D0217816

FEEDS & NUTRITION

(Formerly, Feeds & Nutrition—complete)

by

M. E. ENSMINGER, PH.D.
J. E. OLDFIELD, PH.D.
W. W. HEINEMANN, PH.D.

SECOND EDITION

The Ensminger Publishing Company

648 West Sierra Avenue
Clovis, California 93612
Phone #: 209/299-2263
U.S.A.

FEEDS & NUTRITION

Second Edition

The Ensminger Publishing Company
648 West Sierra Avenue
P.O. Box 429
Clovis, California 93612
Phone #: 209/299-2263
U.S.A.

LIBRARY OF CONGRESS CATALOG CARD NO. 89-083516

ISBN — 0-941218-08-2

Feeds & Nutrition is in several languages

ABOUT THE AUTHORS

The three authors are shown seated in the library of the senior author, Dr. M. E. Ensminger. *Left to right:* Dr. James E. Oldfield, Dr. M. E. Ensminger, and Dr. Wilton W. Heinemann.

• ***About Dr. M. E. Ensminger* (Center)**—Dr. M. E. Ensminger is President of Agriservices Foundation, Clovis, California, a nonprofit foundation serving world agriculture in the area of World Food, Hunger and Malnutrition. Also, he is Adjunct Professor, California State University—Fresno; Adjunct Professor, The University of Arizona—Tucson; Distinguished Professor, University of Wisconsin—River Falls; Honorary Professor, Huazhong Agricultural College—Wuhan, People's Republic of China; and collaborator, U.S. Department of Agriculture.

Dr. Ensminger grew up on a Missouri Farm; completed B.S. and M.S. degrees at the University of Missouri, and the Ph.D. at the University of Minnesota; served on the staffs of the University of Massachusetts, the University of Minnesota, and Washington State University; served as Consultant, General Electric Company, Nucleonics Department; and served as the first President of the American Society of Agricultural Consultants.

Among Dr. Ensminger's honors and awards are: Distinguished Teacher Award, American Society of Animal Science; Washington State University named and dedicated the *Ensminger Beef Cattle Research Center,* in recognition of his contributions to the University; and an oil portrait of him was placed in the 300-year-old gallery of the famed Saddle and Sirloin Club, which is recognized as the highest honor that can be bestowed on anyone in the livestock industry.

Dr. Ensminger founded the International Stockmen's School, which he directed for 40 years. He has lectured and/or conducted seminars in more than 60 countries, including giving 5 invitational lectures before the Chinese Academy of Science and conducting the largest Seminar in China since the revolution. Dr. Ensminger is the author of more than 500 scientific and popular articles, bulletins and columns; and the author or co-author of 19 books, which are in several languages and used all over the world. The whole world is his classroom!

• ***About Dr. James E. Oldfield* (Left)**—Dr. James E. Oldfield is Director, Nutrition Research Institute, and Professor Emeritus of Animal Science, Oregon State University—Corvallis.

Dr. Oldfield was born in Victoria, B.C., Canada. He received the B.S.A. and M.S.A. degrees from the University of British Columbia; and the Ph.D. from Oregon State University. During World War II, he served in the Canadian Army, and was decorated with the Military Cross.

Dr. Oldfield has conducted nutrition research and/or been involved in nutrition problems pertaining to all animal species, including fur and laboratory animals. He has served in numerous scientific capacities, including: member of the Committee on Animal Nutrition, National Research Council; member, National Technical Advisory Committee on Agricultural Uses of Water, U.S. Department of Interior; member, Animal Nutrition Research Council; Director, Council of Agricultural Science and Technology (CAST); member, Nutrition Study Section, Division of Research Grants, National Institutes of Health; and Consultant to the Office of Economic Cooperation and Development, Ankara, Turkey. Dr. Oldfield is a world-renowned authority on selenium, on which subject he has given invitational lectures, participated in symposia, and/or served on committees throughout the United States, and in Canada, New Zealand, Australia, and China.

Among Dr. Oldfield's numerous professional recognitions are the Morrison Award, American Society of Animal Science; Fulbright Research Scholar, Massey University, New Zealand; and President, American Society of Animal Science.

Dr. Oldfield has published 142 technical papers and journal articles, and 92 reports and popular articles.

• ***About Dr. Wilton W. Heinemann* (Right)**—Dr. Wilton W. Heinemann, whose expertise is animal nutrition and pasture/range management, is Professor Emeritus, Washington State University, where he had a long and distinguished career. He completed B.S. and M.S. degrees at Washington State University, and the Ph.D. degree at Oregon State University. Dr. Heinemann is a Fellow in both the American Association for the Advancement of Science and the American Society of Animal Science.

Dr. Heinemann has served as Consultant/Nutritionist to the Bureau of Fisheries and Wildlife, U.S. Department of the Interior; the National Hay Association; the U.S. Feed Grains Council; Battelle Northwest; and to farmers, ranchers and agribusinesses, worldwide.

Dr. Heinemann has presented invitational papers in the U.K., Brazil, Finland, Australia, and the USSR; participated in the Nordic Congress of Agricultural Scientists, Helsinki, and in the Symposium on Ruminant Nutrition, Cambridge University, England; and has twice given invitational lectures before the Polish Academy of Science.

Dr. Heinemann has conducted research and/or been involved in problems pertaining to all animal species, including fish and wildlife. He is the author of 151 technical papers and journal articles, 158 popular articles and reports, and the co-author of a textbook. Dr. Heinemann is an international authority on animal nutrition, and on pasture and range management.

Lest we forget!

Dedicated to the memory of two immortals

W. A. Henry
(1850–1932)

F. B. Morrison
(1887–1958)

authors of

Feeds and Feeding

1898, 1st edition—1956, 22nd edition

(From 1898 to 1910, Henry was sole author;
from 1910 to 1929, Henry and Morrison were coauthors;
from 1929 to 1956, Morrison was sole author.)

PREFACE TO THE SECOND EDITION

Feeds & Nutrition

Throughout history, hunger and malnutrition have been the major cause of disease, morbidity, and death. Pellagra, scurvy, and beriberi decimated armies, ships' crews, and nations; they even reshaped the course of history. But the cause and cure for these diseases was not understood until the 20th century when the biological approach in experiments—the use of laboratory animals (largely white rats and mice, guinea pigs, chickens, pigeons, and dogs)—was ushered in. Their diets were made up of relatively pure nutrients (proteins, carbohydrates, fats, and minerals)—using casein or albumen, lard, and pure carbohydrate such as dextrin. Deficiencies followed, then it was discovered that dramatic cures resulted when minute amounts of a mineral or a vitamin were added. Using the biological approach, in 1913 McCollum and Davis at the University of Wisconsin, and Osborne and Mendel at the Connecticut Experiment Station, working independently, discovered vitamin A. It is noteworthy that these researchers were driven by the desire for scientific discovery—not money; Margaret Davis, a young biologist, volunteered to do the rat work for Dr. McCollum without salary. It was my happy privilege to witness, and even have a small part in, many of the nutritional discoveries that followed.

In my lifetime, various minerals were found to be required by animals and people: phosphorus, in 1918; copper, in 1925; magnesium, manganese, and molybdenum, in 1931; zinc, in 1934; and cobalt, in 1935. But it was not until 1948 that it was discovered that cobalt functions as a component of vitamin B-12. After developing isolation equipment to shield animals fed ultrapure diets from contamination by minute amounts of mineral elements in the environment, the essentiality of selenium was discovered in 1957, and of chromium in 1959. Then, as recently as 1972, it was found that fluorine and silicon are essential.

As a student working my way through the University of Missouri at 25 cents per hour, beginning in 1926, it was my good fortune to work with several of the all-time greats in biological research. I cared for some of the animals that were used in the pioneering growth and body composition studies, which spanned 35 years and were conducted by H. J. Waters, T. F. Trowbridge, L. D. Haigh, C. R. Moulton, and A. G. Hogan. Also, Samuel Brody, the brilliant and prodigious worker in growth and energetics, was a friend of mine. Dr. Brody was in his office or laboratory 7 days a week and at all hours of the day and night. He frequently reminded me and other students that scientists often prove what farmers have long known.

As a student at the University of Missouri, I also worked on some of the early vitamin C studies conducted by Dr. A. G. Hogan, pioneer nutritionist. In the process of conducting my experiments, I hand-squeezed loads of oranges at 4:00 a.m. Everything went well until the Columbia, Missouri police suspicioned that I was violating the Prohibition Act and making moonshine liquor. To keep out of jail, one wintry morning I told the police the story of scurvy, and explained that orange juice contained a factor that would prevent the dreaded disease. I narrowly averted a ride in the paddy wagon (police car) that morning, although the police were not fully convinced relative to the difference between making moonshine liquor and conducting a scientific experiment on scurvy.

The comparative slaughter technique for determining energy storage and heat production was first employed by H. H. Mitchell (whom I knew) and coworkers, in 1926, at the University of Illinois. A part of the experimental animals were slaughtered at the beginning and the end of each experiment, with the calorie content of the carcasses analyzed and the increased energy determined by difference in caloric content at the beginning and the end.

In studies conducted from 1935 to 1955, Dr. W. C. Rose (whom I also knew) and coworkers at the University of Illinois were the first to determine the essentiality of amino acids and the minimum requirements of each. Using rats, Rose found that 10 different amino acids must be supplied in adequate amounts in the food to support the normal growth of young rats; and using his graduate students in nitrogen balance studies, Rose found that only 8 of these amino acids were essential for the maintenance of nitrogen equilibrium in fully grown young men. Subsequently, it was shown that a ninth amino acid (histidine) is essential for human infants; and, in long-term studies, it was found to be essential for adults, too.

In the early 1900s, there were around 100,000 cases of pellagra per year in the United States, and up to 10,000 deaths. A similar disease in dogs was called *blacktongue*. In 1937, C. A. Elvehjem (a biochemist, and a friend of mine, who later became president of the University of Wisconsin) and his colleagues at the University of Wisconsin, in experiments with dogs, identified the antiblacktongue factor as nicotinic acid (niacin) and nicotinamide (niacinamide).

Yes, I was there! I had a ringside seat as I witnessed the discovery of many of the nutrients essential for animals and humans.

But the past is prologue! Genetic wizardry by gene splicing is giving rise to a major scientific revolution called biotechnology and spawning many new developments exceeding our fondest dreams. Biotechnology will involve every facet of animal production from breeding and feeding to the finished product, including the genetic makeup of animals and the feeds they eat; the digestion, physiology, stress tolerance, disease resistance, and efficiency of production of animals; the composition, quality, and quantity of products produced; along with the production of large quantities of drugs and chemicals. While some aspects of biotechnology are decades away from commercial production, others are near, and still others are here now.

But advanced technology calls for advanced animal adaptation, welfare, and environmental control. We need to breed and select animals adapted to an artificially-made environment—animals that not only survive, but thrive, under the conditions in which they are kept. We need to heed the warnings of endangered animals, endangered people, and an endangered planet—presaged by increased pollution, the greenhouse effect, acid rain, depletion of the ozone layer, and destruction of rain forests.

As the senior author of this book, I wish to acknowledge the great contribution of my two co-authors, Dr. James E. Oldfield and Dr. Wilton W. Heinemann, along with their patience with, and understanding of, my idiosyncracies. Audrey Ensminger shepherded the manuscript from beginning to end; Joan Wright deciphered my hieroglyphics and typed the manuscript; Lynn Wright set the type and paged up the copy; Tom Phillips prepared more than 40 original paintings specially for this book; Ran Guang Liang prepared the cover design; and Margo Williams did the traditional art work and prepared the camera-ready copy. Also, I am grateful to Verla Rape, Program Analyst, Office of the Administrator, Economic Research Service, U.S. Department of Agriculture, who gave liberally of her time and talents in providing much of the recent source material for updating. Additionally, at appropriate places in the book, due acknowledgment and appreciation is expressed to all those who reviewed portions of the manuscript or responded so liberally to my call for illustrations and information. Without the help of all these fine folks, the task could not have been completed.

Although books pertaining to animal nutrition abound, it is my fond hope that *Feeds & Nutrition* will be a classic, and that it will give a big assist to ushering animals and people of a biotechnological era into the 21st century and make all of our dreams come true, faster and more abundantly.

Clovis, California
1990

M. E. Ensminger

FEEDS & NUTRITION

CONTENTS

TOM PHILLIPS

F&N PACKING CO.
1
Nutrition

SECTION I

NUTRITION

Feeding had its beginning as an art, the foundations of which were animal instinct and a blend of the caretaker's fads, foibles, and trade secrets. Then came science, founded on chemistry, physics, physiology, and bacteriology.

For many years, the keepers of the herds and flocks were responsible for the very considerable progress made in the art of feeding. They were intensely practical, never overlooking the utility value or the market requirements. No animal met with their favor unless it was earned by meat upon the back, milk in the pail, weight and quality of wool, pounds gained for pounds of feed consumed, draft or speed ability, or some other performance of practical value. In time, scientists teamed up with caretakers to improve the feeding of animals, slowly, but surely, evolving with the science called *nutrition.*

The successful merger of the art and the science of feeding—the joining of feeds and nutrition—ushered in a new era in animal agriculture. In the process, it also stimulated increased interest in human nutrition, where the requirements are similar. This gave rise to the statement that, "We are gradually learning to feed our children as well as our animals." Many maladies which had long plagued both humankind and beast were traced to dietary deficiencies, imbalances, and toxicities. Rectifying these nutritional problems improved the health and performance of all creatures, including humans.

But the past is prologue! The final chapter of the 20th century, which is now being written, may well be the most revolutionary and dramatic of all in animal nutrition.

Section I covers the fundamentals of nutrition, with separate chapters devoted to each of the following:

Chapter 1, Food and Animals—a global perspective
Chapter 2, Principles of Nutrition
Chapter 3, Digestion/Absorption
Chapter 4, Nutrients/Metabolism
Chapter 5, Nutritional Disorders/Toxins

NUTRITION, the preceding two-page spread, is from an original painting by the noted artist, Tom Phillips, prepared specially for this book. It portrays the artist's conception of what the first five chapters of *Feeds & Nutrition* are all about—animal feeds and human foods derived from plants which have their tops in the sun and their roots in the soil, and farmers and ranchers producing bountiful and nutritious foods for all the people, many of whom live in distant cities.

Chapter 1

Original painting by Tom Phillips

FOOD AND ANIMALS —a global perspective

Contents Page

Back of animals are feeds; and back of the feeds are soil resources, spring rains, and the energy of the sun. With the aid of science, technology, and animals, farmers and ranchers combine these to produce a tasty platter of meat and eggs for the table, cream for the peaches, butter for the biscuits, and cheese for the macaroni—all derived from the sun through the process known as photosynthesis.

Fig. 1-1. Without photosynthesis, there would be no oxygen, no plants, no food, no animals, and no people.

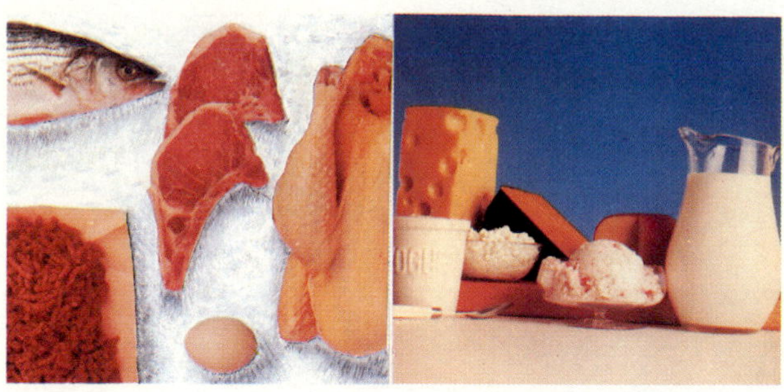

Fig. 1-2. Animal products are good—and good for you. (Courtesy, USDA)

But animal products are far more than just very tempting and delicious foods! From a nutrition standpoint, foods of animal origin contribute certain essentials to the American diet; they supply 35% of the energy, 70% of the protein—along with the essential amino acids, 80% of the calcium, 60% of the phosphorus, and significant amounts of the other minerals and vitamins needed in the human diet. It is noteworthy, too, that animal products contain vitamin B-12, which does not occur in plant foods, and that they are a rich source of iron, the availability of which is twice as high as in plants.

ALL FLESH IS GRASS!

Life on earth is dependent upon photosynthesis. Without it, there would be no oxygen, no plants, no feed, no food, no animals, and no people.

As fossil fuels (coal, oil, shale, and petroleum)—the stored photosynthates of previous millennia—become exhausted, the biblical statement, "all flesh is grass" (Isaiah 40:6), comes alive again. The focus is on photosynthesis. Plants, using solar energy, are by far the most important, and the only renewable, energy-producing method,[1] the only basic food-manufacturing process in the world; and the only major source of oxygen in the earth's atmosphere. Even the chemical and electrical energy used in the brain cells of man is the product of sunlight and the chlorophyll of green plants. Thus, in an era of world food shortages, it is inevitable that the entrapment of solar energy through photosynthesis will, in the long run, prove more valuable than all the underground fossil fuels—for when the latter are gone, they are gone forever.

PHOTOSYNTHESIS AND RUMINANTS

Photosynthesis is the process by which the chlorophyll-containing cells in green plants capture the energy of the sun and convert it into chemical energy; it's the process through which plants synthesize and store organic compounds, especially carbohydrates, from inorganic compounds—carbon dioxide, water, and minerals, with the simultaneous release of oxygen.

Ruminants, which include cattle, sheep, goats, and water buffalo, are even-toed, hoofed animals that ruminate (regurgitate and chew a cud) and have a complex four-compartment stomach characterized by much storage space and microbial fermentation and adapted to the effective use of high-fiber feeds and the manufacture of B-complex vitamins and essential amino acids.

Photosynthesis and ruminants team up to provide every ounce of food that we eat, every breath of oxygen that we inhale, and a very large portion of all the B-complex vitamins and all the essential amino acids that we require. Without photosynthesis and ruminants, there would be no plant and animal life on earth—and no human race. But the nature of photosynthesis and ruminants remained elusive for centuries.

For many years, the *humus theory* prevailed; scholars believed that green plants derived all their nourishment from the organic materials of the soil. Finally, in about 1630, Jean van Helmont, a Belgian physician, performed a revealing experiment which proved this belief false. He placed exactly 200 lb of completely dried soil into a vessel; planted a 5-lb willow shoot in the container; and added rainwater, but no fertilizer, to the soil at regular intervals. At the end of 5 years, he removed the willow tree, now grown large; carefully scraped the soil from the roots and returned it to the container; dried the willow tree and weighed it; and dried and weighed the soil. He found that the willow tree weighed 169 lb and 3 oz—a gain of approximately 164 lb during the 5-year period, and that the soil weighed 199 lb and 14 oz—indicating a loss of only 2 oz during the same period. Van Helmont concluded that the willow tree built its substance from water alone. This classic experiment proved the falsity of the humus theory, even though van Helmont's explanation was only partially correct, since he did not know of the role of carbon dioxide in the synthesis of organic matter by plants. Further studies followed, but more than two centuries elapsed after van Helmont's experiment before the concept of chemical energy developed sufficiently to permit the discovery (in 1845) that light energy from the sun is stored as chemical energy in products formed during photosynthesis. Subsequent studies added to the store of knowledge. Today, scientists know that green plants fashion most of their solid materials from the water of the soil and the carbon dioxide of the air through the process of photosynthesis, and that green plants absorb relatively small amounts of solid substances (mineral nutrients) from the soil.

Photosynthesis is dependent upon the presence of chlorophyll, a green pigment which develops in plants soon after they emerge from the soil. Chlorophyll is a chemical catalyst—it stimulates and makes possible certain chemical reactions without becoming involved in the reaction itself. By drawing upon the energy of the sun, it can convert inorganic molecules, carbon dioxide (CO_2) and water (H_2O), into an energy-rich organic molecule such as glucose ($C_6H_{12}O_6$), and at the same time release free oxygen (O_2). It transforms solar energy into a form that can be used by plants, animals, and humans. Because of this capability, chlorophyll has been referred to as the link between nonliving and living matter, or the pathway through which nonliving elements may become part of living matter.

[1]Certain types of microorganisms, termed chemoautotrophs, get their energy from inorganic compounds, but aside from this minor exception, the energy that runs the life support systems of the biosphere comes from photosynthesis.

Through the photosynthetic process, it is estimated that more than a billion tons of carbon per day are converted from inorganic carbon dioxide (CO_2) to organic sugars ($C_6H_{12}O_6$— glucose), which can then be converted into other carbohydrates, fats, and proteins—the three main groups of organic materials of living matter.

Photosynthesis is a series of many complex chemical reactions, involving the following two stages (see Fig. 1-3):

Stage 1—The water molecule (H_2O) is split into hydrogen (H) and oxygen (O); and oxygen, the necessary gas for breathing of animals, is released into the atmosphere. Hydrogen is combined with certain organic compounds to keep it available for use in the second step of photosynthesis. Chlorophyll and light are involved in this stage.

Stage 2—Carbon dioxide (CO_2) combines with released hydrogen to form the simple sugar (glucose) and water. This reaction is energized (powered) by ATP (adenosine triphosphate), a stored source of energy. Neither chlorophyll nor light is involved in this stage.

The chemical reactions through which chlorophyll converts the energy of solar light to energy in organic compounds is one of nature's best-kept secrets. Scientists have not been able to unlock it, as they have so many of life's processes. Moreover, photosynthesis is limited to plants; animals store energy in their products—meat, milk, and eggs—but they must depend upon plants to manufacture it. Additional facts pertinent to an understanding of photosynthesis follow:

1. During the earth's very long geological past, green plants, growing in warm climates in the presence of more carbon dioxide than the atmosphere now contains, grew faster than they were consumed. As a result, vast quantities of carbon, in the form of organic matter now represented by the fossil fuels (coal, oil, shale, and petroleum), accumulated beneath the earth's surface. The combustion of these fuels provides much of the energy now used in homes, factories, and transportation.

2. Photosynthesis is an energy-requiring process, which uses sunlight as the source of energy. Hence, it can occur only when light shines upon green plant tissues.

3. Plant species and genetics (the inherited set of directions) determine whether a plant will form high or low levels of specific proteins, carbohydrates, minerals, and vitamins. For example, alfalfa always contains more calcium than corn even though they grow side by side.

4. Environmental factors—including the amount of sunlight, the temperature of the air and of the soil, the humidity of the air, and the moisture content of the soil—may also have an important bearing on the concentration of nutrients in a plant. The impact of environmental factors on plant nutrients is of concern to the producer and the nutritionist, as evidenced by the following examples: (a) The amount of vitamin C in a ripening tomato is primarily controlled by the amount of sunlight that strikes the tomato; (b) during cool, cloudy weather some grasses may accumulate high levels of nitrate; and (c) the effects of environment on plant composition may be so pronounced that certain nutritional diseases of animals occur much more frequently in some years than in others, even on the same pastures.

5. Physiological factors of plants—health and maturity—also exert an effect on the rate of photosynthesis.

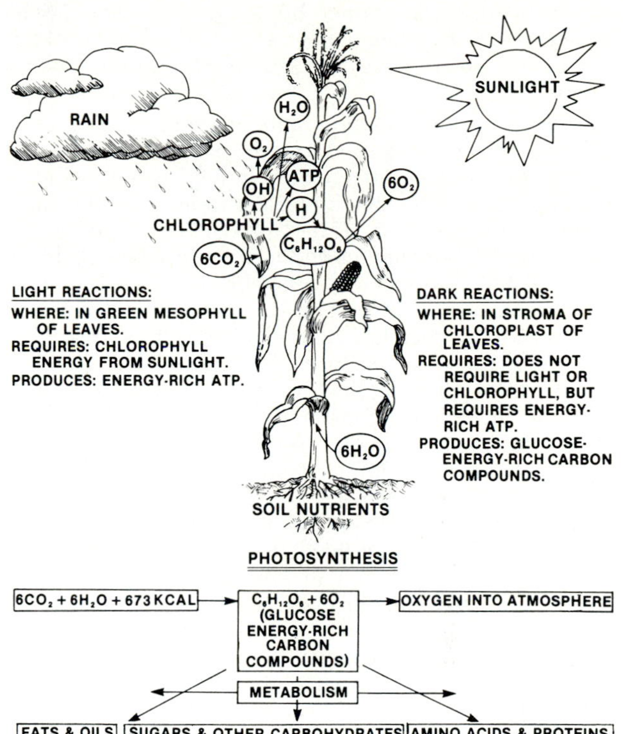

Fig. 1-3. Photosynthesis fixes energy. Diagrammatic summary of (1) photosynthesis, and (2) the metabolic formation of organic compounds from the simple sugars. This diagram shows the following:

1. Carbon dioxide gas from the air enters the green mesophyll cells of plant leaves.
2. Plants take up oxygen from the air for some of their metabolic processes and release oxygen back to the air from other metabolic processes.
3. Plants take up water and essential elements from the soil.
4. The energy essential to photosynthesis is absorbed by chlorophyll and supplied by sunlight.
5. For a net input of 6 molecules of carbon dioxide and 6 molecules of water, there is a net output of 1 molecule of sugar and 6 molecules of oxygen.
6. The process is divided into light and dark reactions, with the light reactions building up the energy-rich ATP required for the dark reactions.
7. In the process, 673 Calories (kcal) of energy are used.
8. The sugar (glucose) manufactured in photosynthesis may be converted into fats and oils, sugars and other carbohydrates, and amino acids and proteins.

From the foregoing, it is apparent that the concentration in plants of the different nutrients required by animals and humans is controlled by several processes that depend on the fertility of the soil, the genetics of the plant, and the environment in which it grows. Any one of these factors may affect the level of different essential nutrients or of toxic substances in feeds and foods.

Although photosynthesis is vital to life itself, it is very inefficient in capturing the potentially available energy. Of energy that leaves the sun in a path toward the earth, only about half reaches the ground. The other half is absorbed or reflected in the atmosphere. Most of that which reaches the ground is dissipated immediately as heat or is used to evaporate water in another important process for making life possible. Only about 2% of the earthbound energy from the sun actually reaches green plants, and only half of this amount (1%) is transformed by photosynthesis to energy storage in organic compounds. Moreover, only 5% of this plant-captured energy is fixed in a form suitable as food for people.

With such a small portion of the potentially useful solar energy actually being used to form plant tissue, it would appear that some better understanding of the action of chlorophyll should make it possible to increase the effectiveness of the process. Three approaches are suggested: (1) increasing the amount of photosynthesis on earth; (2) manipulating plants for increased efficiency of solar energy conversion; and (3) converting a greater percentage of total energy fixed as chemical energy in plants (the other 95%) into a form available to humans. Ruminants are the solution to the latter approach; they can convert energy from such humanly inedible plant materials as grass, cornstalks, and straw into food for humans. (See Fig. 1-4.) Also, it is noteworthy that animals do not require fuel to graze the land and recover the energy that is stored in the grass. Moreover, they are completely recyclable; they produce a new crop each year and perpetuate themselves through their offspring. It would appear, therefore, that there is more potential for solving the future food problems of the world by manipulating plants for increased solar energy conversion and by using ruminants to make more plant energy available to people than from all the genetic and cultural methods combined.

CONSERVE ENERGY

Population growth and food production technology are now creating feed, food, and energy stresses of unprecedented scale and urgency—threatening human existence. It's a case of too many people nibbling away at natural resources faster than the earth can combine the energy of the sun, the rains of the heavens, and the minerals of the soil, to produce food.

Fossil fuels are like a bank account. There is nothing wrong with drawing upon either of them, but neither is inexhaustible. It is highly imprudent not to be aware of big withdrawals and not to cover them. Within a short span of a few years, the world made the transition from a positive energy balance based upon the capture of the energy of the sun via green plants, crops, and forests to an imbalance, or even a negative balance, by resorting primarily to the bank of trapped sun energy of fossil fuels that had accumulated over millions of years.

Modern, mechanized feed and food production requires an extra input of fuel, which is mostly of fossil origin. This auxiliary energy is expended in endless ways to improve agricultural productivity; it is used for drainage and irrigation, clearing of forest land, seedbed preparation, weed and pest control, fertilization and efficient harvesting. In addition to production, as such, there are two other important steps in the feed-food line as it moves from the producer to the consumer; namely, processing and marketing, both of which require higher energy inputs than to produce the food on the farm. (See Table 1-1.)

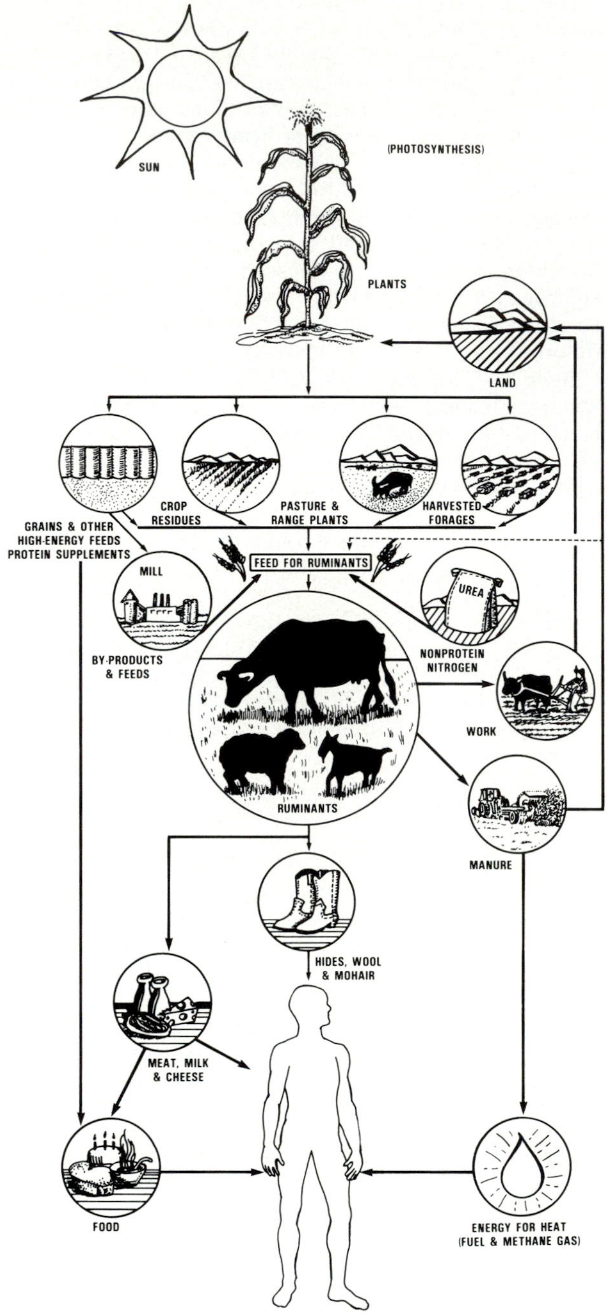

Fig. 1-4. The sun-plant-land-ruminant-human relationship. Ruminants step up energy and manufacture B-vitamins and essential amino acids. Their feed comes from plants which have their tops in the sun and their roots in the soil. Hence, we have the nutrition cycle as a whole—from the sun and the soil, through the plant, thence to the ruminant (and human) and back to the soil again.

Table 1-1 points up the increasing drain that modern food production is putting on the energy supply. In 1980, U.S. farms put in 2.8 calories of fuel per calorie of food grown, 3.1 times more than the on-farm energy input in 1940.

Table 1–1
MODERN FOOD PRODUCTION IS INEFFICIENT IN ENERGY UTILIZATION—THE STORY FROM PRODUCER TO CONSUMER[1]

Year	On the Farm	Food Processing	Marketing and Home Cooking	Total/Person /Year
1940[2]				
Million kcal	0.9	2.2	2.1	5.2
Percent	18.0	42.0	40.0	100.0
1980[3]				
Million kcal	2.8	4.7	4.6	12.1[4]
Percent	23.0	39.0	4.6	100.0
Increase, times, 1940-1980	3.1	2.1	2.2	2.3

[1]Energy in million kcal per capita to produce one million kcal of food in the U.S.
[2]Values from Borgstrom, G., "The Price of a Tractor," *Ceres*, FAO of the U.N., Rome, Italy, Nov.-Dec., 1974, p. 18, Table 3.
[3]Authors' estimate based on several reports detailing trends in energy usage.
[4]This means that in 1980 it required 12.1 million kcal to produce 1 million kcal of food for each person, a daily consumption of 2,740 kcal (1,000,000 ÷ 365 = 2,740).

Fig. 1–5. An Oriental wet rice peasant, using animal power (water buffalo), expends only 1 calorie of energy to produce each 50 calories of food. By comparison, the average U.S. farmer, using mechanical power (tractors), expends 2.5 calories of fuel energy to produce 1 calorie of food. (Courtesy, International Bank for Reconstruction and Development, Washington, D.C.)

Table 1–1 also shows that, in the United States in 1980, a total of 12.1 calories were used in the production, food processing, and marketing-cooking for every calorie of food consumed, with a percentage distribution of the total cost of energy at each step from producer to consumer as follows: on the farm, 23%; food processing, 39%; and marketing and home cooking, 38%. In 1940, it took only 5.2 calories—slightly less than half the 1980 figure—to get 1 calorie of food on the table. It's noteworthy, too, that more energy is required for food processing and marketing/home cooking than for growing the product; and that, from 1940 to 1980, the on-the-farm energy requirement increased by 3.1 times, in comparison with an increase of 2.1 and 2.2 times for each of the other steps—processing and marketing/home cooking.

Prior to the advent of machines and fuel in crop production, 1 calorie of energy input on the farm produced about 16 calories of food energy. Today, on the average, U.S. farms put in about 2.8 calories of fuel per calorie of food grown; hence, to produce a daily intake of 3,000 calories of edible food from cultivated crops may require 8,400 calories of energy from fossil fuels—an exhaustible source. It's more surprising yet—and thought-provoking—to know that, even today in the poorer or developing countries, it takes only 1 calorie to produce each 10 calories of food consumed. The Oriental wet rice peasant uses only 1 unit of energy to produce 50 units of food energy. This gives the Orientals a favorable position among the major powers as the energy crisis worsens.

Modern intensive farming has markedly increased crop yields per acre and per man-hour—by as much as 50- to 100-fold. But this has been done at the cost of large inputs of fuel. Today, for a surprising number of cropping systems, a 10- to 50-fold increase in the energy output merely doubles or triples the food energy. Thus, the law of diminishing returns prevails.

Scarce and high-priced fossil fuels have spurred a search for conserving stored energy and for increased energy production through photosynthesis. Higher productivity of the agriculture of tomorrow must be achieved through ingenious approaches in order to reverse the present lopsided energy balance. In obtaining increased feed and food yields, we must consider how many calories of energy are required to produce each calorie of feed or food. We must remember that photosynthesis does not deplete fossil fuels. We must remember, too, that grazing animals do not require fuel outside of their own body use to harvest the energy and other nutrients of grass (solar energy converted into chemical energy by grass), a renewable source. It follows that ruminants, which utilize grazing land, offer the best means of stepping up and storing energy for humans.

Energy may also be conserved by lessening waste. Pests cause an estimated 30% annual crop loss in the worldwide potential production of crops, livestock, and forests.[2] Every part of our feed, food, and fiber supply is vulnerable to pest attack, including marine life, wild and domestic animals, field crops, horticultural crops, and wild plants. Obviously, reducing these losses would conserve energy and increase the supply of feed, food, and fiber.

FOOD FOR THE 21st CENTURY

As the ghost of hunger, foretold by the English clergyman Thomas Robert Malthus in 1798, stalks the world, the focus is on animals. During periods of food scarcity, it is inevitable that some will suggest that grain be diverted from livestock and poultry feeding—that they will challenge the efficiency of animals in converting feed to food and the place of animals in the economical production of human food. Animal agriculture will be on trial. Increasingly, the charge will be made that much of the world goes hungry because of the substitution of meat, milk, and eggs for direct grain consumption. A response to this accusation requires that animal agriculturalists substitute knowledge for moral indignation. To this end, the important sections that follow are presented.

[2]Ennis, Jr., W. B., W. M. Dowler, and W. Klassen, "Crop Production to Increase Food Supplies," *Science*, Vol. 188, No. 4188, May 9, 1975, pp. 593–598.

Who Shall Eat?

An appalling 500 million people in the world suffer from hunger and malnutrition. More shocking yet, it is estimated that the troubled 21st century will open with 1.3 billion malnourished people. For these starving millions around the globe, life is little more than a heartrending journey to an end which refuses to arrive soon enough to stop their suffering. For the most part, the world's hungry and malnourished are grain eaters.

Cereal grain is the most important single component of the world's food supply, accounting for 50% of the food produced in all the globe. It is the major source of food for many of the world's poorest people, supplying 58% of the total calories in the developing countries (see Fig. 1-6).

WHERE THE CALORIES COME FROM

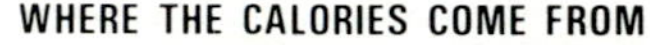

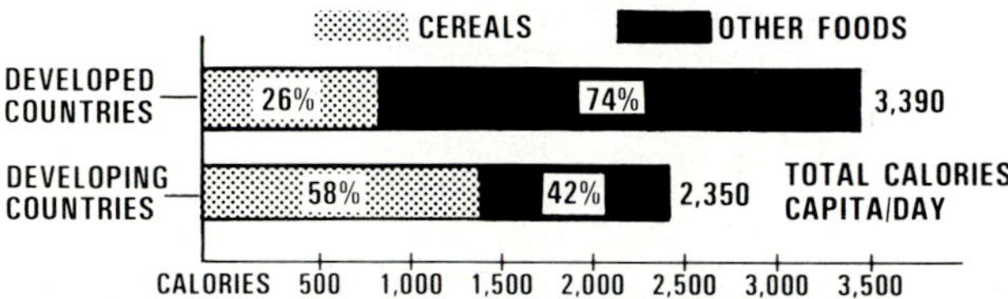

Fig. 1-6. Calories per person per day from cereals vs other foods. In the developed countries, only 26% of the calories comes from direct consumption of cereals, compared with about 58% in developing countries. (*Source:* The Fifth World Food Survey, FAO, United Nations, Rome, Italy, 1985)

However, in many developed countries, more grain is fed to animals than is consumed directly by humans. Under such circumstances, sporadic food shortages and famine in different parts of the world give rise to the following recurring questions:

1. Who should eat grain—people or animals? Shall we have food or feed?
2. Can we have both food and feed?

In attempting to answer these complex questions, those favoring people going on a grain diet often substitute moral indignation for knowledge. The authors' answers are given in the sections that follow.

Favoring Direct Human Grain Consumption

Historically, the people of new and sparsely populated countries have been meat eaters, whereas the population of the older and more densely populated areas have been vegetarians. The latter group has been forced to eliminate most animals and consume plants and grains directly in an effort to avoid famine.

Among the arguments sometimes advanced by those who favor bread alone—the direct human consumption of grain—are the following:

1. **More people can be fed.** Forgetting for a moment the high nutritive value of meats, milk, and eggs, there can be no question that more hunger can be alleviated with a given quantity of grain by completely eliminating animals. About 2,000 lb of concentrates (mostly grain) must be supplied to livestock in order to produce enough meat and other livestock products to support a person for a year, whereas 400 lb of grain (corn, wheat, rice, soybeans, etc.) eaten directly will support a person for the same period of time. Thus, a given quantity of grain eaten directly will feed 5 times as many people as it will if it is first fed to livestock and is eaten indirectly by humans in the form of livestock products. This inefficiency is the result of unavoidable nutrient losses in all animal feeding and the fact that no return is received from that portion of the animal's feed which goes for maintenance (which amounts to approximately ½). This is precisely the reason why the people of the Orient have become vegetarians.

2. **On a feed, calorie, or protein conversion basis, it's not efficient to feed grain to animals and then to consume the livestock products.** This fact is pointed up in Table 1-2 and Figs. 1-7, 1-8, and 1-9.[3]

FEED EFFICIENCY

POUNDS OF FEED REQUIRED TO PRODUCE ONE POUND OF PRODUCT

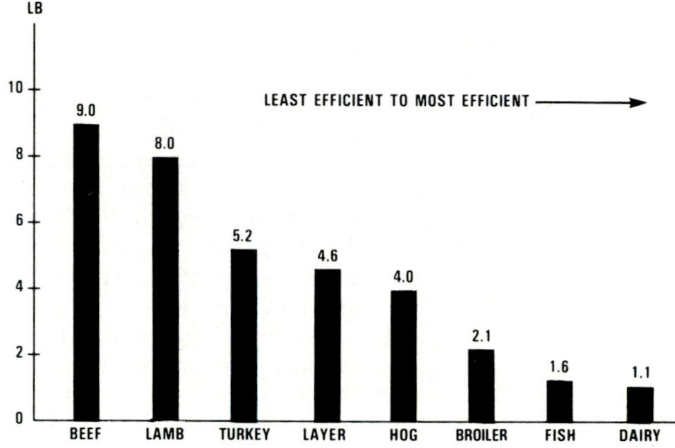

Fig. 1-7. Pounds of feed required to produce 1 lb of product. This shows that it takes 9 lb of feed to produce 1 lb of on-foot beef, whereas it takes only 1.11 lb of feed to produce 1 lb of milk. (Source: Table 1-2 of this chapter)

Thus, in the developing countries, where the population explosion is the greatest, virtually all grain is eaten directly by people; precious little of it is converted to animal products.

As people become more affluent, they actually use more grain, but most of it is converted into animal products, for they consume more meat, milk, and eggs. It is noteworthy, too, that no nation appears to have reached such a level of affluency that its per capita grain requirement has stopped rising.

[3]It could be argued that Table 1-2 makes no provision for the feed used by the sires and dams of these animals—the animals that gave birth to these producers. Others may be critical of using a yearling steer without making provision to get him to the feedlot stage. Finally, it may be contended that any such comparison should be between animals of like age; for example, between broilers and veal calves. Having raised these questions, the authors submit Table 1-2, which in their judgment is as fair a rating on feed to food efficiency as can be made.

TABLE 1–2

FEED TO FOOD EFFICIENCY RATING BY SPECIES OF ANIMALS, RANKED BY PROTEIN CONVERSION EFFICIENCY

(Based on Energy as TDN or DE and Crude Protein in Feed Eaten by Various Kinds of Animals Converted into Calories and Protein Content of Ready-to-Eat Human Food)

		Feed Required to Produce One Production Unit				Dressing Yield		Ready-to-Eat; Yield of Edible Product (meat & fish deboned & after cooking)				Feed Efficiency[4]		Efficiency Rating			
Species	Unit of Production (on foot)	Pounds	TDN[1]	DE[2]	Protein	Percent	Net Left	As % of Raw Product (carcass)	Amount Remaining from One Unit of Production	Calorie[3]	Protein[3]	(lb feed to produce one lb product)		Calorie Efficiency[5]		Protein Efficiency[6]	
		(lb)	(lb)	(kcal)	(lb)	(%)	(lb)	(%)	(lb)	(kcal)	(lb)	(%)	(ratio)	(%)	(ratio)	(%)	(ratio)
Broiler	1 lb chicken	2.1[7]	1.7[8]	3,400	0.21[8]	72[13]	0.72	54[14]	0.39	274	0.11	47.6	2.1:1	8.1	12.4:1	52.4	1.9:1
Fish	1 lb fish	1.6[9]	0.98	1,960	0.57	65[10]	0.65	57[11]	0.37	285	0.27	62.5	1.6:1	14.5	6.9:1	47.6	2.1:1
Dairy cow	1 lb milk	1.11[7]	0.9[8]	1,800	0.1[8]	100	1.0	100	1.0	309	0.037	90.0	1.11:1	17.2	5.8:1	37.0	2.7:1
Turkey	1 lb turkey	5.2[7]	4.21[8]	8,420	0.46[8]	79.7[13]	0.797	57[15]	0.45	446	0.146	19.2	5.2:1	5.3	18.9:1	31.7	3.2:1
Layer	1 lb eggs (8 eggs)	4.6[7]	3.73[8]	7,460	0.41[8]	100	1.0	100[12]	1.0[12]	616	0.106	21.8	4.6:1	8.3	12.1:1	25.9	3.9:1
Hog (birth to market weight)	1 lb pork	4.0[18]	3.2	6,400	0.36	70[16]	0.70	44[17]	0.31	341	0.088	0.25	4.0:1	5.3	18.8:1	24.4	4.1:1
Rabbit	1 lb fryer	3.0[19]	2.20	4,400	0.48	55[19]	0.55	79[19]	0.43	301	0.08	35.7	2.8:1	6.8	14.6:1	16.7	6.0:1
Beef steer (yearling finishing period in feedlot)	1 lb beef	9.0[18]	5.85	11,700	0.90	58[16]	0.58	49[17]	0.28	342	0.085	11.1	9.0:1	2.9	34.2:1	9.4	10.6:1
Lamb (finishing period in feedlot)	1 lb lamb	8.0[18]	4.96	9,920	0.86	47[16]	0.47	40[17]	0.19	225	0.052	12.5	8.0:1	2.3	44.1:1	6.0	16.5:1

[1]TDN pounds computed by multiplying pounds feed (column to left) times percent TDN in normal rations. Normal ration percent TDN taken from M. E. Ensminger's books and rations, except for the following: dairy cow, layer, broiler, and turkey from *Agricultural Statistics 1974,* p. 358, Table 518. Fish based on averages recommended by Michigan and Minnesota Stations and U.S. Fish and Wildlife Service.

[2]Digestible Energy (DE) in this column given in kcal, which is 1 Calorie (written with a capital C), or 1,000 calories (written with a small c). Kilocalories computed from TDN values in column to immediate left as follows: 1 lb TDN = 2,000 kcal.

[3]From *Lessons on Meat,* National Live Stock and Meat Board, 1965.

[4]Feed efficiency as used herein is based on pounds of feed required to produce 1 lb of product. Given in both percent and ratio.

[5]Kilocalories in ready-to-eat food = kilocalories in feed consumed, converted to percentage. Loss = kcal in feed ÷ kcal in product.

[6]Protein in ready-to-eat food = protein in feed consumed, converted to percentage. Loss = pounds protein in feed ÷ pounds protein in product.

[7]*Agricultural Statistics 1974,* p. 358, Table 518. Pounds feed per unit of production is expressed in equivalent feeding value of corn.

[8]Pounds feed (column No. 2) per unit of production (column No. 1) is expressed in equivalent feeding value of corn. Therefore, the values for corn were used in arriving at these computations. No. 2 corn values are TDN, 81%; protein, 8.9%. Hence, for the dairy cow 81% × 1.11 = 0.9 lb TDN; and 8.9% × 1.11 = 0.1 lb protein.

[9]Data from report by Dr. Phillip J. Schaible, Michigan State University, *Feedstuffs,* April 15, 1967.

[10]*Industrial Fishery Technology,* edited by Maurice E. Stansby, Reinhold Pub. Corp., 1963, Ch. 26, Table 26-1.

[11]*Ibid.* Reports that "Dressed fish averages about 73% flesh, 21% bone, and 6% skin." In limited experiments conducted by A. Ensminger, it was found that there was a 22% cooking loss on filet of sole. Hence, these values—73% flesh from dressed fish, minus 22% cooking losses—give 57% yield of edible fish after cooking, as a percent of the raw, dressed product.

[12]Calories and protein computed basis per egg; hence, the values herein are 100% and 1.0 lb, respectively.

[13]*Marketing Poultry Products,* 5th Ed., by E. W. Benjamin et al., John Wiley & Sons, 1960, p. 147.

[14]*Factors Affecting Poultry Meat Yields,* University of Minnesota Sta. Bull. 476, 1964, p. 29, Table 11 (fricassee).

[15]*Ibid.* Page 28, Table 10.

[16]Ensminger, M. E., *The Stockman's Handbook,* 6th Ed., Sec. XII.

[17]Allowance made for both cutting and cooking losses following dressing. Thus, values are on a cooked, ready-to-eat basis of lean and marbled meat, exclusive of bone, gristle, and fat. Values provided by National Live Stock and Meat Board (personal communication of June 5, 1967, from Dr. Wm. C. Sherman, Director, Nutrition Research, to the senior author), and based on data from *The Nutritive Value of Cooked Meat,* by Ruth M. Leverton and George V. Odell, Misc. Pub. MP-49, Appendix C, March 1958.

[18]Estimates by the authors.

[19]Based on information in *Commercial Rabbit Raising,* Ag. Hdbk. No. 309, USDA, 1966, and *A Handbook on Rabbit Raising,* by H. M. Butterfield, Washington State University Ext. Bull. No. 411.

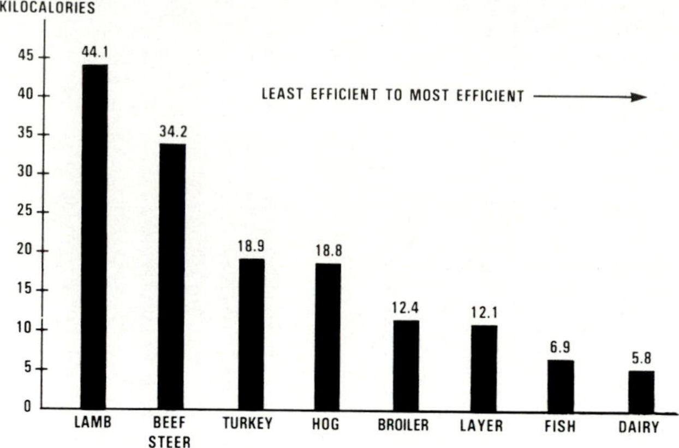

Fig. 1-8. Kilocalories in feed required to produce 1 kcal of product. This shows that it takes 44.1 kcal in feed to produce 1 kcal in lamb, whereas only 5.8 kcal in feed will produce 1 kcal in milk. (Source: Table 1-2 of this chapter)

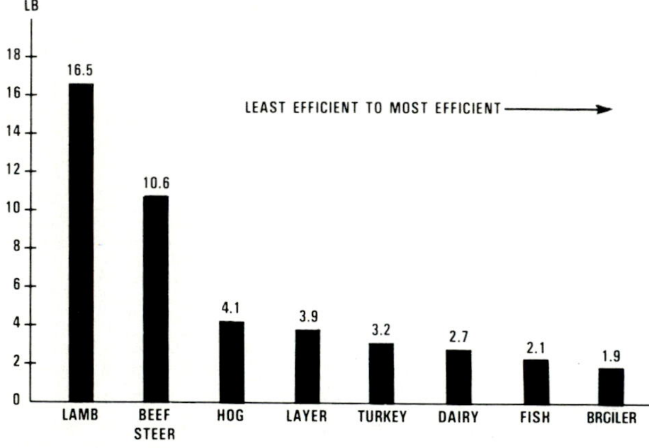

Fig. 1-9. Pounds of feed protein required to produce 1 lb of product protein. This shows that it takes 16.5 lb of feed protein to produce 1 lb of lamb protein, whereas only 1.9 lb of feed protein will produce 1 lb of broiler protein. (Source: Table 1-2. See column headed "Protein Efficiency.") **NOTE WELL:** Where ruminants are supplied with a nonprotein nitrogen supplement, such as urea, they may yield far more product protein than is supplied to them in the feed because they are capable of manufacturing protein from NPN.

Favoring An Animal Agriculture

Among some social reformists, the charge persists that much of the world goes hungry because of the substitution of meat, milk, and eggs for direct grain consumption. A response to this accusation requires far more than a simple denial.

The following facts are presented in favor of sharing grain with animals, then consuming the animal products:

1. **Animals provide needed power.** A century ago, muscles provided 94% of the world's energy needs; coal, oil, and waterpower provided the other 6%. Today, the situation is reversed in the developed nations. They now obtain 94% of their energy needs from coal, oil, natural gas, and waterpower, and only 6% from the muscle power of people and animals. However, in the developing nations, cattle, water buffalo, and horses still provide much of the agricultural power. In this capacity, they contribute to human food supply from plant sources. Such draft animals are a part of the agricultural scene of Asia, Africa, the Near East, Latin America, and parts of Europe; areas characterized by small farms (for example, India's farms average only 6.4 acres), low incomes, abundance of manpower, and lack of capital. But animals have certain advantages. They can be fueled on roughages to produce power, a most important consideration in time of energy shortage; and both cattle and water buffalo are "triple threat animals"—they're used for work, milk, and meat. Also, when it comes to tilling wet, muddy rice paddies, water buffalo are without a peer; and, under adverse conditions, they will outproduce cattle in power, milk yield, and butterfat.

Although the general trend in the world is toward more and more mechanization, animals will continue to provide most of the agricultural power for the small farm food crop agriculture in many of the developing countries.

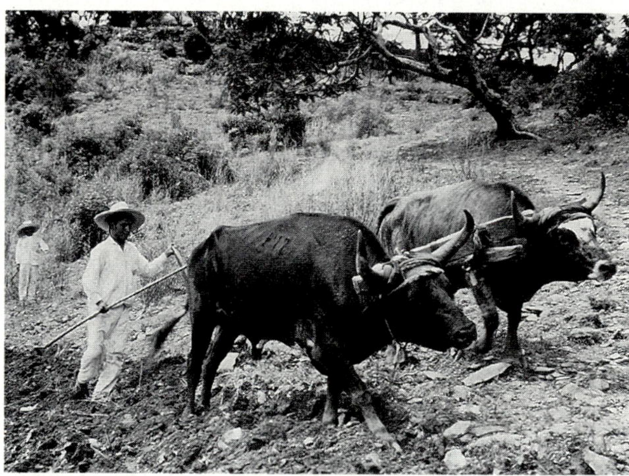

Fig. 1-10. Oxen pulling a stick (one-handled) plow. Draft animals are a part of the agricultural scene in most of the developing countries of the world. (By Burton Holmes, from Ewing Galloway)

2. **Animals provide needed nutrients.** Man cannot live by bread alone. The validity of this statement is generally recognized. Experiments and experiences give abundant evidence that animal products are far more than "empty" calories; they also provide all the essential amino acids (including lysine and methionine in which vegetable sources are deficient), minerals and vitamins, along with digestibility and palatability. This is important for there are two ways to starve to death: (1) lack of food, or (2) lack of one or more essential nutrients.

It is estimated that the average American gets the percentages of food nutrients shown in Fig. 1-11 and Table 1-3 from animal products.

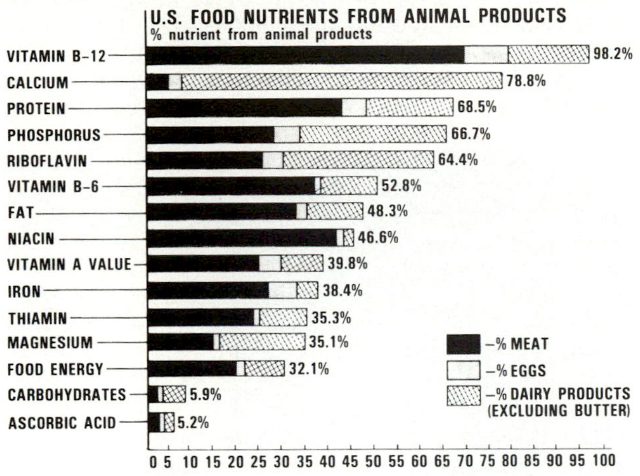

Fig. 1-11. Percentage of food nutrients contributed by animal products of the total nutrient supply in the U.S. (From: *Agricultural Statistics*, USDA)

Foods of animal origin (meat, milk, eggs, and their various by-products) are especially important in the American diet; they provide ⅔ of the total protein, ⅓ of the total energy, ⅘ of the calcium, ⅔ of the phosphorus, and significant amounts of the other minerals, and vitamins needed in the human diet. Note, too, that animal products provide practically all of the Vitamin B-12, which does not occur in plant foods—only in animal sources, and fermentation products. Also, it is noteworthy that the availability of iron in meat is twice as high as in plants.

About ⅔ of the world's protein supply is provided from plant sources, ⅓ from animal sources. (See Fig. 1-12.) The Food and Agriculture Organization of the United Nations reports that the world's diet needs animal protein in amounts equivalent to ⅓ of the total protein requirements. Thus, there should be ample animal protein, *provided* it is equally distributed. But it isn't. The people in the developed countries have 5 times as much high-quality animal protein per person as the people living in the developing countries. The gap between total protein (animal and vegetable combined) is not as wide (96.4 g vs 57.4 g per person per day in the developed and developing countries, respectively).

The most important role of animal protein is to correct the amino acid deficiencies of the cereal proteins, which supply about ⅔ of the total protein intake, and which are notably deficient in the amino acid, lysine. The latter deficiency can also be filled by soybean meal, fish, protein concentrates and isolates, synthetic lysine, or high-lysine corn. But such products have neither the natural balance in amino acids nor the appetite appeal of animal protein.

TABLE 1-3
FOOD NUTRIENTS: PERCENTAGE OF TOTAL CONTRIBUTED BY LIVESTOCK AND POULTRY PRODUCTS[1]

	Food Energy	Protein	Fat	Carbohydrates	Calcium	Phosphorus	Iron	Magnesium	Vitamin A Value	Thiamin	Riboflavin	Niacin	Vitamin B-6	Vitamin B-12	Ascorbic Acid
	(%)	(%)	(%)	(%)	(%)	(%)	(%)	(%)	(%)	(%)	(%)	(%)	(%)	(%)	(%)
Meat, fish & poultry	20.1	42.4	34.0	0.1	4.1	27.9	30.7	13.7	21.0	26.0	22.2	45.2	40.0	70.0	1.9
Eggs	1.8	4.8	2.6	0.1	2.3	5.1	5.1	1.2	5.6	1.9	4.9	0.1	2.0	8.5	0
Dairy products, excluding butter	10.2	21.3	11.7	5.7	72.4	33.7	2.6	20.2	13.2	7.4	37.3	1.3	10.8	19.7	3.3
Total	32.1	68.5	48.3	5.9	78.8	66.7	38.4	35.1	39.8	35.3	64.4	46.6	52.8	98.2	5.2

[1]*Agricultural Statistics*, USDA.

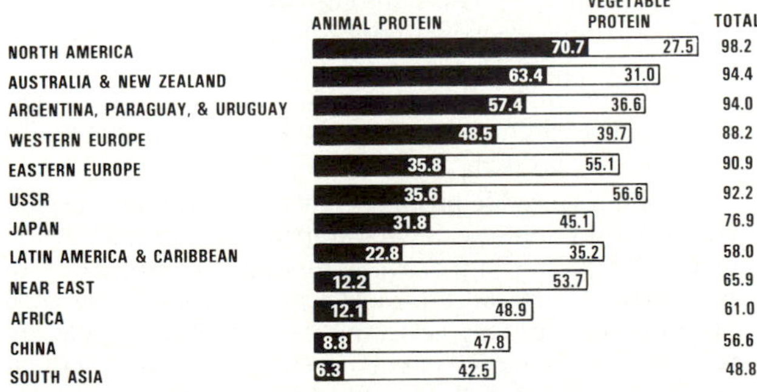

Fig. 1-12. Average grams protein consumption per person per day, with a breakdown into animal and vegetable protein, by geographic areas and countries. (From: *Ceres*, FAO/UN, Vol. 8, No. 3)

As soon as people get enough calories—as they achieve higher incomes, as they approach affluency—they start turning away from a starch-oriented diet to one based on animal protein. This has happened in the United States, Canada, New Zealand, Sweden, and Japan. The affluent do not necessarily eat more animal protein products for nutritional reasons. Rather, they consume more meat, milk, eggs, and fish because they like them—and because they derive a rich enjoyment and satisfaction from them.

3. **Animals produce protein of higher value than plants.** Proteins from animal sources (meat, milk, and eggs) have a higher value than proteins from plant sources because they have every amino acid needed for growth, including lysine, tryptophan, and methionine, which are deficient in vegetable sources. Also, they are an excellent source of zinc,

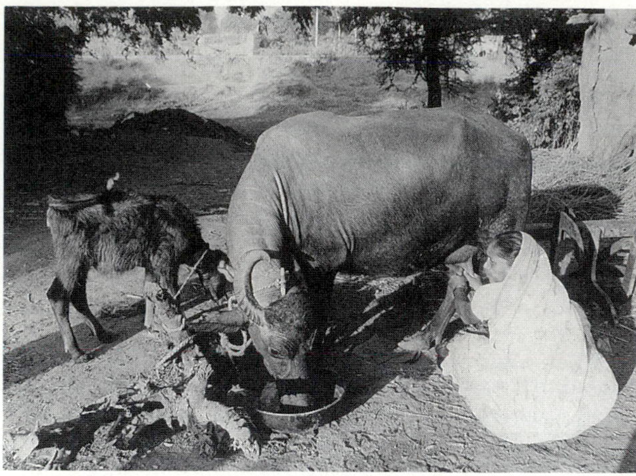

Fig. 1-13. A woman milking a water buffalo in India. Because of the large proportion of vegetarians in India (35 to 40%), milk is by far the nation's most important animal protein food—and a source of essential amino acids. More than 50% of the milk produced in India is buffalo milk. In comparison with cow's milk, buffalo milk is higher in fat content (7.5% vs 4%) and sells at a higher price. (Courtesy, FAO, Rome, Italy)

iron, and many other trace minerals. Thus, animals improve protein quality by converting lower quality plant protein to high-quality, balanced animal protein.

The high value of animal protein becomes apparent when it is realized that one 3½-oz serving of cooked, lean beef is equal to 14 oz of cooked dried beans.

Sometimes people are prone to compare cereal grains and animal proteins, pound for pound, without considering the digestibility of each. The fallacy of such reasoning is pointed up in the example that follows.

A ton of corn (dent, No. 2) contains 8.9% cereal protein, or 178 lb (2,000 × .089 = 178) of available protein. That same ton of corn will finish a 700-lb milk-fed and grass-raised steer to Standard grade and weight of 1,100 lb. Following slaughter, this steer will yield 660 lb of carcass containing 19.4% beef protein, or 128 lb of available protein. Offhand, this appears to favor people eating grain directly—178 vs 128 lb, or 50 lb more protein.

But it's net protein utilization (which is a measurement of both digestibility and biological value of a protein) by the human body that counts. Hence, one further step is necessary.

In the net protein utilization (NPU) scale, eggs rank at the top, with a score of 94%, followed by cows' milk (whole) at 82%. Beef protein has a score of 73%, whereas corn scores only 53%.[4] This means that a much higher proportion of the beef protein than of the corn protein is assimilated by the human body.

Thus, 178 lb (the available protein in a ton of corn) × 53 (net protein utilization of corn) = 94.34 lb of protein actually available to humans who eat the cereal (corn). By using the same arithmetic, the beef story is: 128 lb (the available protein in the beef carcass) × 73 (net protein utilization of beef) = 93.44 lb of protein actually available to people who eat the beef. That makes it a standoff—less than 1 lb difference in net protein between consuming the corn directly or putting it through a steer. But, given the choice, most people would rather eat beef.

4. **Ruminants convert nonprotein nitrogen to protein.** Ruminant animals, by their ability to fix nitrogen through bacterial action, can use nonprotein nitrogen, like urea, to produce protein for humans in the form of meat and milk (see Fig. 1-14). This fact should be recognized when comparing the feed efficiency of ruminants and nonruminants.

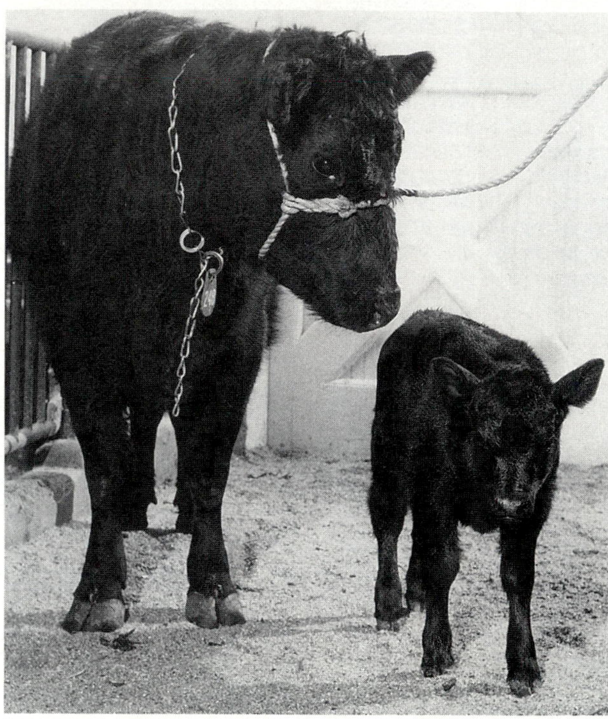

Fig. 1-14. This cow is believed to be the first in the world to have grown, conceived, and given birth to a healthy calf when fed since weaning (7 months) on a protein-free diet that contained urea as the only source of nitrogen. The cow weighed 930 lb and the calf 61 lb at the time of birth. (Courtesy, Robert R. Oltjen, U.S. Meat Animal Research Center, Clay Center, Nebr.)

5. **Animals step up the protein content and quality of foods.** Grains, such as corn, are much lower in protein content in cereal form than after conversion into meat, milk, and eggs. On a dry basis, the protein contents of selected products are corn, 10.45%; beef (Choice grade, total edible, trimmed to retail level, raw), 30.7%; milk, 26.4%; and eggs, 47.0%.[5] Also, animals increase the quality (*i.e.,* biological value) of the protein.

6. **Animals provide products that meet consumer preferences.** Most people who can afford to do so eat a portion of their food in the form of livestock products simply because of preference—because they like them. Thus, when they have the money, they consume meat, milk, and eggs. In turn, when price ratios between grain and livestock products are favorable, more grain is fed to more animals.

[4]Briggs, G. M. and D. H. Calloway, *Nutrition and Physical Fitness,* 1984, p. 80, Table 4-2, from FAO Nutrition Studies.

[5]*Composition of Foods,* Ag. Hdbk. No. 8, Agricultural Research Service, USDA.

7. **Much of the world's land is not cultivatable.** More land throughout the world can, and will be, brought under cultivation and used for crop production. But, like the western range of the United States, vast acreages throughout the world—including arid and semiarid grazing lands; and brush, forest, cutover, and swamplands—are unsuited to the production of bread grains or any other type of farming; their highest and best use is, and will remain, for grazing and forest.

Fig. 1-15. Texas Longhorns in Wyoming. Vast areas throughout the world, such as this rough terrain, are not suited to cultivation. Hence, their only use is for grazing or forest. (Courtesy, Darol Dickinson, Dickinson Ranch, Calhan, CO 80808)

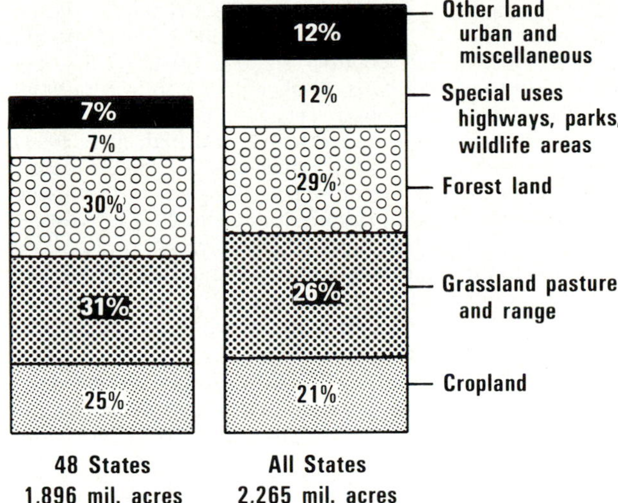

Fig. 1-16. Major uses of land. (From: Ag. Hdbk. No. 652, USDA, p. 16, Chart 41)

Despite grains being relatively plentiful in the United States, forages provide the bulk of animal feeds; pastures and other roughages account for 93.8% of the total feed of sheep and goats, 84.5% of the feed of beef cattle, 58.7% of the feed of dairy cattle, and 61.7% of the feed of all livestock.[6]

Even feedlot cattle consume relatively little grain in total. Generally speaking, feeder cattle, raised on milk and grass and that are to be grain fed, are put into the feedlot at weights of around 600 to 700 lb, to be fed to weights of about 1,050 lb. This means that they attain 60 to 65% of their weight gain before entering the feedlot. In the feedlot, it takes 9 lb of feed to make 1 lb of gain, with 6 lb of this

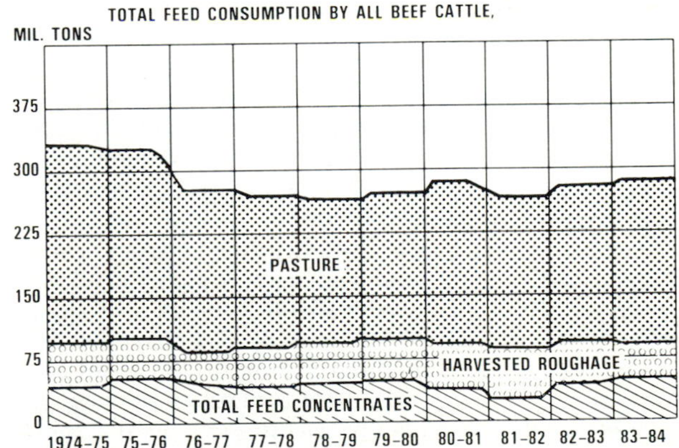

Fig. 1-17. Forages provide most of the feed of beef cattle. (Provided to the authors by the USDA)

The enormous productivity of the U.S. grasslands becomes apparent from the following figures: Every 23.8 lb of usable forage (grass, shrubs, and other plants) eaten by a ewe-lamb combination will produce about 1 lb of lamb; every 14.3 lb of usable forage eaten by a cow-calf combination will yield about 1 lb of calf; and every 10 lb of forage eaten by a calf will produce about 1 lb of calf.

8. **Forages provide most of the feed for livestock.** Pastures and other roughages—feeds not suitable for human consumption—provide most of the feed for livestock, especially for ruminants, throughout the world. Fortunately, the uniqueness of the ruminant's stomach permits it to consume forages, and, through bacterial synthesis, to convert such inedible (to humans) roughages into high-quality proteins—meat and milk. Hence, cattle and sheep manufacture human food from nonedible forage crops. Additionally, they serve as the primary means of storing (on the hoof, without refrigeration) such forage from one season to the next.

[6]Unpublished data provided to the authors by Commodity Economics Division, Economic Research Service, USDA.

consisting of grains and by-product feeds and 3 lb of roughage. Assuming a feeding period of 140 days and a gain of 450 lb in the lot, the total market weight (1,050 lb) would represent 2.57 lb of feed grain expended for each pound of gain (450 × 6 = 2,700; then, 2,700 ÷ 1,050 = 2.57). So, on a birth-to-market basis, it takes only 2 to 3 lb of grain per pound of weight gained. Less grain is consumed during those times when grains are scarce and high in price, at which times cattle are grazed longer, then kept in the feedlot a shorter time. For example, had the steer in the above example been kept on pasture longer, had he been short fed for 90 to 100 days (instead of 140 days), and had he been fed to the same weight, but marketed at Select grade instead of Choice, each pound of on-foot weight would have required only about 1.8 lb of grain, which is comparable to the feed efficiency of broilers.

A Choice steer weighing 1,050 lb on foot will produce 454 lb of salable beef (processed, cut, and trimmed). As noted above, this steer can be finished on 2,700 lb of grain. This means that it requires slightly less than 6 lb of grain to produce a pound of beef for the retail meat counter (2,700 ÷ 454 = 5.95).

Of course, not all beef is grain fed; about 40% of it is strictly grass fed.

9. **Food and feed grains are not synonymous.** Animals do not compete to any appreciable extent with the hungry people of the world for food grains, such as rice or wheat. Instead, they eat feed grains and by-product feeds—such as field corn, grain sorghum, barley, oats, milling by-products, distillery wastes, and fruit and vegetable wastes—for which there is little or no demand for human use in most countries, plus forages and grasses—fibrous stuff that people cannot eat. For example, in the United States only 3% of the corn—the major animal feed grain—is used for human food. Also, it is noteworthy that the feed grains which the United States ships overseas are used almost entirely for livestock and poultry production abroad.

10. **Ruminants utilize low-quality roughages.** Cattle, sheep, and goats efficiently utilize large quantities of coarse, high-cellulose roughages, including crop residues, straw, and coarse low-grade hays. Such products are indigestible by humans, but from 30 to 80% of the cellulose material is digested by ruminants.

Of all U.S. crop residues, the residue of corn (cornstalks and husklage) is produced in greatest abundance and offers the greatest potential for expansion in cow numbers. In 1985, 75,134,000 acres of corn, yielding 118.0 bushels per acre, were harvested in this country.

For the most part, over and above the grain, approximately 2¾ tons of dry matter produced per acre (40 to 50% of the energy value of the total corn plant) were left to rot in the field. That was 207 million tons of potential cow feed wasted, enough to winter 104 million dry pregnant cows consuming an average of 33 lb of corn refuse per head per day during a 4-month period. Mature cows are physiologically well adapted to utilizing such roughage. Moreover, when corn residue is used to the maximum as cow feed, acreage which would otherwise be used to pasture the herd is liberated to produce more corn and other crops. Also, there are many other crop residues which, if properly utilized, could increase the 104 million head figure given above.

Fig. 1-18. Cattle can utilize efficiently large quantities of coarse, humanly inedible roughages, like cornstalks. This shows cows feeding on corn residue which had been harvested by mechanical means. (Courtesy, Iowa State University, Ames)

11. **Animals utilize by-products.** Animals provide a practical outlet for a host of by-product feeds derived from plants and animals, which are not suited for human consumption. Some of these residues (or wastes) have been used for animal feeds for so long, and so extensively, that they are commonly classed as feed ingredients, along with such things as the cereal grains, without reference to their by-product origin. Most of these processing residues have little or no value as a source of nutrients for human consumption. Among such by-products are corncobs, cottonseed hulls, gin trash, oilseed meals, beet pulp, citrus pulp, molasses (cane, beet, citrus, and wood), wood by-products, rice bran and hulls, wheat milling by-products, and fruit, nut, and vegetable refuse. It is estimated that each year ruminants convert more than 9 million tons of by-products into human food.

Fig. 1-19. Chinese hogs, Kwang Tung Province, in China. Their ration consisted of two by-products—rice millfeed and bagasse (the pith of sugarcane), along with water hyacinth—all of which the pigs ate with relish. In China, swine utilize millions of tons of otherwise wasted crop residues and by-products. (Photo by Audrey Ensminger)

12. **Animals provide elasticity and stability to grain production.** Livestock feeding provides a large and flexible outlet for the year-to-year changes in grain supplies. When there is a large production of grain, more can be fed to livestock, with the animals carried to heavier weights and higher finish. On the other hand, when grain supplies are low, herds and flocks can be maintained by reducing the grain that is fed and by increasing the grasses and roughages in the ration. Thus, when grains are in short supply, fewer slaughter cattle are grain fed—more are grass finished. In the years ahead, depending on future grain supplies and prices, it is predicted that less than 55% of the U.S. domestic beef supply will come from feedlot cattle, in comparison with the 61% of U.S. slaughter cattle that were grain fed in 1982. Also, during periods of high-priced grains, heavier feeder cattle will go into feedlots, and they will be fed for a shorter period on less grain and more roughage than when grains are more abundant and cheaper.

In the future, animals will increasingly be "roughage burners," with the proportion of grain to roughage determined by grain supplies and prices.

Producers of beef cattle, dairy cattle, and sheep will more and more rely upon the ability of the ruminant to convert coarse forage, grass, and by-product feeds, along with a minimum of grain, into palatable and nutritious food for human consumption, thereby competing less for humanly edible grains. The longtime trend in animal feeding will be back to roughages; increasingly, flesh will be grass.

13. **Animals provide medicinal and other products.** Animals are not processed for meat alone. They are the source of hundreds of important by-products, including some 100 medicines such as insulin, epinephrine, thyroxin, estrogen, cortisone, and ACTH, along with a multitude of products used in making everything from candles to cosmetics; without which the health and life-style of many people would be altered.

14. **Animals are an effective method of food storage.** In many countries, there are no facilities for storing or transporting crops. Animals may be fed crops in productive years, store food nutrients until needed, and transport themselves to market.

15. **Animals maintain soil fertility.** Animals provide manure for the fields, a fact which was often forgotten during the era when chemical fertilizers were relatively abundant and cheap. One ton of average manure contains about 10 lb of nitrogen, 5 lb of P_2O_5, and 10 lb of K_2O. On the assumption that nitrogen retails at 25¢/lb, P_2O_5 at 20¢/lb, and K_2O at 10¢/lb, then a ton of manure is worth about $4.50/ton.

The energy crisis prompted concern that farmers would not have sufficient chemical fertilizers at reasonable prices in the years ahead. Since nitrogenous fertilizers are oil- and petroleum-based, there is cause for concern. As a result, a growing number of American farmers are returning to organic farming; they are using more manure—the unwanted barnyard centerpiece of the past 30 years, and they are discovering that they are just as good reapers of the land and far better stewards of the soil.

It is noteworthy that China has kept its soils productive for thousands of years, primarily through the use of night soil (human waste) and every other kind of manure, applied to the land in primitive, but effective, fashion. Every Chinese

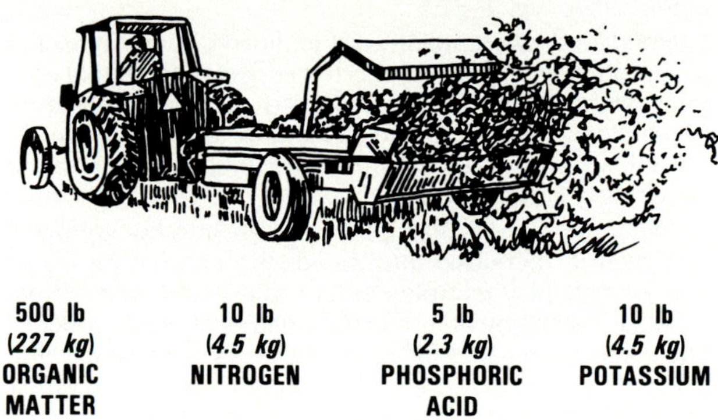

Fig. 1-20. The contents of 1 ton of average manure.

peasant recites the following teaching: "The more pigs, the more manure; and the more manure, the more grain." Indeed, animal manure is very valuable in China; it is carefully conserved and added to the land. Manure is used as a way in which to increase yields of farmland already under cultivation.

WORLD WITHOUT END—WITH ANIMALS

World population, which is increasing by 76 million per year, is outrunning acres of cropland (see Fig. 1-21). So, food supplies have become more dependent upon increased productivity. But just as Middle East oil is being depleted, so too are soils. Reversing this alarming trend calls for an increased animal agriculture worldwide.

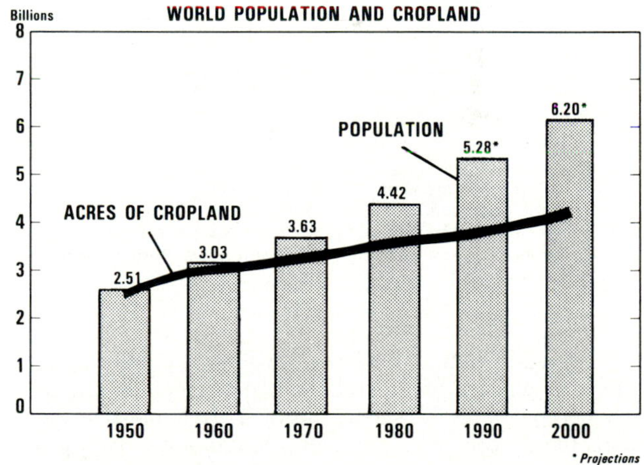

Fig. 1-21. World population is outrunning available cropland. This shows the cropland/population relationship since 1950, along with projections for the year 2000. (World population figures from Worldwatch Institute. World cropland figures from *World Agriculture Outlook and Situation Report,* USDA, Economic Research Service, WAS-36, cover page, June 1984)

World population reached 5 billion in 1987. Population growth during the 1990s is expected to be below the rate of the previous decades—around 1.6% per year, compared to 1.7% in the 1980s, and 1.9% in the 1950s and 1960s. In the year 2000, world population is expected to reach 6.2 billion. (See Fig. 1-21.)

World food consumption is determined by population and the amount eaten per person. It is expected to double over the next three decades, led by greater per capita consumption linked to rising incomes, changing tastes—preference for animal products, and improved food supplies in the developing countries.

Practicality dictates that in the years ahead a hungry world will meet its increased food needs through having plants and animals play complementary roles—and with animal products complementing the deficiencies of plant products. The virtue, even necessity, of using the plant-animal relationship is illustrated in Fig. 1-22. As shown, crops vary in their return of captured solar energy per unit of cultural energy input. Grazing land is highly efficient in the capture of solar energy—requiring little energy for a high return. Hay and silage rank second in energy return, followed by feed grains and oil crops. For the most part, however, these efficient capturers of solar energy do not store the energy in a form available to humans. It follows that ruminants which can utilize grazing land, hay, and silage (not suitable for human consumption) and convert them into meat and milk, are essential.

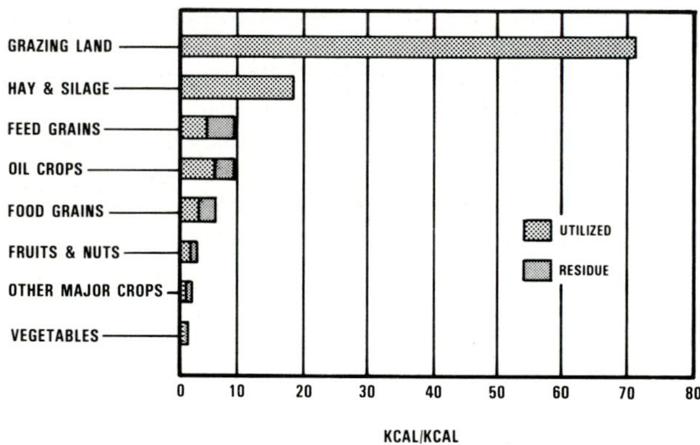

Fig. 1-22. Energy output per unit of cultural energy input (kcal/kcal) for production of food, feed, and fiber crops. (Adapted by the authors from *American Society of Agricultural Engineers,* St. Joseph, Mich., paper No. 75-7505, p. 10, Fig. 5, prepared by L. F. Nelson, W. C. Burrows, and F. C. Stickler, Deere & Company, Moline, Ill.)

For a world without end, the developing countries need a massive infusion of research, technology, and education—self-help programs, with emphasis on a plant-animal relationship. Other approaches serve only to prolong and aggravate the current disparities.

Specific methods for improving the world food situation include the following:

1. Curb population growth.
2. Increase farm prices and profits.
3. Bring more arable land under cultivation.
4. Develop more irrigation.
5. Increase crop yields.
6. Improve pastures and ranges.
7. Feed more roughage and less grain.
8. Produce leaner meats.
9. Develop more efficient animals.
10. Control diseases and parasites.
11. Improve and increase protein sources.
12. Tap the sea for more food.
13. Increase fish farming.
14. Conserve energy.
15. Control pollution.
16. Establish grain reserves.
17. Lessen food waste caused by pests.
18. Increase scientific exchange between countries.
19. Increase research, education, and extension.

QUESTIONS FOR STUDY AND DISCUSSION

1. What is the difference between the art and the science of nutrition?

2. Why is it important that nutritionists meet the feed and nutritional needs of each of the following, now and in the future?
   a. The needs of both great and small operators.
   b. The needs of animals bred for higher production.
   c. The needs of animals in confinement.
   d. The need for the maximum utilization of forages and by-product feeds.
   e. The need for greater efficiency in feeds and nutrition.
   f. The need to control pollution.

3. List the essential nutrients, and give the percentage of each, contributed to the American diet by animal products.

4. Define photosynthesis, and explain the process.

5. Why is photosynthesis classed as the most vital of all chemical reactions on earth?

6. Discuss the efficiency of photosynthesis from the standpoint of capturing the potentially available energy of the sun.

7. What is the most practical approach for converting a greater proportion of the total energy of plants (the other 95%) into a form available to humans?

8. Discuss the potential for solving the world food problems of the future through manipulating plants for increased solar energy conversion.

9. What is a ruminant?

10. Why and how are ruminants so important in lessening world food shortages, hunger, and malnutrition?

11. Why and how has modern food production become so inefficient in energy utilization?

12. How may we conserve energy?

13. Who was Thomas Robert Malthus? What, and when, did he prophesy relative to world population and food?

14. Why is it important that livestock producers answer by more than simple denial the charge that the world goes hungry because of animals?

15. In the developed countries, little more than one-third of the calories come from direct consumption of cereals, compared with 62% in the developing countries. Why this difference?

16. List and discuss the factors favoring direct human grain consumption.

17. Table 1-2 shows that on a calorie or protein conversion basis it is not efficient to feed grain to animals and then consume the livestock products. Evaluate this table.

18. Beef and lamb are the least efficient of all animals in (a) feed conversion, (b) energy conversion, and (c) protein conversion. Why not eliminate them entirely?

19. List and discuss the factors favoring an animal agriculture sharing grain with animals.

20. Why retain so much animal power in the developing countries, rather than go entirely to mechanical power?

21. What nutrients can best be obtained from animal proteins, rather than from plant proteins?

22. Fig. 1-14 pictures a cow and a calf that were produced on a protein-free diet. What's unique and significant about this?

23. World population is outrunning world cropland. What can be done to check or lessen this situation?

24. How and why should plants and animals play complementary roles in lessening world food shortages, hunger and malnutrition in the years ahead?

25. Rank crops in their return of captured solar energy per unit of cultural energy input.

26. How can world food shortages be lessened through increased research, technology, and education?

27. Will world food production be adequate to meet world demand to the year 2000 A.D.?

28. What solution do you propose for the world food problem?

SELECTED REFERENCES

Title of Publication	Author(s)	Publisher
Agricultural Commodity Projections, 1970-1980, Vols. 1 and 2		Food and Agriculture Organization of the United Nations, Rome, Italy, 1971
Animal Agriculture and the World's Food Supply	R. W. Phillips	Institute of Agriculture, University of Minnesota, St. Paul, Minn.
Animal Health Yearbook, 1974		Food and Agriculture Organization of the United Nations, Rome, Italy, 1975
Animal Production and World Food Needs, Spec. Pub. 12	H. DeGraff *et al.*	College of Agriculture, University of Illinois, Urbana, Ill., 1968
Animal Products in the Future Food Supply	J. E. Oldfield	*National Forum* (the Phi Kappa Phi Journal), 1979, Vol. 69, p. 21
By Bread Alone	L. R. Brown E. P. Eckholm	Praeger Publishers, New York, N.Y., 1975
Effect of Feeding Ruminants Non-Protein Nitrogen as the Only Nitrogen Source	R. J. Oltjen	*Journal of Animal Science,* 1969, Vol. 28, p. 673
Energy and Protein Requirements: Report of a Joint FAO/WHO Ad Hoc Expert Committee		Food and Agriculture Organization, and World Health Organization, United Nations, Rome, Italy, 1973
FAO Commodity Review and Outlook 1974–1975		Food and Agriculture Organization of the United Nations, Rome, Italy, 1975
Food and Animals—a global perspective	M. E. Ensminger	Agriservices Foundation, Clovis, Calif., 1975
Food and Nutrition, Vol. 1, No. 1		Food and Agriculture Organization of the United Nations, Rome, Italy, 1975
Foods From Animals	G. C. Smith, Chairman Task Force	Council for Agricultural Science and Technology, Ames, Iowa, Report No. 82, March 1980
Handbook on Human Nutritional Requirements	R. Passmore B. M. Nicol M. N. Rao	Food and Agriculture Organization of the United Nations, Rome, Italy, 1974
Man, Food, and Nutrition	Ed. by M. Rechcigl, Jr.	CRC Press, Cleveland, Ohio, 1975
Production Yearbook 1974, Vols. 28–1 and 28–2		Food and Agriculture Organization of the United Nations, Rome, Italy, 1975
Role of Animals in the World Food Situation, The	Staff	The Rockefeller Foundation, New York, N.Y., 1975

(Continued)

(Continued)

Title of Publication	Author(s)	Publisher
Role of Livestock in Food Production, The	T. C. Byerly	*Journal of Animal Science,* 1966, Vol. 25, No. 2, p. 552
Ruminants as Food Producers	J. E. Oldfield *et al.*	Council for Agricultural Science and Technology, Ames, Iowa, Publication No. 4, 1975
Science, Vol. 188, No. 4188		American Association for the Advancement of Science, Washington, D.C., 1975
State of Food and Agriculture, The: 1973 and 1974		Food and Agriculture Organization of the United Nations, Rome, Italy, 1973 and 1974
Statistical Yearbook 1973, 25th Edition		United Nations, New York, N.Y., 1974
Strategy for the Conquest of Hunger: Proceedings of a Symposium		The Rockefeller Foundation, New York, N.Y., 1968
Strategy for Plenty, A		Food and Agriculture Organization of the United Nations, Rome, Italy, 1970
United Nations World Food Conference: Assessment of the World Food Situation, Present and Future		United Nations, Rome, Italy, 1974
United Nations World Food Conference: The World Food Problem; Proposals for National and International Action		United Nations, Rome, Italy, 1974
Working Papers: Conference on International Development Strategies for the Sahel		The Rockefeller Foundation, New York, N.Y., 1975
Working Papers: Food Production and the Energy Dilemma	R. W. Cummings, Jr.	The Rockefeller Foundation, New York, N.Y., 1974
Working Papers: Perspectives on Aquaculture		The Rockefeller Foundation, New York, N.Y., 1974
World Animal Review, No. 4		Food and Agriculture Organization of the United Nations, Rome, Italy, 1972
World Food Production, Demand, and Trade	L. L. Blakeslee E. O. Heady C. F. Framingham	Iowa State University Press, Ames, Iowa, 1973
World Food Situation and Prospects to 1985, The		Economic Research Service, USDA, Washington, D.C., 1974
Yearbook of International Trade Statistics 1972-1973		United Nations, New York, N.Y., 1974

Fig. 1-23. The dairy cow, known as the foster mother of the human race, converts animal feed to human food. These Jersey cows are manufacturing milk from grass. (Courtesy, the American Jersey Cattle Club, Columbus, Ohio)

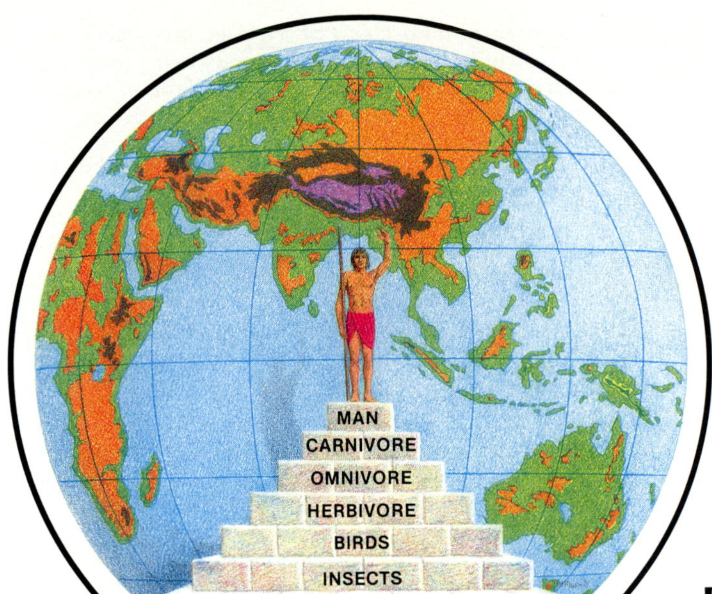

The Biotic Pyramid

Original painting by Tom Phillips

PRINCIPLES OF NUTRITION

Contents *Page*

A massive purebred bull standing belly deep in straw with a manger full of feed in front of him is the result of two forces—heredity and environment. If turned out on the range, an identical twin to the placid bull would present an entirely different appearance. By the same token, optimum nutrition could never make a champion out of a bull with scrub ancestry. But it might well be added that "fat and hair will cover up a multitude of sins." Thus, animals inherit certain genetic possibilites, but how well these potentialities develop depends upon the environment to which they are subjected; and the most important influence in the environment is the nutrition. In turn, all feed comes directly or indirectly from plants which have their tops in the sun and their roots in the soil. Hence, we have the nutrition cycle as a whole—from the sun and soil, through the plant, thence to the animal, and back to the soil again.

From the above, it may be concluded that the terms *nutrition* and *nutritionist* involve more than just feeding. *Nutrition is the science of feeds as they relate to the requirements, production, and health of animals. Nutritionist is a person who is trained in and able to apply knowledge of feeds to the requirements, production, and health of animals.* Nutrition begins with a knowledge of the fertility of the soil and the composition of plants; and it includes the ingestion of feed, the liberation of energy, the elimination of wastes, and all the syntheses essential for maintenance, growth, reproduction, egg production, lactation, fattening (fitting), wool and mohair production, and/or work (running).

A good understanding of nutrition is important because animals and humans are dependent upon food nutrients for the processes of life. This chapter will be limited to an elucidation of the principles of nutrition, including an historical perspective, body composition, classification of nutrients, and functions of nutrients. Separate chapters are devoted to digestion and absorption; nutrients—metabolism; and nutritional diseases—toxins.

PERSPECTIVE OF NUTRITION

Of the basic needs of animals and humans, none is more important than food. The ingestive behavior in mammals begins at birth—with suckling. All animals are impelled by hunger and their desire to survive. Starving animals and people will prey upon each other—as cannibals do.

The desire of primitive people to safeguard their food supply for times when hunting was poor and to have their food close at hand caused them to plant seeds near the campsite, to store roots and grains for winter use, and to tame and confine (domesticate) animals.

The primary purpose of keeping animals is to transform feeds into substances usable by humans—meat, milk, eggs, wool, mohair, and work (running). But the conversion of feed to these uses must be done efficiently and economically. To do this, the principles of nutrition must be applied; and they must be augmented by superior breeding, good health, and competent management.

Like other sciences, nutrition does not stand alone. It draws heavily on the basic findings of chemistry, biochemistry, physics, microbiology, physiology, medicine, genetics, mathematics, endocrinology, and, most recently, animal behavior and cellular biology. In turn, it also contributes richly to each of these fields of scientific investigation.

It is noteworthy that much of our knowledge of human nutrition came the animal route. Human nutritionists often use animal subjects for nutrition studies, because people usually prove too time-consuming, too costly, too difficult to control, and perhaps undesirable from a medical standpoint.

The sections that follow are devoted to the historical. In nutrition, more than in any other science, history is most important. In the final analysis, animals and food are not only inseparable from history—they're part of it. Without them, there would be no history—no humankind.

MILESTONES IN NUTRITION

For thousands of years, the quest for food has shaped the course of history. It dictated population growth, urban expansion, and migration and settlement of new lands. It profoundly influenced economic, social, and political theory. It prompted early sailing and the discovery of new worlds, widened the horizons of commerce, caused wars of dominion, and played no small role in the creation of empires.

Food has played an important role in many things. In religion, it helped to define the separateness of one creed from another by means of dietary taboos. In science, the prehistoric cook's discoveries of the effect of heat applied to raw materials laid the foundations on which much of early chemistry was based. In technology, the waterwheel, first used in the milling of grain, achieved immense industrial importance. In wars, battles were sometimes postponed until the harvest had been gathered in, and the well-fed armies usually defeated the hungry ones. Without food, medicine as we know it today might never have been born, for until early in the 18th century health treatments were largely based on dietary theories.

Obviously, all the significant milestones in a subject as broad as nutrition cannot be presented in this chapter because of space limitations. Obviously, too, the milestones are a progressive series of developments, rather than unrelated individual discoveries. In reality, they are overlapping eras. The intention in the sections that follow is to reveal (1) the general development of nutrition as a field of scientific study, and (2) the primary research emphasis that characterized each era. To this end, the history of nutrition has been aptly divided into eight eras: (1) naturalistic era, (2) chemical-energy era, (3) mineral era, (4) vitamin-biological era, (5) feeding standard era, (6) amino acid era, (7) feed additive and implant era, and (8) biotechnology era.

Naturalistic Era

The naturalistic, or prescientific, era of nutrition was characterized by a fascinating maze of sage philosophy, taboos, bizarre superstitions, and religious precepts.

Even today, meat for the Jewish trade—known as *kosher meat*—is slaughtered and processed according to ancient biblical laws, called *Kashruth,* dating from the days of Moses, more than 3,000 years ago. The Hebrew religion holds that God issued these instructions directly to Moses, who, in turn, transmitted them to the Jewish people while they were wandering in the wilderness near Mount Sinai.

Prior to the 18th century, little of truly scientific nature was accomplished in the field of nutrition. Although the ancient Greek philosophers were interested in science, logical reasoning—rather than experimentation—was the Greek way.

Hippocrates (460 to 357 B.C.), known as the *Father of Medicine,* was the first great physician to indicate an interest in nutrition. The following are among his famous aphorisms: *"Children produce more heat and need more food than adults"*; and *"Persons who are naturally very fat are apt to die earlier than those who are slender."* (The latter has nagged the consciences of many of us ever since.)

Chemical-Energy Era

The great French chemist Antoine Laurent Lavoisier (1743 to 1794) is credited with being the founder of the science of nutrition. Experimenting with guinea pigs enclosed in a chamber that he had constructed, and using the thermometer and balance, Lavoisier measured body heat loss, oxygen consumed, and carbon dioxide expired. He concluded that respiration is a combustion process similar to that occurring when substances are burned outside the body. Further, he was able to show that heat production in the animal body is directly related to oxygen consumption. Lavoisier compared animal heat to that produced by a lamp or candle.

Fig. 2-1. Lavoisier and his beauteous wife. (Courtesy, the Rockefeller Institute)

But the achievements that led to the modern era of nutrition did not bring honor to Lavoisier during his lifetime. Outside his laboratory, he was a public tax collector and landed gentry. During the French Revolution, he was convicted of collecting taxes on water contained in tobacco—a criminal offense, punishable by death. Lavoisier was guillotined at age 51. His close friend, Lagrange, who witnessed the execution, said of him, "It took but a second to cut off his head; yet, a hundred years will not suffice to produce one like it."

Despite some technical inaccuracies in Lavoisier's work and in his interpretation of the results, subsequent refinement in instrumentation and in scientific thought has added little to the basic concepts derived from his experiments. Chemistry and physiology, in which Lavoisier pioneered, served as the foundation upon which the science of nutrition was established. But almost 100 years elapsed following Lavoisier's classical studies before carbohydrates, fats, and proteins were identified as the source of energy for the animal body.

Mineral Era

The occurrence of goiter in both animals and people is as old as antiquity. In 1822, Boussingault, while traveling in South America, observed that the villagers who used salt containing iodine were free from goiter, but that those who used plain salt were afflicted. Thereupon, he advocated the use of iodine at low levels to prevent the malady.

In 1838, Berzelius, noted Swedish chemist, concluded that the iron in hemoglobin made it possible for the blood to absorb much oxygen. As a result of a series of studies initiated in 1925, Hart and his associates, at the University of Wisconsin, reported that white rats developed anemia if fed a milk diet, and that the anemia was not cured by iron salts alone. However, the ash of lettuce, which contained copper, was an effective cure. Mere traces of copper added to pure iron salts were found to be adequate for the cure of iron-deficiency anemia.

The mineral structure of bones and teeth was recognized by early workers. In 1840, Charles J. Chossat, a physician in Geneva, Switzerland, won a prize for showing that a diet of wheat and water must be supplemented with calcium for the bones of a growing pigeon to develop. In 1843, the importance of calcium in the diet was further attested by J. B. Boussingault, a Frenchman, who performed calcium-balance studies with animals.

The general belief at the beginning of the 20th century was that sodium chloride, calcium, phosphorus, and probably iron were the important minerals in animal nutrition. Little attention was given to other elements known to be present in tissues in relatively small amounts. The importance of trace elements in animal nutrition was more readily accepted following the discovery of vitamins.

Vitamin-Biological Era

Until early in the 1900s, if a ration contained proteins, fats, carbohydrates, and minerals—together with a certain amount of fiber—it was considered to be a complete diet. True enough, the disease known as beriberi, having been known to the Chinese as early as 2600 B.C., made its appearance in the rice-eating districts of the orient when milling machinery was introduced from the West; and scurvy was long known to occur among sailors fed salt meat and biscuits. However, for centuries these diseases were thought to be due to toxic substances in the digestive tract caused by pathogenic organisms, rather than food deficiencies; and more time elapsed before the discovery of vitamins.

Largely through the trial-and-error method, it was discovered that specific foods were helpful in the treatment of certain of these maladies. In 1747, Lind, a British naval surgeon, showed that the juice of citrus fruits was a cure for scurvy. In 1881, Lunin had come to the conclusion that certain foods, such as milk, contain small quantities of unknown substances essential to life, in addition to their principal ingredients. In 1897, Eijkman, a Dutch physician working in the East Indies, produced and cured at will the disease beriberi (polyneuritis) in hens, simply by changing the diet of unpolished rice to milled rice, or the reverse. At a very early date, the Chinese used a concoction rich in vitamin A as a remedy for night blindness. Also, cod-liver oil was used in treating or preventing rickets long before anything was known about the cause of the disease.

The significance of these observations relative to diet, however, was not fully appreciated until scientists found it desirable in many types of investigations to use the biological approach, with purified diets to supplement chemical analyses in measuring the value of feeds. These rations were made up of relatively pure nutrients—proteins, carbohydrates, fats, and minerals—from which the unidentified substances were largely excluded. With these purified rations, all investigators shared a common experience—the animals limited to such diets not only failed to thrive, but they even failed to survive if the investigations were continued for any length of time. At first, many investigators explained such failures on the basis of unpalatability and monotony of the rations. Finally, it was realized that these purified rations were lacking in certain substances, minute in amount, and the identity of which was unknown to science. These substances were essential for the maintenance of health and life itself and the efficient utilization of the main ingredients of the food. With these findings, a new era of science was ushered in. The modern approach to nutrition was born.

Casimir Funk, a Polish scientist working in London, discovered evidence of one such substance in the outer hulls of rice, in 1912. He called it a "vital amine" because he (1) believed it to be necessary for life, and (2) found that it was an amine, chemically. This substance, later identified as thiamin, or vitamin B-1, gave rise to the general terminology *vitamine,* later shortened to *vitamin* as we know it today.

Acceptance of the vitamin theory was not immediate, however. It remained for the independent discovery, in 1913, of vitamin A by McCollum and Davis, of the University of Wisconsin, and Osborn and Mendel, of Yale University. These researchers observed that animals which received diets containing lard instead of butterfat would cease to grow and eventually develop an inflammation of the eye. When either butterfat or cod-liver oil was used, the condition was corrected. These discoveries led to the recognition of *fat-soluble A* and *water-soluble B,* as the factors were then designated. But it was not until the decade 1930 to 1940 that the majority of the vitamins were identified, isolated from feeds, and synthesized in the laboratory.

The known existence of vitamins, therefore, dates only to 1912. Since that time, the growth of the vitamin family, the isolation and determination of many vitamins, the partial solution of the puzzle of vitamin functions in the body, the discovery of the amazing therapeutic value of minute quantities of these vitamins in the cure of deficiency diseases, the numerous determinations of feed (food) composition with respect to vitamins, and the synthesis of most of them, have had a profound effect in nutrition.

Feeding Standard Era

Feeding standards are tables showing the amounts of one or more nutrients needed by different species of animals for different purposes. They serve as guides in balancing rations and feeding practices. Most feeding standards are expressed in (1) quantities of nutrients required per day, and/or (2) percent of the ration; the first type is used where animals are provided a given amount of feed during a 24-hour period, and the second is used where animals are provided a ration without limitation on the time in which it is consumed.

The first feeding standard was developed in 1810, by Thaer, a German scientist. He took meadow hay as his standard, compared the extractable nutrients of other feeds to it, then assigned them *hay values.* In 1859, Grouven, another German scientist, made use of analyses of protein, fat, and carbohydrate to formulate the first feeding standard for farm animals. In 1864, the great German scientist, Wolff, devised a standard based on digestible nutrients obtained from feeding trials. In 1897, Lehmann, of Germany, modified Wolff's standards. Other systems followed; among them, the total digestible nutrient system by Henry, of the University of Wisconsin, published in the first edition of his book, *Feeds and Feeding,* in 1898; the starch values of Kellner, a German scientist, in 1907; the Scandinavian feed unit system (Woll, 1912); the dairy cow standards of Haecker, of Minnesota, in 1914; the net energy values of Armsby, of Pennsylvania, in 1915; the productive units developed by Mollgaard, of Denmark, in 1939; and the productive energy values computed by Fraps of the Texas Station, in 1937 and 1941.

Today, the most widely used feeding standards in the United States are those published by the National Academy of Sciences. In England, similar standards are issued by the Agricultural Research Council (ARC). Other countries have similar bodies which make recommendations on the nutritive requirements of animals.

In the United States, the TDN system is gradually giving way to other energy evaluation systems, particularly net energy. England uses metabolizable energy (ME), adjusted according to the efficiency with which a feedstuff or diet is used for a particular purpose. Other European standards are based on starch equivalents, Scandinavian feed units, and other methods of evaluation.

Amino Acid Era

Among the more recent outstanding achievements in nutrition was the identification of the essential amino acids by William C. Rose and co-workers of the University of

Illinois. In 1930, they initiated a brilliant series of studies, using a new technique, out of which evolved specific information relative to the amino acids that must be present in feed. By the use of diets otherwise designed to be adequate for the normal growth of rats, in which the sole source of nitrogen was supplied by amino acids, the effect of the addition or deletion of each of the amino acids was studied. Simultaneously, the doctoral students conducting these investigations confirmed the results on themselves. They tested each amino acid at different levels and determined the amount needed for optimal utilization of dietary protein. As a result of these investigations, the Illinois workers were able to classify 10 amino acids as essential dietary constituents and the others as nonessential.

Feed Additive and Implant Era

In the 1950s and 1960s, a new era of livestock production was ushered in when it was found that antibiotics and stilbestrol would increase both rate and efficiency of gain of animals. They were hailed as the "wonder drugs" of our time. Soon the race was on, and other feed additives and implants followed.

In 1949, Jukes, of Lederle Laboratories, found that antibiotics were something new to be added to livestock feeds—that they would improve growth rate and feed efficiency.

In 1952, Burroughs, of Iowa State University, reported that the feeding of diethylstilbestrol (DES) would lower feed usage and increase weight gains of finishing cattle. In 1954, stilbestrol was approved by the Food and Drug Administration (FDA) for use in cattle finishing rations; and in 1956, FDA approved the use of stilbestrol implants for finishing steers. (**NOTE:** In 1948, Andrews of Purdue University reported on the effect of implanted stilbestrol, but no practical application of the finding was made at the time.)

Today, more than 1,000 drugs are approved by FDA for use by livestock and poultry producers, and 80% of all animals raised for food receive some animal drugs during their lifetime. When used properly, these drugs enable livestock producers to provide safe and wholesome meat, eggs, and milk to consumers at lower costs than would otherwise be possible.

Among the newer additives and implants now being tested or used are various antimicrobials (including antibiotics, antibacterial agents, and antifungal agents), hormones or hormonelike substances, enzymes, and ionophores. Additional products are evolving, some of them made commercially possible by recombinant DNA (gene splicing) technology.

(See Chapter 13, Feed Supplements/Additives/Implants for definitions of, and further information relative to, additives and implants.)

USE ADDITIVES AND IMPLANTS SAFELY

Since the beginning of recorded history, people have been concerned about the purity and safety of their food and drink. In 1202, King John of England proclaimed the first English food law, the Assize of Bread, which prohibited adulteration of bread with such ingredients as ground peas or beans. Regulation of food in the United States dates from 1784. Federal controls over the drug supply started in 1848.

As the list of additives and implants grows, so also do the concerns of consumers, augmented by government regulatory bodies and laws. In 1958, the food additive amendment, better known as the Delaney Clause, was passed by the U.S. Congress. This clause gave rise to the policy of *zero tolerance*—that no substance can be used as an additive or implant if it has been shown to produce cancer in either humans or animals.

The key to using animal drugs safely is to read and follow the label directions. Everything that a livestock producer needs to know about approved species, dosage, methods of administration, and withdrawal time is on the label.

More than half of the FDA-approved drug products require pre-slaughter withdrawal times or milk-discard periods. The FDA has found that approximately 90% of residue violations are caused by producers failing to withdraw drugs from animals soon enough. Any illegal residues found in market livestock, milk, or eggs can result in marketing delays and even lead to condemnation of the shipment.

Biotechnology Era

The development of gene-splicing (also known as recombinant DNA) ushered in a new era. On May 23, 1977, scientists at the University of California-San Francisco reported a major breakthrough as a result of altering genes—turning ordinary bacteria into factories capable of producing insulin, a valuable hormone previously extracted at slaughter from pigs, sheep, and cattle, so essential to the survival of diabetics. The feat gave rise to a major scientific revolution, called *biotechnology.*

Biotechnology is the aspect of technology concerned with the application of an array of biological and engineering tools, from gene splicing to manipulating cells, tissues, and genes so as to control certain characteristics. It involves every facet of animal production, from breeding and feeding to finished products, including the genetic makeup of animals; the feeds they eat; their digestion, physiology, stress tolerance, disease resistance, and efficiency of production; the composition, quality, and quantity of products produced; along with the production of large quantities of drugs and chemicals. Potential uses number into the hundreds, perhaps thousands—they're endless. While some aspects are decades away from commercial production, others are near, and still others are here now.

The entire food team—producers, processors, and consumers—stands to benefit from biotechnology, from the greater abundance of high-quality products produced more efficiently. However, progress will be slowed by regulatory and social problems; and some producers and processors will falter and fail along the way. Nevertheless, with the opening of the 21st century, it is predicted that the magnitude of change wrought by biotechnology will rival that of mechanization, which doubled total American agricultural production and increased output per man hour 20-fold.

Summary of Nutritional Milestones

Obviously, all of the significant milestones in a subject as broad as nutrition cannot be summarized in this section because of space limitations. Obviously, too, the milestones were a progressive series of developments, rather than unrelated individual discoveries. All researchers contributed to the kaleidoscope of knowledge that we have today, whether it was to prove or disprove a theory, or to add just one more piece to the puzzle. Neither can the dates given always be pinpointed because some research projects spanned several years; others were completed on a certain date but not reported or published until later; and still others led to the wrong conclusion, which, with further study, was changed.

A chronology of the history of nutrition is given in Table 2-1.

TABLE 2-1
SUMMARY OF NUTRITIONAL MILESTONES

Date	Nutritional Milestones
B.C. 3000	A Chinese document described goiter and recommended that afflicted people eat seaweed and burnt sponge, which are good sources of iodine.
2600	Beriberi, which affects the nervous system, was known to the Chinese as early as 2600 B.C., but its cause remained elusive for centuries.
1550	Scurvy and diabetes were described by the Egyptians on medical papyrus rolls (the first writing paper, made by the Egyptians from the papyrus plant as early as 2400 B.C.), discovered in Thebes by George Moritz Ebers, a German Egyptologist and novelist, who, in 1874, edited it in a romantic historical novel on medicine which he titled *Papyrus Ebers.*
1000	The Chinese, in their herbal medicine, used a concoction now thought to have been rich in vitamin A as a remedy for night blindness; and cod-liver oil was used in treating or preventing rickets long before anything was known about the cause of the disease.
1100 to 500	In the Old Testament, which was written over a long period, reference is made to scurvy.
460 to 364	Hippocrates, the Father of Medicine, advocated liver as a cure for night blindness, and described the symptoms of scurvy.
A.D. 1250	Jean Sire de Joinville, the French chronicler, accompanied Louis IX of France to Cyprus and Egypt. In 1309, he completed in final form the *History of Saint Louis,* an account of the Crusade, in which he told of a disease "which attacked the mouth and legs" (obviously scurvy).
1497	Vasco de Gama, a Portugese navigator, sailed around the Cape of Good Hope to establish the first European trading colony on the coast of India. One hundred of his crew of 160 men perished with scurvy on the voyage.
1535	In Canada, during the winter, Jacques Cartier, the daring explorer who laid claim to Canada for France, recorded in his log that the lives of many of his men dying of scurvy were saved when they learned from the Indians that a brew made from the tips of pine or spruce trees cured and prevented the malady. (It is now known that the brew contained vitamin C.)
1564	Ronsseus, a Dutch physician, was the first to recommend oranges as the antiscorbutic for sailors. Unfortunately, it was to be over 200 years before this addition to the diet of sailors became mandatory.
1601	Sea Captain James Lancaster recorded that lemon juice protected against scurvy. Thus, some of the sailing ships were freed from this scourge.
About 1650	Leeuwenhoek pioneered in using the microscope to describe the appearance of starch granules.
1730	A Spanish physician, Gaspar Casal, was the first to describe pellagra, which appeared soon after the introduction of corn (maize) into Europe.
1747 to 1753	James Lind, a British naval surgeon, and the first great experimental nutritionist, tested six remedies on 12 sailors who had scurvy, and found that oranges and lemons were curative. His classical studies, the results of which were published in *A Treatise of Scurvy,* 1753, are generally credited as being the first experiments to show that an essential food element can prevent a deficiency disease. But another 50 years elapsed before the British Navy required rations of lemons or limes on sailing vessels, which resulted in the sailors being nicknamed "limeys."
1771	Francesco Frapoli, an Italian physician, named the disease pellagra (*pelle,* for skin; and *agra,* for reddened and rough).
1768 to 1775	On two historic voyages, each of three years' duration, from 1768 to 1771 and from 1772 to 1775, British Captain James Cook, avoided scurvy—hitherto the scourge of long sea voyages. He had his ship stocked with concentrated slabs of thick brown vegetable soup and barrels of sauerkraut. Of the sauerkraut he said: "It is not only a wholesome vegetable food, but, in my judgment, highly antiscorbutic, and spoils not by keeping." In addition, he sent seamen ashore at every port visited to gather all sorts of fresh fruits and green vegetables (including grasses), which the crew prepared, served, and ate. As a result, not one of the crew died from scurvy.
1774	Priestly discovered *dephlogisticated air,* later renamed *oxygen.*
1783	The great French chemist, Antoine Laurent Lavoisier, the Father of Nutrition, published the resuslts of his experiments on respiratory metabolism. He used guinea pigs to measure body heat, oxygen consumed, and carbon dioxide expired, and concluded that the human body was like a little furnace using food to produce heat and energy. Further, he was able to show that heat production in the animal body is directly related to oxygen consumption. Lavoisier compared animal heat to that produced by a lamp or candle.
1791	Dr. George Fordyce of England was the first to use a control group in an experiment with chickens. He found that they required a calcareous (calcium) supplement in order to produce healthy eggs, that did not break easily in the nest.
1807	Sodium, potassium, calcium, magnesium, sulfur, and boron were all isolated by Humphry Davy.
1810	Thaer, a German scientist, developed the first feeding standard, based on "hay values." Cystic oxide (cystine) was isolated from urinary calculi by Wollaston. This heralded the discovery of the first amino acid.
1821	John Gorham of Harvard University did the first chemical analysis of an American food—he analyzed Indian corn, or maize.
1822	Jean Baptiste Boussingault, a Frenchman, while traveling in South America, observed that the villagers who used salt containing iodine were free from goiter, but that those who used plain salt were afflicted.
1824	Cod-liver oil, long known as a folk medicine, was found to be important in the treatment of rickets. But the remedy lost favor with the medical profession because physicians could not explain its action.
1836	J. B. Boussingault (1802 to 1887) founded scientific agriculture—he studied it both in the field and in the laboratory. He (1) demonstrated that some plants take up nitrogen from the air and add it to the soil; (2) studied the digestion and absorption of foods; (3) did work on the mineral requirement for bone growth; (4) developed the first method to distinguish between the nitrogen of urea and of the ammonium salts in urine; (5) analyzed a curative salt from South America and proved that it was iodine that cured goiter; (6) proved that ordinary salt, potassium, calcium, and phosphate were all indispensable nutrients; (7) was the first to use animals and birds, along with the techniques of the chemical laboratory, for experiments; (8) was a clear thinker and an enthusiastic, industrious worker; and (9) made many significant contributions to the history of nutrition.

(Continued)

TABLE 2-1 *(Continued)*

Date	Nutritional Milestones
1838	Berzelius, the noted Swedish chemist, observed that the iron in hemoglobin made it possible for the blood to absorb oxygen.
1840s	The German chemist, von Liebig, who is considered to be the Father of Organic Chemistry, and his co-workers improved the methods of food analysis and suggested the designations of *carbohydrate, fat,* and *protein* for the major organic nutrients.
1843	J. B. Boussingault performed calcium-balance studies with animals and showed the importance of calcium. A controversy developed among scientists over whether the body could convert carbohydrates to fat. Von Liebig showed that some animals produced more fat than they ingested, thus, he was convinced that the body could take carbohydrates and make fat. Finally, Boussingault and Persoz, who were originally convinced that he was wrong, proved, with repeated experiments, that von Liebig *was* correct.
1844 to 1846	N. T. Gobley isolated a substance from egg yolk, which he called *lecithin* (from the Greek, *lekithos,* meaning *egg yolk*). This substance contained glycerol, fatty acids, phosphorus, and nitrogen.
1846	Justus von Liebig produced and named the amino acid, tyrosine.
1849	Strecker, a German chemist, isolated a compound from hog bile, to which he subsequently (in 1862) applied the name *choline* (after *chole,* the Greek word for bile).
1850s	Henneberg and Stohmann, of Germany, developed the system of proximate analysis, which is still used in the analysis of feeds today.
1856	C. Bernard discovered the formation and storage of glycogen in the liver. Pasteur began his experiments that marked the beginning of the study of bacteria, and their role in the scheme of nutrition and health.
1866 to 1867	Baeger and Wurtz, working independently, determined the correct structure of choline and carried out the first synthesis of it. But the compound did not attract the attention of nutrition investigators at the time.
1867	Nicotinic acid was first discovered and named when Huber, a German chemist, prepared it from the nicotine of tobacco, but it lay idle on the shelf for another 70 years. No one suspicioned that it was a cure for pellagra.
1869	DNA (deoxyribonucleic acid) was discovered by F. Miescher.
1873	Van Lent was the first to conclude that the type of diet had something to do with the origin of beriberi. By reducing the ration of rice in the diet of the sailors in the Dutch Navy, he was able to eradicate beriberi almost entirely.
1881	Nicholas Lunin, while a student of Von Bunge at the Univeristy of Dorpat, concluded that certain foods, such as milk, contain small quantities of unknown substances essential to life. He discovered that a purified diet of protein, fat, and carbohydrate would not maintain life.
1882	Kanehiro Takaki, Director-General of the Japanese Navy, cured beriberi in sailors of the Japanese Navy by giving them less rice and more meat and milk, but he concluded that the higher protein content of the diet was responsible for the cure.
1883	J. Kjeldahl developed the method for determining the nitrogen; hence, the protein content of substances. This method is still used today.
1884	Hubl developed a method for determining the iodine number of fats. The German physiologist, Rubner, utilizing data from dietary studies of Voit, calculated the heat of combustion of carbohydrates, fats, and proteins. He corrected for the incomplete metabolism of protein in the body by subtracting the heat of combustion of the urinary product, urea.
1890	Palm, an English physician, observed that where sunshine was abundant, rickets was rare, but where the sun seldom shone, rickets was common.
1890s	Wilbur Atwater, the American nutritionist, who had worked for a while with Voit and Rubner, refined the caloric conversion factors of Rubner by correcting them for the percent losses that occurred in digestion. This required many feeding trials with animals and people in which both the foods consumed and the excretory products were analyzed.
1891	C. Voit, a pupil of von Liebig, was one of the most successful investigators of metabolism. One of his studies showed that glucose, levulose, cane sugar, and maltose were all converted into glycogen in the liver, but that lactose and galactose were not of much value.
1896	Wilbur Atwater published the first extensive table of food values. Atwater approximated that on a typical American diet each gram of carbohydrate, fat, and protein would yield 4, 9, and 4 Calories (kcal), respectively.
1897	Christiaan Eijkman, a Dutch physician assigned to a prison hospital in the East Indies, observed beriberi among the inmates and sought the answer through experiments with chickens. To save money, he fed the birds scraps, mostly polished rice, from the patients' meals. The chickens developed a bad nerve ailment, which resulted in paralysis. Later, the director of the hospital withdrew his permission to use the scraps, so Dr. Eijkman had to buy unmilled rice for the chickens. All of them promptly recovered. This led to the first experimental work that proved a nutritional deficiency. Dr. Eijkman observed that the chickens' ailment was similar to beriberi, but he erroneously attributed it to (1) too much starch from the rice, and (2) the presence of a nerve poison in the endosperm of rice, which the outer layers of unmilled rice neutralized.
1901	Dr. G. Grijns continued the work of Eijkman on beriberi, but he correctly interpreted the results as a deficiency of an essential ingredient.
1900s	Pellagra reached epidemic proportions in southern United States, where the diet was based primarily on corn, which is extremely low in both available niacin and in tryptophan. In 1915, 10,000 people died with the disease; and in 1917 to 1918, there were 200,000 cases of pellagra in this country. In 1914, the U.S. Public Health Service sent a team under the direction of Dr. Joseph Goldberger, a physician-researcher, to study the cause of, and hopefully find a cure for, pellagra. In his studies, Goldberger proved that the disease was caused by a dietary deficiency, and not an infection or toxin.
1902	Emil Fisher determined the chemical structure of amino acids, the building blocks of protein; and he determined the nature of the chemical bond—the peptide bond—holding the amino acids together.
1905	C. A. Pekelharing, a German professor, observed that milk had some unknown substance in minute quantity that would maintain life, but his discovery went largely unnoticed.
1906	Sir Frederick G. Hopkins, of Cambridge University, England, the most advanced thinker in nutrition of his time, determined by careful experiments that rats became sick and died on a purified diet; but that the addition to the diet of either (1) a small amount of milk, or (2) an alcoholic extract of milk solids or of certain dried vegetables enabled the animals to live and grow; however, the ash of either milk or vegetables was ineffective. Thus, Hopkins showed that the essential unknowns which existed in certain foods were organic in nature rather than inorganic (minerals). He called these substances *accessory food factors.*

(Continued)

TABLE 2-1 *(Continued)*

Date	Nutritional Milestones
1907	E. V. McCollum, an American scientist and one of the century's nutritional giants, tried to make purified diets more palatable, because he thought that the problem was unpalatability, but he only proved once again that a purified diet would not maintain life. Holst and Frolich, of Norway, produced scurvy experimentally in guinea pigs by feeding them a diet deficient in foods containing ascorbic acid.
1909	T. B. Osborne and L. B. Mendel, at the Connecticut Experiment Station, demonstrated, by the use of protein-free whey, that there was considerable difference between proteins, that some were complete and some were incomplete. Their work established the quantitative biological values for proteins.
1911	Casimir Funk, a Polish biochemist at the Lister Institute, London, England, isolated crystals with B-complex activity. He coined the name *vitamine* from the Latin *vita* or life, plus *amine,* and applied it to the antiberiberi substance. Later, it was shown that these substances did not contain an amine, so eventually the "e" was dropped.
1912	Funk isolated nicotinic acid from rice polishings while attempting to isolate the antiberiberi vitamin; and, that same year, Suzuki, in Japan, isolated nicotinic acid from rice bran. But both researchers lost interest in the acid when they found it ineffective in curing beriberi. So nicotinic acid was again left on the shelf unobserved.
1913	Working independently, McCollum and Davis at the University of Wisconsin, and Osborne and Mendel at the Connecticut Experiment Station, demonstrated the presence of an essential dietary substance in fatty foods, such as butterfat and egg yolk (Wisconsin), and cod-liver oil (Connecticut). These researchers believed that only one factor, which they called fat-soluble A, was needed to supplement the purified diets.
1914	Dr. Goldberger visited an orphanage in Jackson, Mississippi, and noted that 68 of the 211 children suffered from pellagra, but that none of the employees in the institution had ever contracted the disease. This caused him to doubt the prevailing theory that pellagra was a communicable disease. He concluded that the absence of pellagra in the better-fed employees was due to the presence of meat and milk in their diets; additionally, he found that some of the orphans who were free of the disease were able to pilfer some milk and meat from the limited supply.
1915	Dr. Goldberger continued his research with prisoners from one of the penitentiaries. Out of the 12 volunteers, five developed pellagra, which he cured by the addition of yeast to the diet.
1916	Dr. Elmer V. McCollum, of the University of Wisconsin, designated the concentrate that cured beriberi as water-soluble B, thinking that it was one factor only. Most medical men refused to accept Goldberger's finding for curing pellagra. So, Goldberger and his faithful disciple, Dr. G. A. Wheeler, made themselves the first subjects of a series of experiments designed to prove to the doubters that pellagra could not be contagious. In April, both of them swabbed their throats with secretions obtained from the nose and throat of pellagra patients, but they did not contract the disease. In addition, a group of 21 men and 1 woman took capsules containing blood, feces, and urine from pellagra patients, but none contracted the disease.
1917	E. V. McCollum showed that xerophthalmia, or night blindness, is due to lack of fat-soluble A.
1918	Sir Edward Mellanby of England, working with puppies, demonstrated that rickets was a nutritional deficiency disease. He cured them with cod-liver oil, but incorrectly attributed the cure to the newly discovered fat-soluble A.
1919	R. J. Williams discovered the nutritive significance of pantothenic acid in the proliferation and fermentative activity of yeast cells.
1921	Banting and Best, physiologists, at the University of Toronto, obtained insulin from the pancreas of dogs. The insulin which they obtained cured the diabetes of depancreatized dogs. Following many tests on dogs, the hormone was administered to a male diabetic human, who experienced a remarkable recovery. Banting was awarded the 1923 Nobel Prize for Medicine for the discovery, but Best was not even considered for the award because he held only an undergraduate degree at the time. However, Banting shared the money with him.
1922	McCollum, while at Johns Hopkins University, found that, after destruction of all the vitamin A in cod-liver oil (oxidation, by passing heated air through cod-liver oil), it still retained its rickets-preventing potency. This proved the existence of a second fat-soluble vitamin, carried in liver oils and certain other fats, which he called the *calcium-depositing vitamin.* It is of interest to note that, though McCollum discovered the existence of vitamin D, he did not call it by this name until after this designation was in common use by others. Evans and Bishop of the University of California, discovered that a fat-soluble dietary factor (then called *factor X*) in lettuce and wheat germ was essential for successful reproduction in rats. For many years, what subsequently became known as *vitamin E* was known as the *antisterility vitamin,* because of its effect on the fertility of rats.
1924	Sure, of the University of Arkansas, named the antisterility factor found in wheat germ and lettuce, *vitamin E.* The mystery of how sunlight could prevent rickets was partially solved. Dr. Harry Steenbock and Dr. A. Hess, working independently, showed that the antirachitic activity could be produced in foods and in animals by ultraviolet light. The process, known as the Steenbock Irradiation Process was patented by Steenbock of the University of Wisconsin, with the royalties assigned to the Wisconsin Alumni Research Foundation.
1925	E. B. Hart, who together with Stephen Babcock, helped to make the University of Wisconsin world famous, reported that white rats developed anemia if fed a milk diet, and that the anemia was not cured by iron salts. However, the ash of lettuce, which contained copper, was an effective cure. Mere traces of copper added to pure iron salts were found to be adequate for the cure of iron-deficiency anemia. G. H. Whipple showed that liver was a great benefit in blood regeneration of dogs rendered anemic by bleeding.
1926	Minot and Murphy of the Harvard Medical School reported that feeding large amounts of raw liver restored the normal level of red blood cells in cases of human pernicious anemia. For this discovery, they shared a Nobel Prize with G. H. Whipple. B. C. P. Jansen and W. P. Donath, in Holland, isolated the antiberiberi vitamin, which, at first, was called *aneurin* because of its specific action on the nervous system. Goldberger and Wheeler showed that pellagra in humans and blacktongue in dogs were similar. They confirmed it by curing the blacktongue when they fed yeast to the dogs. They designated this preventive and curative factor as *P-P* (pellagra-preventive); others designated it *vitamin G* for Goldberger. Jansen and Donath isolated vitamin B-1 (thiamin) from rice polishings. It took over 30 years of research to isolate this vitamin. Mitchell and co-workers, at the University of Illinois, first employed the comparative slaughter method of determining net energy (energy storage and heat production).
1928	A Hungarian scientist, Albert Szent-Gyorgy, working in Hopkins' laboratory at Cambridge University, England, isolated a substance from the ox adrenal glands, oranges, and cabbage leaves, which he called *hexuronic acid* (eventually to be called *ascorbic acid*); but he did not test it for antiscorbutic effect. Dr. George Oswald Burr, University of California, Berkeley, published a paper in which he proved that fats were essential in the diet of rats. Subsequently (in 1929), at the University of Minnesota (to which he transferred), he and Mildred M. (Mrs.) Burr found that the essential fat was linoleic acid.

(Continued)

TABLE 2–1 *(Continued)*

Date	Nutritional Milestones
1929	Professor Carl Peter Hendrik Dam, biochemist at the University of Copenhagen, Denmark, observed that certain experimental diets produced fatal hemorrhages in chicks. Bleeding could be prevented by giving a variety of foodstuffs, especially alfalfa (lucerne) and fishmeal. Further, it was found that the active principle in these materials could be extracted with ether; thus, a new fat-soluble factor (later called vitamin K) was discovered. Sir Frederick G. Hopkins shared the Nobel Prize in Medicine with Christiaan Eijkman for their work with the vitamins. W. B. Castle of Harvard showed that pernicious anemia could be controlled by feeding patients beef incubated in normal gastric juice, although neither the beef nor the gastric juice alone was effective. Thus, he found that there is an *extrinsic factor* in food, and an *intrinsic factor* in normal gastric secretion; which, given together, cause red bood cell formation in pernicious anemia.
1930	William C. Rose and co-workers of the University of Illinois used rats to study the essentiality of each amino acid. Later (1935 to 1955), Rose conducted studies on humans to determine how much of each amino acid is needed. These experiments proved to be brilliant additions to the field of nutrition.
1931	P. Karrer, a Swiss biochemist, determined the chemical formula for vitamin A—the first vitamin to have its chemcial structure determined. For this, and for his work with riboflavin, he received the Nobel Prize.
1932	Dr. C. H. Best, codiscoverer of insulin, made the discovery that choline is important to nutrition. Further research proved that choline has a definite relationship in the metabolism of fats. Best and his co-workers at the University of Toronto reported that choline prevented fatty livers in rats fed high fat diets. Windaus of Germany and Askew of England isolated crystals of pure vitamin D_2 (ergocalciferol) from irradiated ergosterol. Charles Glen King and W. A. Waugh, working at the University of Pittsburgh, isolated a crystalline material from lemon juice that possessed antiscorbutic activity in quinea pigs. This marked the discovery and isolation of vitamin C, a deficiency of which caused the centuries-old scourge of scurvy.
1933	R. J. Williams fractionated a compound from yeast and called it *pantothenic acid.* Vitamin C was synthesized by Richstein, a Swiss scientist. Kuhn, working at the University of Heidelberg, isolated pure riboflavin (vitamin B-2) from milk, and demonstrated its similarity to that which Warburg and Christian had isolated.
1934	Dr. Paul Gyorgy found a cure for the severe dermatitis which resulted when Goldberger and Lillie tried to produce pellagra. But the curative compound of the extract was neither thiamin, niacin, nor riboflavin, but a substance which he called *vitamin B-6.*
1935	Kuhn, at the University of Heidelberg, and the Swiss researcher, P. Karrer, both accomplished the synthesis of riboflavin (vitamin B–2). Karrer named it *riboflavin* because it was found to have a pentose side chain—ribitol (similar to the sugar, ribose)—attached to a flavinlike compound. Dam named the antibleeding factor *koagulation vitamin* (the Danish word for *coagulation*), which was later shortened to *vitamin K* (from the first letter of koagulation). W. C. Rose of the University of Illinois discovered threonine (1935 to 1936), thereby (1) paving the way for carefully controlled studies with diets made of synthetic amino acids, and (2) making it possible to learn the dispensability and indispensability of the various amino acids for different species of animals and the role of amino acid imbalance in nutrition. Most of this work was done with rats. But the doctoral students conducting these investigations confirmed the results on themselves; they tested each amino acid at different levels and determined the amount needed for optimum utilization of dietary protein.
1936	Szent-Gyorgy, who won a Nobel Prize in Medicine for his work with vitamin C, isolated a material from citrus rind that he called *citrin,* which he showed consisted of a mixture of flavonoids. His initial test of the new substance with scorbutic guinea pigs seemed to indicate that, in combination with ascorbic acid, it was effective in strengthening the body's smallest blood vessels, the capillaries, in addition to curing scurvy. In 1938, Szent-Gyorgy reported that subsequent tests failed to confirm the results of his earlier experiments. Evans and co-workers isolated crystalline vitamin E from wheat germ oil and named it *tocopherol,* from the Greek words *tokos* (offspring) and *pherein* (to bear), meaning *to bear offspring.* Robert R. Williams, an American, determined the structure of and synthesized the antiberiberi vitamin, which he named *thiamine* (B–1) because it contains sulfur (from *thio,* meaning *sulfur-containing*) and an amine group. Subsequently, the "e" was dropped. Samuel Lepkovsky, Jukes, and Krause, at the University of California, isolated vitamin B–6, or pyridoxine. Almost simultaneously, Keresztesy and Stevens, and Gyorgy reported its isolation. And in 1938, Kuhn and Wendt in Germany, and Ichiba and Michi in Japan, independently isolated vitamin B–6. Kogl and Tonnis of Germany isolated a crystalline substance from boiled yolks of duck eggs, which they called *biotin,* since they believed it to be identical to the *bios* factor needed for yeast growth. The two German scientists, Warburg and Christian, showed that nicotinamide was an essential component of the hydrogen transport system in the form of nicotinamide adenine dinucleotide (NAD).
1937	Paul Gyorgy found that a substance, which he named *vitamin H,* would prevent the pathological condition that resulted from feeding rats and chicks raw egg white. Later, he demonstrated that vitamin H and biotin were one and the same thing. Dr. Conrad Elvehjem, who spent most of his career at the University of Wisconsin, where he eventually became president, discovered that niacin (as either nicotinic acid or nicotinic acid amide, isolated from liver) cured blacktongue in dogs. Shortly thereafter, it was established that niacin was effective in the prevention and treatment of pellagra, and that it was a dietary essential for humans, monkeys, pigs, chickens and other species. Dr. Elvehjem also did much work on the role of copper and iron in nutrition, and amino acid interrelationships.
1938	P. Karrer and his associates succeeded in synthesizing alpha-tocopherol (vitamin E), and in proving its biological activity. Five different laboratories, working independently, isolated vitamin B–6 (pyridoxine) in crystalline form, but credit for obtaining the first crystals is generally given to Lepkovsky of the University of California.
1939	Dam and Karrer isolated vitamin K in pure form; and, that same year, Almquist and Klose synthesized vitamin K. In 1943, Dam shared the Nobel Prize in Physiology and Medicine with Edward Doisy who had studied the chemical properties of vitamin K. Williams isolated pantothenic acid (vitamin B–3) from liver. A group of scientists from Merck & Co. discovered the structure of vitamin B–6 (pyridoxine), and S. A. Harris and K. Folkers synthesized the vitamin.
1940	D. W. Woolley, one of the famous Wisconsin group, isolated inositol from liver. Sure and Gyorgy-Goldblatt, working independently, reported that choline is essential for growth of rats, thereby indicating its vitamin nature. Pantothenic acid was synthesized by R. J. Williams and two other laboratories, all working independently. Thus, many scientists were studying the previous research papers and investigations into the vitamins were at a peak. D. W. Woolley, at the University of Wisconsin, demonstrated that inositol could prevent alopecia (patchy-hair, or baldness) in mice. E. Nielson, Oleson, and Elvehjem reported the concentration and fractionation of para-aminobenzoic acid.

(Continued)

TABLE 2-1 *(Continued)*

Date	Nutritional Milestones
1941	The name *folic acid,* the forerunner of the term *folacin,* was suggested by Mitchell, Snell, and Williams, of Texas, for the bacterial growth factor that was found in spinach and known to be widely distributed in green leafy plants. The word *folic* is derived from the Latin word *folium,* meaning *foliage* or *leaf.* Link and co-workers at the University of Wisconsin discovered dicoumarol—an antagonist or antimetabolite of vitamin K.
1942	V. du Vigneaud and associates, at Cornell, suggested the correct structural formula for biotin based on a study of its degradation products. Snell discovered the vitamin B-6 activity of two closely related substances in natural products, which he named pyridoxal and pyridoxamine.
1943	Harris and his co-workers of Merck & Co. synthesized biotin.
1945	Angier and co-workers isolated and synthesized folacin (folic acid). That same year, Dr. Tom Spies showed that folic acid was effective in the treatment of megaloblastic anemia of pregnancy and of tropical sprue. Willard Krehl and his associates at the University of Wisconsin finally solved the mystery of pellagra when they discovered that tryptophan is a precursor of niacin, thereby explaining two things: (1) why milk, which is low in niacin but high in tryptophan, will prevent or cure pellagra; and (2) why, in earlier concepts, protein deficiency was often related to pellagra. A. I. Virtanen, Finnish biochemist, won the 1945 Nobel Prize in Chemistry for researches and inventions in agriculture and nutrition. He discovered the AIV method, named from his initials, for preserving silage by acidification.
1947	Isler, working in Switzerland, synthesized vitamin A.
1948	Rickes and co-workers of Merck & Co., New Jersey, and Smith and Parker of Glaxo Laboratories, in England, isolated from liver concentrate a crystalline, red pigment, which they called *vitamin B-12.* R. West of Columbia University, New York, showed that injections of vitamin B-12 induced a dramatic beneficial response in patients with pernicious anemia.
1952	R. B. Woodward of Harvard accomplished the first total synthesis of a form of vitamin D (in this case vitamin D_3). He was awarded the Nobel Prize in Chemistry in 1965 for this and other similar achievements.
1953	Francis Crick and James Watson described the structure of the DNA molecule.
1955	The structure of vitamin B-12, cyanocobalamin, was determined by Dorothy Hodgkin and co-workers, at Oxford. In 1964, Hodgkin was awarded the Nobel Prize for her work. Woodward's group at Harvard synthesized vitamin B-12, using a very complicated and expensive procedure. Fortunately, soon thereafter, it was found that it could be produced from cultures of certain bacteria and fungi grown in large tanks containing special media.
1957	Selenium was discovered to be an essential element by Klaus Schwarz, a German medical researcher, who, at the time, was a visiting scientist at the U.S. National Institutes of Health (NIH). In 1951, Schwarz reported that the American type of brewers' yeast contained an unidentified *Factor 3* which apparently acted, along with vitamin E and sulfur-containing amino acids, in protecting the liver against damage due to certain types of diets. In 1957, he and co-worker, C. M. Foltz, reported that Factor 3 contained selenium; thereby making the discovery that selenium is an essential element. In 1958, Oregon State University workers showed that white muscle disease, a myopathy of young ruminants, was caused by a selenium deficiency. Other studies by the group at NIH, and by Scott and his co-workers in the Department of Poultry Husbandry at Cornell, showed that the addition of selenium salts to the diets of chicks prevented certain disorders which resulted from vitamin E deficiencies. Finally, in 1973, Rotruck and co-workers at the University of Wisconsin reported that selenium acted as a cofactor for a recently-discovered enzyme (glutathione peroxidase) which breaks down toxic peroxides, most of which are formed from the oxidation of polyunsaturated fats. Thus, the link between selenium and vitamin E was established.
1959	Chromium was found to be an essential element by W. Mertz and K. Schwarz, German medical scientists working at the NIH. They (1) made the important discovery that the feeding of chromium salts were necessary in the diet for normal metabolism of blood glucose in the rat, and (2) announced the specificity of chromium as the glucose tolerance factor (GTF).
1967	Dr. George Wald of Harvard University was awarded the Nobel Prize in Medicine for clarifying the role of vitamin A in vision.
1969	Prostaglandins ushered in a new era of research. Although a group of lipid compounds—20-carbon fatty acids with various functional groups and configurations—were first extracted from human semen and sheep seminal vesicles in the early 1930s, and named prostaglandins in 1935, little attention was paid to their physiological action at the time. Pharris and co-workers of the Upjohn Company reported that prostaglandin F_2 alpha would regress the corpus luteum in rats, following which prostaglandin research increased manyfold. Prostaglandins are a class of hormonelike compounds, made in various tissues of the body from arachidonic acid (and other derivatives of linoleic acid), which are important in the regulation of such diverse reactions as gastric secretion, pancreatic functions, release of pituitary hormones, smooth muscle metabolism, and control of blood pressure. Several prostaglandins have been identified.
1970	Dr. Norman Borlaug, an American agricultural scientist employed by the Rockefeller Foundation and known as the *Father of the Green Revolution,* was awarded the Nobel Peace Prize for breeding new high-yielding, short-strawed varieties of wheat. A related development was the development of special varieties of rice by scientists at the International Rice Research Institute in the Philippines.
1977	Gene splicing, also known as recombinant DNA technique, ushered in a new era of genetic engineering, when scientists at the University of California-San Francisco reported a major breakthrough as a result of altering genes—turning ordinary bacteria into factories capable of producing insulin, a valuable hormone essential to the survival of diabetics. But this was only the beginning! Building upon the present knowledge and understanding of the nucleic acids DNA and RNA, the *Biotechnology Era* arrived.

The Search Goes On

Modern nutritionists move ahead in their search. Food groups—carbohydrates, fats, proteins, minerals, vitamins, and water—do not go far enough in any sophisticated study of nutrient requirements. Corn, alfalfa, soybean meal, and bone meal never get to the cells and tissues. But the nutrient chemicals—more than 40 of them, including amino acids, minerals, vitamins—do reach the body cells and tissues and are essential to their function. These are the ABCs of modern nutrition. Today, there is increasing research emphasis on the microscopic and subcellular components of the living system, and there is more and more fragmentation and specialization. First we had the animal nutritionists. Next came the ruminant and nonruminant nutritionists. Now these are being replaced by species specialists, but with further breakdown and study of a class of nutrients, or even of an individual nutrient. So, the scientific domain of the individual nutritionist has become increasingly narrow. Too often, this has been done at the expense of understanding the animal as a whole. Too often, there has been little attention to the comparative aspects with other animal species. Too often, modern nutritionists may be likened to an automotive engineer who knows all about the production of energy by combustion of fossil fuels; he even knows about each of the car systems—the fuel system, the transmission system, the brake system, etc.—but who cannot put the automobile together and drive it. Too often, in this day of specialization, the nutritionist has neglected the team approach, with scientists of many disciplines working together.

Despite the many advances that typify nutrition, the following are among the unknowns in nutrition, or among the nutrition-related areas that need improvement, which merit the best efforts of nutritionists:

1. **We need to know more about the nutrient requirements of animals.** We know that nutrients are needed in definite amounts; that excesses may cause toxicity and needless expense, that insufficiencies may result in deficiency diseases, and that imbalances can be very harmful. But neither nutrient requirement tables nor nutrition labeling guarantee superior nutrition.

We need to know if we have identified all of the required nutrients, and we need to know more about the availability of the nutrients and the way that they interact with each other. We need to know more about the unidentified factors, and we need to know if more of the minerals in the periodic table should be listed as requirements. We need to recognize that nutrient availability may be more important than composition.

2. **We need to know more about the net energy system.** As currently measured, net energy represents an animal's predicted response to a given mixture of feedstuffs under an incompletely defined set of environmental conditions. Further refinement is needed.

3. **We need to feed animals more forages and less grain in the future.** As population increases, the demand for grain increases. This, coupled with the increasing cost of fossil fuels, will elevate grain prices to a point where a transition in animal feeding practices will be dictated largely by economics. Shifts to greater reliance on forage in livestock rations and less grain will take place in the future. Ruminants can make this transition easily. Increasingly, they will be "roughage burners," transforming huge quantities of fibrous materials and wastes into high-quality protein found in meat and milk.

4. **We need to know more about animal behavior and environment.** As the art of husbandry evolved into science, people's fluency with the language of their beasts declined; and human-made environments became increasingly artificial until animals were viewed primarily as economic, biological conversion mechanisms. Behavior received little attention; the push was on for quantity and quality of meat, milk, eggs, fiber, and power of animals. But many abnormal animal behaviors evolved to plague those who raise them, including cannibalism, loss of appetite, stereotyped movements, poor mothering, overaggressiveness, dullness, degenerate sexual behavior, tail biting, cribbing, and a host of other behavior disorders. Today, the situation is being righted. Attention is being given to animal space requirements, heat and vapor production, and environmental control. The modern nutritionist must (a) formulate rations for different environments; and (b) develop nutritional programs for periods of stress, such as weaning, shipping, fatigue, illness, and abrupt temperature and weather changes. The development of nutritional programs for periods of stress should receive as much attention as the medical treatment of ills.

5. **We need to conserve energy.** Population growth and feed-food production technology are creating energy stresses of unprecedented scale and urgency—threatening human existence. In an era of world food shortages, the nutritionist is a key person in the entrapment of solar energy and in the conservation of energy as evidenced by the following:

   a. **Photosynthesis fixes energy.** Plants, using solar energy, are by far the most important, and the only renewable, energy-producing method; the only basic feed-food-manufacturing process in the world; and the only major source of oxygen in the earth's atmosphere. Thus, in an era of world food shortages, it is inevitable that the entrapment of solar energy through photosynthesis will, in the long run, prove more valuable than all the underground fossil fuels—for when the latter are gone, they are gone forever.

   b. **Animals concentrate energy.** Crops vary in their return of captured solar energy per unit of cultural energy input. Pasture, hay, and silage are highly efficient in the capture of solar energy—requiring little input of energy for a high return.

   It follows that ruminants, which utilize grazing land, hay, and silage (feeds not suitable for human consumption), offer the best means of concentrating and storing energy for people.

   c. **Crop residues contain energy.** Crop residues left in the field, above and below the ground surface, may well contain four to five times more energy than is harvested. Increasingly, this potential source of added feed will be utilized by animals.

6. **We need to use agricultural chemicals and drugs with discretion.** Sometimes choices must be made; for example, between malaria-carrying mosquitoes and some fish, or between hordes of grasshoppers and the crops that they devour, or between the use of antibiotics and hormones as feed additives and higher consumer costs. In the era ahead, food shortages will increasingly concern all people—producers

and consumers alike. If the world is to "waste not, want not," it must use more chemicals and drugs in the future, not fewer. This merely underscores the need for (a) careful testing through properly designed experiments of all products prior to use, (b) complying with Federal and state laws, and (c) accurate labeling and use of products. (See section headed "Use Additives and Implants Safely.")

7. **We need to control pollution.** Today, there is worldwide awakening to the problem of pollution of the environment (air, water, and soil) and its effects on human health and other forms of life. Much of this pollution was ushered in with industry. But some of this concern stems from the sudden increase of animals in confinement and the disposal of waste. Certainly, there have been abuses of the environment; and there is no argument that such neglect should be rectified in a sound, orderly manner. Nutritionists can, and must, contribute mightily to the control of pollution, and they must do so with a minimum of disruption of the economy and without lowering the standard of living.

BODY COMPOSITION

Nutrition encompasses the various chemical and physiological reactions which change feed elements into body elements. It follows that knowledge of body composition is useful in understanding animal response to nutrition.

In 1843, Lawes and Gilbert, famed English scientists, initiated at the Rothamsted Station the pioneering and laborious task of analyzing the entire bodies of farm animals—studies which extended for over half a century. In 1895, H. J. Walters, of the University of Missouri, began body composition work—experiments which spanned 35 years. (As a student working his way through college, the senior author of this book was privileged to care for the animals used in these studies during the latter years of the experiment.) Other studies followed by certain of the experiment stations and the U.S. Department of Agriculture. From these data, the figures given in Tables 2-2 and 2-3 have been assembled to provide a picture of the gross composition of animals of different species at different ages and in different nutritional states.

TABLE 2-2
RANGE IN CHEMICAL COMPOSITION OF BODIES OF CATTLE, SHEEP, AND PIGS[1]

Species	Number in Group	Range in Body Composition of Ingesta-Free (empty) Body			
		Water	Fat	Protein	Ash
		%	%	%	%
Cattle	256	39.8–77.6	1.8–44.6	12.4–20.6	3.0–6.1
Sheep	221	39.6–73.8	4.9–46.6	10.7–19.5	1.7–5.8
Pigs	714	30.7–80.8	1.1–61.5	8.3–19.6	1.3–5.6

[1]Reid, J. T., *et al.*, "Some Peculiarities in the Body Composition of Animals," *Body Composition in Animals and Man,* National Academy of Sciences, 1968, p. 20, Table 1. Pertinent information about the data follows:

Cattle: Of the 256 animals, 139 were beef type and 117 were dairy type. Seven purebreds and 5 crossbred combinations were represented. Ages ranged from 1 to 4,860 days (13.3 yr). Sex was designated on 249 animals, of which 135 were males and 114 were females.

Sheep: The study included 4 purebred breeds and 2 crossbred populations, representing a considerable range in body conformation, maturing rate, and mature size. All animals were male castrates ranging from 90 to 895 days (2.5 yr) of age and from 26.4 lb to 147.4 lb in weight.

Pigs: The pigs represented 8 distinct breeds and 1 crossbred group, ranging in age from 1 to 923 days (2.5 yr). Of the 714 pigs, sex was identified for 248, of which 153 were male castrates and 95 were females.

TABLE 2-3
BODY COMPOSITION OF ANIMALS, AT DIFFERENT WEIGHTS AND AGES, INGESTA-FREE (EMPTY) BASIS[1]

Species	Age or Status	Weight		Water	Fat	Protein	Ash
		(lb)	*(kg)*	(%)	(%)	(%)	(%)
Cattle:							
Calf	Newborn	70	*31.8*	74.4	2.5	19.0	4.1
Calf	Weanling	450	*204.1*	69.0	9.0	18.0	4.0
Steer	Feeder	650	*294.8*	60.3	18.0	17.2	4.5
Steer	Choice grade	1,050	*476.3*	53.5	26.0	17.0	3.5
Steer	Very fat	1,500	*680.4*	40.0	41.0	16.0	3.0
Cow	Breeding condition	1,100	*499.0*	60.0	18.0	17.5	4.5
Sheep:							
Lamb	Newborn	9	*4.1*	72.8	2.0	20.2	5.0
Lamb	Feeder	65	*29.5*	63.9	17.0	15.7	3.4
Lamb	Fat	100	*45.4*	53.2	29.0	15.0	2.8
Lamb	Very fat	125	*56.7*	39.0	44.0	14.4	2.6
Pig:							
Pig	Newborn	3	*1.4*	74.0	2.0	19.0	5.0
Pig	Weanling	30	*13.6*	70.0	9.0	17.5	3.5
Pig	Growing-finishing	100	*45.4*	66.8	16.2	14.9	3.1
Pig	Market	220	*100.0*	50.0	34.4	13.0	2.6
Pig	Very fat	300	*136.1*	42.5	43.5	12.0	2.0
Poultry:							
Chick	Newly hatched	0.090	*0.04*	78.8	4.0	15.3	1.9
Broiler . . .	Market	3.5	*1.6*	65.7	12.2	18.4	3.7
Hen	Layer	4.5	*2.0*	59.6	20.0	17.0	3.4
Turkey . . .	8 weeks	5.1	*2.3*	70.1	4.5	20.9	4.5
Turkey . . .	Market	18.0	8.2	59.4	18.4	19.1	3.1
Rabbit:							
Rabbit . . .	Market	8	*3.6*	69.0	3.5	9.0	5.0
Horse:							
Foal	Newborn	110	*49.9*	73.0	2.0	20.0	5.0
Foal	Weanling	400	*181.4*	69.0	9.0	18.0	4.0
Horse	Mature	1,050	*476.3*	62.0	17.0	17.0	4.0
Man:							
Baby	Newborn	8	*3.6*	76.0	9.2	12.0	2.8
Man	Mature	150	*68.0*	59.0	18.0	18.0	5.0

[1]Prepared by the authors from numerous sources.

Tables 2-2 and 2-3 show that there is a wide range in the body composition of animals according to age and nutritional state (degree of fatness). Yet, based on these tables, together with other studies, the following conclusions relative to body composition may be drawn:

1. **Water.** On a percentage basis, the water content shows a marked decrease with advancing age, maturity, and fatness. In cattle, for example, the water content from conception to market weight and finish changes as follows: embryo soon after conception, 95%; newborn calf, 74%; 450-lb weaner, 69%; yearling feeder steer, 60%; and slaughter steer, grading Choice, 53.5%.

2. **Fat.** The percentage of fat normally increases with growth and fattening. In swine, for example, the body of a newborn pig contains only about 2.0% fat, whereas the body of a very fat 300-lb pig may run as high as 43.5% fat.

It is recognized, of course, that the amount of fat is materially affected by the feed intake. The Missouri workers reported, for example, that the body of a thin feeder steer contains 18% fat, whereas an overweight very fat steer may contain 41%.

3. **Fat and water.** As the percentage of fat increases, the percentage of water decreases.

4. **Protein.** The percentage of protein remains rather constant during growth, but decreases as the animal fattens.

On the average, there are 3 to 4 lb of water per 1 lb of protein in the body.

5. **Ash.** The percentage of ash shows the least change. However, it decreases as animals fatten because fat tissue contains less minerals than lean tissue.

6. **Composition of body gain.** The data presented in Tables 2–2 and 2–3 clearly indicate that gain in weight tells nothing about the composition of gain.

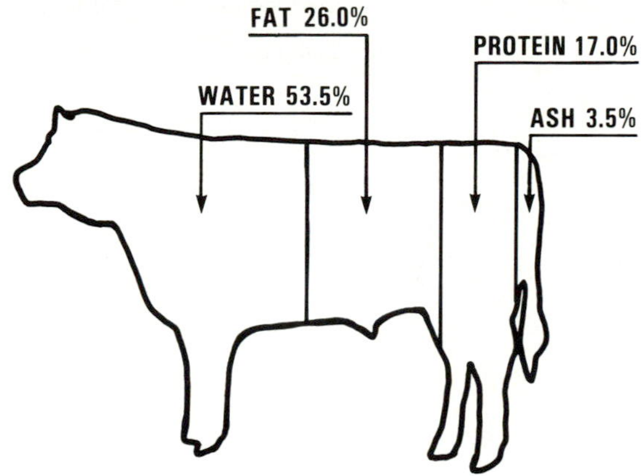

Fig. 2–2. Composition of a 1,050-lb steer, grading Choice.

Composition of gain is affected by feed intake and rate of gain. When feed intake is limited, the rate of gain is depressed and the composition of this gain is primarily protein along with the concomitant deposition of water. As feed (energy) intake is increased, daily weight gain increases correspondingly, but the proportion of protein in this gain decreases and the proportion of fat increases.

Recent research at the Ohio Agricultural Research Development Center and Michigan State University indicates that, in typical feeder steers, the normal upper limit of protein gain is achieved at a daily empty body weight gain of 2 to 2.2 lb. This limit can be increased by hormonal implants, but at gains of about 3.0 lb or higher, the gain above the upper level of protein deposition is mostly fat.

From the above, the following deductions and practical applications relative to composition of gain may be made:

a. Rate of gain is determined primarily by total feed energy intake. But it may be depressed by protein deficiency.

b. In young growing cattle, energy intake above needs for daily protein deposition will result in fat deposition. In older cattle, once the maximum protein content (lean) has been reached, most of the energy intake is used for fattening.

c. In young cattle, maximal protein gain can be achieved without excess fat deposition by providing rations of low energy density, which is usually achieved by feeding high roughage grower rations. As cattle get heavier, they are switched to finishing (high energy density) rations.

d. Feeding high energy density (finishing) rations during the growing phase will result in cattle that reach the desired finish quite early in relation to their mature weight, without the total potential protein (lean) gain being expressed. Further feeding of such rations will result in maximum carcass protein deposition, but the carcass will also be overly finished.

The feeding of grower type, or lower energy density rations such as corn silage, will result in the maximum total carcass protein gain, but a longer feeding period is required to obtain the desired amount of finish.

e. Biologically, it would appear desirable to provide for a rate of gain during the growing period that maximizes carcass lean deposition without concomitant excess fattening, followed by providing extra feed energy to encourage rapid finishing.

7. **Variation between organs and tissues.** The chemical composition of the body varies widely between organs and tissues and is more or less localized according to function. Thus, water is an essential of every part of the body, but the percentage composition varies greatly in different body parts; blood plasma contains 90 to 92% water, muscle 72 to 78%, bone 45%, and the enamel of the teeth only 5%. Proteins are the principal constituents, other than water, of muscles, tendons, and connective tissues. Most of the fat is localized under the skin, near the kidneys, and around the intestines. But it is also present in the muscles (known as marbling in a carcass), bones, and elsewhere.

Tables 2–2 and 2–3 do not reveal the very small amount of carbohydrates (mostly glucose and glycogen) present in the bodies of animals and found principally in the liver, muscles, and blood. Although these carbohydrates are very important in animal nutrition, they account for less than 1% of the body composition. It is noteworthy, too, that the carbohydrate content is one of the fundamental differences between the composition of plants and animals. In animals, the walls of the body cells are made chiefly of protein, whereas in plants they are composed of cellulose and other carbohydrates. Also, in plants most of the reserve food is stored as starch, another carbohydrate, whereas in animals nearly all the reserve is stored in the form of fat.

8. **Species comparison.** The following species differences are noteworthy:

a. The bodies of very fat pigs contain more fat and less water than cattle and sheep in comparable condition.

b. Because of their smaller skeletons, the bodies of pigs contain less ash than those of cattle or sheep.

c. Cattle are higher in protein than pigs of comparable age, weight, and finish. At normal market stage, broilers are higher in protein than four-footed animals.

CLASSIFICATION OF NUTRIENTS

Animals do not utilize feeds as such. Rather, they use those portions of feeds called *nutrients* that are released by digestion, then absorbed into the body fluids and tissues. *Nutrients are those substances, usually obtained from feeds, which can be used by the animal when made available in a suitable form to its cells, organs, and tissues.* They include carbohydrates, fats, proteins, minerals, vitamins, and water. More correctly speaking, the term *nutrients* refers to the more than 40 nutrient chemicals, including amino acids, minerals, and vitamins.

Knowledge of the basic functions of the nutrients in the

animal body, and of the interrelationships between various nutrients and other metabolites within the cells of the animal, is necessary before one can make practical scientific use of the principles of nutrition. To this end, the rest of this chapter is appropriately devoted to a discussion of the "Functions of Nutrients." At the outset, however, it will facilitate the understanding of functions if feed nutrients are first classified. For this reason, Table 2-4 is presented.

TABLE 2-4
CLASSIFICATION OF NUTRIENTS BY ANALYSIS

ORGANIC

- **Carbohydrates**
  - **Nitrogen-Free Extract**
    - **Monosaccharides**: glucose, fructose, galactose, mannose
    - **Disaccharides**: sucrose, lactose, maltose
    - **Polysaccharides**: starch
  - **Crude Fiber**
    - **Polysaccharides (insoluble)**: cellulose, hemicellulose
    - **Lignin**
- **Lipids**
  - **Simple**
    - **Fatty Acids**: palmitic, oleic, stearic, linoleic, arachidonic
    - **Sterols**: cholesterol, ergosterol
  - **Compound**
    - **Triglycerides**: esters of glycerol and fatty acids, such as butterfat, lard, beef tallow, and soybean oil.
    - **Phospholipids**: lecithin, cephalin, sphingomyelin
    - **Waxes**: esters of a fatty acid and long-chain alcohol. Thus, palmitic acid + myricyl alcohol = $C_{30}H_{33}OH$ in beeswax.
- **Nitrogenous**
  - **Protein**
    - **Essential Amino Acids**: arginine, histidine, isoleucine, leucine, lysine, methionine, phenylalanine, threonine, tryptophan, valine
    - **Semiessential Amino Acids**: cysteine, cystine (synthesized from methionine), tyrosine (synthesized from phenylalanine)
    - **Nonessential Amino Acids**: alanine, asparagine, aspartic acid, glutamic acid, glutamine, glycine[1], hydroxyproline, proline[2], serine[1]
  - **Nonprotein**: amines, free amino acids, urea
- **Vitamins**
  - **Fat Soluble**: A (carotene), D_2, D_3, E, K
  - **Water Soluble**
    - B complex: choline, folic acid (folacin), inositol, nicotinic acid (niacin), pantothenic acid, para-aminobenzoic acid, B-1 (thiamin), B-2 (riboflavin), B-6 (pyridoxine), B-12, biotin
    - C (ascorbic acid)

INORGANIC

- **Minerals**
  - **Essential Minerals**[3]
    - **Macro**: calcium (Ca), chlorine (Cl), magnesium (Mg), phosphorus (P), potassium (K), sodium (Na), sulfur (S)
    - **Micro**: chromium (Cr), cobalt (Co), copper (Cu), fluorine (F), iodine (I), iron (Fe), manganese (Mn), molybdenum (Mo), selenium (Se), silicon (Si), zinc (Zn)
  - **Possible Essential Minerals**: barium (Ba), bromine (Br), nickel (Ni), strontium (Sr), tin (Sn), vanadium (V)

[1]Under some conditions, glycine or serine synthesis may not be sufficient for most rapid growth; either glycine or serine may need to be supplied in the diet.
[2]When diets composed of crystalline amino acids are used, proline may be necessary to achieve maximum growth.
[3]Required by at least one animal species.

FUNCTIONS OF NUTRIENTS

Of the feed consumed, a portion is digested and absorbed for use by the animal. The remaining undigested portion is excreted and constitutes the major portion of the feces. Nutrients from the digested feed are used for a number of different body processes, the exact usage varying with the species, class, age, and productivity of the animal. All animals use a portion of their absorbed nutrients to carry on essential functions, such as body metabolism and maintaining body temperature and the replacement and repair of body cells and tissues. These uses of nutrients are referred to as *maintenance.* That portion of digested feed used for growth, fattening, or the production of milk, eggs, wool, and work is known as *production requirements.* Another portion of the nutrients is used for the development of the fetus and is referred to as *reproduction requirements.*

Based on the quantity of nutrients needed daily for different purposes, nutrient demands may be classed as high, low, variable, or intermediate.

Requirements for milk and egg production are considered *high-demand uses,* whereas wool is a *low-demand use.* Work, which may be strenuous for limited periods of time (as in racing), and the last stages of pregnancy have *variable requirements.* Growth and fattening may be classed as intermediate in nutrient demands. Each of these needs will be discussed in more detail.

Maintenance

Animals, unlike machines, are never idle. They use nutrients to keep their bodies functioning every hour of every day, even when they are not being used for production or work.

Maintenance requirements may be defined as the combination of nutrients which are needed by the animal to keep its body functioning without any gain or loss in body weight or any productive activity. Although these requirements are relatively simple, they are essential for life itself. A mature animal must have (1) heat to maintain body temperature, (2) sufficient energy to keep vital body processes functional, (3) energy for minimal movement, and (4) the necessary nutrients to repair damaged cells and tissues and to replace those which have become nonfunctional. Thus, energy is the primary nutritive need for maintenance. Even though the quantity of other nutrients required for maintenance is relatively small, it is necessary to have a balance of the essential proteins, minerals, and vitamins.

No matter how quietly an animal may be lying in a stall or in a pasture, it requires a certain amount of fuel and other nutrients. The least amount on which it can exist is called its *basal maintenance requirement.* With the exception of horses, most animals require about 9% more fuel (calories) when standing than when lying, and even more is needed when they walk or run. This explains why it is desirable, for economic reasons, that finishing animals eat, then lie down as much as possible.

There are only a few times in the normal life of an animal when only the maintenance requirement needs to be met. Such a status is closely approached by mature males not in service; by mature, dry, nonpregnant females; and by idle horses. Nevertheless, maintenance is the standard bench mark or reference point for evaluating nutritional needs.

Although nutrient needs are minimal during maintenance, it is noteworthy that ⅓ to ½ of the feed consumed by animals as a whole is used to meet the maintenance requirement (see Fig. 2-3). Of course, on an individual basis, the higher the production, the smaller the proportion of nutrients needed for maintenance.

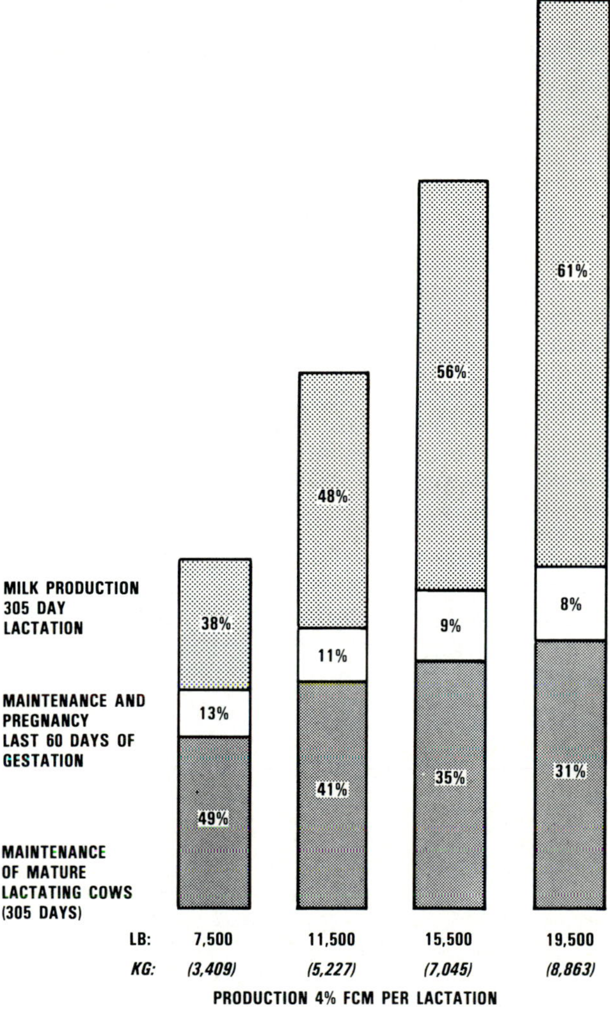

Fig. 2-3. Relative proportions of feed used by a 1,430-lb (*650 kg*) cow for (1) maintenance, (2) maintenance and pregnancy, and (3) milk production, at levels of 7,500, 11,500, 15,500, and 19,500 lb 4% FCM milk. Note that the percentage of feed used for maintenance decreases as production increases.

FACTORS AFFECTING MAINTENANCE REQUIREMENTS

Even though maintenance requirements might be considered an expression of the nonproduction needs of an animal, there are many factors which affect the amount of nutrients necessary for this vital function; among them, (1) exercise, (2) weather, (3) stress, (4) health, (5) body size, (6) temperament, (7) individual variation, (8) level of production, and (9) lactation. The first four are *external factors*—they are subject to control to some degree through management and facilities. The others are *internal factors*—they are part of the animal itself. Both external and internal factors influence requirements according to their intensity. For example, the colder or hotter it gets from the most comfortable (optimum) temperature, the greater will be the maintenance requirements.

Exercise

Movement of animals might be classed as work, except that in the normal use of the term maintenance, all activities which are not considered productive are grouped under maintenance requirements. With this concept, animals in a confined or restricted lot have a smaller maintenance requirement than those in a pasture or on the range. In less productive ranges and pastures, the amount of forage and the type of terrain (topography) determine the amount of energy which it is necessary to expend in order to harvest the feed. The more sparse the forage, the farther an animal must walk each day to eat its fill. The location of water and the distance animals must travel to get a drink also affect daily energy needs. Climbing hillsides and mountains on summer ranges increases the daily maintenance requirements, compared with a similar forage stand on level ground.

Fig. 2-4. Rough terrain and sparse vegetation make for a higher maintenance requirement.

Exercise lots are often provided for confined animals to help keep their muscles in better condition and to reduce injury to feet and legs. Even though the overall benefits may be good, the additional movement of animals increases the energy requirements for maintenance.

Weather

Weather affects the maintenance requirements of animals. For example, under ideal October weather conditions in Missouri, a horse may require 14 lb of a 60% TDN ration daily, whereas in the same area and doing the same work, the same horse may require 16 lb daily of the same feed in July and August and 20 lb in the winter. The good caretaker senses this situation, and changes the feed allowance accordingly.

The maintenance requirements of animals increase as temperature, humidity, and air movements depart from the comfort zone. Likewise, the heat loss from animals is affected by these three items.

The critical temperature is that temperature at which the heat created by digestion and body metabolism just equals that which the animal dissipates by convection, evaporation, radiation, and conduction. The comfort zone is the range in temperature within which the animal may perform with little or no discomfort. At temperatures below the comfort zone, additional nutrients need to be converted to heat to keep the body warm; and at temperatures above the comfort zone, nutrients are needed to help keep the animal cool. *The optimum temperature is the temperature at which the animal responds most favorably, as determined or measured by maximum rate of gain or production, feed efficiency, and/or reproduction.*

The critical temperature, comfort zone, and optimum temperature vary with different species, ages, breeds, and the physiological and productive status of animals. The species differences result primarily from the kinds of thermoregulatory mechanism provided by nature, such as type of coat (hair, wool, feathers), sweat glands, etc. Thus, hogs, which have a light coat of hair, are very sensitive to extremes of heat and cold. On the other hand, nature gave cattle an assist through growing more hair for winter and shedding hair for summer, with the result that they can withstand higher and lower temperatures than hogs.

The critical temperature varies according to age, too. For example, the comfort zone of newborn lambs is 75° to 80°F, whereas the comfort zone of mature sheep is 45° to 75°F.

Fig. 2-5. A band of sheep making use of a tree shelter during a blizzard on the range. (Courtesy, Charles Belden, Pitchfork, Wyo.)

There are also breed differences, which make it possible to select animals well adapted to specific environments. For example, haired sheep (devoid of wool) are adapted to the desert, fat-tailed sheep do well in arid zones, *Bos indicus* (Zebu) types of cattle thrive in tropical areas, and *Bos taurus* cattle (European breeds) do well in temperate zones. The Santa Gertrudis breed of cattle, which evolved from a Brahman X Shorthorn cross, is intermediate between its parent breeds in heat tolerance. The Jersey and Brown Swiss breeds of dairy cattle will maintain their milk production in hot summer temperatures better than Holsteins or other dairy breeds.

Animals that consume large quantities of roughage produce more heat during digestion; hence, they have a different critical temperature than the same animals on a high-concentrate ration. Because of this, experienced cattle feeders decrease the roughage and increase the concentrate of finishing cattle during the hot summer months.

Stresses at both high and low temperatures are increased with high humidity. The cooling effect of evaporating sweat is minimized, and the respired air has less cooling effect. As humidity of the air increases, discomfort at any temperature increases, and nutrient utilization decreases.

Air movement (wind) results in body heat being removed at a more rapid rate than when there is no wind. In warm weather, air movement may make the animal more comfortable; but in cold weather, it adds to the stress of temperature. At low temperatures, the nutrients required to maintain body temperature are increased as the wind velocity increases. In addition to the wind, a drafty condition where the wind passes through small openings directly onto some portion or all of the animal body will usually be more detrimental to comfort and nutrient utilization than the wind itself.

While stock raisers have always been concerned with the effect of weather on animals, this concern has taken on a new dimension with rising costs of feed, labor, land, and financing. There was need to minimize maintenance requirements and maximize the nutrients available for production. This prompted interest in building design. Naturally ventilated buildings, consisting mainly of shells to protect animals from rain and snow, are still used for most animal housing.

Fig. 2-6. Environmentally controlled dairy barn. (Courtesy, Babson Bros. Co., Oak Brook, Ill.)

But environmentally controlled buildings (in which the air temperature, relative humidity, and air velocity are regulated) are rather common in poultry and swine housing, and on the increase for other classes of livestock—especially dairy cattle. They make for the ultimate in feed efficiency, along with a saving in labor and land.

In hot climates, increased use is being made of shades for the purpose of enhancing animal comfort and minimizing the maintenance requirements. Also, studies with lactating dairy cows reveal that putting only the head in an air-conditioned chamber, with the rest of the body left exposed to the heat, will increase production and feed efficiency.

Stress

Stress of any kind increases maintenance requirements. Stress is affected by temperament, excitement, presence of strangers, fatigue, number of animals together, changing corral and corral mates, previous training, previous nutrition, breed, age, and management.

Race and show horses are always under stress; and the greater the speed and the more tired they become, the greater the stress. Also, the greater the stress, the more exacting the nutritive requirements. Thus, the ration of race and show horses should be scientifically formulated, rather than based on fads, foibles, and trade secrets.

Animals can be prepared in such manner as to reduce stress. For example, if calves are properly preconditioned (started on feed, vaccinated, treated for parasites, etc.) prior to weaning, the stress of subsequent weaning and movement to a feedlot will be minimized.

Health

Although animal deaths take a tremendous toll, even greater economic losses—hidden losses—result from poor feed efficiency due to diseases and parasites. It is difficult to assess accurately the effect of animal ill health in terms of the added nutrients necessary for maintenance, for it is affected by the kind and severity of the health problem, the nutritional state of the animal, and the stamina of the individual. Nevertheless, animals that are emaciated give ample evidence of wasted feed.

Body Size

Body size was once used as the basis for determining the maintenance requirements of animals. However, it proved to be inaccurate for such a determination, because it did not represent the relative metabolic activity. Body surface was found to be a better reference point for metabolism than body weight. It was formerly expressed as ⅔ power of body weight ($W^{.67}$). More precise studies, however, have shown that the most accurate measure of metabolic size of an animal is $W^{.75}$, a factor which is generally used by most scientists. In practical terms, this means that a smaller animal has a higher rate of metabolism per pound of body weight than a larger animal. For example, a 1,200-lb animal has less than twice the maintenance requirements of an animal weighing half that amount (600 lb).

Fig. 2-7. On a per pound body weight basis, the small Shetland Pony has a higher rate of metabolism than the big Shire. (Courtesy, Iowa State University, Ames)

Temperament

Nervous animals require more nutrients for maintenance than docile animals. Some animals are naturally nervous; others may be nervous because of the way in which they have been handled. Estrus (heat) is a natural condition; nevertheless, it usually induces excitement and increases nervousness. The presence of strangers and the changing of corrals and corral mates result in induced nervousness. Either natural or induced nervousness has much the same effect on nutrient utilization—both increase it.

Individual Variation

The maintenance requirement varies among individual animals, just as it does in people. Some animals utilize their feed more efficiently than others. A "hard keeper" will require considerably more feed than an "easy keeper." This, of course, is the basis for production testing animals as a means of selecting those which utilize their feed most efficiently.

Racehorses are the *prima donnas* of the animal world; most of them are temperamental, and no two of them can be fed alike. They vary in rapidity of eating, in the quantity of feed that they will consume, in the proportion of concentrate to roughage that they will take, and in response to different caretakers. Thus, for best results, they must be fed as individuals. Recognition of this fact sets apart great trainers from the "also-rans."

Level of Production

Animals with accelerated rates of production tend to have higher maintenance requirements than those with lower levels of production. For convenience reasons, most feeding standards make provision for this increased maintenance as part of the production requirements; it simplifies calculations to use a single maintenance value for all production levels.

Lactation

The maintenance requirements of lactating females of all species are higher than those of dry, nonpregnant females. For example, the maintenance requirement of lactating cows is approximately 10% higher than that of dry, nonpregnant cows. This is attributed, in part, to (1) the increased secretion of thyroxin in lactating cows, which is reflected in a faster heartbeat—increased basal metabolic rate; and (2) the approximately 400 lb of blood which must be pumped through the udder to supply the raw material for 1 lb of milk.[1] (Hence, a cow giving 80 lb of milk in a day must pump 32,000 lb, or 16 tons, of blood through her udder daily.)

Other Factors Affecting Maintenance Requirements

Other factors that affect the maintenance requirements follow:

1. **Shearing sheep and goats.** In cool weather, the maintenance requirements increase at shearing due to decreased insulation. In hot weather, however, shearing may decrease maintenance requirements.
2. **Gestation.** Bearing in mind that the maintenance requirements of gestation do not include fetal growth, the lowered activity of pregnant females generally leads to a lower maintenance requirement.
3. **Mature size of breed.** Larger breeds grow more rapidly than smaller breeds; hence, they have a higher maintenance requirement.
4. **Sex.** Young males gain more rapidly and have a higher maintenance requirement than young females.

Growth

Growth may be defined as the increase in size of bones, muscles, internal organs, and other parts of the body. It is the normal process before birth, and after birth until the animal reaches its full, mature size. Growth is influenced primarily by nutrient intake. The nutritive requirements become increasingly acute when young animals are under forced production, such as when heifers are bred to calve as 2-year-olds and horses are raced as twos.

Growth is the very foundation of animal production. Young cattle, sheep, swine, poultry, rabbits, and other types

[1]Van Sant, W. R., *A Milker's Manual,* The University of Arizona, Bull. A-37, 1965, p. 15.

of meat animals will not make the most economical finishing gains unless they have been raised to be thrifty and vigorous. Likewise, breeding females may have their reproductive ability seriously impaired if they have been improperly grown. Nor can one expect the most satisfactory yields of milk from dairy cows or production of eggs from laying hens unless they were well developed during their growing period. Horses cannot perform the maximum amount of work, and running horses do not possess the desired speed and endurance, if their growth was stunted or if their skeletons were improperly formed by inadequate rations during growth.

Generally speaking, organs vital for the maintenance of life—e.g., the brain, which coordinates body activities, and the gut, upon which the rest of the postnatal growth depends —are early developing; and the commercially more valuable parts, such as muscle, fat, and udder, develop later. However, not all gut is early developing; for example, the growth and functioning of the ruminant stomach is delayed.

MEASURES OF GROWTH

Knowledge of normal growth and development is useful for a variety of purposes. From a nutrition standpoint, growth curves are used primarily as standards against which to gauge the adequacy of nutrient allowances. In fact, such curves are often the entire basis for the allowances set down in dietary and feeding standards. Also, they provide a basis for comparisons of breeding groups and serve as a reference point from which to establish breeding and management objectives. Economically, growth is important, for young gains are cheap gains. This is generally so because in comparison with older animals, young animals (1) consume more feed according to size, (2) use a smaller proportion of their feed for maintenance, and (3) form relatively more muscle tissue which has a lower caloric value than fat (Table 2-3).

Various methods of measuring growth have evolved, the most common of which follow, along with the advantages and disadvantages of each:

1. **Body weight.** This is the most widely used measure of growth from birth to maturity. It may be measured by (a) weight gain per unit of time (pounds gain per day, per month, or per year); (b) weight per day of age; (c) percent of birth weight, mature weight, or weight during prior period; or (d) cumulative weight gain to any given age. Its chief *advantage* is that it is simple to apply—only scales for weighing are necessary. Its *disadvantage,* especially in meat animals, is that it does not tell anything about the composition of gain—there is no indication of skeletal growth, or of carcass yield or quality. Moreover, heavier weight may be due to higher finish, or more fat. Also, an animal may remain at constant body weight while simultaneously losing water or fat, but growing in skeleton.

2. **Body measurements.** Numerous body measurements may be, and are, used. Heart girth measurements are commonly made on replacement dairy heifers; sometimes, height at withers is included, also. Normal horse measurements are height at withers and circumference of girth and bone. Other body measurements that are sometimes used for different species are length from nose to tailhead, length from crown to rump, and width of hips.

3. **Combination body weight and body measurement.** In practical livestock husbandry, a combination of body weight plus a measurement such as height at withers is often of greater value as a guide to feed allowance than either measurement alone. Thus, such a standard—weight-for-height-for-age—makes it possible to distinguish between well-grown individuals and those that are heavy because of excessive fat.

4. **Feed efficiency.** Correctly speaking, this is not a measure of growth as such. However, it is an indication, because fast gainers are usually more efficient. Moreover, feed efficiency does necessitate that body weight be taken. This value is usually expressed by the conversion factor—pounds of feed eaten per pound of gain in body weight. Because of the nature of the growth process, this measurement has the following weaknesses: (a) In the earlier stages of growth, a larger proportion of the liveweight is made up of gut contents and offal than later in life; and (b) deposition of protein (muscle) is accompanied by 3 to 4 parts of water to 1 of protein, whereas fat (mainly laid down in later stages) is not accompanied by water and is higher in caloric value than muscle tissue.

FACTORS AFFECTING NUTRITIVE NEEDS FOR GROWTH

The nutritive needs for growth vary with age, breed, sex, rate of growth, and health.

Age

In comparison with older animals, young animals generally (1) consume more feed per unit of body weight; (2) utilize feed more efficiently, in pounds of feed eaten per pound of body gain; (3) have a higher requirement for protein, energy, vitamins, and minerals per unit of body weight; (4) require a more concentrated and more easily digested diet; and (5) are more subject to nutritional deficiencies.

The digestive systems of newly hatched birds and of newborn mammals have been nonfunctional and are relatively undeveloped. For this reason, the type of ration is particularly important during this period of functional and physical development of the digestive tract and should include a balance of all available nutrients in a readily available form. Young birds (chicks, poults, ducklings, and goslings) should be fed a highly nutritious diet. The first food ingested by young mammals should be colostrum, the initial secretion of the mammary glands following parturition. In addition to being highly nutritious and easily digested, colostrum contains a large supply of globulins and other proteins that provide nutrients as well as a temporary supply of antibodies which greatly increase resistance to many diseases. As the young mammal grows, its capacity to use other types of feeds increases. This transition is most dramatic in the young ruminant, which develops functional microbial fermentation soon after birth and is able to consume a ration with considerable fiber by the time it is four to six months old.

After an initial adjustment period (which may be prolonged if the feed, environment, or disease level are unfavorable), the rate of gain of young animals and birds is very rapid when measured as a percentage of body weight. Table 2-5 shows growth rate as measured by the days needed to double birth weight and the number of months needed to reach 50% of mature body weight. As shown, the long juvenile period of humans is unique.

TABLE 2-5
DAYS NEEDED TO DOUBLE BIRTH WEIGHT AND MONTHS NEEDED TO REACH 50% OF MATURE BODY WEIGHT, FOR SEVERAL SPECIES OF ANIMALS

Species	Days to Double Birth Weight	Months to Reach 50% Mature Weight
Human	150	115–145
Cattle	47–70	12–22
Horse	60	8–9
Goat	22	4–6
Sheep	15	3–5
Swine	14	11
Rabbit	6	2–5
Rat	6	1½–3
Chicken	5	2
Turkey	5	4
Duck	4	1½

Typical growth patterns in body weight gains of four breeds of dairy heifers are shown in Fig. 2–8.

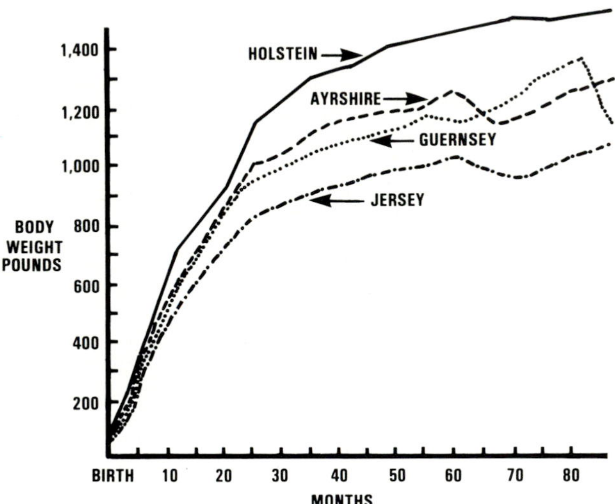

Fig. 2–8. Body weight of four breeds of dairy females from birth to 84 months of age. (Adapted by the authors from Neb. Ag. Exp. Sta. Bull. 179, by H. P. Davis and I. L. Hathaway)

Fig. 2–8, along with studies by Brody (*Bioenergetics and Growth,* by S. Brody) and others, show that growth in dairy heifers, as measured by gain in body weight, is most rapid to about 2 years of age, followed by a more gradual rate of gain. A similar pattern of rapid weight gains early in life occurs in all species.

The typical growth pattern of horses in height at withers and weight is shown in Fig. 2–9.

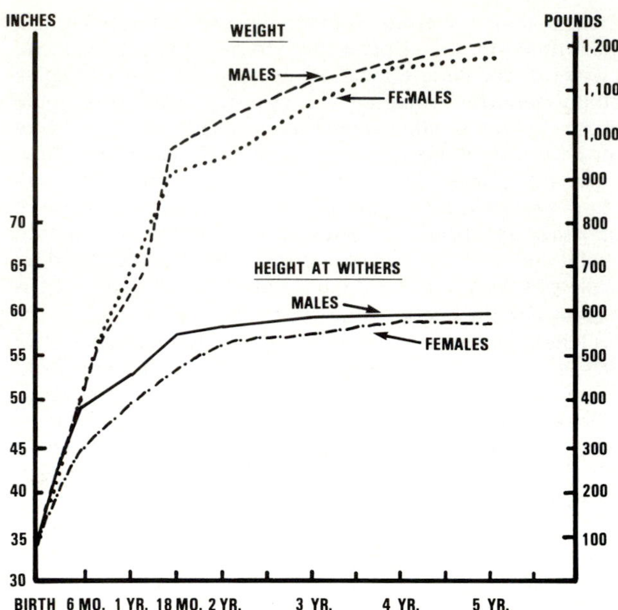

Fig. 2–9. Growth curve (height at withers and weight) of Quarter Horse males and females from birth to maturity. (Based on data from *A Study of Growth and Development in the Quarter Horse,* Bull. No. 546, Louisiana State University, Baton Rouge)

Fig. 2–9 shows that Quarter Horses grow rapidly in weight until about 1 year of age and in height to about 18 months of age, followed by more gradual growth to 5 years of age. Also, it is noteworthy that females reached maturity (tapered off in both weight and height) faster than males. In a study (*Bioenergetics and Growth,* by S. Brody) involving 4 breeds of dairy cows, Brody found a similar pattern of rapid growth in height at the withers to about 2 years of age, followed by a period of slow growth to about 5 years of age. A like growth pattern applies to all species and breeds.

Breed

Larger breeds of all species grow more rapidly than smaller breeds and have a higher nutrient requirement.

The Wisconsin Station studied the feed efficiency of three breeds of beef cattle—Angus, Charolais, and Hereford—cattle varying in size and growth. They concluded that (1) at the same weight (but not at the same degree of finish) faster gaining cattle (like the Charolais) are more efficient than smaller and slower gaining cattle; and (2) feed efficiency does not differ greatly among cattle of different sizes when they are fed to the same grade or degree of finish.[2]

Sex

Growth studies involving young animals of both sexes, and of all species, reveal the following: (1) Males gain more

[2]Brungart, W. H., "Efficiency and Profit Differences of Angus, Charolais, and Hereford Cattle Varying in Size and Growth," *Research Report,* R 2400, University of Wisconsin-Madison, May 1972, p. 2.

rapidly than females and have a higher feed requirement; (2) uncastrated males use feed more efficiently for body weight gains than females, because of the higher water and protein content and the lower fat content of the increased body weight; (3) mature average size is larger in males than in females; and (4) females reach maturity faster than males (see Fig. 2–9).

Rate of Growth

In recent years, the accent in meat production has been on forced production and marketing at an early age; and in breeding animals, it has been on early reproduction. Achieving each of these goals has involved improved nutrition, and, generally speaking, rapid gains and profits have been on the same side of the ledger. Today, broilers may reach market weight in 8 weeks, with notable efficiencies in feed utilization as low as 1 lb of feed per 1 lb of bird. Heifers now calve at 24 months of age, or less, and more and more ewe lambs are being bred to lamb at about 12 months of age.

Rapid gains call for more nutrients. In turn, this necessitates high-energy, palatable, well-balanced rations. For the most part, fast gains are efficient gains; when animals grow at maximum rates, they require fewer nutrients and fewer pounds of feed per pound of gain.

Despite the above factors favoring rapid gains, the following cautions are noteworthy:

1. **Fast gains may be fat gains.** Fast gaining cattle may deposit more fat, according to a study conducted by the Univeristy of Wisconsin, involving Angus, Charolais, and Hereford cattle.[3] The Wisconsin researchers concluded that weight per day of age (WPDA) is not the best measure of growth in cattle, because, within a breed and at a given age (15 months of age in this study), carcasses from cattle with the higher WPDA tend to be fatter and contain a smaller amount of muscle.
2. **Fleshy feeder cattle may not be desirable.** Feeder cattle should not be permitted to gain too rapidly (get too fleshy), because (a) they may reach market finish before they attain desirable market weight, and (b) they will gain more slowly during this finishing period.
3. **Rapid gains may be uneconomical.** When grains are scarce and high in price, greater net returns may accrue from the use of a high-roughage ration and the resulting slower gains.
4. **Rapid gains may impair reproduction in females.** Rapid-gaining young females may become excessively fat, with the result that difficult parturition and heavy birth losses may be encountered, apparently as a result of excess fat decreasing the size of the birth canal.

COMPENSATORY GROWTH

Compensatory growth is making up for a bad start in life. It is common practice for stocker cattle to be "roughed through" the winter as cheaply as possible, with limited daily gains. Then, in the spring, the animals are turned to lush spring pasture or put in a feedlot on a high-energy ration. Animals so managed exhibit the phenomenon of compensatory growth; that is, on a high-energy diet they gain faster and more efficiently than similar cattle that were fed more liberally during the wintering period. Feedlot operators were quick to sense this phenomenon, and to take advantage of it. This is the chief reason for the popularity of "Okie-type" cattle—animals whose growth has been held back to less than their genetic potential. When fed more liberally, they exhibit a surge in growth rate and feed efficiency. Large compensatory growth usually indicates that someone (the stocker operator) has lost money while someone else (the feeder) has made money. It is noteworthy that Holsteins and the larger exotics should never be handled so as to exhibit compensatory gains. If they're held back in the winter, they're too heavy when they finish.

[3] *Ibid.*

Health

Ill health—diseases and parasites—results in lack of thrift and poor development in young stock. When the causative factor is severe, growth may be stunted. Feed is always too costly to waste. Besides, the full productive potential of animals is needed.

Reproduction

Being born (or hatched) and born alive are the first and most important requisites of livestock production, for if animals fail to reproduce, the breeder is soon out of business. A "mating of the gods," involving the greatest genes in the world, is of no value unless these genes result in (1) the successful joining of the sperm and egg, and (2) the birth of live offspring. Stock raisers acknowledge that young crop percentage is the biggest single factor affecting profit. Despite this undeniable fact, it has been estimated that 20 to 50% of all matings are infertile, that 25% of all cows culled from dairy herds are removed because of reproductive inefficiency, that the overall average U.S. calf crop of all cattle (beef and dairy combined) is only 88%—the other 12% abort or are sterile, that 5% of all ewes are sterile, that 15% of all sows bred fail to produce litters, and that only 50% of all mares bred actually produce foals.

Even more nutritional problems with developing embryos are encountered in chickens and turkeys. This may be related to maternal turnover. For example, cattle and horses produce birth-weights each year which are from 7% to 10% of maternal liveweight, whereas a value of about 700% is realized with chickens. Although it could be argued that mammals (such as cattle and horses) subsequently provide milk as a maternal output, we have much greater control over this process, and, in problem situations, we can always supplement the milk or wean the offspring. With chickens and turkeys, however, the "package" (egg) is delivered at laying, and as yet we have no way of controlling embryo nutrition during incubation. Considering the very large maternal turnover of chickens and turkeys, and the lack of control of embryo nutrition during incubation, it is not too surprising that nutritional problems with developing embryos sometimes occur.

Certainly, there are many causes of reproductive failure, but scientists are agreed that nutritional inadequacies play a major role.

Fig. 2–10. "Bonus Calf," an original painting by artist Tom Phillips, 3333 17th St., San Francisco, CA 94110

NUTRITIONAL FACTORS AFFECTING REPRODUCTION

Many factors affect reproduction. For convenience, herein they have been grouped under the following headings: (1) liberal feeding for early sexual maturity, (2) flushing, and (3) nutritional reproductive failure.

Liberal Feeding for Early Sexual Maturity

Rising costs make it imperative, for economic reasons, that breeders get females in reproduction as early in life as possible. This calls for calving 2-year-old heifers, instead of threes; for breeding ewes to lamb as yearlings, and for the early breeding of females of all other species.

It is now known that liberal feeding makes for early sexual maturity. This was confirmed in a longtime experiment conducted at Cornell University, in which 3 groups of Holstein heifers were raised from birth to first calving on 3 different nutritive levels—62% (low), 100% (medium), and 146% (high)—of the standard amount of total digestible nutrients.[4]

[4]Reid, J. T., *et al., Causes and Prevention of Reproductive Failures in Dairy Cattle: IV.* "Effect of Plane of Nutrition During Early Life on Growth, Reproduction, Production, Health, and Longevity of Holstein Cows"; I. "Birth to Fifth Calving," Feb. 1964.

An extremely important finding in the Cornell study was that, regardless of the plane of nutrition, the heifers (Holsteins) came into heat at about 600 lb body weight, showing that size and weight, not age, determine the time of sexual maturity (see Fig. 2–11). This points up the importance of liberal feeding for early breeding.

For heifers of both beef and dairy breeds to breed at 15 months of age and calve at 24 months of age, following weaning the larger size breeds should gain from 1.4 to 1.8 lb per day, and the smaller breeds should gain 1.0 to 1.4 lb per day.

Flushing

Flushing is the practice of having females gain in weight just prior to breeding. The purpose of flushing is to increase the number of ova shed during estrus. Although it is not likely that all the benefits ascribed to flushing will be fully realized under all conditions, the general feeling persists that the practice will result in a 15 to 30% increase in lamb and pig crops, and that females of all species will breed both earlier and more nearly at the same time. Hence, it follows that the offspring will be earlier and more uniform in age and size.

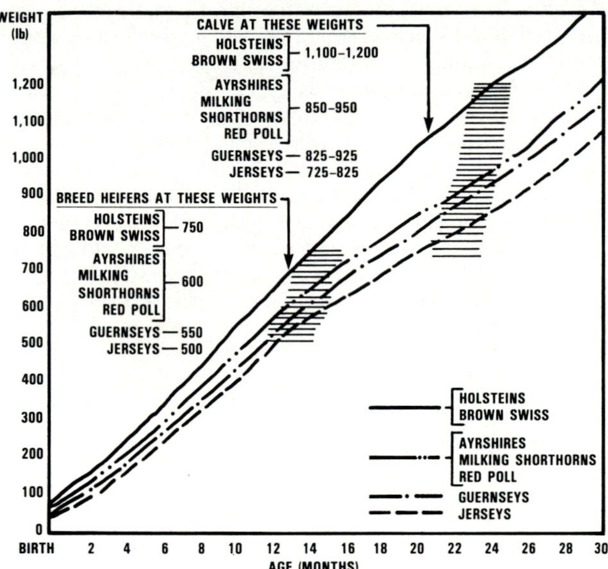

Fig. 2-11. Heifers that are well grown can be bred at the size and age shown, so as to calve at the size and age shown. (Adapted by the authors from Cornell University Ag. Exp. Sta. Bull. 987, Feb. 1964)

Mature females appear to respond better to flushing than virgin females. Also, flushing may be more beneficial early and late in the breeding season than during the peak when the ovulation rate is highest.

Fat females will not respond to flushing. Instead, they should be conditioned for breeding by stepping up the exercise.

Flushing is accomplished by feeding females more liberally 2 to 3 weeks prior to breeding and continuing into the breeding season, so that they are gaining in weight at the time of mating. It may be accomplished by increasing the grain allowance; or in the case of ruminants and mares, by turning them to a fresh, luxuriant pasture.

Soon after breeding, females bred for the first time and dry females should be put back on limited rations. Continuation of a high level of nutrition after breeding will result in a higher embryo mortality.

Nutritional Reproductive Failure

Because livestock producers largely determine their own destiny when it comes to feeding, it is important that they know the causes of reproductive failure and how to rectify them.

A review of the literature clearly points to three reproductive difficulties: (1) a small number of females in heat and bred early in the breeding season, (2) the low conception rate at first service, and (3) the excessive losses at birth or within the first two weeks of age.

Research gives ample evidence that the real cause of most reproductive failure is a deficiency of one or more nutrients just before or immediately following parturition—nutritive deficiencies during the critical period when life begins—a deficiency of energy, protein, minerals, and/or vitamins. Based on an extensive review of the voluminous literature, the authors evolved with the following summary of the nutritional causes of reproductive failures in mammals:

1. **Overfeeding or underfeeding.** Overfeeding, accompanied by extremely high condition, or underfeeding, accompanied by emaciated and run-down condition, usually result in temporary sterility. Overfat females often experience birth difficulties. Excessive thinness results in low birth weights and weak young.
2. **Energy.** A low level of energy during the last third of pregnancy and immediately following parturition will have a marked effect on rebreeding—fewer females will come in heat at the beginning of the breeding season, and fewer will conceive.
3. **Protein.** A low level of protein during gestation results in lowered reproduction, lighter birth weights, and delayed heat following parturition.
4. **Phosphorus.** Low phosphorus will markedly decrease the number of young born.
5. **Iodine.** A deficiency of iodine willl cause impaired reproduction; weak or dead offspring at birth; big-necked (goiterous) calves, lambs, and foals; and hairless pigs.
6. **Vitamin A.** Low vitamin A will result in the birth of weak, malformed, partially blind, or dead young.

With all mammalian species, most of the growth of the fetus occurs during the last third of pregnancy. Additionally, the female must store body reserves during pregnancy, for the demands for milk production are generally greater than can be supplied by the ration fed during early lactation. Hence, the nutrient requirements are very critical during this period, especially for young pregnant females.

It is also known that the ration exerts a powerful effect on sperm production and semen quality. Too fat a condition can lead to temporary or permanent sterility. Moreover, there is abundant evidence that greater fertility of herd sires exists under conditions where a well-balanced ration is provided.

Egg Production

Egg production involves feeding for number of eggs, egg quality, hatchability, and control of molt and broodiness.

The nutritive needs for commercial egg production include those for maintenance of the birds, growth of pullet layers, and the formation of eggs. The nutritive requirements are greater for birds with an inherited capacity for high egg production than for those that lay only a few eggs. The standard-weight egg contains about 95 calories of gross energy, 7.5 grams of crude protein, and 2 grams of calcium.

With poultry, the development and hatching of a fertile egg constitute reproduction. As with mammals, the nutritive requirements of poultry breeders (including chickens, turkeys, ducks, geese, etc.) are more rigorous than those for commerical laying. For high hatchability and good development of young, breeders require greater amounts of vitamins A, D, E, B-12, riboflavin, pantothenic acid, niacin, and of the mineral manganese. Birds intended as breeders should be started on special breeder rations at least a month before hatching eggs are to be saved.

Lactation

Simply stated, milk production is a by-product of the reproductive process.

The lactation requirements of females of all mammalian species for moderate to heavy milk production are much more rigorous than the maintenance or pregnancy requirements. For example, it is estimated that at the peak of lactation the net energy requirements of ewes suckling twins are about three times the maintenance requirement. In fact, the nutritive needs for milk production are exceeded only by sustained, heavy, muscular exercise—like racing. Fortunately, heavy milking females, like dairy cows, can store body reserves of certain nutrients before and during the pregnancy period, to be drawn upon following parturition. Here is how this phenomenon works: When a cow is properly fed before and during pregnancy, certain nutrient deposits are made in the body. Then, during lactation when the demands are greater than can be obtained from the feed, the female draws from the stored body reserves. Thus, both calcium and phosphorus can be stored in the bones, then withdrawn during early lactation when milk production is at its peak. Of course, if there hasn't been proper body storage, something must "give"—and that something will be the mother, for nature ordained that growth of the fetus, and the lactation that follows, shall take priority over the maternal requirements. Hence, when there is a nutrient deficiency, the female's body will be deprived, or even stunted if she is young, before the developing fetus or milk production will be materially affected.

The nutrient needs for lactation depend on the amount and composition of milk secreted. Some idea of the enormity of these requirements, as well as the efficiency of milk secretion, becomes apparent when it is realized that a dairy cow that produces 14,500 lb of milk during 1 year manufactures 523 lb of milk fat, 674 lb of milk sugar, 477 lb of milk protein, and 109 lb of minerals and vitamins, or a total of over 1,783 lb of food. That's equivalent to the carcass weight produced by 2½ steers in 18 months time. Also, it is noteworthy that the cow remains alive and can repeat the production again and again, whereas the steers must be slaughtered or "spent." It is noteworthy, too, that the cow needs additional nutrients for body maintenance (which are about the same as those of one steer), for development of the unborn calf if she is pregnant, and for growth if she is a young cow.

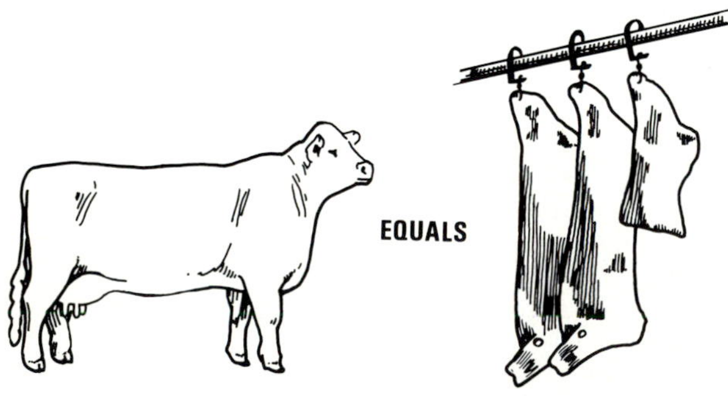

Fig. 2-12. It takes 2½ steers 18 months' time (for each steer) to produce as much carcass weight as one cow produces in milk in 1 year. And the cow remains alive to do it all over again!

Although high-producing cows require more total feed than low producers, they utilize proportionately more nutrients for milk production (Fig. 2-3), and generally they return more net income over feed cost.

One of the most drastic changes in the life cycle of mammals is that which occurs at freshening, when a female suddenly makes the transition from nonlactating to lactating. In high-producing dairy cows, clinical milk fever is an indication of a very sudden drop in the level of calcium in the blood serum and of putting calcium into the milk. If the reversal process is too drastic, the animal may be thrown into a blood calcium deficiency (hypocalcemia), followed by slowing of the heart and eventual coma. Although this problem is most common among high-producing dairy cows, it occasionally occurs in sheep and in other mammals.

Finishing/Fitting

Finishing is what the name implies—the laying on of fat, especially in the tissues of the abdominal cavity and in the connective tissues just under the skin and between the muscles. It is the normal feeding practice followed prior to slaughter, for the purpose of improving the flavor, tenderness, and quality of meat, better to meet consumer demands. Generally speaking, the higher the degree of finish, the higher the dressing percentage and the lower the protein (red meat) content. Also, it takes more nutrients to produce a pound of fat than a pound of lean; hence, excess finish is wasteful and undesirable.

Fattening is usually achieved through the use of high-energy feeds, carbohydrates, and fats—a liberal allowance of grains. However, due to world food shortages, the long-time trend is to incorporate more roughages in finishing rations for ruminants. Such rations are lower in net energy and produce smaller gains than high-concentrate rations, but they may make for more net returns when feed grains are scarce and high in price.

The objective of the livestock producer is to finish animals to the degree of fleshing and carcass weight desired by the consumer, at a maximum of profit for his efforts. Computers have given large operators a big assist in meeting these goals.

Fitting is the conditioning of animals, usually for show or sale, through careful feeding, grooming, and exercising, to enhance their bloom and attractiveness. Fitting animals for show or sale involves the application of similar principles and practices to those followed in fattening (finishing) livestock for market. Animals intended for show or sale should be fed so as to achieve a certain amount of finish or bloom, but they should not be too fat. In general, most fitting rations are similar to the rations used in commercial fattening operations for animals of like species and comparable ages, except that they are usually higher in protein content; experienced caretakers feel that they get more bloom by use of high-protein rations. Also, it is common practice to feed a palatable milk replacer to young animals that are being fitted for show or sale.

Fitting horses for show or sale involves conditioning plus hardening and exercise. Such animals should be worked daily, in harness or under saddle. Fleshing should be obtained without sacrificing action or soundness or without causing filling of the legs and hocks. Thus, they must be carefully fed, groomed, and exercised to bring them to proper bloom.

Wool and Mohair

Wool and mohair are high-protein products. Also, it is noteworthy that they are especially rich in the sulfur-containing amino acid, cystine. But the latter requirement is usually amply met by the cystine of feeds or by methionine—another amino acid which is also rather widely distributed in feeds, as well as being derived from rumen synthesis.

A lack of energy in the ration of sheep will result in lighter fleeces and lower quality of wool, including breaks, or tender spots, in the fiber. A protein deficiency will also make for lighter fleeces.

Copper-deficient sheep produce "steely" wool, lacking in crimp, tensile strength, affinity for dyes, and elasticity. With a severe deficiency, the wool of black sheep is depigmented.

Work/Running

Fig. 2-13. Work being performed by a hunter. The nutritive requirements of such equine athletes are very exacting. (Courtesy, the American Morgan Horse Association, West Moreland, N.Y.)

In the United States, the function of work or running is limited to horses. But, in certain parts of the world, oxen, water buffalo, camels, reindeer, dogs, and other animals are the chief sources of power.

Racehorses, and other horses used for such purposes as hunting, jumping, cutting, or roping, are equine athletes, whose nutritive requirements are very exacting. And the younger the animal, the more intense the use, and the greater the stress, the higher the level of nutrition needed in order to develop and maintain sound legs and build a strong frame and body. It would appear that their nutritive requirements are not unlike human athletes—for example, college football teams and participants in the Olympics—who are required to eat at special training tables, supervised by expert nutritionists; and who are fed generous quantities of meats and other high-protein foods, along with diets high in energy, minerals, and vitamins. Further credence to this analogy is lent by the estimate that, when racing, the energy requirements of a horse are 100 times greater than at rest. Thus, rations for racehorses, and other horses in heavy use, should be rich in available energy, high in protein, and fortified with minerals and vitamins—with all nutrients in proper balance.

For draft horses not in reproduction, the energy needs are met primarily by carbohydrates and fats, nutrients that can be provided in the form of grain. Adequate protein is necessary to maintain, repair, and replace the muscles used in work. The mineral and vitamin requirements for work horses are practically the same as for comparable idle horses —except for the greater need for Na and Cl to offset the salt losses that accompany increased perspiration.

Special care must be exercised to avoid azoturia (Monday morning disease) in horses, a condition which most often develops when a working horse is overfed on an idle day, and which can be prevented by reducing the concentrate allowance on idle days. The disease is commonly referred to as Monday morning disease because it usually occurs after a horse has been idle for a day or two, as over the weekend. The condition appears to be caused by an accumulation of glycogen, a carbohydrate storage compound, in the musculature. When the animal is exercised, the glycogen is broken down for energy and large amounts of lactic acid are formed and accumulated in the muscles. This lactic acid destroys the muscle cells, releasing myoglobin from the cells in the system. The myoglobin produces the dark reddish-brown colored urine, which characterizes the malady. Normally, the acid is carried off by the vascular system and does no damage. It is the unusually large storage of available carbohydrates, and the inability of the blood to "wash" away the wastes formed quickly enough when they are produced in the muscles, that results in muscle fiber damage. The kidneys are not built to cleanse the blood of myoglobin, and kidney damage can result from these by-products of muscle damage.

QUESTIONS FOR STUDY AND DISCUSSION

1. Which is the most important in animal production, heredity or nutrition?

2. Define the terms *nutrition* and *nutritionist.*

3. What is the primary reason for keeping animals?

4. Why has much of our knowledge of human nutrition come via the animal route?

5. How has the quest for food shaped the course of history?

6. Why was medicine prior to the 18th Century largely based on dietary theories?

7. Discuss the leading nutrition developments in each of the following eras, along with the impact of each:
   a. Naturalistic era
   b. Chemical-energy era
   c. Mineral era
   d. Vitamin-biological era
   e. Feeding standard era
   f. Amino acid era
   g. Feed additive and implant era
   h. Biotechnology era

8. What's the purpose of pre-slaughter withdrawal times or milk-discard periods for animals receiving certain drugs? How may livestock producers be certain that they have used drugs safely?

9. Identify and discuss the five nutritional milestones that have had the greatest longtime impact.

10. How will each member of the food team—producers, processors, and consumers—benefit from biotechnology?

11. As the scientific domain of the nutritionist has become increasingly narrow, frequently such individuals (a) lack understanding of animals as a whole, and (b) pay little attention to the comparative aspects with other animal species. Is this good or bad? If it is bad, how may it be rectified?

12. How, and why, should we improve our store of knowledge relative to the following nutrition-related areas:
   a. Nutrient requirements
   b. Net energy system
   c. Proportion of forage to grain
   d. Behavior and environment of animals
   e. Conservation of energy
   f. Pollution control

13. Why is knowledge of body composition important to the nutritionist? What is the effect of each of the following on body composition:
   a. Age
   b. Feed intake
   c. Rate of gain
   d. Grower type rations
   e. Species

Generally speaking, experienced cattle feeders feed grower rations during the growing (stocker) phase, then, as cattle get heavier, they switch to high-energy finishing rations. Discuss the practicality of such feeding programs.

14. On the average, how many pounds of water per pound of protein are found in the animal body?

15. Classify feed nutrients. What is the significance of such a classification?

16. What is meant by the "functions of nutrients"?

17. Discuss the nutrient needs for each of the following body functions, with appropriate application to different species:
   a. Maintenance
   b. Growth
   c. Reproduction
   d. Egg production
   e. Lactation
   f. Fattening/fitting
   g. Wool and mohair
   h. Work/running

18. List and discuss the factors affecting the nutritional requirements for each of the following:
   a. Maintenance
   b. Growth
   c. Reproduction

19. Why is weight gain not the best measure of growth in meat animals? What measure(s) would you substitute for it in meat animals?

20. What is compensatory growth? How does this phenomenon affect (a) the cow-calf producer, and (b) the cattle feeder?

21. Compare the reproductive problems encountered in mammals and poultry. Discuss which is the easiest to improve.

22. What is the justification for liberal feeding of heifers for early sexual maturity?

23. What is flushing? How may females of each species be flushed?

24. Give and discuss some examples of nutritional reproductive failure.

25. What is the difference(s) between (a) finishing animals for market, and (b) fitting animals for show or sale?

26. Why are the nutritive requirements of a 2-year-old racehorse so critical?

SELECTED REFERENCES

Title of Publication	Author(s)	Publisher
Animal Growth and Nutrition	E. S. Hafez I. A. Dyer	Lea & Febiger, Philadelphia, Penn., 1969
Animal Nutrition, 3rd Edition	P. McDonald R. A. Edwards J. F. D. Greenhalgh	Longman, London and New York, 1981
Animal Nutrition, 7th Edition	L. A. Maynard J. K. Loosli H. F. Hintz R. G. Warner	McGraw-Hill Book Company, New York, N.Y., 1979
Animal Science, 8th Edition	M. E. Ensminger	The Interstate Printers & Publishers, Inc., Danville, Ill., 1983
Applied Animal Feeding and Nutrition	M. H. Jurgens	Kendall/Hunt Publishing Company, Dubuque, Iowa, 1973
Applied Animal Nutrition	E. W. Crampton L. E. Harris	W. H. Freeman and Co. Publishers, San Francisco, Calif., 1969
Basic Animal Nutrition and Feeding, Second Edition	D. C. Church W. G. Pond	John Wiley and Sons, New York, 1982
Bioenergetics and Growth	S. Brody	Reinhold Publishing Corp., New York, N.Y., 1945
Body Composition in Animals and Man, Pub. 1598	National Research Council	National Academy of Sciences, Washington, D.C., 1968
Feeds and Feeding	A. Cullison	Reston Publishing Company, Inc., Reston, Va., 1975
Feeds and Feeding, 22nd Edition	F. B. Morrison	The Morrison Publishing Company, Ithaca, N.Y., 1956
Feeds and Feeding, Abridged	F. B. Morrison	The Morrison Publishing Company, Ithaca, N.Y., 1958
Feeds for Livestock, Poultry and Pets	M. H. Gutcho	Noyes Data Corporation, Park Ridge, N.J., 1973
Fire of Life, The	M. Kleiber	John Wiley & Sons, Inc., New York, N.Y., 1961
Food and Animals—a global perspective	M. E. Ensminger	Agriservices Foundation, Clovis, Calif., 1975
Fundamentals of Nutrition	E. W. Crampton L. E. Lloyd	W. H. Freeman and Co. Publishers, San Francisco, Calif., 1959
Manual of Clinical Nutrition	R. S. Goodhart M. G. Wohl	Lea & Febiger, Philadelphia, Penn., 1964
Mineral Nutrition of Livestock, The, 2nd Edition	E. J. Underwood	Food and Agriculture Organization of the United Nations, Rome, Italy, 1981
Nutrition of Animals of Agricultural Importance, Parts 1 and 2	Ed. by D. Cuthbertson	Pergamon Press, London, England, 1969
Proceedings of the Fifth International Congress on Nutrition	National Research Council	Waverly Press, Inc., Baltimore, Md., 1961
Science of Animals That Serve Mankind, The	J. R. Campbell R. T. Marshall	McGraw-Hill Book Company, New York, N.Y., 1969
Science of Providing Milk For Man, The	J. R. Campbell R. T. Marshall	McGraw-Hill Book Company, New York, N.Y., 1975
Stockman's Handbook, The, 5th Edition	M. E. Ensminger	The Interstate Printers & Publishers, Inc., Danville, Ill., 1978
Vitamins, The: Chemistry, Physiology, Pathology, Methods, Vols. I-III, 2nd Edition	Ed. by W. H. Sebrell, Jr., R. S. Harris	Academic Press, Inc., New York, N.Y., 1967, 1968, 1971
Vitamins in Feeds for Livestock	F. C. Aitken R. G. Hankin	Commonwealth Agricultural Bureaux, Farnham Royal, Bucks, England, 1970
Vitamins and Hormones, Vol. XV	Ed. by R. S. Harris, G. F. Marrian, K. V. Thimann	Academic Press, Inc., New York, N.Y., 1957

Fig. 2-14. Nutrition is the science of feeds as they relate to the requirements, production, and health of animals. (Courtesy: (1) J. C. Allen & Son, West Lafayette, Ind.; (2) USDA; (3) Kildangen Stud, Monasterevin, Ireland; (4) USDA; (5) Sperry New Holland, New Holland, Pa.; and (6) J. C. Allen & Son, West Lafayette, Ind.)

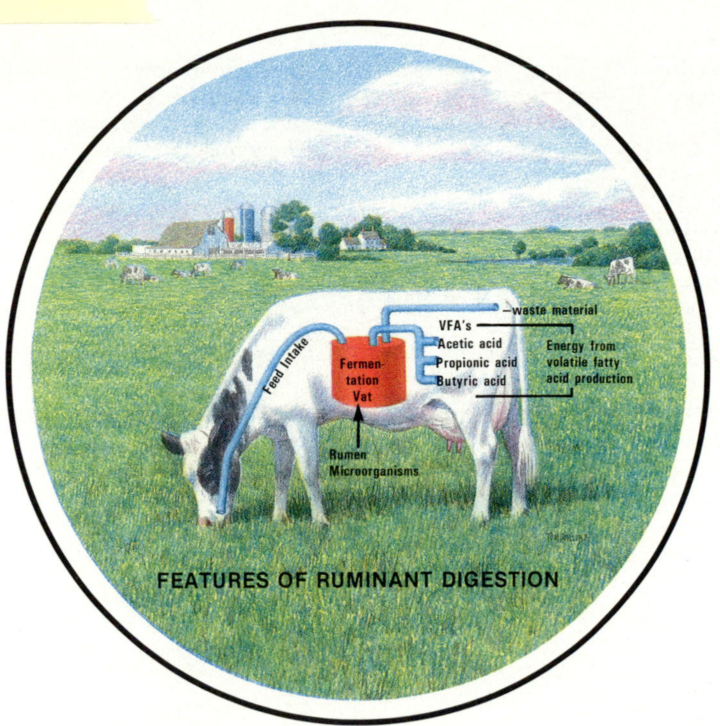

Original painting by Tom Phillips

Chapter 3

DIGESTION/ ABSORPTION

Contents Page

In nature, the primary purpose of grains and other seeds is to provide a means of propagating plants—not to supply nutrients to animals that consume them. Likewise, eggs—a universally used food and feed—provide a medium from which the developing embryo can draw vital nutrients. As a result, numerous differences can be observed in the anatomy and physiology of digestion in various livestock species which enable them to utilize feedstuffs efficiently. These differences can be observed through two modes of adaptation: (1) behavioral, and (2) anatomic. Thus, animals that consume large amounts of fibrous feeds, such as hays, differ behaviorally and anatomically from animals that consume meat and other easily digested feeds. Therefore, in order to maximize the use of feed, the caretaker and nutritionist must have an understanding of these differences in digestion and absorption.

When a feed is evaluated by chemical analysis alone, there is a good possibility that this evaluation is not a true indicator of its value to livestock. No matter how accurate the chemical analysis, it tells us little with regard to how the feed will react when subjected to the physiological phenomena occurring in the animal. Nor does it tell us anything about palatability. If a feed is extremely high in nutrients but is refused by animals, its value is nil. Likewise, once the feed is ingested, the animal must be able to digest and absorb the nutrients efficiently.

HUNGER AND APPETITE

Hunger is the physiological desire for feed following a period of fasting.

Appetite is a learned or habitual response to the presence of feed.

An animal that is extremely hungry may not have an appetite for a type of feed that it deems undesirable. Conversely, if the feed is of a desirable nature, an animal may have an appetite for it in spite of the fact that it is not hungry.

Fig. 3-1. "Hunger," an original painting by artist Tom Phillips, 3333 17th St., San Francisco, CA 94110

Hypothalamic Control of Appetite

The hypothalamus (derived from the terms *hypo* meaning below, and *thalamus,* a region of the brain)—a structure in the ventral region of the diencephalon—has been implicated as one of the major control centers of appetite regulation. Within the hypothalamus, certain areas can be differentiated. Two of these are of particular importance in the regulation of appetite. The first area is that of the lateral hypothalamus. It is commonly called the *feeding center,* because, upon stimulation of this region, the animal commences to eat whether or not it is hungry. If this area is damaged, the animal loses all desire to eat and will eventually starve. The ventro-medial area of the hypothalamus functions as the *satiety center.* Stimulation of this region will depress appetite. If the ventro-medial nuclei are destroyed, there is no inhibition of feed intake, and the animal will have an uncontrollable appetite. It is believed that there is a chronic activity in the lateral hypothalamus which is kept in check by the inhibitory influence of the ventro-medial area.

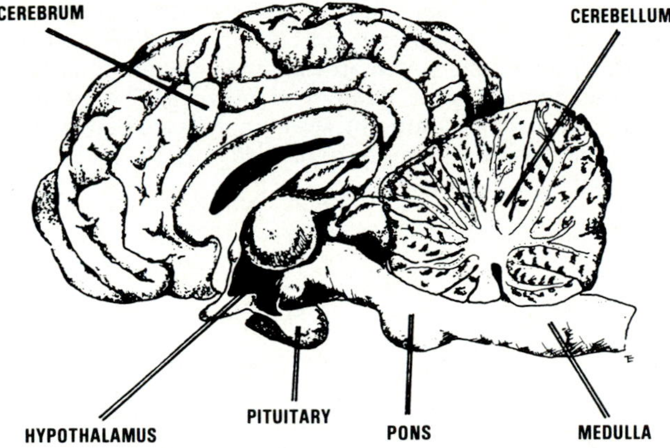

Fig. 3-2. Brain of the horse showing location of important parts, including hypothalamus.

Several theories have been advanced as to the exact physiological mechanism which triggers the hypothalamus to tell the animal when to eat. While each theory has its merits, there is no conclusive proof in favor of any one of them. In the long run, it seems probable that a combination of a number of factors will provide the answer.

The two theories concerning the hypothalamic control of appetite that have received the most attention are (1) the chemostatic hypothesis, and (2) the thermostatic hypothesis. The chemostatic hypothesis reasons that the hypothalamus is sensitive to circulating blood nutrient levels, such as sugar or lipid. When these levels become too low, the hypothalamus sends signals to begin feeding. Once the blood nutrient level is elevated, stimuli from the feeding center are inhibited and the animal feels full. The second theory of appetite control, the thermostatic hypothesis, theorizes that the hypothalamus plays an important role in heat regulation within the body, and that a decrease in hypothalamic temperature will induce feeding.

Gastric Influences on Appetite Regulation

Appetite in ruminants has been shown to be sensitive to volatile fatty acid levels in the rumen. When acetate is injected into the dorsal rumen, feed consumption is reduced. Researchers believe there are receptors in this area which are sensitive to the acetate levels produced in the rumen, thereby influencing feed intake to a limited degree. Rumen propionate levels affect feed intake, but this receptor system may be distinctly different from the one involving acetate. When the gastrointestinal tract is distended, there is normally a cessation of feeding. Thus, the actual physical limit of the digestive system has a direct influence on appetite. If an animal eats bulky or succulent feed, it may become satiated, even though it does not fulfill its energy requirements.

TYPES OF FEEDING BEHAVIOR

In general, the feeding behavior of animals is correlated with the various anatomical adaptions of the gastrointestinal tract. It is logical to assume that animals which consume feed that can be easily digested and absorbed would have gastrointestinal tracts that are smaller and simpler in structure than those animals which utilize feeds that are complex in chemical composition.

Based on the kind of feed eaten, animals are classified as follows:

1. **Carnivores.** These are the flesh eaters. They feed mostly on the flesh of other species. The organization of the digestive tract is relatively simple. When a wild carnivore kills an animal, it will first consume the contents of the gastrointestinal tract which contains partially digested and digested plant material. However, the great bulk of the diet consists of meat, and it is this type of feed that the gastrointestinal tract is best equipped to handle. The feed that is consumed is, for the most part, composed of fat and protein—both of which are readily digested and absorbed. For this reason, the length of the digestive tract is short. Examples of this type of animal are mink, dogs, and cats.
2. **Herbivores.** These are vegetarians. They depend entirely upon plants for their feed supply. Because plant material is difficult to digest, anatomical adaptations enabling the animal to digest the material more efficiently have evolved. The gastrointestinal tract in these animals tends to be long, and morphological changes in the structure of the gastric and/or cecal-colon areas enable these animals to digest and utilize plant material efficiently. Examples of herbivores are cattle, goats, sheep, and horses.
3. **Omnivores.** These consume both flesh and plants. Anatomically, these animals represent an intermediate stage in the types of the gastrointestinal tract. The digestive tracts of omnivores are generally longer than those of carnivores but shorter than those of herbivores. Plant material can be utilized by omnivores, but the efficiency of utilization does not approach that of herbivorous animals. Humans and swine are examples of omnivores.

ANATOMY OF THE DIGESTIVE SYSTEM

The digestive system, or alimentary canal, consists of a tube which courses internally from the lips to the anus. It is specialized in regions called the mouth, esophagus, stomach, small intestine, cecum, large intestine, rectum, and anus. Attached to the tube are the liver and the pancreas, which provide essential secretory products for digestion.

There are major differences in the anatomy and physiology of the organs of the digestive tract of the various animal species. These differences are of great nutritional significance, as they affect the nature of the digestive processes—hence, the kind of feed that can be utilized. Based on the anatomy of the digestive tract, animals may be grouped as (1) nonruminants, (2) ruminants, and (3) avian. An understanding of the types of digestive systems and the kind of feed eaten is essential to the intelligent feeding of each class of animal.

Nonruminant Digestive System

These animals possess stomachs that are relatively simple in structure. This group can be subdivided with respect to the functionality of the cecum and colon.

MONOGASTRIC (SIMPLE-STOMACHED) DIGESTIVE SYSTEM

These animals have the simplest of all digestive systems. It consists of the mouth and associated glands, esophagus, stomach, small intestine, large intestine, pancreas, and liver. This is the type of gastrointestinal tract found in the pig, dog, mink, fish, monkey, and humans. It is characterized by limited capacity and limited microbial action and fiber digestion. It follows that these animals are better adapted to the use of concentrated feeds, such as grains and meat products, than to the use of large quantities of roughages.

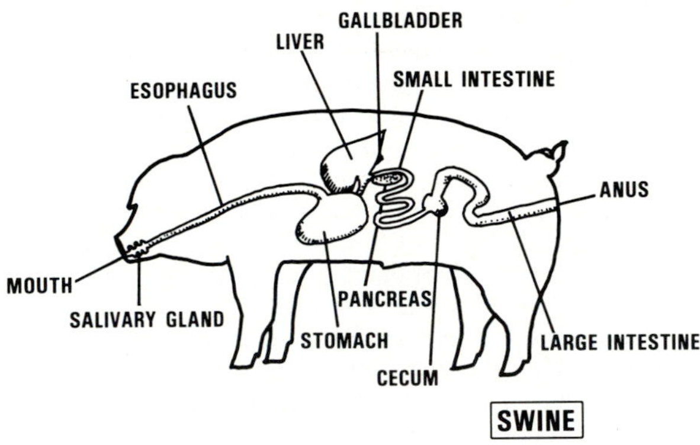

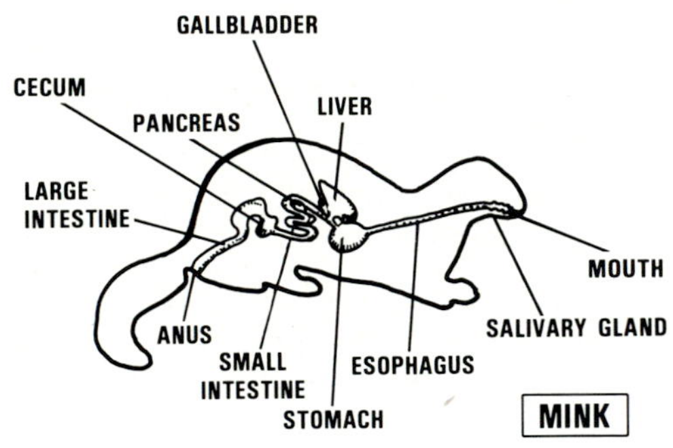

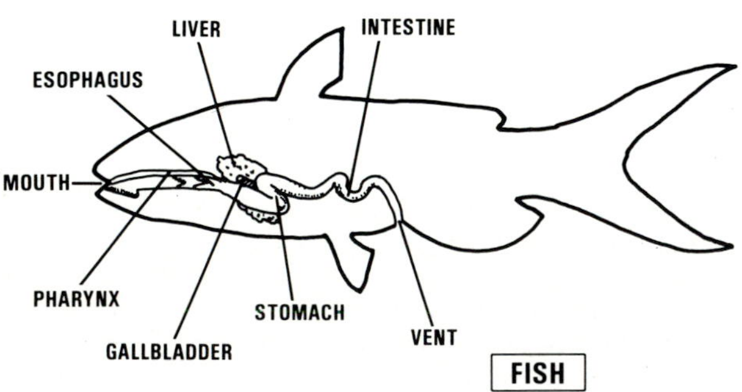

Fig. 3-3. The digestive systems of representative monogastrics with nonfunctional cecums: swine, mink, and catfish.

FUNCTIONAL CECUM DIGESTIVE SYSTEM (Nonruminant Herbivore)

In this type of digestive system—as represented by the horse, rabbit, guinea pig, and hamster—the cecum and colon are extremely large and contain a large population of microorganisms which are capable of digesting fiber as well as synthesizing a number of vitamins. Thus, from a practical feeding standpoint, these animals are between (1) the monogastric or simple-stomached animals, and (2) the ruminants.

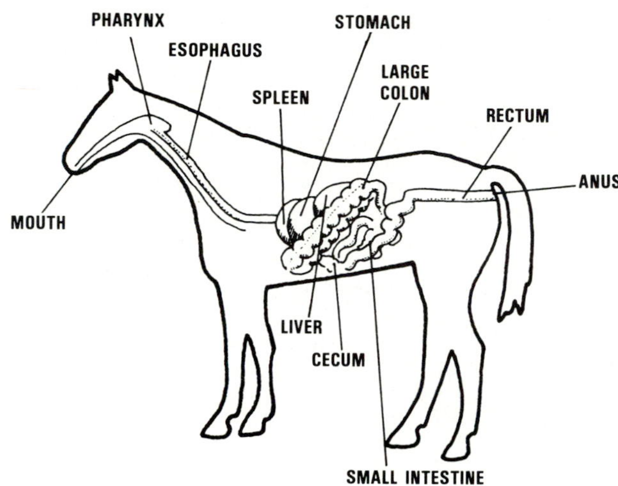

Fig. 3-4. The digestive system of the horse—a nonruminant herbivore with a functional cecum and colon.

Ruminant Digestive System

Cattle, sheep, and goats are ruminants. They differ from monogastric animals in the following important ways:

1. **Mouth.** Ruminants have no upper incisor or canine teeth. Thus, they depend on the upper dental pad and lower incisors, along with the lips and tongue, for the prehension of feed.

2. **Four stomach compartments.** Ruminants possess four stomach compartments—rumen, reticulum, omasum, and abomasum (true stomach)—whereas monogastrics have one. Such a digestive system makes for two primary nutritional differences between ruminants and simple-stomached animals:

   a. **More space.** They have the necessary space for processing large quantities of bulky forages to provide their nutrients. The cow, for example, when compared to the human on a proportion-to-weight basis, has about nine times the digestive tract capacity.

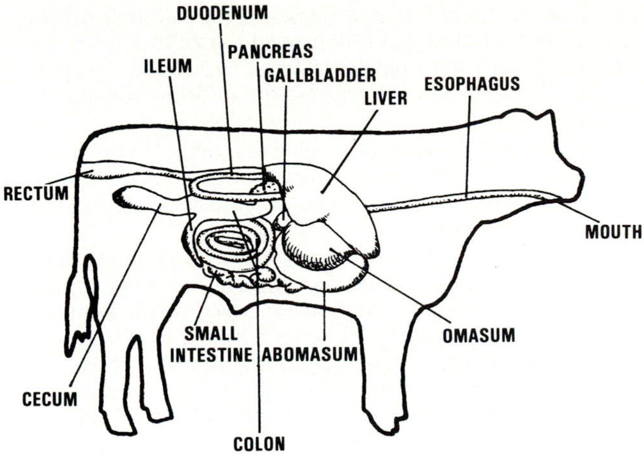

Fig. 3-5. The digestive system of the bovine—a ruminant.

b. **More microorganisms.** The rumen provides a desirable environment for an enormous population of microorganisms. Typical counts of rumen bacteria range from 25 to 80 billion/ml, and typical counts of protozoa range from 200,000 to 500,000/ml. The number of rumen bacteria varies according to the nature of the diet, feeding regimen, time of sampling after feeding, species differences, individual animal differences, season, availability of green feed, and the presence or absence of ciliate protozoa.

Rumen microorganisms serve two important functions:

(1) They make it possible for ruminants to utilize roughage—to digest the fiber therein. They break down the cellulose and pentosans of feeds into usable organic acids, chiefly acetic, propionic, and butyric acid—commonly called the volatile fatty acids (VFA). These VFAs are largely absorbed through the rumen wall and provide the ruminant 60 to 80% of its energy needs. Microbial digestion is of great practical importance in the nutrition of ruminants; it is the fundamental reason why they can be maintained chiefly on roughages.

(2) In exchange for their rumen-housing privileges, the microbes synthesize nutrients for their host, in a true symbiotic relationship. Rumen microbes synthesize, or manufacture, all the B-complex vitamins and all the essential amino acids. The latter can even be made from nonprotein nitrogen compounds (NPN), such as urea or ammoniated products, or from proteins that are deficient in one or more of the amino acids. Finally, the microorganisms give their life to their host in payment for food and shelter, being digested farther along in the gastrointestinal tract.

NOTE WELL: Although rumen microbes synthesize all the B-complex vitamins and all the essential amino acids, it has been shown that they do not supply them in adequate amounts for the maximum growth or milk production of ruminants.

3. **Rumination.** A placid cow lying under a tree slowly chewing her cud conveys a special sense of contentment symbolic of the tranquility of the countryside. But this activity, or phenomenon, which is peculiar to ruminants, is of great practical significance.

During rumination, the animal regurgitates and rechews a soft mass of coarse feed particles, called a *bolus.* Each bolus is chewed for about a minute, then swallowed again. Ruminants may spend 8 hours or more per day in rumination, the amount of time varying according to the nature of the diet. Coarse, fibrous diets result in more time ruminating. Rechewing does not improve digestibility. Rather, rumination has an important bearing on the amount of feed the animal can eat and utilize. Feed particle size must be reduced to allow passage of the material from the rumen. Because high-quality forages contain less fiber than low-quality forages, they require much less rechewing and pass out of the rumen at a faster rate; hence, they allow the cow to eat more. The rumination process also stimulates the salivary glands in the mouth. These produce large amounts of saliva which facilitate food passage through the digestive tract. Saliva also contains buffer salts which resist changes in pH of the rumen contents and protect against acidosis.

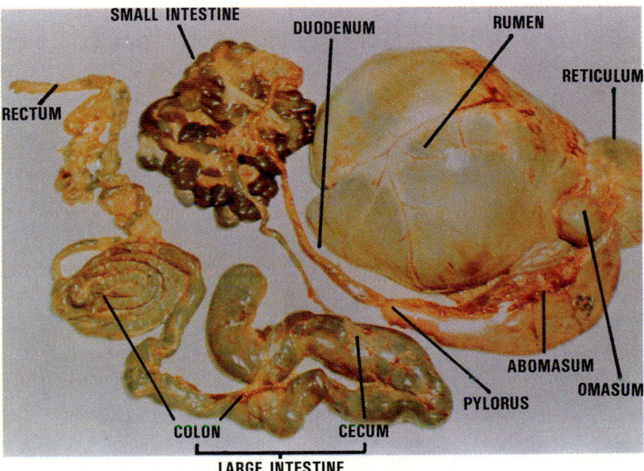

Fig. 3-6. The digestive tract of the sheep. The digestive organs of all ruminants look very similar and function in much the same manner. (Courtesy, CSIRO, Australia)

4. **Eructation (belching of gas).** Substantially more gas is produced in digestion by ruminants than by simple-stomached animals. The microbial fermentation in the rumen results in the production of large amounts of gases (primarily CO_2 and methane) which must be eliminated; otherwise, bloat results. Normally, these gases are expelled quite freely by eructation (belching) and, to a lesser extent, by absorption into the blood draining from the rumen, from which they are eliminated through exhaled air from the lungs.

(Also see Chapter 11, section on "Protein and Amino Acids For Ruminants.")

ESOPHAGEAL GROOVE (RETICULAR GROOVE)

The anatomical peculiarities of the newborn ruminant are: (1) The reticulum, rumen, and omasum are relatively underdeveloped; (2) the esophageal groove, which is formed by two heavy muscular folds, or lips, and which extends from the lower end of the esophagus into the omasum, can convey material from the esophagus directly to the abomasum; and (3) nursing stimulates reflex closure of the esophageal groove and causes the milk consumed to bypass the rumen and reticulum and most of the omasum, thereby escaping bacterial fermentation, and to pass directly into the last compartment of the gastric region—the abomasum.

As the young ruminant grows older, it ingests increasingly larger quantities of solid feed which, in turn, stimulates the growth and development of the reticulum and rumen. Also, through its contact with the environment and other ruminants, the young ruminant becomes naturally inoculated with microorganisms which benefit digestion. At this stage, reflex to milk diminishes and the animal functions as a mature ruminant. Age and type of feed are the main factors associated with loss of response to suckled milk. Normally, most of this transition occurs in the young ruminant by 2 months of age. Thus, at birth the first three compartments (reticulum, rumen, and omasum) of the new born calf represent less than 30% of the total somach capacity; by 2 months of age, they represent 70% of the total stomach capacity; and at maturity, they account for 93% of the total stomach capacity. The greatest development is in the rumen, which accounts for 80% of the total stomach capacity of the mature cow. (Also see subsequent section in this chapter on "Newborn Ruminants.")

COMPARISON OF HORSE AND RUMINANT STOMACHS

The nonruminant animal with a functional cecum and colon, such as the horse, is somewhat similar to the ruminant in that it uses a microbial population to digest fiber. But, several differences between these types of digestive systems can be noted; among them, the following:

1. The ruminant stomach has four compartments, whereas the horse stomach has one.
2. The stomach capacity of the horse is much smaller—less than 5 gal for the mature horse as opposed to about 66 gal for the mature cow. Because of its small stomach, feeding a horse too much roughage may cause impaction, followed by labored breathing and quick tiring. Actually, the horse's stomach is designed for almost constant intake of small quantities of feed, rather than large amounts at any one time.
3. The cecum (the horse's fermentation vat) and colon are located on the distal end of the gut, a fact that is of far greater significance than their size. They follow the small intestine, with the result that the ingesta pass from the ileum directly to the cecum and then to the large intestine. By contrast, the anatomical arrangement of the cow is such that the ingesta pass from the rumen (the cow's fermentation vat) through the other compartments of the stomach to the small intestine, thence to the large intestine.
4. At the time of feeding, the ingesta pass through the horse's stomach very rapidly—so much so that feed eaten at the beginning of the meal passes to the intestine before the last part of the meal is completed. Without feed, the horse's stomach will empty completely in 24 hours, whereas it takes about 72 hours (3 times as long) for the ruminant's stomach to empty.

The anatomical and physiological differences in the horse are of great importance nutritionally, for the following reasons:

1. There is less microbial activity in the horse than in the ruminant. As a result—
    a. The horse does not break down more than about 30% of the cellulose of feed, whereas the ruminant breaks down 60 to 70%. Hence, horses cannot handle as much roughage as ruminants. Therefore, higher quality (lower cellulose content) forages should be fed to horses.
    b. The horse synthesizes only limited amounts of proteins, B vitamins, and vitamin K, whereas the ruminant synthesizes sizable quantities of these. Thus, the addition of B vitamins to the ration of the horse (along with vitamins A and D, which are dietary essentials and not synthesized in the digestive tract of any class of farm animals) is good insurance, especially when horses are under stress and high-quality feeds are not being fed.
2. The efficacy of absorption of nutrients synthesized by the microorganisms in the cecum and colon is questioned, since the synthesis takes place after the most highly absorptive part of the digestive tract has been passed.

It follows that, in comparison with the cow, the horse should be fed less roughage, higher quality proteins (and not nonprotein nitrogenous products such as urea), and added B vitamins and vitamin K. These facts lead the authors to the conclusion that the nutritive requirements of the horse more nearly parallel those of the pig than of the cow.

Avian Digestive System

The digestive system of poultry differs considerably from that of other monogastric animals. Birds have no teeth; hence, there is no chewing. The esophagus empties directly into the crop, where the feed is stored and soaked. From the crop, the feed passes to the proventriculus (or glandular stomach), the thick-walled organ immediately in front of the

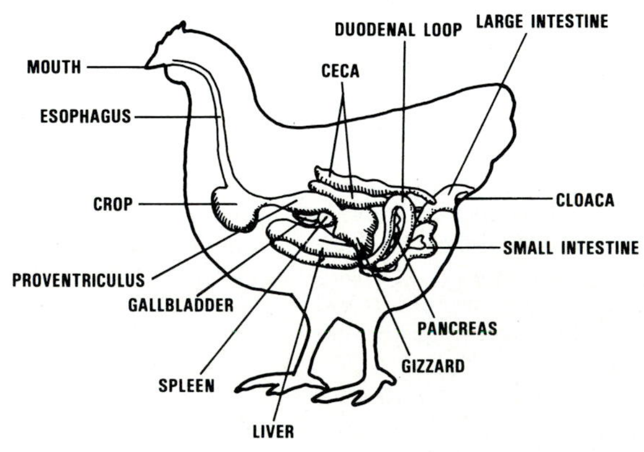

Fig. 3-7. The digestive system of the hen.

gizzard. Here it is stored temporarily while digestive juices are copiously secreted and mixed with it. Thence, it passes to the gizzard, a very muscular organ, which normally contains stones or grit, where it is crushed and ground. Then the feed moves through the small intestine, the ceca, and the large intestine to the cloaca.

Digestion in the fowl is rapid. It requires only about 2½ hours in the laying hen, and 8 to 12 hours in the nonlaying hen, for the feed to pass from the mouth to the cloaca.

CAPACITY OF THE DIGESTIVE TRACT

Due to the anatomic adaptations of the various livestock species, the relative importance of the various digestive organs is reflected in their respective capacities. In ruminants, the stomach has the largest capacity, while in the pig the small intestine and the stomach have about the same capacity. In the horse, the cecum and colon are the largest segments of the gastrointestinal tract. (See Table 3-1.)

TABLE 3-1
PARTS AND AVERAGE CAPACITIES OF DIGESTIVE TRACTS OF SELECTED ANIMALS[1]

	Animal Species									
	Cattle		Sheep or Goat		Horse		Pig		Humans	
	(gal)	*(liter)*	(gal)	*(liter)*	(gal)	*(liter)*	(gal)	*(liter)*	(gal)	*(liter)*
Gastric compartment:										
Rumen (paunch)	53.4	*202.4*	6.2	*23.5*						
Reticulum (honeycomb)	2.0	*7.6*	0.5	*1.9*						
Omasum (manyplies)	5.0	*18.9*	0.2	*0.8*						
Abomasum (true stomach)	6.1	*23.1*	0.9	*3.4*	4.8	*18.2*	2.1	*7.9*	0.3	*1.1*
Subtotal	66.5	*252.0*	7.8	*29.5*	4.8	*18.2*	2.1	*7.9*	0.3	*1.1*
Small intestine	17.4	*65.9*	2.4	*9.1*	16.9	*64.0*	2.4	*9.1*	1.0	*3.8*
Cecum	2.6	*9.8*	0.3	*1.1*	8.9	*33.7*	0.4	*1.5*		
Large intestine	7.4	*28.0*	1.2	*4.5*	25.4	*96.1*	2.3	*8.7*	0.3	*1.1*
Total	93.9	*355.9*	11.7	*44.2*	56.0	*212.0*	7.2	*27.2*	1.6	*6.0*

[1]Adapted by the authors from Swenson, M. J. (ed): *Dukes' Physiology of Domestic Animals,* 9th ed., 1977, by Cornell University, by permission of Cornell University Press.

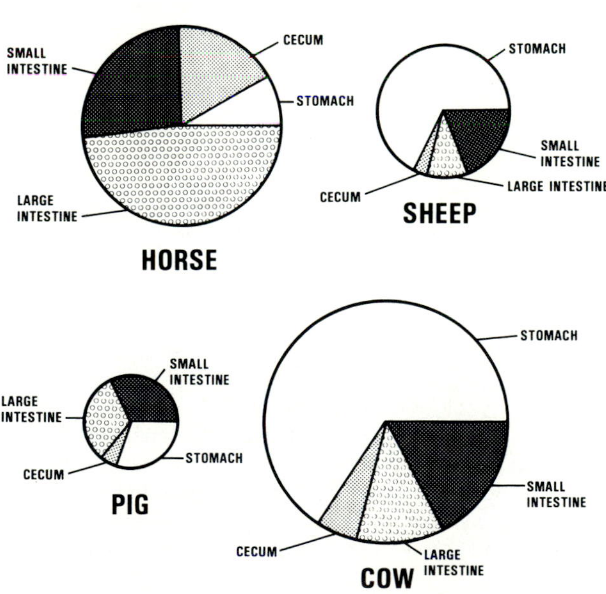

Fig. 3-8. Relative capacities of digestive tract segments.

Fig. 3-8, clearly illustrates the importance of the cecum and large intestine in digestion in the herbivorous nonruminant.

PROCESS OF DIGESTION

Digestion, taken in a narrow sense of the word, can be defined as the process whereby proteins, fats, and complex carbohydrates are broken down into units that are of small enough size to be absorbed through the gut wall, into the animal body, proper. This process is accomplished by both chemical and physical processes.

Enzymes are organic catalysts produced by certain cells within the body which speed biochemical reactions at ordinary body temperatures without being used up in the process. Enzymatic activity is responsible for most of the chemical changes occurring in feeds as they move through the digestive tract. A summary of the enzymes involved in the digestive process of farm animals is presented in Table 3-2.

Many of the digestive enzymes are stored in an inactive form. When they are in the inactive form, they are called *zymogens* or *proenzymes.* Once secreted into a favorable environment for digestion, generally governed by pH, these inactive enzymes "turn on" and perform their specific digestive function.

TABLE 3-2
DIGESTIVE PROCESSES IN FARM ANIMALS AND POULTRY

Region	Secretion (Secreted by)	Enzyme	Enzyme Acts On, or Function	End Product of Digestion	Comments
Mouth	Saliva (salivary glands).	Amylase (ptyalin). Maltase.	Starch, dextrins. Maltose.	Maltose and dextrins. Glucose.	Saliva lubricates food. In ruminants, the buffer salts of saliva help control acidity in the stomach.
Crop (birds)	Mucus.		Lubricates and softens food.		
Rumen		Enzymes from microorganisms.	Cellulose, polysaccharides, starches, sugars, fats, proteins (urea).	Volatile fatty acids. Microbial protein. B vitamins. Vitamin K.	
Stomach (abomasum) in animals; proventriculus in birds	Gastric juice and acids (chiefly HCl)(walls of stomach). Mucus.	Pepsin. Lipase (in carnivores). Amylase.	Protein. Fat.	Proteoses, polypeptides, peptides. Higher fatty acids and glycerol. Coating of stomach lining and lubrication of food.	
Nursing animals.	Gastric juice (walls of stomach).	Renin.	Milk protein (casein).	Coagulates milk protein.	
Gizzard (birds)			Grinding.	Ground foods. Reduced particle size.	
Duodenum (small intestine)	Pancreatic juice (pancreas).	Trypsin. Chymotrypsin. Amylopsin (amylase). Steapsin (lipase). Carboxypeptidase. Collagenase. Cholesterol esterase.	Proteins, proteoses, peptones, and peptides. Starch, dextrins. Fats. Peptides. Collagen. Cholesterol.	Peptones, peptides. Amino acids. Maltose, dextrins. Higher fatty acids and glycerol. Amino acids and peptides. Peptides. Cholesterol esterified with fatty acids.	Low in ruminants.
	Bile (liver).		Fats.	Emulsion of fats (soap, glycerol).	
Small intestine	Intestinal juice (secreted by intestinal wall).	Peptidase (erepsin). Sucrase (invertase). Maltase. Lactase. Polynucleotidase.	Peptides. Sucrose. Maltose Lactose. Nucleic acid.	Amino acids and dipeptides. Glucose and fructose. Glucose. Glucose and galactose. Mononucleotides.	Very low in ruminants. Low in ruminants. High in young mammals.
Large intestine (cecum and colon)		Cellulase from microorganisms.	Cellulose, polysaccharides, starches, sugars.	Volatile fatty acids. Microbial protein. B vitamins. Vitamin K.	

PROCESS OF ABSORPTION

Once the various nutrients have been adequately digested, several modes of absorption can occur. These modes are dependent on the chemical nature of the nutrient and the site of absorption. Virtually no absorption takes place before the feed enters the stomach. In the ruminant, the microflora digest a large proportion of the feed, and considerable absorption takes place in the rumen. In the nonruminant species, very little absorption occurs in the stomach. Rather, the primary site for the absorption of most nutrients is in the small intestine. In nonruminants with a functional cecum and colon, some proteins and amino acids may be absorbed in the colon. For the most part, absorption in the large intestine is restricted to water and electrolytes.

Mechanisms of Absorption

Several mechanisms of nutrient absorption are found in animals. The particular mechanism to be used is dependent on the physical size of the particle, the chemical properties of the nutrient, and the site of absorption. The three basic mechanisms of absorption are diffusion, active transport, and pinocytosis.

DIFFUSION

Diffusion is the mechanism whereby the molecules of a solvent expand to fill all of the available volume of a given area. If two solutions are separated by a permeable membrane, and one solution has either a higher chemical or electrical concentration than the other, diffusion will occur until there is a uniform concentration throughout both solutions.

Osmosis is a form of diffusion whereby there is a migration of solvent molecules through a membrane into a solution of higher concentration which is impermeable to that membrane. The effect of the osmotic pressure of a solution relative to that of plasma is called *tonicity.* If the solution is of a higher concentration than that of plasma, it is referred to as a *hypertonic* solution. Conversely, if the solution is of a lower concentration relative to plasma, it is *hypotonic*. (See Fig. 3–9.)

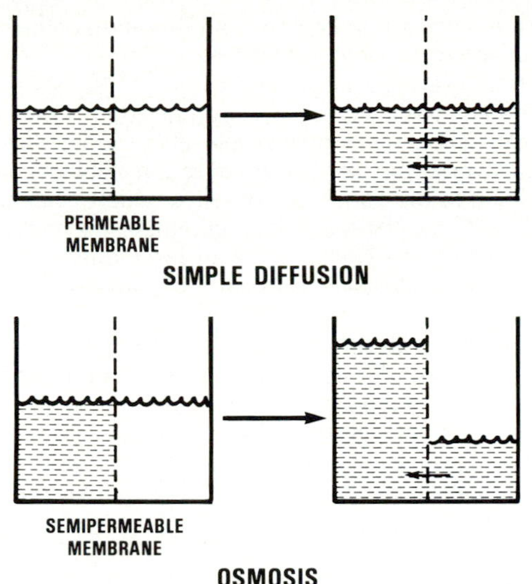

Fig. 3–9. Simple diffusion occurs when a permeable membrane separates two solutions of differing concentrations. There is movement across both sides of the membrane until an equilibrium of concentration is reached. Osmosis occurs when two solutions of unequal concentrations are separated by a semipermeable membrane. Movement is unilateral.

The rate of diffusion depends on the size, shape, charge, and polarity of the particle that is to be absorbed.

ACTIVE TRANSPORT

In the process of diffusion, no energy is expended. In many cases, though, a nutrient must cross a membrane against a concentration gradient (chemical or electrical). This process—called *active transport*—requires the expenditure of a certain amount of energy. Diffusion is a balancing of the solution concentrations on both sides of a membrane, while active transport works against a concentration gradient.

The exact mechanism as to how active transport is achieved has not been elucidated, but there is strong evidence indicating the involvement of a carrier system. The nutrient to be absorbed is believed to combine with a specific membrane-bound carrier with the resulting complex moving the nutrient across the membrane. Once the nutrient passes through the membrane, the carrier and the nutrient dissociate.

In physiological systems, sodium ions are generally found in higher concentrations outside of the cell. This has led to the suggestion that active transport is involved in what is known as the *sodium pump*. There has been speculation, with good evidence, that the sodium ion carrier can combine with either glucose or amino acids to form complexes which are drawn into the cell along with sodium ions. Once across the cellular membrane, the complex dissociates, and the sodium is pumped out of the cell.

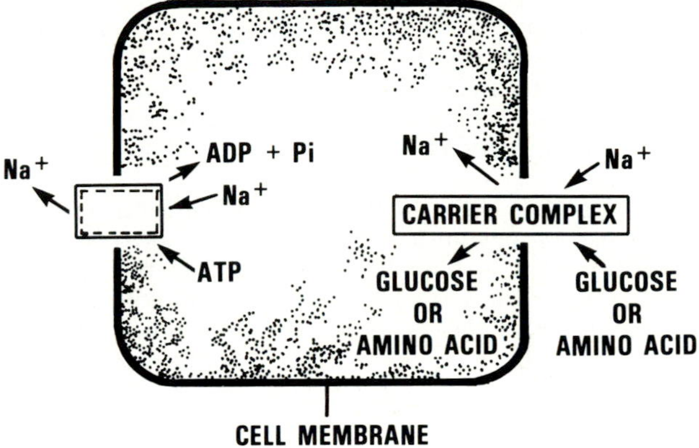

Fig. 3–10. Active transport occurs when there is movement against a concentration gradient, thus necessitating the expenditure of energy. A carrier system utilizing Na^+ has been implicated in the active transport of glucose and some amino acids.

PINOCYTOSIS

Pinocytosis ("cell drinking") is the process by which materials are taken into the cell through an invagination and subsequent dissolving of a part of the cell membrane. This process enables the cell to absorb certain lipids and proteins intact—a critical factor in the newborn animal which must absorb antibodies from colostrum. As can be seen in Fig. 3–11, the material to be absorbed comes into contact with the cell membrane. The membrane then invaginates

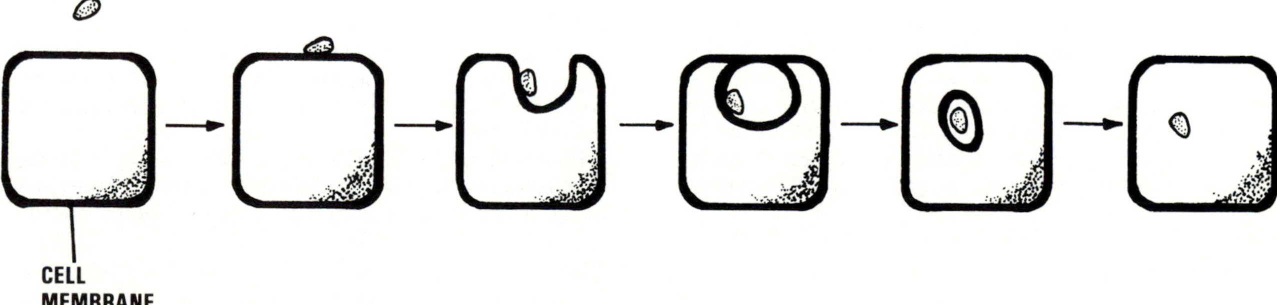

Fig. 3–11. Pinocytosis (cell drinking) involves the invagination of the cell membrane around the material to be absorbed. Once the material has been completely engulfed, the membrane surrounding the material disintegrates and the material is incorporated into the cell.

to surround the material. Once the material has been completely surrounded, the membrane fuses, and the invaginated section of the membrane is dissolved by lysosomal enzymes. Absorptive cells in the small intestine are capable of using this mechanism.

Nutrient Carriers

When the nutrients have been digested and absorbed, they must be transported to tissues that either have an immediate demand for them or can store them for later use. Lymph and blood are the primary transport media for the nutrients which have been absorbed.

LYMPH

Within the mucosal membrane of the intestinal tract, there is a capillary network of lymph vessels. Cholesterol, water, long-chain fatty acids, and some proteins are picked up by this system and transported through a series of larger vessels which ultimately empty into the venous system anterior to the heart.

The immune system of the newborn animal is not well developed. Therefore, it is essential that it receive colostrum from its mother as soon as possible. Through this intake of colostrum, antibodies are passed from the mother to the newborn, imparting a certain degree of immunity to stress and disease which will last for the first critical days of the young animal's life. Many of these antibodies are absorbed intact in the newborn and transported via the lymphatic system.

BLOOD

Most of the low molecular weight products of digestion are absorbed and transported by the blood. These nutrients include water, salts, glycerol, amino acids, short-chain fatty acids, monosaccharides, and certain vitamins. These materials are absorbed into the capillary system of the intestine. The capillary network drains into the venous system, eventually entering the portal vein of the liver. From the liver, the nutrients then travel through the hepatic veins which, in turn, enter the main systemic vein—the vena cava.

PHYSIOLOGY OF DIGESTION

Although there are numerous anatomical differences among the domestic livestock species, the principles of digestion and absorption are quite similar on a physiological basis. In order to compare these processes in the various animals, the discussion of the physiology of digestion will be divided into sections dealing with the regions of digestion instead of individual organs. These regions are the oral region, pharyngeal and esophageal region, gastric region, pancreatic region, hepatic region, and intestinal and cecum-colon region.

Oral Region

Three physical processes occur in the oral region of animals—prehension, mastication, and the initiation of deglutition.

Prehension can be defined as the act of bringing food into the mouth. Numerous modes of prehension can be found in animals. Animals, such as the raccoon and humans, use their forelimbs to bring the food to the oral cavity, while many other types of animals rely on structures of the mouth, such as the tongue, lips, and teeth.

Mastication is the act of chewing food. Most animals chew their food immediately following prehension. One notable exception is the fowl which swallows its food whole. Mastication involves the physical grinding and tearing of the food in addition to the admixture of saliva which lubricates the food as well as initiates a limited amount of enzymatic digestion. *Food that has been masticated and formed into a small compact ball for passage down the digestive tract is called a bolus.*

Deglutition is the act of swallowing. This process involves both voluntary and involuntary reflexes. Upon completion of mastication, the bolus is lifted by the tongue and moved to the back of the mouth. The bolus passes through the pharynx, causing a temporary inhibition of respiration by the reflex closure of the larynx, and finally down the esophagus to the gastric region.

TEETH

The teeth serve primarily as a mechanical aid for mastication. By tearing and grinding the food, they provide a means whereby a large surface area is created which can be exposed effectively to the digestive fluids of the tract. There are four types of teeth, each serving a specialized function: incisors, canines, premolars, and molars. The teeth on the front of the jaw are called incisors and are used for the tearing and slicing of food. Moving progressively to the back of the jaw, the next teeth are called canines. These teeth—sometimes called the eye teeth, or tusks—are also used for tearing. Ruminants do not have these teeth. Following the canines are the premolars and molars, both types of which are used for grinding. Generally grinding can occur on only one side of the jaw at a time.

By observing the dentition patterns of animals, we can get an idea as to their respective eating habits. Table 3-3 shows these patterns for the livestock species to be discussed throughout this book. It is of interest to compare ruminant animals with swine and mink. Ruminants, being entirely herbivorous, do not require many teeth for the tearing of food; hence, there are no canine teeth in ruminants, and the incisors are found only on the lower jaw where they are used for shearing forages in prehension. On the other hand, swine and mink are known to consume meat and animal products; and there is a need for a tearing and shearing action to facilitate digestion. To accommodate this need, incisors and canine teeth represent a large proportion of the overall dentition.

Poultry do not have teeth. Rather, they swallow their food whole, and the mechanical breakdown of food takes place in the gizzard, aided by the presence of stones or grit.

TONGUE

In many species of domestic animals, the tongue is the primary structure for prehension. In the cow, the tongue is elongated and covered with rough papillae, making it

TABLE 3-3
DENTITION PATTERNS OF ADULT DOMESTIC ANIMALS[1]

	Incisors	Canines	Premolars	Molars
Herbivores:				
Sheep	0/8	0/0	6/6	6/6
Goats	0/8	0/0	6/6	6/6
Cattle	0/8	0/0	6/6	6/6
Horses	6/6	2/2	6–8/6	6/6
Rabbits	4/2	0/0	6/4	4–6/6
Omnivores:				
Swine	6/6	2/2	8/8	6/6
Humans	4/4	2/2	4/4	6/6
Carnivores:				
Mink	6/6	2/2	6/6	2/4

[1]The top figure refers to the dentition pattern of the upper jaw, while the bottom figure refers to the lower jaw.

adapted to wrapping around grass and other forages. The cow then brings the forage into the oral cavity where it is sheared by the movement of the incisors against the dental pad.

Throughout the process of mastication, the tongue serves a threefold purpose. First, movement of the tongue transports the feed to the various areas of the mouth to be torn and ground. While doing this, the tongue is also mixing the feed with the various secretions of the mouth, ultimately forming a bolus. Secondly, the presence of taste buds on the tongue provides a neurological control for feed selection and intake. If the feed is bitter or unpalatable, impulses from the taste buds signal the animal to stop eating. Conversely, a desirable taste stimulates appetite. In the cow, the tip of the tongue contains a large number of taste buds while the middle portion has very few. The back portion of the tongue contains the highest density of taste buds. In the chicken, there are very few taste buds; and what few there are lie on the back area of the tongue. Finally, the tongue initiates the process of deglutition. When the bolus has been adequately prepared, the tongue moves it to the back of the mouth where neural receptors are stimulated, and swallowing commences.

SALIVARY GLANDS

The salivary glands represent a network of accessory structures which are essential to digestion. Three pairs of salivary glands are of primary importance—parotid, submaxillary, and sublingual. Fig. 3-12 illustrates the location of these glands.

Saliva, the secretion from these glands, is highly variable in chemical composition. Two basic types of saliva are produced. The first type is extremely thick, being rich in the glycoprotein *mucin.* In this type of saliva, there is very little enzymatic activity. The second type of saliva is serous in composition; that is, it is watery and thin, containing various proteins and enzymes but little mucin. Depending on the particular species of animal, the salivary glands secrete different mixtures of these two kinds of saliva. Saliva from the parotid glands tends to be serous in composition, while the submaxillary glands normally secrete a mixture of the two. In the horse, cow, pig, and dog, the sublingual glands secrete a mixture of the two types; but in humans and rodents, this pair of glands secretes saliva that is primarily mucous in composition.

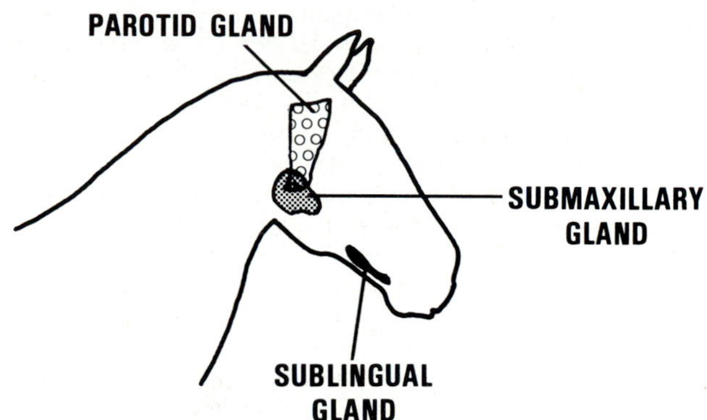

Fig. 3-12. Location of the main salivary glands.

The uses of saliva in digestion are manyfold, including the following:

1. **Lubricant.** These secretions act as aids in mastication, the formation of the bolus, and swallowing. Without this moisture, swallowing would be extremely difficult in most animals.
2. **Enzymatic activity.** The enzyme alpha-amylase (ptyalin) is found in serous saliva of nonruminants. It acts to break alpha 1, 4 glucosidic linkages in starch and glycogen.
3. **Buffering capacity.** A large quantity of bicarbonate is secreted in saliva, thus serving as a buffer in the ingesta.
4. **Nutrients for rumen microorganisms.** Saliva contains considerable amounts of urea, mucin, phosphorus, magnesium, and chloride—all of which can be readily utilized by the bacteria and protozoa in the rumen.
5. **Prevention of frothing.** Gas can accumulate in the rumen and cause serious bloating if the eructation process is impaired. Saliva—acting as a surfactant—helps to prevent these problems.
6. **Taste.** Saliva solubilizes a number of the chemicals in the feed which, once in solution, can be detected by the taste buds.
7. **Protection.** The membranes within the mouth must be kept moist in order to remain viable. Saliva provides one means by which this is accomplished.

Pharyngeal and Esophageal Region

The pharynx is the structure which controls the passage of air and feed. In this organ, the openings of the mouth, esophagus, posterior nares, Eustachian tubes, and larynx come together. During the act of swallowing, the opening into the larynx is reflexly closed by the arytenoid cartilages; and the epiglottis is passively folded over the opening of the larynx. This forces food into the esophagus, thus preventing it from passing into the respiratory tract.

The esophagus is a muscular tube extending from the pharynx to the cardia of the stomach. The musculature and innervation of the esophagus are such that peristalic waves move the bolus. *Peristalsis is the coordinated contraction and relaxation of smooth muscles creating an unidirectional movement which pushes the bolus through the digestive tract.* In nonruminant animals, peristalsis normally moves from the mouth to the stomach. Belching and vomiting, reverse peristalsis, are usually dysfunctions of normal digestion. Reverse peristalsis is a normal process in ruminants as eructation and rumination are essential for digestion. Vomiting is rare in horses; if it does occur, most of the material passes out through the nostrils.

In fish, the esophagus is an extremely distensible organ. Generally, if the fish can get an object in its mouth, the esophagus will accommodate its passage down the digestive tract. Herbivorous fish have no differentiated structure resembling the stomach; essentially, the esophagus extends all the way to the small intestine.

At the junction of the cervical segment and the thoracic segment of the esophagus in birds, there is a differentiated outpouching of the esophagus called the crop. If the bird has been starved, feed will bypass this structure and go directly to the proventriculus and gizzard. As feeding progresses, the crop begins to fill and acts as a storage organ. In the crop, there is limited digestion due to the presence of salivary amylase mixed in the food and a small amount of fermentation. Limited absorption of glucose and volatile fatty acids in the crop has been demonstrated in some birds.

Gastric Region

The primary differences in digestion among domestic livestock can be traced to the specialized development of the gastric region. Variations in the structure of the stomach are reflected in the nonruminant, ruminant, and avian species.

NONRUMINANT ANIMALS

In nonruminant animals, the shape of the distended stomach, being relatively simple, resembles that of a kidney bean. Fig. 3-13 illustrates the various external areas of the stomach.

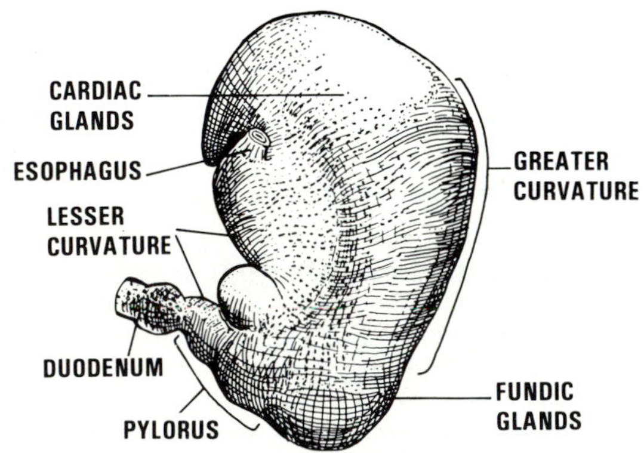

Fig. 3-13. Anatomy of the nonruminant stomach.

Two specialized types of cells in the stomach provide the gastric secretions needed for the initial stages of digestion. The *parietal cells,* located in the fundic region, secrete hydrochloric acid. Hydrochloric acid hydrolyzes a limited amount of protein, but its main function is to establish an acid environment conducive to the activity of certain hormones and enzymes. The second cell type is called the *chief cell* or *peptic cell.* These cells secrete the enzyme pepsinogen. When pepsinogen is secreted in an acid environment (pH of 1.6 to 3.2), the proenzyme is activated forming pepsin—an enzyme that hydrolyzes certain peptide bonds.

Mucus is secreted by cells in the stomach mucosa. This secretion provides protection for the lining of the stomach. If there is a malfunction in the secretion of mucus, the stomach can digest itself—resulting in ulceration.

Two types of motility have been observed in the stomach. The first type is that of *peristalsis* whereby food is moved toward the duodenum of the small intestine. *Tonic contractions,* the second type of motility, churn and knead the ingesta to ensure thorough mixing, but do not propel the food from one end of the stomach to the other.

Feed ingested by carnivores is passed through the stomach at an extremely fast rate (on the order of a couple of hours), while ingesta of herbivores and omnivores tend to remain in the stomach for relatively long periods, sometimes in excess of 24 hours. The rate of passage of feed depends largely on the nutrient composition of the ration. Carbohydrates pass through the stomach faster than either proteins or fats, with proteins being intermediate in rate of passage. Water can pass directly through to the small intestine, spending very little time in the stomach.

In addition to pepsin, gelatinase and rennin are two enzymes secreted in the stomach. Gelatinase liquifies gelatin, and rennin coagulates milk. The latter enzyme is not found in adult humans but is extremely important in the nutrition of all young mammals.

Intrinsic factor, a protein necessary for proper absorption of vitamin B-12, is produced in the parietal cells of the stomach of humans, pigs, rats, and guinea pigs; but it is not produced by ruminants or dogs. If there is a malfunction in the production of this protein, a condition called pernicious anemia results. Quite often, people suffering from this condition ingest amounts of dietary vitamin B-12 that are normally considered adequate, but the vitamin has no means of being absorbed—thus, pernicious anemia develops.

RUMINANT ANIMALS

The ruminant has been described as having four stomachs. In reality, the ruminant possesses a complex stomach consisting of four morphologically distinct compartments. These compartments are reticulum, rumen, omasum, and abomasum.

Scientists have developed highly advanced methods of fermentation technology in which microorganisms are used to produce food and health products. However, the natural fermentation engineering found in the stomach of the ruminant is unexcelled as a culture system for bacteria, protozoa, and fungi.

Reticulum and Rumen

The reticulum and rumen are closely related as to physiological function and are often discussed together. The esophagus empties into the atrium ventriculi, a convex area formed by both the rumen and reticulum. In the adult ruminant, the rumen is an extremely large compartment lined with a large number of papillae that increase the surface area for the churning of digested material and absorption. The reticulum, a structure which has an interior that very much resembles a honeycomb, acts as a collection compartment for foreign objects as well as an organ for digestion.

Ingested feed passes into these two compartments and is digested thoroughly through the action of various microorganisms (bacteria and protozoa) present in the rumen. The rumen, in effect, is a large physiological fermentation vat.

The microbes of the rumen digest carbohydrates to produce carbon dioxide and volatile fatty acids. Although a number of volatile fatty acids are produced, the vast majority of these end products are acetate, propionate, and butyrate. These products are then absorbed from the rumen and supply much of the energy required by the animal. Quite often, when high-concentrate rations are used, large quantities of lactic acid are produced and the pH of the rumen falls. Since most of the bacteria in the rumen are pH sensitive, any dramatic shift in pH will alter the proportions of the various types of microorganisms. When the ruminal pH drops too low, the animal goes off feed—a symptom of acute digestive problems.

Lipids are degraded by the ruminal microbes to fatty acids and glycerol. Glycerol is then primarily converted to propionate, with the long-chain fatty acids passing down to the small intestine for absorption.

Very few dietary proteins escape the degradation process of the rumen. The degree to which dietary protein is degraded is dependent on its solubility. A highly soluble protein will be rapidly degraded while a highly insoluble protein will probably leave the rumen relatively intact. Most dietary proteins are metabolized by the microorganisms and are incorporated as microbial protein. The microbes can also use simpler forms of nonprotein nitrogen, such as urea, to make protein. The microbes are then passed down the tract, which, through their own degradation, provide protein for the animal. With the degradation of the various dietary proteins, ammonia is produced in the rumen which can then either be absorbed through the rumen wall or provide nitrogenous precursors for the synthesis of bacterial protein. If the ration is high in sugars and starches, ammonia concentration is depressed.

Vitamin K and the B-complex vitamins are all synthesized by the ruminal microbes. Therefore, no supplementation of these vitamins is necessary in the adult ruminants. The young ruminant must obtain these vitamins from exogenous sources, but milk generally supplies the young animal's needs. Vitamin C is synthesized on the tissue level in ruminants as well as most nonruminants. Humans are one of the notable exceptions.

Throughout the fermentation process in the rumen, large quantities of various gases are produced and expelled through eructation (belching). Methane and carbon dioxide are the two gases produced most abundantly in the rumen. It has been estimated that methane constitutes 30 to 40% of the total rumen gas volume, and carbon dioxide 20 to 65%.

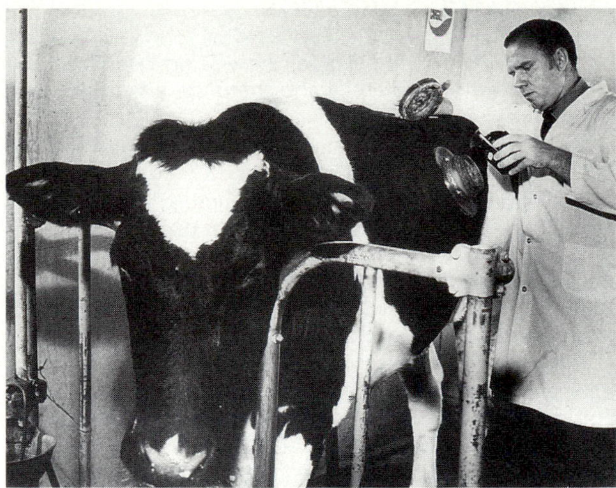

Fig. 3-14. A rumen fistula, a port (opening) that can be opened and closed at will. The feedstuff(s) being studied is placed in porous polyester bags, which are inserted in the rumen. The bags are held by a fishline, suspended in the rumen fluids, so that the microbes digest the contents. The undigested portion of the feedstuff remains in the bag for washing and analysis. The difference in analysis of the feed before and after digestion represents the portion digested. (Courtesy, *Feed Management,* Watt Publishing Co., Mount Morris, Ill.)

Omasum

The omasum, or manyplies, is the next compartment for digestion. It contains numerous laminae (tissue leaves) that help grind ingesta. The exact physiological function of this compartment has not been fully elucidated, but many researchers feel that it serves to absorb water in addition to its function of grinding ingesta.

Abomasum

This compartment is analogous to the stomach of the nonruminant. It is the only compartment of the gastric region of the ruminant containing digestive glands. Digestive processes of this compartment are very similar to those of the stomach in the nonruminant.

Newborn Ruminants

When a calf is born, the rumen is small and the fourth stomach is by far the largest of the compartments. Thus, digestion in the young calf is more like that of a simple-stomached animal than that of a ruminant. The milk which the calf normally consumes bypasses the first two compartments by way of the esophageal groove and goes almost directly to the fourth stomach in which the rennin and other compounds for the digestion of milk are produced. If the calf gulps too rapidly, or gorges itself, the milk may go into the rumen where it is not digested properly and may upset the calf's digestive system. As the calf nibbles at hay, small amounts of material get into the rumen. When certain bacteria become established, the rumen develops and the calf gradually becomes a full-fledged ruminant. (Also see earlier section in this chapter on "Esophageal Groove [Reticular Groove].")

AVIAN GASTRIC DIGESTION

Birds have no teeth; hence, there is no chewing. The esophagus empties directly into the crop, where the feed is stored and soaked.

Gastric digestion in birds occurs in two separate and distinct organs—the proventriculus and the gizzard.

The proventriculus is a small organ, through which ingested feed passes rapidly. Its main function is that of gastric fluid secretion. The fluids secreted by the proventriculus are very similar to those in the stomach of the nonruminant, containing both pepsin and hydrochloric acid. Very little churning and mixing of feed occur in this organ.

The function of the gizzard is the mechanical action of mixing and grinding the feed. Because the bird has no teeth and swallows its feed whole, this muscular organ—sometimes called the "hen's teeth"—acts primarily as an organ for *mastication*. Here, fluids secreted by the proventriculus are mixed in the ingesta during grinding. Grit, such as small pieces of granite, is often added to poultry rations to increase the digestibility of the whole grains or grains with a minimal amount of processing. Grit stimulates motility in the gizzard and provides additional surface for grinding. When feed is provided in mash form, the benefits of grit are minimal.

Pancreatic Region

The pancreatic region involves the pancreas and the pancreatic duct—a duct leading from the pancreas to the small intestine.

The pancreas, an accessory organ of digestion, is a glandular structure that plays an essential role in the digestive physiology of animals. The pancreas—being both an endocrine and exocrine gland—serves two physiologically distinct functions. The endocrine function is that of the secretion of the hormones, insulin and glucagon. The exocrine function deals with the production and secretion of fluids that are necessary for digestion within the small intestine.

The digestive fluid produced by the pancreas is clear and alkaline and consists of two phases—an aqueous phase and an organic phase. Being rich in bicarbonate, the aqueous phase serves primarily to neutralize the highly acid chyme produced in the stomach and passed on to the small intestine. In the organic phase, enzymes produced in the acinar cells of the pancreas are transported to the duodenum. These enzymes are stored in granules in the pancreas and are secreted from the cells through the process of emeiocytosis (cell vomiting). This process is sometimes called reverse pinocytosis because the granule fuses with the cell membrane, followed by a breakdown of the membrane and the evacuation of the granule. A listing of the composition of pancreatic fluids is found in Table 3-4.

Many of the pancreatic enzymes are stored and secreted in an inactive form to be activated at the site of digestion. Trypsinogen is a proteolytic enzyme that is activated in the small intestine by the enterokinase, an enzyme secreted from the intestinal mucosa. When activated, trypsinogen becomes trypsin. Trypsin, in turn, can then activate chymotrypsinogen to chymotrypsin.

The nucleases, lipases, and pancreatic amylase are secreted in their active form. Many of the enzymes require a specific environment before they will function. For example, amylase requires a pH of about 6.9 and the presence of inorganic ions before it will digest complex carbohydrates.

TABLE 3-4
COMPOSITION OF PANCREATIC SECRETIONS

Item	Function
Proteolytic enzymes:	Splits proteins into peptides and amino acids.
Trypsinogen	
Chymotrypsinogen A	
Chymotrypsinogen B	
Procarboxypeptidase A	
Procarboxypeptidase B	
Collagenase	Breakdown of collagen.
Lipolytic enzymes:	Breakdown of lipids.
Prophosphorolipase A	
Pancreatic lipase	
Cholesterol esterase	Esterification of cholesterol.
Nucleolytic enzymes:	Breakdown of nucleic acids.
Ribonuclease	
Deoxyribonuclease	
Amylolytic enzymes:	Breakdown of starches.
Pancreatic amylase	
Cations:	Buffers; cofactors; osmotic regulators.
Sodium	
Potassium	
Calcium	
Magnesium	
Anions:	Buffers; osmotic regulators.
HCO_3	
Cl^-	
$SO_4^=$	
$HPO_4^=$	
Proteins:	Buffers.
Albumin	
Globulin	

Hepatic Region

The hepatic region incorporates the liver, gallbladder, and bile duct.

In addition to the pancreas and salivary glands, the liver is an indispensable accessory organ of the gastrointestinal tract. From the stomach and small intestine, most of the absorbed nutrients travel through the portal vein to the liver—the largest gland in the body. The liver not only plays an important part in nutrient metabolism and storage, but also forms bile, a fluid essential for lipid absorption in the small intestine. The numerous physiological functions of the liver follow:

1. Secretion of bile.
2. Detoxification of harmful compounds.
3. Metabolism of proteins, carbohydrates, and lipids.
4. Storage of vitamins.
5. Storage of carbohydrates.
6. Destruction of red blood cells.
7. Formation of plasma proteins.
8. Inactivation of polypeptide hormones.
9. Urea formation.

The primary role of the liver in digestion and absorption is the production of bile. Bile facilitates the solubilization and absorption of dietary fats and also aids in the excretion of certain waste products such as cholesterol and by-products of hemoglobin degradation. The greenish color of bile is due to the end products of red blood cell destruction—biliverdin and bilirubin. Bile contains a number of salts resulting from the combination of sodium and potassium with bile acids. There are four types of bile acids: cholic acid, deoxycholic acid, chenodeoxycholic acid, and lithocholic acid. These salts combine with lipids in the small intestine to form micelles. *Micelles are colloidal complexes of monoglycerides and insoluble fatty acids that have been emulsified and solubilized for absorption.* When the micelle has been formed, the lipid can be digested and the resulting products (fatty acids and glycerol) can cross the mucosal barrier of the small intestine and enter the lymphatic system. Bile salts, however, do not travel with the lipid; rather, they are recycled into the enterohepatic circulation.

The volume of bile production is highly variable. An animal that has been starved produces little bile. Conversely, an animal that is fed a high-fat ration will produce substantial quantities in order to keep up with absorptive requirements. Generally, the volume of bile is dependent on (1) blood flow, (2) nutritive state of the animal, (3) type of ration being fed, and (4) the enterohepatic bile salt circulation.

In many animals, the gallbladder is the storage site for bile. Several species of livestock and animals, however, do not have gallbladders; among them, horses, rats, gophers, deer, elk, moose, giraffes, camels, elephants, pigeons, and doves.

Intestinal and Cecum-Colon Region

The small intestine is divided anatomically into three sections—duodenum, jejunum, and ileum. The first segment, the duodenum, originates at the pyloric sphincter of the stomach and is closely attached to the body wall by a short mesentery. Both bile and pancreatic fluids are emptied into this segment. The next section is the jejunum. There is no clear demarcation between the jejunum and the ileum, but it is arbitrarily defined as the free border of the ileocecal fold.

Throughout the luminal surface of the small intestine lies an extensive network of fingerlike projections called villi. In the human, there are about 20 to 40 of these projections per square millimeter of intestine, each one being from 0.5 to 1.0 mm long. Each villus contains a lymph vessel called a lacteal and a series of capillary vessels. On the surface of the villi are a great number of microvilli which provide further surface area for absorption.

Three types of motility can be observed in the small intestine. The first type is called *pendular motion*. These waves do not advance down the intestine. Rather, they are merely a localized shortening and lengthening of the intestine which produces a mixing action. *Segmentation contractions* are the second type of intestinal motility. These intestinal movements are ringlike contractions at regular intervals which periodically relax, whereupon the area that had been previously relaxed contracts. This type of motility provides a means of mixing in addition to the pendular contractions. *Peristalsis,* a form of motility that has been previously discussed, is the third type of intestinal motility, providing a means for movement of chyme (intestinal contents) down the tract.

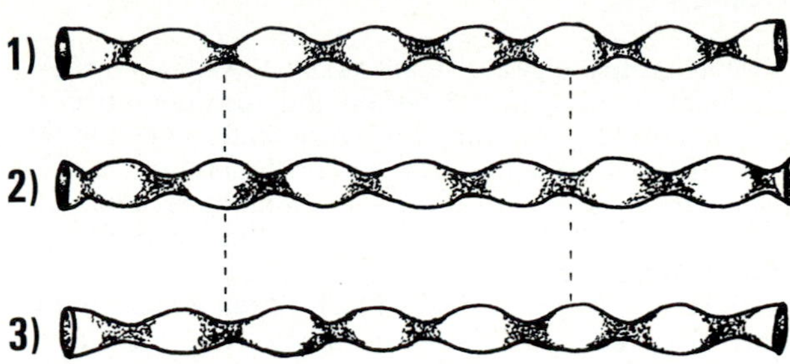

Fig. 3-15. Segmentation waves provide a mixing and churning action by alternately relaxing and contracting in localized areas.

DIGESTION AND ABSORPTION IN THE SMALL INTESTINE

In nonruminant animals, the small intestine is the primary site of both digestion and absorption. While considerable digestion takes place in the gastric region of the ruminant, the small intestine is still of paramount importance for digestion and absorption, especially for lipids and proteins.

Carbohydrates

The digestion and absorption of most carbohydrates, in nonruminant animals, occurs in the small intestine. Here, such enzymes as amylase, sucrase, maltase, and lactase split carbohydrates into monosaccharides, whereupon absorption takes place. The region of the greatest absorption of sugars is in the jejunum. Glucose and galactose are absorbed through an active transport mechanism. Sodium ion concentration within the intestinal contents has been shown to be critical in this mechanism. A high Na^+ concentration will facilitate rapid absorption of these sugars while a low Na^+ concentration will reduce the rate of absorption. Some pentoses and hexoses are absorbed through diffusion—a process considerably slower than that of active transport.

Lipids

Lipids are digested and absorbed primarily in the upper part of the small intestine, but considerable absorption can take place as far down as the ileum. When lipids, emulsified by bile salts, come into contact with the various lipases that are found in the duodenum, they are broken down into monoglycerides and fatty acids. Short-chain fatty acids are then absorbed directly into the mucosa of the small intestine and are transported to the portal circulation. Monoglycerides and insoluble fatty acids are emulsified by bile salts, forming micelles. By attaching to the surface of epithelial cells, the micelles enable these components to be absorbed into the mucosal cells. Once inside these cells, the long-chain fatty acids are reesterified to form triglycerides. Triglycerides then combine with cholesterol, lipoproteins, and phospholipids to form chylomicrons—minute fat droplets. The chylomicrons are then passed into the lymphatic circulatory system.

Proteins

While protein digestion is initiated in the stomach of nonruminant animals, most digestion and absorption occur in the small intestine. Numerous pancreatic and intestinal enzymes split proteins into their constituent amino acids, which are subsequently absorbed. In humans, it has been estimated that 50% of the digested protein comes from the diet, 25% from the proteins in the digestive fluids, and the remaining 25% from sloughed cells of the gastrointestinal tract. The rate of turnover of mucosal intestinal cells is extremely rapid—1 to 3 days—thereby giving an excellent source of recyclable protein. In ruminants, it has been estimated that up to 80% of the ingested protein is converted to microbial protein.

Amino acid absorption is not clearly understood; but an active transport mechanism involving Na^+, similar to that of glucose absorption, has been implicated. Amino acids are rapidly absorbed in the duodenal and jejunal segments, but are poorly absorbed in the ileum. A limited amount of absorption of small polypeptides can occur, especially in newborn animals. This mechanism of absorption, pinocytotic in nature, facilitates the passage of antibodies from the colostrum of the mother to her young.

Minerals and Vitamins

Mineral absorption occurs throughout the small and large intestine, with the rate of absorption depending on a number of factors; among them, pH and carriers. Numerous mechanisms of mineral absorption have been elucidated. Many minerals, for example iron and sodium, require active transport systems. Others, such as calcium, utilize both carrier proteins and diffusion mechanisms. Additional information on specific minerals can be found in Chapter 4, Nutrients/Metabolism.

Most of the vitamins are absorbed in the upper portion of the intestine, with the exception of vitamin B-12 which is absorbed in the ileum. Water-soluble vitamins are rapidly absorbed, but the absorption of fat-soluble vitamins relies heavily on the fat absorption mechanisms which are generally slow.

CECUM AND COLON

The small intestine terminates at the ileocecal valve—a sphincter that controls the flow of ingesta from the small intestine into the cecum and large intestine. This structure prevents the backflow of ingesta into the small intestine.

The cecum and colon in mammals are composed of several layers of muscle. There is a circular layer of muscle that forms the basic tube of the colon and facilitates movement. In addition to this layer of muscle, there are three strips of longitudinal muscle which form the *taenia coli.* These strips form a series of pouches or sacculations throughout the colon which are called *haustrae.* Ingesta are held in these saclike structures to facilitate the removal of water. Subsequently, the feces generally take on the shape of the haustrae. Numerous mucous-secreting goblet cells can be found in the colon, but villi, such as the type that are found in the small intestine, are absent.

Three types of motility can be observed in the colon: (1) haustral contractions, (2) massive peristalsis, and (3) reverse peristalsis.

1. **Haustral contractions.** This type of motility creates a mixing action of the ingesta; hence, the absorption of water from the material is facilitated. These contractions are localized in the various portions of the colon with no coordinated wave movement traveling along the organ.
2. **Massive peristalsis.** These waves are slow, strong movements that propel the digesta down the colon.
3. **Reverse peristalsis.** Reverse peristalsis is a type of movement that aids in the mixing of the digesta as well as the absorption of nutrients. By moving material proximally, the colon has another opportunity to absorb what was missed the first time.

At the proximal end of the colon is a blind sac called the cecum. In carnivores, the cecum is of little importance because most of the ingested material has been digested prior to entering the colon. Water and electrolytes are absorbed in the colon as well as a number of water-soluble vitamins and vitamin K, which are produced in limited amounts by the microflora of the colon.

The cecum and colon are rather large and well-developed in nonruminant herbivores, such as the rabbit and the horse. In these animals, considerable fermentation can take place in the cecum and colon, which, together, act somewhat like a rumen. Fermentation products are synthesized and absorbed. When one considers that these are nonruminant animals, considerable fiber and cellulose digestion can be achieved. Physiologically, this anatomic adaptation can never be as efficient as the ruminant process of digestion. Ruminal contents are passed into the small intestine—the primary site of digestion and absorption, whereas products which are broken down in the cecum of a nonruminant herbivore are transported down the colon where absorption is severely limited. Rabbits and rats can compensate for this fact through the behavioral adaptation of coprophagy. *Coprophagy is the ingestion of fecal material.* Through this feeding practice, the rabbit gives the digestive tract another chance for additional digestion and absorption. Despite this arrangement, the rabbit is not as efficient in digestion as the ruminant.

The cecum of the ruminant is not very well developed and plays a rather insignificant role in digestion. There is some absorption of volatile fatty acids in the ruminant cecum, and considerable amounts of water and electrolytes are absorbed in the colon. It has been estimated that in sheep 70% of the water in the digesta is absorbed between the mouth and the duodenum. In the colon alone, about 19% of the water is absorbed.

There are two blind sacs (ceca) in the fowl, in which a limited amount of bacterial activity and subsequent nutrient absorption have been observed. The large intestine is extremely short and is very similar in structure to the small intestine. It is generally believed that the large intestine in the fowl does not play any significant role in digestion.

GASTROINTESTINAL HORMONES

A hormone can be defined as a chemical released by a specific area of the body that is transported to another region within the animal where it elicits a physiological response. A number of hormones have been isolated and characterized from the gastrointestinal tract. Gastrointestinal endocrinology is an extremely recent area of study; and new hormones are being found and identified chemically. Table 3-5, Gastrointestinal Hormones, lists the gastrointestinal hormones, along with their respective site of production, mechanism of release, and physiological function.

TABLE 3-5
GASTROINTESTINAL HORMONES

Hormone	Origin	Mechanism of Release	Physiological Function
Enterocrinin[1]	Duodenum.	Presence of chyme.	Increases secretion of intestinal fluids.
Enterogastrone	Duodenum.	Presence of fats.	Inhibits gastric acid secretion and motility.
Cholecystokinin-Pancreozymin	Duodenum.	Presence of fats and products of protein digestion.	Contraction of gallbladder and secretion of pancreatic enzymes.
Gastrin	Antral portion of gastric mucosa; pancreatic islets.	Distension of stomach; presence of proteins and polypeptides; alcohol; stimulation of vagus nerve.	Stimulates gastric acid and pepsin secretion; stimulates gastric motility.
Secretin	Duodenum.	Presence of acid and protein.	Stimulates secretion of aqueous pancreatic fluid (high in bicarbonate).
Villikinin[1]	Duodenum.	Presence of chyme.	Increases contractions of villi.
Glucagonlike[1] **immunoreactive factor (GLI)**	Wall of small intestine.		Stimulates insulin secretion.

[1]Isolation and chemical structure of these hormones have not been elucidated.

NEUROLOGICAL CONTROL OF THE GASTROINTESTINAL TRACT

The nervous system can be divided into two anatomical systems—the somatic nervous system, and the autonomic nervous system. The somatic nervous system enables the body to adapt to stimuli from the external environment. Various stimuli, such as touch, are perceived by specialized receptors within this system, and the body responds accordingly. The autonomic system involves the maintenance of homeostasis—the internal environment of the body. This is the system that controls the gastrointestinal tract.

The autonomic system can be further divided into the *sympathetic autonomic nervous system* and the *parasympathetic autonomic nervous system*. The sympathetic system is generally associated with the traditional "fight or flight" response, and the parasympathetic system is usually associated with routine integration of normal activity.

When the sympathetic system is stimulated, there is a need for large amounts of blood in peripheral tissues, such as skeletal muscle. In order to accommodate this need, blood is shunted from the gastrointestinal tract, resulting in reduced digestive activity. For the most part, salivation ceases, and the mouth becomes dry. Secretions from the digestive glands are inhibited as well as peristalsis throughout the tract. The various sphincters of the gastrointestinal tract contract in response to sympathetic stimulation.

Stimulation of the parasympathetic system induces increased gastrointestinal activity. Generally, the parasympathetic system is stimulatory to the gastrointestinal system during rest and normal activity.

With the knowledge of the action of the sympathetic and parasympathetic autonomic systems, one can understand the action of certain drugs. When acute diarrhea is encountered, sympathetic-type drugs or parasympathetic depressant drugs are often used. Drugs that act as parasympathetic stimulators are frequently used as laxatives.

FACTORS AFFECTING DIGESTION AND ABSORPTION

In all animals, both extrinsic and intrinsic influences alter the efficiency of digestion and absorption. Feed given to a particular animal in one form may not be as digestible as the same feed in another form. Numerous factors concerning the feed itself will influence digestion and absorption. The autoregulation of the digestive system can influence digestion, especially when there are severe dysfunctions.

Influences of Feed Composition and Preparation

When planning a nutrition program for livestock, the producer must understand the digestive and absorptive processes of the particular species being fed. If these processes are understood, a feeding program can be planned so as to maximize the digestibility of the feed.

RATE OF PASSAGE OF FEED

If a certain type of feed is known to pass quickly through an animal, it is likely that digestion will not be very efficient because the feed will not have adequate exposure to the digestive enzymes. Likewise, if nonruminant animals having little storage area for ingested feed are fed only once or twice a day, they will ingest large quantities of feed at a single feeding. This also limits the relative exposure of the feed to the digestive enzymes; hence, a nonruminant fed on a free-choice regimen would utilize feed more efficiently.

FEED PROCESSING

Processing of grains, such as grinding, does not markedly increase the digestibility of the feed in animals that masticate their feed thoroughly. On the other hand, processing can offer great advantages when feeding grains to animals that swallow their feed without much chewing. Unbroken seed coats are not easily digested, and unprocessed grains can pass through the tract undigested. By removing or cracking this seed coat, digestive enzymes can readily degrade the feed.

Grinding is a useful process in the feeding of very young or very old animals. In the young animals, dentition is rudimentary and the digestive tract has not been well developed. Old animals frequently have worn or missing teeth, and grinding substitutes for mastication.

Cooking feeds has been shown, in certain cases, to increase digestibility. But the process involves additional labor and expense; so, one must weigh the economic advantages and disadvantages carefully before processing. In some cases, such as with soybeans, cooking is necessary before the feed can be fed to poultry and other species, due to the presence of metabolic inhibitors. More information regarding the use of cooking in processing can be found in Chapter 14, Feed Processing.

DIGESTION AND ABSORPTION

It is possible to process certain protein sources in such a way that they escape microbial destruction in the rumen. Then, the protein passes down to the small intestine where it is digested and absorbed. This preserves the integrity of the amino acid composition of high-quality protein. Such protein is known as protected or by-pass protein.

By definition, protected or by-pass protein is feed protein which escapes digestion in the rumen, but is digested in the small intestine. Among the feed processing methods used to prepare protected protein are: (1) *natural protection* (corn gluten meal, distillers dried grains, dehydrated alfalfa meal, and animal by-products), (2) heat denaturation, (3) aldehyde treatment, (4) combined heat and aldehyde treatment, (5) lipid coating—treatment with certain tannins, and (6) complexing with bentonite clay. Even though protein may be protected, performance is not necessarily guaranteed. For improved performance, the by-pass protein must contain amino acids that complement those of the microbe-synthesized protein, with the mixture absorbed by the animal balanced. But, eventually, by-pass products will be developed that will either cut cost or increase production—or both.

LEVEL OF FEED INTAKE

As feed is offered to animals in excessive amounts, digestibility decreases. It has been observed in ruminants that the offering of feed at twice the maintenance level can reduce dry matter digestibility by 1 to 2 percentage units. In swine, this trend is also observed, but the magnitude of decreased digestibility is not as great.

COMPOSITION OF FEED

Feeds vary considerably in composition and digestibility. One good estimate of digestibility in nonruminant animals can be obtained by the content of crude fiber. Fiber is extremely undigestible in nonruminants due to the absence of the enzymes necessary to break down the complex cell walls of the feed. Therefore, a feed high in fiber is probably highly undigestible. Roughages are high in fiber; hence, their use is limited in rations for nonruminants. Conversely, the microorganisms of the rumen synthesize enzymes that digest the components of plant cell walls, and roughages are well utilized by ruminants. For additional information on the effects of nutrient composition on digestibility, see Chapter 4, Nutrients/Metabolism.

CHELATES

Chelates are formed when certain organic molecules combine with metallic ions to form cyclic compounds. These chelated complexes possess different solubility characteristics than the unbound metallic ion. By this binding, certain minerals may be more or less readily absorbed in the gastrointestinal tract. Several naturally occurring chelating agents are chlorophylls, cytochromes, hemoglobin, ascorbic acid, vitamin B-12, and some amino acids. The most commonly used synthetic chelating agent is EDTA (ethylenediamine tetraacetic acid)—a compound used in some poultry rations to improve the availability of certain minerals.

PHYTIC ACID

Phytic acid is a hexaphosphoric acid ester of inositol. When the acid form is combined with a cation to form a salt, the compound is referred to as phytin. More than 50% of the phosphorus in mature seeds is in the form of phytin. Numerous studies have shown that animals vary greatly in their ability to absorb phytins. Cattle and sheep have little trouble breaking down phytins and absorbing the phosphorus since their rumen flora produce phytase. In the dog and humans, phytic acid may combine with calcium, thus making the calcium less available for absorption.

OXALATES

Oxalic acid, a compound present in certain leafy plants, may interfere with calcium absorption. The acid combines with calcium to form insoluble calcium oxalate which is less available for absorption. Spinach has a high oxalic acid content, which may subsequently tie up substantial portions of its calcium. The magnitude of this problem is debatable as certain species of livestock have been shown to metabolize oxalic acid in such a way as to render it harmless.

Dysfunctions of the Digestive Tract

The digestive tract, though relatively simple in structure, is complex in function. In order for proper digestion and absorption to take place, various humoral and neural mechanisms coordinate the movement of ingesta throughout the tract. If anything goes wrong either physically or chemically in the digestive organs, the entire integrated process can break down, causing malabsorption syndromes or physical changes in the tract.

RETICULITIS

The condition when the collection of foreign objects irritates or punctures the reticulum is called reticulitis, or hardware disease. Ruminants are grazers by nature and are sometimes indiscriminant in their selection of feed. Often they will consume nails, pieces of wire, and other foreign objects. Due to the motility patterns of the gastric region, this paraphernalia tends to accumulate in the reticulum; and the presence of these sharp objects can pose serious problems, especially if the reticulum should be punctured. Magnets may be given ruminants to hold the metal objects and prevent them from harming the animal.

RUMINAL ACIDOSIS

When ruminants are switched from a high-roughage ration to a high-grain ration too rapidly, ruminal acidosis and atony may occur. The microflora of the rumen do not have time to adapt to the new ration, resulting in serious digestive disturbances. The pH of the rumen contents falls, and lactic acid production increases dramatically. Many of the microorganisms cannot live in this environment resulting in radical changes in the bacterial and protozoal populations. Rumen motility, then, becomes static. Afflicted animals show signs of weakness, diarrhea, and abdominal discomfort, and in many cases die.

ACUTE TYMPANY (Bloat)

Quite often, when ruminants are first introduced into the feedlot or lush, legume pastures, bloat problems arise. Normally, the eructation process allows for the expulsion of gases that are produced in the rumen. In cases of bloat, frothing (the trapping of gas in the ingesta) prevents this process and intraruminal pressure builds, thereupon distending the left side of the abdomen until the animal is thrown off feed and goes down due to pain and the buildup of toxic metabolites. A number of theories of the causes of bloat have been postulated, and several treatments for bloat have proven effective. Additional information on bloat may be found in Chapter 5, Nutritional Disorders/Toxins.

PARAKERATOSIS OF THE RUMEN

When rations are fed in a highly processed form—for example, pelleted—morphological changes can occur in the epithelium of the rumen. The papillae of the rumen become enlarged and hardened. Lesions can also be found. This condition is known as parakeratosis.

COLIC

Improper feeding, working, or watering of horses can cause colic—a condition whereby the horse has excruciating abdominal pain. Depending on the type of colic, the symptoms include distended abdomen, increased intestinal rumbling, violent rolling and kicking, profuse sweating, constipation, and refusal of feed and water. Additional information on the treatment and prevention of colic can be found in Chapter 5, Nutritional Disorders/Toxins.

DIARRHEA (Scours)

Diarrhea is a common digestive ailment, especially in young animals. Severe dehydration and acidosis due to the loss of sodium, potassium, bicarbonate, and water can result in death. Even in mild cases of scours, costly setbacks in the growth of animals result. There are numerous causes of scours. Several of these are overfeeding, proliferation of pathogenic microorganisms, lack of bulk in the ration, abrupt changes in the rations, the feeding of spoiled hay or silage, and the inclusion of excessive levels of laxative types of feeds.

QUESTIONS FOR STUDY AND DISCUSSION

1. What is the difference between hunger and appetite?
2. What happens to an animal that is stimulated in the ventro-medial hypothalamus?
3. Discuss the glucostatic theory of appetite control.
4. What are hunger pains?
5. Define: (a) carnivore; (b) herbivore; (c) omnivore.
6. Discuss the differences in the digestive systems of the monogastric and the ruminant.
7. What is the function of rumination?
8. Explain how the esophageal groove functions in the young ruminant, and tell of its importance from a nutrition standpoint.
9. Discuss the differences in the digestive physiology between the cow and the horse.
10. How does the digestive system of poultry differ from that of other monogastrics?
11. Why is knowledge of the capacities of the digestive tracts of importance?
12. Discuss the importance of diffusion, active transport, and pinocytosis.
13. Where is rennin produced?
14. Into what two monosaccharides does sucrase split sucrose?
15. Discuss the role of lymph in the digestive process.
16. Define: (a) prehension; (b) mastication; (c) deglutition.
17. Explain why it is possible to get an idea as to the eating habits of animals by looking at dentition patterns.
18. List the various functions of the tongue.
19. What are the three main salivary glands and where are they located?
20. Discuss the role of saliva in digestion.
21. What type of gastric cell secretes hydrochloric acid? What is the function of hydrochloric acid in digestion?
22. What is peristalsis?
23. Discuss the role of the intrinsic factor.
24. Discuss the anatomy and function of each of the four stomach compartments of the ruminant: reticulum, rumen, omasum, and abomasum.
25. What are the two main gases produced in the rumen?
26. Discuss the changes that take place in the stomach of a newborn ruminant.
27. What is the function of the crop in birds?
28. Discuss the functions of the avian gastric organs.
29. Why is grit sometimes fed to birds?
30. List the physiological functions of the liver.

31. On what factors is the volume of bile dependent?

32. Describe the various types of motility in the small intestine.

33. How are fats digested and absorbed in the small intestine?

34. What are haustrae?

35. Of what benefit is the practice of coprophagy in rabbits?

36. Discuss the roles of the various gastrointestinal hormones.

37. Why are sympathetic-type drugs used in the treatment of diarrhea (scours)?

38. Discuss the influences of feed composition and preparation on digestion and absorption.

39. Discuss the following relative to protected or by-pass proteins: (a) definition; (b) how they function in ruminant digestion; (c) why they are important in ruminant nutrition; and (d) their present and future importance.

SELECTED REFERENCES

Title of Publication	Author(s)	Publisher
Anatomy of the Domestic Animals, The	S. Sisson J. D. Grossman	W. B. Saunders Co., Philadelphia, Pa., 1953
Anatomy and Physiology of Farm Animals	R. D. Frandson	Lea & Febiger, Philadelphia, Pa., 1965
Animal Nutrition	L. A. Maynard, *et al.*	McGraw-Hill Book Co. Inc., New York, N.Y., 1979
Digestive Physiology and Nutrition of the Ruminant	Ed. by D. Lewis	Butterworth, Inc., Washington, D.C., 1961
Digestive Physiology and Nutrition of Ruminants	D. C. Church	O and B Books, Corvallis, Ore., 1975
Duke's Physiology of Domestic Animals	Ed. by M. J. Swenson	Cornell University Press, Ithaca, N.Y., 1970
Nutrition of Animals of Agricultural Importance, Part 1. The Science of Nutrition of Farm Livestock	Ed. by D. Cuthbertson	Pergamon Press, New York, N.Y., 1969
Physiology of Digestion and Metabolism in the Ruminant	Ed. by A. T. Phillipson	Oriel Press, Ltd., Newcastle, England, 1970
Physiology of Digestion in the Ruminants	Ed. by R. W. Dougherty	Butterworth, Inc., Washington, D.C., 1965
Recent Developments in Ruminant Nutrition	W. Haresign D. J. A. Cole	Butterworth, Inc., London, England, 1981
Review of Medical Physiology	W. F. Ganong	Lange Medical Publication, Los Altos, Calif., 1975
Ruminants as Food Producers	J. E. Oldfield, *et al.*	CAST, Ames, Iowa, 1975

Fig. 3–16. Layers must digest and absorb their feed in order to produce eggs. In order to maximize the use of feeds by all species, producers and nutritionists must have an understanding of digestion and absorption. (Courtesy, *Poultry World*, London, England)

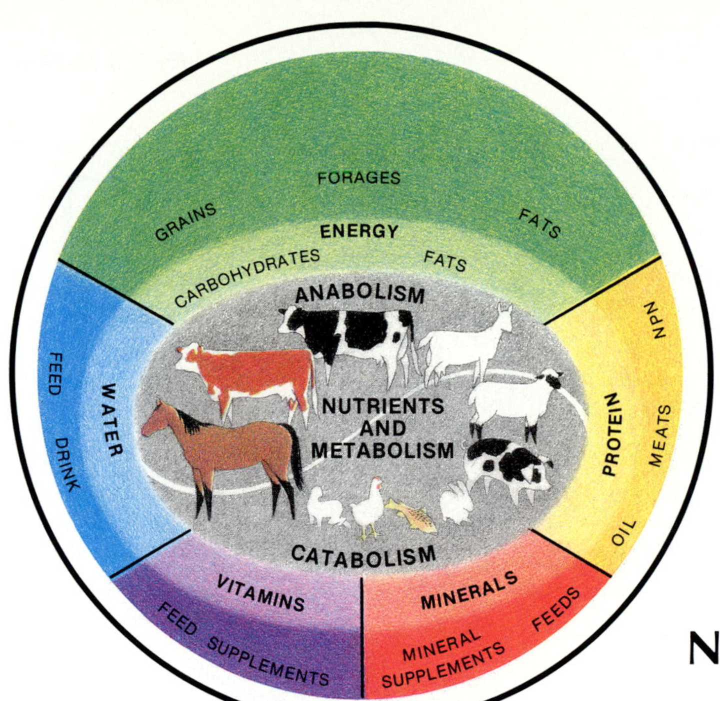

Original painting by Tom Phillips

Chapter

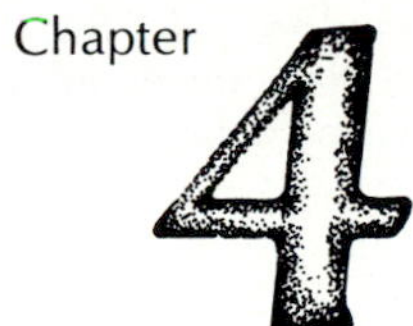

NUTRIENTS/METABOLISM

Contents *Page*

Contents *Page*

Corn, alfalfa, soybean meal, and grass never get to the cells and tissues of animals. But their nutrient chemicals—more than 40 of them, including amino acids, minerals, and vitamins—do reach the body cells and tissues and are essential to their life.

Once feeds are eaten and digested, the nutrients are absorbed into the blood and distributed to the cells of the body. Still more chemical changes are required before the nutrients can be put to work in the body, by transforming them into energy or structural material. Thus, nutrients—carbohydrates, fats, proteins, minerals, vitamins, and water—are subjected to various chemical reactions. These chemical reations occur on the cellular and subcellular level. The sum of all these chemical reactions is termed *metabolism.* It has two phases: *catabolism* and *anabolism.* Pertinent definitions follow.

Nutrients are the chemical substances found in feed materials that can be used, and are necessary, for the maintenance, production, and health of animals.

Metabolism is the process by which physiological changes occur in the body through chemical reactions.

Catabolism is the oxidative breakdown of nutrients, liberating energy (exergonic reaction) which is used to fulfill the body's immediate demands.

Anabolism is the process by which nutrient molecules are used as building blocks for the synthesis of complex molecules. Anabolic reactions are endergonic—that is, they require the input of energy into the system.

Knowledge of what nutrients are needed by animals, and how they are used (metabolized) is the basis of modern nutrition.

CELLS—FUNCTIONAL UNITS OF NUTRITION AND METABOLISM

Nutrition is achieved in cells; they are the functional units of nutrition and metabolism. In the cells, the metabolism of carbohydrates, fats, and amino acids (proteins) takes place. In the cells, compounds are built for use in the body (anabolism), and compounds are broken down into simpler units (catabolism) for new uses or for excretion if not useful. The release of energy by some substances and the acceptance of it by others occurs in the cells, providing energy for use whenever and wherever it may be needed.

Each of the many cells is specialized in some way to carry out the functions of the organism (body) of which it is a part. Grouped together, cells serving the same general function form a tissue; thus, there are muscle, nerve, epithelial, and connective tissues. Structural units, made up of two or more tissues, serving a specific function or functions, are organs; for example, the heart, kidneys, and lungs.

Just as the body has organs, the cells of the body contain organelles (little organs) which are also involved in nutrition and metabolism. The two major parts of the cell are the nucleus and the protoplasm surrounding it, called *cytoplasm* (see Fig. 4-1). The nucleus serves as the center of the cell, controlling its functions; the cytoplasm performs the cell's metabolic activities.

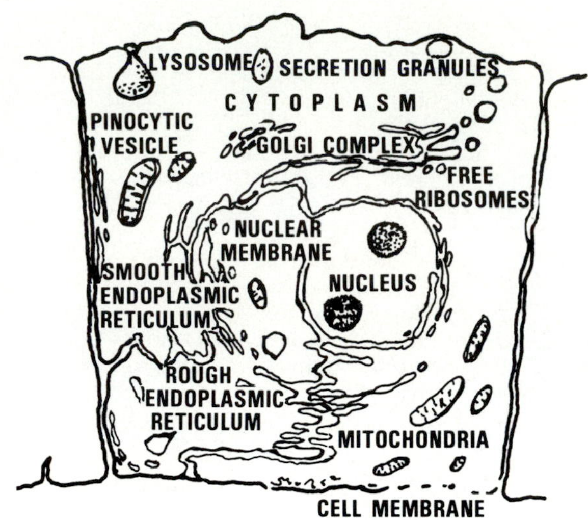

Fig. 4-1. A typical cell showing organelles (little organs) associated with cell nutrition and metabolism.

In the nucleus, the pattern of each of the different proteins is present as deoxyribonucleic acid (DNA). The information stored in DNA is put into action by transcribing it into ribonucleic acid (RNA), which directs the actual protein synthesis in the ribosomes. This process of transcription is the key to nutrition.

Membranes subdivide the cell into compartments and regulate the passage of substances into, out of, and within the cell. The membranes include those of the endoplasmic reticulum, the mitochondria, the golgi body, the lysosomes, the nucleus, and the cell itself. The cell wall separates the cell from the external environment and selectively controls the rate of movement of nutrient and waste material into and out of the cell.

All components for the formation of nutrients in cells come from feeds. Genes determine which ones the body can synthesize and which it cannot.

Nucleic Acids (DNA and RNA)

Nucleic acids were so named because they were originally isolated from cell nuclei. They are the carriers and mediators of genetic information, of which there are 2 types: *deoxyribonucleic acid (DNA)* and *ribonucleic acid (RNA).* The 2 types of nucleic acids, DNA and RNA, differ in that DNA has 1 less oxygen molecule (on carbon atom number 2) in its component sugar ribose, and is a double strand.

Every cell of the body contains the same amount of DNA with the exception of the sperm and egg cells. It is the DNA of the chromosomes in the nuclei of cells that carries the coded master plans for all of the inherited characteristics—size, shape, and orderly development from conception to birth to death. DNA is different for each species, even for each individual within a species. These differences consist of minor rearrangements of sequences among the nitrogenous bases, which constitute a code containing all the information on the heritable characteristics of cells, tissues, organs, and individuals.

The messages carried by DNA are put into action in the cells by the other nucleic acid, RNA. To do this, DNA serves as a template (as the pattern or guide) for the formation of RNA. The genetic message is coded by the sequence of purine and pyrimidine bases attached to the *backbone* of the DNA structure—a long chain of the sugar deoxyribose and phosphoric acid. Purine bases in the DNA include adenine and guanine, while pyrimidine bases include cytosine and thymine. One molecule of DNA may contain 500 million bases. The *backbone* of RNA is also a sugar, the sugar ribose, plus phosphoric acid. However, in RNA the pyrimidine base thymine is replaced by uracil, another pyrimidine. RNA molecules are considerably smaller than DNA, containing from less than a hundred to hundreds of bases—not millions.

In the 1860s, the Austrian monk, Gregor Mendel, experimenting with garden peas in his monastery garden, showed that heredity could be explained in terms of simple mathematical ratios. Although Mendel was not able to explain how hereditary characteristics are transmitted to offspring, he was observing nucleic acids in action. An explanation of Mendel's work was not generally available until 1916, when Thomas Hunt Morgan and his group at Columbia University, using the fruit fly, demonstrated that genes on the chromosomes carried hereditary information from parent to offspring. In 1933, Morgan won a Nobel Prize for his work. Before 1950, George W. Beadle and Edward L. Tatum demonstrated that genes were long-chain polymers in which proteins are linked with nucleic acids. They received a Nobel Prize in 1958. Next, Alfred E. Mirsky and V. G. Allfrey of the Rockefeller Institute showed that DNA was necessary for the manufacture of RNA in the cell and for the buildup of proteins in the nucleus. Then, Erwin Chargoff noted the important relationship between the bases—that adenine and thymine occur in equal amounts, as do cytosine and guanine. Armed with this information and the x-ray diffraction studies of Maurice Wilkins of King's College in London, James Watson, an American biochemist, and Francis Crick, an English physical chemist, set out to determine the actual structure of DNA. Working together, they explored many possible structures, out of which they finally concluded that DNA was composed of two polynucleotide chains joined by bonds between the bases and wrapped around each other in the form of a double helix (see Fig. 4-2). In 1962, those three scientists—Wilkins, Watson, and Crick—shared the Nobel Prize for Physiology and Medicine for their work. But the story did not end! Building upon the present knowledge and understanding of the nucleic acids, DNA and RNA, scientists are making such concepts as genetic engineering and cloning into realities. Thus, humanity is coming closer and closer to understanding the very essence of life—DNA.

SYNTHESIS OF DNA AND RNA

Each time a cell divides, a new DNA is made, while within most cells RNA is continually being synthesized and broken down into its components. When the purines and pyrimidines are combined with the sugars ribose or deoxyribose, the resultant compound is referred to as a *nucleoside*. *Nucleotides* are nucleosides that are esterified (acid and alcohol combination) with phosphoric acid. When a series of nucleotides are joined together, nucleic acids are formed. The secret of the code is a process called base pairing. During the formation of new DNA or RNA, the same bases always pair off—guanine with cytosine, adenine with thymine, and adenine with uracil in RNA. This ensures that the code is duplicated time and time again. During synthesis, the double helix splits, pulling the base pairs apart. New bases line up with the proper partner and form another sugar-phosphoric acid *backbone* (see Fig. 4-2).

Purines, pyrimidines, ribose, and deoxyribose may be synthesized in the body from other compounds, or they may be recycled. Phosphorus, for the formation of phosphoric acid, is a required dietary mineral, and it is employed in a variety of other compounds besides DNA and RNA.

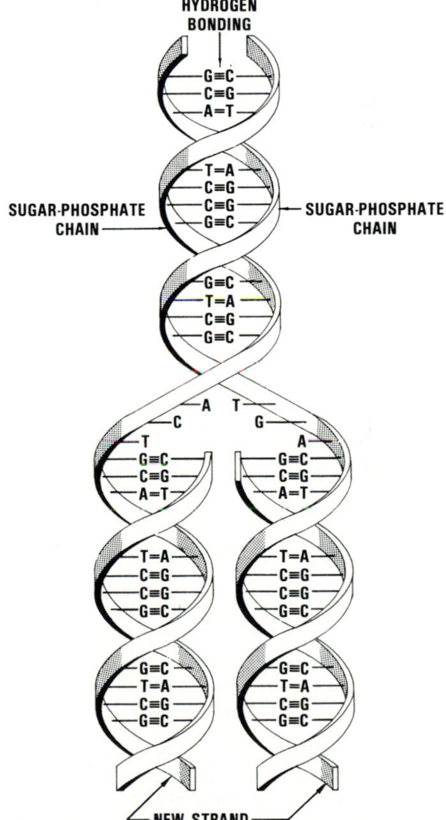

Fig. 4-2. The spiral structure of deoxyribonucleic acid, or DNA—the basic building block of life on earth. It's a double helix (a double spiral structure), with the sugar (deoxyribose)-phosphate (phosphoric acid) *backbone* represented by the 2 spiral ribbons. Connecting the *backbone* are 4 nitrogenous bases (a base is the nonacid part of a salt): adenine (A) paired with thymine (T), and guanine (G) paired with cytosine (C); with the parallel spiral ribbons held together by hydrogen bonding between these base pairs.

GENETIC ENGINEERING (RECOMBINANT DNA)

Genetic engineering has been going on for thousands of years, ever since the human race began raising crops and domestic livestock. From that remote day forward, the most productive plants and animals were held back for breeding while the least productive were eaten. That's a rudimentary form of genetic engineering!

But the development of gene-splicing (also known as recombinant DNA [see Fig. 4-3]) ushered in a new era of genetic engineering. On May 23, 1977, scientists at the University of California-San Francisco reported a major breakthrough as a result of altering genes—turning ordinary bacteria into factories capable of producing insulin, a valuable hormone previously extracted at slaughter from pigs, sheep, and cattle, so essential to the survival of diabetics. This feat opened the door to further genetic engineering or splicing. Already, this genetic wizardry has been used in transplanting into bacteria (and recently into yeast cells) genes responsible for the synthesis of many critical biochemicals in addition to insulin.

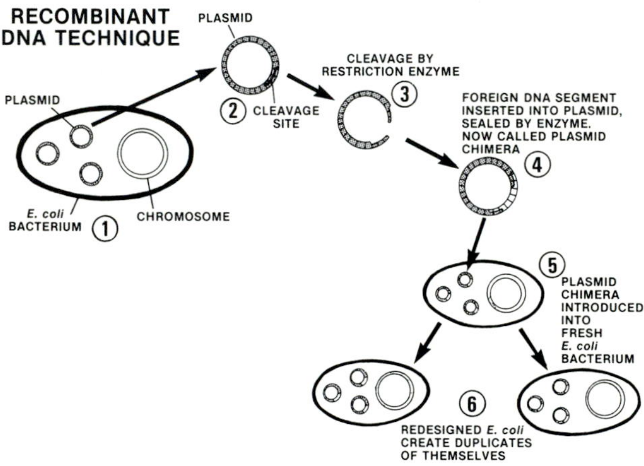

Fig. 4-3. Redesigning *E. coli,* common bacteria of animal intestines. The steps:

1. The scientist places the bacterium in a test tube with a detergent. This dissolves the microbe's outer membrane, causing its DNA strands to spill out.
2. The plasmids (the closed loops), which have only a few genes, are separated from the chromosomal DNA in a centrifuge.
3. The plasmids are placed in a solution with a chemical catalyst called a restriction enzyme, which cuts through the plasmids' DNA strips at specific points.
4. The opened plasmid loops are then mixed in a solution with genes—also removed by the use of restriction enzymes—from the DNA of a plant, animal, bacterium, or virus. In the solution is another enzyme called a DNA ligase, which cements the foreign gene into place in the opening of the plasmids. These new loops of DNA are called plasmid chimeras because, like the chimera—the mythical lion-goat-serpent after which they are named—they contain the components of more than one organism.
5. The chimeras are placed in a cold solution of calcium chloride containing normal *E. coli* bacteria. Then the solution is suddenly heated, at which time the membranes of the *E. coli* become permeable, allowing the plasmid chimeras to pass through and become a part of the microbe's new genetic structure.
6. When the redesigned *E. coli* reproduce, they create duplicates of themselves, new plasmids—and DNA sequences—and all.

BIOTECHNOLOGY

Genetic engineering by gene splicing gave rise to a major scientific revolution, called *biotechnology.*

By definition, *biotechnology is concerned with the application of an array of biological and engineering tools, from gene splicing, to manipulating cells, tissues, and genes so as to control certain characteristics.* It involves every facet of animal production, from breeding and feeding to finished products, including the genetic makeup of animals; the feeds they eat; their digestion, physiology, stress tolerance, disease resistance, and efficiency of production; the composition, quality, and quantity of products produced; along with the production of large quantities of drugs and chemicals. Potential uses are endless. While some aspects of biotechnology are decades away from commercial production, others are near, and still others are here now. Among the major biotechnology breakthroughs visualized, being researched, or reality, are the following:

1. Redesigning animals by gene manipulation and cloning, so that they are more stress tolerant and feed efficient.
2. Modifying ruminant microorganisms to increase their efficiency in digesting forages, from the current approximately 50% to a future 80%.
3. Changing the growth pattern and production of animals by using somatotropin (growth hormone) (a) to increase the milk production of dairy cows by 20 to 40%; and (b) to produce pigs that gain 15 to 18% faster, require 30% less feed per 100 lb gain, and are leaner.
4. Producing safer and more effective vaccines; already, new vaccines have been developed for use against foot-and-mouth disease and pseudorabies, and others will follow.
5. Altering the composition of meat, milk, and eggs.
6. Producing super crops that have built-in defenses to diseases and insects; that are nitrogen-fixing and environmentally adapted; and that yield higher compositions of essential amino acids, starches, low-cholesterol oils, minerals, and vitamins.
7. Developing accurate and simple techniques for the determination of sex, pregnancy, and estrus.
8. Producing multiple births at will.
9. Processing foods that are more attractive, more nutritious, more flavorsome, and longer lasting.
10. Converting wastes into ethanol and high-protein animal feeds.

The entire food team—producers, processors, and consumers—stands to benefit from biotechnology, from the greater abundance of high-quality products produced more efficiently. However, progress will be slowed by regulatory and social problems; and some producers and processors will falter and fail along the way. Nevertheless, with the opening of the 21st century, it is predicted that the magnitude of change wrought by biotechnology will rival that of mechanization, which doubled total American agricultural production and increased output per man hour 20-fold. *Biotechnology will favor those who are there first with the most of the best.*

NUTRIENTS AND THEIR METABOLISM

The diet contains carbohydrates, fats, and proteins. Although each of these nutrients has specific functions in maintaining a normal body, all of them can be used to provide the number one requirement—energy. Hence, the discussion of their metabolism centers first, on their catabolism, and second, on their anabolism. Although they are not used for energy, minerals, vitamins, and water are essential to the metabolic processes.

The sections that follow pertain to the metabolism of nutrients—to all the chemical changes that take place after the end products of the digestion of carbohydrates, fats, and proteins, along with the minerals, vitamins, and water, are absorbed into the body. (See Fig. 4-4.)

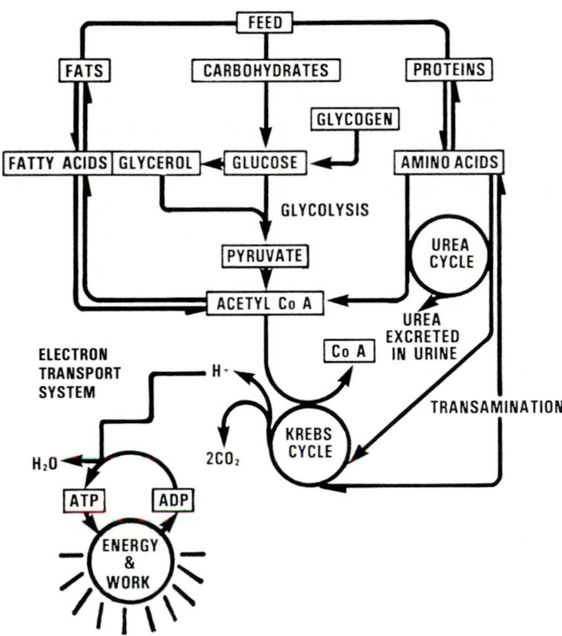

Fig. 4-4. Schematic diagram of the metabolism of nutrients. Note the metabolic interrelationships of carbohydrates, fats, and proteins.

ENERGY (CARBOHYDRATES AND LIPIDS)

Energy is required for practically all life processes—for the action of the heart, maintenance of blood pressure and muscle tone, transmission of nerve impulses, ion transport across membranes, reabsorption in the kidneys, protein and fat synthesis, the secretion of milk, and the production of eggs, wool, and power.

A deficiency of energy is manifested by slow or stunted growth, body tissue losses, and/or lowered production of meat, milk, eggs, fiber, or power, rather than by specific signs such as those which characterize many mineral and vitamin deficiencies. For this reason, energy deficiencies often go undetected and unrectified for extended periods of time.

It is common knowledge that a ration must contain carbohydrates, fats, and proteins. Although each of these has specific functions in maintaining a normal body, all of them can be used to provide energy for maintenance, for work, or for finishing. From the standpoint of supplying the normal energy needs of animals, however, the carbohydrates are by far the most important, more of them being consumed than any other compound, whereas the fats are next in importance for energy purposes. Carbohydrates are usually more abundant and cheaper, and most of them are very easily digested, absorbed, and transformed into body fat. Also, carbohydrate feeds may be more easily stored than fats in warm weather and for longer periods of time. Feeds high in fat content are likely to become rancid, and rancid feed is unpalatable, if not actually injurious in some instances.

Carbohydrates

The carbohydrates are organic compounds composed of carbon, hydrogen, and oxygen. This group includes the sugars, starch, cellulose, gums, and related substances. Although few carbohydrates occur in animal tissues (glucose and glycogen are exceptions), they form the largest part of the feed supply of animals. Carbohydrates make up 75% of the dry weight of the plant world, upon which animal life primarily depends.

The carbohydrates in plants are produced by photosynthesis, the most important chemical reaction in nature. The radiant energy of the sun is captured by the chlorophyll of plants and changed to chemical energy, which, in turn, supports the formation of glucose from carbon dioxide and water—an endergonic process. (See Fig. 4-5.)

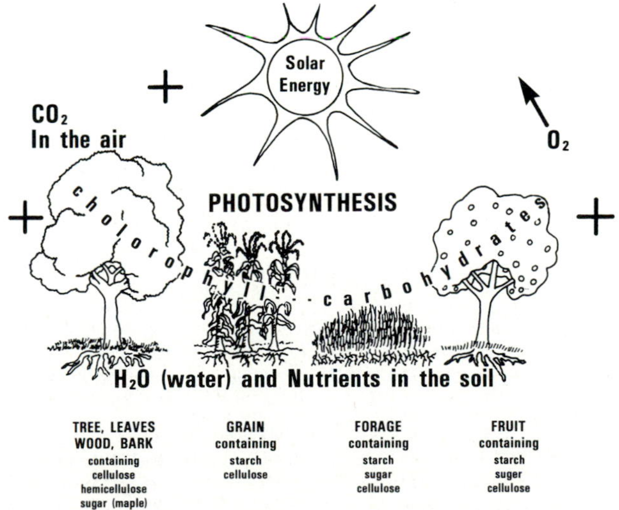

Fig. 4-5. The production of carbohydrates in plants by photosynthesis.

There are many intermediary reactions, but the overall process of glucose formation is as follows:

$$6CO_2 + 6H_2O + \text{energy from sun} = C_6H_{12}O_6 \text{ (glucose)} + 6O_2$$

Carbohydrates form the woody framework of plants as well as the chief reserve food stored in seeds, roots, and tubers. For animals, carbohydrates serve as a source of heat, energy, or bulk; and any excess of them is stored in the body as fat. But there is no specific requirement for any individual carbohydrate compound.

CLASSIFICATION AND CHARACTERISTICS OF CARBOHYDRATES

Nature ordained that carbohydrates foster plant existence and propagation; their use as animal feed is purely incidental. Thus, the more soluble forms serve in plant systems in energy transformation and for tissue synthesis; the less soluble forms, such as starch, serve as reserve energy; while the relatively insoluble fractions, such as cellulose and hemicellulose, form plant structural material.

Carbohydrates are very abundant in plants, occurring in a wide variety of forms. Table 4–1 (p. 73) lists the types of carbohydrates that are commonly involved in nutrition and summarizes their characteristics. It is noteworthy that the members of a given subgroup have the same empirical formula, yet they have different structural formulas and exhibit different physical properties.

Monosaccharides (1-Sugar Units)

Monosaccharides, or simple sugars, are seldom found free in nature. Rather, they constitute the building blocks of more complex carbohydrate molecules; hence, they are produced in the body during the metabolic process. These sugars are classified as to the number of carbon atoms within the molecule. For example, a triose is a sugar containing 3 carbons (dihydroxyacetone and glyceraldehyde are trioses), and a tetrose is a 4-carbon molecule (erythrose is a tetrose). While monosaccharides of various lengths are integral to the metabolism of carbohydrates, the pentoses (5-carbon sugars) and hexoses (6-carbon sugars) are of paramount importance. Only two of the monosaccharides, glucose and fructose, occur to any extent in free form in nature.

PENTOSES (5-CARBON SUGARS)

Very limited amounts of this type of sugar are found in a free form in plants. Generally, pentoses are polymerized (combined chemically into larger molecules) to form pentosans.

Ribose, the most plentiful pentose, is found in every living animal cell. It occurs in a number of compounds which play crucial roles in metabolism; e.g., ATP, ADP, riboflavin, and RNA. In its reduced form, 2-deoxy D-ribose, it is found in DNA. In addition, D-ribose (aldose) and D-xylulose (ketose) participate in the phosphogluconate pathway.

Arabans and xylans make up a significant part of hemicellulose. Xylose can be produced by the hydrolysis of a number of roughages and woody material, including hay, straw, oat hulls, many woods, and corncobs. When polymerized, xylose forms the pentosan xylan. Arabinose is a 5-carbon sugar found in large amounts in gums. When a number of arabinose molecules are linked together, the pentosan, araban, is formed.

HEXOSES (6-CARBON SUGARS)

Four hexoses are found in physiological systems: fructose, glucose, galactose, and mannose. Glucose, fructose, and galactose are the most common simple sugars in feed ingredients. Also, it is noteworthy that fructose and galactose are converted readily to glucose in the animal body and are, therefore, available to body metabolism as glucose.

Fructose, which is the sugar found in fruits and honey, is the sweetest of the simple sugars. Galactose is generally not found free in feeds, but is produced from lactose (milk sugar), then changed to glucose for energy. The reaction is reversible; so, during lactation, glucose may be reconverted to galactose for use in milk production.

Glucose is a moderately sweet sugar. It is not found predominantly as preformed natural glucose in feeds. Rather, it is mainly formed in the animal body from starch or cellulose digestion. In animal metabolism, all other types of sugar are eventually converted by the body to glucose. Then, glucose is the form in which sugar circulates in the bloodstream to provide the body's major fuel source for oxidation to produce energy.

Oligosaccharides (2- to 10-Sugar Units)

The prefix oligo means "few." Oligosaccharides contain 2 to 10 monosaccharides chemically bonded. Most oligosaccharides result from the partial breakdown of polysaccharides. The three most important oligosaccharides are all disaccharides—lactose, maltose, and sucrose.

DISACCHARIDES (2-SUGAR UNITS)

Disaccharides are compound sugars composed of 2 monosaccharides linked together. Both cellobiose and maltose contain 2 molecules of glucose, but they differ in the manner in which the glucose units are joined. Lactose, which is found only in milk, contains 1 molecule of galactose and 1 molecule of glucose. Sucrose contains 1 molecule of fructose and 1 molecule of glucose.

The three main disaccharides of physiologic importance are lactose, maltose, and sucrose.

Lactose is the sugar in milk. It is formed in the body from glucose to supply the carbohydrate component of milk during lactation. It is the least sweet of the disaccharides; it's about one-sixth as sweet as sucrose.

Maltose occurs in malt products of starch hydrolysis and in germinating cereal grains.

Sucrose, or common table sugar, is the most prevalent dietary disaccharide. It is found in cane and beet sugar, sorghum molasses, and carrots.

TABLE 4–1
CARBOHYDRATES

Classification	Ring (Cyclic) Structure	Chief Feed Sources	End Products of Digestion	Nutritional Functions
Sugars (water soluble):				
Monosaccharides (single sugar unit)				
Trioses ($C_3H_6O_3$)	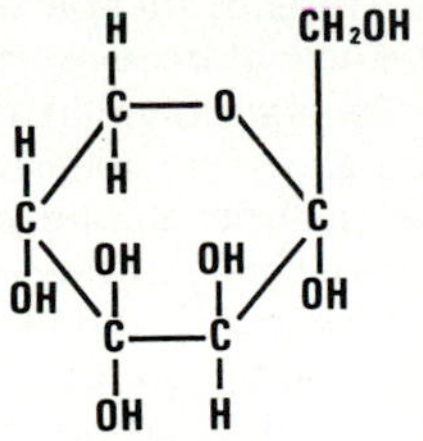			
Dihydroxyacetone .... Ketotriose		Glucose (fruit and plant sap).	Produced by oxidation of glucose.	An intermediate in the metabolism of glucose.
Glyceraldehyde .... Aldotriose				
Tetroses ($C_4H_8O_4$) Erythrose				
Pentoses ($C_5H_{10}O_5$)		Free pentoses constitute a very small part of animal diets.		
Arabinose		Gums such as gum arabic.	Arabinose.	Component of every living animal cell. It occurs in a number of compounds which play crucial roles in metabolism; e.g., ATP, ADP, riboflavin, and RNA. Its reduced form, 2-deoxy D-ribose, is found in DNA.
Ribose	Fructose	Formed through metabolic processes.	Ribose.	
Xylose	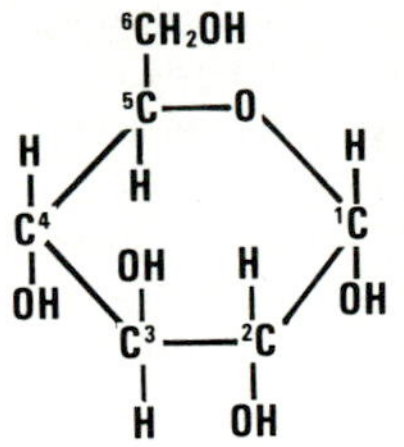	Produced by hydrolysis of hay, straw, oat hulls, many woods, and corn cobs.	Xylose.	
Xylulose		Xylulose is a ketose sugar formed from xylose or D-arabitol.	In all mammals, gulonic acid may be oxidized to L-xylulose.	Xylulose plays a role in carbohydrate metabolism.
Hexoses ($C_6H_{12}O_6$)		A large group of sugars, with a significant role in nutrition.		
Fructose .... [FRUCTOSE]		Fruits; hydrolysis of sucrose from cane sugar.	Fructose.	Changed to glucose in the liver and intestine to serve as a body fuel.
Galactose .... [GALACTOSE]		Component of milk sugar.	Galactose.	Changed to glucose in the liver; cell fuel; synthesized in mammary gland to make lactose of milk; constituent of glycolipids and glycoproteins.
Glucose .... [GLUCOSE]	Glucose	Fruits; hydrolysis of starch, cane sugar, maltose, and lactose.	Glucose.	Body "sugar"; blood and tissue fluids; cell fuel.
Mannose .... [MANNOSE]	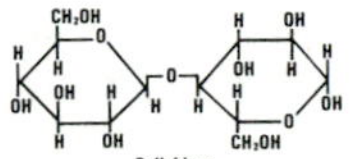	Hydrolysis of plant mannosans and gums; legumes.	Mannose.	Component of polysaccharide of albumins, globulins, mucoproteins, glycoproteins.
Oligosaccharides (2 to 10 sugar units)				
Disaccharides ($C_{12}H_{22}O_{11}$)		Glucose polymers.		
Cellobiose	Cellobiose	Cellobiose does not occur in free form in nature; occurs only as a component of glucose.		
Lactose .... [GALACTOSE] [GLUCOSE]	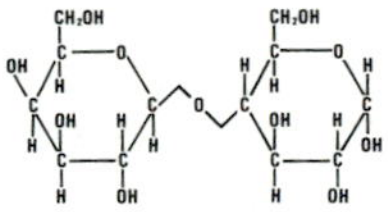	Milk and milk products.	Glucose and galactose.	Hydrolyzed to glucose and galactose, body fuel, constituent of milk production during lactation.
Maltose .... [GLUCOSE] [GLUCOSE]	Lactose	Starch by the action of malt, obtained from the malting of barley.	Glucose.	Hydrolyzed to D-glucose, basic body fuel and metabolite, fermentable.
Sucrose .... [FRUCTOSE] [GLUCOSE]		Cane and beet sugars, molasses.	Glucose and fructose.	Hydrolyzed to glucose and fructose, body fuel.
Trisaccharides ($C_{18}H_{32}O_{16}$) Raffinose	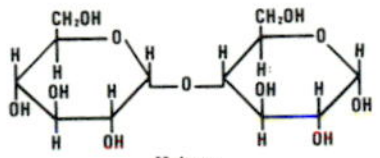	Legume seeds; sugar beets.		
Tetrasaccharides ($C_{24}H_{42}O_{21}$) Stachyose	Maltose	Legume seeds.		
Pentasaccharides ($C_{30}H_{52}O_{26}$) Verbascose		Legume seeds.		
Nonsugars:	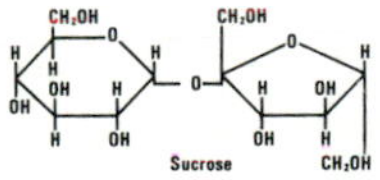			
Polysaccharides (Glycan, >10 sugar units)	Sucrose			
Homoglycans (single sugar units)				
Pentosans $(C_5H_8O_4)_n$[1]				
Arabans (Arabinose)				
Xylans (Xylose)				
Hexosans $(C_6H_{10}O_5)_n$[1]				
Fructans				
Inulin	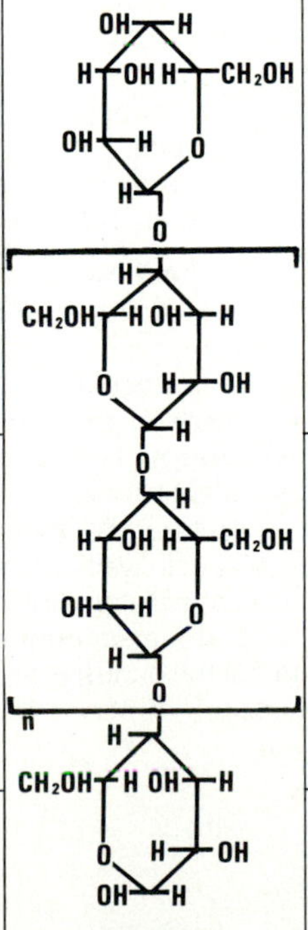	Jerusalem artichokes.	Fructose.	
Levan		A variety of plants.	Sucrose.	Reserve plant food material.
Galactans		Seeds of alfalfa, clovers, and trefoil.	Glucose.	
Glucans				
Cellulose, β-linked (glucose)		Stalks and leaves of plants, hulls of seeds.	Glucose, acetic acid.	Provide energy, hold water; reduce elevated colonic intraluminal pressure; bind zinc.
Dextrins, α-linked (glucose)		Starch of grains subjected to hydrolysis or action of heat.	Glucose.	Provide energy for animal needs.
Glycogen, α-linked (glucose)		Meat by-products, marine by-products.	Glucose.	Provide energy for animal needs.
Starch, α-linked (glucose) ...... 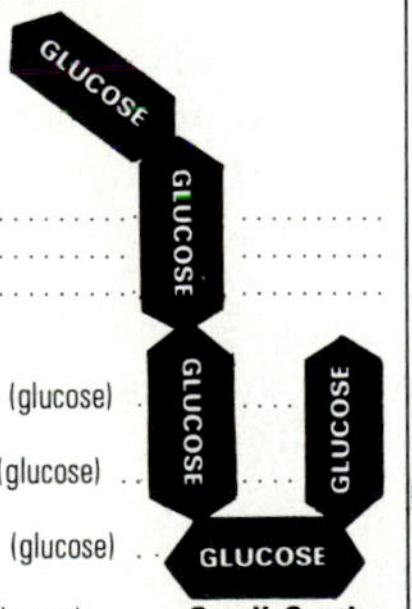 **Small Section of Starch Molecule**		Grains, rhizomes, and seeds; shoots, stems, and leaves of plants.	Glucose.	Provide energy for animal needs.
Mannans		Palm seeds.	Glucose.	
Heteroglycans (2–6 different kinds of sugar units)				
Gums		Secretions of plants; seaweeds.	Monosaccharides and uronic acids.	Slow gastric emptying; provide fermentable material for colonic bacteria with production of gas and volatile fatty acids; binds bile acids.
Hemicelluloses (β-linked)		Cell wall plant material.	A number of monosaccharides, including glucose, xylose, mannose, arabinose, and galactose; acetic acid.	Provide energy, hold water, and increase fece bulk; reduce elevated colonic pressure; bind bile acids.
Mucilages		Plant secretions and seeds.		
Mucopolysaccharides		Animal connective tissue.	Amino sugars.	Provide some rigidity to animal tissues.
Pectins (α-linked)		Citrus fruits, sugar beets, apples.	Galacturonic acid, galactose, arabinose, rhamnose, and other sugars.	High water-holding capacity. Used to reduce diarrhea in calves.
Specialized compounds				
Chitin		Exoskeleton of insects and crustaceans.		
Lignin (not a carbohydrate)	Cellulose	Woody part of plants such as woods, cobs, and hulls; and in the fibrous parts of roots, stems, and leaves.		Antioxidants; bind bile acids and metals.

[1]The "n" indicates any number of sugar units greater than 10.

TRISACCHARIDES

Trisaccharides, 3-sugar polymers, are not abundant in nature, but 2 trisaccharides—raffinose and melezitose—are found in limited amounts in certain plants. Raffinose, which is found in high concentrations in sugar beets, consists of glucose, fructose, and galactose. Melezitose, a component of sap in some coniferous plants, contains 2 molecules of glucose and 1 of fructose.

TETRASACCHARIDES

Stachylose is a tetrasaccharide composed of 2 molecules of galactose, 1 molecule of glucose, and 1 molecule of fructose. Beans produce gas in the digestive tract—flatulence—due to microbial action on stachylose and raffinose which are not split into monosaccharides by enzymes of the digestive tract. Maltotetrose is formed during the digestion or breakdown of starch. It consists of 4 molecules of glucose, which the digestive enzymes of the body can separate.

PENTASACCHARIDES

Raffinose with two or more molecules of D-galactose becomes verbascose, a pentasaccharide.

Polysaccharides (Multiple Sugars)

Polysaccharides are large sugar complexes that contain repeating sequences of simple sugars. They are used in plants for either carbohydrate storage or structural support.

The polysaccharides are divided into homoglycans, those having only 1 glucose unit, and heteroglycans, those having 2 to 6 different glucose units (see Table 4-1). Further, the homoglycans are divided into those that yield pentoses (pentosans) and those that yield hexoses (hexosans).

HOMOGLYCANS

The homoglycans, which are characterized by having a single glucose unit, embrace the pentosans and the hexosans.

Pentosans $(C_5H_8O_4)_n$[1]

The pentosans may compose up to one-fifth of all the complex carbohydrates in hay. Pentosans yield pentoses, such as arabinose and xylose, upon complete hydrolysis with acid. They are not degraded by mammalian enzymes, but they are rapidly hydrolyzed by the microbial and fungal enzymes of ruminants.

When pentosans are boiled with hydrochloric acid, an aldehyde, furfural, is produced. This reaction is used in the commerical production of furfural from oat hulls and corn cobs.

[1]The "n" indicates any number of sugar units greater than 10.

Hexosans

Hexosans are polysaccharide sugars which contain hexoses as their respective repeating units.

Several polymers of glucose are found in physiological systems. The difference among the various compounds results from the linkages between the glucose molecules. These chief polymers of glucose are (1) cellulose, (2) dextrin, (3) glycogen, and (4) starch. Quantitatively, the hexosans provide the greatest amount of energy for animal needs of any of the carbohydrates in nature.

1. **Cellulose.** Cellulose is, by far, the most abundant polysaccharide in nature—composing nearly 50% of the total organic carbon. It is a straight-chain polymer which is extremely resistant to acid and alkaline hydrolysis. Monogastric animals lack the necessary enzymes to cleave the linkages of glucose molecules in cellulose. Hence, they are poor users of fibrous plants. The microorganisms in the rumen of ruminants contain the enzyme cellulase; hence, feeds that are high in cellulose can be utilized effectively by ruminants.

2. **Dextrin.** Dextrins are metabolic by-products of starch digestion. They are more soluble than starches and are found in relatively high concentrations in seeds.

3. **Glycogen.** Glycogen (animal starch) is the primary storage source of glucose in the blood of animals. It is stored in the liver and muscle tissue. The glycogen in the liver may account for as much as 10% of the weight of that organ. As needed, the glycogen in the liver is changed to glucose.

4. **Starch.** Starch is the most important low-fiber polysaccharide found in plants, particularly in grains, legumes, and tubers. Corn, millet, rice, rye, and wheat—the important cereal grains—contain as much as 70% of this carbohydrate, whereas the dried seeds of leguminous plants (such as beans and peas) average about 40%. Starch is also found in small quantities in all parts of the plant such as shoots, stems, and leaves.

Starch is composed of units of glucose, of which there are two types: (1) amylose, a straight-chained structure of repeating glucose molecules; and (2) amylopectin, a highly-branched compound. Starches from different plant sources vary in the ratio of amylose and amylopectin, and in their microscopic physical structure. Starches are generally highly digestible by animals. However, tuber starch, such as found in the potato, is an exception; its granules are quite resistant to rupture, with the result that it must be cooked before being fed to pigs and poultry.

5. **Other hexosans.** In this category, the following carbohydrates are of importance: fructans, galactans, and mannans.

Fructans occur as reserve material in roots, stems, leaves, and seeds of a variety of plants. They can be divided into two groups: (1) the levan group; and (2) the inulin group. Jerusalem atrichoke is rich in inulin.

Galactans and mannans are polysaccharides which occur in the cell walls of plants. The seeds of many legumes, including alfalfa, clovers, and trefoil, contain galactans. Mannan is the main component of the cell walls of palm seeds and of the endosperm of nuts from the South American tague tree.

HETEROGLYCANS

The following heteroglycans (polysaccharides with 2 to 6 different glucose units) are important in animal nutrition: gums, hemicelluloses, mucilages, mucopolysaccharides, and pectins.

1. **Gums.** Plant gums are viscous fluids which are present in all plants, and which are especially high in seaweeds. Also, gums are often released by wounds in plants, and they may arise as natural exudate from bark and leaves.

2. **Hemicellulose.** Hemicellulose, which is the principal component of plant cell walls, is the cell wall fraction most closely allied with lignin. It is a complex, heterogeneous mixture of a number of different polymers of monosaccharides including arabinose, galactose, mannose, xylose, other sugars, and uronic acids.

Hemicellulose is more subject to chemical degradation than cellulose. It is soluble in mild alkali; and it can be hydrolyzed by a relatively mild acid treatment.

3. **Mucilages.** Mucilages are found in certain plants and seeds and function as water-holding compounds which protect against dessication. They are highly variable in composition. Well-known examples of mucilages are: linseed mucilage and agar obtained from red seaweeds.

4. **Mucopolysaccharides.** Mucopolysaccharides, which are widespread in animal connective tissue, are highly viscous and readily form gels. They are synthesized primarily from the amino sugars D-glucosamine and D-galactosamine, together with several uronic acids. They provide some rigidity to animal tissues.

5. **Pectins.** Pectins are found primarily in the spaces between plant cell walls, but they also infiltrate the cell wall itself. They consist of galacturonic acid, galactose, and arabinose units. Pectic substances are particularly abundant in citrus fruits, sugar beets, and apples. Pectin can be extracted with hot or cold water and will form a gel, as in jelly and jam. Because of its water-holding capacity, it is often used as a treatment for diarrhea in calves.

Specialized Compounds

A number of complex mixed polysaccharides are found in nature, many of them serving structural or protective functions; among them, chitin and lignin.

1. **Chitin.** Chitin forms the thick material of the hard exoskeletons of insects and crustaceans. It is composed of N-acetyl-D-glucosamine.

2. **Lignin.** Lignin is not a carbohydrate. But, because of its association with carbohydrates, it is appropriately discussed with them. Its structure is not well understood; its form may vary widely from plant to plant; and its carbon-to-carbon linkages are very resistant to chemical degradation. Lignin is found in the woody parts of plants such as hardwoods (which contain the most lignin of any plant), cobs, and hulls, and in the fibrous portion of roots, stems, and leaves. Its content increases as plants mature, and its chemical linkage, especially with hemicellulose and cellulose, markedly reduces the digestibility of the latter. Alkali treatment of highly lignified forage, such as straw, breaks the hemicellulose-lignin bond, improving digestibility of the hemicellulose without destroying the lignin.

CARBOHYDRATE COMPONENTS WITHIN FEEDS

From a feeding standpoint, carbohydrates are components of the nitrogen-free extract (NFE) and the crude fiber fractions of proximate analysis. The nitrogen-free extract includes the more soluble, and, therefore, the more digestible, carbohydrates—such as the starches, sugars, hemicelluloses, and some cellulose and pentosan. Also, NFE contains a limited amount of lignin. Crude fiber is the woody portion of plants (or feeds) which is not dissolved by weak acids and alkalies. Therefore, crude fiber—including cellulose, hemicellulose, and lignin—is digested less easily.

Availability of Fiber

The ability of animals to utilize roughages—to digest the fiber therein—depends chiefly on microbial action. This is confined largely to the first three compartments of the stomach of ruminants, to the cecum and colon of the horse, and to a lesser extent the large intestine of all animals. This bacterial digestion breaks down the cellulose and pentosans of feeds into usable organic acids (chiefly acetic, propionic, and butyric acid). These volatile fatty acids are absorbed directly through the ruminal wall, and furnish much of the energy required for maintenance. The fiber of growing pasture grass, fresh or dried, is more digestible than the fiber of most hay. Likewise, the fiber of early-cut hay is more digestible than that of hay cut in the late-bloom or seed stages. The difference is due to both chemical and physical structure, especially to the presence of certain encrusting substances (notably lignin) which are deposited in the cell wall with age. Young stock of all classes, finishing steers and lambs, high-producing dairy cows, swine, poultry, and horses must have rations in which a large part of the carbohydrate content of the ration is low in fiber but high in the form of nitrogen-free extract. On the other hand, a considerable amount of fiber or bulk in the ration is considered desirable for mature breeding animals of all classes of livestock, especially when high condition is not desired. Likewise, with young animals being developed for breeding purposes, the increased fiber will tend to develop more growth and not so much fat.

To promote good muscle tone and activity in the gastrointestinal tract, one must feed a certain amount of coarse roughage to all classes of farm animals, except to swine and poultry.

CATABOLISM OF CARBOHYDRATES

The primary use for carbohydrates is to produce energy. In the presence of oxygen, glucose will produce a maximum of 686 kcal/mole (*180.2 g*), as seen in the following equation:

$$C_6H_{12}O_6 + 6O_2 \rightarrow 6CO_2 + 6H_2O + 686\ \text{kcal/mole (heat)}$$

This is a combustion or burning reaction. Hence, the expression of *burning off calories* can be understood.

No physiological system can approach 100% efficiency in the production of energy from the catabolism of a nutrient. Likewise, the body cannot produce 686 kcal net from one mole (*180.2 g*) of glucose. Nevertheless, a great deal of energy is produced from the catabolism of glucose by a series of enzymatic reactions which permit the orderly transfer of energy from glucose to energy-rich compounds:

• **ATP—energy currency**—For metabolism to permit the orderly transfer of energy from feeds to processes requiring energy, there must be a common carrier of energy—something cells can use to exchange goods for services. The energy currency—the primary mechanism by which energy is captured and stored—is a compound called adenosine triphosphate, often abbreviated ATP. Just as pennies are small units of a dollar, ATP is a small unit of energy.

ATP is a compound formed from the purine adenine, the 5-carbon sugar ribose, and 3 phosphate molecules as shown in Fig. 4-6. Two of the phosphates are joined by high-energy bonds.

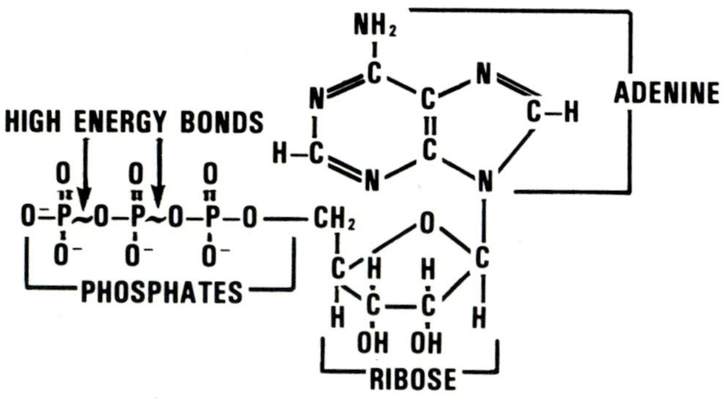

Fig. 4-6. Structure of adenosine triphosphate.

When ATP is formed from ADP (adenosine diphosphate) by the addition of another phosphate, 8 kcal/mole must be put into the system—like a deposit in a bank account. The ATP then acts as a high-energy storage compound for the energy produced in the catabolism of various nutrients. When energy is required for body functions, ATP is broken down to ADP by releasing 1 phosphate and 8 kcal/mole of energy, from the rupture of the high energy bond.

Releasing all of the energy contained in carbohydrates (fats and proteins, too) in one step would result in lost and wasted heat energy. Instead, energy is released in steps. The body utilizes a series of biological oxidations to form ATP. *Oxidation refers to the loss of electrons (or hydrogen) by a compound.* Conversely, *reduction refers to the acceptance of electrons (or hydrogen) by a compound.* Chemical energy is released as the lost electrons flow down energy gradients of an elevated state to the next lower hydrogen (electron) acceptor or carrier. Several coenzymes act as hydrogen (H^+) acceptors, many of which involve vitamins; for example, niacin and riboflavin. Compounds such as nicotinamide adenine dinucleotide (NAD^+), nicotinamide adenine dinucleotide phosphate ($NADP^+$), and flavin adenine dinucleotide (FAD) can accept and transfer hydrogen (or electrons) through a cytochrome system known as *the electron transport system.* Cytochromes are pigmented proteins similar to hemoglobin. In the electron transport system, electrons cascade from one intermediate carrier molecule—cytochrome—to the next lower in line, like water over a small waterfall. For every 2 electrons that tumble down the cytochrome system, enough energy is released to form 3 ATP molecules by adding a high-energy phosphate to ADP. The ultimate hydrogen acceptor in the system is oxygen; hence, water (H_2O) is formed. This is metabolic water. Production of ATP through this system is known as oxidative phosphorylation. Oxygen is required. It should be noted that when oxidation is initiated and NAD^+ carries the hydrogen ($NADH + H^+$) there is a net production of 3 ATP molecules. Only 2 ATPs are yielded, however, when hydrogen is transferred by FAD ($FADH_2$).

Fig. 4-7 illustrates the cytochrome system, and the formation of ATP as hydrogen drops to different energy levels.

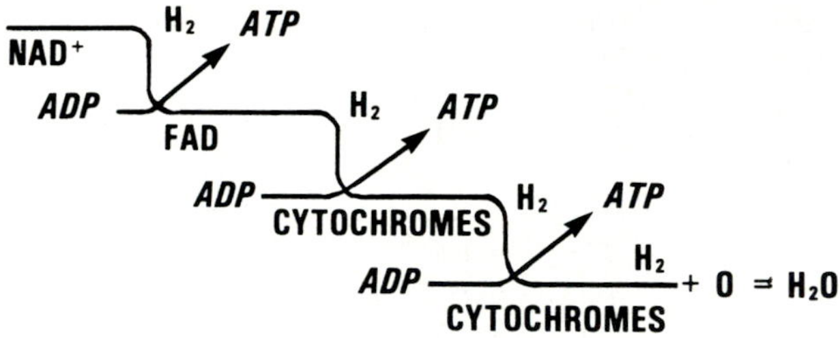

Fig. 4-7. The formation of adenosine triphosphate (ATP) from adenosine diphosphate (ADP) as hydrogen tumbles down a chain of hydrogen carriers—the electron transport system. NAD^+ stands for nicotinamide adenine dinucleotide, the FAD stands for flavin adenine dinucleotide—both hydrogen carriers similar to the cytochromes.

• **Glycolysis**—Before glucose can be used by the cells it must first enter them from the blood. Entry into the cells depends upon the hormone insulin from the pancreas. Without insulin, glucose stays in the blood where it may create the condition known as *diabetes mellitus.* Once inside the cell, the initial steps in the catabolism of glucose begin with a process called glycolysis—meaning glucose breakdown. This pathway of chemical changes is referred to as the Embden-Meyerhof or glycolytic pathway. It is a series of reactions that are oxygen independent—no free or molecular oxygen needs to be involved; hence, these reactions are said to be anaerobic. The first reaction of glucose is that of providing activation energy—energy to get glucose "over the hump" before proceeding "downhill." This is accomplished by a reaction in which ATP donates a high-energy phosphate yielding a new compound called glucose-6-phosphate.

Next, glucose atoms are slightly rearranged to yield fructose-6-phosphate. Then another phosphate is added by a donation from ATP. The compound is now called fructose-1, 6-phosphate. Another enzyme splits this 6-carbon compound into two 3-carbon compounds each containing a phosphate; one compound is glyceraldehyde-3-phosphate, and the other is dihydroxyacetone phosphate. Dihydroxyacetone phosphate can be converted to glyceraldehyde-3-phosphate which receives another phosphate after two hydrogen atoms are removed via $NADH + H^+$. When oxygen is present (aerobic), this hydrogen enters the electron transport system and yields three ATP. Next, the compound with two phosphates—1,3-diphosphoglyceric acid—loses a phosphate thus forming ATP from ADP and converting the compound to 3-phosphoglyceric acid. Eventually, the other phosphate is lost to convert ADP to another ATP and pyruvic acid is formed. Because each glucose provides two 3-carbon compounds for glycolysis, the net gain in ATP molecules to this point is two. Two ATP were used to add phosphate to the 6-carbon compounds while two were gained for each 3-carbon compound converted to pyruvic acid.

Lactic acid is produced only under anaerobic conditions such as in exercise, when oxygen cannot be delivered fast enough by the blood. Since the conversion of pyruvic acid to lactic acid requires the input of two hydrogen atoms, the NADH + H^+ generated earlier in glycolysis can be reconverted to NAD^+, thereby allowing glycolysis to continue. Normally, NADH + H^+ would go to the electron transport system, but this requires oxygen. Lactic acid is a dead end. In order for the body to use it, it must be reconverted to pyruvic acid, and in some cases glucose.

The whole scheme of glycolysis is presented in Fig. 4-8.

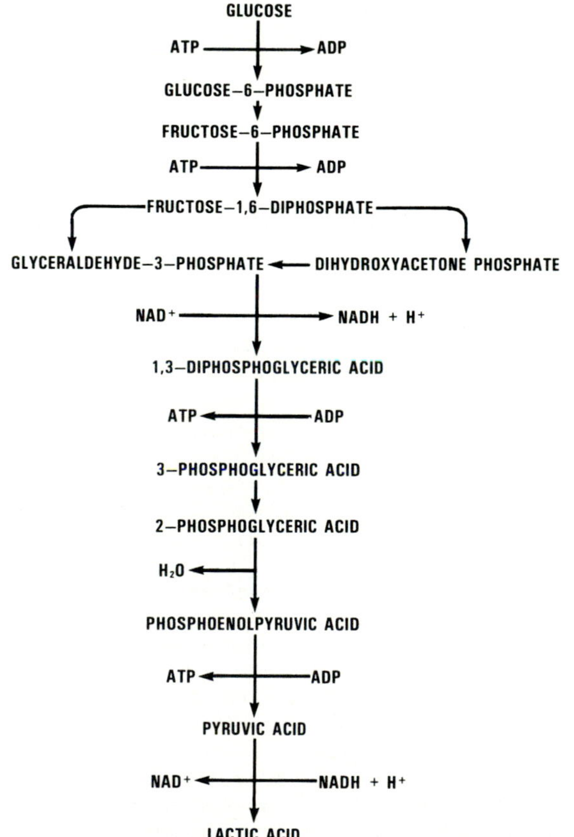

Fig. 4-8. Glycolytic or Embden-Meyerhof pathway.

The following items are important features of glycolysis:

1. Two ATP are netted for each glucose.
2. The system allows for the anaerobic breakdown of glucose.
3. It provides a system whereby energy from glucose stored as glycogen can be made readily available in muscle even when oxygen is in short supply during exercise.
4. Intermediates for the synthesis of other nutrients, such as glycerol and amino acids, are provided.
5. In the step whereby glyceraldehyde-3-phosphate is converted to 1, 3-diphosphoglyceric acid, 2 moles of NADH + H^+ are produced which subsequently can be oxidized in the electron transport system under aerobic conditions to produce 6 additional ATP.

Through more chemical reactions, more energy can be derived from glucose which has reached the point of pyruvic acid. These additional reactions take place in a cellular structure appropriately called the powerhouse of the cell. These are the mitochondria.

• **Mitochondria**—These are watermelon-shaped microscopic structures within the cells of the body. The number and distribution of mitochondria within a cell depend upon the energy requirement of the cell. Those cells with large energy requirements possess a large number of mitochondria. The mitochondria contain the electron transport system. Moreover, they contain the enzymes necessary for the further metabolism of pyruvic acid, and other chemical transformations. It is their unique internal structure of cristae—a series of baffles—that allows for very efficient production of ATP.

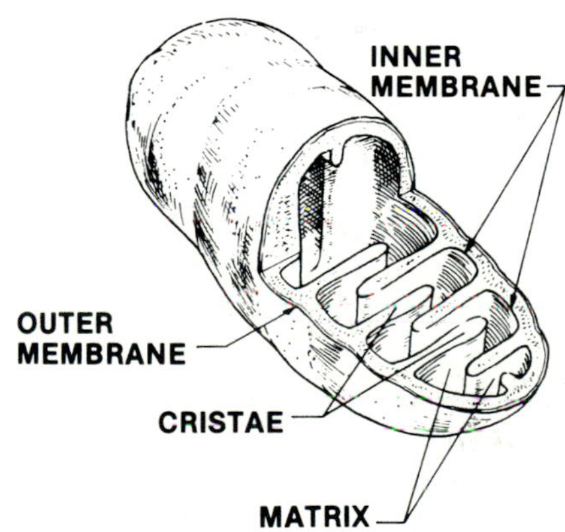

Fig. 4-9. Microscopic view of a mitochondrion—powerhouse of the cell.

Pyruvic acid enters the mitochondria of the cell and begins another series of reactions variously referred to as the Krebs cycle, the tricarboxylic acid (TCA)cycle, or the citric acid cycle. First, pyruvic acid is converted to acetyl coenzyme A (acetyl CoA) by a loss of a carbon atom as carbon dioxide (CO_2) and 2 hydrogen atoms. This conversion yields 3 ATP, since the hydrogen atoms enter the electron transport system via NADH + H^+. Acetyl CoA is then condensed with oxaloacetic acid to form citric acid, thus initiating the Krebs cycle. As the cycle progresses, more electrons (hydrogens) are transferred to different coenzymes, which, in turn, enter the electron transport chain producing ATP. One complete turn of the cycle (1) produces 12 ATP and the loss of 2 carbons as carbon dioxides (CO_2) from citrate, and (2) restores the oxaloacetic acid; completing the cycle. Since 2 pyruvic acid molecules can be formed from 1 glucose molecule, 24 ATP are produced via the Krebs cycle. Thus, 38 ATP are produced from 1 molecule of glucose: 8 from glycolysis, 6 in the conversion of pyruvic acid to acetyl CoA, and 24 from the Krebs cycle. Each pass of the 2-carbon acetyl CoA through the Krebs cycle results in the formation of carbon dioxide (CO_2) and water (H_2O), the end products of catabolism.

Fig. 4-10 is a simplified version of the Krebs cycle, named after Hans Krebs, the scientist who discovered it.

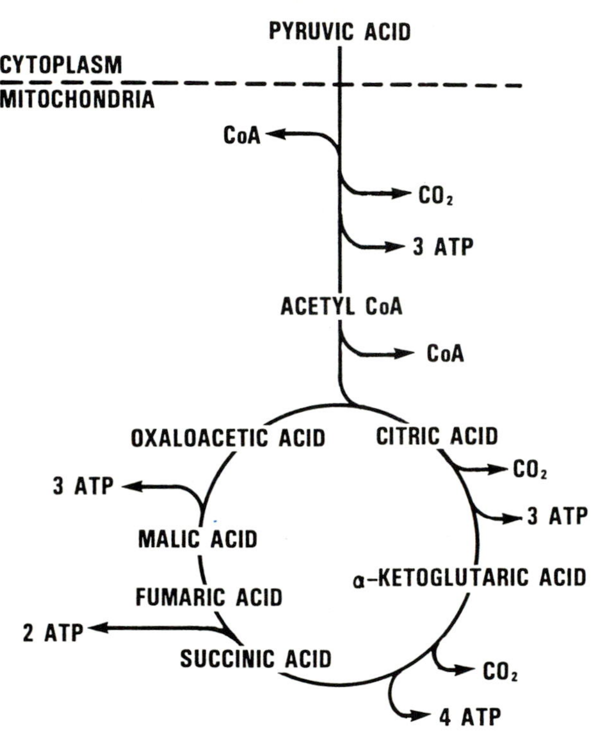

Fig. 4-10. The Krebs cycle.

• **Energy efficiency**—One ATP molecule will yield about 8 Calories (kcal). Since 38 ATP are derived from the complete oxidation of glucose, 304 Calories (kcal) of energy are formed (38 × 8 = 304). Under ideal conditions, the oxidation of glucose yields 686 Calories (kcal). Hence, the body is 44% efficient in converting glucose to energy (304 ÷ 686 × 100 = 44). Diesel engines have a maximum efficiency of about 40%, while many gasoline engines only convert about 25% of their fuel energy into work. So, the body burns its fuel quite efficiently.

• **Hexosemonophosphate pathway (pentose shunt)**—Not all glucose enters the glycolytic pathway (Fig. 4-8). A small amount continually enters another pathway known as the pentose shunt or hexosemonophosphate (HMP) pathway (Fig. 4-11). The functions of this pathway are twofold: (1) to provide some energy in the form of nicotinamide adenine dinucleotide phosphate-reduced (NADPH), another hydrogen carrier, which will function in the formation of fatty acids; and (2) to provide the 5-carbon sugar ribose for use in the nucleic acids DNA and RNA, or nucleotide coenzymes such as ATP. Products of this pathway eventually re-enter the glycolytic pathway.

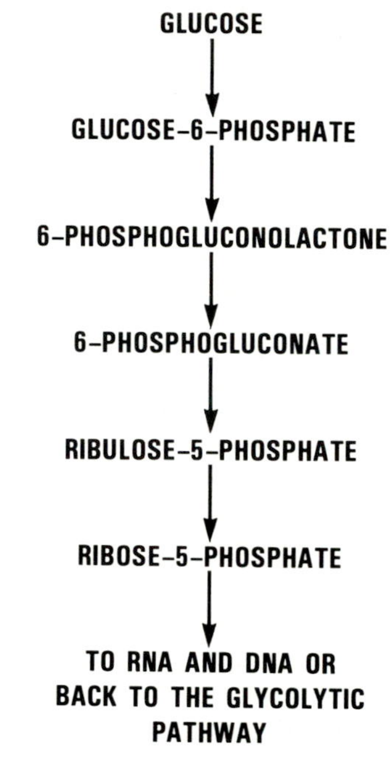

Fig. 4-11. The pentose shunt or hexosemonophosphate pathway.

ANABOLISM OF CARBOHYDRATES

Two processes involved in carbohydrate metabolism may be considered anabolic in that they require energy: (1) storage of glucose as glycogen; and (2) the formation of glucose from noncarbohydrate sources, or gluconeogenesis.

• **Glycogen**—Upon absorption into the bloodstream, glucose travels to the liver where it is converted into glycogen. Glycogen is composed of numerous glucose molecules joined together; hence, glycogen is sometimes referred to as animal starch. Each time a glucose molecule is linked with the glycogen already in the liver, two high-energy phosphate bonds are required. Then, as glucose is needed to maintain the blood glucose levels, glycogen is broken down to release glucose.

Some glycogen is manufactured and stored in the muscles. However, glucose derived from this glycogen is available only for use by the muscle and not the general circulation.

Lactic acid may leave the muscle and be reconverted to glucose in the liver and released into the blood. This process requires 6 ATPs for each glucose molecule, and is referred to as the lactic acid cycle or Cori cycle (Fig. 4-12). A husband and wife team, Carl and Gerty Cori, described it; and they were jointly awarded the Nobel Prize in 1947.

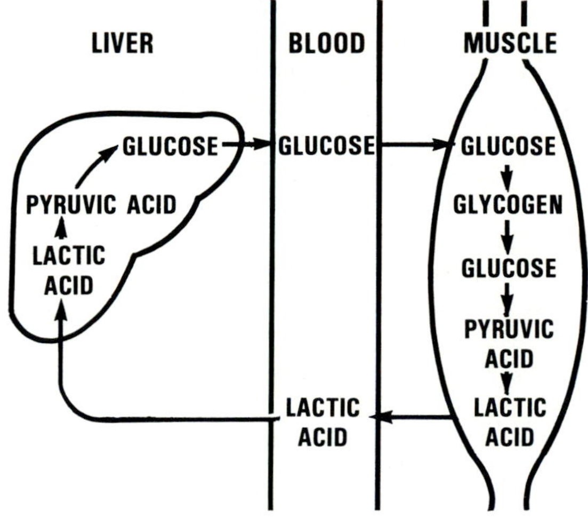

Fig. 4-12. The lactic acid or Cori cycle.

• **Gluconeogenesis**—Gluconeogenesis is the formation of glucose from nutrients other than carbohydrates. This pathway is extremely important when there is an insufficiency of dietary carbohydrates. The liver is the main site of gluconeogenesis. Amino acids are the primary precursors in gluconeogenesis. However, amino acids must first have their amino group (NH_2) removed by a process called deamination, following which the remaining carbon skeleton can enter the Krebs cycle and, essentially, be worked backwards through glycolysis to glucose. All of the amino acids except leucine are glycogenic—capable of forming glucose. Production of glucose by this means is possible because most reactions catalyzed by enzymes are reversible. There is, however, one exception. Pyruvic acid in the mitochondria of cells cannot be directly converted to phosphoenolpyruvic acid. Rather, it is transformed into oxaloacetic acid and malic acid. Malic acid is then able to cross the mitochrondrial membrane into the cytoplasm where it is reconverted to oxaloacetic acid. Oxaloacetic acid is then converted to phosphoenolpyruvic acid due to the presence of a specific enzyme in the cytoplasm. Once this is accomplished, the process closely resembles the reverse of the glycolytic pathway. (See Figs. 4-8 and 4-10, pp. 77 and 78.) The whole process requires an expenditure of energy.

Table 4-2 indicates where each amino acid fits into the metabolic scheme following deamination.

TABLE 4-2
GLUCONEOGENESIS FROM AMINO ACIDS

Amino Acid	Compounds Formed Following Deamination	Metabolic Fate[1]
Alanine	Pyruvic acid	Glycogenic
Arginine	Alpha-ketoglutaric acid	Glycogenic
Aspartate	Fumaric acid	Glycogenic
Cysteine	Pyruvic acid	Glycogenic
Glutamate	Alpha-ketoglutaric acid	Glycogenic
Glycine	Pyruvic acid	Glycogenic
Histidine	Alpha-ketoglutaric acid, beta-ketoglutaric acid	Glycogenic
Proline	Alpha-ketoglutaric acid	Glycogenic
Hydroxyproline	Alpha-ketoglutaric acid	Glycogenic
Methionine	Succinyl CoA	Glycogenic
Serine	Pyruvic acid	Glycogenic
Threonine	Succinyl CoA	Glycogenic
Valine	Succinyl CoA	Glycogenic
Isoleucine	Acetyl CoA, succinyl CoA	Glycogenic and ketogenic
Lysine	Alpha-ketoglutaric acid, acetyl CoA, acetoacetyl CoA	Glycogenic and ketogenic
Phenylalanine	Fumaric acid, acetyl CoA	Glycogenic and ketogenic
Tryptophan	Succinyl CoA, acetyl CoA	Glycogenic and ketogenic
Leucine	Acetyl CoA, acetoacetic acid	Ketogenic

[1]Glycogenic forms glucose while ketogenic forms ketone bodies—acetoacetic acid, beta-hydroxybutyric acid, and acetone.

Fats and Other Lipids

Lipids is an all-embracing term referring to compounds that are soluble in chloroform, benzene, petroleum, or ether. It includes fats, oils, waxes, sterols, and complex compounds such as phospholipids and sphingolipids.

Lipids, like carbohydrates, contain the three elements—carbon, hydrogen, and oxygen.

Not all lipids are fats, but all fats are lipids; so, the two terms are used interchangeably. Fats and oils are differentiated on the basis of melting points; fats are solid at room temperature, while oils are liquid. As livestock feeds, fats and oils function much like carbohydrates in that they serve as a source of heat and energy and for the formation of fat. Because of the larger proportion of carbon and hydrogen, however, fats and oils liberate more energy than carbohydrates when digested, furnishing on oxidation approximately 2.25 times as much heat or energy per pound as do the carbohydrates. A smaller quantity of fat is required, therefore, to serve the same function. Common belief to the contrary, animals can tolerate a rather high-fat content in the ration. As evidence of this, sucklings normally handle a relatively large amount of fat, for milk contains 25 to 40% of this nutrient on a dry matter basis. Also, except for the soft pork problem, no apparent difficulty is encountered in feeding hogs a rather high-fat ration, such as results when large quantities of peanuts or soybeans are fed.

A small amount of fats in the ration is desirable, as these fats are the carriers of the fat-soluble vitamins. Fortunately, normal farm rations contain ample quantities of these nutrients.

CLASSIFICATION OF LIPIDS

Lipids are often classified into three major groups: (1) simple lipids; (2) compound lipids; and (3) derived lipids. When fatty acids are esterified with alcohols, simple lipids result. If compounds such as choline or serine are esterified to alcohols in addition to fatty acids, compound lipids result. The third type of lipid, derived lipids, results from the hydrolysis or enzymatic breakdown of simple and compound lipids. Table 4–3 classifies lipids and provides some examples and characteristics of each of them.

TABLE 4–3
CLASSIFICATION OF LIPIDS

Type of Lipid	Example	Chemistry	General Comments
Simple lipids:			
Neutral fats	Triglycerides (triacylglycerols).	Esters of fatty acids with glycerol; ratio of 3 fatty acids to 1 glycerol.	Most abundant lipids in nature. Mixed triglycerides (those in which at least 2 fatty acids are different) account for 98% of the fats in feeds and over 90% of fat in the body.
Waxes	Beeswax.	Esters of fatty acids with high-molecular-weight alcohols other than glycerol. This group includes the esters of cholesterol, vitamin A, and vitamin D.	More important in commerce than in animal nutrition; occur widely in cuticle of leaves and fruit.
Compound lipids:			
Phospholipids	Lecithins. Cephalins. Lipositols.	Compounds of neutral fat, a phosphoric acid, and a nitrogenous base (choline, ethanolamine, or serine); water-soluble.	Lecithins are largest group of phospholipids. Lecithin may be obtained from egg yolks or soybeans.
Glycolipids	Cerebrosides. Gangliosides.	Sugar-(carbohydrate)-containing fatty acids plus nitrogen.	Sugar can be glucose or galactose; found in nervous tissue; component of cell membrane.
Lipoproteins	Chylomicrons. Very low density lipoproteins (VLDL). Low density lipoproteins (LDL). High density lipoproteins (HDL).	They all contain protein, triglycerides, phospholipids, and cholesterol; but in varying amounts.	The lipoproteins, synthesized in the liver, are composed of about ¼ to ⅓ protein, with the remainder lipids. Means of transporting lipids in the blood.
Derived lipids:			
Fatty acids	Palmitic acid. Oleic acid. Stearic acid. Linoleic acid.	Generally have one acid group (COOH); may be saturated, or unsaturated—contain 1 or more double bonds.	In most cases, there is an even number of carbon atoms in the naturally-occurring fatty acids. There are few odd-numbered carbon atom fatty acids in nature. Release of fatty acids from triglyceride releases glycerol.
Steroids	Cholesterol. Ergosterol. Cortisol. Bile acids. Vitamin D. Androgens, estrogens, and progesterone.	Derivatives of the perhydrocyclopentan-o-penanthrene nucleus (chemical structure is a series of rings).	One of the most studied classes of lipids. Collectively many of these are referred to as steriod hormones—hormones of the adrenal gland, testes, and ovaries.
Hydrocarbons	Terpenes.	Compounds of hydrogen and carbon only.	Includes a series of oils (such as camphor), resin acids, and plant pigments. Beta-carotene is an example of an important terpene.

CHARACTERISTICS OF FATS

Triglycerides—the combination of 3 fatty acid molecules and a glycerol molecule—account for about 98% of the fats in feed and over 90% of the fat in the animal body. The remainder of the diet and body fats is comprised primarily of phospholipids and cholesterol. Since triglycerides and their component fatty acids are so abundant in the diet and body, the discussion which follows centers around fatty acids and triglycerides.

Fatty Acids

Fatty acids are key components of lipids. They are called acids because of the carboxyl organic acid group (COOH) which they contain. Their degree of saturation and the length of their carbon chain determine many of the physical characteristics of lipids. Numerous triglycerides exist due to the variety of fatty acids which may bind with glycerol. The properties of fatty acids depend on their chemical characteristics, which follow.

• **Saturation**—This refers to the ratio of hydrogen atoms to carbon atoms. The backbone of the fatty acid consists of a chain of carbon atoms joined by chemical bonds. When a single bond joins each pair of carbon atoms, carbon atoms within the chain have 2 hydrogen atoms joined to them and carbons at the end of the chain have 3 hydrogens. When carbon atoms are joined by double bonds, the carbon atoms within the chain are able to have only 1 hydrogen bound to them. Therefore, saturated fatty acids contain all possible hydrogen and no double bond between carbon atoms. Unsaturated fatty acids contain at least one double bond within the carbon chain (monounsaturated) or 2 or more double bonds within the carbon chain (polyunsaturated). Therefore, unsaturated fatty acids contain the same number of carbon atoms, but fewer hydrogen atoms than their saturated counterparts. Fig. 4-13 illustrates the concept of saturated and unsaturated.

SATURATED

```
   O H H H H H H H H H H H H H H H H H
   ‖ | | | | | | | | | | | | | | | | |
HO-C-C-C-C-C-C-C-C-C-C-C-C-C-C-C-C-C-C-H
     | | | | | | | | | | | | | | | | |
     H H H H H H H H H H H H H H H H H
```

(STEARIC ACID, $C_{18}H_{36}O_2$)

MONOUNSATURATED

```
   O H H H H H H H   H H H H H H H H
   ‖ | | | | | | |   | | | | | | | |
HO-C-C-C-C-C-C-C-C=C-C-C-C-C-C-C-C-C-H
     | | | | | | | | | | | | | | | |
     H H H H H H H H H H H H H H H H
```

(OLEIC ACID, $C_{18}H_{34}O_2$)

POLYUNSATURATED

```
   O H H H H H H H     H     H H H H H
   ‖ | | | | | | |     |     | | | | |
HO-C-C-C-C-C-C-C-C=C-C-C=C-C-C-C-C-C-H
     | | | | | | | | | | | | | | | |
     H H H H H H H H H H H H H H H H
```

(LINOLEIC ACID, $C_{18}H_{32}O_2$)

Fig. 4-13. Three fatty acids all composed of 18 carbons but different degrees of saturation or unsaturation. The = indicates a double bond and C stands for carbon, H for hydrogen, and O for oxygen.

Most fatty acids in nature contain an even number of carbon atoms. The nomenclature is such that the following suffixes are used to describe the degree of unsaturation:

1. Anoic—no double bond.
2. Enoic—one double bond.
3. Dienoic—two double bonds.
4. Trienoic—three double bonds.
5. Tetraenoic—four double bonds.
6. Pentaenoic—five double bonds.

• **Iodine number**—Unsaturated fat readily unites with iodine; 2 atoms of this element will add to each double bond. Thus, in experimental work, the number of grams of iodine absorbed by 100 g of fat—the iodine number—is an excellent criterion of the degree of unsaturation. In the past, the iodine test was commonly used when studying the soft pork problem—a problem caused when pigs are fattened on feeds rich in unsaturated fats, such as peanuts or soybeans. At the present time, the chief measure used in such determinations is the refractive index, as determined by a refractometer.

Fatty acids that are unsaturated have the ability to take up oxygen or certain other chemicals. This presents both advantages and disadvantages. The value of linseed oil and varnish is due to their high content of unsaturated fatty acids, by virtue of which oxygen is absorbed when they are exposed to air, resulting in a tough, resistant coating. On the other hand, because of their unsaturation, these fats often become rancid through oxidation, resulting in disagreeable flavors and odors which lessen their desirablility as feeds. Moreover, oxidative rancidity in fats results in formation of unstable compounds called peroxides, which can destroy certain essential nutrients in the diet.

• **Rancidity**—This is the oxidation (decomposition) primarily of unsaturated fatty acids resulting in disagreeable flavors and odors in fats and oils. This process occurs slowly and spontaneously, and may be accelerated by light, heat, and certain minerals. Rancidity may be prevented through proper storage and/or the addition of antioxidants such as BHA (butylated hydroxyanisole). Some fats are naturally protected from oxidation due to the presence of vitamin E. Hydrogenation of fats (adding hydrogen to unsaturated fatty acids) increases their hardness and also lessens the threat of rancidity. This process has been used to improve the keeping qualities of vegetable shortenings and lard.

• **Hydrogenation (hardening)**—This process adds hydrogen to the double bonds of unsaturated fatty acids. It may be accomplished with hydrogen gas in the presence of a nickel catalyst. It also occurs in the rumen as a result of microbial activity. The result of hydrogenation is a harder fat because adding hydrogen increases the melting temperature. It may be used on animal or vegetable fats to produce fats with a desired hardness. Many vegetable oils are converted into a solid or semisolid form for use in shortenings and margarines. Hydrogenation is also known as hardening.

Hydrogenation has a drawback in that it converts the naturally-occurring *cis* fatty acids to *trans* fatty acids. The prefixes *cis* and *trans* refer to the orientation of the atoms around the double bond. The *trans* form of essential fatty acids does *not* function as an essential fatty acid in the body. Also, some researchers have found that (1) *trans* fatty acids are not as effective as their *cis* analogs in lowering blood cholesterol, and (2) fats rich in *trans* fatty acids appear to promote atherosclerosis.

The content of *trans* fatty acids generally increases with the extent to which a vegetable oil has been hydrogenated. For example, hard sticks of vegetable oil margarines may contain from 24 to 35% of *trans* acids, whereas lightly hydrogenated liquid oils usually contain 5% or less.

• **Carbon chain length**—Another variable factor in the makeup of fatty acid molecules is the number of carbon atoms. Fatty acids are designated as having (1) short chains when the number of carbon atoms is 6 or less, (2) medium chains where there are 8 or 10 carbon atoms, and (3) long chains when there are 12 or more carbon atoms.

Together, the degree of saturation and the length of the carbon chain influence the melting point of fats as shown in Table 4–4. Short chain fatty acids tend to be more volatile, and acetic, propionic, and butyric acids are collectively called volatile fatty acids (VFA). There is a steady rise in melting point as the chain lengths increase. However, as the number of double bonds increases, the melting point decreases.

TABLE 4–4
THE CHEMISTRY OF SOME FATTY ACIDS

Name	Structural Formula	Chain Length	Melting Point		Example of Source
		(no. C atoms)	(°F)	(°C)	
Saturated:					
Acetic	CH_3CHOOH	2	—	—	Vinegar.
Propionic	CH_3CH_2COOH	3	—	—	Dairy products.
Butyric	$CH_3(CH_2)_2COOH$	4	19	*-7*	Coconut butter.
Caproic	$CH_3(CH_2)_4COOH$	6	18	*-8*	Coconut butter.
Caprylic	$CH_3(CH_2)_6COOH$	8	62	*16*	Coconut butter.
Capric	$CH_3(CH_2)_8COOH$	10	88	*31*	Coconut butter.
Lauric	$CH_3(CH_2)_{10}COOH$	12	112	*44*	Coconut, palm.
Myristic	$CH_3(CH_2)_{12}COOH$	14	129	*54*	Coconut butter, whale blubber.
Palmitic[1]	$CH_3(CH_2)_{14}COOH$	16	146	*63*	Palm, beef tallow, butter, lard, cottonseed oil.
Stearic[1]	$CH_3(CH_2)_{16}COOH$	18	157	*70*	Beef tallow, butter, lard.
Arachidic	$CH_3(CH_2)_{18}COOH$	20	170	*76*	Peanut oil.
Lignoceric	$CH_3(CH_2)_{22}COOH$	24	187	*86*	Beechwood tar.
Monounsaturated:					
Palmitoleic[1]	$CH_3(CH_2)_5CH=CH(CH_2)_7COOH$	16	31	—	Menhaden (fish), chicken, beef tallow.
Oleic[1]	$CH_3(CH_2)_7CH=CH(CH_2)_7COOH$	18	56	*13*	Olive oil, peanut oil, egg.
Polyunsaturated:					
Linoleic[1]	$CH_3(CH_2)_4CH=CHCH_2CH=CH(CH_2)_7COOH$	18	23	*-5*	Safflower, soybean oil, corn oil.
Linolenic	$CH_3CH_2CH=CHCH_2CH=CHCH_2CH=CH(CH_2)_7COOH$	18	12	*-11*	Linseed oil, soybean oil.
Arachidonic[1]	$CH_3(CH_2)_4CH=CHCH_2CH=CHCH_2CH=CHCH_2CH=CH(CH_2)_3COOH$	20	-57	*-50*	Liver, egg.

[1]Most abundant in animal lipids.

• **Saponification**—The combination of a fatty acid with an alkali, such as potassium or sodium hydroxide, forms soap. This reaction is called saponification. Besides forming soap, it is a method of evaluating the average length of the carbon chain in the fatty acids which constitute a fat. The test is performed by reacting fats with potassium hydroxide. The saponification number, or value, is the number of milligrams of potassium hydroxide required for the complete saponification of 1 g of the fat. A high saponification value signifies a short chain length and vice versa.

Saponification may also occur in the alkaline medium of the intestine. For example, calcium may combine with free fatty acids.

• **Emulsification**—Fats (oils) and water do not stay mixed, but often it is desirable for them to do so. Therefore, fats are often emulsified. Minute droplets of fats or oils are evenly distributed throughout a water-based solution. Emulsions are essential for the digestion, absorption, and transport of fats in the body. Emulsifying agents used to create emulsions include some fatlike and fat-derived substances such as monoglycerides (glycerol with one fatty acid), diglycerides (glycerol with two fatty acids), lecithin, and the bile salts.

ESSENTIAL FATTY ACIDS

There is evidence that linoleic, linolenic, and arachidonic acids are dietary essentials. Deficiency symptoms of these fatty acids have been observed in mice, poultry, swine, dogs, guinea pigs, and human infants. Depending on the type of animal, numerous manifestations of these deficiencies are seen; among them, dermatitis, reduced growth, increased water consumption and retention, impaired reproduction, and increases in metabolic rate. Two functions of the essential fatty acids have been postulated: (1) precursors of prostaglandins, and (2) structural components of cells. The evidence for the first theory is conclusive. The second theory has substantial support inasmuch as the essential fatty acids are in highest concentrations in phospholipids, a type of lipid that plays an important role in the structural integrity of the cell.

OTHER LIPIDS

Other important lipids include the phospholipids, lipoproteins, and cholesterol. All cells contain phospholipids. They are structural compounds found in cell membranes and in the blood. The brain, nerves, and liver contain particularly high levels. Lecithin is one of the most abundant phospholipids in the diet and the body. Phospholipids are powerful emulsifying agents. Lipoproteins are the primary vehicle for lipid transport in the blood. There are four main types: chylomicrons; very low density lipoproteins (VLDL); low density lipoproteins (LDL); and high density lipoproteins (HDL). Cholesterol is derived from the diet or synthesized in the body. It is necessary for the formation of hormones, bile salts, and vitamin D.

Metabolism of Fats

In monogastrics, fats are digested in the small intestine, primarily as a result of action (1) of bile which emulsifies the fat, thus greatly increasing the surface area, and (2) of pancreatic lipase, an enzyme which hydrolyzes fatty acids from the glycerol molecule. Some diglycerides are absorbed, but most of the absorption is as monoglycerides and fatty acids. Most of the longer chain fatty acids are absorbed by lacteals into the lymph system which enters the blood stream just before the vena cava vein enters the heart.

In the rumen, many of the nutrients are transformed by the microorganisms into volatile fatty acids—primarily acetate, propionate, and butyrate. These products are then absorbed through the rumen wall and used as energy sources. Acetate is converted to acetyl CoA, whereupon it enters the Krebs cycle. Since 2 ATP are needed in this transformation, the net energy yield is 10 ATP. Propionate is not only a product of rumen fermentation, but it is also the final product of the β-oxidation of odd-carbon fatty acids. Upon absorption, it is converted to propionyl CoA. Propionyl CoA is then converted to succinyl CoA through a series of reactions. Succinyl CoA then enters the Krebs cycle and a net of 18 ATP is produced from propionate. Butyric acid, a 4-carbon fatty acid, undergoes β-oxidation resulting eventually in the formation of 27 ATP.

After absorption as a fatty acid or monoglyceride, triglycerides are resynthesized in the mucosal tissues of the gut. The fats are then transported to the various tissues, particularly the liver, where they are (1) used in synthesis of various compounds required by the body, (2) metabolized as a source of energy, or (3) stored in the tissues (as fat deposits).

CATABOLISM OF FATS

When fats are hydrolyzed, glycerol and fatty acids are released and subsequently catabolized separately. Glycerol is converted to glycerol-3-phosphate, then to dihydroxyacetone phosphate. Dihydroxyacetone phosphate can be readily converted to 3-phosphoglyceraldehyde which enters the last part of the glycolytic pathway and eventually enters into the Krebs cycle. (See Fig. 4-10.) Thus, it yields an amount of energy similar to that of one-half of the glucose molecule passing through glycolysis and the Krebs cycle.

The catabolism of fatty acids occurs in the mitochondria through a systematic process called beta-oxidation, whereby 2-carbon fragments are successively chopped from the fatty acid molecule to form acetyl CoA which then enters the Krebs cycle. The term lipolysis indicates mobilization of fats and their oxidation. Carnitine, a vitaminlike substance, facilitates the transport of fatty acids across the mitochondrial membrane. The first step in the beta-oxidation of a fatty acid is the addition of coenzyme A (CoA) to the end of the molecule. This requires energy, but not nearly so much energy as is subsequently generated. For example, the 16-carbon fatty acid palmitic acid can be oxidized seven times. Each time a 2-carbon fragment—acetyl CoA—is released from a fatty acid, 5 ATP are formed. However, 1 ATP is consumed in activating the fatty acid with CoA. Then, each acetyl CoA enters the Krebs cycle and produces 12 ATP. The energy balance sheet is as follows:

			ATP
Palmitic acid	→	Palmitoyl CoA	-1
Palmitoyl CoA	→	8 Acetyl CoA (7 × 5 ATP)	35
8 Acetyl CoA	→	$H_2O + 18\ CO_2$ (8 × 12 ATP)	96
Net			130

Each mole of ATP contains about 8 Calories (kcal). Therefore, 1 mole of palmitic acid yields 1,040 Calories (kcal), which demonstrates that fats are high-energy compounds. Comparing this value to that obtained by burning fat in a laboratory bomb calorimeter, the efficiency of the biological burning of palmitic acid is about 42%—not bad for any machine.

Acetyl CoA is a common point—a "crossroad"—for many of the biochemical reactions of the body. It can be used for building other substances including (1) new fatty acids and cholesterol, (2) the formation of the ketone body, acetoacetic acid, and (3) the nerve transmitter substance, acetylcholine. It is formed during the catabolism of fatty acids, glucose, and amino acids; hence, excesses of these nutrients can be stored as fat. As already pointed out, acetyl CoA can enter the Krebs cycle. However, entry into this cycle is dependent upon continued availability of carbohydrates.

• **Ketosis**—If acetyl CoA—the end product of β-oxidation—is to enter the Krebs cycle, there must be a parallel catabolism of carbohydrates occurring simultaneously. Pyruvate, a product of carbohydrate catabolism, is needed to produce oxaloacetic acid. Oxaloacetic acid must then condense with acetyl CoA in order to produce citric acid. If there is insufficient oxaloacetic acid in the cell to keep the Krebs cycle functioning efficiently, the acetyl CoA is converted to acetoacetic acid, β-hydroxy butyric acid, and acetone. These ketone bodies accumulate in the blood; and if the condition goes unchecked, acidosis will occur, often resulting in coma and death. This condition is seen in both sheep and cattle. It usually strikes ewes just before lambing, whereas cows are usually affected within the first 6 weeks after calving. In this disorder, ketone bodies pervade the blood, urine, and milk, imparting a peculiar sweetish chloroformlike odor of acetone. Glucose injections provide a quick, effective treatment for animals that are suffering from the condition. Also, propylene glycol or sodium propionate may be given orally (see Chapter 5, Nutritional Disorders/Toxins, Table 5-1, Nutritional Diseases and Ailments).

ANABOLISM OF FATS

Saturated fatty acids and monounsaturated fatty acids are rapidly and abundantly formed from acetyl CoA. Hence, any nutrient capable of yielding acetyl CoA during the metabolic processes can potentially contribute to fatty acid synthesis—a process called lipogenesis. While a limited amount of fat synthesis occurs in the mitochondria of the cell, most synthesis takes place outside the mitochondria via enzymes in the cytoplasm when supplemented with energy (ATP), carbon dioxide (CO_2), hydrogen, and the mineral magnesium (Mg^+). The 2-carbon compound, acetyl CoA, is the primary building block. It is first carboxylated (organic acid group COOH added) to form malonyl CoA, and then it is transferred to a carrier protein—acyl-carrier protein (ACP). Malonyl-ACP then combines with acetyl CoA via acetyl ACP to form acetoacetyl-ACP which then has hydrogen added by NADPH generated in the pentose shunt to convert to butyryl-ACP. More and more acetyl CoA and hydrogen may be added to this molecule until the chain is 16 carbons long—palmitic acid. All fatty acids formed as described above are then transported back into the mitochondria where 3 fatty acids are joined—esterified—with 1 glycerol molecule, thus forming triglyceride fats. The glycerol is derived from the glycolytic pathway. Fatty acids may also be elongated in the mitochondria.

Acetyl CoA may also be used to synthesize cholesterol in many tissues, but the liver is the major site of synthesis.

Measuring and Expressing Energy Value of Feedstuffs

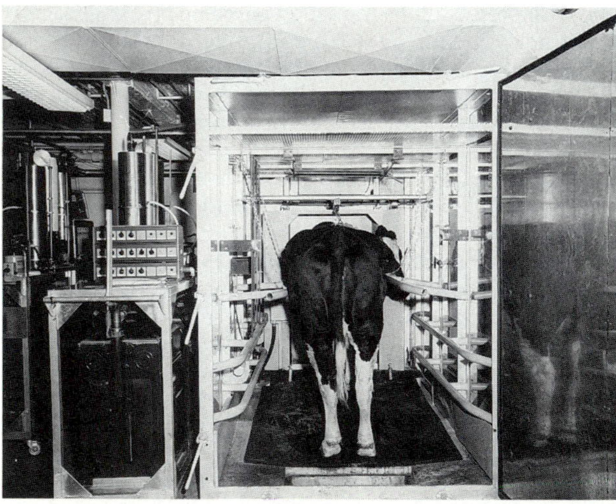

Fig. 4-14. A Holstein cow in an open circuit respiration chamber, used for measuring the energy value of a feedstuff. The unit to the left of the chamber is used to measure the respiratory exchange of the cow (air flow, temperature, relative humidity, gas samples, and animal activity). (Courtesy, Energy Metabolism Lab., USDA, Beltsville, Md.)

One nutrient cannot be considered as more important than another, because all essential nutrients must be present in adequate amounts if efficient production is to be maintained. Yet, historically, feedstuffs have been compared or evaluated primarily on their ability to supply energy to animals. This is understandable because (1) energy is required in larger amounts than anything else, (2) energy is most often the limiting factor in livestock production, and (3) energy is the major cost associated with feeding animals.

Cereal grains are higher in available energy than roughages. Although grains usually cost more on a weight basis than roughages, they are often a cheaper source of energy.

Our understanding of energy metabolism has increased through the years. With this added knowledge, changes have come in both the methods and terms used to express the energy value of feeds.

(Also see Chapter 15, Feed Analysis/Feed Evaluation, Section headed "Types of Feed Evaluation Trials.")

ENERGY DEFINITIONS AND CONVERSIONS

Some pertinent definitions and conversions of energy terms follow:

• **Calorie (cal)**—*The amount of energy as heat required to raise the temperature of 1 g of water 1 °C* (precisely from 14.5 °C to 15.5 °C). It is equivalent to 4.184 joules. Although not preferred, it is also called a *small calorie* and so designated by being spelled with a lower case "c." **NOTE WELL**: In popular writings, the term calorie is frequently used erroneously for the kilocalorie (1,000 small calories).

• **Kilocalorie (kcal)**—*the amount of energy as heat required to raise the temperature of 1 kg of water 1 °C* (from 14.5 °C to 15.5 °C). Equivalent to 1,000 calories. In human nutrition, it is referred to as a kilogram calorie or as a *large Calorie* and is so designated by being spelled with a capital "C" to distinguish it from the *small calorie.*

• **Megacalorie (Mcal)**—*Equivalent to 1,000 kilocalories or 1,000,000 calories.* Also, referred to as a *therm,* but the term megacalorie is preferred.

• **British Thermal Unit (Btu)**—*The amount of energy as heat required to raise 1 lb of water 1 °F;* equivalent to 252 calories. This term is seldom used in animal nutrition.

• **Joule**—*A proposed international unit* (4.184J = 1 calorie) *for expressing mechanical, chemical, or electrical energy, as well as the concept of heat.* In the future, energy requirements and feed values will likely be expressed by this unit.

• **Converting TDN to Mcal**—One pound of TDN = 2.0 Mcal or 2,000 kcal. It is recognized, however, that the roughage component in a ration affects its energy value. Thus, when converting all-roughage rations from TDN to calories, some scientists figure that 1 lb of TDN = 1,500 kcal, instead of 2,000.

• **Hay equivalent (HE)**—*This is the energy equivalent of 1 ton of hay which, on the average, contains 800 Mcal of net energy.* With an Animal Unit Month (AUM) being equivalent to 320 Mcal of net energy, 2.5 AUM are required to furnish the same amount of energy as 1 ton of hay.

ENERGY SYSTEMS

Broadly speaking, two methods of measuring energy are employed in this country—the total digestible nutrient system (TDN), and the calorie system. Each system has its advantages and advocates. But, more and more feedstuffs are being evaluated in calories.

Total Digestible Nutrients (TDN)

Total digestible nutrients (TDN) is the sum of the digestible protein, digestible fiber, digestible nitrogen-free extract, and digestible fat × 2.25. It has been the most extensively used measure for energy in the United States.

Back of TDN values are the following steps:

1. **Digestibility.** The digestibility of a particular feed for a specific species is determined by a digestion trial.

2. **Computation of digestible nutrients.** Digestible nutrients are computed by multiplying the percentage of each nutrient in the feed (protein, fiber, nitrogen-free extract [NFE], and fat) by its digestion coefficient (percentage digestibility). The result is expressed as digestible protein, digestible fiber, digestible NFE, and digestible fat. For example, if No. 2 corn contains 8.9% protein of which 77% is digestible, the percentage of digestible protein is 6.9.

3. **Computation of total digestible nutrients (TDN).** The TDN is computed by use of the following formula:

% TDN = % DCP + % DCF + % DNFE + (% DEE × 2.25)

where DCP = digestible crude protein; DCF = digestible crude fiber; DNFE = digestible nitrogen-free extract; and DEE = digestible ether extract.

TDN is ordinarily expressed as a percent of the ration or in units of weight (lb or kg), not as a caloric figure.

The main **advantage** of the TDN system is that it has been used for a very long time and many people are acquainted with it.

The main **disadvantages** of the TDN system are:

1. It is really a misnomer, because TDN is not an actual total of the digestible nutrients in a feed. It does not include the digestible mineral matter (such as salt, limestone, and defluorinated phosphate—all of which are digestible); and the digestible fat is multiplied by the factor 2.25 before being included in the TDN figure, because its energy value is higher than carbohydrates and protein. As a result of multiplying fat by the factor 2.25, feeds high in fat will sometimes exceed 100 in percentage TDN (a pure fat with a coefficient of digestibility of 100% would have a theoretical TDN value of 225% [100% × 2.25]).

2. It is an empirical formula based upon chemical determinations that are not related to actual metabolism of the animal.

3. It is expressed as a percent or in weight (lb or kg), whereas energy is expressed in calories.

4. It takes into consideration only digestive losses; it does not take into account other important losses, such as losses in the urine, gases, and increased heat production (heat increment).

5. It overevaluates roughages in relation to concentrates when fed for high rates of production, due to the higher heat loss per pound of TDN in high-fiber feeds.

Because of these several limitations, in the United States the TDN system is gradually being replaced by other energy evaluation systems, particularly net energy for ruminants and metabolizable energy for nonruminants. However, due to the voluminous TDN data on many feeds and long-standing tradition, it will continue to be used by many people for a long time to come.

Calorie System

Calories are used to express the energy value of feedstuffs. *One calorie* (always written with a small "c") *is the amount of heat required to raise the temperature of 1 g of water 1 °C* (precisely from14.5 °C to 15.5 °C).

To measure this heat energy, an instrument known as the bomb calorimeter is used, in which the feed (or other substance) to be tested is placed and burned in the presence of oxygen.

Through various digestive and metabolic processes, much of the energy in feed is dissipated as it passes through the animal's digestive system. About 60% of the total combustible energy in grain and about 80% of the total combustible energy in roughage is lost as feces, urine, gases, and heat. These losses are illustrated in Figs. 4–15 and 4–16.

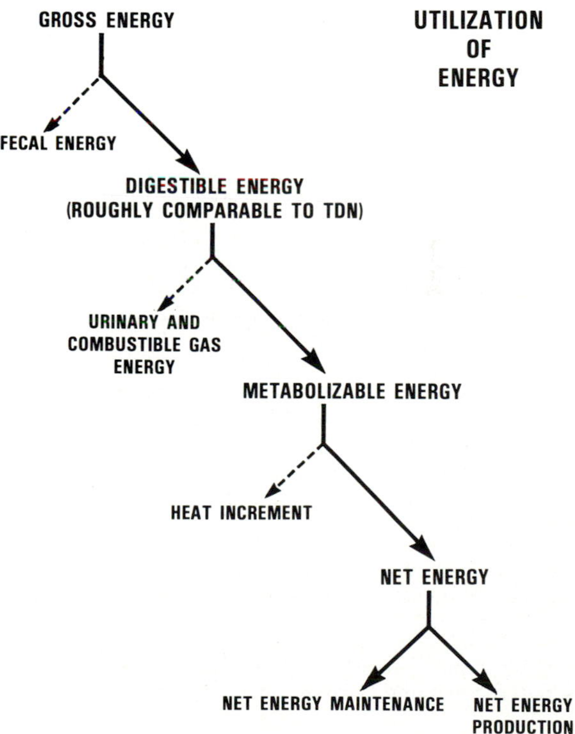

Fig. 4–15. Utilization of energy.

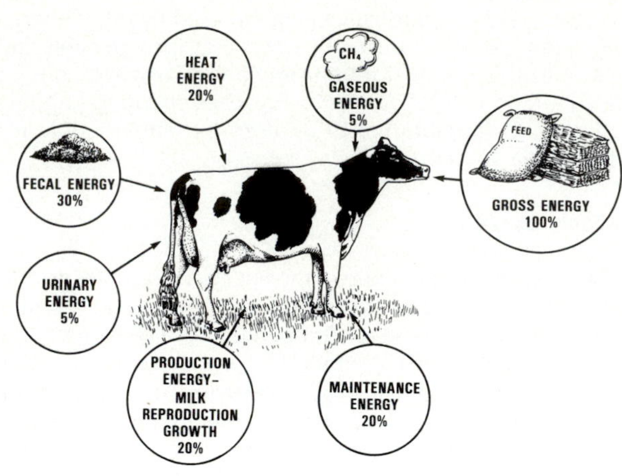

Fig. 4–16. Energy utilization by a lactating cow showing average partition of feed energy by the animal.

As shown in Figs. 4–15 and 4–16, energy losses occur in the digestion and metabolism of feed. Measures that are used to express animal requirements and the energy content of feeds differ primarily in the digestive and metabolic losses that are included in their determination. Thus, the following terms are used to express the energy value of feeds:

- **Gross energy (GE)**—*Gross energy represents the total combustible energy in a feedstuff.* It does not differ greatly among feeds, except for those high in fat. For example, 1 lb of corncobs contains about the same amount of GE as 1 lb of shelled corn. Therefore, GE does little to describe the useful energy in feeds for finishing animals.

- **Digestible energy (DE)**—*Digestible energy is that portion of the GE in a feed that is not excreted in the feces.*

- **Metabolizable energy (ME)**—*Metabolizable energy represents that portion of the GE that is not lost in the feces, urine, and gas.* Although ME more accurately describes the useful energy in the feed than does GE or DE, it does not take into account the energy lost as heat.

- **Net energy (NE)**—*Net energy represents the energy fraction in a feed that is left after the fecal, urinary, gas, and heat losses are deducted from the GE.* The net energy, because of its greater accuracy, is being used increasingly in ration formulations, especially in computerized formulations for large operations.

Although net energy is a more precise measure of the real value of the feed than other energy values, it is much more difficult to determine.

Two systems of net energy evaluation are presently being used. Lofgreen and Garrett [2] developed a system whereby the net energy requirements are listed as dictated by physiological functions—for example, net energy for maintenance (NE_m) and net energy for gain (NE_g). Also, Moe and Flatt [3] developed a net energy system that compares the physiological function to that of lactation through the use of regression analysis. This value, $NE_{lactation}$, is applicable for all physiological functions.

NITROGENOUS FEEDS

The nitrogenous fraction of feeds is generally broken down into two parts—protein and nonprotein. Compounds such as amines, free amino acids, purines, pyrimidines, and urea are included in the nonprotein part.

Nonprotein Nitrogen

When feeds are analyzed by the proximate analysis method, crude protein is determined by multiplying the nitrogen content of the feed by 6.25. This factor is used because protein is approximately 16% nitrogen (100% ÷ 16% = 6.25). This assumes that all of the nitrogen in feed is in the form of protein, but in many cases some of the nitrogen is derived from purines and pyrimidines and other nonprotein nitrogen sources. Ruminal microorganisms have the ability to convert certain nonprotein nitrogen feeds into microbial protein which can subsequently be used by the host animal. Hence, feeds such as urea and ammoniated by-products are commonly incorporated in ruminant rations.

(Also see Chapter 11, section on "Nonprotein Nitrogen (NPN) Feedstuffs.")

PURINES AND PYRIMIDINES

Purines and pyrimidines are cyclic compounds containing carbon, hydrogen, nitrogen, and oxygen. They provide the backbone for a number of compounds that are essential to metabolism.

Structure

In physiological systems, adenine and guanine are the primary purines while cytosine, uracil, and thymine constitute the primary pyrimidines. Their respective structures are given in Fig. 4–17.

When these compounds are combined with ribose or deoxyribose, the resultant structure is referred to as a *nucleoside. Nucleotides are nucleosides that are esterified with phosphoric acid.* When a series of nucleotides are joined together, nucleic acids are formed.

[2]Lofgreen, G. P., and W. N. Garrett, "A System for Expressing Net Energy Requirements and Feed Values for Growing and Finishing Beef Cattle," *Journal of Animal Science*, Vol. 27, 1968, p. 793.

[3]Moe, P. W., and W. P. Flatt, "Net Energy of Feedstuffs for Lactation," *Journal of Dairy Science*, Vol. 52, 1969, p. 928.

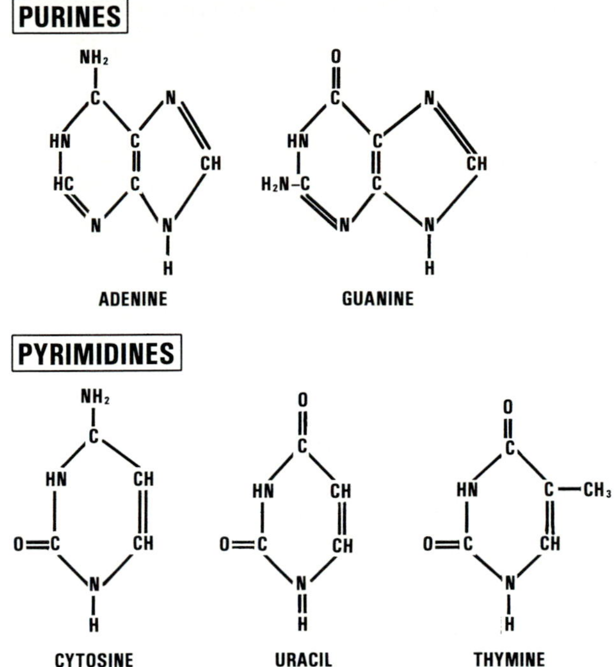

Fig. 4-17. Structures of the primary purines and pyrimidines.

Synthesis

When nucleic acids within the ration are digested, the resultant purines and pyrimidines are absorbed. Most purines and pyrimidines, though, are synthesized from amino acids with the primary site of synthesis being the liver. Upon catabolism of the various nucleic acids, purines and pyrimidines can be either used again or broken down to CO_2 and NH_3 or uric acid.

Functions

Purines and pyrimidines are components of a number of compounds which are required in metabolism. These compounds are (1) coenzymes, (2) high-energy compounds, (3) deoxyribonucleic aicd (DNA), and (4) ribonucleic acid (RNA).

COENZYMES

Adenine is a constituent of such coenzymes as FAD, NAD, NADP, and coenzyme A. Without these coenzymes, energy metabolism would probably grind to a halt. Uridine diphosphate glucose is used in glycogen synthesis.

HIGH-ENERGY COMPOUNDS

As has been previously discussed, ATP is the primary source for the storage of large amounts of energy that are produced in energy metabolism. When energy is required by the body, ATP provides this stored energy, becoming ADP in the process.

UREA AND AMMONIATED FEEDS

It has long been recognized that animals in which there is ruminal fermentation can subsist very well on protein of extremely low quality and that these animals can incorporate nonprotein nitrogen into microbial protein. When dietary protein and nonprotein nitrogen are metabolized by the ruminal microorganisms, ammonia is produced which is, in turn, combined with carbon chains to form the amino acids which are incorporated into microbial protein. In the rumen, there is a constant turnover of microorganisms. Many of these microorganisms are passed down the digestive tract along with ingesta and are subsequently digested and absorbed in the small intestine (see Fig. 4-18, p. 88). When compared with casein, microbial protein has nearly the same biological value, although it is slightly lower in methionine. More recent studies on the need for sulfur and cobalt by rumen microorganisms have accounted for some of the differences in microbial protein utilization.

Urea is a form of nonprotein nitrogen which can be used in the synthesis of protein via microbial fermentation. It is usually the least costly of the nonprotein nitrogen compounds. Its chief disadvantage is that it should be fed along with some readily available carbohydrate and, even then, occasional incidences of toxicity may occur.

The common methods of feeding urea are: urea mixed with concentrates; urea liquid supplements; urea salt blocks; urea mixed with silages; urea added to dry roughages; and slow-release urea products.

In monogastric species, the only microorganisms that can convert urea to protein are found in the lower intestinal tract at a point where absorption of amino acids, peptides, and proteins is rather low or nonexistent. Research with pigs, poultry, horses, and other species indicates only slight utilization of nitrogen from urea.

Nonprotein nitrogen can also be supplied through the ammoniation of certain feedstuffs. Ammoniated molasses represents a logical combination of a nitrogen source and a carbohydrate source. Beet pulp, citrus pulp, rice hulls, and cottonseed hulls are protein sources when ammoniated.

Anhydrous ammonia (NH_3) may be added to forages. Since 1970, the ammoniation of low-quality forages has increased dramatically throughout North America, with the anhydrous ammonia added to silages, and to dry forages in bales, stacks and barns. The ammonia provides an economical source of nitrogen (protein equivalent) which can be utilized by ruminants.

When nonprotein nitrogen sources are fed to ruminants, the rate of ammonia release must be carefully monitored in order to prevent toxicities. For this reason, only restricted levels of NPN are used in ruminant rations. The following toxicity preventive/control measures are recommended: (1) Do not treat hay with more than 1.0 to 1.5% ammonia by weight, and apply the ammonia evenly to the forage; (2) do not feed ammoniated hay alone—feed some grain with it; and (3) if toxicity signs develop when feeding ammoniated forage, discontinue feeding the forage immediately.

(Also see Chapter 5, Nutritional Disorders/Toxins, Table 5-3, Potential Poisons—Urea Toxicity [Ammonia Toxicity]; and Chapter 11, Protein Supplements, the several sections under "Nonprotein Nitrogen (NPN) Feedstuffs.")

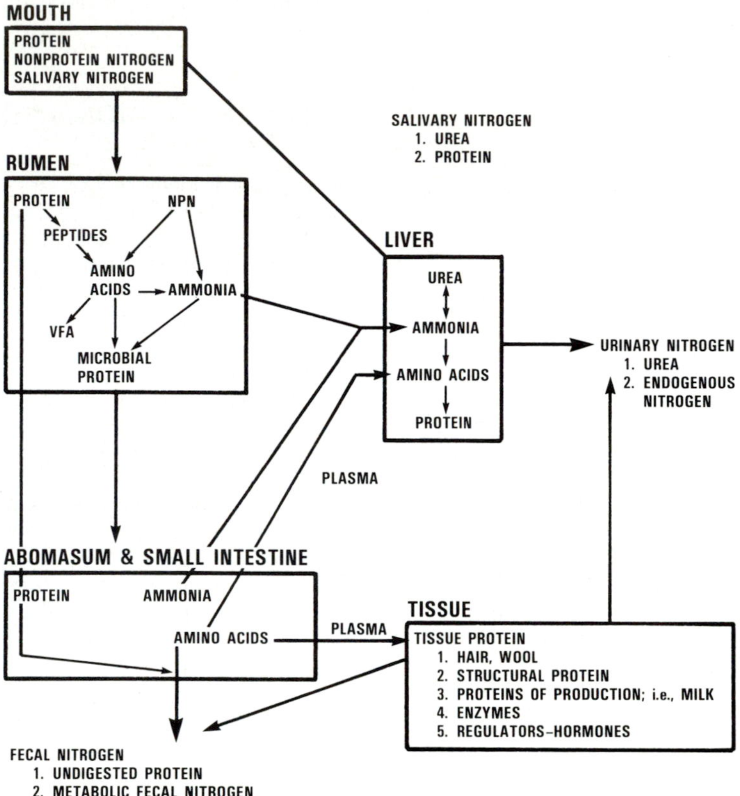

Fig. 4-18. Nitrogen metabolism in the ruminant.

Proteins

Chemically, proteins are complex organic compounds made up chiefly of amino acids. For each different protein there are specific amino acids and a specific number of amino acids which are joined in a specific order. Since amino acids always contain carbon, hydrogen, oxygen, and nitrogen, so do proteins. Moreover, the presence of nitrogen provides a tool for chemically estimating the amount of protein in a tissue, feed, or some other substance. Crude protein is routinely determined by the Kjeldahl process, which involves finding the nitrogen content and multiplying the result by 6.25, since the nitrogen content of all protein averages about 16% (100 ÷ 16 = 6.25). In addition, proteins usually contain sulfur and frequently phosphorus. Proteins are essential in all plant and animal life as components of the active protoplasm of each living cell.

In plants, the protein is largely concentrated in the actively growing portions, especially the leaves and seeds. Plants also have the ability to synthesize their own proteins from such relatively simple soil and air compounds and carbon dioxide, water, nitrates, and sulfates, using energy from the sun. Thus, plants, together with some bacteria which are able to synthesize these products, are the original sources of all proteins.

Proteins in animals are much more widely distributed than in plants. Thus, the proteins of the body are primary constituents of many structural and protective tissues—such as bone, ligaments, hair, hooves, skin, and the soft tissues which include the organs and muscles. The total protein content of animal bodies ranges from about 10% in very fat, mature animals to 20% in thin, young animals. By way of further contrast, it is also interesting to note that, except for the bacterial action in the rumen of ruminants, animals lack the ability of the plant to synthesize proteins from simple materials. They must depend upon plants or other animals as a source of dietary protein.

Animals of all ages and kinds require adequate amounts of protein of suitable quality. The protein requirements for growth, reproduction, and lactation are the greatest and most critical. Actually, the need for protein in the ration is really a need for amino acids.

Each protein has a distinctive function in the animal body, ranging from protection of the body surface (skin, hair) to defense against invading organisms. Structurally, proteins have important functions as components of muscle, cell membranes, skin, hair, and hooves. Metabolically important proteins are the blood serum proteins, enzymes, hormones, and immune bodies—all of which have important specialized functions in the body.

CLASSIFICATION OF PROTEINS

Proteins occur in nature in a number of forms, each one possessing unique chemical properties. Based on chemical composition, proteins may be divided into 2 main categories: (1) simple, and (2) conjugated. Simple proteins consist of only amino acids or their derivatives, while conjugated proteins are joined to various nonprotein substances. Then these 2 main categories are further subdivided. Table 4-5 lists the various categories of proteins along with their distinguishing characteristics.

TABLE 4-5
CLASSIFICATION OF SOME COMMON PROTEINS

Type	Chemical Properties	General Comments
Simple proteins:		
Albuminoids (scleroproteins)	Insoluble in water; highly resistant to enzymatic digestion; some become gelatinous upon boiling in water or dilute acids or bases.	Includes collagen, elastin, and keratin; common in supporting tissues; sometimes referred to as fibrous protein.
Albumins	Readily soluble in water; coagulate upon heating.	Present in egg, milk, and serum.
Globulins	Low solubility in water; solubility increases with the addition of neutral salts; coagulates upon heating.	Abundant in nature; examples are serum globulins, muscle globulins, and numerous plant globulins.
Glutelins	Insoluble in water; soluble in dilute acids or bases.	Abundant in cereal grains; an example is wheat gluten.
Prolamins	Insoluble in water, absolute alcohol, or neutral solvents; soluble in 80% ethanol.	Zein in corn and gliadin in wheat are prolamins.
Conjugated proteins:		
Chromoproteins	Combination of a protein and a pigmented (colored) substance.	Common example is hemoglobin —hematin and protein.
Lecithoproteins	Combination of protein and lecithin.	Found in fiber of clotted blood and vitellin of egg.
Lipoproteins	Water-soluble combination of fat and protein.	A vehicle for the transport of fat in the blood; all contain triglycerides, cholesterol, and phospholipids in varying proportions.
Metalloproteins	Proteins that are complexed with metals.	One example is transferrin, a metalloprotein that can bind with copper, iron, and zinc. Various enzymes contain minerals.
Mucoproteins or glycoproteins	Contain carbohydrates such as mannose and galactose.	Examples are mucin from the mucous secretions which act as protectants and lubricants in many parts of the body.
Nucleoproteins	Combination of proteins and nucleic acids.	Present in germs of seeds and glandular tissue.
Phosphoproteins	Compounds containing protein and phosphorus in a form other than phospholipid or nucleic acid.	Casein in milk and ovovitellin in eggs, are examples.

A third category—*derived proteins* may be added to the two main categories. Essentially, derived proteins are the product of digestion and are fragments of various sizes. From largest to smallest, in terms of the number of amino acids, derived proteins are proteoses, peptones, polypeptides, and peptides.

Proteins may also be classified according to their structure which is very important to their function in the body. Some proteins are round to ellipsoidal and are called *globular proteins.* These include *enzymes, protein hormones, hemoglobin* and *globulins.* Other proteins form long chains bound together in a parallel fashion, and are called *fibrous proteins.* These include *collagen, elastin,* and *keratin*—the proteins of connective tissue, elastic tissue, and hair. Many of the fibrous proteins are not very digestible, and if they are, they are poor-quality proteins.

AMINO ACIDS

The basic structural components of protein are amino acids. Although more than 200 naturally occurring compounds have been classed as amino acids, most proteins contain about 20 of the amino acids shown in Fig. 4-19 (p. 90). Many of the amino acids can be synthesized within the body. These are called nonessential amino acids or dispensable amino acids. If the body cannot synthesize sufficient amounts of certain amino acids to carry out physiological functions, they must be provided in the ration; hence, they are referred to as essential or indispensable amino acids. Actually, it is not entirely correct to say that all indispensable amino acids need to be provided in the diet; the requirement is for the preformed carbon skeleton of the indispensable amino acids, except in the case of lysine and threonine.

The necessity of each amino acid in the diet of the experimental rat has been thoroughly tested, but less is known about the requirements of large animals or even the human. According to our present knowledge, based largely on work with the rat, the following division of amino acids as essential and nonessential seems proper:

Essential (indispensable)	Nonessential (dispensable)
Arginine	Alanine
Histidine	Asparagine
Isoleucine	Aspartic acid
Leucine	Cysteine
Lysine	Cystine
Methionine (may be replaced in part by cystine)	Glutamic acid
	Glutamine
	Glycine
Phenylalanine	Hydroxyproline
Threonine	Proline
Tryptophan	Serine
Valine	Tyrosine

Arginine is regarded as essential for animals, whereas it is not for humans; most young mammals cannot synthesize it in sufficient amounts to meet their needs for growth.

In practical animal nutrition, the amino acids most likely to be deficient are lysine, methionine, and tryptophan. This stems from the fact that the cereal grains, which are primary

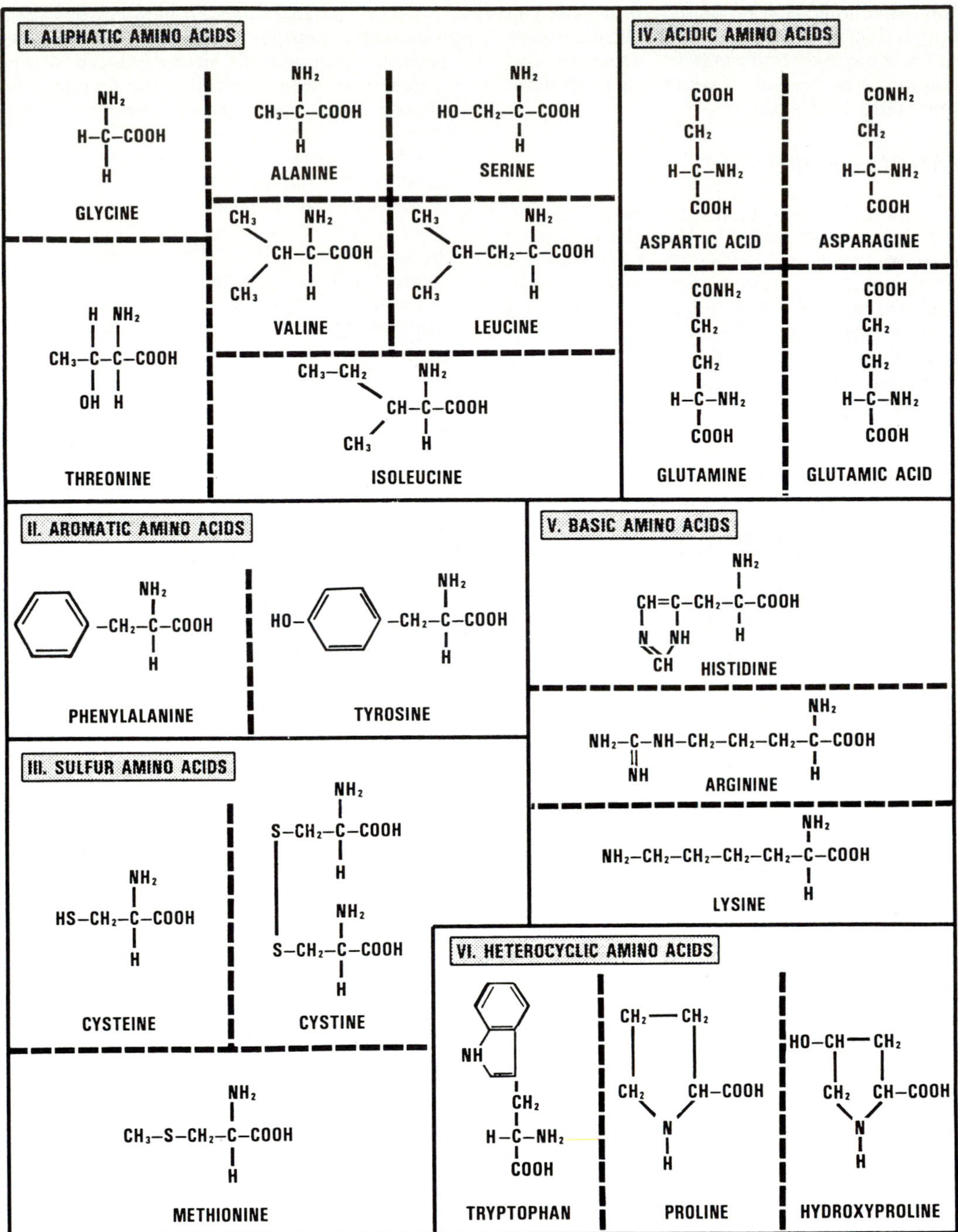

Fig. 4-19. Structures of the amino acids.

energy feeds, are quite low in these amino acids. So, it follows that rations based on a high percentage of these grains usually require supplementation with proteins which contain higher levels of these amino acids.

Fortunately, the amino acid content of proteins from various sources differs. Thus, the deficiencies of one protein may be improved by combining it with another, and the mixture of the two proteins often will have a higher feeding value than either one alone. It is for this reason that a considerable variety of feeds in the ration is usually recommended for monogastric animals.

The feed proteins are broken down into amino acids by digestion. They are then absorbed and distributed by the bloodstream to the body cells, which rebuild these amino acids into body proteins.

The various amino acids are then systematically joined

together to form peptides and proteins. The amino nitrogen of one amino acid will combine with the carboxyl carbon atom of another amino acid to form a peptide linkage. When several of these junctions occur, the resulting molecule is called a protein. (See Fig. 4–20.)

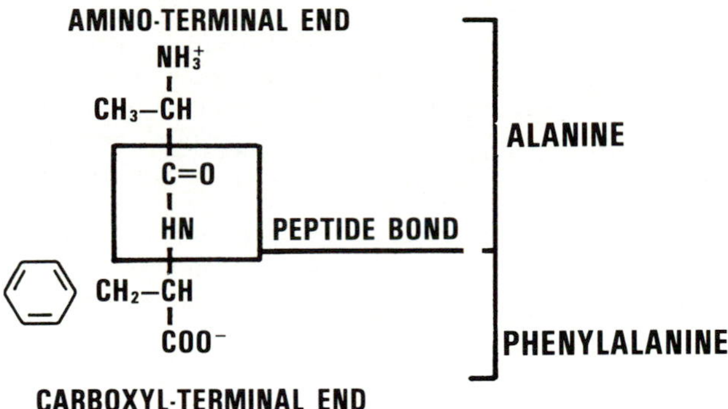

Fig. 4–20. A peptide bond between 2 amino acids.

CATABOLISM OF PROTEINS

Proteins are the primary sources of carbon skeletons utilized in gluconeogenesis. In order that proteins might enter energy metabolic pathways, they must first be deaminated (have their amino [NH_2] group removed). Therefore, oxidative deamination (see Fig. 4–21) provides a mechanism whereby ammonia and a carbon skeleton that can be used for energy are produced.

OXIDATIVE DEAMINATION

$$R\text{—}\underset{NH_2}{\underset{|}{CH}}\text{—}COOH + NAD^+ + H_2O \rightleftharpoons R\text{—}\underset{O}{\underset{\|}{C}}\text{—}COOH + NH_4^+ + NADH$$

Fig. 4–21. Process of oxidative deamination.

Following deamination, the carbon skeletons, depending upon their structure, can enter the Krebs cycle as pyruvic acid, alpha-ketoglutaric acid, succinic acid, fumaric acid, or oxaloacetic acid and be completely oxidized to carbon dioxide and water, yielding ATP. (See Table 4–2.) Transamination—the transfer of the amino group (NH_2) to synthesize new amino acids—also forms carbon skeletons which can enter the Krebs cycle.

Most deaminated amino acids can be used to synthesize glucose by working backwards through glycolysis. This process is called gluconeogenesis, and amino acids capable of this are called gluconeogenic or glycogenic. Alanine is an important gluconeogenic amino acid, which is formed in the muscles by transamination. When it is deaminated in the liver, it becomes pyruvic acid which the liver converts to glucose. Leucine is the only amino acid incapable of gluconeogenesis.

• **Urea**—The fate of the amino group (NH_2) split off from amino acids presents a special metabolic problem. Fortunately, the body has developed the urea cycle (Fig. 4–22) for handling this substance. In mammals, most of the NH_2 is converted to urea in the liver for excretion by the kidneys. To facilitate elimination, one mole of ammonia (NH_3) combines with one mole of carbon dioxide (CO_2). This compound is then phosphorylated to produce carbamyl phosphate. Carbamyl phosphate then combines with ornithine to form citrulline—an intermediate in the urea cycle. Citrulline is then converted to arginine. Urea splits off from arginine, producing ornithine, and the cycle is completed. The kidneys remove urea from the blood and excrete it in the urine. In birds, the NH_2 is excreted as uric acid.

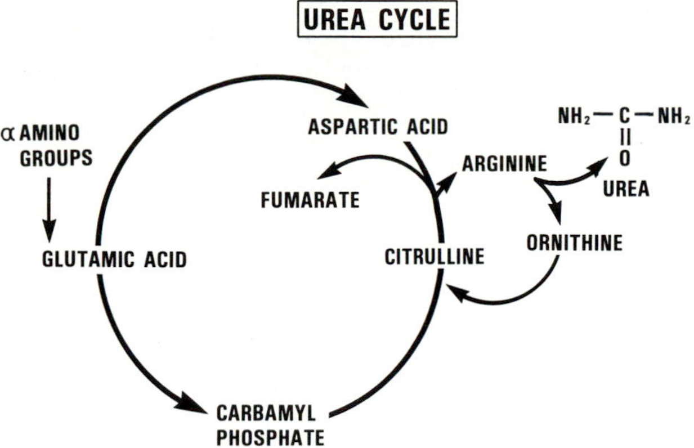

Fig. 4–22. Urea cycle (nitrogen excretion).

SYNTHESIS OF AMINO ACIDS

Monogastric animals have the ability to synthesize a limited number of amino acids; hence, the term nonessential amino acids. Transamination (Fig. 4–23) is one metabolic process which makes this possible. Carbon skeletons are produced through various intermediates of carbohydrate metabolism; and a new amino acid can be produced when an amino group is transferred from an amino acid to the carbon skeleton. The deaminated molecule can then be used as an energy source.

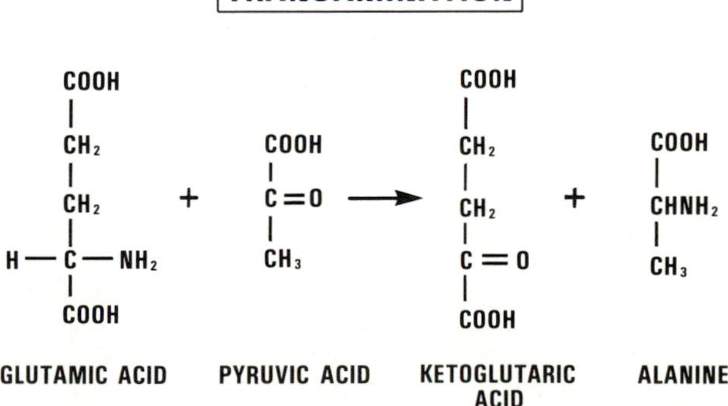

Fig. 4–23. Process of transamination.

Ruminants (and some other herbivores) do not require dietary amino acids to the same extent as monogastrics. This is so because the vast majority of the dietary proteins are degraded and used as precursors for microbial protein. Thus, on a dietary level there is no differentiation between essential and nonessential amino acids in ruminants. However, on the cellular level, the ruminant needs the same amino acids as the nonruminant and possesses the same mechanisms for the interconversion of amino acids as the nonruminant animal.

There is some evidence, however, that high-producing animals, particularly dairy cows, may not receive optimal amounts of lysine or methionine and may benefit from dietary supplementation with them at peak production periods.

ANABOLISM OF PROTEINS

Protein synthesis is not a random process whereby a number of amino acids are joined together; rather, it is a detailed predetermined procedure. Within the cell, DNA serves as the information center concerning the sequences of the various proteins to be synthesized in the cell. When DNA is decoded, amino acids are linked to form a specific protein which has its own particular physiological function.

In order for a protein to be synthesized, all of its constituent amino acids must be available. If one amino acid is missing, the synthesis procedure is halted. When a particular amino acid is deficient, it is referred to as a limiting amino acid because it limits the synthesis of protein. This is the reason why protein quality is so important in the nutrition of monogastrics. High-quality proteins, upon digestion, provide balanced supplies of the various amino acids which can subsequently be absorbed as precursors for protein synthesis. Certain feeds are noted for being deficient in particular amino acids—for example, corn is known to be deficient in lysine. When these feeds are used, the livestock producer must make sure that other feeds are incorporated into the diet to compensate for the deficiencies of the particular amino acid(s).

Protein synthesis involves a series of reactions which are specific for *each* protein. A four-step outline of this procedure follows:

1. Messenger RNA (ribonucleic acid) transcribes the sequence *message* from DNA to form a template (a pattern or guide) for protein synthesis. The sequences of the nucleotides in the DNA are the keys to the sequence pattern forming this template. This is possible because the nucleotides always pair off in the following manner: adenine (A) and thymine (T); adenine (A) and uracil (U); and guanine (G) and cytosine (C). Triplets of the purine (adenine and guanine) and pyrimidine (cytosine, thymine, and uracil) bases in the DNA form *codons* which correspond to specific amino acids and are signals which control protein synthesis. For example, the codon AGG (adenine, guanine, and guanine) signals the incorporation of arginine.

2. A specific transfer RNA (tRNA) combines with each respective amino acid to form an aminoacyl complex. This reaction requires the expenditure of energy. There is at least one specific RNA for most of the 22 amino acids.

3. The initiation of protein synthesis occurs when the ribosome of the rough endoplasmic reticulum (site of protein synthesis) recognizes a codon specific for initiation. This first amino acid will be the amino (NH_2) terminal end of the protein. This amino acid is formulated to prevent the amino group of the amino acid from being incorporated in a peptide bond. Upon completion of the protein, the N-formyl group is cleaved from the protein.

4. Once synthesis has been initiated, the protein is elongated through a series of successive additions of amino acids as determined by the messenger RNA template. Each new amino acid is linked to the next by the formation of a peptide bond. Eventually the procedure will be terminated when a codon specific for terminating protein synthesis is reached. Thereupon, the protein splits off from the ribosomes.

Each sequence of amino acids is a different protein. Hence, different proteins are able to accomplish different functions in the body. With 22 amino acids, the different arrangements possible are endless, yielding a variety of proteins. For example, egg albumin, a small protein, contains approximately 288 amino acid units. Thus, if one assumes that there are about 20 different amino acids in the albumin molecule, then mathematical calculations show that the possible arrangements of this number of amino acids are in excess of 10^{300}. (For comparison, one million is equal to 10^6.)

Since the DNA of each cell carries the master plan for forming proteins, genetic mutations often are manifested by disorders in protein metabolism. Many metabolic defects—inborn errors of metabolism—are the result of a missing or modified enzyme or other protein. The misplacement or ommission of just one amino acid in a protein molecule can result in a nonfunctional protein. Therefore, when proteins are formed three requirements must be met: (1) the proper amino acids, (2) the proper number of amino acids, and (3) the proper order of the amino acid in the chain forming the protein. Meeting these requirements allows the formation of proteins which are very specific and which give tissues their unique form, function, and character.

BIOLOGICAL VALUE OF PROTEIN

Most chemical and microbiological tests for nutrient substances give information about the total amount of a nutrient present in a particular feedstuff or ration. However, they tell nothing about the digestibility and utilization of the feedstuff or ration in the digestive tract of the animal. Hence, biological tests directly involving animals are required to establish the true usefulness of feed in supplying the nutrient needs of animals. These biological tests are particularly important in evaluating protein. (They are also important in evaluating energy-yielding nutrients such as carbohydrates and fats.)

The biological value of a protein is the precentage of the digestible protein of a feed or feed mixture which is usable as a protein by the animal. It can be determined by a balance experiment in which a measured intake of protein is compared to the measured undigested protein in the feces of the animal. Thus, the biological value of a protein is a reflection of the kinds and amounts of amino acids available to the animal after digestion. If the amino acids available to the animal closely match those needed for body protein formation, the biological value of the protein is high. If, on the other hand, there are excesses of certain amino acids and deficiencies of other amino acids as a result of digestion, the biological value of the protein is low because of the increased number of amino acids which must be excreted via the kidney.

The biological values of animal proteins are generally much higher than those of plant proteins. For example, the biological value of whole milk is 85, compared to whole corn at 60. Because of this situation, the nutritional value of plant proteins (corn, beans, etc.) is substantially increased by feeding with them animal products rich in the amino acids in which plant proteins are deficient.

Other measures of protein quality are protein efficiency ratio (PER) and the net protein ratio (NPR). *The PER is, by definition, the number of grams of body weight gain of an animal per unit (lb or kg) of protein consumed. The NPR measures efficiency of growth by comparing body nitrogen resulting from feeding a test protein with that resulting from feeding a comparable group of animals a protein-free diet for the same period of time.*

Most feeds contain some protein, but the amount and quality of the protein varies considerably from one feed to another and the digestibility of the protein also varies. Legumes are particularly noted for high-quality protein—especially in ruminant rations. Young grasses may also be quite high in protein. When protein is deficient, it is necessary to consider other sources of protein. Both protein quality and quantity should be considered in the feeding of nonruminant, monogastric animals.

One of the most common supplemental sources of protein for farm animals is meal residue resulting from the extraction of oil from seeds of certain plants—primarily cotton, soybean, flax, safflower, and sunflower. Other excellent sources of high-quality proteins, although often more expensive than the plant sources, are the animal and fish meals—liver meal, fish meals (many different kinds of fish meals are available as single types of fish or blends of several), and offal (viscera and scraps of slaughtered animals). Meat meal and tankage are variable in nutritional quality, depending on the proportion of gelatin they contain, as well as the conditions used in cooking and drying them. Mill by-products, distillers' and brewers' by-products, and more recently, hydrolyzed feather meal, are high-protein by-products which are used to a lesser extent.

Processing of feeds may have an effect on the quality or availability of proteins. Heated forages show a decreased digestibility of proteins. Normally, such proteins are 60 to 70% digestible, compared to as low as 10% digestibility for hay or silage which has turned dark brown or black from heating. On the other hand, some protein feeds which are heated may be improved in digestibility for some monogastric animals. Heating cottonseed meal may not materially affect the utilization of protein, but it will destroy the gossypol which is toxic to some animals. Beans which contain the enzyme urease are heated when they are mixed with rations containing urea. If not, the urea will be converted to ammonia prior to reaching the rumen.

Rich feed sources of protein are shown in Fig. 4–24.

PROTEIN SOURCES

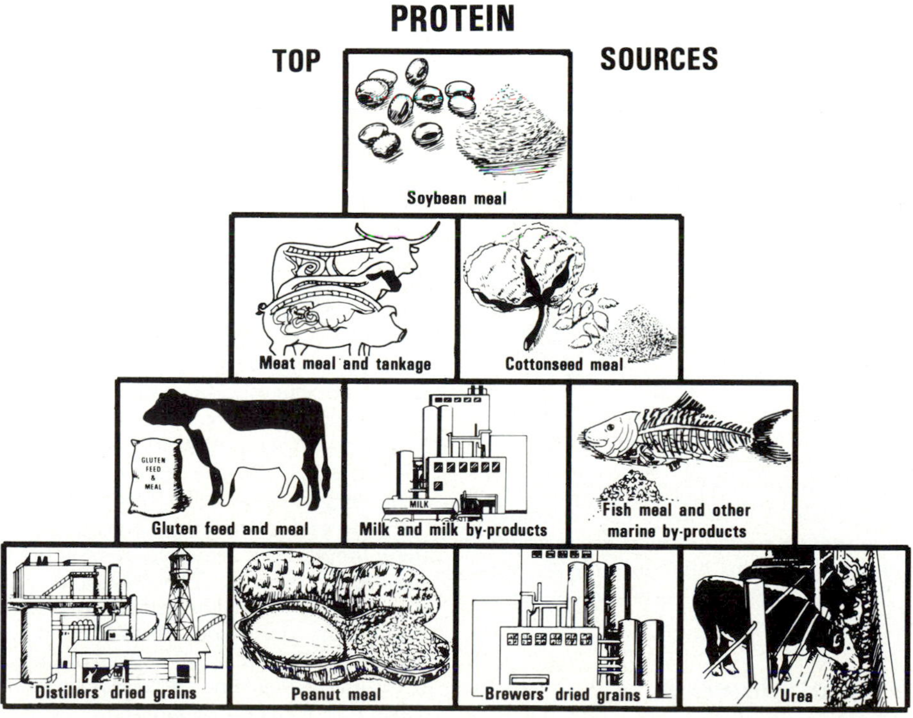

Fig. 4–24. Top sources of protein. No claim is made that all the best sources of protein are listed. Neither are the sources ranked. The evaluation of a protein depends on several factors, including amino acid composition, availability, and price.

MINERALS

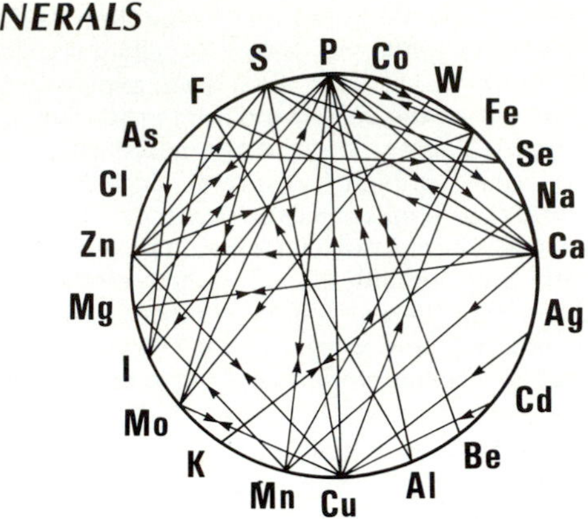

Fig. 4-25. Mineral interaction chart, showing the interrelationship of minerals. The importance of such relationships is evidenced by the following:

1. A great excess of dietary calcium and/or phosphorus interferes with the absorption of both minerals, and increases the excretion of the lesser mineral.
2. Excess magnesium upsets calcium metabolism.
3. Excess zinc interferes with copper metabolism.
4. Copper is required for the proper utilization of iron, but excess copper can markedly depress iron absorption.

And there are others! The maze of connecting lines in this figure shows the relation of each mineral to other minerals.

In nutrition, the term *mineral* denotes certain chemical elements which are found in the ash after a food or a body tissue is burned.

For a good understanding of minerals, it is very important that consideration be given to the following facts pertaining to them:

1. **The soil/plant/animal relationship of minerals.** There is a direct and most important relationship between the content and availability of mineral elements in the soil and the mineral composition of plants (forages and grains). Sometimes the concentration of an essential mineral in the soil is so low that plants growing on it will not contain enough of that mineral to meet the dietary requirements of animals. At other times, plants may contain such high concentrations of a certain mineral(s) that they are toxic to the animals that eat them. Such soil/plant/animal relationships are particularly important in trace elements.

2. **The interrelationship of minerals.** The interrelationship (Fig. 4-25) of minerals greatly complicates the task of determining minimum dietary requirements and tolerances for many elements; and the degree of interaction varies greatly. In addition to the interaction of mineral elements with other mineral elements, there are interactions (1) between mineral elements and the organic components of the diet, and (2) with factors other than nutrients.

3. **The role of minerals.** In addition to some general functions in which several minerals are involved, each essential mineral has at least one specific role. Although the metabolism of minerals does not produce energy, minerals are involved in many of the reactions of the body which comprise metabolism.

History/Discovery of Minerals

It is noteworthy that certain ancient peoples recognized and treated mineral deficiencies, although they did not understand the bases of their treatments. For example, a Chinese document dated about 3000 B.C. described goiter and recommended that afflicted people eat seaweed and burnt sponge, which are good sources of the trace element iodine. Another deficiency disease, anemia, was treated in ancient Greece (around the time of Hippocrates, or about the 4th century, B.C.) by giving the patient iron-containing water in which heated swords had been quenched. However, the effects of such treatments were often unpredictable, because there were no means of identifying or measuring the quantity of the active ingredients in the various medicinal substances.

The breakthroughs in our understanding of the functions of minerals in the body did not come for many centuries because laboratories for research were not developed until the Renaissance; although the medieval alchemists appear to have invented some of the techniques and tools of chemistry in their futile efforts to change base metals into gold.

The great French chemist Lavoisier, who is credited with being the founder of the science of nutrition, predicted in 1799 that such elements as sodium and potassium would soon be discovered because he believed that they were present in certain mineral compounds which were then known as *earths.* Sure enough, within a few years, the British chemist Davy discovered not only sodium and potassium, but also calcium, sulfur, magnesium, and chlorine. Davy's discoveries were so highly regarded by the French Academy that in 1806 they awarded him their new Volta medal, even though France and England were at war when the award was presented.

The pace of progress quickened as the noted Swedish chemist Berzelius added to our knowledge of minerals by (1) reporting his analysis of the calcium and phosphorus content of bone in 1801, and (2) concluding in 1838 that the iron in hemoglobin made it possible for the blood to absorb much oxygen. Similar contributions were made by the French chemist Boussingault, who (1) noted in 1822 that South American villagers who used salt containing iodine were protected from goiter which affected those who used plain salt; and (2) showed by means of animal feeding trials the necessity of providing dietary calcium and iron. Calcium studies were also conducted by Chossat, a physician in Switzerland, who won a prize in 1840 for his demonstration that the addition of calcium carbonate to a diet of wheat and water improved bone growth in pigeons.

Another half century elapsed before the value of iodine became widely accepted; thanks to (1) the discovery by the German biochemist Baumann in 1895 that the thyroid gland contained iodine, and (2) the demonstrations by the American medical scientist Marine and his coworkers between 1907 and 1918 that the administration of minute amounts of iodine prevented goiter in animals and in school children.

By the end of the 19th century only about one-third of the minerals now accepted as essential were known to be required in the diet. This state of affairs existed in spite of the many nutritional investigations which were conducted during that century because the need for vitamins had not yet been discovered. Hence, it was difficult to differentiate between the deficiency diseases due to lack of minerals and those due to lack of vitamins.

The first half of the 20th century was marked by finding that various minerals were required by animals and people, as indicated by the dates of the discoveries which follow: phosphorus, 1918; copper, 1925; magnesium, manganese, and molybdenum, 1931; zinc, 1934; and cobalt, 1935. However, there was often uncertainty as to the metabolic roles of various minerals, even though they were known to be essential. For example, it was not until 1948 that it was established that cobalt functions as a component of vitamin B-12. So, the two decades between the 1930s and the 1950s were marked by feverish activity aimed at learning how each of the newly discovered essential elements acted in the body.

The most recent chapter in this story began when Schwarz, a medical scientist who had emigrated to the United States from Germany, and his co-workers discovered the essentiality of selenium in 1957, and of chromium in 1959. They then developed isolation equipment for shielding their animals and their ultrapure diets from contamination by minute amounts of elements in the environment. Their painstaking work paid off; in 1972, they were able to show that fluorine and silicon were essential, also. It is ironic that only a short time ago all of these four trace elements—selenium, chromium, fluorine, and silicon—were considered to be only unwanted toxic contaminants of feeds, water, and air.

Much work remains to be done in the areas of (1) testing whether other elements might be essential, (2) defining the limits of safe and toxic doses for those already known, and (3) determining how the various elements interact with each other in the animal body.

Definitions/Classification/Functions

Minerals are inorganic elements, frequently found as salts with either inorganic elements or organic compounds. Their availability—and often their metabolic functions—are related to the form in which they are found. For example, in the presence of oxalates and citrates, calcium cannot be absorbed. When combined with phytin, phosphorus is available to some animals but not to others. Chelating agents have a selective attraction with various mineral elements, releasing one mineral element for another for which it has a greater attraction—sometimes creating deficiencies of an element which is present in otherwise adequate amounts.

In the proximate feed analysis, feed samples are burned in a furnace in order to destroy all organic matter, thus leaving only the minerals, or what is termed *ash*. At best, this analysis is a very crude measure of the mineral content in feed, because it (1) does not indicate what minerals are present; (2) does not include the volatile minerals such as iodine, chlorine, and selenium; (3) weighs the minerals as oxides or carbonates, with the weight of these elements included with the minerals; (4) does not indicate anything about the availability or form in which the minerals appear in the feed; (5) provides no information of the relative amounts of each mineral; and (6) ignores the importance of those minerals needed in such small quantities that they are not really contributing much to the overall ash content of feeds. In fact, the ash analysis is used quite often in forages to estimate the amount of dust and soil that has been harvested with the feed. It is also used by some feed inspectors to monitor the amount of *filler* in feedstuffs to which excessive amounts of limestone or other earthy materials are sometimes added.

Eighteen mineral elements are known to be required by at least some animal species. They can be divided into two groups based upon the quantity required in the ration.

- **Major or macrominerals**—These elements are required in amounts ranging from a few tenths of a gram to one or more grams per day.
- **Trace or microminerals**—These elements are required in minute quantities, ranging from a millionth of a gram (microgram) to a thousandth of a gram (milligram) per day.

The terms—major/macromineral and trace/micromineral—do not imply any lesser role for the latter group; rather, they represent quantity designations based on the amounts needed by animals.

The two groups, based on the amounts needed in the ration, follow:

Major/Macrominerals	Trace/Microminerals	
Salt (sodium & chlorine, NaCl)	Chromium (Cr)	Molybdenum (Mo)
Calcium (Ca)	Cobalt (Co)	Selenium (Se)
Phosphorus (P)	Copper (Cu)	Silicon (Si)
Magnesium (Mg)	Fluorine (F)	Zinc (Zn)
Potassium (K)	Iodine (I)	
Sulfur (S)	Iron (Fe)	
	Manganese (Mn)	

The percentages of the principal mineral constituents of the body are indicated by the following data, showing the average analyses of 18 steers of varying ages exclusive of the contents of the digestive tract.[4]

Element	Percent	Element	Percent
Calcium	1.33	Chlorine	0.11
Phosphorus	0.74	Magnesium	0.04
Potassium	0.19	Iron	0.01
Sodium	0.16		
Sulfur	0.15	Total	2.73

The dominant position of calcium and phosphorus in the above data becomes apparent when it is realized that calcium accounts for 49% of the total mineral; phosphorus, 27%; and all other minerals, 24%.

The general functions of minerals are as follows:

1. Give rigidity and strength to the skeletal structure.
2. Serve as constituents of the organic compounds, such as protein and lipid, which make up the muscles, organs, blood cells, and other soft tissues of the body.
3. Activate enzyme systems.
4. Control fluid balance—osmotic pressure and excretion.
5. Regulate acid-base balance.
6. Exert characteristic effects on the irritability of muscles and nerves.
7. Engage in mineral-vitamin relationships.

A summary of each mineral as presented is given in Table 4-6, Animal Mineral Chart, with the minerals listed alphabetically within each of two classifications: (1) Major/Macrominerals, or (2) Trace/Microminerals.

[4]Hogan, A. G., and J. L. Nierman, *Studies of Animal Nutrition—VI, The Distribution of the Mineral Elements in the Animal Body as Influenced by Age and Condition,* Missouri Ag. Exp. Res. Bull. No. 107.

TABLE 4-6
ANIMAL MINERAL CHART

MAJOR/MACROMINERALS:

CALCIUM (Ca)

Calcium made the difference!

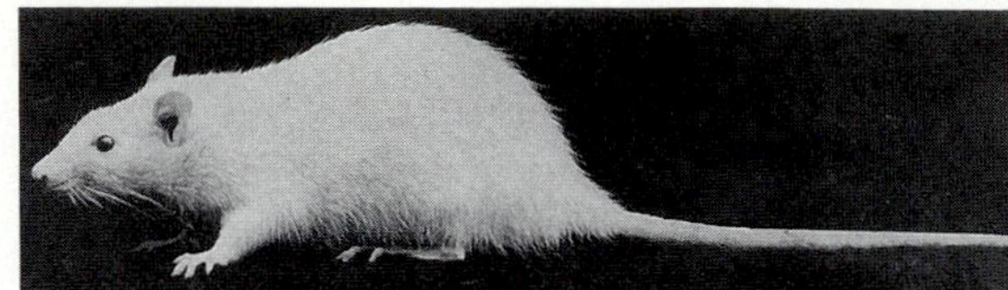

Fig. 4-26. *Left:* Rat on calcium-deficient diet. *Right:* Rat that received plenty of calcium. (Adapted from USDA sources.)

All animals need calcium, which constitutes about 2% of the body weight. Calcium gives strength and structure to bones and teeth, controls the heartbeat, has a role in the transmission of nerve impulses, is related to muscle contraction, is necessary for blood clotting, and activates a number of the enzymes, including lipase—a fat-splitting enzyme.

Farm animals are more likely to suffer from a lack of calcium and phosphorus than from any of the other minerals except common salt.

History:

Calcium was among the first nutrients known to be essential in the diet. As early as 1842, Chossat, a Frenchman, showed experimentally that pigeons developed poor bone on a diet low in calcium. When fed wheat alone, the birds died after 10 months; and, on autopsy, the bones were found very much depleted. Calcium carbonate prevented the trouble. In later studies, Chossat also used chickens, rabbits, frogs, eels, lizards, and turtles.

Absorption/Metabolism/Excretion:

Absorption—Calcium salts are more soluble in acid solution; hence, absorption occurs largely in the upper part (proximal part) of the small intestine, where the food contents are still somewhat acidic following digestion in the stomach.

Not all the calcium in feed becomes available to the body. Normally, depending on the intake, only 20 to 30% of the calcium in the average diet is absorbed from the intestinal tract and taken into the bloodstream. Calcium absorption is dependent upon the calcium needs of the body, the type of feed, and the amount of calcium ingested.

Metabolism—Normally, the small intestine acts as an effective control and prevents an excess of calcium from being absorbed. Body requirement is the major factor governing the amount of calcium that will be absorbed.

After absorption through the intestinal wall, most of the calcium is stored in the bones, especially the spongy bones (*trabeculae*), from which it is withdrawn in time of need. But calcium cannot always be withdrawn rapidly enough to prevent some alteration in the metabolism, such as tetany. Most investigators measure calcium depletion from the long bones; for example, the femur (thigh bone). But, calcium is most readily mobilized from the jaw, a fact that is commonly overlooked in the diagnosis of calcium deficiency. If the amount of calcium in the blood is above normal levels, it can be deposited in the bone for future use or excreted. Deposition and mobilization of calcium are under hormonal control.

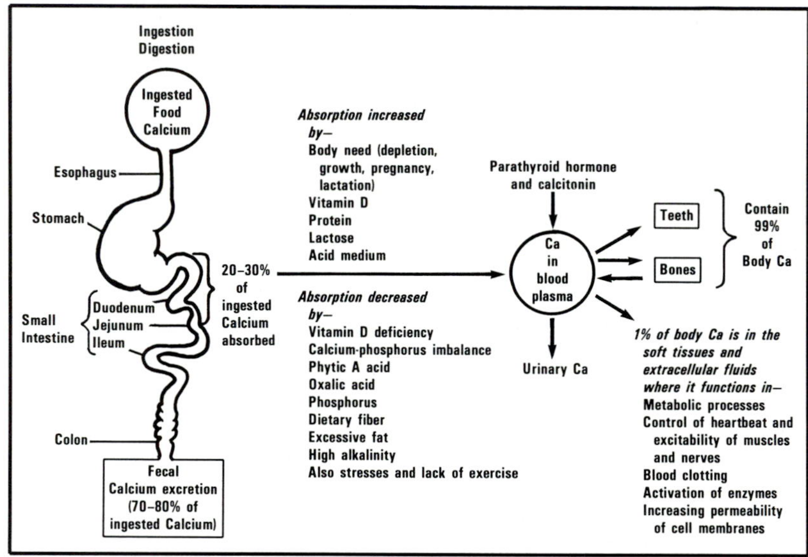

Fig. 4-27. Calcium utilization. Note that healthy animals absorb only 20 to 30% of the calcium contained in their feed and that 70 to 80% is excreted in the feces. Note, too, the factors that increase and decrease absorption.

Excretion—Most of the absorbed calcium that the body does not need is excreted in the urine, although some of it is dispersed in the feces and sweat.

Functions:

The primary fuction of calcium is to build the bones and teeth, and to maintain the bones.

Other functions are:

1. Blood clotting. When calcium is not available, as is the case when it is tied up with citrate or oxalate, clotting does not take place. Thus, when an animal is subjected to surgery or hemorrhaging, a calcium deficiency may seriously impair healing.

2. Muscle contraction and relaxation, especially the heartbeat. An increase in blood calcium causes the heart to beat faster. Conversely, as it decreases, the heart beats at a slower rate.

3. Nerve transmission.

4. Cell wall permeability.

5. Enzyme activation.

(Continued)

TABLE 4-6 *(Continued)*

Calcium (Ca) (Continued)

6. Secretion of a number of hormones and hormone-releasing factors.
7. Milk production and egg shell formation.

Deficiency Symptoms/Toxicity:

Fig. 4-28. *Left:* Pig with severe rickets due to calcium deficiency. *Right:* Healthy pig. (Courtesy, Ohio Agricultural Research and Development Center, Wooster, Ohio)

Deficiency Symptoms—The most dramatic deficiency symptoms are manifested in the bones and teeth of the young, evidenced in—

1. Stunting of growth.
2. Poor quality bones and teeth.
3. Malformation of bones—rickets.

The clinical manifestations of calcium related diseases are—

1. Rickets in young, characterized by (a) enlarged joints of the long bones, (b) soft bones that bend out of shape due to the weight and activity of the animal, and (c) the animal walking in a distorted manner, with the back humped up in the midsection.
2. Osteomalacia, the adult counterpart of rickets.
3. Osteoporosis, a condition of too little bone, resulting when bone resorption exceeds bone formation.
4. Drop in milk production. The calcium content of milk can vary only within narrow limits, below which milk production will be decreased to the amount that can be produced with the available calcium.
5. Drop in egg production, thin- or soft-shelled eggs, and lowered hatchability.
6. Kidney stones, which may form when dietary calcium is excessive.
7. Tetany (hypocalcemia). Normally, the blood calcium is controlled by hormones (see Fig. 4-27). But a severe drop in blood calcium can result in tetany. Soon after calving, cows may show a reduction in their blood calcium of sufficient magnitude to cause hypocalcemic tetany, commonly called *milk fever.* The heart rate is slowed so much that the cow loses control of her legs and will, if untreated, go into a coma and die. An intravenous injection of a suitable calcium solution will usually get the cow back on her feet within a few minutes, provided her condition has not gone too far for recovery.

Toxicity—Normally, the small intestine prevents excess calcium from being absorbed. However, a breakdown of this control may raise the level of calcium in the blood and lead to calcification of the kidneys and other internal organs.

High calcium intake may cause excess secretion of calcitonin and such bone abnormalities as osteopetrosis (dense bone). Also, there may be deposition in some of the soft tissues, particularly the tendons which connect muscles to bones.

High calcium intakes have also been reported to cause kidney stones, although this usually involves more than merely high blood calcium.

Major Interrelationships:

Excess calcium intake may reduce the absorption of magnesium, iron, iodine, manganese, zinc, and copper, especially when the intake of one of these minerals is borderline.

Excess calcium reduces the absorption and utilization of zinc and causes parakeratosis.

Excess magnesium decreases calcium absorption, replaces calcium in the bone, and increases calcium excretion.

• **Calcium:phosphorus ratio and vitamin D**—The proportional relationship that exists between calcium and phosphorus is known as the *calcium:phosphorus ratio.*

When considering the calcium and phosphorus requirements of livestock, it is important to realize that the proper utilization of these minerals by the body is dependent upon three factors: (1) an adequate supply of calcium and phosphorus in an available form, (2) a suitable ratio between them (somewhere between 1 to 2 parts of calcium to 1 part of phosphorus), and (3) sufficient vitamin D to make possible the assimilation and utilization of the calcium and phosphorus.

Generally speaking, nutritionists have advocated a calcium-phosphorus ratio somewhere between 1 to 2 parts of calcium to 1 part of phosphorus. However, there is much evidence indicating that calcium-phosphorus ratios of 1:1 to 2:1 for nonruminants (hogs and horses) and 1:1 to 7:1 for ruminants are satisfactory. But ratios below 1:1 may be disastrous.

If plenty of vitamin D is present (provided either by sunlight or through the ration), the ratio of calcium to phosphorus becomes less critical. Likewise, less vitamin D is needed when there is a desirable calcium:phosphorus ratio.

(Continued)

TABLE 4-6 *(Continued)*

Calcium (Ca) (Continued)

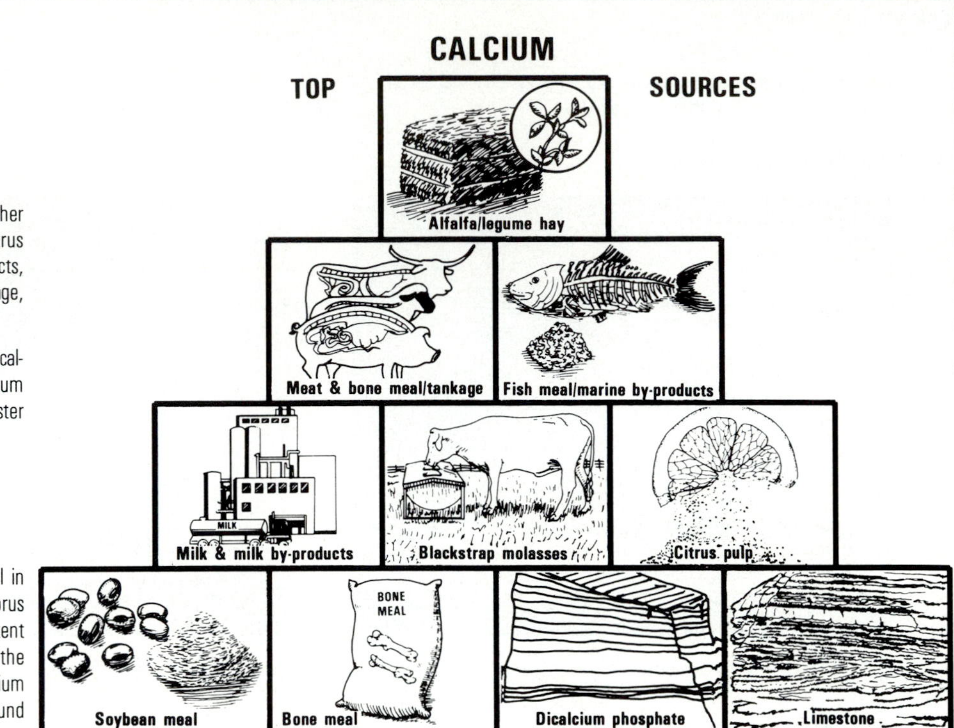

Fig. 4-29. Commonly used, top feed and supplement sources of calcium. No claim is made that all the top sources of calcium are listed. Neither are the sources ranked according to amount of calcium supplied nor is price considered.

Sources for Animals:

Rich feed sources—Alfalfa and other legume forages, blackstrap molasses, citrus pulp, fish meal and other marine by-products, meat and bone meal, meat and bone tankage, milk and milk by-products, and rape.

Supplemental sources—Bone meal, calcium gluconate, calcium lactate, dicalcium phosphate, dolomite, kelp, limestone, oyster shells.

Comments:

Calcium is the most abundant mineral in the body. Together, calcium and phosphorus comprise about 70% of the mineral content of the animal body and from ⅓ to ½ the minerals of milk. About 99% of the calcium and over 80% of the phosphorus are found in the bones and teeth.

CHLORINE (Cl)

Chlorine, which is a strong-smelling, greenish-yellow gas, is never found in nature as an individual element. But its compounds, such as sodium chloride (common salt), are widespread.

The blood contains 0.25% chlorine, 0.22% sodium, and 0.02 to 0.22% potassium; thus, the chlorine content is higher than that of any other mineral in the blood.

History:

Carl Wilhelm Scheele, a Swedish chemist, discovered chlorine in 1774 by treating hydrochloric acid with manganese dioxide. In 1810, Sir Humphry Davy, the English chemist, determined that chlorine was an element (further, he showed that muriatic acid itself contains only hydrogen and the new element, chlorine), which he named *chloros,* a Greek word meaning greenish-yellow. Chlorine functions in animals in the form of salts, or chlorides.

Absorption/Metabolism/Excretion:

Absorption—Chloride from feeds and from gastric juice is absorbed chiefly from the small intestine.

Metabolism—During digestion some of the chloride of the blood is used for the formation of hydrochloric acid in the gastric glands and is secreted into the stomach where it functions temporarily with the gastric enzymes and is then reabsorbed into the blood stream with other nutrients.

The highest concentrations of body chloride are found in the gastric juice and in the cerebrospinal fluid.

Excretion—Excessive chloride in the diet is excreted via the urine. Additional losses occur in sweating, vomiting, and diarrhea.

Functions:

Chloride (Cl) plays a major role in the regulation of osmotic pressure, water balance, and acid-base balance (chloride shift).

It is required for the production of hydrochloric acid of the stomach; this acid is necessary for the proper absorption of vitamin B-12 and iron, for the activation of the enzyme that breaks down protein, and for suppressing the growth of microorganisms that enter the stomach with feed and water.

Deficiency Symptoms/Toxicity:

Deficiency Symptoms—Severe deficiencies of chloride may result in alkalosis (an excess of alkali in the blood), characterized by slow and shallow breathing, listlessness, muscle cramps, loss of appetite, and, occasionally, by convulsions.

Depressed growth rate.

Chicks fed chloride-deficient diets exhibit nervous symptoms induced by sudden noise.

Toxicity—Excess chloride is not likely, particularly when animals have sufficient water.

(Continued)

TABLE 4–6 *(Continued)*

Chlorine (Cl) (Continued)

Major Interrelationships:

Loss of chloride generally parallels that of sodium. When sodium chloride intake is restricted, the chloride level in the urine falls, followed by a drop in tissue chloride levels. Increased losses of sodium that occur with sweating or diarrhea result in concurrent losses of chloride.

Chloride also aids in the conservation of potassium.

Sources for Animals:

Rich feed sources—Alfalfa and other legumes, barley straw, beet molasses, blackstrap molasses, fish meal and other marine by-products, grass-legume silage, meat meal, meat tankage, milk and milk by-products, oat straw, turnips, vetch.

Supplemental sources—Potassium chloride, sodium chloride (table salt).

Comments:

In practice, sodium and chlorine are supplied together as common table salt.

The body's requirement for chlorine is approximately half that of sodium.

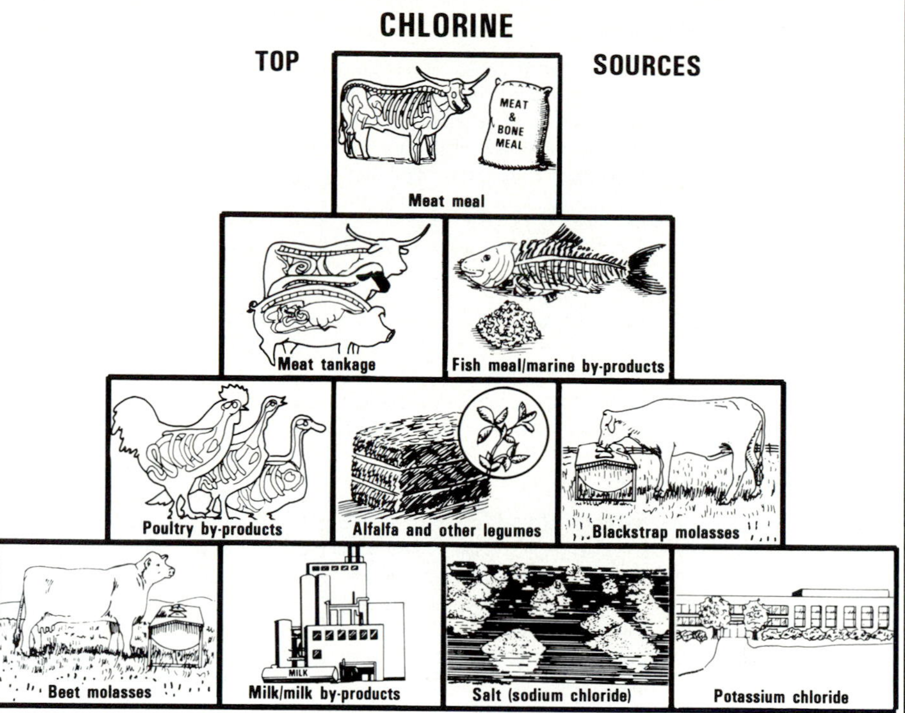

Fig. 4–30. Commonly used, top feed and supplement sources of chloride. No claim is made that all the top sources of chloride are listed. Neither are the sources ranked according to amount of chloride supplied nor is price considered.

MAGNESIUM (Mg)

Magnesium made the difference!

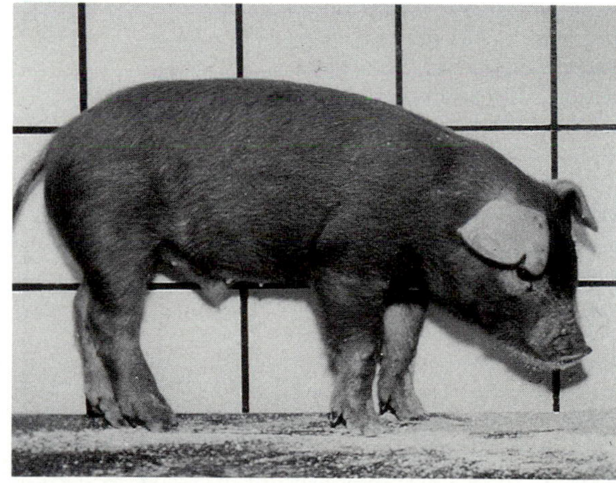

Fig. 4–31. *Left:* Magnesium-deficient pig, fed 70 ppm magnesium for 3 weeks. Note extreme leg weakness, arched back, and general unthriftiness. *Right:* Healthy pig, fed 413 ppm of magnesium for 3 weeks. (Courtesy, Purdue University, Lafayette, Ind.)

Magnesium is found in most tissues of the body. About half of it is found in bone, with the remaining half in the soft tissues and body fluids. It appears to be closely associated with calcium in many functions especially in the formation of bone.

History:

Centuries ago, the ancient Romans claimed that *magnesia alba* (white magnesium salts from the district of Magnesia in Greece, after which the element was eventually named) cured many ailments. But, it was not until 1808 that Sir Humphry Davy, a British chemist, announced that he had isolated the element, magnesium.

In 1926, LeRoy, in France, using mice, first proved that magnesium is an essential nutrient for animals. Subsequently, McCollum and co-workers, in the U.S., described several magnesium deficiency signs in rats and dogs, including magnesium tetany, a form of convulsions in which the nerves and muscles are affected.

(Continued)

TABLE 4–6 *(Continued)*

Magnesium (Mg) (Continued)

Absorption/Metabolism/Excretion:

From 30 to 50% of the average daily intake of magnesium is absorbed from the small intestine. Almost all of the magnesium in the feces represents unabsorbed dietary magnesium. Its absorption is interfered with by high intake of calcium, phosphate, oxalic acid, phytate (whole grain) and poorly digested fats (long chain saturated fatty acids). Its absorption is enhanced by protein, lactose (milk sugar), vitamin D, growth hormone, and antibiotics.

Magnesium is generally reabsorbed in the kidneys, thus minimizing the loss of body reserves.

The main route of excretion of magnesium is the urine. Aldosterone, a hormone secreted by the adrenal gland, helps regulate the rate of magnesium excretion through the kidneys.

Functions:

Constituent of bones and teeth.

Essential element of cellular metabolism, often as an activator of enzymes involved in phosphorylated compounds and of high-energy phosphate transfer of ADP and ATP.

Involved in activating certain peptidases in protein digestion.

Relaxes nerve impulses, functioning antagonistically to calcium which is stimulatory.

Serves as a ruminant alkalizer and buffer, thereby improving the butterfat levels and milk production of dairy cows.

Deficiency Symptoms/Toxicity:

Deficiency Symptoms—A deficiency of magnesium may cause grass tetany (grass staggers) in cattle (and sometimes in sheep), characterized by twitching of muscles (usually of head and neck), head held high, accelerated respiration, high temperature, grinding of the teeth, and abundant salivation. Slight stimulus may precipitate a crash to the ground, and finally death. Grass tetany generally occurs within the first 2 weeks after animals are turned out on new pasture growth, either in spring or fall.

Grass tetany is produced when the level of magnesium in the blood drops dramatically, but it differs from a simple magnesium deficiency in that the blood calcium may also be low. Suckling calves and pigs can also encounter magnesium deficiencies because they are sometimes unable to meet their needs through milk alone.

Treatment of magnesium deficiencies by intravenous injections of magnesium induces the excretion of excessive amounts of calcium in the urine. When magnesium is fed in large quantities, this calcium excretion is not observed, probably due to the regulation of absorption of magnesium through highly selective mechanisms.

Fig. 4–32. Cow with grass tetany (grass staggers). (Courtesy, Watt Publishing Co., Mount Morris, Ill.)

Toxicity—Spontaneous toxicities have not been reported, but intravenous injections of magnesium salts (such as magnesium sulfate—Epsom Salts) will cause the heart to enter a stage of sustained contraction (tetany) resulting in death.

Major Interrelationships:

When magnesium intake is extremely low, calcium is sometimes deposited in soft tissues forming calcified lesions.

An excess of magnesium upsets calcium and phosphorus metabolism. Also, with rations adequate in magnesium and other nutrients, increasing the calcium and/or phosphorus results in magnesium deficiency.

Magnesium activates many enzyme systems, particularly those concerned with transferring phosphate from ATP and ADP. Magnesium is also capable of inactivating certain enzymes, and is known to be a component of at least one enzyme.

Sources for Animals:

Rich feed sources—Blackstrap molasses, cottonseed (whole), crab cannery residue, distillers' solubles, flax seed, legume forages, meat meal with bone, meat tankage with bone, rapeseed meal, rice bran, safflower

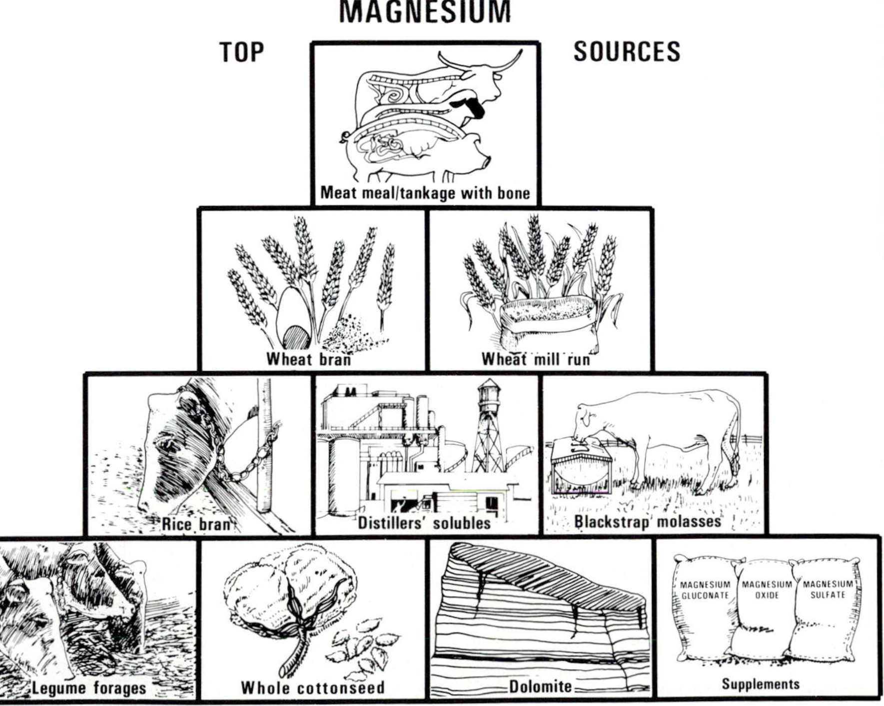

Fig. 4–33. Commonly used, top feed and supplement sources of magnesium. No claim is made that all the top sources of magnesium are listed. Neither are the sources ranked according to amount of magnesium supplied nor is price considered.

(Continued)

TABLE 4-6 *(Continued)*

Magnesium (Mg) (Continued)

meal, sesame meal, shrimp cannery residue, wheat bran, wheat mill run.

Supplemental sources—Dolomite; kelp; magnesium gluconate, magnesium oxide or magnesium sulfate (mixed with other minerals or with the concentrate portion of the feed); rice polishings.

Comments:

Deficiencies of magnesium may be encountered in mature cattle, suckling calves, and pigs.

PHOSPHORUS (P)

Phosphorus made the difference!

Fig. 4-34. *Left:* Phosphorus-deficient cow. *Right:* Same cow following phosphorus supplementation for 4 weeks. Phosphorus improved this cow's condition and increased milk production by 33%. (Courtesy, The National Fertilizer Assn.)

Phosphorus is closely associated with calcium in nutrition—to the extent that a deficiency or an overabundance of one may very likely interfere with the proper utilization of the other. Also, both calcium and phosphorus occur in some of the same major feed sources, both function in the major task of building bones and teeth, both are related to vitamin D in the absorption process, both are regulated metabolically by the parathyroid hormone and calcitonin, and both exist in the blood serum in a definite ratio to each other.

History:

Phosphorus was first identified in urine by Hennig Brand, a German alchemist, in 1669. It created much interest because, in the unnatural free form, it glowed in the dark and took fire spontaneously upon exposure to the air. Evenutally, the name phosphorus (from the Greek for *light-bringing*) was appropriated to this element. Fortunately, phosphorus exists in nature only in combined forms, usually with calcium, in such sources as bone and rock phosphates.

Absorption/Metabolism/Excretion:

Absorption—Phosphorus is more efficiently absorbed than calcium; 70% of the ingested phosphorus is absorbed and 30% is excreted

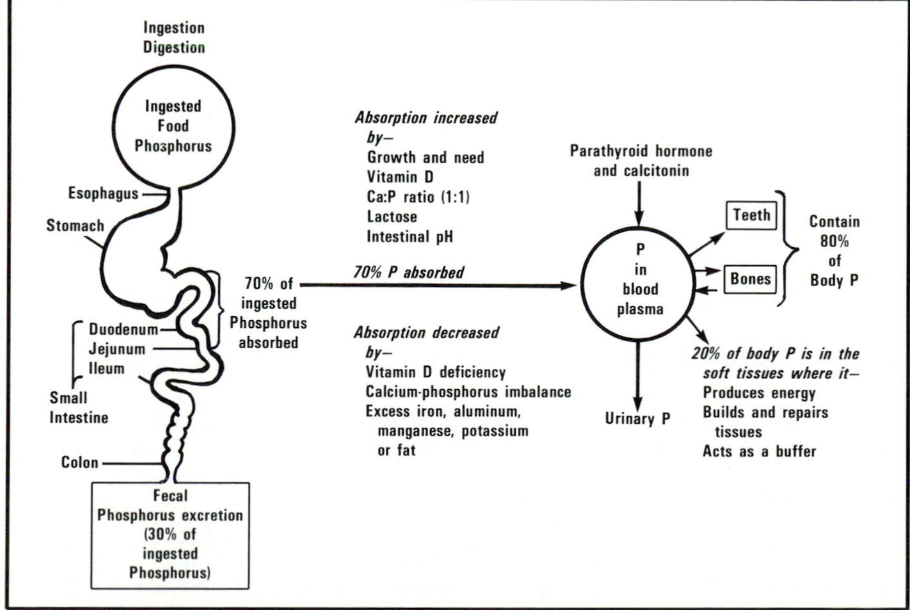

Fig. 4-35. Phosphorus utilization. Note that healthy animals absorb 70% of the phosphorus in their feed, and that 30% is excreted in the feces. Note, too, the factors that increase and decrease absorption.

(Continued)

TABLE 4–6 *(Continued)*

Phosphorus (P) (Continued)

in the feces, whereas only 20 to 30% of the ingested calcium is absorbed and 70 to 80% is excreted in the feces.

Phosphorus is absorbed in both organic and inorganic forms. In many plants, phosphorus is present in the form of phytic acid, a form which is absorbed with varying efficiency depending on the species of animal.

Phosphorus is absorbed chiefly in the upper small intestine. The amount absorbed is dependent on several factors, such as source, calcium:phosphorus ratio, intestinal pH, lactose intake, and dietary levels of calcium, phosphorus, vitamin D, iron, aluminum, manganese, potassium, and fat. As is the case for most nutrients, the greater the need, the more efficient the absorption. Absorption increases, although not proportionally, with increased intake.

Metabolism—Phosphorus absorbed from the intestine is circulated through the body and is readily withdrawn from the blood for use by the bones and teeth during periods of growth. Some incorporation into the bones occurs at all ages. It may be withdrawn from bones to maintain normal blood plasma levels during periods of dietary deprivation.

The level of plasma phosphorus, along with calcium, is regulated by the parathyroid hormone and thyrocalcitonin (calcitonin) and is inversely related to the blood calcium level.

Excretion—The kidneys provide the main excretory mechanism for regulation of the serum phosphorus level, although some phosphorus is excreted in the feces. All of the plasma inorganic phosphate is filtered through the renal glomeruli. If the serum phosphorus level falls, the renal tubules return more phosphorus to the blood; if the serum phosphorus level rises, the renal tubules excrete more. Also, when the ration lacks sufficient phosphorus, the renal tubules conserve phosphorus by returning it to the blood.

Functions:

The primary functions of phosphorus are:

Bone formation and maintenance.

Development of teeth.

Milk secretion.

Building muscle tissue.

Component of nucleic acids (RNA and DNA), which are important in genetic transmission and control of cellular metabolism.

Maintenance of osmotic and acid-base balance.

Also, phosphorus is important in many metabolic functions, especially—

1. Energy utilization.
2. Phospholipid formation.
3. Amino acid metabolism; protein formation.
4. Enzyme systems.

Deficiency Symptoms/Toxicity:

Deficiency Symptoms—General weakness, loss of appetite, depraved appetite (pica), muscle weakness, demineralization of bone, loss of calcium, breeding problems, and reduced egg production in poultry.

Severe and prolonged deficiencies of phosphorus may be manifested by rickets, osteomalacia, and other phosphorus related diseases.

Another deficiency symptom is the excretion of blood in the urine, a condition referred to as nutritional redwater. The pigments of blood turn black in the urine and are quite distinctive.

Toxicity—There is no known phosphorus toxicity *per se.* However, excess phosphate consumption may decrease the absorption of calcium and cause hypocalcemia (a deficiency of calcium in the blood). Also, when phosphorus is high in relation to calcium, urinary calculi may be formed, especially in ruminants.

Fig. 4–36. Phosphorus deficiency in a pig. Note crippled condition due to soft bones. (Courtesy, School of Veterinary Medicine, Auburn University, Auburn, Ala.)

Major Interrelationships:

Ratio of calcium to phosphorus is important; somewhere between 1–2 parts of calcium to 1 part of phosphorus is ideal, although wider ratios are tolerated, especially by ruminants. Also, sufficient vitamin D is necessary for phosphorus assimilation and utilization. (See Calcium, section on "Major Interrelationships.")

Monogastrics utilize phytate phosphorus poorly.

Excess calcium, magnesium, iron, and aluminum cause a decrease in phosphorus absorption.

In ruminants, excess phosphorus in relation to calcium is likely to cause calculi.

Sources for Animals:

Rich feed sources—Cereal grains and their by-products (notably wheat bran), cottonseed meal, distillers solubles, fish meal and other marine by-products, peanut meal, rape seed meal, rice bran, safflower meal, sesame meal, soybean meal, whey.

In some feeds of plant origin, phosphorus is tied up as phytic acid and is unavailable for absorption by some animals, particularly nonruminants. Half to two-thirds of the phosphorus in mature seeds and their by-products is in this form. Thus, despite the fact that chemical analysis may show the diet to contain adequate phosphorus, some animals—unable to use this form of phosphorus—may not be receiving adequate amounts. Therefore, supplementation should be provided. (Also see Chapter 26, Feeding Rabbits, section on Minerals, Calcium and Phosphorus.)

(Continued)

TABLE 4-6 *(Continued)*

Phosphorus (P) (Continued)

Supplemental sources—Ammonium phosphate, bone meal, calcium phosphate, colloidal clay, dicalcium phosphate, monosodium phosphate, phosphoric acid, defluorinated rock phosphate.

Comments:

Phosphorus comprises about 25% of the total mineral matter in the body.

Eighty percent of the phosphorus is in the bones and teeth in inorganic combination with calcium.

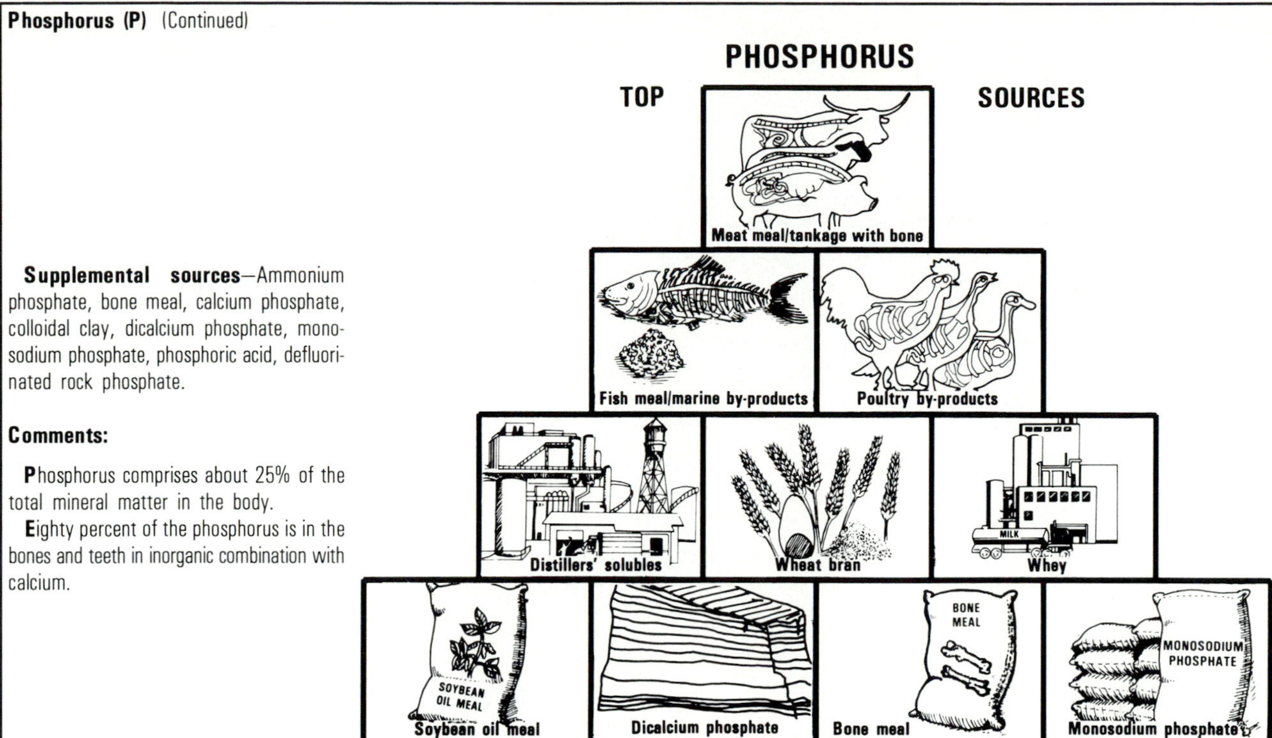

Fig. 4-37. Commonly used, top feed and supplement sources of phosphorus. No claim is made that all the top sources of phosphate are listed. Neither are the sources ranked according to amount of phosphate supplied nor is price considered.

POTASSIUM (K)

Potassium made the difference!

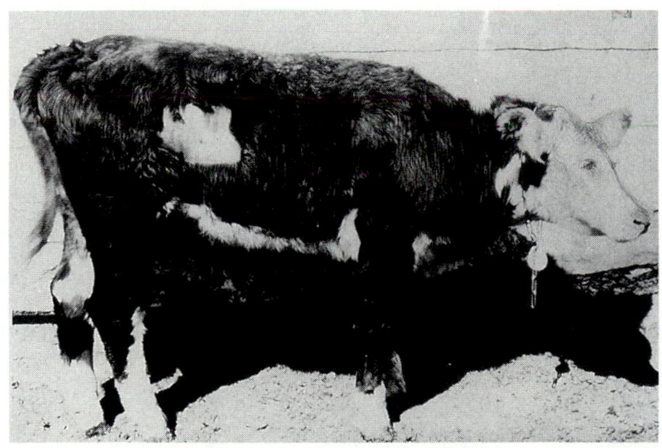

Fig. 4-38. *Left:* Potassium deficient steer fed 0.27% K; lost 125 lb (*57 kg*) in 110 days. *Right:* Healthy steer that received adequate potassium. (*Left photo:* Courtesy, University of Manitoba, Winnipeg, Canada)

Potassium is the third most abundant element in the body, after calcium and phosphorus; and it is present in twice the concentration of sodium.

Potassium constitutes about 5% of the mineral content of the body. It is the primary cation of intracellular (within the cells) fluids. Approximately 98% of the total body potassium is located intracellularly, where its concentration is 30 or more times that of the extracellular (between the cells) fluid. The concentration of sodium in blood plasma is much higher than potassium. On the other hand, the potassium concentration in muscle tissue is many times higher than sodium.

History:

In 1807, a brilliant young English chemist, Sir Humphry Davy, isolated the metal which he named *potassium,* and gave it the chemical symbol K, from *Kalium,* the Latinized version of the Arabic word for *alkali.* However, more than 100 years elapsed following discovery before McCollum, in 1938—using the rat, obtained positive proof that potassium is an essential nutrient, although this had been suggested earlier.

(Continued)

TABLE 4-6 *(Continued)*

Potassium (K) (Continued)

Absorption/Metabolism/Excretion:

Absorption of dietary potassium is very efficient; more than 90% of ingested potassium is absorbed. Most of the absorption occurs in the small intestine.

Although the digestive juices contain relatively large amounts of potassium, most of it is reabsorbed and the loss in the feces is small.

The kidneys provide the major regulatory mechanism for maintaining potassium balance; and relatively wide variations in intake are not reflected in fluctuations in plasma concentration. Aldosterone, and adrenal hormone, stimulates potassium excretion.

Excessive potassium buildup may result from kidney failure or from severe lack of fluid.

Functions:

Involved in the maintenance of proper acid-base balance and the transfer of nutrients in and out of individual cells.

Relaxes the heart muscle—action opposite to that of calcium, which is stimulatory.

Required for the secretion of insulin by the pancreas, in enzyme reactions involving the phosphorylation of creatine, in carbohydrate metabolism, and in protein synthesis.

Deficiency Symptoms/Toxicity:

Deficiency symptoms—Potassium deficiencies are rare but can occasionally occur in the drylot finishing of cattle or sheep when high-concentrate rations are used. Deficiency symptoms are characterized by growth retardation, unsteady gait, general muscle weakness, pica (depraved appetite), diarrhea, distention of the abdomen, emaciation (loss of flesh), hypertrophy (enlargement) of the heart and kidneys, and eventually death.

Toxicity—Excessive levels of potassium interfere with magnesium absorption and utilization. Toxicity from excessive intake is unlikely except (1) when water intake is restricted or water is saline, or (2) when the kidneys are not functioning properly.

Major Interrelationships:

A magnesium deficiency results in failure to retain potassium; hence, it may lead to a potassium deficiency.

Excessive levels of potassium interfere with magnesium absorption. Also, large excesses of potassium may slow the heart to a standstill when the kidneys are unable to excrete the surplus in the urine.

Because sodium and potassium must be in balance, excessive use of salt depletes the body's potassium supply.

The need for potassium is increased when there is growth or deposition of lean tissue; and potassium is lost whenever muscle is broken down owing to starvation, protein deficiency, or injury.

Sources for Animals:

Rich feed sources—Beet molasses, beet tops, blackstrap molasses, carrots, forages, soybean seed by-products, tomato pomace, whey.

Feeds of plant origin, especially forages, are extremely good sources of potassium. For this reason, potassium is not generally added to feeds for herbivores.

Supplemental sources—Kelp, potassium chloride, potassium gluconate, yeast (brewers', torula).

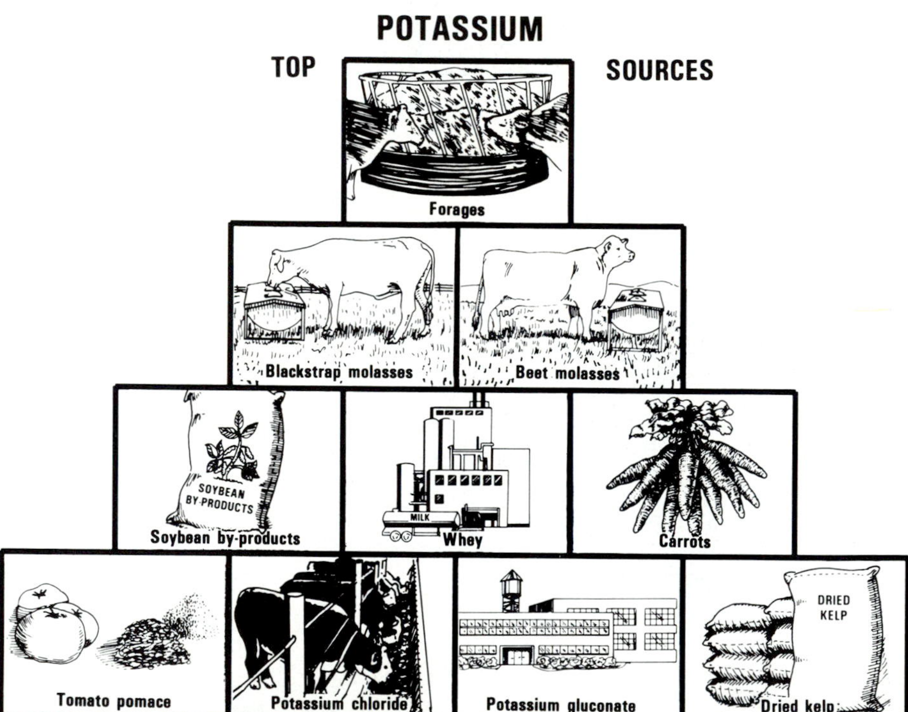

Fig. 4-39. Commonly used, top feed and supplement sources of potassium. No claim is made that all the top sources of potassium are listed. Neither are the sources ranked according to amount of potassium supplied nor is price considered.

Comments:

Potassium deficiency may occur in drylot finishing cattle or sheep fed a high-concentrate ration.

SODIUM (Na)

Most of the sodium in the diet is in the form of sodium chloride (NaCl), *salt,* a white granular substance used to season and preserve feed and food. It is created by combining the elements sodium (Na), a soft, silvery-white metal, and chlorine (Cl), a poisonous, yellow gas; neither of which is desirable for use alone.

The body contains approximately 0.2% sodium.

(Continued)

TABLE 4–6 *(Continued)*

Sodium (Na) (Continued)

History:

Compounds of sodium were known and used extensively during ancient times, but the element was not isolated until 1807, when Sir Humphry Davy, an English chemist, used electricity (in a process called electrolysis) to extract the pure metal from sodium hydroxide. Although sodium had long been believed to be a dietary essential, final proof was not obtained until 1918 when Osborne and Mendel conducted laboratory animal experiments.

Absorption/Metabolism/Excretion:

Virtually all sodium ingested in the diet is readily absorbed from the gut, following which it is carried by the blood to the kidneys, where it is filtered out and returned to the blood in amounts needed to maintain the levels required by the body.

Excess sodium, which usually amounts to 90 to 95% of ingested sodium, is, for the most part, excreted by the kidneys as chloride and phosphate and controlled by aldosterone, a hormone of the adrenal cortex.

The levels of sodium in the urine reflect the dietary intake; if there is high intake of sodium the rate of excretion is high, if there is low intake the excretion rate is low.

Functions:

Major cation in osmotic pressure and acid-base balance in body fluids, upon which depends the transfer of nutrients to the cells and the removal of waste materials and the maintenance of water balance among the tissues.

As a constituent of pancreatic juice, bile, sweat, and tears.

Associated with muscle contraction and nerve functions.

Plays a specific role in the absorption of carbohydrates.

Deficiency Symptoms/Toxicity:

Deficiency Symptoms—Loss of appetite.

Reduced growth and efficiency of feed utilization in growing animals, reduced milk production and weight loss in adults.

Lowered reproduction (infertility in males, and delayed sexual maturity in females).

Craving for sodium, evidenced by such things as drinking urine.

In laying hens, a deficiency of sodium results in lowered production, loss of weight, and cannibalism.

Toxicity—Salt may be toxic when (1) a high intake is accompanied by a restriction of water, or (2) the body is adapted to a chronic low-salt diet. Salt toxicity readily occurs in ruminants, in which it is characterized by a staggering gait, blindness, and other nervous disorders.

Excess sodium results in hypertension.

Major Interrelationships:

Sodium, potassium, and chlorine are closely related metabolically. They serve a vital function in controlling osmotic pressures and acid-base equilibrium; and they play important roles in water metabolism. Both sodium and potassium ions occur in the body chiefly in close association with the chloride ion; therefore, a sodium or potassium deficiency is rarely found in the absence of a chlorine deficiency.

Sources for Animals:

Rich feed sources—Beet molasses, beet pulp with molasses, blackstrap molasses, buttermilk, cracklings (swine), distillers' solubles, fish meal and other marine by-products, meat meal/tankage and other meat by-products, poultry by-product meal, poultry feathers, skimmed milk, whey.

Supplemental sources—Kelp, monosodium glutamate, salt (sodium chloride).

Comments:

Most of the salt in the diet is in the form of sodium chloride (NaCl), fed free-choice or added to the ration at a level of 0.25 to 0.50%.

Deficiencies of salt may occur in animals not provided supplemental salt when on pasture or range.

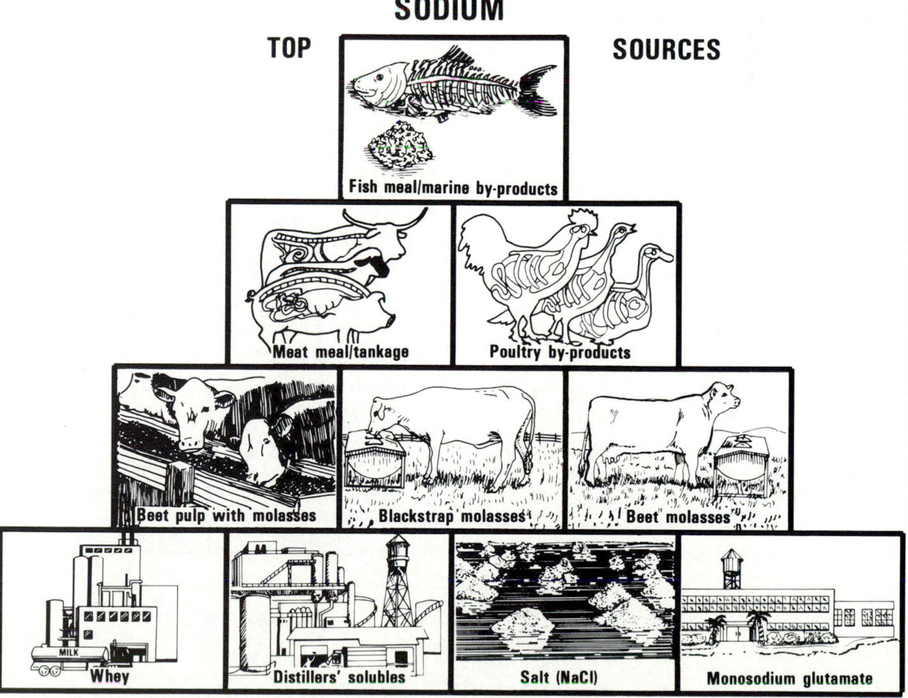

Fig. 4–40. Commonly used, top feed and supplement sources of sodium. No claim is made that all the top sources of sodium are listed. Neither are the sources ranked according to amount of sodium supplied nor is price considered.

(Continued)

TABLE 4-6 *(Continued)*

SULFUR (S)

Sulfur is found in every cell of the body and is essential for life itself, mostly as a component of three important amino acids—cystine, cysteine, and methionine. Also, it is a part of two vitamins—thiamin and biotin, and it is present in saliva and bile, and in the hormone, insulin. Approximately 0.15% of the body weight and 10% of the mineral content of the body are sulfur.

History:

The name is derived from the Latin word *sulphurum.* Sulfur has been used since ancient times. It was often called brimstone (burning stone); and ignited sulfur is mentioned in the earlier records of many countries as having been used in religious ceremonies and for purifying (fumigating) buildings. The early medical books of Dioscorides of Greece and Pliny the Elder mention sulfur; and the Romans used it in medicine and in warfare. Alchemists recognized sulfur as a mineral substance that could be melted and burned. It was first classified by Antoine Lavoisier in 1777.

Absorption/Metabolism/Excretion:

Absorption—The small intestine is the major site of sulfur absorption.

Metabolism—During digestion, the sulfur-containing amino acids are split off from protein and taken into the portal circulation. Sulfur is stored in every cell of the body, with the highest concentrations found in the hair, skin, and hooves.

Excretion—Excess sulfur is excreted in the urine and in the feces. About 85 to 90% of the sulfur excreted in the urine is in the organic form, derived almost entirely from the metabolism of the sulfur amino acids. Since inorganic sulfates are poorly absorbed, it follows that the fecal excretion of sulfur is about equal to the inorganic sulfur content of the diet.

Functions:

As a component of the sulfur-containing amino acids methionine, cystine, and cysteine.

As a component of biotin, sulfur is important in fat metabolism.

As a component of thiamin and insulin, it is important in carbohydrate metabolism.

As a component of coenzyme A, it is important in energy metabolism.

As a component of certain complex carbohydrates, it is important in various connective tissues.

As a component of insulin and glutathione, it regulates energy metabolism.

As a converter of toxic substances to nontoxic forms, it rids the body (via excretion) of such toxic substances as phenols and cresols.

As a primary component of hair, wool, and feathers.

Deficiency Symptoms/Toxicity:

Deficiency Symptoms—Retarded growth, primarily due to not meeting the sulfur amino acid requirement for protein synthesis.

Sheep fed nonprotein nitrogen to replace protein without sulfur supplementation may show reduced wool growth (wool contains approximately 4% sulfur).

Toxicity—The production of hydrogen sulfide—a highly toxic gas—by rumen microorganisms, can pose problems when excessive amounts of sulfur are fed.

Major Interrelationships:

Sulfur is related to the amino acids methionine, cystine, and cysteine, and to biotin, thiamin, insulin, coenzyme A, certain complex carbohydrates, and glutathione.

Sources for Animals:

Monogastric animals should be provided sulfur-containing proteins, for they have little or no ability to utilize elemental sulfur for amino acid synthesis. Ruminants and horses (and probably rabbits) may be provided sulfur in protein, as elemental sulfur, or as sulfate sulfur, from which they can synthesize amino acids through the action of the microorganisms.

Rich feed sources—Alfalfa, barley malt sprouts, beet molasses, blackstrap molasses, blood meal, corn cobs, corn gluten meal, cottonseed meal, distillers' grains, feather meal, fish meal and other marine by-

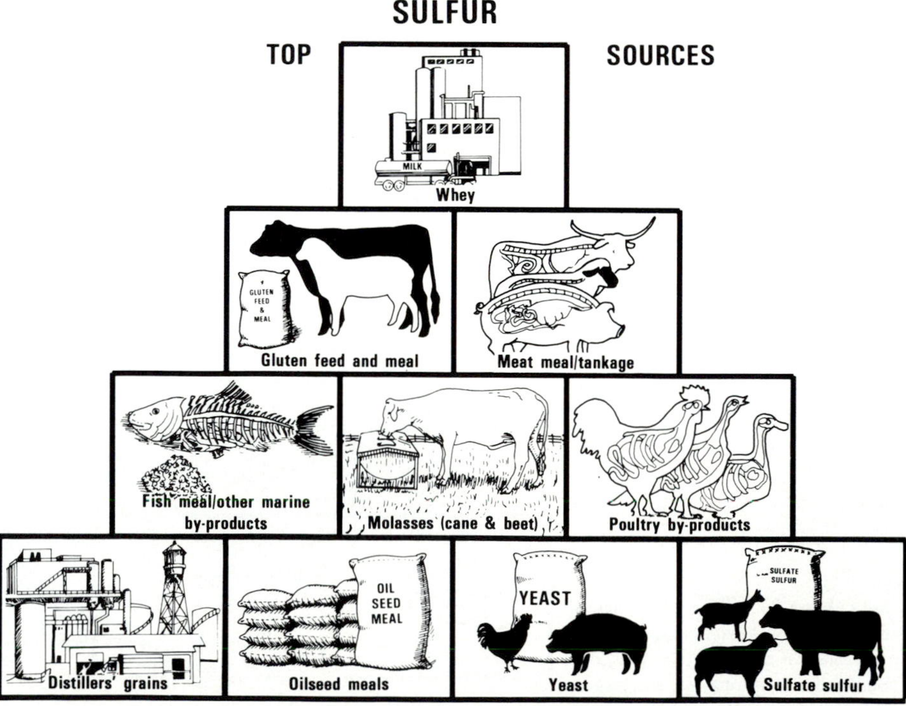

Fig. 4-41. Commonly used, top feed and supplement sources of sulfur. No claim is made that all the top sources of sulfur are listed. Neither are the sources ranked according to amount of sulfur supplied nor is price considered.

(Continued)

TABLE 4-6 *(Continued)*

Sulfur (S) (Continued)

products, linseed meal, meat meal, meat tankage, poultry by-product meal, soybean meal, wheat gluten meal, whey.

Supplemental sources—Elemental sulfur, sulfate sulfur, yeast (brewers', torula).

Comments:

Sulfur requirements are primarily those involving amino acid nutrition.

Ruminants fed urea as a source of protein nitrogen may benefit from supplemental sulfur. It is frequently recommended that the nitrogen-sulfur ratio of ruminant rations should be approximately 10:1.

TRACE/MICROMINERALS:

CHROMIUM (Cr)

The mention of this chemical element makes most people think of the *chrome* plating on the bumpers and body trim of their automobiles. However, it was recently discovered that the shiny metal may also exist in forms which function as (1) an essential element, (2) a hormone, (3) a vitamin, and (4) a poison.

History:

Chromium was discovered by the French chemist Vauquel in 1797, while he was studying the properties of crocoite, an ore which is rich in lead chromate. Its common name of chrome was derived from the Greek work *chroma,* which means color, because the element is present in many different colored compounds. These compounds have long been used as pigments in dyeing and in the tanning of leather. In the early 1900s, chromium became an important ingredient of corrosion-resistant metals—a use which has increased to the present.

It was not until 1959 that the medical scientists W. Mertz and K. Schwarz—who came to the United States from Germany—discovered that the feeding of chromium salts corrected the abnormal metabolism of sugar in rats which resulted from the feeding of diets based upon torula yeast. Later work by these researchers, and by H. Schroeder of the Dartmouth Medical School, established chromium as a cofactor with insulin, necessary for normal glucose utilization and for growth and longevity in rats and mice. (Dr. Schwarz has long studied the nutritional effects of various types of yeasts; in 1957, he had discovered that selenium was present as a vital factor in American-type brewers' yeast.) Shortly thereafter, it was found that the inorganic salts of chromium were utilized poorly, compared to an organically bound form of chromium present in brewers' yeast which was utilized well by both animals and humans. The chromium-containing substance from yeast was named the *glucose tolerance factor* (GTF), because it sometimes restored the metabolism of sugars to normal when diabeticlike tendencies were present.

Absorption/Metabolism/Excretion:

Absorption—Studies have shown that only about 1%, or less, of the dietary intake of inorganic chromium is absorbed, whereas as much as 10 to 25% of the glucose tolerance factor (GTF-chromium) may be absorbed.

Metabolism—Animal studies have shown that a greater proportion of a given dose of chromium is stored in the liver when it is supplied as GTF than when it is supplied as an inorganic salt. Furthermore, chromium in the liver has been shown to have GTF activity. Next to the liver, the kidneys appear to be one of the best sources of GTF-chromium.

Various stressful conditions, such as malnutrition and loss of blood, have long been known to impair the body's utilization of sugars. Likewise, it seems that the chromium needs of the body become more critical under such conditions. For example, studies on animals have shown that the effects of (1) low-protein diets accompanied by controlled exercise, and (2) the withdrawal of measured amounts of blood were more severe in chromium-deficient groups than in those given supplements containing the mineral.

Excretion—The predominant route of excretion of endogenous chromium is via the urine.

Functions:

Component of the glucose tolerance factor (GTF), which enhances the effect of insulin.

Activator of certain enzymes, most of which are involved in the production of energy from carbohydrates, fats, and proteins.

Stabilizer of nucleic acids (DNA and RNA).

Stimulation of synthesis of fatty acids and cholesterol in the liver.

Deficiency Symptoms/Toxicity:

Deficiency Symptoms—Impaired glucose tolerance, which may be accompanied by high blood sugar and the spilling of sugar in the urine.

Disturbance in lipid and protein metabolism.

Toxicity—Chromium is seldom toxic because (1) only small amounts are present in most feeds, (2) the body utilizes it poorly, and (3) there is a wide margin of safety between helpful and harmful doses.

NOTE WELL: Excesses of inorganic chromium are much more toxic than similar amounts of GTF-chromium.

Major Interrelationships:

The following interrelationships are pertinent:

1. Chromium functions best in the body when it is in the form of the GTF.
2. Diets rich in carbohydrates may cause the supply of GTF-chromium to be depleted.
3. Inorganic chromium is utilized, but less efficiently than that in GTF.

(Continued)

TABLE 4–6 *(Continued)*

Chromium (Cr) (Continued)

4. The absorption of chromium is impeded by oxalates and phytates.
5. Zinc and vanadium antagonize the effects of chromium.

Sources for Animals:

Rich feed sources—Blackstrap molasses, corn, liver meal, milk, potatoes, poultry by-product meal, vegetable oils, wheat, wheat bran.

NOTE WELL: The content and/or availability of chromium in feeds may be affected by (1) the chromium content of the soil, (2) the processing of grain, (3) the refining of molasses, and (4) fermentation.

There is no evidence that practical animal rations need to be supplemented with chromium.

Supplemental sources—Brewers' yeast, dried liver.

Comments:

The importance of Cr in glucose metabolism in animals other than the rat and the human has not been established to date.

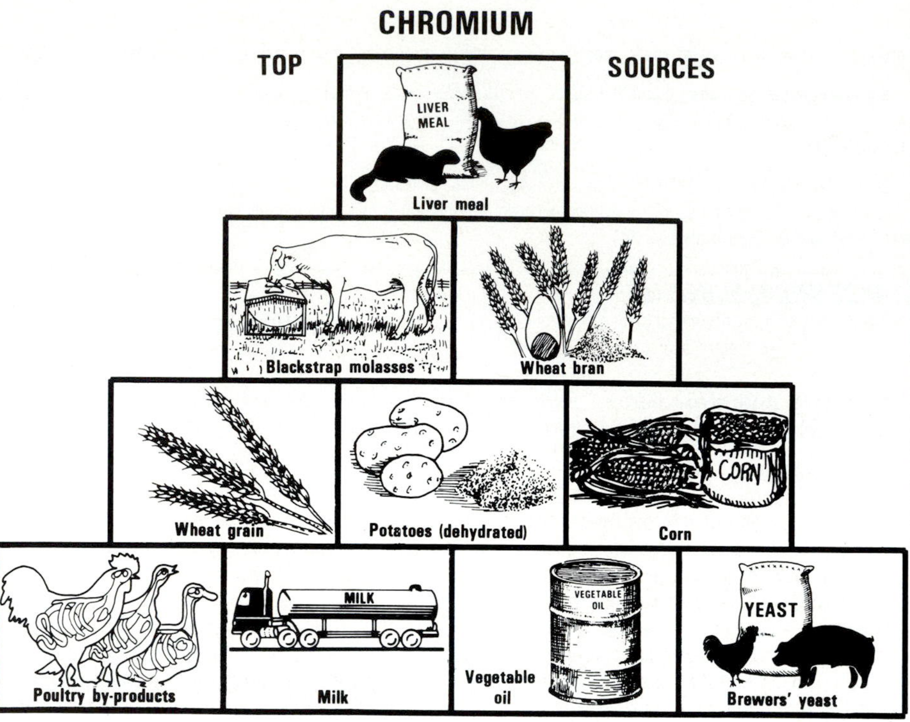

Fig. 4–42. Commonly used, top feed and supplement sources of chromium. No claim is made that all the top sources of chromium are listed. Neither are the sources ranked according to amount of chromium supplied nor is price considered.

COBALT (Co)

Cobalt made the difference!

Fig. 4–43. *Left:* Lamb fed cobalt-deficient ration containing 0.05 ppm cobalt. Lamb weighed 48.5 lb (*22 kg*) at the end of the experimental period. Note the emaciation and loss of wool. *Right:* Control lamb that received the same ration as the lamb on the left plus a daily allowance of 0.1 mg of cobalt as cobalt sulfate. This lamb weighted 92.5 lb (*42 kg*). (Courtesy, Cornell University, Ithaca, N.Y.)

This element is an essential constituent of vitamin B–12. No other function of cobalt has been established.

History:

The word cobalt is derived from the German word *kobold,* meaning *goblin* or *mischievous spirit.* The term originated in the 16th century, when arsenic-containing cobalt ores were dug up in the silver mines of the Harz Mountains. Believing that the ores contained copper, miners heated them and were injured by the toxic arsenic trioxide vapors that were released. These evils were attributed to the goblin or kobold. George Brandt, a Swedish chemist, first isolated the element in 1742, although cobalt had been used for centuries for its blue color in decorative glass and pottery.

The discovery, in 1948, that vitamin B–12 (cyanocobalamin) contains 4% cobalt (Co) proved this element to be an essential nutrient for humans. It is noteworthy that cobalt's essential role in ruminant animal nutrition was known much earlier. In 1935, E. J. Underwood and other Australian scientists discovered that lack of cobalt, resulting from its deficiency in the soil and thus in the herbage grazed, produced a *wasting disease* in animals. Much earlier, and long before the cause was known, livestock producers in different areas of the world learned that this peculiar malady could be prevented and/or cured by transferring animals from "sick" to "healthy" areas.

(Continued)

TABLE 4-6 *(Continued)*

Cobalt (Co) (Continued)

Absorption/Metabolism/Excretion:

As a component of vitamin B-12, cobalt requires the intrinsic factor—a compound secreted in the stomach—for absorption (see Chapter 3, Digestion/Absorption). If this factor is not present, a vitamin B-12 deficiency will result. Since cobalt from the body is excreted through the bile, it can be readily reabsorbed to minimize losses.

Excessive cobalt is excreted primarily in the urine.

Functions:

The only known function of cobalt is that of an integral part of vitamin B-12, an essential factor in the formation of red blood cells.

When cobalt is given to animals which utilize microbial fermentation, the element is used for the synthesis of vitamin B-12; hence, ruminants do not need supplemental sources of the vitamin provided cobalt is supplied. However, most nonruminants and very young ruminants do not have a need for cobalt, *per se,* but do require dietary sources of vitamin B-12.

Deficiency Symptoms/Toxicity:

Deficiency Symptoms—About 0.1 ppm of cobalt in the feed is adequate, with deficiency symptoms appearing when the level drops to the range of 0.04 to 0.07 ppm or lower. Deficiency of cobalt in cattle and sheep produces symptoms similar to a deficiency of vitamin B-12—rough hair coat, scaliness of the skin, absence of estrus, abortion, low milk production, loss of appetite, rapid loss of weight, emaciation, and anemia. Continued deficiency may eventually cause death. The difference between life and death in a 1,000-lb steer can be only 1.0 mg of cobalt. Recovery is quite rapid with the introduction of cobalt into the rumen. Feeds grown in areas where there is adequate cobalt will sustain normal rumen synthesis.

The disease called *salt sick* in Florida is due to cobalt deficiency associated with copper deficiency.

In different parts of the world, cobalt deficiency is known as Denmark disease, coast disease, enzootic marasmus, bush sickness, wasting disease, Nakuritis, and pining disease.

Toxicity—Cobalt can be toxic when intake is excessively high, but the margin of safety is wide and absorption from the intestinal tract is low. Cobalt is stored in the liver, kidneys, adrenal glands, and bones. Cattle may be less tolerant to excessive cobalt intake than sheep, although cobalt toxicities in both species have been reported under practical conditions.

Major Interrelationships:

Cobalt is a component of vitamin B-12.

Sources for Animals:

Rich feed sources—Alfalfa hay/meal, beans (kidney, pinto), beet molasses, blackstrap molasses, corn, corn silage, cottonseed meal, cotton seeds, fish meal (white fish), grass-legume silage, meat meal, orchardgrass hay, peas (field), poultry by-product meal, rice bran, rice hulls, rice groats, safflower meal, sorghum grain, soybean meal, vetch hay, wheat grain, wheat middlings. **NOTE WELL:** The level of cobalt in plant feeds depends on the soil levels from which they come.

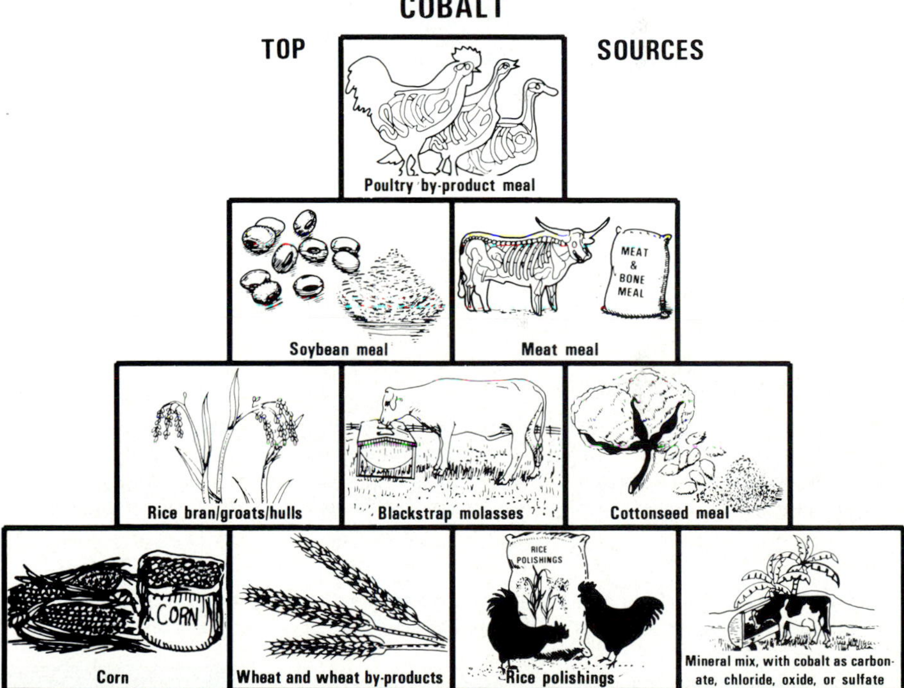

Fig. 4-44. Commonly used, top feed and supplement sources of cobalt. No claim is made that all the top sources of cobalt are listed. Neither are the sources ranked according to amount of cobalt supplied nor is price considered.

Supplemental sources—Cobalt is a component of vitamin B-12. For this reason it must be supplied in ruminant rations to enable the microorganisms to synthesize the vitamin. For nonruminants, cobalt is not added to rations because vitamin B-12 is generally supplied in association with animal protein, or in the vitamin supplement.

The common cobalt supplements are: cobaltized mineral mixture made by adding cobalt at the rate of 0.2 oz/100 lb of salt as cobalt carbonate, cobalt chloride, cobalt oxide, or cobalt sulfate. Also, several good cobalt-containing commercial minerals are on the market. Other cobalt-containing supplements are: brewers' yeast and rice polishings.

Grazing animals may be given pellets composed of cobalt oxide and iron administered orally with a balling gun. The pellets lodge in the rumen and are gradually dissolved over a period of months.

(Continued)

TABLE 4-6 *(Continued)*

Cobalt (Co) (Continued)

Comments:

Cobalt-deficient areas have been reported in Australia, New Zealand, western Canada, and in the U.S. in the states of Florida, Massachusetts, Michigan, New Hampshire, New York, North Carolina, Pennsylvania, South Carolina, and Wisconsin. (See Chapter 5, Nutritional Disorders/Toxins, Fig. 5-30.)

COPPER (Cu)

Copper made the difference!

Fig. 4-45. *Left:* Copper-deficient pig. *Right:* Healthy pig that received a similar diet, with adequate copper. (Courtesy, The Hormel Institute, Austin, Minn.)

Copper and iron are mutually involved in the formation of hemoglobin—the red pigment in blood which carries oxygen.

History:

Copper was discovered and first used by neolithic man during the late Stone Age. The exact date of this discovery will probably never be known, but it is believed to have been about 8000 B.C. The late Bronze Age (3000 to 1000 B.C.) takes its name from the use during this period of bronze, an alloy of copper and tin.

As a result of a series of studies beginning in 1925 (and reported in 1928), Hart and associates at the University of Wisconsin discovered that a small amount of copper is necessary, along with iron, for hemoglobin formation. Today, copper is considered an essential nutrient for all vertebrates and some lower animal species.

Absorption/Metabolism/Excretion:

Absorption—The site of copper absorption varies with the species of animal, but it is generally greatest in the small intestine. Of the intake, about 30% is absorbed.

Metabolism—After absorption in the intestine, copper reaches the bloodstream where most of it (80% or more) becomes bound in ceruloplasmin, a protein (globulin-copper) complex. The remainder is loosely bound to albumen and transported to various tissues.

Excretion—Copper is excreted through the bile for elimination in the feces, but much of this copper is reabsorbed.

Functions:

Facilitates the absorption of iron from the intestinal tract and releases it from storage in the liver and the reticuloendothelial system.

Essential for the formation of hemoglobin, although it is not a part of hemoglobin as such.

Constituent of several enzyme systems.

Development and maintenance of the vascular and skeletal structures (blood vessels, tendons, bones).

Structure and functioning of the central nervous system. Required for normal pigmentation of the hair and wool.

Component of important copper-containing proteins.

Reproduction (fertility).

Deficiency Symptoms/Toxicity:

Deficiency Symptoms—Even though the requirement of copper is very small, many factors relating to its absorption and utilization have created copper deficiencies.

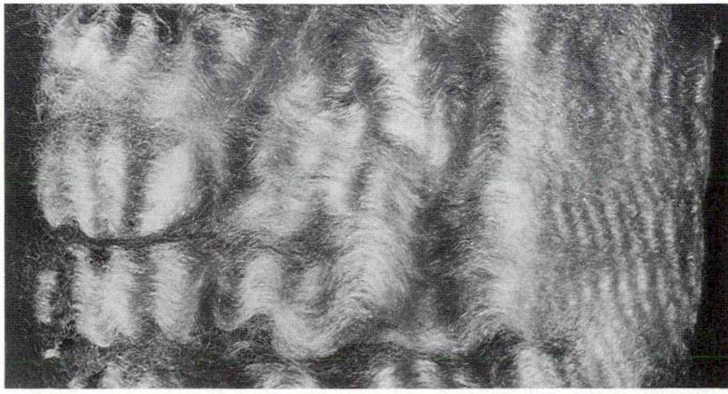

Fig. 4-46. The left ⅔ of this sample was produced by a sheep on a copper-deficient ration, resulting in hair-like or "steely" wool. Then copper was added to the sheep's diet, and normal, well-crimped wool (right) was produced. (Courtesy, Washington State University, Pullman)

(Continued)

TABLE 4-6 *(Continued)*

Copper (Cu) (Continued)

A condition called *swayback* (enzootic ataxia) in newborn lambs has been associated with copper deficiency. Since in most cases the amount of copper in the feed is adequate, either a faulty absorption mechanism or excessive excretion of copper is thought to be the cause. *Falling disease* in animals of various parts of Australia is characterized by staggering, falling, and sudden death. *Peat scours* in New Zealand and *teartness* in England are copper deficiencies created by high molybdenum intake. Sheep consuming too little copper not only have a change in the color of black fleeces, but the wool becomes coarse and straight, similar to hair, and is referred to as *stringy wool* or *steely wool*. Other deficiency symptoms include a fading hair coat, lameness, swelling of the joints, diarrhea, fragility of the bones, a nervous condition known as ataxia, and nutritional anemia, commonly called "salt sick." Deficiencies may be observed in many animals, but most reports have involved cattle and sheep.

Toxicity—Excess copper is toxic; it accumulates in the liver, and death may result. However, there are wide species differences in tolerance to excess copper. Horses are very tolerant, whereas sheep are very sensitive. The maximum tolerable levels for growing animals, according to the National Academy of Sciences, are: horse, 800 ppm; chicken, 300 ppm; swine, 250 ppm; cattle, 100 ppm; and sheep, 25 ppm.

Major Interrelationships:

Copper, along with certain vitamins, is involved in iron metabolism.

Dietary excesses of cadmium, calcium, iron, lead, molybdenum plus sulfur, silver, and zinc reduce the utilization of copper. There should not be less than 2 parts copper to 1 part molybdenum.

An excess of molybdenum in the presence of sulfate causes a condition which can be cured by administering copper.

In high-molybdenum areas, the copper level for horses and cattle should be about 5 times higher than normal.

Sources for Animals:

The amount of copper available in feeds is highly dependent on the copper content of the soil. Soils in Florida and the Coastal Plain are generally considered to be copper-deficient. Also, copper absorption and utilization are affected by many other nutrients—for example, molybdenum, calcium, mercury, cadmium, zinc, and iron. Therefore, trace mineralized salt containing copper sulfate or copper carbonate should be given to animals when feed is suspected of being either low in copper or high in antagonistic nutrients.

Rich feed sources—Blackstrap molasses, brewers' grains, citrus molasses, corn gluten feed, corn gluten meal, cotton seeds, crab cannery residue meal, distillers' grains, distillers' solubles, fish solubles, grass hays, liver meal, meat tankage, millet, safflower meal, sardine meal, shark meal, sorghum gluten meal, soybean meal, Sudangrass hay, sunflower seeds, tomato pomace, wheat bran, wheat germ, whey. **NOTE WELL:** The copper content of plants reflects the soil on which they were grown.

Supplemental sources—Brewers' yeast, copper carbonate, copper sulfate, trace mineralized salt containing copper carbonate or copper sulfate.

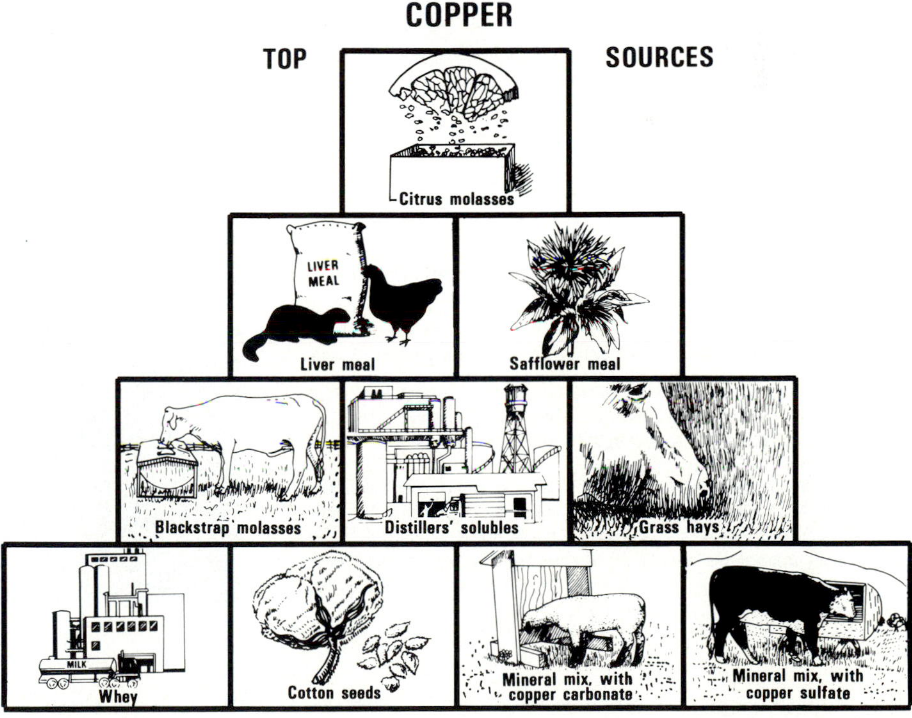

Fig. 4-47. Commonly used, top feed and supplement sources of copper. No claim is made that all the top sources of copper are listed. Neither are the sources ranked according to amount of copper supplied nor is price considered.

Comments:

The amount of copper needed in the diet varies with different individuals, and is very interdependent on the level of other minerals in the feed. When there are no interfering elements in the ration, 5 to 8 ppm of copper in the diet are adequate for most animals.

A variable store of copper is located in the liver and spleen.

Milk is low in copper; hence, young animals raised almost exclusively on milk may develop anemia.

Copper deficiencies are common in parts of Australia and New Zealand and in southern U.S.

(Continued)

TABLE 4-6 *(Continued)*

FLUORINE (F)

Fluorine made the difference!

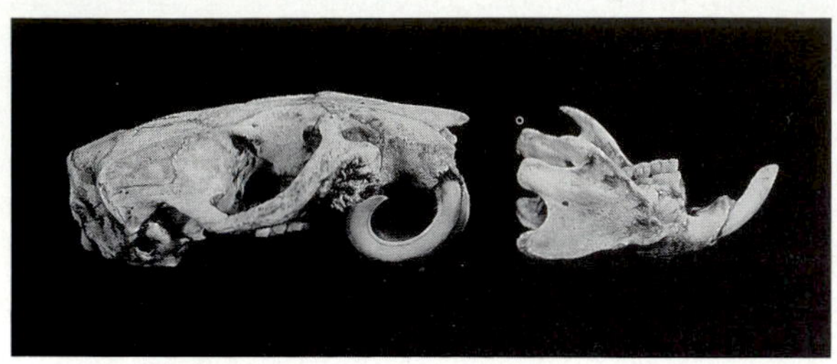

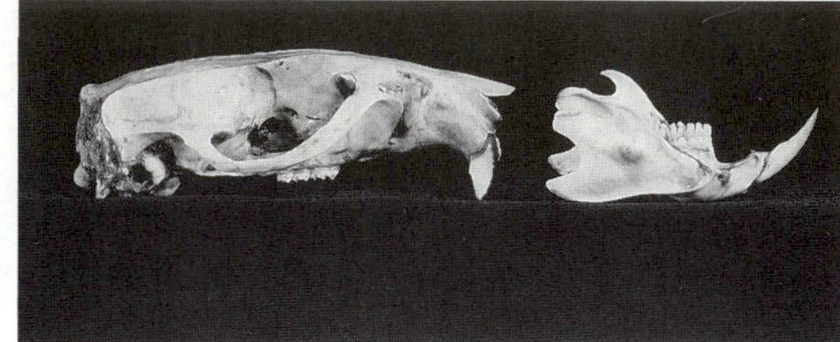

Fig. 4-48. Skulls of rats fed diets with and without fluorine. *Left:* Thickened skull with elongated and chalky incisors caused by excessive fluorine intake. *Right:* Normal skull. (Courtesy, Cornell University, Ithaca, N.Y.)

Fluorine is present in small, but widely varying, concentrations in practically all soils, water supplies, plants, and animals. It is, therefore, a constituent of all normal rations. Also, fluorine is one of the atmospheric contanimants from industries which use certain metal ores, coal, or earthy phosphates; and phosphate fertilizers usually contain fluorides which may become contaminants in the vegetation fed to animals.

In small amounts, fluorine helps develop strong bones and teeth, but in excessive amounts bones become porous and soft and teeth become mottled and easily worn down.

History:

Fluorine was first isolated in 1886 by Henri Moissan, a Frenchman, who obtained it by the electrolysis of anhydrous hydrogen fluroide containing dissolved potassium fluoride. The name fluorine is derived from the Latin *fluo,* meaning to flow, because from early times it was used as a flux in metallurgy.

Absorption/Metabolism/Excretion:

A large proportion (about 90%) of ingested fluorine is normally absorbed, primarily from the small intestine.

That portion of absorbed fluorine which is not taken up by the bones and teeth (50% or more of the fluorine absorbed) is excreted mainly in the urine (although small amounts are excreted in the sweat and the feces), with the result that the level of fluoride in the blood plasma is quite constant.

Functions:

Constitutes .02 to .05% of the bones and teeth. Necessary for sound bones and teeth.

Deficiency Symptoms/Toxicity:

Deficiency Symptoms—Excesses of fluorine are of more concern than deficiencies.

Toxicity—Deformed teeth and bones; softening, mottling, and irregular wear of the teeth; roughened hair coat; delayed maturity; and less efficient utilization of feed. The degree of mottling of the teeth depends upon the level of fluorine intake, species, and individual susceptibility.

Some reduction in toxicity may be obtained by the addition of calcium, aluminum, or fat to the ration, as these elements seem to reduce the absorption of fluorine.

Amounts of fluorine in excess of 20 to 40 ppm of the dry matter of the diet (depending on the species of animal, age, and rate of production) may show a progressively severe toxicity. Also, consuming water containing in excess of 2 to 8 ppm (or more) of fluoride over a long period of time can produce toxicity (fluorosis).

Fluorine is a cumulative poison; hence, chronic fluoride toxicity, known as fluorosis, may not be noticed for sometime.

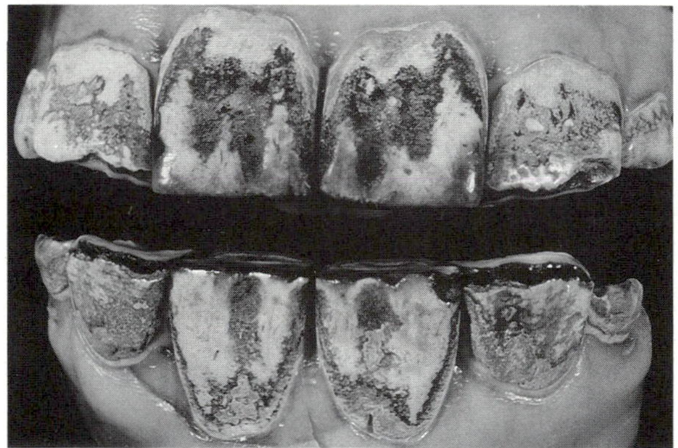

Fig. 4-49. Fluorosis evidenced in the teeth of a horse. Note mottling and irregular wear. (Courtesy, Utah State University, Logan)

Major Interrelationships:

Large amounts of dietary calcium, aluminum, or fat will lower the absorption of fluorine.

Sources for Animals:

Fluorine is found in many feeds, but fish meal and other marine by-products, and kelp, are the richest feed sources.

No need to supplement livestock with fluorine has been demonstrated. Should such supplementation be necessary, 1 ppm of fluorine in the drinking water should suffice.

Comments:

The largest high fluorine area in the U.S. is the West Texas Panhandle. Also, the soils in some volcanic areas of the world contain large amounts of fluorine, with the result that feeds grown in such areas may contain 2 to 3 times more fluorine than feeds grown elsewhere.

(Continued)

TABLE 4-6 *(Continued)*

IODINE (I)

Iodine made the difference!

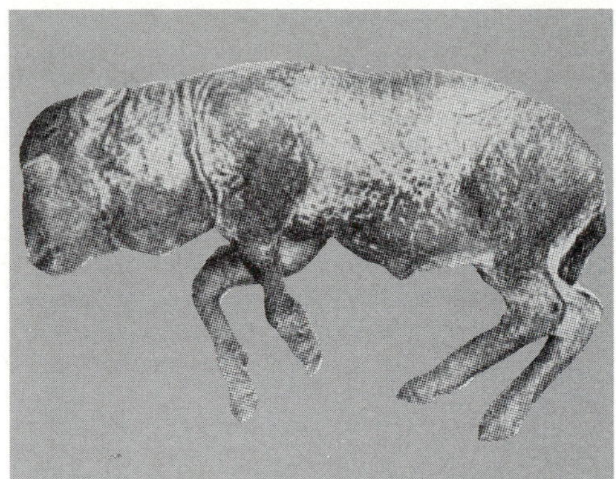

Fig. 4-50. *Left:* Iodine deficiency, exemplified by a woolless, goitered (big-necked) stillborn lamb. *Right:* Healthy newborn lamb. (*Left photo:* Courtesy, Montana State University, Bozeman. *Right photo:* Courtesy, *The Sheepman's Production Handbook,* Denver, Colo.)

Iodine is recognized as an essential nutrient for all animal species. Most of the iodine in the animal body is in the thyroid gland, where it is an integral component of the thyroid hormones, thyroxin and triiodothyronine, both of which have important metabolic roles.

One of the factors affecting the output of thyroid hormones by the thyroid gland is iodine availability. In the absence of sufficient iodine, the gland attempts to compensate for the deficiency by increasing its secretory activity, and this causes the gland to enlarge. This condition is known as simple, or endemic, goiter.

History:

Iodine was the first nutrient to be recognized as essential for animals and humans. As early as 3000 B.C., the Chinese treated goiter by feeding seaweed and burnt sponge. Also, Hippocrates (460 to 370 B.C.), a Greek physician, used the same treatment for enlarged thyroid glands. The name *iodine* is derived from the Greek word *iodes,* meaning violet color, from the color of the fumes of iodine.

In 1811, Bernard Courtois, a French chemist, discovered iodine in seaweed and described some of its basic properties. Five years later, potassium hydriodate was introduced by Prout as a treatment for goiter. However, the widespread appearance of goiter continued in much of the world for many years.

In 1914, Kendall, at the Mayo Clinic, in Minnesota, reported the isolation of a crystalline compound containing 65% iodine from the thyroid gland and named it thyroxin. From this discovery and from other studies, the inclusion of iodine in the diets of animals and man led to a greater reduction of goiter in the U.S. and other developed countries of the world. Kendall received the Nobel Prize for his work on thyroxin and other hormones.

Absorption/Metabolism/Excretion:

Iodine absorption is very efficient (near 100%). Although it is absorbed throughout the entire gastrointestinal tract, most absorption takes place in the small intestine. Following absorption, iodine takes two main pathways within the body. Approximately 30% is removed by the thyroid gland and used for the synthesis of the thyroid hormones; most of the remainder is excreted in the urine, although small amounts are lost in the feces and sweat.

Functions:

The sole function of iodine is making the iodine-containing hormones, thyroxin and triiodothyronine, secreted by the thyroid gland, which regulate the rate of oxidation within the cells; and in so doing influence growth, the functioning of the nervous and muscle tissues, circulatory activity, and the metabolism of all nutrients.

When increased amounts of the thyroid hormones are noted in the blood, the rate of metabolism increases, thus increasing the rate of production (such as milk and eggs). If dietary intake of nutrients is not sufficient for the increased needs, body stores of nutrients will be mobilized with the effect of decreasing body weight. Eventually emaciation and health problems occur if the dietary deficiencies are prolonged.

Also, iodine is sometimes used as an antibacterial agent to control low-grade infections.

Fig. 4-51. Iodine toxicity. This shows hyperparathyroidism, commonly called big head disease, in an 8-year-old Quarter Horse mare, caused by feeding too much iodine. (Courtesy, Texas A&M Univeristy, College Station)

Deficiency Symptoms/Toxicity:

Deficiency Symptoms—With an inadequate dietary intake of iodine, the thyroid gland cannot continue to produce its hormones. Under the stimulus of a thyroid-stimulating hormone from the pituitary gland, the thyroid gland enlarges in an attempt to respond to the needs of the body. Even birds and fish develop enlarged thyroid glands when fed iodine-deficient rations. The enlarged thyroid gland is called a *goiter.* It may be found in adult animals, but it is generally more commonly found in newborn animals from mothers which have been fed an iodine-deficient diet. In addition to the enlarged thyroid gland (called "big neck" in calves),

(Continued)

TABLE 4–6 *(Continued)*

Iodine (I) (Continued)

other symptoms of iodine deficiency include reduced rate of growth, dry skin and brittle hair, resorption of fetuses, abortions, stillbirths, abnormal estrus in females, and reduced libido and poor semen quality in males. Iodine-deficient newborn pigs are hairless. Wool is sparse and of poor quality in iodine-deficient sheep. Foals may be born dead or be so weak at birth that they cannot stand and suckle.

Treatment of iodine deficiencies may not be successful if the effects on the thyroid and other tissues have progressed too far. Prevention is far better and easier since iodine is readily available either in the inorganic form or the organic form, both of which are suitable as ration supplements. The use of iodized salt as part of the ration is the most economical means of supplementing rations where feeds are low in iodine. The intake necessary to prevent goiter development is on the order of 0.002 to 0.004 mg per kg of body weight. The addition of 0.0076% iodine to salt is the usual method of assuring adequate iodine intake. Iodine is stabilized in iodized salt, which prevents wastage.

Toxicity—Long-term intake of large excesses of iodine may disturb the utilization of iodine by the thyriod gland and result in toxicity.

Marked species differences exist in tolerance to, and toxicity symptoms of, high intakes of iodine. In horses, there may be hyperparathyroidism, commonly called big-head disease. In rabbits, there is increased prenatal mortality. In poultry, there is reduced egg production and poor hatchability. In mammals, abortions may occur.

Major Interrelationships:

Certain feeds (especially plants of the cabbage familly—cabbage, rapeseed, and turnips) contain goitrogens, which interfere with the use of thyroxin and may produce goiter. Fortunately, an adequate supply of iodine inhibits or prevents this.

Jointly occurring deficiencies of iodine and vitamin A are likely to cause a more severe thyroid disorder than lack of iodine alone.

Long-term chronic intake of large amounts of iodine reduces thyroid uptake of iodine.

Sources for Animals:

Rich feed sources—Alfalfa meal, blackstrap molasses, brewers' grains, corn gluten feed, distillers' solubles, fish meal and other marine by-products, forages grown on iodine-rich soils, meat meal with bone, oats and oat by-products, poultry by-product meal, sorghum grain, soybean meal, wheat and wheat by-products, whey.

Iodinated casein—thyroprotein—is sometimes fed to animals to increase production. Ducks may be fed thyroprotein to improve feathering and increase rate of gain. Dairy cattle and dairy goats have been fed thyroprotein to increase milk yields. In lactating animals, its effectiveness is somewhat limited—being most effective when it is fed for short periods during the declining phase of lactation. When it is used, feed intake will be increased. *CAUTION:* The use of thyroprotein must be very carefully monitored.

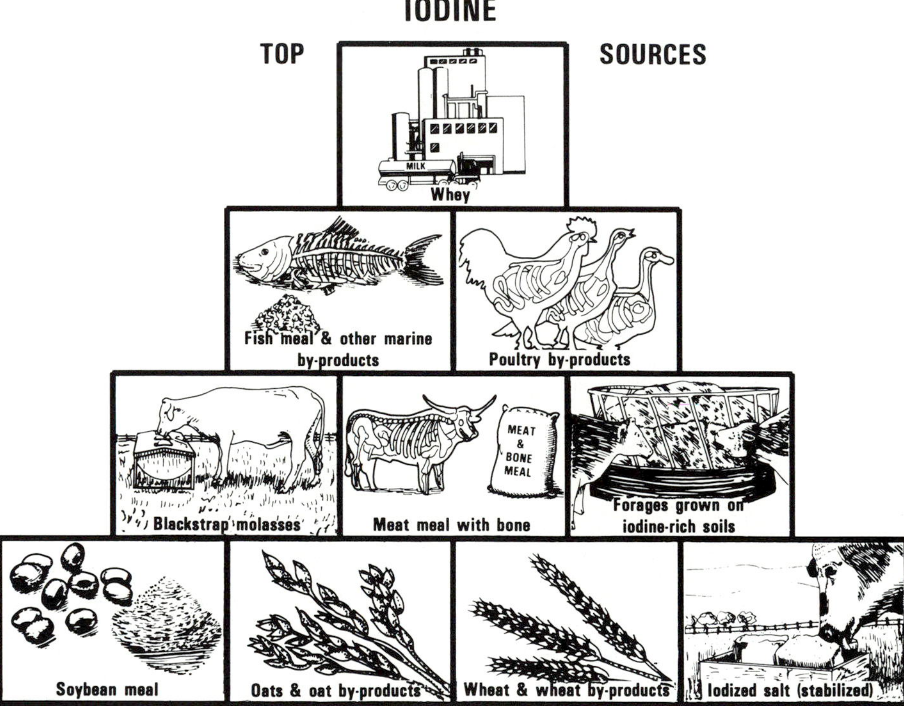

Fig. 4-52. Commonly used, top feed and supplement sources of iodine. No claim is made that all the top sources of iodine are listed. Neither are the sources ranked according to amount of iodine supplied nor is price considered.

Supplemental sources—Calcium iodate, ethylenediamine dehydroiodide (EDDI), stabilized iodized salt containing 0.01% potassium iodide (0.0076% iodine), kelp (dried), yeast (brewers', torula).

CAUTION: Iodized salt should never be used as a feed governor as too much iodine may be ingested and toxicity will result.

Comments:

Enlargement of the thyroid gland (goiter) is nature's way of trying to make enough thyroxin when there is insufficient iodine in the feed.

A mature animal body contains less than 0.00004% iodine.

Iodine deficiencies are worldwide, occurring wherever feeds and foods are grown on iodine-poor soils containing insufficient iodine to meet body needs. The highest incidence has been observed in the Alps, the Pyrenees, the Himalayas, the Thames Valley of England, certain regions of New Zealand, a number of Central and South American countries, and the Great Lakes and Pacific Northwest regions of the U.S.

(Continued)

TABLE 4-6 *(Continued)*

IRON (Fe)

Iron made the difference!

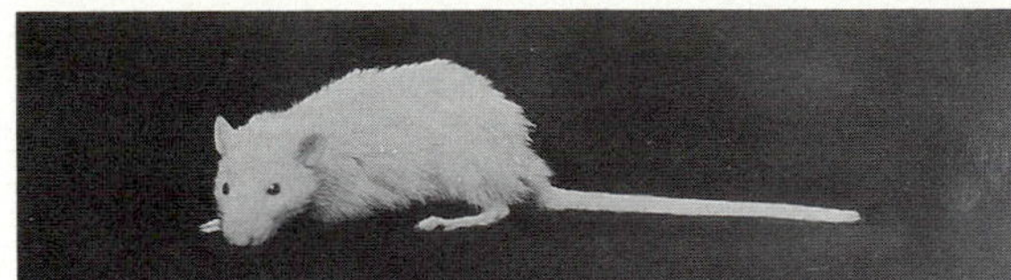

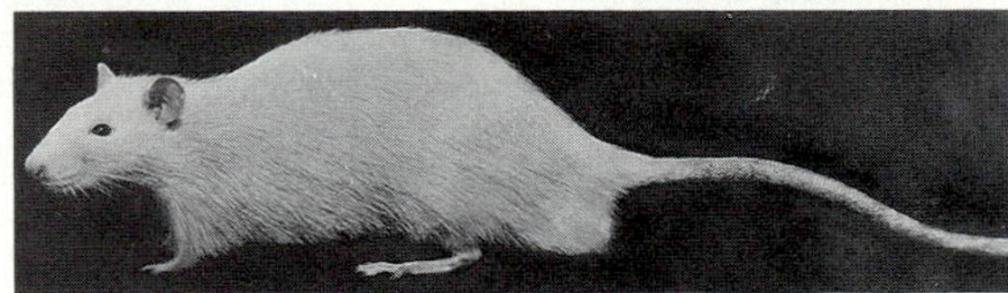

Fig. 4-53. *Left:* This rat did not have enough iron. It had pale ears and tail. At 8 months old, it weighed only 109 g. *Right:* This rat had plenty of iron. Its fur was sleek and its blood had 3 times as much hemoglobin as the rat on the left. Though only 5½ months old, it weighed 325 g. (Courtesy, USDA)

The body contains only about 0.004% iron of which about 70% is present in hemoglobin, the iron-containing pigment of the red blood cells that transports oxygen. Most of the remainder (about 30%) is present as reserve stores in the liver, spleen, and bone marrow. Despite the very small amount in the body, iron is one of the most important elements in nutrition and is of fundamental importance to life. It is a component of hemoglobin and myoglobin (muscle hemoglobin), and of the enzymes, cytochromes, catalases, and peroxidases. As part of these heme complexes and metalloenzymes, it serves important functions in oxygen transport and cellular respiration.

The red blood cells and the pigment within are broken down and replaced about every 120 days, but the liberated iron is not excreted; most of it is utilized to form new hemoglobin.

History:

Elemental iron has been known since prehistoric times. Although how early humans first learned to extract the element from its ores is still debated, scientists are fairly certain that early, highly-prized samples of iron were obtained from meteors. Several references to "the metal of heaven" (thought to be iron) have been found in ancient writings. By approximately 1200 B.C., iron was being obtained from its ores; this achievement marked the beginning of the *Iron Age.*

The early Greeks were aware of the health-imparting properties of iron. Ever since, it has been a favorite health tonic (for better or worse). As early as the 17th century in England, iron was found to be a specific treatment for anemia in humans. In 1867, Boussingault, the French chemist, obtained experimental evidence of the essential nature of iron in nutrition.

Absorption/Metabolism/Excretion:

Absorption—The greatest absorption of iron occurs in the upper part of the small intestine—in the duodenum and jejunum, although a small amount of absorption takes place from the stomach and throughout the whole of the small intestine. A value of 10% is usually taken as the total precentage of iron absorbed from mixed feeds.

It is noteworthy that there are two forms of feed iron—heme (organic) and nonheme (inorganic). Of the two, heme is absorbed from feed more efficiently than inorganic iron and is independent of vitamin C or iron-binding chelating agents. Although the proportion of heme iron in animal tissues varies, it amounts to about ⅓ of the total iron in all animal tissues—including meat, liver, poultry, and fish. The remaining ⅔ of the iron in animal tissues and all the iron of vegetable products are treated as nonheme iron.

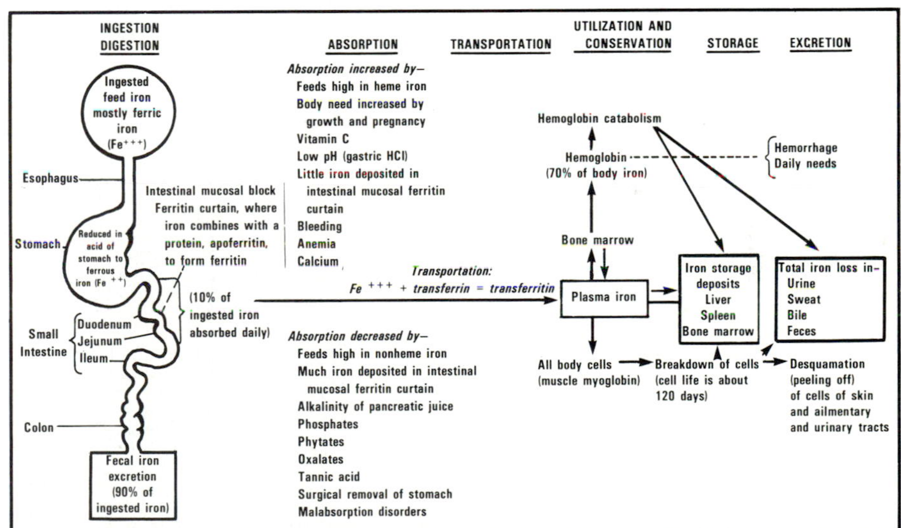

Fig. 4-54. Iron absorption, metabolism, and excretion. Note that healthy animals absorb only about 10% of the iron in feed, and that about 90% is excreted in the feces. Note, too, the factors that increase and decrease absorption.

Iron absorption is controlled by the intestinal mucosal block—the ferritin curtain, the exact mechanism of which is not known. It is increased (1) by feeds high in heme iron; (2) by body needs—increased by growth, menstruation, and pregnancy; (3) by the presence of vitamin C (ascorbic acid) and gastric HCl, which convert the iron from the ferric (Fe^{+++}) to the ferrous (Fe^{++}) state; (4) when little iron is deposited in the intestinal mucosal ferritin curtain; (5) when there is increased *hemoglobin* synthesis—for example, following *hemorrhages* (bleeding), or as a result of anemia or *hemopoetic* abnormalities; and (6) by the presence of calcium. Iron absorption is impaired (1) by feeds high in nonheme iron; (2) when much iron is deposited in the intestinal mucosal ferritin curtain; (3) by excess phosphates, phytates, oxalates, and tannic acid, all of which form insoluble compounds that are not readily absorbed.

Metabolism—Mucosal ferritin delivers ferrous iron (Fe^{++}) to the portal blood system. Thence iron is converted back to the ferric (Fe^{+++}) state by oxidation. As ferric iron, it combines with protein (transferrin), forming a combination known as transferritin. In this form, iron is transported to the bone marrow where it may be incorporated into newly synthesized hemoglobin molecules; or, alternatively, it may be stored in the liver, spleen, and bone marrow, where it combines with

(Continued)

TABLE 4-6 *(Continued)*

Iron (Fe) (Continued)

a protein and is deposited as ferritin.

Excretion—Absorbed iron is lost only by desquamation (shedding or peeling off of cells) from the alimentary, urinary, and respiratory tracts, and by skin and hair losses. The bulk of ingested iron (about 90%) is excreted in the feces. Only negligible amounts of iron are excreted in the urine. The body conserves and reuses iron once it has been absorbed.

Functions:

Iron (heme) combines with protein (globin) to make hemoglobin, the iron-containing compound in red blood cells; so, iron transports oxygen to all cells and tissues of the body as part of the hemoglobin molecule. Myoglobin—a pigmented compound similar to hemoglobin but smaller in size—is found in muscle. It contains iron which has the ability to accept oxygen in much the same manner as hemoglobin. In addition to its major role in the blood transport of oxygen, iron is also an essential part of a variety of enzymes. Some of these include cytochrome oxidase, catalase, and peroxidase. It also functions as an enzyme activator for the enzyme, arginase.

Deficiency Symptoms/Toxicity:

Deficiency Symptoms—Iron-deficiency anemia, characterized by (1) smaller than normal number of red cells and less than normal amount of hemoglobin, and (2) paleness of mucous membranes. Other deficiency symptoms include depraved appetite (pica), diarrhea, loss of appetite for normal feeds, labored breathing, rough hair coat, reduced rate of growth, and an increased susceptibility to stress and disease. Generally this condition is not a problem with animals given a normal ration, but it can be a problem for animals during their suckling period. Most newborn animals have sufficient iron to carry them through the suckling period if milk is the only nourishment they get during that period. The newborn pig, however, does not have such a reserve of iron and is very susceptible to anemia if it is not given supplemental iron. Premature births, multiple births, and low-iron diets fed to females prior to parturition increase the potential problem of iron deficiency of the newborn. Attempts to increase the iron transfer to developing fetuses or to increase the amount of iron in milk, by feeding the mothers feed high in iron, have not been successful.

Fig. 4-55. Suckling pig with nutritional anemia, caused by lack of iron, characterized by swelling about the head and paleness of the mucous membranes around the eyes and mouth. (Courtesy, College of Veterinary Medicine, University of Illinois, Urbana)

Toxicity—Free iron ions are very toxic. So, the iron molecule is always transported in combination with protein; 2 atoms of ferric iron are bound to 1 molecule of beta globulin protein called *transferrin*—and the combination forms *transferritin*. When the level of iron ions exceeds the binding capacity of the transferrin, iron toxemia occurs. Normally, the amount of iron in plasma is sufficient to bind only ⅓ of the transferrin—the remaining ⅔ represents the unbound reserve.

Major Interrelationships:

Iron is associated with hemoglobin and various enzymes. However, the production of hemoglobin in the body also requires protein, copper, vitamin C, vitamin B-6 (pyridoxine), folic acid, and vitamin B-12. An excess of iron in the diet can tie up phosphorus in an insoluble iron-phosphate complex, thereby creating a deficiency of phosphorus.

Copper is required for proper iron metabolism.

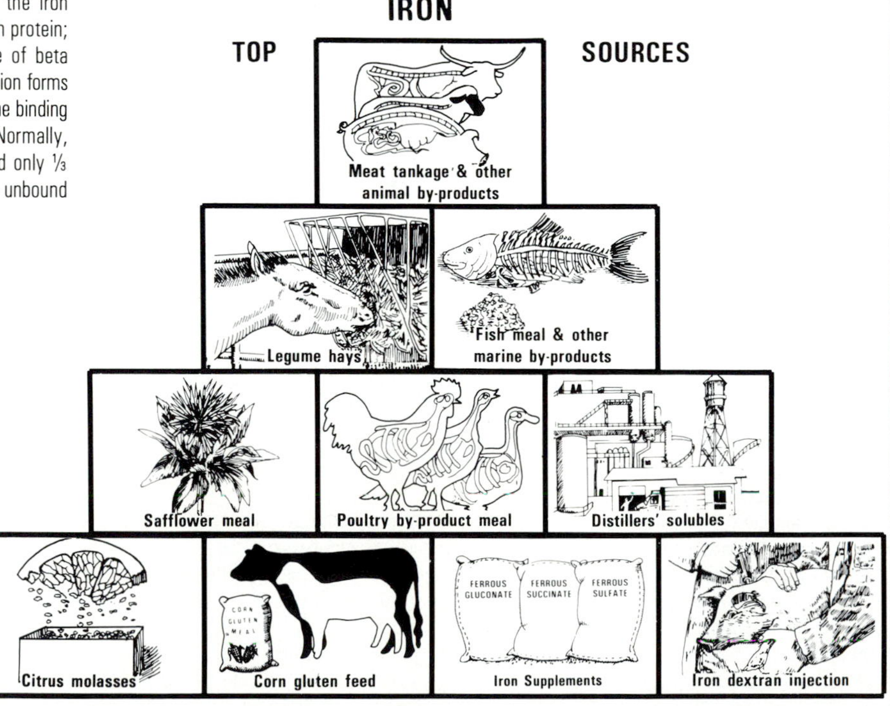

Fig. 4-56. Commonly used, top feed and supplement sources of iron. No claim is made that all the top sources of iron are listed. Neither are the sources ranked according to amount of iron supplied nor is price considered.

Sources for Animals:

Rich feed sources—Blackstrap molasses, blood meal, cereal grains, citrus molasses, copra meal, corn gluten feed, distillers' solubles, fish meal and other marine by-products,

(Continued)

TABLE 4–6 *(Continued)*

Iron (Fe) (Continued)

leafy portions of plants, legume hays, legume seeds, meat meal/tankage and other meat by-products, oil seed meals, pineapple cannery residue, poultry by-product meal, safflower meal, sugar beet tops, tomato pomace, vetch hay.

Supplemental sources—Ferrous gluconate, ferrous succinate, or ferrous sulfate given orally; iron dextran injection; iron fumarate; iron peptonate; trace mineralized salt.

Comments:

Young animals are born with a store of iron, largely in their livers. But, milk is low in iron. So, when young animals are continued on milk, particularly under confined conditions and with little or no supplemental feed, nutritional anemia will likely develop.

The iron requirements in pregnancy increase appreciably as the pregnancy progresses and the fetus increases in size. Likewise, considerable amounts of iron are required by laying hens, because eggs contain large quantities of this element. Extremely large losses of iron are incurred whenever there is extensive hemorrhaging.

MANGANESE (Mn)

Manganese made the difference!

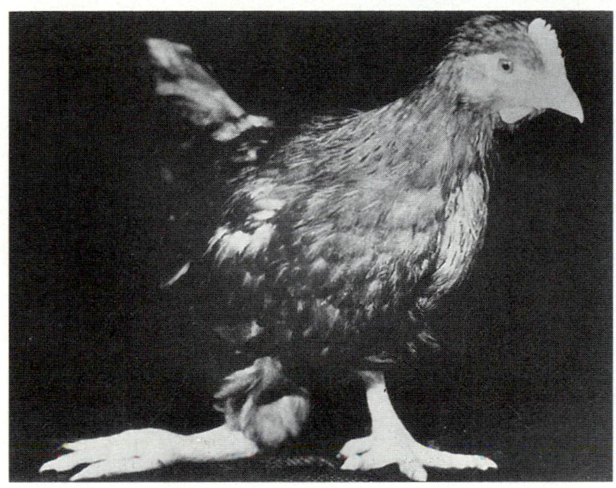

Fig. 4–57. *Left:* Manganese-deficient chicken, showing a bone disease known as perosis or slipped tendon. *Right:* Healthy chicken that received adequate manganese. (*Left photo:* Courtesy, Dept. of Poultry Science, Cornell University, Ithaca, N.Y. *Right photo:* Courtesy, *Poultry Tribune,* Mount Morris, Ill.)

In nutrition, manganese is an essential element for many animal species. It is an activator of several enzyme systems involved in protein and energy metabolism and in the formation of mucopolysaccharides.

History:

Manganese was first recognized as an element in 1774 by the famed Swedish chemist, Carl W. Scheele. It was isolated in the same year by his co-worker, Johann G. Ghan. The name *manganese* is a corrupted form of the Latin word for a form of magnetic stone, magnesia.

About 95% of the world's annual production of manganese is used by the steel industry. Also, manganese is essential for plant growth, and it is found in trace amounts in higher animals, where it activates many of the enzymes involved in metabolic processes.

In 1931, University of Wisconsin researchers reported that manganese is a dietary essential for growth of rats. Later, it was shown to be essential for poultry, swine, guinea pigs, cattle, and other animals.

Absorption/Metabolism/Excretion:

Absorption—Manganese is rather poorly absorbed, primarily in the small intestine. In the average diet about 45% of the ingested manganese is absorbed, and 55% is excreted in the feces. Absorption can be depressed when excessive amounts of calcium, phosphorus, or iron are consumed.

Metabolism—Following absorption, manganese is loosely bound to a protein and transported as transmanganin. The bones, and to a lesser extent the liver, muscles, and skin, serve as storage sites.

The concentration of manganese in the various body tissues is quite stable under normal conditions, a phenomenon attributed to well-controlled excretion rather than regulated absorption.

Excretion—Manganese is mainly eliminated from the body in the feces as a constituent of bile, but much of this is again reabsorbed, indicating an effective body conservation. Very little manganese is excreted in the urine.

Functions:

Formation of bone and growth of other connective tissues.

Blood clotting.

(Continued)

TABLE 4-6 *(Continued)*

Manganese (Mn) (Continued)

Insulin action.

Cholesterol synthesis.

Activator of various enzymes in the metabolism of carbohydrates, fats, proteins, and nucleic acids, (DNA and RNA).

Deficiency Symptoms/Toxicity:

Deficiency Symptoms—Poor growth; lameness, shortening and bowing of the legs, and enlarged joints.

In calves, knuckling over, congenital malformations in the newborn, abnormal formation of bone and cartilage, and impaired glucose tolerance.

In poultry, slipped tendons (perosis), retraction of the head of chicks, decreased egg production, and poor hatchability.

In mammals, delayed estrus, poor conception, and decreased litter size in females, and decreased libido and abnormal spermatogenesis in males.

In pigs, crooked legs and enlarged hocks.

In rabbits, crooked front legs.

Toxicity—Managanese is not toxic to animals in moderate excesses.

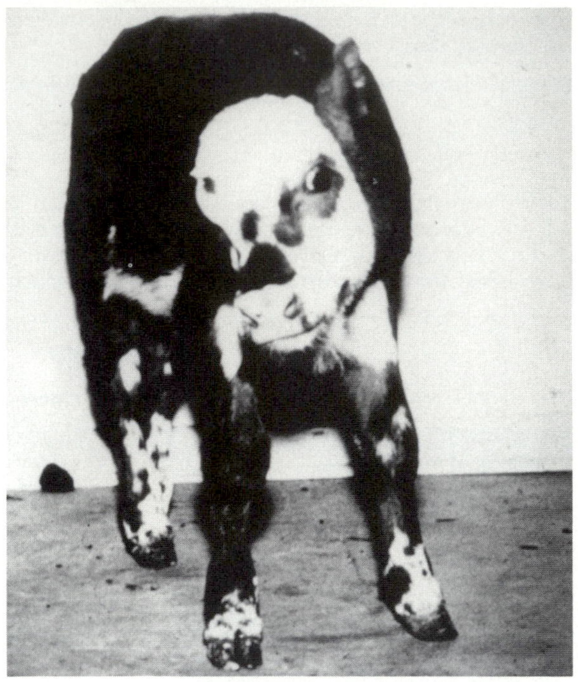

Fig. 4-58. "Crooked calf," resulting from manganese deficiency. Born to a cow receiving a ration with 15 ppm Mn. Typical symptoms: enlarged joints, *knuckled over* pasterns, twisted forelimbs, reduced length and breaking strength of humerus, diminished serum alkaline phosphatase, and low Mn levels in bone, liver, and other tissues. (Courtesy, The American Institute of Nutrition, Bethesda, Md.)

Major Interrelationships:

Manganese interacts with other nutrients. Excess calcium and phosphorus interfere with the absorption of manganese; the functions of manganese, copper, zinc, and iron may be interchangeable in certain enzyme systems; and manganese and vitamin K work together in the promotion of blood clotting.

Sources for Animals:

Rich feed sources—Blackstrap molasses, corn fodder, cottonseed hulls, crab cannery residue, grass/legume hays, rice/rice by-products, salt grass (fresh), shark meal, sorghum fodder, sorghum silage, wheat/wheat by-products.

Supplemental sources—Kelp (dried), manganese gluconate, rice polish, trace mineralized salt containing 0.25% (or more) manganese.

Comments:

The manganese content of plants is dependent on soil content. It is noteworthy, however, that plants grown on alkaline soils may be abnormally low in manganese.

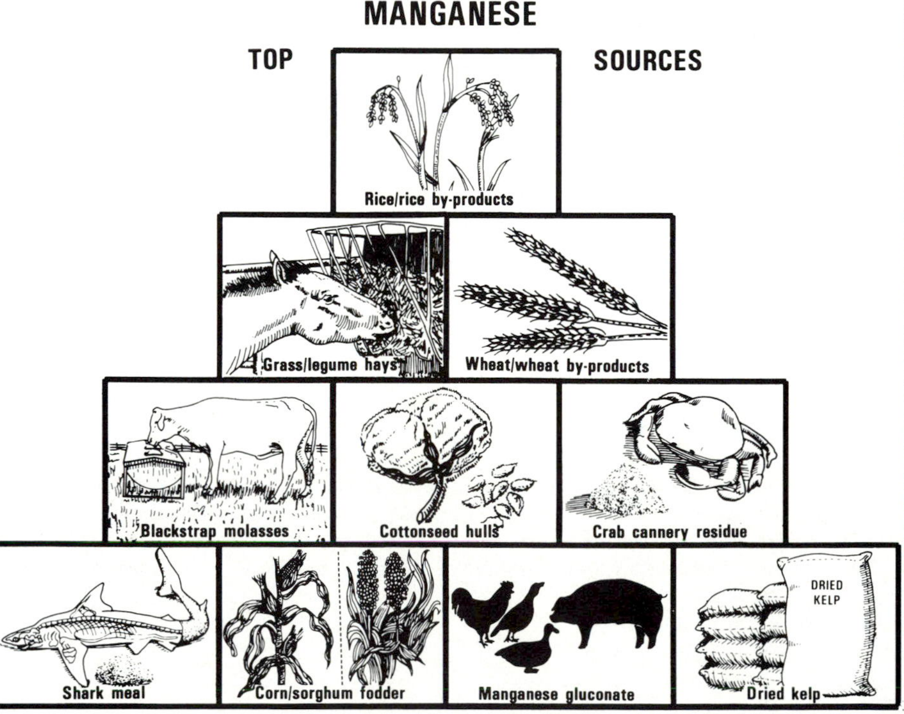

Fig. 4-59. Commonly used, top feed and supplement sources of manganese. No claim is made that all the top sources of manganese are listed. Neither are the sources ranked according to amount of manganese supplied nor is price considered.

(Continued)

TABLE 4-6 *(Continued)*

MOLYBDENUM (Mo)

Molybdenum made the difference!

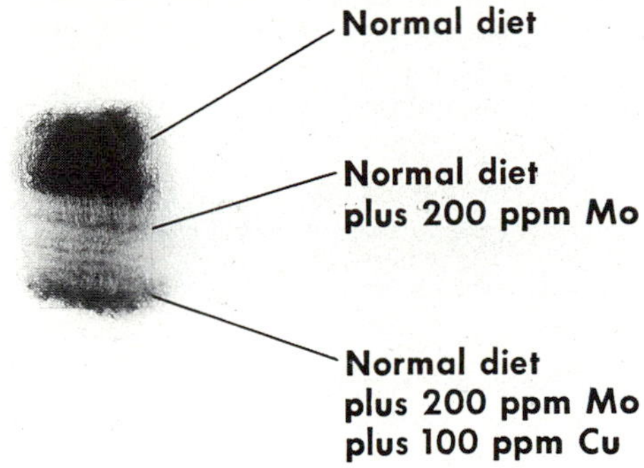

Normal diet

Normal diet
plus 200 ppm Mo

Normal diet
plus 200 ppm Mo
plus 100 ppm Cu

Fig. 4-60. This shows the interrelationship of molybdenum and copper on a black sheep. Note effect of (1) "normal diet" on outer fleece growth (top black band); and (2) "normal diet plus 200 ppm Mo" (center white band); and (3) "normal diet plus 200 ppm Mo plus 100 ppm Cu"—new growth (bottom black band). (Courtesy, University of California, Davis)

The essential role of molybdenum in plants is well known. In legumes, molybdenum aids in nitrogen fixation. Evidence that molybdenum is an essential trace element is based on the fact that it is part of the molecular structure of two enzymes, xanthine oxidase (involved in the oxidation of xanthine to uric acid) and aldehyde oxidase (involved in the oxidation of aldehydes to carboxylic acids), and that diets low in molybdenum adversely affect growth in small animals.

Molybdenum is important in the ration in small amounts (about 2 ppm), but it is detrimental in excessive quantities (about 20 ppm).

History:

In 1778, Karl Scheele of Sweden recognized molybdenite as a distinct ore of a new element. Then, in 1782, P. J. Hjelm obtained the metal by reducing the oxide with carbon and called it molybdenum. The name molybdenum is derived from the Greek *molybdos,* meaning lead.

Molybdenum was long known as essential for the growth of all higher plants. Then, in 1953, it was found in an essential enzyme, xanthine dehydrogenase.

Absorption/Metabolism/Excretion:

Molybdenum is readily absorbed as molybdate in the small intestine, although some absorption occurs throughout the intestinal tract. The rate of absorption can be dramatically depressed by the presence of sulfates in feeds. There is little retention of this element except in the liver, adrenals, kidneys, and bones. Molybdenum is excreted rapidly in the urine, and in limited amounts via the bile and feces.

Functions:

Molybdenum functions as a component of three different enzyme systems which are involved in the matabolism of carbohydrates, fats, proteins, sulfur-containing amino acids, nucleic acids (DNA and RNA), and iron.

Molybdenum is a component of xanthine oxidase, an enzyme which is essential in the formation of uric acid. Iron is also an important component of this enzyme. Excess nitrogen in birds is excreted in the form of uric acid, which is analagous to urea in mammals. Hence, birds require considerably more molybdenum than do mammals.

A growth-promoting effect of molybdenum has been observed in growing lambs. Although the basis for this action is not known, it has been speculated that molybdenum has a stimulating effect on cellulose-degrading microorganisms in the rumen.

Molybdenum is a component of the enamel of teeth.

Deficiency Symptoms/Toxicity:

Deficiency symptoms—Naturally occurring deficiency in animals is not known, *unless* utilization of the mineral is interfered with by excesses of copper and/or sulfate.

Molybdenum-deficient animals are especially susceptible to the toxic effects of bisulfite, characterized by breathing difficulties and neurological disorders.

Toxicity—Toxicities of molybdenum are of greater practical concern than deficiencies.

Severe molybdenum toxicity in animals (molybdenosis), particularly cattle, occurs throughout the world wherever pastures are grown on high-molybdenum soils. In England, molybdenum toxicity in cattle is referred to as *teartness.* The symptoms include severe scours, loss of weight, decreased production, fading of hair color, and other symptoms of copper deficiency.

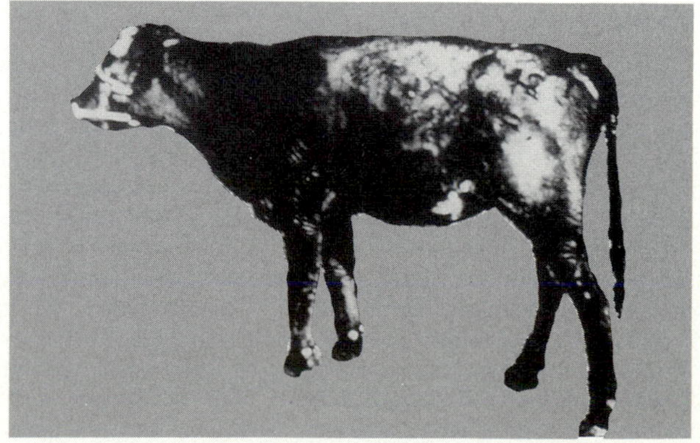

Fig. 4-61. Molybdenum toxicity. This shows an Angus steer with bleached hair caused by high molybdenum. (Courtesy, American Institute of Nutrition, Bethesda, Md.)

(Continued)

TABLE 4-6 *(Continued)*

Molybdenum (Mo) (Continued)

Major Interrelationships:

The utilization of molybdenum is reduced by excess copper, sulfate, and tungsten (an environmental contaminant). Also, even a moderate excess of molybdenum causes significant urinary loss of copper; and toxicity due to excess copper is counteracted by molybdenum.

Molybdenum is related to uric acid formation in poultry and microbial action in ruminants.

Molybdenum as a toxic mineral affects cattle and sheep grazing pastures grown on soils high in molybdenum content.

Sources for Animals:

The concentration of molybdenum in feed varies considerably, depending on the soil in which it is grown.

High feed sources—Most of the dietary molybdenum of animals is derived from the following feeds: grass/legume pastures and hays, meat by-products, whole grains (barley, corn, oats, sorghum grain, wheat), wheat germ, yeast.

Supplemental sources—No molybdenum supplementation of normal rations is necessary.

Comments:

In cattle, a relationship exists between molybdenum, copper, and sulfur. Excess molybdenum will cause copper deficiency. However, when the sulfate content of the diet is increased, the symptoms of toxicity are avoided inasmuch as the excretion of molybdenum is increased.

SELENIUM (Se)

Selenium made the difference!

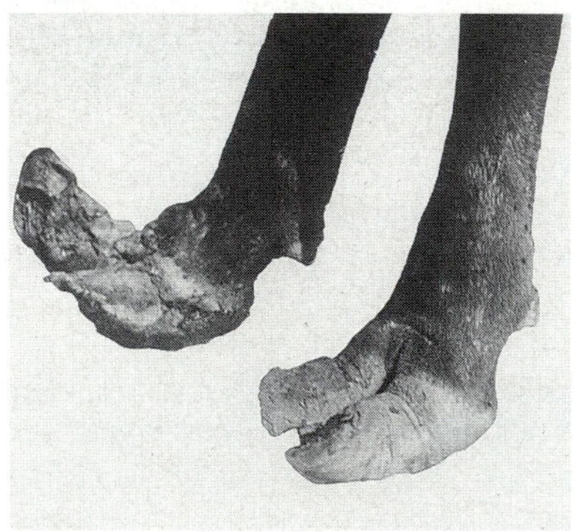

Fig. 4-62. *Left:* Hind feet of cow afflicted with selenium toxicity. *Right:* Normal hind feet of cow. (Courtesy, The American Institute of Nutrition, Bethesda, Md.)

Selenium is named after *selene,* the moon goddess. This comparatively rare element has recently been found to be essential. Fortunately, it is needed only in minute amounts, because (1) only traces are present in most feeds, and (2) poisoning may result when the dietary level of this element is 0.0003% or greater. However, it seems that adequate dietary selenium is one of the keys to the maintenance of health under stressful conditions such as (1) prematurity at birth, (2) protein-energy malnutrition, and (3) the tissue disorders which accompany aging.

History:

In 1817, the Swedish chemist Berzelius discovered the element selenium while testing the residue which remained after sulfur had been burned to make sulfuric acid.

During the 1940s, German scientists tested the European type of brewers' yeast for use as a protein supplement and found that it sometimes produced a liver disease in rats which was prevented by feeding wheat germ or other sources of vitamin E. Then, in 1951, a German medical researcher, Schwarz, who at the time was a visiting scientist at the U.S. National Institutes of Health (NIH), discovered that the American type of brewers' yeast contained an unidentified *Factor 3* which apparently acted along with vitamin E and sulfur-containing amino acids in protecting the liver against damage due to certain types of diets. Schwarz stayed in the U.S. to continue his research, and, in 1957, he and co-worker, Foltz, reported that Factor 3 contained selenium.

Other studies by the group at NIH, and by Scott and his co-workers in the Department of Poultry Husbandry at Cornell, showed that the addition of selenium salts to the diets of chicks prevented certain disorders which resulted from vitamin E deficiencies. The pace of discovery quickened as Oldfield (coauthor of this book) and other researchers showed that selenium protected calves and lambs against white muscle disease (nutritional muscular dystrophy). Finally, in 1973, Rotruck and his co-workers at the University of Wisconsin reported that selenium acted as a cofactor for a recently-discovered enzyme (glutathione peroxidase) which breaks down toxic peroxides, most of which are formed from the oxidation of polyunsaturated fats. Hence, the link between selenium and vitamin E was shown; with selenium participating in the breaking down of these highly toxic compounds, and with vitamin E preventing their formation. Recent research suggests that there are other roles for this essential element; so, new and exciting chapters may be added to the selenium story in the years ahead.

(Continued)

TABLE 4-6 *(Continued)*

Selenium (Se) (Continued)

Absorption/Metabolism/Excretion:

Ingested selenium is absorbed in the intestine, mainly in the duodenum. Thence, it is bound to a protein and transported in the blood to the tissues, where it is incorporated into tissue protein as selenocysteine and selenomethionine; in the latter process, selenium replaces the sulfur in the amino acids cysteine and methionine. Excretion of selenium is largely by way of the kidneys, although small amounts are excreted in the feces and in sweat.

Many factors may affect dietary selenium; some enhancing it, others reducing it.

Functions:

Component of the enzyme glutathione peroxidase, the metabolic role of which is to protect against oxidation of polyunsaturated fatty acids and resultant tissue damage.

Protecting tissues from certain poisonous substances, such as arsenic, cadmium, and mercury.

Interrelation with vitamin E—they spare each other, and with the sulfur-containing amino acids. (Also see vitamin E section on "Functions.")

Deficiency Symptoms/Toxicity:

Deficiency symptoms—In calves, the condition is called *white muscle disease.* In lambs, it is called *stiff lamb disease.*

In pigs, the condition is called *liver necrosis.*

In poultry, young chickens develop *exudative diathesis,* which is characterized by the walls of capillaries becoming highly permeable.

(Also see Chapter 5, Table 5-1, Nutritional Diseases and Ailments—White Muscle Disease.)

The selenium content of plant and animal products is affected by the selenium content of the soil and animal feed, respectively. The best means of detecting selenium-deficient pastures or other feeds is to draw a blood sample and have it analyzed by a laboratory experienced in testing for this element. In cattle, blood selenium concentrations above 0.08 ppm are generally regarded as adequate, and less than 0.04 ppm as deficient.

Fig. 4-63. Selenium toxicity in lamb, showing congenital deformity traceable to selenium injury during fetal development. (Courtesy, The University of Wyoming, Cheyenne)

Toxicity—Animals consuming forage or grain produced on seleniferous soils develop blind staggers or alkali disease, characterized by emaciation, loss of hair, soreness and sloughing of hoofs, lameness, anemia, excess salivation, grinding of the teeth, blindness, paralysis, and death.

In poultry and other birds, egg production and hatchability are reduced and deformities are common, including lack of eyes and deformed wings and feet.

Soils containing more than 0.5 ppm of selenium are potentially dangerous for livestock feed production, and diets containing more than 10 ppm will produce observable toxicity problems. Arsenic added to the diet will reduce some of the toxic effects of selenium, but care must be exercised in adding arsenic, since arsenic itself is toxic.

Major Interrelationships:

The major selenium interrelationships are:

1. The functions of selenium are closely related to those of vitamin E and the sulfur-containing amino acids.
2. Selenium protects against the toxic effects of arsenic, cadmium, copper, mercury, and silver. Likewise, these elements counteract the toxic effects of selenium.
3. A diet high in protein or high in sulfate provides some protection against selenium poisoning.
4. Diets rich in polyunsaturated fatty acids but poor in vitamin E may raise the requirements for selenium. Details relative to the selenium-vitamin E relationship follow:

During the 1950s, an interrelationship between selenium and vitamin E was established. It was found that selenium prevented exudative diathesis in vitamin E-deficient chicks and liver necrosis in vitamin E-deficient rats. Subsequent research demonstrated that both selenium and vitamin E protect the cell from the detrimental effects of peroxidation, but each takes a distinctly different approach to the problem. Selenium functions throughout the cytoplasm to destroy peroxides, while vitamin E is present in the membrane components of the cell and prevents peroxide formation. This explains why the biological need for each nutrient can be offset, at least partially although not totally, by the other.

(Continued)

TABLE 4-6 *(Continued)*

Selenium (Se) (Continued)

Sources for Animals:

Rich feed sources—Barley malt sprouts, blackstrap molasses, blood meal, cereal grains (barley, oats, rye, sorghum, wheat), corn distillers' grains, corn distillers' solubles, corn germ meal, corn gluten meal, feather (poultry) meal, fish meal and other marine by-products, meat by-products, oilseed meals, peas (field), poultry by-product meal, rape meal, wheat bran, wheat gluten, wheat grain screenings, wheat middlings, wheat red dog, wheat shorts.

Supplemental sources—Sodium selenate, sodium selenite.

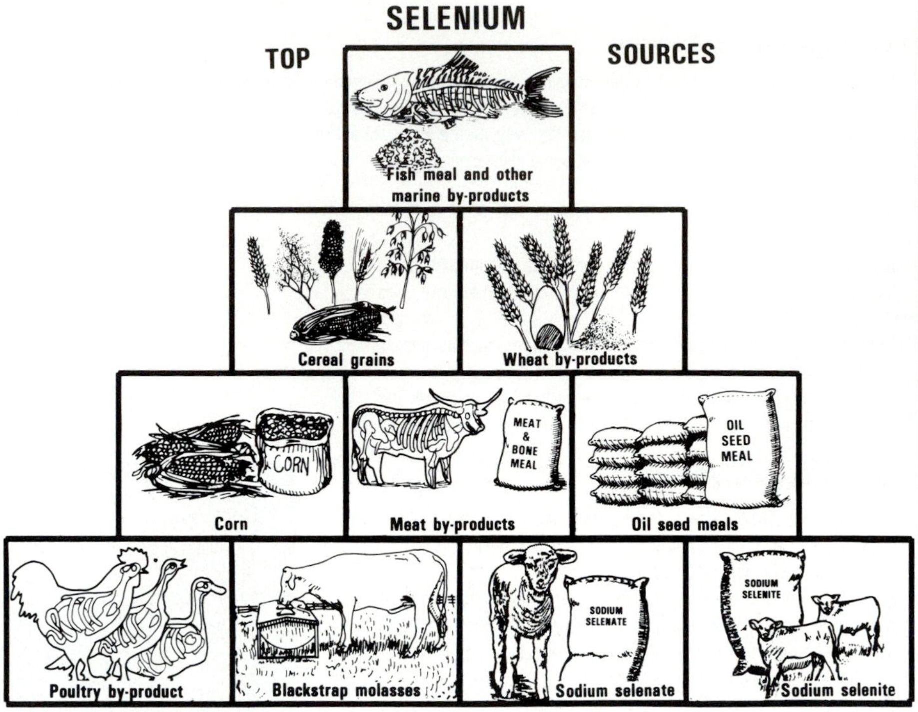

Fig. 4-64. Commonly used, top feed and supplement sources of selenium. No claim is made that all the top sources of selenium are listed. Neither are the sources ranked according to amount of selenium supplied nor is price considered.

Comments:

As a result of extensive experiments, the Food and Drug Administration (FDA) approved the use of selenium for livestock as follows:

1. In 1974, FDA approved the addition of selenium as either sodium selenite or sodium selenate at the rate of 0.1 part per million (ppm) to complete feed (total ration) for swine, growing chickens to 16 weeks of age, breeder hens producing hatching eggs, and nonfood animals; and at the rate of 0.2 ppm in complete rations for turkeys.
2. In 1979, FDA amended the food additive regulations for selenium to include beef cattle, dairy cattle, and sheep; with the addition of selenium approved at the rate of 0.1 ppm of complete feed, along with stipulations on how to incorporate it into the ration.
3. In 1987, FDA provided for (1) an increase in the maximum allowance of selenium in complete feeds for cattle (beef and dairy), sheep, swine, chickens, turkeys, and ducks from 0.1 ppm to 0.3 ppm; (2) a proportional increase in the limit feeding (feed supplements and salt-mineral mixtures) consumption rates for sheep and beef cattle to 0.7 mg and 3 mg per head per day, respectively; and (3) a proportional increase in the selenium fortification levels for salt-mineral mixtures for sheep and cattle to 90 and 120 ppm, respectively.

Since excess selenium is highly toxic, regulations and guidelines should be carefully followed.

SILICON (Si)

Silicon is one of the most abundant elements on Earth—present in large amounts in soils and plants. The highest concentrations in animal tissues are found in the skin and its appendages; for example, the ash of feathers is more than 70% silicon.

History:

In 1823, Jons J. Berzelius, the Swedish chemist, discovered the element. The name *silicon* is derived from the Latin *silex* or *silicis,* meaning *flint*—appropriately indicating its hardness.

In 1972, Dr. E. Carlisle, nutritionist at the University of California, Los Angeles, reported that the trace element silicon (Si) is needed in microgram amounts for normal growth and skeletal development of chicks and rats.

Absorption/Metabolism/Excretion:

Silicon is absorbed readily. Even over a wide range of intake, the concentration in the blood remains relatively constant—not more than 1 mg per 100 ml. It is excreted easily—via both the feces and urine.

Functions:

Necessary for normal growth and skeletal development of the chick and rat.

Deficiency Symptoms/Toxicity:

Deficiency Symptoms—Deficiencies in chicks and rats are characterized by growth retardation and skeletal alterations and deformities, especially of the skull.

Toxicity—High levels of silicon in the diet of farm animals may cause the silicon of the urine to be deposited in the kidneys, bladder, or urethra to form calculi. However, the condition is not simply due to high dietary intakes of silicon as attempts to produce calculi by adding silicon to rations of ruminants have not been successful.

(Continued)

TABLE 4-6 *(Continued)*

Silicon (Si) (Continued)

Major Interrelationships:

Silicon appears to take part in the synthesis of mucopolysaccharides and is a component of the mucopolysaccharide-protein complexes of connective tissue.

Sources for Animals:

Rich feed sources—Meat by-products, whole grains.

Supplemental sources—Not needed.

Comments:

On purified diets, the addition of Si has increased the growth rate of chicks and rats.

ZINC (Zn)

Zinc made the difference!

Fig. 4-65. *Left:* Zinc-deficient pig. *Right:* Control pig that received adequate zinc in the ration. (Courtesy, Dept. of Animal Science, University of Illinois, Urbana)

Zinc is an essential element for animals.

It is widely distributed throughout the body, but the highest concentrations are found in the skin (the skin contains 20% of the total body zinc), hair, hoofs, and eyes. Traces occur in the liver, bones, and blood. Also, it is a constituent of enzymes involved in most major metabolic pathways.

History:

Zinc was first shown to be biologically important more than 100 years ago when it was found to be needed for the growth of certain bacteria. In the 1920s, it was demonstrated to be required for the growth of experimental rats.

Absorption/Metabolism/Excretion:

Absorption—Zinc is poorly absorbed; less than 10% of dietary zinc is taken into the body, primarily in the duodenum. It appears that metallic zinc and zinc in its carbonate, sulfate, and oxide forms are all absorbed equally well. Large amounts of calcium, phytic acid, or copper inhibit zinc absorption. Cadmium appears to be a zinc antimetabolite.

Metabolism—After zinc is absorbed in the small intestine, it combines with plasma proteins for transport to the tissues. Relatively large amounts of zinc are deposited in bones, but these stores do not move into rapid equilibrium with the rest of the organism. The body pool of biologically available zinc appears to be small and to have a rapid turnover, as evidenced by the prompt appearance of deficiency signs in experimental animals.

Excretion—Most of the zinc derived from metabolic processes is excreted in the intestine—in pancreatic, intestinal, and bile secretions. Only small amounts are excreted in the urine.

Functions:

Needed for normal skin, bones, hair, and feathers.

As a component of several different enzyme systems which are involved in digestion and respiration.

Required for the transfer of carbon dioxide in red blood cells; for proper calcification of bones; for the synthesis and metabolism of proteins and nucleic acids; for the development and functioning of reproductive organs; for wound and burn healing; and for the functioning of insulin.

Zinc imparts *bloom* to the hair coat.

Deficiency Symptoms/Toxicity:

Deficiency Symptoms—Loss of appetite and stunted growth, bone problems, improper development of the male gonads, severe delays in wound healing, and an impairment of glucose tolerance.

(Continued)

TABLE 4-6 *(Continued)*

Zinc (Zn) (Continued)

Poor hair or feather development; slipping of wool.
Rough and thickened skin in various animal species, known as parakeratosis.
In pregnant animals, experimental zinc deficiency has resulted in malformation and behavioral disturbances in offspring.

Toxicity—Toxicity of zinc is characterized by anemia, depressed growth, stiffness, hemorrhages in bone joints, bone resorption, depraved appetite, and, in severe cases, death. The anemia appears to result from an interference with iron and copper utilization because addition of these two elements can overcome the anemia caused by excessive zinc.

Fig. 4-66. Zinc deficiency in a calf, showing parakeratosis on the head and neck. (Courtesy, The American Institute of Nutrition, Bethesda, Md.)

Major Interrelationships:

Zinc is involved in many relationships: in the metabolism of carbohydrates, fats, proteins, and nucleic acids; in interference with the utilization of copper, iron, and other trace minerals, when there are excess dietary levels of zinc; in protection against the toxic effects of cadmium, when there is ample dietary zinc; in reduced absorption, when there are high dietary levels of calcium, phosphorus, and copper Also, there is indication that added dietary zinc may partially alleviate lead toxicity in some species.

Sources for Animals:

Rich feed sources—Algae, brewers' grains, cereal grains, citrus molasses, corn germ meal, corn gluten feed, cotton seeds, distillers solubles, fish meal and other marine by-products, meat by-products, poultry by-product meal, poultry feather meal, oatmeal, peanut meal, safflower meal, sesame seeds, sunflower seeds, wheat by-products. Plant levels of zinc are highly variable, depending on growing conditions.

Supplemental sources—Wheat germ, yeast (torula), zinc carbonate, zinc sulfate.

Comments:

Zinc availability is adversely affected by phytates (found in whole grains and beans), high calcium, oxalates, high fiber, copper, and EDTA (an additive).

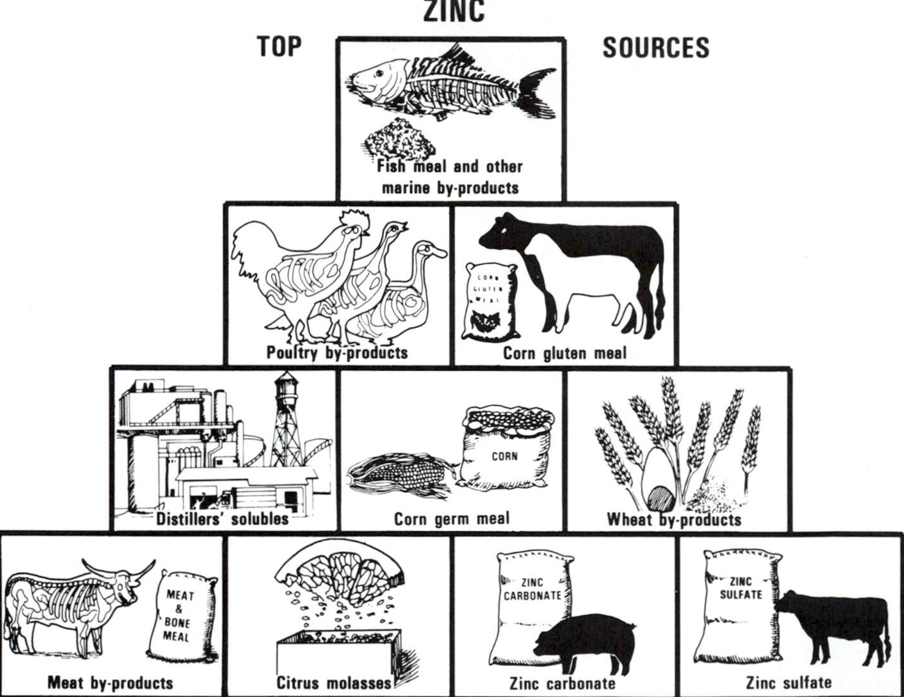

Fig. 4-67. Commonly used, top feed and supplement sources of zinc. No claim is made that all the top sources of zinc are listed. Neither are the sources ranked according to amount of zinc supplied nor is price considered.

VITAMINS

Fig. 4-68. This shows an artist's dramatic impression of how British sailors, who ate little except salt meat and biscuits when on long voyages 200 years ago, suddenly collapsed and died of scurvy. (Reproduced with permission of *Nutrition Today*, Annapolis, Md.)

For proper physiological functions, the animal body requires some 40 to 50 dietary essentials, of which 16 are vitamins.

Throughout history, vitamin deficiencies have been a major cause of disease, morbidity, and death. Pellagra, scurvy, and beriberi decimated armies, ships' crews, and nations; they even reshaped the course of history. The importance of dietary factors in the genesis of diseases became recognized in the 18th century. But the significance of these observations was not fully understood until early in the 20th century, when scientists found it desirable in many types of investigations to use the biological approach—the use of laboratory animals (largely white, albino rats and mice; guinea pigs; and chicks) fed purified diets using pure protein such as casein or albumin, pure fat such as lard, and pure carbohydrate such as dextrin, plus minerals to supplement chemical analyses, in measuring the value of food. These diets were composed of relatively pure nutrients (proteins, carbohydrates, fats, and minerals) from which the unidentified factors were largely excluded. With these purified diets, all researchers shared a common experience—the animals not only failed to thrive, but they even failed to survive if the investigations were continued for any length of time. At first, many investigators explained such failures on the basis of unpalatability and monotony of diets. Finally, it was realized that these purified diets were lacking in certain factors, minute in amounts, the identities of which were unknown to science. These factors were essential for the efficient utilization of the main ingredients of the food and for the maintenance of health and life itself. The discovery, synthesis, and commercial production of vitamins followed. With these developments, the vitamin era of science was ushered in and the modern approach to nutrition was born.

History/Discovery of Vitamins

Until the early 1900s, if a diet contained proteins, fats, carbohydrates, minerals, and water, it was considered to be complete. True enough, the disease known as beriberi made its appearance in the rice-eating districts of the Orient when milling machinery was introduced from the West, having been known to the Chinese as early as 2600 B.C.; and scurvy was long known to occur among sailors fed salt meat and biscuits. However, for centuries these diseases were thought to be due to toxic substances in the digestive tract caused by pathogenic organisms, rather than food deficiencies; and more time elapsed before the discovery of vitamins.

Largely through the trial-and-error method, it was discovered that specific foods were helpful in the treatment of certain of these maladies. Hippocrates (460–377 B.C.), the Greek medical doctor, advocated liver as a cure for night blindness 400 years before the birth of Christ. At a very early date, the Chinese also used a concoction rich in vitamin A as a remedy for night blindness; and cod-liver oil was used in treating or preventing rickets long before anything was known about the cause of the disease. In 1747, James Lind, a British naval surgeon, in a study involving 12 sailors with scurvy on board the ship "Salisbury," showed that the juice of citrus fruits was a preventive and cure for the disease. Nicholas Lunin, as early as 1881, while a student of von Bunge at the University of Dorpat, had come to the conclusion that certain foods, such as milk, contain, beside the principal ingredients, small quantities of unknown substances essential to life. In 1882, Kanehiro Takaki, Director-General of the Japanese Navy, greatly reduced the number of beriberi cases among naval crews by adding meat and evaporated milk to their diet of rice. In 1897, Christiaan Eijkman, a Dutch medical officer, working in Java, had satisfied himself that the disease beriberi was due to the continued consumption of a diet of polished rice.

In 1912, Dr. Casimir Funk, a 28-year-old Polish biochemist working in London, coined the word *vitamine*. Funk postulated, as others had before him, that beriberi, scurvy, pellagra, and possibly rickets, were caused by a lack in the diet of "special substances which are of the nature of organic bases, which we will call vitamines." Presumably, the name vitamines alluded to the fact that they were vital to life, and that they were chemically of the nature of amines (nitrogen-containing). The name caught the popular fancy and persisted, despite the fact that the chemical assumption was not common to all such vital substances, with the result that the "e" was dropped in 1920; hence, the word *vitamin*. In 1922, Funk's book entitled *The Vitamins* was published.

The actual existence of vitamins has been known only since 1912, when Funk, the Polish biochemist working in London, coined the word. But it was much later before it was possible to see or touch any of them in pure form. Previously, they were merely mysterious invisible *little things* known only by their effects. In fact, most of the present fundamental knowledge relative to the vitamin content of both human foods and animal feeds was obtained through measuring their potency in promoting growth or in curing certain disease conditions in animals—a most difficult and tedious method. For the most part, small laboratory animals were used, especially rats, guinea pigs, pigeons, and chicks.

Today, there are 16 known vitamins. Additionally, there are at least nine other vitaminlike substances that have been proposed. But it is unlikely that all of them are distinct essentials. Yet, the probability that there are still undiscovered vitamins is recognized.

A chronological summary of the discovery/isolation and synthesis of the various vitamins is given in Table 4–7.

TABLE 4–7
CHRONOLOGY OF THE DISCOVERY/ISOLATION AND SYNTHESIS OF VITAMINS

Year	Discovery/Isolation	Synthesis
1849[1]	Choline	
1866–67		Choline
1913	Vitamin A[2]	
1926	Thiamin (B–1)	
1929	Vitamin K[3]	
1931	Vitamin A[2]	
1932	Vitamin C (Ascorbic acid)	
	Vitamin D_2	
1933	Riboflavin (B–2)	Vitamin C (Ascorbic acid)
1935		Riboflavin (B–2)
1936	Biotin	Thiamin (B–1)
	Vitamin E	
1937	Niacin (Nicotinic acid)	
1938	Vitamin B–6 (pyridoxine)	Vitamin E
1939	Pantothenic acid (B–3)	Vitamin B–6 (pyridoxine)
	Vitamin K[3]	Vitamin K[3]
1940		Pantothenic acid (B–3)
1943		Biotin
1945	Folacin (Folic acid)	Folacin (Folic acid)
1947		Vitamin A[2]
1948	Vitamin B–12	
1952		Vitamin D_3
1955		Vitamin B–12

[1]In 1844 and 1846, Gobley isolated a substance from egg yolk, which he called lecithin. In 1849, Strecker isolated a compound from hog bile, to which he subsequently (in 1862) applied the name *choline*.

[2]A more detailed chronology of vitamin A is: in 1913, it was discovered, independently, by McCollum and Davis of the University of Wisconsin, and Osborne and Mendel of the Connecticut Experiment Station; in 1931, its chemical formula was determined by P. Karrer, a Swiss researcher; and, in 1947, it was synthesized by Isler, working in Switzerland.

[3]Vitamin K was discovered in 1929, isolated in 1939, and synthesized in 1939.

Definitions/Classification/Functions

Vitamins are organic substances that are essential in small amounts for the health, growth, reproduction, and maintenance of one or more animal species, which must be included in the diet since they either (1) cannot be synthesized at all, or (2) cannot be synthesized in sufficient quantities in the body.

Each vitamin performs a specific function; hence, one cannot replace, or act for, another. In general, the body cannot synthesize them, at least in large enough amounts to meet its needs.

There is no universal agreement on the nomenclature of the vitamins. But the modern tendency is to use the chemical name, particularly in describing members of the B complex. In this book, the most common designations are used.

Today, vitamins are generally classed as (1) fat-soluble, (2) water-soluble, including the vitamin B complex, and (3) vitaminlike substances.

• **Fat-Soluble *vs* Water-Soluble Vitamins**—Many phenomena of vitamin nutrition are related to solubility—vitamins are soluble in either fat or water. Consequently, it is important that nutritionists (1) be well informed about solubility differences in vitamins and (2) make use of such differences in programs and practices. Based on solubility, vitamins may be grouped as follows:

The Fat-Soluble Vitamins	The Water-Soluble Vitamins
Vitamin A	Biotin
Vitamin D	Choline
Vitamin E	Folacin (folic acid)
Vitamin K	Inositol
	Niacin (nicotinic acid, nicotinamide)
	Pantothenic acid (vitamin B–3)
	Para-aminobenzoic acid (PABA)
	Riboflavin (vitamin B–2)
	Thiamin (vitamin B–1)
	Vitamin B–6 (pyridoxine, pyridoxal, pyridoxamine)
	Vitamin B–12 (cobalamins)
	Vitamin C (ascorbic acid, dehydro-ascorbic acid)

It is noteworthy that vitamin C is the only member of the water-soluble group that is not a member of the B family.

The two groups of vitamins exhibit the following several differences that distinguish them both chemically and biologically:

The fat-soluble vitamins contain only carbon, hydrogen, and oxygen, whereas the water-soluble B vitamins contain these three elements plus nitrogen and occasionally sulfur.

Vitamins originate primarily in plant tissues; with the exceptions of vitamins C and D, they are present in the animal tissues only if an animal consumes feed containing them or harbors microorganisms that synthesize them. Fat-soluble vitamins can occur in plant tissue in the form of a provitamin (or precursor of a vitamin), which can be converted into a vitamin in the animal body. Also, the B vitamins are universally distributed in all living tissues, whereas the fat-soluble vitamins are completely absent from some.

The fat-soluble vitamins are stored in appreciable quantities in the body, whereas the water-soluble vitamins are not. Any of the fat-soluble vitamins can be stored wherever fat is deposited; and the greater the intake, the greater the storage. By contrast, the water-soluble B vitamins are not stored in any appreciable amount. Moreover, the large amounts of water which pass through the body daily tend to carry out the water-soluble vitamins, thereby depleting the supply. Hence, they should be supplied in the diet on a daily basis. However, because all living cells contain all the B vitamins, and because the body conserves nutrients that are in short supply by using them only in vital reactions, deficiency symptoms do not appear immediately following their removal from the diet.

• **Vitamin B Complex**—With the exception of vitamin C, all of the water-soluble vitamins can be grouped together under the name vitamin B complex.

Under good practical conditions, the rations of farm animals usually contain adequate quantities of each of the several vitamins. However, deficiencies may occur during periods (1) of extended drought or in other conditions of restriction in diet, (2) when production is being forced, (3) when large quantities of highly refined feeds are being fed, or (4) when low-quality forages are utilized. Also, deficiencies may occur as a result of lack of availability of vitamins or because of the presence of antimetabolites. Both are important concepts. For example, analyses show corn to be adequate in niacin. Yet, due either to an antimetabolite or unavailability, there may be niacin deficiencies when corn is fed—deficiencies that can be remedied by niacin supplementation.

The absence of one or more vitamins in the ration may lead to a failure in growth or reproduction, or to characteristic disorders known as deficiency diseases. In severe cases, death may follow. Although the occasional deficiency symptoms are the most striking result of vitamin deficiencies, it must be emphasized that, in practice, mild deficiencies probably cause higher total economic losses than do severe deficiencies. It is relatively uncommon for a ration, or diet, to contain so little of a vitamin that obvious symptoms of a deficiency occur. When one such case does appear, it is reasonable to suppose that there must be several cases that are too mild to produce characteristic symptoms but which are sufficiently severe to lower the state of health and the efficiency of production. It is also to be emphasized that different species of animals vary in their needs for the vitamins. Further, not all animals suffer from the same deficiency diseases; thus, humans, monkeys, and guinea pigs react severely to the absence of vitamin C in the ration, whereas domestic pigs, fowl, and ruminants are unaffected.

Initially, vitamin nutrition was dependent upon incorporating into the ration feeds that were known to contain a high natural content of the needed vitamins. But this proved to be unsatisfactory, for the vitamin content of feeds varies considerably according to soil, climate conditions, and curing and storing. Through the years, however, methods for the laboratory synthesis of various vitamins were developed. So, today, most of the vitamins are available in pure crystalline form, and at prices that make them economical for supplementation of livestock.

The omission of a single vitamin from the diet of a species that requires it will produce specific deficiency symptoms. Many of the vitamins function as coenzymes (metabolic catalysts); others have no such role, but perform certain other essential functions.

Each of the vitamins is as much a distinct chemical compound as is cane sugar, for example. All of them contain carbon, hydrogen, and oxygen. In addition, all of the B vitamins except inositol contain nitrogen. Certain of the B vitamins also contain one or more of the mineral elements in their molecules. Even when added to the diet in very small amounts, vitamins are extraordinarily potent.

A summary of each vitamin is presented in Table 4–8, Animal Vitamin Chart. In reviewing this table, it must be remembered that single, uncomplicated vitamin deficiencies are the exception rather than the rule; multiple deficiencies are altogether too common, making diagnosis difficult even for the trained observer.

Nutritious feeds being consumed by contented finishing cattle. Knowledge of what nutrients are needed by animals and how they are used (metabolized) is the basis of modern nutrition. (Original painting by artist Tom Phillips, 3333 17th St., San Francisco, CA 94110)

TABLE 4–8
ANIMAL VITAMIN CHART

FAT-SOLUBLE VITAMINS:

VITAMIN A

Vitamin A made the difference!

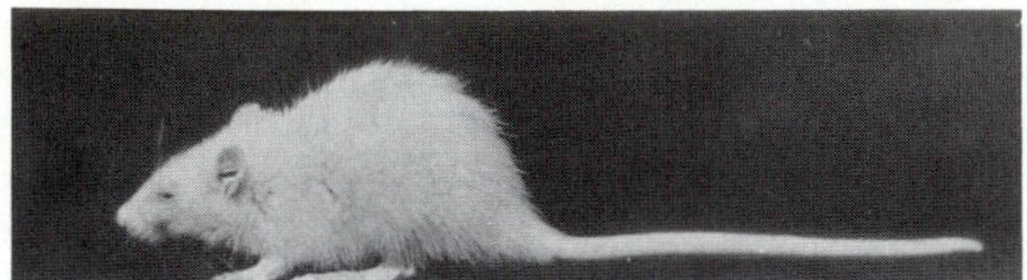

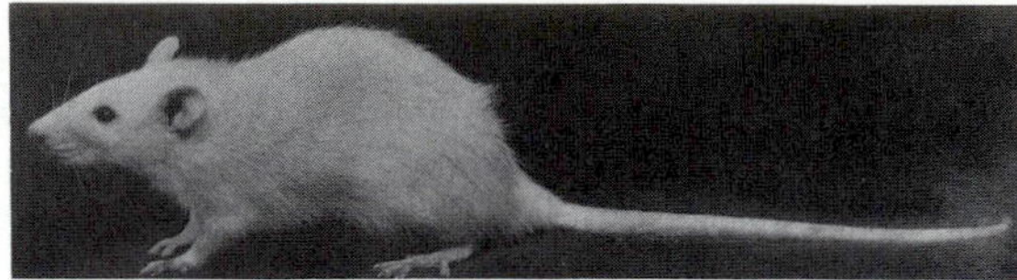

Fig. 4-69. Two rats from same litter, 11 weeks old. *Left:* This rat had no vitamin A. Note the infected eye, rough fur, and sick appearance. It weighed only 53 g. *Right:* This rat had plenty of vitamin A. It has bright eyes, sleek fur, and appears alert and vigorous. It weighed 123 g. (Courtesy, USDA)

Vitamin A is probably the most important of all vitamins, if any vitamin can be singled out and ranked. More than any other vitamin, deficiencies of vitamin A are still widespread throughout most developing countries of the world and involve both animals and people.

Vitamin A is required by all animals. It is strictly a product of animal metabolism—present in all species of mammals, birds, and fish, no vitamin A being found in plants. The counterpart in plants is known as carotene, which is the precursor of vitamin A. Because the animal body can transform carotene into vitamin A, this compound is often spoken of as *provitamin A.*

Carotene, which derives its name from the carrot, from which it was first isolated over 100 years ago, is the yellow-colored, fat-soluble substance that gives the characteristic color to carrots and many other vegetables, to several fruits, and to yellow corn.

Thus, the ultimate source of all vitamin A is the carotenes which are synthesized by plants.

History:

At a very early date, the Chinese used a concoction rich in vitamin A as a remedy for night blindness; and, in ancient Greece, Hippocrates prescribed various forms of liver, a good source of preformed vitamin A, as a treatment for night blindness. But it remained for laboratory experiments conducted in the early 1900s to identify vitamin A as a distinct nutrient. A chronology follows.

In 1913, vitamin A was discovered by Elmer V. McCollum and Marguerite Davis of the University of Wisconsin, and by Thomas B. Osborne and Lafayette B. Mendel of the Connecticut Experiment Station. Working independently, each research team demonstrated the presence of an essential dietary substance in fatty foods. McCollum and Davis found it in butterfat and egg yolks; Osborne and Mendel discovered it in cod-liver oil. These researchers believed that only one factor, which they called fat-soluble A, was needed to supplement purified diets. They described the condition as the "type of nutritive deficiency exemplified in the form of an infectious eye disease prevalent in animals inappropriately fed." In 1915, McCollum and Davis also noted that a deficiency of fat-soluble A caused night blindness. (It is noteworthy that Miss Marguerite Davis, a young biologist who had just obtained her bachelor's degree from the University of California, volunteered to do the rat work for McCollum without salary.)

In 1919, Steenbock and co-workers at the University of Wisconsin noted that an unknown substance, present in sweet potatoes, carrots, and corn, later to be identified as carotene, would support normal growth and reproduction.

In 1920, the English scientist Drummond proposed that this compound be called vitamin A.

In 1930, Moore, of England, demonstrated that this provitamin A activity was beta-carotene.

In 1931, P. Karrer, a Swiss researcher, isolated the active substance in halibut-liver oil and determined the chemical formula for vitamin A—the first vitamin to have its chemical structure determined. For this, and for his work with riboflavin, he received the Nobel Prize. However, vitamin A was not produced (by Karrer) in crystalline form from fish liver oils until 1937, and it did not become available in synthetic form (by Isler working in Switzerland) until 1947. Today, pure, inexpensive synthetic forms of vitamin A are readily available.

Chemistry/Metabolism/Properties:

Chemistry—*Vitamin A* may be a misleading term because it sounds as if only one chemical compound has vitamin A activity. Actually, several forms of vitamin A exist, with each possessing different degrees of activity. Two forms, retinol and dehydroretinol, are of primary importance.

Retinol (formerly called vitamin A), which is found as an ester (retinyl palmitate) in ocean fish oils and fats, and in liver, butterfat, and egg yolk, is biologically active as an alcohol, an aldehyde, and an acid. The alcohol, the most common form (see Fig. 4–70 for formula), is usually referred to as retinol, the aldehyde form as retinal or retinene, and the acid form as retinoic acid.

Dehydroretinol, or vitamin A_2, differs from retinol in that (1) it has an extra double bond, and (2) it has about 40% of the biological value (activity). It is found only in freshwater fish and in birds that eat these fish, hence, it is of limited importance.

Today, the general term *vitamin A* is used for both retinol and dehydroretinol.

In addition to the actual forms of vitamin A, related compounds, known as *carotenes,* are found in several fruits and vegetables. Carotene is also called (1) provitamin A, because it can be converted to vitamin A in the body; and (2) precursor of vitamin A, because it precedes vitamin A. At least 10 of the

VITAMIN A

H_3C H_3C CH_3 CH_3
C H H H H H H
H_2C C–C=C–C=C–C=C–C=C–CH_2OH
H_2C C–CH_3
CH_2

Fig. 4–70. The structure of vitamin A (retinol).

(Continued)

TABLE 4–8 *(Continued)*

Vitamin A (Continued)

carotenoids found in plants can be converted with varying efficiencies into vitamin A. Four of these carotenoids—alpha-carotene, beta-carotene, gamma-carotene, and cryptoxanthine (the main carotenoid of corn)—are of particular importance due to their provitamin A activity. Of the four, *beta-carotene* (see Fig. 4–71) has the highest vitamin A activity and provides about ⅔ of the vitamin A necessary for animal nutrition.

β-CAROTENE

Fig. 4–71. The structure of beta-carotene.

• **Conversion of beta-carotene to vitamin A**—Different species of animals convert β-carotene to vitamin A with varying degrees of efficiency. The conversion rate of the rat has been used as the standard value, with 1 mg of β-carotene equal to 1,667 IU of vitamin A. Based on this standard, the comparative efficiency of each animal species is as shown in the table entitled "Conversion of Beta-Carotene to Vitamin A for Different Species."

CONVERSION OF BETA-CAROTENE TO VITAMIN A FOR DIFFERENT SPECIES[1]

Species	Conversion of mg of Beta-Carotene to IU of Vitamin A		IU of Vitamin A Activity (calculated from carotene)
	(mg)	(IU)	(%)
Standard (rat)	1	1,667	100
Cattle	1	400	24.0
Dairy cattle	1	400	24.0
Sheep	1	400–500	24.0–30.0
Swine	1	500	30.0
Horses			
Growth	1	555	33.3
Pregnancy	1	333	20.0
Poultry	1	1,667	100
Mink	Carotene not utilized		—
Humans	1	556	33.3

[1]Adapted from the *Atlas of Nutritional Data on United States and Canadian Feeds,* NRC-National Academy of Sciences, 1972, p. XVI, Table 6.

Metabolism—Feeds supply vitamin A in the form of vitamin A, vitamin A esters, and carotenes. Almost no absorption of vitamin A occurs from the stomach. In the small intestine, vitamin A and beta-carotene are emulsified with bile salts and products of fat digestion and absorbed in the intestinal mucosa. Here, much of the conversion of beta-carotene to vitamin A (retinol) takes place. There are wide differences in species and individuals as to how well they utilize the carotenoids. Their absorption is affected by several factors, including the presence in the small intestine of bile, dietary fat, and antioxidants. Bile aids emulsification; fat must be absorbed simultaneously; and antioxidants, such as alpha-tocopherol and lecithin, decrease the oxidation of carotene. Also, the presence of enough protein of good quality enhances the conversion of carotene to vitamin A— a matter of great importance in developing countries where protein is limited in both quantity and quality.

The absorption of vitamin A is adversely affected by the presence of mineral oil or parasites in the intestinal tract.

In the blood, vitamin A esters are transported in association with a retinol-binding protein, whereas the carotenoids are associated with the lipid-binding protein. Storage of vitamin A is largely in the liver, but small amounts are also stored in the lungs, body fat, and kidneys.

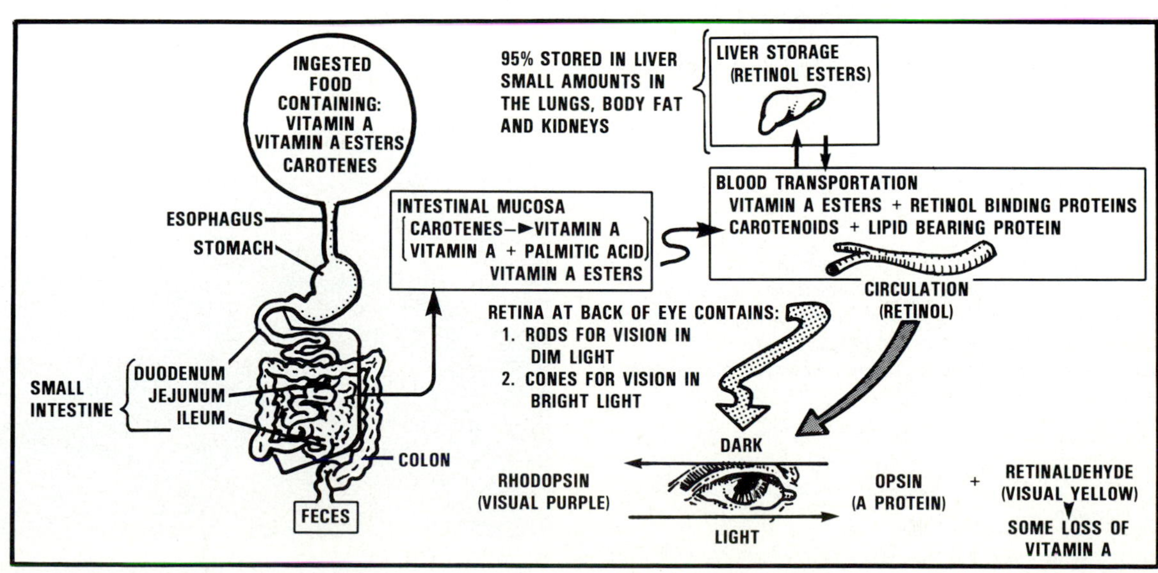

Fig. 4–72. Metabolism of vitamin A and its role in vision in dim light.

(Continued)

TABLE 4–8 *(Continued)*

Vitamin A (Continued)

From the liver, vitamin A enters the bloodstream as a free alcohol, whereupon it travels to the tissues for use.

No vitamin A is excreted in the urine because it is not water-soluble, but considerable unabsorbed carotene is normally found in the feces.

Properties—Vitamin A (retinol) is an almost colorless (pale yellow) fat-soluble substance. It is insoluble in water. Although the esters of vitamin A are relatively stable compounds, the alcohol, aldehyde, and acid forms are rapidly destroyed by oxidation when they are exposed to air and light. Since vitamin A occurs in the stable form (the ester) in most feeds, normal processing procedures do not destroy much vitamin A activity.

The carotenoid pigments (which are referred to as carotene in this discussion) are a deep red color, but in solution they are bright yellow or orange yellow. They impart the color to carrots, sweet potatoes, squash, pumpkins, and yellow corn. As a general rule of thumb, the more intense the pigmentation in feeds, the higher the provitamin A content. Green forages, particularly dark green and leafy ones, are also rich in provitamin A. They contain carotenoids, although their color is masked by that of the green pigment, chlorophyll. The properties of solubility and stability of carotene and carotenoid pigments are similar to those of vitamin A.

Measurement/Assay:

The assay of vitamin A is accomplished by two basic methods: biological, or chemical. The bioassay procedure is based on a biological response such as growth of rats or chicks deficient in vitamin A. It measures the total vitamin A, including provitamin A, present. But, because of the difficulties and time factor in bioassays, chemical assays are usually used.

Dietary allowances of vitamin A are stated in terms of either International Units (IU) or United States Pharmacopeia (USP) units, which are equal. An International Unit of vitamin A is defined on the basis of rat studies as equal to 0.344 mcg of crystalline retinylacetate (which is equivalent to 0.3 mcg of retinol, or to 0.6 mcg of beta-carotene). These standards are based on experiments that show that in rats only about 50% of the beta-carotene is converted to vitamin A.

In order to quantify vitamin A values within the metric system, international agencies have now introduced the biological equivalent of 1 microgram (1 mcg = 1/1000 milligram) of retinol as the standard. This is known as the *Retinol Equivalent* (RE). The RE system of measurement takes into account the amount of absorption of the carotenes as well as the degree of conversion to vitamin A; hence, it is more precise than the IU system.

In terms of International Units, beta-carotene (by weight) is ½ as active, and the other provitamin A carotenoids are about ¼ as active, as retinol. Also, retinol is completely absorbed by the intestine, but only about ⅓ of the intake of provitamin A carotenoids is absorbed. Of the absorbed carotenoids, only ½ of the beta-carotene and ¼ of the other provitamin A carotenoids are converted to retinol. It follows that beta-carotene is only ⅙ as active, and the other carotenoids 1/12 as active, as retinol.

Species Most Affected:

All animals, including poultry.

Functions:

Helps maintain normal vision in dim light—prevents night blindness.

Prevents xerophthalmia, an eye condition which may lead to blindness in extreme vitamin A deficiency.

Essential for body growth.

Necessary for normal bone growth.

Necessary for normal tooth development.

Helps keep the epithelial tissues of the skin, and of the lining of the nose, throat, respiratory and digestive systems, and genitourinary tract, healthy, and free of infection.

New information suggests that vitamin A (1) acts in a coenzyme role, as for instance in the formation of intermediates in glycoprotein synthesis; and (2) functions like steroid hormones, with a role in the cell nuclei, leading to tissue differentiation.

Other functions of vitamin A: necessary for (1) thyroxin formation and prevention of goiter; (2) protein synthesis; and (3) synthesis of corticosterone from cholesterol, and the normal synthesis of glycogen.

Deficiency Symptoms/Toxicity:

Deficiency Symptoms—Night blindness (nyctalopia), xerosis, and xerophthalmia.

Stunted growth.

Slowed bone growth, abnormal bone shape, and paralysis.

Unsound teeth, characterized by abnormal enamel, pits, and decay.

Rough, dry, scaly skin; increased abscesses in ears, mouth, and/or salivary glands; increased diarrhea and kidney and bladder stones.

Reproductive disorders, including poor conception, abnormal embryonic growth, placental injury, and death of the fetus.

Toxicity—Toxicity of vitamin A is characterized by loss of appetite, poor vision, excessive irritability, loss of hair, dryness and flaking of the skin (with itching), swelling over the long bones, diarrhea, nausea, and enlargement of the liver and spleen.

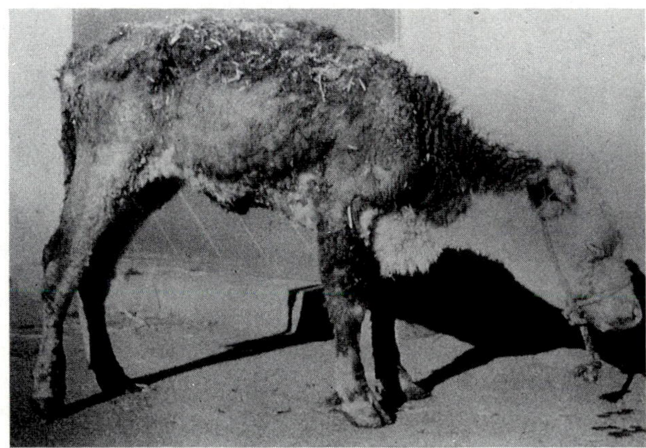

Fig. 4–73. Vitamin A deficiency in a calf. Note excessive watering of the eye (lacrimation), lethargic appearance, diarrhea, and signs of breathing difficulty. (Photo by Chas. Pfizer Co. Print courtesy of D. C. Church)

(Continued)

TABLE 4-8 *(Continued)*

Vitamin A (Continued)

Sources for Animals:

Rich feed sources—Vitamin A can be provided as the synthetic vitamin or as the precursor, carotene. Rich sources of carotene follow: carrots, corn (yellow), grass and/or legume hay/meal (green and leafy; not over 1 year old), grass silages, milk (whole), pastures (immature and green), peas (green and yellow), pumpkins, squash, sweet potatoes.

Supplemental sources—Cod and other fish oils, synthetic vitamin A.

In recent years, many livestock producers have followed the practice of injecting animals with a vitamin A concentrate intramuscularly. This injection provides adequate vitamin A reserves for periods of 3 to 6 months. It is given during periods when the vitamin A intake is low, as in late winter and early spring when the reserves from the previous summer on pasture have been depleted. In many operations, newborn calves are given an injection of vitamin A, usually in combination with vitamins D and E. These injections are inexpensive and readily available. The main disadvantage is that animals must be handled individually, which is more difficult in large commercial operations than in smaller family-operated units.

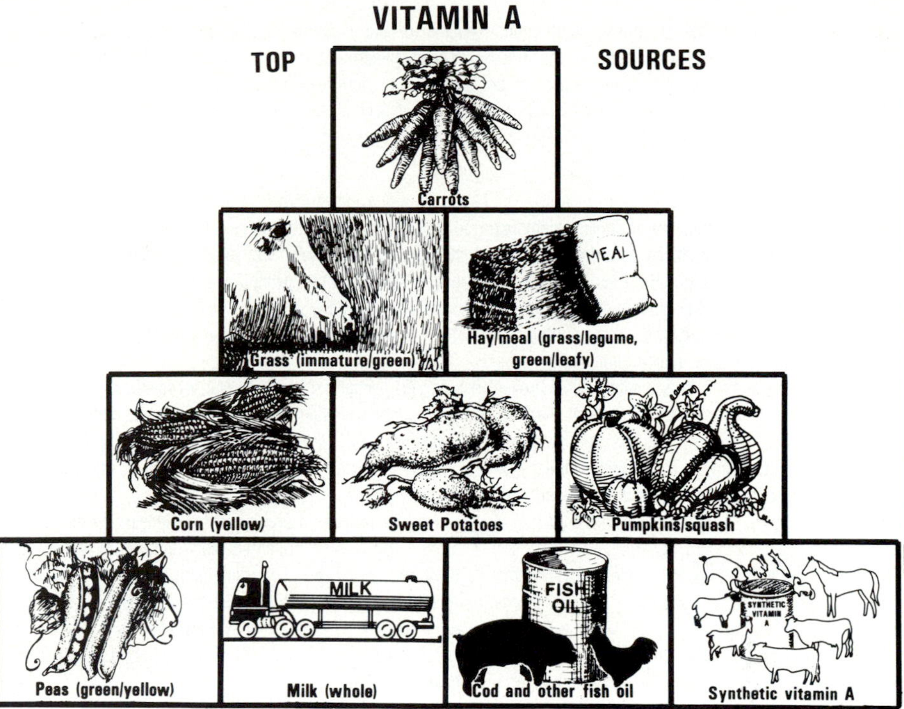

Fig. 4-74. Commonly used, top feed and supplement sources of vitamin A (carotene). No claim is made that all the top sources of vitamin A (carotene) are listed. Neither are the sources ranked according to amount of vitamin A (carotene) supplied nor is price considered.

Comments:

The forms of vitamin A are: alcohol (retinol), ester (retinyl palmitate), aldehyde (retinal or retinene), and acid (retinoic acid).

Retinol, retinyl palmitate, and retinal are readily converted from one form to another, but retinoic acid cannot be converted to other forms. Retinoic acid fulfills some of the functions of vitamin A, but it does not function in the visual cycle.

Animals are able to store considerable vitamin A, but because of their greater requirements and less storage, young animals suffer from a deficiency much sooner than those that are mature.

Both carotene and vitamin A are readily destroyed by oxidation, thus resulting in considerable losses in processing and storing (as in making or storing of hay). There is indication that vitamin A suppresses vitamin E in feeds.

VITAMIN D

Vitamin D made the difference!

Fig. 4-75. *Left:* Calf with rickets caused by a lack of vitamin D in the ration, characterized by stiffness of joints, arched back, and easily broken bones. *Right:* Healthy calf that received plenty of vitamin D. (*Left photo:* Courtesy, Michigan State University, East Lansing. *Right photo:* Courtesy, American Breeders Service, a Division of W. R. Grace and Company, DeForest, Wisc.)

(Continued)

TABLE 4-8 *(Continued)*

Vitamin D (Continued)

The importance of vitamin D—the sunshine vitamin—lies in its role in regulating calcium and phosphorus metabolism. Vitamin D promotes intestinal absorption of calcium and phosphorus and influences the process of bone mineralization. In the absence of vitamin D, mineralization of bone matrix is impaired, resulting in rickets in young animals and osteomalacia, or brittle bones, in mature animals.

Vitamin D is unique among the vitamins in two respects: (1) it occurs naturally in only a few common feeds, and (2) it can be formed in the body by exposure of the skin to ultraviolet rays of the sun—light of short wavelength and high frequency; hence, it is known as the *sunshine vitamin.*

History:

In 1918, Sir Edward Mellanby of England demonstrated that rickets was a nutritional deficiency disease. He produced rickets in puppies, then cured it by giving them cod-liver oil. But Mellanby incorrectly attributed the cure to the newly discovered fat-soluble vitamin A.

In 1922, McCollum at Johns Hopkins University found that, after destruction of all the vitamin A in cod-liver oil (by oxidation, passing heated air through cod-liver oil), it still retained its ricket-preventing potency. This proved the existence of a second fat-soluble vitamin, carried in liver oils and certain other fats, which he called *calcium depositing vitamin.* It is of interest to note that, though McCollum discovered the existence of vitamin D, he did not call it by this name until after this designation was in common use by others.

In 1924, the mystery of how sunlight could prevent rickets was partially solved. Dr. Harry Steenbock of the University of Wisconsin and Dr. A. Hess of Columbia University, working independently, showed that the antirachitic activity could be produced in feeds/foods and in animals by ultraviolet light. The process, known as the Steenbock Irradiation Process was patented by Steenbock, with the royalties assigned to the Wisconsin Alumni Research Foundation of the University of Wisconsin. Subsequent research disclosed that it was certain sterols in feeds/foods and animal tissues that acquired antirachitic activity upon being irradiated. Before irradiation, the sterols were not protective against rickets.

By the late 1920s, it had been established that rickets could be prevented and cured by exposure to direct sunlight (ever since, vitamin D has been popularly called *the sunshine vitamin*), by irradiation with ultraviolet light, by feeding irradiated feed/food, or by feeding cod-liver oil. Later, the natural vitamin D of fish-liver oils was identified as the same substance that is produced in the skin by irradiation.

In 1932, crystals of pure vitamin D_2 (ergocalciferol) were isolated from irradiated ergosterol by Windaus of Germany and Askew of England; and in 1936, crystals of pure vitamin D_3 (cholecalciferol) were isolated from tuna liver oil by Brockman of Germany.

In 1952, the first total synthesis of a form of vitamin D (in this case vitamin D_3) was accomplished by R. B. Woodward of Harvard. He was awarded the Nobel Prize in Chemistry in 1956 for this and other similar achievements.

Chemistry/Metabolism/Properties:

Although about 10 sterol compounds with vitamin D activity have been identified, only 2 of these, known as provitamins D or precursors, are of practical importance today from the standpoint of their occurrence in feeds—ergocalciferol (vitamin D_2) and cholecalciferol (vitamin D_3); the name cholecalciferol of the latter is a reflection of its cholesterol precursor. Because these substances are closely related chemically, the term vitamin D is used collectively to indicate the group of substances that show this vitamin activity.

Chemistry—Fig. 4–76 shows the structures of vitamins D_2 and D_3.

Ultraviolet irradiation of the two provitamins—ergosterol and 7-dehydrocholesterol—will produce vitamins D_2 and D_3, respectively. Ergosterol is found in plants (in yeasts and fungi), whereas 7-dehydrocholesterol is found in fish-liver oils and in the skin of humans and other animals. Therefore, people or animals that are exposed to sunlight for extended periods of time do not need dietary supplementation of vitamin D. Both forms of vitamin D, D_2 and D_3, have equal activity for people and most other mammalian species. But chickens, turkeys, and other birds are exceptions—they utilize vitamin D_3 more efficiently than vitamin D_2.

Metabolism—Vitamin D is unique in that humans and other animals normally obtain it from two sources; formation in the skin, and by mouth. The steps involved in the metabolism of vitamin D follow:

1. **Formation of vitamin D_3 in the skin and its movement into the circulation.** The unique mechanism for the synthesis, storage, and slow, steady release of vitamin D_3 from the skin into the circulation is shown in Fig. 4–77.

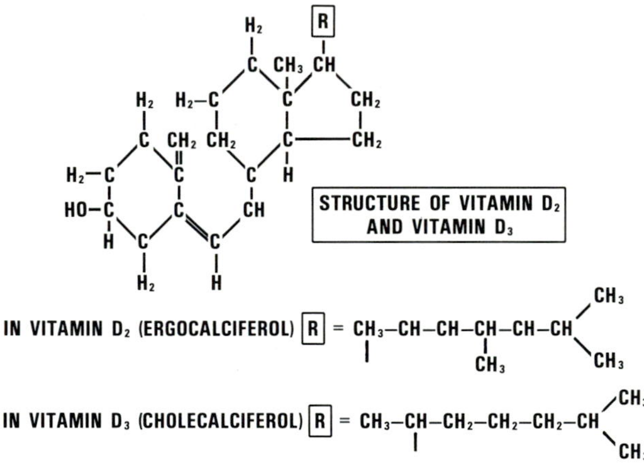

Fig. 4–76. Structure of vitamin D_2 and vitamin D_3.

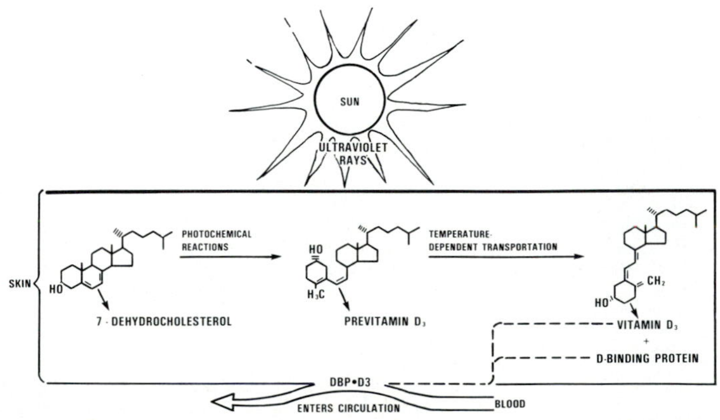

Fig. 4–77. Diagram showing the sequence of steps in the formation of vitamin D_3 in the skin and its transport into the circulation.

(Continued)

TABLE 4–8 *(Continued)*

Vitamin D (Continued)

When the skin is exposed to the ultraviolet radiation of sunlight, part of the store of 7-dehydrocholesterol undergoes a photochemical reaction in the epidermis and the dermis and forms provitamin D_3. Once provitamin D_3 is formed in the skin, it undergoes a slow temperature-dependent transformation to vitamin D_3, which takes at least 3 days to complete. Then, the vitamin D-binding protein transports D_3 from the skin into the circulation.

2. **Absorption.** Vitamin D taken by mouth is absorbed with fats from the small intestine (from the jejunum and ileum), with the aid of bile. Vitamin D formed in the skin by irradiation of the provitamin present there is absorbed directly into the circulatory system.

3. **Utilization.** Cholecalciferol—obtained either from the diet or from the irradiation of the skin—is transported by a specific vitamin D carrier protein (a globulin) to the liver where it is converted to 25-hydroxycholecalciferol (25-OH-D_3).[1] From the liver, 25-OH-D_3 is transported to the kidneys where it is converted to 1,25-$(OH)_2$-D_3, the most active form of vitamin D in increasing calcium absorption, bone calcium mobilization, and increased intestinal phosphate absorption. The active compound, 1,25-$(OH)_2$-D_3, functions as a hormone, since it is a vital substance made in the body tissues (the kidneys) and transported in the blood to cells within target tissues. This physiologically active form of vitamin D_3 is then either transported to its various sites of action or converted to its metabolite forms of 24,25-dihydroxycholecalciferol or 1,24,25-trihydroxycholecalciferol.

Although most of the research on vitamin D metabolism has been conducted on cholecalciferol, studies by DeLuca on ergocalciferol indicate that it is metabolized similarly to cholecalciferol; that it is changed to a similar active metabolite in the liver—25-hydroxyergocalciferol (25-OH-D_2).

4. **Storage.** The major storage sites of vitamin D are the fatty tissues and skeletal muscles. Some of it is also found in the liver, brain, lungs, spleen, bones, and skin. But body storage of vitamin D is much more limited than the storage of vitamin A.

5. **Excretion.** The main pathway of excretion of vitamin D is by way of the bile into the small intestine, thence the feces. Less than 4% of the intake is excreted in the urine.

Properties—Pure D vitamins are white, crystalline, odorless substances that are soluble in both fats and fat solvents (such as ether, choloroform, acetone, and alcohol). They are insoluble in water, and they are resistant to heat, oxidation, acid, and alkali.

Although the precursors are activated by ultraviolet light, excessive irradiation results in the formation of slightly toxic compounds that have no antirachitic activity.

Measurement/Assay:

Vitamin D potency is expressed in International Units (IU) and United States Pharmacopeia (USP) Units, which are equal. *One IU, or one USP, of vitamin D is defined as the activity of 0.025 mcg of pure crystalline vitamin D_3 (cholecalciferol).*

The ultraviolet light absorption property of vitamin D may be used for the assay of pure preparations free of irrelevant absorption. But it does not distinguish between vitamin D_2 and D_3.

Both vitamin D_2 and D_3 give a yellow-orange color with antimony trichloride. This color reaction forms the basis of the USP XVIII method for vitamin D. Since color reactions are subject to interferences from many sources, they should be limited to high potency pharmaceutical preparations or fortified feeds. Combinations of column and thin-layer chromatographic purification steps with the antimony trichloride reaction have been successfully used.

Gas-liquid chromatography provides a means of combining qualitative and quantitative assays.

Feed samples, which contain vitamin D in very low concentrations, are not usually assayed by chemical methods. For these substances, bioassays are the only means available for the assessment of vitamin D activity. Rats and chicks are used as test animals; rats respond equally well to D_2 and D_3, whereas chicks respond only to D_3. The assays measure the alleviation (curative test) or the development (prophylactic test) of vitamin D deficiency in terms of the degree of rickets produced.

A bioassay method, known as the *line test,* uses stained longitudinal sections of the distal end of radius bones to evaluate calcification. Usually the test animal is the rat, although the chick must be used if the vitamin D activity of a sample intended for poultry nutrition is to be determined. Young rats from mothers having a deficient supply of vitamin D are kept on a rachitogenic diet so that no calcification occurs in the ends of the long bones. When a test material is fed to these vitamin D-deficient rats, its value as a source of vitamin D is measured by the amount that must be fed for 7 to 10 days to produce a good calcium line (line test) in the ends of the long bones. Standard cod-liver oil is fed to a similar group of animals and is used as a basis of comparison.

Species Most Affected:

All farm animals, including poultry.

Functions:

Aids in the assimilation and utilization of calcium and phosphorus.
Promotes growth and mineralization of the bones, including the bones of the fetus.
Promotes sound teeth.
Maintains normal level of citrate in the blood.
Protects against the loss of amino acids through the kidneys.
Functions as a hormone, as the active compound of vitamin D (1,25-$(OH)_2$-D_3).

Deficiency Symptoms/Toxicity:

Deficiency Symptoms—Rickets in young, characterized by enlarged joints, bowed legs, knocked knees, and beaded ribs.
Osteomalacia in adults, in which the bones soften, become distorted, and fracture easily.
Tetany, characterized by muscle twitching, convulsions, and low serum calcium.
Chicks: Reduced growth, soft bones (rickets), leg deformities.
Hens: Poor eggshells and lowered hatchability.

Toxicity—Excessive vitamin D may cause hypercalcemia (increased intestinal absorption, leading to elevated blood calcium levels), characterized by loss of appetite, excessive thirst, nausea, vomiting, irritability, weakness, constipation alternating with bouts of diarrhea, retarded growth in young animals, and weight loss in adults.

[1]Identified by DeLuca and co-workers of the University of Wisconsin. First reported in *Proc. Nat. Acad. Sci.,* 60:1503, 1968.

(Continued)

TABLE 4-8 *(Continued)*

Vitamin D (Continued)

Sources for Animals:

Ultraviolet light from the sun is by far the most accessible and cheapest means of supplying vitamin D. The precursor of vitamin D_2—ergosterol—is present in many forages. Grains and their by-products are generally extremely low in vitamin D. Eggs are good sources of vitamin D. Milk is quite variable in content, with the result that most commercial milk processors routinely fortify their products with this vitamin. Fish-liver oils are generally rich sources of vitamin D. Synthetic vitamin D is commercially available at relatively low cost.

Rich feed sources—Vitamin D occurs naturally in only a few feeds. Sun-cured hays are the best natural/practical source. Corn leaves, fodder and stover; and barley, oat, and wheat hays are good sources. Less common feed sources are beet pulp, and cocoa shells (sun-dried).

Supplemental sources—Vitamin D_2 (irradiated ergosterol), the plant form; vitamin D_3, the animal form; cod and certain other fish-liver oils; irradiated yeast.

VITAMIN D
TOP SOURCES
Alfalfa leaf meal sun-cured
Alfalfa hay sun-cured
Grass-legume hay sun-cured
Cocoa shells sun-cured
Corn leaves
Oat hay
Vitamin D_2
Vitamin D_3
Irradiated yeast
Cod and certain other fish oils

Fig. 4-78. Commonly used, top feed and supplement sources of vitamin D. No claim is made that all the top sources of vitamin D are listed. Neither are the sources ranked according to amount of vitamin D supplied nor is price considered.

Comments:

When animals are exposed sufficiently to direct sunlight, the ultraviolet light in the sunlight penetrates the skin and produces vitamin D_3 from traces of 7-dehydrocholesterol in the tissues.

Tissue storage is very limited.

The vitamin D requirement is less when a proper balance of calcium and phosphorus exists.

VITAMIN E

Vitamin E made the difference!

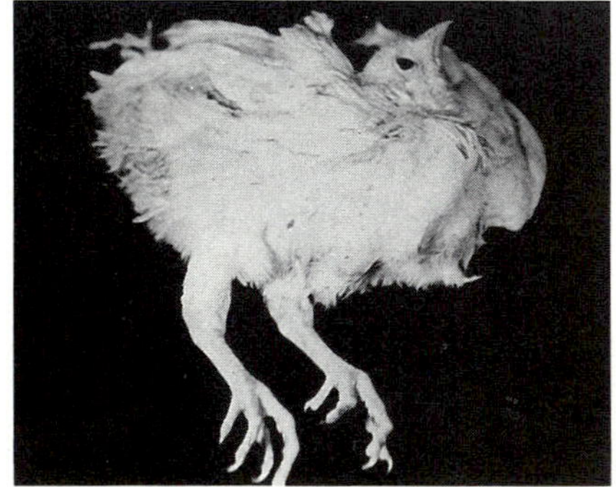

Fig. 4-79. *Left:* Vitamin E-deficient chick with nutritional encephalomalacia (crazy chick disease), due to lack of vitamin E. Note the head retraction and loss of control of legs. *Right:* Normal chick which received adequate vitamin E. (*Left photo:* Courtesy, Dept. of Poultry Science, Cornell University, Ithaca, N.Y.

Today, vitamin E is recognized as an essential nutrient for higher animals.

History:

In 1922, Evans and Bishop, of the University of California, discovered that a fat-soluble dietary factor (then called *factor X*) in lettuce and wheat germ was essential for successful reproduction in rats. In 1924, Sure, of the University of Arkansas, named the factor vitamin E. In 1936, Evans and co-workers isolated crystalline vitamin E from wheat germ oil and named it tocopherol, from the Greek words *tokos* (offspring) and *pherein* (to bear), meaning *to bear offspring.*

(Continued)

TABLE 4-8 *(Continued)*

Vitamin E (Continued)

In 1938, the vitamin was first synthesized by the Swiss chemist, Karrer.

Chemistry/Metabolism/Properties:

Chemistry—Eight tocopherols and tocotrienols with vitamin E activity, collectively called vitamin E, have been identified. Alpha-tocopherol (see Fig. 4-80) has by far the greatest vitamin E activity; the other tocopherols have biological activities ranging from 1 to 50% that of alpha-tocopherol.

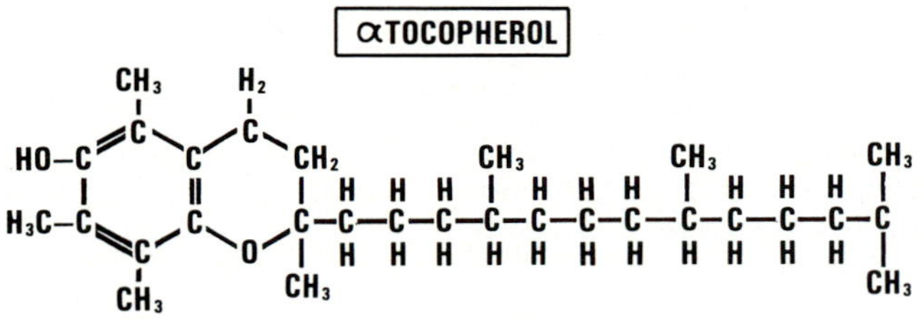

Metabolism—A discussion of the absorption, transportation, storage, and excretion of vitamin E follows:

1. **Absorption.** As with other fat-soluble vitamins, the presence of both bile and fat are required for the proper absorption of vitamin E. Absorption takes place in the small intestine, where 20 to 30% of the intake passes through the intestinal wall into the lymph.

αTOCOPHEROL

Fig. 4-80. Structure of alpha-tocopherol.

2. **Transportation.** Once absorbed, vitamin E is transported attached to the beta-lipoprotein fraction of the blood. The predominant form of vitamin E in both the plasma and red cells is alpha-tocopherol; gamma-tocopherol accounts for most of the remainder.

3. **Storage.** Adipose (fatty) tissue, liver, and muscle are the major storage sites, although small amounts of vitamin E are stored in most body tissues. Relatively high amounts are found in the adrenal and pituitary glands, heart, lungs, testes, and uterus. Vitamin E is deposited in the fat of tissues and is mobilized from them with fat.

4. **Excretion.** The major pathway of excretion of vitamin E is by way of the feces, although a small amount is excreted in the urine.

Properties—The tocopherols and tocotrienols are light yellow, viscous oils, soluble in alcohol and fat solvents, but insoluble in water. They are stable to acids and heat, but they are destroyed upon exposure to oxygen, ultraviolet light, alkali, and iron and lead salts. Their ability to take up oxygen gives them important antioxidant properties.

Currently, the bulk of commercial vitamin E is synthetic dl-alpha-tocopherol acetate, an ester of alpha-tocopherol, a form more stable to heat and oxidation than the free alcohol form, but with the same vitamin E activity. This stable form of alpha-tocopherol is added to vitamin preparations, various medicinal products, foods, and animal feeds. It is changed to the active vitamin by metabolic processes in the animal.

Measurement/Assay:

By agreement of international committees, it is now preferred to use *milligrams of alpha-tocopherol equivalents* as a summation term for all vitamin E activity. In this book, however, alpha-tocopherol equivalents will be used because (1) feed composition tables generally give values in IU, and (2) IU is still used for labeling most feed products.

A colorimetric method is the most commonly used assay method for estimating vitamin E activity. Additionally, spectrofluorometric and gas-liquid chromatographic methods are sometimes used. Animal tests are available, but are now rarely used.

Species Most Affected:

Cattle, sheep, goats, horses, poultry, and perhaps certain other animals.

Functions:

As an antioxidant which (1) retards rancidification of fats in plant sources and in the digestive tracts of animals, and (2) protects body cells from toxic substances formed from the oxidation of unsaturated fatty acids. As a powerful antioxidant, vitamin E readily oxidizes itself (it combines with oxygen), thereby minimizing the destruction by oxidation of unsaturated fatty acids and vitamin A in the intestinal tract and in the tissues.

As an essential factor for the integrity of red blood cells.

As an agent essential to cellular respiration, primarily in heart and skeletal muscle tissues.

As a regulator in the synthesis of DNA, vitamin C, and coenzyme Q.

As a sparer of selenium.

• **Interrelationship of vitamin E and selenium**—During the 1950s, an interrelationship between vitamin E and the element selenium was established. It was found that selenium prevented exudative diathesis in vitamin E-deficient chicks and liver necrosis in vitamin E-deficient rats. Subsequent research demonstrated that both selenium and vitamin E protect the cell from the detrimental effects of peroxidation, but each takes a distinctly different approach to the problem. Vitamin E is present in the membrane components of the cell and prevents free-radical formation, while selenium functions throughout the cytoplasm to destroy peroxides. This explains why selenium will correct some deficiency symptoms of vitamin E, but not others.

Deficiency Symptoms/Toxicity:

Deficiency Symptoms—Innumerable vitamin E deficiency symptoms have been demonstrated in animals; and they are highly variable from species to species.

A deficiency of vitamin E affects the normal functioning of muscle tissue. In most animals, both skeletal and heart muscles are affected by such deficiencies. Both selenium and vitamin E appear to be able to prevent these dysfunctions. In young goats and sheep, the skeletal muscle degenerates, giving rise to the disease called *stiff lamb disease.* In calves, the counterpart is called *white muscle disease.* In poultry, guinea pigs, and rabbits, muscular dysfunctions associated with vitamin E deficiency are referred to as *nutritional muscular deficiency.* Myopathy of the gizzard has been reported in turkeys.

Poultry are extremely sensitive to vitamin E deficiencies. In addition to muscular degeneration, deficiencies of vitamin E produce exudative diathesis (a leaking of the capillaries), loss of blood protein, encephalomalacia, steatitis, and reproductive failure in both the male and the female.

(Continued)

TABLE 4–8 *(Continued)*

Vitamin E (Continued)

Vitamin E has often been touted for increasing reproductive ability. But in cases where adequate amounts of vitamin E are being fed, excessive amounts will not be beneficial. A deficiency of vitamin E has been known to cause sterility in rats, guinea pigs, and hamsters. Problems with fetal development can occur in animals when vitamin E is inadequate.

Steatitis, or "yellow fat," has been observed in vitamin E-deficient mink, pigs, and chicks. Necrosis of the liver occurs in rats and pigs on vitamin E-deficient rations. Kidney degeneration is observed in rats, monkeys, and mink. Selenium can prevent degenerative problems, but not steatitis.

Toxicity—Vitamin E is relatively nontoxic. Excess intakes are excreted in the feces.

Sources for Animals:

Rich feed sources—Alfalfa leaf meal, alfalfa meal, barley and barley by-products, brewers' grains, corn and corn by-products, cottonseed meal, distillers' grains, distillers' solubles, hays (early cut, green), oats and oat by-products, pastures (green), rice and rice by-products, soybeans, sugar beet leaves, wheat and wheat by-products.

Supplemental sources—Rice polishings, wheat germ meal, and synthetic dl-alpha-tocopherol acetate.

Comments:

Vitamin E is widely distributed in all natural feeds.

Utilization of vitamin E is dependent on adequate selenium.

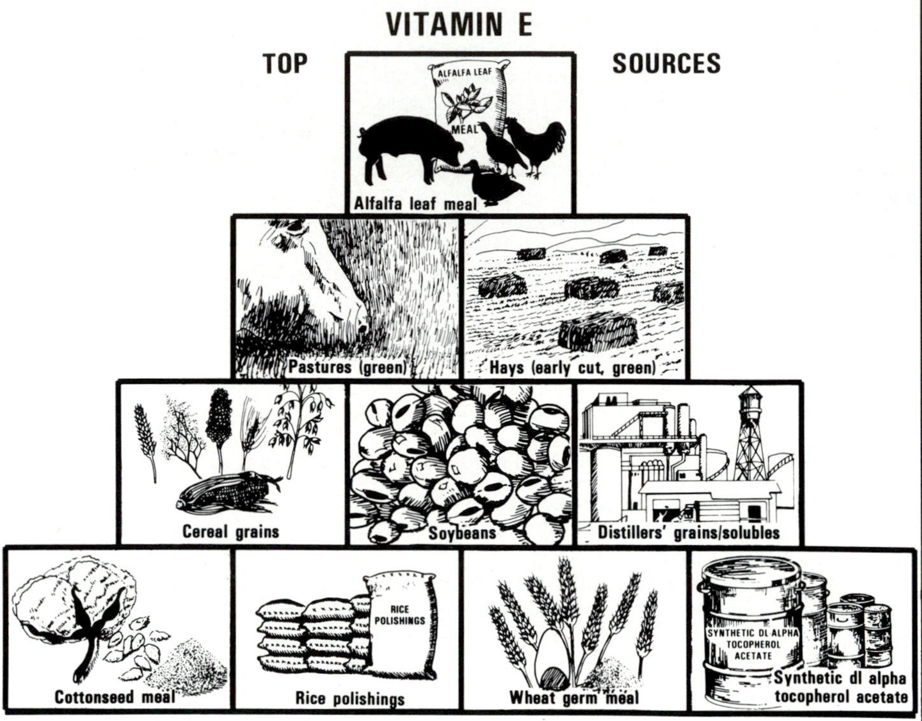

Fig. 4–81. Commonly used, top feed and supplement sources of vitamin E. No claim is made that all the top sources of vitamin E are listed. Neither are the sources ranked according to amount of vitamin E supplied nor is price considered.

VITAMIN K

Vitamin K, known as the *antihemorrhagic vitamin,* is necessary for the synthesis of prothrombin and other blood clotting factors in the liver. Presently, the term vitamin K is used to describe a chemical group of quinone compounds, rather than a single entity, which have characteristic antihemorrhagic effects.

History:

In 1929, Professor Carl Peter Hendrik Dam, biochemist at the University of Copenhagen, Denmark, observed that certain experimental diets produced fatal hemorrhages in chicks. Bleeding could be prevented by giving a variety of feedstuffs, especially alfalfa (lucerne) and fishmeal. Further, it was found that the active principle in these materials could be extracted with ether; thus, a new essential fat-soluble factor was *discovered.* In 1935, Dam named it the koagulation vitamin (the Danish word for coagulation), from which the shortened term vitamin K (from the first letter of koagulation) was derived. In 1939, Dam and Karrer isolated vitamin K in pure form; and, that same year, Almquist and Klose *synthesized* vitamin K. In 1934, Dam received the Nobel Prize in Physiology and Medicine for his brilliant work.

In 1941, Link and co-workers at the University of Wisconsin discovered dicoumarol—an antimetabolite of vitamin K.

Chemistry/Metabolism/Properties:

Chemistry—A number of chemical compounds possessing vitamin K activity have been isolated or synthesized. There are two naturally occurring forms of vitamin K; vitamin K_1 (phylloquinone or phytylmenaquinone) which occurs only in green plants, and K_2 (menaquinone or multiprenyl-menaquinone), which is synthesized by many microorganisms, including bacteria in the intestinal tracts. Additionally, several synthetic compounds have been prepared that possess vitamin K activity, the best known of which is menadione (2-methyl,1,4-naphthoquinone), formerly known as K_3. Menadione, which is converted to vitamin K_2 in the body, is 2 to 3 times as potent as K_1 and K_2. The structural formulas of K_1, K_2, and menadione, all of which are called vitamin K, are shown in Fig. 4–82.

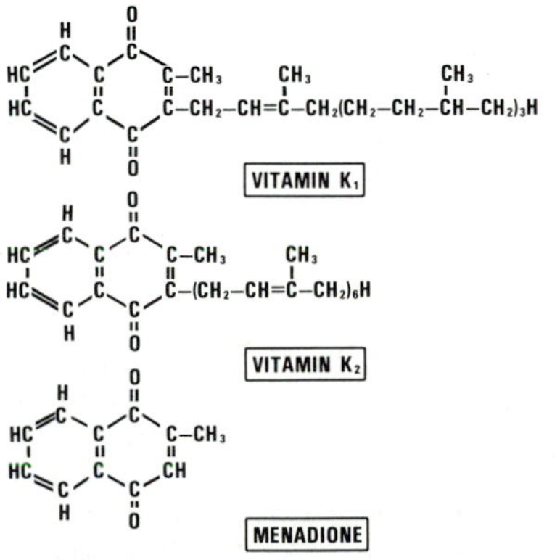

VITAMIN K_1: side chain $C{-}CH_2{-}CH{=}C(CH_3){-}CH_2(CH_2{-}CH_2{-}CH(CH_3){-}CH_2)_3H$; $C{-}CH_3$

VITAMIN K_2: side chain $C{-}(CH_2{-}CH{=}C(CH_3){-}CH_2)_6H$; $C{-}CH_3$

MENADIONE: $C{-}CH_3$; CH

Fig. 4–82. Structures of vitamin K_1, vitamin K_2, and menadione.

(Continued)

TABLE 4-8 *(Continued)*

Vitamin K (Continued)

Metabolism—Normally, vitamin K_1 is taken into the body in the diet, vitamin K_2 is synthesized in the intestine, and menadione is taken as a vitamin supplement.

1. **Intestinal synthesis.** Vitamin K_2 is synthesized by the normal bacteria in the small intestine and the colon; hence, an adequate supply is generally present.

The intestinal synthesis of vitamin K reduces the food dietary requirements for the vitamin in mammals (but not birds; birds have such a short intestinal tract and harbor so few microorganisms that they require a dietary source of vitamin K), although it appears that little of the vitamin K produced in the lower gut is absorbed. It is noteworthy, however, that animals that practice coprophagy, such as the rabbit, can utilize much of the vitamin K that is eliminated in the feces.

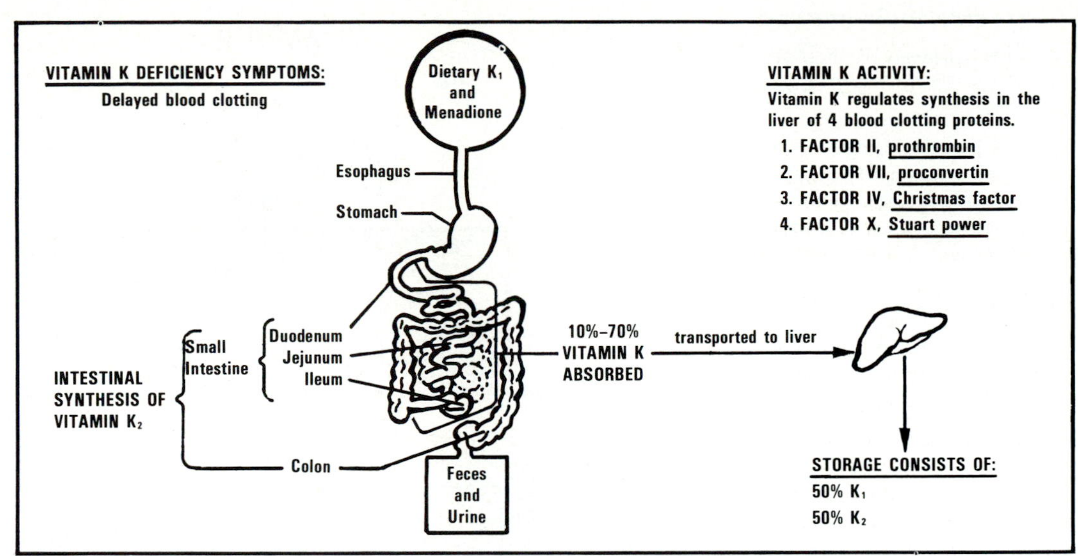

Fig. 4-83. Vitamin K utilization.

2. **Absorption.** Since the natural vitamins K (K_1 and K_2) are fat soluble, they require bile and pancreatic juice in the intestine for maximum absorption. By contrast, some of the synthetic vitamin K compounds are water soluble and more easily absorbed. Absorption takes place mainly in the upper part of the small intestine. Normally, 10 to 70% of the vitamin K in the intestines is absorbed. However, there is uncertainty as to how much of the vitamin K_2 that is synthesized in the colon is absorbed, based on the fact that absorption of nutrients in general from the large intestine appears to be limited because of the nature of the epithelial lining.

3. **Transportation.** Vitamin K passes unchanged from the small intestine into the lymph system. Thence, it is carried to the thoracic duct, where it enters the bloodstream. In the blood, it is attached to beta-lipoproteins and transported to the liver and other tissues.

4. **Metabolism.** Whether the K vitamins' functions are unchanged or are transformed to other metabolically active forms has not been determined. It is known that menadione must be converted to K_2 for it to be biologically active.

5. **Storage.** Vitamin K is stored only in small amounts. Modest amounts are stored in the liver, with the skin and muscle following in concentration.

6. **Excretion.** Excess vitamin K is excreted in the feces and urine.

Properties—The naturally occurring forms of vitamin K are yellow oils; the synthetic forms are yellow crystalline powders. All K vitamins are resistant to heat and moisture, but they are destroyed on exposure to acid, alkali, oxidizing agents, and light—particularly ultraviolet light.

Measurement/Assay:

Vitamin K can be measured in micrograms of the pure synthetic compound (menadione), and the vitamin K activity of other substances can be expressed in similar terms.

The potency of low-concentration samples, such as occurs in most feeds, is commonly determined by bioassay, using young chicks, and is based on the minimum dose that will maintain the normal coagulation of the blood at the end of 1 month. In pure solutions, vitamin K may be assayed by U.V. spectrophotometry or colorimetric methods. Other accepted techniques are the oxidimetric assay after catalytic reduction to the hydroquinone and the polarographic determination.

Species Most Affected:

All species, but ruminants have the advantage of microbial synthesis.

Functions:

Vitamin K controls blood coagulation; recent research suggests that it acts in some way to convert precursor proteins to the active blood clotting factors. Vitamin K is essential for the synthesis in the liver of four blood-clotting proteins: (1) Factor II, prothrombin; (2) Factor VII, proconvertin; (3) Factor IX, Christmas factor; (4) Factor X, Stuart-Power. The mechanism of blood clotting is shown in Fig. 4-84.

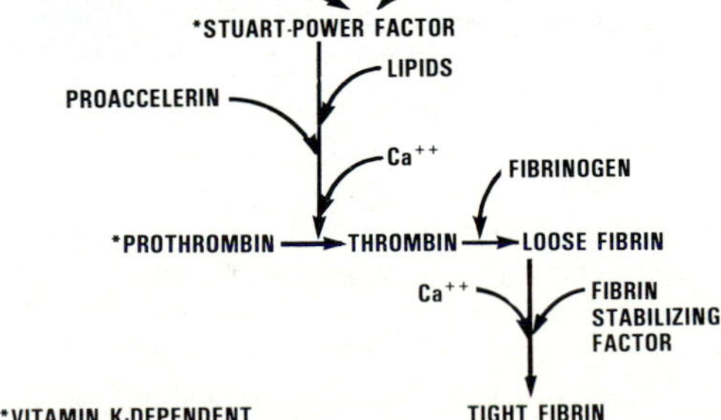

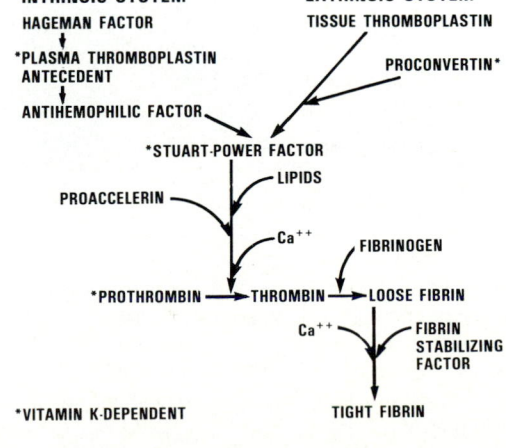

Fig. 4-84. Mechanism of blood clotting.

(Continued)

TABLE 4–8 *(Continued)*

Vitamin K (Continued)

Deficiency Symptoms/Toxicity:

Deficiency Symptoms—Prolonged blood clotting time, generalized hemorrhages, and death in severe cases.

Toxicity—The natural forms of vitamins K_1 and K_2 have not produced toxicity even when given in large amounts. However, synthetic menadione and its various derivatives have produced toxic symptoms in rats.

Sources for Animals:

In general, this factor is widely distributed in normal farm rations. Also, all classes of farm animals synthesize it.

Rich feed sources—Alfalfa meal, barley (whole), corn (whole), fish meal, hays (well cured), milk, pastures (green), peas (green), sorghum grain, soybean meal, wheat.

Supplemental sources—Menadione (vitamin K_3).

Comments:

Well known antagonists of vitamin K are dicoumarol and warfarin.

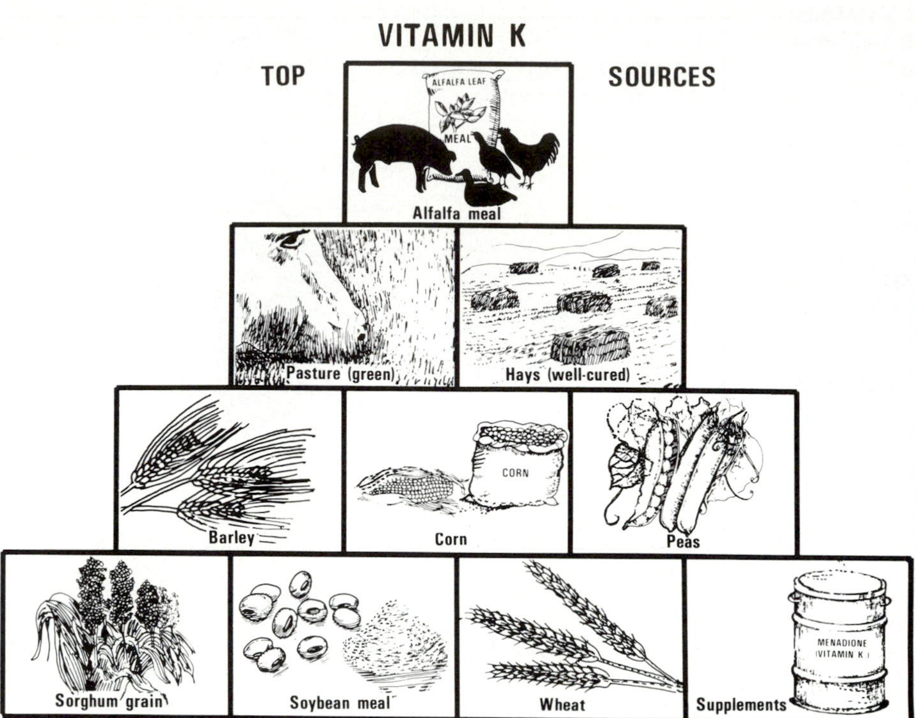

Fig. 4–85. Commonly used, top feed and supplement sources of vitamin K. No claim is made that all the top sources of vitamin K are listed. Neither are the sources ranked according to amount of vitamin K supplied nor is price considered.

WATER-SOLUBLE VITAMINS:

BIOTIN

Biotin made the difference!

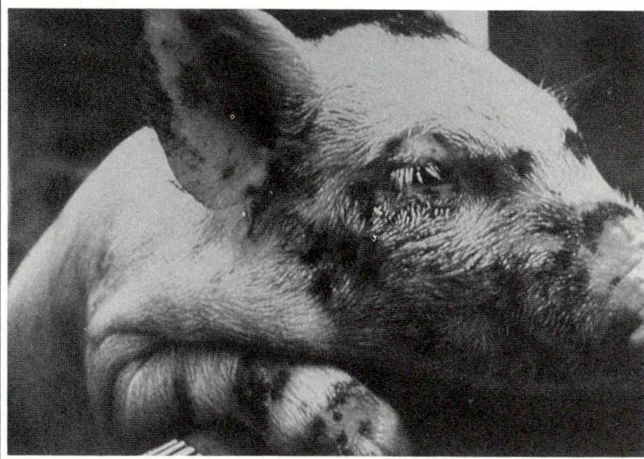

Fig. 4–86. *Left:* Biotin-deficient pig. Note brown crusts on the skin. *Right:* Same pig after biotin supplementation by injection. No crusts on skin. (Courtesy, Roche Chemical Division, Hoffman-LaRoche, Inc., Nutley, N.J.)

Biotin, a member of the vitamin B complex, is a water-soluble, sulfur-containing vitamin that is widely distributed in nature and essential for the health of many animal species. It plays an important role in the metabolism of carbohydrates, fats, and proteins.

History:

The history of biotin is the history of the merging of investigations, a chronological record of which follows:

In 1901, Wildiers expressed the belief that yeast required for its nutrition an organic substance which he called *bios.*

In 1916, Bateman reported that raw egg white had a detrimental effect on rats, but that the raw egg white was rendered *innocuous* by heat coagulation.

(Continued)

TABLE 4–8 *(Continued)*

Biotin (Continued)

In 1927, Boas in England reported that feeding raw egg white to rats produced a dermatitis.

In 1933, Allison and co-workers isolated a nitrogen-fixing bacteria in legume nodules, which they named coenzyme R.

In 1936, Kogl and Tonnis of Germany isolated a crystalline substance from boiled yolks of duck eggs, which they called *biotin* because they believed it to be identical to the *bios* factor needed for yeast growth.

In 1937, Gyorgy, a Hungarian scientist, found that a substance which he named vitamin H, would prevent the pathological condition that resulted from feeding rats and chicks raw egg white.

In 1940, Gyorgy and associates obtained conclusive experimental evidence that coenzyme R, biotin, and vitamin H were the same substance.

In 1942, du Vigneaud and associates, at Cornell, suggested the correct structural formula for biotin based on a study of its degradation products.

In 1943, Harris and co-workers of Merck and Co. synthesized biotin.

Looking back, it is now known that the deficiency (once called egg-white injury) occurs when the biotin in food combines with a factor in the protein of uncooked egg white (called avidin, because of its avidity for biotin); and that when egg white is cooked, avidin is inactivated. This also explains why, in early studies, liver and yeast offered protection against egg-white injury. (Both of them contain sufficiently large amounts of biotin to saturate the avidin completely and leave a surplus of biotin available to meet the needs of experimental animals.)

Chemistry/Metabolism/Properties:

Chemistry—Biotin, like thiamin, is a sulfur-containing vitamin. It is a cyclic derivative of urea with an attached thiophene ring. Its structure is given in Fig. 4–87.

Metabolism—Biotin is absorbed primarily from the upper part of the small intestine. However, avidin, a protein found in raw egg white, binds biotin and prevents its absorption from the intestinal tract. Fortunately, cooking inactivates avidin so that it no longer has the ability to bind biotin.

A considerable amount of biotin is synthesized by intestinal bacteria, as evidenced by the fact that more biotin is excreted in the urine and feces than is ingested. But synthesis in the gut may occur too late in the intestinal passage to be absorbed well and play much of a direct role as a biotin source. Also, several variables affect the microbial synthesis in the intestines, including the carbohydrate source of the diet (starch, glucose, sucrose, etc.), the presence of other B vitamins, and the presence or absence of antimicrobial drugs and antibiotics.

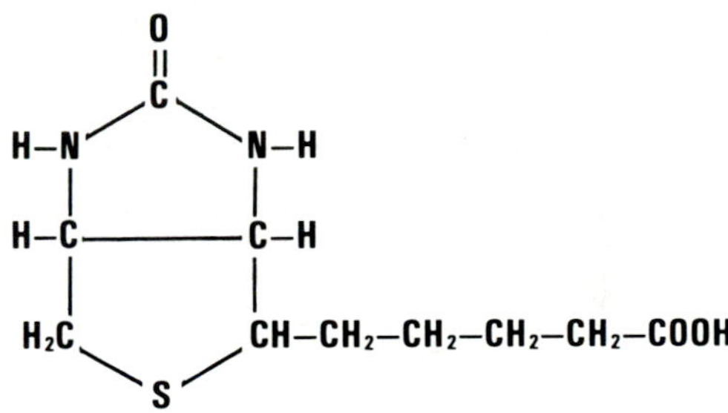

Fig. 4–87. Structure of biotin.

Following absorption, biotin enters the portal circulation. It is stored primarily in the liver and kidneys, although all cells contain some biotin.

Excretion is mainly in the urine. Only traces of biotin are secreted in milk.

Determination of biotin levels in the blood, and urinary excretion levels of biotin, both provide evidence of biotin status.

Properties—Biotin is a colorless, odorless, crystalline substance. It is readily soluble in hot water, but only slightly soluble in cold water; and it is stable to heat. It is destroyed by strong acids and alkalis and by oxidizing agents. Also, it is gradually destroyed by ultraviolet light.

Measurement/Assay:

No International Units have been defined for the biological activity of biotin.

Analytical results are generally expressed in terms of weight units of pure d-biotin. Purified, high-potency solutions of biotin may be assayed by a photometric method based on splitting of an avidin dye complex by biotin. Also, gas-liquid chromatography is sometimes used. However, the microbiological assay is the method of choice for biotin assay of feeds. The biotin content of feeds can be determined by using the rat or the chick. Biological assays with rats or chicks are more reliable than the other methods because of determining availability; for example, it has been shown that the biotin in milo, oats, and wheat is less available to the chick than the biotin in corn.

Species Most Affected:

Required by all species.

Functions:

Biotin is required for many reactions in the metabolism of carbohydrates, fats, and proteins. It functions as a coenzyme mainly in decarboxylation-carboxylation and in deamination.

Biotin serves as a coenzyme for transferring CO_2 from one compound to another (for decarboxylation—the removal of carbon dioxide; and for carboxylation—the addition of carbon dioxide), as shown in Fig. 4–88. Numerous decarboxylation and carboxylation reactions are involved in carbohydrate, fat, and protien metabolism; among them, the following:

1. Interconversion of pyruvate and oxaloacetate. The formation of oxaloacetate is important because it is the starting point of the tricarboxylic acid cycle (TCA), known also as the Krebs cycle, in which the potential energy of nutrients (ATP) is released for use by the body.

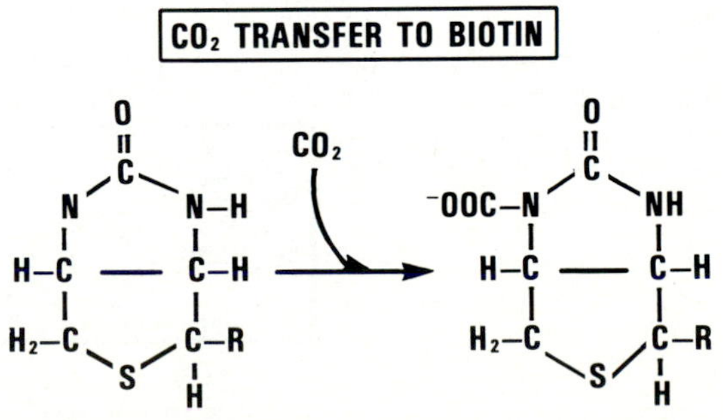

CO_2 TRANSFER TO BIOTIN

Fig. 4–88. Transfer of CO_2 by biotin.

(Continued)

TABLE 4-8 *(Continued)*

Biotin (Continued)

2. Interconversion of succinate and propionate.
3. Conversion of acetyl CoA to malonyl CoA, the first step in the formation of long chain fatty acids (fat synthesis).
4. Formation of purines, essential part of DNA and RNA, and for protein synthesis.
5. Conversion of ornithine to citrulline, an important reaction in the formation of urea.

Biotin also serves as a coenzyme for deamination (removal of $-NH_2$) reactions that are necessary for the production of energy from certain amino acids (at least of aspartic acid, serine, threonine); for amino acids to be used as a source of energy, they must first be deaminated—the amino group must be split off.

Deficiency Symptoms/Toxicity:

Raw egg white contains an antivitamin, a protein called *avidin,* which binds to biotin and renders it unavailable to animals. However, avidin is denatured when eggs are cooked, thus making biotin available to the animal.

In all animals, a deficiency of biotin will depress growth and cause a loss of hair and/or a dermatitis condition.

Deficiency Symptoms—Pigs exhibit spasticity of the hind legs, cracks in the feet, and a dermatitis. There is also lowered efficiency of feed utilization.

Chicks and turkey poults show dermatitis and perosis.

In hens, egg production is not affected, but hatchability is severely reduced.

In mink, biotin deficiency interferes with normal pigmentation of the fur.

Toxicity—There are no known toxic effects.

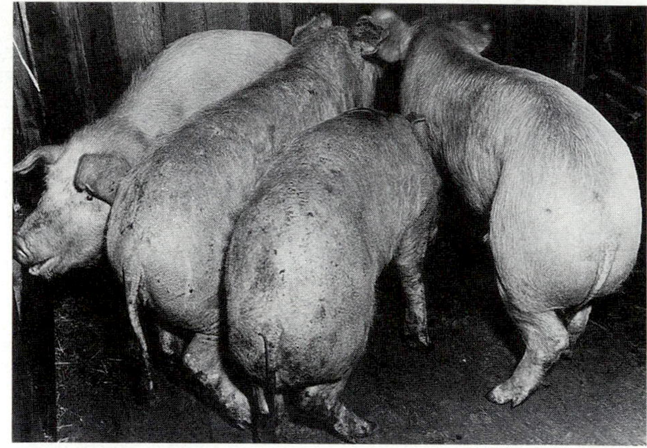

Fig. 4–89. Biotin deficiency in swine. The two pigs in the middle were fed a biotin-deficient ration; note their loss of hair, dry/scaly skin, and signs of lameness. (Courtesy, Washington State University)

Sources for Animals:

Rich feed sources—Alfalfa meal, barley malt sprouts, blackstrap molasses, brewers' grains, corn gluten feed, cottonseed meal, distillers' grains, distillers' solubles, fish solubles, liver meal, milk (whole, skim), peanut meal, rice bran, safflower meal, sorghum grain, soybean meal, soybean seeds, wheat bran, wheat mill run, whey.

Supplemental sources—Alfalfa leaf meal (dehydrated), rice polishings, synthetic biotin, yeast (brewers', torula).

Comments:

Biotin is closely related metabolically to folacin, pantothenic acid, and vitamin B–12.

Ordinary farm rations probably contain ample biotin, or farm animals synthesize all they need. However, the presence of interfering factors, like avidin, in diets may cause biotin deficiency.

Considerable synthesis of biotin occurs through intestinal fermentation. Sulfa drugs can kill these intestinal organisms; hence, when such drugs are used, the biotin allowance should be increased.

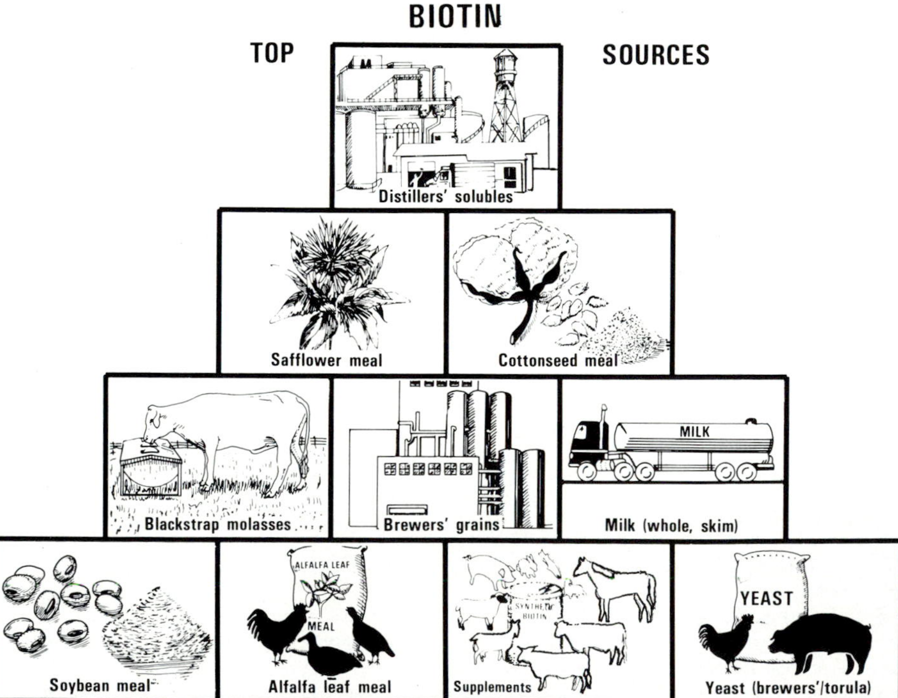

Fig. 4–90. Commonly used, top feed and supplement sources of biotin. No claim is made that all the top sources of biotin are listed. Neither are the sources ranked according to amount of biotin supplied nor is price considered.

(Continued)

TABLE 4-8 *(Continued)*

CHOLINE

Choline made the difference!

Fig. 4-91. *Left:* Choline-deficient pig. *Right:* Pig that received adequate choline. (Courtesy, University of Illinois, Urbana)

Choline, which is a key part of the constituent of lecithin, is vital for the prevention of fatty livers, the transmitting of nerve impulses, and the metabolism of fat. The classification of choline as a vitamin is debated, however, because it does not meet all the criteria for vitamins, especially those of the B vitamins. One reason for this questioning stems from the fact that choline is utilized in much larger quantities than the other B vitamins. Additionally, the body itself can synthesize considerable amounts of choline, thus reducing the need for dietary supplementation. Specific deficiency symptoms have not been observed in humans. Of course, where to draw the line between an essential nutritional factor with vitamin-like activity and a vitamin is no easier than setting any boundary in nature.

History:

In 1844 and 1846, N. T. Gobley isolated a substance from egg yolk, which he called *lecithin* (from the Greek, *lekithos,* egg yolk. Choline is a basic constituent of lecithin).

In 1849, Strecker, a German chemist, isolated a compound from hog bile, to which he subsequently (in 1862) applied the name *choline* (after *chole,* the Greek word for bile).

In 1866–67, Baeger and Wurtz, working independently, determined the correct structure of choline and carried out the first synthesis of it. But the compound did not attract the attention of nutrition investigators at the time.

In 1932, Best and co-workers at Toronto University reported that choline prevented fatty livers in rats fed high-fat diets.

In 1940, Sure and Gyorgy-Goldblatt, working independently, reported that choline is essential for growth of rats, thereby indicating its vitamin nature.

By 1942, the vitamin nature of choline was fully confirmed by many other workers, who used the rat, chicken, and turkey as experimental animals.

Chemistry/Metabolism/Properties:

Chemistry—Structurally, choline ($C_5H_{15}NO_2$) is a relatively simple molecule, containing three methyl groups (CH_3–)(see Fig. 4-92).

Metabolism—Choline is absorbed from the small intestine.

Properties—It is a colorlesss, bitter-tasting, water-soluble white syrup that takes up water rapidly on exposure to air (hygroscopic) and readily forms more stable crystalline salts with acids such as choline chloride or choline bitartrate. It is fairly stable to heat and storage, but unstable to strong alkali. It exists in all feeds in which phospholipids occur liberally.

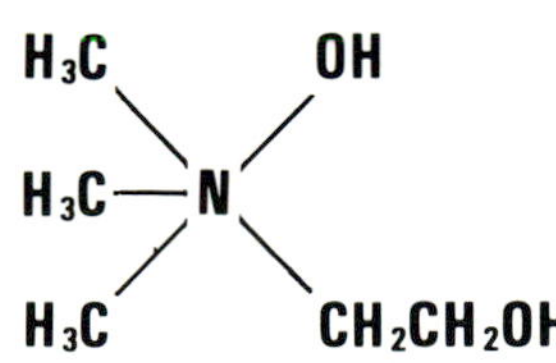

Fig. 4-92. Structure of choline.

Measurement/Assay:

The activity of choline and choline chloride is expressed in grams and milligrams of the chemically pure substances.

Choline content of feeds is usually determined by a colorimetric method or by microbiological assay. Recent assay techniques include: fluorometric enzyme assay, enzymatic radioisotopic assay, and gas chromatography.

Species Most Affected:

Swine, rats, and poultry.

Functions:

Choline has several important functions; it is vital for the prevention of fatty livers, the transmitting of nerve impulses, and the metabolism of fat.

1. It prevents fatty livers through the transport and metabolism of fats. Without choline, fatty deposits build up inside the liver, blocking its function and throwing the whole body into a state of ill health.

(Continued)

TABLE 4-8 *(Continued)*

Choline (Continued)

2. It is needed for nerve transmission. Choline combines with acetate to form acetylcholine, a substance which is needed to jump the gap between nerve cells so that impulses can be transmitted.

3. It serves as a source of labile methyl groups which facilitate metabolism by a phenomenon known as *transmethylation.* Choline can be replaced as a donor of labile methyl groups either by a structurally related compound called betaine, or by the amino acid methionine. In the presence of excess methionine, choline can be formed, a process whereby methyl groups are transferred from methionine to form choline. Likewise, methyl groups from choline can be used to produce methionine from homocysteine. This interchange of methyl groups is referred to as *transmethylation.*

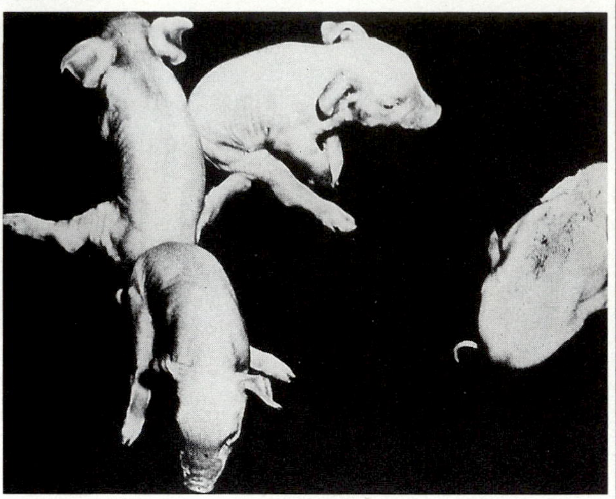

Fig. 4-93. Choline-deficient pigs. Note the spraddled legs. This condition, which was produced on a purified ration, was prevented by choline supplementation. (Courtesy, Washington State University, Pullman)

Deficiency Symptoms/Toxicity:

Deficiency Symptoms—Poor growth and fatty livers are the deficiency symptoms in most species except chickens and turkeys. Chickens and turkeys develop slipped tendons (perosis). In young rats, choline deficiency produces hemorrhagic lesions in the kidneys and other organs. Choline deficiency in swine causes an abnormal gait in growing pigs and reproductive failure in adult females.

Toxicity—No toxic effects have been observed.

Sources for Animals:

The choline content of normal feeds is sufficient. With a high-protein ration, enough choline is synthesized from certain precursors and amino acids. Deficiency symptoms are more likely as the protein content is lowered.

Rich feed sources—Alfalfa meal, blackstrap molasses, brewers' grains, buttermilk (dried), cereal grains (barley, corn, oats, rice, sorghum, wheat), cotton seed meal, distillers' grains, distillers' solubles, fish meal and other marine by-products, hominy, meat meal/tankage and other meat by-products, milk (fresh, skimmed), oats, potatoes, poultry by-product meal, rapeseed meal, rice bran, safflower meal, soybean meal, soybean seeds, wheat bran, whey.

Supplemental sources—Synthetic choline and choline derivatives, rice polish, soybean, lecithin, wheat germ, yeast (brewers', torula).

Also, the body manufactures choline from methionine with the aid of folacin and vitamin B-12. So, the needs for choline can be supplied in two ways: (1) by dietary choline, and/or (2) by body synthesis through transmethylation.

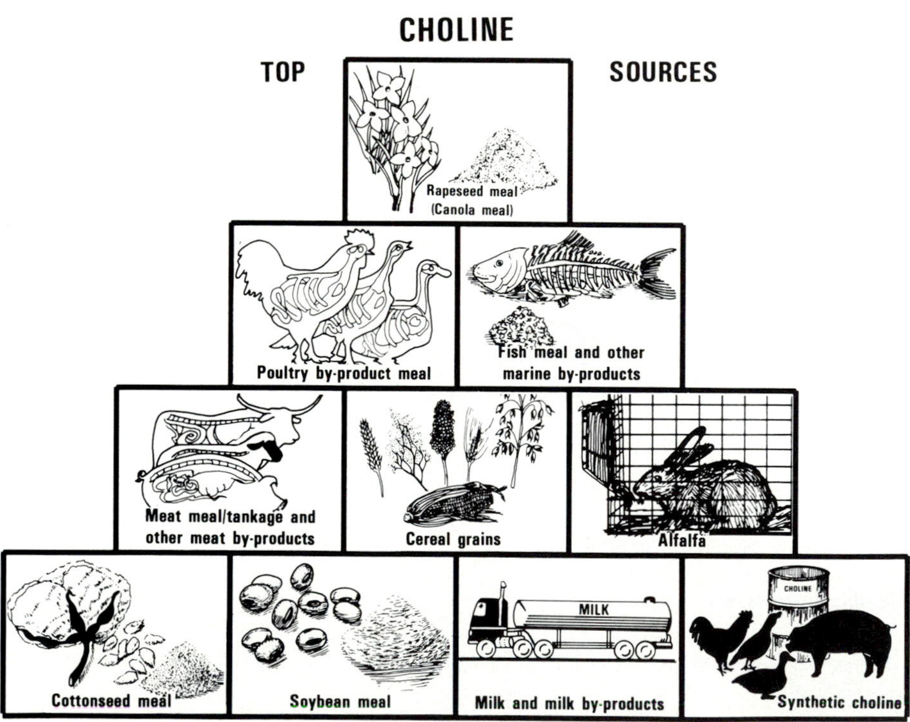

Fig. 4-94. Commonly used, top feed and supplement sources of choline. No claim is made that all the top sources of choline are listed. Neither are the sources ranked according to amount of choline supplied nor is price considered.

Comments:

It is noteworthy that choline has been known for a very long time. It was isolated in 1849, named in 1862, and synthesized in 1866–67. But the compound did not attract the attention of nutrition investigators at the time.

(Continued)

TABLE 4-8 *(Continued)*

FOLACIN (FOLIC ACID)

There is no single compound vitamin with the name *folacin;* rather, the term *folacin* is used to designate folic acid (pteroylmonoglutamic acid, or PGA) and a group of closely related substances which are essential for all vertebrates, for normal growth and reproduction, for the prevention of blood disorders, for important biochemical mechanisms within each cell, and for the prevention of a variety of deficiency symptoms in different species.

The research work related to folacin is considered to be one of the most complicated chapters in the story of the B-complex vitamins.

History:

Long before folacin was isolated or synthesized, its deficiency symptoms had been described in animals, humans, and microorganisms.

In 1941, the name *folic acid,* the forerunner of the term *folacin,* was suggested by Mitchell, Snell, and Williams, of Texas, for the growth factor for bacteria that were found in spinach and known to be widely distributed in green leafy plants. The word *folic* is derived from the Latin word *folium,* meaning foliage or leaf.

In 1945, Angier and co-workers isolated and synthesized folic acid.

Through the years, several names other than folacin and folic acid have been given to this vitamin, including: Wills' factor, pteroylmonoglutamic acid, antianemia factor PGA, vitamin M, vitamin B_c, SLR factor, factor R, factor U, vitamin U, *Lactobacillus casei* factor; citrovoran factor (CF); yeast Norite eluate factor, vitamin B-9, vitamin B-10, and vitamin B-11.

Chemistry/Metabolism/Properties:

Chemistry—This vitamin exists in several different forms in nature, making up the folacin group of compounds. All forms have similar activity when fed to higher animals, but they have widely different activities as growth factors for microorganisms. The structural formula of pure folic acid (pteroylmonoglutamic acid) is given in Fig. 4-95. If the parent molecule (consisting of pteridine, para-aminobenzoic acid, and glutamic acid) is broken, nutritional acitivity is lost. The biologically active form of folacin is a reduction product called tetrahydrofolic acid.

FOLIC ACID

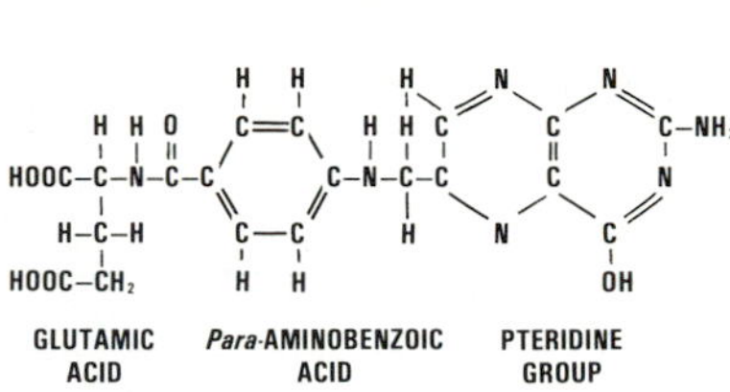

Fig. 4-95. Structure of folic acid (pteroylmonoglutamic acid).

Metabolism—Wide variation exists in the absorption of folates from different feeds, with a low of about 10% and a high of about 80%.

Folate is absorbed by active transport and by diffusion, mainly in the upper part of the small intestine, although some is absorbed along the entire length of the small intestine; glucose, ascorbic acid, and some antibiotics facilitate its absorption.

Folate, bound to protein, is transported in the blood to the liver. There it is methylated and carried to the bone marrow cells, the maturing red blood cells, and perhaps to other cells. Methyl-folate seems to be the chief form of the vitamin in body tissues. Serum levels of folacin range from 7 to 16 nanograms per milliliter of serum. The total body stores of folate normally range between 5 and 12 mg, about half of which is in the liver.

Some folate is excreted in the bile, as well as in the urine.

Properties—Folic acid is a bright yellow crystalline powder, slightly soluble in water, unstable in acid solution, relatively unstable to heat, and rather easily destroyed upon exposure to light.

Measurement/Assay:

Folic acid is measured in micrograms or nanograms (ng = millimicrograms).

Folic acid in feed is usually assayed either biologically by chick or rat growth tests, or microbiologically with *L. casei* or another suitable microorganism. Larger quantities of folacin activity may be measured chemically, fluorometrically, or by paper and thin-layer chromatography.

Species Most Affected:

All animals and birds may be affected.

Functions:

In the body, folic acid is changed to at least five active enzyme forms, the parent form of which is tetrohydrofolic acid. Folacin coenzymes are responsible for the following important functions:

1. The formation of purines and pyrimidines which, in turn, are needed for the synthesis of the nucleic acids DNA and RNA, vital to all cell nuclei. This explains the important role of folacin in cell division and reproduction.
2. The formation of heme, the iron-containing protein in hemoglobin.
3. The interconversion of the 3-carbon amino acid serine from the 2-carbon amino acid glycine.
4. The formation of the amino acids tyrosine from phenylalanine and glutamic acid from histidine.
5. The formation of the amino acid methionine from homocysteine
6. The synthesis of choline from ethanolamine.
7. The conversion of nicotinamide to N-methylnicotinamide, one of the metabolites of niacin that is excreted in the urine.

Deficiency Symptoms/Toxicity:

Deficiency Symptoms—Megaloblastic anemia (of young) and macrocytic anemia (of pregnancy). In chicks, growth is retarded, and depigmentation occurs in colored feathers. Although synthesis occurs in the rumen, newborn lambs require a dietary supply. In humans and dogs, deficiency of folacin is characterized by a sore, red, smooth tongue (glossitis), disturbances of the digestive tract (diarrhea), and poor growth.

Toxicity—Normally, excess folic acid has no adverse effect.

(Continued)

TABLE 4–8 *(Continued)*

Folalcin (Folic Acid) (Continued)

Sources for Animals:

Rich feed sources—Alfalfa hay/meal, cottonseed meal, distillers' grains, distillers' solubles, fish meal, linseed meal, meat meal/tankage and other meat by-products, milk (skimmed), peanut meal, potatoes (dehydrated), rice bran, rye (grain), safflower meal, soybean meal, wheat and wheat by-products, whey.

Supplemental sources—Synthetic folic acid (pteroylglutamic acid, or PGA), wheat germ, yeast (brewers', torula).

Comments:

Folic acid is widely distributed in both plants and animals. It was given this name because of the abundance of the factor in plant leaves.

Ascorbic acid, vitamin B–12, and vitamin B–6 are essential for the activity of the folacin coenzymes in many of their metabolic processes; again and again, pointing up the interdependence of various vitamins.

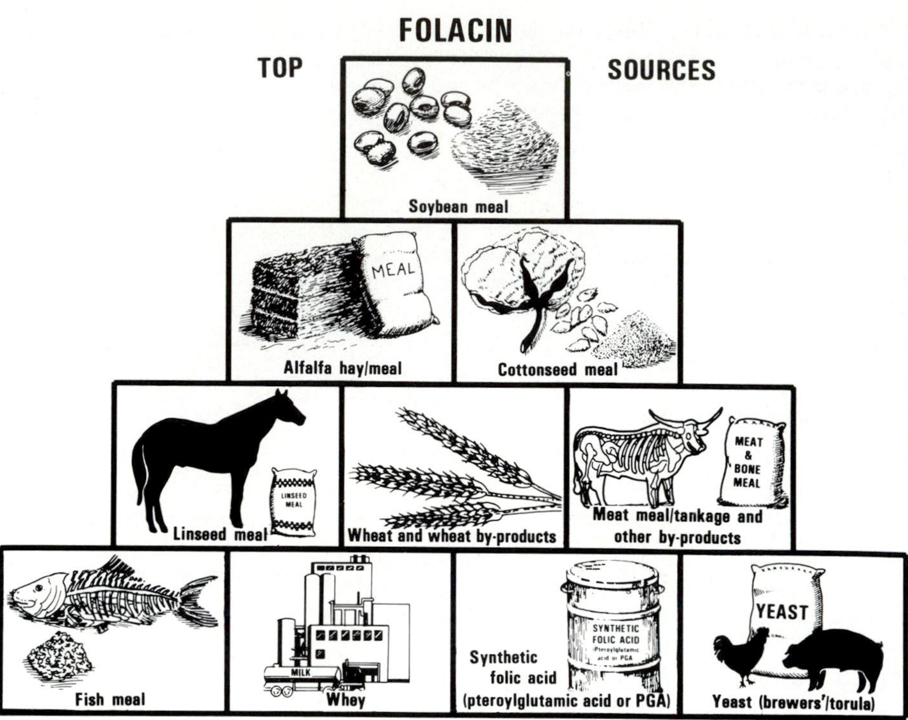

Fig. 4-96. Commonly used, top feed and supplement sources of folacin. No claim is made that all the top sources of folacin are listed. Neither are the sources ranked according to amount of folacin supplied nor is price considered.

INOSITOL

Inositol, which has been known as a chemical compound since 1850, is widely distributed in feeds and closely related to glucose. It was first commonly called *muscle sugar* and given the name *inositol* from two Greek roots: *inos,* meaning sinews; and *-ose,* the suffix for sugars.

Animal experiments conducted in the 1940s indicated that inositol was an essential nutritional factor and led many investigators to group it with the B vitamins. Today, its classification as a vitamin is disputed; more properly perhaps, it should be classified as an essential nutrient, rather than a vitamin, for certain species of bacteria and animals. Nevertheless, listing inositol among the B vitamins persists in some books, catalogs, and diet-ingredient lists, and on some labels.

History:

In 1940, Woolley at the University of Wisconsin demonstrated that inositol could prevent alopecia (patchy-hair, or baldness, condition) in mice. Later studies demonstrated that rats fed inositol-deficient diets developed a denuded area around the eyes that imparted a curious *spectacled-eye* appearance. Research has also indicated a need for dietary inositol for chicks, swine, hamsters, and guinea pigs.

Chemistry/Metabolism/Properties:

Chemistry—Inositol is a cyclic 6-carbon compound with 6 hydroxy groups, closely related to glucose. It exists in nine forms, but only myo-inositol demonstrates any biological activity (see Fig. 4–97). The hexaphosphoric acid ester of inositol is phytic acid, a compound that ties up phosphorus, thus making the phosphorus unavailable for absorption in animals.

Metabolism—In addition to feed sources of inositol, it is synthesized with the cells. Myo-inositol is present in relatively large amounts in the cells of practically all animals and plants.

In animal cells, it occurs as a component of phospholipids, substances containing phosphorus, fatty acids, and nitrogenous bases. In plant cells, it is found as phytic acid, an organic acid that binds calcium, iron, and zinc in an insoluble complex and interferes with their absorption.

Inositol is stored largely in the brain, heart muscle, and skeletal muscles.

Small amounts of inositol are normally excreted in the urine.

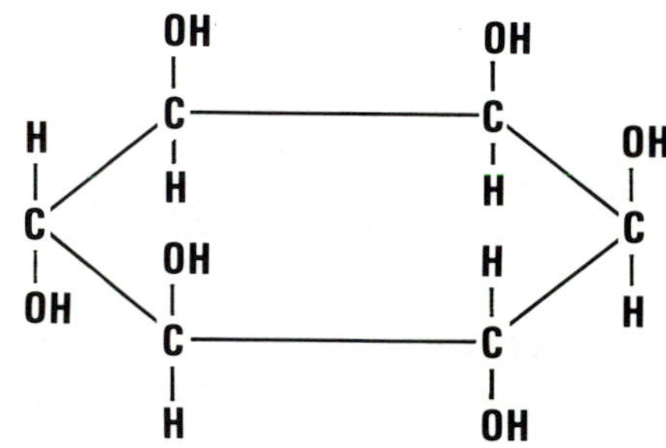

Fig. 4-97. Structure of myo-inositol ($C_6H_{12}O_6$).

(Continued)

TABLE 4-8 *(Continued)*

Inositol (Continued)

Properties—Inositol is a colorless, water-soluble, sweet-tasting crystalline material. It can withstand acids, alkalies, and heat.

Measurement/Assay:

Inositol is measured in milligrams.

Formerly, myo-inositol analysis was by microbiological method only, based on growth of certain yeasts. Later, a time-consuming chemical method became available. Today, microbiological and chemical assays are giving way to chromatographic methods and enzyme assays.

Species Most Affected:

Chicks, fish (several species), guinea pigs, hamsters, swine, rats, mice.

Functions:

The functions of inositol are not completely understood, but the following roles have been suggested:

1. It has a lipotropic effect (an affinity for fat, like choline). In this role, inositol aids in the metabolism of fats and helps reduce blood cholesterol.
2. In combination with choline, inositol prevents the fatty hardening of arteries and protects the heart.
3. It appears to be a precursor of the phosphoinosities, which are found in various body tissues, especially in the brain.

Deficiency Symptoms/Toxicity:

Deficiency Symptoms—Myo-inositol is a *growth factor* for certain yeasts and bacteria, and for several lower organisms up to and including several species of fish.

Earlier experiments indicated that a deficiency of inositol caused retarded growth and loss of hair in young mice, and loss of hair around the eyes (spectacled-eyes) in rats. But these symptoms are now being questioned because the experimental diets fed were partially deficient in certain other vitamins.

Toxicity—There is no known toxicity of inositol.

Sources for Animals:

Rich feed sources—Blackstrap molasses, cereal grains (barley, corn, oats, rice, sorghum, wheat), citrus meal, legumes, liver meal, meat meal/tankage, milk, rice bran, wheat bran.

Supplemental sources—Synthetic inositol, wheat germ, yeast.

Comments:

There is no evidence that animals cannot synthesize all the inositol needed by the body. So, its classification as a vitamin is disputed.

Inositol is widely distributed in animal feeds.

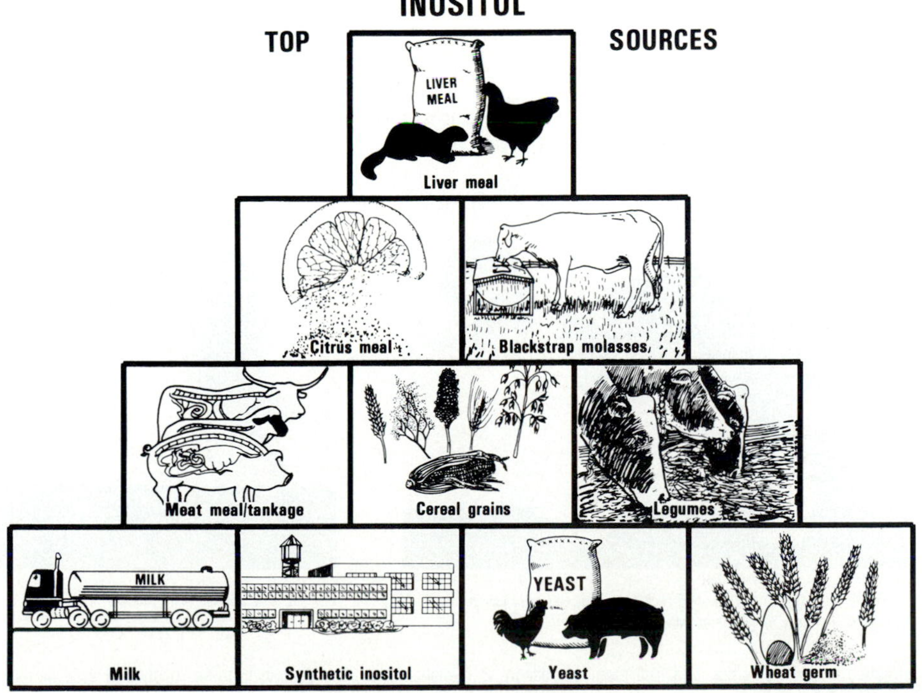

Fig. 4-98. Commonly used, top feed and supplement sources of inositol. No claim is made that all the top sources of inositol are listed. Neither are the sources ranked according to amount of inositol supplied nor is price considered.

(Continued)

TABLE 4-8 *(Continued)*

NIACIN (NICOTINIC ACID, NICOTINAMIDE)

Niacin made the difference!

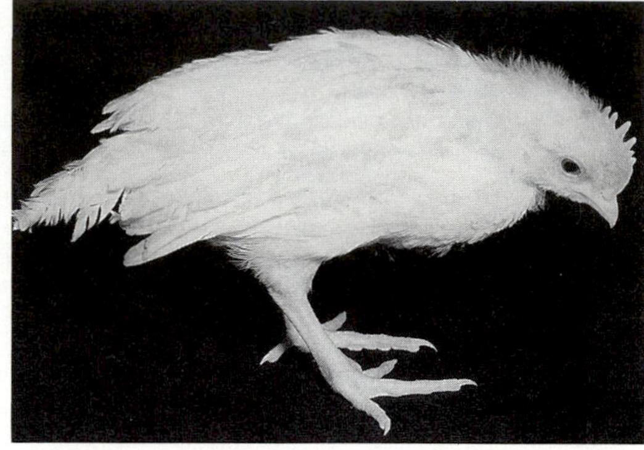

Fig. 4-99. *Left:* Chick on niacin-deficient diet. Note poor growth and abnormal feathering. *Right:* Chick that received plenty of niacin. (Courtesy, University of Wisconsin, Madison)

Niacin, a member of the B complex, is a collective term which includes nicotinic acid and nicotinamide, both natural forms of the vitamin with equal niacin activity. In the body, they are active as nicotinamide adenine dinucleotide (NAD) and nicotinamide adenine dinucleotide phosphate (NADP) and serve as coenzymes, often in partnership with thiamin and riboflavin coenzymes, to produce energy within the cells, precisely when needed and in the amount necessary.

The discovery of the role of niacin as a vitamin of the B group was the result of humanities' age-old struggle against pellagra. The disease was first described by the physician Gaspar Casal in Spain in 1730, soon after the introduction of corn (maize) into Europe; and it was given the name pellagra (*pelle,* for skin; and *agra* for sour) by physician Francesco Frapoli in Italy in 1771.

Pellagra spread with the spread and cultivation of corn and its use in the human diet. In the 19th century, it was common in almost all of the European and African countries bordering on the Mediterranean Sea; and it later spread to other African countries. Pellagra has long been present in both North and South America, but it reached epidemic proportions in southern United States after the Civil War, which left poverty in its wake, as a result of which many of the poor subsisted almost entirely on corn. Outbreaks of the disease were so widespread and severe that most physicians considered the cause to be either an infectious agent or a toxic substance present in spoiled corn.

History:

Nicotinic acid was first discovered and named in 1867, when Huber, a German chemist, prepared it from the nicotine of tobacco. But for the next 70 years it remained idle on the chemists' shelves because no one thought of it, even remotely, as a cure for pellagra. In the meantime, thousands of people died from the disease.

In the early 1900s, pellagra reached epidemic proportions in southern United States, where the diet was based primarily on corn, which is extremely low in both available niacin and tryptophan. In 1915, 10,000 people died of the disease; and, in 1917–18, there were 200,000 cases of pellagra in this country.

In 1914, the U.S. Public Health Service dispatched a team under the direction of Dr. Joseph Goldberger, a physician-researcher, to study the cause of, and hopefuly find a cure for, pellagra. In a series of studies initiated in 1914 and continuing throughout the 1920s, Goldberger proved that the disease was caused by a dietary deficiency, and not an infection or toxin.

In 1926, Goldberger and Wheeler showed that pellagra in people and blacktongue in dogs were similar. Further, they confirmed that dogs with blacktongue could be cured by yeast. Goldberger and Wheeler designated this preventive and curative factor, present in certain foods, as P-P (pellagra-preventive); others designated it vitamin G for Goldberger.

In 1937, Dr. Conrad Elvehjem, and co-workers at the University of Wisconsin, discovered that niacin (as either nicotinic acid or nicotinic acid amide, which he isolated from liver) cured blacktongue in dogs, a condition recognized as similar to pellagra in humans. (See Fig. 4-100.) Shortly thereafter, several investigators found that niacin was effective in the prevention and treatment of pellagra in humans. Soon, the vitamin became recognized as a dietary essential for humans, monkeys, pigs, chickens, and other species.

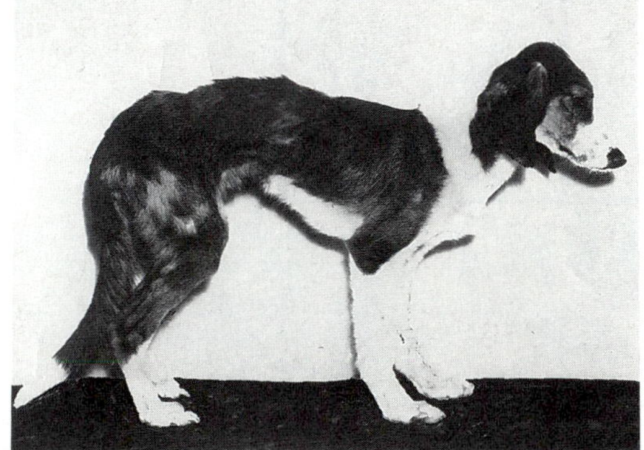

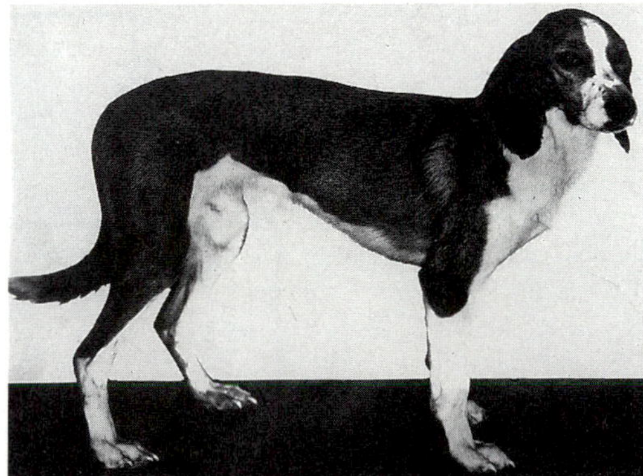

Fig. 4-100. *Left:* A dog that had been fed a diet extremely low in niacin. *Right:* The same dog after having been fed meat—a good source of niacin—for only 2 weeks. (Courtesy, University of Wisconsin, Madison)

In 1945, Willard Krehl and his associates at the University of Wisconsin finally solved another mystery in the story of pellagra prevention when they discovered

(Continued)

TABLE 4-8 *(Continued)*

Niacin (Nicotinic Acid, Nicotinamide) (Continued)

that tryptophan is a precursor of niacin, thereby explaining two things: (1) why milk, which is low in niacin but high in tryptophan, will prevent or cure pellagra; and (2) why, in earlier concepts, protein deficiency was often related to pellagra—for without protein there could be no tryptophan (the precursor of niacin). Corn is low in tryptophan whereas meat contains both tryptophan and niacin.

In 1971, the name *niacin* was adopted by the American Institute of Nutrition and international agencies for all forms of the vitamin.

Today, pellagra is rare in the United States. Even in Latin America and Mexico, where many people eat large amounts of corn, pellagra is seldom seen. This is because of their common practice of soaking the corn in lime, which makes the niacin present in the corn in the bound form (niacytin) more available to the body. This probably explains why Mexicans who eat tortillas are relatively free of pellagra. In making tortillas, the pre-Columbian civilizations of Mexico (Aztec, Mayan, Toltec) devised a procedure to treat corn flour with lime water (alkali) before cooking in order to improve the plastic properties of the dough; and, presumably unbeknown to them, the lime water treatment also freed the niacin from the niacytin and made it fully available to the body tissues.

Africa is the only continent in which pellagra is still a public health problem.

Chemistry/Metabolism/Properties:

The chemistry, metabolism, and properties of nicotinic acid and nicotinamide follow:

Chemistry—The structure of nicotinic acid and nicotinamide are shown in Fig. 4-101.

Metabolism—Niacin is readily absorbed from the small intestine into the portal blood circulation and taken to the liver. There it is converted to the coenzyme nicotinamide adenine dinucleotide (NAD). Also, some NAD is synthesized in the liver from tryptophan. NAD formed in the liver is broken down, releasing nicotinamide, which is excreted into the general circulation. This nicotinamide and the niacin that was not metabolized in the liver are carried in the blood to other body tissues, where they are utilized for the synthesis of niacin-containing coenzymes.

NICOTINIC ACID **NICOTINAMIDE**

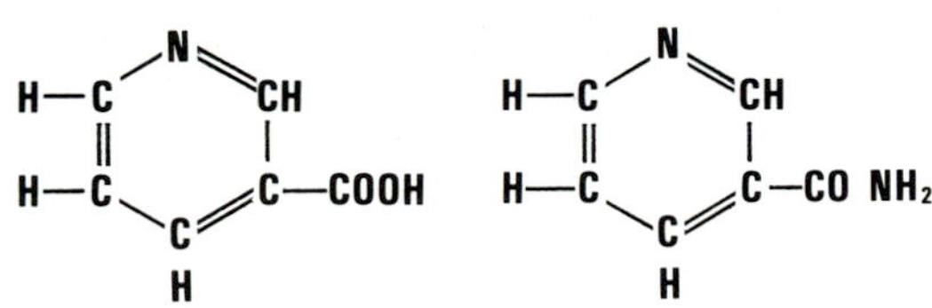

Fig. 4-101. The formulas of nicotinic acid and nicotinamide reveal that the compounds are derivatives of pyrimidine.

Niacin is found in the body tissues largely as part of two important coenzymes, nicotinamide adenine dinucleotide (NAD) and nicotinamide adenine dinucleotide phosphate (NADP); together, NAD and NADP are known as the pyridine nucleotides. The structure of NAD is given in Fig. 4-102.

NAD is composed of nicotinamide, adenine, 2 molecules of ribose, and 2 molecules of phosphate. NADP is similar in structure except it contains 3 phosphate groups.

Little niacin is stored in the body. Most of the excess is methylated and excreted in the urine, principally as N-methylnicotinamide and N-methyl pyridine (in about equal quantities). Also, small amounts of nicotinic acid and niacinamide are excreted in the urine. With a low niacin intake, there is a low level of metabolite excretion in any form.

Properties—Nicotinic acid appears as colorless needlelike crystals with a bitter taste, whereas nicotinamide is a white powder when crystallized. Both are soluble in water (with the amide being more soluble than the acid form) and are not destroyed by acid, alkali, light, oxidation, or heat.

Nicotinic acid is easily converted to nicotinamide in the body. In large amounts, nicotinic acid acts as a mild vasodilator (as a mild dilator of blood vessels), causing flushing of the face, increased skin temperature, and dizziness. Since nicotinamide does not cause these unpleasant reactions, its use is preferred in therapeutic preparations.

NICOTINAMIDE ADENINE DINUCLEOTIDE (NAD⁺)

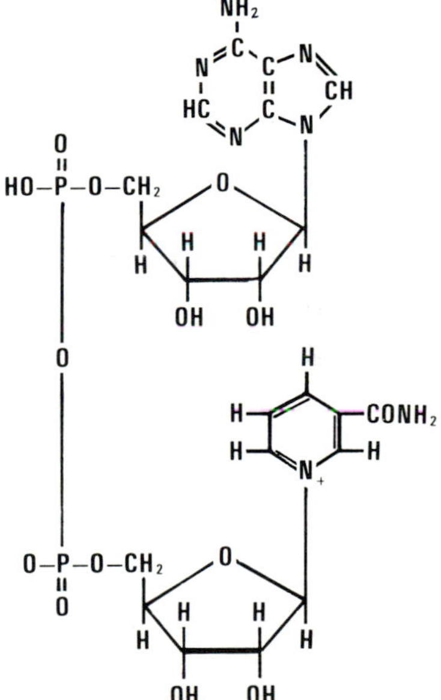

Fig. 4-102. Structure of NAD.

Measurement/Assay:

Niacin in feeds and niacin requirements are expressed in milligrams of the pure chemical substance.

Chemical and microbiological methods for niacin assay are now generally used rather than animal assays.

The biological vitamin activity of new compounds can be assayed by the curative dog test (black-tongue disease) or by growth test with chicks and rats.

Species Most Affected:

Niacin is a dietary essential for pigs, chickens, monkeys, and humans.

Apparently, it is synthesized in the digestive tract of ruminants (sheep and cattle) and the horse.

Functions:

The principal role of niacin is as a constituent of two important coenzymes in the body: nicotinamide adenine dinucleotide (NAD) and nicotinamide adenine dinucleotide phosphate (NADP). These coenzymes function in many important enzyme systems that are necessary for cell respiration. They are involved in the release of energy from carbohydrates, fats, and protein. They function in biological oxidation-reduction systems by virtue of their ability to serve as hydrogen-transfer agents.

NAD and NADP are also involved in the synthesis of fatty acids, protein, and DNA.

Also, niacin is thought to (1) have a specific effect on growth, and (2) reduce the levels of blood cholesterol.

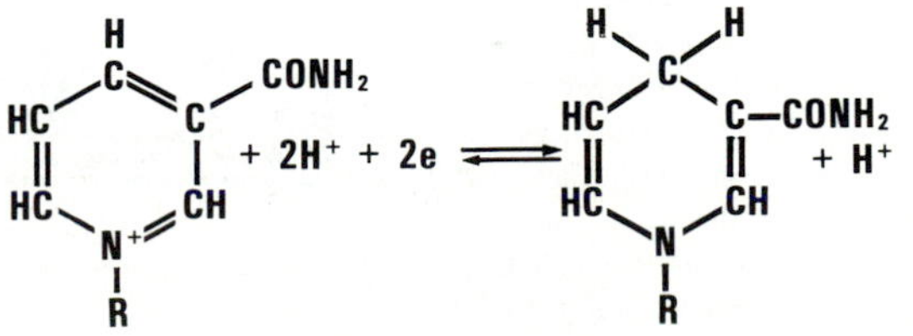

$$NAD^+ + 2H^+ \rightleftharpoons NADH + H$$

Fig. 4-103. Hydrogen acceptor function of nicotinamide containing coenzymes. R = adenine dinucleotide (= NAD): = adenine dinucleotide phosphate (= NADP).

(Continued)

TABLE 4-8 *(Continued)*

Niacin (Nicotinic Acid, Nicotinamide) (Continued)

Deficiency Symptoms/Toxicity:

Deficiency Symptoms—Reduced growth and appetite.

Swine exhibit diarrhea, vomiting, dermatitis, unthriftiness, and ulcerated intestine.

Chicks show poor feathering, scaly dermatitis, and sometimes, a *spectacled eye.*

Dogs show a darkening of the tongue (black-tongue) and mouth lesions.

Humans develop pellagra, characterized by a bright red tongue, mouth lesions, anorexia, and nausea.

Toxicity—Ingestion of large amounts of niacin may result in vascular dilation, or *flushing* of the skin, itching, liver damage, elevated blood glucose, elevated blood enzymes, and/or peptic ulcer.

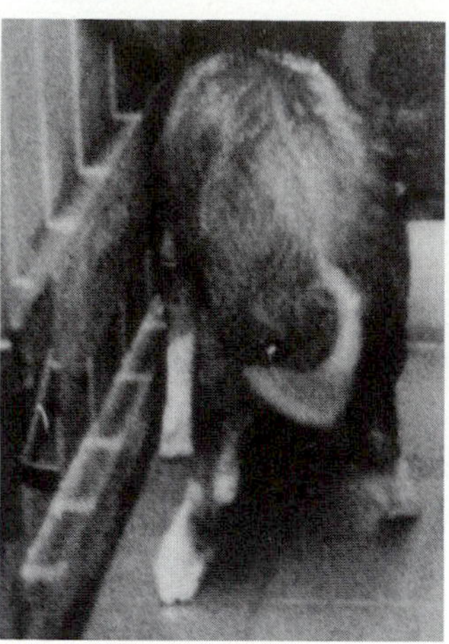

Fig. 4-104. *Left:* Niacin-deficient pig, showing retarded growth. *Right:* Healthy pig that received plenty of niacin. The ages of the 2 pigs were approximately the same, and the rations were identical except that niacin was left out of the ration of the stunted pig on the left. (Courtesy, *Journal of Nutrition* and the University of Utah, Salt Lake City)

Sources for Animals:

Rich feed sources—Barley, corn gluten, distillers' grains, distillers' solubles, fish meal and other marine by-products, grass (immature, dehydrated),mature, dehydrated), liver meal, rice bran, sorghum gluten, wheat and wheat by-products.

Supplemental sources—Nicotinamide (synthetic), nicotinic acid (synthetic), rice polishings, yeast (brewers', torula)

NOTE: When estimating the total amount of niacin supplied in the ration, the tryptophan content of the feeds should be considered, also. Note, too, that most feeds that are rich in animal protein are also rich in tryptophan.

Comments:

Niacin is the most stable of the B complex vitamins.

Niacin present in most cereal grains is not available to the pig and other simple-stomached animals.

Niacin can be synthesized in the body from surplus tryptophan.

Mature ruminants do not need dietary niacin under most conditions because of synthesis by rumen microflora. However, high producing dairy cows may require supplemental niacin for maximum performance.

Born of centuries of experience, (1) in Mexico, corn has long been treated with lime water before making tortillas; and (2) the Hopi Indians of Arizona have long roasted sweet corn in hot ashes. Both practices liberate the nicotinic acid in corn.

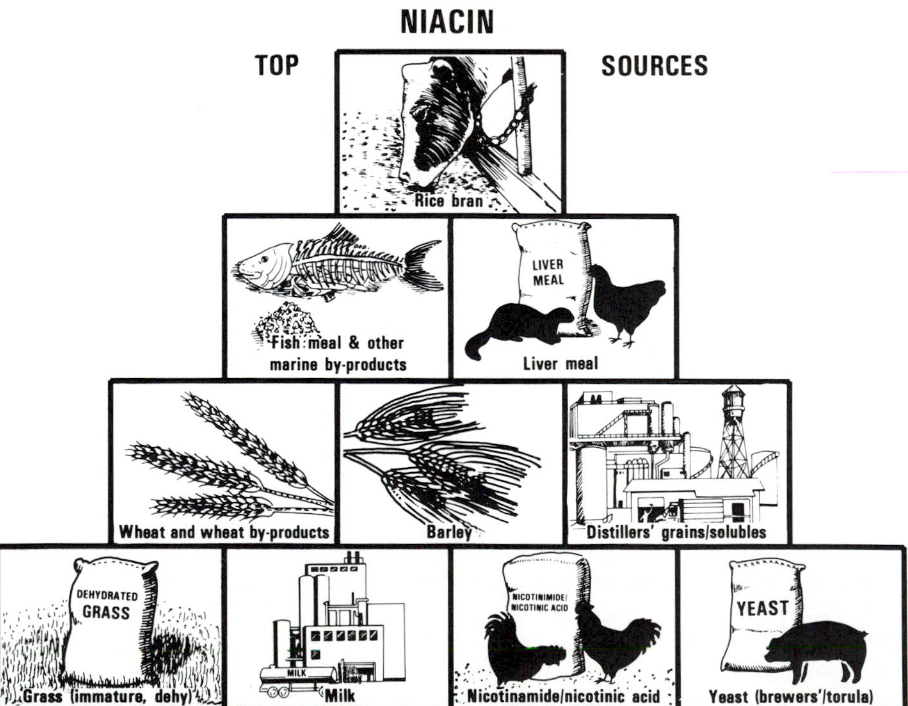

Fig. 4-105. Commonly used, top feed and supplement sources of niacin. No claim is made that all the top sources of niacin are listed. Neither are the sources ranked according to amount of niacin supplied nor is price considered.

(Continued)

TABLE 4-8 *(Continued)*

PANTOTHENIC ACID (VITAMIN B-3)

Pantothenic acid made the difference!

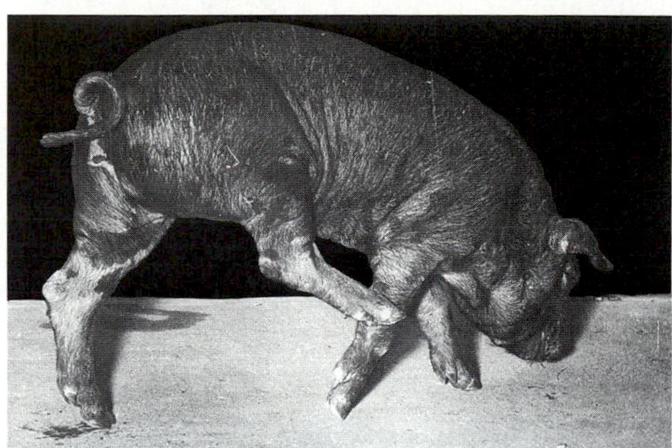

Fig. 4-106. *Left:* Pantothenic acid-deficient pig, showing typical symptoms. Note goose-stepping gait. *Right:* Healthy pig that received plenty of pantothenic acid. (*Left photo:* Courtesy, Michigan State University, East Lansing)

Pantothenic acid, a member of the vitamin B complex, is a dietary essential for animals and humans; and, as an important constituent of coenzyme A (CoA), it plays a key role in energy metabolism.

In recognition of its wide distribution in feeds and foods, the name *pantothenic acid,* derived from the Greek word *pantothen*—meaning everywhere, was first given to it in 1933, by R. J. Williams, then of Oregon State University, later at the University of Texas.

History:

The existence of pantothenic acid and its nutritive significance for the proliferation and fermentative activity of yeast cells first emerged from the studies of R. J. Williams, beginning in 1919. In 1933, Williams fractionated this compound from yeast and called it pantothenic acid; and in 1939, he isolated pantothenic acid from liver. In 1939, T. H. Jukes concluded that the antidermatitis factor isolated from liver was one and the same thing as the factor found in yeast. In 1940, pantothenic acid was synthesized by Williams and scientists at two other laboratories, all working independently. Also, in 1940, pantothenic acid received widespread attention as a possible preventive for gray hair, since it had been observed that the black hair of a rat would turn gray when the animal was deprived of the vitamin; but subsequent studies did not reveal any such benefits to accrue to humans. In 1946, Lipmann and co-workers showed that coenzyme A is essential for acetylation reactions in the body; and in 1950, reports from the same laboratory showed pantothenic acid to be a constituent of coenzyme A.

Chemistry/Metabolism/Properties:

Chemistry—Pantothenic acid is composed of pantoic acid and the amino acid beta-alanine, as shown in Fig. 4-107.

Metabolism—Pantothenic acid, like the other B vitamins, is readily absorbed through the mucosa of the small intestine and enters the portal circulation. Within the tissues, most of the pantothenic acid is used in the synthesis of coenzyme A (CoA); but a significant amount found in the cells is bound to a protein in a compound known as acyl carrier protein (ACP).

Pantothenic acid is present in all living tissue, with high concentrations in the liver and kidney. (Large amounts of CoA are found in the liver, with lesser amounts in the adrenal glands.)

The vitamin is excreted from the body by way of the kidneys.

PANTOTHENIC ACID

```
      H   CH3  OH  O    H   H   H
      |    |    |  ||   |   |   |
HO—C—C—C—C—N—C—C—COOH
      |    |    |           |   |
      H   CH3   H           H   H
```

Fig. 4-107. Structure of pantothenic acid.

Properties—In pure form, pantothenic acid is a viscous yellow oil, soluble in water, quite stable in neutral solutions, but destroyed by acid, alkali, and prolonged exposure to dry heat. Calcium pantothenate, the form in which it is commercially available, is a white, odorless, bitter, crystalline substance, which is water-soluble and quite stable. (Pantothenic acid is also available as the sodium salt.)

Measurement/Assay:

The activity of pantothenic acid is expressed in grams and milligrams of the chemically pure substance.

Pantothenic acid content may be determined by chemical methods (including gas chromotography), microbiologic procedures, and the chick and rat bioassay.

Species Most Affected:

Rats, dogs, pigs, chickens, and turkeys. Synthesized in the rumen of cattle and sheep; perhaps the horse also synthesizes it.

(Continued)

TABLE 4-8 *(Continued)*

Pantothenic Acid (Vitamin B-3) (Continued)

Functions:

Pantothenic acid functions in the body as part of two enzymes—coenzyme A (CoA) and acyl carrier protein (ACP).

Coenzyme A is one of the most important substances in body metabolism. The way in which pantothenic acid is incorporated in the molecule of coenzyme A is shown in Fig. 4-108.

CoA functions in the following important reactions:

1. The metabolic processes by which carbohydrates, fats, and proteins are broken down and energy is released.
2. The formation of acetylcholine, a substance of importance in transmitting nerve impulses.
3. The synthesis of porphyrin, a precursor of heme, of importance in hemoglobin synthesis.
4. The synthesis of cholesterol and other sterols.
5. The steroid hormones formed by the adrenal and sex glands.
6. The maintenance of normal blood sugar, and the formation of antibodies.
7. The excretion of sulfonamide drugs.

ACP, along with CoA, is required by the cells in the biosynthesis (the building up) of fatty acids. (CoA, without ACP, is involved in the breakdown of fatty acids.)

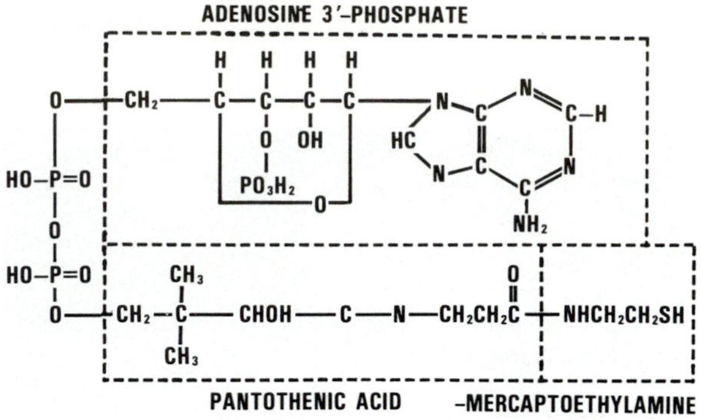

ADENOSINE 3'-PHOSPHATE

PANTOTHENIC ACID —MERCAPTOETHYLAMINE

Fig. 4-108. Structure of CoA.

Deficiency Symptoms/Toxicity:

Deficiency Symptoms—All species exhibit reduced growth, loss of hair, and enteritis.

Mature ruminants synthesize pantothenic acid in the rumen. Signs of deficiency in calves are rough coat, dermatitis, anorexia, and loss of hair around eyes.

Pigs develop a neurological disorder that causes them to walk with a *goose-stepping* gait.

Chicks show dermatitis and embryonic death.

Dogs vomit and develop fatty infiltration of the liver.

Toxicity—Pantothenic acid is relatively nontoxic. Surplus quantities are readily excreted.

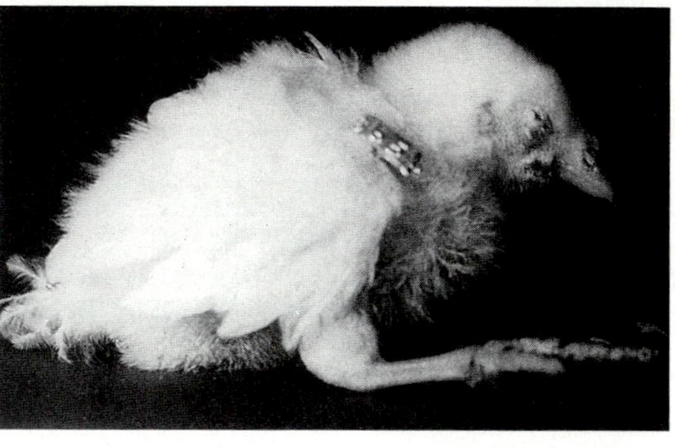

Fig. 4-109. Chick showing an advanced stage of pantothenic acid deficiency. Note the lesions at the corners of the mouth and on the eyelids and feet. (Courtesy, Dept. of Poultry Science, Cornell University, Ithaca, N.Y.)

Sources for Animals:

Rich feed sources—Alfalfa leaves, alfalfa meal, blackstrap molasses, buttermilk, distillers' solubles, fish solubles, liver meal, milk, peanut meal, potatoes (dehydrated), peas (seeds), rice bran, safflower meal, skim milk, wheat bran, wheat red dog, whey.

Supplemental sources—Calcium pantothenate (synthetic), rice polish, yeast (brewers', torula).

Comments:

Grain is very deficient in pantothenic acid.

Of all B vitamins, pantothenic acid is most likely to be deficient under drylot conditions.

Pantothenic acid is commonly added to commercial poultry and swine rations.

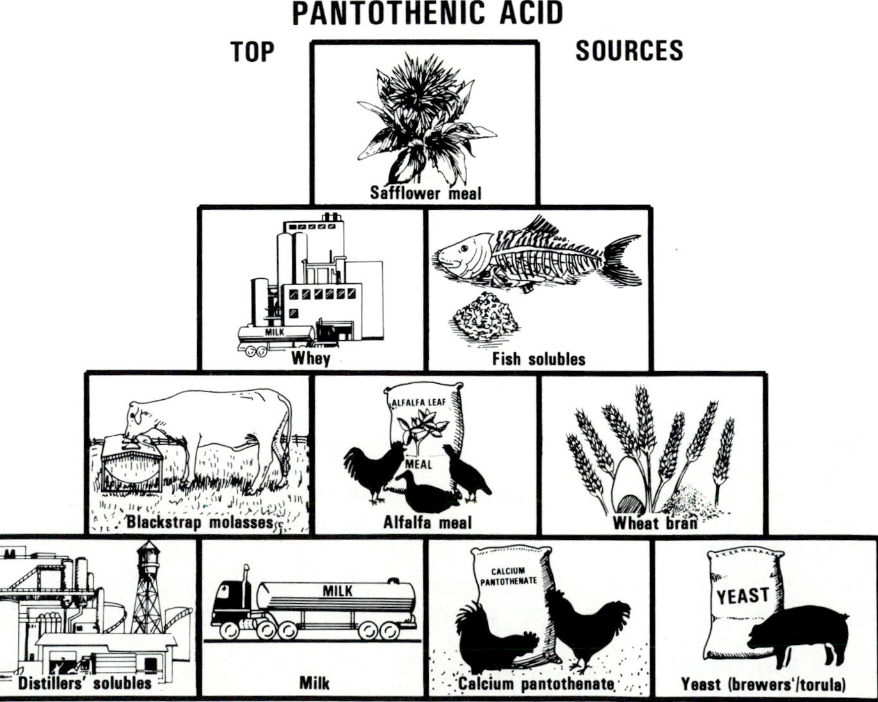

Fig. 4-110. Commonly used, top feed and supplement sources of pantothenic acid. No claim is made that all the top sources of pantothenic acid are listed. Neither are the sources ranked according to amount of pantothenic acid supplied nor is price considered.

(Continued)

TABLE 4–8 *(Continued)*

PARA-AMINOBENZOIC ACID (PABA)

Para-aminobenzoic acid (PABA) is a consitituent of many feeds, which is sometimes listed with the B vitamins.

In addition to having activity as a growth factor for certain bacteria, PABA has considerable folacin activity when fed to deficient animals in which intestinal synthesis of folacin takes place. For example, for rats and mice, it can completely replace the need for a dietary source of folacin.

For higher animals, PABA is an essential part of the folacin molecule. But, it has no vitamin activity in animals receiving ample folacin, it can be produced in the intestinal tract by microbial synthesis, and it is not required in the ration; hence, it can no longer be considered a vitamin, contrary to its listing in many vitamin preparations on the market.

History:

PABA was first identified as an essential nutrient for certain microorganisms. Later, it was shown to act as an antigray hair factor in rats and mice (but not people) and as a growth-promoting factor in chicks.

Chemistry/Metabolism/Properties:

Chemistry—The chemical structure of PABA is given in Fig. 4–111. The chemical structure of PABA is very similar to some of the sulfonamides which explains why it can conteract inhibition of microbial growth by these drugs.

Metabolism—The body manufactures its own PABA if conditions in the intestines are favorable.

Properties—PABA is a yellow, crystalline, slightly water-soluble substance.

Para-AMINOBENZOIC ACID

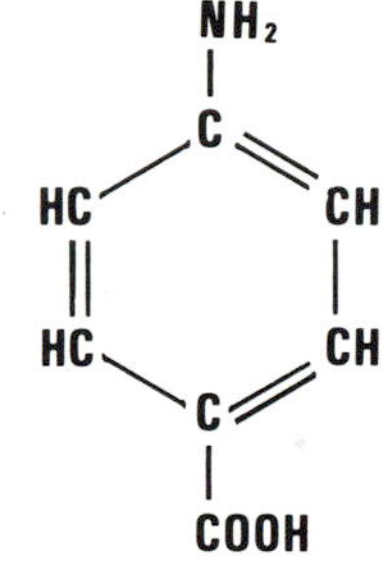

Fig. 4–111. Structure of para-aminobenzoic acid.

Measurement/Assay:

The activity of PABA, and of its sodium and potassium salts, is ordinarily expressed in grams of chemically pure substances.

Species Most Affected:

Essential growth factor for certain microorganisms.

Functions:

For higher animals, PABA functions as an essential part of the folacin molecule.

Deficiency Symptoms/Toxicity:

Deficiency Symptoms—Not demonstrated in animals.

Toxicity—PABA is not known to be toxic to animals. But continued high doses may result in nausea and vomiting.

Sources for Animals:

Rich feed sources—Feed composition tables do not list PABA, but the following feeds are generally recognized as rich sources: Blackstrap molasses, eggs (dried), fish meal, liver meal, peanut meal, soybean meal.

Supplemental sources—Lecithin, para-aminobenzoic acid (synthetic), wheat germ, yeast (brewers').

Comments:

In addition to having activity as a growth factor for certain bacteria, PABA has considerable folacin activity when fed to deficient animals in which intestinal synthesis of folacin takes place. For example, in rats and mice, it can completely replace the dietary source of folacin.

PARA-AMINOBENZOIC ACID
TOP SOURCES

Fish meal
Soybean meal
Peanut meal
Blackstrap molasses
LIVER MEAL
Liver meal
Eggs (dried)
SOYBEAN LECITHIN
Soybean lecithin
Wheat germ
PARA AMINOBENZOIC ACID (Synthetic)
Para-aminobenzoic acid (synthetic)
Yeast (brewers')

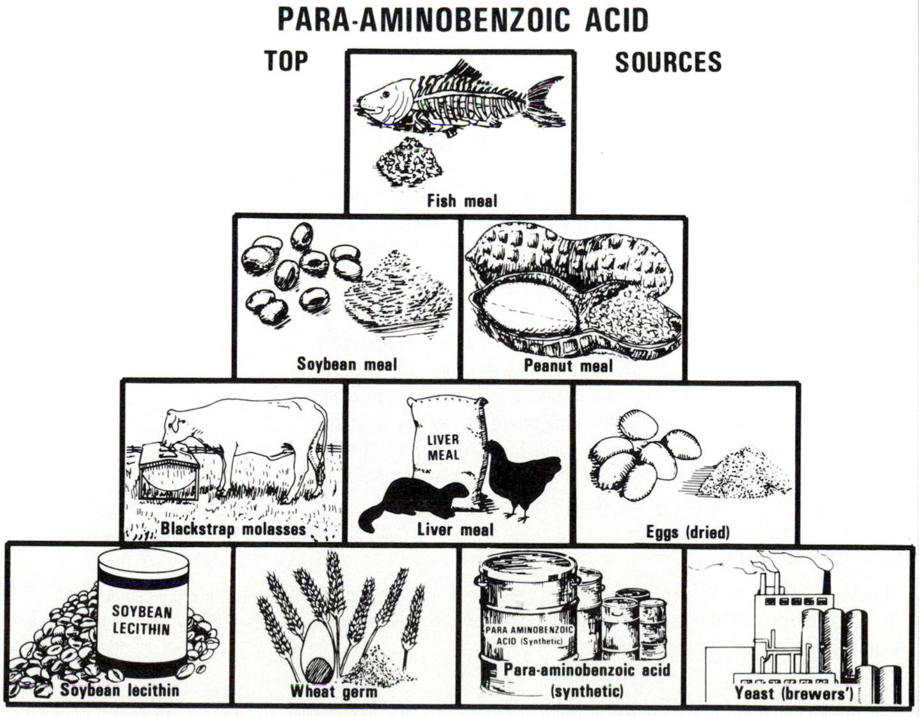

Fig. 4–112. Commonly used, top feed and supplement sources of para-aminobenzoic acid. No claim is made that all the top sources of para-aminobenzoic acid are listed. Neither are the sources ranked according to amount of para-aminobenzoic acid supplied nor is price considered.

(Continued)

TABLE 4–8 *(Continued)*

RIBOFLAVIN (VITAMIN B-2)

Riboflavin made the difference!

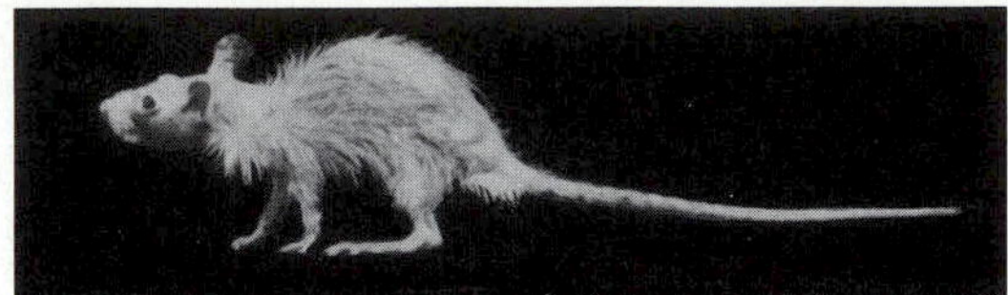
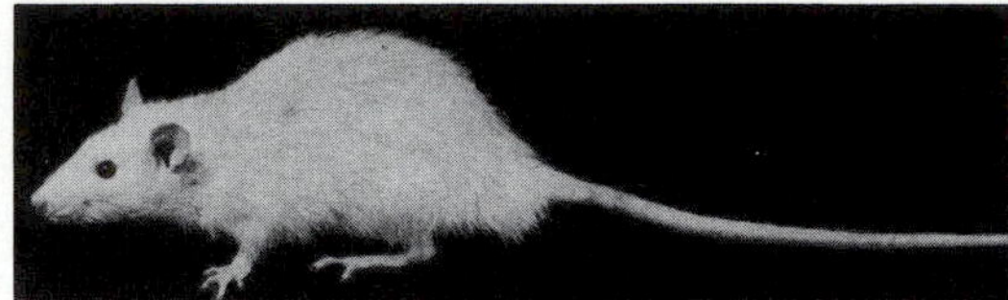

Fig. 4–113. Same rat before (left) and after (right). *Left:* Rat shown at 28 weeks of age, without receiving any riboflavin. It soon became sick; it lost hair, especially about the head; and it weighed only 63 g. *Right:* The same rat 6 weeks later, after receiving feed rich in riboflavin. It recovered its fine fur and weighed 169 g. (Courtesy, USDA)

Riboflavin is present in virtually all living cells; and, like niacin, it has an essential role in the oxidative mechanisms in the cells.

As early as 1879, the existence of a yellow-green fluorescent pigment in milk whey was recognized. Subsequently, other workers found this pigment in widely varying sources. This pigment, which possessed fluorescent properties, was called *flavin.* But, at the time, the biological significance of the pigment was not understood.

By 1928, it became evident that what had been called vitamin B was not a single vitamin. Numerous investigators found that a growth-promoting substance remained after heat had destroyed the beriberi-preventive factor (thiamin, or vitamin B–1) in yeast. This unknown substance was called *vitamin G* by U.S. research workers and *vitamin B-2* by British scientists. At the time, it was thought to be only one vitamin; later, it was found that the heat-stable fraction was composed of several vitamins.

History:

The first serious attempts to isolate the long-known and widely-distributed fluorescent pigment was undertaken by workers in Germany and Switzerland in the early 1930s. In 1932, two German scientists, Warburg and Christian, isolated a *yellow enzyme,* part of which was later identified as flavin mononucleotide (FMN)—riboflavin phosphate. In the following year (1933), Kuhn, working at the University of Heidelberg, isolated pure riboflavin from milk; and, in 1935, he elucidated the structure and synthesized the vitamin. Independently, Swiss researchers, Karrer and co-workers accomplished the same feat that year (1935). Karrer[2] named it riboflavin, because it was found to have a pentose side chain—ribitol (similar to the sugar, ribose)—attached to a flavinlike compound; and, in 1952, the name was adopted by the Commission on Biochemical Nomenclature.

Riboflavin is widely distributed in both plant and animal tissues. It is formed by all higher plants, chiefly in the green leaves. Microbial synthesis of riboflavin occurs in ruminants. The bacteria in the intestinal tract may be a variable, but undependable, source of riboflavin for monogastrics; so, they must rely on feed for their riboflavin.

Chemistry/Metabolism/Properties:

Chemistry—Chemically, riboflavin is composed of an alloxazine ring linked to an alcohol derived from the pentose sugar ribose (see Fig. 4–114).

Metabolism—Riboflavin is absorbed in the upper part of the small intestine by passive diffusion, which controls the amount of the vitamin taken up by the cells of the intestinal mucosa. It is phosphorylated in the intestinal wall and carried by the blood to the tissue where it may occur as the phosphate or as a flavoprotein.

The body has limited capacity for storing riboflavin, although higher concentrations are found in the liver and kidneys than in other tissues. So, day-to-day tissue needs must be supplied by the diet or by continuing bacterial synthesis, as in ruminants. Excretion is primarily via the urine, with the amount excreted related to uptake. When the intake is high, urinary excretion is high; when the intake is low, excretion is low. Some riboflavin is excreted in the feces. All mammals secrete riboflavin in their milk.

Properties—In pure form, riboflavin exists as fine orange-yellow crystals, which are bitter tasting and practically odorless. In water solutions, it imparts a greenish-yellow fluorescence. It is sparingly soluble in water. It is heat stable in neutral or acid solutions, but it may be destroyed by heating in alkaline solutions. It is easily destroyed by light, especially ultraviolet light; for example, it may be destroyed by sunlight. Because of its heat stability and limited water solubility, very little riboflavin is lost in processing feeds.

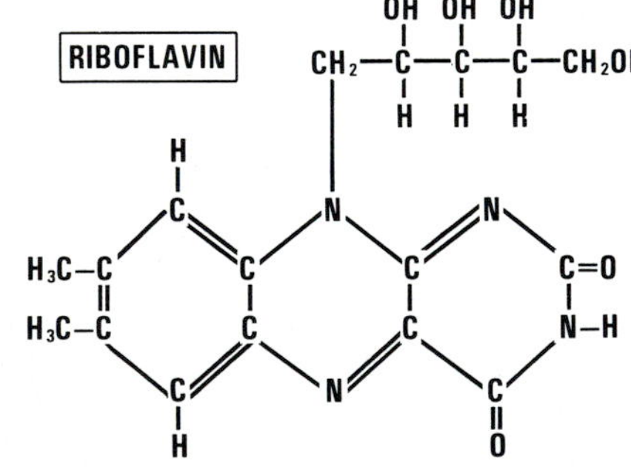

Fig. 4–114. Structure of riboflavin.

Measurement/Assay:

Riboflavin is measured in terms of the metric weight of pure riboflavin; requirements are expressed in milligrams, and feed content in milligrams or micrograms.

Although the growth of rats and chicks may occasionally be used to assay riboflavin in mixed diets, the biological method of assay has been generally superseded by microbiological and chemical methods.

[2]Karrer, who died in 1971, was also the first to synthesize carotene. In 1937, he was awarded the Nobel Prize for his important discoveries in nutrition.

(Continued)

TABLE 4-8 *(Continued)*

Riboflavin (Vitamin B-2) (Continued)

Species Most Affected:

Riboflavin is thought to be required by all animals, but deficiency symptoms are not observed in ruminants, due to rumen synthesis. Deficiency symptoms are noted in poultry, swine, and horses.

Functions:

Riboflavin functions as part of a group of enzymes called flavoproteins. *Flavin mononucleotide* (FMN) and *flavin adenine dinucleotide* (FAD) operate at vital reaction points in the respiratory chains of cellular metabolism. The structure of these two compounds is shown in Fig. 4-115.

FMN and FAD function as coenzymes in a number of different flavoprotein systems. They play a major role with thiamin- and niacin-containing enzymes in a long chain of oxidation-reduction reactions by which energy is released. In the process, hydrogen is transferred from one compound to another until it finally combines with oxygen to form water. Thus, riboflavin functions in the metabolism of amino acids, fatty acids, and carbohydrates. During this process, energy is released gradually and made available to the cell. In addition, riboflavin, through its role in activating pyridoxine (vitamin B-6), is necessary for the formation of niacin from the amino acid tryptophan. Also, riboflavin is thought to be (1) a component of the retinal pigment of the eye; (2) involved in the functioning of the adrenal gland; and (3) required for the production of corticosteroids in the adrenal cortex.

Deficiency Symptoms/Toxicity:

Deficiency Symptoms—Retarded growth in most species, with a wide variety of other symptoms somewhat variable with the species. Periodic ophthalmia (moon blindness) in horses; reproductive failure in the sow, and slow growth, anemia, diarrhea, unthrifty appearance, eye opacities, and an abnormal gait in the young pig; and curled-toe paralysis in birds.

Toxicity—There is no known toxicity of riboflavin.

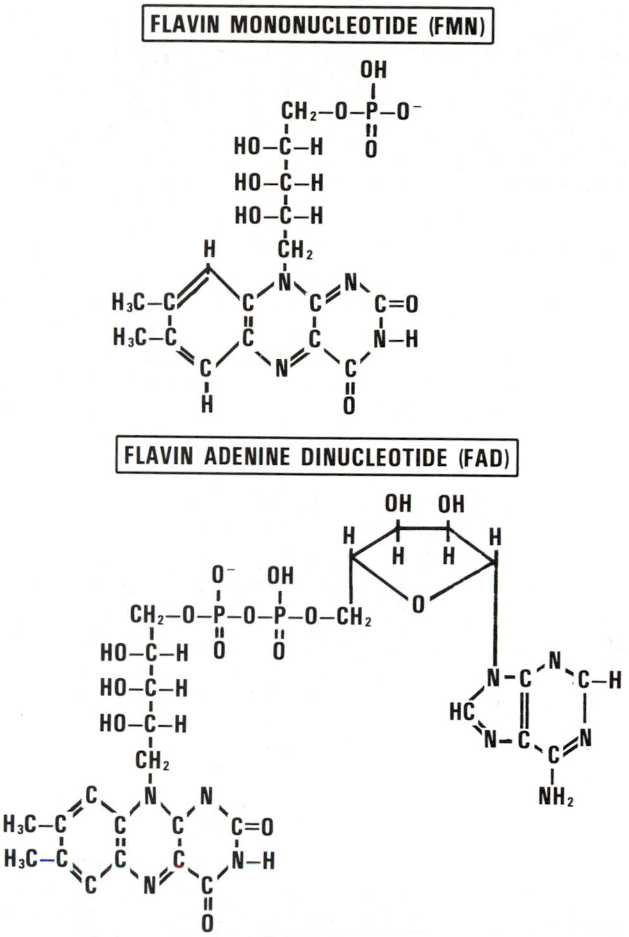

FLAVIN MONONUCLEOTIDE (FMN)

FLAVIN ADENINE DINUCLEOTIDE (FAD)

Fig. 4-115. The structure of FMN and FAD.

Fig. 4-116. Riboflavin deficiency. All pigs from this sow that received a riboflavin-deficient ration during gestation were either born dead or died within 48 hours. (Courtesy, Washington State University, Pullman)

(Continued)

TABLE 4-8 *(Continued)*

Riboflavin (Vitamin B-2) (Continued)

Sources for Animals:

Rich feed sources—Alfalfa hay/leaves/meal, alsike clover hay, birdsfoot trefoil hay, buttermilk, distillers' solubles, fish solubles, grass (immature, green), ladino clover hay, liver meal, milk, poultry by-product meal, red clover hay, skim milk, sweet clover (fresh), wheat hay, white clover (fresh), whey.

Supplemental sources—Riboflavin (synthetic), yeast.

Comments:

Body storage of riboflavin is very limited; so, day-to-day needs must be supplied in the ration, except for ruminants, where microbial synthesis occurs.

Two properties of riboflavin may account for major losses: (1) it is destroyed by light; and (2) it is destroyed by heat in an alkaline solution.

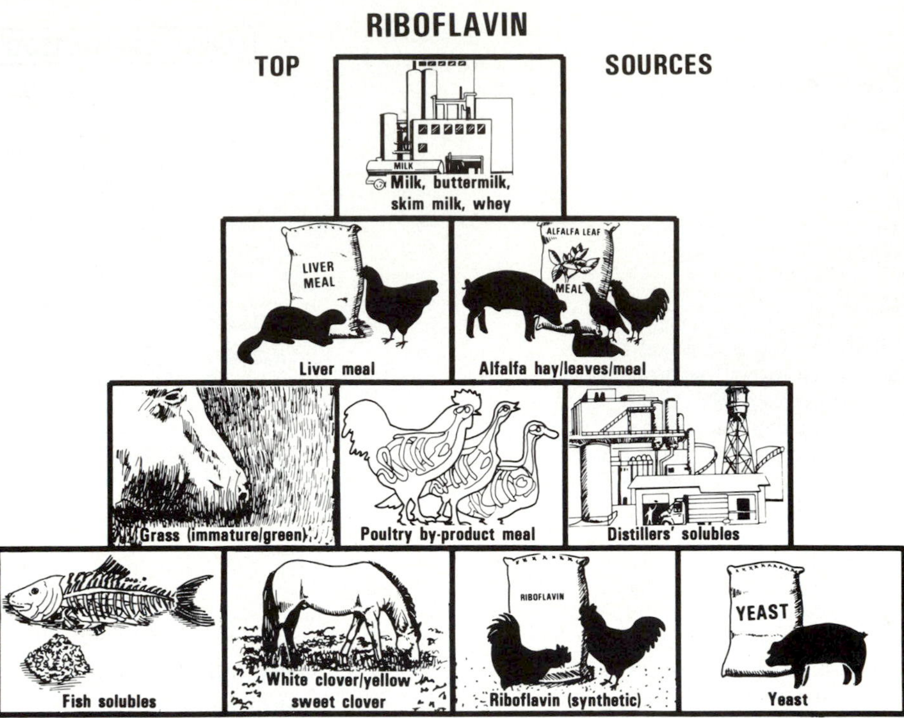

Fig. 4-117. Commonly used, top feed and supplement sources of riboflavin. No claim is made that all the top sources of riboflavin are listed. Neither are the sources ranked according to amount of riboflavin supplied nor is price considered.

THIAMIN (VITAMIN B-1)

Thiamin made the difference!

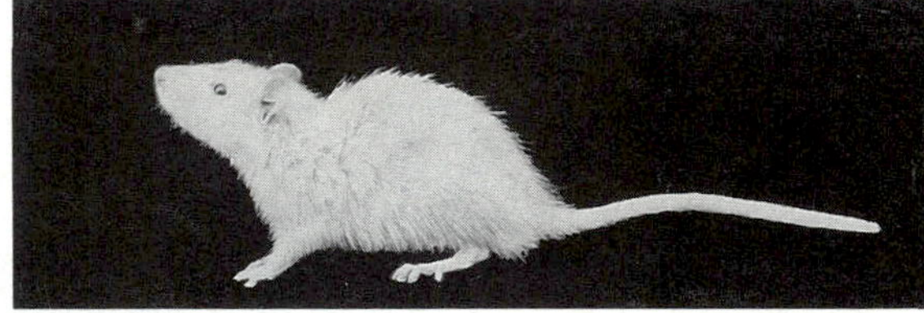

Fig. 4-118. Same rat before (left) and after (right). *Left:* This rat, shown at 24 weeks of age, had practically no thiamin. It lost the ability to coordinate its muscles. *Right:* The same rat 24 hours later, after receiving food rich in thiamin—fully recovered. The thiamin treatment is one of the most dramatic cures in medicine. (Courtesy, USDA)

Thiamin (vitamin B-1)—the antiberiberi, or antineuritic, or antipolyneuritis vitamin—was the first of the B-complex vitamins to be obtained in pure form; hence, the name B-1, a name proposed by the British in 1927. Various other names were used for short periods along the way, including antineuritic factor, antiberiberi factor, water-soluble B, aneurin, and simply vitamin B.

The mystery of beriberi, an ancient disease among rice-eating peoples in the East, was eventually unraveled as a deficiency disease caused by lack of thiamin.

Thiamin is required by all species of animals. They must have a dietary source, unless it is synthesized for them by microorganisms in the digestive tract, as in the case of ruminants.

History:

The history of thiamin begins with a study of the age-old disease beriberi.

Beriberi, which affects the nervous system was known to the Chinese as early as 2600 B.C. But the cause of beriberi in humans, and of polyneuritis—the counterpart in poultry, remained elusive for centuries.

Recognition of the cause of beriberi, and its possible cure by a better diet, is a landmark in the history of nutrition, a chronological record of which follows:

In 1873, Van Lent was the first to conclude that the type of diet had something to do with the origin of beriberi, based on reducing the amount of rice in the diet of sailors in the Dutch navy.

In 1882, Kanehiro Takaki, a Japanese medical officer, cured beriberi in sailors of the Japanese navy by giving them less rice and more meat and milk.

In 1897, Christiaan Eijkman, a Dutch physician, working in a military hospital in the East Indies, produced polyneuritis, a condition resembling beriberi, in chickens, pigeons, and ducks by feeding polished rice; which he wrongfully attributed to too much starch.

(Continued)

TABLE 4-8 *(Continued)*

Thiamin (Vitamin B-1) (Continued)

In 1901, G. Grijns, another Dutch physician, who continued the work of Eijkman at the same hospital, concluded that beriberi in birds and humans resulted from the lack in the diet of an essential nutrient.

In 1912, Casimir Funk, working at the Lister Institute in London, coined the term *vitamine* and applied it to the antiberiberi substance.

In 1916, Elmer V. McCollum of the University of Wisconsin, designated the concentrate that cured beriberi as *water-soluble B.*

In 1926, B. C. P. Jansen and W. P. Donath, in Holland, isolated the antiberiberi vitamin, which, at first, was called aneurin because of its specific antineuritic action.

In 1936, Robert R. Williams, an American biochemist, determined the structure, synthesized it, and gave it the name *thiamine* because it contains *sulfur* (from *thio,* meaning sulfur-containing) and an amine group. Subsequently, the "e" was dropped and the spelling *thiamin* became to be preferred.

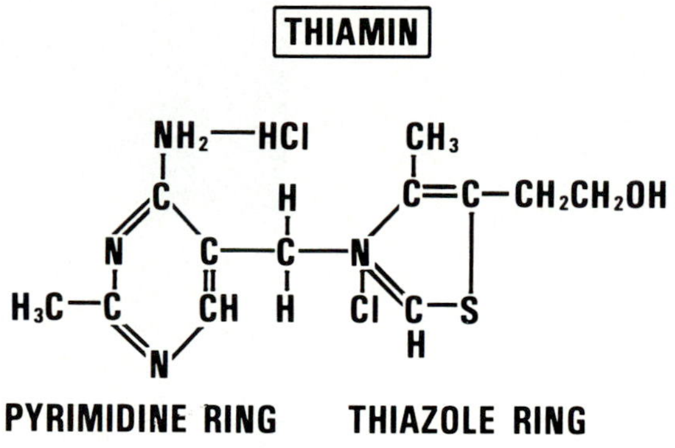

THIAMIN

PYRIMIDINE RING THIAZOLE RING

Fig. 4-119. Structure of thiamin hydrochloride, the white, crystalline, stable form in which thiamin is usually marketed.

Chemistry/Metabolism/Properties:

Chemistry—Thiamin is made up of carbon, hydrogen, oxygen, nitrogen, and sulfur (see Fig. 4-119). It consists of a molecule of pyrimidine and a molecule of thiazole linked by a methylene bridge.

Metabolism—The thiamin ingested in feed is available (1) in the free form, or (2) bound as thiamin pyrophosphate (also called thiamin diphosphate), or (3) in a protein-phosphate complex. The bound forms are split in the digestive tract, following which absorption takes place principally in the upper part of the small intestine where the reaction is acid.

Following absorption, thiamin is transported to the liver where it is phosphorylated under the action of ATP to form the coenzyme thiamin diphosphate (formerly called thiamin pyrophosphate or cocarboxylase) (see Fig. 4-120); although this phosphorylation occurs rapidly in the liver, it is noteworthy that all nucleated cells appear to be capable of bringing about this conversion.

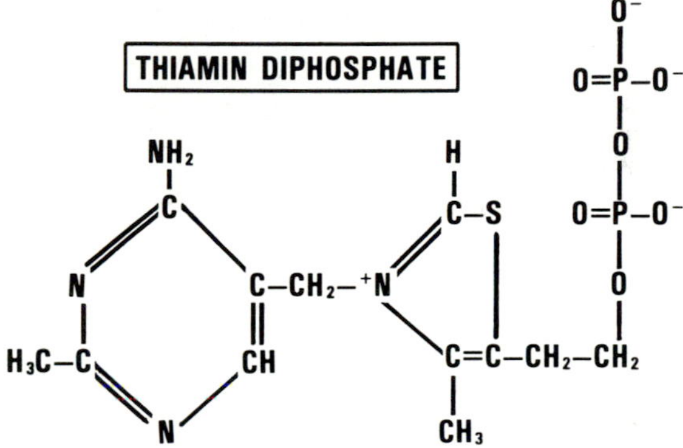

Fig. 4-120. Structure of thiamin diphosphate.

Thiamin is the least stored of all the vitamins. If the diet is deficient, tissues are depleted of their normal content of the vitamin in 1 to 2 weeks; so, fresh supplies are needed regularly to provide for maintenance of tissue levels. Body tissues take up only as much thiamin as they need; with the need increased by metabolic demand (fever, increased muscular activity, pregnancy, and lactation) or by composition of the diet (carbohydrate increases the need for thiamin, while fat and protein spare thiamin). Because thiamin is water soluble, most of the vitamin not required for day-to-day use is excreted in the urine. This means that the body needs a regular supply, and that unneeded intakes are wasted.

Properties—Synthetic thiamin is usually marketed as thiamin hydrochloride, which is more stable than the free vitamin. It is a crystalline white powder, with a faint yeastlike odor and a salty nutlike taste. It is stable when dry but readily soluble in water, slightly soluble in alcohol, and insoluble in fat solvents. Thiamin mononitrate is more stable to heat than thiamin hydrochloride.

Derivatives of thiamin, thiamin propyl disulfide and thiamin tetrahydrofurfural disulfide, have been synthesized. These products are recommended for oral administration when there is evidence of thiamin deficiency, because they are absorbed more rapidly than thiamin hydrochloride.

Measurement/Assay:

The thiamin content of feeds is expressed in milligrams or micrograms. It is usually determined by rapid chemical or microbiological methods, which have largely replaced the older bioassay methods in which chicks, pigeons, and rats were used.

Species Most Affected:

All animals must have a dietary source, unless there is rumen synthesis, as in cattle and sheep.

Functions:

As a coenzyme in energy metabolism. Without thiamin, there could be no energy. As a coenzyme in the conversion of glucose to fats—the process called transketolation (keto-carrying).

In the functioning of the peripheral nerves. In this role, it has value in the treatment of beriberi.

(Continued)

TABLE 4-8 *(Continued)*

Thiamin (Vitamin B–1) (Continued)

Indirect functions in the body, including (1) maintenance of normal appetite, (2) the tone of the muscles, and (3) a healthy mental attitude.

Deficiency Symptoms/Toxicity:

Deficiency Symptoms—Reduction in appetite (anorexia) and loss in weight.

Cardiovascular disturbances.

Lowered body temperature.

Chicks: Polyneuritis (retraction of the head).

Hens: Lowered egg production.

Mink and foxes: Chastek paralysis.

Humans: Beriberi.

Toxicity—There are no known toxic effects from thiamin.

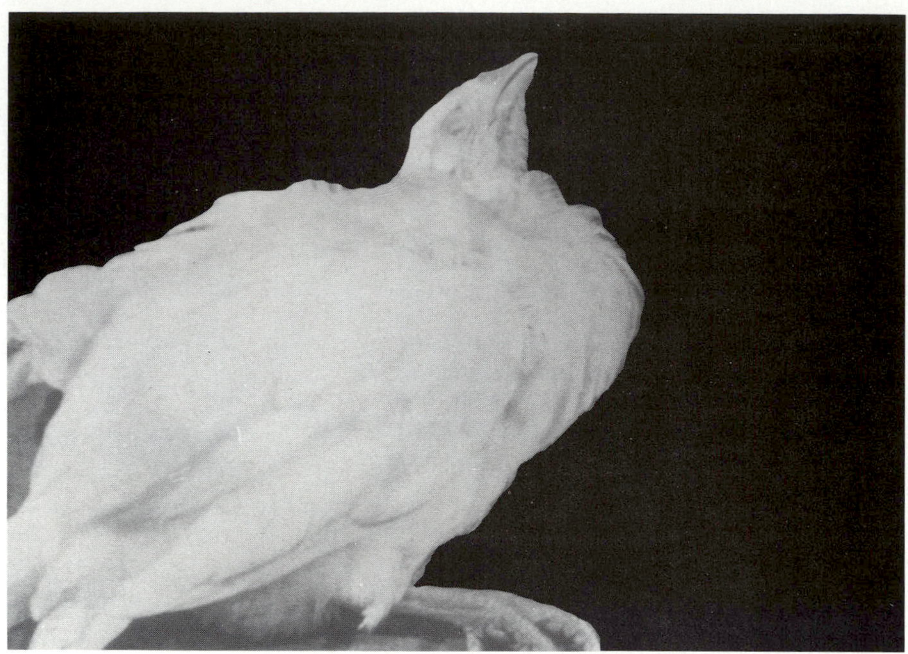

Fig. 4–121. Thiamin-deficient chick, showing polyneuritis (inflammation of the nerves), the counterpart of beriberi in man. Note the characteristic head retraction. (Courtesy, Dept. of Poultry Science, University of Wisconsin, Madison)

Sources for Animals:

Rich feed sources—Barley malt sprouts, birdsfoot trefoil hay, cotton seeds/cottonseed meal, cowpea seeds, distillers' solubles, fish solubles, hominy, linseed meal, millet, navy bean seeds, oat grain/groats, peanut meal, pinto bean seeds, rice bran, safflower meal, soybean seeds/meal, sunflower seeds.

Supplemental sources—Rice polish, thiamin hydrochloride, thiamin mononitrate, wheat germ meal, yeast (brewers', torula).

Comments:

Certain raw fish and seafood—particularly carp, herring, clams, and shrimp—contain the enzyme thiaminase, which inactivates the thiamin molecule by splitting it into two parts. This effect has been seen in mink and fox fed 10 to 25% levels of certain raw fish, giving rise to a thiamin deficiency disease known as Chastek paralysis. This action can be prevented by cooking the fish prior to feeding, thereby destroying the thiaminase.

Fats exhibit a thiamin-sparing effect.

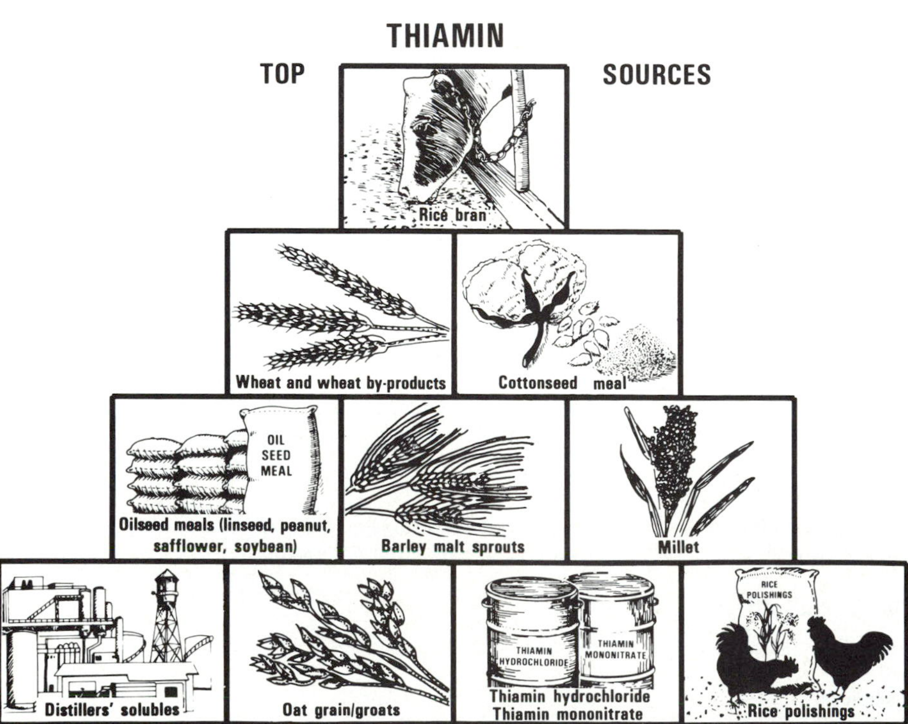

Fig. 4–122. Commonly used, top feed and supplement sources of thiamin. No claim is made that all the top sources of thiamin are listed. Neither are the sources ranked according to amount of thiamin supplied nor is price considered.

(Continued)

TABLE 4-8 *(Continued)*

VITAMIN B-6 (PYRIDOXINE, PYRIDOXAL, PYRIDOXAMINE)

Vitamin B-6 made the difference!

Fig. 4-123. *Left:* Vitamin B-6 deficient chick, showing retarded growth and abnormal feathering. *Right:* Healthy chick that received adequate vitamin B-6 in a similar diet. (Courtesy, University of Georgia, Athens)

By official action of the Society of Biological Chemists and the American Institute of Nutrition, vitamin B-6 is now the approved collective name for 3 closely related naturally occurring compounds with potential vitamin B-6 activity: pyridoxine, pyridoxal, and pyridoxamine. Pyridoxine is found largely in plant products, whereas the pyridoxal and pyridoxamine forms occur primarily in animal products. In rats, the biological activity of the 3 compounds is equal if given parenterally (injected intramuscularly or intravenously).

The need for vitamin B-6 was first demonstrated in rats, but it is now established that it is also a dietary essential for the pig, chick, dog, human, and other species, including microorganisms.

History:

In 1926, Goldberger and Lillie conducted an experiment designed to produce pellagra in rats. A severe dermatitis resulted, which was believed to be analogous to pellagra. In 1934, the Hungarian scientist, Gyorgy, produced a cure for this condition with an extract from yeast; but the curative compound of the extract was neither thiamin, niacin, nor riboflavin, but a substance which he called *vitamin B-6.* In 1938, five different laboratories, working independently, isolated the vitamin in crystalline form, but credit for obtaining the first crystals is generally given to Lepkovsky of the University of California. In 1939, Stiller *et al.,* of Merck, along with Kuhn, a German scientist, synthesized the compound.

In 1942, Snell discovered the vitamin B-6 activity of two closely related substances in natural products, which he named pyridoxal and pyridoxamine. Umbreit followed in 1945 with a report of the coenzyme functions of the vitamin in phosphate forms.

Chemistry/Metabolism/Properties:

Chemistry—Vitamin B-6 is found in feeds in three forms which are readily interconvertible—pyridoxine, pyridoxal, and pyridoxamine. Also vitamin B-6 is found in physiological systems in the forms of pyridoxal phosphate and pyridoxamine phosphate. The structural formulas of the three naturally occurring free forms of these compounds are given in Fig. 4-124.

VITAMIN B-6 – THREE FORMS

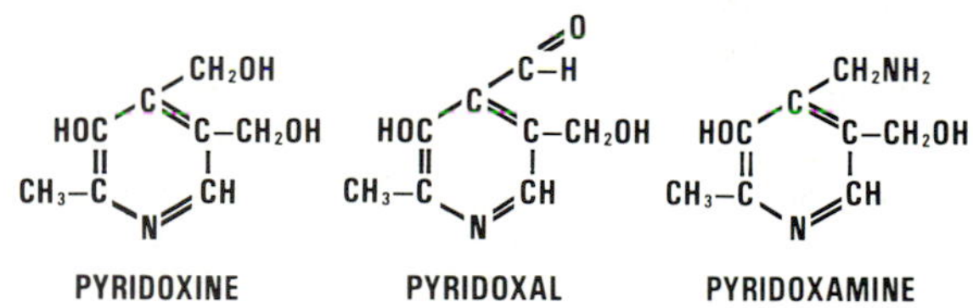

Fig. 4-124. Structure of pyridoxine, pyridoxal, and pyridoxamine.

Metabolism—In the free form, vitamin B-6 is absorbed rapidly from the upper part of the small intestine, thence it enters the body by the portal vein. It is present in most body tissues, with a high concentration in the liver. It is secreted into milk and excreted primarily via the urine. Measurement of B-6 in the urine is the assay method used in nutrition surveys.

Properties—Vitamin B-6 is readily soluble in water, quite stable to heat and acid, but easily destroyed by oxidation and exposure to alkali and ultraviolet light. All three forms are white crystalline substances.

Of the three forms, pyridoxine is more resistant to feed processing and storage conditions and probably represents the principal form in feed products.

Measurement/Assay:

No international standard or unit system of vitamin B-6 is in current usage. Analytical results are expressed in weight units of pyridoxine hydrochloride. One milligram of pyridoxine hydrochloride is equivalent to 0.82 mg pyridoxine or 0.81 mg pyridoxal, or 0.82 mg pyridoxamine.

The vitamin B-6 content in feeds and tissues is determined by microbiological assay, chemical methods, and animal bioassays. The animal bioassays, using either the rat or chick are time consuming, expensive, and variable; therefore, they have been generally replaced by microbiological and chemical methods.

Species Most Affected:

B-6 is a dietary essential for the rat, pig, chick, and dog. It is synthesized in the rumen of cattle and sheep and perhaps in the cecum of the horse; thus, no deficiency symptoms in these species have been reported.

(Continued)

TABLE 4-8 *(Continued)*

Vitamin B-6 (Pyridoxine, Pyridoxal, Pyridoxamine) (Continued)

Functions:

Vitamin B-6, in its coenzyme forms, usually as pyridoxal phosphate but sometimes as pyridoxamine phosphate, is involved in a large number of physiological functions, particularly:

1. In protein (nitrogen) metabolism, including—(a) transamination (see Fig. 4-125), (b) decarboxylation, (c) deamination, (d) transsulfuration, (e) tryptophan conversion to nicotinic acid, (f) hemoglobin formation, (g) absorption of amino acids.
2. In carbohydrate and fat metabolism, including—(a) the conversion of glycogen to glucose-1-phosphate, (b) the conversion of linoleic acid to arachidonic acid.
3. In clinical problems, including—(a) anemia that is iron-resistant, (b) kidney stones, (c) physiologic demands in pregnancy.

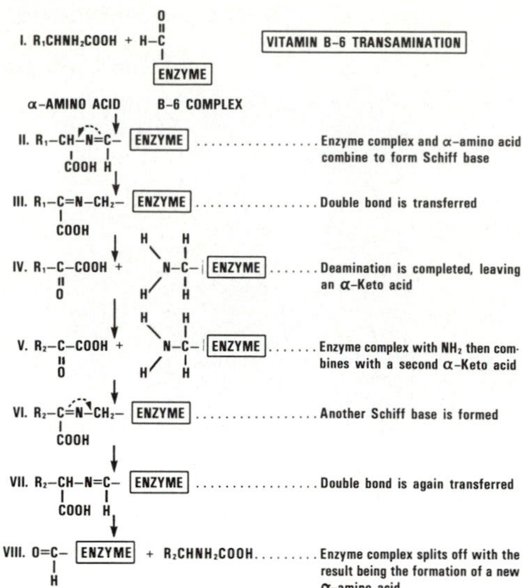

Fig. 4-125. Vitamin B-6 in transamination.

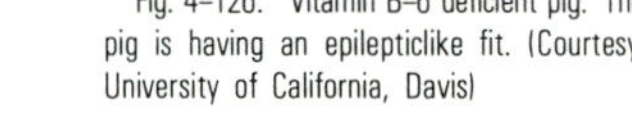
Fig. 4-126. Vitamin B-6 deficient pig. This pig is having an epilepticlike fit. (Courtesy, University of California, Davis)

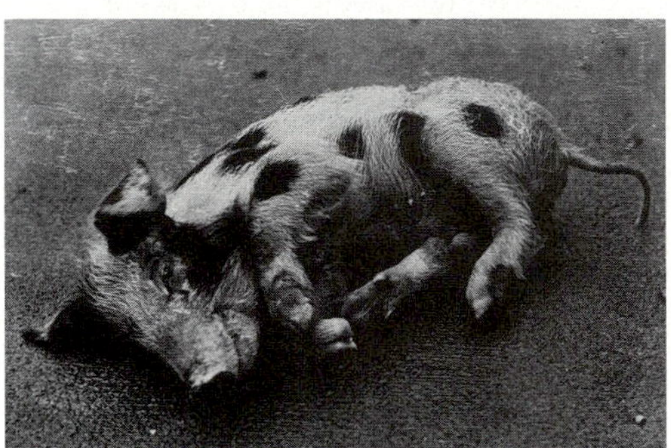

Deficiency Symptoms/Toxicity:

Deficiency Symptoms—All species exhibit convulsions.
Pigs show anorexia and poor growth.
Chicks show retarded growth and abnormal feathering.
Hens show lowered egg laying and hatchability.
Rats develop a specific dermatitis (acrodynia).

Toxicity—B-6 is relatively nontoxic.

Sources for Animals:

Normally, animal rations contain adequate vitamin B-6.

Rich feed sources—Alfalfa hay/meal, barley malt sprouts, corn gluten feed, corn gluten meal, distillers' solubles, fish meal, fish solubles, hominy, linseed meal, meat meal/tankage, pastures (green), potatoes (dehy), rice bran, wheat germ meal, wheat middlings, wheat mill run.

Supplemental sources—Rice polish, pyridoxine hydrochloride, wheat germ, yeast (brewers', torula).

Comments:

In rats, the three forms of vitamin B-6 have an equal activity.

VITAMIN B-6
TOP SOURCES
Safflower meal
Fish meal solubles
Pastures (green)
Meat meal/tankage
Wheat and wheat by-products
Alfalfa hay/meal
MEAL
Soybean meal
Corn gluten feed/meal
CORN GLUTEN MEAL
Pyridoxine hydrochloride
PYRIDOXINE HYDROCHLORIDE
Yeast (brewers'/torula)
YEAST

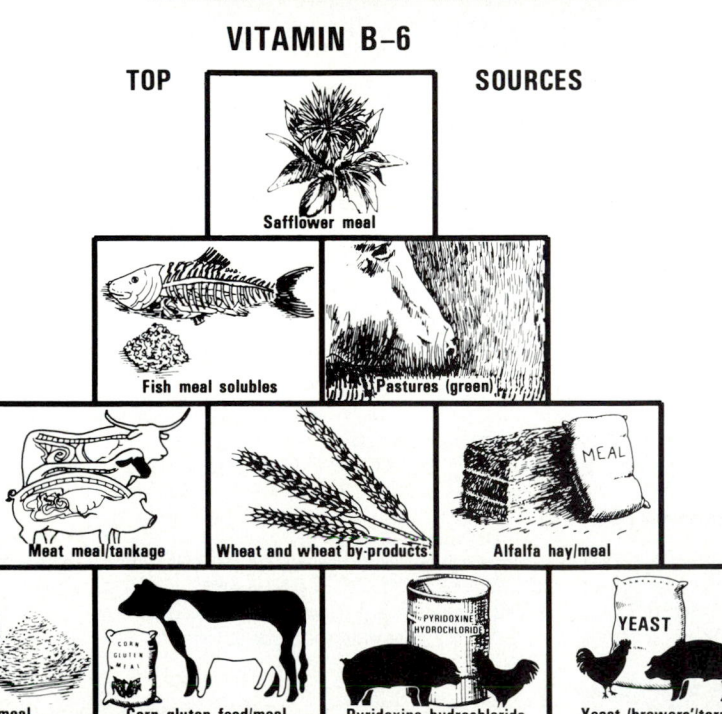

Fig. 4-127. Commonly used, top feed and supplement sources of vitamin B-6. No claim is made that all the top sources of vitamin B-6 are listed. Neither are the sources ranked according to amount of vitamin B-6 supplied nor is price considered.

(Continued)

TABLE 4-8 *(Continued)*

VITAMIN B-12 (COBALAMINS)

Vitamin B-12 made the difference!

Fig. 4-128. The smaller chick at the left and his bigger companion at 3½ weeks old. *Left:* The small chick, fed a ration deficient in vitamin B-12, weighed 157 g. *Right:* The larger chick, fed the same ration plus vitamin B-12, weighed 280 g. (Courtesy, Merck and Co., Rahway, N.J.)

Vitamin B-12, like so many other members of the B complex, is not a single substance; rather, it consists of several closely related compounds with similar activity. The term *cobalamins* is applied to this group of substances because all of them contain cobalt. Vitamin B-12, which is the most active member, is cyanocobalamin, named after the cyanide ion in the molecule. Other chemically related compounds known to have vitamin B-12 activity include hydroxocobalamin, nitritocobalamin, and thiocyanate cobalamin.

Unlike any other vitamin, higher plants cannot synthesize vitamin B-12, although it can be synthesized by microorganisms and animals.

History:

The chronological history of vitamin B-12 as it relates to animals follows:

1. In 1948, two groups of researchers working independently, Rickes and co-workers of Merck and Co., Inc., of New Jersey, and Smith and Parker of Glaxo Laboratories in England, isolated from a liver concentrate a crystalline, red pigment, which they called *vitamin B-12.*

2. In 1955, the structure of vitamin B-12 (cyanocobalamin) was determined by Dorothy Hodgkin and co-workers, at Oxford. Later (1964), Hodgkin was awarded the Nobel Prize.

3. In 1955, Woodward's group at Harvard synthesized vitamin B-12 using a very complicated and expensive procedure. Fortunately, soon thereafter, it was found that highly active vitamin B-12 concentrates can be produced from cultures of certain bacteria and fungi grown in large tanks containing special media; and this remains the main method of commercial production.

Chemistry/Metabolism/Properties:

Chemistry—Vitamin B-12 is the largest and the most complex of all vitamin molecules. The main part of the molecule consists of a porphyrin ring containing cobalt as the central element. A cyanide ion is also attached to the cobalt atom. This cyanide group can be replaced by such anions as hydroxy, bromo, chloro, sulfato, or nitro groups, and the molecule will remain biologically active.

The structure of vitamin B-12 ($C_{63}H_{90}O_{14}N_{14}PCo$) is shown in Fig. 4-129.

Vitamin B-12 occurs as a protein complex in animal proteins. The ultimate source, however, is the microorganisms in the gastrointestinal tract of the herbivorous animals. Such microorganisms are found in large amounts in the rumen (the first stomach) of cows, sheep, and other ruminants. For synthesis by the microorganisms to occur, cobalt must be supplied in the feed so that it can be incorporated in the molecule. If the ration is deficient, vitamin B-12 will not be synthesized.

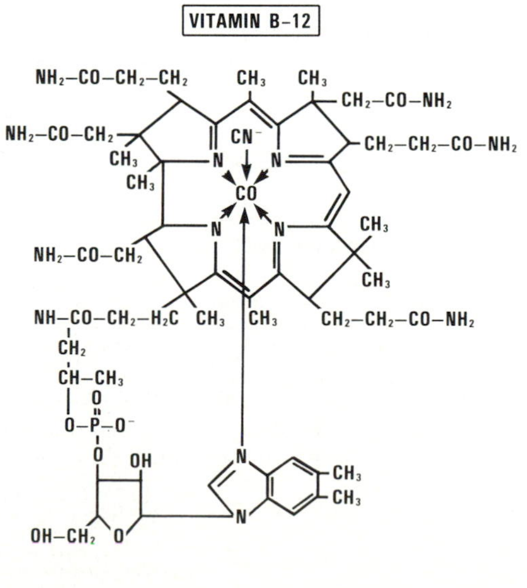

VITAMIN B-12

Fig. 4-129. Structure of vitamin B-12.

(Continued)

TABLE 4-8 *(Continued)*

Vitamin B-12 (Cobalamins) (Continued)

Metabolism—It is noteworthy (1) that vitamin B-12 is the only vitamin that requires a specific gastrointestinal tract secretion for its absorption (*intrinsic factor*); and (2) that the absorption of vitamin B-12 in the small intestine requires about 3 hours (compared to seconds for most other water-soluble vitamins).

The liver is the principal site of storage of vitamin B-12.

Vitamin B-12 is excreted by way of the kidneys and in the bile.

Properties—The deep-red needlelike crystals are slightly soluble in water, stable to heat, but destroyed by light and by strong acid or alkaline solutions.

Vitamin B-12 is remarkably potent. It has a biologic activity 11,000 times that of the standard liver concentrate formerly used in the treatment of pernicious anemia.

Measurement/Assay:

No International Units have been defined for the biological activity of vitamin B-12. However, pure cobalamin can be used as a standard substance. Vitamin B-12 is measured in micrograms or picograms (pg = micromicrograms).

High potency preparations of vitamin B-12 are usually assayed by spectrophotometry. Also, vitamin B-12 may be assayed colorimetrically or fluorometrically. Some assays involve measurement of cobalt. However, feed sources are usually assayed for vitamin B-12 by either (1) the microbiological method, or (2) the biological method using chicks or rats.

Species Most Affected:

Swine, rats, poultry, and humans.

Ruminants synthesize B-12 unless cobalt is deficient.

Functions:

In the body, vitamin B-12 functions in two coenzyme forms: coenzyme B-12, and methyl B-12.

Vitamin B-12 coenzymes perform the following physiological roles at the cellular level, especially in the cells of the bone marrow, nervous tissue, and gastrointestinal tract: (1) red blood cell formation and control of pernicous anemia; (2) maintenance of nerve tissue; (3) carbohydrate, fat, and protein metabolism; (4) synthesis or transfer of single carbon units; and (5) biosynthesis of methyl groups ($-CH_3$), and in reduction reactions such as the conversion of disulfide (S-S) to the sulfhydryl group (-SH).

The role of B-12 is closely interrelated with other vitamins—among them choline, folic acid, and pantothenic acid, acting primarily in the synthesis of labile methyl groups. Lipid metabolism involves vitamin B-12 in the conversion of methylmalonyl CoA to succinyl CoA.

Deficiency Symptoms/Toxicity:

Deficiency Symptoms—All animals show retarded growth.

Pigs show uncoordinated hind leg movements; and there's reproductive failure in sows.

Eggs from B-12-deficient hens fail to hatch.

Birth defects—hydroencephaly and bone and eye disorders—sometimes occur in newborn rats born to vitamin B-12 deficient mothers.

Toxicity—No toxic effects of vitamin B-12 are known.

Fig. 4-130. *Top:* Vitamin B-12-deficient pig. Note rough hair coat and dermatitis. *Bottom:* Healthy pig that received adequate vitamin B-12. (*Top photo:* Courtesy, Iowa State University, Ames)

(Continued)

TABLE 4–8 *(Continued)*

Vitamin B–12 (Cobalamins) (Continued)

Sources for Animals:

Plants cannot manufacture vitamin B–12, so they do not contain appreciable quantities of it. Vitamin B–12 can be made available to the animal through exogenous sources (food) and endogenous sources (gastrointestinal microorganisms). Most animal products are high in vitamin B–12. Synthetic sources of vitamin B–12 are commercially available. Practical sources of B–12 follow:

Rich feed sources—Fish meal and other marine by-products, liver meal, meat meal/tankage and other meat by-products, milk, poultry by-product meal, whey.

Supplemental sources—Cobalamin, of which there are at least three active forms produced by microbial growth and yeast.

Comments:

It is noteworthy (1) that vitamin B–12 is the only vitamin that requires a specific gastrointestinal tract secretion for its absorption (intrinsic factor); and (2) that the absorption of vitamin B–12 in the small intestine requires about 3 hours (compared to seconds for most of the other water-soluble vitamins).

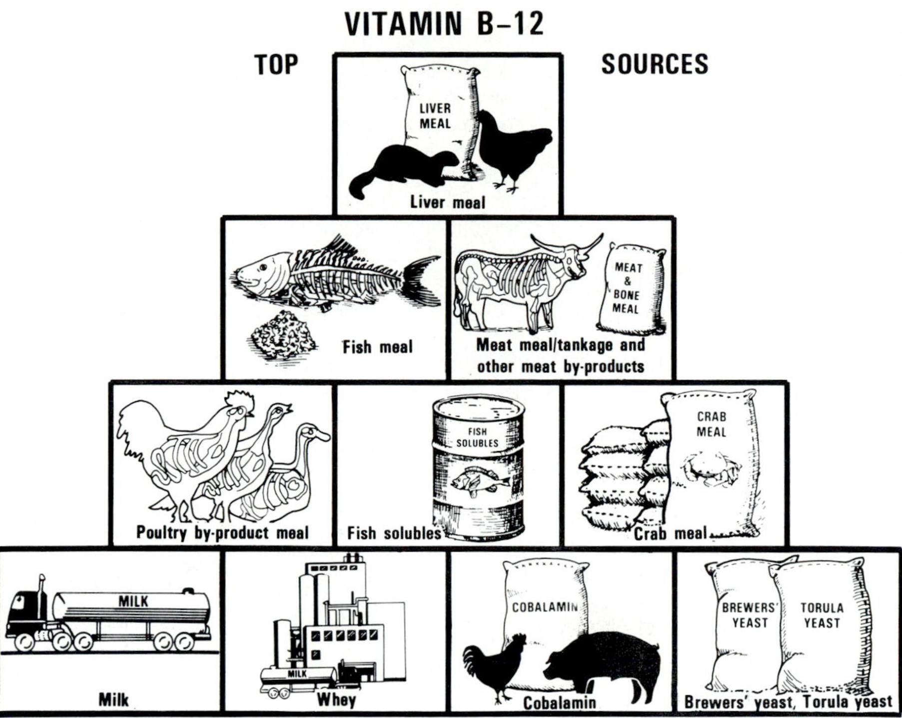

Fig. 4–131. Commonly used, top feed and supplement sources of vitamin B–12. No claim is made that all the top sources of vitamin B–12 are listed. Neither are the sources ranked according to amount of vitamin B–12 supplied nor is price considered.

VITAMIN C (ASCORBIC ACID, DEHYDROASCORBIC ACID)

Vitamin C made the difference!

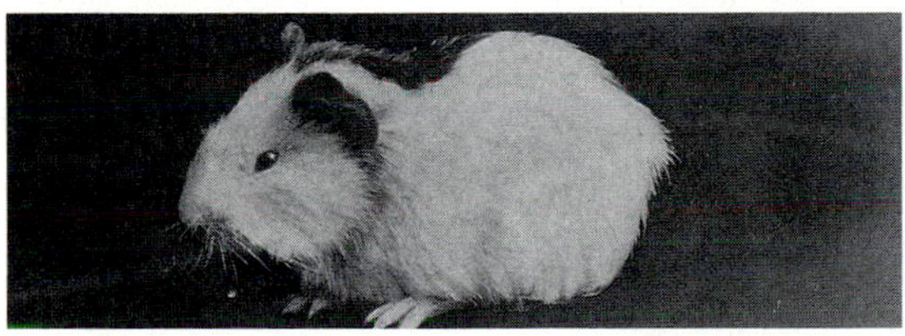

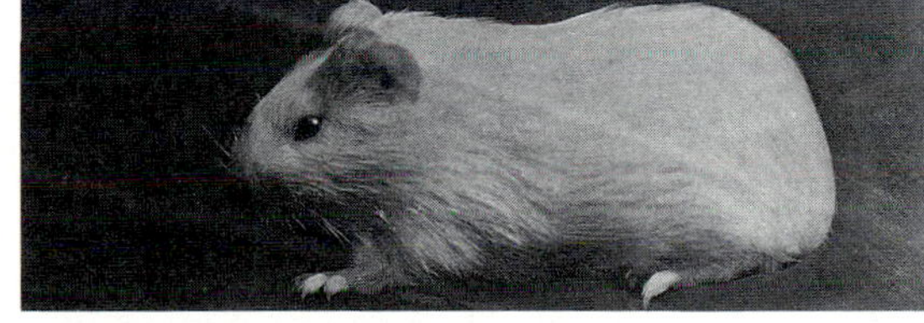

Fig. 4–132. *Left:* This guinea pig had no ascorbic acid and developed scurvy. Note the crouched position due to sore joints. *Right:* This guinea pig had plenty of ascorbic acid. It was healthy and alert, and its fur was sleek and fine. (Courtesy, USDA)

Vitamin C—also called *ascorbic acid, dehydroascorbic acid, hexuronic acid,* and the *antiscorbutic vitamin*—is the very important substance, first found in citrus fruits, which prevents scurvy, one of the oldest scourges of humanity. All animal species appear to require vitamin C, but a *dietary need* is limited to humans, guinea pigs, monkeys, fruit bats, certain fish, and perhaps certain reptiles. These species lack the enzyme L-gulonolactone oxidase which is necessary for vitamin C synthesis from 6-carbon sugars.

History:

Scurvy, now known to be caused by a severe deficiency of vitamin C, was a dread disease in ancient times. It was once common among sailors who ate little except bread and salt meat while on long voyages.

The historical incidence and conquest of scurvy constitute one of the most thrilling chapters in the development of nutrition as a science. A chronological summary of the saga of scurvy and vitamin C follows:

As early as 1550 B.C., scurvy was described by the Egyptians on medical papyrus rolls (the first writing paper, made by the Egyptians from the papyrus plant).

In the Old Testament (which was written over a long period of time, thought to be from 1100 B.C. to 500 B.C.), reference is made to this disease.

About 450 B.C., Hippocrates, the Greek "father of medicine" described the symptoms of the malady—gangrene of the gums, loss of teeth, and painful legs in soldiers.

In 1248–54, Jean Sire de Joinville, the French chronicler, accompanied Louis IX of France to Cyprus and Egypt. In 1309, he completed, in final form, the *History of Saint Louis,* an account of the Crusade, in which he told of a disease (scurvy) "which attacked the mouth and legs."

In 1497, when Vasco da Gama, Portuguese navigator, sailed around the Cape of Good Hope and established the first European trading colony on the coast of Malabar in India; 100 of his crew of 160 men perished of scurvy on the voyage.

During the winter of 1535 in Canada, Jacques Cartier, the daring explorer who laid claim to Canada for France, recorded in his log that the lives of many of his men dying of scurvy were saved *almost overnight,* when they learned from the Indians that drinking a "brew" made from the growing tips of pine or

(Continued)

TABLE 4–8 *(Continued)*

Vitamin C (Ascorbic Acid, Dehydroascorbic Acid) (Continued)

spruce trees cured and prevented the malady. (It is now known that the "brew" contained vitamin C.)

In the 15th and 16th centuries, scurvy was a scourge throughout Europe, so much so that medical doctors wondered if all diseases might stem from it.

In 1600–1603, Captain James Lancaster, an English navigator, recorded that on the long voyage to the East Indies he kept his crew hearty merely by the addition of a mandatory "three spoonsful of lemon juice every morning."

In 1747, James Lind, an English naval surgeon, tested six remedies on 12 sailors who had scurvy and found that oranges and lemons were curative. His classical studies, the results of which were published in 1753, are generally credited as being the first experiments to show that an essential food element can prevent a deficiency disease. But another 50 years elapsed before the British Navy required rations of lemons or limes on sailing vessels.

On two historic voyages, each of 3-years duration, from 1768 to 1771 and from 1772 to 1775, British Captain James Cook avoided scurvy—hitherto the scourge of long sea voyages. He had his ship stocked with concentrated slabs of thick brown vegetable soup and barrels of sauerkraut. Of the sauerkraut he said: "It is not only a wholesome vegetable food, but, in my judgment, highly antiscorbutic, and spoils not by keeping." In addition, he sent seamen ashore at every port visited to gather all sorts of fresh fruits and green vegetables (including grasses), which the crew prepared, served, and ate. As a result, not one of the crew died from scurvy.

In 1795 (1 year after Lind's death), by Admiralty Order, the British Royal Navy began providing 1 oz of lime juice daily in every sailor's food ration; from this date forward, British sailors were stuck with the nickname "limeys."

In 1907, Holst and Frolich, of Norway, produced scurvy experimentally in guinea pigs by feeding them a diet deficient in foods containing ascorbic acid.

In 1928, Szent-Gyorgy, an Hungarian scientist, working in Hopkins' laboratory at Cambridge University, in England, isolated a substance from the ox adrenal glands, oranges, and cabbage leaves, which he called *hexuronic acid;* but he did not test it for antiscorbutic effect.

In 1932, Charles Glen King and W. A. Waugh, at the University of Pittsburgh, isolated from lemon juice a crystalline material that possessed antiscorbutic activity in guinea pigs; this marked the discovery of *vitamin C,* a deficiency of which caused the centuries-old scourge of scurvy.

In 1933, vitamin C was synthesized by Reichstein, a Swiss scientist.

In 1938, *ascorbic acid* was officially accepted as the chemical name for vitamin C.

Chemistry/Metabolism/Properties:

Chemistry—Ascorbic acid is a compound of relatively simple structure, closely related to the monosaccharide sugars. It is synthesized from glucose and other simple sugars by plants and by most animal species (see Fig. 4–133).

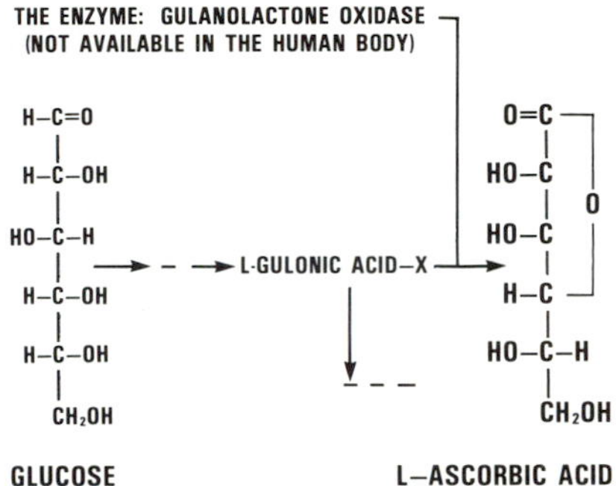

Fig. 4–133. Metabolic relation of glucose to ascorbic acid. In humans, monkeys, guinea pigs, fruit-eating bats, and red-vented bulbul birds, the absence of oxidase prevents this reaction, making the intake of preformed ascorbic acid in food necessary.

Humans, monkeys, guinea pigs, fruit-eating bats, and red-vented bulbul birds (the latter two are native to India), cannot make the conversion from glucose to ascorbic acid, because these species lack the necessary enzyme (oxidase). Scurvy, then, in these species, can really be classed as a disease of distant genetic origin—an inherited metabolic error; a defect in carbohydrate metabolism due to the lack of an enzyme, which, in turn, results from the lack of a specific gene.

Two forms of vitamin C occur in nature, ascorbic acid (the reduced form), and dehydroascorbic acid (the oxidized form).[3] Their structural formulas are shown in Fig. 4–134.

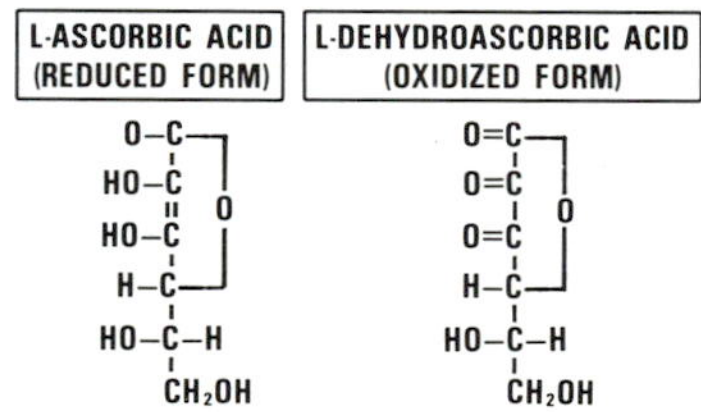

Fig. 4–134. Structural formulas of vitamin C.

Metabolism—Pertinent facts about the absorption, storage, and excretion of vitamin C follow:

1. **Absorption.** Vitamin C is readily and rapidly absorbed from the upper part of the small intestine into the circulatory system. Thence, it is taken up by the tissues.

2. **Storage.** Unlike the majority of the water-soluble vitamins, limited stores of vitamin C are held in the body.

Ascorbic acid is easily oxidized to dehydroascorbic acid, which is just as easily reduced back to ascorbic acid. However, dehydroascorbic acid may be irreversibly oxidized, particularly in the presence of alkali, to diketogulonic acid which has no antiscorbutic activity (see Fig. 4–135).

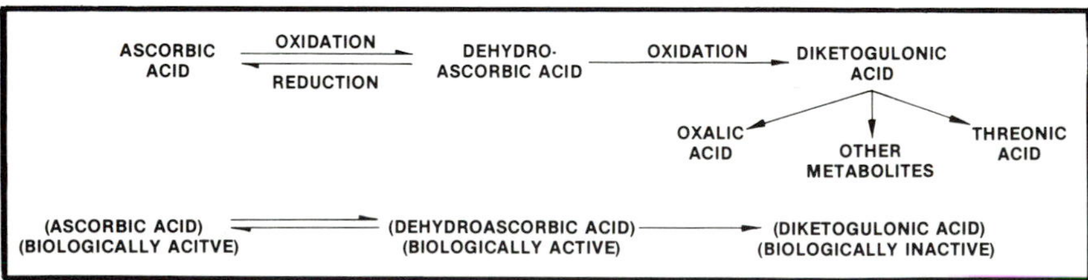

Fig. 4–135. Relationship of ascorbic, dehydroascorbic, and diketogulonic acids.

3. **Excretion.** Vitamin C is largely excreted in the urine, with the amount excreted controlled by the kidney tubules.

Properties—Ascorbic acid is a white, odorless, crystalline powder, which is quite stable when dry.

Of all the vitamins, ascorbic acid is the most unstable when in solution. It is highly soluble in water, but not in fat. The oxidation (destruction) of ascorbic acid is accelerated by air, heat, light, alkalies, oxidative enzymes, and traces of copper and iron.

[3]Actually, several chemical compounds have vitamin C activity. So, it is now recommended that the term *vitamin C* be used as the combined name of all compounds having the biological activity of ascorbic acid, and that the terms *ascorbic acid* and *dehydroascorbic acid* be used only when specific reference to them is made.

(Continued)

TABLE 4-8 *(Continued)*

Vitamin C (Ascorbic Acid, Dehydroascorbic Acid) (Continued)

Measurement/Assay:

The concentration of ascorbic acid in tissues and feeds is expressed in milligrams. One I.U. is the activity of 0.05 mg of ascorbic acid.

Ascorbic acid is generally determined by chemical assay, accomplished by taking advantage of its reducing properties.

For bioassay work, guinea pigs are the preferred experimental animal because of their susceptibility to a deficiency of vitamin C. Thus, they are still used for demonstration of deficiency of the vitamin and to make comparative assays.

Species Most Affected:

Dietary need is limited to humans, guinea pigs, monkeys, fruit-eating bats, and bulbul birds. Vitamin C is probably required by other species but synthesized in the body.

Functions:

Formation and maintenance of collagen, the substance that binds body cells together. So, vitamin C makes for more rapid and sound healing of wounds and burns.

Metabolism of the amino acids tyrosine and tryptophan.

Absorption and movement of iron.

Metabolism of fats and lipids, and cholesterol control.

Sound teeth and bones.

Strong capillary walls and healthy blood vessels.

Metabolism of folic acid.

As a general antioxidant.

Vitamin C requirements have been observed to increase in periods of stress. Catfish and trout require dietary sources of vitamin C when they are raised in extremely crowded conditions. This theory of stress has been carried over to human nutrition. In 1970, Linus Pauling, the two-time winner of the Nobel Prize, published a book, *Vitamin C and the Common Cold,* which was to create considerable furor in nutrition and biochemistry circles. Pauling suggests that the daily ingestion of .25 g to 5 g (1 to 2 g optimal) of vitamin C will in many cases lower the incidence rate or severity of the common cold. To date, there is little convincing evidence to support Pauling's claims. But the debate over vitamin C and the common cold will likely stir emotions for years to come.

Deficiency Symptoms/Toxicity:

Deficiency Symptoms—Scurvy: In humans and monkeys, this deficiency is characterized by swollen, bleeding, ulcerated gums; loose teeth; malformed and big joints; fragility of the capillaries with resulting hemorrhages throughout the body; large bruises; big joints, such as the knees and hips, due to bleeding into the joint cavity; anemia; degeneration of muscle fibers, including those of the heart; and tendency of old wounds to become red and break open. Sudden death from severe internal hemorrhage and heart failure is always a danger.

Toxicity—Adverse effects in humans reported of intakes in excess of 8 g per day (more than 100 times the recommended allowance) include: nausea, abdominal cramps, and diarrhea; absorption of excessive amounts of iron; destruction of red blood cells; increased mobilization of bone minerals; interference with anticoagulant therapy; formation of kidney and bladder stones; inactivation of vitamin B-12; rise in plasma cholesterol; and possible dependence upon large doses of vitamin C.

Sources for Animals:

Rich feed sources—Citrus pulp, Irish potatoes, liver meal, pastures (green), sweet potatoes, tomato pulp.

Supplemental sources—Acerola cherry (*camu-camu*), rose hips, vitamin C (ascorbic acid).

Comments:

Of all the vitamins, ascorbic acid is the most unstable. It is easily destroyed during storage and processing; and it is water-soluble, easily oxidized, and attacked by enzymes.

Ordinary rations and body synthesis provide adequate vitamin C for farm animals and poultry.

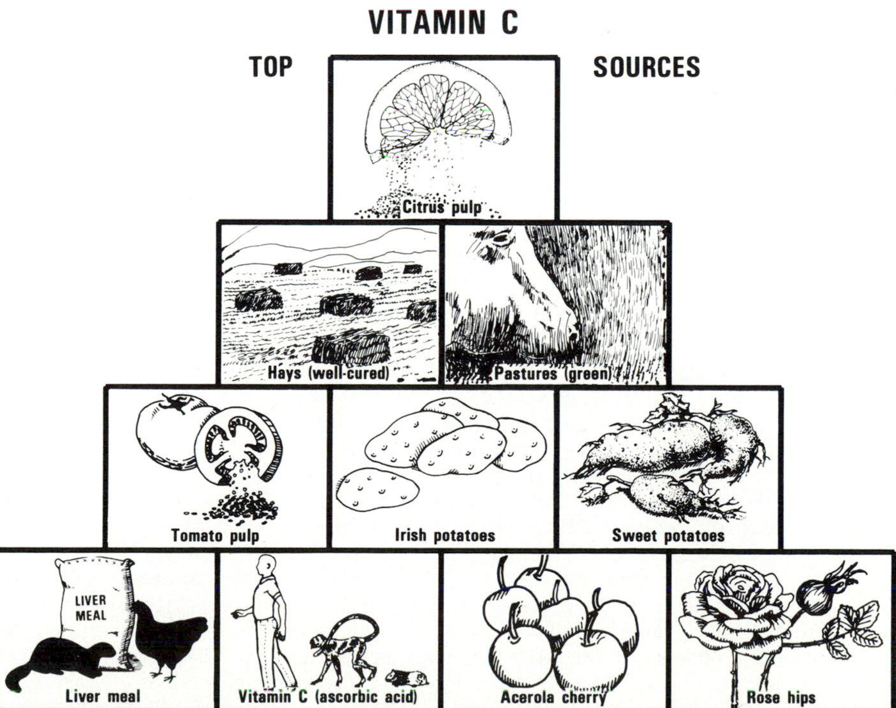

Fig. 4-136. Commonly used, top feed and supplement sources of vitamin C. No claim is made that all the top sources of vitamin C are listed. Neither are the sources ranked according to amount of vitamin C supplied nor is price considered.

Vitaminlike Substances

Certain substances, although not considered true vitamins, closely resemble vitamins in their activity, and are sometimes classed with the B-complex vitamins. They're commonly referred to as *vitaminlike substances.*

The nutritional status and biological role(s) of each of the vitaminlike substances require further clarification; indeed, some of them are very controversial. For a substance to be classified as a vitamin requires proof (1) that it is an essential and indispensable feed constituent, and (2) that deficiency syndromes are known which can be cured specifically by its administration. Both of these prerequisites are not met by the vitaminlike substances. Nevertheless, the authors list them for informational purposes, since in certain specific situations they can be useful. Additionally, cognizance is taken of the fact that there is usually a considerable time lag between the scientific validation and the acceptance of a vitamin, essential nutrient, and/or treatment.

The vitaminlike substances are listed and described in the sections which follow.

BIOFLAVONOIDS (VITAMIN P)

Bioflavonoids, also known as vitamin P, are a group of natural pigments in vegetables, fruits, flowers, and grains. They appear as companions of natural vitamin C, but they are not present in synthetic vitamin C. Because of the basic yellow color of most of them, they are called flavonoids, after the Latin, *flavus,* for yellow. (Some of these substances are naturally occurring dyes.) To date, about 800 different flavonoids have been identified, of which more than 30 are in the genus *Citrus* alone. Three of the better known bioflavonoids are *hesperidin, naringin,* and *rutin.*

- **Functions**—Bioflavonoids are promoted chiefly because of their function in preventing capillary fragility and supporting normal permeability. This is important because all body cells depend on the capillaries to bring them everything they need and to take away wastes, this is important.

The mechanism by which bioflavonoids exert their claimed influence on capillary fragility and permeability is not fully understood. It is known that capillary breakage is characteristic of the vitamin C deficiency disease, scurvy; and that vitamin C has a vital role in maintaining capillary health. Since bioflavonoids and vitamin C are found together in nature, it is conjectured that they function together in increasing the strength of the capillaries and regulating their permeability. These actions help prevent hemorrhages and ruptures in the capillaries and connective tissues and build a protective barrier against infection.

Although flavonoids are not classified as vitamins for higher animal life because of lack of proof that they are essential feed constituents, and that deficiency symptoms can be cured by their administration, there are strong indications that they are essential for lower stages of animal life—butterflies, silkworm larvae, crickets, and some species of beetles. Moreover, there are indications that similar essential effects may exist in higher animals and humans, too. For example, rutin has been shown to exert a growth-promoting action on young rats and on bacteria. Some authorities believe that a vitaminlike effect of flavonoids may manifest itself only under stress conditions. Moreover, flavonoids have been shown to exhibit a great number of specific effects contributing to good health in mammals.

In addition to their reputed effect on capillary fragility and permeability and on health, it is claimed that flavonoids function as follows:

1. **They are active antioxidant compounds in feed,** ranking second only to the fat-soluble tocopherols in this regard.
2. **They possess a metal-chelating capacity;** and they affect the activity of enzymes and membranes.
3. **They have a synergistic effect on ascorbic acid (vitamin C),** and they appear to stabilize ascorbic acid in tissues.
4. **They possess a bacteriostatic and/or antibiotic effect,** which is sufficiently high to account for measurable anti-infectious properties in normal daily feed.
5. **They possess anticarcinogenic activity in two ways;** a cytostatic effect against malignant cells (stopping or inhibiting the growth of cells), and a biochemical protection of the cell from damage by carcinogenic substances.

- **Deficiency symptoms**—Symptoms of bioflavonoid deficiency are closely related to those of a vitamin C deficiency. The tendency to bleed (hemorrhage) and bruise easily are especially claimed.
- **Sources**—Animals are unable to synthesize bioflavonoids. The top sources are: citrus pulp, rose hips, and buckwheat leaves.

CARNITINE (VITAMIN B_T)

Carnitine, a vital coenzyme in animal tissues and involved in fat metabolism, is a vitaminlike substance that has received much attention recently. It is similar to a vitamin with the exception that under normal conditions higher animals synthesize their total requirement within their bodies; hence, it is unnecessary to supply this substance in feed on a daily basis. However, some studies suggest (1) that synthesis of body carnitine may be inadequate for some individuals, and (2) that a number of diseases alter levels of carnitine in body fluids and tissues. This prompts two questions: (1)Are carnitine needs met adquately by body synthesis to assure good health; and (2) what role, if any, does carnitine play in certain diseases?

- **Functions**—Carnitine plays an important role in fat metabolism and energy production in mammals. It functions as follows:

1. **Transport and oxidation of fatty acids.** Carnitine plays an important role in the oxidation of fatty acids by facilitating their transport across the mitochondrial membranes.
2. **Fat synthesis.** Although this role is controversial, carnitine appears to be involved in transporting acetyl groups back to the cytoplasm for fatty acid synthesis.
3. **Ketone body utilization.** Carnitine stimulates acetoacetate oxidation; thus, it may play a role in ketone body utilization.

- **Deficiency symptoms**—If carnitine is to be considered an essential dietary nutrient, it must be possible to show that a lack in the diet results in a reproducible deficiency disease in animals. To date, carnitine has not met this criterion.
- **Sources**—Generally speaking, carnitine is high in meat meal/tankage and other meat by-products, milk, whey, and yeast (brewers' and torula).

The lower level of carnitine in plant feeds in comparison

with animal feeds is explainable on the basis that plant materials are most likely deficient in the essential amino acids, lysine and methionine, precursors of carnitine.

Higher animals appear to be able to synthesize their total needs within the body. But the mechanism of carnitine synthesis in animals is unknown.

COENZYME Q (UBIQUINONE)

Coenzyme Q, or ubiquinone, is a collective name for a number of ubiquinones—lipidlike compounds that are chemically somewhat similar to vitamin E.

Coenzyme Q is found in most living cells where it is concentrated in the mitochondria. Because it is synthesized in the cells, it cannot be considered a true vitamin.

• **Functions**—Coenzyme Q functions in the respiratory chain in which energy is released from the energy-yielding nutrients as ATP.

There is evidence that specific ubiquinones function in the remission/prevention of some of the symptoms of vitamin E deficiency.

• **Sources**—Quinones occur widely in aerobic organisms, from bacteria to higher plants and animals. Because they are synthesized in the body, they cannot be considered a true vitamin.

LAETRILE (VITAMIN B-17, AMYGDALIN, NITRILOSIDES)

Laetrile, or amygdalin, is a natural substance obtained from apricot kernels, which its advocates claim to have cancer preventive and controlling effects. Dr. Ernest Krebs, Sr., who was the first to use laetrile therapeutically in this country, considered it to be an essential vitamin and called it vitamin B-17.

Laetrile is manufactured and used legally for the prevention and treatment of cancer in about 20 countries throughout the world, including Belgium, Germany, Italy, Mexico, and the Philippines.

• **Functions**—Nitrilosides, including apricot kernels and bitter almonds, supply the body a low, but steady, level of hydrogen cyanide (HCN). Humans and other mammals have an enzyme, rhodanase, which converts the cyanide to thiocyanate.

There are at least three theories as to how the HCN interferes with tumor growth, but all theories relative to the action of nitrilosides recognize the steady low-level supply of HCN as the active agent.

• **Deficiency symptoms**—The advocates claim that prolonged deficiency of laetrile, or amygdalin, may lead to lowered resistance to malignancies.

• **Sources**—A concentration of 2 to 3% laetrile is found in the whole kernels of most fruits, including apricots, apples, cherries, peaches, plums, and nectarines.

LIPOIC ACID

Lipoic acid is a fat-soluble, sulfur-containing substance. It is not a true vitamin because it can be synthesized in the body and is not necessary in the diet of animals. However, it functions in the same manner as many of the B-complex vitamins.

• **Functions**—Lipoic acid functions as a coenzyme. It is essential, together with the thiamin-containing enzyme, thiamin pyrophosphatase (TPP), for reactions in carbohydrate metabolism which convert pyruvic acid to acetyl-coenzyme A. Lipoic acid, which has two sulfur bonds of high-energy potential, combines with TPP to reduce pyruvate to active acetate, thereby sending it into the final energy cycle. It joins the intermediary products of protein and fat metabolism in the Krebs Cycle in the reactions involved in producing energy from these nutrients. A metal ion (magnesium or calcium) is involved in this oxidative decarboxylation, along with lipoic acid and four vitamins: thiamin, pantothenic acid, niacin, and riboflavin. Thus, this underscores the concept of the interdependent relationships among the vitamins.

• **Deficiency symptoms**—No characteristic deficiency symptoms have been produced.

• **Sources**—Lipoic acid is found in many feeds. Yeast and liver are rich sources.

ORTIC ACID (VITAMIN B-13)

It is possible that ortic acid, or the so-called vitamin B-13, may be a growth promotant and a preventative of certain disorders. At this time, however, it is not known whether it plays an essential role in an otherwise adequate diet; hence, this presentation is for two purposes; (1) informational, and (2) stimulation of research.

This compound was first obtained from distillers' solubles.

• **Functions**—Ortic acid (B-13) has been found to stimulate the growth of rats, chicks, and pigs under certain conditions.

Ortic acid is utilized by the body in the metabolism of folic acid and vitamin B-12. Also, it appears to aid the replacement or restoration of some cells.

• **Deficiency symptoms**—Deficiency symptoms have not been proved. But it is believed that a deficiency may lead to liver disorders, cell degeneration, and premature aging.

• **Sources of ortic acid**—Ortic acid is found in such natural sources as distillers' solubles, whey, soured or curdled milk, and root crops. Also, this nutrient is available in supplement form as calcium orotate.

PANGAMIC ACID (VITAMIN B-15)

Pangamic acid (also known as vitamin B-15, calcium pangamate, and dimethylglycene) is a chemical substance of an organic nature. It is sometimes classed as a vitamin or vitaminlike substance. Obviously, however, pangamic acid does not meet the classical definition of a vitamin, which is: *A substance, organic in nature, necessary in the diet in small amounts to sustain life, in the absence of which a specific deficiency disease develops.* Proponents counter with the argument that: "It's not essential, strictly speaking, but it's helpful biologically under many circumstances."

In 1951, Krebs *et al.* reported the presence of a water-soluble factor in apricot kernels, which they subsequently isolated in crystalline form from rice bran and polishings. Later, it was extracted from brewers' yeast, cattle blood, and horse liver. The name, *pangamic acid* (*pan,* meaning universal; *gamic,* meaning seed) was applied to the substance to connote its seeming universal presence in seeds; and it was assigned the fifteenth position in the vitamin B series by its discoverers.

The claim that pangamic acid is a B vitamin is based primarily on its presence in B-vitamin rich feeds, and on the broad spectrum of physiological functions attributed to it. However, it is not known whether animals have the capacity to synthesize pangamic acid, and no specific disease can be attributed exclusively to a deficiency of the substance. Thus, the designation of pangamic acid as a vitamin is not accepted by most U.S. scientists.

• **Functions**—Numerous functions have been attributed to pangamic acid; among them, the following:

1. **Stimulation of transmethylation reactions.** Pangamic acid possesses methyl groups ($-CH_3$) which are capable of being transferred from one compound to another within the body, resulting in the stimulation of creatine synthesis in muscle and heart tissues.

2. **Stimulation of oxygen uptake.** Pangamic acid stimulates tissue oxygen uptake, thereby helping to prevent the condition called *hypoxia*—an insufficient supply of oxygen in living tissue, especially in heart and other muscles.

3. **Inhibition of fatty liver formation.** Oral administration or injection of pangamic acid to rats and rabbits exerts a protective effect against fatty infiltration of the liver induced by starvation, protein-free diets, anesthetics, carbon tetrachloride, or cholesterol.

4. **Adaptation to increased physical activity.** Pangamic acid may enable animals to adapt to increased exercise. After periods of enforced exercise, animals previously treated with pangamic acid have been reported to demonstrate a better maintenance of oxidative metabolism and energy levels than untreated controls; moreover, these effects persist for several days.

Some horse trainers have been known to administer pangamic acid to racehorses because of its reputed, but undocumented, quality of enabling animals to run faster and tire less.

5. **Control of blood cholesterol levels.** In some cases, the administration of pangamic acid may cause a fall in both cholesterol biosynthesis and blood levels.

• **Deficiency symptoms**—There is indication that a deficiency of pangamic acid may cause fatigue, hypoxia, heart disease, and glandular and nervous disorders.

• **Sources**—The best natural sources of pangamic acid are sunflower seed, pumpkin seed, yeast, liver, rice, whole cereal grains, apricot kernels, and other seeds. Evidence suggests its occurence wherever B-complex vitamins are found in natural feeds.

WATER

Fig. 4-137. "Water," an original painting by artist Tom Phillips, 3333 17th St., San Francisco, CA 94110)

Chemically, water is the combination of 2 gases—hydrogen (H) and oxygen (O)—which are joined in the ratio of 2 hydrogen atoms to 1 oxygen (as H_2O). It is the most abundant chemical substance, and it performs endless functions in its three forms—liquid, solid, gas.

Water is vital to nutrition. It is the solvent wherein the metabolic reactions of the body take place. Also, as a solvent, water carries (1) the nutrients which are subjected to cellular metabolism, and (2) the waste products of metabolism. Also, it serves to disperse the heat generated by the metabolic reactions. In many of the metabolic reactions water is either added or subtracted. Subtracted water is termed metabolic water. The addition of water is termed hydrolysis.

Animals can survive for a longer period without feed than they can without water. Only oxygen is more important to animal life. Fortunately, under most conditions, water can be readily provided in abundance—and at little cost. In addition to what animals drink, water is found in all feeds, ranging from about 10% in air-dry feeds to over 80% in fresh green forage.

Water is one of the largest single constituents of the animal body, varying in amount from 40% in fat hogs to 80% in newborn pigs, 50% in a 1,000-lb steer to 70% in a newborn calf, and 50% in a fat lamb to 80% in a newborn lamb. In general, the percentage of water in the bodies of animals varies with species, condition, and age. The younger the animal, the more water it contains. Also, the fatter the animal, the lower the water content. Thus, as an animal matures, it requires proportionately less water on a weight basis because it consumes less feed per unit of weight and the water content of the body is being replaced by fat. This accounts for the fact that gains in older animals are more costly nutritionally than those in younger animals.

Water performs the following important functions in animals:

1. It is necessary to the life and shape of every cell and is a constituent of every body fluid.

2. It acts as a carrier for various substances, serving as a medium in which nourishment is carried to the cells and waste products are removed therefrom.

3. It assists with temperature regulation in the body, cooling the animal by evaporation from the skin as perspiration.

4. It is necessary for many important chemical reactions of digestion and metabolism.

5. It lubricates the joints, as a constituent of the synovial fluid; it acts as a water cushion for the nervous system, in the cerebrospinal fluid; it transports sound, in the perilymph in the ear; and it is involved with sight and provides a lubricant for the eye.

6. It acts as a solvent for a number of chemicals which can subsequently be detected by taste buds.

7. It aids in gas exchange in respiration by keeping the alveoli of the lungs moist.

The total body water involved in all of these functions is contained in two major compartments in the body: (1) the extracellular water outside the cells (about 20% of the body weight), and (2) the intracellular water inside each cell (about 45% of the body weight).

Deficits or excesses of more than a few percent of the total body water are incompatible with health, and large deficits, of about 20% of the body weight, lead to death. Under normal circumstances, thirst ensures that water intake meets or exceeds the requirement for water.

The specific water requirements of each class of animal receive further consideration in the section devoted to the respective species. In general, however, under practical conditions, the needs for water can best be taken care of by allowing the animals free access to plenty of clean, fresh water at all times.

Water Balance

In healthy animals, total body water remains reasonably constant. An increase or decrease in water intake brings about an appropriate increase or decrease in water output to maintain the balance. Water enters the body as a liquid, and as a component of feed—including metabolic water derived from the breakdown of feed. Water is lost from the body (1) by the skin as perspiration, (2) by the lungs as water vapor in expired air, (3) by the kidneys as urine, and (4) by the intestine in the feces. Therefore, under normal conditions the intake of water from the various sources is approximately equal to output of water by the various routes.

If water is withheld, the following compensatory mechanisms are initiated to provide enough water for maintenance:

1. Urine excretion is reduced as is the water content of the feces.

2. The animal oxidizes much of its tissue reserves to provide metabolic water. This results in a loss of body weight.

3. Animals become sedentary, seeking shade whenever possible to reduce the loss of water from surface evaporation and sweating.

4. There is a reduction in feed consumption except for feeds that are high in moisture.

If adequate water supplies are not made available to the animal, the blood eventually thickens, resulting in a decreased ability to transport nutrients and waste products. Animals can lose most of their fat, about half of their protein, and many other constituents of their body tissues, but a loss of one-tenth of the water in the body is lethal.

Water Sources

Water can be obtained from three sources: (1) drinking water, (2) water contained in feed, and (3) metabolic water derived from the breakdown of carbohydrates, fats, and proteins.

DRINKING WATER

Many livestock producers tend to underestimate the value of water. It is not enough to make water available. Quality, as dictated by the levels of pollutants, is as important as availability. With an increase in management of livestock in confinement, it becomes increasingly necessary to consider the use of water from approved, inspected sources, since the water is also used by the people working with the animals as well as for the washing of utensils used in the operation. It is advisable for the livestock manager to check local codes and regulations on water quality.

The pollution of streams is of increasing concern to our government and is monitored by the Environmental Protection Agency (EPA). Because of their regulations, it may become necessary to alter some practices in livestock operations, relating to accessibility of water to animals and subsequent waste removal procedures. Table 4-9 lists the criteria for water quality as governed by pollution levels.

TABLE 4-9
CRITERIA FOR WATER QUALITY FOR LIVESTOCK[1]

Pollutants	Limit
	(ppm)
Elements:	
Aluminum	5
Arsenic	0.2
Beryllium	—
Boron	5
Cadmium	0.05
Chromium	1
Cobalt	1
Copper	0.5
Fluorine	2
Iron	—
Lead	0.1
Manganese	—
Mercury	0.001
Molybdenum	—
Nitrate	100
Nitrite	10
Salts, soluble	300
Selenium	0.05
Vanadium	1
Zinc	25
	(ml)
Coliform bacteria:	
Fecal	1,000/100
Total	5,000/100

[1]Adapted by the authors from *Proposed Criteria for Water Quality*, Vol. 1, U.S. Environmental Protection Agency.

A particularly troublesome problem of drinking water found on ranges is that of salinity. Dissolved salts can affect the intake of water; but animals can eventually become conditioned to the salinity if it is not excessive. Production can also be affected by the use of salty water. Most livestock species can tolerate a total dissolved solid concentration of 15,000 to 17,000 ppm; but at these levels, production will, in all likelihood, be dramatically affected.

Nitrate pollution has received special emphasis recently as part of the environmental protection program. Quantities as small as 100 to 200 ppm can be dangerous while 3,000 ppm is potentially lethal.

The amount of water required by the various types of livestock varies considerably, as evidenced in Table 4-10. Several factors can affect the amount of water a particular animal will consume: (1) age, (2) body weight, (3) production, (4) weather (heat and humidity), and (5) type of ration. The intensity of production dramatically affects the water requirement. A steer fed a maintenance ration will consume approximately 35 lb of water daily, whereas a steer fed a fattening ration will consume double this quantity. A cow that is not in lactation will have a daily water intake of about 90 lb. When she produces 20 to 50 lb of milk, this figure will increase to about 160 lb. When milk production reaches 80 lb per day, water intake will be near 200 lb.

WATER IN FEEDS

The water content of feeds is extremely important, especially for animals which do not have ready access to drinking water. The term succulence refers to the property of feeds whereby the composition of the feed is relatively high in water compared to the amount of dry matter. When water is scarce, succulent feeds become extremely important for production. In the case of the dairy cow in heavy production, however, the use of succulent feeds is curtailed because the intake of these feeds may severely limit the intake of adequate amounts of dry matter needed for milk production.

Water on the surface of the plant, such as dew, can serve as an important source for some animals. On arid ranges, cattle, goats, and sheep may rely heavily on this type of water; but this supply is rarely sufficient to meet the needs of livestock.

METABOLIC WATER

Metabolic water is produced from the catabolism of nutrients. When 100 g of carbohydrates are oxidized, 60 g of water are produced. The oxidation of proteins yields 42 g of water for every 100 g of protein. Fats can be said to be "wetter than water." For every 100 g of fat that are oxidized, nearly 110 g of water are produced. However, there are some losses of water in the oxidation of both proteins and fats. Water must be used to excrete nitrogen in the deamination process of protein—thus lowering the net availability of water. In fact, it requires more water to excrete nitrogen as urea than is formed in the deamination process. The oxidation of fats requires increased respiration. Water is lost from the lungs during this increased respiration, and the net yield of water produced from fat is less than that from the oxidation of carbohydrates.

Water Excretion

Water can be eliminated from the body through three routes: (1) urine, (2) feces, and (3) insensible perspiration.

Urine provides a means whereby water-soluble products of metabolism can be excreted. These substances can be end products of nutrient catabolism or products of the various detoxification and protective mechanisms of the body. Generally, when rations are high in protein or mineral content, urine flow is increased.

The amount of water lost in the feces is highly dependent on the species. Fecal pellets of sheep contain much less water than the feces of cattle. Likewise, when production demands are high, as in lactation, water can be a limiting nutrient, thereby necessitating conservation efforts within the animal. When cattle are first placed on succulent, fresh pasture or green chop, the feces become very watery—a reflection of the increased water intake.

Insensible perspiration is the loss of water through the lungs and skin. Animals that do not have sweat glands rely on the cooling effect of panting. During the exhalation process of respiration, water vapor is lost from the lungs. Water is also lost through the skin as sweat in order to act as a temperature regulator.

TABLE 4–10
DAILY WATER CONSUMPTION BY LIVESTOCK

Species	Age	Body Weight	Condition	Water Consumption[1]
	(weeks)	(lb)		(gal)
Cattle	4	112	Growing	1.3– 1.5
	8	152	Growing	1.6– 2.0
	12	204	Growing	2.3– 2.5
	16	263	Growing	3.1– 3.5
	20	327	Growing	4.0– 4.5
	26	416	Growing	4.5– 6.0
	60	779	Growing	6.0– 8.0
	84	1,023	Pregnant	8.0–10.0
	1–2 yr	1,000–1,200	Fattening	8.0– 9.0
	2–8 yr	1,200–1,600	Lactating	10.0–25.0
	2–8 yr	1,200–1,600	Grazing	4.5– 9.0
Sheep		20	Growing	0.5
		50	Growing	0.4
		150–200	Grazing	0.5–1.5
		150–200	Grazing, salty feeds	2.1
		150–200	Hay and grain	0.1–0.8
		150–200	Good pasture	less than 0.5
Swine		30	Growing	0.3–1.0
		60– 80	Growing	0.7–1.2
		80–125	Growing	1.0–2.0
		200–400	Maintenance	1.5–3.5
		200–400	Pregnant	4.0–5.0
		200–400	Lactating	5.0–6.5
Chickens	1– 3		Growing	0.5–2.0
(per 100	3– 6		Growing	1.5–3.0
birds)	6–10		Growing	3.0–4.0
	9–13		Growing	4.0–5.0
	Mature		Nonlaying hens	5.0
	Mature		Laying	5.0–7.5
	Mature (90 °F)		Laying	9.0
Turkeys	1– 3			1.1– 2.6
(per 100	4– 7			4.0– 8.5
birds)	9–13			9.0–14.5
	15–19			16.7–17.0
	21–26			13.5–15.0
	Mature			17.0
Horses	Mature			12.0

[1]A gallon of water weighs 8.3 lb.

QUESTIONS FOR STUDY AND DISCUSSION

1. Define and discuss each of the following terms: *nutrients, metabolism, catabolism,* and *anabolism.*

2. Why are cells referred to as functional units of nutrition and metabolism?

3. Discuss the impact of the research work of Wilkins, Watson, and Crick, for which they shared the Nobel Prize for Physiology and Medicine, in 1962, on nucleic acids (DNA and RNA), and in turn, on nutrition and metabolism.

4. Discuss the synthesis of DNA and RNA. Why is the structure of DNA called a double helix?

5. What is gene-splicing?

6. What is the present and potential impact of genetic engineering (recombinant DNA) on nutrition and metabolism?

7. Define biotechnology.

8. Has biotechnology given rise to a major scientific revolution? Justify your answer by citing some examples.

9. Discuss the role of each of the following nutrients in meeting the energy requirements of the body: carbohydrates, fats, proteins, minerals, vitamins, and water.

10. Discuss each of the following pertaining to energy for the body: (a) what energy is required for, (b) the symptoms of energy deficiency, and (c) the role of carbohydrates, fats, and proteins in providing energy.

11. Discuss the following aspects of carbohydrates: (a) what they are, (b) their importance in feeds, and (c) how they are produced by photosynthesis.

12. Give the distinguishing characteristics of each of the following types of carbohydrates, and list the most important subgroups under each of them: (a) monosaccharides, (b) oligosaccharides, (c) polysaccharides, and (d) specialized compounds. (See Table 4–1 and the narrative sections that follow it.)

13. Classify and describe each of the following carbohydrates, and give (a) the chief feed sources, (b) the end products of digestion, and (c) the nutritional functions of each of them: fructose, galactose, glucose, mannose, lactose, maltose, sucrose, cellulose, dextrins, glycogen, starch, gums, hemicelluloses, pectins, and lignin. (See Table 4–1 and the narrative sections that follow it.)

14. Discuss the availability and utilization of fiber as related to (a) ruminants, (b) stage of plant maturity, and (c) classes and ages of animals.

15. Of what importance is ATP in carbohydrate metabolism?

16. Define (a) oxidation, (b) reduction, and (c) the electron transport system.

17. Briefly outline glycolysis—the aerobic oxidation of glucose. What's the glycolytic or Embden-Meyerhof pathway?

18. Briefly outline the Krebs Cycle.

19. Compare the efficiency of the body in converting glucose to energy with the efficiency of diesel and gasoline engines in converting their fuel energy into work.

20. What are the functions of the pentose shunt?

21. Discusss the anabolism of carbohydrates.

22. How may fats and oils be differentiated? When digested, why do fats and oils liberate more heat than carbohydrates?

23. What are the three major groups of lipids?

24. Discuss the importance of each of the following chemical characteristics as related to fatty acids: (a) saturation, (b) iodine number, (c) rancidity, (d) hydrogenation, (e) carbon chain length, (f) saponification, and (g) emulsification.

25. Name the essential fatty acids, and describe their deficiency symptoms and primary functions.

26. Discuss the (a) digestion, (b) catabolism, and (c) anabolism of fats.

27. Discuss the cause, symptoms, and treatment of acidosis (ketosis) in sheep and cattle.

28. What's the Total Digestible Nutrients (TDN) system? What are its main (a) advantages, and (b) disadvantages?

29. What's the Calorie System? Diagram the utilization of energy, showing the losses along the way.

30. Compare the advantages and disadvantages of the Total Digestible Nutrient and Net Energy Systems of expressing energy requirements. Which one do you favor? Support your answer.

31. What are the functions of purines and pyrimidines?

32. Discuss nucleic acids, including their synthesis and involvement in genetic engineering.

33. Discuss the utilization of urea and ammoniated feeds by ruminants.

34. Chemically, (a) what are proteins, and (b) how may proteins be classified?

35. What is meant by (a) essential and (b) nonessential amino acids? List the amino acids in each of these categories.

36. Discuss the (a) catabolism and (b) anabolism of proteins.

37. Define and discuss the biological value of protein.

38. Discuss the history and discovery of minerals.

39. Define: (a) minerals, (b) major or macrominerals, and (c) trace or microminerals. List the 18 required minerals in either of two groups—major or trace.

40. Why is the ash fraction of the proximate analysis considered to be only a crude measurement of the mineral content of feeds?

41. List the general functions of minerals.

42. What are (a) the functions and (b) the deficiency symptoms of calcium.

43. Discuss the relationship of calcium, phosphorus, and vitamin D.

44. Explain the interrelationship between chloride and sodium.

45. What function does chromium perform as a component of the glucose tolerance factor (GTF)?

46. What is the function of cobalt in animal nutrition?

47. How does a copper deficiency affect the color and crimp of the fleece of sheep?

48. How may excess fluorine be evidenced in the teeth?

49. Explain how iodine functions in the body. Why should iodized salt never be used as a feed governor when self-feeding?

50. Explain how the ferritin curtain controls iron absorption. How does iron function in the body?

51. What is the cause and treatment of grass tetany?

52. What is perosis?

53. Discuss the effect of molybdenum and copper on black sheep.

54. Describe phosphorus deficiency symptoms.

55. Why is a potassium deficiency most commonly evidenced in drylot finishing cattle and sheep?

56. Why is selenium sometimes correctly referred to as "an essential poison"?

57. What animal species require silicon? What are its functions?

58. What are the interrelationships between sodium, potassium, and chlorine? How is sodium usually provided?

59. Should elemental sulfur be fed to monogastric animals? State your reasons.

60. List the functions of zinc.

61. Summarize the history and discovery of vitamins.

62. Define (a) vitamins, (b) the fat-soluble vitamins, (c) the water-soluble vitamins, and (d) the vitamin B complex. List the vitamins in either of 2 groups—the fat-soluble vitamins, and the water-soluble vitamins.

63. Formerly, vitamin nutrition was dependent upon incorporating into the ration feeds that were known to contain a high natural content of the needed vitamins. Today, vitamin deficiencies may be averted by using vitamin supplements. What caused this transition?

64. Discuss (a) the relationship between vitamin A and carotene, (b) the forms of vitamin A, and (c) the role of vitamin A in vision in dim light (night blindness).

65. For what 3 closely related naturally occurring compounds with potential vitamin B-6 activity is vitamin B-6 now the approved collective name? What one vitamin B-6 deficiency symptom do all species share?

66. What is unique about B-12 from the standpoints of (a) its natural synthesis by plants/animals, (b) its requiring a specific gastrointestinal tract secretion (intrinsic factor) for its absorption, and (c) its rate of absorption?

67. What factor in uncooked eggs will produce biotin deficiency in animals? How may this be averted?

68. Why cannot humans, monkeys, guinea pigs, fruit-eating bats, and red-vented bulbul birds convert glucose to ascorbic acid? Describe the deficiency symptoms of scurvy. Is there sufficient experimental evidence to warrant people taking massive doses of vitamin C to lower the incidence and severity of the common cold?

69. The classification of choline as a vitamin is debated because it does not meet all the criteria for vitamins. What criteria does it not meet? Describe choline deficiency symptoms.

70. Discuss the following pertaining to vitamin D: (a) two ways in which it is unique among vitamins, (b) the comparative activity of vitamins D_2 and D_3 for people and birds, (c) the formation of vitamin D_3 in the skin, and (d) the characteristics of the vitamin D deficiency diseases known as rickets and osteomalacia.

71. Explain the interrelationship between vitamin E and selenium. Describe the vitamin E deficiency diseases known as stiff lamb disease and white muscle disease.

72. Describe folacin deficiency symptoms. List the leading supplemental sources of folacin.

73. Should inositol be classified as a vitamin? Justify your answer.

74. What is the relationship between vitamin K and dicoumarol? How is dicoumarol used?

75. Describe niacin deficiency symptoms. Explain how the following historical practices may have prevented pellagra, unknowingly: (a) in Mexico, treating corn with lime water before making tortillas; and (b) in Arizona, the Hopi Indians roasting sweet corn in hot ashes.

76. Describe pantothenic acid deficiency symptoms in the pig and the chick. Of all the B vitamins, pantothenic acid is most likely to be deficient under drylot conditions. Why is this so?

77. Para-aminobenzoic acid can completely replace a certain vitamin. Name it.

78. Why has riboflavin deficiency not been observed in ruminants? Summarize riboflavin deficiency symptoms, and list rich feed sources of riboflavin.

79. Summarize the history of thiamin, beginning with the study of the age-old disease beriberi. Describe the thiamin deficiency symptoms.

80. Define vitaminlike substances. Why are these substances so controversial? List some vitaminlike substances and summarize the functions of each of them.

81. Chemically, what is water? What is the comparative importance of feed and water?

82. List the functions of water in animals.

83. How does water enter the body; and how does it leave the body?

84. If water is withheld from an animal, what compensatory mechanisms are initiated to provide enough water for body maintenance?

85. List the 3 sources through which animals may obtain water.

86. Explain how fats are "wetter than water."

87. What are the 3 ways through which water is eliminated from the body?

SELECTED REFERENCES

Title of Publication	Author(s)	Publisher
Animal Growth and Nutrition	Ed. by E. S. Hafez, I. A. Dyer	Lea & Febiger, Philadelphia, Pa., 1969
Animal Nutrition, 2nd Edition	P. McDdonald R. A. Edwards J. F. Greenhalgh	Oliver and Boyd, Edinburgh, Scotland, 1973
Animal Nutrition, 7th Edition	L. A. Maynard, *et al.*	McGraw-Hill Book Company, New York, N.Y., 1979
Biochemistry	A. L. Lehninger	Worth Publishers, Inc., New York, N.Y., 1970
Bioenergetics and Growth	S. Brody	Reinhold Publishing Company, New York, N.Y., 1945
Comparative Nutrition of Man and Domestic Animals	H. H. Mitchell	Academic Press, New York, N.Y., 1961
Dukes' Physiology of Domestic Animals, 8th Edition	Ed. by M. J. Swenson	Cornell University Press, Ithaca, N.Y., 1970
Energy Metabolism of Ruminants, The	K. L. Blaxter	Hutchinson Scientific and Technical, London, England, 1962
Fats in Animal Nutrition	J. Wiseman	Butterworth, London, 1984
Fat-Soluble Vitamins, The	Ed. by H. F. DeLuca, J. W.Suttic	University of Wisconsin Press, Madison, Wisc., 1969
Fat-Soluble Vitamins, Vol. 9 of the International Encyclopedia of Food and Nutrition	Ed. by R. A. Morton	Pergamon Press, New York, N.Y., 1970
Feeds and Feeding	A. Cullison	Reston Publishing Co., Inc., Reston, Va., 1979
Foods & Nutrition Encyclopedia, Vols. 1, 2	A. H. Ensmsinger, *et al.*	Pegus Press, 648 W. Sierra Ave., Clovis, CA 93612, 1983
Fundamentals of Nutrition, 2nd Edition	L. E. Lloyd E. W. Crampton B. E. McDonald	W. H. Freeman and Company, San Francisco, Calif., 1978

(Continued)

(Continued)

Title of Publication	Author(s)	Publisher
Growth, Nutrition, and Metabolism of Cells in Culture, Vol. III	Ed. by G. H. Rothblat, V. J. Cristofalo	Academic Press, New York, N.Y., 1977
Guide to the Vitamins—Their Role in Health and Disease, A	J. Marks	Medical and Technical Publishing Company, Ltd., Lancaster, England
Handbook of Vitamins and Hormones	R. J. Kutsky	Van Nostrand Reinhold Company, New York, N.Y., 1973
Livestock Feeds and Feeding, 2nd Edition	D. C. Church	O & B Books, Inc., Corvallis, Ore., 1984
Metabolism	Compiled & Ed. by P. U. Altman, D. S. Dittmer	Federation of American Societies for Experimental Biology, Bethesda, Md., 1968
Mineral Metabolism, Vols. 1, 2, and 3	Ed. by C. L. Comar, F. Bronner	Academic Press, New York, N.Y. Vol. 1, Part A, 1960 Vol. 2, Part A, 1964 Vol. 2, Part B, 1962 Vol. 3, Calcium Physiology, 1969
Mineral Nutrition of Animals	V. I. Georgievskii B. N. Annenkov V. I. Samokhin	Butterworth, London, 1981
Mineral Nutrition of Livestock, The	E. J. Underwood	F.A.O., Rome, Italy, 1966
Nitrogen and Energy Nutrition of Ruminants	R. L. Shirley	Academic Press, Inc., Orlando, Fla., 1986
Nutrients and Toxic Substances in Water for Livestock and Poultry	National Research Council	National Academy of Sciences, Washington, D.C., 1974
Nutrition of Animals of Agricultural Importance, Parts I and II	Ed. by D. Cuthbertson	Pergamon Press, Oxford, England, 1986
Nutrition of the Chicken	M. L. Scott M. C. Nesheim R. J. Young	M. L. Scott and Associates, Ithaca, N.Y., 1976
Physiology of Digestion and Metabolism in the Ruminant	Ed. by A. T. Phillipson	Oriel Press, Ltd., Newcastle Upon Tyne, England, 1970
Principles of Biochemistry, 6th Edition	A. White, *et al.*	McGraw-Hill Book Company, New York, N.Y., 1978
Protein Contribution of Feedstuffs for Ruminants	E. L. Miller I. H. Pike A. J. H. Vanes	Butterworth, London, 1982
Recent Advances in Animal Nutrition—1985	W. Haresign D. J. A. Cole	Butterworth, London, 1985
Review of Medical Physiology, 7th Edition	W. F. Ganong	Lange Medical Publications, Los Altos, Calif., 1975
Review of Physiological Chemistry, 15th Edition	H. A. Harper	Lange Medical Publications, Los Altos, Calif., 1973
Science of Nutrition of Farm Livestock, The, Vol. 17 of the International Encyclopedia of Food and Nutrition	Ed. by D. Cuthbertson	Pergamon Press, New York, N.Y., 1969
Trace Elements in Human and Animal Nutrition	E. J. Underwood	Academic Press, New York, N.Y., 1974
Trace Minerals in Agriculture	V. Sauchelli	Van Nostrand Reinhold Company, New York, N.Y., 1969
Vitamin Manual		Upjohn Company, Kalamazoo, Mich., 1957
Vitamins, The, Vols. I through V	Ed. by W. H. Sebrell, Jr., R. S. Harris	Academic Press, New York, N.Y., 1967
Vitamins, The, Vols. VI and VII	Ed. by P. Gyorgy, W. N. Pearson	Academic Press, New York, N.Y., 1967

Chapter

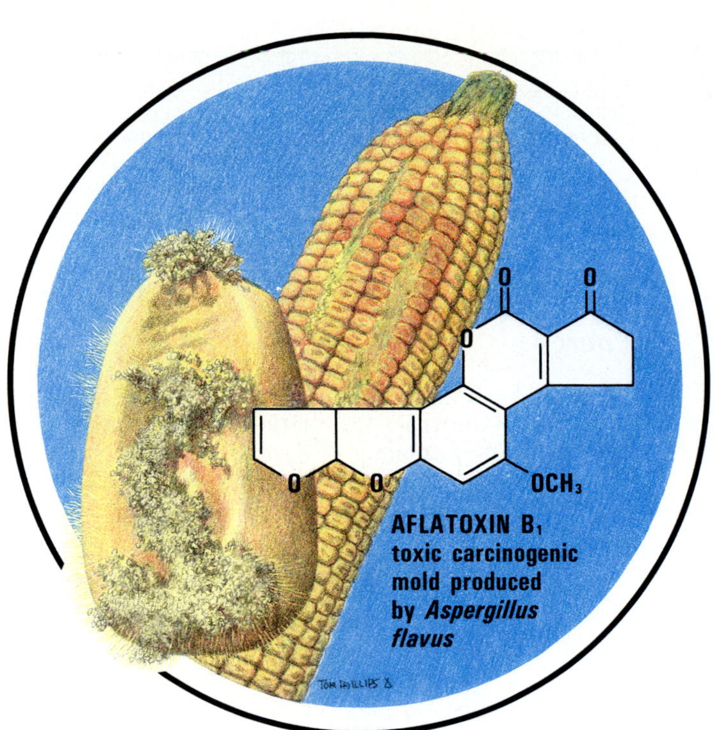

Original painting by Tom Phillips

NUTRITIONAL DISORDERS/TOXINS

Contents Page

Nutritional disorders may be caused by nutritional deficiencies and imbalances, by diseases and parasites, or by nutrition-related problems. Toxicity may be caused by poisonous plants, or by a host of agricultural chemicals and drugs.

NUTRITIONAL DEFICIENCIES AND IMBALANCES

Nutritional deficiency diseases may be brought about by (1) too little feed, (2) rations that are too low in one or more nutrients, (3) imbalance of nutrients, or (4) presence of antinutritional factors, such as antimetabolites or antivitamins.

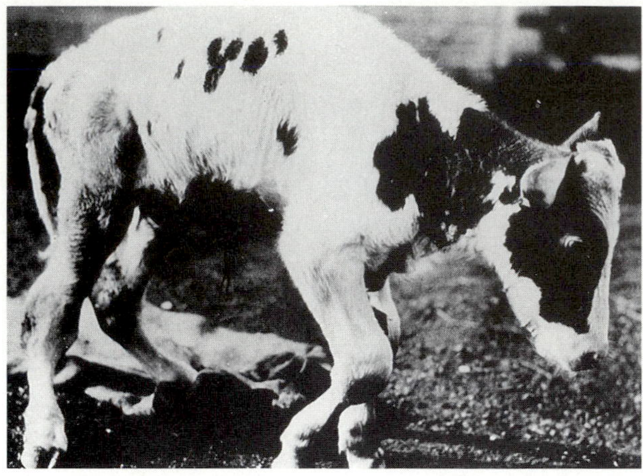

Fig. 5-1. Calf with severe rickets. Note the emaciation, humping of back, swelling of joints, knuckling of pasterns, and bowing of legs. Rickets may be caused by a lack of calcium, phosphorus, or vitamin D, or by an incorrect ratio of the two minerals. (Courtesy, Michigan State University, East Lansing)

Forced production (such as very high milk yields and marketing animals at early ages) and the feeding of forages and grains which may be produced on leached and depleted soils have created many problems in nutrition. This condition has been further aggravated through the increased confinement of stock, many animals being confined to corrals or buildings all or a large part of the year so that they have little or no opportunity to select their own diet. Under these unnatural conditions, nutritional diseases and ailments, along with abnormal behavior, have become increasingly common.

Animals have rather narrow tolerances between what they need and what will cause problems. Whether nutrients are needed in small or large amounts, most of them have an upper tolerance level beyond which excesses will prevent the animals from performing normally, or may even cause death. The range between that which is necessary and that which can be tolerated varies among species and nutrients. Selenium, for example, which is needed in very minute quantities by most species is quite toxic when fed above a certain level. Water, on the other hand, which is needed by all animals, can cause problems if consumed in excess, but the range between what is needed and what will be troublesome is very great.

Although the cause, prevention, and treatment of most nutritional diseases and ailments are known, they continue to reduce profits in the livestock industry simply because the available knowledge is not put into practice. Moreover, those widespread nutritional deficiencies which are not of sufficient proportions to produce clear-cut deficiency symptoms cause even greater economic losses because they go unnoticed and unrectified. It is important, therefore, that livestock producers and those who counsel with them be able to recognize subacute nutritional deficiencies, as well as the more obvious symptoms of acute deficiencies or toxicities.

It is often difficult to isolate deficiencies or toxicities due to the deficiency of a single nutrient. It is one thing to control animals experimentally and to isolate a single factor, and quite another to distinguish it under field conditions where there may be many complicating factors. When an animal goes off feed for whatever reason, several nutrient deficiencies may enter into the symptoms that appear. What caused the animal to go off feed, rather than the fact that it is off feed, is the important thing in returning it to normal. This is why it is so important that caretakers detect the early symptoms of deficiencies or toxicities, thereby making it possible for them to rectify the causes.

Energy

Many animals throughout the world are underfed all or some part of the year. Thus, lack of sufficient energy is probably the most common deficiency suffered by animals, although it is frequently complicated by a concomitant shortage of protein and other nutrients.

Fig. 5-2. Lack of energy!

Restricted rations often occur during periods of drought, when pastures and ranges are overstocked, or when winter rations are skimpy. Fortunately, during such times of restricted energy intake, animals may have nutritive reserves —mainly body fat—upon which they may draw. Although they may survive for a considerable period of time under these conditions, there is an inevitable loss in body weight; and, varying with the degree of energy shortage, there may be slowing or cessation of growth, failure to conceive, and increased mortality. Low energy intake also commonly results in increased deaths from toxic plants and from lowered resistance to diseases and parasites.

During times of energy shortage when animals are withdrawing stored energy—mostly from body fat—the mobilized fat may not be completely metabolized, with the result that ketosis (incompletely metabolized fatty acids) develops (see Table 5-1). Many high-producing animals are mildly ketotic, because it is nearly impossible for them to consume sufficient energy during maximum production. Mild cases of ketosis may not affect production or health. But animals with severe ketosis go off feed, which further aggravates the malady by lowering energy intake still more.

During cold weather, some energy—perhaps as much as 20% of the maintenance requirements—must be converted to heat to keep the body warm. Tables of nutrient requirements do not usually make provision for this added energy. The amount of added energy needed is proportional to the decrease in temperature below the comfort zone. It is further increased by humidity and wind. The combination of temperature, humidity, and wind is commonly referred to as the *chill factor.*

(Also see Chapter 4, section on "Energy," for more complete information relative to energy deficiencies and utilization.)

Protein

The protein allowance for animals, regardless of age or system of production, should be ample to replace the daily breakdown of the tissues of the body, including muscles, blood, hair, and hooves. Protein needs are increased by growth and gestation-lactation.

Depressed appetite is the primary symptom of protein deficiency in livestock rations. Going off feed may, in turn, lead to an inadequate intake of energy; hence, protein deficiency and energy deficiency often occur together.

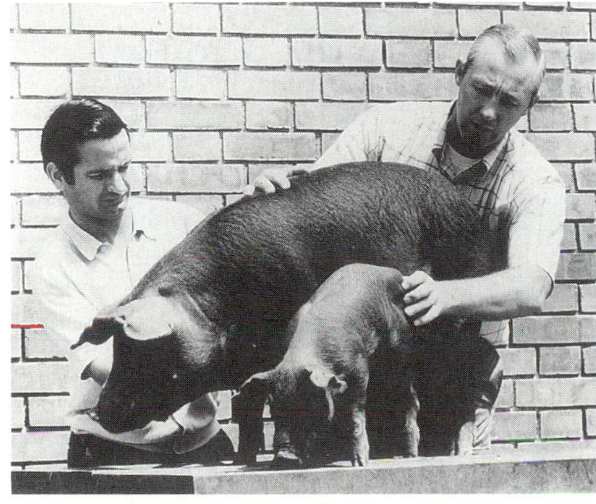

Fig. 5-3. Lysine made the difference! These littermate pigs were started on test at weaning, at a weight of 20 lb. The only difference in their ration was the kind of corn. The big pig (left) received high-lysine (opaque-2) corn; the little pig (right) got regular corn. During the 130-day trial, the pig fed opaque-2 gained a respectable 73.2 lb, whereas the pig eating ordinary corn gained only 6.6 lb. (Courtesy, The Rockefeller Foundation, New York, N.Y.)

Other symptoms of protein deficiency are loss of weight, poor growth, irregular or delayed estrus, and reduced milk production.

Because of synthesis of essential amino acids by microorganisms, the quality of protein (or balance of essential amino acids) is of less importance in feeding ruminants than in feeding nonruminant animals. Thus, the latter must not only have sufficient amounts of available protein, but the protein must contain a balance of essential amino acids. When one or more of the essential amino acids is not available, the remaining amino acids cannot be used effectively, and may be converted to energy.

(Also see Chapter 4, Section on "Proteins," for more complete information relative to protein deficiencies and quality.)

Minerals

Animal bodies contain small amounts—only 2 to 5%—of inorganic elements, called minerals. But these constituents play a vital role in animal nutrition. They furnish structural materials for bones and teeth. Additionally, as constituents of the soft tissues, the blood, the fluids of the body, and certain of the secretions, they regulate many of the vital processes.

Fig. 5-4. Ewe showing goiter (big neck) due to a deficiency of one mineral —iodine. (Courtesy, Montana State University, Bozeman)

Although acute mineral deficiency diseases and actual death losses are relatively rare, inadequate supplies of any one of the 18 essential mineral elements may result in lack of thrift, poor gains, inefficient feed utilization, lowered reproduction, and decreased production of meat, milk, eggs, wool, or work. Only when the mineral deficiency reaches such proportions that it results in excess emaciation, reproductive failure, or death is it likely to be detected.

With minerals, it is also very important that consideration be given to both nutrient relationships and availability.

(Also see Chapter 4, section on "Minerals," for more complete information relative to mineral deficiencies, relationships, and availabilities.)

Vitamins

Vitamin deficiencies in animals may occur as a result of lack of availability of vitamins, or because of the presence of antinutritional factors. Corn is an example of the latter concept. Analyses show corn to be adequate in niacin. Yet, due either to an antimetabolite or unavailability, there may be a niacin deficiency when corn is fed—a deficiency which can be rectified by niacin supplementation.

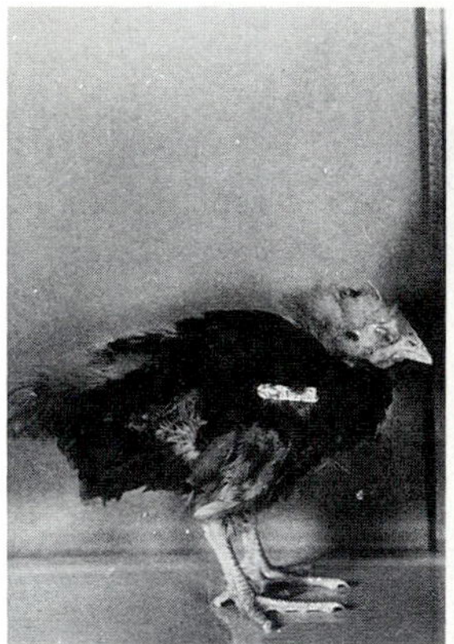

Fig. 5-5. Vitamin A made the difference! *Left:* Vitamin A-deficient chick, showing poor growth, watery eyes, unsteady gait, and ruffled appearance. *Right:* Chick that received the same basal ration, plus plenty of vitamin A. (Courtesy, University of Maryland, College Park)

The fat-soluble vitamins (A, D, E, and K) are stored in appreciable quantities in the body, whereas the water-soluble vitamins are not. Thus, vitamin A and/or carotene may be stored by an animal in its liver and fatty tissue in sufficient quantities to meet its requirements for a period of 6 months or longer. By contrast, the large amounts of water which pass through most animals daily tend to carry out the water-soluble vitamins of the body, thereby depleting the supply. Thus, they must be supplied in the ration on a day-to-day basis for those animals having a simple stomach in which microbial synthesis is limited.

(Also see Chapter 4, section on "Vitamins," for more complete information relative to vitamin deficiencies and storage.)

Multiple Nutritional Deficiencies

Multiple nutritional deficiencies are altogether too common, making diagnosis difficult even for the trained observer. Among the nonspecific indicators of multiple nutritional deficiencies are the following: loss of appetite; depraved appetite; unthrifty appearance; failure to grow; rough coat; rough, scaly skin; lowered production of meat, milk, eggs, wool, or work; impaired reproduction (or egg hatchability); weakness; lack of coordination; depression; difficult breathing; increased pulse rate; and a host of others.

Pica

Pica, a perverted or depraved appetite—a craving for substances not ordinarily considered feed—may indicate a nutritional deficiency. Animal species differ in their preference for foreign material. Cattle have been known to ingest cloth, leather, pieces of metal, wood, stone, and carcass material such as bone and hide. Horses may chew bones or eat dirt or sand. Sheep may eat dirt, wool, and bones.

Fig. 5-6. Bone chewing by cattle is a common sign of phosphorus deficiency. (From *Tex. Sta. Bull.* 344, courtesy of The Fertilizer Institute, Washington, D.C.)

Classically, pica is associated with a phosphorus deficiency. However, lambs may eat dirt to soothe abdominal pain caused by enterotoxemia (overeating disease). In all species, pica may occur as a result of a deficiency of protein, fiber, minerals, and/or vitamins. In addition, boredom, especially in stabled animals, may lead to excessive licking and chewing.

If pica is of nutritional origin, the deficiency or imbalance should be identified through the aid of blood analyses, and corrected.

NUTRITIONAL DISEASES AND AILMENTS

Nutritional diseases, which afflict both animals and humans, provide a dramatic and vivid way in which to relate the story of undernutrition and malnutrition.

It is noteworthy that nutritional deficiency areas throughout the world generally affect all species within the area, and that the manifestations of most deficiency diseases are similar regardless of species. For example, soils of an iodine-deficient area will produce iodine-deficient crops and result in many big-necked (goiterous) animals and people.

Table 5-1 lists the most common nutritional diseases and ailments of animals and summarizes pertinent information relative to each of them.

TABLE 5-1
NUTRITIONAL DISEASES AND AILMENTS

ACETONEMIA

(See KETOSIS.)

ACIDOSIS (LACTIC ACIDOSIS)

A metabolic disease of cattle and sheep.

Species Affected:

Cattle, especially feedlot cattle; and sheep, especially feedlot lambs.

Cause:

Acidosis is caused by an increase in lactic acid-producing bacteria and the rapid production of lactic acid (both the d- and l-forms). It commonly occurs when there is a sudden shift from a high-roughage to a high-concentrate ration. However, cattle maintained on high-energy rations may be in a marginal state of acidosis due to the formation of lactic acid by the rumen flora. Thus, ingredient changes, poor mixing of grain in the ration, or faulty feeding can produce acute acidosis.

Symptoms and Signs (or age group most affected):

Marginal acidosis is characterized by poor performance and inconsistent feed ingestion. If ingredient changes or erratic feeding persist, acute acidosis may result, creating laminitis—and eventually "ski shoe" cattle (founder). In severe cases, the rumen becomes immobilized, followed by increased pulse and respiration rate, variable rectal temperature, sunken eyes, loss of dermal elasticity (dehydration), staggering, coma, and death.

Distribution and Losses Caused By:

Acidosis occurs wherever beef or dairy cattle and lambs are fed, especially when consuming high-concentrate rations.
The annual loss from acidosis has been estimated at about 1% of the production.

Treatment:

Different treatments have been used with varying degrees of success; among them: (1) removing rumen contents and replacement by contents of an animal on a normal ration; (2) feeding a high level of an antibiotic to suppress lactic acid-producing bacteria; (3) drenching (or intravenous injection) with a solution of sodium bicarbonate to restore the acid-base balance; (4) administering intramuscularly antihistamines and cortical steroids daily for each of several days to help prevent intoxication and laminitis; and/or (5) backing the cattle down on both amount and kind of feed (lessening the total amount of the ration, and returning to a higher forage mix).

Control:

Acidosis is best controled by (1) avoiding accidental access of cattle to large amounts of concentrates, (2) changing gradually and stepwise from a low to a high proportion of concentrate in the ration, and (3) adding buffer salts, such as sodium bicarbonate, to the ration.

Prevention:

Prevention consists of starting animals on a high-roughage ration and gradually reducing the roughage and increasing the grain; avoiding erratic feeding; and avoiding abrupt ration changes.

Remarks:

A feedlot history of deliberate or accidental starting of animals on high-energy feeds, or of sudden ration changes, helps establish the correct diagnosis.
A rapid field test can be used to diagnose and differentiate between rumen acidosis and urea poisoning. Samples can be collected by stomach tube or postmortem collection. In general, a rumen content pH of 5.0 or less is indicative of rumen acidosis; a rumen content pH greater than 7.5 is indicative of urea or NPN toxicosis.

ALKALI DISEASE

(See Table 5-3, POTENTIAL POISONS, SELENIUM TOXICITY.)

ANEMIA, NUTRITIONAL

Species Affected:

All warm-blooded animals, including humans.

Cause:

Commonly an iron deficiency, but it may be caused by a deficiency of copper, cobalt, and/or certain vitamins.
The baby pig is born with a total of about 40 mg of iron in the body. With an iron requirement of about 7 mg daily, it is apparent that without supplemental iron, body stores will not last very long.

(Continued)

TABLE 5-1 *(Continued)*

ANEMIA, NUTRITIONAL (Continued)

Fig. 5-7. Anemia, caused by an iron deficiency, characterized by listlessness, rough hair coat, and wrinkled skin. (Courtesy, University of Florida, Gainesville)

Sow's milk is a good source of all nutrients the baby pig is known to require with the exception of iron.

Symptoms and Signs (or age group most affected):

Loss of appetite, progressive emaciation, and death.
Most prevalent in suckling young.
Pigs show listlessness, rough hair coat, wrinkled skins, drooping ears and tails, pale membranes around the mouth and eyes, labored breathing, and a swollen condition about the head and shoulders.

Distribution and Losses Caused By:

Worldwide.
Losses consist of slow and inefficient gains, and deaths.

Treatment:

Provide sources of the nutrient or nutrients, the deficiency of which is known to cause the condition. If iron deficiency is indicated, iron may be given by injection, in organic combination (iron dextran).

Control:

When nutritional anemia is encountered, it can usually be brought under control by supplying dietary sources of the deficient nutrient(s).

Prevention:

Supply dietary sources of iron, copper, cobalt, and certain vitamins (especially folacin, riboflavin, and vitamin B-6).
Keep confinement of suckling animals to a minimum and provide dry feeds at an early age.
Anemia in pigs can be prevented by providing supplemental iron in one of the following forms:

1. Inject intramuscularly 100 to 200 mg of iron from iron dextran into baby pigs at 2 to 3 days of age. If pigs remain in confinement and do not have access to creep feed at an early age, a second injection at 2 to 3 weeks of age is desirable. Injection is the method of choice, for it assures that every pig receives its requirement.
2. Orally administer iron dextran in a liquid or a solid preparation. To ensure daily intake by all pigs, it is important to have a preparation that is palatable and readily consumed. Also, placement of the oral preparation at the right location in the creep area is most important.
3. Give the pigs iron tablets or paste at 2 to 3 days of age. Repeat the treatment every 7 to 10 days until the pigs are eating the creep ration adequately. If pills are given, it is important to see that the pigs swallow them and not spit them out.
4. Place clean soil in the farrowing pen daily. Soil should not be contaminated with parasite eggs and other disease organisms. Iron sulfate can be sprinkled over the soil.
5. Swab sow's udder daily with a solution of 1 lb ferrous sulfate dissolved in 1 gal of warm water.
6. Provide pigs with access to a creep feed by the time they are 10 days old.

Remarks:

Anemia is a condition in which the blood is either deficient in quality or quantity. (A deficient quality refers to a deficiency in hemoglobin and/or red cells.)
Levels of iron in dry feed are generally believed to be ample, since feeds contain 40 to 400 mg/lb.
(**A**lso see Chapter 4, Table 4-6, ANIMAL MINERAL CHART, IRON.)

APHOSPHOROSIS

Species Affected:

Cattle; sheep to a lesser extent.

Cause:

Low available phosphorus in feed.

Symptoms and Signs (or age group most affected):

Decreased growth rate; inefficient feed utilization; depraved appetite—chewing bones, wood, hair, rags, and other objects; stiff joints and fragile bones. Breeding problems and a high incidence of milk fever in dairy cattle.

(Continued)

TABLE 5–1 *(Continued)*

APHOSPHOROSIS (Continued)

Fig. 5–8. Aphosphorosis, or phosphorus deficiency, in feedlot cattle. *Left:* Steer fed a ration consisting of wet beet pulp, alfalfa hay, and beet molasses, which contained 0.12% phosphorus. *Right:* Steer received the same ration as the steer on the left plus 1/10 lb of steamed bone meal daily, which brought the phosphorus content up to 0.18% and provided an average intake of 17 g of phosphorus daily. (Courtesy, Purdue University, West Lafayette, Ind.)

Phosphorus deficiency in cattle results in decreased growth rate; inefficient feed utilization; depraved appetite—chewing bones, wood, hair, rags; and stiff joints and fragile bones. In dairy cattle, there are breeding problems and a high incidence of milk fever. In sheep, there may be a knock-kneed condition.

Distribution and Losses Caused By:

Worldwide. Southwestern U.S.

Treatment:

Intravenous drench with suitable phosphorus solution.
Add sources of phosphorus, such as dicalcium phosphate or steamed bone meal, to the ration.

Control:

Controlled by feeding sources of phosphorus, either free-choice or added to the ration.

Prevention:

Feed phosphorus salts in feed and/or as a mineral supplement (free-choice).
Keep the calcium-phosphorus ratio within the range 4:1 to 1:1.

Remarks:

Generally caused by lack of phosphorus in the pasture. Phosphorus fertilizing may help.
(**A**lso see Chapter 4, Table 4-6, ANIMAL MINERAL CHART, PHOSPHORUS.)

AZOTURIA (HEMOGLOBINURIA, MONDAY MORNING DISEASE, BLACKWATER)

Species Affected:

Horses.

Cause:

Sudden exercise, following a day or two of rest during which time the horse has been on full feed, resulting in partial spasm or "tie-up." Azoturia is caused by an abnormal amount of glycogen being stored in the muscle. As the glycogen breaks down, lactic acid is formed, which builds up in the muscle causing severe muscle destruction and the release of myoglobin which manifests itself as partial spasm or "tie-up" and wine-colored urine.

Symptoms and Signs (or age group most affected):

Symptoms usually develop 15 to 60 minutes after the beginning of exercise. Azoturia is characterized by profuse sweating, elevated temperature and pulse, wine-colored urine (caused by the release of myoglobin—the red pigment in muscle tissue), tight (cramping) and sore loin hindquarter muscles—they're "tied up" due to semi-paralysis, stiff gait, reluctance to move due to pain, and knuckling over of the hind pasterns. Finally, the animal may assume a sitting position and, eventually, fall prostrate on its side. The breath and urine may have a peculiar odor.

(Continued)

TABLE 5-1 *(Continued)*

AZOTURIA (HEMOGLOBINURIA, MONDAY MORNING DISEASE, BLACKWATER) (Continued)

Distribution and Losses Caused By:

Worldwide, but the disease is seldom seen in horses at pasture and rarely in horses at constant work.

Treatment:

Absolute rest and quiet. While awaiting the veterinarian, apply heated cloths or blankets, or hot-water bottles to the swollen and hardened muscles, *but don't try to move the horse—don't take the horse back to the barn.* Keep it on its feet if possible, even if you have to use a sling.

The veterinarian should determine treatment. In mild cases, treatment may consist of the use of a tranquilizer or a sedative. In severe cases, the veterinarian may use (1) muscle relaxers or (2) sodium bicarbonate in solution to readjust the acid balance in the muscles.

Control:

When trouble is encountered, decrease the concentrate ration and increase the exercise on idle days.

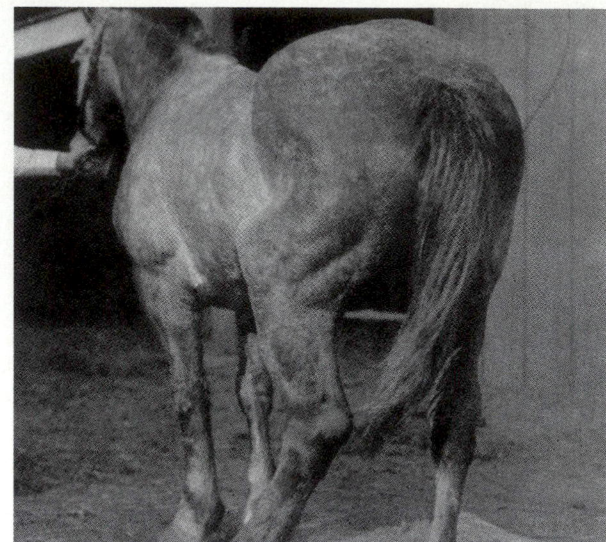

Fig. 5-9. Horse with Azoturia, evidencing sore loin and hindquarters—it is "tied up." (Courtesy, Pitman-Moore, Indianapolis, Ind.)

Prevention:

Restrict the grain ration, increase good quality roughage, and provide daily exercise when the animal is idle. Give a wet bran mash the evening before an idle day or turn the idle horses to pasture.

Some believe that a diuretic (a drug which will increase the flow of urine) will prevent the tie-up syndrome. This is a common treatment of racehorses.

Others feel that increased B vitamins will prevent the lactic acid buildup.

Remarks:

The chances of recovery are good for horses that remain standing, are not forced to move after the signs are noticed, and whose pulse returns to normal within 24 hours.

Azoturia and colic have some similar symptoms; hence, there is danger of misdiagnosis and the wrong treatment. Walking, a standard part of colic treatment, is the worst thing to do when a horse has azoturia.

BABY PIG SHAKES

(See HYPOGLYCEMIA.)

BLOAT—FEEDLOT

Species Affected:

All ruminants.

Cause:

Bloat is an excessive accumulation of gas in the rumen and reticulum of ruminants. High-concentrate rations, especially when finely ground, increase numbers of slime-producing bacteria in rumen. Slime traps fermentation gas and produces bloat. Both frothy and free gas bloat occur in feedlot bloat.

Genetic tendency or physiological abnormality.

Symptoms and Signs (or age group most affected):

Symptoms same as pasture bloat (see "Bloat—Pasture" which follows).

Occurs when cattle or sheep have been fed high-concentrate, low-roughage rations for approximately 60 days or longer.

Distribution and Losses Caused By:

A survey of Kansas feedlots showed the following losses from bloat: 0.1% died of bloat; 0.2% bloated severely; and 0.6% bloated mildly to moderately, with animal performance affected adversely.

Treatment:

Reduce intraruminal pressure as quickly as possible. This may be done by means of a large stomach tube, although this method is usually disappointing in foamy bloat.

Administer a defoaming agent immediately, such as (1) 1 pint of corn oil, peanut oil, or soybean oil; or (2) poloxalene administered according to the manufacturer's directions.

(Continued)

TABLE 5-1 *(Continued)*

BLOAT—FEEDLOT (Continued)

As a last resort, a trocar and cannula can be inserted on the left side of the animal at the center of a triangle formed by the backbone, the hipbone, and the last rib. The trocar is removed, but the cannula should stay in place until all gases have dissipated. If a trocar and cannula are not available, a knife may be used in emergencies.

Control:

If feasible, increase proportion of nonlegume roughage in ration. However, good-quality legume hay may increase incidence of feedlot bloat. In this case, poloxalene or oxytetracycline are effective preventives when used according to manufacturers' directions.

Prevention:

(1) Use poloxalene (Bloat Guard) or oxytetracycline (Terramycin and Neo-Terramycin) according to manufacturers' directions; and (2) proper management.

Remarks:

Feedlot bloat may occur during any month of the year; however, it is more common during hot, humid weather.

Two products are cleared by FDA for bloat control; namely, poloxalene (trade name, Bloat Guard) and oxytetracycline (trade names, Terramycin and Neo-Terramycin).

BLOAT—PASTURE (LEGUME BLOAT)

Species Affected:

All ruminants.

Cause:

Bloat is caused by the inability of the animal to get rid of ruminal gas. Lack of scabrous (rough) material in the rumen to stimulate eructation (belching), along with the formation of heavy foam bubbles, seems to be the main cause of pasture bloat. Pasture bloat is most common on immature, rapidly growing legumes and on wheat pasture. Pasture bloat is a frothy bloat caused by interaction of several factors—plant, animal, and microbial. Soluble plant proteins and the presence of saponins play a prominent role in permitting stable foam formation.

Heavy applications of urea fertilizer on pastures may also induce bloat.

Animals that will bloat on any feed are known as chronic bloaters. These animals, in which there may be a genetic tendency, are unable to eructate (belch) fermentation gases because of some physiological abnormality.

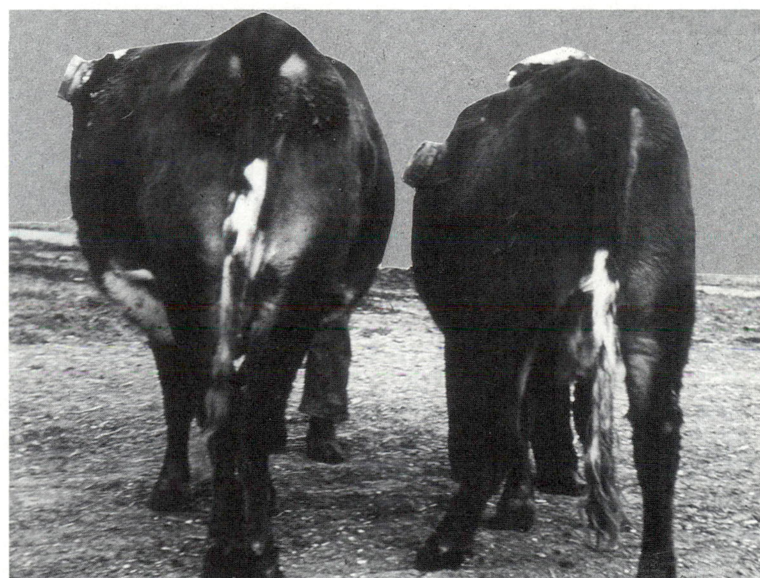

Fig. 5-10. Identical twins. *Left:* Bloated animal showing distention of the paunch on the left side in front of the hipbone. *Right:* Twin mate showing no bloating. (Courtesy, Kansas State University, Manhattan)

Symptoms and Signs (or age group most affected):

First observed as a distention of the paunch on the left side in front of the hipbone. This is followed by distention of the right side, protrusion of the anus, respiratory distress, cyanosis (bluish coloration) of the tongue, struggling, and death if not treated. The entire period of time from when a ruminant enters a pasture until death occurs can be as short as a half hour.

Distribution and Losses Caused By:

Widespread, although some areas appear to have more bloat than others.

It often results in death.

Bloat causes annual losses in beef and dairy cattle of more than $100 million from reduced weight gain and lower milk production.

Treatment:

Time permitting, severe cases of bloat should be treated by a veterinarian. The use of a stomach tube, carefully inserted, is usually very helpful in eliminating gases. Puncturing of the paunch with a trocar and cannula should be a last resort. A knife may be used in emergencies.

Mild cases may be home treated by (1) keeping the animal on its feet and moving; and (2) drenching cattle either with (a) 1 pint of corn oil, or soybean oil; or (b) 1–2 oz poloxalene.

(Continued)

TABLE 5–1 *(Continued)*

BLOAT—PASTURE (LEGUME BLOAT) (Continued)

Control:

When there is a high incidence of bloat, it may be desirable to change the feed.

Where legume bloat is encountered, use poloxalene (Bloat Guard), oxytetracycline (antibiotic), or polyoxyethylene (23) lauryl ether (Laureth–23/Enproal Bloat Blox), according to the respective manufacturers' directions.

Prevention:

The incidence is lessened by (1) avoiding straight legume pastures and immature legumes, (2) feeding a coarse grass hay prior to turning onto lush pasture, (3) feeding dry forage along with pasture, (4) avoiding a rapid fill from an empty start, (5) keeping animals continuously on pasture after they are once turned out, (6) keeping salt and water conveniently accessible at all times, (7) avoiding frosted pastures, or (8) using poloxalene (Bloat Guard), oxytetracycline (Terramycin or Neo-Terramycin), or Laureth–23 (Enproal Bloat Blox) according to manufacturers' directions, including placing blocks containing these antifoaming agents in various parts of the pasture.

Oils and fats have been used successfully to control bloat in Australia and New Zealand.

Remarks:

Legume or cereal pastures, or alfalfa hay, appear to be associated with a higher incidence of bloat than any other feeds.

Legume pastures are particularly hazardous when immature, when moist, after a light rain or dew.

COBALT DEFICIENCY

(See SALT SICK.)

COLIC

Species Affected:

Horses.

Cause:

Internal parasites are the number one cause of colic; additional causes are improper feeding, working, or watering. There are more than 70 different things that can cause colic.

Symptoms and Signs (or age group most affected):

Severe pain, usually in the abdomen; and depending on the type of colic, other symptoms are: the horse looking at his belly, distended abdomen, increased intestinal rumbling, violent rolling and pawing, profuse sweating, constipation, and refusal of feed and water.

Distribution and Losses Caused By:

Worldwide.

Colic is the most common ailment among horses and is the leading cause of death. Livestock insurance companies report about ⅓ of all deaths of insured horses can be attributed to colic.

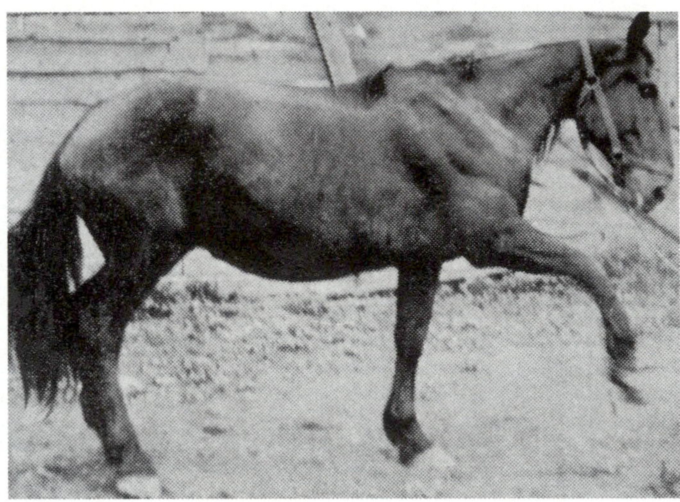

Fig. 5–11. Horse with colic, evidencing severe pain, but being taken for a slow walk. (Courtesy, Pitman-Moore, Indianapolis, Ind.)

Treatment:

Call a veterinarian. To avoid danger of inflicting self-injury, (1) place the animal in a large, well-bedded stable, or (2) take the animal for a slow walk. Do not give the horse any type of drug, unless so advised by the veterinarian when telephoned. Painkillers may cover up symptoms which are vital for the veterinarian in making an accurate diagnosis.

Depending on the diagnosis, the veterinarian may use one or more of the following: sedatives; laxatives, such as mineral oil; drugs; or surgery. The surgeon may avoid recurrence of twists and displacements of the horse's colon by attaching it to other organs or the abdominal wall, thereby deliberately creating adhesions which prevent further twisting.

In the late 1980s, Colorado State University scientists developed a "scorecard" method of evaluating colic (1) to help differentiate between a colicking horse that needs surgery and one that should be treated medically, and (2) to predict how likely a horse in each category is to survive.

Control:

Follow a good management program, including parasite control.

Feed, work, and water horses properly.

Prevention:

Proper feeding (including adequate roughage), working, watering, and parasite control.

TABLE 5-1 *(Continued)*

COLIC (Continued)

Remarks:

The word colic is not specific. There are many syndromes that can result in colic, and not all of them are gastrointestinal. For example, blood worms can cause colic due to damaging the walls of blood vessels, and mares with uterine torsions exhibit colic pain as do horses with urinary stones in the bladder. So, the first thing is to determine what is causing the animal to exhibit colic symptoms.

CROOKED CALVES

Species Affected:

Calves.

Cause:

Crooked calves may be caused by either—

1. Manganese deficiency, or
2. Lupine consumption. Certain alkaloids, which are found in lupines, are responsible for the disease.

Symptoms and Signs (or age group most affected):

Calves born with crooked necks and legs.

Distribution and Losses Caused By:

Manganese deficiency occurs in northwestern U.S.

Treatment:

There is no treatment.

Fig. 5-12. Manganese deficiency in a young calf, resulting in the crooked calf syndrome. Note weak legs and knuckling over. (Courtesy, Washington State University, Pullman)

Control:

Provide manganese where needed. Avoid pasturing pregnant cows on pasture or range containing lupines.

Prevention:

Feed manganese; 30 ppm of total feed. Either manganese carbonate or manganese sulfate may be used.

Remarks:

Crooked calf disease due to manganese deficiency can be eliminated almost completely by providing manganese where needed.
(Also see Chapter 4, Table 4-6, ANIMAL MINERAL CHART, MANGANESE)

DOWNER COW SYNDROME

Species Affected:

Cattle.

Cause:

The downer cow syndrome commonly develops following a malady that causes recumbency (lying down); for example, following anesthesia, arthritis, calving paralysis, cold stress and inadequate nutrition, copper metabolism interacted with other compounds and elements, exhaustion, grass tetany, ketosis, mastitis, metritis, milk fever (parturient paresis), protein/energy malnutrition, toxicity, or trauma.

Generally, downer cows do not have forelimb problems, but the hindlimbs are paralyzed and they are unable to rise. Experimental evidence indicates that, in less than 6 hours, the pressure on the sciatic nerve and muscles in the caudal thigh of a downer cow can cause hindlimb paralysis.

(Continued)

TABLE 5–1 *(Continued)*

DOWNER COW SYNDROME (Continued)

Fig. 5–13. *Left:* An unusual "downer cow." After being down for 7 weeks, with the cause unknown, she walked again. Not only that, she gave birth to a fine heifer calf. *Right:* Same cow 13 months later, along with her 8-month-old heifer calf and her caretaker, Mr. Clair Pollard. Her treatment: TLC (tender loving care), along with plenty of water and alfalfa hay. (Photos by Mrs. A. H. Ensminger)

Symptoms and Signs (or age group most affected):

Cows go down and are weak and reluctant or unable to rise voluntarily.

Distribution and Losses Caused By:

The downer cow syndrome may be found wherever there are cattle. In afflicted beef herds, up to 10% of the cows may go down.

Treatment:

There is little experimental support for any of the treatments that have been suggested. Good nursing is the best treatment.

Move animal to an area with adequate bedding and footing; turn animal from side to side frequently, if it does not turn alone; and provide adequate nursing and feeding. If recumbent for 2 weeks without improvement, the animal will likely suffer nonreversible muscle and nerve damage, due to pressure, and not get up. On big cows, hip lifters may do more harm than good by adding to the muscle damage. Slings are less likely to cause additional injury, but they are difficult to use. Currently, a British firm is manufacturing an inflatable air bag that gradually lifts a cow onto her feet so that she can be treated.

Control:

Beef cattle: Evaluate the condition of the herd in the fall. Separate the first calf heifers and the thinner cows and give them special feeds.

Dairy cows: Feed cows properly, especially those that are likely candidates for milk fever.

Prevention:

Beef cattle: When supplemental feeding is indicated, feed 5 to 6 lb corn and 1 to 2 lb soybean meal/head/day.

Dairy cows: Recognize the earliest signs of milk fever so that the cows can be treated before they become recumbent.

Remarks:

"Downer cow" is not a good term because it lumps together all cows that remain incumbent for whatever reason.

ENCEPHALOMALACIA (CRAZY CHICK DISEASE)

Species Affected:

Chicken.

Cause:

Vitamin E deficiency. In addition to causing encephalomalacia, a deficiency of vitamin E may cause two other classical deficiency disorders; namely, exudative diathesis or muscular dystrophy. Also, it is noteworthy that synthetic antioxidants can prevent encephalomalacia.

(Continued)

TABLE 5–1 *(Continued)*

ENCEPHALOMALACIA (CRAZY CHICK DISEASE) (Continued)

Symptoms and Signs (or age group most affected):

Chicks: Head retraction and loss of control of legs.
Adults: Poor reproductive performance; prolonged vitamin E deficiency results in permanent sterility in the male and reproductive failure in the female.

Distribution and Losses Caused By:

Encephalomalacia occurs with diets borderline in vitamin E that also contain polyunsaturated fats, such as cod-liver oil or soybean oil, that are oxidized and become rancid.

Treatment:

No treatment of afflicted birds is effective.

Control:

Add vitamin E or another suitable antioxidant compound to the diet.

Prevention:

Rations of growing chicks and breeding hens are usually supplemented with a source of vitamin E or a suitable antioxidant.

Remarks:

In growing chicks, a deficiency of vitamin E can result in (1) encephalomalacia, (2) exudative diathesis, or (3) muscular dystrophy. Under farm conditions in the U.S., encephalomalacia is the major vitamin E deficiency disease found in growing chicks.
(Also see Chapter 4, Table 4-8, ANIMAL VITAMIN CHART, VITAMIN E.)

ENTEROTOXEMIA (OVEREATING DISEASE, PULPY-KIDNEY DISEASE)

Species Affected:

Sheep; less frequently goats, and rarely cattle.

Cause:

Clostridium perfringens type D. However, predisposing factors are essential; the most common of these are overconsumption of high-energy feeds, an abundant milk supply, and lush pastures. Under such conditions, the *Clostridium perfringens* bacteria grow rapidly and produce a powerful toxin.

Symptoms and Signs (or age group most affected):

Sudden death; frequently a lamb is found dead in the field or feedlot without having shown any previous signs of illness. Quite often it is the biggest lamb with the biggest appetite. The disease develops rapidly; and the animal becomes weaker and weaker and shows nervous disturbances such as circling, butting, or throwing the head from side to side or backwards. Finally, the animal collapses and may go into convulsions before dying. Enterotoxemia can be confirmed by laboratory tests if necropsy is performed shortly after death.

Distribution and Losses Caused By:

Worldwide.
The death rate is a minimum of 1%, with an average of 3 to 4% in unvaccinated feedlot lambs. In explosive outbreaks, losses range from 10 to 40%.

Treatment:

None.

Control:

The method of control depends on the age of the lambs, the frequency with which the disease occurs, and the method of husbandry.
When an outbreak in feeder lambs occurs, for several days (1) increase the amount of roughage in the ration, and (2) add 200 g of chlortetracycline per ton of feed.
When an outbreak in nursing lambs occurs, injection of all susceptible lambs with enterotoxemia antiserum will provide protection for about 14 to 21 days, at which time the lambs can be vaccinated.

Prevention:

Along with proper feeding, vaccinate. Ewes should be vaccinated with type C and D toxoid. Lambs should be vaccinated with type D only.

(Continued)

TABLE 5-1 *(Continued)*

ENTEROTOXEMIA (OVEREATING DISEASE, PULPY-KIDNEY DISEASE) (Continued)

Nursing lambs can be protected by vaccinating the ewes for enterotoxemia. For best results, previously unvaccinated ewes should be vaccinated with type C and D toxoid twice before lambing. The 2 doses should be spaced at least 1 month apart, with the second dose given 2 to 4 weeks before start of lambing. Ewes which have been vaccinated previously need only one booster shot before lambing. Vaccinating the ewes prior to lambing ensures that the lambs will receive colostral protection for 2 to 3 weeks, following which the lambs should be vaccinated at 4 to 6 weeks of age.

Feedlot lambs can be protected by giving them one dose of type D toxoid soon after their arrival in the feedlot. It takes about 10 days after vaccination for immunity to develop. Sometimes revaccination with the toxoid or bacterin (a booster shot) is required 2 to 4 weeks following the first vaccination.

Remarks:

The disease is caused by bacteria, but it is triggered by high-energy rations or excellent pastures.

FESCUE FOOT (FESCUE TOXICOSIS)

Species Affected:

Cattle, sheep, and horses.

Cause:

A fungus (endophyte), *Acremonium coenophialum,* which lives in the leaves, stems, and seeds of tall fescue, without adversely affecting the fescue plant.

Symptoms and Signs (or age group most affected):

The symptoms vary. Some animals show no apparent lameness, whereas others show varying degrees of sloughing (necrosis) on the ends of their tails. Mild fescue toxicosis is characterized by poor conception rates, low pasture gains, and depressed milk production.

The most common symptoms in horses, in the order of their occurrence, are a decrease or absence of milk production, prolonged gestation, abortion, and thickened placenta.

Fescue toxicity is more common in animals suffering from malnutrition and/or parasitism.

Distribution and Losses Caused By:

Fescue foot has occurred in the U.S., Australia, New Zealand, and Italy. In the U.S., it has been reported in California, Colorado, Florida, Kentucky, Missouri, and Tennessee.

Tall fescue is currently grown on about 35 million acres in the U.S.

The Mississippi Station reports that studies extending over several years show that each 10% fungus infection in a fescue pasture will lower the daily gains of cattle by about 10%.

University of Kentucky researchers report that dairy cows grazing 70% infected fescue produced an average of 11.2 lb less milk per day than cows grazing noninfected fescue.

Treatment:

There is no effective medication. Cattle usually recover if removed from fescue pasture or fescue hay.

Control:

Until, and unless, scientists find a way to remove the toxic factor(s) from fescue, the best control consists of good management, proper nutrition of animals, early detection of symptoms, and/or destroying toxic pastures and reseeding with endophyte-free seed.

Prevention:

The seeding of fungus-free fescue seed is the best way to prevent fescue foot. Also, interseeding alfalfa, clovers, or other grasses into fescue stands dilutes the amount of fescue eaten and helps to reduce the toxic effects.

Also, tall fescue selections low in, or free of, this endophytic fungus are evolving.

In areas where fescue toxicity is a problem, gestating mares should be removed from fescue pasture the last 2 or 3 months of pregnancy.

Remarks:

Most cases of fescue toxicity occur among cattle that graze pure stands of fescue during late fall and winter; and most toxic stands of fescue pasture are several years old.

FOUNDER (LAMINITIS)

Species Affected:

Horses, cattle, sheep, goats.

(Continued)

TABLE 5–1 *(Continued)*

FOUNDER (LAMINITIS) (Continued)

Cause:

A variety of causes have been recognized, including (1) overeating and too rapid increase in the ration (grain founder), (2) digestive disturbances (enterotoxemia), (3) retained afterbirth (foal founder), (4) lush pastures (grass founder), and (5) concussion (road founder).

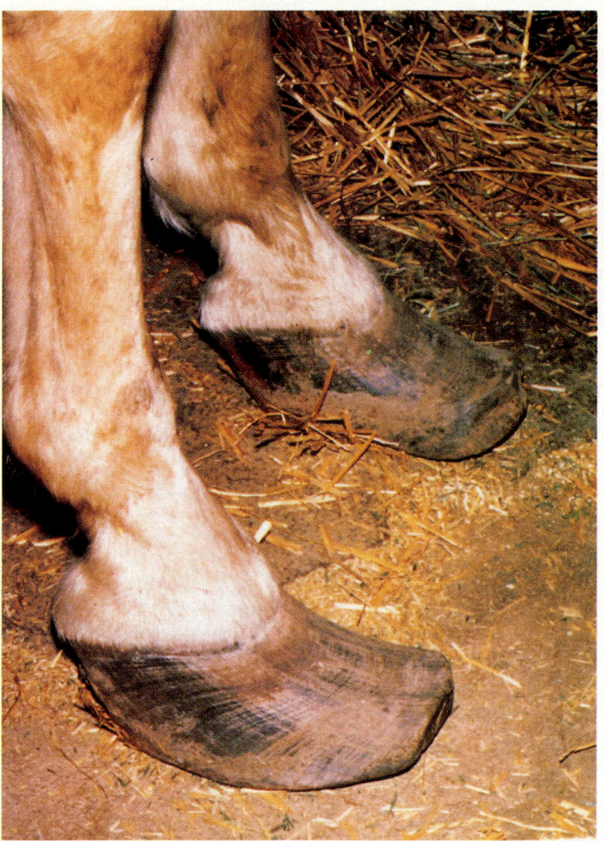

Fig. 5–14. "Snowshoe feet" of a horse due to chronic founder. (Courtesy, Colorado State University, Ft. Collins)

Symptoms and Signs (or age group most affected):

Extreme pain, fever (103°–106°F), and reluctance to move—the animal appears to be "walking on eggs." If neglected, it causes an acute or chronic degeneration of the joining of the sensitive and insensitive laminae of the foot; and, if the degeneration is severe, the coffin bone may rotate and come through the bottom of the foot.

Distribution and Losses Caused By:

Worldwide.

Actual death losses from founder are not very frequent, but animal usefulness may be severely affected.

Treatment:

There is no widely accepted, standard method of treating founder. If known, the condition(s) that caused the problem should be alleviated. Treatment of acute horse founder usually involves one or more of the following procedures and medications:

1. *Mineral oil.* To aid passage of the excessive feed consumed and prevent further absorption of lactic acid and endotoxin into the bloodstream. One gallon should be given via stomach tube. **NOTE WELL**: A purgative should not be employed in cases involving pneumonia or parturient laminitis of mares.
2. *Analgesics (pain-killers).* To obtain pain relief.
3. *Injectable antihistamines.* To provide anti-inflammatory effects.
4. *Antibiotics.* To combat the subsequent formation of endotoxin which destroys tissue.
5. *Sodium bicarbonate (baking soda).* To neutralize the acidic toxicity of these inflammatory products.
6. *Temporarily deadening the nerve supply to the feet.* To alleviate pain and assist in restoring blood supply by allowing the horse to be walked.
7. *Water soaks.* To stimulate and massage the blood supply to the feet in an effort to open up previously constricted blood vessels, but there is disagreement as to whether the water should be warm or cold.
8. *Wraps applied to the affected feet.* To cushion the painful sole soreness which is evident in the toe region due to coffin bone rotation following breakdown of the laminae.

Other treatments for founder that are sometimes used, with varying degrees of reported success, are cortisones; and methionine, a sulfur containing amino acid.

NOTE WELL: Due to the complexity of laminitis and its far-reaching effect on many of the horse's internal systems, the treatments described above may not be effective in controlling founder. If the disease is not diagnosed and treated early enough, the coffin bone rotation and damage to the associated hoof wall structure can become irreversible.

Treatment of chronic founder consists of attempting to restore the normal alignment of the rotated coffin bone by lowering the heels, removing excess toe, and protecting the dropped sole. This may be accomplished by a competent farrier through proper trimming and perhaps by using leather pads or a steel-plate shoe. Also, soft acrylic plastics are sometimes applied to the sole to replace the injured hoof area and help realign the rotated coffin bone. Because of the tendency to refounder, weight control is extremely important and overfeeding should be avoided.

Control:

Alleviate the causes.

Prevention:

Avoid (1) overeating, (2) overdrinking (especially when hot), and/or (3) inflammation of the uterus following parturition.

Veterinary attention should be given if mares retain the after-birth longer than 12 hours.

Careful management practices related to grain feeding will prevent many cases of founder in cattle, sheep, and goats.

Remarks:

Unless foundered animals are quite valuable, it is usually desirable to dispose of them following a case of severe founder.

Swine do not founder because they can unload their stomachs by vomiting.

(Continued)

TABLE 5-1 *(Continued)*

GOITER

(See IODINE DEFICIENCY.)

GOSSYPOL TOXICITY

Species Affected:

Monogastrics—swine, poultry.
Ruminants seldom affected, since rumen microorganisms can detoxify gossypol.

Cause:

Gossypol, a toxic yellow pigment contained in the glands of most cottonseed, but which is absent in the new glandless selections of cottonseed.

Symptoms and Signs (or age group most affected):

Swine: Difficult or labored breathing, loss of appetite, retarded growth, lack of thrift, weakness, and death.
Poultry: When fed to laying hens, high-gossypol will cause egg yolk discoloration in stored eggs.

Distribution and Losses Caused By:

Gossypol toxicity is seldom a problem in the U.S., due to (1) inactivating the gossypol of cottonseed in processing, and (2) enlightened knowledge relative to the use and limitations of cottonseed and cottonseed meal.

Fig. 5-15. A pig showing gossypol toxicity, from eating too much cottonseed meal high in gossypol. This pig died soon after this picture was taken. (Courtesy, University of Arkansas, Fayetteville)

Treatment:

Remove the cause; alleviate or limit the gossypol in the feed.

Control:

Swine: Limit the cottonseed meal content of swine rations (especially rations for growing swine) to ½ the protein supplement of the ration. At this level, it is unlikely that the total ration will contain more than 0.01% free gossypol. However, if the level of free gossypol is higher than 0.01%, then iron as ferrous sulfate ($FeSO_4$) should be added in a 1:1 weight ration up to a maximum of 500 ppm of added iron in the total ration. When used in this manner, cottonseed meal can be an economical and satisfactory protein supplement for swine.
Chickens: Limit cottonseed meal to a free gossypol level below 0.03%, and add iron salts.

Prevention:

Gossypol toxicity can be alleviated in three ways: (1) limiting the amount of free gossypol, (2) adding iron salts, or (3) using glandless cottonseed and cottonseed products.

Remarks:

During the processing of regular cottonseed, the "cooking" or moist heat treatment causes changes to take place which largely inactivate the gossypol. Also, methods have been devised to extract the gossypol.

GRASS TETANY (HYPOMAGNESEMIC TETANY, GRASS STAGGERS, WINTER TETANY. In Europe, it's called* FOG FEVER.*)

Species Affected:

Cattle (beef and dairy; in the U.S., tetany is more common in beef herds than in dairy herds). Sometimes sheep and goats, in which the disease occurs under essentially the same conditions and has the same clinical signs as in cattle.

Cause:

Grass tetany is a nutritional disease caused by an inadequate level of magnesium (Mg) in the blood. It most commonly occurs among lactating animals grazing rapidly growing, lush spring pastures containing less than 0.2% magnesium and more than 3% potassium and 4% nitrogen (25% protein). Forage that is high in potassium and nitrogen should have a magnesium content of at least 0.25%. Such low magnesium pastures are most commonly encountered during the first two weeks of the pasture season, although somewhat later in the season outbreaks have been reported during rainy and foggy weather. Sometimes tetany is a problem when cattle are allowed to overgraze a field, then moved abruptly to a field of new lush growth. Small grain pastures (wheat/rye/oats/barley) are especially troublesome. Also, the disease may occur when animals are fed poor quality hay, straw, or corn stover—feeds that are low in magnesium. It is not common on legume pasture or in animals wintered on legume hay. (Legumes may contain twice the magnesium concentration of grasses grown on the same soil.)

(Continued)

TABLE 5–1 *(Continued)*

GRASS TETANY (HYPOMAGNESMIC TETANY, GRASS STAGGERS, WINTER TETANY. In Europe, it's called *FOG FEVER.*) (Continued)

Several factors adversely influence magnesium metabolism in cattle and may "trigger" grass tetany; among them, drastic fluctuations in spring temperatures, prolonged cloudy weather, organic acid content of plants, hormonal status of the animal, level of higher fatty acids in plants, energy intake of the animal, and additional stress—such as a dog chasing animals, parasites, or a cold rain.

Grass tetany is most likely to occur on pasture plants grown on soils that are low in available magnesium and high in available potassium. If calcium is low as well as magnesium, the hazard of tetany is even greater. Many state soil-testing laboratories provide information on the danger of tetany on pastures, and can recommend corrective fertilization or dolomitic liming (which contains magnesium). Also, the historical record of grass tetany in an area or on a specific pasture is important.

Symptoms and Signs (or age group most affected):

The initial signs of magnesium deficiency include nervousness, attentive ears, and decreased milk yield; signs which, to the experienced and observing caretaker, indicate the need for immediate preventive measures—before the animals become sick tomorrow. In more severe cases, affected animals may avoid the rest of the herd, walk with a stiff gait, lose their appetite, and urinate frequently. They are nervous, have staring eyes, and keep their head and ears in an erect position. Also, they stagger; have a twitching skin, especially on the face, ears, and flanks; and lie down and get up frequently. Animals may be irritable and behave aggressively; they may even charge or fight persons in the immediate area. After a time (as long as 2 to 3 days), extreme excitement and violent convulsions may develop. Animals lie flat on their sides, the fore legs pedal periodically, saliva flows freely, breathing is labored, and the heart pounds rapidly. If treatment is not given at this stage, animals usually die during or after a convulsion. The various symptoms of animals suffering from grass tetany indicate that the nervous system controlling both voluntary and involuntary muscles is affected. Quite often, clinical signs are not observed, and the only evidence is a dead animal.

Chronic grass tetany is generally slow to develop and muscular affection may be limited to twitching, a clumsy walk or exaggerated motions, but convulsions may occur if animals are driven or handled roughly.

Older cows are more susceptible to grass tetany than those with their first or second calves, because of lowered magnesium stores and decreased absorption efficiency. Also, the disease is most likely to strike beef cows during early lactation, especially those with high levels of milk production. Dry cows and bulls are seldom affected.

Normal plasma magnesium levels range from 1.8 to 2.0 mg/100 ml; values below 1.0 to 1.2 mg/100 ml are indicative of magnesium deficiency. However, not all cattle with low plasma magnesium develop tetany. Also, plasma magnesium levels in affected animals may return to almost normal during the convulsive stage. So, diagnosis can not be based on blood tests alone. Since the kidneys apparently start conserving magnesium when the serum level reaches about 1.8 mg/100 ml, one of the better diagnostic aids to indicate grass tetany is low urinary magnesium.

Grass tetany should not be confused with nitrate toxicity or calcium deficiency. In nitrate toxicity, the blood is brown and there is a grayish to brownish discoloration of white areas on the skin and on nonpigmented mucous membranes of the mouth, nose, eyes, and vulva. In calcium deficiency, animals may be sluggish rather than nervous as they are when they have magnesium deficiency. Also, animals may have a calcium and magnesium deficiency at the same time, thus masking the signs of magnesium deficiency; this happens in wheat pasture poisoning, in which the animals may be deficient in both calcium and magnesium.

Distribution and Losses Caused By:

Grass tetany is a worldwide problem, with occurrence sporadic and unpredictable for any given area. It is generally considered to be the leading cause of cattle deaths in the U.S., killing an estimated 1 to 3% of the cattle in temperate regions.

Treatment:

Treatment of tetany cases can be successful if given early and without excessive handling of the affected animals. Chance of recovery is slight if treatment is delayed 8–12 hours; so, call the veterinarian immediately.

Under range conditions, 200 cubic centimeters (cc) of a sterile, saturated solution of magnesium sulfate (Epsom Salts) injected under the animal's skin (inject only 50 cc at any one place on the animal) places a high level of magnesium in the blood in 15 minutes.

Some veterinarians use intravenous injections of chloral hydrate or magnesium sulfate to calm excited animals, then follow with a calcium-magnesium gluconate solution. If the animal again goes into convulsions, a second dose of calcium-magnesium gluconate solution may be required. Intravenous injections should be administered slowly (allow about 15 minutes for a 500 cc bottle) by a trained person because there is a danger of heart failure if they are given too rapidly.

An enema of 60 g (*2 oz*) of magnesium chloride ($MgCl_2 \bullet 6H_2O$) in 10 oz of water is helpful. The enema may be given with an esophageal or oral calf feeder with the probe inserted 10 in. into the anus. Magnesium is absorbed through the walls of the large intestine and the lower bowel.

Oral administration of magnesium to sick animals, in place of intravenous injections or enemas, has not been effective because too much time is required for the magnesium to reach that part of the GI tract where it can be absorbed.

Herd treatment of the animals that are not down may involve adding magnesium sulfate (Epsom Salts) or magnesium acetate or chloride to the drinking water. Some diarrhea may occur, but this is not reason for concern. To be effective, the treated tanks should be the only source of drinking water. **NOTE WELL**: Production will be lowered by this treatment due to lowered consumption of water.

Follow-up treatment may involve removing all animals from the tetany-producing pasture and feeding alfalfa hay (plus concentrates if necessary). Additionally, each animal should consume 30 g of magnesium daily for 1 to 2 weeks, preferably through a highly palatable supplement; force-feeding should be resorted to if necessary.

Cattle that get tetany are likely to get it again later in the season or in later years; they are usually the high producers.

NOTE WELL: "Downer cows" should be turned daily—and more frequently if possible. (Also see Downer Cow Syndrome.)

• **Toxicity**—Magnesium toxicity does not occur in cattle fed normal rations. Supplemental levels of magnesium of 170 to 350 g have resulted in deleterious effects. Maximum tolerable levels have been established as 0.4% of the ration by the National Research Council. Feeding toxic levels has resulted in anorexia (loss of appetite), reduced performance, and occasional diarrhea. Also, cattle experiencing toxicity may exhibit lack of reflexes and respiration depression.

Control:

Commonly used feedstuffs vary widely in magnesium concentration and availability. The magnesium content of most cereal grains runs between 0.12 and 0.18%. Protein supplements of animal origin are low in magnesium, while those of plant origin usually contain 0.3 to 0.6%. Fat is not as beneficial as carbohydrate as a source

(Continued)

TABLE 5-1 *(Continued)*

GRASS TETANY (HYPOMAGNESMIC TETANY, GRASS STAGGERS, WINTER TETANY. In Europe, it's called *FOG FEVER.*) (Continued)

of energy under tetany conditions, as fat tends to tie up calcium and magnesium in the digestive tract, rendering them less available to the animal. The magnesium content of forages varies greatly; normally, the legumes contain more than the grasses. Magnesium availability increases with plant maturity. Magnesium fertilization usually increases plant magnesium content. The inclusion of energy or protein supplements in high-magnesium mineral supplements will help to overcome palatability problems.

Prevention:

Prevention of grass tetany is always preferred to treatment. Prevention consists of providing magnesium daily throughout the high-risk period, because very little of it is stored in the body. **NOTE WELL**: Crash feeding programs begun after tetany appears in a herd are usually not adequate to stop the disease. A magnesium supplement should be started 30 days before grass tetany is usually observed in the area in order to get the animals accustomed to it. Since magnesium oxide or sulfate are not very palatable, cattle may not consume sufficient of them.

Meeting the magnesium requirements of beef cows calls for providing 10 g of magnesium daily for the dry cows, and 20 to 25 g daily for cows suckling calves. For dairy cows, 30 g of magnesium per day is recommended. For calves, 4 to 8 g per day is needed, depending on their ages.

Lactating ewes and does, just after parturition, which is the most tetany-susceptible period, should receive about 3 g of magnesium per day.

High levels of aluminum, potassium, phosphorus, or calcium decrease the efficiency of magnesium absorption and/or utilization; so, in areas where the levels of these elements are high, the magnesium allowance should be increased to overcome their antagonistic effect.

Normally, animals on pasture during the summer and early fall months receive an adequate supply of magnesium from the grasses on which they feed. However, during the late fall, winter, and spring months, many pastures are magnesium deficient. To prevent grass tetany during these months, cattle, sheep, and goats on pasture should receive a magnesium-rich feed in addition to pasture and/or have ready access to a magnesium mineral supplement.

One of the following high-magnesium feeds is commonly used:

1. Alfalfa, or other legume hay; 20 lb of average alfalfa hay will provide 30 g of magnesium. Additionally, the legume hay provides increased energy.
2. For self-feeding (with adequate available water and forage containing 10% protein), 65% ground grain (corn, barley, or grain sorghum), 20% magnesium oxide, and 15% iodized salt. Since early green grass is high in protein, a supplement containing cereal grain is preferred to a supplement containing an oilseed protein. When consumed at the rate of ½ lb/head/day, such a mix will provide 27 g of magnesium daily.
3. For self-feeding (with adequate available water and low-protein forage), 65% cottonseed meal or soybean meal and 35% magnesium oxide. When consumed at the rate of ⅓ lb/head/day, such a mix will provide 31 g of magnesium daily.
4. For hand-feeding of supplements, use (2) or (3) above, but omit the salt in (2). With the salt omitted, each ½ lb of mix No. 2 will provide 32 g of magnesium daily.
5. A liquid molasses supplement fortified with 4% magnesium sulfate[1] (80 lb of magnesium sulfate per ton of liquid molasses). When consumed at the rate of 2 lb/head/day, this will provide 7.2 g of magnesium.

Several good sources of inorganic magnesium may be used to supplement cattle; among them, those listed in the table that follows:

Magnesium Content of Various Magnesium Salts

Name	% Mg	Pounds of Mineral Required to Supply the Same Amount of Mg as 30 lb of MgO
Magnesium oxide (Magnesia)(available in light and heavy grades; heavy is more stable and easier to mix)	60.32	30.0
Magnesium hydroxide	41.69	43.4
Magnesium carbonate (Magnesite)	28.8	62.8
Magnesium carbonate hydroxide	27.0	67.0
Magnesium sulfate (Epsom Salts)	20.2	89.6
Potassium magnesium sulfate (Langbelinite)	11.6	156.0
Magnesium acetate (Cromosan)	11.34	159.6

Magnesium from dolomitic limestone is less readily available to cattle, sheep, and goats than some other salts. Any of the following mineral supplements are satisfactory:

1. A mineral mix made by mixing ⅓ each magnesium oxide, iodized salt, and either soybean oil meal or cottonseed meal. This mix should be made available as the only mineral. Each ⅓ lb of this mix will provide 30 g of magnesium.
2. A mineral mix made by mixing 30% magnesium oxide, 30% iodized salt, 30% bone meal, and 10% dried molasses. Each ⅓ lb of this mix will provide 27 g of magnesium.
3. A mineral mix made by mixing ⅔ (66⅔%) magnesium oxide and ⅓ (33⅓%) salt as the only source of salt is effective. Each ⅙ lb of this mix will provide 30 g of magnesium.
4. A commercial high-magnesium supplement, in blocks or mineral-salt mixtures which usually contain molasses, grain, and/or some other material to make them more palatable to animals. Generally, these are formulated to be fed at the rate of ½ to 1½ lb/head/day, and to provide 10 to 15 g of magnesium daily.

NOTE WELL: Blocks or mineral mixes usually give the best results if no additional salt is provided. Since the desire for salt varies among animals and with seasons, a high-salt mixture may not provide the required level of magnesium consumption. It should be emphasized that cattle must consume adequate magnesium on a regular basis. When using a supplement, one should pay particular attention to (1) the percentage of magnesium it contains and (2) the daily intake of the supplement. From these two factors the daily intake of magnesium can be determined and compared to the animal's requirement. The intake should be checked frequently as magnesium salts are generally unpalatable. The intake may be increased by adding grain or cottonseed meal/soybean meal.

[1]Although magnesium sulfate contains only 20% Mg, it is commonly used as a magnesium additive to a molasses supplement because of its solubility. Magnesium oxide will not remain dispersed in liquid feed without the aid of suspending agents.

(Continued)

TABLE 5–1 *(Continued)*

GRASS TETANY (HYPOMAGNESMIC TETANY, GRASS STAGGERS, WINTER TETANY. In Europe, it's called *FOG FEVER.*) (Continued)

On farms and ranches with a history of grass tetany, free-choice feeding of a mineral mix to insure a daily intake of 25 g of magnesium, with about half the daily intake coming from the natural magnesium in the pasture or other feed and half from the magnesium-containing mineral supplement, will usually provide protection from the development of grass tetany.

In high-risk situations—such as cows near calving grazing lush spring grass, highly fertilized with nitrogen or potassium, or both—the total magnesium requirement should be provided in the supplement. The reasons: (1) cows near calving are approaching lactation, when the magnesium demands are the highest; (2) nitrogen and potassium are antagonistic to magnesium; and (3) magnesium availability in early spring grass is low—besides, cattle cannot consume enough such grass to meet their energy requirements. A free-choice or hand-fed grain or protein-mineral supplement, providing 35 g of magnesium daily, will usually prevent the occurrence of grass tetany in such high-risk situations. If the magnesium supplement is self-fed, it should be located for easy access to the cattle, especially near water and shade where cattle tend to congregate and loiter. If the magnesium supplement is hand-fed, it is important that all animals have access to feeder space.

Where grass tetany is particularly troublesome, as an additional preventive measure, consideration should be given to (1) applying magnesium fertilizer and dolomitic limestone, or (2) dusting the pastures with magnesium oxide (MgO). Also, generally speaking, such pastures may be grazed without hazard by steers. But, before fertilizing or dusting, or shifting from a cow-calf program to a steer program, the counsel and advice of the local Farm Advisor/County Extension Agent should be sought.

Remarks:

Affected animals may be aggressive on getting up. So, watch out!
(Also see Chapter 4, Table 4-6, ANIMAL MINERAL CHART, MAGNESIUM.)

HEAVES

Species Affected:

Horses, mules.

Cause:

Exact cause unknown, but it is known that the condition is often associated with (1) the feeding of damaged, dusty, or moldy hay; and/or (2) the use of dusty bedding or paddocks.
It often follows severe respiratory infection such as strangles.
Probably an allergy.

Symptoms and Signs (or age group most affected):

Difficulty in forcing air out of the lungs resulting in a jerking of flanks (double flank action) and coughing. The nostrils are often slightly dilated, and there is a nasal discharge.
Heaves in horses is similar to emphysema in people.

Distribution and Losses Caused By:

Worldwide.
Losses are negligible.

Treatment:

Antihistamine granules can be administered in feed to control coughing due to lung congestion.

Control:

Affected animals are less bothered if turned to pasture, if used only at light work, if fed an all-pelleted ration, or if the hay is sprinkled lightly with water.

Prevention:

Avoid the use of damaged feeds.
Feed an all-pelleted ration, thereby alleviating dust.

Remarks:

Basically, heaves is a rupture of some of the alveoli in the lungs, the specific cause of which is unknown.

HYPOGLYCEMIA (BABY PIG SHAKES)

Species Affected:

Swine.

(Continued)

TABLE 5–1 *(Continued)*

HYPOGLYCEMIA (BABY PIG SHAKES) (Continued)

Cause:

Low blood-sugar level is characteristic of the trouble, but the cause of the low blood sugar is unknown.

Predisposition to piglet hypoglycemia occurs from any disease of the sow which decreases or inhibits milk production or let down.

Symptoms and Signs (or age group most affected):

Shivering, weakness, failure to nurse, with no evidence of scouring. If disturbed, the pigs emit a weak, crying squeal. Hair becomes erect and rough, and the heart action slow and feeble. Without treatment, death usually occurs in 24 to 36 hours after the first symptoms appear.

Confined to baby pigs only.

Distribution and Losses Caused By:

Worldwide.

Hypoglycemia accounts for 15 to 25% of total piglet mortality.

Fig. 5–16. Hypoglycemia in pig. Lack of milk during the first few days of life accompanies this condition; chilling speeds the process. Note the erect hair coat. (Courtesy, College of Veterinary Medicine, University of Illinois, Champaign-Urbana)

Treatment:

Provide heat lamps for pigs.

At earliest symptoms either (1) force feed at frequent intervals a mixture of 1 part corn syrup diluted with 2 parts of water, or (2) give intraperitoneal injections of 5% sterile glucose solution every 4-6 hours. Oxytocin may be administered to the sow to promote milk let down.

Consult the veterinarian.

Control:

Apparently not contagious.

Prevention:

Adequate rations and good care and management of the gestating sows may lessen the incidence of the disease.

Be sure there is adequate milk for baby pigs during first days of life.

Remarks:

One of the hazards of hypoglycemia is that the milk flow of the sow will not be stimulated or may even cease due to the inactivity of the affected pigs. In the latter case, the pigs may have to be either transferred to a foster mother or hand-fed.

IODINE DEFICIENCY (GOITER, BIG NECK)

Species Affected:

All farm animals and humans.

Cause:

A failure of the body to obtain sufficient iodine from which the thyroid gland can form thyroxin (an iodine-containing compound).

Symptoms and Signs (or age group most affected):

Goiter (big neck, which is a swelling under the chin) is the most characteristic symptom of iodine deficiency in calves, lambs, kids, and humans. Also, there may be reproductive failure and weak offspring that fail to survive. Pigs may be born hairless and show edema of the shoulders and neck. Foals may be born weak.

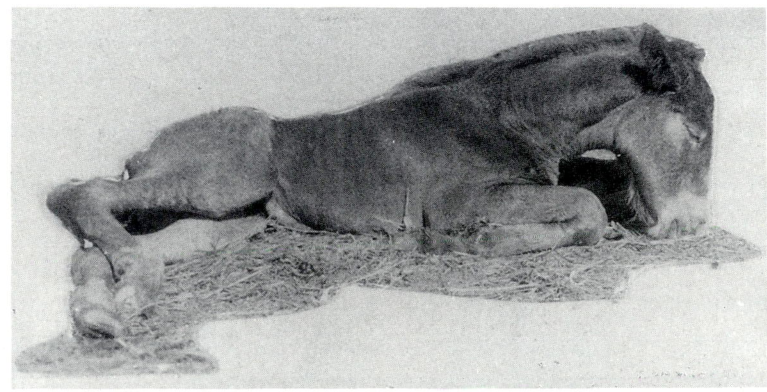

Fig. 5–17. Weak newborn foal due to iodine deficiency of the mare during pregnancy. (Courtesy, Washington State University, Pullman)

(Continued)

TABLE 5-1 *(Continued)*

IODINE DEFICIENCY (GOITER, BIG NECK) (Continued)

Distribution and Losses Caused By:

Iodine deficiencies occur worldwide; wherever feeds are grown on iodine-poor soil containing insufficient iodine to meet animal needs. The highest incidence has been observed in the Alps, the Pyrenees, the Himalayas, the Thames Valley of England, certain regions of New Zealand, a number of Central and South American countries, and the Great Lakes and Pacific Northwest regions of the U.S. Fig. 5-18 shows the goiter areas of the world.

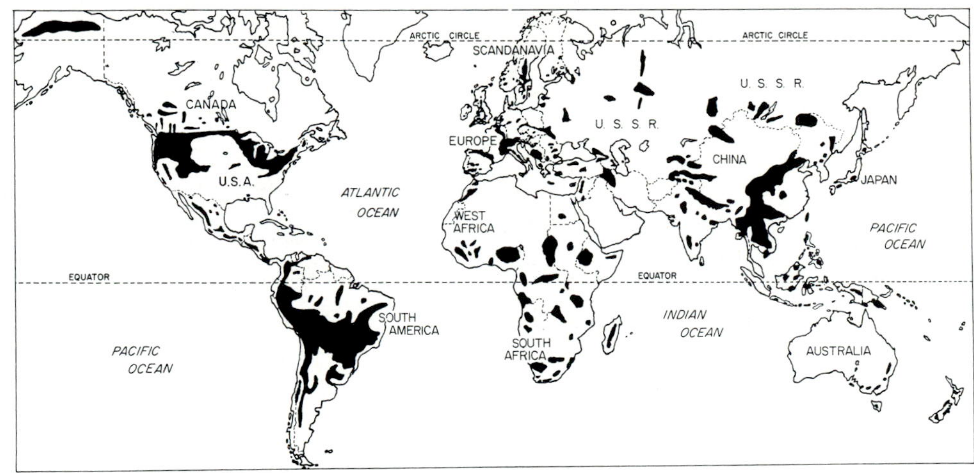

Fig. 5-18. Goiter areas of the world. (Map prepared by the authors on the basis of information from the World Health Organization, Geneva, Switzerland)

Treatment:

Occasionally borderline cases may survive; in these the moderate thyroid enlargement disappears in a few weeks.
Once the iodine deficiency symptoms appear, no treatment is very effective.

Control:

At the first signs of iodine deficiency, stabilized iodized salt should be fed to all farm animals.

Prevention:

In iodine-deficient areas, feed stabilized iodized salt containing 0.01% potassium iodide to all farm animals throughout the year.
Organic forms of iodide are also suitable sources of iodine, but they are usually more costly.

Remarks:

The enlarged thyroid gland (goiter) is nature's way of attempting to make sufficient thyroid hormone, thyroxin, under conditions when an iodine deficiency exists.
Mares fed excess iodine (48 mg or more) during late gestation will produce foals with hyperplastic goiter. Some mares will also develop goiter.
(**A**lso see Chapter 4, Table 4-6, ANIMAL MINERAL CHART, IODINE.)

KETOSIS (ALSO KNOWN AS ACETONEMIA IN CATTLE AND PREGNANCY DISEASE IN SHEEP)

Species Affected:

Cattle, sheep, goats.

Cause:

A metabolic disorder of nutritional origin, characterized by hypoglycemia (low blood sugar). If the increased nutrient requirements are not met by more feed during the high-demand periods (in cows, 1-6 weeks after calving; in ewes, 2 weeks before lambing), the animal must draw on body fat reserves. If this is done too rapidly, and without adequate carbohydrates in the ration, ketosis follows.

Symptoms and Signs (or age group most affected):

In cows, ketosis or acetonemia is usually observed 2 to 6 weeks after calving. Affected animals show loss of appetite and condition, a marked decline in milk production, and the production of a peculiar sweetish chloroform-like odor of acetone that may be present in the milk and urine and pervade the barn. A positive diagnosis can be made by testing the milk or urine for the presence of ketones.

In ewes and goats, ketosis or pregnancy disease generally strikes during the last 2 weeks of pregnancy. Usually, affected ewes are carrying twins or triplets. Symptoms include going off feed suddenly, grinding of teeth, dullness, weakness, frequent urination, trembling when exercised, and blindness—with the final stage being complete collapse, followed by death in 90% of the cases. In dairy goats, lactation ketosis, which is similar to the ketosis of dairy cows, may be observed in high milk producers following kidding.

Fig. 5-19. Ewe with ketosis, or pregnancy disease. (Courtesy, College of Veterinary Medicine, University of Illinois, Champaign-Urbana)

(Continued)

TABLE 5-1 *(Continued)*

KETOSIS (ALSO KNOWN AS ACETONEMIA IN CATTLE AND PREGNANCY DISEASE IN SHEEP) (Continued)

Distribution and Losses Caused By:

Worldwide.

Ketosis or acetonemia affects cattle throughout the U.S.

Ketosis or pregnancy disease in sheep affects farm flocks more than range bands, the losses in the former sometimes being as high as 25%.

Treatment:

Cattle: ½–1 lb of either propylene glycol or sodium propionate daily, with the dose divided into 2 treatments for 5–10 days. Put treatment in grain if cow is eating; otherwise, give as drench.

Intravenous injection of glucose solution and glucocorticoids (to increase blood sugar levels temporarily) as well as the oral administration of propylene glycol. Numerous other treatments are sometimes used.

Sheep and goats before parturition: 3 to 4 oz of propylene glycol, given orally 3 times daily. Cesarean section early in the course of the disease usually leads to recovery and, if near term, the offspring may be saved.

Dairy goats after kidding: 6 to 8 oz of propylene glycol, given orally twice daily. Severe cases may be aided by intravenous injections of 50% dextrose solution. Cortocosteroid injection may be used in conjunction with either propylene glycol or dextrose solution in does that have kidded.

Control:

Cows: Maintain relatively high-energy intake before calving; increase energy intake substantially after calving.

Ewes: Avoid obesity in early pregnancy. Feed grains rather liberally the last 6 weeks of pregnancy.

Prevention:

Cows: The incidence of ketosis can be lessened by (1) avoiding excessively fat cows at calving; (2) increasing the level of concentrates gradually after calving; (3) feeding good-guality hay in preference to high-silage rations after calving, and avoiding abrupt changes in roughage; (4) feeding adequate proteins, minerals, and vitamins; and (5) providing comfort, exercise, and ventilation. In problem herds, feeding ¼ lb daily of propylene glycol or sodium propionate may be helpful.

Sheep and goats: Feed more hay and ½ to 1 lb of grain beginning a month before parturition. Good management is important, too, including exercise, freedom from parasites, and avoiding stress.

Remarks:

The clinical findings are similar in the case of affected cattle and sheep, but it usually strikes ewes just before lambing, whereas cows are usually affected within the first 2–6 weeks after calving.

LIVER ABSCESSES

Species Affected:

Cattle, especially feedlot cattle.

Cause:

Fusobacterium necrophorum, the same bacteria that causes foot rot, appears to be the major organism responsible for liver abscesses.

High concentrate (low roughage) finishing rations predispose cattle to a high incidence of liver abscesses.

Symptoms and Signs (or age group most affected):

Liver abscesses generally go undetected until cattle are slaughtered, at which time they appear as "walled-off" areas filled with pus. However, reduced feed intake and gain near the end of the feeding period may be indicative.

Distribution and Losses Caused By:

Liver abscesses occur in all countries where intensive beef production is practiced and are common in the Corn Belt and western U.S. and in western Canada.

The USDA records show that 12% of the beef livers condemned in the U.S. are due to abscesses.

The incidence of liver abscesses is highest in feedlot cattle, where an estimated 18 to 20% of the livers are affected. For a particular lot of feedlot cattle, liver abscesses may range from 1 to 90%.

Since the liver of a 1,000-lb steer weighs approximately 11 lb, liver condemnation represents a considerable monetary loss, but the loss from reduced feed efficiency and gains may be even greater.

Treatment:

Treatment of an acute liver abscess in feedlot cattle should be left to the veterinarian, who may administer (1) sulfapyridine, or (2) an antibiotic.

Control:

The low level feeding of certain antibiotics during the finishing period will markedly reduce the number of liver abscesses.

(Continued)

TABLE 5-1 *(Continued)*

LIVER ABSCESSES (Continued)

Prevention:

The incidence of liver abscesses in feedlot cattle can be reduced, but not entirely eliminated, by (1) changing from high-roughage to high-concentrate rations gradually, and (2) feeding an antibiotic (commonly chlortetracycline or tylosin) at the daily rate of 70 mg/animal.

Remarks:

Liver abscesses, as indicated by the name, are single or multiple abscesses of the liver observed at slaughter. Usually a liver abscess consists of a central mass of necrotic liver surrounded by pus and a wall of connective tissue. At slaughter, livers affected with abscesses are condemned for human food.

MILK FEVER (PARTURIENT PARESIS, HYPOCALCEMIA)

Species Affected:

Cattle, sheep, goats.

Milk fever, a metabolic disease, is similar in cows, ewes, and does. So, the discussion that follows pertaining to cows also applies to sheep and goats, with treatment and prevention adjusted for size of animal.

Cause:

Low blood calcium concentration. The name *milk fever* is a misnomer, because the animal does not have a fever.

Initiation of lactation places a severe strain on the calcium balance of the cow due to the amount of calcium secreted in the milk. All cows are slightly hypocalcemic at the time of calving, but some become so hypocalcemic that clinical signs of milk fever develop.

Fig. 5-20. Milk fever in Jersey cow, showing characteristic position of head—turned back over shoulder. (Courtesy, Washington State University, Pullman)

Symptoms and Signs (or age group most affected):

Commonly occurs within 3 days after calving and in high-producing cows. Rarely occurs at first calving. First symptoms are loss of appetite, constipation, and general depression. This is followed by nervousness and finally collapse and complete loss of consciousness. The head is usually turned back toward the flank.

The incidence of the disease increases with the age of the cow and is highest in Guernsey and Jersey breeds.

Distribution and Losses Caused By:

A common widespread disease of dairy cows. It is estimated that more than 8% of all dairy cows are stricken by milk fever; occasionally, with up to 80% affected in a single herd.

Losses are not great, although untreated animals are likely to die.

Causes estimated average annual losses in dairy cattle (including milk) of $100 million.

Treatment:

Milk fever should be regarded as an emergency and the affected animal treated as soon as possible.

The standard treatment is intravenous infusion of calcium borogluconate as soon as the first signs appear. *CAUTION:* Overdoses of calcium salts can result in acute heart damage; so, dose level should be carefully calculated, and intravenous administration should be performed slowly.

Cows that are already down will usually stand up within 1 to 2 hours following treatment.

Control:

(See Prevention.)

Prevention:

Each of the following measures will lessen the incidence of milk fever:

1. *Low calcium during the dry period.* Feeding low calcium (less than 100 g/day)—high phosphorus (more than 40 g/day) rations during the dry period is important. High calcium levels in dry cow rations aggravate the problem. So, feeding a low calcium ration (less than 0.1 lb/day) before calving has shown promise of preventing milk fever.

2. *Calcium shock treatment.* Feed low calcium, high phosphorus rations containing only 15 to 20 g of calcium per day for 2 weeks prior to the expected calving date, rather than a restricted, but somewhat higher, calcium intake during the entire dry period. This creates a mild calcium deficiency which stimulates production of the biologically active form of vitamin D in the animal's body. In turn, this form of vitamin D stimulates the bone and gut to supply more calcium and phosphorus. As a result, when the greater demand for calcium and phosphorus occurs at calving, the bone and gut are already activated and are able to meet the increased demands for calcium. Thus, milk fever is avoided.

3. *Calcium-phosphorus ratio and amounts.* Balancing cow rations to contain 0.5% calcium and 0.25% phosphorus on a dry matter basis will limit the incidence of milk fever.

(Continued)

TABLE 5-1 *(Continued)*

MILK FEVER (PARTURIENT PARESIS, HYPOCALCEMIA) (Continued)

4. *High vitamin D.* Feeding massive doses of 20 million I.U. of vitamin D/cow/day starting about 5 days before calving and continuing through the first day postpartum, with a maximum dosage period of 7 days, has been effective in controlling milk fever. However, difficulty in predicting calving dates accurately has reduced the effectivemess of this treatment under practical conditions.

5. *Avoid excessive fatness.* Excessive fatness, or any other conditions that reduce feed intake at calving, tends to cause more milk fever.

Remarks:

The name *milk fever* is a misnomer, because the disease is not accompanied by fever, the temperature really being below normal.

MOON BLINDNESS

(See PERIODIC OPHTHALMIA.)

NIGHT BLINDNESS (NYCTALOPIA)

Species Affected:

All farm animals and humans.

Cause:

Deficiency of vitamin A.

Ability to see in dim light depends upon the rate of resynthesis of rhodopsin (visual purple) in the retina of the eye. When vitamin A is deficient, rhodopsin resynthesis is impaired, resulting in a lessened ability to see in dim light, commonly known as night blindness.

Symptoms and Signs (or age group most affected):

The deficiency first manifests itself as a slow, dark adaptation, then progresses to total night blindness. All animal species may be affected.

Distribution and Losses Caused By:

Worldwide. Especially prevalent where one of the following conditions frequently prevails: (1) extended drought, and/or (2) winter feeding on bleached grass cured on the stalk or on bleached hay.

Treatment:

Treatment consists in correcting the dietrary deficiencies.

The vitamin A requirement and allowances vary (1) according to species, and (2) within species according to age, weight, and reproductive status. The chapters in this book devoted to each animal species present the recommended vitamin A requirements and allowances.

Control:

(See Treatment and Prevention.)

Prevention:

Provide good sources of carotene (provitamin A) through green, leafy hays, silage, green pasture, and/or yellow corn.

Add stabilized vitamin A to ration or inject slow-release vitamin A intramuscularly.

Remarks:

Vitamin A can be provided as the synthetic vitamin or as the precursor, carotene.

(Also see Chapter 4, TABLE 4-8, ANIMAL VITAMIN CHART, VITAMIN A.)

OSTEOMALACIA

(An adult form of rickets)

Species Affected:

All species.

Cause:

Inadequate phosphorus; sometimes inadequate calcium; incorrect calcium:phosphorus ratio; or lack of vitamin D in confined animals.

(Continued)

TABLE 5-1 *(Continued)*

OSTEOMALACIA (Continued)

Symptoms and Signs (or age group most affected):

Phosphorus deficiency symptoms are depraved appetite (gnawing on bones, wood, or other objects, or eating dirt); and lack of appetite, stiffness of joints, failure to breed regularly, decreased milk production, and an emaciated appearance.

Calcium deficiency symptoms are fragile bones, reproductive failures, and lowered lactations.

Mature animals most affected. Most acute cases occur during pregnancy and lactation.

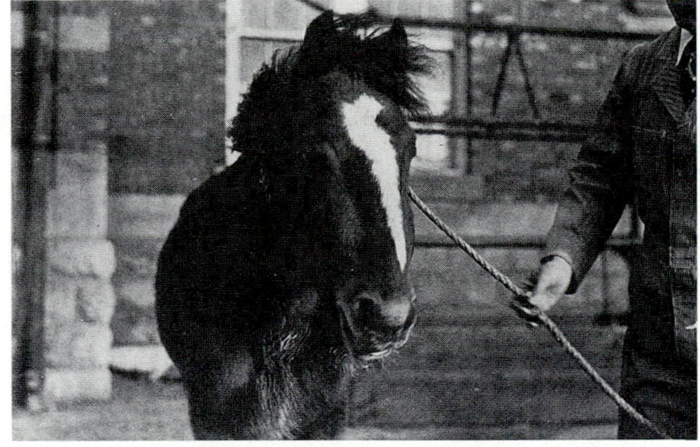

Fig. 5-21. Osteomalacia of the facial bones in a mature Hackney. (Courtesy, College of Veterinary Medicine, University of Illinois, Champaign-Urbana)

Distribution and Losses Caused By:

Southwestern U.S. is classed as a phosphorus-deficient area, whereas calcium-deficient areas have been reported in parts of Fla., La., Nebr., Va., and W. Va.

Treatment:

Select natural feeds that contain sufficient quantities of calcium and phosphorus.

Feed a special mineral supplement or supplements.

If this disease is far advanced, treatment will not be successful.

Control:

(See Treatment.)

Prevention:

Feed balanced rations, and, if necessary, allow animals free access to a suitable phosphorus and calcium supplement.

Increase the calcium and phosphorus content of feed by fertilizing the soils.

Remarks:

Calcium deficiencies are much less frequent than phosphorus deficiencies in cattle, sheep, and horses.

Calcium deficiencies are fairly common in swine because grains, which are their chief feed, are low in this mineral.

PARAKERATOSIS (GREASY SKIN DISEASE)

Species Affected:

Swine. A somewhat similar clinical condition has been reported in cattle and sheep.

Cause:

Low zinc level, or high calcium level in the ration—above 0.8%.

Symptoms and Signs (or age group most affected):

The disease is characterized by a mangy appearance, reduced appetite and growth rate, diarrhea, and vomiting. It generally affects 6- to 16-week-old pigs.

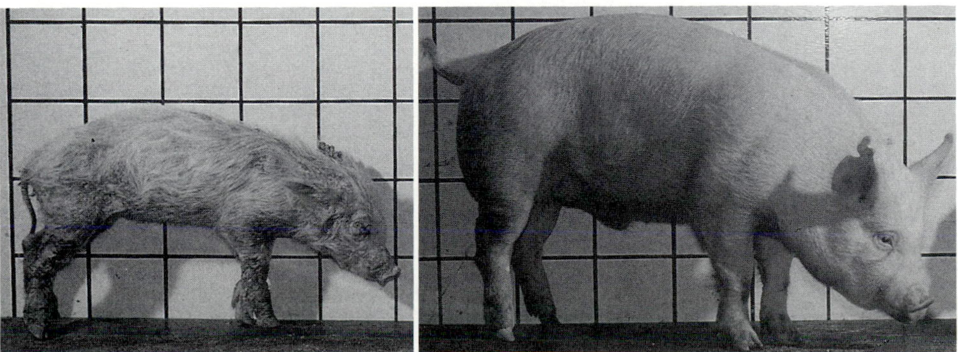

Fig. 5-22. Zinc made the difference! *Left:* Pig with parakeratosis due to zinc deficiency. This pig received only 17 ppm of zinc and gained only 3 lb in 74 days. *Right:* This pig received the same ration as the pig on the left, except that the ration contained 67 ppm of zinc. This pig gained 111 lb in 74 days. (Courtesy, Purdue University, West Lafayette, Ind.)

(Continued)

TABLE 5-1 *(Continued)*

PARAKERATOSIS (GREASY SKIN DISEASE) (Continued)

Distribution and Losses Caused By:

Mortality is not high; but economic loss occurs due to reduced gains and lowered feed efficiency.

Treatment:

The amount of zinc to use in treatment will vary with the type of protein fed, along with the following dietary factors: phosphorus, cadmium, magnesium, copper, molybdenum, cobalt, iron, amino acids, chelating agents, and vitamin D. (See Chapter 23, Feeding Swine.)

Control:

This is a metabolic problem; hence, it is not contagious.

Prevention:

Meet the zinc requirements for swine, as given in Chapter 23, Feeding Swine.

Remarks:

Excess calcium reduces the absorption and utilization of zinc. In swine, this frequently causes parakeratosis.
Zinc carbonate, sulfate, and oxide are effective sources of zinc in prevention or treatment.

PERIODIC OPHTHALMIA (MOON BLINDNESS, EQUINE RECURRENT UVEITIS [ERU])

Species Affected:

Horses, mules, asses.

Cause:

There are many theories as to the cause of periodic ophthalmia; among them: (1) lack of riboflavin; (2) an autoimmune reaction; (3) an allergic reaction; (4) genetics; (5) leptospirosis, brucellosis, and strangles; (6) parasitic infections; and (7) viral, fungal, chlamydial and mycoplasmal infections. The most prevalent current thinking is that Periodic Ophthalmia is an immune-mediated disease of varying etiology.

Symptoms and Signs (or age group most affected):

Periods of cloudy vision, in one or both eyes, which may last for a few days to a week or longer and then clear up; but it recurs at intervals (approximately as often as the new moon; hence, the name *moon blindness*), eventually culminating in blindness in one or both eyes.

Distribution and Losses Caused By:

In many parts of the world. It is the most common cause of blindness in the horse.
In the U.S., it occurs most frequently in the northeastern U.S., where an estimated 10 to 20% of the horses are afflicted.

Treatment:

The main objectives in treatment are to preserve vision, to control pain and inflammation, and to reduce the development of permanent scarring in the eye. In cases in which the cause is determined, specific therapy can be implemented. In all cases, the treatment is symptomatic and aimed at controlling the inflammation. Antibiotics, corticosteroids, and/or atropine administered promptly, under the direction of a veterinarian, may be helpful in some cases.
Immediately (1) change to greener hay or grass, and (2) add riboflavin to the ration at the rate of 40 mg/animal/day.

Control:

There is no sure control, because of the elusive causes of the disease.

Prevention:

Feed green grass, or well-cured green, leafy hay; or add riboflavin to the ration at the rate of 40 mg/animal/day.

Remarks:

This disease has been known to exist for at least 2,000 years.
Because of the recurrent nature of the disease, it should be considered an unsoundness in the horse.

(Continued)

TABLE 5–1 *(Continued)*

PEROSIS (SLIPPED TENDON)

Species Affected:

Chickens, turkeys, ducks, geese, pheasants, grouse, and quail.

Cause:

Manganese deficiency. Except for the animal by-products, most of the feedstuffs used in poultry rations do not contain sufficient manganese to meet the requirements of poultry.

Perosis is made more severe by excessive amounts of calcium and phosphorus in the ration, and by rearing birds on wire or slotted floors.

Perosis may also be caused by a deficiency of biotin or choline.

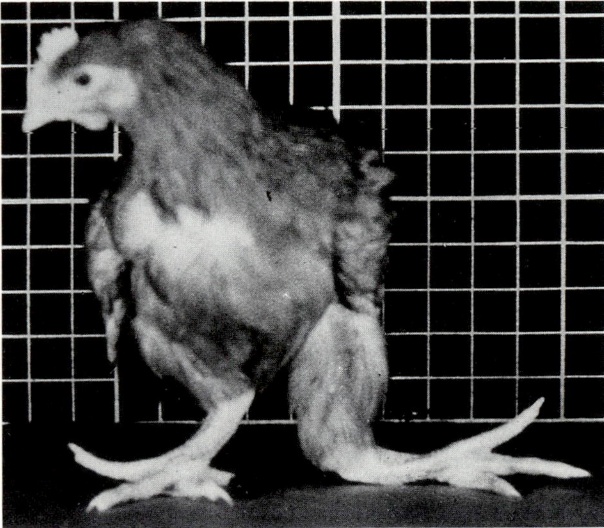

Fig. 5–23. Manganese-deficient chicken, showing a bone disease known as perosis or slipped tendon. (Courtesy, Virginia Poultry Federation, Harrisonburg, Va.)

Symptoms and Signs (or age group most affected):

Perosis affects young chickens and the young of other birds.

It is characterized by a deformity of the hock joint. The hock joint becomes enlarged and the gastrocnemius tendon at this location slips from the condyle, twisting the shank to one side. One or both legs may be affected.

Distribution and Losses Caused By:

Perosis is seldom seen in large commercial poultry operations, where rations are scientifically formulated. However, it is sometimes seen in farm flocks.

Treatment:

After the deformity has occurred, it cannot be corrected by feeding an adequate ration.

Control:

Provide sufficient manganese in the ration, along with sufficient biotin and choline.

Prevention:

Since most poultry rations composed largely of grains tend to be deficient in manganese, prevention consists in routinely adding this element to poultry rations in such forms as manganous chloride, manganous carbonate, manganese dioxide, or manganese sulfate. Dietary levels of 30 to 50 g of manganese per ton of feed should be sufficient in most poultry rations.

Remarks:

(Also see Chapter 4, Table 4–6, ANIMAL MINERAL CHART, MANGANESE.)

PHOTOSENSITIZATION

Species Affected:

Cattle, horses, sheep, swine.

Cause:

The sensitization of light-colored skin to sunlight. Some feeds, forages and certain medicines contain substances which may sensitize the skin (primary photosensitization). In other cases, products of metabolism, which normally would be removed from the body, accumulate because of faulty liver function (hepatogenous photosensitization). Primary photosensitization usually occurs in the spring when plants are lush, green, and growing rapidly. St. John's Wort (Klamath weed) and buckwheat are two of the most common sources of photosensitizing substances. Also, rape, kleingrass, kale, trefoil, alfalfa, alsike clover, Swedish clover, lamb's tongue, and plantain have been associated with photosensitization at one time or another.

(Continued)

TABLE 5-1 *(Continued)*

PHOTOSENSITIZATION (Continued)

Symptoms and Signs (or age group most affected):

The signs of the disease are essentially those of severe sunburn. The lesions are confined to white, or lightly pigmented, exposed areas of skin. The muzzle, eyes, face, and light areas over the back are usually affected first. Areas of the belly and udder, which are exposed to the sun when the animal lies down, may also be affected. The earliest signs are redness and swelling of the skin. Later, tissue fluids ooze from the affected areas and crusting of the skin occurs, with resultant matting of the hair. In severe cases, the eyelids and nostrils may be swollen closed. In extreme cases, sloughing of the skin and gangrene result.

Distribution and Losses Caused By:

Photosensitization occurs worldwide. Death loss is higher in sheep than in cattle. But the monetary losses from the afflicted living are considerable, including loss in weight, damaged udders and teats, screwworms, secondary infections, eye damage, and stunted offspring.

Fig. 5-24. Photosensitization in a cow, showing (1) swollen eyelids and nostrils, and (2) crusting of the skin and matting of the hair. (Courtesy, USDA-ARS Poisonous Plant Research Laboratory, Logan, Utah)

Treatment:

Discontinue the forage that's causing the trouble and keep the animal out of sunlight. In some cases, treatment of local lesions is warranted. In severe cases, supportive treatment, such as intravenous fluids, antibiotics, and other special medicines may be directed by the veterinarian.

Control:

Control of photosensitization can be achieved by preventing access to offending plants and keeping animals in the shade.

Prevention:

Good range and pasture management practices generally prevent the problem of photosensitization.

Remarks:

Photosensitization should be differentiated from sunburn in which the white or lightly pigmented skin of a normal animal becomes inflamed following overexposure to ultra-violet rays.

POLIOENCEPHALOMALACIA (CEREBROCORTICAL NECROSIS, POLIO)

Species Affected:

Cattle, sheep, goats, deer.

Cause:

Although the cause is not known, it appears to be due to a deficiency. One hypothesis with considerable substantiation is that an acute thiamin deficiency is an important factor in the cause as evidenced by a favorable response to thiamin injection. Yet, the reasons why a thiamin deficiency should exist are not understood, because the thiamin intake appears to be more than adequate. One theory is that abnormally high concentrations of thiaminase enzyme from unusual plants or microflora destroy the vitamin before absorption takes place.

Symptoms and Signs (or age group most affected):

Sudden deaths in animals. Sick animals are excitable, incoordinated, and have impaired vision. On driving, these animals go down into convulsions.

Affects feedlot and pasture cattle 3 months to 2 years of age.

In sheep, the incidence is highest in feedlot lambs 5 to 8 months of age. However, outbreaks of the disease may occur in farm flocks on pasture, especially in lambs following changing from overgrazed to lush pasture.

In goats, it may strike suckling young on pasture.

(Continued)

TABLE 5-1 *(Continued)*

POLIOENCEPHALOMALACIA (CEREBROCORTICAL NECROSIS, POLIO) (Continued)

Distribution and Losses Caused By:

Most common in feedlot animals.

The disease occurs worldwide.

In sheep, the morbidity rate may range from a few cases up to 10% of the flock, and 50% of the affected animals may die.

Treatment:

Treatment consists of the IV or IM administration of thiamin at a dosage of 1 to 2 mg/lb. Twice daily treatment may be necessary for 2 days.

Rapidity of recovery relates directly to the speed of disease recognition and institution of thiamin treatment. Good nursing will help.

Usually, animals severely affected for more than 24 hours cannot be expected to respond well to treatment.

Control:

Dietary cereal content should be decreased and additional good quality roughage supplied for a period of 5 days prior to a gradual return to higher energy rations.

Prevention:

Until the cause is discovered, little can be done to prevent the disease, except to provide a good ration.

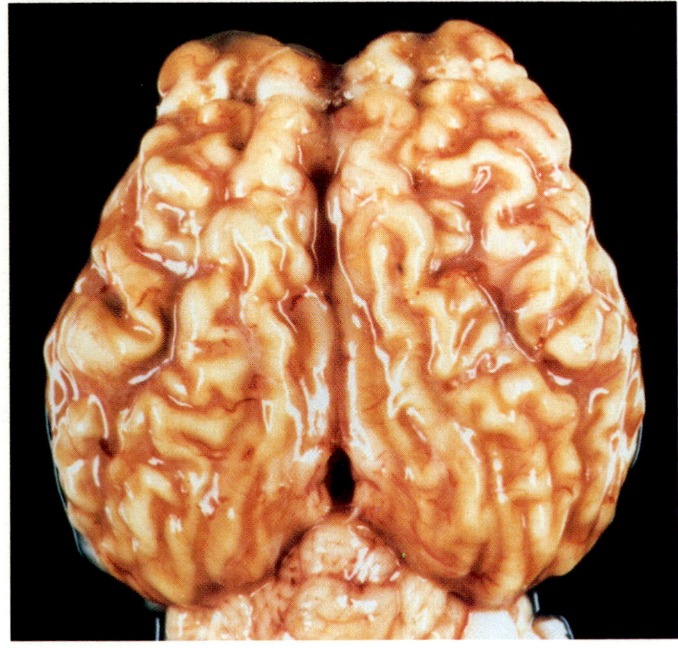

Fig. 5-25. Soft, swollen, yellowish cerebral gyri in brain of cow afflicted with polioencephalomalacia. (Courtesy, College of Veterinary Medicine, University of Florida, Gainesville)

Remarks:

Ruminants normally derive adequate thiamin from symbiotic ruminal activity; the inadequacy is thought to be a result of intraruminal thiamin destruction either by the enzymes of microbes or other dietary sources.

(**A**lso see Chapter 4, Table 4-8, ANIMAL VITAMIN CHART, THIAMIN.)

POLYNEURITIS

Species Affected:

Chickens.

Cause:

Deficiency of thiamin (vitamin B–1), which is required by poultry for metabolism of carbohydrates.

Symptoms and Signs (or age group most affected):

Loss of appetite, followed by loss in weight, ruffled feathers, leg weakness, and unsteady gait.

Head retraction.

Nervous disorders, culminating in paralysis of the peripheral nerves.

Distribution and Losses Caused By:

Thiamin is found in abundance in whole grains which make up the major part of most poultry rations; hence, a deficiency is not observed under practical conditions.

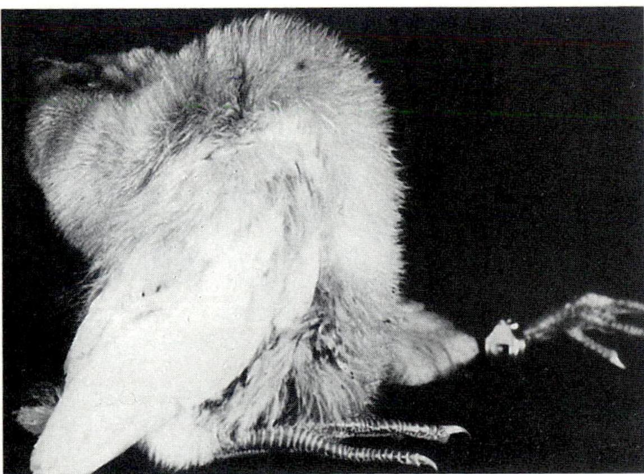

Fig. 5-26. Polyneuritis in a thiamin-deficient chick. Muscle paralysis causes extended legs and retraction of the head. (Courtesy, Cornell University, Ithaca, N.Y.)

Treatment:

Chickens suffering from a thiamin deficiency respond in a matter of a few hours to oral administration of the vitamin. Since thiamin deficiency causes extreme loss of appetite, supplementing the feed with the vitamin is not a reliable treatment until after chickens have recovered from acute deficiency by oral administration; hence, injection of chickens suffering from thiamin deficiency is necessary.

Control:

Most of the feedstuffs used in feeding poultry contain more than adequate quantities of thiamin; hence, control of this disease is not a problem.

(Continued)

TABLE 5-1 *(Continued)*

POLYNEURITIS (Continued)

Prevention:

Special sources of thiamin are not normally added to poultry rations.

Remarks:

Thiamin is the least stored of all the vitamins. So, it should be supplied regularly in the ration.
Today, the classic disease, polyneuritis, is primarily of historic interest; it seldom occurs in practice because of the common fortification of rations with thiamin. (Also see Chapter 4, Table 4-8, ANIMAL VITAMIN CHART, THIAMIN.)

PREGNANCY DISEASE IN SHEEP

(See Ketosis.)

PROTEIN POISONING

Species Affected:

Horses in particular, but claims of protein poisoning are also made relative to other classes of farm animals.

Cause:

High levels of protein incriminated. Some horses appear to be allergic to certain proteins or to excesses of specific amino acids.

Symptoms and Signs (or age group most affected):

"Protein bumps" over the body—an allergic reaction.

Distribution and Losses Caused By:

Evidence to support claims of "protein poisoning" are lacking. Extremely high protein rations may be laxative, but no toxic effects have been noted.

Treatment:

Lower protein content of the ration.

Control:

Economics generally control the level of protein feeding. High-protein feeds are more expensive than high-energy feeds, with the result that there is temptation to feed too little of them.

Prevention:

Do not feed excessive levels of protein.

Remarks:

There is no proof that heavy feeding of high-protein feeds to horses is harmful, provided (1) the ration is balanced in other respects, (2) the animal's kidneys are normal and healthy (a large excess of protein in terms of body needs increases the work of the kidneys for the excretion of the urea), (3) any ration change to a high-protein feed is made gradually, as is recommended in any change in feed, and (4) there is adequate exercise and normal metabolism.

PULMONARY EMPHYSEMA (BOVINE PULMONARY EMPHYSEMA, COW ASTHMA)

Species Affected:

Cattle.

Cause:

The condition can be traced to the quantities of the amino acid, tryptophan, consumed when cattle are pastured on lush, rapidly growing forage plants. The change from dry feed to lush pasture produces conditions favorable to an abnormal growth of clostridial organisms in the rumen. These organisms help convert tryptophan to 3-methylindole (3-Mi). When large quantities of 3-Mi are absorbed into the blood stream, pulmonary emphysema may result.
Also, the disease may be caused by pneumonia, allergic reactions to lungworm larvae or inhaled fungal organisms, or inhalation of irritating gases.

(Continued)

TABLE 5-1 *(Continued)*

PULMONARY EMPHYSEMA (BOVINE PULMONARY EMPHYSEMA, COW ASTHMA) (Continued)

Symptoms and Signs (or age group most affected):

Difficult breathing. In severe cases, the common signs include panting, difficulty in exhaling air from the lungs, coughing, excessive salivation, reluctance to move, extreme weakness, and rapid loss in condition. Death may occur within a few hours after onset of the disease.

Distribution and Losses Caused By:

The disease occurs throughout the world. In affected herds, up to 20% of the cattle may develop emphysema and as many as 10% may die.

Treatment:

No specific medications are available. The recommended treatment consists of—

1. Removing the animals from lush pasture and placing them on hay.
2. Injecting antihistamines, steroids and other compounds to lessen the respiratory distress.
3. Using antibiotics and sulfonamides to prevent secondary bacterial infections.

Control:

The removal of cattle from lush pasture and placing them in a drylot and feeding hay will control the disease.

Prevention:

Measures that will prevent pulmonary emphysema include—

1. Removing cattle from summer range before feed becomes too dry.
2. Making a gradual transition from summer range to lush pasture.
3. Continuing to feed hay or straw while the cattle are on pasture.

Remarks:

Pulmonary emphysema is primarily a nutritional disease, although there are other causes.

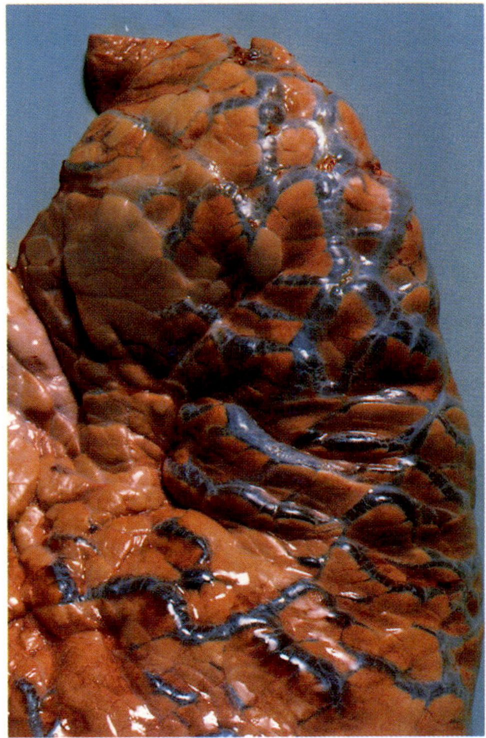

Fig. 5-27. Lungs from a cow afflicted with pulmonary emphysema. The lungs are greatly enlarged, firm, and edematous. The lungs do not collapse normally, and they are pinkish gray in color. (Courtesy, College of Veterinary Medicine, University of Tennessee, Knoxville)

RICKETS

Species Affected:

All young farm animals and young humans.

Cause:

Lack of either calcium, phosphorus, or vitamin D; or an incorrect ratio of the 2 minerals.

In housed animals, vitamin D deficiency is not uncommon; grazing animals are more likely to be phosphorus-deficient.

Symptoms and Signs (or age group most affected):

Enlargement of the knee and hock joints, and the animal may exhibit great pain when moving. Irregular bulges (beaded ribs) at juncture of ribs with breastbone, and bowed legs.

Rickets is a disease of young animals—calves, foals, pigs, lambs, kids, pups, and chicks.

Poultry: Bones of growing birds become soft and rubbery.

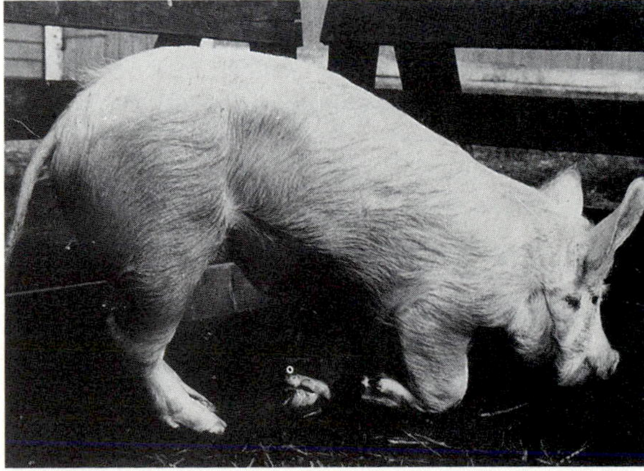

Fig. 5-28. Rickets (advanced case) caused by a deficiency of vitamin D. The pig was fed indoors, without exposure to sunlight. Because of leg abnormalities, it was unable to walk. (Courtesy, University of Saskatchewan, Saskatoon, Saskatchewan, Canada)

(Continued)

TABLE 5-1 *(Continued)*

RICKETS *(Continued)*

Distribution and Losses Caused By:

Worldwide.
It is seldom fatal, but it can be severely debilitating and economically disastrous.

Treatment:

If the disease has not advanced too far, treatment may be successful by supplying adequate amounts of vitamin D, calcium, and phosphorus, and/or adjusting the ratio of calcium to phosphorus.

Control:

Control of rickets is usually achieved by providing a balanced ration, with special consideration given to calcium, phosphorus, and vitamin D.

Prevention:

Provide (1) sufficient calcium, phosphorus, and vitamin D, and (2) a correct ratio of the 2 minerals. Vitamin D_3, rather than D_2, is required by the chicken.

Remarks:

Rickets is characterized by a failure of growing bone to ossify, or harden, properly.
Hens fed rations deficient in vitamin D lay eggs with progressively thinner shells until production ceases.

SALT (SODIUM CHLORIDE) DEFICIENCY

Species Affected:

All farm animals and humans.

Cause:

Lack of salt (sodium chloride).

Symptoms and Signs (or age group most affected):

Loss of appetite, retarded growth, loss of weight, a rough coat, lowered production of milk, and a ravenous appetite for salt.
All age groups affected after they have been weaned from milk.
Horses: Horses are most likely to develop signs of sodium chloride (salt) deficiency when worked hard in hot weather, since perspiration and urinary losses are appreciable. When deprived of salt, horses tire easily, stop sweating, and exhibit muscle spasms.
Poultry: Cannibalism and poor growth.

Distribution and Losses Caused By:

Worldwide, especially among grass-eating animals.

Treatment:

Salt-starved animals should be gradually accustomed to salt, slowly increasing the hand-fed allowance until the animals may be safely allowed free access to it.

Control:

(See Treatment and Prevention.)

Prevention:

Provide plenty of salt at all times, preferably by free-choice feeding.

Remarks:

Common salt is one of the most essential minerals for grass-eating animals, and one of the easiest and cheapest to provide.
Excessive salt intake can result in toxicity if animals are deprived of water (see Table 5-3, POTENTIAL POISONS, SALT POISONING).

SALT SICK (COBALT DEFICIENCY)

Species Affected:

Cattle, sheep, goats.

(Continued)

TABLE 5-1 *(Continued)*

SALT SICK (COBALT DEFICIENCY) (Continued)

Cause:

Cobalt deficiency.
In Florida, cobalt deficiency is associated with copper deficiency.

Symptoms and Signs (or age group most affected):

Loss of appetite, emaciation, depraved appetite, scaliness of skin, rough hair coat, listlessness, and lack of thrift.

Fig. 5-29. Cobalt deficiency. *Left:* A heifer suffering from cobalt deficiency. Anemia, loss of appetite, and roughness of hair coat characterize the malady. *Right:* Illustrates the remarkable recovery in the same animal brought about by the administration of cobalt. (Courtesy, Michigan State University, East Lansing)

Distribution and Losses Caused By:

Cobalt deficiency is widespread. In different parts of the world, it is known as *Denmark disease, coast disease, enzootic marasmus, bush sickness, wasting disease, Nakuritis,* and *pining disease.* Fig. 5-30 shows cobalt-deficient areas of the U.S.

Treatment:

Provide 0.2–0.5 oz cobalt salt/100 lb of salt—or feed a suitable trace mineral supplement. Injection of cobalt salts is not satisfactory, since ruminal action is needed to form vitamin B-12, the active form of cobalt.

Control:

Provide adequate cobalt in the ration—about 0.1 ppm. Deficiency symptoms appear when the level drops to the range of 0.04 to 0.07 ppm or lower.

Prevention:

Mix 0.2–0.5 oz of cobalt chloride, cobalt sulfate, or cobalt carbonate/100 lb of either (1) salt, or (2) whatever mineral mix is being used.

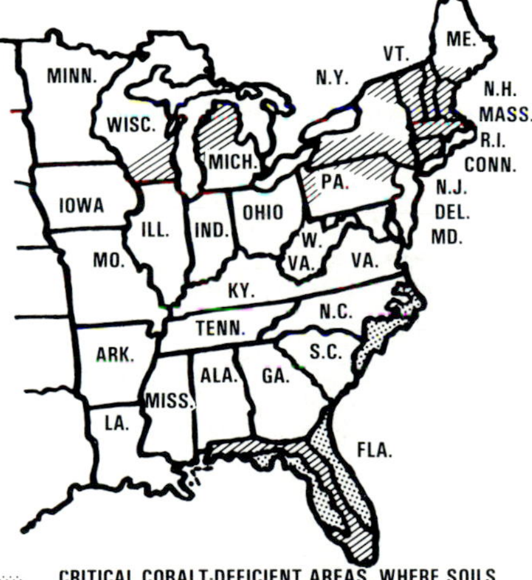

Fig. 5-30. Cobalt-deficient areas in eastern U.S., resulting from its deficiency in the soil and thus in the herbage produced thereon.

Remarks:

Cobalt is needed especially for rumen microbial synthesis of vitamin B-12. Nonruminants must be fed preformed vitamin B-12.
(Also see Chapter 4, Table 4-6, ANIMAL MINERAL CHART, COBALT.)

STIFF-LAMB DISEASE (WHITE MUSCLE DISEASE)

(See White Muscle Disease.)

(Continued)

TABLE 5-1 *(Continued)*

SWEET CLOVER DISEASE

Species Affected:

Cattle; rarely affects sheep or horses.

Cause:

Usually produced only by moldy or spoiled sweet clover hay or silage.
In moldy or spoiled sweet clover hay, the harmless natural coumarins are converted to dicoumarol, which interferes with vitamin K in blood clotting.

Symptoms and Signs (or age group most affected):

Loss of clotting power of the blood. As a result, blood forms soft swellings beneath the skin on different parts of the body. Serious or fatal bleeding may occur at time of dehorning, castration, parturition, or following injury.
All ages affected. A newborn animal may also have the condition at birth.

Fig. 5-31. Sweet clover disease in calf. Note the collection of blood at the point of the left shoulder. (Courtesy, College of Veterinary Medicine, University of Illinois, Champaign-Urbana)

Distribution and Losses Caused By:

Wherever sweet clover is grown and cured for hay.

Treatment:

Remove the offending materials and administer menadione (vitamin K_3).
The veterinarian usually gives the affected animal an injection of plasma or whole blood from a normal animal that was not fed on the same feed.

Control:

When a case of sweet clover disease is observed in the herd, either (1) discontinue feeding the damaged product, or (2) alternate it with a better-quality hay—especially alfalfa.

Prevention:

Properly cure any sweet clover hay or ensilage.
Cultivars of sweet clover that are low in coumarin content, and hence safe to feed, have been developed.

Remarks:

The disease has also been produced from feeding moldy lespedeza hay and from sweet clover pasture.
(Also see Chapter 4, Table 4-8, ANIMAL VITAMIN CHART, VITAMIN K.)

URINARY CALCULI (GRAVEL, STONES, WATER BELLY, UROLITHIASIS)

Species Affected:

Cattle, sheep, goats, horses, mink, and humans.

Cause:

The precipitation of various salts, usually inorganic, in the urine, frequently associated with rations high in cereal grains or grazing on the silica-rich soils of the northwest plains of Canada and the U.S. However, not all causative factors are known.
Experiments and experiences have shown a higher incidence of urinary calculi when there is (1) a high potassium intake, (2) a high phosphorus-low calcium ratio (from the standpoint of preventing urinary calculi, the Ca:P ratio should be about 2:1), (3) a high-silica content in the ration, or a high proportion of high-silica grains and forages, such as native grasses, wheat straw, sugar beet leaves or pulp, sorghums, and cottonseed meal. High dosages of diethylstilbestrol or a deficiency of vitamin A may be contributing factors.

Fig. 5-32. Lamb suffering from urinary calculi. (Courtesy, Washington State University, Pullman)

Symptoms and Signs (or age group most affected):

Frequent attempts to urinate, dribbling or stoppage of the urine, pain and renal colic.

(Continued)

TABLE 5–1 *(Continued)*

URINARY CALCULI (GRAVEL, STONES, WATER BELLY, UROLITHIASIS) (Continued)

Usually only males affected; females are able to pass the concretions.
Bladder may rupture, with death following. Otherwise, uremic poisoning may set in.
Urinary calculi is one of the most important diseases in feedlot cattle and sheep, particularly in steers and wethers on full feed.

Distribution and Losses Caused By:

Worldwide. The economic loss may be considerable, since calculi formation frequently comes near the end of the feeding period.
Affected animals seldom recover completely.

Treatment:

(1) Add ammonium chloride at the rate of 1 oz (lambs) or 1¼–1½ oz (cattle) per head daily—or 50–60% more ammonium sulfate; (2) increase the phosphorus content of the ration (pasture) so that it equals the calcium content (by adding monosodium phosphate); (3) increase salt content of ration to 3 to 4% so as to increase water consumption (too much salt may lower feed intake); (4) incorporate 20% alfalfa in the ration; (5) administer muscle relaxants to help the passage of calculi from the bladder; or (6) surgically remove the calculi.
In cattle, surgical removal of the calculi is the most effective treatment, with the stone(s) removed at the point of blockage. In steers, the urethra may be bisected and brought to the outside of the body to bypass the constricted portion of the tract. After a short time to eliminate any tissue residue of urine, such animals are marketable.
In sheep, amputation of the urethral process is simple and allows the immediate passage of urine.
In horses, bladder calculi must be removed surgically.
Careful observation of susceptible animals by an experienced person several times daily will allow early detection and more successful treatment.

Control:

If severe outbreaks of urinary calculi occur in finishing steers or lambs, it is usually well to dispose of them if they are carrying acceptable finish.
Increase water consumption by including 3 to 4% sodium chloride (salt) in the ration.
The addition of a broad-spectrum antibiotic to the ration has been useful in controlling urinary calculi in some cases.

Prevention:

The basis of prevention is identification of the chemical composition of the calculi so that appropriate steps can be taken to reduce the concentration of the particular chemical in the urine by increasing urine volume, eliminating infection, changing urine pH, or altering the metabolism with drugs.
Good feed and management appear to lessen the incidence.
Delayed castration (castration of bull calves at 4–5 mo. of age) and high-salt rations for feedlot cattle (1–3% salt in the grain ration, using the upper limits in the winter months) in order to induce more water consumption are effective preventive measures.
Avoid (1) a high potassium or phosphorus intake, (2) an incorrect Ca:P ratio, or (3) an excessive amount of beet pulp or grain sorghum in the ration.

Remarks:

Calculi are stonelike concretions in the urinary tract which almost always originate in the kidneys. These stones block the passage of urine, resulting in the condition commonly referred to as *water belly.*
The mineral deposits may be of variable sizes, shapes, and composition. In cattle, the phosphatic type predominates under feedlot conditions and the silicate type occurs most frequently in range cattle.
According to researchers at Canada's Lethbridge, Alberta, research station, the incidence of calculi in calves grazing native pasture is 10 times higher than in calves grazing Russian wild ryegrass.

WEAK CALF SYNDROME

Species Affected:

Cattle.

Cause:

Deficiencies of protein and other essential nutrients and/or weather stress.

Symptoms and Signs (or age group most affected):

Calves show severe depression, general weakness, arched back, red crusted muzzle, diarrhea, and inability to stand and nurse. Calves from first calf heifers are more frequently affected than calves from older cows.

Distribution and Losses Caused By:

Weak calf syndrome was first described as a distinct entity in the Bitterroot Valley of Montana in 1963. It is common to northwestern U.S.
Weak calf syndrome has resulted in calf mortality as high as 48% in some Pacific Northwest herds. In addition to calf losses, another economic loss is in the reproductive failure of cows.

(Continued)

TABLE 5-1 *(Continued)*

WEAK CALF SYNDROME (Continued)

Treatment:

Improve the nutrition and minimize weather stress.

Control:

Control consists in providing an adequate ration and shelter where necessary.

Prevention:

The following preventive measures will minimize weak calf syndrome:
1. Feed and manage pregnant females so that they are gaining weight prior to calving.
2. Separate first-calf heifers and older and thinner cows from the herd and feed and manage them separately.
3. Feed adequate energy, protein, minerals, and vitamins.
4. Control internal and external parasites.

Remarks:

Weak Calf Syndrome actually results from the nutritional deficiency of the pregnant cow, along with weather stress.

WHITE MUSCLE DISEASE (MUSCULAR DYSTROPHY; in sheep, STIFF-LAMB DISEASE)

Species Affected:

Calves, lambs, and foals. In lambs, it is commonly referred to as stiff-lamb disease.

Cause:

Selenium deficiency, due to the continuous consumption of a ration containing less than 0.02 ppm selenium.

Symptoms and Signs (or age group most affected):

In calves, white muscle disease is characterized by lameness or inability to stand, and heart failure. It most commonly affects calves 2 to 4 months of age.

In lambs, the symptoms and signs are: A stiff, stilted way of moving, chiefly in the hind legs, although the front legs and shoulders may be involved. The back is usually humped or "roached." Lambs that live are usually stunted. Young, rapidly growing lambs are especially susceptible.

It seems that more calves than lambs or foals develop heart damage, which may be fatal, especially if subjected to unusual exercise. Affected calves and lambs show similar pathological lesions—whitish areas or streaks in the heart and other muscles.

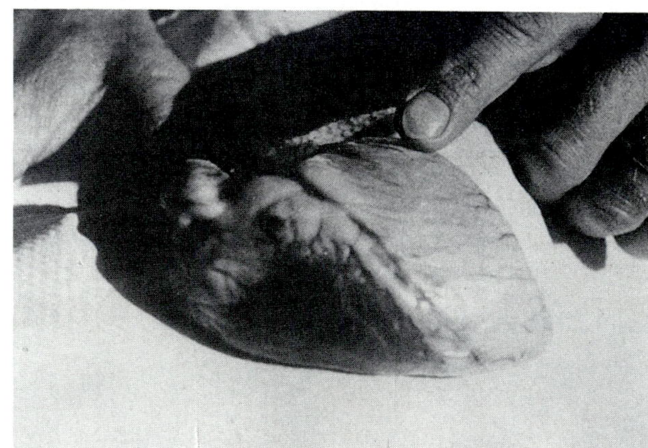

Fig. 5-33. White muscle disease in a calf. *Left:* Shows the generalized weakness of muscles, lameness, and difficulty in locomotion of an afflicted calf. Calf is about 3 months old. *Right:* Shows abnormal white areas in the heart muscles of a 6-week-old calf afflicted with white muscle disease. (Courtesy, Oregon State University, Corvallis)

Distribution and Losses Caused By:

Geographically, white muscle disease has been reported in Australia, Canada, Finland, Italy, Japan, New Zealand, Norway, Scotland, South Africa, Sweden, U.S., U.S.S.R., and Yugoslavia.

In the U.S., the disease is widely distributed, but the severity is greatest on the two coasts. Fig. 5-57 shows low, variable and adequate or high selenium areas in the U.S.

Economic losses result from the deaths of severely affected calves and lambs, the unthriftiness of survivors, and the cost of preventive programs.

Death losses range up to 50%, with an average of 15%; and the mortality of untreated animals may reach 80%.

(Continued)

TABLE 5-1 *(Continued)*

WHITE MUSCLE DISEASE (MUSCULAR DYSTROPHY; in sheep, STIFF-LAMB DISEASE) (Continued)
Treatment:
Affected animals should receive early treatment—the intramuscular injection of sodium selenite/vitamin E in aqueous solution at the rate of 0.25 mg Se per pound of body weight. This may be repeated in 2 weeks, but should not exceed 4 doses. **NOTE WELL**: Federal law restricts injectable Se to the order of a licensed veterinarian. Do not use within 30 days of slaughter.
Control:
Control consists in meeting the selenium requirements by (1) supplementing the ration of cows or ewes during the last ⅓ of pregnancy and the first part of lactation with selenium in the form of sodium selenite at the rate of 0.3 ppm dry matter; or (2) injecting intramuscularly each cow or ewe 1 month before parturition with approved levels of selenium/vitamin E preparation.
Prevention:
Add selenium to the ration. In 1987, FDA approved the addition of 0.3 ppm selenium to the complete feed of cattle, sheep, swine, and poultry. **A**lso, mineral mixes containing selenium are available for free-choice feeding in areas of known selenium deficiency.
Remarks:
White muscle disease is most common (1) in rapidly growing calves and lambs, and (2) in calves and lambs on lush pastures on selenium-deficient soils. **B**ecause of muscle failure, severely affected young animals die from starvation or heart failure. **Wh**ite muscle disease cannot be produced in calves on vitamin E-free rations unless the rations are high in unsaturated fats. (**A**lso see Chapter 4, Table 4-6, ANIMAL MINERAL CHART, SELENIUM.)

NUTRITION AND DISEASE/PARASITE INTERACTION

In general, well-nourished animals are more resistant to bacterial, viral, and parasitic diseases than those poorly nourished. This is attributed to better body tissue integrity, more antibody production, more immunity to disease, greater detoxifying ability, increased blood regeneration, and other factors. Also, it is recognized that proper nutrition, which heads the list of what consititutes *good nursing,* is essential for fast recovery from diseases and the ravages of parasites.

It is estimated that, annually, diseases and parasites in the United States (1) decrease animal productivity by 15 to 20%, and (2) make for losses aggregating $10 billion; and that nutrition has some involvement in 85% of the veterinary cases. In the developing countries, diseases and parasites take an even greater toll—they decrease animal productivity by 30 to 40%.

A nutrition program can be fully effective only if the animals are healthy. The converse is equally true. Also, the higher the productivity level, the higher the nutritional and health requirements.

Some noteworthy interactions of nutrition and diseases/parasites follow:

1. Some nutrients (for example, vitamin A) are important in keeping the epithelial tissue (the skin and mucous membranes), the body's first line of defense, in a healthy condition.

2. Protein, certain B-complex vitamins, and some trace minerals and other nutrients are essential for the production of antibodies and phagocytes, which serve as defenders against infectious agents that enter the body through one of the openings or through the skin.

3. Adequate nutrition is essential for an animal to respond properly to a vaccination. Although many factors affect the effectiveness of a vaccination, studies are showing that the good nutritional health of the animal is one of the most important requisites for a vaccination to induce an immune response.

4. Certain diseases increase the need for various nutrients by animals. This may be due to reduced appetite resulting in inadequate nutrient intake, vomiting or diarrhea resulting in a loss of nutrients from the intestinal tract, fever, decreased absorption or utilization of nutrients, or other causes.

5. Certain minerals, vitamins, and proteins (amino acids) are required at higher than normal levels in order to produce maximum immune response and resistance to diseases. Thus, the decline of immunoglobulins in the milk, followed by early weaning of calves, pigs, and lambs necessitate superior nutrition in order to avoid outbreaks of diarrhea (scours).

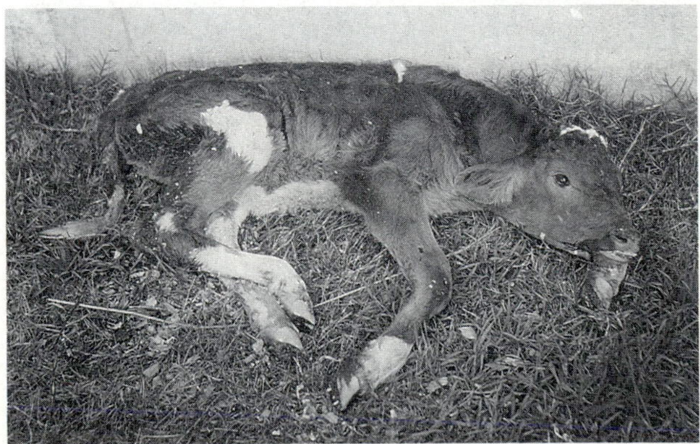

Fig. 5-34. Calf with severe scours. Scours may be (1) nutritional/management related (caused by overfeeding, irregular feeding, use of unclean utensils, too rapid changes in feed, or exposure to drafts and cold, damp floors), or (2) infections; and it is sometimes difficult to distinguish between the two causes. (Courtesy, North Carolina State University, Raleigh)

6. Many extra nutrients are needed for the repair and restoration of tissues, red blood cells, vital organs, and other parts of the body destroyed by diseases and parasites.

7. Diseases and parasites that cause diarrhea (scours) or vomiting decrease intestinal absorption of nutrients and cause electrolyte loss and dehydration; hence, extra nutrients and electrolytes may be beneficial in treating these conditions.

8. Diseases and parasites that reduce appetite and decrease total feed intake increase the need for higher levels of the nutrients in the feed consumed as a means of ensuring that the total daily nutrient needs will be met.

9. Parasites that cause severe damage to the digestive tract result in impairment of absorption of a number of essential nutrients. Thus, the protozoan parasites, particularly the coccidia, produce profound effects on the digestive physiology of animals.

10. Ketosis increases the need for niacin by dairy cows.

11. Certain nutrients are required at higher than normal levels during stress produced by such things as uncomfortable environmental conditions, vaccinations, crowding, loud noises, debeaking birds, and castrating and dehorning animals. Thus, it is noteworthy that shipping fever (bovine respiratory disease complex), which causes estimated losses of $500 million annually, is most frequently associated with animals in which resistance has been lowered due to change in weather and feed, overcrowding, hard driving, lack of rest, and improper shelter that accompany shipping.

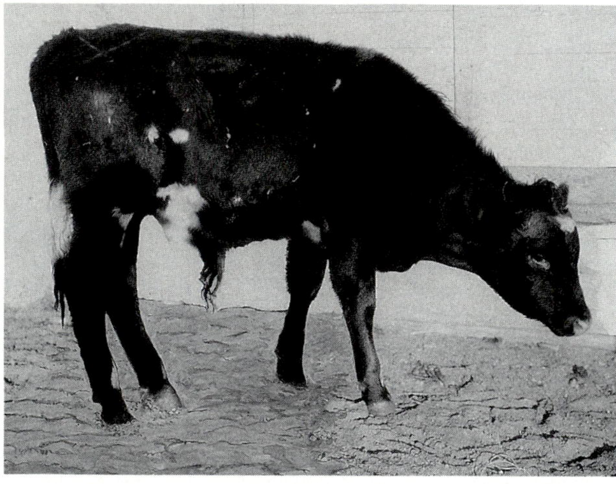

Fig. 5-35. Calf with shipping fever (bovine respiratory disease complex), most commonly associated with animals whose resistance has been lowered due to the stresses of travel. (Courtesy, USDA)

12. Well-fed animals have fewer parasites. For example, there is experimental evidence, substantiated by practical observation, that milk and its by-products are helpful in holding in check some of the internal parasites of swine. Milk-fed pigs make more rapid gains and have fewer parasites in the digestive tract than pigs not receiving milk.

13. Infectious diseases and parasites increase feed costs and lower feed efficiency.

The above list clearly shows that good nutrition is the first requisite of a disease and parasite prevention and control program. It also points up (1) that further intensive studies on these interrelationships hold great potential for discoveries of new ways to improve resistance to diseases and parasites, and (2) that there is a continuing need for cooperative effort between nutritionists and veterinarians.

NUTRITION RELATED DISORDERS

Not all noncontagious diseases are nutritionally responsive. Some of them are due to physical factors. Others are caused by faulty management. Regardless of the nature or the cause, however, all of them have an impact on the nutritional well-being of the affected animal; hence, they are nutrition related. Several of the most important of these disorders are discussed in the sections that follow.

Choking

Occasionally, feeds become lodged in the esophagus of cattle or horses, causing them to choke.

Cattle may choke on such feeds as beets, potatoes, apples, hay cubes, or ears of corn. Afflicted animals drool saliva from the mouth, make frequent attempts at swallowing, bloat rapidly due to closure of the outlet for gas from the stomach, and switch the tail. If the obstruction is in the region of the neck, it can be felt from the outside. Treatment consists in an attempt to work the object into the mouth, rather than force it into the stomach. A speculum should be applied to hold the mouth open, then the hand should be inserted into the animal's throat to grasp and remove the obstacle. (*CAUTION:* Beware of the hazard of a bitten hand or arm.) If this fails, a stomach tube or rubber hose lubricated with water or oil and pushed down the esophagus usually will free the object; but care must be taken not to damage the lining of the esophagus. Should there be marked bloating, the stomach may have to be punctured.

Horses choke most frequently from bolting (eating too rapidly) their grain, although they may choke on hay cubes, ears of corn, potatoes, and apples. Afflicted animals become excited, squeal, and thrust the head forward. Treatment consists in controlling the pain with sedatives, confining the animal, and allowing access to water but not feed. Passage of a stomach tube to the obstruction and repeated pumping and siphoning may relieve grain choke. As a last resort, the obstruction may be gently pushed into the stomach with a large stomach tube or rubber hose.

Minimizing choking in both cattle and horses consists in avoiding, to the extent practical, feeds that are most likely to cause choking. Such feeds as potatoes, apples, and roots are less apt to cause trouble if they are sliced or chopped. When feeding potatoes to cattle, choking can be materially lessened by forcing the animals to eat them from the ground level with their heads down. This can be accomplished by having a cable or pole arrangement at the top of their necks to keep their heads down. Choking can be lessened in gluttonous horses (horses that eat their feed too rapidly—that bolt their feed) by putting into the grain box several smooth stones about the size of a baseball, thereby slowing their eating.

Displaced Abomasum

This disorder is being diagnosed with increasing frequency in dairy cows.

Normally, the abomasum is located on the right side of the rumen of the cow and rests on the floor of the abdomen. It is attached at the front end to the omasum and at the back end to the small intestine. The mid-portion of the abomasum is not held rigidly in place by other structures; however, the weight of the material within it usually keeps it in place on the floor of the abdomen. Sometimes when the metabolism of the stomach is abnormal, the abomasum slides under the rumen, becomes partially filled with gas, and rises on the left side between the rumen and the body wall. When this happens, the afflicted cow goes off feed and exhibits many symptoms similar to ketosis. In fact, secondary ketosis may result from displaced abomasum because of decreased feed intake, thereby inducing a ketotic state. However, an experienced veterinarian can differentiate between displaced abomasum and primary ketosis.

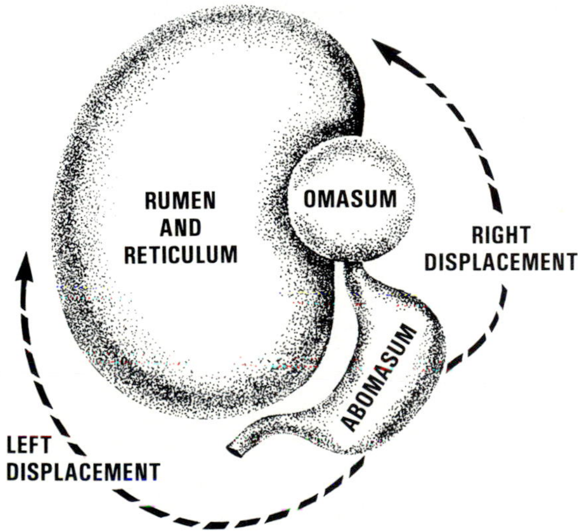

Fig. 5-36. Normally, the abomasum rests near the belly floor and to the right of the rumen. In most displaced abomasums, it shifts to the left.

Several treatments for displaced abomasum have been used with variable success; most commonly, the following two:

1. Rolling the cow onto her back and getting the abomasum back into normal position. Frequently, the problem recurs following this treatment.

2. Performing rather simple surgery, involving the veterinarian suturing the bottom of the rumen wall to the body wall, thereby preventing the abomasum from slipping between the two. When properly done, this usually provides a permanent cure. The operation can be done using local anesthesia. Frequently, cows so treated do not miss any milkings.

The cause of displaced abomasum is not known; however, the following theories have been advanced:

1. High-concentrate rations and inadequate bulk, as a result of which the rumen pulls away from the body wall.

2. Reduced ruminal contents for various reasons, such as cows going off feed, resulting in a gap between the rumen and the body floor.

When displaced abomasum is a problem, the incorporation of more roughage in the ration is recommended. A rule of thumb is that at least half of the dry matter should be in the form of roughage-type feeds, such as hay, silage, or pasture.

Hardware Disease

The term *hardware disease* (traumatic gastritis) is used to describe the condition that results from swallowing foreign materials, usually metal (nails, wire, screws, and pins). Cattle are involved more than other classes of animals; however, cases have been reported in horses and goats. In most cases, the metal is found in the reticulum (second stomach). However, the material may puncture the reticulum lining and pass into the body cavity. Some objects may pierce the diaphragm and enter the thoracic cavity, thence work their way to the heart or lungs, where they may cause serious damage or death.

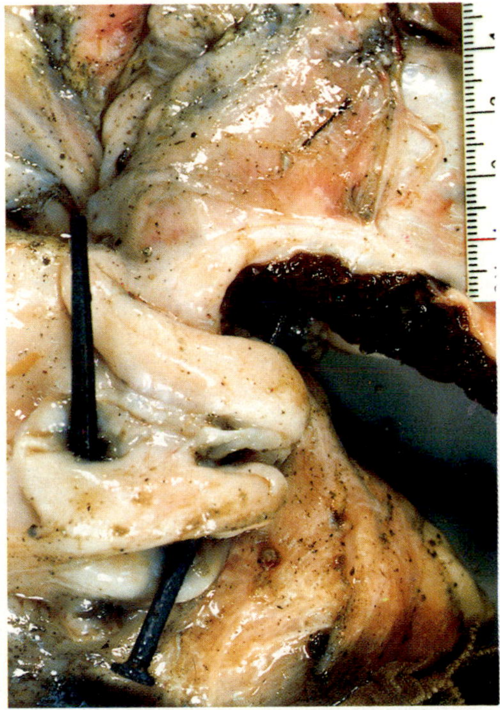

Fig. 5-37. Hardware disease. Note the nail puncturing the reticulum. (Courtesy, College of Veterinary Medicine, University of Tennessee, Knoxville)

Nearly 7,000 cattle are condemned each year by the Federal Meat Inspection Service as unfit for food because of hardware disease. Clinical reports indicate that the problem is increasing due to the use of more chopped feeds and more contamination. Sharp objects will injure the lining of the stomach and cause infection and inflammation, a condition known as traumatic gastritis.

Hardware disease is a problem in cattle because of their eating habits and stomach arrangement. The usual source of metals is the feed. The animals eat rapidly and are not able to sort foreign objects from their feed.

The most common symptoms of hardware disease are loss of appetite and digestive disturbance; slow and stiff movement and arched back; elbows that bow outward; decreased rumen movement and chewing; tendency to stand with the front feet elevated so as to lessen the pressure of the viscera on the inflamed area; rise in body temperature; and swellings under the jaw, at the brisket, and at the hock joints. Bulls may be reluctant to mate.

Prevention consists in avoiding foreign objects getting into the feed through good management. Also, it is recommended that strong magnets be installed, in keeping with the manufacturer's directions, (1) at the outlets of mechanical silo unloaders, and (2) in feed processing equipment.

Magnets may also be permanently placed in the cow's second stomach, for the purpose of holding objects that have not penetrated the stomach wall. However, the only sure cure for traumatic gastritis is veterinary surgery. Surgery will be successful only if performed before the condition has progressed to the point that damage has been done to the heart or other organs.

Immunoglobulin Deficiency

Immunoglobulin, found in colostrum, is the means by which most newborn farm mammals acquire passive immunity against pathogenic diseases and infections.

Deficiencies in immunoglobulins may result (1) when the young fail to nurse properly following birth, (2) when the mother is exhausted or sick after giving birth, (3) when the mother's colostrum contains low levels of specific antibodies, (4) when the caretaker fails to give colostrum to newborn, and/or (5) when other feeds are consumed prior to colostrum. Also, immunoglobulin deficiencies may exist in milk from first-calf heifers and in milk from cows that haven't dried up or have had less than a 30-day dry period. Older cows have been exposed to a wider range of diseases than first-calf heifers and therefore produce more immunoglobulins against them. So, if livability problems are encountered, colostrum from older cows should be fed to calves from heifers or from cows that haven't dried up or have had too short a dry period.

The highest absorption of intact immunoglobulins directly into the lymphatic system (thoracic duct) occurs immediately after birth (within 15 to 30 minutes) and during the first 6 hours after birth, but by 24 hours most of the absorption capacity is lost. Hence, for maximum protection against infection, newborn mammals should be fed colostrum very early in life.

Where the immunoglobulin phenomenon exists, as in the calf, the intestinal villi of the newborn are able to absorb the globulins by pinocytosis (engulfing). This enables those species which do not normally obtain adequate immune protection through placental transfer to acquire instant immunity by ingesting colostrum high in immunoglobulins. Aside from this unique stituation, protein must be digested.

(Also see Colostrum.)

Impaction In Horses

Impaction is a form of colic caused by obstruction of the cecum or colon by fibrous feeds. It is usually caused by feeding horses large amounts of straw, cornstalks, or other coarse, high-fiber feeds, along with lack of water intake. Distention of the large colon with gas causes acute abdominal pain. The usual signs described by colic are seen.

Medical treatments, which should be administered by a veterinarian, meet with variable success. Mineral oil and magnesium sulfate are popular laxatives. Dioctyl calcium sulfosuccinate may be used to penetrate the impacted mass. Antiferments and oral antibiotics such as neomycin are helpful in preventing gas formation. Surgery should be the last resort.

Sand Colic/Impaction

Cattle on sandy soils, especially when consuming large amounts of crop residues or root crops, may ingest so much sand that pathological amounts collect in the rumen and the abomasum. By the time signs are diagnostic, the condition is usually beyond medical therapy or surgery.

Horses may consume too much sand when grazing following dust storms or as a result of pulling up plants when grazing on sandy pasture. This may cause acute or chronic enteritis, commonly called *sand colic.* The amount of sand in the dung is indicative of *sanding.* Affected horses are sluggish, dull, and listless; and they have a harsh, dry coat. They lose condition, become weak, and, finally, are unable to rise. Treatment consists in removing the animals from sandy grazing areas and providing high quality laxative feeds.

"Washy" Pastures

Early spring pastures are high in protein and often low in energy; usually, they are described as "washy," meaning that their consumption results in watery feces.

In any season, the sudden turning of animals to lush pastures will result in "looseness." Thus, they should not be turned to pasture too early, and the adjustment to lush pastures should be gradual, preferably augmented by the feeding of some dry forage, such as grass hay or straw.

Water Deprivation

Increasingly, animals are being fenced away from natural water sources and confined, with water provided by automatic, mechanical devices. As a result, water deprivation can, and does, occur for numerous reasons, including mechanical failure of waterers, frozen pipes, broken faucets, communication failures, irresponsible employees, and shock due to an electric short. Also, dehydration may become a critically important problem in animals with diarrhea, during heavy exercise, or when they are unable or unwilling to eat or drink.

Normally, the water content of animals is approximately two-thirds of the body weight. (See Chapter 2, Table 2-3, Body Composition of Animals.) However, it decreases with advancing age and fatness.

Dehydration, or loss of body water, occurs when fluid losses exceed intake. This can have serious consequences, because of the loss of essential body salts called electrolytes, the most important of which are sodium, chlorine, and potassium.

Animals differ markedly in their ability to conserve water and withstand water deprivation. In a hot environment, dehydration of about 12% is fatal to humans. Some desert animals, such as the camel, can tolerate a more severe dehydration than humans without suffering an explosive heat rise. Also, donkeys and mules are more tolerant of water deprivation than horses. Animals deprived of water refuse to consume feed in the early stages of dehydration and will not eat dry feed until after they have had a drink.

When dehydration exists, prompt provision of water is the treatment of choice, but large amounts of water should not be given at one time. *CAUTION:* Water dehydration, accompanied by concentration of sodium and other ions in the brain cells, may cause fatal cerebral edema if thirsty animals are suddenly given all the fresh water that they will drink.

If the dehydrated animal(s) is unable to drink, intravenous fluids must be administered. However, if the administered fluids are of an inappropriate composition or given too rapidly, little of the fluid will be retained. This points up the need for studies designed to develop a basic understanding of fluid and electrolyte loss and replacement in animals of various species, ages, and uses, when deprived of water under different environmental conditions.

POISONOUS PLANTS

Fig. 5-38. Lupine toxicity. Calf showing abnormalities due to ingestion by dam of lupines during 40th to 70th days of gestation. (Courtesy, The American Institute of Nutrition, Bethesda, Md.)

Poisonous plants have been known since time immemorial. Biblical literature alludes to the poisonous properties of certain plants, and history records that hemlock (a poison made from the plant from which it takes its name) was administered by the Greeks to Socrates and other state prisoners.

No section of the United States is entirely free of poisonous plants, for there are hundreds of them. But the heaviest livestock losses from them occur on the western ranges because (1) there has been less cultivation and destruction of poisonous plants in range areas, and (2) the frequent overgrazing on some of the western ranges has resulted in the elimination of some of the more nutritious and desirable plants, and these have been replaced by increased numbers of the less desirable and poisonous species. It is estimated that poisonous plants account for 8 to 10% of all range animal losses each year, and even more in some areas. It is further estimated that poisonous plants cause average annual losses of beef cattle, sheep, goats, and horses of more than $200 million in the 17 western states. Additionally, toxic plants contribute to indirect losses, such as reduced weight gains; reduced calving, lambing, kidding, and foaling percentages; chronic illness; birth defects; fencing; abandoned ranges; and supplemental feeding.

Many plants contain substances which, if consumed in amounts above safe levels, may be toxic. These substances may be found in the leaves, stems, fruits, seeds, roots, and/or tubers. The type of substance, stages of highest concentration, destruction by heat or other processing, type and age of animals consuming the plant, and the amount of storage or excretion of the toxic substance, may influence the severity of toxicity.

Common Poisonous Plants

The list of poisonous plants is so extensive that no attempt is made herein to describe them in detail. Nevertheless, both producers and veterinarians should have a working knowledge of the principal poisonous species in the area in which they operate. The common poisonous plants of the intermountain ranges to which cattle and/or sheep are susceptible at certain times of the grazing season are listed in Table 5-2.

TABLE 5-2
TYPES OF RANGE ANIMALS SUSCEPTIBLE TO POISONOUS PLANTS AT DEFINITE SEASONS

Poisonous to Cattle	Time of Year	Poisonous to Sheep	Time of Year	Poisonous to Cattle and Sheep	Time of Year
Low larkspur	Spring	Death camas	Spring	Broomweed	Spring and summer
Oak	Spring and fall	Greasewood	Fall	Chokecherry	Spring
Tall larkspur	Early summer and early fall	Horsebrush	Spring	Copperweed	Summer
Timber milk vetch	Spring and summer	Rubberweed	Summer	Desert parsley	Spring
Water hemlock	Spring	Sneezeweed	Summer	Halogeton	All year
				Loco	All year
				Lupine	Summer and fall
				Milkweed	Summer
				Veratrum	Summer

Diagnosing Plant Poisoning

The diagnosis of plant poisoning in animals is not an easy or precise procedure. Any case of sudden illness or death with no apparent cause is commonly considered to be a poisoning. This may not always be correct. When large numbers of animals are suddenly affected, however, a suspicion of poisoning is justified until it has been proven otherwise.

Symptoms or signs induced by eating poisonous plants may include sudden death; transitory illness; general body weakness; disturbance of the central nervous, vascular, and endocrine systems; photosensitization; frequent urination; diarrhea; bloating; chronic debilitation and death; embryonic death; fetal death; abortion; extensive liver necrosis and/or cirrhosis; edema and/or abdominal dropsy; tumor growths in tissues; congenital deformities; metabolic deficiencies; and physical injury.

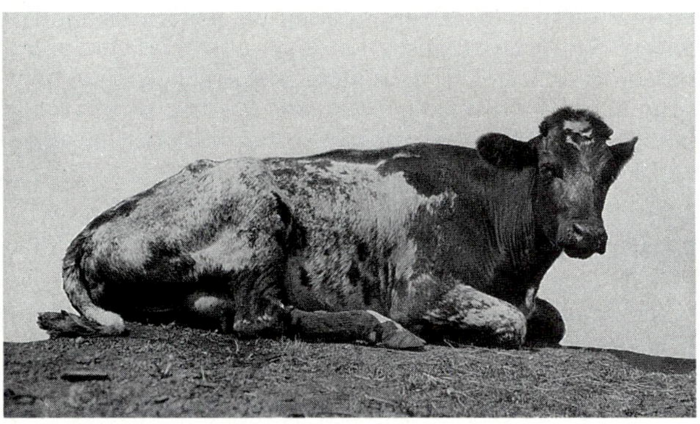

Fig. 5-39. Cow with white snakeroot poisoning. Marked weakness results in the "trembles" characteristic of this condition. (Courtesy, College of Veterinary Medicine, University of Illinois, Champaign-Urbana)

No general set of symptoms and signs *per se* provides all the information necessary to make an irrefutable diagnosis of plant poisoning. Nevertheless, a careful description of the toxic signs coupled with information pertaining to available plants provides a meaningful basis for a tentative diagnosis. Additional information essential to a poisonous plant diagnosis includes (1) type of feed, site grazed, and availability of water; (2) identification and relative abundance of all poisonous plants available to animals; (3) amount and stage of growth of the various poisonous plants being grazed; (4) the toxicity and palatability of the plants in relation to their stage of growth; (5) time from eating plants until onset of toxic signs; (6) species, age, and sex of animals affected; (7) clinical signs of toxic reactions; (8) chemical analysis of plants; and (9) a careful evaluation of all the information relative to the etiology of the disease.

Why Animals Eat Poisonous Plants

A frequently asked question is: Why do animals eat poisonous plants? The answer is not simple, but among the reasons are the following: (1) total lack of sufficient palatable forage—the animals are hungry; (2) decrease in palatability and nutrients of mature, weathered range grasses, with the result that poisonous plants become more appealing, comparatively speaking; (3) insufficient spring grass; (4) rain, melting snow, and heavy dew may enhance the palatability of some poisonous plants; and (5) going without water too long, which results in a reduction in feed intake, then, after watering, they develop a ravenous appetite and eat anything in sight—including less palatable poisonous plants. Poisonous plants vary in palatability—between species, and within species, at different stages of growth. For example, poison hemlock is never palatable and is eaten only as a last resort—when palatable forage is not available. Locoweed and black nightshade are eaten at any stage of growth or when mixed with hay. Others, such as lupines, horsebrush, and death camas may be eaten only at certain stages of growth. Still others, such as milk vetch, larkspur, and halogeton, are highly palatable to livestock at any and all times, with the result that if they're present animals will seek them out and there will be losses. Then, too, certain plants are poisonous to cattle but not to sheep (and vice versa), as shown in Table 5-2.

Treatment of Plant-Poisoned Animals

Unfortunately, plant-poisoned animals are not generally discovered in sufficient time to prevent loss. Thus, prevention is decidedly superior to treatment.

When trouble is encountered, the owner or caretaker should *promptly* call a veterinarian. In the meantime, the animal should be (1) placed where adequate care and treatment can be given, (2) protected from excessive heat and cold, and (3) allowed to eat only feeds known to be safe.

The veterinarian may determine the kind of poisonous plant involved (1) by observing the symptoms, and/or (2) by finding out exactly what poisonous plant was eaten through looking over the pasture and/or hay and identifying leaves or other plant parts found in the animal's digestive tract at the time of autopsy.

It is to be emphasized, however, that many poisoned animals that would have recovered had they been left undisturbed, have been killed by attempts to administer home remedies by well-meaning but untrained persons.

Preventing Losses from Poisonous Plants

With poisonous plants, the emphasis should be on prevention of losses rather than on treatment, no matter how successful the latter. The following are effective preventive measures:

1. **Follow good pasture or range management.** This is necessary in order to improve the quality of the pasture or range. Usually, plant poisoning is nature's sign of a "sick" pasture or range, resulting from misuse. When a sufficient supply of desirable forage is available, poisonous plants may not be eaten, for they are usually less palatable. On the other hand, when overgrazing reduces the available supply of the more palatable and safe vegetation, animals may, through sheer hunger, consume the toxic plants.

2. **Know the poisonous plants common to the area.** This can usually be accomplished through (a) studying drawings, photographs, and/or descriptions; (b) checking with local authorities; or (c) sending 2 or 3 fresh whole suspect plants

(if possible, include the roots, stems, leaves, flowers, and seeds) to the state agricultural college—first wrapping the plants in several thicknesses of newspaper.

By knowing the poisonous plants common to the area, it will be possible—

• To avoid areas heavily infested with poisonous plants which, due to concentration and overgrazing, usually include waterholes, salt grounds, bed grounds, and trails.

• To control and eradicate the poisonous plants effectively, by mechanical or chemical means (as recommended by local authorities) or by fencing off.

• To recognize more surely and readily the particular kind of plant poisoning when it strikes, for time is important.

• To know what first aid, if any, to apply, especially when death is imminent or where a veterinarian is not readily available.

• To graze with a class of livestock not harmed by the particular poisonous plant or plants. Many plants seriously poisonous to one kind of livestock are not poisonous to another, at least under practical conditions (see Table 5-2).

• To shift the grazing season to a time when the plant is not dangerous. That is, some plants are poisonous at certain seasons of the year, but comparatively harmless at other seasons (see Table 5-2).

• To avoid cutting poison-infested meadows for hay when it is known that the dried cured plant is poisonous. Some plants are poisonous in either green or dry form, whereas others are harmless when dry. When poisonous plants (or seeds) become mixed with hay (or grain), it is difficult for animals to separate the safe from the toxic material.

3. **Know the symptoms that generally indicate plant poisoning.** Such knowledge makes it possible to take early action.

4. **Avoid turning animals on pasture in very early spring.** Nature has ordained most poisonous plants as early growers —earlier than the desirable forage. For this reason, as well as from the standpoint of desirable pasture management, animals should not be turned to pasture in the early spring before the usual forage has become plentiful.

5. **Provide supplemental feed during droughts, after plants become mature, and after early frost.** Otherwise, hungry animals may eat poisonous plants in an effort to survive.

6. **Avoid turning out very hungry animals where there are poisonous plants.** This caution especially applies to animals that have been in corrals for branding or other reasons; that have been recently shipped or trailed long distances; or that have been wintered on dry forage. First feed the animals to satisfy their hunger or allow a fill on an area known to be free from poisonous plants.

7. **Avoid driving animals too fast when trailing.** On long drives, either allow them to graze along the way or stop frequently and provide supplemental feed.

8. **Remove promptly all animals from infested areas when plant poisoning strikes.** Hopefully, this will check further losses.

9. **Treat promptly, preferably by a veterinarian.** Rapid and proper treatment may save some animals.

AGRICULTURAL CHEMICALS AND DRUGS

In the everyday pursuit of modern agriculture, more and more chemicals and drugs are being used. Hand in hand with this development, there has been increased public concern over the use of these products, for fear of poisoning human food.

Chemicals and drugs must be used with discretion, especially those designed to kill some living organism. But sometimes choices must be made; for example, between malaria-carrying mosquitoes and some fish, or between hordes of locusts and grasshoppers and the crops that they devour. This merely underscores the need for (1) careful testing through properly designed experiments of all products prior to use, (2) conforming with Federal and state laws, and (3) accurate labeling and use of products. Additionally, food producers need to relate the miracle of agriculture. They need to tell that back of food and clothing are agricultural chemicals and drugs—herbicides, insecticides, pesticides, disease control materials, feed additives, fertilizers, and many others. They need to show how these products are as indispensable to modern food production as tractors, trucks, hybrid seeds, and improved livestock.

In an era of food shortages, losses of feed, food, and fiber will increasingly concern all people, producers and consumers alike. If the world is to "waste not, want not," it must either (1) use more agricultural chemicals and drugs in the future, not fewer; or (2) vastly improve biological control methods, such as the use of predatory insects, resistant plants, and sterilization of insects. Here are just a few wanton wastes awaiting the proper use of agricultural chemicals and drugs:

1. Pests cause an estimated 30 to 40% annual loss in the worldwide potential production of crops, livestock, and forests. Every part of our feed, fiber, and food supply is vulnerable to pest attack, including wild and domestic animals, marine life, field crops, horticultural crops, and wild plants. Obviously, if these losses could be prevented, or reduced, world food supplies could be increased by a potential 30 to 40%.

2. Weeds reduce yields and quality by competing with crops for nutrients, water, light, and space. Also, they may poison livestock, interfere with harvesting, and slow the flow of water for irrigation and drainage. In the United States, some 2,000 species of weeds and brush cost farmers and ranchers $8 billion annually, with reductions in quantity and quality of crops heading the list.

3. Insects devour growing crops, attack animals, lower yields and quality, and attack food products in storage and during transport. Also, insects harbor and transmit diseases to plants, animals, and people. In the United States alone, approximately 10,000 species of insects are destructive enough to be called "enemies"; and about 4/5 of them are injurious to animals and crops.

4. Each year, every rat consumes feed, damages additional feed, destroys property, and spreads disease, costing an estimated average of $13. Thus, the yearly keep on U.S. rats totals nearly $2 billion.

Indeed, through the proper use of agricultural chemicals—herbicides, insecticides, pesticides, rodenticides, and others —food and fiber losses could be reduced substantially—perhaps by as much as 30 to 40%. The net result would be an increase of 10 to 15% in the world food supply, with no new land required. In no other way can the hunger gap be filled so quickly and at so little cost.

Farmers have an obligation to produce more foods and fibers efficiently, to reduce production costs, and to increase their income. In turn, consumers—all—benefit from agricultural chemicals and drugs which produce more products at less cost.

When properly used, agricultural chemicals are an important adjunct to providing feed for animals and food for people. However, improper use can result in toxicoses. Moreover, certain chemicals can accumulate in the body fat of animals, and be found in the meat or secreted in milk.

The vast majority of agricultural chemicals and drugs have been properly used. Of course, it shouldn't be too surprising that a few have been improperly used when it is realized that there are approximately 300,000 trade name products on the market.

When chemical poisoning or drug misuse happens, it can be both devastating and perplexing. Usually, the causative agent can be diagnosed after an investigation of the environment and the feed. However, few poisons can be diagnosed with certainty by clinical signs alone. When trouble is encountered, the producer should promptly call a veterinarian if animals are involved, or a medical doctor if people are involved.

A voluminous amount of information is available on the deleterious effects of poisonous chemicals and drugs. Because of space limitations, only a few of the more important ones will be covered in Table 5-3.

Potential Poisons

A poison is a substance which in sufficient quantities and/or over a period of time kills or harms living things. Toxic substances are chemical substances that may present an unreasonable risk of injury to health or to the environment. Many poisons are called toxins. The study of poisons is called *toxicology.* The discussion that follows and Table 5-3 pertain primarily to feed-related poisons that may be eaten by animals. For most of these, there is both a safe level and a poisonous level; and the severity of the effect depends upon (1) the amount taken, (2) the period of time over which the substance is taken (certain poisons are cumulative), and (3) the age and physical condition of the animal. This lends credence to the toxicological adage: "Only the dose makes the poison."

There are more than 4 million chemical compounds, of which more than 60,000 are commercially produced; and about 1,000 new ones are introduced each year. Some of these make their way into feed and water. With the growth and use of chemicals, feed supplies are subject to contamination from or treatment with chemicals in the course of growing, fertilizing, harvesting, processing, and storing. In addition to manufactured chemical poisons, there are many naturally occurring poisonous substances.

Farmers know that unless they follow state and federal regulations they risk having their products condemned and seized, or refused by food processors. Nevertheless, the economics dictate that new products be used as soon as they prove useful and are approved. On the other hand, food faddists may feel that they are being poisoned; wildlife conservationists may be concerned over possible damage to songbirds and other animals; beekeepers become unhappy if insecticides kill honeybees; and public health agencies are concerned about contamination of soil, water, and food supplies. Thus, great care should be exercised in handling chemicals and drugs; the labels on the containers should be read and heeded carefully, and partly used packages and empty containers should not be left where animals have access to them.

When poisoning happens, it can be both devastating and perplexing. No part of veterinary diagnostics is as difficult and complex as toxicology. First, what compound is being tested for out of the many thousands known? Second, detecting trace levels, such as parts per billion, of pesticides and other chemicals in feed and water by low level residue analysis can be as difficult as understanding them.

Table 5-3 lists the most common potentially toxic substances, both synthesized and naturally occurring, and presents pertinent facts pertaining to each.

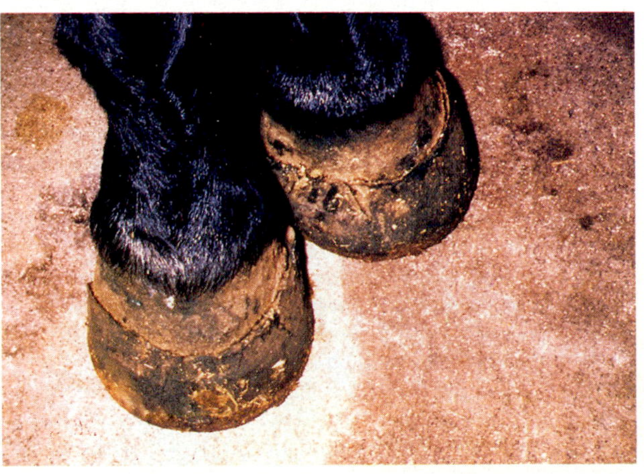

Fig. 5-40. For some substances, such as selenium and copper, a little is good but more may be poisonous. *Left:* Selenium toxicity resulting in severe hoof damage of a horse—note the horizontal cracks. (Courtesy, Colorado State University, Ft. Collins) *Right:* Copper toxicity evidenced by gunmetal-colored kidneys. (Courtesy, University of Tennessee, Knoxville)

TABLE 5-3
POTENTIAL POISONS

ACORN POISONING

It is caused by the tannin in oak buds, leaves, and acorns.

Source:

The oak buds, leaves, and acorns produced by the oak tree, *Quercus*. More than 60 species of oak have been identified in North America, and all should be considered potentially toxic. Cattle are poisoned in the spring by oak leaves and buds and in the fall by acorns. Young leaves and green acorns are more toxic than their mature counterparts. The tannin content diminishes as they mature. Most episodes of acorn poisoning occur soon after the acorns fall. Acorn shells are the major source of tannic acid.

Species Affected:

Cattle, although it occasionally occurs in horses, sheep, and swine.

Symptoms and Signs (or age group most affected):

Loss of weight, rough hair coat, loss of appetite, thin nasal discharge that may become red-tinged, excessive urination, and scouring. Calves that are severely affected will show signs of severe abdominal pain and appear bloated; often they will be found dead before any signs are noted.

Distribution and Losses Caused By:

Acorn poisoning may occur wherever oak trees grow, especially when there is a shortage of pasture.

Treatment:

Animals suspected of acorn poisoning should (1) be removed from pastures containing acorns, and (2) given a mild laxative, such as the magnesium hydroxide products, to purge the digestive tract and remove the digestive material as soon as possible. However, treatment is not very effective.

Prevention:

Prevention consists in not allowing cattle access to oak acorns or buds. Partial protection can be provided cows on oak-containing pastures by supplying supplemental feed, especially when there is a shortage of pasture.

Remarks:

Swine thrive and fatten on acorns, while cattle and calves often sicken and die from eating quantities of nuts.

ARSENIC (As) POISONING

Source:

Arsenic used to control insects and weeds, and to defoliate crops.
Overdosing of phenylarsonic compounds as feed additives to swine and poultry for growth and disease control.

Species Affected:

All farm animals.

Symptoms and Signs (or age group most affected):

The onset is sudden; characterized by groaning, restlessness, rapid breathing, muscular incoordination, blindness, and photosensitization. Death in 3–4 hours to a few weeks, depending on amount of arsenic consumed.

Necropsy reveals severe hemmorrhagic inflammation of the stomach and intestines, with perhaps areas of erosion on mucous membranes.

Distribution and Losses Caused By:

Arsenic has long been a leading cause of chemical poisoning.

Treatment:

Handled by the veterinarian. If caught in time, first remove the material from the animal. Sodium thiosulfate may be used, and supportive treatment may be indicated.

Fig. 5-41. Steer evidencing great pain from arsenic poisoning. This animal was accidentally poisoned by eating bran treated with arsenic for grasshopper bait. (Courtesy, College of Veterinary Medicine, University of Illinois, Champaign-Urbana)

(Continued)

TABLE 5-3 *(Continued)*

ARSENIC (As) POISONING (Continued)

British Anti-Lewisite (Dimercaprol) is a specific antidote for some forms of arsenic poisoning.

Prevention:

Keep animals away from arsenic.

Remarks:

Accumulation of arsenic in soils may sharply decrease crop growth and yields, but it is not hazardous to animals or humans that eat plants grown in these fields, provided they do not eat the foilage of the plants.

In spite of the recognized toxicity of many forms of arsenic, various arsenicals have been used in the practice of medicine. Also, animal feeds have been supplemented with growth-promoting organic arsenicals for many years. Another curious feature of arsenic biochemistry is the ability of the element partially to counteract the toxic effects of excess selenium.

AVOCADO POISONING

Source:

Avocado leaves, fruit, or bark. Even wilted leaves blown into a pasture during a storm are reported to cause poisoning.

Species Affected:

Horses, cattle, goats.

Some animals react to avocado poisoning, others do not.

Symptoms and Signs (or age group most affected):

Swelling around the mouth and head, which sometimes extends to the neck and chest; inflammation (swelling) of the mammary glands or udder; depression; loss of appetite; and perhaps colic. The swellings resemble snakebite.

Distribution and Losses Caused By:

The poison occurs in areas where avocados are grown.

Afflicted horses are uncomfortable, but they generally recover.

Treatment:

The veterinarian will probably use antibiotics and corticosteroids to control infection and inflammation.

Prevention:

Keep animals away from avocados.

Remarks:

Nothing can be done to restimulate late lactation if it stops.

BLACK WALNUT TOXICOSIS

It is thought to be due to the toxic compound *jugolene,* contained in black walnut trees.

Source:

Black walnut (*Juglans nigra*) shavings or sawdust used as bedding.

Species Affected:

Horses.

Symptoms and Signs (or age group most affected):

Founder (laminitis) occurs within 12 to 24 hours after horses have been exposed to black walnut shavings or bedding. Apparently, only skin contact is necessary; the material need not be eaten.

The temperature, pulse, and respiration are elevated; both front and hind legs may be affected, with edema from the knees and hocks downward; lameness may be severe or non-existent; some animals show sensitivity to hoof testers, especially over the toes; and the more severely affected horses may have to be destroyed because of severe founder. The horse's age may also be a factor; foals may be unaffected while their dams develop founder.

(Continued)

TABLE 5-3 *(Continued)*

BLACK WALNUT TOXICOSIS (Continued)

Distribution and Losses Caused By:

Black walnut toxicosis has been reported when black walnut shavings and sawdust have been used as bedding for horses. One-third or more of the horses bedded on black walnut materials may be affected. Usually, the condition is diagnosed promptly and the black walnut bedding removed, with the result that few animals must be destroyed because of founder.

Treatment:

Remove *all* black walnut bedding. Call the veterinarian, who may administer medical treatment.

Prevention:

Do not use black walnut shavings or sawdust for bedding horses.

Remarks:

Exposure of black walnut shavings or sawdust to the air appears to diminish their toxic effect. Nevertheless, the use of this material for horse bedding is not recommended.

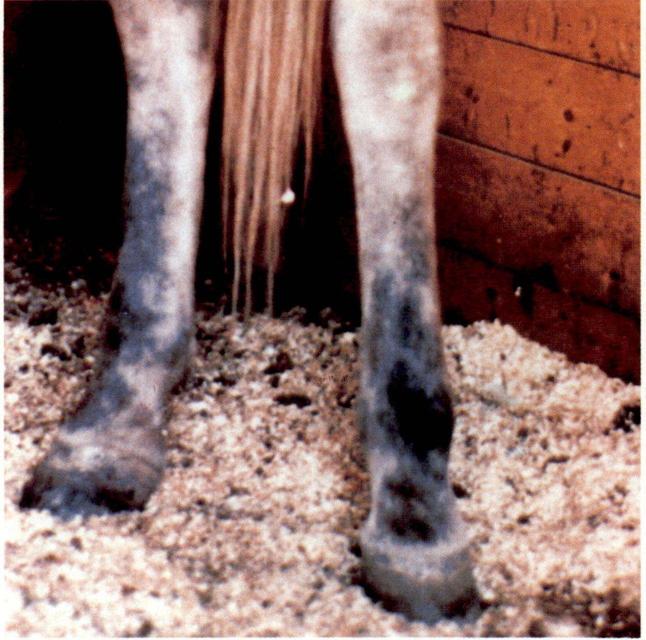

Fig. 5-42. Black walnut toxicosis, showing characteristic lower leg edema, caused by the use of black walnut bedding. (Courtesy, Colorado State University, Ft. Collins)

BLISTER BEETLE POISONING (CANTHARIDIN)

Source:

The three-striped blister beetles (*Epicauta* spp), which contain in their tissues the poisonous substance cantharidin. The poison is very irritating to the digestive tract and causes severe kidney damage.

Adult blister beetles feed on the leaves of alfalfa. This habit, along with the tendency of adults to swarm when feeding, results in the inclusion of beetles in bales of alfalfa. The problem of blister beetles in alfalfa hay has increased in recent years with the use of the hay conditioner or crimper, which prevents escape of the beetles at time of cutting. In the past, when hay was mowed and left to be raked and baled, the beetles would leave the hay before baling.

Fig. 5-43. Three-striped blister beetles, which are very toxic and found in southwestern U.S. (Courtesy, Colorado State University, Ft. Collins)

Species Affected:

Horses, although cattle, sheep, and goats may be affected.

Symptoms and Signs (or age group most affected):

Large doses may cause shock and death within a few hours. Smaller doses commonly produce the following symptoms and signs: abdominal pain, fever, severe depression, increase in respiratory and pulse rate, and muscle tremors. Animals demonstrate the symptoms of colic by lying down and rising with abdominal straining, discomfort, and sweating. Horses that survive more than 24 hours may exhibit signs of frequent urinary straining while voiding only small amounts of urine. The urine may be blood tinged and contain blood clots. Frequently, affected horses immerse their lips and muzzle in water without drinking, trying to wash away the burning chemical.

A positive diagnosis can be made by finding the beetles in alfalfa hay.

(Continued)

TABLE 5-3 *(Continued)*

BLISTER BEETLE POISONING (CANTHARIDIN) (Continued)

Distribution and Losses Caused By:

Blister beetle poisoning is most common in the South and Southwestern U.S.
The incidence of reported blister beetle poisoning is low.

Treatment:

No specific treatment is known. Supportive treatment includes the use of mineral oil or activated charcoal to protect the intestinal tract and prevent absorption of the cantharidin. Analgesics and steroids may be helpful in controlling pain and shock. Fluid therapy with balanced electrolyte solutions should also be instituted to combat dehydration and shock. Treatment of blister beetle toxicity can be successful if the diagnosis is made and the treatment is begun early in the course of the poisoning.

Prevention:

The following preventive measures are suggested:
1. Know your alfalfa supplier and inquire as to what precautions were taken to avoid blister beetles in the hay.
2. Inspect the hay (flake by flake) before feeding for the presence of blister beetles.
3. Call the veterinarian immediately if symptoms appear.

Remarks:

The blister beetle is considered a beneficial insect because its larva feed on grasshopper larva.

BOTULISM

Botulism is a poisoning caused by ingestion of feed containing *Clostridium botulinum,* which organism proliferates in decomposing animal matter and sometimes in plant material. The toxins formed from these bacteria are the most potent poisons known; botulism Type A—the most lethal—is 10,000 times as deadly as cobra venom and millions of times more potent than strychnine or cyanide.

Source:

Ingestion of toxin in feed. The usual source of this toxin is decaying carcasses or vegetable materials such as decaying grass, hay, grain, or spoiled silage. In horses, the most common predisposing cause is moldy silage, haylage, or hay.

Species Affected:

All animals are subject to botulism. But it occurs most frequently in poultry, horses, and mink; it is relatively infrequent in cattle; and it is uncommon in other species.

Symptoms and Signs (or age group most affected):

The signs of botulism are associated with the paralysis of muscles, including motor paralysis, disturbed vision, difficulty in chewing and swallowing, and generalized progessive weakness. Death is usually due to respiratory or cardiac paralysis. Type B *Clostridium botulism* appears to cause "staggers," or the "shaker foal syndrome," in young equines.

Signs usually appear 3 to 7 days after animals gain access to toxic material, although rare, mild cases caused by the ingestion of small amounts of toxin may recover.

Distribution and Losses Caused By:

Botulism occurs worldwide.
The incidence in animals is not known with accuracy. Losses are particularly high in waterfowl; an estimated 10,000 to 50,000 birds are lost in most years, with losses reaching 1,000,000 during bad outbreaks in western U.S. Most botulism in cattle occurs in South Africa, where phosphorus-deficient cattle chew bones (and flesh) that they find on the range. Botulism in sheep has been encountered in Australia, where protein-deficient sheep eat the carcasses of rabbits and other small animals that they find on the range. Usually, botulism in mink is caused from eating meat or fish containing the toxin.

Treatment:

Botulism antitoxin has been used for treatment with varying degrees of success, depending upon the type of toxin involved and the animal species. Treatment of ducks and mink is often successful. In cattle, however, the treatment is rarely used.

Prevention:

Prevention consists of correcting dietary deficiencies and the proper disposal of all carcasses. Immunization of cattle with toxoid has been successful in South Africa and in Australia. Also, effective toxoids are available for immunizing mink and foals.

Remarks:

Botulism was first described as food poisoning of humans in Germany in 1817.
The toxin blocks transmission of the neuromuscular junctions.

(Continued)

TABLE 5-3 *(Continued)*

COAL TAR POISONING (PITCH, CLAY PIGEON POISONING, CRESOLS, CRUDE CREOSOTE)

Source:

Expended clay pigeons; roofing material, certain types of tar paper, and plumbers' pitch.

Species Affected:

Coal tar poisoning is confined almost entirely to hogs.

Fifteen grams of clay pigeons consumed over a 5-day period will kill swine.

Symptoms and Signs (or age group most affected):

An acute, highly fatal disease (the first sign may be several dead pigs), characterized clinically by depression, and pathologically by striking liver lesions.

Anemia and jaundice.

Distribution and Losses Caused By:

Wherever there is access to coal tar.

Fig. 5-44. Liver from a pig afflicted with coal tar posioning. Note the striking liver lesions, characterized by red and yellow mottling and deep red dots about the size of a pinhead. (Courtesy, College of Veterinary Medicine, University of Tennessee, Knoxville)

Treatment:

No known treatment.

Prevention:

Do not allow animals access to pitch-containing or coal tar-containing products.

Remarks:

Pastures containing clay pigeons are dangerous for years; deaths have been reported 35 years after area was used for trap-shooting.

COPPER (Cu) POISONING

Source:

The (1) ingestion of large amounts of copper salts at one time, or (2) the continued ingestion of small amounts of copper over a long period.

Acute copper toxicity is seen when animals (1) accidentally ingest large amounts of copper such as copper sulfate from a foot bath, or (2) graze in orchards that have been sprayed with copper-containing chemicals. Toxicity has also occurred when sheep have access to cattle or swine mineral mixtures containing high levels of copper, or when sheep drink water from farm ponds treated with copper sulfate to control algae. Poisoning may result from the following intakes of copper per lb of body weight: Sheep and young calves, 10 to 50 mg; mature cattle, 100 to 400 mg. Excess copper accumulates in the liver.

Chronic copper poisoning occurs in sheep and cattle with daily intakes of 1.5 mg of copper per lb body weight. Cases of such toxicity have been reported from ingesting forages (such as pastures contaminated by smelter fumes), pellets, and mineral mixes containing high levels of copper; and from the feeding of seed grain which has been treated with antifungal agents containing copper.

Copper is sometimes used as a feed additive for swine at concentrations of 125 to 250 ppm; but levels greater than 250 ppm are hazardous.

The amount of copper required to cause copper poisoning varies with the concurrent intake of molybdenum, sulfur, and possibly zinc and iron.

Species Affected:

Copper poisoning commonly occurs only in sheep and cattle, with sheep being more susceptible than cattle. Monogastric animals are more resistant to copper toxicity than ruminants.

(Continued)

TABLE 5-3 *(Continued)*

COPPER (Cu) POISONING (Continued)

Symptoms and Signs (or age group most affected):

Most animals that develop signs of copper poisoning die within 1 to 2 days. Poisoning is characterized by severe gastroenteritis, marked by abdominal pain, nausea, salivation, and diarrhea. Monogastric animals may vomit. Feces and vomitus are bluish green. Chronic copper poisoning is manifested by loss of appetite, thirst, jaundice, dark brown urine, and marked depression.

Fig. 5-45. Copper toxicity signs. Left to right: Kidney, liver, and urine. Note the gunmetal-colored kidneys and liver, and the port-wine-colored urine. (Courtesy, College of Veterinary Medicine, University of Tennessee, Knoxville)

Distribution and Losses Caused By:

Sporadic outbreaks of copper poisoning occur in many circumstances. In both acute and chronic cases, the mortality rate approximates 100%.

Treatment:

In acute cases, gastrointestinal sedatives and symptomatic treatment for shock are recommended. Calcium versenate and penicillamine may be useful if administered in the early stages of the disease. Daily oral treatment with 100 g of ammonium molybdate and 1 g anyhdrous sulfate appears to prevent death of lambs known to have ingested toxic amounts of copper. But, generally speaking, treatment for copper posioning is not very effective.

Prevention:

When chronic copper poisoning appears probable, providing additional molybdenum in the diet will be effective as a preventive. Molybdenum exerts a protective effect by inhibiting copper absorption from the gut. As a standard preventive practice, the ration of sheep should not contain over 10 ppm of copper.

Remarks:

A particular level of dietary copper can induce either symptoms of deficiency or toxicity depending on the concurrent intake of molybdenum, sulfur, and possibly other elements.

ERGOT POISONING (ERGOTISM)

(caused by the parasitic fungus *Claviceps purpurea*)

Source:

The parasitic fungus replaces the seed in the heads of grasses and cereal grains, in which it appears as a purplish-black, hard banana-shaped dense mass from ¼ to ¾ in. long.

Most common in rye, wheat, wild rye, bromegrass, and dallisgrass.

Species Affected:

Cattle, sheep, swine, horses, and humans.

Symptoms and Signs (or age group most affected):

Acute ergot poisoning, caused by large quantities eaten at one time, may produce paralysis of the limbs and tongue, disturbance of the gastrointestinal tract, and abortion.

It is a cumulative poison; hence, poisoning may develop from lesser quantities eaten over a long period of time.

Chronic poisoning produces gangrene of the extremities, with subsequent sloughing off of hoofs, ears, and tail.

Delirium, spasms, and paralysis may occur before death.

Distribution and Losses Caused By:

Ergot is found throughout the world. However, seldom is sufficient of it ingested to cause poisoning.

(Continued)

TABLE 5-3 *(Continued)*

ERGOT POISONING (ERGOTISM) (Continued)

Treatment:

If noticed in time, stricken animals may recover if taken off the affected feed.

Tannin used as a drench is an antidote, and sedations such as chloral hydrate, may be given to nervous animals.

Prevention:

Never feed heavily ergot-infested hay or grain.

Remarks:

Poultry are more tolerant of ergot than other animals.

Grain containing 0.05% ergot will reduce gain and feed efficiency of finishing cattle.

Six different alkaloids are involved in ergot poisoning.

Grain containing even small amounts of ergot should not be fed to pregnant and lactating sows.

Fig. 5–46. A mixture of ergot sclerotia and wheat kernels as they appeared after harvest. (Courtesy, University of Idaho, Moscow)

FLUORINE (F) POISONING (FLUOROSIS)

Source:

Ingesting excessive quantities of fluorine through either the feed, water, or air; or some combination of these.

Toxic quantities of fluorides occur naturally in certain raw rock phosphates and in the superphosphates produced from them, and in partially defluorinated phosphates and in phosphatic limestones.

Species Affected:

All farm animals, poultry, fish, and humans.

Symptoms and Signs (or age group most affected):

Abnormal teeth (especially mottled enamel) and bones, roughened hair coat, stiffness of joints, loss of appetite, emaciation, reduction in milk flow, diarrhea, delayed maturity, and salt hunger (see tabulated symptoms below).

In most practical feeding situations, problems with fluorine have been associated with cattle. This is largely due to the fact that cattle have lower tolerances to fluorine than other classes of livestock.

SYMPTOMS OF FLUORINE TOXICITY IN CATTLE[1]

Symptom	Total Fluorine in Ration (ppm)			
	20–30	30–40	40–50	>50
Mottling of teeth[2]	Yes	Yes	Yes	Yes
Decreased growth of enamel[2]	No	No	Yes	Yes
Lameness	No	No	No	Yes
Decreased milk production	No	No	No	Yes

[1]Adapted by the authors from *Effects of Fluorides in Animals,* National Academy of Sciences, Washington, D.C.

[2]Occurs only when fluoride is present during the formative period of the tooth.

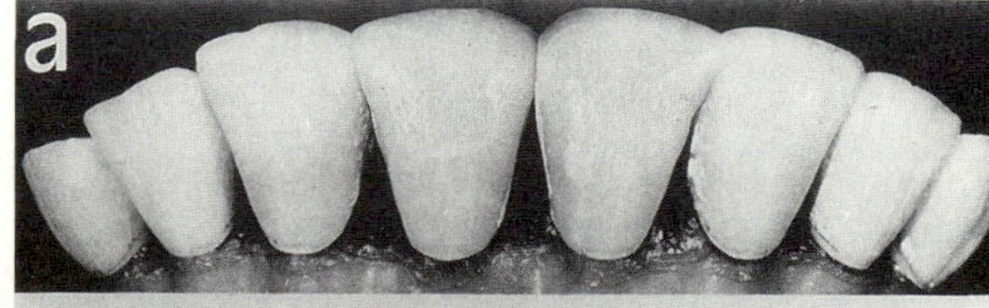

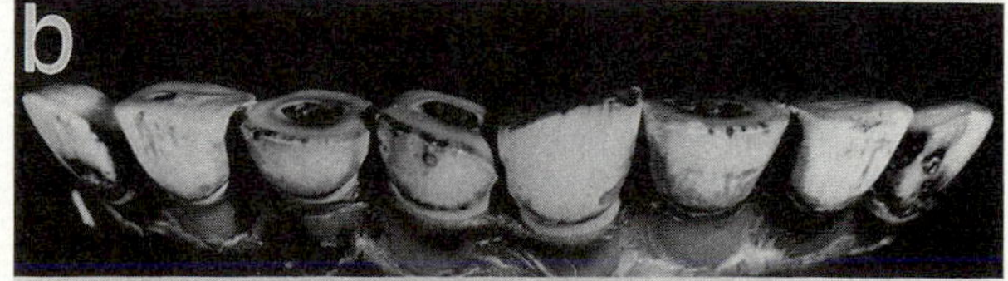

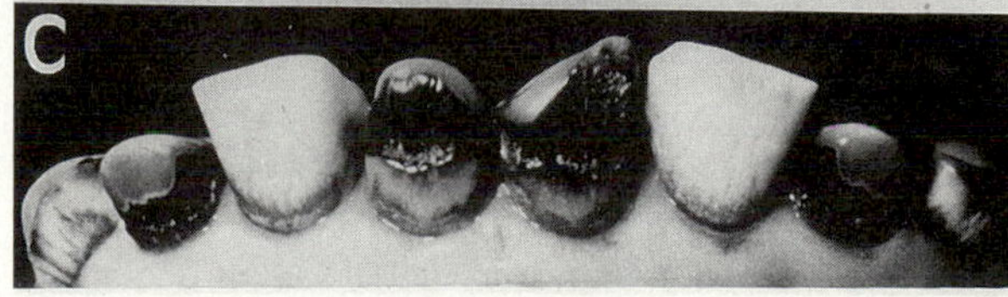

Fig. 5–47. Fluorosis evidenced in teeth of cow. A, normal teeth. B, fluorosis caused by continuous ingestion of high fluorine levels. C, fluorosis caused by intermittent ingestion of high fluoride levels. (Courtesy, Utah State University, Logan)

(Continued)

TABLE 5-3 *(Continued)*

FLUORINE (F) POISONING (FLUOROSIS) (Continued)

Distribution and Losses Caused By:

The water in parts of Arkansas, California, South Carolina, and Texas has been reported to contain excessive fluorine. The largest high-fluorine area in the U.S. is the West Texas Panhandle. Occasionally, throughout the U.S. high-fluorine, raw-rock phosphates are used in mineral mixtures.

Areas near smeltering or metal-production industries which heat ores or burn high-fluoride coal may be a problem.

Treatment:

Any damage may be permanent, but animals which have not developed severe symptoms may be helped to some extent, if the source of excess flourine is eliminated. Some reduction in toxicity may be obtained by the addition of calcium, aluminum, or fat into the ration, as these elements seem to reduce the absorption of fluorine.

Prevention:

Avoid the use of feeds, water, or mineral supplements containing excessive fluorine (see Table, previous page).

The maximal safe level of fluorine in the total dry ration is as follows: Beef and dairy cattle, 50 to 100 ppm; sheep and swine, 100 to 200 ppm; and chickens, 300 to 400 ppm.

Remarks:

Fluorine is a cumulative poisoning.

Undefluorinated rock phosphate often contains 3.5 to 4.0% (35,000 to 40,000 ppm) of fluorine.

Phosphate clays (soft phosphates) are usually too high in fluorine to be used safely unless defluorinated.

(Also see Chapter 4, Table 4-6, ANIMAL MINERAL CHART, FLUORINE.)

HYPERVITAMINOSIS (VITAMIN OVERDOSES)

Poison:

Excess vitamin A (hypervitaminosis A); excess vitamin D (hypervitaminosis D); or excess menadione (K_3)(hypervitaminosis K).

Source:

Massive doses, from either intentional or accidental ration supplementation, of vitamins A, D (especially vitamin D_3 [cholecalciferol]), or menadione (K_3) may result in hypervitaminosis. Excessive intake of vitamin A and vitamin D may produce toxicities because (1) these vitamins are fat-soluble, (2) small amounts of them have strong effects, and (3) they tend to accumulate in the liver. Excess intake of synthetic menadione (formerly called K_3) has produced toxic symptoms in rats.

Toxic problems occur most frequently with fat-soluble vitamins, which are stored in the animal body.

Toxicities from excessive intakes of vitamins E, K_1 and K_2, vitamin C, and the B-complex vitamins are rare.

Species Affected:

All species, including people.

Symptoms and Signs (or age group most affected):

Symptoms of vitamin A toxicity are: loss of appetite, poor vision, excessive irritability, loss of hair, dryness and flaking of the skin (with rubbing), bone abnormalities, and swelling over the long bones, and diarrhea. **NOTE WELL**: Massive intakes of carotene are not harmful because they are not converted to vitamin A rapidly enough to cause toxicity. The excess carotene will merely produce a yellow coloring of the skin, which disappears when the intake is reduced.

Symptoms of vitamin D toxicity are: hypercalcemia (increased intestinal calcium absorption, leading to elevated blood calcium levels and bone changes), loss of appetite, excessive thirst, irritability, weakness, constipation alternating with bouts of diarrhea, retarded growth in young animals, and weight loss in adults. If massive doses of vitamin D are continued, widespread calcification of the soft tissues may ultimately prove fatal.

Symptoms of menadione (vitamin K_3) toxicity are: depression in growth, anemia, methemoglobinaria (the presence of methemoglobin [oxygen in firm union with iron (Fe^{+++}) in the urine]), and porphyrinuria (the excretion of porphyrin [a pigment in the urine]).

Distribution and Losses Caused By:

Hypervitaminosis A, D, or menadione (K_3) are not common under natural conditions; hence, toxicities are not widely distributed. However, where synthetic vitamins are added to rations, there is always the hazard of abnormally large amounts being ingested, intentionally or unintentionally.

Treatment:

Remove the cause—the overdose of the particular vitamin (vitamins A, D, or menadione).

Prevention:

Avoid an overdose of vitamins A, D, or menadione (K_3). Large overdoses (megadoses) of any vitamin in the form of supplements should be under the supervision of a nutritionist.

(Continued)

TABLE 5-3 *(Continued)*

HYPERVITAMINOSIS (VITAMIN OVERDOSES) (Continued)

Remarks:

The concurrent administration of large quantities of vitamin A with potentially toxic levels of vitamin D will reduce the toxicity of the latter.
(Also see Chapter 4, Table 4-8, ANIMAL VITAMIN CHART: VITAMIN A, VITAMIN D, VITAMIN K.)

LEAD (Pb) POISONING

Source:

Lead is discharged into the air from auto exhaust fumes and other sources.
Lead pollution of feed and food crops as a result of lead being deposited on the leaves and other edible portions of the plant by direct fallout.
Inhaling airborne lead.
Lead may get into feed or food and water from contact with lead pipes, lead utensils, discharged storage batteries, old lead base paint, or used motor oil.

Species Affected:

All animals, but cattle, sheep, and mink are especially susceptible.
Swine, goats, and chickens are relatively resistant.
There is no evidence that lead constitutes a health problem of fish in the U.S.

Symptoms and Signs (or age group most affected):

Symptoms develop rapidly in young animals, but slowly in mature animals.
Loss of appetite and evidence of gastroenteritis.
Feces may become very dark gray and be tinged with blood.
Salivation, champing of the jaws, frenzy, blindness, convulsions, coma, and death.
Mature animals usually have diarrhea and show incoordination, especially in the hind limbs, and prostration.

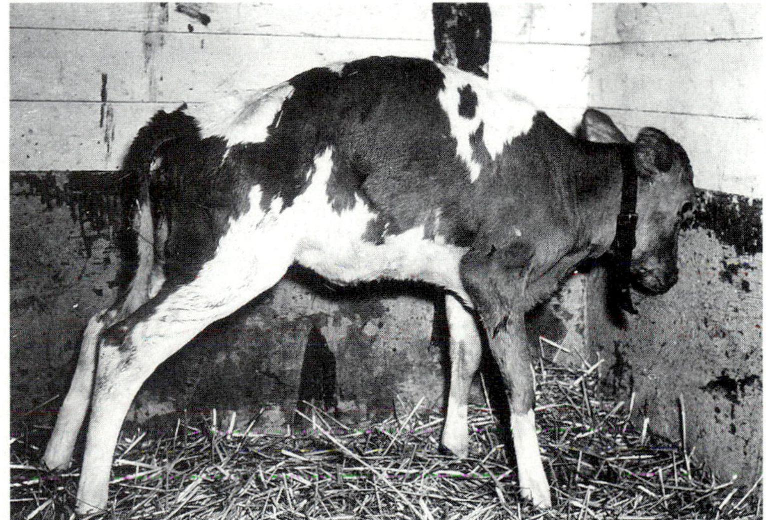

Fig. 5-48. Lead poisoning. Calf shows evidence of gastroenteritis and incoordination of the hind legs. (Courtesy, College of Veterinary Medicine, University of Illinois, Champaign-Urbana)

Distribution and Losses Caused By:

At one time, lead was a component of sprays used to control insects and plant diseases and of paints. But, leaded paint is now banned by law.
Lead poisoning is the most frequently diagnosed poisoning of domestic animals. However, its occurrence appears to be declining because lead is no longer used in paint and there is a growing tendency toward the use of nonleaded gasoline.

Treatment:

If damage to tissue has been extensive, treatment is of little value; in any event it should be handled by a veterinarian.
Magnesium sulfate (Epsom Salts) may be employed to remove any lead remaining in the digestive tract.
Calcium disodium EDTA should be given for several days to absorb lead from the tissues.

Prevention:

Avoid sources of lead.

Remarks:

Lead poisoning is cumulative.
When incorporated in the soil, nearly all the lead is converted into forms that are available to plants. Any lead taken up by plant roots tends to stay in the roots, rather than move up to the top of the plant.
Lead poisoning can be diagnosed by positively analyzing the blood tissue for lead content.

MERCURY (Hg) POISONING

Source:

Mercury is discharged into air and water from industrial operations and is used in herbicide and fungicide treatments.

(Continued)

TABLE 5-3 *(Continued)*

MERCURY (Hg) POISONING (Continued)

Consumption of seed grains treated with fungicides that contain mercury, for the control of fungus diseases of oats, wheat, barley,and flax.

Mercury poisoning has occurred where mercury from industrial plants has been discharged into water and then accumulated in fish and shellfish. Fish and shellfish concentrate high levels of methylmercury from the small quantities in the waterways.

Species Affected:

All animals, but especially cattle and hogs.

Symptoms and Signs (or age group most affected):

Gastrointestinal, renal, and nervous disturbances; but impossible, on basis of symptoms to differentiate mercury from other poisons.

Case history of animals consuming mercury-treated grains should be considered strong circumstantial evidence.

Distribution and Losses Caused By:

Mercury poisoning is not common in farm animals. It usually occurs when, through ignorance or negligence, mercury-treated grain is fed to animals.

Treatment:

Usually, treatment is not satisfactory.

Protein (milk, eggs, blood serum) may reduce G.I. absorption.

Sodium thiosulfate may be used.

Selenium, in carefully-controlled levels, appears helpful.

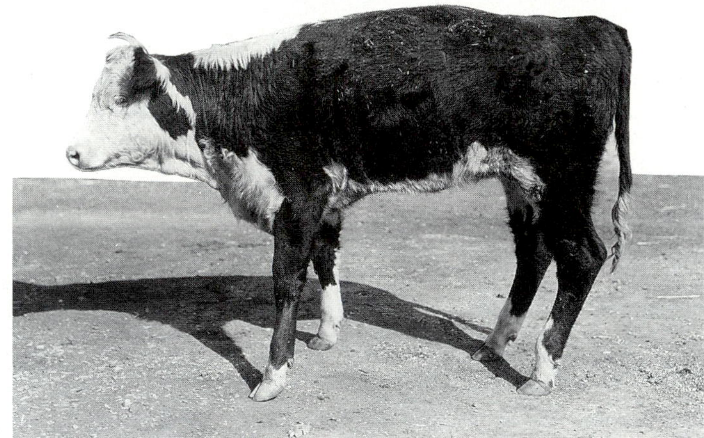

Fig. 5-49. Heifer poisoned on seed corn that had been treated with a mercury compound. Note stiffness and weakness. (Courtesy, College of Veterinary Medicine, University of Illinois, Champaign-Urbana)

Prevention:

Do not feed livestock seed grains treated with a mercury-containing fungicide, which is usually dyed bright pink.

Surplus of treated grain should be burned and the ash buried deep in the ground. Corn may be detreated legally under very strict conditions, by washing or roasting.

Remarks:

Ultimate diagnosis depends upon demonstrating the presence of mercury in the tissues, especially in the kidneys and liver.

Food and Drug Administration prohibits use of mercury-treated grain for feed or food.

Mercury is a cumulative poison.

Grain crops produced from mercury-treated seed and crops produced on soils treated with mercury herbicides have not been found to contain harmful concentrations of this element.

MOLYBDENUM TOXICITY (MOLYBDENOSIS)

Source:

Toxicity can show up at as little as 6 ppm, depending on the amount of copper available.

Species Affected:

Ruminants, especially calves and cows in milk.

Symptoms and Signs (or age group most affected):

Toxic levels of molybdenum interfere with copper metabolism, thus increasing the copper requirement and producing typical copper deficiency symptoms.

The toxicity signs in cattle are: diarrhea (scouring), loss of appetite, anemia, lack of coordination, bone malformation, and depigmentation of hair.

In sheep, there is depigmentation of the wool and loss of crimp.

Distribution and Losses Caused By:

Canada (Manitoba), England, and in California, Florida, Nevada, and other areas of the U.S.

Treatment:

For cattle 1 yr or older, 1 g of copper sulfate per head daily, added to the feed, will usually cure symptoms of molybdenum toxicity. For calves up to 1 year of age, add ½ g of copper sulfate to the feed daily.

(Continued)

TABLE 5-3 *(Continued)*

MOLYBDENUM (Mo) TOXICITY (MOLYBDENOSIS) (Continued)

Prevention:

Where the molybdenum content of the forage is below 5 ppm, add 1% copper sulfate to the salt.

Where the molybdenum content of the forage is above 5 ppm, add 2 to 5% copper sulfate to the salt, depending on the level of the molybdenum.

Remarks:

When feeds are high in sulfate, toxic symptoms will be produced on lower levels of molybdenum and, conversely, higher levels of molybdenum can be tolerated with low levels of sulfate.

MYCOTOXINS (TOXIN-PRODUCING FUNGI OR MOLDS)

Source:

Aflatoxin (most studied of the group) associated with peanuts, brazil nuts, silage, corn and most other cereals, hay, and grasses. The mold can produce toxic compounds on virtually any feed/food (even synthetic) that will support growth.

Aflatoxin is actually a group of toxins which are similar in chemical structure, the principal ones of which are aflatoxin B_1, aflatoxin B_2, aflatoxin G_1, and aflatoxin G_2. Aflatoxin B_1 is usually found in greatest abundance and is the most toxic form of aflatoxin.

While aflatoxin appears to cause most of the problem, it is not the only mycotoxin to be feared. Other mycotoxins are being studied, especially ochratoxin and T_2 (trichothecenes).

Species Affected:

Cattle, chickens, ducklings, goats, horses, mink, pheasants, sheep, swine, trout, turkeys, and humans.

In all species, the young are far more susceptible than mature animals.

Generally, ruminants appear to tolerate higher levels of mycotoxins and longer periods of intake than simple-stomached animals.

Symptoms and Signs (or age group most affected):

Mold affects animals in a variety of ways, from decreased production to sudden death. Usually, the first sign is the loss of appetite and weight.

A few animals will abort, and an occasional animal will die.

Aflatoxin exposure has been associated with liver cancer in children, and with trout hepatoma.

With high intakes of mycotoxin, or with the several types of molds, any one or a combination of the following symptoms may develop: liver damage, hyperkeratosis, a typical interstitial pneumonia, bloody slimy scours, arched back, dry gangrene at the end of the tail or top of hoof, hemorrhagic hepatitis, renal damage, lameness and/or swollen legs. In swine, an estrogenic mycotoxin produced by *Fusarium graminearum* in corn produces swelling of the vagina and possibly mammary development in gilts and preputial enlargement in boars.

Generally, aflatoxin does not affect animals when the level in the ration is below 100 ppb (0.1 ppm). Typically, aflatoxin residues are short-lived in animals that don't develop liver lesions. In swine, levels of 500 ppb (0.5 ppm) have been eliminated from the tissues within 4 days.

There are many mycotic diseases. Only Aflatoxicosis (the response of animals to aflatoxin) is herewith summarized:

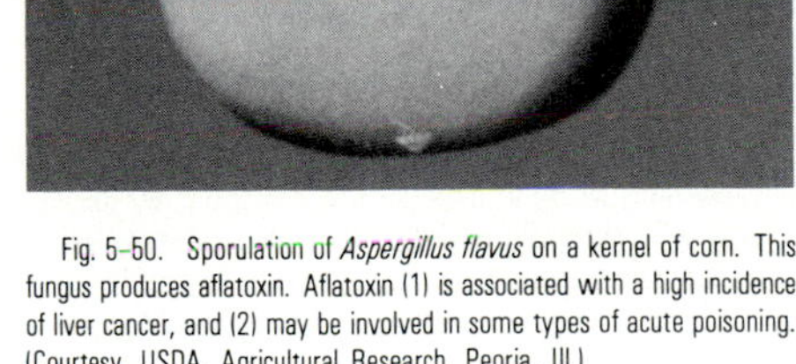

Fig. 5-50. Sporulation of *Aspergillus flavus* on a kernel of corn. This fungus produces aflatoxin. Aflatoxin (1) is associated with a high incidence of liver cancer, and (2) may be involved in some types of acute poisoning. (Courtesy, USDA, Agricultural Research, Peoria, Ill.)

Distribution and Losses Caused By:

Widely distributed throughout the world.

In addition to the effect of mycotoxins on the animal's health, milk and eggs are contaminated by the residues or mycotoxins, or by their metabolic products.

Class of Animal	Level of Aflatoxin in Feed	Effects on Animals
Cattle:	10–20 ppm	Jaundice, hemorrhage, liver necrosis. Death in 1–2 weeks.
Calves	0.2 ppm	Fed 2–4 weeks will cause reduced weight gain and impaired blood coagulation.
Dairy	2–4 ppm	Off feed, reduced milk production.
Yearling Steers	0.7 ppm	Multiple doses (3–4) months may lead to liver damage, reduced rate of gain, and death.
Poultry:		
Broilers	5–10 ppm	Liver necrosis, hemorrhage, death.
	2–5 ppm	Impaired blood coagulation, reduced growth and feed efficiency.
	1.5–2.5 ppm	Decreased gain and feed efficiency.
Layers	2.0 ppm	Reduced egg production.
Ducks	0.3 ppm	Liver damage, death.
Turkeys	0.25 ppm	Decreased weight gain and impaired immunity to disease.
Swine	10–20 ppm	Single exposure, lethal. Hemorrhage, acute hepatitis.
	2–4 ppm	Lethal. Multiple doses required.
	0.8–2.0 ppm	Subacute, and may be lethal. Liver necrosis, jaundice, fibrosis (formation of fibrous tissue), hemorrhages.
	0.2–0.5 ppm	Reduced gain, impaired immunity to disease.

(Continued)

TABLE 5–3 *(Continued)*

MYCOTOXINS (TOXIN-PRODUCING FUNGI OR MOLDS) (Continued)

Treatment:

Remove the source of the mold.
Animals suffering from molds frequently respond to vitamin B injections.
Iron therapy may be helpful, since hemorrhaging is a frequent problem.

Prevention:

The prime cause of aflatoxin is moisture, which favors mold growth; hence, proper harvesting, drying, and storage are important factors in lessening contamination and toxin production.
Propionic and acetic acids, and sodium propionate, will inhibit mold growth; hence, their use in preserving high-moisture grains is encouraged.
Increasingly, as a preventive measure, grains and processed feeds are being tested for mycotoxins and required to meet acceptable levels.

Remarks:

Certain molds produce toxins, or mycotoxins.
Aflatoxin has been clearly shown to be a carcinogen (tumor producing).
Ultraviolet irradiation and anhydrous ammonia under pressure will reduce the toxicity of aflatoxins and, if continued long enough, will deactivate them entirely.
Not all mold products are harmful. For example, zeranol is being commercially produced as a growth-promotant hormone for cattle.
Food and Drug Administration regulations do not permit feeding more than 100 ppb aflatoxins in the total ration of livestock and poultry.
(Also see Chapter 13, section on "Mold [Fungi] Inhibitors.")

NITRATE/NITRITE POISONING (OAT HAY POISONING, CORNSTALK POISONING)

Acute nitrate/nitrite poisoning is caused by the presence of nitrite in the blood at a level sufficient to cause anoxia (internal suffocation). Nitrate (NO_3) can be reduced to nitrite (NO_2) by microorganisms in the gastrointestinal tract, especially in the rumen, at a rate which overwhelms the body's defense system. Nitrite combines with the hemoglobin of the red blood cells to form methemoglobin, which cannot transport oxygen to the body tissues.

Source:

Nitrate is a naturally occurring form of nitrogen and a desirable part of our environment, found in most soils and in a number of fertilizers. Under normal conditions, plants use nitrates and other nitrogen compounds to form plant proteins. However, when nitrate concentrations are excessive and/or out of place, as may happen when normal growth is altered by the environment (such as drought) protein formation may be slowed and the nitgrogen may remain in the plants as non-protein nitrogen—nitrates, nitrites, amides, free amino acids, and peptides. The nitrate is of special concern because of its potential toxicity when excessive amounts are ingested.

The three principle sources of nitrate for animals and humans—plants, water, and air—are interrelated. The sources, along with their relationships, are depicted in Fig. 5–51. As shown, nitrogen may be (1) fixed by microorganisms in ruminant animals and in legumes (natural nitrogen fixation), (2) formed when plant residues, animal manures, and human wastes decompose, or (3) added to soils as nitrogen fertilizer.

In order to maintain production, when nitrogen is removed from the cycle, it must either be (1) returned so that nature can reuse it, or (2) added as chemical fertilizer nitrogen. The cyclic nature of nitrogen in the environment is depicted in Fig. 5–52. The nitrogen cycle includes various changes from elemental atmospheric nitrogen to inorganic, to organic, and back to inorganic forms.

Crop production has been increased by making more nitrogen and other nutrients available to crops. But this practice also increases the chance of getting the nitrogen cycle out of balance. Even when proper fertility and crop selection decisions have been made, changes in the environment—the rainfall, temperature, sunlight, and shifting of seasons—can still alter the nitrate concentration of crops and the water supply. Nitrate poisoning of animals and people can result from ingesting plants or water high in nitrate content, commonly one or more of the following sources:

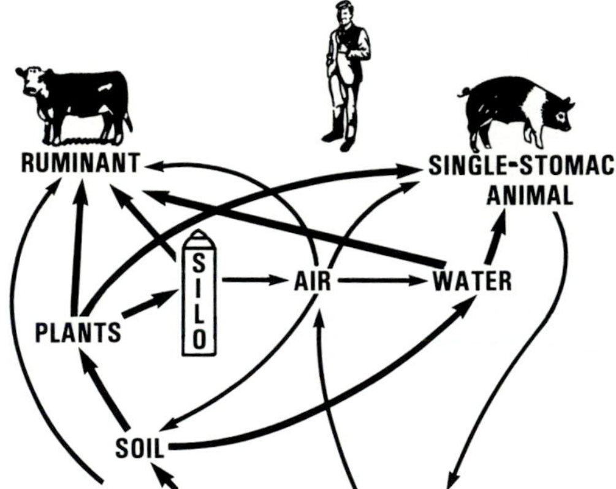

Fig. 5–51. Nitrates in relation to animals and humans.

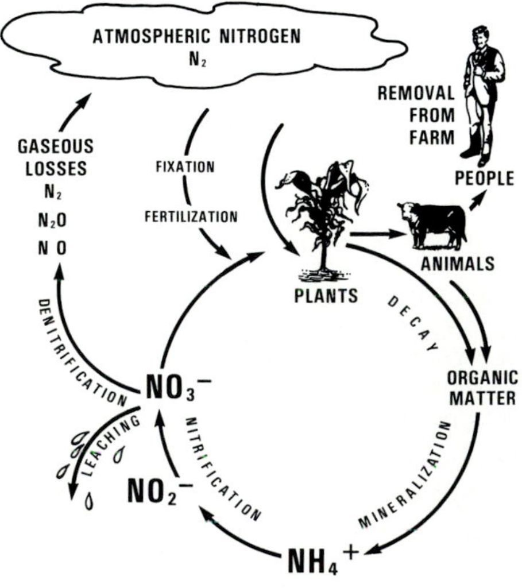

Fig. 5–52. The nitrogen cycle.

1. Forages (vegetative part) of most grain crops (oats, wheat, barley, rye, corn, sorghum), Sudangrass, and numerous weeds, especially (a) when under stress such as drought, insufficient sunlight, or after spraying with weed killer (herbicide); or (b) following heavy nitrate fertilization of soils (commercial, green manure crop, barnyard manure). Some nitrate may be formed after forage is stacked.

(Continued)

TABLE 5–3 *(Continued)*

NITRATE/NITRITE POISONING (OAT HAY POISONING, CORNSTALK POISONING) (Continued)

2. Inorganic nitrate or nitrite salts, or fertilizer left where animals have access to them, or where they may be mistaken for salt.
3. Pond or shallow well water into which surface runoff from barnyard or well-fertilized soil may drain.

Species Affected:

Primarily cattle. Sheep. Horses. Ruminants are most susceptible because of the conversion of nitrates to nitrites by the microorganisms in the rumen.

Symptoms and Signs (or age group most affected):

Accelerated respiration and pulse rate; diarrhea; frequent urination; loss of appetite; general weakness; trembling and staggering gait; frothing from mouth; lowered milk production; abortion; blue color of the mucous membrane, muzzle, and udder due to lack of oxygen in the blood; death within 4½–9 hours after consuming nitrates.

A rapid and accurate diagnosis of nitrate poisoning may be made by examining blood. Normal blood is red and becomes brighter when exposed to air, whereas blood from cows toxic with nitrates is a brown color due to formation of the methemoglobin. Nitrates oxidize ferrous hemoglobin (oxyhemoglobin) to ferric hemoglobin (methemoglobin) which is not an efficient oxygen transporter. The animal essentially suffocates for lack of oxygen in tissues. When ¾ of the oxyhemoglobin is converted to methemoglobin, the animal will die.

Distribution and Losses Caused By:

Excessive nitrate content of feeds is an increasingly important cause of poisoning in farm animals, due primarily to more and more high nitrogen fertilization. But nitrate toxicity is not new, having been reported as early as 1850, and having occurred in semiarid regions of this and other countries for years.

Treatment:

Death usually occurs so suddenly that treatment is not possible, and few treated animals recover.

The most common treatment is a 4% solution of methylene blue (in a 5% glucose or a 1.8% sodium sulfate solution) administered by a veterinarian intravenously at the rate of 100 cc/1,000 lb liveweight.

Prevention:

More than 0.9% nitrate (dry basis) may be considered as potentially toxic. Feed should be analyzed when in question, by using a simple test to detect the presence of nitrates (qualitative); if present, follow with a quantitative test to determine how much is present.

Nitrate poisoning may be reduced by (1) feeding high levels of grains and other high-energy feeds (molasses) and vitamin A, (2) limiting the amount of high-nitrate feeds, (3) ensiling forages which are high in nitrates (fermentation reduces some nitrates to gas, but care must be taken to avoid nitric oxide and nitrogen dioxide released in early stages of fermentation), and avoid feeding until 3–4 weeks in storage.

Remarks:

The nitrate form of nitrogen does not appear to cause the actual toxicity. During digestion, the nitrate is reduced to nitrite, a far more toxic form (10–15 times more toxic than nitrates). In cows and sheep, this conversion takes place in the rumen (paunch); in horses, in the cecum.

Lethal dose varies with (1) nutritional state, size and type of animal; and (2) the consumption of feed other than nitrate-containing material.

Methods of reporting nitrates (dry basis) in rations in relation to death losses follow:

	Potentially Lethal Levels	
	(%)	**(ppm)**
Nitrate (NO_3)	over 0.9	9,000
Nitrate nitrogen (NO_3N)	over 0.21	2,100
Potassium nitrate (KNO_3)	over 1.5	15,000

OAT HAY POISONING

(See NITRATE POISONING.)

ORGANOPHOSPHORUS COMPOUNDS

A large number of these compounds have been developed for animal and plant protection. When properly used, they are not toxic. When improperly used, they may be toxic. Triaryl phosphate, a component of lubricants, has been implicated in toxicity.

Source:

A number of well-known compounds used for animal protection; among them, malathion, ronnel, coumaphos, trichlorfon, and ruelene. Triaryl phosphate is a component of lubricants.

(Continued)

TABLE 5–3 *(Continued)*

ORGANOPHOSPHORUS COMPOUNDS (Continued)

Species Affected:

All animals.

Symptoms and Signs (or age group most affected):

The organophosphorus compounds vary greatly in toxic effect, and each has its own characteristics in regard to tissue storage and excretion of milk. Experimentally, several organophosphorus compounds have produced delayed neurotoxicity. The most common signs are excess salivation, shortness of breath, signs of abdominal pain, a staggering gait, diarrhea, and, occasionally, convulsions.

Distribution and Losses Caused By:

Organophosphorus toxicity may occur wherever these compounds are not used properly.

Treatment:

Organophosphorus insecticide poisoning can usually be treated successfully with atropine sulfate, using the following dosages initially: cattle, 30 mg/100 lb of body weight; sheep, 50 mg/100 lb; and horses, 6.5 mg/100 lb. The veterinarian may prescribe treatment for removal of the poison from the animal. Rest and quiet surroundings are indicated.

Prevention:

Prevention consists in using all organophosphorus compounds in keeping with the directions on the label.

Remarks:

The course of organophosphorus toxicity is influenced principally by the dosage.

PENTACHLOROPHENOL POISONING (PENTA, OR PCP)

Poison:

Pentachlorophenol (Penta or PCP), a wood preservative, which is also used as a herbicide, fungicide, molluscocallide, and insecticide. **NOTE WELL**: Commercial Pentas vary considerably in purity. Use only Pentas that are low in dioxin, the toxic substance.

Source:

Penta-treated wood, such as barns, telephone poles, corrals, pens, fence posts, and feed bunks. Do not allow animals to lick PCP-treated wood or eat feed that has been in contact with treated wood. Also, avoid inhalation and skin contact.

Penta-treated foliage is not dangerous to animals when the herbicide has been used correctly.

Species Affected:

All animals.

Symptoms and Signs (or age group most affected):

The most common symptoms of pentachlorophenol poisoning are: nervousness, restlessness, rapid pulse and respiratory rates, weakness, muscle tremors and chronic convulsions, followed by death. It is an intense irritant to the skin and mucous membranes. Animals fed in troughs made of lumber treated with Penta may show salivation and irritation of the oral mucosa.

Distribution and Losses Caused By:

Losses have occurred wherever impure Penta has been used improperly.

Treatment:

No effective treatment or antidote is known.

Prevention:

Comply with the label directions of the Environmental Protection Agency (EPA)/manufacturer, including: (1) selling to and using by certified applicators (or persons under their direct supervision); (2) using only for those purposes covered by the certified applicators' certification; (3) consumers washing their hands after working with treated wood, wearing dust masks and eye goggles when sawing or machining treated wood, and disposing of treated wood by burial or trash collection rather than by burning.

Do not (1) allow animal feed to come in contact with Penta; (2) permit animals to lick, chew, or rub on recently treated wood; and (3) provide outside lots and plenty of ventiliation for animals placed in freshly Penta-treated buildings.

Exercise the following precautions when filling wood silos that have been treated with Penta: steam clean and either paint with a non-toxic latex paint or line with plastic.

(Continued)

TABLE 5-3 *(Continued)*

PENTACHLOROPHENOL POISONING (PENTA, OR PCP) (Continued)

Remarks:

Penta has been used in the U.S. since the 1930s.

When purified grade Penta is used in keeping with the directions on the label, poison problems are unlikely.

NOTE WELL: Farmers and ranchers should check with their local County Extension Agent, or other local authority, relative to the use of wood preservatives (penta, creosote, zinc chloride, or other wood preservatives), making certain that they are approved and properly used.

PESTICIDE POISONING

Poison:

Pesticides are chemicals used to destroy, prevent, or control pests, but they can also be toxic (poisonous) to animals, people, and plants.

Pests can be classified into 6 main groups: (1) insects (plus mites, ticks, and spiders); (2) snails and slugs; (3) vertebrates, including rats, mice, and certain birds (starling, linnets, English sparrows, crows, and blackbirds); (4) weeds; (5) plant diseases; and (6) nematodes.

Source:

Pesticides are chemicals. When properly used, they are beneficial; when improperly used, they may be hazards.

Pesticide poisoning may be caused by either (1) sudden exposure to lethal quantities, or (2) as a result of repeated exposure to nonlethal quantities (chronic poisoning) during a protracted period of time.

Species Affected:

Animals, people, and plants.

Many pesticides are so highly toxic that very small quantities can kill an animal and exposure to a sufficient amount of almost any pesticide can make an animal ill. Even fairly safe pesticides can irritate the skin, eyes, nose, or mouth.

Some pesticides produce unplanned and undesirable "side effects," particularly when they are not used properly. Among such effects are: reduction of beneficial species; drift; wildlife losses; honeybee and other pollenating insect losses; and pollution of air, soil, water, and vegetation.

Symptoms and Signs (or age group most affected):

Knowing something of the toxicity and symptoms or signs of poisoning of each type of pesticide may result in getting medical advice quickly and in saving a life. The overall symptoms of pesticide poisoning are:

1. Organophosphates—most toxic; they injure the nervous system.
2. Carbamates—safer than the organophosphates; they produce the same symptoms as the organophosphates, but they respond more easily to treatment.
3. Fumigants—cause poor coordination and confusion.
4. Plant-derived pesticides—some are very toxic; pyrethrum may cause allergic reaction, rotenone may irritate the respiratory tract, nicotine is a fast-acting nerve poison.

Distribution and Losses Caused By:

In every pursuit of modern agriculture, more and more pesticides are being used. They are the first line of defense against pests that affect human health and well-being and attack livestock, crops, and structures.

Pesticides are used to control many of the estimated 10,000 species of harmful insects; more than 160 bacteria, 250 viruses, and 8,000 fungi known to cause plant diseases; 2,000 species of weeds and brush; and 150 million rats.

Treatment:

Each label contains a "Statement of Practical Treatment." Read it before using a pesticide.

Prevention:

The first and most important precaution to observe when using any pesticide is to read and heed the directions on the label. In the event of an accident, the label becomes extremely important in remedial measures.

For public protection, all chemicals are rigidly controlled by federal laws. Each one is required to be registered by the Environmental Protection Agency before it can be sold in the U.S., and each one is issued a tolerance for residues that may result from its use on food or feed crops.

Remarks:

The Environmental Protection Agency (EPA) administers the following acts:

1. The Federal Insecticide, Fungicide, and Rodenticide Act (FIFRA) as amended in 1972. Every pesticide must be registered, and commercial applicators must be certfied—showing that they know the safe and correct way to use them.
2. The Toxic Substances Control Act (TOSCA) of 1976, which regulates chemicals that are not presently covered under other federal acts. It might more properly be described as the "chemical health and environmental regulation act."

U.S. consumers benefit greatly from the proper use of pesticides. They enjoy the world's most abundant food supply with quality and variety second to none; and they spend less than 20% of their income for food, compared with the 50% or more that people in many other countries spend for food.

(Continued)

TABLE 5-3 *(Continued)*

PINE NEEDLE ABORTION

Source:

Pine needles or slash of *Pinus ponderosa* (ponderosa pine; western yellow pine).

Species Affected:

Cattle.

Symptoms and Signs (or age group most affected):

Pregnant cows, free of brucellosis, abort, especially during the last 3 months of pregnancy.

Pine needle abortion usually appears 1 to 3 days after pregnant cattle have eaten the needles or buds. Abortions will continue for up to 2 weeks though cattle are removed from the needles.

At calving, excessive hemorrhaging; retained placenta; septic metritis, often followed by peritonitis.

If cow is affected near parturition, calf may be born normal, but weak.

Fig. 5-53. Cow aborting, as a result of eating pine needles. (Courtesy, USDA-ARS Poisonous Plant Research Laboratory, Logan, Utah)

Distribution and Losses Caused By:

Wherever ponderosa pine trees are found; in the Northern Plains, Rocky Mountain, and Pacific Northwest regions of the U.S. and in western Canada.

Not all pregnant cows will abort after eating pine needles, but the disease has been known to affect as much as 50% of the herd.

Treatment:

None known.

Prevention:

Keep pregnant cows away from yellow pine, especially during the latter part of gestation.

Maintain the cow herd on the recommended level of nutrition and avoid stress conditions as much as possible.

Remarks:

Cows will eat pine needles even when well fed.

Generally, cattle eat pine needles or buds only under the following circumstances:

1. Sudden weather changes causing animals to seek shelter under ponderosa pine where they will eat the needles.
2. Severe storms which place large quantities of the needles on the ground.
3. Animals concentrated near ponderosa pine.
4. The cattle are hungry.
5. Hay is fed on the ground beneath the trees.
6. The cattle are changed to unfamiliar or poor quality feed.

POLYBROMINATED BIPHENYLS (PBBs)

These are fire retardants which may cause cancer when taken into the feed/food supply.

Source:

Until 1973, PBBs were used without known incident of toxicity, primarily to coat building roofs and to treat clothing for resistance to flammability. Then, a Michigan chemical company, short on bags for their commercial fire retardant, packaged some of it in bags normally used, and labeled, for magnesium oxide, which was being used as a dairy supplement. Some of the toxic fire retardant found its way into dairy rations.

All of the 1973 Michigan toxicity problem was traced to livestock feed which became contaminated with PBB when the fire retardant was shipped by mistake to a feed manufacturer.

Species Affected:

Humans and animals—including poultry, fish, mink, wildlife.

(Continued)

TABLE 5-3 *(Continued)*

POLYBROMINATED BIPHENYLS (PBBs) (Continued)

Symptoms and Signs (or age group most affected):

Loss of appetite, the secretion of excess tears, salivation, depression, abortion, loss of weight, diarrhea, and thickening and wrinkling of the hide.

Distribution and Losses Caused By:

In 1973, the accidental contamination of animal feeds exposed many people in Michigan to PBB in dairy products and other foods.

The Michigan incident eventually led to the slaughter and burial of nearly 25,000 cattle, 3,500 hogs, and 1.5 million chickens; and the disposal of about 5 million eggs and tons of milk, butter, cheese, and feed.

Treatment:

Follow the prescribed treatment of the veterinarian.

Prevention:

Avoid human error such as substituting PBB for a livestock feed supplement. Avoid harmfully contaminated feed/food.

Remarks:

PBBs are long-term, low-level contaminants, very stable and resistant to decay.

POLYCHLORINATED BIPHENYLS (PCBs)

These are chlorinated hydrocarbons which may cause cancer when taken into the feed/food supply.

Source:

The polychlorinated biphenyls (PCBs), a group of chlorinated hydrocarbons, were first used in 1930.

PCBs were produced in huge quantities for use as insulation for electrical equipment, and as plasticizers in paints, adhesives, and caulking compounds. Their chief virtue: they are fire resistant; hence, they lessen fires. Not until tens of millions of pounds of these products were produced and released into the environment was there any realization of their toxicity and persistency.

In 1966, scientist Soren Jensen in Sweden detected PCB as an environmental contaminant.

Since 1967, monitoring efforts in this country have revealed PCBs in milk, eggs, meat, fish, fats and oils, and cereal products.

Despite limited restrictions imposed by industry in the early 1970s to reduce the production and use of PCBs and the halting of production in 1977, high levels continue to persist in the Great Lakes and many other major waters across the nation.

Species Affected:

Cattle, chickens, ducks, fish, humans, mink, wildlife.

PCBs have been found in the tissues of humans and in the milk of nursing mothers.

Symptoms and Signs (or age group most affected):

The clinical signs of PCB toxicity are:

In chicks, labored breathing, abdominal edema, enlarged heart, swollen kidneys, and internal hemorrhage.

In hens, lowered or prevented hatchability.

In mink, PCB in the ration has caused reproduction failure.

Distribution and Losses Caused By:

PCBs are widespread. They have, directly and indirectly, found their way into animal feeds and animal food products through water, paints, sealants used on the interior of silos, heat transfer fluids, and plastic and cardboard food-packaging materials.

PCB constitutes the first example of a widespread environmental hazard from an industrial chemical not classed as a pesticide. The wide usage and the persistence indicate that PCB toxicity will be of considerable concern for many years.

The following U.S. incidents dramatically illustrate the potential health hazard and economic harm that can be caused by PCBs: (1) the discharging of PCBs into the Hudson River with waste water from capacitor plants from 1946 to 1976, and the contamination of Hudson River fish; (2) in 1977, a fire in an animal-feed warehouse in Puerto Rico contaminated fish meal, which was subsequently fed to poultry and resulted in the destruction of about 400,000 chickens and millions of eggs in the U.S.; and (3) in 1979, the fat used to produce meat and bone meal feeds in a meat-packing plant in Billings, Montana, contaminated by PCBs from a damaged transformer, resulted (to October 1979) in the destruction of 600,000 to 700,000 chickens, several hundred thousand eggs, and 16,000 lb of fresh pork.

Treatment:

Animals afflicted with PCB should be treated by a veterinarian.

(Continued)

TABLE 5-3 *(Continued)*

POLYCHLORINATED BIPHENALS (PCBs) (Continued)

Prevention:

Comply with the law; do not use PCBs.
Avoid harmfully contaminated feed.

Remarks:

Although the production of PCBs was halted in 1977, and the importing of PCBs was banned Jaunary 1, 1979, the chemicals had been widely used for 40 years; and they are exceptionally long-lived. So, it is expected that these persistent chemicals will be around for many years to come.

Most other countries that manufacture PCBs, including the U.S.S.R., still haven't passed laws restricting their use.

PCBs have been widely used in dielectric fluids in capacitors and transformers, hydraulic fluids, and heat transfer fluids. Also, they have more than 50 minor uses, including plasticizers and solvents in adhesives, printing ink, sealants, moisture retardants, paints, and pesticide carriers.

PCB will cause cancer in laboratory animals (rats, mice, and rhesus monkeys). It is not known if it will cause cancer in humans. More study is needed to gauge its effects on the ecological food chain and on human health.

When fed coho salmon from Lake Michigan with 10–15 ppm PCB, mink in Wisconsin stopped reproducing or their kits died.

In 1977, polar bears at the very top of the Arctic food chain showed PCB levels of up to 8 ppm in their fatty tissue. This indicates that PCBs are spread throughout the biosphere.

PRUSSIC ACID (HCN) POISONING

The toxic material is hydrocyanic acid—HCN.

Source:

Most outbreaks of hydrocyanic acid poisoning are caused by the ingestion of plants which contain cyanogenetic glucosides. In this form, the acid is non-toxic but it may be liberated from the organic complex by the action of an enzyme which may also be present in the same plant or in another plant, or by the activity of rumen microorganisms. Among the common plants that may produce prussic acid poisoning are: Johnsongrass, Sudangrass, sorghum, wild black cherry, chokeberry, and arrow grass.

Some clovers, particularly white clover (*Trifolium repens*) and members of the *Brassica* genus (mustard) as well as plants of the flax family may also contain large amounts of the poison. A high soil nitrogen level increases the hydrocyanic acid content of the plants, especially when soil phosphorus levels are low. The prussic acid content is increased by heavy nitrate fertilization, excessive irrigation, wilting, trampling, and plant diseases. Very young, rapidly growing plants also contain greater quantities of the glycoside than more mature plants. Freezing does not ordinarily increase glycoside content in these plants but it does tend to increase the quantity of free hydrocyanic acid in the plants, thus resulting in a temporary increase in toxicity. Spraying cyanogenetic plants with plant herbicides apparently also increases the toxic hazard.

Species Affected:

All animals. However, ruminants are more susceptible to HCN poisoning from plants than nonruminants, because the rumen microflora and pH encourage greater glucoside breakdown than occurs in nonruminants.

Symptoms and Signs (or age group most affected):

If extremely large doses are consumed rapidly, generalized spasms develop and animals die within a few minutes. If small doses are consumed over a longer period, the more common clinical signs may be seen. The onset of this form is sudden and characterized by slobbering or frothing at the mouth and the gradual increase in the breathing rate. Within 5 to 15 minutes, this is developed to the point of open mouth breathing. Pulse becomes rapid and weak, muscle twitching occurs early, and progresses into generalized spasms just before death. Most animals stagger around and fall considerably before they go down. The mucous membranes may be bright red in color but become blue near death. Death from respiratory paralysis occurs during the convulsions. Bright red blood often passes from the nostrils and the mouth near the time of death. The course is rapid and does not usually exceed 30–45 minutes. A high percentage of animals that live for 2 hours after the onset of signs generally will recover. Diagnosis of prussic acid poisoning is made on the basis of history, cherry red color of the blood, signs and symptoms, post-mortem findings, and the demonstration of the presence of hydrocyanic acid in the stomach contents.

Distribution and Losses Caused By:

Prussic acid poisoning is worldwide. Approximately 1,000 plant species in 250 genera are known to contain HCN. Although reliable figures are not available, the losses are large.

Treatment:

Often when prussic acid poisoning is first noted, the animal is dead before there is time to administer an antidote. Animals which have not shown much evidence of toxicity may be injected intravenously with a mixture of sodium nitrite and sodium thiosulfate. The dose rates are 3 g of sodium nitrite and 15 g sodium thiosulfate in 200 ml water for cattle; for sheep, 1 g sodium nitrite and 2.5 g sodium thiosulfate in 50 ml water. Treatment may have to be repeated because of further liberation of hydrocyanic acid. Prussic acid poison is not cumulative; hence, on being removed from the feed, those animals not showing evidence of being poisoned will likely not be affected adversely.

(Continued)

TABLE 5-3 *(Continued)*

PRUSSIC ACID (HCN) POISONING (Continued)

Prevention:

Efforts to prevent prussic acid poisoning must be directed toward preventing the use of cyanogenetic plants for grazing. The risk of cyanide poisoning may also be decreased by the heavy feeding of ground grains before the animal is turned out to graze. Carbohydrates tend to inhibit the action of the enzyme that hydrolyzes the glucoside. Hungry cattle should not be allowed access to toxic plants, especially cultivated sorghum species when they are immature, wilted, frost bitten or growing rapidly after a stage of retarded growth. Plants of the sorghum family should be in flower before they are grazed or chopped to be fed green. If there is doubt as to the toxicity of a field, plants should be tested.

Observation of the following precautions will lessen prussic acid poisoning:

1. Allow 18 to 24 in. of growth before grazing Sudangrass. Sudangrass-sorghum crosses should be 25 to 30 in. high.
2. Do not graze frost-injured Sudangrass.
3. Do not allow hungry cattle to feed on pasture which is just recovering from a dry soil moisture condition and/or frost.
4. Watch out for Johnsongrass or sorghum which would be growing in Sudangrass fields. They contain higher concentrations of prussic acid (HCN) than Sudangrass.
5. Do not turn hungry animals into very short succulent growth.
6. Consider tissue as safe if it contains less than 500 ppm HCN, critical at 500 to 750 ppm, and dangerous above 750.

Remarks:

Care must be taken to distinguish between nitrate and HCN poisoning, because the treatment would cause death if the animal suffered from nitrate poisoning.

Sorghum, sorghum hybrids, and Sudangrasses should not be used for horse pasture; they are capable of producing a glycoside which converts to free cyanide in the horse and causes cystitis.

Piper Sudangrass is low in HCN.

SALMONELLOSIS

This is a toxic disease of all animals caused by many species of salmonellae.

Source:

Infected animals shed the bacteria in their feces, which are consumed by other animals in contaminated feedstuffs. Rodents and birds are also sources of infection.

Species Affected:

All animals, including humans.

Symptoms and Signs (or age group most affected):

Diarrhea, depression, dehydration, and fever. Often blood is seen in the feces. In pigs, a dark red-to-purple discoloration of the skin is common, especially of the ears and ventral abdomen. Nervous signs may appear in calves and pigs.

Distribution and Losses Caused By:

The disease is worldwide, and the incidence is increasing with the intensification of livestock production.

Usually 10 to 80% of the animals in a group are affected. Death losses vary with the severity of the infection and the treatment given.

Treatment:

Fluid therapy to correct acid-base balance and dehydration, and antibacterial drugs (broad-spectrum antibiotics, nitrofurans, or ampicillin); under the direction of a veterinarian.

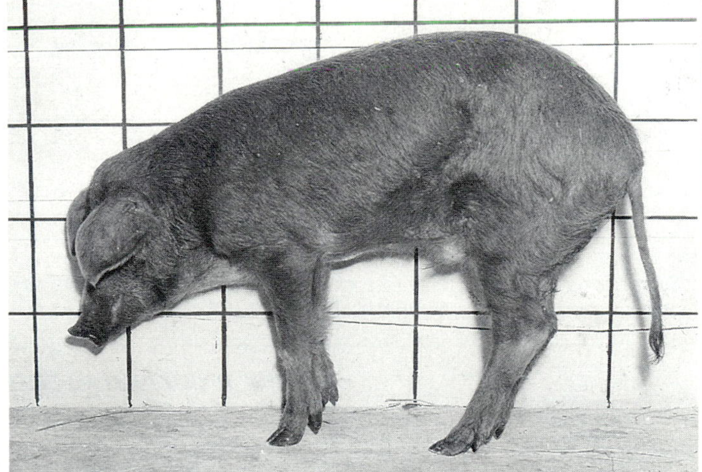

Fig. 5-54. Pig with salmonellosis. Note signs of marked depression and weakness. (Courtesy, Department of Veterinary Pathology and Hygiene, College of Veterinary Medicine, University of Illinois, Champaign-Urbana)

Prevention:

Quarantine new animals and avoid contaminated feed. Practice good sanitation. Clean and disinfect animal quarters, followed by drying. Control rodents and birds. Reduce stress, because stress triggers the disease. **NOTE WELL**: Producers could do away with salmonellosis by separating manure from animals. Also, salmonella in human foods can be destroyed by irradiation and proper cooking. **NOTE WELL**: Before fresh poultry can be irradiated, FDA must approve the treatment; and before it will be irradiated, consumers must be willing to buy the treated product.

Remarks:

Salmonellosis is named for the American bacteriologist and veterinarian, Daniel E. Salmon, who first isolated the organism in 1885.

Animals in close proximity to one another, such as veal calves or confined cattle, are of greater risk than isolated animals.

(Also see Chapter 13, section on Antibiotics, subsection headed "Safety And Future Of Antibiotics As Feed Additives.")

(Continued)

TABLE 5-3 *(Continued)*

SALT POISONING (NaCl—SODIUM CHLORIDE)

Salt may be toxic when excessive quantities are ingested and water intake is limited. "Salt poisoning" is a misnomer because the condition usually occurs with concomitant water deprivation.

Source:

Brine from cured meats; wet salt.
Where large amounts of brine or salt have been mixed in hog slop.
When excess salt is fed following salt starvation.
When salt is improperly used to govern self-feeding of concentrate.

Species Affected:

All farm animals, but swine and sheep are most frequently affected.

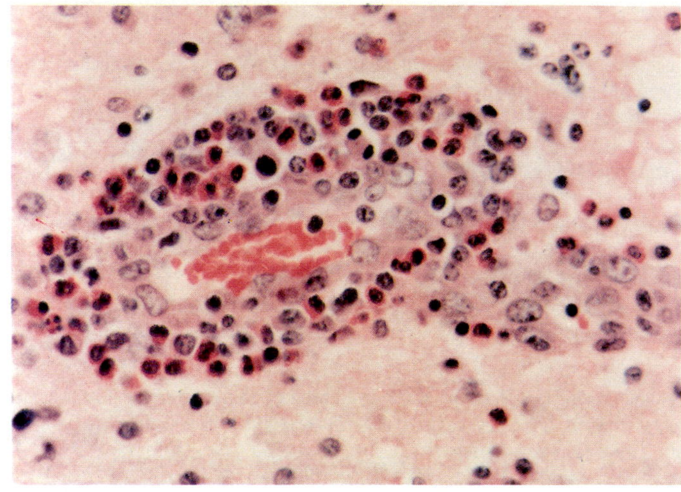

Fig. 5-55. Photomicrograph of the brain of a salt-poisoned pig, showing edema and lesions of the brain. (Courtesy, College of Veterinary Medicine, University of Tennessee, Knoxville)

Symptoms and Signs (or age group most affected):

Sudden onset—1–2 hours after ingesting salt; extreme nervousness; muscle twitching and fine tremors; much weaving, wobbling, staggering and circling; blindness; weakness; normal temperature, rapid but weak pulse,and very rapid and shallow breathing; diarrhea; death from a few hours up to 48 hours. Convulsions seldom occur, except in pigs.

Distribution and Losses Caused By:

Salt poisoning is relatively rare except in pigs.

Treatment:

Provide large quantities of fresh water to affected animals. Those unable to drink should be given water via stomach tube, by a veterinarian.

Prevention:

If animals have not had salt for a long time, they should first be hand-fed salt, gradually increasing daily allowance until they leave a little in the mineral box; then self-feed.

Remarks:

Indians and pioneers handed down many legendary stories about huge numbers of wild animals that killed themselves simply by gorging at a newly found salt lick after having been salt-starved for long periods of time.
Salt poisoning in pigs and poultry does not occur unless there is water deprivation. Even normal salt concentration may be toxic if water intake is low. (Also see Chapter 4, Table 4-6, ANIMAL MINERAL CHART, SODIUM.)

SELENIUM TOXICITY (ALKALI DISEASE, BLIND STAGGERS)

Source:

Chronic selenium poisoning, commonly called *alkali disease,* may result when animals consume forages and grains containing 5 to 40 ppm selenium.
Acute selenium poisoning, commonly called *blind staggers,* may result when animals consume high selenium content plants known as "selenium accumulators." Two of the accumulators in North America are: *Astragalas* and *Stanleya.* Also, cases of selenium toxicity have been caused by animals breaking into bags and consuming selenium supplements containing high levels of selenium.

Species Affected:

All farm animals and people. Young animals are especially susceptible. **NOTE WELL**: Confirmed cases of selenium toxicity in humans are rare because (1) the foods and beverages which are consumed by people are not likely to contain excesses of the element, and (2) well-nourished people are protected by metabolic processes that convert selenium into harmless substances which are excreted in the urine or in the breath. Nevertheless, a few cases of poisoning have occurred under unusual circumstances such as (1) very high levels of the element in the drinking water, or (2) the presence of malnutrition, parasitic infestation, or other factors which may make people highly susceptible to selenium toxicity.

(Continued)

TABLE 5-3 *(Continued)*

SELENIUM TOXICITY (ALKALI DISEASE, BLIND STAGGERS) (Continued)

Symptoms and Signs (or age group most affected):

In chronic selenium poisoning (alkali disease), there is a loss of hair from the mane and tail of horses, a loss of hair from the tail of cattle, and a general loss of hair in swine. The hoofs slough off, lameness occurs, feed consumption decreases, and death may occur from starvation. Chronic selenium poisoning in poultry and other birds is characterized by reduced egg production and hatchability, and by deformities in young, including lack of eyes and deformed wings and feet.

In acute selenium poisoning (blind staggers), vision is impaired. In cattle, blind staggers is manifested in 3 stages. In stage 1, there is a tendency to wander, the vision is poor, and there is a loss of appetite. In stage 2, the wandering increases, the vision becomes poorer, and the front legs become weak. In stage 3, the throat and tongue become paralyzed, the temperature is subnormal, and death follows from respiratory failure. In sheep, the 3 stages are not as clearly differentiated as in cattle.

Fig. 5-56. Selenium toxicity in cow grazing on forage produced on alkali soil containing excessive selenium. Note emaciated condition, curvature of back, and deformed hoofs. (Courtesy, Wyo. Ag. Exp. Sta., Laramie, Wyo.)

Distribution and Losses Caused By:

Most of the selenium poisoning in livestock occurs in Colorado, Nebraska, South Dakota, and Wyoming. Also, high selenium has been reported in Alberta, Saskatchewan, and Manitoba in Canada; in Mexico, Ireland, Israel, and China; in northern Queensland in Australia; and in parts of South America.

Losses result primarily from the failure to thrive on forages grown on seleniferous soils. But some deaths occur from the chronic and acute forms of the disease.

Fig. 5-57 shows the selenium level of U.S. soils and feeds. Note the deficient and high regions of the country.

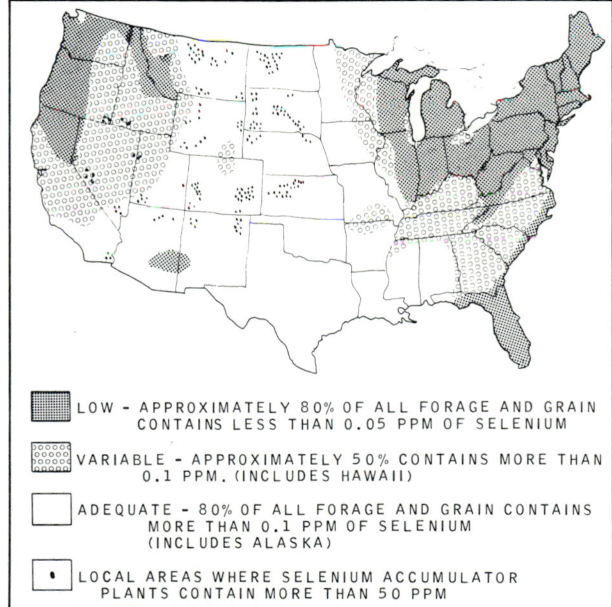

Fig. 5-57. A geographic distribution of low, variable and adequate or high selenium areas in the U.S. (Source: Kubota, J. and W. H. Allaway, "Geographic Distribution of Trace Element Problems," *Micronutrients in Agriculture,* Soil Science Society of America)

Treatment:

The effect of chronic selenium toxicity may be reduced by feeding a high-protein ration, by the use of trace amounts of arsenic compounds, and/or by the oral administration of such compounds as naphthalene and bromobenzene; with such treatments under the direction of a veterinarian or nutritionist.

There is no known treatment for acute selenium poisoning.

Prevention:

Soils which contain more than 0.5 ppm selenium are potentially dangerous. Chronic toxicity is caused by rations containing as little as 8.5 ppm of selenium. In swine, levels as low as 10 ppm have been found to lower the conception rate and result in a higher percentage of small, weak, and dead pigs at birth. Acute cases of poisoning have been reported from levels of 500 to 1,000 ppm. So, prevention consists in the cautious use of soils and crops exceeding these levels. Some of these areas may be used by pasture rotation and the use of supplemental feeds.

The maximum and toxic levels of selenium for farm animals are:

	Maximum Total Recommended By FDA	Toxic Level	
	(mg/head/day)	(ppm in feed)	(mg/head/day)
Beef cattle	1.0	10–30	100–300
Dairy cattle	2.0	3–5	30–60
Sheep	0.23	3–20	7–50
Swine		5–10	8–16
Chickens		2	
All species	2.0 (or 2 ppm)[1]		

[1]Sugggested maximum tolerable level for all species (NRC, 1980).

Remarks:

Although selenium is highly toxic in overdoses, it is essential in trace amounts to maintain life and to prevent such conditions as muscular dystrophies in many species and exudative diathesis in chicks.

The toxic effects of selenium were observed long before the existence of this element was known. In his travels in the mountains of western China at the end of the 13th century, Marco Polo recorded that consuming certain forages caused the hoofs of grazing animals to drop off.

(Also see Chapter 4, Table 4-6, ANIMAL MINERAL CHART, SELENIUM.)

(Continued)

TABLE 5-3 *(Continued)*

SULFUR POISONING

Source:

Excessive doses of elemental sulfur (flowers of sulfur).

Species Affected:

All animals, but most commonly cattle and sheep.

Symptoms and Signs (or age group most affected):

Dullness, abdominal pain, muscle twitching, diarrhea, dehydration, and a strong odor of hydrogen sulfide on the breath. The animals go down, are short of breath, develop convulsions, and die in a coma.

Lime sulfur, which is a complex of sulfides, may cause skin irritation, discomfort, or blistering.

Distribution and Losses Caused By:

Sulfur toxicity is not common. However, potential toxicity exists wherever sulfur is used as a tonic, or to control parasites, or to restrict the feed by lambs and thus reduce the incidence of enterotoxemia.

Treatment:

Treatment consists in deleting the sulfur.

Prevention:

Prevention consists in not overdosing with sulfur. The feeding of 3 oz per head to cattle has been fatal, and the minimum lethal dose of a sulfur-protein concentrate for sheep is estimated to be 4.5 g per lb.

Remarks:

It appears that sulfur is most toxic when fed in rations that are high in protein.

TRIARYL PHOSPHATE

(See ORGANOPHOSPHORUS COMPOUNDS)

UREA TOXICITY (AMMONIA TOXICITY/BOVINE BONKERS/CRAZY CATTLE DISEASE)

Ammonia is the actual toxic agent in urea poisoning.

Source:

Primarily urea, which dominates the market. But other nonprotein nitrogen (NPN) products, such as ammonium salts and ammoniated feeds, which are used in ruminant rations.

Urea is hydrolyzed by the urease activity of the rumen microorganisms with the production of ammonia as follows:

$$\left.\begin{matrix} NH_2 \\ \\ NH_2 \end{matrix}\right\rangle \underset{\text{(urea)}}{C{=}O} + \underset{\text{(water)}}{H_2O} \xrightarrow[\text{urease}]{\text{microbial}} \underset{\text{(ammonia)}}{2NH_3} + \underset{\text{(carbon dioxide)}}{CO_2}$$

When urea is fed at excessive levels, large amounts of ammonia are liberated in the rumen. Eventually, the pH of the ruminal fluid increases, thus facilitating the passage of ammonia across the rumen wall. If the levels of ammonia absorbed are greater than the ability of the liver to convert ammonia to urea, ammonia accumulates in the blood. If blood ammonia levels reach toxic levels (80 mg per 100 ml), the animal shows signs of acute ammonia poisoning.

Species Affected:

Primarily ruminants, for little nonprotein nitrogen is used by nonruminants. There is hazard of ammonia toxicity (urea toxicity) when NPN is fed to young ruminants and young equines, due to their limited bacterial action.

Symptoms and Signs (or age group most affected):

The animal shows signs of nervousness, excessive salivation, muscular tremors, respiratory difficulty, and tetanic spasms. Death occurs within ½ to 2½ hours.

(Continued)

TABLE 5-3 *(Continued)*

UREA TOXICITY (AMMONIA TOXICITY) (Continued)

Distribution and Losses Caused By:

Urea or ammonia toxicity should never occur in practice, if feeds are thoroughly mixed and total intakes are moderate. Errors in formulation and improper mixing of urea with other ration ingredients are probably the major factors causing urea toxicity in the feeding of ruminants.

Treatment:

An effective treatment for urea toxicity of cattle, if applied before tetanic spasms occur, consists in administering, immediately, 5 to 10 gal of cold water orally. A gallon of either dilute acetic acid or vinegar given with cold water is more effective than cold water alone.

Prevention:

Prevention of urea toxicity consists in alleviating, or lessening, the following predisposing factors:

1. Poor mixing of feed.
2. Errors in ration formulation.
3. Inadequate period of adaptation.
4. Low intake of water.
5. Feeding urea in conjunction with poor quality roughages.
6. Low feed intake prior to feeding urea.
7. Treating hay with more than 1.5% anhydrous ammonia, and treating it unevenly.

Remarks:

Urea is less effective in young ruminants in which the rumen is not fully functional.

Urea and other NPN sources are not used to any appreciable extent for swine or poultry. The mature horse can utilize limited NPN (such as access to protein blocks containing urea); however, the efficiency of nitrogen utilization from NPN is considerably less than that of nitrogen from intake protein.

(Also see Chapter 11, Protein Supplements, sections on "Urea" and "Ammoniated Products.")

VITAMIN A TOXICITY

(See HYPERVITAMINOSIS.)

VITAMIN D TOXICITY

(See HYPERVITAMINOSIS.)

VITAMIN K (MENADIONE) TOXICITY

(See HYPERVITAMINOSIS.)

Diagnosing and Treating Livestock Poisoning

It is often difficult to make a definite diagnosis of an animal poisoning. Clinical signs are not usually specific, and all signs are not always seen in every poisoned animal. However, in a herd of poisoned animals, every sign or toxic effect will likely be seen in some animal. The recommended procedure for making a diagnosis of the cause of poisoning follows:

1. Check on the accessibility of a poisonous substance. A highly toxic substance may or may not be hazardous to livestock, depending on whether the animals could conceivably have come into contact with it.
2. Study the clinical signs. This may be difficult, especially with possible combinations of toxins or infectious agents.
3. Use a few test animals in a feeding trial.
4. Make a pathologic examination of the animal's internal organs and tissues.
5. Chemically analyze the feed, water, and animal tissues for the presence of suspected toxin. It is necessary to have enough information so that certain poisons or groups of poisons can be suspected, because the analytical methods are quite specific and certain tissues are required.
6. Use a specific antidote (where available) for the suspected poison. If it alleviates the clinical signs, it gives evidence of the cause.

The principles of treatment are directed toward accomplishing the following:

1. Preventing injury and controlling convulsions with a sedative, usually a barbiturate.
2. Relieving pain by use of chemical analgesics.
3. Removing or neutralizing the poison by—
   a. Washing off any surface poison.
   b. Using gastric lavage with activated charcoal for absorbing toxins in the stomach.
   c. Using cathartics to help fecal elimination of unabsorbed toxins.
   d. Using diuretics to help urinary elimination of absorbed toxins.
   e. Performing a rumenotomy for physical removal of unabsorbed toxins.
   f. Using a specific antidote, if available.
4. Maintaining the vital signs of respiratory, circulatory, and renal functions by physical or chemical resuscitation, fluid therapy, etc.

5. Observing the animal for further treatment needs, because the toxin may continue to be absorbed from the skin, gut, or respiratory system of the animal.

National Animal Poison Control Center

Established in 1978 and maintained at the University of Illinois, Urbana-Champaign, the National Animal Poison Control Center hotline number is: *217/333-3611*. Recognizing that accidents don't wait for business hours, the Center is open 24 hours a day, every day of the week. The toxicology group is staffed to answer questions about known or suspected cases of poisoning or chemical contaminations involving any species of animal. It is not intended to replace local veterinarians or state toxicology laboratories, but to complement them.

The toxicologists at the Center constantly update their files on chemicals, feed additives, human and veterinary drugs, pesticides, environmental contaminants, and plant and mold toxins. Their comprehensive file of information contains comparative species toxicity data, product ingredients, and recommended therapeutic and decontamination measures. The goal is a computer database containing 200,000 entries to facilitate quick and accurate responses to all types of poisoning/contamination incidents and inquiries.

Many times a proper treatment regimen can be recommended over the telephone. When telephone consultation is inadequate or the problem is of major proportions, a team of veterinary specialists can arrive at the scene of a toxic or contamination problem within a short time.

The cost of an investigation varies according to distance traveled, personnel time, and laboratory services required. Where consultation over the telephone is adequate, there is no charge to the veterinarian or producer.

HAIR ANALYSIS

Minerals are deposited in hair as it grows; and, in theory, the hair reflects the mineral status of an animal or person at the time of hair growth. Scientific analytical methods, such as atomic absorption spectometry, neutron activation analysis, and x-ray fluorescence spectometry, are sensitive enough to detect the levels of even trace minerals in hair samples. Because hair samples are easily and painlessly obtainable, and because hair samples are stable and store easily, there is considerable interest in the use of hair as a diagnostic tool for mineral deficiencies and/or toxicities. However, like many other diagnostic tests, hair analysis is only a tool to complement other tests and observations. For the reasons which follow, nutritional status cannot be assessed solely on the basis of hair analysis:

1. The relationship between the concentration of a mineral or of a vitamin in the hair and that in other body tissues is unknown.

2. Laboratory results depend upon the proper treatment of the hair sample. Hair samples must be washed properly to remove outside contamination from environmental sources, but washing leaches some of the deposited minerals out of the hair.

3. Sampling procedures vary. Mineral levels at the ends of hair may be different from that near the hide, or mineral levels of hair may vary according to the part of the body from which the hair sample is taken.

4. Factors other than ration influence the mineral content of hair. Such factors include sex, age, and hair color. For example, red hair contains more iron than hair of other colors. Also, sweat contains trace minerals.

Clearly, some scientific findings demonstrate that if dietary intakes of certain minerals (chromium, copper, iron, manganese, and selenium) are extremely low, if the ration includes toxic minerals (arsenic, cadmium, lead, and mercury), hair analysis can detect these changes. Furthermore, some disorders such as anemia are reported to change the mineral levels of the hair.

Hair analysis is a growing industry, and some laboratories encourage individuals to submit hair samples to assess their nutritional status by multiple mineral hair analysis. Some of these laboratories then prescribe minerals and vitamins to correct the metabolic imbalances, or deficiencies, discovered by the analysis. Not surprisingly, some of these laboratories also sell minerals and vitamins.

In summary, while hair analysis has been used to detect certain types of heavy metal poisoning (e.g. lead, arsenic, mercury), its value in determining nutritional status remains to be established. There are a number of limitations to hair analysis, both in terms of analytical procedures and in interpretation of the results.

GOOD LIVESTOCK REQUIRE GOOD SOILS

Fig. 5-58. Good farmers/ranchers, good livestock, good pastures and crops, and good soils go together. (Courtesy, *Holstein World*, Sandy Creek, N.Y.)

Proper animal nutrition is obtained by feeding hays, silages, root crops, grains, and pastures grown on fertile soils. Good livestock require good soils.

Present research, together with practical observation, points to the fact that the mere evaluation of crop yields in terms of tons of forage or bushels of grain produced per acre is not enough. Neither does a standard feed analysis (a proximate analysis) tell the whole story. Rather, there is a direct and most important relationship between the fertility of the soil and the composition of the plant. Sometimes the concentration of an essential mineral in the soil is so low that plants growing on it will not contain enough of that mineral to meet the dietary requirements of animals. Under such circumstances, adding the deficient element to the soil may help the animals meet their requirements. At other times, feed crop plants may contain such high concentrations of certain minerals that they are toxic to the animals that eat them.

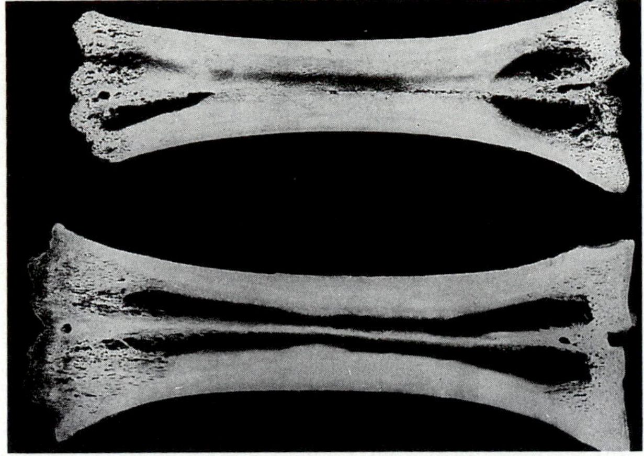

Fig. 5-59. Soil nutrients made the difference! Split bones from two calves of similar breeding and age. Small, fragile, pitted bone (top) obtained from calf pastured on belly-deep grass grown on highly weathered soil low in mineral content. Big, rugged, strong bone (bottom) from calf grown on moderately weathered, but highly mineralized soil. (Courtesy, University of Missouri, Columbia)

The land surface of the earth is covered with many kinds of soil. Some soils naturally contain an abundant supply of most of the elements needed by both plants and animals. Others may have an abundant supply of most required elements and yet be deficient in one or more essentials. For example, southwestern United States is known as a phosphorus-deficient area; northwestern United States and the Great Lakes area are iodine-deficient areas, and southeastern United States is a cobalt-deficient area.

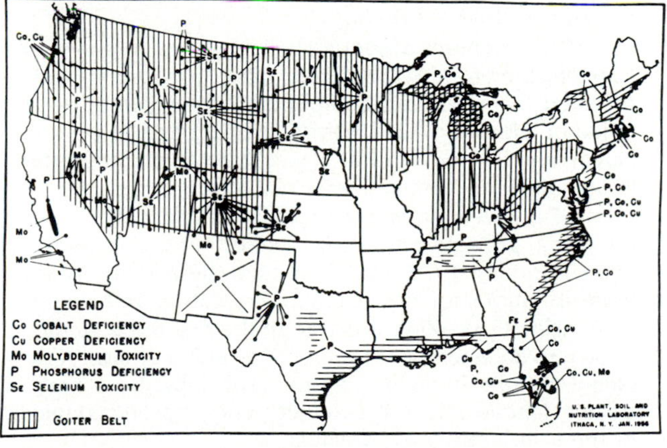

Fig. 5-60. Mineral deficiency areas of the U.S. and the excess selenium area of the northern and central Great Plains. (Courtesy, USDA)

Also, differences between plant species in their tendency to accumulate different elements are often important in determining the mineral status of animals that eat these plants. For example, in many places in the United States such forages as alfalfa and clover contain adequate levels of cobalt for cattle and sheep, whereas grass species in the same fields or pastures do not contain enough cobalt to meet the requirements of these animals. In places in the Great Lakes area, some native species of vetch accumulate toxic levels of selenium, yet farm crops and pasture grasses growing in the same field will contain substantially lower, and generally nontoxic, levels of this element.

At every step in the chain from soils to animals, the essential mineral elements interact with other elements, and these interactions may profoundly affect the availability of essential elements to plants or animals or the amount of the essential element required for normal growth or metabolic function. For example, a high level of soluble iron in the soil may depress the solubility of phosphorus and cause plants, and the animals that feed upon the plants, to suffer from phosphorus deficiency. The availability of zinc to animals may be depressed if the ration is high in calcium; and high levels of molybdenum may interfere with copper metabolism in animals. These and other interactions must be considered in assessing whether a given soil will supply plants with needed nutrients, and, in turn, whether plants will supply the animals that consume them with needed nutrients.

Thus, the transfer of essential nutrient elements from soils to plants, thence to animals, is a complicated process.

Some of the pathways and relationships of soil elements pertinent to the nutrition of animals follow:

1. **Calcium (Ca).** The genetic nature of the plant is very important in determining its calcium composition. Thus, the legumes are able to withdraw more calcium from the soil than grasses. It follows that red clover grown on the low-calcium soils of the northeastern United States contains more calcium than grasses grown on the high-calcium soils of the West.

A deficiency of calcium has been observed where cattle graze on pastures produced on acid, low-calcium soils and containing grasses but no legumes. Also, a calcium deficiency may occur when farm animals are fed high cereal grain rations, because grains are generally low in calcium.

2. **Cobalt (Co).** Cobalt moves from the soil into plants. When plants are eaten by ruminants, the cobalt is combined with other materials in the rumen to form vitamin B-12. Areas of low-cobalt soils in the United States occur in the New England states and along the South Atlantic Coastal Plain.

Adding cobalt to soils, either as cobalt sulfate or as cobaltized superphosphate, can be used to increase the level of cobalt in plants and prevent cobalt deficiency in cattle and sheep.

3. **Copper (Cu).** Copper is required by both plants and animals. In some parts of Australia, livestock production was impossible until copper fertilizers were used on pastures.

If the molybdenum concentration in the forage is high, extra amounts of copper will be needed to prevent deficiency. Also, with high molybdenum, higher levels of copper can be tolerated without danger of toxicity.

4. **Fluorine (F).** In animals and humans, low levels of fluorides have beneficial effects on teeth and bone structure.

But excessive levels of certain fluorine compounds—from fumes and dusts emitted from industrial plants, and from high levels of fluoride in water—are very toxic to both plants and animals.

Where increased fluoride intake is desirable to prevent dental caries in humans, carefully controlled direct additions to the drinking water is more promising than adding fluorides to soils that produce food crops.

5. **Iodine (I).** The relationship between the iodine levels of the soil and the incidence of iodine deficiency (goiter) in animals and people is well known. Iodine is not required by plants. Nevertheless, if iodine is present in the soil, it is taken up by most plants and moves on into diets of animals (and humans) in forms that are effective in preventing goiter. However, there are important differences among plant species, and even among varieties of the same species, in their tendency to take up iodine from the soil.

Many of the iodine-deficient areas of the world have been identified and mapped. Thus, northwestern United States and the Great Lakes region are well-known iodine-deficient areas.

The use of iodized salt is such an effective and low-cost way of supplying this element to animals and people that there is little need to include iodine in fertilizers.

6. **Iron (Fe).** Iron deficiency is a serious problem in crop production in certain areas, and some nutritionists consider iron-deficiency anemia to be the most frequently observed mineral deficiency in young pigs and in people. Nevertheless, iron fertilization of soils is not likely to be effective in decreasing the incidence of iron deficiency in pigs and people. Actually, most soils contain plenty of iron; however, it may not be soluble. Some soils that are red from iron compounds contain too little available iron for normal plant growth.

Among farm animals, iron deficiency is most common in young pigs raised in confinement. Injecting the baby pigs with iron compounds and painting the sow's udder with iron compounds are the accepted methods to prevent iron deficiency on modern hog farms.

7. **Magnesium (Mg).** The level of magnesium in plants is affected by soil content, plant species, and the presence of potassium. The leguminous plants (alfalfa, clover, beans, and peas) usually contain more magnesium than the grasses, corn, and other nonleguminous plants. Also, a very high level of available potassium in the soil interferes with the uptake of magnesium by plants; hence, magnesium deficiency in plants is often found when they are grown on soils that are very high in available potassium.

The most dramatic indication of acute magnesium deficiency in farm animals is that of grass tetany (grass staggers) in cattle (see Table 5-1). Not all factors causing grass tetany are fully understood. However, it is known that the grasses eaten by afflicted animals are usually low in magnesium and high in potassium.

8. **Manganese (Mn).** Manganese is required by both plants and animals. The uptake of manganese in feed and food plants is more dependent on the acidity or alkalinity of the soil than on the amount of manganese used in fertilizers.

Chickens are subject to manganese deficiency, especially when fed high corn rations. Manganese deficiency causes the deformity of the legs known as *perosis,* or *slipped tendon.* Because of this situation, commercial poultry rations are nearly always supplemented with extra manganese.

9. **Molybdenum (Mo).** Molybdenum is required in very low amounts by both plants and animals, but deficiencies of molybdenum are infrequent in either. However, nutrient imbalances involving molybdenum and copper have caused serious problems in cattle and sheep.

Molybdenum toxicity is actually a molybdenum-induced copper deficiency. The symptoms of molybdenum toxicity are identical to those of copper deficiency and include fading of the hair and diarrhea. It may be prevented by copper supplementation or injection.

10. **Phosphorus (P).** When phosphorus is added to soils deficient in available forms of this element, increased crop and pasture yields usually follow. Sometimes the phosphorus concentration in the crop is increased, but this is not always the case.

Washington State University produced alfalfa hay on soils containing two different levels of phosphorus, then fed it to rabbits.[1] The rabbits in the low-phosphorus group made slower growth, required more matings per conception, and had more fragile bones than the high-phosphorus group. There is reason to believe that soil nutrients affect all animals similarly—in growth, conception, and soundness of bone; but more experimental work on this subject is needed.

11. **Potassium (K).** Potassium is required by most plants and animals. Most soils contain an abundance of total potassium, but the level of available, or soluble, forms of potassium is frequently inadequate. The role of potassium fertilizers is to increase crop yields rather than to improve the nutritional quality of the crop produced.

12. **Selenium (Se).** Selenium is not required by plants, but it is required in very small amounts by animals and birds. The dividing line between selenium requirement and selenium toxicity is very narrow, and excess selenium produces toxicity.

In large areas of the United States, the soils contain little available selenium. Crops produced in these areas are low in selenium, and selenium deficiency in livestock is a serious problem. On the other hand, the soils in some areas of the Great Plains and Rocky Mountain states are rich in readily available selenium with the result that the plants produced thereon are so high in selenium that they are poisonous to the animals that eat them.

In 1957, selenium was found to be essential in preventing liver degeneration of laboratory rats. Subsequent studies revealed that selenium compounds, either added to the ration or injected, would prevent certain diseases of lambs, calves, and chicks. Further, it was shown that rations high in vitamin E necessitated less selenium.

In 1974, FDA approved the addition of selenium to rations for swine, growing chickens, breeder hens, and nonfood animals; and in 1979, FDA approved the addition of selenium in rations for beef cattle, dairy cattle, and sheep.

13. **Zinc (Zn).** Zinc was one of the first trace elements known to be essential for both plants and animals. A deficiency of zinc in swine rations, especially those high in calcium (above 0.8% Ca), produces a dry, cracked condition of the skin, known as *parakeratosis.*

14. **Other elements.** In addition to the elements already discussed, several others may be beneficial to plants and/or animals. Some of these are relatively abundant; others await more research.

Boron (B) is required by plants, but not known to be required by animals.

[1]Heinemann, W. W., *et al., Wash. Ag. Exp. Sta. Tech. Bull 24.*

Chlorine (Cl) is required by both plants and animals, but chlorine deficiencies under practical conditions are relatively rare. The addition of salt (sodium chloride) to animal and human diets has been a common practice for centuries. Salt accomplishes two things: (1) it improves flavor; and (2) it ensures an adequate intake of chlorine.

Chromium (Cr) is one of the most recent elements to be added to those required by animals and people. It appears that certain compounds of chromium activate insulin during sugar metabolism. Chromium is not required by plants.

Silicon (Si) is an essential element for both plants and animals. However, it is one of the most abundant elements of the earth's crust and a major component of most soils.

Sodium (Na) is required by all animals. Adding salt (sodium chloride) to animal and human diets is the oldest known dietary supplementation practice.

Sulfur (S) is an important component of most proteins and an essential element of all plants and animals. Ruminants can use inorganic sulfur in their rations, because the microorganisms in the rumen can convert the inorganic sulfur into sulfur amino acids. However, pig and chicken rations are usually fortified with sulfur-containing amino acids.

Nickel (Ni), strontium (Sr), tin (Sn), and vanadium (V) have been shown to have some beneficial effects on laboratory animals. But there is no evidence of any relationship between the levels of these elements in soils and plants and the nutrition of animals and humans.

Soil Testing

Soil testing is the most important single guide to the profitable application of fertilizer and lime. When properly done, it provides a reliable basis for planning the fertility program for each field.

Traditionally, soil testing has been used to decide how much lime and fertilizer to apply. Today, it is important to determine where fertilizer should not be used for two reasons: (1) with high fertilizer prices, no more fertilizer should be applied than is needed; and (2) with increased concern about the environment, soil tests are a logical tool to determine areas where adequate or excessive fertilization has taken place.

If soils tests are to be of value, the samples must be taken properly and the results interpreted correctly. The local county agent (farm advisor) can provide instructions on how to take soil samples and recommend where to send them. Some land grant colleges run soil analyses at nominal cost.

ORGANIC FARMING

The U.S. Department of Agriculture's definition of organic farming follows:

> *"Organic farming is a production system which avoids or largely excludes the use of synthetically compounded fertilizers, pesticides, growth regulators, and livestock feed additives. To the maximum extent feasible, organic farming systems rely upon crop rotations, crop residues, animal manures, legumes, green legumes, off-farm organic wastes, mechanical cultivation, mineral-bearing rocks, and aspects of biological pest control to maintain soil productivity and tilth, to supply plant nutrients, and to control insects, weeds, and other pests."*[2]

Often, the meanings for *organic, natural,* and *health,* which people imply or conjure up in their minds, are misleading, harmful, and tend to polarize people. Frequently, it is proclaimed that such products are safer and more nutritious than conventionally grown and marketed foods. The growing interest of consumers in the safety and nutritional quality of the American diet is a welcome development. Regrettably, however, much of this interest has been colored by alarmists who state or imply that the American food supply is unsafe or somehow inadequate to meet our nutritional needs.

The FDA has taken no position on use of the terms *organic, natural,* and *health* in food labeling, because the terms are often used loosely and interchangeably. The Federal Trade Commission (FTC) in its proposed Food Advertising Rule would prohibit use of the words *organic* and *natural* in food advertising because of concern about the ability of consumers to understand the terms in the conflicting and confusing ways they are used.

- **Feed/food quality**—Undeniably, many benefits accrue from organic farming, not the least of which is the valuable exercise that the practicing advocates get from growing their own organically produced vegetables. However, no laboratory test or animal feeding trial can distinguish between crops fertilized with inorganic fertilizers and those fertilized with organic fertilizers. Moreover, experiments designed to compare the levels of different essential nutrients in crops produced with organic fertilizers against those produced with comparable amounts of nutrients supplied as inorganic materials have shown little difference, with the advantage in favor of the inorganic as often as the organic. In the few experiments in which the plants produced under the two systems have been fed to test animals, the small differences noted in animal growth have not consistently favored either the organic or inorganic sources of the nutrients.

The above results are as one would expect, based on the function of plants in the food chain—to convert inorganic compounds to organic compounds. If organic materials containing the nutrients are incorporated in the soil, the microorganisms in the soil must first break down the organic matter into inorganic forms. Inorganic ions of the essential nutrients are then taken up by the plant roots and manufactured into new organic materials. In the plants, and in the bodies of animals, these essential nutrient elements have the same effect, regardless of whether they were added to the soil in the form of organic or inorganic fertilizers.

- **Environmental control**—An important reason for adding organic materials to agricultural and garden soils is that this practice can be used to recycle the organic material without damaging the environment. On the other hand, too much organic matter can be added to the soil. Problems from excessive application of organic materials are generally confined to fields in close proximity to large cattle feedlots or poultry operations. Under such circumstances, nitrate toxicity and grass tetany have been serious problems where pastures have received excessive applications of manure.

---

[2]*Report and Recommendations on Organic Farming,* USDA, 1980.

• **Feed/food quantity**—The use of chemical fertilizers has been partly responsible for the abundance of food available. If all farmers were to adopt organic methods, there would be a decline in productivity as shown in Table 5-4.

TABLE 5-4
ESTIMATED NATIONAL AVERAGE CROP YIELDS UNDER CONVENTIONAL AND ORGANIC FARMING[1]

Crop	Bushels per Acre	
	Conventional	Organic
Corn	98	49
Wheat	43	20
Soybeans	40	20
Other grains	57	17

[1] *Organic and Conventional Farming Compared*, Council for Agricultural Science and Technology (CAST), Report No. 84, October 1980, p. 24, Table 6.

In a 1980 report issued by the Council for Agricultural Science and Technology (CAST), it was estimated that if organic farming were widely adopted the cost of food would increase because the total production from the land under cultivation would decrease. Those now practicing organic methods realize their yields are less, but they may receive higher prices for their goods because they sell to a specialty market.

• **Some inorganic fertilizers and mineral supplements are essential**—It is noteworthy that some nutritional problems can be corrected only by inorganic fertilizers or mineral supplements. For example, iodine deficiency in animals or people cannot be corrected by organic fertilizers, unless, of course, they contain kelp (seaweed) or marine by-products, good sources of iodine. The cobalt deficiency that plagued the cattle of the colonists in New Hampshire did not respond to organic fertilizers that contained little cobalt.

• **Summary**—Undoubtedly, world food production of the future will make use of a combination of both organic and inorganic fertilizers, with the nature and proportions of the combination for different farms and for different countries dependent on their access to fossil fuels, the availability and price of fertilizers, their soils, their food production requirements, their environmental control problems, and many other factors. Regardless of the combination of organic and inorganic fertilizers used, feed and food plants of adequate nutritional quality can be produced.

QUESTIONS FOR STUDY AND DISCUSSION

1. What are the primary differences between nutritional deficiencies brought about by (a) too little feed, and (b) rations too low in one or more nutrients? Which is the more serious?

2. What are the symptoms (a) of an energy deficiency, and (b) of a protein deficiency?

3. Discuss calcium-phosphorus ratio and vitamin D, from the standpoints of an adequate supply of each of them and a suitable ratio between calcium and phosphorus.

4. Analyses show corn to be adequate in niacin. Yet, there may be niacin deficiencies when corn is fed—deficiencies which can be rectified by niacin supplementation. How do you explain this?

5. What is pica? Discuss the pica/behavior of cattle suffering from phosphorus deficiency and of lambs suffering from enterotoxemia.

6. List what you consider to be the 10 most important nutritional diseases and ailments. Summarize pertinent information relative to each of the 5 nutritional diseases and ailments which you feel cause the greatest economic loss, using the following headings: Species affected; Cause; Symptoms and Signs; Distribution and Losses caused by; Treatment; Control; and Prevention.

7. List and discuss 5 important interactions of nutrition and diseases/parasites.

8. Discuss the cause and importance of each of the following nutrition related disorders: choking, displaced abomasum, hardware disease, immunoglobulin deficiency, and water deprivation.

9. Discuss the importance of each of the following points pertaining to poisonous plants:
   a. Economic importance
   b. Diagnosis
   c. Treatment
   d. Prevention

10. Why do the heaviest livestock losses from poisonous plants occur on the western ranges?

11. Have agricultural chemicals been good or bad? Justify your answer.

12. What is a poison? When it is suspected that an animal has been poisoned, what should the owner or caretaker do?

13. List what you consider to be the 10 most important potential animal poisons. Summarize pertinent information relative to each of the 5 potential poisons which you feel cause the greatest economic loss, using the following headings: Source; Species affected; Symptoms and Signs; Distribution and Losses caused by; Treatment; and Prevention.

14. Outline the recommended procedure for making a diagnosis of the cause of poisoning.

15. What is the purpose of the National Animal Poison Control Center? Where is it located?

16. What is your evaluation of hair analysis? Would you recommend that it be used for a determination of each of the following: heavy metal poisoning, mineral deficiencies, vitamin deficiencies?

17. Identify and discuss the major mineral deficiency areas of the United States.

18. Discuss the transfer of each of the following essential elements from soils to plants, thence to animals:
   a. Calcium
   b. Cobalt
   c. Copper
   d. Iodine
   e. Phosphorus
   f. Selenium

19. Why should soils be tested? How would you go about getting a soil test?

20. Do you advocate organic farming, as opposed to chemical farming? Justify your answer.

SELECTED REFERENCES

Title of Publication	Author(s)	Publisher
Aflatoxin: Scientific Background, Control and Implications	Ed. by L. A. Goldblatt	Academic Press, New York, N.Y., 1969
Animal Health Livestock and Pets. 1984 Yearbook of Agriculture	Norvan L. Meyer, Chairman of the Committee	U.S. Department of Agriculture, Washington, D.C., 1984
Animal Nutrition, 7th Edition	L. A. Maynard J. K. Loosli	McGraw-Hill Book Company, New York, N.Y., 1979
Animal Science, 8th Edition	M. E. Ensminger	The Interstate Printers & Publishers, Inc., Danville, Ill., 1983
Applied Animal Feeding and Nutrition: An-Outline	M. H. Jurgens	Kendall/Hunt Publishing Company, Dubuque, Iowa, 1973
Basic Animal Nutrition and Feeding	D. C. Church W. G. Pond	D. C. Church, O & B Books, Corvallis, Ore., 1974
Crops in Peace and War: Yearbook of Agriculture	Ed. by A. Stefferud	U. S. Department of Agriculture, Washington, D.C., 1950–51
Diseases of Poultry, 7th Edition	M. S. Hofstad, *et al.*	Iowa State University Press, Ames, Iowa, 1978
Diseases of Sheep	R. Jensen B. L. Swift	Lea & Febiger, Philadelphia, Pa., 1982
Diseases of Swine, 5th Edition	A. D. Leman, *et al.*	Iowa State University Press, Ames, Iowa, 1981
Effects of Soils and Fertilizers on Human and Animal Nutrition, The, Ag. Info. Bull, No. 378	W. H. Allaway	U. S. Department of Agriculture, in cooperation with Cornell University Agricultural Experiment Station, Washington, D.C., 1975
Feeds and Feeding, 2nd Edition	A. Cullison	Reston Publishing Company, Inc., Reston, Va., 1979
Feeds and Feeding, 22nd Edition	F. B. Morrison	The Morrison Publishing Company, Ithaca, N.Y., 1956
Health Issues Related to Chemicals in the Environment	A. L. Craigmill, Chairman of the Committee	Council for Agricultural Science and Technology (CAST), Ames, Iowa, 1987
Keeping Livestock Healthy: Yearbook of Agriculture 1942	Ed. by G. Hambidge	U. S. Department of Agriculture, Washington, D.C., 1942
Livestock Feeds and Feeding, 2nd Edition	D. C. Church	O & B Books, Inc., Corvallis, Ore., 1984
Livestock-Poisoning Plants of Arizona	E. M. Schmutz B. N. Freeman R. E. Reed	The University of Arizona Press, Tucson, Ariz., 1968
Merck Veterinary Manual, The, 6th Edition	C. H. Fraser, Editor	Merck & Co., Inc., Rahway, N.J., 1986
Organic and Conventional Farming Compared	S. R. Aldrich, *et al.*	CAST, Ames, Iowa, 1980
Pasture and Range Plants		Phillips Petroleum Company, Bartlesville, Okla., 1963
Plants for Man, 2nd Edition	R. W. Schery	Prentice-Hall, Inc., Englewood Cliffs, N.J., 1972
Plant World, The, 4th Edition	H. J. Fuller Z. B. Carothers	Holt, Rinehart and Winston, Inc., New York, N.Y., 1963
Range Development and Improvements	J. F. Vallentine	Brigham Young University Press, Provo, Utah, 1971
Soil Fertility and Fertilizers, 2nd Edition	S. L. Tisdale W. L. Nelson	The Macmillan Company, New York, N.Y., 1966
Soil Sense	G. E. Hull, *et al.*	U. S. Department of Agriculture, Washington, D.C., 1973
Soil: The Yearbook of Agriculture 1957	Ed. by A. Stefferud	U. S. Department of Agriculture, Washington, D.C., 1957
Stockman's Handbook, The, 6th Edition	M. E. Ensminger	The Interstate Printers & Publishers, Inc., Danville, Ill., 1983
Weeds, 2nd Edition	W. C. Muenscher	The Macmillan Company, New York, N.Y., 1955
Western Fertilizer Handbook, 4th Edition	Ed. by E. J. Shaw	Soil Improvement Committee, California Fertilizer Assn., Sacramento, Calif., 1965

TOM PHILLIPS

2
Feeds

SECTION II

FEEDS

Section II pertains to feeds in the broadest sense of the word. It includes natural and artificial products that are provided to animals for purposes of (1) sustaining them; (2) increasing production and/or efficiency, providing flavor, adding color, reducing stress; and/or (3) enhancing palatability, bulk, or the preservation of feeds.

Pasture and range forages, hays, silages, grains, and other high-energy feeds, protein supplements, and a host of by-product feeds and residues are the basis of successful livestock production. About two-thirds of the feeds consumed by animals are not suited to human comsumption. Many of them are produced on lands not adapted to the growth of bread grains or gardens.

Most feeds come directly or indirectly from plants that have their roots in the soil. Thus, we have the cycle as a whole—from the soil, through the plant, thence to the animal and back to the soil again.

The primary purpose of animals is to convert feeds into food, clothing, power, and recreation—hopefully, at a profit to the producer and with benefit to the consumer.

Chapter 6 presents a panoramic view of feeds and feedstuffs. It is followed by a chapter devoted to each of the major types of feedstuffs. Then, Chapters 13 to 16—supplements, processing, analysis/evaluation, and buying feeds—are pertinent to all feeds.

FEEDS, the preceding two-page spread, is from an original painting by the noted artist, Tom Phillips, prepared specially for this book. It portrays the artist's conception of what Chapters 6 to 16 are all about—the growing, harvesting, processing, and feeding of a great array of feeds. All flesh is grass!

Original painting by Tom Phillips

Chapter 

TYPES AND ROLES OF FEEDSTUFFS

Contents Page

Contents Page

Feeds are naturally occurring ingredients or materials fed to animals for the purpose of sustaining them. In many cases, the term feed connotes complete feeds, rations, or diets. The term feedstuff is generally synonymous with feed, except for one difference. *A feedstuff is any product, of natural or artificial origin, that has nutritional value in the ration when properly prepared.* Frequently, nonnutritive products are included in the ration for such purposes as increasing production and efficiency, providing flavor, adding color, reducing stress, or for reasons related to palatability, bulk, or preserving feeds. Hence, compounds such as butylated hydroxytoluene (commonly abbreviated as BHT, an antioxidant) and vitamin A acetate (a synthetic vitamin) can be considered feedstuffs but not feeds.

Different species of livestock utilize various feeds with different efficiency. That is, what may be a high-energy feed to a ruminant may not be so classified for a nonruminant due to the varying ability of the respective animals to digest, absorb, and utilize the nutrients within the feed. Thus, the livestock producer must be knowledgeable relative to the specific application of many types of feeds.

Chapter 6, Types and Roles of Feedstuffs, is an umbrella type of chapter, designed to present an overall view of feeds and feedstuffs. Detailed discussions of various feeds and feedstuffs are presented in subsequent chapters of this book under the following classifications: Pasture and Range Forages (Chapter 7); Hay (Chapter 8); Silage/Haylage/High-Moisture Grain (Chapter 9); Grains/High-Energy Feeds (Chapter 10); Protein Supplements (Chapter 11); By-product Feeds/Crop Residues (Chapter 12); Feed Supplements/Additives/Implants (Chapter 13); Feed Processing (Chapter 14); Feed Analysis/Feed Evaluation (Chapter 15); and Buying Feeds/Commercial Feeds/Feed Laws (Chapter 16). Additionally, a glossary of feedstuffs and feed composition tables are presented in Chapter 30 and Section V, respectively.

A complete listing of all feedstuffs fed to livestock would be extremely long and diverse—ranging from acacia to yucca. More than 2,000 different products have been classified as animal feeds, exclusive of variety, grade, and stage of maturity differences within individual feeds. In the future, new advances in plant breeding will increase this number, thereby providing feeds which yield higher and have greater adaptability to the environment, and which more nearly meet the nutritive needs of animals. Recent pressures to recycle industrial wastes have ushered in a new era for the feed industry. Thus, within a few years, the vast majoity of animal feeds will include one or more by-product feeds.

Along with the production of new feeds, new chemicals, antibiotics, and growth-promoting factors are currently being developed and extensively tested for safety and efficacy. The list of feed additives grows longer—each new product offering ways of improving rate and/or efficiency of production.

The relative importance of the principal U.S. livestock feeds is shown in Fig. 6-1.

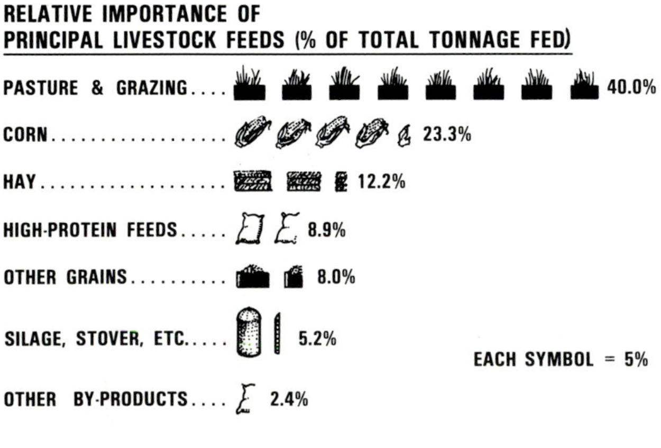

RELATIVE IMPORTANCE OF PRINCIPAL LIVESTOCK FEEDS (% OF TOTAL TONNAGE FED)

PASTURE & GRAZING	40.0%
CORN	23.3%
HAY	12.2%
HIGH-PROTEIN FEEDS	8.9%
OTHER GRAINS	8.0%
SILAGE, STOVER, ETC.	5.2%
OTHER BY-PRODUCTS	2.4%

EACH SYMBOL = 5%

Fig. 6-1. These principal livestock feeds are converted into meat, eggs, milk, and wool. (Source of data: USDA, Economic Research Service)

CLASSIFICATION OF FEEDSTUFFS

In general, feedstuffs may be classified into one of the following categories: (1) forages, (2) concentrates, (3) complete feeds, (4) premixes, (5) by-product feeds and crop residues, (6) specialty feeds, and (7) supplements, additives, and implants. Regardless of the criteria used to classify them, there are always some feedstuffs that do not fit into any one category or which fit the criteria of more than one class. However, livestock producers need not be too concerned relative to the precise classification of each feedstuff. Rather, they should familiarize themselves with the characteristics, along with the conditions that affect the nutritive value of each available feed as it pertains to the species of animal they are feeding and their particular feeding program.

National Research Council Classification and Nomenclature

Under the auspices of the National Research Council (NRC), a system of classifying and naming feeds in a detailed, systematic manner was developed to meet the following needs:

1. To provide a means of describing feeds accurately and in detail. By establishing uniform nomenclature, all farmers and feed manufacturers should, theoretically, know what others are talking about.
2. To facilitate computer formulation of rations.
3. To provide an international classification and listing of feedstuffs.

The NRC classification of feedstuffs follows:

1. **Dry roughages and forages**
   - Hay
     - Legume
     - Nonlegume
   - Straw
   - Fodder
   - Stover
   - Other feeds with greater than 18% fiber
     - Hulls
     - Shells
2. **Pasture, range plants, and green forages**
3. **Silages**
   - Corn
   - Legume
   - Grass
4. **Energy or basal feeds**
   - Cereal grains
   - Mill by-products
   - Fruits
   - Nuts
   - Roots
5. **Protein supplements**
   - Animal
   - Marine
   - Avian
   - Plant
6. **Mineral supplements**
7. **Vitamin supplements**

8. **Nonnutritive additives**
   Antibiotics
   Coloring materials
   Flavors
   Hormones
   Medicants

For the purpose of standardizing the nomenclature of feedstuffs, the NRC established the following order of feedstuff description: (1) origin; (2) species, variety, or kind; (3) part eaten; (4) processes and treatment; (5) stage of maturity; (6) cutting or crop; (7) grade or quality designations; and (8) classification. It is apparent that the name of each feed so classified provides considerable information about it. For this reason, many feeds which are closely related have many listings. For example, there are over 400 entries under alfalfa in the *Atlas of Nutritional Data on United States and Canadian Feeds.* These large numbers of entries often make it difficult to locate a specific feed composition. Feed nomenclature is constantly being reviewed and revised to meet the needs of the feed and livestock industries; and, in the future, it will likely be simplified.

ECONOMIC IMPORTANCE OF FEEDS FOR LIVESTOCK

Feed costs generally account for 12 to 15% of the total cost of agricultural production (see Table 6-1). This represents the second greatest single item of farm production expenses, being exceeded only by replacement of capital equipment. Thus, feed is very important from an economic standpoint.

TABLE 6-1
EXPENSES: FARM PRODUCTION EXPENSES, UNITED STATES[1]

Year	Feed Purchased	Livestock Purchased	Seed Purchased[2]	Fertilizer and Lime	Repairs and Operation of Capital Items[3]	Depreciation and Other Consumption of Farm Capital[4]	Hired Labor[5]	Taxes on Farm Property[6]	Interest on Farm Mortgage Debt[7]	Net Rent to Nonoperator Landlords	Miscellaneous[8]	Total Production Expenses
	(million dollars)											
1971	8,049	5,123	1,072	2,654	4,707	7,416	4,341	2,704	1,905	2,028	7,107	47,107
1975	12,907	4,954	2,138	6,660	7,806	12,354	6,586	3,193	3,317	4,024	11,104	75,043
1980	20,971	10,670	3,220	9,491	15,527	21,474	9,293	3,891	7,544	6,075	24,983	133,139
1985	19,588	8,991	3,369	7,258	14,034	21,101	10,347	4,423	9,878	7,387	29,729	136,105

[1]*Agricultural Statistics 1986,* USDA, p. 414, Table 587.
[2]Includes bulbs, plants, and trees.
[3]Includes expenditures for repairs and maintenance of farm buildings and other land improvements, petroleum fuel and oil, other motor vehicle operation, and repairs on other machinery.
[4]Estimated outlay necessary, at current prices, for the replacement of capital equipment that has been used up during the year.
[5]Includes contract labor expenses and hired labor cash wages, perquisites, and Social Security taxes paid by employers.
[6]Includes taxes levied against farm real estate and farm personal property.
[7]Interest charges payable during the calendar year on outstanding farm-mortgage debt.
[8]For 1980 and 1985, includes interest on nonreal estate debt, pesticides, electricity, machine hire and custom work, marketing and storage, transportation, livestock supplies and equipment, livestock health and breeding services and supplies, livestock feeding, production fees, tools and shop equipment, net insurance, uninsured damages paid, and other management expenses.

As shown in Table 6-2, in 1975, 66% of the feed consumed by livestock and poultry consisted of forage (harvested roughage and pasture) and 34% consisted of concentrates. Subsequently, the trend has been to feed slightly more concentrates and slightly less harvested roughage and pasture.

TABLE 6-2
FEED CONSUMED BY LIVESTOCK AND POULTRY, BY TYPE OF FEED, WITH QUANTITY EXPRESSED IN EQUIVALENT FEEDING VALUE OF CORN[1]

Year Beginning October	Concentrates		Harvested Roughage		Pasture		Total
	(million tons)	(%)	(million tons)	(%)	(million tons)	(%)	(million tons)
1975	173	34	96	19	243	47	512
1980	193	37	102	19	232	44	527
1981	197	38	93	18	231	44	521
1982	203	39	107	20	213	41	523
1983	197	38	100	20	215	42	512
1984	206	40	82	16	227	44	515
1985	210	41	80	16	222	43	512

[1]*Agricultural Statistics, 1986,* USDA, p. 58, Table 77.

While a wide variety of feedstuffs is currently being used, relatively few of these products make up the great bulk of the U.S. supply, as shown in Table 6-3. From these figures, it is possible to ascertain the relative importance of the various feeds. For example, more alfalfa hay is used for feed than all other types of hays combined. Corn and soybean oil meal are the chief energy and protein sources, respectively.

TABLE 6-3
ANIMAL FEEDS CONSUMED IN THE UNITED STATES[1]

	Acreage Harvested	Use for Feed		Yield per Acre
	(1,000 acres)	(1,000 tons)		(tons)
Hay:	61.4	143,800		
Alfalfa	26.8	90,048		3.36
All other hay	34.6	53,752		1.75
Silage:				
Corn	7.5	104,000		13.9
Sorghum	0.6	6,000		10.6
Other	1.0	10,500		12.5
		(mil. tons)	(mil. bu)	(bu)
Grains:				
Corn	71.9	115,276	4,117	106.7
Sorghum	15.4	15,316	547	56.4
Barley	11.2	7,296	304	53.4
Oats	8.2	6,928	433	58.0
High-Protein:				
Oilseed meals:				
Soybean		19,501		
Cottonseed		1,782		
Sunflower		470		
Linseed		127		
Peanut		110		
Animal proteins:				
Tankage and meat meal		1,567		
Fish meal and solubles		251		
Feather meal		130		
Dried milk products		112		
Grain protein feeds:				
Gluten feed and meal		1,160		
Distillers' dried grains		551		
Brewers' dried grains		298		
Miscellaneous feeds:				
Wheat millfeeds		4,795		
By-products from hominy, oats, etc.		2,646		
Molasses		2,646		
Alfalfa meals		992		
Rice millfeeds		678		
Fats		644		
Urea[2]		265		

[1]USDA, Economic Research Service, estimates for the feed year 1984.

[2]Estimate for the feed year 1986-87 from *Feed Management,* June, 1987, p. 18.

Due to the efficiency of the American farmer, the United States is the largest exporter of agricultural commodities in the world—worth approximately $31.2 billion in 1985. Considerable amounts of feed grains are exported, with corn far in the lead (see Fig. 6-2). Of the $31.2 billion total farm product exports in 1985, feed grains accounted for $6.8 billion (with corn alone accounting for $5.8 billion of it), wheat and products for $4.5 billion, and oilseeds and products for $6.2 billion. These 3 commodities accounted for 53% of the total exports of farm products that year.

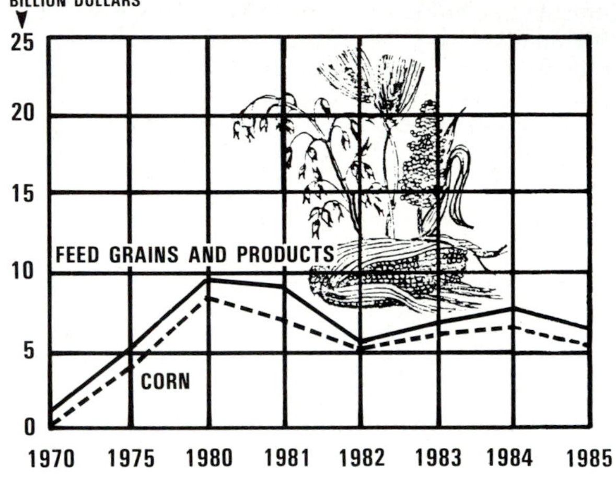

Fig. 6-2. Importance of feed grains as exports. Feed grains include corn, grain sorghum, barley, oats, and their products. Corn is also listed separately in order to show that it dominates feed grain exports. (From: *Statistical Abstracts of the United States 1987,* p. 796, No. 1408.)

FORAGES

Forage is defined as vegetable material in a fresh, dried, or ensiled state (pasture, hay, or silage) which is fed to livestock. In the dry state, forages average more than 18% fiber. The term roughage is often used interchangeably with forage, although roughage usually implies a coarser, bulkier feed than forage. Forages are extremely important feeds for ruminants and nonruminant herbivores, not only for the nutrients they provide but also for the stimulatory effects of forages on the muscle tone and activity of the gastrointestinal tract. Because swine and poultry have very limited capacities to utilize forages, the role of forages in their nutritional program is, for the most part, minor.

From a feeding standpoint, the following general characteristics of forages are pertinent, although some well-known forages can be cited as an exception to each characteristic (for example, on a dry basis well-eared corn silage runs 18% crude fiber, but the TDN is high—about 70%):

1. **Bulk.** They are bulky feeds with a low weight per unit of volume.

2. **Fiber and energy.** They contain more than 18% crude fiber, and they are lower in energy than the concentrates.

3. **Digestibility.** They are generally lower in digestibility than concentrates, due to the lignin content.

Fig. 6-3. Forages that are indigestible by humans being utilized by ruminants. *Left:* Cattle in Madagascar eating fodder. *Right:* Sheep in Tunisia eating spineless cacti. (Courtesy, FAO, Rome, Italy)

4. **Minerals.** They are generally higher in calcium, potassium, and trace minerals than most concentrates; but phosphorus content is apt to be moderate to low.

5. **Vitamins.** They are higher in fat-soluble vitamins than most concentrates. Legumes are good sources of B vitamins.

6. **Protein.** They are variable in protein content. Legumes may run 20% or more crude protein, whereas other forages, such as straws, may have only 3 to 4% crude protein.

From an overall nutritional standpoint, forages may range from very good nutrient sources (such as lush young grass, legumes, and high-quality silage) to very poor feeds (such as straws, hulls, and some browse). Nevertheless, all of them can be used advantageously, provided (1) they are properly prepared and supplemented, and (2) the feeder uses judgment in selecting the species and class of animal to which the particular roughage is fed.

Forage is the natural feed of all herbivorous animals, including ruminants and horses. Swine can survive solely on forage, but their productivity thereon is generally too low to be economical.

Some form of forage is generally included in all balanced rations for ruminants—even in high-concentrate rations. Research and practical feedlot management have repeatedly shown that the addition of small amounts of forage (3 to 15%) to high-energy cattle and sheep finishing rations increases performance and reduces digestive disturbances. Forages appear to stimulate fermentation and promote good muscle tone and epithelial growth in the rumen. In many cases when a ration is 100% concentrate, parakeratosis of the rumen and ruminal atony develop.

Nonruminant herbivores can utilize forages reasonably well, although not as efficiently as ruminants. This is due to the fact that the enlarged cecum—the fermentation vat of the nonruminants—is located distally to the small intestine (the primary organ of absorption). Thus, the digested fiber in the nonruminant herbivore is not exposed to as great an absorptive surface as in ruminants. Rabbits practice coprophagy (eating of feces), resulting in increased forage utilization; but the efficiency of digestion and absorption is still lower than in ruminants. Swine have neither a rumen nor an enlarged cecum and are, therefore, relatively poor utilizers of forages. However, forages can provide an effective means of reducing the feed costs for swine fed maintenance or low-intensity production rations, such as brood sow rations.

Forages include, but are not limited to, pastures, hays, and silages. Table 6-4 shows that roughages account for 61.7% of all U.S. livestock feeds. Of course, the proportion of forages to concentrate consumption varies widely according to relative price and the class of animal. As shown in Table 6-4, sheep and goats head the list of forage consumers, with 93.8% of their total feed coming therefrom, including pasture. Beef cattle obtain 84.5% of their feed from forages. Poultry consume only a negligible amount of forage.

TABLE 6-4
PERCENTAGE OF FEED FOR DIFFERENT CLASSES OF U.S. LIVESTOCK DERIVED FROM (1) CONCENTRATES, AND (2) FORAGES, INCLUDING PASTURE[1]

Class of Animal	Concentrates	Roughages
	(%)	(%)
Beef cattle	15.5	84.5
Dairy cattle	41.3	58.7
Sheep and goats	6.2	93.8
Swine	95.7	4.3
Horses and mules	27.0	73.0
Poultry	100.0	0.0
All livestock	38.3	61.7

[1]USDA, Economic Research Service. Data for the feed year 1983-1984.

Pasture and Range Forages

A pasture is an area of land on which there is a growth of forage that animals may graze.

Broadly speaking, pastures are of two types: (1) cultivated pasture, and (2) native pasture. *Cultivated pastures are those which either receive in excess of 20 in. of rainfall annually or are irrigated.* They include the seeded (cultivated) pastures of the Corn Belt, the South, and the East, and the

Fig. 6-4. Pastures provide 44% of the feed consumed by U.S. livestock and poultry. Much of the area grazed is land that cannot be cropped. However, animals enable this land to make a significant contribution to the U.S. food supply. This shows a fine herd of Jersey cows on pasture. (Courtesy, American Jersey Cattle Club, Columbus, Ohio)

irrigated areas, and smaller scattered moderate- to high-rainfall areas throughout the West. Thus, cultivated pastures generally produce high yields and good-quality forages. *Native pastures include those which receive less than 20 in. of rainfall annually.* These pastures allow only a limited stocking rate, and much of the available forage is of low quality.

Pastures may also be classified as (1) permanent, (2) rotation, and (3) temporary and supplemental, depending on the topography of the land, the climate, and the cultivation program of the operation.

Numerous advantages accrue from using pasture and range forages for livestock—especially ruminants. But as with all types of feeds, certain disadvantages must also be recognized and weighed against the advantages. In rangelands and other relatively unproductive types of land, the stock grower generally has little choice as to the use of the land; hence, it becomes pasture.

The following are among the **advantages** of pasture and range forages:

1. They lessen feed cost. The amount of grain and protein supplements fed to livestock—both of which tend to be expensive—can be dramatically reduced when pasture and range forages are used wisely. Good pasture will produce 200 to 400 lb of beef per acre—superior pasture will produce even more. When a good pasture program is used in swine production, the cost of feeding brood sows can be reduced by 50%.
2. Pasture forages lessen the hazard of nutritional deficiencies. Well-managed pastures and ranges provide good sources of high-quality protein, certain vitamins and minerals, and unknown factors.
3. The threat of communicable diseases is reduced. When animals are reared in confinement, one sick animal poses a potential threat to the entire group. Animals on pasture have less close contact with each other; hence, disease problems are generally minimal.
4. Pastures require less capital for buildings and equipment than confinement production. Generally, all one needs is fenced land, a source of water, simple shelter to protect the animals from the elements, and a good stand of forage.
5. The need for highly developed skills in management are generally not as critical in grazing programs as in confinement production.
6. Pastures make for improved soil conservation on rolling land.
7. Pastures and ranges provide a fairly uniform supply of feed throughout the entire season, provided they are properly cared for by avoiding overgrazing and keeping them relatively free from trash plants.
8. Animals get valuable exercise by grazing on pasture. This is particularly important for breeding animals.
9. A good pasture program permits maximum use of land that is not suited for crop production.

The **disadvantages** of pasture and range forages are:

1. Some land used for pasture may bring a higher return from other uses.
2. Some range areas require large amounts of land to support each animal. Hence, it is difficult to check the animals; and rounding them up usually takes considerable time and labor.
3. The nutritive value of the forage is directly related to the composition of the soils. Hence, a soil that is low in certain minerals produces forages that are low in those minerals. Conversely, toxicities can result in animals grazing crops grown on soils with high concentrations of certain trace minerals: for example, selenium.

(Also see Chapter 7, Pasture and Range Forages.)

GREEN CHOP (SOILING; ZERO GRAZING)

Green chop, or soilage or zero grazing, is fresh herbage cut and chopped in the field and fed to animals in confinement. This type of forage is most frequently fed to lactating dairy cattle. By using green chop, losses in moisture, color, and nutrients are minimized. However, this type of forage is labor intensive because the feed must be harvested daily.

Also, the crop must be harvested at the optimum stage of maturity for best results, thus presenting limitations for extended use.

(Also see Chapter 7, Pasture and Range Forages, section on "Green Chop.")

Hay

Fig. 6-5. Haystacks. Hay provides 12% of the feed consumed by U.S. livestock. (Courtesy, USDA)

Hay is forage harvested during the growing period and preserved by drying for later use in animal feeding. Hay provides the most permanent form of feed storage. While there is some loss of nutrients over a period of time, properly cured hay can be stored for years with little danger of spoiling. It is used as one of the primary feeds for ruminants and nonruminant herbivores. Also, good legume hay is frequently used as a vitamin and mineral supplement in swine and poultry rations.

The primary objective of haymaking is to lower the moisture content of the forage to a stage whereby enzymatic and microbial degradation in the plant is inhibited. This occurs when the moisture content is reduced to 15 to 20%. As forages mature, they contain more dry matter, but they also become less digestible. Thus, hay should be harvested when there is a balance between dry matter content and digestible nutrients.

Hay varies more in nutritive value than any other feed, primarily because of (1) differences in the crop from which it is made, (2) stage of maturity at which it is cut, (3) handling, and (4) possible weather damage during curing. Average-quality hay will run 25 to 35% crude fiber and 45 to 55% TDN.

The **advantages** of hay are:

1. It is the best form for long-term storage of forages. Silage can be stored for extended periods, but once the silo is opened, the storage life of the feed decreases.
2. It is an excellent source of certain vitamins and minerals. For example, poultry producers routinely use alfalfa meal as a source of vitamin A and for its pigment-producing value.
3. When animals are fed high-concentrate rations, hay can be added to facilitate digestion and prevent digestive disturbances.
4. Once hay has been harvested and properly packaged, it is easy to handle and feed.

But there are several **disadvantages** to hay; among them, the following:

1. Considerable labor and expensive equipment are needed to harvest it.
2. The physical process of harvesting hay creates considerable losses of material—for example, leaf shattering.
3. Throughout the curing process, dry matter and some nutrients are lost, with the magnitude of the losses determined largely by the weather (rain, air velocity, temperature).
4. If hay is not cured for a sufficiently long period, spontaneous combustion of the stored forage can occur.
5. Quite often, land that is used for the production of hay is also suited for more intensive crop production.

(Also see Chapter 8, Hay.)

Silages and Haylages

Silages and haylages are fermented forages stored under anaerobic (without air) conditions in a silo. Silos are designed so that anaerobic conditions prevail during the storage of high-moisture feedstuffs. These anaerobic conditions provide an environment whereby microorganisms can ferment the soluble carbohydrates of the feeds, thereby producing lactic acid, volatile fatty acids, and dicarboxylic acids. With the buildup of these end products of fermentation, the pH of the stored feed falls to a range of 3.8 to 5.0. The microorganisms in the feeds are sensitive to pH; and once these high hydrogen concentrations are achieved, microbial growth is inhibited. At this stage, lactic acid represents 8 to 12% of the feed.

So long as the ensiled feed is kept in anaerobic conditions, it can be stored for several years. However, once the silo is opened and silage is removed, it should be emptied within 12 months since exposure to air will speed up the deterioration of the feed.

Fig. 6-6. Silage stored in upright silos on a dairy farm. (Courtesy, USDA)

SILAGE

Usually, the forages made into silage are green crops, or dry crops to which moisture has been added, preferably stored at a level of 60 to 65% moisture. Corn is by far the most extensively used silage, followed by sorghum. A rule of thumb used by many producers to determine what crops can be ensiled is: Crops that are palatable and nutritious to animals as pasture, freshly harvested feed, or dry forage are suitable for ensiling.

Silages are very palatable, and they allow the farmer to get maximum yields per acre because the entire aerial part of the plant is used. The primary problems with silage are the high initial costs of constructing the silo and the increased managerial skills required of the producer—for example, knowing the procedure of harvesting, loading, and unloading the silage, and the problems inherent in feeding silage.

(Also see Chapter 9, Silage/Haylage/High-Moisture Grain.)

HAYLAGE

Haylage is low-moisture silage that is made from grasses and/or legumes that are wilted to 40 to 60% moisture content before ensiling. It is generally produced where the climate is too cool and the growing season too short for corn or sorghum silage. Haylage is higher in protein and carotene, but lower in TDN and vitamin D, than corn and sorghum silage. Animals will ingest more dry matter when offered haylage than when fed corn silage.

For haylage to be properly ensiled, it is very important that as much air as possible be excluded from the silo. Haylage is more difficult to ensile than conventional silage materials, because it is difficult to pack.

(Also see Chapter 9, Silage/Haylage/High-Moisture Grain.)

CONCENTRATES

Concentrates are feeds that are high in nitrogen-free extract and TDN and low in crude fiber (less than 18%). These feeds can be either high or low in protein content. Cereal grains, oil meals, and by-products of the milling industry—for example, corn gluten—are classified as concentrates. Concentrates can be broken down into (1) carbonaceous feeds, and (2) nitrogenous feeds.

Concentrates represented about 38.3% of the feeds consumed by all livestock in the United States in 1983–1984. While ruminant animals are fed considerable amounts of concentrate feeds—especially when in heavy production, monogastrics, such as poultry and swine, are fed almost entirely on concentrates.

The economic impact of concentrate feeds extends well beyond the farm level. Numerous jobs are involved in the processing and handling of these feeds. Additionally, feed grain exports give the United States a big assist in achieving a favorable balance of trade in the international market system.

Fig. 6–7 shows the tonnage of feed concentrates fed to U.S. livestock and poultry from 1975 to 1985. This figure points up (1) that corn is by far the most important livestock feed, (2) that by-products rank second, and (3) that wheat and rye are normally of minor importance as livestock feeds. In 1984, the following quantities of concentrates were fed (in million short tons): feed grains (corn, sorghum, oats, and barley), 145.3; wheat and rye, 8.2; by-products, 36.6; and total concentrates (including by-products), 190.1.

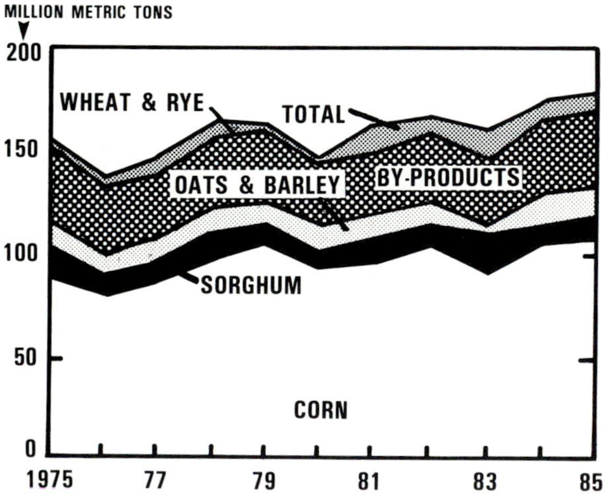

Fig. 6–7. Tonnages of concentrates fed to livestock and poultry. (From *1986 Handbook of Agricultural Charts,* USDA, p. 104, Chart 257)

Since concentrates are relatively low in fiber, most of them are highly digestible. Grains contain large amounts of starch that are easily digested and absorbed in the rumen of ruminants and in the small intestines of ruminants and monogastrics. The protein in most grains and protein concentrates is also highly digestible.

Grains and High-Energy Feeds (Carbonaceous Feeds)

High-energy feeds are feeds used primarily for their energy content. In most cases, high-energy feeds contain less than 20% protein and 18% crude fiber. However, many protein supplements can also be classified as high-energy feeds.

In addition to grains, numerous other feedstuffs are used as high-energy feeds. Price, availability, and nutrient composition determine which of these feeds are used and at what level.

GRAINS

Fig. 6–8. Corn, leading U.S. feed grain, provides 22% of the feed consumed by the nation's livestock and poultry. (Courtesy, USDA)

Grains are the seeds from cereal plants. They constitute the bulk of the high-energy feeds. Some contain as much as 85% carbohydrate (starch) and 6% fat. Most harvested feed grains have relatively little moisture, about 10%, and are not as variable in composition as forages. Corn—representing more than 60% of the total tonnage of concentrate feed—is, by far, the most widely used high-energy feed in the United States. In 1984, more than 8 times as much corn was fed to livestock as grain sorghums—the second most widely used grain.

Fig. 6–9. A collection of the types of corn found in the western world. Corn is by far the most widely used feed grain in the U.S. Over 115.3 million short tons of corn were fed in the U.S. in 1984. (Courtesy, USDA)

Except for the drought year of 1983, the acreage of harvested feed grains has remained relatively constant since 1975 (see Fig. 6–10), but the total production of feed grains has, with some yearly fluctuations, steadily increased. This reflects the higher yields per acre resulting from better agronomic practices and improved varieties of grains.

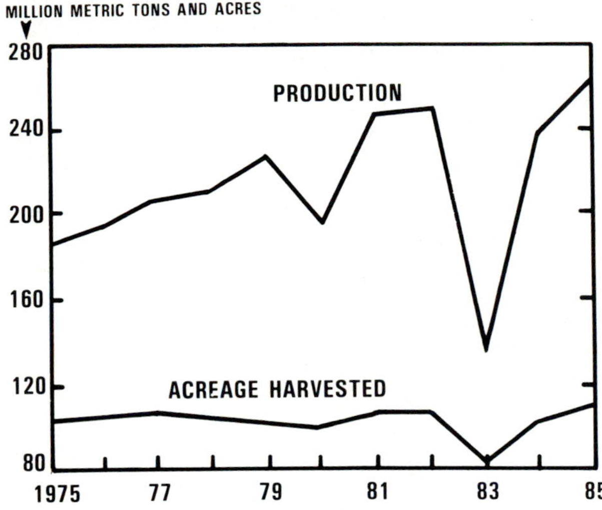

Fig. 6–10. Production and acreage of grain for use as feed. (From *1986 Handbook of Agricultural Charts,* USDA, p. 104, Chart 256)

HIGH-MOISTURE GRAIN

High-moisture grain is grain that contains 22 to 40% moisture. It includes considerable quantities of (1) ear corn, (2) shelled corn, and (3) small grains. Ensiling or acid treatment alleviates costs for drying and reduces the risks of spoilage due to molding or heating.

Corn is the most widely used high-moisture grain, but barley, milo, and other grains can also be used effectively. Although grain is ensiled whole, it is often ground or rolled before being fed, to increase digestibility.

By using high-moisture grain systems, the producer can harvest at an early date—often avoiding unfavorable weather thereby. Through this system, field and storage losses are minimized.

(Also see Chapter 9, Silage/Haylage/High-Moisture Grain.)

FATS AND OILS

Fats and oils are the most potent feed energy source—having 2.25 times as much energy as carbohydrates. By incorporating fat into livestock rations, the producer can increase the caloric density of the ration, control dustiness, reduce wear on mixing equipment, provide a protective agent for some micronutrients, and increase palatability of the feed.

Animal and vegetable fats seem to be almost equally effective additions to rations; thus, selection should be determined solely by comparative price—based on energy content. Ordinarily, animal fats are much cheaper than such vegetable oils as soybean oil or cottonseed oil. Vegetable oils are generally priced out of the animal feed market, for use in margarine, and for use in paint and other industrial purposes.

Several different fat products are used as animal feed; among them, acidulated soap stock (foots), tallows, greases (white and yellow), blended feeding fat, house grease, brown grease, sewer grease, and modified yellow grease. Each of them should be stabilized with an antioxidant and should be bought by specifications and guarantees.

Also, in recent years, the price of whole cottonseed has been favorable enough to include in cattle feed, for both dairy and beef animals.

(Also see Chapter 10, sections on "Cottonseed" and "Fats and Oils.")

FRUITS, ROOTS, AND NUTS

If a livestock producer is located in an area where fruits, root crops, or nuts are grown, a limited amount of these feedstuffs or of their by-products may be incorporated in the ration. Since fruits and roots contain considerable quantities of water, their respective feeding values on an as-fed basis are low. However, on a moisture-free basis, these feeds are quite comparable in TDN to the conventional feed grains. Quite often, these feeds are used so as to provide variety, as well as for economy reasons. Nuts are seldom used in livestock rations unless they are off-quality.

(Also see Chapter 7, section on "Turnips For Fall/Winter Grazing.")

MILL BY-PRODUCT FEEDS

Since many of the cereal grains are processed routinely, mill by-products constitute an additional source of feedstuffs. By-products such as wheat mill run, corn gluten, and bran are used in all types of livestock feeds. Many of these feedstuffs are also excellent sources of protein, in addition to their high-energy content.

(Also see Chapter 10, Grains/High-Energy Feeds.)

LIQUIDS AND SEMILIQUIDS

Liquid and semiliquid products are often incorporated in livestock feeds. Most of these feedstuffs are by-products of some industry—for example, molasses, distillers' solubles, fish solubles, and corn fermentation solubles. Of these feedstuffs, molasses (including cane or blackstrap, beet, citrus, and wood molasses) is the most common; about 2.7 million tons are fed annually in the United States.

According to the American Feed Industry Association, nearly 25% of the nation's feedlot animals received a liquid feed supplement in 1986.

(Also see Chapter 10, section on "Molasses.")

Protein Supplements

Protein supplements are feedstuffs containing more than 20% protein or protein equivalent. Protein levels in livestock rations are extremely important, especially for young animals and animals in high production. Muscle growth, egg production, wool and hair growth, lactation, and gestation all require considerable quantities of protein because the products of these types of production are largely protein in composition.

The use of high-protein feeds is increasing. Fig. 6–11 shows the high-protein feeds fed to U.S. livestock and poultry in soybean meal equivalents from 1975 to 1985. The following points are pertinent to this figure: "other oilseed meals" includes cottonseed, linseed, peanut, and sunflower meals; "animal/marine proteins" includes tankage/meat meal, fish meal and solubles, dried milk products, and feather meal; and "grain proteins" include gluten feed and meal, distillers' grains, and brewers' grains. The figure illustrates that soybean meal is by far the most important high-protein feed. In 1984, in million short tons, the following quantities of different proteins were fed: soybean meal, 19.8; all oil meals, 21.8; animal protein, 2.7; grain protein 1.2; and total of all high-protein feeds, 25.8.

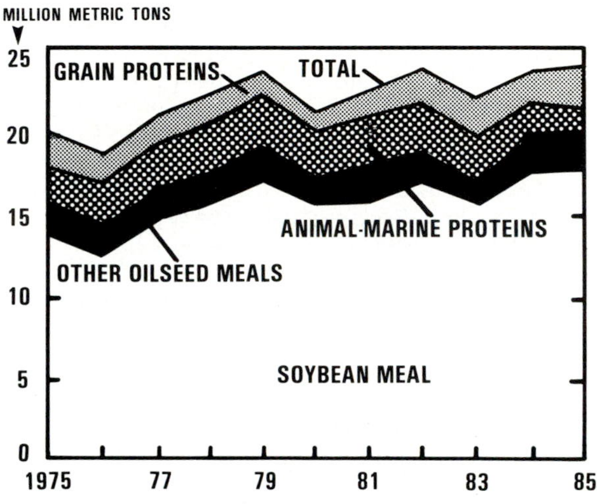

Fig. 6-11. Importance of high-protein feeds. (From *1986 Handbook of Agricultural Charts,* USDA, p. 104, Chart 258)

Fig. 6-12. Unprocessed soybeans. Soybean meal is the most widely used protein supplement in livestock rations. (Courtesy, USDA)

High-protein feeds are usually named and classified according to their origin and method of processing. On the basis of origin, they are usually grouped into the following categories: (1) plant, (2) animal (mammalian, avian, and marine), (3) nonprotein nitrogen (NPN), and (4) single-cell protein. (Also see Chapter 11, Protein Supplements.)

PLANT PROTEINS

This group includes the common oilseed by-products—soybean meal, cottonseed meal, linseed meal, peanut meal, safflower meal, sunflower meal, rapeseed meal, and coconut (or copra) meal. They vary in protein content and feeding value, depending on the seed from which they are produced, the amount of hull and/or seed coat included, and the method of oil extraction used.

Soybean meal is, by far, the most widely used protein supplement for livestock, with an estimated 19.8 million tons used in 1984. Approximately 2.0 million tons of other oilseed meals (including cottonseed, linseed, peanut, and sunflower meals) were fed to livestock in 1984.

ANIMAL PROTEINS

Animal protein supplements are generally high-quality protein feeds which are derived from inedible tissues from meatpacking or rendering plants, from surplus milk or milk products, and from marine sources. They include proteins from meat, fish, poultry, eggs, milk, and their processing by-products. Marine protein supplements—such as fish meal—are generally considered to be better quality feeds than the other animal protein feedstuffs because they contain an excellent balance of amino acids and are good sources of minerals and vitamins. In the past, hog and poultry rations almost always contained some type of animal protein source to provide an "unknown factor(s)." With the discovery and commerical availability of vitamin B-12, high-protein feeds of animal origin became less essential for swine and poultry.

Not all animal proteins are of high quality. For example, feather meal, a by-product from poultry processing, runs about 85% protein; but this protein is very poorly digested by monogastrics and must be hydrolyzed for good utilization. Even then, not more than 3 to 5% should be used in swine rations.

Although most animal protein supplements are excellent feeds, the following concerns are inherent in the utilization of these feedstuffs:

1. **Susceptibility to autoxidation.** Some animal protein supplements contain large amounts of fat and are, therefore, vulnerable to autoxidation and rancidity.
2. **Sources of bacterial contamination.** Animal protein feedstuffs provide excellent media for the growth of bacteria. To prevent this contamination, many of the products must be processed and stored in such a way as to prevent bacterial growth.
3. **Cost.** Most animal sources are more costly than either plant protein or nonprotein nitrogen.

NONPROTEIN NITROGEN (NPN)

Protein quality is less important with ruminants and nonruminant herbivores than in omnivores or carnivores because of microbial synthesis in the gut. As a result, certain nonprotein nitrogen (NPN) sources may be substituted for all or much of the supplemental protein required in most ruminant rations, provided such rations are adequate in minerals and readily available carbohydrates. Among such products are urea, ammoniated molasses, ammoniated beet pulp, ammoniated cottonseed meal, ammoniated citrus pulp, and ammoniated rice hulls. In recent years, the use of liquid protein supplements containing NPN as the protein source has dramatically increased. Also, slow-release nonprotein nitrogen products have come on the market.

(Also see Chapter 11, Protein Supplements.)

SINGLE-CELL PROTEIN (SCP)

Some single-cell protein types, such as yeast, algae, and bacteria, can be useful sources of protein and vitamins for animal feeding. The safety of these feeds depends on the organisms selected, the quality of substrate used, and the conditions of growth. Of course, yeast and bacteria have been used for centuries in the baking, brewing, and distilling industries; in making cheeses and other fermented foods; and in the storage and preservation of foods.

(Also see Chapter 11, Protein Supplements, section on "Single-Cell Protein.")

COMPLETE FEEDS

Complete feeds are prepared products that provide all the nutrients, along with roughage, required to support the form of animal production for which they were designed.

Examples of complete feeds are a 17% layer mash or a 15% lactating dairy ration.

Today, most poultry and swine are fed complete (all-in-one) rations. However, most cattle, sheep, and horses are fed roughages and concentrates separately; yet, most experiments and experiences have not shown any difference between mixed rations and feeding roughages and concentrates separately insofar as efficiency and production are concerned.

PREMIXES

Premixes are concentrated mixes that provide one or more micronutrients and/or specialty products to larger mixtures of feed ingredients.

Premixes are generally commercially prepared and consist of minerals, vitamins, and perhaps other additives, along with a carrier, formulated for blending with a larger mix. Rather than purchasing individual micronutrients (including minerals, vitamins, and other additives), then mixing the ration from the ground up, many commercial feed companies use premixes. Also, farmers may use premixes to blend with local or homegrown grains.

BY-PRODUCT FEEDS AND CROP RESIDUES

By-product feeds are roughages and concentrates other than the primary product from plant and animal processing and from industrial manufacturing. The term crop residue refers to that part of a crop which remains following harvesting. All processing industries produce some by-products which can be considered of secondary, or no value, compared to the primary product. Frequently, these by-products can provide cheap alternatives to other feeds. On many farms, crop residues, which were originally thought to be of little value, are now considered valuable feedstuffs.

(Also see Chapter 12, By-product Feeds/Crop Residues.)

Plant By-products

Plant by-products can be divided into two categories: (1) high-roughage by-products, and (2) high-energy by-products. The high-roughage by-products, which are generally low to moderate in energy and are fed to ruminants that can utilize the fiber, include cottonseed hulls, rice hulls, peanut hulls, soybean hulls, nut shells, sugarcane and bagasse, peelings, vines, and corncobs. Included in the high-energy by-products are molasses, grain milling by-products, and dried beet pulp.

Fig. 6-13. Cows grazing cornstalks (corn stover), the residue of corn, which is available in great abundance. (Courtesy, Ron Baker, C & B Livestock, Inc., Hermiston, Ore.)

Also, numerous crop residues produced on farms can be used as feeds for livestock; among them, the following:

1. **Stover (fodder).** It consists of mature, cured stalks from which the seeds have been removed.
2. **Chaff.** This includes glumes, husks, or other seed covering, together with other plant parts, which are separated from the seeds during threshing or processing.
3. **Straw.** This is the plant residue remaining after separation of seeds in threshing (including chaff).
4. **Stalklage.** This is all of the residue remaining after corn is harvested with a picker or combine.
5. **Husklage.** This is forage discharged from the rear of the combine when harvesting corn. Husks, cobs, and some grain are found in the residue.

Animal By-products

Virtually everything that goes into a slaughterhouse can be processed for some purpose. Feathers, bones, connective tissues, organs, blood, meat scraps, and hoofs are used as either protein, vitamin, or mineral supplements.

In recent years, considerable attention has been devoted to the feeding of livestock manure and litter. In the past, feedlots and large poultry operations were often hard pressed to find ways of disposing of the mountainous volumes of manure and litter (manure with absorbant material). Today, these producers are turning this material—once thought to be of value only as fertilizer—into valuable feed.

(Also see Chapter 17, Animal Behavior/Environment.)

Industrial By-products

Many feedstuffs are being derived from industrial by-products. Thus, many products which were once thought of as wastes are now considered valuable livestock feeds.

FERMENTATION BY-PRODUCTS

Both the brewing industry (beer and ale) and the distilling industry (liquor) produce by-products which are of high nutritive value to livestock. Brewers' dried yeast is 100% yeast solids and is rich in B complex vitamins, proteins, minerals, and unidentified factors. Brewers' dried grains contain about 65% TDN and 21% digestible protein. Distillers' solubles—a syrup produced in the distillation process—is high in protein, energy, linoleic acid, and unidentified factors. Likewise, distillers' dried grains are good sources of energy and protein.

BY-PRODUCTS FROM THE WOOD AND PAPER INDUSTRY

Many of the by-products from the wood and paper industry can be effectively used in limited amounts by ruminants and nonruminant herbivores—animals that can digest fiber effectively. Torula yeast and wood molasses are two commonly used by-products derived from the processing of wood.

BAKERY BY-PRODUCTS

Bakeries process many types of cereals, and the resultant by-products make excellent and highly digestible high-energy feeds.

MUNICIPAL GARBAGE

Vast quantities of foods unfit for human consumption, as well as garbage resulting from the wastage of food, pose serious disposal problems in many urban areas. One of the solutions to the problem is the cooking of garbage, so as to alleviate the hazard of trichinosis, and feeding it to hogs. All states now have laws requiring that commercial garbage be cooked.

SPECIALTY FEEDS

The term *specialty feeds* as used herein refers to unusual or uncommon feeds, or feeds that are used for a specific purpose or under special circumstances. The following specialty feeds are discussed herein:

Milk Replacers
Drought Area Feeds
Weeds

Milk Replacers

Milk replacers are formulated feeds designed to replace the mother's milk of young mammals during the critical, early suckling or milk-feeding stage of life. Milk replacers are available for calves, lambs, pigs, foals, and other animals.

Although scientists have not yet learned how to formulate a synthetic product that will alleviate the necessity of colostrum, in certain other respects they have been able to improve upon nature's product, milk. For example, it has long been known that milk is deficient in iron and copper, thus resulting in anemia in suckling young if proper precautions are not taken. In addition to correcting these deficiencies, milk replacers are fortified with minerals, vitamins, and antibiotics.

A good commercial milk replacer should contain (1) 22–25% protein, preferably derived from milk products (whey, casein, or nonfat dry milk, although about 25% of the milk protein can be replaced by a modified soybean protein); (2) 10 to 20% fat (preferably derived from animal fats, although homogenized soy lecithin is a very acceptable fat source), the higher fat level tends to reduce the severity of diarrhea and provide additional energy for growth; (3) carbohydrates from lactose (milk sugar) and dextrose, and *not* from starch and sucrose (table sugar); and (4) the essential minerals and vitamins. Acidified milk replacers for free-choice calf feeding systems, which were developed and tested in Europe, are gaining increased acceptance in large U.S. commercial operations.

Most milk replacers are formulated to be mixed and fed like normal whole milk. They should be diluted, mixed, and fed according to the directions of the manufacturer.

From the standpoint of livestock producers, synthetic milk is of interest in raising orphaned or early weaned animals of each class of livestock. For the dairy producers, it is generally more profitable to sell the whole milk and purchase a high-quality milk replacer for the young calves. Also, it is a valuable adjunct in certain disease control programs, especially those diseases that may be transmitted from dam to offspring; and, in some cases, it makes it practical to retain in production those valuable females which, due to injury or disease to the udder, cannot suckle their young.

The economics of using milk replacer instead of whole milk for calves can be determined by the following simple rule: 25 lb of milk replacer should not cost more than the selling price of the milk that it replaces. Thus, if 25 lb of milk replacer replaces 225 lb of whole milk selling at $12/100 lb, 25 lb of milk replacer has an equivalent value of $27 (2.25 × $12 = $27).

The raising of early-weaned or orphaned young of mammalian animals will be simplified if they have first received colostrum. During the first few days, it is generally best to feed the young mammal from a bottle with a rubber nipple. Later, it should be taught to drink from a suitable receptacle. It is important that all feeding utensils be kept absolutely clean and sanitary (cleaned and sterilized each time) and that feeding be at regular intervals. Also, young animals should be given grain (and high-quality, fine stemmed hay in the case of ruminants and foals) at the earliest possible time.

Drought Area Feeds

In drought-stricken areas, finding enough feed is an age-old problem. Grain and by-product feeds can be shipped from surplus to deficit areas with relative ease. Silage cannot be shipped. Long hay can be shipped, but at great expense because of its bulk. Less desirable local feeds, such as weeds, prickly pear, and yucca, may be used.

The following alternative feeds have been used for ruminants and horses during droughts in different parts of the world:

• **Mixed feeds**—Three mixed rations that have been used successfully for cattle in drought areas, and which should be satisfactory for sheep and other ruminants, follow:

1. **Southern United States Silage Mix.** The formula:

Ingredient	Pounds
Soybean mill feed .	800
Cottonseed hulls . .	600
Ground corn	500
Cottonseed meal . .	100
Total	2,000

This silage mix has a calculated crude protein content of 11%.

2. **Denmark Straw Mix.** The formula:

Ingredient	Pounds
Barley straw (chopped)	1,000
Molasses	500
Beet pulp	435
Urea	20
Ammonium phosphate	20
Minerals	25
Total	2,000

This Danish Straw Mix has a calculated crude protein content of 9%.

3. **British Straw Mix.** The formula:

Ingredient	Pounds
Barley straw (chopped)	1,200
Soybean meal (48%)	300
Molasses	260
Barley	200
Minerals	40
Total	2,000

This British Straw Mix has a calculated crude protein content of 13%.

The following points are pertinent to the three mixed extender rations above:

1. Each ration combines a low-quality (poorly digested) roughage with a supply of highly digestible fiber and readily available carbohydrates.

2. The rations work best when prepared and fed as a mix, rather than some of the ingredients being fed separately.

3. These are *extenders.* So, whenever possible, they should be fed along with some good forage; the good forage should not be fed until it is all gone, followed by dependence upon extenders alone.

4. These extender rations should be fortified with needed minerals and vitamins.

5. Any one of the three extender mixed rations can be adapted and used for lactating cows by adding locally available grain and by-products. *Example:* A producer wishes to formulate a ration for a group of lactating cows with an average body weight of 1,300 lb, producing 60 lb of 3.8% milk, and with a dry matter intake of 46 lb per day. *The nutrient requirements for these cows may be met by the following daily ration:* 10 lb of good alfalfa hay, 12 lb of any one of the extenders, and 28 to 30 lb of grain and by-products; with the actual composition of the grain and by-products, particularly of the protein, determined by the available forage. Thus, for lactating cows in a drought area, a suitable ration can be formulated by combining a poorly digested roughage with highly digestible fiber, adding a readily available energy source, some protein and minerals, and feeding the ration as a mix.

• **Forage substitutes**—When hay is scarce and high in price, it may be replaced (preferably, partially only) by such forages as cereal straws, corn stalks, alfalfa straw, beet tops, bean straw, soybean hulls, cottonseed hulls, or poultry litter; adding protein, minerals, and vitamins as necessary to meet the nutrient requirements of the animals being fed.

• **Weeds and other unusual feeds**—During droughts, certain weeds and such unusual feeds as cactus (prickly pear) with the spines burned off, and yucca chopped or shredded, may be fed. (See section on "Weeds" in this chapter.) Before feeding any weeds, the veterinarian and/or county extension agent should be consulted.

• **Substituting grain for hay**—Half or more of the energy requirement of most rations may be furnished by substituting about 5 lb of grain for 8 lb of hay. (See Chapter 8, Hay, section headed "Stretching the Hay Supply.")

Weeds

Weeds may be defined as plants growing where they are not wanted and interfering with desired land use.

Despite the above degrading definition, it is noteworthy that many of today's most useful plants were once considered weeds, and that a number of weeds furnish forage for animals, especially cattle and sheep, when more palatable plants are scarce. This is frequently the case in late autumn or during droughts, when the more desirable vegetation is dried up but some of the deep-rooted perennial weeds may still be green.

When young and tender, some weeds, such as Russian thistles, pigweeds, sunflowers, ragweeds, lamb's-quarter, wild oats, and others, may be used for silage.

It is noteworthy that pigweed, which is the common name for *Amaranthus,* is used as a protein supplement by people in many protein-poor developing nations. Depending on the variety, pigweed plants may contain up to 28% protein and the seeds about 13% protein; and the protein is of high quality. Pigweeds are also a good source of calcium, iron, vitamin A, and vitamin C.

Precautions are necessary when weeds are harvested and used as feeds. For example, foxtail must be harvested and used before the bloom stage because the awns of the seeds are harmful to the mouth of the consuming animal. Also, some weeds are unpalatable, others are toxic.

NOTE WELL: Before feeding any weeds, the veterinarian and/or county extension agent should be consulted.

SUPPLEMENTS, ADDITIVES, AND IMPLANTS

A feed supplement is a feedstuff that is mixed with a primary grain and/or roughage to provide all the nutrients required to support the form of production for which it is intended.

An additive is a substance of nonnutritive nature which when added to feed will improve feed efficiency and/or production of animals.

An implant is a substance that is implanted into the body for the purpose of growth promotion or controlling some physiological function. They are generally slow releasing in action and often contain a substance that would be destroyed before absorption if included in the feed.

(Also see Chapter 13, Feed Supplements/Additives/Implants.)

Mineral Supplements

Mineral supplements are rich sources of one or more of the inorganic elements needed to perform certain essential body functions.

Generally, the macrominerals of concern are NaCl (common salt), Ca, P, Mg, and sometimes S; and the trace elements that may be deficient are Cu, Fe, I, Mn, Zn, Co, and Se. Most feeds provide minerals in addition to basic organic nutrients, although fat and urea are marked exceptions. Nevertheless, most rations require more concentrated sources of one or more mineral elements.

Needed mineral supplements may be either home-mixed or provided by a commercial mineral product. Commercial mineral mixes are mixed by manufacturers who specialize in the mineral business, either handling minerals alone or in combination with a feed business. Because mineral mixes have become more complicated with the recognition of the importance of trace elements and interrelationships, and because most farmers and ranchers do not have the equipment with which to mix minerals properly, commercial minerals are finding a place of increasing importance in all livestock feeding.

Minerals may, in most cases, be either incorporated in the ration or self-fed.

Vitamin Supplements

Vitamin supplements are rich synthetic or natural feed sources of one or more of the complex organic compounds, called vitamins, that are required in minute amounts by animals for normal growth, production, reproduction, and/or health.

Formerly, a wide variety of feed ingredients was added to livestock rations for their vitamin content. But it was found that the vitamin concentration of feedstuffs varied tremendously, being affected by plant species and part (leaf, stalk, or seed), harvesting, storing, and processing. Generally speaking, vitamins are easily destroyed by heat, sunlight, oxidation, and mold growth. So, today, nutritionists rely on vitamin supplements, which in many cases are chemically pure sources that need to be used only in very minute amounts. In modern feed formulation, premixes often represent the commonsense approach to providing vitamins.

For adult ruminants, vitamins A, D, and E are of concern, with A being the one most likely to be deficient. Under ordinary circumstances, ruminants synthesize adequate B vitamins, and vitamins C and K. Unless they are kept indoors, they usually receive sufficient exposure from direct sunlight to meet their needs for vitamin D.

Because of the greater prevalence of confinement feeding, along with limited gastrointestinal synthesis, swine are more apt to suffer from vitamin deficiencies than ruminants. Under practical conditions, special consideration should be given to the need for supplementing swine rations with vitamins A, D, E, riboflavin, niacin, pantothenic acid, B–12, and choline.

Vitamins A, D, (D_3 for poultry), B–12, and riboflavin are commonly low in poultry rations. Also, it is in the nature of good insurance to add the following vitamins to poultry rations, as they may be deficient: E and K, and the rest of the B vitamins (in addition to vitamin B–12 and riboflavin, already mentioned).

Because only limited amounts of water-soluble vitamins can be stored, they should be fed regularly in monogastric livestock rations in adequate amounts.

Feed Additives and Implants

These nonnutritive products are used to improve the rate and/or efficiency of production of animals, prevent certain diseases, or preserve feeds. But there is no evidence of a nutritional deficiency when they are omitted from a ration.

Most animal scientists and livestock producers agree that antibiotics and hormones stand out as the two nutritional discoveries since 1949 that have had the greatest impact on the livestock industry. In 1949, it was discovered that antibiotics were something new to be added to livestock feeds. Then, in 1954, stilbestrol was approved by the Food and Drug Administration for use in cattle finishing rations. In each case, a new era in livestock feeding was ushered in—comparable to the vitamin era which was born in 1912—and more feed additives and implants followed. Today, it is estimated that 75% of the nation's growing-finishing cattle, lambs, and pigs receive some form of feed additives or implants.

Some glowing reports to the contrary, there is no evidence to indicate that the use of these additives or implants can or will alleviate the need for vigilant sanitation, improved nutrition, and superior management. Instead, with the unfolding and applying of scientific information relative to these promotants, the producer will be able to achieve still greater efficiency of production. Also, practical producers should weigh the benefits of each one against its cost.

ANTIBIOTICS

Antibiotics are chemical substances which are produced by living organisms (molds, bacteria, or green plants) and which have bateriostatic or bactericidal properties.

Commercially produced antibiotics are thought of as substances produced by moldlike organisms grown in nutrient solutions in large steel tanks. Hundreds of antibiotics are known—including commercially produced bacitracin (and zinc bacitracin), chlortetracycline (Aureomycin), erythromycin, hygromycin, neomycin, oleandomycin, oxytetracycline (Terramycin), penicillin, streptomycin, and tylosin (Tylan).

• **Role of antibiotics**—Evidence indicates that antibiotics improve rate and efficiency of gain through the following actions:

1. They reduce the incidence of subclinical levels of bacterial infections in the digestive tract.
2. They stimulate appetite and have a nutrient-sparing effect.
3. They stimulate certain enzyme systems.

In addition to improvement in rate and efficiency of gains, antibiotics usually (1) reduce the incidence of diarrhea in young animals, especially in young mammals deprived of colostrum; (2) lessen the tendency of animals to go off feed; (3) reduce enterotoxemia and death loss of lambs fed high-grain rations; and (4) lower the incidence of abscessed livers of cattle fed high-grain rations.

GROWTH STIMULANTS AND IMPLANTS

Most livestock producers are familiar with, or have used, one or more of the growth stimulants or implants. In 1954, diethylstilbestrol (DES)—a synthetic female hormone, and not a nutrient—was approved by the Food and Drug Administration for use in cattle finishing rations; and 2 years later, in 1956, the Food and Drug Administration approved the use of DES implants for steers. Subsequently, DES was shown to be a carcinogen and banned as feed or implants in 1979. Nevertheless, DES, along with antibiotics, ushered in the Feed Additive and Implant Era. (See Chapter 2, Principles of Nutrition, section on "Feed Additive and Implant Era.")

Today, feed additives and implants constitute a diverse group. More than 1,000 drugs are approved by the Food and Drug Administration (FDA) for use by livestock and poultry producers. Although many different products are used as growth stimulants, and implants, antibiotics and hormones (or hormonelike products) are most common. The recommended growth stimulants and implants for each animal species, along with pertinent details relative to their use, are presented in separate chapters devoted to each class of animal, Chapters 19 to 28 of this book.

OTHER ADDITIVES

A number of other additives are used from time to time and for specific purposes; among them—

1. Anthelmintics
2. Antioxidants
3. Arsenicals
4. Bloat control products
5. Buffers
6. Chemotherapeutics
7. Drugs
8. Enzymes
9. Flavoring agents.
10. Grit
11. Hormones
12. Ionophores
13. Isoacids (IsoPlus)
14. Mold (fungi) inhibitors
15. Pellet binders
16. Probiotics
17. Roughage substitutes
18. Tranquilizers

(Also see Chapter 13, Feed Supplements/Additives/Implants.)

WET FEEDS (HIGH-MOISTURE FEEDS)

Wet feeds, or high-moisture feeds, can be defined as those containing more than 20% water, or less than 80% dry matter (DM). Some common categories of wet feeds are:

Feed	Dry Matter
	(%)
Fresh forages	15–30
Haylage	45–60
High-moisture grain	60–78
Milk, fresh	9–13
Molasses	60–80
Roots	9–30
Silages	25–40
Wet by-products	10–25

Wet feeds may have the following **advantages**: (1) if produced locally, they may be inexpensive; (2) the moisture in the ration lessens dust and facilitates mixing; and (3) they improve palatability.

On the other hand, wet feeds may have the following **disadvantages**: (1) the price that is being paid for water may make the feed expensive when computed on an equivalent dry matter basis; (2) long hauls of wet feeds are expensive, because it costs to transport water; (3) high-producing animals may not be able to consume sufficient wet feeds to meet their requirements for dry matter, energy, protein, and other nutrients; (4) the dry matter content of wet feeds may differ from load to load; and (5) wet feeds cannot be exposed to the air long because of the production of molds.

So, when the livestock producer is considering the use of wet feeds, the advantages must be weighed against the disadvantages and cost should be computed on the basis of dry matter delivered as feed to the animals.

(Also see Chapter 9, section on "Buying and Selling High-Moisture Grain"; and Chapter 16, "Moisture is Important.")

QUESTIONS FOR STUDY AND DISCUSSION

1. Differentiate between the two terms, feeds and feedstuffs.

2. How do you account for the fact that pasture or grazing and corn rank first and second, respectively, in relative importance as livestock feeds?

3. Describe the NRC system of naming and classifying feeds. What are its advantages and disadvantages?

4. Feed constitutes the second largest single expense in agricultural production, being exceeded only by replacement of capital equipment. As a percentage of total expenses, are feed costs showing an increasing or decreasing percentage? Do you think the trend will continue, if there is a trend? Justify your answer.

5. Since 1975, the trend has been to feed slightly more concentrates and slightly less harvested roughage and pasture. How do you account for this?

6. How do you account for the fact that corn and soybean oil meal are the chief animal feed energy and protein sources, respectively?

7. Why is the American farmer the largest exporter of agricultural commodities in the world?

8. Define the term roughage. What is the role of this type of feedstuff in monogastric rations? What are the characteristics of roughages?

9. How do you account for the fact that sheep and goats head the list for forage consumers whereas poultry consume the least forage?

10. Differentiate between cultivated and native pasture.

11. What are the advantages and disadvantages of pasture and range forages?

12. What is green chop?

13. Define the term hay. What type of hay is predominantly used in the United States?

14. What are the advantages and disadvantages of growing and harvesting hay?

15. What are the basic principles of the ensiling of forages?

16. What is the difference between silage and haylage? Which would you prefer?

17. Define concentrate. How important is the feed grain industry? What are the recent trends of acreage harvested and yields?

18. What is meant by the terms (a) carbonaceous feeds, and (b) nitrogenous feeds?

19. What are grains? What are some of the criteria for evaluating the nutritive value of grains?

20. What is high-moisture grain? What advantages accrue from a high moisture grain system?

21. List the practical functions of added fat in livestock feed.

22. Discuss the role of protein quality in feeds for the feeding of ruminants and nonruminants.

23. List the various classes of high-protein feeds and give examples of each.

24. Why has there been an increased interest recently in using urea in rations for ruminants?

25. Why are slow-release NPN products advantageous?

26. What is single-cell protein?

27. What is a complete feed? What is a premix?

28. What is the difference between stalklage and husklage; between straw and chaff?

29. For esthetic reasons, some consumers balk at purchasing products that came from animals that have been fed certain by-products and residues. How would you reassure them that their fears are unfounded?

30. What are milk replacers? How are they used?

31. Discuss the importance of developing and using special feeds and rations during a severe drought.

32. Under what conditions are weeds sometimes fed?

33. Which minerals are most likely to be needed as supplements in livestock rations?

34. Why did the practice of selecting feed ingredients for their vitamin content give way to the use of vitamin supplements?

35. List and discuss the various roles of antibiotics.

36. Why are hormones used so extensively for ruminants?

SELECTED REFERENCES

Title of Publication	Author(s)	Publisher
Applied Animal Feeding and Nutrition	M. H. Jurgens	Kendall/Hunt Publishing Company, Dubuque, Iowa, 1972
Applied Animal Nutrition, 2nd Edition	E. W. Crampton L. E. Harris	W. H. Freeman and Co., San Francisco, Calif., 1969
Digestive Physiology and Nutrition of Ruminants, Vol. 3	D. C. Church	D. C. Church, Corvallis, Ore., 1975
Feed Formulations, 3rd Edition	T. W. Perry	The Interstate Printers & Publishers, Inc., Danville, Ill., 1982
Feeds and Feeding, 2nd Edition	A. Cullison	Reston Publishing Co., Reston, Va., 1979
Feeds and Feeding, 22nd Edition	F. B. Morrison	The Morrison Publishing Company, Ithaca, N.Y., 1956
Feeds and Feeding, Abridged	F. B. Morrison	The Morrison Publishing Company, Ithaca, N.Y., 1956
Feeds for Livestock, Poultry and Pets	M. H. Gutcho	Noyes Data Corporation, Park Ridge, N.J., 1973
Handbook of Feedstuffs, The	R. Seiden W. H. Pfander	Springer Publishing Company, Inc., New York, N.Y., 1957
Livestock Feeds and Feeding, 2nd Edition	D. C. Church	O & B Books, Inc., Corvallis, Ore.
Stockmen's Handbook, The, 6th Edition	M. E. Ensminger	The Interstate Printers & Publishers, Inc., Danville, Ill., 1983

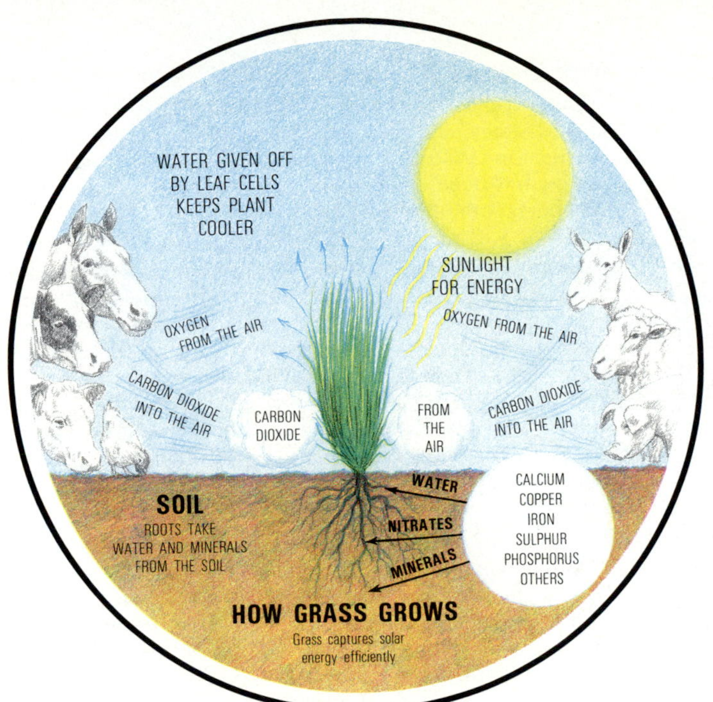

Original painting by Tom Phillips

Chapter 

PASTURE AND RANGE FORAGES[1]

Contents — *Page*

Part I—Pasture

Part II—The Western Range

Part III—Multiple Use/Conservation of Land

[1]The authors gratefully acknowledge the helpful suggestions of the following authorities who reviewed this entire chapter: J. E. Baylor, Ph.D., Professor of Agronomy, the Pennsylvania State University, State College; R. A. Forsberg, Ph.D., *et al.*, Department of Agronomy, University of Wisconsin-Madison; S. C. Fransen, Ph.D., Forage Agronomist, Western Washington Research and Extension Center, Washington State University, Puyallup; C. S. Hoveland, Ph.D., Professor of Agronomy, College of Agriculture, The University of Georgia, Athens; D. A. Miller, Ph.D., Professor of Plant Breeding and Genetics, College of Agriculture, University of Illinois at Urbana-Champaign; and W. E. McMurphy, Ph.D., Department of Agronomy, Oklahoma State University, Stillwater. Other authorities reviewed a section or sections, for which credit is given.

Fig. 7-1. "All flesh is grass"! (Isaiah 40:6) Hereford breeding herd on pasture. (Courtesy, American Hereford Assn., Kansas City, Mo.)

From time immemorial, the keepers of herds and flocks have recognized that productive grasslands require good care—and that they must not be overgrazed. Abraham and Lot, patriarchs of the Old Testament, were the first grassland management specialists of recorded history. The story of their dispute over the water holes in Canaan reads with all the freshness of a page from the early history of the United States. Canaan was a wide open country, and just as subject to overgrazing and water depletion as any grazing area in America. So, Abraham said to his kinsman: ..."Let there be no strife...between me and thee"..."Is not the whole land before thee?..." Thereupon, these two stockmen separated, each trailing his herd in a different direction (Genesis 13:5-12). So it was, and ever shall be! However, through the years, overgrazing has been lessened by advancing soil, plant, and animal technology.

Grass is the largest and most remunerative crop in the United States, and the cornerstone of successful livestock production. Also, and most important, no method of harvesting has been devised which is as cheap as that which can be accomplished by animals.

As the ever-increasing human population of the world consumes a higher proportion of grains and seeds directly, there will be increased reliance on grass for meat, milk, and wool production. In this connection, it is noteworthy that petroleum is not needed to make wool, and that animals do not require fuel to graze the land and recover the energy that is stored in the grass. Noteworthy, too, is the fact that animals are completely recyclable; they produce a new crop each year and perpetuate themselves through their offspring. But it takes thousands of years to create coal, oil, and natural gas; and when they're gone, they're gone forever.

It is also noteworthy that soil conservation—erosion control—is accomplished more effectively by pasture crops than by any other means. The American Forage and Grassland Council is authority for the statement that, "A good stand of grasses and legumes is over 300 times as effective in saving soil, and six times as effective in reducing water runoff, as a clean tilled crop on the same kind of land."

Grassland agriculture, better than any other type of agriculture, will continue in the face of economic and social changes to conserve the land and ensure a food supply of the desired quantity, variety, and quality. At its best, it calls for an interdisciplinary approach—for knowledge and application of soil, plant, and animal sciences. This joint focus characterizes the great livestock areas of the world.

From the standpoint of organization, this chapter is presented in three parts: Part I—Pasture; Part II—The Western Range; and Part III—Multiple Use/Conservation of Land.

Part I—**Pasture**

The term *pasture* is of Latin origin, from the word *pastus,* meaning *an area of land on which there is a growth of forage that animals may graze.* Additional definitions of pasture-related terms follow.

Rangeland, which is of American origin, *refers to western lands on which the native (range) vegetation is predominantly grasses, grasslike plants, forbs, or shrubs suitable for grazing or browsing.*

Grassland includes both pasture and rangeland; it denotes all land on which animals are grazed, with the exception of annually seeded row crops.

Forage refers to the vegetative portion of plants in a fresh, dried, or ensiled state, which is fed to livestock as grass, hay, or silage.

Grazing refers to the process by which animals harvest their own feed (grass, legume, browse, and/or forbs) from its growing condition.

The importance of pasture, rangeland, and forage in the United States is evidenced by the facts (1) that 29.1% of the total land area is devoted to pasture and rangeland (Fig. 7-2)—more than any other land use; and (2) that 61.7% of the feed for all livestock is derived from forage (Table 6-4).

Pastures vary greatly in quality, depending on type (plant species), soil, growing conditions, and stage of maturity. Mature grasses, especially those that are leached and bleached, are low in palatability, digestibility, protein, and carotene, and in some of the minerals. Grasses are usually adequate in calcium, magnesium, and potassium, but they are apt to be borderline or deficient in phosphorus, and they may be low in some of the trace minerals.

Also, pastures vary in quality as a result of management. Poor pasture management results in poor pasture utilization and often in stand loss. Since plant growth rates vary, animal stocking rates should be adjusted in keeping therewith. Too close grazing during the growing season will lower yield and stand persistence. Overgrazing during periods of low rainfall will reduce animal performance and photosynthesis and result in forage plants lacking vigor because of weakened root systems.

So, pastures should no longer be taken for granted. Most grazing areas can be improved by seeding new and better varieties of grasses and legumes, by fertilizing, and by management, including scientifically controlled grazing, avoiding overgrazing by both domestic livestock and wild animals, and supplemental feeding.

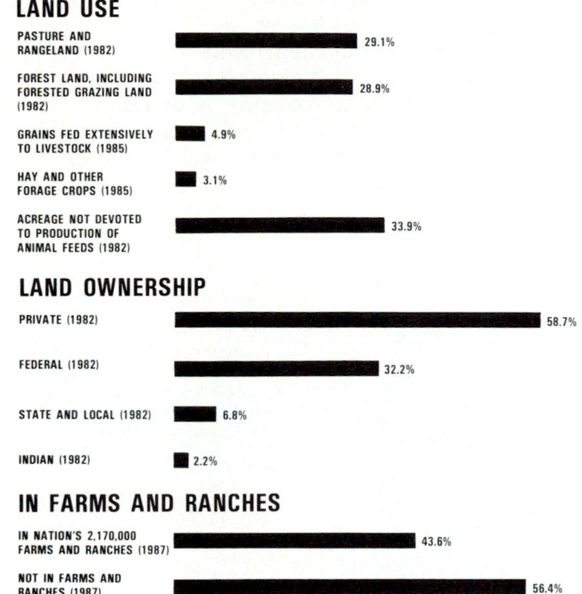

Fig. 7-2. Land use, ownership, and number of farms and ranches of the United States, including Alaska and Hawaii; with the year of the data given in parentheses. It is noteworthy (1) that 29.1% of the total U.S. land area is devoted to pasture and rangeland, exclusive of forested grazing land; (2) that 37.1% of the total U.S. land is devoted to the production of livestock feed (pasture and rangeland, hay and other forage crops, and grain); (3) that 58.7% of the U.S. land is privately owned vs 41.2% publicly owned; and (4) that there were 2,170,000 U.S. farms and ranches. (*Source:* Land use from *Agricultural Statistics 1986,* p. 373, Table 539; p. 388, Table 558; and p. 390, Table 559; and land ownership from *Statistical Abstracts of the United States 1987,* p. 182, No. 318.)

Again and again, scientists and practical farmers and ranchers have demonstrated that the following desired goals in pasture production are well within the realm of possibility:

- To produce higher yields of palatable and nutritious forage.
- To extend the grazing season from as early in the spring to as late in the fall as possible.
- To provide a fairly uniform supply of feed throughout the entire season.

At the outset, it should be recognized that no one plant embodies all the desirable characteristics necessary to meet the above goals. None of them will grow year-round, or during extremely cold or dry weather. Each of them has a period of peak growth which must be conserved for periods of little growth. Consequently, the progressive producer will find it desirable (1) to grow more than one species, and (2) to plan pastures for each season of the year. In general, a combination of permanent, rotation, and temporary pastures —accompanied by scientific management—will best achieve these ends.

Beef cattle, dairy cattle, and sheep compete with each other for many of the grazing areas of the United States. Horses also relish the same kind of pastures; and breeding swine may utilize high quality pastures advantageously. Dried forage, rich in xanthophyll and carotene, makes an excellent poultry feed, especially in providing the pigmentation needed to make broilers customer-appealing.

The value and use of pasture and other forages for each species is covered in the respective chapters devoted to each class of livestock, Chapters 19 through 25.

CLASSES OF PASTURE

Broadly speaking, all U.S. pastures may be classified as either (1) seeded pastures, or (2) native pastures (see Fig. 7-3). Although no sharp line of demarcation exists between the two groups, seeded pastures include those which either receive more than approximately 20 in. of rainfall annually or are irrigated. They are the seeded (cultivated) pastures of the Corn Belt, the South, the East, and the irrigated areas, and smaller and scattered moderate to high rainfall areas throughout the West. Seven grass species—Bahiagrass, Bermudagrass, orchardgrass, reed canarygrass, ryegrass, smooth bromegrass, and tall fescue—account for the major portion of seeded forage grass production in the United States.

The native pastures include those range pastures which receive less than 20 in. of rainfall annually. Their vegetative cover, known as native plants, consists of adapted plants developed by natural selection that have existed in the area for many years, and that were not intentionally introduced.

In addition to the seeded pastures, and the native pastures, there is a considerable acreage of nonirrigated, hill-land pastures on the west coast of Washington, Oregon, and California which do not fit under either of these traditional classifications. The annual rainfall of the area varies from 12 to 14 in. in California to more than 100 in. in coastal Oregon and Washington. More and more of these brushy hill lands are being developed into productive seeded pastures. The higher rainfall areas are being seeded to subterranean clover and perennial ryegrass, with tall fescue included where beef cattle are more important than sheep. In the drier areas, ryegrass, Hardinggrass, and some annual grasses are being seeded. There are an estimated 25 million acres of this brushy hill land, once classified as range, which have potential for development as improved or tame pasture.

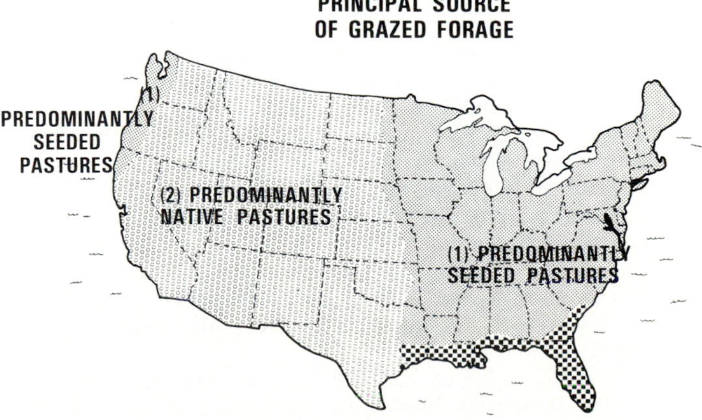

Fig. 7-3. The two major U.S. pasture areas—(1) seeded, and (2) native (range)—about equally divide the 48 contiguous states into east and west halves. (Courtesy, USDA)

Pasture may be further classified as—

1. **Permanent pastures.** Those which, with proper care, last for many years. They are most commonly found on land that cannot be used profitably for cultivated crops, mainly because of topography, moisture, or fertility. The vast majority of the farms of the United States have one or more permanent pastures, and most range areas come under this classification.

2. **Semipermanent or rotation pastures.** Those that are used as a part of the established crop rotation. These are seeded pastures that are generally used for 2 to 7 years before plowing.

3. **Temporary and supplemental pastures.** Those that are used for a short period; and they are usually annuals, such as Sudangrass, sorghum, millet, rye, barley, wheat, oats, ryegrass, arrowleaf clover, crimson clover, ball clover, or rape. They are generally seeded for the purpose of providing supplemental grazing during the season when the permanent or rotation pastures are relatively unproductive.

Pasture plants are classed as (1) grasses, (2) legumes, (3) browse, and (4) forbs. The definition of each of these terms follows:

1. **Grass.** *Botanically, any plant of the family, Gramineae.* In grassland agriculture, grass refers to the forage species of *Gramineae* when either grown alone or with a legume.

2. **Legume.** *Plants, such as alfalfa and the clovers, that obtain nitrogen through bacteria that live in their roots are known as legumes.* The nitrogen fixation aspect of legumes will be of increasing interest as energy sources become more scarce and costly.

3. **Browse.** *The edible parts of woody vegetation, such as leaves, stems, and twigs from bushes.*

4. **Forbs.** *Nongrasslike range herbs which animals eat (forbs are generally called weeds by western livestock producers).*

PASTURES FOR CATTLE, SHEEP, AND HORSES

The economic importance of pastures for beef cattle, dairy cattle, sheep, and horses has been demonstrated in many experiments and on thousands of livestock farms and ranches.

On Corn Belt beef farms, brood cows normally obtain practically all their feed from pasture for 200 days of the year. While pastures furnish about 55 to 60% of the total feed for the whole year, the cost of pasture is only about ⅓ of the annual feed bill; hence, pasture is a good buy. In the southern part of the United States, year-round grazing can be achieved.

As larger numbers of dairy cattle are concentrated on smaller acreages, and as milk production per cow increases, dairy producers depend less on pasture and more on other feeds.

Sheep are able to utilize the various grasses, legumes, weeds, herbs, and browse that grow on millions of acres of cultivated and uncultivated lands in this and other countries. This characteristic, plus their gregarious or flocking instinct, has made sheep raising a frontier industry throughout the world.

Although suburban horse owners are finding it increasingly difficult to provide pasture, horses compete with cattle and sheep for grazing areas.[2]

Further facts pertinent to pastures for beef cattle, dairy cattle, sheep, and horses follow:

1. **Beef cattle.** It is estimated that 84.5% of the total feed supply of all U.S. beef cattle is derived from forage (see Table 6-4); in season, this means pasture. Good pasture alone will produce 200 to 400 lb of beef per acre annually (in weight of calves weaned, or in added weight of older cattle); superior pastures will do much better.

Fig. 7-4. More than four-fifths of the total feed supply of U.S. beef cattle is derived from forage. (Courtesy, American Polled Hereford Assn., Kansas City, Mo.)

Generally speaking, cattle can be finished more cheaply on pasture than in the drylot because (a) less labor is required, for the animals do their own harvesting; (b) grass is the cheapest of all roughages; (c) less expensive protein supplement is required; (d) the animals scatter their own droppings, which aids in maintaining fertility, and alleviates hauling it; and (e) fewer buildings and less equipment are necessary.

• **Custom cattle grazing (pasture leasing)**—*Custom cattle grazing is the grazing of cattle for a fee without taking ownership of the animals.* This practice has long existed in such noted tall-grass areas as the Flint Hills of Kansas, the Sand Hills of Nebraska, and the Osage Pastures of Oklahoma, where ranch owners grow out cattle, primarily yearling steers, owned by Corn Belt Feeders. Gains in weight depend upon the condition of the range, the length of the grazing season, the age and condition of the cattle, and the management. Daily gains of 2 lb on large-framed, thin yearling steers are not uncommon.

---

[2]The pasture recommendations made in this section for cattle and sheep are equally applicable to horses, except for Sudangrass and sorghum-Sudangrass hybrids, which may cause cystitis—a fatal inflammation of the bladder in horses. Hence, *Sudangrass should not be grazed by horses.*

With the tightening of credit in recent years, the concept spread to other areas, especially to the South. Also, increasing numbers of calves (instead of yearlings) and of cows-calves are being custom grazed.

Fig. 7-5. Yearling steers grazing tall-grass in the Sand Hills of Nebraska. (Courtesy, Panhandle Research Station, University of Nebraska, Scottsbluff)

Custom grazing contracts should be in writing, with the terms and conditions detailed. Generally the contract calls for the custom grazer to guarantee acreage, care, salt, and water. The contract should also detail the kind of cattle (stockers/cows-calves), the incoming weight of stockers, weighing conditions, death losses, vet costs, length of the grazing period, time of payment, and the treatment of mortgages and liens.

In some cases, custom grazers charge a flat fee per head per day for the season. In other contracts, the charge is based on per pound of gain. Also, some are using an incentive basis, in which the custom grazer is given a base rate for a modest gain (for example, 1 lb/day), then a bonus payment is made for increases above this figure.

In the late 1980s, custom grazers in the Flint Hills of Kansas, the Sand Hills of Nebraska and the Osage Pastures of Oklahoma were charging from $40 to $60 per head for the season for stockers and from $75 to $110 per cow-calf for the season.

(Also see Chapter 19, section on "Stocker and Grower Contracts.")

2. **Dairy cattle.** It is estimated that 59% of the nation's milk is produced from forages (see Table 6-4). Good pasture alone will provide cows with sufficient nutrients for body maintenance and for the production of about 20 lb of milk daily; superior pasture will provide for maintenance and more than 40 lb of milk daily.

As larger numbers of lactating cows are concentrated on smaller acreages and as milk production per cow increases, dairy producers depend less on pasture and more on other feeds. A major reason for this is the inability of high-producing cows to consume enough feed to supply their energy requirement when pasture is their main feed source. The physical form and volume of pasture fill the rumen to capacity before nutrient needs of high producers are fulfilled. However, pastures continue to be practical in smaller herds and for heifers and dry cows in both large and small herds.

The general trend among dairy producers, however, is a gradual reduction in use of pasture and increased dependence on stored feeds and green chop as herd size and milk production per cow increase.

Fig. 7-6. More than half of the U.S. milk supply is produced on pasture. (Courtesy, American Milking Shorthorn Assn., Springfield, Mo.)

3. **Sheep.** It is estimated that 94% of the total feed supply of all U.S. sheep is derived from forage (see Table 6-4); for the most part, this means pasture. No other class of farm animals is so well adapted to the utilization of maximum quantities of pasture as sheep. They are unique in that the vast majority of the young are marketed as milk-fed animals directly off grass.

Fig. 7-7. Ninety-four percent of the total feed supply of U.S. sheep is derived from forage. (Courtesy, William Rueter, Gateway Star Rt., Madras, Ore.)

4. **Horses.** When idle, horses do well on pasture and other forages as the only feed. Even when working, they can use some pasture, with the amount depending on the degree and severity of the work. Most pleasure horses are kept in pasture paddocks when not in use, and they receive supplemental grain and hay when at work.

Do not use Sudangrass for horses.

Fig. 7–8. Good pastures make for a good start in equine life. (Courtesy, University of Kentucky, Lexington)

Adapted and/or Common Grasses and Legumes of U.S.

The specific grass or grass-legume mixture will vary from area to area, according to differences in soil, temperature, and rainfall. Fig. 7–9 shows the 10 generally recognized U.S. pasture areas; and Chart 7–1 shows the best adapted and/or most common grasses and legumes for each of these areas.

In using Fig. 7–9 and Chart 7–1, bear in mind that many species of forages have wide geographic adaptation, but varieties often have rather specific adaptation. Thus, alfalfa, for example, is represented by many varieties which give this species adaptability to nearly all states. Variety then, within species, makes many forages adapted to widely varying climate and geographic areas. The county agricultural agent or state agricultural college can furnish recommendations for the area that they serve.

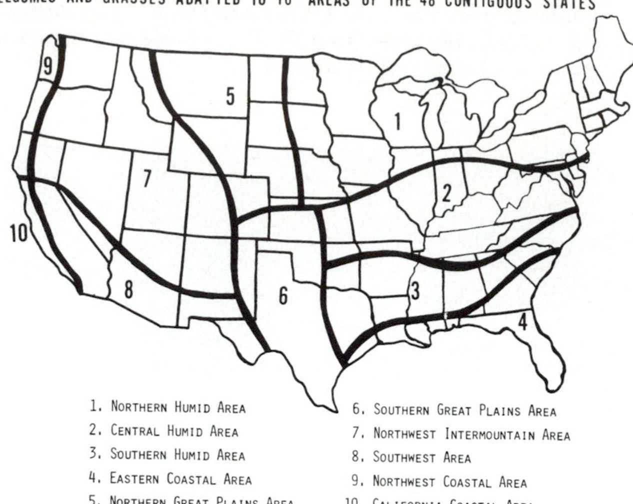

Fig. 7–9. The 10 generally recognized U.S. pasture areas.

For more specific and individual farm recommendations, farmers and ranchers are urged to seek the advice of local authorities or to write to their state agricultural college.

The following points are pertinent to the recommendations given in Chart 7–1:

1. **Fertilizer rates.** Because of the high price of fertilizer, along with concern relative to possible pollution of groundwater, fertilizer rates should be based on soil test values. Although the practice of soil testing is increasing, from authoritative sources the authors have determined that of the nation's currently chemically fertilized pastures (not including lime), only an estimated 15 to 20% of the applications are made on the basis of soil tests. *Soil tests are urged.*

After legumes have been lost from a grass-legume sward, it is recommended that they be reestablished in the grass sod. If the latter is not feasible, an annual nitrogen application at the rate of 60 to 100 lb of actual nitrogen per acre per year should be applied in split applications or increments.

Where a center pivot irrigation system is used, nitrogen is usually applied in increments, 3 to 4 times during the growing season.

2. **Varieties.** The best guide for the varietal selection of grasses and legumes from the numerous varieties available is the use of certified seed of an adapted variety.

CHART 7-1
ADAPTED GRASSES AND LEGUMES (INCLUDING BROWSE AND FORBS) FOR CATTLE, SHEEP, AND HORSE PASTURES, BY 10 GEOGRAPHICAL AREAS OF THE UNITED STATES (SEE FIG. 7-9 FOR GEOGRAPHICAL AREAS)[1]

	Areas of the United States									
	1	2	3	4	5	6	7	8	9	10
Grasses, shrubs, forbs:										
Alfileria (filaree)								x	x	
Bahiagrass (a paspalum)			x	x						
Beardgrass (a bluestem)								x		
Bentgrass	x	x								
Bermudagrass		x	x	x		x		x		x
Bluegrass	x	x			x		x	x		
Bluestem	x	x			x	x		x		
Bristlegrass (a millet)								x		
Bromegrass	x	x			x		x	x	x	
Buckwheat (wild)								x		
Buffalograss					x	x				
Buffelgrass						x				
Chamiza (fourwing saltbush)								x		
Cottontop								x		
Curly mesquite (a Hilaria)						x		x		
Dallisgrass (a paspalum)			x	x						
Dropseed						x		x		
Fescue, tall	x	x	x				x	x	x	x
Foxtail							x		x	
Galleta (a Hilaria)						x		x		
Grama grass	x				x	x		x		
Hardinggrass									x	x
Indiangrass	x	x			x	x				
Indian ricegrass								x		
Indianwheat								x		
Johnsongrass (a sorghum)			x	x		x				
Junegrass					x	x		x		
Kleingrass						x				
Lovegrass	x	x				x		x		
Mesquite (vine; a panicum)						x		x		
Millet	x	x	x	x		x				
Mormon tea (ephedra, jointfir)								x		
Muhly								x		
Needlegrass (needle-and-thread)					x		x			
Oatgrass									x	
Oats	x	x	x	x	x	x		x	x	x
Orchardgrass	x	x	x	x	x		x	x	x	x
Pangola digitgrass			x	x						
Panicgrass (a panicum)						x		x		
Paragrass (malojillo)				x						
Pea bush								x		
Pearlmillet		x	x	x		x				
Ratany								x		
Redtop	x						x		x	
Reed canarygrass	x	x			x		x		x	
Rescuegrass			x	x					x	

	Areas of the United States									
	1	2	3	4	5	6	7	8	9	10
Rhodesgrass			x	x						
Rye	x	x	x	x	x	x		x	x	x
Ryegrass, annual		x	x	x		x			x	
Ryegrass, perennial	x	x	x						x	
Sacaton								x		
St. Augustine grass			x							
Sorghum-Sudan hybrids	x	x	x	x	x	x	x			
Stargrass			x							
Sudangrass	x	x	x	x	x	x	x	x	x	x
Switchgrass (a panicum)	x	x	x		x	x				
Three-awn (wiregrass)						x		x		
Timothy	x	x					x		x	
Tobosa (a Hilaria)						x				
Wheat	x	x	x	x	x	x	x	x	x	x
Wheatgrass					x	x	x	x		
Wild-rye					x	x	x			
Winterfat (white sage)								x		
Wintergrass, Texas						x				
Legumes:										
Alfalfa (lucerne)	x	x	x	x	x	x	x	x	x	x
Alyceclover			x	x						
Black medic (yellow trefoil)			x			x		x		
Bur-clover		x						x		x
Cicer milkvetch	x				x		x			
Clover, alsike	x	x			x		x	x	x	
Clover, arrowleaf		x	x							
Clover, crimson		x	x							
Clover, Hubam (white sweet clover)	x	x						x	x	
Clover, Kura	x	x			x		x		x	
Clover, Ladino	x	x	x	x			x	x	x	x
Clover, prairie						x		x		
Cover, red	x	x	x	x			x	x	x	
Clover, strawberry					x		x	x		x
Clover, subterranean			x	x				x	x	x
Clover, white	x	x	x	x			x	x	x	x
Cowpeas			x	x						
Crownvetch	x	x								
Field pea			x	x			x			
Hairy indigo				x						
Lespedeza (annual)		x	x	x						
Lespedeza (perennial, sericea)		x	x	x						
Peas (flat)									x	
Soybeans	x	x	x	x			x			
Sweet clover	x	x			x	x	x	x	x	
Trefoil, birdsfoot	x	x	x				x		x	x
Velvet beans			x	x						
Vetch		x	x	x	x	x		x	x	

[1]Authoritative recommendations for this chart were made by the following agronomists: J. E. Baylor, Ph.D., Professor Emeritus of Agronomy Extension, The Pennsylvania State University, State College; R. A. Forsberg, Ph.D., *et al.*, Department of Agronomy, University of Wisconsin-Madison; J. R. Forwood, Ph.D., Research Agronomist, USDA-ARS, University of Missosuri, Columbia; S. C. Fransen, Ph.D., Forage Agronomist, Western Washington Research and Extension Center, Washington State University, Puyallup; C. S. Hoveland, Ph.D., Professor of Agronomy, Department of Agronomy, The University of Georgia, Athens; W. E. McMurphy, Ph.D., Professor, Department of Agronomy, Oklahoma State University, Stillwater; D. A. Miller, Ph.D., Professor, Department of Agronomy, University of Illinois, Urbana-Champaign; R. R. Smith, Ph.D., Professor, Department of Agronomy, University of Wisconsin, Madison.

PASTURES FOR SWINE

Fig. 7–10. Sows and litters on pasture. Good pasture can reduce the concentrate requirement of sows and pigs by 20%. (Courtesy, Hampshire Swine Registry, Peoria, Ill.)

Prior to 1950, pastures were considered essential for successful swine production. Subsequently, the importance of pastures for swine declined with increased knowledge of nutrition and escalated land and labor costs. These forces resulted in more and more confinement rearing, accompanied by increased labor efficiency and less use of pastures.

Today, only 4.3% of U.S. swine feed is derived from forage, including pasture (see Table 6–4). Yet, hogs, especially gestating sows, will often yield greater return from an acre of good pasture than any other class of farm animal.

Good pasture can reduce the concentrate requirements for the breeding herd by 75%, for sows and pigs by 20%, and for growing-finishing pigs by 15%.

In a 3-year study, the Alabama Agricultural Experiment Station compared a rotational grazing system vs drylot management for gestating sows. The perennial pasture consisted of orchardgrass and crimson clover, along with a summer annual pasture of millet and a winter annual pasture of rye, ryegrass, and clover. The forage grazing program replaced up to 67% of the drylot feed needed for sows. On a dollar basis, the saving by grazing gestating sows amounted to as much as $40 per sow per year over drylot feeding.[3]

The most common forages used in U.S. swine production are alfalfa and ladino pastures and dehydrated alfalfa meal. But many other pastures are used, including other clovers, lespedeza, birdsfoot trefoil, rape, winter rye, and certain grasses. In addition to suitability for swine, when choosing forage species and cultivars, consideration should be given to adaptation.

(Also see Chapter 23, Feeding Swine, section on "Pastures."

[3]Prince, T. J., A. L. Stephenson, and J. T. Eason, *Highlights,* Alabama Agricultural Experiment Station, Fall 1986, p. 5.

SEEDING AND MANAGEMENT OF SUBHUMID, HUMID, AND IRRIGATED PASTURES

This section, and the subsections under it, has reference to those pastures which either receive above approximately 20 in. of rainfall annually or are irrigated. This includes the pastures of the Corn Belt, the South, the East, and the irrigated valleys and smaller, scattered moderate-to-high-rainfall areas throughout the West.

Fig. 7–11. Hereford breeding herd on clover. (Courtesy, American Hereford Assn., Kansas City, Mo.)

Establishing a New Pasture

The following practices are usually adhered to in successfully establishing a new pasture in the subhumid, humid, and irrigated areas:

1. **The species and cultivars are selected.** The selection of species and cultivars to seed should receive first priority. One species or a specific combination best fits the needs of a certain climate, soil, and intended use. A single species may be adequate for one pasture while several species may be best for another.

Chart 7–1 gives the general recommendations of competent specialists residing in different areas of the United States. For more specific recommendations for a particular farm or ranch, the livestock producer should consult such local authorities as the county extension agent or successful neighbors.

Where grass-legume mixtures are to be grown, a 50–50 mixture is satisfactory for most purposes and conditions.

2. **The soil is tested and limed/fertilized.** Following testing of the soil, it should be limed and/or fertilized according to needs. In humid regions where periodic application of lime is needed, it is best to work lime into the soil considerably in advance of seeding. Phosphorus levels are especially critical when establishing a new seeding. When the soil is deficient in nitrogen, the application of a small amount of nitrogen at seeding time may be helpful.

A thin, uniform mulch of manure is especially valuable in establishing a new seeding.

3. **High-quality seed is purchased.** The seed should be of good quality, of high germination and purity as indicated on the tag, and free of noxious weeds. Also, proof of origin is of prime importance when an imported variety is obtained. Certified seed carries more assurance of being high quality than noncertified seed, and gives proof of its origin much as a registration certificate does for a purebred animal.

4. **Scarified legume seed is used.** In the purchase of certain legume seed, it is important that it be scarified, which breaks the seed coat and allows faster moisture penetration —thus assuring quicker and more uniform germination and a better stand the first year.

5. **Legume seed is inoculated.** Since legumes can use nitrogen from the air provided they are inoculated with the proper bacteria, it is important that legume seed be inoculated. This is traditionally done at planting time by coating the seed with a water-based slurry consisting of a commercial preparation of bacteria mixed with a peat carrier enriched with sugars, gums, and complex polysaccharides to provide nutrition, adhesion, and protection. The expiration date is given on the container. It is important that the seed not be inoculated more than a few hours before seeding because these nitrogen-fixing bacteria are easily killed by drying, heat, sunlight, chemical seed treatment, or by direct contact with acid fertilizers.

In many localities where the legume species has been grown regularly for a number of years, inoculation of seed is not necessary because the soil contains a sufficiently large population of nodule bacteria to fulfill crop requirements.

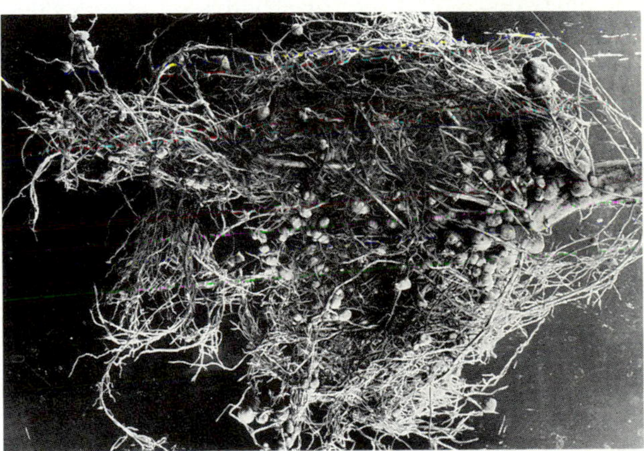

Fig. 7-12. Nodules (small bumps) containing nitrogen-fixing bacteria on a well-inoculated soybean root. (Courtesy, USDA)

6. **A good seedbed is prepared.** A good seedbed is free from weeds, fine-textured, firm, and moist.

Weeds are usually destroyed by growing row crops or a small grain the year preceding seeding to pasture and by cultivating frequently following the harvesting of this crop.

There are many different ways in which to prepare a good seedbed. Perhaps as good a method as any consists in (a) plowing as far in advance of seeding as possible, (b) discing, (c) harrowing one or more times to level the field and smooth the surface, and (d) cultipacking or rolling. A properly prepared seedbed should be so firm that one barely leaves a footprint when walking across it; the firmer the better from the standpoint of moisture conservation of small seeds.

7. **The seeding operation is timed and carried out properly.** The seeding time will vary, being determined primarily by the area and by the species or mixture used. Cool-season species should be seeded early enough, in the spring or fall, for seedling development so they can withstand the stress of hot summers and/or avoid winter injury. Conversely, warm-season species may rot before germination if planted before critical soil temperatures are reached in the spring.

The actual seeding operation may be (a) by broadcast, with a whirlwind seeder or by hand; (b) by cultipacker, consisting of two corrugated rollers with seed-metering boxes; or (c) by drilling with one of several types of conventional seeders over a band of fertilizer ⅕ to ⅗ in. deep. Drilling is the preferred method, for it ensures more uniform placement of seed in both depth and amount of seed per acre and results in a more uniform stand.

Seeding rates are dependent on many factors. Normally, only about a third of sown seed produces seedlings, and only half of the seedlings survive the first year. So, high seeding rates appear to be justified. Moreover, high seeding rates reduce weed invasion and increase yields the first year.

Since most grass and legume seeds are very small, they should not be covered deeply. A good rule is that they should not be covered more than 4 or 5 times the diameter of the seed; usually this means not more than ⅙ to ½ in. deep.

8. **A companion (nurse crop) or a preemergence herbicide may be used.** The value of planting a "companion" or nurse crop—usually consisting of annuals—with new seed crops is controversial. The **advantages** are (a) it furnishes a crop of value while the new seeding is being established, (b) it lessens erosion, and (c) it reduces the weed population. The **disadvantages** are (a) it may retard the growth of the seedlings for whose protection it is grown, and (b) it may rob the new seedlings of so much moisture and light that it kills them during dry spells unless the companion crop is harvested early as pasture, hay, or silage.

Preemergence herbicides may replace the traditional companion crop where clear seedings (not grass-legume mixtures) are involved. Usually these result in higher forage yields in the seeding year than can be obtained when a companion crop is used.

Renovating An Old Pasture

Fig. 7-13. No-till seeder in operation, overseeding a pasture in Caroline County, Maryland. (Courtesy, University of Maryland, College Park)

Renovation is the improvement of pasture by partial or complete destruction of a sod, plus liming, fertilizing, weed control, and seeding as required to establish desired species without an intervening crop. Extension of the grazing season by sod-seeding small grains or other winter annual species into permanent warm-season grass pastures was started in the early 1950s. About the same time, effective herbicides and sod-seeding machines evolved, making no-till pasture renovation a practical way to improve many grasslands. Today, both winter and summer annual species are used to increase yields and extend the grazing season.

Production from renovated permanent pasture may be increased 2- to 5-fold, depending on the soil characteristics and the condition of the sod.

The practice of sod-seeding works best with cool-season annuals seeded into warm-season perennial grasses. Winter annuals, such as small grains (rye, wheat, or oats), sod-seeded into Bermudagrass swards work especially well since there is minimum competition between the two species.

Successful seeding of summer annuals into perennial cool-season swards can also provide improved forage distribution.

Permanent pastures may be renovated by the following methods:

1. **Complete seedbed preparation with reseeding.** Complete seedbed preparation followed by reseeding has been superior to sod-seeding or overseeding in several areas of the North where perennial grass weeds in the sod are difficult to subdue. The establishment of a legume is markedly improved by plowing as compared to tilling with a field cultivator. Depending on soil conditions, the use of an intervening row crop for one year may or may not be desirable to reduce weed population and competition from the sod, and to help defray costs.

2. **Sod-seeding or overseeding (no-till seeding).** In humid regions, reseeding of legumes and/or grasses into thin permanent pasture sods by sod-seeding or overseeding is often preferred to complete seedbed preparations. The basic requirements for successful sod-seeding are: the partial or complete destruction of the existing plants (including invading weeds) by close grazing and the use of herbicides, thereby lessening top growth, surface resistance, and competition; liming when necessary; fertilizing; proper seed placement; and good pasture management following seeding.

In the North, sod-seedings are made in the spring at a time coinciding with the maximum growth period of grasses in the old sod.

In the South, sod-seeding is most successful when introduced legumes and grasses are seeded at a time that does not coincide with the maximum growth period of the existing sward. So, seedings are generally made in late summer or early fall. Winter annuals or perennial grasses and legumes can be incorporated into the sod to provide winter grazing. Overseeding with pelleted seed of legumes such as white clover and alfalfa from an airplane has also been successful in parts of the humid region.

While most renovations using sod-seeding have been made on humid pastures of eastern United States, the technique has also been effectively used on some western rangelands.

3. **Fertilizing and liming.** Large areas of permanent pasture have adequate cover but are low in productivity because of lack of fertility. Fertilization can be an economical method of increasing the productivity of such pastures.

Legume-grass pastures with more than 50% legumes usually do not respond to N fertilization. However, such pastures usually respond to lime, P, and K; the lime and P-K fertilizer encourage legume growth and increase the N supply. Also, such fertilization of legume-grass enables earlier spring grazing and extends the pasture season.

Grass pastures and grass-legume pastures with less than 50% legumes respond to N fertilization. However, adequate pH, P, and K soil levels must be maintained to achieve top results from N. High N fertilization (120 to 240 lb N/acre) of tall-growing grasses gives greater response than high N fertilization of short-growing grasses such as Kentucky bluegrass. In addition to increasing yields and carrying capacity, N fertilization of grass pasture encourages growth earlier in the spring and sustains production later in the fall.

Factors Affecting Value of Pasture

Many factors affect the value of pasture, including (1) soil and fertilizer, (2) plant species, (3) stage of maturity, (4) rate of growth and season of year, and (5) grazing.

• **Soil and fertilizer**—Soil and fertilizer affect the growth and composition of pasture crops. Many experiments have been conducted to determine the effect of soil fertility and fertilizer application on pasture. A brief summary of the benefits that generally accrue from pasture fertilization follows:

1. **Increased yields.** The chief benefit to accrue from applying fertilizer to pasture is an increase in yield.

2. **Increased proportion of legumes.** In grass-legume pastures, proper fertilizing can make for an increase in the proportion of legumes; in turn, this increases the protein, calcium, phosphorus, and vitamin content of the mixture.

3. **Extended grazing season.** Properly fertilized pasture plants grow over a longer period of time than those on infertile soils; they begin growth earlier in the spring, and they continue growth later in the fall.

4. **Increased protein and palatability.** The protein content of young, immature nonlegume pasture is increased appreciably by nitrogenous fertilization, unless, of course, there is already plenty of nitrogen in the soil. This increase may be sufficient to add materially to the feeding value of grass pasture. Also, nitrogen fertilization increases the palatability of grasses.

5. **Increased calcium and phosphorus.** Calcium-deficient soils affect pastures in two ways: (a) The percentage of calcium in nonlegume crops is considerably reduced, and (b) the legume crop, if present, will not thrive.

On phosphorus-deficient soils, the phosphorus content of grasses may drop so low that phosphorus deficiency in animals is produced, unless a phosphorus supplement is provided. According to research workers at the Texas Station, pasture forage having less than 0.15% phosphorus on a dry matter basis will not supply enough of this element to meet safely the requirements of beef cattle grazed thereon without supplemental feed. The phosphorus content of legumes is less affected by a deficiency of soil phosphorus than that of nonlegumes. However, most legumes do not thrive or yield well on phosphorus-deficient soils.

• **Plant species**—Plant species affect the feeding value of pasture. Generally speaking, legumes contain a higher percentage of protein and calcium than nonlegumes. Also, there are marked differences between different kinds of pasture plants as growth advances. For example, bromegrass retains its palatability and nutritive value over a longer period than most grasses. By contrast, reed canarygrass is readily eaten when young, but becomes woody, high in alkaloids, and unpalatable with maturity. Most legumes retain their palatability and nutritive value as they mature better than most grasses. An exception to the latter rule is *lespedeza sericea,* which becomes bitter and distasteful with maturity due to the accumulation of tannin in the plants. However, plant breeders have developed *sericea* that is low in tannin, thereby overcoming this problem to a considerable extent.

• **Stage of maturity**—Many livestock producers are not aware of the great differences in nutritive value between young, immature pasture and the same plants when they are mature or even at the usual hay stage. These wide differences are shown in Section V—Composition of Feeds, and in Fig. 7-14.

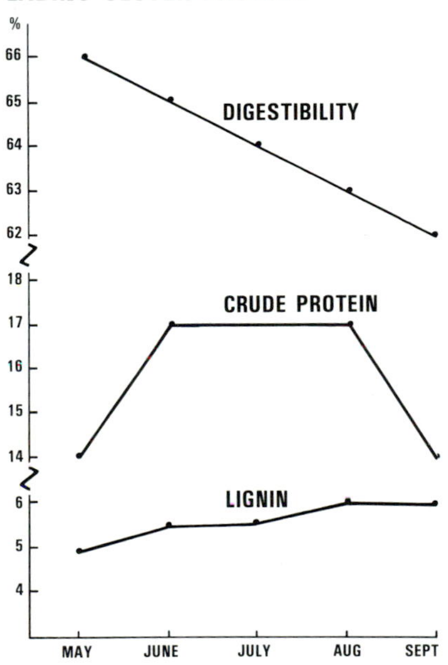

Fig. 7-14. Changes in digestibility, crude protein, and lignin content of orchardgrass *(Dactylis glomerata)*/ladino clover *(Trifolium repens)* pasture in Tennessee, during a 5-year period, from May to September. **Note well**: Digestibility decreased with maturity; it decreased steadily from 66% in May to 62% in September. Protein decreased with maturity; it dropped from 17% in June to 14% in September. Lignin increased with maturity; it increased from 4.8% in May to 5.8% in September. (*Source of data:* Lane, C. D., K. M. Barth, and J. B. McLaren, *Journal of Animal Science,* Vol. 34, No. 2, 1972, p. 351.)

Stage of maturity affects pasture composition as follows:

1. **Protein decreases with maturity.** Young pastures are much richer in protein, on a dry basis, than the same plants at maturity. Young actively growing grasses may run 16 to 20% protein content, or better; and young legumes, such as alfalfa and clover, are even higher—20 to 25% protein or more. Very mature grasses may contain as little as 2.0% protein. The protein content of grasses decreases greatly as they head out, largely because of (a) the accumulation of carbohydrates, and (b) the smaller proportion of leaves on the plants.

The protein content of legumes decreases with maturity, but to a lesser extent than in grasses.

The above underscores the need for producers to take advantage of the high-protein content of young pastures when formulating rations so as to avoid overfeeding costly protein supplements.

2. **Fiber and lignin increase with maturity.** Young plants are lower in fiber and lignin than mature plants with the result that they are more tender and more digestible. On a dry matter basis, young grass-legume pasture will range up to 68% total digestible nutrients, in comparison with 51% for clover and timothy of good quality cut at the normal hay stage. As straight grass becomes mature, the digestibility and nutritive value decrease even more. If it is then left to bleach and weather on the stalk, it will resemble poor-grade straw in composition and nutritive value.

To a considerable degree, the decrease in digestibility as plants mature is due to an increase in lignification, which lowers digestibility, even by ruminants.

3. **Calcium and phosphorus decrease with maturity.** The calcium content of plants decreases with maturity. However, the percentage change in calcium is much less than occurs in phosphorus.

Young grasses or legumes grown on phosphorus-rich soils usually contain 0.25%, or more, phosphorus. Although phosphorus in pasture decreases with maturity, there is generally plenty left for livestock, unless (a) it is produced on phosphorus-deficient soils, or (b) it is left to cure on the stalk, which makes for weathering and bleaching. Early spring grass (April) may contain five times more phosphorus than leached and weathered grass in the winter (March).

4. **Vitamin value decreases with maturity.** Actively growing green parts of plants are high in carotene. Likewise, such plants are usually rich in most of the B complex vitamins, in vitamin E, in ascorbic acid, and in certain unidentified factors. But the content of vitamins, especially of carotene, decreases as plants mature. When pasture plants are left to cure on the stalk, and to leach and weather, carotene disappears rapidly.

• **Rate of growth and season of year**—Rapidly growing grass is usually rich in protein and in other nutrients on a dry basis. It is important, therefore, that pasture plants be properly fertilized and managed so that they keep growing and be prevented from heading out.

Grass is usually higher in protein and other nutrients early in the spring than later in the season. If plant growth is sharply checked in the summer—due to drought, hot weather, and/or lack of available plant food—the protein content and the digestibility will be lower than that of grass at the same stage of maturity earlier in the season.

If pasture resumes growth after the fall rains come, it may be nearly as high in protein and other nutrients as spring growth.

Also, it is noteworthy that warm season perennial grasses generally have a lower percentage of protein and are less desirable than cool season perennial grasses.

• **Grazing**—When pastures are grazed closely throughout the season, the total yield of dry matter is usually 30 to 50% less than when they are allowed to grow to the normal hay stage. This is due to the smaller leaf surface and lowered photosynthesis. This explains why rotational, strip, and green chop grazing usually yield more than close continuous grazing.

The effect of frequent grazing will depend on the kind of plants. The yield of tall-growing plants—such as timothy, orchardgrass, alfalfa, and the erect clovers—is reduced much more than that of low-growing spreading plants, such as bluegrass, Bermudagrass, and white clover.

In contrast to the lowering of the yield of dry matter, frequent grazing usually results in greater total production of protein for the season than when the crop is cut for hay. Also, because immature plants are lower in fiber and more digestible than mature plants, the yield of total digestible nutrients is not reduced as much by frequent grazing as the dry matter yield—dry matter production is lowered by 30 to 50%, whereas digestibility is lowered only by 25 to 40%.

Also, it is noteworthy that plenty of available forage results in selective grazing—with the animals picking and choosing the leaves and finer parts of stems, which are more tender and more nutritious, and rejecting the coarser, stemmy parts. Thus, the portion consumed under such circumstances may differ appreciably from the chemical composition of the entire plant.

Pasture Management

Many good pastures have been established only to be lost through careless management. Good pasture management in the subhumid, humid, and irrigated areas involves the following practices:

1. **Controlled grazing.** Nothing contributes more to good pasture management than controlled grazing. At its best, it embraces the following:

   a. **Protecting first-year seedings.** First-year seedings should be grazed lightly or not at all in order that they may get a good start. Where practical, instead of grazing, it is preferable to mow a new first seeding about 3 in. above the ground and to utilize it as hay or silage, provided there is sufficient growth to so justify.

   b. **Shifting the location of salt, shade, and water.** Where portable salt containers are used, more uniform grazing and scattering of the droppings may be obtained simply by the practice of shifting the location of the salt to the less grazed areas of the pasture. Where possible and practical, the shade and the water should be shifted likewise.

   c. **Deferred spring grazing.** Allow 6 to 8 in. of growth before turning out to pasture in the spring, thus giving grass a needed start. Anyway, the early spring growth of pastures is high in moisture and washy.

   d. **Avoiding close late fall grazing.** Pastures that are closely grazed late in the fall start late in the spring. With most pastures, 3 to 5 in. of growth should be left for winter cover. An exception to this close grazing rule should be made where winter annual clovers are to be seeded, or are expected to volunteer, on Bermudagrass or bahiagrass. Under such circumstances, close grazing or mowing (1 to 2 in.) of the Bermudagrass or bahiagrass is recommended. Likewise, in the Pacific Northwest, perennial ryegrass must be grazed hard in the summer in order for subterranean clover seeds to germinate and become established in the autumn.

   e. **Avoiding overgrazing.** Never graze more closely than 2 to 3 in. during the pasture season. Continued close grazing reduces the yield, weakens the plants, allows weeds to invade, and increases soil erosion. The use of temporary and supplemental pastures, such as Sudan, may "spell off" regular pastures through seasons of drought and other pasture shortages and thus alleviate overgrazing.

   f. **Avoiding undergrazing.** Undergrazing seeded pastures should be avoided, because (1) mature forage is unpalatable and of low nutritive value; (2) tall-growing grasses may drive out low-growing plants such as white clover due to shading; and (3) weeds, brush, and coarse grasses are more apt to gain a foothold when the pasture is grazed insufficiently. It is a good rule, therefore, to graze the pasture fairly close at least once each year.

2. **Clipping pastures and controlling weeds.** Pastures should be clipped at such intervals as necessary to control weeds (and brush) and to get rid of uneaten clumps and other unpalatable coarse growth left after incomplete grazing. Good grazing management will reduce the amount of clipping needed. Pastures that are grazed continuously may be clipped at or just preceding the usual haymaking time; rotated pastures may be clipped at the close of the grazing period. Weeds and brush may also be controlled by chemicals and burning.

3. **Topdressing.** Like animals, for best results, grasses and legumes must be fed properly throughout a lifetime. It is not sufficient that they be fertilized (and limed if necessary) only at or prior to seeding time. In addition, in most areas it is desirable and profitable to topdress pastures with fertilizer annually, and, at less frequent intervals, with reinforced manure and lime. Such treatments should be based on soil tests, and are usually applied in the spring or fall.

4. **Scattering droppings.** The droppings should be scattered at the end of each grazing season in order to prevent animals from leaving ungrazed clumps and to distribute the droppings over a larger area. This can best be done by the use of a brush harrow or chain harrow.

5. **Grazing by more than one kind of animal.** Grazing by two or more species of animals makes for more uniform pasture utilization and fewer weeds and parasites, provided the area is not overstocked. Different kinds of livestock have different habits of grazing; they show preference for different plants and they graze to different heights. For example, sheep consume shorter and finer forages and more forbs than cattle.

6. **Irrigating where practical and feasible.** Where irrigation is practical and feasible, it alleviates the necessity of depending on natural precipitation.

Extending the Grazing Season

In practically all U.S. pasture areas, the grazing season can be extended by grazing earlier in the spring and later in the fall/winter, thereby lessening the amount of stored feed needed for winter.

Fig. 7–15 illustrates in graphic form the growth period of each of the common pasture plants of area 2 (see Fig. 7–9 for areas), the northern part of the humid South. As shown, by selecting the proper combination of crops, pastures for each month of the year are assured. A similar chart for each area of the country can be developed.

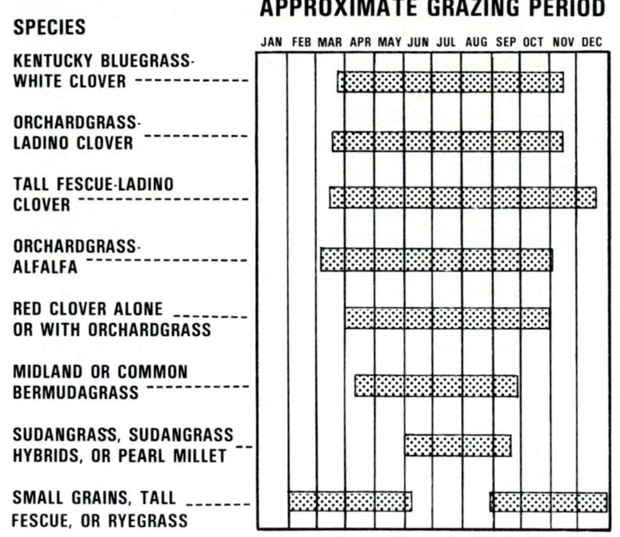

Fig. 7-15. Approximate grazing period of common pasture crops in area 2 (see Fig. 7-9 for areas), the northern part of the humid South. As shown, by selecting the proper crops year-round grazing can be achieved.

Some practical ways of extending the grazing season in each of the 10 recognized pasture areas shown in Fig. 7-9 follow.

• **Area 1, Northern Humid Area (the lake states and northeastern United States)**—For supplemental summer pasture, (1) Sudangrass, sorghum-Sudangrass hybrids, or millet may be grown; or (2) switchgrass or Caucasian bluestem provide good mid-summer grazing in pasture systems that include tall grasses fertilized with nitrogen.

For fall pasture, brassica crops such as turnips, rape (canola), swedes, or fodder beets may be grown.

Where Bermudagrass is grown, rye, barley, oats, or winter wheat may be sod-seeded into the Bermudagrass sod to extend the grazing season into the cooler months.

• **Area 2, Central Humid Area (the northern part of the southern humid area)**—For supplemental summer pasture, Sudangrass, sorghum-Sudangrass hybrids, or millet may be grown.

For fall, winter, and early spring pasture, cereal grains (rye, oats, winter wheat, or barley), tall fescue, or ryegrass may be grown.

For year-round pasture, tall fescue may be interseeded into Bermuda sod.

• **Area 3, Southern Humid Area (southern United States)**—For fall, winter, and early spring grazing, fall-sown small grain (oats, barley, wheat, or rye) or perennial tall fescue may be grown.

For summer grazing, Sudangrass, sorghum-Sudangrass, or millet may be grown.

Also, to extend the grazing season into the winter months, Bermuda sod may be overseeded with small grain (oats, barely, wheat, or rye) or legumes (arrowleaf or crimson clover, vetch, or winter peas).

• **Area 4, Eastern Coastal Area (southeastern United States)**—For supplemental summer grazing, annual lespedeza, soybeans, sorghum-Sudangrass hybrids, or millet may be used.

For winter grazing, warm-season perennial grass (such as Bermudagrass) may be fall sod-seeded to annual ryegrass, a small grain, crimson clover, or arrowleaf clover.

• **Area 5, Northern Great Plains Area (central and northern Great Plains)**—For early spring grazing, crested wheatgrass may be used.

For hot-season grazing when cool-season grasses become dormant, switchgrass or Sudangrass may be used.

For fall grazing, Russian wildrye is especially good because of its high protein content.

• **Area 6, Southern Great Plains area (the southern Great Plains and the southwest)**—In the irrigated areas, alfalfa, the sorghums, or a number of grass species may be grown. Among the grass species that may be grown under irrigation in this area are Bermudagrass, tall fescue, tall and intermediate wheatgrasses, ryegrass, blue panicgrass, Hardinggrass, and kleingrass. Orchardgrass, smooth bromegrass and reed canarygrass may be grown in the northern part of the region.

For fall, winter, and spring grazing, Bermudagrass and Old World bluestems may be overseeded with a small grain or a cool-season annual.

For winter pasture in the eastern part of the area, small grain pasture (wheat, oats, barley, or rye) may be grazed.

• **Area 7, Northwest Intermountain Arèa (the intermountain area and Alaska)**—For range reseeding in this area, crested wheatgrass is the most widely used species.

In Alaska, smooth bromegrass or timothy may be seeded on tilled croplands, or winter rye may be used for early spring grazing.

• **Area 8, Southwest Area (the southwest)**—For summer grazing, Sudangrass, Sudangrass-sorghums, or annual ryegrass may be used.

For winter and early spring grazing, Bermudagrass or sod may be overseeded with small grain (oats, barley, wheat, or rye) or a legume (crimson clover, vetch, or winter peas).

For augmenting winter range, perennial irrigated pastures may be used.

• **Area 9, Northwest Coastal Area (northern Pacific Coast)**—For summer grazing, Sudangrass, Sudangrass-sorghum hybrids, fescue, ryegrass, subterranean clover, or rape may be grown.

For winter feed, turnips or kale may be grown.

• **Area 10, California Coastal Area (southern Pacific Coast)**—For summer grazing, Sudangrass, Sudan-sorghum hybrids, or annual ryegrass may be grown.

For fall grazing in northern California, ladino clover or subterranean clover may be used.

For fall grazing in central and southern California, alfalfa may be used.

In addition to lengthening the grazing season through the selection of species, it may be extended as follows:

1. **By obtaining earlier spring pastures.** This can be accomplished by avoiding grazing too late in the fall and by applying nitrogen fertilizer in the fall or early spring. Nitrogen fertilizers will often stimulate the growth of grass so that it will be ready for grazing 10 days to 2 weeks earlier than unfertilized pastures.

2. **By saving fall growth for winter grazing.** For example, tall fescue that has been cut for winter feed can be used to stockpile regrowth in August through October for grazing from November into the winter. Also, often the regrowth after hay harvest may be utilized.

3. **By using crop residues.** Following harvest, cornstalks will provide 2 months of winter grazing for dry cows and ewes. In addition to extending the grazing season, this utilizes feed which would otherwise be wasted or unmarketable.

TURNIPS FOR FALL/WINTER GRAZING

Turnips are a popular fall/winter forage for cattle and sheep in some northern areas, especially in the Pacific Northwest. They are commonly planted in July following harvest of early-maturing crops such as sweet corn, early potatoes, or small grains. Grazing usually begins in October or early November. Two varieties, Barive Cow Turnips and Purple Top White Globes, are used as forage.

The factors that usually make turnips a profitable fall/winter crop are:

1. They grow rapidly in cool weather, producing high yields in 90 days (4 tons of dry matter per acre for Purple Tops; 2.5 tons of dry matter per acre for Cow Turnips).
2. They are highly palatable (100% of the leaves utilized; 78% of the roots utilized).
3. They have excellent nutritional value. On a moisture-free basis, they are high in energy (roots 86% TDN; leaves 69% TDN), and they are good in protein (roots 9–12% crude protein, and leaves 12–19% crude protein).

Turnip fields normally provide up to 350 steer days of grazing per acre. In stocker operations, weight gains of 1.5 to 2.0 lb per head per day may be expected on 600–700 lb steers. A roughage supplement of 2–3 lb of straw or hay per animal daily is recommended; it will improve digestion and slow down the passage of the turnips through the digestive tract.

Cattle generally remain healthy while grazing turnips, but occasionally they develop polioencephalomalacia, pulmonary emphysema, bloat, or hemolytic anemia; and sometimes they choke. Pregnant ewes grazing turnips without iodine supplementation may give birth to lambs with goiters and low survivability. These diseases may be lessened, or prevented, by the following dietary management:

1. Place animals on a high-quality diet 2–3 weeks prior to grazing turnips, thereby preparing the rumen microorganisms to digest turnips.
2. Immunize against enterotoxemia.
3. Avoid abrupt change to turnips.
4. Supplement the turnips with straw, hay, or other roughage.
5. Strip-graze turnips (small strips, moved weekly).
6. Feed an iodized salt-trace mineral mix at all times.

SEEDED PASTURE GRAZING SYSTEMS

Several systems of grazing management have been successfully applied to semipermanent and supplementary pastures. Generally speaking, the more intensive the system of management on such pastures, the higher the yield of forage and of livestock products.

It is noteworthy that pasture grazing systems have been changed/adapted by both researchers and farmers; and with such changes/adaptations, they have been given different names. Nevertheless, under whatever name, the basic types of rotation grazing, intensive grazing, creep grazing, strip grazing, and green chop are covered in the sections that follow.

(Also, see the section on Range Systems in this chapter. The principles involved in grazing seeded pastures and western ranges are similar. However, the application differs because western range pastures are generally much larger and have lower rainfall.)

Continuous Grazing

The name identifies the practice. *Continuous grazing is the uninterrupted grazing of a specific pasture by livestock throughout a year or grazing season.* It can be successful provided variable stocking is practiced, with some adjustment in animal numbers to reduce the severity of under- or overgrazing.

The **advantages** of continuous grazing as compared to rotational grazing are (1) lower costs for fencing and watering facilities, because there are fewer animals when they are confined to one pasture; and (2) fewer management decisions when animals are not moved from pasture to pasture.

The **limitations** of continuous grazing are (1) animal numbers are seldom flexible; and (2) pastures must be stocked lighter than desired when forage growth is maximal to avoid overgrazing during periods of minimal forage growth.

Rotation Grazing

Rotation grazing is that system in which two or more pastures are grazed and rested in a planned sequence. In this system, pastures are divided and fenced into two or more pastures.

Rotation grazing involves the concept of *time* as a management variable for either the grazing period or the pasture regrowth interval of each pasture. Duration of grazing and rest generally are governed by herbage growth rate, which depends primarily on the time of year, moisture, sod fertility, and species. The number of animals grazing each pasture may be fixed or variable. Experimental studies comparing continuous and rotation grazing have given inconsistent results. In general, however, rotation grazing is favored in temperate regions and continuous grazing is favored in the tropics.

The **advantages** of a rotation grazing system are:

1. It permits the farmer to match grazing more adequately to the growth habit of the forage species, condition of the pasture, and animal needs than does continuous grazing.
2. It improves stand persistence. Plants are given recovery periods during the growing season for more or less unhampered development of tillers, and leaves. This is essential to replenish root reserves. It is the only system of grazing which enables the tall-growing legumes and grasses to survive.
3. It increases carrying capacity. Greater amounts of feed nutrients can be removed in the form of herbage with reduced losses due to trampling, fouling, and herbage death and decay.
4. Equalization of grazing is encouraged. It prevents overgrazing and undergrazing, and results in maintaining a better balance of the legumes and grasses. Also, both the palatable and the inferior species will be grazed more nearly the same.

5. It provides more nutritious herbage since the herbage is at the most ideal pasture stage. It will be high in protein and low in crude fiber. The protein content will average 10 to 20% compared with 6 to 10% for more mature grasses.

6. It prevents the grasses from heading out, particularly if the mower is used each time the animals are shifted to new pasture. This allows new growth to come back uniformly and keeps it more palatable.

7. It helps control livestock parasites.

8. It makes it easier to harvest surplus forages as hay or silage.

The **limitations** of rotation grazing are:

1. It requires a higher input of capital and management than continuous grazing.

2. There is a continuous day-to-day decline in the quality of the available forage. At first turn-on, animals have access to leafy, high quality forage, but the quality of the forage gets poorer and poorer during the grazing period.

• **Conventional (rectangular) system**—This is a short duration system, similar in principle to the Savory (cell) system, but different in design; the Savory system generally uses the wagonwheel design, whereas the conventional (rectangular) system uses retangular pastures and an alley. Like the Savory system, it was first used in the West, but it is finding increased use in seeded pastures in the humid area. (See section on "Short Duration Grazing Systems" later in this chapter.)

• **Savory (cell) system**—Although the Savory (cell) system, which is a short duration grazing system, started, and was first popularized, in the West, it is now finding increased use in seeded pastures in the humid area. (See section on "Short Duration Grazing Systems" later in this chapter.)

Intensive Grazing

Several ingenious intensive grazing systems have evolved. All of them are designed to provide the maximum of high quality forage, to utilize the highest quality pastures for the highest producing animals, and/or to make more profit. They call for maximum spring and early summer grazing, and for using the best quality pastures for dairy cows, cows suckling calves, ewes with twin lambs, does with kids, or mares with foals.

• **First and second grazers**—This grazing system involves two herds: first grazers, and second grazers. Here is how it may work with dairy cows: High producing lactating cows, which have a high energy requirement, may be first grazers, they are allowed to graze the higher-quality (leafy) portion of pasture No. 1, following which they are moved to pasture No. 2—a fresh pasture. Dry cows, which have a low energy requirement, may be second grazers; they are turned onto pasture No. 1 immediately following the removal of the high producers. This progression is continued through all pastures, then the cycle is repeated. Also, it may involve three or more groups of animals. The groups may consist of any animal species or class. For example, the first grazers may consist of beef cows and suckling calves, and the second grazers of stocker steers; or the first grazers may be ewes with twin lambs, and the second grazers gestating ewes.

The chief **advantage** of the system of first and second grazers is the enhanced productivity of the first grazers. The main **limitations** are the necessity of maintaining (1) two groups of animals of different productivity levels, and (2) balanced stocking rates and pasture sizes.

• **Intensive early season stocking**—This grazing system calls for heavy stocking (perhaps twice the average year-round carrying capacity of the pasture) in the spring and early summer, when the pasture is of highest quality and most productive. Here is how it works with stocker cattle: Double the normal stocking rate in the spring and early summer. Then, around July 1, either sell half or more of the herd or place them in a feedlot.

The **advantages** of this system are (1) more pounds of beef per acre, (2) lower interest charges, because of owning the cattle for a shorter period of time, and (3) higher net returns. The main **limitation** is the lack of flexibility relative to removal of half the herd; they must either be sold or moved into the feedlot as scheduled—in mid-summer.

Creep Grazing

Creep grazing is a system of grazing nursing young on a high-quality pasture(s) (a grass-legume mixture, all legumes, or high-quality annuals) separated from their dams. This system may be used for cows and calves, ewes and lambs, goats and kids, and mares and foals. Creep grazing may be accomplished as follows:

1. **Allowing young to forward graze ahead of their dams, then following with their dams later.** This is similar in principle to "first and second grazers," with the young having access to the choicest, most succulent pasture(s) without competition from their dams. It may be accomplished by either of the following methods:

a. Confining the dams and young in one field for a period, but allowing the young to enter a nearby, choice-quality pasture(s) through a creep opening large enough for the young but small enough to keep the dams out. The dams are kept on each pasture in the rotation until all forage is well utilized, then moved to a new pasture. This same pattern is continued through all pastures in the rotation.

b. Keeping the dams and young together in part of a field, but allowing the young without their dams to forward graze the rest of the field. This is accomplished by positioning an electric wire high enough to allow young to pass under but low enough to keep the dams out.

2. **Keeping the dams and young on a base pasture, and providing an additional creep pasture for the young.** The dams and young are kept on a base pasture (or pastures if rotational). In addition, the young are given access to a high quality creep pasture(s) through a creep opening.

This system works best by using a two-pasture rotation for the young, with these creep pastures seeded to perennial legumes or summer annuals, although they may have the same species as the base pasture. Once the young have selectively grazed the No. 1 pasture, the dams are also turned on it, and both dams and young are allowed to graze it together until it is well utilized. Then, the No. 1 pasture is closed, the dams are returned to the base pasture, and the creep is opened to the No. 2 pasture for the young.

Limited studies in creep grazing calves indicate that as much as ½ lb extra weight gain per day may be obtained by creep grazing.

Strip Grazing

In this system, animals are allowed access to a strip which may be large enough for several days of grazing or small enough for one-half day to one-day of grazing. Heavy stocking rates of upwards to 50 animal units per acre are used by fencing each strip with a movable electric fence. The **advantages** claimed for this method are:

1. Increased utilization of herbage, with wastage reduced to 10 to 20%.
2. Increased meat and milk yields per acre up to 25%.
3. Improved stability of meat and milk yield because the nutritive value of the pasturage consumed is quite constant.
4. Improved utilization of the available forage. Less herbage is soiled by dung, urine, and treading. Under strip grazing, animals are quieter and settle down quickly for steady grazing rather than roaming about and tramping forage.
5. Increased animal units maintained on a given area, although animal productivity may not be increased.

Green Chop

Green chop is fresh herbage that is cut and chopped in the field, then transported and fed to animals in confinement.

Green chop, which is also called soilage or zero grazing, consists of growing a succession of forage crops, harvesting them with mechanized equipment, and hauling the green feed to the animals rather than allowing the animals to harvest their own forage. Historically, cutting green forage and hauling it to animals developed where land and forage were scarce and labor plentiful. Under this system of feeding, each animal unit has to be supplied with upwards of 150 lb of green forage daily, depending upon its succulence.

Green chop minimizes the loss of moisture, color, nutrients, and wastage. Alfalfa, ladino clover, orchardgrass, bromegrass, grass-legume mixtures, Sudangrass, corn, sorghum, soybeans, and cereal grains are sometimes used in this manner. With tall-growing crops, more feed value may be realized from a given area than can be obtained by conventional pasturing. However, green chop requires special equipment and harvesting every day. Also, there are harvesting problems in wet weather.

Fig. 7-16. Harvesting forage as green chop. (Courtesy, Deere & Company, Moline, Ill.)

Most green chop is fed to lactating dairy cows, usually in combination with hay or silage because the total intake tends to be greater. Green chop has increased with herd size, with more intensive form of dairying, with drylotting of cows, and with high grain prices. Also, the use of green chop has been facilitated by the greater mechanization present on larger and more modern dairy farms.

IRRIGATED PASTURES

Fig. 7-17. Holstein cows on irrigated pasture. (Courtesy, Holstein Assn., Brattleboro, Vt.)

The irrigated land in conterminous United States fell from a high of 50.2 million acres in 1978 to 44.7 million in 1984, primarily due to the rise in energy costs and the fall in crop and livestock prices.[4]

Irrigated pastures provide forage of high quality at a relatively low cost, often on land unsuited to other crops. Both perennial and annual irrigated pastures are important feed crops.

Successful pasture irrigation involves special decision making relative to (1) irrigation—the method, frequency, and amount of irrigating, and the removal of excess water; and (2) the kind and amount of fertilizer.

• **Method of water application**—Three basic methods of irrigating pastures are practiced: (1) border-flood, (2) furrow, and (3) sprinkler. Additionally, variations or combinations of the first two methods are sometimes used. The choice of the method for any given field should be determined by soil type, topography, water supply, and funds available for irrigation development.

The border-flood method is adapted where a large head of water is available and the land is level or requires only minor soil movement. When properly used, there is no runoff and efficiency of water utilization is high.

Furrow irrigation utilizes rills or corrugations about 4 in. wide and 4 in. deep, usually spaced 24 to 36 in. apart. Runoff and erosion are minimized by regulating the water flow through the furrows. Effective and economical furrow irrigation is limited to (1) better soils, (2) slopes that do not exceed 2%, and furrow lengths under 400 ft.

[4]*1986 Agricultural Chartbook,* USDA, Agricultural Handbook No. 663, p. 27, Chart 68.

Sprinkler irrigation may be preferable to the other methods (1) on land that is not level enough for surface irrigation or where the cost of leveling would be prohibitive; (2) on soils of variable texture where the amount and frequency of application can be adjusted to the water-holding capacity of the soil; or (3) where water cost is high or the supply of water is limited. Sprinklers are on the increase throughout the United States, with laborsaving, center-pivot, and wheel-move systems making hand-move systems obsolete.

• **Frequency and amount of irrigation**—Many pasture plants, especially the clovers, are shallow-rooted and require more frequent and lighter irrigations than deep-rooted plants. The Washington Station reports that the highest yields per acre from an orchardgrass-ladino clover pasture can be obtained with a summer irrigation frequency of 7 to 11 days, rather than at less frequent intervals, and that more frequent irrigations also give the highest proportion of clover. Of course, the time of irrigation is often determined by the availability of water. Such restriction may seriously limit pasture production. In important irrigated areas, county extension agents are usually knowledgeable relative to the proper time to irrigate specific crops in the particular area; hence, they should be consulted when developing a schedule.

Water penetrates into the soil of a well-established pasture more rapidly than in the same soil with arable crops. For this reason, a much larger head of water is usually necessary to obtain good distribution throughout the root zone. The objective at each irrigation should be to apply enough water to moisten the root zone.

• **Excess water**—Excess water in irrigated pastures is caused by either (1) overirrigation and the inability of the excess water to drain from the soil, or (2) subsurface drainage from adjacent and higher land. Surface drains are necessary to remove excess irrigation. Drainage is particularly important wherever there is danger of salt accumulation. Deep drainage ditches spaced at proper intervals help to remove these excess salts and control the level of the water table. In some areas, drainage ditches must be augmented by tile drainage in order to keep the salt content below levels that are harmful to plants.

• **Fertilization**—irrigated pastures require high soil fertility to be productive. The kind and amount of fertilizer should be determined by the level of the productivity desired, and the role of the legumes in the mixture. Production levels of irrigated pastures are increased more by N fertilizer than by other fertilizer elements, with responses also obtained from P and K where soils are deficient in these elements. Nitrogen stimulates grass growth, whereas P increases the legume component. Nitrogen can be supplied by either fertilizer or innoculated legumes. When legumes are a major component of pasture, economic returns from applied N, measured in increased animal production may not be obtained.

Part II—**The Western Range**[5]

The opening up of the western range accompanied the establishment of the Jesuit missions in what is now Arizona, New Mexico, and Texas in the period from 1670 to 1690. But livestock grazing did not spread throughout the West until the 19th century, with the annexation of Texas, the Southwest, and Oregon. Then, in 1848, came the magic news—*gold* had been discovered in California. The race was on! The trickle of adventurers turned into a flood. Trails and trading posts came alive; and there was need for cattle for meat.

The growth of the industrial East and the subsequent development and extension of the railroads provided the necessary stimulus for further expansion of the range cattle and sheep industries. The grass supply of the vast ranges seemed unlimited, and the region was regarded as a permanent paradise for cattle and sheep. About 1880, the lure of the grass bonanza fired the imagination of investors, big and little. Cowboys, sheepmen, lawyers, farmers, merchants, laborers, and bankers—many of them English and Scotch investors in great companies—rushed to seek their fortunes on the open ranges. Cattle barons and sheep kings reached a "high water mark." The number of cattle and sheep increased dramatically, joining the wild animals that were already there. Soon the range was overstocked. Regulations were few, and the guiding philosophy was that the grass belonged to the cattleman or sheepman who got there "firstest with the mostest." The cattle-sheep feuds waxed hot. Deadlines were set up and enforced by the side that could muster the greatest strength.

Suddenly, it became apparent that greed had taken its toll! The supply of tall grass was exhausted. Even more tragic, the winter of 1886–87 was ferocious. Few owners had made provisions for winter feed. Cattle and sheep perished by the thousands. In some herds, 85% of the animals starved to death. This marked the beginning of the end of the large livestock companies, the death knell of the open range, the gradual growth of smaller operations, and increased attention to range management, winter feed, and shelter.

In the mid-1890s, H. L. Bentley, a special agent with the U.S. Department of Agriculture, initiated the first range experiments concerning correct stocking rates and range seeding. Government intervention into grazing problems on the western range began in 1898, when the Department of Interior granted grazing permits to limit the number of livestock on federal lands. In 1905, the Forest Service was set up in the Department of Agriculture and the process of forage allotment was initiated. Between 1910 to 1915, Arthur Sampson conducted the first grazing system experiments in Oregon; he reasoned that deferment of grazing until seed maturity would allow for seedling establishment and replenishment of carbohydrate reserves. Because the price of cattle was high during World War I, a period of very severe overgrazing took place between 1915 and 1920. National Forest ranges began to improve again after World War I because scientific range management practices and controlled grazing were once again implemented. However, other rangelands in the West continued to deteriorate.

The discipline of range management developed in the 1920s. By 1925, some 15 colleges were offering courses in range management.

The biggest event of the 1930s in range management was

[5]The authors are very grateful to the following specialists who reviewed this section: R. M. Williamson, Director of Range Management, Forest Service U.S. Department of Agriculture, Washington, D.C.; E. Moris, Assistantto the Director and Chief, Office of Public Affairs, Bureau of Land Management, U.S. Department of the Interior, Washington, D.C.

the passage of the Taylor Grazing Act of 1934, which placed the administration of remaining public lands under the Grazing Service, which later became the Bureau of Land Management. In 1933, the Soil Erosion Service was formed because of alarm over the drought in the Great Plains. In 1948, the Society for Range Management was organized.

During the 1960s, the multiple use concept of range management on federal lands developed. Wildlife, water, and recreation became recognized as important range products as well as red meat. Previously, range research and management had been geared towards producing forage for livestock.

Toward the end of the 1970s and into the present, concern over the world population explosion generated renewed interest in using public rangeland for livestock production. Simultaneously, energy costs spiraled, and it was recognized that lower energy inputs are required to produce red meat from rangeland than cropland. In addition, there was an awareness that range forage can only be converted by grazing animals into products usable by people. Scientific range management on private land accelerated during this period and replaced much of the frontier spirit and the romantic, adventurous life of the cowboy and sheep herder.

Fig. 7-18. On the range. (Courtesy, American Angus Assn., St. Joseph, Mo.)

RANGE AREA

Range refers to large, naturally vegetated, mostly unfenced areas where animals harvest a rather sparse growth of grasses, legumes, and other edible parts of forbs and browse.

Various geographical divisions are assumed in referring to the western range area—the native pasture area. Sometimes reference is made to the 17 range states, embracing a land area of approximately 1.16 billion acres. At other times this larger division is broken down, chiefly on the basis of topography, into (1) the Great Plains area (the 6 states of Kansas, Nebraska, North Dakota, Oklahoma, South Dakota, and Texas); and (2) the 11 western states (Arizona, California, Colorado, Idaho, Montana, Nevada, New Mexico, Oregon, Utah, Washington, and Wyoming).

Almost half (47.6%) of the land area in the 11 western states is federally owned and administered. Domestic livestock graze on 89% of this area. Federal land is estimated to supply 17% of all grazing resources in the region.

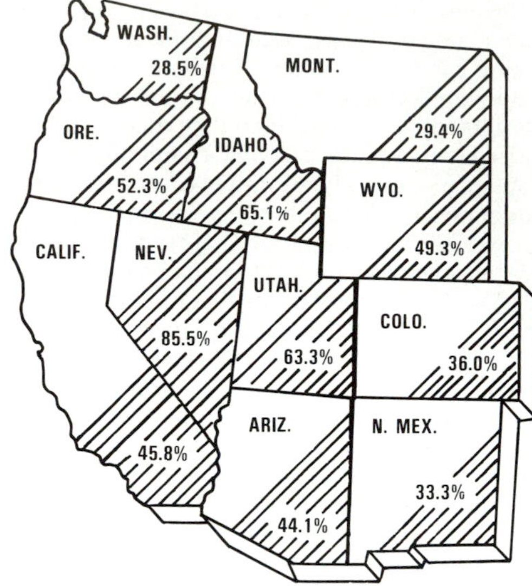

Fig. 7-19. A map showing the 11 western states and the proportion of land in each of these states that is owned by the U.S. Government. (Source: Public Land Statistics, 1984, Vol. 169, August 1985, p. 10.)

Because of the magnitude of the range livestock industry and the fact that it is a highly specialized type of operation, considerable discussion will be devoted to the range area and the care and management of cattle and sheep in the range method.

The carrying capacity of much of the western range is low, and little of it provides yearlong grazing. Moreover, variation in vegetative types, climate, and topography in the range country is accompanied by great diversity in the seasonal use made of it. As a result, rangelands are usually grazed during different times of the year, and the herds and flocks migrate with the season, moving to the mountains and higher elevations in summer and returning to the lower ranges in winter.

From the standpoint of vegetation and utilization by livestock, ranges differ from cultivated pastures as follows:

1. **They are less productive.** Generally, their productive capacity is lower. This is as one would expect, for they are largely made up of the residue remaining after the usable agricultural lands have been taken up. Also, plant growth on rangelands frequently is limited by low and undependable rainfall (even drought), short growing seasons, shallow or rocky soil, alkali or salt accumulations, and steep topography. Under such conditions, forage plants are usually less resistant to grazing damage than those growing under a more favorable environment.

2. **They are more likely to progress to less palatable plants.** Range vegetation consists of a mixture of native and introduced plants, varying greatly in palatability, nutritive value, and productive ability. Grazing animals select the most palatable plants first. Thus, unless careful management is practiced, the best plants are crowded out through a combination of grazing injury and competition from the ungrazed, low-value plants. Continued poor management can result in good forage plants being almost completely replaced by low-value annual, weedy, or shrubby vegetation, or left denuded and subject to severe erosion.

3. **They are more difficult to restore when depleted.** Once a range becomes depleted, it is a slow process to rebuild it. Plowing and drilling are impractical on most rangelands; thus, very often the only feasible way of restoring a range to good condition is to stock it conservatively and manage it well.

4. **They often serve multiple uses.** Rangelands often have other uses in addition to grazing values. Among such uses are water production, timber production, mineral production, wildlife production, and recreation (camping, hiking, picnicking, etc.).

Thus, many people, in addition to the livestock producer, have an interest in the grazing management practiced on ranges. This is part of the justification given for Federal Government ownership of large tracts of rangeland.

RANGE NUTRIENT DEFICIENCIES

Nutrient deficiencies are rather common on the range. Many soils are deficient in certain nutrients, which affect the plants, and, in turn, the animals feeding on them. During droughts, and early or late in the season, forage is in short supply, limiting energy and other nutrients. Early spring pastures are washy. Later in the season they become leached or bleached—they increase in fiber and decrease in protein, phosphorus, and carotene.

• **Energy deficiencies**—Hunger, due to lack of feed, is the most common deficiency of range livestock. The most important requirement is sufficient feed for body maintenance. Over and above this, surplus energy is used for growth or fattening.

With bulky, low-quality roughages—such as range grass cured on the stalk, animals cannot consume sufficient quantities to meet their energy needs. The younger the animal, the more acute the problem. Under these circumstances, the low-energy intake is met by breaking down of body tissues. This results in loss of weight and condition and lack of growth.

In breeding animals, low-energy intake affects reproduction. Cows take longer to come in heat and require more services per conception, thus reducing the calf crop. Also, calves born from energy-deficient cows are lightweight at birth. Supplemental feeding is the practical way to eliminate energy deficiencies.

• **Protein deficiencies**—The protein intake of beef cattle must be adequate to develop muscle (meat) and replace worn body tissues. The protein need is most critical in young calves and in gestating-lactating cows.

Mature, weathered native range plants are almost always deficient in protein—sometimes containing as little as 3% (see Fig. 7-20). A deficiency of protein results in depressed appetite, poor growth, loss of weight, reduced milk production, irregular heat periods, and lowered calf crops.

Because protein supplements ordinarily cost more per ton than grains, the temptation is to feed too little of them. When grazing mature, weathered forage, cows should be fed a protein supplement.

• **Mineral deficiencies**—Phosphorus deficiencies are rather common on the range (see Fig. 7-21). A severe phosphorus deficiency results in depraved appetite, emaciation, retarded growth and development, failure to breed regularly, lowered calf crop, lowered milk production, and high death losses.

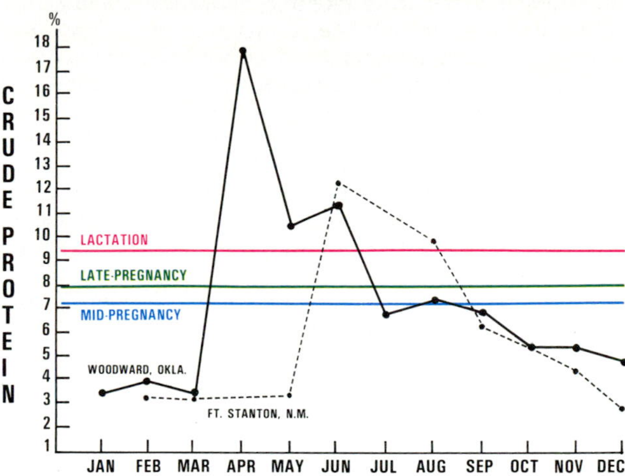

Fig. 7-20. Crude protein content (moisture-free basis) of blue grama *(Bouteloua gracilis)* grass. *Woodward, Okla. data from:* USDA Tech. Bull. 943, 1947, p. 53, Table 12, by Savage, D. A. and V. G. Heller. *Ft. Stanton, N. Mex. data from:* New Mexico Ag. Exp. Sta. Bull. 662, 1978, p. 22, Table 1, by Pieper, R. D., *et al.* **NOTE:** The peaks in the protein content of the vegetation in the two areas reflect the periods of greatest rainfall. Thus, approximately 60% of the annual precipitation of the Woodward, Okla. area occurs in April, May, and June, whereas 60% of the annual precipitation of the Ft. Stanton, N. Mex. area occurs in July, August, and September.

Superimposed on the chart are the crude protein requirements in a moisture-free ration of a 1,100-lb cow during mid-pregnancy, late-pregnancy, and lactation, respectively (taken from Table 19-2 of this book). Note that, in both areas, there is a protein deficiency except during periods of highest rainfall. (Chart provided by Ted McCollum, Ph.D., Animal Science Department, Oklahoma State University, Stillwater.)

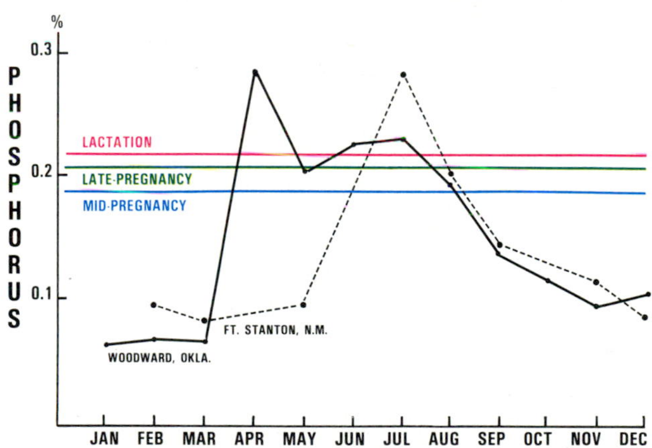

Fig. 7-21. Phosphorus content (moisture-free basis) of blue grama *(Bouteloua gracilis). Woodward, Okla. data from:* USDA Tech Bull. 943, 1947, p. 52, Table 11, by Savage, D. A. and V. G. Heller. *Ft. Stanton, N. Mex. data from:* New Mexico Ag. Exp. Sta. Bull. 662, 1978, p. 28, Table 7, by Pieper, R. D., *et al.* **NOTE:** The peaks in the phosphorus content of the vegetation in the two areas reflect the periods of greatest rainfall. Thus, approximately 60% of the annual precipitation of the Woodward, Okla. area occurs in April, May, and June, whereas 60% of the annual precipitation of the Ft. Stanton, N. Mex. area occurs in July, August, and September.

Superimposed on the chart are the phosphorus requirements in a moisture-free ration of a 1,100-lb cow during mid-pregnancy, late-pregnancy, and lactation, respectively (taken from Table 19-2 of this book). Note that, in both areas, there is a phosphorus deficiency except during periods of highest rainfall. (Chart provided by Ted McCollum, Ph.D., Animal Science Department, Oklahoma State University, Stillwater.)

• **Vitamin deficiencies**—Range plants are frequently low in carotene. A severe deficiency of carotene (vitamin A) may result in low conception rate, a small calf crop, many calves weak or stillborn, with some calves born blind, more cows with retained afterbirth, low gains, greater susceptibility to calf scours, and more respiratory troubles.

Severe vitamin A deficiency in bulls may result in decreased sexual activity and lowered semen quality.

When grazing dry range longer than 4 to 6 months, it is recommended that a supplement containing 20,000 to 30,000 IU of vitamin A per pound be fed to brood cows and bulls.

Range cattle usually receive sufficient vitamin D from exposure to direct sunlight or from sun-cured forages.

Range Livestock Supplementation

Energy, protein, phosphorus, and vitamin A are the major nutrients limiting the performance of range livestock.

Energy supplementation may be advantageous when range forage is in short supply, as during a drought or heavy snowfall. Also, energy supplementation may reduce nitrite toxicity problems and improve protein/energy ratios of livestock grazing lush, high protein pastures. Livestock perform best when energy (particularly grain) is provided on a daily basis.

Protein is the major supplement cost of most ranches. Economically, protein supplementation of livestock is most advantageous when the crude protein content of range forage drops below 6 to 7%. Protein supplements can be provided to livestock on an every-other-day or every-third-day basis without affecting their performance. Salt or fat may be used to limit the consumption of cottonseed meal or other protein supplements.

Phosphorus supplementation during periods of forage dormancy seems justified. Also, the routine inclusion of essential microminerals in salt blocks or mineral mixes is a good practice.

Vitamin A supplementation is recommended when livestock must be maintained for over 4 months without access to green grass or browse.

(Also see Chapter 11, Protein Supplements, Chapter 13, Feed Supplements/Additives/Implants; Chapter 19, Feeding Beef Cattle, Part II Feeding Breeding Beef Cattle, section on "Range and Pasture Feeding"; Chapter 21, Feeding Sheep, section on "Feeding Range Sheep"; and Chapter 22, Feeding Goats, "Part III Feeding Angora and Spanish [Meat] Goats.")

RANGE MANAGEMENT CONSIDERATIONS

Good range management may be achieved if an inventory or analysis is made of the forage resources and all contributing factors, followed by a sound plan of management based upon the analysis. Consideration should be given to such factors as proper stocking rate, and safe degree of use, season of use, kind of livestock, condition and trend of forage, soil stability, system of use, improvements needed, etc.

Stocking Rate

The key to successful long-term operation of rangeland lies in making (1) a reliable determination of the land that is suitable or adaptable to grazing use over a long period of time; (2) a realistic estimate of grazing capacity for this land; and (3) a flexible stocking rate, even within a single season, followed by (a) application of proper stocking intensity, and (b) frequent observations to determine the effect of the stocking rate upon changes in condition of the forage cover. Too light stocking wastes forage, while too heavy stocking results in a change of forage plant cover from an abundance of valuable forage plants to an abundance of less valuable, or worthless, plants.

Of course, the stocking rate for any given unit may vary widely from year to year, and within a given season, depending on the forage production as affected by weather and other factors. For this reason, stocking should either be adjusted to forage yield each year, and within season, or set at a constant rate that will assure a sustained yield of most valuable forage plants (constant stocking at about 25% below average capacity will usually achieve the latter).

Recognition must also be given to the fact that animals do not graze uniformly over a range unit—that certain areas are more attractive to them. Consequently, some areas produce most of the grazed forage, while others may go practically unused. Cattle tend to congregate on fairly level creek bottoms, ridge tops, and around water and shade, whereas sheep prefer hillsides. If herded, sheep can be moved rather uniformly over a range. If not herded, they tend to congregate in the portion of the pasture in the direction of the prevailing winds, and to spot graze more than cattle. For the purpose of determining grazing capacity, the key areas—those rather extensive parts of the range which are most heavily grazed—must be given greatest consideration. If preferred or key areas are maintained in good condition, the whole unit will generally remain in good condition. Conversely, if key areas are allowed to deteriorate, the grazing capacity of the whole unit will be endangered.

The following rule of thumb, applied to the more heavily grazed key areas, may be used in arriving at the proper stocking rate: "Use half and save half, and the half you save will grow bigger and bigger." The rule refers to half the weight, which is concentrated at the bottom of the plant, and not to half the height. Thus, when the 50% rule of thumb is applied to bluebunch wheatgrass, a common range plant, it means that about 75% of the bunches have been grazed to an average stubble height of about 4 in., and the remaining 25% of the plants left relatively ungrazed.

The scientific basis for this rule is one of the great miracles of nature: Plants manufacture food in their green leaves through the use of solar enengy. They do not produce food in their roots; the roots merely pull water and minerals from the soil. The leaves take in carbon dioxide gas from the air through tiny pores. Using solar energy, they recombine the carbon with hydrogen and oxygen to make sugars and starches. The sugars then combine with minerals from the soil to make fibers, proteins, plant oils, and fats. The plant uses these sugars, starches, proteins, oils, and fats to grow and reproduce. Removing too many leaves will retard forage production and damage the plant's root system. Generally, when up to 50% of the plant is grazed, root growth continues unimpaired.

Grazing capacity determinations are relatively complex and require careful study over a period of several years. They are arrived at most simply and accurately by observing soil stability conditions and changes in plant cover. If the best plants are being destroyed and soil movement is observed, numbers of animals or season of use should be reduced; conversely, if excessive forage remains at the end of the grazing season, numbers should be slowly increased until a balance is struck.

In arriving at grazing capacity, it is generally wise to seek assistance from qualified range technicians, who need to know:

1. The potential of the particular range.
2. The present state of the vegetation as it relates to potential on each site.
3. The alternative methods of changing present conditions to meet management objectives, including such things as flexible stocking, seeding, and brush control, fences, watering, and trails.

The commonly used terms for describing range condition are (1) excellent, (2) good, (3) fair, and (4) poor. If the range is covered almost entirely with high-value forage plants which are producing near maximum for the site and has a stable soil, it is classified as being in excellent condition. If the best plants are scarce and the soil is exposed and unstable, it is classified as being in poor condition. Good and fair classifications are intermediate. The trend in condition is also important; if the range condition is improving, the trend is upward, and *vice versa*. Actually, the range condition reflects the kind of management practiced in the past.

Season of Use

A prime requisite of successful management for both cattle and sheep is that there shall be as nearly year-around grazing as possible and that both the animals and the range shall thrive. In some areas, especially in the southwestern Great Plains areas, these conditions are met without necessitating extensive migration of animals. The winter climate is mild, and the native forages cure well on the stalk, thus providing nutritious dry feed at times when green vegetation is not available. Generally speaking, however, most of the cattle and sheep from such areas are marketed via the feeder route rather than as grass-fat slaughter animals.

In general, the most desirable management, both from the standpoint of the animals and the vegetation, consists of the proper seasonal use of the range. Although there is wide variety in the customs and requirements for seasonal use of the range—because of the spread in climate, topography, and vegetative types included in the vast expanse of range country—seasonal-use ranges are usually placed in four major classes: (1) spring-fall, (2) winter, (3) spring-fall-winter, and (4) summer.

Because a range band of sheep can be moved and herded on unenclosed areas with greater ease than a herd of cattle and because investigations in range livestock management have been conducted more extensively with sheep, greater seasonal use of ranges is made with sheep. On the other hand, the more progressive cattle producers are finding ways and means of adopting many of the same methods.

Despite the value of yearlong grazing, it is recognized that the prevalence of severe winters in some parts of the West preclude winter grazing except to a limited degree, and stock must be fed during at least a part of the winter season. Where these conditions prevail, cattle and sheep are usually wintered in the irrigated valleys, close to the feed supply, especially a supply of alfalfa or meadow hay.

Some pertinent points in determining the proper season to use the range follow:

1. **Elevation.** Generally speaking, vegetative development is delayed 10 to 15 days by each 1,000-ft increase in elevation. Also, severe storms occur later in the spring and earlier in the fall at higher altitudes than at lower ones.
2. **Availability of water.** Certain desert areas are so poorly watered that only the occurrence of winter snows makes their use practical.
3. **Early forage "washy."** Early spring forage is extremely "washy," and may be incapable of supporting stock. Spring grazing should be delayed until the plants are developed enough to meet the nutritive needs of animals.
4. **Soil tramping.** Soil tramping may be serious in early spring. In order to avoid plant damage and soil compaction, grazing should be delayed until the soil is firm.
5. **Poisonous plants grow early.** Most poisonous plants are very early growers and cause their greatest damage when animals are turned out too early. Larkspur, which affects cattle, and death camas, which affects sheep, are two examples. Poisoning losses from these two plants are usually negligible if stock are detained until the best forage plants have made suitable growth.
6. **Winter range should be saved.** If stock are allowed to remain on winter ranges too long after spring growth begins, the next winter's feed will be reduced, because the forage produced on these ranges grows mainly during spring and early summer.

Kind of Livestock

Sheep and cattle share in the utilization of the western range. In fact, some ranges are simultaneously grazed by these two kinds of animals. This dual system of grazing is practical and beneficial provided the grazing capacity for each is properly adjusted so that the major forage plants are properly used, and that, at intervals, a careful determination is made of condition and trend of soil and forage. Some ranges, especially in Texas, are grazed by three classes of animals—cattle, sheep, and goats—with the goats controlling the brush without adversely affecting environment.

When sheep and/or goats are added to a cattle range (or *vice versa*), the increased numbers should not result in a total which exceeds the previous animal units of a single species by more than 10%; otherwise, overgrazing will likely result.

Actually, economic factors—often unrelated to range characteristics—probably have the greatest influence on the selection and popularity of kinds of livestock. The kind which the operator feels will return the greatest net profit is selected, and the choice changes with changing times. Nevertheless, range characteristics may be so specific as to favor one kind of livestock to the point that other kinds would be produced under handicap. Among such range characteristics which should be considered in choosing the kind of livestock are:

1. **Poisonous plants.** The presence of certain poisonous plant species may limit the use of the range to one kind or

mals come to the center, or hub, for water and minerals, they can be moved, between pastures by opening and closing gates. Producers using the Savory system generally use electric fences in order to reduce costs. The Savory system is designed to lessen movement stress. However, it can add to nutritional stress if (1) stock are held too long in pastures, (2) too low stock density is combined with too fast a move, or (3) accelerated grazing with rests become inadequate.

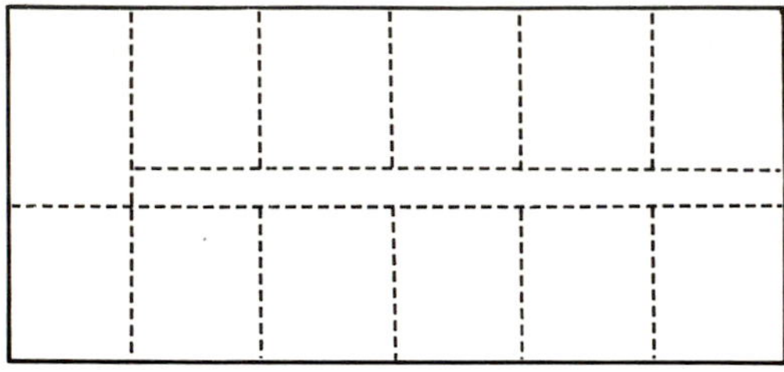

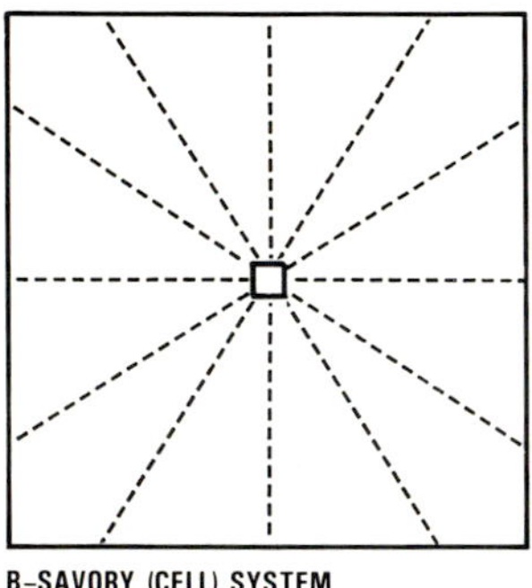

Fig. 7-22. Two basic layouts (designs) for short duration grazing: A, the conventional (rectangular) system; and B, the Savory (cell) system.

RANGE IMPROVEMENT METHODS

The warning signals of a range that is on the downgrade and that is in need of improvement are:

1. Desirable forage plants "going out" and being replaced by undesirable ones; the number of young, inferior plant species increasing.
2. Thinning of perennial grass cover, with the grass tufts breaking down and dying; and an increase in annual plants and perennial weeds. The poorer the condition of the range, the more rapidly this process takes place.
3. Weakened vitality of the important forage plants as shown by pale color and reduced height and yield in period favorable to good growth.
4. Increased soil erosion, by wind and/or water.
5. Excessive trampling damage.

The above warning signals have appeared in various intensities over part of the western range area of the United States.

There is no quick, easy, and inexpensive method by which poor ranges can be improved. However, one or more of the following methods should be employed.

Conservative Stocking

Usually controlled stocking and natural range reseeding are accomplished by employing one or more of the following practices: (1) rotation-deferred grazing; (2) rest-rotation grazing; (3) a lighter continuous grazing load; or (4) a shorter season of use, such as may be accomplished by using supplemental pastures.

With seeded areas in humid regions, it is generally recommended that pastures be grazed fairly close at least once each year and that the plants not be permitted to reseed. In continuously grazed range areas, however, good management consists of limiting the number of animals to the point where at least one-fourth of the seed from the better forage plants is permitted to ripen annually. This requires frequent and careful inspection to make sure that the better forage species are not progessively being eliminated.

Distribution of Animals on the Range

Next to the proper rate of stocking and proper seasonal use, distribution of the animals on the range is the most important feature in range management. Proper distribution of animals is reflected in more even utilization of the forage. This assignment is more difficult with cattle than with sheep, especially on rough mountainous land. Cattle have a strong tendency to utilize the flatter areas and to congregate around watering places. Also, sheep are usually herded.

Better distribution of animals on the range may be accomplished through (1) the grazing system (concentrating livestock into large herds and rotating the herds through 2 or more pastures tends to improve grazing distribution for each pasture); (2) fencing; (3) riding the range (or herding); (4) providing sufficient watering places (under ideal conditions, the distance between water in rough country should not exceed ½ mile for cattle or 1 to 2 miles for sheep and goats; in level country, 1 to 1½ miles for cattle or 2 to 3 miles for sheep and goats—water hauling on both cattle and sheep ranges is increasing); (5) systematically locating salt blocks or salt containers away from watering areas in underused areas; (6) building trails into inaccessible parts of the range; and (7) controlling livestock pests such as grubs and flies, which cause animals to congregate and seek protection.

Range Reseeding

Where improper management and overstocking have seriously reduced the quality of the forage and the grazing capacity of the range, some method of reseeding may be the only logical alternative.

Where considerable of the better forage plants remain, natural regeneration is preferred. The latter is accomplished by managing the grazing season so as to favor the propagation of the remaining desirable native forage and by controlling low-value brush and other competing vegetation. Often the recovery process can be speeded up by controlling most undesirable vegetation through the use of herbicides. In western United States, this process is especially successful on rangeland dominated by sagebrush and wyethia where a residual stand of native grasses is present as an understory. There are also other weed types which respond well to chemical treatment.

Where most of the desirable plants have been destroyed and the soils are suitable, artificial reseeding is advocated—even though it is expensive and subject to failure.

Fig. 7-23. No-till seeding on the range. (Courtesy, Montana State University, Bozeman)

It is recognized (1) that only a relatively small proportion of the western range can be seeded, (2) that seeding is not a satisfactory substitute for good management practices needed to prevent further destruction of forage plants, (3) that if seeding is necessary, some practice(s) followed in the past has been faulty, and (4) that seeded areas require a high level of management to maintain them.

For successful seeding, the following rules are usually adhered to:

1. Select a site where success can reasonably be achieved. Rainfall, soil, topography, and other site factors must be suitable for seeding. Otherwise, the practice is foredoomed.
2. Prepare a firm, weed-free seedbed. Light soils receiving limited rainfall are notoriously difficult to firm into a satisfactory seedbed. Also, seedlings of perennial grasses commonly used in range reseeding are weak competitors with weeds.
3. Select a species or mixture of species adapted to the climate and other features of the area being planted and fitted to the seasonal forage needs of the particular ranch. Keep mixtures simple; single species are often best.
4. Plant just prior to the season of most dependable moisture and growing conditions.
5. Cover the seed, but do not get it too deep. The best depth for most range grasses is ¼ to ¾ in.; depth regulators on drills are advisable. In drier areas, deep furrow drills are usually more successful.
6. Protect the seeding from grazing use until the plants have become well established. This may require from 8 months to 3 years—depending on species planted, weather conditions, etc.
7. Fence the seeding or otherwise provide a means of managing it. Without adequate control, animals congregate on the new seedings and destroy them through overuse.

Brush and Weed Control

Brush and weeds compete with both native and introduced forage plant species for water, nutrients, and light. Their control is particularly difficult on rangelands, because they are not normally cultivated or rotated with other crops.

Brush is a primary problem on U.S. rangelands. An estimated 320 million acres of grazing land are predominantly brush. Mesquite, juniper, and sagebrush make for the major brush problems.

Brush can be controlled by the following methods:

1. **Chemical methods.** This involves applying select herbicides to the soil or the plant by ground equipment or by aircraft. Success depends on applying the right herbicide at the right rate and at the right time. The county extension agent or other specialist should be consulted for the latest recommendations, and the directions on the label of the product being used should be followed.
2. **Mechanical methods.** Where there is a dense stand of brush and few or no desirable grasses, mechanical methods may be most effective. Mowing, axing, root plowing, chaining and bulldozing have proven quite effective.
3. **Prescribed burning.** Fire is the oldest method of controlling unwanted woody plants, and it is still used. Prescribed burning refers to burning that is done at the end of the dormant season or at the time the desirable grasses are beginning spring growth; this suppresses certain undesirable plants, thereby giving an improved environment for the more desirable plants. In addition to doing prescribed burning at the proper time, the following factors should be considered: fuel for the fire, favorable weather conditions, safety, and air pollution.
4. **Proper grazing management.** Proper grazing management avoids overgrazing and assures healthy, vigorous forage, in a healthy ecosystem, thereby reducing brush problems.

Fig. 7-24. Brush control would increase the carrying capacity of this range. (Courtesy, American Shorthorn Assn., Omaha, Nebr.)

To be effective, brush control must be followed by proper grazing management. During the first growing season following treatment, grazing should be deferred/limited in order to allow desirable grasses to become established. Reseeding of these areas may be necessary where a natural seed source of the desirable forage plants is not available.

Weeds—the unwanted plants of pastures and ranges—are undesirable because (1) They compete with higher quality plants which might otherwise grow; (2) they are often less palatable and less well consumed than desirable species, with the result they are undergrazed; (3) they may impart undesirable flavors and quality to the food product produced by the grazing livestock; (4) they often contain toxic substances which may impair the usefulness of, or even kill, the animals which consume them; (5) they may inflict mechanical injury; (6) they may shade the desirable plants and interfere with their establishment and maintenance in the stand; and (7) they usually give an unsightly appearance to the pasture.

Weeds may be controlled in the following ways, singly or in combination: (1) biologically, through the use of insects, diseases, and other organisms, including higher animals, as control agents; (2) tillage, either as an idle field or as a row crop which permits cultivation; (3) roguing, or individual weeding, if the plants are scattered and not found in large numbers; (4) mowing the aftermath following grazing, so that the plants which the animals do not consume are unable to go to seed or otherwise have an advantage over those plants which were better utilized; (5) chemically, by treating either the plants or the soil to prevent seed germination or to kill the plant itself; or (6) plowing up the pasture and cropping it in some other way for one or more years while the weeds are controlled, and then reseeding it to pasture. Each of these methods has certain advantages over the others. The first step in any weed control program consists of identifying the nature of the weed problem. Next, using this information, along with a specialist's advice, one can determine how weeds can be controlled. Many counties have a weed control program, with provision to help by spraying roadways, fencelines, and even concentrated spots in fields if the problem has the potential to spread to adjoining fields.

• **Biological control of brush and weeds**—Interest in the biological control of brush and weeds is increasing. The classical example of the biological control of an undesirable species was the practical elimination of prickly pear cactus (*Opuntia* spp) on 30,000,000 acres of grazing land in Queensland, Australia, by using a moth introduced from Agrentina in 1925. In the United States, insects have been used to control St. Johnswort, lantana, puncturevine, and knapweed. The use of sheep and goats is also a form of biological control. Sheep will eat leafy spurge, tansy ragwort, tall larkspur, and spotted knapweed. Goats are effective for keeping some species of chaparral and other brushes under control.

GRAZING PUBLICLY OWNED LANDS

The ownership of U.S. land is summarized in Table 7-1.

About one-third of U.S. public lands are in Alaska. Because of its remoteness and northern location, land development has been slow in this state. As a result, the Federal Government still owns almost 67% of all the lands in Alaska.

The other two-thirds of the public lands are located in the 48 contiguous states, but are not evenly distributed across the country. About 93% of these federal lands outside Alaska are in the 11 western states.

Today, in the 11 western public land states, the Federal Government owns and administers approximately 320 million acres on which grazing is allowed. At one time or another during the year, domestic cattle and sheep graze on about half of these public lands. More of the public lands are used for this purpose than for any other economic activity. In 1983, lands in the 11 western states administered by the Bureau of Land Management and the U.S. Forest Service provided grazing all or part of the year for an estimated 4.5 million cattle and horses and 6.7 million sheep and goats, or a total of 11.2 million head of all classes, or a total of 17.6 million animal unit months.

TABLE 7-1
OWNERSHIP OF U.S. LAND (50 STATES)[1]

Ownership	Area		Percentage of Total
	(million acres)	*(million ha)*	(%)
Private ownership	1,329	*538*	58.7
Indian land	51	*20.6*	2.2
Public ownership	885	*358.3*	39.1
Federal	730	*295.5*	32.2
State and local governments	155	*62.7*	6.8

[1]*Statistical Abstract of the United States,* 1987, p. 182, Table 318.

Agencies Administering Public Lands

Because much of the grazing land that ranchers rely upon to maintain their cattle and sheep enterprises is built up into operating units by leasing or by obtaining use permits from several federal and state agencies, private corporations, and individuals, it is imperative that the owner have a working knowledge of the most important of these agencies. Some range operators are placed in the position of using range rented from as many as six landlords—either private, state, and/or federal.

The bulk of federal land is administered by the following six agencies: the Bureau of Land Management, the U.S. Forest Service, the Bureau of Indian Affairs, the Department of Defense, the National Park Service, and the Bureau of Reclamation. The largest land area from the standpoint of grazing permits and utilization of grazing areas by animals is administered by the first three of these agencies; hence, each of these three agencies is discussed at this point, followed by pertinent information relative to state and local government-owned lands, and railroad-owned lands.

1. **Bureau of Land Management.** The Bureau of Land Management of the U.S. Department of the Interior administers more than 40% of all federal lands. More than one-third of the land it manages is in Alaska. The remainder is almost entirely in the 11 western states.

From the standpoint of the livestock producers, the most important function of the Bureau of Land Management is its administration of the grazing district established under the Taylor Grazing Act of 1934 and of the unreserved public land situated outside of these districts which are subject to grazing lease under Section 15 of the Act. This federal act and its amendments authorize the withdrawal[6] of public domain from homestead entry and its organization into

[6]On May 28, 1954, a bill was signed by President Eisenhower lifting the 142 million acre limitation on public domain lands that can be included in Taylor Grazing Act districts.

grazing districts administered by the Department of the Interior. Also, this legislation, as amended, allows the Bureau of Land Management to administer state and privately owned lands under a cooperative arrangement.

In 1984, the Bureau of Land Mangement had 52 grazing districts, operating in the 11 western states and totaling 157.3 million acres of public lands. In these districts, 12,000 operators were granted privileges to graze 3,973,000 head of livestock for an average of about 5 months each year. These operators paid the United States, as grazing fees for this range use, a total of $12,396,000. In addition to this livestock use, in 1984 the grazing districts supported, for approximately 5 months of the year, an estimated 1.8 million big game animals, of which approximately 1.1 million were deer.

In addition to, and outside of, the grazing districts, in 1984 the Bureau of Land Management supervised 17.7 million acres of public domain in the western states, most of which was leased to 7,300 livestock producers for 651,956 head of livestock for about 5 months. These operators paid rentals in the amount of $2,049,000 for the use of these lands.

Each district is administered by a District Manager, who is a technically trained employee of the Bureau of Land Management. The District Manager is responsible to the state bureau office for the proper use, management, and welfare of the public land resources of his district. In turn, the state office is responsible to the Director's office in Washington, D.C.

Grazing privileges are allocated to individual operators, associations, and corporations on the basis of (1) priority of use; (2) ownership or control of base property dependent on grazing district land for forage during certain seasons of the year, or control of permanent water needed to graze district land; (3) proximity of base property to public lands outside home ranch to the grazing district; and (4) adequate property to supply the feed needed along with grazing privileges, to maintain throughout the year the livestock permitted on public range. All of these lands are subject to classification and disposal under Sections 7 and 14 of the Taylor Grazing Act, for any higher use or other appropriate purpose. Grazing privileges may, therefore, be cancelled whenever such lands are determined to be more suitable for other purposes.

A fee is charged for grazing privileges. In 1989, the basic fee was equivalent to $2.29 per animal unit month (AUM). An AUM is the equivalent of the grazing of a mature cow, 5 sheep, or 1 horse, for 1 month.

The Taylor Grazing Act has been responsible for many changes, not all of which have been popular. Some livestock producers complain about the loss of their ranges; others tell of increased costs; and there are those who resent government controls, and, above all, the confusion which results from dealing with several agencies. Without doubt, many of these criticisms are justified, and some errors in administration should be rectified; but those who would be fair are agreed that the ranges as a whole have improved under the supervision of the Bureau of Land Management and that further improvements are in the offing.

2. **U.S Forest Service.** Almost one-fourth of the federal lands are administered by the Forest Service. Over 100 million acres of the national forests are used for grazing under a system of permits issued to local farmers and ranchers by the Forest Service of the U.S. Department of Agriculture. In 1985, about 1,482,000 mature sheep/goats and 1,469,000 mature cattle and horses (mostly cattle), owned by over 16,000 paid permit operators, were grazed on national forests for some part of the year. In addition, there were many calves and lambs for which no fee is charged and additional stock that were grazed under free permits to local settlers.

The forest service issues term grazing permits and annual permits. Among other things, the permit prescribes the boundaries of the range which they may use, the maximum number of animals allowed, the season in which grazing is permitted, and the expiration date of term permits.

Temporary permits may be waived back to the government when permittees sell livestock or base property. Then, the purchaser of the permitted livestock or base property may apply for and be issued a permit if qualified.

The requisites in order to qualify for a term permit are:

a. U.S. citizenship.

b. Ownership. The ownership of both the livestock and commensurate ranch property.

A term grazing permit is not a property right. Rather, it is approved for the exclusive use and benefit of the person to whom it is issued. Permits may be revoked in whole or in part for a clearly established violation of the terms of the permit, the regulations upon which it is based, or the instructions of forest officers issued thereunder.

A ranger administers the grazing use on each National Forest Ranger District. Several districts (usually 3 to 6 or more) comprise a national forest. A forest supervisor, with his staff, administers the national forest. Several national forests, under the direction of a regional forester and staff, comprise a forest service region. The Chief administers the Forest Service from Washington, D.C., under the supervision of the Secretary of Agriculture.

Local farmers and ranchers act in an advisory capacity in reviewing allotment management plans and the use of range betterment funds. About 800 such livestock associations are recognized and in operation.

Forest Service grazing fees are based on a formula which takes into account livestock prices over the past 10 years, the quality of forage on the allotment, and the cost of ranch operation. In 1989, average charges were $1.86 per animal unit month (AUM); or $1.86 for a mature cow or horse, or for 5 sheep, for a month. The use resulted in the payment of $9,039,100 in grazing fees in 1985.

Although shortcomings exist in the management of the national forests, it is generally agreed that these ranges have been vastly improved under the administration of the Forest Service. Many of them now approach the quality that existed in their virgin state. Perhaps the most heated arguments between livestock producers and the Forest Service arise over the relative importance attached to the multiple use of big game and other wildlife, recreation, etc. For example, it was estimated that in 1974 there were 4.6 million big game animals (85% of which were deer) in the national forests. As would be expected, these wild animals compete with domestic animals for use of the range, thus creating a most difficult problem.

3. **Bureau of Indian Affairs.** Most Indian lands, comprising 51 million acres, are really not public lands. Rather, these lands are held in trust for the benefit or use of the Indians and are merely administered by the Bureau of Indian

Affairs of the Department of the Interior. Because over 80% of Indian lands are in the range area of the West, they are suited primarily to livestock. Thus, it is noteworthy that the sale of livestock and animal by-products regularly accounts for ⅔ of the total Indian agricultural income. Although the Indians themselves own most of the stock that graze these lands, animals owned by non-Indians utilize ¼ of the Indian lands devoted to grazing. Provision for such use is handled under lease agreement jointly approved by the Indian owners and the Bureau of Indian Affairs.

Many of the Indian lands have suffered serious vegetative depletion, but a concerted effort is now being made to decrease livestock numbers in keeping with available feed supplies and to improve the quality of animals produced. However, overstocking continues to be a difficult problem on the Navajo, Hopi, and Papago Reservations.

4. **State and local government-owned lands.** A total of 134 million acres are owned by state and local governments. For the most part, the management of these areas is diverse and confused, each state and local government having established different regulations relative to the lands under its ownership. In general, however, such lands are operated on a stipulated lease arrangement. On many such areas, range depletion has been severe.

5. **Railroad-owned lands.** Recognizing that the main deterrent to rapid settlement and development of the West was lack of adequate transportation facilities, the Federal Government very early encouraged the construction and westward extension of the railroads by means of large grants of land. It was intended that the railroads should sell or otherwise utilize these lands in financing their costs of construction. These initial grants, totaling 94,355,739 acres, consisted of alternate sections extending in a checkerboard fashion for a distance of from 10 to 40 miles on each side of the right-of-way. Today, less than 20 million acres of these lands are held by railroads. Many of these holdings are leased to livestock producers; but because of inconvenience, past abuses, or other reasons, some of these lands are considered worthless for grazing. In general, railroad lease agreements do not restrict the number of stock to be grazed or the season during which the land may be so used.

Part III—**Multiple Use/Conservation of Land**

The multiple use of publicly-owned lands evolved in response to public pressure. The multiple use of privately-owned lands followed, in response to economic pressure—the need to increase net returns. Soil and water conservation evolved on both public and private lands in recognition that they are national resources that should be preserved for posterity.

The sections that follow pertain to the multiple use/conservation aspects of both public and private lands—to both introduced (seeded) pastures and native (range) pastures.

MULTIPLE-USE CONCEPT

Multiple-use of land is the management of all the various resources of lands, both public and private, so that they are utilized in combination. With federal lands, multiple use is determined by the needs of people; and management decisions are made publicly. With private lands, multiple use is based on their most profitable use and management decisions are made privately by owners/managers. Important multiple uses include livestock grazing, mining, national heritage preservation, occupancy, recreation, water, wildlife, and wood/timber production.

The multiple use concept developed as a compromise relative to the use of public lands; it evolved as a result of attempting to placate individuals and groups who wish to have the land used for purposes which they consider desirable or to prevent others from using the land for purposes which they consider to be undesirable.

MULTISPECIES GRAZING

Grazing two or more species of livestock together or separately on the same land unit in a single growing season is known as multispecies grazing. Research indicates that multispecies grazing contributes to better and more uniform forage use and higher economic returns from livestock.

Multispecies grazing evolved in regions with diverse vegetation types and suitable climates. Grazing by a mix of domestic and wild animals can often result in more efficient use of forage and browse, more total animal gains, and a more vigorous plant community. While multispecies grazing is a common management practice on rangelands of the West, it is much less commonly practiced on the pasture lands of eastern United States.

Western rangelands are characterized by vast diversity in elevation, precipitation, temperature, and other climatic factors. These differences make for a multitude of range cites dispersed among several major vegetation or habitat types. It follows that great potential exists on these lands for multispecies grazing by livestock and wildlife to maintain forage production and species diversity.

Where multispecies grazing is practiced, cattle and sheep dominate. In the Southwest, goats are sometimes a component. Goats are without a peer in rough, unimproved areas and as browsers. Sheep prefer steeper terrain and eat more shrubs and forbs than cattle. Cattle stick to the more gentle slopes and prefer grasses. So, multispecies grazing can result in more complete and uniform utilization of multiplant species pastures and greater animal production. However, predators and labor problems have caused decreased sheep and goat numbers. In turn, this has resulted in lower income and the deterioration of many ranges, due to the invasion of undesirable plants such as bushy species.

In the past, wildlife has generally been incidental. Now, and in the future, economic pressures dictate that wildlife be an integral part of multiple land use.

WILDLIFE

Fig. 7-25. Deer at home on the range. Good range management is good for wildlife. (Courtesy, USDA-ARS, Cheyenne, Wyo.)

Wild animals and birds are becoming more valuable to today's landowner. Higher livestock production costs and demand for outdoor recreation have prompted practical landowners to seek means of increasing income by providing game for hunters. In some areas, wildlife income exceeds livestock income. This has caused landowners to include wildlife in farm and ranch planning.

There is a close association between kinds of plants and animals present in the habitat. Also, livestock numbers, domestic species, and grazing patterns can be manipulated to enhance wildlife habitat and maintain wildlife populations. Through proper habitat management, the farmer and rancher can maintain healthy, abundant wildlife populations. To accomplish this, land managers must place wildlife high on their priority list and consider wildlife in overall farm/ranch planning.

• **Kinds of wildlife**—Many kinds of wild animals and birds live on pastures and ranges. Identifying the kinds is necessary because management will vary for different species. Deer, for example, need browse, forbs, and grasses for feed, and timber and brushy areas for cover. Quail feed on weed seeds, nuts, and seeds of certain grasses and shrubs; and they prefer a mixture of wooded and open areas with small plots of low shrubs and vines for cover. Normally, management will involve meeting the needs of several different kinds of animals and birds.

• **Numbers of wild game**—Nationwide, in 1984 there were an estimated 10 to 11 million deer, 700 thousand pronghorn antelope, and 500 thousand elk.[7] Additionally, in 1984, in U.S. National Forests and Grasslands, there were an estimated 231,339 wild turkeys, 106,480 bear, 26,246 moose, 25,783 javelina (peccary), 25,152 mountain goats, 20,753 wild sheep, 11,091 mountain lion, 3,480 wild boar, 1,206 wolf, 323 caribou, and 206 bison.[8] Also, in 1984, hunters harvested 391,000 deer and 109,000 other big game in National Forests.[9]

[7]Oldfield, J. E., Chairman, *et al.*, *Forages,* Council for Agricultural Science and Technology, Ames, Iowa, 1986, p. 30.

[8]*Wildlife and Fish Habitat Management in the Forest Service,* USDA, 1985, pp. 52-53.

[9]*Agricultural Statistics 1986,* USDA, p. 492, Table 680.

In addition to the big game animals listed above, there are many species of small game animals. Also, there are numerous species of game birds, including quail, partridge, and pheasants.

• **Wildlife management**—Wildlife can exist in harmony with livestock operations provided (1) wildlife needs and species are inventoried and included in the management plan, and (2) the following management aspects prevail:

1. **Grazing system.** For proper use of vegetation, it is important that the grazing system allow livestock and wildlife to harvest the forage without overgrazing. This protects the quality of the forages, provides wildlife cover, and prevents erosion. Grazing systems where livestock are concentrated and rotated between pastures reduces some of the competition with wildlife.

2. **Revegetation.** Some grasslands do not have adequate plants to meet the needs of livestock, wildlife, and erosion control. Such areas may be in need of reseeding. When reseeding, consideration should be given to including plants that have special value for wildlife.

3. **Water.** Water is as important to wildlife as it is to livestock. Reliable and well distributed supplies of water should be provided and maintained.

4. **Brush control.** Controlling brush can improve grasslands for wildlife and livestock. But it must be done properly and in harmony with other conservation practices. If poorly planned and not followed with good grassland management, it can harm the habitat for wildlife. One method of brush control is prescribed burning. If this method is used, a burning plan should be developed to meet the objectives of the owner/manager. Patterned brush control, or leaving strips or mottes of brush in pastures, increases the edge effect and enhances wildlife habitat for many species.

SOIL EROSION CONTROL

Soil erosion control is any management plan to reduce soil and water losses.

Soil erosion is a natural occurrence. However, it may be increased by activities that disturb the natural balance of the pasture or range ecosystem. Poor grazing management is a major cause of erosion.

A raindrop that hits bare soil has a different effect from one that falls on a plant or litter. A racing raindrop smashes against bare soil with great force, splashing water and soil particles and packing the surface soil together; it seals pores, with the result that little water goes into the soil and runoff occurs. By contrast, when a raindrop hits a plant or litter, its force is broken and it trickles into the soil. Grasses are very effective in catching and holding moisture.

When plant cover is reduced by poor management and the distance between plants allows wind to reach the soil, wind erosion may result.

The amount of plant cover on the soil surface at the time of a rain or wind storm is the primary factor in preventing erosion. Both the bulk, or total weight, of cover and the distribution over the surface are important in reducing erosion.

The primary method of controlling soil erosion is plant cover. Other practices to control erosion on grasslands include brush control, deferred grazing, reseeding, and mechanical land treatments. Among the latter are terracing, contour furrowing, pitting, small dams, and diversions.

The combination of practices used for erosion control will gradually result in better production of grass, improved condition of the range, and a better water supply for domestic animals and wild game. Also, and most important, it will lessen the sedimentation due to soil erosion, which (1) reduces channel capacity and reservoir storage, and (2) results in increased flooding and reduced water supply.

WATER

Fig. 7-26. "The Western Range," an original painting by artist Tom Phillips, 3333 17th St., San Francisco, CA 94110.

Water is often a limiting factor in pasture productivity, affecting forage production and/or drinking water for grazing animals.

The water cycle is the never-ending movement of water from clouds to soil, through plants, and back to clouds again. The cycle begins when precipitation strikes the land and ends when the water leaves the land either through runoff or evaporation. During the intervening time, a livestock producer should store as much water as possible, in the soil and in reservoirs. The shortage of water over much of the West is particularly important. In addition to limiting livestock production, lack of water may limit stream flow for fish, cultivated crops, and industries.

There are various types of stock water developments. These include *natural* water supplies such as lakes, ponds, streams, springs, and seeps, and *made* developments such as wells, reservoirs, dugouts, sand tanks, and catchment basins. A combination of two or more types of water development is often more advantageous than one type only.

QUESTIONS FOR STUDY AND DISCUSSION

1. In a period of world food shortages, what characteristcs of an animal agriculture will favor its survival?

2. What is the difference between the terms pasture and forage?

3. Give facts and figures pointing up the magnitude and importance of pastures in the United States.

4. What are the primary differences between (a) seeded pastures, and (b) native pastures?

5. What are the primary differences between (a) permanent pastures, (b) semipermanent or rotation pastures, and (c) temporary and supplemental pastures?

6. Define each of the following terms: (a) grass, (b) legume, (c) browse, and (d) forbs.

7. Discuss the economic importance of pastures for each of the following classes of livestock: (a) beef cattle, (b) dairy cattle, (c) sheep, (d) swine, (e) horses, and (f) poultry.

8. Define and describe custom cattle grazing.

9. How would you go about determining what grass and/or legume to seed on a particular farm or ranch?

10. How may a livestock producer use Chart 7-1, Adapted Grasses and Legumes?

11. Why should pasture fertilizer rates be based on soil tests?

12. Why are swine producers using less and less pasture? Are pastures outmoded in modern swine production?

13. List the practices that are usually adhered to in successfully establishing a new pasture in the subhumid, humid, and irrigated areas.

14. Define "pasture renovation." List and discuss three different methods by which permanent pastures may be renovated.

15. Discuss how each of the following factors affects the value of pasture: (a) soil and fertilizer, (b) plant species, (c) stage of maturity, (d) rate of growth and season of year, and (e) grazing.

16. List and discuss some of the important practices involved in good pasture management of subhumid, humid, and irrigated areas.

17. Why and how should a farmer/rancher attempt to extend the grazing season?

18. Discuss and compare each of the following grazing systems: (a) continuous grazing, (b) rotation grazing, (c) conventional (rectangular) short duration grazing, (d) Savory (cell) short duration grazing, (e) first and second grazers, (f) intensive early season stocking, (g) creep grazing.

19. Define and compare strip grazing and green chop.

20. Is there a need and a place for more irrigated pastures in the United States? Discuss some of the special decisions that must be made if pasture irrigation is to be successful.

21. Discuss the history of the western range.

22. Why is so much of the range area of the West publicly owned and unenclosed? Is it good or bad to have so much public domain?

23. From the standpoint of vegetation and utilization by livestock, how do ranges differ from cultivated pastures?

24. List and discuss common range nutrient deficiencies. How are supplements of these nutrients usually provided?

25. Discuss the scientific basis for proper stocking rate on the western range.

26. Discuss the seasonal use of western ranges.

27. List and discuss the range characteristics which may affect the choice of the kind of livestock.

28. Discuss and compare each of the following range grazing systems: (a) continuous grazing, (b) rotation grazing, (c) deferred-rotation grazing, and (d) short duration grazing.

29. What are the warning signals of a range that is on the downgrade and in need of improvement?

30. List and discuss the methods that may be employed to improve a range.

31. Discuss each of the following types of ownership of U.S. land: (a) private ownership, (b) Indian land, (c) public ownership.

32. Discuss the role of each of the agencies administering public lands.

33. Some environmentalists are agitating for a ban of grazing rights of public lands. What are the pros and cons for such action, and what is your recommendation?

34. Define multiple-use of land. How and why did this concept evolve?

35. Define multispecies grazing. Where and how may it contribute to better and more uniform forage use and higher economic returns?

36. What forces have caused land owners to include wildlife in farm and ranch planning?

37. List and discuss the requisites for wildlife to exist in harmony with livestock operations.

38. Discuss the soil erosion control aspects of pasture and range management.

39. Discuss the water aspects of pasture and range management.

SELECTED REFERENCES

Title of Publication	Author(s)	Publisher
Commercial Beef Cattle Production, 2nd Edition	Ed. by C. C. O'Mary, I. A. Dyer	Lea & Febiger, Philadelphia, Pa., 1978
Crop Production, 5th Edition	R. J. Delorit L. J. Greub H. L. Ahlgren	Prentice-Hall, Inc., Englewood Cliffs, N.J., 1984
Forage and Pasture Crops	W. A. Wheeler	D. Van Nostrand Company, Inc., New York, N.Y., 1950
Forage Conservation in the 80's	Ed. by C. Thomas	British Grassland Society, Hurley, Maidenhead, Berks, SL 65 LR, United Kingdom, 1979
Forage Legumes for Energy-efficient Animal Production	Ed. by R. F. Barnes, *et al.*	USDA, Agricultural Research Service, 1985
Forages: Resources for the Future	J. E. Oldfield, Chairman, Task Force	Council of Agricultural Science and Technology, 1986
Forages, The Science of Grassland Agriculture, 4th Edition	M. E. Heath R. F. Barnes D. S. Metcalfe	The Iowa State University Press, Ames, Iowa, 1985
Grass, The Yearbook of Agriculture 1948	U.S. Department of Agriculture	U.S. Government Printing Office, Washington, D.C., 1948
Intensive Grazing Management	B. Smith	The Graziers Hui, Kamuela, Hawaii
Livestock Husbandry on Range and Pasture	A. W. Sampson	John Wiley & Sons, Inc., New York, N.Y., 1928
Manual of the Grasses of the United States, 2nd Edition	A. S. Hitchcock, Rev. by A. Chase	U.S. Government Printing Office, Washington, D.C., 1950
One Third of the Nation's Land	Public Land Law Review Commission	U.S. Government Printing Office, Washington, D.C., 1970
Pasture Book, The	W. R. Thompson	W. R. Thompson, State College, Miss., 1950
Pasture and Range Plants	Phillips Petroleum Company	Phillips Petroleum Company, Bartlesville, Okla., 1963
Practical Grassland Management	B. W. Allred, Ed. by H. M. Phillips	*Sheep and Goat Raiser Magazine,* San Angelo, Tex., 1950
Problems and Practices of American Cattlemen, Wash. Ag. Exp. Sta. Bull. 562	M. E. Ensminger M. W. Galgan W. L. Slocum	Washington State University, Pullman, Wash., 1955
Profitable Pasture Management	R. A. Chessmore	The Interstate Printers & Publishers, Inc., Danville, Ill., 1979
Range Development and Improvements	J. F. Valentine	Brigham Young University Press, Provo, Utah, 1979
Range Management, Principles and Practices	A. W. Sampson	John Wiley & Sons, Inc., New York, N.Y., 1952
Ruminants as Food Producers	J. E. Oldfield, Chairman, Task Force	Council for Agricultural Science and Technology, Ames, Iowa, 1975
Stockman's Handbook, The, 6th Edition	M. E. Ensminger	The Interstate Printers and Publishers, Inc., Danville, Ill., 1983
Veld and Pasture Management in South Africa	N. M. Tainton	Shuter & Shooter (pty) Ltd., Grey's Inn, 230 Church Street, Pietermaritzburg, South Africa, 1984
Western Range Livestock Industry, The	M. Clawson	McGraw-Hill Book Company, Inc., New York, N.Y., 1950

Original painting by Tom Phillips

Chapter 

HAY[1]

Contents	Page

[1]The authors gratefully acknowledge the helpful suggestions of the following eminent authorities who reviewed this chapter: J. E. Baylor, Ph.D., Professor of Agronomy, The Pennsylvania State University, State College; R. A. Forsberg, Ph.D., *et al.*, Department of Agronomy, University of Wisconsin-Madison; and S. C. Fransen, Ph.D., Forage Agronomist, Western Washington Research and Extension Center, Washington State University, Puyallup.

Fig. 8–1. "Winter In Wyoming," an original painting by artist Tom Phillips, 3333 17th St., San Francisco, CA 94110. Hay is still fed like this on some western snow-covered ranges.

Hay is forage harvested during the growing period and preserved by drying for subsequent use. Hays are made from legumes, grasses, and cereal crops. It is the most important harvested forage fed to livestock, and it ranks third among all livestock feeds, being exceeded only by pasture and corn. Hay is primarily a cattle, sheep, and horse feed, although alfalfa (especially ground, dehydrated alfalfa) may be included in swine and poultry rations. Average-quality hay runs 25 to 35% crude fiber and 45 to 55% TDN on an as-fed basis, whereas such concentrates as corn and wheat contain approximately 2 to 3% fiber and 80% TDN.

The object of haymaking is to (1) harvest the crop at the optimum stage of maturity which will provide the maximum yield of nutrients per acre without damage to the next crop, and (2) cure the crop properly by lowering the water content of the green herbage from 65 to 85% to 20% or less.

Drying, or making hay, is the most common method of preserving forage for storage, primarily because it is relatively easy to handle. It can be stored or transported long, chopped, pelleted, cubed, or packaged into various types and sizes of bales. Modern equipment and chemicals hasten drying time; and automated systems facilitate handling.

The great capacity and specialized functions of the rumen allow cattle and sheep to use hay, and other forages, in large amounts. Bacteria and protozoa in the rumen break down and make available to the host animal part of the nutrients in cellulose or fibrous material.

In addition to the nutrients that it contains, and to its value in providing feed throughout the year, hay has other values. Dry feed is essential for the proper functioning of the digestive tract; it acts as a stimulant in moving the feed through the intestines, and it maintains the proper conditions in the rumen for the microbial action which plays such a vital role in the digestion of the fibrous portions of feeds. Hay is often used as a supplement to "washy" pastures and succulent silages. Also, it speeds along the development of the rumen function of the young ruminant, lessens the incidence of displaced abomasum in cattle, and prevents a lowering of the fat content of the milk of lactating cows (unless it is finely ground). Also, and most important, good-quality hay is a hedge against high-concentrate prices, for when the price of such feeds increases disproportionately, increased amounts of hay may be fed and concentrates may be decreased, with a higher net return to the producer.

Despite its several advantages, hay has some shortcomings. It varies in nutrient content and palatability more than any other feed, because of differences in the (1) crops from which it is made, (2) stage of cutting, (3) handling, and (4) weather damage during curing. Not even ruminants can consume enough hay alone to meet the demands of high production; for example, when fed hay alone, dairy cows will produce only 50 to 70% as much milk as they would when fed a ration consisting of 50% concentrates. Also, fiber is poorly digested by monogastric animals, with the result that hay serves primarily as a source of minerals and vitamins for swine and poultry.

An estimated 80% of all hay is fed on the farms or ranches on which it is produced, rather than being purchased. It is important, therefore, that producers know how to produce good hay, as well as how to feed it, for most of them determine their own destiny from the standpoint of quality. For this reason, this chapter covers hay from production to feeding.

HISTORY OF HAY

Fig. 8–2. Haymaking has gone modern! It's no longer a backbreaking, pitchfork job. This shows a pick-up baler with a bale loader. (Courtesy, Deere & Company, Moline, Ill.)

Haymaking evolved with the domestication of animals, for from that remote day forward caretakers assumed responsibility for storing feed for them for use during times of scarcity. More than 2,000 years ago, the Roman agricultural writer, Columella, described haymaking as "throwing forage loosely together for a few days to heat and concoct itself and then cool before putting into the mow." But another 20 centuries were to pass before haymaking changed materially. As recently as 1850, it was cut with a scythe; and pitchfork haymaking persisted into the present century. Beginning about 1940, scientists and engineers pooled their efforts to transform roughages into high-quality hay. Haymaking went modern, with automated one-operator pick-up balers, field choppers, cubing machines, and other modern equipment, replacing the backbreaking, labor-intensive methods of old.

Automated haymaking and surplus grain were ushered in together. At the close of World War II, U.S. grain bins bulged. This spawned the era of high-energy, low-forage rations. Then, suddenly, in the early 1970s, there were world food shortages. The 20-year grain-feeding binge in the United States began reversing itself. Now, and in the future, more and more grain will be used for direct human consumption. Animals (especially cattle and sheep) will increasingly be "roughage burners."

MAGNITUDE AND IMPORTANCE OF HAY

The importance of the nation's hay crop is attested to by the fact that the total area devoted to hay in the United States exceeds 60 million acres, the total production averages about 150 million tons, and the annual crop is worth approximately $10 billion—it is worth more than any other crop except corn and soybeans. On an air-dry tonnage basis, about 3 times as much hay is produced as silage.

Despite the importance of hay, no other feed crop suffers a higher loss of nutrients from the time it is cut to the time it is fed. During the curing process, the quality and feeding value of hay decreases rapidly by rain, sun bleaching, raking, handling when too dry, and storing with too much moisture. Studies by the U.S. Department of Agriculture revealed that the following losses accrued in field-cured, second-cut alfalfa hay from the time of cutting to the time of feeding: leaves, 35%; dry matter, 20%; and proteins, 29% (see Fig. 8–3).

The longer hay remains in the field until it is dry enough to store, the greater the nutrient losses (Fig. 8–3). These losses have been estimated to have a feeding value of more than a billion dollars annually.

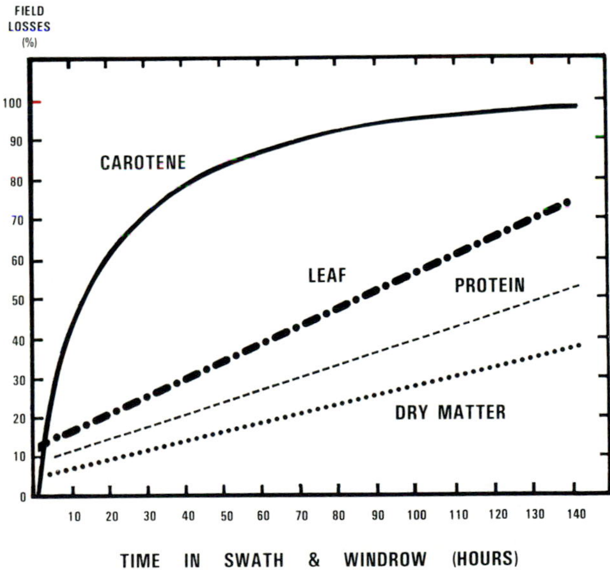

Fig. 8–3. Losses in sun-curing alfalfa hay as related to time in the field to reduce moisture to a safe level for storage. (Adapted by the authors from USDA data.)

Hay As an Energy Source

Hay is an important source of energy for cattle, sheep, and horses. Table 8–1 (p. 300) shows the percentage of total energy (TDN) intake provided to these species by hay and other kinds of feeds.

TABLE 8-1
PERCENTAGE OF ENERGY SUPPLIED BY HAY AND OTHER KINDS OF FEEDS[1]

Animal	Concentrates	Hay	Other Harvested Forages	Pasture	All Forage	Total
	(%)	(%)	(%)	(%)	(%)	(%)
Lactating cows . .	37.9	23.1	19.4	19.6	62.1	100
Other dairy cows	19.4	29.0	5.9	45.7	80.6	100
Finishing beef cattle	69.8	16.3	8.7	5.2	30.2	100
Other beef cattle	8.7	15.5	4.1	71.7	91.3	100
Sheep and goats	10.4	4.7	3.1	81.8	89.6	100
Horses and mules	20.6	18.3	10.2	50.9	79.4	100

[1]Based on USDA data. From paper entitled, "Hay Production, Preservation and Quality," by J. E. Baylor, The Pennsylvania State University, *Beef Cattle Science Handbook,* Vol. 13, p. 199, published by Agriservices Foundation, edited by M. E. Ensminger.

As shown in Table 8-1, hay is a more important source of energy for dairy cows than for any other class of farm animal. But it is also an important feed source for beef cattle and horses. It is noteworthy, too, that about one-half the total hay tonnage produced in North America is fed to dairy cattle, while beef cattle consume almost one-third of all hay produced. As increasing quantities of concentrates go to feed the world's hungry people, it is expected that livestock producers will depend even more on hay to meet a larger percentage of the total feed needs of ruminants.

Comparative Value of Hay

In ruminant rations, hay is primarily a source of energy, but the legumes also serve as a source of protein. For swine and poultry, ground hay (especially alfalfa) is fed primarily as a source of minerals and vitamins.

TABLE 8-2
COMPARATIVE ECONOMICS OF THE NUTRIENTS IN CORN GRAIN, CORN SILAGE, AND ALFALFA HAY OF THREE QUALITIES[1]

Item	Corn Grain	Corn Silage	Alfalfa Hay		
			Mature	Mid-Bloom	Early Bloom
Analyses, DM basis, %					
Dry matter (DM)	88.0	33.0	90.0	90.0	90.0
Crude protein (CP)	10.1	8.1	12.9	17.0	18.0
Total Digestible Nutrients (TDN)	90.0	70.0	50.0	58.0	60.0
Value of 100 lb DM, $					
CP value[2]	1.72	1.38	2.19	2.89	3.06
TDN value[3]	5.04	3.92	2.80	3.25	3.36
Total value	6.76	5.30	4.99	6.13	6.42
Total value/acre,[2] [3] $					
16 tons silage or 100 bu grain	333.00	560.00	—	—	—
21 tons silage or 150 bu grain	500.00	735.00	—	—	—
5 tons hay	—	—	449.00	552.00	578.00
8 tons hay	—	—	719.00	883.00	924.00

[1]Adapted by the authors from: *Haymaker's Handbook,* by J. E. Baylor, Professor of Agronomy, The Pennsylvania State University, and M. A. Balas, New Holland, Inc., published by Ford New Holland, Inc., New Holland, Pa., 1987, p. 140, Table 17.1.

[2]44% soybean meal used as a standard protein source, priced at $150 per ton or $.17 per pound of crude protein.

[3]Corn grain used as a standard TDN source, priced at $2.50 per bushel or $.056 per pound of TDN.

Table 8-2 shows the dry matter (DM), crude protein (CP), and total digestible nutrients (TDN) in 100 lb of dry matter from corn grain, corn silage, and three different types of alfalfa hay (mature, midbloom, and early bloom). Note, too, that this table compares these crops on a per acre basis.

Of course, the economic comparisons in Table 8-2 are valid only at the stated prices of soybean meal and corn. However, in most practical feeding situations the comparisons are meaningful. It is noteworthy that, in terms of the economic value of the energy and protein provided by the different feeds listed in Table 8-2, early bloom alfalfa hay had a value nearly 95% (6.42 ÷ 6.76 × 100) that of corn grain. Note, too, that in terms of energy produced per acre, corn silage leads all other feeds.

Hay As a Grain Replacement

In the future, livestock producers will increasingly rely upon the ability of ruminants to convert coarse forage, grass, and by-product feeds, along with a minimum of concentrate, into food for human consumption, thereby competing less for humanly edible grains.

The University of Arizona found that 60% alfalfa in a grower calf ration was converted as well as one containing only 15% roughage with the remainder consisting of high-priced concentrate or grain. The study involved steer calves averaging 400 lb at weaning, and a grow-out period of 84 days. Actual gains were 2.84 lb per head daily for the calves fed the 60% alfalfa ration vs 2.34 lb for those fed the ration containing 15% roughage. Feed conversion was practically the same—5.63 to 1 for the high roughage, and 5.60 lb of feed per pound gain for the control animals.

The U.S. Department of Agriculture conducted a study of all-forage rations for finishing cattle, the results of which are given in Table 8-3.

TABLE 8-3
FEEDLOT PERFORMANCE AND CARCASS EVALUATION OF STEERS FED ALL-FORAGE VS ALL-CONCENTRATE RATIONS[1]

Item	All-Forage Ration[2]	All-Concentrate Ration[2]
Average daily feed intake . . . . (lb)	23.3	16.0
. . . . *(kg)*	*10.59*	*7.26*
Average daily feed intake in % of body weight	3.23	2.15
Average daily gain . . . . (lb)	2.3	2.8
. . . . *(kg)*	*1.05*	*1.27*
Feed-gain ratio	10.06	5.71
Average carcass grade	Low Choice	Medium Choice
Dressing percentage . . . . (%)	55.4	59.9
Marbling score	Abundant	Abundant
Rib eye area . . . . (sq in.)	11.0	10.6
. . . . *(sq cm)*	*71.1*	*68.5*
Fat over rib eye . . . . (in.)	.37	.67
. . . . *(mm)*	*9.4*	*17.0*
Taste panel evaluation[3]	7.6	7.2

[1]Oltjen, R. R., T. S. Rumsey, and P. A. Putnam, "All-Forage Diets for Fattening Beef Cattle," *Journal of Animal Science,* Vol. 32, No. 2, 1971, pp. 327-333.

[2]Corn grain provided 90% of the all-concentrate ration; pelleted alfalfa provided 98% of the all-forage ration.

[3]Overall desirability rated on a scale of 1 to 9, with 9 being the most desirable.

As a result of the experiment summarized in Table 8–3, the U.S. Department of Agriculture researchers concluded that (1) beef cattle of an acceptable quality were produced on a pelleted, all-forage ration; (2) steers on an all-forage ration had to be fed a month longer than those on the all-concentrate ration; (3) the all-forage-fed steers consumed about 95% as much metabolizable energy and were about 86% as efficient converters of it to body weight gains as were the all-concentrate steers; (4) the forage-fed steers had only 55% as much fat over the rib eye as did the all-concentrate-fed steers; and (5) there was a 4.5% difference in dressing percentage in favor of the animals receiving all-concentrate ration. Based on this study, the following conclusion may be drawn: Since cattle fed high-roughage rations normally have lower dressing percentages, forages must be cheaper than grain in order for the feeder to obtain the same net return; this situation usually exists relative to pastures, but it doesn't always apply to dry forages.

The Georgia Station compared (1) a conventional high-energy ration, composed of 72.8% corn, 20% peanut hulls, and 7.2% protein supplement; and (2) an all-forage ration consisting of Bermudagrass pellets (in drylot) and small grain pastures (in season), for finishing calves and yearlings.[2] The results of this study are reported in Table 8–4.

TABLE 8–4
FEEDLOT PERFORMANCE AND CARCASS CHARACTERISTICS OF CALVES AND YEARLINGS FINISHED ON A HIGH-ENERGY RATION AND AN ALL-FORAGE RATION

Item	Calves				Yearlings			
	High-Energy		All-Forage		High-Energy		All-Forage	
	(lb)	*(kg)*	(lb)	*(kg)*	(lb)	*(kg)*	(lb)	*(kg)*
Average initial weight	488	*222*	492	*224*	670	*305*	672	*306*
Average final weight	1,042	*474*	1,053	*479*	1,165	*530*	1,147	*521*
Average daily gain	2.84	*1.29*	2.33	*1.06*	3.07	*1.40*	2.31	*1.05*

Yearling steers fed the high-energy ration gained 3.07 lb daily vs 2.31 lb for the steers fed the all-forage ration. Average daily gains of steer calves and yearlings were similar when fed the all-forage ration, but average gain for yearlings was 8% faster than for calves when fed the high-energy ration. Steers fed the high-energy ration dressed out about 7.5% more carcass, and had more marbling, higher yield grades, and more fat covering over the rib eye than steers fed the all-forage ration. Under the conditions of these trials, finishing steers of both age groups were more profitable when an all-forage ration was fed.

KINDS OF HAY

Fig. 8–4. Alfalfa (lucerne), known as the "Queen of the Forages," accounts for 57% of the U.S. hay crop—more than all other kinds of hay combined. (Courtesy, Deere & Company, Moline, Ill.)

[2]Utley, P. R., R. E. Hellwig, and W. C. McCormick, "Finishing Beef Cattle for Slaughter on All-forage Diets," *Journal of Animal Science*, Vol. 40, No. 6, 1975, p. 1034.

Although there are favorite hays, a great variety of legumes, grasses, and cereal crops can be, and are, successfully used for hay. In terms of total tonnage produced annually, alfalfa (or lucerne), the "Queen of the Forages," accounts for approximately 57% of the U.S. hay crop. Many different kinds of hay make up the other 43% of the nation's hay supply; among them, other perennial legumes, cool season grasses, warm season grasses, cereal hays, summer annuals, and annual legumes. (See Table 8–5.)

The kind of forage grown should be determined by soil type, soil drainage, soil pH, topography, climatic conditions, preferred use, and the animals to which it will be fed. Also, more and more farmers are coming to appreciate the flexibility afforded by growing varieties of grasses and legumes that may be used three ways: for pasture (see Chapter 7, Pasture and Range Forages), for hay, or for silage (see Chapter 9, Silage/Haylage/High-Moisture Grain). With such an arrangement, surplus pasture may be converted into hay, or, if the weather is not favorable for haymaking, the crop can be ensiled.

Generally speaking, legumes should be used as hay crops wherever they are adapted, either alone or in combination with grass(es). There is one possible exception to this recommendation—where horses are involved, sometimes a good-quality grass hay may be preferable.

Whenever feasible, it is recommended that a legume be grown for hay, for the reasons that, in comparison with grasses, legumes are (1) higher in protein, vitamins, and minerals; (2) higher yielding; and (3) nitrogen-fixing when inoculated, because the bacteria (rhizobia) on their roots take free atmospheric nitrogen from the air. However, a mixture of grasses and legumes is often preferred for reasons of palatability, ease in curing, erosion control, and lessening bloat.

Table 8–5 (p.302) summarizes the adaptation and use of a number of species grown for hay. Note that soil drainage and pH are important. In addition to the adaptation and use of various species, consideration should be given to (1) variety, because some varieties may be better suited than others to a specific soil and management; and (2) quality of seed.

TABLE 8-5
ADAPTATION AND USE OF SOME COMMON FORAGE SPECIES GROWN FOR HAY[1]

Species	Estimated Life of Stand	Approx. Yield Pure Stand[2]	Soil Adaptation	Preferred Use	Preferred Associated Species for Mixtures
	(yrs)	(tons/acre)			
Legumes:					
Alfalfa	3-5	4-7	Deep, well drained, pH 6.5 or above.	Hay, silage, pasture.	Tall, cool season grasses and perennial ryegrass.
Red clover	1-2	3½-5	Medium to well drained, pH 6.0 or above.	Hay, silage, pasture.	Timothy, bluegrass, tall fescue.
Birdsfoot trefoil	3-5	2½-4	Somewhat poorly to well drained, pH 5.5 or above.	Hay, silage, pasture.	Timothy, Kentucky bluegrass, perennial ryegrass, reed canarygrass.
Alsike clover	1-2	1-2	Tolerates wet, acid soils, pH 6.0 or above.	Pasture, silage, hay.	Red clover, timothy.
Cool season grasses:					
Perennial ryegrass	3-4	2-4	Moist to well drained, pH 6.0 or above.	Pasture, silage, hay.	Alone, or with a mix of either grass or legume.
Orchardgrass	3+	3-5	Medium to well drained, pH 6.0 or above.	Pasture, silage, hay.	Alone or with alfalfa or ladino clover.
Smooth bromegrass	3+	3-4	Fertile, well drained, pH 6.0 or above.	Hay, silage, pasture.	Alone, or with alfalfa.
Timothy	2-3	3-4	Medium to well drained, pH 6.0 or above.	Hay, silage, pasture.	Alone, or with red clover.
Reed canarygrass	3+	3½-6	Widely adapted from wet to droughty, pH 5.5 or above.	Pasture, silage, hay.	Alone, or with a legume.
Tall fescue	3+	3-5	Widely adapted shallow to deep soil, pH 5.5 or above.	Pasture, silage, hay.	Alone, or with lespedeza.
Warm season grasses:					
Switchgrass	4+	5-6	Moderately deep, from somewhat dry to poorly drained, and tolerant to low fertility.	Pasture, hay.	Alone.
Big bluestem	4+	4-5	Moderately deep, from somewhat dry to poorly drained, and tolerant to low fertility.	Pasture, hay.	Alone.
Bermudagrass	4+	5-8	Moderate to well drained and fairly heavy soil, pH 5.5 or above.	Hay, silage, pasture.	Alone, or with crimson or arrowleaf clover.
Cereal hays:					
Barley	Annual	1-3	Well drained, pH 7.0 or above.	Grain, hay, silage.	Alone.
Oats	Annual	1-3	Well drained, pH 5.5 or above.	Grain, hay, pasture, silage.	Alone, or with field peas.
Rye	Annual	1-3	Sandy, acid, infertile, pH 5.5 or above.	Grain, pasture.	Alone, or with vetch or crimson clover.
Triticale	Annual	1-3	Droughty and poor soil, pH 5.5 or above.	Grain, pasture, hay.	Alone.
Wheat	Annual	1-3	Fertile, well drained, pH 6.0 or above.	Grain, pasture, hay.	Alone, or with vetch, crimson clover or Austrian peas.
Summer annuals:					
Millet	Annual	1-3	Poor soil, pH 7.0 or above.	Hay, silage.	Alone.
Sorghum hybrids	Annual	2-4	Fertile, well drained, drought resistant, pH 7.0 or above.	Grain, silage, green chop, pasture, hay.	Alone.
Sudangrass	Annual	2-4	Fertile, well drained, drought resistant, pH 7.0 or above.	Pasture, silage, green chop, hay.	Alone, or with soybeans or cowpeas.
Annual legumes:					
Cowpeas	Annual	2-4	Poorer soils, pH 6.0 or above.	Pasture, green manure, hay, seed.	Alone, or with corn or sorghum.
Soybeans	Annual	2-5	Fertile soil, pH 6.0 or above.	Seed, hay.	Alone, or with millet, sorghum or Sudangrass.
Vetch	Annual	2-2½	Poor soil, pH 6.5 or above.	Hay, pasture, soil improvement.	Alone, or with small grain.

[1]Adapted by the authors from: *Haymaker's Handbook*, by J. E. Baylor, Professor of Agronomy, The Pennsylvania State University, and M. A. Balas, New Holland, Inc., published by Ford New Holland, Inc., New Holland, Pa., 1987, pp. 17-18, Table 3.2.

[2]These are short tons (2,000 lb). To convert to metric tons/acre, multiply by 0.907; then, to obtain metric tons/hectare, multiply by 2.47.

Weeds and Other Potential Hay Crops

In periods of serious drought, many other plants are used for feed. Russian thistle was used to keep cattle alive during the drought of the 1930s. Even when harvested at a very early stage, it is a better feed when ensiled than when dried for hay. Its abrasive surface is less irritating to the mouths of livestock when softened as silage.

Pigweeds, sow thistles, and other weeds have been used as livestock feeds in emergencies. Even though their yield is low and their nutrient content is poor, they will sustain life.

In using weeds and other unusual crops for livestock feed in times of emergency, it is always advisable to obtain authoritative information relative to toxic substances in them and the stage at which it is best to harvest them for feeding.

HAY QUALITY

Hay quality is the degree of excellence, or the productive worth, that hay possesses. It refers to the nutritive value of hay. For hay to be of superior quality, it must be high in four factors: (1) nutrients, (2) palatability (intake), (3) digestibility, and (4) efficiency of utilization.

The most accurate method of determining hay quality involves live animal experiments on the farm or ranch where the forage is to be fed. However, this is often too costly, slow, and impractical. Therefore, forage value is predicted by visual inspection, chemical analysis, and/or new methods such as near infrared analysis.

Importance of Hay Quality

Hay is feed. Thus, as with any feed, it's the end results from feeding hay—the value as determined by animals—that count. Generally speaking, livestock producers recognize that the feeding value of hay varies according to quality. However, it is doubtful that they realize just how much returns in production—in meat, milk, wool, reproduction, and speed and endurance—are affected by quality.

The quality of hay greatly affects its consumption. High-quality forage is more digestible and passes through the digestive tract more rapidly than low-quality forage; hence, animals will consume more of it.

Feeding trials at the University of Wisconsin, Madison, showed, conclusively, the effect of hay quality on milk production. Alfalfa hay at four stages of maturity—prebloom, early bloom, midbloom, and full bloom—was fed to lactating cows. Table 8-6 summarizes the results. It gives for the four different hays the crude protein (CP), neutral detergent fiber (NDF), and acid detergent fiber (ADF); the digestible dry matter (DDM); the dry matter intake (DMI); and the 4% fat corrected milk (FCM) produced. Note that with maturity of the hay the crude protein, DDM, DMI, and milk production decreased, while the NDF and ADF increased. The increase in NDF and ADF with maturity is as expected, because NDF is inversely correlated with intake, whereas ADF is highly correlated with digestibility.

TABLE 8-6
EFFECT OF QUALITY OF ALFALFA HAY ON PERFORMANCE OF LACTATING COWS[1]

Stage of Harvest	Composition			DDM	DMI	4% FCM
	CP	NDF	ADF			
	(%)	(%)	(%)	(%)	(% BW[2])	(lb/day[3])
Prebloom	21.1	40.5	30.2	62.7	2.08	87.1
Early bloom ...	18.9	42.0	33.0	61.6	1.97	77.2
Midbloom	14.7	52.5	38.0	54.8	1.48	66.2
Full bloom	16.3	59.5	45.9	52.9	1.42	64.7

[1]Kawas, J. R., N. A. Jorgensen, A. R. Hardie, and J. L. Danelon, *Journal of Dairy Science,* Abstract, Supplement 1, 1983, Vol. 66, p. 181; and Kawas, J. R., N. A. Jorgensen, and D. A. Rohwede, *Proceedings Wisconsin Forage Council Eighth Forage and Use Symposium,* 1984, p. 21, Tables 3 and 4.

[2]80% hay and 20% concentrate (DM basis).

[3]46% hay and 54% concentrate (DM basis).

At the Illinois Station, test cows ate 2.21 lb of early cut hay per 100 lb of body weight compared to 1.6 lb of late cut hay. This meant that cows fed late cut hay ate 9 lb less per day than cows fed hay cut in the bud stage. To make up this nutritional difference in intake, the cows fed the late cut hay required 7 lb more grain daily for each cow. Hence, a 100-cow herd would require 700 lb more grain daily as a penalty for late cutting.

In feeding trials with lactating cows, Cornell workers compared alfalfa hay cut at two different stages of maturity—early bloom vs late bloom. They found that, in comparison with the late cut hay, the cows that were fed the early cut hay consumed 7 lb more of it per head per day and it was 16% more digestible and produced 12 lb more milk per day (see Fig. 8-5).

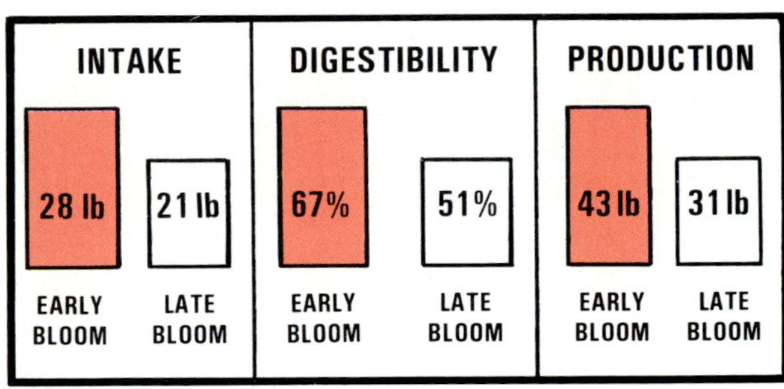

Fig. 8-5. Comparison of alfalfa hay cut at two different stages of maturity when fed to lactating cows (based on Cornell data).

Cornell University workers also compared the economic impact of feeding 5 different qualities of hay to a 120-cow dairy herd. In this study, half the dry matter was corn silage and the other half was hay. Everything else was held constant except for the hay quality. The results are reported in Table 8-7. Note that, as hay quality declined, the increase in purchased feed costs spiraled. The cost of purchased feed for the 120-cow dairy herd fed grass hay, the lowest hay quality, was $79.04 more per day, or $28,849.60 more per year, than for the cows fed the early cut legume. As dramatic as these results were, they do not tell the whole story. The higher quality hays also resulted in increased production and greater returns.

TABLE 8-7
EFFECT OF 5 DIFFERENT QUALITIES OF HAY ON THE DAILY FEED COST OF A 120-COW DAIRY HERD[1]

Hay Quality		Forage Fed	Concentrate Purchased	Purchased Feed Cost
Description	Percent Protein			
	(%)	(tons DM)	(lb)	($)
Early cut legume	21	2.17	531	94.74
Legume	18	2.07	733	117.62
Mixed, mainly legume	15.5	1.91	977	138.68
Mixed, mainly grass ..	12	1.84	1,158	160.84
Grass	10	1.77	1,278	173.78

[1]Adapted by the authors from: *Haymaker's Handbook,* by J. E. Baylor, Professor of Agronomy, The Pennsylvania State University, and M. A. Balas, New Holland, Inc., published by Ford New Holland, Inc., New Holland, Pa., 1987, p. 66, Table 9.2.

The Wisconsin, Illinois, and two Cornell University studies previously referred to confirm what experienced dairy producers know—lactating cows simply cannot consume sufficient amounts of low-quality forage to permit optimum production.

The importance of hay quality in beef cattle gains was pointed up in studies conducted by the Tennessee Station (Table 8-8). Note that, with advancing maturity, the composition (quality) of the hay declined sharply; the crude protein decreased from 13.8 to 7.6%, and the digestibility decreased from 68 to 56%. With the decreased quality, (1) it required more hay per pound of gain, and (2) the cattle gains plummeted from 1.39 to 0.42 lb/head/day.

TABLE 8-8
EFFECT OF QUALITY OF FESCUE HAY ON CATTLE GAINS[1]

Stage of Harvest	Lb of Hay Harvested/Acre (1st cutting)	Composition		Intake/Animal/ Day (DM Basis)	Lb of Hay/ Lb of Gain	Gain/Head/Day
		Crude Protein	Digestibility			
	(lb)	(%)	(%)	(lb)	(lb)	(lb)
Late boot to head (cut May 3)	1,334	13.8	68	13.0	10.1	1.39
Early bloom—10% shedding pollen (cut May 14)	1,838	10.2	66	11.7	13.5	0.97
Early milk—seed forming (cut May 25)	2,823	7.6	56	8.6	22.5	0.42

[1]Adapted by the authors from: *Southern Regional Beef Cow-Calf Handbook, SR 5004,* "Quality Hay Production," by J. Burns, Extension Forage Specialist, University of Tennessee, and J. K. Evans and G. Lacefield, Extension Forage Specialists, University of Kentucky. Holstein heifers weighing 500 lb were used in the experiment and were fed salt and water plus hay. The research was conducted at the University of Tennessee by M. Montgomery, B. Bearden, and K. Barth.

Thus, experiments and experiences show that, in addition to the low-nutrient content that characterizes poor-quality hay, a more serious loss may follow from feeding it. Studies show that part of the poor results obtained from feeding low-quality hay can be attributed to its failure to support maximum microflora in the rumen, with the result that the digestibility of the crude fiber suffers. Hand in hand with the decline in microflora activity, forage consumption goes down. Of course, if animals won't eat feed, it won't do them any good.

Visual Inspection

Although not as reliable as chemical analysis, most hay is still bought and sold on the basis of visual appraisal.

Fortunately, hay quality and value can be estimated by certain characteristics. It is important, therefore, that those who grow hay, and those who buy and sell hay, be acquainted with the recognizable characteristics of hay which indicate high palatability and nutrient content. If in doubt, the animals will tell you, for they like and thrive on high-quality hay.

The factors to look for in high-quality hay are:

1. **Species of plants.** Determine what plants are present and the proportion of each. Hay with a high percentage of legume is usually higher in feed value than pure grass hay.

2. **State of maturity when cut.** Plants should not be in full bloom, nor should they have formed seeds. Early cut hay assures the maximum content of protein, minerals, and vitamins, and the highest digestibility.

3. **Percentage of leaves present.** Leaves are the part of the plant of highest quality; hence, a high proportion of leaves relative to stems is indicative of high quality.

4. **Green color.** A bright green color indicates (a) minimum of bleaching and leaching losses of carotene and other nutrients, and (b) palatability.

Fig. 8-6. James Tappan, Arizona Dairy Co., checking the quality of new-mown alfalfa hay. (Courtesy, James Tappan, Arizona Dairy Co., Higley, Ariz.)

5. **Aroma and fragrance.** High-quality hay has a pleasing, fragrant aroma. Moldy smells are undesirable.

6. **Stemminess.** Large stiff, woody stems make for low acceptability and quality. High-quality hay is fine stemmed and pliable.

7. **Foreign material.** High-quality hay is free from such foreign materials as weeds, stubble, sticks, dirt, etc.

8. **Condition.** Hay that has been cured and stored properly does not contain excess moisture, is not in layers or chunks due to excess moisture or heating, is not moldy, and is not dry and brittle.

Guidelines for sensory evaluation of (1) legumes, (2) grasses, and (3) grass-legume mixtures are given in Table 8-9, Hay Scorecard.

TABLE 8-9
HAY SCORECARD[1]

Factor	Hay Type		
	Legumes	Grasses	Grass Legume Mixtures
	Point Score		
Leafiness. Legume hay should contain 40%, or more, leaves.	25		15
Color/aroma. Hay should be green, bright, and have a pleasant aroma.	25	30	25
Softness and pliability of stems. These qualities are indicative of harvest at early stage of maturity, and of high nutrients, palatability, and consumption.	15	30	20
Freedom from foreign material. Few weeds and little trash, no toxic substances, and minimum waste.	15	20	20
Condition. Hay cut, cured, and stored properly, with the result that it is free from dust and mold.	20	20	20
Totals	100	100	100

[1]Adapted by the authors from *Harvesting Quality Hay,* Circular R-624, North Dakota State University, Fargo.

Chemical Analysis

Visual estimates of hay quality are of value and should be used, but the most precise way to determine the nutrient value of hay is through chemical analysis. Analyses are not infallible, however; a Pennsylvania study revealed errors of as much as 5% in crude protein and 9% in TDN (energy) content of a forage, with evaluations made by trained individuals.

Visual inspection of hay is still needed. For example, hay cut at the right stage of maturity can become low-quality hay by poor hay-making practices and conditions, which only visual inspection can detect. Also, visual inspection is needed for (1) weed detection and color, (2) predicting palatability, and (3) detecting the effects of mold, rain damage, and brittleness. So, chemical analyses should supplement, but not replace, visual inspection. Also, it is recognized that any method of determining hay quality by means of chemical analyses is of value only if it is related to feeding value.

Most livestock and hay producers are aware that hay harvested at an early stage of maturity is high in protein and low in fiber (cellulose), although hay yields at immature stages are low. But the magnitude of the variation is usually greater than suspected. Thus, both chemical composition and yields must be considered in a practical management system. The highest yield of protein and of most of the other important chemical constituents is obtained at near the $\frac{1}{10}$ bloom stage of growth. Although the yield of hay may continue to increase between $\frac{1}{10}$ bloom and full bloom, it is due largely to an increase in the yield of cellulose (Fig. 8-7).

As shown in Fig. 8-7, the yield of protein is greatest when alfalfa is harvested at the first flower ($\frac{1}{10}$ bloom) stage of maturity.

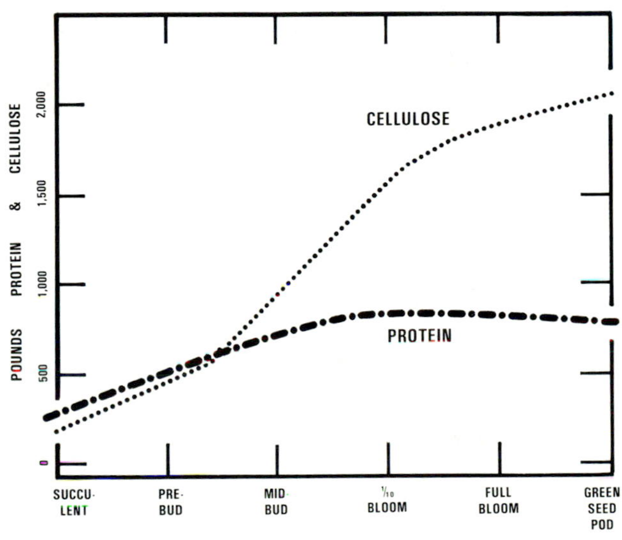

Fig. 8-7. Yield of crude protein and cellulose per acre from first cutting alfalfa. (*From:* Rohweder, D. A. and D. Smith, *Establishing and Managing Alfalfa,* Pub. A 1751, University of Wisconsin, 1982, p. 12.)

Research has generally shown a good relationship between the chemical composition of hay and its feeding value. As a result, a growing number of states now have laboratories where, at a nominal charge, a quick determination can be made of the chemical composition of hay. As hay matures, protein decreases (pounds of protein per acre decreased from 935 lb in early cut hay to 605 lb in late cut hay, according to a Cornell study) and fiber increases. Likewise, weathering lowers the protein and raises the fiber content, since soluble nutrients are washed out by rain and leaves are lost during harvest. It is also noteworthy that palatability is negatively correlated with crude fiber levels—the higher the fiber content, the lower the palatability; this is important, for if the animals won't eat it, they can't produce. Cornell investigators found that cows ate 2 lb more of the early-cut hay per day than of late-cut hay.

But protein and fiber content don't tell the whole story! The concentration of each essential nutrient decreases with maturity (Table 8–10).

TABLE 8–10
CHEMICAL COMPOSITION OF FIRST CUTTING ALFALFA CUT AT FOUR STAGES OF MATURITY[1]

Constituent		Prebud	Bud to Midbud	First Flower to 1/10 Bloom	Near Full Bloom
Experiment A					
IVDDM[2]	(%)	67.7	61.4	56.0	53.0
Nitrate (NO_3)	(%)	0.11	0.07	0.04	0.06
Crude protein	(%)	26.5	23.3	17.9	15.8
Crude fiber	(%)	21.0	27.9	34.0	35.4
Cellulose	(%)	24.8	30.5	34.6	36.2
Ether extract (fat)	(%)	1.8	1.5	1.5	1.4
Total ash	(%)	11.5	8.7	8.8	7.8
Calcium	(%)	1.48	1.47	1.41	1.28
Phosphorus	(%)	0.40	0.34	0.28	0.24
Potassium	(%)	2.16	1.61	1.42	1.10
Iron	(ppm)	310	184	154	153
Manganese	(ppm)	58	50	52	46
Copper	(ppm)	7.8	6.8	5.8	6.0
Cobalt	(ppm)	0.11	0.09	0.10	0.09
Zinc	(ppm)	31	29	23	24
Carotene[3]	(ppm)	228	224	208	105
pH of tissue[3]		5.58	5.46	5.65	5.50
Experiment B					
TNC[4]	(%)	15.7	13.6	11.1	13.3
Starch	(%)	4.8	6.8	6.0	6.2
Sucrose	(%)	5.6	2.7	2.4	2.7
Glucose	(%)	2.6	2.1	1.3	2.1
Fructose	(%)	2.7	2.0	1.4	2.3
Experiment C					
Crude protein	(%)	28.60	23.69	17.90	15.31
Alanine	(%)	1.28	1.07	0.80	0.81
Arginine	(%)	1.27	0.94	0.78	0.74
Aspartic acid	(%)	3.28	—	1.85	1.88
Glutamic acid	(%)	2.20	2.07	1.42	1.53
Glycine	(%)	1.14	0.94	0.76	0.77
Histidine	(%)	0.55	0.42	0.37	0.35
Isoleucine	(%)	1.07	0.88	0.69	0.68
Leucine	(%)	1.82	1.53	1.20	1.17
Lysine	(%)	1.41	1.10	0.90	0.87
Methionine	(%)	0.40	0.30	0.21	0.22
Phenylalanine	(%)	1.19	0.93	0.81	0.75
Proline	(%)	1.00	0.89	0.75	0.69
Serine	(%)	1.08	0.86	0.74	0.71
Threonine	(%)	1.20	—	0.78	0.61
Tyrosine	(%)	0.82	0.62	0.54	0.51
Valine	(%)	1.35	1.08	0.97	0.86
Experiment D					
Total-N	(%)	4.54	3.21	2.89	2.08
Protein-N	(%)	3.11	2.38	2.13	1.56
Non-protein-N	(%)	1.43	0.84	0.76	0.52
Ammonium-N	(%)	0.009	0.007	0.006	0.004
Nitrate-N	(%)	0.053	0.034	0.016	0.013
Free alpha-amino-N	(%)	0.243	0.194	0.137	0.103
Asparagine-N	(%)	0.178	0.148	0.110	0.089
Glutamine-N	(%)	0.009	0.006	0.006	0.004

[1]Summarized and adapted from studies conducted by the Wisconsin Agricultural Experiment Station, Madison, Wisc. (*From:* University of Wisconsin Circular A 2539.)

[2]*In vitro* digestible dry matter.

[3]Analyses on fresh tissue.

[4]Total nonstructural carbohydrates.

The concentration of *in vitro* digestible dry matter (IVDDM), crude protein, amino acids, carotene, and minerals decrease, while fiber increases (Experiment A and C). Total nonstructural carbohydrates which are highly digestible, decrease to early bloom (Experiment B). Protein N (nitrogen) is the largest N component in alfalfa (Experiment D), making up over three-fourths of the total N. Nitrate N is highest at prebud, but is not found at toxic levels. Also, countless research studies have shown that animals consume much larger quantities of early-cut forage than of late-cut forage, resulting in greater yields of livestock products. The best time to harvest is when both yield and quality are high.

NDF, ADF, and NIRS Analyses

Since 1865, fibrous materials traditionally have been analyzed by the *Weende proximate analysis* method. Although this method is still widely used, it is often supplemented with additional analyses. In a series of reports beginning in 1963,[3] Peter J. Van Soest proposed the *detergent analysis system,* better to evaluate the feeding value of fibrous materials; by using detergents, he separated the sample into two fibrous fractions: (1) a neutral detergent fibrous fraction (NDF), and (2) an acid detergent fibrous fraction (ADF). Further, he reported that, in comparison with traditional proximate analysis, NDF provided a better estimate of dry matter intake (consumption) by animals, and ADF provided a better estimate of the *in vivo* (inside the animal) dry matter digestibility. Van Soest partitioned forages into the following two fibrous fractions:

1. **Neutral detergent fiber (NDF),** which is the cell wall material or plant structure in feed, is comprised of hemicellulose, cellulose, lignin, lignified N, and insoluble ash. This constituent is insoluble in neutral detergent and is only partially available to animals. The lower the NDF percentage, the more the animal will eat; it is inversely related to voluntary intake (consumption). Thus, a low percentage of NDF is desirable.

   The NDF content is also positively correlated with eating time and rumination, and this may be related to the rate of particle size reduction. Additionally, NDF is related to the proper function and health of the rumen; roughage value (the amount, source, and physical form of dietary NDF) is associated with chewing time, saliva flow (buffering and pH in the rumen), rumen fermentation patterns, milk fat test, and total energy output.

2. **Acid detergent fiber (ADF),** which is the highly indigestible plant material in a forage, is comprised of cellulose, lignin, and insoluble ash. This constituent is insoluble in acid detergent. ADF differs from crude fiber in that it contains silica. Silica and lignin in plants are associated with low *in vivo* (in animal) digestibility. The lower the ADF, the more feed an animal can digest. Thus, a low ADF percentage is desirable.

[3]Goldring, H. K. and P. J. Van Soest, *Forage Fiber Analyses,* Agriculture Handbook No. 379, ARS, USDA, 1970, p. 20 on which 12 papers are listed.

But chemical determinations are slow! The laboratory determination of crude protein, neutral detergent fiber, acid detergent fiber, *in vitro* (in a test tube or other artificial environment) dry matter digestibility, mineral, and vitamin analyses may take up to 2 weeks. This prompted the search for a more rapid, yet reliable, method of assessing the nutritive value of hay. In 1976, Norris, *et al.*, indicated (*Journal of Animal Science,* 47:747–759) that a relatively new procedure, known as *near infrared reflectance spectroscopy,* which had been applied successfully to grain quality evaluation, could be used for predicting forage quality, also. Additional research confirmed the initial findings of Norris and developed data bases and procedures for quickly analyzing hays.

The near infrared reflectance spectroscopy (NIRS) is a nonconsumptive instrumental method for fast, accurate, and precise evaluation of the chemical composition and associated feeding value attributes of forages and other feedstuffs. The instrument, known as a *near infrared analyzer,* produces infrared radiation over a given range of wavelengths and this radiation is focused onto the sample being tested. Because of the chemical structure of the sample material, certain combinations of infrared wavelengths are reflected and certain combinations are absorbed for each chemical characteristic tested, *e.g.,* energy values, crude protein, digestibility, minerals, NDF, and ADF. By using a system of filters and detectors, the instrument senses these reflected wavelengths and passes this information on to the computer. The computer sorts out the appropriate wavelength combinations and their relative magnitudes for each chemical characteristic and transforms these data into percentages.

The near infrared reflectance measures hay quality by comparing the energy reflected back from a hay sample with computerized standards established by conventional laboratory analysis of a large number of reference samples.

By using a portable near infrared analyzer in a mobile van, samples at hay markets can be evaluated in minutes, and hay can be bought and sold on the basis of known quality. Also, tests made at the farm by mobile units can provide accurate information needed to make decisions about balancing rations. Additionally, forage researchers are using NIRS to select experimental plant lines for improved nutrient quality.

The NIRS method of analysis has four main advantages: speed, simplicity of sample preparation, multiplicity of analysis with one operation, and nonconsumption of the sample (it can be analyzed again by the same or another procedure). With the NIRS method of analysis, it is possible to take a sample from a truckload of hay and provide, in less than 3 minutes, an analysis for crude protein, NDF, ADF, dry matter, lignin, and *in vitro* dry matter digestibility.

The chief disadvantages of the NIRS method are instrumentation requirements and costs, dependence on calibration procedures, complexity in the choice of data treatment, and lack of sensitivity for minor constituents.

On a voluntary basis, NIRS technology is being used currently and successfully in the NIRS Forage Research Project Network; and either NIRS or wet chemistry tests may be used in the laboratory certification program operated by the National Alfalfa Hay Test Association in conjunction with the American Forage and Grassland Council (AFGC) and the National Hay Association.

Correct Sampling Necessary

No forage test is any better than the sample taken. Stated differently, a chemical analysis is valid only to the extent that the sample analyzed represents the lot of hay under consideration. Thus, the most important single step in determining the chemical composition of hay is sampling. No matter how accurate the chemical analysis, a poor sampling technique can easily invalidate the results and lead to an erroneous conclusion.

It is difficult to obtain a representative, meaningful sample of forage because of its bulky nature and the variability within a given lot of hay as compared to most other crops. For the sample to be representative of a given lot of hay, it should have been taken from the same field (or at least it should have been produced under the same cultural conditions—same irrigation and fertility conditions, for example), same cutting, same stage of maturity, and all of it should have been baled within a 48-hour period using only one harvesting method. With conventional, rectangular bales, at least 20 bales should be sampled at random, by probing every third bale, for example. The probe, or core sampler, should be at least ⅜ in. in diameter. The center of either end of a rectangular bale may be probed by inserting the probe at a right angle to the face of the bale and to a depth of 12 to 18 in. The hay testing laboratory should be consulted relative to sampling large round bales and stacks.

Hay samples should be placed in a closed plastic bag or freezer carton; otherwise, the moisture content will not be meaningful.

For further instructions on correct sampling, and to determine if the state has a laboratory for analyzing forage samples, the hay grower or buyer should see the local county extension agent or write to the state college of agriculture.

What Tests to Make

In modern haymaking, marketing, and feeding, proximate feed analysis no longer suffices. In order to predict the level of animal performance (milk, rate of gain) that may be obtained from hay, the following chemical components are commonly determined: crude protein (CP), neutral detergent fiber (NDF), acid detergent fiber (ADF), *in vitro* dry matter digestibility (IVDMD), crude fiber (CF), lignin, moisture, calcium, and phosphorus.

Moisture and crude protein are the most commonly analyzed components. NDF measures cell wall portion and is used to predict intake (consumption); and CF, ADF, and IVDMD give good estimates of the digestibility of the forage.

Other analyses which are often useful in evaluating hay are carotene and certain trace minerals. There are times when the amount of vitamins and amino acids in the feed might be useful, particularly when hay is used as a supplement in nonruminant rations. These analyses are costly to run, however, and it may not be economically feasible to run very many of them.

Evaluating Test Results

Test results can best be evaluated by comparing them with some standard. The testing laboratory may provide such information, possibly along with recommendations for applying the test results in balancing rations. For convenience, average crude protein and crude fiber values of some common hay crops are given in Table 8–11.

TABLE 8–11
APPROXIMATE CHEMICAL COMPOSITION (MOISTURE-FREE) OF VARIOUS SUN-CURED HAYS

Kind of Hay	Crude Protein		Crude Fiber	
	Average	Range	Average	Range
	(%)	(%)	(%)	(%)
Alfalfa	16.0	12.0–24.0	28.0	22.0–39.0
Bermudagrass	10.0	7.0–15.0	33.0	28.0–37.0
Bromegrass	10.5	6.0–15.0	28.0	24.0–31.0
Ladino clover	18.5	16.0–21.5	22.0	18.5–23.0
Red clover	12.0	10.5–18.5	27.0	18.0–34.0
Lespedeza	13.0	11.5–14.5	27.0	22.5–32.5
Oat hay	5.0	4.0– 6.0	28.0	26.0–32.0
Orchardgrass	8.1	6.0–14.0	30.0	26.0–31.0
Soybean	14.5	9.0–16.5	28.0	20.5–41.0
Timothy	6.5	5.5– 9.5	30.0	28.0–31.5
Sudangrass	8.8	6.5–11.0	28.0	26.0–30.5

If a chemically analyzed sample runs higher in protein and lower in fiber than the average figures given in Table 8–11, it means that the sample is better than average quality hay; conversely, if it is lower in protein and higher in fiber, the sample tested is below average quality. Of course, producers should not settle for average protein or fiber content, for, on the whole, the vast majority of the U.S. hay crop is of low quality. For this reason, one should strive for the upper figures of the range given in Table 8–11. Certainly, poor-quality hay can be fed, and, under certain circumstances, it may even be economical and quite satisfactory—for example, for dry cows. However, when buying poor hay, the purchase price should be lowered accordingly; and the feed analysis should also be used as a basis for balancing the ration. By the same token, it is usually good business to pay a premium for high-quality hay. Some producers are very wisely applying an escalator principle to hay purchases. They may pay $1.00 to $1.50 per ton for each 1% of protein above an agreed-upon figure; or they may dock the price by a corresponding amount if the content is lower. For example, if a vendor guarantees to deliver alfalfa with 15% crude protein and it is agreed that a $1.50 per ton premium will be paid for each 1% protein in excess of this figure, a $4.50 per ton premium would be added for alfalfa running 18% crude protein.

In some cases, hay is also purchased on the basis of moisture content; and the price drops as the moisture content increases—thereby discouraging selling water. For example, if 15% moisture is agreed upon, hay running 18% moisture would be docked 3% in price. Thus, if the base price of hay is $75.00 per ton, the price would be lowered to $72.75 per ton ($75.00 × 3% = $2.25; then, $75.00 − $2.25 = $72.75).

Others use crude fiber in the same manner—paying a premium for hay of low fiber content. Still others apply more complicated formulas to arrive at the dollar value of hay.

Thus, a chemical test provides informed appraisal of hay values. Except for actual feeding trials, it is the best method of evaluating hay quality presently available. With this information at hand, livestock producers can do a better job of feeding animals and realizing higher profits. Additionally, an analysis and payment on an incentive quality basis should result in the following benefits: (1) Compensate the grower for extra effort in keeping hay quality high; (2) induce more hay growers to produce high-quality hay, rather than just tonnage; and (3) provide livestock producers with better-quality feed, which, in turn, will result in greater feed efficiency.

Supply and demand will still regulate price. But, in the end, hay will be upgraded; it will be cut at an earlier stage of maturity and put up with greater care.

MAKING QUALITY HAY

The object of haymaking is to (1) harvest the crop at the optimum stage of maturity which will provide the maximum yield of nutrients per acre without damage to the next crop; and (2) cure properly, which involves lowering the water content of the green herbage from 65 to 85% moisture to 20% or less.

Hay quality begins with the soil and ends with the manger, with many intermediate factors affecting it along the line. Once forage is cut, opportunities to increase nutrient content are over; from that point on, quality can only be preserved.

There is no one best haymaking method or kind of equipment. These must necessarily vary with the size of the operation, the kind of hay, the climate of the area, the individual farm or ranch conditions and buildings, and the available labor and machinery and their cost. Yet, the principles of good haymaking and the objectives sought are the same everywhere.

About 80% of all harvested forage is now baled. Only 10% is stored as loose hay and cubes, and the remaining 10% is stored as hay crop silage.

Growing Forage

Growing forage for hay has long been neglected. Average yields per acre are still well under one-half their potential. Little more than 1 acre of hay in 10 is fertilized on a regular basis; and the precious few acres that are fertilized get an average of only 12 to 15 lb per acre—a paltry amount compared to corn, which receives an average of about 200 lb of fertilizer per acre.

The steps in growing quality hay are:

1. **Match crop to soil.** Some forage crops will do better on certain soils than others. So, the crop should be fitted to the soil.

2. **Choose quality seed, and proven varieties and mixtures.** Most grass and legume seeds are extremely small and contain very little stored food material. Thus, it is important that good seed be used.

Also, recommended varieties should be selected. If there is a choice, consider simple grass-legume mixtures. Research has shown that in many areas and over a period of years such mixtures are frequently higher yielding and more persistent than legumes grown alone. Mixtures are also easier to harvest and cure as hay.

3. **Lime and fertilize.** To obtain top yields of hay, lime the soil (if needed) according to soil test, then use the right kinds and amounts of fertilizer for the species and soil. Legumes, such as alfalfa, need a nearly neutral soil to promote growth of nodule-forming bacteria. Fertilization should be based on soil test.

4. **Get good stand.** A good stand is important. In this connection, it is noteworthy that it takes about one-half million successfully established individual forage plants per acre for a productive stand.

5. **Irrigate where practical.** In arid and semiarid regions, irrigation is a must for hay production. Whether it will pay to irrigate hay crops in humid and semihumid areas will depend on the water supply, irrigation costs, and returns. Census figures show that more than 14 million acres of U.S. hay and pasturelands are irrigated—that's about 1 in 3 of all irrigated acres.

6. **Control insects and diseases.** Forage insects destroy a lot of hay. But they can be controlled by using a proper combination of insecticides, cultural control practices (such as timely cutting), and biological control agents.

Haying Equipment

Mechanization of hay harvesting started with the use of the barn hayfork in 1864, followed by the hay-loader in 1874 and the side-delivery rake in 1893. But the greatest milestone in haymaking was the automatic pick-up baler in 1940.

In addition to saving labor, modern harvesting equipment improves the quality of hay. For example, conditioned hay dries in ⅓ to ½ the time required for nonconditioned hay. Artificial curing of hay, using either natural or heated air, can reduce the field exposure time of hay an additional 20 to 50%.

The largest remaining problems of hay handling are (1) its removal from storage, and (2) its mechanical feeding to animals.

A brief rundown on today's haying tools follows:

1. **Mowers.** Simplicity, high speeds, and greater widths continue to make the conventional cutterbar the most common mechanism used in forage harvesting.

2. **Mower-conditioners.** These combine cutting and conditioning in one operation.

Fig. 8–8. Mower-conditioner. (Courtesy, Deere & Company, Moline, Ill.)

3. **Disc mowers/disc mower conditioners.** These mowers were developed for tough cutting conditions that often plugged sicklebar cutters. The addition of intermeshing rubber rolls to the disc mower permits conditioning the crop when cutting.

4. **Forage mat machine.** This newly developed machine mows alfalfa, shreds it, presses the shredded leaves and stems into a ¼-in.-thick mat, and spreads it on the stubble to dry—all in one operation. The forage mat machine squeezes out a lot of moisture followed by the forage drying quickly and uniformly.

5. **Windrowers.** Multipurpose, self-propelled windrowers provide up to 16 ft of cutting capacity. Most models feature hay conditioners as standard or optional equipment.

6. **Rakes.** Side-delivery rakes are still widely used to make hay crop windrows. Parallel bar rakes, which are most popular today, reduce impact between rake teeth and hay by moving hay from the outside of the swath to the windrow in less than 13 ft of travel.

7. **Tedders, rake-tedders.** In humid regions, making hay can be a problem because of the constant threat of showers and moisture from the morning dew. To help speed drying time, tedders have been developed to ted or fluff the hay crop either from a swath or windrow. Rake-tedders do a tedding operation in addition to raking.

8. **Balers.** Automatic pick-up balers package most hay today.

Balers that package small, rectangular bales, weighing 60 to 140 lb, continue to be common on farms and ranches that produce hay for their own use.

Balers that make large square bales are designed for custom operators or hay growers who have large volumes of hay or straw. Bale size is 3′ × 4′ × 8′ long, and bale weight ranges from 1,000 to 1,500 lb. Large square bales are suitable packages for long distance transportation.

Balers that produce large round bales, weighing from 850 to 2,000 lb, are increasing in popularity, especially with cow-calf operators, and even with some dairy producers.

Fig. 8–9. Large round bales averaging 1,200 lb weight. (Courtesy, McArthur Farms, Inc., Okeechobee, Fla.)

9. **Bale handlers.** Mechanization took the backache out of bale handling. Today, the following types of sophisticated hay harvesting and handling equipment are available: bale throwers that throw bales into trailer wagons and eliminate lifting; bale conveyors for putting bales where desired;

automatic bale wagons, which pick up bales from the field, load, transport, and stack them; and stack retrievers for transporting bales from storage to where they will be fed.

10. **Stack machines.** These are hydraulically operated field machines that compress long hay into stacks, weighing from 1 to 6 tons.

Harvesting at Proper Stage

Whether the crop is a grass or a legume, or a combination, the stage of maturity of the plants at the time of harvest affects digestibility, yield, and feeding value (see Fig. 8–10). Young, immature plants are high in protein and low in fiber or lignin. As hay crops mature, feeding value goes down and fiber content increases. Digestibility of the forage (TDN) declines about 0.5% each day cutting is delayed beyond the early bloom stage (Fig. 8–10); and the intake of forage decreases during this same period at more than 0.5% each day. Thus, in total, the feeding value of forage drops more than 1% for each day's delay after early bloom.

Forage dry weight yields increase until midbloom to late bloom stages (Fig. 8–10). Timothy and bromegrass fully headed, and red clover and alfalfa at full bloom, will give maximum yield of dry matter. However, maximum feeding value of first cutting forage is reached at least 10 days before the time of maximum dry weight yield (Fig. 8–10).

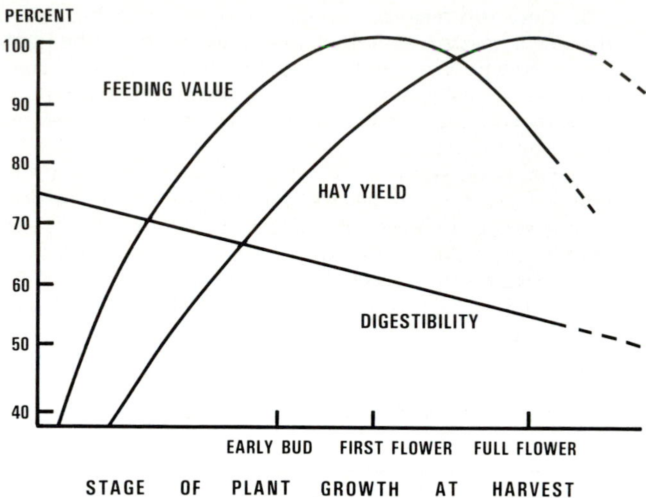

Fig. 8–10. Effect of advancing maturity on the feeding value, yield, and digestibility of alfalfa hay. Note that maximum feeding value per acre is reached 10 to 15 days before maximum yields.

Table 8–12 shows the primary changes in chemical constituents of alfalfa (a legume) and timothy (a grass) as affected by maturation. Note that, for both species, the DE, TDN, CP, lysine, Ca, and P decrease with maturity.

TABLE 8–12
EFFECT OF STAGE OF MATURITY AT HARVEST OF ALFALFA AND TIMOTHY ON COMPOSITION OF HAY[1]

Forage Crop	Stage of Maturity	DE[2]	TDN[3]	CP[4]	Lysine	Ca	P
		(Mcal/kg)	(%)	(%)	(%)	(%)	(%)
Alfalfa	Prebloom	2.78	63	19.4	1.10	2.10	0.34
	First flower[5]	2.42	56	17.9	0.94	1.75	0.28
	Midbloom	2.29	52	16.0	0.90	1.50	0.25
	Full bloom	2.16	49	15.0	0.64	1.29	0.25
Timothy	Prehead	2.20	50	11.5	—	0.50	0.25
	Head	1.98	45	9.0	—	0.41	0.19

[1]Adapted by the authors from: Rohweder, D. A. and R. Antoniewicz, *Alfalfa, the high quality hay for horses,* produced by Certified Alfalfa Seed Council, Inc., Woodland, Calif., p. 6, Table 1.

[2]Digestible energy. Divide by 2.2 to obtain Mcal/lb.

[3]Total digestible nutrients. Improved harvest techniques can increase values 10%.

[4]Crude protein.

[5]First flower to 1/10 bloom.

As shown in Fig. 8–10 and Table 8–12, advancing stage of maturity at harvest affects the chemical composition of hay adversely. More importantly, it has a direct and dramatic effect on animal production, as reported in Table 8–13. When alfalfa hay that was harvested at four different stages of maturity was fed to yearling steers, the animals that were fed the very mature alfalfa (seed stage) made less than half the average daily gains and total gains of the animals that were fed early harvested alfalfa hay (bud stage). Moreover, they required more than twice as much hay per 100 lb gain.

TABLE 8–13
EFFECT OF STAGE OF MATURITY OF ALFALFA HAY AT HARVEST ON GAINS OF YEARLING STEERS[1]

Stage of Maturity	Average Daily Gains	Total Gains Per Steer	Feed/ 100 lb Gain
	(lb)	(lb)	(lb)
Bud	1.07	96	959
1/10 bloom	0.76	69	1,351
Full bloom	0.63	58	1,600
Seed	0.48	44	2,144

[1]Rohweder, D. A., *Maintaining Forage Stands for Efficient Production,* University of Wisconsin, Madison, A 2907, 1978, p. 9, Table 12; adapted from a study conducted by Kansas State University, with steers that had an initial weight of 440 lb.

In Wisconsin, first-growth alfalfa-bromegrass hay was harvested each year for 3 years at four stages of maturity. Dry matter digestibility, digestible energy, milk production, and animal gains all declined sharply with increasing maturity of the hay (Table 8–14).

TABLE 8–14
EFFECT OF STAGE OF MATURITY AT HARVEST OF ALFALFA-BROMEGRASS ON MILK AND MEAT PRODUCTION[1]

Stage of Maturity	Dry Matter Digestibility[2]	Digestible Energy[3]	4% Milk Production[4]	Lamb Gains[5]
	(%)	(kcal/g)	(lb/day)	(lb/day)
Vegetative	71.4	3.20	45	0.38
First flower	64.6	2.86	29	0.21
Full bloom	58.0	2.54	15	0.15
Green seed pod	55.2	2.43	4	0.05

[1]Adapted by the authors from University of Wisconsin data. The alfalfa-brome was first cutting, made at Arlington, Wisc.; and the data are an average of 3 years, except for lamb gains which are one year only.

[2]Animal digestion trial data, values similar to total digestible nutrients.

[3]Animal energy digestibility × forage gross energy.

[4]Estimated for a 1,200-lb cow fed hay alone.

[5]Lambs fed hay alone.

Stage of maturity also affects the vitamin content of hay. Carotene (precursor of vitamin A) and the B vitamins decrease as plants mature. Vitamin D content is the one exception—it increases as the forage is sun-cured.

Everything considered, there is a loss of about 1% in nutrient value for each day that hay harvest is delayed beyond the late vegetative stage of growth.

Table 8–15 gives guidelines relative to the proper forage-harvesting stage for maximum protein and minimum fiber.

TABLE 8–15
HAY CUTTING GUIDE

Kind of Hay	When to Cut
Alfalfa	Bud stage for first cutting; 1/10 bloom for second and later cuttings.
Alsike clover	Early bloom to ½ bloom stage.
Bermuda	When 16–18 in. tall, before lodging.
Birdsfoot trefoil	First flower to full bloom.
Bromegrass	Heads emerging.
Cowpeas	When pods are ½ to fully matured.
Crested wheatgrass	When the plants begin to head.
Crimson clover	From early bloom to ½ bloom.
Fescue	Boot to early head stage.
Grass-legume mixtures	When the legume is at the proper stage.
Johnsongrass, millet, Sudangrass, sorghum hybrids	40 in. height or early boot stage, whichever comes first.
Ladino clover	Few blooms to full bloom.
Lespedeza, annual	Early blossom.
Orchardgrass	Boot to early head stage.
Red clover	Late bud.
Sericea	When 12–15 in. high.
Small grains (oats, barley, wheat)	Boot stage to early dough stage.
Soybeans	Mid-to-full bloom and before bottom leaves begin to fall.
Sweet clover	Bud to very early flowering stage.
Timothy	Boot to early head stage.

RETAINING ADEQUATE PLANT FOOD RESERVES

Cutting or grazing plants when carbohydrate reserves are low may leave too little energy available to support new growth; *e.g.*, continued cutting at immature stages of growth will eventually exhaust the plant and weaken it to the point of death. Plants weakened by too early, too close, or too frequent cutting or grazing are more susceptible to winter injury, drought, and disease. Usually, the closer to maturity that plants are cut or grazed, the higher the stored food reserves will be and the easier it is to maintain vigor for productivity. However, as emphasized in the preceding section, harvesting (by cutting or grazing) forage at a young growth stage makes for high feed value. So, a compromise must be reached between cutting or grazing forages for maximum quality of the forage and maximum vigor of the plants.

• **Plant food reserves and storage**—Perennial forages store energy as total nonstructural carbohydrates (also called carbohydrate reserves or total available carbohydrates). These food reserves are used by plants mainly to develop cold hardiness, to live during winter, and to initiate growth after each cutting or grazing.

The place of storage of the reserve food varies with the plant. In alfalfa and birdsfoot trefoil, it is stored in the roots. In orchardgrass and tall fescue, it is stored in the leaf sheaths and stem bases; in timothy, it's in the swelled basal stem internodes called haplodorms; and in reed canarygrass and smooth bromegrass, it's in the underground creeping stems called rhizomes.

• **Reserves supply food for first spring growth**—To illustrate the importance of management as related to plant food storage, the growth pattern of alfalfa will be detailed in this section.

When alfalfa growth starts in the spring, food reserves in the roots and crowns are used to start new top growth from small crown buds or underground stems. Depletion of the food reserves continues until the plant is 6 to 8 in. tall. By this time, food is being synthesized in the leaves more rapidly than it is being used, and some food storage begins. Storage continues and reaches its highest level in the roots when alfalfa is about full bloom. The changes that occur in root energy reserves and dry matter yield in alfalfa during one growth period are illustrated in Fig. 8–11.

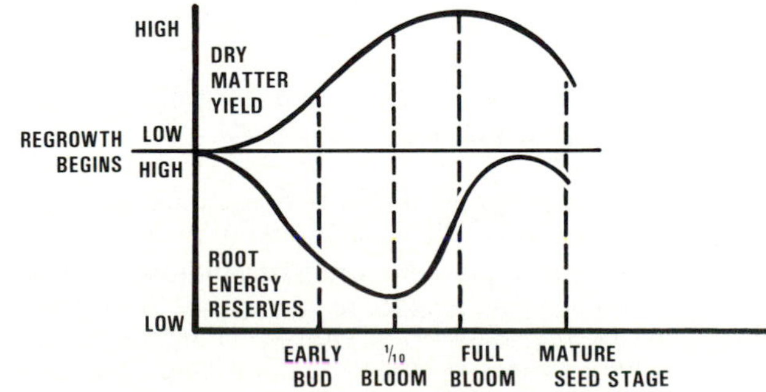

Fig. 8–11. Changes in root energy reserves and dry matter yield of alfalfa during one growth period of the crop. Note the changes by stages: early bud; 1/10 bloom; full bloom; mature seed stage. Note, too, that root reserves decrease with dry matter increases, with the use of root reserves preceding dry matter yields. (Adapted by the authors from *Haymaker's Handbook*, by J. E. Baylor, Professor of Agronomy, The Pennsylvania State University, and M. A. Balas, New Holland, Inc., published by Ford New Holland, Inc., New Holland, Pa., 1987, p. 121.)

• **Regrowth**—After the first spring cutting of alfalfa is harvested, preferably when the flower buds have formed, the process of food reserve depletion and renewal is repeated for the new growth. Although root reserves are not at the highest level at the bud stage, plants can maintain a satisfactory level of reserves if harvest of later cuttings is delayed until early bloom and careful management is practiced in the fall. Also, the practice of cutting at early growth stages should be rotated among alfalfa fields.

• **Fall management important**—The period before the first frost is usually critical to forage survival—especially with alfalfa, because at this time a large supply of food reserves is a "must" to assure winter survival and vigorous spring growth.

Harvesting shortly before a killing frost weakens plants and causes winter injury. The reason: Alfalfa cut in late September or early October in the northern states may produce 8 to 10 in. of top growth that is killed by frost before restorage of reserve has occurred. When this happens, plants may go into winter in a weakened condition. So, under average conditions, the 4 to 6 weeks before the first killing frost is considered the *critical period.* However, some scientists now feel that the rule of not cutting during the critical period is more flexible than many people believe, particularly when winter dormant and multiple pest resistant varieties are used and a high level of fertility is maintained. On the other hand, some Canadian scientists report reduced stands and yields following a cutting during the critical period.

So, the question: What is the best system to follow relative to fall cutting or grazing? *The answer:* To play it safe, either (1) harvest early enough to give the plants sufficient time to build up reserves before frost, or (2) delay harvest until near or after a killing frost so that there is little or no regrowth after cutting. *CAUTION:* If a late harvest is made in a cold area, leave at least 4 in. of stubble to help plants trap snow which acts as an insulator and shields plants from cold weather, and lessens heaving.

Cutting and Field Curing Hay

Proper cutting and field curing of hay embraces all the steps from cutting to ready-for-packaging or storing. In modern hay making, it includes (1) cutting, curing in the swath and windrow, and raking or cocking; (2) reducing moisture content, while minimizing shattering, bleaching and fermenting; (3) reducing rain damage; and (4) considering chemical and preserving agents.

CUTTING/CURING IN THE SWATH AND WINDROW/RAKING OR COCKING

The common steps, methods, and equipment used in cutting and field curing hay are as follows:

1. **Cutting and curing in the swath or windrow.** Cutting, followed by curing in the swath or windrow, is the first step in haymaking, regardless of the subsequent method or type of equipment employed.

Any one of several types of mowers may be used, for all of them are designed to get the hay down. The most important thing is that the hay be cut at the proper stage of maturity (as noted in Table 8-15).

The following points are also pertinent to cutting hay and curing it in the swath or windrow:

a. **Direction of mowing.** It is highly desirable to mow in the same direction as will be traveled in raking and in picking up the hay in subsequent operations.

b. **Time of the day to mow.** Some opinions to the contrary, quality of hay is only slightly affected by either the time of day at which the forage is cut (the proportion of sugar does vary with the time of day), or the presence of dew.

c. **Mechanical hay conditioning.** Mechanical conditioning reduces drying time and field losses each by about 50%. These machines are designed so that slow-drying stems are split, cracked, crushed, or broken as they pass through rollers or knives. As a result, the stems dry at about the same rate as the leaves; and there is more uniform curing and less leaf loss. The three types of hay conditioning machinery are:

(1) **The crusher.** It has two flat counter-rotating rolls under spring tension which actually crush stems as the hay passes through.

(2) **The crimper.** It has two corrugated rollers which crack the stems every 2 or 3 in. as the hay passes through.

(3) **The flail harvester (or chopper).** It has loose pointed knives on a horizontal cylinder. These knives chop the hay in about 6-in. lengths and also crack the stems vertically.

All new units both mow and condition the hay with the same power transmission system, thereby necessitating that the field be gone over only once.

The flail harvester cuts the hay and conditions it in one operation, or it will also condition hay that is in the windrow. It operates somewhat like a silage cutter with a blower tube and downspout which places the hay back on the ground in a windrow.

Although hay conditioners have considerable merit, farmers and ranchers with smaller hay acreages may not be able to justify the added cost of the equipment.

d. **Length of swath curing.** Hay dries more rapidly in the swath than in the windrow, even if the windrow is small and fluffy. Therefore, it should be left to cure in the swath as long as is possible without damaging it; until the forage is wilted but before there is danger of the leaves shattering and/or of excessive bleaching and loss of carotene. At this stage, the moisture content will be about 40%. The point at which swath curing is sufficient should be carefully determined, because, without a conditioner, the leaves become dry and brittle long before the stems are cured, especially on legumes.

No definite period of time for swath curing can be assigned as it will vary according to the tonnage of hay per acre, temperature, sunshine, wind, and atmospheric humidity. Instead, the time should be determined entirely by the condition of the hay in the swath.

Sometimes curing in the swath is speeded up by using swath fluffers; machines which pick up the forage, lift it to a standing position, and then release it to fall loosely onto the stubble—thereby fluffing the hay.

Where it seldom rains, sometimes hay is not cured in the swath at all. Under such conditions, it may be cut and windrowed immediately. For the latter purpose, swathers—self-propelled mowers that cut and windrow hay in one simultaneous operation—may be used.

2. **Raking.** After the hay has wilted sufficiently in the swath, but while it is still tough and the leaves will not shatter, it should be windrowed. For this assignment, the side-delivery rake is preferred to the dump rake. The side-delivery rake rolls hay into fluffy, cylindrical windrows, which allows for good circulation of air; whereas dump rakes produce large windrows which are apt to remain damp underneath and bleach excessively on top. Where the hay crop is exceedingly heavy, the size windrow can be kept desirably small by limiting the width raked into each windrow.

If considerable shattering appears probable, it may be desirable to do the raking early in the morning when the dew makes the hay a bit tough.

Where windrowed hay is rained on, wait until the top half dries out, and then turn it upside down with the side-delivery rake (the use of the tedder for rewindrowing is not recommended because of excessive shattering).

3. **Cocking.** Formerly, well-made cocks, often adorned by hay caps, were considered a necessary part of good haymaking. However, this practice has greatly decreased, due primarily to higher labor costs and the advent of modern haymaking machinery. Today, the cocking of hay is confined almost entirely to use (a) in hot, arid regions where the leaves shatter if the hay is left in the swath or windrow for any appreciable length of time, and (b) as an emergency measure in order to protect hay when a storm is imminent.

REDUCING MOISTURE CONTENT/SHATTERING/ BLEACHING AND FERMENTING

Proper curing ensures that (1) the hay can be stored safely without heating excessively or becoming moldy; and (2) the maximum leafiness, green color, aroma, nutrient value, and palatability shall be retained. To the end that these desired objectives may be achieved, the following information is pertinent:

1. **Moisture content.** Freshly cut forage contains 75 to 80% moisture, whereas the maximum moisture content for safe hay storage is as follows:

For loose hay—25% moisture.

For baled hay—20 to 22% moisture (the lower figure for larger bales).

For chopped hay—18 to 20% moisture.

For cubes—16 to 17% moisture.

Hay of a higher moisture content than indicated should not be stored because (a) its value may be greatly lowered due to mold or to nutrient losses accompanying fermentations, and (b) of the ever-present danger of spontaneous combustion and a costly fire.

Two rule-of-thumb methods used by farmers in determining when hay is dry enough for storage are:

a. **The twist method.** Twist a wisp of the hay in the hand. If the stems are slightly brittle and there is no evidence of moisture on the twisted stems, the hay can be stored safely.

b. **The scrape method.** Scrape the outside of the stems with the thumbnail or a knife. If the epidermis can be peeled from the stem, the hay is not sufficiently cured. If the epidermis does not peel off, the hay is usually dry enough to stack or put in the mow.

2. **Shattering losses.** Legume forages contain a larger proportion of leaves than do grasses, but unfortunately, the fine, thin legume leaves dry out more rapidly than the coarse stems to which they are attached. This results in considerable shattering losses, unless great care is taken. The importance of this condition is readily apparent when it is realized that in alfalfa, for example, 50% of the total weight of the plant is contained in the leaves, but the leaves contain 70% of the protein and 90% of the carotene content of the entire plant.

In field curing hay, losses from leaf shattering range from 2 to 5% for grass hay and 3 to 39% for legume hays, with as much as 15 to 20% for legume hays field cured under the most favorable conditions. Based on extensive experiments with field-cured alfalfa hay, the U.S. Department of Agriculture reported that leaf losses averaged 38.5% when none of the hay was wet; 47.3% when the hay was wet by 2 showers; and 74.5% when the hay was wet by 3 showers—and milk production per acre was 19.7% less per acre when cows were fed rain damaged, field-cured alfalfa hay in comparison with field-cured hay without damage by rain.

3. **Bleaching and fermenting losses.** In general, the carotene or provitamin A content of freshly cured hay is proportional to the greenness. With severe bleaching, more than 90% of the vitamin A potency may be destroyed.

Even under the best of conditions, there is an unavoidable loss through fermentation, especially losses in sugars, starch, and carotene. With good weather and proper curing methods, however, these losses will not be excessive.

REDUCING RAIN DAMAGE

The leaching losses from rain are less severe soon after mowing, but increase in severity as curing progresses. Also, repeated showers are more damaging than one heavy rain. Experimental studies have revealed that damaging rains may lower the feeding value of hay by one-fourth to one-third, or even more with severe exposure.

The effects of rain on hay quality depend on the amount of rainfall and drying conditions (see Table 8–16, next page). In a Wisconsin study, alfalfa and red clover were subjected to various rain treatments during field hay drying. The effects of rain damage on CP concentrations were small. The concentration of CP in both species decreased when harvest was delayed to full-late bloom compared with harvesting at an earlier state; based on these data, the reduction in CP due to a delay in harvest was larger than any change due to rain damage. The effects of rain on IVDMD were greater than the effects on CP, and differed with drying conditions and amount of rain. The loss in IVDMD due to rain under poor drying conditions was from 73 to 57% for the alfalfa harvested at early maturity and from 62 to 39% for the alfalfa harvested at late maturity.

Losses from weather damage may be reduced (1) by using haymaking equipment that reduces the field drying time, (2) by understanding and using existing weather aides, and (3) by using proven chemical conditioning and preserving agents.

TABLE 8-16
EFFECT OF RAIN DAMAGE ON ALFALFA AND RED CLOVER HAY[1]

Legume/ Harvesting Stage	Good Drying Conditions		Poor Drying Conditions	
	No Rain	Rain	No Rain	Rain
	% Crude Protein			
Alfalfa:				
Early to late bud . . .	19.9	20.3	26.3	24.6
Late bloom	14.9	14.9	18.1	13.9
Red clover:				
Late bud to first flower	22.4	22.1	23.8	26.5
Full to late bloom . .	15.0	14.9	15.4	14.9
	% Digestibility (% IVDMD)[2]			
Alfalfa:				
Early to late bud . . .	67	65	73	57
Late bloom	60	59	62	39
Red clover:				
Late bud to first flower	75	72	68	47
Full to late bloom . .	67	63	62	49
	% NDF Concentration			
Alfalfa:				
Early to late bud . . .	41	42	32	45
Late bloom	49	49	42	64
Red clover:				
Late bud to first flower	31	37	30	44
Full to late bloom . .	42	45	40	55

[1]Collins, M., *Proceedings Wisconsin Forage Council, Eighth Forage Production and Use Symposium*, "The Effects of Rain Damage on Yield and Quality of Field Cured Hay," 1984, p. 98, Tables 3 and 4.

[2]*In vitro* dry matter disappearance.

CONSIDERING CHEMICAL CONDITIONING AND PRESERVING AGENTS

Chemical hay drying agents and preservatives are giving haymakers a big assist in lessening hay making losses and improving hay quality. The big advantage of these products is that they speed up the haymaking process and reduce exposure to weather damage. In comparison with no treatment, the use of a desiccant, or drying agent, along with mechanical conditioning, can reduce the moisture content by an additional 2 to 10% during a 24-hour period. Adding a preservative to hay that is in the 25 to 35% moisture range will allow it to be baled and stored without undue heating.

• **Chemical conditioners**—Chemical drying agents, which are sprayed on the crop at mowing time, break down the waxy cutin layer on the wall of the stem and allow moisture to escape, thereby promoting faster drying time, with the drying rate of the stems approaching that of the leaves. Several chemicals can be, and are, used for conditioning, including potassium carbonate, sodium carbonate, and sodium silicate. Also, methyl esters of fats, vegetable oils, or animal fats have been mixed with potassium carbonate in an attempt to increase the effectiveness of chemical conditioning.

Chemical conditioners are effective on legumes such as alfalfa, birdsfoot trefoil, and red clover, but, generally, they are not effective on grasses. Although they will reduce drying time on all cuttings of legumes, they are most effective on second and third cuttings and least effective on first and late autumn cuttings. This situation is attributed to the fact that conditioners work best when drying conditions are best (in the summer), and that first cutting has heavier yields and heavier swaths than later cuttings—conditions that hamper drying, because the moisture movement inside the swath is inhibited.

Studies show that drying agents are more effective as an addition to, and not as a substitute for, mechanical conditioners. The chemical of choice is applied at the time of cutting, by either of two techniques: (1) a spray boom mounted ahead of the reel; or (2) spray nozzles are mounted behind the reel, but in front of the conditioning rollers so that the rollers help distribute the spray.

The recommended application rates range from 5.7 to 8.5 lb of the chemical powder per ton of hay, mixed and applied with 15 to 30 gal of water per acre in order to assure good coverage; at a cost of $5.00 to $10.00 per ton for the chemical in the late 1980s.

Additional costs are involved for labor and equipment. This prompts the question: Does chemical conditioning pay? The answer depends on the area and season. Michigan studies indicate that it pays for second and third cuttings, but not for first and last.

CAUTION: Since the crop is standing when desiccants are applied and won't be cut until a few milliseconds later, the EPA requires that some desiccants must be labeled as *pesticides.* So, before using a chemical conditioner, the haymaker should check with local regulatory authorities for their interpretation of this point.

• **Preserving agents**—Under normal conditions, for safe baling a moisture content of 20% or less is a must. Studies in many states have shown that, if properly treated with an adequate amount of the right preservative, alfalfa hay can be baled at 25 to 30% moisture, thereby speeding harvesting and lessening losses significantly. Preservatives act as fungicides and inhibit the growth and reproduction of the microorganisms that cause heating and molding in wet hay. A brief description of each of the common preservatives follows:

1. **Organic acids.** Propionic acid is the organic acid of choice. It is sometimes mixed with acetic acid, inorganic acids, formaldehyde, water, flavoring ingredients, and/or antioxidants. But, to be most effective, organic acid formulations should have at least 60% propionic acid.

To be effective, the organic acid(s) must be applied at the proper rate, depending on the moisture content of the hay; and it must be uniformly distributed throughout the hay mass. Table 8-17 gives the recommended rates of actual propionic acid to be applied at baling for different moisture levels.

TABLE 8-17
RECOMMENDED RATE OF APPLICATION OF PROPIONIC ACID TO HAY[1]

High Moisture Level	Rate, Dry Weight Basis	Actual Acid/Ton Hay
(%)	(%)	(lb)
20–25	0.5	10
26–30	1.0	20
31–35	1.5	30

[1]Always follow the manufacturer's directions.

To assure proper coverage, an applicator with a proper displacement pump and appropriately placed nozzles is needed. To date, preservatives have been more effective on conventional, rectangular bales than on large round bales. The application of perservatives to large round bales is difficult because of the way hay is rolled in a round bale rather than compacted in a conventional bale chamber.

Using 1% propionic acid as a preservative, Michigan State University studies showed (1) that it was very effective in lowering dry matter losses in each of four stages: field, baling, storage, and feeding; and (2) that it was effective with rectangular bales, round bales, and stacks.[4] (See Table 8–18.)

TABLE 8–18
HOW HAY-HANDLING METHODS COMPARE[1]

	Rectangular Bales		Round Bales		Stacks	
	1% Acid	No Acid	1% Acid	No Acid	1% Acid	No Acid
Harvest moisture	27	17	25	16	27	15
Type of dry matter loss						
Field	7	13	7	13	7	13
Baling	3	4	9	10	7	13
Storage	5	8	10	12	11	17
Feeding	0	5	6	14	6	16
Total	15	30	32	49	31	60

[1]100% propionic acid was used.

Note, however, that the Michigan studies revealed that greater storage and feeding losses occurred in round bales and stacks than in rectangular bales.

Organic acid preservatives are *acids;* hence, they can cause skin irritation and eye damage. Thus, rubber gloves and protective goggles must be used when handling these products. Some corrosion of equipment can be expected. Another disadvantage is that treated hay is usually somewhat bleached, thereby making it less attractive in the market place.

2. **Anhydrous ammonia.** When properly applied, anhydrous ammonia will stop bacteria and mold growth; and when applied to poor quality hay, it has the added advantages of increasing protein and digestibility. However, as a preservative it's not as effective as propionic acid. Also, unless large round bales are covered and/or contain less than 28% moisture, too much of the ammonia escapes. Excessive amounts of ammonia may cause animal disorders. (See Chapter 5, Nutritional Disorders/Toxins, section on "Nutrition Related Disorders.")

Where high-quality alfalfa hay is involved, which is already high in protein and digestibility, it's doubtful that the added expense of using ammonia can be justified.

3. **Bacterial inoculants.** Claims are made that most bacterial inoculants on the market will produce lactic acid, which acts as a fungicide and inhibits mold growth. More experimental work is needed, substantiating the effectiveness of bacterial inoculants as hay preservatives.

Value of Different Cuttings of Alfalfa Hay

In many areas, there is a decided preference among livestock producers in favor of a certain cutting of alfalfa hay.

Generally, first cutting alfalfa hay is coarser stemmed and less leafy than later cuttings, and therefore of somewhat lower feeding value when the different cuttings are equally well cured. Also, the weather is often less favorable for curing the first cutting. Yet, dairy producers participating in the Wisconsin Forage Analysis Superbowl Contest, sponsored by the Wisconsin Forage and Grassland Council, report (*Holstein World,* March 10, 1986, p. 90) that they obtain higher milk production from feeding higher fiber first cutting alfalfa than from lower fiber subsequent cuttings. For unknown reasons, this appears to contradict the expected performance based on average chemical analysis, from feeding different cuttings of alfalfa hay. As shown in Fig. 8–12, on the average, each successive cutting is lower in crude fiber and higher in crude protein.

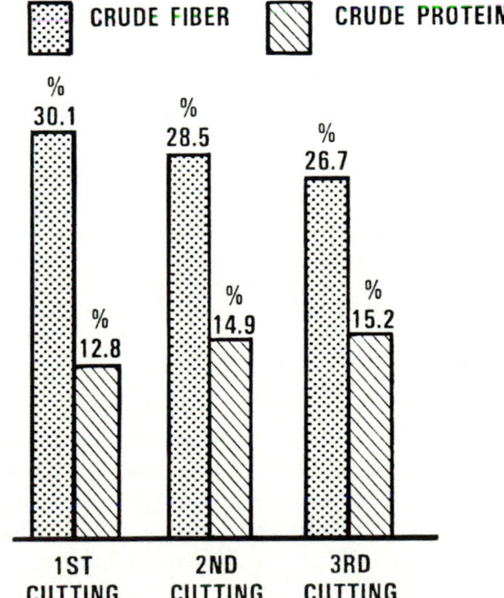

Fig. 8–12. Each successive cutting of alfalfa hay tends to be higher in nutritive value. Note that each successive cutting is lower in fiber and higher in protein. (Adapted by the authors from *Atlas of Nutritional Data on United States and Canadian Feeds,* NRC-National Academy of Sciences, 1972.)

[4]Thomas, J. W., Michigan State University, Lansing, "Hay Crop Losses Can Vary Greatly," *Hoard's Dairyman,* June 10, 1986, p. 559.

Nitrogenation of Low-Quality Hay

Nitrogenation is the combining with nitrogen. Low-quality hay may be combined with ammonia or urea.

• **Ammoniation of low-quality hay**—In the late 1970s and the early 1980s, the ammoniation of low-quality forages increased greatly throughout North America, primarily as a means of providing an economical source of nitrogen (protein equivalent) for ruminants.

During the 3 years 1983–1986, Texas A&M researchers treated mature Coastal Bermudagrass hay, in large round bales, with anhydrous ammonia by weight, then fed the treated and untreated hays, with and without shelled corn, to beef cattle. Generally, the anhydrous ammonia treatment increased the crude protein of the mature Bermudagrass by 2.5 to 3.5% and increased the *in vitro* digestibility by 10 to 14%. The results are reported in Tables 8–19 and 8–20; Table 8–19 is a summary of forage quality, and Table 8–20 is a summary of the cattle feeding studies.[5]

TABLE 8–19
FORAGE QUALITY OF UNTREATED AND AMMONIA TREATED COASTAL BERMUDAGRASS HAY

	CP[1]			IVDMD[2]		
	83–84	84–85	85–86	83–84	84–85	85–86
	(%)	(%)	(%)	(%)	(%)	(%)
Untreated	7.02	8.10	7.14	50.28	49.92	47.23
Treated	11.02	11.57	11.55	64.12	59.46	60.24

[1]Average crude protein of samples taken from each bale immediately prior to feeding.
[2]Average *in vitro* dry matter digestibility of samples taken from each bale immediately prior to feeding.

TABLE 8–20
HAY DISAPPEARANCE AND AVERAGE DAILY GAINS (ADG) OF THE THREE YEARS OF FEEDING STUDIES UTILIZING UNTREATED AND TREATED (ANHYDROUS AMMONIA TREATMENT) MATURE COASTAL BERMUDAGRASS HAY

	Hay Disappearance[1]			ADG		
Hay Feeding Regime	83–84	84–85	85–86	83–84	84–85	85–86
	(lb)	(lb)	(lb)	(lb)	(lb)	(lb)
Untreated alone	—	14.8	20.1	—	0.76	0.85
Treated alone	—	16.4	21.4	—	1.17	1.03
Untreated plus supplement[2]	15.0	12.7	16.7	0.50	1.60	1.40
Treated plus supplement[2]	17.3	13.9	20.8	1.35	1.57	1.60

[1]Calculated sum of hay consumed plus hay wasted per head per day.
[2]In 1983–84, supplement was 2 lb whole shell corn. In 1984–85 and 1985–86, supplement was 2.4 lb whole shell corn plus 1.7 lb guar meal. Both supplements were fed daily.

The Texas researchers concluded: (1) that the small differences in gains between cattle fed treated hay, with and without corn supplement, may not be large enough to pay for the treatment of the hay; and (2) that in the 1985–86 study, 3 of the 5 heifers receiving only ammonia-treated hay exhibited *crazy cow syndrome.* **NOTE WELL:** The 3% added anhydrous ammonia by weight used in the Texas A&M trials was higher than the 1.0 to 1.5% added anhydrous ammonia by weight that the authors recommend elsewhere in this book. (See Chapter 11, Protein Supplements, section on "Ammoniated Products"; and Chapter 5, Nutritional Disorders/Toxins, Table 5–3, section on "Urea Toxicity [Ammonia Toxicity/Bovine Bonkers/Crazy Cattle Disease].)

CAUTION: Anhydrous ammonia can be hazardous to people and animals if not handled properly. So, (1) use proper equipment, (2) take good care of equipment, and (3) follow safe practices.

• **Adding urea to low-quality hay**—Experimentally, University of Missouri scientists nitrogenated low-quality hay with urea; by injecting it into the bale, or by spraying it upon the hay at the time the hay was baled, at the rate of 2 to 4% of the dry weight of the hay. In some cases, the crude protein content was doubled, from 6 to 12%. In comparison with liquid ammonia, urea is easier to store and handle and may be applied at baling. Also, it is less hazardous to the applicator and to the animals consuming the treated hay.

[5]Ocumpaugh, W. R. and G. L. Williams, Texas A&M University, College Station. Reported in *Feedstuffs,* May 25, 1987, p. 11.

Haymaking Systems

In haymaking, the term *system* refers to a team of processes and machines that does the work from field through feeding, saves crop nutrients, reduces manpower requirements, and eliminates drudgery. When each step is mechanized, it must be matched; otherwise, workers and machines end up waiting.

In recent years, automation has had great impact on haymaking. Some haymaking systems are completely mechanized from field to feeding.

There is no one best haymaking system for all conditions. Nevertheless, all good systems are fast, make handling easy, save labor and nutrients, and increase profits. Baling is the most popular hay-handling system in North America.

LONG, LOOSE HAY

The acreage harvested as long, loose hay has declined sharply in recent years, especially in the humid areas, because (1) of high labor cost, and (2) long hay is too bulky for mechanized feeding. Nevertheless, long, loose hay is still popular in many western areas where specialized handling equipment is used. Moreover, some of the newer systems of handling and self-feeding loose hay show promise.

The two common methods of handling long hay are:

1. **Loading with hay-loader directly from windrows.** In this method, cured hay is loaded on a truck or wagon directly from the windrow by means of a hay-loader. Usually the hay is then transported to a barn or stack where it is unloaded by fork or sling and moved away by hand. Sometimes it is chopped into the barn or other storage area.

The hay-loader method requires less investment in equipment than the pick-up baler or field chopper; but, unless the acreage is small, the cost of handling on a per ton basis may be excessive because of the relatively high labor requirement. For these reasons, the use of the hay-loader is usually confined to smaller hay operations.

2. **Hauling cured hay from windrows or cocks with buck rakes, sweep rakes, or sled.** In the West, much of the hay is cured in windrows or cocks and then transported by buck rakes, sweep rakes, or sleds to field stacks where it is stacked by hay stackers or other large mechanical devices. Then

after going through a sweat in the stack, the hay is fed out as loose hay, or, if intended for market, it is baled.

Without doubt, this method results in the production of the highest percentage of good-quality hay of any known method, primarily because (a) more latitude is permissible in the moisture content when stacking than when baling or chopping, (b) the practice predominates in an area which normally has good haying weather, and (c) the method is prevalent on farms and ranches where haymaking is frequently a major enterprise and where the operators have the know-how to produce good hay.

In a somewhat modified form this method of haying has spread from the West to other sections of the United States. In the latter areas, where hay is generally grown for winter feed rather than for sale, it is usually transported to barns or sheds from nearby fields by means of a buck rake. It is then unloaded with a hay sling, grapple fork, or blower. Where limited distances and small acreages are involved, this is probably the cheapest method of handling hay, because a minimum of equipment and labor is required.

CHOPPED HAY

Chopped dry hay fits into some feeding systems, particularly in the West.

For safe storage, the moisture of chopped hay should not exceed 18 to 20%.

Two common methods of chopping freshly cured hay follow:

1. **Field chopping cured hay directly from the windrow.** In this method, a field chopper gathers the cured hay from the windrow, chops it, and blows it into a truck or trailer. The chopped hay is then blown into the barn, stack, or other storage area.

Some pertinent facts relative to field chopping cured hay directly from the windrow are:

a. **Equipment cost.** The equipment cost, on a per ton basis—for the chopper, the equipment for hauling, and the blower for unloading—is apt to be high unless a considerable acreage of hay is involved, or the operator can also use this equipment for corn or sorghum silage.

b. **Moisture content.** The moisture content must be lower than is permissible for baling—from 18 to 20%.

c. **Convenience in feeding.** Chopped hay is often more convenient to feed, since it can be handled mechanically.

d. **Feeding value and wastage.** Chopping does not increase the feeding value of the hay, but it may make for less wastage, especially where low-quality hay is involved.

e. **Length of cut.** A 1½- to 2-in. cut is recommended. Finer chopping requires more power, increases the tendency to heat in storage, and makes for dusty and unpalatable feed. Regardless of the length of cut, repeated blower action, first at the chopper and later at the barn, pulverizes the leaves and aggravates the dust nuisance.

f. **Storage space.** Chopped hay requires only one-half to one-third as much storage space as long hay. This has the advantage of effecting a saving in space, but the disadvantage of requiring caution in order to prevent overloading mows.

2. **Chopping into the barn or other storage area.** In this method, the cured hay is generally hauled from the windrow or cock to the barn or other storage area where it is chopped by a hay chopper or silage cutter and blown directly into the storage area. This method is slower and requires more labor than where field-cured hay is chopped directly from the windrow, but less expensive equipment is necessary.

PACKAGED HAY

Great strides have been made in hay packaging in recent years, characterized by the advent of round bales, large rectangular bales, loaflike stacks, and cubes.

Although large round bales and compressed stacks are better adapted to outside storage than small round bales, unrestricted access at feeding will result in excess wastage. Large rectangular bales are suitable for commercial marketing and long distance hauling.

Compressing hay into cubes and pellets makes for many advantages; among them, (1) completely mechanized handling; (2) high density, with more economical transportation and storage; (3) easier self-feeding and higher intake by animals; and (4) lower feeding losses.

Bales

The following choices of bales are available:

1. **Rectangular bales.** Conventional, small, rectangular packages (often called *square bales*), weighing from 60 to 140 lb, were produced by the first baling machines; and they are still popular on farms and ranches that produce hay for their own use.

Fig. 8-13. Conventional rectangular baler in operation. (Courtesy, Deere & Company, Moline, Ill.)

Today, hay is also being packaged in large rectangular bales, weighing from 1,000 to 1,500 lb, designed for custom operators or hay growers with large volumes of hay or straw.

2. **Large round bales.** Many makes and models of large round balers are on the market. All of them can be classified

by the method of rolling the hay into a bale. One method is to pick up the hay from the windrow and roll it in a chamber, between a series of belts or chains. The other method is to roll the windrow on the ground, similar to rolling up a carpet. Both methods produce a rounded shape which gives weather protection. Large round bales range in weight from 850 to 2,000 lb, depending on the make of equipment.

Fig. 8-14. Large round bales being transported on a New Holland hay hauler. (Courtesy, McArthur Farms, Inc., Okeechobee, Fla.)

Most rolled bales are moved twice: (a) The intial move from the field to a storage area prevents damage to the crop stand, reduces spoilage losses in the bale, and makes rolled bale feeding more convenient; and (b) the final move from the storage area to the feeding area.

If not handled properly, big round bales can be dangerous. Three types of bale handling hazards have surfaced: (a) a downhill roll; (b) tractor mounted front-end loaders carrying such a heavy, bulky load have a tendency to upset if not properly counterbalanced; and (c) a bale can roll out of the loader bucket back down the loader arm onto the operator when a loader handling a round bale is raised to full height.

Stacks

These are loaf-shaped (one system makes a circular stack), mechanically pressed haystacks. Long, loose hay is blown into a wagon and pressed down by a hydraulically operated canopy roof. Stacks range in size from 7 to 10 ft wide, 8 to 22 ft long, and 8 to 11 ft high, and weigh from 1 to 6 tons.

Fig. 8-15. Loaflike stacked hay; machine made, 1-ton size, sheds rain and snow. (Courtesy, USDA)

The stack system saves labor and permits nearly the same latitudes as loose hay, with the efficiencies of mechanization.

Limitations of the system are (1) Investment costs are high, (2) heavy rain or snow may seep down through the stacks if they are not formed properly, and (3) they are not efficiently transported over long distances. Nevertheless, the principles involved are good; hence, some of the problems will likely be overcome with more experience.

Cubes (Wafers)

Fig. 8-16. Hay cuber in operation, making cubes 1¼ in. square and 2 to 3 in. long. (Courtesy, Deere & Company, Moline, Ill.)

Field cubers are machines that move across hayfields, pick up windrows of forage, and produce dense, high-quality forage cubes or wafers. Stationary cubers are used to produce similar cubes from loose haystacks or bales.

The following are the major problems which remain to be overcome:

1. Lowering cost.
2. Lessening equipment weights and power requirements.
3. Adapting them for use in humid areas.

Nevertheless, the benefits derived from field cubing of hay are great. It (1) simplifies haymaking, (2) lessens transportation and storage space—cubed or wafered forages weigh between 45 and 55 lb per cubic foot (baled hay density is 8 to 15 lb), (3) reduces labor, (4) makes automatic hay feeding possible, (5) decreases nutrient losses, (6) eliminates dust, and (7) makes for increased feed comsumption, gains, and feed efficiency. Also, with cubing, the spread between high- and low-quality roughage is narrowed; that is, within reason the poorer the quality of the roughage, the greater the advantage from cubing or wafering. The latter is so because such preparation assures complete consumption of the roughage.

Hay cubes are of special interest to dairy producers because they have the advantages of pellets, without their disadvantages. Like finely ground forage that is pelleted, cube-feeding can be readily automated; and, in comparison with long hay, there is less transportation and storage cost. Besides, cubes will not lower the fat content of the milk as much as pellets when appreciable quantities of them are fed.

The Utah Station researchers compared cubed alfalfa hay with each (1) baled alfalfa, and (2) corn silage.[6] These studies showed that consumption of feed is higher when alfalfa hay is fed in cubed than in baled form, thereby increasing the potential for milk production. In one study, cubes resulted in a slight lowering in milk fat percentage; in the other study, cubes did not affect milk fat percentage.

Pellets

Pelleted forages are finely ground, then condensed. The **advantages** of pellets are:

1. Pelleted feeds are less bulky than any other hay package (pelleted roughage requires one-fifth to one-third as much space as is required by the same roughage in loose or chopped form), and are easier to store and handle—thus lessening transportation, building, and labor costs. For these reasons, it is particularly advantageous to use pelleted feeds where storage space is limited and feed must be transported considerable distances, conditions which frequently characterize small enterprises.
2. Pelleting prevents animals from selectively refusing ingredients likely to be high in certain dietary essentials; each bite is a balanced feed.
3. Pelleting practically eliminates wastage. Since animals frequently waste up to 20% of long hay, less pelleted feed is required. Wastage of conventional feed is highest where low-quality hay is fed and/or feed containers are poorly designed.
4. Pelleting eliminates dustiness and lessens heaves.

The biggest deterrent to increased pelleting at the present time is the difficulty of processing chopped forage coarse enough so that it will not cause digestive disturbances. A minimum of a ¼-in. chop is recommended.

ARTIFICIAL DRYING

The use of forced air, either heated or unheated, for final drying is the most dependable way in which to preserve quality in hay. The application of heat provides faster drying, saves more leaves, and reduces losses of nutrients. Yet, artificial drying is on the decline in the United States because of the added cost in heating the air and the increased labor.

Hay-drying equipment permits handling while hay is green and tough enough to withstand mechanized processes without excessive leaf loss, and it minimizes weather damage. The most common methods of artificial drying are (1) mow curing, (2) artificial dehydrators, and (3) wagon driers.

Mow Curing

Mow curing, or drying, refers to the practice of curing partially dried hay—either long, chopped, or baled—in barn mows equipped with ventilation systems through which either unheated or heated air is forced. The pertinent facts relative to mow curing are:

1. **Equipment.** Although numerous variations in the design of mow-drying equipment exist, generally speaking the equipment consists of one of the following two types:
   a. A blower or fan (usually powered with an electric motor, but with provision for tractor operation in case of power failure) which forces air through a main duct or flue and out through a system of lateral ducts.
   b. A slotted floor built above the regular mow floor, with provision under this slotted floor for forced air circulation up through the hay.
   Usually the air is not heated, but in some systems oil furnaces and other heating units are used to provide warm air and to hasten drying.

   For mow-drying equipment plans and specifications, farmers should contact their local county extension agent or write to their state agricultural college.
2. **Moisture content of forage.** The following moisture conditions usually prevail in mow drying:
   a. Newly mown forage has a moisture content of 75 to 80%.
   b. When placed in the mow for further curing, the moisture content should not exceed 35 to 40%, which means that it is about half cured.
   c. For safe storage the moisture content of hay must be reduced to between 20 and 25%.

   When weather conditions are favorable, it is recommended that field curing be prolonged so as to lower the moisture content under the 35 to 40% figure indicated. This will lessen the weight of water that must be handled, speed up barn drying, and reduce the cost of operating the drying equipment. Naturally, field drying should not extend to the point that there is much loss of leaves by shattering.
3. **Mow drying time.** With unheated air, mow curing will require from 7 to 14 days' time; with heated air considerably less time is necessary—the hay dries from the bottom up. When the top is dry, the curing is completed.
4. **Quality of mow-dried hay.** Generally mow-dried hay is greener, leafier, and of a higher grade than similar hay that is field cured. This is particularly true in those himid areas which consistently have poor haymaking weather. Of course, during ideal haymaking weather, it is possible to make about as good quality hay by field curing as can be made by mow curing.

   Based on a limited number of experiments and practical observations, it may be concluded that there is no appreciable difference in either the palatability or the feeding value per ton of mow-cured and field-cured hay, provided they are of comparable quality. Naturally, weather damaged field-cured hay is inferior to good-quality mow-cured hay.
5. **Cost.** Although higher quality hay can generally be produced in mow curing than in field curing, the cost is also greater. The higher cost in mow curing is due to (a) the cost of the mow-drying equipment, (b) the cost of operating the equipment, (c) the added labor required in handling heavy forage of high-moisture content, and (d) the higher storage cost per ton, because in mow curing the hay cannot be piled too high over the ducts at any one time (it is recommended that not more than 4 to 8 ft of uncured hay be placed in the mow at any one time) without decreasing the all-important air circulation.

---

[6]Anderson, M. J., *et al.*, "Cubed Versus Baled Alfalfa for Dairy Cows," *Journal of Dairy Science*, Vol. 58, No. 1, 1975, p. 72.

Because of the relatively high cost involved in mow curing hay, in areas where poor haymaking weather generally prevails and field curing is hazardous, farmers may well consider the desirability of making silage.

Artificial Dehydrators

Artificial dehydrating refers to a process in which forage is taken from the field as soon as it is cut (or in some instances after wilting), put through a hay chopper or silage cutter, and dried in large driers of various types, following which it is finely ground. For the most part, this method of curing is limited to large commercial operations which process early cut alfalfa (or its leaves) and other legume and/or grass crops chiefly as a supplement for swine and poultry. Occasionally, artificially dehydrated hay is produced for other classes of animals, especially in those areas which rarely have good haymaking weather.

Other pertinent facts relative to artificial dehydrating of hay are:

1. **Drier.** The most popular type of artificial dehydrator in use in this country is one that uses a high initial heat (1,200° to 1,400°F), and which is heated by gas or oil. In a good drier, the forage does not get sufficiently hot to be damaged, primarily because of the cooling effects produced when the water evaporates from the plant tissues, and because the forage is in contact with the hot air for only a few minutes.

   Any burning of the leafy portions of the forage indicates that the temperature was too high or that the forage was overdried.

2. **Nutritive value of artificially dehydrated hay.** Generally, artificially dehydrated forage is of high nutritive value, for few leaves are lost and the maximum content of protein, carotene, and riboflavin is retained. But the protein may be slightly less digestible due to the effect of the heat of dehydration, and the vitamin D content is low for the reason that in the curing process the hay is not exposed to the vitamin D-imparting ultraviolet rays of the sunlight.

3. **Cost.** Due primarily to high cost of equipment and fuel, artificial dehydration is usually considered practical only in commercial operations where large tonnages are processed. Even then, the artificially dehydrated forage must command a premium price over field-cured hay in order to justify the higher investment in equipment, the higher cost in moving the heavier high-moisture forage from field to drier, and in operating the dehydrator.

The producer is well justified in paying a premium price for high-quality dehydrated forage for use in swine and poultry rations. Except in special circumstances or as a partial grain replacement, however, it is seldom economical to feed artificially dehydrated hay to cattle or sheep. For ruminants, it is generally more practical to make silage or to resort to mow drying in those areas commonly having poor haying weather. This is especially true where the forage is to be fed on the farm on which it is produced.

Wagon Dryers

Wagon dryers were developed to reduce the high labor requirements of batch or platform drying. Essentially, they are batch dryers on wheels.

Some wagons are covered; they are known as *covered wagons* because of the outward appearance of the ballooned cover on the wagons during drying. The cover consists of a durable, lightweight material that won't leak water or air. Heat is conveyed under the cover. Other wagons are open and adapted to drying in a shed. In both types, the drying wagons are loaded in the field directly behind the baler, taken to the drying system where they are connected to a main air duct, and the hay is dried. After drying, they are taken to the storage place and unloaded. This procedure eliminates the unloading and reloading of wagons required in batch or platform drying.

Drying costs money, both for equipment and heat. Of course, the cost will vary according to (1) the initial moisture content of the forage, (2) the cost of electricity, gas, or fuel oil, and (3) how much the temperature of the outside air must be raised to bring it to at least 140°F. In any case, it will pay to let the sun do as much of the drying as practical without risking weather damage or shattering.

Although good-quality hay can be produced by wagon drying, it requires more labor and handling than most other systems of haymaking.

Storing

Good hay should never be poorly stored. Naturally, the type of storage will vary from area to area. In the more arid sections where little rainfall comes during the fall and early winter, a good stack of loose or baled hay may provide entirely satisfactory storage. On the other hand, in high-rainfall areas, more expensive waterproof storage should be provided. At and between these two extremes, hay may be and is successfully stored in many different ways in different sections of the country.

In the West, a considerable amount of hay is chopped (either at the time of gathering from the windrow or adjacent to the stack) and stack stored. Most such stacks are round, and are built by sliding a snow fence toward the top as the stack is built. Generally these stacks are rounded off at the top and left uncovered. The advantages claimed for stack storage of chopped hay are (1) minimum labor in haymaking, (2) minimum stack storage space and spoilage, and (3) ease of feeding.

Where different kinds and qualities of hay are produced or purchased, each kind and each quality should be stored in such manner that it will be accessible when needed. Otherwise, it may not be convenient to provide for variety in feeding and to feed some of the low-quality along with some of the high-quality hay.

• **Round bale losses/storage**—Prior to 1970, most hay was packaged in rectangular bales, and in humid areas it was stored inside. With the advent of large round bales weighing 850 to 2,000 lb, most of them were left uncovered in the field or in a fence row.

The nutritional losses from bales stored outside and unprotected depend on the storage area (it should be high and dry), the amount of rainfall during the storage period, the hay type and condition when baled, the bale shape and density, and the method of feeding; and the monetary value of the losses depends on the price of hay and the nutritional loss. Legume hays do not form as tight a thatch as grass hays; consequently, they are not as weather-resistant.

Table 8-21 shows the range of dry matter losses that can be expected in large round bales stored outside and unprotected. The lower value of each range represents well-formed bales located in areas with low rainfall (less than 25 in. per year) and low relative humidity. The higher values are for areas with high rainfall (greater than 40 in. per year) and high relative humidity.

TABLE 8-21
DRY MATTER LOSS IN LARGE ROUND BALES STORED OUTSIDE AND UNPROTECTED

Forage	Storage Period	
	Up To 9 Months	12 To 18 Months
	(% loss)	(% loss)
Alfalfa	6–24	16–50
Grass hay	4– 8	10–14
Sorghum X Sudan	6–12	10–24
Cereal hay	10–24	20–40

Not all the deterioration that occurs is the result of rain falling on the bale. Moisture movement at the bottom of the bale, where it contacts the ground, can also cause considerable loss. Dry matter losses can be reduced by as much as 10% if the bales are stored on high ground on a well-drained site; set on racks or pallets, fence posts, railroad ties, or a 3-in. base of rock; and spaced 12 to 28 in. between bales.

Large round bales stored outside and covered with plastic or canvas bonnets (or caps), sleeves, or bale bags sustain much less loss than unprotected bales.

Also, hay losses vary with the method of feeding. When animals have free access to hay, they trample and stand on it, with the result that losses may be as high as 40 to 50%. By using a barrier (a feeding rack, panels, feeding wagons, or gates), losses can be cut to 5 to 10%.

Fig. 8-17. Round bales can be fed with a minimal of waste if proper equipment is used such as this round bale feeder for sheep. (Courtesy, Shalom Valley Sheep Equipment, Inc., Gary, S.D.)

ADDITIVES FOR HAY

Farmers in many countries of the world have traditionally added about 20 lb of salt per ton of new hay at the time of stacking or putting it into the hay mow in the belief that the salt would prevent the hay from molding and heating. Carefully controlled experiments have failed to substantiate claims that salt will prevent excess heating or sweating; nor has it prevented spontaneous combustion of hay. However, when salt is used in moderate amounts, it may improve the color, aroma, and palatability of poor-quality hay. It is recognized, too, that much higher levels of salt—quantities sufficiently high to harm animals—may prevent mold.

Over the years a number of products, both liquids and powders, said to preserve hay have become commercially available. While claims have been made that there will be no heating or molding when these materials are used, the results have been highly variable.

Recent studies of several state experiment stations have shown that propionic acid or a combination of propionic and other organic acids can be used successfully to preserve baled or stacked hay stored at 35% or less moisture. Anhydrous ammonia and ammonium isobutyrate have also been found to be effective in preventing heating and preserving the quality of high-moisture hay. (See earlier section in this chapter on "Considering Chemical Conditioning and Preserving Agents.")

SPONTANEOUS COMBUSTION

Wet hay ferments and generates heat. Sometimes this results in spontaneous combustion and fire, usually about a month to 6 weeks after storing. Here are the facts:

1. **Symptoms of heating.** The warning signals are hay that feels hot to the hands, strong burning odor, and visible vapor.

2. **Temperature of hot hay.** Hot spots may be located by probing the hay with a steel rod. Then the temperature of the hot spots may be tested with a thermometer (a dairy thermometer or other type) attached to a wire and dropped down a pipe. If the hay is over 140°F, it should be checked periodically during the day. If the hay is 160°F, it should be checked hourly. If the hay is 180°F, there are apt to be fire pockets, and it should be removed from a barn.

3. **Cooling hay.** Hay that is heating may be cooled by discharging through pipes into the hot areas either dry ice or liquid carbon dioxide.

4. **Removing hot hay.** When a fire is imminent and hot hay must be removed, it is important to have plenty of help on hand including the fire department. Then the hay should be removed cautiously and without wetting unless necessary.

5. **Precautions.** Never walk on hay that is heating—place planks over it, and do not breathe hot and noxious fumes.

BUYING AND SELLING HAY

Historically, most hay has been fed on the farms where it was produced. But this practice is changing. Today, about 29 million tons, or about 25% of the U.S. production, with a cash value of over $2 billion, is sold off the farm. The Northwestern states of Washington, Idaho, and Oregon, along with California and Texas, are the leading states in total value of hay sold. In California alone, the annual value of hay sold exceeded $425 million in the late 1980s. In four states—California, Arizona, New Mexico, and Washington—some 40% or more of the hay produced is sold. In California, nearly 70% of the hay produced is sold.

New hay markets are developing. As dairy producers, beef cow-calf producers, cattle feeders, and sheep feeders become more specialized, they prefer to grow animals and rely on other specialists to grow hay. Also, more high-quality hay is needed for expanding horse numbers. Mushroom growers are becoming an important market for hay, too; they require a high-energy hay, although it can be moldy.

Additionally, considerable hay is exported. In order to save space and cut down transportation costs, this generally requires a special hay package—pellets, wafers, or high-density bales.

Hay Sources

The five basic sources of hay in the United States are:

1. **Hay dealers or brokers.** Hay dealers or brokers are important suppliers of hay. In California, 90% of the hay is marketed through this channel. For the nation as a whole, dealers and brokers market 5 to 10% of the hay produced. They purchase hay from the producer and sell it to the consumer. Other hay marketing firms match hay sellers and buyers, based on a detailed description of the hay, and/or samples and tests; and charge a commission for their services, which may also include trucking the hay.

2. **Neighbor to neighbor.** This is the oldest market channel for hay, and it is still quite common. This is a disorganized market without any particular pricing structure. Hay is purchased on visual inspection.

3. **Associations of cooperatives.** These organizations vary in size, from very small to very large. The San Joaquin Hay Producers Association in California is one of the largest, grossing over $10 million annually. Such associations normally purchase hay for cash and store it or move it to the consumer. This is a high-risk operation, because there are no futures markets that enable hedging protection.

4. **Auctions.** Hay auctions are trading centers where hay is sold by public bidding to the buyer who offers the highest price. Hay is sold by visual inspection, and the price is determined by supply and demand.

5. **Contract.** This is an agreement between a hay producer and a livestock producer to supply hay of a specified quality at a prior agreed-upon price. It assures both the buyer and the seller an orderly market. Such eventualities as weather-damaged hay should be covered in the contract.

How Hay is Priced

Traditionally, most hay has been bought and sold by using the ancient art of bartering, based on visual evaluation.

Several states have long had programs for selling and buying hay on the basis of analysis, usually involving crude protein and moisture. Since 1958, California, which is the largest hay marketing state in the nation, has used a modified crude fiber analysis to estimate TDN. While the California test is satisfactory for alfalfa, it is not good for mixed hays or grasses.

Nevada growers use crude protein and acid detergent fiber (ADF) tests for hay pricing.

Minnesota forage scientists have incorporated relative feed value (RFV) into the pricing scheme of legume and grass hays. They determine a substitution value, called Dollar Feed Value ($FV), based on local corn and soybean meal market prices, which they incorporate into the relative value. Such a method allows for consideration of intake caculated from neutral detergent fiber (NDF) in addition to protein and digestibility of TDN.

But there is need to eliminate the confusion caused by variation between states in the chemical components considered and the laboratory bases of determining them. The goal is a uniform, nationwide system that transcends state boundaries and is based on chemical analysis and visual appraisal that will project the feeding value of hay.

Federal Grades of Hay and Straw of Historic Interest

Today, the federal grades of hay and straw are of historic interest only. They were created by the Agricultural Marketing Act of 1946, and they were terminated March 14, 1988. At the time the grades were discontinued, fewer than 50 requests per year for their use were being received by the U.S. Department of Agriculture.

The historic federal grades of hay were based entirely on subjective factors—leafiness, color, and presence or absence of foreign material. There were no indicators of nutrient content and projected animal productivity, such as crude protein, fiber, NDF, or ADF.

Thus, the demise of the 42-year-old federal grades of hay resulted because they did not adequately reflect feeding value. Progress had bypassed them.

Futuristic Hay Evaluating and Pricing

Recognizing the problems of hay marketing, the American Forage and Grassland Council (AFGC) formed, in 1972, a Hay Marketing Task Force, made up of representatives of the AFGC, industry, and the National Hay Association, and charged them with the following primary responsibilities: (1) to develop improved measurements for reflecting the feeding value of hay, and (2) to develop improved standards (grades) for marketing hay.

- **Proposed chemical measures of hay quality**—The first challenge for the committee was to research and select the best practical methods of chemical analysis suitable for the widest range of hay species.

Following careful study, three chemical values were selected as best suited to measure hay quality: crude protein (CP), neutral detergent fiber (NDF), and acid detergent fiber (ADF). Studies showed that NDF is good for predicting intake, or the amount of a given forage an animal can consume, while ADF is the best predictor of forage dry matter digestibility. All three analyses can be run in most forage testing labs.

Hand in hand with evolving with chemical measures of hay quality, the Hay Marketing Task Force discovered the potential of infrared reflectance analysis as an accurate and rapid means of measuring forage feed quality. (See earlier section in this chapter entitled, "NDF, ADF, and NIRS Analyses.")

- **Proposed hay standards (grades)**—Using the above analyses, the committee developed new proposed hay standards, including five hay grades and one sample grade for all legumes and grasses. But, for various reasons, these proposed standards were not accepted by the Federal Grain Inspection Service.

However, nine states, including Minnesota, Oregon, and Wisconsin, did implement the proposed AFGC standards in their hay marketing systems. (See Table 8-22 for the Wisconsin adaptation.) Note that Table 8-22 lists seven grades, and that the grades are based on maturity and species composition. Note, too, that legumes rank highest in the chemical components used in the grades, followed by legume-grass mixtures, grasses, and heavily weathered forage.

TABLE 8-22
MARKET HAY GRADES FOR LEGUMES, LEGUME-GRASS MIXTURES, AND GRASSES[1]

Grade	Description[2]	CP	NDF	ADF	DDM	DMI[3]	DDMI	RFV
		(%)	(%)	(%)	(%)	$gm/Wkg^{0.75}$	$gm/Wkg^{0.75}$	(%)
Prime	Leg.-Pre.B1.	>19	<39	<30	>65.5	>143	>93.5	>143
1	Leg.-EB1., 20% grass-V.	17–19	40–46	31–35	62–65	134–143	82–93	126–143
2	Leg.-MB1., 30% grass-EH.	14–16	47–53	36–40	58–61	128–133	74–81	113–126
3	Leg.-FB1., 40% grass-Head	11–13	53–60	40–42	56–57	113–127	64–73[4]	97–113[4]
4	Leg.-FB1., 50% grass-Head	8–10	61–65	43–45	53–55	106–112	55–63	86–97
Fair	Grass-head, and/or rain damaged	<8	>65	>46	<53	<105	<55	<86
6	Sample	—	—	—	—	—	—	—

[1]Description and DDM adopted by National Alfalfa Hay Quality Committee.

[2]Prebloom, EB1. = early bloom, MB1. = midbloom, FB1. = mid- to full bloom, V. = vegetative, EH. = early head.

[3]DMI for sheep and goats = >82, 76–81, 72–75, 63–71, 52–62, and <56 for grades Prime through Fair, respectively.

[4]Reference hay mid- to full bloom alfalfa (Lema and Kawas and Jorgensen) DDM = 54.2, DMI = 120.2 $gm/Wk^{0.75}$, DDMI = 65.2 $gm/Wk^{0.75}$, and RFV = 100%.

In addition to leading to the development of the proposed hay grades listed in Table 8–22 the earlier effort of the Hay Marketing Task Force prompted the formation of the National Alfalfa Hay Quality Committee and the development of national alfalfa hay quality standards, with the intent of having uniform testing throughout the United States so that both buyer and seller can receive accurate and interpretable results on any given lot of hay. While these standards are based on a minimum test of alfalfa hay for dry matter (DM), crude protein (CP), acid detergent fiber (ADF), and estimated digestible dry matter (EDDM) calculated from the ADF, provision is also included for a description sheet of visual characteristics.

Procedures for the national alfalfa hay quality standards are carefully spelled out and include: (1) specific sampling procedures using an improved hay core sampler, (2) suggested visual factors to be used with chemical analysis to describe the sample, (3) approved testing procedures using either NIRS or wet chemistry, (4) development of an acceptance of acid detergent fiber (ADF) to predict estimated digestible dry matter, and (5) a voluntary laboratory certification program operated by the National Alfalfa Hay Test Association in conjunction with AFGC and the National Hay Association. See Table 8–23 for the estimated digestibility, intake, and relative feeding values of the grades given in Table 8–22.

TABLE 8-23
ESTIMATED DIGESTIBILITY, INTAKE AND RELATIVE FEED VALUES FOR THE MARKET HAY GRADES GIVEN IN TABLE 8-22

Grade	DDM[1]	DMI[2] Sheep	DMI[3] Cattle	DDMI[4] Maintenance	DDMI[5] 3 × Maint.	Relative Feed Value[6]
	(%)	$g/kgW^{0.75}$	$g/kgW^{0.75}$	$g/kgW^{0.75}$	$g/kgW^{0.75}$	(%)
Prime	>65	>82	>143	>93	>89	>132
1	62–64	76–81	133–142	83–92	78–88	118–132
2	58–61	72–75	122–132	71–82	67–77	101–117
3	56–57	63–71	110–122	62–70	59–66	88–100
4	54–55	56–62	98–109	53–61	50–58	75–87
5	<53	<56	<98	<52	<49	<75
6	—	—	—	—	—	—

[1]DDM = digestible dry matter, *in vivo*, = 102 + .008 CP% − .382 ADF% − $4.63\sqrt{ADF\%}$.

[2]DMI = dry matter intake sheep = 96.4 − .0003 CP% − .0482 NDF% − .0085 $NDF\%^2$.

[3]DMI = dry matter intake cattle = (DMI sheep × 1.75).

[4]DDMI = digestible dry matter intake cattle at maintenance level of intake = (DDM × DMI cattle) ÷ 100.

[5]DDMI for dairy cattle based on intake 3 × maintenance = (DDM × .95) • (DMI cattle) ÷ 100.

[6]Relative feed value = DDMI maintenance × 1.6.

• **Hay marketing summary**—At today's feed prices, hay (especially alfalfa) is usually a profitable crop to grow, both for direct utilization through livestock and as a cash crop, *provided* it is of high quality and is priced at its true value for feed. However, there is general agreement that if hay is to take its rightful place in the marketplace as a major source of protein and energy, many of the traditional selling arts must be replaced by developments of modern science.

But there are a number of obstacles which must be overcome before the marketing of hay on the basis of analysis will be widely adopted; among them, the following:

1. Keeping the cost of analysis nominal.
2. Developing simple and practical sampling techniques.
3. Reducing the time lag between sampling and getting the results. Hay dealers report that much of their hay is purchased one day and sold the next. As a result, they do not have time to take advantage of a laboratory analysis. Therefore, hay is not evaluated properly by traditional chemical analysis unless this is done by the producer in advance of selling.
4. Developing standards and grades based on feed value.
5. Arriving at a satisfactory pricing structure.

With increased marketing of hay, these problems will likely be resolved. Also, there will evolve (1) a more uniform method of analyzing hay for quality, and (2) a system of pricing consistent with the quality of the product sold.

Other developments are necessary if hay is to become a leading cash crop, including:

1. **More markets.** There is need for more markets—such as auctions, dealers, and associations—to provide a ready market outlet for hay growers, and to attract both ample supplies of hay and buyers.
2. **Price protection for the dealer while he has possession of the hay,** There is need for something similar to the futures market that will permit hedging protection.
3. **More hay grower storage facilities.** There is need for additional grower storage so as to spread hay marketing over a longer period and avoid harvest market gluts that depress prices.
4. **Better hay market information.** There is need for hay growers and hay buyers to be better informed relative to supply, demand, and going prices.

Freedom From Toxic Residues

With the emphasis on residues in foods, it is important that hay be free from those residues which are prohibited. If meat or milk (or products derived from them) are found to have residues, the blame cannot be shifted to the hay grower by the livestock producer, unless there is a clear-cut case of fraudulent representation. The best assurance of freedom from such residue rests with the integrity of the hay growers, or those who represent them in selling their hay.

It is important that livestock producers and hay growers be well informed relative to (1) the chemicals which are banned, and (2) the conditions of application for those chemicals which are still permitted. In this way, disastrous financial effects from confiscation of market animals or products can be averted. So, *read the label*. Use proper pesticides and observe minimum days from last application to harvest. Wear protective rubber gloves, respirators, and coveralls when applying pesticides, especially those that are highly toxic. Wash before eating.

(Also see Chapter 5, Nutritional Disorders/Toxins, section on "Agricultural Chemicals and Drugs.")

Hay Shrinkage

Hay buyers should figure hay shrinkage closely. Here's why: If a ton of hay containing 90% dry matter is bought for $75, 1,800 lb of dry matter have been purchased at this price. However, if the $75 per ton hay contains 80% dry matter, only 1,600 lb of dry matter have been purchased for this same price. Purchase of the high-moisture, 80% dry matter hay has resulted in a loss of 200 lb of hay, or ⅑ of the dry matter, worth $8.33 (⅑ × $75). Thus, if the 90% dry matter hay is worth $75 per ton, then the 80% dry matter hay is worth only $66.67 ($75 − $8.33) per ton. If 1,000 tons of hay are involved, that's a loss of $8,330 (1,000 × $8.33).

In addition to moisture losses, newly harvested hay may be expected to lose about 5% weight from going through the sweat.

Hidden Hay Costs

When buying and feeding hay, all costs and losses—handling, shrinkage and wastage, grinding costs and losses, insurance, interest, and storage—in getting it from the point of purchase to the feed bunk should be taken into consideration. The example that follows will illustrate this situation.

Which is the better buy—baled alfalfa hay at $75 per ton roadside or sun-cured alfalfa pellets at $117.50? Simple arithmetic would indicate that baled hay is the cheaper, by $42.50 per ton. But, what are the facts?

Most researchers report that pellets have a higher feeding value for beef cattle and sheep than long or chopped hay. But they warn that the increased rate and efficiency of gain may not be sufficient to cover the cost of pelleting, running $12 to $16 per ton.

Unfortunately, most people fail to account for all the costs and losses inherent in getting baled hay from roadside to feed bunk. Generally speaking, they do not make allowance for hauling costs, storage, shrinkage and wastage, chopping costs and losses, insurance, interest, and breakdown losses in time and equipment due to hardware in the bales—not to mention the fuss and muss of handling baled or chopped hay.

A comparison of baled alfalfa hay vs alfalfa pellets, based on 1988 figures in the San Joaquin Valley of California, reveals the figures shown in Table 8-24.

Since it is net returns that count, in each case (bales vs pellets) one must consider (1) cost delivered to the feed bunk, and (2) added returns from pelleting in terms of more gains and/or less feed. In the above example, when all the hidden losses and costs were added to baled hay, the price in the bunk was $117.50 per ton—the same as for sun-cured alfalfa pellets.

Actual feeding trials conducted by Illinois, California, Washington, and Okalhoma researchers—with roughage constituting 80 to 100% of the rations—show that in terms of increased rate and efficiency of gains, each ton of pellets is worth $20.00 per ton more than a ton of chopped forage.

TABLE 8-24
HIDDEN HAY COSTS

	Baled Alfalfa Hay	Sun-cured Alfalfa Pellets
	(per ton)	(per ton)
Cost in the feed bunk:		
Price at roadside	$ 75.00	
Hauling and stacking	7.50	
Moisture, shrinkage, wastage, and spoilage	7.50	
Chopping or grinding costs	13.00	
Grinding losses	4.50	
Insurance	3.00	
Interest (6 months)	5.00	
Storage cost	2.00	
Total cost in the feed bunk	$117.50	$117.50
Added returns from pelleting:[1]		
Value of improved rate of gain		5.00
Value of improved feed conversion		15.00
Total		$ 20.00

[1]Computations by the authors, based on feeding trials conducted by the Illinois, California, Washington, and Oklahoma Experiment Stations.

In this example, pellets were the best buy, by $20.00 per ton—the amount of the added feeding value, since baled and pelleted hay cost the same when delivered to the bunk.

In one Illinois Station test, weaned steer calves were fed timothy-alfalfa either (1) long (baled), (2) chopped, or (3) pelleted. Gains were increased 175% by pelleting. Even with an allowance of $10.00 per ton for pelleting, feed cost per pound of gain was 3.7¢ cheaper for calves fed the pellets.

On the average, cattle fed high-roughage rations (above 80%) will eat about ⅓ more pellets than long or chopped hay, make about 50 to 75% faster daily gains, and require 200 to 250 lb less feed per 100-lb gain.

Of course, producers should apply their own cost figures to baled hay and pellets in order to determine which is the better buy for them. The important thing is that all costs be accounted for—that in computing the cost of baled hay there be added such hidden costs and losses as handling, shrinkage and wastage, grinding costs and losses, insurance, interest, and storage. Additionally, allowance should be made for the added feeding value of the pellets. Also, the age and grade of cattle, other available feeds and prices, and starter vs finishing rations must be considered.

This same method of cost analysis can be applied to hay cubes, or wafers, or to any other alternate method of haymaking.

HAY FEEDING FUNDAMENTALS

Feeding is the end of the line for hay. No matter how carefully it has been grown, harvested, and stored, all that has gone before can be dissipated if it is improperly fed—unless hay feeding fundamentals are observed.

Monogastric animals, including swine and poultry, must eat a large percentage of grains and other concentrates and depend almost entirely on digestive enzymes to break down these compounds. But ruminants, with their four stomach compartments and the help of microorganisms, can subsist largely, or entirely, on bulky, high-fiber forages which, because of their low energy per unit weight of dry matter, must be consumed in large quantities to supply their nutrient needs. The horse, because of its greatly enlarged cecum and large intestine, can utilize quantities of hay intermediate between simple-stomached and ruminant animals.

The economics of the situation—the relative price of forage and grain—call for greater emphasis on forage accompanied by less grain feeding. With greater quantities of forage incorporated in rations, it is expected that performance—the production of meat and milk—will decrease. However, maximum net returns, rather than just maximum production, will be the primary objective.

Increasingly, forage testing will be used in two ways: (1) to purchase hay on a quality basis, and (2) to balance rations more precisely.

Maximum Use of Homegrown Forages

Although increased quantities of hay will be bought, especially by large dairies and feedlots, most forages will continue to be homegrown for the following reasons: (1) Hay is bulky and costly to transport, unless it is cubed or pelleted; (2) hay crops have an important place in most crop rotations; and (3) hay crops can be grown on wasteland not suited to other crop production.

Legume Vs Grass Hays

With the higher cost of both protein supplements and grains, legume hays have a particularly valuable place in rations. Legumes are also a rich source of many minerals and vitamins.

Hay Preparation

Hay is fed as long hay or in processed form. The common methods of processing are chopping, grinding, cubing, and pelleting.

Fig. 8-18. Cows eating chopped alfalfa hay in a mixed ration. (Courtesy, James Tappan, Arizona Dairy Co., Higley, Ariz.)

Ruminants need some roughage. Hay fed early in life will develop the calf's rumen.

Many serious problems in high-producing dairy herds have been traced to lack of hay in the ration; among them, increased incidence of ketosis and displaced abomasums. In addition to these maladies attributable to no-hay rations, it is noteworthy that milk fat percent can be as much as 1% less on an all-silage and concentrate ration. The explanation: The amount and length of hay in the ration affects cud chewing time and percent milk fat. Cud chewing time is the biological response that indicates several desirable factors—rumen muscle tone and function, the supply of saliva going into the rumen to buffer rumen acids, and rumen fermentation with the production of acetic acid. It is estimated that about 15 minutes of cud chewing time is needed per pound of dry matter eaten.

Finishing cattle also require a minimum amount of roughage factor for normal rumen function. Feedlot cattle fed all-concentrate rations show a response to small amounts of roughage factor. As little as 1 lb per day of some low-quality roughage, such as alfalfa stems, rice hulls, or sorghum straw, has given improvement in efficiency ranging up to 15% of all-concentrate rations. To meet the *roughage factor* need, different roughage substitutes have been developed and used with varying degrees of success.

Different Qualities of Hay May Be Used

The type of ration which will be least costly and result in satisfactory performance will differ according to species, level of performance, reproductive status, age, etc. For example, the nutritive needs of a dry, pregnant beef cow are much lower than those of a high-producing dairy cow. Thus, a low-quality hay may be quite satisfactory for wintering a beef cow without calf at side, whereas high-producing, lactating cows should always have high-quality hay. Also, high-quality hay is important for swine and poultry. Where forage is incorporated in monogastric rations, high-quality dehydrated alfalfa is most commonly used. High-quality hay is also essential for horses.

Hay Waste and Refusal

In a recent Texas study involving 10 different hay feeding racks, the cows at the best conventional feeder still wasted 14% of their hay. Dairy producers have commonly accepted 10% refusal as normal.

High-priced feeds and smaller margins are causing livestock producers to scrutinize hay losses, and to do something about them. Chopping hay and/or adding molasses will lessen wastage. But feeding high-quality hay is the best way in which to lessen waste and refusal.

Supplementing the Hay Ration

Hay is generally lower in energy and higher in fiber than most grains and concentrates. Legume hays have a high-calcium content, but they vary in available phosphorus. If sun-cured properly, they are high in carotene and vitamin D, along with many of the B vitamins. A supplement should supply the nutrients that are most likely to be lacking in the hay; thus, supplements for alfalfa hay should be high in energy and phosphorus and low in fiber. Also, carotene or vitamin A should be provided if the hay has been bleached or turned brown. Salt is lacking in hays and other natural feedstuffs and should be provided as a supplement, along with trace minerals that are deficient in the local area.

Stretching the Hay Supply

When hay is scarce and high in price, the supply of it for ruminants and horses may be stretched. As the amount of hay fed is reduced, it must be replaced with other feeds so that the total ration is still balanced and fulfills all the nutrient requirements.

Most grains contain 75 to 80% TDN, while most medium- to good-quality hays contain 45 to 50% TDN. Hence, as a general rule of thumb, about 5 lb of grain equal 8 lb of hay, provided they are of comparable quality. Thus, it follows that if corn can be bought for $100.00 per ton, hay should be bought at $62.50 per ton, or less.

If the price of hay is less than five-eighths the price of grain, relatively more of it should be fed; whereas, if the price of hay is higher than this, relatively more grain will make for cheaper production.

When hay is scarce and high in price, the hay supply for ruminants and horses can be stretched as follows:

1. Feed only ½ to ⅔ the normal ration of hay, but be on the alert for digestive disturbances. With lactating cows, too little hay or other forage, or hay that is chopped less than 0.25 in. in length, will result in low milk fat test and cow health problems. A good rule of thumb for the dairy producer to follow is to feed at least 1.5% of the body weight of the cow daily as hay, or hay equivalent from other forages. For a 1,400-lb cow, this calls for a minimum of 21 lb of hay or hay equivalent. Under conditions of extreme shortages, hay consumption can probably be lowered to the 5 to 10 lb per cow range, but there is not general agreement on the minimum amount of hay necessary to maintain maximum feed intake and milk production.
2. Replace 1 lb of hay with 3 lb of silage. The basis for this substitution: hay is usually 90% dry matter, whereas silage runs about 30% dry matter (70% moisture); hence, $90\% \div 30\% = 3$.
3. Replace 1 lb of hay with 4 lb of green chop. The basis for this substitution: green chop generally runs 22.5% dry matter (77.5% moisture); hence, $90\% \div 22.5\% = 4$.
4. Replace each 2 lb of hay deleted with 1 lb of grain.
5. Make the maximum use of such feeds as cottonseed hulls, corncobs, straw, and grass aftermath in the ration for (a) all but 5% of the alfalfa (or other legume) hay of grower rations; and (b) all of the "hottest" finishing ration, adding such supplementary proteins, minerals, and vitamins as necessary to balance the ration.
6. Get finishing cattle and lambs, and animals being fitted for show or sale, on high-concentrate rations as expeditiously as possible. In cattle, eliminate the stocker feeding period—get weaned calves on full feed as quickly as possible.
7. Provide such supplementary proteins, minerals, and vitamins as necessary. This is especially important with gestating-lactating females or young, growing animals. This may be accomplished by (a) feeding some legume, either

hay or silage, and/or (b) adding suitable protein, mineral, and vitamin supplements. For example, pregnant cows that are in medium to good condition can be wintered satisfactorily on 12 to 20 lb of straw or other low-quality roughage, plus 1 lb of oilseed cake or meal (or equivalent protein supplement), or on straw plus 4 to 5 lb of alfalfa or other legume hay. Unless cows have had good green pasture in the fall, and consequently have a store of vitamin A in their bodies, alfalfa pellets or a vitamin A supplement should be fed. In addition, straw-fed cattle should always have access to a mineral supplement high in calcium.

QUESTIONS FOR STUDY AND DISCUSSION

1. Why is making hay the most common method of preserving forage for storage?

2. Discuss the history of haymaking.

3. Tell of the magnitude and importance of the U.S. hay crop.

4. Discuss hay from each of the following standpoints:
   a. As an energy source.
   b. Camparable value to corn grain and corn silage.
   c. As a grain replacement.

5. Why is alfalfa known as "Queen of the Forages"?

6. What factors should be considered when determining the kind of forage to grow?

7. For hay to be of superior quality, it must be high in four factors. What are these factors?

8. Cite experimental evidence of the importance of hay quality.

9. What visual/sensory factors indicate high quality hay? Why should visual inspection supplement chemical analyses?

10. Discuss the per acre yield of crude protein and cellulose from first cutting alfalfa harvested at different stages of maturity.

11. Discuss the effect of harvesting alfalfa at different stages of maturity on the following:
    a. *In vitro* digestible dry matter (IVDDM), CP, amino acids, carotene, minerals, and fiber.
    b. Total nonstructional carbohydrates.
    c. Protein N and nitrate N.

12. What is the difference between neutral detergent fiber (NDF) and acid detergent fiber (ADF)? In what ways are they superior to proximate analysis?

13. What is near infrared reflectance spectroscopy (NIRS)? How is it used in forage analyses? What are its advantages and disadvantages?

14. Discuss each of the following points with reference to determining hay quality by chemical composition: (a) proper sampling, (b) what tests to make, (c) evaluating the test results, and (d) using the results in buying hay and balancing rations.

15. List and discuss each of the steps in growing quality hay.

16. Give a brief rundown on today's haying tools.

17. Describe the proper forage harvesting stage of alfalfa. How does harvesting stage of alfalfa affect chemical composition, gains of yearling steers, milk production, and lamb gains?

18. Why are plant food reserves important? Usually, the closer to maturity that plants are cut or grazed, the higher the stored food reserves. However, harvesting at a young growth stage makes for high feed value. So, when should alfalfa be harvested during the spring, the summer, and the fall?

19. Discuss modern hay cutting, curing in the swath and windrow, and raking or cocking.

20. Discuss reducing moisture content, shattering, bleaching and fermenting in modern haymaking.

21. Discuss the effect of rain damage on *in vitro* dry matter disappearance (IVDMD) and on crude protein (CP) of alfalfa and red clover.

22. What chemical hay drying agents and preservatives are being used? How and why are they being used?

23. Is there any basis for livestock producers favoring a certain cutting of alfalfa hay?

24. What products may be used in the nitrogenation of low-quality hays? What are the advantages and hazards of such treatments?

25. Describe and discuss the place of each of the following haymaking systems: long hay, chopped hay, packaged hay, and artificial drying.

26. What prompted the development of big round bales, large rectangular bales, and small mechanical stacks?

27. How do cubes and pellets compare?

28. How may the losses of large round bales due to weather and feeding be lessened?

29. What is meant by spontaneous combustion? What causes it in hay?

30. Discuss each of the following factors pertaining to buying and selling hay: (a) the amount, proportion, and value of the U.S. hay crop sold off the farm where it is produced, (b) hay sources, and (c) how hay was priced prior to the development of the near infrared reflectance spectroscopy.

31. Why were the federal grades of hay and straw that were created by the Agricultural Marketing Act of 1946 so little used, then finally terminated in 1988?

32. What are the recommendations of the American Forage and Grassland Council "Hay Marketing Task Force" and the National Alfalfa Hay Test Association relative to (a) determining hay quality by infrared (NIRS) or wet chemistry, along with visual inspection; and (b) hay standards (grades)?

33. When buying and selling hay, of what importance are the following: (a) freedom from toxic residues, (b) hay shrinkage, and (c) hidden hay costs?

34. Discuss (a) hay preparation, (b) hay feeding systems, and (c) feeding hay packages.

35. What factors determine the proportion of hay to concentrate in a ration?

36. Cite proof of the following statement: Ruminants need hay. What desirable factors does cud chewing time indicate? How much chewing time is needed?

37. For what animals may low-quality hay be used? For what animals is high-quality hay essential?

38. Should livestock producers accept a 10% hay waste and refusal as normal? If not, what can they do to lessen it?

39. When hay is high in price and scarce, as may happen after a long, hard winter and much snow in the northern part of the United States or during a severe drought, how can producers *stretch* their hay supply?

SELECTED REFERENCES

Title of Publication	Author(s)	Publisher
Crop Production, 5th Edition	R. J. Delorit L. J. Greub H. L. Ahlgren	Prentice-Hall, Inc., Englewood Cliffs, N.J., 1984
Crop Quality, Storage, and Utilization	C. S. Hoveland, Editor	American Society of Agronomy, Crop Science Society of America, Madison, Wisc., 1980
Crops in Peace and War: Yearbook of Agriculture 1950–1951		U.S. Department of Agriculture, Washington, D.C.
Effect of Processing on the Nutritional Value of Feeds	National Research Council	National Academy of Sciences, Washington, D.C., 1973
Forage and Pasture Crops	W. A. Wheeler	Van Nostrand Reinhold Company, New York, N.Y., 1950
Forage Conservation in the 80's	C. Thomas, Editor	British Grassland Society, The Grassland Research Institute, Hurley, Maidenhead, Berkshire, UK, 1980
Forage Crops	D. A. Miller	McGraw-Hill Book Company, Inc., New York, N.Y., 1984
Forages, 4th Edition	H. D. Hughes M. E. Heath D. S. Metcalf	The Iowa State College Press, Ames, Iowa, 1985
Forages: Resources for the Future	J. E. Oldfield, Chairman of Task Force	Council for Agricultural Science and Technology, Ames, Iowa, 1986
Grass to Milk	C. P. McMeekan	The New Zealand Dairy Exporter, Wellington, New Zealand, 1964
Grass: yearbook of Agriculture 1948		U.S. Department of Agriculture, Washington, D.C.
Haymaker's Handbook	M. A. Bales J. E. Baylor	New Holland, Inc., New Holland, Pa., 1987
Manual of the Grasses of the United States, 2nd Edition	A. S. Hitchcock; rev. by A. Chase	U.S. Department of Agriculture, Washington, D.C.
Pasture Book, The	W. R. Thompson	W. R. Thompson, State College, Miss., 1950
Plants for Man, 2nd Edition	R. W. Schery	Prentice-Hall, Inc., Englewood Cliffs, N.J., 1972
Practical Grassland Management	B. W. Allred	The Interstate Printers & Publishers, Inc., Danville, Ill., 1952
Stockman's Handbook, The, 6th Edition	M. E. Ensminger	The Interstate Printers & Publishers, Inc., Danville, Ill., 1983

Original painting by Tom Phillips

Chapter 

SILAGE/HAYLAGE/HIGH-MOISTURE GRAIN[1]

Contents Page

Contents Page

[1]The authors gratefully acknowledge the helpful suggestions of the following authorities who reviewed this entire chapter: J. E. Baylor, Ph.D., Professor of Agronomy, Pennsylvania State University, State College; R. A. Forsberg, Ph.D., *et al.,* Department of Agronomy, University of Wisconsin-Madison; S. C. Fransen, Ph.D., Forage Agronomist, Western Washington Research and Extension Center, Washington State University, Puyallup; D. A. Miller, Ph.D., Professor of Plant Breeding and Genetics, College of Agriculture, University of Illinois at Urbana-Champaign; and W. E. Murphy, Ph.D., Department of Agronomy, Oklahoma State University, Stillwater.

Fig. 9-1. Silage has gone modern! On this large family-owned dairy in the Hudson River Valley of Eastern New York, the oxygen-limiting silos are used to store haylage, corn silage, and high-moisture corn for lactating cows and young dairy stock. (Courtesy, A. O. Smith Harvestore Products, Arlington Heights, Ill.)

Silage may be defined as fermented forage plants. It is a very old method of preserving feed. Columbus found that the American Indians used pits or trenches in which to store their grain, and, centuries earlier in the Old World, silos were used as a means of preserving both grain and green forage. The Frenchman, Auguste Goffart, preserved forage in a silo in 1865. The discovery was so highly regarded that the French government awarded him the Cross of the Legion of Honor. The first tower silo built in the United States by white man is said to have been erected by F. Morris in Maryland in 1876.

Silage making is one of the 3 common methods of utilizing forage crops, the other 2 methods being grazing and haying. Grazing is the least expensive of the 3 methods, but it is seasonal in nature. In the spring and early summer, forage plants generally grow faster than they can be utilized by normal grazing; and they become dormant in cold weather.

The surplus forage produced during the growing season may be preserved for feeding during the winter months and other periods of pasture scarcity by haymaking. But weather conditions are not always favorable to haymaking. Ensiling, on the other hand, can be done in inclement weather. Also, it has the added virtues of succulence and of preserving a higher proportion of the nutrients of the plant than can be accomplished in haymaking.

Silage is primarily a beef and dairy feed, where it is used as part or the only roughage in the ration. It is also a good sheep feed. Sometimes it is fed to brood sows. Very little silage is fed to horses.

The importance of silage in this country is attested to by the fact that more than 170 million tons are made annually, of which more than 100 million tons are corn silage. Further, there is ample evidence that silage making is on the increase. It is estimated that 2,000 to 3,000 silos are constructed in the United States each year,[2] with tower silos increasing more rapidly than other types. Most silage is fed on the farms or ranches on which it is produced, rather than being purchased. It is important, therefore, that livestock producers know how to produce good silage, as well as how to feed it, for most of them determine their own destiny from the standpoint of quality. For this reason, this chapter covers silage from production to feeding.

[2]Estimate made by International Silo Assn., Inc., 410 North Michigan Ave., Chicago, IL 60611, in a personal communication to the senior author. The 2,000 to 3,000 estimate includes moved and rebuilt structures and is exclusive of baled and bagged silage.

Fig. 9-2. Silage is a popular feed for lactating dairy cows. (Courtesy, Holstein-Friesian Assn. of America, Brattleboro, Vt.)

THE ENSILING PROCESS

The ensiling process refers to the changes which take place when forage or feed with sufficient moisture to cause fermentation is stored in a silo in the absence of air. Many different kinds of plants and plant products can be preserved in this way. Sauerkraut, for example, is the silage form of cabbage.

The ensiling process is governed by the interaction of three factors: (1) the chemical composition of the plant material placed in the silo, (2) the amount of air entrapped or allowed to enter the mass, and (3) the activity of the bacterial population.

The entire ensiling process requires 2 to 3 weeks, during which time the following phases of varying intensity occur:

1. **Aerobic phase (with air).** This is the respiration phase. In this phase, the living plant cells of the forage continue to respire, or take up oxygen, and the plant enzymes and aerobic bacteria use the readily available carbohydrates to produce heat, water, and carbon dioxide.

The oxygen supply is usually exhausted in 4 to 5 hours, but carbon dioxide continues to accumulate for about 48 hours; the temperature of the ensiled material increases over a period of about 15 days, but seldom exceeds 85 to 90°F, then decreases gradually. At this point, anaerobic conditions prevail.

2. **Anaerobic phase (without air).** This is the "pickling" phase. When the available oxygen of the entrapped air has been consumed, anaerobic bacteria—chiefly acid-forming and proteolytic—multiply at a prodigious rate. Simultaneously, the molds and the yeasts die, but continue in a

minor way to provide enzyme systems which produce alcohol and other end products.

The combined anaerobic activity produces the following changes: (a) The complex carbohydrates and sugars (especially the sugars) are broken down into lactic acid (the acid in sour milk), some acetic acid (the acid in vinegar), and a small amount of other acids and alcohols; (b) small quantities of the proteins are broken down into ammonia, amino acids, amines, and amides; and (c) the acidity finally reaches a point when the bacteria themselves are killed, and the silage-making process is completed. At this stage, ideally the lactic acid concentration is equivalent to 4 to 10% of the dry matter.

3. **Stable phase.** When a pH of 4.2 or less is reached,[3] silage is stable and may be kept for years if air is excluded. Drier silages stabilize at a higher pH.

Yeasts, followed by molds, will again become active if silage is exposed to air—with the opening of the silo, or through leaks or air pockets.

These changes in the ensiling process and the quality of the silage produced can be altered in a number of ways; by moisture content, legume content, stage of maturity, length of cut, speed of filling, distribution in the silo, and amount of packing.

Some additional and pertinent facts about the ensiling process follow:

1. **Microbiology.** A desirable fermentation is dependent upon the proliferation and activity of two types of lactic acid-producing bacteria: (a) homofermentative bacteria, the principal end product of which is lactic acid; and (b) heterofermentative bacteria, which produce lactic acid, acetate, sorbitol, and ethanol.

The type of fermentation and keeping quality of silage may be affected by the activity of bacteria belonging to genera *Clostridium,* which are involved in spoilage, characterized by high pH, high water soluble nitrogen content, and high volatile nitrogen content. The activity of *Clostridium* can be minimized by (a) reduction of forage moisture under 70%, and reduction of the pH to less than 4.2.

Sometimes yeasts are a problem as oxygen becomes available, following opening of the silo. The activity of yeasts is influenced by the availability of carbohydrates, oxygen, and pH. Fungal infestation may follow yeast activity. This is reason for concern because some fungi are capable of producing toxins that can have undesirable effects on animals consuming the mold-infested silages.

2. **Optimum pH.** The formation of a pH of 3.5 to 4.5 is the key to good wilted silage preservation, because it will prevent the growth of bacteria, including those which cause rotting and putrefaction. Excellent low-moisture silage (40 to 60% moisture) is frequently made in the pH range of 4.0 to 4.5, and even up to 5.0.

3. **Sugar content of forage.** When cut at the proper stage of maturity, corn and sorghum forage possess just the right amount of sugar for the production of good silage. On the other hand, if cut when too immature, these crops, being high in sugar and water, may cause excessive acid formation and result in unpalatable silage.

Today, high sugar content corn is being grown for silage. It makes very high-quality feed, although yield and quality are not superior to conventional corn made into silage.

4. **Moisture of grass/legume silage.** In making good-quality grass/legume (hay crop) silages, either the forage should be wilted below 70% moisture content or a preservative should be added. Two theories have been advanced for this necessity; namely, (a) that such forages are low in sugar, and (b) that the physical conditions are not suitable for the growth of the desirable acid-producing organisms. Without this precaution, high-moisture or grass-legume silages form a cold fermentation containing high levels of butyric acid, which is undesirable.

• **Browning reaction/Maillard reaction**—Lowering of the moisture content without excluding air may lead to undesirable effects on silage. It may cause high temperature in the silo and damage the protein and energy value of low-moisture silage as a result of the nonenzymatic browning reaction, also termed the *Maillard reaction,* evidenced by a tobacco-brown or black color and a caramelized or tobacco odor. In this reaction, heat causes carbohydrates to combine with protein to produce an insoluble product, decreasing the digestibility of both protein and energy sources. The loss of feeding value depends on the degree of the heat damage. Determination of heat damage to protein can be assessed by determining the residual insoluble N in acid detergent fiber.

ECONOMICS/ADVANTAGES/DISADVANTAGES OF SILAGE

Storing feed as silage often makes it possible to get more forage preserved from fewer acres. As a result, there is evidence that silage making in the United States is increasing. During the past 20 years, the total tonnage of corn silage, which comprises about ¾ of all the silage fed in the United States, has increased by approximately 25%. Corn, sorghum, and grass/legume silage production have benefited from improved technology, including higher yielding silage crops, improved harvesting and handling machinery, and lower cost of storage. But silage has many other **advantages;** among them, the following:

1. It makes it possible to increase the livestock carrying capacity of a farm or ranch. Thus, corn, the chief U.S. silage crop, (a) yields more total digestible nutrients per acre than most other forage crops, and (b) has 30 to 50% higher feeding value as silage than when fed as grain and stover.

2. It retains a higher proportion of the nutrients of plants than can be accomplished by haymaking, even if the weather is satisfactory for the latter, chiefly because shattering and bleaching losses are held to a minimum. Thus, ensiling grass preserves 85% or more of the feed value of the crop, whereas haymaking under the best of conditions will preserve only 80%, and under poor conditions only 50 to 60%.

3. It is feasible to produce top quality hay crop silage during times of inclement weather when it would normally be impossible to cure the forage crop properly as hay.

[3]The pH refers to the degree of acidity of the silage. Acidity or alkalinity is measured on a scale of 0 to 14, with 7 being neutral. Numbers that are below 7 indicate an acid condition, with the degree of acidity increasing as the numbers get smaller. Numbers that are above 7 indicate an alkaline condition, with the degree of alkalinity increasing as the numbers get larger. The preservation of silage involves only pH values that are on the acid side of the scale.

4. It is the most economical form in which the whole stalk of corn or sorghum can be processed and stored.

5. It requires less storage space per pound of dry matter than dry hay, even when the latter is baled or chopped. A cubic foot of silage contains about 3 times more dry weight of feed than a cubic foot of long hay stored in the mow.

6. It practically eliminates the danger of loss by fire if stored within the recommended moisture range.

7. It is the most satisfactory and economical way in which to preserve a number of by-product feeds.

8. It makes it possible to remove forage crops from the land earlier than would otherwise be possible.

9. It is one of the best methods of controlling the European corn borer since the removal of cornstalks is required in making silage.

10. It helps to control weeds, which are often spread through hay or fodder.

11. It is the cheapest form in which a good succulent winter feed can be provided on most farms and ranches.

12. It is a better source of protein and of certain vitamins, especially carotene, and perhaps some of the unknown factors, than dried forage.

13. It is a very palatable feed and slightly laxative in nature.

14. It makes for less waste, the entire plant being consumed, which is an important consideration with coarse, stemmy forages.

15. It is without a peer from the standpoint of longtime storage, holding the feeding value of protein, carbohydrates, and carotene better than any other method of preservation, and providing a desirable backlog against drought or any other crop failure.

16. It may be completely mechanized as a feeding system, thereby eliminating much labor and time.

17. It offers many advantages over pasture, including (a) no fencing required, (b) approximately one-third more forage from the same acreage, (c) harvesting at optimum maturity, (d) more uniform quality, (e) little or no bloat, (f) closer observation of animals that are confined to a lot or corral, (g) reduced damage to the growing sward, and (h) lessened topsoil loss as a result of alleviating the hoof action of grazing animals.

Some of the **disadvantages** of silage are:

1. It requires a silo or storage structure and other special equipment, for best results. In comparison with the simpler methods of field curing and storing hay, this may mean higher costs for the small operator.

2. It contains considerably less vitamin D than sun-cured hay.

3. It necessitates that 2 to 3 times as much tonnage be handled as when the same forage is dried for hay, due to the high water content.

4. It incurs an added expenditure when preservatives are necessary.

5. It lessens the amount of organic material returned to the soil, which is needed in some soil types.

SILOS

Silage may be stored in almost any kind of container. The main requisites of a good silo, regardless of kind, are:

1. That its size be in keeping with the number and kind of animals to be fed daily, the length of the feeding period, and the amount of forage available for ensiling.

2. That it exclude air from the stored material, including entrance of air around the doors of tower silos.

3. That the sidewalls be straight and smooth in order to prevent the formation of air pockets and allow for unimpeded packing.

4. That it be of adequate depth, thereby making for better packing and less surface area to total mass exposed.

5. That it be properly reinforced. This point is especially important where direct cut grass silage is made, because it exerts from ½ to 2½ times as much pressure on the walls as does corn silage. Thus, tower silos which are originally built for corn or sorghum silage but which are to be filled with wet grass silage should be either (a) reinforced with extra bands placed around the lower part to strengthen the walls if an inspection reveals that the existing strength is not adequate, or (b) not filled to more than half capacity.

6. That adequate provision be made for the escape of surplus juices, either by a drain or by a gravel bottom.

7. That it be conveniently located and accessible in all kinds of weather, from the standpoint of both filling and feeding.

Silos may be classified according to the five basic methods used for processing forages. Each method is associated with the shape and material of the structure, which also influences the efficiency of preserving the silage. Also, the different shaped structures are adapted to different methods of filling and unloading. Within each classification there are many variations of each type depending upon the manufacturer.

The kind of silo and the choice of construction material should be determined primarily by economics and the suitability to the particular needs of the farm or ranch.

Silos may be classified as follows:

- I. Conventional Upright (Tower) Silos
  1. Concrete stave
  2. Galvanized steel
  3. Wood stave
  4. Monolithic concrete (poured in place)
  5. Tile block
  6. Brick
- II. Gastight (Oxygen-Limiting) Silos
  1. Glass-lined structures
  2. Concrete stave
  3. Galvanized steel
  4. Monolithic concrete
- III. Pit Silos
- IV. Horizontal Silos
  1. Trench silos (belowground level)
  2. Bunker silos (aboveground level)
- V. Temporary Silos
  1. Enclosed stack silos
  2. Open stack silos
  3. Modified trench-stack silos
  4. Plastic silos

Some pertinent information relative to each main kind of silo is given in the discussion which follows, but it is not within the scope of this book to give detailed silo plans and specifications. The latter may be obtained from local authorities, from silo manufacturers, or by writing to the state agricultural college.

Conventional Upright (Tower) Silos

The upright or tower silo, which is sometimes referred to as the "watch tower of prosperity," is a cylinder built aboveground. Its round shape withstands pressure well and is adapted to good packing.

The tower silo is a permanent farm structure, and, as such, should be constructed to withstand long usage. Although tower silos are usually handy, they are, in their initial cost, generally the most expensive of all types. However, they have the following advantages: (1) durability, (2) minimum top and side spoilage, (3) convenience for feeding during periods of inclement weather, and (4) well adapted to automation (loading and unloading machinery).

Fig. 9-3. Upright (tower) concrete stave silos. (Courtesy, USDA)

Recent developments in construction of tower silos have been made in (1) bottom unloaders, with elimination of doors; (2) center-core unloaders, with elimination of most of the doors; (3) top-unloaders; and (4) large diameter features (24 to 30 ft in diameter). These new developments are also available in gastight (oxygen-limiting) silos.

Gastight (Oxygen-Limiting) Silos

These silos resemble conventional tower silos, but they are more expensive because of their construction.

Sealed silos are designed for storage of wilted or even overwilted forage with as little as 40 to 55% moisture content or for the storage of high-moisture grain containing 22 to 40% moisture. These structures may be partly filled on widely separated dates, provided they are sealed between fillings. Packing and tramping of forage is not necessary or recommended although distribution is desirable.

Fig. 9-4. Chopped forage being transferred from a forage truck into a blower which elevates it through a pipe to the top of, and into, the silo. (Courtesy, Deere & Company, Moline, Ill.)

Practically all outside air is kept out of the oxygen-limiting silo, and carbon dioxide formed during fermentation is kept in. A plastic breather bag and a pressure relief valve located in the top of some of these structures compensate for differences in inside and outside pressures without allowing outside air to contact the forage. Before each filling, the plastic breather bag should be checked for holes and flaws, the pressure relief valves should be inspected, the structure should be checked for leaks, door seals should be inspected, and unloaders should be checked for wear in order to prevent malfunctioning during unloading.

There is not sufficient oxygen inside a gastight (oxygen-limiting) silo to sustain life. So, never enter a filled or partially filled sealed silo without proper life supporting equipment.

Pit Silos

The pit silo is shaped like the tower silo, but inverted into the ground. It resembles a well or cistern. The walls of a pit silo may or may not be lined. Where the water table is low enough that the silo will not fill with water, such as in semiarid areas, the pit silo is very satisfactory.

In comparison with tower silos, pit silos have the following advantages: (1) they are never damaged by storm or fire, (2) they require less reinforcing, (3) they minimize silage losses because they do not have doors, and (4) they avoid frozen silage. But they have the following disadvantages: (1) they are dangerous, due to the frequent presence of suffocating carbon dioxide gas, and (2) considerable work is involved in removing the silage, despite the development of a number of hoist devices.

Before entering a pit silo, it is recommended that a lighted cigarette lighter, candle, or lantern be lowered into the silo. If the flame goes out, assume that the pit is dangerous to enter and replenish the carbon dioxide gas with fresh air before entering. This precaution also applies to upright silos, particularly when first filled.

Horizontal Silos

Only two types of horizontal silos will be discussed herein; trench silos and bunker silos (or horizontal surface silos), both of which may be adapted to self-feeding.

Fig. 9-5. Unloading and packing corn silage in a horizontal silo. (Courtesy, USDA)

TRENCH SILOS

The trench silo is a horizontal, trenchlike structure that can be built quickly and at low cost. It is most popular in areas where the weather is not too severe and where there is good drainage. The walls of a trench silo may or may not be lined, but for making good silage they should always be smooth; and there may or may not be a floor. A trench silo should be wider at the top than at the bottom, and the bottom should slope away from one end in order that excess juices will drain off if material with too high moisture content is ensiled.

Fig. 9-6. Front-end loader moving alfalfa silage from trench silo to truck. (Courtesy, James Tappan, Arizona Dairy Co., Higley, Ariz.)

Trench silos have the **advantages** of (1) low initial cost, (2) low cost of filling machinery, for a blower is not necessary, (3) relative freedom from freezing, and (4) ease of construction. The chief **disadvantages** of trench silos in comparison with tower silos are the (1) larger area to seal, (2) higher spoilage losses, and (3) inconvenience in feeding during inclement weather. Because of shallowness, throughout the filling process the forage should be packed very thoroughly in a trench silo by driving a tractor back and forth over it, with the tractor kept as level as possible to prevent tipping over. When filling is completed, the top should be carefully sealed (1) by 3 to 6 in. of limestone or soil, with or without a seeding of rye or winter wheat (where birds are a problem, they may eat the grain before it sprouts; hence, grain seeding will not work); (2) by wet straw, poor-quality hay, marsh grass, or sawdust; (3) by a mixture of soil and straw; (4) by waterproof paper lapped about 12 in. at the joints and covered with soil or straw; or (5) by polyethylene, plastic, aluminum, or other materials, properly weighted down, commonly by used tires.

BUNKER OR SELF-FEEDER SILOS

As a laborsaving measure, some operators are now constructing horizontal silos aboveground (or slightly recessed) —usually with concrete floors, and side walls of wood, concrete, or other materials—and self-feeding silage to cattle by making use of either a feeding fence or an electrified pipe suspended 30 to 48 in. from the floor of the silo.

In the first of these methods, the fence (or gate, hurdle, or stanchion) through which the animals put their heads, is placed at the end (or side) of the silo and moved toward the silage as it is eaten. Feed wastes are reduced if the bottom section of the gate or hurdle is solid and the top slopes away from the silo (about 7½ in. out at the top, on gates 5 ft high), so that the animals are forced to eat down toward the bottom of the stack.

The principle of the electrified pipe is similar to that of the feeding fence—both are designed to self-feed silage with minimum wastage. However, in comparison with the fence, the pipe is much lighter and easier to move, and not so easily broken.

Most of those who have self-fed successfully from bunker silos recommend the following:

1. That there be about 6 in. of feeder space per animal (cattle unrestricted feeding time).
2. That the silage not be piled higher than 6 ft.
3. That the silage be loosened in front of the fence or pipe.
4. That there be once-a-day policing (a) to push uneaten good silage back up against the stack face, and (b) to push the spoiled material and manure to the rear of the silo.

Temporary Silos

Several kinds of above ground temporary silos are used. Generally, this kind of storage is used to meet emergencies, to supplement permanent silos, or to ensile such by-product feeds as cannery refuse, pea vines, and beet tops or pulp. Above ground temporary silos are low in cost, can be erected at short notice, require no special foundation, and can be set up on almost any level site convenient for filling and feeding.

The amount of spoilage in aboveground temporary silos can be kept to a minimum by having straight sides, considerable height, proper packing, and protection with fiber-reinforced paper, plastic, or other suitable material. Also, the use of propionic acid, formic acid, or other effective organic acids, applied to harvested materials at the time of chopping, is very effective in reducing spoilage in temporary silos. It is important that stacked material treated with organic acid be covered with plastic to prevent dilution of the acid by rain and snow.

The spoilage on the sides of temporary silos will vary from 4 to 20 in., with greater spoilage in grass silage than in easier-fermenting corn and sorghum silage.

Most above ground temporary silos can be classed as belonging to one of the following four kinds:

ENCLOSED STACK SILOS

These are built entirely aboveground, without trenches or holes. They are upright, are generally circular, and are enclosed by snow or picket fences, poles, wooden staves, heavy woven wire, or other materials. Most of them are lined with tar paper, plastic, or tough fiber-reinforced paper made especially for the purpose. Because of the relatively weak walls of these silos, their height should not be greater than twice their diameter unless poles are set at four to six points around their circumference and tied together at the top.

OPEN STACK SILOS

These are similar to enclosed stack silos, except that no supports or walls are used. As would be expected, greater spoilage is encountered in the open stack than in the enclosed stack, because of the greater evaporation and spoilage which accompanies the exposed sides. Less spoilage, percentage-wise, occurs in stacks of considerable size—stacks that contain 500 to 1,000 tons or more silage—than in smaller stacks.

MODIFIED TRENCH-STACK SILOS

This silo, which is intermediate between a trench and stack silo, is adapted to areas where the ground-water level is high. It is constructed by excavating a shallow trench 12 to 18 in. deep, by piling the excavated earth on either side of the trench to support the silage and to keep out surface water, by packing silage thoroughly in and over the trench to a height of 10 to 15 ft and by covering the stack with any one of the materials recommended for covering the trench silo (see trench silo). The modified trench-stack silo is designed to give greater protection and less spoilage than can be accomplished by open or enclosed stacks. Also, this type of silo is easier to feed from than a trench silo.

PLASTIC SILOS

Plastic (polyethylene) is now available for use as temporary silos, and for use as covers for trench, bunker, and tower silos, and as silo liners. If not punctured, it is nearly airtight. Plastic thicknesses range from 4 to 9 mils. The thicker grades have better tear and puncture resistance, and low permeability by both air and moisture; however, they cost more and are difficult to tie tightly. Thinner grade plastics are less costly, more pliable, and easier to seal.

The two common types of plastic silos are: (1) enclosed plastic bag or tube silos, and (2) round bale plastic covered silage.

Enclosed Plastic Bag or Tube Silos

These temporary silos are made of heavy plastic in the form of a tube into which forage is forced by a special machine (much like stuffing sausage). The machine needed to pack the tube is generally rented or owned cooperatively. The filled structure is 8 ft in diameter and about 100 ft long. Preservation of silage is excellent provided the ends are kept sealed and the plastic is not torn or damaged by rodents or other animals. To remove or self-feed the silage, the plastic is cut and folded back at one end to expose as much silage as needed each day. The plastic cannot be re-used.

Fig. 9-7. Plastic tube silos. (Courtesy, Montana State University, Bozeman)

Round Bale Plastic Covered Silage

The most common methods for using plastic material to produce round bale silage are:

1. **Individual bags.** Bags come in various lengths, diameters, and thicknesses. A tractor-mounted spear device is needed to lift the bale while applying the bag. Then the bale is placed in storage position before it is tied off. If possible, the bales should be stacked in cord-wood fashion to reduce exposed surface area. Then, a plastic cover over the entire stack may reduce storage damage.

2. **Plastic tubes.** These consist of several round bales stuffed by a machine into a long plastic tube which is then sealed at both ends. The filled plastic tube resembles an "Enclosed Plastic Bag or Tube Silo" described earlier, except that it consists of a row of round bales covered with plastic rather than long, continuous sausage-type silage material. Plastic tubes can be effective and time-saving, but the multiple bales stored in one package tend to increase the loss if the bag is torn, punctured, or opened for feeding. However, the tube can be easily tied off into one-bale (or more) segments for feeding.

3. **Sheet plastic.** Several round bales can be stacked under two sheets of plastic, with the plastic ends on the ground covered with soil, sand, or other effective sealing procedure. The hazard with this type of storage is that there are more possibilities for air leaks to develop, which may result in a large number of bales being spoiled.

Fig. 9-8. Several round bales of silage stacked under sheets of plastic. (Courtesy, Sperry New Holland, New Holland, Pa.)

ADVANTAGES AND DISADVANTAGES OF ROUND BALE PLASTIC COVERED SILAGE

Round bale silage may serve as a supplement to, rather than a replacement for, other stored forages on most livestock farms. Some **advantages** are:

1. It doesn't require silo structures.
2. Hay-making equipment may be used to harvest it.
3. When silo capacity is lacking because of a surplus of forage, round bale silage can offer an effective method of storing excess forages.

Fig. 9-9. Round bale silage properly fed in a steel rack, thereby alleviating the energy for chopping and lessening the labor for feeding. (Courtesy, Sperry New Holland, New Holland, Pa.)

4. The round bale silage system can be used to save a mowed field of hay when an anticipated rainstorm or extremely high humidity interfere with proper hay curing.
5. Round bale silage saves about one-third of the harvesting energy plus the fuel required for chopping silage.
6. Round baled silage can be self-fed if properly presented, thereby saving both labor and fuel by not requiring daily silage feeding.

But there are **disadvantages**; among them, the following:

1. Conditions associated with round bale silage are not optimum for fermentation.
2. Extreme care must be taken to eliminate air leaks.
3. The system requires prompt handling and storage of bales.
4. Machines for lifting and moving heavy, high-moisture bales must be available.
5. Plastic bags, storage tubes, or plastic sheets to cover group-stacked bales must be purchased.
6. Plastic is easily damaged, which can result in forage losses greater than in conventional silos.

Generally speaking, users of baled silage have liked the concept, but (1) not the machinery for bagging it, and (2) not the sometimes poor fermentation. Recently, automatic equipment has been developed that individually shrink-wraps big bales with tough plastic and provides an airtight seal; and additives/preservatives have been successfully added to large round bales.

How to Determine the Size Silo to Build

The size of silo to build should be determined by needs. With tower type and pit silos, this means (1) that the diameter should be determined by quantity of silage to be fed daily, and (2) that the height (depth in a pit silo) should be determined by the length of the silage feeding period. Similar consideration should be accorded trench silos.

SIZE OF TOWER SILO

If the diameter is too great, the silage will be exposed too long before it is fed; and, unless a quantity is discarded each day, spoiled silage will be fed.

The minimum recommended rate of removal of silage varies with the temperature. In most sections of the United States, it is desirable that a minimum of 1½ in. of silage be removed from tower silos daily during the winter feeding period, with the quantity increased to a minimum of 3 in. when summer feeding is practiced. Of course, the total daily silage consumption on any given farm or ranch will be determined by (1) the class and size of animals, (2) the number of animals, and (3) the rate of silage feeding. (Some suggestions on how much silage to feed are given in the chapters devoted to feeding each species of livestock.)

Silo height should be determined primarily by the length of the intended feeding period. In general, however, the height should not be less than twice, nor more than three and one-half times the diameter. The greater the depth, the greater the unit capacity. Extreme height is to be avoided because (1) of the excessive power required to elevate the cut silage material, and (2) of the heavier construction material required. Also, it is noteworthy that, with silos of large diameters, more labor is required in carrying the silage to the silo door if silage is manually removed.

Table 9-1 may be used as a guide in computing the proper diameter of tower silo for any given farm or ranch.

TABLE 9-1
MAXIMUM DIAMETER OF TOWER SILO TO BUILD IF SILAGE IS TO BE KEPT FRESH

Inches of Silage Removed Daily	Total Silage Removed Daily With An Inside Silo Diameter Of:											
	10 ft		12 ft		14 ft		16 ft		18 ft		20 ft	
	(lb)	*(kg)*	(lb)	*(kg)*	(lb)	*(kg)*	(lb)	*(kg)*	(lb)	*(kg)*	(lb)	*(kg)*
Summer: 3 in. *(7.6 cm)* daily will remove[1]	786	*357*	1,312	*596*	1,539	*700*	2,010	*914*	2,545	*1,157*	3,142	*1,428*
Winter: 1½ in. *(3.8 cm)* daily will remove[1]	393	*179*	656	*298*	770	*350*	1,005	*457*	1,272	*578*	1,571	*714*

[1]The lb*(kg)* listed in each of the columns are approximations based on an average constant weight of 40 lb of silage per cu ft *(18.2 kg/.028 m³)*. Low-moisture silages to 40–55% moisture content will weigh somewhat less than 40 lb per cu ft *(18.2 kg/.028 m³)*.

Fig. 9-10 shows capacities of tower silos of different heights and diameters. It is based on well-eared corn silage harvested in the early dent stage, cut in ¼-in. lengths, well-tramped when filled, and with the silo refilled once after settling for a day.

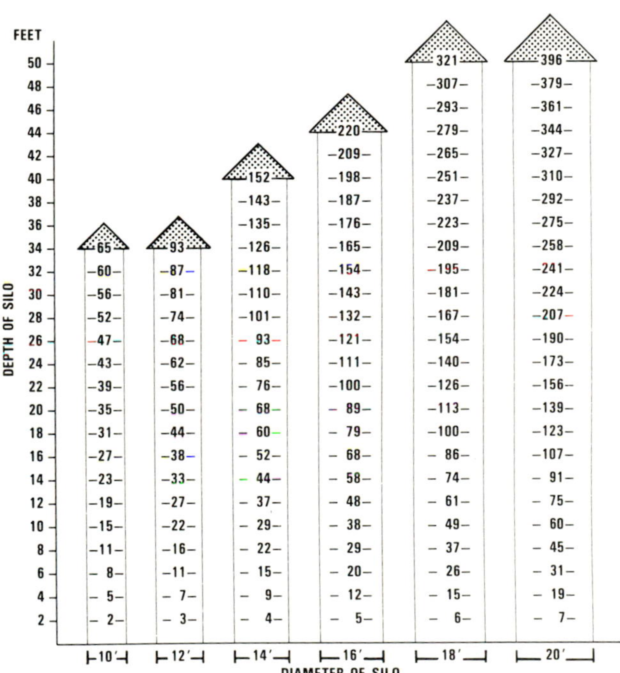

Fig. 9-10. Capacity in tons of settled corn silage in tower silos of varying sizes. See Table 9-4 for tabular material.

Fig. 9-10 can be adapted for corn silage of different stages of maturity and grain content, and for other kinds of silage, by applying the rules of thumb given in Table 9-2.

The following example will serve to illustrate how to determine the size tower silo to build:

Over a period of years, a farmer plans to winter 34 head of 425-lb stocker calves on a ration of corn silage and protein supplement. There is a 240-day wintering period. No increase in the herd is planned. What size tower silo should be built?

The answer is obtained as follows:

TABLE 9-2
EFFECT OF KIND OF SILAGE ON WEIGHT

Kind Of Silage	Changes To Be Made In The Number Of Tons Shown In Table 9-4
1. For corn silage ensiled when less mature than usual	Add 5–10%.
2. For corn ensiled when dry or overripe	Deduct 5–10%.
3. For corn very rich in grain	Add 5–10%.
4. For corn with very little grain	Deduct 5–10%.
5. For sorghum silage	Use the same weights as used for corn silage of comparable grain and maturity.
6. For sunflower silage	Add 5–10%.
7. For grass silage	Add 10–15%.[1]

[1]For this reason, a stronger structure is necessary where grass silage is stored.

1. First, here are the silage requirements:

a. Stocker calves weighing 425 lb on a ration of corn silage and protein supplement should receive about 30 lb of silage per head daily.

b. 34 × 30 = 1,020 lb of silage required daily for the 34 calves.

c. 1,020 × 240 = 244,800 lb, or 122.4 tons, of silage required for the 240-day wintering period for the 34 calves.

2. Next, here is the size silo to build:

a. Table 9-1 shows that in order to remove 1,005 lb of silage daily (which is only slightly less than the 1,020 lb needed daily), with 1½ in. removed from the top of the silo each day, the diameter of the silo should not be greater than 16 ft.

b. Fig. 9-10 can now be used as a guide in determining both the proper height (or depth) and diameter of the silo. Fig. 9-10 shows that a silo 16 ft in diameter and 27 ft high will hold 127 tons of silage, which would allow for 4.6 tons spoilage in excess of the required 122.4 tons. However, the height of a silo should not be less than twice the diameter. It appears best, therefore, to plan on a 14-ft diameter silo. As noted in Fig. 9-10, 34 ft of settled silage in a 14-ft diameter silo will provide 126 tons of silage, which would allow for 3.6 tons spoilage in excess of the required 122.4 tons. To allow for settling, an additional 4 to 6 ft should be added to the height, thus making a 38- to 40-ft height.

c. The size of silo to build to meet the needs outlined in this example, therefore, is one that is 14 ft in diameter and 38 to 40 ft high. Sufficient additional height, usually 4 to 6 ft, should be added to provide the necessary space required for the silage unloader.

shown in Table 9-5. Losses in trench and open stack silos are also influenced by depth; less surface is exposed in deeper silos.

TABLE 9-5
ESTIMATED (1) AVERAGE, AND (2) RANGE OF SILAGE STORAGE LOSSES

Type Of Silo	Percent Of Loss	
	Average	Range
	(%)	(%)
Gastight upright	5	1-10
Conventional upright	6	2-12
Horizontal (trench)	15	8-25
Open stack	20	12-30

Losses in the silo are of four types: (1) surface or top spoilage, (2) seepage, (3) gaseous, and (4) heating (browning reaction and spontaneous combustion).

Surface or top spoilage losses of 20% or more may occur in stack silos and in any uncovered bunk, trench, or pit silo. These losses can be reduced by the use of suitable protection, such as a plastic cover.

Seepage losses can be high in high-moisture silage stored in upright silos. The higher the silo, the greater the pressure and the higher the losses through seepage. The seepage carries soluble feed nutrients with it. Horizontal silos have less seepage loss than upright (tower) silos because of lower vertical pressure. Seepage losses can be reduced by wilting forages to less than 65% moisture before ensiling.

Gaseous losses are unavoidable so long as the plant material respires and there is subsequent fermentation. However, these losses can be minimized by avoiding entry of air into the silo, by having the pH decline rapidly, and by encouraging favorable fermentations.

Lowering the moisture without excluding the air may lead to heat damage, known as the browning reaction or Maillard reaction. (See "Browning reaction/Maillard reaction" in the earlier section of this chapter headed "The Ensiling Process.")

Spontaneous ignitions sometimes occur in low-moisture silage (haylage). For such losses to occur, there must be a build-up of temperature to the combustion point in the silo mass, combined with a low transfer of heat. These fires are very difficult, and usually impossible, to extinquish. The addition of water may build up pressure and lead to an explosion. Most silo fires should be allowed to burn.

KINDS OF SILAGE

A great variety of crops can be and are made into silage. Generally speaking, crops that are palatable and nutritious to animals as pasture, as green chop, or as dry forage also make palatable and nutritious silage. Likewise, crops that are unpalatable and unnutritious as pasture, as green chop, or as dry forage make unpalatable and unnutritious silage.

Most silage in the United States is made from either corn or sorghum, with corn silage far in the lead—over 16 times as much corn silage as sorghum silage is made. In 1985, 102.6 million tons of corn silage and 6.3 million tons of sorghum silage were produced in the United States. At the present time, it is estimated that 65% of the nation's silage is made from corn and sorghum and 35% from grasses, legumes, and other feeds. In addition to the kinds of silage already mentioned, silage is made from sunflowers, the small grains, sugar beet tops, crop residues, wastes from food processing (sweet corn, green beans, green peas), root crops, and various vegetable residues.

Many of the leading silage crops are listed in Table 9-6, which also includes an estimate of the expected and potential production from these crops, along with the normal stage of maturity at harvest. The average composition of the silages produced from several of these crops is given in Table 9-7.

TABLE 9-6
LEADING SILAGE CROPS[1]

Crop	Expected Production[2]			Potential Production[3]			Stage For Ensiling
	Dry Matter Yield	Digestible Protein	TDN	Dry Matter Yield	Digestible Protein	TDN	
	(tons/A)	(lb/A)	(lb/A)	(tons/A)	(lb/A)	(lb/A)	
Corn	6.0	600	8,100	10.5	1,100	14,300	Hard dough or early glaze
Grain sorghum	5.5	550	7,200	9.0	900	11,700	Soft to medium dough
Forage sorghum	6.0	600	6,900	9.0	900	10,400	Soft to medium dough
Sorghum-Sudangrass hybrids	4.5	540	5,000	8.0	960	8,800	Early bloom
Sudangrass	3.5	420	3,900	6.0	720	6,600	Early bloom
Alfalfa	5.0	1,500	6,500	9.0	2,700	11,700	Bud stage, first cut; early bloom, other cuts
Legume-grass mixtures	4.5	900	5,400	8.0	1,600	9,600	Legumes, early bloom
Small grains	4.0	560	4,800	7.0	980	8,400	Boot to early head

[1]From paper entitled "Silage Production, Preservation, and Quality," by John E. Baylor, The Pennsylvania State University, *Dairy Science Handbook*, Vol. 9, 1976, p. 224, published by Agriservices Foundation, edited by M. E. Ensminger. To convert to metric: tons/A = 0.907 metric ton/0.4 hectare; lb/A = 0.454 kg/0.4 hectare.

[2]Expected production—Figures presented are better than average yields for the country. They are yields which might be expected for average or better farmers who are following recommended practices.

[3]Potential production—Figures presented are production yields attainable where good soils, excellent growing conditions, and top management are present. These silage yields equivalents have been produced on farms and experiment stations prior to this time.

TABLE 9-7
COMPOSITION OF VARIOUS SILAGES

Type Of Silage	Analyses On A Dry Matter Basis			
	Crude Protein	TDN	Ca	P
	(%)	(%)	(%)	(%)
Corn	8.3	68.0	0.31	0.27
Grain sorghum	7.9	55.0	0.34	0.19
Forage sorghum	9.2	57.9	0.30	0.24
Oats	10.0	57.0	0.47	0.33
Alfalfa	17.4	59.0	1.75	0.27

Corn and Sorghum Silage

For the United States as a whole, corn ranks first in importance as a silage crop. Generally more total digestible nutrients can be obtained from an acre of corn as silage—which will yield from 5 to 25 tons of forage per acre, with an average of about 14 tons—than can be obtained from an acre of any other crop. Also, corn ensiles easily without the aid of a preservative, and keeps almost indefinitely in a good silo, is highly palatable, is well adapted to mechanized feeding, and may be fed with little waste.

There are four kinds of corn silage; namely,

1. **The whole corn plant.** When at the peak of its nutritive value and right for ensiling, the whole corn plant contains 1½ times the nutrients of the ripened grain that the plant would have yielded. Also, in corn silage made from the whole crop more than 90% of the nutrients produced are saved.

Fig. 9-11. Harvesting corn silage for cattle in open-fronted housing, in a midwest beef operation. (Courtesy, A. O. Smith Harvestore Products, Arlington Heights, Ill.)

2. **Ear corn silage.** The ensiled ears contain up to 68% of the nutrients of the entire corn plant.

3. **Corn stover silage.** The forage remaining after harvesting a grain crop. This accounts for about one-third of the total nutritive value of the crop.

4. **Shelled-corn silage.** This consists of the kernels only. At 70% dry matter (30 % moisture), shelled-corn silage contains 61 to 66% of the nutrients in the whole crop (also see "High-Moisture Grain" later in this section).

The sorghums are more dependable and higher yielding than corn in certain areas, particularly in unirrigated, and relatively dry areas, of western and southwestern United States. Sorghum for silage is harvested with the same equipment as is used for corn silage. It should not be harvested for silage until the heads are soft to medium dough stage. Harvesting at this stage provides the highest yields of total feed material, enhances preservation, and makes silage that has good palatability.

Fig. 9-12. Harvesting grain sorghum for silage on a southeast Kansas dairy farm. (Courtesy, A. O. Smith Harvestore Products, Arlington Heights, Ill.)

On a dry-matter basis, corn silage contains an average of 8.3% crude protein, 68.0% total digestible nutrients, 0.31% Ca, and 0.27% P. Grain sorghum silage contains less protein and TDN than corn silage. Grass/legume silages contain more protein and less TDN than corn silage. The carotene content of corn silage is variable, but on the low side.

Corn and Sorghum Residue Silage

Corn and sorghum residues—the forages that remain after harvesting a grain crop of corn or sorghum—may be used as cattle feed three ways: (1) grazed, (2) harvested (stacked or baled) and fed dry, or (3) ensiled and fed as silage. The discussion that follows will be limited to ensiling corn and sorghum residue.

When ensiled, cornstalks (stover) produce a product known as corn stover silage or cornstalk silage. When stalks are processed as silage, the use of a forage harvester equipped with a screen or a recutter-blower at the silo is necessary in order to chop the material finely. Fine chopping will ensure good packing and improve consumption by avoiding selectivity.

Where corn stover silage is made, the residue should be harvested as soon as possible after the grain is taken off, before the residue loses any moisture. At that time, the grain moisture will generally be under 30% and the refuse moisture will be above 48%. In an airtight silo, 40 to 45% moisture is very satisfactory for ensiling. In an unsealed or bunker silo, the moisture content should be 48 to 55% for proper lactic acid formation. Water may be added at the silo if necessary. As a precaution, some authorities recommend the addition of 56 lb of corn meal (or other finely ground grain) per ton of corn stover silage, as a means of providing readily-fermentable carbohydrates from which acids will form and act as a preservative. With husklage, the latter precaution is not necessary since there is sufficient grain remaining in the husk and cob.

Purdue University researchers fed corn stover silage supplemented with varying increments of energy for growing dairy heifers.[5] Twenty-eight yearling Holstein heifers were randomly assigned to each of 4 corn stover silage rations to determine the level of supplemental energy required to obtain growth response of 1.3 to 1.4 lb per day as recommended by the National Research Council (NRC). The 4 stover silage complete rations contained 0.37, 0.42, 0.47, and 0.52 Mcal estimated net energy (ENE) per pound of complete feed. Additions of increasing increments of ENE produced a linear increase in daily body weight gains of 0.75, 0.95, 1.28, and 1.5 lb for the 4 rations, respectively. Total dry matter and ENE intake increased linearly with increasing ENE concentration of the rations. The investigators concluded that, under the conditions of their study, a total ENE content of corn stover silage ration of approximately 0.5 Mcal per pound ration could have produced the growth response recommended by NRC for dairy heifers.

The biggest deterrent to harvesting stalklage, in either dry or ensiled form, is the cost—primarily for equipment. Rather than own such expensive equipment, which is used for a short period only, custom harvesting of stalklage is likely cheaper for most operators.

Husklage—the forage discharged from the rear of a corn combine, and consisting of the husks, cobs, and any grain carried through the combine—may also be ensiled. Ensiling husklage, along with recutting and adding water, results in increased cow consumption and less rejection of cobs.

Like corn, sorghum stover may either be grazed or harvested and stored either as dry feed or silage. Because the sorghum plant stays green late in the fall, good sorghum stover silage can be made without additional water.

Grass/Legume (Hay Crop) Silage

Grass/legume (hay crop) silage refers to silage made from any of the green crops which might otherwise be grazed or dried and made into hay. This includes grasses (such as timothy or fescues), legumes (such as alfalfa or clovers), grass-legume mixtures, and cereal grains (such as oats).

Grass/legume silage can be produced in areas where the climate is too cool and the growing season too short for corn or sorghum silage.

Although grass and legume crops have been ensiled in Europe for hundreds of years, the practice did not become widely used in the United States until the 1930s. At that time, interest in hay crops for silage increased as a result of farmers (1) becoming aware of the field losses that occur in haymaking, (2) being provided with the information necessary to make high-quality silages from grasses and legumes, and (3) having access to field choppers, which facilitated making silage from hay crops.

The following are the most important **advantages** of grass/legume silage:

1. It minimizes field, harvest, and storage losses of grass/legume forages. (See Fig. 9–13.)

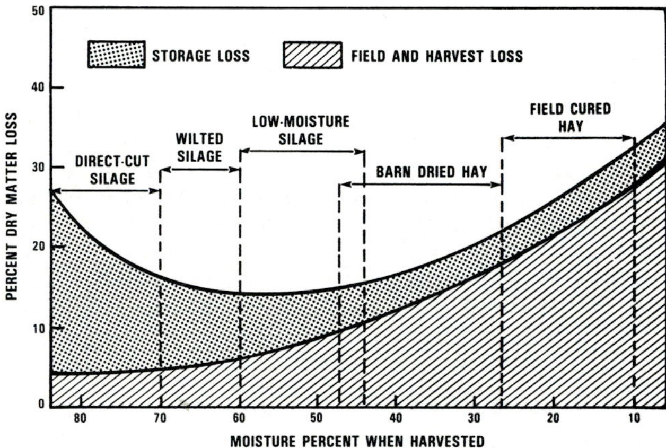

Fig. 9–13. Estimated total field, harvest, and storage loss when grass/legume forages are harvested at different moisture levels by alternative harvesting methods. (Courtesy, C. R. Hoglund, Professor Emeritus, Michigan State University, East Lansing)

2. It minimizes the dependence on favorable weather to harvest the crop.
3. It can be harvested with modern, efficient machinery and stored in large-volume structures.
4. It requires less supplemental feed than corn silage.
5. It can be handled and fed by mechanized methods, thereby reducing the labor requirements.
6. It kills weed seeds as a result of the fermentation.

Although grass/legume silage has important advantages, it also has the following **disadvantages**:

1. The initial investment costs for machinery, storage units, and feeding facilities are very high.
2. Silage-making machinery and storage and feeding facilities are highly specialized with the result that they have limited use for other purposes.
3. Inadequate fermentation may occur under certain conditions, resulting in poor-quality feed.
4. Storage and feeding losses may be high under poor management.
5. Like all silages, grass/legume silage is heavy, bulky, and costly to transport, thus its off-farm market value is limited.
6. Existing upright silos may not be in sufficiently good condition to store grass/legume silage.
7. Once grass/legume silage is removed from the silo, it must be fed within 12 to 24 hours to alleviate spoilage. Except for an oxygen-limiting silo, once feeding begins during warm weather, it is necessary to feed a minimum of 3 to 4 in. off the exposed surface daily to prevent spoilage in the silo.

[5]Colenbrander, V. F., *et al.*, "Corn Stover Silage Supplemented With Varying Increments of Energy for Growing Dairy Heifers," *Journal of Animal Science*, Vol. 33, No. 6, 1971, p. 1306.

Grass/legume silages are of three kinds based on moisture level:

1. Direct-cut silage, 70% moisture or above.
2. Wilted silage, 60 to 70% moisture.
3. Low-moisture silage (or haylage), 40 to 60% moisture.

DIRECT-CUT SILAGE

Direct-cut grass/legume silage is forage that is harvested and stored without field-drying, usually containing more than 70% moisture. Although direct-cut ensiling is the standard practice with mature corn and sorghum, it is not recommended for grass/legume crops because of (1) the difficulty in getting good preservation due to the high moisture content, and (2) the increased nutrient losses due to seepage.

The higher the moisture content of grass/legume forage the more critical the need for a low pH to obtain good preservation. Direct-cut silage is subject to excessive deterioration of protein, undesirable odor, excess loss of nutrients due to seepage, and deterioration of concrete stave silos. Also, when fed as the major forage, direct-cut silage usually results in lower consumption of dry matter and consequently lower animal production than wilted or low-moisture silage.

When making direct-cut silage, the following directions should be observed: (1) harvest at the proper stage of maturity; (2) avoid cutting when the forage is wet with dew or rain; (3) provide drainage for excess juice; (4) add an additive or preservative (coarsely ground cereal grain will absorb some of the excess moisture in addition to providing sugars for fermentation); (5) distribute evenly in the silo and pack thoroughly; and (6) cover with plastic or suitable material.

WILTED SILAGE

Today, a high percentage of grass/legume forage is dried to some degree prior to ensiling. Wilting gets rid of some water in the field, so less weight is handled. Also, in comparison with direct-cut silage, odor and seepage problems are reduced and additives or preservatives are usually not needed.

Fig. 9–14. Alfalfa forage wilted to 65% moisture being field chopped for ensiling. (Courtesy, James Tappan, Arizona Dairy Co., Higley, Ariz.)

Authorities generally agree on the following rules for making good wilted silage: (1) harvest at the proper stage of maturity; (2) allow the forage to wilt in the swath and/or windrow until the moisture reaches about 65%, which may take from 1 to 4 hours, or longer, depending on the weather, but which may be expedited by the use of a forage conditioner; (3) use a short cut (about ⅜ in.) on the forage harvester; (4) fill the silo rapidly and continuously; (5) distribute evenly and pack thoroughly; and (6) top off with 2 ft of unwilted material and cover with plastic or other suitable cover. No additive or preservative is needed with properly wilted silage, although it may be added if desired.

LOW-MOISTURE SILAGE (HAYLAGE)

This method of making silage is not new, contrary to common belief. It was developed many years ago at the Bacteriological Station at Crema, Italy.

Low-moisture grass/legume silage (or haylage), containing 40 to 60% moisture, is made with limited bacterial growth and fermentation. The term *oatlage* is sometimes used specifically to indicate low-moisture silage made from oats.

Fermentation is of minor concern in making low-moisture silage since little acid is produced and pH is not a useful criterion of quality. The most important factor is the establishment and maintenance of air-free conditions through fine chopping, rapid filling, and a good silo. Because of the difficulty of maintaining air-free conditions, stacks, bunkers, and trench silos are seldom used for storing low-moisture silage. Infiltration of air into the silage mass will result in growth of yeasts and molds and increase in temperature. Temperatures about 95°F for a few days will cause certain proteins to combine chemically with carbohydrates to form a product that is indigestible, termed *bound protein*. When more than 12% of the total protein is in the "bound" form, the silage has undergone excessive heating.

Properly made and stored low-moisture silage has a pleasant aroma and is a palatable, high-quality feed. Animals usually receive more dry matter and net feed value in low-moisture silage than in wilted silage made from the same cut. Low-moisture silage is increasing in popularity, especially as a dairy feed. It may be fed like wilted silage, with adjustment for difference in moisture content.

In a Wisconsin study,[6] low-moisture alfalfa-grass silage (haylage) was compared with wilted alfalfa-grass silage plus second crop hay in rations for dairy cows. Both low-moisture silage and wilted silage were stored in oxygen-limiting silos. Total dry matter consumption was higher for the low-moisture silage ration ($P \leq .01$) and higher milk and fat corrected milk (FCM) production ($P \leq .01$) resulted. There was no significant difference in milk fat percentage. Forage dry matter consumption for cows fed low-moisture silage was 32.5 lb per day vs 30.5 lb for wilted silage plus hay. The FCM per day per cow fed the low-moisture silage treatment was 45.6 lb vs 42.2 lb for cows fed the wilted silage plus hay treatment. Total dry matter required per pound of FCM produced was similar for both rations. No detrimental effects were noted in the health of cows fed either forage ration.

[6]Larsen, H. J., E. L. Jensen, and R. F. Johannes, "A Comparison of Haylage and Wilted Grass Silage Plus Hay in the Dairy Diet," *Research Report*, The University of Wisconsin-Madison, August 1971.

Workers at the New Hampshire Station compared hay crop silage and hay for lactating cows.[7] Holstein cows were divided into 3 groups for a 51-month continuous trial to study feed intake, production, profit, and dry cow responses when 43 to 44% of the forage dry matter was from either (1) wilted hay crop silage, (2) silage and hay (50:50), or (3) hay, all from the same field. The remainder of the forage for all cows was urea-treated corn silage. Common concentrate was fed to all cows *ad libitum* from calving to peak of lactation, then according to requirements until the end of lactation. Forages were fed *ad libitum* during lactation. Dry cows received a small amount of concentrate plus forage. The forage intake of hay was higher than the intake of hay crop silage, but this did not result in significantly more milk or income over feed cost. It was concluded that the decision to harvest the hay crop as wilted silage or as hay depends on economics of earlier harvest and earlier regrowth with silage, and on its adaptability to blended complete rations rather than on differences in profitability or production response.

In addition to excluding air by providing an airtight silo and by thorough packing, the following directions should be observed to make the best quality low-moisture silage: (1) harvest at the proper stage of maturity; (2) wilt in the swath and/or windrow until the moisture reaches 40 to 60% (35 to 40% for a bunker silo), with the required time determined by the weather, but which may be expedited by the use of a forage conditioner; (3) chop short, a ¼ in. cut is best; (4) fill silo rapidly and continuously; (5) add an additive or preservative if desired; (6) distribute silage evenly in the silo; (7) apply a top seal of forage containing 65 to 70% moisture, level, and tramp to remove air; and (8) crown the center slightly and cover with a plastic silo cap.

Corn or Sorghum Silage Vs Grass/Legume Silage

Frequently livestock producers are confronted with choosing between corn or sorghum silage and grass/legume silage. Under these circumstances, the following facts are pertinent:

1. Where adapted, corn or sorghum will generally produce a greater tonnage of feed per acre than grass/legume silage.
2. Good-quality corn or sorghum silage can be made more consistently and with greater ease than good-quality grass/legume silage.
3. Corn or sorghum silage may be more palatable than grass/legume silage, even when the latter is carefully preserved.
4. Grass/legume silage is generally higher in protein and carotene but lower in total digestible nutrients and vitamin D (wilted grass/legume silage is higher in vitamin D than unwilted) than corn or sorghum silage (generally, grass/legume silage contains about 90% as much TDN as corn silage, but it will equal corn silage in TDN when 150 lb of grain per ton have been added). Thus, grass/legume silage generally requires the addition to the ration of less protein supplement but more total concentrates than corn or sorghum silage. This would indicate that corn or sorghum silage would be slightly preferable to grass/legume silage in high-forage finishing rations for beef cattle and sheep, whereas grass/legume silage would be preferable in high-forage rations for dairy animals and young beef cattle and sheep.
5. Grass/legume silage can be produced in areas where the climate is too cool and the growing season too short for corn or sorghum silage.
6. The production of grass/legume silage will result in less soil washing than the production of corn or sorghum silage on lands subject to erosion.
7. Grass/legume silage will freeze next to the silo wall more than corn or sorghum silage, especially if it is ensiled when too wet (unwilted).
8. The silo can be kept working full time by using both grass/legume and corn or sorghum silage; ensiling the first cutting of grass/legume silage for summer feeding, and ensiling corn or sorghum silage for winter feeding.

Other Silage Crops

In the Northwest and North Central states, where the weather is cool and the growing season is short, sunflowers are sometimes grown for silage. Although they yield and ensile well, sunflower silage is neither as palatable nor as nutritious as corn, sorghum, or grass silage. Pound for pound, sunflower silage is about 80 to 85% as valuable as corn silage.

Throughout the United States, a great array of by-product feeds are ensiled, especially in the less expensive and temporary types of silos. Among such by-products are grain chaff, pea and bean vines, beet tops and pulp, sunflower hulls and chaff, potatoes, cannery refuse, cull and surplus fruits and vegetables, pulp and trimming wastes from market vegetables and fruits, wet brewers' and distillers' grains, almond hulls, and poultry litter. Sometimes Russian-thistles and other weeds are ensiled.

When potatoes, which contain about 80% moisture, are ensiled for cattle, it is recommended either (1) that 20 to 25 lb of dry hay, straw, or chaff be run through the ensilage cutter with each 100 lb of potatoes, or (2) that 1 ton of corn or sorghum silage be chopped with each 500 lb of potatoes. Frozen and sprouted potatoes should not be ensiled. Potato silage intended for swine should be made from cooked or steamed potatoes ensiled alone in a shallow pit or silo. Potato processing wastes (cull potatoes, off-flavor french fries and chips, etc.) can be ensiled in the same manner as unprocessed potatoes.

Either of the methods recommended for ensiling potatoes for cattle is equally adapted for the preservation of other high-moisture crops, such as apples, beets, pears, tomatoes, cauliflower, broccoli, kale, and trimming wastes from market vegetables—provided the added forage is in proportion to their respective moisture contents.

Cabbage, rape, and turnips should not be ensiled, as they make unsatisfactory, watery, foul-smelling silage.

(Also see Chapter 12, By-product Feeds/Crop Residues.)

[7]Holter, J. B., W. E. Urban, Jr., and H. A. Davis, "Haycrop Silage Versus Hay in a Mixed Ration for Lactating Cows," *Journal of Dairy Science*, Vol. 59, No. 6, 1976, p. 1087.

Combining Crops for Silage

Sometimes, in order to lower the moisture content, to alleviate the necessity of a preservative, and to assure better quality silage, forages of high sugar content are combined with forages of low sugar content. Thus, excellent silage can be made by mixing 1 ton of sorghum forage with each 3 tons of grass/legume silage material, or a ton of corn forage with each ton of grass/legume forage material (less sorghum forage is necessary than corn forage, because of the higher sugar content of the former).

At times such combination silage crops are even grown together; for example, corn and soybeans; millet or Sudangrass; and soybeans, oats, and peas.

A major difficulty in combining ensiling crops is that it is almost impossible to synchronize the stage of maturity of different crops so that they reach maximum yield and nutrient level at the same time.

Rain-Damaged Hay Silage

Partly cured hay that has been rained upon, but is not moldy, may be salvaged as silage (although it will not be of high quality), provided it is finely chopped, distributed evenly, and packed in the silo thoroughly enough to squeeze out the air. It is recommended that it be placed in the bottom of the silo, and, preferably, that alternate loads of a green crop be mixed with it. Otherwise, satisfactory packing can be obtained by putting a few loads of greener-than-ordinary material on top of it.

Frosted Crop Silage

Sometimes corn, sorghum, sunflowers, small grains, beans, and other crops, which may or may not have been intended for silage, are frosted before they reach the silage cutting stage. Corn that has been frosted before reaching maturity is commonly known as *soft corn*. Such frosted crops may be salvaged as silage. They should be cut at recommended moisture contents and ensiled according to directions. If they are too dry, water should be added.

Frosted crops, especially frosted sorghum, may be high in cyanide (HCN). (See *CAUTION* under the section on "Drought Stricken Crop Silage.")

Drought Stricken Crop Silage

Sometimes corn or sorghum, or other crops, are drought stricken to the extent that little or no grain will be produced. Such crops may be harvested for silage and used as an energy source for ruminants. They should be cut and ensiled like any other silage crop. If they are too dry, water should be added.

Drought stricken crop silage may be used in the same manner as any other low-energy source. It is well-suited for wintering breeding beef cattle and stockers, for backgrounding finishing cattle—to approximately 850-lb weights, and for dry dairy cows.

CAUTION: Danger of cyanide toxicity is much greater from sorghum than from corn. Drought stricken plants can accumulate cyanogenetic glycoside which hydrolizes to form free cyanide (HCN). The danger is increased when crops are grown on heavily nitrogen-fertilized soils or if any of the following have occurred: frosting, wilting, trampling, or hail. Any combination of these conditions can lead to a dangerous build-up or release of cyanide.

SILAGE ADDITIVES AND PRESERVATIVES

Silage additives are products that provide supplemental nutrients which enhance the feeding value of silage.

Silage preservatives are products that enhance the keeping qualities of silage.

High-quality silage can be made without the use of additives or preservatives if good material is started with and all proven good practices are followed. But there are times when the ensiled material is either too wet or too dry; does not contain sufficient fermentable carbohydrates; is deficient in certain nutrients; is lacking in palatability; and/or the proven good practices cannot be followed. Under such circumstances, silage additives or preservatives may reduce silage losses and/or improve the feeding value of the silage.

Additives or preservatives should not be used as a substitute for a good silo or for proper chopping, packing, and sealing. Normally, additives or preservatives are neither needed nor added to corn or sorghum silages. But additives or preservatives may be very helpful if a grass/legume forage with over 70% moisture is ensiled.

A number of materials are available to incorporate in silage, with claims made that they will enhance the nutrient value, the preservation of the nutrients, and/or the palatability of the silage. *The bottom line when using any silage additive or preservative is how much it improves animal performance and net profit; and not whether it merely makes silage look better and smell better.*

Thorough testing of these materials would necessitate that each of them be used at several levels, with many kinds of silage, with each forage at various moisture contents, and under different storage conditions. Obviously, it would be highly impractical, if not impossible, to carry out such an extensive study. However, there is sufficient understanding of the process of silage formation, the requirements for the preservation of silage nutrients, and the mode of action of the ingredients used in various additives to make sound decisions as to whether they might be economically worthwhile. Also, some experimental testing has been done with certain additives.

In order to be effective, an additive or preservative should serve one or more of the following purposes:

1. Add nutrients.
2. Provide fermentable carbohydrates.
3. Furnish additional acids to increase acid conditions.
4. Inhibit undesirable types of bacteria and molds.
5. Reduce the amount of oxygen present, directly or indirectly.
6. Reduce the moisture content of the silage.
7. Absorb some nutrients which might otherwise be lost in seepage.

Four types of additives or preservatives are used in silage making: (1) feed additives, (2) acids, (3) fermentation aids, and (4) preservatives.

Feed Additives

Feed additives may be used to provide a readily available source of carbohydrates for fermentation into lactic acid, to reduce the moisture content, to provide needed nutrients, and/or to enhance palatability.

Feedstuffs used as silage additives include:

1. Corn-and-cob meal, ground corn, barley, or oats; applied in amounts varying from 100 to 300 lb per ton, depending on the moisture content of the crop.
2. Beet pulp, citrus pulp, chopped corncobs, or chopped hay to reduce seepage losses from the silo if moisture is high.
3. Molasses, either liquid or dehydrated, at rates of 40 to 80 lb per ton of green forage.
4. Dried whey, a product of the dairy industry, applied at the rate of 30 to 300 lb per ton, as a source of fermentable carbohydrate, protein, and minerals.
5. Nonprotein nitrogen (NPN) products such as urea and anhydrous ammonia.
6. Ground limestone.

• **Grain and other feed ingredients**—Silage made from legumes or grasses may be improved under certain conditions by the addition of ground grain (corn, wheat, or barley), ground ear corn, beet pulp, citrus meal, citrus pulp, or other appropriate feed ingredient. The ground material will provide a readily available source of carbohydrates (sugar and starch) for bacterial fermentation and the production of acids; increase the feeding value of the silage, because 75 to 85% of the feeding value of the grain will be retained if the silo excludes air properly; and likely improve palatability. If the primary concern is to reduce the moisture content of the silage, cheaper materials, such as ground corncobs, cottonseed hulls, oat hulls, or chopped straw or hay, may be more appropriate than ground grain. Dry, finely ground corn cobs will absorb nearly 200 lb of water per 100 lb of the cob material; dried beet pulp will absorb even more.

When green forage is ensiled at a proper moisture content, there is usually no advantage to adding grain for the purpose of preservation or palatability.

These materials may be added by feeding them into the blower from a properly adjusted hopper attachment.

• **Molasses (including cane or blackstrap, beet, corn, and citrus molasses)**—Some green forages, such as legumes and certain grasses, are rather low in sugar content. Hence, adding molasses, which is high in sugar, may increase lactic and acetic acid production and improve silage quality and preservation. Also, molasses improves the palatability of silage and increases its nutritive value. For legumes, about 80 lb of molasses per ton are generally used, and for grasses about 40 lb per ton (molasses weighs 12 lb/gal). Additions of much less than these amounts, as an ingredient in mixed preservatives, is of little value. Much of the feeding value of the molasses is retained in the silage under good storage conditions and where there is no seepage loss.

Molasses may be added in either liquid or dehydrated form as the forage enters the blower.

When a grass and/or legume is wilted to 50 to 60% moisture content and adequately protected from air, an excellent feed with a good aroma and keeping quality can be obtained without the addition of molasses (sugar). Neither is the addition of sugar needed for corn silage.

• **Whey**—Experiments and experiences indicate that adding dried whey to alfalfa silage or haylage slightly improves the quality and digestibility. There is also indication that adding dried whey to urea-treated corn silage may help reduce nitrogen losses from and improve the feeding value of the silage.

South Dakota Station workers studied corn silages containing either (1) 0.5% urea plus 1% dried whey, or (2) 0.5% urea; in a lactating trial, in a feeding trial with heifers 4 to 8 months of age, and in a digestion trial.[8] Adding 1% dried whey to urea-treated corn silage improved the feeding value of silage. This was evidenced by a 6.5% increase in milk production by lactating cows, a 7% increase in weight gains by growing heifers, a 2 to 3% increase in dry matter digestibility, and a 16% increase in apparent nitrogen digestibility by steers. While these improvements in feeding value are not large, they are sufficient to make adding dried whey to corn silage an economically sound practice under most farm conditions.

• **Urea, ammonia, and other NPN products**—Urea, ammonia, and other NPN products can be added to corn or sorghum silage at the time of ensiling as a source of nonprotein nitrogen.

Urea increases the crude protein content of the silage and the amount of lactic and acetic acids produced. The addition of 10 lb of urea per ton of ensiled corn material will make for the following approximate increases on a dry matter basis: the crude protein from 8.3 to 12.3%, the lactic acid from 4.2 to 5.4%, and the acetic acid from 0.9 to 1.2%. Since the amount of nonprotein nitrogen that can be converted to microbial protein by the organisms in the rumen is limited, no more than 10 lb of urea should be added to a ton of ensiled corn material. The urea can be added by spreading it over the top of each load of chopped corn or it can be added to the chopped corn through the blower by commercially manufactured metering equipment.

More recently, ammonia-containing materials have been added to corn silage as a source of nonprotein nitrogen, including ammonia-water solutions, ammonia-mineral solutions, ammonia-mineral-molasses solutions, anhydrous ammonia gas, and cold flow ammonia. For dairy cattle, 5 lb of actual nitrogen (about 6 lb of ammonia) may be added per ton of wet silage. Ammonia treated corn silage has been found to contain increased concentrations of true protein, lactic acid, and acetic acid. Also, it may have a higher pH and be more stable than untreated silage when exposed to air. Special equipment is required to add ammonia or ammonia-containing materials.

• **What NPN-silage experiments show**—Interest in adding nonprotein nitrogen (NPN) to corn silage has prompted a number of experiments.

Michigan Station researchers compared corn silages (34 to 37% dry matter) ensiled with (1) no additive; (2) 0.5% urea; (3) 0.75% urea; and (4) 0.75% urea + 0.17% $CaSO_4$

[8]Schingoethe, D. J., and G. L. Beardsley, "Feeding Value of Corn Silage Containing Added urea and Dried Whey," *Journal of Dairy Science*, Vol. 58, No. 2, 1975, p. 196.

when fed as the only forage to lactating cows averaging 64.5 lb milk per day at the time treatments began.[9] Concentrates containing 8, 12, or 18% crude protein were fed at 1 lb per 3 lb of milk. The 0.5% urea silage was fed with each of 3 protein concentrates—8, 12, and 18%. The other 3 silages were fed with 8% concentrate only. Cows fed low-protein rations (corn plus corn silage) consumed less feed and produced less milk than those fed rations to which 0.5 or 0.75% urea was added to the corn silage at ensiling time. Highest milk yields were obtained from the ration in which approximately 50% of the supplemental nitrogen came from urea added to corn silage and the remainder from soybean meal. Because of the slightly higher milk yields and the larger savings in the price of the concentrates fed, income over feed costs was about 10% greater for the urea-fed group than for the group fed all of its supplemental nitrogen as soybean meal. These studies clearly demonstrate a beneficial response from adding urea to corn silage at ensiling time.

The Michigan Station workers also compared the feeding value of corn silage treated with an ammonia solution at ensiling with that of urea-treated and control silages in dairy cattle rations.[10] On all-silage rations with no protein supplement, heifers ate more of the ammoniated than of the control silage. Lactating cows fed ammonia- and urea-treated silages showed higher milk yields than cows fed negative control rations with no added nitrogen. No significant differences in production were noted for cows fed control, urea-treated, or ammoniated silages at equal dietary nitrogen. This study demonstrated that the addition of an ammonia solution to corn silage results in a high-quality feed enriched with NPN which is equal to urea-treated silage in supporting milk yields of lactating cows. The lower cost of manufacturing ammonia than urea increases its potential as a silage additive. The Michigan data suggest that the ammonia solution exerts a buffering action which increases protein and lactic acid concentrations in ensiled material.

• **Sulfur may be needed**—When corn silage reinforced with urea or ammonia is the major source of protein fed to ruminants, there may be inadequate sulfur for the rumen organisms to manufacture their own protein. In such cases, there is some experimental evidence that the addition of sulfur to achieve a nitrogen-sulfur ratio of less than 15:1 improves both the growth and milk production of cattle. The most practical way to provide additional sulfur is to add gypsum (calcium sulfate, $CaSO_4 \bullet 2H_2O$) at the rate of 1.8 lb per ton of silage.

• **Advantages and disadvantages of adding NPN**—Before adding NPN to corn silage, one should weigh carefully the advantages and disadvantages to accrue therefrom.

The **advantages** of adding NPN to corn silage are:

1. It is possible to raise the crude protein content by 4% on a dry basis—from 8.3 to 12.3% by the addition of 10 lb of urea per ton of ensiled corn material.
2. It (NPN) is effectively utilized by rumen bacteria.
3. It will not likely make for any intake or refusal problems (as can happen with urea in grain mixtures).
4. It alleviates or lessens the need for a protein supplement to balance rations.
5. It makes corn silage more stable. (There is less secondary fermentation or heating after removal from the silo.)

The **disadvantages** of adding NPN to corn silage are:

1. Seepage and ammonia gas losses occur. Even under ideal conditions, about 10% of the urea may be lost; and under average conditions, up to 30% of the urea may seep away or be lost as ammonia gas.
2. The expense of adding NPN occurs at ensiling time, thereby tying up capital.
3. Feed intake can be lowered if corn silage is too dry.
4. It necessitates care because ammonia gas, which is dangerous and causes irritation, can accumulate in the silo.
5. It lessens flexibility of feeding because the same amount of urea is contained in all the silage.
6. The monetary investment in NPN could be tied up in the silo for an extended period of time before any return on the investment.

(Also see Chapter 11, Protein Supplements, section on "Urea.")

• **Limestone**—Limestone (calcium carbonate) may be added at a level of 0.5 to 1.0% to corn silage to increase acid production. It neutralizes some of the initial acids as they are formed, allowing the lactic acid bacteria to perform longer and to produce more desirable acids.

Research has not shown any consistent increase in the nutritive value of silage treated with limestone.

The addition of limestone at ensiling time raises the naturally low calcium content of corn silage—a fact which should be considered when balancing rations. For lactating dairy cows fed high-alfalfa rations, this would be disadvantageous; but it can be rectified by increasing the phosphorus of the ration. Yet, experiments have shown that dairy cattle do not respond to the addition of limestone to silage. For most beef cattle rations, however, the added calcium is a virtue because (1) when high corn silage rations are fed, the added calcium is needed, and (2) when silage is limited, little calcium is added.

Acids

Both inorganic (mineral) and organic acids may be used as additives. Mineral acids lower the pH immediately, while organic acids have a limited effect on lowering pH. Both mineral and organic acids limit microbial growth and help to stabilize silage.

The use of inorganic acids, such as hydrochloric, sulfuric and phosphoric, for forage preservation was pioneered by A. I. Virtanen, Finnish biochemist, in the 1920s. He discovered the AIV method, named from his initials, for preserving silage by acidification, for which he was awarded the Nobel Prize in Chemistry in 1945. His work was highly regarded in that area of the world because hay drying was difficult and dairying was, and still is, important.

• **Inorganic acids**—Inorganic acids (hydrochloric acid, sulfuric acid, phosphoric acid) have been used as silage preservatives, almost entirely in Europe, in connection with the

[9]Huber, J. T., and J. W. Thomas, "Urea-Treated Corn Silage in Low Protein Rations for Lactating Cows," *Journal of Dairy Science,* Vol. 54, No. 2, 1971, p. 224.

[10]Huber, J. T., and O. P. Santana, "Ammonia-Treated Corn Silage for Dairy Cattle," *Journal of Dairy Science,* Vol. 55, No. 4, 1972, p. 489.

ensiling of high-moisture material. These acids substitute for the acids produced by bacterial action. However, they are very corrosive, causing problems in their application, including problems with the silo walls and silage handling equipment. Of the three acids, phosphoric is preferred because (1) it is less corrosive than sulfuric acid or hydrochloric acid, (2) it may enhance the phosphorus content of the silage, and (3) it increases the residual manure value from the silage. But phosphoric acid may introduce a problem of proper calcium-phosphorus ratio. This can result in some abnormal conditions and unsatisfactory performance in the animals to which it is fed.

When mineral acid preservatives are used, it is recommended that ground limestone or some other form of calcium or sodium carbonate be fed to animals at the rate of approximately 1 oz for each 10 lb of silage in order to neutralize the acid and prevent undesirable effects.

In general, the use of mineral acid preservatives is not considered as desirable as the use of molasses or grain, because (1) they produce a more sour and less palatable silage; (2) they may damage clothing, machinery, and/or masonry silo walls, due to their corrosiveness; and (3) they do not add to the nutrient value of the silage except by enhancing the preservation of carotene.

In general, the use of mineral acids has more disadvantages than advantages.

• **Organic acids**—Propionic, acetic, lactic, citric, and formic acids are included in this group. They are used in a manner similar to inorganic acids, but they are much less corrosive and not so difficult to handle, although precautions must be taken.

Organic acids will enhance the preservation of forage without the loss of palatability. Also, they serve as mold inhibitors. Even so, like all additives, they cost money; hence, the economics of using them in making silage must be considered. When an organic acid is used, the following guidelines should be observed:

1. Add 1% of the acid to the forage in the field at the time of harvest or at the chopper.
2. Limit the presence of oxygen by using a sound, well-built silo.
3. Prevent dilution of organic acid treated silage by rain or snow by covering it with plastic when it is stored outside or in a temporary silo.

It appears that organic acids will find their greatest use in the preservation of high-moisture grain. (See section on "High-Moisture Grain.")

Fermentation Aids

This group includes bacterial cultures, yeast cultures, and enzyme supplements. Controlled experiments support the claims made for some of these products, but not all of them. So, they should be purchased only from reputable sources that have valid research data to support the claims made for them.

• **Bacterial cultures**—Silage additives containing cultures of acid-forming bacteria *(Lactobacillus)* are on the market. The basis for including these is to provide an inoculum or to increase the numbers of these bacteria and ensure rapid fermentation.

There are two schools of thought relative to the value of bacterial cultures as silage additives. The advocates claim that these products increase the dry matter, energy, and protein of the silage. Others report that the addition of such cultures is unnecessary and of questionable value because: (1) There are always sufficient numbers of these bacteria present on the ensiled material to bring about the proper fermentation; (2) the number of live bacteria present in these preparations cannot be guaranteed with accuracy; and (3) the number of bacteria provided through such additives is insignificant in comparison with numbers already present on the ensiled material.

• **Yeast cultures**—Yeast cultures have also been included in certain silage additives. However, yeasts will sometimes grow in silage without an inoculum being added. When this happens, the silage has a yeasty odor and taste, which is considered undesirable. Yeast does have nutritional value, but because of the small quantity involved in additives, the contribution in this regard is minimal.

• **Enzymes**—Cultures of molds, or of molds with other microorganisms, are sometimes added to silage to provide a source of enzymes. It is claimed that these enzymes improve the nutritive value of the silage by increasing its digestibility or digestible nutrient content. Although the enzyme activity of these preparations has not been measured experimentally, no doubt they vary considerably from batch to batch. Further, the quantity of enzymes added is insignificant compared to those already present in the silage.

Preservatives

This group includes antibiotics, salt, and sterilants. These products preserve silage by inhibiting microbial action or undesirable fermentations. All of them are of questionable value if air is properly excluded from the silage. If air is not excluded, they must be added at very high levels in order to be effective.

• **Antibiotics**—Theoretically, antibiotics could preserve silage by selective action; by inhibiting undesirable microbial activity while allowing the desirable organisms to develop. So far, the results have been inconsistent.

• **Salt (NaCl)**—At an appropriate level, salt inhibits certain microorganisms without preventing the action of bacteria which produce the desirable acids.

• **Sterilants**—This group includes sulfur dioxide, sodium diacetate, sodium metabisulfite (sodium sulfite), sodium benzoate, and sodium nitrate. Sodium propionate and other organic acids have also been used as preservatives, because of their mold inhibiting properties. Each of these products appears to reduce carotene losses, improve the odor of silage, and/or lessen the production of toxic gases. But their effect on palatability is variable. The cost and inconvenience of application of these products may not justify their advantages.

Silage Additive and Preservative Considerations

A variety of silage preservatives is presently on the market; and, no doubt, new ones will follow. Many of these products have been inadequately tested. Yet, farmers and ranchers are often in the position of having to decide whether an additive shall be used. In addition to understanding the silage-forming process and how different additives function, they should consider the following:

1. Additives and preservatives will not substitute for the proper exclusion of air.
2. Additives and preservatives do not produce new nutrients in silage, although they may aid in retention of those already present.
3. Additives and preservatives that add nutrients to the silage will be partially lost with any spoilage or seepage.
4. The cost of an additive or preservative is usually high in relation to the value of the silage.
5. Chemical analyses are of very limited use in evaluating silage additives and preservatives.

Silage Additive and Preservative Recommendations

When added to silage, the following materials will increase the amount of nutrients it contains:

1. Grain or grain by-products will increase total digestible nutrients and dry matter.
2. Molasses will increase the total digestible nutrients (TDN, or energy) and may improve fermentation in legumes and certain grasses.
3. Urea or other NPN products will increase the nitrogen (crude protein).
4. Limestone will increase the calcium content.

Most of the arguments center around the use of the non-nutrient silage additives. A review of the literature indicates varying degrees of success from the use of such products. Some reports show positive effects from them while others show no effect. The most important consideration is whether the improved quantity and quality of forage from the use of an additive or preservative will offset its cost and application.

At the outset, it should be recognized that no silage additive or preservative can rectify mistakes that were made earlier—prior to incorporating the product. Neither can feeding more grain compensate for poor-quality silage.

Additives or preservatives are not essential to good silage formation when conditions of moisture and storage are right. Yet, under special circumstances they can be recommended for use. For example, molasses, grain, or grain by-products might be a wise addition to silage when conditions do not allow for proper wilting prior to ensiling, or when an *all-in-one* silage is being made. Urea may be an appropriate addition to an *all-in-one* silage or where increasing the protein content of the silage will simplify its feeding. It is doubtful that there is any justification for adding limestone unless this is a convenient method of calcium supplementation. The economy of most nutritive additives of this type depends largely on how well their nutrients are retained in the silage and the use made of them in balancing the rations.

When forages are stored at the proper moisture content, and when air is properly excluded, nutrient losses are low and a good-quality silage forms. Additives such as lactic acid bacteria, mold inhibitors, antibiotics, salt, enzymes, yeast cultures, and mineral acids, can, therefore, do little if anything to improve the preservation of the silage or its feeding value. When high-moisture material is ensiled, grain is superior to any of these additives. When air is not properly excluded, none of these additives will correct the large fermentation and spoilage losses.

In short, there is no substitute for good management of forage crops for silage, with proper control of such factors as stage of maturity at harvest, harvesting methods, moisture content, fineness of chopping, distribution and packing, and exclusion of air.

In order to assess the value of a silage additive or preservative, it is recommended that the following criteria be applied:

1. Does the product lower the ensiling temperature?
2. Does the product increase aerobic stability?
3. Does the product increase dry matter and nutrient recovery from the silo?
4. Does the product improve feed value and animal performance, particularly when silage is a major ingredient of the ration?
5. Does the product make for sufficient benefits to offset costs and give a return on investment?

HARVESTING METHODS AND MACHINERY

There is no one best silage-making method or kind of equipment. These must necessarily vary with the kind of forage, the kind of silo, the size of operation, and the labor and machinery cost.

Three principal kinds of machines are used for harvesting silage; namely, field forage harvesters, row-crop binders, and stationary silo fillers.

Field forage harvesters, which were first developed around 1936, are more widely used for harvesting silage than any other type of equipment. They tend to be concentrated in those states where the production of silage is most important.

Fig. 9-15. Harvesting corn for silage with a self-propelled forage harvester. (Courtesy, Deere & Company, Moline, Ill.)

With different attachments, field forage harvesters can be used to harvest row crops for silage, grass silage as a standing crop or from the windrow, and hay from the windrow. Also, they can be used to harvest straw and other kinds of forage. Field choppers can even be adapted for grinding and blowing high-moisture cob corn for ensiling. With appropriate attachments, a modern field harvester can be used to harvest all major ensiled crops. Thus, with a minimum of complementary machinery, a forage harvester can be the major piece of equipment in providing a completely ensiled ration for beef cattle, dairy cattle, or sheep. But such equipment is expensive and may or may not be economical where a small operation is involved.

Field chopped forage is generally transported on wagons equipped with mechanical unloading devices or by means of dump trucks. Blowers and conveyors are used in filling both tower and horizontal silos. Frequently, trench silos are filled by dumping over the sides.

The use of row-crop binders reached a peak in 1942, following which they declined. Reduction in the numbers of these machines reflects the increased use of cornpickers, field forage harvesters, and grain combines. In addition to being used to harvest silage, row-crop binders are also used to harvest corn for grain, corn for fodder, and sorghum as bundle feed.

Beginning in 1951, the use of stationary silo fillers declined markedly. Today, few of them are used, primarily because of their high labor requirement.

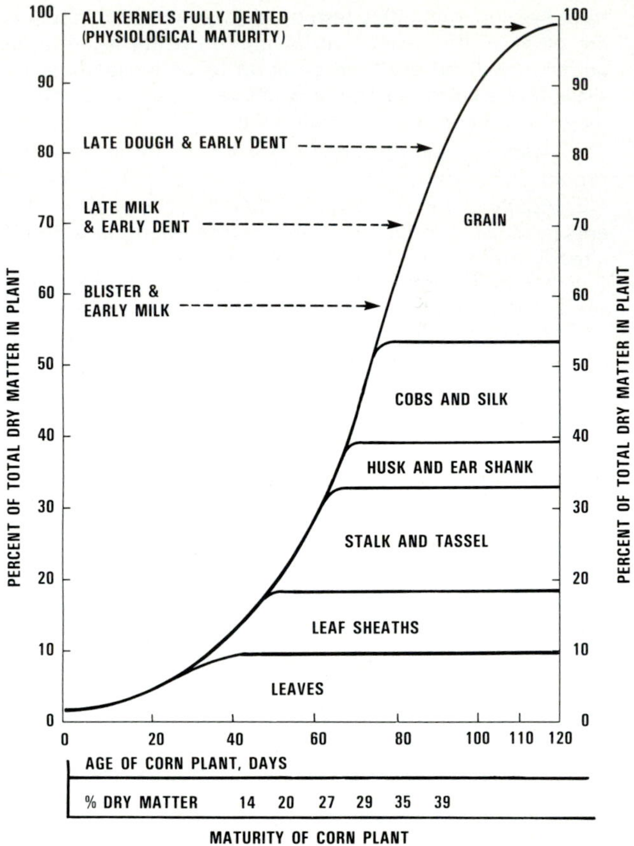

Fig. 9-16. Effect of maturity of corn plant on total dry matter accumulation.

Fig. 9-17. Black layer near the tip of the kernel indicates that the grain is physiologically mature and ready for the silo.

HOW TO MAKE GOOD SILAGE

In addition to using a sound silo of proper size, those who make good silage generally harvest at the proper stage of maturity, cut to proper length, control the moisture content, add an additive or preservative when needed, fill rapidly, distribute forage uniformly in the silo, and seal or top-off the silo. Each of these factors will be discussed.

Harvest at Proper Stage of Maturity

Harvesting at the proper stage of maturity assures the maximum yield and nutrient content.

Fig. 9-16 shows the effect of stage of maturity of the corn plant on total dry matter accumulation.

The *black layer test* can be applied quickly and easily to determine when to harvest corn for maximum yield and nutrient quality (see Fig. 9-17).

When the grain reaches physiological maturity, several layers of cells near the tip of the kernel turn black, forming the *black layer.* This layer can be detected by removing several kernels from the middle of the ear, then splitting them lengthwise or just cutting off the tip, and looking for the black layer near the tip. If the black layer is present, the grain is physiologically mature and ready for the silo.

At the black layer stage, the grains are usually dented and glazed, the lower 4 to 6 leaves of the corn plant are brown, and the plant contains 60 to 67% moisture. It can be cut 3 to 4 weeks past this stage with very little loss in dry matter or in feeding value.

Sorghum should be cut for silage when the seeds are hard.

Grass silage forages (grasses, legumes, and cereal crops) should be cut at the same stage at which they would make the best hay. (See Table 8-15 in Chapter 8, Hay.)

The New Hampshire Station investigators harvested and ensiled corn at four stages of maturity: soft dough, medium-hard dough, early dent, and glazed-frosted.[11] When fed to steers, digestibility, nutritive value, and *ad libitum* intake were highest for silage harvested at the medium-dough stage.

[11]Colovos, N. F., "Digestibility, Nutritive Value, and Intake of Ensiled Corn Plant *(Zea mays)* in Cattle and Sheep," *Journal of Animal Science,* Vol. 30, No. 5, 1970, p. 819.

The Tennessee Station workers also harvested corn for silage at 4 stages of maturity which they described as late milk, early dough, late dough, and mealy endosperm.[12] For the last stage, the corn was permitted to stand until there was approximately 35% moisture in the grain, and the endosperm was mealy. As high as 50% of the leaves were fired and essentially all the color was gone from the husks. These corn silages were fed to slaughter heifers in 3 successive years. Results indicated that corn plants harvested for silage at the late milk, early dough, and late dough stages did not affect the average daily gains of slaughter heifers. However, there was a significant reduction in average daily gain at the mealy endosperm stage.

In further studies, the Tennessee researchers did not detect any significant difference in the milk production of Jersey cows fed corn silage at 3 different stages of maturity.[13] They concluded that selecting a high grain yielding corn silage and harvesting the crop when the dry matter content is between 30 and 35% to prevent seepage and spoilage losses may be as important as the physiological stage of maturity to determine the time of harvest.

Cut to Proper Length

The length of the cut sections affects the packing and, hence, the quality of the silage. Also, the proper length of cut varies with the crop and the moisture content. Thus, for corn and sorghum crops, forage harvesters should be set to make a theoretical cut of ¼ to ⅜ in. If the knives are sharp and set up to the cutter bar, this will result in about 15% of the particles being 1½ in. and over, 25% of the particles being ¾ to 1½ in., and 60% being ⅛ to ¾ in. in length. Such a combination of particle size is necessary for high-quality feed. Grass silages should be more finely chopped than corn or sorghum silage. Also, wilted and dry forage and forage with hollow stems should be chopped more finely than forage of high-moisture content, thus permitting more thorough packing and eliminating most air pockets.

Control the Moisture Content

Moisture content is one of the most important factors in determining quality of silage. Experimental work and practical experience have indicated that 60 to 67% is the best moisture content for most crops to be ensiled. However, low-moisture silage of 40 to 60% moisture is now being preserved successfully in either oxygen-limiting silos, or tall conventional silos that are properly topped off with heavy, wet forage or sealed with a plastic cover.

Forage containing more than 60 to 67% moisture (1) is heavier and more costly to handle than is necessary; (2) is apt to produce slimy, putrid silage, due to the presence of butyric and other undesirable acids; (3) will have excessive seepage of the juices and some loss of nutrients, except carotene, from the silo; (4) will result in excessive deterioration in the silo walls due to the high acidity; and (5) will exert higher pressure on the silo walls—for the greater the moisture content, the greater the pressure.

If corn and sorghum are harvested at the stage recommended, their moisture content will be right. However, freshly cut grass and/or legume forage contains 75 to 80% moisture, which means that for proper ensiling its moisture content must be lowered by 10 to 15%.

HOW TO LOWER THE MOISTURE CONTENT

The moisture content of silage material may be lowered by any one or a combination of the following methods: by conditioning and/or wilting, by adding dry hay or straw, by combining with corn or sorghum silage, or by adding a dry additive/preservative of grain, dried molasses, or dried by-products of citrus or beets.

Conditioning-Wilting

This method is particularly applicable to the making of grass silage. Conditioning and/or wilting of grass silage increases the percentage of sugar in the forage, lessens the leakage of juice from the silo, lessens the pressure on the silo walls, and decreases the destructive action of the acids on the silo walls.

The needed 10 to 15% reduction in the moisture content of grass silage material can be accomplished by wilting for 3 to 4 hours on a good drying day and up to 1 day or longer in slow drying weather.

The combination of conditioning and wilting is the method most commonly followed today. Excellent equipment is available for conditioning.

Excess drying should be avoided, as it will result in the forage becoming too dry for proper ensiling.

Adding Dry Hay or Straw

The moisture content of any wet silage material can be lowered effectively by mixing dry hay or straw with it at the time of filling. Thus, during poor wilting weather, the moisture content of grass forage can be brought within the desired range by adding 5 to 20% hay or straw. Also, this is the standard method of lowering the moisture content when it is desired to ensile such high-moisture products as potatoes. (See "Other Silage Crops.")

Conditioning and wilting is the preferred method of lowering the moisture content of grass silage, rather than adding dry hay or straw, which reduce digestibility and energy content.

Combining With Corn or Sorghum Silage

Sometimes the moisture content can be lowered sufficiently by merely mixing high-water content crops with low-water content crops. (See sections entitled "Other Silage Crops" and "Combining Crops for Silage.") Simultaneously, usually more desirable bacterial action can be assured by this procedure.

[12]Chamberlain, C. C., "Effect of Maturity of Corn Silage at Harvest on the Performance of Feeder Heifers," *Journal of Animal Science*, Vol. 33, No. 1, 1971, p. 161.

[13]Montgomery, M. J., *et al.*, "Effect of Maturity of Corn on Silage Quality and Milk Production," *Journal of Dairy Science*, Vol. 57, No. 6, 1974, p. 698.

Adding a Dry Additive/Preservative

Such dry additives/preservatives as ground grain, corn-and-cob meal, dried molasses, and dried citrus meal, citrus pulp, or beet pulp will reduce the moisture content of freshly cut and unwilted forage, and, in turn, lessen the leakage (or seepage) from the silo.

HOW TO INCREASE THE MOISTURE CONTENT

Drier material may be used for silage by cutting shorter and packing more thoroughly. If necessary, water should be added or the dry material should be mixed with very green, freshly cut material by alternating loads.

HOW TO DETERMINE THE MOISTURE CONTENT

Some methods of determining moisture content follow:

1. **The grab test (or squeeze method).** This test consists in taking a handful of the chopped forage and giving it a good hard squeeze for about 30 seconds. Then opening the hand slowly, noting the condition of the ball of forage in the hand, and referring to Fig. 9–18.

1.
Juice runs freely or shows between the fingers. The crop contains 75 to 85% moisture and is too wet to make high-quality silage without treatment. Silages made from crops in this condition will lose large quantities of juice. When possible, wilt these crops. If they must be ensiled without wilting, use an effective chemical preservative (not all of them are effective) or 200 lb of ground grain per ton of crop.

2.
The ball holds its shape—the hand is moist. The crop contains 68 to 75% moisture. Some juices will escape from tower silos. Additional wilting in the field is desirable. Where this is not done, use a chemical preservative or 150 lb of ground grain per ton of crop, or layer with wilted crops. Odors will be strong without some treatment.

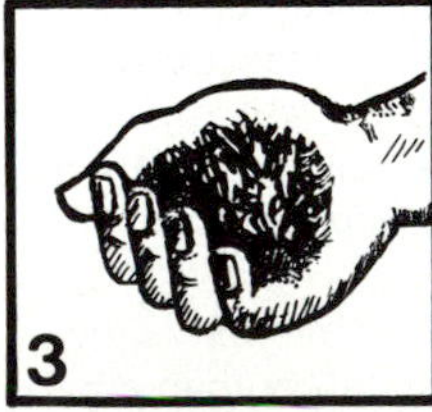

3.
The ball expands slowly—no dampness appears on the hand. The crop contains 60 to 67% moisture. This is the best condition for ensiling legumes without treatment.

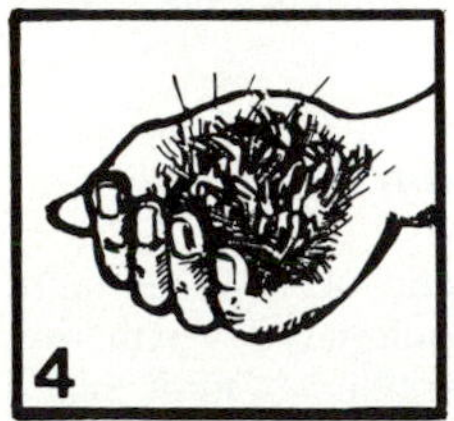

4.
The ball springs out in the opening hand. The crop contains less than 60% moisture. Only very young crops wilted to this condition can be safely ensiled. Others are likely to mold in the silo unless layered with wet crops or placed in gastight silos.

Fig. 9–18. The grab test.

2. **The twist method.** Before chopping, the forage should be so well wilted that the stems may be twisted without breaking, but the limp leaves should show no signs of dryness. This test cannot be used for such coarse crops as sweet clover.

3. **The oven-drying method.** If in doubt or until more experience is obtained, the moisture content of a sample of any kind of forage may be obtained in about an hour's time by the following procedure:

a. Weigh an empty tray on kitchen, bathroom, dairy barn, or other scales.
b. Spread some of the forage in a thin layer on the tray.
c. Weigh the tray and forage.
d. Subtract "a" from "c," in order to obtain the weight of the green forage sample.
e. Place the tray and sample in a preheated oven at 275°F. If the oven doesn't have a vent, leave the door slightly open.
f. When the forage seems to be dry, weigh it again. Return it to the oven and reweigh at 5 to 10 minute intervals until the weight is constant. Record final weight of the tray and dry forage.
g. Subtract "a" from "f," in order to obtain the weight of the dry forage sample.
h. Subtract "g" from "d," in order to obtain the weight of the water in the sample.
i. Divide "h" by "d" (and multiply by 100), in order to obtain the percent of moisture in the forage.

4. **Other methods.** The heated-oil method and certain patented devices (as forced air dryers) may be used for moisture determination. Recently, several electronic testers which give instantaneous moisture readings of forages have come on the market. Also, microwave ovens can be used for moisture determination, provided the forage is spread in a thin layer to prevent burning. The latter two methods are expensive, but fast.

Fill Rapidly

Once silo filling is started, it should be rapid, so as to avoid spoilage before the silo is filled and sealed. Generally speaking, a silo should be filled in 2 days or less.

Distribute Forage Uniformly in the Silo

In order to avoid the presence of air pockets and spoilage, it is essential that any kind of chopped forage be distributed uniformly in the silo and that it be packed well. Proper silage distribution is obtained by keeping the material nearly level or slightly higher at the center. Silage distributing equipment is available for keeping the material in an upright silo level. These devices are very helpful, especially in silos of 14-ft or larger diameters.

Where corn, sorghum, and sunflower forage is harvested at a green, immature stage and cut into short lengths, tramping in an upright silo will not be necessary; but uniform distribution is very important. The only filling precaution under these conditions is to see that the top is carefully leveled and well packed and covered whenever filling is completed.

Grass silage (especially when wilted), hollow-stemmed forages, and forages that have matured or dried beyond the best silage stage should always be trampled well, especially near the wall.

Packing in a trench silo should be obtained by use of a tractor.

Seal or Top-Off the Silo

Sealing or topping-off is necessary in order to avoid excess spoilage, especially with grass silage, which tends to dry on the surface and to shrink away from the silo walls. This may be accomplished by carrying out one or more of the following procedures:

1. Leveling off the top and thoroughly trampling the last few feet, especially near the walls.
2. Topping-off the silo with 2 to 3 loads of wetter material.
3. Covering the top with plastic cut to fit the silo diameter and turned up against the silo wall a distance of 5 to 8 in.

FEEDING VALUE AND ECONOMY OF SILAGE

Fig. 9-19. Silage stored in oxygen-limiting silos and conveyor-delivered to cows on the John Wardlow dairy in Mississippi. (Courtesy, A. O. Smith Harvestore Products, Arlington Heights, Ill.)

A common rule of thumb is that 3 lb of 70% moisture grass silage or 2 lb of 40% haylage are equivalent to 1 lb of hay of similar kind and quality; a difference due primarily to the higher water content of silage or haylage. Suggested practical rations for different classes of livestock in which silage is incorporated, usually in combination with hay or some other dry forage, are given in this book in the chapters devoted to each species.

Many factors enter into any figures which propose to show the comparative economy of silage vs dry forages; among them, (1) the comparative yield of total digestible nutrients per acre, (2) the cost per ton for preserving and storing, (3) the relative nutrient and feeding value, (4) the distribution of labor, (5) the control of weeds, (6) the kind of haymaking weather, (7) the hazard of curing so much hay without it becoming too mature, (8) the price per ton, and (9) the machinery and efficiency of the methods used in each method.

CHARACTERISTICS OF GOOD SILAGE

Fig. 9-20. Guernsey cows eating round bale alfalfa silage with relish, from an enclosed steel rack. (Courtesy, Sperry New Holland, New Holland, Pa.)

In order to make good-quality silage, producers need to know what constitutes silage quality. They need to be acquainted with those recognizable characteristics of silage which indicate high palatability and nutrient content. These are:

1. **Odor.** It has a "clean," rather pleasing acid odor, in contrast to the foul or objectionable odor of poor silage.
2. **Taste.** The taste is pleasing, not bitter or sharp.
3. **Absence of mold and rot.** There is no visible mold, and it is not musty or slimy.
4. **Moisture and color.** It is uniform in moisture and color. Very high-moisture silage is likely to be dark colored, slimy textured, and have a disagreeable odor. Generally, green or brownish silage is good; tobacco brown, dark brown, caramelized, or charred silage indicate excessive heat; and black silage is rotten and should not be fed.
5. **Animal acceptance.** Animals like and thrive on good silage.

SILAGE POINTERS

Some additional pointers which may be of value to the farmer or rancher who is making or feeding silage follow.

Coating the Silo

Since wet grass silage has a somewhat more corrosive action on concrete (monolithic) than does corn or sorghum silage, it may be desirable to apply a protective coating to the inside of concrete silos, whether of solid concrete or of stave construction. The problem is to find an effective, nontoxic, and economical coating. For information on the latest recommendations, the farmer or rancher should contact the local county extension agent, cement dealer, or write to the state college of agriculture.

Nutrient Losses in Seepage

Seepage losses vary with the moisture content, depth of silage, distribution of the silage, and the amount of nutrients in the seepage. They may be as high as 14% of the dry matter stored.

The nutrient losses vary, but generally they are in proportion to the run-off. The nutrients lost in seepage from a 100-ton silo may equal the nutrients in ¾ ton or more of hay.

Exposure to Air

Spoilage begins the moment silage is exposed to the air. Therefore, once the silo is opened for use, feed should be removed daily. In the wintertime, a minimum of 1½ in. of silage should be removed daily from a tower silo; in the summertime, 3 in.

Also, it should be realized that spoilage is likely to occur on the surface of the ensiled material if more than 2 days elapse between filling periods.

Removal of Silage from Silo

In the past, the common method of feeding silage was by hand—with a fork. In the present era of bigness and automation, the removal and feeding of silage is being completely automated.

It is possible to achieve complete push button controlled feeding in an upright silo. With horizontal silos, the silage may be handled with a scoop, and sometimes a mechanical unloading wagon or truck—depending on distance from feedlot to silo.

Self-feeding from both tower and horizontal silos can be achieved, but this requires more management on the part of the operator.

Moldy Silage

Moldy silage may be harmful. Any spoiled material that causes animals to go off feed, or that upsets the metabolic processes, should not be fed.

Some conditions cause certain molds to produce toxins. The toxins are called *mycotoxins* and the effects of the toxins on animals are called *mycotoxicoses.*

Mature ruminants appear to tolerate higher levels of mycotoxins than young ruminants, monogastric animals, or horses.

One way in which to determine the potential toxicity of moldy silage is to feed it to some less valuable animals for at least 2 weeks. Observe the animals daily for signs of toxicity—such as reduced gain and going off feed. If no toxic effects are noticed, it is probably safe to feed the suspect silage to other animals. If ill effects are noticed, switch them to other feed immediately and dispose of the suspect silage by spreading it on the land and plowing it under.

Recently, it has been determined that farmers may become afflicted with mycotoxicoses, characterized by severe congestion of the respiratory tract, high fever, and sometimes complete immobility. Thus, prevention of mold growth and protection from molds is essential from the standpoint of the people who feed moldy silage, as well as the animals that consume it.

Silage for Summer Feeding

Some farmers and ranchers, especially dairy producers, use silage effectively as a summer feed. This practice is especially desirable in areas where pastures dry up during the hot, dry months. It appears that more and more dairy producers will change to year-around silage feeding in confinement.

Effect of Silage on Milk Odor and Flavor

Silage sometimes affects the flavor and odor of milk, especially when ensiled too wet. This effect may be somewhat more pronounced with some silages than with others. The dairy producer will do well, therefore, to feed all silages after, rather than before, milking.

Dangerous Silage Gases

Two types of toxic gases may be formed when making silage: (1) carbon dioxide (CO_2), and/or (2) nitrogen dioxide (NO_2).

Carbon dioxide forms soon after filling begins and continues until fermentation stops. It is a colorless, suffocating gas, which is heavier than air and tends to collect in low places.

Under drought conditions, corn, sorghum, and other grass species may accumulate higher than normal levels of nitrates. When ensiled, nitrates are converted to nitrites, then nitrites are converted to nitrogen oxide by bacteria and plant cells. As the nitrogen oxide comes in contact with air, it is oxidized to form nitrogen dioxide, a reddish brown-colored gas, which is heavier than air. This gas is highly toxic to both humans and farm animals.

Precautions against hazards caused by silage gases include (1) operating the blower for a 15-minute period if it is still connected, (2) swinging a piece of canvas, a tree branch, or a burlap bag vigorously so as to agitate the air and dilute gases that may be present, or (3) taking proper life support equipment when entering an oxygen-limiting, or sealed, silo. Also, adequate provision for ventilation of the silo through the roof is essential.

A victim of silo gas should be moved into fresh air immediately, and artificial respiration should be applied. A physician should be called immediately.

HIGH-MOISTURE GRAIN

High-moisture grain refers to grain that is harvested at a moisture level of 22 to 40% and stored without drying.

Interest in storing and feeding high-moisture grain was prompted by the shift toward more field shelling of corn, instead of picking ear corn, because much shelled corn must be dried for safe storage whereas ear corn can be safely stored at moisture contents up to 24% without drying. More recently, interest in high-moisture grain increased with high energy cost for drying. It takes approximately 1 gal of propane fuel to dry 4½ to 6 bu of corn, or 1 kilowatt hour of electricity to dry 10 to 12 bu of corn, with conventional high temperature drying to reduce the moisture content of wet grain 10 percentage units (*i.e.,* to dry from 25% down to 15% moisture).

Fig. 9-21. High-moisture corn being stored in one of the cement-lined trench storages at the Farr Feeders, Inc., cattle feedlot northeast of Greeley, Colorado. The high-moisture corn is stored in large quantities (60,000 tons) to provide a year's continuous supply of feed. Note truck unloading high-moisture corn at bottom of trench and tractor packing it at top of the silo. (Courtesy, William D. Farr, Farr Feeders, Inc., Greeley, Colo.)

The **advantages** of high-moisture grain in comparison with dried grain are:

1. It alleviates the high energy cost for drying grain.
2. It lessens field losses at harvest time.
3. It permits harvesting earlier, at higher moisture content, and usually during more desirable weather. As a result, it releases land for fall plowing in the North and for fall seeding of a second crop in the South.
4. It makes it practical to use later maturing, higher yielding varieties of corn and sorghum in the northern areas which often have early frost.
5. It usually requires less investment in processing equipment.
6. It improves the feeding value of grain for beef and dairy cattle.

The **disadvantages** of high-moisture grain in comparison with dried grain are:

1. It requires a large inventory of high-moisture grain, which may increase capital requirements.
2. It limits market flexibility for the grain, since it must be fed to livestock.
3. It may result in higher storage losses than for dry grain if proper ensiling or acid-treatment is not followed.
4. It may freeze in the bunk in the winter, and flies may be a problem in the summer.

Fig. 9-22. High-moisture shelled corn pouring from a combine into a grain truck. By properly storing the moist grain, the cost and labor of artificial drying wil be alleviated. (Courtesy, A. O. Smith Harvestore Products, Arlington Heights, Ill.)

Today, considerable quantities of high-moisture ear corn, shelled corn, sorghum (milo), and small grains (wheat, barley, and oats), containing about 30% moisture, are stored and fed. Some high-moisture grains are planned for and intended. Others are the result of happenstances, such as crops planted late, early frost damage, or harvesting when wet.

(Also see Chapter 14, Feed Processing, section on "High-Moisture Grain.")

Harvesting

A number of field equipment combinations may be used to harvest high-moisture grains. Harvest cost plus product quality should be considered in selecting the most desirable combination.

High-moisture grain should be harvested when it reaches physiological maturity and the moisture is 22 to 40%; with ear corn, sorghum (milo), and small grains higher in moisture than corn at harvest time. At this stage, grains yield the maximum available nutrients per acre and preservation conditions are best. If grain is harvested when too wet and immature, dry matter yield per acre will be less. If the grain is too dry, mold will be high and fermentation will be low; thus, grain containing less than 22% moisture should be reconstituted by adding water.

Storage of High-Moisture Grain

There are three basic storage methods for high-moisture grains: (1) ensiling in sealed (airtight) storage, (2) ensiling in nonsealed storage, and (3) preservation with an organic acid or ammonia.

SEALED (AIRTIGHT) STORAGE

High-moisture grain may be stored in an oxygen-limiting silo. It is not necessary to crack or grind grains before storing in this manner. Another advantage of sealed storage is that it can be unloaded from the bottom.

Sealed storage is the most popular method of storing high-moisture grains even though greater initial capital investment is required. This type of storage eliminates the 2 to 5% spoilage loss normally associated with unsealed storage.

Fig. 9-23. High-moisture grain (corn) coming straight from the combine and going directly into oxygen-limited storage. (Courtesy, A. O. Smith Harvestore Products, Arlington Heights, Ill.)

UNSEALED STORAGE

Two types of unsealed high-moisture grain storage are used; (1) conventional upright silos made of concrete or steel, which are structurally adequate for storage of grass silage; or (2) horizontal silos. The grain should be ground into the storage unit, then firmly packed; otherwise, spoilage will result. With upright silos, about a 3 in. layer off the ground, high-moisture grain should be removed from the exposed surface each day during mild weather; with horizontal silos, 4 in. should be removed daily. Greater amounts should be removed from both types of storage units during warm weather to prevent spoilage.

PRESERVATION WITH AN ORGANIC ACID OR AMMONIA

Storage involving the treatment of high-moisture grain with an organic acid or ammonia to inhibit mold or spoilage is favored by many farmers because it alleviates artificial drying or the necessity to store in an airtight silo. Several different organic acids may be used—propionic, acetic, isobutyric, formic, benzoic, or a combination of these acids—but the most commonly used acids are propionic or propionic-acetic acid mixtures, marketed under various trade names. Anhydrous ammonia and other gaseous mixtures are also effective.

• **Mode of action of organic acids**—High-moisture grain should be treated promptly after harvesting to avoid heating. When properly applied, propionic acid and other acid-type grain preservatives will kill most fungi and other micro-organisms present on the outside of the treated grain. As the acid moves into the seed, it kills the embryo, thus nearly eliminating respiration and enzymatic activity. These actions prevent heating and retain nutrient content. While the exact mode of action is not known, the acid preservatives continue to inhibit growth of molds. The effect is probably due in part to the lower pH created by the acid; however, not all products which depress pH inhibit mold. To be effective, the treatment must provide continuing protection against mold and other microbial growth.

• **Amount of organic acid to apply**—The amount of acid required to treat high-moisture grain depends on moisture content of the grain, length of storage desired, and the temperature. The recommended rate of application of propionic acid to high-moisture grain to provide protection for 1 year is shown in Table 9-8. These rates are for corn, but they are suitable for other high-moisture grains.

TABLE 9-8
AMOUNT OF 100% PROPIONIC ACID REQUIRED FOR 1 YEAR OF STORAGE[1]

% Of Moisture Content Of Grain	% By Weight	Lb (*0.45 Kg*) Per Wet Bushel[2] (*35.2 l*)	Lb (*0.45 Kg*) Per Ton (*0.907 Metric Ton*)	Gal (*3.78 l*) Per Ton (*0.907 Metric Ton*)
16–18	0.50	0.28	10	1.3
20	0.75	0.42	15	1.8
25	1.00	0.56	20	2.4
30	1.25	0.70	25	3.0
35	1.50	0.84	30	3.6

[1]The amounts of acid listed are for long term storage (1 year). For storage periods of 6 months or less, the amount of acid used could be reduced by ½.

[2]56 lb (*25.5 kg*) of high-moisture corn.

• **Application guidelines**—

1. Check grain for moisture content so selected rate of acid application meets requirement for preservation.
2. Treat the grain immediately after harvesting so as to eliminate heating and mold development.
3. Make sure that all the grain is coated with the acid and that the application rate is correct.
4. Treat outdoors if possible; otherwise, provide adequate ventilation.
5. After treatment, flush equipment with water or untreated grain to prevent corrosion.
6. Observe safety guidelines at all times.

• **Safe handling of grain preservatives**—When preservatives are applied, the following safety precautions should be observed by the user:

1. Follow the manufacturer's instructions.
2. Avoid storage of acids with fuels, lubricants, and pesticides. Store acids *only* in original container, tightly closed, with the bungs upright.

3. Use organic acids *only* on grain destined for animal feed.

4. If grain is treated in a building, adequate ventilation should be provided.

5. Acid should not be allowed to come in contact with skin or eyes. Protective gloves, goggles, respirators, or a face shield and protective clothing should be worn by the person applying it when there is a risk of contact.

6. A supply of water should be available to wash away any acid coming in contact with skin or eyes.

7. When contact occurs, drench and remove contaminated clothing immediately. Contaminated clothing should be washed before wearing.

8. Avoid breathing fumes during filling or when entering a freshly treated grain bin a day or two after treatment.

9. Organic acids are flammable so care should be exercised to avoid fire. Avoid smoking or having an open flame or heating device in treatment areas. Keep the acid away from any ignition source and maintain good ventilation in areas when it is being applied.

• **Storage facility guidelines**—Nearly any weatherproofed facility can be used to store treated grains, with the precaution that the acid is very corrosive to steel, especially if galvanized. Concrete is also subject to corrosion, but to a lesser degree than steel. Among the facilities that may be, and are, used are the following:

1. Metal bins and buildings coated with acid-resistant paint, or linseed oil, or covered with polyethylene plastic.

2. Concrete silos or bins coated with acid-resistant paint or linseed oil, or covered with polyethylene plastic.

3. Wooden bins, cribs, and buildings. Except for nailhead corrosion, wooden bins or cribs are not affected by acids.

4. Quonset-type buildings, provided grain is not stored on a dirt floor and the metal is protected by acid-resistant paint, or linseed oil, or covered with polyethylene plastic.

Acid-treated grain stored in makeshift bins, piles, or bunkers should be provided with covers that will prevent moisture from entering the storage area.

• **Handling and feeding**—

1. Since the acid is absorbed by the grain, protection is provided after removing from storage. Thus, treated grain can be mixed with dry grains and other feeds.

2. Treated grains can be transported. They have been successfully moved from one silo to another. However, movement from a large diameter to a smaller diameter silo is suggested for better packing.

3. Acid-treated high-moisture grain is readily accepted by cattle. In studies at the University of Wisconsin, dairy cows offered propionic acid-treated high-moisture ground ear corn consumed more dry matter than those offered similar ensiled corn or dried corn that had received no propionic acid treatment.

• **Advantages and disadvantages of acid-treated grain**—In comparison with ensiled high-moisture grain, acid-treatment has the following advantages: (1) Removal from the silo is eliminated; (2) it can be stored in a barn or other temporary storage facility; (3) treated grain can be transported over long distances; and (4) large batches of a ration which include acid-treated high-moisture grain can be mixed without risk of spoilage.

The major disadvantages to acid-treated grain are: (1) It must be fed to livestock—it cannot be marketed for any other purpose; (2) it will not germinate, so it cannot be used for seed; (3) acids are corrosive to metal or concrete storage facilities; and (4) organic acids are costly, so for grain with over 30% moisture they may be uneconomical.

Feeding Value of High-Moisture Grain

The feeding value of high-moisture grain is equal or slightly superior to that of dry grain, with some variation according to class and productivity of livestock.

Beef cattle: Improvement is greater and more consistent with high-moisture ear corn than with high-moisture shelled corn. A summary of 14 experiments showed that high-moisture ear corn increased gains by 3% and feed efficiency by 10% over dried ear corn. Studies have also shown 3 to 5% improvement in the value of the dry matter in high-moisture shelled corn for cattle. Rate of gain has been similar for cattle fed dry or high-moisture shelled corn.

High-moisture storage improves the feeding value of sorghum (milo) more than it does corn for cattle. High-moisture harvested milo increases daily gains slightly (0 to 2%) and improves feed efficiency from 6 to 10% over dry sorghum grain for beef cattle.

High-moisture storage does not improve the feeding value of wheat for cattle.

Grain is stored whole in gastight silos, but it should be rolled or ground when removed for feeding. Grain should be ground or rolled when it is stored in horizontal or conventional upright silos in order for it to pack tightly and exclude air.

(Also see Chapter 19, Feeding Beef Cattle, section on "High-Moisture Grain.")

Dairy cows: Research studies have shown that the feeding value of properly ensiled or acid-treated high-moisture corn is equal to that of dry corn for lactating dairy cows; the milk yield and feed intake were similar. However, when high-moisture shelled corn supplies more than 50% of the total ration dry matter, depressed milk fat percentage may occur, with inadequate fiber given as the probable cause.

Some form of processing (e.g., rolling or grinding) of high-moisture corn improves utilization.Whole kernels appearing in feces indicate incomplete digestion. Processed corn is higher in digestible, metabolizable, and net energy for dairy cows than rations containing whole shelled corn.

Swine: When compared on an equal dry matter basis and in mixed rations, rate of gain and feed efficiency are essentially the same for hogs fed high-moisture or dry grains. Free-choice feeding of high-moisture grain may be used successfully for hogs weighing more than 60 lb provided proper intake of the protein supplement relative to grain intake is assured. But free-choice feeding is not recommended for pigs weighing under 60 lb.

There is no advantage to grinding or cracking high-moisture corn for growing-finishing pigs other than for mixing purposes. However, grinding or cracking increases the feeding value of high-moisture sorghum (milo), barley, and wheat.

Reconstituted Grain

Reconstituted grain is mature grain to which water has been added to raise the moisture content to 25 to 30% for storage.

Reconstituted sorghum (milo) stored whole in an airtight silo, then processed at feeding time and fed to finishing cattle, will improve rate of gain slightly and feed efficiency by 12 to 15%. However, if sorghum is ground before reconstituting, there is little improvement in feed value over dry milo for cattle. Reconstituted whole corn fed to finishing cattle will not improve rate of gain, but will improve feed efficiency by about 4.5%.

(Also see Chapter 14, Feed Processing, section on "Reconstituted Grain.")

Buying and Selling High-Moisture Grain

Most high-moisture grain is grown by livestock producers for livestock feed, or grown as a cash crop and sold at harvest to livestock producers. High-moisture grain is bought/sold at a lower figure than dry grain because of the water content. Also a greater quantity of it must be fed because of the water content. In most cases, the buying and selling transactions of high-moisture grains are based on an 87% dry matter and 13% moisture basis.

(Also see Chapter 16, Buying Feeds/Commercial Feeds/ Feed Laws, section on "Moisture is Important.")

QUESTIONS FOR STUDY AND DISCUSSION

1. What is silage? How important is it as a U.S. livestock feed?

2. Since most silage is fed on the farms or ranches on which it is produced, how important is it that the farmer or rancher know how to produce it?

3. Define the *ensiling process*. Describe each of the 3 phases of the ensiling process. What is the browning reaction/Maillard reaction in silage?

4. Discuss the economics of silage production. List both the advantages and disadvantages of silage.

5. Classify and describe different kinds of silos. What are the advantages and disadvantages of each kind of silo?

6. List the advantages and the disadvantages of round bale plastic covered silage.

7. What factors determine the size silo to build of each (a) a tower type, and (b) a trench type?

8. Under what circumstances might it be important to determine the amount of silage in a silo after part of it has been fed? How would you go about estimating the weight of silage remaining in each (a) a tower silo, and (b) a trench silo?

9. List and discuss the importance of each of the 4 primary types of silage storage losses.

10. What accounts for corn ranking first in importance as a silage crop? Under what circumstances might it be desirable for a farmer to grow sorghum for silage rather than corn?

11. What is grass/legume (hay crop) silage? What are the advantages and the disadvantages of grass/legume (hay crop) silage?

12. What are the distinguishing characteristics of (a) direct-cut silage, (b) wilted silage, and (c) low-moisture silage (haylage)?

13. Give directions for making top-quality low-moisture silage.

14. Under what circumstances should a farmer or rancher make grass (hay crop) silage rather than corn or sorghum silage? Nutritionally, how do corn/sorghum silage and grass/legume silage compare?

15. Under what circumstances might it be desirable to make silage from each of the following types of products: (a) by-product feeds, (b) rain-damaged hay, (c) frosted crop, or (d) drought-stricken crop?

16. What are: (a) a silage additive, and (b) a silage preservative? List 4 types of additives or preservatives, and give examples and describe the role of each type.

17. Why, and under what circumstances, might nonprotein nitrogen (NPN) additives be mixed with silage?

18. What criteria for assessing the value of a silage additive or preservative would you recommend?

19. Why have silage harvesting methods and machinery changed so greatly in recent years, with field forage harvesters becoming more widely used than any other type of equipment, and with stationary silo fillers declining markedly?

20. Discuss the importance of each of the following practices from the standpoint of making good silage: (a) harvest at the proper stage of maturity, (b) cut to proper length, (c) control the moisture content, (d) fill rapidly, (e) distribute forage uniformly in the silo, and (f) seal or top-off the silo.

21. Discuss the feeding value and economy of silage.

22. What are the easily recognized characteristics of silage of high feeding value?

23. What types of toxic gases may be formed when making silage; what causes them; and what precautions may be taken against the hazards caused by them?

24. What is meant by high-moisture grain? How and why is it produced? Why has interest in high-moisture grain increased with higher energy cost?

25. What are the advantages and disadvantages of high-moisture grain in comparison with dried grain?

26. List and discuss the 3 basic storage methods for high-moisture grains.

27. Discuss the feeding value of high-moisture grain for beef cattle, dairy cows, and swine.

28. What is reconstituted grain? Under what circumstances might it be desirable to reconstitute grain?

29. What factors will determine whether a particular crop should be utilized as pasture, hay, or silage?

SELECTED REFERENCES

Title of Publication	Author(s)	Publisher
Crop Production	R. J. Delorit L. J. Greub H. L. Ahlgren	Prentice-Hall, Inc., Englewood Cliffs, N.J., 1984
Crop Quality, Storage, and Utilization	C. S. Hoveland, Editor	American Society of Agronomy, Crop Science Society of America, Madison, Wisc., 1980
Crops in Peace and War: Yearbook of Agriculture 1950–1951		U.S. Department of Agriculture, Washington, D.C.
Forage Conservation in the 80s	Ed. by C. Thomas	British Grassland Society, Hurley, Maidenhead, Berks, United Kingdom, 1979
Forage Crops	D. A. Miller	McGraw-Hill Book Company, Inc., New York, N.Y. 1984
Forages, The Science of Grassland Agriculture, 3rd Edition	M. E. Heath R. F. Barnes D. S. Metcalfe	The Iowa State University Press, Ames, Iowa, 1985
Forages: Resources of the Future	J. E. Oldfield, Chairman of Task Force	Council for Agricultural Science and Technology, Ames, Iowa, 1986
Grass to Milk	C. P. McMeekan	The New Zealand Dairy Exporter, Wellington, New Zealand, 1964
Grass: Yearbook of Agriculture 1948		U.S. Department of Agriculture, Washington, D.C.
Haymaker's Handbook	M. A. Bales J. E. Baylor	New Holland, Inc., New Holland, Pa., 1987
Manual of the Grasses of the United States, 2nd Edition	A. S. Hitchcock; rev. by A. Chase	U.S. Department of Agriculture, Washington, D.C.
Pasture Book, The	W. R. Thompson	W. R. Thompson, State College, Miss., 1950
Plants for Man, 2nd Edition	R. W. Schery	Prentice-Hall, Inc., Englewood Cliffs, N.J., 1972
Practical Grassland Management	B. W. Allred	The Interstate Printers & Publishers, Inc., Danville, Ill., 1952
Silage Additives USA	K. Bolsen	Chalcombe Publications, Marlow, Bottom, Marlow, Bucks, Great Britain, 1985
Stockman's Handbook, The, 6th Edition	M. E. Ensminger	The Interstate Printers & Publishers, Inc., Danville, Ill., 1983

Fig. 9-24. A modern dairy. (Courtesy, Holstein-Friesian Assn. of America, Brattleboro, Vt.)

Fig. 9-25. Feeding silage by a mechanical (auger) feeder. (Courtesy, Massey-Ferguson)

Original painting by Tom Phillips

Chapter 10

GRAINS/HIGH-ENERGY FEEDS[1]

Contents Page

[1]The authors gratefully acknowledge the helpful suggestions of the following authorities who reviewed all or parts of, and furnished material for, this chapter: J. D. Axtel, Ph.D., Professor of Agronomy, Department of Agronomy, Purdue University, West Lafayette, Ind.; R. A. Forsberg, Ph.D., *et al.*, Department of Agronomy, University of Wisconsin, Madison; R. G. Fulcher, Ph.D., Grain Quality Laboratory, Plant Research Centre, Agriculture Canada, Ottawa, Ontario; R. C. Hoseney, Ph.D., Department of Grain Science and Industry, Kansas State University, Manhattan; D. A. Miller, Ph. D., Professor, Department of Agronomy, University of Illinois, Urbana-Champaign; M. W. Phillips, Ph.D., Head, Department of Agronomy, Purdue University, West Lafayette, Ind.; Y. Pomeranz, Ph.D., Director, U.S. Grain Marketing Research Laboratory, Agricultural Research Service, USDA, Manhattan, Kan.; C. M. Wilson, Ph.D., Research Chemist and Professor of Plant Physiology, Agricultural Research Service, USDA, Urbana, Ill.; and V. L. Youngs, Ph.D., Research Food Technologist, Agricultural Research Service, USDA, Fargo, N.D.

All physiological processes require energy. Before any animal can reach maximum performance, energy must be expended to maintain routine body functions, such as blood pressure, body temperature regulation, respiration, transmission of nerve impulses, muscle tone, and metabolic processes. Only when these essential functions are provided can the animal synthesize tissues and secretions which we commonly call *production*. All production parameters—lactation, growth, pregnancy, egg laying—require substantial quantities of energy. To maximize production, the livestock producer must use a high-energy ration—that is, a ration with a high caloric density and digestibility. When low-energy rations are used—for example, high-roughage feeds—the animal cannot physically ingest sufficient feed to fill its energy needs for production. Hence, there is a suppression of production potential. A high-energy ration properly complemented with the other essential nutrients allows the animal to fulfill its energy requirements with the ingestion of considerably less feed.

Concentrates are feeds that are high in nitrogen-free extract and TDN and low in crude fiber (less than 18%). They can be broken down into two classes—(1) carbonaceous feeds, and (2) nitrogenous feeds. The feeds discussed in this chapter belong to the first category, as their main function is to supply energy. Sugars, various types of polysaccharides, fats, and oils—all consisting primarily of carbon, hydrogen, and oxygen—provide the necessary materials for energy production.

Of all the energy sources used in livestock and poultry rations, cereal grains fill the most important role, providing the bulk of the energy for animals as well as substantial portions of the protein.

Although grains are the most widely used high-energy feed, by-products of grain milling, fats, oils, fruits, nuts, roots, and specialized feeds such as molasses, often provide excellent alternatives. While nitrogenous concentrates are used primarily for the synthesis of protein in the body, a limited amount of the protein in these feeds can be used for energy. Nitrogenous concentrate feeds are discussed in Chapter 11, Protein Supplements.

Grains and high-energy feeds, being low in fiber, are generally considered to be highly digestible by all types of livestock. Since poultry, swine, and other nonruminants which do not have functional ceca cannot digest fiber efficiently, the bulk of their respective rations consists of concentrate feeds—even for maintenance and low production. Ruminants and herbivorous nonruminants, on the other hand, can obtain a large proportion of their energy needs through the consumption of forages. However, when production demands are increased in these animals, forages are generally partially replaced by concentrate feeds which have higher caloric densities and digestibilities. For example, many finishing rations for beef cattle contain more than 85% concentrate.

GRAINS

Grains are seeds from cereal plants—members of the grass family, Gramineae. In addition to the production of grains which contain large quantities of carbohydrates, the entire plant can be harvested prior to grain maturation and used for forage—pasture, hay, or silage. The primary use of cereal plants, however, is through the utilization of the highly digestible seeds.

Fig. 10-1. Corn, the grain on which much of American livestock production is based. (Courtesy, USDA)

All cereal crops are annuals. Depending on the type and variety of grain crop, they may be either (1) winter annuals—planted in the fall and harvested in the spring, or (2) summer annuals—seeded in the spring and harvested in the late summer or early fall.

Such by-products of harvested grains as chaff, stover, and straw can be utilized as low-quality forages for ruminant animals. Moreover, many of the grains are milled or processed in some manner, thereby creating additional by-products which can be fed to livestock with varying degrees of success. These by-products range from high-energy, high-protein wheat germ to relatively low-quality, coarse oat hulls.

Corn, oats, barley, and sorghums are the primary grains fed to livestock and poultry; hence, they are known as feed grains or coarse grains. The United States produces 41% of the world's corn, 30% of the grain sorghum, and more than 25% of all feed grains. Rice and wheat are consumed primarily by humans, but many of their milling by-products are fed to livestock. Additionally, wheat is sometimes fed to livestock when the price becomes competitive with the more widely used feed grains, or when it is contaminated or damaged by insects or weather. Barley and rye are fed to livestock in limited amounts, but a large proportion of these two grains is used in the brewing and distilling industries. Millet, emmer, spelt, and triticale are used occasionally, but the impact of these grains on the feed industry is very limited. With the exception of rye, grains are very palatable to livestock.

The utilization of grains by humans and animals varies from country to country. For example, most U.S. wheat is used for human food; in Europe, however, where there is a surplus of wheat, large quantities of it are fed to livestock. In Mexico, corn is considered as a food grain, where it is ground and made into tortillas or gruels, while sorghum is the primary feed grain. Although sorghum is almost exclusively a feed grain in the United States, it is the primary food grain for people in a number of African countries and in parts of India, Pakistan, and China.

Table 10–1 gives the yearly breakdown of the amounts of the main feed grains fed to livestock and poultry in the United States. Corn is, by far, the most widely used feed grain. Sorghum ranks second.

TABLE 10–1
FEED GRAINS FED TO LIVESTOCK AND POULTRY (MILLION TONS)[1]

Grain	Year 1965	1970	1975	1980	1985
Corn	94.1	100.6	99.6	115.9	114.7
Sorghum	15.9	19.0	14.3	8.6	18.6
Barley & Oats	10.8	19.0	13.2	11.1	14.0
Total	126.8	138.6	127.1	135.6	147.3

[1]*Agricultural Statistics,* USDA.

Table 10–2 gives a breakdown of the specific purposes for which U.S. feed grains (corn, sorghum, oats, and barley) are used. The amount of feed grain used for food, alcohol, industrial purposes, and seed for replanting has been increasing over the last 5 years at a slow, steady rate. On the other hand, the amount of feed grain exported has been decreasing. During this same period of time, the amount of feed grain fed to livestock in the United States has greatly increased.

TABLE 10–2
USAGE OF FEED GRAINS[1]

Year[2]	Domestic Use: Feed (million metric tons)	Feed (% of total disappearance)	Food, Alcohol, Industrial, and Seed (million metric tons)	Food, Alcohol, Industrial, and Seed (% of total disappearance)	Total (million metric tons)	Total (% of total disappearance)	Exports (million metric tons)	Exports (% of total disappearance)	Total Disappearance (million metric tons)
1982–1983	139.0	63.2	26.5	12.0	166.9	75.9	53.0	24.1	219.9
1983–1984	119.7	54.4	28.3	12.9	149.5	67.9	56.6	25.7	206.1
1984–1985	131.1	59.6	30.6	13.9	163.2	74.2	56.6	25.7	219.8
1985–1986	134.9	61.3	33.5	15.2	169.9	77.3	36.6	16.6	206.5
1986–1987	145.5	66.2	34.2	15.6	181.0	82.3	46.3	21.1	227.3
1987–1988	146.9	66.8	36.5	16.6	183.4	83.4	51.7	23.5	235.0

[1]Feed grains include corn, sorghum, barley, and oats. Adapted by the authors from *Feed Situation and Outlook Report,* USDA, Nov. 1987.

[2]The marketing year for corn and sorghum begins Oct. 1; June 1 for oats and barley.

Fig. 10–2. Sorghum, the second ranking feed grain in the U.S., ready for harvest. (Courtesy, National Grain Sorghum Producers Assn., Abernathy, Tex.)

Grains provide the livestock producer an excellent source of highly digestible energy for animals that are either on a high level of production or unable to utilize forages effectively. However, several problems are inherent in the use of grains; among them, the following:

1. **In ruminant animals, high-concentrate rations may cause digestive disturbances, such as acidosis and parakeratosis of the rumen.** Ruminants need some *roughage factor* or *scratch factor* to stimulate the rumen papillae. In some cases, this is accomplished by rolling or coarse grinding the grain. Most cattle feedlot operators include 10 to 15% levels of some roughage source in finishing rations.

2. **Some grains must be processed before they can be fed.** The need for processing is primarily governed by the type of grain and the particular animal being fed. For example, grain should be ground when fed to very young or very old animals and animals which do not thoroughly masticate their feed.

3. **Grains are more costly than most fibrous feeds on a weight basis.** However, when comparing the costs of grain to roughage, the energy content, digestibility, and other nutrients must be considered on a per unit feed basis. Thus, a relatively expensive grain containing large amounts of highly digestible energy may in reality be a better buy than a low-cost, low-quality roughage.

4. **Grains are extremely deficient in calcium and certain vitamins.** Most grains contain less than 0.1% calcium. Ade-

TABLE
CHARACTERISTICS AND FEEDING

Grains	Lb per bu	Ruminant/ Swine	Energy Content[1]						Protein Content[1]				Value (lb for lb) as % of Corn (with Corn=100)[2]
			TDN		Digestible Energy				Crude Protein		Digestible Protein		
			%		A-F		M-F		%		%		
			A-F	M-F	Mcal per		Mcal per		A-F	M-F	A-F	M-F	
					(lb)	*(kg)*	(lb)	*(kg)*					
Barley *(Hordeum vulgare)*	48	Ruminant	**75**	85	**1.6**	***3.4***	1.8	*3.9*	**11.7**	13.2	**8.8**	10.0	Cattle 90. Sheep and goats 90. Swine 90–95. Horses 110.[2] Poultry 80–85. Rabbits 110.[2]
		Swine	**70**	79	**1.4**	***3.1***	1.6	*3.5*	**11.7**	13.2	**9.6**	10.8	
Corn *(Zea mays)*	56	Ruminant	**80**	91	**1.6**	***3.6***	1.9	*4.1*	**9.9**	11.2	**8.1**	9.2	100. Horses 115.[2] Rabbits 125.[2]
		Swine	**79**	90	**1.5**	***3.3***	1.7	*3.8*	**9.9**	11.2	**7.3**	8.2	
Emmer *(Triticum dicoccum)*	40, but variable	Ruminant	**72**	80	**1.4**	***3.1***	1.5	*3.4*	**11.7**	12.9	**9.4**	10.3	Cattle 70–90.
		Swine	**67**	74	**1.3**	***3.0***	1.5	*3.3*	**11.7**	12.9	**8.9**	9.8	
Millet *(panicum milaceum)* Also called hog millet, proso millet, broomcorn millet.	56	Ruminant	**61**	68	**1.5**	***3.4***	1.7	*3.8*	**12.1**	13.5	**6.8**	7.5	Swine 85–90. Poultry 95–100.
		Swine	**66**	73	**1.3**	***2.9***	1.5	*3.2*	**12.1**	13.5	**8.8**	9.8	
Oats *(Avena sativa)*	32	Ruminant	**69**	77	**1.4**	***3.0***	1.5	*3.4*	**11.9**	13.3	**9.2**	10.3	Cattle 70–90. Sheep and goats 80. Swine 70–80. Horses 100.[2] Poultry 70–80. Rabbits 100.[2]
		Swine	**65**	72	**1.3**	***2.8***	1.4	*3.2*	**11.9**	13.3	**9.7**	10.9	
Rice *(Oryza sativa)*	45	Ruminant	**68**	76	**1.4**	***3.0***	1.5	*3.4*	**7.5**	8.4	**4.0**	4.5	Cattle 80. Sheep and goats 55–75. Swine 80–85. Horses 115.[2] Poultry 80–85.
		Swine	**72**	81	**1.5**	***3.3***	1.7	*3.7*	**7.5**	8.4	**5.1**	5.7	
Rye *(Secale cereale)*	56	Ruminant	**73**	84	**1.4**	***3.1***	1.6	*3.6*	**12.0**	13.8	**8.4**	9.6	Cattle 96. Sheep and goats 83–87. Swine 90. Horses 115.[2] Rabbits 100.[2] Poultry 90.
		Swine	**75**	86	**1.5**	***3.2***	1.7	*3.7*	**12.0**	13.8	**9.1**	10.4	

quate amounts of phosphorus are generally present in grain, but the calcium to phosphorus ratio is highly unbalanced. Additionally, grains are deficient in certain vitamins; for example, vitamin A is low in all grains except fresh yellow corn.

Pertinent information relative to the primary grains fed to livestock is given in Table 10–3. A summary of grains not discussed elsewhere in this chapter is given in Chapter 30, Glossary of Feedstuffs.

10–3
VALUE OF CEREAL GRAINS (See footnotes at end of table.)

Special Characteristics and Importance	Livestock Use	Cultural Characteristics/ Geographic Adaptation	Other Comments
Barley contains more protein, lysine, and fiber than corn. Palatable. New hull-less types have higher value but lower yield. One of oldest crops known to humanity. Fourth highest tonnage of grain crops in U.S.	Beef—excellent. Dairy—excellent. Sheep—good. Swine—high in fiber. Poultry—high in fiber. Horses—with other grains. Should be ground or otherwise processed for all species.	Adapted to cool areas. Northern Great Plains to California. High N in soil may produce excess straw causing lodging (should not follow legumes in rotation). Can be used as a winter or summer annual.	Much of the domestic barley is used for brewing—about 25% of total production. Malting barley is usually higher in energy and lower in protein. Used as nurse crop for grasses and legumes. Sometimes seeded with oats for mixed grain. Should purchase No. 2 quality or better for feed.
High in starch. Low in protein, especially deficient in the amino acid lysine. Low in fiber. Yellow corn is only grain with appreciable carotene—most lost in long storage. Very palatable. Most acreage and tonnage of all grains in U.S. Normally, corn constitutes about 81% of all the grains fed to U.S. livestock. China is the second leading producer of corn.	Sometimes considered as *heavy* feed. Used by all animals. Cattle and sheep are often fed ground ear corn (grain + cobs). About 25% of the total production is fed to hogs. For horses, mix with bulky feed to avoid colic. Limit when fed to breeding stock to avoid excess fat.	Does best in warm areas with ample moisture, but less well in hot southern states. Needs warm nights for best growth. Grown primarily in north central states but expanding over wider area with new hybrids.	New high-lysine corn, Opaque–2, is nutritionally superior to other varieties, but yields less protein and feed per acre. Nutritive sweeteners from corn have taken over 50% of the U.S. sweetener market. When converted to alcohol, each bushel of corn produces 2.5 gal of alcohol and 16.5 lb (*7.5 kg*) of by-product.
Close relative of wheat. Minor grain crop. Less yield than other feed grains.	Use similar to barley and oats.	Hulls not removed in threshing. Looks more like barley than wheat. Limited climatic adaptation. Grown mostly in the Dakotas.	
In comparison with sorghum, millet is equal as an energy source and higher in lysine.	Millet may be used as a feed for livestock and poultry, or as birdseed. Should always be ground or rolled except for poultry.	Northern Great Plains in areas where the growing season is too short for sorghums. Low water requirement. Does best on well-drained, fertile soil.	Millet is grown throughout the world, especially in semi-arid regions.
Bulky feed, 30% hulls. Palatable, considered as an excellent conditioning feed. Higher protein than most grains; and the best balanced in amino acids of the cereal grains. Newer hull-less types higher in energy, comparable to wheat and corn. Fifth in acreage of grains in U.S.	Excellent for all livestock and especially for horses and breeding stock of most species. Often used as nurse crop for grasses and legumes. Sometimes seeded with other cereals for mixed grains. Reduces cannibalism in poultry.	Most in north central states but grown in most states. Yields less feed per acre than most cereal grains—otherwise, would be more widely used. Short season crop. Grows well in cool weather.	Oat straw used for bedding and as supplement for cattle fed high-concentrate ration. Too much N in soil increases straw height and chance for lodging. Do not follow legume in rotation.
One of most important grain crops in the world—mostly used for human food.	Should be ground or cracked before feeding. Used to only a limited extent in animal feeds.	Requires warm, long growing season, with an abundance of water.	
One of least important of U.S. grain crops. Used primarily for pasture, hay, silage, and erosion control.	High protein. Lower palatability than most grains. Especially liked by sheep. Needs to be ground for all animals but sheep. Laxative if fed in high amounts.	Mostly in north central states, but some in most states. Grows on soils and under conditions not tolerated by most other grains. Most of rye in U.S. is winter variety. Extremely cold resistant.	Ergot contamination may be toxic to animals.

(Continued)

TABLE 10-3

Grains	Lb per bu	Ruminant/ Swine	Energy Content[1]						Protein Content[1]				Value (lb for lb) as % of Corn (with Corn=100)[2]
			TDN		Digestible Energy				Crude Protein		Digestible Protein		
			%		A-F		M-F		%		%		
			A-F	M-F	Mcal per		Mcal per		A-F	M-F	A-F	M-F	
					(lb)	*(kg)*	(lb)	*(kg)*					
Sorghums *(Sorghum vulgare)*	56	Ruminant	**67**	75	**1.3**	***3.0***	1.5	*3.3*	**11.5**	12.8	**7.1**	7.9	Cattle 90–95. Sheep and goats 85–100. Swine 95. Horses 110–115.[2] Poultry 100. Rabbits 125.[2]
		Swine	–	–	**1.6**	***3.5***	1.7	*3.8*	**11.5**	12.8	**8.7**	9.6	
Spelt *(Triticum spelta)*	40, but variable	Ruminant	**68**	76	**1.4**	***3.0***	1.5	*3.4*	**12.0**	13.3	**8.0**	8.9	Cattle 70–90. Swine 65–80.
		Swine	**67**	74	**1.3**	***2.9***	1.5	*3.3*	**12.0**	13.3	**9.1**	10.1	
Triticale (a hybrid of durum wheat and rye)	48–50	Ruminant	**75**	84	**1.4**	***3.2***	1.6	*3.6*	**15.4**	17.3	**11.1**	12.5	Swine 80–90. Poultry 80–90.
		Swine	–	–	**1.5**	***3.2***	1.6	*3.6*	**15.4**	17.3	**12.2**	13.7	
Wheat *(Triticum vulgare)*	60	Ruminant	**77**	87	**1.5**	***3.4***	1.7	*3.8*	**13.1**	14.7	**10.5**	11.7	Cattle 100–105. Sheep and goats 100–110. Swine 100–105. Horses 115.[2] Poultry 90–95. Rabbits 120.[2]
		Swine	**79**	89	**1.5**	***3.4***	1.7	*3.8*	**13.1**	14.7	**11.1**	12.4	

[1]Values obtained from Table V-1A of this book.

[2]The feed substitution values given in this column for each species correspond to the feed substitution values given in the chapters of this book devoted to each of the respective species. For horses and rabbits, oats is used as the base feed and given a value of 100.

Structure of Grain

Grain develops from the ovary and its ovule after fertilization by pollen. The structures of the flowers of the various cereal plants differ, and these differences are reflected in the structure of the individual kernel.

In corn, the male and female structures are found in separate inflorescences (flower structures) on the plant. The male inflorescence, commonly referred to as the tassel, is located at the top of the cornstalk. Pollen is shed from the tassels and subsequently comes into contact with the female inflorescence, thereby producing grain. The female inflorescence contains a central rachis, commonly referred to as the cob. The cob contains series of rows of sessile spikelets and is enclosed by overlapping bracts (husks). The silks which are found on the corncob structure are the stigmas—the pollen-receiving organs. As seen in Fig. 10-3 each spikelet contains two flowers—one fertile and one sterile.

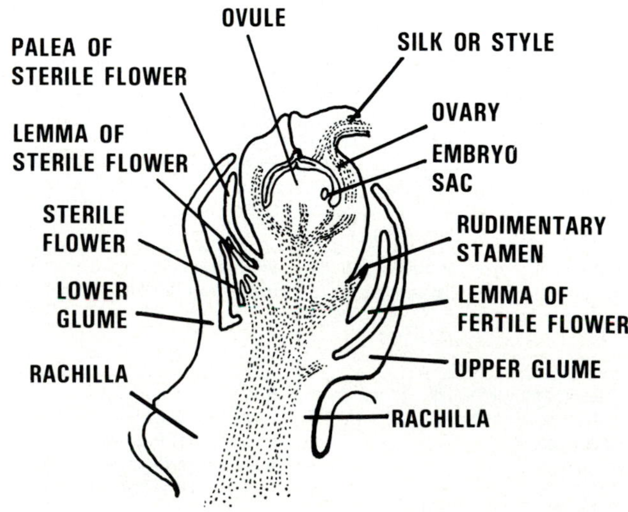

Fig. 10-3. Structure of a corn spikelet.

(Continued)

Special Characteristics and Importance	Livestock Use	Cultural Characteristics/ Geographic Adaptation	Other Comments
Increasing in importance. Sometimes fed as whole head but not to swine or poultry because of fiber. Considerable variation in protein and starch exists between varieties of sorghums.	About 98% of the sorghum grain used in the U.S. goes into livestock and poultry rations. Sorghum constitutes 6 to 8% of all the concentrates fed to livestock and poultry. Should be processed except for sheep. For cattle, heat and/or high-moisture treatment, followed by rolling, improves sorghum more than corn.	The species *Sorghum vulgare* encompasses the grain sorghums, broomcorn, Sudangrass, and tall sorghums. Grown primarily in central and south plains states. Grows best on fertile, sandy loam soils.	Use of hybrids adds potential for future use of sorghums and their adaptation to other regions of the country. In the past, there has been some prejudice against sorghum due to the poor performance by animals attributed to the high tannin content of the grain. However, most of the varieties of sorghums presently grown in the U.S. are low in tannin.
Grain is of no value for milling. Use principally as a feed for livestock.	The hull remains attached to the kernel. So, the feeding value of spelt is similar to oats.	Very little grown in the U.S., mostly in the East.	In 1987, Ohio State University released the first spelt variety ever developed in the U.S.
Studies are in progress to improve yield and nutrient value. Triticale is higher in protein and in the essential amino acids than corn.	Grain may be used more for human food with higher cost making it difficult to use for livestock until there is a greater supply than needed for human food.	Hardiness of rye. Has some characteristics of both wheat and rye.	Triticale is a hybrid cereal created from a cross between durum wheat *(Triticum)* and rye *(Secale)*, followed by doubling the chromosomes in the hybrid. Triticale has created much interest because of having higher quality protein and of being more winter hardy than wheat. Triticale is subject to ergot.
Most important grain crop in the world. Mostly used for human food. Second only to corn in the U.S. in total acreage and production.	Not as suitable as some other feed grains because of digestive upsets which may develop. Usually best not to exceed 50% of the concentrate mix. Should be cracked or coarsely ground for most animals. Keep coarse to avoid lack of palatability.	Grown in all states. Types and varieties vary by location. Greatest acreage in the Central Plains and north central states.	Wheat usually not used very extensively in animal rations because of high cost. Most wheat for animal feed is of the soft types which are not as valuable for milling.

Fig. 10-4. Dent corn. At maturation, the crown region contracts, giving the kernel a characteristic indentation. (Courtesy, USDA)

The grain from barley, oats, wheat, rice, and sorghum develops from flowers which contain the ovary, three stamens, and two scalelike lodicules. These structures are surrounded by a pair of bracts called the lemma (found on the dorsal side of the flower) and the palea (found on the ventral side of the flower).

In rye and wheat, the lemma and palea are loosely attached to the grain. During threshing, these particles are separated from the grain and constitute what is known as chaff.

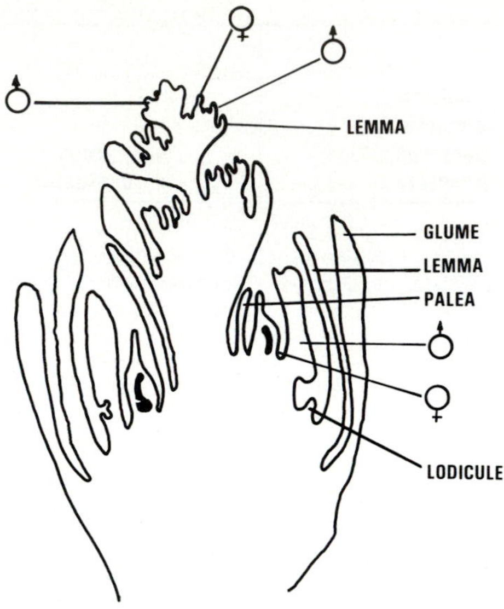

Fig. 10-5. Structure of the wheat flower.

Barley, oats, and rice retain their lemma and palea during threshing, thus giving rise to structures called husks or hulls. In barley, the lemma and palea fuse with the grain. In oats, the lemma and palea do not fuse with the kernel but enclose and adhere tightly to the entire grain. This hull structure can be removed during processing, resulting in dehulled oat grains called groats. Rice hulls are removed during processing.

Fig. 10-6. Barley heads. Barley retains its lemma and palea during threshing; hence, it is called a hulled grain. (Courtesy, USDA)

Individual kernels of grain are called caryopses. Grains which contain husks (oats, barley, and rice) are called covered caryopses, whereas grains lacking husks (corn, wheat, rye, and sorghum) are referred to as naked caryopses. Each kernel (caryopsis), exclusive of the husk, is composed of two main parts—pericarp (fruit coat) and seed.

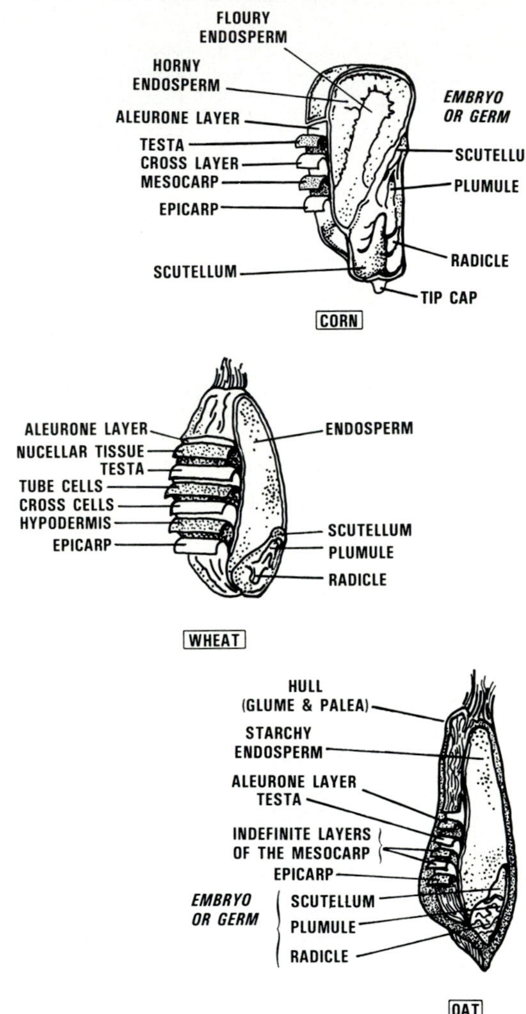

Fig. 10-7. Structure of corn, wheat, and oat caryopses.

PERICARP

The pericarp consists of two layers. The outer layer contains the epidermis and the hydroderm, collectively referred to as the beeswing. The cells of the epidermis are thin-walled and rectangular. The hydroderm is highly variable in thickness. These components of the outer layer are oriented lengthwise in the grain. The inner layer of the pericarp contains cross cells and tube cells. The cross cells are oriented transversely along the grain. Throughout the ripening process of grain, the innermost layer of the pericarp becomes distorted and torn, thus giving a tubelike appearance—hence, tube cells.

SEED

The seed portion of grain can be divided into three parts: (1) seed coat (testa) and hyaline layer (nucellar layer), (2) endosperm, (3) germ (embryo). When grain is processed in such a way that the germ and starch endosperm are removed, the composite of the remaining parts of the seed and the pericarp is called bran.

Seed Coat (Testa) and Hyaline Layer (Nucellar Layer)

The seed coat (testa) is generally either one or two layers thick. There is very little cellular structure in this region. Likewise, the hyaline layer lacks any cellular structure but, rather, acts as an embryo sac.

Endosperm

The endosperm of grain can be divided into two parts—the aleurone and the starchy endosperm.

ALEURONE

The aleurone surrounds the starchy endosperm of grain but does not encompass the scutellum of the embryo. The cells of the aleurone are thick-walled, cuboidal, and rich in oil, niacin, and mineral matter. Phytic acid is also produced in rather large quantities in this region.

The number of layers of cells in this region varies according to the type of grain. Wheat, rye, oats, and sorghum, generally have only one layer of cells in the aleurone. Depending on the particular variety, corn can contain anywhere from 1 to 6 layers; barley, 2 to 4; and rice, 2 to 6.

STARCHY ENDOSPERM

The starchy endosperm portion of grains contains thin-walled cells that are highly variable in shape, size, and contents. Pentosans are found in large amounts in the cell walls of this region, but starch and protein make up most of the cell contents. Starch is found primarily in the form of granules, with protein filling the intergranular spaces.

In wheat, the cells adjacent to the aleurone are relatively higher in protein and lower in starch than the rest of the starchy endosperm.

The concentration of starch in corn depends largely on the type of corn and the area of the kernel being analyzed. The endosperm of corn is divided into two regions—the crown (area opposite that of the embryo) and the horny region (area next to the embryo). The crown region contains loosely packed starch granules with little protein. In the horny region of the yellow varieties of corn are layers of proteinaceous material with starch granules interspersed. Hence, the protein content of the horny region is much higher than that of the crown region (about twice as high). Additionally, the oil content of the horny endosperm is high. In some varieties of corn, the crown region contracts during maturation, thereby creating an indentation in the kernel forming dent corn (see Fig. 10–4).

Embryo (Germ)

The embryo (germ) of the seed consists essentially of an immature, undeveloped plant, surrounded by a quantity of stored food for its early nourishment and a protective seed coat.

Between the embryo proper and the endosperm is an organ called the scutellum. Upon germination, food reserves in the endosperm are mobilized and passed on to the embryo by the scutellum. The plumule of the embryo gives rise to the growing bud, and the radicle to the root system.

Grain Milling and Milling By-products

All grains can be fed to livestock with varying efficiencies in their intact state. However, routine processing, such as grinding, can often improve the feeding value of the grains. A detailed discussion on the processing of grains for livestock is given in Chapter 14, Feed Processing. Additionally, most of the grains are milled in some manner for the preparation of foods for human consumption. In these milling processes, a number of by-products are produced which are generally considered to be of little value to humans but which can be and are used extensively as livestock feeds.

GRAIN MILLING

The milling procedures of the grains discussed in the sections that follow illustrate the ways that by-product feeds for livestock are obtained. However, it is recognized that no two mills (no two corn mills, for example) are ever exactly alike—in the sequence, identity, or placement of machinery. Nevertheless, the accompanying figures show in a general way the basic steps involved in grain milling to obtain food for people and feed for farm animals.

Corn Milling

Fig. 10–8. Corn ears. (Courtesy, Watt Publishing Co., Mount Morris, Ill.)

Between 5 and 6% of the corn grown in the United States is processed by wet milling for the production of starch, sugar, and corn oil. Slightly less corn, about 3.5% is dry milled in the production of flour, oil, and breakfast cereals. By-products from both of these processes are fed to livestock.

WET MILLING

As evident in Fig. 10-9, the wet milling of corn makes valuable use of each constituent of the corn kernel. In addition to starch and the various products derived from it, along with highly valued oil from the germ, 25 to 30% of the corn used by the wet-milling industry goes into the production of livestock feeds. Details relative to the wet-milling process follow.

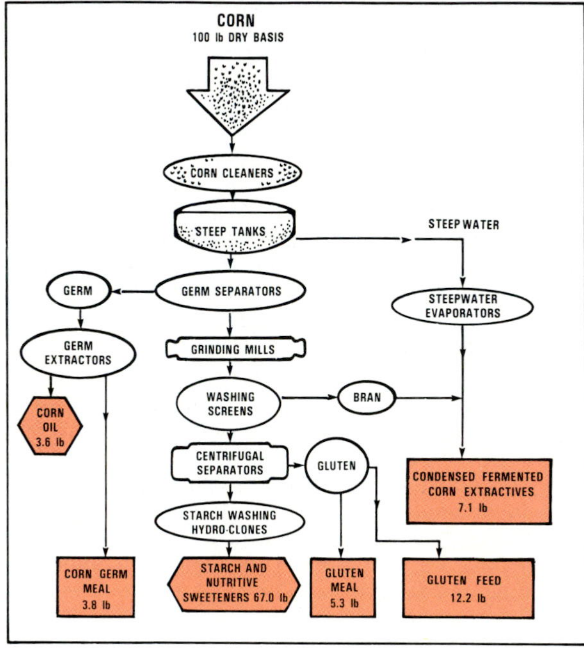

Fig. 10-9. Schematic outline of wet milling of corn.

Before storage, shelled corn is initially cleaned to remove extraneous material such as pieces of cob, foreign seeds, stray metal, fine dirt, or light, unwanted material. When the corn is taken from storage, it is cleaned again and steeped in a series of tanks where, at regular intervals, a solution of sulfurous acid water is recirculated. This procedure, which takes anywhere from 28 to 48 hours, disintegrates the protein that binds the starch within the kernel, thereby softening the kernel and facilitating grinding. The steep water which is removed from this process contains about 6% solids, of which about 35 to 45% consist of protein. The steep water then undergoes evaporation, whereupon the moisture level is reduced to 45 to 65%. This high-protein extract may be used as a nutrient for microorganisms in the production of enzymes, antibiotics, and other fermentation products. The major portion, however, is combined with fiber and gluten in the production of animal feed ingredients.

After steeping, the extracted corn is degermed by coarse grinding, yielding a pulpy material which contains the germ, hull, starch, and gluten. This material is then passed through a liquid cyclone which separates the germ from the rest of the material. Oil is then extracted from the germ through hydraulic, expeller, or solvent processes. The extracted germ is dried and sold as corn germ meal for animal feed.

Following the separation of the germ, the remaining materials containing starch, hulls, and gluten pass through a series of screens which separate the various fibrous fractions from starch. The fibrous fractions are dried and sold as mill by-products to the animal feed industry.

The water slurry of starch and gluten is separated by centrifugation. Typical operations yield a gluten stream containing over 60% protein, while the starch stream is over 99% starch. The gluten is dried and sold as gluten meal (60% protein) or it may be used as an ingredient in corn gluten feed (21% protein).

The white, nearly-pure starch slurry is further washed to remove small quantities of solubles, dewatered using filters or centrifuges, and dried to produce common starch (unmodified).

Various modified or derivatized starches may be produced by treating the slurry of washed starch with chemicals or enzymes. After treatment, the products are recovered by filtration or centrifugation and the starch is dried.

Thus, the wet milling of corn produces four major feed ingredients: corn gluten feed, corn gluten meal, corn germ meal, and condensed fermented corn extractives or corn steep liquor. Additionally, three further products used in feed manufacturing are obtained by wet millers: hydrol, a starch molasses; wet bran; and dried steep liquor concentrate.

DRY MILLING

The dry milling of corn is generally less involved than that of wet milling. Two basic processes—degerming and nondegerming—are used extensively.

In the degerming process, the hull, germ, and endosperm are separated before milling. This process is used for the production of grits, flakes, meal, flour, oil, and feeds.

The entire kernel is ground intact in the nondegerming process of dry milling. The resulting product is an oily flour which is subsequently used in baked products. Although some hulls and germ sift out in this process, the quantity of by-products is too insignificant to be considered as a useful source of animal feed.

Sorghum Milling

Fig. 10-10. Sorghum grain, headed and ready for harvest. (Courtesy, National Grain Sorghum Producers Assn., Abernathy, Tex.)

Sorghum is processed in much the same way as corn. It is used in the wet milling and dry milling industries as well as in the fermentation industries. The primary products of the wet milling of sorghum are starch, edible oil, and gluten feed. The primary products of dry milling are low-protein flour and feed by-product.

Wheat Milling

Fig. 10–11. A field of wheat ready for harvest. (Courtesy, USDA)

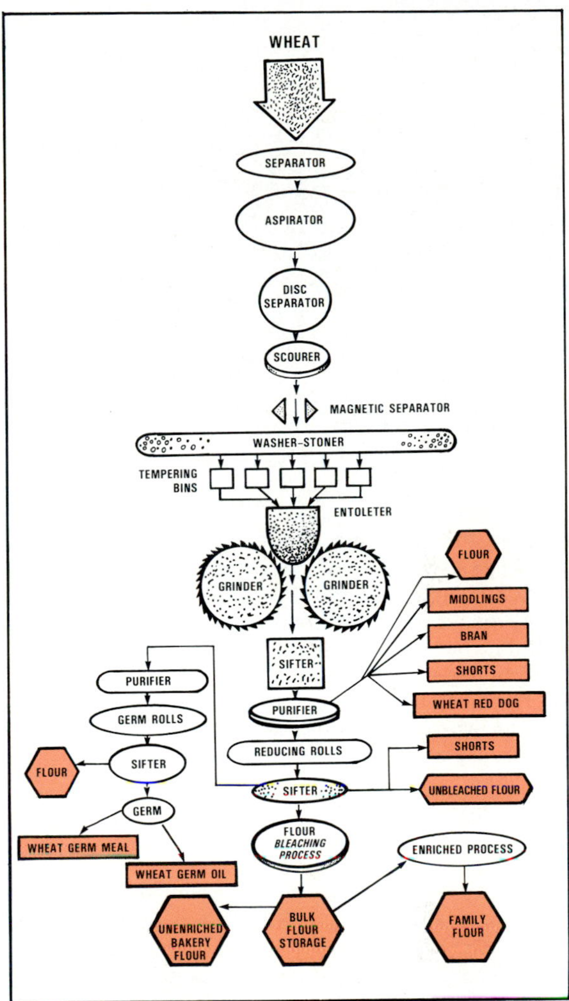

Fig. 10–12. Schematic outline of milling of wheat.

In the United States, wheat is cultivated primarily as a food grain for human consumption. As a result, most of the wheat fed to livestock is in the form of mill by-products. Worldwide, however, an average of about 19% of the global wheat use is for animal feed.

The initial steps in the milling of wheat involve the cleaning and sifting of the grain, as illustrated in Fig. 10–12. The separator is a series of screens which removes stones, sticks, and other foreign material. From the separator, the wheat passes through an aspirator where jets of air remove many of the light impurities, such as dust and chaff. The next step involves a disc separator. The surfaces of the discs are indented so that wheat is caught, but larger and smaller particles, such as foreign grains, are rejected.

After passing through the disc separator, the wheat is scoured to remove the beard from the individual grains and passes through a magnetic separator to the washer-stoner, a machine which utilizes high speed rotors and water to remove stones.

Prior to grinding, the cleaned wheat is tempered, to facilitate the removal of the bran from the endosperm. The tempered wheat then passes into an Entoleter machine which breaks and removes unsound wheat. The sound wheat then passes through a series of grinders, sifters, and purifiers to separate the various parts of the grain. The latter process is repeated over and over again—sifters, purifiers, reducing rolls—until the maximum amount of flour is separated. Generally, millers remove about 72% of the wheat kernel for wheat flour, and the other 28% goes into the production of livestock feeds, primarily wheat middlings, wheat bran, wheat shorts, wheat red dog, wheat screenings, wheat germ meal, and wheat germ oil.

Rye Milling

The same basic type of machinery is used in milling both rye and wheat; and the primary objectives are the same—to produce a granular or powdery product by pulverizing the seed. The final milled product is free of bran and germ.

Fig. 10-13. Rye crop ready for harvest. (Courtesy, USDA)

However, the following differences between the milling procedures of wheat and rye should be noted:

1. Rye milling does not utilize purifiers.
2. In milling rye, the break rolls are used to make as much flour and as little middlings as possible whereas in wheat milling the converse is true.
3. The cleaning process with rye is much more difficult than with wheat. Rye must be initially separated by grain size because of its high variability in size.

Rice Milling

Fig. 10-14. Rice growing in flooded lowland. Rice requires a lot of water. (Courtesy, USDA)

Rice, like wheat, has traditionally been one of the staple grains for human food. However, several rice milling by-products unsuitable for humans can be effectively fed to livestock.

The purpose of milling rice is to separate the outer portions from the inner endosperm with a minimum of breakage. In modern mills, the rough rice passes through several processes in the mill: cleaning, (sometimes parboiling), hulling, pearling, polishing, and grading. (See Fig. 10-15.)

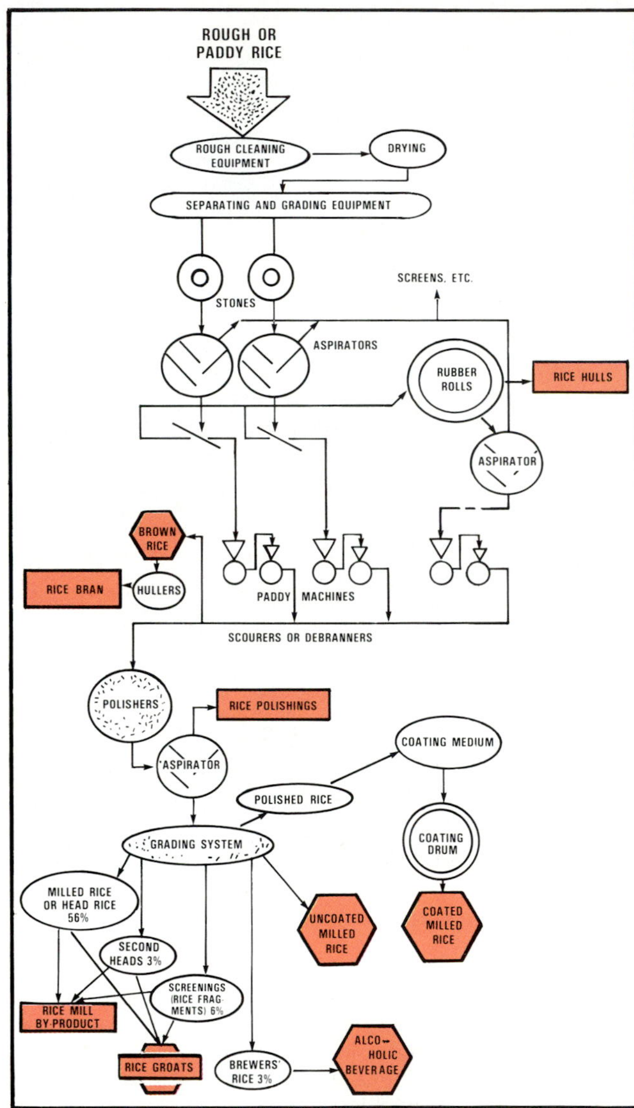

Fig. 10-15. Schematic outline of milling of rice.

Dried rice that still has its hull is referred to as *rough* or *paddy* rice. Following the removal of foreign material by means of a fan and screen separator and a magnetic separator, the rough rice passes into an awning machine where the awns are removed from the grain. The grain then passes through a series of shellers, and the hulled grain is separated from the unhulled in the paddy machine—a complex sifter.

At this stage, the hulled rice is referred to as *brown* rice. This brown rice undergoes scouring in a series of hullers to remove the outer bran and inner bran (polishings).

Throughout the entire milling process, a number of kernels are unavoidably broken. These broken kernels are sorted by size and used as by-product feeds. The largest particles, generally whole or ¾ kernels, are referred to as head rice. Broken rice consisting of ¾ to ⅓ size grains are termed second heads. Screenings—⅓ to ¼ size—are the next type of broken rice. The smallest fragments are used for brewing; hence, the term brewers' rice.

The following by-products of rice milling are used as feed ingredients: rice hulls, rice bran, and rice polishings. Additionally, the broken rice (milled rice or head rice, second heads, and screening) is generally used as livestock feed—as rice groats, or as rice mill by-product. Rice groats, which is rice with the hulls removed, may be used as either human food or livestock feed. Rice mill by-product is the total mixture obtained from the milling of rice (hulls, bran, polishings, and broken rice grains). Brewers' rice is occasionally used as animal feed, although it is used chiefly in the brewing industry.

Oat Milling

Fig. 10-16. Oats. The loose-fitting flumes surrounding the grain are easily removed during the threshing, leaving the oat grain with hulls. Once these hulls are removed and the oat groat is exposed, the fiber content is decreased dramatically. (Courtesy, USDA)

Cleaned oats are initially roasted to 212°F for 1 hour. This process reduces the moisture of the grain to about 6% and weakens the hulls to facilitate removal. After cooling, the oats are separated by size, and the hulls, which account for about 30% of the weight of the grain, are subsequently removed by passing the kernels through 2 large circular milling stones. The milled oats are then screened to separate the groats (hulless grain), hulls, and broken grains. The hulless grain can then be rolled or milled into a number of attractive products, including oatmeal.

Barley Milling and Malting

Fig. 10-17. Barley head. (Courtesy, USDA)

The most important uses of barley are for human food, beverages, and livestock feed. Normally, about 51% of the U.S. barley supply is used for food, alcohol, and seed, and 49% is used for feed. Additional barley by-product feeds are obtained from milling and from malting/brewing/distilling. (See Fig. 10-18, page 376.)

Barley can be milled through the usual milling procedures used for the other cereal grains. However, barley is also used extensively in the malting and brewing/distilling industries and can be processed in a manner unique to traditional grain processing. In the malting process, clean, graded barley is wetted and germinated under carefully controlled conditions to minimize the loss of weight due to respiration.

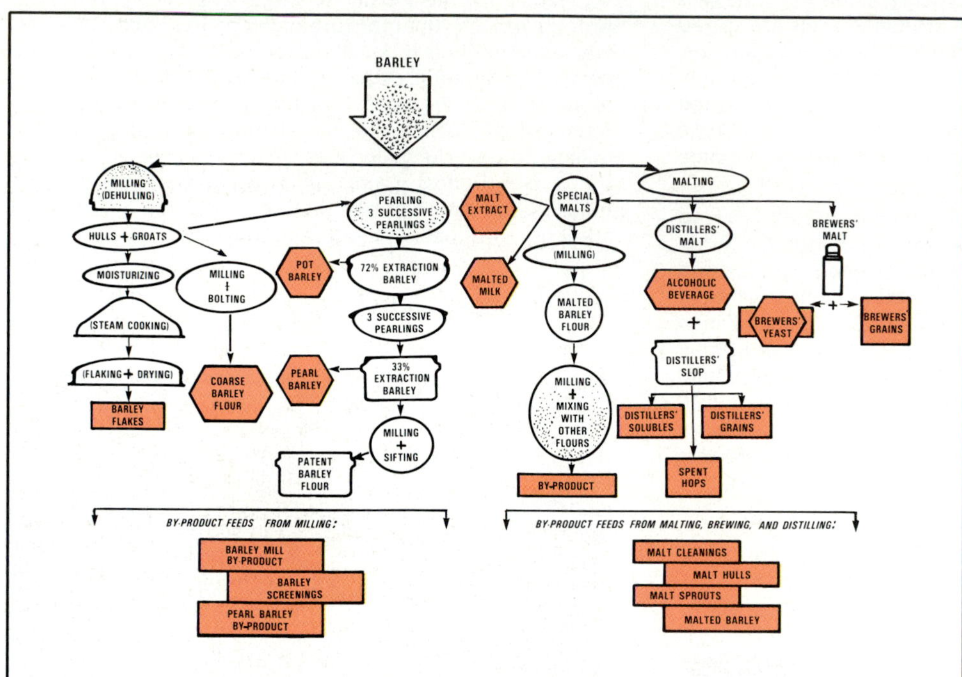

Fig. 10–18. Schematic outline of malting of barley.

Once germinated, the grain is dried to stop growth and to produce a storable product. Following drying, the malt sprouts are removed and subsequently used as a by-product feed. The resulting malt is then used in the brewing and distilling industries and in the specialty food industry.

Malting, brewing, and distilling result in the following valuable by-product feeds for livestock: malt cleanings, malt hulls, malt sprouts, malted barley, brewers' grains, distillers' grains, distillers' solubles, spent hops, and brewers' yeast.

MILLING BY-PRODUCTS

The feed industry got its beginning from the flour milling industry, as an outlet for milling by-products.

Quite often, the terms used to describe the various cereal by-products are confusing. Table 10–4, Guide to Cereal By-product Terms, lists these terms with a brief description as to what part of the grain is included, from what types of grain they are derived, and their relative feeding value.

TABLE 10–4
GUIDE TO CEREAL BY-PRODUCT TERMS (See footnote at end of table.)

			Nutritive Value												
			Digestible Energy								Crude Protein		Crude Fiber		
			Ruminant				Swine								
			A-F		M-F		A-F		M-F		%		%		
Milling By-product[1]	Description	Cereals	Mcal per		Mcal per		Mcal per		Mcal per		A-F	M-F	A-F	M-F	Comments
			(lb)	*(kg)*	(lb)	*(kg)*	(lb)	*(kg)*	(lb)	*(kg)*					
Bran	Outer coarse coat (pericarp) of grain separated during processing.	Corn	**1.4**	***3.0***	1.5	*3.4*	**1.5**	***3.2***	1.6	*3.6*	**8.4**	9.4	**9.6**	10.8	There are many brans. Corn, rice, and wheat are most common. Laxative in action.
		Rice	**1.1**	***2.4***	1.2	*2.7*	**1.5**	***3.3***	1.6	*3.6*	**13.0**	14.3	**11.9**	13.1	
		Wheat	**1.3**	***2.8***	1.4	*3.1*	**1.1**	***2.5***	1.6	*3.5*	**15.5**	17.5	**10.0**	11.2	
Flour	Soft, finely ground and bolted meal from the milling of cereal grains and other seeds. Consists primarily of gluten and starch of the endosperm.	Rye	**1.5**	***3.3***	1.7	*3.7*	**1.5**	***3.4***	1.7	*3.8*	**10.4**	11.7	**0.5**	0.6	There are many flours. Only the major ones are listed herein. Wheat flour must have less than 1.5% crude fiber.
		Wheat	**1.6**	***3.5***	1.8	*3.9*	**1.6**	***3.6***	1.9	*4.1*	**13.7**	15.5	**0.9**	1.0	
Germ meal	Ground germ (embryo) of the seed.	Corn germ meal	**1.4**	***3.1***	1.5	*3.4*	**1.6**	***3.5***	1.7	*3.8*	**20.7**	22.6	**12.2**	13.3	Wheat germ meal must contain at least 25% crude protein and 7% crude fat.
		Wheat germ meal	**1.7**	***3.7***	1.9	*4.1*	**1.7**	***3.8***	2.0	*4.3*	**24.4**	27.6	**3.1**	3.5	
Gluten (feed and meal)	Substance remaining after extraction of starch and germ in the manufacture of starch and syrup.	Corn gluten feed	**1.4**	***3.2***	1.6	*3.5*	**1.4**	***3.0***	1.5	*3.4*	**23.0**	25.6	**8.7**	9.7	Gluten feed and gluten meal are fed to all livestock and poultry. Corn gluten provides a high level of bypass protein for cattle.
		Corn gluten meal	—	—	—	—	—	—	—	—	**60.8**	67.5	**1.8**	2.0	
		Sorghum gluten meal	**1.6**	***3.5***	1.8	*3.9*	**1.6**	***3.5***	1.8	*3.9*	**44.4**	49.2	**4.9**	5.4	

(Continued)

TABLE 10-4 *(Continued)*

Milling By-product[1]	Description	Cereals	Nutritive Value												Comments
			Digestible Energy								Crude Protein		Crude Fiber		
			Ruminant				Swine								
			A-F		M-F		A-F		M-F		%		%		
			Mcal per		Mcal per		Mcal per		Mcal per		A-F	M-F	A-F	M-F	
			(lb)	*(kg)*	(lb)	*(kg)*	(lb)	*(kg)*	(lb)	*(kg)*					
Grain screenings	Small imperfect grains, weed seeds, and other foreign material of value as a feed that is separated through the cleaning of grain with a screen.	All grains	**1.2**	***2.7***	1.4	*3.0*	**1.5**	***3.3***	1.7	*3.7*	**12.1**	13.4	**12.0**	13.4	Quality varies according to percentage of weed seeds and other foreign material. Should be finely ground in order to kill noxious weed seeds.
Groats	Grain from which the hulls have been removed.	Oat groats	**1.7**	***3.8***	1.9	*4.2*	**1.4**	***3.1***	1.6	*3.5*	**15.8**	17.6	**2.5**	2.8	Improved feeding value over whole grain with hulls.
		Rice groats	**1.6**	***3.5***	1.8	*4.0*	**1.7**	***3.7***	1.9	*4.2*	**7.0**	7.9	**0.4**	0.4	
Hominy feed	A mixture of corn bran, corn germ, and some of the starchy portion, produced in the manufacture of pearl hominy, hominy grits, and table meal.	Corn	**1.7**	***3.7***	1.9	*4.1*	**1.6**	***3.6***	1.8	*4.0*	**10.3**	11.4	**4.8**	5.3	Must contain not less than 4% crude fat.
Hulls	Outer covering of grain.	Oat hulls	**0.6**	***1.4***	0.7	*1.5*	—	—	—	—	**3.7**	4.0	**30.9**	33.4	Obtained in milling breakfast cereal or in groating oats or rice.
		Rice hulls	**0.3**	***0.6***	0.3	*0.7*	—	—	—	—	**3.0**	3.2	**38.9**	42.2	
Malt sprouts	The rootlets and sprouts, along with some of the malt hulls, obtained from malted barley.	Barley	**1.3**	***2.9***	1.4	*3.1*	**1.2**	***2.7***	1.3	*2.9*	**22.9**	24.6	**14.2**	15.3	Barley malt sprouts must contain not less than 24% crude protein.
Meal, oat	Rolled oat groats.	Oat	**1.7**	***3.8***	1.9	*4.2*	**1.6**	***3.6***	1.8	*4.0*	**14.8**	16.3	**3.6**	4.0	There are 2 types: cereal oat meal, and feeding oat meal. Oat meal must contain less than 4% fiber.
Middlings	A by-product of the flour milling industry, consisting of bran, shorts, germ, flour, and some of the offal from the "tail of the mill."	Rye middlings	**1.5**	***3.3***	1.7	*3.7*	**1.3**	***3.0***	1.5	*3.3*	**17.1**	19.1	**5.2**	5.8	Deficient in calcium, carotene, and vitamin D. Wheat middlings cannot contain more than 9.5% crude fiber. Rye middlings cannot contain more than 8.5% crude fiber.
		Wheat middlings	**1.4**	***3.1***	1.6	*3.5*	**1.3**	***2.9***	1.5	*3.3*	**16.4**	18.5	**7.7**	8.7	
Mill run (mill by-product or pollards)	State in which a material comes from the mill, usually ungraded and having no definite specifications. It consists of bran, shorts, germ, flour, and the offal from the "tail of the mill."	Rye mill run	**1.3**	***2.9***	1.5	*3.2*	**1.3**	***3.0***	1.5	*3.4*	**15.6**	17.6	**4.9**	5.5	Grain sorghum mill feed must contain more than 5% crude fat and less than 6% crude fiber. Oat mill by-product cannot contain more than 25% crude fiber. Rice mill by-product cannot contain more than 32% crude fiber. Rye mill run cannot contain more than 9.5% crude fiber. Wheat mill run cannot contain more than 9.5% crude fiber.
		Wheat mill run	**1.5**	***3.2***	1.6	*3.6*	**1.4**	***3.2***	1.6	*3.5*	**15.1**	16.7	**8.2**	9.1	
Polishings	By-product of rice, consisting of a fine residue that accumulates as rice kernels are polished.	Rice	**1.6**	***3.5***	1.7	*3.8*	**1.7**	***3.7***	1.9	*4.1*	**12.0**	13.3	**3.2**	3.5	Good source of thiamin.
Red dog	By-product of milling wheat for flour. Consists of the offal from the "tail of the mill," along with fine particles of wheat bran, wheat germ, and wheat flour.	Wheat	**1.5**	***3.2***	1.7	*3.7*	**1.4**	***3.1***	1.6	*3.6*	**15.6**	17.6	**2.9**	3.3	Cannot contain more than 4% crude fiber.
Shorts	A by-product of flour milling consisting of a mixture of small particles of bran, germ, flour, and the offal from the "tail of the mill."	Wheat	**1.5**	***3.2***	1.7	*3.7*	**1.4**	***3.1***	1.6	*3.5*	**16.5**	18.7	**6.4**	7.2	Cannot contain more than 7% crude fiber.

[1]Cereal by-products recognized by the Association of American Feed Control Officials.

Nutrient Composition of Grain

Grains are fed to livestock primarily for their high-energy content. Of the grains most widely used in livestock rations, corn contains the most energy (see Table 10-5). Although corn has the highest energy value, it is the grain with the lowest crude protein content (9 to 11%). The rest of the feed grains contain from about 12.8 to 14.7% crude protein with wheat having the highest amount of crude protein.

TABLE 10-5
COMPOSITION OF TYPICAL FEED GRAINS[1]

Grain	Digestible Energy (Mcal/lb moisture-free grain)		Crude Protein
	Ruminants	Swine	(%, dry basis)
Corn, No. 2	1.80	1.81	10.2
Barley	1.75	1.58	13.2
Wheat, all analyses	1.73	1.73	14.7
Rye	1.62	1.68	13.8
Oats	1.53	1.43	13.3
Sorghum	1.49	1.74	12.8

[1]*From:* This book, Table V-1A, Energy Feeds.

Whole grains contain large amounts of digestible energy, as illustrated in Table 10-6. Many of the by-products from these grains contain as much or more energy than the whole grain from which they are derived. Of particular note is the fact that whole oats contain about 90% of the digestible energy of groats—oats with the hulls removed.

TABLE 10-6
COMPARISON OF ENERGY VALUES OF GRAINS AND THEIR MILL PRODUCTS FED TO SWINE (AS-FED BASIS)[1]

Grain	Digestible Energy (kcal/lb)
Barley:	
Whole	1,396
Malt sprouts	1,219
Corn:	
Grain, No. 2	1,586
Hominy	1,620
Oats:	
Grain	1,278
Groats	1,410
Meal	1,641
Wheat:	
Grain, all analyses	1,544
Bran	1,119
Middlings	1,321
Shorts	1,413

[1]*From:* This book, Table V-1A, Energy Feeds.

CARBOHYDRATES

About 83% of the dry matter of wheat, corn, rye, barley, sorghum, millet, and rice consists of carbohydrates. In oats, this figure is roughly 79%. The carbohydrate fraction of grain is composed primarily of pentosans, starch, dextrins, sugars, cellulose, and hemicellulose.

Starch is the most abundant type of carbohydrate in grain. About 60% of the entire wheat grain (as-fed basis) and 70% of its endosperm are starch. Two forms of starch are found in grain—amylose and amylopectin. Amylose is a straight chained polymer of glucose, whereas amylopectin is a highly branched chain polymer of glucose. In wheat, about ¼ of the starch is in the form of amylose while the remainder is amylopectin. Starch units are stored in granules found throughout the endosperm. The size and shape of these granules tend to vary according to the type of cereal. Simple structured granules are found in corn, wheat, rye, barley, and sorghum. In rice, these granules are compound. For the most part, oats contain compound granules, although the simple type can be found, also.

Fiber is extremely low in grains—ranging from about 0.5% in polished rice to about 12% in whole oats. Once the hulls are removed from oats, the crude fiber content of the grain is lowered dramatically. Thus, grains serve as valuable, highly digestible feeds for nonruminants as well as ruminants.

LIPIDS

Wheat, rye, barley, and rice contain about 1 to 2% lipid (fatty) material. The lipid content of sorghum is approximately 3%; whereas corn, whole oats, and millet contain from 4 to 6%. When the hulls are removed from oats, the resulting grain (groats) contains in excess of 7% lipid.

Wheat and corn germ have a rather large lipid fraction, containing 10 to 35%, respectively. However, other cereal by-products tend to contain substantially lower amounts. Wheat germ, bran, and endosperm contain lipids in concentrations of 6 to 10%, 3 to 5%, and 0.8 to 1.5%, respectively. Corn bran contains less than 1% lipid.

For the most part, the fatty acids in cereals are unsaturated, abounding in both oleic and linoleic acid. However, relatively large amounts of the saturated fatty acid, palmitic acid, can be found in grains (ranging from 11.5% of total fatty acids in barley to about 25% of total fatty acids in millet). Linoleic acid composes more than 50% of the total fatty acids in wheat, barley, rye, corn, and millet, and more than 40% in oats and sorghum. Only about ⅓ of the fatty acids in rice is linoleic acid. Oleic acid ranges from about 11% of total fatty acids in wheat to about 50% in rice.

PROTEIN

Although whole grains are not used as protein feeds, the fact that they are incorporated in livestock rations in large amounts means that a sizable amount of protein is provided from them.

Protein is found in all parts of the grain, but the embryo, scutellum, and aleurone layer contain considerably higher protein concentrations than the endosperm, pericarp, and testa. Table 10-7 shows the protein distribution in corn and wheat. Most of the protein in the respective grains is located in the endosperm; but it must be remembered that the endosperm is the largest segment of the grain kernel, 79.6% in corn and 82.5% in wheat. In both types of grain, the embryo and the scutellum have high concentrations of protein. Of particular note is the extremely low-protein content of the pericarp in each of the 2 grain types.

TABLE 10–7
PROTEIN DISTRIBUTION IN CORN AND WHEAT[1]

Part of the Kernel	Proportion of the Kernel		Protein Content		Proportion of the Total Protein of Kernel	
	Corn	Wheat	Corn	Wheat	Corn	Wheat
	(%)	(%)	(%)	(%)	(%)	(%)
Pericarp	6.5	8.0	3.0	4.4	2.2	4.0
Aleurone layer	2.2	7.0	19.2	19.7	4.7	15.5
Endosperm (outer)	3.9	12.5	27.7	13.7	11.9	19.4
Endosperm (middle)	58.1	12.5	7.5	8.8	48.2	12.4
Endosperm (inner)	17.6	57.5	5.6	6.2	10.9	40.7
Embryo	1.1	1.0	26.5	33.3	3.2	3.5
Scutellum	10.6	1.5	16.0	26.7	18.9	4.5

[1]Adapted by the authors from Hinton, J. J., "Distribution of Protein in the Maize Kernel in Comparison with That in Wheat," *Cereal Chem.* 30:441, 1953.

In nonruminant rations, the essential amino acid composition of the various feed ingredients warrants careful consideration. Since grains and grain products generally compose the bulk of these rations, their respective essential amino acid profiles can determine what type of protein supplement must be added to the ration. Table 10–8 gives a comparison of the amino acid compositions of corn grain, hominy feed, gluten feed, and gluten meal; oat grain, middlings, and groats; sorghum grain; and wheat grain, bran, middlings, and shorts. Amino acid data are often highly variable; hence, the important considerations of this table are the trends of amino acid profiles. For example, all the grains and their by-products, especially corn, are extremely low in tryptophan and lysine. Methionine—one of the sulfur amino acids—is also low in grains and their by-products. A large amount of data suggest that lysine is the first limiting amino acid in most nonruminant feeds. The grains are notably low in lysine. Additionally, many of the protein supplements of plant origin are low in lysine, thereby magnifying the problem. (See Chapter 11, Protein Supplements.)

TABLE 10–8
PROTEIN AND ESSENTIAL AMINO ACIDS IN BARLEY, CORN, OATS, SORGHUM, AND WHEAT, MOISTURE-FREE BASIS[1]

	Barley Grain	Corn				Oats			Sorghum Grain	Wheat			
		Grain No. 2	Hominy Feed	Gluten Feed	Gluten Meal	Grain	Middlings (Feeding Oat Meal)	Groats		Grain (All Analyses)	Bran	Middlings	Shorts
	◄—— (% of dry matter) ——►												
Protein	13.2	10.2	11.4	25.6	47.3	13.3	16.3	17.6	12.8	14.7	17.5	18.5	18.7
Amino acid:													
Arginine	0.58	0.52	0.52	0.87	1.53	0.80	0.89	0.99	0.43	0.69	0.96	1.10	1.35
Histidine	0.28	0.23	0.22	0.68	1.06	0.19	0.33	0.31	0.26	0.34	0.37	0.45	0.50
Isoleucine	0.52	0.46	0.44	0.98	2.46	0.54	0.61	0.60	0.47	0.55	0.62	0.77	0.64
Leucine	0.85	1.15	0.94	2.44	8.08	0.97	1.16	1.16	1.63	1.01	1.00	1.25	1.22
Lysine	0.45	0.22	0.42	0.71	0.88	0.45	0.58	0.61	0.29	0.44	0.61	0.77	0.90
Methionine	0.18	0.13	0.18	0.41	1.13	0.20	0.23	0.23	0.15	0.20	0.19	0.21	0.32
Phenylalanine	0.65	0.52	0.36	0.90	3.12	0.64	0.76	0.78	0.62	0.69	0.56	0.75	0.75
Threonine	0.41	0.40	0.44	0.87	1.56	0.43	0.53	0.49	0.40	0.45	0.45	0.64	0.67
Tryptophan	0.17	0.10	0.12	0.17	0.23	0.17	0.22	0.57	0.10	0.17	0.28	0.22	0.26
Valine	0.64	0.40	0.55	1.22	2.44	0.69	0.81	0.83	0.56	0.69	0.76	0.90	0.92

[1]*From:* This book, Table V–2, Composition of Feeds.

High-Lysine Corn (Opaque–2)

It has been known for many years that corn, the world's third most important human food after rice and wheat, is nutritionally inadequate. In 1914, researchers at the Connecticut Agricultural Experiment Station induced starvation in laboratory rats by feeding them generous helpings of corn. Further, it was found that rats could be restored to health by supplementing the high-corn diet with two protein fractions—the amino acids lysine and tryptophan.

Although normal corn contains about 10% protein, half of the protein consists of zein, which is especially low in lysine and tryptophan, essential amino acids that the nonruminant cannot manufacture and must get from feed.

This deficiency of corn shows up in people wherever corn is a major source—if not the only source—of protein in the diet. Known by the name, *kwashiorkor,* this nutritional deficiency disease is the leading cause of mortality among infants and children in many parts of the world.

For years, plant scientists assayed the world's corn varieties one by one, looking for a strain with more nutritionally balanced protein. Finally, in 1963, a Purdue University team headed by biochemist Edwin T. Mertz analyzed an odd lot of corn characterized by soft, floury endosperm inside an opaque, chalk-white kernel. The Purdue scientists found that the opaque characteristic of corn, which had been noted for years without exciting much scientific interest, is associated with a recessive gene that replaces some of the kernel's nutritionally deficient zein with needed lysine and tryptophan. The mutant, labeled Opaque–2, was twice as high in lysine and tryptophan as ordinary corn. Later, another high-lysine strain was found, which was named Floury–2.

Although the nutritional value of the high-lysine corn is recognized, two major hurdles between research discovery and application must yet be overcome: (1) The mutant gene is linked to Opaque–2's soft, floury kernel, which is both light in weight and vulnerable to pest attacks, producing lower yields for farmers; and (2) Opaque–2 has not been accepted by the majority of consumers.

• **High protein/high amino acid corn**—In 1989, University of Minnesota researchers announced the discovery of a gene in corn that controls the level of protein produced, which can be used to produce corn with 3% more protein and 20% more methionine and lysine than normal corn, without lowering the yields or producing soft kernels. A patent application on the process is pending.

MINERALS

Grains generally contain more minerals than forages. However, all grains (especially corn) are extremely low in calcium but fair to good sources of phosphorus. Therefore, special attention must be given to mineral supplementation of rations that are predominantly grain.

Naked caryopses and grains with their coverings removed contain most of their minerals in the forms of phosphates and sulfates. The primary cations are potassium, magnesium, and calcium. Substantial portions of the phosphorus are bound in the form of phytic acid—a form of phosphorus that is of little value to certain livestock.

The hull portions of rice, barley, and oats are excellent sources of minerals, yielding ash contents of 22.6, 6.0 and 5.2%, respectively. However, the majority of this ash portion is silica.

VITAMINS

Vitamin A activity is low in all cereal grains except fresh yellow corn. Oats and wheat germ contain extremely high amounts of vitamin E.

Thiamin, riboflavin, pantothenic acid, and pyridoxine are found throughout the grain kernel. However, thiamin is found in relatively high concentrations in the embryo and scutellum. Niacin is found in relatively high concentrations in wheat, corn, and rice. However, in corn, most of the niacin is in the form of niocytin—a form of niacin that is biologically unavailable.

Effect of Stage of Maturity on the Nutritive Value of Grains

As grains mature, there is a steady decrease in moisture content paralleled by an increase in carbohydrate content. Workers at the Minnesota Agricultural Experiment Station demonstrated the effect of stage of maturity on the nutritive value of corn. Tables 10–9 to 10–11 give the results of their study. As can be seen in Table 10–9, dry matter content increased dramatically as the grain matured. Likewise, carbohydrate parameters (nitrogen-free extract and starch) showed increases. Fat content also increased from 3 to 4.9%. As corn matured, crude protein, crude fiber, ash, and cell wall constituents decreased. Because corn is used for its energy content, the drop in crude protein is not that important; especially when, on an as-fed basis, corn in the early milk stage contains only 3.5% crude protein.

TABLE 10–9
PROXIMATE, STARCH, CELL WALL CONSTITUENTS, AND GROSS ENERGY COMPOSITIONS OF CORN GRAIN AT FOUR STAGES OF MATURITY[1]

Nutrient	Concentration in Dry Matter			
	Early Milk	Early Dough	Mid-dent	Mature
	(%)	(%)	(%)	(%)
Dry matter	20.9	35.7	55.5	76.6
Crude protein	16.6	12.5	10.7	10.9
Ether extract	3.0	4.0	4.8	4.9
Crude fiber	5.4	3.3	2.5	2.1
Ash	2.8	2.3	1.7	1.5
NFE	72.2	77.9	80.3	80.6
Starch	47.4	55.0	58.7	63.7
CWC[2]	27.7	24.6	16.3	13.9
	(kcal/kg)	(kcal/kg)	(kcal/kg)	(kcal/kg)
Gross energy	4,560	4,540	4,590	4,580

[1]Adapted by the authors from Thorton, J. H., R. D. Goodrich, and J. C. Meiske, "Corn Maturity. I. Composition of Corn Grain of Various Maturities and Test Weights," *Journal of Animal Science*, Vol. 29, 1969, pp. 977–982.

[2]Cell wall constituents.

On a dry matter basis, there was a decreasing trend in each amino acid concentration as the corn matured (Table 10–10); but when concentration in protein was considered, there was relatively little change among the amino acids due to stage of maturity.

TABLE 10–10
AMINO ACID COMPOSITION OF CORN GRAIN AT FOUR STAGES OF MATURITY[1]

Amino Acid	Concentration in Dry Matter				Concentration in Protein			
	Early Milk	Early Dough	Mid-dent	Mature	Early Milk	Early Dough	Mid-dent	Mature
	(%)	(%)	(%)	(%)	(%)	(%)	(%)	(%)
Alanine	1.18	0.91	0.79	0.79	7.62	7.49	7.40	7.34
Arginine	0.61	0.49	0.46	0.41	3.94	5.02	4.31	3.81
Aspartic acid	1.54	0.97	0.79	0.75	9.95	7.98	7.40	6.96
½ cystine[2]	0.10	0.08	0.07	0.07	0.65	0.66	0.66	0.65
Glutamic acid	2.61	2.09	1.88	1.97	16.86	17.20	17.62	18.29
Glycine	0.57	0.45	0.40	0.39	3.68	3.70	3.75	3.62
Histidine	0.34	0.28	0.27	0.26	2.20	2.30	2.53	2.41
Isoleucine	0.58	0.45	0.40	0.41	3.75	3.70	3.75	3.81
Leucine	1.55	1.35	1.27	1.30	10.01	11.11	11.90	12.07
Lysine	0.56	0.39	0.33	0.31	3.62	3.21	3.09	2.88
Methionine	0.23	0.21	0.17	0.17	1.49	1.73	1.59	1.58
Phenylalanine	0.67	0.55	0.49	0.53	4.33	4.53	4.59	4.92
Proline	1.04	0.99	0.93	0.94	6.72	8.15	8.72	8.73
Serine	0.66	0.53	0.48	0.53	4.26	4.36	4.50	4.92
Threonine	0.55	0.41	0.37	0.37	3.55	3.37	3.47	3.44
Tyrosine	0.39	0.30	0.21	0.24	2.52	2.47	1.97	2.23
Valine	0.79	0.61	0.54	0.52	5.10	5.02	5.06	4.83

[1]Adapted by the authors from Thornton, J. H., R. D. Goodrich, and J. C. Meiske, "Corn Maturity. I. Composition of Corn Grain of Various Nutrients and Test Weights," *Journal of Animal Science*, Vol. 29, 1969, pp. 977–982.

[2]The cysteic acid or *S*-carboxymethylcysteine content of a protein does not indicate whether both cysteine and cystine are present in the protein or only one of the two. Thus, the number of residues of cysteic acid or *S*-carboxymethylcysteine is referred to as the half-cystine content. To determine the proportion of cysteine and cystine in a protein, the thiol content can be estimated by a variety of methods, *e.g.*, reaction with Ellman's reagent, which indicates the cysteine content. If the total half-cystine and thiol (cysteine) contents are known, the amount of cystine in the molecule can be calculated.

The Minnesota investigators also made interesting observations relative to the mineral composition of corn (Table 10–11). When concentration in dry matter was used as a basis of comparison, both phosphorus and potassium showed dramatic declines as the corn matured. On the other hand, when the concentration in ash was used for comparison, potassium showed a dramatic decline, but phosphorus showed a dramatic increase. Magnesium content closely paralleled phosphorus. As the grain matured, trace minerals constituted larger proportions of the ash concentration.

TABLE 10–11
MINERAL COMPOSITION DATA OF CORN GRAIN AT FOUR STAGES OF MATURITY[1]

Mineral	Concentration in Dry Matter				Concentration in Ash			
	Early Milk	Early Dough	Mid-dent	Mature	Early Milk	Early Dough	Mid-dent	Mature
	(%)	(%)	(%)	(%)	(%)	(%)	(%)	(%)
Sodium (Na)	0.02	0.04	–	0.02	0.85	1.77	–	1.02
Calcium (Ca)	0.02	0.01	0.01	0.01	0.74	0.43	0.46	0.41
Magnesium (Mg)	0.19	0.16	0.16	0.15	6.74	6.72	9.08	10.14
Phosphorus (P)	0.47	0.40	0.37	0.35	16.67	17.24	21.26	23.81
Potassium (K)	1.31	0.90	0.58	0.46	46.45	38.79	33.33	31.29
	(ppm)	(ppm)	(ppm)	(ppm)	(ppm)	(ppm)	(ppm)	(ppm)
Copper (Cu)	8.0	6.1	6.0	5.3	285	264	348	359
Iron (Fe)	50.5	47.5	53.4	52.6	1,792	2,047	3,070	3,579
Manganese (Mn)	12.7	8.9	9.4	8.4	449	383	542	573
Molybdenum (Mo)	0.5	0.5	0.5	0.4	19	20	26	28
Zinc (Zn)	54.0	47.9	44.6	42.4	1,914	2,064	2,561	2,887

[1]Adapted by the authors from Thornton, J. H., R. D. Goodrich, and J. C. Meiske, "Corn Maturity. I. Composition of Corn Grain of Various Maturities and Test Weights," *Journal of Animal Science*, Vol. 29, 1969, pp. 977–982.

Grain Standards

Grains, like all feeds, vary in quality. To assist grain producers, buyers/sellers, and users in recognizing this variability, the government has established a set of standards for the grading of grains. Grain grading is done by the Federal Grain Inspection Service (FGIS), on a per hour charge basis. Grades are based on the physical and biological factors present in a sample. In general, the criteria for grading are (1) test weights per bushel, (2) moisture content, (3) foreign material and other grains, (4) broken and damaged kernels, and (5) discoloration. Also, it is important that a visual inspection be made for mold and/or fungi. There are additional grading criteria for some grains. In all cases, the highest grade is U.S. No. 1 and the lowest is U.S. Sample grade.

Table 10–12, p. 382, gives the standards by which barley, corn, flaxseed, oats, rye, sorghum, soybeans, and wheat are graded.

As is true of most standards, not everyone approves of the current grain standards; and they are subject to change. The chief criticisms of the present grain standards are:

1. Their enforcement is not rigid enough.
2. They need to define or describe quality more than by a grade number.
3. They do not provide bonuses (incentives) to encourage the production and sale of high-quality grain. Yet, others counter by saying that there is price incentive now—that quality is reflected in the prices of buying contracts.
4. They do not reflect consumer concerns relative to insect infestations; they do not force food processors to avoid grain purchases which have insect infestations or have been fumigated for insects.
5. They do not guarantee shipments of grain that meet the standards of nations to which U.S. grains are exported.
6. They are lower than the grain standards of two of our chief export competitors, notably, Australia and Canada.

Fig. 10–19. Yellow grain sorghum. Under U.S. federal grades, there are 4 classes of sorghum according to color—yellow, brown, mixed sorghum, and white. (Courtesy, National Grain Assn., Abernathy, Tex.)

TABLE 10-12
UNITED STATES STANDARDS FOR GRAINS[1]

Grain	Grade	Minimum Test Weight per Bushel	Minimum Sound Grain	Maximum Limits of: Moisture	Foreign Material	Other Grains	Broken Grain	Heat Damaged Kernels	Total Damaged Kernels	Discolored Grain
		(lb)	(%)	(%)	(%)	(%)	(%)	(%)	(%)	(%)
Barley[2]	U.S. No. 1	47.0	97	—	1.0	—	5.0	0.2	2.0	0.5
	U.S. No. 2	45.0	94	—	2.0	—	10.0	0.3	4.0	1.0
	U.S. No. 3	43.0	90	—	3.0	—	15.0	0.5	6.0	2.0
	U.S. No. 4	40.0	80	—	4.0	—	20.0	1.0	8.0	5.0
	U.S. No. 5	36.0	70	—	6.0	—	30.0	3.0	10.0	10.0
	U.S. Sample grade	U.S. Sample grade shall include barley of the class barley which does not come within the grade requirements of any of the grades from U.S. No. 1 to U.S. No. 5, inclusive; or which contains more than 16.0% of moisture; or which contains stones; or which is musty, or sour, or heating; or which has any commercially objectionable foreign odor except of smut or garlic; or which contains a quantity of smut so great that any one or more of the grade requirements cannot be applied accurately; or which is otherwise of distinctly low quality.								
Malting and blue malting barley[3][4]	U.S. No. 1	47.0	97	—	1.0	2.0	4.0	—	2.0	0.5
	U.S. No. 2	45.0	94	—	2.0	3.0	6.0	—	3.0	1.0
	U.S. No. 3	43.0	90	—	3.0	5.0	8.0	—	4.0	2.0
Western barley	U.S. No. 1	—	98	—	0.5	1.0[5]	3.0	0.1	—	0.5
	U.S. No. 2	—	96	—	1.0	2.0[5]	6.0	0.2	—	1.0
	U.S. No. 3	—	93	—	2.0	3.0[5]	10.0	0.3	—	2.0
	U.S. No. 4	—	88	—	3.0	5.0[5]	15.0	0.5	—	5.0
	U.S. No. 5	—	80	—	4.0	10.0[5]	25.0	1.0	—	10.0
	U.S. Sample grade	U.S. Sample grade shall include barley of the class western barley which does not come within the grade requirements of any of the grades from U.S. No. 1 to U.S. No. 5, inclusive; or which contains more than 15% of moisture; or which contains stones; or which is musty, or sour, or heating; or which has any commercially objectionable foreign odor except of smut or garlic; or which contains a quantity of smut so great that any one or more of the grade requirements cannot be applied accurately; or which contains the seeds of wild brome grasses of a character and in a quantity sufficient to cause the grain to be of low quality for feeding purposes; or which is otherwise of distinctly low quality.								
Corn	U.S. No. 1	56.0	—	14.0	—[6]	—	—[6]	0.1	3.0	—
	U.S. No. 2	54.0	—	15.5	—[6]	—	—[6]	0.2	5.0	—
	U.S. No. 3	52.0	—	17.5	—[6]	—	—[6]	0.5	7.0	—
	U.S. No. 4	49.0	—	20.0	—[6]	—	—[6]	1.0	10.0	—
	U.S. No. 5	46.0	—	23.0	—[6]	—	—[6]	3.0	15.0	—
	U.S. Sample grade	U.S. Sample grade shall be corn which does not meet the requirements for any of the grades from U.S. No. 1 to U.S. No. 5, inclusive; or which contains stones; or which is musty, or sour, or heating; or which has any commercially objectionable foreign odor; or which is otherwise of distinctly low quality.								
Flaxseed (linseed)	U.S. No. 1	49.0	—	—	—	—	—	0.2	10.0	—
	U.S. No. 2	47.0	—	—	—	—	—	0.5	15.0	—
	U.S. Sample grade	U.S. Sample grade shall be flaxseed which does not meet the requirements for grade U.S. No. 1 or U.S. No. 2; or which contains more than 9.5% of moisture; or which contains castor beans *(Ricinus communis)*, crotalaria seeds (*Crotalaria* spp), stones, unknown foreign substances, or commonly recognized harmful or toxic substances; or which is musty, sour, or heating; or which has any commercially objectionable foreign odor; or which is otherwise of distinctly low quality.								
Oats[7][8]	U.S. No. 1	36.0	97	—	2.0	2.0[5]	—	0.1	—	—
	U.S. No. 2	33.0	94	—	3.0	3.0[5]	—	0.3	—	—
	U.S. No. 3	30.0	90	—	4.0	5.0[5]	—	1.0	—	—
	U.S. No. 4	27.0	80	—	5.0	10.0[5]	—	3.0	—	—
	U.S. Sample grade	U.S. Sample grade shall be oats which— (a) do not meet the requirements for the grades U.S. Nos. 1, 2, 3, or 4, (b) contain more than 7 stones which have an aggregate weight in excess of 0.2% of the sample weight or more than 2 crotalaria seeds (*Crotalaria* spp) per 1,000 g of oats or more than 16% of moisture, (c) have a musty, sour, or commercially objectionable foreign odor (except smut or garlic odor), or (d) are heating or otherwise of distinctly low quality.								
Rye[9][10]	U.S. No. 1	56.0	—	—	3.0	—	—	0.1	2.0	—
	U.S. No. 2	54.0	—	—	6.0	—	—	0.2	4.0	—
	U.S. No. 3	52.0	—	—	10.0	—	—	0.5	7.0	—
	U.S. No. 4	49.0	—	—	10.0	—	—	3.0	15.0	—
	U.S. Sample grade	U.S. Sample grade shall include rye which does not come within the requirements of any of the grades from U.S. No. 1 to U.S. No. 4, inclusive; or which contains more than 16% of moisture; or which contains inseparable stones and/or cinders; or which is musty, or sour, or heating, or hot; or which has any commercially objectionable foreign odor except of smut or garlic; or which contains a quantity of smut so great that any one or more of the grade requirements cannot be applied accurately; or which is otherwise of distinctly low quality.								

[1]Adapted by the authors from *The Official United States Standards for Grain*, USDA, Dec. 1975. *(Continued)*

[2]One additional grade requirement is used for barley—thin barley (%). The maximum limits of thin barley for U.S. grades Nos. 1, 2, 3, 4, and 5 are 10, 15, 25, 35, and 75%, respectively.

[3]One additional grade requirement is used for malting and blue malting barley—thin barley (%). The maximum limits of thin barley for U.S. grades Nos. 1, 2, and 3 are 7, 10, and 15%, respectively.

TABLE 10–12 *(Continued)*

Grain	Grade	Minimum Test Weight per Bushel	Minimum Sound Grain	Maximum Limits of: Moisture	Foreign Material	Other Grains	Broken Grain	Heat Damaged Kernels	Total Damaged Kernels	Discolored Grain
		(lb)	(%)	(%)	(%)	(%)	(%)	(%)	(%)	(%)
Sorghum[11]	U.S. No. 1	57.0	—	13.0	—[12]	—[12]	—[12]	0.2	2.0	—
	U.S. No. 2	55.0	—	14.0	—[12]	—[12]	—[12]	0.5	5.0	—
	U.S. No. 3	53.0	—	15.0	—[12]	—[12]	—[12]	1.0	10.0	—
	U.S. No. 4	51.0	—	18.0	—[12]	—[12]	—[12]	3.0	15.0	—
	U.S. Sample grade	U.S. Sample grade shall be sorghum which— (a) does not meet the requirements for the grades U.S. Nos. 1, 2, 3, or 4, (b) contains more than 7 stones which have an aggregate weight in excess of 0.2% of the sample weight or more than 2 crotalaria seeds (*Crotalaria* spp) per 1,000 g of sorghum, (c) has a musty, sour, or commercially objectionable foreign odor (except smut odor), or (d) is badly weathered, heating, or distinctly low quality.								
Soybeans[13 14]	U.S. No. 1	56.0	—	13.0	1.0	—	10.5[15]	0.2	2.0	1.0[16]
	U.S. No. 2	54.0	—	14.0	2.0	—	20.0[15]	0.5	3.0	2.0[16]
	U.S. No. 3	52.0	—	16.0	3.0	—	30.0[15]	1.0	5.0	5.0[16]
	U.S. No. 4	49.0	—	18.0	5.0	—	40.0[15]	3.0	8.0	10.0[16]
	U.S. Sample grade	U.S. Sample grade shall be soybeans which do not meet the requirements for any of the grades from U.S. No. 1 to U.S. No. 4, inclusive; or which are musty, sour, or heating; or which have any commercially objectionable foreign odor; or which contain stones; or which are otherwise of distinctly low quality.								
Wheat[17] (hard red spring or white club wheat)	U.S. No. 1	58.0	—	—	0.5	3.0[17]	3.0[18]	0.1	2.0	—
	U.S. No. 2	57.0	—	—	1.0	5.0[17]	5.0[18]	0.2	4.0	—
	U.S. No. 3	55.0	—	—	2.0	10.0[17]	8.0[18]	0.5	7.0	—
	U.S. No. 4	53.0	—	—	3.0	10.0[17]	12.0[18]	1.0	10.0	—
	U.S. No. 5	50.0	—	—	5.0	10.0[17]	20.0[18]	3.0	15.0	—
	U.S. Sample grade	U.S. Sample grade shall be wheat which does not meet the requirements for any of the grades from U.S. No. 1 to U.S. No. 5, inclusive; or which contains more than 2 crotalaria seeds (*Crotalaria* spp) in 1,000 g of grain, or contains castor beans *(Ricinus communis)*, stones, broken glass, animal filth, an unknown foreign substance(s), or a commonly recognized harmful or toxic substance(s); or which is musty, sour, or heating; or which has any commercially objectionable foreign odor except of smut or garlic; or which contains a quantity of smut so great that any one or more of the grade requirements cannot be applied accurately; or which is otherwise of distinctly low quality.								
Wheat[17] (all other classes and subclasses)	U.S. No. 1	60.0	—	—	0.5	3.0[17]	3.0[18]	0.1	2.0	—
	U.S. No. 2	58.0	—	—	1.0	5.0[17]	5.0[18]	0.2	4.0	—
	U.S. No. 3	56.0	—	—	2.0	10.0[17]	8.0[18]	0.5	7.0	—
	U.S. No. 4	54.0	—	—	3.0	10.0[17]	12.0[18]	1.0	10.0	—
	U.S. No. 5	51.0	—	—	5.0	10.0[17]	20.0[18]	3.0	15.0	—
	U.S. Sample grade	U.S. Sample grade shall be wheat which does not meet the requirements for any of the grades from U.S. No. 1 to U.S. No. 5, inclusive; or which contains more than 2 crotalaria seeds (*Crotalaria* spp) in 1,000 g of grain, or contains castor beans *(Ricinus communis)*, stones, broken glass, animal filth, an unknown foreign substance(s), or a commonly recognized harmful or toxic substance(s); or which is musty, sour, or heating; or which has any commercially objectionable foreign odor except of smut or garlic; or which contains a quantity of smut so great that any one or more of the grade requirements cannot be applied accurately; or which is otherwise of distinctly low quality.								

[4]Barley which does not meet the requirements of any of the grades U.S. No. 1 to U.S. No. 3, inclusive, for the subclasses malting barley and blue malting barley shall be classified and graded according to the grade requirements for the subclass barley.

[5]Wild oats.

[6]In the grading of corn, the criteria of broken kernels and foreign material are combined. Therefore, corn of grades U.S. Nos. 1, 2, 3, 4, and 5 shall have no more than 2, 3, 4, 5, and 7% broken corn and foreign material, respectively.

[7]Oats that are slightly weathered shall be graded no higher than U.S. No. 3.

[8]Oats that are badly stained or materially weathered shall be graded not higher than U.S. No. 4.

[9]One additional grade requirement is used for rye—foreign matter other than wheat (%). The maximum limits for rye grading U.S. Nos. 1, 2, 3, and 4 are 1, 2, 4, and 6%, respectively.

[10]The rye in grade U.S. No. 1 may contain not more than 10%, in grade U.S. No. 2 not more than 15%, and in grade U.S. No. 3 not more than 25% of *thin* rye, which *thin* rye shall consist of rye and other matter that will pass readily through a sieve 0.032 in. thick with perforations 0.064 by 0.375 in.

[11]Under the U.S. Grades and Standards, there are four classes of sorghum: yellow, brown, mixed sorghum, and white. Sorghum which is distinctly discolored shall not be graded higher than U.S. No. 3.

[12]In the grading of sorghum, the criteria of broken kernels, foreign material, and other grains are combined. Therefore, sorghum of grades U.S. Nos. 1, 2, 3, and 4 shall have no more than 4, 8, 12, and 15% broken kernels, foreign material, and other grains, respectively.

[13]Soybeans which are purple mottled or stained shall be graded no higher than U.S. No. 3.

[14]Soybeans which are materially weathered shall be graded no higher than U.S. No. 4.

[15]Splits.

[16]Brown, black, and/or bicolored soybeans in yellow or green soybeans.

[17]Other grains = wheat of other classes (total). Red durum wheat of any grade may contain not more than 10% of wheat of other classes. An additional grading criterion is used in wheat—contrasting classes. Grades U.S. Nos. 1, 2, 3, 4, and 5 may not exceed 1, 2, 3, 10, and 10% of contrasting classes, respectively.

[18]Includes shrunken kernels.

NOTE WELL: Effective May 1, 1988, the standards permit the presence of fewer insects before the grain is discounted or termed *infested.* Samples of corn, sorghum, oats, barley, or soybeans are discounted if they have 2 or more live weevils, or 1 live weevil and 5 or more other live insects injurious to stored grain, or 10 or more other live insects. Wheat, rye, or triticale samples will be penalized for 2 or more live weevils, or 1 live weevil and 1 or more live insects injurious to stored grain, or 2 or more live insects injurious to stored grain.

Grain Storage

When grain is properly stored in well-designed facilities, losses due to spoilage and contamination can be held to a minimum. However, when little attention is given to the construction of the facilities or the condition of the grain to be stored, serious economic consequences can, and in all likelihood will, result.

Fig. 10-20. A steel bin for farm grain storage. (Courtesy, Western Illinois University, Macomb)

When storing grain, special attention must be given to moisture, temperature, and to damage from molds, insects, and rodents.

MOISTURE

When grains are stored at moisture levels less than 12%, the growth of most microorganisms is kept in check. If the moisture level is reduced to 10%, development of most insects will be arrested. However, when moisture level is decreased, especially to levels below 12%, incidence of grain breakage is markedly increased.

Sprouting can occur in stored grains when moisture levels exceed 30%. In addition to the alteration of the physical form of the grain during sprouting, heat is generated in the stored grain, thus creating a potential fire hazard.

(Also see Chapter 14, section on "Drying [Dehydrating].")

TEMPERATURE

As the temperature of storage is decreased, the number of problems associated with storage likewise decreases. For this reason, small grain bins painted white or completely shaded are generally better than large bins that are unpainted or exposed to the sun.

When storage temperatures are decreased to about 40°F, mites do not develop. Temperatures below 60° and 32°F inhibit growth of insects and fungi, respectively. However, the effects of temperature are closely interrelated with moisture in the grain and air movements throughout the storage bins.

BIOLOGICAL SOURCES OF GRAIN DAMAGE DURING STORAGE

The actual physical destruction of stored grain by moisture and temperature themselves can be great, but the losses due to biological factors—such as mold growth and insect and rodent damage—can be devastating. Quite often, the careful control of moisture and temperature can reduce these losses. However, additional precautions involving the construction of the storage facilities and the treatment of stored grains with pesticides may be needed.

Because grain is an excellent source of nutrients for livestock, it follows that it is also relished by insects and rodents. In cases of excessive moisture and poor storage management, grain provides an excellent growth medium for toxin-producing microorganisms that can create morbidity or mortality in livestock which consume the contaminated feed. *CAUTION:* For additonal assistance in choosing an insecticide or rodenticide, consult the local County Extension Agent; and always follow the manufacturer's label directions. *Seed that is dyed bright pink or reddish purple has been chemically treated to control insects and other pests, or molds, and is for planting purposes only; so, do not use for animal feed or human food.*

(See Chapter 5, Table 5-3, Potential Poisons.)

Insects and Mites

Infestations by insects and mites is most prevalent when the stored grain is in poor condition—for example, when a large proportion of the kernels are broken. Grain is subject to exposure to these pests at all stages of handling—harvesting, storing, transporting, processing, and packaging. Hence, it is imperative that precautionary measures be employed throughout handling.

Two basic methods of controlling insects and mites are used: (1) fumigation, and (2) residual contact pesticides. Fumigants are particularly effective because they can diffuse throughout the intergranular spaces of the stored grain. Many of the fumigants that are used to kill pests are equally effective against humans; hence, it is imperative that only well-trained personnel undertake this procedure. Two factors should be considered when fumigants are used: (1) effective lethal concentration in the air, and (2) length of exposure time. Residual contact pesticides are used to kill pests present in the grain and to prevent further infestation. Little or no specialized equipment is necessary for the application of these pesticides, and the dangers to the applicator are not as great as with fumigants. Additionally, these pesticides retain their potency for several months. Five types of residual contact pesticides are used: (1) dusts, (2) wettable powders, (3) emulsions, (4) aerosols, and (5) smoke.

Both temperature and moisture affect the effectiveness of pesticides. High temperatures in stored grains can reduce the amount of adsorption and absorption of fumigants and increase the rate of metabolism of the insecticide by the grain. On the other hand, extremely low temperatures reduce the effectiveness of fumigants. As moisture in grain increases, the rate of penetration of the pesticide into the grain will be reduced.

When pesticides of any type are used, the applicator should follow the directions and specifications on the label.

If too little pesticide is used, infestations will be neither eliminated nor prevented. If too much pesticide is used, serious consequences from residues can result.

Rodents

It has been estimated that rats consume 10% of their body weight in feed per day. In addition to the loss of feed, contamination from the feces and urine of rats constitutes further damage to grain. The United Nations Food and Agriculture Organization estimates that rats eat or contaminate 42.5 million tons of the world's grains each year—enough to feed 200 million people. For these reasons, rodenticides are commonly used where grain is stored.

The factors to be considered when selecting a rodenticide are (1) toxicity (for example, dosage levels), (2) acceptance, (3) reacceptance, (4) development of tolerances, (5) hazards to other animals, and (6) duration of potency. It is important to follow the label directions of the rodenticide carefully. Dosages recommended by the manufacturer are generally of sufficient strength to kill rodents with above-average tolerances. By using more than the recommended dosage level, the user may be decreasing the acceptability of the poison to the pest as well as increasing the risk of poisoning animals for which the poison is not intended.

Microbial Contamination

Both molds and bacteria can create serious problems in feeds. Molds will actively develop in grains stored at moisture contents in excess of 14.5%. Under certain temperature and moisture conditions, molds can produce mycotoxins which, if incorporated in feeds, can have an adverse effect on animals, *i.e.*, reduce weight gains, cause sickness, or even result in death.

Fig. 10-21. A mixture of ergot, a parasitic fungus (see black, banana-shaped masses), and wheat kernels as they appeared after harvest. (Courtesy, University of Idaho, Moscow)

The aflatoxin-producing fungi on feed grains in storage can be controlled, without affecting the feeding value for livestock and poultry, by treatment with an organic acid (propionic, acetic, or isobutyric; or a mixture of these). If applied properly to the grain as it is augered into the bins, organic acids will prevent growth of *A. flavus*. However, they will not remove any aflatoxins which formed within the grain before the fungus was killed.

Toxin-concentrated grain can be detoxified to levels below 20 ppb by using anhydrous ammonia. **NOTE WELL:** Ammonia is not registered by the Food and Drug Administration for use on grains to be shipped out of state, but it may be used on grains remaining in the state.

Animal feeds can be an important link in the *Salmonella Cycle*. Salmonella bacteria are responsible for feed-borne infections in animals and food-borne infections in humans.

Salmonella contamination can be prevented by good sanitation. This involves controlling rodents and birds and avoiding the contamination of grains by manure.

In all treatments, always follow manufacturer's directions.

(Also see Chapter 5, Table 5-3, under "Mycotoxins," "Pesticide Poisoning," and "Salmonellosis.")

OTHER HIGH-ENERGY FEEDS

Although feed grains and their milling by-products comprise the vast majority of the energy feeds, numerous other feeds are routinely used to supply energy to livestock. Seeds from plants other than *Graminae* can be used effectively (for example, beans). Fats provide an extremely concentrated source of energy. Molasses is a liquid energy feed that is highly palatable and digestible. When the price and availability are advantageous, roots, tubers, and certain other by-product feeds are fed to livestock.

Other Seeds

Seeds from plants other than cereal grains are used in livestock feeds when they are readily available and when the price is right. Legume seeds, or pulses, such as soybeans and peanuts, and whole cottonseed are used for their energy content in addition to protein. Many types of seeds are by-products of cash crop enterprises, representing culls of processing or marketing. On occasion, a surplus of a certain seed generally used for human consumption may reduce the cost to a level where it becomes economically feasible to incorporate it in livestock feeds. For this reason, livestock producers should become familiar with the feed substitution tables found in the feeding chapters devoted to each class of livestock.

Sprouted Grains

Sprouted grains are seeds that, following stimulation by water, air, and temperature, have germinated, or started to grow. Adverse moisture during harvest can cause grain to sprout in the head. Sorghum, barley, and wheat are especially affected, with the result that large quantities of these energy feeds become unsuited for milling purposes and become available for livestock feeding as sprouted grain.

U.S. Grade Standards discount grain on the basis of percent of sprouted kernels. For example, wheat showing more than 2% sprouted kernels is classified as sprouted wheat; and the grade is lowered with increased sprouting until, at 15%, the grain is classified as Sample grade. Likewise, in commercial trading channels, sorghum with 15% or more sprouted kernels is classed as Sample grade; and this, along with its light test weight per bushel, makes for a depressed price.

When sprouting drops the price of grain, the nutritive value of the sprouted grain should be evaluated to establish a fair price to both grain growers and livestock feeders. Fortunately, the nutritional value of sprouted grain is generally good, despite its dismal appearance. Also, it is noteworthy that the metabolic changes occurring in a grain kernel during germination (sprouting) are similar in many respects to what occurs in high-moisture grain or in reconstituted grain. It follows that experiments and experiences generally show that the sprouting of grain does not significantly affect its feeding value for cattle, sheep, swine, and poultry.

The Idaho Agricultural Experiment Station staff studied the value of sprouted wheat for growing-finishing hogs when the proportion of sprouted kernels represented 10, 20, and 30%, respectively, of the ration. Average daily gains were not affected, but more feed per pound of gain was required due to the lower energy value of the sprouted wheat. The Idaho researchers reported the following reduction in the energy value of the sprouted wheat for swine as compared to normal unsprouted wheat:[2]

Proportion of Sprouted Wheat Kernels	Energy Value of Sprouted Wheat Relative to Unsprouted Wheat For Swine
(%)	(%)
20	92.5
40	87.2
60	85.6

The Idaho researchers also conducted trials with cattle and chicks, feeding wheat that was classified as 60% sprouted. The rate and efficiency of gains of steers and chicks was not affected by the sprouted wheat that was fed.

North Dakota State University researchers reported that sprouted barley gave pig performance comparable to that of unsprouted barley. Kansas State University workers fed sprouted sorghum to growing-finishing pigs without significantly reducing daily gains, although slightly more feed was required per pound of gain.

It should be recognized that percentage of sprouted kernels alone is not an adequate measure of the nutritional value of sprouted grain for livestock and should not be the only criterion for discounting the price of sprouted grain. Substantial sprouting (involving most kernels and long sprouts) will reduce the energy available in the kernels, and require more feed per pound of gain. Also, when buying or selling sprouted grain the following additional factors should be considered: (1) the possible presence of molds and toxins, (2) the high moisture content of sprouted grains, and (3) storage problems.

[2]Eide, W. D., *et al., Feeding Value of Sprouted Grain,* Circular AS-647, North Dakota State University, Fargo, 1978. (**Note:** This North Dakota publication reported the Idaho study. M.E.E.)

Cottonseed

About 5 million tons of cottonseed are produced in the United States annually.

In recent years, the price of whole cottonseed has been favorable enough to include it in cattle feed, for both dairy and beef animals. Previously, the oil was extracted and used in human foods and industrial oils, leaving cottonseed meal and cottonseed hulls for animal feeds.

Fig. 10–22. Whole cottonseed, much of which is now being used as a high-energy feed for ruminants. (Courtesy, National Cotton Council of America, Memphis, Tenn.)

Whole cottonseed is very high in energy (95% TDN for ruminants, on a moisture-free [M-F] basis); high in protein (24% crude protein on M-F basis); high in phosphorus (0.76% M-F); and high in fiber (21.4% M-F); and it requires no feed processing.

Whole cottonseed is especially valuable as a hot weather feed for lactating dairy cows and finishing beef cattle because of its low heat increment, which means that less incidental heat is produced during its metabolism than from carbohydrates and proteins.

• **Whole cottonseed for lactating dairy cows**—Whole cottonseed is an important and rather unique feed for lactating cows, because it combines high energy and high-crude fiber with relatively high crude protein. In about 80% of the experimental feeding trials, whole cottonseed increased fat of milk about 0.3%. This increases milk values about 50¢ per cwt, which almost pays for the whole cottonseed. However, whole cottonseed tends to depress the protein fraction of solids-not-fat (SNF). So, where SNF is a milk price factor, lower protein in the milk may defeat the benefits of a high butterfat content.

Cottonseed can be fed to lactating cows "as-is," requiring no grinding or pelleting; and either linted or delinted seed may be used. It can be easily top-dressed over silage or green chop, mixed with other grains for bulk feeders, or processed with other grains for pelleted parlor mixes.

The amount of whole cottonseed to feed should be determined by comparative feed prices, cow weights, and the other ration components. For large animals, like Holsteins, most producers feed a minimum of 4 lb and a maximum of 7 lb per cow daily. There are a few reports favoring feeding up to 12 to 14 lb per cow daily. Also, some producers feed large amounts of whole cottonseed to the highest producers, a modest amount to average producers, and none at all to low producers.

- **Whole cottonseed for beef cattle**—Beef producers consider whole cottonseed a good buy when 100 lb of it costs less than the combined cost of 35 lb of cottonseed meal and 65 lb of sorghum. Pertinent points relative to the feeding of whole cottonseed to beef cattle follow:

1. Whole cottonseed may constitute up to 20% of the ration of finishing cattle. Because of the high oil content of cottonseed, weaned calves should not receive more than 4 lb per head daily, yearlings not more than 6 lb daily, and more mature animals not receive more than 7 lb daily.
2. As with any new feed, cattle may have to be enticed to eat cottonseed, which can be accomplished by topdressing it with molasses or other palatable feed.

The following cautions should be observed when feeding whole cottonseed.

1. Cottonseed can combust spontaneously if stored too wet and stacked too high, so the moisture level should not exceed 14%, and it should be stored in a flat bulk container.
2. Aspergillus mold, which produces aflatoxins, grows in any moist feed, but cottonseed is more likely to grow aflatoxins than many other feeds. This emphasizes that cottonseed must be stored and kept dry.
3. Whole cottonseed containing more than 1.0% gossypol can be toxic when fed to young calves under 4 months of age or to monogastric animals.

NOTE WELL: (1) In mature ruminants, gossypol is detoxified by the action of the rumen contents; and (2) the new glandless varieties of cottonseed do not contain gossypol.

In 1988, the National Cottonseed Products Association developed quality standards/trading rules for whole cottonseed, with descriptions and specifications for prime feed grade cottonseed, delinted prime feed grade cottonseed, and feed grade cottonseed—off quality.

Fats and Oils

Feeding of fats was prompted in an effort to find a profitable outlet for surplus vegetable oils and packinghouse fats. For the most part, fats were formerly used for soapmaking, but they are not used extensively in manufacturing detergents. Thus, with the rise in the use of detergents in recent years, they became a competitive energy feed. Hand in hand with these circumstances, the use of fat in feeds was enhanced by (1) recognition of the effect of high-energy feeds on efficiency of feed utilization; (2) discovery of the role of amino acids in improving the energy utilization by animals; and (3) discovery of the role of antioxidants in maintaining the quality of feed grade fat and other renderers' products.

In 1984, 1,443,000,000 lb of fats and oils were used in animal feeds in the United States, 93% of which were tallow and grease.

Fats and oils are high-energy sources. They enable animals to meet a high-energy requirement with less feed, because they contain 2.25 times as much energy as carbohydrates. For ruminants, the energy value of fats in relation to carbohydrates is higher than the normal 2.25 to 1 for monogastric animals; for cattle feed, fat is worth about 3.2 times the energy value of corn.

Common belief to the contrary, animals can tolerate a rather high-fat content in the ration. As evidence of this, it is noteworthy that sucklings normally handle a relatively large amount of fat, for milk contains 25 to 40% of this nutrient on a dry matter basis. Also, except for the soft pork problem, no apparent difficulty is encountered in feeding hogs a rather high-fat ration, such as results when large quantities of peanuts or soybeans are fed.

Fig. 10-23. Feed fats affect body fats. The soft lard sample (left) came from hogs fed a ration containing whole soybeans. Both samples of lard had been exposed to room temperature, 70°F, for 2 hours prior to photographing. (Courtesy, University of Illinois, Urbana)

A small amount of fat in the ration is desirable, as these fats are the carriers of the fat-soluble vitamins. Also, there is evidence that some species (humans, swine, rats, and dogs) require certain fatty acids. Although the fatty acid requirement of farm animals has not been determined, it is thought that ordinary farm rations contain ample quantities of these nutrients.

Because animal and vegetable fats seem to be equally effective additions to most livestock rations, with the exception of those for lactating cows, selection should be determined solely by comparative price. Ordinarily, animal fats are much cheaper than such vegetable fats as soybean or cottonseed oil.

There are many different feeding fats. Also, the terminology and standards differ. So, fats should be bought by specifications and guarantees. The quality of fat is based on free fatty acid content, moisture, insolubles, unsaponifiable matter, color, odor, and titer (temperature developed as a result of the heat of crystallation during cooling). Fats are designated as tallows or greases based on the titer value. The American Association of Feed Control Officials lists the following types of fats: animal fat; vegetable fat, or oil; hydrolyzed fat, or oil; fat product; corn endosperm oil; vegetable oil refinery lipid; corn syrup insolubles; beef fat; pork fat; poultry fat; and tall oil fatty acids (product obtained from making pulp from pine trees). The California Cattle Feeders Association uses the following more specific designations:

1. **Acidulated soap stocks (foots).** This end product is derived from vegetable or animal sources which have been washed free of mineral acids.
2. **Tallows.** This end product consists of rendered, clean, filtered animal fats. The melting point of tallows is greater than 104°F.
3. **Greases (white and yellow).** This end product is rendered, clean, filtered animal fats, and/or restaurant greases. The melting point of greases is less than 104°F.
4. **Blended feeding fat.** This end product is composed of rendered animal fat and grease, restaurant grease, vegetable oil, or acidulated soap stock in any combination.
5. **Other materials.** House grease, brown grease, sewer grease, modified yellow grease, and other low-grade materials are sometimes used.

Generally, vegetable oils are used for human consumption; hence, their cost is usually prohibitive to the livestock producer. However, vegetable oils can be used for livestock when the following circumstances prevail:

1. When the size of the individual purchase is small, the costs of plant oils can sometimes be justified.
2. When animal fats are unavailable or scarce, crude vegetable oils are sometimes used.
3. When edible vegetable oils become off-color, their marketability for human consumption decreases; and the prices may be subsequently reduced to a level at which the livestock producer can afford to use them.

Fat serves the following functions when added to livestock rations: (1) increases the caloric density of the ration; (2) controls dust in feed processing; (3) lessens the wear and tear on feed mixing equipment; (4) facilitates pelleting of feeds; (5) increases palatability; (6) helps to homogenize and stabilize certain feed additives, especially those of a very fine particle size; and (7) enhances the digestibility of the other feed ingredients of the ration. Fats also provide needed caloric density in the ration for high-producing animals, such as high producing dairy cows in early lactation and horses in endurance trials. During hot and cold weather, fats provide added energy without overloading the digestive system. Results of cattle feedlot trials have also indicated that fat will improve the feeding value of wheat as a feed grain, making it more economically comparable to corn and sorghum.

Fats are used to some degree in all types of livestock operations. Of the livestock feeds, milk replacers, used in feeding suckling animals, contain the most fat—ranging from 15 to 30%.

The following general guidelines relative to the use of fats and oils in animal feeds are based on experiments and experiences:

1. Add supplemental fat when it is desired to increase energy density, lower heat increment during hot weather, and increase feed efficiency.
2. Make such adjustments as necessary in other dietary ingredients when feeding fats because other nutrients may be decreased as a result of the higher energy level and possibly decreased nutrient intake of the ration.
3. Exercise extreme care to prevent contamination of milk or meat from such items as pesticides, chlorinated hydrocarbons, and PCBs, which are fat-soluble and may be introduced with blended fats.

Additionally, the following species recommendations are pertinent when adding fats to rations:

- **Fat for beef cattle**—It is recommended that 2 to 5% fat be added to high concentrate cattle finishing rations in which milo, barley, and/or wheat are the chief grain sources. When corn is the major grain, the addition of fat is less effective than with small grains, since corn contains approximately 4% fat compared to 1 to 1½% fat in other feed grains.
- **Fat for dairy cows**—It is recommended that 1 to 1½ lb of blended fat be fed daily to high producing dairy cows in early lactation and/or during hot weather, to stimulate milk production and improve feed efficiency; and that the maximum content of the fat in the ration not exceed 6% of the dry matter content. But not all fats perform alike when fed to lactating cows! Most unsaturated types of fats such as corn oil, soybean oil, safflower oil, and cottonseed oil have been shown to depress fiber digestibility followed by lowering of the butterfat (BF) percentage and the protein fraction of the solid-not-fat (SNF). So, it is recommended that lactating cows be fed either (1) a blend of more-saturated animal fat and vegetable oils and soapstocks, or (2) whole cottonseed or raw or cooked soybeans, which break down and release fats slowly.
- **Fat for horses**—It is recommended that 5 to 10% fat be added to the ration of horses subjected to intense and prolonged exercise and stress, such as endurance trials, to increase energy intake and improve performance.
- **Fat for swine**—It is recommended that approximately 5% fat be added to brood sow rations during late gestation to increase pig survival at birth, milk yield of sows, and rebreeding conception.

  It is recommended that 3 to 7% fat be added to pig starter rations for early-weaned pigs (pigs weaned at about 3 weeks of age) to improve daily gains and feed efficiency.
- **Fat for poultry**—It is recommended that 4 to 6% fat be added to broiler and turkey rations to increase growth rate, and to layer rations to increase egg production.

One of the more exciting nutritional developments is the so-called rumen protected fat system. In this system, the fat is emulsified with a protein. Then the protein is treated with formaldehyde to harden it. The product cannot be digested in the rumen due to the formaldehyde linkage on the protein, but the pH of the abomasum is such that this linkage is broken, with the result that the protein in fat will then be digested in the small intestine (similar to that of the monogastric animal). Studies have shown that the utilization of fat by the ruminant can be improved by the use of this system. Also, and most significant, by the use of the protein protected fat system, it is possible to alter the ratio of the saturated to the unsaturated fatty acids in beef depot fat. The same is true of butterfat. A beef fat which contains high levels of unsaturated fatty acids is especially attractive to people with certain types of heart conditions. Thus, in the future, we may see some beef cattle fed rations containing protein protected fat for the express purpose of producing Choice beef in which the fat contains a high level of unsaturated fatty acids.

(Also see Chapter 6, section on "Fats and Oils"; Chapter 11, section on "Protein Bypass (Protected Protein, Escaped Protein)"; Chapter 14, sections on "Fat Added," and "Slow-Release and Rumen Bypass Treatments"; Chapter 19, section on "Fats/Cottonseed"; and Chapter 25, section on "Fats.")

Molasses

Molasses (including cane or blackstrap, beet, citrus, wood, and starch molasses) is extremely palatable and an excellent source of energy. Approximately 638 million gal or 3,745,265 tons, of molasses of all kinds are used annually in the United States. About 73%, or 2,734,044 tons, is consumed in feeds and the balance is used for industrial purposes.

Cane and beet molasses are by-products of the manufacture of sugar from sugarcane and sugar beets, respectively. Citrus molasses is produced from the juice of citrus wastes. Wood molasses is a by-product of the manufacture of paper, fiberboard, and pure cellulose from wood; it is an extract from the more souble carbohydrates of the wood material. Starch molasses is a by-product of the manufacture of dextrose from starch derived from corn or grain sorghums in which the starch is hydrolyzed by use of enzymes and/or acid. Cane or blackstrap is, by far, the most extensively used type of molasses. The different types of molasses are available in both liquid and dehydrated forms.

In addition to its use as an energy feed, molasses is used in the following ways: (1) as an appetizer, (2) to reduce the dustiness of a ration, (3) as a binder for pelleting, (4) to stimulate rumen microbial activity, (5) to supply unidentified nutrient factors, (6) in the case of cane molasses, to provide trace minerals, and (7) to provide a carrier for nonprotein nitrogen and vitamins in liquid supplements.

In ruminant rations, molasses is restricted to the level of 10 to 15% of the ration since it is most efficiently utilized when it does not exceed these levels. Excessive amounts of molasses (greater than 15%) will cause the feed to become messy and unmanageable as well as create digestive disturbances. Poultry are rather sensitive to molasses as excessive levels cause diarrhea. Hence, although molasses occasionally constitutes as much as 10% of poultry rations, levels are generally restricted to from 2 to 5%.

The quality of molasses is determined by its sugar content, which is expressed by the term, *Brix*. Brix is determined by measuring the specific gravity of molasses. After the specific gravity has been obtained, the value is applied to a conversion table from which the level of sucrose (or degrees Brix) can be determined. As sugar content increases, degrees Brix likewise increases. Since molasses also contains lipids, protein, inorganic salts, waxes, gums, and other material, the Brix classification can often be misleading, because each of these contaminants has an influence on the specific gravity of the solution. However, degrees Brix does give a relatively accurate indication as to the sugar content of molasses and is, therefore, a good means of determining quality. A summary of the minimum specifications for molasses is presented in Table 10-13.

TABLE 10-13
SPECIFICATIONS FOR MOLASSES

Type	Degrees Brix	Total Sugars as Invert	Weight per Gallon
	(minimum)	(minimum %)	(lb)
Beet	79.5	48	11.75
Cane	79.5	43	11.75
Citrus	71.0	45	11.29

(Also see Chapter 14, Feed Processing, section on "Molasses Added.")

CANE MOLASSES

As early as 1900, sugar planters in Louisiana fed cane molasses in long, open troughs to mules, cows, and hogs. The animals did their own mixing, alternating between a mouthful of hay, corn, oats, and molasses, and washing the molasses down with a drink of muddy water from a nearby bayou. Feeding was a sticky, messy business, and attracted swarms of flies. Even so, molasses was fed simply because it had to be disposed of, and the animals liked it. It couldn't be sold for the price of the barrel container, and there was still the matter of freight charges. Gradually, the feeding value of the product was appreciated, and, by 1914, the demand exceeded the domestic supply. Shipments were brought in from Cuba. Today, a large quantity of this carbohydrate feed is produced in the southern states, and an additional tonnage is imported.

Cane molasses weighs about 11.7 lb per gallon. One hundred pounds of cane molasses will yield about 50 to 55 lb of highly digestible sugars and 26 lb of water. Cane molasses contains an extremely low amount of protein (about 4.3% as-fed basis). In fact, large amounts of molasses in feed can depress protein utilization of other feeds. Molasses is deficient in thiamin, riboflavin, vitamin A, and vitamin D, but it is rich in niacin and pantothenic acid. Although cane molasses is low in phosphorus, it is an excellent source of other minerals. It may contain up to 10% ash.

Cane molasses can be used most effectively by adding small amounts to poor-quality roughages for ruminants.

BEET MOLASSES

The use of beet molasses is generally restricted to the western states which produce most of this country's sugar beets. When properly used, this type of molasses has the same feeding value as cane molasses. However, care should be taken when using beet molasses because it has highly laxative properties due to its high-mineral content. Total sugar content generally varies from about 48 to 53%. Protein values are higher than in cane molasses (6.6% vs 4.3%).

CITRUS MOLASSES

Citrus molasses contains about 41 to 43% total sugars—substantially less than either cane or beet molasses. Moisture content is also higher, ranging from 27 to 30%. Protein content is about 5.8% as-fed basis, which is intermediate between cane and beet molasses.

WOOD MOLASSES

This type of molasses is little used in the feed industry. The energy content of wood molasses is high since it contains at least 55% total carbohydrates. The protein content is very low, only 0.6% on an as-fed basis. Use of this feed has been limited to beef cattle rations, for the most part.

Roots and Tubers

A root crop consists of the fleshy subterranean parts of a harvested plant—for example, carrots and beets. Tubers are short, thickened, fleshy stems, or terminal portions of stems or rhizomes, that are usually formed underground—for example, peanuts and potatoes.

Root and tuber crops have traditionally been used more extensively as livestock feeds in Europe than in this country. They yield relatively large amounts of nutrients per acre;

but the cost of the labor needed to harvest these crops has been prohibitive from a livestock feeding standpoint. Another limitation in their use is their high moisture content. However, in the production of roots and tubers for human consumption, considerable wastes become available which can be fed to livestock as by-product feeds.

Due to the large amounts of moisture in these feeds, only limited amounts are fed to high-producing animals which require considerable amounts of dry matter. Ingestion of highly succulent feeds decreases the amount of space in the stomach and small intestine, thus restricting the amount of dry matter that can be consumed.

When considered on a dry matter basis, roots and tubers are relatively good sources of energy (see Table 10-14). However, they generally have limited amounts of protein and, except for carrots and sweet potatoes, contain very little carotene. Calcium and vitamin D levels are also extremely low. Best results are secured in most animals when roots are limited to a replacement of not more than one-fourth of the grain ration.

TABLE 10-14
DIGESTIBLE ENERGY VALUES OF VARIOUS ROOTS AND TUBERS[1]

Feed	Dry Matter	Digestible Energy			
		Ruminants		Swine	
		As-Fed	Moisture-Free	As-Fed	Moisture-Free
	(%)	(Mcal/lb)	Mcal/lb)	(kcal/lb)	(kcal/lb)
Carrots	11	0.19	1.69	208	1,805
Cassava	32	0.51	1.58	511	1,576
Chufa	27	0.39	1.47	480	1,812
Mangels	11	0.18	1.59	181	1,652
Potatoes	24	0.38	1.62	398	1,695
Sugar beets	20	0.33	1.67	355	1,786
Sweet potatoes	33	0.52	1.59	530	1,610
Turnips	9	0.16	1.70	146	1,594

[1]*From:* This book, Table V-1A, Energy Feeds.

Some of the most widely used root and tuber crops in livestock feeds are Irish potatoes, sweet potatoes, chufas, cassava, beets, mangels, carrots, and turnips.

• **Potatoes (Irish potatoes, *Solanum tuberosum*)**—In the United States, the culls of Irish potatoes are fed to livestock. In limited amounts and in properly balanced rations, they are a satisfactory energy feed for beef cattle, dairy cows, sheep, swine, horses, and poultry. They are rich in starch, fair in protein, and deficient in vitamins A and D—lacks which are made good when they are fed with well cured legume hay. About 400 to 450 lb of potatoes are equivalent in energy value to 100 lb of cereal grains. Stated in another way, potatoes are worth 22 to 25% as much as grain.

Irish potatoes should be cooked (steamed or boiled, preferably in salt water to increase palatability, with any cooking water discarded) for swine or poultry, but this is not necessary for other species.

Animals should be accustomed to potatoes gradually, as they are not very palatable. Also, raw potatoes should not be fed in too large amounts, as they may cause scours.

Sometimes cattle choke on potatoes, but this risk is minimized if the potatoes are fed from low troughs, with a cable or bar located so as to keep their heads down.

The sprouts of potatoes contain an alkaloid, solanin, which is toxic. Thus, long sprouts should be removed prior to feeding. Frozen potatoes should never be fed.

Potatoes may be ensiled (see Chapter 9, Silage/Haylage/High-Moisture Grain, section on "Other Silage Crops"). Potato silage is eaten readily by animals and is approximately equal to corn silage in value per ton.

Potatoes can be dehydrated, or dried, to make potato meal. This is a satisfactory substitute for 20 to 25% of the concentrate of beef cattle, dairy cattle, or sheep. If heated sufficiently in the drying process and cooked thoroughly, 10 to 20% potato meal can be incorporated in swine and poultry rations. Potato meal is equal to grain when substituted for part of it.

Dried potato pulp, or potato pomace, a by-product of starch production from potatoes, is slightly higher than potato meal in total digestible nutrients. It is palatable and nearly equal to hominy feed for dairy cows when forming 20% of the concentrate mixture.

Some by-products of processing potatoes for human food, including peelings, cull french fries, and discarded potato chips, are fed to animals kept in close proximity to the potato processing plants. These by-products may be fed as-is provided they are properly stored or ensiled.

• **Sweet potatoes *(Ipomea batatas)***—Cull or unmarketable sweet potatoes are available for feeding. Also, in the South, some sweet potatoes are grown especially for livestock feed. Pigs are often turned into the field to harvest the crop.

Sweet potatoes are high in dry matter for a root crop—averaging 31.8%, rich in starch, and high in carotene. But they are low in protein, calcium, and phosphorus. For best results, they should not replace more than half of the grain in a ration; and the ration should be properly supplemented with protein, minerals, and vitamins. It requires about 4.3 lb of sweet potatoes to equal 1 lb of grain and other concentrates for pigs in drylot, and about 4.9 lb of potatoes to equal 1 lb of corn when the crop is hogged-down. Cooking increases the value of sweet potatoes for swine.

Sweet potatoes produce a hard pork, but pigs fed high levels of them tend to be paunchy and have a low dressing percentage.

Sweet potato meal, or dried sweet potato, is worth 90 to 100% as much as corn for beef cattle, dairy cattle, and sheep. But it is not palatable to horses, and it is least useful for swine and poultry. Also, the cost of dehydrating is rather high.

Satisfactory silage may be made from chopped sweet potatoes, without a preservative.

Green sweet potato vines are a nutritious forage. Sometimes animals are turned into a sweet potato field to graze on the vines after the crop is harvested. Wilted vines make a good silage, without a preservative. Sweet potato vine silage is equal to corn silage for finishing cattle or dairy cows.

• **Turnips**—Turnips are a popular fall/winter forage for cattle and sheep in some northern areas. Generally, they are harvested by grazing.

(See Chapter 7, section on "Turnips for Fall/Winter Grazing." Also see the feed substitution tables in the chapters devoted to each animal species.)

Fig. 10–24. Ewes strip-grazing turnips. (Courtesy, Purdue University, Lafayette, Ind.)

Processing Industry By-product Feeds as Energy Sources

Numerous by-products from various processing industries are used as energy feeds. Spent grains from the brewing and distilling industries are excellent sources of energy and other nutrients. The dairy processing industry produces numerous high-energy by-products, such as whey. Bakery residue provides a good source of highly digestible energy. Citrus pulp and beet pulp are relatively good energy feeds. These by-product feeds, along with several other types of by-products, are discussed in Chapter 12, By-product Feeds/Crop Residues. Additional information on the individual by-product feeds may be found in Chapter 30, Glossary of Feedstuffs.

QUESTIONS FOR STUDY AND DISCUSSION

1. What are concentrates? How are they classified?

2. Give your reasons for agreeing with or challenging the following statement: "Of all the energy sources used in livestock and poultry rations, cereal grains fill the most important role."

3. Define the term *grain*. In the U.S., what grains are grown primarily for food for humans; what grains are produced chiefly as feed for livestock?

4. How do you explain the following variations in the utilization of grains in other countries, in comparison with the U.S.: (a) In Europe, large quantities of wheat are fed to animals; (b) in Mexico, corn is favored as a food grain for humans; and (c) in Africa, sorghum is the primary human food grain?

5. Has the amount of grain fed to livestock in the U.S. increased or decreased in recent years? What are the reasons for this trend?

6. Give the pertinent characteristics and feeding value of each of the following cereal grains: barley, corn, oats, and sorghum under the following headings: (a) TDN, (b) protein, (c) livestock use, and (d) cultural characteristics.

7. List the problems that may be encountered by producers in the use of grains.

8. Diagram and label the flower structures of corn or wheat. How are the grains formed?

9. The seed portion of grain can be divided into four parts. List them.

10. Why does a kernel of dent corn have an indentation?

11. Describe and/or diagram the wet milling of corn.

12. Name the valuable by-products commonly used as livestock and poultry feeds which result from the milling of each of the following grains: corn, wheat, rice, and barley.

13. Give a brief, but pertinent, description, including the nutritive value, of each of the following cereal by-products: bran, gluten feed and meal, groats, malt sprouts, middlings, mill run, polishings, and shorts.

14. Rank the commonly used feed grains in decreasing order of available energy content.

15. Approximately how much of the dry matter of grains consists of carbohydrate? Of what is the carbohydrate fraction of grain composed?

16. Which grains contain the highest amount of lipid material? What part of the grain contains the most oil?

17. Are grain oils high in saturated fatty acids or in polyunsaturated fatty acids?

18. Which part of corn grain contains the highest concentration of protein?

19. What amino acids are found in very low concentrations in the cereal grains and their by-products? Are these essential amino acids?

20. Why is the nutritional deficiency disease kwashiorkor so prevalent where corn or sorghum is the leading cereal grain consumed by people?

21. What is Opaque–2 corn? What are its advantages? Why hasn't it been more widely accepted?

22. Discuss the calcium and phosphorus content, and the phosphorus availability, of cereal grains.

23. Niacin is found in relatively high concentrations in the aleurone of corn, yet corn is known to be a producer of pellagra (niacin deficiency disease). How can this be explained?

24. Briefly outline what changes in chemical composition occur as corn matures.

25. What basic criteria are used in the grading of grains?

26. What chief criticisms are being made of the current grain standards?

27. What level of insect infestation results in wheat and corn being termed *infested* and discounted in price? (See the *note well* footnote at the bottom of Table 10–12.)

28. Discuss the various factors that must be dealt with in the storage of grain.

29. Discuss the feeding value of sprouted grain of each of the following: sorghum, barley, wheat.

30. Why has whole cottonseed evolved as a high-energy feed in recent years? Discuss the feeding value of whole cottonseed for beef cattle and for dairy cattle.

31. List the various types of fat products that are used as feeds.

32. What functions does the addition of fats into feeds serve? What are some of the disadvantages?

33. Discuss the feeding of fats to each of the following species: beef cattle, dairy cattle, horses, swine, and poultry.

34. Explain the rumen protected fat system. How may this phenomenon be used?

35. Discuss molasses from the standpoints of kinds, advantages, level in the ration, and quality.

36. Differentiate between a root crop and a tuber crop. Discuss the feeding value, species to which fed and method of feeding potatoes, sweet potatoes, and turnips.

SELECTED REFERENCES

Title of Publication	Author(s)	Publisher
Animal Feeds	M. Gutcho	Noyes Data Corporation, Park Ridge, N.J., 1970
Association of American Feed Control Officials Incorporated, Official Publication	Association of American Feed Control Officials, Inc.	Association of American Feed Control Officials, Inc., Annual
Atlas of Nutritrional Data on United States and Canadian Feeds	E. W. Crampton, Chairman, Subcommittee on Feed Composition	National Academy of Sciences, Washington, D.C., 1971
Cereal Processing and Digestion	U.S. Feed Grains Council, USDA Foreign Agricultural Service	U.S. Feed Grains Council, London, England, 1972
Cereal Science	S. A. Matz	Avi Publishing Company, Westport, Conn., 1969
Chemistry and Technology of Cereals as Food and Feed, The	S. A. Metz	Avi Publishing Company, Westport, Conn., 1959
Composition of Cereal Grains and Forages, Pub. 585	National Research Council	National Academy of Sciences, Washington, D.C. 1958
Crop Production, 5th Edition	R. J. Delorit L. J. Greub H. L. Ahlgren	Prentice-Hall, Inc., Englewood Cliffs, N.J., 1984
Digestibility and Amino Acid Availability in Cereals and Oilseeds	Ed. by J. W. Finley D. T. Hopkins	American Association of Cereal Chemists, Inc., St. Paul, Minn., 1985
Effect of Processing on the Nutritional Value of Feeds	National Research Council	National Academy of Sciences, Washington, D.C. 1973
Energy for World Agriculture	B. A. Stout	Food and Agriculture Organization of the United Nations, Rome, Italy, 1979
Energy Use and Production in Agriculture	B. A. Stout, Chairman, and Task Force Members	Council for Agricultural Science and Technology, Ames, Iowa, 1984
Feed Manufacturing Technology III	R. R. McEllhiney Tech. Ed.	American Feed Industry Association, Inc., Arlington, Va., 1985
Feeds and Feeding, 22nd Edition	F. B. Morrison	The Morrison Publishing Company, Ithaca, N.Y., 1956
Grain Storage: Part of a System	R. N. Sinka W. E. Muir	Avi Publishing Company, Westport, Conn., 1973
Grain Storage: The Role of Fungi in Quality Loss	C. M. Christensen H. H. Kaufmann	University of Minnesota Press, Minneapolis, Minn., 1969
Storage of Cereal Grains and Their Products	Ed. by C. M. Christensen	American Association of Cereal Chemists, St. Paul, Minn., 1974
Technology of Cereals	N. L. Kent	Pergamon Press, Oxford, England, 1975
Tropical Feeds	B. Gohl	Food and Agriculture Organization of the United Nations, Rome, Italy, 1975
United States-Canadian Tables of Feed Composition, 3rd revision	J. H. Conrad, Chairman, Subcommittee on Feed Composition	National Academy Press, Washington, D.C., 1982

Chapter 11

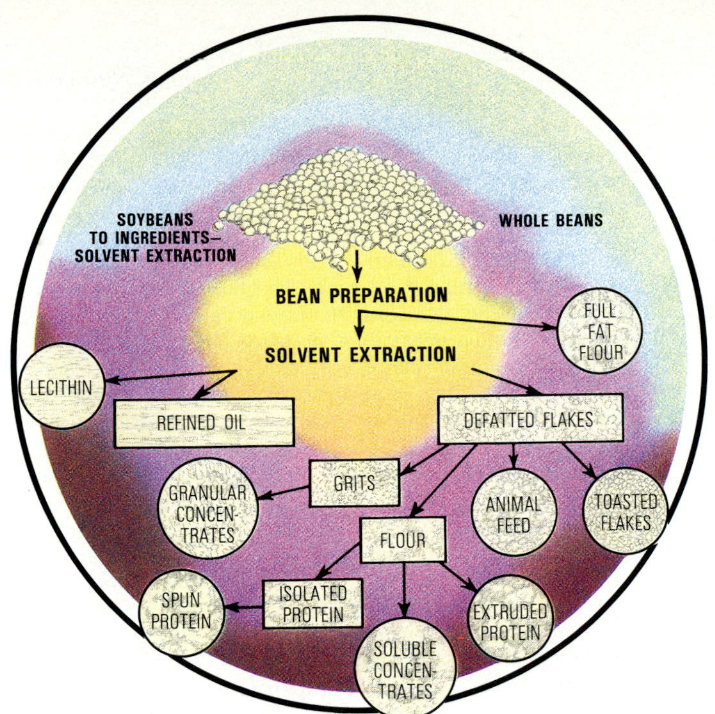

Original painting by Tom Phillips

PROTEIN SUPPLEMENTS

Contents	Page

The following definitions are pertinent to the discussion that follows in this chapter:

Proteins are complex organic compounds made up chiefly of combinations of amino acids present in characteristic proportions and arrangements for each specific protein; they always contain carbon, hydrogen, oxygen and nitrogen, and, in addition, usually sulfur and frequently phosphorus. Proteins are essential in all plant and animal life as components of the active protoplasm of each living cell. In plants, the percentage of protein is high in the young growing plant and falls as the plant matures. In animals, muscle, skin, hair, feathers, wool, and horns/hoofs/nails contain protein. Plant proteins differ from each other and from animal proteins; and each animal species has its own specific proteins, and a given animal has thousands of different proteins in its various tissues. Moreover, each protein has a specific function in the animal body (or other biological organism), ranging from protection of the body surface (skin, hair) to defense against invading organisms. Chemically, it has been determined that all proteins contain about 16% nitrogen; so, crude protein is determined by finding the nitrogen content, then multiplying the result by 6.25 ($100 \div 16 = 6.25$).

Amino acids are nitrogen-containing compounds that constitute the "building blocks," or units, from which more complex proteins are formed; they contain both amino (NH_2) and carboxyl (COOH) groups. At least 22 amino acids are found in proteins; and they occur in different combinations to form an almost limitless number of proteins. For monogastric animals and very young ruminants (preruminants), the amino acids that make up proteins are really the essential nutrients, rather than the protein molecule itself. For ruminant species, the dietary need is (1) to nourish the microorganisms, and (2) to have an adequate supply of digestible essential amino acids in the gut.

Protein quality is a term used to describe the amino acid balance of protein. A protein is said to be of good quality when it contains all the essential amino acids in proper proportions and amounts needed by a specific animal; and it is said to be poor quality when it is deficient in either content or balance of essential amino acids. Protein quality is much less important in ruminants than in nonruminants. In the rumen, a high proportion of the dietary proteins are hydrolyzed by rumen microbes to amino acids, many of which are further degraded to organic acids, ammonia, and carbon dioxide. If conditions are optimal, the free ammonia and carbon chains in rumen fluid are utilized by bacteria to synthesize new amino acids. The bacteria, in turn, may be ingested by protozoa which go through the same cycle of degradation and resynthesis of proteins. Ammonia that is not resynthesized is lost in the urine, while bacterial, protozoal, and undegraded dietary proteins pass into the intestinal tract where they are digested and the amino acids absorbed. Bacterial and protozoal proteins are generally lower in biological value than the better animal proteins and soybean meal, but of higher quality than most plant proteins. Thus, the tendency of rumen synthesis is to lower the quality of very high-quality proteins and to upgrade low-quality dietary proteins.

Nonprotein nitrogen (NPN) is nitrogen which comes from other than protein sources, but which may be used by ruminants in building proteins. NPN feed sources include urea, anhydrous ammonia, biuret, ammonium sulfate, monoammonium phosphate, ammoniated molasses, ammoniated beet pulp, ammoniated cottonseed meal or hulls, and ammoniated rice hulls.

Protein equivalent is a term indicating the total nitrogenous contribution of a substance in comparison with the nitrogen content of protein; for example, the nonprotein nitrogen (NPN) compound urea contains approximately 45% nitrogen and has a protein equivalent of 281% ($6.25 \times 45\%$).

Protein supplements are feedstuffs that contain more than 20% protein or protein equivalent, and which are commonly used to improve the protein quantity and/or quality of the basal feed(s). Ideally, a protein supplement should provide the amino acids needed to match up the amino acids of the basal feed so that, combined, they meet the essential amino acid needs of the animal to which they are fed, without either a surplus or a deficiency of essential amino acids.

For more than a century, proteins and their structural units, the amino acids, have been studied and recognized as important dietary constituents. Yet, prior to 1890, no one was concerned about adding them to livestock rations. The flour mills in Minneapolis dumped wheat bran into the Mississippi River, because nobody wanted to buy it. Cottonseed meal was used as a fertilizer, if used at all. Most of the linseed meal was shipped to Europe. Soybeans were little known outside the Orient, and tankage had not been processed.

Beginning about 1900, scientists discovered that the kind or quality of protein in livestock feeds was of tremendous importance, thus ushering in the golden era in nutrition. Soon the race was on for protein-rich feeds. Many of the by-products that once polluted the streams of the nation were in unprecedented demand, first as protein supplements, and later because of the recognized added vitamin value of many of them.

Protein in plants is largely concentrated in the actively growing portions, especially the leaves and seeds. Plants, especially legumes, possess the ability to synthesize their own proteins from such relatively simple soil and air compounds as carbon dioxide, water, nitrates, and sulfates (see Chapter 1 for more details on the photosynthetic process). Thus, plants, together with some bacteria which are able to synthesize these products, are the original sources of all

Fig. 11-1. Soybeans, source of soybean meal, America's leading protein supplement. As with most plants, the protein is generally concentrated in the leaves and seeds. (Courtesy, American Soybean Assn., St. Louis, Mo.)

proteins. It follows that the solution to world protein shortage is to increase the acreage of legume crops and the number of ruminant animals.

Proteins in animals are much more widely distributed than in plants. The proteins of the animal body are primary constituents of many structural and protective tissues—such as bones, ligaments, hair, hooves, skin, and the soft tissues which include the organs and muscles. The total protein content of animal bodies ranges from about 10% in very fat, mature animals to 20% in thin, young animals (see Chapter 2, Tables 2-2 and 2-3). By way of further contrast, it is noteworthy that, except for the bacterial action in the rumen, animals lack the ability of plants to synthesize proteins from inorganic materials. Hence, they must depend upon plants or other animals as a source of dietary protein. In brief, except for the high-quality proteins synthesized by the microorganisms in the paunch of ruminants, animals must have amino acids, or complete proteins, present in the ration. Animals of all ages and kinds require adequate amounts of dietary protein of suitable quality for maintenance, growth, finishing, reproduction, work, and wool production.

Proteins are found in most of the feeds commonly fed to animals. The amount, digestibility, and balance of essential amino acids, of the protein are important factors that must be considered in balancing rations. In general, animal proteins are superior to plant proteins for monogastric animals (including humans). For example, zein (a corn protein) is a low-quality or unbalanced protein, being deficient in the essential amino acids lysine and tryptophan. On the other hand, aminal proteins are excellent sources of lysine, and many (especially milk and eggs) are abundant in tryptophan. But not all animal proteins are superior and not all plant proteins are inferior; gelatins and soybeans are exceptions. Gelatin, an animal protein, is lacking or low in several amino acids. Soybean protein, a plant form, has good amounts of the essential amino acids.

Fortunately, the amino acid content of proteins varies from various plant and animal sources. Thus, the deficiencies of one protein source may be improved by combining it with another; and the mixture of the two proteins will often have a higher feeding value than either one alone. It is for this reason that a combination of protein feeds in the ration is usually recommended when the person formulating the ration does not have access to specific amino acid values of the feeds to be used or to supplements of individual amino acids.

In the past, it was common practice to use several sources of protein in nonruminant rations so that their respective amino acid profiles would complement each other. Today, the use of the computer in formulating rations and the increased availability of amino acid supplements, such as methionine hydroxy analog (MHA), allow nutritionists to formulate complete rations with a minimum number of protein feeds. The computer can rapidly determine what specific amino acids must be added and at what levels. Thus, the trend is toward fewer protein feed sources, properly supplemented with specific amino acids. A detailed discussion on the advantages and disadvantages to computer formulation is found in Chapter 18, Feeding Standards/Ration Formulation.

Also, it is noteworthy that researchers are bolstering plant protein sources by increasing, through genetic means, both the content and the nutritional quality of the protein in cereal grain. High-lysine corn is one such product of research.

(Also see Chapter 4, section on "Nitrogenous Feeds"; Chapter 6, section on "Protein Supplements"; Chapter 10, section on "High-Lysine Corn—Opaque 2, or O_2"; Chapter 13, sections on "Protein Supplements" and "Amino Acid Supplements.")

PROTEIN REQUIREMENTS VS AMINO ACID REQUIREMENTS

Many feeding standards list requirements for protein as determined by age, weight, sex, and production. However, the validity of listing total protein requirements has recently come under considerable scrutiny. Proteins are not always digestible or available to the animal. Certain processing methods, especially those involving heat, may reduce the digestibility or availability of proteins in feeds. In many cases, it is necessary to heat certain feeds to destroy a toxic substance, even when it is known that the protein value will be reduced. On the other hand, some proteins are made more readily available by processing. Another weakness of the total protein concept is the fact that it does not take into consideration protein quality—that is, the amino acid composition of feeds or their availability.

(Also see the section on "Protein Assessment Systems" at the end of this chapter and in Chapter 4, Nutrients/ Metabolism.)

Protein and Amino Acids for Nonruminants

Amino acid availability is critical in the nutrition of nonruminants because they do not have the ability to synthesize certain amino acids. Therefore, in recent years nonruminant nutritionists have concentrated on establishing specific amino acid requirements for animals rather than establishing protein levels. A deficiency, and sometimes an excess, of a particular amino acid can severely affect production even though total protein levels appear to be adequate.

Protein and Amino Acids for Ruminants

Amino acids in the rations of ruminants is more important than formerly thought. It has been shown that the protein produced by ruminal synthesis does not supply all the amino acids in quality and quantity needed for maximum growth or peak milk production of ruminants. Moreover, it has been found that protein efficiency can be increased by protecting protein from the degradation of the microbes in the rumen and increasing the escape of the protein from the rumen to the intestine where it is digested and absorbed. Because protein is a costly ingredient, it follows that increasing its efficiency reduces cost since less of it is necessary. This technology of manipulating the quantity of dietary protein rumen fermentation, thereby increasing the supply of protein (amino acids) in the small intestine, is known as *protein bypass* (*protected* protein, *escaped* protein).

PROTEIN BYPASS (PROTECTED PROTEIN, ESCAPED PROTEIN)

Fig. 11-2 is a schematic summary of protein utilization by cattle (and other ruminants) and is useful in understanding where and how bypass (protected) proteins may be beneficial in ruminant rations. Most ruminant rations contain both true protein and nonprotein nitrogen (NPN). Essentially, all the NPN fraction is converted to ammonia in the rumen.

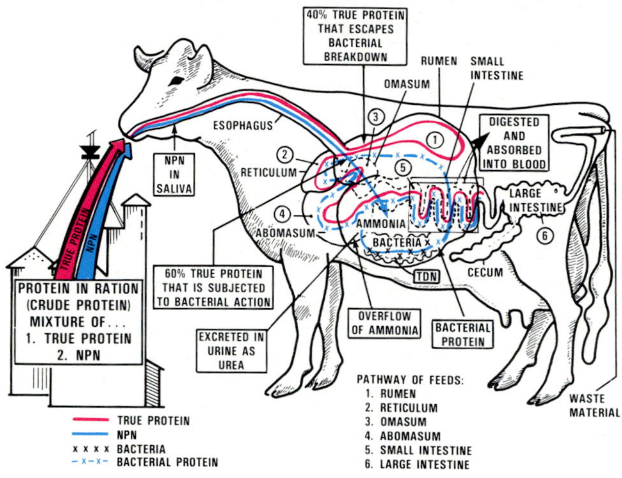

Fig. 11-2. Schematic summary of protein utilization by ruminants, using the cow as an example.

Ruminant bacteria can use various sources of nitrogen, energy (derived from fermentation), and minerals for growth. Lack of any of these factors can limit bacterial growth because requirements are interrelated. It is estimated that of the bacterial species normally present in the rumen, 80% can use ammonia as the sole source of nitrogen for growth, 26% require ammonia absolutely, and 55% can use either amino acids or ammonia. The supply of ammonia can be inadequate when either the intake of protein or the ruminal degradation of protein is low. Ammonia deficiency in the rumen reduces the efficiency of bacterial growth and may reduce the rate and extent of digestion of feed in the rumen, which, in turn, may reduce feed intake.

In addition to the microbial protein synthesized in the rumen, the animal has available another source of amino acids via the true protein in the feed which escaped breakdown in the rumen and arrived at the small intestine. On the average, as indicated in Fig. 11-2, approximately 40% of the true protein escapes degradation in the rumen. Under most production situations, the microbial protein synthesized plus the escape of dietary protein is adequate to meet the animal's protein requirement. However, there are practical production instances, such as rapidly growing, young ruminants and ruminants at peak lactation, in which microbial protein plus the normal dietary protein escaping ruminal breakdown may not be adequate to provide for optimal growth and feed efficiency. Present technology is being geared to manipulate the quantity of dietary protein rumen fermentation or breakdown, thereby increasing the supply of protein (amino acids) to the small intestine. In addition to increasing the amount of bypass protein that escapes degradation in the rumen, two other requisites are important: (1) The ruminant must be able to digest the bypass protein in the small intestine, and (2) the bypass protein must provide the amino acids that the animal needs.

The following definitions are pertinent to the discussion that follows:

Bypass protein (protected protein, escaped protein) is the feed protein which escapes digestion in the rumen and which passes into the lower digestive tract where it is digested and absorbed. Because all bypass protein resides in the rumen for a period of time, the term *bypass protein* is not factual; the terms *protected protein* and *escaped protein* are more accurate.

Degradability of protein refers to the rate and extent to which a feed protein may be broken down in the rumen. This is important for two reasons: (1) Breakdown in the rumen supplies the microbes with peptides, amino acids, and ammonia; and (2) the nondegraded portion is swept out of the rumen to the abomasum and small intestine for possible breakdown there and absorption as peptides and amino acids. The degradability of protein is affected by a number of factors, including: (1) solubility of the various nitrogen fractions in the feed, (2) the particle size of the ingested material (small particles pass out of the rumen more quickly), and (3) the level of feeding (a high level of feeding pushes feed through the rumen more rapidly than a low level of feeding). Also, the ionophores (lasalocid [Bovatec] and monensin [Rumensin]), are very effective in decreasing the breakdown of natural protein by rumen bacteria (enhancing rumen bypass). (See Chapter 13, Section on "Ionophores.")

Urea (a source of non-protein N) is 100% degraded. Casein, a very high-quality protein, is 90% degraded. However, fish meal, which is also a high-quality protein, is not degraded to any great extent in the rumen. Upon passing into the intestinal tract, where it is degraded rather completely, it supplies the animal with a great abundance of amino acids, such as lysine and methionine, which otherwise might be limiting.

CAUTION: Merely increasing the amount of protein escaping degradation in the rumen may not result in improved animal performance for either or both of the following reasons:

1. It may not be digested in the small intestine. For example, feather meal, unless hydrolized in processing, escapes rumen degradation but is poorly digested in the small intestine.

2. It may not provide the amino acids which the animal needs for the intended function.

Solubility of protein is an estimate of the protein that breaks down rapidly in the rumen. If a protein is extremely insoluble, it is less likely to be degraded. For example, zein, a protein found in corn, is extremely insoluble, with 40 to 60% escaping ruminal degradation. On the other hand, casein (milk protein) is highly soluble and is almost totally degraded in the rumen.

Microbial protein is protein produced by the microbes in the rumen. This involves a wide variety of microbe types including many species of anaerobic bacteria, protozoa, and even fungi, along with ammonia or, in some cases, peptides or amino acids. Ammonia may be provided in many ways, but it must be available for the microbes when needed and at the levels needed. The common recommendation in the use of NPN is that it can provide up to one-third of the total protein.

With ruminant (cud-chewing) animals, the amino acid dietary need is complicated. Swallowed food goes into the rumen (the largest compartment of the stomach) where it is partly digested by bacteria. It is then regurgitated to the mouth to be rechewed and reswallowed. During this process, the great bulk of dietary protein fed to ruminants is transformed into microbial protein in the rumen, with the result that the amino acids originally in the feed are not the ones actually used by the ruminant animal. But a portion of the dietary protein escapes (bypasses) rumen bacterial action, and, along with the microbes, passes to the lower gastrointestinal tract. So, the ruminant has two sources of protein: microbial protein and bypass protein, with the former predominating.

Until recent years, it was thought that ruminants were capable of producing the essential amino acids needed for microbial protein in sufficient amounts to meet all their needs, including those of rapid growth or peak milk production. This was likely true until improved genetics, nutrition, and management resulted in production increases demanding more protein (amino acids) than microbial synthesis could produce. Hand in hand with this new awareness, it was found that microbes were often wasteful. Much of the nitrogen liberated in the degradation of dietary protein is synthesized into nonprotein products. For example, it has been estimated that 20% of the nitrogen found in the crude protein fraction of ruminal microorganisms is actually nitrogen from nucleic acids. This fraction of the crude protein is essentially wasted as the nucleic acids are, for the most part, degraded to form urea in the postruminal section of the gastrointestinal tract. Additionally, the true protein fraction from ruminal microorganisms is not as digestible as the dietary protein in many cases.

• **Rumen bypass**—Ferguson, of Australia, opened up a new and exciting area of research designed to enhance rumen bypass. He demonstrated that formaldehyde-treated protein was protected from degradation in the rumen but was digested and absorbed in the abomasum and small intestine of ruminants.[1]

Also, in the 1960s and 1970s cumulative evidence indicated that the existing theories on crude protein were not entirely sound. Scientists in several countries reported that certain feed ingredients supported better production than others in rapidly growing and peak lactating ruminants even though the experimental rations were isonitrogenous.

During the early 1970s, too, several researchers infused proteins and amino acids directly into the abomasum (thereby escaping rumen degradation) and noted significant increases in production. From this, it was concluded that some high-producing ruminants may be operating on suboptimal levels of certain amino acids, and that if certain amino acids are deficient in the microbial protein, maximum production cannot be attained. Further, it was concluded that if these amino acids could be supplied, production will increase, and that, quantitatively, ruminants have the same essential amino acid requirements as nonruminants on a tissue level.

• **Practicality of protein bypass**—Because protein is a costly ingredient, increasing efficiency of its utilization is important. Increasing bypass protein can reduce cost because less protein is fed. This is accomplished by (1) maximizing the amounts of less expensive nonprotein nitrogen (NPN) as a source of ammonia for microbial synthesis in the rumen, and minimizing the amount of rumen degradable protein for microbial protein; and (2) maximizing the bypassing (escaping degradation) of the more expensive protein from the rumen to the intestines, where it is digested. For high-producing cows in early lactation, this calls for combining NPN and protected protein. Thus, the ammonia that is needed by the microbes in the rumen may be supplied in varying amounts by cheaper urea rather than entirely by expensive protein. Young, of Southern Illinois University at Carbondale, reported that, in studies with beef cattle, he was able to reduce the amount of supplemental soybean fed from 1.5 to 0.75 lb per head daily when the soybean was protected with 0.4% formaldehyde.[2] Young cautioned, however, that, when applying this protein-saving technique, only the preformed protein level is lowered. The microbial need for nitrogen (urea or other NPN sources *plus* some true protein) must still be provided, along with mineral, vitamin, and other feed additive levels.

Dr. R. L. Preston, Texas Tech University, Lubbock, Texas, suggests the following rule in order to improve the predictability of animal response when fed various feeds: If the crude protein (CP) level in the final ration exceeds 10% of the dry matter, all CP above this amount should be bypass protein. Thus, if the final ration is to contain 16% CP, 6 of the 16 percentage units, or 37.5% of the CP, should be in the form of bypass protein. Once these relationships have been better quantified, crude protein requirements may be lowered, especially at higher protein levels.

• **Solubilities of protein sources**—Protein solubility is an indicator of the rapidity with which protein breaks down in the rumen. This characteristic is important because the soluble proteins are more easily degraded by rumen microbes than insoluble ones. This early breakdown leaves less protein to be digested in the abomasum and small intestine, where it is of most use to high-producing animals. For example, Clark states that maximum milk yield of lactating cows can usually be attained if the ration contains 10 to 25% soluble nitrogen.[3] At this level, the amounts of soluble nitrogen provide adequate ammonia for maximum bacterial growth while still allowing large quantities of dietary protein to escape degradation in the rumen.

But solubility must be considered along with other factors. For example, ensiled and fermented feeds may have increased amounts of soluble N compared to unfermented feedstuffs, but this soluble N may not be available in the rumen.

---

[1]Ferguson, K. A., J. A. Hensley, and P. J. Reis, "The Effect of Protecting Dietary Protein From Microbial Degradation in the Rumen," *Australian Journal of Science*, Vol. 30, 1967, p. 215.

[2]Young, A. W., "Protected Proteins in Beef Cattle Diets," *Feed Management*, Dec. 1980, p. 31.

[3]Clark, J. H., "Protein Nutrition of the High-Producing Cow," *Feed Management*, Feb. 1981, p. 36.

Bypass protein values of many feeds have not been determined. However, Table 11-1 shows the percentage bypass protein of some common feeds. Note that several of the by-product feeds and dehydrated alfalfa are high in bypass protein.

Normally, the amount of protein that escapes ruminal degradation is dependent largely on the length of time the protein remains in the rumen. One can speed the rate of flow through the rumen by increasing dietary intake or frequency of feeding, or by reducing the feed particle size. Also, if large quantities of salt are fed, there is an increased water intake resulting in a faster rate of flow through the rumen. Thus, speeding feeds through the rumen ultimately increases the amount of dietary protein that reaches the lower gut.

TABLE 11-1
BYPASS PROTEIN (PERCENT OF UNDEGRADED PROTEIN) IN COMMON FEEDS[1]

Feed[2,3]	Bypass Protein (%)
Alfalfa, fresh	20
Alfalfa, artificially dehydrated, 17% protein[4]	59
Alfalfa hay, early bloom	18
Alfalfa hay, midbloom	22
Alfalfa hay, full bloom	28
Alfalfa hay, mature	35
Alfalfa cubes	35
Alfalfa stems	40
Alfalfa silage[4]	23
Alfalfa silage, wilted	22
Bakery product, dried	20
Barley[4]	27
Barley, flaked[4]	67
Barley silage	25
Barley silage, mature	35
Beet pulp, wet	30
Beet pulp, dried	35
Beet pulp, wet with molasses	25
Beet pulp, dried with molasses	25
Blood meal	82
Brewers' grains, wet	57
Brewers' grains, dried[4]	49
Coconut meal[4]	63
Corn, yellow dent[4]	52

Feed[2,3]	Bypass Protein (%)
Corn, high moisture[4]	80
Corn, steam flaked[4]	68
Corn and cob meal	50
Corn cobs	50
Corn gluten feed[4]	25
Corn gluten meal[4]	55
Corn silage, milk stage	25
Corn silage, mature, well eared	40
Corn whole plant, pelleted	45
Corn fodder	45
Cottonseed meal, screw-pressed, 41% protein[4]	50
Cottonseed meal, prepressed*	36
Cottonseed meal, solvent, 41% protein[4]	41
Cottonseed hulls	40
Distillers' grain, barley	60
Distillers' grain, corn	57
Distillers' grain, with solubles	47
Distillers' stillage, corn	65
Feathermeal, hydrolyzed[4]	71
Fescue, Kentucky 31, fresh	30
Fescue, Kentucky 31, hay, early bloom	30
Fescue, Kentucky 31, hay, mature	35

Feed[2,3]	Bypass Protein (%)
Fish meal[4]	60
Grass silage[4]	29
Linseed meal, solvent[4]	35
Meat and bone meal[4]	49
Meat meal	63
Oats[4]	17
Oat groats	25
Oat middlings	20
Oat silage	25
Orchardgrass, fresh, immature	25
Orchardgrass hay	30
Peanut meal, solvent[4]	25
Peas[4]	22
Rapeseed meal, solvent (canola meal)[4]	28
Rye[4]	19
Ryegrass, fresh[4]	48
Sorghum grain (milo), ground	60
Sorghum grain (milo), flaked	50
Sorghum silage	50
Soybeans, whole[4]	26
Soybean meal, solvent, 44% protein	26
Soybean meal, solvent, 49% protein	23
Soybean meal, dried 248°F *(120°C)*[4]	59

Feed[2,3]	Bypass Protein (%)
Soybean meal, dried 266°F *(130°C)*[4]	71
Soybean meal, dried 284°F *(140°C)*[4]	82
Soybean meal, HCHO (formaldehyde treatment)[4]	80
Sunflower meal, solvent[4]	26
Sunflower meal, with hulls	40
Timothy, fresh, pre-bloom	20
Timothy hay, early bloom	25
Timothy hay, full bloom	35
Timothy silage	25
Triticale	25
Wheat[4]	22
Wheat, hard	35
Wheat, soft	35
Wheat, sprouted	20
Wheat bran[4]	29
Wheat middlings[4]	21
Wheat mill run	20
Wheat shorts	20
Wheat straw	80
Wheat straw, ammoniated	25

[1]In addition to the bypass protein values given in this table, more complete compositions of these and other feeds are given in Section V of this book.

[2]Feeds without an asterisk obtained from *Feedstuffs,* Oct. 12, 1987, pp. 18–26, by Dr. R. L. Preston, Animal Science Department, Texas Tech University, Lubbock; with the permission of Dr. Preston and *Feedstuffs.*

[3]Feeds followed by an asterisk obtained from *Ruminant Nitrogen Usage,* National Research Council, Washington, D.C., 1985, p. 33, Table 6.

[4]Adapted by the authors from *Nutrient Requirements of Dairy Cattle,* 6th rev. ed., NRC, National Academy Press, 1988, pp. 113–115, Table 7-3.

• **Methods of protecting proteins from ruminant degradation**—Scientists have been developing ways of manipulating ruminant protein digestion to make more of the amino acids in the feed available to the animal after bacterial breakdown in the rumen. This technique is known as bypass proteins or protected proteins. Such proteins actually bypass the rumen, and their amino acids are not broken down by bacteria. Among the methods of protecting proteins from ruminant degradation are the following:

1. **Naturally protected proteins.** A few proteins are naturally protected by heat during normal processing and have good bypass characteristics. These include corn gluten meal, brewers' grains (wet or dry), distillers' grains (wet or dry), extruded or roasted soybeans, extruded soybean meal, meat meal, blood meal, fish meal, and hydrolyzed feather meal. Normally, solvent extracted soybean meal contains 28% bypass protein; by comparison, a modified expeller processed soybean meal contains 65% bypass protein.

2. **Heat and pressure treatment.** Heating protein feeds, usually under pressure, creates cross-linkages of free amino groups within and between protein molecules. In addition to altering the solubility properties of the protein, these cross-linkages reduce the surface of the protein that comes into contact with enzymes, thereby blocking the site of enzymatic attack. *Caution:* Overheating of protein may produce enzymatic browning which renders the protein indigestible in both the rumen and small intestine.

3. **Treatment of protein with tannins.** The mechanism of this treatment is not clearly understood, but it has been hypothesized that the tannins form hydrogen bonds with proteins, thereby protecting the protein from bacterial degradation. As the treated protein travels down the digestive tract, the acidity of the fluids is altered and the tannin-protein complex dissociates.

4. **Treatment of protein with formaldehyde or other aldehydes, with or without heat.** This is the most widely used technique of protein protection. The aldehyde reacts

with free amino groups and N-terminal groups to form Schiff bases and cross-linkages between protein chains. This process decreases the solubility of the protein and protects it from bacterial degradation. Once the treated protein enters into a highly acid environment, the reaction is reversed, and the protein is degraded. Combined aldehyde and heat treatment is also being used.

5. **Lipid (fat) treatment.** In this system, fat is emulsified with a protein, then the protein is treated with formaldehyde. The product cannot be digested in the rumen due to the formaldehyde linkage on the protein. However, the pH of the abomasum is such that the formaldehyde linkage of the protein is broken, with the result that the product is then digested in the same manner as in monogastric animals. Also, and most significant, by use of the protein protected fat system, it is possible to alter the ratio of the saturated to the unsaturated fatty acids in beef fat and in butterfat.

Lipid coating of protein with certain tannins is also being used.

6. **Complexing with bentonite clay.** Bentonite, which is a naturally occurring substance commonly used in feeds as an anticaking agent and pelleting aid, is being used for protecting protein.

7. **Encapsulation of amino acids.** This involves coating (protecting) specific amino acids for ruminal bypass. Among the materials that have been used as encapsulating agents are kaolin, tristearin, a pH sensitive copolymer, cellulose propionate-3-morpholino butyrate, and imidamine polymers. Methionine and lysine are considered to be among the first limiting amino acids in most rations.

8. **Use of amino acid analogs.** If these analogs are to be used successfully in ruminant rations, they must be able to escape ruminal degradation and yet be absorbed in a biologically active form. One analog that has produced variable, but promising, results is methionine hydroxy analog, or MHA. MHA is less soluble in ruminal fluids than methionine. As a result, much of the MHA escapes degradation in the rumen.

9. **Increase the microbial metabolism in the rumen.** It has been speculated that another way to improve protein utilization is to increase the rumen bacteria population. To accomplish this the following conditions would need to exist: (1) an adequate source of energy for maximum utilization of ammonia; and (2) an environment in the rumen that stimulates bacterial growth, especially a pH ranging from 5.6 to 6.7.

10. **Adding ionophores.** Bovatec (lasalocid) and Rumensin (monensin), two ionophores, are effective (a) in decreasing rumen degradation of natural protein and (b) in increasing rumen bypass—perhaps by as much as 25%. (See Chapter 13, section on "Ionophores.")

11. **Non-enzymatic browning.** Through a process known as non-enzymatic browning, University of Nebraska researchers have increased the bypass content of soybean meal by as much as 2½ times. The process involves mixing the soybean meal, or other protein source, with xylose sugar, then heating the mixture to 200 to 250°F. A patent has been applied for on the process.

- **How to measure the degree of protection**—Several methods of measuring protein degradability are being used; among them, (1) determining solubility values, obtained by extraction with hot or cold water, rumen fluid, buffers, alkaline or acid solutions (The disadvantage to this technique: Solubility is not synonymous with degradability); (2) using duodenally cannulated animals to obtain estimates of undegraded protein leaving the rumen (the disadvantages of this technique: two markers are needed—flow and microbial, the animal is surgically altered, intakes are seldom normal, and the metabolizability of the protein is not measured); and (3) measuring animal performance (the disadvantages of this technique: the protein sources may supply rumen degradable protein or other nutrients which affect animal performance, and animal variation makes measurement difficult).
- **Commercial bypass proteins**—Many commercial feed companies have developed brand name protected protein products, some of which are patented. Most of these high bypass supplements consist of a combination of protein and urea, with naturally protected feeds serving as the primary sources of bypass protein.
- **Future of bypass protein and amino acids**—The future of bypass protein is bright, but much fine tuning of the concept must come, through experiments and experiences.

Protein degradability is quite variable (1) among feedstuffs, and (2) within feedstuffs—depending on the amount and kind of bacteria in the rumen and the length of time the protein stays in the rumen.

Protein bypass produces rather dramatic responses in young, rapidly growing ruminants and during peak lactation. Researchers believe that it may prove useful in lightweight, stressed calves, newly-arrived feedlot cattle, bulls, and cattle with large mature size; all of which may have protein requirements that may not be met from conventional microbial protein synthesis. It is doubtful that protected proteins will be beneficial to finishing cattle weighing more than 750 to 800 lb which are fed high-grain rations.

There is reason to believe that protected proteins will be as effective for sheep as for cattle. Ohio researchers have found that milk production is increased 27% in ewes nursing twins when blood meal replaced soybean meal as the protein supplement in a corn silage based ration; and that weight gains of each twin lamb increased 24% by replacing 7.7% soybean meal with 5.4% fish meal in the ration.[4]

As with most feed ingredients, the future use of protected proteins and amino acids will be dictated by economics. If supplement costs can be reduced by feeding less protein or amino acids in a protected form and supplying cheaper NPN for the microbes, protected proteins and amino acids may become major feed ingredients.

Eventually, bypass products will be developed that will either cut costs or increase production—or both.

(Also see Chapter 3, section on "Ruminant Digestive System.")

IMPORTANCE OF PROTEIN FEEDS

Since protein feeds are usually among the more costly components of animal rations, it is important to provide enough protein for the animal to perform its assigned function, but to avoid feeding more than is necessary. The primary functions of protein feeds are to supply (1) those

[4]"Bypass Protein to Boost Milk Output in Sheep," *Feed Management*, Vol. 37, No. 11, 1986, p. 47.

amino acids not provided in adequate amounts by the cereal portion of nonruminant rations, or (2) nitrogen precursors of microbial protein in the case of ruminants. Some animals are equipped to utilize a large amount of protein for energy; for example, mink and fish may obtain more than half of their energy from protein. But usually there are more economical sources of energy than protein.

Protein supplements may be further categorized according to source of origin as (1) plant proteins, (2) animal proteins, (3) nonprotein nitrogen, and (4) single-cell (microbial) proteins.

Plants provide 79% of the protein feeds used in livestock rations (see Table 6–3). Most protein feeds of plant origin consist of processed oilseeds. Mill products generally make up the remainder of the plant protein feeds. Many protein feeds of animal origin are derived from sources that are considered unsuitable for human consumption. Also, animal protein feeds are generally more expensive than those of plant origin. It is noteworthy that in the 14-year period from 1970 to 1984 (see Table 11–2), the kinds and amounts of protein feeds used in livestock rations remained relatively constant.

PLANT PROTEINS

The bulk of the protein for ruminants (cattle, sheep, goats), whose requirements for specific amino acids are met by microbial fermentation of the material in the forestomach (rumen), comes from plant sources.

The protein content of plants varies considerably from one type to another. Even within the same plant, there is considerable variation from one stage of maturity to another or from one part of the plant to another. Proteins in plants are

TABLE
COMMERCIAL FEEDS FED TO LIVESTOCK IN THE

Year Beginning October	Oilseed Cake and Meal						Animal Proteins			
	Soybean[3]	Cottonseed	Linseed	Peanut	Sunflower	Total	Tankage and Meat Meal	Fish Meal and Solubles	Dried Milk[4]	Total
1970	13,467	1,693	258	173	99	15,690	2,039	609	260	2,908
1975	15,612	1,266	94	313	119	17,404	2,001	508	147	2,656
1980	17,591	1,636	103	94	44	19,468	2,458	379	104	2,941
1984	19,480	1,732	120	122	338	21,792	2,889	228	98	2,987

[1]From USDA.

[2]Other mill products that are not listed include screenings, hominy, and oat feeds.

[3]Includes use in edible soy products and shipment to U.S. territories.

[4]Includes commercial dried milk products and noncommercial milk products.

primarily associated with the tissues that are actively metabolizing, such as the leaves, centers of growth, and the seeds which must have the potential to metabolize nutrients at certain critical stages of their development.

Even though they are not especially high in protein by comparison with other feedstuffs, the vegetative portions of many plants supply an extremely large portion of the protein in the total ration of livestock, simply because these feeds are consumed in large quantities. Needed protein not provided in these feeds is commonly obtained from one or more of the oilseed by-products—soybean meal, cottonseed meal, linseed meal, peanut meal, safflower meal, sunflower meal, rapeseed meal, or coconut (copra) meal. The protein content and feeding value of these products vary according to the seed from which they are produced, the geographical area in which they are grown, the amount of hull and/or seed coat included, and the method of oil extraction used. Sometimes, the unprocessed seed is used to provide both a source of protein and a concentrated source of energy. The unprocessed oil-bearing seeds are especially high in energy because of the oil that they contain.

Additional plant proteins are obtained as by-products from grain milling, brewing and distilling, and starch production. Most of these industries use the starch in grains and seeds, then dispose of the residue, which contains a large portion of the protein of the original plant seed.

Oilseed Meals

Several rich oil-bearing seeds are produced for vegetable oils for human food (oleomargarine, shortenings, and salad oil), and for paints and other industrial purposes. In processing these seeds, protein-rich products of great value as

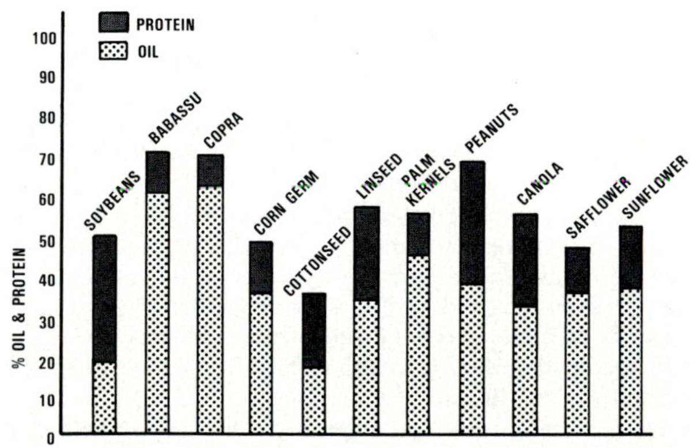

Fig. 11–3. The oil and protein content of commonly used protein supplements. (Adapted by the authors from USDA sources.)

livestock feeds are obtained. Among such high-protein feeds are soybean meal, coconut meal, cottonseed meal, linseed (flax) meal, peanut meal, canola meal (rapeseed meal), safflower meal, sesame meal, and sunflower seed meal.

Oil is extracted from these seeds by one of the following basic processes or modifications thereof: solvent extraction, hydraulic extraction, or expeller extraction. For purpose of illustration, all three processes are discussed relative to soybeans, since soybeans are the most widely used oilseed in the United States, and since the principles involved in processing all oilseeds are essentially the same. However, it should be noted that very little soybean meal is produced today via hydraulic or expeller extraction.

Although crude protein determination is the most frequently used chemical assay for comparing oilseed meals, the real quality measurement goes far beyond; there is more to oilseed meals than just protein. The availability and digestibility of amino acids, the concentration of minerals and vitamins, and the level of moisture, fiber, and urease can all affect oilseed meal performance in a livestock ration. Also, the feeding value of oilseed meals varies according to the method of extraction. Oilseeds vary in oil content, both between varieties and within a variety. Meals produced by mechanical extraction of the oil from the seed contain more fat and fiber and a lower percentage of protein than those produced by solvent extraction (hexane).

SOYBEAN MEAL

Soybean meal, processed from an oil-bearing seed that originated in the Orient, has the highest nutritive value of any plant protein source. It is now the most widely used protein supplement in the United States.

11-2
UNITED STATES (1,000 SHORT TONS)[1]

Mill Products[2]								
Wheat Millfeeds	Gluten Feed and Meal	Rice Millfeeds	Brewers' Dried Grains	Distillers' Dried Grains	Dried and Molasses Beet Pulp	Alfalfa Meal	Total	Total Commercial Feeds
4,499	1,237	436	361	382	1,509	1,584	10,008	28,606
4,933	1,477	547	321	400	1,861	1,569	11,108	31,168
5,114	839	778	320	499	1,310	1,096	9,956	32,365
5,556	2,065	607	156	967	1,207	891	11,449	36,228

Processing of soybeans, and the production of soybean oil and meal, began in a small way in the early twenties; but the real impetus came in 1928 when the American Milling Company agreed to purchase on contract all soybeans produced from 50,000 acres in Illinois at a guaranteed minimum of $1.37 per bushel. Following this, soybean acreage continued to expand, and with it the production of soybean meal. Soybean meal passed cottonseed meal in tonnage during World War II when soybean acreage was increased greatly to provide needed oil for use in human foods and for industrial purposes. Today, the United States produces about 55% of the world's soybeans.

Fig. 11-4. Iowa soybean field in August. (Courtesy, American Soybean Assn., St. Louis, Mo.)

Fig. 11-5. Close-up of soybean pods. (Courtesy, American Soybean Assn., St. Louis, Mo.)

Fig. 11-6. Harvesting soybeans. (Courtesy, Deere & Company, Moline, Ill.)

Processing Soybean Meal

Raw whole soybeans are seldom fed to livestock. Rather, they are processed and utilized as either oil meal or whole, heat processed beans. Today, most soybeans are fed in oil meal form. But, as a result of increased processing costs, the use of heat processed whole soybeans may play an increasing role in livestock nutrition. Whole soybeans must be heated to deactivate the antinutritional factors of the seeds; but once properly heated, the whole seed provides a valuable source of energy and protein. Many of the problems associated with feeding this high-oil feed, such as soft pork, have been overcome by adjustments in feed formulations. (See Chapter 23, Feeding Swine, section headed "Full-Fat Soybeans.")

Fig. 11-7. A soybean processing plant in Iowa. (Courtesy, American Soybean Assn., St. Louis, Mo.)

In the past, oil was extracted by the solvent, hydraulic, and expeller processes (see Fig. 11-8). But, today, almost all soybeans are solvent extracted. Soybean meal normally contains 41, 44, 48, or 50% protein, depending on the amount of hull removed. Because of its well balanced amino acid profile, the protein of soybean meal is of better quality than other protein-rich supplements of plant origin. However, it is low in calcium, phosphorus, carotene, and vitamin D.

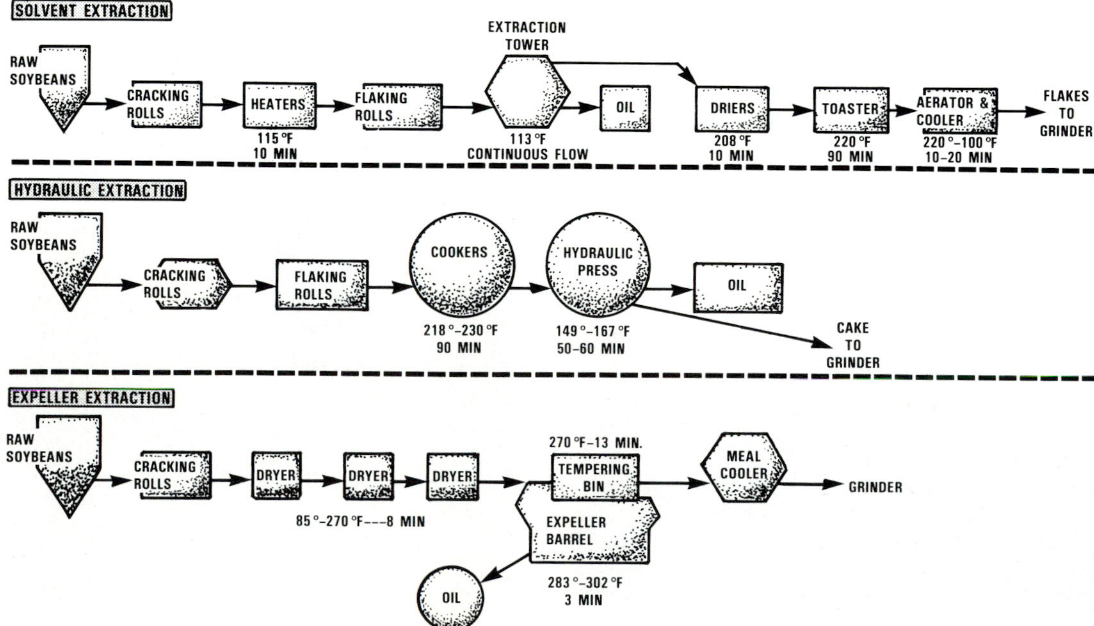

Fig. 11-8. Processing soybean oil meal, contrasting the three methods.

SOLVENT EXTRACTION

In this method, the soybeans are first cracked, then heated to 140°F for about 10 minutes. After the cracked seeds are heated, they proceed through a series of grinding rollers where they are flaked. The flakes are allowed to cool to about 113°F and are then conveyed to the extraction equipment where the oil is removed by the solvent—hexane.

The extracted flakes then proceed to driers where the solvent is completely volatilized and removed. From the drier, the flakes are conveyed to a toaster, thence they are cooled and ground.

HYDRAULIC EXTRACTION

In the hydraulic extraction procedure, raw soybeans are cracked, ground, and flaked. The flakes (called meats) are then transported to cookers where they are exposed to both dry and steam heat. The cooking stage takes about 90 minutes.

After cooking, the meats are formed into cakes and wrapped in heavy cloth whereupon they are placed in hydraulic presses for the mechanical extraction of the oil. This procedure takes about 1 hour. Following extraction, the extracted cakes are ground. Hydraulic press cake may have 5 to 8% residual oil.

Since this form of extraction is labor intensive and inefficient in the removal of oil, very little meal is produced this way.

EXPELLER EXTRACTION

In this type of extraction, raw soybeans are initially cracked and dried to about 2% moisture. The dried soybeans are then transported hot to a steam jacketed tempering device which is directly above the expeller apparatus. The tempering apparatus stirs the cracked soybeans for about 10 to 15 minutes so that the seeds are heated uniformly.

From the tempering bin, the soybeans are fed into an expeller barrel (screw presses). A central revolving worm shaft creates pressure within the expeller barrel thereupon extracting the oil from the ground soybeans. The extracted soybeans leave the expeller in the form of flakes, which are subsequently ground.

The expeller process tends to extract less oil than the solvent process; consequently, it is now used less frequently. Generally, expeller extracted soybean meal contains 4 to 5% oil while solvent extracted soybean meal contains less than 1%. However, expeller processed soybean meal is higher in bypass protein than solvent extracted meal.

CANOLA MEAL (RAPESEED MEAL)

Canola was created from specially selected rapeseed by Canadian plant scientists in the 1970s. The old rapeseed was high in glucosinolate compounds, which, when fed to animals at high levels, made for palatability problems and lowered performance due to goitrogenic action. Canola changed this. The new canola is low in glucosinolates in the meal, and low in erucic acid (a long-chain fatty acid) in the oil. Canola is grown mainly in Canada, but it is increasing in the United States.

Fig. 11-9. Canola (rapeseed) plant in flower. (Courtesy, Canola Council of Canada, Winnipeg, Manitoba, Canada)

Canola meal averages about 36% crude protein and its amino acids compare favorably with soybean meal. When the price is favorable, canola meal may be used as a protein

Fig. 11-10. Cattle feeding on canola meal, a good protein supplement. (Courtesy, Canola Council of Canada, Winnipeg, Manitoba, Canada)

supplement for all classes of livestock and poultry. It may be used at maximum levels of 20% in most rations, but the amount in the total ration should be limited to about 10% for layers and breeding ducks, 15% for breeding turkeys, and 12% for young pigs and breeding swine.

CAUTION: The lowering of the erucic acid and glucosinolate in rapeseed (canola) has proceeded at different rates in different countries. But the change is nearly complete in Canada, the leading nation, which produces about 20% of the world crop.

COCONUT MEAL (COPRA MEAL)

This is a by-product from the production of oil from the dried meats of coconuts. The oil is generally extracted by either (1) the hydraulic process, or (2) the expeller process. Coconut meal averages about 21% protein content. The quality of the protein is not high; hence, its use should be restricted in nonruminant rations. If copra meal is fed to nonruminants, it should be supplemented with the amino acids lysine and methionine.

The lipid component of copra meal is very low in unsaturated fatty acids. Hence, the feeding of copra meal produces firm body fat in swine. Also, dairy producers use copra meal to produce a pleasant flavored, rather hard (highly saturated) butterfat. The maximum recommended level in dairy rations is 3.3 to 6.5 lb per day. Higher amounts tend to produce tallowy butter.

No copra meal is produced in the United States and imports have practically ceased since 1984.

COTTONSEED MEAL

According to historical records, oil was first successfully extracted from cottonseed in a New Orleans plant in 1885. However, due to strong prejudice, most of the meal continued to be sold as fertilizer at low prices until sometime after 1900. Stock raisers who had ventured to feed the product attributed blindness and abortion to its use. Eventually, the experiment stations came to the rescue of the manufacturers and demonstrated that cottonseed meal would give excellent results if fed in properly balanced rations.

Fig. 11-11. Cottonseed meal is derived from the cotton plant. This shows a mature cotton boll with lint attached to the seed, ready for harvest. The oil is extracted from the seed, leaving the meal and hulls for livestock feeds. (Courtesy, National Cotton Council of America, Memphis, Tenn.)

Among the U.S. oilseed meals, cottonseed meal ranks second in tonnage to soybean meal. The processing steps in making cottonseed meal are as follows: (1) cleaning the seeds; (2) dehulling the seeds; (3) crushing the kernels; (4) extracting the oil by either (a) mechanical (or screw press), (b) solvent, or (c) partially mechanically extracted and then solvent extracted process; and (5) grinding the remaining residue or cake, thus forming cottonseed meal. Today, 80% of all processed cottonseed is solvent or combined mechanical-solvent extracted.

The protein content of cottonseed meal varies from about 22% in meal made from undecorticated seed to 50% in flour made from seed from which the hulls have been removed completely. Thus, by screening out the residual hulls, which are low in protein and high in fiber, the processor is able to make a cottonseed meal of the protein content desired—usually 36, 41, 44, and 48%. The protein content of cottonseed meal varies with the geographical location in which it was grown. Meals manufactured from cottonseed produced on the West Coast generally contain higher protein levels than those produced throughout the rest of the United States.

For monogastric animals, cottonseed meal is low in lysine and tryptophan and deficient in vitamin D, carotene (vitamin A value), and calcium. Also, it contains a toxic substance known as gossypol, varying in amounts with the seed and the processing. But, it is rich in phosphorus.

Today, glandless cottonseed, free of gossypol, is being grown and improved. Someday, meal made from it may replace conventional cottonseed meal in nonruminant rations and alleviate (1) many of the restrictions as to levels of meal, and (2) the need to add iron to tie up free gossypol. University of California researchers report (1) that glandless cottonseed meal caused no problems in egg production, hatchability, or egg yolk discoloration in laying hens; and (2) that glandless cottonseed meal supplemented with 0.3% or more lysine produced growth in broilers not statistically different from that obtained by feeding a practical starting ration.[5]

Today, considerable whole cottonseed, which averages about 21.8% crude protein and 87% TDN for ruminants, is being fed to ruminants, as both a protein and energy feed source (see Chapter 10, section on "Cottonseed").

LINSEED MEAL

While the feeding value of linseed meal was recognized prior to 1900, most of the output was shipped to Europe. As late as 1917, 331,000 tons out of the 500,000 tons produced in this country were sold abroad. Today, less than 100,000 tons of linseed meal is fed annually in the United States.

Linseed meal is a by-product of flax, a fiber plant which antedates recorded history. In this country, most of the flax is produced as a cash crop for oil from the seed and the resulting by-product, linseed meal. Practically none of the U.S. flax crop is grown for fiber, for it is more economical to import it from those countries where cheaper labor is available.

Most of the nation's flax is produced in North Dakota, South Dakota, and Minnesota. Normally, an additional

[5]Ryan, J. R., *et al.*, "Glandless Cottonseed Meal for Laying and Breeding Hens and Broiler Chicks," *Poultry Science*, Vol. 65, No. 5, May 1986, p. 949.

quantity of seed is imported and processed in U.S. plants.

The oil is extracted from the seed by either of two processes: (1) the mechanical process, or (2) the solvent process. Horse caretakers prefer old process linseed meal, because it is palatable, and it imparts glossiness to the hair coat due to the higher oil content.

Linseed meal is the finely ground residue (known as cake, chips, or flakes) remaining after the oil extraction. It averages about 35% protein content (33 to 37%), about half as much as tankage. For swine, the proteins of linseed meal do not effectively supplement the deficiencies of the cereal grains; since linseed meal is also low in the amino acids lysine and tryptophan. Also, linseed meal is lacking in carotene and vitamin D, and is only fair in calcium and the B vitamins. It is laxative; hence, it may be used to regulate the bowels.

LUPINE

Lupine is an ancient legume crop in which there is new interest. Sweet (or white) lupine produces less harmful alkaloids than the other approximately 100 varieties. Also, it grows well in marginal soils and tolerates both drought and cold temperature; conditions not suited to soybeans. The beans may be fed to animals as a protein supplement without heating.

PEANUT MEAL, AND PEANUT MEAL AND HULLS

Fig. 11-12. A Georgia farmer inspecting his crop of peanuts. (Courtesy, USDA)

Peanut meal, a by-product of the peanut industry, is ground peanut cake, the product which remains after the extraction of part of the oil of peanuts by pressure or solvents. It is a palatable, high-quality vegetable protein supplement used extensively in livestock and poultry feeds. Peanut meal ranges from 41 to 50% protein and from 4.5 to 8% fat. It is low in methionine, lysine, and tryptophan, and low in calcium, carotene, and vitamin D.

Peanut meal and hulls is ground peanut meal with added hulls, or the ground by-product remaining after extraction of part of the oil from whole or unshelled peanuts. Since about one-fourth of peanut meal and hulls consists of peanut hulls, it is high in fiber, averaging about 22.5%.

Since peanut meal tends to become rancid when held too long—especially in warm, moist climates—it should not be stored longer than 6 weeks in the summer or 2 to 3 months in the winter.

SAFFLOWER MEAL

India produces about half of the world safflower crop. In the United States, it is a minor crop—only about 100,000 tons of seed being produced annually.

A large proportion of the safflower seed is composed of hull—about 40%. Once the oil is removed from the seeds, the resulting product contains about 60% hulls and 18 to 22% protein. Various means have been tried to reduce this high-hull content. Most meals contain seeds with part of the hull removed, thereby yielding a product containing 35 to 50% protein and 10 to 15% fiber.

Dehulled (decorticated) safflower meal can be fed to swine and poultry, but it should be supplemented with soybean or meat meal to provide adequate levels of lysine and the sulfur-containing amino acids. High-fiber safflower meal (above 15%) should be fed only to ruminants.

SESAME MEAL

Worldwide, sesame ranks sixth among oilseed crops in the production of edible oilseeds and tenth in tonnage of vegetable oils. But, little sesame is grown in the United States despite the fact that it is one of the oldest cultivated oilseeds. The oil meal is produced from the entire seed. Solvent extraction yields higher protein (45%) but lower fat levels (1%) than either the screw press or hydraulic methods which produce meals containing about 38% protein and 5 to 11% oil.

SUNFLOWER SEED MEAL

The Russians pioneered in plant breeding research which resulted in sunflower seed which yielded more than 40% oil, and, in addition, increased the seed yield per acre. The improved Russian cultivars were brought to the United States in the 1960s.

Worldwide, sunflowers rank second as an oilseed crop, being exceeded only by soybeans. Twenty-eight percent of the world sunflower seed is produced in the U.S.S.R. In

Fig. 11-13. The Cargill crushing facility at Riverside, North Dakota is flanked by a field of sunflowers in full bloom. (Courtesy, National Sunflower Assn., Bismarck, N.D.)

the United States, about 3 million acres of sunflowers are grown annually, with North Dakota the leading state, followed by South Dakota and Minnesota.

Sunflower meal—a relative newcomer to the commercial oilseed industry in the United States—is rapidly gaining acceptance as a high-quality source of plant protein. It competes with other oilseed meals. The oil meal varies considerably in nutrient content, depending on the extraction process and whether the seeds are dehulled. Meal from prepressed solvent extraction of dehulled seeds contains about 44% protein, as opposed to 28% for whole seeds. The hulls are not easily removed, with the result that sunflower meal is high in fiber, ranging from 15 to 24%. Meal produced from seed that has not been dehulled has a high fiber content; hence, it should be fed only to ruminants. When sunflower meal is used in nonruminant feeds, it should be combined with high-lysine supplements, such as meat scrap or fish meal.

FEEDING VALUE OF OILSEED MEALS

The feeding value of oilseed meals varies with the composition of the seed and the method of oil extraction. In general, meals produced by mechanical extraction of the oil from the seed contain more oil and fiber and a lower percentage of protein than those produced by solvent extraction using hexane.

TABLE 11-3
NUTRITIONAL CONSIDERATIONS OF OILSEED MEALS

Meal	Form of Extraction	Feeding Value[1]					Antinutritional Factors	Comments
		Cattle	Sheep	Swine	Poultry	Horses		
Soybean meal	Solvent Expeller Hydraulic	***100***	***100***	***100***	***100***	***100***	Urease and a trypsin inhibitor may be present in meals prepared under low temperature.	Most valuable source of vegetable protein. Highly palatable.
Canola (rapeseed) meal	Prepressed solvent Direct solvent Expeller	88	88	85–90	80		Goitrogenic glycosides.	Ruminants are less susceptible to toxic effects. Feeding values to left are for low glucosinolate rapeseed meals.
Coconut (copra) meal	Mechanical Hydraulic	90–100	90–100	50	50	90–100		Used in dairy rations to increase milk fat and produce firm butter. Very low in lysine. Too high in fiber for widespread use in nonruminant rations.
Cottonseed meal	Solvent Mechanical Prepressed solvent	100	100	85	85	100	Gossypol; cyclopropenoid fatty acids.	Excellent source for ruminants. Very low in lysine. Free gossypol levels should not exceed 0.03% in growing chicken rations. Glandless cottonseed meal (gossypol-free) holds considerable promise for the future.
Linseed meal	Solvent Mechanical	95	90	80	80	100	Prussic acid may pose problems when the meal is manufactured under low temperatures.	Reputation for producing bloom on the coats of livestock. Laxative due to oil content.
Peanut meal	Solvent Mechanical	100	100	95	95	100		High-protein quality. Palatable.
Safflower meal	Solvent Mechanical	40–45	40–45	45–50	45–50			Undecorticated meal is used for ruminants only. Not very palatable.
Sesame meal	Solvent Mechanical	90–95	90–95	90–95	95–100	90–95		Mildly laxative. May produce soft butter or pork if fed in large amounts. Rich in methionine and arginine.
Sunflower meal	Solvent Mechanical	95–100	100	90–95	95–100	100		Decorticated meal can be effectively fed to all livestock. Undecorticated should be restricted to ruminants.

[1]Comparison of relative feeding value pound for pound with soybean meal, 41% protein (100). The oilseed meals used for comparison are: (1) canola (rapeseed) meal, 37% protein; (2) coconut meal, 21% protein; (3) cottonseed meal, 41% protein; (4) linseed meal, 35% protein; (5) peanut meal, 40% protein; (6) safflower meal with hulls, 20% protein; (7) sesame meal, 40% protein; and (8) sunflower meal, 39% protein.

TABLE
ESSENTIAL AMINO ACID CONTENT OF SOME OILSEED MEALS

Meal	Process	Crude Protein	Arginine		Glycine		Histidine		Isoleucine	
		(%)	(% DM)	(% CP)	(% DM)	(% CP)	(% DM)	(% CP)	(% DM)	(% CP)
Coconut (copra) meal	Solvent	22.8	2.48	10.8	1.19	5.2	—	—	—	—
Cottonseed meal, 41% protein	Solvent	45.2	4.66	10.3	2.31	5.1	1.22	2.7	1.70	3.8
	Prepressed solvent	46.1	5.11	11.1	1.89	4.1	1.22	2.6	1.48	3.2
	Mechanical	44.0	4.36	9.9	2.22	5.0	1.12	2.5	1.62	3.7
Peanut meal	Solvent	52.9	5.00	9.5	2.51	4.7	1.01	1.9	1.83	3.5
	Mechanical	49.0	4.51	9.2	—	—	0.94	1.9	1.89	3.9
Rapeseed meal	Prepressed solvent	44.0	2.42	5.5	2.11	4.8	1.19	2.7	1.58	3.6
Safflower meal	Solvent	46.5	4.09	8.8	2.65	5.7	1.11	2.4	1.88	4.0
	Mechanical	46.8	4.98	10.6	2.71	5.8	—	—	—	—
Sesame meal	Mechanical	47.7	4.96	10.4	4.19	8.8	1.25	2.6	2.12	4.4
Soybean meal, 41% protein	Solvent	44.3	3.21	7.2	1.82	4.1	1.10	2.5	1.91	4.3

Soybean meal is, by far, the most widely used oilseed meal in the United States. For this reason, it is generally used as the standard to which all the others are compared (see Table 11-3). Many of the oilseed meals compare favorably with soybean meal, but their use is generally restricted because of (1) limited supply, (2) antinutritional factors (such as gossypol in cottonseed meal), or (3) lack of palatability.

For information pertaining to the uses of oilseed meals in specific livestock rations, the reader is referred to the appropriate chapters of Section III—Feeding.

The quality of various oilseed proteins is compared in Table 11-4. Soybean protein is the best in all measurements of quality, with sunflower protein following closely.

TABLE 11-4
MEASUREMENTS OF OILSEED PROTEIN QUALITY[1]

Oilseed	Biological Value (BV)	Net Protein Utilization (NPU)	Protein Efficiency Ratio (PER)
Soybeans	73	61	2.32
Sunflower	70	58	2.10
Cottonseed	67	53	2.25
Peanuts	55	43	1.05

[1]Adapted by the authors from *Improvement of Protein Nutriture* by D. M. Hegsted, NRC-National Academy of Sciences, 1974.

Table 11-5 lists the essential amino acid profile of several of the most widely used oilseed meals. In all of the meals, the level of the sulfur amino acid methionine is low relative to the other amino acids. Sesame meal contains considerably more methionine than the other oilseed meals.

Corn Gluten Feed and Corn Gluten Meal

Corn gluten is a mixture of bran (the high-fiber shell, or hull, of the kernel) and steepwater solubles (a molasseslike material derived from the water used to soak the corn during wet milling). More precisely, corn gluten feed and corn gluten meal may be defined as follows:

Corn gluten feed is that part of corn that remains after the extraction of the larger portion of the starch, gluten, and germ by the process employed in the wet milling manufacture of corn starch or corn syrup. Corn gluten feed may or may not contain fermented corn extractives and/or corn germ meal. It contains about 23% protein.

Corn gluten meal is the dried residue from corn after the removal of the larger part of the starch and germ, and the separation of the bran by the process employed in the wet milling manufacture of corn starch or syrup, or by enzymatic treatment of the endosperm. It may contain fermented corn extractives and/or corn germ meal. Corn gluten meal averages about 43% crude protein.

Corn gluten feed is an old product with a new look. The recent switch of the soft drink industry to corn sweeteners (high fructose) has made it abundant. But corn gluten feed has been around a very long time. A hundred years ago (in the 1880s), corn gluten feed was considered to be a worthless waste product of starch manufacture; a corn refining plant in Buffalo, New York dumped it into a nearby canal. In time, however, its value as a dairy feed became known.

Both corn gluten feed and corn gluten meal are by-products of the wet milling of corn (see Chapter 10, section on "Wet Milling"). In the dry form, corn gluten feed contains about 23% protein and has a bulky consistency due to its content of corn bran (hulls). Corn gluten feed is also available in wet form.

Corn gluten meal, which contains much less bran and is less bulky, averages about 43% crude protein. Both products are low in lysine and tryptophan. When made from yellow corn, as nearly all corn gluten feed and corn gluten meal are, they are quite rich in carotene and in the yellow pigment xanthophyll. The presence of xanthophyll makes corn gluten feed and corn gluten meal valued ingredients in poultry rations because it produces the yellow skin color, which is prized in dressed poultry.

Today, dairy cows and poultry consume most of the gluten feed and meal. Because of its bulk, gluten feed is better suited for dairy cows than for poultry. Beef cattle, sheep, and swine use lesser amounts of both feeds. Corn gluten feed and corn gluten meal are of special value as ruminant protein supplements because of their good protein bypass (escape) characteristics. Most of the U.S. annual production of more than 5 million tons of corn gluten feed and corn gluten meal is marketed in mixed commercial feeds as a substitute protein source for higher priced protein feeds.

Pulse Proteins

Pulses are the seeds of leguminous plants. They are used primarily for human consumption, but they can be fed to livestock effectively when the price is right. Although there are over 13,000 species within the family *Leguminosae,* only about 20 species are used for food and/or feed. A listing of various types of pulses, along with the protein content of each, is given in Table 11-6. Soybeans and peanuts are pulses, but they are used almost entirely as oilseed meals in livestock rations.

11-5
EXPRESSED AS A PERCENTAGE OF DRY MATTER (DM) AND CRUDE PROTEIN (CP)

Leucine		Lysine		Methionine		Phenylalanine		Threonine		Tryptophan		Valine	
(% DM)	(% CP)	(% DM)	(% CP)	(% DM)	(% CP)	(% DM)	(% CP)	(% DM)	(% CP)	(% DM)	(% CP)	(% DM)	(% CP)
—	—	0.58	2.5	0.36	—	—	—	—	—	0.22	1.6	—	—
2.69	6.0	1.85	4.1	0.66	1.5	2.35	5.2	1.52	3.4	0.60	1.3	2.80	6.2
—	—	1.90	4.1	0.58	1.3	2.47	5.4	1.47	3.2	0.52	1.1	2.09	4.5
2.47	5.6	1.68	3.8	0.61	1.4	2.20	5.0	1.39	3.2	0.53	1.2	2.07	4.7
2.83	5.3	1.75	3.3	0.45	0.9	2.15	4.1	1.18	2.2	0.51	1.0	1.94	3.7
2.91	5.9	1.48	3.0	0.56	1.1	2.17	4.4	1.30	2.7	0.53	1.1	2.11	4.3
2.95	6.7	2.33	5.3	0.84	1.9	1.67	3.8	1.85	4.2	0.53	1.2	2.11	4.8
2.89	6.2	1.44	3.1	0.76	1.6	2.05	4.4	1.49	3.2	0.66	1.4	2.54	5.5
—	—	1.44	3.1	0.76	1.6	—	—	0.88	1.9	0.67	1.4	—	—
3.34	7.0	1.35	2.8	1.47	3.1	2.30	4.8	1.72	3.6	0.77	1.6	2.54	5.3
3.36	7.6	2.49	5.6	0.50	1.1	2.12	4.8	1.67	3.8	0.72	1.6	1.95	4.4

All of the pulses contain components which possess antinutritional properties. Fortunately, processing procedures, such as cooking, germination, and fermentation, can reduce the risks of feeding pulses to livestock. Among the chemical factors that can create problems in feeding pulses are protease inhibitors, goitrogens, cyanogens, antivitamins, metal-binding factors, lathyrogens, and phytohemagglutins (see Table 11-7).

TABLE 11-6
PULSE PROTEINS[1]

Common Name	Latin Name	Crude Protein	Areas Where It Is Grown	General Comments
		(%)		
Beans—includes common bean, dry bean, snap bean, kidney bean, navy bean, mung bean, lima bean	*Phaseolus* spp	20–28	In the U.S., beans are grown primarily in the north central and northwestern regions.	Many of these beans contain components which are harmful if they are not processed properly. (See Table 11-7.)
Chickpea	*Cicer arietinum*	19–21	Grown in subtropics and in the cool season of the tropics.	Chickpea is extremely drought resistant. Good source of lysine, but deficient in tryptophan, methionine, and cystine. Can be used in swine rations uncooked. Cooking appears to improve utilization by poultry.
Cowpeas	*Vigna sinensis*	18–29	Grown primarily in Africa; very little is produced in the U.S.	Cooking or germinating seeds improves feeding value greatly.
Field bean (horse or broad bean)	*Vicia faba*	24–31	Grown throughout the temperate and subtropical regions of the world.	Deficient in methionine, but relatively high in cystine.
Field pea	*Pisum sativum*	22–29	Grown in temperate regions or as hill crops or crop for the cool season in tropical and subtropical regions.	Highly palatable feed for all types of livestock. When price of peas and cereals is equal, peas can be substituted for grain. Can be used in swine rations uncooked.
Peanut	*Arachis hypogaea*	26–30	U.S. Cotton Belt, and in humid subtropical areas.	Generally used for human consumption. Excessive quantities of whole peanuts fed to hogs will cause soft pork.
Pigeonpea	*Cajanus cajan*	20–23	Grown in subtropical and tropical areas.	Pigeonpea is extremely drought resistant. Very deficient in tryptophan, methionine, and lysine. Must be cooked before feeding.
Soybeans	*Glycine max*	39–45	Corn Belt, and northern part of Cotton Belt.	Used primarily for production of vegetable oil and oilseed meal. Should be cooked before feeding to nonruminants.

[1]Adapted by the authors from *Alternative Sources of Protein for Animal Production,* NRC-National Academy of Sciences, 1973, pp. 87–118.

TABLE 11-7
ANTINUTRITIONAL FACTORS OF PULSE PROTEINS

Antinutritional Factors	Mode of Action	Method of Detoxification	Pulses Containing the Factors
Antivitamin factors	These factors render certain vitamins physiologically inactive.	Cooking. Supplementation with vitamins.	Soybeans contain a rachitogenic factor and an antivitamin B-12 factor. Kidney beans contain an antagonist to vitamin E.
Cyanogens	Upon hydrolysis, these compounds release hydrogen cyanide.	Cooking volatilizes HCN, thus reducting risks of feeding.	All legumes contain some cyanogens. Lima beans contain extremely large quantities.
Goitrogens	Enlargement of the thyroid in the rat and chick.	Administration of iodide. Heat treatment in some cases.	Soybeans. Peanuts. Plants of the cabbage family (for example, cabbage, kale, turnips, cauliflower, Brussell sprouts, rapeseed, and mustard seed).
Lathyrogens	Nervous disorders and weakness in humans.	Soaking and heat treatment.	Peas of the genus *Lathyrus* (for example sweet peas).
Metal-binding factors	These factors decrease the availability of certain minerals.	Heat treatment. Addition of chelating agents.	Peas are known to have a factor that ties up zinc. Soybeans have been shown to contain factors that can tie up several trace minerals (for example, copper, iron, manganese, and zinc).
Phytohemagglutinins	Agglutinates red blood cells.	Heat treatment.	Found in all leguminous seeds.
Protease inhibitors	Combines with trypsin to form an inactive complex. Causes hypertrophy of the pancreas. Counteracts feedback inhibition of pancreatic enzyme secretion by trypsin.	Heat treatment: for soybean, autoclaving at 15 lb/sq in. for 15–20 minutes; extrusion cooking; soaking, followed by steaming at atmospheric pressure. Germination. Fermentation.	All legumes contain trypsin inhibitors to some extent.

Leaf Protein Concentrate (LPC)

In recent years, considerable attention has been focused on the extraction of protein from green, leafy plants. Green leaves are among the best sources of protein in feeds. To date, most of the research on leaf protein concentrate has centered around the processing of alfalfa; but there is a great potential for LPC development from other sources—for example, vegetable packinghouse by-products and field wastes.

The advantages of using alfalfa for a protein resource are numerous; among them, (1) It is a perennial crop that can be harvested several times in one growing season; and (2) being a legume, it has the ability to convert nitrogen into protein. Hence, alfalfa yields more protein per acre than any other crop.

Ruminants can utilize unprocessed alfalfa extremely well, due to microbial digestion of large amounts of fiber characteristic of all forages. However, if nonruminants are to utilize the protein in alfalfa efficiently, processing is required. It has been well documented that the leaves of alfalfa contain more protein than the stems, and that the contents of the cells in both the stems and the leaves contain more protein than the cell walls. Unfortunately, when alfalfa is sun-cured for hay, losses of 15 to 40% of the dry matter are incurred, most of which are leaves. If alfalfa is ensiled, many of the water-soluble nutrients are lost due to leaching. It would appear, then, that the most efficient means of preserving the nutrients in alfalfa would be to process the alfalfa immediately after cutting to reduce field losses. Leaves could then be separated from the stems, thereby yielding a high-protein leaf meal. The stems could then be fed as green chop, ensiled or dehydrated to form a medium-protein dry roughage. Unfortunately, this has not been found economically feasible on a large-scale basis. Some leaf protein concentrate is commercially available, but due to the high cost of processing, little of it is produced; and most of that which is currently produced is used primarily as a supplement to provide high content protein (more than 40% crude protein), vitamins, xanthophylls, and unidentified factors. Until new developments lower the cost of processing, leaf protein concentrate will remain a protein supplement of minor importance.

One recent development showing promise for the future is the process whereby juice is obtained from the pressing of freshly cut alfalfa. About 25 to 50% of the protein in the leaves can be found in this juice. The high-protein liquid concentrate can then be fed to animals in its crude form or coagulated and used as a solid protein supplement. One such supplement that is currently available is marketed under the trade name of X-Pro. The extract can be coagulated and dried, forming a solid product that is 50% protein.

ANIMAL PROTEINS

Protein supplements of animal origin are derived from (1) meat packing and rendering operations, (2) poultry and poultry processing, (3) milk and milk processing, and (4) fish and fish processing. Before the discovery of vitamin B-12, it was generally considered necessary to include one or more of these protein supplements in the rations of hogs and chickens. With the discovery and increased availability of synthetic vitamin B-12, high-protein feeds of animal origin have become less essential, although they are still included to some extent in rations for most monogastric animals.

With improvements in the protein quality of plants, such as the development of high-lysine corn, the use of meat proteins may decline in the future. The cost of such proteins will need to be more competitive than they are at the present time if they are to be included in any major quantity in animal rations. Blending several proteins with complementary balances of amino acids and supplying more concentrated sources of individual amino acids may also be a factor affecting the future role of animal proteins in animal and poultry rations.

Many protein supplements of animal origin are difficult to process and store without some spoilage and nutrient loss. If they cannot be dried, they must usually be refrigerated. If not heated to destroy disease-producing (pathogenic) bacteria, they may be a source of infection. On the other hand, protein availability will be reduced and some nutrients lost if the feed is heated excessively.

Meat Packing By-products

Although the meat or flesh of animals is the primary object of slaughtering, modern meat packing plants process numerous and valuable by-products, including protein-rich livestock feeds.

In the early days of the meat packing industry, the only salvaged by-products were hides, wool, tallow, and tongue. The remainder of the offal was usually either carted away and dumped into a river or burned or buried. In some instances, packers even paid for having the offal taken away. Later, during the 1870s and 1880s, some of these by-products were recovered, dried, and utilized as fertilizers.

In the early 1890s, some of the selected meat residues from lard-rendering—known as cracklings—were sold locally for chicken feed. Finally, about 1900, poultry feeders on the Pacific Coast, principally at Petaluma, California, started using dried tankage in poultry rations. Almost simultaneously, this new feed was utilized in hog feeding, with experiment stations leading the way. Thus was born a new era in the utilization of packinghouse by-products, and a new era in livestock feeding.

Fig. 11-14. Modern equipment used by Iowa Beef Processors for producing 50% meat and bone meal. (Courtesy, IBP, Dakota City, Nebr.)

MEAT AND MEAT BY-PRODUCTS

Meat is defined by the Association of American Feed Control Officials *as the clean, wholesome flesh derived from slaughtered animals and is limited to that part of striate muscle which is skeletal or which is found in the tongue, diaphragm, heart, or esophagus.*

Meat by-products are the nonrendered, clean, wholesome parts, other than meat. This feed classification includes blood, bones, brains, intestines, kidneys, lungs, spleens, and stomachs, but it does not include hair, hooves, or teeth.

TANKAGE AND MEAT MEAL

In 1984, 2.9 million tons of tankage and meat meal were fed to livestock. Tankage and meat meal are made from the trimmings that originate on the killing floor, inedible parts and organs, cleaned entrails, fetuses, residues from the production of fats, and certain condemned carcasses and parts of carcasses. In order to be used in animal feeds, carcasses condemned as unfit for human food because of antibiotic or biological residues, must meet the standards of the Food Safety and Inspection Service.

The end products, and the methods of processing each, are:

1. **Meat meal tankage (tankage)** is produced by the older wet-rendering method, in which all of the material is cooked by steam under pressure in large closed tanks; hence, the derivation of the name tankage. After cooking, the fat is skimmed off, the soupy liquid drained off, and the remaining residue pressed to remove as much of the fat and water as possible. The soupy liquid is then evaporated until it becomes gluey, at which state it is called *stick.* The stick is added to the pressed residue, following which the mixture is dried and ground.

According to the Association of American Feed Control Officials, the raw materials from which tankage is made should not contain hair, hoof, horn, manure, stomach contents, and hide trimmings, except in such traces as might occur unavoidably in good factory practice.

The level of protein in tankage (generally 60%) is standardized during manufacture by the addition of enough blood to raise the total protein to the desired level. The protein content must be designated.

The label shall include guarantees for minimum crude protein, minimum crude fat, maximum crude fiber, minimum phosphorus (P), and minimum and maximum calcium (Ca). If the product bears a name descriptive of its kind, composition or origin, it must correspond thereto. The calcium level shall not exceed the actual level of phosphorus by more than 2.2 times.

In addition to tanking under live steam (as previously described), tankage may also be made by the dry-rendering method (a description of which follows), or by mixing products containing both wet-rendered and dry-rendered materials.

2. **Meat meal** is produced by the newer and more efficient dry-rendering method, in which all of the material is cooked in its own grease by dry heat in open steam-jacketed drums until the moisture has evaporated. Then as much of the fat as possible is removed by draining off and the solid residue is passed through a screw press. Next the dry residue is granulated or ground into a meal. This product is lighter colored and does not have as strong an odor as wet-rendered tankage.

According to the Association of American Feed Control Officials, the raw materials from which meat meal is made should not contain hair, hoof, horn, hide trimmings, blood meal, manure, and stomach contents, except in such traces as might occur unavoidably in good factory practice.

The level of protein of meat meal (generally 50 to 55%) was originally established because the normal proportions of raw materials available for rendering resulted in a product which, after being pressed and ground, contained approximately 50% protein. The protein content of meat meal is adjusted up or down by raising or lowering the quantity of bone and fat in the raw material. The protein content must be designated.

Tankage and meat meal are widely used as protein supplements. Generally, their proteins are of excellent quality, effectively correcting the deficiencies in the proteins of the cereal grains. They are also rich in minerals (especially calcium and phosphorus) and good sources of riboflavin, niacin, and vitamin B-12. However, they are lacking in vitamins A and D and pantothenic acid.

The protein contents of meat meal is somewhat lower than that of meat meal tankage; yet, the two products are about equal in feeding value, probably due to the greater digestibility and high nutritive value of the protein in the dry-rendered product, since it is subjected to less heat.

Protein content alone is not an infallible criterion of the feeding value of tankage or meat meal, for they vary considerably according to the kind of raw material from which they are produced. Further, there is strong suspicion in many quarters that these products are not as nutritious as they used to be, chiefly because the modern meat packing plant has developed more remunerative outlets for some materials that formerly went into tankage or meat meal. Today, the heart, liver, brains, kidneys, tongue, cheek meat, tail, pigs' feet, sweetbreads, and tripe—former components of tankage or meat meal—are sold over the counter as *variety meats* or *fancy meats.* With this change, there has been a very natural tendency to incorporate in tankage a higher proportion of gristle, connective tissue, bone, stick, and blood meal—all materials of lower nutritive value than meat and glands. It would appear, therefore, that it is timely that some reevaluation be made of the feeding value of present-day tankage. Also, it is important that the practical livestock producer (1) purchase tankage or other meat meal from a reputable source, and (2) mix tankage with other protein supplements, especially when confinement feeding is practiced.

The label shall include guarantees for minimum crude protein, minimum crude fat, maximum crude fiber, minimum phosphorus (P), and minimum and maximum calcium (Ca). If the product bears a name descriptive of its kind, composition, or origin, it must correspond thereto. The calcium level shall not exceed the actual level of phosphorus by more than 2.2 times.

MEAT AND BONE MEAL TANKAGE/MEAT AND BONE MEAL

When, because of added bone, tankage or meat meal contains more than 4.4% phosphorus, the word *bone* must be inserted in the name; and they must be designated, according to the method of processing, as either (1) meat and bone meal tankage, or (2) meat and bone meal. Thus, when such high-phosphorus products are prepared by the older wet-rendering method, they are known as *meat and bone meal tankage.* Likewise, when products in excess of 4.4% phosphorus are prepared by the newer dry-rendering method, they are designated as *meat and bone meal.*

The protein content of meat and bone meal tankage and meat and bone meal must be designated. Generally, they contain less protein than tankage or meat meal, usually 45 to 50%. Except for noting this difference, the swine feeding recommendations given relative to tankage and meat meal are equally applicable to meat and bone meal tankage and meat and bone meal.

The requirements for *meat and bone meal tankage* and *meat and bone meal* are the same: The label shall include guarantees for minimum crude protein, minimum crude fat, maximum crude fiber, minimum phosphorus, and minimum and maximum calcium. If it bears a name descriptive of its kind, composition, or origin, it must correspond thereto. It shall contain a minimum of 4.0% phosphorus, and the calcium level shall not be more than 2.2 times the actual phosphorus level.

BLOOD MEAL

Although the Germans had long incorporated blood in livestock rations, in this country there was strong prejudice against the practice until quite recently. Further, manufacturing difficulties were encountered, due chiefly to inadequate drying methods. It is not surprising, therefore, to learn that, along about 1900, the use of dried blood as a feed ingredient was tried but found wanting. Later, the process was perfected.

Today, dried blood meal may be, and is, prepared by any one of three processes: (1) spray drying, (2) cooker drying, or (3) flash drying. Their market designations and descriptions follow:

1. **Spray dried blood meal.** This is prepared by a two-step operation: (a) moisture is first removed by a low temperature evaporator under vacuum until it contains approximately 30% solids, and (b) it is then spray dried. Spray dried blood must have a maximum of 8% moisture and minimum of 85% crude protein.
2. **Cooker dried blood meal.** This is produced by drying blood in a convential cooker. It usually has a dark black color and is rather insoluble in water. It must have 85% crude protein.
3. **Flash dried blood meal.** This is processed in two steps as follows: (a) a large portion of the moisture (water) is usually removed by a mechanical dewatering process or by condensing by cooking to a semisolid state; and (b) the semisolid blood mass is then transferred to a rapid drying facility where the more tightly bound water is rapidly removed. The minimum biological activity of lysine in flash dried blood shall be 80%. It must have 80% crude protein.

Blood meal is low in the essential amino acid isoleucine and low in calcium and phosphorus.

Blood meal ranks very high as an escape (bypass) protein for ruminants (see Table 11-1). For this reason, it is finding an increasing place as a ruminant feedstuff.

OTHER ANIMAL BY-PRODUCTS

Meat meal tankage, meat meal, meat and bone meal tankage, and meat and bone meal are the leading meat packing by-products. Additional meat packing by-products follow:

1. **Animal by-product meal.** This is the dried product from animal tissue prepared by tanking with live steam or dry rendering. The protein content must be designated.

   This ingredient definition is intended to cover those individual rendered animal tissue products that cannot meet the criteria as set forth elsewhere in this section. This ingredient is not intended to be used to label a mixture of animal tissue products.
2. **Animal digest.** This product results from chemical and/or enzymatic hydrolysis of clean and undecomposed animal tissue, exclusive of hair, horns, teeth, hooves, and feathers; suitable for animal feed.
3. **Animal liver meal.** This is obtained by drying and grinding liver from slaughtered mammals.
4. **Blood protein.** This is produced by quick freezing and/or transporting in a chilled state, clean, fresh, whole or dewatered animal blood.
5. **Fleshings hydrolysate.** This product is obtained by acid hydrolysis of the flesh from fresh or salted hides. It is defatted, strained, and neutralized. If evaporated to 50% solids, it shall be designated "Condensed Fleshings Hydrolysate." It must have a minimum crude protein and a maximum salt guarantee.
6. **Glandular meal and extracted glandular meal.** This product is obtained by drying liver and other glandular tissues from slaughtered mammals. When a significant portion of the water-soluble material has been removed, it may be called *Extracted Glandular Meal.* These products contain proteins of high quality and are rich sources of both the fat-soluble and water-soluble vitamins.
7. **Hair, hydrolyzed.** Hydrolyzed hair is prepared from clean, undecomposed hair by heat and pressure. At least 80% of the crude protein content must be digestible by the pepsin digestibility method.

   Although this product contains high levels of protein—about 95%—it is extremely deficient in the amino acids methionine, lysine, and tryptophan, thereby dramatically reducing its feeding value.
8. **Leather meal, hydrolyzed.** This product is produced from leather scrap that is treated with steam under pressure. It must not contain more than 10% moisture, not less than 60% crude protein, not more than 6% crude fiber, not more than 27.75% chromium, and with not less than 80% of its crude protein digestible by the pepsin digestibility method.
9. **Meat solubles, dried.** Dried meat solubles are prepared through the drying of the defatted water obtained in the steaming or hot water extraction of meat and/or organs. It must contain at least 70% crude protein.
10. **Unborn calf carcass (cattle fetus).** This is obtained from whole unborn fetuses taken from slaughtered cows at government inspected slaughter plants. The product is produced by grinding whole unborn fetuses, exclusive of hides, following which it is denatured and fresh frozen.

Poultry Wastes

By-product feedstuffs are derived from all segments of the poultry industry—from hatching all the way through processing for market; and they come from the broiler and turkey segments of the industry as well as from egg production. Centralization of these industries into large units, with enough volume of wastes to make it feasible to process the potential feeds, has opened new markets for what was previously a disposal problem. Certain precautions have had to be included to make the products most useful and safe, but considerable amounts of poultry products are currently being included in rations for various animals, both ruminants and monogastrics.

The three most extensively used high-protein by-products of the poultry industry are hatchery by-product, poultry by-products, and poultry feathers. Cull birds, unsalable eggs, eggshells, and slaughter wastes are also used in animal feeds.

In addition to the by-products of the poultry processing industry, poultry manure and litter are now being processed to produce a palatable, high-protein feed. A detailed discussion of this phase of by-product utilization is presented in Chapter 12, By-product Feeds/Crop Residues.

EGGSHELL MEAL

This is a mixture of eggshells, shell membranes, and egg content obtained by drying the residue of an egg-breaking plant in a dehydrator to an end product temperature of 180°F. It must be designated according to its protein and calcium content.

FEATHER MEAL

Feathers, a by-product that is nearly all protein, can be used in rations after they are hydrolyzed with heat and pressure to make the proteins available. *Hydrolyzed feather meal is defined as the product resulting from the treatment under pressure of clean, undecomposed feathers from slaughtered poultry, free of additives, and/or accelerators.* Not less than 75% of its crude protein content must be digestible by the pepsin digestibility method. Although hydrolyzed feather meal is high in protein (from 85 to 90%), it is rather low in nutritional value, being low in the amino acids histidine, lysine, methionine, and tryptophan. Those amino acids which are present are readily available.

Ruminants can utilize feather meal rather well; and benefit from its high bypass protein. Levels up to 10% have been used successfully in concentrated feeds for dairy cattle. Feather meal should be included in ruminant feeds gradually, since sudden addition of it may decrease feed intake.

Because of the deficiencies of several amino acids, care must be used when incorporating feather meal into nonruminant feed. The addition of fish meal, meat meal, or blood meal tends to complement feather meal and facilitates its use. In practice, feather meal rarely exceeds 5% of the ration for nonruminants.

HYDROLYZED POULTRY BY-PRODUCTS AGGREGATE

This is the product resulting from heat treatment under pressure of all by-products of slaughtered poultry, clean and undecomposed, including such parts as heads, feet, undeveloped eggs, intestines, feathers, and blood.

POULTRY BY-PRODUCTS

This must consist of non-rendered clean parts of carcasses of slaughtered poultry such as heads, feet, and viscera, free from fecal content and foreign matter.

POULTRY BY-PRODUCT MEAL

This consists of the ground, dry-rendered or wet-rendered clean parts of carcasses of slaughtered poultry—such as heads, feet, undeveloped eggs, and viscera—free from feathers, fecal content and foreign matter except for such trace amounts as are unavoidable in good factory practice. It must contain not more than 16% ash and not more than 4% acid-insoluble ash.

Because of the heads and feet, poultry by-products are lower in nutritional value than the flesh of animals, including poultry. The biological value of the proteins is lower than the other animal proteins and the better plant proteins. They may be successfully used in animal rations, however, provided they are not the sole source of proteins.

The label shall include guarantees for minimum crude protein, minimum crude fat, maximum crude fiber, minimum phosphorus, and minimum and maximum calcium. The calcium level shall not exceed the actual phosphorus level by more than 2.2 times.

POULTRY HATCHERY BY-PRODUCT

Of the various by-products from poultry, the most valuable is *hatchery by-product* consisting of eggshells, infertile and unhatched eggs, and culled chicks which have been cooked, dried, and ground, with or without removal of part of the fat. This product deteriorates quite rapidly if not cooled promptly—a factor which is true of most poultry, fish, and meat products.

Dairy Products

Skimmed milk and buttermilk have long been used on, or in close proximity to, the farms where they are produced. However, in their liquid form it is economically impossible to ship them long distances or to store them; and it is difficult to maintain sanitary feeding conditions in using them.

Along about 1910, processes were developed for drying buttermilk. Soon thereafter, special plants were built for dehydrating buttermilk, and the process was extended to skimmed milk and whey. Beginning about 1915, dried milk by-products were incorporated in commercial poultry feeds. In 1984, 98,000 tons (dry basis) of dairy products were fed to livestock.

The superior nutritive values of milk by-products are due to their high-quality proteins, vitamins, a good mineral balance, and the beneficial effect of the milk sugar, lactose. In addition, these products are palatable and highly digestible. They are an ideal feed for young animals and for balancing the deficiencies of the cereal grains. The chief limitation to their wider use is price, together with the perishability and bulkiness of the liquid products.

Although whole milk is an excellent feed—worth about twice as much as skimmed milk—it is usually too expensive to feed. In general, liquid milk products contain 10% dry matter; semisolid milk products, 30% dry matter; and dried milk products, 90% dry matter.

BUTTERMILK

Buttermilk and skimmed milk have approximately the same composition and feeding value, providing the buttermilk has not been diluted by the addition of churn washings. Accordingly, it may be considered that (1) 15 lb of buttermilk will replace 1 lb of tankage or meat meal, or (2) 6 lb of buttermilk will replace 1 lb of complete feed. Other buttermilk products are:

1. **Buttermilk, condensed.** This is made by evaporating buttermilk to about ⅓ of its original weight. It contains 27% minimum total solids, 0.055% minimum milk fat for each percent of total solids, and 0.14% maximum ash for each percent of total solids

One pound of condensed buttermilk is worth about 3 lb of liquid buttermilk. But it requires about 3 lb of condensed buttermilk to equal 1 lb of dried milk. Another basis of evaluating condensed buttermilk is that, pound for pound, it is worth approximately ½ as much as tankage.

2. **Buttermilk, dried.** This is the residue obtained by drying buttermilk. It must contain less than 8% moisture, less than 13% ash, and more than 5% milk fat. One pound of dried buttermilk has about the same composition and feeding value as 10 lb of liquid buttermilk.

CHEESE RIND (CHEESE MEAL)

This is a by-product from the manufacture of processed cheese, consisting of the cheese trimmings from which most of the fat has been removed. It contains around 50% protein and 9% fat.

DAIRY FOOD BY-PRODUCTS

These products are derived from the collection of solids contained in the washwater from normal processing and packaging of various food manufacturing plants. Dairy products are the primary source, but non-dairy products may occasionally constitute a minor amount of the total volume. Dairy food by-products should be fed at levels less than 25% of the animal's total dry matter intake.

MILK PROTEIN PRODUCTS

The five milk protein products currently available are:

1. **Albumin, dried milk.** This product is produced by drying the coagulated protein residue separated from whey. It contains at least 75% crude protein on a moisture-free basis.
2. **Casein.** This is the solid residue that remains after the acid or rennet coagulation of defatted milk. It contains at least 80% crude protein.
3. **Casein, dried hydrolyzed.** This is the residue obtained by drying the water-soluble product resulting from the enzymatic digestion of casein. It contains at least 74% crude protein.
4. **Milk protein, dried.** This product is obtained by drying the coagulated protein residue resulting from the controlled coprecipitation of casein, lactalbumin, and minor milk proteins from defatted (skim) milk.
5. **Whey protein concentrate, dried.** This is the product obtained by removal or separation of water, lactose, and/or minerals from whey by ultrafiltration, dehydration, or other process. It must contain a minimum of 25% crude protein.

Although the quality and quantity of protein from milk protein products are excellent, these sources of protein are generally too expensive to be used routinely as major protein sources in livestock feeds.

MILK, WHOLE DRIED

This is the residue left following the drying of milk. It contains at least 26% milk fat but no more than 8% moisture. This product is generally too expensive to feed to livestock.

SKIMMED MILK

Because of the removal of the fat, skimmed milk has little vitamin A value. However, in comparison with whole milk, it is higher in content of protein, milk sugar, and minerals. Like all milk products, skimmed milk is low in vitamin D and iron.

Skimmed milk is the best single protein supplement for swine. It is especially valuable for young pigs prior to and immediately after weaning. The addition of either pasture or a choice legume hay will supplement skimmed milk with the needed vitamins A and D.

Skimmed milk should be fed consistently sweet or sour, because abrupt changes are apt to produce digestive disturbances. Where a choice is possible, fresh skimmed milk is recommended.

The amount of skimmed milk to feed will vary according to (1) the available supply, (2) the relative price of feeds, (3) the kind of grain ration fed, and (4) whether pasture is available.

The feeding value of skimmed milk varies with the age of the animal, the type of ration, the price of other feeds, and the amount of milk fed. Naturally, it has a higher value per pound when fed in limited amounts than when an excess is fed. On the other hand, when the supply of milk is abundant and cheap, a larger proportion may be fed profitably—especially when grain is scarce and high in price. Roughly, it can be figured that (1) 15 lb of skimmed milk will replace 1 lb of tankage or meat meal, or (2) 6 lb of skimmed milk will replace 1 lb of complete feed.

In addition to liquid skimmed milk, the following processed dairy products are valuable feeds:

1. **Skimmed milk, condensed.** This is the residue obtained by evaporating defatted milk. It contains 27% minimum total solids.
2. **Skimmed milk, condensed cultured.** This is the residue obtained by evaporating lactic acid bacteria cultured defatted milk. It contains 27% minimum total solids.
3. **Skimmed milk, dried.** As the name indicates, this product is dehydrated skimmed milk. It contains less than

8% moisture, and averages 32 to 35% protein. One pound of dried skimmed milk has about the same composition and feeding value as 10 lb of liquid skimmed milk. Although dried skimmed milk is an excellent swine feed, it is generally too high priced to be economical for this purpose except for limited use in pig starter rations.

Prior to World War II, dried skimmed milk was the most widely used dried milk product included in feeds. During and since the war, however, much of it has been marketed as a human food.

4. **Skimmed milk, dried cultured.** This is the residue obtained by drying lactic acid bacteria cultured defatted milk. It contains less than 8% moisture.

WHEY

This is the product obtained as a fluid by separating the cheese coagulum from milk, cream, or skimmed milk and from which a portion of the milk fat may have been removed. The following whey products, each involving different processing, are available:

1. **Whey, condensed.** This is the product obtained by partially removing water from whey. Minimum percent solids must be declared on the label.

2. **Whey, condensed hydrolyzed.** This is the residue obtained by evaporating lactase enzyme hydrolyzed whey. It contains 50% minimum total solids and 0.3% minimum total glucose and galactose for each percent total solids.

3. **Whey, cultured condensed.** This is the product obtained by partially removing water from whey which has been cultured. The minimum percent of solids must be declared on the label.

4. **Whey, dried.** This is the product obtained by removing water from whey. It contains less than 11% protein and less than 61% lactose.

5. **Whey, dried hydrolyzed.** This is the residue obtained by drying lactase enzyme hydrolyzed whey. It contains a minimum of 30% glucose and galactose.

6. **Whey, fermented ammoniated condensed.** This product is produced by the *Lactobacillus bulgaricus* fermentation of whey, followed by the addition of ammonia. It contains 35 to 55% crude protein of which not more than 42% is equivalent crude protein from non-protein nitrogen. Fermented ammoniated condensed whey is a suitable protein supplement for finishing cattle and lactating dairy cows.

7. **Whey product, condensed.** This is the product obtained by partially removing water from whey from which a portion of the lactose, protein, and/or minerals have been removed. The minimum percent of solids, crude protein, and lactose, and the maximum percent ash must be declared on the label.

8. **Whey product, dried.** This is the product obtained by drying whey from which a portion of the lactose, protein, and/or minerals have been removed. The minimum percent of crude protein and lactose and the maximum percent of ash must be declared on the label.

9. **Whey solubles, condensed.** This is the product obtained by concentrating the whey residue after removal of whey protein, with or without partial removal of lactose.

10. **Whey solubles, condensed modified.** This is the product obtained by concentrating the whey residue after removal of whey protein and partial removal of lactose, and modifying the sugar content so that there is a minimum of 0.3% nonlactose carbohydrate for each percent solids.

11. **Whey solubles, dried.** This product is obtained by drying the whey residue after removal of whey protein, with or without partial removal of lactose. Minimum percent crude protein and lactose and maximum percent ash must be declared on the label.

Marine By-products

Fish and marine by-products are most commonly used in swine, poultry, and mink rations. Increasingly, they are being used as a ruminant feedstuff to supply a dietary protein of proper amino acid balance which can survive ruminal degradation (see Table 11-1).

FISH PROCESSING

Processing of fish can be a difficult task. Many fish contain large amounts of fat which, if not properly processed, can create numerous problems with the processing machinery (see Table 11-8).

TABLE 11-8
CLASSIFICATION OF RAW MARINE PRODUCTS[1]

Classification	Composition Limits		Types of Fish
	Oil	Protein	
	(%)	(%)	
Low oil—low protein	less than 5	less than 15	oysters clams
Low oil—high protein	less than 5	15-20	carp cod crab flounder haddock hake mullet ocean perch pollack rockfishes scallop shrimp whiting
Low oil—very high protein ..	less than 5	over 20	halibut tuna
High oil—high protein	over 5	15-20	anchovy herring mackerel salmon sardine

[1]Adapted by the authors from *Tropical Feeds,* FAO-UN, 1975.

There are several methods of manufacturing fish products, each adaptable to a particular type of fish. These follow:

- **Simple direct drying**—Most fish containing little fat can be dried through this process. The raw material is cut extensively in a disintegrator and passed through a flame dryer. Once the fish material passes through the dryer, it is ground, cooled, and bagged.
- **Recirculation drying**—When whole fish are heated, large quantities of water mixed with lipids and other materials are

liberated. Since this emulsion can create processing problems through clumping of the material subsequently clogging the machinery, material that has been partially dried is recirculated with the incoming material to dilute the concentration of the gluey water.

• **Traditional reduction processing**—Fig. 11-15 shows a flowsheet of this type of processing. Material is carried to a cooker via a screw conveyer. The cooking stage of this type of processing is the most critical because the fish must be properly coagulated if pressing is to be efficient. If the material is improperly cooked, the resulting presscake will not have a uniform texture due to the presence of oil which should have been previously extracted. Before the cooked material is pressed, it passes through a screen to remove much of the fine sludge which is produced during cooking. After screening, the material is passed through presses to remove moisture, oil, and dissolved solids. The solid material formed in this stage is called the presscake. It can contain as little as 50% moisture and 5% fat. The presscake is then pulverized to increase surface area for drying and subsequently dried, ground, and bagged. The liquid extracted during pressing is then separated into oil and gluewater.

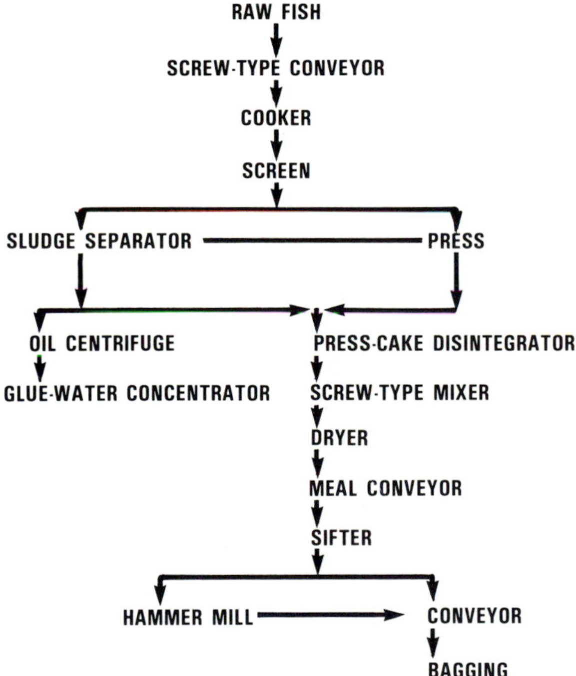

Fig. 11-15. Flow diagram of fish reduction processing.

When the glue water is condensed so that it contains about 50% total solids, it is marketed under the name "fish solubles," which is an excellent source of B vitamins and water-soluble proteins. Sometimes, this gluewater is added back to the meal to form what is known as *whole meal.*

• **Other processing methods**—Most fish used for feed are processed by the preceding methods. However, numerous other methods are being investigated; among them, (1) dry rendering, (2) solvent extraction, (3) ultrasonic extraction, and (4) solubilization processing.

The processing of fish meal and solubles varies widely, reflecting fish catch and competing feed prices. In 1970, 609,000 tons were processed; in 1984, the tonnage dropped to 228,000 tons.

MARINE FEEDS

Marine feeds consist of the by-product waste from processing fish, along with unmarketable species of fish and other marine animals.

On a worldwide basis, some 25 million metric tons of fish are processed into fish meal for animal feed each year, but less than ½ million metric tons of this amount is consumed in the United States. Nevertheless, fish by-products are generally considered by livestock producers to be excellent sources of nutrients. Proteins, vitamins, and minerals are all readily available in most fish products.

Fish Meal

Fish meal—a by-product of the fisheries industry—consists of dried, ground whole fish or fish cuttings—either or both—with or without the extraction of part of the oil. If it contains more than 3% salt, the salt content must be a part of the brand name. In no case shall the salt content exceed 7%.

The feeding value of fish meal varies somewhat, according to:

1. **The method of drying.** It may be either vacuum, steam, or flame dried. The older flame drying method exposes the product to a higher temperature, which makes the proteins less digestible and destroys some of the vitamins.

2. **The type of raw material used.** It may be made from the offal produced in fish packing or canning factories, or from the whole fish with or without extraction of part of the oil.

Fish meal made from offal containing a large proportion of heads is less desirable because of the lower quality and digestibility of the proteins. Although few feeding comparisons have been made between the different kinds of fish meals, it is apparent that all of them are satisfactory when properly processed raw materials of good quality and moderate fat content are used. A high-fat content may impart a fishy taste to eggs, meat, and milk.[6] Also, such meal is apt to become rancid in storage.

It is of interest to swine and poultry producers to know the sources of the commonly used fish meals. These are:

1. **Herring meal.** This is a high-grade product produced in the Pacific Northwest and Alaska.

2. **Menhaden fish meal.** Over 90% of the fish meal produced in the U.S. is made from menhaden (a very fat fish not suited for human food) caught in the Atlantic primarily for their body oil. The meal is the dried residue after most of the oil has been extracted.

[6]Research workers at the Indiana Station (Vestal, *et al., Journal of Animal Science,* Vol. 4, No. 1, 1945) found that the addition to the ration of 0.5 and 1.5% fish oil produced a fishy flavor in pork, which was more pronounced in the roasts and bacon than in the chops.

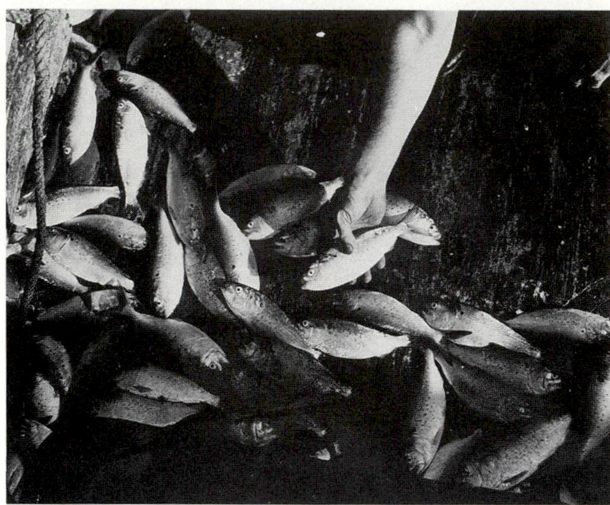

Fig. 11-16. Menhaden are a fat fish unsuitable for human consumption. They are caught primarily for their body oil, but the dried residue after oil extraction provides an excellent high-quality protein feed for livestock. (Courtesy, National Fisheries Institute, Inc., Washington, D.C.)

3. **Salmon meal.** This is a by-product of the salmon canning industry in the Pacific Northwest and in Alaska.
4. **Sardine meal or pilchard meal.** This is made from sardine canning waste and from the whole fish, principally on the West Coast.
5. **White fish meal.** White fish meal, which ranks second to menhaden in U.S. production, is a by-product from fisheries making cod and haddock products for human food. Its proteins are generally of high quality.

Fish meal should be purchased from a reputable company on the basis of protein content. It varies in protein content from 57 to 77%, depending on the kind of fish from which it is made; and it must not contain over 10% moisture. When of comparable quality, fish meal is superior to tankage or meat meal as a protein supplement for nonruminant animals. The protein of a good-quality fish meal is 92 to 95% digestible. If fish meal is poorly processed or improperly stored, the digestibility of protein decreases dramatically. Since fish meals are cooked, there is danger that certain amino acids—notably lysine, cystine, tryptophan, and histidine—will be denatured, but these losses are minimized when proper processing techniques are used.

Fish meals containing high levels of fat are considered to be low quality. If they are incorporated into feeds, they tend to impart a fishy flavor to the animal products. Also, problems of rancidity are greater in high-fat fish meals.

Fish meal is an excellent source of minerals. Calcium and phosphorus are especially abundant, being present in the amounts of 3 to 6% and 1.5 to 3.0%, respectively. Many of the trace minerals, especially iodine, required by livestock can be supplied in part by fish meal.

Fish meal is not a particularly good source of fat-soluble vitamins, which are lost during the extraction of oil, but a fair amount of the B vitamins remain. However, fish meal is one of the richest sources of vitamin B-12 and unidentified growth factors.

Historically, in the United States, fish meal was used largely in poultry feeds, where it supplied unidentified growth factors. But, in the 1980s, the use of fish meal changed significantly. The three most important new uses for fish meal are: aquaculture feeds, early weaned pig feeds, and ruminant bypass feeds.

Other Marine By-products

Other valuable marine by-products follow:

1. **Fish by-products.** This product consists of non-rendered, clean undecomposed portions of fish which result from fish processing.
2. **Fish digest residue.** This is the clean, dried, undecomposed residue (bones, scales, undigested solids) of the digest resulting from the enzyme hydrolysis of producing fish protein. It must be designated according to its protein, calcium, and phosphorus contents.
3. **Fish liver and glandular meal.** This marine by-product is obtained by drying the entire viscera of fish. The Association of American Feed Control Officials (AAFCO) specifies that it must contain at least 18 mg of riboflavin per pound, and that at least 50% of its dry weight must consist of livers.
4. **Fish protein concentrate.** Fish protein concentrate is prepared through the solvent extraction processing of clean, undecomposed whole fish or fish cuttings. The product cannot contain more than 10% moisture and must contain at least 70% protein, according to AAFCO. The solvent residues must conform to the rules as set forth by the Food Additive Regulations.
5. **Fish protein digest, condensed.** This is the condensed enzymatic digest of clean undecomposed whole fish or fish cuttings. The product must be free of bones, scales, and undigested solids, with or without the extraction of part of the oil.
6. **Fish protein digest, dried.** This is the dried enzymatic digest of clean undecomposed whole fish or fish cuttings. The product must be free of bones, scales, and undigested solids, with or without the extraction of part of the oil. It must contain not less than 80% protein and not more than 10% moisture.
7. **Fish residue meal.** This is the dried residue from the manufacture of glue from nonoily fish. If the product contains more than 3% salt, the amount of salt must be included in the product name. However, salt content must never exceed 7%.
8. **Fish solubles, condensed.** This is a semisolid by-product obtained by evaporating the liquid remaining from the steam rendering of fish, chiefly sardines, menhaden, and redfish. The gluewater that comes from processsing contains about 5% total solids. After this liquid is evaporated or condensed, it contains about 50% total solids. Condensed fish solubles, which must contain a minimum of 30% crude protein, are a rich source of the B vitamins and unidentified factors. They are particularly rich in pantothenic acid, niacin, and vitamin B-12.
9. **Fish solubles, dried.** Dried fish solubles are obtained by dehydrating the gluewater of fish processing. It must contain at least 60% crude protein.

10. **Liquified seafood (fish) waste.** This is a relatively new protein source, which consists of the waste from the processing of fish; it may include the heads, guts, scales, and the skeleton of fish filleted for market, along with unmarketable species of fish and other marine animals inadvertently netted while fishing for other species. Most of this waste is dumped at sea from factory ships or in other locations from land-based processing plants. However, some of it is ground and frozen, then used as pet, mink and fox food.

Europe has pioneered in the development of *fish silage*, prepared by grinding the fish waste, then adding an acid, usually formic acid, or a combination of sulfuric and formic acids for oily fish. Well prepared fish silage can be kept up to 2 years. Liquified seafood waste has been successfully used for swine, poultry, early-weaned calves, and sheep.

11. **Shrimp meal and crab meal.** Along the coastal regions of the United States, large quantities of potential feedstuffs are produced by the shellfish industry. By-products from the shrimp and crab industries are presently being converted to high-protein feeds.

Shrimp meal is the ground, dried waste of shrimp processing. It may consist of the head, shell, and/or whole shrimp. The provisions relative to salt content are the same as those for fish meal. It is either steam dried or sun dried, with the former method being preferable. On the average, it contains 32% protein and 18% mineral.

Crab meal, the by-product of the crab industry, is composed of the shell, viscera, and flesh. Mineral content is exceedingly high, about 40%, and protein content is above 25% (generally about 30%). A rule of thumb for feeding crab meal is that 1.6 lb of crab meal can replace 1 lb of fish meal.

Both shrimp and crab meal are relatively high in chitin, a glycoprotein of low digestibility, which makes up much of the shell.

12. **Unprocessed fish and fish scraps.** The feeding of unprocessed fish is limited primarily to mink and other fur animals. Both freshwater and saltwater fish are fed to these carnivores. Care should be taken in the feeding of raw fish because many of these fish contain antimetabolites that can cause disease problems, such as cotton fur, steatitis, and Chastek paralysis in mink. For more information concerning the use of unprocessed fish as feed, the reader is referred to Chapter 27, Feeding Mink.

NONPROTEIN NITROGEN (NPN) FEEDSTUFFS

Feedstuffs which contain nitrogen in a form other than proteins or peptides are termed nonprotein nitrogen (NPN). Since microorganisms in the rumen of ruminant animals degrade dietary protein to synthesize microbial protein, it follows that if one feeds readily available carbohydrate and nonprotein nitrogen sources, both precursors of amino acids, microbial protein can be successfully synthesized.

Although urea dominates the market, other nonprotein nitrogen (NPN) sources, such as ammonium salts, and ammoniated by-products, have been and are being used successfully in ruminant rations. Some of the nonprotein nitrogen sources are listed in Table 11-9.

Table 11-9
SOME NONPROTEIN NITROGEN SOURCES FOR RUMINANTS

	Formula	Nitrogen Content	Protein Equivalent[1]
		(%)	(%)
Ammonium acetate	$CH_3CO_2NH_4$	18	112
Ammonium bicarbonate	NH_4HCO_3	18	112
Ammonium carbamate	$NH_2CO_2NH_4$	36	225
Ammonium lactate	$CH_3CHOHCO_2NH_4$	13	81
Biuret	$NH_2CONHCONH_2$	35	219
Urea, pure	$(NH_2)_2CO$	46.7	292
Urea, feed grade[2]	—	42-45	262-281

[1]Nitrogen × 6.25.

[2]Feed grade urea is diluted with varying amounts of materials to prevent lumps forming.

(Also see Chapter 4, section on "Nonprotein Nitrogen.")

Urea

The structural formula of urea is given in Fig. 11-17.

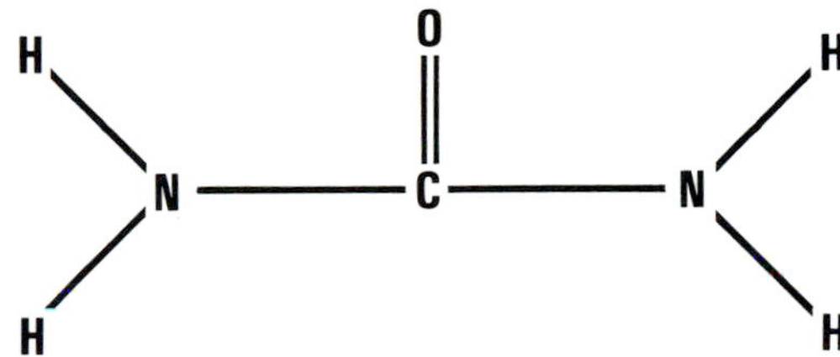

Fig. 11-17. Urea.

Crude protein is determined by multiplying the nitrogen content of a feedstuff by 6.25. Thus, feed grade urea (45% nitrogen) would have a crude protein equivalent of 281%. However, urea contains no precursors for the formation of a carbon skeleton for protein. Therefore, some readily available carbohydrate must be fed simultaneously to maximize protein production.

For specific recommendations as to the proper levels of urea or other NPN sources that can be used in livestock feeds, the reader is referred to the appropriate chapters of Section III in this book.

HISTORY OF UREA AS A LIVESTOCK FEEDSTUFF

The interest in the chemistry of urea dates back more than 200 years, when Rouelle first identified the compound in urine in 1773. Various workers at the time began to test the physiological roles and implications of urea, and by 1823, two French scientists, Prevost and Dumas, had determined that urea was removed from the blood by the kidneys and hypothesized that it was formed in the liver. The following year, Proust determined the empirical structure of urea. In 1828, Wohler demonstrated the synthesis of urea from inorganic compounds.

During the latter part of the 19th century, scientists began to study the role of ruminal microorganisms in digestion. By

1891, German researchers Zuntz and Hagemann had clearly shown that the microorganisms in the rumen could convert nonprotein nitrogenous compounds into essential nutrients. The role of nonprotein nitrogen in ruminant nutrition was subsequently studied and investigated throughout the first 30 years of the 20th century—primarily by German workers. By 1913, the Germans had developed a way of manufacturing ammonia directly from atmospheric nitrogen. By 1924, Morgen and his German co-workers had demonstrated that 30 to 40% of the protein in the ration of a sheep could be replaced by urea.

The use of urea in ruminant rations was generally viewed with skepticism in the United States and Great Britain through-out the 1920s and 1930s. This resistance began to weaken in the late thirties. By then, a distinct difference in protein nutrition between ruminants and monogastric animals had been established. Unlike protein nutrition in most nonruminants, the emphasis of protein in ruminant rations had been placed on quantity, not quality. While this newborn interest in urea was being generated in the United States, Europeans were already using 10,000 tons of urea per year in sheep and cattle feeds by the mid-thirties.

In 1949, Loosli at Cornell, using a nearly protein-free purified diet with lambs, produced specific evidence (1) that microbial action in the rumen can synthesize from urea and associated carbohydrates all the 10 amino acids required for rat growth, and (2) that 3 to 10 times as much of each amino acid was excreted as was fed.[7]

The ability of ruminants to use rations containing large quantities of nonprotein nitrogen was dramatically illustrated in an experiment conducted by the Finnish Nobel laureate A. I. Virtanen.[8] Lactating cows averaged over 8,800 lb of milk per year despite the fact they were fed a purified ration containing no preformed protein. Urea and ammonium salts were the only nitrogen sources, and purified carbohydrates were the only energy sources. While these cows were managed under carefully controlled experimental conditions, and their production level was low in comparison with today's high producers, the study demonstrated conclusively the qualitative value of nonprotein nitrogen. Additionally, the Finland study showed that when 20% of the dietary nitrogen came from protein, milk production increased. Thus, depending entirely upon microbial protein is possible, but it is neither efficient nor practical.

Thus, the use of urea in ruminant rations was catapulted to the forefront of protein nutrition. Today, urea is routinely used in ruminant rations in Europe and the United States. It is estimated that about 265,000 tons of urea are fed annually in the United States (see Table 6-3).

CONVERSION OF UREA TO PROTEIN

Our current understanding of the pathway of dietary NPN metabolism by rumen microbes is shown in Fig. 11-18.

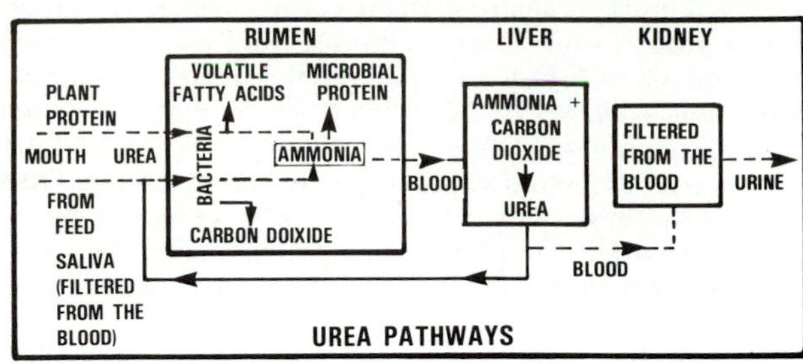

Fig. 11-18. The urea pathway.

When urea is fed to ruminants, it is first broken down in the rumen into ammonia (NH_3) and carbon dioxide by an enzyme—*urease*—produced by certain microorganisms. Paralleling the catabolism of urea in the rumen, carbohydrates (chiefly starch from cereal grains) are degraded by other microorganisms to produce volatile fatty acids (acetic, propionic, and butyric), which are used as an energy source.

In their multiplication and growth, rumen microorganisms use the ammonia released from the breakdown of protein and nonprotein nitrogen compounds (like urea), along with carbon skeletons and reduced sulfur, to form amino acids which are, in turn, manufactured into microbial protein.

For ruminants to use urea, the microorganisms in the rumen must be able to convert the ammonia released from the urea into microbial protein. As shown in Fig. 11-18, the ammonia released from the urea can go via two pathways: (1) It can be made into microbial protein, or (2) it can be absorbed through the rumen wall into the blood stream, which carries it to the liver. The liver detoxifies ammonia by reconverting it to urea to be excreted in the urine. Some of the urea is recycled in the rumen, however, through saliva and by absorption from the blood through the rumen wall. If ammonia escapes the rumen too rapidly, the capacity of the liver is exceeded and ammonia spills into the main blood system. High levels of ammonia circulating in the blood can cause toxicity or even death. Thus, if urea is to be an effective source of true protein for ruminants, conditions in the rumen must be favorable for microorganisms to use the ammonia from urea before it escapes the rumen. So, whenever ammonia is overflowing the rumen and bacteria already have sufficient ammonia for metabolism, additional urea in the ration is of no value in meeting the protein needs of ruminants and may be harmful. This means that, for efficient microbial protein production, ammonia must be available when needed and at the levels needed.

The amount of urea that can be used by rumen microorganisms will depend upon the number of microbes and how rapidly they are growing and whether ammonia and other essential nutrients are available when needed. In addition to nitrogen, microbes need minerals, vitamins, and a readily available source of energy for fast growth. Therefore, more ammonia can be utilized when high energy feeds are fed. Cattle and sheep fed high grain rations can make greater use of urea than animals fed low energy roughage rations.

Throughout digestion, microorganisms are passed from the rumen to the more distal digestive organs. In the abomasum and small intestine, these microbes are hydrolyzed

[7]Loosli, J. K., *et al.*, "Synthesis of Amino Acids in the Rumen," *Science*, Vol. 110, 1949, p. 144.

[8]Virtanen, A. I., "Milk Production of Cows on Protein-Free Feed," *Science*, Vol. 153, 1966, p. 1603.

and digested to such a degree that the microbial protein is broken down to free amino acids which can then be absorbed by the host animal. So, once the protein, be it microbial or undegraded true protein, passes into the small intestine, it is digested and absorbed by the ruminant much like that by the nonruminant. Thus, the host animal itself does not utilize urea directly.

UREA FERMENTATION POTENTIAL

Urea fermentation potential (UFP) is a term used to indicate the amount of urea that can be utilized in a ruminant ration.

The UFP system was developed by Iowa State University scientists for the purpose of evaluating the urea fermentation potential of feeds, and, in turn, estimating the amount of urea that can be useful in a ruminant ration. *A positive UFP value of a feed or ration can be defined as the grams of urea per kilogram of feed dry matter consumed (or pounds per 1,000 lb) that can be useful in fermentation by microorganisms in the rumen.* A positive UFP value implies that this quantity of urea feeding is a satisfactory level for achieving maximum or near maximum formation of urea nitrogen into microbial protein.

Calculation of a UFP value involves the amount of fermentable energy present in a feed and the amount of ammonia formed from breakdown of the protein in a feed by rumen fermentation. A feed with a positive UFP value is one that has more fermentable energy present than that needed for transforming the ammonia degraded from its own protein into rumen microbial protein.

For example, corn is given a positive UFP value of 11.8 which means 11.8 g of urea is useful for fermentation by rumen bacteria for each kilogram (1,000 g) of corn dry matter fed. Another way to express this would be that 1.18 lb of urea can be used by bacteria for each 100 lb of corn dry matter fed (1.18%). UFP values for selected feeds are shown in Table 11–10.

TABLE 11–10
UREA FERMENTATION POTENTIAL (UFP)[1]
(VALUES ON DRY MATTER BASIS)

Feed	TDN	Protein	Estimated Protein Degraded In Rumen	UFP[2]
	(%)	(%)	(%)	(%)
Corn	91	10	62	1.18
Milo	80	12.4	52	0.68
Oats	76	13.2	70	–0.47
Corncobs	47	2.8	75	1.0
Corn silage	70	8.1	68	0.64
Fescue hay	62	10.5	85	–0.87
Orchardgrass hay	57	9.7	85	–0.82
Soybean meal	81	51.5	75	–10.7

[1]Table 11–10 was prepared by Iowa State University scientists and was taken from *Agricultural Guide*, "Urea Supplements for Beef Cattle," published by the University of Missouri-Columbia, 1983.

[2]The pounds of urea that can be fermented for 100 lb of dry matter in the feed. To change to g/kg mutiply by 10.

A feed with a negative UFP value is one that has less fermentable energy than needed to transform into microbial protein all the ammonia arising from the breakdown of its protein during rumen fermentation. Urea addition to this type of feed would be of no value in satisfying the metabolizable protein requirements of the ruminant host; it would only add to the surplus nitrogen load in the rumen. An example is fescue hay with a negative UFP value of –0.87. This value indicates that the available energy in fescue hay is not great enough to transform to microbial protein all the nitrogen that is released by the degradation of its own protein in the rumen.

NOTE WELL: The UFP values given in Table 11–10 are only guides for the use of urea in ruminant rations. The protein and energy levels of the feed and other factors change these UFP values.

ESSENTIALS FOR OPTIMUM USE OF UREA

In this section, discussion centers on the proper use of urea. But the same principles apply to all NPN products.

Urea can be successfully and effectively used, or it can be abused. Observance of the following pointers will assure optimum results:

1. **Feed urea only to ruminants.** Never feed urea to nonruminants—swine, poultry, or horses.[9]

2. **Supply adequate energy.** The best utilization of urea is obtained when the ration contains adequate energy, preferably from readily available carbohydrates (grain), because they provide carbon skeletons in addition to energy, which, along with ammonia, are needed for making microbial protein. Generally speaking, urea can be used effectively in high-energy rations (greater than 75% TDN on a dry basis) and when the crude protein content is below 12 or 13%. As an energy source, molasses is good, but its highly fermentable sugars may be used up too quickly. Energy from roughages is made available too slowly in the rumen for good utilization of urea by bacteria.

3. **Provide the necessary minerals.** Because many of the minerals are constituents of key cofactors involved in the production of microbial protein, the mineral requirements of rumen microorganisms must be supplied if they are to make the best use of urea. In addition to the major and minor minerals normally included in ruminant rations, sulfur must be provided in high-urea rations to allow bacteria to synthesize the sulfur-containing amino acids, cystine and methionine. The nitrogen-sulfur ratio should not be wider than 15:1.

4. **Mix well.** Thorough mixing of urea will lessen the hazard of overconsumption by an individual animal and possible toxicity.

5. **Don't feed excessive levels of urea.** Urea should not supply more than ⅓ of the protein equivalent in the total ration. Generally, it should not be fed in excess of 1% of the total ration, or 3% of a concentrate mix, of finishing cattle or lambs or of high-producing dairy cows. Using urea to increase the protein level of high-energy rations containing more than 12 to 13% crude protein on a dry matter basis seldom gives any increase in performance because urea is of no value if the rumen bacteria already have sufficient nitrogen.

[9]Most state laws restrict the use of urea to ruminant rations; hence, it is illegal to add it to swine, poultry, or horse rations. Urea may be toxic to foals.

Also, excess urea in the ration may cause acute or chronic ammonia toxicity.

(For recommended levels of urea for beef cattle, dairy cattle, sheep, and goats, see the respective chapters devoted to each class of animals, Chapters 19, 20, 21, and 22.)

6. **Do not use urea as the sole source of N for (a) young, rapidly growing ruminants or (b) high producing, lactating ruminants.** For maximum production, these animals require more than microbial proteins; in addition, they need some true protein in the feed which will escape breakdown in the rumen and become digested in the small intestine.

7. **Give animals time to adjust to urea.** Animals started on a feed containing urea should be given an initial period of adaptation, the length of which depends on the class of animal and the stage and level of production. Generally 5 to 7 days is adequate—except for newly arrived and heavily stressed animals.

8. **Give heavily stressed, new arrivals time to adjust.** Do not feed urea to newly arriving cattle or sheep, many of which may have been off feed for 2 or 3 days. Preferably, use true protein supplements for the first 2 to 3 weeks.

9. **Feed frequently.** Allowing animals to eat two or more times daily provides a more uniform entry of urea into the rumen and helps prevent a burst of ammonia release that may exceed the capacity of the bacteria to use it. Urea should never be used in a feeding program when feed containing urea is fed infrequently or at irregular intervals, such as when range cubes are fed two or three times weekly.

10. **Enhance urea utilization by proper fermentation, protein degradation, and length of time in the rumen.** Urea utilization is enhanced by the capacity of the rumen fermentation to produce microbial protein; the proper extent of dietary (preformed) protein degradation in the rumen—there should be neither too much nor too little; and moderate microbial protein passage to the abomasum—it should be neither too fast nor too slow.

11. **Include alfalfa meal in urea rations.** As a source of unidentified factors, alfalfa may stimulate microbial synthesis of protein.

12. **Provide vitamin A.** The proper level of vitamin A should be provided, along with such other vitamins as needed and additives as desired.

13. **Include adequate salt for palatability.** The palatability of urea may be enhanced by including 0.5% salt in complete rations and 3.5% salt in protein supplements.

14. **Use a free-flowing urea.** This is important for proper mixing.

15. **Never use urease-containing seeds in urea rations.** Soybeans or raw beans of any kind, lespedeza seed, alfalfa seed, or wild mustard seed should not be used in concentrate mixtures containing urea for the reason that the enzyme urease in them will break down urea rapidly into ammonia and carbon dioxide. The liberated ammonia may be strong enough to be objectionable to animals.

(Also see Chapters 19 and 20, relative to feeding urea to beef cattle and dairy cattle.)

PERTINENT FACTS ABOUT UREA

The following facts are pertinent to the informed use of urea:

1. **Feed grade urea.** Initially, the protein equivalent value of feed grade urea was 42% (nitrogen) times 6.25 (common protein factor), or 262%. Today, more concentrated 45% nitrogen (45 × 6.25 = 281) urea has replaced most of the 42% grade, at a lower unit cost.

2. **Feeding value of urea.** Attempts have been made to equate urea to oil meals by various thumb rules. One such thumb rule is that 1 lb of urea plus 6 lb of corn equals 7 lb of soybean (or cottonseed) meal. This combination of corn and urea supplies as much nitrogen as does soybean meal and, thus, could be considered equal to it in crude protein content. This is true if the rumen microorganisms can convert the urea nitrogen to protein. But this doesn't tell the whole story! Table 11–11 shows the inequalities in mineral content of these two feeds.

TABLE 11–11
COMPARISON OF A CORN:UREA (6:1) MIXTURE WITH SOYBEAN MEAL

Nutrient	Corn:Urea (6:1) as a % of Soybean Meal
TDN	88.9
N	100.0
Ca	13.5
P	38.1
Mg	36.0
K	14.7
S	9.1
Mn	14.6
Co	95.2
Cu	20.0
Fe	13.1
Zn	63.9

As noted in Table 11–11, the corn-urea combination is low in TDN, or energy. Additionally, the combination supplies only 13.5% as much calcium and 38.1% as much phosphorus as does soybean meal; and only in the case of cobalt does the combination supply as high a mineral level as does the soybean meal. Unless these deficiencies are met, poor utilization of the urea-containing ration can be expected. Of course, minerals can be added, but all ingredients cost money.

3. **Urea is best utilized in well-balanced, high-energy rations.** Urea is not well utilized in supplements to low-quality roughages. The explanation is that the carbohydrates in grasses and hays appear to be so slowly available that the bacteria have difficulty in using the energy from roughages to make use of urea in preparing bacterial protein. It is generally held that some preformed protein should be present in the feed, also. Part of this will be provided by the grains, and frequently some oil meals are used in preparing the formula feeds.

Other components of a balanced feed include calcium, phosphorus, iron, copper, cobalt, manganese, iodine, and perhaps zinc, sulfur, and magnesium. The need for these minerals, as well as for vitamin A, will depend upon local conditions with respect to the types of roughages produced and the influence of weather upon the quality of such roughages.

4. **Toxicity.** When urea is fed at excessive levels, large amounts of ammonia are liberated in the rumen. Eventually, the pH of the ruminal fluid increases, thus facilitating the

passage of ammonia across the rumen wall. If the levels of ammonia absorbed are greater than the ability of the liver to convert ammonia to urea, ammonia accumulates in the blood. If blood ammonia levels reach toxic concentrations (1mg/100 ml in cattle), the animal shows signs of acute ammonia poisoning. Cattle become susceptible to urea toxicity under the following conditions:

a. Poor mixing of feed.
b. Errors in ration formulation.
c. Inadequate period of adaptation.
d. Low intake of water.
e. Feeding of urea in conjunction with poor-quality roughages.
f. Low feed intake prior to exposure to feed containing urea.
g. Rations that promote a high pH in ruminal fluid.

Symptoms of ammonia toxicity resemble those observed in strychnine poisoning—tetany, respiration difficulty, bloat, excessive salivation, ataxia, convulsions, and bellowing. If not promptly treated, death will follow in about 30 minutes to 2.5 hours.

The common treatment of ammonia toxicity in cattle consists in drenching with 5 to 8 gal of cold water. This lowers the temperature of the ruminal fluid, thereby inhibiting ureolytic activity while diluting the concentration of ammonia. If about 1 gal of dilute acetic acid—for example, vinegar—is given in addition to the cold water, the basic concentration of the ruminal fluid will become neutralized.

(Also see Chapter 5, Table 5-3, Potential Poisons—Urea Toxicity.)

5. **Palatability.** Although various opinions exist relative to the palatability of urea and urea-containing feeds, most feeders feel that urea is not palatable and, therefore, that feed consumption may be lowered in comparison with rations in which oil meal protein supplements are used. For this reason, care should be exercised in selecting an appetizing urea-supplemented mixture.

In contrast to the above opinion, it should be noted that, occasionally, cattle will consume straight fertilizer urea or amonium nitrate in sufficient amounts to poison themselves.

Sometimes cattle will consume a urea-containing feed for a few days or weeks, then refuse it. This has occurred in drought areas where farmers have tried to extend their roughage supplies by feeding straw and other mineral poor, low quality roughages.

6. **How to compute how much urea is in a feed.** The level of urea in a feed may be determined in the following ways:

a. **Percent of urea in the feed.** When the percent of urea is given, one can calculate the amount of protein furnished by urea by multiplying the percent urea by 281% (the protein equivalent of urea). For example, if a 40% supplement contains 5% urea, then 14% protein is furnished by urea (281% × 5% = 14%). To determine the percent of the total protein furnished by urea, divide the percent of protein as urea by the percent of protein in the supplement (14% ÷ 40% = 35%). In this case, slightly more than ⅓ of the protein in the supplement is furnished by urea.

b. **Percent protein as urea.** When the urea in the supplement is expressed in percent protein as urea, one can determine the amount of urea by dividing this value by 281%. For example, if a 35% protein supplement has 12% protein as urea, it contains 4.3% urea (12 ÷ 281 = 4.3%). Slightly more than ⅓ of the protein in the supplement is furnished by urea (12% ÷ 35% = 34.3%).

METHODS OF FEEDING UREA

Urea can be provided by different methods and systems, with consideration given to the following factors: (1) protein needs of the animal as dictated by type of production; (2) availability of urea; (3) toxicity hazards; (4) cost of processing and mixing; and (5) availability of energy sources and amount of plant protein being used. The common methods of feeding urea follow:

1. **Urea mixed in concentrate.** Most of the urea fed to growing and lactating dairy cattle and to finishing beef cattle is incorporated into the concentrate portion of the ration. It can be supplied in a protein supplement or in the entire concentrate ration. Also, urea may be provided in a complete mixed feed.

Urea can be mixed in feed either as a powder or as an aqueous solution. Both methods are relatively simple and inexpensive. When urea is added as a powder, there is a chance that it will sift through the grain and be unevenly distributed, thus increasing the chance of toxicity. However, if careful mixing procedures are followed, this hazard can be kept to a safe minimum.

2. **Urea liquid supplements.** Liquid supplements combining molasses for energy and urea as a protein precursor are widely used. This type of supplement can also be used as a carrier for micronutrient and nonnutritive additives. It is fed primarily in conjunction with low-quality forages.

The **advantages** of liquid supplements are: (a) They provide a way to assure uniform distribution of urea throughout the supplement; (b) they are adapted to limited feeding—cattle consumption can be regulated by means of a lick-wheel feeder or the addition of a bitter principle in the liquid supplement; (c) they require less labor to feed; and (d) they eliminate dustiness of the ration and loss from blowing.

Fig. 11-19. Lick-wheel in use, for limiting consumption of urea liquid supplement. (Courtesy, *Feed Management,* Watt Publishing Co., Mt. Morris, Ill.)

The **disadvantages** of liquid supplements are: (a) They require special mixing and feeding equipment; (b) they require greater transportation costs; (c) they are not well suited to keeping calcium and certain other nutrients in solution; (d) they may subject urea to degrading, once it is in solution, during prolonged storage; and (e) they tend to be highly corrosive.

3. **Urea salt blocks.** Another simple way of supplying protein precursors to livestock on pasture is through the use of urea in salt licks or blocks. This practice is used extensively where large and/or inaccessible ranges limit the amount of contact that ranchers have with their animals. Numerous combinations of salt and urea have been used.

4. **Urea mixed with silage.** Excellent utilization of urea can be obtained when it is ensiled with cereal grain silage, especially corn silage. If chopped, whole plant corn is being ensiled at 35 to 40% dry matter, urea can be added at a level of .5% of the wet material. This level should increase the crude protein level of the silage on a dry matter basis about 5 percentage points. Urea levels higher than .5% can create palatability problems as well as storage problems. When the silage contains little or no grain, the amount of urea to be added should be reduced.

Silage tends to be variable in moisture, and this variability can affect the benefits of added urea. Hence, one should have a reasonable estimate of the moisture content of the material that is to be ensiled. Likewise, water in silage will create some leaching of the urea. Also, ammonia is produced during the ensiling process, representing an additional loss of urea.

(Also see Chapter 9, section headed "Feed Additives.")

5. **Urea added to dry roughages.** A limited amount of research has been reported on the addition of a urea-molasses mixture to hay. Based on available information, the addition of urea to good-quality hay (over 10% crude protein) is generally not beneficial.

6. **Slow-release urea products.** Several products in which urea is bound in a slow-release complex have been developed in recent years; among them, urea combined with starch from grain, and urea combined with the sugars in molasses through heat and chemical treatment. These products are designed to decrease the solubility of urea in the rumen and thereby slow the release of ammonia. Slow ammonia release, or a more uniform ammonia level in the rumen throughout the day, is desirable, especially for urea used in low-energy rations. Additionally, there should be less danger of urea toxicity from overconsumption with slow-release products.

(Also see Chapter 14, Feed Processing, section headed "Slow-release and Rumen Bypass Treatment.")

Other Nonprotein Nitrogen Feedstuffs

Nonprotein nitrogen sources other than urea have been used to a limited extent. These products can be classified into two categories: (1) ammoniated products, and (2) biuret.

AMMONIATED PRODUCTS

Numerous ammonium salts have been used effectively as sources of nitrogen. Both organic and inorganic salts of ammonia can be utilized efficiently by ruminants; among them, ammonium acetate, ammonium bicarbonate, ammonium butyrate, ammonium carbonate, ammonium chloride, ammonium lactate, ammonium polyphosphate, ammonium propionate, ammonium sulfate, monoammonium and diammonium phosphate, and methionine hydroxy analog (MHA) (the latter is used as a source of methionine for poultry).

A wide range of feedstuffs has been treated with anhydrous ammonia for a variety of reasons. One of the first synthesized protein supplements for cattle and sheep was ammoniated molasses. The ammoniation of beet pulp and citrus pulp followed. Cottonseed meal has been ammoniated for two purposes: (1) to inactivate aflatoxin-contaminated feed, and (2) to provide a supplemental source of nitrogen for livestock. Aflatoxin-contaminated shelled corn has been detoxified and made safe for animals by treatment with anhydrous ammonia. Rice hulls and rice straw have also been ammoniated.

Experiments have demonstrated that the addition of anhydrous ammonia to corn silage can double the protein equivalent and increase the feeding value of the silage dramatically. Huber and Santana, of Michigan State University, found that corn silage treated with ammonia at the rate of .28% compared favorably with silage treated with urea when milk persistency and feed intake of lactating cows were considered. Cows from both the urea-treated silage and the ammonia-treated silage exhibited better milk persistency than cows fed untreated silage. The cows also consumed more of the treated silage than of the untreated silage (see Table 11–12).

TABLE 11–12
MILK YIELDS AND FEED INTAKES OF COWS FED CORN SILAGE TREATED WITH UREA OR AMMONIA[1]

Item	Untreated Silage	Ration: Urea Silage	Ration: Ammonia Silage
Milk persistency (kg/day)	71	98	91
Feed intakes (% of body weight)			
Silage dry matter	1.12	1.51	1.67
Total dry matter	2.36	2.91	2.93

[1]Adapted by the authors from "Ammonia-Treated Corn Silage for Dairy Cattle," by J. T. Huber and O. P. Santana, *Journal of Dairy Science*, Vol. 55, 1972, p. 489.

In the late 1970s and the early 1980s, the ammoniation of low-quality forages increased greatly throughout North America, with the anhydrous ammonia applied to forages in round bales, stacks, and barns. The reasons:

1. The ammonia provides an economical source of nitrogen (protein equivalent) which can be utilized by ruminants.
2. The ammonia treatment increases the digestibility of fiber, thereby increasing the amount of energy that animals can obtain from the forage.
3. The ammonia is inhibitory to many microbes. As a result, treated hay can be baled at a high moisture content (up to 30%) without mold and/or bacterial spoilage, and with reduced weather and harvesting losses.

All went well until 1983, when rare episodes of toxicity in animals consuming certain ammoniated hays was reported.

The afflicted animals behaved abnormally, including running in circles, extreme nervousness, and convulsions. Many animals died, usually from injuries sustained while they were in an excited state. The problem was observed in cows, bulls, yearlings, newborn calves, and even in calves nursing cows fed ammoniated forage. The diagnosis: ammonia toxicity, also known as *bovine bonkers* or *crazy cattle disease*. After several cases were investigated and compared, it was determined that the syndrome resembled what had occurred 30 years earlier, in the 1950s, when ammoniated molasses was first fed to cattle and sheep. Subsequently, methods were developed to produce synthetic proteins that did not cause toxicity, and the problem was largely forgotten.

Ammoniated feedstuffs reported to have caused toxicity include alfalfa hay, barley hay, Bermudagrass hay, bromegrass hay, fescue hay, oat hay, orchardgrass hay, rice straw, sorghum forage, Sudangrass hay, wheat hay, and wheat straw. It is noteworthy that most of the toxic forages appeared to have been treated with high levels of ammonia (3% or more) and/or that the treatment was not evenly applied. It is noteworthy, too, that ammoniated beet pulp, ammoniated citrus pulp, ammoniated cottonseed meal, ammoniated shelled corn, and ammoniated corn silage have not been reported to have caused ammonia toxicity.

Although the specific toxin responsible for the malady has not been positively identified, the following preventive/control measures are recommended:

1. Do not ammoniate high-quality hay that is fed to lactating cows, sheep, or goats.
2. Do not treat hay with more than 1.0 to 1.5% ammonia by weight; and apply the ammonia evenly to the forage.
3. Do not ammoniate hay that contains more than 30% moisture.
4. Do not feed ammoniated hay alone—feed some grain with it.
5. If toxicity signs develop when feeding ammoniated forage, discontinue feeding the forage immediately. Affected animals may return to normal in 1 to 2 days; and the toxic hay will become less toxic with the passing of time.

(Also see Chapter 5, Nutritional Disorders/Toxins, Table 5, Potential Poisons—Urea Toxicity [Ammonia Toxicity].)

BIURET

In the past, biuret ($NH_2CONHCONH_2$) was used as a source of nonprotein nitrogen that liberated its nitrogen slowly, thus making for more efficient use of the nitrogen. In recent years, however, the cost of manufacturing biuret has been so high as to make it uneconomical as a feedstuff. Hence, it is not being produced as a commercial feed supplement at the present time. Because biuret will show up in the milk, it is not approved for use by lactating dairy cows.

SINGLE-CELL (MICROBIAL) PROTEIN (SCP)

Single-cell protein (SCP) is protein obtained from single-cell organisms, such as yeast, bacteria, fungi, and algae, that have been grown on specially prepared growth media. Production of this type of protein can be attained through the fermentation of petroleum derivatives or organic waste or through the culturing of photosynthetic organisms in special illuminated ponds.

Of course, yeast and bacteria have been used for centuries in the baking, brewing, and distilling industries, in making cheese and other fermented foods, and in the storage and preservation of foods. Dried brewers' yeast, a residue from the brewing industry, and torula yeast, resulting from the fermentation of wood residue and other cellulose sources, have been marketed as animal feeds for years.

A wide variety of materials can be used as substrates for the growth of these organisms. Current research deals with the use of products which otherwise would have little or no economic value; among them: straws; fodders, and other low quality cellulose wastes; wood and wood processing wastes; food, cannery, and food processing wastes; residues from alcohol production; and animal excreta.

The potential of single-cell protein as a high-protein source for both humans and livestock is enormous, but many obstacles must be overcome before it becomes widely used. It has been calculated that a single-cell protein fermenter covering .386 square mile could yield enough protein to supply 10% of the world's needs. To put this potential in a different perspective, a 1,000-lb steer will produce about 1 lb of protein per day. One thousand pounds of rapidly growing soybeans will produce 80 lb of protein per day. One thousand pounds of single-cell organisms might well produce up to 50 tons of protein per day. This is not to imply that we have solved the world's protein needs with single-cell protein because serious problems involving palatability, gastrointestinal disturbances, uric acid accumulation, and simple economics must be solved before wide-scale production of single-cell protein becomes a reality. There is a limited amount of single-cell protein on the market in the form of brewers' yeast and torula yeast, but these products are generally too expensive to use as a major protein supplement.

But science and technology have overcome the commercial production of single-cell protein! In 1981, England announced that, following 25 years of research on SCP, they were in full-scale production in the world's largest fermenter. They projected annual production of 50,000 tons of a 70% microbial protein supplement for animals, with the microorganisms feeding on methanol (methyl alcohol).

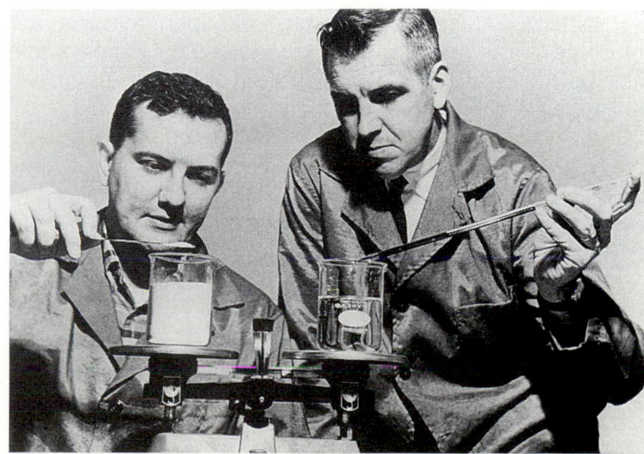

Fig. 11-20. Protein from oil through petroleum microbiology. (Courtesy, FAO, Rome, Italy)

Types of Single-Cell Protein

Single-cell protein can be produced by nonphotosynthetic and photosynthetic organisms. Of the nonphotosynthetic organisms, yeasts are the most popular sources, but bacteria and fungi are also currently being investigated as potential sources. The photosynthetic organisms—algae—are grown in ponds that are illuminated and fortified with simple salts such as carbonates, nitrates, and phosphates.

NONPHOTOSYNTHETIC ORGANISMS

Traditionally, yeasts have been used as sources of vitamins and unidentified factors. They are easily harvested and pose few consumer resistance problems. One disadvantage to the utilization of yeasts as protein sources is that they are deficient in the sulfur amino acids. Fortunately, methionine hydroxy analog (MHA) is a cheap commercial product that can make up for this deficiency.

Bacteria grow very rapidly and tend to have a more balanced amino acid profile than yeasts. Additionally, they generally contain more protein. Since humans will probably avoid this type of protein for esthetic reasons, this type of single-cell protein, in all likelihood, will be developed as livestock feed. Currently, two serious problems exist with the culture of bacterial protein: (1) susceptibility to phages, and (2) high ribonucleic acid content.

Until recently, the fungi have received little attention as single-cell protein. Since fungi can be harvested with relative ease and possess considerable enzymatic activity, they will likely be intensively studied in the near future.

PHOTOSYNTHETIC ORGANISMS

Photosynthetic organisms, such as algae, provide nutrients in relatively large quantities in a limited area. Depending on the type of organism, temperature, and latitude, yields of harvested protein can be attained on the order of 4 to 9 tons per acre yearly.

These organisms, containing from 5 to 15% dry matter, can be fed to livestock, either directly after harvesting or in a dehydrated form. Nutrient composition (moisture-free) tends to be rather variable, with protein ranging from 8 to 75%, carbohydrate from 4 to 40%, lipids from 1 to 86%, and ash from 4 to 45%. The protein obtained from this type of single-cell protein is deficient in the sulfur amino acids. The biological value of algae protein has been estimated to range from 50 to 70.

Using the power of the sun and single-cell algae, Israeli scientists are now turning domestic waste water into valuable animal feed and, at the same time, producing purified water for irrigation—all at minimal cost. The algae are dried, mixed with other ingredients, and fed to poultry, fish, and cattle. The algae production of SCP points the way to a cheap, solar powered system of protein production.

Current Problems Associated with Single-Cell Protein

Although single-cell protein appears to be an excellent alternative source to the protein feeds currently used, several problems must be overcome before it becomes a widely used feedstuff; among them, palatability, digestibility, nucleic acid content, toxins, protein quality, and economics.

- **Palatability**—Microbial cells must be processed to some extent if they are to be palatable. Yeasts are bitter tasting, and algae and bacteria have characteristically unpleasant tastes which can depress feed intake. At the present time, processes making single-cell protein tasteless are being developed so that this source of protein can be used as a supplement to the more conventional feeds.
- **Digestibility**—Ways of making single-cell protein more digestible must be developed if it is to be competitive with the traditional protein feeds. Digestibility among single-cell product sources tends to be extremely variable. If the organisms are not killed prior to being used as a feed, digestibility is dramatically reduced. Certain forms of processing can improve the digestibility of certain single-cell protein products. For example, if algae are cooked prior to use, digestibility can be doubled in many cases. However, processing beyond the killing of yeasts does little to improve digestibility.
- **Nucleic acid content**—Much of the nitrogen found in single-cell organisms is in the form of nucleic acids. When the purines of the nucleic acids are metabolized, uric acid is formed. In humans, uric acid is relatively insoluble with the result that uric deposits accumulate leading to kidney stones or gout. In birds and many mammals, uric acid is further metabolized to allantoin by the enzyme uricase (see Fig. 11-21). Allantoin is much more soluble than uric acid and can be excreted from the body with relative ease. This is one reason why ruminants do not have problems resulting from the digestion of ruminal microorganisms.

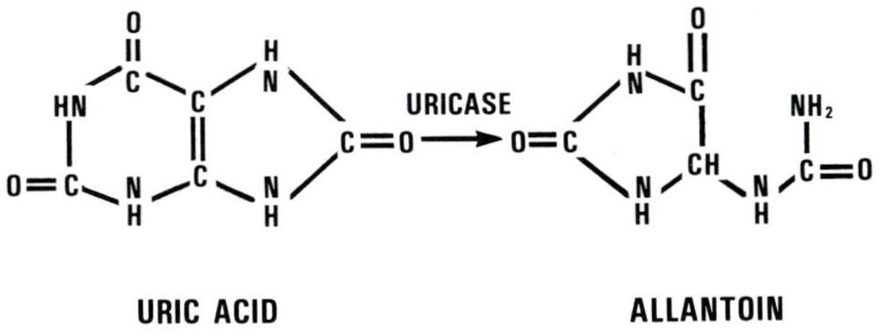

URICASE

URIC ACID → ALLANTOIN

Fig. 11-21. Conversion of uric acid to allantoin.

Although the feeding of SCP to livestock may not produce physiological problems, it will be necessary to demonstrate to the consuming public that problems will not arise from the consumption of animal products produced from the feeding of SCP.

- **Toxicities**—Toxicities arising from the use of single-cell protein can result from two sources: (1) from toxins produced by the microorganisms themselves, and (2) from contaminated microorganisms. The second type of risk is probably the most likely to occur. Much of the single-cell protein will be derived from processes whereby by-products of industry are used as substrates. Thus, the microorganisms will in some ways reflect the chemical composition of the by-product. For example, if chemical residues, such as pesticides, or large amounts of trace minerals are present in the by-product substrate, the microorganism will, in all likelihood, absorb these chemicals. When livestock consume these contaminated microorganisms, toxicities may result.

• **Protein quality**—Research to date has indicated that SCP is deficient in the sulfur-containing amino acids and possibly in isoleucine. While the amino acid profile of SCP is more balanced than that of the cereal grains, it is clearly inferior to the traditional protein supplements. However, this inferiority can often be nullified through the addition of commercially available methionine.

• **Economic problems**—So long as the more traditional sources of protein—such as fish meal, the oilseed meals, and meat products—are readily available at reasonable prices, single-cell protein will remain a relatively obscure alternative. But, unlike the United States, many countries cannot produce sufficient animal protein supplement by conventional agriculture. For them, the production of single-cell protein may be the answer.

As the world's human population increases, there will be an increasing demand for cheap protein. This demand could dry up the sources of the tradional protein feeds for livestock, thereby opening the way for the intensive development of alternative sources of protein.

We are also entering an era of concern for the maintenance of the quality of our environment. This means that a greater emphasis will be placed on the transformation of industrial by-products into usable commodities. Single-cell protein is one way that this challenge can be met.

TYPES AND METHODS OF FEEDING PROTEIN SUPPLEMENTS

There is no one best type (or form) and method of feeding protein supplements for any and all conditions. When mixed feeds (either complete rations or concentrates) are fed, usually the protein ingredient(s) is incorporated in the mix. When the protein supplement is fed separately, such as on the pasture or range or in the corral, any one of the following types and methods of feeding may be used: (1) hand-fed at intervals, (2) liquid protein supplements, (3) protein blocks, (4) range cubes or pellets, or (5) self-fed in salt-feed mixtures.

Hand-feeding at Intervals

Fig. 11-22. Hand-feeding range cubes. (Courtesy, USDA)

The Texas Station compared the following intervals of range feeding: (1) daily feeding, (2) twice weekly feeding, and (3) 3 times weekly feeding; with each group hand-fed. Cottonseed meal was used as the supplement at 2 levels: (1) 14 lb/head/week, and (2) 21 lb/head/week.

As a result of a 4-year study it was reported that—

1. The group fed twice weekly had a slight advantage over feeding daily and 3 times weekly in (a) weight change of cows, (b) percent calf crop weaned, and (c) weaning weight of calves.
2. Twice weekly feeding saved 60% in time and equipment over daily feeding.
3. Twice weekly feeding did not cause any digestive disturbances, even when fed 10½ lb at one time (½ the allowance of the 21 lb/week level). It took these cows about 2 hours to consume their 10½ lb share of cottonseed cake, which gave the timid and the slow eaters an opportunity to get their share.
4. The cows fed twice weekly grazed more widely over their range than those fed daily.

Based on the above study, plus observations and experiences, the authors recommend feeding nonurea range supplement twice weekly, allocating in each of the two feedings one-half as much supplement as would have been fed in a week on a daily feeding basis.

Protein cubes may be scattered on the ground—2 or 3 times a week. This offers a method of checking the animals because they are attracted by the sight or sound of the vehicle when they know that there is something to eat.

Twice weekly feeding has two distinct advantages over the use of salt-feed mixes: (1) It alleviates the cost of using excess salt, which has no nutritive value when so used; and (2) it forces inspection of the herd two times per week, which is as infrequent as is desirable.

Liquid Protein Supplements

Liquid supplements are the fastest growing type of protein supplement in the United States. They are fed primarily to beef cattle (both breeding and finishing cattle) and dairy cattle; they are fed in the corral and on the pasture and range; and they are incorporated in mixed feed, used as a top dressing, and self-fed.

The formulation and manufacture of modern liquid supplements makes use of sophisticated technology involving suspensions, emulsions, and slurries. Molasses may or may not be the major component of the new liquid supplements; if the formulation calls for blended feed fat and/or a buffering agent, some manufacturers can add them. Moreover, today's liquid supplement manufacturers do not have to rely on urea or other nonprotein nitrogen sources as the only source of protein (or protein equivalent)—they can also provide conventional sources of true protein, including rumen bypass products. Minerals and vitamins can be incorporated as needed. So, today's liquid supplements can be tailor-made to meet the nutritional needs of beef and dairy cattle at any age or production level, and in the corral or on the range. This means that the producer may now choose between liquid and dry supplements on the basis of service, cost, convenience, and performance.

When self-fed, consumption may be controlled by use of a *lick tank* and/or by incorporating in the formulation such ingredients as phosphoric acid, beet solubles, and citrus peel liquor.

In the United States, the vast majority of the liquid protein supplements are patented. Also, they are difficult to home mix. As a result, few universities have done experimental work on them. However, Australia has long fed a great deal of liquid supplements on the range, due to its frequent droughts. Also, daily feeding on its vast stations (ranches), many of which stretch over many miles, isn't practical—it requires too much labor and travel. Hence, the Australians are more experienced than we are when it comes to using liquid feeds to supplement grass and to save labor. The senior author of this book has come to know their formulations and techniques, firsthand, in three visitations that he made "down under." Their usual feeding directions are as follows:

1. Do not feed urea-containing supplement to animals that have been without feed for 36 hours until they have had a chance to fill the rumen with grass.
2. Restrict consumption to a desired level (1 to 2 lb/head/day), since some animals tend to overeat. This may be accomplished by a free-turning plastic or wood wheel (a lick wheel) dipped in a tank or other similar equipment, or by adding phosphoric acid to the mix at a level of 1% phosphorus. The phosphoric acid also serves as a source of phosphorus and keeps the molasses free-flowing in cold weather.
3. Allow several days for cattle grazing dry grass to become accustomed to a liquid protein supplement.
4. Never let the animals run out of the liquid supplement once feeding is started, until you discontinue feeding it entirely.

Protein Blocks

Protein blocks are just what the designation implies. They are compressed protein blocks, ranging in weight from 50 to 500 lb each. They are particularly adapted to supplementing cattle on rather inaccessable range when the grass dries.

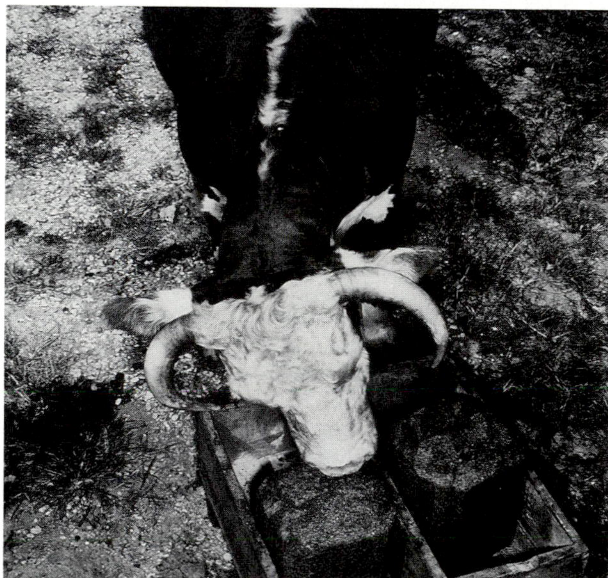

Fig. 11-23. Protein block in use on pasture—a means of lessening the labor attendant to the daily feeding of a protein supplement on pasture or range. (Courtesy, Moorman Mfg. Co., Quincy, Ill.)

Blocks may be placed in grazing areas where cattle have frequent access to them, with one 50-lb block provided to 15 cows. Intake will vary with the feed supply and the type of block. Generally, it is planned to limit feed consumption to about 2 lb/head/day by intake limiters such as hardness of block, salt, and/or fat content.

Range Cubes or Pellets

Traditionally, cattle have been supplemented either once or twice daily on pasture or range. Where this practice is followed, a urea-containing (preferably, a slow-release urea) range cube or pellet, similar to the formulation shown in Table 11-13 may be used. Cubes may be scattered on the ground. The formulations of two range cubes or pellets are given in Tables 11-13 and 11-14; one with urea, the other without.

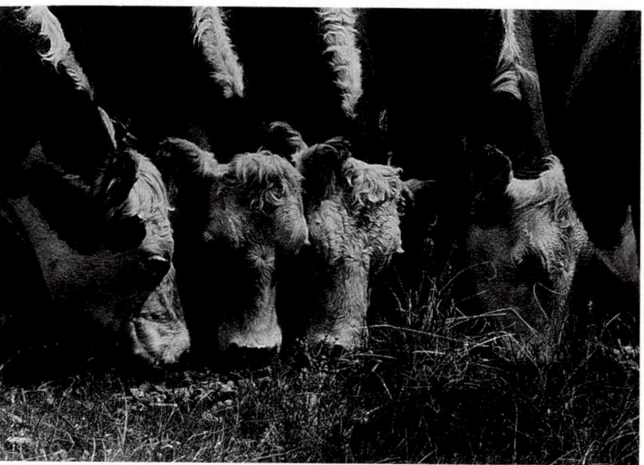

Fig. 11-24. Range cubes fed on pasture or range. Many producers prefer this method of supplementation, primarily for reasons of convenience and to reduce losses from wind blowing. (Courtesy, Ralston Purina Company, St. Louis, Mo.)

TABLE 11-13
RANGE CUBE OR PELLET, WITH UREA
(PREFERABLY A SLOW-RELEASE PRODUCT; AS-FED BASIS)

Ingredient	Percent	Per Ton
	(%)	(lb)
Corn #2 (barley, wheat, oats, and/or milo)	34.7	694
Soybean meal, 44% (cottonseed, linseed[1], and/or peanut meal)	32.5	650
Alfalfa meal, 15%	15.0	300
Molasses, sugarcane	10.0	200
Urea, 45% grade (use slow-release product)	4.0	80
Dicalcium phosphate, or equivalent	2.0	40
Salt	1.0	20
Trace minerals[2]	.5	10
Vitamin A[3] (30,000 IU/g potency)	.3	6
Total	100.0	2,000
Proximate analysis:	(%)	
Crude protein[4]	31.8	
Fat	2.2	
Fiber	6.6	
Calcium	.9	
Phosphorus	.7	
TDN	67.5	

[1]If linseed is used, limit to 6% of the ration.

[2]Trace minerals should be in keeping with Table 19-7. Generally, trace minerals can best be provided by a mineral or feed company, rather than home mixed.

[3]In low-sunshine areas, also add 6 million IU of vitamin D/ton of finished feed.

[4]This includes not more than 11.24% equivalent protein from nonprotein nitrogen; 34.9% of the total protein is furnished by urea (use a slow-release product).

TABLE 11–14
RANGE CUBE OR PELLET, WITHOUT UREA (AS-FED BASIS)

Ingredient	Percent	Per Ton
	(%)	(lb)
Soybean meal, 44% (or cottonseed meal)	72.7	1,454
Alfalfa meal, 15%	15.0	300
Molasses (sugarcane)	8.5	170
Dicalcium phosphate, or equivalent	2.0	40
Salt	1.0	20
Trace minerals[1]	.5	10
Vitamin A[2] (30,000 IU/g potency)	.3	6
Total	100.0	2,000
Proximate analysis:	(%)	
Crude protein	35.9	
Fat	1.2	
Fiber	8.3	
Calcium	1.01	
Phosphorus	.9	
TDN	68.7	

[1]Trace minerals should be in keeping with Table 19–7. Generally, trace minerals can best be provided by a mineral or feed company, rather than home mixed.

[2]In low-sunshine areas, also add 6 million IU of vitamin D/ton of finished feed.

Urea-containing supplements, particularly those containing high levels of urea, should not be fed at long intervals (such as twice weekly) on the range because (1) range forages are relatively low in energy, and (2) urea is extremely soluble and its nitrogen becomes available very quickly in the rumen. Also, it is important that NPN-containing range supplements provide readily available energy and needed minerals.

Self-feeding Salt-Feed Mixtures

The practice of using salt as a governor to limit feed consumption on pasture or range has been around a very long time. It was ushered in as a laborsaving device for cattle and sheep in inaccessible and rough areas.

Two suggested salt-meal supplements follow. Either 41% cottonseed or soybean meal may be used. Neither mix should be pelleted.

Ingredient:	Salt-Meal Mix No. 1	Salt-Meal Mix No. 2
	(lb)	(lb)
Salt	665	499
Meal (either 41% cottonseed or soybean meal)	1,331	1,497
Vitamin A (30,000 IU/g)	4	4
	2,000	2,000
Consumption level:	approx. 1½ lb daily	approx. 2 lb daily
Guarantee:		
Crude protein	min. 27%	min. 30%
Salt	max. 35%	max. 27%
Vitamin A	24,000 IU/lb	24,000 IU/lb

Based on experiments and experiences, the following points are pertinent to self-feeding salt-feed mixtures to range cattle and sheep, to cattle grazing stalk fields, or to cattle that are being grain-finished on pasture:

1. The practice need not be limited to any specific protein supplement or feed.
2. It is best that salt mixes be in meal form, rather than pelleted. If pellets are small and soft, they will work satisfactorily. However, there is always the hazard that they will be hard enough to permit animals to swallow them without the salt being fully effective as an inhibitor, with overeating resulting.
3. The proportion of salt and feed may vary anywhere from 5 to 40% (with 30 to 33⅓% salt content being most common), with the actual intake of feed supplement limited to 1 to 2½ lb daily. By varying the proportion of salt in the mixture, it is possible to hold the consumption of feed supplement to any level desired. In some range areas, a reduction of the salt level from 33⅓ to 24% will increase consumption by about 50%. When a liberal feeding of grain on pasture is desired, 5% salt may be sufficient.[10]
4. The quantity of salt and the proportion of salt to supplement required to govern supplement consumption varies according to (a) the daily rate of feed consumption desired, (b) the age and weight of animals (higher quantities of salt are required in the case of older animals), (c) the fineness of the salt grind (fine grinding lowers the salt requirement), (d) the salinity of the water, (e) the severity of the weather, (f) the quality and quantity of forage, and (g) the length of the feeding period (as animals become accustomed to the mixture, it may be necessary to increase the proportion of salt).
5. It is common practice to prepare the starting feed by mixing 1 lb of salt to 4 lb of feed supplement, and to increase the proportion of salt in the mixture as the animals become accustomed to the feed.
6. It lessens the difficulty in starting animals on a supplement, for sprinkling a little salt on the meal makes it more palatable.
7. It is recommended that animals be hand-fed a week or so before allowing free-choice to a salt-feed mixture; thus getting them on feed gradually.
8. It is necessary to regulate or limit (by hand-feeding for a few days) the supply of salt-feed mixture when it is desired to shift animals from a straight feed supplement (such as cottonseed meal alone) to a salt-feed mixture. Otherwise, hungry animals may consume too much.
9. It is estimated that the practice increases the total salt consumption to 8 to 10 times that required in conventional salt feeding, and doubles or triples the water consumption.
10. If the salt-feed mixture is placed in close proximity to the water supply, it will make for restricted grazing distribution on the range, because of the greater intake of water on a high-salt diet. On the other hand, if the salt-feed mixture is shifted about on the range, it will make for desirable distribution, because of the animals following the feed supply.
11. It reduces the labor required in feeding, promotes more uniform feed consumption (among the greedy and the timid), and permits animals to eat at their leisure with less disturbance during blizzards or cold weather.
12. It lessens the space required for feed equipment (bunks or feeders) to 20% of that required in conventional hand-feeding, but makes it desirable that the feeder be constructed

[10]At the Irrigated Agriculture Research and Extension Center, Prosser, Wash., it was found that 7½% salt limited grain consumption by yearling steers on pasture to 10 to 12 lb daily; 5% salt limited grain consumption to 12 to 14 lb.

so as to protect the mixture from the weather (especially wind and rain).

13. It is equally applicable to feeding during droughts, on dry summer range, and in the winter months.

14. It is commonly believed that under conditions of short feed supply (submaintenance) and relatively inaccessible water supply, animals may consume sufficient salt in this manner to produce toxic effects, especially during the winter months when low temperatures tend to lessen the water intake.

15. The practice of self-feeding salt-feed mixtures is well adapted to inaccessible and rougher areas, where daily feeding is difficult. In no case, however, should it be an excuse to neglect animals, for herds and flocks need to be checked often.

16. It reduces the consumption of minerals other than salt to practically nothing, with the result that correction of mineral deficiencies must be considered.

17. With adequate water, high salt intake has no effect on fertility, calf crop percentage, weaning weight, or bloom of animals.

18. *CAUTION:* Do not use trace-mineralized salt, because it may result in toxicity or mineral imbalances due to excessive intake of certain trace elements.

HIGH-PROTEIN BY-PRODUCT FEEDS

Many industrial by-products can be used as protein supplements. The distilling and brewing industries have done considerable research in the development of grain by-products as feeds for livestock. Likewise, many poultry operations are developing new ways of using poultry wastes as protein supplements. The meat and fish processing industries, discussed earlier in this chapter, have provided a variety of protein supplements. The grain milling industry now sells many of its by-products, such as corn gluten meal and wheat gluten meal, as protein feeds. (For additional information on by-product feeds which can be used as protein supplements, the reader is referred to Chapter 12, By-product Feeds/Crop Residues, and Chapter 30, Glossary of Feedstuffs.)

AMINO ACID SUPPLEMENTS

Amino acid supplements are supplements carrying large amounts of one or more pure amino acids which may be added to a ration to make up for an amino acid deficiency (or deficiencies).

The amino acids, of which there are 22, are the basic components, or building blocks, of proteins. But, according to our present knowledge, based largely on rat work, only the following 10 of these are essential in rations, because animals are able to synthesize sufficient quantities of others: arginine, histidine, isoleucine, leucine, lysine, methionine, phenylalanine, threonine, tryptophan, and valine. But all proteins are not created equal! Plant proteins often contain insufficient quantities of lysine, methionine, cystine, tryptophan, and /or threonine. Since the synthesis of body proteins (muscle, milk, eggs, wool, and hair/feathers/hooves/nails) is an "all or nothing" proposition, if any one of the essential amino acids needed to form the particular protein is deficient, that protein cannot be made.

Generally the first limiting amino acid (*i.e.,* the amino acid in lowest amount relative to need) in grain-based swine rations is lysine; and the second and third most limiting amino acids may be methionine and threonine, with their ranking depending on the grains used. So, normal swine rations may be deficient in lysine, methionine, and threonine.

Due to the large amount of vegetable proteins used, methionine is usually the first limiting amino acid in poultry rations. Also, lysine and cystine are often inadequate in normal poultry feedstuffs. So, most poultry rations call for supplementation with methionine, lysine, and cystine.

Ruminants have the same amino acid requirements as nonruminants at the tissue level. When ruminants are producing at average or low levels, they are capable of meeting any feed deficiencies of amino acids through microbial synthesis. But, high-producing ruminants (young, rapidly growing ruminants, and ruminants in peak milk production) cannot achieve maximum production if certain amino acids are deficient. So, ruminants in high production have need for amino acids in excess of those furnished by microbial synthesis and the normal escape of approximately 40% of the dietary protein from the rumen. The needed limiting amino acids for ruminants may be provided by protected (escape) amino acids or by true proteins.

When formulating rations, therefore, it is necessary not only to provide certain levels of protein, but also to provide specific levels of the essential amino acids. Theoretically, a perfectly balanced protein is one in which the available amino acids meet the requirements of the animal, with neither a surplus nor a deficiency of any amino acid.

Synthetic amino acids are now available for feed supplementation. But, due to their high cost, supplementation to date has been largely limited to methionine in poultry rations and lysine in starter and prestarter pig rations. Beginning in the late 1980s, tryptophan and threonine were marketed in feed grade form and used to a limited extent. But, biotechnology has become more sophisticated, and amino acids have become less expensive to produce. Today, both American and Japanese firms are producing amino acids with the same technology as is used in manufacturing drugs. Microorganisms use corn or molasses as an energy source to produce amino acids through fermentation.

It is noteworthy that 97 lb of corn and 3 lb of lysine can replace 100 lb of soybean meal. As soybean prices rise and/or lysine prices fall, this alternative becomes more attractive.

PROTEIN ASSESSMENT SYSTEMS

The crude protein method of evaluating proteins by the century-old Kjeldahl method, or some minor variation of it, involves finding the nitrogen content of the sample, then multiplying it by 6.25, since the nitrogen content of all protein averages about 16% ($100 \div 16 = 6.25$). However, beginning about 1960, the poultry industry, and to a lesser extent the swine industry, developed rations on an amino acid balance basis rather than on a crude protein basis. Along the way, research evidence began to accumulate indicating that crude protein content was not an accurate evaluation of nitrogenous sources for ruminants. It was found that certain feed ingredients achieved better production than others in ruminants even though the rations were isonitrogenous. This led to the development of several proposed protein assessment systems; among them, (1) the metabolizable protein concept (the quantity of protein digested or amino acids absorbed in the postruminal portion of the digestive tract of ruminants) proposed by Burroughs, Trenkle, and Vetter of Iowa State, in 1972; and (2) the ammonia overflow system as related to feed ingredients developed by Satter and Roffler of Wisconsin, in 1978. Then, adding to the complexity, are the different methods used around the world for evaluating protein sources for ruminants. In England, it's the New Protein System, recommended by the ARC, based on the fact that the use of protein is dependent on the energy intake. In Denmark, it's the new AAT-PBV system, designed to predict both the amino acid supply to the ruminant body (AAT) and the nitrogen supply to the rumen microorganisms. In France, it's the PDI, slightly modified into PAI in Switzerland. In West Germany, it's a new model replacing the use of digestible crude protein and making allowance for complex nitrogen conversions in the alimentary tract. Despite these national differences, or perhaps because of them, in 1985 the European Association of Animal Production initiated serious discussions with the aim of agreeing on standard terminologies and methods of feed assessments in ruminant protein evaluation; with each nation incorporating these into its own preferred method of final evaluation. Protein degradability will be an essential component in the eventual description, and at the outset dacron bags will be involved in its determination. Eventually, however, the dacron bag determination of digestibility will be replaced by a more rapid and less laborious method. Presently, none of the above systems provide a quick means of measuring degradability. It may be concluded, therefore, that, currently, there is no single procedure for determining protein efficiency or biological value for ruminants which is best for all purposes. However, the critical scientist can find methods that will serve the purpose at hand and which will yield reliable results provided they are properly interpreted.

The goal ahead is to develop a protein/carbohydrate evaluation system which predicts well for all types of feeding systems and all types of feed processing.

QUESTIONS FOR STUDY AND DISCUSSION

1. Define each of the following: proteins, amino acids, protein quality, nonprotein nitrogen (NPN), protein equivalent, and protein supplements.

2. Trace the utilization of each (a) true protein, and (b) nonprotein nitrogen (NPN) by ruminants, from consumption to absorption/metabolism/excretion.

3. Define and explain the significance of each of the following: bypass protein, degradability of protein, solubility of protein, and microbial protein.

4. List and discuss each of the methods for protecting proteins from ruminant degradation.

5. Discuss the relative importance of plant and animal protein sources. Which sources have become increasingly more important in recent years?

6. List and give pertinent information relative to each of the commonly used oilseed meals.

7. Outline the three processes of oil extraction of soybeans. What are the advantages and disadvantages of each?

8. What are corn gluten feed and corn gluten meal? Why have these feeds become so abundant in recent years?

9. Define pulses. List five types of pulses and give the relative crude protein content of each. What precautions should be taken when pulses are to be used as either feed or food?

10. Discuss the future of leaf protein concentrates as it pertains to livestock.

11. Thirty years ago, it was commonly believed that animal protein had to be incorporated into nonruminant feeds. Why? Why has this thinking changed?

12. Define, describe the processing, and tell of the importance and use of each of the following animal proteins: tankage, meat meal, meat and bone meal tankage, meat and bone meal, and blood meal.

13. Define, describe the processing, and tell of the importance and use of each of the following poultry wastes: feather meal, hydrolyzed poultry by-product aggregate, poultry by-products, poultry by-product meal, and poultry hatchery by-product.

14. What are the limitations to the use of dairy products? What is whey? Why have so many new whey products evolved in recent years?

15. Why are fish meals becoming of increasing importance as a ruminant feedstuff? What factors cause variation in the feeding value of fish meal? List the two main sources of fish meal.

16. The research on urea by Loosli of Cornell and by Virtanen of Finland had great impact on the subsequent use of urea. Tell about each of these studies, and explain why they were so significant.

17. Discuss the conversion of urea to protein and its subsequent utilization in the ruminant animal.

18. Define fermentation potential. What is the practicality of this concept?

19. List and discuss 10 pointers that should be observed in order to assure optimum results from urea.

20. If a 35% crude protein supplement contains 4% urea, what percent protein is furnished by urea?

21. What method would you use to feed urea to (a) dry dairy cows in a corral, (b) finishing steers in a dry lot, and (c) pregnant cows on an isolated range?

22. What are the advantages of slow-release NPN products?

23. List the precautions that should be observed if hay is ammoniated.

24. Briefly discuss the advantages and disadvantages inherent with single-cell protein.

25. What forms of protein supplements are fed to animals on range and pasture? Discuss the advantages and disadvantages of each pasture/range supplement.

26. Define amino acid supplements. What amino acid supplements are particularly important to each (a) swine, and (b) poultry? Why may they be deficient for these species?

27. Under what circumstances may there be amino acid deficiencies in ruminants? When amino acids are needed for ruminants, how should they be provided if they are to be effective?

28. Why has the century-old Kjeldahl crude protein method of evaluating protein come under considerable scrutiny in recent years? Is there a better and quicker system for protein evaluation?

SELECTED REFERENCES

Title of Publication	Author(s)	Publisher
AFMA Liquid Feed Symposium Proceedings	American Feed Manufacturers Association	AFMA, Chicago, Ill., 1971
Alternative Sources of Protein for Animal Production	National Research Council	National Academy of Sciences, Washington, D.C., D.C., 1973
Applied Animal Nutrition	E. W. Crampton L. E. Harris	W. H. Freeman and Co., San Francisco, Calif., 1969
Association of American Feed Control Officials Incorporated, Official Publication	Association of American Feed ControlOfficials, Inc.	Association of American Feed Control Officials, Inc., Annual
Commercial Chicken Production Manual, 3rd Edition	M. O. North	Avi Publishing Company, Inc., Westport, Conn., 1984
Effect of Processing on the Nutritional Value of Feeds	National Research Council	National Academy of Sciences, Washington, D.C., 1973
Evaluation of Novel Protein Products	A. E. Bender, *et al.*	Pergamon Press, Oxford, England, 1970
Feeds and Feeding, 2nd Edition	A. Cullison	Reston Publishing Company, Inc., Reston, Va., 1979
Feeds and Feeding, 22nd Edition	F. B. Morrison	The Morrison Publishing Company, Ithaca, N.Y., 1956
Feeds and Feeding, Abridged	F. B. Morrison	The Morrison Publishing Company, Ithaca, N.Y., 1956
Fish as Food, Volumes II and III	Ed. by G. Borgstrom	Academic Press, New York, N.Y., 1965
Fundamentals of Nutrition	E. W. Crampton L. E. Lloyd	W. H. Freeman and Co., San Francisco, Calif., 1959
Iowa State University Nutrition Symposium on Proteins		Iowa State University, Ames, Iowa, 1972
Livestock Feeds and Feeding, 2nd Edition	D. C. Church	O & B Books, Inc., Corvallis, Ore., 1984
Nonprotein Nitrogen in the Nutrition of Ruminants	J. K. Loosli I. W. McDonald	Food and Agriculture Organization of the United Nations, Rome, Italy, 1968
Nitrogen and Energy Nutrition of Ruminants	R. L. Shirley	Academic Press, Inc., Orlando, Fla., 1986
Nutrition of the Chicken, 3rd Edition	M. Scott M. C. Nesheims R. J. Young	M. L. Scott and Associates, Ithaca, N.Y., 1982
Poultry Feeds & Nutrition, 2nd Edition	P. J. Schaible	Avi Publishing Co., Inc., Westport, Conn., 1980
Poultry Nutrition, 5th Edition	W. R. Ewing	Ray Ewing Company, Pasadena, Calif., 1963
Processed Plant Protein Foodstuffs	Ed. by A. M. Altschul	Academic Press, Inc., New York, N.Y., 1958
Protein Nutritional Quality of Food and Feeds	Ed. by Mendel Friedman	Marcel Dekker, Inc., New York, N.Y., 1975
Protein Resources and Technology Status and Research Needs		U.S. Government Printing Office, Washington, D.C., 1975
Rumen and Its Microbes, The	R. E. Hungate	Academic Press, Inc., New York, N.Y., 1966
Ruminant Nitrogen Usage	National Research Council	National Academy of Sciences, Washington, D.C., 1985
Single-Cell Protein	Ed. by R. I. Mateles, S. R. Tannenbaum	The M.I.T. Press, Cambridge, Mass., 1968
Swine Production in Temperate and Tropical Environments	W. Pond and J, Maner	W. H. Freeman and Co., San Francisco, Calif., 1974

(Continued)

(Continued)

Title of Publication	Author(s)	Publisher
Urea and Non-Protein Nitrogen in Ruminant Nutrition, 2nd Edition	Ed. by H. J. Stangel	Nitrogen Division, Allied Chemical Corporation, Morristown, N.J., 1963
Urea as a Protein Supplement	Ed. by M. H. Briggs	Pergamon Press, New York, N.Y., 1967
World Protein Resources	A. Jones	John Wiley & Sons, New York, N.Y., 1974

Fig. 11-25. Canola (rape), source of canola meal, in bloom. (Courtesy, Canola Council of Canada, Winnipeg, Canada)

Fig. 11-26. Soybean plant, source of soybean meal, flowering. (Courtesy, American Soybean Assn., St. Louis, Mo.)

Fig. 11-27. Sunflowers, source of sunflower meal. (Courtesy, National Sunflower Assn., Bismark, N.D.)

Fig. 11-28. Assorted texturized soy protein products. (Courtesy, American Soybean Assn., St. Louis, Mo.)

Chapter 12

BY-PRODUCT FEEDS/ CROP RESIDUES[1]

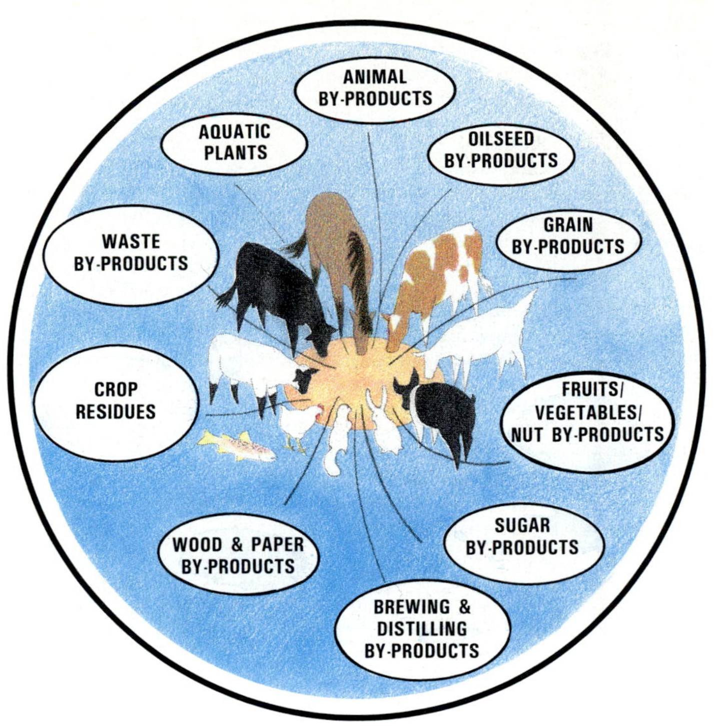

Original painting by Tom Phillips

Contents	Page

(Continued)

[1]The authors gratefully acknowledge the helpful suggestions of the following authority who reviewed this chapter: D. A. Miller, Ph.D., Department of Agronomy, College of Agriculture, University of Illinois, Urbana-Champaign.

Contents (Continued) Page

Contents Page

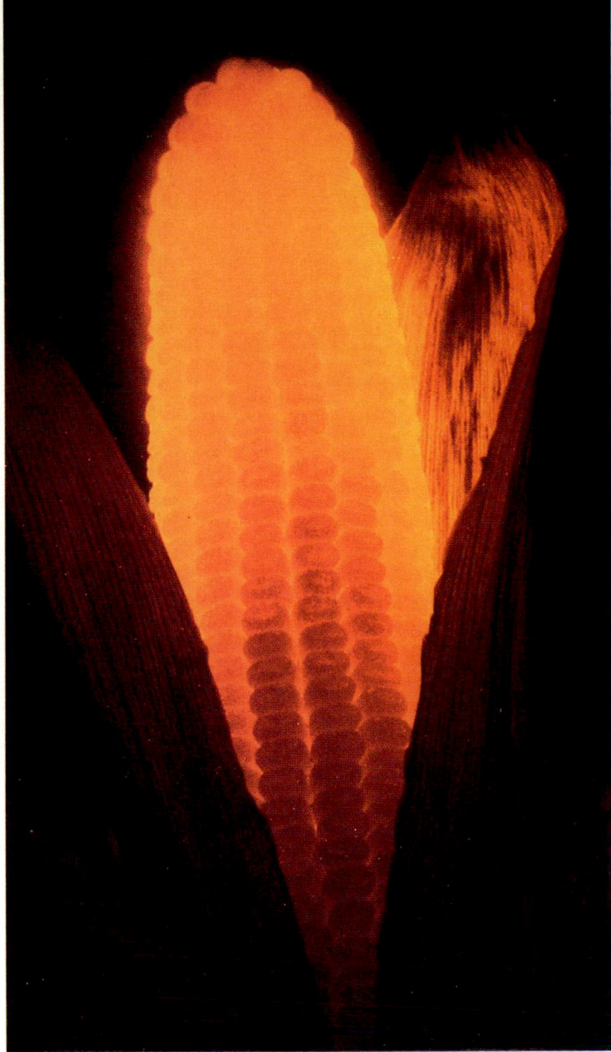

Fig. 12–1. Corn, leading source of U.S. by-product feeds and crop residues. From each bushel of corn processed for sweeteners or ethanol, enough corn oil is produced for 2 lb of margarine and enough by-product feedstuffs are produced to feed 3 broilers to market weight. (Courtesy, Archer Daniels Midland Company, Decatur, Ill.)

Producing and processing animals and plants for food for people and feed for animals result in many by-products and crop residues which can be utilized as livestock feeds.

By-product feeds are concentrates and roughages other than the primary products from animal and plant processing and from industrial manufacturing.

Crop residues are the portions of crops that are normally left in the field following harvesting.

In addition to conventional by-product feeds and crop residues, many unusual and underutilized feeds are discussed in this chapter, including surplus and cull fruits and vegetables, certain weeds that are fed on occasion, wood and paper by-products, animal waste (excreta), and aquatic plants.

Animals provide a practical outlet for a host of products which are not suited for human consumption. Some of these products have been used for animal feeds so long, and so extensively, that they are commonly classed as feed ingredients, along with such things as the cereal grains, without reference to their by-product origin. Such ingredients include both animal and plant materials. More than 36 million tons of these by-product feeds are used as animal feeds each year in the United States. (See Chapter 6, Types and Roles of Feedstuffs, section on "Concentrates.")

In the United States, cereal grain residues are the principal crop residues, although some areas have significant amounts of cotton, soybean, sugar beet and sugarcane residues. Corn stover, wheat straw, and soybean residues account for more than 80% of the estimated 500 million tons of residues produced annually, with corn stover alone making up 50% of the total.

The list of by-product feeds and crop residues is very long; and various classifications of them have evolved. In this chapter, and in Table 12–1, by-product feeds and crop residues are classified in four major categories: animal by-product feeds, plant by-product feeds, waste by-product feeds, and aquatic plants. Regardless of the classification used, there are always some feedstuffs which do not fit well in any one category, or which may fit in more than one category. However, livestock producers need not be too concerned with the precise classification of each feedstuff. Rather, they should familiarize themselves with the characteristics and nutritive value of each feedstuff.

In addition to the by-product feeds and crop residues listed in Table 12–1 and discussed in this chapter, these and other by-product feeds and crop residues are listed in Part V, Composition of Feeds; and the replacement values for many of them are given in the feed substitution tables for each species, in Chapters 19 to 26. Additionally, many by-product feeds and crop residues are listed and discussed in Chapter 10, Grains/High-Energy Feeds; Chapter 11, Protein Supplements; and Chapter 30, Glossary of Feedstuffs. In order to avoid needless repetition in presenting such a great array of by-product feeds and crop residues, cross referencing and indexing are used.

TABLE 12–1

BY-PRODUCT FEEDS/CROP RESIDUES[1] (See footnotes at end of table.)

By-products/ Crop Residues	Description[2]	Moisture Basis: A-F (as-fed) or M-F (moisture-free)	Dry Matter	Crude Fiber	TDN		Digestible Energy				Crude Protein	Ca	P	Feeding Recommendations[3]	Cross References
					Ruminant	Swine	Ruminant Mcal per		Swine kcal per						
			(%)	(%)	(%)	(%)	(lb)	(kg)	(lb)	(kg)	(%)	(%)	(%)		
ANIMAL BY-PRODUCT FEEDS															
DAIRY BY-PRODUCT FEEDS															
Buttermilk, condensed	Condensed buttermilk is made by evaporating buttermilk to about ⅓ of the original weight. Its superior nutritive value is due to its content of high quality proteins, its vitamins, and its good mineral balance, along with its high palatability and digestibility.	**A-F** M-F	**29** 100	**0.1** 0.3	**26** 88	**22** 76	**0.52** 1.77	***1.14*** *3.89*	**442** 1,503	***974*** *3,314*	**10.8** 36.9	**0.44** 1.51	**0.26** 0.89	Condensed buttermilk is an excellent supplement for young pigs and chicks; it is ideal for balancing the deficiencies of the cereal grains.	Chapter 11, under "Dairy Products."
Cheese rind (cheese meal)	Cheese rind is a by-product from the manufacture of processed cheese, obtained by cooking cheese trimmings. It contains about 50% protein and 9% fat.	**A-F** M-F	**85** 100	**0.2** 0.3	**78** 91	**80** 95	**1.69** 2.00	***3.73*** *4.40*	**1,609** 1,897	***3,547*** *4,183*	**45.5** 53.7	**0.98** 1.16	**0.56** 0.66	Cheese rind is an excellent high-protein and high-energy feed for young stock.	Chapter 11, under "Dairy Products."
Milk, whole dried	Dried whole milk is obtained by artificially drying whole milk. The AAFCO specs stipulate that it contain at least 26% milk fat, but not more than 8% moisture.	**A-F** M-F	**95** 100	**0.2** 0.2	**113** 119	**109** 115	**2.27** 2.38	***4.99*** *5.24*	**2,183** 2,290	***4,813*** *5,049*	**25.3** 26.6	**0.89** 0.93	**0.70** 0.74	Most whole dried milk is used for human food. It is generally too expensive to feed to livestock.	Chapter 11, under "Dairy Products."
Skimmed milk, dried	Dried skimmed milk is dehydrated skimmed milk. It contains less than 8% moisture and averages 32 to 35% protein. One pound of dried skimmed milk has about the same composition and feeding value as 10 lb of liquid skimmed milk.	**A-F** M-F	**94** 100	**0.2** 0.2	**80** 85	**86** 92	**1.43** 1.52	***3.15*** *3.35*	**1,758** 1,867	***3,876*** *4,115*	**33.3** 35.4	**1.28** 1.36	**1.02** 1.09	Dried skimmed milk is generally too high priced to feed to livestock, except for limited use in starter pig, chick, and foal rations.	Chapter 11, under "Dairy Products."
Whey, fresh	Fresh whey is the liquid by-product which results from cheese manufacture.	**A-F** M-F	**7** 100	**—** —	**7** 94	**—** —	**0.13** 1.88	***0.29*** *4.15*	**—** —	**—** —	**0.9** 13.2	**0.06** 0.81	**0.05** 0.71	Fresh whey is high in moisture; hence, it is not economical to transport it a great distance. It is sometimes fed as a liquid to cattle and swine in close proximity to the plant where it is produced.	Chapter 11, under "Dairy Products."
Whey, condensed	Condensed whey is the product obtained by partially removing water from whey. Minimum percent solids must be declared on the label.	**A-F** M-F	**54** 100	**0.3** 0.5	**47** 87	**46** 86	**0.94** 1.74	***2.07*** *3.84*	**928** 1,726	***2,046*** *3,806*	**6.9** 12.8	**0.39** 0.72	**0.47** 0.88	Condensed whey contains too much water to transport far. So, it is usually fed to cattle and swine in close proximity to the plant where it is produced.	Chapter 11, under "Dairy Products."
Whey, dehy	Dehydrated whey is the product obtained by artificially dehydrating whey. It must contain not less than 11% protein and not less than 61% lactose.	**A-F** M-F	**93** 100	**0.2** 0.2	**76** 82	**77** 83	**1.51** 1.62	***3.33*** *3.57*	**1,444** 1,549	***3,183*** *3,415*	**13.3** 14.2	**0.86** 0.92	**0.76** 0.82	Most dehydrated whey is incorporated in swine and poultry rations.	Chapter 11, under "Dairy Products."
MARINE BY-PRODUCT FEEDS															
Fish liver meal	Fish liver meal is dried fish liver. Today, most of this product is marketed as "Fish Liver and Glandular Meal," which is obtained by drying the entire viscera of fish. The AAFCO specify that the latter product must contain at least 18 mg of riboflavin per pound, and that at least 50% of its weight must consist of livers.	**A-F** M-F	**93** 100	**1.2** 1.3	**97** 104	**96** 103	**1.94** 2.09	***4.28*** *4.61*	**1,914** 2,062	***4,219*** *4,547*	**62.8** 67.7	**—** —	**—** —	Fish liver meal and fish liver and glandular meal are excellent protein/vitamin/unidentified supplements for use in poultry starter rations, early weaned pig feeds, and aquaculture feeds.	Chapter 11, under "Marine Feeds."

(Continued)

TABLE 12–1 *(Continued)*

By-products/ Crop Residues	Description[2]	Moisture Basis: A-F (as-fed) or M-F (moisture-free)	Dry Matter	Crude Fiber	TDN Ruminant	TDN Swine	Digestible Energy Ruminant Mcal per		Digestible Energy Swine kcal per		Crude Protein	Ca	P	Feeding Recommendations[3]	Cross References
			(%)	(%)	(%)	(%)	(lb)	*(kg)*	(lb)	*(kg)*	(%)	(%)	(%)		
MARINE BY-PRODUCT FEEDS (Continued)															
Fish meal	Fish meal is the clean, dried, ground tissue of undecomposed fish or fish cuttings, either or both, with or without the extraction of part of the oil. The AAFCO specify that fish meal not contain more than 10% moisture, and not more than 7% salt. If fish meal contains between 3 and 7% salt, the amount of salt must constitute part of the brand name.	**A-F** M-F	**92** 100	**0.7** 0.8	**67** 73	**66** 72	**1.34** 1.46	***2.95*** *3.22*	**1,317** 1,438	***2,903*** *3,169*	**64.3** 70.2	**6.63** 7.24	**3.61** 3.94	Traditionally, fish meal has been extensively used in poultry feeds. But, in the 1980s, the use of fish meal changed significantly. The three most important new uses for fish meal are: acquaculture feeds, early weaned pig feeds, and rumen by-pass feeds. The bypass protein as a percent of crude protein of fish meal is 65% (see Chapter 11, Table 11–1).	Chapter 11, under "Marine Feeds."
Fish solubles, condensed	Condensed fish solubles are a semisolid by-product obtained by evaporating the liquid remaining from the steam rendering of fish. Condensed fish solubles, which must contain a minimum of 30% crude protein, are a rich source of the B vitamins and unidentified factors. They are particularly rich in pantothenic acid, niacin, and vitamin B–12.	**A-F** M-F	**50** 100	**0.5** 1.0	**41** 82	**44** 87	**0.85** 1.68	***1.87*** *3.71*	**866** 1,717	***1,909*** *3,784*	**31.5** 62.5	**0.16** 0.32	**0.57** 1.14	Condensed fish solubles are an excellent vitamin supplement. They are recommended for inclusion in young pig and young poultry rations, and in acquaculture feeds.	Chapter 11, under "Marine Feeds."
Fish solubles, dried	Dried fish solubles are obtained by dehydrating the stickwater. It must contain not less than 60% crude protein.	**A-F** M-F	**93** 100	**2.0** 2.1	**77** 83	**66** 71	**1.50** 1.61	***3.30*** *3.56*	**1,467** 1,581	***3,234*** *3,485*	**60.4** 65.1	**0.40** 0.43	**1.27** 1.37	Dried fish solubles are a high quality vitamin supplement, recommended for use in feeds for young pigs and young poultry, and in fish feeds.	Chapter 11, under "Marine Feeds."
Shrimp meal	Shrimp meal is the undecomposed ground dried waste of shrimp, consisting of parts and/or whole shrimp. If it contains more than 3% salt, the amount of salt must constitute part of the brand name; but the salt content cannot exceed 7%.	**A-F** M-F	**90** 100	**14.1** 15.6	**41** 46	**—** —	**0.82** 0.92	***1.82*** *2.02*	**—** —	**—** —	**38.7** 43.0	**10.40** 11.55	**1.85** 2.06	Shrimp meal is recommended for use as a protein and calcium supplement in swine, poultry, and fish feeds.	Chapter 11, under "Marine Feeds."
MEAT ANIMAL BY-PRODUCT FEEDS															
Animal fat, hydrolyzed	Hydrolyzed animal fat is obtained from the fat processing procedures commonly used in edible fat processing or soap making. It consists predominantly of fatty acids and must contain not less than 85% total fatty acids, not more than 6% unsaponifiable matter, and not more than 1% insoluble matter. If an antioxidant is used, the common name or names must be indicated, followed by the word *preservative(s)*.	**A-F** M-F	**99** 100	**—** —	**223** 225	**209** 211	**4.46** 4.50	***9.84*** *9.92*	**4,144** 4,177	***9,135*** *9,210*	**—** —	**—** —	**—** —	Hydrolyzed animal fat may be used in feeds for any species to increase the caloric density. Milk replacers, used in feeding suckling young, contain the most fat—ranging from 15 to 30%.	Chapter 10, under "Fats and Oils."
Animal fat, lard	Lard is pork fat obtained in the commercial processes of rendering or extracting. It consists predominantly of glyceride esters of fatty acids and contains no additions of free fatty acids or other materials obtained from fats. It must contain not less than 90% total fatty acids, not more than 2.5% unsaponifiable matter, and not more than 1% insoluble matter. If an antioxidant is used, the common name or names must be indicated, followed by the word *preservative(s)*.	**A-F** M-F	**99** 100	**—** —	**—** —	**—** —	**—** —	**—** —	**3,669** 3,694	***8,089*** *8,143*	**—** —	**—** —	**—** —	Lard (pork fat) may be used to increase the caloric density of feeds for any species.	Chapter 10, under "Fats and Oils."

(Continued)

TABLE 12–1 *(Continued)*

By-products/ Crop Residues	Description[2]	Moisture Basis: A-F (as-fed) or M-F (moisture-free)	Dry Matter	Crude Fiber	TDN Ruminant	TDN Swine	Digestible Energy Ruminant Mcal per (lb)	Digestible Energy Ruminant Mcal per *(kg)*	Digestible Energy Swine kcal per (lb)	Digestible Energy Swine kcal per *(kg)*	Crude Protein	Ca	P	Feeding Recommendations[3]	Cross References
			(%)	(%)	(%)	(%)	(lb)	*(kg)*	(lb)	*(kg)*	(%)	(%)	(%)		
MEAT ANIMAL BY-PRODUCT FEEDS (Continued)															
Animal tallow	Animal tallow is beef and/or sheep fat obtained in the commercial processes of rendering or extracting. It consists predominantly of glyceride esters of fatty acids and contains no additions of free fatty acids or other materials obtained from fats. It must contain not less than 90% total fatty acids, and not more than 2.5% unsaponifiable matter, and not more than 1% insoluble matter. If an antioxidant is used, the common name or names must be indicated, followed by the word *preservative(s).*	**A-F** M-F	**97** 100	**—** —	**203** 209	**—** —	**4.07** 4.18	***8.97*** *9.22*	**—** —	**—** —	**1.5** 1.6	**—** —	**—** —	Animal tallow (beef and/or sheep fat) may be used to increase the caloric density of feeds for any species.	Chapter 10, under "Fats and Oils."
Blood meal	Blood meal is produced from clean, fresh animal blood. It may be prepared by any one of three processes: (1) spray drying, (2) cooker drying, or (3) flash drying. Blood meal is usually of dark black color and rather insoluble in water. Blood meal is low in the essential amino acid isoleucine and low in calcium and phosphorus.	**A-F** M-F	**91** 100	**1.0** 1.1	**61** 66	**61** 67	**1.20** 1.32	***2.65*** *2.91*	**1,220** 1,336	***2,690*** *2,946*	**80.5** 88.2	**0.29** 0.32	**0.25** 0.28	Blood meal ranks very high as an escape (bypass) protein for ruminants (see Chapter 11, Table 11–1). For this reason, it is finding an increasing place as a ruminant feedstuff. Blood meal is also used as an ingredient in milk replacers, especially for ruminants.	Chapter 11, under "Blood Meal."
Liver meal	Liver meal is dried, ground liver from slaughtered mammals.	**A-F** M-F	**93** 100	**1.4** 1.5	**89** 97	**93** 101	**1.79** 1.93	***3.94*** *4.26*	**1,867** 2,017	***4,116*** *4,446*	**66.1** 71.4	**0.56** 0.61	**1.26** 1.36	Liver meal is used as a high quality vitamin/high protein supplement in feeds for young animals and in fish feeds.	Chapter 11, under "Other Animal By-Products."
Meat meal	Meat meal is the dry-rendered product from mammal tissues, exclusive of blood, hair, hoof, horn, hide trimmings, manure, stomach and rumen contents. The calcium level shall not exceed the level of phosphorus by more than 2.2 times. Meat meal is a high protein feed, with protein of excellent quality. Also, it is rich in calcium and phosphorus and a good source of riboflavin, niacin, and vitamin B–12.	**A-F** M-F	**94** 100	**2.7** 2.9	**67** 71	**64** 68	**1.20** 1.28	***2.64*** *2.81*	**936** 998	***2,064*** *2,200*	**50.7** 54.0	**8.61** 9.18	**4.58** 4.88	Meat meal is a widely used protein supplement, especially in swine and poultry feeds. Meat meal ranks high as a rumen bypass protein (see Chapter 11, Table 11–1).	Chapter 11, under "Tankage and Meat Meal."
Meat meal with bone	Meat meal with bone is the dry-rendered product from mammal tissues, *including bone,* but exclusive of blood, hair, hoof, horn, hide trimmings, manure, stomach and rumen contents. It shall contain a minimum of 4.0% phosphorus and the calcium level shall not be more than 2.2 times the phosphorus level. Meat and bone meal is similar to meat meal, but higher in calcium and phosphorus due to the added bone.	**A-F** M-F	**93** 100	**2.4** 2.6	**66** 71	**68** 73	**1.32** 1.41	***2.91*** *3.11*	**1,028** 1,102	***2,267*** *2,430*	**50.4** 54.0	**10.00** 10.72	**4.94** 5.30	Meat and bone meal is widely used as a protein supplement, especially in swine and poultry feeds. Meat meal with bone ranks high as a rumen bypass protein (see Chapter 11, Table 11–1).	Chapter 11, under "Tankage and Meat Meal."

(Continued)

TABLE 12–1 *(Continued)*

By-products/ Crop Residues	Description[2]	Moisture Basis: A-F (as-fed) or M-F (moisture-free)	Dry Matter	Crude Fiber	TDN		Digestible Energy				Crude Protein	Ca	P	Feeding Recommendations[3]	Cross References
					Ruminant	Swine	Ruminant Mcal per		Swine kcal per						
			(%)	(%)	(%)	(%)	(lb)	*(kg)*	(lb)	*(kg)*	(%)	(%)	(%)		
MEAT ANIMAL BY-PRODUCT FEEDS (Continued)															
Paunch, rumen, dehy	In large processing operations, the rumen content, which may contain in excess of 130 lb of feed, is first pressed to remove a large portion of the liquids. Next, it is dehydrated to reduce the moisture and kill most of the pathogens that may be present. The product is marketed as dehydrated paunch product, which is described as follows: Dehydrated paunch product is composed of the rumen contents of slaughtered cattle, dehydrated at temperatures over 212°F *(100°C)* to a moisture content of 12% or less.	**A-F** M-F	**87** 100	**24.4** 28.0	**51** 58	**42** 48	**1.00** 1.14	***2.20*** *2.51*	— —	— —	**15.3** 17.6	**0.69** 0.79	**0.58** 0.67	Most dehydrated paunch product is used as a feed for mature ruminants. Sometimes, it is used as litter in poultry operations. The liquid obtained from pressing is a good source of B vitamins; so, sometimes it is condensed or dried and used in swine rations.	
Tankage	Tankage, which is also known as meat meal tankage, is the wet-rendered product from mammal tissue exclusive of hair, horn, hide trimmings, manure, stomach and rumen contents. It may contain added blood or blood meal, however, to raise the protein to the desired level. The calcium level shall not exceed the phosphorus level by more than 2.2 times. The protein content of tankage is somewhat higher than meat meal. Tankage is a high protein feed, and rich in riboflavin, niacin, and vitamin B-12.	**A-F** M-F	**92** 100	**1.8** 2.0	**67** 73	**67** 73	**1.34** 1.46	***2.96*** *3.21*	**1,112** 1,207	***2,451*** *2,660*	**60.5** 65.6	**5.87** 6.37	**3.09** 3.36	Tankage is widely used as a protein supplement in swine and poultry feeds.	Chapter 11, under "Tankage and Meat Meal."
Tankage with bone	Tankage with bone, which is also known as meat and bone meal tankage, is the wet-rendered product from mammal tissue, *including bone,* but exclusive of hair, hoof, horn, hide trimmings, manure, stomach and rumen contents. It may contain added blood or blood meal, however, to raise the protein to the desired level. It shall contain a minimum of 4.0% phosphorus and the calcium level shall not be more than 2.2 times the phosphorus level. Tankage with bone is a high protein/high calcium-phosphorus feed, and rich in riboflavin, niacin, and vitamin B-12.	**A-F** M-F	**93** 100	**2.2** 2.4	**63** 68	**68** 73	**1.44** 1.55	***3.17*** *3.42*	**1,382** 1,488	***3,047*** *3,280*	**46.6** 50.2	**11.16** 12.01	**5.41** 5.82	Tankage with bone is widely used as a protein supplement in swine and poultry feeds.	Chapter 11, under "Tankage and Meat Meal."
POULTRY BY-PRODUCT FEEDS															
Feather meal	Feather meal (hydrolyzed poultry feathers) consists of hydrolyzed, dried, ground feathers from slaughtered poultry. Not less than 75% of its crude protein content must be digestible by the pepsin digestibility method. Feather meal is rather low in nutritional value, being low in the amino acids histidine, lysine, methionine, and tryptophan.	**A-F** M-F	**93** 100	**1.4** 1.5	**67** 72	**62** 67	**1.21** 1.30	***2.66*** *2.87*	**1,238** 1,332	***2,729*** *2,936*	**83.8** 90.2	**0.30** 0.33	**0.62** 0.67	Feather meal is used as a high protein feed for mature ruminants. It ranks high as an escape (by-pass) feed for ruminants. Levels up to 10% of the ration may be used for mature ruminants.	Chapter 11, under "Poultry Wastes."

(Continued)

TABLE 12–1 *(Continued)*

By-products/ Crop Residues	Description[2]	Moisture Basis: A-F (as-fed) or M-F (moisture-free)	Dry Matter	Crude Fiber	TDN		Digestible Energy				Crude Protein	Ca	P	Feeding Recommendations[3]	Cross References
					Ruminant	Swine	Ruminant Mcal per		Swine kcal per						
			(%)	(%)	(%)	(%)	(lb)	*(kg)*	(lb)	*(kg)*	(%)	(%)	(%)		
POULTRY BY-PRODUCT FEEDS (Continued)															
Poultry by-product meal	Poultry by-product meal consists of the ground, rendered, clean parts of the carcass of slaughtered poultry, such as necks, feet, undeveloped eggs, and intestines, exclusive of feathers. The calcium level shall not exceed the level of phosphorus by more than 2.2 times.	**A-F** M-F	**94** 100	**2.2** 2.3	**74** 79	**76** 81	**1.44** 1.53	***3.17*** *3.37*	**1,406** 1,499	***3,101*** *3,305*	**61.2** 65.3	**3.97** 4.23	**2.06** 2.20	Poultry by-product meal is widely used as a protein supplement in swine and poultry feeds.	Chapter 11, under "Poultry Wastes."
Poultry fat	Poultry fat (feed grade) is primarily obtained from the tissue of poultry in the commercial process of rendering or extracting. It must contain not less than 90% total fatty acids and not less than 3.0% of unsaponifiables and impurities. It shall have a minimum titer of 33 C. If an antioxidant is used, the common name or names must be indicated, followed by the word *preservative(s).*	**A-F** M-F	**99** 100	**—** —	**188** 189	**196** 198	**3.57** 3.60	***7.87*** *7.94*	**3,750** 3,783	***8,267*** *8,340*	**—** —	**—** —	**—** —	Poultry fat may be used to increase the caloric density of feeds for any species.	Chapter 10, under "Fats and Oils."
PLANT BY-PRODUCT FEEDS															
OILSEED BY-PRODUCT FEEDS (all meals solvent extd)															
Canola (rapeseed) meal	Canola meal (rapeseed meal) is derived from canola, which was created from selected rapeseed by Canadian scientists in the 1970s. In comparison with the old rapeseed, canola is lower in glucosinolates and erucic acid (a long-chain fatty acid).	**A-F** M-F	**90** 100	**13.0** 14.4	**—** —	**—** —	**—** —	**—** —	**—** —	**—** —	**34.0** 37.8	**—** —	**—** —	Canola meal may be used as a protein supplement for all classes of livestock and poultry. It may be used at maximum levels of 20% in most rations, but the amount in the total ration should be limited to about 10% for layers and breeding ducks, 15% for breeding turkeys, and 12% for young pigs and breeding swine.	Chapter 11, under "Canola Meal (Rapeseed Meal)."
Coconut (copra) meal	Coconut meal (also known as copra meal) is the by-product from the production of oil from the dried meats of coconuts.	**A-F** M-F	**91** 100	**14.4** 15.8	**69** 75	**73** 80	**1.37** 1.50	***3.01*** *3.31*	**1,460** 1,602	***3,218*** *3,532*	**21.3** 23.4	**0.17** 0.19	**0.60** 0.66	The quality of the protein in coconut (copra) meal is not high; hence, it should be used primarily in ruminant rations. Coconut meal produces a rather hard (highly saturated) butterfat. The maximum recommended level in dairy rations is 3.3 to 6.5 lb/day. Higher levels produce tallowy butter.	Chapter 11, under "Coconut Meal (Copra Meal)."
Cottonseed, whole	As indicated by the name, this is whole cottonseed, with or without the lint. As shown in the columns to the right, it is very high in energy, high in protein, high in phosphorus, and high in fiber. It does not require processing.	**A-F** M-F	**91** 100	**19.4** 21.4	**87** 95	**71** 79	**1.57** 1.72	***3.46*** *3.80*	**1,430** 1,572	***3,152*** *3,467*	**21.8** 24.0	**0.14** 0.16	**0.69** 0.76	Whole cottonseed is used as a cattle feed, for both beef and dairy animals. Generally, whole cottonseed will increase the fat of milk by about 0.3%, making for an increase in the market value of the milk by about 50¢/cwt. Lactating Holstein cows are normally fed 4 to 7 lb of whole cottonseed/head/day.	Chapter 10, under "Cottonseed."

(Continued)

TABLE 12-1 *(Continued)*

By-products/ Crop Residues	Description[2]	Moisture Basis: A-F (as-fed) or M-F (moisture-free)	Dry Matter	Crude Fiber	TDN		Digestible Energy				Crude Protein	Ca	P	Feeding Recommendations[3]	Cross References
					Ruminant	Swine	Ruminant Mcal per		Swine kcal per						
			(%)	(%)	(%)	(%)	(lb)	*(kg)*	(lb)	*(kg)*	(%)	(%)	(%)		
OILSEED BY-PRODUCT FEEDS (Continued)															
Cottonseed meal	Cottonseed meal is the finely ground flakes of the residue that remains after most of the oil from whole cottonseed has been extracted. It must contain not less than 36% crude protein. Cottonseed meal is low in lysine and tryptophan and deficient in vitamin D, carotene, and calcium. But it is rich in phosphorus. Also, it may contain a toxic substance known as gossypol.	**A-F** M-F	**91** 100	**12.1** 13.4	**68** 75	**61** 67	**1.48** 1.63	***3.27*** *3.60*	**1,209** 1,330	***2,666*** *2,933*	**41.2** 45.4	**0.17** 0.19	**1.11** 1.22	Cottonseed meal is an excellent protein supplement for ruminants, since they do not require lysine and tryptophan and they can tolerate gossypol. However, it must be limited in the rations of monogastric animals, due to gossypol; unless derived from glandless seed.	Chapter 11, under "Cottonseed Meal."
Linseed (flaxseed) Meal	Linseed meal is the ground residue that remains after most of the oil from flaxseed has been extracted. Linseed meal is low in the amino acids lysine and tryptophan. Also, it is lacking in carotene and vitamin D, and only fair in calcium and the B vitamins.	**A-F** M-F	**90** 100	**8.9** 9.9	**70** 78	**67** 74	**1.41** 1.56	***3.10*** *3.43*	**1,336** 1,479	***2,945*** *3,260*	**35.7** 39.6	**0.40** 0.44	**0.82** 0.91	Linseed meal is the protein supplement of choice by horse caretakers. They especially prefer old process linseed meal, because it is palatable, and it imparts glossiness to the hair coat due to the high oil content. For swine and poultry, the amino acid deficiencies of linseed meal do not effectively compensate for the deficiencies of the cereal grains.	Chapter 11, under "Linseed Meal."
Peanut meal	Peanut meal, a by-product of the peanut industry, is ground peanut cake, the product that remains after the extraction of part of the oil of peanuts. It must contain not more than 7% crude fiber. Peanut meal is low in methionine, lysine, and tryptophan, and low in calcium, carotene, and vitamin D.	**A-F** M-F	**93** 100	**7.7** 8.3	**73** 79	**74** 80	**1.43** 1.54	***3.15*** *3.40*	**1,296** 1,401	***2,857*** *3,088*	**49.0** 52.9	**0.36** 0.39	**0.61** 0.66	Peanut meal, which is palatable and contains high quality protein, is used extensively in livestock and poultry feeds. Since peanut meal tends to become rancid when held too long—especially in warm, moist climates—it should not be stored longer than 6 weeks in the summer or 2 to 3 months in the winter.	Chapter 11, under "Peanut Meal, and Peanut Meal and Hulls."
Safflower meal	Safflower meal is the ground residue obtained after extracting oil from the whole safflower seed.	**A-F** M-F	**92** 100	**14.6** 16.0	**66** 72	**56** 62	**1.15** 1.26	***2.53*** *2.77*	**1,508** 1,647	***3,325*** *3,632*	**42.7** 46.7	**0.38** 0.41	**1.08** 1.18	Dehulled (decorticated) safflower meal can be fed to swine and poultry, but it should be supplemented with soybean and meat meal to provide adequate levels of lysine and the sulfur-containing amino acids. High-fiber safflower meal (above 15%) should be fed only to ruminants.	Chapter 11, under "Safflower Meal."
Sesame meal	Sesame meal is obtained by grinding the residue remaining after extraction of most of the oil from whole sesame seed.	**A-F** M-F	**92** 100	**6.8** 7.4	**69** 75	**81** 88	**1.38** 1.50	***3.04*** *3.30*	**1,610** 1,750	***3,549*** *3,858*	**45.0** 48.9	**2.01** 2.18	**1.28** 1.39	Sesame meal is a low-fiber, excellent quality plant protein supplement suitable for all animal species.	Chapter 11, under "Sesame Meal."
Soybean meal	Soybean meal is the product obtained by grinding the flakes which remain after extracting most of the oil from soybeans. It must contain not more than 7% crude fiber. Soybean meal has the highest nutritive value of any plant protein source.	**A-F** M-F	**89** 100	**6.2** 7.0	**76** 85	**75** 84	**1.45** 1.62	***3.19*** *3.57*	**1,565** 1,756	***3,450*** *3,872*	**44.4** 49.8	**0.35** 0.40	**0.64** 0.71	Soybean meal, which is the most widely used protein supplement in the U.S., is very palatable and suitable for all animal species.	Chapter 11, under "Soybean Meal."

(Continued)

TABLE 12–1 *(Continued)*

By-products/ Crop Residues	Description[2]	Moisture Basis: A-F (as-fed) or M-F (moisture-free)	Dry Matter	Crude Fiber	TDN		Digestible Energy				Crude Protein	Ca	P	Feeding Recommendations[3]	Cross References
					Ruminant	Swine	Ruminant Mcal per		Swine kcal per						
			(%)	(%)	(%)	(%)	(lb)	*(kg)*	(lb)	*(kg)*	(%)	(%)	(%)		
OILSEED BY-PRODUCT FEEDS (Continued)															
Sunflower meal	Sunflower meal is obtained by grinding the residue remaining after extraction of most of the oil from whole sunflower seed. It varies considerably in nutrient content, depending on the extraction process and whether the seeds are dehulled.	**A-F** M-F	**93** 100	**11.0** 11.8	**65** 70	**69** 74	**1.22** 1.32	***2.70*** *2.90*	**1,366** 1,469	***3,012*** *3,239*	**46.8** 50.3	**—** —	**—** —	Sunflower meal is a suitable protein supplement for all animal species. Meal produced from seed that was not dehulled has a high fiber content; hence, it should be fed only to ruminants. When sunflower meal is used in monogastric feeds, it should be combined with high-lysine supplements, such as fish meal or meat scraps.	Chapter 11, under "Sunflower Seed Meal."
GRAIN BY-PRODUCT FEEDS															
Corn bran	Corn bran is the outer coating of the corn kernel obtained primarily in the wet milling of corn, with little or none of the starchy part.	**A-F** M-F	**89** 100	**9.6** 10.8	**67** 76	**73** 82	**1.37** 1.54	***3.02*** *3.39*	**1,464** 1,645	***3,227*** *3,626*	**8.4** 9.4	**0.03** 0.04	**0.20** 0.22	Corn bran adds bulk to the ration, and is a laxative.	Chapter 10, Table 10–4.
Corn gluten feed	Corn gluten feed is a major by-product feed that remains after the extraction of most of the starch and germ in the wet milling of corn to produce starch and syrup.	**A-F** M-F	**90** 100	**8.7** 9.7	**75** 83	**75** 84	**1.44** 1.60	***3.17*** *3.52*	**1,375** 1,529	***3,031*** *3,371*	**23.0** 25.6	**0.32** 0.36	**0.74** 0.82	Corn gluten feed is generally not fed to monogastric animals because of its bulkiness, poor quality protein, and unpalatability. It is used extensively for dairy cows. Because corn gluten feed varies widely, depending on the manufacturer, it should be purchased on a guaranteed nutrient basis.	Chapter 10, Table 10–4.
Corn gluten meal	Corn gluten meal is a major by-product feed obtained in the wet milling of corn to produce starch and syrup. It contains more than twice the crude protein content of corn gluten feed.	**A-F** M-F	**90** 100	**1.8** 2.0	**—** —	**—** —	**—** —	**—** —	**—** —	**—** —	**60.8** 67.5	**0.07** 0.08	**0.45** 0.50	Corn gluten meal, like corn gluten feed, is used primarily as a feed for ruminants, especially for dairy cows. However, it is somewhat more valuable because of its higher protein content. Corn gluten meal is an excellent rumen by-pass feed (see Chapter 11, Table 11–1).	Chapter 10, Table 10–4.
Corn oil	Corn oil is the oil from the germ obtained in the wet milling of corn to produce starch and syrup.	**A-F** M-F	**99** 100	**—** —	**203** 205	**208** 210	**3.57** 3.60	***7.86*** *7.94*	**3,421** 3,456	***7,543*** *7,619*	**—** —	**—** —	**—** —	Corn oil may be used to increase the caloric density of feeds of any species. When a small quantity is used in a horse ration, it imparts gloss to the hair.	Chapter 10, under "Fats and Oils."
Hominy feed	Hominy feed is a mixture of corn bran, corn germ, and part of the starchy portion of either white or yellow corn kernels or mixture thereof, produced in the manufacture of pearl hominy, hominy grits, or table meal. It must contain not less than 4% fat.	**A-F** M-F	**90** 100	**4.8** 5.3	**84** 93	**82** 91	**1.69** 1.88	***3.73*** *4.14*	**1,620** 1,796	***3,571*** *3,959*	**10.3** 11.4	**0.05** 0.05	**0.51** 0.57	Hominy feed is used as a grain replacement in the rations of all animal species.	Chapter 10, Table 10–4.
Wheat bran	Wheat bran is the coarse outer covering of the wheat kernel which is separated from wheat in commercial milling. Bran is rich in niacin, vitamin B-1, phosphorus, and iron. A large part of the phosphorus is phytin phosphorus.	**A-F** M-F	**89** 100	**10.0** 11.2	**63** 70	**57** 64	**1.26** 1.42	***2.78*** *3.12*	**1,119** 1,256	***2,466*** *2,768*	**15.5** 17.5	**0.13** 0.14	**1.16** 1.30	Wheat bran is widely used in horse rations because of its bulky nature and laxative effect. Also, it is a favored supplement for use in gestating cow, sheep, and swine rations.	Chapter 10, Table 10–4.

(Continued)

TABLE 12–1 *(Continued)*

By-products/ Crop Residues	Description[2]	Moisture Basis: A-F (as-fed) or M-F (moisture-free)	Dry Matter	Crude Fiber	TDN		Digestible Energy				Crude Protein	Ca	P	Feeding Recommendations[3]	Cross References
					Ruminant	Swine	Ruminant Mcal per		Swine kcal per						
			(%)	(%)	(%)	(%)	(lb)	*(kg)*	(lb)	*(kg)*	(%)	(%)	(%)		
GRAIN BY-PRODUCT FEEDS (Continued)															
Wheat middlings	Wheat middlings, a by-product of the flour milling industry, consist of the fine particles of wheat bran, wheat shorts, wheat germ, wheat flour, and some of the offal from the "tail of the mill." It cannot contain more than 9.5% crude fiber. Middlings are deficient in calcium, carotene, and vitamin D.	**A-F** M-F	**89** 100	**7.7** 8.7	**74** 83	**68** 77	**1.39** 1.57	***3.07*** *3.45*	**1,321** 1,487	***2,912*** *3,279*	**16.4** 18.5	**0.13** 0.15	**0.89** 1.00	Wheat middlings are widely used as a potential grain replacement in rations for all animal species. When fed to monogastrics, middlings are most efficient when used along with an animal protein supplement.	Chapter 10, Table 10–4.
Rice bran	Rice bran is the pericarp (outer covering), or bran, layer and germ of rice, obtained in the milling of edible rice. It is rich in the B-complex vitamins.	**A-F** M-F	**91** 100	**11.9** 13.1	**64** 71	**69** 77	**1.10** 1.21	***2.42*** *2.67*	**1,474** 1,626	***3,250*** *3,586*	**13.0** 14.3	**0.07** 0.08	**1.44** 1.59	Rice bran is best used as a ruminant feed, because of its high fiber content.	Chapter 10, Table 10–4.
Rice polishings	Rice polishings are a by-product of rice milling, obtained in brushing the grain to polish the kernel. It is rich in many of the B-vitamins; it is especially high in thiamin.	**A-F** M-F	**89** 100	**0.4** 0.4	**78** 88	**86** 97	**1.59** 1.80	***3.51*** *3.96*	**1,697** 1,916	***3,741*** *4,223*	**7.0** 7.9	**0.02** 0.03	**0.11** 0.13	Most rice polishings are incorporated in swine and poultry rations, as a rich source of the B-vitamins, especially thiamin. Because of their high fat content, (1) they should be limited in swine rations in order to avoid soft pork, and (2) they tend to become rancid when stored very long.	Chapter 10, Table 10–4.
Screenings, cereal grain	Grain screenings consist of 70% or more of light and broken grains, along with weed seeds, and other foreign material that is separated in the cleaning of grain with a screen. The quality varies according to the percentage of weed seeds and other foreign materials.	**A-F** M-F	**90** 100	**12.0** 13.4	**62** 69	**55** 61	**1.23** 1.37	***2.71*** *3.02*	**1,493** 1,663	***3,291*** *3,666*	**12.1** 13.4	**0.33** 0.37	**0.35** 0.39	Grain screenings should be finely ground in order to destroy noxious weed seeds. Most grain screenings are incorporated in ruminant rations. Normally, screenings are limited to 15 to 20% of feedlot rations and 10% of concentrate mixes in lactating dairy rations.	Chapter 10, Table 10–4.
FRUIT, VEGETABLE, AND NUT BY-PRODUCT FEEDS															
Acorns	Acorns are the nuts of the oak species of trees, of which the white oak group are most widely used in feed. They are high in oil.	**A-F** M-F	**50** 100	**9.3** 18.6	**23** 47	**39** 77	**0.60** 1.21	***1.33*** *2.66*	**771** 1,543	***1,700*** *3,401*	**2.2** 4.4	**—** —	**—** —	Acorns are an important feed for wild animals, and sometimes they are fed to domestic animals. They are high in tannin, which, at high levels, may produce toxicities and result in cows drying up and in deformed calves known as *acorn calves.*	
Apple pomace, wet	Apple pomace, which contains pulp, peels, cores, and cull apples, is the by-product that remains (1) after expressing juice from apples or (2) after canning, drying, and freezing apples. It may be fed fresh (wet), as silage, or dry.	**A-F** M-F	**23** 100	**4.7** 20.7	**17** 74	**17** 74	**0.33** 1.46	***0.73*** *3.21*	**337** 1,483	***744*** *3,270*	**1.3** 5.6	**0.02** 0.10	**0.02** 0.10	Wet apple pomace may be used as a feed for cattle and sheep, for which it is very palatable.	
Apple pomace, dehy	Dried apple pomace is the dried residue obtained by the removal of cider from apples. It is high in fiber and very palatable. Dried apple pomace is recognized as a feedstuff by the AAFCO.	**A-F** M-F	**89** 100	**16.2** 18.2	**60** 68	**69** 77	**1.21** 1.35	***2.66*** *2.98*	**1,380** 1,548	***3,043*** *3,413*	**4.4** 5.0	**0.11** 0.13	**0.10** 0.12	Dried apple pomace is used as a cattle and sheep feed. It is an excellent appetizer when included as a replacement for ¼ to ⅓ of the grain in the ration.	

(Continued)

TABLE 12–1 *(Continued)*

By-products/ Crop Residues	Description[2]	Moisture Basis: A-F (as-fed) or M-F (moisture-free)	Dry Matter	Crude Fiber	TDN Ruminant	TDN Swine	Digestible Energy Ruminant Mcal per		Digestible Energy Swine kcal per		Crude Protein	Ca	P	Feeding Recommendations[3]	Cross References
			(%)	(%)	(%)	(%)	(lb)	*(kg)*	(lb)	*(kg)*	(%)	(%)	(%)		
FRUIT, VEGETABLE, AND NUT BY-PRODUCT FEEDS (Continued)															
Asparagus stem butts, meal	Asparagus meal is the ground, dried meal of asparagus butt ends and broken spears. It is very high in fiber.	**A-F** M-F	**91** 100	**29.0** 31.9	**43** 47	**26** 29	**0.87** 0.95	***1.91*** *2.10*	— —	— —	**14.2** 15.6	— —	— —	Asparagus may be fed to cattle when there is a great surplus or it is of low quality. Because of the high energy cost, it is seldom dehydrated.	
Bananas, dehy	Dehydrated bananas consist of the dried fruit without the peelings. It is palatable and low in fiber and protein.	**A-F** M-F	**86** 100	**1.0** 1.2	**64** 74	**80** 93	**1.33** 1.55	***2.94*** *3.41*	**1,604** 1,859	***3,537*** *4,099*	**3.5** 4.1	— —	— —	Dehydrated bananas are available as a livestock feed if there is a surplus or if the product is of low quality. When used as a feed, it is commonly fed to swine.	
Banana peelings, dehy	Banana peelings are a by-product of processing flaked bananas. After removing the fruit, the peelings may be dried.	**A-F** M-F	**91** 100	**9.5** 10.5	**58** 64	**78** 85	**1.24** 1.36	***2.74*** *3.01*	**1,554** 1,707	***3,425*** *3,762*	**8.6** 9.4	— —	— —	Dehydrated banana peelings may be fed to ruminants.	
Beans, navy, cull	Cull navy beans consist of the broken, small, shriveled, and cull beans, obtained in the normal processing of dried navy beans for human consumption.	**A-F** M-F	**89** 100	**4.4** 5.0	**76** 85	**78** 87	**1.53** 1.71	***3.37*** *3.77*	**1,737** 1,941	***3,830*** *4,280*	**22.9** 25.6	**0.17** 0.19	**0.54** 0.61	Dry cull beans may be ground and replace 15 to 20% of the grain of cattle and sheep. When fed to hogs, they should be cooked and fed with an animal protein.	
Beechnut meal	Beechnut meal is the dried, ground seeds from beech trees.	**A-F** M-F	**87** 100	**23.4** 26.9	**37** 42	— —	**0.74** 0.85	***1.63*** *1.87*	— —	— —	**17.0** 19.5	— —	— —	Beechnut meal is of little importance as a livestock feed. Most beech nuts are gleaned by wild animals and swine.	
Broccoli, fresh, immature	Fresh broccoli consists of the leaves and stems, along with surplus or low quality broccoli.	**A-F** M-F	**11** 100	**1.5** 13.8	— —	— —	— —	— —	— —	— —	**3.6** 33.0	**0.10** 0.92	**0.08** 0.73	Fresh broccoli may be used as a livestock feed. Because of its high moisture and high fiber most of it is fed to ruminants.	
Carrots, roots, fresh	Carrots are a succulent, thickened root, high in carotene—the precursor of vitamin A. They may be available as a result of surpluses, or culls, tips, crowns, and tops may be available as a by-product from dehydrating plants.	**A-F** M-F	**11** 100	**1.1** 9.5	**10** 84	**10** 90	**0.19** 1.69	***0.43*** *3.72*	**208** 1,805	***458*** *3,979*	**1.2** 10.0	**0.05** 0.40	**0.04** 0.35	Carrots are an excellent cattle feed. They are usually fed in bunks or on the ground. A mature cow will eat about 35 lb of carrots daily, along with other feeds. Horses relish carrots as a treat.	
Citrus pulp, dehy	Dried citrus pulp is the peel, seeds, pulp, and cull fruits, left from the canning of citrus juices and other citrus products, which have been dried. It is high in fiber and low in phosphorus.	**A-F** M-F	**87** 100	**13.2** 15.1	**76** 87	**65** 74	**1.40** 1.60	***3.09*** *3.53*	**1,293** 1,479	***2,850*** *3,260*	**12.1** 13.8	**1.66** 1.90	**0.12** 0.14	Most of the dried citrus pulp is fed to ruminants. It is seldom used for monogastric animals because of its high fiber content. A level of 10% citrus pulp in chicken rations will depress growth. Incorporation of 2.5% dried citrus pulp in layer rations may result in off-colored yolks.	
Citrus molasses	Citrus molasses is the partially dehydrated juices obtained from the manufacture of dried citrus pulp. It must contain not less than 45% total sugars expressed as invert; and its density determined by double dilution must not be less than 17.0° Brix. Citrus molasses is used as an energy source.	**A-F** M-F	**67** 100	— —	**51** 75	**54** 80	**1.01** 1.49	***2.22*** *3.29*	**1,084** 1,609	***2,390*** *3,547*	**5.8** 8.5	**1.18** 1.76	**0.09** 0.13	Citrus molasses is used extensively as an energy feed for cattle, despite being bitter tasting. It is usually restricted to about 10% fo the ration.	Chapter 10, under "Citrus Molasses."

(Continued)

TABLE 12–1 *(Continued)*

By-products/ Crop Residues	Description[2]	Moisture Basis: A-F (as-fed) or M-F (moisture-free)	Dry Matter	Crude Fiber	TDN Ruminant	TDN Swine	Digestible Energy Ruminant Mcal per (lb)	Digestible Energy Ruminant Mcal per *(kg)*	Digestible Energy Swine kcal per (lb)	Digestible Energy Swine kcal per *(kg)*	Crude Protein	Ca	P	Feeding Recommendations[3]	Cross References
			(%)	(%)	(%)	(%)	(lb)	*(kg)*	(lb)	*(kg)*	(%)	(%)	(%)		
FRUIT, VEGETABLE, AND NUT BY-PRODUCT FEEDS (Continued)															
Date palm pits	Date palm pits (stones) are the flat seeds that are sometimes removed from the edible fruit, then ground.	**A-F** M-F	**92** 100	**14.4** 15.7	**76** 82	**83** 90	**1.52** 1.65	***3.36*** *3.63*	**1,669** 1,804	***3,680*** *3,977*	**8.3** 9.1	**0.60** 0.65	**0.18** 0.20	Ground date pits may constitute up to 20% of ruminant rations with good results, provided they are properly supplemented with protein or urea.	
Grape pomace, wet (marc)	Wet grape pomace is the refuse from the production of grape juice and wine, consisting of seeds, stems, pulp, and skins. It is low in both energy and protein. With new harvesting and winery techniques, grape pomace containing few or no stems can be produced.	**A-F** M-F	**37** 100	**9.7** 25.9	**12** 32	**—** —	**0.37** 0.98	***0.81*** *2.17*	**—** —	**—** —	**5.2** 13.8	**—** —	**—** —	Grape pomace is of low feeding value. If relatively free of stems, up to 15 to 20% may be used in complete cattle finishing rations.	
Grape pomace, dehy (marc)	Dried grape pomace is the dried refuse from the production of grape juice and wine, consisting of seeds, stems, pulp, and skins. It is high in fiber and low in energy.	**A-F** M-F	**90** 100	**27.9** 30.9	**24** 27	**—** —	**0.48** 0.53	***1.06*** *1.17*	**—** —	**—** —	**12.1** 13.4	**0.52** 0.58	**0.15** 0.17	Dried grape pomace may be fed to cattle and sheep, but it should be limited to 20% of the balanced, total ration.	
Olive fruit without pits, dehy	Dried olive fruit without pits consists of the pulp from which the kernels have been separated, followed by dehydrating. It contains 20 to 25% oil.	**A-F** M-F	**93** 100	**27.9** 29.9	**51** 55	**—** —	**1.02** 1.10	***2.25*** *2.42*	**—** —	**—** —	**10.8** 11.6	**—** —	**—** —	Dried olive fruit without pits may constitute up to 15% of the ration of ruminants and swine without lowering feed efficiency.	
Pear cannery residue, wet	Pear cannery residue is the waste that remains from canning pears. The peels, cores, and screened solids are sometimes used as feed.	**A-F** M-F	**15** 100	**2.6** 17.1	**11** 72	**11** 75	**0.22** 1.45	***0.49*** *3.19*	**227** 1,495	***501*** *3,295*	**0.6** 3.9	**—** —	**—** —	Based on feeding trials, pear cannery residue can replace up to 25% of the molasses and dried beet pulp in the concentrate ration of finishing steers, with the pear cannery residue having a value of 70 to 75% of the molasses and dried beet pulp. Because of the high water content and transportation costs, these wastes are fed fresh by farmers in close proximity to the cannery.	
Peas, cull (split), dry	Dry cull peas are the cull and split peas obtained from the production of yellow, green, and black peas grown for seed.	**A-F** M-F	**90** 100	**23.7** 26.3	**78** 87	**55** 61	**1.34** 1.49	***2.96*** *3.29*	**—** —	**—** —	**17.7** 19.7	**—** —	**—** —	Dry cull peas may be used as a grain replacement and/or protein supplement for cattle, sheep, and hogs. Cull peas may replace up to 40% of the grain mixture in rations of finishing steers with good results. As a swine feed, 2 tons of cull peas are equivalent in feeding value to 1 ton of soybean meal plus 1 ton of grain.	Chapter 11, Table 11–6.
Pineapple bran	Pineapple bran is a by-product of the pineapple canning industry. It consists of the outer shells, and it may also include the cores of pineapples; sometimes, molasses is added. Its composition is similar to dried beet pulp. It is high in fiber and low in protein.	**A-F** M-F	**87** 100	**18.2** 20.9	**64** 73	**61** 70	**1.27** 1.46	***2.81*** *3.23*	**1,223** 1,405	***2,696*** *3,097*	**4.0** 4.6	**0.20** 0.23	**0.11** 0.13	Pineapple bran may be used in the rations of ruminants and horses. Because of its low protein content, supplementation is needed. Levels up to 15 lb per cow daily have been fed in Hawaii with good results.	

(Continued)

TABLE 12-1 *(Continued)*

By-products/ Crop Residues	Description[2]	Moisture Basis: A-F (as-fed) or M-F (moisture-free)	Dry Matter	Crude Fiber	TDN Ruminant	TDN Swine	Digestible Energy Ruminant Mcal per (lb)	Digestible Energy Ruminant Mcal per *(kg)*	Digestible Energy Swine kcal per (lb)	Digestible Energy Swine kcal per *(kg)*	Crude Protein	Ca	P	Feeding Recommendations[3]	Cross References
			(%)	(%)	(%)	(%)	(lb)	*(kg)*	(lb)	*(kg)*	(%)	(%)	(%)		
FRUIT, VEGETABLE, AND NUT BY-PRODUCT FEEDS (Continued)															
Potatoes, cull, fresh	Potatoes, cull and surplus, are high in water (76%), low in fiber, high in carbohydrates, and comparable to whole grains in protein content on a dry basis, and deficient in vitamins A and D.	**A-F** M-F	**24** 100	**0.6** 2.4	**19** 81	**20** 85	**0.38** 1.62	***0.84*** *3.58*	**398** 1,695	***878*** *3,736*	**2.2** 9.3	**0.01** 0.04	**0.06** 0.24	**When** properly processed, potatoes may be fed to all classes of livestock. Fresh potatoes may be fed to ruminants and horses. Choking of cattle can be minimized by feeding whole potatoes from low bunks, under a bar or cable, to keep the animals heads down, or by chopping the potatoes. Potatoes should be cooked for swine and poultry. About 400 to 450 lb of potatoes are equivalent in energy value to 100 lb of cereal grain. **Sprouted** potatoes contain (in the sprouts) the toxic substance, *solanine,* and should not be fed to animals.	Chapter 10, under "Roots and Tubers."
Potatoes, cull, cooked	**Cooked** potatoes are potatoes that have been cooked (steamed or boiled, preferably in salt water to increase palatability, with the cooking water discarded).	**A-F** M-F	**24** 100	**0.7** 3.0	**17** 69	**21** 88	**0.33** 1.38	***0.73*** *3.05*	**416** 1,742	***918*** *3,841*	**2.2** 9.3	**0.01** 0.04	**0.05** 0.22	**Cooked** potatoes may be fed to swine and poultry, but cooking is not necessary for other classes of livestock. **Cooked** potatoes are efficiently used by both breeding and finishing swine. Up to 11–13 lb *(5 to 6 kg)* per day may be fed to mature swine. **Cooked** potatoes may constitute up to 40% of poultry rations.	Chapter 10, under "Roots and Tubers."
Potatoes, dehy, meal	**Dehydrated** potatoes—also called potato meal (if powdered) or potato flakes—are sun dried or artificially dried cull or surplus potatoes.	**A-F** M-F	**91** 100	**2.1** 2.3	**75** 83	**80** 88	**1.42** 1.56	***3.14*** *3.45*	**1,565** 1,718	***3,450*** *3,788*	**8.1** 8.9	**0.07** 0.08	**0.20** 0.22	**Dehydrated** potatoes may replace 20 to 25% of the concentrate of beef cattle, dairy cattle, or sheep rations. If heated sufficiently in the drying process and cooked thoroughly, 10 to 20% can be incorporated in swine and poultry rations. Potato meal is equal to grain, pound for pound, when substituted for part of it.	Chapter 10, under "Roots and Tubers."
Raisins, cull, dehy	**Cull** dehydrated raisins (dried grapes or dried raisins) are grapes that have been dried, either in the sun or artificially.	**A-F** M-F	**87** 100	**10.1** 11.6	**44** 51	**72** 83	**1.11** 1.27	***2.44*** *2.80*	**1,449** 1,664	***3,195*** *3,668*	**6.4** 7.4	— —	— —	**Raisins** may replace up to ⅓ of the grain ration of beef cattle, dairy cattle, sheep, or swine. They have a feeding value of about 85% that of barley.	
Sweet potato, dehy, meal	**Dehydrated** sweet potatoes are sweet potatoes that have been dried, either in the sun or artificially. They are low in fiber, fat, and protein, but high in starch and vitamin A.	**A-F** M-F	**89** 100	**3.7** 4.1	**72** 81	**64** 72	**1.42** 1.58	***3.12*** *3.49*	**1,280** 1,431	***2,823*** *3.156*	**6.4** 7.2	**0.12** 0.13	**0.15** 0.16	**Dehydrated** sweet potatoes are valuable as a partial substitute for grain in rations for cattle and sheep. It is worth 90 to 100% as much as corn for ruminants. But it is not palatable to horses, and it is least useful for swine and poultry.	Chapter 10, under "Roots and Tubers."
Tomato, fruit, fresh	**Fresh** tomatoes consist of cull or surplus tomatoes, which may be available as a livestock feed. On a fresh (as-fed) basis, tomatoes are high in water (94%), and low in fiber, energy, protein, calcium, and phosphorus.	**A-F** M-F	**6** 100	**0.5** 9.1	**5** 86	**5** 77	**0.10** 1.60	***0.21*** *3.53*	**92** 1,549	***204*** *3,416*	**1.0** 16.4	**0.01** 0.17	**0.03** 0.49	**Fresh** tomatoes cannot be transported very far, because of their high water content; and fresh tomatoes should be limited to about 15% of the concentrate ration, because of their low nutritive value. They may be fed to cattle, sheep, or swine.	

(Continued)

TABLE 12–1 *(Continued)*

By-products/ Crop Residues	Description[2]	Moisture Basis: A-F (as-fed) or M-F (moisture-free)	Dry Matter	Crude Fiber	TDN Ruminant	TDN Swine	Digestible Energy Ruminant Mcal per (lb)	Digestible Energy Ruminant Mcal per (kg)	Digestible Energy Swine kcal per (lb)	Digestible Energy Swine kcal per (kg)	Crude Protein	Ca	P	Feeding Recommendations[3]	Cross References
			(%)	(%)	(%)	(%)	(lb)	*(kg)*	(lb)	*(kg)*	(%)	(%)	(%)		
FRUIT, VEGETABLE, AND NUT BY-PRODUCT FEEDS (Continued)															
Tomato pomace, wet	Wet tomato pomace consists of skins, pulp, and crushed seeds that remain after the processing of tomatoes for juice, soup, or catsup. It is high in water (75%).	**A-F** M-F	**25** 100	**8.5** 33.7	**16** 62	— —	**0.31** 1.23	***0.68*** *2.71*	— —	— —	**5.4** 21.5	— —	**0.16** 0.64	Wet tomato may constitute up to 15% of the ration of beef cattle, dairy cattle, sheep, and swine. At higher levels, its bitterness may make for unpalatability.	
Tomato pomace, dehy	Dried tomato pomace is the dried mixture of tomato skins, pulp, and crushed seeds that remain after processing of tomatoes for juice, soup, or catsup. It is high in fiber, high in protein, a good source of some B vitamins, and a fair source of vitamin A.	**A-F** M-F	**92** 100	**25.0** 27.2	**60** 66	**47** 51	**1.21** 1.31	***2.66*** *2.89*	**934** 1,015	***2,058*** *2,238*	**21.0** 22.9	**0.39** 0.43	**0.55** 0.60	Dried tomato pomace may be incorporated in the rations of cattle, sheep, and goats at a level up to 15% of the concentrate mixture with satisfactory results; but it is too high in fiber to be very useful in rations for swine, poultry, and other monogastrics. Dried tomato pomace is somewhat bitter; so, it should be used with well-liked feeds. Some dried and bagged tomato pomace is used in pet food and for fur-bearing animals to prevent diarrhea, with dry pomace constituting 5% of the wet diet.	
Turnip, roots, fresh	Turnips are a root crop, popular as a winter forage in some northern areas, especially for sheep. On a moisture-free basis, they are high in energy and good in protein.	**A-F** M-F	**9** 100	**1.1** 11.5	**8** 85	**7** 80	**0.16** 1.70	***0.34*** *3.74*	**146** 1,594	***322*** *3,513*	**1.2** 13.1	**0.06** 0.64	**0.03** 0.32	Turnips are a popular fall/winter forage grazing crop for cattle and sheep. Stocker cattle, weighing 600 to 700 lb, will make weight gains of 1.5 to 2.0 lb per head per day when grazing turnip fields. A roughage supplement of 2 to 3 lb of straw or hay per day is recommended.	Chapter 7, under "Turnips For Fall/Winter Grazing."
SUGAR BY-PRODUCT FEEDS															
Beet molasses	Beet molasses is a by-product of sugar beets, obtained in the process of manufacturing sugar. It must contain not less than 48% total sugars expressed as invert and its density determined by double dilution must be not less than 79.5° Brix. It is a good energy source, but it is low in protein. Beet molasses is very laxative when used at high levels.	**A-F** M-F	**78** 100	— —	**61** 78	**57** 73	**1.20** 1.54	***2.64*** *3.38*	**1,130** 1,451	***2,491*** *3,198*	**6.6** 8.5	**0.12** 0.16	**0.03** 0.03	Beet molasses may be fed to cattle, sheep, and swine. It may replace up to ¼ to ⅓ of the grain in finishing rations.	Chapter 10, under "Beet Molasses."
Beet pulp, wet	Wet beet pulp results from processing sugar beets. It is palatable, but high in moisture (89%).	**A-F** M-F	**11** 100	**2.3** 21.3	**8** 72	**7** 64	**0.16** 1.44	***0.35*** *3.18*	**140** 1,272	***308*** *2,804*	**1.2** 11.2	**0.10** 0.87	**0.01** 0.10	Wet beet pulp may be used as a primary feed source for finishing cattle near beet sugar processing plants. However, its high moisture content (and consequent high transportation cost) makes it economically unfeasable to feed wet beet pulp very far from processing plants.	
Beet pulp, dried	Dried beet pulp is the dried residue from sugar beets obtained in the manufacture of sugar. It is high in fiber, but the fiber is highly digestible. It is high in calcium and low in phosphorus.	**A-F** M-F	**91** 100	**18.2** 20.1	**67** 74	**67** 73	**1.31** 1.44	***2.89*** *3.18*	**1,305** 1,436	***2,878*** *3.166*	**8.8** 9.7	**0.63** 0.70	**0.09** 0.10	Dried beet pulp may replace up to 45% of the concentrate mix for dairy cows, finishing cattle, and finishing lambs. Beet pulp is an excellent ingredient for maintaining normal milk fat test when dairy cows are fed restricted roughage rations.	

(Continued)

TABLE 12–1 *(Continued)*

By-products/ Crop Residues	Description[2]	Moisture Basis: A-F (as-fed) or M-F (moisture-free)	Dry Matter	Crude Fiber	TDN Rumi-nant	TDN Swine	Digestible Energy Ruminant Mcal per		Digestible Energy Swine kcal per		Crude Pro-tein	Ca	P	Feeding Recommendations[3]	Cross Refer-ences
			(%)	(%)	(%)	(%)	(lb)	*(kg)*	(lb)	*(kg)*	(%)	(%)	(%)		
SUGAR BY-PRODUCT FEEDS (Continued)															
Cane molasses	Cane molasses is a by-product of the manufacture of sucrose from sugar cane. It must contain not less than 43% total sugars expressed as invert. If its moisture content exceeds 27%, its density determined by double dilution must not be less than 79.5° Brix. It is a good energy source but low in protein. Cane molasses is a very palatable feed—an excellent appetizer.	**A-F** M-F	**74** 100	**0.4** 0.5	**60** 81	**56** 76	**1.22** 1.63	***2.68*** *3.60*	**1,135** 1,526	***2,502*** *3,364*	**4.3** 5.8	**0.74** 1.00	**0.08** 0.11	Cane molasses may be fed to any animal species. Normally, it is limited to 5 to 10% of rations. It is commonly restricted to 2 to 3 lb per cow daily in dairy rations, but in Hawaii, where molasses is abundant and low cost, levels as high as 12 lb per cow daily are reported.	Chapter 10, under "Cane Molasses."
Corn molasses	Corn molasses (also known as starch molasses) is a by-product of the manufacture of starch from corn. It must contain not less than 43% reducing sugars expressed as dextrose and not less than 50% total sugars expressed as dextrose; and it shall contain not less than 73% total solids. It is very low in protein.	**A-F** M-F	**73** 100	— —	— —	— —	— —	— —	— —	— —	**0.3** 0.4	— —	— —	Corn molasses is suitable for ruminants, at a level of 5 to 10% of the ration. It may be used in the same manner as cane molasses, but it does not have the sweet odor of cane molasses.	
BREWING AND DISTILLING BY-PRODUCT FEEDS															
Brewers' dried grains	Brewers' dried grains are the dried extracted residue of barley malt alone or in a mixture with other cereal grain or grain products resulting from the manufacture of wort or beer. They have about 80% the energy value of barley grain. Brewers' grains are not as palatable in the dried form as the original grain.	**A-F** M-F	**92** 100	**13.0** 14.1	**65** 71	**66** 71	**1.25** 1.36	***2.76*** *2.99*	**1,045** 1,134	***2,303*** *2,500*	**27.3** 29.6	**0.30** 0.33	**0.51** 0.55	Brewers' grains are used chiefly as a cattle feed; they are used interchangeably with other feeds of similar bulk, fiber, and crude protein content. They are usually limited to 15 to 20% of cattle finishing rations and 25% of the dairy concentrate mix. Brewer's dried grains rank high as a rumen bypass protein (see Chapter 11, Table 11–1).	
Distillers' dried grains	Distillers' dried grains are by-products of the production of distilled liquors from grains. Distillers' dried grains are high in fiber and high in protein. Corn is the most widely used grain in alcohol production. But rye, sorghum, and wheat are sometimes used for alcohol production. Distillers' dried grains are identified by the type of grain from which they are made.	**A-F** M-F	**93** 100	**12.8** 13.8	**78** 84	**82** 89	**1.61** 1.74	***3.54*** *3.85*	**1,637** 1,779	***3,609*** *3,922*	**27.3** 29.5	**0.12** 0.13	**0.54** 0.59	Distillers' dried grains are recommended for use by beef cattle, dairy cattle, and sheep. They may constitute up to 25% of the concentrate mix. Distillers' dried grains rank high as a rumen bypass protein (see Chapter 11, Table 11–1).	Chapter 10, Fig. 10–18.
Distillers' dried grains with solubles	Distillers' dried grains with solubles are by-products of the production of distilled liquors from grains, including solubles of fermentation which are added to the grain before drying. The predominant grain is declared as the first word in the name.	**A-F** M-F	**92** 100	**3.4** 3.7	**68** 75	**75** 81	**1.39** 1.52	***3.07*** *3.34*	**1,493** 1,627	***3,291*** *3,587*	**28.8** 31.3	**0.24** 0.26	**1.35** 1.47	Distillers' dried grains with solubles are used as a protein/B-vitamin supplement for all classes of animals, but especially for ruminants. For swine and poultry, they are usually limited to 5 to 10% of the ration.	Chapter 10, Fig. 10–18.

(Continued)

TABLE 12–1 *(Continued)*

By-products/ Crop Residues	Description[2]	Moisture Basis: A-F (as-fed) or M-F (moisture-free)	Dry Matter	Crude Fiber	TDN Ruminant	TDN Swine	Digestible Energy Ruminant Mcal per		Digestible Energy Swine kcal per		Crude Protein	Ca	P	Feeding Recommendations[3]	Cross References
			(%)	(%)	(%)	(%)	(lb)	*(kg)*	(lb)	*(kg)*	(%)	(%)	(%)		
BREWING AND DISTILLING BY-PRODUCT FEEDS (Continued)															
Malt sprouts, barley, dehy	**D**ried malt sprouts are obtained from malted barley by the removal of the rootlets and sprouts. It may include some of the hulls and other parts of the malt, but it must contain not less than 24% crude protein. Malt sprouts obtained from different grains are identified as barley malt sprouts, rye malt sprouts, or wheat malt sprouts.	**A-F** M-F	**93** 100	**14.2** 15.3	**66** 71	**61** 66	**1.31** 1.41	***2.89*** *3.11*	**1,219** 1,313	***2,688*** *2,894*	**22.9** 24.6	**0.18** 0.19	**0.63** 0.68	**B**ecause only about ½ of the protein is true protein, malt sprouts can be used most effectively in ruminant rations. They have a bitter flavor and can pose some palatability problems if fed at high levels. **D**ried malt sprouts are used chiefly in mixed feeds for dairy cattle, with an upper limit of about 3.0 lb *(1.4 kg)* per head daily.	Chapter 10, Fig. 10–18 and Table 10–4.
Yeast, brewers', dried	**D**ried brewers' yeast is the dried, nonfermentative, nonextracted yeast of the botanical classification *Saccharomyces* resulting as a by-product from the brewing of beer and ale. It must contain not less than 35% crude protein. **B**rewers' dried yeast is an excellent source of highly digestible protein of good quality.	**A-F** M-F	**93** 100	**3.0** 3.2	**73** 78	**70** 76	**1.46** 1.57	***3.21*** *3.45*	**—** —	**—** —	**43.8** 47.1	**0.14** 0.15	**1.36** 1.47	**D**ried brewers' yeast can replace up to 80% of the animal protein portion of swine and poultry rations. However, in swine and poultry rations, it is used primarily as a source of B vitamins and unidentified growth factors.	Chapter 10, Fig. 10–18.
WOOD AND PAPER BY-PRODUCT FEEDS															
Ash leaves	**A**sh leaves are the foilage from trees of the genus *Fraxinus.* Untreated ash leaves have a digestibility of 17%, in comparison with a digestibility of 31 to 37% of untreated aspen leaves. Forest foliage leaves are known as *muka,* a term that originated in the Soviet Union.	**A-F** M-F	**85** 100	**12.7** 14.9	**62** 73	**—** —	**1.24** 1.46	***2.73*** *3.21*	**—** —	**—** —	**12.4** 14.6	**—** —	**—** —	**I**n the Soviet Union, where large quantities of muka are fed (about 100,000 metric tons), muka is fed to cattle, swine, and poultry, primarily as a source of carotene, trace elements, and vitamins.	Table 12–2 of this chapter.
Aspen wood, boiled, ground	**A**spen wood is obtained from trees of the genus *Populus.* Ground aspen wood is composed of the entire tree, including leaves, branches, trunk, and bark. The particle size of the product shall not exceed ⅜ in.	**A-F** M-F	**90** 100	**—** —	**—** —	**—** —	**—** —	**—** —	**—** —	**—** —	**—** —	**—** —	**—** —	**B**oiled ground aspen wood is best suited for use in high-roughage, maintenance type rations, such as for wintering beef cows.	Table 12–2 of this chapter.
Paper waste	**W**aste paper consists of discarded newspaper, office paper, wrapping paper, and cardboard, more than 20 million tons of which are produced annually. The dry-matter digestibility of waste paper by cattle is as follows: newspaper, 30%; wrapping paper and cardboard, 40 to 60%; and high-quality chemical pulp paper about 98%.	**A-F** M-F	**92** 100	**63.4** 68.9	**25** 27	**—** —	**0.92** 1.00	***2.02*** *2.20*	**—** —	**—** —	**0.6** 0.7	**0.09** 0.10	**0.06** 0.07	**W**aste paper may be used as a roughage for ruminants, especially cattle. When levels greater than 12% are incorporated in cattle feed, there may be a reduction in feed intake. **O**ne of the potential hazards of feeding waste paper is the lead in ink. But no toxicities have been reported to date.	
Sawdust	**S**awdust is small fragments of wood made by a saw in cutting. Sawdust of the coniferous species is essentially undigested by rumen microorganisms. Deciduous species, with a few exceptions, are only slightly digested. Aspen is the most digestible; aspen sawdust has a digestibility of about 35%, and aspen bark has a digestibility of about 50%. **S**everal chemical and physical treatments of sawdust are partially effective. But generally the cost of treatment exceeds the increased feeding value of the product.	**A-F** M-F	**90** 100	**71.5** 79.4	**32** 36	**—** —	**0.65** 0.72	***1.43*** *1.59*	**—** —	**—** —	**0.3** 0.3	**—** —	**—** —	**S**awdust is effective as a roughage replacement. Concentrations of 5 to 15% screened sawdust in the ration of beef cattle appear practical. For lactating dairy cows, sawdust may be used as a roughage extender or as a partial roughage substitute in high grain rations; but some long hay in the ration appears to be necessary in order to stabilize feed intake.	Table 12–2 of this chapter.

(Continued)

TABLE 12–1 *(Continued)*

By-products/ Crop Residues	Description[2]	Moisture Basis: A-F (as-fed) or M-F (moisture-free)	Dry Matter	Crude Fiber	TDN		Digestible Energy				Crude Protein	Ca	P	Feeding Recommendations[3]	Cross References
					Ruminant	Swine	Ruminant Mcal per		Swine kcal per						
			(%)	(%)	(%)	(%)	(lb)	*(kg)*	(lb)	*(kg)*	(%)	(%)	(%)		
WOOD AND PAPER BY-PRODUCT FEEDS (Continued)															
Wood molasses	**W**ood molasses is obtained from wood wastes of the timber and lumber industries. The wood is converted to sugar by boiling it under pressure with diluted acid, neutralizing it with alkali, and evaporating it to wood molasses. Up to 200 gal of molasses may be obtained from 1 ton of dry wood waste. Wood molasses is lacking in palatability.	**A-F** M-F	**62** 100	**0.5** 0.9	**52** 83	**51** 82	**1.03** 1.66	***2.27*** *3.66*	**1,033** 1,664	***2,278*** *3,669*	**0.6** 1.0	**1.17** 1.88	**0.05** 0.07	**W**ood molasses can be substituted for cane molasses in ruminant rations, although it is not as palatable. For monogastric animals, it should be limited to upper levels of 2 to 2.5%; higher levels in nonruminant rations may cause digestive disorders.	**C**hapter 10, under "Molasses."
Yeast, torula	**T**orula yeast *(Torulopsis utilis)* is a hardy type of yeast that can be grown on a variety of substances, such as sulfite waste liquor and wood. While brewers' and distillers' yeast are truly by-product feeds, torula yeast is cultured specifically as a feed for livestock. **I**t is usually dried. Dried torula yeast is an excellent source of high quality protein, minerals, and B vitamins. It is tasteless, rather than bitter like brewers' yeast.	**A-F** M-F	**93** 100	**2.5** 2.7	**72** 78	**64** 69	**1.49** 1.60	***3.29*** *3.53*	**1,287** 1,383	***2,837*** *3,049*	**49.6** 53.3	**0.55** 0.59	**1.61** 1.73	**D**ried torula yeast can be included in the mixed feeds for all classes of livestock. However, its high price usually limits its use to providing a supplement to the amino acid and vitamin deficiencies of cereal grains. **T**orula yeast is used primarily in the rations of monogastrics and young ruminants, at a level of 3 to 5% of the total ration.	
CROP RESIDUE FEEDS															
Alfalfa screenings	**A**lfalfa screenings are obtained in cleaning alfalfa seed. They include light and broken seeds, along with weed seeds and other foreign material. Alfalfa screenings should be ground to increase digestibility. They are not very palatable.	**A-F** M-F	**90** 100	**11.3** 12.5	**78** 87	**73** 80	**1.52** 1.69	***3.36*** *3.72*	**1,452** 1,608	***3,202*** *3,546*	**31.0** 34.3	**1.11** 1.23	**0.34** 0.37	**A**lfalfa screenings are commonly used as a protein supplement for ruminants. They are not well suited to monogastrics because of their high fiber content.	
Alfalfa straw	**A**lfalfa straw consists of the stems and leaves remaining after alfalfa is threshed for seed.	**A-F** M-F	**93** 100	**40.6** 43.8	**43** 46	**—** —	**0.89** 0.96	***1.95*** *2.11*	**—** —	**—** —	**9.2** 9.9	**—** —	**0.13** 0.14	**M**ost alfalfa straw is fed to cattle, primarily as a winter roughage for gestating cows. **F**or cattle and sheep, alfalfa straw should be fed with good quality hay at about a 1:1 ratio.	
Almond hulls	**A**lmond hulls are a by-product of the almond nut industry. After the almonds are harvested, the nuts with attached hull are processed through a machine which removes the hull. Almond hulls are very variable in feeding value—from very poor to very good.	**A-F** M-F	**90** 100	**13.5** 14.9	**66** 73	**73** 81	**1.34** 1.48	***2.95*** *3.27*	**1,457** 1,615	***3,213*** *3,560*	**4.1** 4.5	**0.19** 0.21	**0.09** 0.10	**A**lmond hulls are very satisfactory for ruminants when fed in combination with such feeds as barley and alfalfa hay which compensate for the lack of protein in the hulls.	
Apple pomace silage	**A**pple pomace is the by-product obtained (1) after expressing juice from apples for cider or vinegar, or (2) after processing apples for canning, drying, and freezing; consisting of pulp, peels, cores and cull apples. It may be fed fresh, ensiled, or dried. Apple pomace is often mixed with alfalfa or corn for ensiling. Apple pomace silage is a very palatable feed, medium in energy, but very low in protein.	**A-F** M-F	**21** 100	**4.4** 20.6	**15** 68	**16** 74	**0.30** 1.40	***0.66*** *3.10*	**314** 1,479	***691*** *3,261*	**1.6** 7.8	**0.02** 0.10	**0.02** 0.10	**A**pple pomace silage is used primarily as a feed for cattle; in beef cattle finishing rations and in dairy rations. ***Caution:*** (1)pesticide contamination has occasionally been a problem, but is less of a problem now due to stricter control of pesticides; and (2) urea or other nonprotein nitrogen compounds should not be fed with apple pomace because of the possibility of abortions and/or abnormalities of offsprings, the reason for which is unknown.	

(Continued)

TABLE 12–1 *(Continued)*

By-products/ Crop Residues	Description[2]	Moisture Basis: A-F (as-fed) or M-F (moisture-free)	Dry Matter	Crude Fiber	TDN		Digestible Energy				Crude Protein	Ca	P	Feeding Recommendations[3]	Cross References
					Ruminant	Swine	Ruminant Mcal per		Swine kcal per						
			(%)	(%)	(%)	(%)	(lb)	*(kg)*	(lb)	*(kg)*	(%)	(%)	(%)		
CROP RESIDUE FEEDS (Continued)															
Bagasse, sugarcane	**S**ugarcane bagasse, the fibrous residue of sugarcane stalks which remains after the juice is pressed out, is one of the principal by-products of the sugar-making process. **T**reatment of sugarcane bagasse with sodium hydroxide and/or steam pressure dramatically increases its feeding value.	**A-F** M-F	**91** 100	**42.3** 46.5	**43** 48	— —	**0.88** 0.96	***1.93*** *2.12*	— —	— —	**1.4** 1.6	**0.47** 0.51	**0.26** 0.29	**B**agasse may be used in limited amounts as a low quality roughage for beef and dairy cattle. **I**n Hawaii, bagasse is becoming less available as a feedstuff due to its high fuel value—it is burned by the sugar mills to generate heat and electricity.	
Barley straw	**B**arley straw is the product remaining after separating the seed from the mature plants. **B**arley straw is not equal to oat straw in feeding value, but usually it is superior to wheat straw. **T**he feeding value of barley straw can be increased by treatment with calcium hydroxide, sodium hydroxide, or ammonium hydroxide. But the feeding value may not be increased sufficiently to pay for the treatment.	**A-F** M-F	**91** 100	**37.9** 41.5	**43** 47	— —	**0.84** 0.92	***1.86*** *2.03*	— —	— —	**4.0** 4.4	**0.27** 0.30	**0.07** 0.07	**B**arley straw may be fed to cattle on a maintenance ration, such as gestating beef cows, along with a legume forage. **B**arley straw is usually used for bedding unless supplemented with other feeds that supply energy, protein, minerals, and vitamin A.	
Bean straw	**B**ean straw is the product remaining after separating the seeds from the mature plant. It varies widely in quality.	**A-F** M-F	**89** 100	**36.8** 41.3	**43** 49	— —	**0.87** 0.97	***1.91*** *2.15*	— —	— —	7.1 8.0	1.65 1.85	0.13 0.14	**B**ean straw of good quality is satisfactory as part of the roughage for cattle and sheep when fed with good legume hay. It is worth about half as much as alfalfa hay when fed this way. In the West, bean straw is used mainly for wintering livestock and to a limited extent in finishing rations of cattle and sheep.	
Beet top silage	**B**eet top silage is the by-product obtained from the tops of sugar beets, preserved by ensiling. The average production of beet tops is 5 to 6 tons of green weight per acre, of which 40% is in the crowns and 60% is in the leaves. The tops are usually ensiled in trench silos or in stacks.	**A-F** M-F	**25** 100	**3.2** 12.8	**13** 52	— —	**0.26** 1.04	***0.58*** *2.29*	— —	— —	**3.4** 13.4	**0.39** 1.56	**0.07** 0.28	**B**eet top silage is a good, succulent winter feed for cattle and sheep. It is a fairly laxative feed. Mature cattle should not be fed more than 30 lb of beet top silage per day, and mature sheep not more than 3 lb per day.	
Clover, crimson, straw	**C**rimson clover straw consists of the stems and leaves remaining after crimson clover is threshed for seed.	**A-F** M-F	**88** 100	**38.8** 44.2	**40** 46	— —	**0.83** 0.94	***1.82*** *2.07*	— —	— —	**7.5** 8.6	— —	— —	**C**rimson clover straw is a satisfactory winter maintenance feed for cattle and sheep. It is worth about half as much as good clover hay. For best results, it should be fed with a good quality hay at about a 1:1 ratio.	

(Continued)

TABLE 12–1 *(Continued)*

By-products/ Crop Residues	Description[2]	Moisture Basis: A-F (as-fed) or M-F (moisture-free)	Dry Matter	Crude Fiber	TDN		Digestible Energy				Crude Protein	Ca	P	Feeding Recommendations[3]	Cross References
					Ruminant	Swine	Ruminant Mcal per		Swine kcal per						
			(%)	(%)	(%)	(%)	(lb)	*(kg)*	(lb)	*(kg)*	(%)	(%)	(%)		
CROP RESIDUE FEEDS (Continued)															
Corn stover	Corn stover is the mature corn plant from which the ears have been removed.	**A-F** M-F	**85** 100	**29.3** 34.4	**51** 59	**26** 31	**1.01** 1.19	***2.23*** *2.62*	— —	— —	**5.4** 6.4	**0.49** 0.57	**0.08** 0.10	Corn stover may be (1) stored as dry roughage, or (2) grazed in the field. Its highest and best use is for wintering beef cows. The traditional way of utilizing corn stover in the field is to allow the animals to do their own harvesting. However, grazing cornstalks results in considerable wastage. In an open field and winter, 2 acres of cornstalks will carry a pregnant cow for 100 to 120 days.	Section on "Grazing Corn Residues" in this chapter; and Chapter 19, under "Crop Residues and Winter Pastures."
Corn husks	Corn husks (husklage, shucklage) are the outer covering of ears of corn. It is the most digestible of the corn residues, which rank as follows in digestibility, in decending order: husks, leaf, cob, and stalks.	**A-F** M-F	**89** 100	**30.0** 33.9	**55** 62	**29** 32	**1.11** 1.25	***2.44*** *2.75*	— —	— —	**3.3** 3.7	**0.16** 0.18	**0.12** 0.14	Corn husks may be collected from the rear of the combine when harvesting corn, then stored. They are commonly fed to gestating cows, along with a protein supplement. As calving time approaches, a supplementary energy feed should be provided.	Section on "Grazing Corn Residues" in this chapter, and Chapter 19, under "Crop Residues and Winter Pastures."
Corncobs	Corncobs are the fibrous axis of the corn ear to which the kernels are attached, exclusive of the kernels and husk. Corncobs are a low quality roughage comparable to poor hay.	**A-F** M-F	**90** 100	**32.2** 35.8	**44** 50	**28** 32	**0.91** 1.01	***2.00*** *2.23*	— —	— —	**2.8** 3.1	**0.11** 0.12	**0.04** 0.04	Corncobs may be fed to cattle or sheep. Except for feedlot cattle, they should not replace more than half the roughage.	
Cottonseed gin trash	Cottonseed gin trash is composed of fragments of burs and stems, small amounts of immature cottonseed, lint, leaf fragments, and dirt. It is about 90% as valuable as cottonseed hulls in feeding value.	**A-F** M-F	**89** 100	**30.3** 34.1	**41** 46	**24** 26	**0.96** 1.08	***2.11*** *2.37*	— —	— —	**9.1** 10.2	**0.70** 0.78	**0.20** 0.23	Cottonseed gin trash is commonly fed to cattle, beef and dairy. It should be fed at low levels because of limited nutritional value. ***Caution:*** Pesticide contamination has been of concern in recent years, particularly when feeding it to dairy cattle. It should be tested and cleared before feeding it to livestock.	
Cottonseed hulls	Cottonseed hulls are the outer covering of cottonseeds. The hulls are removed from the seed before the oil is extracted. They are a high-fiber, low-protein by-product.	**A-F** M-F	**90** 100	**43.2** 47.8	**42** 47	— —	**0.86** 0.95	***1.90*** *2.10*	— —	— —	**3.8** 4.2	**0.13** 0.15	**0.09** 0.09	Cottonseed hulls may be used as a roughage source for beef cattle, dairy cattle, and sheep. As a roughage, they have about the same energy value as oat straw.	
Lespedeza straw	Lespedeza straw is the residue remaining after separation of the seeds by threshing.	**A-F** M-F	**90** 100	**29.2** 32.4	**44** 49	**38** 43	**0.95** 1.06	***2.10*** *2.33*	— —	— —	**6.8** 7.6	— —	— —	Lespedeza straw is a satisfactory winter maintenance feed for cattle and sheep. It is worth about half as much as good lespedeza hay. For best results, it should be fed with a good quality hay at about a 1:1 ratio.	

(Continued)

TABLE 12–1 *(Continued)*

By-products/ Crop Residues	Description[2]	Moisture Basis: A-F (as-fed) or M-F (moisture-free)	Dry Matter	Crude Fiber	TDN Ruminant	TDN Swine	Digestible Energy Ruminant Mcal per (lb)	Digestible Energy Ruminant Mcal per *(kg)*	Digestible Energy Swine kcal per (lb)	Digestible Energy Swine kcal per *(kg)*	Crude Protein	Ca	P	Feeding Recommendations[3]	Cross References
			(%)	(%)	(%)	(%)	(lb)	*(kg)*	(lb)	*(kg)*	(%)	(%)	(%)		
CROP RESIDUE FEEDS (Continued)															
Mint hay	Mint hay is the forage that remains after mowing mint, distilling off the oil, then drying the plant residue.	**A-F** M-F	**88** 100	**20.5** 23.3	**49** 56	**49** 55	**1.01** 1.15	***2.23*** *2.54*	— —	— —	**12.8** 14.6	**1.50** 1.71	**0.19** 0.22	Mint hay can replace all or nearly all of the legume hay in cattle and sheep rations. Usually, animals tire of mint hay after several months.	
Oat hulls	Oat hulls are the outer covering of oat seeds, obtained in the milling of table cereals or in the groating of oats.	**A-F** M-F	**92** 100	**30.9** 33.4	**32** 35	**27** 29	**0.64** 0.69	***1.41*** *1.52*	— —	— —	**3.7** 4.0	**0.14** 0.15	**0.14** 0.15	Oat hulls have little value as a feedstuff. During extreme feed shortages, they may be used as a small part of the roughage in cattle rations.	
Oat straw	Oat straw is the residue remaining after separation of the seeds by threshing of oats.	**A-F** M-F	**92** 100	**37.2** 40.4	**46** 50	— —	**1.15** 1.24	***2.53*** *2.74*	— —	— —	**4.1** 4.4	**0.22** 0.24	**0.06** 0.07	Oat straw is the most nutritious and palatable of the cereal straws. It may constitute up to half the roughage of breeding and stocker beef cattle, provided the other half consists of a good legume hay.	
Peanut hulls	Peanut hulls are the outer covering of peanut seeds. Peanut hulls are high in fiber, but they have 8% protein on a moisture-free basis.	**A-F** M-F	**91** 100	**57.2** 62.9	**16** 18	— —	**0.33** 0.36	***0.72*** *0.79*	— —	— —	**7.3** 8.0	**0.24** 0.26	**0.06** 0.07	Peanut hulls may constitute up to 10% of cattle and lamb finishing rations, and up to 30% of the rations of gestating and stocker beef cattle.	
Pear cannery refuse silage	Pear cannery refuse silage is the ensiled residue that remains from canning pears, consisting of peels, cores, and screened solids.	**A-F** M-F	**28** 100	**9.0** 32.6	**16** 59	— —	**0.33** 1.19	***0.72*** *2.62*	— —	— —	**3.3** 12.0	— —	— —	Pear cannery refuse silage is a satisfactory feed for cattle and sheep when mixed with feeds that have sufficient dry matter to prevent most of the drainage. Corn stover, hay, straw, and corncobs can be used for the latter purpose. Also, the addition of ground grain or beet pulp, at the time of ensiling will greatly improve the feeding value. Up to 20 lb of pear cannery silage daily can be fed to finishing cattle.	
Pea vine hay	Pea vine hay is the forage that is dried after removing the peas for human food.	**A-F** M-F	**87** 100	**23.7** 27.2	**51** 58	**45** 52	**1.02** 1.17	***2.24*** *2.57*	— —	— —	**11.8** 13.5	— —	— —	Pea vine hay is a very palatable and highly nutritious forage which may be fed to all ruminants. It is equal or superior to alfalfa hay of comparable quality.	
Pea straw	Pea straw is the residue remaining after separation of the seeds by threshing mature peas. With modern combine threshing more leaves are lost (left in the field) than with outmoded stationary threshing machines.	**A-F** M-F	**87** 100	**34.3** 39.5	**43** 49	**26** 30	**0.85** 0.98	***1.88*** *2.16*	— —	— —	**7.8** 8.9	— —	— —	Pea straw is worth about ⅔ as much as a legume hay for breeding, stocker, and finishing cattle. For best results, at least 2 lb of alfalfa hay should be fed daily when pea straw constitutes most of the roughage.	
Pea vine silage	Pea vine silage is the forage that is ensiled after removing the peas for human food. It contains more protein and carotene than corn silage, but less energy.	**A-F** M-F	**25** 100	**7.3** 29.8	**14** 57	— —	**0.28** 1.13	***0.61*** *2.49*	— —	— —	**3.2** 13.1	**0.32** 1.31	**0.06** 0.24	Pea vine silage is an excellent forage for cattle and sheep. In balanced rations, it is worth 90% as much as well-eared corn silage.	
Rice hulls	Rice hulls are the outer covering of rice grains. Rice hulls are high in fiber, low in protein, and high in lignin.	**A-F** M-F	**92** 100	**38.9** 42.2	**11** 12	— —	**0.27** 0.30	***0.60*** *0.65*	— —	— —	**3.0** 3.2	**0.11** 0.12	**0.10** 0.10	Ground rice hulls may be used as a filler during extreme feed shortages; other than this, they have no place in normal feeding programs. They should be used as bedding material.	

(Continued)

TABLE 12–1 *(Continued)*

By-products/ Crop Residues	Description[2]	Moisture Basis: A-F (as-fed) or M-F (moisture-free)	Dry Matter	Crude Fiber	TDN Rumi-nant	TDN Swine	Digestible Energy Ruminant Mcal per		Digestible Energy Swine kcal per		Crude Pro-tein	Ca	P	Feeding Recommendations[3]	Cross Refer-ences
			(%)	(%)	(%)	(%)	(lb)	*(kg)*	(lb)	*(kg)*	(%)	(%)	(%)		
CROP RESIDUE FEEDS (Continued)															
Rice straw	Rice straw is the residue remaining after separation of the seeds by threshing rice. It is very indigestible.	**A-F** M-F	**91** 100	**31.9** 35.1	**40** 44	**12** 13	**0.80** 0.88	***1.76*** *1.94*	— —	— —	**3.9** 4.3	**0.19** 0.21	**0.07** 0.08	In the past, rice straw has been burned after harvest. Treatment with sodium hydroxide, ammonia, and other substances show promise of increasing its feeding value, as well as being a better environmental practice.	
Rye straw	Rye straw is the residue remaining after separation of the seeds by threshing mature rye. It is less palatable and less digestible than oat straw.	**A-F** M-F	**91** 100	**38.3** 42.1	**39** 43	— —	**0.78** 0.86	***1.72*** *1.89*	— —	— —	**2.8** 3.0	**0.22** 0.24	**0.08** 0.09	Rye straw may constitute up to ⅓ of the roughage of breeding and stocker beef cattle, provided the other ⅔ consists of a good legume hay.	
Sorghum stover	Sorghum stover is the stalks and leaves of sorghum after removing the mature heads.	**A-F** M-F	**92** 100	**29.9** 32.6	**47** 51	**27** 29	**0.92** 1.00	***2.02*** *2.21*	— —	— —	**4.4** 4.9	**0.37** 0.40	**0.10** 0.11	Sorghum stover may be (1) grazed in the field, (2) stored as dry roughage, or (3) ensiled. It is commonly fed to beef cattle. ***Caution:*** After harvest, sorghum will send up new shoots if the moisture is favorable. The prussic acid content of these shoots may be toxic to grazing animals. These shoots can be grazed safely 4 to 6 weeks after a hard killing frost.	
Soybean hulls	Soybean hulls are the outer covering of soybean seeds. They are high in fiber and fair in protein content. The hull constitutes 8% of the seed weight.	**A-F** M-F	**91** 100	**36.2** 40.0	**69** 77	**47** 52	**1.20** 1.33	***2.65*** *2.92*	**851** 940	***1,877*** *2,073*	**10.8** 11.9	**0.45** 0.49	**0.19** 0.21	Soybean hulls are incorporated in commercial feeds, primarily. Because of their high fiber, they are best suited for use by lactating cows, growing cattle and sheep. But they are not recommended for use in cattle finishing rations.	
Soybean straw	Soybean straw is the residue remaining after separation of the seeds by threshing mature soybeans. It is a high fiber, low protein, and high lignin product.	**A-F** M-F	**88** 100	**38.9** 44.3	**36** 41	— —	**0.71** 0.81	***1.57*** *1.79*	— —	— —	**4.6** 5.2	**1.40** 1.59	**0.05** 0.06	Soybean straw may be used for wintering gestating beef cows and dry dairy cows. It should be fed with a good quality hay at about a 1:1 ratio.	
Sweet corn cannery refuse silage	Sweet corn cannery refuse silage consists of the ensiled husks and cobs, plus some kernels not satisfactory for canning. It is low in carotene.	**A-F** M-F	**31** 100	**10.1** 32.7	**22** 72	— —	**0.42** 1.35	***0.92*** *2.97*	— —	— —	**2.5** 8.0	**0.10** 0.32	**0.24** 0.77	Sweet corn cannery refuse silage can be fed liberally to gestating beef cows, finishing cattle, and finishing lambs. It is worth 85 to 95% as much as silage made from well-eared field corn.	
Tumbleweed hay	Tumbleweed hay is made by cutting tumblegrass *(Schedonnardus paniculatus)* before maturity. At maturity, this plant breaks away and rolls in the wind; hence, the name *tumbleweed.* It is found primarily in the prairies and plains of the U.S.	**A-F** M-F	**86** 100	**25.1** 29.0	**39** 45	**29** 34	**0.79** 0.91	***1.73*** *2.00*	— —	— —	**10.2** 11.8	**1.31** 1.51	**0.22** 0.25	Tumbleweed hay may be used for maintaining cattle during droughts.	
Tumbleweed silage	Tumbleweed silage is ensiled tumblegrass *(Schedonnardus paniculatus).*	**A-F** M-F	**34** 100	**10.3** 29.8	**14** 42	— —	**0.31** 0.90	***0.69*** *1.99*	— —	— —	**2.5** 7.4	— —	— —	Tumbleweed silage is a satisfactory feed for maintaining cattle during emergency, drought conditions.	
Wheat chaff	Wheat chaff includes the husks, hulls, joints, and small fragments of straw that are separated from the seeds in threshing of wheat. Wheat chaff is bulky, high in fiber, and low in protein.	**A-F** M-F	**93** 100	**30.4** 32.7	**35** 37	**19** 20	**0.76** 0.82	***1.68*** *1.81*	— —	— —	**5.4** 5.9	**0.19** 0.21	**0.08** 0.08	Wheat chaff may be used as a filler and to provide some nutrients for stocker and breeding cattle, and for dry dairy cows. It should be fed with a good legume hay at about a 1:1 ratio.	

(Continued)

TABLE 12–1 *(Continued)*

By-products/ Crop Residues	Description[2]	Moisture Basis: A-F (as-fed) or M-F (moisture-free)	Dry Matter	Crude Fiber	TDN		Digestible Energy				Crude Protein	Ca	P	Feeding Recommendations[3]	Cross References
					Ruminant	Swine	Ruminant Mcal per		Swine kcal per						
			(%)	(%)	(%)	(%)	(lb)	*(kg)*	(lb)	*(kg)*	(%)	(%)	(%)		
CROP RESIDUE FEEDS (Continued)															
Wheat straw	**W**heat straw is the residue remaining after separation of the seeds by threshing of wheat.	**A-F** M-F	**90** 100	**37.4** 41.7	**40** 44	**—** —	**0.86** 0.96	***1.90*** *2.12*	**—** —	**—** —	**3.2** 3.6	**0.16** 0.18	**0.05** 0.05	**W**heat straw may constitute up to ½ the roughage of breeding and stocker beef cattle, provided the other half consists of a good quality legume hay.	
WASTE BY-PRODUCT FEEDS															
Bakery by-product, dried	**B**akery by-product (bakery waste) consists of unsold bread, doughnuts, cakes, and other pastries. It is usually very low in fiber, high in fat, and fair in protein.	**A-F** M-F	**91** 100	**1.2** 1.3	**82** 89	**89** 98	**1.62** 1.78	***3.57*** *3.91*	**1,794** 1,966	***3,954*** *4,335*	**10.1** 11.1	**0.14** 0.15	**0.22** 0.24	**B**akery by-product is an excellent energy feed for cattle. It may constitute most of the maintenance of stocker and breeding cattle, provided it is properly supplemented with minerals and vitamin A. Up to 10% may be included in cattle finishing rations, and up to 15% may be included in lactating cow rations (due to its low fiber content, higher levels may depress milk fat).	
Garbage, boiled, wet	**B**oiled, wet garbage is animal and vegetable waste that has been boiled. It is usually collected by contractors sufficiently often to avoid decomposition, and injurous materials (crockery, glass, metal, string, and similar materials) are sorted out. Best quality garbage comes from restaurants. **B**oiled garbage must be cooked at a temperature high enough to destroy all organisms capable of producing animal diseases.	**A-F** M-F	**23** 100	**0.7** 3.0	**20** 89	**25** 110	**0.42** 1.83	***0.92*** *4.02*	**501** 2,191	***1,105*** *4,831*	**3.6** 15.8	**0.10** 0.44	**0.06** 0.28	**B**oiled garbage may be utilized either as a feed for a sow and pig enterprise or for finishing feeder pigs. A combination of garbage and grain feeding is best.	
Garbage, dehy, ground	**D**ehydrated, ground garbage is garbage that has been artificially dried at a temperature sufficient to destroy all organisms capable of producing animal diseases. It is a good energy feed. **B**ecause of high energy cost, very little garbage is artificially dehydrated.	**A-F** M-F	**92** 100	**2.2** 2.4	**82** 89	**97** 106	**1.65** 1.79	***3.63*** *3.94*	**1,947** 2,113	***4,293*** *4,659*	**16.1** 17.4	**0.42** 0.46	**0.32** 0.35	**B**est results are obtained if pigs weigh at least 100 lb when they are started on garbage. A mineral supplement should be provided. **G**arbage-fed hogs often produce soft pork.	
Manure, cattle, without bedding, dehy	**D**ehydrated cattle manure without bedding (dehydrated cattle waste without bedding) is a mixture of cattle excrements (consisting of undigested feeds plus certain body wastes). It is high in fiber, low in energy, and contains considerable nonprotein nitrogen.	**A-F** M-F	**93** 100	**30.7** 33.0	**44** 47	**—** —	**0.88** 0.94	***1.94*** *2.08*	**1,281** 1,374	***2,825*** *3,029*	**12.0** 12.9	**1.35** 1.45	**1.08** 1.15	**T**he high fiber and considerable nonprotein nitrogen of cattle manure indicates that it is best suited as a ruminant feed. Also, because of its low energy content, it is best adapted for use in maintenance and gestating cattle rations, rather than in lactating and growing rations.	
Manure, poultry, with litter, dehy	**D**ehydrated poultry manure with litter is the type that is generally produced by broiler operations. Poultry litter is the most collectable and most nutritious of all animal wastes. It is fair in energy, but high in protein and minerals.	**A-F** M-F	**86** 100	**15.0** 17.5	**47** 55	**9** 10	**0.99** 1.16	***2.19*** *2.56*	**175** 205	***386*** *451*	**24.6** 28.7	**2.67** 3.12	**1.69** 1.98	**D**ehydrated poultry manure with litter is best fed to beef cattle—breeding cattle, stockers, or finishing cattle.	
Manure, poultry, without litter, dehy	**D**ehydrated poultry manure without litter is the kind that is generally produced in caged layer operations. On a moisture-free basis, it is medium in fiber, fair in energy, and high in protein and minerals.	**A-F** M-F	**90** 100	**12.0** 13.3	**46** 51	**—** —	**0.92** 1.01	***2.02*** *2.24*	**—** —	**—** —	**25.4** 28.2	**8.07** 8.95	**2.22** 2.46	**D**ehydrated poultry manure without litter may be fed successfully to beef cattle, lactating cows, sheep, swine, and poultry.	Table 12–12 in this chapter.

(Continued)

TABLE 12–1 *(Continued)*

By-products/ Crop Residues	Description[2]	Moisture Basis: A-F (as-fed) or M-F (moisture-free)	Dry Matter	Crude Fiber	TDN Ruminant	TDN Swine	Digestible Energy Ruminant Mcal per (lb)	Digestible Energy Ruminant Mcal per (kg)	Digestible Energy Swine kcal per (lb)	Digestible Energy Swine kcal per (kg)	Crude Protein	Ca	P	Feeding Recommendations[3]	Cross References
			(%)	(%)	(%)	(%)	(lb)	*(kg)*	(lb)	*(kg)*	(%)	(%)	(%)		
							AQUATIC PLANTS								
Algae, green *Scenedesmus quadricuada* whole, fan air dried	Algae are primitive aquatic plants that contain chlorophyll and convert the sun's energy into single-cell protein by photosynthesis. Like most aquatic plants, algae are high in moisture. On a dry basis, however, algae are low in fiber, medium in energy, and high in crude protein and minerals. Three species of algae are of interest in producing single-cell protein *(Scenedesmus quadricuada, Chlorella vulgaris,* and *Spirulina maxima).* Only *Scenedesmus quadricuada,* known as green algae, is reported in Table 12–1.	**A-F**	**93**	**6.1**	**52**	**47**	**1.11**	***2.44***	**938**	***2,069***	**46.7**	**1.76**	**2.05**	Algae may be fed to sheep, swine, and chicks. Algae meal should be limited to 10% of swine rations and 6% of poultry rations. Pellets in which 50% of the nitrogen was supplied by algae have been fed to sheep on summer range.	
		M-F	100	6.5	56	50	1.19	*2.62*	1,009	*2,224*	50.2	1.89	2.20		
Seaweed (kelp) *Laminariales* (order) *Fucales* (order) whole, dehy	When botanists speak of seaweed, they usually mean one of the larger brown or red varieties. In this table, the composition of *Laminariales* (order) and *Fucales* (order) are given. Seaweed is low in protein (5.7% on an as-fed basis), and the protein is of low biological value.	**A-F**	**91**	**6.5**	**29**	–	**0.58**	***1.27***	–	–		**2.47**	**0.28**	Dehydrated seaweed may constitute up to 10% of cattle and sheep rations, and up to 6% of swine and poultry rations. Higher levels have been used experimentally in lactating dairy cow and sheep rations.	
		M-F	100	7.1	32	–	0.63	*1.40*	–	–		2.72	0.31		

[1]For more complete compositions of the by-product feeds listed herein, along with compositions of other by-product feeds, see the tables in Section V, Composition of Feeds.

[2]Many of the descriptions given herein were adapted by the authors from the Official Publication of the Association of American Feed Control Officials (AAFCO).

[3]The feeding recommendations given herein were adapted by the authors from many sources, especially from Agricultural Experiment Station Reports and from *Underutilized Resources as Feedstuffs,* NRC, National Academy Press, Wash., D.C., 1983.

ECONOMY OF BY-PRODUCT FEEDS AND CROP RESIDUES

Feed cost, the largest single expense in animal production, may often be lowered by including by-product feeds and crop residues in the ration. But more than cost per pound or per ton should be considered!

When determining the economy of by-product feeds and crop residues, consideration should be given to the cost of the nutrients supplied; transportation costs; storage costs and losses; possible variations in nutrient content because of different milling and processing procedures; palatability; possible toxicity or contamination with pesticides or heavy metals; effect upon the digestibility and utilization of the total ration; and labor costs in feeding. Reduced animal performance and less profit can result from improper feed substitution.

Increasingly, new uses for by-products will compete with their traditional use as livestock feeds. Fuel, plastics, paints, lubricants, road deicers, biodegradable plastic bags, and printing ink are among the products that can be made from crop by-products. Biotechnology will make these alternative uses more viable.

ANIMAL BY-PRODUCT FEEDS

Animal by-products are derived from animals; they include dairy by-products, marine by-products, meat animal by-products, and poultry by-products. As a whole, animal by-products are high quality protein feeds; they generally contain an excellent balance of amino acids and are good sources of minerals and vitamins. However, the following concerns are inherent in the utilization of these feedstuffs:

1. Some animal by-products contain large amounts of fat and are subject to oxidation and rancidity.
2. Animal by-products must be processed and stored properly to prevent bacterial contamination and growth.
3. Most animal by-products are more costly than plant by-products.

Dairy By-product Feeds

Dairy by-products, while excellent nutritionally, furnish a relatively small percentage of the total high-protein feeds consumed by livestock in the United States. In 1984, the several dairy by-products—in liquid, dry, and condensed forms—constituted less than 0.4% of the high-protein feeds fed to animals in the United States.

Milk is processed in numerous ways to provide a wide variety of products, for both human and animal consumption. Generally speaking, milk by-products are of high quality; they are good sources of protein and the essential amino acids, minerals, vitamins, and lactose (milk sugar), and they are palatable and highly digestible. The chief deterrent to their wider use in animal feeds is price.

Some common dairy by-product feeds are listed in Table 12-1 of this chapter and in Section V, Composition of Feeds. Additionally, milk by-products are discussed in Chapter 11, under "Dairy Products."

Marine By-product Feeds

In the beginning, marine wastes were dumped into the sea. Eventually, some of them were dried at high temperatures—usually in an open flame dryer—and used as fertilizers. Then, about 1910, it was discovered that these waste materials from fish canning plants and fish species not used by humans were a desirable protein source for livestock feeding. Experiment stations led the way in determining the feeding value of these new products which are now incorporated extensively in swine and poultry rations.

Although fish meal and solubles are the leading marine by-products, in 1984 they accounted for only 1.0% of the high-protein feeds fed to animals. In addition to several fish by-products, marine products include crab meal and shrimp meal. The common marine by-product feeds are listed in Table 12-1 of this chapter and in Section V, Composition of Feeds. Additional marine by-products are discussed in Chapter 11, under "Marine By-products."

Meat Animal By-product Feeds

In the early days of the meat-packing industry, the only salvaged by-products were hides, wool, tallow, and tongue. The remainder of the offal was usually carted away and dumped into rivers, or burned or buried. In some instances, packers even paid for having the offal taken away.

Finally, in the early 1890s, the selected meat residues from lard rendering—known as cracklings—were sold locally for chicken feed. Then, about 1900, poultry producers in California started using dried tankage in chicken rations. Almost simultaneously, this new product was utilized as a hog feed. Thus was born a new era in the utilization of packinghouse by-products and a new era in livestock feeding.

Fig. 12-2. Cracklings, the residue remaining after removal of the fat from the skin of animals by dry heat. (Courtesy, Iowa Beef Processors, Dakota City, Nebr.)

Today, packinghouses produce numerous by-products; cattle slaughter alone produces approximately 80 by-products which have a great variety of uses. Some of these, such as fats and gelatin, are used as human food; others, like feathers and tankage, are only fed to animals.

Currently, tankage and meat meal provide 6.0% of the total supply of high-protein feeds available for livestock feeding. In 1986, 2,650,000 tons of tankage and meat meal were fed in the United States. The most widely used meat processing and by-product feeds are listed in Table 12–1 of this chapter and in Section V, Composition of Feeds. Additional meat animal by-product feeds are discussed in Chapter 11, under "Meat Packing By-products," and in Chapter 10, under "Fats and Oils."

Poultry By-product Feeds

The by-products of the poultry industry consist of offal, dead birds deemed unfit for human consumption, feathers, blood, and eggs—including cracked eggs, those with blood spots, rots, and those that did not hatch. For many years, the chief problem in utilizing poultry by-products was that of assembling enough material in one place to make the operation profitable. But the growth and centralization of poultry production accomplished this.

Poultry by-products are widely used as protein supplements for four-footed animals and poultry, and in fish hatcheries. Some of the common poultry by-product feeds are listed in Table 12–1 of this chapter and in Section V, Composition of Feeds. Additional poultry by-products are discussed in Chapter 11, in the section on "Poultry Wastes."

PLANT BY-PRODUCT FEEDS

Almost every type of plant that is produced or processed for human food or animal feed yields one or more by-products which can be utilized as feed for animals. Additionally, some plants, such as trees, yield by-products which, when initially considered, seem inedible, but which upon proper processing yield valuable by-product feeds.

In Table 12–1 and the sections that follow, plant by-product feeds are classed and discussed in the following categories: oilseed by-product feeds; grain by-product feeds; fruit, vegetable, and nut by-product feeds; sugar by-product feeds; brewing and distilling by-product feeds; wood and paper by-product feeds; crop residues; waste by-product feeds; and aquatic plants.

The lexicon used in discussing plant by-product feeds and residues is often confusing. For this reason, some of the terms commonly applied to the various by-products from plant processing are defined as follows:

- **Bagasse**—Solid residue remaining after extraction of juice.
- **Cannery residue**—The waste remaining following canning vegetable products for human consumption.
- **Chaff**—Husks, hulls, joints, and small fragments of straw that are separated from seed in threshing of small grains.
- **Corncobs**—Fibrous portion of the fruiting head (grain producing portion) of the corn plant, excluding the husk.
- **Corn husks**—Fibrous outside cover of the fruiting head (grain producing portion) of the corn plant.
- **Crop residue**—Fibrous residue remaining after harvest of the primary product (grain, fruit, *etc.*).
- **Crown**—The top portion of a root from which the aerial part extends. It is generally used with respect to such root crops as beets, carrots, and rutabagas which have a green growth originating from the crown. In most cases, part of the crown is attached to the green portions of the plant following processing of the root.
- **Cull**—This denotes inferior quality. This should not be construed as implying spoiled products; rather, it refers to products which do not meet certain minimum quality standards for marketing.
- **Husk**—Outer covering of kernels or seeds, especially when fibrous.
- **Husklage**—Corncobs and husks.
- ***In vitro***—Outside the living organism in an artificial environment.
- ***In vivo***—Within the living organism.
- **Pit, shell, hull**—These terms are applied to the hard, woody coating which surrounds the seed.
- **Pith**—This is the soft, spongy tissue in the center of certain plant stems.
- **Pod**—Empty seed vessel.
- **Pulp, pomace, marc**—These terms refer to the solid residue, including such materials as skins and seeds, after the extraction of juices from the intact product. Pomace is often used to denote the dried product.
- **Skins, rinds, peel**—These terms refer to the outer covering of a product, which usually provides some protection from excessive loss of moisture and from attack by microorganisms. A *pod* differs from these terms in that it is a vessel which contains the seeds but is not part of the seed itself.
- **Stalk**—Main stem of a herbaceous plant.
- **Stalklage**—Moist, ensiled stalks.
- **Stover**—Stalks and leaves of corn or sorghum after grain harvest.
- **Straw**—Plant residue remaining after removal of the seeds of cereal grains or grasses by threshing.
- **Stubble**—Lower parts of plant stems that remain standing in the field after harvest.

Oilseed By-product Feeds

Several seeds high in oil are produced primarily as sources of vegetable oils for oleomargarine, shortening, and salad oil for human consumption, and for paints and other industrial purposes. In processing these seeds, residual protein-rich products of great value as livestock feeds are obtained. These oilseed by-product feeds are listed in Table 12–1 of this chapter and in Section V, Composition of Feeds. The traditional oil extraction procedures of the oilseeds, and the by-products resulting therefrom, are discussed in Chapter 11, under "Oilseed Meals." Whole cottonseed is discussed in Chapter 10, under "Cottonseed."

Grain By-product Feeds

Fig. 12–3. Cattle eating corn gluten meal, a by-product of the production of starch, oil, and sweeteners or alcohol from corn. The protein and hulls of the corn kernel produce corn gluten feed and corn gluten meal. (Courtesy, Archer Daniels Midland Company, Decatur, Ill.)

Many of the cereal grains are processed to obtain a great array of foods for human consumption and by-product feedstuffs for animals. The common grain by-product feeds are listed in Table 12–1 of this chapter and in Section V, Composition of Feeds. The traditional milling procedures of cereal grains to obtain food for people, and the by-products resulting therefrom, are illustrated and described in Chapter 10, in the section headed, "Grain Milling and Milling By-products." Also, Chapter 10, Table 10–4 lists, describes, and shows the nutritive value of the leading cereal grain by-products.

Fruit, Vegetable, and Nut By-product Feeds

Fruit, vegetable, and nut by-product feeds may come from three sources: (1) cull, unmarketable, or damaged, but wholesome, commodities; (2) residues left in the field; or (3) canning, juicing, or processing wastes. These products can be used successfully in many feeding programs. But problems involving their continued availability, storage, and handling must be considered, because many of them are highly perishable. Most of these by-product feeds are restricted to areas where processing and canning operations are located.

Some of the common fruit, vegetable, and nut by-product feeds are listed in Table 12–1 of this chapter and in Section V, Composition of Feeds. Additional fruit, vegetable, and nut by-products are discussed in the sections that follow.

BEANS, GREEN

Cull green canning beans and their vines, with or without residual beans, may be used for feed. If they are harvested while green and growing, on a dry basis they have a nutritive value comparable to fair quality hay. If the plants have been allowed to dry or have been killed by frost, their value is slightly higher than the cereal straws but lower than poor quality hay crops.

Green beans and cannery wastes, which include cull beans and trimmings from processing, on a dry basis have a slightly higher feeding value than hay crops. They are relatively high in moisture (they contain only about 10% dry matter) and high in fiber; their use had best be restricted to beef cattle, sheep, horses, goats, and rabbits. Green beans are rich sources of vitamins, including carotene. Raw green beans have a slightly bitter taste, but this is not likely to present a problem in livestock feeding.

DATES, CULL

There are only limited data relative to the feeding value of cull dates. However, practical information indicates that a level of 5 to 10% can be used in a complete cattle finishing ration.

GRAPES, CULL

The vast majority of grapes are cultivated for juice, wine, raisins, or fresh table grapes.

Up to 35 lb of fresh table grapes have been fed to cattle daily, without any evidence of digestive disturbance or scouring.

LETTUCE

The feeding value of cull lettuce depends on the water content. Fresh lettuce in the field contains about 93% moisture. Cull leaves contain 85 to 88% moisture. A load of lettuce consists of a mixture of heads and leaves, about half and half. Lettuce is palatable. Cattle will eat the heads first, followed by the leaves. The supply of cull lettuce is not always dependable, even during the normal harvest season.

Because of the high water content, it does not pay to haul lettuce very far. Moreover, lettuce cannot be held longer than 3 to 5 days following harvest, after which it begins to spoil and is of little value.

On a dry basis, lettuce is comparable to low grade oats; having about 70% TDN and 10% crude protein. When fed strip lettuce, cattle will gain about 1 lb per head daily. When 6 to 7 lb of cereal hay (or equivalent) is added, cattle gains may be increased to 1.25 lb per day. With more hay, plus grain and supplement, greater gains can be expected.

ONIONS

Up to 20 lb of cull onion bulbs per head daily have been fed to cattle; and onions have been fed free-choice to sheep. However, onions contain an alkaloid which may cause anemia and toxicity in cattle, horses, and to a lesser extent in sheep. There is no known satisfactory treatment for onion toxicity (poisoning), so feeding them as a major part of the ration is risky and may result in death.

PEACHES, FRESH

Both cling and freestone peaches have been fed to cattle, with the animals eating about 20 lb of clings daily and 30 lb of freestones. The cattle swallowed a few pits, but discarded most of them after removing the pulp. In spite of some spoiling and swallowing of pits, no detrimental effects were noted. The peaches were eaten with relish and there was no scouring.

PEACHES, DRIED

Three pounds daily of dried peaches have been fed to cattle without any detrimental effects. However, when as much as 6 lb of dried peaches were fed daily, the animals scoured, and after 3 days refused to eat the fruit.

PEARS, FRESH

When pears are harvested, some fruit invariably drops to the ground and, if damaged, is considered cull. Also, pears are sometimes damaged in shipment and marketing.

Pears do not appear to be as palatable as peaches; and spoiled pears seem to be more objectionable to livestock than spoiled peaches. However, Western Regional Extension Workers reported that dry cows and 2-year-old heifers consumed an average of about 20 lb of pears daily without noticeable adverse effects.

PEARS, DRIED

A larger quantity of dried pears than of dried peaches can be fed to cattle. When cattle were fed 4.5 lb of dried pears daily, there was no laxative effect and no loss of appetite.

PEAS, GREEN

Cull green canning peas and their vines, with or without residual peas, may be used as ruminant feed. Pea vine silage and pea vine hay are comparable in nutritive value to alfalfa silage and alfalfa hay of comparable quality.

PINEAPPLE BY-PRODUCTS

Several valuable by-product feeds are obtained from pineapples, two of which are described here.

- **Pineapple cannery by-product**—When pineapples are canned, the outer portions and the core are discarded as wastes. These cannery by-products can be fed fresh, dried, or ensiled. The dried product is generally supplemented with about 9% molasses to increase the energy content and enhance palatability. Because the fresh product contains large amounts of water and is highly corrosive, hay or wilted grass is often mixed with it during ensiling. The fresh or dried product can constitute up to 50% of the ration for older pigs, but is not a suitable feed for poultry.

- **Pineapple syrup**—In the production of citric acid from pineapple juice, the extracted juice can be concentrated into a high-energy syrup. Studies have shown that older pigs thrive on a ration of 80% syrup and 20% protein supplement.

(Also see the following sections under Crop Residues: "Pineapple Greenchop," "Pineapple Juice Presscake," "Pineapple Stump Meal," and "Pineapple Tops and Crowns.")

POTATO, FRENCH FRY WASTE

This is a by-product from frozen food processors. It contains more than 5% fat and about 48% dry matter. The dry matter is more than 87% digestible by cattle.

Potato french fry waste is very palatable to both cattle and hogs. On a free-choice basis, hogs will eat about 5 lb of french fry waste for each 1 lb of corn consumed. About 20% of the grain ration of finishing cattle and hogs may be replaced with french fry waste, pound for pound. When french fry waste replaces up to 20% of the corn in swine rations, the fat used in french fry produces desirable hard pork carcasses.

POTATO PROCESS RESIDUE (SLURRY)

Potato process residue is the effluent recovered from potato freezer processors. Because of high moisture content and consequent high transportation cost, potato process residue is stored and fed near the processing plants. Up to about 40% potato process residue may be incorporated in cattle finishing rations without lowering gains.

Fig. 12-4. Cattle eating potato process residue, a high-moisture by-product of the potato processing industry. (Courtesy, Washington State University, Pullman)

PRUNES, FRESH

When cattle are limited to about 15 lb of fresh prunes per head daily, the fruit is eaten readily and there is no scouring. Generally, cattle discard most of the pits after removing the pulp. When more than 15 lb of fresh prunes per animal are fed daily, scouring results and the animals refuse the fruit.

PRUNES, DRIED

Cattle will eat up to 6 lb of dried prunes daily, with only slight scouring resulting. They will not eat the pits.

Sugar By-product Feeds

Although the United States imports a considerable amount of sugar, the nation's sugar industry is large. Two crops—sugarcane and sugar beets—account for the bulk of refined sugar. Several by-product feeds are produced in various stages of processing these two crops for sugar.

SUGARCANE PROCESSING

In the initial steps of processing sugarcane, the cane is crushed by rollers to press out the sweet juices. The juice is then centrifuged and partially condensed. At this stage, the unrefined sucrose product is brown and sticky. Upon refinement of this product, the white, crystalline sucrose is separated from the molasses and brown sugars. This molasses product is referred to as blackstrap or final molasses. A detailed discussion on molasses is found in Chapter 10, Grains/High-Energy Feeds.

After the juice has been extracted from the cane, the remaining by-product is known as bagasse. About 60% of the bagasse is used for fuel in the sugar mills. Being high in fiber, it has a low dry matter digestibility—only about 25%. Additionally, its TDN is extremely low, ranging from 20 to 25%. However, bagasse has been used effectively as a carrier of molasses, the combination of which yields a relatively high-fiber, high-energy feed. "Camola" is a term applied to a mixture of 4 parts bagasse pith to 10 parts cane molasses. "Molascuit" contains 1 part pith to 6.25 parts molasses.

SUGAR BEET PROCESSING

In the production of sugar from sugar beets, the beets are initially shredded into strips or slices (cossettes). The cossettes are then soaked in hot water. The sugar diffuses out of the cossettes producing a *raw juice* containing about 10 to 15% sugar. This juice is then processed in much the same manner as juice from sugarcane.

The resulting beet pulp can be fed wet if used within a short time, or it can be ensiled or dried when long-term storage is desired. Molasses is often added to beet pulp to increase the energy content; and, on occasion, beet pulp is ammoniated to provide a source of nonprotein nitrogen.

Beet tops and crowns are relished by livestock, but since they contain large quantities of oxalic acid, extra calcium must be added to the ration. Additionally, tops from beets grown on soils heavily fertilized with nitrogen are apt to create nitrate poisoning problems in cattle that do not have access to other feed. Beet tops can be fed fresh, dried, or ensiled.

MOLASSES AND OTHER SUGAR BY-PRODUCT FEEDS

Annually, about 638 million gallons, or 3,745,265 tons, of molasses of all kinds are used in the United States, of which about 73%, or 2,734,044 tons, is consumed in feeds or foods and the balance is used for industrial purposes. The United States produces about half of the molasses that it uses; the remainder is imported. Of the U.S. production, 48% is cane molasses and 52% is beet molasses.

Some of the common sugar by-product feeds are listed in Table 12-1 of this chapter and in Table V, Composition of Feeds. Additionally, both cane molasses and beet molasses are discussed in Chapter 10, under "Molasses."

Brewing and Distilling By-product Feeds

Considerable quantities of grains are used in the brewing of beers and ales and in the distilling of liquors. After processing, the remaining by-products can be readily adapted to many feeding programs. In 1984, 298,000 tons of brewers' dried grains and 551,000 tons of distillers' dried grains were fed to livestock. In addition to these feeds, solubles and yeast products from these industries are used in livestock feeds.

Chapter 10, Fig. 10-18, illustrates how malting, brewing, and distilling of barley result in several valuable by-product feeds for livestock.

Some of the common brewing and distilling by-product feeds are listed in Table 12-1 of this chapter and in Section V, Composition of Feeds. Additional brewing and distilling by-product feeds are discussed in the sections that follow.

BREWERS' BY-PRODUCTS

Barley is the primary grain used for the brewing of beers and ales. The initial step of brewing involves the malting of the barley. Special varieties of barley are steeped under carefully controlled conditions to germinate the grains. Upon germination, the enzymes which are activated convert the starch in the grain to dextrin and sugars. After the germination process is completed, the sprouted barley is dried by heating to halt enzymatic activity. The *malt sprouts* and *malt hulls* are then separated from the malted barley, forming two by-products which can be used as feed.

The clean malted barley is mixed with other grains (generally corn or rice) and a flavoring agent, hops, to form a mash. This mash is then cooked in water to enhance further enzymatic activity, and, following cooking, it is separated into liquid and solid fractions. The liquid—called *wort*—undergoes yeast fermentation to form the alcoholic beverage of either beer or ale.

An outline of the brewing process is shown in Fig. 12-5.

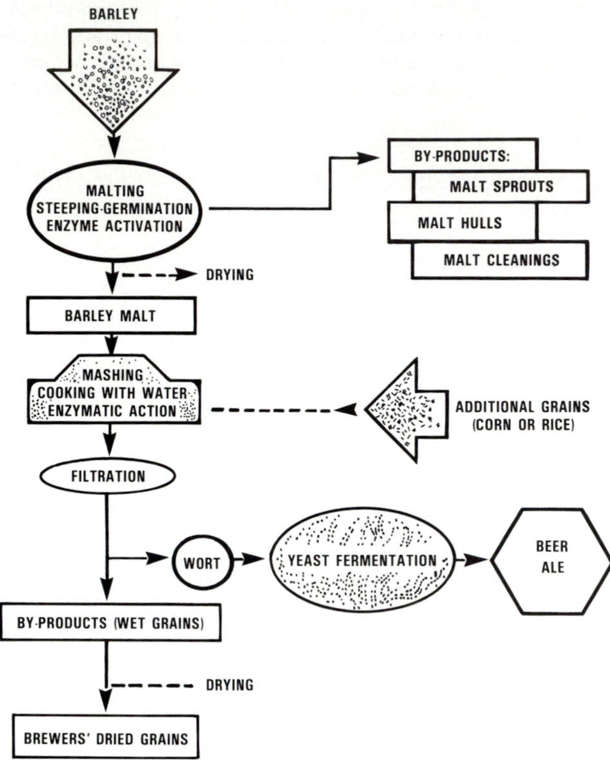

Fig. 12-5. An outline of the brewing of beer and ale, and the production of by-product feeds.

Brewers' Condensed Solubles

This by-product is obtained by condensing the liquids resulting as by-products of the production of beer or wort. It must contain at least 20% total solids, and on a dry matter basis it must contain at least 70% carbohydrates.

Brewers' Dried Yeast

Brewers' dried yeast is the dried nonfermentative, non-extracted yeast of the botanical classification, Saccharomyces, resulting from the brewing of beer and ale. (See Table 12-1, under "Brewing and Distilling By-product Feeds," for further description and for feeding recommendations.)

Brewers' Spent Grains

The extracted residue of barley malt, alone or in mixture with other cereal grains or grain products resulting from the manufacture of wort or beer, is termed *brewers' wet grains.* If the residue is dried, the product is termed *brewers' dried grains.* These products may contain some pulverized spent hops.

Wet grains become rancid rather quickly, so it is necessary to use them immediately or store them where exposure to air is limited. They can be ensiled if long-term storage is desired. The addition of 2 to 3% molasses facilitates fermentation, and good silage will be produced after 4 to 6 weeks of fermentation.

Brewers' grains are rather low in energy and high in fiber; hence, they are not widely used in poultry and swine rations. However, they are used extensively in dairy feeds, primarily as a source of protein. Up to one-third of the concentrate mix of dairy rations can consist of dried brewers' grains.

Hops

Hops are tall vines from which cones are dried and used to impart a distinct flavor to beer and ale. The vines and leaves, crop residues remaining after the harvesting of the cones, can be used as forage. The harvested cones are used extensively in beer and ale production.

When hopped wort is filtered, the residue, called spent hops, can be dried (as dried spent hops) and fed to livestock. Spent hops are often combined with brewers' grains to mask their bitter flavor. Dried spent hops can be added to an otherwise heavy ration as an aid in keeping cattle on feed.

Malt Cleanings

Malt cleanings are by-products produced in cleaning of malted barley or from the recleaning of malt which does not meet the minimum standards for crude protein in malt sprouts.

This by-product contains about 18% crude protein and 15% crude fiber.

Malt Hulls

Barley grain is covered by a hull. During the cleaning of malted barley, the hulls are removed and subsequently used as feed. They contain about 22% crude fiber.

Malt Sprouts (Malt Culms)

Malt sprouts or culms are obtained by removal of the sprouts of malted barley. (See Table 12-1, under "Brewing and Distilling By-product Feeds," for further description and for feeding recommendations.)

DISTILLERS' BY-PRODUCTS[2]

A large number of distilled spirits are produced throughout the world, each characterized by (1) the area of origin, (2) type of material used, (3) preparation of those materials, (4) proportions of materials, (5) fermentation conditions, (6) distillation processes, (7) maturation processes, and (8) mixture techniques. The art of distilling liquors has been largely a localized, traditional process; and many countries have developed their own characteristic spirit—for example,

[2]The section on Distillers' By-products was authoritatively reviewed by R. Hatch, Ph.D., Executive Director, Distillers Feed Research Council, Cincinnati, Ohio.

tequila is associated with Mexico; Scotch Whisky, with Scotland; bourbon, with the United States; and vodka, with the U.S.S.R.

It is necessary to review two basic distillation processes—British and American—because the stages from which by-products are produced vary (see Fig. 12-6).

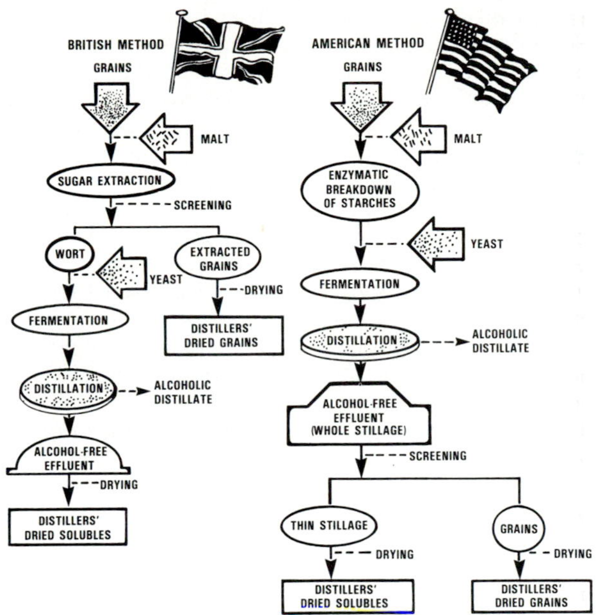

Fig. 12-6. An outline of the British and American processes of distillation, and the production of by-product feeds.

In the British method, grains are mixed with malt to facilitate the breakdown of starch to fermentable sugars. After the malting process, the resulting product is screened to separate the wort from the extracted grains. These grains can then be dried and used as feed. Yeast is added to the wort, and the mixture allowed to ferment. Following fermentation, the wort is distilled to remove the portion of the wort containing the alcohol. The product that remains is essentially alcohol free and can be dried to produce another by-product, *distillers' dried solubles.*

In the American process, there is no screening of the product formed during the enzymatic breakdown of the starches. Rather, the entire mixture is fermented and distilled. The alcohol-free effluent is termed whole stillage. The whole stillage is screened to separate the distillers' grain and the thin stillage. Drying procedures are applied to both products to produce distillers' dried grains from the grain portion and distillers' dried solubles from the thin stillage; or the grains and solubles may be combined prior to drying, then dried to produce distillers' dried grains with solubles.

Distillers' Grains (Draff)

Distillers' grains can be fed fresh (as distillers' wet grains), dried, or ensiled. The dried product is by far the easiest to handle and store, but it is the most expensive.

This product is less palatable to livestock than brewers' grains, but it contains more crude protein (and the protein is superior bypass protein) and less crude fiber. Like brewers' grain, it is fed primarily to dairy cattle as a source of protein. However, it has also been used successfully in sheep and beef rations.

Distillers' dried grains, as officially defined by AAFCO, are the product obtained after the removal of ethyl alcohol by distillation from the yeast fermentation of a grain or a grain mixture by separating the resultant coarse grain fraction of the whole stillage and drying it by the method employed in the grain distilling industry. (See Table 12-1, under "Brewing and Distilling By-product Feeds" for further description and for feeding recommendations.)

Distillers' dried grain with solubles is the product obtained after the removal of ethyl alcohol by distillation from the yeast fermentation of a grain or a grain mixture by condensing and drying at least three-fourths of the solids of the resultant whole stillage by methods employed in the grain distilling industry. (See Table 12-1, under "Brewing and Distilling By-product Feeds" for further description and feeding recommendations.)

Condensed Distillers' Solubles

The by-product feed officially known as *condensed distillers' solubles are a product obtained following the removal of ethyl alcohol by distillation from the yeast fermentation of a grain or a grain mixture by condensing the thin stillage fraction to a semisolid.* As with the other distillers' product, the predominating grain is listed before the feed term.

Condensed distillers' solubles have long been noted as a source of B vitamins and unidentified growth factors for swine and poultry. There have been indications that distillers' solubles may also contain a rumen-stimulating factor that facilitates cellulose digestion.

Stillage

Stillage is the mash from fermentation of grains after removal of alcohol by distillation. With the high cost of gasoline and diesel fuel, there is growing interest in producing alcohol (ethanol) as an alternate energy source. The residue from ethanol production, called stillage, may be used as a livestock feed for all animal species, but because of its high water content, it is best suited as a cattle feed. Whole stillage from corn contains about 92% moisture. Pressed grain (pressed stillage) from fuel alcohol production contains about 60% moisture (32% less than whole stillage). High energy costs make it impractical for many distillers to dry stillage. But wet stillage must be fed within about 3 days in order to prevent spoilage. Also, because of bulk and high water content, wet stillage must be fed in close proximity to the distillery. So, much attention is being given to locating cattle feedlots near ethanol plants.

When converted to alcohol, a bushel of corn will produce 2.5 gal of ethanol and 16.5 lb of stillage by-product.

Grain Distillers' Dried Yeast

Grain distillers' dried yeast is the dried nonfermentative yeast of the botanical classification, Saccharomyces, resulting from the fermentation of grains and yeast which is separated from the mash either before or after distillation. It must contain at least 40% crude protein. This by-product is extremely rich in all of the B complex vitamins except vitamin B-12.

Wood and Paper By-product Feeds[3]

The forest biomass is the world's largest storehouse of carbohydrates. In the past, it has been used primarily for traditional forest products and fuel; and its use for animal feed has been relatively minor, and for the most part limited to browse for goats, emergencies, and experimental work.

Wood residues, such as wood pulp and bark, have been used as energy sources for ruminants during periods of critical feed shortages, but they have never been considered as alternatives for conventional feedstuffs under normal economic conditions. Although more than 1.5 million tons of sulfate and sulfite pulps from spruce, pine, and fir were fed to cattle and horses in the Scandinavian countries during World War II when feed supplies were limited, the feeding of wood pulp to livestock ceased when conventional feedstuffs became available.

The primary structural components of wood are cellulose, hemicellulose, and lignin—all three of which are chemically bound to form a complex called lignocellulose. Cellulose in pure form is almost completely digestible by ruminants. Hemicellulose, composed primarily of 5-carbon sugars, varies in digestibility from about 45 to 90%. Lignin is completely indigestible. As a component of the lignocellulose complex, lignin can dramatically lower the digestibility of cellulose and hemicellulose by preventing cellulose-splitting enzymes from gaining access to the complex. So, whether the carbohydrates contained in wood lignocellulosic residues can be utilized by rumen microbes depends largely on how extensively the lignin-carbohydrate complex can be altered or opened up.

Of the woods tested, all of the coniferous species are essentially undigested by rumen microorganisms. Deciduous species, with a few exceptions, are only slightly digested. Aspen is the most digestible species tested, giving both an *in vitro* and *in vivo* digestibility of about 35%. Aspen bark is about 50% digestible.

Several chemical and physical treatments of woods are partially effective, but the effectiveness of each of them varies according to species. Hardwoods are generally more responsive to treatment than softwoods. Aspen is particularly responsive to treatment. Although several of the treatment methods can be readily adapted to commercial production, the cost of treatment has been the biggest deterrent to their use. Conventional feedstuffs must be relatively high priced before the treatment of wood to produce feed can be profitable. On occasion, such as during a war, the prices of conventional feedstuffs have been high enough to make processing woods into feeds attractive.

Wood residues are primarily an energy source, comparable to average or low-quality hay. But they are low in protein; so, they require protein supplementation. Nutritionally, treated wood residues are best suited for ruminants having low energy requirements, such as pregnant cows and ewes, and dry dairy cows.

Wood is effective as a roughage replacement. Concentrations of 5 to 15% screened sawdust in the ration of beef cattle appear practical. For lactating dairy cows, aspen sawdust may be used as a roughage extender or as a partial roughage substitute in high-grain rations; but some long hay appears to be necessary in the ration to stabilize feed intake.

Some of the pulp and papermaking residues which are already partially delignified, but which have little fiber value for paper manufacture, have excellent potential as feedstuffs. Care must be used in their selection, however, because some of them may contain toxic materials.

Both the research and the commercial production of wood and paper by-product feeds have lagged because of (1) an abundance of traditional feeds available at reasonable prices, (2) the lack of a steady market for these innovative by-product feeds, and (3) the rising cost of the energy and chemicals involved in treatment and processing.

Some common wood and paper by-product feeds are listed in Table 12-1 of this chapter and in Section V, Composition of Feeds. Additional wood and paper by-product feeds are discussed in the sections that follow.

WHOLE-TREE RESIDUE AND FRACTIONS OF WHOLE TREES

The wood products available for processing into livestock feeds consist mainly of tree tops and branches, leaves and needles, twigs, buds, and wood and bark fines.

Aspen has the most potential as a livestock feedstuff. Also, it is the most widespread species in North America; it grows from Mexico to the Arctic Ocean, and from Maine to Alaska. The primary markets for aspen are in the Lake States, northeastern United States and central Canada, where it is used for plywood, sawlogs, and composition board. Residues from these uses are bark (which contains up to 50% wood) and sawdust. Where whole-tree harvesting is practical, the fractions of whole trees, consisting of tree tops and branches, along with twigs, are also available. Part of the residues and fractions of whole trees are burned to provide heat and kiln dry the lumber; the remainder is used in landfills, put back on the land as fertilizer, or sold for livestock bedding, mulch, or fuel. These residues are suitable for processing into feeds, but unless traditional livestock feeds are scarce and high in price, it is not economically practical to do so.

When residues and whole-tree materials are fresh or have a moisture content above 17%, decomposition and spontaneous combustion may be encountered. For this reason, fresh products should not be piled high, should not be covered with canvas or plastic, and should be monitored, at intervals, for internal temperatures.

Nutritive Value

Most untreated woods are quite indigestible. Untreated hardwoods range in digestibility from 2 to 37%, with aspen being most digestible, followed by soft maple and black ash. All untreated softwoods are essentially indigestible.

There is a positive correlation between the digestibility of the bark and the wood of a given species, with the bark being the more digestible. Twigs and buds are more digestible than stemwood because they contain less lignin and

[3]In the preparation of this section on "Wood and Paper By-product Feeds" including most of the research cited, the authors drew heavily from the following source: Fontenot, J. P., Chairman of Subcommittee, National Research Council, "3. Forest Residues," *Underutilized Resources As Animal Feedstuffs*, NRC-National Academy of Sciences, 1983, pp. 69-120.

have a higher percentage of digestible protein and ether extract; which explains why buds and twigs are the preferred browse of goats and a number of wild animals.

The official publication of the American Feed Control Officials lists and describes aspen feedstuff as follows:

> *Ground Whole Aspen and/or parts* is generally recognized as a feed ingredient in cattle rations when used in accordance with good nutritional practices. Ground whole aspen *(Populus tremuloides Michiz* and *Populus gradidentata)* is composed of the entire tree including leaves, branches, trunk, and bark. Ground aspen parts may also include leaves, branches, trunk, and bark. The particle size of the product shall not exceed ⅜ in.

A summary of the nutritional value of aspen as a feedstuff is presented in Table 12-2.

TABLE 12-2
ASPEN AS A FEEDSTUFF

Product	Chemical Composition	Nutrient Utilization	Animal Performance	Comments
Aspen bark	M-F basis: Crude protein, 3% Ether extract, 5–10% Crude fiber, 40–55%	TDN on a dry basis, 36.7%. This compares with a TDN of 45 to 50% for barley, oat, rye, wheat straw. Dry matter digestibility, 30–50%.	Wethers receiving 37 to 53% aspen bark, along with oats and soybean meal, gained 0.9 lb *(0.04 kg)*/day during a 48-day trial. In this same trial, wethers on 68% bark lost 0.9 lb *(0.04 kg)*/day. Ewes fed, for 11 months, 67–70% aspen bark on a dry matter basis, along with dehydrated alfalfa, corn, oats, and soybean meal, gestated, lambed, lactated, and were rebred, with performance comparable to the control group which was fed hay.	
Aspen sawdust	Crude protein, 2%. ADF, more than 60%. Klason lignin, 16–22%.	Dry matter digestibility in a high-roughage ration, 41%. Dry matter digestibility in a high-concentrate ration, 28%.		
Whole aspen tree	Crude protein, 1.2%. ADF, 64.9%.	Dry matter digestibility of whole aspen silage in 80% grain ration, 5%. Dry matter digestibility of whole aspen silage in 40% grain ration, 37%; which is about 80% of that expected with wheat straw.	A series of 3 experiments involving rather large numbers of cattle fed whole aspen tree material was conducted at South Dakota State University. The first 2 trials were with finishing steers. The 3rd trial involved wintering beef cows. The conclusions as a result of these 3 trials: 1. The net energy content of the aspen appeared to be considerably higher when fed to steers receiving high-forage rations than when fed to steers receiving high-grain rations. 2. Whole aspen tree material is best suited for inclusion in high-roughage, maintenance type rations, such as for wintering beef cows.	
Forest foliage (muka) The term *muka* originated in the Soviet Union, where it refers to forest foliage flour or meal used for animal feed. The term is also accepted and used in the U.S. Muka may be derived from softwood needles or hardwood leaves. It is usually dried and ground, and it may be fed in loose form or pelleted.	Crude protein, 5%. It is a good source of carotene and riboflavin.	Dry matter digestibility of aspen muka by goats is 40%. Dry matter digestibility of spruce and fir muka is 43%.	A number of studies in the U.S.S.R., where muka is fed to poultry, swine, and cattle, have indicated that the addition of muka to rations improved animal performance. In North American studies, muka in broiler diets lowered feed efficiency and gains.	The Soviet Union produced 100,000 metric tons of muka in 1975; and, at that time indicated that they planned to double production by 1980. In the U.S.S.R., forest foliage is primarily a source of carotene, trace elements, and vitamins and is used in poultry, swine, and cattle rations. When used, it constitutes 3 to 8% of the ration. Foliage makes up 10% of the weight of mature softwoods and 25% of the weight of young softwoods. Corresponding values for hardwoods are 5 and 15%.

Processing Methods

(Also see Chapter 14, section on "Treatment of High-Cellulose Feeds.")

Certain fractions of forest biomass can be useful in animal rations without treatment; however, many species and fractions should be treated to enhance digestibility. Among the treatments that have been used commercially or experimentally to increase the digestibility of wood by animals are: hydrolysis, treatment with alkali and ammonia, delignification, biochemical degredation, grinding, and irradiation.

HYDROLYSIS

At the present time, three U.S. plants produce wood molasses (hemicellulose extract) from wood as a by-product of the production of hardboard by the wet process. Two of these plants use a 1- to 2-minute high-pressure steam cook after which the steam pressure is suddenly released. The other plant uses a steam-pressurized refiner to defibrate the wood chips. At each plant, the solubilized wood materials are neutralized and concentrated or spray dried for use as animal feed. The combined production of this molasses and dried product is about 68,000 tons per year on a dry matter basis. A typical analysis of this molasses is: 0.5% protein, fat, and fiber; 6% ash; 3% calcium; 0.8% phosphorus; 34% total sugar; and 35% moisture.

The so-called Madison process, using dilute sulfuric acid, is used if additional molasses is needed. More than 40 plants based on a modified Madison process are currently operating in the Soviet Union to make wood sugars for fermentation to yeast for human food and animal feed. Also, some of the wood sugars are fermented to ethanol.

Steaming has been used to increase the digestibility of both wood and straw. The steam treatment of aspen (*Populus* spp) chips at 320°F to 338°F for about 2 hours results in a product that is readily accepted by sheep at up to 60% of their ration and produces normal weight gains and carcass yields. This research led to a process developed by Stake Technology, Ltd., Toronto, to make a steamed product from wood and agricultural residues. In this process, the steaming time is a few minutes following which the pressure is rapidly reduced. Currently, two U.S. plants (one in Maine, the other in Minnesota) are using this process to produce feed from aspen.

ALKALI AND AMMONIA

Various wood species respond differently to sodium hydroxide treatment. As a group, hardwoods are more responsive than softwoods, perhaps due to the higher lignification of the latter.

Experimentally, the dry matter digestibilities of untreated and alkali-treated (NaOH) aspen sawdust have been studied. Aspen yielded dry-matter digestibilities of about 41% for untreated aspen and 52% for the alkali-treated aspen. So, alkali treatment can increase ruminant utilization of aspen wood by approximately 25%, thereby making it equal to medium-quality hay as a source of energy.

Another long-standing approach to upgrading the feeding value of lignocellulosic materials involves treatment with aqueous or gaseous ammonia (NH_3). Experimentally, a number of woods have been treated with ammonia (NH_3) in both liquid and gaseous forms and the results assayed by changes of *in vitro* digestibility. As with alkali treatment, a variable response toward NH_3 treatment is exhibited, which is attributed to the degree of lignification. Digestibility of approximately 46% was obtained for the ammonia-treated aspen, which was about 6% lower than the NaOH-treated aspen. However, this was sufficient to rank the NH_3-treated aspen equivalent to medium-quality hay as a source of energy. As an added benefit, the NH_3 treatment provided some usable nitrogen.

DELIGNIFICATION

Delignification would appear to provide a straightforward solution to increasing the digestibility of wood by-products. As evidence of the feasibility of this approach, it is noteworthy that more than 1.5 million tons of sulfate and sulfite pulps from pine, spruce, and fir were consumed by cows and horses in the Scandinavian countries during World War II. This was a wartime emergency situation, however, and the use of treated wood pulps for animal feeds declined rapidly following the war, with the availability of more conventional feedstuffs.

Complete delignification is costly and is not required to prepare a nutritionally acceptable feedstuff. Current thinking is that the goal should be to obtain a product having an *in vitro* digestibility of about 60%, which is comparable to good-quality hay. Treatment with sodium chlorite is one approach. In acetic acid solution, sodium chlorite exerts a marked specificity for lignin solubilization.

Another possibility for enhancing the availability of wood residue carbohydrates consists in disrupting the lignin-carbohydrate association without removal of either constituent. Under proper conditions, gaseous sulfur dioxide (SO_2) can effect a disruption; the treatment appears to be applicable to both hardwoods and softwoods. The 60 to 65% digestibility of SO_2-treated hardwoods ranks these materials equal to moderate- or high-quality forage. The SO_2-treated softwoods are comparable to low-quality forage.

OTHER PROCESSING METHODS

Experimentally, the following additional processing methods have been tested to increase the digestibility of wood by animals:

- **Biochemical**—Some white-rot fungi are effective, but slow, in removing lignin from wood.

- **Grinding**—While vibratory ball milling for 140 minutes is an effective treatment, the response is quite species-selective; for example, following milling the digestibility ranged from about 80% for aspen to only 20% for red alder; and softwoods were even less responsive than red alder. This selectivity, along with the time and energy costs, imposes restrictions on this method.

- **Irradiation**—Irradiating wood or straw by gamma rays or by high-velocity electrons substantially improves digestibility by rumen organisms. However, a strong species specificity exists; for example, an electron dosage of 10^8 rad results in digestibility of 78% for aspen and only 14% for spruce.

PULPMILL AND PAPERMILL RESIDUES

Some of the many pulp and papermaking residues which are already partially delignified but which have little fiber value for paper manufacture have excellent potential as ruminant feedstuffs. The residues in greatest abundance and with greatest potential as livestock feeds are: (1) fines and screen rejects, (2) sludges, and (3) spent sulfite liquor.

Fines and Screen Rejects

The composition and *in vitro* rumen digestibility of several papermill residues are shown in Table 12-3.

As shown in Table 12-3, groundwood fines yielded digestibility values comparable to that of sawdust of the same species: 37% for aspen and 0% for pine and spruce. All of the screen rejects and chemical pulp fines had digestibilities of more than 40%, and digestibility of two of the pulp fines was more than 70%. It should be noted (although not shown in Table 12-3) that fines and screen rejects are very low in protein and high in acid detergent fiber. Thus, based on *in vitro* dry matter digestibility, any of the screen rejects and chemical pulp fines could serve as useful sources of feed energy for ruminants.

TABLE 12-3
COMPOSITION AND *IN VITRO* RUMEN DIGESTIBILITY OF PAPERMILL RESIDUES (FINES AND SCREEN REJECTS)[1]

Type of Residue	Composition			*In Vitro* Digestibility
	Lignin	Carbohydrate	Ash	
	(%)	(%)	(%)	(%)
Groundwood fines:				
Aspen	21	73	1	37
Southern pine	31	59	1	0
Spruce	31	60	1	0
Screen rejects:				
Aspen sulfite	19	77	2	66
Mixed hardwood, sulfite	24	65	14	54
Mixed hardwood, kraft	25	74	9	44
Chemical pulp fines:				
Mixed hardwood, kraft (bleached)	<1	109	1	95
Aspen sulfite (parenchyma cells)	20	73	2	73
Southern pine, kraft (unbleached)	28	68	4	46

[1]Fontenot, Joseph P., *et al.,* National Research Council subcommittee, *Underutilized Resources as Animal Feedstuffs,* National Academy Press, 1983, p. 108, Table 27. *Original source:* Baker, A. J., *et al.,* 1975, "Wood and Wood-Based Residues in Animal Feeds," p. 75 in *Cellulose Technology Research,* A. F. Turbak, ed., ASC Symposium Series 10, Washington, D.C.: American Chemical Society.

Generally, animals readily accept rations containing up to 75% pulp residues. Experimental feeding trials with both breeding and finishing cattle and sheep have given favorable results.

Some pulp and paper residues are highly digestible and do not require further processing. Presently, pulp fines are being incorporated in livestock rations on a limited basis. Pulp fines containing about 65% water are being fed to dairy and beef cattle in Pennsylvania. The use of additional spent sulfite pulp fines in animal rations awaits further studies to determine (1) how to make the material available without lowering its feeding value or disrupting the mills, and (2) how best to use it near its source.

Sludges

Pulpmills and papermills may produce primary clarifier or lagoon sludges. Most of these by-products are very high in ash, and moderate to high in lignin, both of which diminish their feeding value. Also, because the groundwood mill sludges are mostly groundwood, digestibility is low, although the total carbohydrate content may be high. Moreover, sludge is relatively unpalatable, with the result that feed consumption is low. For these reasons, sludge is the least valuable of the three pulpmill and papermill residues discussed in this section.

Spent Sulfite Liquor

The chemical composition of spent sulfite liquor varies greatly, depending upon wood species, pulping reagent used, and pulping conditions. In the past, much of this product was disposed of by being burned or sewered. Because of its adhesive properties, it has recently been used in the production of linoleum paste, cement, ceramic mixes, and pellet binders for animal feeds.

The ammonium base chemical pulping process produces large quantities of spent sulfite liquor, which, on a moisture-free basis, contains about 25 to 30% carbohydrates and 25% crude protein equivalent (most of which is nonprotein nitrogen); hence, it can be considered a potential feedstuff for ruminants.

Washington State University scientists investigated the feeding value of this waste material for beef cattle. Using a corn-alfalfa-molasses control ration, spent sulfite liquor was substituted for molasses at levels of 8 and 12% of the total ration. The results of the experiment are given in Table 12-4. As indicated, spent sulfite liquor did prove to be a viable alternative feed source, producing as good or better gains than the control ration.

TABLE 12-4
SPENT SULFITE LIQUOR AS A FEED FOR BEEF CATTLE[1]

Item	Control		Spent Sulfite Liquor			
			8%		12%	
	(lb)	*(kg)*	(lb)	*(kg)*	(lb)	*(kg)*
Daily gain	3.57	*1.62*	3.62	*1.64*	3.88	*1.76*
Daily feed intake	22.90	*10.40*	22.90	*10.40*	25.60	*11.60*
Feed:Gain ratio	14.10	*6.41*	13.98	*6.34*	14.53	*6.59*

[1]Adapted by the authors from "Feeding Value and Net Energy for Gain of Spent Sulfite Liquor for Beef Cattle," by F. S. Chang, *et al., Journal of Animal Science,* Vol. 41, 1975, p. 625.

Paper Products

A novel and widely publicized by-product feed for ruminants is paper. It has been estimated that more than 20 million tons of waste paper are produced annually, much of which could be a potential feedstuff for ruminants.

Digestibility of paper is highly variable. The dry matter digestibility of newsprint by cattle is about 30%; wrapping paper and cardboard, 40 to 60%; and high-quality chemical pulp paper about 98%. When high levels of newsprint (greater than 12%) or office paper (15%) are incorporated in cattle feed, there may be a reduction of feed intake.

One of the potential hazards of feeding printed paper is the high level of heavy elements, notably lead, in the ink. While no toxicities have been reported, more research must be conducted to monitor possible metal accumulation in animals before widespread use of such waste paper can be encouraged.

WOOD SUGAR PRODUCTS

Various processes have been developed to break down the long polymers of cellulose and hemicellulose into their constituent simple sugars. Wood molasses is the most widely used of these products.

Cellulose is degraded to glucose by high temperatures and pressures and dilute acids. One process by which this is accomplished is the Madison wood sugar process, in which a dilute sulfuric acid solution is sprayed on chips heated in a digester at 300°F. The temperature is gradually raised to 365°F, then cooled to 280°F and neutralized with lime. The sugar solution is then condensed to form a bitter-tasting syrup which can be substituted for cane molasses in cattle rations. One ton of wood will yield approximately ½ ton of sugar.

Wood molasses is produced from the solubilization of hemicellulose by steam. It can be substituted for cane molasses in ruminant rations. Recent evidence indicates that wood molasses protects a substantial portion of the plant protein in rations from ruminal degradation, so it may take on increasing importance. For nonruminant animals, wood molasses can be used at levels up to 2 to 2.5% of the ration. Excessive levels in nonruminant rations can cause digestion disorders.

TORULA YEAST

Torula yeast *(Torulopsis utilis)* is a hardy type of yeast that can be grown on a variety of substrates, such as sulfite waste liquor and saccharified wood. While brewers' and distillers' yeast are truly by-product feeds, torula yeast is cultured specifically as a feedstuff for livestock. (See Table 12–1, under "Wood and Paper By-product Feeds," for further description and for feeding recommendations.)

WOOD RESIDUES AS ROUGHAGE SUBSTITUTES IN RUMINANT RATIONS

Ruminants generally require a certain level of roughage in their ration. Roughage provides tactile stimulation of the rumen wall and promotes rumination, which in turn increases salivation and supply of buffer for maintenance of rumen pH. In the dairy ration, roughage can help maintain the level of milk fat. In high-grain beef cattle finishing rations, it can lessen the incidence of parakeratosis and liver abscesses. When traditional sources of forage or roughage are in short supply and high priced, wood has been shown to be an effective replacement. Depending upon the other ration ingredients, concentrations of 5 to 15% screened sawdust in the rations of beef cattle appear practical. Oak sawdust or pine sawdust, which are essentially indigestible, are effective as roughage substitutes in cattle finishing rations. For lactating dairy cows, aspen sawdust may be used as a roughage extender or as a partial roughage substitute in high grain rations; some long hay appears to be necessary in the ration to stabilize feed intake.

ANIMAL HEALTH/REGULATORY ASPECTS

Generally, whole-tree or tree residues will not affect the health of livestock adversely. However, they are primarily energy sources, so it is essential that rations containing wood be properly balanced for all the essential nutrients.

Pine needles from ponderosa pine *(Pinus ponderosa)* may cause abortion in cattle. (See Chapter 5, Table 5–3, section on "Pine Needle Abortion.")

Wood residues that have been exposed to chemical treatments, such as in the pulping process, should be carefully examined for residual chemicals. Sludges in particular need careful scrutiny.

The use of wood and wood-derived products in animal feeds requires the consideration of and compliance with state and federal regulations.

Crop Residues

Most of the by-product feeds result from some sort of industrial manufacturing; for example, the by-products from milling grains. Another excellent source of feed is crop residues—parts of plants that are normally left in the field following harvest of the primary crops.

As production costs increase, livestock producers become more interested in the enormous potential for animal production through feeding crop residues.

KIND AND QUANTITY OF RESIDUE PRODUCED

The quantity of crop residues produced may be estimated by multiplying the annual grain production by a grain weight: residue weight ratio. Normally, grain-producing plants produce as much (or more) weight of vegetative material as of grain. The crops, ratios used for conversion from grain to residue, and estimated annual production of residues are given in Table 12–5.

TABLE 12-5
ESTIMATED SUPPLY OF CROP RESIDUES[1]

Crop Source	U.S. Grain Production	World Grain Production	Canadian Grain Production	Ratio Residue/Grain[2]	U.S. Residue	Residue % of Total
	(mil. metric tons)	(mil. metric tons)	(mil. metric tons)		(mil. metric tons)	(%)
Barley	13.3	182.0	14.6	2.0	26.6	6.53
Corn	209.6	476.6	5.9	1.0	209.6	51.48
Cottonseed	4.79	30.63	–	3.0	14.37	3.53
Flax	0.21	2.36	0.90	3.0	0.63	0.15
Oats	5.6	47.5	3.3	1.0	5.6	1.37
Peanuts	1.87	19.99	–	1.5	2.8	0.69
Rice, rough	6.0	466.9	–	1.0	6.0	1.47
Rye	0.5	31.0	0.6	1.0	0.5	0.12
Sorghum	23.8	64.3	–	1.0	23.8	5.84
Soybeans	57.11	97.03	1.01	1.0	57.11	14.03
Sugarbeets[3]	22.9	285.7	0.94	0.14	3.21	0.79
Wheat	56.9	529.7	31.4	1.0	56.9	13.98
Total	402.58	2,233.71	58.65		407.12	

[1]U.S., World, and Canadian grain production from: *World Agricultural Production,* USDA, Foreign Agricultural Service, Circular Series WAP 5-88, May 1988, except for sugarbeets.

[2]Ratio residue/grain from: *Underutilized Resources as Animal Feedstuffs,* National Research Council, National Academy Press, 1983, p. 180, Table 48.

[3]Sugarbeets production from: *1986 FAO Production Yearbook,* Vol. 40, p. 163.

As shown in Table 12-5, corn, wheat, and soybeans account for nearly 80% of the total residues. It follows that the major U.S. corn-, wheat-, and soybean-producing states produce the most residues.

Corn is the most widely produced grain crop in the United States. It usually produces an amount of residue equal to the quantity of grain produced. So, as shown in Table 12-5, the production of 209 million tons of corn in 1986 resulted in 209 million tons of corn residue, which was over ½ the total available crop residue that year.

Wheat, which produces much less tonnage of grain per acre than corn, accounts for about 14% of the residue. About 57 million tons of wheat straw were produced in 1986 (see Table 12-5).

Soybean residue usually provides another 14% of the crop residue (57 million metric tons in 1986) and grain sorghum 6% (23.8 million metric tons in 1986). Other crops account for the remaining residue (see Table 12-5).

More than 400 million metric tons of straws, stalks, and stubble are available in the United States each year (see Table 12-5) and another 60 million tons in Canada. Worldwide, more than 2.2 billion tons of crop residues are produced annually.

In addition to being used as livestock feeds, crop residues may be, and are, used for bedding, soil improvement, and as a substitute for fossil fuels.

ECONOMY OF CROP RESIDUES

Evaluating the nutrient content of crop residues, along with collecting, storing, treating, transporting, and feeding them, is much more difficult than determining the quantities available. Also, when evaluating crop residues for animal feeding, three major questions must be answered: (1) are they available in sufficient quantity to make their use as a feedstuff worthwhile, (2) do they have high enough nutrient content to justify feeding them to livestock, and (3) are they cost competitive? These and other important *field to feed* aspects of crop residues are discussed in the sections that follow.

Harvesting

Because of differences in plant structure, grain harvesting methods, and moisture content, harvesting crop residues may not be easy. Straws from cereal grains are easily collected in dry state behind the combine. Corn and sorghum stovers often are too wet for dry storage, but they can be stored as silage. Soybean residue is difficult to harvest if allowed to drop on the ground behind the combine. Soil contamination during harvest may be a problem with all residues. Some of the residues, such as cottonseed hulls, rice milling by-products, and sugarcane bagasse, which are processed at central locations, have the advantage of being collected and available for treatment.

Nutrient Value of Crop Residues

Almost all crop residues are harvested after the plants reach physiological maturity; so, they are high in cell walls and lignin and low in protein and digestible dry matter. They are a feed energy source, but suitable only for ruminants—beef and dairy cattle, sheep, and goats.

Most of the energy in crop residues is in the lignocellulose complex (lignin, cellulose, and hemicellulose). The amount of the total lignocellulose and of its constituents varies widely with forage species and stage of maturity. Generally, the higher the lignin content the lower the digestibility of the cellulose material. The chemical and physical binding of cellulose and hemicellulose in the cell wall of plants is important; so, the amount of lignin *per se* is not always a good indicator of digestibility. Nutritionally, the lignocellulose complex consists of these fractions: (1) lignin, which is unavailable as an energy source, (2) a digestible energy source that can be utilized by rumen microorganisms, and (3) a fraction which is very resistant to bacterial action in the rumen, but which becomes an energy source after special treatment. The third fraction is of major interest because of the potential additional energy which can be made available in the cellulosic crop residues. In many of the crop residues, this third fraction is large enough that considerable research has been done on treatment to "unlock" its energy.

In addition to lack of available energy, crop residues may be deficient in protein, phosphorus and possibly other minerals, and vitamin A. Also, they are usually bulky and lacking in palatability. So, it is important to provide proper supplementation based on the performance expected of animals grazing or being fed crop residues.

Treating Crop Residues to Increase Digestibility

Crop residues are inefficiently utilized by animals because of the high content and poor digestibility of the fibrous fraction. This poor digestibility is related to the extent of lignification of the cell wall component of these low-quality forages. Although crop residues provide a satisfactory ration for dry gestating animals, they do not provide sufficient energy for either young or lactating ruminants—they simply cannot hold enough of these low-quality roughages to provide adequate energy. This prompts interest in increasing the digestibility of these crop residues.

Two approaches offer the best possibilities for increasing the digestibility of crop residues: (1) manipulation of plant genetics or harvest time to obtain higher quality residues, and (2) physical or chemical treatments to increase intake and digestibility. Among the physical and chemical treatments for delignifying and increasing digestibility are the following:

1. **Grinding and pelleting.** The physical treatment of grinding and pelleting is effective on corn stover, corncobs, and straw; it will increase feed intake by 25%, and improve feed efficiency and rate of gain by 35 and 100%, respectively. Not all crop residues will respond so dramatically to grinding and pelleting, however.

2. **Sodium hydroxide (NaOH).** Treatment with sodium hydroxide is one of the oldest methods of improving digestibility of high fiber crop residues. The main disadvantages of the NaOH treatment are excessive labor, high cost, and the loss of soluble nutrients by washing to remove alkali residues.

3. **Calcium hydroxide (CaOH).** Calcium hydroxide is less caustic than sodium hydroxide, but greater amounts are needed to obtain results equal to sodium hydroxide.

4. **Ammoniation.** The application of ammonia (either as anhydrous gas or aqueous liquid) to crop residues to improve feeding value is effective. Ammonia appears to react in a manner similar to sodium hydroxide. However, the reaction time is much longer; up to 20 days is required for ammoniation vs 24 hours for sodium hydroxide. Also, during treatment the crop residue must be stored in an airtight structure so there will be no loss of ammonia. The two major advantages of the ammonia treatment are: (1) the added nonprotein nitrogen increases the nutrient value of the residue, and (2) no mineral residue remains which might be detrimental to the animals or to the soil to which the manure is added. *CAUTION:* Animals will not eat ammonia-treated residues unless they are aerated or mixed with a fermented feed so that the organic acids neutralize the ammonia.

5. **High pressure steam.** In this system, the fibrous material is moistened and placed in a pressure vessel; steam pressure of 5 to 10 lb/sq in. is applied for a few seconds and then the pressure is released. The high pressure steam treatment of corncobs fed to calves increases digestibility by about 16% and doubles daily gains. Pressure treatment with steam also increases wheat straw digestibility.

The five treatments briefly described have been tried and found successful on corn and small grain residues. Soybean residues are very high in lignin and do not respond well to chemical treatment. Grain sorghum residue responds to sodium hydroxide and calcium hydroxide treatments, but probably not as well as corn residue or straws. Rice hulls are improved by ammoniation, and peanut hulls are improved by treatment with calcium hypochlorite. Chemical treatment of sugarcane bagasse with sodium hydroxide or calcium hydroxide dramatically increases its value as a feed for ruminants. Steam pressure treatment also increases *in vitro* digestibility of bagasse.

Summary: Because of the low nutritional value of crop residues, methods of increasing value by treatment are of interest. Sodium hydroxide has been widely studied. While it is effective, concerns about human safety and sodium residues may limit its ultimate usefulness. Calcium hydroxide and ammonia treatments offer the greatest long-term potential.

The potential of crop residue treatments becomes apparent when it is realized that straw, for example, is only 30 to 40% digestible before treatment. When pressure heated with water, it becomes 50 to 60% digestible; and digestibility increases to 70 to 80% when sodium hydroxide is added prior to cooking. By treating corn husklage and milo residue, workers at the Nebraska Station were able to increase the energy value of these residues to 90% that of corn silage.

Lowering the cost of treating crop residues to increase digestibility is the primary area which must be researched before these procedures can be applied to practical operations.

(Also see Chapter 14, Feed Processing, section headed "Treatment of High-Cellulose Feeds.")

Transporting, Packaging, and Storing

The relatively low value of crop residues in comparison with grains necessitates low-cost harvesting, packaging, and storing. Because of their bulky nature and low market value per unit weight, it is not economical to move them very far. Also, the bulkiness of most crop residues makes packaging difficult.

In the past, the efficient management of the family farm generally included the use of most crop residues, such as straw and stalks, as bedding or feed, or both. However, the development of large volume, high density livestock and poultry units, which may not be located where the feed grains are produced, has created a situation in which there are large quantities of crop residues that are not being efficiently utilized. Moreover, many grain-producing farms no longer have fences or water for livestock; so, animals cannot graze the crop residues unless they are herded and water is hauled.

Most of the crop residues become available in the summer and fall, but the greatest need for them as livestock feeds is in the winter. So, some kind of packaging and storage is desirable, but may not be practical.

Because of the transportation, packaging, and storage costs involved, the increased use of crop residues in large volume, high density livestock operations is unlikely so long as conventional forages for use in high concentrate feeding systems are available at relatively low cost. However, crop residues have high potential as an alternative source of feed energy.

Feeding Systems for Crop Residues

Generally speaking, crop residues may be grazed, processed as dry feed, or made into silage. The important thing to remember is that their relatively low value, in comparison with grains, necessitates low-cost harvesting, storing, and feeding. Also, they must be fed to the right class of animals, and they must be properly supplemented.

The use of low quality crop residues is restricted primarily to ruminants, such as wintering beef cows. In many parts of the United States, cornstalks, sorghum stubble, and cereal grain stubble can be grazed, provided there are fences and an available supply of water. This is the most economical system of utilizing crop residues; there is essentially no harvesting machinery or energy cost, nothing is removed from the soil, and the manure is deposited on the land. However, weather problems are often encountered when winter grazing stalks and stubbles. In the higher rainfall areas of eastern United States, the residues deteriorate rapidly, and muddy fields may prevent continuous grazing. In the colder areas of the western Corn Belt and Plains States, snow cover may prevent grazing at times, with the result that an alternate reserve feed supply is needed.

SOME CROP RESIDUES

Some of the common crop residue feeds are listed in Table 12-1 of this chapter and in Section V, Composition of Feeds. Additionally, crop residue feeds are discussed in Chapter 6, in the section on "Plant By-products"; in Chapter 9, in the sections on "Corn and Sorghum Residue Silage" and "Other Silage Crops"; in Chapter 11, in the section on "Ammoniated Products"; and in Chapter 19, in the sections on "Crop Residues" and "Corn Residues." Some crop residue feeds are discussed in the sections that follow.

Bean, Snap, Cannery Waste

On a dry basis, snap bean cannery waste is higher in protein than corn cannery waste. But it contains only about 10% dry matter. Snap bean cannery waste can replace part of the hay or silage in the rations of ruminants, but because of its high moisture content it should not be fed in large amounts to lactating animals or young stock.

Beet Tops, Fresh

On the average, a 20 ton-per-acre crop of beets will supply about 14 tons of tops. This will provide about 2.5 tons of dry matter, including 560 lb of protein. When included in a balanced ration, this will produce more than 325 lb of beef or lamb gains.

Fresh beet tops may be left in the field and grazed by livestock, but this method of harvesting results in considerable waste. For this reason, they are usually cocked, dehydrated, or ensiled.

Beet Tops, Cocked

Sometimes beet tops are cocked by placing them in 200- to 400-lb stacks. If properly cocked, tops will keep 4 months or longer following harvest.

Beet Tops, Dehydrated

Dehydrated beet tops can be used in about the same manner as dried molasses beet pulp. However, dehydrating is costly, with the result that most beet tops will continue to be fed fresh, as ensilage, or stored in cocks.

Bluegrass Straw

Bluegrass straw is the residue that remains after threshing bluegrass for seed.

In studies at Washington State University, Heinemann found that bluegrass straw had an *in vivo* dry matter digestibility of 57.6%, and that it was more palatable to cattle than wheat straw. He also fed yearling beef steers on corn-beet pulp-alfalfa rations with each of three crop residues: (1) bluegrass straw, (2) wheat screenings, and (3) pelleted and bulk cottonseed hulls. Daily feed intake as a percentage of animal weight and average daily gain was highest for the steers fed bluegrass straw.

Chaff, Cereal Grain

Chaff is material that is separated from grains in the usual commercial cleaning process. It may include hulls, joints, grain, sand, and/or dirt. Wheat chaff averages about 29% fiber and 4.4% protein. Chaff had best be fed to ruminants on maintenance rations, along with a legume hay or a protein/mineral/vitamin supplement.

Clover Straw

Clover straw is the residue remaining after separating the seeds in threshing. It is worth about half as much nutritionally as good clover hay. Clover straw should be fed to ruminants, along with a good quality hay at about a 1:1 ratio.

Corn Residues

Of all crop residues, the residue of corn is produced in greatest abundance and offers the greatest potential for expansion in ruminant numbers.

In 1986, 69,189,000 acres of corn, yielding 119.3 bu per acre, were harvested for grain in this country. For the most part, over and above the grain, about 2¾ tons of dry matter per acre were left to rot in the field. That's 190 million tons of potential feed wasted, enough to winter 96 million dry pregnant cows consuming an average of 33 lb of corn refuse per head daily during a 4-month period. Also, there are many other crop residues, which, if properly ultilized, could increase the 96 million figure given above. Since goats and sheep require even less feed per day, the potential for increasing animal numbers is even greater!

Although corn refuse offers tremendous potential as a feed for ruminants, there are difficulties in harvesting and storing it. But science and technology have teamed up and are working ceaselessly away at solving these problems.

Fig. 12-7. The feeding of crop residues is becoming an important facet of livestock feeding. Here cornstalks are being harvested for feed. (Courtesy, Gehl Company, West Bend, Wisc.)

Broadly speaking, three alternate methods of salvaging corn refuse are being used: (1) grazing, (2) harvesting and dry feeding, and (3) ensiling; with different ways of accomplishing each. The choice of the method should be determined primarily by cost, the proportion of refuse utilized, and how well it meshes with other farm enterprises—for example, in some cases the need for fall plowing will necessitate removal of the material from the land and eliminate grazing as an alternative. Costs and the proportion of the refuse slavaged vary widely, as shown in Table 12-6.

TABLE 12-6
ESTIMATED TOTAL HARVESTING, STORING, AND FEEDING COST PER TON OF CORN FORAGE[1] [2]

Harvesting System	Tons Harvested Per Year		
	100	500	1,500
	($)	($)	($)
Corn silage, upright silo	—	12.63	6.48
Corn silage, bunker silo	31.00	9.07	4.68
Cornstalk silage, bunker	35.12	11.63	7.01
Field-stored packages, self-fed:			
Harvesting all rows:			
Husklage, 4-row combine	7.84	1.87	1.02
Husklage, 6-row combine	7.77	1.76	0.85
Round bale	20.14	6.78	4.69
1-ton stack	15.59	6.65	5.48
3-ton stack	23.78	8.36	5.32
Harvesting 2 of 6 rows:			
Round bale	12.80	4.55	2.72
1-ton stack	14.34	5.25	3.90
3-ton stack	22.50	6.93	4.52
Add cost to move large packages for feeding:			
Round bale	3.64	3.35	3.30
1-ton stack	3.94	2.89	2.73
3-ton stack	6.57	2.39	1.72

[1]Ayres, G. E., "Large-Package Forage Machinery Costs," Iowa State University A.S. Leaflet R222, 1975.

[2]One ton of corn forage = 1 ton of corn silage at 65% moisture, 1 ton of cornstalk silage at 55% moisture, or 1 ton of husklage or baled and stacked cornstalks at 35% moisture.

Table 12-6 shows that the cost of each method of harvesting is lowered as the tonnage increases.

GRAZING CORN RESIDUES

This refers to turning the animals directly into the stalk field—the traditional way of utilizing cornstalks. Letting the animals do their own harvesting is the simplest and least

Fig. 12-8. Cows winter grazing cornstalks—the residue remaining in the field after harvesting the corn grain. (Courtesy, Ron Baker, C & B Livestock, Inc., Hermiston, Ore.)

expensive method devised for utilizing a crop. However, there is considerable wastage, and it is not possible to prolong the winter feeding period. In an open fall and winter, 2 acres of cornstalks will carry a pregnant cow for 100 to 120 days. But the following problems are associated with this method of harvesting:

1. **Selective grazing.** Some animals, such as cows, are selective grazers. They will consume the more palatable portions of corn refuse first, in the following order: corn ears, husks, leaves, and stalk.

2. **Waste.** Only an average of 15 to 25% of the stover is actually used when cattle are given access to grazing areas, with the amount varying from 15 to 40%, depending primarily on weather conditions.

3. **Fencing.** Many cornfields are unfenced; hence, fence must be constructed in order to confine the animals. Also, strip grazing (grazing a part of the field at a time) will improve the utilization of stalks in large fields by making more uniform nutrition available throughout the grazing period. It prevents selective grazing over the entire field, with the result that the animals consume the more palatable portions of the plant first, and leave the bare stalks until last. The fencing problem may be solved economically with modern types of electric fence.

4. **Snow cover; fall plowing; soil puddling; stock water.** In the northern part of the United States, snow cover prevents grazing for part of the winter and necessitates a reserve feed supply.

Another drawback is that grazing prevents fall plowing, a recommended practice on heavy soils. Also, animals may puddle and pack the soil, which lowers crop yields.

Frequently, supplying the herd with drinking water is costly. It may necessitate drilling a well, piping, or hauling water from a distance.

STALKLAGE

Stalklage refers to all the residue remaining after harvesting corn with a combine or picker. It may either be stored dry, or ensiled.

• **Dry stalklage**—Stalklage is more difficult to collect than husklage, and more expensive, since it involves more equipment and another trip across the field. A number of different machines for harvesting stalklage are being used; among them, forage harvesters, balers, stackers with flail pickups, and choppers and stackers. By operating the machine a few inches above the ground to prevent excess soil pickup, a yield of 1 to 3 tons of residue per acre may be obtained, with the moisture content ranging from 20 to 55%, depending on the time of harvest. Stacked or baled cornstalks should be at the low end of this moisture range (20 to 35%) to reduce heating and spoilage.

Cows like dry stover. Self-feeders around a stack make feeding convenient. Leftover material may be used as bedding.

• **Stalklage ensilage (stover silage)**—Stalklage may also be ensiled, producing a product known as corn stover silage or cornstalk silage. When this is done, the use of a forage harvester equipped with a screen or a recutter-blower at the silo is necessary in order to chop the material finely. Fine chopping will ensure good packing and improve consumption by avoiding selectivity.

Where corn stover silage is to be made, the residue should be harvested as soon as possible after the grain is taken off, before it loses any moisture. At that time, the grain moisture will generally be under 30% and the refuse will have about twice the moisture content of the shelled grain. In an airtight silo, 40 to 45% moisture will suffice. In an unsealed or bunker silo, the moisture content should be 48 to 55% for proper lactic acid formation. Water may be added at the silo if necessary. As a precaution, some authorities recommend the addition of 56 lb of corn meal (or other finely ground grain) per ton of corn stover silage, as a means of providing carbohydrates from which acids will form and act as a preservative. With husklage, the latter precaution is not necessary since there is sufficient grain remaining in the husk and cob. Because corn residues are characteristically low in protein—less than 5%—it is possible to increase the protein value of the ensiled material by adding a nonprotein nitrogen source to the corn residue at the time of ensiling. An ammonia-molasses-mineral mix provides an effective means of improving the protein content of corn residues, as well as providing needed energy and minerals. Further information concerning the addition of nonprotein nitrogen to silage is found in Chapter 9, Silage/Haylage/High-Moisture Grain, under "Feed Additives."

The biggest deterrent to harvesting stalklage, in either dry or ensiled form, is the cost—primarily for the equipment. Rather than own such expensive equipment, which is only used for a short period, custom harvesting of stalklage is likely cheapest for most operators. Although custom harvesting may be cheaper, the timeliness of harvest is important for quality and palatability. If too much time elapses before the stalklage is harvested, husks and leaves may be blown away, thereby lowering quality.

HUSKLAGE (SHUCKLAGE)

Husklage is the forage discharged from the rear of a combine when harvesting corn. It consists of the husks, cobs, and any grain carried over the combine, collected in a wagon or straw buncher pulled behind the combine. This operation minimizes labor and does not slow the grain harvest, because the husklage piles can be dumped at the end of the field for supplemental feeding or later pickup by a front-end loader and moved to another location for stacking or ensiling. The moisture content of this material will usually run between 30 and 40%, and the yields will be between 1 and 1½ tons per acre.

The greatest difficulty encountered in feeding husklage dumps at the end of the field is waste. Depending on weather conditions, as much as 50% of the material may be wasted. But wastage of husklage dumps can be materially lessened by controlling access to them.

Stacking of husklage has been satisfactory for some producers.

Ensiling husklage, along with recutting and adding water, results in increased cow consumption and less rejection of cobs.

Since grazing stalk fields is widely practiced in the fall and winter, the feeding of baled, piled, or stacked corn residues in the field permits feeding cows on stalks most of the winter without supplemental feeding of hay or silage. However, a protein supplement is needed. As the latter stage of gestation approaches, supplemental energy feeds should be provided. Some molds may develop in the collected material, but usually they are not sufficient to affect either the feed intake or health of mature cows.

GUIDELINES FOR USING CORN RESIDUES

The following additional information is pertinent to the feeding value and supplementation of corn residues:

1. **Digestibility.** The components of corn residue rank as follows in digestibility, in descending order: remaining grain, husk, leaf, cob, and stalks.
2. **Energy.** Corn residues provide adequate energy for animals on maintenance levels of nutrition, but they must be supplemented with additional energy when fed to animals that are growing or in lactation.
3. **Protein.** The crude protein content of corn stover is low, running about 5.4% on an as-fed basis. The addition of NPN to corn residues can increase the crude protein content of corn stover up to a level of 9.4%.
4. **Minerals.** Phosphorus should be provided to all animals fed corn residue as a major part of the ration. Calcium may be deficient, especially for lactating animals. Also, some of the trace elements may be deficient. Hence, it is recommended that all animals fed high corn refuse rations have free access to a complete mineral mix. A mineral mixture with a Ca:P ratio of 1:1 is recommended for gestating ruminants.
5. **Vitamin A.** Corn residue, along with other drop residues, is extremely low in carotene. Hence, vitamin A must be supplemented.

(Also see Chapter 19, section on "Crop Residues and Winter Pastures.")

Cotton Crop Residues

Cottonseed hulls and gin trash are widely used fibrous by-product feeds derived from the cotton crop.

COTTONSEED HULLS

Cottonseed hulls are one of the most important roughages in the South, especially for cattle. On an as-fed basis, they supply about 42% TDN, which is about as much as is furnished by late-cut grass hay or by oat straw. They are low in protein (3.8%)—and practically none of it is digestible—low in calcium (0.13%), very low in phosphorus (0.09%), and lacking in carotene. To correct these deficiencies when fed to dry pregnant cows, hulls should be supplemented with a daily allowance of either (1) 6 lb of a good-quality legume hay, or (2) 2 lb of a 30 to 40% protein supplement, along with free access to a complete mineral, high in phosphorus unless a phosphorus-rich supplement such as cottonseed meal is fed. If no legume is fed, vitamin A should be fed or injected.

Cottonseed hulls can be fed without further processing—there is no chopping; and they are well liked by cattle, even when fed as the only roughage. In trials with lactating dairy cows, workers at the University of Arizona substituted pelleted or nonpelleted cottonseed hulls for 10, 30, and 50% alfalfa hay cubes in a ration containing 50% alfalfa hay cubes and 50% high-energy concentrate, with no differences due to ration in total milk production or percent fat.

Fig. 12-9. Dairy cows eating cottonseed hulls. (Courtesy, Watt Publishing Co., Mt. Morris, Ill.)

Pelleted hulls are now on the market. In comparison with regular hulls, they are more digestible, require less transportation and storage space—because of their high density—and are easier to handle.

(Also see Table 12-1, under "Crop Residue Feeds.")

GIN TRASH

Gin trash is a by-product of the ginning of cotton. (See Table 12-1, under "Crop Residue Feeds," for description and feeding recommendations.)

Hop Vine Silage

Hop vine silage is a suitable and palatable feed for cattle and sheep. On an as-fed (31% dry matter) basis, it contains about 2% crude protein. Hop vine silage should be fed along with a legume hay.

Pineapple Greenchop

Pineapple greenchop consists of the chopped upper ½ to ¾ of the mature pineapple plant after it is no longer used for fruit production. It is a fresh, succulent, very palatable, medium-quality roughage that is low in digestible protein. Pineapple greenchop is usually allowed to ferment, or ensile, for 3 to 7 days before feeding in order to ensure a more consistent acidity level in the feed and thus avoid potential digestive upsets. In Hawaii, pineapple greenchop is fed to dairy and beef cattle, at a recommended level of 20 to 35 lb of greenchop per cow daily.

Pineapple Juice Presscake (Also Called Juice Plant Pulp, Juice Press Residue, Pressed Pineapple Core, Pineapple Presscake, Belovit Presscake, and Presscake)

This is a high-moisture by-product of the pineapple juice press. Pineapple juice presscake is very acid, so cows should be accustomed to it gradually. Because of this high acidity, it does not undergo normal ensiling fermentation. But it will keep well for about 2 weeks if stacked. Hawaiian dairies feed up to 30 lb of pineapple juice presscake per cow per day. It may be used as a replacement or substitute for pineapple bran or pineapple greenchop; 4 lb of pineapple presscake can replace 1 lb of pineapple bran, or 2 lb of presscake can replace 3 lb of pineapple greenchop.

Pineapple Stump Meal (or Pineapple Stem Meal)

This is a by-product of the production of the proteolytic enzyme, *bromelain,* which is derived from pineapple stumps or stems and used as a meat tenderizer. It is generally available as a semimoist product, in relatively short particle lengths. In Hawaii, it is successfully fed to dairy and beef animals at levels of 5 to 35 lb per cow per day. On a dry-matter basis, it is equal in energy to pineapple bran, but not quite as palatable. Pineapple stump meal may be used as as substitute for, or in addition to, pineapple bran; 2 lb of semimoist stump meal may be substituted for 1 lb of pineapple bran. Because of its intermediate moisture level, it spoils readily if stored for longer than about 1 week.

Pineapple Tops and Crowns

These plant parts can be chopped or ground and fed to ruminants in much the same way as good-quality grass. However, they should not be fed to nonruminants.

Sorghum Residues

Ruminants will make good use of sorghum stover as a winter feed. It can be grazed or harvested and stored either as dry feed or silage. The sorghum plant stays green late in the fall; hence, good sorghum stover silage can be made without additional water. In comparison with corn residue, sorghum residue (1) is less palatable (if given a choice, cows will select corn refuse in preference to sorghum refuse); (2) comprises a lower percentage of the total plant dry matter than corn (40% of the total plant dry matter of sorghum is residue compared with 40 to 50% for corn); and (3) is lower yielding.

After harvesting, sorghum will send up new shoots if moisture permits. The prussic acid (hydrocyanic acid) content of these shoots may be harmful to grazing animals; hence, animal caretakers should be aware of this possible poisoning. These shoots can be grazed safely 4 to 6 days after a killing frost.

(Also see Table 12-1, under "Crop Residues.")

Soybean Residues

The stems and pods of soybean refuse available for feeding yield approximately ¼ ton per acre, with a ratio of stems to pods of about 2:1. The digestibility of stems is low—25 to 35%—due to their high-lignin content (18 to 20% for the stalk portion). The digestibility of pods is much higher, ranging from 58 to 63%. Soybean refuse should be used only as a filler or a high-quality feed stretcher. Cows consuming soybean refuse as a primary energy source have been shown to lose excessive weight.

(Also see Table 12-1, under "Crop Residues.")

Straws and Tailings

These are the chaff and grain behind the combine. In the days of binders and threshing machines, straw stacks, used extensively for winter cattle feed, were commonplace. With the advent of combines, much of the straw was left in the field. During periods of scarce and high-priced hay, straw is frequently used as either a *hay-stretcher* or *hay-replacer.*

Of the common cereal straws, oat straw is the most palatable and nutritious. Barley straw ranks second, and wheat straw is third.

Straw is a bulky feed, and it must be properly supplemented. It is low in protein (on an as-fed basis, wheat straw averages about 3.2% crude protein), low in phosphorus, and low in vitamin A. Dry pregnant cows can be wintered on straw plus a daily allowance of either 5 to 6 lb of good-quality alfalfa hay or 1 to 2 lb of a 30 to 40% protein supplement, along with free access to a high-phosphorus mineral (one containing at least 12% phosphorus). If no legume hay is fed, vitamin A should be fed or injected. When oilseed meals are scarce and high in price, a slow-release nonprotein nitrogen may be used.

The tailings are generally used by farmers and ranchers, either as dry feed or mixed with silage.

In addition to the cereal straws, other low-cost roughages available in certain sections of the United States are lentil straw, field pea straw, bean straw, clover straw, and bluegrass straw.

Sugarcane Strippings (Also Called Sugarcane Trash, or Strip Cane)

This is a by-product of cane sugar. When cane is harvested and sent to the mill with some or all of the leaves attached, the leaves and leaf sheaths are stripped from the stalks. The strippings are suitable as a cattle feedstuff. It is a medium-grade roughage, but it is successfully fed to lactating cows as a fiber source at levels of 5 to 20 lb per cow per day. Strippings do not ensile well, due to insufficient carbohydrate for fermentation and lactic acid production and to poor packing because of bulkiness and dryness. An improved silage can be made by mixing molasses or pineapple greenchop with the strippings. Recommended feeding levels are in the range of 5 to 15 lb per head per day. The protein in strippings is poorly digested and should be disregarded when balancing rations.

Sweet Corn Stover

This is the sweet corn plant without the ears. It may be ensiled, or allowed to cure on the stalk and harvested and fed dry or grazed in the field. If it is ensiled, it should be harvested soon after the ears are removed. Nutritionally, it is comparable to field corn stover, or about half as valuable as well-eared field corn.

WASTE BY-PRODUCT FEEDS

Waste by-products represent a feed resource which, presently, is not being used to its nutritional and economical potential. Four major types of wastes are involved: (1) animal wastes (manure and litter), (2) municipal wastes (sewage sludge and garbage), (3) bakery wastes, and (4) coffee wastes.

Animal wastes refer to a mixture of animal excrements (consisting of undigested feeds plus certain body wastes), with or without bedding materials.

Municipal wastes refer to sewage sludge and garbage, both of which have been receiving increasing attention as feeds in recent years.

Bakery wastes refer to bread and other bakery products, along with dough, which make excellent energy feeds for cattle and swine.

Coffee wastes consist primarily of the pulp, a jellylike substance, and coffee grounds.

Some common waste by-product feeds are listed in Table 12-1 of this chapter and in Section V, Composition of Feeds. Additional waste by-product feeds are discussed in the sections that follow:

Animal Wastes (Manure and Litter)

Historically, animal wastes have been used as fertilizer and feed, although their potential as a feed received little attention until recently.

The utilization of animal wastes as feedstuffs is not new, however. Coprophagy (eating of feces) is normal in rabbits and in many wild and domestic species. In 1914, Osborne and Mendel demonstrated that feeding 1% feces from normally fed rats to rats on a deficient purified diet prevented death.[4] In 1925, Evvard and Henness reported that on the average one pig following 1.9 steers recovered the equivalent of 212 lb of corn during a 120-day feeding period.[5] In 1943, Bohstedt found that cow manure had nutritional value for pigs, in addition to the grain that it contained.[6] In 1956, Fuller reported that hydrolyzed poultry litter was as effective as fish meal in achieving growth from commercial type broiler diets.[7] Most animal wastes are fed to ruminants, especially to beef cattle.

[4]Osborne, T. B. and L. B. Mendel, "The Contribution of bacteria to the feces after feeding diets free from indigestible components," *Journal of Biological Chemistry,* 1914, 18:177.

[5]Evvard, J. M. and K. K. Henness, "An experiment to study hogs following cattle," *Proceedings American Society of Animal Production,* 1925, p. 55.

[6]Bohstedt, G., R. H. Grummer, and O. B. Ross, "Cattle manure and other carriers of B-complex vitamins in rations for pigs," *Journal of Animal Science,* 1943, 2:373.

[7]Fuller, H. L., "The value of poultry by-products as sources of protein and unidentified growth factors in broiler rations," *Poultry Science,* 1956, 35:1143.

PHYSICAL CHARACTERISTICS OF ANIMAL WASTES

An understanding of the physical characteristics of animal wastes is requisite to their collection and best use.

Animal wastes are of two types, solid and liquid. In mammals, urine and feces are voided separately; but in poultry the end products coming from the urinary and gastrointestinal tracts are mixed together and voided via the cloaca. In poultry droppings, the fecal material prevents much of the nitrogen from leaching out, which is the primary reason that poultry manure has a higher nitrogen content than manure from four-footed animals.

It follows that the solid and liquid wastes of mammals are generally collected separately, unless pits are involved, whereas poultry wastes are collected as a semisolid mixture of feces and urine. Because of the different methods of urinary excretion in mammals and poultry, a large part of the urinary nutrients tend to be lost from mammals but retained from poultry. An additional factor preventing loss of nitrogen in poultry waste is the form in which nitrogen is excreted in the urine; it is insoluble uric acid. Wastes containing bedding or floor litter tend to be drier and may contain absorbed urinary waste.

Cattle waste may contain bedding material, such as straw or wood shavings, and may contain appreciable levels of soil if the animals are in a dirt corral. Sheep and horse feces are relatively dry, and usually contain some bedding material. Swine wastes from confinement operations are generally collected in pits, in which the feces and urine are mixed. Poultry wastes are of two types, with or without litter. The waste from caged birds is without litter; generally, it contains shed feathers, spilled feed, and broken eggs in addition to excretory products. The waste from birds kept on a floor system contains litter, such as peanut or rice hulls, wood shavings, sawdust, or straw, along with some feathers and spilled feed. Usually litter comes from broiler and turkey houses.

AMOUNT, COMPOSITION, AND VALUE OF MANURE PRODUCED

Figures relative to the amount of manure produced vary widely, primarily because of differences in (1) year-to-year numbers of animals, (2) estimated amount of bedding, (3) estimated recoverable manure and feces, and (4) estimated dry matter content of the wastes. Cattle (beef and dairy) account for almost 80% of the total manure production, but only about 40% is recoverable because of being voided on pastures and ranges, whereas virtually all of the poultry manure is collectable. Also, manure tonnage estimates may be on the basis of with or without bedding, and on the basis of moisture-free, 15% dry matter, or wet.

The authors' bases and estimated tonnages of animal wastes produced daily and annually follow.

Daily Manure Production and Storage

Table 12-7 shows the daily volume of manure produced by different kinds of animals. These figures can be used to determine storage needs. A rule of thumb for under the floor pits is to figure that the pit will fill at the rate of 1 ft per month. Of course, the amount of the excreta is influenced by the composition of the ration; for example, high-silage rations produce more manure than high-grain rations.

TABLE 12-7
APPROXIMATE DAILY MANURE PRODUCTION, WITHOUT BEDDING[1]

Animal	Cu Ft/Day Solids and Liquids[2]	Gallons/Day[3]
1,000-lb cow	1½	11
1,000-lb steer	1	7½
10 head of sheep	½	4
10 head of hogs:		
50 lb	⅔	5
100 lb	1⅓	10
150 lb	2¼	17
200 lb	2¾	20½
250 lb	3½	26
1,000-lb horse	¾	5½

[1]Adapted by the authors from *Michigan State University Circ. 231.*

[2]There are about 34 cu ft in a ton of manure.

[3]One cubic foot = 7½ gal.

Manure may be stored in a separate tank, in a nearby earthen dam lagoon, or it may be left to accumulate in a pit under slotted floors.

Storage capacity can be computed as follows:

Storage capacity = number of animals × daily manure production × desired storage time in days + extra water.

Example: *80 cows (1,000 lb each) × 1½ × 120 days = 14,400 cu ft: 7½ gal × 14,400 cu ft = 108,000-gal capacity.*

Water should be added to a pit (1) initially, and (2) subsequently to replace evaporation losses from wastes. Thus, if the manure is to be pumped, ⅕ to ⅖ of the storage volume may be needed for the extra water. For irrigation, there should be about 95% water and 5% manure. Water should be kept to a minimum if the manure is to be field spread with a tank wagon.

Generally 3 to 6 months' storage capacity is desirable.

MANURE GASES

When stored inside a building, gases from liquid wastes create a hazard and undesirable odors. Most (95% or more) of the gas produced by manure decomposition is methane, ammonia, hydrogen sulfide, and carbon dioxide. Several have undesirable odors or may cause animal and human toxicity, and some promote corrosion of equipment. Table 12-8 gives some properties of the more abundant gases.

TABLE 12-8
PROPERTIES OF THE MORE ABUNDANT MANURE GASES[1]

Gas	Weight Air = 1	Physiologic Affect	Other Properties
CH_4 (Methane)	½	Anesthetic	Odorless, explosive.
NH_3 (Ammonia)	⅔	Irritant	Strong odor, corrosive.
H_2S (Hydrogen sulfide)	1+	Poison	Rotten-egg odor, corrosive.
CO_2 (Carbon dioxide)	1⅓	Asphyxiant	Odorless, mildly corrosive.

[1]*Beef Housing and Equipment Handbook,* Midwest Plan Service, Iowa State University, Ames, 1968, p. 10.

Animals and people can be killed (asphyxiated) because methane and carbon dioxide displace oxygen.

Most gas problems occur when manure is agitated or when ventilation fans fail.

No one should enter a storage tank, unless (1) the space over the wastes is first ventilated with a fan, (2) another person is standing by to give assistance if needed, or (3) wearing self-contained breathing equipment—the kind used for fire fighting or scuba diving.

It is important that maximum building ventilation be provided when agitating or pumping wastes from a pit. Also, an alarm system (loud bell) to warn of power failures in tightly enclosed buildings is important, because there can be a rapid buildup of gases when forced ventilation ceases.

Yearly Manure Production, Composition, and Value

The quantity, composition, and value of manure produced vary according to species, weight, kind and amount of feed, and kind and amount of bedding. The computations herein are on a fresh manure basis (exclusive of bedding). Fig. 12-10 presents manure production data per year per 1,000 lb animal liveweight. Table 12-9 (p. 478) gives the quantity, composition, and value of manure excreted per year per 1,000 lb liveweight by different species of farm animals. Table 12-10 (p.478) gives the tonnage and value of manure excreted by U.S. livestock by species per year, along with the total tonnage and total value per year.

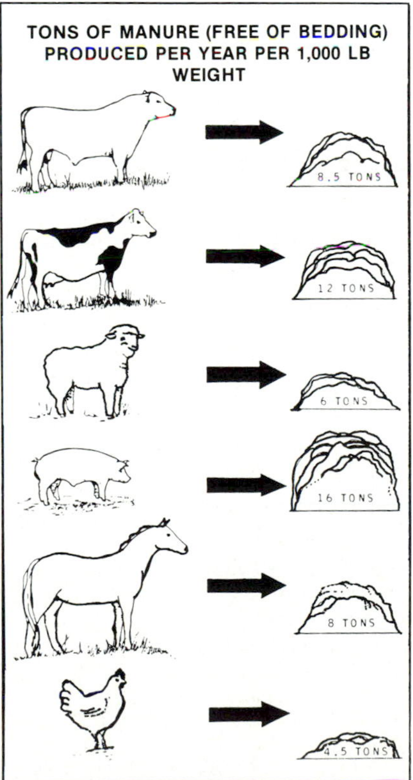

Fig. 12-10. On the average, each class of confined animals produces per year per 1,000 lb weight the tonnages shown. (Drawing by R. F. Johnson)

TABLE 12-9
QUANTITY, COMPOSITION, AND VALUE OF FRESH MANURE
(FREE OF BEDDING) EXCRETED PER YEAR PER 1,000 LB LIVEWEIGHT BY VARIOUS KINDS OF FARM ANIMALS

		Composition and Value of Manure on a Tonnage Basis[2]						
(1) Animal	(2) Tons Excreted/Year/1,000 Lb Liveweight[1]	(3) Excrement	(4) Lb/Ton[3]	(5) Water	(6) N	(7) P_2O_5[4]	(8) K_2O[4]	(9) Value/Ton[5]
			(lb)	(%)	(lb)	(lb)	(lb)	($)
Cow (beef or dairy)	12	Liquid	600	79	11.2	4.6	12.0	6.07
		Solid	1,400					
		Total	2,000					
Steer (finishing cattle)	8.5	Liquid	600	80	14.0	9.2	10.8	8.12
		Solid	1,400					
		Total	2,000					
Sheep	6	Liquid	660	65	28.0	9.6	24.0	13.88
		Solid	1,340					
		Total	2,000					
Swine	16	Liquid	800	75	10.0	6.4	9.1	5.81
		Solid	1,200					
		Total	2,000					
Horse	8	Liquid	400	60	13.8	4.6	14.4	7.11
		Solid	1,600					
		Total	2,000					
Poultry	4.5	Total	2,000	54	31.2	18.4	8.4	15.58

[1] *Manure Is Worth Money—It Deserves Good Care,* University of Illinois Circ. 595, 1953, p.4.

[2] Columns 5, 6, 7, and 8 from *Farm Manures,* University of Kentucky Circ. 593, 1964, p.5, Table 2.

[3] From *Reference Material for 1951 Saddle and Sirloin Essay Contest,* compiled by M. E. Ensminger, p. 43; data from *Fertilizers and Crop Production* by Van Slyke, published by Orange Judd Publishing Co.

[4] P_2O_5 can be converted to phosphorus (P) by dividing the figure given above by 2.29, and K_2O can be converted to potassium (K) by dividing by 1.2.

[5] Calculated on the assumption that nitrogen (N) retails at 29¢, P_2O_5 at 30¢, and K_2O at 12¢ per pound in commercial fertilizers.

TABLE 12-10
TONNAGE AND VALUE OF MANURE (EXCLUSIVE OF BEDDING) EXCRETED BY U.S. LIVESTOCK[1]

Class of Livestock	Number of Animals on Farms[2]	Average Liveweight	Tons Manure Excreted/Year/1,000 Lb Liveweight[3]	Total Manure Production	Total Value of Manure[4]
		(lb)	(tons)	(tons)	($)
Cattle (beef and dairy; including steers)	102,468,000	900	11	1,014,433,200	7,202,475,720
Sheep	10,328,000	100	6	6,196,800	86,011,584
Swine	50,960,000	200	16	163,072,000	947,448,320
Chicken, layers	280,140,000	4.5	4.5	5,672,835	88,382,769
Broilers	4,646,312,000	3.5	4.5	73,179,414	1,140,135,270
Turkeys	207,216,000	22	4.5	20,514,384	319,614,102
Horses	8,519,000	1,000	8	68,152,000	484,560,720
				1,351,220,633	10,268,628,485

[1] In these computations, no provision was made for animals that died or were slaughtered during the year. Rather, it was assumed that their places were taken by younger animals, and that the population of each species was stable throughout the year.

[2] From USDA, *Agricultural Statistics 1987,* except horse numbers from American Horse Council, Inc., Wash., D.C. The cattle and sheep figures are for Jan. 1, 1987; the swine and layer figures are for Dec. 1, 1986; the broiler and turkey figures are number raised during the year; the horse numbers are for 1986. All figures are assumed averages throughout the year.

[3] *Manure Is Worth Money—It Deserves Good Care,* University of Illinois Circ. 595, 1953, p. 4.

[4] Computed on the basis of the value per ton given in the right-hand column of Table 12-9.

The data in Fig. 12-10 and Tables 12-9 and 12-10 are based on animals confined the year-around. Actually, the manure recovered and available to spread where desired is considerably less than indicated because (1) animals are kept on pasture and along roads and lanes much of the year, where the manure is dropped, and (2) losses in weight often run as high as 60% when manure is exposed to the weather for a considerable time.

As shown in Fig. 12-11, about 75% of the nitrogen, 80% of the phosphorus, and 85% of the potassium contained in animal feeds are returned as manure. In addition, about 40% of the organic matter in feeds is excreted as manure. As a rule of thumb, it is commonly estimated that 80% of the total nutrients in feeds are excreted by animals as manure.

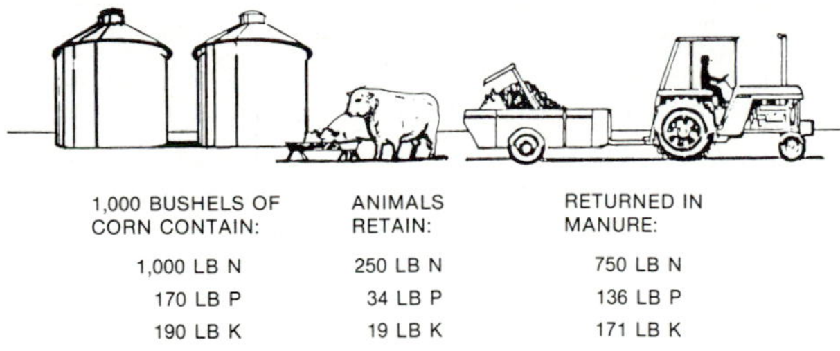

Fig. 12-11. Animals retain about 20% of the nutrients in feed; the rest is excreted in manure.

Naturally, it follows that the manure from well-fed animals is higher in nutrients and worth more than that from poorly fed ones. For example, steer manure produced by finishing cattle liberally fed on nutritious concentrates is more valuable than that produced from cattle wintered on hay.

The urine makes up 20% of the total weight of the excrement of horses, and 40% of that of hogs; these figures represent the 2 extremes in farm animals. Yet the urine, or liquid manure, contains nearly 50% of the nitrogen, 6% of the phosphorus, and 60% of the potassium of average manure; roughly ½ of the total plant food of manure (see Fig. 12-12). Also, it is noteworthy that the nutrients in liquid manure are more readily available to plants than the nutrients in the solid excrement. These are the reasons why it is important to conserve the urine.

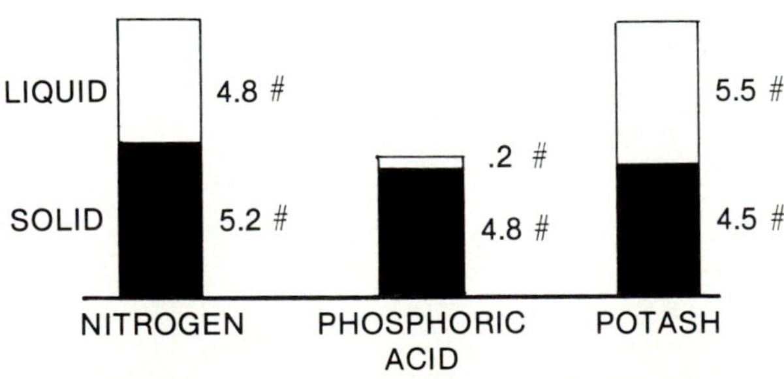

Fig. 12-12. Distribution of plant nutrients between liquid and solid portions of a ton of average farm manure. As noted, the urine contains about half the fertility value of manure.

Currently, we are producing manure (exclusive of bedding) at the rate of 1.35 billion tons annually (see Table 12-10). That is sufficient manure to add nearly ¾ ton each year to every acre of the total land area (1.9 billion acres) of continental United States.

MANURE USES

Historically, manure has been used as a fertilizer. But high feed prices and shortages of fossil fuels have made for new uses. Although it is expected that manure will continue to be used primarily as a fertilizer for many years to come, increasingly it will be recycled and used as a feed and converted into energy. Other manure-based products will continue to evolve, but it is expected that they will be of minor importance.

Manure as a Fertilizer

The actual monetary value of manure can and should be based on (1) equivalent cost of a like amount of commercial fertilizer, and (2) increased crop yields. Table 12-9 gives the equivalent cost of a like amount of commercial fertilizer. Numerous experiments and practical observations have shown the measurable monetary value of manure in increased crop yields.

There was a time when farmers fed livestock to produce manure, to grow more crops, to feed more livestock, to produce more manure. But things changed! The use of chemical fertilizer expanded manyfold; labor costs rose to the point where it was costly to conserve and spread manure on the land; more animals were raised in confinement; and a predominantly urban population didn't appreciate what they referred to as "foul-smelling, fly-breeding stuff." As a result, what to do with manure became a major problem on many livestock establishments.

Based on equivalent fertilizer prices (see Table 12-9, right-hand column), and livestock numbers (Table 12-10), the yearly manure crop is worth $10.3 billion. That's a potential income of $4,905 for each of the nation's 2.1 million farms.

The value of manure as a fertilizer varies according to (1) the class of animals, (2) the kind of feed consumed and the kind of bedding used, (3) the method of handling, (4) the rate and method of application, and (5) the kind of soil and crops on which it is used.

Livestock producers sometimes fail to recognize the value of this barnyard crop because (1) it is produced whether or not it is wanted, and (2) it is available for only the cost of handling.

Manure as a Nonfeed Energy Source

Manure may also serve as a source of nonfeed energy, which, or course, is not new. The pioneers burned dried bison dung, which they dubbed *buffalo chips,* to heat their sod shanties. In this century, methane from manure has been used for power in European farm hamlets when natural gas was hard to get. While the costs of constructing plants to produce energy from manure on a large-scale basis may be high, some energy specialists feel that a prolonged fuel shortage will make such plants economical. India now has about 10,000 anaerobic digestion plants in operation.

Methane, of course, is usable like natural gas. There is nothing new or mysterious about this process. Sanitary engineers have long known that a family of bacteria produces methane when they ferment organic material under strictly anaerobic conditions. (Grandad called it *swamp gas;* his city cousin called it *sewer gas.*) However, it should be added that, due to capital and technical resources needed, for some time to come, the production of methane by anaerobic digestion will likely be limited. If all animal manure were converted to energy, it has been estimated that it could produce energy equal to 10% of the petroleum requirements or 12½% of our natural gas requirements.

Manure as a Feed

Recycling manure as a livestock feed is the most promising of the nonfertilizer uses. Various processing methods are being employed; and some manure is being fed without processing. More and more feedlot manure will be either (1) incorporated in a grower ration, or (2) fed to breeding herds during periods when pasture supplementation is beneficial, with the residues distributed over grazing areas where they would have fertilizing value.

CHEMICAL COMPOSITION OF ANIMAL WASTES

Animal wastes contain several nutrients that are capable of being utilized when the material is recycled by feeding. Nitrogen, which is present in both protein and nonprotein forms, is a major constituent. Available energy is rather low. Fiber and ash are generally high. The high ash indicates that animal wastes are high in minerals; they are especially rich in phosphorus. Additionally, they contain certain vitamins synthesized in the digestive tract. The wastes possessing the highest nutritive value are broiler litter and layer waste.

One characteristic of all animal wastes is variability in composition due to diet regime, kind and amount of bedding, length of time before collecting, and processing method. The main difference in composition between raw and processed wastes is in moisture content; many of the processed wastes are low in moisture.

The high fiber and considerable nonprotein nitrogen of animal wastes indicate that they are best suited for feeding to ruminants, since they possess a digestive tract capable of efficiently utilizing high fiber and nonprotein nitrogen. Also, because of their low energy content, they are best adapted for use in maintenance and gestating rations, rather than in lactating and growing rations.

NUTRIENT UTILIZATION

Animal wastes processed by ensiling, dehydrating, and other methods can be fed successfully to a wide range of animals. But, for best results the rations in which they become a part should be well balanced following their incorporation. Several workers have shown that the inclusion of too high levels of waste in a ration results in an excessive level of fiber and/or minerals, followed by lowered animal performance. Because of this limitation, not more than 10 to 20% waste should be included in high-energy rations, such as in cattle finishing rations. However, much higher levels (up to 80%) can be incorporated in rations of gestating beef cows.

PERFORMANCE OF ANIMALS FED ANIMAL WASTES

When rations containing animal wastes are properly balanced and fed to the right species and class of animals, the performance of animals in growth, or in meat, milk, or egg production, is comparable to animals fed traditional feed ingredients of comparable chemical content.

Cattle Waste

Much of the early research on feeding cattle manure was conducted by Anthony at the Alabama Agricultural Experiment Station; in 1962, he established the feasibility of feeding steer manure to cattle. In a 1970 report, Anthony reappraised the feeding of manure-containing rations to finishing cattle and presented the results of a new experiment designed to determine if cooking the manure improved its feeding value.[8]

Anthony compared a typical corn-based finishing ration to rations containing untreated manure and cooked manure. The results of this experiment are summarized in Table 12–11. Steers fed the basal ration gained faster than the basal plus manure rations, but this difference was not statistically significant.

When feed per unit gain was quantitated, less concentrate was used in the manure treatments. However, this was not statistically significant. Feed digestibility was significantly higher in the rations with manure than the basal ration. It was concluded that (1) manure could effectively be used in steer rations, and (2) cooking did not improve the feeding value of the manure.

TABLE 12–11
RATE OF GAIN, FEED EFFICIENCY, AND FEED DIGESTIBILITY[1]

Treatment	Average Daily Gain		Feed Dry Matter Per Unit of Gain		Feed Digestibility (nylon bag)
			Concentrate	Manure	
	(lb)	*(kg)*			(%)
Basal	2.56	*1.16*	9.20	—	40.36
Untreated manure	2.20	*1.00*	8.52	1.66	46.61
Autoclaved manure	2.18	*.99*	8.75	1.70	45.56

[1]Adapted by the authors from "Feeding Value of Cattle Manure for Cattle," by W. B. Anthony, *Journal of Animal Science,* Vol. 30,1970, p. 274.

Subsequently, several investigators have reported satisfactory performance of cattle when fed cattle waste that was ensiled, treated with sodium hydroxide, or fed fresh. But, in limited experiments, incorporating cattle waste in rations for swine and laying hens has lowered production and required more feed per unit of production, perhaps due to the cattle waste being too high in fiber for monogastric animals.

[8]Anthony, W. B., "Feeding Value of Cattle Manure for Cattle," *Journal of Animal Science,* Feb. 1970, Vol. 30, No. 2, p. 274.

Swine Waste

The practice of fertilizing ponds with manure to promote the growth of fish has been followed extensively in China and Southeast Asia, where swine operations are often coordinated with fish production. The manure produced by the swine is flushed into nearby ponds where it promotes growth of vegetation and microbic animals which are subsequently eaten by the fish. Often the fish are fed solely on pig manure and grass. Hence, the Chinese have provided their own pollution control of pig manure and, at the same time, recycled and used the feed twice—first through hogs, and second through fish.

Limited U.S. studies with swine waste incorporated in cattle rations have not been encouraging; both the rate of gain and feed efficiency have been affected adversely.

Poultry Waste

Nearly 100 million tons of poultry wastes (from layers, broilers, and turkeys) are produced annually (see Table 12–10). Because poultry production is highly intensive, with many birds in a small area, waste disposal is a major problem. Most cage-layer operations produce manure free of litter as the primary form of waste. Broiler operations, generally, produce litter.

On a moisture-free basis, cage-layer manure generally contains 25 to 35% crude protein and minimal fiber, while broiler litter contains somewhat less protein—about 18 to 30% and substantially more fiber due to the presence of absorbent materials.

Poultry litter is the most collectable and the most nutritious of all animal wastes. It follows that many experiments have been conducted with it, involving feeding trials with different species. The results of numerous experiments are summarized in Table 12–12. The mean values for waste-fed animals reported therein were obtained by averaging all levels of feeding poultry wastes in the respective categories, though some of the levels were excessive. As shown in Table 12–12, the performance of animals fed wastes was generally slightly lower than that of the controls that were fed traditional feed ingredients. But, on a dry-matter basis, animal wastes generally make for least cost rations and highest net returns.

TABLE 12–12
PERFORMANCE OF ANIMALS FED RATIONS CONTAINING POULTRY WASTES[1]

Species of Experimental Animal Used	Kind of Poultry Waste Studied	Performance of Experimental Animals		
		Criteria	**Control Group**	**Waste-Fed Group**
Cattle	Dehydrated layer waste	Daily gain, lb Daily feed dry-matter intake, lb Feed/gain ratio	2.35 (*1.07 kg*) 15.82 (*7.19 kg*) 7.81	2.31 (*1.05 kg*) 15.44 (*7.02 kg*) 7.72
Lactating cows	Dehydrated layer waste	Milk yield, lb/day Milk fat, % Milk total solids, %	41.8 (*19.0 kg*) 3.51 12.04	38.94 (*17.7 kg*) 3.63 12.01
Sheep	Dehydrated layer waste	Daily gain, lb Feed/gain ratio	0.42 (*0.19 kg*) 5.52	0.40 (*0.18 kg*) 6.66
Swine	Dehydrated layer waste	Daily gain, lb Feed/gain ratio	1.32 (*0.60 kg*) 4.12	1.14 (*0.52 kg*) 4.82
Growing chicks	Dehydrated layer waste	Daily gain, grams Feed/gain ratio	16.1 2.36	15.7 2.60
Laying hens	Dehydrated layer waste	Egg production, % lay Feed/dozen eggs, lb	71.9 4.18 (*1.90 kg*)	72.8 4.18 (*1.90 kg*)
Cattle	Poultry litter[2]	Daily gain, lb Feed/gain ratio	2.2 (*1.0 kg*) 10.18	1.91 (*0.87 kg*) 11.58

[1]Adapted by the authors from *Unidentified Resources as Animal Feedstuffs*, NRC, National Academy Press, Wash., D.C., 1983, pp. 132–144, Tables 35–41.

[2]Also, dried poultry litter has been fed successfully to dry and lactating dairy cows, to growing and breeding sheep, to growing swine, and to broilers.

Also, dried poultry litter has been fed successfully to dry and lactating dairy cows, to growing and breeding sheep, to growing swine, and to broilers.

PROCESSING ANIMAL WASTES FOR FEED

Several methods have been used to process animal wastes for feed; among them, dehydration, ensiling, pelleting, preparation for liquid feeding, oxidation-ditch aerobic processing, commercial (patented) systems, and the use of wastes as substrates for single-cell protein production.

(Also see Chapter 14, section on "Animal Waste [Manure] Processing.")

Dehydration

When voided, layer waste contains about 75% water. Reduction of the moisture content from 75% to 15% in dehydrators (698 to 1,292°F) requires removal of 1,284 lb of water per ton of dry solids, at an energy cost of $25 to $50/ton of dehydrated material.

The main **advantages** to artificial dehydration are: (1) reducing pathogens to low levels or eliminating them entirely, (2) lessening or removing odors, and (3) facilitating storing and handling.

The chief **disadvantages** of artificial drying are: (1) the high energy cost, and (2) the considerable loss of nitrogen and certain other components due to heating.

It may be concluded that artificial dehydration of animal waste results in excellent products, but the process may not be economically feasible due to high energy costs.

Ensiling

Ensiling of animal waste is a controlled anaerobic fermentation process during which the carbohydrates in the mixture are converted to lactic and other acids. Once sufficient acids are produced, bacterial action ceases and the ensilage is stable. Heat is generated in the process, with an internal temperature of about 77°F achieved. Processing animal wastes by ensiling has the **advantages** of (1) being economical, (2) diminishing the hazards from certain potentially pathogenic organisms, (3) rendering the waste mixture more palatable, and (4) producing a product with a pleasant aroma.

Because of the considerable expense and energy required for drying animal wastes, the trend is toward ensiling. Except for the cost of the silo, ensiling of wastes involves few expenses. Poultry wastes can be ensiled in bunker-type silos as well as oxygen-limiting tower silos.

South Carolina Experiment Station researchers conducted extensive studies on ensiling poultry wastes. They found that manure mixed with forage or litter takes about 6 weeks to ensile adequately. In an experiment designed to test the proper moisture level for ensiling a manure-forage combination, they found that hay and manure ensiled at 44% moisture produced the most desirable combination from a pH standpoint (see Table 12-13). Based on their studies,

TABLE 12-13
PH OF MANURE-FORAGE COMBINATIONS AT VARYING MOISTURE LEVELS[1 2]

Moisture Level	Forage		
	Hay	Peanut Hulls	Straw
(%)			
33	5.91	6.09	6.31
36	5.84	5.79	6.16
39	5.70	5.68	5.92
44	5.52	5.80	5.98
49	5.72	6.05	6.44
54	5.77	6.25	6.77

[1]Adapted by the authors from "Feeding Ensiled Poultry wastes to Cattle," by D. L. Cross, *Proceedings, 2nd Annual North Carolina Poultry Nutrition Conference,* 1975, p. 24.

[2]Ensiled in small-scale silos for 60 days.

the South Carolina Agricultural Experiment Station workers recommend the following practices for ensiling poultry litter in an upright silo:

1. The litter should be ensiled at about 37% moisture. Although maximum fermentation takes place at higher moisture levels, it is difficult to blow wet litter into a tall silo because it clogs the blower pipe. At 37% moisture, there is adequate moisture to promote good fermentation; and blower difficulties are minimized. Bunker or trench silos do not pose any moisture problem and can, therefore, be used to ensile litter of higher moisture content.

2. In order to remove metal objects which commonly get into the litter, a magnet should be included in the ensiling and the feeding systems.

3. The easiest place to add enough water to obtain the desired moisture is in the poultry house. A portable moisture tester can be used to check the moisture content. However, a preliminary check on the moisture content of litter may be obtained by squeezing it; litter first begins to stick together at 35% moisture.

4. A front-end loader can be used to clean out the poultry trucks. This clean-out process facilitates mixing of the litter, thereby evenly distributing the moisture.

Other Processes

Other processes that have been used follow:

- **Pelleting**—Pelleting animal wastes prevents ingredient-sorting by animals. However, the waste must be dried before pelleting; hence, pelleting is costly.
- **Commercial patented systems**—Several commercial (patented) systems have been developed for processing animal wastes for feeding, but details relative to these are proprietary to the companies involved.
- **Substrates for protein production**—The use of wastes as substrates for the production of protein supplements for livestock feeds is feasible, with systems using algae, yeasts, bacteria, and fungi all showing promise.

AAFCO FEED TERMS AND DEFINITIONS

The Association of American Feed Control Officials (AAFCO) has developed and adapted the following official feed terms and feed ingredient definitions for animal wastes:

- **Dried poultry waste**—This is processed poultry waste, composed primarily of feces from commercial poultry which has been processed to remove part or all of the equivalent crude protein, NPN as urea, and/or uric acid; which has been thermally dehydrated to a moisture content not to exceed 15%. It shall contain not less than 11% crude protein, and not more than 15% crude fiber, 30% ash, and 1% feathers.
- **Dried poultry litter**—This is processed poultry waste, composed of a combination of feces from commercial poultry together with litter that was present in the floor production of poultry, which has been artificially dehydrated to a moisture content not to exceed 15%. It shall contain not less than 18% crude protein, and not more than 25% crude fiber, 20% ash, and 4% feathers.
- **Dried ruminant waste**—This is processed ruminant waste, composed primarily of processed ruminant excreta which has been artificially dehydrated to a moisture content not to exceed 15%. It shall contain not less than 12% crude protein, and not more than 40% crude fiber, including straw, wood shavings, or other material, and not more than 30% ash.
- **Dried swine waste**—This is processed swine waste, composed primarily of swine excreta which has been artificially dehydrated to a mositure content not to exceed 15%. It shall contain not less than 20% crude protein, not more than

35% crude fiber, including other material such as straw, wood shavings, or other acceptable bedding material, and not more than 20% ash.

• **Undried processed animal waste product**—This is processed animal waste, composed of excreta, with or without litter, from farm animals or poultry, which may or may not include other feed ingredients, and which contains in excess of 15% feed ingredients, and in excess of 15% moisture. It shall contain no more than 30% combined wood, wood shavings, litter, dirt, sand, rocks, and similar extraneous materials.

• **Processed animal waste derivative**—This is a product resulting from the chemical, physical, or microbiological alteration of an animal waste. Examples of processed animal waste derivatives are composts, yeasts, algae, or other organisms produced from farm animal wastes, or wastes treated with ammonia, formaldehyde, or other chemicals. The specific name of each derivative product must be descriptive, and efficacy and safety data must be submitted and approved before the produce is registered or offered for sale.

ANIMAL AND HUMAN HEALTH/ESTHETIC ASPECTS

There appears to be only minimal risk from feeding wastes that have been processed by ensiling or dehydration. However, spore-forming bacteria are not destroyed by either process.

Copper may present a mineral problem accompanying the recycling of animal wastes; and this occurs primarily in sheep. The problem of excess copper in sheep feeds is well known; and it is not connected solely with the feeding of animal waste.

Drug and chemical residues do not appear to present a major problem as a result of waste feeding. However, as a precautionary measure, it is recommended that waste feeding should be followed by a withdrawal period of at least 15 days when waste-fed animals are intended to provide meat, milk, or eggs for human consumption. Food quality is not affected by feeding waste.

Finally, there is the esthetic consideration. The vast majority of consumers are not knowledgeable about the physiological process of digestion and absorption—especially in the ruminant. To them, the subject of manure does not make for pleasant conversation. Worse yet, manure as a livestock feed is downright repugnant. Nevertheless, the esthetic angle does not appear to be unsurmountable. Many manure feeding trials have included meat quality evaluations. Livestock producers need to make a concerted effort to present facts and results; otherwise, consumer groups may seize the initiative and push for a ban on manure as a livestock feed.

REGULATORY ASPECTS

In December 1980, the U.S. Food and Drug Administration published a document revoking its earlier policy regarding the feeding of animal waste and leaving regulations governing the feeding of animal waste to the states. So, U.S. livestock producers using, or planning to use, animal wastes as feed ingredients should comply with the specific regulations of the state involved. In Canada, waste-feeding is governed by the Foods and Drugs Act.

Future Handling and Uses of Manure

Increasingly, what to do with manure and how to protect the environment will be a major problem for livestock producers. More and more laws and zoning will govern the handling of manure to prevent it from polluting water and air, and to minimize its potential to increase insects, odors, and dust. Responsible livestock producers need to be attuned to the worldwide awakening to the problem of pollution of the environment (air, water, and soil) and its effect on animal and human health, and science and technology need to evolve with new methods of handling manure, and new uses for it, so that animals and humans can continue to live happily together and enhance each other.

Municipal Wastes

As the world's population continues to expand, human waste will also increase—creating a major disposal problem. The U.S. Department of Energy estimated that, on a dry basis, 68 million tons of municipal solid waste were produced in 1975. The livestock industry offers one alternative to deal with this problem. The feeding value of two human waste products—sewage sludge and garbage—have been receiving increasing attention in recent years.

SEWAGE SLUDGE

Wherever there are high concentrations of people, problems arise concerning the safe disposal and treatment of excreted material. Most cities now have elaborate waste treatment plants that process sewage in such a manner that the end product poses no health hazard. But disposal of this end product can pose logistical problems. One solution is to recycle it through livestock.

Activated sludge, a gelatinized mass of microorganisms, is rich in nitrogen and vitamin B-12. However, when fed at high levels, it can depress feed intake and cause diarrhea; so, it is recommended that amounts incorporated in swine, poultry, and ruminant rations not exceed 5, 8, and 8% respectively.

One of the major problems with the feeding of sewage sludge is the danger of chemical contamination and subsequent residues in animals fed this by-product. Heavy metals, polychlorinated biphenyls (PCBs), pesticides, and other hazardous chemicals may be introduced into the sewage system, thereby creating a health risk to both livestock and people. As a result, the use of sewage sludge for animal feed is not considered viable at this time.

GARBAGE

From the remote day of domestication forward, swine have been considered scavengers—often fed table scraps and other wastes. Even today, in most of the developing countries, pigs, and, to a lesser extent other livestock, consume few products that are suitable for human food.

Municipal garbage has long been fed to hogs. But beginning about 1940, the practice declined because of a gradual lowering in the feeding value of garbage, along with other competition for it—notably, its manufacture into lawn, greenhouse, and garden fertilizer. By June, 1960, only 1.85% of the nation's hogs were being fed garbage.

Prior to 1940, the garbage feeder figured that a ton of city garbage would produce 60 to 100 lb of pork; whereas, at the present time, it is estimated that a similar quantity will not produce more than 30 lb of pork.[9] The change in feeding value may be largely attributed to improved refrigeration, the effective use of leftovers, and the change of the general eating habits of humans; for example, the increasing use of frozen and highly processed foods. Institutional, hotel, and restaurant garbage is superior to household garbage.

Garbage may be utilized either as a feed for a sow and pig enterprise or for finishing feeder pigs that are obtained from other sources. Usually, the venture seems most successful when a combination of grain and garbage feeding is practiced.

It is also observed that the most successful garbage feeders use concrete feeding floors, practice rigid sanitation, and take every precaution to prevent diseases and parasites. Unless considerable grain is fed to market hogs, especially after weights are over 100 lb, soft pork and paunchiness will result in garbage-fed hogs.

Swine are more likely to become infected with trichinosis, vesicular exanthema, and certain other diseases when fed raw garbage. For this reason, all states now have laws requiring that commercial garbage be cooked. Also, it is noteworthy that there is no danger of transmitting trichinosis from swine to humans provided pork and pork products are thoroughly cooked.

Because of high energy costs, very little garbage is artificially dehydrated. However, the Association of American Feed Control Officials list "Dehydrated Garbage," which they describe as follows:

> Dehydrated garbage is composed of artificially dried animal and vegetable waste collected sufficiently often that harmful decomposition has not set in, and from which have been separated crockery, glass, metal, string, and similar materials. It must be processed at a temperature sufficient to destroy all organisms capable of producing animal diseases. If part of the grease is removed, it must be designated as "Degreased Dehydrated Garbage."

Recent research has indicated that garbage can be fed successfully to ruminants. While the data are somewhat limited, it appears that garbage may be incorporated into maintenance or grower rations when availability and price permit. As with activated sludge, the level of pesticides, PCBs, and heavy metals must be carefully monitored to avoid possible residues.

(Also see Chapter 6, section on "Municipal Garbage.")

[9]Some Canadian authorities consider 4 lb of heavy garbage to be equivalent to 1 lb of concentrate (*Feeder's Guide and Formulae for Meal Mixtures*, 13th ed., published by the Quebec Provincial Feed Board for April 1959–61).

Bakery Waste

Bakery waste consisting of bread, cookies, cake, crackers, flours, and doughs can be dried and sold as livestock feed under the name *dried bakery product.* Since it is high in digestible fat and carbohydrate, it is often used to replace grain when economically feasible. However, like grains, dried bakery product is low in vitamin A, protein, and minerals. It contains rather large amounts of salt; and if the salt content exceeds 3.5%, the Association of American Feed Control Officials stipulate that the maximum amount of salt must be so labeled in the name of the product. Because of these high levels of salt, poultry should not be fed dried bakery product in excess of 15% of the ration. On the other hand, it can replace all of the grain in swine rations and up to 30% in cattle rations without adversely affecting palatability.

(Also see Chapter 6, section on "Bakery By-products.")

Coffee Wastes

To date, the natural levels of caffeine and tannin in coffee berries (cherries) have restricted the inclusion of the pulp waste in animal feeds to a maximum of 12% of the total ration. However, a new process developed in Costa Rica extracts the noxious elements from the pulp, allowing more coffee waste to be used as animal feed and making the caffeine available for use in soft drinks and drugs.

The coffee bean comprises only 44% of the weight of the average coffee berry. The rest is made up of pulp and a jellylike substance which processors dump in local rivers and elsewhere, creating pollution and health hazards. Annually, an estimated total of 16.9 million tons of coffee wastes (potential livestock feed) are discarded where coffee is grown—in Latin America, Africa, and Asia and Oceania.

After harvesting, the cherries are dried and the covering around the beans (the pulp) is removed by one of two methods: (1) the dry method, and (2) the wet method.

The composition of coffee products follows:

TABLE 12-14
COMPOSITION OF COFFEE PRODUCTS[1]

Product	Dry Matter	Nitrogen Free Extract	Crude Protein
	(%)	(%)	(%)
Pulp, dry method	89.1	39.8	20.4
Coffee grounds	19.7	4.2	13.3
Coffee oil meal	89.8	48.3	17.4

[1]From: *Tropical Feeds,* FAO, Rome, 1975, p. 299. The pulp and coffee ground analyses are from Trinidad; the coffee oil meal composition is from Colombia.

The leaves of the coffee bush are sometimes dried and fed to cattle. They are palatable and nutritionally comparable to medium-quality hay. The dried coffee leaves contain 52.5% nitrogen-free extract and 9.9% crude protein.

AQUATIC PLANTS[10]

The word *aquatic* means of, or pertaining to, water. *Aquatic plants are plants that grow in water.*

Aquatic plants occur throughout the world; in the sea, saltwater marshes, rivers, lakes, and waste-treatment ponds.

To date, these plants have been regarded more as problems than resources; among their adverse effects are: blocking canals and pumps in irrigation projects, interfering with hydroelectric production, wasting water by evapotranspiration, hindering boat traffic, increasing waterborne diseases, impeding drainage, causing flooding, interfering with fish culture and fishing, and harboring mosquito breeding.

Presently, aquatic plants are being recognized as potentially valuable resources for animal feed, human food, and other uses.

There exist almost unlimited opportunities for increased sea farming. The water area of the world is many times greater than the land space; and phenomenal yields of aquatic plants are obtained—as much as 60 tons per acre. Not only that, aquatic farming does not suffer from drought or loss of crop through pests and disease; and aquatic plants require no planting, weeding, or fertilizing. Some scientists predict that soon after the turn of the century the world's sea crops will have to be farmed to ensure the survival of our teeming population.

With one of the world's densest populations and a long coastline, the Japanese already have great expertise in the art of sea farming. Teams of girls, all expert swimmers and skin divers, play a part in this form of cropping, which is also known as aqua-culture. These underwater laborers are specially trained and equipped to carry out the cutting of aquatic plants from cultivated beds off the seashores.

Quantity and Kinds of Aquatic Plants

In 1976, the Food and Agriculture Organization of the United Nations reported that, worldwide, 2,402,000 metric tons (wet weight) of aquatic plants were harvested, most of which was seaweed. Thus, the total tonnage harvested is very small, considering the huge production.

Data are not available on the quantity of aquatic plants other than seaweed produced, but the potential is very great. For example, growth rates of 713 lb dry matter/acre/day have been recorded for water hyacinth. If such growth could be sustained for 6 months a year, that would make for a yield of 64 tons of dry matter per acre. By comparison, on the average, alfalfa yields 4 to 7 tons of hay per acre (see Chapter 8, Hay, Table 8–5), on a 20% moisture basis.

The aquatic plants with greatest potential for feed production are: algae, seaweed, and water hyacinth, with certain other plants like duckweed (of the family *Lemnaceae*) having possibilities.

[10]In the preparation of this section, the authors drew heavily from *Underutilized Resources as Feedstuffs,* NRC, National Academy Press, Washington, D.C., 1983.

Physical Characteristics of Aquatic Plants

The physical characteristics of aquatic plants make for difficulties in harvesting and processing. Algae are small (5 to 15 microns, or 0.000195 to 0.000585 in.) in diameter. Kelp varies in length from 20 ft to 200 ft. Water hyacinth plants are connected by stolons. Duckweed is a tiny free-floating vascular (tubular) plant. Harvested aquatic plants are slippery and tangled, with the result that it is difficult to handle them mechanically.

All aquatic plants contain much water—they are 85 to 95% water.

Algae (Three species are of interest for producing single-cell protein: *Scenedesmus quadricuada, Chlorella vulgaris, Spirulina maxima*)

Algae are primitive plants. They contain chlorophyll and are capable of photosynthesis—converting the sun's energy into food. As part of the food chain, many provide food for fish. Algae vary in size and shape. Some consist of individual microscopic cells while others form flat sheets of narrow filaments or stemlike structures that may be more than 100 ft long. Algae are able to live almost anywhere—salt water, fresh water, hot springs, polar snow, soil, trees, and rocks.

Characteristically, green algae are found in fresh water, though some may dominate tropical marine waters, and some are occasionally found on land. Green algae form an early link in the food chain as food for fish. Furthermore, some species of green algae grown on water or lagoons of waste material provide a potential source of animal and human food. The natives of Lake Chad, Africa, and of Mexico have harvested and eaten algae for years. Algae provide nutrients in relatively large quantities in a limited area, depending on the type of organism, temperature, and latitude. Grown as a crop, 4 to 16 tons of dried algae can be harvested from an acre each year.

These organisms, containing from 5 to 15% dry matter, can be fed to livestock, either directly after harvesting or in a dehydrated form. Nutrient composition (moisture-free) tends to be rather variable, with protein ranging from 8 to 75%, carbohydrate from 4 to 40%, lipids from 1 to 86%, and ash from 4 to 45%. The protein obtained from this type of single-cell protein is deficient in the sulfur amino acids. The biological value of algae protein has been estimated to range from 50 to 70.

Using the power of the sun and single-cell algae, Israeli scientists are now turning domestic waste water into valuable animal feed and, at the same time, producing purified water for irrigation—all at minimal cost. The algae are dried, mixed with other ingredients, and fed to poultry, fish, and cattle.

The algae production of single-cell protein points the way to a cheap, solar powered system of protein production. Additionally, algae provide a means of recycling animal wastes and renovating waste water.

- **Animal performance**—Most algae are dried and ground, and fed as a meal, primarily to swine and poultry. Up to

10% algae meal may be included in swine rations without lowering growth rate or feed conversions. Up to 6% algae meal may be included in poultry rations with good results. Because it is high in xanthophyll, it colors egg yolks most effectively. Algae meal is low in energy; hence, it reduces animal performance when included in the ration at high levels.

Algae have been fed experimentally to rats, chicks, swine, and sheep. Performance of chicks and rats fed algae-supplemented diets was lower than that of comparable animals fed soybean-protein and casein-supplemented diets. Supplementing the algae-containing diets with seven essential amino acids increased growth rate 49% and feed efficiency 47% in chicks, indicating amino acid deficiency in algae. The algae contain substantial quantities of xanthophyll, which can be utilized by the chick for shank and skin pigmentation. Chicks can use up to 10% algae meal in the diet.

Including up to 10% algae in swine rations did not alter rate of gain or feed efficiency. Supplementing lambs on a California summer range with pellets in which 50% of the nitrogen was supplied by algae and 50% by alfalfa resulted in higher weight gains than supplementing only with alfalfa pellets.

Seaweed or Kelp (Laminariales [order], *Fucales* [order])

Any plant that grows in the sea is a seaweed.

Seaweeds are algae, of which there are four principal groups:

Brown algae
Red algae
Green algae
Blue-green algae

When botanists speak of seaweed, they usually mean one of the larger brown or red algae. The seaweeds of cold waters are chiefly brown algae; those of the tropics are mainly red algae.

For centuries, seaweed has been used as a feed for animals and as a food for humans; prized for its minerals and vitamins.

The ancient Chinese thought highly of seaweed and made offerings of it as a food to their ancestors; and, in 800 B.C., mention was made of seaweed in a Chinese book of poetry. In his *Natural History,* in the 1st century A.D., Pliny the Elder referred to the use of seaweed. The Greeks are known to have used seaweed to treat intestinal disorders and to counteract goiter, despite knowing nothing about iodine. On his first voyage to America, Columbus' ships passed through masses of seaweed floating in the Atlantic Ocean.

One of the most confusing things about seaweed is the various names by which it is known; among them, seaweed, agar-agar, algae, carrageenan, dulse, Iceland moss, Irish moss, kelp, laver, rockweed, and sea lettuce. It has been estimated that there are some 2,500 varieties of marine plants; so, seaweed embraces a great variety of plants of many hues, shapes, and sizes. It is noteworthy, too, that seaweed is one of the few members of plant life that has remained unchanged for centuries, since its cultivation and growth is still controlled by the elements—not by humans. Although seaweeds occur at depths of up to 100 ft, certain red algae grow to depths of 600 ft.

The giant seaweed of the Pacific, a type of brown algae, is commonly known as *kelp.* The stems of these huge weeds sometimes grow more than 200 ft long. There are said to be over 800 species of kelp.

• **Composition of seaweed**—The Norwegian Seaweed Institute reports the following proximate analysis of seaweed:

Component	Percent
Protein	5.7
Fat	2.6
Fiber	7.0
Nitrogen-free extract	58.6
Ash	15.4
Moisture	10.7
	100.0

Contrary to claims that are sometimes made, seaweeds are low in protein, and the protein is of very low biological value.

The Norwegian Seaweed Institute has found an assortment of 60 different mineral elements in seaweed, all harnessed from the sea (see Table 12-15). Additionally, seaweed contains carotene, vitamin D, vitamin K, and most of the water-soluble vitamins, including vitamin B-12.

TABLE 12-15
MINERAL ELEMENTS IN DRIED SEAWEED[1]
(AVERAGE ANALYSIS, NORWEGIAN BROWN VARIETY)

Element	Percent	Element	Percent
Ag (silver)	.000004	**Mg** (magnesium)	.213000
Al (aluminum)	.193000	**Mn** (manganese)	.123500
Au (gold)	.000006	**Mo** (molybdenum)	.001592
B (boron)	.019400	**N** (nitrogen)	1.467000
Ba (barium)	.001276	**Na** (sodium)	4.180000
Be (beryllium)	Trace	**Ni** (nickel)	.003500
Bi (bismuth)	Trace	**O** (oxygen)	Undeclared
Br (bromin)	Trace	**Os** (osmium)	Trace
C (carbon)	Undeclared	**P** (phosphorus)	.211000
Ca (calcium)	1.904000	**Pb** (lead)	.000014
Cb (niobium)	Trace	**Pd** (palladium)	Trace
Cd (cadmium)	Trace	**Pl** (platinum)	Trace
Ce (cerium)	Trace	**Ra** (radium)	Trace
Cl (chlorine)	3.680000	**Rb** (rubidium)	.000005
Co (cobalt)	.001227	**Rh** (rhodium)	Trace
Cr (chromium)	Trace	**S** (sulfur)	1.564200
Cs (caesium)	Trace	**Se** (Selenium)	.000043
Cu (copper)	.000635	**Sb** (antimony)	.000142
F (fluorine)	.032650	**Si** (silicon)	.164200
Fe (iron)	.089560	**Sn** (tin)	.000006
Ga (gallium)	Trace	**Sr** (strontium)	.074876
Ge (germanium)	.000006	**Te** (tellurium)	Trace
H (hydrogen)	Undeclared	**Th** (thorium)	Trace
Hg (mercury)	.000190	**Ti** (titanium)	.000012
I (iodine)	.062400	**Tl** (thallium)	.000293
Id (indium)	Trace	**U** (uranium)	.000004
Ir (irridium)	Trace	**V** (vanadium)	.000531
K (potassium)	1.280000	**W** (tungsten)	.000033
La (lantanum)	.000019	**Zn** (zinc)	.003516
Li (lithium)	.000007	**Zr** (zirconium)	Trace

[1]Data from the Norwegian Seaweed Institute, as reported in *Review of Seaweed Research,* Research Series No. 76, Clemson University, 1966.

Seaweed is almost always fed as a dried meal. Seaweed meal of good quality may constitute up to 10% of cattle and sheep rations with good results. The optimum level of seaweed meal in swine and poultry rations appears to be about 6%. Sometimes, seaweed imparts a fishy smell to pork.

The following animal feeding trials with seaweed have been reported: Including 20% seaweed in the concentrate of a calf ration tended to lower the rate of gain, but the difference was not significant statistically. Feeding lactating dairy cows up to 30% seaweed did not alter milk yields. Including up to 15% dried seaweed meal did not affect feed intake or weight gain in sheep.

CAUTION: Dried kelp (seaweed) is so rich in iodine that consumption of a large amount over a prolonged period may be harmful; hence, it should be used as directed by a competent nutritionist.

Water Hyacinth (Eichhornia crassipes)

Water hyacinth is an unattached and free-floating waterplant with the leaves above the surface of the water and the roots in the water. Generally, it is considered a troublesome weed; it multiplies rapidly, clogs lakes, rivers, and ponds; and seriously obstructs traffic on waterways. Also, it is extremely difficult to eradicate. However, water hyacinth has considerable potential as a livestock feed.

In China, water hyacinth is sometimes harvested, boiled, and fed to swine. Because it is a floating plant, it is easily harvested. The fresh plant contains prickly crystals which make it unpalatable; and the high water content of the plant imposes a limitation on the amount of dry matter an animal is capable of ingesting. For these reasons, normally not more than 25% boiled water hyacinth is included in swine rations in China.

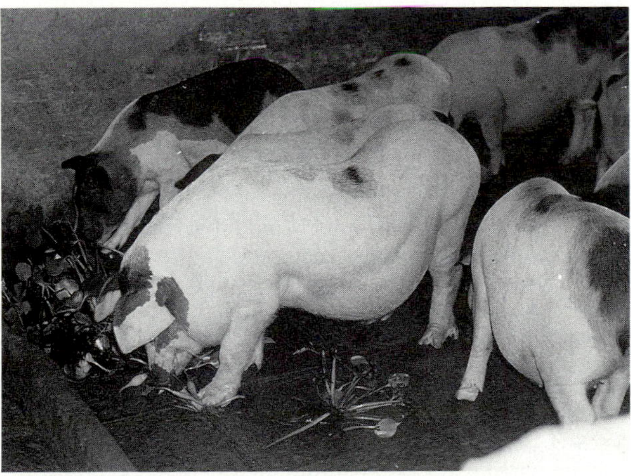

Fig. 12-13. Pigs eating water hyacinth in China. (Photo by A. H. Ensminger)

• **Animal performance**—Boiled water hyacinth is used as a pig feed in Southeast Asia. The hyacinth is chopped and sometimes mixed with other vegetable wastes, such as banana stems, and boiled slowly for a few hours until the ingredients turn into a paste. To this paste, the following ingredients are commonly added: oil cake, rice bran, corn, and salt.

The following results of feeding trials with water hyacinth have been reported: Cattle and sheep readily consumed complete rations containing processed water hyacinth. The dry matter intake of pangolagrass silage by sheep was higher than that of water hyacinth silage.

Other Aquatic Plants

It has been estimated that there are approximately 2,500 varieties of marine plants. Many of these are potential livestock feeds. In addition to algae, seaweed, and water hyacinth, duckweed (small floating aquatic plant of the family *Lemnaceae,* especially of the genus *Lemna*) appears to have possibilities as a livestock feed.

• **Animal performance**—The rate of gain of dairy heifers fed rations containing duckweed was equal to or higher than that of heifers fed control rations. The body weights and shank pigmentation scores of chicks fed a diet containing duckweed were similar to those of chicks fed diets containing alfalfa.

Processing

The harvesting and processing of aquatic plants is different. Since they are located in water, specially adapted harvesting equipment is essential. The high water content at harvest (up to 95%) adds volume and weight, and makes for special problems in processing.

PROCESSING ALGAE

Harvesting and processing algae involves three steps: (1) initial concentration, with chemical precipitation and centrifuging appearing to be the most promising method for this step; (2) dewatering, which is best achieved by using a modified industrial gravity filter; and (3) drying, with minimum exposure to high temperature during drying in order not to lower the quality of the protein.

PROCESSING SEAWEED

The seaweed industry is greatly in need of improved efficiency in harvesting. Worldwide, most harvesting is by hand collection. Following hand harvesting, most seaweed is dried naturally by exposure to the sun and air, with the washed seaweeds spread thinly on portable racks, wooden platforms, or flat rocks. Under favorable drying conditions, 1 to 2 days are sufficient to reduce the moisture to 18 to 20%.

Few successful mechanized methods have been developed, the principal exception being the harvesting of the California *macrocystis* beds.

The giant seaweed (kelp) off the coast of California is mechanically harvested by specially equipped barges, following which it is unloaded by a mechanical fork, artificially dried, then ground. After grinding, the resultant meal is olive green in color. The best of the harvest is reserved for the production of kelp tablets and seaweed powder for human consumption, while the bulk of the crop is marketed for livestock feed.

A relatively new process has been developed for speeding the drying process of seaweed (kelp). It consists of: (1) chopping the plants in a silage cutter, (2) grinding in a vertical shaft hammermill, (3) adding calcium chloride to the mixture, followed by heating and pressing, with removal of 75% of the moisture and 65% of the ash.

PROCESSING WATER HYACINTH

Because water hyacinth is a floating plant, it has been harvested by nets, traditionally. However, it is also being harvested by mechanical harvesters designed for the specific purpose.

Following harvesting, water hyacinth may be processed as follows:

1. Boiled as a feed for pigs.
2. Pressed to reduce the moisture level from about 95% to 65%, then ensiled. The quality of silage may be improved by adding molasses, grain, or rice straw.
3. Pressed to reduce the moisture level from about 95% to about 45%, then sun dried.

Animal and Human Health

Data are not available relative to possible pathogens in water plants. The main concern appears to be in plants harvested in waste-treatment ponds and lagoons. Data on mineral composition do not implicate a serious heavy metal threat.

Too much dried kelp can result in iodine toxicity in both animals and humans.

Regulatory Aspects

There are no specific regulations concerning the feeding of aquatic plants.

QUESTIONS FOR STUDY AND DISCUSSION

1. Define (a) by-product feeds, and (b) crop residues. If by-product feeds and crop residues were not used for animals, would they be used for human food?

2. When determining the economy of by-product feeds and crop residues, what factors in addition to cost per pound or per ton should be considered?

3. List the four major categories of animal by-product feeds; and give the nutritional characteristics of animal by-product feeds as a whole.

4. Describe and give the feeding recommendations pertaining to each of the following dairy by-product feeds: (a) cheese rind, (b) skimmed milk, dried, (c) whey, fresh, and (d) whey, dehydrated.

5. Describe and give the feeding recommendations pertaining to each of the following marine by-product feeds: (a) fish liver meal, (b) fish meal, (c) fish solubles, dried, and (d) shrimp meal.

6. Describe and give the feeding recommendations pertaining to each of the following meat animal by-product feeds: (a) animal fat, hydrolyzed, (b) blood meal, (c) liver meal, (d) meat meal, and (e) tankage.

7. Describe and give the feeding recommendations pertaining to each of the following poultry by-product feeds: (a) feather meal, (b) poultry by-product meal, and (c) poultry fat.

8. Define each of the following plant by-product terms: (a) pulp, (b) pit, (c) cannery residue, (d) cull, (e) bagasse, (f) *in vitro,* (g) *in vivo,* (h) stover, and (i) stubble.

9. Describe and give the feeding recommendations pertaining to each of the following oilseed by-product feeds: (a) canola (rapeseed) meal, (b) cottonseed, whole, (c) cottonseed meal, (d) linseed meal, and (e) soybean meal.

10. Describe and give the feeding recommendations pertaining to each of the following grain by-product feeds: (a) corn gluten feed, (b) hominy feed, (c) wheat bran, (d) rice polishings, and (e) screenings, cereal grain.

11. Describe and give the feeding recommendations pertaining to each of the following fruit, vegetable, and nut by-product feeds: (a) acorns, (b) apple pomace, dehydrated, (c) beans, navy, cull, (d) carrot roots, fresh, (e) citrus pulp, dehydrated, (f) lettuce, (g) pineapple cannery by-product, (h) potatoes, cull, fresh, (i) potato process residue, (j) tomato, pomace, wet, and (k) turnip, roots, fresh.

12. Describe and give the feeding recommendations pertaining to each of the following sugar by-product feeds: (a) beet molasses, (b) beet pulp, wet, (c) beet pulp, dried, and (d) cane molasses.

13. Describe and give the feeding recommendations pertaining to each of the following brewing and distilling by-product feeds: (a) brewers' dried grains, (b) distillers' dried grains, (c) malt sprouts, barley, dehydrated, and (d) yeast, brewers' dried.

14. Although more than 1.5 million tons of sulfate and sulfite pulps from spruce, pine, and fir were fed to cattle and horses in the Scandinavian countries during World War II when feed supplies were limited, the feeding of wood pulp to livestock ceased when conventional feedstuffs became available. Why wasn't the feeding of wood and paper by-products continued in these countries after World War II?

15. Evaluate by chemical composition and animal performance when used as a feed the following wood feedstuffs: (a) aspen bark, aspen sawdust, and whole aspen tree, and (b) forest foliage (muka).

16. Discuss the method and effectiveness of each of the following treatments to enhance the digestibility of wood by animals: (a) hydrolysis, (b) alkali and ammonia, and (c) delignification.

17. Describe and give the feeding recommendations pertaining to each of the following pulpmill and papermill residues: (a) sludges, (b) spent sulfite liquor, (c) waste paper products, (d) wood molasses, and (e) torula yeast.

18. Discuss the following aspects of crop residues: (a) kind and quantity produced, (b) economy of crop residues, (c) harvesting, and (d) nutrient value.

19. Discuss the method and effectiveness of each of the following treatments for delignifying and increasing digestibility of crop residues by animals: (a) grinding and pelleting, (b) sodium hydroxide, (c) calcium hydroxide, (d) ammoniation, and (e) high pressure steam.

20. Discuss the advantages and the disadvantages of each for the following feeding systems for utilizing crop residues: (a) grazing, (b) processing as dry feed, or (c) ensiling.

21. Describe and give the feeding recommendations pertaining to each of the following crop residues: (a) almond hulls, (b) bagasse, sugarcane, (c) beet tops, fresh, (d) corn stover, (e) corncobs, (f) cottonseed hulls, (g) oat straw, (h) pineapple greenchop, (i) rice straw, (j) sorghum stover, (k) soybean hulls, (l) sweet corn cannery refuse silage, and (m) tumbleweed silage.

22. Review the historic use of manure as an animal feed.

23. What is the annual production, composition, and value of manure produced in the United States?

24. Discuss each of the following potential uses of manure: (a) fertilizer, (b) nonfeed energy source, or (c) feed. What is the highest and best use of manure?

25. Discuss the performance of animals in growth, or in meat, milk, or egg production, when fed (a) cattle manure, (b) swine manure, or (c) poultry manure, with and without litter.

26. What are the advantages and the disadvantages of each of the following methods for processing animal wastes: (a) dehydrating, and (b) ensiling.

27. When feeding manure, of what concern are the following aspects: (a) human health, (b) esthetic, and (c) regulatory.

28. What do you forsee relative to the future handling and uses of manure?

29. Why is the feeding of sewage sludge not considered to be viable at this time?

30. How is garbage normally prepared and fed to swine? How may humans alleviate the danger of trichinosis from eating pork derived from garbage-fed swine?

31. What is bakery waste? How may it be used as a feed?

32. Based on chemical analyses, do coffee wastes possess potential feeding value provided their caffeine and tannin contents can be lowered? Justify your answer.

33. What are aquatic plants? Currently, what are their adverse effects; and what are their potentials as a feed resource?

34. Discuss the physical characteristics, quantity, yield, harvesting, storing, and feeding of each (a) algae, (b) seaweed, and (c) water hyacinth.

SELECTED REFERENCES

Title of Publication	Author(s)	Publisher
Agricultural Waste Utilization and Management	Ed. by J. C. Converse	American Society of Agricultural Engineers, St. Joseph, Mo., 1985
American Feed Control Official Publication	Association of American Feed Control Officials	AAFCO, Donald H. James, West Virginia Dept. of Ag., Capital Complex, Guthrie Center, Charleston, W. Va., annual
Animal Feeds	M. Gutcho	Noyes Data Corporation, Park Ridge, N.J., 1970
Animal Waste Utilization on Cropland and Pastureland	Coordinated by C. B. Gilbertson	USDA, 1979
Applied Animal Nutrition	E. W. Crampton L. E. Harris	W. H. Freeman and Co., San Francisco, Calif., 1969
By-Products and Unusual Feedstuffs in Livestock Rations	D. L. Bath, *et al.*	Cooperative Extension, University of California, Davis, 1980
Composition of Concentrate By-Product Feeding Stuffs, Pub. 449	National Research Council	National Academy of Sciences, Washington, D.C., 1956
Crop Residue Management Systems	W. R. Oschwald	American Society of Agronomy, Madison, Wisc., 1979
Farm Animal Manures: an overview of their role in the agricultural environment	J. Azevedo P. R. Stout	University of California, Davis, 1974
Feeds and Feeding, 22nd Edition	F. B. Morrison	The Morrison Publishing Company, Ithaca, N.Y., 1956
Handbook of Feedstuffs, The	R. Seiden W. H. Pfander	Springer Publishing Company, Inc., New York, N.Y., 1957

Title of Publication	Author(s)	Publisher
Processing and Utilization of Animal By-Products	I. Mann	FAO, Rome, Italy, 1967
Stockman's Handbook, The, 6th Edition	M. E. Ensminger	The Interstate Printers & Publishers, Inc., Danville, Ill., 1983
Tropical Feeds	B. Gohl	FAO, Rome, Italy, 1975
Underutilized Resources as Animal Feedstuffs	J. P. Fontenot, Chairman of Committee	NRC, National Academy Press, Washington, D.C., 1983
Washington By-Product Feeds for Livestock	W. W. Heinemann	Washington State University, Cooperative Extension Service, Pullman, Wash., 1978
Working Papers The Rockefeller Foundation The Role of Animals In The World Food Situation	Conference Papers by Participants	The Rockefeller Foundation, New York, N.Y., 1975

Fig. 12-14. Canola and its by-products: *top,* canola flower; *left,* canola seed; *center,* canola oil; *left front,* canola pellets; *right front,* canola meal. (Courtesy, Canola Council of Canada, Winnipeg, Manitoba, Canada)

Fig. 12-15. Wet Citrus pulp. (Courtesy, D. L. Bath, Extension Dairy Nutritionist, University of California, Davis)

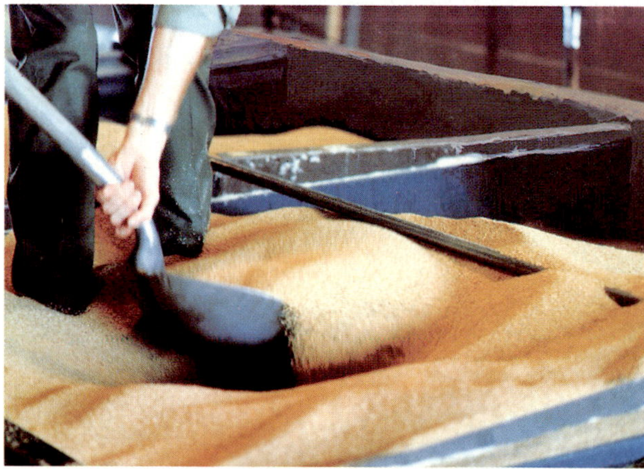

Fig. 12-16. Soybean oil meal. (Courtesy, American Soybean Assn., St. Louis, Mo.)

Fig. 12-17. Cottonseed meal. (Courtesy, Maddox Dairy, Riverdale, Calif.)

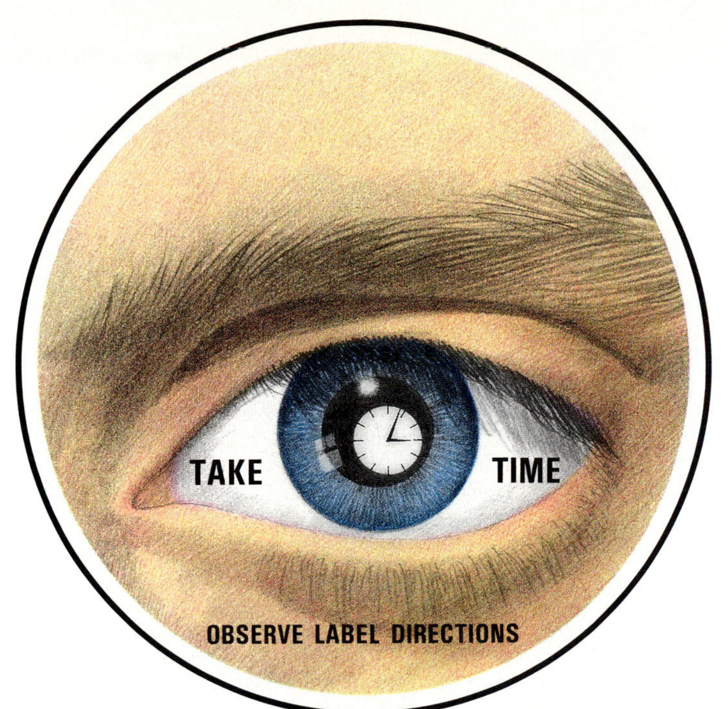
TAKE
TIME
OBSERVE LABEL DIRECTIONS

Original Painting by Tom Phillips

Chapter 13

FEED SUPPLEMENTS/ ADDITIVES/IMPLANTS

Contents — Page

From the remote day of animal domestication forward, people sought ways of improving their performance by using different feeds. Originally, it was not understood why two different feeds might elicit different responses, but the end result was recognized.

Throughout the 19th century, the roles of energy and protein in nutrition were beginning to be elucidated, and the livestock producer was able to differentiate between the feeding values of certain feeds. Then came the discovery of vitamins in the early 1900s, ushering in a new era of nutrition. Along with the expanding knowledge of vitamins, mineral research flourished throughout the first half of the 20th century. Feeding standards evolved, showing the nutrient needs of different species of animals for various purposes. Supplements of micronutrients became available, enabling the producer to provide specific nutrients in small quantities.

Then came the feed additive/implant era. In 1949, Stokstad and Jukes of American Cyanamid Company discovered that antibiotics were something new to be added to livestock feeds. In 1952, Burroughs of Iowa State University announced a major breakthrough in increased feed efficiency and weight gains by feeding the compound diethylstilbestrol (DES). In 1954, DES was approved by the Food and Drug Administration (FDA) for use in cattle finishing rations; and 2 years later, FDA approved the use of stilbesterol implants for steers. Soon the race was on; and other additives and implants followed.

The successful livestock producer uses supplements, additives, and implants to maximize performance; to improve animal health, to increase feed intake and hopefully feed efficiency, and/or to alter some physiological process in the animal that will stimulate production and/or improve the quality of the product.

The term, supplement, refers to feedstuffs that are used to improve the value of basal feeds. Thus, supplements are products that provide an additional nutrient or nutrients. They can be used in large quantities, such as protein supplements, or in extremely small quantities, such as trace minerals and vitamins.

An additive is a substance of non-nutritive nature which when added to feed will improve feed efficiency, production, and/or health of animals. In general, the term *feed additive* refers to a non-nutritive product that affects utilization of the feed or productive performance of the animal.

An implant is a substance that is implanted into the body for the purpose of growth promotion or controlling some physiological function. Compounds used in implants are generally materials (for example peptide hormones) that would be destroyed during digestion. Hence, implants enable the body to absorb the compounds intact. Additionally, implants provide a means of controlled continuous release of chemicals.

Today, feed additives, biologicals, and pharmaceuticals are big business; in 1986, they made for U.S. sales of $2.1 billion. So, in order to justify the use of these products, they must be cost effective; the increased returns from their use should be great enough to offset the cost of the product, plus any additional management expenditure, and allow for a profit.

FOOD AND DRUG ADMINISTRATION (FDA)

The Food and Drug Administration (FDA), which is part of the U.S. Department of Health, Education, and Human Services, is charged with the responsibility of regulating the use and safety of additives and implants. Through the years, it has established rigorous testing policies requiring proof relative to product safety and efficacy. The requirements for new product approval by FDA have become more rigid, and enforcement has become more unrelenting. Thus, in 1962, a drug could be developed, tested, and approved for about $1 million—all in a period of 1 to 2 years. Twenty five years later, development to approval of a new drug took 7 to 8 years and cost $14 to $15 million. It is understandable, therefore, why manufacturers resist the threat of a ban on a product in which they have such an investment.

The first law enacted to protect the food we eat dates back to 1784, in Massachusetts. In the period from 1891 to 1895, laws were passed by the U.S. Congress requiring the inspection of animals for diseases before slaughter. The furor over the quality of meat came to a head in 1906 with the passage of the Meat Inspection Act, which came as a direct result of shocking disclosures of conditions under which meat was processed. On the same day that the Meat Inspection Act was passed, the Food and Drug Act of 1906 was signed. This act was designed to prevent the use of poisonous preservatives and dyes in foods, and to regulate the sales of patent medicines.

In 1927, the Food, Drug, and Insecticide Administration, later to be named the Food and Drug Administration (FDA), was created. In 1938, the Federal Food, Drug, and Cosmetic Act of 1938 was signed, broadening the scope of the Food and Drug Act of 1906 to include (1) coverage of cosmetics and devices; (2) requirements for predistribution clearance for the safety of new drugs; (3) provisions for tolerances for unavoidable poisonous substances; (4) standards of identity, quality, and fill of food containers; (5) authorization for factory inspections; and (6) provision for court injunctions in matters of seizures and possession. With its numerous amendments, the Federal Food, Drug, and Cosmetic Act is probably the most extensive law of its kind in the world.

Since 1960, manufacturers have been required to demonstrate both the efficacy and the safety of proposed new additives before they can be certified for use. The required demonstrations include tests at 3 different levels: (1) acute toxicity test to show the effects of a single lethal dose given to a variety of laboratory animals; (2) short-term (90-day) toxicity studies showing the effects of feeding different concentrations to 2 kinds of laboratory animals; and (3) long-term toxicity studies of 2 years or more to show the effects of lifetime consumption. Tolerance levels are based on the amount of chemical shown to be safe in long-term studies. The margin of safety is extremely wide. For example, if toxicity studies show that the use of 1,000 ppm of a particular chemical is safe, then the FDA will permit maximum use of 10 ppm (1% of the no-effect level). The one exception to this standard is that any substance demonstrated to be carcinogenic (cancer-producing) is not permitted in any amount, if such use leaves detectable residues in the product.

Wherever additives are involved, it is the responsibility

of the producer to comply with the withdrawal periods prior to slaughter or milking, as established by the Food and Drug Administration.

FDA's attitude toward the regulation of animal health biotechnology products is similar to that of any other new drug entity.

Delaney Clause

In 1958, the food additives amendment, better known for its Delaney Clause (named after the congressman who sponsored it), was passed. This bill has proven to be one of the most controversial pieces of legislation ever to affect the American livestock industry. The Delaney Clause states:

> "*Provided,* That no additive shall be deemed safe if it is found to induce cancer when ingested by man or animal, or if it is found, after tests which are appropriate for the evaluation of the safety of food additives, to induce cancer in man or animal. . ."

This clause gave rise to the policy of *zero tolerance*—that is, no substance can be used as a feed additive, even in miniscule amounts, if it has been, in any way, implicated as an inducer of cancer in either human or beast. What, at the time, appeared to be a well-intentioned law aimed at protecting the consumer from potential health hazards proved to be a nightmare for the drug industry and livestock producers. The additive manufacturers must now prove a negative hypothesis which many feel is impossible. That is, an additive must be demonstrated to be 100% noncarcinogenic. Unfortunately, the lawmakers are in a tenuous position because, to repeal the Delaney Clause, they run the risk of being accused of supporting the addition of cancer-causing drugs to our food supply.

In 1973, the FDA banned the use of diethylstilbestrol (DES), a widely used growth promotant in cattle, for the reason that it had proven to be carcinogenic in large doses. Emotions waxed hot! The drug and beef industries brought suit against the FDA; and on January 24, 1974, a U.S. Court of Appeals overruled the ban on DES, thereby granting the feed additive a temporary reprieve. But the ruling of the court did not change the Delaney Amendment. In the meantime, newer and more sophisticated techniques evolved which made it possible to detect the presence of DES residues in meat at lower levels than formerly. So, in 1979, the use of DES as feed or implants for cattle and sheep was banned.

In 1977, the FDA called for a ban on the artificial sweetener, saccharin, citing two Canadian studies showing it to be carcinogenic. Since a ban on this sweetener would have severely affected many weight-conscious consumers, there was an immediate and indignant outcry. Moves to modify the Delaney Clause—to change the controversial *zero tolerance* to *biological zero tolerance*—followed. The emotional nature of the debate that ensued demonstrated how difficult it is to get objective statements and to discover the truth.

Only when those who have chosen sides become informed on the issues, realistic in their demands, and willing to listen to the opposition, will the issues be resolved. The Delaney Clause will, in all likelihood, be modified, but the debate of how safe is *safe* will, in all likelihood, continue without end.

With or without the Delaney Clause, *America's food supply is the safest in the world.*

Use Feed Supplements, Additives, and Implants Safely

The list of supplements, additives, and implants grows longer. So, also do the concerns of producers and consumers.

Producers use drugs to produce products faster, more abundantly, on less feed, and/or of improved quality. Consumers want to be assured of the wholesomeness of their meat, milk, and eggs.

Modern agriculture and a significant share of our present feed/food technology are dependent upon the use of chemicals. Yet our increasing awareness of food safety and environmental quality has resulted in increased scrutiny of many of the chemical compounds and their use patterns. Two primary concerns prevail: (1) to assure that the chemicals used in the production of food and fiber at every level are safe when used properly, and (2) to assure that the benefits of these chemicals in providing adequate food supplies and in protecting the health of consumers not be displaced by an abstract goal of zero risk. These concerns necessitate that those who use such products read and heed the labels.

The answer to avoiding illegal residue violations begins with every package, with the label (and in some products, with the package insert), which is written by the manufacturer and approved by the federal government. Everything that the producer needs to know about approved species, dosage, method of administration, and withdrawal time is on the label or insert. Thus, the key to using additives and implants safely is to read and follow the label directions. Since withdrawal times may change, the producer should read the label instructions on *each* package used.

Concerns about proper use of additives and implants does not stop with the marketing of the animals or products. To guard against possible misuse of products, meat, milk, eggs, and other animal products are routinely monitored by the U.S. Department of Agriculture's Food Safety and Inspection Service for over 40 compounds, including drugs, pesticides, and industrial contaminants.

More than half of the U.S. drug products that FDA has approved for animal use require pre-slaughter drug withdrawal times or a milk-discard period. This is done to protect consumers from harmful drug residues. But not all residues are harmful! That's the reason that withdrawal times are not required on some drugs. They may, in fact be beneficial. For example, the vitamin B-12 contributed to our diet by meat, milk, and eggs is a vital contribution to our health and well-being.

The FDA reports that many residue violations are caused by producers themselves, especially by their failure to withdraw drugs from livestock far enough in advance of marketing products. Additional sources of residue violations which should be guarded against are: (1) contaminated feed storage, mixing, and handling equipment; and (2) the waste products (feces and urine) of treated animals coming in contact with untreated animals.

Summary: The label holds the key to proper livestock drug use. But no matter how clear and accurate the label, it won't communicate anything if it is not read; and it won't prevent illegal residues if it is not followed. The alternatives to reading and heeding the label relative to the proper use of livestock drugs are: (1) the losses to non-complying producers through condemnation of their products, plus possible fines; and (2) the losses to the entire livestock industry as a result of some scare headlines in the news media, followed by the erosion of consumer confidence in the safety and wholesomeness of meat, milk, and eggs, the consequences of which no one in the livestock industry can afford.

FEED SUPPLEMENTS

In the formulation of rations, the energy and fiber feeds which are usually produced on the farm or ranch are the basal feeds. Such feeds are commonly deficient in protein and perhaps in one or more amino acids, and in minerals and vitamins.

A feed supplement is a feed that is used to improve the nutritional value of a basal feed. It follows that feed supplements are usually concentrated sources (1) of protein, or of one or more amino acids; (2) of one or more minerals; (3) of one or more vitamins; or (4) of a protein/mineral/vitamin mix. Additionally, the following special supplements are sometimes used:

1. High fiber supplements, such as soybean hulls or corn bran, both of which are highly digestible, fed to dairy cows on low fiber rations to prevent lowered milk fat content.
2. High energy supplements, such as cereal grains, fat, and whole cottonseed, fed to animals to improve ration energy density and performance.
3. Protein/energy blocks; usually containing some molasses, natural protein and/or urea, minerals, vitamins, in some cases fat, and an intake limiter(s) such as salt; and weighing 50 to 500 lb; used primarily as winter range supplements.
4. Medicated health supplements.

Supplements may be fed (1) undiluted as an addition to other feeds; (2) further diluted and mixed to produce a complete feed; or (3) offered free-choice, with other parts of the ration fed separately. Also, supplements are provided in different forms—in meals, granules, pellets, cubes, blocks (ranging in weight from 50 to 500 lb), liquids (usually with molasses), or salt-limited.

The term *supplement* is descriptive within itself. Thus, in order properly to balance a basal high energy feed, additional protein, minerals, and vitamins are usually needed. Normally, these deficiencies are met by adding ingredients that are richer in the needed nutrient(s) than the basal feed. Consequently, such ingredients are commonly referred to as feed supplements.

Specific nutrient supplements are commercially available, thereby permitting the nutritionist to add small amounts of a specific nutrient or combination of nutrients to a ration without altering the general makeup of the initial formulation.

Normally, the United States produces about 25 million tons of protein supplements, exclusive of urea. But it is estimated that an additional 2 to 5 million tons of these products could be used advantageously if all animals were supplied an adequate amount of protein.

In general, little or no supplementation of specific amino acids is done for ruminants, although some research with methionine hydroxy analog (MHA) has indicated its possible use in the future. However, several amino acids, especially arginine, cystine, lysine, methionine, and tryptophan, warrant careful consideration in feeds for nonruminants; and sometimes individual amino acids or amino acid analogs are added to feed.

Mineral and vitamin supplementation is of paramount importance to *all* livestock rations. Imbalances, deficiencies, or excesses of minerals pose major problems. While toxicities of vitamins are rare, deficiencies are not; and in this era of highly refined scientific feeding, there can be no excuse for these occurrences.

Feed supplements are expensive. However, it may be practical and profitable to add them for one or more of the following reasons:

1. To correct some deficiency in the ration of animals.
2. To serve as a carrier for some additive.
3. To enhance animal health, such as controlling bloat or parasites.
4. To teach young animals to eat, as in creep feeding.
5. To balance dry summer or winter range.
6. To increase the carrying capacity of pasture or range.
7. To stretch the feed supply during a drought.
8. To add fiber to low fiber dairy rations, in order to prevent a drop in milk fat content.
9. To increase the energy density of low energy rations.
10. To gather animals, such as for heat detection.

Protein Supplements

Protein supplements are feedstuffs containing more than 20% protein or protein equivalent. They are obtained from animal, marine, plant, or microbial sources, as well as from nonprotein nitrogen sources such as urea, biuret, or ammoniated products. Additionally, a number of specially designed pasture and range protein supplements are available, many with added energy, minerals, and/or vitamins; and they are used in the following forms and systems that lessen the labor attendant to daily feeding: (1) cubes, hand-fed at intervals, (2) blocks, (3) salt-feed mixtures that are self-fed, and (4) liquid feed supplements that are self-fed.

Most high energy feeds (except for purified products like fat, starch, or sugar) supply some protein, but, except for adult animals during maintenance, they usually do not supply enough to meet total needs. Thus, supplementary sources are commonly needed in the formulation of rations for all species of animals. In particular, protein is a critical nutrient for young, rapidly growing animals and for high-producing dairy cows and layers. Animals cannot (1) develop their genetic potential, (2) produce the maximum of meat, milk, and eggs, or (3) produce the maximum power, animation, and speed in the case of horses unless their rations contain sufficient protein, along with the correct amino acid composition in the case of monogastrics.

Protein supplements are regularly in shorter supply and higher priced than the cereal grains and other high-energy feeds. As a result, the tendency is not to feed sufficient protein supplements to balance rations.

(Also see Chapter 6, Types and Roles of Feedstuffs, Section on "Protein Supplements"; Chapter 11, Protein Supplements; and Chapters 19 to 28, the chapters devoted to the respective classes of livestock.)

AMINO ACID SUPPLEMENTS

For monogastrics and very young ruminants (preruminants), the amino acids that make up proteins are really the essential nutrients, rather than the protein molecule itself. For ruminant species, the dietary need is (1) to nourish the rumen microorganisms, and (2) to have an adequate supply of digestible essential amino acids in the gut. High-producing ruminants (such as high-producing dairy cows) may have higher amino acid needs than can be satisfied by rumen synthesis; so, protein quality is more important under these circumstances than for animals producing at low levels and consuming much less feed.

Of the 22 amino acids, 5 are deemed critical, whereas the others are usually in sufficient supply from the combination of feedstuffs found in the rations of most monogastric species. The critical 5 are: arginine, cystine, lysine, methionine, and tryptophan. When a monogastric animal's ration is low in 1 or more of these 5 amino acids, protein supplements carrying large amounts of pure amino acids may be added to make up the deficiencies. Thus, swine feeds consisting chiefly of corn, wheat, and barley are deficient in the essential amino acid, lysine, and corn is also deficient in tryptophan. Due to the large amounts of vegetable proteins used, along with low levels of animal and fish proteins, poultry rations are most often lacking in methionine. Also, lysine and cystine are often inadequate in normal feedstuffs. So, poultry rations frequently call for the supplementation of DL-methionine; and lysine and cystine are often added, also.

Proper amino acid supplementation calls for meeting the minimum requirements for all amino acids, with little excess of any of them. This is practically impossible; there is some waste. Usually the value of the protein portion of the ration is determined by the limiting amino acid. Excesses of other amino acids are usually of no value if one is limited.

In practice, the amino acid requirements of most animals, including monogastrics, young ruminants, and high-producing dairy cows, are met by proteins from plant and animal sources. Protein supplements that most nearly supply the essential amino acids are known as *high-quality.* Usually it is necessary to choose more than one source of dietary protein, then combine in such a way that the amino acid composition of the mixture meets the requirements of the animal. For example, corn protein is very low in lysine. On the other hand, the proteins of soybean meal, milk solids, whey, and fish meal are much higher in lysine and can be mixed with corn to satisfy the lysine requirement.

Some amino acid characteristics of the major feed ingredients follow:

Barley . . . . Low in tryptophan and lysine
Corn . . . . Low in lysine and tryptophan
Sorghum . . . . Low in lysine
Soybean oil meal . . . Low in methionine, high in lysine
Corn gluten meal . . Low in lysine

(Also see Chapter 11, Protein Supplements.)

Mineral Supplements

The metabolic functions and interrelationships among the minerals are extremely varied and complex. An excessive amount of one mineral can create a deficiency of another. Additionally, several trace minerals have relatively narrow toxicity tolerances. For example, selenium can be legally added to feed to a maximum of 0.3 ppm (part per million), but toxicities can result when animals ingest 10 ppm. Therefore, supplementation with this and several other minerals must be carefully monitored.

Almost all feeds contain at least limited amounts of the various minerals, but these levels are highly variable and often reflect the profile of the soil on which they are grown and the genetic variations among its plant species. In addition to the variability of mineral levels in individual feeds, the mineral requirements of animals are highly variable, depending on such factors as age, size, sex, type of production, and stage of production.

(Also see Chapter 4, Nutrients/Metabolism, Table 4-6, Animal Mineral Chart.)

CLASSES OF MINERAL SUPPLEMENTS

Mineral supplements can be divided into three basic categories: (1) packinghouse by-products, (2) naturally occurring mineral sources, and (3) synthetic mineral compounds. A listing, grouped by these categories, of mineral products recognized by the Association of American Feed Control Officials is given in Table 13-1.

Packinghouse Mineral By-products

Packinghouse mineral by-products are exactly what the term implies. Bones and connective tissue resulting from the processing of various meat products represent an excellent source of quality calcium, phosphorus, and some trace minerals. Various processes—such as steaming, cooking, and precipitating—are applied to these products to (1) alter their relative mineral content, and (2) sterilize and stabilize the products for storage and use.

Of the three classes of mineral supplements, packinghouse by-products are probably the most variable in chemical composition. For this reason, minimums of one or more of their respective nutrients are listed on the label to serve as a guide to the user.

Naturally Occurring Mineral Sources

These are mineral supplements which are obtained from our natural environment and processed in such manner as to render them safe for feeding. Such mineral sources as rock phosphate must be processed in some way to remove contaminants which can be toxic to livestock—in this particular case, fluorine. When phosphates are defluorinated, they are exposed to extremely high temperatures (1,800° to 2,000°F). Subsequently, their chemical composition is altered in such a way that the end product is tricalcium phosphate. As with packinghouse by-products, it is important to read the label carefully to determine the guaranteed levels of the minerals in question.

Synthetic Mineral Compounds

In recent years, chemists have developed processes for synthesizing mineral supplements that are cheap and yet of extremely high purity. Many of these compounds are sold as *Commercial Grade.* This term implies that there are trace amounts of impurities in the product which render it unusable for analytical purposes. This should not concern the buyer as the level of impurity is negligible. Rather, the buyer should be concerned about how much of each mineral is being bought.

When using a synthetic compound, the nutritionist often needs to know the percentages of the respective elements within the compound. If the chemical formula of the compound is known, the percentages can be established quickly.

TABLE 13-1
MINERAL SUPPLEMENTS[1] (See Footnote at end of table.)

Mineral	Source		
	Packinghouse By-products	Natural Sources	Synthetic
Major or Macrominerals:			
Sodium (Na)		Salt (NaCl)	Sodium bicarbonate, $NaHCO_3$ Sodium sulfate, Na_2SO_4 Sodium tripolyphosphate, $Na_5P_3O_{10}$
Chlorine (Cl)		Salt (NaCl) Potassium chloride (KCl)	
Calcium (Ca)	Bone ash Bone charcoal (bone black) Bone phosphate Spent bone charcoal (spent bone black) Steamed bone meal	Calcite Calcium carbonate Chalk, precipitated Chalk rock Dolomite limestone Ground limestone Ground rock phosphate Oystershell flour Shell flour Soft rock phosphate	Calcium carbonate, $CaCO_3$ Calcium carbonate, precipitated Calcium chloride, $CaCl_2$ Calcium gluconate, $C_{12}H_{22}CaO_{14}$ Calcium hydroxide, $Ca(OH)_2$ Calcium oxide, CaO Calcium periodate, $Ca(IO_4)_2$ Calcium sulfate, $CaSO_4$ Defluorinated phosphate Dicalcium phosphate, $CaHPO_4$ Ground rock phosphate, low fluorine Tricalcium phosphate, $Ca_3(PO_4)_2$
Phosphorus (P)	Bone ash Bone charcoal (bone black) Bone phosphate Spent bone charcoal (spent bone black) Steamed bone meal	Ground rock phosphate Ground rock phosphate, low fluorine Soft rock phosphate	Ammonium polyphosphate Defluorinated phosphate Diammonium phosphate, $(NH_4)_2HPO_4$ Monoammonium phosphate, $NH_4H_2PO_4$ Dicalcium phosphate, $CaHPO_4$ Disodium phosphate, Na_2HPO_4 Monocalcium phosphate, $CaH_4(PO_4)_2$ Monosodium phosphate, NaH_2PO_4 Phosphoric acid, H_3PO_4 Phosphate, partially defluorinated Sodium tripolyphosphate, $Na_5P_3O_{10}$ Tricalcium phosphate, $Ca_3(PO_4)_2$
Magnesium (Mg)		Dolomitic limestone Smectite-vermiculite	Magnesium carbonate, $MgCO_3$ Magnesium hydroxide, $Mg(OH)_2$ Magnesium oxide, MgO Magnesium sulfate, $MgSO_4$
Potassium (K)		Smectite-vermiculite Potassium chloride (KCl)	Potassium bicarbonate, $KHCO_3$ Potassium carbonate, K_2CO_3 Potassium iodate, KIO_3 Potassium iodide, KI Potassium sulfate, K_2SO_4
Sulfur (S)			Ammonium sulfate, $(NH_4)_2SO_4$ Calcium sulfate, $CaSO_4$ Potassium sulfate, K_2SO_4 Sodium sulfate, Na_2SO_4 Zinc sulfate, $ZnSO_4$
Trace or Microminerals:			
Cobalt (Co)			Cobalt acetate, $Co(C_2H_3O_2)_2$ Cobalt carbonate, $CoCO_3$ Cobalt chloride, $CoCl_2$ Cobalt oxide, CoO Cobalt sulfate, $CoSO_4$

(Continued)

TABLE 13-1 *(Continued)*

Mineral	Source		
	Packinghouse By-products	Natural Sources	Synthetic
Trace or Microminerals (Continued)			
Copper (Cu)			Copper carbonate, $CuCo_3{\bullet}Cu(OH)_2$ Copper chloride, CuCl or $CuCl_2$ Copper gluconate, $C_{12}H_{22}CuO_{14}$ Copper hydroxide, $Cu(OH)_2$ Copper orthophosphate, $Cu_3(PO_4)_2$ Copper oxide, CuO or CuO_2 Copper sulfate, $CuSO_4$ Cuprous iodide, CuI
Iodine (I)			Calcium iodate, $Ca(IO_3)_2$ Calcium iodobehenate, $Ca(C_{21}H_{42}ICO_2)_2$ Calcium periodate, $Ca(IO_4)_2$ Cuprous iodide, CuI Diiodosalicylic acid, $C_7H_4I_2O_3$ Ethylenediamine dihydroiodide, $C_2H_8N_22HI$ Iodized salt Potassium iodate, KIO_3 Potassium iodide, KI Sodium iodate, $NaIO_3$ Sodium iodide, NaI Thymol iodide, $C_{20}H_{24}I_2O_2$
Iron (Fe)		Smectite-vermiculite	Ferrous fumarate, $C_4H_2FeO_4$ Iron ammonium citrate, $Fe(NH_3)C_6H_8O_7$ Iron carbonate, $FeCO_3$ Iron chloride, $FeCl_3$ Iron gluconate, $C_{12}H_{22}FeO_{14}$ Iron oxide, Fe_2O_3 Iron phosphate, $FePO_4$ Iron pyrophosphate, $Fe_4(P_2O_7)_3$ Iron sulfate, $FeSO_4$ Iron, reduced
Manganese (Mn)			Manganese acetate, $Mn(C_2H_3O_2)_2$ Manganese carbonate, $MnCO_3$ Manganese chloride, $MnCl_2$ Manganese citrate, $Mn_3(C_6H_5O_7)_2$ Manganese gluconate, $C_{12}H_{22}MnO_{14}$ Manganese orthophosphate, $Mn_3(PO_4)_2$ Manganese phosphate, $MnHPO_4$ Manganese sulfate, $MnSO_4$ Manganous oxide, MnO
Selenium (Se)			Sodium Selenate, Na_2SeO_4 Sodium Selenite, Na_2SeO_3
Zinc (Zn)			Zinc acetate, $Zn(C_2H_3O_2)_2$ Zinc carbonate, $ZnCO_3$ Zinc chloride, $ZnCl_2$ Zinc oxide, ZnO Zinc sulfate, $ZnSO_4$

[1]Mineral products accepted by the Association of American Feed Control Officials, Inc.

NEED FOR MINERAL SUPPLEMENTATION

Only the specific minerals that are needed should be provided. Excesses and mineral imbalances should be avoided. Except for substances like fat and urea, most feeds provide some minerals. Nevertheless, many rations require more concentrated sources of one or more of the macro and/or microminerals.

• **Macrominerals**—Of the macrominerals demonstrated to be required by livestock, only salt (sodium chloride), calcium, and phosphorus are routinely added to all livestock rations. Two other macrominerals, magnesium and sulfur, are sometimes added to ruminant rations in specialized cases. Magnesium is sometimes provided in mineral mixes for cattle on pasture in areas where grass tetany is a problem. Sulfur is routinely added to rations containing urea, since urea replaces protein which is normally a source of sulfur. The levels of all the macrominerals should be carefully monitored for nonruminants.

• **Micro, or trace, minerals**—The following 7 trace minerals are common supplements: cobalt, copper, iodine, iron, manganese, selenium, and zinc. All 7 essential microminerals can be added to salt at a cost of 1¾¢ per pound, for the resulting trace mineralized salt. So, multiplying the yearly salt consumption figure in pounds for each class of animals by 1¾¢ gives the cost of protecting them for a whole year.

Even if the animal's ration is not deficient in all seven essential trace minerals, there is no harmful effect from their supplementation because there is a large safety factor between the level needed and the level that will cause a harmful effect. Also, a little extra of the trace minerals is in the nature of good insurance due to variations in the feed ingredients, level of animal productivity, stress, nutrient relationships, and other factors.

Many studies have shown that adding trace minerals to the ration improves animal productivity and profitability. A University of Nebraska study of growing-finishing pigs on pasture showed that a trace mineral expenditure of 4⁷⁄₁₀¢ per pig returned $1.64—a 35-fold increase on the investment.

GUIDELINES FOR MINERAL SUPPLEMENTATION

No single plan can be proposed as being the best for mineral supplementation. Rather, livestock producers must tailor their supplement regimens to encompass the following considerations:

1. **Needs of the particular animal.** Age, sex, weight, and production parameters must all be considered.
2. **Types of feed.** An all- or high-concentrate ration will require a different mineral supplement than an all- or high-roughage ration.
3. **Region from which the feeds were obtained.** The mineral content of the feed will reflect the mineral composition of the soil and the genetic makeup of the specific plant.
4. **Facilities.** If the mineral mix is offered free-choice, containers protected from the elements—*i.e.,* rain and wind—may have to be constructed.

Supplementation of Mixed Feeds

When livestock are fed a mixed feed, totally or in part, the needed minerals are usually incorporated in the ration in keeping with the known requirements. In general, the following recommendations are applicable to the mineral supplementation of mixed feeds.

1. Salt is usually incorporated in the ration at levels of 0.25 to 0.50%. If less salt is added, it can also be made available *ad libitum.*
2. Calcium and phosphorus are added as needed to balance the ration. Numerous calcium and phosphorus supplements are available at reasonable cost; and the wide variety enables producers to select those which fit their particular needs (see Table 13-2).
3. If animals are housed in confinement where they receive little exposure to sunlight, careful attention must be given to providing adequate vitamin D, because vitamin D affects the assimilation and utilization of a number of minerals, especially calcium and phosphorus.
4. When the ration is suspected of being deficient in one or more minerals, a trace mineralized salt or specific minerals should be added to the ration.

Supplementation of Unmixed Feeds or Pasture

When animals are fed an unmixed ration or are on pasture, minerals may be provided as follows:

1. **When animals are on liberal grain feeding.** Provide free access to a 2-compartment mineral box, with (a) trace mineralized salt in one side; and (b) in the other side, a mixture of ⅓ trace mineralized salt (salt included for purposes of palatability), ⅓ dicalcium or defluorinated phosphate or steamed bone meal, and ⅓ ground limestone or oystershell flour.
2. **When animals are primarily on roughage (pasture, hay, and/or silage).** Provide free access to a 2-compartment mineral box, with (a) trace mineralized salt in one side, and (b) in the other side, a mixture of ⅓ trace mineralized salt (salt included for purposes of palatability); and ⅔ dicalcium or defluorinated phosphate or steamed bone meal.

Free Choice Mineral Feeding vs Incorporating Minerals in the Ration

There are several serious flaws behind the rationale of *ad libitum* (free choice) mineral feeding. It is based primarily upon the assumption that animals know their specific needs for minerals. That is to say, if an animal is deficient in calcium, it instinctively knows that it needs to consume the calcium supplement.

In a review paper, R. D. Goodrich *et al.* of the University of Minnesota concluded that there is evidence that rats have the ability to choose certain minerals of which they have a deficiency, but they question that this nutritional wisdom exists in other animals. They state:

> "It appears that acceptability, rather than a true appetite or craving for specific minerals, influences ruminants to consume free choice minerals. When animals are allowed to consume free choice minerals there is much individual variation; some animals consume minerals in excess of their requirements while others consume no minerals."

It appears, therefore, that required minerals should be fed in a complete ration whenever possible. However, when feeding practices make this impossible, a complete mineral mixture which includes salt (sodium chloride) should be available to all animals.

SALT (NaCl)

Salt, which serves both as a condiment and a nutrient, is needed by all classes of animals, but more especially by herbivora (grass-eating animals). Ratios of potassium to sodium may reach 17 to 1 in forage feeds; thus, salt is required to narrow this ratio to counteract the metabolic action of high levels of potassium. It may be provided in granulated, rock, or block form. In general, the form selected is determined by price and availability. It should be pointed out, however, that very hard block and rock salt are difficult for stock to eat, often resulting in sore tongues and inadequate consumption. Also, if there is much competition for the salt block, the more timid animals may not satisfy their requirements.

Both sodium and chlorine are essential for animal life. The blood contains 0.25% chlorine, 0.22% sodium, and 0.02% to 0.22% potassium; thus, the chlorine content is higher than that of any other mineral in the blood. The salt requirements are greatly increased under conditions which cause heavy sweating, thereby resulting in large losses of this mineral from the body. Unless replaced, fatigue will result. For this reason, when engaged in hard work and perspiring profusely, both horses and humans should receive liberal allowances of salt.

The amount of salt required by animals varies with their stage of growth, ration composition, level of production, and the temperature of their environment. Some animals sweat more than others and their salt requirements are reflected proportionately. Animals exposed to heat and those doing heavy work will need more salt than similar animals under unstressed conditions. Additionally, ruminants on pasture need salt to balance the high-potassium, low-calcium content of grass.

Carnivores usually require less supplemental salt than animals maintained largely on a plant-feed diet, because animal tissues and blood have higher salt concentrations than plants. Even with this lesser need, it is wise to make additional salt available to carnivorous animals.

• **Sources**—Salt, as commonly used in feeding, is sodium chloride. It may be mined from salt deposits found in various places of the world, or it may be produced by evaporating salt water, leaving the salt and other nonvolatile substances as a dry residue. Salt can also be refined by recrystallization from the more crude forms found in many of nature's deposits.

Salt can be fed free-choice to cattle, sheep, swine, and horses provided they have not previously been salt starved.

• **Deficiency and toxicity symptoms**—If animals have not been fed salt for a considerable length of time, they may overeat, resulting in digestive disturbances and even death. The Indians and the pioneers of this country handed down legendary stories about the huge number of buffalo and deer that killed themselves by gorging at a newly found "salt lick" after having been salt starved for long periods of time. Salt-starved animals should first be fed salt by hand, with the daily allowance increased gradually until they start leaving a little salt in the mineral box. When this point is reached, a self-feeding regimen may be followed.

In general, deficiencies of salt will produce rather non-specific symptoms. Production drops dramatically, and in some types of livestock, such as poultry, cannibalism results. Animals suffering from a salt deficiency show a depraved appetite (They will chew soil, boards, and other substances.), a reduced rate of growth, infertility in males, delayed sexual maturity of females, lowered production, and a general unthrifty condition. Chicks fed a chlorine-deficient diet are nervous and easily excitable, especially to sudden noises. A condition called nursing sickness is seen in lactating mink. Large amounts of sodium are lost from the female in the milk; and the female becomes extremely weak and emaciated. Death will result unless rapid corrective measures are taken.

Toxicities occur when the body is unable to excrete enough salt to maintain proper water balance. If there is an excessive intake of salt or a malfunction in this excretory mechanism, generalized edema will occur as a result of water retention.

(Also see Chapter 4, Nutrients/Metabolism, Table 4–6, Chlorine and Sodium; Chapter 5, Nutritional Disorders/Toxins, Table 5–1, Salt Deficiency; Chapter 11, Protein Supplements, section on "Self-feeding Salt-Feed Mixtures"; and Chapter 14, Feed Processing, "Self-feeding Governors.")

CALCIUM (Ca) AND PHOSPHORUS (P) SUPPLEMENTS

Table 13–2 lists several sources of calcium and phosphorus and gives the typical analysis of each.

TABLE 13–2
TYPICAL ANALYSIS OF CALCIUM AND PHOSPHORUS SUPPLEMENTS (MOISTURE-FREE BASIS)[1]

Compound	Calcium Content	Phosphorus Content	Sodium Content	Protein Equivalent N × 6.25	Fluorine Content
	(%)	(%)	(%)	(%)	(%)
Calcium compounds:					
Calcium carbonate	38.13	0.04	0.07	–	0
Limestone, ground	37.22	0.22	0.06	–	–
Oystershells, ground	36.27	0.10	0.21	0.7	–
Defluorinated phosphates manufactured from defluorinated phosphoric acid:					
Defluorinated phosphate	32.10	17.13	3.27	–	0.18
Diammonium phosphate	0.52	20.54	0.04	115.5	0.16
Dicalcium phosphate	22.67	19.00	1.61	–	0.10
Monoammonium phosphate	0.39	24.99	0.08	71.0	0.19
Monocalcium phosphate	18.80	21.27	0.06	–	0.14
Defluorinated phosphates manufactured from furnace phosphoric acid:					
Dicalcium phosphate	23.71	19.07	0.08	–	0.19
Disodium phosphate	–	22.32	32.00	–	–
Feed-grade phosphoric acid	0.18	27.84	0.23	–	0.25
Monocalcium phosphate	22.92	23.96	–	–	0.03
Monosodium phosphate	0.04	25.60	19.63	–	–
Sodium tripolyphosphate	–	25.38	31.23	–	0.03
Tricalcium phosphate	31.44	17.34	0.17	–	0.05
High-fluoride phosphates:					
Ground low-fluorine rock phosphate	36.00	14.00	–	–	0.45
Ground rock phosphate, raw	35.00	13.00	0.03	–	3.70
Soft rock phosphate	16.09	9.05	0.10	–	1.21
Packinghouse by-products:					
Bone charcoal (bone black)	34.08	15.85	–	–	–
Bone meal, steamed	27.31	12.40	0.42	19.5	0.07

[1]Adapted by the authors from data provided especially for this book by the former International Feedstuffs Institute, now the Feed Composition Data Bank, National Agricultural Library, USDA, Beltsville, Md.

When calcium alone is needed, ground limestone or ground oyster shells are commonly used. Other calcium supplements are: bone meal, calcium gluconate, calcium lactate, dicalcium phosphate, dolomite, and kelp.

The most common supplemental sources of phosphorus are: ammonium phosphate, bone meal, calcium phosphate, colloidal clay, dicalcium phosphate, monosodium phosphate, phosphoric acid, and defluorinated phosphate.

(Also see Chapter 4, Nutrients/Metabolism, Table 4–6, Animal Mineral Chart.)

Precautions Relative to Calcium and Phosphorus Supplements

Earlier experiments cast considerable doubt on the availability of phosphorus when it was largely in the form of phytin. Although wheat bran is very high in phosphorus, containing 1.32%, there was some question as to its availability since most of it occurred in phytin combination. More recent studies, however, indicate that cattle and sheep can partially utilize phytin phosphorus. They can utilize about 60% of the total phosphorus from most plant sources. It must be emphasized, however, that phosphorus availability depends to a large extent on phosphorus sources, dietary supplies of calcium, and adequate vitamin D. Recent work indicates that high calcium levels enhance the formation of the insoluble calcium phytate, whereas high vitamin D levels aid materially in the utilization of phosphorus in the form of phytin. On the other hand, in the case of humans, swine, and poultry, the evidence seems clear that phytin phosphorus is a less satisfactory source of phosphorus.

During World War II, the shortage of phosphorus feed supplements led to the development of defluorinated phosphates for feeding purposes. Raw, unprocessed rock phosphate usually contains from 3.25 to 4.0% fluorine, whereas steamed bone meal normally contains only 0.05 to 0.10%. Fortunately, through heating at high temperatures under conditions suitable for elimination of fluorine, the excess fluorine of raw rock phosphate can be removed. Such a product is known as defluorinated phosphate.

CHELATES

The word chelate is derived from the Greek *chelae*, meaning a claw or pincerlike organ. A chelate is a cyclic or complex ring structure in which a divalent or multivalent metal atom is held through two or more bonds in a coordination complex. Generally, chelates are chemically more stable than complexes in which the mineral element is held through only one chemical bond.

Chelating agents, such as ethylenediaminetetraacetic acid (EDTA) or preparations involving such natural materials as kelp, are sometimes used to increase the availability and absorption of certain minerals. For example, in chicks zinc absorption has been demonstrated to be enhanced through the addition of EDTA.

Chelating forms of various trace minerals, such as iron, copper, and cobalt, are now being marketed as mineral supplements.

Irrespective of whether one adds specialty chelates, however, chelation is involved in mineral nutrition. Chelating substances are found in most feedstuffs and include such compounds as proteins, amino acids, peptides, polyphosphates, sugars, starches, cellulose, citric acid, oxalic acid, and numerous other organic substances.

Those selling chelated minerals generally recommend a smaller quantity of them (but at a higher price per pound) and extol their "fenced-in" properties.

Many claims have been made for the benefit of chelated minerals: (1) greater physical stability, which reduces the tendency for trace mineral separation in feeds; (2) less oxidation of vitamins and their labile derivatives; and (3) better animal performance and/or higher bioavailability.

Perhaps the most important practical question concerning the use of specialty chelated mineral supplements should center around whether their use will make animal production more or less profitable than using alternative sources (inorganic) of mineral supplements.

Vitamin Supplements

As with mineral supplements, careful consideration must be given to the vitamin supplementation of livestock feeds. While the requirements of vitamins are extremely small in comparison with energy and protein, the omission of a single vitamin from the diet of a species that requires it will produce specific deficiency symptoms, thereby reducing production. Moreover, the cost for vitamin supplementation constitutes a very small fraction of the total feed bill.

It is to be emphasized that subacute deficiencies can exist although the actual deficiency symptoms do not appear. Such borderline deficiencies are both costly and difficult to cope with, going unnoticed and unrectified; yet they may result in poor and expensive gains, impaired reproduction, or depressed production. Also, under farm conditions one will usually not find a single, isolated vitamin deficiency. Instead, deficiencies usually represent a combination of factors, and usually the deficiency symptoms will not be clear-cut.

Formerly, a wide variety of feed ingredients was added to livestock rations for their vitamin content. But it was found that the vitamin concentration of feedstuffs varied tremendously, being affected by plant species and part (leaf, stalk, or seed), harvesting, storing, and processing. Generally speaking, vitamins are easily destroyed by heat, sunlight, oxidation, and mold growth. So, today, nutritionists rely on vitamin supplements, which in many cases are chemically pure sources that need to be used only in very minute amounts. In modern feed formulation, premixes often represent the common sense approach to providing vitamins.

For adult ruminants, vitamins A, D, and E are of concern, with A being the one most likely to be deficient. Under ordinary circumstances, ruminants synthesize adequate B vitamins, and vitamins C and K. However, cobalt—a component of vitamin B–12—must be supplied in adequate amounts. Unless they are kept indoors, they usually receive sufficient exposure from direct sunlight to meet their needs for vitamin D. Young ruminants do not have fully functional rumens, with the result that the B vitamins and vitamin K should be provided in adequate amounts in feed until their rumens have developed.

(Also see Chapter 4, Nutrients/Metabolism, Table 4–8, Animal Vitamin Chart.)

VITAMIN A AND CAROTENE

Vitamin A is required by all farm animals. No vitamin A is synthesized in plants; but carotene, a precursor of vitamin A, is found in varying quantities in plants. Since the animal body transforms carotene into vitamin A, it is often referred to as "provitamin A."

The degree of greenness in a roughage is a good index of its carotene content, provided it has not been stored too long. Early cut, leafy green hays are very high in carotene.

Aside from yellow corn, all cereal grains have little carotene or vitamin A value. Yellow corn has only about one-tenth as much carotene as well-cured hay, and even this small amount deteriorates over a period of time.

The most common supplemental sources of vitamin A are: cod and other fish liver oils and synthetic vitamin A. The latter is more stable and currently more widely used.

When vitamin A deficiency symptoms appear, producers should add a stabilized vitamin A product to the ration. Since these products are very reasonably priced, there is very little excuse for vitamin A deficiencies occurring in well-managed operations. But, it is wasteful to feed more vitamin A than is needed. Also, feeding exceedingly high levels of vitamin A over an extended period of time may cause bone fragility, hyperostosis, and exfoliated rumen epithelium. When fed as directed, the vast majority of livestock feeds won't provide excesses of vitamin A. But when higher product levels than needed are used as a sales gimmick, there is hazard that the caretaker will be guided by the philosophy that—"If a little is good, more is better."

Fortunately, animals are able to store vitamin A, primarily in the liver, during periods of abundance to tide them through periods of scarcity. Thus, animals on green pasture store reserves to help meet their needs during the winter feeding period when their rations may be deficient. Mature animals may be able to store sufficient vitamin A in the liver to last 6 months; young animals store much less. Currently, many producers routinely follow the practice of injecting ruminants with large doses of vitamin A, sometimes in combination with vitamins D and E. Since these vitamins are stored in the body for a relatively long period, the injection provides a means of sustaining animals through periods when forages may be vitamin-deficient.

It is generally believed, though not conclusively proved, that stressed animals have a higher vitamin A requirement than those not under stress. Among such stress factors are racing, showing, fatigue, hot weather, shipping, confinement, excitement, nitrate and mold poisoning, and number of animals run together.

The vitamin A potency (whether due to the vitamin itself, to carotene, or to both) of feeds is usually reported in terms of IU or USP units. These two units of measurement are the same. They are based on the growth response of rats, in which several different levels of the test product are fed to different groups of rats, as a supplement to a vitamin A-free diet which has caused growth to cease. A USP or IU is the vitamin A value for rats of 0.30 microgram of pure vitamin A alcohol, or of 0.60 microgram of pure beta-carotene. The carotene or vitamin A content of feeds is commonly determined by colorimetric or spectroscopic methods.

(Also see Chapter 4, Nutrients/Metabolism, Table 4–8, Animal Vitamin Chart.)

VITAMIN D

As with vitamin A, vitamin D is required by all farm animals and humans.

For four-footed animals (cattle, sheep, swine, and horses), both D_2 (the plant form) and D_3 (the animal form) are equally effective, so there is no need to use some of each. With poultry, however, vitamin D_3 is more active than vitamin D_2, and should, therefore, be used.

The most potent supplemental products are obtained by irradiating plant or animal sterols that are subject to activation. Thus, ergosterol produced by plants is irradiated and sold for human use in a variety of forms. Irradiated animal sterol, activated 7-dehydrocholesterol, is most frequently used in poultry feeds in view of the superior value of the D_3 form of the vitamin for this species. Yeast is rich in ergosterol; thus, its irradiation results in a potent source that is used for farm animals other than poultry.

Young animals sometimes develop rickets because of insufficient vitamin D, calcium, or phosphorus. This condition can be prevented by exposing the animal to as much direct sunlight as possible, by allowing free access to a suitable mineral mixture, and/or providing good-quality sun-cured hay. In confinement operations, and in northern areas that do not have adequate sunshine, producers should provide young stock and lactating females with a vitamin D supplement.

The vitamin D requirement is less when a proper balance of calcium and phosphorus exists in the ration.

With vitamin D, as with vitamin A, there is need for adequacy without harmful excesses. Too much vitamin D may harm an animal. Vitamin D toxicity is characterized by calcification of the blood vessels, heart, and other soft tissues, and by bone abnormalities. Also, there is general weakness and loss of body weight. In general, the level of vitamin D intake that might prove harmful to farm animals is so far above the amounts fed for nutritional purposes that extensive studies on toxicity have not been made. Also, there appears to be species and individual difference in the tolerance to high levels of vitamin D. As a result, evidence of vitamin D toxicity in farm animals is fragmentary and inadequate. It appears, however, that it takes massive doses of vitamin D to produce toxic symptoms in farm animals. As a result of several swine studies, the Nebraska researchers concluded that vitamin D can be toxic to swine when fed at levels of 80,000 IU/lb of ration, or above, while many nutritionists subscribe to the view that the vitamin D_3 levels must be in excess of 100 times the requirement to be toxic to poultry.

Other factors pertinent to vitamin D follow:

1. **Vitamin D, and cholesterol and ergosterol.** Most of the commonly used feeds contain little or no vitamin D; yet when animals are exposed to sunlight, there is no widespread need for special supplements containing this factor. Fortunately, the skins of animals and many feeds contain provitamins in certain forms of cholesterol and ergosterol, respectively, which, through the action of ultraviolet light (light of such short wavelength that it is invisible) from the sun, are converted into vitamin D. These certain forms of cholesterol and ergosterol themselves have no antirachitic effect.

2. **Vitamin D limited in feeds.** Of all the known vitamins, vitamin D has the most limited distribution in common feeds.

Very little of this factor is contained in the cereal grains and their by-products, in roots and tubers, in feeds of animal origin, or in growing pasture grasses. The only important natural sources of vitamin D are sun-cured hay and other roughages. The chief vitamin D-rich concentrates include sun-cured hay, cod-liver and other fish oils, irradiated cholesterol and ergosterol, and irradiated yeast.

As might be suspected from the preceding discussion, artificially dehydrated hay contains little vitamin D.

3. **Effectiveness of sunlight in producing vitamin D.** The effectiveness of sunlight is determined by the lengths and intensity of the ultraviolet rays which reach the body. It is more potent in the tropics than elsewhere, more potent in the summers than in the winter, and more potent at high altitudes. The ultraviolet rays are largely screened out by clothing, window glass, clouds, smoke, or dust. Also, some biochemists theorize that the color of the skin of humans is nature's way of regulating the manufacture of vitamin D—that the dark skin of races near the equator filters out excess ultraviolet light.

(Also see Chapter 4, Nutrients/Metabolism, Table 4–8, Animal Vitamin Chart.)

VITAMIN E

Vitamin E is required by a large number of animal species, but the deficiency signs may differ greatly among species and even within the same species. There is no experimental evidence that vitamin E deficiency will cause reproductive failure in cattle, sheep, and goats. Studies on the relation of vitamin E to reproduction in horses have been contradictory.

Vitamin E is widely distributed in plants; leafy forages (especially alfalfa) are a good source. Grains (particularly their germs) are fair sources of vitamin E. Germ oils, such as wheat germ oil, are rich supplemental sources of vitamin E.

Most practical rations contain liberal quantities of vitamin E, perhaps enough except under conditions of work, stress, or reproduction, or when there is interference with its utilization. Rather than buy and use costly vitamin E concentrates indiscriminately, the producer should only add them to the ration on the advice of a nutritionist or veterinarian.

The requirements for vitamin E are influenced by interrelationships with other essential nutrients—increased by the presence of interfering substances, and spared by the presence of other substances that may be protective or that may assume part of its functions, particularly the trace element selenium.

(Also see Chapter 4, Nutrients/Metabolism, Table 4–8, Animal Vitamin Chart.)

VITAMIN K

When vitamin K is deficient, the blood prothrombin level is decreased, and the coagulation time of the blood is increased. This is the main justification for adding this vitamin to the ration. However, vitamin K is widely distributed in normal farm feeds. Also, it appears that it is synthesized in adequate amounts by the intestinal microflora of ruminants and the horse.

When animals are on medication, additional vitamin K may be warranted as the medication may decrease the population of vitamin K-producing organisms in the gut. Additionally, animals consuming sweet clover may ingest sufficient amounts of the antimetabolite, dicoumarol, to create a vitamin K deficiency.

Menadione (vitamin K_3), the synthetic form of the vitamin, is the most widely used commercial source of vitamin K. Also, the following are rich natural sources: alfalfa meal, barley, corn, fish meal, hays (well-cured), milk, pasture (green), peas (green), sorghum grain, soybean meal, and wheat.

(Also see Chapter 4, Nutrients/Metabolism, Table 4–8, Animal Vitamin Chart.)

B-COMPLEX VITAMINS

In the past, it has been commonly assumed that ruminants do not need dietary supplementation with the B-complex vitamins because the rumen microflora synthesize these vitamins in sufficient quantities to fulfill the host's requirement. However, based on recent research, this assumption has been re-examined. There is now experimental evidence that supplemental sources of three of the B vitamins—thiamin, niacin, and choline—need to be provided to ruminants under some conditions. Animals receiving high-concentrate and low-forage rations (*i.e.* feedlot cattle and sheep and high-producing dairy cattle), or stressed animals such as newly weaned or shipped calves or lambs, may receive some benefit from additions of these vitamins. Normally, grazing cattle do not need supplementation with these vitamins. However, lambs moved from overgrazed to more lush pastures may need thiamin supplementation.

Usually, deficiency symptoms of B vitamins are not observed in ruminants. In the case of niacin and choline, however, a response to dietary additions can be measured in terms of weight gain and feed efficiency. Despite the fact that rumen microbes synthesize thiamin and that whole grains contain it, thiamin deficiencies may develop in cattle and sheep and cause a deficiency disease known as polioencephalomalacia (PEM), a disorder of the central nervous system. PEM is thought to be caused by a severe deficiency of thiamin at the tissue level, created by the presence of the enzyme thiaminase, which destroys the thiamin that is in the feed or manufactured by the rumen organisms. (Also see Chapter 5, Nutritional Disorders/Toxins, Table 5–1, Nutritional Diseases and Ailments—Polioencephalomalacia.)

Unlike ruminants, pigs, mink, fish, and poultry have one stomach compartment and no large cecum. As a result, they do not synthesize enough of certain B vitamins. Consequently, these factors must be provided regularly in the ration in adequate amounts if deficiencies are to be averted. This means that the nutritionist and the caretaker should provide a dietary source of water-soluble B vitamins on a daily basis for simple-stomached animals—pigs, mink, fish, and poultry.

Commercially produced, synthetic sources of each of the B vitamins are available. Also, the following natural feeds are excellent sources of most of the B vitamins: green pastures, green hays, yeast, distillers' solubles, and animal and marine products.

(Also see Chapter 4, Nutrients/Metabolism, Table 4–8, Animal Vitamin Chart.)

VITAMIN C (ASCORBIC ACID, DEHYDROASCORBIC ACID)

A dietary need for vitamin C is generally limited to humans, guinea pigs, monkeys, and certain fruit bats. Hence, there is no need to add this vitamin to most animal rations. However, since it has been demonstrated that some types of fish, such as the catfish, require dietary vitamin C when placed in stressed conditions, it may be advisable to supplement vitamin C for fish in intensive production.

VITAMIN IMBALANCES

Experiments have shown that the amounts needed of certain vitamins may be affected by the supply of another vitamin or of some other nutritive essential. Also, it is known that excess fortification of the animal's diet with certain vitamins may prove more detrimental than helpful. Thus, harmful imbalances should be avoided; and vitamins should be provided on the basis of recommended allowances. Also, when fortifying with vitamins, consideration should be given to the vitamins provided by the ingredients of the normal ration, for it is the total composition of the feed that counts.

UNIDENTIFIED FACTORS

In addition to the vitamins as such, certain unidentified or unknown factors are important in animal nutrition. They are referred to as *unidentified* or *unknown* because they have not yet been isolated or synthesized in the laboratory. Nevertheless, rich sources of these factors and their effects have been well established. A ration that supplies the specific levels of all the known nutrients but which does not supply the unidentified factors may be inadequate for best performance. There is evidence that these production factors exist in dried whey, marine and packinghouse by-products, distillers' solubles, antibiotic fermentation residues, yeasts, alfalfa meal, and certain green forages. There is also evidence that at least one unknown hatchability factor is in fish solubles and green forage. Most of the unidentified factor sources are added to the ration at a level of 1 to 3%.

ADDITIVES, IMPLANTS, AND INJECTIONS

More than 1,000 drug products are approved by the Food and Drug Administration (FDA) for use by livestock and poultry producers. This includes additives, implants, and injectables, along with other drugs that are used to fight diseases and protect animals from infections. Two other statistics which point up the important role of drugs in animal production are: (1) 8 out of every 10 animals raised for food in the United States receive some drugs during their lifetime; and (2) chemicals that regulate growth, modify the rumen's activity, and/or improve feed efficiency increase U.S. meat, milk, and egg production approximately 15% each year. Used properly, these drugs enable livestock producers to provide safe and wholesome meat, eggs, and milk to consumers at lower costs than would otherwise be possible. Used improperly, however, these drugs can be hazardous to consumers.

Unfortunately, some of the drugs used in providing this bountiful supply of food can leave potentially harmful residues. For this reason, more than half of the drug products approved by FDA require drug withdrawal times, in order to protect consumers from drug residues.

Consumers are very much aware of what is in their food. While they enjoy the price and supply benefits of modern food production technology, they want to be assured of the wholesomeness of the food they eat.

Thus, livestock producers have the unenviable task of choosing the right drug(s) to maximize rate and efficiency of production, while, at the same time, observing FDA regulations and protecting the consumer. Under such circumstances, they should carefully analyze all the information presented by each company in support of its product. Also, they should study the results of unbiased experimental work, as reported in both scientific literature and popular articles; and they should sound out reliable users of the product. Finally, in the United States, they must comply with FDA regulations.

When considering the use of a feed additive or implant, the following questions should be asked:

1. **What are the specific needs of the animal?** As intensity of production increases, the need for production stimulants increases.

2. **Does the additive have a withdrawal period or a product discard period?** Quite often a drug will have a withdrawal period requirement before the animal can be slaughtered or before the products produced, such as milk or eggs, can be marketed. If a withdrawal period is required, the producer must have separate storage facilities for unmedicated and medicated feed. Additionally, feed mixing facilities must be thoroughly cleaned after mixing medicated feed. If residues are found in the animal products, the producer is subject to prosecution.

The ultimate guide to using any drug properly is the label or tag. These should be read and followed carefully.

3. **Can the additive be used in combination with other additives?** The FDA has strict regulations as to which additives can be used in combination. If there is any doubt, the producer should contact the county agent or a reputable feed company.

4. **What is the best form of the product to be used?** There are advantages and disadvantages to using feed additives as opposed to implants. Likewise, the active ingredient of one product may be more stable or readily available than that of a competing brand.

5. **What is the cost?** Quite often the true value of an additive cannot be determined until it has been used. The ultimate criterion for evaluation is the increased income over the cost of the additive. If an additive increases production by $1 per head, but costs $1 per head to use, obviously there is no financial benefit from its use.

Today, feed additives and implants constitute about two-thirds of the animal drugs sold. But the development of new products is being stymied by the high cost of research and development, overregulation by the government, and consumer concern about adulteration of foods. However, recombinant DNA (gene splicing) technology may make it possible to produce certain additives and implants more cheaply and in large volume in the future.

Feed additives and implants constitute a diverse group—so diverse that it is difficult to classify all of them as to mode of action or function. For this reason, some are grouped whereas others are merely listed alphabetically in the sections that follow. **NOTE WELL**: Recommended additives or implants for each animal species, along with pertinent details relative to their use, are presented in the separate chapters devoted to each class of animal, Chapters 19 to 28 of this book.

Abortifacients

An abortifacient is a drug or other agent that induces abortion. In the livestock industry, the primary use of abortifacients is to abort feedlot heifers. Prostaglandins and prostaglandin analogues are the abortifacients of choice during the first 150 days of pregnancy; beyond 150 days pregnancy, additional products, such as dexamethasone or estradiol, may be used.

The economic loss that accrues from pregnant feedlot heifers was pointed up by two different studies made and reported by Monfort of Greeley, Colorado.[1] As a result of a 1983 survey of a number of cattle feeders and packers, Monfort reported the following relative to feedlot heifers: (1) a pregnancy rate of 16.5% on incoming feedlot heifers, (2) feeders placed a lower value of $30.32 per head on pregnant vs. open heifers, (3) lower dressing percentage at slaughter of 3.3% on pregnant vs. open heifers, and (4) a cost of $5.29 per head to pregnancy check and attempt an abortion on each incoming pregnant heifer. Additionally, in an actual study of 10,000 head of heifers and 1,000 heiferettes (heiferettes are large heavy heifers possessing nearly the size and development of a mature cow) slaughtered at Monfort's Greeley, Colorado plant in 1983–84, it was found that pregnancy lowered the dressing percentage on heifers by 5.6%, and on heiferettes 6.1%. Thus, the bottom line is that feeding pregnant heifers is costly.

In light of the preceding facts, the following management options may be considered when feeding heifers of unknown pregnancy status:

1. Feed heifers the same as steers, and meet the calving problems (difficult births, and caring for newborn calves) as they occur.
2. Buy only open or spayed heifers, the supply of which is limited.
3. Pregnancy examine all heifers and use an abortive agent on the pregnant ones, according to directions.

By pregnancy testing and the use of an abortifacient, the termination of early pregnancy can be brought about. However, the cost and setback in performance may not justify such action.

(Also see Chapter 19, Feeding Beef Cattle, section on "Spayed Heifers."

Activated Carbon (Charcoal)

In recent years, sporadic cases of pesticide contamination in livestock have been reported. In a number of these cases, activated carbon, or charcoal, has been fed to the contaminated animals to facilitate excretion of the pesticides. It has been suggested that the activated carbon absorbs the pesticide as it is excreted into the digestive tract and recycled in the bile. However, the use of activated carbon is controversial because the success rate has been highly variable.

[1]Bennett, Bill, *Animal Nutrition & Health,* "The Pregnant Feedlot Heifer," May 1985.

Additives That Enchance Market Value

Traditional methods of breeding and selecting animals to meet changing consumer demands take time—lots of time. So, the race is on to speed up this transition, with genetic engineering and additives leading the way. Moreover, it no longer suffices to have additives that stimulate growth and improve feed efficiency. The additives of tomorrow must also improve the composition and quality of the product produced—the meat, milk, and eggs. That's not easy, but the stakes are high! Among the types of product-enhancing additives currently being used or tested experimentally are the following:

- **Xanthophylls and carotenoids in poultry**—Many consumers believe that a deep yellow color of broiler skin/shanks and egg yolks is indicative of top quality. Consequently, the poultry producer may receive a premium price for such cosmetically esthetic products. Xanthophylls or carotenoid additives are commonly used for this purpose. (See Chapter 24, Feeding Poultry, section on "Additives.")
- **More lean/less fat**—Several products for the purpose of producing more muscle/protein and less fat in beef cattle, lambs, hogs, and broilers are being tested.
- **Lower cholesterol**—Consumers are cholesterol conscious; so, scientists are developing and testing products that will lower the cholesterol in meat, milk, and eggs.

The Bottom Line: Most people have strong food preferences; some justified, others fickle. Often it is easier to change a product than a food fad. In recognition of this fact, scientists and livestock producers are researching, discovering, creating, and advancing; working ceaselessly to meet consumer demands in a changing world.

Additives That Physically Aid Digestion

In some types of livestock—notably poultry and ruminants—the physical characteristics of the feed can markedly alter its digestibility. Two products, which have received considerable attention as physical aids to digestion are grit for poultry and roughage substitutes for ruminants.

GRIT

Since poultry do not have teeth to facilitate grinding of feed, most grinding takes place in the thick-muscled gizzard. The more thoroughly feed is ground, the more surface area is provided for digestion and subsequent absorption. Hence, when hard, coarse, or fibrous feeds are fed to poultry, grit is sometimes added to supply additional surface for grinding within the gizzard. Additionally, grit serves to break down ingested feathers and litter which can sometimes lead to gizzard impaction. When mash or finely ground feeds are used, the value of grit is greatly diminished.

Oyster, clam, coquina shells, and limestone are sometimes

used for grit. Being relatively soft and calcareous, they provide a source of calcium as they, too, are ground in the process. Gravel and pebbles have been used successfully as long-lasting sources of grit. Several granite products are available commercially.

ROUGHAGE SUBSTITUTES

Roughages are bulky, coarse feeds which are high in fiber (cellulose and related compounds) and thus low in digestibility; such as hay, straw, silage, wheat bran, corncobs, and cottonseed hulls. In human nutrition, roughage has been a long-standing means of alleviating constipation. In ruminants, roughage promotes rumination and rumen function and the production of milk of normal fat content. The capacity of feed sources of roughage to elicit these effects varies with the composition and physical character of the fiber source. Thus, there must be a balance between lignification and digestible cellulose; for example, one cannot feed plastic hay to a lactating cow and maintain milk fat because acetate-producing and cellulose-digesting bacteria must be maintained. Also, fine grinding and pelleting decrease the value of roughage for ruminants because of the loss of the scratch factor and reduced rumination.

The above statement, which has been well documented by research, clearly shows that ruminants require some form of roughage or scratch factor in their feed in order to maintain a healthy, functioning rumen and produce milk with normal fat content. However, as a source of energy, roughages are sometimes too expensive in comparison with concentrate feeds. Under such circumstances, ruminants may be fed high-concentrate rations.

Polyethylene and oystershell have been tested relative to their respective efficacies as roughage substitutes. While several studies have indicated some improvement from the use of these products, some nutritionists feel that they are of little or no value. The most common practice in feeding high-energy feeds is to incorporate in the ration some source of natural roughage, such as hay, cottonseed hulls, almond hulls, or ground corncobs.

(Also see Chapter 4, section on "Availability of Fiber"; and the section on "Buffers" in this chapter.)

Anthelmintics (Wormers; Vermifuges)

Anthelmintics are drugs used to control worms.

The prevention and control of parasitic infection is one of the quickest, cheapest, and most dependable methods of increasing production with no extra animals, no additional feed, and little more labor. This is important, for, after all, the farmer or rancher bears the brunt of this reduced production, wasted feed, and damaged products.

Knowing what kind of worm(s) is present in an animal is the first requisite to the choice of the proper anthelmintic. Since no one drug is appropriate or economical for all conditions, the next requisite is to select the right one; the one which, when used according to directions, will be most effective and produce a minimum of side effects on the animal treated. Coupled with knowledge of the kind of worm(s) present, an individual assessment of each animal is necessary. Among the factors to consider are age, pregnancy, illnesses and medications, and the method by which the drug is to be administered. Some drugs characteristically put animals off performance for several days after treatment, whereas others have less tendency to do so. Some drugs are unnecessarily harsh or expensive for the problem at hand, whereas a safe inexpensive alternative would be more suitable.

Each livestock establishment should, in cooperation with the local veterinarian, evolve with a parasite control program and schedule. It is recommended that several different wormers be used, and that they be rotated. Also, a schedule of treatments should be prepared, based on knowledge of the life cycles of the various parasites.

From time to time, new vermifuges, or wormers, are approved and old ones banned or dropped. When parasitism is encountered, therefore, it is suggested that the producer obtain from local authorities the current recommendation relative to the choice and concentration of the vermifuge to use. This information can be obtained from a county extension agent, entomologist, veterinarian, or agricultural consultant.

Antibiotics

Antibiotics are substances which are produced by living organisms (molds, bacteria, fungi, or green plants) and which have bacteriostatic or bactericidal properties. They are the most widely used of the microbial drugs.

The newer knowledge of antibiotics dates from the discovery of penicillin by Dr. Alexander Fleming, a British scientist, at St. Mary's Hospital in England, in 1928.[2] Quite by accident, a stray mold spore floated in on the breeze, and landed on a culture plate of bacteria with which Dr. Fleming was working. It inhibited the growth of bacteria. Dr. Fleming correctly interpreted his observation: the possible value of the mold in the treatment of disease, thus ushering in the antibiotic era. However, penicillin did not come into prominence until 10 years later; and it was not until 1944 that streptomycin, the second most widely known of the antibiotics, was discovered by Wakeman, a soil microbiologist, and his colleagues at the New Jersey Station.

Penicillin, streptomycin, and other antibiotics were rushed into use against a long list of human ailments. They were hailed as the *wonder drugs* of our time, but no one thought of them, even remotely, as something new to be added to livestock feeds. Finally, in 1949, quite by accident, while conducting nutrition studies with poultry, Stokstad and Jukes of American Cyanamid Company obtained startling growth responses from feeding a residue from Aureomycin production. Later experiments revealed that the supplement used by Stokstad and Jukes supplied residual amounts of the antibiotic chlortetracycline (Aureomycin). Such was the birth of feeding antibiotics to livestock.

In addition to their use as growth stimulators, antibiotics are used as nutritional stimulants to promote better feed efficiency in ruminants and swine, and to increase egg production, hatchability, and shell quality in poultry. They are also added to feed in substantially higher quantities to remedy pathological problems.

[2]Actually, the presence of antibiotics was known much earlier than the discovery of penicillin, but no commercial use was made of them.

MODE OF ACTION OF ANTIBIOTICS

Numerous theories, each with convincing support, have been hypothesized with respect to the mode of action of growth-stimulating antibiotics.

One fact is well substantiated. Antibiotics are effective in controlling certain environmental stresses, as evidenced by the fact that animals raised under germ-free conditions exhibit no improved responses to them. Additionally, it has been observed that pigs raised in clean pens which had not previously housed other pigs, did not exhibit any improved responses to antibiotic supplementation. These facts indicate that exposure to everyday stresses may reduce performance. However, antibiotics are not a means whereby one can ignore good sanitation and management practices.

The following six mechanisms by which antibiotics increase rate of gain and feed efficiency have been suggested:

1. **Disease control.** There is evidence that antibiotics exert a "disease defense effect" by suppressing microorganisms which might otherwise produce subclinical diseases in the animals. Although the effect of the microorganisms might be so mild that there would be no clinical symptoms of disease, it could still slow the animal's growth.

2. **Nutrient-sparing effect.** When antibiotics are fed, the populations of microorganisms in the animal's digestive tract change. It follows that if the number of microorganisms which manufacture nutrients which the animal can use increases, or the number of microorganisms which compete with the animal for nutrients decreases, the animal will be able to grow more on the same amount of feed. Studies indicate that antibiotics have a sparing effect on some vitamins and amino acids.

3. **Metabolic effect.** According to this theory, there is a "metabolic effect," by which the antibiotic affects the body functions. Since some of the antibiotics used as growth promotants are not absorbed from the digestive tract into the animal's body, they could not act in this way. However, other antibiotics could exert a metabolic effect.

4. **Feed and water intake.** Antibiotics usually increase feed and/or water intake.

5. **Toxic waste products or toxins.** Antibiotics may inhibit the growth of organisms which produce toxic waste products or toxins.

6. **Digestion and absorption.** Antibiotics may improve the digestion and subsequent absorption of certain nutrients. Thus, it has been noted that the intestinal wall of animals fed antibiotics is thinner than the intestinal wall of those not fed antibiotics, which may promote better utilization of nutrients.

More than likely, all six modes of action apply to antibiotics in general, but not to any specific antibiotic—for antibiotics differ. Also, there are dose-related responses. Since antibiotics are most effective in stressful conditions where animals are more likely to be exposed to disease and more susceptible to it, the disease-control effect is probably the most important of the six modes of action.

IMPORTANCE AND USE OF ANTIBIOTICS IN ANIMAL PRODUCTION

Antibiotics are used in the following two ways in livestock and poultry production:

1. **Low levels in feeds.** Low, subtherapeutic doses of antibiotics are included in livestock and poultry feeds to increase growth rate, and/or feed efficiency, and help prevent bacterial diseases. The effects of antibiotics on rate of gain or feed efficiency of each animal species is presented in the respective chapters devoted to each class of livestock.

In addition to improvement in rate and efficiency of gains, low levels of antibiotics usually (1) reduce the incidence of diarrhea in young animals, especially in young mammals deprived of colostrum; (2) lessen the tendency of animals to go off feed; (3) reduce enterotoxemia and death loss of lambs fed high-grain rations; (4) lower the incidence of abscessed livers of cattle fed high-grain rations; (5) control atrophic rhinitis in swine; (6) reduce respiratory diseases, non-specific enteritis (blue comb), and infectious sinusitis of poultry; and (7) lessen death losses in rabbits due to enteritis.

Routine low dose feeding of antibiotics is standard practice in most livestock and poultry production operations in the United States. The vast majority of poultry, swine, beef cattle, and dairy calves raised in this country receive antibiotics in their feed at some point in their lives.

It is noteworthy, too, that the addition of antibiotics to animal feeds began in the early 1950s along with the shift of animal production to larger units, greater concentration of animals, and close confinement. Without the use of antibiotics, major infectious disease problems might have prevented this transition.

2. **High (therapeutic) levels in feeds.** High levels of antibiotics are used to treat diseases, just as they are in human medicine. Used for short periods, high levels have been quite effective in treating anaplasmosis and controlling shipping fever in cattle, bacterial enteritis in swine, and respiratory diseases, diarrheas, fowl cholera, typhoid, and breast blisters in poultry. Also, high levels of antibiotics have been useful for preventing and treating stresses associated with transporting animals and their adjustment to a new environment.

About 6 million lb of antibiotics, at a cost of more than $250 million, are used in animal production in the United States each year. This represents about 45% of the total usage of these drugs. Approximately 80% of the poultry, 75% of the swine, 75% of the dairy calves, and 60% of the beef cattle in the United States are fed antibiotics at some time in their lives. For every dollar livestock producers spend on antibiotic-enriched feed, they get a return of 2 to 4 dollars.

Economically, antibiotics are of benefit to both producers and consumers of animal foods. They reduce the cost of animal production; therefore, they reduce the retail price of meat and poultry (and to a lesser extent, milk and eggs). The Council for Agricultural Science and Technology (CAST) is authority for the statement that banning the use of penicillin and tetracyclines in animal feeds would make for added food costs to U.S. consumers of more than $3.5 billion per year, in the short term, if no substitutes were available.[3]

Most antibiotics are approved for cattle, swine, chickens, and turkeys. Few are approved for sheep, horses, rabbits,

[3]Hays, Virgil W., *et al., Antibiotics In Animal Feeds,* Report No. 88, Council for Agricultural Science and Technology (CAST), Ames, Iowa, 1981.

ducks, pheasants, and quail. No approvals have been given for goats, geese, dogs, and cats. The reason for this situation is the cost of obtaining approval in relation to potential sales.

SAFETY AND FUTURE OF ANTIBIOTICS AS FEED ADDITIVES

Although the practice of low level (subtherapeutic) feeding of antibiotics has been routine for decades, it has recently come under fire. The concern: Are animals serving as the *training ground* for bacteria that, in effect, learn how to disarm the antibiotics that we use against them? Are antibiotic-laced feeds encouraging the growth of antibiotic-resistant bacteria, which can be transmitted to people and cause sickness?

Those favoring low-level feeding of antibiotics (at a dose much lower than is required to treat an illness caused by bacteria) claim (1) that antibiotics are essential to raising healthy animals under today's crowded, confined, stressful conditions; and (2) that banning antibiotic feeding would cost American consumers hundreds of millions of dollars in higher food prices yearly. Opponents counter (1) that human health is being sabotaged by the feeding of antibiotics, and (2) that antibiotics are being substituted for poor livestock sanitation and management.

Scientists generally agree that the use of antibiotics (whether in livestock or people) results in the proliferation of bacteria that are resistant to one or more antibiotics. Currently, for example, bacteria causing everything from diarrhea to sore throat to meningitis in people, can be killed only when antibiotics are given in much higher doses than were originally required, and in some cases they are completely unaffected by the antibiotics that once destroyed them. Since some bacteria are common to both animals and humans, the concern is that disease organisms resistant to treatment by antibiotics will develop in animals and spread to humans. But scientists disagree on the key issue involved in the low-level feeding of antibiotics: Are antibiotic-resistant organisms causing diseases in people as a direct consequence of feeding antibiotics to animals? The answer to the question centers around the ability of certain bacteria to develop resistance to antibiotics and to spread this resistance to other bacteria, a phenomenon known as *episome.* This well-documented ability involves what are termed *resistance factors* (RF). *An episome is an independent genetic element occurring in addition to the normal cell genetic makeup, which can be transmitted to other bacteria, and which can replicate either as an incorporated piece of the host's genome (chromosomal material) or as a separate unit within the bacteria.* This means that an episome (1) can be fused with the chromosomal material of the host bacteria or (2) can be a separate piece of genetic material in the cytoplasm.

Plasmids are episomes which never become part of the host chromosome but rather become a self-replicating unit; they are short segments of DNA which carry genes for resistance to drugs. Plasmids are the major equipment which bacteria use for defending themselves against antibiotics. A set of genes in a plasmid can confer resistance against a whole list of antibiotics. Worse yet, plasmids can be easily passed back and forth among bacteria, even between bacteria of different types. Thus, the ingenuity of one strain may benefit all others.

Circumstantial evidence, without positive proof, that antibiotic-resistant bacteria in animals can lead to disease in humans was reported in the respected *New England Journal of Medicine.*[4] The circumstantial evidence: A particular strain of *Salmonella* appeared to be linked to a herd of cattle that had been fed the antibiotic chlortetracycline. These bacteria, in turn, caused 18 cases of diarrhea in humans, one case of which was fatal. The particular strain of *Salmonella* was resistant to multiple antibiotics. All 18 of the victims studied had handled or eaten ground beef presumed to have originated from the same antibiotic-fed cattle the week before becoming ill. Two-thirds of them had also been taking antibiotics to treat various minor illnesses. The conjecture: *Salmonella,* already resistant to antibiotics as a result of their being hosted by antibiotic-fed cattle, arrived in the intestines of their human hosts where they found that the bacteria normally present had been depleted by the humans' use of antibiotics. With competition reduced, the resistant *Salmonella* easily gained a foothold and established an infection, which proved highly resistant to treatment with an antibiotic. The investigators were unable to obtain samples of the beef in question, but indirect evidence linked the *Salmonella* organisms found in these people to those present in beef cattle. Some scientists have accepted this study as showing transfer of resistant bacteria from animals to people through the meat. Others dispute the conclusion, claiming that the evidence is too indirect.

The New England Journal of Medicine carried a second article on drug-resistant salmonella in the March 5, 1987 issue.[5] Based on a California study, a team of 15 researchers from the Federal Centers for Disease Control, Atlanta, the USDA Food Safety and Inspection Service, the California Department of Health Services, and the Los Angeles County Department of Health Services, concluded "that food animals are a major source of antimicrobial-resistant salmonella infections in humans, and that these infections are associated with antimicrobial use on farms." The study involved tracing drug-resistant salmonella infections in humans to farms where antibiotics were used in dairy cattle that were later processed into ground beef. The researchers found that those who got the illness had used tetracycline and penicillin prior to the onset of the illness and had eaten undercooked hamburger (ground beef) containing a drug resistant strain of salmonella within the week just prior to the start of the illness.

In both *The New England Journal of Medicine* reports summarized above, cattle and beef were implicated as the sources of drug-resistant salmonella. But, in the late 1980s, poultry was accused of being an even greater source of salmonella poisoning of people; it was estimated that about 40% of the poultry meat was contaminated. Although thorough cooking will destroy the microbes, and proper storage following cooking, will keep poultry wholesome, this will not alleviate concerns. So, programs and studies were initiated to minimize the contamination of meat and other food items of resistant salmonella.

---

[4]Holmberg, S. D., *et al,* "Drug-Resistant *Salmonella* From Animals Fed Antimicrobials", *The New England Journal of Medicine,* Vol. 311, No. 10, September 6, 1984, pp. 617–621.

[5]Spika, J. S., *et al.,* "Chloramphenicol-Resistant *Salmonella newport* Traced Through Hamburger to Dairy Farms," *The New England Journal of Medicine,* Vol. 316, No. 10, March 5, 1987, pp. 565–570.

Although the cause is elusive, the fact remains that more and more drug-resistant Salmonella bacteria are infecting animals and moving up through the food chain to humans.

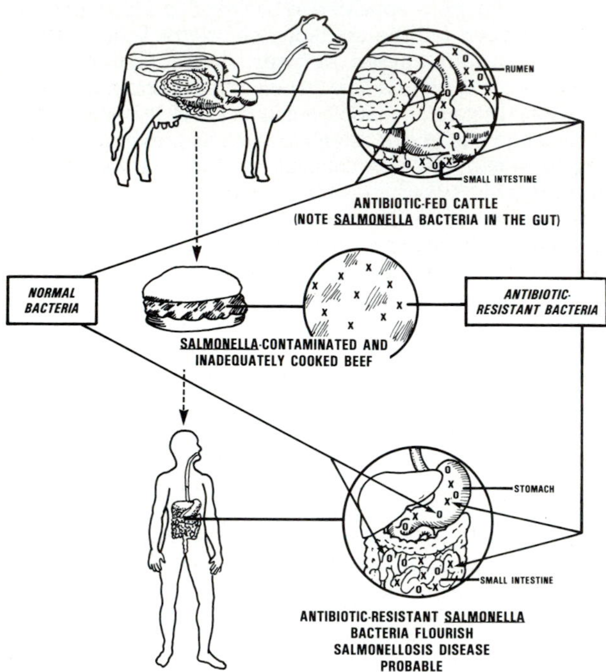

Fig. 13-1. Theory of antibiotic resistance—how it could happen. This shows how antibiotic-resistant *Salmonella* could cause Salmonellosis disease in humans. The steps: (1) antibiotic-fed cattle produce antibiotic-resistant *Salmonella* bacteria; (2) at slaughter, *Salmonella*-contaminated beef may be produced; and (3) when *Salmonella*-contaminated beef is inadequately cooked and consumed by people, antibiotic-resistant *Salmonella* may flourish and produce Salmonellosis disease.

(**NOTE WELL**: The U.S. meat supply is the world's safest. The above figure is for the purpose of illustrating a theory. M.E.E.)

It is noteworthy (1) that in England in the 1960s, the rise of resistant organisms, especially *Salmonella,* led to the removal of therapeutic antibiotics from subtherapeutic use; and (2) that 10 to 25% of the identified cases of *Salmonella* infections in the U.S. are now drug-resistant. There is general agreement that antibiotic use is responsible for the rise in resistant bacteria, but controversy continues over whether antibiotic use in people or animals has contributed most to the environmental pool. The issue is complicated by the facts that both U.S. groups receive the same kinds of antibiotics and each receives about half of the total antibiotics produced.

To date, however, all cases of animal-to-human spread of disease from resistant bacteria have originated in animals in which antibiotic use has been either at therapeutic use or ill-defined. A clear association between low-level (subtherapeutic) use and human disease has not been shown. Moreover, the defenders of the subtherapeutic use of antibiotics for animals charge that the resistance to antibiotics results mostly from their use in human treatment in hospitals and clinics, rather than from antibiotics in animal feeds; then, they add: 'The use of antibiotics in food animals may actually improve human health by decreasing the number of diseased animals and reducing transference of disease organisms to the human population.'

Since there is risk to human health from the use of antibiotics in feed, the careful monitoring and periodic review of the human health impact of the subtherapeutic use of antibiotics in animal feeds is prudent. But decisions should be based on scientific facts rather than political expediency.

In 1977, the federal Food and Drug Administration (FDA) proposed a ban on the use of penicillin and tetracycline at subtherapeutic levels in animal feed, as growth promotants, because of concern that antibiotic-resistant bacteria would spread from animals to humans, rendering penicillin and tetracycline useless in combating human illness. The U.S. Congress stopped the ban and requested the National Academy of Sciences to study the issue. In 1980, the Academy reported that existing studies neither proved nor disproved the human health hazard of antibiotics in feed. Thereupon, Congress called for further studies. More arguments and more petitions followed. Finally, in 1985, after an elapse of 8 years following the ban proposed by FDA, the Secretary of Health and Human Services ruled that the therapeutic feeding of antibiotics to livestock does not pose any "imminent hazard."

But the antibiotic issue won't go away! There will be more petitions, more hearings, more research, and more rulings. Scientific facts are subject to change when new findings indicate the need to revise prevailing views. Hopefully, the question of actual hazard will be decided by scientific facts.

• **Some facts pertinent to the antibiotic issue**—The following facts are pertinent to the issue of antibiotics in feed:

1. **Current federal (FDA) regulations already limit the use of antibiotic feed additives.** In cattle rations, FDA prohibits the use of antibiotics in combination with ionophore feed additives, such as Bovatec and Rumensin. The only antibiotic that can be used in combination with Rumensin is Tylan. Also, no two feed additives can be used in combination unless those antibiotics have been researched for residue problems, which is a lengthy and expensive process.

2. **Use of antibiotics in animal feeds by other countries.** Australia, Canada, Japan, Latin America, and Taiwan, all use antibiotics in animal feeds. In Czechoslovakia, the addition of antibiotics to animal feeds is mandatory. Great Britain divides all antibiotics into two categories: (a) those used as feed additives, and (b) those restricted to prescription by licensed veterinarians. In 1986, Sweden became the first country to ban the use of growth promoting antibiotic additives in animal feeds.

3. ***Old* antibiotics still potent growth promotants.** Poultry scientists at the University of Wisconsin have published papers documenting the continued effectiveness in chicken and turkey feeds of penicillin and tetracycline, two *old* antibiotics that have been around and used since the early 1950s. The effects of the *old* antibiotics are comparable or superior to the *new* antibiotics, although it is known that bacteria resistant to penicillin and the tetracyclines have been present in the environment of chickens and turkeys for more than 30 years.

4. **Antibiotics specific for animals are increasing.** A number of new antibiotics specifically for animals have been developed, and more will come. These products improve livestock performance in rate of gain and/or feed efficiency, but have no medical applications, and, therefore, pose no

health hazard with regard to bacteria becoming resistant to them. Among such antibiotics designed specifically for animals, but which will not cause resistance to antibiotics in humans are virginiamycin, bambermycin, bacitracin, and the ionophores, monensin and lasalocid.

APPROVED ANTIBIOTICS AND ANTIBIOTIC COMBINATIONS

The list of antibiotics and antibiotic combinations approved as feed additives is long—and growing longer; hence, space limitations in this book will not permit a listing of each of them, along with giving their FDA status, use level, and indications of use. For up-to-date and detailed information on this subject, feed manufacturers and livestock producers are referred to the *Feed Additive Compendium,* published annually and up-dated monthly, by The Miller Publishing Company, 12400 Whitewater Drive, Minnetonka, MN 55343.

Antioxidants

Many of the most useful products in the feed industry are readily subject to autoxidation; among them, fat sources, fish by-products, animal-rendered products, essential vitamins, and pigmentation sources. Feeds which are high in unsaturated fatty acids are especially prone to autoxidation and subsequent rancidity.

Most animals will refuse to eat spoiled feed. But when feed is limited, they may consume it, with digestive disturbances resulting in many cases. It has been demonstrated that cows consuming oxidized oils produce milk with an off-flavor. To curb the oxidation of feeds, antioxidants are routinely added to many livestock feeds.

Antioxidants are compounds that prevent oxidative rancidity of polyunsaturated fats. It is important that rancidity of feeds be prevented because it may cause destruction of vitamins A, D, and E, and several of the B-complex vitamins. Also, the breakdown products of rancidity may react with the epsilon amino group of lysine and thereby decrease the protein and energy values of the ration. These effects can be prevented by inclusion in the ration of an effective antioxidant such as ethoxyquin (6-ethoxy-1, 2-dihydro-2, 4-trimethylquinoline), BHT (butylated hydroxytoluene), or BHA (butylated hydroxyanisole); with these products used singly or in combination with each other.

Most antioxidants function by providing the unsaturated chemical bond of a fatty acid with an opportunity to combine with a hydrogen molecule. But this is a numbers game! There must be an overabundance of antioxidants to go around; otherwise, oxygen will take over as the dominant chemical partner with fat.

Vitamin E can serve as an antioxidant both in the feed and in the cell of the animal ingesting the feed. Such antioxidants as ethoxyquin, BHT, or BHA are unable to prevent peroxidation within the cell; consequently, they cannot reduce the dietary requirements for vitamin E.

Oxidation reactions are accelerated by high temperature, light (ultraviolet and blue), ionizing radiation, peroxides (including oxidized fats), lipoxidase enzymes, organic ion catalysts (hemoglobin), and trace metals (copper).

Oxidation reactions are inhibited by refrigeration, exclusion of light, exclusion of oxygen, destruction of enzymes, metal deactivators, and antioxidants.

Arsenicals

Arsenilic acid and sodium arsanilate are the most widely used growth-promoting arsenicals. They are FDA-approved for use alone or in certain drug combinations for chickens, turkeys, and swine. When used alone according to directions, they increase rate of gain and improve feed efficiency of chickens, turkeys, and swine; improve pigmentation in growing chickens and turkeys; increase egg production in layers; and prevent coccidiosis in chickens and dysentery in swine. Arsenic compounds are toxic; hence, they should be used with care and according to manufacturer's directions.

Bloat Control Products

Three products are approved by FDA for bloat control; namely, poloxalene (trade name, Bloat Guard), oxytetracycline (trade names, Terramycin or Neo-Terramycin), or laureth-23 (Enproal Bloat Blox). (See Chapter 5, Table 5-1, Nutritional Diseases and Ailments—Bloat [Feedlot] and Bloat [Pasture]).

Also, two ionophores (lasalocid [trade name, Bovatec] and monensin [trade name, Rumensin]) inhibit gas formation and decrease methane production in the rumen, thereby reducing feedlot bloat. (See section on "Ionophores" in this chapter.)

These products are effective preventives when used according to manufacturers' directions, but they should be accompanied by proper management.

Buffers

Buffers are substances which lessen the change in hydrogen ion concentration produced by adding acids or alkalis to ruminant rations. The counterpart substances for humans are called *antacids.*

When used in beef, dairy, and sheep rations, a buffer chemically maintains a balanced pH in the animal's digestive system. The pH, which is a measure of the level of acid and alkali in a solution, is expressed in numbers from 0 to 14. Acids are on the low end of the scale—below 7; alkalis (bases) are on the upper end—7 or above. A rumen pH between 6.2 and 6.8 is generally considered best for optimum digestion and rumen function in ruminants, although this will vary with feeding regimen and time after feeding.

When the rumen pH drops from a normal of about 6.5 to a range of 5.0 to 6.0, milk fat decreases; and when it goes on down to 4.0, there is acidosis.

Ruminants were developed as forage utilizers. The unique arrangement of four-compartment stomachs enables them to consume fibrous feedstuffs which are digested and used to meet the energy demands of the animal. This is accomplished by a large microbial population that inhabits the rumen. However, when today's ruminants are in high production, especially feedlot cattle and lambs and high-producing dairy cows, they are routinely given high-concentrate and low-forage rations to maximize production. Characteristically, such rations lower the pH of the rumen. The lower ruminal pH of high-concentrate rations can be attributed to decreased saliva secretion, increased energy intake, and increased rate of carbohydrate degradation. The saliva contains bicarbonates and phosphates that buffer the ruminal contents and help maintain the pH between 6.5 and 7.0. The increased feed intake of high-concentrate rations along with more rapid carbohydrate degradation make for more rapid feed conversion to volatile fatty acids in the rumen.

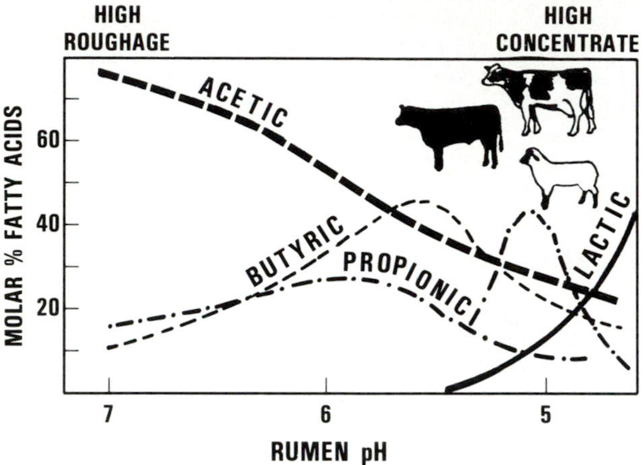

Fig. 13-2. Interrelationships among volatile fatty acids (acetic, butyric, and propionic), lactic acid, and pH during transition from high-roughage to high-concentrate rations. (Source: Adapted by the authors from *Buffers in Ruminant Physiology and Metabolism,* Edited by M. S. Weinberg and A. L. Sheffner, Published by Church & Dwight, Inc., New York, N.Y., 1976, p. 92.)

Fig. 13-2 summarizes the important changes encountered in rumen conditions. Obviously, this chart cannot be applied to any specific case; rather, it portrays the general findings of research. When high-roughage rations are fed, the pH of rumen fluid tends to remain relatively high, with acetic acid as the predominant fatty acid. As more concentrates are fed, propionic acid and butyric acid tend to increase relative to acetic acid. Finally, as the concentrates are increased still further, lactic acid is formed. In some instances, increased propionic acid levels accompany the increased lactic acid concentrations. Presumably, there is an optimum ratio of these acids based upon the ultimate production goal, i.e., milk or meat. Presumably, too, the effect of buffer additions depends on (1) the status of the rumen prior to addition of the buffer and (2) the amount and kind of buffer added.

Reducing the particle size of the feeds by chopping, grinding, and/or pelleting also decreases ruminal pH, primarily because of the reduction in saliva secretion and a more rapid breakdown of feed to volatile fatty acids. Likewise, fermented feeds (silage or haylage) increase the *acid load.*

Also, a sharp drop in pH is noted when animals are switched from high-roughage feeds to high-concentrate feeds without allowing a period of gradual adjustment.

So, high-concentrate rations, smaller feed particle size, fermented feeds, and rapid shifts from high-roughage to high-concentrate rations all lower the pH of the rumen. Since many of the microorganisms in the rumen cannot tolerate low pH concentrations, the normally heterogeneous, balanced population of the microbes becomes skewed, favoring the acidophilic (acid-loving) bacteria. This condition often leads to upsets. However, the addition of feed buffers can prevent dramatic changes of pH in the rumen, thereby stabilizing the microbial population.

It is now known that buffers do more than neutralize acids and alkalis in the rumen. Sodium and potassium bicarbonates and sodium bentonite increase rate of passage of liquids from the rumen. Magnesium oxide regulates pH in the small intestine and increases the uptake of acetate by the mammary gland.

COMMON BUFFERS AND THEIR USES

Many buffer products are available. However, the following six products make up the most common ingredients: sodium bicarbonate, magnesium oxide, sodium bentonite, sodium sesquicarbonate, limestone, and whey. The recommended feeding level for the four most used of these products is given in Table 13-3.

TABLE 13-3
RECOMMENDED FEEDING LEVELS FOR COMMON BUFFERS

Buffer	Percent of total ration[1]	Percent of Grain Mix	Pounds/ cow/day
	(%)	(%)	(lb)
Sodium bicarbonate	0.6–1.0	1.2–2.0	0.3–0.5
Magnesium oxide	0.2–0.35	0.4–0.7	0.1–0.2
Sodium bentonite	2.0–3.0	4.0–6.0	1.0–1.5
Sodium sesquicarbonate	0.6–1.0	1.2–2.0	0.3–0.5

[1]Assumes about 50% forage and 50% grain on a dry matter basis in a total mixed ration. Note that recommended buffer levels are given for either total ration (column 2) or grain mix (column 3).

The observed effects of various buffers upon volatile fatty acid production and upon animal performance are variable. However, it appears that the addition of buffers will improve animal performance provided the appropriate buffer and the right buffer concentration for a particular dietary situation can be determined. A brief summary relative to each of the common buffers follows:

- **Sodium bicarbonate/bicarb/baking soda ($NaHCO_3$)**—Sodium bicarbonate is the major natural buffer in the ruminant's body, being present in saliva, blood, and other body fluids where it acts to stabilize acidity. Commercial sodium bicarbonate, as sold all over the world, is produced synthetically.

Bicarb is the standard buffer in the dairy industry. Properly used, it results in (1) reduction in rumen acidity and improvement in milk fat percentage, especially when the milk fat is depressed (below about 3.3%); (2) higher feed intake and higher milk production; and (3) marked increase in milk production occurring 2–4 weeks after calving.

Potassium bicarbonate is equal to sodium bicarbonate as a buffer.

- **Magnesium oxide (MgO)**—Traditionally, magnesium oxide has been the second most popular rumen buffer. Although its role as a buffer in the rumen and lower gut is not clear, it results in (1) lower rumen acidity, (2) improved feed efficiency, (3) improved fat tests, and (4) increased milk production. Like bicarb, magnesium oxide is most effective for early-lactation cows and when dietary and environmental conditions are causing low fat tests. The ability of magnesium oxide to regulate pH appears to be related to particle size and solubility.

- **Sodium bentonite**—Sodium bentonite is a form of clay. It may lower rumen acidity under some conditions, but its effects are not consistent. Its primary use is in the poultry industry, as a pellet binder.
- **Sodium sesquicarbonate**—This product is a combination of sodium bicarbonate and sodium carbonate. While the value of sesquicarbonate is not as well documented as bicarb, initial research indicates that the two products work equally well.
- **Limestone/calcium carbonate ($CaCO_3$)**—Although limestone is regarded as an excellent and economical source of supplemental calcium, research has shown that it exerts little or no buffering effect in the rumen. It may have some benefit as an intestinal buffer, but it is unlikely that it will counteract rumen acidity. So, supplementing calcium carbonate in excess of the cow's requirement for calcium is not likely to improve efficiency of feed utilization, milk production or milk composition.
- **Whey, partially delactosed**—Partially delactosed whey helps to maintain fat tests near normal levels when dairy cow rations are low in roughage, or when finely ground roughages are fed. Whey is palatable, but it must be fed at about 10% of the total ration in order to be effective. The high cost of the product and difficulties in blending and pelleting are its major drawbacks.

The common uses of buffers follow:

- **Buffers for feedlot cattle and feedlot lambs**—High-grain rations fed to feedlot cattle and lambs make for rapid and efficient weight gains, but they may also result in rumen acidity. Too much acidity in the rumen can depress appetite, reduce feed efficiency, and lower gains.

High-energy grain rations or acidic, ensiled crops (silage or haylage) can cause cattle and sheep to produce more acid than their own supply of sodium bicarbonate can neutralize naturally. A buffer supplement may be the answer. It can help reduce excess acid and maintain the normal balance of acidity and alkalinity.

- **Buffers for dairy cattle**—Studies indicate that buffers can replace a portion of the forage or fiber in the ration of lactating dairy cows even though forage is an essential component thereof. The conditions under which buffers will be of greatest benefit to the dairy cow are: (1) when large amounts of concentrates are fed, as in early lactation; (2) when a fermented forage (such as silage or haylage) is the major or only forage in the ration; (3) when the particle size of the ration has been reduced by chopping, grinding, or pelleting, with the result that it increases the rate of ruminal fermentation and depresses saliva secretion and buffering capacity; (4) when cows are suddenly switched from high-forage to high-concentrate rations; (5) when the milk fat test is low, or (6) when cows are off-feed due to feeding rapidly fermentable rations.

None of the additives will increase butterfat above normal levels. The bicarbonates or magnesium oxide lower palatability and feed consumption, resulting in lower milk production; however, total fat may be greater.

Despite the role and importance of buffers, feeding lactating cows rations with adequate fiber (a minimum of 17% crude fiber or 21% acid-detergent fiber in the ration dry matter) is the best preventive of depressed fat known at the present time.

- **Buffers for poultry**—There is some indication that sodium bicarbonate will reduce the incidence of rough shells and improve eggshell quality during hot weather, but the evidence is not conclusive.

Sodium bentonite is often added to pelleted feed for poultry in order to improve the hardness of the pellets. Additionally, it absorbs water from processed pellets, tends to reduce wet droppings, and improves the growth of young chicks.

The bottom line: Any economic analysis of buffer feeding should take into consideration the net returns based on improved production from feeding buffers and the price of the buffer.

Chemotherapeutics

Chemotherapeutics are organic compounds with bacteriostatic or bactericidal properties similar to those of antibiotics. But, unlike antibiotics, these compounds are produced chemically rather than microbiologically. Table 13-4 lists the common chemotherapeutics and gives the class of animal(s) and use(s) for which each is approved.

As indicated from purusing Table 13-4, the chemotherapeutics are used primarily for disease control. However, arsenicals, carbadox, furaxolidone, and roxarsone are also used as growth promotants and to improve feed conversion efficiency.

The arsenicals and the sulfas, two of the most widely used chemotherapeutics, have been around for a very long time. Pertinent information about each of them not contained in Table 13-4 follows:

- **Arsenicals**—The history of arsenic as a therapeutic agent antedates the time of Christ. In the time of Hippocrates (around 400 B.C.), arsenic sulfide was used in the treatment of ulcers. During the Middle Ages, it became a widely used poison, a reputation that has persisted to the present time. At the beginning of the 20th century, it was extolled as the "magic bullet"—as a cure for syphilis. With the advent and rapid use of antibiotics, the use of arsenic compounds declined.

Various compounds containing arsenic, when used at carefully-controlled levels, have been found to increase rate of gain and improve feed efficiency in chickens, turkeys, and swine; and to prevent coccidiosis in chickens and turkeys and prevent dysentery (bloody scours) in swine.

- **Sulfonamides**—The therapeutic use of the sulfonamides precedes that of antibiotics. The coming of the antibiotics, many of which were more specific in action and easier to administer than the sulfas, markedly reduced the use of the sulfas. In time, however, bacteria emerged that were resistant to many of the antibiotics, which led to a resurgence in the use of the sulfonamides, in feed or water.

All went well until the 1960s and the early 1970s, when the U.S. Department of Agriculture became aware of sulfa residues in pork carcasses. Then, in the late 1980s, a preliminary FDA study showed that massive doses of sulfamethazine caused thyroid tumors in laboratory mice. Then, FDA followed with a warning that sulfamethazine, a drug used subtherapeutically to prevent respiratory problems and as a growth promotant in swine, may be carcinogenic.

TABLE 13-4
CHEMOTHERAPEUTICS[1]

Feed Additive	Animal	Use
Arsenicals	Chickens and turkeys	Increase rate of gain and improve feed efficiency, improve pigmentation, and increase egg production. Prevent coccidiosis.
	Swine	Increase rate of gain and improve feed efficiency. Prevent swine dysentery (bloody scours).
Carbadox	Swine	Increase rate of gain and improve feed efficiency. Control of swine dysentery (bloody scours).
Ethylenediamine dihydriodide (EDDI)[2]	Cattle	Treatment of foot rot and soft tissue lumpy jaw.
	Chickens and turkeys	Removal of mucus from upper respiratory tract following treatment for chronic respiratory disease (air-sac infection).
	Sheep	Prevention of soft tissue lumpy jaw.
	Swine	Control of respiratory difficulties.
Furazolidone	Chickens and turkeys	Stimulate growth and improve feed efficiency. Prevention and treatment of fowl typhoid, paratyphoid and pullorum, chronic respiratory disease, infectious sinusitis, and blue comb. Treatment of blackhead. Prevention of coccidiosis in chickens.
	Swine	Prevention and treatment of bacterial scours and vibrionic (bloody) dysentery.
Nitrofurazone	Chickens and turkeys	Prevention of coccidiosis.
	Mink	For control and treatment of gray diarrhea.
	Swine	Treatment of necrotic enteritis.
Roxarsone	Chickens and turkeys	Increased rate of gain, improved feed efficiency, and improved pigmentation.
	Swine	Increased rate of gain and improved feed efficiency.
Sulfaquinoxalene	Chickens and turkeys	Prevention of coccidiosis and controlling fowl cholera and fowl typhoid.
	Rabbits	Prevention and controlling coccidiosis.
Sulfonamides, *in combination* with antibiotics and other products approved by FDA. There are many different sulfonamides; they differ in absorption and excretion rates, in their availability and active form in the body, and in their potential toxicity. *Except for sulfaquinoxalene for chickens, turkeys, and rabbits, it is illegal to use sulfas alone; they must be used in legal combinations approved by FDA.*	Beef cattle	Controlling respiratory diseases such as shipping fever.
	Chickens	Prevention of coccidiosis, infectious coryza, colibacillosis, and fowl cholera.
	Ducks	Control of bacterial infections, including fowl cholera.
	Rabbits	Prevention of coccidiosis.
	Swine	Prevention and treatment of enteritis, control atrophic rhinitis, reduce cervical abscesses, and control leptospirosis.
	Trout and salmon	Control furunculosis disease.
	Turkeys	Prevention of coccidiosis, fowl cholera, fowl typhoid, and blackhead.

[1]In addition to being used alone, each of the chemotherapeutics is approved for use in certain drug combinations. Chemotherapeutics, and other additives, are subject to FDA approval—and change. So, always read and follow the directions on the label.

[2]FDA considers a maximum of 10 mg of EDDI per head per day appropriate as a source of iodine in the ration of beef and dairy cattle. However, levels above 10 mg per head per day with claims for the prevention or treatment of animal diseases (primarily foot rot, soft tissue, lumpy jaw, and wooden tongue) became unapproved effective September 1, 1986.

Animal products are deemed to be in violation of sulfa residue regulations if levels of 0.1 ppm sulfa are found in the muscle, liver, kidney, eggs, or milk. The 0.1 ppm level is based on toxicological studies with rats and dogs and provides a 2,000-fold safety factor for humans.

Research at the University of Kentucky indicates that as little as 1 g of sulfa per ton of complete feed can cause 100% volatile liver residues and 63% kidney condemnations. This means that as little as ¼ teaspoon of sulfa per ton of complete feed can result in pork carcasses that are in violation.

The sulfonamides are also capable of premise contamination. If excreted in the feces or urine, they may contaminate the premises. Thus, if the pens are not thoroughly cleaned between groups of pigs, untreated animals reared in these pens may show sulfonamide residues.

In order to alleviate sulfa residues in excess of tolerance levels, it is essential (1) that producers observe pre-slaughter withdrawal periods and avoid contaminated premises and animals, (2) that producers use one of the new tests to detect the presence of sulfa in the serum or urine of hogs, and (3) that feed companies avoid cross contamination of sulfa-containing feeds. For the latter reason, feed companies should not use the powder form of sulfa, because it can attach itself more easily to the sides of mixers, bins, augers, and feeders than granulated sulfa products.

CHEMOTHERAPEUTIC CONSIDERATIONS

Producers may choose among chemotherapeutics and among several types of feed mixes with chemotherapeutics (see Table 13-4).

Although many of the antibiotics and chemotherapeutics have similar properties, the expertise of the nutritionist should be sought when increased growth rate and improved feed efficiency is the primary objective, and the veterinarian should be consulted where disease control and/or treatment is involved. Correctly (1) determining the objective and/or problem and (2) matching the additive to the objective and/or problem are the first and most important decisions when choosing a feed additive.

Generally, the chemotherapeutic chosen can be purchased in any of the following types of mixes:

1. A complete mixed feed.
2. A supplement designed to be mixed with a grain source.
3. A separate premix to be added either to complete feeds or supplements.

In addition, some medicated feeds containing chemotherapeutics are formulated to be fed alone for short periods of time.

The type of feed mix/chemotherapeutic used will depend on the method of feeding and the facilities the producer has available.

Whatever the type of feed mix with additive(s) used, consider carefully and determine (1) which chemotherapeutic is needed, and (2) the actual amount of chemotherapeutic in a ton of mixed feed basis. The feed tag must list and show the actual amounts of additive(s). For comparative purposes, these should be converted to amounts of additive(s) per ton of complete feed as fed.

Each producer should have a specific feed additive program. Some may need to feed a certain chemotherapeutic during all stages of growth and development. Others may need to feed a certain chemotherapeutic or combination for a certain period, then change to another additive for additional growth periods.

PROPER USE OF CHEMOTHERAPEUTICS

It is the responsibility of the producer to comply with regulations and use additives properly. To this end, the producer should—

1. **Read and follow directions on the feed tag.** With increasing concern about bacterial resistance and drug residues, it is very important that producers read and follow the directions on the feed tag or label.

2. **Follow withdrawal period.** To avoid residues and insure safe and wholesome animal products, producers must observe appropriate withdrawal periods. Failure to do so will likely result in unacceptable residues and render the producer liable to prosecution under the Food and Drug Act.

Producers should always follow good feeding, sanitation, and disease control programs, and not expect to buy management in a bag of feed with additive(s).

(Also see Chapter 23, Feeding Swine, section on "Additives, • Sulfonamides [Sulfas].")

Copper (Cu)

Copper is widely used as a swine feed growth additive in Europe, where numerous experiments have shown growth responses and increased feed efficiency comparable to those obtained with antibiotics.

A supplemental level of 175 to 200 ppm copper is recommended. No toxicity problems are expected when copper is added at a level up to 250 ppm in a well-mixed, balanced ration adequate in zinc and iron. Levels above 250 ppm may be toxic. As with antibiotics, copper is more effective with younger pigs than older hogs. European reports indicate that copper will increase rate of gain by about 8.0% and decrease feed per unit of gain by about 5.0%.

Copper may not be as effective as antibiotics as a growth promotant or in certain subclinical disease situations. Also, because of its toxic nature, care must be exercised in adding copper.

The increased copper levels in the liver that result if copper feeding is continued to market weight do not appear to present any threat to human health, but they can be reduced by copper withdrawal at 150 lb weight. There appears to be no danger of poisoning the environment due to the copper in the feces.

Copper carbonate, copper chloride, copper oxide, copper sulfate, copper glycinate, and copper methionine are effective sources of copper, with copper sulfate the preferred form. They cost less than antibiotics.

Electrolytes

An electrolyte is a substance which when dissolved in water enables the resulting solution to conduct an electric current. The most common electrolytes in the animal body are salts of such minerals as sodium, potassium, magnesium, calcium, phosphorus, sulfur, and chlorine.

The most commonly used electrolyte solution is 0.9% sodium chloride, which is also called *physiological saline* because it has the same solute strength (tonicity) as the body fluids. Some solutions also contain potassium and magnesium salts because these minerals are also highly essential in the maintenance of a variety of vital functions.

Solutions of electrolytes are administered when it is necessary to replace the mineral salts and water that have been lost under circumstances such as dehydration, diarrhea, hemorrhage, excess urination, and vomiting.

Before starting electrolyte treatment, accurate diagnosis of the disease condition is essential since an effort must be made to stop the processes causing the excessive loss or deficient intake of fluids. Also, consideration must be given to the choice of the electrolyte, the volume of the fluid to be administered, and the route of administration—oral vs. intravenous (IV).

There are many electrolytes on the market. The veterinarian should make the choice, based on prior knowledge of the type of body fluid disturbance likely to be caused by a particular syndrome.

The volume of the electrolyte administered must be adequate, with the veterinarian determining the amount. Up to 7 to 10% of the body weight may be administered over a 24-hour period.

The oral route of administering the electrolyte should be chosen whenever the condition of the animal so permits. The IV route of administration is indicated in life-threatening situations (severe vomiting, diarrhea, impending circulatory failure). Subcutaneous or intraperitoneal administration may be used effectively if circulation is adequate to ensure absorption.

Enzymes

Enzymes are complex protein compounds produced in living cells which cause changes in other substances without being changed themselves. They are organic catalysts.

Normally, the enzymatic output of the digestive system of animals is adequate for maximum digestion of the starches, fats, and proteins. However, repeated experiments in many laboratories with western U.S. barley have shown that the metabolizable energy value of barley for poultry produced in the West under semiarid conditions is improved either by soaking in water before being fed or by the addition of enzyme preparations derived from fungal fermentations. However, little or no improvement in metabolizable energy has been shown from the use of the same fungal enzymes with other common poultry feedstuffs, or even with barley produced in the eastern part of the United States.

Results from feeding enzymes to species other than poultry have been negative to slightly favorable. So, enzymes are little used as feed additives.

Flavoring Agents

If the flavor is not enticing to the animal, the feed will not be eaten; hence, no matter how important the feed may be nutritionally, it won't contribute to the ration unless consumed.

Flavoring agents are feed additives that are designed to increase palatability and feed intake. But not all feed flavors are the same. The International Organization of the Flavour Industry (I.O.F.I.) gives the following breakdown:

1. **Natural flavors.** These substances are obtained from vegetable or animal raw materials by physical methods.
2. **Nature-identical flavors.** These substances are chemically identical to natural products, but they are obtained by synthesis or isolated through a chemical process from natural raw materials.
3. **Artificial flavors.** These substances are not present, or as yet not identified, in a natural product. They are mostly synthesized chemically.

There are four primary taste sensations—sweet, bitter, salty, and sour. Additional feed flavors are the result of the sense of smell. In humans (and perhaps in farm animals, too), smells arouse emotions; they evoke fear, sadness, disgust, longing, love, and/or passion.

Taste and smell play an important role in the sensory apparatus of animals. Most animals will eat only what they can first smell. Then, if it tastes good, they will consume it with relish.

Today, chemists can make chemicals in the laboratory which, alone or in various combinations, can imitate many of the natural feed flavors. In many cases, the synthetic flavors are superior to natural flavors in terms of (1) withstanding processing, (2) cost, (3) availability, and (4) consistent quality. Synthetic flavors may be substances that are prepared in the laboratory but chemically identical to those found in nature, or substances prepared in the laboratory which as yet have not been found to occur in nature but which produce familiar aromas.

Flavor, regardless of whether it is naturally occurring or created from synthetic chemicals, often results from a complex mixture of chemicals in the proper proportion. Hence, a vast number of synthetic compounds can be employed to create endless flavor formulations. Creating flavors is a science.

Chickens have the ability to differentiate between sucrose solutions, for which they show preference, and saccharine solutions which they avoid. Other studies have shown that chickens possess a sense of taste, but little or no ability to smell. Numerous chemical agents are sufficiently objectionable to chickens that they cause a decrease in normal feed consumption. Also, there is evidence that certain feedstuffs are relatively unpalatable to chickens; among them, barley, blood meal, buckwheat, fermentation by-products, and fish solubles. Whether chickens avoid certain feedstuffs on the basis of taste, lack of eye appeal, or because of adverse effects upon metabolism or "sense of well-being" is unknown.

Researchers at the University of Alberta, Canada, reported that two commercial feed flavor additives improved palatability (feed consumption) and performance of starter (young) pigs fed canola meal supplemented rations.[6]

Horses have an acute sense of smell and taste. They will often refuse musty feed when there is so little mold that the caretaker fails to detect it; and foals may back away from a ration that contains cod-liver oil. Also, horses are very fond of apple flavor; many a case of colic has resulted from "old dobbin" making a stolen visit to the apple orchard.

Ruminants prefer sweet compounds. Additionally, cattle and goats respond positively to salts of volatile fatty acids.

Hand in hand with improved animal nutrition and care have come feed palatability problems. Antibiotics, wormers, growth stimulants, and waste products often taste bad. Besides, today's animals are in forced production and stressed. There is a place, therefore, for feed flavors and aromas that will overcome palatability problems, attract animals to feed and keep them on feed, and increase feed consumption and performance. To meet this need, a wide array of feed flavor products is on the market; and they are available in both dry and liquid forms, usually under alluring brand names. However, additional research work is needed in order to establish the true value of flavoring agents in feeds.

Hormone and Hormonelike Compounds

Hormones are chemicals released by a specific area of the body that are transported to another region within the animal where they bring about a physiological response. The word *hormone* comes from the Greek word *hormon,* which means "to spur on, to set in motion, to excite to action." All these phrases are very descriptive of the hormones.

Scientists have identified many of the hormones of the body and have successfully synthesized several hormone or hormonelike compounds which produce the same physiological responses as those from naturally produced hormones. In some cases, this has made it possible to administer these products to obtain a specific response; for example, increased growth, milk production, or meat production.

Each hormone or hormonelike compound elicits a different response. In separate sections that follow, the nutrition-related responses of each of the following products are presented:

1. Somatotropin (growth hormones), including the bovine somatotropin (BST) for dairy cattle and porcine somatotropin (PST) for swine.

---

[6]Baidoo, S. K., M. K. McIntosh, and F. X. Aherne, *Canadian Journal of Animal Science,* 66, 1986, pp. 1039–1049

2. Melengestrol alcetate (MGA).
3. Thyroprotein and goitrogens.

(Also see Chapter 19, section on "Growth Stimulants and Implants.")

SOMATOTROPIN (GROWTH HORMONE)

Initially, this product was known as *growth hormone*. Today, the term *somatotropin* is preferred, because the general public associates the word *hormone* with estrogens, progestins, and testosterone, all of which are steroids, whereas growth hormone or somatotropin is a peptide (protein) hormone.

Somatotropin is secreted naturally by the anterior pituitary gland, located in the skull at the base of the brain of all vertebrates. It is a peptide, and it is species-specific. Thus, bovine somatotropin (BST) and porcine somatotropin (PST) are distinctly different.

Pure somatotropins, obtained from the pituitaries of many species, have been around for many years, but they were far too costly to use routinely. For example, it required the pituitaries of 200 cows to obtain enough bovine growth hormone to treat one cow for one day. However, the "new biology," using recombinant DNA technology (genetic engineering), has made it possible for scientists to identify the gene responsible for producing the bovine growth hormone, isolate it from experimental animals, splice it onto bacteria, and use the altered bacteria to synthesize somatotropins cheaply and in quantity. In turn, research to find new and exciting uses for somatotropins has been accelerated.

Many questions must still be answered before growth hormones become commercially available and widely used, including their long-term effect. Also, a practical method of administering somatotropin to animals must be devised. Since it is a protein, as is insulin, it cannot be taken orally in the feed because it would be digested in the gut; and, unlike commercially-available beef cattle implants (Compudose, Ralgro, Steer-oid and Heifer-oid, and Synovex), a somatotropin implant would break down too quickly. Therefore, a specially protected implant or long-lasting, slow-release injection is required. Finally, as with all such products, FDA approval must precede commercial use of growth hormones.

Currently, scientists are working feverishly away perfecting and testing growth hormones. For dairy cattle, somatotropin is being tested as a stimulator of milk production, and as a growth promotant for replacement heifers. For beef cattle, sheep, swine, and poultry, somatotropin is being tested as a promotant to improve gains and feed efficiency, and, hopefully, produce leaner meat.

More recently, researchers at Michigan State University reported preliminary studies with a growth hormone-releasing factor (GRF), which increased milk production by 23%. GRF is a natural hormone, identical to the hormone produced by the cow. Unlike bovine growth hormone (BST), which adds growth hormone directly to the cow, GRF stimulates the cow's pituitary gland to release more growth hormone, which, in turn, increases milk production.

The full impact of these products on the livestock industry won't be felt until the 1990s—and beyond. It appears, however, that growth hormones will be one of the early significant commercial successes of DNA technology; and that they will usher in a new era in livestock production.

Bovine Somatotropin (BST) for Dairy Cattle

In 1985, Bauman and Eppard of Cornell University, and DeGeeter and Lanza of the Monsanto Company in St. Louis, reported that responses of high-producing dairy cows to long-term treatment with pituitary somatotropin and recombinant somatotropin.[7] Daily injection of graded doses of somatotropin were started 84 days after calving and continued for 188 days. Cows received either 13.5, 27, or 40.5 mg per day of genetically-engineered bovine somatotropin, or 27 mg per day of pituitary-derived bovine somatotropin, while a control group was injected with a placebo. Cows injected with either type of somatotropin produced from 16 to 41% more 3.5% fat-corrected milk (FCM) than the control group. Cows injected with somatotropin consumed more feed on a body weight basis than the control group, but their relative feed energy efficiency was higher. The increased feed efficiency could be accounted for by the smaller proportion of total intake required to meet the cows' maintenance energy requirements.

This report showing the dramatic effect of the hormone on milk production sparked much controversy. Some dairy groups questioned the need or desirability to develop new technology for increasing milk production at a time when there are tremendous milk surpluses, nationally and worldwide; lawsuits were filed to prevent further research with bovine somatotropin and institutions conducting such research were picketed by activist groups. Also, fear was expressed that the commercial application of growth hormone technology would result in the demise of the family dairy farm because of fewer and fewer cows being needed to produce the nation's milk supply. Even some U.S. Congressmen got into the act; an informational House Agriculture Subcommittee hearing on genetically engineered bovine somatotropin was held on Capitol Hill.

Before the commercial use of bovine somatotropin can become a reality, more research is needed; a practical method of administering the product must be developed; and FDA approval must be secured. But, given time, there is every reason to expect that all these requisites will be forthcoming.

Some additional pertinent information relative to bovine somatotropin follows:

• **Historical**—Somatotropin is not new. Cows have been producing it ever since there were cows, and dairy researchers have been studying it since the late 1930s.

In the early days of BST research, which began in the U.S.S.R. about 1937, scientists showed that the growth hormone (then known as pituitary extract) would boost mild production of dairy cows. But the only source of the hormone was the pituitary gland; so, supplies of the product were extremely short and incredibly expensive. But genetic engineering has changed this! Scientists reprogrammed bacteria to produce large quantities of BST.

• **Human safety**—BST is a natural protein produced by all cows, which is always present in milk. It is not a steroid or sex hormone; and it is not related to the steroid hormones which are used as growth promotants. There is no evidence

[7]Bauman, Dale E., *et al.*, Responses of High-Producing Dairy Cows to Long-Term Treatment with Pituitary Somatotropin and Recombinant Somatotropin, *Journal of Dairy Science*, Vol. 68, No. 6, 1985, p. 1352.

that the BST normally found in milk is harmful when consumed by humans. Moreover, milk levels do not increase significantly following treatment. As a peptide, when the BST molecule is consumed by humans, it is broken down and inactivated in the digestive tract in the same manner as any other protein. Both the U.S. Food and Drug Administration and England's Milk Marketing Board have approved the sale of milk from BST-treated cows. Thus, there does not appear to be any negative effect on human health.

• **How to administer**—Because BST is a peptide and would be broken down in the digestive tract, it cannot be fed. But daily injections don't seem practical. (In the experimental stage, it was injected, daily, in the hip area.) So, a long-lasting time-released injection or implant appears to be the answer.

• **Increased feed efficiency and milk production**—In a commercial dairy and over a full lactation, bovine growth hormone may be expected to increase feed efficiency from 10 to 20%, and milk production by 10 to 25%. But scientists still don't know exactly how it works.

• **Growth rate of heifers**—Bovine growth hormone will increase the growth rate of heifers by 8 to 10% and stimulate the development of secretory tissue in the mammary gland.

• **No adverse effect on cows**—It appears that the growth hormone is a safe product to use on dairy animals; treated cows have about the same somatic cell count, disease incidence, post-calving behavior, and rebreeding performance as untreated cows. However, because of their higher production, treated cows will require greater feed intake and more intensive management and observation.

• **Sire evaluation**—Future sires may require two proofs, one developed with and the other developed without hormonally-treated daughters; or correction equations may be developed to equalize bulls regardless of how the proof was developed.

• **Progress can be slowed, but not stopped**—In the future, higher milk yields per cow and fewer dairy cows in the U.S. are inevitable, with or without bovine growth hormone. This will come about as a result of improved genetics, feeding, and management. However, use of the bovine growth hormone will speed the process. In this connection, it is noteworthy that the following developments were products of research: artificial insemination, balancing rations, somatic cell testing, sire proving formulas, new milking systems, and various disease control measures.

• **Progressive dairymen will survive**—The survivors of these changes are likely to be dairymen who adopt technology which increases production efficiency. Milking fewer cows with higher milk yields and higher efficiency appears to be in the best interest of both producers and consumers in the future.

• **Price supports are more responsible for dairy surpluses than new technology**—Government policies which maintain price supports above market-clearing levels and prevent milk prices from fully adjusting to increased supplies are more responsible for dairy surpluses than new dairy technology.

Porcine Somatotropin (PST) for Swine

Porcine somatotropin (PST) is the scientific name for the growth hormone in swine, which is the counterpart of bovine somatotropin (BST) in dairy cattle; and it appears to be equally as effective. Both products are normally produced by the anterior pituitary at the base of the brain. But each is species specific; that is, PST works only on swine, and BST works only on cattle. PST is present in all pigs. It stimulates protein synthesis and growth in most tissues of the body, but it causes breakdown of fat deposits in adipose tissue.

Scientists have known for sometime that providing extra somatotropin to a pig will cause it to grow more rapidly and produce leaner pork. However, it was not possible to take advantage of this knowledge practically, because the only way to get porcine somatotropin was to isolate it from the pituitary glands of slaughtered hogs, which was extremely expensive because of the very low yield of hormone from each hog. But new techniques in molecular genetics (gene splicing, by inserting the pig gene that causes production of somatotropin into the genetic material of the bacteria) make it possible for bacteria to make large quantities of porcine somatotropin. As a result, it is now possible to utilize commercially the positive effects of extra somatotropin. When injected, lactating sows produce more milk and wean heavier pigs; and growing-finishing hogs grow faster and more efficiently, and produce leaner pork. Under field conditions, it is estimated that the use of PST will produce 15 to 20% (1) more rapid gain, (2) greater feed efficiency, and (3) increase in lean muscle.

A major deterrent to the commercial use of PST is lack of an appropriate method of administering the product to swine. Daily injections, as used in early experimental studies, are not practical in commercial hog production. A suitable method of administering PST will be developed; the only question is when and by whom. Also, FDA approval will be necessary.

(Also see Chapter 23, section on "Porcine Somatotropin [PST] for swine.")

MELENGESTROL ACETATE (MGA)

Melengestrol acetate (MGA) a synthetic progestogen hormone, is approved by FDA for use as a feed additive for nonpregnant heifers.

Melengestrol acetate is similar in structure and activity to progesterone, the naturally-occurring hormone of pregnancy. It suppresses estrus (heat) and ovulation and promotes growth.

MGA is fed at a very low level to feedlot heifers—0.25 to 0.40 mg/head/day. On the average, it will increase daily rate of gain by 10% and feed efficiency by 6%. A 48-hour withdrawal period prior to slaughter is required.

MGA also appears promising for use in synchronizing estrus as well as inducing estrus in non-cycling females.

THYROPROTEIN AND GOITROGENS

The primary function of the thyroid gland is the production of thyroxin—a hormone that increases the metabolic rate of animals. Researchers have approached the use of thyroid regulators through two contradictory methods.

The first rationale is that the feeding of thyroxinlike compounds, such as iodinated casein, should increase the metabolic rate and, therefore, enhance growth and production. In the dairy cow, this has proven to be true to a limited degree. Under certain conditions, thyroprotein can stimulate milk production (up to 20% more) of a higher fat content (up to 30% more). However, in order to be effective, the product must be used during a specific period of the lactation. Also, since the cows have an increased metabolic rate, they require additional feed.

The second type of regulator of metabolic rate involves the depression of thyroid activity. Goitrogens, such as thiouracil, depress growth but increase the rate of fattening.

While in theory, and in some cases practice, thyroid regulators appear to have some applications to livestock feeding, their practical use is extremely limited. There is considerable variation among animals with regard to metabolic rate; and the mechanism of the control of the metabolic rate is extremely complex, thereby limiting the potential of these products. Some plants, especially members of the *Prussica* family (*e.g.*, rape), produce natural goitrogens that can interfere with metabolic activity of animals that consume them.

Implants

An implant is a small pellet that is deposited underneath the skin behind the ear of an animal for the purpose of promoting growth.

The idea of implants evolved from watching the response of animals to alfalfa. For years, livestock producers have known that alfalfa produces faster gains and higher production than other feeds with similar nutrient content. This phenomenon prompted researchers to attempt to isolate the factor in alfalfa responsible for stimulating growth and production. It was found that alfalfa contained substances that had estrogenic activity. Subsequent research led to the development of implants that utilized sex hormones as growth promotants.

In 1956, the use of implants was ushered in when the Food and Drug Administration (FDA) approved the use of stilbestrol (diethylstilbestrol, DES) implants for steers. All went well for the next 20 years, with both feeders and consumers greatly benefiting from the increased rate of gain and improved feed efficiency accruing from the use of this product. Then, in the 1970s, it was found that, in large enough quantities, DES can produce cancer in humans and animals; hence, its use was in violation of the Delaney Amendment of the Food and Drug Act. So, on July 6, 1979, the use of DES as implants for cattle and sheep was banned. Subsequently, several new products which produce similar effects with safety have been developed and are now on the market.

Table 13-5 lists the presently available and FDA-approved growth implants, and summarizes how to use them and the results to expect. Except for the most recently FDA-cleared implant, Finaplex (trenbolene acetate [TBA]), all these implants contain the estrogen, estradiol, or an estrogenlike compound. Although their mode of action is not entirely clear, estrogens seem to affect protein synthesis indirectly by altering the animals' endocrine system. Acting primarily on the anterior pituitary gland, estrogen implants increase growth hormone production, which stimulates skeletal muscle growth. Some research indicates that estrogens may also alter the production of other hormones, such as insulin, which might also contribute to additional protein deposition.

Finaplex (TBA) is a synthetic analog of testosterone. It appears to increase growth and protein deposition by acting directly on skeletal muscle and other tissues.

All the products listed in Table 13-5 have been shown to improve rate of gain and feed efficiency. From an economic standpoint, the use of these growth promoting implants can be expected to return \$10 to \$15 for each dollar spent.

(Also see Chapter 19, section on "Growth Stimulants and Implants.")

Pertinent facts not covered in Table 13-5 relative to each of the implants follow:

• **Compudose**—In 1983, this long-lasting implant was approved by FDA for use in steers of any age and weight. The active ingredient in Compudose is estradiol-17 beta, a naturally occurring steroid hormone in all mammalian species. The 1 in. long, 3⁄16 in. diameter, pellet, which contains 24 mg of estradiol, is implanted underneath the skin in the middle of the back side of the ear. Estradiol is then slowly released into the circulatory system and carried to the pituitary gland, where it causes an increase in the growth hormone, which increases protein deposition.

• **Finaplex**—This implant, chemically known as trenbolene acetate (TBA), was cleared by the FDA in 1987, to be used in growing and finishing feedlot steers. It is a synthetic androgen with much greater anabolic potency than testosterone. Although TBA is a new product from the standpoint of commercial use in the United States, it has been used in Europe for more than 10 years. One of the unique properties of TBA appears to be the additive response obtained when it is used in combination with an estrogen. In Europe, implanting steers with both estrogen and TBA often results in 20% greater gains and 12% greater feed efficiency than the controls. TBA seems to increase protein deposition by decreasing the rate of protein degradation.

• **Ralgro**—This implant is derived from a mold produced on corn by the organism, *Gibberella zeae*. The active ingredient is zeranol, an anabolic agent—not a hormone. An anabolic agent enhances retention of nitrogen (*i.e.*, protein deposition) and promotes skeletal growth without an increased deposition of fat. Ralgro improves the rate of gain and feed efficiency of cattle of either sex in the suckling, growing, or finishing stages, in either the feedlot or on pasture.

Ralgro is also approved as an implant for sheep.

• **Steer-oid and Heifer-oid**—These products were developed in Canada. Essentially, Steer-oid, which contains estradiol benzoate and progesterone, is similar to Synovex-S; and Heifer-oid, which contains estradiol benzoate and testosterone, is similar to Synovex-H.

• **Synovex-S, Synovex-H, and Synovex-C**—Synovex-S and Synovex-H are sex-specific implants; "S" is for steers, and "H" is for heifers. Synovex-C, which was approved by FDA in 1984, is for calves of either sex 45 days of age or older. The Synovex implants are a combination of naturally occurring steroid hormones. All these products contain the growth hormone estradiol benzoate. Additionally, Synovex-S (for steers) contains progesterone, Synovex-H (for heifers) contains testosterone, and Synovex-C (for calves) contains progesterone.

TABLE 13–5
IMPLANT SUMMARY[1]

Trade Name	Chemical Name (Active Ingredients)	Approved for Use in	Dosage	Effective Days	Increase in Daily Rate of Gain	Increase in Feed Efficiency	Pre-Slaughter Withdrawal	Comments
			(mg)		(%)	(%)	(days)	
Compudose	Estradiol–17 beta	Steers, birth through feedlot	24	200	10–15	5–10	none required	The name of the product suggests "computed dosage." Recommended dosage is one implant regardless of size of cattle. Compudose is long-lasting; it has an effective life of about 200 days.
Finaplex	Trenbolene acetate (TBA)	Steers, growing and finishing in feedlot	140	120	7	7		TBA is a synthetic analog of testosterone. It gives an additive response when used in combination with an estrogen.
(Also see Chapter 19, Table 19–11, Growth Stimulants and Implants For Cattle.)								
Ralgro	Zeranol	*Cattle*						
		Steers, birth through feedlot	36	90–120	10	5–10	65	Each implant pellet contains 12 milligrams of zeranol.
		Heifers, birth through feedlot	36	90–120	10	5–10	65	
		Suckling calves	36	90–120	10	5–10	65	Calves should be re-implanted every 90–100 days prior to slaughter.
		Sheep						
		Feedlot lambs	12	45–60	10	6	40	Lambs may be re-implanted but must be held 40 days prior to slaughter.
Steer-oid	Estradiol benzoate and progesterone	Steers over 400 lb	220	90–120	10–15	5–10	none required	These products are practically identical to Synovex.
Heifer-oid	Estradiol benzoate and testosterone	Heifers over 400 lb	220	90–120	10	5–10	none required	
Synovex								
Synovex-S	Estradiol benzoate and progesterone	Steers over 400 lb	220	90–120	10–15	5–10	none required	Synovex-S is for steers only.
Synovex-H	Estradiol benzoate and testosterone	Heifers over 400 lb	220	90–120	10	5–10	none required	Synovex-H is for heifers only.
Synovex-C	Estradiol benzoate and progesterone	Steer/heifer calves over 45 days of age	110	90–120	5–10	—	none required	Synovex-C is for calves of either sex. Heifer calves implanted at 45–90 days of age have normal fertility when bred the following spring.

[1]Also, these products are approved by FDA for use in certain combinations.

IMPLANT CONSIDERATIONS

In order that implants shall be used properly and produce the maximum results, those using them should be well informed. To this end, the following considerations are presented:

1. **Be alert to possible FDA changes.** In the past, FDA regulations governing the use of implants have occurred frequently. So, use only an approved product, and watch for possible changes.

2. **Choose implant for specific use.** Select an implant that is approved for the specific class, age, and sex of animal.

3. **Follow label instructions.** When using implants, follow label instructions for implanting technique, level of application, and withdrawal period prior to slaughter.

4. **Lessen stress in implanting.** Experience and experiments show that there is little loss in gains from implanting and reimplanting provided (a) a proper corral, chute, and squeeze are used, and (b) the animals are handled gently.

5. **Implant fast-gaining animals.** The faster animals are gaining, the greater will be the benefits from implanting. For example, there is little advantage from implanting cattle gaining less than 1 lb per day.

6. **Implant suckling calves.** The response of suckling

calves to implanting is influenced by milk production of the dam, pasture conditions, creep feeding, age, and genetic potential. Generally, implanting suckling calves will increase weaning weight by 5 to 10%.

7. **Implant cattle on good pasture.** Implants will usually increase the gains of yearling steers and heifers by 20 to 30 lb during the grazing season, provided the pasture is good enough to make for daily gains of 1 lb or better.

8. **Do not implant breeding bulls or heifers.** If bulls and heifers are to be used for breeding purposes, they should not be implanted. Moreover, implants are not approved by FDA for cattle that are to be used for breeding stock.

Suckling heifers can be implanted once early in life with little adverse effect on subsequent reproduction. However, multiple implants before weaning and implanting after weaning can severely reduce future reproductive performance.

9. **Implant bulls intended for slaughter.** Limited experimental work shows that bulls intended for slaughter may be implanted at birth, then reimplanted subsequently as necessary. Implanted bulls are less responsive than steers; generally, they show some improvement in rate of gain, but little improvement in feed efficiency.

10. **Implant cull cows.** Limited experimental work shows that implanting cull cows will result in increased weight gains, improved feed efficiency, and an increase in the proportion of lean meat in the carcass.

11. **Avoid side effects.** Side effects—raised tailheads, sunken loins, udder development, and riding—usually result from improper dosages, but will be rare if proper implanting technique is used.

12. **Capitalize on more rapid gains, greater feed efficiency, and leaner beef.** Implanted cattle will make more rapid gains, require less feed per pound of gain, and produce leaner carcasses.

13. **Reimplanting may be needed.** Ralgro, Steer-oid, Heifer-oid, and Synovex (S, H, and C) implants are effective for approximately 100 days. So, for continuous response from these implants, cattle should be reimplanted every 100 days, except that the withdrawal of Ralgro 65 days before slaughter must be observed. Compudose has an effective life of about 200 days.

14. **Use implants and feed additives together.** Adding feed additives, such as Rumensin, Bovatec, and MGA, to the ration of implanted cattle increases performance; that is, the combined use of implants and feed additives gives greater performance than when either type of product is used alone.

15. **Be prepared for possible controversies on meat exports.** In 1986, for example, the European Economic Community (EEC) announced that, effective January 1, 1988, the EEC would ban the use of hormones in animals produced by EEC member countries, and on imported meats from other countries where animals receive hormones; with the ban covering Compudose, Ralgro, Steer-oid, Heifer-oid, Synovex S/H/C, and MGA. The Unites States protested the ban as a non-tariff trade barrier, and as being without scientific basis. If put into effect, such a ban would leave U.S. producers two alternatives, neither of which would be good: (1) lose millions of export dollars in sales of red meat products (exclusive of pork), or (2) forego the use of the banned drugs and incur higher production costs. An EEC spokesperson denied that the directive is a subtle trade barrier, and explained that they are merely asking that the United States comply with the same standards as the EEC meat plants. A coalition of U.S. industry groups filed a complaint with the U.S. Trade Representative, under Section 301 of the 1974 federal trade law. But the process under Section 301 could take several years. In the meantime, it is hoped that a settlement may be negotiated.

PROPER IMPLANTATION

If one is to obtain maximum results from using implants, it is essential that the products be properly implanted. Here is how:

1. Restrain the animal properly.
2. Determine which ear you want to implant and adjust the implant instrument so that the needle can be positioned next to and parallel to the ear with the slant side of the needle facing outward. Implant all animals in the same ear to minimize confusion.
3. Select the proper implant site on the back of the ear (Fig. 13–3).

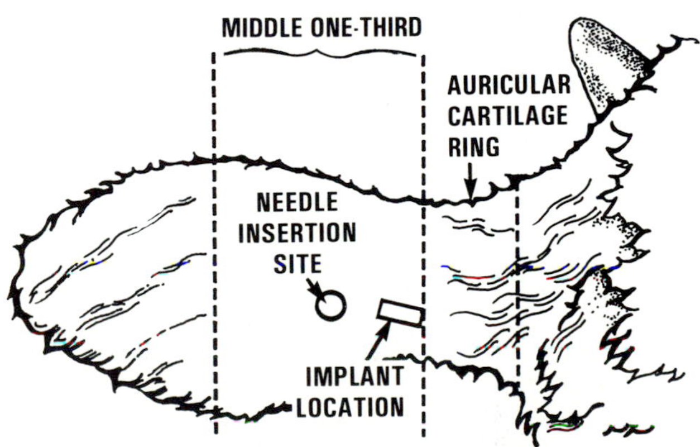

Fig. 13–3. Site of insertion for implants; subcutaneous between the skin and cartilage on the back side of the ear, in the middle third, and not closer to the head than the edge of the annular cartilage ring farthest from the head.

4. Clean the needle and implant site with cotton dipped in alcohol to reduce contamination of the needle wound.
5. Grasp the ear with one hand while the other hand positions the instrument parallel to and nearly flush with the ear. The point of the needle should be against the ear with the beveled part facing you.
6. Use the tip of the needle to prick the skin, lift slightly and completely insert the needle under the skin.
7. Pull the instrument and needle back to create a space for the implant.
8. Depress the plunger of the implant gun and withdraw the needle.
9. Feel the ear for the implant under the skin to ensure that it was inserted properly.

Implantation of growth stimulants is not difficult providing proper restraining equipment is used, but it must be done correctly to maximize benefits. Some of the errors commonly experienced in implanting are:

1. Not restraining the animal properly, with the result that the animal may throw its head when the ear is grasped and

the needle inserted. This can be prevented by using a nose lead, halter, or a headgate equipped with a head and nose bar.

2. Placements of implants in the wrong part of the ear.
3. Crushing of the implant in the ear.
4. Pushing the needle completely through the ear.
5. Hitting and puncturing a blood vessel.
6. Placing the implant in one of the cartilage bands.
7. Depositing the implant in the skin.
8. Failure to follow instructions on proper use of cartridge gun.
9. Sacrificing careful implantation for speed.

Ionophores

Ionophores are feed additives that change the metabolism within the rumen by altering the rumen microflora to favor propionic acid production.

Currently, two ionophores—Bovatec (lasalocid) and Rumensin (monensin)—are approved for cattle. But both products are antibiotics that have been around for a long time. However, they were initially approved as anticoccidial drugs for poultry; monensin was marketed under the trade name Coban, and lasalocid as Avatec. There are at least 76 known polyether ionophores; so, more than likely additional ones will be approved as feed additives.

• **Mode of action**—In order to understand how lasalocid and monensin work, a basic understanding of the biochemistry of carbohydrate metabolism in the rumen is necessary; so, a brief summary follows.

The main end products of the metabolism of carbohydrates by rumen microorganisms are acetic acid, propionic acid, butyric acid, carbon dioxide, and methane. Rations that contain high proportions of forages and other roughage-type feeds favor the production of acetic acid in the rumen. When high-forage rations are fed, the ratios among the three major fatty acids are approximately 50–65% acetate, 18–25% propionate, and 12–20% butyrate. As the proportion of concentrate in the ration is increased, the proportion of acetic acid falls and propionic acid rises. With all-concentrate rations, the proportion of propionic acid may even exceed that of acetic acid. When the percentage of propionate is increased relative to the other fatty acids, a depression in the percentage of milk fat often occurs, accompanied by an increase in body weight as a result of deposition of fat in the cow. In dairy cows, where the value of milk is partially dependent on its fat content, this is bad; in fattening cattle, it is good.

In beef cattle, an increase in propionate is important because propionate tends to be utilized more efficiently by ruminants as an energy source than either acetate or butyrate. Also, a decrease in the production of methane gas by rumen microorganisms is desired as a means of improving feed efficiency, because normally the ruminant loses 7–10% of its feed energy as methane. Additionally, a decrease in the breakdown of natural protein by rumen bacteria, along with a decrease in microbial synthesis is desired, thereby allowing greater quantities of high-quality protein to escape ruminal degradation and decreasing the animal's need for feed protein. The ionophores, lasalocid and monensin, are very effective in altering the rumen microflora; resulting in (1) increasing propionic acid production and decreasing acetic acid, methane, and carbon dioxide production, and (2) decreasing the breakdown of natural protein by rumen bacteria (making for rumen bypass).

During the 1980s, the ionophores were fed to the majority of feedlot cattle. Table 13-6 lists the available and approved ionophores, and summarizes how to use them and the results to expect.

In addition to the Table 13-6 summary, the following facts are pertinent to the use of ionophores:

• **Bovatec (lasalocid)**—In 1982, Bovatec (lasalocid) was approved by the Food and Drug Administration for use in feedlot cattle, both to improve feed efficiency and daily gain. Lasalocid is a polyether antibiotic produced by *Streptomyces lasaliensis.* Research has shown that lasalocid is an effective coccidiostat in both cattle and sheep. Lasalocid is toxic to horses, but higher doses are required than of monensin to cause death.

• **Rumensin (monensin)**—Rumensin, an antibiotic fermentation product produced by a strain of *Streptomyces cinnamonensis,* alters the metabolism within the rumen. Extensive testing has shown Rumensin to be a highly effective improver of feed efficiency in feedlot cattle and in stocker and feeder cattle and beef and dairy replacement heifers on pasture; Rumensin-fed cattle eat less feed per day but gain about the same amount of weight per day as cattle not fed Rumensin. Hence, higher feed efficiency makes for higher profits.

• **Lactic acidosis**—Experiments and experiences show that both lasalocid and monensin lessen the formation of lactic acid in animals fed high-concentrate rations; hence, they lower the incidence of lactic acidosis.

• **Feedlot bloat**—Both lasalocid and monensin inhibit gas formation and decrease methane production in the rumen; hence, they are effective in reducing feedlot bloat.

• **Toxicity to Horses and Swine**—Ionophores can be toxic to horses (and other equines) and swine; so, these species should not have access to either lasalocid or monensin.

• **Ionophores effective for pasture cattle**—The use of ionophores will increase beef and profit per pasture acre. They may be provided to cattle with their mineral, or with the protein or energy supplement. **NOTE**: Ionophores should not be used as a substitute for cattle implants, or vice versa. Ionophores alter fermentation in the rumen, while implants work on the animal's hormonal system. It follows that greater pasture gains may be expected from using ionophores and implants together than from using either product alone.

Isoacids (IsoPlus)

Isoacids provide three branched-chain fatty acids (isobutyric acid, isovaleric acid, and 2-methyl butyric acid) and valeric acid—the same fatty acids that are made by ruminant bacteria and are present naturally in the rumen of cattle. Isoacids are essential for the growth of some rumen organisms that digest fiber. Adding isoacids to the cow's ration results in higher milk production.

The U.S. Food and Drug Administration has approved a blend of isoacids as a feed additive for lactating cows, sold as the calcium salts, under the trade name IsoPlus, manufactured by Eastman Chemical Co., a division of Eastman Kodak.

TABLE 13-6
IONOPHORE SUMMARY[1]

Trade Name	Drug/ Chemical Name	Approved for use in	Approved Use Level	Increase in Daily Rate of Gain	Increase in Feed Efficiency	Disease Control	With-drawal Period	Caution	Comments
				(%)	(%)				
Bovatec (Avatec for poultry).	Lasalocid/ Lasalocid sodium.	Beef cattle.	10–30 g/ton complete feed, fed at rate of 100–360 mg/day; for feed efficiency.		8		none required	Do not allow equines or swine access to lasalocid.	Approved for use in complete mixed feed, complete liquid feed, or free-choice mineral supplement.
			25–30 g/ton complete feed, fed at rate of 250–360 mg lasalocid/day; for feed efficiency and rate of gain.	8	5				Approved for use in heifers in combination with MGA to suppress heat and improve gain/feed efficiency.
		Pasture cattle.	60–200 mg lasalocid/day with drug contained in at least 1 lb of feed.	10					
		Dairy heifers.	60–200 mg lasalocid/head/day.	10					No age or weight restrictions.
		Broilers or friers.	68–113 g/ton.			Prevention of coccidiosis.			
		Sheep.	20–30 g/ton complete feed, fed at rate of 15–70 mg lasalocid/day.			Prevention of coccidiosis.			
Rumensin (Coban for poultry).	Monensin/ Monensin sodium.	Beef cattle in confinement fed for slaughter.	5–30 g/ton complete feed, fed at rate of 50–360 mg monensin/day; for feed efficiency.	—	6–10		none required	Do not allow equines or swine access to monensin.	Approved for complete mixed feed as-fed basis, or complete liquid feed.
		Cattle: slaughter, stocker, feeder, and dairy and beef replacement heifers weighing more than 400 lb on pasture.	25–400 g/ton of feed, fed as directed by the manufacturer; for increased rate of weight gain. 60 g/lb in a feed block, which may be fed free-choice.	10					Approved as a block for stocker and feeder cattle and beef and dairy replacement heifers on pasture. It is not to be fed in conjunction with other protein blocks, and it should not be fed to horses.
		Goats.	20 g/ton in a complete feed.			Prevention of coccidiosis.			To be fed continuously. Cannot be fed to lactating does when milk is used for human consumption.
		Broilers.	90–110 g/ton.			Prevention of coccidiosis.			
		Replacement chickens intended for use as layers.	90–110 g/ton.			Prevention of coccidiosis.			

[1]In addition to being used alone, both Bovatec (lasalocid) and Rumensin (monensin) are FDA approved for use in certain drug combinations.

IsoPlus is marketed as (1) a dry, flowable powder, and (2) a carrier containing protein and minerals.

Pertinent information relative to isoacids as a feed additive to dairy cows follows:

• **Results**—Use of isoacids may boost milk production 8 to 10%, or 4 to 6 lb per day, with little or no increase in feed consumption.

• **Cost**—The feed additive is relatively expensive; it costs 25 to 30¢ per cow per day. A response of 2 to 3 lb more milk per day is needed to break even.

• **A time lag in activity**—A 30- to 60-day time lag occurs from the time isoacids are first fed until an economic response occurs.

• **Not all herds respond**—Generally, response has been positive, with approximately 85% of the herds showing increases in milk. However, about 15% do not improve.

• **Isoacids deliver the greatest response early in the lactation period**—The manufacturer recommends that it be discontinued 220 to 250 days after calving, because the economic response is marginal at this time.

• **Product has strong, disagreeable odor**—The product has an odor that people find disagreeable, but it is masked when incorporated in a mixed feed and it does not carry over in the milk.

• **Greatest response from marginal to below normal protein rations**—The greatest response may be expected from cows fed marginal to low protein rations.

• **The response of first-calf heifers is marginal**—The response of first-calf heifers has been marginal, at about 2.5 lb more milk per day.

• **Mode of action**—The exact mode of action of isoacids by which milk yields increase by 4 to 6 lb per cow per day is not entirely clear. But it appears to be due to enhancing fiber digestion and acetate production, without stimulating insulin secretion.

• **Acceptance by dairy producers**—As is true of all additives, the acceptance by users will be a major factor in determining the future of isoacids. The following two factors may dampen the enthusiasm of some potential users of isoacids: (1) not all dairy cattle benefit from its use, and (2) the delayed returns—the benefits of a profitable response are delayed 30 to 60 days following the use of the additive.

Kelp

Kelp, or seaweed, grows in the sea. Botanically, it is a member of the algae family.

For centuries, kelp has been promoted as a natural feed for animals and food for humans, prized for its minerals and vitamins.

The nutritive value of seaweed is quite variable; it is affected by species, geographic area, season of year, and temperature of water. The Norwegian Seaweed Institute reports an assortment of 60 different mineral elements in seaweed. Additionally, it contains carotene, vitamin D, vitamin K, and most of the water-soluble vitamins, including vitamin B-12.

Kelp is always a rich source of iodine; dried kelp contains more than eight times as much iodine as iodized salt. So, dried kelp may be harmful if fed to animals in large amounts and over a prolonged period.

Kelp has long been promoted for its therapeutic properties for both animals and people, but with few of these claims substantiated by properly conducted and controlled experiments. When used at proper levels, it is an excellent source of certain minerals and vitamins.

KIC (2-ketoisocaproate)

In 1986, Iowa State University applied for patents on the technology of a new substance called "KIC," a natural product, not a drug or hormone. According to the Iowa State Animal Scientist-Developer, the product has the potential to greatly increase profits for livestock producers, promote muscle growth in livestock, and improve the quality of products. KIC (2-ketoisocaproate) is one of several keto-acids derived from dietary protein, which occurs naturally in all animal tissues. The product improves feed efficiency and increases growth rate in broilers, turkeys, sheep, and cattle (but not pigs). Also, it enhances egg production in layers, and milk production and butterfat in dairy cows; it appears to decrease the cholesterol content in meat, milk, and eggs; and it enhances the immune systems of animals. Practical use of KIC awaits commercial production, lowered cost, and more experimental work.

Medicated Feeds

Medication is the administration of remedies for the prevention or healing of disease. It follows that medicated feeds are feeds that contain remedies to prevent or heal disease.

Various routes of administering medications to animals are used. Incorporating coccidiostats, histomonostats, and anthelmintics, and other medicants in feed or water is often the most convenient method. Medications and/or feed additives are also being incorporated in molasses/salt blocks and given to animals free-choice.

Coccidiostats are drugs that are used to prevent coccidiosis. Histomonostats are used to prevent blackhead disease in turkeys. A host of anthelmintic drugs are used to treat cattle, sheep, goats, swine, and horses suffering from a large variety of worm parasites. In each category—coccidiostats, histomonostats, and anthelmintics—a wide variety of drugs sold under many trade names are available. But medicated feeds are not limited to these three groups. Other common feed additives are: ethylenediamine dihydriodide (EDDI) used to treat bovine foot rot, soft tissue, lumpy jaw, and wooden tongue; bloat control products; and ammonium chloride used in the prevention of urinary calculi.

It is the responsibility of the Food and Drug Administration to establish satisfactory safety standards for these drugs, then it is up to livestock producers to comply with these standards.

Despite the convenience of administering drugs in feed or water, their therapeutic effectiveness may be affected by level of consumption. Sick animals or birds have lowered feed consumption. Also, variations in feed consumption are associated with (1) hot and cold weather, (2) body weight, (3) level of production, (4) the energy and fiber content of the feed, and (5) stress. It follows that a 10 to 20% lower feed consumption associated with one of these factors will make for a lower intake of the medicament in the feed by the same amount. The intake of medicated water is even more erratic. Increased water intake during hot weather can spell disaster through overconsumption of water medication.

On the other hand, a given concentration of a drug in water may fail to control a parasite under circumstances where water consumption is very low, as in very cold weather. Many tragedies have resulted from overdosing or underdosing due to carelessness, miscalculation, or failure to consider feed and water intake and the factors that cause it to vary.

The following guidelines may help improve the effectiveness of medicated feeds:

1. Select the proper medication for the specific disease.
2. Use proper dosage; a low level should not be relied upon when a therapeutic level is required.
3. Start medication early; the earlier the start, the better the response.
4. Avoid illegal levels of drugs, illegal combinations of drugs, or combining drugs that are not compatible.
5. Avoid cross-contamination of medicated feeds at the feed mill, in the feed bin, or in handling feed.
6. Observe withdrawal times.
7. Be on the alert for drug resistance, which, sooner or later, is encountered whenever drugs have been widely used for the control of disease, such as coccidiosis control. As resistance develops to older compounds, use a new one—provided a new one is available.
8. Do not try to remedy poor management by feed medication.

CAUTION: Medicated feeds can be a major source of residues, especially of antibiotics and sulfas because of their wide use; 75% of all animals receive one or both of these products during their lives. So, medicated feeds should be used in keeping with the instructions on the label, especially as they pertain to (1) the drug withdrawal period from animals in advance of marketing products, and (2) proper mixing, handling, and storage of medicated feeds.

Mold (Fungi) Inhibitors

Molds are fungi distinguished by the formation of mycelium (a network of filaments or threads), or by spore masses.

Mold inhibitors are substances that prevent the growth of molds.

All feeds are suitable mediums for growth of a wide variety of molds, provided temperature and moisture conditions are favorable.

More than 100 different molds which grow on standing crops or in feeds are known to produce toxins (mycotoxins), and about 20 of these mycotoxins have been associated with diseases in animals or humans.

In recent years, nutritionists have been giving increasing attention to the effects of fungal infestations of feeds. It has been speculated that perhaps many nutritional problems of the past (for example, suspected nutrient deficiencies) were, in fact, caused by feeds contaminated with fungi.

Fungi can affect feed intake and subsequent production through contamination at one or more of four stages in the feeding chain: (1) in the field (preharvest), (2) during storage, (3) at mixing, and (4) in the animal itself. Fungal contamination can pose problems through the production of toxins, alterations of the chemical composition of the feed, or alterations of the metabolic functioning of the animal ingesting or harboring the fungus.

Certain fungi, most notably the organism *Aspergillus flavus,* produce toxins; the toxin produced by *Aspergillus flavus* is known as aflatoxin. Aflatoxin, which has clearly been shown to be a carcinogen (cancer-producing), causes much trouble in livestock. But it is not the only mycotoxin to be feared. Mycotoxins affect all species, especially the young. Generally, ruminants appear to tolerate higher levels of mycotoxins over longer periods of intake than simple-stomached animals. Growing chickens are markedly less susceptible to aflatoxins than ducklings, goslings, pheasants, or turkey poults. Fish are probably one of the most susceptible species of animals to aflatoxin poisoning.

The primary condition conducive to the growth of *A. flavus,* and therefore, the production of aflatoxin, is moisture; hence, proper harvesting, drying, and storage will minimize mold growth and subsequent toxin production. All feedstuffs should be dried below the critical moisture content which permits the growth of molds—approximately 12%. Additionally, mold inhibitors should be added to high-moisture feeds that are exposed to air during storage.

Of all the mold inhibitors currently available at an economical price, propionic acid is the most efficacious. It is Generally Recognized As Safe (GRAS), by FDA, available in liquid or dry form, of low toxicity to animals, and economical for addition to feeds at an effective level. Sorbic acid is also very effective in preventing mold growth. Mold inhibitors should be applied in keeping with the directions of the manufacturer.

The toxicity of aflatoxin-contaminated feed can be reduced when irradiated by ultraviolet light or exposed to anhydrous ammonia under pressure.

(Also see Chapter 5, Table 5-3, Potential Poisons—Mycotoxins [toxin-producing fungi or molds]; Chapter 9, section on "Acid Preservation of High-Moisture Grain"; and Chapter 14, section on "Organic Acids.")

Pellet Binders

Pellet binders are products that enhance the firmness of pellets. Several feed additives are known to produce a marked increase in the firmness of pellets; among them, (1) sodium bentonite (clay), (2) cellulose products from the wood pulp industry, (3) lignin derivatives, and (4) grain industry byproducts. Although bentonite has no nutritive value, several reports indicate that at the level of common usage (2 to 2.5% of the ration) it may even improve the growth and/or feed utilization of animals. Hemicellulose preparations at levels up to 2.5% may serve as good energy sources for ruminants, but lignin has practically no nutritive value.

Molasses or fat are sometimes added to feed as an aid in pelleting, as well as being a concentrated source of energy.

Probiotics (Microbial Enhancers)

Although the term *probiotic* is relatively new, the practice is very old. The first scientific mention of the practice that we now call probiotics came from the Russian scientist Metchnikoff who observed the long life span of many Bulgarian residents who regularly consumed large quantities of yogurt. He attributed their good health as due to reducing the putrefaction products in their large intestines by the bacterium *Bacillus bulgaricus,* later renamed *Lactobacillus bulgaricus,* in the yogurt.

Probiotics are substances that contain desirable gastrointestinal microbial cultures and/or ingredients that enhance the growth of desirable gastrointestinal microbes. They establish a desirable balance of gastrointestinal organisms and/or the substances which contribute toward the balance.

The concept of probiotics is not unlike the use of bacterial cultures as a silage preservative or the inoculation of legume seeds.

As knowledge of the types and functions of microorganisms in the digestive tract unfolded, there evolved with it the concept of inoculating animals with beneficial microorganisms and/or giving them substances to encourage the growth of beneficial microorganisms.

Because the field of probiotics is not an exact science, myths, misinformation, and "merchants of menace" abounded in the formative period in the 1950s. Today, a number of respected firms produce a wide variety of probiotics, with research to back up their claims. However, the *right balance of microflora* is affected by the immediate environment, the type of feed, and the health of the animal. Without doubt, young animals and animals under stress—adverse weather, changes in ration, weaning, transporting, co-mingling, or other stressful situations—will be more likely to respond to probiotics. It follows that the greater the need, the greater the response from probiotics.

Microbial products are available in powder, granular, or liquid form; for use as feed additives, water additives, or drenches. They should be used in accordance with the label of the manufacturer.

Steroids

Steroids are a group of fat-related organic compounds. They include cholesterol, 7-dehydrocholesterol and ergosterol, bile acids, and steroid hormones.

- **Cholesterol**—Cholesterol occurs free or in combination with fatty acids in all animal cells, in blood, and in wool grease (lanolin). Also, it is an important constituent of the brain, where it may form up to 17% of the dry matter. It is synthesized from acetate in the liver; hence, it is not a dietary essential. In recent years, cholesterol has been implicated in atherosclerosis of humans, which involves a thickening of the cell walls due to deposits containing cholesterol.
- **7-dehydrocholesterol**—This is the animal-derived precursor of vitamin D_3, which is produced when the sterol is exposed to ultra-violet light.
- **Ergosterol**—This is the principal plant sterol of importance in animal nutrition. It is important as the precursor of ergocalciferol or vitamin D_2, into which it is converted by ultraviolet irradiation.
- **Bile acids**—These are polar derivatives of cholesterol which are synthesized in the liver. They are either stored in the gall bladder from which they are released after eating, or released continuously from animals having no gall bladder (e.g., horses and rats). They aid in the emulsification of fats.
- **Steroid hormones**—These hormones are complex alcohols synthesized from cholesterol, primarily at the following sites: androgens from the testicles, estrogens from the ovaries, and glucocorticoids and mineral corticoids from the adrenal cortex.

Varying types of steroids have been used to assist human and equine athletes increase muscle mass and eliminate pain from creaky joints. Additionally, steroids are sometimes used on horses to assist in the healing of broken bones, to fight against parasitic anemias, to correct low blood counts, to combat respiratory infections, to increase appetites, and to improve performance of race horses.

Steroids used to increase muscle mass are based on testosterone, the male hormone secreted by the testicles in males, and are known as *anabolic steroids.* Steroids used to treat inflamed joints, along with a variety of other ailments, are based on the hormone secreted by the adrenal gland, and are known as *corticosteroids.* Steroids heavily involved in pregnancy are produced by the ovaries; they include progesterone and estrogen. Estrogens are also used to stimulate muscle growth in meat animals (see earlier section in this chapter on "Implants").

But the use of anabolic steroids may reduce fertility in both stallions and mares. The use of corticosteroids carries the risk of the horse using the injured part more than it should and making the injury worse. Also, on most U.S. race tracks the use of steroids is forbidden within a certain period prior to a race (usually within 48 hours of a race). Because horses may be used for human consumption in a number of European countries, notably in France and Belgium, the use of steroid drugs is banned in the European Economic Community (EEC).

While anabolic steroids and corticosteroids are controversial because of some of their side effects, no such argument surrounds the proper use of progesterone and estrogen.

Tranquilizers

Several drugs such as reserpine, aspirin, ethylene glycol, and other tranquilizers, have been fed to animals for the purpose of quieting and curbing activity.

Tranquilizers are sometimes added to poultry feeds to quiet birds being moved from place to place, to reduce the incidence of cannibalism, or to calm flocks affected with hysteria. But the use of these drugs as feed additives for other animal species is essentially nonexistent.

QUESTIONS FOR STUDY AND DISCUSSION

1. Define each of the following: (a) supplement, (b) additive, and (c) implant.

2. What is the responsibility of the FDA relative to additives and implants?

3. What is the Delaney Clause? How does it affect the feed industry?

4. Why do livestock producers use feed additives and implants?

5. Why is compliance with the feed additive/implant label of importance to both producers and consumers?

6. What is a feed supplement? List the common types of feed supplements, and discuss the importance of each type.

7. Under what circumstances are amino acid supplements needed?

8. Why must considerable caution be exercised when choosing a mineral premix?

9. List the three classes of mineral supplements and give examples of each.

10. What macrominerals and what microminerals are commonly needed as mineral supplements?

11. List the pertinent guidelines for mineral supplementation.

12. Outline the guidelines for a mineral supplementation program for mixed feeds. For unmixed feeds or pasture.

13. What are chelates? Would you recommend the use of chelated mineral supplements instead of alternative sources of inorganic mineral supplements?

14. Why do nutritionists rely on vitamin supplements to meet vitamin needs rather than select feed ingredients for vitamin content?

15. Discuss the need for, and the common supplemental sources of, each of the following vitamins: vitamin A, vitamin D, vitamin E, vitamin K, B-complex vitamins, and vitamin C.

16. Give three examples of products believed to contain unidentified factors.

17. Cite statistics that show the important role of additives, implants, and injections in modern animal production.

18. Should feedlot heifers be aborted? Justify your answer.

19. Why is so much research time and money being devoted to the development of additives that will produce meat (a) with more lean and less fat, and (b) with lower cholesterol?

20. What is the role of roughage or scratch factor for ruminants?

21. What are anthelmintics? Why is a parasite control program and schedule important?

22. What is an antibiotic? Through what mechanisms do they appear to increase rate of gain and feed efficiency?

23. Economically, how do antibiotics benefit both producers and consumers?

24. Discuss the theory of antibiotic resistance and explain how it could happen. Why and how are more and more drug-resistant *Salmonella* bacteria infesting animals and moving up the feed chain to humans?

25. Should the use of penicillin and tetracyclines be discontinued as livestock feeds; are suitable alternatives available? Justify your answer.

26. Discuss the role and importance of each of the following additives: (a) antioxidants, (b) arsenicals, and (c) bloat control products.

27. What are buffers? List and discuss the factors that commonly lower the pH of the rumen.

28. Explain the mode of action and importance of buffers in dairy cattle.

29. What are chemotherapeutics? How do they differ from antibiotics? When chemotherapeutics and antibiotics perform a similar role, what bases should be used in making a choice between them?

30. Discuss the role and importance of each of the following additives: (a) copper, (b) electrolytes, (c) enzymes, and (d) flavoring agents.

31. What are hormones? Discuss the mode of action and the importance of each of the following hormone and hormonelike compounds: (a) growth hormones, including the bovine growth hormone (BGH) for dairy cattle, (b) melengestrol acetate (MGA), and (c) thyroprotein and goitrogens.

32. What are implants? List the presently available and FDA-approved implants and briefly summarize how to use each of them and the results to expect.

33. What are ionophores? List the presently available ionophores, explain the mode of action of each of them, and briefly summarize how to use them and the results to expect.

34. Discuss the role and importance of each of the following types of additives: (a) isoacids (IsoPlus), (b) kelp, (c) KIC (2-ketoisocaproate), (d) medicated feeds, (e) mold (fungi) inhibitors, (f) pellet binders, (g) probiotics, and (h) tranquilizers.

SELECTED REFERENCES

Title of Publication	Author(s)	Publisher
Animal Nutrition, 7th edition	L. A. Maynard, *et al.*	McGraw-Hill Book Co., New York, N.Y., 1979
Antibiotic Feeding, Antibiotic Resistance & Alternatives	D. J. Fagerberg C. L. Quarles	American Hoechst Corp., Somerville, N.J., 1979
Antibiotics in Animal Feeds	V. M. Hays, Chairman Task Force	Council for Agricultural Science and Technology (CAST), Report No. 88, March, 1981
Antibiotics in Nutrition	T. H. Jukes	Medical Encyclopedia, Inc., New York, N.Y., 1955
Applied Animal Nutrition, 2nd Edition	E. W. Crampton L. E. Harris	W. H. Freeman and Co., San Francisco, Calif., 1969
Buffers in Ruminant Physiology and Metabolism	Ed. by M. S. Weinberg, A. L. Sheffner	Church and Dwight Co., Inc., New York, N.Y., 1976

(Continued)

Title of Publication	Author(s)	Publisher
Efficiency in Animal Feeding with Particular Reference to Nonnutritive Feed Additives	Group of University Scientists	Council for Agricultural Science and Technology (CAST), Report No. 22, January, 1974
Feed Additive Compendium	Staff	The Miller Publishing Company, Minnetonka, Minn., Annual with monthly update.
Feed Flavor and Animal Nutrition	T. B. Tribble	Agriaids, Inc., Chicago, Ill., 1962
Feeds and Feeding, 2nd Edition	A. E. Cullison	Reston Publishing Co., Inc., Reston, Va., 1979
Feeds and Feeding, 22nd Edition	F. B. Morrison	The Morrison Publishing Company, Ithaca, N.Y., 1956
Feeds and Feeding, Abridged	F. B. Morrison	The Morrison Publishing Company, Ithaca, N.Y., 1961
Handbook of Feedstuffs, The	R. Seiden W. H. Pfander	Springer Publishing Company, Inc., New York, N.Y., 1957
Hormonal Relationships and Application in the Production of Meat, Milk, and Eggs	National Research Council	National Academy of Sciences, Washington, D.C., 1966
Livestock Feeds and Feeding, 2nd Edition	D. C. Church	O. & B. Books, Inc., Corvallis, Ore., 1984
Merck Veterinary Manual, The	C. M. Fraser, Editor	Merck & Co., Inc., Rahway, N.J., 1986
Nutrition of the Chicken, 3rd Edition	M. L. Scott M. C. Nesheim R. J. Young	M. L. Scott and Associates, Ithaca, N.Y., 1982
Official Publication Assoc. of American Feed Control	E. Miller, Pres. B. J. Sims, Sec.	West Virginia Dept. of Agriculture, Charleston, W. Va., annual
Palatability and Flavor Use in Animal Feeds	H. Bickel, Editor	Verlag Paul Parey, Hamburg & Berlin, Germany, 1980
Stockman's Handbook, The, 6th Edition	M. E. Ensminger	The Interstate Printers and Publishers, Inc., Danville, Ill., 1983
Use of Drugs in Animal Feeds, The, Pub. 1979	National Research Council	National Academy of Sciences, Washington, D.C., 1969

Fig. 13-4. Liquid protein supplement being self-fed. (Courtesy, Watt Publishing Co., Mount Morris, Ill.)

Chapter 14

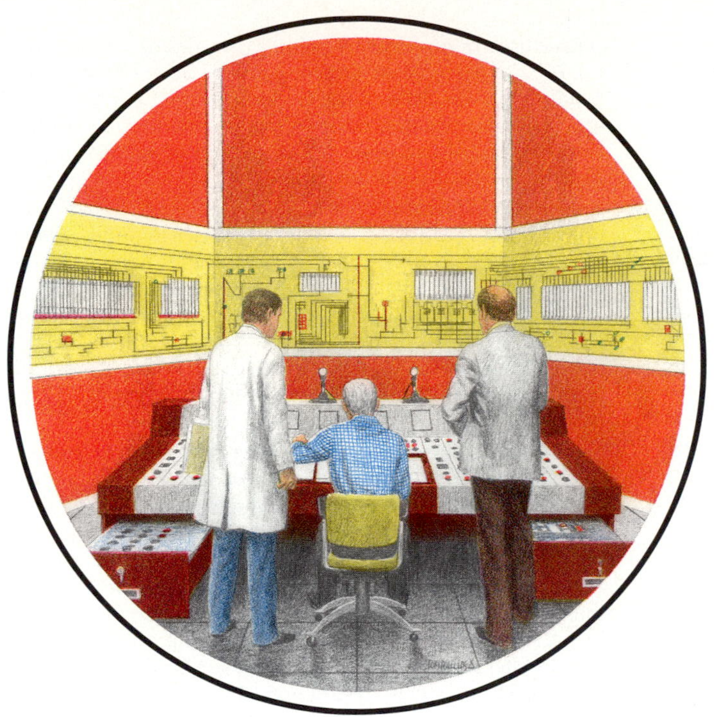

Original painting by Tom Phillips

FEED PROCESSING

Contents	Page

Feed processing refers to the operations necessary to achieve the maximum potential nutritional value of a feedstuff. Practically speaking, it involves changing ingredients in such manner as to maximize their natural value and the net returns from their use.

Feed processing may be physical and/or chemical. Physical changes result from such things as moisture addition or removal, heat, pressure, agglomeration (a clustering together), and particle reduction. Chemical changes may include structural changes in the starch and disrupting the protein matrix, resulting in changes in digestibility and metabolic end products. In some cases, both physical and chemical alterations, known as physiochemical changes, occur simultaneously.

Accelerated rate of ingesta passage and altered site of digestion within the G.I. tract are both likely end results of physiochemical changes in processed grains.

Feed processing has been practiced for a very long time. The "swill barrel" and slop bucket evolved with the domestication of swine. Grinding, crushing, rolling, and soaking of cereal grains for livestock have been practiced by livestock producers for many years. Cattle exhibitors have long cooked grain, and horse fanciers have long used wet bran mashes.

Fig. 14-1. This advanced electronic feed mill, which houses the computerized central control facility and utilizes a continuous flow mixing process, is capable of producing more than a ton of mixed feed per minute for an 80,000 head per year cattle feedlot near Greeley, Colo. The continuous flow system, incorporating the use of augers and weigh belts, ensures ration accuracy within 1% on each ingredient. (Courtesy, Farr Feeders, Inc., Greeley, Colo.)

New technology in feed preparation for feedlot cattle was pioneered by W. H. Hale, of the University of Arizona, beginning about 1963.[1] But much of it is applicable to all ruminants. Feed preparation for swine and poultry has remained relatively simple as compared to the variety of methods available and in use for ruminant feeds. The major change in horse feed preparation has been the increased use of all-pelleted rations (hay and grain combined). Complete, dry pellets for mink that can be dispensed from a self-feeder are only now being realized.

[1]Hale, W. H., *et al.*, "The Effect of Steam Processing Milo and Barley and the Use of Various Levels of Fat with Steam Processed Milo and Barley in High-Concentrate Steer Fattening Rations," *Arizona Cattle Feeders' Day*, May 6, 1965, p. 16.

Feed is a major cost in animal production. It accounts for 70% of the cost of finishing cattle, 55% of the cost of producing milk, 50% of the cost of finishing lambs, and 65 to 75% of the cost of producing pork. Hence, it is economically important that feed be processed in such a manner as to make for maximum efficiency.

Improvements of even 5 to 10% in feed efficiency (feed/unit of production) make for large increases in profit. Assuming a total feed requirement of 3,000 lb per head in feedlot cattle during the finishing period and a feed cost of $4 per 100 lb, 5 and 10% improvements in feed efficiency would result in feed cost savings of $6 and $12 per head, respectively. These would translate to 50 and 100% greater profits per head (assuming an average feeding return of $12 per head).

Feed processing influences the nutritional value of feedstuffs —enhancing some, lowering others. Many of the methods that have been developed for the commercial processing of grains have resulted in decreases in the nutritional value of some of the ingredients. Thus, it is well known that the outer bran portion of a cereal grain is higher in most vitamins than the inner floury portion. Heat treatment of grain results in gelatinization of the starch and denaturing (changing structurally) the protein. Some processes to which feedstuffs are subjected result in destruction of certain vitamins, others enhance the stability of some vitamins, and still others bring about an improvement in availability of vitamins from certain feedstuffs. For example, fine grinding and pelleting of forages tend to increase rate of passage through the gut, which lowers fiber digestibility. However, overall animal response to pelleted forages is usually increased over the same forage fed in long or chopped form, because the slightly lower digestibility is more than offset by increased feed consumption.

Feed processing may also affect handling. For example, hay cubes facilitate automation. Since labor is the second greatest cost item of animal production, it is important that feed processing be evaluated from the standpoint of mechanization.

But feed processing costs money! Thus, recent increases in energy costs have forced producers to take a hard look at the economics of grain processing. Since it is net returns that count, improvements accruing from feed preparation in production and feed efficiency and in laborsaving must be weighed against the cost of processing. Is the processing sufficiently beneficial to cover all costs and return a profit? The economics of processing are well illustrated by the following example: A cattle feeder is considering installation of steam flaking equipment vs feeding whole shelled corn. At best, it can be expected that the steam flaker will improve the feeding value of corn 4%. If shelled corn costs $2.75 per bushel or $4.91 per 100 lb, a 4% improvement would be worth 20¢ per 100 lb or $3.93 per ton. Further investigation into the cost of installing, maintaining, and operating a steam flaker would indicate to the feeder that it would cost about $4.00 per ton, to get $3.93 worth of improvement, or there would be a net loss to the feeding operation of 7¢ per ton as a result of installing the steam flaking equipment. On the other hand, a cattle feeder using ground or dry rolled milo could install steam flaking equipment which would make for an 8% improvement in $3.50-per-100 lb milo. This would represent a 28¢ per 100 lb or a $5.60 per ton improvement for a processing cost of $4.00 per ton, or a net return of $1.60 per ton by switching from grinding or dry rolling to steam flaking. Thus, under the above circumstances, the feeder could not afford to flake corn. However, flaking milo, rather than grinding or dry rolling, would be profitable, because it enhances its value much more than similar processing of corn.

Generally speaking, the higher the level of feeding and the greater the production desired, the more important proper feed preparation becomes. This is so because (1) the higher the level of feeding, the more selective animals become in their eating habits; and (2) in ruminants, digestibility decreases as level of feeding increases, primarily because the feed does not remain in the digestive tract long enough for maximal effect of the various digestive processes.

Also, farm and feedlot differences make for differences in feed preparation. The sheer size of an operation is a major factor in amortizing equipment costs. Thus, greater animal numbers, along with using equipment to capacity, make it easier to amortize expensive equipment. Smaller operations may be able to afford and justify sophisticated feed processing methods through a feed processing cooperative or by having the work done on a custom basis on the farm or at the mill.

Then, too, alternative uses of capital must be considered. The owner must weigh the return from a large investment in processing equipment against a similar investment in more animals or in a confinement slotted-floor facility. Small feeders who have limited capital may opt to omit processing equipment if the estimated return therefrom will likely be less than that which could be obtained if the same money were invested in animals, or in a confinement building, or in other alternative investments.

Finally, energy requirements and costs for processing must be considered. It is inevitable that energy costs will continue to rise. Thus, those processing methods that have the lowest energy requirements will be favored in the years ahead.

PURPOSES OF PROCESSING

The primary reasons for processing feeds are:

1. **To make more profit.** Most people are in business to make money—and livestock producers are people.

Profits in the livestock business may be enhanced by either reducing costs, or increasing production—or both. The profitability of feed preparation is determined by these same criteria.

Feed efficiency can be routinely improved by as much as 10%, and occasionally by as much as 15 to 20% by changing the method of grain processing. It is more difficult to increase production (such as increased rate of gain or milk production) by grain processing. Some processing methods may even improve feed efficiency, but reduce gain. It follows that the relative importance of feed efficiency and rate of gain should be known. As shown in Table 14-1, improving feed efficiency by 10% may have more than 7 times as much impact (1.5¢ vs 10.8¢) on the cost of gain as does improving rate of gain by 10%. However, it should be noted that an increase in rate of gain or feed intake is usually reflected in better feed efficiency.

TABLE 14-1
RELATIVE IMPORTANCE OF IMPROVING RATE OF GAIN AND FEED EFFICIENCY OF CATTLE

	Normal	10% Improvement	Daily Cost Saving/Head
	(lb)	(lb)	($)
Rate of gain	3.00	.30	10% of 15¢ (daily overhead charge/head) = 1.5¢
Feed/lb gain	9.00	.90	.9 lb feed × 3.0 lb gain/day × 4¢/lb feed = 10.8¢

2. **To alter particle size.** Some feeds need to be reduced in size so that they can be consumed, or so that they may be more digestible. In some instances, particle size is increased (agglomerated) by pelleting or cubing, such as (a) when feeding cattle or sheep on the range; (b) when fines need to be alleviated to control waste; (c) when there is selectivity, with the animals consuming the more palatable ingredients and leaving those that are least palatable; or (d) when improved handling efficiency is desired.

3. **To change moisture content.** The moisture content of a feedstuff may need to be changed to make it safer to store, more palatable, more digestible, or to prepare it for other processes. Also, with very high moisture content, such as root crops, the total dry matter intake may be reduced; which may be undesirable when maximum production is desired, but desirable for control of obesity.

For safe storage, grains should be below the following moisture maximums: corn, 14%; grain sorghum, 14%; soybeans, 13%; barley, 14.5%; oats, 14.5%; and rye, 14%. These figures must be modified according to temperature and desired storage time; higher temperatures and longer storage necessitate lower moisture.

Basically, there are two ways to remove moisture from grain: (a) with heat (heated air drying), or (b) without heat (aeration)—or some combination of the two.

For safe storage, the moisture content of hay should not exceed the following maximums: loose hay, 25%; baled hay, 20 to 22% (the lower figure for larger bales); chopped hay, 18 to 20%; and cubes, 16 to 17%.

Water additions to feeds stored as high-moisture grain, silage, or haylage may be desirable. For best results, it is recommended that high-moisture grain be stored at about 30% moisture content, although the moisture may be as high as 40% or as low as 20%. But water should be added when it falls below 26%.

Experimental work and practical experience show that 60 to 67% is the best moisture content of a crop to be ensiled. If the crop is overripe and too dry when cut, or if it becomes overwilted, it will be necessary either to add water to the silo or mix it with green, freshly cut material by alternating loads.

It may be desirable to add water to a finely ground meal mixture at the time of feeding in order to lessen dustiness and increase palatability; and added water may be necessary in preparing feedstuffs for other processes, such as adding moisture in the form of steam to ground feeds that are to be pelleted.

4. **To change density of feed.** The weight per unit volume, or the bulk, of the ration affects total intake; hence, bulky rations which make for "fill" reduce dry matter con-

sumption. For this reason, very bulky feeds are sometimes pelleted or cubed in order to increase energy density and feed consumption. Also, pelleted hay or cubed hay lessen transportation costs and storage space.

On the other hand, horse caretakers favor less dense rations. They prefer that grains be flaked, rather than ground or pelleted, because it makes for a lighter ration and fewer digestive disturbances.

Bulky rations are sometimes prepared for the purpose of limiting energy intake.

5. **To change acceptability (palatability).** In most instances, feeds are processed in such manner as to increase acceptability (palatability) and feed intake. To this end, molasses, flavors, and fats may be added. However, processing may be used for the purpose of decreasing palatability and limiting feed consumption. Salt-feed mixtures are an example of the latter. In limited quantities, salt is very palatable to animals; in excess quantities, it is unpalatable. Thus, by varying the proportion of salt in the mixture, it is possible to hold the consumption of feed supplement to any desired level. This feed processing technique is frequently used for the purpose of limiting the feed consumption of cattle and sheep that are self-fed on pastures and ranges.

6. **To change nutrient content.** When used alone and in their natural state, few feedstuffs meet the nutrient requirements of the animals to which they are fed. Even milk, when fed to young animals over an extended period of time, may be improved by enrichment with certain minerals and vitamins.

It is noteworthy that processing may affect the availability and nutritional value of iron, although this is not the reason for processing. Three examples of increasing iron content in processing follow:

a. The iron content of dried milk products intended for animal feeds runs much higher than the iron content of comparable human food grades due to the generous contribution of iron in the animal feed grades from processing equipment.

b. Nutrition studies in Ethiopia have shown an extremely high iron content in teff, a small seed that constitutes the chief cereal product in the Ethiopian diet. Further investigation revealed that at least three-fourths of this iron was due to contamination with high-iron soil during primitive threshing by treading oxen. However, this iron was mostly in the oxide form and had very low bioavailability.

c. The Bantu consume large quantities of native beer that is brewed in iron containers; and the iron dissolved from these vessels by the acid beer is in a highly available form.

Galvanized equipment contributes zinc to acid solutions much like the situation with iron.

The vitamin D content of ergosterol-containing feeds can be changed through irradiation, with the provitamin ergosterol converted into vitamin D_2.

7. **To increase nutrient availability and digestibility.** Milo that has been dry ground or rolled, and not processed in some more sophisticated manner, has a relative TDN value of only 90% that of corn, and a net energy value of only 80 to 89% that of corn. Yet, chemical composition data indicate that there should be less difference in the feeding value of milo and corn than suggested above. The starch, which represents 70 to 80% of the total dry matter, and perhaps other fractions such as protein, appear to be less available in milo than in other grains. However, some of the newer processing techniques have produced dramatic improvements in the feeding value of milo, with it responding more than other grains. This is attributed to a gelatinization of the starch granules (hydration or rupturing of the complex starch molecules), rendering them more digestible, with the level of gelatinization appearing to be influenced by such factors as steaming time, temperature, grain moisture, roller size and tolerance, processing rate, and variety of milo.

Although feed processing is not known to influence calcium and magnesium utilization, pelleting of feeds increases the utilization of phosphorus by chickens and pigs.

The processing of grains for ruminants enhances availability and digestibility by (1) increasing surface area for greater microbial activity, and (2) giving rumen microorganisms and digestive enzymes easier access to starches and readily utilizable nutrients.

8. **To detoxify or remove undesirable ingredients.** Some feeds may contain toxic substances, the excesses of which will prevent them from performing normally, or even cause death. Table 14-2 lists nine feedstuffs that sometimes cause toxic reactions, along with the inhibitor(s) and the deactivation process.

TABLE 14-2
NATURAL INHIBITORS IN FEEDSTUFFS[1]

Feedstuff	Inhibitor(s)	Deactivation Process
Cottonseed meal	Gossypol: cyclopropene fatty acids	Adding iron salts; rupturing pigment gland
Soybean meal	Trypsin inhibitor: an unidentified factor	Heat; autoclaving
Linseed meal	Crystalline water-soluble substance	Water treatment
Raw fish	Thiaminase	Heat
Alfalfa meal	Saponins; pectin methyl esterase	Limit amount fed
Rye	5-N-Alkyl resorcinols	Limit amount fed
Sweet clover	Dicoumarol	
Wheat germ	Unidentified	Heat
Rapeseed	Isothiocyanate; thyroactive materials	

[1]Stephenson, E. L., "Processing Feeds to Destroy Natural Toxins and Inhibitors," *Effect of Processing on the Nutritional Value of Feeds,* National Academy of Sciences, 1972, p. 68, Table 1.

Fortunately, some of these toxic principles can be removed by processing. Considerable control of gossypol, the yellow pigment of cottonseeds which is toxic to simple-stomached animals, is possible by heating; and the addition of iron salts ($FeSO_4 \cdot 7H_2O$) will protect against egg discoloration. Heating soybeans destroys the factors which inhibit the digestive enzymes, trypsin and chymotrypsin. The toxicity of linseed meal can be removed by adding 2 or 3 parts of water to the meal and allowing it to stand for 12 to 18 hours at a temperature between 72 and 99°F.

As shown in Table 14-2, certain other feeds contain toxic materials. But the amount fed is usually not sufficient to cause difficulty, or the inhibitor(s) can be deactivated by processing as shown. Through heating at high temperatures under certain conditions, the excess fluorine of rock phosphate can be removed.

9. **To improve keeping qualities.** Because feeds are seasonally produced, some of them must be stored for use in the nongrowing season. Usually, this involves the application of some type of preservation.

Feeds carrying more than 14% moisture cannot be stored in bulk. They are likely to mold, and spontaneous combustion may also take place. High-moisture grains may be processed by either drying or chemical treatment (adding an organic acid), or they may be stored in oxygen-limiting silos.

Forages may either be dried to safe storage levels, ranging from 25% moisture in loose hay to 16 to 17% for cubes, or preserved by ensiling (preferably, at 60 to 67% moisture content).

10. **To reduce storage and transportation space and cost.** Sometimes feeds are processed in a certain way in order to reduce storage and transportation space. This is particularly true of forages. Thus, alfalfa hay averages about the following pounds per cubic foot in different forms: loose hay, 4.2; chopped hay, 6.2; and baled hay, 7.5. Cubes and pellets require even less storage space; hence, it is economically feasible to ship cubed or pelleted hay from mainland United States to Hawaii.

11. **To improve mechanization.** Because it is difficult to mechanize the feeding of long hay, pitchfork haymaking persisted in the present century. But, beginning about 1940, scientists and engineers pooled their efforts to automate haymaking and feeding. Hay went modern with one person pick-up balers, field choppers, cubing machines, and other equipment.

Compressing hay into cubes or pellets makes for mechanized handling, high density, with more economical transportation and storage, easier self-feeding, and lower feeding losses. Today, large round bales weighing 1,000 lb and up, and mechanically pressed haystacks weighing from 1 to 6 tons, permit the same latitudes as loose hay, with the added efficiency of mechanization.

Likewise, silage feeding has been automated. It is possible to achieve complete push button controlled feeding in an upright silo. With horizontal silos, silage is handled with a front-end loader, or with a mechanical unloading wagon or truck. Self-feeding of silage is also being achieved.

Fig. 14–2. Ground ear corn being processed in the field—mechanized and laborsaving. (Courtesy, Avco New Idea Farm Equipment, Coldwater, Ohio)

12. **To lessen molds, salmonella, and other harmful substances in feeds.** Sometimes feeds are subjected to a certain process in order to ensure safety and avoid contamination, especially from molds and salmonella.

Molds on feeds and foods have long been a problem. Aflatoxins, part of a larger group called mycotoxins and classified as carcinogens, are toxic substances produced as a result of mold (fungus) growing on grains and other feedstuffs. Fortunately, through processing, much can be done to prevent the formation of molds in feeds, and to reduce the level of aflatoxin or deactivate it in contaminated feeds. Proper harvesting, drying, and storage are important factors in lessening aflatoxin contamination and toxin production. Propionic and acetic acids will inhibit mold growth; hence, they are finding increasing use in preserving high-moisture grains. Treatment with ammonia or ammonium hydroxide will detoxify feeds.

Salmonella, rod-shaped bacteria, the incidence of which is high in meat meal, may be destroyed by pelleting.

13. **To enhance rumen bypass.** Processing, especially heat and pressure treatment, may be used for the purpose of increasing rumen bypass (protected or escape protein).

SELECTING THE PROCESSING METHOD

Many grain processing methods have been developed. Obviously, no single method will be suited to all feeds, all species, and all functions (maintenance, growth, finishing, reproduction, eggs, lactation, work, and wool). Both nutritional and nonnutritional factors should be considered in arriving at a choice.

Nutritional Considerations

The following important nutritional factors should be considered in choosing grain processing methods:

1. **Type of grain.** Grains differ in their response to processing. For example, milo is more responsive to processing than any other grain.

2. **Uniformity and quality of finished product.** Will processing influence the amount of fines, the separation of the components, the palatability, the surface area, and the density or weight per bushel?

3. **Moisture content.** After processing, can the grain be stored or self-fed if desired? Addition of moisture during processing requires that the processed feed be fed soon after treatment; otherwise, it must be dried prior to storage.

Steam heating may add 6 to 8% moisture to milo, while dry heat or pressure may decrease moisture content by a like amount. This point is important to both feed suppliers and custom feeders.

Handling cereals as high-moisture harvested grain will produce nearly the same results as more costly processes; e.g., steam flaking.

Reconstituted (moisture added) grain produces nearly the same results as high-moisture harvested grain.

4. **Percentage of concentrate in the ration.** Grain processing for ruminants gives greater returns when feed intake of grains is high. Thus, cattle fed maintenance rations are not normally fed much grain; hence, the increase in feed efficiency may not return the added processing cost.

5. **Change in structure of the starch.** Milo responds to some of the newer processing techniques (heat provided either by steam, dry heat, microwaves, or pressure) which gelatinize the starch granules and increase digestibility and production of volatile fatty acids.

6. **Feed intake, rate of production, and feed efficiency.** What effect will the processing have upon feed intake, rate of production, and feed efficiency? Will increased performance more than offset any increase in cost?

7. **Effect on health.** Do the animals go on feed and stay on feed satisfactorily? Will the processed feed increase acidosis or other metabolic disorders?

8. **Influence on end product.** Will processing affect the end product—the carcass grade and yield, the milk, wool, eggs, or power?

Nonnutritional Considerations

The following important nonnutritional factors should be considered in choosing the grain processing method:

1. **Time of grain purchase.** Will grain be purchased at harvest time or throughout the year?

2. **Size of operation.** A small operation may not be able to justify a large investment in sophisticated processing equipment.

3. **Effect on hauling costs.** Are transportation costs increased or decreased according to density? Does the process in question increase or decrease the number of tons or loads hauled per day from the mill and/or storage area to the bunks?

4. **Type of ration and kind of operation.** The type of processing should vary according to the type of ration and the kind of operation, with these factors determining the relative merits of such choices as (a) hay cubes vs big round bales, (b) grass silage vs hay, and (c) all-concentrate vs low-concentrate rations.

5. **Capacity of mill.** Some grain processing methods are slower than others. Thus, hourly volume may be a limiting factor when the mill is already operating to capacity if the process is more time-consuming.

6. **Initial investment in equipment.** There is a wide range in the initial cost of processing equipment. Also, the greater the quantity of feed processed, the lower the per ton processing cost. It follows that the larger the operation and the greater the tonnage of feed processed, the greater the initial investment in equipment that can be justified.

7. **Maintenance, repair, and operating cost.** Before buying and installing processing equipment, consideration should be given to the normal maintenance, repair, availability of replacement parts, and operating cost. Some types of processing equipment are more expensive to use than others.

8. **Labor requirements.** Feed processing equipment varies from the standpoints of both (a) hours of labor per ton of feed processed, and (b) the caliber of labor required, with the more complicated types of equipment requiring an engineer or expert mechanic for operation.

9. **Energy requirements.** Increasing energy shortages and rising energy costs necessitate that careful consideration be given to the energy requirements when choosing the feed processing method.

CONCENTRATE PROCESSING METHODS

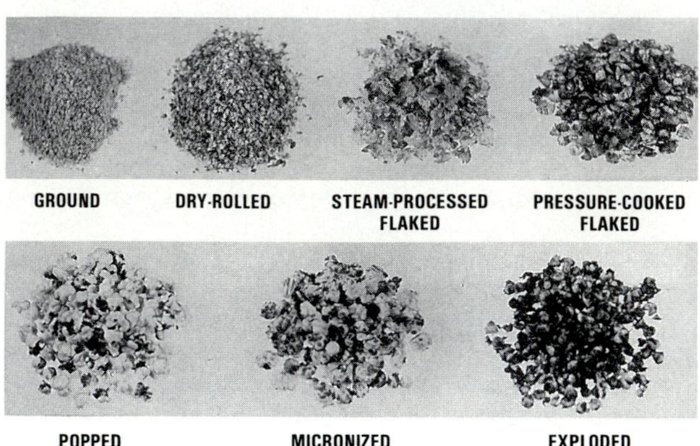

Fig. 14-3. Milo processed by several different methods. (Courtesy, Department of Animal Science, University of Arizona, Tucson)

Most concentrate processing methods have as their primary objective, improvement in starch availability in cereal grains, which, in turn, enhances digestion and feed efficiency. Since 70 to 80% of the dry matter in cereal grains is composed of starch, this is a logical approach. However, the method of accomplishing this is complicated because (1) the type of starch varies among grains, with the starch in some grains more digestible than in others; and (2) availability of starch even varies from one grain variety to another, particularly in milo.

Grain processing gives greater returns when feed intake of grains is high. Animals fed maintenance rations are not normally fed much grain; hence, the increase in feed efficiency may not return the added processing cost.

Several grain processing methods have evolved. A survey of the literature reveals that no process prior to the development and use of steam flaking for feedlot cattle improved performance so dramatically. Thus, most modern processing techniques have been developed in an attempt to obtain similar, and hopefully better, performance than steam flaking—with possibly a reduction in cost. Some are physical, others are chemical; some are dry processing, others are wet processing. Hale listed 18 different methods of processing grain and classified them according to dry and wet processing.[2] It is recognized that any grouping of processing methods cannot be precise, for two or more processing treatments may be involved in a feed; for example, in making pellets grinding is followed by adding heat and moisture, thence pressure. Despite some overlapping, the authors have evolved with the following classification of grain processing methods:

Mechanical alterations
- Dehulling
- Extruding (Gelatinization)
- Grinding
- Rolling

[2]Hale, W. H., "Influence of Processing on the Utilization of Grains (Starch) by Ruminants," *Journal of Animal Science,* Vol. 37, No. 4, 1973, p. 1075.

- Dry rolling (cracking, crushing)
- Steam rolling (crimping, steam crimping)

Heat treatments

- Dry heat processing
  - Micronizing
  - Popping —Jet-sploding
  - Roasting
- Moist heat processing
  - Cooking
  - Exploding
- Flaking
  - Steam flaking
  - Pressure flaking
- Pelleting
  - Crumbling

Moisture alterations

- Bran mash
- Drying (dehydration)
- High-moisture grain (early harvested)
- Reconstituted grain
- Watered feeds

Blocks

Liquid supplements

Fermenting

Hydroponics (sprouted grain)

Unprocessed (whole) corn

Mechanical Alterations

The oldest and most widely used methods for processing grains are those which merely cause physical disruption of the cells by mechanical means. The fact that the more nutritious portions of the grain are surrounded by an outside coating or hull makes it easy to understand how the exposure of these nutrients to the action of digestion processes would increase the utilization of the nutrients. The mechanical methods by which the grain kernel is broken vary, but generally speaking, they involve either shearing, cutting, or mashing. In the milling of grains, there is also the abrasive action to scrub off the outer coats in processes referred to as burring, pearing, polishing, dehulling, and other similar terms.

When any of the dry processing methods are used, it is important that the kernel be broken, but that there be coarseness and relative freedom from fines.

DEHULLING

Dehulling is the process of removing the outer coat of grain, nuts, and some fruits. The hulls are high in fiber and low in digestibility by swine, poultry, and other monogastric animals.

The best known outer coverings of cereal grains are barley hulls, oat hulls, and rice hulls. Today, hulls are combined with other residue from the milling of these cereal grains and marketed under the names of Barley Mill By-Product, Oat Mill By-Product, and Rice Mill By-Product, respectively.

The protein content of such unhulled (undecorticated) oilseeds as soybeans, cottonseeds, peanuts, sunflowers, and safflowers is relatively low. For example, in cottonseed it varies from about 22% in meal made from undecorticated seed to 60% in flour made from seed from which the hulls have been removed completely. Thus, by screening out the residual hulls, which are low in protein and high in fiber, the processor is able to make cottonseed meal of the protein content desired—usually 41, 44, and 50%.

The best known of the oilseed hulls is cottonseed hulls. Today, cottonseed hulls are available in both loose and pelleted forms.

EXTRUDING (GELATINIZATION)

Extruding is a process by which feed is pressed, pushed, or protruded through constrictions under pressure.

Extruding usually involves grinding the grain, followed by heating with steam in order to soften it, then forcing the material through a steel tube by an auger. The softened material is then extruded through cone-shaped holes which are smaller where the feed enters and gradually enlarge where the feed is expelled. The expansion causes disruption, or granulation, of the starch granules. Various factors, including moisture of the grain, influence the character of the final product.

Extruding is an effective way in which to cook and shape raw ingredients into very specialized products. For example, through the choice of ingredients (through the formulation), along with the use of appropriate equipment and proper extrusion procedures in manufacturing, fish feed can be made to sink or float.

In Colorado work with finishing cattle, extruded milo produced results similar to corn processed in various ways. In Kansas trials, milo, processed by dry rolling, high-moisture storage, steam flaking and extruding, produced similar gains, but feed efficiency was 9 and 15% better for the flaked and extruding treatments, respectively, over dry rolled milo. In other Kansas trials, extruding improved feed efficiency by 11 and 15% over dry rolling. Results were similar to steam flaked and high-moisture grain. Tests, however, are too limited in number to draw firm conclusions as to the average improvements in gain and efficiency.

GRINDING

Grinding is that process by which a feedstuff is reduced in particle size by impact, shearing, or attrition. It may change the digestibility of cellulose and protein.

Grinding is the most common, cheapest, and simplest method of feed preparation. It is usually accomplished by means of a hammer mill, which, by impact, reduces the particle size of the grain until it passes through a screen of a certain size. Medium-fine grinding, which can be distinguished by a gritty feeling as some of the feed is rubbed between the fingers, is best. Very fine grinding makes feeds dusty and lowers palatability. However, fine grinding may be desirable (1) when pelleting is to follow, or (2) when grains contain small weed seeds, the viability of which should be destroyed.

Feed manufacturers commonly refer to a feed mixture in which all the ingredients are ground as a mash or meal. Mash is more often used when referring to poultry rations, whereas meal is used when referring to rations for four-footed animals.

Fig. 14-4. Hens eating mash. (Courtesy, Ralston Purina Company, St. Louis, Mo.)

Factors influencing the nature of ground feed are screen size, hammer mill size, power and speed, type of grain, and moisture content of grain.

A major advantage of grinding compared with more sophisticated processing methods is the economic feasibility of having a hammer mill on the farm, or of having a custom grinder come to the farm or ranch periodically to process grain.

Differences in animal performance from ground grains as reported in the literature are partially due to variations in the fineness of grind used in the various experiments.

ROLLING

Rolling refers to the process by which grain is compressed into flat particles by passing it between rollers. The rolling may be accomplished without the addition of water (dry rolling) or after subjecting the grain to steam (steam rolling).

Dry Rolling (Cracking, Crushing)

Dry rolling, which is also called cracking or crushing, refers to passing grain, without steam, between a closely fitted set of steel rollers which are usually grooved on the surface. It breaks the hull and/or seed coat and results in an end product resembling coarsely ground grain. Particle size varies from very small to very coarse and is influenced by roller pressure and groove spacing, moisture content of the grain, and rate of grain flow.

Steam Rolling (Crimping, Steam Crimping)

Steam rolling, which is also called crimping or steam crimping, refers to exposing grain to steam for a short period of time, usually 1 to 8 minutes, followed by rolling. The steam softens the kernel, producing a more intact, crimped-appearing product than that produced by dry rolling, Steam rolling offers little or no advantage in feed efficiency over grinding or dry rolling. However, the particle size and physical form of the steam-rolled grain may improve palatability and animal acceptance in some instances.

The moisture content of steam rolled grain is increased—perhaps by an average of 6%. Thus, when whole grain is exchanged for processed grain on a ton for ton basis, as is frequently done by farmers who have grain processed at the local mill on a custom basis, there may be a loss of 6% of the grain in addition to the cost of processing. If grain is worth $5 per 100 lb, this would add 30¢ per 100 lb, or $6 per ton to the processing cost.

Heat Treatments

It is not known when feeds for animals were first processed by heating. However, it is postulated that humans shared choice tidbits of cooked foods with their favorite animals from the remote day of their domestication forward. Thus, the heat treatment of feeds for animals has been around for a very long time. In recent years, several sophisticated heat treatments have evolved.

Excess heating damages some nutrients, such as the amino acids, and vitamins, whereas proper heating of protein sources (such as soybeans) and of carbohydrate sources (such as cereal grains, potatoes, and beans) results in better availability of nutrients. Heating soybeans destroys the trypsin inhibitor or a possible active protein fraction in raw soybeans, increases the amino acid availability, results in better availability of the fat, and increases metabolizable energy.

Proper heating of cereal grains, such as corn, barley, and milo, will make for partial gelatinization and improve rate and efficiency of gains of cattle.

Mink may suffer from thiamin deficiency when fed certain types of raw fish that contain the enzyme thiaminase; but since thiaminase is heat labile, cooking the fish or dehydrating the fish meal averts the problem. Biotin deficiency can be produced experimentally in different animals by feeding raw egg white; but, avidin, the causative factor in egg white, is destroyed by thorough cooking.

Prolonged heat treatment in processing destroys several of the vitamins. The fat-soluble vitamins and biotin, folacin, pantothenic acid, and thiamin are particularly susceptible to destruction by heat. This is especially true if the material contains high levels of polyunsaturated fat.

Losses of most of the water-soluble vitamins occur during cooking when the cooking water is drained off and not dried with the feed material. Thus, the B vitamin content of fish stick water is much higher than that of fish meal, and the final fish meal dried without the fish-soluble fraction is very low in water-soluble vitamin content.

In general, heat treatments do not improve the nutritional value of most feedstuffs for monogastrics. However, they are the most successful of the newer feed processing techniques for ruminants.

DRY HEAT PROCESSING

Dry heat processing consists of surrounding the feed with dry air. It has the following advantages: (1) The temperature may be changed (turned on, or cut off) rapidly; (2) it does not add moisture to the feed, so it may be used on feeds that are to be stored following processing; and (3) it may remove some of the water from feeds that contain too much moisture for safe storage.

The common methods of processing by dry heat are micronizing, popping, and roasting.

Micronizing

Micronizing is a coined word used to describe a dry heat treatment of grain by microwaves emitted from infrared burners. In micronizing, grain is heated to 300°F by gas-fired infrared generators as it passes along an oscillating steel plate or skillet, thence is dropped into knorling rolls. Micronized grain is not popped. It is reduced to about 7% moisture, then rolled to produce a uniform, stable, dry, free-flowing product. The product has an intact, flakelike appearance, resembling some steam flaked grains. Densities of micronized grain normally range from 18 to 30 lb per bushel, with approximately 25 lb per bushel being recommended. Water is usually added just prior to feeding to adjust to a 10% moisture content.

Micronized grain sorghum compares favorably to steam flaked grain sorghum, from the standpoint of rate and efficiency of gain. However, cost of processing favors the micronizing technique over steam flaking because of a lower initial cost in equipment.

Popping (Jet-sploding)

Popping is the exploding, or puffing, of grain resulting from the rapid application of dry heat. Popping grain for livestock involves the same principle as processing popcorn for people, and the end results are similar. But no experimental work on popping was reported until 1966. Ellis and Carpenter obtained slightly lower gains on 16% less feed per unit of gain when 40% of cracked milo was replaced by popped milo in an all-concentrate mixture for yearling steers.[3]

The principle of popping is based on the use of superhot dry heat, with the grain quickly heated to a temperature of 300° to 310°F (preferably grain with 15 to 20% moisture content) and an exposure of 15 to 120 seconds (depending upon temperature). Rapid heating by dry heat volatizes the internal, natural moisture in the kernel until the pressure is great enough to explode it (to gelatinize and expand, or disrupt, the starch granules), causing the grain to puff out upon reaching atmospheric pressure. Usually not more than 45% of the grain pops, although the jet-sploding method approaches 100% popping. The percentage of grain popped depends primarily on moisture content, temperature, and rate of flow through the machine. The exploded product is then dry rolled.

All grains can be processed by this method, but it appears that it is especially effective in processing sorghum grain.

[3]Ellis, G. F., and J. A. Carpenter, Jr., "Popped Milo in Fattening Rations for Beef Cattle," *Journal of Animal Science,* Vol. 25, p. 594.

Several different popping methods have evolved. One of the most successful is known as jet-sploding, a dry heat processing method. High capacity can be built into this system without difficulty, and quality control is automatic. Also, the jet-sploder can change from one type of grain to another without adjustment. Studies show that jet-sploded milo compares favorably with steam processed and flaked milo.

Densities from various popping methods prior to rolling range from 6 to 15 lb per bushel. Remoisturizing, rolling, or regrinding increase the density.

Some conclusions relative to popping milo, based on experimental studies, are:

1. The resulting product very much resembles ordinary popcorn and has a moisture content of approximately 3%.
2. Popping increases digestibility and efficiency. However, there is disagreement among scientists as to whether the increased digestibility of starch is due to the expansion associated with popping *per se* or due to heat.
3. Popping causes disruption of the starch granule by using natural moisture in the kernel to steam, gelatinize, and expand the starch granules. Rolling and moisturization are usually essential.
4. Popped milo is palatable and very satisfactory for starting cattle on feed, but bulk densities are so light that they sometimes result in severely depressed feed intake and reduced daily gains.
5. Initial investment and operating costs per ton are lower than for steam flaking operations; estimates generally range from $.50 to $1.00 per ton lower.
6. Popped milo requires more storage space than most processing methods, due to its light density. Also, it may create handling problems in bin flow, hang up, bridging, and conveying.

Roasting

Roasting is a simple process of heating feed to the desired temperature in some form of oven for a period of time. It is another method of heat treatment.

The effects of roasting are not fully understood. But it appears to increase the availability of nutrients, possibly as a result of changes in the starch (perhaps gelatinization) due to the heat, along with some effect on the proteins.

Corn and soybeans are the principle feeds that are processed by roasting.

- **Roasting corn**—In roasting corn, the grain is heated to about 300°F. The roasted grain has a pleasant, "nutty" aroma and a puffed, carmelized appearance. Very few of the kernels are actually popped. However, there is some expansion during the roasting process; raw corn weighs 45 lb per cubic foot, whereas the roasted corn weighs only 39 lb per cubic foot. Also, the moisture content of the grain is decreased to 5 to 9%. Purdue University reports that for fattening cattle roasting improves feed efficiency by 10% and increases weight gains by 14% over ground corn.
- **Whole cooked (or roasted) soybeans**—Raw soybeans are poorly utilized by pigs and poultry. Numerous experiments have shown that the nutritive value of soybeans and of soybean oil meal for monogastric animals is much improved by proper cooking. The explanations are that cooking soybeans does the following things: (1) It destroys a substance

known as the *trypsin inhibitor,* which depresses the growth of nonruminants and prevents the action of the protein digestive enzymes, trypsin and erepsin; (2) it greatly increases the availability and value of protein for these animals, through making the sulfur-containing amino acids more readily available; and (3) it improves the palatability of soybeans for monograstics.

Research has shown that whole cooked soybeans can be used to replace soybean meal or other forms of protein supplements in swine rations. Whole cooked beans have little effect on rate of gain, but they usually improve feed efficiency by 5 to 10% due to the higher fat content of the whole soybeans, which makes for a higher energy ration. However, this improvement in feed efficiency may be offset by the lower protein content of whole soybeans; whole cooked beans average about 37% protein, whereas soybean meal usually runs 44%. Also, due to the higher energy of the beans, the protein content of the ration must be 1 to 2% higher than in a soybean meal ration.

In order to process whole soybeans, a large investment in cooking or roasting equipment is required. Such an investment commits the operator to long-term use even though lower prices of soybean meal, relative to the price of whole soybeans, may not justify the practice.

It is noteworthy that hogs fed cooked whole soybeans have a softer carcass than those fed a soybean meal ration. Whether this condition (soft pork) will influence the price packers are willing to pay for live hogs is yet to be determined.

MOIST HEAT PROCESSING

Moist heat processing consists in surrounding the feed by water or steam and (1) cooking either in a conventional vessel or under pressure, or (2) compressing.

The common moist heat processing methods are cooking, exploding, flaking (steam flaking, pressure flaking), pelleting, and crumbling.

Cooking

Cooking is processing by applying heat.

Professional beef cattle fitters have long cooked feed (especially barley) for show cattle. However, the practice declined with the deemphasis of livestock shows, better milking dams of young show cattle, and fewer hard-working caretakers willing to cook grain.

Farmers have long known that potatoes, beans, and soybeans should be cooked for pigs. In the early 1900s, many pork producers used steam cookers to prepare cereal grains and their by-products for hogs; but experiments at many locations showed that this practice (1)is apt to decrease the digestibility of the proteins, even though it may slightly increase the digestibility of the starches, and (2) will not improve performance.

In general, cooking feedstuffs does not improve their nutritional value for monogastrics—swine and poultry. However, soybeans are an exception. Their nutritional value is greatly enhanced by heating.

Garbage is cooked to control *Trichinella,* which causes trichinosis in humans. Actually, raw garbage has a higher feeding value than cooked garbage. Cooking allows for less feed selection on the part of the pig and lowers the digestibility of some of the nutrients, expecially the proteins. On the other hand, swine are more likely to become infested with *Trichinella* and certain other diseases when fed raw garbage. For this reason, all states now have laws requiring that commercial garbage be cooked. The claim is frequently made that the greatest number of cases of trichinosis in humans occurs in communities where garbage is fed to hogs. It should be understood, however, that there is little or no danger in transmitting the disease in this way provided the pork and pork products are thoroughly cooked.

Exploding

Exploding is the swelling of grain, produced by steaming under pressure followed by releasing to the air. This technique involves delivering raw grain into high tensile strength steel "bottles" which hold approximately 200 lb of grain. Live steam is injected into bottles until pressure reaches 250 psi. After about 20 seconds, a valve opens to let the grain escape as expanded balls with the hulls removed. Under the high pressure, moisture is forced into the kernels, which, when released into the air, swell to several times the original size. The product resembles puffed breakfast cereals. Excellent quality control and uniformity of product are possible with this process.

California workers compared exploded milo with steam flaking. The puffed material produced feed intake, gain, and feed efficiency comparable to the best performing flaked grain treatment.

Flaking

Flaking is a modification of steam rolling in which the grain is subjected to steam either for a longer period of time or under pressure. The end product has a distinct and pleasant aroma, resembling cooked cereal. Proper flaking of grains renders the starch fraction more readily available to rumen microorganisms and enzyme degradation than conventional methods of steam or dry rolling.

The flaking process varies according to the grain. The grain that responds the most to flaking is milo. In comparison with dry rolling or grinding, cattle fed flaked milo will gain from 0.25 to 0.5 lb more, or about 10% more, per head per day and require 5 to 10% less feed. In studies with dairy calves, the Arizona Experiment Station researchers found that feed requirements were lowered 10% by steam flaking grain sorghum as compared to steaming and rolling the grain in the older conventional manner, and that there was no difference between steam flaking and pressure flaking. In terms of improvement due to flaking, corn follows milo. Steam processing and flaking of barley and wheat appear to improve gain but not utilization of the grain. This is probably due to improved palatability and intake of the flaked product as compared to dry rolled or ground product. Flaking is the preferred method of processing grains for horses; it produces light, fluffy particles which result in fewer digestive disturbances than any other method of feed preparation.

Flaking is rolling into flat pieces following either (1) steaming at atmospheric pressure, or (2) steaming under pressure.

STEAM FLAKING

This was the first modern technique which markedly increased feed efficiency and rate of gain in the case of milo. This process differs from steam rolling or crimping in that the grain is subjected to steam under atmospheric conditions for a longer period of time, usually 15 to 30 minutes, prior to rolling. Large, heavy roller mills set at near zero tolerance produce a very thin, flat flake which usually weighs from 22 to 28 lb per bushel and contains 16 to 20% moisture. The flaking process causes gelatinization of the starch granules (hydration or rupturing of the complex starch molecule), rendering them more digestible. The degree of flaking and level of gelatinization are influenced by such factors as steaming time, temperature, grain moisture, roller size and tolerance, processing rate, and type and variety of grain.

PRESSURE FLAKING

In pressure flaking, the grain is subjected to steam under pressure for a short time, such as 50 psi for 1 to 2 minutes. A continuous flow cooker is operated by air lock valves to inject and eject grain. Steam is injected into the cooker at the desired pressure (somewhat like a pressure cooker used for food preparation). The grain in the chamber reaches a temperature approaching 300°F. When the grain is expelled from the cooker, it is generally cooled (by use of a cooling and drying tower) to below 200°F and 20% moisture before flaking. In comparison with steam flaking, flakes produced by pressure are less brittle and less subject to fragmenting during the mixing and feeding operation.

Pelleting

Fig. 14-5. Broilers eating bite-sized pellets. (Courtesy, Ralston Purina Company, St. Louis, Mo.)

Pelleting is the agglomerating of feed by compacting and forcing it through die openings by a mechanical process. Pellets can be made into small chunks or cylinders of different diameters, lengths, and degrees of hardness. Large pellets—especially those large enough to be fed on pasture or range—are commonly called range cubes.

Fig. 14-6. Range cubes fed to replacement heifers wintered on low-quality pasture. (Courtesy, Ralston Purina Company, St. Louis, Mo.)

Grains and other concentrates are pelleted for the purposes of (1) facilitating mechanization in handling; (2) eliminating fines and dust, and increasing palatability; (3) alleviating separation of ingredients and sorting; (4) increasing feed density—thereby lessening transportation and labor costs; (5) reducing storage space; (6) making it possible to feed on the ground or in windy areas with little loss; and (7) improving the nutritional value of certain feedstuffs through the instantaneous heat and pressure.

On rations containing a low level of crude fiber, there is no advantage in pelleting feed for beef cattle or swine. However, with more fibrous feeds—especially barley—there is a decided advantage in pelleting feed for swine. Pelleted feeds are popular with horse owners and caretakers; and this includes pelleted concentrates, pelleted hay, and pelleted complete feeds (concentrates and hay combined).

Pelleting broiler and turkey feeds improves performance. But there is no advantage from pelleting layer feeds, with the result that layers are usually fed mash feeds. Pelleting results in marked improvement in the nutritional value for chicks and poults of certain feedstuffs, such as wheat bran, wheat germ meal, dehydrated alfalfa meal, rye, rapeseed oil meal, and field peas. Much of the improvement is still apparent when the pellets are reground and fed as mash; hence, most of the improvement is due either to enhanced nutrient availability or destruction of heat-labile toxins. But pelleting brings about much less enhancement in the value of certain other feedstuffs. High-temperature steam pelleting may even be detrimental in rations that are borderline in protein or in certain critical amino acids.

Pelleting feeds may destroy vitamins A, E, and K, especially if the ration does not contain sufficient antioxidants to prevent the accelerated oxidation of these vitamins under conditions of moisture and high temperature.

Steaming and pelleting may have a beneficial effect on the availability of nicotinic acid and biotin, which are present in bound form in many feedstuffs. Although this possible effect of pelleting has not been adequately investigated, it is noteworthy that the niacin deficiency disease of humans, pellagra, does not exist in Mexico despite their high-corn diet. This is attributed to their treating their corn meal with limewater before making it into tortillas. The limewater releases the nicotinic acid so that it is nutritionally available.

Pelleting could have a beneficial effect on vitamin E nutrition in a ration containing raw soybeans or other raw beans, with the heat of pelleting lessening the amount of vitamin E needed.

The following concentrates may be pelleted: (1) the entire concentrate, (2) the fines only, (3) the protein supplement, and (4) range supplements. Also, concentrates may be combined with a roughage(s) to make a complete feed, and the entire mix may be pelleted.

CRUMBLING

Crumbles are crushed pellets. They are made by crushing pellets into a coarse, granular form. In comparison with pellets, crumbles are preferred by many poultry producers and are better adapted to mechanical feeders. Crumbles retain the heating and density advantages of pellets, but alleviate the sometimes disadvantages of pellets being difficult to chew, swallow, and digest. In comparison with ground feeds, crumbles have the advantage of being dust-free, irregular, and granular.

Moisture Alterations

Water is important in feed preparation and processing. Sometimes the water content of a feed must be altered for proper feed storage, and sometimes it must be changed for feeding purposes.

Some feeds must be stored dry; others must be stored wet. Feeds carrying more than 14% moisture cannot be stored in bulk, for they will likely mold. For safe storage, therefore, grains with higher moisture content must be dried, ensiled, or acid treated.

The moisture content of forages that are to be preserved for ensiling is also of importance, since it affects the ease with which ensiling can be effected. Grass-legume forages must frequently be wilted to reduce moisture content to about 60 to 67%. On the other hand, mature forages often require the addition of water during the ensiling process.

Very dry feeds are often very dusty following grinding or dry rolling. Animals universally dislike dusty feeds; consequently, powdery rations are not eaten well. Dry, dusty rations may be improved by adding small quantities of water, by steaming, or by feeding the product in wet form.

BRAN MASH

A bran mash is steamed wheat bran. It is the traditional feed for use in regulating the bowels of horses on idle days and at such other times as required. The wet mash is prepared by filling a 2- to 2½-gal bucket with wheat bran, pouring enough boiling water over it to make it the consistency of breakfast oatmeal, covering the bucket with a blanket and allowing it to steam until cooked, then feeding it to the horse.

DRYING (DEHYDRATING)

Drying is the removal of moisture by artificial or natural means. To avoid spoilage in storage, grains must be dry enough to prevent the growth of bacteria and molds.

Generally speaking, shelled or threshed grains stored in unventilated bins should not have more than about 14% moisture; preferably, it should not exceed 10 to 12%. Grain may be dried (1) by the use of fuel—artificially; (2) by natural air drying; or (3) by a combination of the two methods. Artificial drying is usually accomplished by running the grain through a heated chamber at a rate that will ensure its being adequately dried when it passes from the drier. The amount of heat and the drying time will vary with the amount of moisture to be removed. The process is expensive.

In addition to cost, prolonged drying at high temperature may adversely affect the feeding value of grain, especially the protein, carotene, and B vitamins. Depressing effects on protein digestibility have been noted, particularly in nonruminant animals. However, experiments show that moderate levels of drying temperatures will not adversely affect performance of swine and ruminants. The adverse effects can be minimized by decreasing exposure time, by faster aeration, and perhaps by nutrient supplementation.

Energy shortages and costs favor delaying harvest until grain is lower in moisture, along with maximum natural air drying. Also, the following alternatives to drying should be considered: (1) the immediate feeding of high-moisture grain, (2) Storing it as high-moisture grain in an oxygen-limiting silo, or (3) treating it with an organic acid(s).

HIGH-MOISTURE GRAIN

High-moisture grain refers to grain that is harvested at a moisture level of 22 to 40% and stored without drying. Optimum conditions for ensiling high-moisture grain appear to be 25 to 32% moisture content. Correctly speaking, high-moisture grain does not involve moisture alteration.

The use of high-moisture grain for finishing cattle was prompted soon after 1900 when early frost terminated the natural maturity of corn in the Corn Belt. Kennedy *et al.* concluded that, on a dry matter basis, soft corn containing 35% moisture was equal to mature corn for finishing steers.[4]

But it remained for Indiana workers to rediscover and popularize high-moisture corn. In 1958, Beeson and Perry reported on a comparison of high-moisture and low-moisture corn for finishing cattle. They found no significant difference in rate of gain, but the cattle fed high-moisture corn showed a substantial improvement in feed conversion when the two feeds were compared on a dry basis.[5]

Similar results have been obtained from high-moisture sorghum grain. The Texas Experiment Station workers harvested and stored sorghum grain with 23 to 32% moisture, then ground it as fed. Cattle fed ground moist grain required 11 to 26% less dry matter from grain than cattle fed ground dry grain.[6]

High-moisture grain may be successfully stored in either of three ways:

1. It may be ensiled (fermented) in an oxygen-limiting silo.
2. It may be ensiled in unsealed storage (in conventional upright silos, or in horizontal silos).

---

[4]Kennedy, W. J., *et al.*, "The Feeding Value of Soft Corn for Beef Production," *Iowa Agriculture Experiment Station Bulletin 75,* 1904.

[5]Beeson, W. M., and T. W. Perry, "The Comparative Feeding Value of High-Moisture Corn and Low-Moisture Corn with Different Feed Additives for Fattening Cattle," *Journal of Animal Science,* Vol. 17, 1958, p. 368.

[6]Riggs, J. K., and D. D. McGintry, "Early Harvested and Reconstituted Sorghum Grain for Cattle," *Journal of Animal Science,* Vol. 31, No. 5, 1970, p. 991.

3. It may be preserved by the addition of an organic acid, most commonly propionic acid (or a mixture of propionic acid with either acetic acid or formic acid), or ammonia.

(Also see Chapter 9, Silage/Haylage/High-Moisture Grain, section on "High-Moisture Grain.")

RECONSTITUTED GRAIN

Reconstituted grain is mature grain that is harvested at the normal moisture level (10 to 14% moisture), following which water is added to bring the moisture level to 25 to 30% and the wet product is stored in a suitable structure for 15 to 21 days prior to feeding. Thus, reconstituted grain involves processing that resembles soaking, and which results in an end product similar to high-moisture grain.

When stored in upright silos, the grain is stored whole, then rolled or ground at the time of removal. Reconstituted grain cannot be satisfactorily stored in horizontal silos as compaction cannot be obtained.

When reconstituting grain, the amount of moisture should be regulated so as to avoid getting too much or too little and reducing the benefits which may be derived. For example, a ton of dry grain normally contains 10 to 12% moisture, or 88 to 90% dry matter. Thus, the dry matter in a ton of dry grain usually totals 1,760 to 1,800 lb. If it is desired to increase the moisture content of this grain to 30%, it would require the addition of 500 to 570 lb of additional water per ton of grain. This means that the high-moisture grain would have the same nutrient value in 2,500 to 2,570 lb as the original grain would have in 2,000 lb. Feeding should be adjusted accordingly.

To make certain that reconstituted grain contains the desired amount of moisture, it is usually advisable to use a simple commercial moisture tester, into which is weighed a given amount of grain, followed by heating, then weighing again. The scales are calibrated in percentage of moisture.

Properly reconstituted milo and steam processed flaked milo give similar results with fattening cattle. Corn is also greatly improved by reconstituting, but there appears to be less advantage from reconstituting barley or wheat. It is noteworthy that, unlike most other methods of processing, no gelatinization of the starch occurs in reconstituted grain, yet the utilization of the starch is similar to that of other processing methods. Also, protein utilization of reconstituted grain is higher than that of other processing methods.

(Also see Chapter 9, Silage/Haylage/High-Moisture Grain, section on "Reconstituted Grain.")

WATERED FEEDS

Water is frequently added to feed, with the amount varying from just enough for dust control to making a slop.

Ground and dry rolled grains, and finely ground alfalfa, tend to be dusty. The palatability of such feeds may be improved by adding a small amount of water at the time of feeding.

- **Soaking**—Sometimes hard grains that are not mechanically processed are soaked for 12 to 24 hours. The soaking softens and swells the grain. Also, dried beet pulp and soybean flakes may be fed in wet form.
- **Liquid and paste feeding**—Liquid feeding usually involves mixing predetermined amounts of feed and water prior to, or at the time of, feeding. When properly used, this method can practically eliminate feed dust in the feeding area and minimize wastage. Ratios of feed and water can be varied to produce a free-flowing liquid or a thick paste.

Some swine producers feed a slop (slurry, gruel, or swill), especially to early weaned pigs and pigs being fitted for show or sale, feeling that they get greater feed consumption and gains thereby.

Blocks

Blocks are compressed packages, generally weighing from 30 to 50 lb each, although high-energy blocks (high in fat content) weighing up to 500 lb are now available. Mineral blocks have been used for a very long time. These were followed by the development of protein blocks, primarily for supplementing cattle on the range and horses on pastures or in corrals. More recently, high-energy blocks evolved.

Fig. 14-7. Block in use on pasture—a means of lessening the labor attendant to the daily feeding of a supplement on pasture or range. (Courtesy, Moorman Manufacturing Co., Quincy, Ill.)

Blocks may be placed in grazing areas where cattle have frequent access to them, with one block provided to 15 mature cattle. Intake will vary with the feed supply and the type of block. Generally, it is planned to limit feed consumption to about 2 lb per head per day by hardness of block and salt and/or fat content.

Range cattle producers use blocks as a means of (1) lessening the labor attendant to the daily feeding of a range supplement, (2) alleviating the loss that accompanies feeding a meal, and (3) distributing cattle on the range.

Liquid Supplements

Liquid supplements are supplements in liquid form. Many of them contain water, molasses, and urea, usually with added trace minerals and vitamins. This is a convenient way of feeding supplements to cattle on pasture or in a corral. Also, liquid supplements are sometimes added to complete ration mixes, either as part of the mix or as a top dressing.

The amount of molasses in most liquid supplements varies from 50 to 70% of the total weight. Most liquid supplements contain ½ to 2% phosphorus, often phosphoric acid. Other compounds that may be present in liquid feed supplements are fat, either animal or vegetable, to increase the amount of energy; alcohols—both ethyl alcohol and propylene glycol are used; and/or a product(s) to govern consumption.

Liquid supplements in a "lick" tank can be offered free-choice. This is a convenient and satisfactory way in which to supply protein, energy, and other nutrients, so long as the cattle do not consume more than they need.

Until the latter 1980s, liquid feeding of swine was considered too messy. However, technological advances, including computerized feeding, resulted in some swine liquid feeding programs being less messy than dry feeding. Additionally, liquid feeding may improve weight gain and feed conversion by as much as 10%.

Fermenting

Two fermentation processes are of practical importance in livestock feeding: (1) ensiling; and (2) improvement of the nutritional value of feeds, either by fermenting the feedstuff itself, or by fermenting other materials that may be used as feed additives to supplement the original feed.

• **Ensiling**—The earliest use of fermentation in animal feeding, and still the most extensive one, involves the ensiling process which takes place when certain feeds with sufficient moisture are stored in a silo in the absence of air. The entire ensiling process requires 2 to 3 weeks, during which time a small amount of oxygen is deleted with aerobic respiration, and anaerobic fermentation occurs.

Ensiling is notable for its versatility. It can be conducted in facilities ranging from simple to sophisticated, and it can be applied to a wide variety of feedstuffs. Its greatest use is in preserving forages, with acetic, lactic, and other lower acids formed. The addition of grains (at the rate of about 150 lb per ton) as preservatives in ensiling forage crops also involves the principle of fermentation. Likewise, when high-moisture grain is stored in an oxygen-limiting silo, it undergoes a fermentation process. Even fish are ensiled by fermentation in the scandinavian countries, by the addition of about 20% starch.

(Also, see Chapter 9, Silage/Haylage/High-Moisture grain.)

• **Improvement of feed nutrient content by fermentation**—The age-old practice of slop-feeding pigs was a continuous-batch fermentation process. Today, more sophisticated and better controlled fermentation techniques are being used; among them, the following:

1. **Proteins and amino acids.** Yeasts, which are a fermentation product, probably have the most favorable characteristics of all microorganisms as major protein sources. Brewers' dried yeast, the most common kind of yeast used in stock feeding, is made from the yeast filtered from beer or ale after the fermentation is completed. Torula-type yeast is made by the fermentation of pentoses, such as the waste sulfite liquor from making paper pulp. In addition to its protein value, yeast carries with it some nutritionally important bonuses, especially the B complex vitamins.

It is frequently advantageous to supplement livestock rations with specific amino acids, rather than intact proteins. Thus, it is noteworthy that fermentations are already operated on a commercial scale to produce lysine and glutamic acid.

2. **Vitamins.** Certain vitamin supplements may be provided by fermentation. Feasible processes have been developed for synthesis of carotene and vitamin A, riboflavin, and vitamin B-12. Also, the B complex vitamins are provided by a multitude of microorganisms.

3. **Antibiotics.** While conducting nutrition studies with poultry in 1949, Jukes and Stokstad of Lederle Laboratory and McGinnis of Washington State University, demonstrated that much of the growth-promoting activity in certain fermentation solubles was due to their antibiotic content. This ushered in the era of feeding antibiotics to livestock.

4. **Enzymes.** Microbiological processes for producing various enzymes are available, and the thought of improving the digestive efficiency of animals by adding the appropriate enzyme(s) to the ration is intriguing. However, the enzymatic output of the digestive system of animals is adequate for maximum digestion of starches, fats, and proteins.

The above discussion identifies some of the contributions that have been made by fermentation processes to the feeding of animals. Beyond these accomplishments there exists an area of significant impact of fermentative processes in converting by-products and waste materials into livestock feeds, and, at the same time, lessening the pollutants in the environment of humans. For example, considerable research is in progress to convert manure from poultry and other animals through bacterial fermentation into animal protein feed. This recycling process could produce much more protein and help solve a pollution problem.

Hydroponics

Hydroponics is the growing of plants with their roots immersed in an aqueous solution containing the essential mineral nutrient salts, instead of soil. This means that the plants are produced with water and chemicals, but without soil.

The Wisconsin Alumni Research Foundation chemically analyzed and compared the composition of oat grain and 5-day oat grass on a dry matter basis (see Table 14-3).

TABLE 14-3
COMPOSITION OF OAT GRAIN AND 5-DAY OAT GRASS, MOISTURE-FREE BASIS[1]

Constituent		Oat Grain	Oat Grass
Dry Matter	(%)	100.00	100.00
Protein	(%)	15.00	21.00
Ether extract (fat)	(%)	4.21	5.20
Nitrogen-free extract	(%)	65.86	42.79
Fiber	(%)	11.71	26.11
Ash	(%)	3.22	3.90
Calcium	(%)	0.063	0.238
Phosphorus	(%)	0.360	0.509
		(mg/kg)	**(mg/kg)**
Carotene[2]		0	39.067
Vitamin E		17.95	48.87
Niacin		7.18	103.96
Riboflavin		1.96	22.29
Thiamin		3.14	12.86
Vitamin C		0	218.3

[1]Analyses by Wisconsin Alumni Research Foundation.

[2]Each mg of beta carotene was considered to be equivalent to 1,556 IU of vitamin A.

As shown in Table 14–3, the 5-day oat grass is a better source than oat grain of calcium, phosphorus, carotene, vitamin E, the B vitamins (riboflavin, thiamin, and niacin), and vitamin C. In addition to these comparative figures, however, the following facts are pertinent:

1. Supplemental quantities of calcium and phosphorus can be provided in many forms at a relatively low cost.
2. Sprouting greatly increases the carotene content; hence, if carotene, or vitamin A, is deficient, sprouted grains are a good supplemental source. However, supplemental vitamin A can be provided in a dry, stabilized form at a low cost.
3. Most rations are adequate in vitamin E. The B vitamins (riboflavin, thiamin, and niacin) are produced by the microorganisms in cattle, sheep, and horses; hence, supplemental quantities of them are not normally needed. Vitamin C is not required in the ration of farm animals.

The Michigan Station researchers made a study of hydroponically-grown oat forage as a feed for dairy cows. They reported that the cost of the hydroponically produced oat forage was over four times that of the original oats or similar grains; and that there was a loss in nutrients during sprouting, a decrease in digestibility of sprouted oats, and no observed increase in milk production when sprouted oats were added to an adequate ration.[7]

Based on studies conducted by the different universities, sprouting results in an average loss of 83% of the dry matter of the oat grain. One study showed a reduction in TDN from 75.7% in the oat grain to 70.2% for the sprouted oats. Also, the digestibility of dry matter, energy, protein, ether extract, and nitrogen-free extract was lower for the sprouted oats than for the oat grain. The composition of hydroponically-grown forage will vary according to the growth stage of the plants, the temperature, the nutrients in the aqueous solution, and several other variables

In arriving at a decision whether to produce feeds hydroponically, consideration should be given to (1) the needs of different classes of livestock for each nutrient, and (2) the cost of supplying these nutrients hydroponically.

Although hydroponically-produced forages are high-quality feeds from the standpoint of certain minerals and vitamins, the need for supplemental quantities of such nutrients in common rations for livestock is questionable.

Without doubt, sprouted grains will give an assist when added to poor rations—and the poorer the ration, the bigger the boost. However, with our present knowledge of nutrition, it should be recognized that balanced rations can be formulated without the added equipment and labor costs of producing forages hydroponically.

Unprocessed (Whole) Corn

Unprocessed (whole) corn refers to shelled corn, the kernels of which have not been broken.

It is generally recognized that young cattle (both beef and dairy animals under 6 months of age) masticate their feed well. Thus, although the digestibility of corn may be increased when it is processed for young bovines, the increased feeding value may not be sufficient to offset the added cost of processing. With the exception of young cattle, it has been assumed that corn should be ground, or otherwise processed, for cattle. Recent experiments at a number of experiment stations have indicated that there are exceptions—that the proportion of concentrate to forage is a factor in determining whether corn should be processed for cattle. Cattle fed dry, whole shelled corn gain an average of 5% faster and require 7% less feed per pound of gain than cattle fed ground, rolled, or crimped corn *when high-concentrate rations are fed.* However, processing appears to have some value for dry shelled corn in rations with 20% or more roughage content or when corn is very dry—less than 12% moisture.

Corn kernels appearing in the feces of cattle fed whole-shelled corn have caused some feeders to think that whole corn is much less digestible than ground corn. Some of the ground grain passes through the digestive tract, too, but it is not as noticeable in the feces as whole grain. Tests at Ohio State University showed very little difference in the digestibility of whole- and ground-shelled corn fed to steers on high-concentrate rations.

Eliminating processing costs is the main advantage of feeding whole corn.

SOME NONNUTRITIVE FEED ADDITIVES

In processing practical livestock feeds, nutritionists may include those nonnutritive feed additives that will improve the ingestion, digestion, protection, absorption, and/or transport of the nutrients to an extent that will increase the nutritive value of the feed and decrease the feed cost for production. No blanket recommendation is possible as to which additives are most useful. Each nutritionist must decide which feed additive(s) is needed under each specific set of circumstances. Among such nonnutritive additives are the following:

1. **Antifungals (mold inhibitors).** Used to prevent harmful molds in feeds and/or in the digestive tract.
2. **Antioxidants.** Used to protect the polyunsaturated fatty acids and the fat-soluble vitamins from destruction by peroxidation.
3. **Enzymes.** Used to improve the digestibility of certain feedstuffs.
4. **Flavoring agents.** Used in an effort to improve the palatability of feed.
5. **Pellet binders.** Used to improve firmness of pellets.

Of course, many other additives than those listed above are incorporated in modern livestock rations; among them, antibiotics, arsenicals, and hormones. These are covered in Chapter 13, Feed Supplements/Additives/Implants, and in the respective chapter devoted to each of the several species; hence, the reader is referred thereto. But a word of caution at this point may be in order. Many additives, such as the hormonelike materials, are extremely potent. So, contamination must be avoided, and withdrawal time (such as a required period prior to slaughter) and other government regulations must be observed. For example, the addition of stilbestrol to a ration for finishing cattle will make for more rapid and efficient gains. Yet, this same additive may make for infertility in mink. Hence, it is important that contamination be avoided, especially where the same feed mixing equipment is used for preparing feeds for both finishing

[7]Report from *Quarterly Bulletin,* Vol. 44, No. 4, Michigan State University, East Lansing, May 1962, pp. 654–665.

cattle and mink. Where such potent additives as stilbestrol are used, the equipment must be thoroughly cleaned before mixing a ration in which the additive will be harmful.

EFFECT OF STORAGE ON FEEDSTUFFS

The extent of the effects of storage on feedstuffs is dependent upon a number of factors, including moisture content, temperature, degree of maturity when harvested, the manner of handling until it is placed in storage, the type of construction of storage bins or containers, and the length of time stored. Also, whole grains generally withstand storage better than the same grains after processing or milling.

It is generally recognized that several of the vitamins are unstable in storage—that they encounter many antagonists in storage, with oxygen heading the list. All the obvious factors contributing to the destruction of vitamins—time, moisture, heat, trace minerals, and low pH—hasten the oxidation of vitamins. This instability of vitamins, along with their recognized very great nutritional importance, prompted many commercial vitamin manufacturers to improve their storage qualities in two ways: (1) by enveloping the vitamin or vitamins in a stable fat or gelatin, forming small beads, thereby preventing most of the vitamins from coming in contact with oxygen until consumed by the animal, and (2) by the use of effective antioxidants that will delay oxidation.

Based on limited work, it may be tentatively concluded that storage affects the nutritive value of feeds as follows:

1. **Loss of carotene.** Carotene content decreases significantly in storage, with the consequent loss of vitamin A activity. Experiments have shown that the loss of carotene in alfalfa meal during storage is only 10% in 6 months at a temperature of $-9°$ to $-15°F$, whereas the loss is 60 to 73% over the same period of storage at room temperature. Also, studies show that the carotene of alfalfa is completely preserved by storage *in vacuo* or under nitrogen.
2. **Trace minerals destroy vitamins.** Of the many ingredients present in feeds or vitamin premixes, trace mineral elements probably have the greatest effect on vitamin losses. When heat and moisture are encountered, vitamin destruction is accelerated; and in the presence of some trace minerals, the rate of loss is compounded.
3. **Destruction by light.** Riboflavin, pyridoxine, and ascorbic acid are readily destroyed by light. It is important, therefore, that premixes of feeds containing these vitamins be stored in a dark place.
4. **Vitamins A, D, and E lowered.** Vitamins A, D, and E are unstable in storage.
5. **Thiamin (B-1) little affected.** Thiamin (B-1) content is little changed by longtime storage under favorable conditions.
6. **Fats deteriorate.** Fats deteriorate, with the formation of free fatty acids, which may affect palatability. These effects can be prevented by proper use of an effective antioxidant, like ethoxyquin or butylated hydroxytoluene (BHT).
7. **Proteins may deteriorate.** Proteins may deteriorate, especially if grains are placed in storage bins immediately following harvesting.
8. **Insects may destroy.** Unless proper steps are taken to control insects through sanitation and the use of insecticides, insects may destroy feedstuffs.

In view of the above, adequate overage of the various vitamins should be provided in all feeds or premixes subjected to processing and/or time and temperature storage conditions conducive to rapid loss of vitamins. Also, trace minerals should be used with discretion, light should be subdued, antioxidants should be used, and insects should be eliminated.

FORAGE PROCESSING METHODS

In recent years, researchers and feeders have been much interested in improved processing of grains. But little study has been made of forage preparation, except from the standpoint of mechanizing and ease in mixing. With the increased competition of grain for human consumption around the world, it is expected that roughage preparation will assume greater importance.

Before discussing each of the common methods of forage preparation, the following generalizations are pertinent to all of them:

1. Most forages are roughages; and ruminants need roughages. Biologically, roughages are coarse feeds that stimulate or require chewing activity and influence the passage of residues through the gastrointestinal tract. Roughage implies coarse texture and high fiber. However, if high-fiber feed is finely ground, it is no longer a roughage even though the fiber concentration is unchanged; the fine grinding of forage results in decreased chewing activity.
2. In preparing forages, avoid processing those (a) with high moisture, which may heat and produce spontaneous combustion, and (b) in which there are foreign objects (wire and other hardware) which animals may not be able to avoid, and which may generate sparks and ignite a fire during processing (grinding-chopping, conveying, or mixing).
3. Processing forages does increase cost from $2 to $10 a ton, depending on the method of processing. Therefore, livestock producers should apply their own cost figures, then determine which processing method would be most profitable for them. The important thing is that all costs be accounted for. For example, in computing the cost of baled hay, with which most processing methods are compared, such added *hidden* costs as losses in handling, shrinkage and wastage, grinding costs and losses, insurance, interest, and storage must be considered. Also, the age and grade of the animals, other available feeds and prices, and starter vs finishing rations must be considered.
4. Processed forages result in the forced feeding of the entire plant, including stems which may be of low nutritional value. With high-producing animals, this may be a disadvantage.
5. The preparation of forages does not increase the value of the initial product.

The common methods of forage preparation are chopping, grinding, shredding, cubing (wafering), drying, ensiling, and pelleting.

Chopping, Grinding, or Shredding

Chopping, grinding, or shredding result in forages divided into smaller particles; but they differ in how they section it, and in the size of the particles. In comparison with a similar forage fed in long form, a forage subjected to any one of these three processes (1) is easier to handle and mechanize, (2) can be stored in a smaller area at less cost, (3) is fed with less feed refusal and waste, and (4) may make for slightly greater production.

Low-quality, coarse forages are usually improved more from chopping than high-quality, fine forages. This should not be construed as license to make poor-quality forage, then improve it by processing. Rather, processing makes for less waste, and perhaps some improvement in digestiblity, but it does nothing to improve the nutrient content.

• **Chopping**—This refers to cutting forage not less than 2 in. in length. (The 2 in. refers to the set of the choppers. Some of the material will be cut longer than this, and some shorter).

Fig. 14-8. Power-bale feeder moving bales into portable on-the-farm mill for chopping and mixing. (Courtesy, Gehl Company, West Bend, Wisc.)

Chopping has the disadvantage of being dusty. Also, there may be considerable leaf loss, or shattering, in field chopping because the hay must be drier than when it is baled or put up as long hay.

(Also see Chapter 8, Hay, section on "Chopped Hay.")

• **Grinding**—This refers to processing forage less than 1 in. in length. Usually grinding is accomplished by means of a hammer mill, in which the forage is beaten by revolving metal hammers until it is small enough in size to pass through the screen placed in the grinder. Generally, screens with holes ¼ in. or larger are used so as to avoid pulverizing the hay. Chopping to a length of less than 1 in. is also referred to as grinding, even without the hammer mill treatment.

Fine grinding is more costly than coarse chopping; hence, it is less appealing from a practical standpoint. Yet, fine grinding is sometimes desirable when the material (either sun-cured or dehydrated) is to be incorporated in the rations of swine or poultry. Ground forages are less digestible for ruminants because they pass through the paunch more rapidly, with only limited bacterial action. When finely ground hay is fed to lactating cows, the fermentation in the rumen produces less acetic acid and more propionic acid than when coarse forage is fed, and, in turn, this results in the fat content of the milk being substantially reduced.

When it is advantageous to use ground hay in a ration, the addition of molasses, fat, or water will lessen the dustiness and reduce the air pollution by nutrients. Some commercial mills spray a small amount of liquid fat on bales of hay just before they enter the grinder. Fat is easier to work with in a grinder or mixer than molasses, for the latter has a tendency to be sticky and "gum up" the equipment.

• **Shredding**—This process is similar to chopping, except shredding tends to separate the stems longitudinally rather than cut them crosswise. Coarse forages, such as fodder and stover, are better suited to shredding than to chopping and grinding. In some ways, it may be superior to chopping, because of exposing more of the inner part of the stem to fermenting bacteria in the rumen, thereby increasing the likelihood of better digestion. Shredding necessitates that hay be as dry as when it is chopped (10% or less); hence, it may result in as much leaf shattering as chopping. Shredding appears to have a more desirable image than chopping, with the result that hay chopping is sometimes referred to as shredding, when it is really nothing more than conventional chopping.

Cubing[8] *(Wafering)*

When applied to forages, the term cubing (wafering) refers to the practice of compressing long or coarsely cut hay into cubes about 1¼ in. square and 2 in. long, with a bulk density of 30 to 32 lb per cubic foot. They do not necessitate fine grinding, and they facilitate automation in both haymaking and feeding. Cubing costs about $5 per ton more than baling.

This method of haymaking is increasing, because it offers most of the advantages of pelleted forages, with few of the disadvantages. Because cubed forage is relatively coarse, it lowers milk fat percentage only slightly, if at all.

It is noteworthy, however, that horses occasionally choke when fed cubes.

(Also see Chapter 8, Hay, section on "Cubes [Wafers].")

Drying

For safe storage, the moisture of hay must be lowered to the following levels: loose hay, 25%; baled hay, 20 to 22% (the lower figure for larger bales); field chopped hay, 18 to 20%; and cubes (wafers), 16 to 17%. These figures must be modified according to temperature; higher temperatures necessitate lower moisture.

Generally, hay moisture is lowered by field curing. However, artificial drying—artificial dehydrators, mow curing, and wagon dryers—may be used during times of inclement weather or when very high-quality forage is desired.

Artificial dehydrating refers to that process in which forage is taken from the field as soon as it is cut (or in some instances after wilting), put through a hay chopper or silage cutter, and dried in large rotating drum driers of different types. For the most part, this method of drying is limited to processing forage for swine and poultry feeds. The most popular type of artificial dehydrator in use in this country is one that uses

[8]There is overlapping in the use of the word *cube*. The compressed long or coarsely cut hay packages about 1¼ in. square and 2 in. long are known as cubes. Also, pellets that are large enough to be fed on pasture or range are commonly called cubes.

a high initial heat (1,200° to 1,400°F), and which is usually heated by natural gas or fuel oil. Due to high equipment and fuel cost, along with the added cost of moving heavy high-moisture forage from field to dryer and in operating the dehydrator, artificially dried forages must command a premium price over field-cured forage.

(Also see Chapter 8, Hay, section on "Reducing Moisture Content/Shattering/Bleaching and Fermenting.")

Ensiling

Ensiling refers to the changes which take place when forage or feed with sufficient moisture to allow fermentation is stored in a silo in the absence of air. The entire ensiling process requires 2 to 3 weeks, during which time a small amount of oxygen is deleted with aerobic respiration, and anaerobic fermentation occurs.

Ensiling is notable for its versatility. It can be conducted in facilities ranging from simple to sophisticated, and it can be applied to a wide variety of feedstuffs. Its greatest use is in preserving forages, with acetic, lactic, and other of the lower acids formed. The addition of grains (at the rate of about 150 lb per ton) as preservatives in ensiling forage crops also involves the principle of fermentation. Likewise, when high-moisture grain is stored in an oxygen-limiting silo, it undergoes a fermentation process.

Pelleting[9]

When applied to forages, the term pelleting refers to the process of forcing ground forage (usually with some added moisture) through a thick steel die and compressing it into a circular or rectangular mass which is cut at predetermined lengths. They can be formed into shapes of varying thickness, length, and hardness. The larger shapes, commonly fed to cattle and sheep on the range, are referred to as cubes.

Binding agents are sometimes added to feedstuffs to regulate the hardness of pellets, especially forage pellets which bind less than concentrates.

The two biggest deterrents to pelleting forages are (1) fine grinding, and (2) cost. From the standpoint of the animal, pelleted forages should be chopped coarsely in order to allow for optimum cellulose digestion in the rumen and to alleviate the incidence of bloat. As a rule of thumb, one would be on the safe side if the forage were not chopped more finely than silage. Also, there is a cost factor; processors charge up to $10 per ton for an all-forage pellet. Of course, the increased cost of pellets should be appraised against their increased value.

Since the pelleting of forages usually involves quite fine grinding prior to the pelleting process, and since finely ground feeds cause lowered butterfat, it follows that pelleted forage is not suitable for lactating cows.

Because of the high cost of pellet mills and their lack of mobility, pelleting is largely confined to commercial feed companies and very large operations, who manufacture sufficient tonnage to make pelleting economically practical.

Pelleted feeds for horses (hay, concentrate, and complete feed) have become fairly popular in recent years.

On the average, cattle fed high-roughage (above 80% roughage) or all-roughage rations will eat about ⅓ more pellets than long or chopped hay, make about ½ to ¾ lb faster daily gains, and require 2 to 2½ lb less feed per pound of gain. Also, it is recognized that low-quality roughages are improved most by pelleting.

Cottonseed hulls, one of the most important roughages in the South—especially for cattle—are now on the market in pelleted form. In comparison with regular hulls, they are more digestible, require less transportation and storage space—because of their greater density, and are easier to handle.

The practice of pelleting forages will likely increase for the following reasons:

1. They are less bulky and easier to store and handle than any other form of forage, thus lessening storage and labor costs.
2. Pelleting forages prevents animals from selectivity, such as eating the leaves and leaving the stems.
3. Pelleting decreases wastage of relatively unpalatable forages, such as ground alfalfa.
4. Pelleting roughages increases intake by 50% or more. Larger responses in intake are associated with poor-quality roughage, high-roughage rations, and younger cattle.
5. Pelleting of roughages improves utilization of digestible energy, partially because processing causes a higher percentage of the roughage to be digested postruminally.

Both cubing (wafering) and pelleting forages will (1) simplify haymaking, (2) lessen transportation costs and storage space, (3) reduce labor, (4) make automatic hay feeding feasible, (5) decrease nutrient losses, and (6) eliminate dust.

With cubing or pelleting, the spread between high- and low-quality roughage is narrowed; that is, the poorer the quality of the roughage, the greater the improvement from cubing or pelleting. This is so because such preparation assures complete consumption of the roughage. Also, cubing or pelleting, especially the latter, tends to speed the passage of roughage through the digestive system.

(Also see Chapter 8, Hay, section on "Pellets.")

MISCELLANEOUS PROCESSING METHODS

There is hardly any limit to the number of processing methods—some old, others new. Some preserve quality, others increase consumption and lessen labor, and still others change the chemical composition and feeding value. In addition to the processing methods already covered, several miscellaneous, but important, methods are discussed in the sections that follow.

Ammoniation

Ammonium salts and anhydrous ammonia (gas or liquid) have been used for ammoniating feeds that contain high levels of carbohydrates and low levels of nitrogen. Among such ammoniated feeds are: citrus pulp, beet pulp, molasses, sugarcane bagasse, and rice hulls. Also, low quality roughages may be ammoniated.

(Also see Chapter 11, Protein Supplements, section on "Ammoniated Products.")

[9]Pellets may refer to (a) the entire concentrate in pellet form, (b) the fines of the concentrate in pellet form, which are usually added back to the grain for feeding, (c) the forage in pellet form, (d) the protein supplement, or (e) the range supplements in pellet form.

Animal Waste (Manure) Processing

Animal waste (manure) has nutritive value for ruminants because these animals are capable of utilizing nonprotein nitrogen and fiber. So, proper processing is important.

Broiler and layer litter has been successfully used as an ingredient of cattle feed for many years. However, wastes from all species may be, and are, used. Among the methods employed to process animal wastes prior to feeding are: deep-stacking, ensiling (fermentation), dehydration, and pelleting. The two most common and practical methods of processing are:

1. **Deep-stacking.** In this method, the litter is deep-stacked for several weeks, during which it generates temperatures of 160°F or higher, which render it free of any potentially pathogenic microorganisms that might be present. (Pathogenic bacteria do not grow at temperatures over 80°F, and they are killed at 145°F in a matter of minutes.) It follows that there have been no documented animal health problems associated with feeding broiler or layer litter processed in this manner.

2. **Ensiling (fermentation).** Ensiling is a controlled fermentation process during which carbohydrates in the mixture are converted to lactic, acetic, and other acids. Once sufficient acids are produced, bacterial action ceases and the ensilage is stable. During the fermentation, heat is generated, thereby diminishing the hazard from certain pathogenic organisms that might be present.

The nutritive value of the ensilage is improved by blending the waste with other feed ingredients (such as a cereal grain or corn forage) prior to ensiling and adjusting the moisture content to about 40%.

Residues from medical drugs and mineral supplements are affected differently by ensiling; some are unaffected by fermentation, others are largely removed. However, no disease problems have been reported from poultry wastes in practical rations for beef cattle, dairy cattle, or sheep.

Dehydration and pelleting of animal wastes are excellent processing methods as such. However, current energy costs make them uneconomical.

In December 1980, the U.S. Food and Drug Administration published a document leaving regulation of feeding animal waste to the individual states.

Fat Added

Typical livestock rations contain relatively small quantities of fat. Although most animals require minimal amounts of certain fatty acids in their diets, these minimals are low and easily met by normal rations. Nevertheless, fat serves the following practical functions when added to livestock rations:

1. **It increases the caloric density of the ration.** Since fat contains approximately 2¼ times as much energy as soluble carbohydrate, it is possible to increase the energy content with little increase in the bulk of the ration. Thus, with the same feed intake, energy intake is higher. Since animals tend to eat until a definite caloric intake is reached, they consume less total weight when fed high-fat rations.

2. **It improves palatability.** Fats contribute a great deal to the palatability of feeds for all classes of animals. Even in the case of poultry, where taste is not a measurable factor, supplemental fat tends to overcome problems associated with acceptability of feed particles that are too fine.

3. **It facilitates absorption of vitamins A and D and provides fatty acids.** A small amount of fat is necessary for the absorption and movement of vitamins A and D from the digestive tract into the bloodstream. Also, the animal body requires a small amount of unsaturated fatty acids; for the most part, this means linoleic acid. Apparently, all species require linoleic acid, though sufficient of it is likely synthesized by rumen bacteria so that a dietary requirement may not be necessary for the adult ruminant.

4. **It delays the sensation of hunger.** Fats require a longer period of time to pass through the stomach than carbohydrates or proteins. For this reason, they delay the sensation of hunger, a characteristic which, in human diets, is sometimes referred to as "sticking to the ribs."

5. **It controls dust.** The addition of 1 to 2% fat materially lessens the dust involved in feed processing. Also, it is well known that dusty rations are not consumed readily by animals; hence, fat enhances consumption.

6. **It lubricates feed processing equipment.** This is important, because it increases throughput and decreases wear.

7. **It improves handling qualities.** The addition of fat makes for fewer fines and lessens troubles with nozzles or valves in storage and handling equipment.

However, the following problems are inherent in the incorporation of fats in feeds:

1. **Animal fats tend to solidify in cold weather.** This can cause a lumping of the feed.

2. **Fats can coat and clog mixing and distribution equipment.** Because fats tend to build up in the equipment, routine cleaning is a necessity.

3. **High levels of fat in pelleted feeds can cause soft pellets.** For this reason, a high-fat feed must be carefully formulated and processed; or the fat can be applied after pelleting.

4. **Fats can become rancid.** Antioxidants or cold storage must be used to protect fats from autoxidation. These precautions create additional expenses which must be considered in the overall cost of the feed.

If the price is favorable, fat may be added to rations at the following levels: for swine and poultry, 5 to 10%; and for cattle, 2 to 6%. Higher levels of fat usually result in drastically lowered feed consumption. When fed at the levels recommended above, the energy value of fat is approximately 2¼ times that of the grains. When corn is the major grain, fat additions can be expected to be less useful than with the small grains. This is understandable when it is realized that corn contains approximately 4% fat as compared to 1 to 1½% for the other feed grains.

Animal and vegetable fats seem to be equally effective additions to rations, provided they are stable and nonreactive (noncorrosive); hence, selection should be determined solely by comparative price.

The degree of unsaturation of fats is important. Unsaturated oils (e.g., cod-liver oil, corn oil, soybean oil, sunflower oil, and linseed oil) increase the requirements for vitamin A, vitamin E, vitamin D, vitamin K, and biotin. This is especially true if these oils are allowed to undergo oxidative rancidity in the ration or are in the process of peroxidation when consumed by the animal. These effects can be prevented by adding an effective antioxidant, such as ethoxyquin.

Usually, digestibility decreases with saturation. Also, there is a relationship between the degree of unsaturation of the fat consumed and the type of body fat formed, especially in single-stomached animals. Thus, when fed high levels of unsaturated fats, such as peanuts and soybeans, hogs produce soft pork.

Adding fats also poses the problems of how to add them, how to blend them in the mixture, and how to form relatively small pellets that will not crumble; problems which can best be solved by a specialist with expertise in feed processing.

(Also see Chapter 10, Section on "Fats and Oils"; and Chapter 13, section on "Mold (Fungi) Inhibitors.")

Freezing

A considerable amount of mink food, consisting of meat and fish, is frozen. Freezing inhibits bacterial growth and slows the enzymatic processes which can destroy the product.

Proper freezing is the key to good mink nutrition. Fresh meat and fish should be chilled immediately, quick frozen at − 10°F, and stored at 0°F. Rapid freezing makes for small ice crystals, which thaw without lowering quality. To facilitate rapid freezing and thawing, meat and fish packages should not exceed 4 to 5 in. in thickness; and to prevent freezer burn (dry out during freezing), moistureproof bags should be used. Once feeds have been thawed, they should be fed immediately; and they should never be refrozen. On the average, frozen products should not be stored longer than 6 months.

Irradiation

For many years, it was known that both ultraviolet light (from the sun) and cod-liver oil had identical effects in healing of rickets. In 1924, Steenbock of Wisconsin and Hess of Columbia University, independently announced that certain food materials could be made antirachitic by exposing them to ultraviolet light.

Upon irradiation, ergosterol, a plant sterol, yields ergocalciferol, commonly known as vitamin D_2.

The ultraviolet radiation in sunlight serves as a source of radiant energy necessary to convert 7-dehydrocholesterol (an animal sterol stored beneath the skin surface) into biologically active vitamin D_3.

Vitamin D_2, the plant form of the vitamin, and vitamin D_3, the animal form, have the same antirachitic value for the rat, dog, pig, ruminant, and human, but vitamin D_3 is more active for poultry.

Sun-cured hay is a good natural source of vitamin D for four-footed animals, but not for poultry. Fortified fish oils and irradiated sterols are good sources of vitamin D for both four-footed animals and poultry.

Molasses Added

Molasses (including cane or blackstrap, beet, citrus, wood and starch molasses) is extensively used as a livestock feed. When used at levels of 5 to 15% of the ration, it has about ¾ the energy value of corn. However, molasses has added values as an appetizer, to reduce dustiness of a ration, as a binder for pelleting, to stimulate rumen microbial activity, and as a source of unidentified factors. Also, cane molasses is a good source of certain trace minerals.

In hot, humid areas, molasses should be limited to 5% of the ration; otherwise, mold may develop. Where mustiness is a problem, it may be controlled by adding calcium propionate to the feed according to the manufacturer's directions.

When it is desired to add molasses to a ration, small- or medium-sized farmers or feed mixers must resolve the problem of how to get the molasses blended with the dry ingredients, and how to mix it. From a practical standpoint, they must determine whether they can afford an efficient machine that will blend the molasses fairly homogeneously.

(Also see Chapter 10, Grains—High-Energy Feeds, section on "Molasses.")

Organic Acids

The proper use of organic acids provides another way in which to preserve high-moisture grains. The organic acid treatment involves the application of 1 to 1½% acid (*i.e.,* propionic, acetic, formic, ammonium isobutyric, etc.) at time of harvest, followed by storage in a pile. Commercial applicators have been developed whereby the grain flows into a hopper and is sprayed with acid during spiral action of exposed auger flights. The acid flow rate is adjustable and is dependent on auger throughput and moisture content of the grain. The acid is absorbed into the kernel surface, giving it a shiny or glazed appearance, and becomes absorbed into the kernel. Properly treated high-moisture corn has the same appearance after 18 months as it has immediately after treatment.

Acid treatment inhibits the growth of molds and bacteria. Research has shown that propionic acid alone or a mixture of 75% acetic and 25% propionic acid are quite effective. Acetic acid should not be used alone. Limited research indicates that sodium propionate, formalin, ammonium isobutyrate and citric acids have been successful, as well as combinations of propionic acid and formic acid or formalin.

Experimental studies indicate that acid-treated grain has approximately the same feeding value as high-moisture grain stored in an oxygen-limiting silo. Also, it alleviates the cost of drying. Thus, organic acid treatment of grain may be a practical way in which to preserve high-moisture grains.

(Also see Chapter 9, Silage/Haylage/High-Moisture Grain.)

Preservatives

A preservative is a material added at the time of mixing or storing to enhance the keeping qualities of a feed.

From time immemorial, a major concern of keepers of herds and flocks has been that of preserving feeds over a period of time and in a suitable condition, thereby making it possible to store them in times of abundance for use in times of scarcity. Various methods of preserving feeds have been used, including drying, oxygen-limiting silos (and smaller containers), freezing, and organic acids. Several of these preservatives are covered earlier in this chapter. A brief description of hay and silage preservatives is in order, however.

• **Hay preservatives**—Preservatives are available commercially which can be applied to hay. Usually the directions (1) recommend the addition of 1 to 3 lb of these products for each ton of damp hay, and (2) claim that there will be no heating or molding.

More experimental work is needed relative to chemical hay preservatives. But, available data indicate that propionic acid is the hay preservative of choice. Missouri workers report that moist hay (28% moisture at baling) treated with an organic acid was equivalent in digestion to dry hay, whereas hay baled moist and not treated had significantly lower digestibility than treated or control hays.

Anhydrous ammonia is one of the most recent materials being studied as a hay preservative. In Indiana trials, applying this material at the rate of 1.0% to hay baled at 30% moisture successfully prevented molding, heating, and quality deterioration.

The Maryland Station found propionic acid and ammonium isobutyrate equally effective in preventing heating and preserving the quality of high-moisture hay. They recommend an application rate of 1.5% for preserving hay of approximately 30% moisture.

(Also see Chapter 8, Hay, section on "Additives for Hay.")

• **Silage preservatives**—Two types of additives have generally been used in silage making: (1) feed additives, and (2) chemical additives.

Feed additives supply a readily available source of carbohydrates for bacterial fermentaion of the silage. Some feed additives, such as corn-and-cob meal, when mixed with high-moisture forages, also absorb water and help to reduce run-off. When used as preservatives, approximately 75 to 85% of the feed nutrients added may be recovered as feed.

A large number of chemical additives have been used in silage making, with variable results. These are fully discussed in Chapter 9, Silage/Haylage/High-Moisture Grain, section on "Silages, Additives, and Preservatives"; hence, the reader is referred thereto.

Self-feeding Governors

The commonly used self-feeding governors are (1) bulky, fibrous feeds; (2) salt-feed or fat-feed mixtures; (3) fat content of block; and (4) liquid supplements.

• **Bulky, fibrous feeds**—Bulk can be used as a self-feeding governor. This consists in adding to the bulkiness of the ration, such as can be achieved by increasing the amount of chopped hay and lessening the concentrate. Actually, this is a way in which to lower the energy content of the ration. Since an animal can hold only so much, it is an effective control of feed intake.

• **Salt-feed mixtures**—The practice of using salt as a governor to limit feed consumption on pasture or range has been used for a very long time. It was ushered in as a labor-saving device for cattle and sheep in inaccessible and rough areas. Today, salt-feed mixtures are used in either meal or block form.

• **Fat content of block**—Since animals tend to eat until a certain caloric intake is reached, they consume less total weight when fed high-fat rations. Thus, pounds of feed consumed can be governed by the amount of fat in a block. It is noteworthy, too, that fat serves as a needed feed nutrient, whereas consuming more salt than required (as happens when a salt-feed mixture is used) makes for a waste of salt.

• **Liquid supplements**—When self-fed, the consumption of liquid supplements is generally controlled by (1) the use of a lick tank, and/or (2) incorporating in the formulation phosphoric acid, beet solubles, and/or citrus peel liquor.

(Also see Chapter 11, section on "Types and Methods of Feeding Protein Supplements.")

Slow-Release and Rumen Bypass Treatments

Two feed processing techniques—slow-release nonprotein nitrogen, and rumen bypass protein—are designed to delay digestion.

• **Slow-release nonprotein nitrogen**—Among the slow-release nonprotein nitrogen products that liberate nitrogen slowly are a combination of urea and gelatinized starch, and urea combined with gelatinized corn.

• **Rumen bypass**—This refers to bypass protein (also known as protected or escaped protein) in feed that escapes digestion in the rumen and passes into the lower digestive tract where it is digested and absorbed. Feed processors have developed treatments through which the bypass proteins in certain feeds can be increased; among them, heat and pressure treatment, treatment with tannins, treatment with formaldehyde or other aldehydes, lipid (fat) treatment, complexing with bentonite clay, use of amino acid analogs, increasing microbial metabolism in the rumen, and adding ionophores.

(Also see Chapter 11, section on "Protein Bypass (Protected Protein, Escaped Protein.")

Treatment of High-Cellulose Feeds

High feed prices and more stringent burning regulations have spurred research to find a practical method of improving the feeding value of several high-cellulose products, such as rice, wheat, barley and oat straws; bagasse; tree bark; corncobs; gin trash; newspaper; and seed hulls.

In their natural state, these products make poor feedstuffs because lignin or silica, or a combination of the two, (1) encrust the energy-rich carbohydrates, cellulose, and hemicellulose; and (2) keep the microbes in the ruminant's stomach from breaking them down to release energy.

The answer to this problem lies in some treatment that opens up the fibers enough to permit increased digestion in the rumen. Several methods of chemical and/or physical treatment are being investigated; among them, alkali treatment, ammoniation, hydrogen peroxide treatment, and high pressure steam.

• **Alkali treatment**—Sodium hydroxide is the common alkali treatment of high-cellulose products (crop residues), although calcium hydroxide and potassium hydroxide are sometimes used. The effectiveness of alkali treatment depends on the residue or waste being treated and the technique employed. Treatment level ranges from about 2 to 10% of the chemical based on the total dry matter content of the material being treated; and treatment time is about 24 hours. There is indication that mild heat can be used to reduce chemical levels. The results from alkali treatment vary substantially, but based on presently available information, the following deductions can be made:

1. On the basis of efficacy of the treatment, the cereal straws rank as follows: wheat straw, barley straw, and oat straw.
2. It can (a) increase the rate of passage of indigestible material through the digestive tract, and (b) improve the intake of low-quality roughage by up to 50%.
3. It increases potential digestion of cell walls.
4. It increases digestibility of dry matter or energy up to 10%; and in a high-straw ration it may even be greater.
5. Improvements in intake and digestibility may be small when treated straws constitute 50% or less of the ration.
6. Through its heating effect, it depresses the nitrogen digestibility by ruminants.
7. Because energy availability in the rumen is enhanced by alkali treatment, supplementation of treated roughages with more extensively degraded protein sources is usually beneficial.

It is noteworthy that the alkali soaking process was first used on straw in Germany in 1919, during World War I when there was a critical shortage of livestock feeds.

• **Ammoniation treatment**—This method involves placing the high-cellulose products (crop residues) in an air-tight enclosure (such as black plastic) and adding either anhydrous gas or liquid ammonia. It is important to add the correct amount of ammonia; too much makes for unnecessary expense, and too little makes for poor quality feed. Optimum treatment level appears to be 3.0 to 3.5% anhydrous ammonia based on the total dry matter content of the material being treated; and optimum treatment time is about 20 days vs 24 hours for the sodium hydroxide treatment. The two major advantages of using ammonia in comparison with the alkali treatment are: (1) it adds nonprotein nitrogen to the product, and (2) no mineral residue remains that might be detrimental to the animal or to the soil to which the manure is added. Ammoniation produces the following benefits:

1. It increases the crude protein equivalent by 3 to 10%.
2. It increases the digestible energy (TDN) by 3 to 23%.
3. It increases animal intake by 20 to 27%.
4. It prevents molding of crop residues and high-moisture forages.

CAUTIONS relative to ammoniation: Anhydrous ammonia in liquid form is very toxic to the skin and eyes; so, when a person's skin/eyes come in contact with anhydrous ammonia, it should be flushed away with water immediately; otherwise, serious injury may result. Also, ammonia can be flammable and even explosive; so, never smoke or light a flame near it.

Although the ammoniation treatment of low-quality forages will, at low cost, increase both protein equivalent and fiber digestibility, the toxicity problem cannot be ignored.

Ammonia toxicity (characterized by hyperexcitability, circling, convulsions, and some deaths) occurs among some cattle and sheep receiving ammoniated feeds and among some young suckling lactating mothers fed ammoniated feeds. Ohio researchers reported that 3 out of 9 sheep (33%) fed ammoniated hay (4% ammonia on a dry basis) developed signs of toxicity, and 2 of them (22%) died; and that 4 out of 5 calves (80%) that received milk from cows fed ammoniated hay (5% ammonia on a dry basis) developed signs of toxicity, and 1 calf (20%) died.[10]

(Also see Chapter 4, section headed "Urea and Ammoniated Feeds"; Chapter 11, section headed "Ammoniated Products"; and Chapter 5, Table 5-3, Potential Poisons—Urea [Ammonia Toxicity].)

• **Hydrogen peroxide**—In this treatment, the crop residue (corn stalks, wheat straw, or other crop residue) is shredded and mixed with a 1% solution of hydrogen peroxide. The pH level of the mixture is brought up to 11.5 with an alkali material. The crop residue literally disintegrates into a mushy substance, which is rinsed off; following which it can be dried or handled wet, in the same way as corn gluten feed. When dried, the material is light and flaky. It can be pelleted to increase its density and ease of handling. Limited experiments with cattle and sheep show that corn stalks and wheat straw treated with hydrogen peroxide has feed value higher than corn silage. At this stage, the process requires too much water to be practical and is too expensive; but more than likely these matters will be overcome. The chief researcher has applied for a patent on the process.

• **High pressure steaming**—High pressure steaming, with or without added chemicals, has been used to a limited extent in treating crop residues and wood. Aspen, which is the most digestible of the woods, has been shown to reach a digestibility of 56.6% after steaming for 2 hours at 165°C, with the treated product readily accepted by sheep at up to 60% of the ration and producing normal weight gains and carcass yields. The treating of wood products is limited because: (1) the cost of treatment is high, (2) conventional feedstuffs need to be relatively high priced before treated wood residues can compete in the marketplace, and (3) the lack of a steady market for the treated products.

Decisions as to type and amount of chemicals and treatment system must be based on evaluation of processing costs in relation to the value of the finished feedstuff.

(Also see Chapter 12, By-product Feeds/Crop Residues, section headed "Treating Crop Residues to Increase Digestibility.")

COMPLETE (ALL-IN-ONE) RATIONS

Most experiments and experiences have not shown any difference between mixed rations and the feeding of roughage and concentrates separately insofar as efficiency and production are concerned. However, a mixed ration has the following advantages:

1. It makes for greater efficiency in feeding and lessens the sorting at the feed bunk.
2. When the roughage is relatively unpalatable, a mixed ration forces consumption.

---

[10]Weiss, W. P., *et al.*, Etiology of Ammoniated Hay Toxicosis, *Journal of Animal Science,* Vol. 63, No. 2, 1986, p. 525.

3. When it is desired to limit concentrate consumption, mixing with the roughage is desirable.

4. A mixed ration makes it easier to get animals on full feed.

Thus, each feeder must decide on the matter of mixed feed vs feeding roughage and concentrate separately, with relative costs and other factors considered. Most large cattle and sheep feedlots use completely mixed rations. Also, the trend is toward complete feeds for both dairy cows and swine, primarily because such complete feeds (1) lend themselves better to automation, and (2) provide better control of nutrient intake.

• **All-pelleted rations (grain and forage combined)**—Increasingly, complete pelleted rations are being used for horses, swine, and fish. Among the virtues ascribed to all-pelleted rations are (1) They prevent selective eating—if properly formulated, each mouthful is a balanced diet; (2) they alleviate waste; (3) they eliminate dust (thereby lessening heaves in horses); (4) they lessen labor and equipment; (5) they lessen storage; and (6) they facilitate automation.

CHOICE OF PROCESSING METHOD

The choice of a processing method is highly dependent on the feedstuff to be fed. It is clear that a given processing technique may be very desirable for one grain, but quite detrimental to another. Corn may be fed without any processing, but not milo. Pressure treating appears to be desirable for milo, but harmful to wheat.

Comparison of grain processing techniques is difficult because there are a number of interactions between processing technique and roughage level or type of ration fed. For example, data from Ohio State University have shown that whole shelled corn was superior to crimped corn in very low-roughage rations, whereas crimped corn was clearly superior in high-roughage rations.

Interactions cause results on new grain processes to be biased by the kind of control ration fed. Consequently, feeders should consider only those comparative tests which involve rations and conditions very similar to those they intend to use in their own feeding program.

FARM PROCESSED FEED

There are two alternative sources of most feeds and rations—home mixed vs commercially mixed; and the able manager will choose wisely between the two.

The value of farm-grown grains—plus the cost of ingredients which need to be purchased to balance the ration, and the cost of processing and mixing—as compared to the cost of commercial ready-mixed feeds laid down on the farm, should determine whether it is best to mix feeds at home or depend on ready-mixed feeds.

Although there is nothing about the mixing of feeds which is beyond the capacity of the intelligent farmer or rancher, under many conditions a commercially mixed feed supplied by a reputable dealer may be the most economical and the least irksome, especially when many ingredients and additives are involved, such as in young stock and poultry rations. The commercial dealer has the distinct advantages of (1) purchase of feeds in quantity lots, making possible price advantages, (2) economical and controlled mixing, and (3) the hiring of scientifically trained personnel for use in determining the rations.

Modern feed mixing is much more complicated than in the era when hog rations consisted primarily of corn and tankage. Today, micronutrients are added, and this necessitates both sophisticated equipment and skillful mixing. For example, a recommended swine ration may call for the addition of 1 mg of riboflavin per pound of mixed ration. When one considers that a grain of wheat weighs 60 mg, some perspective of the mixing problem comes into focus.

Fig. 14-9. On-the-farm portable grinder-mixer in operation. (Courtesy, Sperry New Holland, New Holland, Pa.)

Baby chick rations are even more exacting. With a consumption of only 2 to 3 oz of feed per day, it is important that the chick get all the trace amounts of the various required nutrients in that quantity. This necessitates that the feed be blended so well that a tablespoon from one sack will be almost identical in chemical composition to a tablespoon taken from the same batch 20 bags later.

Because of these several advantages, commercial feeds are finding a place of increasing importance in American agriculture. Nevertheless, practical considerations favor the use of much homegrown feed. Many feed manufacturers formulate supplements for this particular purpose. For example, a supplement may be prepared for a corn-soybean ration for hogs.

FEED PROCESSING TABLE

Table 14-4 is a summary of pertinent information relative to the preparation of feeds for each class of livestock.

TABLE 14–4
PREPARATION OF FEEDS

Class of Animal	Concentrates	Forages	Comments
Beef cattle	**E**xtruding, flaking, micronizing, popping, roasting, or high-moisture grain—with choice determined by cost—are preferable, especially for full-fed animals on a high-grain ration. But such equipment is costly to purchase and operate; hence, a large-volume operation is required to cover fixed costs. **D**ry or steam roll or grind coarsely for most beef cattle, especially those not full fed high-grain rations and those in smaller operations. **O**n high-concentrate rations (those with 80% or more concentrate), whole corn need not be processed. **G**rain (except for very hard seeds) need not be processed for calves under 6 months of age, for young calves masticate feed thoroughly. **C**ubes (large pellets) preferred for feeding on pasture or range. **P**rofessional caretakers often cook feed (especially barley) for show cattle to increase palatability.	**L**ong hay is satisfactory for most cattle other than commercial feedlot operations. **C**hopped (2″ length), cubed (wafered), or pelleted forage should be used (1) in commercial feedlots or when quality of hay is poor; and (2) in all cattle operations from standpoints of ease of handling and lessening wastage. **S**hredding fodders and stovers (corn or milo) makes them easier to handle and lessens waste.	**F**ine grinding grain increases incidence of hyperkeratosis (ruminal parakeratosis) in feedlot cattle. **D**ry or steam rolling or coarse grinding of grains are of about equal value for most beef cattle. Either method is just as satisfactory as more expensive methods (like flaking) when grain intake is relatively low. **C**hopping or pelleting low-quality hay is more advantageous than chopping high-quality hay. **F**inely ground hay not recommended, as it decreases digestibility.
Dairy cattle	**G**rinding is the simplest and the most widely used grain processing method for dairy cattle. Cracking, steam rolling, and pelleting are the other popular procedures. **E**xploding, extruding, flaking, micronizing, popping, roasting, or high-moisture grain are preferable for high-producing lactating cows, but they are not widely used. **D**ry or steam roll or grind coarsely for dry cows, young stock, and low-producing cows. **G**rain (except for very hard seeds) may be fed whole to young calves under 6 months of age.	**L**ong hay or cubes. Cubes lend themselves to automation; and lower milk fat percentage only slightly, if at all.	**B**utterfat is depressed unless the ration contains some threshold level of coarse material. **F**inely ground or pelleted roughage will result in reduced rumen acetate production and lower milk fat percentage.
Sheep and goats	**P**rocessing grains not necessary unless seeds are hard (like sorghum or millet) or the teeth are poor. **H**ard seeds (like sorghum) may be prepared by exploding, extruding, flaking, micronizing, popping, roasting, or high-moisture grain, with cost determining the choice. **P**ellets are increasingly being used by lamb feeders. **C**ubes or pellets preferred for feeding on pasture or range. **P**rofessional shepherds prefer flaked grain for show sheep, as the ration is lighter and there are fewer digestive disturbances.	**C**hop (2″ in length), pellet, or cube. **M**any lamb feeders are using all-pelleted rations (hay and grain combined).	**S**heep and goats masticate grain more thoroughly than cattle, with the result that feed preparation for them is of less value than for cattle. **A** high incidence of parakeratosis—a degeneration of the rumen papilla—appears to result from feeding pellets, especially when low forage-high concentrate pellets are used. Hence, breeding sheep should not be fed pellets for extended periods without any long or chopped forage.
Swine	**C**orn, barley, grain sorghum, and oats should be finely ground for swine. Medium to coarse grinding is best for wheat, because fine grinding makes it pasty and less palatable. **P**elleting corn-soybean rations generally improves feed utilization and increases rate of gain by at least 4–5%. **C**ook Irish potatoes, beans, soybeans, and garbage. **C**ooking (except for the feeds listed above), soaking, or fermenting are not of value when swine are on full feed. **L**iquid and paste feeding give inconsistent results in feed consumption and rate of gain; hence, they should be evaluated on the basis of a mechanical means of dispensing feed. However, slop (slurry or gruel) is desirable for early weaned pigs, and perhaps for pigs being fitted for show or sale. **H**igh-moisture corn does not result in any improvement of efficiency for swine; hence, the value of high-moisture corn as compared to regular corn should be computed on a dry matter basis.	**A**lfalfa (or other legume) that is to be incorporated in mixed feeds should be ground. **R**ations containing considerable amounts of fiber are improved by pelleting because of increased consumption, improved carbohydrate digestibility, and reduced sorting and wastage compared to meal rations.	**F**ine grinding will cause some bridging in self-feeders. Also, finely ground feed is associated with increased incidence of stomach ulcers in swine.
Poultry	**G**rains for poultry are prepared in 3 forms—mash, pellets, crumbles. 1. *Mash:* grind medium fine. 2. *Pellets:* composed of mash feeds that are pelleted. Birds usually consume more of a pelleted ration than the same ration in mash form. Usually there is more cannibalism with pellets than with mash or crumbles. 3. *Crumbles:* produced by rolling pellets.	**G**rind hay that is to be included in poultry feeds.	**P**roper heating of protein sources will result in better availability of nutrients if temperature and time of heating are not excessive. **H**eating soybean meal destroys trypsin inhibitor and possibly other factors which limit protein digestion especially in growing chicks. **M**ethionine addition to heated meals increases chick growth.
Horses	**F**laking is the preferred method of grain preparation for horses; it makes for a light ration and few digestive disturbances. **F**or horses with good teeth, the value of oats is increased only 5% by processing.	**E**ither feed long hay or an all-pelleted ration (grain and hay combined).	**C**ubes (wafers) sometimes cause horses to choke. **H**orses are very sensitive to moldy feed. **H**orses should not be fed dusty feed, because of the hazard of heaves.
Rabbits	**N**utritionally complete pellets containing 50–60% concentrate and 40–50% roughage, ⅛–3/16″ diameter and ⅛–¼″ long, of two types: 1. Production rations, high in protein and energy. 2. Maintenance rations, medium in protein and energy. **G**rains may be fed whole, but to maximize digestibility oats and barley should be rolled and corn should be cracked. **P**rotein supplements should be in cake, pellet, or crumble form.	**R**oughages should be cut to lengths of 3″ and fed in a rack.	**R**abbits are very sensitive to moldy and dusty feeds, so they should always be fed clean, high-quality feeds. **P**rotein supplements in mash form are not desirable, because they may settle out from the rest of the feed and be wasted. **U**ncut roughage will result in wastage.

(Continued)

TABLE 14-4 *(Continued)*

Class of Animal	Concentrates	Forages	Comments
Mink	Traditionally, mink rations are high in moisture (60–70%) and are fed as a semisolid mass (with a consistency not unlike that of a hamburger) on the wire mesh constituting the top of the pen. **Fresh or frozen animal products and by-products from:** 1. Horses (now in limited supply). 2. Meat-packing houses. 3. Poultry processing. And many other animal products. **Fresh or frozen fish and fish by-products.** **Processed grains and cereal by-products:** 1. From breakfast food industry. *Limit to*—10–15% in breeding diets. 10–25% in production diets. 2. Supplemental cereals (same as No. 1 above) plus skim milk, wheat germ meal, alfalfa meal, yeast, and meat and fish meals. *Limit to*—15–20% in breeding diets. 15–35% in production diets. 3. Dried bakery products. *Limit to*—15–20% of diet. **Vegetables:** *Limit to*—8% of diet. **Completely dehydrated mixed mink feeds.** Such feeds are now available in bags. They do not require refrigeration. The ultimate goal of the large mink producer, which is now being realized, is to supply the complete diet as dry pellets that can be dispensed from a self-feeder.		Fresh meat and fish should be chilled immediately, put in moistureproof bags not to exceed 4–5" thick, quick frozen at −10°F (*−23°C*), stored at 0°F (*−18°C*), and not held more than 6 months. Dehydration is an alternative to freezing for longtime storage, but protein quality is reduced. Fish containing thiaminase should be cooked. Cooking grain improves digestibility of carbohydrates but reduces protein digestibility. However, there may be diarrhea when fed raw.
Fish	Fish diets may be either wet or dry. *Wet diets:* 1. Natural diets from the immediate environment. No feed preparation involved. 2. Artificial diets consisting of various organs, meats, and by-products from animals, poultry, and fish. These products must be refrigerated to prevent spoilage. *Dry diets:* Fed in pellet form, with the size of the pellet altered according to the size of the fish. Pellets are of 2 kinds: 1. Pellets that *float* (they're spongelike) for surface feeders, like rainbow trout. 2. Pellets that *sink* for bottom feeders, like brown trout. Eels are fed mixed diets in paste form.		Some researchers recommend that at least 50% of the ration of carp consist of natural ingredients. Maximum production of natural feeds usually calls for fertilization with organic or inorganic fertilizers. Catfish ponds should be fertilized to establish an optimum level of plankton growth. Fish products should be heated to destroy thiaminase. Normally, catfish are bottom feeders, but they can be taught to feed at the surface.

Fig. 14-10. Turkey eating pellets. (Courtesy, Ralston Purina Company, St. Louis, Mo.)

QUESTIONS FOR STUDY AND DISCUSSION

1. Why did much of the recent technology in feed preparation evolve with cattle feedlots following World War II?

2. Cite examples of feed processing lowering the nutritional value of feedstuffs.

3. Why does the importance of proper feed preparation become increasingly important with higher levels of feeding and greater production?

4. How does the size of feeding operation affect the choice of feed preparation?

5. List and discuss what you consider to be the six most important reasons for processing feeds.

6. List and discuss what you consider to be the four most important nutritional factors in choosing the grain processing method.

7. List and discuss what you consider to be the four most important nonnutritional factors in choosing the grain processing method.

8. Briefly describe each of the following processing methods:

Extruding	Steam flaking
Popping	Pelleting
Roasting	Crumbling
Cooking	

9. For each of two cattle finishing operations, one involving 300 steers and the other 10,000 steers, justify your choice of a processing method between grinding vs steam flaking of milo. Would your choice of a processing method be changed were you processing corn instead of milo?

10. What does cooking do for soybeans? Why is there concern lest the feeding of whole soybeans produce soft pork?

11. Are beef cattle fitters justified in cooking barley for animals being fitted for show?

12. Why do poultry producers tend to favor crumbles?

13. How is a bran mash prepared and used for horses? Why is a bran mash used?

14. How will energy shortage affect the choice between the following preparation and processing methods of corn harvested at 30% moisture?

a. Storage in a silo
b. Artificially dried
c. The use of an organic acid

15. Can the use of hydroponics be justified for race and show horses?

16. Why are antioxidants incorporated in feeds?

17. Do flavoring agents really increase the palatability of feeds, or do they merely impart to caretakers the grand feeling of showering the maximum amount of love and affection on their animals?

18. Explain why forage should not be finely ground when fed to lactating dairy cows. Why do dairy operators favor cubed (wafered) forage over pellets for lactating cows?

19. How may a farmer-feeder add fat to a ration? How may molasses be added?

20. Of what practical value is irradiation in the production of vitamin D?

21. Under what circumstances would you use a preservative for each (a) hay, and (b) silage? Which preservative(s) would you use?

22. Under what circumstances would you use self-feeding governors, such as salt or fat, on the western range?

23. What makes rumen bypass such an exciting area for research?

24. What's the practicality and future of improving the feeding value of high-cellulose feeds through alkali treatment, ammoniation treatment, hydrogen peroxide treatment, and high pressure steaming?

25. Discuss the difference between mixed rations vs feeding roughage and concentrate separately.

26. Under what circumstances would you recommend commercially prepared feed? Under what conditions would you recommend home mixing?

SELECTED REFERENCES

Title of Publication	Author(s)	Publisher
Animal Nutrition, 2nd Edition	P. McDonald R. A. Edwards J. F. D. Greenhalgh	Oliver & Boyd, Edinburgh, Scotland, 1973
Applied Animal Nutrition	E. W. Crampton	W. H. Freeman, San Francisco, Calif., 1956
Beef-Forage Notebook		Cooperative Extension Service, University of Nebraska, Lincoln, Nebr.
Cereal Processing and Digestion	Collected Papers	U.S. Feed Grain Council, London, England, 1972
Cornell Beef Production Reference Manual		Cooperative Extension, Cornell University, Ithaca, N.Y.
Effect of Processing on the Nutritional Value of Feeds	Proceedings of a Symposium	National Academy of Sciences, Washington, D.C., 1973
Feed Formulations, 2nd Edition	T. W. Perry	The Interstate Printers & Publishers, Inc., Danville, Ill., 1982
Feed Manufacturing Technology	R. R. McEllhiney, Editor	American Feed Manufacturing Association, Inc., Arlington, Va., 1985
Feed Mixers Handbook	R. M. Sherwood	The Interstate Printers & Publishers, Inc., Danville, Ill., 1951
Feedlot, The	G. B. Thompson C. C. O'Mary	Lea & Febiger, Philadelphia, Pa., 1983
Feeds and Feeding, 22nd Edition	F. B. Morrison	The Morrison Publishing Company, Ithaca, N.Y., 1956
Fundamentals of Nutrition	E. W. Crampton L. E. Lloyd	W. H. Freeman and Company, San Francisco, Calif., 1959
Nutrition of the Chicken, 3rd Edition	M. L. Scott M. C. Nesheim R. J. Young	M. L. Scott & Associates, Ithaca, N.Y., 1982
South Dakota Beef Cattle Handbook		Cooperative Extension Service, South Dakota State University, Brookings, S.D.
Texas Cattle Feeders Handbook		Texas Agricultural Extension Service, Texas A&M University, College Station, Tex.
Underutilized Resources As Animal Feedstuffs	J. P. Fontenot, Chairman of Committee	National Academy Press, Washington, D.C., 1983

Chapter 15

Original painting by Tom Phillips

FEED ANALYSIS/FEED EVALUATION

Contents | Page

At first glance, a feed may appear to be nutritious and of high quality; but unless it is systematically analyzed—through physical, chemical, and/or biological means—there is no way that one can be sure of its true value to livestock. Two hays may look alike, yet one may contain 12% protein and the other 18%. Thus, chemical analysis of the 2 hays can be justified. But the chemical composition of feed is not enough! Experiments must be designed to determine the availability of the nutrients in the feed to the animal. This involves the digestibility of the nutrients within an ingredient. A feed that is high in nutrients, but low in digestibility, is merely a filler and of little value to livestock.

The most accurate method of determining feed value would involve live animal trials on the farm or ranch where the feed is to be fed. However, this is too costly, slow, and impractical. Therefore, laboratory methods have been developed to predict feed value, and new ways are evolving.

The first record of a systematic approach to measuring nutritive quality can be traced to the early 1800s. Gradually, it became possible to measure many of the nutritional qualities of feeds, but some nutritional characteristics have escaped the chemists' efforts even to the present time. Indeed, some properties of feeds, such as palatability, may never be determined in the laboratory, but may always remain in the domain of the animals themselves.

At the turn of the present century, someone analyzed coal and found that its energy content was similar to that of some of the better animal feeds. Yet, animals could not utilize the energy of coal. Even today, many of the procedures we use to measure nutritive value of feeds are merely rough approximations. For example, the most commonly used procedure to measure protein consists of quantitating the amount of nitrogen within the feed, then multiplying the value derived therefrom by an adjustment factor. The primary assumption of this technique, therefore, is that all of the nitrogen is in the form of protein, which is possible but not probable.

The science of nutrition came into its own during the 20th century. The isolation and identification of growth-promoting factors and nutrients, such as the vitamins, have enabled the producer greatly to increase the efficiency of production. The development of the basic science of biochemistry has paralleled that of the more applied nutritional sciences. Many of the analytical techniques developed in biochemical laboratories have recently been adapted by nutritionists for the analysis and evaluation of feeds. New and sophisticated techniques are now being used in the feed industry which can detect certain compounds in the range of parts per billion (mcg/kg of tissue or feed). Today, many analytical procedures have been improved through automation, thereby reducing the cost per sample.

In addition to the increased sensitivity of new techniques in the quantitative determination of nutrients, new procedures are now being used to monitor contaminants, such as pesticides and chemical pollutants, in feed. These developments have come as a result of pressures applied by the government, the producer, and the consumer to ensure the quality and safety of food products. Thus, many commercial feed companies are now routinely monitoring feed quality to protect themselves from damages incurred from the sale of mislabeled feed or feed contaminated with toxic materials.

IMPORTANCE OF MONITORING FEED QUALITY

Profit is the ultimate criterion of success in any livestock operation, and the cost of nutrients is an important factor in determining success. The feed composition values presented in this and other books merely represent an averaging of an accumulation of data concerning the nutritive value of feeds. Considerable variation is inherent in the nutrient content of different samples of feeds. Thus, the successful producer must recognize the value of a well-planned feed analysis program.

An example of the variability of feedstuffs is presented in Table 15-1. Investigators at North Carolina State University summarized the quality of various feed ingredients used within the state. Table 15-1 shows the variation that they found in the composition of samples of 48.5% crude protein soybean meal. If the price of soybean meal is $180 per ton, the meal containing 41.88% protein (the low extreme) costs 21.5¢ per pound of protein. The meal containing 54.86% protein (the upper extreme) costs 16.4¢ per pound of protein. Thus, even though the buyer is theoretically buying the same product, the meal at the upper extreme of the range actually costs only 76% as much as the meal of the lower extreme on a cost per pound of protein basis.

TABLE 15-1
VARIATION IN THE COMPOSITION OF 48.5% PROTEIN SOYBEAN MEAL NORTH CAROLINA[1]

Nutrient	Number of Samples	Average Composition	Low	High
		(%)	(%)	(%)
Moisture	1,423	11.86	1.22	15.17
Protein	1,425	48.69	41.88	54.86
Fat	128	2.08	0.68	4.70
Fiber	1,424	3.20	2.10	33.00

[1]Adapted by the authors from "Ingredient Quality Analysis and Reporting," by D. W. Murphy and J. B. Ward, *Feedstuffs*, June 7, 1976.

Variations in composition of the magnitude shown in Table 15-1 are unusual today, because of improved methods of analysis and computerized blending equipment. However, nutrient variations of 10 to 15% are normal; due to differences from farm-to-farm and year-to-year; due to differences in samplings, varieties, soil fertility, weather, maturity of harvest, and storage; and due to differences in laboratory testing techniques, and differences in reporting results—some report the results on an as-fed basis whereas others report on a moisture-free basis. Because of such spread in feed compositions, the actual analysis of a given lot of feed should be obtained and used wherever practical. In most cases, however, there is insufficient time to sample and analyze a given lot of feed, and/or the lot of feed may be too small to justify the expense. So, feed composition tables based on large numbers of samples tested, serve well as the bases on which (1) to conduct buying and selling transactions, and (2) to formulate rations to meet the requirement of animals.

HOW TO TAKE FEED SAMPLES

An analysis of a feed is only as good as the sample it represents. If the sample is not representative of the entire batch, the usefulness of the evaluation is limited, no matter how extensively the sample is analyzed. Feed tends to be highly variable in composition. Thus, one bale of hay may be very different in feeding value from another bale immediately next to it. For this reason, several samples are taken from various representative areas, then the samples are thoroughly mixed together to form a representative sample of the entire feed; and an aliquot (part of a whole) is taken therefrom so that the final analysis will represent an overall average of the feed being sampled.

All feed samples should be put in plastic containers or insulated bags and immediately sent for analysis. If, for some reason, there is a delay in submitting samples that have high moisture content, such as silages, they should be frozen until analyzed. The date, place of sampling, and an identification number should be supplied with each sample.

Quite often, handlers may want to send split samples for analysis. In this procedure, the final sample is carefully split in half and sent as two separate samples. This gives evidence of the accuracy of the laboratory analysis. While this procedure increases cost, it does test the reliability of the laboratory. If the reported results of the split samples differ materially, the producer can deduce that one or both of the feed analyses were faulty.

Keeping feed samples can be very helpful should any feed-related problems arise. So, a 1- to 2-lb sample of feed from each lot should be placed in a plastic bag, properly labeled, and stored in a cool, dry place for 30 to 60 days. If no problems arise during that period and the feed has not become moldy, it can be fed.

Mixed Feeds

Feeds that are mixed in either a horizontal or a vertical mixer can be sampled with relative ease. When it is certain that the feed is thoroughly mixed, samples can be taken periodically as the feed comes down the chute. Samples should be taken at random intervals.

Since the various ingredients in mixed feeds tend to separate out according to size and hygroscopic properties, accurate sampling of mixed feeds can sometimes be difficult. In many cases, an analysis profile of the individual ingredients used in the mixed feed is more accurate, assuming care is taken in the weighing and mixing of the feed.

Grains

Grain samples are generally obtained with a grain probe. A minimum of five cores should be taken from various well separated places in the truck or bin. Grain obtained from these cores should be mixed thoroughly, thence a sample (about 1 lb) obtained from the mixture should be bagged and labeled.

Silage

Silage is difficult to sample for the following reasons:

1. It is highly variable in moisture content.

2. It is usually harvested at different times; hence, the silage within a silo will vary in stage of maturity.

3. It is difficult to obtain a sample from various locations throughout the silo, due to the physical structure of the silo.

To obtain a representative sample of silage from a tower silo that is equipped with a mechanical unloader, the person collecting the sample should catch at least 12 handfuls of silage. It is best to let the unloader run until fresh, clean silage is available. When a pit silo is to be sampled, the sampler should collect about 20 handfuls of silage from various areas of the freshly cut face. These samples should then be thoroughly mixed together; and a single sample should be taken from the mixture. This procedure should be repeated every third or fourth face cut in order to take into account variation within the silo. All samples of silage must be placed in an airtight plastic bag and either sent immediately for analysis or frozen.

Hay

With standard hay bales, core samples should be taken from the ends of bales. With large round bales, 2 to 3 core samples should be taken from each side. It is advisable to take random samples of about 5% of the bales.

If loose or chopped hay is to be analyzed, core samples must be taken from numerous locations of the stack, thence the samples should be mixed and sealed in a plastic bag.

Pasture

When sampling a pasture, it is advisable to move through the pasture in a "Z" or "X" pattern. At predetermined intervals, for example every 50 paces, a 12- by 12-in. sample is cut at mowing height. If there is very little forage at this predetermined sampling point, cut what is available and move on. Do not move off the path to cut a more densely populated area. Upon completion of the walking pattern, mix the cut forage thoroughly and bag a sample for analysis.

FEED ANALYSIS

Feeds are analyzed by physical, chemical, or biological procedures. Although physical evaluation may be the least accurate, it provides a quick and easy means of obtaining considerable information about the overall quality of a feed. The chemical procedure is more accurate than a physical evaluation, but it takes time. The biological method necessitates considerable time and expense, and the results are often variable, but it helps to assess the availability of the feed nutrients to animals.

Physical Evaluation of Feedstuffs

In order to produce or buy superior feeds, producers need to know what consititutes feed quality, and how to recognize it. They need to be familiar with those recognizable characteristics of feeds which indicate high palatability and nutrient content. If in doubt, observation of the animals consuming the feed will tell them, for livestock prefer and thrive on high-quality feed.

The physical evaluation of feedstuffs, especially forages, is based largely on visual and smell appeal. Does it look good and smell good?

• **Characteristics of good hay**—The easily recognizable characteristics of hay of high feeding value are:

1. It is made from plants cut at an early stage of maturity, thus assuring the maximum content of protein, minerals and vitamins, and the highest digestibility.
2. It is leafy, thus giving assurance of high content of protein and other nutrients.
3. It is bright green in color, indicating proper curing, a high carotene or provitamin A content, and palatability.
4. It is free from foreign material, such as weeds, dirt, and stubble.
5. It is free from mold and dust.
6. It is fine stemmed and pliable—not coarse, stiff, and woody. This is particularly true when comparing first cut hay with subsequent cuttings. Later cuttings tend to be less fibrous.
7. It has a pleasing, fragrant aroma; it smells good enough to eat. (Also see Chapter 8, Hay.)

• **Characteristics of good silage**—The easily recognized characteristics of silage of high feeding value are:

1. It has a *clean,* rather pleasing lactic acid odor, in contrast to the foul or objectionable butyric acid odor of poor silage.
2. It has a pleasing taste—it is not bitter or sour.
3. It is not moldy, musty, or slimy.
4. It is uniform in moisture and color. Generally, green or brownish silage is good; tobacco brown or dark brown silage indicates excessive heat; and black silage is spoiled and should not be fed.

(Also see Chapter 9, Silage/Haylage/High-Moisture Grain.)

• **Characteristics of good grains and other concentrates**—The easily recognizable physical characteristics of good grains and other concentrates are:

1. Seeds are not split or cracked.
2. Seeds are of low moisture content—generally containing about 88% dry matter.
3. Seeds have a good color, characteristic of the species.
4. Concentrates and seeds are free from mold.
5. Concentrates and seeds are free from rodent and insect damage.
6. Concentrates and seeds are free from foreign material, such as iron filings.
7. Concentrates and seeds are free from rancid odor.

(Also see Chapter 10, Grains/High-Energy Feeds.)

MICROSCOPIC EXAMINATION

Many laboratories use microscopes to aid in feed evaluation. Through the use of this instrument, a trained technician can identify what ingredients are in the feed mix, along with the quantities of each. Also, the microscopist can detect adulteration and variation in quality more quickly and economically than with any other known technique. Weed seeds, foreign objects—such as iron filings and rodent excreta—mold, and damaged feed (for example, split grains), can all be observed under the microscope.

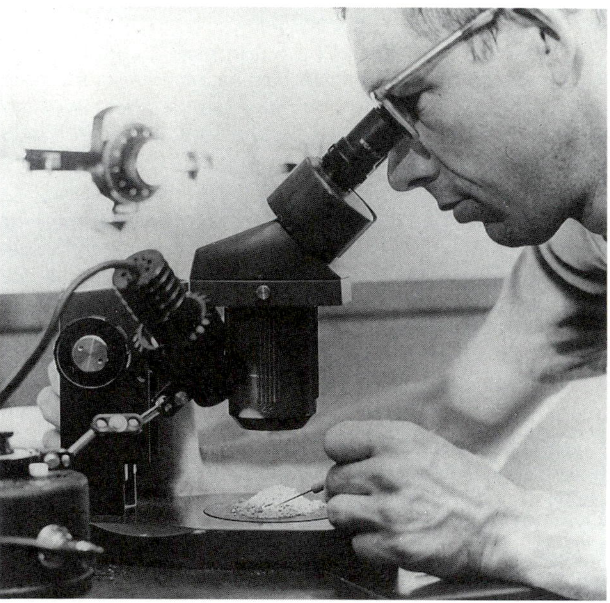

Fig. 15-1. Microscopic examination of feed. (Courtesy, Ralston Purina Company, St. Louis, Mo.)

Proximate Analysis

For more than 100 years, feeds have been analyzed by a method developed by 2 scientists, Henneberg and Stohmann, at the Weende Experiment Station in Germany. This method is called the proximate analysis, or the Weende System, of feed analysis. Feeds are evaluated in terms of 6 components: (1) moisture, (2) ash, (3) crude protein, (4) ether extract, (5) crude fiber, and (6) nitrogen-free extract (see Table 15-2).

Today, feeds are being analyzed routinely through highly sophisticated chemical procedures. Many agricultural experiment stations, as well as most large feed companies, have facilities to analyze feeds for both the prevention and diagnosis of nutritional problems.

A chemical analysis gives a solid foundation on which to start in the evaluation of feeds. Thus, feed composition tables serve as a basis for ration formulation and for feed purchasing and merchandising. Commercially prepared feeds are required by state law to be labeled with a list of

TABLE 15-2
THE FRACTIONS OF PROXIMATE ANALYSIS

Fraction	Procedure[1]	Major Components
1. Moisture (dry matter).	**H**eat sample to constant weight at temperature just above boiling point of water. Loss in weight equals water.	**W**ater and any volatile compounds (100% − H_2O = DM%).
2. Ash (mineral matter).	**B**urn at 500° to 600°C for 2 hours.	**M**ineral elements.
3. Crude protein (protein averages 16% N; hence, N × 6.25 = crude protein).	**D**etermine nitrogen by Kjeldahl process.	**P**roteins, amino acids, nonprotein nitrogen.
4. Ether extract (fat).	**E**xtraction with diethyl ether.	**F**ats, oils, waxes, resins, pigments.
5. Crude fiber (CF).[2]	**R**esidue after boiling in weak acid and weak alkali.	**C**ellulose, hemicellulose, lignin.
6. Nitrogen-free extract (NFE).[2]	**R**emainder; *i.e.,* 100 minus sum of the other fractions.	**S**tarch, sugars, some cellulose, hemicellulose, and lignin.

[1]Each procedure can be applied to a separate sample, of standard weight, of the feedstuff to be analyzed; or a single sample can be used to determine dry matter, crude fat, and crude fiber. In the latter case, separate samples would be run for ash and crude protein.

[2]Carbohydrates (CHO = CF + NFE).

ingredients and a guaranteed analysis. Although state laws vary slightly, most of them require that the feed label (tag) show in percent the minimum crude protein and fat; and maximum crude fiber and ash. Some feed labels also include maximum salt, minimum TDN, and/or minimum calcium and phosphorus. These figures are the buyer's assurance that the feed contains the minimal amounts of the higher cost items—protein and fat; and not more than the stipulated amounts of the lower cost, and less valuable, items—the crude fiber and ash.

Fig. 15-2. Weighing samples of feed for analyses. (Courtesy, International Multifoods, Minneapolis, Minn.)

MOISTURE

In order to compare the nutritional value of feeds, the first step is to determine how much water is contained in them. Many high-moisture products, such as beets, compare very favorably with corn and other more traditional feedstuffs on a dry matter basis; but the high amounts of water contained in them tend to restrict the dry matter intake of livestock. Due to this variation in moisture of many of the feeds used in livestock production, feeding standards are now being converted to a moisture-free (dry matter) basis. **NOTE WELL:** The feed composition tables in Part V of this book give compositions on both as-fed (A-F) and moisture-free (M-F) bases. Thus, fresh alfalfa pasture, all analyses, is listed as having 26% dry matter. So, the moisture is determined by subtracting the dry matter from 100 (100 − 26 = 74% moisture content).

The determination of moisture of various compounds or feeds requires the use of different techniques. The appropriate method is selected by the analyst based on such characteristics as: (1) presence of volatile matter, (2) possibility of browning of some ingredients, (3) need for low temperatures and vacuum, and (4) the presence of some compounds which might be chemically altered during drying—for example, sugars.

Moisture is now being determined in five ways:

1. **Oven dried.** According to the official methods of the Association of Official Analytical Chemists (AOAC), samples are heated at 275°F to a constant weight. The loss in weight represents the amount of water contained in the feed. This method is not exactly correct for the determination of water content, since many short chain fatty acids and organic acids become volatilized and are lost in addition to the evaporation of water.

2. **Vacuum dried.** Samples can be dried *in vacuo* at lower temperatures. The boiling point of water is lowered when the samples are placed in a vacuum. Therefore, vacuum ovens are sometimes used to minimize the loss of compounds other than water.

3. **Distillation with toluene.** The dry matter of samples having large quantities of volatile acids and bases can be determined through distillation with toluene.

4. **Freeze drying.** This method of dry matter determination is receiving increasing attention. The freeze drier basically consists of heated shelves surrounded by a refrigerated condenser. Samples are frozen prior to freeze drying. After the frozen sample is placed in the freeze drier, the apparatus is evacuated to an extremely low pressure. Under these conditions, sublimation takes place. That is, the frozen water crystals within the sample pass directly into a vapor phase without first becoming a liquid. This technique prevents the loss of many of the volatile organic compounds in the sample.

5. **Rapid moisture testing equipment.** Quite often, farmers or feeders need to know the moisture content of a feed immediately. For example, if they are buying high-moisture grain, they may not be able to wait for some chemical laboratory to run a moisture analysis. For this purpose, several types of rapid moisture testing equipment have been developed and are available at reasonable prices. One type of moisture analyzer is an apparatus that comes in two parts—a drier and a small scale. The farmer or feeder weighs

out a portion of feed as specified by the operating directions. The drier consists of a heating element and a fan. The sample is then placed on a screen which allows hot air to pass through the feed sample and dry it until a constant weight is attained (generally about 5 minutes). After drying, the sample is removed from the drier and immediately weighed to determine the loss of water. Although this method of determining moisture content is not extremely accurate, it does permit the farmer to get a rapid and reasonably good indication of the dry matter content of feed. A second type of moisture analyzer measures the conductance or resistance of the feed, and the value derived therefrom is compared to a calibration chart. When this type of apparatus is used, it should be periodically checked and calibrated against a standard of which the exact moisture is known.

ASH

The ash fraction of the proximate analysis represents the inorganic constituents of the feed. Samples in porcelain crucibles are placed in a muffle furnace and ignited at temperatures in excess of 500 to 600°C.

The residue that is left after burning is termed *ash*. In plants, ash composition is highly variable since soil conditions determine the makeup of this fraction.

CRUDE PROTEIN (CP)

Protein, on the average, contains about 16% nitrogen. Thus, theoretically, if we know the nitrogen content of feed, we can estimate the amount of protein that it contains by multiplying the nitrogen content by 6.25 (100% ÷ 16%).

The commonly used procedure for determining the nitrogen content of feeds is called the *Kjeldahl process,* after the discoverer, the Danish chemist, Johan Kjeldahl. The organic matter of the sample is destroyed by digestion with sulfuric acid. The nitrogen is then converted to an ammonia compound which is quantitatively released as ammonia during alkaline distillation. The precise amount of ammonia is titrated against a standard solution; and the figure therefrom derived is converted to nitrogen content, and finally to protein.

It should be remembered that the figure derived from the analysis is rather crude. This procedure involves the following two basic assumptions which are generally applicable to feeds:

1. **Protein contains approximately 16% nitrogen.** In reality, this merely represents an average. Some feeds contain protein that averages more than 16% nitrogen, whereas others contain protein with lesser amounts of nitrogen.
2. **All nitrogen is in the form of protein.** Practically speaking, this may be true. But there are a number of other compounds that may be present in feeds that contain nitrogen; for example, yeasts contain a large quantity of nitrogen as nucleic acids.

• **Heat damaged protein (unavailable protein)**—Unless carefully controlled, the heating that occurs in many commercial feed processes can injure protein quality. Special care must be exercised to avoid heat damage when drying fish meal and milk products, when removing oil from oil-bearing seeds by the expeller process, and when baling, stacking, or barn-storing hay high in moisture. Severely heat-damaged protein is indigestible. The amount of indigestible protein in a feed can be estimated by measuring the nitrogen content of the acid detergent fiber, then multiplying it by 6.25 (ADF − N × 6.25).

• **Available crude protein**—Crude protein minus heat-damaged protein equals available crude protein. This is the protein figure that should be used when balancing rations, if there is heat-damaged protein in the feed.

ETHER EXTRACT (FAT)

Many people refer to this fraction as the fat portion of the sample. This tends to be an oversimplification as the ether extract also contains organic acids, oils, pigments, alcohols, and the fat-soluble vitamins. Many of the complex lipids, such as phospholipids, are not completely extracted in this procedure.

The procedure is exactly what the name implies. The sample is continuously extracted with ether, using a specially designed apparatus. After the extraction is completed, the ether solvent is evaporated and the residue that remains constitutes the ether extract.

CRUDE FIBER (CF)

The crude fiber fraction is an indicator of the relative indigestibility and bulkiness of the sample. The feed sample is first boiled in dilute acid and then in dilute alkali to simulate the digestive action of gastric secretions. The residue of the sample that remains undigested after these boiling procedures is then weighed and ashed. The difference between the initial residue weight and the ashed weight indicates the amount of fiber present in the sample.

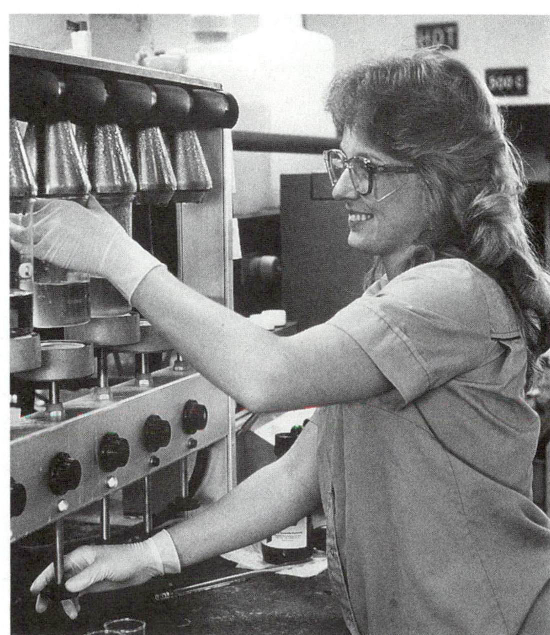

Fig. 15-3. Crude fiber analysis. (Courtesy, International Multifoods, Minneapolis, Minn.)

After being used more than 100 years (its development dating back to 1865), crude fiber (CF) is declining in popularity as a negative measure of feed quality; negative because, typically, digestibility, and thus energy value, of a feed decreases as crude fiber percentage increases. But this isn't always true, especially in ruminants; they are able to digest much of the cellulose and hemicellulose in feeds and are only limited in their ability to utilize crude fiber by lignification and the capacities of their digestive tracts. So, crude fiber is only a rough approximation of differences in the digestibility of feeds. For example, the fiber of immature plants is mostly cellulose, which is highly digestible by ruminants. By contrast, the fiber of mature plants contains large amounts of lignin, which is indigestible. Another major problem is that variable amounts of hemicellulose and lignin end up in other chemical fractions (*i.e.*, in the ether extract and ash). This is important because lignin is essentially indigestible whereas hemicellulose is partially digestible. Hence, CF does not accurately measure all of the lignin. Nevertheless, CF has remained in the scheme of feedstuff analysis because of its requirement for the determination of TDN.

In recent years, a system developed by Van Soest, of Cornell, for evaluating the digestible fraction of fiber has increased in usage. (See later section in this chapter on "Neutral Detergent Fiber [NDF] and Acid Detergent Fiber [ADF].")

NITROGEN-FREE EXTRACT (NFE)

The nitrogen-free extract (NFE) fraction in the proximate analysis program is determined by the following formula:

NFE = 100 − (% moisture + % crude fiber + % ash + % ether extract + % crude protein)

This fraction, represents a catchall for the organic material for which there is no specific analysis. The vast majority of components in this fraction are carbohydrates (starch, fructans, pectins, cellulose, hemicellulose, and lignin), but other substances, such as pigments, organic acids, and water-soluble vitamins, are also present.

USEFULNESS OF PROXIMATE ANALYSIS

As with all analytical techniques, there are advantages and disadvantages to the use of the proximate analysis in the evaluation of feeds.

The **advantages** of the system should not be minimized. They are:

1. **Most laboratories are equipped to run this type of analysis.** Expensive and sophisticated equipment is not needed.
2. **It provides a good general evaluation of the usefulness of the feed to animals.** A feed that is high in crude fiber will probably be inferior in feeding value to one that is very low in crude fiber. Likewise, a feed having a high quantity of ether extract is likely to be a high-energy feed.
3. **The total digestible nutrient (TDN) system of feeding standards is based on the proximate analysis system.** Today, many nutritionists are switching to the net energy system of feeding standards, but the TDN system will likely remain in use, decreasingly, for some time to come.
4. **Most of the data available on feed composition to date are reported in terms of proximate analysis.**

Some of the **disadvantages** of the proximate analysis are:

1. **The system does not define the individual nutrients of the feed.** Rather, the fractions represent mixtures of the various nutrients.
2. **It is not precisely accurate.** Adjustments are used in the calculations of several components. Crude protein and crude fiber are rough estimates of their respective fractions.
3. **The procedure is time-consuming.** There is little adaptability to automation in proximate analysis. Many of the fractionations involve several weighings of samples and other procedures which must be done by the laboratory technician.
4. **It does not tell how much indigestible material there is in a feed.** Unfortunately, the acid-alkali treatment dissolves much of the plant lignin, a plant skeletal substance that no animal can digest, making it impossible to predict accurately how much indigestible matter there is in the feed. The method overestimates the nutritive value of some feeds, underestimates that of others, and fails to indicate how the constituents of the indigestible residue are related to each other or what function some of them perform in digestion.
5. **It does not go far enough.** Proximate analysis does not provide any information relative to palatability, texture, toxicity, digestive disturbances, the associative effect of feedstuffs, or nutritional availability. Neither does it tell anything about the soil on which the feed was grown, despite the fact that soils high in molybdenum and selenium affect the composition of the feeds produced. Other similar soil-plant-animal relationships exist and are important. Thus, further steps need to be taken to evaluate a feed.

It is likely that the use of proximate analysis will decline in the future as new techniques become popular but, for the present, it offers a means of evaluating feed.

Neutral Detergent Fiber (NDF) and Acid Detergent Fiber (ADF)

Fiber can have at least three definitions:

1. In the plant, fiber refers to the structural components that form the plant cell wall, providing the plant with rigidity and protection.
2. In the chemical laboratory, fiber is the analytically measured constituents of cell walls—lignin, cellulose, and hemicellulose.
3. In the cow, fiber (a) requires chewing to reduce particle size and facilitate passage through the digestive system, unless it is ground; (b) occupies space as it moves slowly through the digestive tract; and (c) contains lignin, cellulose, and hemicellulose.

Lignin is indigestible and reduces the digestibility of the other feed components; it is the primary factor limiting forage fiber digestibility. Cellulose and hemicellulose have slow digestion rates compared to other nutrients such as starch, protein, and fat. In the traditional proximate analysis system, crude fiber (CF) and nitrogen-free extract (NFE) together represent the carbohydrate fraction. Crude fiber measures the feed residue that is insoluble after successive boiling in dilute alkali and acid; it is supposed to estimate the indigestible portion of the feed. NFE is intended to represent the more readily digestible carbohydrate fraction. Unfortunately, the proximate analysis system fails to separate carbohydrates according to their nutritive value; in some instances, NFE is less digestible than crude fiber because some lignin is included in the NFE fraction. On the other hand, crude fiber fails to recover completely the plant cell wall components (lignin, cellulose, and hemicellulose) which represent the less digestible portions of feeds. Since a large portion of crude fiber may be digested (cellulose is highly digestible, and hemicellulose is partially digestible), proximate analysis does not provide adequate distinction between the digestible and the indigestible portions of feeds.

• **Van Soest analysis**—The inadequacies of proximate analysis gave rise to the detergent system of feed analysis for estimating energy content of forages, developed by Peter J. Van Soest at the USDA Beltsville National Research facility, in the 1960s. By using detergents, the Van Soest system separates fibrous feeds into two fractions: a neutral detergent fibrous fraction; and an acid detergent fibrous fraction (see Fig. 15-4).

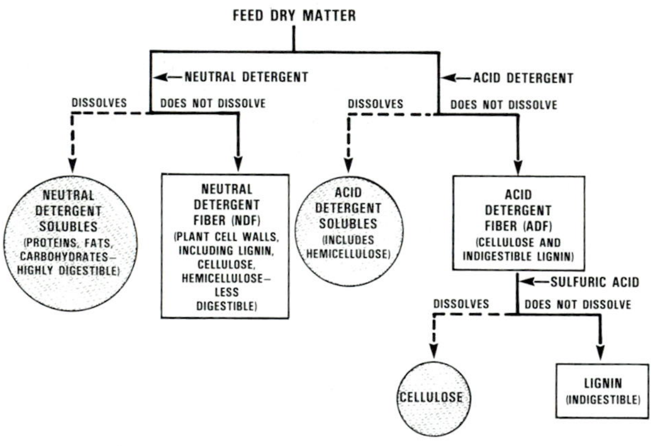

Fig. 15-4. Van Soest method of feed analysis.

• **Neutral detergent fiber (NDF)**—Neutral detergents are used to separate the feed into two fractions: (1) neutral detergent solubles, representing the highly digestible portion of the feed and consisting of proteins, fats, and carbohydrates (along with nonprotein nitrogen, pectin, and soluble materials); and (2) neutral detergent fiber (NDF), representing the less digestible portion of the feed, consisting of plant cell walls, including lignin, cellulose, and hemicellulose.

NDF is closely related to feed intake because it contains all the fiber components that occupy space in the rumen and are slowly digested. Thus, the lower the NDF percentage, the more the animal will eat; it is inversely related to voluntary feed consumption. Hence, a low percentage of NDF is desirable. It is noteworthy that milk production of lactating cows is more highly correlated with the NDF portion of the ration than with the ADF in the ration.

• **Acid detergent fiber (ADF)**—Acid detergent solutions are used to separate the feed into two fractions: (1) acid detergent solubles, containing the more readily digestible hemicellulose; and (2) acid detergent fiber (ADF), representing the less digestible portion of the feed and consisting of lignin (indigestible) and cellulose (digestible).

ADF is an indicator of forage digestibility because it contains a high proportion of lignin which is the indigestible fiber fraction. NDF will always be a higher number than ADF because ADF does not contain hemicellulose. The lower the ADF, the more feed an animal can digest. Hence, a low ADF percentage is desirable.

• **Acid detergent lignin (ADL)**—Sulfuric acid may be used further to separate the ADF into (1) cellulose, which is digestible; and (2) lignin, which is indigestible.

(Also see Chapter 8, section on "NDF, ADF, and NIRS Analyses.")

Bomb Calorimetry

When compounds are burned completely in the presence of oxygen, the resulting heat is referred to as gross energy or heat of combustion. The bomb calorimeter is an instrument used to determine the gross energy of feed, waste products from feed (for example, feces and urine), and tissues.

It should be reiterated that *the calorie is defined as the amount of heat required to raise the temperature of 1 g of water 1 °C (precisely from 14.5 to 15.5 °C).*

Briefly stated, the procedure is as follows: An electric wire is brought in contact with the material being tested, so that it can be ignited by remote control; 2,000 g of water are poured around the bomb; 25 to 30 atmospheres of oxygen are forced into the bomb; the material is ignited; the heat given off from the burned material warms the water; and a thermometer registers the change in temperature of the water.

A correction factor is determined for the calorimeter through the use of a standard—benzoic acid. This factor is referred to as the hydrothermal equivalent of the calorimeter. Two additional correction factors are adjusted for in the determination of gross energy—(1) oxidation of the fuse wire, and (2) acid production from the combustion, as corrected for by titration with a sodium carbonate standard. The hydrothermal equivalent of the bomb calorimeter can be derived from the following equation:

$$\text{HE (cal/degree)} = \frac{\text{Weight of benzoic acid sample} \times \text{calories per gram of benzoic acid} + \text{length of fuse wire burned} \times \text{cal/cm of fuse wire} + \text{ml } Na_2CO_3}{\text{(final temperature} - \text{initial temperature)}}$$

From these factors, the gross energy (cal/g) of a feed can be calculated using the following formula:

$$\frac{(T_f - T_I)\,HE - (L_w \times cal/cm) - ml\ Na_2CO_3}{weight\ of\ sample}$$

Where T_f = final temperature
T_I = initial temperature
HE = hydrothermal equivalent
L_w = length of wire

Example:

A 1.1 g sample is burned. The initial temperature was 20.50°C and the final temperature 22.73°C. The hydrothermal equivalent of the bomb calorimeter is 2,400 cal per degree C.

The correction for the wire burn is this: 7.5 cm of wire burned with the correction of 2.3 cal/cm wire. The total amount of Na_2CO_3 titrated is 7.5 ml.

Therefore:

Gross energy (cal/g) =

$$\frac{(22.73 - 20.50)\,2{,}400 - (7.5 \times 2.3) - 7.5 = 5{,}352 - 17.25 - 7.5}{1.1} = 4{,}843.0$$

It is noteworthy that the determination of the heat of combustion with a bomb calorimeter is not as difficult or time-consuming as the chemical analyses used in arriving at TDN values.

Chromatography

In 1903, Tswett, a Russian botanist, first described his attempts to separate colored substances; hence, the origin of the term chromatography. Today, many of the compounds that are separated and identified by chromatographic techniques are colorless; but new refinements in these techniques enable the feed evaluator to quantitate extremely minute amounts of many compounds.

Chromatography separates compounds through the use of two phases—a stationary or fixed phase and a mobile phase. The differences between the various chromatographic techniques lie in the variation of the materials used in these phases. The stationary phase can be either solid or liquid material, while the mobile phase or *carrier* can be either gas or liquid in nature. The various types of chromatography are listed in Table 15-3.

TABLE 15-3
TYPES OF CHROMATOGRAPHY

Stationary Phase	Mobile Phase	Chromatographic Procedure
Solid	Liquid	Thin layer chromatography. Ion exchange chromatography. Gel chromatography. Absorption-column chromatography.
Liquid	Gas	Gas-liquid chromatography. Capillary-column chromatography.
Solid	Gas	Gas-solid chromatography.
Liquid	Liquid	Paper chromatography. Partition chromatography. Zone electrophoresis.

Fig. 15-5. Liquid chromatograph being used to test the amino acid and vitamin content of feeds. (Courtesy, International Multifoods, Minneapolis, Minn.)

Numerous materials from feeds, such as proteins, amino acids, sugars, fatty acids, minerals, and many other components, are routinely identified and quantitated through the chromatographic technique. In addition to nutrient analysis, chromatography can be adapted to the detection of drug residues, hormones, pesticides, and other feed contaminants.

The **advantages** of chromatography are numerous; among them, the following:

1. **Extreme sensitivity.** Many compounds can be quantitated to parts per billion (mcg/kg of sample).
2. **Inexpensive.** Many chromatographic techniques can be adapted by almost any laboratory. Relatively little equipment is needed for many chromatographic procedures.
3. **Rapid.** In techniques such as thin layer chromatography, a large number of samples can be done simultaneously in a relatively short period of time.
4. **Readily adaptable.** Chromatography can be readily adapted to almost any chemical compound.

With its advantages, there are several **disadvantages,** such as:

1. **Complexity of operation.** Many of the newer techniques involve sophisticated equipment which require the operator to be familiar with the theory and the "art" of chromatography.
2. **Sample preparation.** Most samples require considerable preparation before they can be chromatographed. This may involve such procedures as extraction, hydrolysis, and/or evaporation.
3. **Sample size.** Samples must be small in order to be chromatographed. This means that the sampling procedure must be carefully planned to promote a truly representative sample.

Colorimetry and Spectrophotometry

Colorimetry and spectrophotometry are chemical analyses whereupon light is passed through solutions to yield information about the concentration of certain compounds. A particular wavelength of light is passed through the samples, and the amount of light absorbed by the sample gives an indication of the concentration of the compound being tested. Colorimetry differs from spectrophotometry in that colorimetry is useful for measuring wavelengths in the visible region of the light spectrum whereas spectrophotometry utilizes wavelengths in the ultraviolet, visible, and infrared regions of the spectrum.

The analytical procedures for many nutrients and drugs involve either colorimetry or spectrophotometry. Vitamin A is a good example of a colorimetric procedure. The standard assay for vitamin A determination involves the treatment of the sample with antimony trichloride. A deep blue colored solution is produced, the intensity of which is dependent on the amount of vitamin A in the sample. The solution of unknown concentration is measured in the colorimeter and compared to a series of standards of known concentrations. Spectrophotometric assays are essentially the same as colorimetric assays except the researcher has a more versatile machine with which to work.

The atomic absorption spectrophotometer is one of the most widely used instruments for mineral analysis, having the ability to detect many minerals at concentrations less than 1 part per billion (1 mcg/kg of sample). In addition to its high sensitivity, this machine is readily adaptable to automation, thus presenting the chemist a rapid, accurate method of feed analysis. The atomic absorption spectrophotometer works on a slightly different principle from the regular spectrophotometer. The main principle behind this machine is that when certain compounds (for example, minerals) are volatilized, they emit light of a characteristic wavelength. The machine is calibrated to detect this light.

Protein and Amino Acid Analyses

In the past, the Kjeldahl procedure was considered to be the most efficient way to quantitate protein content in feed. In recent years, several new approaches to protein determination have been developed.

When colorimetric techniques for the determination of total protein nitrogen are used, a certain degree of automation can be designed, thereby offering distinct advantages over the Kjeldahl procedure. One such system has been developed by scientists at the University of Missouri.[1] In this semiautomated system, feed samples are digested and analyzed using an automated ammonia-perchlorite-salicylate analysis system. This system has proven to be highly sensitive in the determination of pepsin-insoluble nitrogen in forages.

Since the Kjeldahl procedure measures the total amount of nitrogen in a compound, one can obtain only a rough indication of the true protein content. Thus, many animal nutritionists and biochemists are developing techniques that yield estimates of the total true protein. A few of these procedures follow:

1. **Biuret assay.** This colorimetric assay compares the unknown to a standard curve of known protein concentrations using a biuret reagent.
2. **Lowry assay.** This colorimetric assay is based on the presence of tyrosine and tryptophan in protein.
3. **Turbidity measurements.** This assay utilizes turbidity measurements after precipitation has been accomplished in a controlled manner.
4. **Peptide bond method.** This spectrophotometric assay is based on the fact that peptide bonds absorb light in the 195 to 225 region of the spectrum.
5. **Warburg-Christian assay.** This spectrophotometric assay utilizes ultraviolet absorption analysis after all nonprotein material has been removed by fractionation or dialysis.

In rations for monogastric animals, the amino acid composition of the feed must be considered. While most procedures for determining the amino acid profiles of feeds are automated, considerable time and labor is still involved. In order to determine the amino acid profile of a feed, the protein within the feed must be completely hydrolyzed into its constituent amino acids. In many cases 6N HC1 is used. Unfortunately, this and many other hydrolysis procedures have the following problems:

1. Some amino acids are destroyed in the process.
2. Some amino acids are chemically altered during hydrolysis.
3. Some of the peptides remain incompletely hydrolyzed, thereupon tying up amino acids which should be quantitated. Following the hydrolysis of the protein, the resulting amino acids are commonly separated by ion-exchange chromatography and subsequently quantitated by colorimetric techniques or by gas-liquid chromatography.

Regression equations are available for estimating the amino acid levels of selected feed ingredients. Although these equations are not a replacement for other procedures for determining amino acid profiles of feeds, they do provide a better estimate of amino acid content for formulation purposes than can be calculated from simple ratios of protein alone.

(Also see Chapter 4, section on "Biological Value of Protein.")

Biological Analysis

Quite often, biological assays are used in the analysis of micronutrients in feeds. There are two basic types of biological assays—(1) microbiological assays, and (2) the use of nutrient-deficient animals.

Biological assays tend to be laborious and time-consuming. Large numbers of samples are needed to produce statistically reliable results, and quite often data obtained from these assays are highly variable. The assay utilizing nutrient-deficient animals is particularly cumbersome because (1) the animals should be of the same sex and of approximately the same age and weight, and (2) time is required to induce deficient conditions in these animals.

(Also see Chapter 4, section on "Biological Value of Protein.")

[1]Wall, L. L., C. W. Gehrke, and R. A. Smith, *Semiautomated Method for Total Protein Nitrogen in Feeds—Applications,* Experiment Station Chemical Laboratories, Columbia, Mo.

MICROBIOLOGICAL ASSAY

In microbiological assay, a microorganism is selected that is known to require the nutrient in question. Therefore, if the nutrient is unavailable, the selected microorganism will not grow. The growth medium is prepared so that it is nutritionally complete except for the nutrient to be tested. Graded levels of the nutrient are then added to the media and a growth response curve is prepared. The sample to be assayed can then be tested and compared to the growth response curve to determine the concentration of the nutrient. Many of the micronutrients, such as the B complex vitamins, are assayed in this manner.

NUTRIENT-DEFICIENT ANIMALS

In this type of assay, experimental animals, such as the rat and the chick, are fed diets deficient in a specific nutrient. Growth response curves are developed by the feeding of known amounts of the nutrient to some of the deficient animals. Other deficient animals are given the product to be assayed, and their responses are compared to the growth curves. In addition, the evaluator can observe changes in specific tissues as various levels of the specific nutrients are supplied.

An excellent example of this type of assay is the bioassay for vitamin A. Young female rats are initially fed vitamin A-deficient diets to deplete their reserves. The feed to be assayed and graded levels of known concentrations of vitamin A are fed to the vitamin A-deficient rats. Three parameters are then measured to determine the vitamin A content of the feed:

1. Growth response.

2. Concentration of vitamin A in the liver of the rats from the various treatments.

3. Examinationn of the degree of cornification of the vaginal epithelium. In a vitamin A deficiency, the lining of the vagina undergoes cornification.

Another example of this type of assay is the bioassay of vitamin D. When vitamin D supplements are assayed, chicks are used because of the great differential in bioactivity between vitamin D_2 and vitamin D_3. Unlike mammals which can use either form equally well, the chick cannot utilize vitamin D_2 nearly as well as vitamin D_3. In this assay, young chicks are fed a rachitogenic (rickets-producing) ration for 21 days. Known quantities of vitamin D are given to some chicks to establish a standard curve. Other chicks are fed the supplement to be tested. Following the feeding period, the chicks are sacrificed, and their tibia bones are dried and ashed. The ash content of the bones of the chicks in the test rations are then compared to the standard curves derived from the data of the chicks fed known concentrations of vitamin D; and the vitamin D content of the unknown is derived. This type of assay has been shown to be superior to most chemical analyses for vitamin D.

Near Infrared Reflectance Spectroscopy (NIRS)

Chemical determinations are slow! This prompted the search for a more rapid, yet reliable, method of evaluating feeds and led to the use of the near infrared reflectance spectroscopy (NIRS).

The infrared technique, discovered in 1954, introduced for the analysis of grains and seeds in 1965, perfected in the 70s, and used in analyzing forages and other feedstuffs today, is based on the fact that near infrared light is absorbed variously by different molecular bonds. NIRS is a method of quantitative analysis, which comes under the category of applied analytical chemistry. When the instrument, known as an *infrared analyzer,* directs different wavelengths of infrared light at a sample of material, information can be gathered about the chemicals in the sample. The instrument passes this information on to a computer. In turn, the computer can automatically determine the percentage of nutrients in the sample. In a hay sample, for example, NIRS can determine the percentage of moisture, crude protein, calcium, phosphorus, sulfur, magnesium, potassium, neutral detergent fiber (NDF), and acid detergent fiber (ADF).

Today, modern NIRS technology is widely used in the feed, grain, and food industries. Instruments capable of measuring starch, oil, sugar, fiber, moisture, and protein are commercially available; and research has shown that many other applications are possible, such as the determination of amino acids and mineral constituents. Also, NIRS is being used to detect impurities. Furthermore, it is quicker and easier than other tests, and it doesn't use caustic chemicals like sulfuric acid or sodium hydroxide.

(Also see Chapter 8, section on "NDF, ADF, and NIRS Analyses.")

Molds or Fungi Assays

Certain types of molds or fungi have caused death losses among various animal species for many years. A very damaging mold is *Aspergillus flavus,* which produces aflatoxin.

The three basic tests for molds are mold counts, culture and identification of the organisms, and assay for the mycotoxins that may have been liberated in the growth of the mold. Of the three tests, assay for mycotoxins is preferred since the toxic effects are known for some of them. Biological assays for the toxin are available in which ducklings or poults are used. Chemical assays are also available, using fluorescent or chromatographic techniques. A practical test consists in feeding some of the moldy feed to 2 or 3 less valuable animals and observing their health and production.

(Also see Chapter 5, Table 5-3, "Mycotoxins.")

TECHNIQUES UTILIZED IN FEED EVALUATION

Two commonly used techniques that have been developed to aid in feed evaluation are the surgical techniques of cannulation and fistulation of the digestive systems of animals and the use of indicators.

Cannulation and Fistulation (Nylon Bag Technique)

Quite often in the evaluation of feeds and their utilization in the body, it is necessary to obtain or introduce samples into various parts of the digestive tract. Through surgical procedures, permanent openings can be made in the body which give the researcher ready access to the desired segment of the gastrointestinal tract.

Cannulation and fistulation techniques enable the researcher to (1) determine flow rates, (2) collect digestive fluids, (3) infuse materials in various parts of the digestive tract, and (4) observe movement of the organs.

CANNULATION

Cannulation is a procedure whereby a cannula (tube) is inserted in the cavity of an organ and subsequently brought to the exterior of the body. Probably the most common and simple form of cannulation is the cannulation of veins—quite often the jugular vein. A hollow needle is inserted in the vein to be cannulated and a small tube is passed through the needle into the vein. Once the tube is in the vein, the needle can be either left in place or removed. An anticoagulant, such as heparin, is used to flush the tube; and blood samples can then be taken periodically without having to puncture the vein each time. When the cannula is not in use, a small wire is placed in the lumen of the tube to prevent blood clots from stopping up the cannula.

Quite often, a re-entry cannula is used in the investigation of digestion in the intestine. After the intestine is cut, a cannula is inserted and brought to the exterior of the body. It is then secured to the skin and reinserted into the body cavity, thus forming a loop on the outside of the body. Finally, the other piece of the cut intestine is attached to the cannula. This type of cannulation does not impede the flow of ingesta but does give the researcher ready access to intestinal fluids with a minimum of trauma to the animal.

- **Mobile bag technique**—Scientists at the University of Alberta, Edmonton, Canada, have adapted the nylon bag technique to swine, in a procedure which they refer to as the *mobile bag technique.* It consists of the following procedure: Approximately a 1 g sample of the feedstuff is weighed and placed in a small nylon bag. Then, the bag is inserted in the pig's duodenum via a duodenal cannula. The bag travels through the intestinal tract and is voided with the feces 36 to 48 hours following insertion in the duodenum.

FISTULATION

Fistulation is a procedure in which an opening is made in a hollow organ with the edges of this opening exteriorized by being sewn to the body wall. A plug is then attached to the opening to prevent leakage of the contents from within the organ and to maintain an anaerobic environment. When a sample is to be taken from the organ, the plug is removed, thereby giving the investigator ready access to the interior of the organ. A fistula differs from a cannula in that a fistula is a permanent opening in the body wall and the organ while a cannula is a tube which leads to the interior of the body.

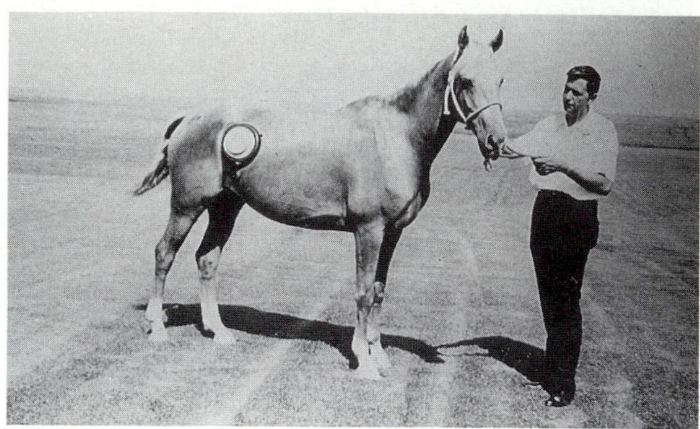

Fig. 15-6. A satisfactory fistulation technique in the horse has been developed. It consists in placing a large plastic window (4 in. in diameter) through the flank into the cecum. (Courtesy, Morris Animal Foundation, Denver, Colo.)

Rumen fistulas are commonly employed in the study of ruminant digestion. By fistulating the rumen, researchers can clearly observe the movements of the rumen and collect ingesta for analysis or for use in *in vitro* digestion trials. One such adaptation of fistulation is a procedure developed by Heinemann and VanKeuren at the Washington Station. In this procedure, the bags containing samples are then attached to a ¾- × 8-in. plastic stick with a lead weight attached to one end; and the entire apparatus is inserted in the rumen of a fistulated steer and imbedded in the ingesta. Here the forage samples are exposed to the ruminal fluids and microorganisms for digestion. At predetermined intervals, the sticks are removed; and forage samples are placed in nylon bags measuring 2 × 4½ in. The nylon bags plus digested samples are dried, weighed, and analyzed to determine how much of the sample has disappeared. This technique offers advantages to the traditional techniques of *in vitro* digestion trials where the feed samples are digested in test tubes under conditions which simulate the environment of the rumen. By using the nylon bag technique, the researchers can determine digestibility parameters in a natural environment, thereby eliminating many factors which can inherently affect the results of an *in vitro* trial involving an artificial rumen.

Indicators

Fig. 15-7. A steer with a harness and bag for the collection of feces. (Courtesy, University of California)

Traditionally, indicators, or tracers, have been used to study rate of passage, forage intake, and digestibility of feed. Numerous compounds have been used for this purpose; among them, chromic oxide, ferric oxide, various dyes, silica, lignin, and chromogen (a pigment found in plants). Polyethylene glycol has been used to trace fluid changes in the digestive tract. Indicators can be broken down into two classes: (1) internal indicators, those inert portions of the plant themselves (for example, lignin, silica, and plant chromogen); and (2) external indicators, inert chemicals which are fed to animals in addition to the feed (for example, chromic oxide and certain dyes).

The properties of a good indicator are:

1. It should not alter the digestive process.
2. It should be palatable.
3. It and the feed should move through the digestive tract together.
4. It should be highly recoverable—almost 100%. The indicator must not be absorbable or degradable.

Digestion trials tend to be laborious and cumbersome since a total collection of the feces is necessary to determine the digestibility of the feed. Indicators provide a quick, indirect way of determining the digestibility of feed. The following equations are used in the determination of digestibility through the use of indicators:

$$\text{Digestion coefficient of dry matter} = 100 - 100\left[\frac{\%\text{ indicator in dry matter of feed}}{\%\text{ indicator in dry matter of feces}}\right]$$

$$\text{Digestion coefficient of a particular nutrient} = 100 - 100\left[\frac{\%\text{ indicator in feed} \times \%\text{ nutrient in feces}}{\%\text{ indicator in feces} \times \%\text{ nutrient in feed}}\right]$$

Indicators are used to estimate forage intake and digestibility in grazing animals. Since the animals are not in confinement, it is often difficult to determine how much feed is consumed. By placing a feces collection bag on the animal (see Fig. 15–7), the researcher can determine feed intake through the use of internal indicators. The total collection of fecal material is weighed and the dry matter content determined. Samples of the forages being grazed are analyzed to determine the content of the internal indicator (*i.e.*, lignin). The fecal material is then analyzed for the content of the internal indicator, and from this information the amount of dry matter consumed can be calculated as follows:

$$\text{Dry matter consumed} = \frac{\text{fecal output (dry basis)} \times \text{internal indicator in feces (dry basis)}}{\%\text{ indicator in forage (dry basis)}}$$

Digestibility of various nutrients in the forage can be determined by the following equation:

$$\text{Apparent digestibility coefficient (\%)} = 100 - 100\left[\frac{\%\text{ internal indicator of forage} \times \%\text{ nutrient in feces}}{\%\text{ internal indicator in feces} \times \%\text{ nutrient in forage}}\right]$$

Through the use of external indicators, it is possible to estimate the total amount of fecal dry matter without having to worry about total collections. All the researcher has to do is obtain grab samples—that is, small samples of fecal material taken directly from the rectum. From the information of how much external indicator was fed and how much was present in the feces, the fecal dry matter output can be calculated from the following equation:

$$\text{Fecal dry matter output} = \frac{\text{weight of external indicator fed}}{\%\text{ external indicator in fecal dry matter}}$$

TYPES OF FEED EVALUATION TRIALS

Once the proper techniques have been developed for feed analysis, it is necessary to set up trials to evaluate the feed. This can be done *in vivo* (through the use of animals) or *in vitro* (through the use of artificial means).

Feeds can be evaluated in a number of ways, with the choice of the type of trial determined by the desires of the operator and the following factors:

1. **Resources available.** The evaluator should decide whether to use *in vivo* or *in vitro* procedures. If animals are involved, the number to be used and the facilities to handle them should be carefully planned. Also, labor and analytical equipment must be considered.
2. **Costs.** Many trials involve considerable labor and equipment, as well as expensive animals. The price of one cow can buy quite a few mice, sheep, or even test tubes if an *in vitro* trial can be used.
3. **Time.** Some feed evaluation trials can last for weeks, whereas others can be done in a matter of hours.
4. **Statistical design.** Ultimately, the evaluator must produce statistically sound results. To obtain these results, adequate replication of treatments must be planned. There are many variations of statistical designs; and valid results can be obtained from relatively few animals if the trial is set up correctly. If small animals, such as chicks, or *in vitro* techniques are used, replication of treatments poses few problems.

Digestion and Metabolism Trials

Animals are not able to extract all the nutrients present in feeds. The actual value of ingested nutrients is dependent upon the extent to which they are metabolized. The first consideration here is digestibility, since undigested nutrients are not incorporated into the body proper.

A digestion trial is made by determining the percentage of each nutrient in the feed through chemical analysis; giving the feed to the test animal for a preliminary period (usually 7 to 10 days), so that all residues of former feeds will pass out of the digestive tract; giving weighed amounts of the feed during the test period (7 to 10 days); collecting, weighing, and analyzing the feces; determining the difference between the amount of the nutrient fed and the amount found in the feces; and computing the percentage of each nutrient digested. The latter figure is known as the *digestion coefficient* for that nutrient in the feed.

Various techniques and equipment may be used to make the fecal collections; among them, a specially designed digestion stall (see Fig. 15-8); collection harness and bag; markers (such as carmine, ferric oxide, chromic oxide, or soot), fed with the ration at the beginning and the end of the collection period; and indicators of an inert reference subject.

It is important that the urine and feces be collected separately, since only the feces are analyzed in a digestion trial. Anatomical problems prevent the use of female animals in most digestion trials, since it is difficult to prevent the contamination of their fecal samples with urine.

The digestibility of a feedstuff is affected by a number of factors, including:

1. **Species of animal.** Ruminants can digest fibrous feeds better than nonruminants.
2. **Physical form of the feed.** Quite often processing can alter the digestibility of feeds.
3. **Variation among individual animals.** Some animals have more efficient digestive systems than others.
4. **Frequency of feeding.** Digestibility tends to increase with frequency of feeding.
5. **Composition of the feed.** Such factors as plane of nutrition, adequacy of vitamins and micronutrients, and palatablility can alter the digestibility of feed.
6. **Adaptability to the ration.** Since much of the digestion within the ruminant is due to microbial action, any changes in the makeup of the microbial population will influence digestibility.

Nutrient Balance Trials

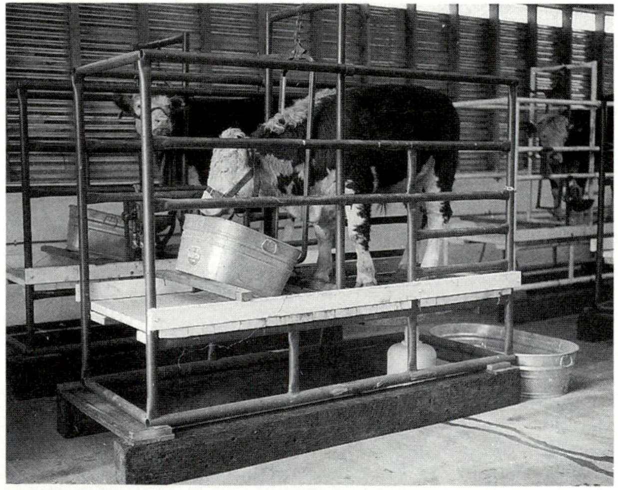

Fig. 15-8. Steers in a nutrient balance trial. Note that the stalls are designed so that fecal material and urine are collected separately. (Courtesy W. W. Heinemann, Washington State University, Prosser)

Nutrient balance trials are experiments which attempt to account for all losses of a particular nutrient. If an animal is losing more of the nutrient being tested (for example, energy) than it is ingesting, the animal is said to be in a negative balance. Conversely, if the animal is ingesting more of the nutrient than it is losing, it is said to be in a positive balance. A negative balance indicates that catabolism of stored nutrients in tissues is occurring to maintain the animal. Positive balances indicate that the excess amounts of the particular nutrient are being incorporated into body tissues and fluids.

Balance trials can be run for any nutrient, but the two most widely used types of balance trials are for the measurement of protein and energy.

Nitrogen Balance Trials

Nitrogen balance trials are used to determine the availability of protein in feeds. Fig. 15-9 illustrates the organization and breakdown of the various components in a nitrogen balance trial.

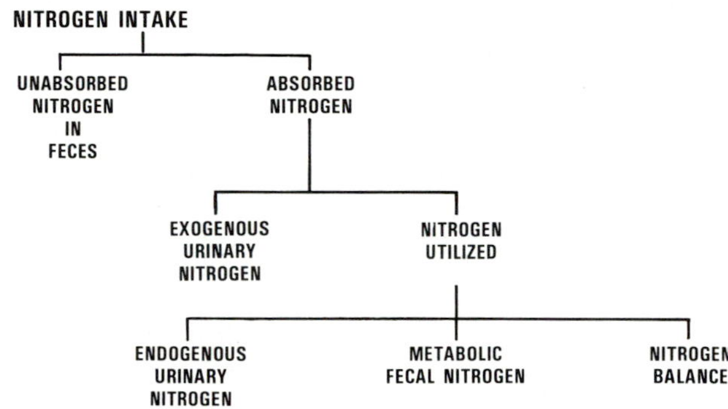

Fig. 15-9. Format of the nitrogen balance scheme.

Both feces and urine are collected in this type of experiment. Since metabolic processes involve considerable amounts of nitrogen, much of the nitrogen found in the feces and urine comes from sources other than ingested feed. To correct for this fact, feed is withheld from animals so that their bodies enter into a catabolic condition. The nitrogen excreted in the urine and feces during this period represents the nitrogen lost through metabolic processes. This urine production represents endogenous urinary nitrogen. Likewise, the fecal nitrogen excreted during the fasting period is called metabolic fecal nitrogen. These values are then subtracted from the nitrogen values obtained in the collection of the urine and feces during the feeding stage of the balance trial. Exogenous urinary nitrogen and fecal nitrogen from feed are determined by the following equations:

1. **Exogenous urinary nitrogen = total urinary nitrogen − endogenous urinary nitrogen.**
2. **Fecal nitrogen from feed = total fecal nitrogen − *metabolic fecal nitrogen.***

Through the use of the nitrogen balance scheme, the *biological value* (BV) of a feedstuff can be determined. *Biological value can be defined as the percent of absorbed nitrogen retained.* Mathematically, it is expressed as:

$$\mathbf{BV} = 100 \times \left[\frac{\text{nitrogen intake} - (\text{total fecal nitrogen} - \text{metabolic fecal nitrogen}) - (\text{total urinary nitrogen} - \text{endogenous urinary nitrogen})}{\text{nitrogen intake} - (\text{total fecal nitrogen} - \text{metabolic fecal nitrogen})} \right]$$

Another method of evaluating protein sources is the calculation of the protein efficiency ratio (PER). Animals are weighed prior to commencement of the feeding period and upon completion of the feeding period. The gain in body weight is then computed as a ratio with the amount of protein consumed, as follows:

$$\text{PER} = \frac{\text{gain in body weight}}{\text{protein intake}}$$

PER is dependent on the level of protein fed. That is, at a dietary level of 10% protein, the experimenter may get a completely different PER than if a 20% protein diet was fed. Therefore, PER values are usually related to a standard; or the value is obtained by taking the maximum obtained through the use of several levels.

A third method of protein evaluation is the determination of net protein utilization (NPU), by the following equation:

$$\text{NPU} = \frac{B - B_n}{I} \times 100$$

Where B = nitrogen content of the bodies of the animals given the diet containing the protein source to be tested.

B_n = nitrogen content of the bodies of the animals given a diet excluding nitrogen.

I = nitrogen intake of the group fed the test protein.

Through the use of NPU, it is possible to determine biological value in the following manner:

$$\text{Biological value} = \frac{\text{NPU}}{\text{digestibility}}$$

Energy Balance Trials

The gross energy of feed is the total energy content of the feed as determined by bomb calorimetry. When feed is consumed, the energy contained therein can be partitioned into fecal energy and apparent digestible energy. The energy content of feces is quantitated and subtracted from the gross energy to give apparent digestible energy. Since some of the energy in the feces represents energy derived from metabolic end products, the digestible energy term must be labeled apparent digestible energy. True digestible energy is the gross energy of a feed minus fecal energy from feed alone. The fecal energy from feed alone can be quantitated by subtracting the fecal energy of a fasting animal from the fecal energy of the same animal when it is properly fed. Most researchers, however, feel that apparent digestible energy is a relatively accurate and simple means of evaluating the energy content of feeds.

Apparent digestible energy is then partitioned into metabolizable energy, urinary energy, and gas loss (for example, methane). Urine and gases are collected, and the energy values derived from these two sources are substracted from the apparent digestible energy value to yield metabolizable energy.

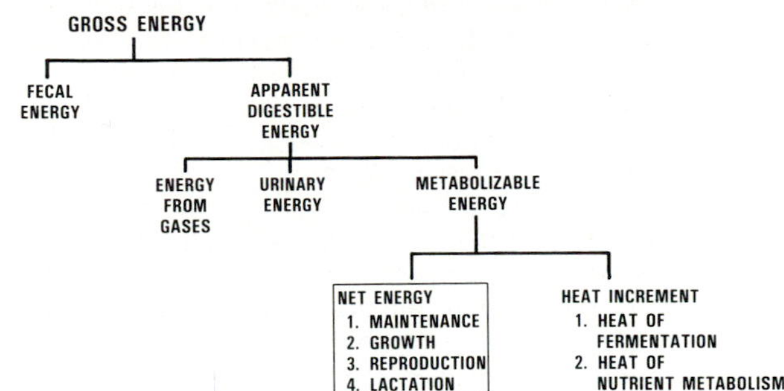

Fig. 15-10. Format of the energy balance scheme.

Metabolizable energy can then be broken down into heat increment and net energy. Heat increment is the term used to describe the heat produced in the processes of nutrient metabolism and fermentation. Net energy is the energy remaining for maintenance, growth, and production. When the values for heat increment and net energy for maintenance are combined, we can obtain an indication of the heat production in the body.

On a mathematical basis, the energy balance scheme can be analyzed in the following manner:

Apparent digestible energy = (gross energy) – (fecal energy)

Metabolizable energy = (apparent digestible energy) – (urinary energy) – (energy from gas losses)

Net energy = (metabolizable energy) – (heat increment)

Heat production = (heat increment) + (net energy for maintenance)

• **True metabolizable energy**—Some of the leading research centers of Europe and Canada are now using the term True Metabolizable Energy (TME) in the evaluation of feed for poultry. TME is metabolizable energy that has been corrected for metabolic and endogenous losses. It takes cognizance of the fact that the feces contain both (1) undigested feed, and (2) metabolic and endogenous products.

Calorimetric Systems

Measurement of animal heat began 200 years ago, when Lavoisier and LaPlace, the great French scientists, enclosed a guinea pig in a chamber surrounded by ice. The amount of ice melted by the heat of the animal multiplied by the latent heat of the ice indicated the heat given off by the guinea pig, thus giving rise to direct calorimetry. They also noted that the melting of the ice was directly related to the amount of carbon dioxide given off by the animal. This was the basis of indirect calorimetry. Since that time, more sophisticated types of equipment for measuring body heat have been developed, but the basic principles remain the same.

DIRECT CALORIMETRY

Armsby, while Director of the Agriculture Experiment Station at Pennsylvania State College in the late 1800s, felt that there was a great need for more sophisticated techniques in the evaluation of feeds. At the time, feeds were evaluated solely on their digestibility. So, together with two other Pennsylvania State professors, Fries and Osmond, Armsby constructed a respiration calorimeter capable of directly measuring heat production in large animals. It took six people at any one time to operate the machine, plus an additional person to handle the animal being studied. Through the use of this apparatus, Armsby became one of the foremost authorities in the research of energy utilization in animals. Today, newer and cheaper methods of measuring energy utilization in animals have rendered the machine obsolete; but Armsby's calorimeter still stands, now housed in a museum commemorating the monumental research which he conducted.

Fig. 15-11. Dr. R. W. Swift, former head of the Department of Animal Nutrition at Pennsylvania State University, is shown operating the respiration calorimeter built by H. P. Armsby. (Courtesy, Pennsylvania State University, University Park)

In direct calorimetry, the animal is confined to a well-insulated chamber and the heat losses (by radiation, convection, and conduction from the body surface; by evaporation of water from the skin and lungs; and by excretion of urine and feces) are measured either by (1) the increase in temperature of a known volume of water, or (2) electrical current generated as heat passes across thermocouples (gradient layer calorimetry). Brody[2] estimated that about one-fourth of the heat lost by the body resulted from moisture vaporization. The remaining three-fourths of the body heat is lost by radiation, conduction, and convection.

[2]Brody, S. B., *Bioenergetics and Growth*, Reinhold Publishing Company, New York, N.Y., 1945.

Direct calorimetry has been useful in the determination of the heat increment fraction of energy utilization. This parameter includes the heat of fermentation and the heat of nutrient metabolism. The heat produced by feeding a particular ration is first measured. Then, the intake of that feed is increased and the heat produced by the animal is measured. The increase in heat due to the increase in feed intake is subsequently termed heat increment.

Direct calorimetry is the most accurate method of measuring the heat production of animals, but it is costly and arduous. The machine itself is very expensive to build and operate; and considerable labor is involved in controlling the animal, running the machine, and analyzing the results. For these reasons, indirect calorimetry is usually the method of choice for the evaluation of energy in feeds.

INDIRECT CALORIMETRY

In indirect calorimetry, heat of production is calculated from measurement of the respiratory exchange—the O_2 consumption, and usually the CO_2 production—of the animal. This method is based on the fact that O_2 consumption and CO_2 production are closely related to heat production. The ratio of carbon dioxide produced to oxygen consumed, which is known as the respiratory quotient (RQ), is distinctive for each compound; hence, it served to indicate the type of nutrient being metabolized. Respiratory quotient can be determined through the following equation:

$$\mathbf{RQ} = \frac{\mathbf{CO_2\ produced}}{\mathbf{O_2\ consumed}}$$

Therefore, if we look at the equation for the catabolism of carbohydrates, we can calculate a respiratory quotient.

1. $\mathbf{C_6H_{12}O_6 + 6\ O_2 \rightarrow 6\ CO_2 + 6\ H_2O}$ **+ heat**
2. $\mathbf{RQ = \frac{6\ CO_2}{6\ O_2} = 1}$

The RQ for fat would be less than 1, as seen in the following example:

1. **Palmitic acid** $\mathbf{(C_{16}H_{32}O_2) + 23\ O_2 \rightarrow}$
   $\mathbf{16\ CO_2 + 16\ H_2O}$ **+ heat**
2. $\mathbf{RQ = \frac{16\ CO_2}{23\ O_2} = .70}$

Most mixed fats have RQ values of about .7, while short-chain fatty acids have higher values (about .8). Proteins have values intermediate between carbohydrates and fats—about .81.

It should be noted that in stress conditions, the RQ value can be greater than one. This can occur when there is hyperventilation where large quantities of carbon dioxide are exhaled while an oxygen debt exists in the body. Metabolic acidosis creates an excess exhalation of carbon dioxide as a compensatory mechanism.

The total heat production of the animal may be computed by either (1) measuring RQ along with oxygen, then making readings from tables; or (2) using a single equation relating heat production to the respiratory exchange.

Comparative measurements of direct and indirect calorimetry reveal that the two methods give results that are in close agreement.

The actual measurement of the respiratory exchange of animals may be accomplished by several different types of apparatus; among them, (1) the open circuit gravimetric system, (2) the open circuit chamber system involving gas analysis, (3) the open circuit mask system, and (4) the closed circuit mask system (spirometer).

1. **Open circuit gravimetric system.** This system is by far the simplest and easiest to set up, maintain, and operate. Also, it is readily adaptable to measure heat production from small animals (poultry, mice, rats, and rabbits). The needed materials are an enclosed holding chamber, scales, water and carbon dioxide absorbants, and an air pump. A schematic diagram of the setup is presented in Fig. 15-12.

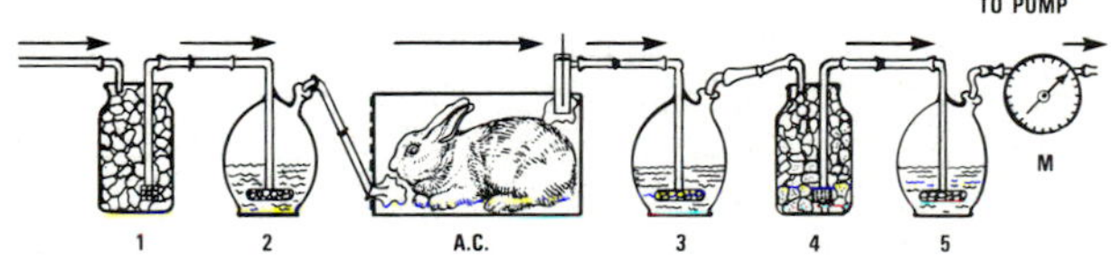

Fig. 15-12. Haldane respiration apparatus for the open circuit gravimetric system of indirect calorimetry: A.C., animal chamber. M, meter for measuring rate of ventilation. Bottles 1 and 4 contain soda lime (or caustic alkali) for absorption of carbon dioxide. Bottles 2, 3, and 5 contain sulfuric acid for absorption of water. Air entering the animal chamber is freed of carbon dioxide and moisture by passage through bottles 1 and 2. The animal gives off carbon dioxide and moisture, and these are collected in the bottles of the outgoing chain. Bottle 5 is necessary because soda lime gives off moisture. The gain in weight of bottles 4 and 5 represents the carbon dioxide production. The gain in weight of bottles 3, 4, and 5 minus the loss in weight of the animal and chamber represents the oxygen consumption.

2. **Open circuit chamber system.** This system of indirect calorimetry is very similar to the open circuit gravimetric system with a few exceptions. The open circuit chamber system enables researchers to use large animals. Since the chamber is too large to allow for accurate weighing procedures, and since large quantities of CO_2 are being exhaled, modifications of the open circuit gravimetric system are employed.

The composition of the air that is pumped into the chamber is carefully analyzed. The air circulation within the chamber is measured with a gas flow meter; and at given periods, air is sampled and analyzed to determine the decrease of oxygen and the increase of CO_2.

Fig. 15-13. A cow in an open circuit respiration chamber. The gas meter to the left of the chamber is used to measure the respiratory exchange of the cow. These data, plus the gas composition, provide the information needed to calculate the heat production (HP) of the cow. The HP of an animal consuming feed in a thermoneutral environment is composed of the heat increment (heat of fermentation plus heat of nutrient metabolism) plus heat used for maintenance (basal metabolism plus voluntary activity). (Courtesy, USDA)

3. **Open circuit mask system.** This type of indirect calorimetry is commonly used in human research. The subject is told to exhale in a mask for a short period, and all of the exhaled air is collected in a bag or spirometer for analysis. Because large animals exhale large amounts of air, a special collection and sampling apparatus is utilized.

4. **Closed circuit mask system (spirometer).** This system is commonly used in both human and animal research. Doctors frequently use this method of indirect calorimetry to measure the basal metabolic rate of humans. The subject, either human or animal, breathes into a mask which is connected to a spirometer—an apparatus which is filled with oxygen. A one-way valve enables the subject to inhale the oxygen from within the bell of the spirometer, and the rate of oxygen consumption is then recorded on a kymograph. Once the rate of oxygen consumption has been determined, the researcher can calculate the utilization of energy through the use of a number of correction factors for sex, weight, and age of the subject, as well as atmospheric temperature and barometric pressure.

Comparative Slaughter Method of Determining Net Energy

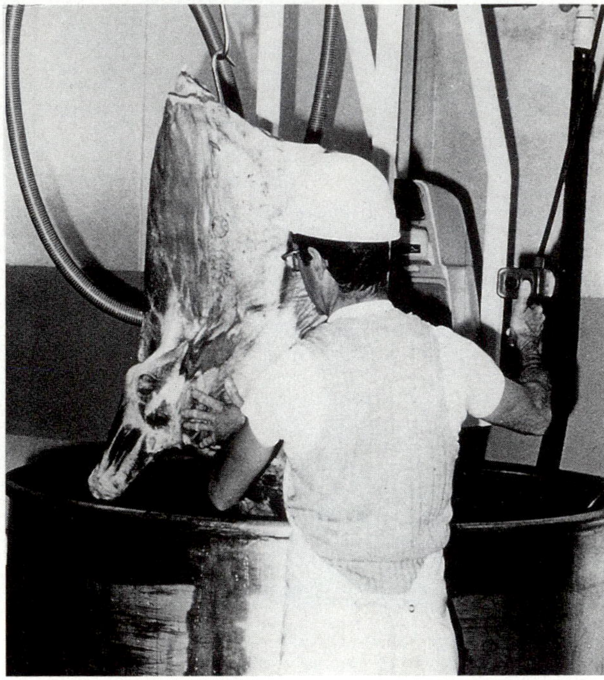

Fig. 15-14. This shows the dipping procedure being used to obtain carcass density from which specific gravity is computed, which, in turn, is used to estimate carcass fat content. By use of an initial and a final slaughter group of animals in each feeding trial, the energy gain can be measured, giving a more accurate measure of the true feed value than does just liveweight gain. Further, the method provides a measure of energy content without grinding and chemical analysis of the carcass. (Courtesy, University of California, Davis)

The comparative slaughter method of determining net energy (energy storage and heat production) is an old technique with a new look. it was first employed by Mitchell and coworkers, of the Illinois Agricultural Experiment Station, in 1926. But, for the most part, it was discontinued, because chemically analyzing the body (carcass) is slow, tedious, and expensive. Then, in 1959, Lofgreen and Garrett, of the California Agricultural Experiment Station, reported an ingenious modification of the comparative slaughter technique. By making use of established relationships between carcass density and the composition of the animal, they were able to estimate the energy content of the carcass without analyzing the body chemically. The density of the carcass is determined by weighing in water—using a dipping procedure (see Fig. 15-14), which can be done quickly and without affecting the sale value of the carcass.

The comparative slaughter method is especially well suited to studies involving growing and fattening animals—cattle, lambs, hogs, and broilers, in which the amount of energy stored in the carcass can be measured. However, it is not adapted to use with dairy animals.

The comparative slaughter method requires relatively large numbers of animals. A random sample, or check group, is selected and slaughtered at the beginning of an experiment to determine the initial body composition. Then, at the close of the experiment, the remaining animals are slaughtered and analyzed. The difference in the calorie content of the two groups represents the energy storage or gain, which is a far more accurate measure of the true value of feed than liveweight gain.

The comparative slaughter method of feed evaluation is unique in the following respects:

1. It provides a relatively inexpensive way in which to determine net energy values.
2. The animals can be kept under more natural conditions, similar to those found in a commercial enterprise.
3. Feeds can be assigned NE_m (net energy for maintenance) and NE_g (net energy for gain) values in keeping with the efficiency of metabolizable energy utilization for these different physiological processes.

The modified comparative slaughter technique has had a major impact on feed evaluation and ration formulation throughout the United States.

In Vitro Digestion Determination

Many laboratories are using *in vitro* digestion techniques to estimate the digestibility of various feeds. Some investigators refer to this as the artificial rumen evaluation. Ruminal fluid is obtained by either a stomach pump or collection through a rumen fistula and strained through cheesecloth. Test tubes containing buffers and the feed samples are inoculated with the fluid that contains, in theory, a representative sample of the microflora of the rumen. The samples are then maintained at the body temperature of the animal in a carbon dioxide environment (anaerobic). In this environment, microbes digest the feed samples.

There are many **advantages** to this type of digestion trial; among them, the following:

1. **Only small amounts of sample material are needed for evaluation.** If a new forage is being tested, the researcher can fully utilize what may be a limited amount of material with which to work.
2. **A large number of samples can be done at one time.** This increases the number of replications per treatment, thereby increasing the statistical accuracy of the results.
3. **Very little equipment is needed; and what equipment is used can be obtained at little cost.** This reduces the cost of analysis.
4. **The results of *in vitro* testing are highly correlated with *in vivo* results.** There are exceptions, but, in general, *in vitro* analysis will yield reliable results as to the *relative* value of the feed.

The conditions under which *in vitro* analyses are carried out are extremely variable. There are many modifications of this type of procedure, and digestibility values obtained from one laboratory will more than likely be different from those obtained at another laboratory. The strength of this procedure lies in the ability to compare feeds—especially when the feeds are from the same class; for example, a comparison of the digestibility values for two types of alfalfa.

Many factors are involved in this type of procedure, which explains why results tend to vary from laboratory to laboratory. The results of the comparisons of feeds should not vary though. If forage "A" is more digestible than forage "B" in one laboratory, the same results should be observed in most of the laboratories. Among the factors which can create variations in this procedure are:

1. Temperature of incubation.
2. Ruminal fluid and inoculant.
3. Buffers used.
4. Degree of agitation used throughout the incubation period.
5. Length of the incubation period.
6. Form of feed tested (degree of processing).
7. Variation of respective components in mixed feeds.

Feeding Trials

Each method of evaluating feedstuffs, discussed earlier in this chapter, has a place and is valuable. But none of them takes into consideration all the factors which determine the true value of any feed for a particular class of livestock. The "court of last judgment" for determining the true value of a feedstuff is the animal. How well do animals eat the feed? How does it affect their health and well-being? How are they producing? Answers to these questions call for feeding the ingredient or ration under controlled conditions to the particular class of livestock.

The U.S. Department of Agriculture and the state experiment stations have conducted numerous experiments to determine the feeding value of specific feeds and of different rations, for each class of livestock. The results of these studies have paid handsome dividends.

Generally speaking, it is in the best interest of producers that experiment stations continue to assume responsibility for the majority of research, including the evaluation of feedstuffs. In comparison with private industry, they have more trained research personnel; generally their studies involve greater accuracy and more controls; and their results are unbiased and unquestioned. But even experiment stations need to bear in mind that adequate animal numbers are important; that there are species and individual animal differences; that feeds differ widely; and that the results must be repeatable.

CONDUCTING APPLIED FEEDLOT TESTS

When carefully conducted and properly interpreted and used, feedlot trials can be a valuable adjunct in the operation of a large feedlot. Among their virtues, the feedlot operator can study area and feed differences. Among their limitations, usually most applied feedlot trials can afford fewer controls than most university-conducted experiments thereby resulting in less accuracy. For the latter reason, most of them should be looked upon as applied tests or demonstrations *per se,* rather than carefully controlled, basic experiments; terminology which does not detract from their value, but which does place them in proper perspective.

The number of pens which a feedlot should devote to test work will vary according to the size of the operation and the number of treatments planned at one time.

There should always be a minimum of 2 lots for controls, plus 2 lots for each treatment evaluated. Generally, the 2 control lots should be fed the standard feedlot ration, and each treatment to be evaluated should be given to 2 lots.

The local county extension agent should be invited to participate in the test; and will usually welcome the opportunity.

The following procedure is recommended in conducting tests in commercial cattle feedlots:

1. **Cattle.** The animals should be of uniform breeding, background, age and weight, and of the same sex. Use cattle owned by the operator, rather than custom-fed animals.

2. **Number per lot.** Ten head if individually weighed; 20 to 40 head, or more, if group weighed.

3. **Randomization.** Gate or chute cut; one per treatment, or not more than five at a time.

4. **Identity.** Preferably (a) apply a different brand to each lot, and (b) individually identify each animal with duplicate numbers—one in each ear. For the latter, use plastic ear tags, the numbers on which can be easily read at a distance.

5. **Variables.** Have as few variables in each treatment group as possible. Let us suppose, for example, that in a certain feedlot, cattle are given a standard ration without implants. But there is a choice of two implant products that claim to promote feed efficiency and rate of gain. So, the owner would like to determine (a) if either product does, in fact, promote feed efficiency and rate of gain; and (b) if so, which product is better. The design would be as follows:

	Lot	Treatment
Control	1	Standard ration, no implants
	2	Standard ration, no implants
Treatment 1, Implant A	3	Standard ration + Implant A
	4	Standard ration + Implant A
Treatment 2, Implant B	5	Standard ration + Implant B
	6	Standard ration + Implant B

6. **Adjustment period.** After sorting cattle into test lots, allow a minimum adjustment period of 7 days, during which the animals should be individually tagged and handled as necessary, and gradually accustomed to their new rations. In case of sickness, a longer adjustment period may be necessary—sometimes as much as 2 to 4 weeks.

7. **Weighing conditions.** Keep off feed and water overnight, then weigh the next morning. Weigh (preferably using a self-recording beam, so as to eliminate the human error) pens in the same order and at the same time each morning when (a) initiating the experiment, (b) at 28-day intervals, and (c) at the close of the test.

Also, weigh and record the amount of feed given to each lot of cattle. In some experiments, it is best to limit all lots (both controls and treatments) to the level of the lot consuming the least, although this will vary according to the treatment being evaluated.

8. **Carcass data.** Sell, or have animals custom slaughtered, with the stipulation that the slaughter plant provide individual (according to individual ear tags) (a) carcass weight and yield, and (b) Federal grade. If slaughter data cannot be obtained on all cattle, get it for as many as possible and of the same number from each lot.

9. **Summarize results.** At the end of the trial, summarize the results, using as criteria (a) rate of gain, (b) feed efficiency, and (c) carcass results.

10. **Determine the application.** If both lots of a given treatment are considerably better than the controls, decide (a) whether to repeat the test, or (b) adopt and use the new treatment throughout the feedlot. The first course of action should be taken under the following circumstances: (1) when there is wide variation in response within the treatments; (2) when the responses are consideraly less than expected; (3) when unusual circumstances prevail—for example, disease problems; and (4) when the experimenter has any doubts as to the validity of the results. If the new treatment becomes the standard, continue with it until a new and superior treatment evolves, based on new trials.

PREDICTING PROTEIN AND ENERGY VALUES BY REGRESSION EQUATIONS

The National Research Council (NRC) has utilized regression equations developed by different researchers to convert various nutrient parameters into more readily usable values. Through the collection of considerable amounts of reported data, equations have been developed to obtain estimates for various animals of digestible protein from crude protein (see Table 15-4).

TABLE 15-4
EQUATIONS USED TO ESTIMATE DIGESTIBLE PROTEIN (Y) FROM CRUDE PROTEIN (X)[1]

Animal Species	Type of Feed	Equation
Cattle	Dry forages or roughages.	Y = 0.886X - 3.06
	Pastures and green feeds.	Y = 0.850X - 2.11
	Silages.	Y = 0.908X - 3.77
	Energy feeds.	Y = 0.918X - 3.98
Sheep	Dry forages or roughages.	Y = 0.897X - 1.33
	Pastures and green feeds.	Y = 0.932X - 3.01
	Silages.	Y = 0.908X - 3.77
	Energy feeds.	Y = 0.916X - 2.76
Goats	Dry forages, roughages, pastures, green feeds.	Y = 0.933X - 3.44
	Silages.	Y = 0.908X - 3.77
	Energy feeds.	Y = 0.916X - 2.76
Horses	Dry forages, roughages, pastures, green feeds.	Y = 0.849X - 2.47
	Silages.	Y = 0.908X - 3.77
	Energy feeds.	Y = 0.916X - 2.76
Rabbits	Dry forages, roughages, pastures, green feeds.	Y = 0.722X - 1.33

[1]Formulas, courtesy of L. E. Harris, Utah State University, Logan.

The livestock industry is gradually changing from the TDN system of energy evaluation to the calorimetric systems—for example, digestible energy, metabolizable energy, and the net energy systems. Unfortunately, calorimetric values for only a small number of feeds have been determined experimentally because of the relative newness of these approaches. Hence, it has been necessary to develop regression equations that allow one to interconvert TDN values and calorimetric values. Thus, many of the calorimetric values given to feeds are empirical in nature—that is, they have not been determined by experiments designed for that parameter *per se*. Therefore, it is necessary to use these regression equations until enough data are available to establish reliable estimates of the energy parameters. For finishing cattle, sheep, and horses, metabolizable energy values can be derived from digestible energy and crude protein values through the use of the following equation:

$$\text{ME (kcal/kg)} = \text{DE (kcal/kg)} \times \frac{96 - (0.202 \times \text{crude protein \%})}{100}$$

Total digestible energy estimates for ruminants can be calculated from either digestible energy or metabolizable energy from the following equations:

$$1.\ \text{TDN (\%)} = \frac{\text{DE (kcal/kg)}}{4{,}409} \times 100$$

$$2.\ \text{TDN (\%)} = \frac{\text{ME (kcal/kg)}}{3{,}616} \times 100$$

Lofgreen and Garrett[3] have established equations whereby net energy for maintenance and growth can be estimated. These equations follow:

$$\text{Log F} = 2.2577 - 0.2213\ (\text{metabolizable energy [kcal/kg]})$$

$$\text{NE}_{\text{maintenance}}\ (\text{Mcal/kg}) = 77 \div \text{F}$$

$$\text{NE}_{\text{growth}}\ (\text{Mcal/kg}) = 2.54 - 0.0314\ \text{F}$$

The net energy system of Moe and Flatt[4] for lactating cattle is based on the following equation:

$$\text{NE}_{\text{lactation}} = -0.77 + 0.84(\text{digestible energy [Mcal/kg dry matter]})$$

FEED GRADES

Grades have been established for many of the commonly marketed grains and other feeds. These grades allow for a more uniform marketing system whereby purchasers of feeds can get a reasonable idea of the quality of the feed—even though they may not have seen it.

Federal grades for grains are based on weight per bushel, moisture content, percentage of damaged grains, and amount of foreign material.

On a voluntary basis, near infrared reflectance spectroscopy (NIRS) is being used currently in the NIRS Forage Research Project Network; and either NIRS or wet chemistry tests may be used in the laboratory certification program operated by the National Alfalfa Hay Test Association in conjunction with the American Forage Grassland Council and the National Hay Association.

(Also see Chapter 8, sections on "Hay Quality" and "Buying and Selling Hay"; and Chapter 10, section on "Grain Standards.")

[3]Lofgreen, G. P., and W. N. Garrett, "A System for Expressing Net Energy Requirements and Feed Values for Growing and Finishing Beef Cattle," *Journal of Animal Science*, 27:793-806, 1968.

[4]Moe, P. W., and W. P. Flatt, "Net Energy Value of Feedstuffs for Lactation," *Journal of Dairy Science*, 52:928 (Abstr.), 1969.

QUESTIONS FOR STUDY AND DISCUSSION

1. What is the most accurate method of determining feed value? Why isn't it used more?

2. Of what importance is a feed evaluation program?

3. What causes such a wide variation in the composition of the same kind of feed; for example, in 48.5% protein soybean oil meal?

4. Why must care be taken in the sampling of feeds?

5. Why should feed samples be kept for 30 to 60 days?

6. How would you go about obtaining a representative sample of each: silage, baled hay, and pasture grass?

7. What are the qualities to look for in the physical evaluation of each of the following feedstuffs: hay, silage, and grains and other concentrates?

8. What evaluations of feed can a trained microscopist make?

9. The Proximate Analysis, or Weende, system of feed analysis divides feeds into several fractions. What are they?

10. For routine moisture determination of grain, which procedure would you choose? Why?

11. What is heat damaged protein? Why should it be considered when balancing rations?

12. Why is the determination of crude fiber by the proximate analysis declining in usefulness?

13. List the advantages and disadvantages of proximate analysis. Do you feel that the disadvantages outweigh the advantages? Why?

14. Outline the Van Soest procedure of fiber analysis and identify the fractions which are digestible by (a) nonruminants, and (b) ruminants.

15. Briefly describe energy determination by use of the bomb calorimeter.

16. Briefly discuss the principles of chromatography.

17. Why is spectrophotometry a good technique in the determination of vitamins in tissues and feed?

18. What are the two basic types of biological assays? Discuss their relative merits.

19. Explain how near infrared reflectance spectroscopy (NIRS) works. How is it being applied in feed analysis?

20. Why should livestock producers be concerned about molds or fungi in feed? Identify the most common damaging mold.

21. Define cannulation and fistulation. How are these techniques useful in the evaluation of feeds?

22. List six indicators commonly used in nutrition research. What are the properties of a good indicator?

23. What factors should be considered when setting up a feed evaluation trial?

24. What factors may affect the digestibility of a feedstuff?

25. How does one determine endogenous fecal nitrogen and metabolic fecal nitrogen?

26. Differentiate between the following terms used in protein evaluation: biological value, protein efficiency ratio, and net protein utilization.

27. Outline the energy balance scheme. Which energy value do you think gives the best indication as to the energy needs of the animal? Why?

28. Why is indirect calorimetry used preferentially over direct calorimetry?

29. Discuss the differences in the setup and adaptability of the various types of indirect calorimetry.

30. Describe the comparative slaughter method of determining net energy.

31. Briefly describe *in vitro* fermentation trials. What are the advantages of this type of digestion trial?

32. Why is it important to have feeding trials conducted by the U.S. Department of Agriculture and the state agricultural experiment stations?

33. Outline the recommended procedure for conducting feeding tests in commercial cattle feedlots.

34. Of what value are regression equations for predicting protein and energy values?

SELECTED REFERENCES

Title of Publication	Author(s)	Publisher
Animal Nutrition, 7th Edition	L. A. Maynard, *et al.*	McGraw-Hill Book Company, New York, N.Y., 1979
Animal Nutrition, 2nd Edition	P. McDonald, *et al.*	Oliver and Boyd, Edinburgh, Scotland, 1973
Applied Animal Nutrition, 2nd Edition	E. W. Crampton L. E. Harris	W. H. Freeman and Co., San Francisco, Calif., 1969
Bioenergetics and Growth	S. Brody	Reinhold Publishing Co., New York, N.Y., 1945
Evaluation of Feeds Through Digestibility Experiments, The	B. H. Schneider W. P. Flatt	University of Georgia Press, Athens, Ga., 1975
Feed Manufacturing Technology III	Ed. by R. R. McEllhiney	American Feed Industry Assn., Inc., Arlington, Va., 1985

(Continued)

(Continued)

Title of Publication	Author(s)	Publisher
Feeds and Feeding, 22nd Edition	F. B. Morrison	The Morrison Publishing Company, Ithaca, N.Y., 1956
Feeds and Feeding, Abridged	F. B. Morrison	The Morrison Publishing Company, Ithaca, N.Y., 1956
Livestock Feeds and Feeding	D. C. Church	O & B Books, Inc., Corvallis, Ore., 1984
Microscopic-Analytical Methods in Food and Drug Control, Food and Drug Tech. Bull. No. 1	U.S. Department of HEW, Food and Drug Administration	Superintendent of Documents, Washington, DC 20402-9325
Near-Infrared Technology in the Agricultural and Food Industries	Ed. by P. Williams K. Morris	American Assn. of Cereal Chemists, Inc., St. Paul, Minn., 1987
Nutritional Ecology of the Ruminant	P. J. Van Soest	Comstock Publishing Associates, a division of Cornell University Press, Ithaca and London, 1982
Nutrition Research Techniques for Domestic and Wild Animals, Vol. 1	L. E. Harris	L. E. Harris, Publisher, Logan, Utah, 1970
Official Methods of Analysis of the Association of Official Analytical Chemists, 12th Edition	Ed. by W. Horwitz	Association of Official Analytical Chemists, Washington, D.C., 1975
Official Publication American Feed Control Officials, Inc.	Staff	West Virginia Dept. of Agriculture, Charleston, W. Va., annual

Fig. 15-15. Probing a load of soybeans to obtain a sample for moisture determination. (Courtesy, American Soybean Assn., St. Louis, Mo.)

Fig. 15-16. Animals used in a feeding trail to evaluate molasses vs corn on orchardgrass pasture for finishing cattle. Note the molasses-fed steers licking molasses off each other. (Courtesy, USDA)

Fig. 15-17. Evaluating high-moisture corn that has been stored in an oxygen-limiting silo, which has undergone mild fermentation. Since the fermentation takes place with little oxygen, there is no darkening of kernels. (Courtesy, A. O. Harvestore Products, Inc., Naperville, Ill.)

Fig. 15-18. Measuring thickness of soybean flakes. (Courtesy, American Soybean Association, St. Louis, Mo.)

Chapter 16

Original painting by Tom Phillips

BUYING FEEDS/ COMMERCIAL FEEDS/ FEED LAWS

Contents Page

Providing feeds for, and feeding, livestock and poulty are important parts of today's agriculture, involving farmers and ranchers, feed and food processors, and commercial feed industries. Feed is the major item of expense in producing animals; it generally accounts for 60% or more of the cost of production. About 28% of the grains fed to animals are used on the farms where grown; the other 72% moves through commercial channels—it is bought and sold.

In the crop year 1984–85, the livestock and poultry industries consumed 509.5 million tons of feed, 5% more than the 485.3 million tons fed in 1965–66 (see Table 16–1). The quantity of concentrates fed increased 27%, while roughage consumption declined 6%, reflecting higher concentrate rations.

TABLE 16-1
KINDS AND QUANTITIES OF FEED CONSUMED BY LIVESTOCK AND POULTRY, FEEDING YEARS 1965-66 and 1984-85[1]

Feed Materials	1965-66 Feeding Year	Percent Of Total	1984-85 Feeding Year	Percent Of Total
	(million tons)	(%)	(million tons)	(%)
Grains:				
Corn	81.5	16	115.2	22.6
Wheat and rye	3.0	1	13.7	2.7
Other feed grains	32.1	7	40.0	7.9
Protein feeds	31.7	6	28.7	5.6
By-product feeds	11.5	3	13.3	2.6
Total concentrates	160.4	33	203.8	40.0
Hay	49.4	10	59.6	11.7
Other harvested roughages	26.3	5	20.0	3.9
Pasture	249.1	52	226.1	44.4
Total roughage	324.8	67	305.7	60.0
Total, all feeds	485.3	100	509.5	100.0

[1]Measured in feed units (corn equivalents). From *1987 Fact Book of U.S. Agriculture,* p. 19, Table 8.

In 1984–85, pasture, hay, and other roughage provided 60% of the total tonnage of feed used, while concentrates furnished 40%.

Some significant shifts (not shown in Table 16–1) occurred in the feed consuming animal species from 1965 to 1985. Poultry accounted for 22% of the grain consuming animal units in 1965–66, compared with 26% of the total in 1984–85, reflecting the expanding poultry industry.

Increasing quantities of livestock and poultry are produced on big and intensive livestock operations, along with increased confinement; resulting in more feeds being purchased, rather than homegrown, and more feeds being milled at or near the feeding location.

The economic impact of purchased feed is clearly reflected in the following figures: In 1975, $12.9 billion worth of feed was purchased by the nation's farms, ranches, and feedlots; in 1985, the feed purchase bill totaled $18 billion (see Fig. 16–1), representing 13.5% of the farm expenditures that year (see Fig. 16–2).

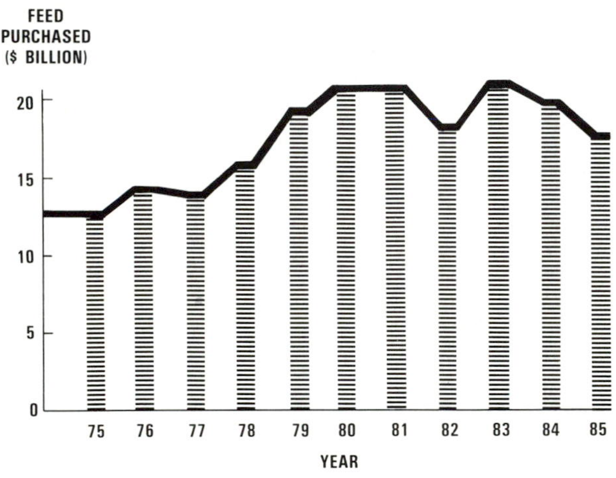

Fig. 16–1. Feed expenditures for 1975–1985. (*Agricultural Statistics 1987*, USDA, p. 414, Table 585)

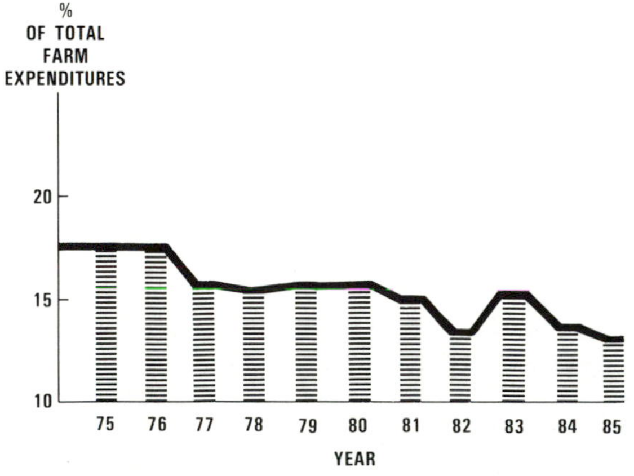

Fig. 16–2. Feed costs as a percent of total farm expenditures, 1975–1985. (*Agricultural Statistics 1987*, USDA, p. 414, Table 585)

HOMEGROWN VS PURCHASED FEEDS

A major factor in determining whether feeds shall be homegrown or purchased is the system of farming, three broad systems of which are practiced in the United States: (1) crop farming, (2) livestock farming, and (3) crop and livestock farming combined. Each system may be defined as follows:

1. **Crop farming.** *The system in which crops are grown for feed, food, clothing, with the income derived primarily from the sale of cash crops.*
2. **Livestock farming.** *The system in which animals are produced, with all or part of the feed bought, and with the income derived primarily from the sale of animals, meat, milk, eggs, and/or wool.*
3. **Crop and livestock farming combined.** *The system that combines significant amounts of crop and livestock farming, with the income derived both from the sale of cash crops and animals and their by-products.*

Over a period of time, farmers and ranchers do those things which are most profitable to them, with some modification according to personal preference. Thus, profitability, with modification by personal preference, largely determines the system of farming. It follows that the choice between homegrown and purchased feeds for the strictly livestock farm or the crop and livestock farm combination is largely based on profitability.

Usually a combination of several factors determines the profits to accrue from the use of homegrown or purchased feeds. Except for very large and intensive livestock operations, net returns are generally highest on those farms or ranches that produce most of their pasture, hay, and/or silage. However, on large specialized operations, such as big cattle feedlot, dairy, and poultry operations, where land and labor are limited and expensive, highest net returns may accrue from the purchase of all feeds.

Generally speaking, part of or all animal feeds are purchased under the following circumstances:

1. When land, water, and/or labor are limited.
2. Where the soil and climate of the farm are not suited to the production of a desired feed(s).
3. When feeds of comparable quality can be bought on a delivered basis more cheaply than they can be produced.
4. Where storage facilities are limited, with the result that feeds produced on the farm cannot be held for yearlong use.
5. When season-long financing for crop production cannot be secured, but when it is possible to "pay as you go," such as a dairy producer using his monthly paycheck to buy feeds.
6. Where homegrown feeds need to be supplemented by certain nutrients, such as protein, minerals, and/or vitamins. This need exists on almost all livestock operations, for few of them produce all the nutrients needed for a balanced ration.
7. When special feeds are desired, such as molasses and fats, which are not farm-produced.

8. Where a purchased feed can be substituted for a homegrown feed on an equal feeding value basis, but with the purchased feed acquired at a lower cost than the selling price of the homegrown product.

9. Where it is more profitable to use available capital, management, and labor for maximum animal production. This frequently applies to large operations among cattle and sheep feedlots, dairies, confinement swine units, and poultry operations, where profits may be maximized by increasing livestock output rather than spending time and capital on raising crops.

10. Where the operator is not knowledgeable relative to crop production.

Generally speaking, forages are more apt to be homegrown than grains, primarily due to their greater bulkiness and higher cost of transportation. However, increased pelleting and cubing have made it economically feasible to transport forages greater distances.

BUYING FEEDS

Buying feeds is an integral part of modern livestock production. Moreover, the trend to purchase feeds, rather than to grow them on location, will continue. In a broad sense, modern sophisticated buying involves knowledgeable buyers, futures trading, consideration of feed substitutions, volume buying, storage, capital outlay, and how to determine the best buy in feeds.

What the Livestock Producer/Feed Buyer Should Know

The vast majority of feeds used on the farms and ranches, and in the feedlots, of America, is purchased by producers—by practical operators who subsequently feed it. So, no one has a greater incentive to purchase wisely and well than they do. However, altogether too often they think only in terms of cost per bushel, per hundred-weight, or per ton. Too often they merely consider the feed analysis or the guarantee based on averages, oblivious to such quality-affecting factors as variety, soil, and weather. Too often buying means haggling with the seller over the price of a feed, rather than considering the nutrient quality. Too often when buying feeds these practical operators are competing with highly trained professionals with great expertise in buying—they are competing with buyers representing commercial feed companies or brokers. For these reasons, more and more livestock producers are purchasing formula feeds, rather than separate ingredients. By so doing, they are acquiring a service—ration formulation—as well as a feed.

Of course, price of feed is very important. For this reason, a sizable section in this chapter is devoted to "Best Buy in Feeds." Additionally, a number of complex and interrelated factors should be considered by the producer-buyer when purchasing feeds. Successful feed buying necessitates knowledge of all the factors that affect net returns, from the time a deal is made to buy the feed until the end product is marketed. Today, sophisticated livestock producers-feed buyers need to know the following:

1. **The nutritive requirements of animals.** The class of animal, age, sex, weight, type of production, and environment influence the more than 40 nutrients required by animals; hence, all these factors must be considered when buying feeds and/or feedstuffs.

2. **The language of the feed trade.** A mark of distinction of feed buyers is that they know feed terms and speak the language. For example, buyers should know that copra meal and coconut meal are synonymous. Likewise, they should be familiar with the various processing methods and the advantages and limitations of each. If given the option to buy dry rolled or micronized milo at the same price, they should know which to choose. Also,with by-product feedstuffs increasing in importance, it behooves buyers to have a working knowledge of the processing methods used in their preparation.

3. **The production and economic trends.** Shrewd buyers follow price trends. They know the current annual production of feedstuffs, and they know the carry-over. Likewise, they know that prices are generally lowest during the harvest season.

4. **The business aspects.** Sophisticated feed buyers know the business aspects, including sources of credit, interest rates, contracts, futures, and possible tax savings to accrue from purchasing feeds before the end of the year.

5. **The different feed grades and quality classifications.** Buyers should know the common feed grades and have a reasonably clear understanding of the specifications of each grade. Thus, they should know the grades of corn and what each grade implies.

6. **The restrictive use of certain feedstuffs.** Buyers should be knowledgeable relative to the numerous factors that may restrict the use of a particular feedstuff. For example, they should know (a) that rye is unpalatable if it constitutes more than ⅓ of the grain ration of finishing cattle; (b) that molasses should be limited to 5% of rations intended for storage in hot, humid areas—otherwise, molds may develop; (c) that unless cottonseed meal is low in gossypol, it should be kept out of the rations of layers because of the effect on egg quality, but that cottonseed meal may be used in broiler diets if the gossypol and noxious fatty acids are minimized; (d) that peanuts should not constitute more than 10% of the ration of growing-finishing hogs if the soft pork problem is to be averted.

7. **The associative, or additive, effects of certain feedstuffs.** In some experiments, some feedstuffs have been more digestible when combined with certain other ingredients than when included with others. This is known as *associative effect* or *additive effect.* Practical application of this phenomenon often occurs when different protein feeds are used in combination, with their respective amino acid composition complementing each other.

8. **The origin of the feed ingredients.** Buyers should be cognizant of the direct and important relationship between the fertility of the soil and the feed that it produces. If all the feed comes from a particular area, this can be very

important. Among the nutrient elements that may be transferred from soil to plants, thence to animals, in a complicated process, are calcium, cobalt, copper, fluorine, iodine, iron, magnesium, manganese, molybdenum, phosphorus, potassium, and selenium. The relationship between the iodine levels of the soil and the incidence of iodine deficiency (goiter) in animals is well known; hence, buyers should know the iodine-deficient areas. Selenium is required in very small amounts by animals and birds, but the dividing line between selenium requirement and selenium toxicity is very narrow. In large areas of the United States, the soils contain little available selenium. Crops produced in these areas are low in selenium, and selenium deficiency in livestock is a serious problem. On the other hand, the soils in some areas of the Great Plains and Rocky Mountain states are high in readily available selenium, with the result that the plants produced thereon are so high in selenium that they are toxic to the animals that eat them.

The feeding value of some feeds is affected by climate. Thus, northern grown oats run heavier than oats grown in the South; and Pacific Coast barley is higher in hull content and lower in protein and energy than barley produced in the Great Plains.

9. **The local potential to grow certain feeds.** In some cases, it may be economically feasible to grow a portion of the ration either on the farm or under contract with local farmers, rather than buy feed(s) that is imported into the area from a distance.

10. **The longtime availability of certain feeds.** Buyers should determine the continuous availability of feedstuffs. Although ration changes can be made, many animals are sensitive to radical changes. Also, it may be costly to change processing methods and feeding systems. For example, roasting equipment for whole soybeans requires a large investment; hence, it would not be wise to tie up capital unless there is reasonable certainty that soybeans will be available at favorable prices for a period of time. Likewise, liquid feeding on pasture requires special equipment; hence, there should be reasonable assurance of the availability of molasses at favorable prices over an extended period before switching from another method of pasture supplementation.

11. **The moisture content of the feed ingredients.** Of the cereals, corn commonly varies most in water content. For this reason, values are given in Part V—Composition of Feeds, for dent corn of the various federal grades, which are based largely on the moisture content.

The significance of water content of feed becomes obvious in the following example: A cattle feeder purchased grain on a 12% moisture basis, but actually received 16% moisture grain. Since 50,000 tons of grain were used during the year and since it had 4% extra moisture, the feeder bought 2,000 tons of water. At $75 per ton, that's $150,000. Thus, this feeder paid about 30¢ per gallon for water.

Such roughages as corn fodder and corn stover also vary widely in water content; they may range from over 50% in wet seasons to 10% or less in dry seasons. Corn fodder with only 10% water supplies 80% more nutrients per 100 lb than corn fodder that contains 50% water.

When using total digestible nutrients (TDN) as a measure of energy value, some of the high-moisture tubers show almost the same feeding value per unit of their dry matter content as the cereal grains (see Table 16–2).

TABLE 16–2
ENERGY VALUE OF TUBERS COMPARED TO CORN ON A MOISTURE-FREE BASIS

Feed	Water	Dry Matter	Energy Value (TDN)	
			As-Fed	Moisture-Free Basis
	(%)	(%)	(%)	(%)
Corn, grain	10	90	80	90
Barley, grain	10	90	77	85
Melons, whole	94	6	5	80
Potatoes, tubers	79	21	18	85
Apples, fruit	82	18	13	74

As shown, moisture-free becomes a common denominator for the comparison for feeds, particularly as to energy value; but this applies to other nutrients, also.

12. **The transportation cost of feeds.** Once feed is purchased, costs are incurred in transporting it. Major factors determining transportation costs are distance, bulk density, and ease of handling. Thus a Hawaiian dairy producer can transport alfalfa cubes from mainland U.S. at a lower cost than baled alfalfa.

13. **The storage capabilities of feeds.** Some feeds require more costly storage facilities than others. For example, raw meat for mink must be refrigerated. The moisture content of a feedstuff must be at a safe storage level; otherwise, moisture must either be removed, or special storage facilities, such as airtight silos, must be provided. For safe storage, grains should be below 14% moisture, and loose hay should not exceed 25% moisture. Additionally, the sophisticated livestock producer-feed buyer needs to know how much feeding value is lost by feeds in storage.

14. **The characteristic shrinkage of feeds.** Buyers should figure shrinkage closely. Here's why: If a ton of hay containing 90% dry matter is bought for $50, 1,800 lb of dry matter have been purchased at this price. However, if the $50 per ton hay contains 80% dry matter, only 1,600 lb of dry matter have been purchased for this same price. Purchase of the high-moisture, 80% dry matter hay has resulted in a loss of 200 lb of hay, or ⅑ of the dry matter, worth $5.56 (⅑ × $50). Thus, if the 90% dry matter hay is worth $50 per ton, the 80% dry matter hay is worth only $44.44 ($50 − $5.56) per ton. If 1,000 tons of hay are involved, that's a loss of $5,560 (1,000 × $5.56).

In addition to moisture losses in storage, newly harvested hay may be expected to shrink about 5% of its weight from going through the sweat.

15. **The risks.** Certain risks are always inherent where a perishable product is involved. Thus, wet hay ferments and generates heat in storage, with the result that there is the hazard of spontaneous combustion and fire, usually about a month or 6 weeks after storage.

16. **The processing that will be involved.** Before buying feed ingredients, consideration should be given to the necessary processing, the available processing equipment, and the cost of processing—all of which add to the cost of the feed at the time it is placed before the animal. For example, whole shelled corn need not be processed when fed to cattle on high-roughage rations, but milo must always be processed. Pressure treating appears to be desirable for milo, but harmful to wheat.

Small operators may not be able to afford sophisticated feed processing equipment. Instead, for practical reasons, they may have to grind or dry roll grains.

In addition to equipment costs, labor costs are a part of feed preparation. Labor is required for the operation of the processing machinery, for bagging the feed, etc.

17. **The truth about certain feedstuffs affecting the product produced.** When buying hay for dairy cows, the buyer must avoid weeds that impart a strong flavor to milk, such as wild onions, skunk cabbage, some members of the mustard family, bitterweed, ragweed, and others.

Likewise, when buying feeds for lactating cows, the informed dairy producer knows that finely ground pelleted forage will lower the fat content of milk, whereas cubed hay has little effect on fat content.

18. **The facts about toxic residues.** Buyers need to be well informed relative to chemicals that are banned, and the conditions of application of those chemicals that are still permitted. In this way, disastrous financial effects from confiscation of market milk or animals can be averted. The buyer must be ever aware of the fact that most pesticides have *zero* tolerance in foods, and that the producer cannot shift blame to the grower for any residues found in meat or milk unless there is a clear-cut case of fraudulent representation.

19. **The government regulations.** The buyer must keep abreast of the current government regulations pertaining to the incorporation of feed additives, especially as to levels, drug combinations, and withdrawals. Of course, reliable commercial feed manufacturers provide this information on the label; hence, the livestock producer should read and follow the label with care.

20. **The impact of foreign feed purchases.** Livestock producers-feed buyers need to understand the effect of large foreign purchases of feed ingredients on domestic feed costs.

OTHER FEED REQUISITES

In addition to the considerations already noted, it is important that all feeds—both bought and homegrown—meet the following requisites:

1. **Palatability.** If animals don't eat it, they won't produce; and if they don't eat enough, feed efficiency will be poor. The relationship of feed consumption to feed efficiency becomes clear when it is realized that the maintenance requirement of an animal producing at a low rate represents a much greater percent of the total feed required than for an animal producing at a more rapid rate.

Palatability is the result of the following factors sensed by the animal in locating and consuming feed: appearance, odor, taste, texture, temperature, and, in some cases, auditory properties of the feed (like the sound of pigs eating corn). These factors are affected by the physical and chemical nature of the feed.

2. **Variety.** Some variety in the ration is desirable, particularly from the standpoints of assuring (a) increased palatability, and (b) balance of nutrients—for example, all the essential amino acids.

3. **Digestive disturbances.** Bloat, colic, scours, and constipation are the bane of all feeders. The choice of feeds can give a big assist in minimizing such disturbances. For example, bloat in cattle and colic in horses can be lessened by avoiding lush or frosted pastures; scours can be lessened by proper feeding; and constipation can be corrected by feeding alfalfa, wheat bran, linseed meal, or molasses.

4. **Bulk.** The amount of bulk in the ration will vary. Ruminants can consume bulkier feeds than monogastric animals; lactating dairy cows require a certain amount of bulk, or fiber, to maintain milk fat production; the younger the animal, the less bulk; and the higher the production desired, the less bulky the ration. Also, the relative cost of feeds—concentrates vs roughages—will influence the relative amount of bulk in the rations.

5. **Poisonous plants and feeds.** Poisonous plants and feeds should be avoided. The livestock producer should know and avoid the poisonous plants common to the area. Also, the following poisons should be avoided: hydrocyanic acid (prussic acid), ergot, scabbed grain, smut on grain, spoiled or moldy feed, botulism, aflatoxins, selenium poisoning, nitrate poisoning, lead poisoning, and mercury poisoning.

Futures Trading in Feed[1]

Futures trading is not new. It is a well-accepted, century-old procedure used in many commodities, for reducing risks, protecting profits, stabilizing prices, and smoothing out the flow of merchandise.

Agricultural prices are highly unstable, due to the perishable nature of farm commodities, seasonal production, the continually changing yield expectations resulting from fluctuating weather conditions, changing international demand, and other factors. Futures trading allows producers, processors, and merchandizers to provide a continuous supply of agricultural goods to consumers at smaller overall profit margins than would otherwise be possible for commodities with such volatile prices. Normal competitive pressures force the buyer to pass on most of the benefits of futures trading to the farmer and the consumer. The farmer is paid higher prices, and the goods are sold to the consumer at lower prices than would be possible without an insurance mechanism to protect against price fluctuations.

A commodity exchange is a place where buyers and sellers meet on an organized market and transact business on paper, without the physical presence of the commodity. The exchange neither buys nor sells; rather, it provides the facilities, establishes rules, serves as a clearinghouse, holds the margin money deposited by both buyers and sellers, and guarantees delivery on all contracts. Buyers and sellers either trade on their own account or are represented by brokerage firms.

• **History of the Chicago Board of Trade**—In the middle of the 19th century, Chicago was the commercial hub of the Midwest. The Plains' farmers took their grains into Chicago each year. In many cases, upon arrival in Chicago, they found themselves at the mercy of the buyers, because little storage space was available for all of the grain converging in the city. So, they hauled their grain from buyer to

[1]The entire section on "Futures Trading in Feed" was reviewed by the following authorities: P. J. Catania, Vice President, Chicago Board of Trade, La Salle at Jackson, Chicago, Ill.; and W. J. Brodsky, President, Chicago Mercantile Exchange, 30 South Wacker Drive, Chicago, Ill.)

buyer in the hope of securing an acceptable offer. Grain that could not be sold was either dumped onto the road or into Lake Michigan.

In this chaotic period, cash contracts were often broken and prices fluctuated wildly. To bring order out of this chaos, 82 Chicago businessmen founded the Chicago Board of Trade in 1846. Their stated objectives were: to establish a commercial exchange; to promote uniformity in the customs and usage of merchants; to install principles of justice and equity to trade; to facilitate speedy adjustments of business disputes; to acquire and disseminate valuable commercial and economic information; and to secure for its members the benefits of cooperation in the furtherance of their ligitimate pursuits. From these concepts, the Chicago Board of Trade has grown to be the largest commodity exchange in the world and has served as a model for the establishment of other exchanges.

Fig. 16-3. Action in the marketplace. The grain trading floor of the Chicago Board of Trade. (Courtesy, Chicago Board of Trade)

• **Characteristics of futures markets**—The unique characteristic of all futures markets is that trading is in terms of contracts for delivery in the future, rather than the immediate transfer of the physical commodity. *A futures contract is a standardized, transferable agreement to buy (take delivery of) or to sell (make delivery of) a specific amount and type of commodity at a future date up to 30 months ahead, at a price established at the time of trading.* In practice, very few contracts are held to delivery date; the vast majority of them are cancelled by offsetting transactions made before the delivery date.

• **Feeds traded on the Chicago Board of Trade**—Since feed represents such a large proportion of the cost of certain types of livestock production, it may be wise to set the price months in advance whenever possible. **NOTE WELL:** Because feed costs represent approximately 80% of the cost of finishing cattle, exclusive of the purchase price of the feeder cattle, cattle feedlot owners are used in the example that follows. But the same principles apply to any feed-using livestock operation, regardless of the class of animals.

Usually, feed can be bought most advantageously at harvest time. Thus, cattle feedlot owners who have adequate storage and finances generally buy their main feed ingredients at that time. By so doing, they can project with reasonable accuracy what it will cost them to feed cattle. Futures permit the cattle feeder to accomplish the same thing without actually taking delivery on the feed and incurring storage costs and risks of physical deterioration. Cattle feeders can use such futures to protect against increases in feed prices. Table 16-3 lists the feeds that are traded on the Chicago Board of Trade.

TABLE 16-3
FEEDS TRADED ON THE CHICAGO BOARD OF TRADE

Grain	Ticker Symbol	Contract Size		Minimum Fluctuation Per Contract	Daily Price Limit Per Contract
		(bushels)	(metric tons)	(per bushel)	(per bushel)
Wheat	W	5,000	136.08	\$12.50 (¼¢)	\$1,000 (20¢)
Corn	C	5,000	127.01	\$12.50 (¼¢)	\$ 500 (10¢)
Oats	O	5,000	72.57	\$12.50 (¼¢)	\$ 500 (10¢)
Soybeans	S	5,000	136.08	\$12.50 (¼¢)	\$1,500 (30¢)
Soybean meal	SM	—	90.72[1]	\$10.00	\$1,000

[1]90.72 metric tons = 100 short tons.

• **Traders**—Traders of commodities are classified as either hedgers or speculators.

Hedgers are farmers, processors, or merchandizers who sell a futures contract when a commodity is produced or bought, and who buy an equivalent contract at the market price when a commodity is sold; thus, hedging is a form of insurance. These traders seek to lock in profits or to minimize risk at adverse price movements.

Speculators seek to assume risks for the sake of profit. They provide necessary liquidity to offset the trading of hedgers.

Hedgers want to protect their positions in the grain markets and as such the use of futures is considered to be a marketing tool. Through the use of futures, the farmer, the grain elevator, or the grain processor can settle on a price well before harvest if so inclined.

Hedging, *per se*, is the act of taking equal and opposite positions in the cash and futures markets. Through this practice, cash prices may fall but the futures position will insure an adequate price. If cash prices rise, the hedger gains when selling the grain, but loses on the futures market.

The key to the concept of hedging between cash and futures is *basis*. *Basis is the difference between the cash price at a particular location and the futures price established at the commodity exchange.* Some of the factors affecting basis include:

1. The overall supply and demand of the commodity.
2. The overall supply and demand of substitute commodities and comparable prices.
3. Geographic disparities in supply and demand.
4. Transportation pricing structures.
5. Storage space available.
6. Carrying costs.

A strong basis is a narrow difference between cash and future prices. A weak basis represents a wide difference in prices and generally indicates an over-supply situation.

Ideally, prices of cash grain and futures should move simultaneously in parallel patterns, but this is rarely the case. However, when basis trends are plotted over time, the pattern becomes more predictable.

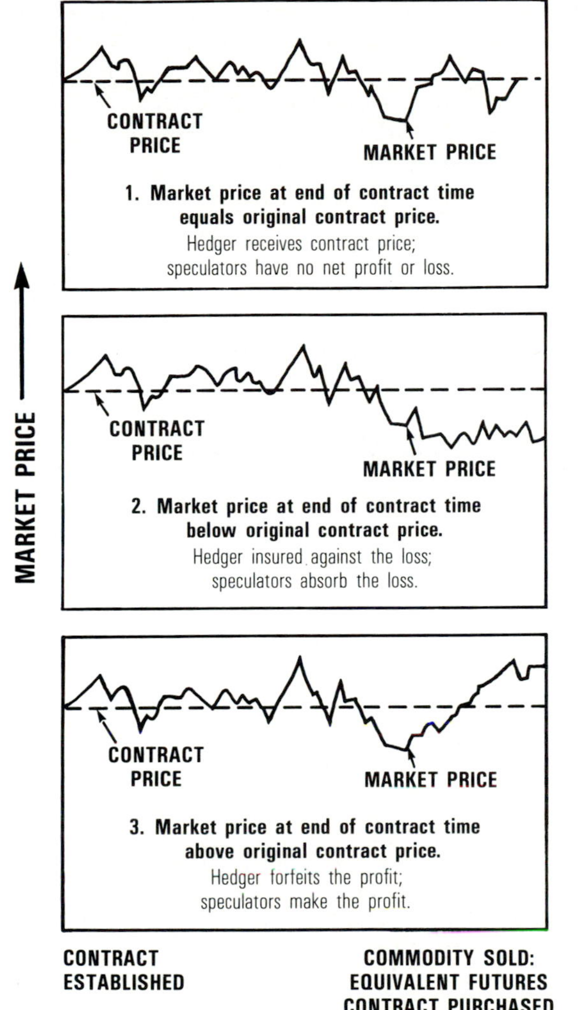

Fig. 16-4. Illustrations of three futures contracts. (Adapted by the authors from: Wisner, R. N., "Back to the Futures," *Science of Food and Agriculture*, Iowa State University, Ames, January 1988)

The components of basis include transportation costs and carrying charges (the charges for services and storage). Once hedgers become familiar with basis patterns, they are ready to protect their interests. Hedgers can be classified as either selling hedgers or buying hedgers. The selling hedgers protect the value of existing inventory and earn, if possible, a stronger return. Here the hedgers own the actual commodity and protect their price of selling, commonly called a short hedge (see example 1).

A buying hedge is used by those who purchase commodities, *i.e.*, elevators, processors, livestock feeders, or exporters. Through futures they can contract to lock in a price for their raw materials. By taking a buying hedge (long hedge), the first transaction involves a purchase rather than a sale (see example 2).

Example 1: Short hedge (selling hedge)

A farmer needs a total of $2.25 per bushel of corn to cover costs and lock-in a reasonable profit. The local basis is 10 ¢ under the Chicago prices; so, the farmer wishes to hedge 10,000 bu, the equivalent of 2 contracts.

Thus, the farmer follows the Board of Trade quotations until on July 1, December corn is $2.35. Thereupon, the farmer sells 2 contracts for December delivery to cover the requirement of $2.25/bu + $.10/bu basis.

Cash	Futures
July 1 owns 10,000 bu price objective of $2.25/bu.	July 1 sells 2 contracts at $2.35/bu

As time for delivery approaches, the futures price drops to $1.85/bu and the local cash price drops to $1.75/bu. So, the farmer sells cash corn at $1.75 and buys back futures contracts at $1.85. The net result is no loss or gain in the total transaction with the farmer receiving in the long run the objective price of $2.25/bu.

Cash, Final	Futures, Final
sells 10,000 bu at $1.75/bu. Result: loss of 50¢/bu.	buys 2 contracts at $1.85/bu. Result: gain of 50¢/bu.

Example 2: Long hedge (buying hedge)

A soybean meal exporter receives an order on April 1 for 1,000 tons of soybean meal to be shipped in 3 months at a certain price. Thereupon, the exporter establishes the order price on the basis of the cash price of the day the deal is concluded, even though the meal will not be bought until shortly before delivery is due. The exporter needs to lock in a profit and secure protection from upward prices. So, 10 futures contracts (1,000 tons) are bought for August delivery to protect the position.

Cash	Futures
April 1 sells 1,000 tons of soybean meal at $150/ton.	April 1 buys 10 contracts (1,000 tons) of August soybean meal at $140/ton.

Assuming that the exporter must buy the cash soybean meal July 15 to fulfill the original contract and that both the cash and futures price of soybean meal has risen $10/ton, the exporter loses nothing on the net transaction.

Cash	Futures
July 15 buys 1,000 tons of soybean meal at $160/ton. Result: loss of $10/ton.	July 15 sells 10 contracts of August soybean meal at $150/ton. Result: gain of $10/ton.

Net result: Futures transaction covered cash market risk.

Without hedging, the exporter would have had to pay $10,000 more to fulfill the contract than was received in the original transaction. However, if the prices had fallen comparably, the exporter would have made an additional $10,000, based on the agreement to sell at the higher price. Hedging, however, allowed the exporter the opportunity to manage the risk and know with some degree of certainty the outcome of the transaction at the time of agreeing to the sale.

Feed Substitutions

Successful livestock producers are keen students of values. They recognize that feeds of similar nutritive properties can and should be interchanged in the ration as price relationships warrant, thereby making it possible at all times to obtain a balanced ration at the lowest cost.

In arriving at feed substitutions, two primary factors besides cost, chemical composition, and feeding value should be considered—namely, palatability and quality of product produced.

Special feed substitution tables have been prepared for each animal species. In order to facilitate their use, they are presented in the feeding chapters devoted to each class of livestock, Chapters 19 through 26.

Best Buy in Feeds

Feed prices vary widely. For profitable production, therefore, feeds with similar nutritive properties should be interchanged as price relationships warrant.

Purchase of feed nutrients at least cost, coupled with knowledge of the nutritive needs of animals and production results, provides a sound basis for arriving at the best buy in feeds.

Two different methods of arriving at the best nutrient buy in feeds are: (1) the computer method, and (2) the cost per unit of nutrients. The computer method is presented in Chapter 18, in the section on "Computer Method." The cost per unit of nutrients is presented in the section that follows.

COST PER UNIT OF NUTRIENTS

One method of arriving at the best buy in feeds is to compute and compare the cost per unit of nutrients, based on feed composition. Where a chemical analysis of a specific feed is not available, feed composition tables, such as those in Section V of this book, may be used as good indicators. Thus, feed composition tables may serve as a basis of feed purchasing and merchandizing, as well as for ration formulation.

The use of the cost per unit of nutrients method can best be illustrated by the examples that follow:

- **Cost per pound of protein and TDN**—If 44% protein (crude) soybean meal is selling at $9.88 per 100 lb whereas 35% protein (crude) linseed meal sells for $6.25 per 100 lb, which is the better buy? Divide $9.88 by 44 to get 22.4¢ per pound of crude protein for the soybean meal. Then divide $6.25 by 35 and get 17.8¢ per pound of crude protein for the linseed meal. Thus, at these prices linseed meal is the better buy—by 4.6¢ (22.4 − 17.8 = 4.6) per pound of crude protein.

When buying energy feed, one can compare the cost per pound of total digestible nutrients (TDN). For example, if corn is priced at $3.63 per 100 lb and has a TDN of 91%, divide $3.63 by 91 and the result is 3.99¢ per pound of TDN. If milo with 86% TDN sells for $3.25 per 100 lb, divide $3.25 by 86, and the price is 3.78¢ per pound of TDN. Thus milo would be the better buy by 0.21¢ (3.99 − 3.78 = 0.21) per pound of TDN.

- **Cost per pound of phosphorus**—When buying a mineral supplement, the livestock producer may also check price against value received. For example, let's assume that the main need is for phosphorus, and that we wish to compare minerals, which we shall call brands "X" and "Y." Brand "X" contains 12% phosphorus and sells at $340 per ton or $17/cwt, whereas brand "Y" contains 10% phosphorus and sells at $320 per ton or $16/cwt. Which is the better buy?

COMPARATIVE VALUE OF BRANDS "X" AND "Y"
(based on phosphorus content alone)

Brand	Phosphorus	Price/cwt	Cost/lb Phosphorus
	(%)	($)	($)
"X"	12	17.00	1.42
"Y"	10	16.00	1.60

Hence, brand "X" is the better buy, even though it costs $1 more per hundred, or $20 more per ton.

One other thing is important when buying minerals. As a usual thing, the more scientifically formulated mineral mixes will have plus values in terms of trace mineral needs and balance.

- **Factors other than price affect feeding value of feeds**—Of course, it is recognized that many other factors affect the actual feeding value of each feed, such as (1) palatability, (2) grade of feed, (3) preparation of feed, (4) ingredients with which each feed is combined, and (5) quantities of each feed fed. It follows that, from the standpoint of the producer, the most important measurement of a feed's usefulness is in terms of *net returns,* rather than cost per bag or cost per ton. To a swine producer, for example, cost per pound or per ton of feed, and pounds of feed required to produce a pound of pork, are important only as they reflect or affect the cost per unit of pork produced. Thus, if the cost of a growing-finishing ration is 6¢ a pound and 4 lb of the ration are required to produce 1 lb body weight, then the feed cost per pound of body weight can be arrived at by multiplying the above figures (6 × 4), which gives a feed cost of 24¢ per pound of pork. Obviously when rations are compared, the ration that produces a pound of pork at the lowest total feed cost is the most desirable from an economic point of view.

Moisture is Important

When buying grains, feeders should never lose sight of how much water they may be purchasing. Table 16-4 illustrates the relative value (dry matter purchased) when paying for corn on a 15.5% moisture basis while actually receiving corn of another moisture content. Thus, if feeders were receiving 19% moisture corn and paying for 15.5% moisture, they would receive only 95.86% of the dry matter for which they paid. On the other hand, if corn is delivered with 7% moisture, while paying on a 15.5% moisture basis, feeders would receive 110.06% of that for which they paid.

TABLE 16-4
RELATIVE VALUE OF U.S. NO. 2 CORN (15.5% MOISTURE) AS AFECTED BY CHANGES IN MOISTURE[1]

Moisture	DM Basis Multiplier	Moisture	DM Basis Multiplier
(%)		(%)	
0	1.1834	19	.9586
1	1.1716	20	.9467
2	1.1598	21	.9349
3	1.1479	22	.9231
4	1.1361	23	.9112
5	1.1243	24	.8994
6	1.1124	25	.8876
7	1.1006	26	.8757
8	1.0888	27	.8639
9	1.0769	28	.8521
10	1.0651	29	.8402
11	1.0533	30	.8284
12	1.0414	31	.8166
13	1.0296	32	.8047
14	1.0178	33	.7929
15	1.0059	34	.7811
16	.9941	35	.7691
17	.9822	36	.7574
18	.9704		

[1]If 15.5% moisture corn is the purchase basis, it will require 1.1834 units of purchase base corn to make 1 unit of 100% dry matter base corn.

So, when high-moisture grain is bought, it should be purchased at a lower figure than dry grain. Likewise, a greater quantity of it must be fed to compensate for the higher moisture content. Table 16-5 shows how the ration needs to be changed when high-moisture corn is included, as well as the dollar value of corn at different moisture contents.

Fig. 16-5. High-moisture corn stored in a gastight (oxygen-limiting) silo for feeding growing-finishing hogs in confinement. The steel bin in front contains the pelleted supplement. (Courtesy, Iowa State University, Ames)

TABLE 16-5
COMPARATIVE VALUE OF CORN OF DIFFERENT MOISTURE CONTENTS

Moisture	Dry Matter	Lb to Equal 100 Lb Dry Corn	Lb to Equal 1 Bu Dry Shelled Corn	Lb to Equal 1 Bu Dry Ear Corn	Estimated Feed Value of Dry Corn	Estimated Dollar Value of Dry Corn
(%)	(%)	(lb)	(lb)	(lb)	(%)	($)
15	85	100	56	70	100	2.75
20	80	106	60	74	94	2.58
25	75	113	63	79	88	2.42
30	70	121	68	85	82	2.25
35	65	131	73	92	76	2.09
40	60	142	79	99	71	1.95

The computations that follow are pertinent to the use of high-moisture corn.

To estimate feed value in percent of dry corn:

Dry corn (15% m) has 85% dry matter.
High-moisture corn (30%) has 70% dry matter.
Therefore, the 30% high-moisture corn has $^{70}/_{85} \times 100 = 82\%$ as much energy feed value.

To estimate dollar value of high-moisture corn:

Dry corn (15% m) costs $2.75/bu.
High-moisture (30%) corn has 82% as much energy feed value.
Therefore, it should cost no more than: $2.75 × .82 = $2.25/bu.

To estimate amount of high-moisture corn to substitute for dry corn:

Total ration mix	=	2,000 lb
Dry (15%) corn	=	1,500 lb
Other ingredients	=	500 lb

How much high-moisture (30%) corn should replace the dry corn?

121 lb of 30% moisture corn = 100 lb dry corn.
Therefore, 1,500 lb × 1.21 = 1,815 lb of 30% moisture corn would be needed to replace the 1,500 lb of dry corn.

High-moisture (30%) corn	=	1,815 lb
Other ingredients in mix	=	500 lb
Total ration mix	=	2,315 lb

Thus, due to the moisture, the mix with the high-moisture corn is about one-seventh less nutritious than the dry corn mix, and it is necessary to feed about one-seventh more of this mix than when feeding the dry corn mix.

(Also see Chapter 6, section on "High-Moisture Feeds.")

HOME-MIXED VS COMMERCIAL FEEDS

The value of farm-grown grains—plus the cost of ingredients which need to be purchased in order to balance the ration, and the cost of grinding and mixing—as compared to the cost of commercial ready-mixed feeds laid down on the farm, should determine whether it is best to mix feeds at home or depend on commercial feeds. Of course, the ultimate criterion for choosing between home-mixed and commercial feeds is which program will make for maximum returns to producers for their labor, management, and capital. Generally speaking, the use of commercial feeds makes it possible for the producer to have more animals and concentrate on production, whereas home mixing restricts animal numbers and necessitates that part of the time and capital be devoted to feed formulating and manufacturing.

Fig. 16-6. Home processing feed center for swine unit, consisting of three bins: *large bin,* corn storage; *middle-sized bin,* soybean meal; and *small bin,* mineral-vitamin premix. (Courtesy, Iowa State University, Ames)

The producer has the following options from which to choose for home mixing feeds:

1. Purchase of a commercially prepared protein supplement (likely reinforced with minerals and vitamins), which may be blended with local or homegrown grain.
2. Purchase of a commercially prepared mineral-vitamin premix which may be mixed with an oil meal, and then blended with local or homegrown grain.
3. Purchase of individual ingredients (including minerals and vitamins) and mixing the feed "from the ground up."

Many farmers and ranchers are faced with the choice of home-mixing vs the purchase of commercially prepared feed. Even though the economics may favor the use of homegrown feeds, the following searching questions should be asked before launching a home processing feed operation:

1. Do I have the necessary equipment to process my feed efficiently, effectively, and without segregation?
2. Do I have a reliable, cost-competitive, quality source of the ingredients which I must buy?
3. Do I need to get FDA approval to add certain medications; must my feed processing operation meet FDA inspection?
4. Do I have the necessary facilities in which to store bulk ingredients and/or finished feed?
5. Do I have the expertise to go it alone without the help of my commercial feed representative?

COMMERCIAL FEEDS

Commercial feeds are feeds that are mixed by commercial feed manufacturers who specialize in the business, instead of being home-mixed.

In the United States, commercial feeds had their beginning as horse *tonics, conditioners, potions,* and *cure-alls.* Claims were made for increased growth and development, improved breeding, more speed, and increased stamina. The feeding directions called for a cup or for 3 to 4 Tbsp per horse daily.

In 1860, the first of these horse tonics was shipped to the United States from London. It contained beans, barley, flax seed, Peruvian bark, and quinine tonic. The price: $14.00/100 lb, at a time when oats were selling at $1.40/100 lb.

Pelleted feeds were introduced in Europe in 1870, during the Franco-Prussian War, when a compressed product was needed to save space. The German armies used pelleted feeds extensively during World Wars I and II, because they required only one-fifth as much transportation and storage space as loose hay and bulk grain.

One of the most famous products in the United States in the early 1900s was the International Stock Food Tonic. The developer and promoter was W. W. Savage, owner of the great Standardbred horse, Dan Patch, holder of the mile pacing record from 1906 to 1933 and the idol of his day.

The advertisements for the tonic claimed that Dan Patch was fed the tonic every day and included a picture of the great horse suitable for framing.

But tonics were not for horses only! The labels on these secret and magic formulas were expanded to include cattle, sheep, swine, and poultry.

Not all commercial horse feeds of the Dan Patch era were tonics. Numerous high-quality feeds were also available. Typical commercial horse feeds during the period 1915 to 1930 consisted of corn, oats, barley, alfalfa, wheat bran, and molasses. During the 1930s, calcium and vitamins were added to commercial feeds.

Today, the commercial feed business is a complex and highly respected industry.

Fig. 16-7. Complete feed automated to hog finishing pens. In recent years, there has been an increasing trend toward the use of complete mixed rations for growing-finishing hogs and baby pigs. (Courtesy, University of Illinois, Urbana)

Importance and Nature of Commercial Feeds

The importance and the nature of commercial feeds in the United States is attested to by the following statistics for the year 1984, based on a survey made by the USDA:[2] A total of 109.6 million tons of commercial feeds were produced, which represented 21.5% of the 509.5 million tons of feed consumed that year. Of the commercial production, 87% was primary feed (which the researchers defined as a mixed feed, but to which a premix is sometimes added at the rate of less than 100 lb/ton) and 13% consisted of secondary feed (which the researchers defined as feed to which a formula feed supplement is added at a rate of 100 lb per ton or more). Primary feed production had the following breakdown by type of livestock feed: poultry, 35.5%; dairy, 22%; beef and sheep, 21.2%; hogs, 15%; and others, 6.5%. Also, the survey revealed that 51.7% of the primary feed was pelleted. Corporations owned 58.5% of all U.S. feed mills and produced 71.1% of the commercial feed; farmer cooperatives owned 28.5% of the feed mills and produced 21.8% of the feed tonnage; and the remaining 13% of the mills that produced 7.1% of the feed were privately owned.

Unfortunately, the terms identifying the types or classes of commercial feeds are not standardized; that is, there are no universally accepted definitions of kinds of commercial feeds. What may be a concentrate to one person may be a supplement to another. So, the authors have evolved with the definitions that follow, which are a good consensus of the terms presently used and/or understood in the feed industry, and which will serve as the basis for discussion of commercial feeds in this book.

- **Complete feed**—*A complete feed is a ready-to-use feed which is nutritionally balanced and requires no additional ingredients, except it may or may not contain processed hay and/or other roughage.*

- **Supplement**—*A supplement is a formula feed which contains a substantial portion of the protein, minerals, and vitamins required in the final ration.* When grains are available locally at a reasonable price, a supplement is commonly bought and mixed with the grains. Thus, the grains provide the primary source of the energy, while the supplement completes the formulation with sources of protein, minerals, vitamins, and other performance-enhancing products. Normally, a supplement is used at the rate of 200 lb or more per ton of finished feed.

- **Base mix**—*A base mix differs from a supplement in that it provides less of the animal's protein requirements.* Where grain(s) and a protein source(s) are available locally at a reasonable price, a base mix may be added to the grain and protein source to complement the protein quality and to provide a source of nutrients and performance-enchancing compounds not found in the grain and the protein source. Normally, the base mix is added to the grain and high-protein source at the rate of 100 lb or more per ton to balance the ration.

- **Premix**—*A combination of one or more trace minerals, vitamins, and/or performance-enhancing compounds with a carrier.* A premix may be used at a level of 2 to 100 lb per ton of finished feed, although the vast majority of premixes are used at levels of 2 to 10 lb per ton of finished feed.

- **Specialty feeds**—*Specialty feeds refer to a number of products that are produced by feed manufacturers for special purposes, including such products as milk replacers and range cubes.*

The various types and classes of commercial feeds can be designed to take into account specific performance parameters, condition of the animals, and need for medication. The determination of what type of class of feed to use depends on a variety of factors, including kind of animals fed, level of performance desired, and availability and cost of ingredients. For example, poultry rations tend to be rather complex; so, most poultry feed is sold as a complete feed.

[2]The statistics presented, and the terms used, herein were taken from a nationwide survey made by the USDA's Economic Research Service (ERS) in 1984, to which there was more than a 40% response. The report was written by ERS economists Mark Ash, William Lin, and Mae Dean Johnson and reported in *Feedstuffs,* March 14 and March 21 issues, 1988.

Many swine producers also produce corn; hence, they may home-mix the corn with a supplement. Other swine producers may grow corn and have access to soybean meal that is processed nearby at a reasonable cost; for them, a base mix may be most practical. A horse owner who has access to hay, oats, and linseed meal may need a premix.

The commercial feed manufacturer has the distinct advantages of (1) purchasing feed in quantity lots, making possible price advantages; (2) using computers for purchasing and least-cost formulating; (3) having the knowledge to manufacture medicated feeds; (4) having the knowledge and the facilities to manufacture specialty feeds, such as milk replacers; (5) processing and mixing economy and control; (6) hiring scientifically trained personnel for use in determining the rations; and (7) controlling quality. Most producers have neither the know-how nor the quantity of business to provide these services on their own. Because of these several advantages, commercial feeds are finding a place of increasing importance in livestock feeding.

Generally, representatives of commercial feed companies have been good feed information sources and teachers. In a survey which *Hoard's Dairyman* conducted in Virginia to determine where dairy producers get their information, it was found that feed companies were most used—by a wide margin. A total of 54% of the respondents obtained their information from the feed company. Other sources, by rank, were: the producer (self), 24%; private consultant, 23%; agricultural extension, 19%; veterinarian, 15%; and none, 9%.[3] **NOTE:** The numbers do not add up to 100% because some respondents used more than one source of information.

In summary, it may be said that there exist two good alternative sources of most feeds and rations—home-mixed or commercial—and the able manager will choose wisely between them, or use some of each.

Types of Commercial Feed Formulations

Commercial feed can be classified under three types of feed formulations—closed formulas, open formulas, and custom formulas.

CLOSED FORMULAS

Closed formula feeds are commercial feed preparations that do not list the proportions of the respective ingredients. Instead, they either (1) list the ingredients on the feed tag in decreasing order of incorporation, or (2) use group terms such as animal proteins, *etc.* In either case, however, feed buyers are unable to tell how much of each feedstuff they are buying. They can obtain a rough idea of the quality of feed by the specific types of feed ingredients.

It is understandable that feed companies prefer closed formulas. They make sizable investments in the research and development of each feed formula; hence, they are reluctant to make the formula public because (1) competitors could then mix the same feed without having to test the product, and (2) they would not likely get a return on their investment.

[3]*Hoard's Dairyman,* May 10, 1988, pp. 452–453.

OPEN FORMULAS

Open formula feeds are those that have a statement on the tag that tells how many pounds of each ingredient are incorporated in the mix. By knowing the amount of each ingredient in the ration, the buyer can calculate the amount of TDN and digestible protein in the ration through the use of feed composition tables. While these types of feeds may benefit the purchaser, the harsh realities of a highly competitive free market system discourage this practice. Additionally, merely knowing the amount of each ingredient in the ration is not the total answer, because no specifications of quality of the ingredients are listed.

Some feed cooperatives do have open formulas. Therefore, buyers have an opportunity to see exactly what feedstuffs they are buying—a fact that may make it easier to decide which feed to buy. By knowing what is in the feed, the buyer is better able to evaluate the claims of the manufacturer as to the merits of the product.

CUSTOM FORMULAS

Custom formula feeds are those that are mixed according to the specifications of the purchaser. Through the use of these feeds, livestock producers can tailor their feeding programs to the specfic needs of their animals. In many cases where drugs are to be added, the feed must be registered with the FDA; hence, these drugs in question are generally not added to custom feeds.

In order for custom formulations to be successful, feed buyers must be well versed relative to the principles of ration formulation because the feed manufacturer is not responsible for the "completeness" of the feed—the feed manufacturer is responsible for only the quality of the ingredients and the actual mixing.

When this type of feed formula is used, it is suggested that the feed tag or invoice contain the following information:

1. Name and address of the feed manufacturer.
2. Name and address of the purchaser.
3. Date of delivery.
4. Net weights of the various feed ingredients.
5. Directions for use, especially when drugs or feed additives are incorporated into the feed.
6. Precautionary statements, such as withdrawal periods or restrictions for use if feed additives are used.

How to Select Commercial Feeds

There is a difference in commercial feeds! That is, there is a difference from the standpoint of what livestock producers can purchase with their feed dollars. Smart operators will know how to determine what constitutes the best in commercial feeds for their specific needs. They will not rely solely on the appearance or aroma of the feed, nor on the sales person.

REPUTATION OF THE MANUFACTURER

The reputation of the manufacturer can be determined by (1) conferring with other producers who have used the particular products, and (2) checking on whether the commercial feed under consideration has a good record for meeting its guarantees. The latter can be determined by reading the bulletins and reports prepared by the state department in charge of monitoring feed quality and enforcing feed laws.

Quite often, a feed that costs a little more will be of a higher quality than its competition. Therefore, the increased costs can be justified by the increased performance of the animals to which the ration is fed.

Many of the larger manufacturers offer services other than selling feed, including the consulting services of well-trained and experienced staff and excellent publications.

SPECIFIC NEEDS OF THE ANIMALS

Feed requirements vary according to (1) the class, age, and productivity of the animals; and (2) whether the animals are fed primarily for maintenance, growth, finishing (or show-ring fitting), reproduction, lactation, or work (running). The wise producer will buy different formula feeds for different needs.

Feeding livestock has become a sophisticated and complicated process. Feed manufacturers have extensive resources with which to formulate and test rations for different needs. As a result, most manufacturers have a large selection of feeds—one of which should be applicable to the needs of the producer. It is essential that the producer make clear to the feed sales person the needs of the animals to be fed.

LABELING OF FEEDS

Most states require that feeds carry labels guaranteeing the ingredients and the chemical makeup of the feed, along with directions for its use. The feed tag should contain the following information:

1. **Net weight of the feed.**

2. **Brand name and product name.** The brand name refers to any work, name, symbol, or logo which identifies the feed of a distributor and distinguishes it from the feeds of other manufacturers. The product name identifies the specific use for the feed.

3. **Guaranteed analysis.** Most feed labels give minimum and/or maximum guarantees of certain nutrients within the feed. Laws vary from state to state as to what analyses must be guaranteed, but most, if not all of the following analyses are generally listed: (a) dry matter; (b) crude protein; (c) crude fat; (d) crude fiber; (e) ash; (f) nitrogen-free extract; (g) calcium, phosphorus, and salt; and (h) other nutrients. Generally, feeds with more protein and fat and less fiber indicate quality. Most labels list minimum values of crude protein and crude fat and maximum values of crude fiber and nonprotein nitrogen when used. Both maximum and minimum guarantees are listed for calcium, phosphorus, and salt on most feed tags.

A high-fiber content often indicates a low feeding value. In ruminant feeds, for example, if the fiber content is less than 8%, the feed may be considered top quality; if the fiber is more than 8% but less than 12%, the feed may be considered as medium quality; while feeds containing more than 12% fiber should be scrutinized carefully. Many feeds are high in fiber simply because they contain liberal amounts of alfalfa; yet they may be perfectly good feeds for the purpose intended. On the other hand, if oat hulls and similar types of high-fiber ingredients are responsible for the high-fiber content of the feed, the quality should be questioned. The latter type of fiber is poorly digested and does not provide the nutrients required to stimulate the digestion of the fiber in roughages.

Nonruminant feeds—feeds for swine and poultry, for example—generally contain negligible amounts of fiber. But they are higher in fats and micronutrients than ruminant feeds.

4. **Listing of ingredients.** The feed tag lists the ingredients of the feed in descending order of quantity used. While the exact quantities of the ingredients are not generally given, the buyer can obtain a rough idea as to the composition of the feed.

5. **Directions for use.** If the feed is to be used for a specific purpose, directions may be given on the label. They should specify what kind of animal and for what particular purpose the feed was formulated. For example, the label of a chicken starter ration may indicate that it shoud be used from age day one to 5 weeks.

6. **Name and mailing address of the manufacturer.** Manufacturers are responsible for the quality of their feed. Any failure to meet guarantees or any contamination problems incurred with the feed makes manufacturers liable for penalties and damages incurred from their feed. For this reason, manufacturers should be identified clearly on their product.

7. **Warnings.** When drugs have been added to commercial feed, the feed label must clearly indicate that it is medicated. Many states require that the name of the drug with a listing of the quantity of active ingredients added be stated on the feed tag. Also, the purpose of medication, any restrictions of use, and withdrawal period for the medicated feed should be so stated.

Fig. 16–8 shows a typical feed tag.

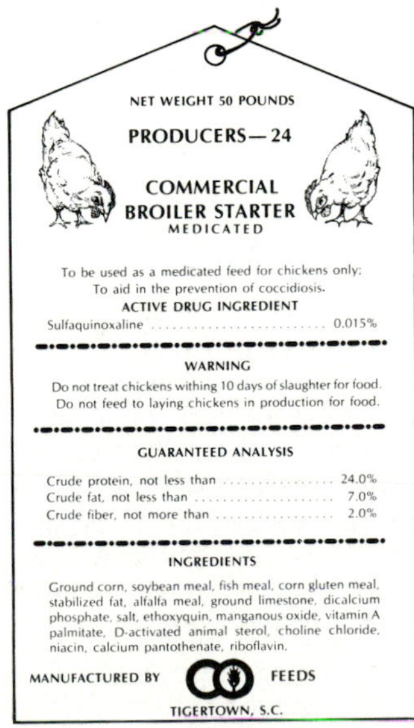
NET WEIGHT 50 POUNDS

PRODUCERS—24

COMMERCIAL
BROILER STARTER
MEDICATED

To be used as a medicated feed for chickens only;
To aid in the prevention of coccidiosis.

ACTIVE DRUG INGREDIENT

Sulfaquinoxaline 0.015%

WARNING

Do not treat chickens withing 10 days of slaughter for food.
Do not feed to laying chickens in production for food.

GUARANTEED ANALYSIS

Crude protein, not less than 24.0%
Crude fat, not less than 7.0%
Crude fiber, not more than 2.0%

INGREDIENTS

Ground corn, soybean meal, fish meal, corn gluten meal, stabilized fat, alfalfa meal, ground limestone, dicalcium phosphate, salt, ethoxyquin, manganous oxide, vitamin A palmitate, D-activated animal sterol, choline chloride, niacin, calcium pantothenate, riboflavin.

MANUFACTURED BY FEEDS
TIGERTOWN, S.C.

Fig. 16–8. Example of a feed tag.

FLEXIBLE FORMULAS

Feeds with flexible formulas are generally the best buy. This is because the price of feed ingredients in different source feeds varies considerably from time to time. Thus, good manufacturers, having access to least-cost computer programs, will shift their formulas as prices change, in order to give the producers the most for their money. This is as it should be, for (1) there is no one best ration; and (2) if substitutions are made wisely, the price of the feed can be kept down and the feeder will continue to get equally good results.

Cooperatives

Cooperatives are business organizations that are formed, operated, and financed by patron members. The members themselves are responsible for the decisions of the organization and subsequently share the risks and reap the benefits. For example, a group of livestock feeders may decide to form a cooperative for the procurement of feeds.

The nationwide survey of the feed industry conducted by the U.S. Department of Agriculture in 1984 (see footnote No. 3) revealed that farmer cooperatives owned 28.5% of the nation's feed mills and produced 21.8% of the feed tonnage.

Cooperatives enjoy substantial tax advantages. Among their additional advantages are pride of ownership, collective bargaining, influence in decision making, and sharing in any profits.

The key to a successful cooperative is strong and able leadership. While management is responsible to the members of the cooperative, it must be willing to make decisions. Failure to make prompt decisions ultimately leads to failure of the entire organization.

To form a cooperative, members must invest considerable capital. This means that they take certain risks. Although they share in profits, they must also assume losses.

FEED QUALITY

Feed quality is the degree of excellence of a feed.

Quality control is a system of assuring the maintenance of standards by checking.

Feed quality and quality control are important in both home-mixed and commercial feeds. High-quality feed must be palatable—the animals must like and eat it. Also, finished (mixed) feeds must meet the nutritional needs of the animals to which they are fed, without deficiencies.

Feed quality begins with the selection of feed ingredients. High quality mixed feed cannot be produced from poor quality feed ingredients. To assure high quality feed ingredients, they should be subjected to the following tests and meet high standards in each test:

1. They should be physically evaluated for quality; including color, smell, taste, texture, and microscopic examination.
2. They should be tested by proximate analysis and/or other methods for components; and roughages should be tested for NDF and ADF.
3. They should be examined for rodent contamination, salmonella bacteria, molds and mycotoxins, and undesirable chemicals such as PCBs and pesticides.

Quality control involves—

1. Buying (or producing) feed ingredients that meet high standards.
2. Inspecting and sampling ingredients at receiving.
3. Monitoring processing, including keeping inventory of drugs, flushing out the feed system after preparing certain medicated feeds, isolating some ingredients, and keeping records of ingredients and finished feeds.
4. Providing finished feed storage and transportation, with rigid sanitation and pest control.

FEED LAWS

The U.S. Food and Drug Act was passed in 1906, giving the Federal Government authority to regulate and inspect feeds shipped in interstate commerce. Additional controls were authorized in the Food, Drug, and Cosmetic Act of 1938. In addition to the Federal laws, nearly all states have laws regulating the sale of commercial feeds. These benefit both producers and reputable feed manufacturers. In most states the laws require that every brand of commercial feed sold in the state be licensed, and that the chemical composition be guaranteed.

Since most commercial feeds are closed formulas, it is necessary to sample and analyze the feed periodically to ensure that it is fulfilling its guarantees. Generally, samples of each commercial feed are taken periodically and analyzed chemically in the state's laboratory. Additionally, skilled microscopists examine the sample to ascertain that the ingredients present are the same as those guaranteed. Flagrant violations on the latter point may be prosecuted as it represents willful mislabeling.

With the great volume of feeds that are manufactured, the procedure of sampling and analyzing feeds can be an enormous task for regulatory agencies. Various types of sampling are used by different states. Some state regulatory officials routinely sample all manufactured feeds as a quality control and preventative measure for avoiding any problems involving feeds. Most states use a system of random sampling, supplemented by additional testing when it appears that there may be a problem—a procedure that is less accurate but also less cumbersome in quality control.

Results of these examinations are generally published annually by the state department in charge of such regulatory work. Usually, the publication of the guarantee alongside any "short-changing" is sufficient to cause any manufacturer promptly to rectify the situation, for such public information soon becomes known to both users and competitors.

In order to standardize the laws concerning the licensing and regulation of feed manufacturers, the Association of American Feed Control Officials has drawn up a sample document from which the states can model their feed laws. A copy of the "Uniform State Feed Bill" may be obtained from the following address: Donald James, Department of Agriculture, State Capital Bldg., Charleston, WV 25305.

(Also see Chapter 13, section on "Food and Drug Administration (FDA).")

QUESTIONS FOR STUDY AND DISCUSSION

1. Why do feed costs often determine the success or failure of a livestock operation?

2. Why did the use of concentrates increase while the use of roughages declined during the 20-year period 1965 to 85?

3. Poultry accounted for 22% of the grain consuming animal units in 1965-66, compared with 26% of the total in 1984-85. What accounted for this increase in poultry feeding?

4. Fig. 16-1 shows that total feed expenditures for feed trended upward from 1975-1985, while Fig. 16-2 shows the feed costs as a percent of total farm expenditures trended downward from the mid-1970s to the mid-1980s. How and why did total feed expenditures trend upward at a time when feed costs as a percent of total farm expenditures trended downward?

5. On a livestock farm or on a crop and livestock farm combined, what circumstances will determine whether feeds will be homegrown or purchased?

6. List and discuss 10 factors about which producers-feed buyers need to be knowledgeable in order to be successful.

7. When buying feed, how important are the following factors: palatability, variety, digestive disturbances, bulk, residues, and poisonous plants?

8. Define commodity exchange futures contracts, hedgers, speculators, and basis.

9. Under what circumstances would you recommend that a livestock producer use futures in corn and/or soybean meal?

10. What precautions should be taken when using a feed substitution table?

11. A producer has the opportunity to select from among the following feeds, to be used for supplying energy:

   a. Milo—80% TDN at $3.40 cwt
   b. Corn—89% TDN at $3.75 cwt
   c. Barley—83% TDN at $3.50 cwt

   Which is the best buy as determined by cost per unit of TDN?

12. Formerly, feeds were evaluated largely on the basis of energy and/or protein. Computer technology makes it possible to evaluate feeds on the basis of a number of nutrients simultaneously. Does this fact justify the purchase of a computer on a medium-sized livestock farm?

13. List five factors other than price that affect the feeding value of each feed.

14. Why is moisture important when buying feeds and balancing rations?

15. What factors should determine whether it is best to mix feeds at home or depend on commercial feeds?

16. What are commercial feeds? Discuss the early history of commercial feeds in the United States.

17. Discuss the magnitude of U.S. commercial feed industry. Rank primary feed production according to type of livestock to which it is fed.

18. Define the following terms: complete feed, supplement, base mix, premix, and specialty feed.

19. List the main advantages of commercial feed manufacturers in comparison with home-mixing.

20. Differentiate between closed, open, and custom formula feeds.

21. How important are the following when selecting commercial feeds: (a) the reputation of the manufacturer, and (b) the specific needs of the animals.

22. What information can be found on a feed tag?

23. Discuss the advantages and disadvantages of cooperatives vs private feed companies.

24. Define (a) feed quality, and (b) quality control.

25. How may a commercial feed company assure high quality feed ingredients?

26. What four steps are involved in feed quality control?

27. Why is it important for the government, both state and federal, to monitor the commercial feed industry? How is it done?

SELECTED REFERENCES

Title of Publication	Author(s)	Publisher
Association of American Feed Control Officials, Incorporated, Official Publication	Association of American Feed Control Officials, Inc.	Association of American Feed Control Officials, Inc., Annual
Feed Additive Compendium	Ed. by D. Natz	The Miller Publishing Company, Minneapolis, Minn., Annual
Feed Formulations	T. W. Perry	The Interstate Printers & Publishers, Inc., Danville, Ill., 1975
Feed Manufacturing Technology III	R. R. McEllhiney, tech. ed.	American Feed Industry Association, Inc., Arlington, Va., 1985
Feeds and Feeding, 22nd Edtion	F. B. Morrison	The Morrison Publishing Company, Ithaca, N.Y., 1956
Livestock Feeds and Feeding	D. C. Church	O & B Books, Inc., Corvallis, Ore., 1984
Stockman's Handbook, The, 6th Edition	M. E. Ensminger	The Interstate Printers & Publishers, Inc., Danville, Ill., 1985

3
Feeding

SECTION III

FEEDING

Animals inherit certain genetic possibilities, but how well these potentialities develop depends upon the environment to which they are subjected; and the most important influence in the environment is the feed.

Also, feeding is important from an economic standpoint; it is the major item of expense in producing livestock. Normally, it accounts for approximately the following proportions of the cost of livestock production: finishing cattle, 70%; feedlot lambs, 50%; pork, 65 to 75%; milk production, 55%; and poultry, 55 to 75%, with the production of eggs toward the lower side of this range and the production of broilers and turkeys toward the upper side. It is important, therefore, that feeding practices be as efficient and economical as possible. To this end, livestock producers should endeavor to provide rations that are both satisfactory and inexpensive—rations that make for maximum production of a quality product per unit of feed consumed and for maximum net returns.

Section III is devoted to the art and the science of livestock feeding. In recognition of the effect of artificial environments on animal nutrition, the innovative Chapter 17, Animal Behavior/Environment, is presented. Chapter 18, Feeding Standards/Ration Formulation, tells how feeding programs may be tailored to specific operations. Then, Chapters 19 to 28 embrace an in-depth discussion of the feeding of each species.

OLD MACDONALD'S FARM, the preceding two-page spread, is from an original painting by the noted artist, Tom Phillips, prepared specially for this book. It portrays the artist's conception of OLD MACDONALD'S FARM, immortalized in the Old English Folk Song. And on that FARM he had a menagerie of chatty animals: some cows that *MOO*ed; some sheep that *BAA*ed; some pigs that *OINK*ed; some horses that *NAY*ed; some chickens that *CLUCK*ed; some turkeys that *GOBBLE*d; and some ducks that *QUACK*ed.

Those of us who grew up on such a farm learned barnyard talk. The animals kept up a running conversation with us as we gave a handful of hay to a still-hungry cow, treated a favorite horse to an apple, took a last look at a litter of newborn pigs, closed the gate on the sheep corral, and shut the poultry house door. It was a sign language, but it spoke louder than words; it told us how the animals felt and what they wanted. Every movement and every sound conveyed a message of well-being, distress, or disease. Lack of interest, dull eyes, sluggishness, rough coat, poor appetite, and/or abnormal feces or urine spelled trouble. Alertness, stretching on rising, yawning, vocalizing, eating with relish, and frisking were good omens and told us that all was well in the barnyard.

Artist Tom Phillips' *OLD MACDONALD'S FARM* captures the marvels and mysteries of animal behavior and brings on a bad case of nostalgia among those of us who remember *the good old days.*

Animal behavior and
environment are closely
related—so closely that they are
best discussed together.

Original painting by Tom Phillips

Chapter 17

ANIMAL BEHAVIOR/ ENVIRONMENT[1]

Contents Page

(Continued)

[1]The authors gratefully acknowledge the helpful suggestions of the following eminent scientists who reviewed this chapter: J. V. Craig, Ph.D., Department of Animal Sciences and Industry, Kansas State University, Manhattan; R. Kilgour, D.Phil., Ruakura Animal Research Centre, Ministry of Agriculture and Fisheries, Private Bag, Hamilton, New Zealand; J. J. McGlone, Ph.D., Department of Animal Science, Texas Tech University, Lubbock; and W. R. Stricklin, Ph.D., Department of Animal Sciences, The University of Maryland, College Park. Additionally, the authors are grateful to S. E. Curtis, Ph.D., Professor of Animal Sciences, University of Illinois, Urbana, for arranging for one of his former students J. J. McGlone, to serve as a reviewer. Also, the senior author obtained many helpful ideas from the authoritative books by Drs. Craig, Curtis, and Kilgour, which are listed in the *Selected References* section at the end of this chapter.

Contents (Continued) *Page*

Fig. 17-1. A serious case! Knowledge of behavioral norms is necessary in order to detect and treat abnormal situations—especially illness. (Courtesy, The Bettmann Archive)

"Wherefore, come one, O young husbandman
Learn the culture proper to each kind."

Human life, animallike and stagnant for a million years, came alive with the domestication of animals a mere 11,000 years ago. Ever since, admonitions like the preceding quote by Virgil, the great Roman poet, have characterized the relationship of caretakers with their herds and flocks. But as the art of husbandry evolved into science, caretakers' fluency with the language of their beasts declined, and human-made environments became increasingly artificial, with emphasis on economic and biological conversion mechanisms. Today, ethologists (those who study animal behavior) are assisting practical caretakers, scientifically.

Presently, there is great interest in animal behavior and environment. Those who grew up around animals and dealt with them in practical ways already have accumulated substantial workaday knowledge about their reaction to certain stimuli or their environment. But those who are less familiar with them may need to acquaint themselves with animal behavior in a people-shaped environment, better to feed and care for them, and in order to recognize the signs when all is not well.

Those of us who remember the last round of the barn at night learned the marvels and mysteries of barnyard talk. The animals kept up a running conversation with us as we gave a handful of hay to one still-hungry cow, treated a favorite horse to an apple, took a last look at a litter of newborn pigs, closed the gate on the sheep corral, and shut the poultry house door. It was a sign language, but it spoke louder than words; it told us how the animals felt and what they wanted. Every movement and every sound conveyed a message of well-being, distress, or disease. Lack of interest, dull eyes, sluggishness, rough coat, poor appetite, and/or abnormal feces or urine spelled trouble. Alertness, stretching on rising, yawning, vocalizing, eating with relish, and frisking were good omens and told us that all was well in the barnyard. Whether we come from farms and hold affectionate remembrances of our animal companions or are city-bred and know them secondhand, animal behavior is one of the marvels and mysteries of life.

Neolithic (New Stone Age) man showed some understanding of animal behavior from the remote day of their domestication, beginning about 11,000 years ago. It required knowledge of basic behavior patterns to capture, confine, and herd animals, as did the subsequent breeding, feeding, watering, and sheltering them. Without this understanding, domestication would have failed and animals might not have survived.

Caretakers were little concerned with environment so long as animals roamed over pastures and ranges. But concentration of animals in smaller spaces and concentration of people in urban areas changed all this. It produced profound changes in the environment, many of which are unfavorable to the quality of both human and animal life. Husbandry that reduces labor, land, and housing costs often results in physical and social conditions that increase behavioral problems. Some stressors acting on animals kept as sources of meat, milk, eggs, fiber, and power are unavoidable. Nevertheless, means of reducing behavioral stress may be needed so that decreased labor, land, and housing costs are not offset by losses in productivity and profits. Also, for humanitarian reasons the stress of animals serving people should be minimal. So, today, the keepers of herds and flocks are giving attention to environmental control, involving space requirements, light, air temperature, humidity, air velocity, wet bedding, ammonia buildup, dust, odors, and manure disposal.

The relationship between animal behavior and environment is evidenced by tail biting in swine, which is sometimes referred to as an anticomfort syndrome, because any part of the pig's environment that makes it uncomfortable may lead to tail biting. Among such environmental factors causing this abnormality are malnutrition; lack of space in confinement; abrasive floors and lack of bedding; improper temperature, high humidity, and poor ventilation; inadequate feed and water space; lack of uniformity within pens; infestation by parasites; or just plain boredom.

The domestication and intensive/confinement systems of animal production imply sort of a contract. Livestock producers, in fulfilling their obligation to the animals that serve them, must protect them from the elements, parasites, and diseases, and furnish them comfortable quarters and balanced rations.

ANIMAL BEHAVIOR

Animal behavior is the reaction of animals to certain stimuli, or the manner in which they react to their environment. Stated simply, *behavior is the movements animals make.* It embraces much more than locomotion; it includes the movements animals make when feeding, when breathing, and when mating. Also, it involves such things as pricking up the ears or making a sound.

The study of animal behavior began with people's first attempts to draw conclusions and make predictions from their observations of the creatures around them. However, the subject did not receive the serious attention of scientists until the middle of the 19th century, when Darwin's theory of evolution placed great emphasis on the concept of progressively improved adaptation, which is largely accomplished through its behavior.

Fig. 17-2. "Defending his harem" shows horses on an open range behaving naturally. (An original painting by artist Tom Phillips, 3333 17th St., San Francisco, CA 94110)

Most scientists agree that natural selection favors animal behaviors that increase chances of survival and reproduction, and that information gained from wild and feral relatives provides clues about the behaviors of domestic animals.

The term *ethology* is now used to describe *the study of the behavior of animals.* More recently, animal behavior studies have come to be termed *applied ethology,* which covers such things as feeding, care and management, stress, transporting, and welfare. In this chapter, emphasis will be placed on feeds and feed-related aspects, although it is recognized that there is interaction between all factors affecting animals.

Through the years, behavior has received less attention than the quantity and quality of the meat, milk, eggs, fiber, and power produced by animals. But modern breeding, feeding, and management have brought renewed interest in behavior, especially as a factor in obtaining maximum production and efficiency.

How Animals Behave

Animals behave in a great variety of ways, differing by species and environments.

Our ancestors had practical knowledge, gained firsthand or learned from others with experience, on how animals behaved. In primitive societies, knowledge of how animals behave was necessary when hunting or when capturing animals for domestication. Later, domesticated animals were maintained in an open pastoral environment, with freedom of movement, where they behaved under relatively natural conditions.

Then came the industrial age! Cheap labor and cheap land were no longer available, and marginally productive animals could no longer be afforded. Intensive production, confinement, and artificial environments evolved. Today, mother-care of young is being removed at earlier and earlier ages; and human care is being replaced by mechanical devices. Animals are segregated by age, size, and sex. This calls for a special fit between the behavior of animals and their artificial environments. Ethology, the scientific study of animal behavior, coupled with physiological evaluations of stress, will make for innovative management and housing methods in the decades to come.

BEHAVIORAL SYSTEMS

Some behavioral systems or patterns are better developed in certain species than in others. Ingestive and sexual behavior systems have been most extensively studied because of their importance commercially. Nevertheless, most animals exhibit the following nine general functions or behavioral systems:[2]

1. **Ingestive (eating and drinking) behavior.** This is characteristic of animals of all species and all ages. Without feed and water, animals cannot survive.

Fig. 17-3. Ingestive (eating) behavior begins with nursing. (Courtesy, J. V. Craig, Department of Animal Sciences and Industry, Kansas State University, Manhattan)

2. **Allelomimetic behavior.** This is mutual mimicking behavior. Thus, when one member of a group does something, another tends to do the same thing; and because others are doing it, the original individual continues.

Fig. 17-4. Allelomimetic (mimicking) behavior exhibited by chickens. Feed consumption is facilitated by the presence of feeding companions. (Courtesy, Ralston Purina Co., St. Louis, Mo.)

[2]Adapted by the authors from: Scott, J. P., *Animal Behavior,* 2nd ed., The University of Chicago Press, Chicago, Ill., 1972.

3. **Eliminative behavior (defecation and urination).** Nature ordained that if animals eat, they must eliminate wastes—they produce manure (feces and urine). Properly handled, manure is an asset; improperly handled, it may be a pollution problem.

Fig. 17-5. Properly handled, manure can contribute to a sustainable agriculture. This shows manure being spread on corn stubble. (Courtesy, Deere & Company, Moline, Ill.)

4. **Gregarious behavior—this refers to the flocking, or herding, of certain species.** It is closely related to allelomimetic behavior. If animals imitate each other, they must stay together. If they stay together as a mobile unit, they use either allelomimetic behavior or leader-follower relationship to do so.

Fig. 17-6. Gregarious (flocking) behavior exhibited by a range band of sheep. (Courtesy, USDA)

5. **Sexual behavior.** This involves courtship and mating. It is largely controlled by hormones. Each animal species has a special pattern of sexual behavior.

Fig. 17-7. "Courtin' Time in Texas." Sexual behavior. (An original painting by artist Tom Phillips, 3333 17th St., San Francisco, CA 94110)

6. **Care-giving and care-seeking (mother-young) behavior.** The care-giving behavior among domestic animals is largely confined to females, where it is usually described as *maternal.* The care-seeking behavior, which extends until the young are weaned, is normal for young animals. Expressions of care-giving and care-seeking vary widely among different species of farm animals.

Fig. 17-8. "Madonna." Care-giving and care-seeking (mother-young) behavior. (An original painting by artist Tom Phillips, 3333 17th St., San Francisco, CA 94110)

7. **Agonistic behavior (combat).** This type of behavior includes fighting, flight, submission, and other related reactions associated with conflict. Among all species of farm mammals, males are more likely to fight than females. Nevertheless, females may exhibit fighting behavior under certain conditions. Castrated males are usually quite passive, which indicates that hormones (especially testosterone), are involved in this type of behavior.

Fig. 17-9. Agonistic (combat) behavior exhibited by two pigs. (Courtesy, J. V. Craig, Department of Animal Sciences and Industry, Kansas State University, Manhattan)

8. **Investigative behavior.** All animals are curious and have a tendency to explore their environment. Investigation takes place through seeing, hearing, smelling, tasting, and touching. Whenever an animal is introduced into a new area, its first reaction is to explore it.

Fig. 17-10. Investigative behavior evidenced by a curious pig. (Courtesy, USDA)

9. **Shelter-seeking behavior.** This is the type of behavior which causes all species of animals to seek shelter—protection from the sun, wind, rain and snow, insects, and predators.

Fig. 17-11. Shelter-seeking behavior, showing cattle in a ravine, seeking protection from the storm. Note that they are grouped together and facing away from the storm. (Courtesy, American Hereford Assn., Kansas City, Mo.)

Of the nine behavioral systems, the following four involve feeding, either directly or indirectly; hence, they are crucial to animal production and are detailed by species in Table 17-1. (See pages 600-601.)

1. Ingestive behavior (eating and drinking).
2. Eliminative behavior (defecation and urination).
3. Allelomimetic behavior (mimicking).
4. Gregarious behavior (flocking).

TABLE
HOW ANIMALS

Behavior	Cattle (Beef and Dairy)	Sheep
Ingestive behavior (eating and drinking):		
This type of behavior includes eating and drinking; hence, it is characteristic of animals of all species and all ages. It is very important because animals cannot live without feed and water. *Rumination is the act of chewing the cud, characteristic of herbivorous animals with split hoofs—cattle, sheep, and goats.* It involves regurgitation of ingesta from the reticulo-rumen, swallowing of regurgitated liquids, remastication of the solids accompanied by reinsalivation, and reswallowing of the bolus. The first ingestive behavior trait, common to all young mammals, is suckling. Each species has its own particular method of ingesting feed.	The natural feeding (grazing) position of cattle is heads down. In this position, they produce more saliva; and saliva aids digestion. When grazing, cattle wrap their tongues around grass, then jerk their heads forward so that the vegetation is cut off by the lower incisor teeth. (There are no upper incisor teeth; only the thick hard dental pad.) When grazing, cattle also move their heads from side to side. This movement gives them a continuous view of their entire surroundings, an essential for wild cattle in an environment containing dangerous predators. It is important that artificial feeding devices and arrangements not depart too far from this natural pattern. Rumination occupies about 8 hours of the cow's time each day. In addition, the harvesting or grazing time may take another 8 hours. This means that a cow may be active for a 16-hour day. The Iowa Station reported that steers in lots on self-feeders spent 12 hours per day lying down, and that this time was unaffected by shelter or season. When the cow regurgitates, a soft mass of coarse feed particles, called a bolus, passes from the rumen through the esophagus in a fraction of a second. She chews each bolus for about 1 minute, then swallows the entire mass again. Originally, it was thought that the regrinding which occurred during rechewing helped the digestion by exposing a greater surface area to fiber-digesting microflora. But recent studies indicate that rechewing does not improve digestibility. Instead, rumination has an important effect on the amount of feed the animal can utilize. Feed particle size must be reduced to allow passage of the material from the rumen. It follows that high-quality forages require much less rechewing and pass out of the rumen at a faster rate; hence, they allow a cow to eat more. This concept is very important to the production of beef and milk because a cow will eat only as much coarse material as she can grind up by ruminating.	Sheep graze very much like cattle, but their cleft upper lip allows them to graze vegetation closer to the ground. As in cattle, the incisors are in the lower jaw only. The total grazing time of sheep ranges from 10 to 12 hours per day, which is normally divided into two periods and highly correlated with sunrise and sunset. Sheep are very selective grazers; they will consistently select plants higher in protein and lower in crude fiber than can be obtained by harvesting the forage. In general, the selectivity is proportional to the amount of herbage available; thus, when the feed supply is short, sheep become less discriminating. Sheep also form preferences for plant species based on previous experience. Sheep are one of the most drought-resistant of domestic animals. In some cases, they graze without access to water, relying on dew, snow, and/or moisture in plants. Goats graze in the same manner as cattle and sheep, but they are very fond of browse—the young shoots of shrubs and trees.
Eliminative behavior (defecation and urination):		
In recent years, elimination has become a most important phenomenon, and pollution has become a dirty word. Nevertheless, nature ordained that if animals eat, they must eliminate.	Cattle deposit their feces in a random fashion. Although cows can defecate while walking, with the result that their feces are scattered, generally they deposit their "chips" in neat piles. Most cows elevate their tails and hump up to urinate, whereas bulls are inclined to stand squarely on all "fours."	The eliminative behavior in sheep is very similar to that of cattle. However, they differ from cattle in that ewes usually assume a squat position when they urinate, and their feces are relatively dry and pelleted.
A full understanding of the eliminative behavior will make for improved animal building design and give a big assist in handling manure. Right off, it should be recognized that the eliminative behavior in farm animals tends to follow the general pattern of their wild ancestors; but it can be influenced by the method of management.		
Allelomimetic behavior (mimicking):		
Allelomimetic behavior is mutual mimicking behavior. Thus, when one animal in a group commences to eat, this triggers a response for others to do likewise. In the wild state, this trait was advantageous in detecting the enemy, and in providing protection therefrom. In wolves and coyotes, this behavior is important in attacks on prey, since a pack working together is much more likely to be successful than when working alone.	Cows moving across a pasture toward a milking barn often display allelomimetic behavior. One cow starts toward the barn, and the others follow. Because the rest of the herd is following, the first cow proceeds on. Because of stimulating and competing with each other, there is usually higher per steer feed consumption among a group of steers than by one steer alone. Thus, one steer penned alone may consume "X" pounds of feed per day. However, when he is placed with other steers, his intake may be "X + Y" pounds. But, of course, the feed consumption advantage can be nullified when animals are placed together too closely, with the result that the agonistic behavior comes into play.	Sheep walk, run, graze, and bed down together. Sheep graze when they observe others in adjacent paddocks doing so.
Gregarious behavior (flocking):		
Gregarious behavior refers to the flocking, or herding, of certain species. It is closely related to allelomimetic behavior. If animals imitate each other, they must stay together. If they stay together as a mobile group, they must use allelomimetic behavior to do so. All such behavior arises out of the process of social attachment. Gregarious behavior differs among species.	Cattle tend to roam in groups of various sizes when a large herd is placed on a pasture or range. However, there is usually considerable space between the members of the herd. Moreover, on close observation it is evident that there are several small groups within a herd, each ranging from 3–5 head.	The gregarious, or flocking, behavior is particularly strong in sheep. Moreover, it is more evident in some breeds than in others. The Merino, and animals carrying Merino breeding, are noted for their flocking behavior. This makes it possible to herd them on the range.
	It is noteworthy that the gregarious instinct of sheep diminishes to some extent when they are placed within fenced holdings, instead of herded. As a result, those who handle western range bands do not try to switch back and forth from fenced range to herding, for the reason that the band becomes unmanageable from the standpoint of herding once they have been in a fenced holding for an extended period of time. Packers use the gregarious behavior of sheep by having an old goat, appropriately called a *Judas,* lead sheep to slaughter. A well-trained Judas will lead group after group to slaughter.	

17–1
BEHAVE

Swine	Horses	Poultry
Swine possess incisor teeth in both the upper and lower jaws; hence, they bite off grass or take a mouthful of grain, then chew and swallow it. **P**igs have a single stomach, whereas ruminants have a 4-compartment stomach. **B**y nature, pigs love to root. If given the opportunity, they will stick their noses into the ground and lift forward and upward, moving earth out of the way and exposing earthworms, grubs, and roots.	**T**he mobile upper lip of the horse is used in gathering grass and other feed, in the same way that a cow uses her tongue or an elephant uses his trunk. The upper lip is very sensitive. **T**he long, strong, and roughened teeth of the horse are well suited to grinding common feeds. **A** foal may have difficulty in getting its head to the ground for grazing, because, in the evolutionary process, the proportionate elongation of the legs and head was not always perfect. As a result, a long-legged foal with a short head and neck may have to stand with the front legs spread apart in order to reach the ground. **W**hen snow covers the pasture or range, horses will paw with a front foot (either right or left) through the snow and clear an area so that they can reach the grass. **T**he horse has a blind spot in front of its nose. Hence, it cannot see the feed as it is eaten. **H**orses rarely browse; they will not eat the leaves of trees or shrubs provided grass is available. **H**orses prefer grazing in an open area, where they can watch for their enemies. Also, they prefer young, tender grass to coarse-stemmed plants, a preference which often causes them to overgraze certain areas. **T**he horse evolved to graze small amounts almost continuously, rather than large amounts infrequently. This explains why horses do better if fed small amounts—often. **I**n areas where water is scarce, wild horses usually graze 6–8 miles from water and come to water every other night. **A** thirsty horse may lower its head in the water deep enough to cover the nostrils.	**C**hickens and turkeys ingest their feed by pecking; ducks scoop their feed with their broad, soft bills. Except for geese (which graze on grass), poultry do not eat much forage. **C**hicks do not peck much until their second day after hatching, presumably due to ingestion of the yolk sac during hatching. Normal pecking experience requires some light. **I**nitially, chicks peck and ingest both nutritive and non-nutritive substances. When foraging for food, a mother hen will cluck to her chicks and call them each time a choice morsel is found. All the chicks wil come running to participate in the find.
If given an opportunity, pigs are of very clean habits. They like to keep their nesting area clean and dry. Hence, they usually deposit their feces and urine away from the feeding and resting areas. By proper building design, it is possible to facilitate feeding, resting, and defecating in separate areas.	**H**orses tend to deposit their feces in certain locations, such as along well-traveled paths, like those leading to waterholes. Hence, if given the opportunity, they often return to these locations for defecation. **S**tallions are much more prone to deposit droppings on the same old mound than mares or geldings. Mares and geldings are inclined to use the border of their defecating area, with the result that they enlarge it each time. **T**here is disagreement whether a feral or range stallion marks the outside boundary of his home range with his feces, thereby staking it out for himself and his harem of mares, with most horse owners believing and most ethologists disbelieving.	**E**xcept when on the roosts at night, chickens and turkeys deposit their excreta at random.
Swine exhibit allelomimetic behavior in their eating habits. Thus, when one pig eats, there is a tendency for the rest to join it. As a result, pigs in a group usually average higher feed consumption than one pig alone.	**A** timid horse will follow behind the pack, in order not to be left behind. **W**hen kept alone, high-strung racehorses may become nervous and fail to eat properly. To alleviate this situation, a companion—known as a mascot, is often provided. All sorts of mascots are used—a goat, a sheep, a chicken, a duck, or a pony.	**P**ecking and feeding are facilitated by social stimulation; *e.g.,* the presence of a feeding companion. **S**ocial facilitation of feeding also occurs in adult chickens. For example, when a solitary hen eats from a large portion of feed until satisfied, she will immediately resume eating if a second hungry hen is introduced and starts consuming some of the feed.
In the wild state, swine roved through the forest in herds. Usually these wild groups consisted of 1 to 4 females, along with their young of the year and their yearlings. Adult males join these groups during the mating season, but range separately during the rest of the year. **U**nder domestication, swine retain their gregarious nature. However, it has been altered a great deal. Today, hogs are usually confined to a very limited area. Also, under domestication, they have lost most of their ferocity and are usually gentle and easily handled.	**I**n the wild state, horses ran in bands; thus, they were gregarious by nature. These bands seldom consisted of more than 40 animals, and always there was a stallion in each group. **U**nder domestication, horses show definite preferences for their herdmates; they will even avoid certain horses in the herd. In the draft horse era, animals that were worked together usually stayed together when they were turned to pasture.	**C**hickens, turkeys, ducks, and geese tend to flock together. Of course, under domestication, humans have interfered with normal flocking of both chickens and turkeys. Even under domestication, however, ducks and geese exhibit their gregarious or flocking nature as they walk Indian file, one behind the other.

Goats. The biggest difference between sheep and goats in feeding behavior is the greater amount of browsing done by goats. Goats have unique preferences for shrubs and tree leaves. But, for maximum performance it is necessary to provide supplemental feed when pastures are poor. Moreover, lactating dairy does should be fed high quality rations which are very similar to lactating dairy cow rations.

• **Control of feed intake**—Sheep adapt their feed intake to environmental temperature, as if appetite is a thermoregulatory mechanism. Lambs grow most rapidly when the temperature is 41° to 77°F. Also, high roughage rations have been found to be beneficial to lambs when the environmental temperature is below 41°F, but detrimental when it is above 77°F.

Goats. There is lack of experimental work on the control of feed intake of goats. Without doubt, the same factors that control the feed intake of sheep and cattle govern the feed intake of goats.

• **Drinking (watering)**—Sheep get water by drinking, and from snow, dew, and feed. The amount of water they consume voluntarily is affected by temperature, rainfall, covering of snow and dew, age, breed, stage of production, number of lambs carried or level of milk production, wool covering, rate of respiration, frequency of watering, kind and amount of feed, and exercise. On the average, mature animals consume approximately 1 gal of water per day, while feeder lambs require about half this amount. However, when foraging on grasses and other feeds of high-moisture content, sheep may go for weeks without drinking water. This condition often prevails on desert ranges in the early spring and on many of the mountain ranges during the summer months.

Goats. Goats are among the most efficient domestic animals in the use of water, approaching the camel in the low rate of water turnover per unit of body weight. Nevertheless, water is important for goats, and the amount required depends on how much is needed for the maintenance of normal water balance and for levels of production. Thus, the daily range of free water intake may be from zero to several quarts. A safe general recommendation for goats, as for other animals, is to provide them with all the clean water that they will drink.

• **Fostering (grafting/adopting)**—Very frequently, there is need to have ewes foster (adopt) one or more lambs as a result of (1) a ewe dying at lambing, (2) a ewe not having milk, (3) a ewe disowning a lamb, (4) a ewe having too many lambs and not enough teats, or (5) an old ewe having twins.

Fostering alien lambs on ewes is not too difficult provided it is done soon after birth (preferably within the first 8 hours), it is accompanied by a deception in the sense of smell, and it is done in close confinement. Shepherds commonly use one of the following deception-of-smell approaches: (1) covering the lamb with the skin (pelt) of the lamb that died at birth, (2) smearing mucus from the fetus or lost lamb on the alien lamb and on the nose of the foster mother, or (3) milking some of the foster mother's milk on the rump of the alien lamb and smearing some of it on the ewe's nose. If these methods fail and the ewe persists in fighting the grafted lamb away, blindfold her so that she cannot see the lamb or give her a tranquilizer. As a last resort, and when all other methods have failed, tie a dog in the adjoining pen; sometimes, this will cause latent maternal instincts to rise to a surprising degree.

Goats. The reasons for fostering kids, and the technique used in accomplishing it, are similar to what has been detailed for fostering lambs.

• **Cross fostering goats and sheep**—Fostering lambs on does and fostering kids on ewes is a well known practice. It will work, but behavioral differences of the species are evidenced. When a doe fosters a lamb along with her kid, the lamb trails her constantly and nurses frequently; as a result, the lamb soon outgrows the kid co-twin that spends a lot of time lying out. When a ewe fosters a kid, the ewe keeps losing the kid because it lies out, rather than follow her; as a result, the ewe becomes frustrated and stressed.

• **Handling**—Sheep are easy to handle in groups. The handling facilities should be designed so that sheep can see other sheep at all times, and so that there will be a good flow of animals. A cutting chute, or dodge gate, should be used for sorting sheep. By flaring the top out, both ewes and lambs can be sorted in the same chute. The dodge gate should swing from one side of the chute to the other to permit cutting into separate pens. In addition to making for greater convenience and reducing labor, a cutting chute saves much disturbance and running of animals when working over the flock or band.

Goats. Goats may be handled in pens, yards, and chutes similar in arrangement and design to the handling facilities used for sheep, except fences and chutes must be much higher; otherwise, goats will jump over them. Fences and chutes should be 54 in. high for handling does and 60 to 66 in. high for handling bucks.

• **Preparing and shipping sheep and goats to market**—The feed-related considerations for preparing and shipping cattle to market also apply to the marketing of sheep and goats. So, see the earlier section on "Beef and Dairy Cattle Behavior, *Preparing for and shipping to market.*"

SWINE BEHAVIOR

A knowledge of swine behavior is necessary in order to feed, care for, and manage them successfully. Pertinent feed-related behavioral characteristics of swine follow:

• **Grazing**—Feral pigs forage. They eat a wide variety of plants, roots, tubers, grasses, seeds, buds, leaves, insects, frogs, snakes, eggs, birds, and rodents. Their daily activity is governed by light and temperature. They arise from their sleeping area and start foraging at daybreak, and the warmer the environment the earlier. Toward midday, they move to water, where they drink and wallow according to the environmental temperature. In mid-aftenoon, they resume foraging, moving to the furthest point on their range. In the late afternoon, they return to their sleeping area where they settle for the night.

Domestic pigs in many parts of the world still forage for much of their subsistence. However, in the United States, many pigs are raised from birth to market in total confinement; nevertheless, some producers use pastures advantageously for breeding animals—for gestating sows and herd boars.

Domestic pigs on pasture forage very much like their feral counterparts; they graze early and late in the day, and they rest in midday. However, the amount of time spent in actual grazing depends directly on the amount of feed provided by the caretaker. When provided little supplemental feed on pasture, domestic swine will graze 8 to 9 hours per day. But when they are provided most of their feed needs, they spend 19 to 20 hours of each day resting and sleeping, and they forage, without working very hard at it, for only 4 to 5 hours daily.

• **Feeding behavior**—The feeding behavior of pigs is characterized by social facilitation; the sound of a self-feeder lid or of another pig eating corn, causes other pigs to begin eating, also.

Pigs are adapted to any of the following feeding systems: self-feeding, free-choice feeding; on-floor feeding; liquid feeding; and limited feeding. When self-fed grain and protein supplements in separate compartments, it is important that the feeds be of equal palatability; otherwise, pigs will consume too much of one of the feeds and too little of the other.

• **Control of feed intake**—The amount of feed eaten is affected by palatability, competition, bulk, interval of feeding, and temperature.

Feed consumption is affected by several palatability factors, including appearance, odor, taste, texture, temperature, and in some cases by the auditory properties of the feed (like the sound of pigs eating corn).

Competition—group size and interaction—affects feed consumption. However, the physical presence of another pig within the same pen is not required so long as other pigs are nearby. Danish swine testing station results indicate that gains in weight are essentially the same for an individually fed pig in close proximity to other pigs and for group-fed pigs. Energy consumption can be lessened by feeding considerable bulky, fibrous feeds such as silage, haylage, or alfalfa. The amount of feed consumed may be controlled by interval feeding: (1) by varying the interval from every other day to twice a week, (2) by varying the length of time that the animals are left on self-feeders (from 2 to 8 hours), or (3) by hand-feeding.

• **Drinking**—Hand-fed pigs generally eat all their feed, then drink water afterwards. Self-fed pigs alternate between eating and drinking. Pigs learn to manipulate various kinds of mechanical self-watering devices. The watering device must be durable, as pigs will rub, nuzzle, and chew on it. Proper placement is also important; it should be located near the dunging area, and it should not be a source of bruises. Nipples should be located so that pigs have to reach up when drinking.

• **Nursing**—Most pigs stand within two minutes after birth and start sucking within 3 to 150 minutes after birth. Sows appear to identify their own piglets primarily by the sense of smell, although they seem to form bonds slowly.

During the first day of life, baby pigs generally show a strong preference for the front teats. Soon, this preferred position becomes established, resulting in what is known as the "teat order." The front teats are longer, spaced further apart, and have more milk than the back teats. So, the front-teat sucking pigs grow faster and usually maintain their social ranking in the litter. Nursing is frequent (every 50 to 60 minutes), with each nursing lasting 2 to 6 minutes.

• **Fostering**—Pigs may be moved from sow to sow during the first two days after farrowing without being rejected. So, compared to other species, fostering of pigs is relatively easy. Thus, when a number of sows have farrowed about the same time, litter size can be adjusted by taking pigs from large litters and fostering them on sows with small litters and adequate milk. Sows are less tolerant of alien pigs that are more than two days old. However, they are more receptive to adopting older foreign pigs if a large number are introduced simultaneously, especially if the environment is changed at the same time.

Where sows farrow about the same time and are on pasture, litters may mix and cross-suckle from birth. There are few acceptance problems, but the caretaker should make sure that the pigs are properly distributed among sows; otherwise, there will be wide variation in weaning weights.

• **Handling**—Handling methods can affect growth and reproductive performance of hogs more than housing or confinement, according to a study conducted by Paul Hemsworth, *et al.*, at the Animal Research Institute, Victoria, Australia. The Australian study was designed to measure the pig's response to humans, with a comparison of (1) gentle (positive) and (2) rough (negative) treatments. The gentle treatment involved either patting or stroking the pigs. The rough treatments consisted of either slapping or briefly shocking the hogs with a battery-operated prod.[3]

The results:

1. Young pigs handled gently for 10 weeks (from 8 to 18 weeks of age) gained over 9 lb more than pigs handled roughly.
2. Gilts handled gently had an 88% conception rate compared to 33% for those handled roughly.
3. Boars handled gently had a fully coordinated mating response 32 days sooner than those handled roughly.
4. Sows handled gently weaned 4 more pigs per sow than sows handled roughly.

Hemsworth's admonition: Select caretakers who like pigs and get rid of those who do not.

• **Porcine stress syndrome (PSS)**—Some pigs stressed by handling, crowded transportation, or sudden environmental change react adversely; are usually associated with low-quality, or pale, soft, and exudative (PSE) pork; and may even succumb. These conditions commonly occur in rapidly growing, heavily muscled pigs. In the 1960s, Dr. David Topel, Iowa State University, coined the term *porcine stress syndrome* (PSS) to identify it.

The Livestock Conservation Institute is authority for the statement that the U.S. pork industry loses $38 to $39 million annually from PSS; $32 million from pale, soft, exudative (PSE) pork, and $6 to $7 million from deaths.

Authorities estimate that up to 8 to 10% of U.S. hogs have severe PSE. PSE pork is caused by an interaction of many factors, the most important of which are: genetic, breed,

[3]Hemsworth, Paul, seminar presentation at Winnipeg, Manitoba, Canada, as reported by Dave Wilkins, *Hog Guide*, May, 1988.

strain within a breed, weather fluctuations, fighting, transportation, time off feed, lack of rest prior to stunning, and handling in the stunning chute. PSE pork is characterized by a rapid breakdown of glycogen which produces lactic acid and lowers muscle pH. Glycogen is rapidly broken down when hogs become excited, with some genetic lines more excitable and susceptible than others.

To reduce PSS losses, as well as other shipping losses, the following cautions should be observed when preparing and shipping hogs to market:

1. Do not feed for a minimum of 8 to 12 hours before shipment.
2. Handle them gently on the farm.
3. Eliminate the use of electric prods and clubs.
4. Do not mix different pens of hogs when trucking.
5. Keep hogs cool by sprinkling when it is hot.
6. Truck hogs properly: Do not overload; haul at night or in the early morning when it is cool; bed with wet sand or shavings when it is hot, and with straw when it is cold; and load and unload trucks promptly.
7. Use packing plant chutes and crowd pens that are designed for gentle handling.
8. Rest hogs 1 to 4 hours at the packing plant prior to slaughter.
9. Reduce stunning and bleeding time, and the interval between stunning and bleeding.

HORSE BEHAVIOR

Fig. 17-17. Even when grazing, the eyes of a horse allow it to see what's going on all around it—to the front, the side, and the back—all at the same time. By turning very slightly to the left, the horse shown above was able to keep an eye on the photographer who was standing behind it. (Courtesy, Wild Horse Research Farm, Porterville, Calif.)

A knowledge of horse behavior is necessary in order to feed, care for, and manage them successfully. Pertinent feed-related behavioral characteristics of horses follow.

• **Grazing**—When grasses are available, horses prefer to feed on them. They manipulate the plants with their mobile and very sensitive upper lip, nip the blades off with their incisors, and use their tongue to ingest the material. When grazing, foals spread their forelegs in order to compensate for their long legs and short necks.

The amount of time spent in grazing depends on the quality and quantity of forage available, along with such factors as temperature, precipitation, and pests. But grazing time averages about 12 hours per day, although horses may graze up to 18 hours per day when feed is scarce. Horses are selective grazers and will eat a wide range of pasture plants, herbs, shrubs, and weeds.

Horses are notorious for overgrazing certain areas and for defecating and urinating in other areas. Moreover, defecating and urinating rituals are an important means of communicating on the range. Stallions routinely defecate on stud piles, but they are more likely to urinate on the excrement of mares. The defecating behavior of horses probably evolved for two reasons:

1. As a means of stallions marking their area or territory. Apparently, such markings serve to warn rival stallions that they are encroaching on the territory of another stallion.
2. To provide some protection from infestation by internal parasites. Because many internal parasites are spread from one horse to another by grazing pastures contaminated with parasite eggs from the droppings of infested horses, it is conjectured that wild horses may have reduced the opportunity for the spread of internal parasites by defecating in certain areas and grazing in others.

Fig. 17-18. "Marking his range," painting by artist Tom Phillips. (Courtesy, Roscoe E. Dean, M.D., Washington Springs, S.D.)

• **Feeding behavior**—Stall-fed horses learn to anticipate their feed; so, they should be fed regularly and at the same time each day. Usually the concentrate ration is fed first. Little time is spent eating concentrate feeds, but they need time in which to eat bulky roughages more leisurely.

If given the opportunity, horses prefer to eat small amounts of feed throughout the day. When kept on pasture, they are able to eat in this manner. When kept in stalls and small paddocks, however, they must eat at the convenience of the caretaker. If fed concentrates *ad libitum,* they usually overeat and get digestive disturbances.

Horses practice coprophagy—the eating of feces. Young foals will eat the feces of mares. Adult horses will eat their own feces or the feces of other horses.

• **Control of feed intake**—When given all the feed that they will consume, mature horses will generally eat an amount equivalent to about 2.5% of their body weight daily. Growing foals and lactating mares eat more heartily—they will consume up to 3% of their body weight.

The horse is not adapted to self-feeding of high-energy rations. Overeating may result in two consequences: If done suddenly, it may cause founder (laminitis); if prolonged, it will likely result in obesity.

Because the horse has rather limited digestive capacity, the amount of concentrate must be increased and the roughage decreased when the energy needs rise with the greater amount, severity, and speed of work. Additional factors affecting the quantity of grain and hay required by horses are: (1) the individuality and temperament of the horse; (2) the age, size, and condition of the animal; and (3) the weather. The total allowance of concentrate and hay should be within the range of 2.0 to 2.5 lb daily per 100 lb liveweight. No grain should be left from one feeding to the next, and all edible forage should be cleaned up at the end of the day.

• **Drinking**—Przewalski Horses, the wild species discovered by the Russian explorer in Mongolia, sometimes went for 2 to 3 days without water. Feral horses walk to a water hole once each, day, and on days of extreme heat they usually return a second time.

Domestic horses drink 10 to 12 gal of water daily, the amount varying according to weather, amount of work and sweating, ration fed, and size of horse.

Free access to water is desirable. When this is not possible, horses should be watered at approximately the same times daily. Opinions vary among caretakers as to the proper times and methods of watering horses. However, most caretakers agree that water may be given before, during, and after feeding.

Frequent, small waterings between feedings are desirable during warm weather or when the animal is being put to hard use. Horses should not be allowed to drink heavily when they are hot, because they may founder; and they should not be allowed to drink heavily just before being put to work.

Automatic waterers are the modern way to provide water for stall-fed horses at all times, as nature intended.

• **Nursing**—Care-giving behavior toward the foal is displayed by the mare from the time of birth; it seems to be triggered by the taste and smell of the allantoic fluid.

The mare generally licks the newborn foal, starting at the head. A strong, healthy foal will usually be up and ready to nurse within 30 minutes to 2 hours after birth. Some mares facilitate the first nursing by maintaining a position which makes it easy for the foal to locate the teats and suckle. Others resent the foal's activity around their sensitive udders. Occasionally, an awkward foal needs the assistance and guidance of the caretaker during its first time to nurse; such assistance should consist of coaxing, forcing is useless. The very weak foal should be given the mare's first milk even if it must be drawn in a bottle and fed by nipple for a time or two. The colostrum contains antibodies that temporarily protect the foal against certain infections, especially those of the digestive tract. Because of this, it is important that foals nurse within 10 to 12 hours after birth. Also, colostrum serves as a natural purgative, removing fecal matter (meconium) that has accumulated in the digestive tract. Suckling is frequent early in life; studies have recorded 3 to 4 feedings per hour during the first week of life. But frequency of nursing gradually decreases until it occurs about once every 2 hours at weaning time. A mare is very protective of her foal; she dissuades it from other herd members, she threatens intruders, and she shelters it by keeping it alongside, when standing and moving. Foals are usually weaned at about 6 months of age. Natural weaning takes place at about 9 months of age.

• **Fostering (adopting)**—Occasionally a mare dies during or immediately after parturition, leaving an orphan foal to be raised. Also, there are times when a mare rejects her foal or fails to give sufficient milk to sustain it. Sometimes there are twins. In such cases, it may be necessary to resort to other milk supplies. The problem will be simplified if the foal has received colostrum from its mother.

If at all possible, the foal should be fostered on another mare. Some horse breeding establishments follow the practice of breeding a mare that is a good milk producer, but whose foal is expected to be of little value. Her own foal is either destroyed or raised on a bottle, and the mare is used as a foster mother, or nurse mare, for a more valuable foal. Generally foals can be fostered on other mares for up to 2 to 4 days after birth. If there is resistance, a *foster gate* may be used to protect the orphan foal. (See Chapter 25, Fig. 25–23 for a picture of a foster gate in use.)

• **Handling**—A round pen is a valuable handling facility. It is the best place in which to conduct training lessons: restraint, sacking out, longeing, ground driving, saddling, ponying, and first rides. The size and construction of the round pen depends somewhat on its intended uses. With planning, a design can suit a variety of needs. A *breaking pen* that is built solely for initial gentling and first rides should be about 35 ft in diameter with solid 7 ft walls to assure maximum control in unpredictable training situations. A *training pen* that is built for routine longeing, driving, or riding should be 66 ft in diameter to allow the horses enough space for balanced movement.

Horses are big, strong, and fleet; hence, it is best that they want to do something, rather than be forced. Also, frequent and improper use of such artificial aids as whips, spurs, reins, and bits makes them less effective; worse yet, it will make for a mean horse. So, proper and modern handling calls for knowledge of horse behavior and the judicious use of rewards (a praising voice, a gentle stroking with the hand, and/or a lump of sugar or a carrot) and punishment (spur or whip).

Handling should begin soon after birth, with the caretaker speaking to the foal softly and gently caressing it about the neck, as a result of which the foal will habituate better and be more self-confident when older. Subsequent training should progress to more difficult tasks. The foal should become accustomed to the halter, to being tied, and to being led. The yearling should be given the following intermediate training: to respond to "whoa" and its name; to stand when hobbled; to become accustomed to the saddle blanket; to get used to the saddle; to drive, turn, stop, and back up; to flex its neck and set its head; to respond to the bosal; and to being ridden, and to leg pressure. The 2-year-old should be given the following advanced training: to respond to the aids; to pivot (if a western horse); and to make a sliding stop (if a western horse).

The good caretaker who has followed a program of training and educating the foal from soon after birth has already eliminated the word *breaking.* To such a trainer, the saddling and/or harnessing of the young horse is merely another step in the training program, which is carried out with ease and satisfaction.

When handling horses, always try kindness first; pat the animal and speak to it reassuringly. If this fails, and the horse is uncontrollable, try one of the following techniques as a last resort:

1. If the horse is unruly, especially if it is tossing its head about, apply a twitch to the upper lip.
2. If the horse is very excited and is about to break away, either (a) blindfold it, or (b) dash a bucket of water in its face. Usually, the animal will calm down following such treatment.
3. If the horse will not move or is kicking, grab its tail and push it over its back. In this position, the horse cannot kick, but can be pushed along.

• **Transporting horses**— Horses are transported via trailer, van, truck, rail, boat, and plane. Regardless of the method of movement, for long hauls the following feed-related considerations should be accorded when preparing and transporting horses:

1. **Feed lightly.** Allow horses only a half feed of grain before they are loaded for shipment and at the first feed after they reach their destination. In transit, horses should be fed alfalfa or other good quality hay to keep the bowels open, but no concentrates should be fed. When in transit, commercial hay nets or homemade burlap containers may be used to hold the hay, but they should not be placed too high.
2. **Water liberally.** When transporting horses, give them all the clean, fresh water they will drink at frequent intervals unless the weather is extremely hot and there is danger of gorging. A tiny bit of molasses may be added to each pail of water, beginning about a week before the horses are shipped, and the addition of molasses to the water may be continued in transit. This avoids any taste change in the water.
3. **Make nurse stops.** Nurse stops should be made at about 3–hour intervals when mares and foals are transported together.
4. **Be calm when loading and unloading.** When loading and unloading horses, always be patient and never show anger. Try kindness first. If it fails, it may be necessary to resort to twitching, blindfolding, dashing a bucket of water in the face, or pushing the horse's tail over its back.
5. **Drive carefully.** Drive at a moderate, constant speed as distinguished from fast or jerky driving, which causes added stress and tiring.

POULTRY BEHAVIOR

A knowledge of behavior and environment is necessary in order to feed, care for, and manage poultry successfully. Pertinent feed-related behavioral characteristics of poultry follow:

• **Feeding and drinking behavior (ingestive behavior)**— This type of behavior includes eating and drinking. Each species of poultry has its own particular method of ingesting feed. Chickens and turkeys ingest their feed by pecking; ducks scoop their feed with their broad, soft bills. Except for geese, which graze grass, poultry do not eat much forage.

Chickens. Chicks do not peck much until their second day after hatching, presumably due to ingesting the yolk sac during hatching. Normal pecking experience requires some light. Initially, chicks peck and ingest both nutritive and non-nutritive substances.

Chickens prefer crumbles; scratch vigorously when grain is scattered in the litter; require increased feeding space with growth; peck and feed more in the presence of a feeding companion; and peck more when there are sounds associated with feeding, such as made by finger taps or a pecking model. After feeding, a dominant bird may return when its inferiors begin active feeding, thereby increasing its consumption and reducing the feed consumption of those in the lowest rank.

The number of birds feeding at any given time is influenced by dominance relations, hunger, and feeding space.

Like other animals, chickens are able to survive much longer without feed than without water; the loss of 10% of body water will cause serious disorders, and the loss of 20% will cause death. Poultry should have free access to clean, fresh water at all times. A general rule is that chickens will drink approximately twice as much water by weight as the feed they consume. The amount of water birds will consume is affected by (1) age, (2) body weight, (3) production, (4) weather (temperature and humidity), and (5) type of ration. Also, the intensity of production dramatically affects the water requirement. The water intake of chicks can be increased slightly by adding blue food coloring to the water supply.

Turkeys. Turkeys swallow feed without the necessity of raising the head. But drinking is accomplished differently; after getting a beak full of water, the bird raises its head, extends the beak upward and repeats several rapid closures of the beak.

Newly hatched poults peck indiscriminately at bright spots and small objects that contrast with the background. Thus, they may be encouraged to consume feed and water by placing brightly colored marbles in the feed and waterers. In a few days, this indiscriminate eating changes to selective eating and drinking.

Low-fiber, high-energy pelleted rations make for maximum feed efficiency and growth of turkeys. But, because such rations are consumed rapidly and the birds have more idle time, they predispose the flock to feather picking and cannibalism.

Ducks. Ducks exhibit their gregarious or flocking nature, and they follow each other.

Fig. 17-19. In addition to being gregarious, ducks walk Indian file, one behind the other. (Courtesy, California Polytechnic University, San Luis Obispo)

Mallard ducks employ several feeding methods, including (1) dabbling (as the birds swim or walk through mud puddles or along shorelines) of the bill along the water surface or in mud which achieves a straining of planktonic organisms and mud-dwelling invertebrates; (2) feeding from the bottom of ponds; and (3) snapping flying insects by rapid closing of the bill.

Drinking by ducks involves a distinctive movement sequence; the bill is dipped in water and then lifted slightly above the horizontal.

Quail. The ingestive behavior of the quail is characterized by (1) sharp increases in feed intake shortly after the onset of morning light and during the period of approximately 3 hours prior to darkness; (2) preference for sweet and sour tastes, and avoidance of salt and bitter tastes; and (3) the ability to tolerate moderate levels of salt in the drinking water for prolonged periods, although they normally avoid a salt solution.

• **Control of feed intake**—The feed intake of poultry is governed primarily by the energy concentration of the feed, although increasing the bulkiness of the ration tends to decrease feed consumption.

High energy layer rations (1350 to 1400 kcal/pound) in moderate to cool climates generally give the most economical results because they produce marked improvements in efficiency of feed utilization as compared to lower energy rations; and the improved efficiency of higher energy diets usually more than offsets the higher cost of these feeds.

Broilers consume somewhat more energy as the energy concentration of the diet increases. Also, high-energy diets tend to produce fat chickens while low-energy diets produce lean chickens.

But feed intake of chickens is affected by many factors in addition to the energy content and bulkiness of the diet. Temperature is a big factor; during very cold weather, layers may consume 20 to 30% more calories than would be needed in a moderate environment. Other factors regulating the amount of feed eaten by chickens are: genetics, size, sex, age, feather cover, activity, feed palatability, quality of feed ingredients, water consumption, type of housing, body fat content, and degree of stress.

• **Handling**—Birds are excitable; so, they should be handled gently at all times. The critical stages when gentle handling is especially important follow.

When moving ready-to-lay pullets. The manner in which pullets are loaded, transported, and placed in the cages on the buyer/layer ranch will significantly affect their subsequent performance. Thus, the contract should specify who is to deliver the birds, and that all equipment be thoroughly sanitized. Delivery instructions should be sufficiently detailed to include method and time of loading, type of hauling equipment to be used, number of birds per crate or cage, the route of travel, and arrival time. The owner or manager should be on hand to supervise the distribution of the pullets into their laying cages.

When marketing broilers. Most broilers are marketed when they are between 7 and 8 weeks of age. For the most part, marketing involves moving the birds from the house(s) in which they are produced to the processing plant.

Improper handling of broilers immediately prior to and during shipment will result in excess bruises, deaths, and lowered quality. Such losses may be minimized by preparing, catching, and transporting the birds as follows:

1. Discontinuing grit feeding at least 2 weeks prior to marketing. (Usually grit is not fed after 5 weeks of age.)
2. Avoiding exciting the birds on the day prior to catching, as it will cause them to hit the feeders and waterers and inflict bruises upon themselves.
3. Letting the feeders become empty about 2 hours before the catching crew arrives; and removing or elevating feeding and watering equipment to prevent bruises during catching. (Water should be removed just before the birds are caught; too early removal will result in excess dehydration.)
4. Catching and loading the birds properly, by: (a) using an experienced crew; (b) working under a dim blue light at night; (c) corralling them in small groups (of about 200 birds) to prevent smothering and undue injury; (d) grasping them by the shanks, with no more than 4 or 5 being carried at a time; (e) placing them in the crates gently; and (f) handling the crates carefully, preferably on pallets that can be moved easily with a hoist.
5. Driving loaded truck carefully, slowing down on turns and avoiding sudden stops.
6. Protecting the in-transit birds from extremes in weather. In cold weather, cover the truck to prevent chilling, as the latter will result in poor bleeding and downgrading of carcasses. In hot weather, protect against overheating in shipment by using open crates and avoiding lengthy stops enroute.
7. Unloading the crates and removing the birds gently at the processing plant; and putting them under cover in an area that is adequately ventilated and comfortable.

Shrinkage, or weight loss from the time feed and water are removed until the birds are weighed at the processing plant, varies according to temperature and length of time involved; it ranges from 2% for a 3-hour period to 6% for a 15-hour period.

Broiler condemnations average about 2.5%, with great variation according to season (condemnations during the winter are about twice as high as during the summer) and between processing plants. Condemnations may be due to many things; respiratory ailments, diseases of the leukosis complex, and bruises. Sometimes bruises result in downgrading or in only partial condemnation; for example, a severe breast or wing bruise may be cut out or off as unacceptable for human consumption. Bruises are responsible for over half of the downgrading. All too often, they are due to mishandling of the birds during catching, loading, and transporting to the processing plant. Improper bleeding, bruising, eviscerating, and over-scalding in processing also result in condemnations. Broiler growers and processors should strive to minimize such condemnations, because they represent a direct monetary loss.

When moving turkey poults to the range. When poults are about 8 weeks of age, they are ready to be moved from the brooder house to the range, semiconfinement sheds, or confinement buildings. Movement to the range is most critical and should be carefully planned. Two things should be checked before moving poults to the range: (1) the 5-day or 1-week weather forecast, and (2) the readiness of equipment. Avoid moving poults when the weather is threatening. Cool, wet, and rainy weather places a severe stress on poults that have been newly moved to the range. It is best to move in the early morning, as this allows poults plenty of time to locate water, feed, and shelter before nightfall. Never move in the heat of the day, because this is hard on both birds and handlers. Some growers move about one-third of their poults the first morning, then skip a day or two and move the remainder of the flock. Large, wire-enclosed four-wheel trailers are best for moving turkeys to ranges. Moving trailers may be constructed so that the birds can be loaded and unloaded by driving or herding. This system will require less labor and reduce injuries to poults.

Fig. 17-20. Turkeys on the range, with sunflowers in the background. (Courtesy, J. C. Allen & Son, West Lafayette, Ind.)

When driving and catching turkeys. Turkeys of all ages can be easily driven from place to place with the aid of light poles or sticks several feet long. With a pole in each hand, one person is able to control a good-size flock. Driving entails much less labor than crating and hauling, and, if distances are not too great, it is easier on the birds. Dogs can be used for driving turkeys, but they must be well trained and gentle and the turkeys must be accustomed to them.

Fig. 17-21. Turkeys being driven up a ramp for trucking to market. (Courtesy, *Turkey World*, Mt. Morris, Ill.)

For catching turkeys, a darkened room is best. In the dark or semidarkness, turkeys can be picked up by both legs without confusion or injury. For this reason, market turkeys sometimes are loaded at night.

Portable catching chutes, preferably with a conveyor, can be used for catching and loading range turkeys.

Metal hooks of ¼ in. round iron for weighing turkeys can be made up in several sizes. If hooks are the right sizes and the bird's legs are properly placed in them, the legs will not be injured. However, mature heavy turkeys should not be left hanging for more than 3 or 4 minutes, and it is better to pick up turkeys by both legs in a dark room rather than catch them with a hook.

When marketing ducks and geese. Pekin ducks are generally marketed at 6.5 to 7.5 lb liveweight, at 7 to 8 weeks of age. Muscovies are not ready for market until 10 to 17 weeks of age.

Most geese are marketed in the fall or winter, when they are 5 to 6 months old and weigh from 11 to 15 lb (depending on the strain and breed).

Feed should be withheld from the birds 8 to 12 hours before slaughter, but water may be provided up to the time of killing. Clean, uncrowded rearing facilities will help to prevent bruising, cutting, and other factors that cause poor market acceptability.

Market ducks or geese can be transported in crates or herded into trailers for delivery to the slaughterhouse.

Anamalous (Abnormal) Animal Behavior

Some behaviors that are useful in natural environments may be maladaptive to the individual animal or to a group of animals in artificial environments. Also, we have learned from studies of captured wild animals that when the amount and quality, including variability, of the surroundings of an animal are reduced, there is increased probability that abnormal behaviors will develop.

It is recognized that confinement of animals makes for limited space, and that this often leads to unfavorable changes in habitat and social interactions for which the species have become adapted and best suited over thousands of years of evolution. This is due to a genetic time lag; the keepers of domestic herds and flocks have altered the environment faster than the genetic makeup of animals. As a result, abnormal behaviors have evolved in domestic animals in confinement to plague those who raise them.

Homosexual behavior in all species is common where adult animals of one sex are confined together. Intersucking sometimes occurs in young calves, lambs, and pigs that are reared artificially. Pica, the eating of unnatural material(s), occurs in several species, but it is more prevalent in cattle and horses. Other abnormal behaviors in different species that frequently develop among domestic animals are listed and described in the sections that follow.

ABNORMAL CATTLE BEHAVIOR

Abnormal behavior in cattle may take many forms, including those that follow.

- **Kicker**—Milk cows may kick because they are in pain or frightened, or because they have been mistreated.

- **"Mean bull" complex**—There are inherited differences in the temperaments of cattle. Nevertheless, constant stress can change the temperament of an animal. Thus, when a bull is kept for hand-mating in a corral by which the cow herd passes each day, cows in heat stimulate his sexual behavior. Since he cannot respond naturally by mating, he may become a mean bull.

- **Pica**—The eating of unnatural material(s) is called *pica.* Cattle may develop pica (consumption of dirt, hair, bones, and/or feces) due to boredom, nutritional deficiencies, or physiological stress. Even after these conditions have been rectified, it may be disconcerting to find that pica persists among certain animals—perhaps due to habit.

ABNORMAL SHEEP BEHAVIOR

Abnormal behavior of sheep may take many forms, including wood or metal bar chewing, head banging, and wool-pulling. Only the latter is of economic importance.

- **Wool-pulling**—This abnormal trait may occur among sheep, especially animals kept in confinement and fed a finely-processed or pelleted ration and limited roughage. It decreases the value of the fleece; and it may have a permanent, adverse effect because coarse, black fibers replace the wool fibers that have been pulled.

ABNORMAL SWINE BEHAVIOR

Abnormal behavior in swine may take many forms,including those that follow.

- **Fighting (agonistic behavior)**—Fighting is a vicious act of aggression, which invites immediate retaliation from the victim. It occurs among pigs as a result of (1) establishing *teat order* (the front teats give more milk—they are worth fighting for); (2) aggression around the feeder; (3) tail biting; (4) mixing pigs together, as a means of establishing dominance hierarchy; and (5) bringing together sexually mature animals, especially boars. The following factors appear to increase the incidence of fighting: physical discomfort, frustration, and high light intensity. Limited experimental work indicates that spraying *androstenone,* a chemical extracted from pork fat which has a scent like a boar, will stop fighting among pigs, but the product has not been approved by FDA for preventing swine fighting. Also, researchers in Sweden recently presented impressive experimental evidence showing the effectiveness of *amperozide* in reducing agonistic behavior of newly grouped pigs.

- **Intersucking**—Newly weaned pigs may nuzzle their penmates' abdomens, mimicking early stages of nursing. Persistent nuzzling may cause ulcers and necrosis (destruction of the tissue).

- **Savage sow syndrome**—Occasionally, a sow becomes savage during or soon after parturition, at which time she may injure, kill, or even eat some or all of her pigs. The cause of this behavior is unknown, but it seems to be more prevalent among gilts farrowing their first litter or where sows are maladjusted to their farrowing quarters. Where this syndrome is encountered, it is recommended (1) that sows be moved to their farrowing quarters several days ahead of parturition, thereby giving them an opportunity to adjust to the new environment before the pigs come; and (2) that the attendant remove each pig just as soon as it is born, keep the pigs away until the sow has settled down (usually soon after parturition), then return the entire litter to the sow. Some producers have found tranquilizers helpful.

- **Tail biting**—Tail biting usually accompanies close confinement. Bad environment, poor nutrition, inadequate management, and a host of other factors appear to increase tail biting. Docking most of the tail soon after birth is the most effective means of preventing tail biting in swine. Odor-masking sprays sometime appear to be effective in reducing fighting and tail biting, especially when pigs are regrouped or moved to a different pen. Some producers have provided different objects for pigs to bite and play with, such as chains or old automobile tires suspended from the ceiling, and bowling balls. Initially, these are of some benefit, but the pigs seem to become habituated to their novelty quickly.

It is recommended that tail docking be a part of the regular management program, with the tails docked at the same time that the needle teeth are cut, when the pigs are about 3 days old. The side cutting-type pliers will work for both jobs, but tails will bleed less when they're cut with a dull blade. Emasculators and poultry debeakers also work well.

To dock the tail, clean it first, then cut it ½ to ¾ in. from its base, lifting it gently so as not to stretch the skin. The skin won't heal over the end bone as rapidly if you pull the tail away from the body. Don't cut the tail shorter than ½ in. because it will cause excessive bleeding and slow healing.

ABNORMAL HORSE BEHAVIOR (VICES)

Domestication and confinement of horses have spawned many abnormal behaviors, which are commonly referred to as *vices* (bad habits). Horses have more abnormal behaviors than any other species; not because they are naturally bad, but because there are a lot of spoiled horses—horses that have received too much tender, loving care (including lumps of sugar and carrots) and too little discipline.

• **Barn sour**—This refers to a horse that refuses to leave the barn—the horse refuses to leave home, friends, security, and feed. There are no easy cures for barn-sour horses; and each individual is different. But the most used, and the most successful, treatment for barn-sour horses consists of giving a bit of feed along the trail, with the feed given further and further from the barn.

• **Biting**—This vice is acquired as a result of incompetent handling. Generally, it is started in either of two ways: (1) The horse has been accustomed to treats (such as a lump of sugar), and nips as an expression of disappointment when there is no treat (treats should always be placed in the feed manger); or (2) as a result of rubbing the horse's nose while petting it (never rub the nose; it teaches a horse to bite).

• **Bolting feed**—Bolting feed is the name given to the habit of eating too fast (gulping the feed down without chewing). This condition can be controlled by spreading the concentrate thinly over the bottom of a large grain box, so that the horse cannot get a large mouthful; by adding chopped hay to the grain ration; or by placing some large, round stones, as big or bigger than baseballs, in the feed box.

• **Charging (attacking)**—This refers to a deliberate attack on a person, with the horse's mouth wide open. In the beginning stage, this vice can usually be corrected. The technique is to discipline without inflicting pain. The horse must be taught to obey, but it must have confidence in the handler. If mature stallions have had this vice for a long time, it is difficult, and perhaps impossible, to break.

• **Cribbing (wind-sucker, stump-sucker)**—A horse that has the vice of biting or setting the teeth against some object, such as the manger or a post, while sucking air is known as a cribber. This causes a bloated appearance and hard keeping; and such horses are more subject to colic. The common remedy for a cribber is a *cribbing strap* buckled around the neck in such a way that it will compress the larynx when the head is flexed, but not cause any discomfort when the horse is not indulging in the vice. A surgical operation to relieve cribbing has been developed and used with some success.

Fig. 17-22. A cribber (wind-sucker, or stump-sucker) in action. This is the vice of biting or setting the teeth against some object, such as a post or manger, while sucking in air. (Courtesy, Dr. George H. Waring, Department of Zoology, Southern Illinois Univeristy, Carbondale)

• **Eating bedding**—Sometimes gluttonous horses eat their bedding. This is undesirable because (1) most bedding materials are low in nutritional value, and (2) feces soiled bedding will likely add to the parasite problem. This habit can be prevented by muzzling the horse.

• **Halter pulling**—A confirmed halter puller breaks halters and lead ropes (straps) as it pulls back, then escapes. Either of two methods of tying will likely break the habit: (1) Tie a strong rope that the horse cannot break around the throatlatch, using a bowline knot so that the rope cannot slip and choke the horse; or (2) run a strong rope around the chest just back of the withers (some prefer to run it around the back in the area of the rear flanks), using a bowline knot. After a few struggles and attempts to break away, the horse will give up.

• **Handler aversion**—As an aversion to handlers, some horses display aggressive vices such as bucking, or objection to catching, harnessing, saddling, and grooming. Flight responses occur in the form of backing, balking, bolting, or running away. Most of these vices originate with incompetent handling. Nevertheless, they may be difficult to cope with or to correct, especially in older animals.

• **Kicking**—Two types of kickers are encountered: (1) the kicker that kicks the stall wall or door, and (2) the kicker that kicks people. The stall kicker kicks for no other excuse or satisfaction than to strike something and make a noise (kicking hollow tile makes a loud noise). Padding the stall will stop some stall kickers. Also, a chain or stick strapped to the back of the leg will usually break the habit; when the horse kicks, the chain or stick strikes the leg.

A horse that kicks people is dangerous. In the formative stage, usually the vice can be eliminated by prompt attention. But it is difficult to correct a confirmed and seasoned kicker.

• **Pawing**—This refers to a horse that digs at the stall floor with its front feet. Heavy rubber mats on the stall floor and under the bedding will discourage stall digging.

• **Pica**—This refers to a depraved appetite, or the eating of unnatural materials. This type of behavior is most common among horses that are kept in stalls and fed highly concentrated rations.

• **Rearing**—Rearing is a very dangerous vice. When the horse rears up, the flailing forelegs can inflict injury on the handler. Such horses can usually be corrected by proper use of a lead shank or whip.

A horse that rears while being ridden should be handled by an experienced trainer, who can usually find the cause and correct the vice.

• **Shying**—Shying at unfamiliar objects makes a horse dangerous to ride. The only solution is, patiently and gently, to take the horse over new trails and in new surroundings, again and again until there is no more shyness.

• **Stall walking**—This is a stereotypic movement about the stall. A mascot, e.g. a goat, may calm a stall walker.

• **Striking**—Striking with the front feet is a dangerous vice, because the handler is always vulnerable. The handler should always stay at the side of such a horse, never in front of it. Each time that a horse attempts to strike, it should be punished with a war bridle or whip.

• **Tail rubbing**—This refers to persistent rubbing of the tail against the side of the stall or other objects, resulting in the loss of hair and an unsightly tail. Tail rubbing is a common vice of Saddlebred horses that wear tail sets. Also, the presence of parasites may cause animals to acquire this vice. Installation of a tail board (a 2″ × 12″ shelf that runs around the stall at a height just above the point of the horse's buttock), or an electric wire similarly placed, may be necessary to break an animal of this habit.

• **Weaving**—This is a rhythmical swaying back and forth while standing at the stall door. The prevention and cure are exercise, ample room, and freedom from stress. A mascot, e.g. a goat, may calm a weaver.

• **Wood chewing**—This is the chewing of wood, usually a wood manger or a board fence. It is generally caused by (1) boredom, (2) nutritional inadequacies, or (3) psychological stress and habit. There is only one foolproof way in which to prevent wood chewing; to have no wood on which horses can chew—to use metal, or other similar materials, for barns and fences. But wood chewing can be lessened, although it cannot be entirely prevented, through one or more of the following practices:

1. Stepping up the exercise.
2. Feeding three times daily, rather than the normal two times; without increasing the total daily allowance.
3. Spreading out the feed in a large feed container and placing a few large, smooth stones about the size of a baseball in the feed container, thereby making the horse work longer and harder to obtain its feed.
4. Providing 2 to 4 lb of straw or coarse grass hay per animal per day, thereby giving the horse something to nibble on.

ABNORMAL CHICKEN BEHAVIOR

With the restriction, or confinement, of flocks, many abnormal behaviors evolved, including those that follow.

• **Cannibalism**—This is the most common abnormal behavior observed in chickens in confinement. It may be encountered among birds of all ages. Many types of cannibalism occur, with the following being most common:

1. **Toe picking.** This type of cannibalism is most commonly seen in baby chicks. It may be brought on by hunger.
2. **Vent picking.** Picking of the vent, or of the area below the vent, is the most severe form of cannibalism. This type is generally seen in pullet flocks in high production. Predisposing factors are prolapse or tearing of the tissues caused by the passage of a very large egg.
3. **Head picking.** This type of cannibalism usually follows injuries to the comb or wattles caused by freezing or by fighting.

Cannibalism appears to be brought on by boredom and too much light, accentuated by deficiencies in management and nutrition. The best way to control cannibalism is by debeaking and dubbing.

Debeaking is the removal of a portion of the upper (and often a lesser portion of the lower) beak of the fowl. Many broiler producers, and some egg producers, have their chicks debeaked at the hatchery. Additionally, layers can be debeaked at the time of placement in the laying house.

Dubbing is the removal of the comb, and in some cases the wattles of chickens. It may be done to either males or females. Generally, dubbing is done when chicks are a day old, using curved manicure scissors. At this age, very little bleeding or discomfort is noted.

• **Egg eating**—Egg eating is a costly vice. It is usually predisposed by factors favoring egg breakage, including insufficient nests, insufficient nesting material, not collecting eggs frequently enough, and soft-shelled or thin-shelled eggs. Prevention consists in alleviating these conditions. Once the egg eating habit has started, it is very difficult to stop and it usually spreads quickly throughout the flock. If the birds have not been debeaked, this should be done immediately. Also, nests should be darkened and eggs should be collected frequently.

• **Feather pecking**—The term *feather pecking* is used to describe feather loss and hemorrhaging of skin in chickens.

• **Hysteria**—Sometimes, excessive fright occurs among growing pullets or layers. With floor-housed birds, it may take the form of flight and piling up in a corner(s) of the house, with the result that many birds may be suffocated. Caged layers may attempt to fly, resulting in injuries to wings and legs or broken necks. Ordinarily, an episode of hysteria only lasts about a minute, but the losses can be devastating. The cause of hysteria is unknown. However, it appears to be triggered by such, things as loud noise, quick movements, and sharp changes in light intensity.

• **Polydipsia**—This refers to excess water drinking. It may occur in caged birds which, due to boredom, play excessively with drinking nipples. Polydipsia leads to regurgitation of food and water.

ABNORMAL TURKEY BEHAVIOR

Abnormal behavior in turkeys may take many forms, including those that follow.

- **Blueback**—Blueback is a permanent, dark discoloration of the skin on the back and sometimes the sides and breast of turkeys with dark plumage, but not turkeys with white plumage. It is caused by feather picking, followed by exposure to sunlight. Blueback may result from overcrowding in the brooder, keeping the poults on the sun porch too long, or lack of sufficient fiber in the ration.

- **Feather picking (cannibalism)**—Feather picking is a mild form of cannibalism to which turkeys may become addicted, especially during the growing period. It results in unsightly appearance, more trouble from pinfeathers when the birds are marketed, and blueback in varieties with dark plumage. It can develop into flesh picking and become serious enough to retard growth. Feather picking usually reaches serious proportions only in turkeys raised in confinement.

Feather picking and cannibalism of turkeys raised in confinement can be prevented almost completely by debeaking poults, preferably by the hatchery when poults are 1 day old.

Additional management practices that tend to prevent feather picking include: (1)avoiding overcrowding in confinement rearing quarters, (2) feeding an adequate diet, (3) feeding pelleted feed rather than loose mash, and (4) avoiding confinement of turkeys to roosts or other closely restricted quarters, particularly in the early morning.

Applied Animal Behavior

The presentation to this point has been for the purpose of understanding animal behavior. The sections that follow give some practical applications of animal behavior.

BREEDING FOR ADAPTATION

The wide variety of livestock in different parts of the world reflects a continuous process of natural and artificial selection which has resulted in the survival of animals well adapted to climate and other environmental factors. Changes in the physical structure of species is dependent upon (1) the ability of animals to mutate and/or respond to selection pressure (natural or artificial), and (2) the effect of environmental pressure on the animal which results in a survival of the fittest. Among the examples of adaptation to environment are haired sheep (devoid of wool) in desert areas, fat-tailed sheep in arid zones, *Bos indicus* (Zebu) types of cattle in tropical areas, and *Bos taurus* cattle in temperate zones. Such adaptations relate to survival of the animals, but they do not necessarily entail maximum productivity of food. European cattle usually have much higher yields of milk and propensities for rapid growth than the breeds native to Africa or India. It is understandable, therefore, why there have been many attempts to introduce improved European livestock into countries in which the productivity of native stock is low. But there are many problems in breed replacement, with the result that a large number of experimental introductions of new breeds have not been successful. Tropical Africa provides an example. Because of disease problems, poor resistance to high temperatures, and limited feed supplies, many of the attempts made by former colonial powers to improve the output of native stock by replacing them with the European breeds failed. Breed replacement or a crossbreeding system might seem to be a simple panacea for low productivity. However, unless associated with special provisions for subsequent importation of breeding stock and simultaneous improvement of the nutritional, parasitological, disease, and husbandry environment of the crossbreds, it is not likely to succeed.

Selection should be from among animals kept in an environment similar to that in which it is expected that their offspring shall perform—this requisite applies to animals brought in from another herd, either foundation or replacement animals. For example, animals that are going into a range herd should be selected from animals handled under range conditions, rather than from among stall-fed animals.

Animals can be changed through heredity and selection. For example, in Israel, which has one of the highest average milk yields per cow of any nation in the world, the individual distance between cows approaches contiguousness in some herds; this is due to Israel's having selected intensively for docility for 25 years. In other words, the animals are literally touching each other, with no agonistic or dominance-type response. This allows them to concentrate their animals even more than they had previously, thereby giving them a higher productivity per unit area. The only problem reported by Israel is that estrus, or heat, in animals in close proximity is difficult to detect.

Another example of breeding and selecting for a behavioral characteristic is milk production. When milkers first modified the milking technique and went from hand milking to machine milking, producers assumed the necessity of hand stripping. In recent years, many producers have given up hand stripping and selected cows capable of producing large quantities of milk by being milked by machine, without either stimulation before milking or stripping afterwards. This has been a selection for a behavioral characteristic—a low threshold to the milking stimulus. As a result, today a very large number of cows, especially within the Holstein-Friesian breed, will let down their milk effectively with no other stimulation than having the milking machine applied.

QUICK ADAPTATION—EARLY TRAINING

We need to breed and select animals that adapt quickly to an artificially-made environment—animals that not only survive, but thrive, under the conditions imposed upon them.

Also, early training and experience are extremely important. In general, as with humans, young animals learn more quickly and easily than adults; hence, advance preparation for adult life will pay handsome dividends. The optimum time for such training varies according to species.

Stress can be reduced or avoided entirely if animals proceed through a graduated sequence of events leading to an otherwise noxious experience. Preconditioning of cattle is an application of this principle to production practices. If calves are properly preconditioned (e.g., started on feed, vaccinated, treated for parasites, etc.) prior to weaning, the stress of subsequent weaning and movement to a feedlot is minimized.

MANURE ELIMINATION

Body waste is a major concern; although unavoidable, it is expensive and time-consuming to handle, and it may create a major pollution problem. But manure handling can be facilitated by an understanding and application of eliminative behavior.

Cattle are indiscriminate eliminators. Even so, this trait can be used effectively. For example, if cattle are fed at the same time each day, feed is released from the rumen into the true stomach regularly, causing a gastro-colic reflex. When this happens and cattle are put under slight stress, they defecate. Knowing this, cattle can be moved to the defecating area at the right time.

Pigs can be trained to defecate in a particular area, separate and apart from their sleeping and resting areas, by: (1) keeping defecating areas cool and resting areas warm, as pigs like to defecate where it is cool and rest where it is warm; (2) locating the water near the defecating area, as a wet floor is conducive to elimination; (3) having the defecating area 1–2 in. below the feeding and resting areas seems to direct the pigs to where the defecating should occur; (4) wetting down the defecating area; and (5) feeding on the floor of the eating and sleeping areas for a few days will encourage pigs to keep these areas clean and to defecate elsewhere.

COMPANIONSHIP

Companionship in animals is of great practical importance. Except for the cat, all domestic animals are highly social and have constant need for companionship.

If not too crowded, placing animals together sometimes accomplishes two things: (1) greater feed consumption, due to the competition among them (allelomimetic behavior); and (2) a quieting effect. For example, putting a barrow with a high-strung boar usually has a quieting effect.

The best known animal companionship pertains to high-strung racehorses and stallions, in which all sorts of companions are used—a goat, a sheep, a chicken, a duck, or a pony. Such companions are commonly referred to as "mascots." The great Stymie, Thoroughbred winner of $918,485, became attached to a hen of nondescript breeding who came to dinner one day and never left.

The expression "to get his goat" was born of the common custom of having goats as mascots for racehorses. Back in the days when skulduggery was as important as form in winning races, the grooms of one stable sometimes plotted to kidnap the goat mascot of a rival's horse. By "getting the goat" of a favorite, they cleaned up by betting against a horse that was odds-on to win, but too upset to run at its best.

CONTROLLING ANIMAL BEHAVIORS

Many abnormal behaviors can be lessened or alleviated rather easily. The primary ones are listed, along with causes (if known) and methods of controlling them, under the earlier section headed, "Anamalous (Abnormal) Animal, Behavior," with a section thereunder devoted to each species.

ANIMAL ENVIRONMENT[4]

An animal is the result of two forces—heredity and environment. Heredity has already made its contribution at the time of fertilization, but environment works ceaselessly away until death. Since most animal traits are only 30 to 50% heritable, the expresssion of the rest (more than 50%) depends on the quality of all of the components of the environment. Thus, it is very important that the keepers of herds and flocks have enlightened knowledge of, and apply expert management to, animal environment.

Environment may be defined as all the conditions, circumstances, and influences surrounding and affecting the growth, development, and production of animals. The most important influences in the environment are the feed and quarters (space and shelter).

The branch of science concerned with the relation of living things to their environment and to each other is known as ecology.

Through the years, the domesticated animals best suited to a particular environment survived, and those that were poorly adapted either moved to a more favorable environment or perished. During the past two centuries, livestock producers have made great strides in the selection and propagation of animals suited to a particular environment, and during the past 50 years they have made progress in modifying the environment for the benefit of their animals and themselves.

It is becoming increasingly difficult to define environment, because scientists continue to discover important new environmental factors. Primitive people recognized that the sun and fire provided both heat and light, that body heat could be conserved by draping the body with animal skins, and that trees and caves provided protection from the weather. Today, it is recognized that these, along with a host of other environmental factors, affect animals and people.

The keepers of herds and flocks were little concerned with the effect of environment on animals so long as they grazed on pastures or ranges. But rising feed, land, and labor costs, along with the concentration of animals into smaller spaces, changed all this. Today, most layers and broilers are maintained in confinement throughout life; many layers are kept in cages, and essentially all broilers are on litter floors. Turkeys are shifting rapidly from range to confinement. Water is important for ducks, but even with ducks the trend is toward higher population densities and more confinement. Many swine are raised partially or totally in confinement; and confinement production is increasing with beef cattle, dairy cattle, and sheep.

Among animals, environmental control involves space requirements, light, air temperature, relative humidity, air velocity, wet bedding, ammonia buildup, dust, odors, and manure disposal, along with proper feed and water. Control or modification of these factors offers possibilities for improving animal performance. Although there is still much

[4]In the preparation of this section, and the sub-sections under it, the authors adapted considerable material, including experimental results, from the following source: Ames, D. R., Chairman, *et al., Effect of Environment on Nutrient Requirements of Domestic Animals,* NRC, National Academy Press, Washington, D.C., 1981.

to be learned about environmental control, the gap between awareness and application is becoming smaller. Research on animal environment has lagged, primarily because it requires a melding of several disciplines—nutrition, physiology, genetics, engineering, and climatology. Those engaged in such studies are known as ecologists.

Samuel Brody, brilliant pioneer in the field of environmental physiology, growth, and energetics, at the Missouri Agricultural Experiment Station from 1921 to 1956, frequently reminded his students that scientists often prove what farmers have long known. This idea is well illustrated by the practical observation that when exposed to cold weather farm animals tend to increase their heat production by increasing the amount of feed consumed. Thus, cattle increase forage consumption during cold weather, thereby taking advantage of the warming effect of the heat increment of the feed. They also tend to increase their tolerance to cold by increasing their deposition of an insulating subdermal fat layer. Also, farm animals may increase heat production by shivering and conserve heat by huddling together.

Effect of Environmental Factors on Animals

The effect of environment on dairy cattle was clearly demonstrated in an experiment in New Zealand. It involved the selection of 20 calves from low-producing herds and 20 calves from high-producing herds. All of them were sired artificially by outstanding bulls. The 40 head were assembled at the Ruakura Experiment Station, raised and milked together for the first lactation. Under these conditions, no significant difference between the production of the 2 groups was observed. Then, they were sent back to the respective herds from whence they came, whereupon their production was comparable to that of the cows with which they were being milked. Then, for a second time, they were returned to the Ruakura Experiment Station, where again there was no significant difference in their production. The Ruakura Station then went one step further; they confirmed these results by using identical twins, with both twins milked at the Ruakura Station, and then later divided between high- and low-producing herds for subsequent lactation.

The New Zealand experiment underscores the importance of environment. No matter how good the genetics, a good environment is essential to obtain high production.

Also, selection of genetically superior individuals to be parents of the next generation is hampered by environmental factors that tend to mask the actual breeding values of individuals being selected. Thus, the contribution of these environmental factors to the total phenotypic variation should be minimized before estimating the genetic parameters.

The environmental factors affecting animals vary. The factors known to predominate in determining sheep productivity under arid conditions are year of birth, age of dam, type of birth and rearing, sex, and breed.[5]

In collaboration with the U.S. Department of Agriculture, Cornell University scientists studied the effect of summer weather on performance of Holstein cows in 3 stages of lactation. They reported that, for all stages of lactation, 9% of the variation in milk yield, 13% in milk fat, 5% in feed intake, and 65% in rectal temperature were attributable to weather conditions.[6]

We now know that controlled environment must embrace far more than an air-conditioned chamber, along with ample feed and water. The producer needs to be concerned more with the natural habitat of animals. Nature ordained that they do more than eat, sleep, and reproduce. For example, studies on the behavior of swine show that they spend much of their day in active investigative behavior, primarily rooting and manipulating their environment. When free ranging, pigs may spend 40% of their day resting, 35% investigating novel surroundings, 15% eating, and 10% in other activities. What happens when pigs are confined in a building on slotted floors? How is the nervous energy dissipated that would normally be used to satisfy the drives for investigating and rooting? Evidently, environmental deficiencies are manifested by tail biting, gastric ulcers, poor maternal care and loss of young, or other physiological functions resulting in a sudden death syndrome or tissue degeneration.

Preventing disorders by merely cutting off the tails of pigs to alleviate tail biting, debeaking poultry to prevent cannibalism, and using choke collars on horses to inhibit cribbing, is not unlike trying to control malaria fever in humans by the use of drugs without getting rid of mosquitoes. Rather, we need to recognize these disorders for what they are—warning signals that conditions are not right. Correcting the cause of the disorder is the best solution. Unfortunately, this is not usually the easiest. Correcting the cause may involve trying to emulate the natural conditions of the species, such as altering space per animal and group size, providing training and experience at opportune times, promoting exercise, and gradually changing rations. Over the long pull, selection provides a major answer to correcting confinement and other behavioral problems; we need to breed animals adapted to people-made environments.

The following factors are of special importance in any discussion of animal environment:

1. Feed
2. Water
3. Weather
4. Facilities
5. Health
6. Stress

FEED/ENVIRONMENTAL INTERACTIONS

Animals may be affected by either (1) too little or too much feed, (2) rations that are too low in one or more nutrients, (3) an imbalance between certain nutrients, or (4) objection to the physical form of the ration—for example, it may be ground too finely.

[5]Eltawil, E. A., *et al.*, "Evaluation of Environmental Factors Affecting Birth, Weaning and Yearling Traits in Navajo Sheep," *Journal of Animal Science*, Vol. 31, No. 5, Nov. 1970, p. 823.

[6]Maust, L. E., R. E. McDowell, and N. W. Hoover, "Effect of Summer Weather on Performance of Holstein Cows in Three Stages of Lactation," *Journal of Dairy Science*, Vol. 55, No. 8, Aug. 1972, p. 1133.

Forced production (such as growth, milk production, and racing 2-year-old horses) and the feeding of forages and grains which are often produced on leached and depleted soils have created many problems in nutrition. These conditions have been further aggravated through the increased confinement of animals, many animals being confined to stalls or lots all or a large part of the year. Under these unnatural conditions, nutritional diseases and ailments have become increasingly common.

Also, nutritional reproductive failures plague livestock operations. Generally speaking, energy supply tends to be more limiting than protein in reproduction. The level and kind of feed before and after parturition will determine how many females will show heat and conceive. After giving birth, feed requirements increase tremendously because of milk production; hence, a female suckling young needs approximately 50% greater feed allowance than during the pregnancy period. Otherwise,she will suffer a serious loss in weight, and she may fail to come in heat and conceive. This basic fact, along with other pertinent findings, was confirmed by researchers at the Montana Agricultural Experiment Station. Based on 12 years research at the Havre and Miles City Stations, they concluded that beef cattle size and milk production should be tailored to fit the environment. Big size and more milk are not better unless the range forage supply is better. The best sized cow is one that fits the range conditions. Small cows do best on poor range because they can usually get 100% of their daily feed requirement for maintenance and milk production, whereas big cows on a poor range are borderline hungry all the time. Also, cows that give a lot of milk must have a good range; otherwise, they are stressed by lack of feed; and their fertility rate and calf crops drop. So, cow size and milk production should match their environment.

The next question is whether a breeding program can make maximum progress under conditions of suboptimal nutrition (such as is often found under some farm and range conditions). One school of thought is that selection for such factors as body form and growth rate in animals can be most effective only under nutritive conditions promoting the near maximum development of those characters of which the animal is capable. The other school of thought is that genetic differences affecting usefulness under suboptimal conditions will be expressed under such suboptimal conditions, and that differences observed under forced conditions may not be correlated with real utility under less favorable conditions. Those favoring the latter thinking argue, therefore, that the production and selection of breeding animals for the range should be under typical range conditions and that the animals should not be highly fitted in a box stall.

In general, the results of a 10-year experiment conducted by the senior author and his colleagues at Washington State University, designed to study the effect of plane of nutrition on meat animal improvement, support the contention that selection of breeding animals should be carried on under the same environmental conditions as those under which commercial animals are produced.[7]

[7]Fowler, S. H., and M. E. Ensminger, *Relationship of Plane of Nutrition to the Improvement of Swine for Meat Production Through Selection*, Tech. Bull. 34, Washington State University, Pullman, 1961.

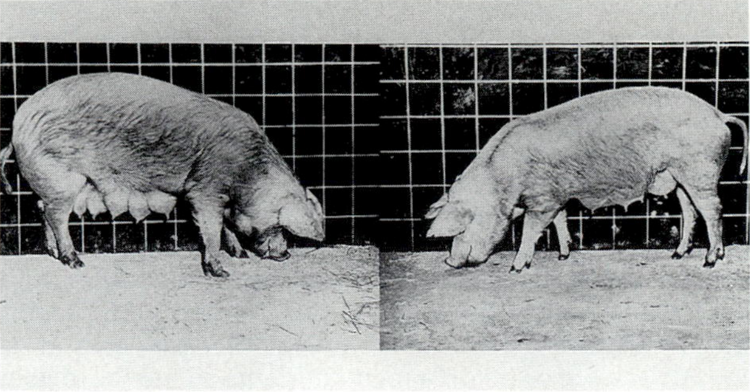

Fig. 17-23. Feed made the difference! The 2 sows are of the same age and breeding, but the sow shown at left received all she could eat from birth; whereas the gaunt sow shown at right was limited to 70% of the ration consumed by the better-fed animal. (Courtesy, Washington State University, Pullman)

The following additional feed/environmental factors are pertinent:

• **Appetite/intake**—Animals control their feed intake through a combination of the following mechanisms:

1. **The hypothalmus.** The two primary theories pertaining to hypothalmic control are: (a) the chemostatic hypothesis, which reasons that when blood nutrient levels, such as sugar and lipid, become too low, the hypothalmus sends signals to begin feeding; and (b) the thermostatic hypothesis, which theorizes that a decrease in hypothalmic temperature will induce feeding.

(Also see Chapter 3, section on "Hunger and Appetite.")

2. **The volatile fatty acids.** The appetite in ruminants has been shown to be sensitive to the volatile fatty acid levels in the rumen. Thus, increased acetate and propionate result in satiety and reduced feed consumption.

(Also see Chapter 3, section on "Gastric Influences on Appetite Regulation.")

3. **The physical size of the digestive tract.** Generally, animals eat to satisfy their energy requirements. However, if the caloric density of the ration is low, as in the case of certain roughages, the bulk may limit the animal's ability to hold sufficient of the feed to meet its energy needs.

4. **The thermostatic control.** According to the thermostatic theory, in hot environments animals reduce metabolic rate by reducing feed intake; and in cold environments animals increase metabolic rate by increasing feed intake.

5. **The fiber content of the ration.** Low-fiber high-energy rations stimulate animal growth and production in hot environments, whereas high-fiber rations increase the heat increment and keep the body warm in cold weather.

6. **Disease.** Feed intake is reduced quickly during most metabolic diseases such as pregnancy toxemia, acetonemia, D-lactic acidosis, ketosis, or bloat. Also, most gastrointestinal disorders of either infectious or parasitic origin as well as many systemic diseases result in decreased feed intake.

• **Calving time affected by feeding time**—Limited research indicates that partial calving control can be achieved by night feeding (feeding at 5:00 p.m. to 10:00 p.m. starting about 2 to 4 weeks before calving), with about 15-35% more daytime calving than when pregnant cows are fed in the morning, between 8:00 and 9:00 a.m. The reason that night feeding causes daytime calving has not been established.

(Also see Chapter 19, section on "Calving Control [Daytime Calving].")

• **Compensatory growth**—Compensatory growth is a cattle feeding term, which refers to increased growth rate in one time period as a result of growth restriction imposed during an earlier time period. This phenomenon is evidenced by stocker cattle that are roughed through the winter with limited daily weight gains, followed by feeding more liberally on a higher energy ration in the spring, when they are turned to lush pasture or put in a feedlot. Cattle so managed gain faster and more efficiently than similar cattle that are fed more liberally during the winter stocker stage.

(Also see Chapter 19, section on "Compensatory Growth.")

• **Competition**—When one animal eats (or grazes) or drinks, others may be stimulated to do likewise, even when they are not hungry or thirsty. This behavior is called social facilitation (competition). It follows that group-fed animals generally consume more feed per animal than individually-fed animals, due to competition. However, if there is fighting (agonistic behavior), the feed consumption of subordinates is likely to be reduced.

• **Familiarity of feed**—When given a choice, animals usually continue eating those feeds with which they are familiar. However, by gradually adding new and novel feeds, they learn to accept them (animals adapt). Also, early exposure of young animals to a feed may result in ready consumption of the feed later in life. When grazing, cattle and sheep develop preferences for certain pasture plants, with the result that they may not readily accept supplementary concentrates on pasture, and they usually have even greater difficulty in changing from grazing to eating dry forages and concentrates in confinement.

• **Feed as a reward**—Feed rewards are particularly common with cattle and horses.

Feed rewards may be used in gentling cattle as follows: Place up to 12 cattle in a relatively small pen where you can walk among and stay fairly close to them. Then, (1) carry a small bucket half-full of pellets, and shake it to attract the attention of the animals; (2) hold one pellet between your thumb and forefinger, and feed each animal that will take a pellet from you, one by one; and (3) repeat this procedure daily, or more often, until the animals get sufficiently gentle that they can be hand-fed and back-rubbed; and (4) *never, never* touch the animals on the forehead or horns, as this will teach them to butt. This procedure substitutes the application of animal behavior and human patience for rough handling.

Satisfying a horse's liking for a lump of sugar or a carrot can be most effective. To be effective, rewards must not be given promiscuously, only when deserved; for example, after carrying out a training command.

• **Frequency of feeding**—Except for swine, all species of animals fed for maximum production seem to respond to frequent feeding or to self-feeding (most poultry are fed on a free-choice basis), along with regularity of feeding. Thus, when fed three or more times daily (or self-fed), rather than twice daily, beef cattle, feeder lambs, and broilers make more rapid gains, lactating dairy cows produce more milk, and layers produce more eggs; and their increased production requires less feed per unit of product (meat, milk, or eggs) produced. However, unless self-fed, more frequent feeding generally makes for increased labor; so, the practicality of increased frequency of feeding should be determined by net returns—the value of the increased product, less the cost of producing it.

There is little evidence that frequency of feeding affects the rate of gain and feed efficiency of swine. Also, animals fed limited rations, such as dry beef cows, dry dairy cows, dry ewes, and gestating sows, may be fed less frequently than the traditional twice daily. Dry cows and ewes on the range may be fed as infrequently as three times a week without appreciably affecting performance.

• **Flushing**—*Flushing is the practice of having females gain in condition 2 to 3 weeks before breeding.* This may be accomplished by grain feeding, or cows and ewes may be turned to more lush pasture or range. Under most circumstances, the following amounts per head daily of a suitable concentrate are added to the ration that the females were receiving prior to flushing: cows, 2 to 5 lb; ewes, 1 to 2 lb; and sows, about 2 lb. Flushing fat animals is not effective. Immediately after breeding, females should be returned to normal rations.

Although it is not likely that all the benefits ascribed to flushing will be fully realized under all conditions, the general feeling persists that the practice will result in (1) more eggs being shed, (2) the females coming in heat more promptly, (3) more certain and prompt conception—with the young arriving more nearly at the same time, and (4) a 15 to 30% increase in lamb and pig crops.

• **Hair growth**—Hair growth on cattle and horses is greatly affected by three environmental factors—feed, temperature, and health. Cattle and horses fed maintenance and submaintenance rations do not shed as early in the spring as animals on a higher nutritional plane. In cold areas and during the winter months, long, shaggy coats of hair are nature's way of protecting animals against excessive energy loss. Sick animals tend to retain their winter hair coats longer than normal.

Fig. 17-24. Hair growth of Hereford cattle is nature's way of protecting them against excessive energy loss in the winter months. (Courtesy, American Hereford Assn., Kansas City, Mo.)

• **Milk composition affected by feed**—Some feeds reduce the fat content of milk. Among such feeds are cod-liver oil and other fish oils, early spring (lush) pasture, and pearl millet. Also, too small an amount of roughage, fine grinding of forage, or heated starch will lower the butterfat content of milk. On the other hand, such feeds as whole cottonseed, soybeans, and coconut oil increase the fat content of milk.

The amount of fat-soluble vitamins A, D, and E in milk are influenced by the amount of these particular vitamins in the ration; and in the case of vitamin D, exposure to sunlight is a factor, also.

• **Nutrient deficiencies**—The nutrient requirement tables (feeding standards) presented for each species in this book list values for animals under conditions presumed to be relatively free from environmental stress and for animals expected to perform near their genetic potential. In practice, environmental factors are not always ideal. Stresses are produced by weather (temperature, humidity, air movement, and solar radiation); diseases and parasites; surgery, dehorning, and castrating; altitude; sound; density and confinement; and pollution. As a result, animal performance often falls short of genetic potential. Animal shelters and housing are intended to eliminate or moderate the impact of some of the environment, but if they are poorly designed, they may create a new array of stresses with which the animal must contend.

So, although nutrient requirement tables are excellent and needed guides, nutrient deficiencies may result even when they are followed because the environment in which the animals are produced can modify the requirements. For this reason, along with making provision for variations in feed composition, *nutrient allowances,* which provide margins of safety, are presented in this book for each species, in addition to the requirements.

• **Overfeeding**—Too much feed is wasteful. Besides, it creates a health hazard; there is usually lowered reproduction in breeding animals, and a higher incidence of digestive disturbances (acidosis, bloat, founder, and scours)—and even death. Animals that suffer from mild digestive disturbances are commonly referred to as *off feed.*

• **Palatability**—*Palatability refers to the combination of factors that result in a feed being well liked and eaten with relish.* If animals don't eat their feed, they won't produce; and if they don't eat enough feed, production efficiency will be poor. Only an animal can assess the palatability of a feed. Palatability is the result of the following factors: taste, appearance, odor, texture, temperature, and in some cases auditory properties of the feed (like the sound of pigs eating corn).

• **Preconditioning**—*Preconditioning, which is a beef cattle term, is a way of preparing calves to withstand the stress and rigors of leaving their mothers, learning to eat new kinds of feeds, and shipping from the farm or ranch where they were raised to markets, feedlots, or other farms or ranches.* To the cow-calf producer (the seller), it is a program of management (including castration, dehorning, and weaning), nutrition, and immunization. To the feedlot operator (the buyer), preconditioning is a way in which to prepare calves to fit into the program and to minimize costly and unnecessary procedures. Economically, complete preconditioning is difficult to justify for both the cow-calf producer and the cattle feeder. But limited creep feeding plus presale vaccination may be economically feasible.

(Also see Chapter 19, section on "Preconditioning.")

• **Regularity of feeding**—Animals are creatures of habit; hence, they should be fed at regular times each day.

• **Selective grazing**—Different species of animals have different habits of grazing; they show preference for different plants and graze to different heights. Cattle are less selective in their grazing habits than other animal species; they will eat many kinds of vegetation, and at all stages of maturity. Sheep consume shorter and finer forages and more forbs than cattle; and their cleft upper lip allows them to graze vegetation close to the ground. Goats are fond of browse—the young shoots of shrubs and trees. Swine prefer legume pastures and love to root in the soil. Horses prefer young, tender grass to coarse-stemmed plants, a preference which often causes them to overgraze cetain areas. Except for geese, which graze grass, poultry do not eat much forage. Because of selective and preference grazing, grazing by two or more classes of animals makes for more uniform pasture utilization and fewer weeds and parasites, provided the area is not overstocked.

• **Stress**—Stress should be considered when formulating rations. All environmental stressors, whether physiological, immunological, or behavioral in nature, require energy expenditure on the part of the animal. Stress especially increases the animal's need for metabolizable energy, which, in turn, affects the optimal ratio of metabolizable energy to protein and other nutrients. So, when under extreme stress, nutrients will be diverted to maintenance, which is high priority, with the result that production, reproduction, and disease resistance will be reduced. So, it is important that rations be formulated to meet stressful conditions.

• **Underfeeding**—Too little feed results in slow and stunted growth of young stock; in loss of weight, poor condition, and excessive fatigue of mature animals; and in poor reproduction, failure of some females to show heat, more services per conception, lowered young crop, and light birth weights.

WATER/ENVIRONMENTAL INTERACTIONS

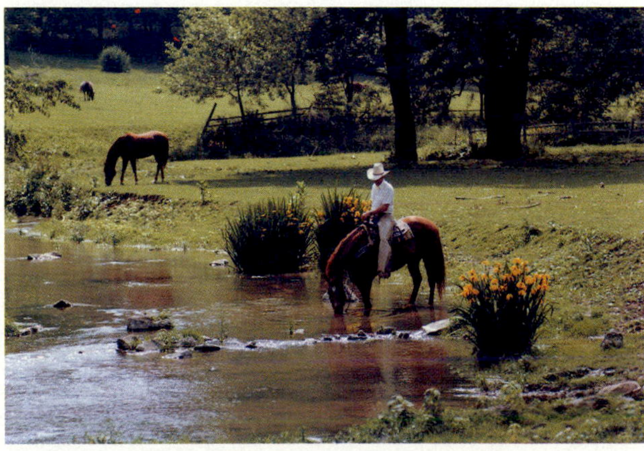

Fig. 17-25. Clear, cool mountain water. (Courtesy, American Quarter Horse Assn., Amarillo, Tex.)

Animals can survive for a longer period without feed than without water. Water is one of the largest constituents in the animal body, ranging from 40% in very fat, mature animals to 80% in newborn animals. Deficits or excesses of more than a few percent of the total body water are incompatable with health, and large deficits of about 20% of the body weight lead to death.

The total water requirement of animals varies primarily with the weather (temperature and humidity); feed (kind and amount); the species, age, and weight of animal; and the physiological state. The need for water increases with increased intakes of protein and salt, and with increased milk production of lactating animals. Water quality is also important, especially with respect to the content of salts and toxic compounds.

It is generally recognized that animals consume more water in summer than in winter. Based on 5 summer and 4 winter trials, the Iowa Agricultural Experiment Station reported that yearling cattle consumed an average of 8.5 gal per day in summer vs 5 gal per day in winter.

Pertinent details relative to important water/environmental interactions follow.

(Also see Chapter 4, section on "Water"; and see Chapter 4, Table 4-10, for daily water consumption.)

• **Air temperature**[8]—Numerous experiments have shown a very positive correlation between water intake and ambient temperature. Some species differences are noted:

Cattle. Under controlled temperature conditions, it has been demonstrated that cattle tend to increase water intake as temperature rises above 81°F. Below this point, water consumption is largely a function of dry matter intake.

British investigators reported that at daily temperatures around 46°F, water intake of lactating cows was significantly correlated with daily milk yield and dry matter content of the forage, but was not significantly related either to air temperature or relative humidity.

Sheep. The relation of drinking water intake to ambient temperature for sheep appears to parallel that for cattle. The water intake of growing-finishing lambs at 32 to 50°F was 2.0 lb/lb dry matter (DM) consumed; at 59 to 68°F, it was 2.5 lb/lb DM; and above 68°F, it was 3.0 lb/lb DM.

Swine. Experiments under controlled temperature conditions have shown an inconsistent relationship between ambient temperature and water intake of swine.

Poultry. As ambient temperatures rise, chickens consume increasing amounts of water; when compared to water consumption at 70°F, intake at 90°F is 2.0-fold, and intake at 99°F is 2.5-fold. The hen drinks water at a ratio of 2 to 3 g of water per gram of feed. Water availability for poultry is important for survival under heat stress.

• **Feed**—The water content of feeds ranges from about 10% in air-dry feeds to more than 80% in fresh, green forage. Feeds containing more than 20% water are known as *wet feeds.*

The water content of feeds is especially important for animals which do not have ready access to drinking water. Also, the water on the surface of plants, such as dew, may serve as an important source for cattle, sheep, and goats on arid ranges, but this supply is rarely sufficient to meet their needs.

• **Frequency of watering**—Home range behavior exists among most wild and feral animals, with each home range including at least one watering hole—generally a stream, spring, lake, or pond. The frequency of watering wild and feral animals, as well as of domestic animals on extensive pastures or ranges, is determined primarily by temperature and humidity—the higher the temperature and humidity, the more frequent the watering. However, under average conditions, the frequency of watering of different species is about as follows: Cattle, 1 to 4 times per day. Sheep, 1 time per day, although when grazing desert ranges in early spring, they may go for weeks without drinking water. Goats, 1 time per day, although goats approach camels in their water requirements. When hand-fed, pigs generally eat all their feed, then drink; when self-fed, they alternate between eating and drinking. Horses, usually drink once each day, but during extreme heat they may return a second time. Poultry drink frequently, alternating between eating and drinking. The frequency of watering of cattle and sheep decreases as distance to water increases.

Under practical conditions, the frequency of watering is best determined by the animals, by allowing them access to clean, fresh water at all times.

• **Physiological state**—The physiological state of animals dramatically affects the water requirements. A steer fed a maintenance ration will consume approximately 35 lb of water daily, whereas a steer fed a fattening ration will double this quantity. A dry cow will drink about 90 lb of water daily; during the last 4 months of pregnancy, she will consume 30% more water than when dry and open; when she produces 20 to 50 lb of milk, the daily water consumption will increase to about 160 lb; and when she produces 80 lb of milk per day, water intake will be near 200 lb. Young calves generally drink 1¼ to 1½ times more water per pound of dry matter consumed than older cattle.

The pounds of water/pound dry matter consumed by pregnant ewes increase from about 2.0 in the first month of pregnancy to 4.3 in the 5th month. Ewes carrying twins will consume over twice the amount of water of nonpregnant ewes; and those carrying single lambs will consume 138% more water than nonpregnant ewes.

The daily water intake of sows is estimated to be about as follows: nonpregnant sows, 11 lb; pregnant sows, 11 to 18 lb; and lactating sows, 33 to 44 lb. Weanling pigs will consume approximately 20 lb of water daily per 100 lb of body weight, whereas hogs near market weight will consume only 7 lb per 100 lb of body weight.

• **Quality of water**—Quality of water is determined by the level of dissolved substances, or pollutants. Water quality is almost as important as availability.

A particularly troublesome problem of water found on ranges is salinity. Most species can tolerate a total dissolved solid concentration of 15,000 to 17,000 ppm, but at these levels production will be affected.

[8]Adapted by the authors from: Ames, D. R., Chairman Subcommittee on Environmental Stress, et al., *Effect of Environment on Nutrient Requirements of Domestic Animals,* NRC, National Academy Press, Washington, D.C., 1981, pp. 44-50.

Nitrate pollution may also be a problem. Quantities as small as 100 to 200 ppm can be dangerous, while 3,000 ppm is potentially lethal.

• **Species differences**—The amount of water consumed varies by species.

The water consumption per pound of dry matter consumed may be as much as 40% more for cattle than for sheep. Over the temperature range of 1 to 81°F, the estimated requirements for cattle are 3.5 to 5.5 lb water/pound dry matter consumed, whereas in about the same temperature range sheep need only 2.0 to 3.0 lb water/pound dry matter consumed. The estimated water needs for swine are near those for sheep, 2.0 to 2.5 lb/pound dry matter.

(Also see Chapter 4, Table 4–10, showing the daily water consumption of different species, along with the effect of age, body weight, and condition.)

• **Water excretion**—Water is excreted from the body through three routes: (1) urine, (2) feces, and (3) evaporation from the body surface and respiratory tract.

Urine provides a means whereby water-soluble products of metabolism can be excreted. Generally, when rations are high in protein or mineral content, urine flow is increased.

The amount of water lost in the feces is highly dependent on the animal species and the ration. Sheep, goat, and horse feces contain only 60 to 65% water and are relatively dry, whereas cattle feces contain 75 to 85% water. When cattle are first placed on early spring pasture, the feces become very watery—a reflection of the increased water intake.

Water loss from the respiratory tract is very variable, depending on humidity and respiration rate. Expired air is over 90% saturated; hence, when the humidity is low, respiratory losses are high. Conversely, respiratory losses are low when inhaled air is near saturation. When respiration rate is increased in response to high temperature or other behavioral stimulus, the rate of respiratory water loss is increased.

There are large differences among species in the importance of sweating, with animals ranked in descending order as follows: horses, donkeys, cattle, buffalos, goats, sheep, and swine.

Swine and poultry depend more on the respiratory than the skin route for water loss. In addition to panting, if given the opportunity swine wallow in water and mud to cool themselves.

• **Water sources**—The water needs of animals are filled from three major sources: (1) drinking water, (2) water contained in feed, and (3) metabolic water produced by oxidation of carbohydrates, fats, and proteins.

The water contained in or on feed is very variable; it may range from a low of 5% in dry grains to more than 80% in lush, young grasses. In addition, the amount of dew or precipitation on the grass at the time of grazing is subject to wide fluctuations. In the case of swine and poultry, the water in feed accounts for about 10% of the total feed intake.

• **Water temperature**—Findings on the effect of temperature on water intake are variable. However, cows in a cold environment generally drink more water when it is heated, but cooling water in a warm environment has no effect. Unlike cattle, heating water for sheep in a cold environment does not influence consumption.

WEATHER/ENVIRONMENTAL INTERACTIONS

Webster defines weather as a state of the atmosphere with respect to heat or cold, wetness or dryness, calm or storm, clearness and cloudiness.

Extreme weather can cause wide fluctuations in animal performance. The difference in weather impact from one year to the next, and between areas of the country, causes difficulty in making a realistic analysis of buildings and management techniques used to reduce weather stress.

The research data clearly show that winter shelters and summer shades improve production and feed efficiency. The issue is clouded only because the additional costs incurred by shelters have frequently exceeded the benefits gained by the improved performance, particularly in those areas with less severe weather and climate.

The animal kingdom may be divided into two kinds of animals: (1) *poikilotherms,* or cold-blooded animals, in which the body temperature fluctuates according to the environmental temperature, and (2) the *homeotherms,* or warm-blooded animals, which maintain almost constant body temperature despite wide fluctuations of environmental conditions. The giant dinosaurs which once dominated the earth were poikilotherms, as are the lizzards of today that resemble them. Since farm animals are homeotherms, they require a delicate balance between the heat produced within the animal, the heat or cold gained from the environment, and the heat lost by the animal to the environment. Thus, hot and cold weather are of great importance to them. This becomes evident when it is realized that the annual atmospheric temperature range in Minnesota may be more than 150°F (−40 to 110°F), yet the body temperatures of cattle, sheep, and horses wintering outdoors are constant within a few degrees.

The maintenance requirement of animals increases as temperature, humidity, and air movement depart from the comfort zone. Likewise, the heat loss from animals is affected by these three factors. Animals adapt to weather as follows.

In cold weather, the heating mechanisms are employed, including (1) increased insulation from growth of hair and more subcutaneous fat; (2) increase in thyroid activity; (3) seeking protective shelter and warming solar radiations (the animals sun themselves); (4) huddling together; (5) consumption of more feed, which increases the heat increment and warms animals; and (6) increasing activity. The most important animal body heating mechanisms are amount of feed consumed and body activity, which are also evidenced in people. For example, after skiing in bitter cold weather, a skier feels comfortable after eating a beefsteak; and during a marathon race, a runner may feel quite warm when the temperature is near freezing (30°F).

In hot weather, the cooling mechanisms are employed, including (1) moisture vaporization (from the skin and lungs), (2) avoidance of the heating solar radiation (the animals seek shade), (3) depression of thyroid activity, and (4) loafing (including lessening the production of meat, milk, and eggs, since they increase heat production).

Thermoneutral Zone (Comfort Zone)

Fig. 17–26 and the definitions that follow it are pertinent to an understanding of thermal zones.

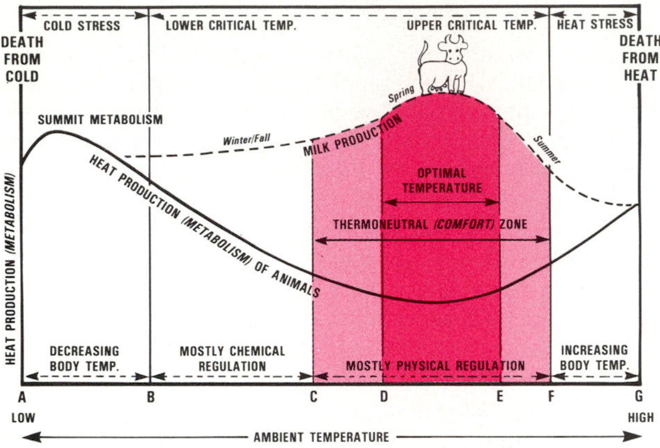

Fig. 17–26. Diagram showing (1) the influence of thermal zones and temperature on homeotherms (warm-blooded animals), and (2) the peak of milk yields in the spring, followed by the summer slump due to high (hot) summer temperature and lignification of forage.

In Fig. 17–26, *heat production (metabolism)* is plotted against *ambient temperature* to depict the relationship between chemical and physical heat regulation. Note, too, the broad range of accommodation to low (cool) temperatures in contrast to the restricted range of accommodation to high (warm) temperatures. Definitions of terms pertaining to Fig. 17–26 follow.

• ***Thermoneutral (comfort) zone*** *(C to F) is the range in temperature within which the animal may perform with little discomfort, and in which physical temperature regulation is employed.*

• ***Optimum temperature*** *(D to E) is the temperature at which the animal responds most favorably, as determined by maximum production (gains, milk, wool, work, eggs) and feed efficiency.*

• ***Lower critical temperature*** *(B) is the low point of the cold temperature beyond which the animal cannot maintain normal body temperature.* The chemical temperature regulation is employed in the zone below C. When the environmental temperature reaches point B, the chemical-regulating mechanism is no longer able to cope with the cold, and the body temperature drops, followed by death. The French physiologist, Giaja, used the term *summit metabolism* (maximum sustained heat production) to indicate the point beyond which a decrease in ambient temperature causes the homeothermic mechanisms to break down, resulting in a decline in both heat production and body temperature and eventually in death of the animal.

• ***Upper critical temperature*** *(F) is the high point on the range of the comfort zone, beyond which animals are heat stressed and physical regulation comes into play to cool them.*

The cow produces the maximum yield of milk during the spring when the temperature is optimum (D to E), and the minimum yield in the summer when it is hot (F to G).

Two facts about Fig. 17–26 should be noted: (1) The animal's cool range is wider than its hot range, which is as nature ordained it—the animal has more body regulators which it can call upon in a cool environment than in a hot environment; and (2) the comfort zone may not be best for the highest productivity, because productive processes involve a *heat increment* which is not easily dissipated in a warm environment, so a cooler environment is more stimulating to high productivity and activity than a warmer environment.

At temperatures immediately below optimum, but still within the comfort zone, there is a *cool* zone (see Fig. 17–26) in which animals invoke mechanisms to conserve body heat, including postural adjustments, changes in hair or feathers, and vasoconstriction of peripheral blood vessels. When the evnironmental temperature goes below the *lower critical temperature* (see Fig. 17–26), the animal is *cold stressed* and chemical regulation comes into play. As the environmental temperature continues to drop, the homeothermic mechanisms fail, resulting in a decline in both heat production and body temperatue, followed by the tissues freezing and eventually death of the animal.

At temperatures immediately above optimum, but still within the comfort zone, there is a *warm* zone (see Fig. 17–26) in which animals invoke limited thermoregulatory reactions, including decreasing tissue insulation by vasodilation and increasing effective surface area by changing posture (for example, hogs sleep stretched out full length so as to expose the maximum body surface to the air) to facilitate rate of heat loss. When the environmental temperature goes above the *upper critical temperature* (see Fig. 17–26), the animal is *heat stressed* and employs physical regulation, especially evaporative heat loss mechanisms such as sweating and panting; when these become insufficient to cool the body, death follows. Also, the immediate response of animals to heat stress is reduced feed intake, to attempt to bring heat production in line with heat dissipation capabilities. Higher producing animals with greater metabolic heat from product synthesis tend to be more susceptible to heat stress than lower producing animals. Evaporation of moisture from the skin surface or respiratory tract is the primary mechanism used by animals to lose excess body heat in a hot environment; this mechanism is limited by air vapor pressure but enhanced by air movement.

The comfort zone, optimum temperature, and both upper and lower critical temperatures vary with different species, breeds, ages, body sizes, physiological and productive status, acclimatization, feed consumed (kind and amount), the activity of the animal, and the opportunity for evaporative cooling.

The temperature varies according to age, too. For example, the comfort zone of newborn lambs is 75 to 80°F, whereas the comfort zone of mature sheep is 45 to 75°F.

Animals that consume large quantities of roughage or high-protein feeds produce more heat during digestion; hence, they have a different critical temperature than the same animals fed a high-concentrate, moderate-protein ration. Because of this, experienced cattle feeders decrease the roughage and increase the concentrate of finishing cattle during the hot summer months.

Stresses at both high and low temperatures are increased with high humidity. The cooling effect of evaporating sweat is minimized and the respired air has less of a cooling effect. As humidity of the air increases, discomfort at any temperature, and nutrient utilization, decrease proportionately.

Air movement (wind) results in body heat being removed at a more rapid rate than when there is no wind. In warm weather, air movement may make the animal more comfortable, but in cold weather it adds to the stress temperature. At low temperatures, the nutrients required to maintain body temperature are increased as the wind velocity increases. In addition to the wind, a drafty condition where the wind passes through small openings directly onto some portion or all of the animal body will usually be more detrimental to comfort and nutrient utilization than the wind itself.

Adaptation, Acclimation, Acclimatization, and Habituation of Species/Breeds to the Environment

Every discipline has developed its own vocabulary. The study of adaptation/environment is no exception. So, the following definitions are pertinent to a discussion of this subject:

Adaptation refers to the adjustment of animals to changes in their environment.

Acclimation refers to the short-term (over days or weeks) response of animals to their immediate environment.

Acclimatization refers to evolutionary changes of a species to a changed environment which may be passed on to succeeding generations.

Habituation is the act or process of making animals familiar with, or accustomed to, a new environment through use or experience.

Species differences in response to environmental factors result primarily from the kind of thermoregulatory mechanism provided by nature, such as type of coat (hair, wool, feathers), and sweat glands. Thus, hogs, which have a light coat of hair, are very sensitive to extremes of heat and cold. On the other hand, nature gave cattle an assist through growing more hair for winter and shedding hair for summer, with the result that they can withstand higher and lower temperatures than hogs. The long-haired, shaggy yak of Tibet and the wooly Scotch Highland cattle of Scotland are as cold tolerant as the arctic-dwelling caribou and reindeer.

Fig. 17–27. Long and wooly haired Scotch Highland cattle equipped by nature to withstand very cold weather. (Courtesy, American Scotch Highland Breeders Assn., Edgemont, S.D.)

From time to time, American buffalo *(Bison bison)* and domestic beef cattle *(Bos taurus)* have been crossed to obtain a more hardy beast than cattle. The most publicized early work of this type was the development of the Cattalo (bison x domestic cattle), the initial cross for which was made at the Dominion Experiment station, at Scott, Saskatchewan, in Canada, in 1915.

Fig. 17–28. Cattalo (¼ buffalo, ¾ domestic cattle) cow. The initial Cattalo breeding experiment was started by the Dominion Experimental Station, Scott, Saskatchewan, Canada, in 1915. The foundation herd consisted of 16 female and 4 male hybrids. (Courtesy, Research Station, Canada Department of Agriculture, Lethbridge, Alberta, Canada)

Male fertility and female reproductive rate have remained a problem in Cattalo. Although unquestionably hardy, bison x cattle crosses can be outperformed in nearly all environments by the currently available cattle breeds or crosses; and management procedures.

Also, there are breed differences, which make it possible to select animals well adapted to specific environments. Thus, the breeds of cattle that originated in the British Isles and Northern Europe are cold tolerant, whereas the Indian-evolved Zebu, or Brahman, cattle are heat tolerant. The long-fibered Black-faced Highland sheep are cold tolerant; the haired sheep are suited to hot, desert areas; and the fat-tailed sheep are adapted to arid conditions. The Shetland Pony, native to the Shetland Isles, not more than 400 miles from the Arctic Circle, evolved in the rigors of the northland climate and on sparse vegetation, which imparted that hardiness for which the stocky breed is famed. The long-legged donkey is adapted to hot, desert areas.

In recent years, attempts have been made to combine the heat tolerance characteristics of tropical breeds with the high productive capacity of European stock. The best known of these planned beef breeds is the Santa Gertrudis, developed on the famed King Ranch of Texas, in the early 1900s, which carry approximately five-eighths Shorthorn and three-eighths Brahman breeding.

Animals have the ability to become acclimated to high temperature. Thus, two trials of 46-day-old broilers were used in an experiment at Mississippi State University. One group was acclimated to a diurnal cycle of 75–95–75°F for a 4–day period, while the controls were kept at a constant 70°F. During the fourth day, both groups of broilers were exposed to 106°F temperature for 210 minutes. Acclimated broilers had significantly lower heat stress mortality. At the end of Trial 1, total mortality was 30% for the controls and 10% for the acclimated broilers. In Trial 2, mortality was 60% and 10% for the control and acclimated broilers, respectively.[9]

The Missouri Station workers conducted classical studies designed to show breed differences between Shorthorn, Santa Gertrudis (five-eighths Shorthorn and three-eighths Brahman), and Brahman cattle.[10] The animals were housed in *climatic chambers,* in which the temperature, humidity, and air movements were regulated as desired. The ability of representatives of the different breeds to withstand different temperatures was then determined by studying the respiration rate and body temperature, the feed consumption, and the productivity in growth, milk, beef, etc. Dr. Samuel Brody, who directed the study, reported the following pertinent points:

1. The most comfortable temperature for the Shorthorns was in the range of 30 to 60°F, while for the Brahmans it was 50 to 80°F, and for the Santa Gertrudis it was intermediate between the ideal temperatures for the two parent breeds.

2. The Brahman cattle could tolerate more heat—they could withstand higher temperature better than Shorthorns, whereas the Santa Gertrudis approached the Brahman in heat tolerance.

3. The Shorthorn cattle could tolerate more cold—they could withstand a lower temperature better than the other two breeds, while the Santa Gertrudis were more cold-tolerant than Brahman cattle.

Dr. Brody attributed the higher heat tolerance of Brahman cattle to their lower heat production, greater surface area (their loose skin) per unit weight, shorter hair, and "other body temperature-regulating mechanisms not visually apparent."

A brief summary of the known species and breed differences follows.

BEEF CATTLE

Beef cattle are kept in many climatic regions, and, except for confinement production systems, are largely exposed to the prevailing weather conditions. In intensive beef production systems such as feedlots with shelters or confinement barns, there may be some modulation and protection from the weather, but other stress factors may be generated that have an adverse effect on animal performance, including mud, dust, gaseous contaminants, and/or crowding.

[9]May, J. D., J. W. Deaton, and S. L. Branton, "Body Temperature of Acclimated Broilers During Exposure to High Temperature," *Poultry Science,* Vol. 66 (2), 1987, pp. 378–380.

[10]Brody, Samuel, *Climate Physiology of Cattle,* Jour. Series No. 1607, Mo. Ag. Exp. Sta.

Experimental work indicates that cold and heat affect beef cattle as follows:

• **Cold weather tolerance**—Beef cattle are quite tolerant to cold weather. Nevertheless, winter cold stress can be compensated for by either, or both, of the following ways: (1) providing shelters that will create an environment which will enable animals to maintain the same production with a given amount of energy intake, or (2) increasing the energy intake which will better enable animals to withstand cold stress conditions.

1. **Voluntary feed intake.** In general, voluntary feed intake increases when ambient temperatures decrease and decreases as ambient temperature increases. The beef cattle breeds of European origin have a relatively small surface area with respect to body mass and are adapted to cold climates.

2. **Maintenance requirement.** The lower critical temperature for beef cattle is lower than for other domestic animals; thus, direct cold stress of beef cattle is generally not a practical problem except in areas with extremely cold winters or in the chilling effects of moisture and wind on young stock or animals with poor resistance to cold. Increased heat production during cold stress requires an immediate utilization of energy from either the ration or tissue reserves.

Increasing the percentage of roughage in cattle rations results in slight increase in heat production because of increased heat increment associated with feeding. So, for cattle on maintenance rations in cold climates an increase of about 25 to 30% in roughage during cold stress is advantageous, with the amount of the increase determined by the chill factor and the animal's hair coat and age.

3. **Finishing requirement.** Although it is usually advantageous to increase the roughage level of beef cattle on maintenance rations during cold stress, this does not apply to finishing cattle. Roughages produce more heat per unit of digestible energy, but they do not produce more heat per pound of dry matter. Thus, more energy will be generated from 1 lb of grain than from 1 lb of roughage. So, to combat cold stress in finishing cattle, it is best to continue feeding the normal finishing ration, but increase the daily allowance. Nevertheless, roughage feeding does contribute to warming cattle due to the dissipation of heat increment energy.

4. **Lower critical temperature.** The estimated lower critical temperatures for different classes of beef cattle are:

a. For finishing cattle gaining 3.3 lb daily, −22°F or even colder.

b. For pregnant cows, 14 to −4°F.

• **Hot weather tolerance**—The heat stress of beef cattle varies considerably among animals depending on breed, acclimatization, ration, level of productivity, and daily (diurnal) fluctuations in radiant heat load.

The Zebu breeds, such as the Brahman, have a large amount of surface area in relation to body mass (enlarged by the loose and pendulous skin and the hump) and are well adapted to hot climates. Also, there is indication that cattle with Zebu breeding tolerate protein deficiencies better than the English-derived breeds.

Fig. 17-29. Brahman cattle are heat tolerant, as evidenced by this herd on summer pasture on the Texas Gulf Coast. (Courtesy, American Brahman Breeders Assn., Houston, Tex.)

1. **Voluntary feed intake.** In general, voluntary feed intake decreases as ambient temperature increases and increases as ambient temperature decreases.

Voluntary feed intake of cattle is also affected by breeds. For example, the breeds of cattle (both beef and dairy) developed in temperate regions are generally more productive than their tropical counterparts. However, they are poorly tolerant to heat; so, when they are exported to the tropics, their performance may be no better, and is frequently worse, than that of the native breeds. This is largely because their productivity in temperate regions is due to their high feed intake, which, in a hot climate, they are unable to sustain.

2. **Maintenance requirement.** During severe heat stress, maintenance requirements increase due to the energy required for increased panting and alterations in tissue metabolism because of increased tissue temperature.

3. **Finishing requirement.** Heat hurts finishing cattle gains more than cold! Two factors affect the response of finishing cattle to heat stress: (1) the amount of insulation (hair and external fat), and (2) the heat given off by digestion. With hot temperature and humidity, feed intake may be lowered up to 30%. Heat stress is reduced if accompanied by low humidity and/or relief through more comfortable nighttime conditions. Drop in feed consumption due to elevated temperature and humidity is more severe on high-roughage rations than on low-roughage rations.

4. **Higher critical temperature.** When the minimum temperature is 81°F, the daily weight gain and feed consumption of finishing cattle drop. Increasing the percentage of roughage in cattle rations results in a slight increase in heat production because of the increased increment of roughage. Under hot environmental conditions, this increased heat production can result in reduced voluntary feed intake. So, it is advantageous to feed rations of low roughage content during hot weather. Rapid and wide daily temperature fluctuations, as occur in the spring and fall, tend to reduce feed intake of cattle and cause more health problems, especially more pneumonia.

DAIRY CATTLE

The efficiency of performance of dairy cattle of all ages is influenced by both low-temperature and high-temperature conditions, with high temperature producing the greatest stress. For lactating cows, the range of temperature for highest efficiency of energy utilization is about 55 to 65°F; however, significant changes in feed intake will not usually occur within the range of 41 to 77°F. Above 77°F or below 41°F, appetite will be influenced by the thermal environment.

Dairy cows are generally kept (1) on pasture with full exposure to the weather, (2) in drylot with access to a constructed shelter, or (3) confined within a barn or shelter with or without environmental control.

• **Cold weather tolerance**—It is practical to increase the feed allowance of lactating cows when environmental temperature is below 32°F.

With decreasing temperatures below 50°F, Jersey cows show a faster decrease in milk production than Holstein cows.

1. **Voluntary feed intake.** Using 64 to 68°F as a base point of 100%, feed intake of lactating cows increases with decreasing temperature, reaching about 150% at −4°F.

2. **Maintenance requirement.** When the temperature decreases from 64 to 68°F down to −4°F, the estimated dry matter intake of lactating cows will increase from about 40 to 45 lb. But, due to rising maintenance requirements, less feed intake will leave less ME available for milk synthesis with the result that milk yield will drop from about 59 to 44 lb.

3. **Calves need more fat in winter.** During cold weather, calves in cold housing (especially in hutches) need a high fat milk replacer and some forage. The milk replacer should have a minimum of 20% fat. If whole milk is being fed instead of a replacer, calves get sufficient fat; 3.5% butterfat milk is equivalent to 29% fat on a dry matter basis.

4. **Lower critical temperature.** When constant environmental temperature drops below 23°F, the dry matter intake of lactating cows increases, but the milk yield decreases.

• **Hot weather tolerance**—Heat stress in dairy herds during hot weather is a serious problem. In heat stressed cows, feed intake drops, milk production may decrease as much as 30%, only 10 to 20% of the inseminations may result in normal pregnancies, general health problems are more frequent, and death losses increase.

Rations high in grain and low in forage (fiber) cause less heat stress in lactating cows because of lower heat of digestion. Milk fat is usually depressed and digestive disorders increase during hot weather when forage intake is limited; however, a buffer will help control rumen acidity, prevent lower fat tests, and improve digestibility.

Since total heat load is derived from metabolic heat production plus heat from the environment, high-producing cows react more negatively to rising temperatures than low producers. So, the natural way for the high-producing cow to react to heat stress is to eat less feed and give less milk.

Evaporation by sweating and panting is the most effective natural way for a cow to lose heat. Several methods may be used by dairy producers in the summer to minimize heat stress, the most common and effective of which are as follows:

1. Provide plenty of clean, fresh drinking water.
2. Reduce roughage (fiber), increase energy (by adding up to 6% fat, or by adding whole oil seeds such as whole cottonseed), and increase protein and minerals to compensate for reduced total feed intake. However, excess protein over requirements should be avoided, because it will contribute to heat build-up. Add 1 to 1½% of a buffer to help control rumen acidity, improve digestibility, and prevent lower fat tests.
3. Increase the potassium (to 1.5%), the sodium (to 0.5–0.6%), and the magnesium (to 0.3%) in the total diet.
4. Increase the vitamin A level to about 150,000 units per cow per day for high-producing cows.
5. Feed more frequently, and increase the proportion of feed fed at night when it is cooler.
6. Add water to dry feeds to reduce dust and increase palatability.
7. Cover feeding areas and provide shades.
8. Provide fans, sprinklers, misters, and/or evaporative coolers, properly engineered and located.
9. Prevent environmental mastitis by providing a clean, cool, dry environment.

The Jersey and Brown Swiss breeds of dairy cattle will maintain their milk production in hot summer temperatures better than Holsteins or other dairy breeds.

When managed traditionally, cows calving in the fall or winter produce more milk during the total lactation period than those coming fresh in spring or summer. This is because spring and summer calvers peak in milk yield during hot weather.

1. **Voluntary feed intake.** At daytime temperatures above 86°F, feed intake of lactating cows declines because of behavioral changes directed toward conservation of body heat, and also because summer forages are often less palatable.

2. **Maintenance requirement.** When the temperature increases from 86 to 104°F, the estimated dry matter intake of lactating cows will decline from about 37 to 22 lb and expected daily milk yield will drop from 50 to 26 lb.

3. **Higher critical temperature.** When constant environmental temperature exceeds 77 to 86°F, the dry matter intake and milk yield of cows fall.

Summary Relative to Dairy Cows. It is practical to increase the feed allowance of lactating cows when environmental temperature is below 32°F, but lessening heat stress is more difficult and costly.

Environmental modifications to alleviate or reduce the stress of cold or hot conditions of lactating cows offer promise as an alternative to higher intakes of feed, but caution must be exercised to ensure that the increased milk production and feed efficiency are sufficient to compensate for the cost of environmental modification.

SHEEP

Sheep are more tolerant of climatic extremes than other farm animals. They are unique because of the potentially large insulatory value of the fleece.

Lambs grow most rapidly when the environmental temperature is between 41 and 77°F. High-roughage rations are beneficial to lambs when environmental temperature is below 41°F, but detrimental when it is above 77°F.

• **Cold weather tolerance**—The cold tolerance of sheep depends largely on the amount of external insulation provided by the fleece. But it is also affected by feed intake; wetting of the fleece, which reduces its insulatory value; and wind.

Cold weather, especially if accompanied by rain and wind, increases mortality, due to pneumonia, of newly shorn sheep.

1. **Voluntary feed intake.** Results of several experiments clearly show that voluntary feed intake is increased above the thermoneutral (comfort) zone during mild cold, but that a limit in voluntary intake was reached before animals were severely cold stressed.

2. **Maintenance requirement.** When exposed to ambient temperatures below the cool zone, energy expenditures must compensate for increased energy loss. Wind, rain, and wind and rain combined have been shown to increase further the rate of heat loss of sheep exposed to cold. From a practical standpoint, increased maintenance energy for sheep during cold weather can best be achieved, at lowest cost and with highest heat increment, by increasing the roughage and reducing the concentrates.

Confinement rearing of sheep has been estimated to reduce energy requirements by 30% below range-reared animals.

3. **Lower critical temperature (LCT).** The lower critical temperature for sheep depends largely on the amount of external insulation provided by the fleece. Shorn growing lambs fed a maintenance ration have a LCT of about 55°F, whereas LCT is estimated as low as −4°F in similar animals with fleeces. Experiments also show that the LCT is influenced by fasting, by wetting of the fleece, and by wind.

• **Hot weather tolerance**—High temperatures have a marked effect on fertility, embryo survival, and fetal development. Experiments show that air temperatures higher than 100°F for periods of 3 months or longer almost eliminate reproduction in sheep, and that constant temperatures of 90°F reduced fertility by 50%, with no embryo survival. When ewes were exposed to a continuous 90°F temperature on the day of breeding, none of the embryos survived. When ewes were exposed to 90°F temperature one day after breeding, 70% of the embryos were lost. Exposure to heat 3, 5, and 8 days after breeding had progressively less effect on embryo survival; and very little embryo loss occurred when ewes were exposed to heat later than 25 days after mating. All these results were obtained when the relative humidity was constant at 60 to 65%. With lower relative humidity, heat stress may not be as harmful. Heat stress during gestation also has an adverse effect on fetal development, resulting in significantly smaller lambs at birth.

Overeating in extremely warm weather may increase body temperature. Also, excessively fat ewes are more susceptible to heat stress than ewes in moderate breeding condition.

In hot areas, ewes should not be bred until late fall, when the weather is cooler. Also, heat stress in ewes may be lessened by shearing, keeping them in shaded areas or cool buildings during the day, and pasturing only at night. In range bands, where ewes cannot be housed or shaded during the day, they should not be sheared. Body temperature in direct sunlight will be lower in sheep with 1 or 2 in. of wool than in freshly shorn sheep, as the fleece serves as an insulator against both heat and cold.

During warm weather, shorn suckling lambs make about $\frac{1}{10}$ lb higher daily gains than their unshorn counterparts.

North Carolina State University scientists are attempting to produce sheep that are better adapted to warm climates than the existing breeds. To this end, they have crossed tropical Barbados blackbelly sheep, which have hair instead of wool, on Dorset sheep.

1. **Voluntary feed intake.** Exposure to heat reduces voluntary feed intake with sheep as it does with other species, with depression most severe when the rations contain high levels of roughage.

2. **Maintenance requirement.** During heat exposure, the energy requirements increase because of energy expended during panting, sweat gland activity, and the calorigenic effect of hormones. For relief of heat stress of growing-finishing lambs, the heat increment of the rations should be lowered by reducing the roughage while increasing the net energy. For example, in hot weather, rations might contain 40% roughage and 60% concentrate (with fat added), instead of 60% roughage and 40% concentrate (with no fat).

3. **Higher critical temperature.** Sheep will survive acute periods of 104 to 140° F, but successful growth and reproduction at these temperatures have not been demonstrated and are unlikely. Wool (length of fleece), wind, and humidity are of major importance in determining survival during heat.

SWINE

The relations between practical thermal environments, feed intake, and performance have been studied for swine, more than any other species of farm animals except chickens.

There is consistent evidence that feed efficiency decreases during extremely cold and extremely hot environmental conditions.

Swine are sensitive to changes in the effective thermal environment. Size is highly correlated with fatness and therefore related to rate of heat loss. Consequently, responses of swine to specific environmental conditions are largely dependent on size.

It is generally easier and less costly to protect animals from a cold environment than from a hot environment. The cold weather protection can be provided by enclosing the building to conserve the animal's sensible body heat and by supplying supplementary heat. Supplying supplemental heat is normally less expensive than removing heat mechanically from the thermal environment surrounding the animal.

Cold stress and heat stress will be considered separately.

• **Cold weather tolerance**—Young pigs up to 8 weeks of age are adversely affected by a cold environment. During the first week of life, the nursery temperature should be at 85° F, thereafter, it should be reduced weekly until it is 79° F during the fifth week (see Chapter 23, Table 23-29). From the fifth week to the eighth week of age, the optimum temperature is around 75° F.

Early weaned pigs have trouble maintaining a positive body-energy balance in a cold climate and mortality is often high. Also, cold-stressed pigs are more susceptible to infectious diseases; resistance to both bacterial and viral diseases has been shown to be lower in pigs during cold weather. The incidence of diarrhea in baby pigs is twice as high in pigs subjected to 50° F temperature as compared to 82° F. Also, the growth rate per day of early weaned pigs is nearly twice as high at 82° F as at 50° F.

Increased heat loss during cold stress reduces gains of growing-finishing pigs unless compensation is made by increasing rate of feed intake.

A Missouri Agricultural Experiment Station study showed that gilts kept in cool chambers (65° F) reached puberty 3 weeks earlier than those kept at 92° F.

Alfalfa hay will assist sows in tolerating cold stress. When a sow eats alfalfa, about 90% of the metabolizable energy in it is converted to heat. So, 1½ lb of alfalfa per sow daily will provide as much heat as 1 lb of a corn-based ration.

1. **Voluntary feed intake.** Under practical conditions, it is estimated that the extra feed requirement for pigs under cold stress averages in the range of 17 to 22 g/° F of coldness per day for the growing-finishing period when body weight ranges from 44 to 220 lb.

2. **Maintenance requirement.** When self-fed, a pig normally consumes feed at a rate of 2 to 4 times the maintenance level. As a consequence, its heat production is higher and its lower critical temperature is lower than when metabolizable energy is at the maintenance level.

In cold stress, extra feed is needed to keep the body warm. A pig weighing 44 lb should consume additional feed at the rate of 7 g/day/° F coldness, and a pig weighing 222 lb should consume additional feed at the rate of 19 g/day/° F.

3. **Lower critical temperature (LCT).** The lower critical temperature of pigs in a group on a well-insulated floor averages about 55 to 57° F in the growing period and 50 to 52° F during finishing. Slotted floors and wet floors cause heat losses and change the critical temperatures. The LCT of *ad libitum*-fed pigs on partly wet, slotted floor will be about 64 to 66° F during the growing phase and 57 to 59° F for finishing hogs.

Daily weight gain is depressed by 8 g per ° F of coldness when feed intake remains constant.

Three additional factors related to LCT are important: (1) Individual pigs consume different amounts of feed, and those with a relatively low feed intake have higher LCT and therefore are especially vulnerable to cold stress; (2) lean-type hogs are more sensitive to cold stress than fat-type hogs; and (3) high vapor pressure may have a negative effect on performance. Also, pigs in groups huddle in cold environments, with the result that their heat-loss rate is less and their critical temperature is lower than that of pigs held singly.

• **Hot weather tolerance**—Heavy hogs are affected more dramatically by hot weather than by cold. During heat stress, rate of feed intake is depressed, resulting in lowered performance.

Most investigators have found the temperature range of 64 to 70°F optimum for growing-finishing pigs. Above approximately 79°F, swine are heat stressed, although the temperature at which heat stress occurs, as well as its severity, are affected by (1) vapor pressure (since the pig must rely on evaporative heat loss in heat stress); and (2) weight of hogs (heavier hogs are more sensitive to heat stress than lighter ones).

Because they have few sweat glands, hogs have to breathe faster and faster under hot weather conditions in an attempt to remove their surplus body heat by evaporation from the moist surfaces of the lungs and air passageways.

In hot areas with low humidities, environmental coolers incorporated in the ventilation systems of swine buildings are effective. Also, sprinkling and wallows are effective in relieving heat stress.

The farrowing house presents a special problem in temperature control because, ideally, two different temperatures are needed—heat for the baby pigs, and cooling for the sows. When farrowing crates are used, this may be solved by (1) providing warmer floor temperatures, supplemental radiant heat, and bedding for pigs in the creep areas, and (2) providing snout coolers for the sows.

1. **Voluntary feed intake.** It is estimated that pigs eat about 33 to 55 g less feed each day per °F heat stress (at 90°F as opposed to 70°F), and that this decline in feed intake will result in a reduction in daily gain of 19 to 32 g/°F.

McGlone reported, in a personal communication to the senior author, that heat stressed sows at Texas Technology University, Lubbock, benefited from drippers; they ate more feed, weaned heavier litters, and lost less weight. Also, the drippers were more effective on relatively warm plastic coated metal floors than on cooler concrete slats. This study underscored two important facts about the environment: (1) it includes more than one important component, and (2) its components interact.

2. **Maintenance requirement.** Based on observation, a swine caretaker would describe the upper critical temperature as *(1) that point at which a pig with dry skin can maintain maximal rate of heat loss, and (2) that point at which there is a rise in the frequency of respiration.*

Table 17-2 shows decreases in feed intake and gain for swine of different weights as a result of heat stress. As noted, heavier pigs are more sensitive to heat stress than lighter ones.

TABLE 17-2
DECREASE IN FEED INTAKE AND RATE OF GAIN DURING HEAT STRESS[1]

Temperature Increase		Weight of Hogs	Decreases			
			Feed Intake		Gain	
(°F)	*(°C)*	(lb)	(g/°F)	*(g/°C)*	(g/°F)	*(g/°C)*
71.6–100.4	*22–38*	29.7	4	*8*	2	*4*
71.6–100.4	*22–38*	77	18	*33*	4	*8*
71.6–100.4	*22–38*	143	23	*41*	7	*12*

[1]Ames, D. R., *et al., Effect of Environment On Nutrient Requirements of Domestic Animals,* NRC, National Academy Press, Washington, D.C., 1981, p. 106, Table 34.

Based on the averages of several sets of data, growing-finishing hogs (consuming about 6.2 lb of feed daily at thermoneutrality) will decrease (1) their feed consumption about 22 g daily per °F of heat stress, and (2) their daily gain by 10 to 20 g.

It is logical that adding about 5% fat to swine rations may be advantageous during heat stress, because fat has a lower heat increment than either carbohydrate or protein. In addition, fat has a high caloric density that helps offset lowered feed intake during heat stress. Also, it is important that the crude protein and amino acids of heat stressed swine be provided in accordance with the daily requirements and not fed as a constant percentage of the ration.

3. **Higher critical temperature.** Although there are both variables and disagreements, perhaps a temperature of 79°F may be considered an acceptable figure above which hogs become heat stressed. Of course, not all hogs will be heat stressed. Heat stress can be lessened by providing shade, increasing air velocity, using wallows, foggers/misters, and taking other measures; in turn, these measures will influence feed intake and daily gain.

HORSES

Wild horses were little affected by their environment as long as they roamed pastures and ranges. But the domestication of horses and their confinement into smaller spaces has introduced a whole set of environmental influences.

Properly designed horse barns and other shelters, shades, insulation, ventilation, and air conditioning can be used to approach the most suitable environment. Naturally, the investment in environmental control facilities must be balanced against the expected increased returns; and there is a point beyond which further expenditures for environmental control will not increase returns sufficiently to justify the added cost. This point of diminishing returns will differ among sections of the country, quality of the horses (the more valuable the animals, the higher the expenditures for environmental control can be), and operators; and labor and feed costs will enter into the picture, also.

Environmentally controlled buildings are costly to construct, but they make for the ultimate in animal comfort, health, and efficiency of feed utilization. Also, they lend themselves to automation, which results in a saving in labor; and, because of minimizing space requirements, they effect a saving in land cost. Today, environmental control is rather common in poultry and swine housing, and it is on the increase for horses.

Environmental control is of particular importance in horse barn construction, because many horses spend the majority of their lives in stalls—for example, race and show horses may be confined as much as 95% of the time.

A temperature range of 50 to 75°F is considered the comfort zone for the horse, with 55°F considered optimum. Until a new born foal is dry, it should be warmed to 75 to 80°F, which can be accomplished by means of a heat lamp.

Unfortunately, few experiments have been conducted on the effect of temperature on the well being of horses.

• **Cold weather tolerance**—Horses are not very sensitive to cold. In the wild state, they developed a long, shaggy coat of hair in the wintertime, sought shelter from storms under trees and in the valleys, and pawed the snow to obtain forage. Like cattle, horses face away from the direction of a severe storm.

Although clipping and daily grooming make the horse more attractive, they remove the animal's natural protection against cold. So, when the weather is cold, idle horses should be blanketed.

1. **Voluntary feed intake when cold.** Feed intake is influenced by weather. For example,under ideal (optimum) weather conditions in October, a horse may eat 14 lb of 60% TDN feed daily, whereas in the same area the horse may eat 20 lb in the winter. Thus, feed consumption is increased with lower than optimum temperature.

2. **Maintenance/cool temperature energy partition.** A mature horse must have heat to maintain body temperature, sufficient energy to cover the internal work of the body and the minimum movement of the animal, and a small amount of proteins, minerals, and vitamins for the repair of body tissues. Normally, about half of all the feed consumed by horses is used in meeting the body maintenance requirements. The harder horses work and the more they eat, the smaller the proportion of the ration that is used for maintenance. For example, during brief periods when racing, horses use up to 100 times the energy utilized at rest, with the result that a much smaller proportion of their energy is used for maintenance than when idle.

In cold weather, horses eat more to keep warm, with the result that a higher proportion of their feed is used for maintenance.

3. **Lower critical temperature.** The lower critical temperature of horses is not known. Their shaggy coats of hair and a layer of fat under the skin, along with a windbreak and feed, equip them to withstand very low temperatures. If the environment becomes too cold, the demand for heat production may exceed the animal's metabolic capacity, and the animal may die of hypothermia (excessively low body temperature).

• **Hot weather tolerance**—The horse is very tolerant to hot weather provided (1) plenty of salt and water are provided, and (2) it is not allowed to gorge on water when hot.

The salt requirements vary with work and temperature. Normally; the horse needs about 2 oz of salt daily, or less than 1 lb per week. However, when at hard work during hot weather—conditions accompanied by profuse perspiration and consequent loss of salt in the sweat—even greater quantities may be required. The white, encrusted sides of a horse after work are evidence of the large amount of salt drawn from the body through sweat (2 g of salt/pound of sweat). Horses at moderate work may lose 50 to 60 g of salt in the sweat and 35 g in the urine daily. Unless this salt is replaced, the animal will soon exhibit signs of excessive fatigue.

Frequent, small waterings between feedings are desirable during warm weather or when the horse is being put to hard use. Horses should not be allowed to drink heavily when they are hot, because they may founder; and they should not be allowed to drink heavily just before being put to work.

1. **Voluntary feed intake when hot.** Feed consumption is reduced as the ambient temperature rises. Thus, the nutrients provided in the ration are subject to variation because they vary with the consumption of the ration. For example, if the horse eats half as much feed as recommended, the percent of nutrients (proteins, minerals, vitamins) must be doubled in order that an adequate amount of each nutrient will be consumed daily.

2. **Maintenance/warm temperature energy partition.** Warmer temperatures reduce basal metabolic rate and maintenance energy requirement. Also, both body weight and degree of work affect the feed required by horses.

3. **Higher critical temperature.** Although the higher critical temperature of horses is not known, when extremely hot they employ physical means of cooling, including profuse sweating, increased blood flow to the skin and underlying tissue, decreased feed consumption, and decreased activity. Also, horses shed their winter coats with the coming of warm weather. Horses have more defense against cold than against heat.

• **Heat stroke (sun stroke)**—When being exercised very vigorously during hot, humid weather, horses may suffer from heat stroke. The symptoms: rapid breathing and heartbeat, weakness, dry skin, and no sweating. First aid consists of immediately lowering the body temperature by placing the horse in a cool place and spraying or sponging its head, lower side of the neck, and legs with cold water. In extreme cases, ice packs may be used and ice may be placed in the rectum. It may be necessary for the veterinarian to restore lost body fluids (the electrolytes).

POULTRY

The most conspicuous thermoregulatory behavior of birds is migration to warmer or cooler areas. But domestic poultry are poorly adapted to flying, so they are dependent on their natural thermoregulatory mechanisms plus the environmental modifications provided by their caretakers.

Many environmental factors have an impact on the production of poultry meat and eggs, including temperature, humidity, light (length of day and intensity), altitude (air pressure and partial pressures of oxygen and carbon dioxide), wind velocity (air movement), solar energy, quality of air and water, and density of population. The discussion that follows will deal primarily with one environmental factor only—temperature.

Birds are warm-blooded animals—homeotherms. Hence, a bird must maintain a constant body temperature for normal physiological functions. If a bird is in a cold environment, the body must produce heat through metabolic processes and use insulatory mechanisms to keep the heat from escaping. If a bird is in a hot environment, the body must give off heat and utilize cooling mechanisms.

The relatively high deep body temperature (chickens 106.7°F), the absence of sweat glands, and the very effective insulation provided by the plumage distinguish the thermoregulatory physiology of poultry.

In general, during cold weather, the feed conversion efficiency of layers and broilers usually falls, and during very cold weather, egg laying and growth rate are reduced, also. In hot environments, layers and broilers reduce feed intake, growth rate, and feed conversion efficiency.

Research data clearly show that shelters to shield poultry from the weather almost always improve production and feed conversion efficiency. However, the additional costs incurred by environmentally controlled buildings have sometimes exceeded the benefits gained by improved performance.

Tests conducted by the U.S. Department of Agriculture showed that hens kept at 55°F laid at a rate of 75% and consumed 3.5 lb of feed for each pound of eggs, whereas those maintained at 85°F laid at a rate of only 50% and consumed 4 lb for each pound of eggs, while those kept at 23°F laid at a rate of only 26% and ate 12.3 lb of feed for each pound of eggs. Experiments conducted at the University of Connecticut showed a 12.5% increase in feed efficiency in broilers grown in a 75°F house compared to those grown at 45°F.

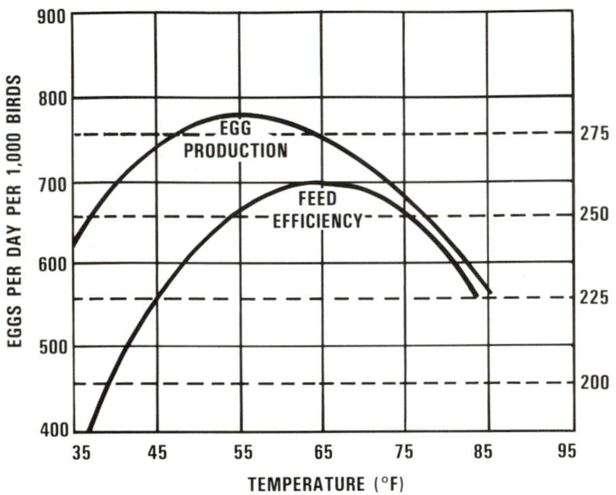

Fig. 17-30. Egg production (upper curve) and feed efficiency (lower curve) at various house temperatures. (Courtesy, USDA)

Nutritionally, the perfect environment for poultry would be the ambient conditions that maximize gain in weight or egg output at the least expenditure of nutrients. Unfortunately, the perfect environment may not be economically feasible. So, a compromise must be reached. The latter is known as the *optimum environment,* in which both production and cost are considered. Thus, if a low ambient temperature (a cool temperature) reduces productivity, providing a higher environmental temperature will be practical if the increased productivity equals or exceeds the expenditure required to supply it. Conversely, a moderately higher ambient temperature (not in the critical zone) may result in the birds expending less of their metabolizable energy for maintaining a constant body temperature and increase feed efficiency. This is so because a smaller proportion of the feed is used for maintenance. For example, when receiving a ration containing 1,386 kcal/pound of metabolizable energy, White Leghorn hens require about 100 g of feed/hen/day, or 124 g of feed/egg, in a cool climate; in comparison with about 90 g of feed/hen/day, or 115 g of feed/egg, in a warm climate.

Breeds differ in their energy requirements for maintenance, due primarily to differences in body weights. Thus, in a moderate environmental temperature and when receiving well balanced rations, White Leghorn hens require approximately 300 to 320 kcal of metabolizable energy per hen per day. Broiler breeder hens, because of their heavier weight and higher energy requirement for maintenance, require approximately 400 to 450 kcal of ME per hen per day. Rhode Island Reds, New Hampshire Reds, and various heavy crossbreeds, with hens that weigh about midway between White Leghorns and broiler breeders, have energy requirements that are about halfway between the 300 for Leghorns and the 420 for broiler breeders, or about 360 kcal per hen per day.

The comfort zones and optimum temperatures for poultry are as shown in Table 17-3; lower temperatures produce cold stress and higher temperatures produce heat stress. If acclimated, however, poultry can withstand sudden short-term cold or hot stresses without adverse effects. For example, laying hens require only approximately 7 days to adjust to a shift from cyclic cold to cyclic hot; and turkeys adapt in 8 to 14 days when shifted from cold to hot environments.

TABLE 17-3
COMFORT ZONE AND OPTIMUM TEMPERATURE OF POULTRY

Poultry	Comfort Zone		Optimum Temperature	
	(°F)	*(°C)*	(°F)	*(°C)*
Chicks	95-70[1]	*35-21*[1]	94-85	*34-29*
Broilers	95-70[2]	*35-21*[2]	70	*21*
Layers	50-75	*10-24*	55-70	*13-21*
Turkeys	100[3]	*38*[3]	60-70[3]	*16-21*
Ducks	90-70[4]	*32-21*[4]	60-70[4]	*16-21*

[1]Chicks: 95°F for the first week, with a reduction of 5°F weekly until 70°F is reached the 6th week.

[2]Same as chicks to 6th week. Then, 70°F until marketed.

[3]For the first 2 weeks, the temperature for white poults should be about 100°F, following which it may be reduced about 5°F each week until heat is no longer needed. The optimum temperature for mature turkeys is 60 to 70°F.

[4]For the 1st week, the brooder temperature of ducklings should be kept at about 90°F, then it should be reduced by 5°F per week during the next 4 weeks. The optimum temperature of adult ducks is 60 to 70°F.

• **Cold weather stress**—Birds are very well adapted to cold, due primarily to their highly efficient insulation provided by feathers. Adult chickens have been known to survive temperatures as low as −58°F for 1 hour, and pigeons have survived temperatures as low as −148°F for 1 hour. When a chicken is exposed to cold stress, several compensatory mechanisms are used: It reduces its surface area, and hence its heat loss, by "hunching." An additional reduction in heat loss, amounting to 12%, may be achieved by tucking the head under the wing. A still further saving of 20 to 50% can be made if the chicken sits rather than stands, thereby reducing the heat loss from the unfeathered legs and feet. Also, birds shiver in response to cold, which increases the metabolic rate of the body to produce additional heat.

1. **Voluntary feed intake when cold.** Feed consumption is increased as the temperature lowers. Also, an improvement in growth performance of broilers may accompany a decline in ambient temperature as a result of an increase in feed consumption.

2. **Maintenance/cool temperature energy partition.** Animals kept for productive purposes must be fed to maintain life, even when they are not producing. Thus, a considerable amount of the feed consumed by all classes of poultry must be used for maintenance.

The maintenance requirement for energy includes the need for basal metabolism and normal activity. The basal metabolism is the minimum energy expenditure or heat production under conditions when the influence of feed, environmental temperature, and voluntary activity are removed. The basal heat production varies with the size of the bird; in general, as size increases, basal heat production per unit of body weight decreases. The minimum heat production of day-old chicks is about 5.5 small calories per gram of live weight per hour, whereas the figure for adult hens is about half this.

The energy required for activity varies considerably, but is usually estimated to be about 50% of the basal metabolism.

Both body weight and rate of egg production affect the feed required by laying hens. It requires considerably less feed to maintain a 4-lb hen than a 5- or 6-lb hen; and the feed required to maintain a non-laying hen is more than half that needed for full production. So, to produce a dozen eggs, high production is more efficient than low production, because less maintenance feed is required per egg. The same principle applies to broiler production. Maintenance is prolonged in slow-growing birds. So, broilers that reach market weight in 8 weeks require considerably less feed per unit weight than those requiring 10 weeks to reach the same weight.

Because of the abililty of warm-blooded animals, including poultry, to maintain a constant body temperature that is normally several degrees above the environmental temperature, the animal is constantly losing heat to its surroundings. This loss of heat means a loss of energy that must be supplied in the feed. Since heat production must be equal to heat loss if the animal is to maintain its normal temperature, this means that there is a rapid increase in the metabolic rate whenever the environmental temperature falls below the critical point. Thus, the maintenance energy required at a low environmental temperature is greater than that required at a more comfortable (warmer) temperature.

When nutrient intake is greatly shifted by environmental influence on feed intake, an adverse effect on productivity (eggs or growth) should be alleviated by adjusting nutrient density to compensate for the altered intake of feed. However, environmental temperatures within the range of 39 to 88°F do not appear to affect nutrient requirements for protein, lysine, or vitamin A, as measured by egg production or growth.

3. **Lower critical temperature.** When laying hens are on full feed, the excess heat available for maintaining the normal body temperature is adequate for air temperature above 40°F. When such hens are active during the day, and are producing additional heat due to muscle activity, the critical temperature is lowered to 15 to 20°F. Below this temperature, they must draw on stored energy or feed nutrients for the heat energy necessary to maintain normal body temperature, and egg production will be lowered. Combs begin to freeze at above 6°F in dry air, and at 9 to 10°F in moist air.

The low lethal temperature for baby chicks ranges from 62°F on the day of hatch to about 67°F at 2 weeks of age.

• **Hot weather tolerance**—Like other farm animals, poultry have more defenses against cold than against heat.

Poultry respond to heat stress by eating less and drinking more, resting more, and by spreading their wings so that air can circulate past the less insulated undersurface of the wings. Also, they dissipate heat through the comb and wattle areas, which are highly vascularized and exposed. However, the primary cooling mechanism is respiratory, accompanied by opening the mouth and panting. When air is inhaled, heat is emitted from the nasal membranes and other areas of the respiratory tract. This heat is then lost when the air is exhaled.

Heat is also transferred from the inner body to the surface through conduction, convection, and countercurrent heat exchange.

In conduction, heat is transferred from molecule to molecule in the body and is eventually lost when the heat reaches the outermost portions. Conduction is increased by the birds through crouching to the ground and compressing the breast feathers to facilitate heat loss to the soil.

In convection, heat is transferred from heat-producing tissues to the blood, which then travels to the skin, resulting in an increased skin temperature and a loss of heat to the atmosphere.

In countercurrent exchange, nature makes use of the fact that arterial blood is warmer than venous blood. Thus, when blood reaches the limbs, some heat is lost to the environment and additional heat is given up to the venous blood returning from the limbs.

High environmental temperatures cause problems in all types of poultry production. Feed consumption and efficiency; egg production, size, and quality; growth rate; hatchability; and mortality are adversely affected under conditions of severe heat stress.

The primary techniques for reducing heat-related stress of poultry follow:

Provide heat protection. This usually consists of one or more of the following heat protection systems: insulation; not crowding; whitewashing or painting the roof a light color; roof sprinkling; fans; in-house fogging; and pad and fan evaporative cooling. Have back-up motors and related ventilation equipment readily available.

Provide cool water. Have plenty of clean, fresh, cool water readily accessible at all times.

Adjust ration. Increase nutrient density by adding 3 to 5% fats, protein (amino acids), minerals, and vitamins so that when heat stressed birds eat less total feed they will get the same quantity of nutrients daily. **NOTE:** For each 1.8°F above 80°F, the feed intake of layers will decrease an average of 1.5 to 2%. At 90°F, there may be a decline of up to 30% in feed intake.

Consider (1) adding potassium chloride to the drinking water, (2) using carbonated water, and/or (3) adding vitamin C to the ration; as determined by further and favorable experimental work with these additives. The physiology basis for carbonated water is that panting may reduce the carbon dioxide concentration in the blood to a level at which the birds are no longer stimulated to breathe; hence, carbonated water may help them to continue panting by restoring carbon dioxide.

Shift feeding schedule. During very hot weather, the birds should be fed early in the morning and at night.

Fast broilers. Fasting (feed depriving) the birds 3 to 24 hours before maximum heat stress is expected will dramatically increase survival time in acute heat stress. *The reason:* When birds are digesting feed, they are producing heat. The extra energy from metabolizing the feed can raise the body temperature by 1 to 2°F, which may make the difference between survival and death.

Acclimate the birds to heat. Experimental work in both England and the United States has shown that acclimation to heat stress is feasible and effective in both layers and broilers. Broilers exposed to high temperatures (95 to 100°F) during the first 1 to 3 weeks of life withstand acute heat stress later in life, with the subsequent losses during heat stress reduced 50% or more. (**NOTE:** In one broiler study, the birds were acclimated just 3 days prior to the heat stress.) Poultry producers need to make a judgment as to the practicality and the protection derived from acclimation.

1. **Voluntary feed intake when hot.** Feed consumption is reduced as the ambient temperature rises. However, a decline in feed intake may or may not affect production (egg production and egg quality of layers, or growth of broilers), depending on how drastically feed intake is reduced and for how long.

2. **Maintenance/warm temperature energy partition.** In hot evnironments, a drop in egg production by layers, or in growth by broilers, is partially due to lower energy intake. High fat or high fat-high density rations alleviate to some extent the weight loss of broilers at high temperatures. Warmer temperatures reduce basal metabolic rate and maintenance energy, the latter by an estimated 4% with each 1.8°F rise above thermoneutrality.

3. **Higher critical temperature.** The critical temperature of hens is about 85°F. At higher temperatures, egg size decreases, eggshells become thinner, and production is reduced. Death losses from heat prostration are often severe when the temperature rises above this point, with survival closely related to the availability of drinking water and the humidity. Tests at the U.S. Department of Agriculture, Beltsville Research Center showed that, when no method of cooling was provided, hens were able to survive a temperature of 90°F, provided the relative humidity did not exceed 75%. At 95°F, they survived at humidities below 60%, but at 100°F, they survived only if the humidity was 30% or lower.

The high lethal body temperature of the baby chick is 117°F.

Above the critical temperature, in which the surrounding air temperature equals or exceeds the skin temperature so that heat can no longer be dissipated by radiation, the bird employs physical means of cooling, especially more panting and increased vaporization, decreased feed consumption, increased drinking, and decreased activity. But this cannot continue for long, and the fowl dies from heat exhaustion.

Turkeys

There are species differences between turkeys and chickens in sensitivity to ambient environmental temperatures and in their performance characteristics. Turkey poults must be kept warmer than baby chicks, but poults are very sensitive to high temperatures and may die from hot weather stress. Turkeys of all weights show an abrupt drop in feed conversion efficiency at temperatures above 86°F. Also, the growth responses of toms and hens are significantly different.

FACILITIES/ENVIRONMENTAL INTERACTIONS[11]

Optimum facility environments can only provide the means for animals to express their full genetic potential of production, but they do not compensate for poor management, health problems, or improper rations.

Research has shown that animals are more productive and feed-efficient when raised in an ideal environment. The primary reason for having facilities, therefore, is to modify the environment. Proper barns and other shelters, shades, sprinklers, insulation, ventilation, heating, air conditioning, and lighting can be used to approach the desired environment. Also, increasing attention needs to be given to other stress sources such as space requirements, and the grouping of animals as affected by class, age, size, and sex.

The principal scientific and practical criteria for decision making relative to the facilities for animals in modern, intensive operations is the productivity and cost of production of animals, which can be achieved only by healthy animals under minimal stress. So, the investment in environmental control facilities is usually balanced against the expected increased returns.

Recommended Environmental Controls

Temperature, humidity, and ventilation recommendations for different classes of livestock are given in Table 17-4, Recommended Environmental Control. This table will be helpful in obtaining a satisfactory environment in confinement livestock buildings, which require careful planning and design.

Facilities for Beef Cattle

Most beef-production systems involve the maximum use of pastures and ranges in season. Calves nurse their mothers and are weaned at about 7 months of age at weights of 400 to 500 lb. After weaning, most young cattle are grazed on pastures or ranges, or wintered on high-forage rations, to a weight of about 700 lb. Then, those destined for slaughter usually are fed a high-energy finishing ration in a feedlot for 2 to 5 months and marketed at about 1,050 lb.

Traditional management is still practiced on the nation's rangelands, whereas more intensive practices are commonplace where land is more valuable and feed more abundant. Greater attention to the environment, nutrition, and genetic improvement, however, has made for major changes in the facilities and management of beef cattle.

- **Space requirements**—In feedlots, the recommended space is 200 sq ft per animal in an unpaved lot and 20 sq ft per animal in complete confinement units.

- **Shelter and shades**—Most experiments show that providing winter shelter for confined feedlot cattle makes for higher gains and greater feed efficiency. Providing summer shade in hot climates usually improves performance, but it is not always profitable because of the initial cost of shades plus maintenance.

[11]In the preparation of this section, the authors adapted considerable material, including experimental results from: *Scientific Aspects Of The Welfare Of Food Animals,* F. H. Baker, task force chairman, CAST, Ames, Iowa, 1981.

TABLE 17-4
RECOMMENDED ENVIRONMENTAL CONTROL

Class of Animal	Temperature: Comfort Zone		Temperature: Optimum		Acceptable Humidity	Commonly Used Ventilation Rates[1]: Basis		Winter[2]		Summer[2]		Drinking Water: Winter		Drinking Water: Summer	
	(°F)	*(°C)*	(°F)	*(°C)*	(%)	(lb)	*(kg)*	(cfm)	*(m³/minute)*	(cfm)	*(m³/minute)*	(°F)	*(°C)*	(°F)	*(°C)*
Beef cow	40–70	*5–21*	50–60	*10–15*	50–75	1,000	*454*	100	*2.8*	200	*5.7*	35–37	*1.7–2.8*	60–75	*15–24*
Steer, enclosed building on slotted floor	40–70	*5–21*	50–60	*10–15*	50–75	1,000	*454*	100	*2.1–2.3*	200	*14.2*	35–37	*1.7–2.8*	60–75	*15–24*
Dairy cow	40–70	*5–21*	50–60	*10–15*	50–75	1,000	*454*	100	*2.8*	200	*5.7*	35–37	*1.7–2.8*	60–75	*15–24*
Dairy calves	50–75	*10–24*	65	*17*	—	per 100 lb *(45 kg)*		10	—	25	—	—	—	—	—
Sheep:															
Ewe	45–75	*7–24*	55	*13*	50–75	—	—	20–25	*0.6–0.7*	40–50	*1.1–1.4*	35–37	*1.7–2.8*	60–75	*15–24*
Feeder lamb	40–70	*5–21*	50–60	*10–15*	50–75	—	—	15	*0.3*	30	*0.65*	35–37	*1.7–2.8*	60–75	*15–24*
Newborn lamb	75–80	*24–27*	—	—	—	—	—	—	—	—	—	—	—	—	—
Swine:															
Sow, farrowing house	60–70	*15–20*	65	*17*	60–85	Sow & litter		80	*1.4*	210	*2.8*	35–37	*1.7–2.8*	60–75	*15–24*
Newborn pigs[3] (pig level)	79–85	*27–32*	85	*29*	60–85	—	—	—	—	—	—	—	—	—	—
Growing-finishing hogs	60–65	*15–17*	60	*15*	60–85	125	*57*	15	*0.7*	75	*2.1*	35–37	*1.7–2.8*	60–75	*15–24*
Horse	45–75	*7–24*	55	*13*	50–75	1,000	*454*	60	*1.7*	160	*4.5*	35–37	*1.7–2.8*	60–75	*15–24*
Newborn foal	75–80	*24–27*	—	—	—	—	—	—	—	—	—	—	—	—	—
Poultry:															
Layers	50–75	*10–24*	62–68	*17–20*	50–75	per bird		2	—	5	—	50	*10*	60–75	*15–24*
Broilers	85–95 (baby chicks)	*21–27*	70	*24*	50–75	per lb body weight		½	—	1	—	50	*10*	60–75	*15–24*
Turkeys	95–100 (beginning poults)	*35–38*	70	*24*	—	per lb body weight		½	—	1	—	50	*10*	60–75	*15–24*

[1]Generally two different ventilating systems are provided; one for winter, and an additional one for summer. Hence, as shown in this table, the winter ventilating system in a beef cow barn should be designed to provide 100 cfm (cubic feet/minute) *(2.8 m³/minute [cubic meters/minute])* for each 1,000-lb *(454-kg)* cow. Then, the summer system should be designed to provide an added 100 cfm *(2.8 m³/minute)*, thereby providing a total of 200 cfm *(5.7 m³/minute)* for summer ventilation.

In practice, in many buildings, added summer ventilation is provided by opening (1) barn doors, and (2) high-up hinged walls.

[2]The ventilation rates for winter and summer are designed to meet the needs in the column headed *Basis*. For example, the commonly used winter ventilation rate of a beef cow weighing 1,000 lb *(454 kg)* is 100 cubic feet per minute (cfm) *(2.8 cubic meters per minute [m³/min.])*. Approximately ¼ of the winter rate should be provided continuously for moisture removal.

[3]The nursery temperature at pig level should be 85°F *(20°C)* for the first week following birth, then decrease to 79°F *(26°C)* at the fifth week. See Chapter 23, Feeding Swine, Table 23-29, for additional baby pig guidelines.

• **Social problems**—When new animals are introduced into a herd or when herds are mixed, social relationships must be established among animals. In feedlots especially, steers may ride each other in a phenomenon called "bulling." The cause is unknown, but it is desirable to remove "bullers" and keep them together in a separate lot to minimize losses due to injuries, exhaustion, and reduced weight gains.

Facilities for Dairy Cattle

In the United States, the traditional scenes of contented dairy cows grazing in pastures have been largely replaced by cows kept in confinement, provided shelters, and with a high degree of mechanization for feed and waste handling.

• **Housing and space needs**—Methods of housing and caring for dairy cows have changed. During the first half of the 20th century, most cows were confined by stanchions, where they rested and were fed and milked. Since about 1950, other methods of housing, designed to give more freedom to the cows, have become common, including comfort stalls, pen or loafing barns, and free-stall housing. It is recommended that free stalls be 45 to 48 in. wide and 78 to 90 in. long, depending on cow size. Comfort stalls range in size from 42 X 66 in. to 49 X 84 in. It is unclear what constitutes crowding in dairy cattle corrals; when studied over a period of 4 weeks and given either 100 or 25 sq ft of space per cow, there was no difference in milk production.

• **Temperature**—Lactating cows are considerably more sensitive to high than to low environmental temperatures; hence, milk production in hot weather can be increased by providing cows with shades and other methods of cooling.

• **Dairy calves**—Because the dairy cow turnover rate is about 25% annually, many of the heifer calves are ultimately returned to the herd as lactating cow replacements Nearly all of the bull calves born, except those of exceptionally high genetic merit, are fed for veal or are castrated and fed for dairy beef. Heifers not retained as replacements are also marketed as veal or dairy beef.

In the United States, dairy calf mortality averages an appalling 15 to 20%; hence, there is much need for improvement.

Usually, dairy calves are separated from their mothers at 1 to 3 days of age. Early separation of the calves improves their protection from certain diseases, makes possible more careful monitoring of udder diseases of cows, and increases the economic return from selling the milk.

1. **Feeding.** It is very important that newborn calves consume colostrum, the first milk in the udder of the cow after calving. In addition to being nutritious, colostrum contains

antibodies that protect the calf against disease. Colostrum must be consumed within the first 24 hours of a calf's life because the intestine loses its ability to absorb antibodies about that time.

After receiving colostrum, calves to be used as herd replacements are usually fed whole milk or milk replacers for the first 4 to 8 weeks of their lives.

2. **Housing.** Calves are housed in many types of facilities including warm and cold stall types, raised slotted floors and conventional floors, outside cold and inside cold pens, outside hutches, and tied to the wall in the alley behind the cows in old fashioned wood stanchion barns. Any of these types may be satisfactory provided the following principles are observed: (1) They should be constructed so that they can be sanitized easily, (2) they should assure adequate ventilation, and (3) calves should be kept in individual rather than community pens during their early life to lessen the chance of disease spread and to prevent the tendency of calves to suck and lick each other.

Fig. 17-31. Dairy calves housed in individual pens to lessen the chances of disease spread and prevent the tendency of calves to suck and lick each other. (Courtesy, James Tappan, Arizona Dairy Co., Higley, Ariz.)

• **Veal calves**—*A veal calf is defined as a young bovine animal, usually not over 4 months of age, that has subsisted largely on milk or milk replacers, thus making the color of the lean meat light, grayish pink.* The majority of veal calves are of dairy breeding, consisting of bull calves and heifer calves not retained as replacements; few calves of beef breeding are marketed as veal.

Many of the husbandry methods for raising veal calves are similar to those followed in raising herd replacements. The primary difference is: Veal calves are fed for more rapid weight gains than herd replacements; thus, they are fed at higher levels of whole milk or of higher-energy milk replacer than herd-replacement calves.

Veal calves are usually grown and marketed at 300 to 400 lb weight and about 12 weeks of age. Because of market demand, producers receive a premium if the lean meat of veal has the light grayish pink color (due to reduced muscle myoglobin), characteristic of feeding milk, which is naturally low in iron. Some producers attempt to enhance the grayish pink color of veal calves by restricting iron intake (by feeding low iron milk replacers) and restricting the exercise. Research has shown that, irrespective of the iron status of calves at birth, a dietary iron concentration of 11.4 to 13.6 mg per pound of dry matter in milk substitute diets is sufficient for the well being of veal calves and produces desirable grayish pink carcasses, with or without exercise.

Facilities for Sheep

Sheep are produced in relatively large bands in western range areas and in small flocks in the farming regions. These units (groups) vary in size from farm flocks ranging from a few head to 500 head or more, to range bands with as many as 3,500 head.

Where pasture or range conditions are good, the sheep producer strives to produce grass-finished lambs at weaning. When the forage is not good, lambs are usually sold as feeders for further finishing in feedlots.

• **Space for feeder lambs**—Feeder lambs confined in barns or sheds require 4 to 6 sq ft per head; in paved lots, they require about 16 sq ft per head; and in a dirt lot they require 25 to 30 sq ft per head.

• **Lights**—(See section on "Modifications of Facilities—Lights.")

• **Predation**—The small size and defenselessness of sheep make them subject to predation when not protected by fences or guards (humans, dogs, or other animals). The most common predator of sheep is the coyote. The major methods of controlling sheep predators are guns, trapping, poisons, repellants (lights, propane guns), guard dogs, and special electric fences.

Facilities for Swine

The shift to fewer and larger swine enterprises has been rapid. The larger operators generally incorporate labor saving practices and reduce environmental stress on the animals by confining them in environmentally controlled housing. Swine are highly sensitive to stress and disease; over which well managed operations improve control, and with it improve production efficiency.

As with other animals, the investment in environmental control facilities must be balanced against the expected increased returns; and there is a point beyond which further expenditures for environmental control will not increase returns sufficiently to justify the added cost.

• **Space requirements**—Studies have shown that when the available space per pig is too small the rate of gain is decreased and aggressive behavior is increased.

Fig. 17-32. Through behavioral research, scientists have discovered creative ways to help pigs better to utilize confined space. The pig in the center is hiding its head, which reduced fighting by 40%. (Courtesy, J. J. McGlone, Texas Tech University, Lubbock)

The authors recommend the minimum floor space shown in Table 17-5 for growing-finishing hogs.

TABLE 17-5
MINIMUM FLOOR SPACE FOR GROWING-FINISHING HOGS

Pig Weight		Recommended Floor Space Per Pig			
		Solid or Partially Slotted Floors		Slotted Floors	
(lb)	*(kg)*	(sq ft)	*(sq m)*	(sq ft)	*(sq m)*
25- 40	*11.4- 18.2*	4.0	*0.65*	3.0	*0.28*
40- 75	*18.2- 34.1*	5.5	*0.51*	3.5	*0.33*
75-125	*34.1- 56.8*	8.5	*0.79*	5.5	*0.51*
125-240	*56.8-109.0*	12.0	*1.12*	8.0	*0.74*

• **Lights**—(See section on "Modifications of Facilities—Lights.")

• **Grouping**—Although there is little experimental work upon which to base intelligent recommendations relative to proper grouping or separating of hogs of different sexes, ages, and sizes, observant producers recognize that this is a most important part of management. The following practices are generally followed and advocated by successful producers:

1. **Replacement gilts.** Gilts to be retained for the breeding herd should be separated from market hogs at 4 to 5 months of age.

2. **Pregnant gilts and sows.** Gilts and sows should be kept in separate groups during the gestation period, unless they are self-fed a bulky ration.

3. **Boars of different ages.** Junior and mature boars should not be run together. Boars of the same age or size can be run together during the off-breeding season.

4. **Adjusting size of litter.** Where possible, the size of litters should be adjusted to the number of functioning teats or nursing ability of the sow. Transferring pigs from sow to sow should be done as early as possible; 3 to 4 days after farrowing is usually the maximum length of time that this can be done, unless the odor of the pigs is masked, when it may be possible to transfer at a later date.

5. **Running sows and litters together.** Pigs should be about 2 weeks old before placing sows and litters together, although small groups may be put together as early as 1 week. The age difference between such litters should not be more than 1 week in a central farrowing house or 2 weeks on pasture. Not more than 4 sows and litters should be grouped together in a central farrowing house; and not more than 6 on pasture.

6. **Creep feeding.** A maximum of 40 pigs per creep may be allowed.

7. **Early weaning.** In early weaning, not over 10 pigs should be placed together up to 3 weeks of age; 20 may be placed together at 3 to 4 weeks of age; and 25 at 5 weeks of age.

8. **Pigs of different weights.** Growing-finishing pigs of widely varying weights should not be run together. It is recommended that the range in weight should not exceed 20% above or below the average.

• **Flooring**—Basically, there are two types of floors for swine, solid and slotted. As the name implies, solid floors are solid. Slotted floors have slots through which the feces and urine pass to a storage area below or nearby. Slotted floors save space, eliminate the need for bedding, reduce labor requirements, improve sanitation and disease management, and lead to more efficient waste handling. However, slotted floors make for more critical environmental conditions and may increase foot problems if not properly designed.

Fig. 17-33. Clean pigs on a sanitary concrete slotted floor. (Courtesy, Iowa State University, Ames)

• **Farrowing stalls or crates**—*Farrowing stalls or crates are structures in which sows are housed when giving birth to litters of pigs.* They are widely used because they prevent sows from lying on some of their pigs and killing them. Also, they facilitate the use of heat lamps, electric heating pads, and other means of making the baby pigs comfortable.

Generally, sows and litters are moved from farrowing stalls to other quarters about 2 weeks after farrowing.

Fig. 17-34. Electric heating pads in a farrowing crate used for baby pig lying areas. (Courtesy, Iowa State University, Ames)

Facilities for Horses

The comfort of horses is a function of temperature, humidity, and air movement.

All horse facilities should modify winter and summer temperatures; protect the animals from rain, snow, sun, and wind; and minimize stress.

Based on currently available information, the authors recommend that horse facilities incorporate the following environmental conditions:

• **Temperature**—A range of 45 to 75°F is satisfactory, with 55°F considered optimum. Until a newborn foal is dry, it should be warmed to 75 to 80°F. This can be achieved with a heat lamp.

• **Humidity**—A range of 50 to 75% relative humidity is acceptable, with 60% preferred.

• **Insulation**—This need will vary from area to area. Where a wide spread between summer and winter temperature exists, and where horses are confined much of the time, proper insulation is of prime importance. Under such circumstances, for moisture control in winter and temperature control in summer, horse barns should have at least 2 in. of insulation on the ceiling, under the roof, and the sidewalls should be insulated, also.

• **Ventilation**—The prime function of the winter ventilation system is to control moisture, whereas the summer ventilation system is primarily for temperature control. If air in horse barns is supplied at a rate sufficient to control moisture—that is, to keep the inside relative humidity in winter below 75%—then this will usually provide the needed fresh air, help suppress odors, and prevent an ammonia buildup. Also the barn should be free from drafts. In a properly ventilated barn, the ventilation system should provide 60 cu ft per minute (cfm) for each 1,000 lb of horse in winter and 160 cfm per 1,000 lb of horse in summer. In warm weather, satisfactory ventilation usually can be achieved by opening barn doors and by installing hinged doors or panels near the ceiling that swing open. Then, on extremely hot or quiet days, the natural ventilating system may be augmented with the fan ventilating system.

The design of the barn, and the existing weather temperature of the area, will determine the best type of ventilating system to use. Also, the requirements for summer and winter are so different that it is best to use two different ventilating systems—one for winter, and the other for summer.

A professional engineer should be engaged to design the ventilating system. Generally, summer exhaust fans should be placed high, and winter exhaust fans low, but in any case drafts on horses should be avoided.

• **Light**—Windows should be provided in the ratio of 1 sq ft for each 30 sq ft of floor area. They should be protected from horses and screened to keep flies out. Additionally, artificial light should be provided for the convenience of the caretaker. One 60-watt bulb, properly recessed and protected, in each stall, plus lighting in the aisle, should suffice.

(Also see section on "Modifications of Facilities—Lights.")

Facilities for Poultry

The poultry industry moved rapidly into the use of environmentally controlled housing and intensive production practices. Today, it is the most efficient of the animal industries. This transition has resulted in widespread use of cages, increased population density, beak trimming, dubbing (partial removal) of combs, and complete automation of feeding, watering, and egg gathering. Economic considerations have dictated much of this change.

Improvements have been made in poultry production methods, including nutrition, control of diseases and parasites, and protection against predators, extreme temperatures, and storms. These improvements have greatly enhanced the well-being of poultry as evidenced by lessened mortality during the first year of life. In the wild, chickens normally experienced up to 85% mortality in the first year. Early domestic chickens, which were partially protected, experienced up to 50% mortality the first year. Today, there is less than 12% mortality in the first year of life.

LAYER FACILITIES

Basically, layer confinement houses are either floor system or cage system.

• **Floor vs cage systems**—The floor system is the older of the two systems. It consists of litter covering the entire floor, with each floor housing a single flock. The individual birds can move about so that each bird can make physical contact with all the other birds in the entire flock. Thus, hens housed in a floor system must adjust to a much more extensive social system than hens in a cage system. Weaker or submissive birds can miss out on feeding in floor systems; and hens housed in floor systems are subject to suffocation from mass hysteria or panic caused by sudden noise or other disturbance, whereas those in cages are not.

In the cage system, the flock is divided into many subflocks. Each subflock (often three to six birds) may contact each other, but they are isolated from the birds in all the other cages.

Texas A&M Agricultural Research workers made extensive studies of stress in caged vs floor hens. They reported that caged hens were less stressed than floor pen hens; as measured by production rates (76.3 vs 73.9), weight gains, feed efficiency, egg and egg shell weights, and plasma corticosterone levels.

Broiler houses continue to be mostly of floor management design. The floor management systems are more labor intensive, but the costs of equipment are less. In layer houses, additional benefits of cages are the elimination of "floor eggs" and litter handling, which stimulated the earlier change to cages.

Fig. 17-35. Broilers on floor, with litter covering the entire floor. (Courtesy, USDA)

More and more pullet growing enterprises are changing to cages, primarily because of the benefit of growing the young birds in conditions similar to the cage laying houses to which they will be moved, thereby minimizing the transition trauma.

Fig. 17-36. Replacement pullets in cages. By growing pullets in cages, the stress of transferring them to cages in the laying house is minimized. (Courtesy, J. V. Craig, Department of Animal Sciences and Industry, Kansas State University, Manhattan)

• **Layers/cages—historical**[12]—In the transformation from living in the jungle to living in cages, the hen has been subjected to radical facility (habitat) changes, most, but not all, of which have improved her well-being. The red jungle hen, *Gallus gallus,* roosted in thick bamboo, nested on the ground, established a strong pecking order, competed with other species for food, and suffered high mortality from predators and adverse weather. Beginning with the first U.S. settlement at Jamestown in 1607 and continuing until 1900, chickens roamed the farmstead, perched on roosts in a poultry house at night, and were tenderly cared for by the farmer's wife, who fed them on table scraps and grain from the crib. Little attention was given to sanitation, rations were not balanced, and diseases (both communicable and nutritional) and parasites were common—often wiping out an entire flock. Before World War II, small farm flocks were kept in houses with roosts, nests, litter to scratch in, and access to outdoor areas in warm weather. Beginning in 1940–1945, the hen's genetic background, life-style, and environment were greatly changed. The numerous breeds of the exhibition-small farm flock era were replaced almost entirely by three layer breeds; White Leghorns for white egg production, and Rhode Island Reds and New Hampshires for brown egg production; artificial incubation and brooding replaced the old setting hen; poultry flocks became larger and means were sought to reduce housing and labor costs per hen. To meet the needs of the time, the floor system evolved, with deep litter covering the entire floor, which was cleaned once or twice per year, and each floor housing a single flock. As the flock size increased, so did crowding, diseases, cannibalism, submissive hen problems, and suffocation from mass hysteria. Next came row upon row of individual cages stacked above each other, with wastes mechanically removed from beneath. All went well until it was found that hens kept in individual cages suffered excessively from "cage fatigue," a malady characterized by fragile bones and inability to stand. To alleviate cage fatigue,

Fig. 17-37. Layers in limited-multiple cages stacked above each other, and with manure removed from beneath. (Courtesy, J. V. Craig, Department of Animal Sciences and Industry, Kansas State University, Manhattan)

[12]In the preparation of this section, the authors adapted material from: Craig, J. V., "Behavioral and Genetic Adaptation of Laying Hens to High Density Environments," *Bio Science,* Vol. 32, No. 1, January, 1982.

producers put more than one hen in a cage. As an added bonus, muiltiple-cages also reduced the per hen cost for labor, housing, and equipment. But as multiple-cages became larger and accommodated more birds, other problems evolved to plague egg producers, including cannibalism, excessive feather loss, and reduction in egg production signaling too many birds together and too little space per bird.

There is no easy solution to the housing of layers! Nothing would be gained by turning back the pages of time and reverting, for each system along the way had its problems—only the nature of the problems differed. Jungle hens had to fend for themselves, without the help of caretakers or the protection from the weather by buildings; the farm flock was plagued by lack of sanitation and poor nutrition; the floor system, with the birds eating and sleeping in their own discharges, spawned diseases and parasites.

The following approaches merit additional research and practical application:

1. Selecting and breeding birds for better adaptation to the intended facilities.
2. Improving cage design and space. Among the new concepts being tried are plastic coated cage floors, abrasive strips to control claw growth, and horizontal bars replacing vertical bars for easy bird removal.
3. Debeaking and declawing as methods of coping with the stresses involved.

• **Layers/cages—present status**—At present, nearly all table eggs in the United States are produced by hens housed in cages of 3 to 6 birds.

Experiments show that decreasing the space allowance in cages to less than 0.5 sq ft per layer generally decreases egg production, reduces feed consumption, lowers body weight gain, and increases mortality. The needed space allowance may be influenced by the shape as well as the size of the cages. Greater width across the front is more important than greater depth, which is attributed to (1) greater availability of feed and water, and (2) reduced face-to-face contact of the birds. A feed space of 4 lineal inches per hen is recommended.

The European Common Market countries have established minimum space requirements of 70 sq in. per caged hen, whereas a minimum of 48 sq in. is commonly accepted in the United States.

In recent studies conducted at the Cornell Agricultural Experiment Station, researchers found that, in high density cages, low dominance rank of laying hens results in reduced feeding activity and reduced egg production. In low density cages, however, no association between dominance rank and egg production was noted.[13]

Different genetic strains of layers differ significantly in their environmental requirements. For example, 0.33 sq ft per bird may be adequate for some strains of layers presently available. The newer strains, along with improved cage design, have greatly increased the production per unit of space.

[13]Cunningham, D. L., "Investigating Behavior, Performance, and Well-Being of Hens In Cages," *Poultry Science*, Nov., 1986, pp. 12-27.

With a given space allowance per bird, the production of eggs per hen declines as the number of hens per cage increases, presumably as a consequence of the greater contact, communication, and activity in the larger cages. The most successful cage holds four birds or fewer.

The main disadvantages of housing layers in cages are (1) increased bone fragility, and (2) a fatty condition of the liver.

• **Lights**—(See section on "Modification of Facilities—Lights.")

• **Temperature for layers**—The ideal temperature for laying hens is 62 to 68°F. High temperatures result in thinner egg shells. High humidity accentuates the effect of high temperature.

The optimum temperature for day-old chicks is 86 to 95°F. Generally, chicks are started at 95°F, and the temperature is lowered daily until a room temperature of 68 to 72°F is reached in about 3 weeks.

BROILER FACILITIES

Most broiler chickens are produced in floor systems.

In the commercial broiler industry, the current space allowance is generally 0.8 sq ft per bird, and in some environmentally controlled houses as low as 0.7 sq ft per bird. Growth of broilers is impaired with a space allowance of 0.5 sq ft per bird.

Growing broilers in cages has been hindered by the development of leg weakness and breast blisters. Perhaps new genetic strains that perform well in cages will evolve in the future. The advantages of cages would be a decrease in ammonia fumes, fewer disease problems, reduced labor, and greater numbers of broilers per house because of tiering cages.

Most broilers make little use of perches until they are forced to do so by a shortage of floor space.

• **Lights**—(See section on "Modifications of Facilities—Lights.")

• **Temperature for broilers**—From the third through the seventh week of age, a temperature of not to exceed 80°F is recommended.

TURKEY FACILITIES

Both breeder hens and growing turkeys are produced on floors. Turkeys experience leg problems when they stand on wire in cages. Production of turkey breeder hens is reduced and mortality is greater in cage systems than in floor systems.

Turkey breeder hens of a large strain need a minimum of 4½ sq ft of floor space, but the turkey hatching egg industry usually allows 5 sq ft or more per breeder hen.

Growing turkeys of medium weight require 1 sq ft per bird to 10 weeks of age and 1.6 sq ft per bird from 10 to 14 weeks of age. Turkeys of a large strain grow satisfactorily when allowed a range between 1½ and 3 sq ft per bird up to 8 weeks of age. The optimum floor space for large turkeys appears to be about 3 sq ft per bird for hens 15 to 23 weeks of age and 4½ sq ft for males through 23 weeks of age.

• **Lights**—(See section on "Modifications of Facilities—Lights.")

• **Temperature for turkeys**—A temperature of not to exceed 80°F is recommended.

Modifications of Facilities

In recent years, there has been a trend to modify the environmental control facilities as much as possible; among such modifications designed for maximum animal comfort and efficiency of production are fans, floors, lights, shades, sprinklers/sprayers/foggers, ventilation, wallows, and windbreaks.

FANS

Generally, the ventilation system of environmentally controlled buildings involves the use of fans, to maintain oxygen; keep carbon dioxide levels low; remove dust, moisture, and ammonia from the building; and maintain suitable temperature. Rate of production and feed efficiency are lowered when animals must endure excess stresses, such as temperatures appreciably below or above the comfort zone. Cold animals eat more and gain less, while heat stressed animals eat less and gain less.

There are two basic types of fans: (1) exhaust fans, and (2) pressure or intake fans.

In an exhaust system, fans force air out of the building, which creates a partial vacuum. Then, air comes in through intake openings to equalize pressure. Where fresh air inlets appear uniformly around the building, fresh air distribution should occur uniformly.

In a pressure fan system, fans draw fresh air from the outside and build up enough pressure to push stale, moisture-laden air out through exhaust ports and any other openings.

Fan capacity is important. Fans are rated in cubic feet per minute (cfm) of air they move. By selecting and using different numbers and sizes of fans, the needed variable rate, or flexibility, of ventilation can be achieved. The amount of ventilation required depends upon (1) the inside-of-building temperature desired, (2) the outside air temperature, (3) the relative humidity, (4) the number and size of animals in the building, (5) the amount of insulation, and (6) the size of the building.

Most large swine and poultry producers use fans in confinement buildings to insure adequate air circulation during warm weather and to control the amount of air change in cold weather. Also, workers at the University of California are testing a new concept involving fan/water misted air for cooling cattle in unenclosed areas. Details relative to fan-cooling of cattle, swine, and poultry follow:

1. **Cattle.** The Agricultural Experiment Station of the University of California is testing the cooling of cattle by a large fan up to 20 ft in diameter, powered by a 2 horsepower electric motor, operated at a speed of about 40 revolutions per minute, suspended or mounted on a pedestal or widely spread legs 10 ft or higher above ground level, with a mister injecting a fine spray of water into the airstream just below the fan. This fan/water mister is designed for cooling feedlot cattle and for lactating dairy cows in wash pens and under free-stall roofs and shades.
2. **Swine.** A farrowing house maintained at 70°F and 75% relative humidity while the outside air temperature is 30°F and 75% relative humidity requires a ventilation rate of approximately 25 cubic feet of air per minute (cfm) to remove the 1 lb of moisture produced per hour by a sow and litter.
3. **Poultry.** Fans are being used to cool both broiler and layer houses. A layer house requires fan capacity of approximately 1 cu ft per minute (cfm) per 1 lb bird. The rule of thumb is: 3 cfm per bird for light breeds, 4 cfm per bird for medium breeds, and 5 cfm per bird for heavier birds. However, many commercial poultry systems provide 7 cfm per bird for added safety.

Fig. 17-38. Fans used to cool a brooder house. (Courtesy, California Polytechnic University, San Luis Obispo)

A knowledgeable animal scientist or agricultural engineer should be consulted when designing and installing a fan system for ventilation. If in doubt about the fans and air movement, the comfort of the animals, the litter condition, and the odor buildup will tell you.

In environmentally controlled buildings, provision should be made for emergency ventilation in the event of an electric power failure. This can be done by either providing sufficient auxiliary electric power to operate fans, or by installing doors in the sides of the building which can be opened to allow natural airflow and thus prevent smothering the animals if the fan ventilation system fails.

Fans may also be effective in open-sided houses. When the outside temperature is unusually high, they should be placed in the house so as to blow the air lengthwise of the building. High-speed fans are more effective than low-speed fans.

FLOORS

Most livestock producers feel that a perfect flooring material has not yet been developed, as each of the existing types has certain disadvantages. Rough wood floors furnish good traction for animals and are warm to lie upon; but they are absorbent and unsanitary. They also lack durability and often harbor rats and other rodents. Concrete floors are durable, impervious, easily drained, and sanitary; but they are rigid and without resilient qualities, slippery when wet, and cold to lie upon. Clay floors are noiseless and springy, and afford a firm natural footing unless wet; but they are difficult to keep clean and level.

Basically, there are two types of floors: solid floors and slotted floors.

• **Solid floors**—Solid floors for animals may be constructed of numerous materials—including clay, clay with a concrete border, plank, concrete, concrete with board surfacing, cork brick, creosoted wooden blocks, cinders, or various combinations of these materials. Regardless of the type of flooring material, for a good dry bed there should be a combination of surface and subsurface drainage, together with a cover provided by a suitable absorbent litter.

• **Slotted floors**—*Slotted floors are floors with slots through which the feces and urine pass to a storage area below or nearby.* Slats may be used for the entire floor area, or only part of it. Such floors are not new; they have been used in Europe for over 200 years. More and more slotted floors are being used for swine and poultry in this country, and there is increased interest in using them for cattle and sheep.

Fig. 17-39. Broiler breeder hens on raised, plastic slotted floor. (Courtesy, Leo S. Jensen, Department of Poultry Science, University of Georgia, Athens)

The main **advantages** of slotted floors are (1) they facilitate automation and save labor; (2) they lessen or eliminate bedding; (3) they facilitate handling of manure; (4) they necessitate less space per animal; (5) they require less land; (6) they increase sanitation; (7) they lessen mud, dust, odor, and fly problems; and (8) they lessen pollution.

The chief **disadvantages** of slotted floors are (1) higher initial cost than conventional solid floors, (2) less flexibility in the use of the building, (3) any spilled feed is lost through the slots, (4) animals raised on slotted floors resist being driven over a solid floor, and (5) environmental conditions become more critical.

Slats may be made of concrete, metal (steel or aluminum), or wood.

LIGHTS

In modern livestock operations, both the length of day and the intensity of light may be altered artificially.

The number of hours of light in the day affects the initiation of the normal breeding season of ewes and mares, both of which are seasonal breeders; and it regulates the reproductive function (egg-laying) of poultry.

The ratio of hours of daylight to darkness throughout the year acts on nerves in the region of the pituitary gland, and stimulates or inhibits the release of the follicle-stimulating hormone (FSH). Lengthening the daylight hours activates the pituitary, and causes it to release increasing amounts of the FSH which stimulates ovarian function. Thus, sometime after the daylight period increases, the estrous cycle begins in ewes and mares, and egg-laying begins in poultry. Artificial lighting will accomplish the same thing as daylight.

• **Sheep**—Normally, ewes come in heat during the late summer or early fall, though there are both area and breed differences. The breeding season is usually restricted to about 4 months.

Ewes generally begin cycling when the number of daylight hours drops below 14. This is the reason that most breeds of sheep come into heat during the fall months. To initiate estrus, however, it appears that the shorter days must be preceded by longer days.

• **Swine**—Traditionally, diurnal (daily cycle) lighting is used for swine. However, there is limited experimental evidence that the light-dark cycle that they prefer is more nearly hourly than diurnal.

• **Horses**—Normally, the natural breeding season of mares begins in March and extends to late July or August.

Artificial lighting of broodmares enables breeders to bring mares in season about 6 weeks earlier than normal. By the use of the artificial light technique, a mare that would normally conceive on March 15 may get in foal sometime in January. By avoiding the necessity of skipping a year due to late breeding, this technique may actually result in obtaining two additional foals during the lifetime of a mare.

The procedure consists in using a 200-watt light bulb in a box stall to extend the hours of light to 16 hours daily. By beginning the light treatment of mares about December 1, they may be bred the latter part of January.

Slight adjustments in the schedule will need to be made in different locations, depending upon the sunrise and sunset times of the particular area.

• **Layers**—Artificial light in the laying house should give a 16-hour day. If natural daylight is longer, artificial light should maintain the longest daylight period, although a light regimen in excess of 17 hours is of doubtful value. Use intensity of 1 footcandle at bird height. (Approximately a 40-watt bulb with reflector every 12 ft, 7 ft above the floor in floor systems.) Never decrease day length or light intensity during the laying period. A photoelectric cell may be used in connection with a time clock to turn lights off at dawn and on at twilight.

• **Replacement pullets**—During the growing period, pullets should not be exposed to increasing amounts of light prior to 21 weeks of age if best egg production is to be attained. Rather, short or decreasing light during the growing period is desirable.

The following lighting regimen is recommended for pullets: 0 to 3 weeks, 1.0 footcandle, 20 hours of light and 4 hours of darkness; 3 to 12 weeks of age, 0.1 to 0.5 footcandle, 16 hours of light and 8 hours of darkness; 12 to 21 weeks of age, 0.5 footcandle, continue 16 hours of light and 8 hours of darkness to 16 weeks of age, followed by decreasing light to 8 hours of light and 16 hours of darkness at 21 weeks of age.

Long or increasing light should begin at 21 weeks of age, with the light increased at the rate of 15 minutes per week until 17 hours, then maintained.

One 25-watt bulb with reflector every 12 ft, 7 ft above the floor will give approximately 0.5 footcandle intensity.

• **Broilers**—The recommended broiler lighting in light-tight houses is: 0 to 5 days, 3.5 footcandle continuous light; 6 days to market, 0.35 footcandle, 23 hours of continuous light and 1 hour of darkness. Alternatively, intermittent light, consisting of 1 hour of dim light (feeding time) to 3 hours of darkness (resting time), may be used from 6 days to marketing.

Recommended lighting for broilers in open-sided houses is: 0 to 2 days, 3.5 footcandle of continuous light; 2 days to market, 0.5 footcandle, 23 hours of continuous light and 1 hour of darkness.

• **Turkeys**—The recommended lighting for market turkeys is: 0 to 2 weeks, 10.0 to 15.0 footcandle, continuous light; 2 weeks to market, 0.5 to 1.5 footcandle, 16 hours of continuous light and 8 hours of darkness.

The recommended turkey breeder lighting is the same as for market turkeys to 29 weeks of age. After 29 weeks, 5.0 to 7.0 footcandle, 14 hours light and 10 hours darkness in spring and summer (or 8 hours of light and 16 hours of darkness in the fall).

• **Ducks, breeders**—Mature ducks may be brought into production in the winter by providing a 14-hour day (14 hours of light and 10 hours of darkness).

• **Other poultry**—Pheasants and chukars appear to have a short day requirement. Coturnix (quail) are unique, for they may be sexually mature by 6 weeks of age if exposed to day lengths in excess of 12 hours, in comparison with the 21 to 29 weeks for chickens and turkeys, respectively.

SHADES

Heat stressed animals are poor doers! They eat less and gain less. So, providing adequate shade to protect animals from hot sun is among the more important and widely used devices for improving their environment in hot climates.

During the hot summer months, grazing animals and animals in corrals often show signs of heat stress from midmorning to late afternoon, during which time they will seek shades.

• **Beef cattle shades**—Numerous studies show that the voluntary feed intake of beef cattle is depressed at high temperatures, but that intake is depressed less when shades or cooling devices are available and when low fiber rations are fed. At 95°F, cattle on full feed consume 10 to 35% less feed than at the comfort zone of 59 to 77°F; and cattle fed maintenance rations show 5 to 20% feed depression when subjected to similar temperatures.

Tests conducted by the University of California, at the Imperial Valley Field Station, showed that it required 200 to 300 lb more feed to make 100 lb gain during midsummer without shade than with shade. Based on work conducted at the Yuma, Arizona Station, where summer temperatures of 110°F are not uncommon, University of Arizona workers showed that good shades can increase feedlot gains by 20 to 25% and improve feed efficiency by 14 to 20%.

Fig. 17-40. Shades protecting beef cattle from the hot sun. (Courtesy, Carl Stevenson, Red Rock Feeding Co., Red Rock, Ariz.)

Shades should be 12 ft or more high; provide 20 to 25 sq ft per animal; and be oriented north and south, so that the sun will shine under the shade early in the morning and late in the evening. Shades may be run east-west in the hot deserts of the Southwest to take fullest advantage of the cooler north sky.

Fans, sprinklers, and other cooling devices increase the effectiveness of shades. At Yuma, Arizona, where August temperatures reach daily highs of 110°F, researchers reported that yearling steers in pens equipped with shades and evaporative coolers gained at the rate of 3.01 lb per head daily, compared with 2.66 lb daily for steers equipped with shades and sprinklers, and 2.62 lb per day in the control pens—shades only.

University of Missouri researchers found that Shorthorn calves gained about ½ lb more per head per day and required 0.49 fewer lb of TDN per pound of gain at 50°F than at 80°F, and that heat-tolerant Brahmans were more efficient at 80°F than at 50°F; thus, there was a marked breed difference in the comfort zone. At the University of California Station in the Imperial Valley, Herefords gained 0.36 lb more per head per day when they had access to shelter cooled 7°F below the high outside temperature.

• **Dairy cattle shades**—Lactating cows under continuous heat stress begin to show a decline in feed intake at 77 to 81°F, with a marked decline occurring at 86°F. At 104°F, intake is usually no more than 60% of the 64 to 68°F level. Also, the higher milk producing animals with greater metabolic heat (from milk synthesis) tend to be more susceptible to heat stress.

Environmental modifications, such as shades and cooling devices, can be used to increase the feed intake of lactating cows during the hot summer months.

The specifications for, and the orientation of, dairy cattle shades are the same as those given for beef cattle.

• **Sheep shades**—Studies of temperature-feed intake interactions of sheep are limited, mainly because the vast majority of sheep are kept under extensive grazing conditions. However, the exposure of sheep, especially if unshorn, to hot, summer temperatures reduces voluntary feed intake just as it does in other species.

When exposed to summer temperatures above 86°F, shorn lambs show a marked decrease in average daily gains (ADG) and feed efficiency (gain/feed). One study showed that, when the temperature rose from 86 to 95°F, the ADG of shorn lambs fell from 0.24 to 0.09 lb and the feed efficiency decreased from 0.08 to 0.04. Without doubt, the modification of such high temperatures by the use of shades would be effective.

The most satisfactory sheep shades (1) provide 4 to 6 sq ft of shade per animal, (2) are 4 to 8 ft high, (3) are located with a north-south placement because they are drier—the sun can get underneath them to dry out the manure and urine, and (4) are open all around, so as to permit maximum air movement. Portable shades may be used.

• **Swine shades**—Younger and lighter pigs are less subject to heat stress than older and heavier hogs. Thus, the feed intake of pigs at 8 weeks of age may be higher at 77°F than at 68°F. Most investigators have found the range 64–70°F optimal for growing-finishing pigs. At higher temperatures, pigs eat less and gain less. But, by providing shades, wallows, and increasing air velocity, heat load can be lessened, which, in turn, will influence feed intake and daily gain.

• **Horse shades**—A shade, either trees or constructed, should be provided for horses that are in the hot sun.

The most satisfactory constructed horse shades are (1) oriented with a north-south placement, (2) at least 12 to 15 ft in height (in addition to being cooler, high shades allow a mounted rider to pass under), and (3) open all around.

• **Poultry shades**—The environmental controls for poultry have been perfected to a higher degree than for any other domesticated species.

Environmental modifications for today's U.S. commercial poultry are far more complete and sophisticated than merely providing shades; they include temperature, humidity, light (length of day and intensity), altitude (air pressure and partial pressures of oxygen and carbon dioxide), wind velocity (air movement), solar energy, quality of air and water, and density of population.

EVAPORATIVE COOLERS

Evaporative coolers for heat removal are especially effective in low humidity areas. They may be used for cooling buildings, and for cooling animals—especially dairy cattle, swine, and poultry.

Fig. 17-41. Lactating cows under a shade and evaporative coolers. (Courtesy, James Tappan, Arizona Dairy Co., Higley, Ariz.)

• **Roof sprinklers**—Open-sided houses are still much in evidence, although the construction of completely enclosed, environmentally controlled units is increasing for swine and poultry.

As long as there is some air movement, open-sided houses work well in warm weather, but in hot weather a warm breeze may stress the animals severely. The buildup of heat within the building is rapid, and at inside temperatures of 95°F or above, animals are greatly stressed. When this condition prevails, the animals may be comforted and losses minimized by sprinkling the roof of the house with circulating microjet sprinklers installed at the point of the roof.

• **Hog sprinklers**—Hogs have very few sweat glands; so, they cannot cool themselves by perspiration. Instead, they breathe faster and faster when they are very hot in an attempt to remove their surplus body heat by evaporation from the wetted surfaces of the lungs and air passageways. Under such circumstances, sprinklers are effective in relieving heat stress. One study showed that, when subjected to a 100°F air temperature, the respiration rate of hogs was within the range of 140 to 150 per minute. When sprinkled, their respiration rates dropped dramatically about 50% within 20 to 40 minutes. Furthemore, when the sprinkling was continued at either 40- or 80-minute intervals, the respiration rates remained in the range of 40 per minute.

Where sprinklers are used for cooling hogs, (1) provide 1 nozzle per 25 to 30 hogs, (2) station the nozzles 4 to 6 ft from the floor or ground and about 8 ft apart, (3) spray about 0.09 gal of water per hour per pig, (4) control by timer set at 2 minutes on in each hour and set thermostat at 75°F, and (5) provide a fine in-line filter. Sprinklers should be limited to concrete lots or sandy soils where mud holes do not develop, and to the manure collection area of the pens.

• **Hog wallows**—In the South, mature breeding animals and finishing hogs kept in the open (not housed) need a wallow during the hot, summer months. Instead of permitting an unsanitary mud wallow, successful swine producers construct hog wallows. This equipment, which maybe either movable or fixed, will keep the animals cool and clean and make for faster and more economical gains. The wallow should be near to shade, but in no case should shade be built directly over the wallow, because such an arrangement will encourage all the hogs to lie in the water all day. The size of wallow to build will depend upon the number and size of animals. Up to 50 growing-finishing pigs can be accommodated per 100 sq ft of wallow provided shade or shelter is nearby.

Fig. 17–42. Sows wallowing in mud to relieve heat stress. A constructed (usually concrete) wallow is more sanitary, and is recommended. (Courtesy, J. V. Craig, Department of Animal Sciences and Industry, Kansas State University, Manhattan)

• **Poultry foggers**—Foggers emit a fine mist of water. By placing them over the birds, the mist keeps the birds wet, and helps to keep them cool.

Death losses from heat prostration in open-sided poultry houses that are not closed and environmentally controlled are often severe in the humid sections of the country, and in the hot, dry areas where maximum daily temperatures may range from 105 to 115°F, or higher. Survival under such conditions is closely related to the availability of drinking water and to the persistence with which birds consume it. Losses can also be reduced by intermittent spraying of the birds, using fine mist foggers. Foggers are especially helpful in dry areas. A very small amount of water is used, only about 1 gal per hour for each nozzle, but the fine mist promotes evaporative cooling. When humidity is high, cold water can be used in the foggers to lower the body temperature by contact.

VENTILATION

Ventilation is the act of causing the movement of air through buildings with the objective of supplanting foul air with fresh air containing needed oxygen.

Ventilation is important only when animals are housed in crowded quarters. Contrary to common opinion, when a feeling of discomfort is noticed, it is the result of oxygen starvation rather than carbon dioxide poisoning.

The amount of moisture in the air is important. When improper ventilation prevents proper evaporation, the moisture content of the air increases. If humidity rises too high, interfering with heat elimination, heat stroke may ensue. Moist air is generally a more favorable medium for the existence of microorganisms, thus lending itself to the transmission of contagious diseases. When one animal is infected with a contagious disease and is closely housed with others, an epidemic will usually follow. The air may also pick up various noxious gases, such as ammonia from decomposing urine, which may cause irritation to the sensitive membranes of the mouth, eyes, nose, and respiratory tract.

Ventilation is measured in cubic feet per minute (cfm). The required ventilation differs according to species of animal, size of animal, and outside temperature.

There is continuing interest among livestock producers in naturally ventilated buildings as opposed to environmentally controlled buildings, primarily because (1) most of the livestock shelters of the world rely on the free flow of air through the building for ventilation, and (2) of their significantly lower construction and operating costs.

Naturally ventilated buildings can be successfully used for most livestock housing; among them, (1) free-stall housing for dairy cattle, (2) loafing or bedded pack barns for dairy, beef, or sheep, (3) swine-finishing buildings, (4) calf barns, and (5) poultry houses.

Fig. 17–43. Open-air, free-stall housing dairy cows, with milking parlor and maternity barn in the background. (Courtesy, Babson Bros. Co., Oak Brook, Ill.)

Naturally ventilated buildings are mainly a shell to protect animals from rain and snow, and to protect the building contents (e.g. grain, hay, etc.). Inside temperature in the winter will often be within 3 to 10°F of outside temperatures. Thus, such buildings are often referred to as cold confinement livestock buildings.

A naturally ventilated building has a continuous opening at the high point (normally the ridge) of the building for air exhaust and continuous openings or inlets along the long sidewalls of the building for fresh air. The size of these openings is based on rules of thumb or experience. Air entering along the sidewalls (normally under the eaves) of the building is warmed by the heat from the animals in the building and picks up moisture as it rises toward the ridge, where it escapes.

During warm weather, the building should serve mainly to keep rain out and act as a sunshade. Large continuous openings in the sidewalls allow summer breezes to blow through the building.

Typical naturally ventilated buildings can be divided into two types:

Fig. 17-44. Naturally ventilated, confinement facilities for market turkeys. (Courtesy, *Turkey World,* Mount Morris, Ill.)

1. **Open front.** These buildings have one long side completely open at least one-half the height of the sidewall. The open side faces away from the direction of prevailing winter winds.

2. **Enclosed.** These buildings have all sides closed but provide continuous eave openings and large doors or vent panels for summer conditions. Enclosed naturally ventilated buildings offer more protection from wind and precipitation.

WINDBREAKS

Natural shelters for beef cattle and horses may consist of hills and valleys, timber, and other natural windbreaks. If natural windbreaks are adequate, it may avoid the necessity of constructing shelter.

The vast majority of feedlot cattle are fed in open pens, without shelter. Some are provided wind protection by means of a board fence (see Fig. 17-45).

Fig. 17-45. Feedlot windbreak and cable fence. Upright fence boards are spiked to horizontal nail ties (2″ lumber) for a windbreak in mounded feedlot. Windbreak, supported by treated poles, is vertical to the ground so it spans curved ground without leaning. Cable fence in foreground is braced with pipes. Pipes have "feet" welded to them that are spiked to the posts. Cables pass through holes drilled in posts. A cable brace with turnbuckle is anchored diagonally from gatepost to second post. (Photo by A. M. Wettach, Mount Pleasant, Iowa)

HEALTH/ENVIRONMENTAL INTERACTIONS

Health is the state of complete well-being, and not merely the absence of disease.

Environment embraces the forces and conditions, both physical and biological, that (1) surround animals, and (2) interact with heredity to determine behavior, growth, and development.

Disease is defined as any departure from the state of health.

Parasites are organisms living in, on, or at the expense of another living organism.

Feed, air quality, lighting, noise, other animals, and weather are among the many factors that constitute an animal's environment. Extremes or alterations in the environment may subject an animal to stress; and stress may affect health and lead to more diseases and parasites.

The importance of good animal health is underscored by the following statistics: It is estimated that animal diseases and parasites in the United States (1) decrease animal productivity by 15 to 20%, and (2) make for annual losses of $10 billion. Further, there is evidence that nutrition has some involvement in 85% of the veterinary cases. In the developing countries, diseases and parasites take an even greater toll—they decrease animal productivity by 30 to 40%.

Some important health/environmental interactions not covered elsewhere in this book are discussed in the sections that follow.

(Also see Chapter 5, Nutritional Disorders/Toxins.)

Altitude

High altitude not only affects environment it is part of it. In extremely high altitudes, animal life is sparse; and few species are well adapted. Thus, the llama is well adapted to the Andes Mountains, and the Yak is well adapted to Tibet—the roof of the world. At such high altitudes, people and animals breathe more rapidly in order to obtain the necessary oxygen. Thus, the environmental effect of altitude is important and warrants further study.

Antibody Production

An antibody is a protein substance (a modified type of blood-serum globulin) developed or synthesized by lymphoid tissue of the body in response to an antigenic stimulus.

Antigens may be (1) components of certain drugs or feed; (2) infectious microorganisms or parasites; (3) substances from the environment, such as chemicals, dusts, pollen grains, etc.; or (4) tissues of the body itself.

Each antigen elicits production of a specific antibody. In disease defense, the animal must have an encounter with the pathogen (antigen) before a specific antibody is developed in its blood.

Normally, antibodies react with antigens and render them harmless. However, certain antibodies may attack body tissue. This abnormal condition is called autoimmunity. Generally, repeated exposure to a specific type of antigen increases the rate at which antibodies against the substance are produced.

Antibody production may be impaired under such conditions as (1) malnutrition, (2) oversecretion of stress hormones, (3) advanced aging, or (4) inherited inability to produce certain antibodies.

Disease Defense

Pathogens affecting animals are ever present in the environment. But, in order to produce disease, they must overcome the body's first line of defense—they must first gain entrance to the animal by one of the body openings or through the skin. Then, they usually multiply and attack the tissues. To accomplish this, they must be sufficiently powerful (virulent) to overcome the defenses of the animal body. The defenses of the animal body vary and may be weak or entirely lacking, especially under conditions of a low nutritional plane and poor management and sanitation practices.

Pathogens commonly gain entrance into the body through one or more of the following channels:

1. Respiratory tract
2. Digestive tract
3. Genital tract, especially during mating or parturition
4. Wounds
5. Mucous membranes of the eye, e.g., pinkeye and leptospirosis (the latter may be acquired when the urine of an infected animal gets into the eye)
6. Teat canal, especially in lactating females
7. Navel cord in the newborn
8. Contaminated syringes or surgical instruments
9. Insect bites

If the pathogens gain entrance to the animal body, they are usually subjected to one or more of the following reactive defenses; inflammatory reactions, febrile reactions, or immune reactions.

• **Inflammatory reactions**—Inflammatory reactions are characterized by the following four cardinal signs: (1) an increased blood supply (redness), (2) increased temperature of the part (heat), (3) swelling of the part (edema), and (4) increased sensitivity (tenderness and pain).

• **Febrile reactions**—Febrile reactions are characterized by (1) an overall increase in body temperature, and (2) an increase in metabolic activity.

• **Immune reactions**—Immune reactions are characterized by the ability of animals to resist and/or overcome disease either through (1) natural (inherited) immunity, or (2) acquired immunity. Immunity is very important; hence, further elucidation of its protective mechanisms against disease follows:

The animal body is remarkably equipped to fight disease. Chief among this equipment are large white blood cells, called phagocytes, which are able to overcome many invading organisms.

The body also has the ability, when properly stimulated by a given organism or toxin, to produce antibodies and/or antitoxins. When an animal has enough antibodies to overcome particular disease-producing organisms, it is said to be immune to that disease.

When immunity to a disease is inherited, it is referred to as a *natural immunity.* For example, when sheep are exposed to hog cholera they never contract the disease because they have a type of natural immunity referred to as species immunity. Likewise, humans are naturally immune to Texas fever. Algerian sheep are said to be highly resistant to anthrax; this constitutes a type of natural immunity called racial immunity.

Acquired immunity or resistance is either *active* or *passive.* When the animal is stimulated in such manner as to cause it to produce antibodies, it is said to have *acquired active immunity.* On the other hand, if an animal is injected with the antibodies (or immune bodies) produced by an actively immunized animal, it is referred to as an *acquired passive immunity.* Such immunity is usually conferred by the injection of blood serum from immunized animals, the serum carrying with it the substances by which the protection is conferred. Passive immunization confers immunity upon its injection, but the immunity disappears within 3 to 6 weeks.

In active immunity, resistance is not developed until after 1 or 2 weeks; but it is far more lasting, for the animal apparently keeps on manufacturing antibodies. It can be said, therefore, that active immunity has a great advantage. There are exceptions, however—for example, when a disease must be checked immediately as in a virulent outbreak of swine erysipelas, when immune serum from actively immunized horses is injected.

It is noteworthy that young suckling mammals obtain a passive immunity from the colostrum that they obtain from their mothers following birth.

• **Vaccination**—*Vaccination may be defined as the injection of some agent (such as a bacteria or vaccine) into an animal for the purpose of preventing disease.*

In regions where a disease appears season after season, it is recommended that healthy susceptible animals be vaccinated before being exposed and before there is a disease outbreak. This practice is recommended not only because it takes time to produce an active immunity but also because some animals may be about to be infected with the disease. The delay of vaccination until there is a disease outbreak may increase the seriousness of the infection. In addition, a new outbreak will "reseed" the premises with the infective agent.

In vaccination, the object is to produce in the animal a reaction that in some cases is a mild form of the disease.

It is a mistake, however, to depend on vaccination alone for disease prevention. One should always ensure its success by the removal of all interfering adverse conditions. It may also be said that varying degrees of immunity or resistance result when animals are actively immunized. Individual animals vary widely in their response to similar vaccinations. Heredity also plays a part in the determination of the level of resistance. In addition, nutritional and management practices play important parts in degrees of resistance displayed by animals.

COLOSTRAL DEFENSE

Colostrum is the first milk secreted by mammalian females following parturition.

Newborn mammals are unable to produce antibodies within their own bodies for some time after birth; they acquire these antibodies from their mothers either while in the uterus before birth or through colostrum after birth. Newborn calves, lambs, kids, pigs, and foals do not acquire passive immunity while *in utero*; so, the transfer of immunoglobulins via colostrum is of special importance to them. However, they should receive colostrum during the first 12 to 24 hours after birth if they are to acquire passive immunity. After about 24 hours following birth, gut closure occurs, following which the newborn animal digests these proteins, which then lose their immunization properties. Apparently, this results from the newborn not being able to absorb the large protein molecule.

Orphan calves may receive colostrum from any fresh cow or from frozen colostrum that has been stored. Also, calves can absorb antibodies in ewe or mare colostrum, but certain diseases are species specific; hence, colostrum from another species may not afford the desired protection.

(Also see Chapter 5, section on "Immunoglobulin Deficiency.")

IMMUNE SUPPRESSION (ALTERED IMMUNE RESPONSE)

People blush when they are embarrassed, and their hearts race when they are frightened—reactions that show the linkage of the brain and body. Recent studies in both humans and animals link the brain to the body's immune system—the complex array of organs, glands, and cells that comprise the body's principal mechanism for repelling invaders.

Studies show that people's emotions have a great impact on their health—that people stressed by bereavement (such as the loss of a loved one) or by loneliness (social isolation) suffer high incidences of disease and mortality. Also, research has shown that chronic stress causes the adrenal gland to pump increased amounts of corticosteroids into the bloodstream, and that these chemical messengers inhibit immune action.

Although animals do not blush, they react to stress. Thus, for many years, livestock producers have tried to minimize stress among their animals by providing a clean sanitary environment, and by keeping them warm in the winter and cool in the summer. Yet, animals still experience unavoidable and harmful stresses, especially during weaning and shipping. Following such stressful times, outbreaks of animal diseases like shipping fever, coccidiosis, scours, and transmissible gastroenteritis are more common.

Stress suppresses an animal's immune system and leaves it more vulnerable to bacterial and viral infections and all other immune disoders. Here is how: When an animal is stressed, the secretion of a hormone, called corticosteroid (cortisol, corticosterone), by the adrenal gland, is increased. The main purpose of the corticosteroids is to increase the animal's immediate chances for survival by producing more energy to withstand the current stressful condition. They are the frontline warriors. However, corticosteroids also decrease immunity and suppress the subsequent ability of the animal to fight infection. For example, corticosteroids help calves survive the stress of weaning at the expense of a lowered abiliy to withstand infection. So, at the end of weaning, the calves are left in a state of weakness to fight disease. As a result, a few days later, the animals may succumb to shipping fever, a disease caused by agents in the environment.

Many stressors increase corticosteroid production and alter the animal's immune system by depressing antibody production and leucocyte levels in the blood, both of which are important in fighting diseases. The most common stressors that can alter animals' immune systems are: heat, cold, crowding, mixing, weaning, fatigue, limit-feeding, noise, and restraint.

If the stress is brief, detrimental effects are not generally observed. However, if the response to stress exceeds a few hours, the catabolic effects cause reduced feed efficiency, growth, product production, and disease resistance.

Based on the above concept, blood corticosterone levels may be used as one of the measures of stress.

Heaves

Heaves are an environmentally-related disease of horses and mules. The disease is associated with (1) the feeding of damaged, dusty, or moldy hay; and/or (2) the use of dusty bedding or paddocks. Heaves are characterized by difficulty in forcing air out of the lungs, resulting in jerking of the flanks and coughing. Heaves in equines is similar to emphysema in people. Control consists of providing an environment that is as nearly dust-free as possible.

(Also see Chapter 5, Table 5-1, section on "Heaves.")

Mastitis

Mastitis is an infectious inflammation or irritation in the udder which interferes with the normal flow of milk and/or its quality.

Mastitis is the most common and most costly disease affecting dairy cows. According to the National Mastitis Council, nearly 40% of all dairy cows have some form of mastitis. Monetary losses from mastitis range from $90 to $250 per infected cow per year, depending on the severity of the infection.

It has been said that producers themselves are responsible directly, or indirectly, for 90% of their mastitis troubles; however, most producers implicate their milking machines. The three main routes through which mastitis comes are: (1) dirty, or poorly adjusted, milking equipment; (2) poor milking practices; and (3) injuries to cows because of their surroundings—their environment.

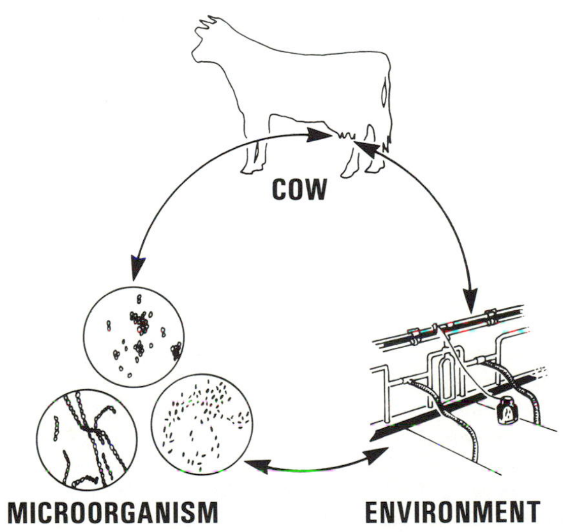

Fig. 17-46. Interrelationship of the three major factors involved in bovine mastitis.

Several species or groups of microorganisms may cause mastitis but over 95% of all cases are caused by the following species of streptococci and staphylococci: *Streptococcus agalactiae, Streptococcus dysgalactiae, Streptococcus uberis,* and *Staphylococcus aureus.* Infection with any of these organisms is usually chronic, with flare-ups recurring frequently. No amount of drugs given to cows today can prevent another attack next month under the same conditions.

Dairy producers themselves are unwittingly setting the stage for mastitis flare-ups in their herds, by providing ideal conditions for it—poor milking practices, poor milking equipment, and improper surroundings. By rectifying these shortcomings—through managed milking and sanitation—producers can reduce or eliminate mastitis.

Pests and Pesticides

Pests are counterproductive—they make for an adverse environment and cause an estimated 30% annual loss in the worldwide potential production of crops, livestock, and forests.[14] Every part of our feed, food, and fiber supply—including that from wild and domestic animals, marine life, field crops, horticultural crops, and wild plants—is vulnerable to pest attack. Obviously, if these losses could be prevented, or reduced, world food supplies would be increased.

This means that a worldwide annual loss of 30% potential food productivity occurs despite the use of advanced farming technology and mechanized agriculture. This also means that in many of the developing countries losses greatly exceed this figure.

Among the pests that cause high losses are: (1) plant diseases, caused by more than 160 bacteria, 250 viruses, and 8,000 fungi; (2) insects, of which 10,000 species in the United States are destructive; (3) weeds and brush of which some 2,000 species in the United States cause losses in crops and animals; (4) rats, mice, gophers, and other rodents, which consume and damage feed, destroy property, and spread disease; and (5) certain birds, such as starlings, which consume and contaminate feed and spread many diseases.

Fig. 17-47. Face-fly avoidance behavior of dairy cows, showing rosette-like formation with heads directed medially. (Courtesy, E. Schmidtmann, Livestock Insects Laboratory, Agricultural Environment Quality Institute, Beltsville, Md.)

Although science and technology have been the great multipliers in increasing our food supply, potential food supplies are still destroyed by the ravages of pests. For example, in the high-rainfall belt of tropical Africa, the dreaded tsetse fly has kept large areas out of agricultural production. It is estimated that if *trypanosomiasis,* a disease borne by the tsetse fly, were brought under control the savannah pastures of the tsetse fly infested area would carry a cattle population of 120 million head, which is more than the cattle population of the United States.

[14]Ennis, W. B., Jr., W. M. Dowler, and W. Klassen, "Crop Production to Increase Food Supplies," *Science,* Vol. 188, No. 4188, May 9, 1975, pp. 593-598. The authors are staff scientists on the National Program Staff, Agricultural Research Service, U.S. Department of Agriculture, Beltsville, Md.

• **Pesticides**—*A pesticide is any substance that is used to control pests.* Pesticides are an integral part of modern agricultural production and contribute greatly to the quality of food, clothing, and forest products that we enjoy. Also, they protect our health from disease and vermin. Pesticides have been condemned, however, for polluting the environment, and in some cases for posing human health hazards. Unfortunately, opinions relative to pesticides tend to become polarized. A report by the National Research Council summarized the situation as follows:

> "Users of pesticides fear that they will be regulated to the point where pests cannot be effectively controlled, with concomitant losses of food while opponents of the use of pesticides fear that people are being poisoned and that irreversible damage is being done to the environment."[15]

No pest control system is perfect; and new pests keep evolving. So, research and development on a wide variety of fronts should be continued. We need to develop safer and more effective pesticides, both chemcial and nonchemical. In the meantime, there is need for prudence and patience.

Pollution

Anything that defiles, desecrates, or makes impure or unclean the surroundings pollutes the environment and can have a detrimental effect on animal health and performance. Thus, gases, odorous vapors, and dust particles from animal wastes (feces and urine) in buildings directly affect the quality of the environment. Muddy lots and stray electrical voltage may also pollute the environment. For healthy and productive animals, each of these pollutants must be maintained at an acceptable level.

Fig. 17-48. Pollution control. The runoff from the cattle feedlots (left) on the Carl Embry farm, in Arkansas, flows into a basin (right). Solids remain in the basin, from which they are removed from time to time, and liquids are pumped onto surrounding fields. (Courtesy, USDA, Soil Conservation Service)

[15]*Pesticide Decision Making,* Vol. VII of the Analytical Studies for the U.S. EPA, NRC, National Academy Press, 1978, pp. 14–15.

DUST

Dust may be defined as a mixture of small particles of different sizes of dry matter.

Dust is a contributing factor to both animal and human health, especially with respect to respiratory diseases. Thus, it should be considered a significant contaminant that adversely affects environmental quality of animal houses and feedlots.

Dust may be present in significant amounts both inside and outside buildings.

Cattle and sheep feedlot dust is both organic, from excreta, and inorganic, from soil. Sprinkling and increased animal density (resulting in more moisture from urine) are the most effective methods of preventing feedlot dust in dry climates or during dry seasons. In swine houses and horse barns, most of the dust comes from the feed. In poultry houses, the dust contains a considerable amount of feather and skin debris, along with particles from the feed and litter. Dust in animal buildings may also carry microbes, gases, and vapors.

GASES AND ODORS FROM ANIMAL WASTES

In closed buildings, gases arise from manure decomposition, respiratory excretion, and fuel burning heaters. Some gases are poisonous.

Odors arise predominantly from manure decomposition. They consist mostly of hydrocarbons containing sulfur or nitrogen. Most states use a nuisance theory to control odor. However, in advance of designing and constructing animal facilities, it is important to know how much abatement equipment is necessary to prevent the neighbors from being offended.

The animal's respiratory system is in continuous contact with the environmental air; hence, if the air contains excessive pollutants, respiratory diseases may occur.

The principal gases generated by stored manure are carbon dioxide, ammonia, hydrogen sulfide, and methane. Of these, ammonia and hydrogen sulfide are odorous. These two gases, along with methane, also provide an explosive hazard when mixed with sufficient air. Ventilation for adequate moisture and temperature control in a confined livestock building is usually sufficient to keep noxious gases from reaching harmful or lethal concentrations unless the ventilation system fails or the liquid manure in the underfloor storage is agitated vigorously. A good indication of the effectiveness of a livestock housing ventilation system may be obtained if certain odors are detected when entering the building.

• **Ammonia (NH_3)**—Ammonia is released from fresh manure and during aerobic decomposition of it. Eyes begin to burn at 25 to 30 ppm, and inflammation in the eyes of chickens occurs at 50 ppm. Concentrations of 200 ppm cause animals to lose appetite, salivate excessively, and sneeze, without loss of feed efficiency. A concentration of 5,000 ppm is dangerous to humans.

• **Carbon dioxide (CO_2)**—Carbon dioxide can increase substantially with animal respiration and manure decomposition, thereby contributing to an oxygen deficiency in the building. Concentrations of carbon dioxide up to 10% can be tolerated by animals and humans if exposure is not for more than an hour.

• **Hydrogen sulfide (H_2S)**—Hydrogen sulfide (the gas with the rotten egg odor) is the most toxic gas in liquid manure systems. It is produced by anaerobic action. In a mixture with air, it will burn with a bluish flame and it can explode violently. Animals subjected to levels of 20 ppm develop fear of light, nervousness, and loss of appetite. At higher levels, symptoms include diarrhea. During agitation, levels of 800 to 1,000 ppm may occur, which may be fatal to animals. With adequate ventilation, levels well below 20 ppm can be maintained except during agitation of liquid manure.

• **Methane**—Methane is flammable and hazardous as an explosive gas when mixed with air in a concentration of 5 to 15%. Being lighter than air, it rises to, and accumulates near, the top of tight enclosures. Ruminants exhale small amounts of methane, but most of it comes from manure decomposition. At high concentrations, it can be asphyxiating.

The most important requisite for controlling noxious gases in a livestock building is continuous ventilation to exhaust the gases and odors, replacing them with fresh outside air. A breakdown in the ventilation system, due to power failure or some other cause, can result in loss of animals due to noxious gases, as well as by heat prostration and asphyxiation. The normal ventilation rate for control of temperature and humidity is sufficient for maintaining levels of gases not harmful to animals except during agitation or pumping of liquid manure from storage.

(Also see Chapter 12, section on "Animal Wastes [Manure and Litter].")

MUDDY LOTS

Muddy lots often plague beef cow-calf producers, dairy producers, and cattle feedlot operations, especially during the winter months. Mud increases calf scours and other diseases in newborn calves and reduces gains and feed efficiency in feedlot cattle.

California Agricultural Experiment Station studies show that mud can reduce finishing cattle gains by as much as 10 to 35%, and increase the feed required per pound of gain by a like amount. Thus, it is important that the problem be minimized, especially in high rainfall areas. Good drainage is the first essential. This should be assured at the time the lot is located and constructed.

Mounds 6 to 12 ft high, preferably perpendicular to the feed bunk, will provide finishing cattle a dry place on which to lie. Concrete aprons 10 X 12 ft wide and sloping 1 in. per foot along the bunk will provide them with solid footing on which to stand and feed. Also, lessening of cattle density during the winter months—fewer animals per lot—is an effective method of controlling the mud problem. Thus, many feedlots plan to feed fewer cattle during the muddy season.

To cope with mud and alleviate calf scours, cow-calf operators should move the cows to a clean pasture during the calving season. If no pasture is available, dirt mounds and a coarse straw bedding are recommended.

STRAY VOLTAGE (TINGLE VOLTAGE, NEUTRAL-TO-EARTH VOLTAGE)

Stray electrical voltage has caused serious problems on many dairy farms—affecting animal behavior and lowering milk production, although it may affect other animal species also. Contrary to popular belief, stray voltage is not new; it is as old as electricity itself. However, it has become a problem on many farms recently for two reasons: (1) There is more electrical load on today's farms; and (2) in the last 20 years we have used more equipment grounding for safety purposes.

Stray voltage is excessive voltage between two animal contact points. The conditions that cause stray voltage are, electrically, quite simple: If sufficient voltage is present, it may force a current through any available conductor, including a cow's body. Cows are good conductors because of their body design (the length from mouth to front and rear legs); cows bridge the gaps between electrically grounded objects and "true earth." The cow doesn't feel the voltage as such; she feels the tingling current running through her body.

People seldom feel the current for several reasons. Usually, caretakers wear rubber-soled shoes when in the barn, whereas the bare-footed cow stands on concrete that is often wet. Also, humans have only two legs instead of four like the cow, and human's legs touch the floor near the same vicinity.

• **Sources of stray voltage**—Any electrical condition which creates large enough voltage between any two animal contact points may create a stray voltage problem. The source of stray voltage may be either "on-farm" or "off-farm."

On-farm voltage problems stem from defective equipment, faulty wiring, bad connections, or having several 120–volt motors on the same line. On-farm stray voltage can be minimized by maintaining good electrical wiring systems that meet the requirements of the National Electrical Code. Also, properly balanced 120–volt circuits and conversion of larger 120–volt meters to 240 volts will reduce the effect of secondary neutral voltage drops at the farm service entrance.

Off-farm voltage comes onto the farm through the electrical supplier's lines. Voltage will vary with the load and the natural grounding ability of the area. As usage increases, so may stray voltage. Heavier loads are seen at milking time and in the fall when grain dryers may be running on many farms.

• **Signs of stray voltage**—One or more of the following signs may indicate that stray voltage exists in a dairy:

1. **Cows reluctant to enter the parlor.** When cows are subjected to stray voltages in the parlor stalls, they soon become reluctant to enter the parlor.

2. **Cows nervous in parlor.** Cows often dance or step around almost constantly while in the milking parlor.

3. **Uneven milk let-down and milk-out.** When milk let-down and milk-out are uneven, more machine stripping is required and longer milking time becomes apparent.

4. **Increased mastitis.** When milk-out is incomplete, more mastitis is likely to occur; all that is required is the presence of infectious bacteria. In turn, this will result in increased somatic count.

5. **Reduced feed intake in the parlor.** If cows encounter stray voltage while eating from the grain feeders, a reluctance to eat and reduced feed intake usually follow.

6. **Reluctance to drink water.** If stray voltage reaches the cows in stall barns through the water supply or metal drinking cups, the animals soon become reluctant to drink.

7. **Lowered milk production.** Each of the symptoms listed above is associated with stress and reduced feed intake, followed by a drop in daily milk production.

But detection of stray voltage is not easy! Other factors such as mistreatment of animals, milking machine problems, disease, sanitation, and nutritional disorders can create problems which manifest themselves in the seven symptoms listed above.

• **Use voltmeter to monitor voltage**—The only sure method to determine if significant stray voltage is present is to have a qualified person perform a stray voltage survey, using approved equipment and monitoring the voltage through one, and preferably two, milkings. Point to point measurements between cow contact points will determine if the voltage is actually getting to the cow. Generally, stray voltage is not constant throughout the day; so, readings should be taken over a long period.

Most milking machine company representatives, many power supplier employees, some milking equipment dealers, and some veterinarians and county extension agents have equipped themselves with suitable voltmeters and are prepared to lend assistance. *Someone familiar with electrical systems, wiring, and equipment should be present when measurements are made.*

• **Install neutral isolators**—Neutral isolators may be installed to eliminate off-farm sources of stray voltage. They must be installed by a licensed electrician and must meet the code requirements.

As a result of a study of 395 Minnesota dairy farms on which neutral isolators had been installed, Minnesota Agricultural Experiment Station workers reported that the isolators increased milk production by 700 lb per cow over a 12-month period. Today, neutral isolators are commonly used to eliminate off-farm sources of stray voltage.

• **Establish equipotential planes**—This method of control can be applied to existing facilities or new facilities. But it must be expertly engineered and installed. Details follow.

1. **Existing facilities.** In an existing facility, this may be accomplished by cutting slots in the concrete floor, placing bare copper wire in the slots, and grounding the wires to the neutral at the electrical service panel. If all animal contact points are bonded together, a voltage differential cannot develop. However, if a neutral-to-earth voltage exists on the farm, the animals may react as they enter onto the equipotential plane. So, to reduce the extent of this voltage differential a gradual voltage gradient must be established between the remote floor or earth and the equipotential plane. This may be accomplished by running a bare copper wire in the concrete out through the entry doors of the milking parlor a distance of 20 ft with the wire embedded deeper (to a final depth of 24 in.) as it becomes more distant from the equipotential plane.

2. **New facilities.** An equipotential plane should be established in all new milking facilities and other areas where animals make contact directly with grounded equipment. The equipotential plane should consist of 6–10 gauge steel mesh with a grid dimension of 6″ X 6″. Sheets of reinforcement mesh should be welded together and welded to any steel stanchions embedded into the concrete. This mesh must be bonded to the neutral of the electrical service panel. Here again, transitions on and off the plane are critical.

Porcine Stress Syndrome (PSS)

Swine with porcine stress syndrome (PSS) cannot cope with stressful situations. When present, the disorder is usually associated with heavily muscled animals and results in sudden unexplained death losses. Animals afflicted with PSS often show signs of nervousness and may have muscle tremors indicated by a rapid tremor of the tail. When exposed to a stressful situation such as a change in surroundings, a sudden change in the weather, vaccination, castration, estrus or mating, they often become overly excited and develop blotches on their skin, along with muscle rigidity, followed by rapid labored breathing. Their body temperature also rises and they show signs of heat stress even in cold weather. Death losses from PSS usually occur when hogs are being sorted and delivered for slaughter, with the highest losses occurring during the summer months. Further details follow:

1. **Cause.** Although the true cause of PSS is not known, it appears to be genetic—inherited as a recessive. It afflicts all breeds; it is common in animals with superior muscling; and its symptoms are triggered by stress. When exposed to stress, PSS pigs undergo a very rapid depletion of their muscle energy stores (glycogen) and a corresponding increase in lactic acid in both the muscle and blood. Normal pigs can remove the lactic acid from the muscle and blood fast enough to prevent build-up, but PSS pigs cannot. The excess lactic acid results in acidosis, which, in turn, leads to the production of pale, soft, exudative (PSE) pork at slaughter in affected animals.

2. **Tests.** Two different tests may be made for the presence of PSS for the purpose of evaluating animals for the breeding herd:

   a. **The creatine phosphokinase (CPK) test.** This involves obtaining a sample of blood and analyzing it for CPK, a serum enzyme that is abnormally high in PSS swine.

   b. **The halothane test.** This involves anesthetizing the animals with halothane, to which PSS swine respond by showing extreme muscle rigidity within 3 minutes.

3. **Relation of PSS to pork quality.** Most pale, soft, and exudative (PSE) pork is the end result of PSS. The two problems are closely related. But not all PSS pigs produce PSE pork; if there is little stress, the pork may be normal.

4. **Preslaughter handling and prevention of PSE pork.** Some environmental conditions may be comfortable to a normal animal, but stressful to a PSS pig. Moreover, it is impossible to handle pigs under practical conditions without imposing some stress. Nevertheless, the amount of PSE pork may be lessened by following good management practices when preparing and shipping hogs to market.

(Also see the earlier discussion in this chapter under the section headed "Swine behavior," subsection headed "Porcine stress syndrome [PSS].")

Ulcers

Gastric ulcers are of special importance among swine and foals, and abomasal ulcers appear to be increasing in mature cattle and young calves.

• **Swine ulcers**—Stomach ulcers in growing-finishing pigs are widespread throughout the world, with the incidence higher in England and Sweden than in the United States. Finely ground feeds and stresses (caused by such things as long duration of shipment, crowding, and fasting) increase the incidence. The economic loss to the industry is restricted mainly to deaths among affected pigs, as growth rate and feed efficiency are not adversely affected.

• **Foal ulcers**—In a 5-year study conducted at the University of Florida, gastric and/or duodenal ulcers were found in 25% (129) of the 511 foals that were presented for necropsy. A single cause of foal ulcer was not found, although stress and diet seemed to be especially important. At the University of Kentucky's Department of Veterinary Science, where gastric ulcers in foals have been studied for 15 years, the 3 main predisposing factors causing the gastric ulcer syndrome in foals, in order of incidence, were: (1) excessive milk from the mare, (2) stress on the foal, and (3) drug-induced ulcers. The mares that produced the most milk tended to have the most foals with ulcers—the foals over-indulged in milk. Stress-related ulcers were caused by sudden weather changes, especially the onset of hot weather, by mares coming in heat and being bred, and by weaning. Drug-induced ulcers tended to follow the drug treatments of foals for diseases, especially diarrhea and pneumonia.

STRESS/ENVIRONMENTAL INTERACTION

"A merry heart doeth good like a medicine;
but a broken spirit drieth the bones."
Proverbs 17:22

Stress, is defined by Hans Selye, M.D., the world's leading authority on stress, as *the nonspecific response of the body to any demand.*

The authors use the term stress to indicate an environmental condition that is adverse to an animal's well-being, either external (nutritional, weather, social) or internal (disease, parasites).

Stresses of many kinds affect animals; among them, cold stress, heat stress, drafts, poor ventilation, excitement, presence of strangers, fatigue, mixing animals, number of animals together, space, changing corral and corral mates, weaning, previous nutrition, hunger, thirst, poor sanitation, disease, parasites, surgical operations, injury, and management.

Race and show horses are always under stress; and the greater the speed and the more tired they become, the greater the stress. Also, the greater the stress, the more exacting the nutritive requirements. Thus, the ration of race and show horses should be scientifically formulated.

Animals can be prepared, or adapted to the environment, in such a manner as to reduce stress. For example, if calves are properly *preconditioned* (started on feed, vaccinated, treated for parasites, etc.) prior to weaning, the stress of subsequent weaning and movement to a feedlot will be minimized.

In the life of an animal, some stresses are normal, and they may even be beneficial—they can stimulate favorable action on the part of an individual. Thus, we need to differentiate between stress and distress. Distress—not being able to adapt—is responsible for the harmful effects. The trick is to manage stress so that it doesn't become distress and cause damage and to recognize the warning signals of distress. For example, Texas Agricultural Experiment Station workers recently reported that added vitamin C, in either the feed or water, may reduce many of the health hazards associated with various kinds of stress to chickens such as hot weather, interaction with other birds in crowded conditions, and exposure to diseases.

The principal criteria used to evaluate, or measure, the well-being or stress of people are: increased blood pressure, increased muscle tension, body temperature, rapid heart rate, rapid breathing, and altered endocrine gland function. In the whole scheme, the nervous system and the endocrine system are intimately involved in the response to stress and the effects of stress.

The principal criteria used to evaluate, or measure, the well-being or stress of animals are: growth rate or production, efficiency of feed use, efficiency of reproduction, body temperature, pulse rate, breathing rate, mortality, and morbidity. Other signs of animal well-being, any departure from which constitutes a warning signal, are: contentment, alertness, eating with relish (and cudding by ruminants), sleek coat and pliable and elastic skin, bright eyes and pink eye membranes, and normal feces and urine.

Stress is unavoidable. Wild animals were often subjected to great stress; there were no caretakers to modify their weather, often their range was overgrazed, and sometimes malnutrition, predators, diseases, and parasites took a tremendous toll.

Domestic animals are subjected to different stresses than their wild ancestors, especially to more restricted areas and greater animal density. However, in order to be profitable, their stresses must be minimal.

Bleeders—A Stress/Environmental Interaction of Equines

Stress triggers different syndromes in different species and in different individuals. When subjected to heavy exercise, such as in racing, horses are prone to a condition known as *bleeders.*

It has been estimated that 70 to 80% of Thoroughbred racehorses are bleeders. In a recent major U.S. race, it was reliably reported that 5 horses in the field of 11 ran with the aid of Lasix, a diuretic used to combat bleeding. So, bleeders are a serious problem!

Bleeders are horses afflicted by blood flowing from their nostrils or bronchial tubes, but originating in the lungs, following strenuous exercise. But the problem is not confined to Thoroughbreds, nor is it limited to racehorses. It also afflicts Quarter Horses and Standardbreds, and it afflicts horses when they are subjected to maximum stress by strenuous physical activity of whatever kind; for example, in endurance rides. But it does not affect racing dogs or humans who race competitively. The species dissimilarity is attributed to the difference in the horse's anatomy. The horse has a sloping diaphragm and is primarily an abdominal breather, inhaling by movement of the diaphragm. This type of breathing appears to create stress in the equine lung.

Bleeders among horses racing or training have been recognized for at least 300 years. One of the earliest reported bleeders was a horse named Bleeding Childers, so named because of his frequent bleeding from the nostrils. Retired to stud because of training and racing difficulties associated with bleeding, the horse was tactfully renamed Bartlet's Childers and went on to gain fame as the grandsire of Eclipse, England's undefeated racing champion of the 18th century. The story goes that another famous English Thoroughbred stallion, Herod, raced in the Great Subscription Purse at York "but broke a blood vessel and was beaten off."

Until recently, one of the major difficulties in determining the source of blood in bleeders has been the lack of an adequate diagnostic instrument to examine safely the horse's upper respiratory tract. The advent of the flexible fiberoptic endoscope revolutionized diagnosis.

Current indications are that the blood originates in the lung tissue, and not within the nasal cavity. Recent findings indicate that the bleeding is caused by ruptured capillaries (pulmonary hemorrhaging) rather than nosebleed; and that, although a large number of horses bleed from the lungs following racing, only a small percentage actually show external evidence of blood at the nostrils—it is swallowed into the gastrointestinal tract.

Bleeding appears to affect the performance of racehorses. But the cause and cure remain elusive. The condition is exercise-induced; and there appears to be a higher frequency of bleeders among older horses. When bleeding is excessive, the most common treatment is to discontinue training and racing temporarily or permanently.

While scientists work to find the cause and an effective treatment of bleeders, among the remedies, mostly highly secretive, attempted by distraught horse owners and caretakers, with little documented success, are copper bands aroung the tail, coins in water buckets, bloodletting, and a variety of folk remedies. Based on studies to date, supplementation with vitamin K, vitamin C, or bioflavinoids does not appear to be beneficial. Furosemide, popularly known by the trade name Lasix, which reduces (but does not prevent) pulmonary edema almost instantly, was approved by the FDA as an equine medication in 1967, and is a regulated raceday medication permitted in a majority of states. According to researchers at Washington State University, Lasix promotes the clotting of blood.

Research studies indicate that bleeders are a stress/environmental interaction of equines induced by heavy exercise, but the findings do not signal any restriction in racing horses; because galloping is a very natural behavior in horses, inherited from their wild ancestors who escaped from the attacks of their predators by flight.

Effect of Animals on the Environment

The effect of environment on animals was presented in earlier sections. The effect of animals on the environment will be presented in the sections that follow.

A sustainable animal agriculture must (1) make proper use of the natural resources—air, water, land, and energy; (2) preserve environmental quality; (3) control animal diseases that are transmissible to humans; and (4) produce nutritious and wholesome foods—meat, milk, and eggs—in greater quantity and more efficiently.

POLLUTION OF THE ENVIRONMENT

Pollution is the issue of the decade. Everything that defiles, desecrates, or makes impure or unclean atmosphere, streams, or land areas must be controlled.

• **Manure**—In animal agriculture, we need to give particular attention to the pollution caused by manure. One cow produces as much waste as 16 humans. Hence, with 20,000 steers in a feedlot, the disposal problem is equal to a city of 320,000 people. In addition to being used as a fertilizer, manure is now being recycled as a livestock feed and serving as a source of energy (methane gas).

Of course, there is no one best manure management system for all situations. But, one way or another, science and technology must evolve with ways of disposing of 1.5 billion tons of manure annually; and this must be accomplished without polluting streams or the atmosphere or being offensive to the neighbors.

We must ever be mindful that life, beauty, wealth, and progress depend upon how wisely people use nature's gifts —the soil, the water, the air, the minerals, and the plant and animal life.

If not managed properly, animals may produce the following pollutants in troublesome quantities: manure, gasses/odors, dust, and flies/other insects. Also, they may pollute water supplies.

(Also see Chapter 12, section on "Animal Wastes [Manure and Litter].")

• **Gases/odors**—Gases in confinement buildings may arise from the animals, and/or from the bacterial decomposition of stored excreta.

Carbon dioxide (CO_2) and methane (CH_4) are the primary constituents of air expired by animals. Carbon dioxide is the gas produced in greatest quantity; it is one of the byproducts of animal energy metabolism and is expired via air exchange in the lungs. Ruminants also expire smaller quantities of methane (CH_4). A cow may produce up to 300 qt of carbon dioxide per day but only 50 qt of methane. The remaining contaminant gases found within livestock buildings are produced as a result of the decomposition of animal waste.

The manure (feces and urine) produced in confinement operations and stored inside buildings may produce gases and undesirable odors. The principal gases are: ammonia, carbon dioxide, hydrogen sulfide, and methane. Ammonia is odorous, may become explosive when mixed with sufficient air, and is dangerous to humans at a concentration above 5,000 ppm. Carbon dioxide is odorless, and can kill (asphyxiate) people at high levels because it displaces oxygen. Hydrogen sulfide is odorous (rotten-egg odor),may become explosive when mixed with air, is very irritating to the eyes and respiratory tract of humans at low concentrations, and will cause immediate unconsciousness and death due to respiratory paralysis at concentrations of 100 ppm. Methane is odorless, may become explosive when mixed with air, and can be asphyxiating at high concentrations.

• **Dust**—Dust may be present both inside and outside of buildings. Inside swine and poultry buildings,most of the dust comes from the feed, animals (skin debris, hair, feathers), and bedding; and it is not particularly harmful to humans, except for the microbes, gases, and vapors that it may carry, and for possible allergic reactions that it may cause. Dust problems in buildings are minimal when there is proper ventilation and sanitation. Dust derived from dirt feedlots and corrals may contain silica from the soil which can cause fibrous scar tissue in the lungs of humans. Dust in feedlots in dry climates may be effectively controlled by sprinkling with fence-line sprinklers or water trucks, by increased animal density (accompanied by increased in-pen urination [moisture]), and by manure removal. Generally, if the surface moisture of feedlots is maintained in excess of 20%, dust problems will be minimal.

• **Flies and other insects**—Most people recognize flies as a public health nuisance. Moreover, the public associates flies and other insects around food as evidence that things may not be as clean or sanitary as they should be, which may or may not be true.

Also flies and other insects originating at animal operations and annoying nearby homeowners and others are increasingly causing lawsuits to be filed against the owners of the animal operations based on the "nuisance law." Legal decisions frequently instruct livestock producers to control such insect populations or risk being driven out of business.

Pesticides provide livestock producers with a means of combating pests that attack animals. However, improper use of agricultural chemicals to control flies and other insects harbored by animals can be toxic to people. Moreover, certain chemicals can accumulate in the body fat of animals and be found in the meat or milk. Because of possible injurious effects of certain chemicals on humans, new methods of insect control are evolving, with biological control and integrated pest management leading the way.

• **Water pollution**—In recent years, there has been worldwide awakening to the pollution of water and its effect on human health and other forms of life. Much of this concern stemmed from the amount of manure produced by the sudden increase of animals in confinement, without adquate attention to disposal. The major concerns in the 1960s, now largely rectified, were: nitrates and "blue babies," and suffocating fish.

1. **Nitrates and "blue babies."** Nitrates are a compound form of nitrogen, usually found in the soil, which may be derived from a breakdown of fertilizers, animal wastes, human wastes, crop residues, and sewage sludges. Plants utilize nitrates in the growth process to obtain nitrogen. However, when the amount of nitrogen contained within nitrates exceeds 10 parts per million (ppm) in underground water, there is a chance that the excess nitrates may cause methemoglobinemia, a disorder in human babies, commonly known as *blue babies*. The major concern, therefore, is that the subsurface water from a cattle feedlot, or other similar animal facility, might lead to a nitrate buildup and cause blue babies. However, research evidence indicates that there is little nitrate buildup under active feedlots or runoff holding ponds.

2. **Suffocating fish.** In the 1960s, before feedlot control runoff measures were instituted, the runoff from many of the feedlots in the U.S. Great Plains areas carried organic material that has a high oxygen demand. This runoff sludge was carried down the streams, using up available oxygen, and causing fish to suffocate. The average pollutional loading on a stream was relatively minor, but the effect of a few hours without oxygen was spectacular. Cattle raisers reacted swiftly by constructing runoff control facilities in large commercial feedlots.

But fish in different parts of the world are still being suffocated by excess manure! In 1988, the Irish government announced a $4 million project designed to clean up lake Lough Sheelin, a former trout angler's paradise in central Ireland in which the fish were destroyed by excess manure from pig farms in the surrounding area. The pollution cleanup program involves the construction of five anaerobic digesters around the lake and the production of methane gas (for energy) from the manure.

Pollution of Pastures and Ranges

Little pollution potential exists from pasture or range systems with proper animal densities or numbers, or where grazing areas are rotated.

Grazing influences the environment on federal lands. Under poor range management, the environment is affected adversely; under good range management, such as exists on most ranges today, grazing actually improves the environment.

The environmental effects of grazing depend upon the kind of range, the intensity of grazing, and the kind of management employed to control livestock on the range. It is generally recognized that unregulated heavy grazing results in loss of desirable forage plants, increased runoff and erosion, and other indications of range deterioration. On the other hand, planned seasonal grazing and controlled animal distribution foster rapid vegetational growth. Most grazing experiments have shown that ranges may be improved more rapidly under proper grazing management than with no grazing at all.

There is no evidence that well-managed grazing of domestic livestock is incompatible with a high-quality environment. But there is ample evidence that managed grazing by livestock enhances certain uses and that poor management detracts from them. Properly managed grazing is a reasonable and beneficial use of the range.

Ecologists report that good range management will support more wildlife than the wilderness. This explains why big game numbers on federal lands have increased during recent years, and why wildlife production is an increasingly important use of rangelands.

Fig. 17–49. A well-managed range supporting good cattle and an abundance of wildlife (note the deer and wild turkeys in the distance). This is Infante Ranch, Coahuila, Mexico, owned by Sr. Guillermo Osuna, as portrayed in a painting by artist Tom Phillips. Courtesy, Sr. Guillermo Osuna, Coahuila, Mexico)

Ranges actually improve while being properly utilized by domestic livestock. The benefits which accrue to the range include increased vegetation cover, improved plant species composition, improved soil fertility and soil structure, and greater retention of high-quality water. When cattle and sheep go, rank underbrush takes over, and fire becomes a real hazard.

Both upland game birds and big game animals are benefited by grazing that promotes good cover for mating sites and enhances food supply and other habitat requirements.

On ranges with mixed types of vegetation, herbaceous species increase and browse species decline when grazed only by game. The converse is true when cattle graze the land. The combined grazing by two groups of animals maintains a better balance of browse species—preferred by game animals, and of herbaceous species, preferred by cattle and sheep.

Heavy livestock grazing is beneficial to irrigated pastures used by geese and other migratory waterfowl. Unless the vegetation is closely cropped, these areas are unattractive to the birds.

Thus, livestock grazing of the public lands is contributing to improved wildlife habitat conditions and increased numbers of game animals. Range development programs, particularly livestock water developments, have made more public land usable by game animals and are partly responsible for the vast increase in game numbers over the years.

On many grass-shrub ranges, livestock grazing reduces the danger of fire by preventing a buildup of dry grass, which is highly flammable.

Ranges properly grazed by hoofed animals produce safe water. Counts of fecal coliform organisms, as indicators of water pollution by warm-blooded animals, relate more closely to the quantity of the fecal material than to the kind of animal. Investigations have shown that the count of harmful bacteria in streams is not greater in areas grazed by livestock than in areas grazed by wild animals alone, and that modern livestock grazing has little effect upon the chemical and physical quality of the water.

It is noteworthy, too, that few western ranges are ever in a stable, natural condition, whether or not they are grazed by domestic animals. Rather, most of them are in a stage of vegetational development following disturbances by such phenomena as drought, flood, avalanche, frost, or fire. Also, cyclic phenomena, such as large numbers of deer, rodent epidemics, or insect plagues, temporarily change the natural ecosystems. Thus, an absolutely stable rangeland is seldom attained or maintained.

Significantly, the greatest diversity of animal and plant species and the highest rates of reproduction occur when the landscape supports many stages of ecosystem development. Fire, grazing, and drought stimulate plants and animals to new growth. Each stage of vegetational development is more productive of certain animal species than of others.

Ecologists recognize (1) that forage is a renewable natural resource, which regrows each year and is wasted unless it is utilized annually; (2) that grazing on Federal rangelands helps to keep the natural environmental systems active and productive; (3) that we cannot allow overgrazing by domestic livestock, bison, deer, or wild horses; and (4) that grazing must be scientifically controlled and responsive to the needs of all users.

TRANSMITTING DISEASES AND PARASITES TO HUMANS

The progress of humans, from cave to condominium, has been greatly influenced by animals. From the remote day of their domestication forward, the most advanced civilizations of the day have been the keepers of herds and flocks. Although progress walks Indian file behind animals, certain diseases follow. Many of these diseases are transmitted through meat, milk or eggs; others are transmitted through close contact with animals, contact with their excreta, or contact with their products—such as hides, wool, or hair; still others are carried from animals to humans by insect vectors.

This group of shared animal/human diseases is known as *zoonoses.* It is important to realize that many zoonotic diseases cannot be prevented or controlled in people except through their control in animals. The common zoonotic diseases transmitted from animals to humans include the following:

1. **African sleeping sickness.** This disease, caused by microscopic parasites called trypanosomes, is confined to Africa, wherever *tsetse flies* (the vectors) occur. At least 50 million African people in 38 tsetse-infested countries live with the threat.
2. **Anthrax.** This acute, infectious disease affects all warm-blooded creatures, animals and humans, throughout the world.
3. **Brucellosis.** This serious disease, which causes abortion, affects cattle, sheep (but rarely), goats, swine, and people, worldwide.
4. **Leptospirosis.** This worldwide disease is transferable between cattle, sheep, goats, swine, horses, dogs, foxes, rats and other rodents, and humans. Leptospirosis is spread from animal to animal, and by infected urine, waterholes, and milk.
5. **Q fever.** This disease affects cattle, goats, sheep, and humans. The causative organism is *Coxiella burnetti,* of which ticks are the most important vector. Prevention consists in avoiding ticks, pasteurizing milk, and vaccination.
6. **Rabies.** This acute, infectious disease affects all warm-blooded animals and humans. It is caused by a filtrable virus that is usually carried into a bite wound by a rabid animal, especially dogs. All dogs should be immunized against it. Complete eradication will be difficult to achieve because of the reservoir of infection in wild animals.
7. **Trichinosis.** This is a parasitic disease of humans, which is contracted primarily by consuming infested pork, eaten raw or imperfectly cooked. It also infects swine, without causing specific symptoms.
8. **Tuberculosis.** This chronic infectious disease affects all animals and humans, worldwide. Periodic testing and removal of reactors is the only effective method of control.
9. **Tularemia.** This disease affects cattle, sheep, swine, horses, many wild animals, and humans. The primary modes of transmission to humans include ticks or other anthropods, handling of infected animals, and the inhaling of dust or vapor containing *F. tularensis.*

• **Diseases for which animals serve as passive carriers**—Animals may host the spores of several pathogenic organisms in their intestinal tracts, without exhibiting any symptoms of disease. Among such diseases are the following:

1. **Tetanus.** *Clostridium tetani,* the tetanus organism, is commonly found in the intestines of herbivorous animals (cattle and horses) and to a lesser extent in humans. Manure, and soil that has been fertilized with manure, are prime sources of these spores. It follows that people working around horses and stables are more likely to be carriers than those engaged in other occupations.

   In humans and farm animals (especially horses, sheep, and goats), tetanus often follows a deep puncture wound, such as may be inflicted by a nail, a firecracker, or a gunshot. The deep puncture provides both an entry (broken skin) and anaerobic conditions for growth of the organism.
2. **Gas gangrene** results from the introduction of spores through the broken skin. The clostridia organisms, especially *C. perfringens,* are frequently found in the intestinal tract of most farm animals and humans.
3. **Botulism.** Botulism in people and animals is caused by the ingestion of food in which the organism *C. botulinum* has produced toxins. Among farm animals, horses are the most susceptible to the disease. They develop botulism from eating moldy or spoiled forages or grains in which botulism toxin has been produced. Poultry are also quite susceptible to the disease.

• **Regulations relative to animal disease control**—Where human health is involved, animal diseases are much too important to be entrusted to individual action. In the United States, therefore, certain regulatory activities in animal disease control are under the supervision of the U.S. Department of Health and Human Services and the U.S. Department of Agriculture.

SAFETY OF PRODUCTS

America's food supply is the safest in the world! Nevertheless, there is need for constant vigilance and improvement, especially in animal products which are subject to all the hazards of other foods (spoilage, pesticides, toxicities), plus being capable of transmitting, or serving as passive carriers, of certain diseases to humans.

In colonial times, the livestock producer slaughtered animals and processed meats, milked the cows, and gathered the eggs; then, delivered the products door-to-door to urban customers. If the products were not acceptable (spoiled meat, sour milk, cracked eggs), the matter was resolved quickly and on the spot, or the producer lost a customer. Today, the public expects the livestock team—farmers, processors, and retailers—to provide wholesome and safe products free from disease agents, toxic subsances, and pesticide and drug residues.

Uptake of pesticides by animals, leading to residues in animal products, can result either from direct application of pesticides to animals or from animals ingesting feeds carrying pesticide residues. Drug residues are caused by (1) producers failing to withdraw drugs from livestock far enough in advance of marketing products; (2) contaminated feed storage, mixing, and handling equipment; and/or (3) the wastes (feces and urine) of treated animals coming in contact with untreated animals. *Reading and following the directions on the label is the key to safe pesticide and drug use.*

Because the welfare of the nation is dependent upon the health of its people, animal (and other) products are carefully monitored by various government agencies to assure consumers that they are wholesome and safe; and because of recognizing the importance of consumers in the safety of their food, the private sector may do additional testing. The agencies most responsible for this important work are:

1. The U.S. Department of Health and Human Services, including the following agencies: the Center for Disease Control, the Food and Drug Administration (FDA), and the National Institute of Health.

2. The U.S. Department of Agriculture, including the following agencies: the Agricultural Research Service, the Animal and Plant Health Inspection Service, the Cooperative State Research Service, the Federal Extension Service, the Labeling and Registration Section, and the Veterinary Service Division.

3. State and Local Government Agencies.

4. International organizations engaged in health and/or nutrition activities, including the World Health Organization (WHO), and Food and Agriculture Organization (FAO).

5. Private industry groups such as the National Livestock and Meat Board and the National Dairy Council.

6. Professional organizations, including dentists, dietitians, doctors, health educators, nurses, and public health workers.

7. Food processors and retailers. For example, in 1988 Lucky Stores, Inc., the largest food handler in California, initiated a testing program in cooperation with the California Department of Food and Agriculture (CDFA) to have the CDFA check their produce for pesticide residues on a regular basis, as a way of assuring their customers that the produce that they buy in Lucky Stores is completely safe. Some food handlers are using private laboratories to conduct similar tests.

(Also see Chapter 13, sections on "Food and Drug Administration [FDA]" and "Use Feed Supplements, Additives, and Implants Safely.")

Pollution Laws and Regulations[16]

Invoking an old law (the Refuse Act of 1899, which gave the Corps of Engineers control over runoff or seepage into any stream which flows into navigable waters), the U.S. Environmental Protection Agency (EPA) launched a program to control water pollution by requiring that all cattle feedlots which had 1,000 head or more the previous year must apply for a permit by July 1, 1971. The states followed suit; although differing in their regulations, all of them increased legal pressures for clean water and air. Then followed the Federal Water Pollution Control Act Amendments, enacted by Congress in 1971, charging the EPA with developing a broad national program to eliminate water pollution.

The current federal requirements, which are rather broad, follow:

1. **Who must apply.** Owners/operators of animal feeding facilities with more than 1,000 animal units. Animal units are computed as follows: multiply number of slaughter and feeder cattle by 1.0; multiply number of mature dairy cattle by 1.4; multiply number of swine weighing over 55 lb by 0.4; multiply the number of sheep by 0.1; and multiply the number of horses by 2.0. (See Table 17-6, footnote 1, for what constitutes 1,000 animal units.)

2. **Definition of *animal feeding operation*.** *An animal feeding operation is defined as a lot or facility where animals have been, or will be, stabled or confined and fed for a total of 45 days or more in any 12-month period and where crops, vegetation, forage growth, or post-harvest residue are not sustained over any portion of the lot or facility during the normal growing season.*

3. **Who does not need a permit.** Permits are required only if there is a discharge to a waterway. Totally enclosed units without pollutant discharge do not require a permit regardless of size. No permit is required if all pollutants are recycled or spread on land. Runoff control ponds or filter strips can be employed to prevent discharges.

4. **How to apply.** Fill out an Application Form For Permit To Discharge Wastewater, and submit as directed. Forms may be obtained from the offices of EPA and state environmental agencies, the county agent, or the Soil Conservation Service district offices. Then, either the federal EPA or the state agency will make an on-the-site inspection. They will draft a proposed permit, put it on public notice, and give the applicant and the public 30 days to comment on it. Then, if there are no protests, the federal discharge permit will be issued.

[16]This section was authoritatively reviewed and updated by Lawrence J. Jensen, U.S. Environmental Protection Agency, Washington, D.C.

TABLE 17-6
SUMMARY OF REGULATIONS

Feedlots with 1,000 or More Animal Units[1]	Feedlots with Less than 1,000 but with 300 or More Animal Units[2]	Feedlots with Less than 300 Animal Units
Permit required for all feedlots with discharges[3] of pollutants.	Permit required if feedlot— 1. Discharges[3] pollutants through an unnatural conveyance, or 2. Discharges[3] pollutants into waters passing through or coming into direct contact with animals in the confined area. Feedlots subject to case-by-case designation requiring an individual permit only after on-site inspection and notice to the owner or operator.	No permit required unless— 1. Feedlot discharges pollutants through an unnatural conveyance, or 2. Feedlot discharges pollutants into waters passing through or coming into direct contact with the animals in the confined area, and 3. After on-site inspection, written notice is transmitted to the owner or operator.

[1]More than 1,000 feeder or slaughter cattle, 700 mature dairy cows (milked or dry), 2,500 swine weighing over 55 lb *(24.9 kg)*, 500 horses, 10,000 sheep or lambs, 55,000 turkeys, 100,000 laying hens or broilers with continuous overflow watering, 30,000 laying hens or broilers with liquid manure handling, 5,000 ducks; or any combination of these animals adding up to 1,000 animal units.

[2]More than 300 slaughter or feeder cattle, 200 mature dairy cows (milked or dry), 750 swine weighing over 55 lb *(24.9 kg)*, 150 horses, 3,000 sheep, 16,500 turkeys, 30,000 laying hens or broilers with continuous overflow watering, 9,000 laying hens or broilers with liquid manure handling, 1,500 ducks; or any combination of these animals adding up to 300 animal units.

[3]Feedlot not subject to requirement to obtain permit if discharge occurs only in the event of a 25-year, 24-hour storm event.

Before constructing an *animal feeding operation*, the owner should become familiar with both state and federal regulations. The state regulations can be obtained from the state water board. They differ from state to state, but most states require a catch basin (detention pond) sufficient to contain the runoff from a storm of the magnitude of the largest rainfall during a 48-hour period of the most recent 10 years. A feedlot may minimize runoff by locating near the top of the slope and, if necessary, by using diversion embankments to divert runoff from other areas.

Cattle feedlots located near centers of population are having an increasing number of complaints lodged against them because of manure, dust, and odor. Lawsuits, based on the nuisance law, have been filed against them.

Sustainable Agriculture

Endangered species—and more! Today, it is endangered planet, endangered people and animals, and endangered agriculture. Among the deluge of warnings of environmental catastrophes are:

• Pollution-caused warming of the atmosphere, known as the *greenhouse effect*, threatening weather changes that could render large areas of the planet unproductive and uninhabitable.

• Toxic and radioactive wastes and dumped garbage that could poison drinking water and despoil the land.

• Chemical pollution that is depleting the atmosphere's protective ozone layer.

• Slashing and burning of tropical rain forests, driving thousands of species to extinction, increasing the amount of carbon dioxide in the atmosphere, and contributing to the greenhouse effect that warms the earth.

Is ¼ lb hamburger worth ½ ton of Brazil's rain forest? Is 67 sq ft of rain forest (an area about the size of one small kitchen) too much to pay for 1 hamburger? Should we form cattle pastures to produce hamburgers in the Amazon, or should we retain the rain forest and the natural environment? These and other similar questions are being asked too little and too late to preserve much of the great tropical rain forest of the Amazon and its environment. It took nature thousands of years to form the rain forest, but it took a mere 25 years for people to destroy much of it. And when a rain forest is gone, it is gone forever![17]

Although less dramatic, the Amazon rain forest story has been, or is being, repeated all over the world in the form of the greenhouse effect, toxicities, polluted streams, and/or other harbingers of threats to our environment. Too long we have managed our nonrenewable resources like there is no tomorrow! Now, the situation is being righted. Worldwide, environmental quality and economic efficiency are in vogue. In the United States, this movement is called *Sustainable Agriculture.*

Sustainable agriculture is often described as farming that is ecologically sound and economically viable. It may be high or low input, large scale or small scale, a single crop or diversified farm, and use either organic or conventional inputs and practices. Obviously, the actual practices will differ from farm to farm. A definition follows.

A sustainable agriculture is farming with reduced off-farm purchased inputs of pesticides, herbicides, and fertilizers, along with reduced negative impact on natural resources and improved environmental quality and economic efficiency, while producing and distributing abundant, nutritious, affordable, high-quality foods and fibers for American and world markets.

The development of improved crops, cropping systems, irrigation, farm management, and marketing will be needed to make farms more profitable and sustainable. Typically, such farms will rely more on biological resources and management than on nonrenewable inputs of energy and chemicals. The foundation of a sustainable farm system is a comprehensive understanding of the land, the farm resources and operations, and potential short- and long-term markets.

[17]Uhl, C. and G. Parker, "Is a One-Quarter Pound Hamburger Worth A Half-Ton of Rain Forest?," *Interciencir*, 1986, Sept.-Oct., Vol. II, No. 5, p. 213.

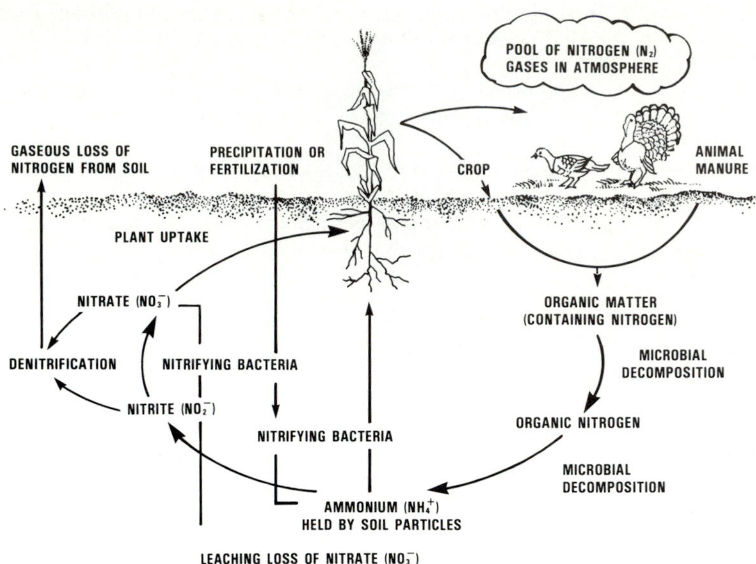

Fig. 17-50. At its best, a sustainable agriculture enhances the *cycle of nature:* Manure is applied to soils and decomposed by microbes; complex protein in manure is broken down to release nitrogen as ammonia (NH_4^+); aerobic microbes convert ammonium nitrogen to nitrite (NO_2^-), thence to nitrate (NO_3^-) nitrogen; nitrate is either (1) taken up by plants and built back into protein compounds; (2) leached downward when the soil is saturated—contaminating surface and groundwater if excessive nitrogen has been applied to the soil; and/or (3) released into the atmosphere when soils are wet for extended periods of time and the absence of air causes anaerobic microbes to convert the nitrate nitrogen to gaseous form. **NOTE:** Nitrates may be derived from a breakdown of fertilizers, animal wastes, human wastes, crop residues, and sewage sludges.

In the Soviet Union, the counterpart to United States' *Sustainable Agriculture* is called *Perestroika;* and the movement is being spearheaded by a newly established "State Committee for Nature Conservation," charged with the responsibility of protecting their natural resources—land, waters, forests, and other national assets. The Committee Chairman, Mr. Fyodor Morgun, is quoted in *Soviet Life* as saying:

> Consideration is being given to (imposing) fines for the discharge of pollutants. ..... There can be no profit in the destruction of nature. ..... We intend to introduce the experience of all humankind in the development of ecologically clean, low-waste or waste-free technologies, nonhazardous types of energy, wholesome foodstuffs, and so on. ..... We have been entrusted with full responsibility for the protection of nature and for the organization of rational use and reproduction of natural resources in the U.S.S.R. ..... Perestroika in nature conservation is a must.[18]

Many of the practices advocated under a sustainable agriculture are not new; they involve such timeless agricultural practices as soil erosion control, the protection of groundwater, the use of legumes as a source of nitrogen, biological insect and weed control, and the use of pastures as a primary feed source.

[18] "Perestroika In Nature Conservation Is Essential," an interview with Fydor Morgun, reported in *Soviet Life,* Nov., 1988, No. 11, p. 16.

ANIMAL WELFARE/ANIMAL RIGHTS

In recent years, the behavior and environment of animals in confinement have come under increased scrutiny of animal welfare/animal rights groups all over the world. For example, in 1987 Sweden passed legislation designed (1) to phase out of layer cages as soon as a viable alternative can be found; (2) to discontinue the use of sow stalls and farrowing crates; (3) to provide more space and straw bedding for slaughter hogs; and (4) to forbid the use of genetic engineering, growth hormones, and other drugs on farm animals except for veterinary therapy. Also, the law provides for fining and imprisoning violators.

Animal welfarists see many modern practices as unnatural, and not conducive to the welfare of animals. In general, they construe animal welfare as the well-being, health, and happiness of animals; and they believe that certain intensive production systems are cruel and should be outlawed. The animal rightists go further; they maintain that humans are animals, too, and that all animals should be accorded the same moral protection. They contend that animals have essential physical and behavioral requirements, which, if denied, lead to privation, stress, and suffering; and they conclude that all animals have the right to live.

Livestock producers know that abuse of animals in intensive/confinement systems leads to lowered production and income—a case in which decency and profits are on the same side of the ledger. They recognize that husbandry that reduces labor and housing costs often results in physical and social conditions that increase animal problems. Nevertheless, means of reducing behavioral and environmental stress are needed so that decreased labor and housing costs are not offset by losses in productivity. The welfarists/rightists counter with the claim that the evaluation of animal welfare must be based on more than productivity; they believe that there should be behavioral, physiological, and environmental evidence of well being, too. And so the arguments go!

But wild animals were often more severely stressed than domesticated animals. They didn't have caretakers to store feed for winter or to irrigate during droughts; to provide protection against storms, extreme temperatures, and predators; and to control diseases and parasites. Often survival was grim business. In America, the entire horse population died out during the Pleistocene Epoch. Fossil remains prove that members of the horse family roamed the plains of America (especially the area that is now known as the Great Plains of the United States) during most of tertiary period, beginning about 58 million years ago. Yet no horses were present on this continent when Columbus discovered America in 1492. Why they perished, only a few thousand years before, is still one of the unexplained mysteries. As the disappearance was so complete and so sudden, many scientists believe that it must have been caused by some contagious disease or some fatal parasite. Others feel that perhaps it was due to multiple causes, including (1) climatic changes, (2) competition, and/or failure to adapt. Regardless of why horses disappeared, it is known that conditions were favorable to them at the time of their reestablishment by the Spanish conquistadores about 500 years ago.

To all animal caretakers, the principles and application of animal behavior and environment depend on understanding; and on recognizing that they should provide as comfortable an environment as feasible for their animals, for both humanitarian and economic reasons. This requires that attention be paid to environmental factors that influence the behavioral welfare of their animals as well as their physical comfort, with emphasis on the two most important influences of all in animal behavior and environment—feed and confinement.

Animal welfare issues tend to increase with urbanization. Moreover, fewer and fewer urbanites have farm backgrounds. As a result, the animal welfare gap between town and country widens. Also, both the news media and the legislators are increasingly informed from urban centers. It follows that the urban views that are propounded will have greater and greater impact in the years ahead.

QUESTIONS FOR STUDY AND DISCUSSION

1. Why did caretakers' fluency with the language of animals decline, and why did environments become increasingly artificial, as the art of husbandry evolved into science?

2. Why was it necessary for Neolithic (New Stone Age) man to have knowledge of animal behavior in order to capture and domesticate wild animals?

3. How did concentration of animals in smaller spaces and concentration of people in urban areas prompt interest in the environment?

4. Cite evidence that animal behavior and environment are closely related.

5. Define the term *animal behavior.*

6. How may information gained from wild and feral relatives provide clues about the behavior of domestic animals?

7. Define the term *applied ethology.*

8. List nine general functions of behavioral systems that most animals exhibit.

9. Discuss the following behavioral systems as they pertain to each major farm animal species:
   a. Ingestive behavior
   b. Eliminative behavior
   c. Allelomimetic behavior
   d. Gregarious behavior

10. Define the following terms: *social behavior* and *social organization.*

11. Describe the social organization in your choice of any two of the following species: cattle (beef and dairy), sheep, goats, swine, horses, poultry.

12. How is the "peck order" of chickens established?

13. Explain the difference between dominance and leader-follower.

14. Discuss the importance of people-animal relationships.

15. Describe each of the following feed-related normal behavioral characteristics of animals:
   a. Grazing, ruminating, and loafing of beef cattle
   b. Sucking behavior of dairy calves
   c. Fostering of sheep
   d. Feeding behavior of goats
   e. Handling of swine, including pigs afflicted with porcine stress syndrome
   f. Transporting horses
   g. Control of feed intake of poultry

16. Describe each of the following anamalous (abnormal) animal behaviors, and tell what you would do to rectify each condition:
   a. Kicker in milk cow
   b. Pica in beef cattle
   c. Wool-pulling in sheep
   d. Fighting (agonistic behavior) in pigs
   e. Tail-biting in pigs
   f. Bolting feed in horses
   g. Wood chewing in horses
   h. Cannibalism in chickens
   i. Feather picking in chickens and turkeys

17. List and discuss the significance of one example of the practical application of animal behavior in each of the following categories:
   a. Breeding for adaptation
   b. Quick adaptation-early training
   c. Manure elimination
   d. Companionship
   e. Controlling animal behavior

18. Define the following terms: *environment* and *ecology.*

19. An animal is the result of two forces—heredity and environment. Which is the more important?

20. Describe the classic New Zealand experiment showing the effect of environment on dairy cattle. How did it underscore the importance of environment?

21. Discuss how each of the following feed/environmental interactions affects animals:

a. Size and milk production of beef cattle and the quality of the range
b. Compensatory growth of beef cattle
c. Flushing
d. Milk composition
e. Palatability
f. Preconditioning of beef cattle
g. Stress

22. Discuss how each of the following water/environmental interactions affects animals:

a. Air temperature
b. Physiological state
c. Quality of water

23. Define each of the following terms:

a. Thermoneutral zone (comfort zone)
b. Optimum temperature
c. Lower critical temperature
d. Upper critical temperature

24. Discuss how each of the following weather/environmental interactions affects animals:

a. Cold weather animal heating mechanisms
b. Hot weather animal cooling mechanisms
c. Adaptation, acclimation, acclimatization, and habituation of species/breeds to the environment, including pertinent facts about the buffalo X domestic cattle cross made in Canada and the classic beef cattle study of breed differences (Shorthorn, Santa Gertrudis, Brahman) conducted by Dr. Samuel Brody at the University of Missouri
d. Finishing requirement of beef cattle
e. Hot weather tolerance of dairy cattle
f. Voluntary feed intake of sheep in hot weather
g. Feeding alfalfa hay to sows in cold weather
h. Salt requirement of horses in hot weather
i. Hot weather tolerance of poultry, including reducing heat stress by dissipating body heat, adjusting ration, adding vitamin C, fasting broilers, and acclimating birds

25. Discuss how each of the following facilities/environmental interactions affects animals:

a. Shelter and shades for beef cattle
b. Housing and feeding of dairy and veal calves
c. Space of feeder lambs
d. Grouping of hogs of different sexes, ages, and sizes
e. Ventilation of horse barns
f. Cages for layers
g. Space allowance of broilers

26. Discuss the impact on animal comfort and efficiency of production of the modification of each of the following types of facilities:

a. Fans for each species
b. Slotted floors for swine and poultry
c. Lights for controlling reproduction in sheep, horses, and poultry
d. Shades for each species
e. Evaporative coolers for each species
f. Ventilation for each species
g. Windbreaks for beef cattle

27. Define each of the following terms: *health, disease, parasite.*

28. Discuss how each of the following health/environmental interactions affects animals:

a. Disease defense
b. Colostral defense
c. Immune suppression (altered immune response)
d. Mastitis
e. Pests and pesticides
f. Pollution, including dust, gasses/odors, muddy lots, and stray voltage
g. Porcine stress syndrome (PSS)
h. Ulcers

29. Define the term *stress;* and give examples of six common kinds of animal stresses.

30. What criteria are commonly used to evaluate, or measure, the well being or stress (a) of people and (b) of animals?

31. Define the term *bleeders* as applied to horses; and discuss the magnitude and concerns of this malady.

32. Discuss in detail the effect of animals on the environment from the standpoints of—

a. Pollution of the environment
b. Transmitting diseases and parasites to humans
c. Safety of animal products

33. Summarize the current requirements of the pollution laws and regulations.

34. Define the term *sustainable agriculture.* Are the following harbingers really dangerous threats to our environment; if so, what can and should be done about each of them?

a. The conversion of the Amazon rain forest to cattle pastures
b. Toxicities
c. Polluted streams

35. Define the terms *animal welfare* and *animal rights.*

36. How do the stresses of wild and feral animals compare with those of domestic animals?

37. What more can livestock producers do to reduce behavioral and environmental stress of animals without increasing costs and/or lowering productivity to the point at which consumers will no longer benefit from an abundance of animal products?

SELECTED REFERENCES

Title of Publication	Author(s)	Publisher
Agricultural Waste Utilization and Management, Proceedings	F. J. Humenik, Chairman Executive Committee	American Society of Agricultural Engineers, St. Joseph, Mich., 1985
Agriculture and Groundwater Quality	P. F. Pratt, Chairman	Council for Agricultural Sciences and Technology, Ames, Iowa, 1985
Animal Behavior, 2nd Edition	J. P. Scott	The University of Chicago Press, Chicago, Ill., 1972
Animal Behavior, 3rd Edition	V. G. Dethier E. Stellar	Prentice-Hall, Inc., Englewood Cliffs, N.J., 1970
Animal Behavior	N. Tinbergen	Time Incorporated, New York, N.Y., 1965
Animal Behavior and Its Application	D. V. Ellis	Lewis Publishers, Inc., Chelsea, Mich., 1986
Animal Behavior in Laboratory and Field	A. W. Stokes	W. H. Freeman and Company, San Francisco, Calif., 1968
Animal Behavior, The Marvels of	T. B. Allen, Editor	National Geographic Society, Washington, D.C., 1972
Animal Behaviour: A Syntheses of Ethology and Comparative Psychology	R. A. Hinde	McGraw-Hill Publishing Co., New York, N.Y., 1970
Applied Animal Ethology	W. R. Stricklin, Guest Editor	Elsevier Science Publishers B. V., Amsterdam, The Netherlands, Vol. II, No. 4, Feb., 1984
Behavior of Domestic Animals, The, 3rd Edition	E. S. E. Hafez, Editor	The Williams and Wilkens Company, Baltimore, Md., 1975
Bibliography of Livestock Waste Management	J. R. Miner D. Bundy G. Christenbury	Office of Research and Monitoring, U.S. Environmental Protection Agency, Washington, D.C., 1972
Biology of Stress In Farm Animals: an integrative approach	P. R. Wiepkema P. W. M. van Adrichem	Kluwer Academic Publishers, Hingham, Mass., 1987
Brazil's Imperiled Rain Forest, Rondonia's Settlers Invade	W. S. Ellis	*National Geographic,* Vol. 174, No. 6, pp. 772–799, Dec., 1988
Development and Evolution of Behavior	Ed. by L. R. Aronson, *et al.*	W. H. Freeman and Company, San Francisco, Calif., 1970
Domestic Animal Behavior	J. V. Craig	Prentice-Hall, Inc., Englewood Cliffs, N.J., 1981
Effect of Environment on Nutrient Requirements of Animals	D. R. Ames, Chairman	NRC, National Academy Press, Washington, D.C., 1981
Effects of Air Temperature, Air Humidity, and Air Movement On Heat Loss From the Pig	T. Kamada I. Notsuki	Proc. 3rd AAAP Anim. Sci. Cong. 2:1174, Seoul, S. Korea, 1985
Environmental and Functional Engineering of Agricultural Buildings	H. J. Barre L. L. Sammet G. L. Nelson	Van Nostrand Reinhold Co., New York, N.Y., 1988
Environmental Biology	P. L. Altman D. S. Dittmer	Federation of American Societies for Experimental Biology, Bethesda, Md., 1966
Environmental Control for Agricultural Buildings	M. L. Esmay J. E. Dixon	The AVI Publishing Company, Inc., Westport, Conn., 1986
Environmental Management in Animal Agriculture	S. E. Curtis	Animal Environment Services, Mahomet, Ill., 1981
Ethology of Free-Ranging Domestic Animals	G. W. Arnold M. L. Dudzinski	Elsevier Scientific Publishing Company, Amsterdam, The Netherlands, 1978
Ethology, The Biology of Behavior, 2nd Edition	I. Eibl-Eibesfeldt	Holt, Rinehart and Winston, New York, N.Y., 1975
Farm Animal Manures: an overview of their role in the agricultural environment	J. Azevedo P. R. Stout	Agricultural Publications, University of Califorina, Berkeley, Calif., 1974
Guide to Environmental Research on Animals, A	R. G. Yeck, Chairman	NRC, National Academy of Science, Washington, D.C., 1971
Health Issues Related to Chemicals in the Environment: A Scientific Perspective	A. L. Craigmill, Chairman	Council for Agricultural Sciences and Technology, Ames, Iowa, 1987
Impact of Stress, The, Proceedings	Ed. by R. E. Moreng, J. R. Herbertson	Colorado State University, Ft. Collins, Colo., 1986

(Continued)

(Continued)

Title of Publication	Author(s)	Publisher
Introduction to Animal Behavior, An,: ethology's first century, 2nd Edition	P. H. Klopfer	Prentice-Hall, Inc., Englewood Cliffs, N.J., 1974
Kinships of Animals and Man	A. H. Morgan	McGraw-Hill Book Company, Inc., New York, N.Y., 1955
Livestock Behaviour, a practical guide	R. Kilgour C. Dalton	Westview Press, Boulder, Colo., 1984
Livestock Environment, Proceedings, Second International Livestock Environment Symposium	D. S. Bundy, Planning Chairman	American Society of Agricultural Engineers, St. Joseph, Mich., 1982
Mechanisms of Animal Behavior	P. Marler W. J. Hamilton, III	John Wiley & Sons, New York, N.Y., 1966
Organic Farming: current technology and its role in a sustainable agriculture	D. M. Kral, Editor	American Society of Agronomy, Madison, Wisc., *1984*
Our Friendly Animals and Whence They Came	K. P. Schmidt	M. A. Donohue & Co., Chicago, Ill., 1938
Portraits in the Wild	C. Moss	Houghton Mifflin Company, Boston, Mass., 1975
Poultry Welfare, Proceedings	R. M. Wegner, Editor	German Branch, The World Poultry Science Association, Federated Agricultural Research Centre, Braunschweig-Volkenrode, Germany, 1985
Principles of Animal Behavior	W. N. Tavolga	Harper & Row, New York, N.Y., 1969
Principles of Animal Environment	M. L. Esmay	The AVI Publishing Company, Inc., Westport, Conn., 1978
Readings in Animal Behavior	T. E. McGill, Editor	Holt, Rinehart and Winston, New York, N.Y., 1973
Safe and Effective Use of Pesticides, The	P. J. Marer	University of California, Publications, Oakland, Calif., 1988
Scientific Aspects of the Welfare of Food Animals	F. H. Baker, Chairman	Council for Agricultural Science and Technology, Ames, Iowa, 1981
Social Hierarchy and Dominance	Edited by M. W. Schein	Dowden, Hutchinson & Ross, Inc., Stroudsburg, Pa., 1975
Social Space for Domestic Animals	R. Zayan, Editor	Kluwer Academic Publishers, Hingham, Mass., 1985
Social Structure in Farm Animals	G. J. Syme L. A. Syme	Elsevier Scientific Publishing Co., Amsterdam, The Netherlands, 1979
Stray Voltage: Proceedings of the National Stray Voltage Symposium	O. M. Majerus R. O Martin R. A. Peterson	American Society of Agricultural Engineers, St. Joseph, Mich., 1984
Stray Voltages in Agriculture, Proceedings	H. J. Hansen L. J. Endahl, Co-Chairman	American Dairy Science Association, Champaign, Ill., 1983
Stress Physiology in Livestock Vol. 1, *Basic Principles* Vol. 2, *Ungulates* Vol. 3, *Poultry*	M. K. Yousef	CRC Press, Inc., Boca Raton, Fla., 1985
Structures and Environment Handbook		Midwest Plan Service, Iowa State University, Ames, Iowa, 1972
Utilization, Treatment, and Disposal of Waste on Land, Proceedings	E. C. A. Runge, President of Society	Soil Science Society of America, Inc., Madison, Wisc., 1986
Wild Animals in Captivity	H. Hediger	Dover Publications, Inc., New York, N.Y., 1964

Chapter 18

Original painting by Tom Phillips

FEEDING STANDARDS/ RATION FORMULATION[1]

Contents Page

[1]The authors gratefully acknowledge the authoritative review accorded this chapter by L. M. Larsen, Ph.D., Consultant, Nutri-Systems, 426 E. Shields, Fresno, CA 93704.

Fig. 18-1. Weather affects the feed requirements! More energy is required when it is cold. ("The first step," original painting by artist Tom Phillips, 3333 17th Street, San Francisco, CA 94110)

During the last half of the 20th century, livestock production has been transformed from a secondary source of farm income to a highly sophisticated technological industry. Pigs are no longer scavengers, and chickens are no longer tenderly cared for by the farmer's wife and fed on table scraps and unaccounted for grain from the crib. With the expanding knowledge of the nutrient requirements of animals, new and refined feeding standards have been developed, thereby allowing producers to tailor their feeding program to their particular operations.

Hand in hand with the revision of feeding standards, came new analytical techniques whereby feedstuffs can be rapidy and accurately analyzed to determine their nutritive values. Thus, with an accurate estimate of what the nutritive requirements of their animals are and the nutrient value of their feeds, producers are now able to formulate balanced rations which result in efficient utilization of feed.

In the past, producers balanced rations by hand calculations, often using long and tedious trial-and-error methods. But in the past two decades, American industries have adapted computers to almost every conceivable task, and the livestock industry is no exception. Today, computers are used to formulate rations utilizing a wide variety of feeds; and the days of pencil-pushing formulations by big operations and commercial feed companies are, for the most part, only a memory. However, sophisticated livestock producers realize that this new computerized technology is a tool, which must be used wisely, and that they must clearly understand the principles on which computer calculations are based.

FEEDING STANDARDS

Feeding standards are tables listing the amounts of one or more nutrients required by different species of animals for specific productive functions, such as maintenance, growth, finishing, lactation, work, wool, or mohair. They are necessary guides in balancing rations. Most feeding standards are expressed in either (1) quantities of nutrients required per day, and/or (2) concentration in the ration; the first type is used where animals are provided a given amount of a feed during a 24–hour period, and the second is used where animals are provided a ration without limitation on the time in which it is consumed.

Fig. 18–2. Angora goats are fed for mohair production. (Courtesy, Dr. Gary C. Smith, Texas A&M University, College Station)

Today, the most widely used feeding standards in the United States are those published by the National Research Council (NRC) of the National Academy of Sciences. In England, similar standards are issued by the Agricultural Research Council (ARC). Other countries have similar bodies which make recommendations relative to the nutritive requirements of animals.

In the United States, the older TDN system is gradually giving way to newer energy evaluation systems, particularly net energy. England uses metabolizable energy (ME), adjusted according to the efficiency with which a feedstuff or ration is used for a particular purpose. Other European standards are based on starch equivalents, Scandinavian feed units, and other methods.

History of Feeding Standards

The first feeding standard was developed by Thaer, a German scientist, in 1810. He used meadow hay as his standard, then compared the extractable nutrients of other feeds to it and assigned them "hay values." Since the standard was somewhat subjective in listing these relative values, it proved to be rather controversial; nevertheless, it did succeed in laying the ground work for subsequent research.

In 1859, Grouven, a Dutch scientist, made use of analyses of protein, fat, and carbohydrate to formulate the first feeding standard for farm animals. This system was based on the total protein, carbohydrate, and ether extract of various feedstuffs; but it did not attempt to adjust for digestibility of these nutrients.

In 1864, the great German scientist Wolff, published the first standards based on digestible nutrients as determined from feeding trials. He attempted to set up standards to meet the needs of the animal with a minimum of waste of any nutrient. He continued to revise these feeding standards throughout the latter half of the 19th century, publishing revised standards annually in the Mentzel-Lengerke Agricultural Calendar from 1864 to 1896. By 1874, the United States had become interested in these feeding standards, largely through the teaching of the American educator, W. O. Atwater, a professor at Wesleyan University. Subsequently, in 1880, Armsby, of the Pennsylvania Station, published standards in his book, *Manual of Cattle Feeding.*

One of the main criticisms of the Wolff system was that it was based on the amount of organic material contained in the feed, rather than on the dry matter. Since feeds differ greatly in organic matter, this variability was reflected in Wolff's standards. In 1897, the Mentzel-Lengerke Agricultural Calendar was prepared by Lehmann, another German scientist. In this publication, he revised Wolff's standards so that they were presented on a dry matter basis, giving rise to what were known as the Wolff-Lehmann feeding standards. These standards were used widely in the first part of the 20th century.

In 1898, W. A. Henry, of the University of Wisconsin, published the first edition of *Feeds and Feeding,* a book which became the livestock producer's bible on feeding for the first three-quarters of the 20th century. Until 1910, Dr. Henry was the sole author. At that time, Frank B. Morrison became junior author; and it remained *Feeds and Feeding* by Henry and Morrison until 1929. From that date until 1956, with the completion of the twenty-second edition, Morrison was the sole author. In 1915, with the printing of the fifteenth edition of *Feeds and Feeding,* Henry and Morrison introduced their feeding standards based on dry matter, digestible protein, and total digestible nutrients. The twenty-second edition of this monumental work, published in 1956, is frequently referred to in this book.

Throughout the first 40 years of the 20th century, numerous other feeding standards evolved; among them, the starch values of Kellner, a German scientist, in 1907; the Scandinavian feed unit system (Woll, 1912); the dairy cow standards of Haecker, of Minnesota, in 1914; the net energy values of Armsby, of Pennsylvania, in 1915; the productive feed units developed by Mollgaard of Denmark, in 1939; and the productive energy values computed by Fraps of the Texas Station (1937, 1941).

In 1944, the National Research Council (NRC) of the National Academy of Sciences established a series of feeding standards for poultry, swine, dairy cattle, beef cattle, sheep, and horses. Initially, these standards were listed as feed allowances—that is, standards with a margin of safety. As more information became available relative to the specific nutritive needs of animals, the NRC developed more precise nutrient requirements. From time to time, the NRC revises these feeding standards in keeping with new information and changing feeding practices.

National Research Council Feeding Standards

In 1942, the Committee on Animal Nutrition of the National Research Council (NRC) of the National Academy of Sciences undertook the task of setting forth the quantitative needs of all the recognized nutrients for each species of farm animals. Following the earlier lead of the human dietary standards, prepared by the Food and Nutrition Board, initially the NRC feeding standards were *feed allowances,* so called because the recommended nutrients were set higher than the determined requirements in order to provide margins of safety. However, in 1953, the Committee on Animal Nutrition decided not to provide such margins of safety in future recommendations; rather, the policy was changed to provide intakes considered adequate for normal growth, health, and production, based on the average needs of groups of animals to achieve these results. With the transition, the NRC reports were designated *nutrient requirements.*

Today, the NRC recommended nutrient requirements for each species of farm animals are the most authoritative feeding standards in the United States. Periodically, a specific committee, composed of outstanding researchers who have worked extensively with the class of animal whose requirements are being reviewed, revises the nutrient requirements of each species for different functions. Thus, the nutritive needs of each type of livestock are dealt with separately and in depth.

The National Research Council (NRC) has established feeding standards for beef cattle, dairy cattle, sheep, goats, swine, poultry, horses, rabbits, mink and foxes, coldwater fishes, warmwater fishes and shellfishes, dogs, and cats.

The NRC standards pertaining to the classes of animals covered in this book are reproduced in the respective chapters pertaining to each species, in Chapters 19 to 28. In order to make for greater convenience and enhance the usefulness of the standards, the authors adapted and standardized the formats and included both U.S. customary and metric values.

Feeding Standards of Other Countries

The Weende, or proximate analysis, system of evaluating feedstuffs as the basis of feeding standards is widely used throughout the European countries. However, the several systems of expressing energy that evolved in Europe differ from the TDN and net energy systems common to the United States. Because of this, considerable confusion and disagreement results when attempting to compare the standards of different countries. Four of the most widely used energy systems of Europe are:

1. **Starch equivalent (SE).** This system is expressed as the amount of starch that produces as much fat as 100 kg (220 lb) of the feed being used.

2. **Russian oat unit (OU).** The oat unit is based on the amount of oats required to produce as much fat as 100 kg of the feed being used.

3. **Scandinavian feed unit (FE, from the term Foderenhet).** This system uses barley as the reference feed and refers to milk production instead of fattening.

4. **Modified feed unit or French feed unit (FE_c or UF).** This system should not be confused with the Scandinavian feed unit system as the conversion factors are different.

Table 18-1 illustrates how these energy standards relate to the United States TDN system. It should be noted that these systems use true protein values, whereas the TDN system uses crude protein.

TABLE 18-1
DETERMINATION OF ENERGY UNITS

Energy Units[1]	Unit of the Digestible Nutrient	Digestible Protein	Digestible Ether Extract	Digestible Crude Fiber	Digestible N-Free Extract	Further Procedure
		Multiply by				
TDN	%	1 (crude protein)	2.25	1	1	Sum up
SE	Weight, units, or %	0.94 (true protein)	2.41, 2.12, or 1.91[2]	1	1	Sum up and multiply by the "availability"
FE_c	g/kg	0.94 (true protein)	2.41, 2.12, or 1.91[2]	1	1	Sum up, multiply by the "availability," and divide by 700
OU	g/kg	0.94 (true protein)	2.41, 2.12, or 1.91[2]	1	1	Sum up, multiply by the "availability," and divide by 600
FE	g/kg	1.43 (true protein)	2.41, 2.12, or 1.91[2]	1	1	Sum up, multiply by the "availability," and divide by 750

[1]TDN = total digestible nutrients; SE = starch equivalent; FE_c = modified or French feed unit; OU = Russian oat unit; FE = Scandinavian feed unit.

[2]Digestible ether extract of oily seeds, cakes, and feeds of animal origin should be multiplied by 2.41; that of leguminous seeds, cereal grains, and their by-products by 2.12; and that of hays, straws, chaffs, green fodders, silages, roots, and tubers by 1.91.

Limitations of Feeding Standards

Although feeding standards are almost indispensable guides for meeting the nutritive needs of animals, numerous factors are not considered in setting the standards, especially their economy. For example, dairy producers are interested in obtaining that level of milk production which will make for the largest net returns in light of current feed costs and the market price of milk. Moreover, feeding standards tell nothing about the palatability, physical nature, or possible digestive disturbances associated with a ration. Neither do they give consideration to individual animal differences, management differences, and the effects of such stresses as weather, disease, parasitism, and surgery (e.g., dehorning and castrating). Thus, there are many variables that alter the nutrient needs and utilization of animals—variables that are difficult to include quantitatively in feeding standards, even when feed quality is well known. The experiences of the ration formulator and the feeder are invaluable in adjusting for such variables.

RATION FORMULATION

To supply all the needs for maintenance, growth, finishing, reproduction, lactation, work (or running), egg production and/or wool production, the different classes of animals must receive sufficient feed to furnish the necessary quantity of energy, proteins, minerals, vitamins, and water. Perhaps under certain conditions nonnutritive feed additives may be desirable, although they are not essential. A ration that meets all these needs is said to be balanced. More specifically, by definition, *a balanced ration is one which provides an animal the proper proportions and amounts of all the required nutrients for a period of 24 hours.*[2]

Consideration of Ration Ingredients

When rations are formulated, feeds are initially divided into three categories: (1) concentrates, (2) roughages, and (3) supplements. Additionally, when formulating rations for lactating cows, consideration should be given to the amount and kind of fiber.

Rations can be formulated whereby (1) all components are mixed together to form a complete feed, or (2) each component is considered as a separate entity. In the latter case, each feed is fed separately; nevertheless, when considered collectively, a balanced ration results. Most rations for nonruminants without functional cecums are formulated by the first method. On the other hand, many ruminants and nonruminant herbivores are fed concentrates and roughages separately, with vitamin and mineral supplements free-choice. In the latter type of feeding program, careful consideration must be given to how much of each class of feed is offered in order to prevent animals from eating too much concentrate and not enough roughage, or *vice versa.*

CONCENTRATES

Most of the protein and energy of the ration is supplied by the concentrate feeds. For ruminants in heavy production (*i.e.,* finishing, or lactation) and almost all nonruminants, the concentrate portion of the ration constitutes most of the dry matter intake.

The processing of grains and other concentrate feeds can dramatically increase or decrease their digestibility. For this reason, in the formulation of livestock rations, careful consideration must be given to the physical form of the various feedstuffs.

Additionally, the relative capacity of animals to digest and absorb such feeds as fats and molasses must be given careful attention.

Amino acid composition of feedstuffs and protein digestibility are especially critical for animals with nonfunctional cecums. Hence, many livestock producers routinely use more than one high-protein feed so that the amino acid deficiencies of one ingredient can be corrected by another. In nonruminant animal rations, levels of nonprotein nitrogen must be monitored to avoid ammonia toxicity.

ROUGHAGES

The amount of roughage incorporated in a ration depends largely upon intensity of production of the particular animal being fed. Nonruminants without functional cecums utilize very little roughage. On the other hand, ruminants and animals with functional cecums are generally given at least a small amount of roughage to maintain healthy, functional gastrointestinal activity.

Animals with nonfunctional cecums, such as swine, can utilize limited amounts of forage. When swine are fed at maintenance levels, forages can be included in the formulation to reduce costs. Some forage, such as high-quality alfalfa, is frequently added to swine and poultry rations for its high vitamin content.

Ruminants and animals with functional cecums can be maintained relatively easily on high-roughage rations. In these cases, a micronutrient supplement containing minerals and vitamins may be all that is additionally required. As production pressures increase, roughage is replaced by concentrate feed. The concentrate can be fed separate and apart from the roughage or mixed with the roughage, along with any supplements, to make a complete feed. Beef cattle have been successfully fed finishing rations with no roughage. However, when on high-concentrate rations, they still need a little roughage to supply the *scratch factor* to promote good ruminal activity. This roughage may be supplied by either (1) high-quality forages which contribute substantial amounts of nutrients to the ration, thereby reducing the need for expensive concentrates, or (2) low-quality forages, such as corncobs or cottonseed hulls, which add a scratch factor to the ration along with limited nutrients.

(Also see Chapter 19, Feeding Beef Cattle, Part IV, section headed "All-Concentrate and High-Concentrate Rations.")

SUPPLEMENTS

When rations are formulated, supplements are generally considered after the macronutrients (for example, protein and energy) have been balanced. The ration is then checked for any deficiencies or imbalances of micronutrients; and supplements are added to correct the deficiencies. The supplements can be in the form of either a premix (combination of many micronutrients) or individual micronutrients (for example, lysine). The producer may also want to incorporate some feed additives in the ration.

NOTE: Instead of using the term *supplement* in a broad sense, commercial feed manufacturers commonly use three terms: supplement, base mix, and premix. These terms are defined in Chapter 16, in the section headed "Importance and Nature of Commercial Feeds."

AMOUNT AND KIND OF FORAGE TO FEED LACTATING COWS (ADF AND NDF)

The amount and kind of fiber should be considered when formulating rations for lactating cows. The forage should constitute a minimum of 40% of the total dry matter of the

[2]Although Webster defines the noun *ration* as *the amount of food (feed) supplied to an animal for a definite period, usually for a day,* to most livestock producers the word implies the feeds fed to an animal or animals, without limitation to the time in which they are consumed. In this and other sections of *Feeds and Nutrition,* the authors accede to the common usage of the word, rather than to dictionary correctness.

ration and account for an intake of approximately 1.5% of the body weight daily. The acid detergent fiber (ADF) should constitute 19% of the ration dry matter, increased to 21% during the first 3 weeks of lactation. The neutral detergent fiber (NDF) should constitute 25% of the ration dry matter, increased to 28% during the first 3 weeks of lactation. However, these general guidelines may need to be modified due to either the source of roughage or its physical form. The NDF and ADF feeding recommendations are given in Chapter 20, Feeding Dairy Cattle, Table 20–5.

(Also see Chapter 8, section headed "NDF, ADF, and NIRS Analyses"; and Chapter 15, sections headed "Neutral Detergent Fiber [NDF] and Acid Detergent Fiber [ADF]," and "Near Infrared Reflectance Spectroscopy [NIRS].")

Fig. 18–3. Lactating cows require fiber for ruminal fermentation and maintaining milk fat percentage. This shows lactating cows eating a mixed ration containing alfalfa, rolled barley, whole cottonseed, distillers' grains, molasses, minerals, and vitamins. (Courtesy, James Tappan, Arizona Dairy Co., Higley, Ariz.)

Health Considerations in Ration Formulation

All animals are susceptible to health-related problems when radical changes are made in the composition of their feed. Therefore, it is recommended that any great changes in feed composition be done gradually over a period of time. Chapter 5, Table 5–1, includes a number of animal disorders, or diseases, attributable to the feeding regimen.

Also, a number of toxic substances may be found in feeds. Chapter 5, Table 5–3, includes a number of these potential poisons.

Whether caused by faulty management or undetected poisons, feed-related disorders are costly and should be guarded against when formulating rations.

How to Balance Rations

When in confinement, animals have access only to the feed provided by the caretaker. Therefore, it is important to provide balanced rations.

Fig. 18–4. Animals in confinement must rely totally on the feed supplied by the producer; hence, any errors in ration formulation may result in lowered or uneconomical production. (Courtesy, Leo S. Jensen, Ph.D., Department of Poultry Science, The University of Georgia, Experiment)

Suggested rations for different classes of livestock are given in Chapters 19 to 28. Generally these rations will suffice, but it is recognized that rations should vary with ingredient availability and cost, and that many times they should be formulated to meet the conditions of a specific farm or ranch or the practices common to an area.

Good livestock producers should know how to balance rations. They should be able to select and buy feeds with informed appraisal; to check on how well their manufacturers, dealers, or consultants are meeting their needs; and to evaluate the results.

Ration formulation consists of combining feeds that will be eaten in the amount needed to supply the daily nutrient requirements of the animal. This may be accomplished by the methods presented later in this chapter, but first the following pointers are necessary:

1. In computing rations, more than simple arithmetic should be considered, for no set of figures can substitute for experience and livestock intuition. Formulating rations is both an art and a science—the art comes from animal know-how, experience, and keen observation; the science is largely founded on mathematics, chemistry, physiology, bacteriology, and nutrition. Both are essential for success.

2. Before attempting to balance a ration, the following major points should be considered:

a. **Availability and cost of the different feed ingredients.** Preferably, cost of ingredients should be based on delivery after processing—because delivery and processing costs are quite variable.

b. **Moisture content.** When considering costs and balancing rations, feed should be placed on a comparable moisture basis; usually, either *as-fed* or *moisture-free.* This is especially important in the case of high-moisture grain or silage.

c. **Composition of the feeds under consideration.** Feed composition tables *(book values),* or average analyses, should be considered only as guides, because of wide variations in the composition of feeds. For example, the protein and moisture contents of sorghum, hay, and silages are quite variable. Whenever possible, especially with large operations, it is best to take a representative sample of each major feed ingredient and have a chemical analysis made of it for the more common constituents—protein, fat, fiber, nitrogen-free extract, and moisture; and often calcium, phosphorus, and carotene. Such ingredients as oil meals, and prepared supplements, which must meet specific standards, need not be analyzed so often, except as quality-control measures.

Despite the recognized value of a chemical analysis, it is not the total answer. It does not provide information on the availability of nutrients to the animal; it does not tell anything about the associated effect of feedstuffs—for example, the apparent way in which beet pulp enhances the value of ground milo; and it does not tell anything about taste, palatability, texture, or undesirable physiological effects, such as bloat and laxative effect. Nevertheless, a chemical analysis does give a sound basis on which to start the evaluation of feeds. Also, with chemical analysis at hand, and bearing in mind that it's the composition of the total feed (the finished ration) that counts, the person formulating the ration can more intelligently determine the quantity of protein to buy, and the kind and amounts of minerals and vitamins to add.

d. **Quality of feed.** Numerous factors determine the quality of feed, including—

- **Stage of harvesting**—For example, early cut forages tend to be of higher quality than those that are mature.
- **Freedom from contamination**—Contamination from foreign substances such as dirt, sticks, and rocks can reduce feed quality, as can aflatoxins, pesticide residues, and a variety of chemicals.
- **Uniformity**—Does the feed come from one particular area or does it represent a blend from several sources?
- **Length of storage**—When feed is stored for extended periods, some of its quality is lost due to its exposure to the elements. This is particularly true with forages and fats/oils.

e. **Degree of processing of the feed.** Often, the value of feed can be either increased or decreased by processing. For example, heating some types of grains makes them more readily digestible to livestock and increases their feeding value.

f. **Soil analysis.** If the origin of a given feed ingredient is known, a soil analysis or knowledge of the soils of the area can be very helpful; for example, (1) the phosphorus content of soils affects plant composition, (2) soils high in molybdenum or selenium affect the composition of the feeds produced, (3) iodine- and cobalt-deficient areas are important in animal nutrition, and (4) other similar soil-plant-animal relationships exist.

g. **Nutrient requirements and allowances.** These should be known for the particular class of animals for which a ration is to be formulated. Also, it must be recognized that nutrient requirements and allowances must be changed from time to time, as a result of new experimental findings.

3. In addition to providing a proper quantity of feed and to meeting the nutritive requirements, a well-balanced and satisfactory ration should be:

a. **Palatable and digestible.**

b. **Economical.** Generally speaking, this calls for the maximum use of feeds available in the area, especially forages.

c. **Suited to the unique needs of the species involved.** Thus, for ruminant animals, the ration should nourish both rumen bacteria and the animal. This necessitates providing adequate rumen degradable protein along with sufficient amino acids post-ruminally.

d. **One that will enhance, rather than impair, the quality of the product (meat, milk, eggs, or wool) produced.**

4. In addition to considering changes in availability of feeds and feed prices, ration formulation should be altered at stages to correspond to changes in the animal life cycle, weight, and productivity.

STEPS IN RATION FORMULATION

The ideal ration is one that will maximize production at the lowest cost. A costly ration may produce phenomenal gains in livestock, but the cost per unit of production may make the ration economically infeasible. Likewise, the cheapest ration is not always the best since it may not allow for a satisfactory level of production.

Therefore, the cost per unit of production is the ultimate determinant of what constitutes the best ration at any given time. Awareness of this fact separates successful producers from marginal or unsuccessful ones.

The following four steps should be taken in an orderly fashion in order to formulate an economical ration:

1. **Find and list the nutrient requirements and/or allowances for the specific animal to be fed.** It should be remembered that nutrient requirements generally represent the minimum quantity of the nutrients that should be incorporated while allowances take into consideration a margin of safety. Factors to be considered in determining the nutrient requirements of animals are:

a. Age.
b. Sex.
c. Body size.
d. Type of production. Is the animal being fed for maintenance, growth, finishing, reproduction, lactation, egg production, wool production, or work?
e. Intensity of production. Is the growing animal gaining 0, 1, 2, or 3 lb per day? Is the lactating animal at the peak of milk production?

Fig. 18–5. The nutrient requirements of a new born foal differ from those of a mature horse. ("Nosing him up," original painting by artist Tom Phillips, 3333 17th Street, San Francisco, CA 94110)

2. **Determine what feeds are available and list their respective nutrient compositions.** In rations for ruminants, dry matter, protein, energy, phosphorus, calcium, and vitamin A are the factors that are generally considered in ration formulation. Additional minerals are generally supplied either as free-choice salt mix or as a premix incorporated in the ration. Animals in confinement may need some vitamin D supplementation. In rations for nonruminants, one must also ensure that adequate amounts of the essential amino acids, essential fatty acids, vitamins D and E, B complex vitamins, and minerals are supplied. Because of these many considerations, it is easy to see why many large producers are using the computer as an aid in ration formulation.

3. **Determine the cost of the feed ingredients under consideration.** Not only should the cost of the feed be considered, but also the cost of processing, transportation, and storage. Some feeds require antioxidants and/or refrigeration to prevent spoilage. Others lose nutritive value when stored for extended periods.

4. **Consider the limitations of the various feed ingredients and formulate the most economical ration.** Remember that the ultimate goal is to formulate a ration that minimizes the cost per unit of production.

ADJUSTING MOISTURE CONTENT

A careful feeder must constantly monitor the moisture content when purchasing feeds, and consider the effect of moisture on nutritional quality control. Most good feeders will readjust feeding formulas whenever moisture in a leading ingredient changes more than 2 or 3%.

Moisture changes may cause imbalances, as pointed up in the following example:

> Let's assume that a cattle feeder is using a ration which has as one of its main ingredients corn silage with 68% moisture content, and that this ration requires 1.9% protein supplement on an *as-fed* basis. Now, assume that the moisture of the silage suddenly decreased to 55%, and with it the necessary supplement to balance the ration increased to 2.62%. Obviously, if the feeder did not adjust the feeding formula, a serious shortage of protein could result. In this case, the cattle would receive only 72.5% as much supplement as they should have since the mixing formula was not recalculated.

The simplest way to avoid errors in ration formulation is to formulate on a 100% dry matter basis.

The multipliers in Table 18–2 may be used to convert feeds of various moisture contents to a 100% dry matter basis.

TABLE 18–2
CORRECTION FACTORS TO USE WHEN CONVERTING FEEDS OF VARIOUS MOISTURE CONTENTS TO A 100% DRY MATTER BASIS (0% MOISTURE)

Percent Moisture	100% DM Basis Multiplier	Percent Moisture	100% DM Basis Multiplier	Percent Moisture	100% DM Basis Multiplier
0	1.0000	29	1.4084	58	2.3809
1	1.0101	30	1.4285	59	2.4390
2	1.0204	31	1.4492	60	2.5000
3	1.0309	32	1.4705	61	2.5641
4	1.0416	33	1.4925	62	2.6315
5	1.0526	34	1.5151	63	2.7020
6	1.0638	35	1.5384	64	2.7777
7	1.0752	36	1.5625	65	2.8571
8	1.0869	37	1.5873	66	2.9411
9	1.0989	38	1.6129	67	3.0303
10	1.1111	39	1.6393	68	3.1250
11	1.1235	40	1.6666	69	3.2258
12	1.1363	41	1.6949	70	3.3333
13	1.1494	42	1.7241	71	3.4482
14	1.1627	43	1.7543	72	3.5714
15	1.1765	44	1.7857	73	3.7037
16	1.1904	45	1.8181	74	3.8461
17	1.2048	46	1.8518	75	4.0000
18	1.2195	47	1.8867	76	4.1666
19	1.2345	48	1.9231	77	4.3478
20	1.2500	49	1.9607	78	4.5454
21	1.2658	50	2.0000	79	4.7619
22	1.2820	51	2.0408	80	5.0000
23	1.2987	52	2.0833	81	5.2631
24	1.3157	53	2.1276	82	5.5555
25	1.3333	54	2.1739	83	5.8824
26	1.3513	55	2.2222	84	6.2500
27	1.3698	56	2.2727	85	6.6666
28	1.3889	57	2.3255		

The majority of feed composition tables are listed on an "as-fed" basis, while most of the National Research Council nutrient requirement tables are on either an "approximate 90% dry matter" or "moisture-free basis." Since feeds contain varying amounts of dry matter, it would be much simpler, and more accurate, if both feed composition and nutrient requirement tables were on a dry basis. In order to facilitate ration formulation, the authors list both the "as-fed" and "moisture-free" contents of feeds in Section V—Composition of Feeds.

The significance of water content of feeds becomes obvious in the examples given in Table 18-3. When using total digestible nutrients (TDN) as a measure of energy, the two high-moisture feeds, carrots and milk, have a higher energy value than oats on a moisture-free basis. The same principle applies to other nutrients, also.

TABLE 18-3
COMPARATIVE ENERGY VALUE OF THREE FEEDS ON (1) AS-FED, AND (2) MOISTURE-FREE BASIS

Feed	Water	Dry Matter	Energy Value (TDN) As-Fed	Energy Value (TDN) Moisture-Free Basis
	(%)	(%)	(%)	(%)
Oats, grain	11	89	69	77
Carrots, roots	84	16	12	78
Milk	88	12	16	128

Mink diets probably have the highest moisture content of any livestock feed—generally 60 to 65% moisture—and are, therefore, used for the purpose of illustrating the formulas for adjusting moisture contents.

Conversion of Rations from an As-Fed Basis to a Moisture-Free Basis

To convert as-fed rations to a moisture-free basis, the following formulas can be used:

Formula 1

When the diet is listed on an as-fed basis, and the producer wishes to compare the content of the various ingredients with the requirements on a moisture-free basis, the equation is as follows:

$$\text{\% nutrient in dry diet (total)} = \frac{\text{\% nutrient in wet diet (total)}}{\text{\% dry matter in diet (total)}} \times 100$$

For example, a diet containing 34% dry matter and 7% protein on a wet basis becomes a 20.6% protein diet on a moisture-free basis.

Formula 2

If the dry matter content of the ingredient, the percentage of the ingredient in the wet diet, and the percent dry matter wanted in the diet are known, it is possible to calculate the amount of the ingredient in the diet on a moisture-free basis as follows:

$$\text{Amount of ingredient in dry diet} = \frac{\text{\% of ingredient in wet diet}}{\text{\% dry matter wanted in diet}} \times \text{\% dry matter of ingredient}$$

Therefore, if a 34% dry matter diet is desired, and if an ingredient containing 25% dry matter is incorporated at a level of 30% of the wet diet, the ingredient constitutes 22% of the moisture-free weight in the diet.

Formula 3

If the producer wants to change the amounts of the ingredients from an as-fed basis to a moisture-free basis, the following equation should be used:

$$\text{Parts on a wet basis} = \text{\% ingredient in wet diet} \times \text{\% dry matter of the ingredient}$$

This calculation should be done for each ingredient and the products added. Each product should then be divided by the sum of the products.

Thus, the diet listed below would break down in the following manner:

	Wet Diet	Dry Matter	Parts	Dry Basis
	(%)	(%)		(%)
Animal by-products	10	20	200	6.3
Blood	5	29	145	4.6
Dry cereal	20	91	1,820	57.6
Horsemeat	25	24	600	19.0
Liver	10	33	330	10.5
Miscellaneous feeds	6.4	10	64	2.0
Water	23.6	0	0	0.0
Total			3,159	100.0

Conversion of Rations from a Moisture-Free Basis to an As-Fed Basis

To convert the components of a dry diet to a wet diet having a given percent of dry matter, one can use the following equation:

$$\text{Parts of ingredient in wet diet} = \frac{\text{\% ingredient in dry diet} \times \text{\% dry matter desired in diet}}{\text{\% dry matter in ingredient}}$$

The total number of parts should be summed and water added to make 100 parts.

The dry diet that follows would break down as shown to make a diet of 31% dry matter:

	Dry Diet	Dry Matter	Dry Matter Desired For Diet	Parts	Wet Diet
	(%)	(%)	(%)		(%)
Animal by-products	3	5	31	18.6	18.6
Dry cereal	50	89	31	17.4	17.4
Horsemeat	15	24	31	19.4	19.4
Liver	15	29	31	16.0	16.0
Miscellaneous feeds	3	10	31	9.3	9.3
Total				80.7	

Add 19.3 parts water to get the desired moisture level (100 − 80.7).

(Also see Chapter 16, Section headed "Moisture is Important.")

METHODS OF FORMULATING RATIONS

In the sections that follow, five different methods of ration formulation are presented: (1) the square method, (2) the trial-and-error method, (3) the simultaneous equation method, (4) the 2 × 2 matrix method, and (5) the computer method. In today's world, some of these are primarily of historic interest, but they present a progressive development of ration formulation technology. Despite the sometimes confusing mechanics of each system, if done properly, the end result of all 5 methods is the same—a ration that provides the desired allowance of nutrients in correct proportions economically (or at least cost), but, more important, so as to achieve the greatest net returns—for it is net profit, rather than cost, that counts. Since feed usually represents the greatest cost item in livestock production, the importance of balanced rations is evident.

An exercise in ration formulation follows for purposes of illustrating the application of each of these five methods:

1. **Square method,** applied to a swine ration.
2. **Trial-and-error method,** applied to a lactating cow ration.
3. **Simultaneous equation method,** applied to poultry and sheep rations.
4. **Two × two matrix method,** applied to sheep and horse rations.
5. **Computer method,** applied to a lactating cow ration.

Fig. 18-6. These pigs which are in a nursery, require more than 40 different nutrients. Formulation of such rations by computer is fast and accurate. (Courtesy, Iowa State University, Ames)

Square (or Pearson Square) Method

The square method is a simple, direct, and easy way in which to figure proportions between two ingredients. It permits quick substitution of feed ingredients in keeping with market fluctuations, without disturbing the protein content.

In balancing rations by the square method, it is recognized that one specific nutrient alone receives major consideration. Correctly speaking, therefore, it is a method of balancing one nutrient requirement, with no consideration given to the other nutritive requirements.

To compute rations by the square method, or by any other method, it is first necessary to have available both feeding standards (see the nutrient requirement tables in the respective chapters devoted to each class of livestock, Chapters 19 to 28) and feed composition tables (Section V—Composition of Feeds).

The following example shows how to use the square method in formulating a swine ration:

Example. *A swine producer has 40-lb pigs to which it is desired to feed a 16% protein ration until they reach 120 lb weight. Corn containing 8.9% protein is on hand. A 36% protein supplement, which is reinforced with minerals and vitamins, can be bought. What percent of the ration should consist of corn and of the 36% protein supplement?*

Step by step, the procedure in balancing this ration is as follows:

1. Draw a square, and place the number 16 (desired protein level) in the center.

2. At the upper left-hand corner of the square, write *protein supplement* and its protein content (36); at the lower left-hand corner, write *corn* and its protein content (8.9).

3. Subtract diagonally across the square (the smaller number from the larger number), and record the difference at the corners on the right-hand side (36 − 16 = 20; 16 − 8.9 = 7.1). The number at the upper right-hand corner gives the parts of concentrate by weight, and the number at the lower right-hand corner gives the parts of corn by weight to make a ration with 16% protein.

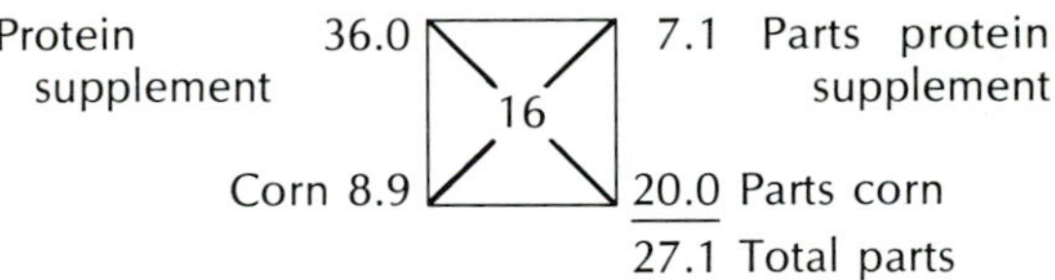

4. To determine what percent of the ration would be corn, divide the parts of corn by the total parts and multiply by 100: 20.0 ÷ 27.1 × 100 = 73.8% corn. The remainder, 26.2%, would be supplement.

Trial-and-Error Method

In the example that follows, the trial-and-error method is used, with consideration given to energy and protein. Also, crude protein rather than digestible protein is used because (1) this is what feed manufacturers want to know as they plan feed formulas, and (2) this is what livestock producers see on the tag when they purchase feed. In most mixed feeds, approximately 80% of the total protein is digestible.

Example. *Let's assume that a dairy producer has a 1,433-lb cow producing 65 lb of milk testing 4.0% fat. The producer is feeding 14 lb of alfalfa hay and 40 lb of corn silage per day. Corn, oats, and soybean meal are available. What concentrate mix should the producer use to meet the needs of this lactating cow, from the standpoint of energy and protein?*

The available feeds have approximately the following composition (as-fed basis):

	TDN	Crude Protein
	(%)	(%)
Alfalfa hay, all analyses	51.0	16.0
Corn silage, all analyses	18.0	2.2
Corn, all analyses	80.0	9.9
Oats, all analyses	69.0	11.9
Soybean meal, solv extd, 44%	76.0	44.4

Here are the steps in balancing this ration:

Step 1. The daily TDN and crude protein requirements of this cow (1,433 lb body weight, 65 lb of milk testing 4% fat) are:[3]

Requirements of cow for—

	TDN	Crude Protein	
	(lb)	(lb)	*(g)*
Maintenance	9.94	0.94	*428*
Milk production	20.9	5.87	*2,665*
Total	30.84	6.81	*3,093*

Step 2. The forage (14 lb alfalfa hay, 40 lb corn silage) is supplying:

	TDN	Crude Protein
	(lb)	(lb)
Alfalfa hay, 14 lb	7.14	2.24
Corn silage, 40 lb	7.20	0.88
Total from forage	14.34	3.12

Step 3. Remainder, to be supplied by concentrate:

TDN	Crude Protein
(lb)	(lb)
16.5	3.69

Step 4. Let's try out (that's why it is called the *trial-and-error method*) a grain mix of 700 lb corn, 280 lb oats, 10 lb monosodium phosphate, and 10 lb salt, and determine the amounts of TDN and crude protein in 1,000 lb of the grain mix:

	TDN	Crude Protein
	(lb)	(lb)
Corn, 700 lb	560.0	69.30
Oats, 280 lb	193.2	33.3
Monosodium phosphate, 10 lb	—	—
Salt, 10 lb	—	—
Total	753.2	102.60
or in percent	75.3%	10.3%

Step 5. Divide the TDN needed from concentrate (16.5 lb) by the percent TDN in the mixture (75.3%). Thus, feeding 21.9 lb of the concentrate will meet the energy needs.

Step 6. Will this level of grain mix (21.9 lb) also meet the crude protein needs? By multiplying the pounds of concentrate mixture by the percent crude protein (21.9 × 10.3%), we find that the proposed concentrate would supply 2.26 lb of crude protein, whereas 3.69 lb are needed. Therefore, a high-protein supplement must be substituted for some of the homegrown grain.

Step 7. Let's substitute 175 lb of soybean meal for 175 lb of corn. Hence, the concentrate mix as now proposed will consist of:

	TDN	Crude Protein
	(lb)	(lb)
Corn, 525 lb	420.0	52.0
Oats, 280 lb	193.2	33.3
Soybean meal, 175 lb	133.0	77.7
Monosodium phosphate, 10 lb	—	—
Salt, 10 lb	—	—
Total	746.2	163.0
or in percent	74.6%	16.3%

Step 8. By referring back to Step 3, we can divide the pounds of TDN and crude protein needed from the concentrate, by the percentage of TDN and crude protein found in the grain mix in Step 7. We find that 16.5 ÷ .746 = 22.1 lb needed to supply 16.5 lb TDN; and 3.69 ÷ .163 = 22.63 lb needed to supply 3.69 lb crude protein. Thus, we find that the following ration will supply the needed TDN (with a slight overage) and crude protein for a 1,433-lb lactating cow producing 65 lb of milk testing 4% fat:

	TDN	Crude Protein
	(lb)	(lb)
Alfalfa hay, 14 lb	7.1	2.2
Corn silage, 40 lb	7.2	0.9
Concentrate mix (Steps 7 & 8), 22.63 lb	16.9	3.7
Total	31.2	6.8

In many sections of the country, especially in grain-deficient areas and on highly specialized dairies where little or no grain is grown, the dairy producer may find it most economical to purchase a commercial dairy feed to augment the roughage that is being fed.

Simultaneous Equation Method[4]

It is possible to formulate rations involving two sources and one nutrient quickly through the solving of simultaneous equations:

Example 1. *A poultry producer has on hand No. 2 corn containing 8.9% protein (as-fed basis). A 40% protein supplement, which is reinforced with minerals and vitamins can be bought. A ration containing 18% protein is desired.*

[3]From Chapter 20, Table 20-3, "Daily Nutrient Requirements of Lactating and Pregnant Dairy Cattle."

[4]The computations for this section were made by Dr. T. L. Cross, Instructor, Agricultural and Research Economics, Oregon State University, Corvallis.)

Step by step, the procedure in balancing this ration is as follows:

Step 1. Let X = amount of corn to be used in 100 lb of mixed feed, and Y = amount of 40% protein supplement to be used in 100 lb of mixed feed. We know that the corn contains 8.9% protein and the protein supplement 40% and that the ration should contain 18% protein. Therefore, the equation we must solve is as follows:

.089X + .40Y = 18 (lb of protein in 100 lb of feed)

Step 2. In order to solve for two unknowns (X and Y), we must create a "dummy equation." This can be done in the following manner:

X + Y = 100 lb of feed

Step 3. We must now multiply our dummy equation by .089 in order that the X term will cancel out with the original equation. Therefore:

X + Y = 100 becomes **.089X + .089Y = 8.9**

Step 4. Subtracting our new dummy equation from the original equation, we can solve for Y as shown below:

Original equation:	**.089X + .40 Y = 18**
Dummy equation:	**− .089X − .089Y = − 8.9**
	.000X + .311Y = 9.1

$$Y = \frac{9.1}{.311}$$ **or 29.26 lb of 40% protein supplement per 100 lb of feed or 29.26%**

Step 5. We can now substitute our newly acquired value for Y in the original equation and solve for X, thus:

X = 100 − 29.26
= 70.74 lb of corn per 100 lb of feed or 70.74%

It is also possible to use simultaneous equations in the formulation of rations involving 2 feed sources and 2 nutrients. Since many formulations involve solving for more than 1 nutritional parameter, this method has many advantages.

Example 2. *A sheep feeder is feeding 88-lb finishing lambs. The feeder has access to a concentrate containing 89% dry matter and 80% TDN (as-fed). Alfalfa hay, containing 93% dry matter and 54% TDN (as-fed), is also available. How much of each feed should be fed?*

From Chapter 21, Table 21-1, Daily Nutrient Requirements of Sheep, we find that an 88-lb lamb requires 3.9 lb of dry matter and 2.7 lb of TDN daily.

Step by step, the procedure in balancing this ration is as follows:

Step 1. Let X = amount of alfalfa hay to be fed and Y = amount of concentrate to be fed. Since the lamb requires 2.7 lb of TDN and 3.9 lb of dry matter, our equations will be:

.54X + .80Y = 2.7 TDN equation
.93X + .89Y = 3.9 dry matter equation

Step 2. If we compare the coefficients for alfalfa in both equations, we find that an adjustment factor of .58 $\frac{.54}{.93}$ is needed to balance the dry matter equation with the TDN equation in order that we may solve for Y. If we multiply the dry matter equation by .58, we arrive at the following:

[.58(.93X + .89Y) = .58(3.9)] equals the equation .54X + .52Y = 2.26

Step 3. With the dry matter equation adjusted so that the X terms of the dry matter and TDN equations cancel out when substracted, we can solve for Y in the following manner:

.54X + .80Y = 2.70 TDN equation
− (.54X + .52Y = 2.26) dry matter equation

.00X + .28Y = .44

$$Y = \frac{.44}{.28}$$

Y = 1.57 lb of concentrate should be fed

Step 4. Now that we know the value of Y, we can substitute it in either original equation to solve for X.

.54X + .80(1.57) = 2.70
.54X + 1.26 = 2.70
.54X = 1.44
X = 2.67 lb of alfalfa should be fed

If we substitute X and Y in our original equations, we can check our accuracy.

.54(2.67) + .80(1.57) = 2.7 TDN equation
.93(2.67) + .89(1.57) = 3.9 dry matter equation

Two X Two Matrix Method

In addition to the traditional algebraic method of solving simultaneous equations, matrix algebra provides an alternative which some people find easier and quicker. The 2 X 2 matrix provides a quick and accurate way of solving for 2 nutritional parameters—such as energy and protein—through the use of 2 ingredients.

A matrix is a mathematical array which allows for the solution of unknowns through the use of a series of equations. Consider the 2 equations:

$$a_1X + b_1Y = c_1$$
$$a_2X + b_2Y = c_2$$

Let us assume that X represents one type of feed, and that Y represents another type. In order to solve for X and Y, we can set up a 2 X 2 matrix using their respective coefficients: c_1 and c_2 could represent 2 nutrient levels we want (for example, energy and protein). These equations can be written in matrix notation as:

$$\begin{bmatrix} a_1 & b_1 \\ a_2 & b_2 \end{bmatrix} \begin{bmatrix} X \\ Y \end{bmatrix} = \begin{bmatrix} c_1 \\ c_2 \end{bmatrix}$$

The matrix would consist of 2 rows and 2 columns. In order to solve for X and Y, we must find the determinant of the matrix. The determinant is established as follows:

$$\begin{bmatrix} a_1 & b_1 \\ a_2 & b_2 \end{bmatrix} = a_1b_2 - a_2b_1$$

If the matrix is $\begin{bmatrix} 1 & 2 \\ 3 & 4 \end{bmatrix}$, the determinant would be

$$\begin{vmatrix} 1 & 2 \\ 3 & 4 \end{vmatrix} = (1 \times 4) - (3 \times 2) = 4 - 6 = -2$$

Note that a determinant of a square matrix is enclosed by straight vertical lines, and a square matrix is enclosed by curved lines. Through a series of derivations from the original 2 equations, the unknowns can be solved in the following manner:

$$X = \frac{\begin{bmatrix} c_1 & b_1 \\ c_2 & b_2 \end{bmatrix}}{\begin{bmatrix} a_1 & b_1 \\ a_2 & b_2 \end{bmatrix}} \text{ or } \frac{(c_1b_2 - c_2b_1)}{(a_1b_2 - a_2b_1)}$$

$$Y = \frac{\begin{bmatrix} a_1 & c_1 \\ a_2 & c_2 \end{bmatrix}}{\begin{bmatrix} a_1 & b_1 \\ a_2 & b_2 \end{bmatrix}} \text{ or } \frac{(a_1c_2 - a_2c_1)}{(a_1b_2 - a_2b_1)}$$

Using the same example as in example 2 of the simultaneous equation section (sheep ration), the 2 X 2 matrix method will arrive at the same answer.

Example 1. *A sheep feeder is feeding 88-lb finishing lambs. The feeder has access to a concentrate containing 89% dry matter and 80% TDN (as-fed). Alfalfa hay, containing 93% dry matter and 54% TDN (as-fed), is also available. How much of each feed should be fed?*

From Chapter 21, Table 21-1, Daily Nutrient Requirements of Sheep, we find that an 88-lb finishing lamb requires 3.9 lb of dry matter and 2.7 lb of TDN daily.

Step 1. To balance this ration by the matrix method, we will proceed as follows: Let X = amount of alfalfa hay to be fed, and Y = amount of supplement to be fed. Therefore, our equations will be:

$$.54X + .80Y = 2.7 \text{ TDN equation}$$
$$.93X + .89Y = 3.9 \text{ dry matter equation}$$

Step 2. From these equations, we can set up the following 2 X 2 matrix:

$$\begin{bmatrix} a_1 & b_1 \\ a_2 & b_2 \end{bmatrix} = \begin{bmatrix} .54 & .80 \\ .93 & .89 \end{bmatrix} \text{ and } \begin{bmatrix} c_1 \\ c_2 \end{bmatrix} = \begin{bmatrix} 2.7 \\ 3.9 \end{bmatrix}$$

Step 3. Once we have set up our matrices, we can solve for X and Y by calculating the determinants, as shown below:

$$X = \frac{[2.7(.89) - 3.9(.80)]}{[.54(.89) - .93(.80)]} = \frac{2.403 - 3.120}{.481 - .744} = \frac{-.717}{-.263} = 2.73$$

We need to feed 2.73 lb of alfalfa.

$$Y = \frac{[.54(3.9) - .93(2.7)]}{[.54(.89) - .93(.80)]} = \frac{2.106 - 2.511}{.481 - .744} = \frac{-.405}{-.263} = 1.54$$

We need to feed 1.54 lb of supplement.

These figures coincide with those calculated by the traditional form of solving simultaneous equations (X = 2.67 and Y = 1.57). The slight variation in amounts of alfalfa and supplement to be fed determined in the simultaneous equation method and the matrix method can be attributed to rounded error.

Example 2. *A horse breeder is feeding a lactating mare (first 3 months lactation) 14 lb of mid-bloom timothy hay a day (moisture-free basis). The breeder realizes that a supplement is required and plans to use a mixture of soybean meal (49%, without hulls, solvent extracted) and No. 2 corn, which are readily available. The supplement is formulated in the following manner:*

Step 1. The breeder finds that the lactating mare requires 23.36 Mcal of digestible energy and 2.46 lb of crude protein daily. Fourteen pounds of timothy hay (moisture-free) should supply 14.84 Mcal and 1.36 lb of protein daily (Section V—Composition of Feeds). Therefore,

	Digestible Energy	Protein
	(Mcal)	(lb)
Requirement	23.36	2.46
Subtract amount supplied by hay . . .	−14.84	−1.36
Needed in supplement	8.52	1.10

Step 2. Section V—Composition of Feeds shows that the corn (No. 2 grade, moisture-free) has 1.27 Mcal/lb and 10.2% protein. Section V—Composition of Feeds shows a protein value of 54.6% for the soybean meal (moisture-free), but does not list a DE value for the soybean meal; so, based on available data, the authors estimated the soybean meal DE value at 1.72 Mcal/lb. Let x = dry matter supplied by meal, and y = dry matter supplied by corn. Therefore,

$$1.72X + 1.27Y = 10.90 \text{ energy equation}$$
$$.546X + .102Y = 1.42 \text{ protein equation}$$

Step 3.

$$\begin{bmatrix} a_1 & b_1 \\ a_2 & b_2 \end{bmatrix} = \begin{bmatrix} 1.72 & 1.27 \\ .546 & .102 \end{bmatrix} \text{ and } \begin{bmatrix} c_1 \\ c_2 \end{bmatrix} = \begin{bmatrix} 10.90 \\ 1.42 \end{bmatrix}$$

Step 4.

$$X = \frac{\begin{bmatrix} 10.90 & 1.27 \\ 1.42 & .102 \end{bmatrix}}{\begin{bmatrix} 1.72 & 1.27 \\ .546 & .102 \end{bmatrix}} = \frac{-.692}{-.518} = \text{1.34 lb of soybean meal (moisture-free) per day}$$

$$Y = \frac{\begin{bmatrix} 1.72 & 10.90 \\ .546 & 1.42 \end{bmatrix}}{\begin{bmatrix} 1.72 & 1.27 \\ .546 & .102 \end{bmatrix}} = \frac{-3.509}{-.518} = \text{6.77 lb of corn (moisture-free) per day}$$

In least-cost formulations, matrix algebra is used by the computer, but the matrices that are used are much larger than 2 X 2 and are far more complicated. The 2 X 2 matrix offers a rapid means of calculating a simple ration using 2 feeds to fulfill 2 nutrients.

Computer Methods[5]

Most large livestock establishments and feed companies now use computers in ration formulation. Also, many of the state universities, through their Federal-State Extension Services, are offering ration balancing computer services to farmers within their respective states on a charge basis. Consulting nutritionists are available throughout the United States and provide computer services, as well as other services. With the recent advent of the low cost personal computer, this powerful technology is available to almost everyone.

Fig. 18–7. Today, most poultry rations, like the ration for this layer operation in Maryland, are formulated with the use of a computer. (Courtesy, University of Maryland, College Park)

Despite their sophistication, there is nothing magical or mysterious about the use of computers in ration balancing. Their primary advantages are accuracy and speed of computation. In addition, computer programs (software) used in ration balancing provide a means of organizing needed information in a logical and systematic manner. The computer should be viewed as an extension of the knowledge and skills of the formulator.

At this time, there is no "pushbutton system" of feed formulation available. The degree of success realized is very dependent on the management of data put into the computer, and on the evaluation of the resulting formulations that the computer generates. In the hands of experienced users, the computer enables the producer and nutritionist to be more precise in carrying out ration formulation.

Two basic approaches to ration formulation are practiced with computers:

1. Trial-and-error formulation.
2. Linear programming (LP).

TRIAL-AND-ERROR FORMULATION WITH THE COMPUTER

For a discussion of the trial-and-error method of ration balancing, see the earlier section in this chapter headed "Trial-and-Error Method." Many ration balancing software programs written for the computer allow for trial-and-error ration balancing. Feed mill nutritionists frequently use this technique to enter into the computer rations that are given to them by other nutritionists or by a producer. The objective in this case is to confirm the nutrient values for the ration based on the specific ingredients used by the feed manufacturer. In many cases, these rations are not to be altered without permission. In other cases, the number of ingredients for a specific ration may be limited so that the trial-and-error technique is just as fast as using linear programming to arrive at the desired nutrient levels in the ration.

> **NOTE:** It does not take specialized computer software to use the trial-and-error method. Spreadsheet (or Financial Spreadsheet) programs, for instance, organize data into rows and columns. Information, such as nutrient values for a feedstuff, may be entered into data cells (see Fig. 18–8). Simple and complex arithmetic operations can be controlled by the user to the extent that rather large trial-and-error method rations can be programmed and run.
>
> Spreadsheets have been developed with specific microcomputers in mind; and there are a great number of them on the market. Detail on their choice and use may be obtained from specialized books listed in the Selected References, at the end of this chapter.

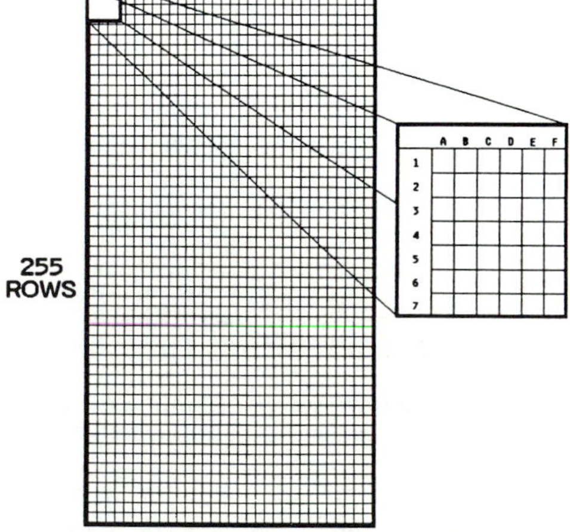

Fig. 18–8. Graphic representation of a spreadsheet (from Lane, R. J. and T. L. Cross, *Spreadsheet Applications for Animal Nutrition and Feeding,* Reston Publishing Co., Inc., Reston, Va., 1985).

---

[5]This section on "Computer Methods" was prepared especially for this book by L. M. Larsen, Ph.D., Consultant, Nutri-Systems, 426 E. Shields, Fresno, CA 93704.

LINEAR PROGRAMMING (LP)

The most common technique for computer formulation of rations is the linear programming (LP) technique. At times, this is referred to as *least cost* ration formulation. This designation results from the fact that most LP techniques for ration formulation have as their objective *minimization of cost.* A few LP programs are in use that solve for *maximization of income over feed costs.* Regardless, the livestock producer and nutritionist should always keep in mind that maximizing net profit is the only true objective of most ration formulations. A skilled user of the LP system will control ration quality by writing specifications that lead to rations that will maximize profit.

Briefly described, the LP program is a mathematical technique in which a large number of simultaneous equations are solved in such a way as to meet the minimum and maximum levels of nutrients and levels of feedstuffs specified by the user at the lowest possible cost. It is not necessary to understand the inner workings of the computer program to use LP, though it does take experience to use it to good advantage and to avoid certain pitfalls. The most common pitfalls are incorrectly entered or missing data and the specification of minimums and maximums that cannot be met with the feedstuffs available. The latter is called an infeasible solution. When an infeasible solution is encountered, the user must determine (1) if this is due to incorrect or missing data, or (2) if the specifications must be relaxed.

PROCEDURE FOR USE OF LINEAR PROGRAMMING (LP)

Before using the LP approach to ration formulation, the user should become familiar with the specific software package to be used. (See later section on "Selection of Computer Software and Hardware for Ration Formulation.") It is also desirable to study the LP technique as applied to feed formulation. After users are familiar with LP and their computer software, they are ready to begin using the computer for ration formulation by LP. It must first be understood that all data entered into the computer is directed to files. In most cases, these files are located on disks, or perhaps on tapes. Currently, most computers use keyboards and CRT (cathode ray tube) displays for entry of data. The necessary data files are generally created in steps as follows:

1. **Enter names of available feed ingredients, and the cost of each.** It is necessary that all of the available feeds be listed along with the unit cost. It matters little if the formulator uses cost per ton, cost per hundred weight (cwt), or cost per pound, but the same method of cost input must be used for all feeds. The computer software may call for a specific form for entering costs.

2. **Enter nutrient values for feeds.** Tables of feed composition using average or typical values, like those in Section V of this book, may be used, but, because of the wide variation in the composition of feeds, chemical analysis of a representative sample of each lot of feed is more precise and should be used if available. This is especially true of forages, in which composition may be affected in a major way by cultural conditions and stage of maturity.

3. **Enter ration specifications.** Ration specifications are generally broken into two parts: (1) Nutrient limits, and (2) Ingredient limits. In each case, the formulator specifies either a lower limit and/or an upper limit for each item. If no specification for the particular item is desired, it may be specified as zero (0) or left blank, depending on the circumstances. It is also appropriate to list feedstuffs available, but not currently on hand (with an upper limit of zero). Most LP solutions will then tell the user the highest cost at which such feeds would enter the solution if allowed to do so. Ratios between nutrients (such as a calcium/phosphorus ratio) or feedstuffs (corn/barley ratio) may also be specified in most LP software packages. The experienced formulator usually deals with palatability or feedstuff quality considerations by setting an upper limit on the amounts of problem feeds or a lower limit on feeds that contribute a positive quality to the ration. Nonnutritive attributes, such as bulk density, may also be programmed into the LP system. The LP technique is a very flexible and powerful ration balancing tool.

> **NOTE:** Important additional items to consider when creating ration specifications are upper limits on the use of nonprotein nitrogen (or urea) and limits on the usage of feed additives like drugs, feed flavors, and the like.

Fig. 18–9 illustrates, by means of a worksheet, a logical method of organizing the restrictions for a ration.

LEAST-COST FORMULATION WORKSHEET

Specifications	Ingredient A	Ingredient B	Ingredient C	Restrictions
Cost				Minimize
Total weight				1,000 lb
Crude protein				133 lb
Digestible protein				100 lb
Ether extract				25 to 80 lb
Net energy lactation				900 Mcal
Calcium				5 to 10 lb
Phosphorus				7 lb
Vitamin A equivalent				35,000
Vitamin D				60,000
Limits on ingredients				
Minimum				
Maximum				

Fig. 18–9. Sample worksheet for a least-cost formulation. The first column lists the specifications. The various feedstuffs to be considered are listed in the succeeding columns with their respective costs and nutritive values. The last column lists the restrictions desired on the final formulation.

4. **Submit all of the above information to the matrix building and solving portion of the LP software package.** Matrix building and solving are generally accomplished automatically by the computer software once the specifications have been entered into the computer. Mathematically, the procedure involves the solution of a complex algebraic problem, with an answer being derived in seconds or minutes. Using the LP program, the computer produces a mix that will meet the desired specifications at the lowest possible cost.

5. **Examine the solution provided by the computer software.** The end result should be feasible, both from a mathematical standpoint and from a nutritional standpoint. The feedstuff mixture should be acceptable to the animals for which it is intended. In most cases the first solution provided to the user is not acceptable. Repeat runs may be necessary to obtain the best solution.

Figure 18–10 is a computer printout of an LP solution. The various columns of the report have been numbered for indentification. Similar columns have been given the same number. The three sections of the report are each identified with a Roman numeral.

ANYCO GRAIN AND MILLING
P. C. 3CX 1234
ANYTOWN, USA 90909 (123) 456-7890
* FEASIBLE* RATION: L507 LAYER MASH, 17 PCT.

16-JAN-89 (7) ID. NO. 29706

COSTS....	$/CWT	$/TON
BASE LP	8.069	161.379
BATCH	8.08	161.60

	(1)	(2)	(3)	(4)	(5) LIMITS	(6)	(7)	(8) STABLE COST	(9) RANGE	(10) COST PER	(11) UNIT	(12) EFFECTIVE	(13) RANGE
	#	INGREDIENT	AMOUNT	BATCH	LOWER	UPPER	COST	LOWER	UPPER	DECREASE	INCREASE	DECREASE	INCREASE
I	4	GROUND CORN	33.333	665.00	10.000		6.40	6.163	6.470	0.0007	0.0024	25.050	64.496
	6	GROUND MILO	33.848	675.00			6.00	5.933	6.218	0.0022	0.0006	0.000	43.256
	10	WHEAT MILLRUN	2.085	40.00		15.000	6.00	4.339	6.280	0.0028	0.0116	0.000	28.642
	19	SOYBEAN MEAL,47.5	16.117	320.00			14.95	11.036	26.372	0.1142	0.0391	14.804	18.805
	20	MEAT SCRAP,50	6.000	120.00		6.000	15.03		20.133	0.0510	-0.0510	3.938	7.540
	21	DICAL PHOS(22CA/18P)	0.396	10.00			17.00	0.412	139.266	1.2227	0.1659	0.273	2.033
	23	LIMESTONE	7.584	150.00			1.45	0.000	12.895	0.1144	0.0309	7.533	7.843
	24	SALT, PLAIN	0.250	5.00	0.250	0.250	2.35	0.000		-0.0393	0.0393	0.000	0.930
	27	POULTRY PREMIX	0.250	5.00	0.250	0.250	56.37	0.000		-0.5795	0.5795	0.000	0.930
	33	DL-METHIONINE,99	0.086	1.73			157.00	76.279	540.275	3.8327	0.8072	0.067	0.094
	37	SELENIUM, 90.8	0.050	1.00	0.050	0.050	17.50	0.000		-0.1908	0.1908	0.000	0.730
		TOTALS	100.000	1992.73	REQUESTED BATCH WEIGHT IS 2000.00 POUNDS								

	(1)	(2)	(3)	(5) LIMITS	(6)	(10) COST PER	(11) UNIT	(12) EFFECTIVE	(13) RANGE
	#	NUTRIENT	AMOUNT	LOWER	UPPER	DECREASE	INCREASE	DECREASE	INCREASE
II	1	WEIGHT	100.000	100.000	100.000	0.0158	-0.0158	99.320	104.209
	8	CRUDE PROTEIN	17.000	17.000		-0.2428	0.2428	15.864	18.136
	12	CRUDE FAT	2.930	0.000		0.0540	0.0532	2.822	5.748
	13	CRUDE FIBER	2.308		4.000	0.1370	0.0969	2.250	2.368
	14	ASH	11.963	0.000		0.6566	0.0333	11.907	12.203
	15	CALCIUM	3.600	3.600	3.700	-0.0798	0.0798	2.001	3.858
	16	PHOSPHORUS	0.671	0.650		0.5568	0.9380	0.661	0.961
	17	AVAIL. PHOSPHORUS	0.470	0.470		-0.9095	0.9095	0.448	0.769
	31	M. E. (POULTRY)/LB	1270.000	1270.000		-0.0035	0.0035	1203.931	1280.669
	47	LYSINE	0.798	0.700		20.0361	1.6941	0.686	0.801
	48	METHIONINE	0.350	0.350		-1.3590	1.3590	0.331	1.798
	49	METHIONINE + CYSTINE	0.619	0.600		3.5320	1.3803	0.617	2.034
	61	LINOLEIC ACID	1.134	1.000		0.0609	0.1073	1.038	2.531
	62	XANTHOPHYLL /LB	3.000	3.000		-0.0264	0.0264	1.570	5.805

	(1)	(2) ** NOT USED **	(3)	(14) RELATIVE	(6) UPPER	(11) COST/UNIT	(13) INCREASE
	#	INGREDIENT	COST	WORTH	LIMIT	INCREASE	RANGE
III	8	GROUND BARLEY	6.500	5.782		0.0072	6.531
	12	ALFALFA, DEHY.,17	8.000	7.374	2.500	0.0063	0.932
	31	VEGETABLE FAT	17.200	11.914		0.0529	2.835
	32	CANE MOLASSES	4.450	2.384	0.000	0.0207	1.738

Fig. 18–10. Example Leghorn layer ration processed by computer linear programming. See text for explanation of marked () columns. (Courtesy, Nutri-Systems, Fresno, Calif.)

An explanation of the information contained in each column of Fig. 18–10 follows:

Column (1)—Ingredient and nutrient numbers.

Column (2)—Ingredient and nutrient names.

Column (3)—Solution *amounts* given in percentage for feed ingredients (Section I) and nutrients (Section II).

Column (4)—The percentage solution for ingredients has been converted to a "ton" *batch* using prespecified rounding factors for each ingredient. (The batch totals 1,992.73 lb, rather than an exact 2,000 lb because of the rounding requirements.)

Columns (5 & 6)—*Lower* and *upper limits* specified for each ingredient (Section I) and each nutrient (Section II).

Column (7)—Ingredient *costs* in dollars per hundred weight (cwt).

Columns (8 & 9)—The *stable cost range (lower)* gives the feed cost below which the present optimal solution would no longer be valid. Similarly, the *stable upper* gives the feed cost above which the present solution would no longer be valid. The *stable cost* figures let the LP user know when it is desirable to reprocess the ration.

Columns (10 & 11)—The *cost per unit decrease* and *increase* values indicate how much the cost of the ration would be changed if either an ingredient or nutrient is increased or decreased by one unit in the percent solution. A positive value means that cost would be increased and a negative value means that cost would be decreased.

Columns (12 & 13)—The *effective range decrease* and *increase* values are related to the *stable cost range* columns and the *cost per unit decrease/increase* columns. The values delineate the limits over which the *stable cost* and *unit decrease/increase* columns are applicable. (An example: If the cost of ground corn decreases to $6.163/cwt [Column 7], then the usage of corn would increase to 64.496% [Column 13]. Of course, there would be changes in the usage of other ingredients as corn increases in amount.)

Column (14)—Section III contains information about the *ingredients not used* in the solution. The *relative worth* column indicates the cost at which each of these ingredients would enter the solution.

6. **Reformulate with LP at periodic intervals.** Changes in ingredient costs, in ingredient availability, and in the needs of animals dictate the need for reprocessing the ration. The good formulator monitors all these items on a regular basis. It is also critical to evaluate the feeding results to confirm that production goals and cost objectives are being met with the ration. Computers don't feed animals—people do!

SELECTION OF COMPUTER SOFTWARE AND HARDWARE FOR RATION FORMULATION

Numerous companies market computer software for ration formulation. The software varies from the very simple and straight-forward to very complex packages intended for large feed manufacturers. The latter packages include applications for formula costing, inventory control, control of usage of ingredients in limited availability, production of feed tags, etc. Most software is intended for use on a single computer model or at least a certain family of computers (IBM-PC tm, for example). It is therefore most desirable to select the software desired before purchasing the computer hardware. Computer type and size of memory and disk drive storage capacity must meet the criteria of the software developer or the software may not be usable.

Directories which list software by application are a good place to start looking. Other sources of information are feed and livestock trade publication advertisements, university personnel, and nutritionists who use feed blending software. Nutritionists are a good source of information as to how well a certain software package performs.

Ration formulation software may be generalized so that it can be made applicable to all species of animals or it may be designed with the unique requirements of specific species such as poultry, dairy cattle, etc. When the software has been designed for a certain species, it may incorporate tables of nutrient requirements and tables of typical feedstuffs and their nutrient values. This can save the user time, but it does not mean that the software will run itself without the judgment of the user. No one has yet developed software that will anticipate all the conditions under which livestock will be fed. Computers are not able to assess all aspects of ingredient quality, environment, and animal management. The judgment of the producer and formulator must be imposed on the computer software. Look for the freedom to make changes as needed. When in doubt, seek advice from those with experience.

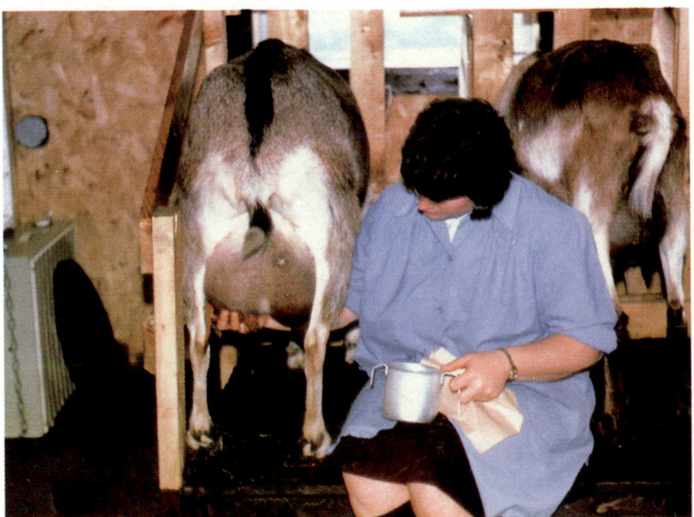

Fig. 18-11. The final determinant of a successfully formulated ration is the cost per unit of production. In the case of dairy goats, it is the cost per pound, gallon, or hundredweight of milk. The udder of this doe is being prepared for milking. (Courtesy, University of New Hampshire, Penacook)

USE OF THE FLEXIBLE FORMULA

The flexible formula is a ration formulation that allows for the substitution of various feeds on the basis of price and availability. The overall formula does not change. The only changes that take place are substitutions of feeds—for example oats for corn—within the formula to supply the same nutrient levels.

The following procedure may be used in setting up a flexible formula:

1. The nutrient requirements for the animals to be fed should be obtained from feeding standards, and should be listed.

2. A chart should be set up with four divisions: (1) energy feeds, (2) protein feeds, (3) mineral supplement, and (4) vitamin supplement. Within each division, subdivisions can be listed—for example, under protein feeds, two types of protein feeds, plant and animal, can be listed (see Table 18-4).

Fig. 18-12. Feed prices fluctuate widely. So, flexible rations formulas, providing for feed substitutions, make for greatest net returns. Thus, sheep, and other animals, may be, and are, fed any of the cereal grains and/or their by-products, provided they are properly processed and incorporated in a balanced ration. (Courtesy, American Hampshire Sheep Assn., Ashland, Mo.)

TABLE 18-4
EXAMPLE OF A FLEXIBLE FORMULA CHART

Nutrient Classification of Feed	Subdivisions	Feeds for Utilization	Restrictions for Use		Amount to be Incorporated
			(Min.)	(Max.)	(lb/100lb feed)
Energy (Average 10% protein)	Grains	Corn Wheat Milo			86.7
	By-products	Wheat mill-run Corn gluten feed			
Protein (Average 45% protein)	Plant (Average protein)	Soybean meal Cottonseed meal Linseed meal			11.8 Total (7.06 plant and 4.73 animal)
	Animal (Average 60% protein)	Fish meal Tankage Meat meal rendered			
Vitamin		Premix			.01
Mineral		Premix			1.5
			Total		100.0

3. The proportion of the nutrient divisions must be established. This can be accomplished by balancing the various divisions for protein or any other feed parameter that is to be balanced. For example, a feed manufacturer wants to establish a flexible formula for a 14% protein feed. The manufacturer has premixes for vitamins and minerals and wants to incorporate them at the levels of 0.01% and 1.5%, respectively. This then means that about 98.5% of the ration can be interchanged to fit protein and price specifications. If the energy sources average 10% protein and the protein feeds average 45% protein, the manufacturer can balance the classes of feeds using the square method. Since 98.5% of the 14% protein ration is to contain all of the proteins, this portion of the ration must contain 14.2% $\left[\frac{14}{.985}\right]$ protein.

Step 1. Energy feeds 10% — 14.2 — 30.8 = 88%
Protein feeds 45% — 4.2 = 12%
35.0

Step 2. Energy feeds to be incorporated in the ration.

100 × .88 × .985 = 86.7%

Protein feeds to be incorporated in the ration.

100 × .12 × .985 = 11.8%

4. Once the main nutrient classifications have been established, it is possible to subdivide the amounts of the various feedstuffs to be used. For example, we can divide the protein portions into plant and animal sources using the square method again.

Step 1. We know that 11.8% of the ration will be protein feeds and that the average protein level of this fraction will be 45%. Therefore, 100 × .118 × .45 = 5.31 lb of protein will be supplied by this factor for every 100 lb of feed.

Step 2. Plant protein 42% — 5.31 — 54.7 = 59.8%
Animal protein 60% — 36.7 = 40.1%
91.4

Step 3. Plant proteins to be incorporated into the ration.

100 × .598 × .118 = 7.06%.

Animal proteins to be incorporated into the ration.

100 × .401 × .118 = 4.73%.

5. Once the proportions of the nutrient classifications have been established, the feed manufacturer can then place minimum or maximum restrictions on the individual feeds within each classification. As the price fluctuates, feeds can be readily substituted for each other within their own restrictions for use. The feed substitution tables in the chapters devoted to the respective livestock species, Chapters 19 through 28, can be used as guides for setting restrictions.

Flexible formulas are extremely useful in ration formulation. No one ration is best and, if substitutions are made wisely, the prices of feed can be minimized and the feeder will continually obtain good results. Common sense is an invaluable tool when using flexible formulas because the feed manufacturer must always keep in mind the factors which cannot always be quantitated, such as palatability and bulkiness of feed.

FORMULATION WORKSHEET

When formulating rations, it is advisable to record the ration on a worksheet similar to that in Fig. 18–13. This worksheet is merely an example of the format that should be used. A similar sheet can be developed for micronutrient composition of premixes for minerals and vitamins as well as for amino acids. In modern feed formulation practice, computer printouts provide convenient worksheets. The worksheet serves three purposes:

1. It provides a means of reviewing and double checking the calculations used to formulate the ration. If there is a gross error, it will become obvious when listed on the worksheet.

2. It can be used to organize mixing procedures. It is vital that the person mixing feed be able to refer to a worksheet on which can be recorded what has been mixed and what mixing order should be followed.

3. The worksheet can be filed for future reference. If any questions should arise when the feed is fed, the worksheet provides an orderly record of the content of the feed and its mixing.

Each ration should be assigned a number for future reference, and the date of formulation and/or mixing should be recorded. In addition to listing the feed ingredients and their respective amounts, the nutrient requirements to be fulfilled by the ration should be listed on the worksheet immediately below the totals of various components of the feed. By subtracting the totals contained in the feed from the nutrient requirements, the person formulating the ration can then determine if there are any severe excesses or deficiencies in the ration.

MACRONUTRIENT WORKSHEET

Ration Number: **Date:**

			Proximate Analysis					Minerals and Vitamins		
Ingredient	✓ if mixed	Amount	Crude Fiber	Ether Extract (Fat)	N-Free Extract	Crude Protein	Energy	Calcium (Ca)	Phosphorus (P)	Vitamin A
		(lb)	(lb)	(lb)	(lb)	(lb)	TDN = lb ME = Mcal NE = Mcal	(lb)	(lb)	(IU)
TOTAL										
NUTRIENT REQUIREMENTS										
NUTRIENT BALANCE (Total–Nutrient Requirements)										

Fig. 18–13. Formulation worksheet for macronutrients.

HOW TO APPLY THE NET ENERGY METHOD

In order to apply the net energy method to the feeding of livestock, the following net energy values must be available:

1. A table showing the net energy requirements of the particular class of animal. Chapter 19, Table 19-3, Daily Nutrient Requirements of Growing and Finishing Cattle, shows the net energy requirements for growing-finishing beef cattle (in megacalories [Mcal] per animal per day), with a breakdown into steers, bulls, and heifers.

2. A table showing the nutrient composition of feeds, with the net energy of each feed partitioned into energy used for body maintenance and for gain; thus, the net energy values in megacalories (Mcal) per unit (lb or kg) are needed for each feed for maintenance (NE_m) and for gain (NE_g)(see Section V—Composition of Feeds).

The two examples that follow will show how to apply the net energy method. In the first example, net energy values of feeds are used to calculate the number of pounds of a given ration that a steer would need to consume to make a specified daily gain. In the second example, net energy is used to predict average daily gain based on consuming a certain number of pounds of a specified ration. Bear in mind that the ration in both cases (in these examples, the ration in Table 18-5) must be balanced for protein, minerals, and vitamins, in order for these net energy values to have validity for calculating daily consumption and predicting average daily gain.

Example 1. *Using net energy values to calculate the number of pounds of the ration that must be consumed to produce a specific gain*—How many calories would a 770-lb medium-frame steer calf need to consume to gain 2.6 lb daily?

Step 1. Calculate the net energy for maintenance (NE_m) and gain (NE_g) values for a pound of the ration shown in Table 18-5.

By referring to Section V—Composition of Feeds, it is determined that l lb of the Table 18-5 ration supplies 0.7620 megacalories of net energy for maintenance (Mcal NE_m) and 0.5015 megacalories of net energy for gain (Mcal NE_g).

Step 2. From Chapter 19, Table 19-3 of this book, we find that the requirement for a 770-lb medium-frame steer calf to gain 2.6 lb daily is as follows:

	Mcal/day
NE_m	6.24
NE_g	5.50

Step 3. Pounds of feed to meet the daily maintenance requirement:

6.24 Mcal ÷ .7620 Mcal = 8.19 lb

Step 4. Pounds of feed to meet the requirement for 2.6 lb daily gain:

5.50 Mcal ÷ .5015 Mcal = 10.97 lb

Step 5. Total pounds of feed that the steer calf must eat daily to gain 2.6 lb:

8.19 lb + 10.97 lb = 19.16 lb

Example 2. *Using net energy to predict the average daily gain of a 770-lb medium-frame steer calf that is consuming a certain number of pounds of a specified ration*—Let's assume that we have a 770-lb steer that is consuming 18 lb of the ration shown in Table 18-5. What daily gain should be expected?

TABLE 18-5
RATION FOR FINISHING CATTLE

Ration Ingredient			Composition of Ingredients (as-fed basis) NE_m[1]		Ration Supplies NE_m[1]	Composition of Ingredients (as-fed basis) NE_g[2]		Ration Supplies NE_g[2]
	(lb)	*(kg)*	(Mcal/lb)[3]	*(Mcal/kg)*[3]	(Mcal)[3]	(Mcal/lb)[3]	*(Mcal/kg)*[3]	(Mcal)[3]
Shelled corn, all analyses	68.60	*31.14*	0.86	*1.90*	59.00[4]	0.59	*1.89*	40.47[5]
Soybean meal (solvent), 44%	4.00	*1.82*	0.79	*1.74*	3.16	0.53	*1.17*	2.12
Alfalfa hay (mid-bloom)	27.00	*12.26*	0.52	*1.15*	14.04	0.28	*0.62*	7.56
Salt	0.40	*0.18*	—	—	—	—	—	—
Total	100.00	*45.4*	—	—	76.20	—	—	50.15

[1]NE_m = net energy for maintenance. [2]NE_g = net energy for gain. [3]Mcal stands for megacalorie. [4]68.60 lb × 0.86 = 59.00 [5]68.60 lb × 0.59 Mcal = 40.47

Step 1. Pounds of feed to meet the daily maintenance requirement = 8.19 lb (see prior example).

Step 2. Pounds of feed left for gain:

18 lb − 8.19 lb = 9.81 lb

Step 3. Mcal of NE_g supplied by remaining feed:

9.81 lb × .5015 Mcal = 4.92 Mcal

Step 4. Daily gain expected from 4.92 Mcal of NE_g (Table 19-3):

4.51 Mcal produces 2.2 lb gain

Therefore, 4.92 Mcal will produce 2.40 lb daily gain

$$\left[\frac{4.92\ (2.2)}{4.51}\right]$$

QUESTIONS FOR STUDY AND DISCUSSION

1. What are feeding standards? How are they expressed?

2. Briefly trace the development of feeding standards.

3. What are the National Research Council feeding standards? Of what value are they? How do the NRC feeding standards compare with the feeding standards of other countries?

4. What are the limitations of feeding standards?

5. Define "balanced ration."

6. Several factors should be considered before a ration is balanced. List them. Why should special consideration be given to the amount and kind of fiber, including ADF and NDF, for lactating cows?

7. What major points should be considered before attempting to balance a ration?

8. Outline the four steps involved in formulating an economical ration.

9. In order to feed livestock efficiently and economically, one must understand thoroughly the nutrients furnished by the available feeds, the extent to which livestock can utilize each feed, and the actual feeding value of these feeds. This can be accomplished only through careful and thorough study of the different feeds.

The following exercises are designed to acquaint the student with the feeds commonly fed to ruminants.

a. **A study of available forages.** It is well known that a relatively large portion of the feed consumed by ruminants is used in meeting the energy needs. Thus, a convenient and reasonably accurate way of determining which forages are most economical under the conditions existing in a particular area at any given time is to compute the cost at which each of the available forages furnishes 100 lb of total digestible nutrients. This is a measure of the economy with which the various feeds furnish fuel or energy. Refer to Section V—Composition of Feeds in this book for analyses, and obtain prices of available forages locally. Then fill out the following table of "Available Roughages":

AVAILABLE ROUGHAGES (AS-FED)

Feed	Farm Price Per Ton	TDN Per 100 Lb	Cost Per 100 Lb TDN	Total Protein Per 100 Lb	Calcium Content	Phos-phorus Content	Carotene Mg Per Lb
					(%)	(%)	

b. **A study of available grains and by-products.** At least one of the cereal grains is grown in every section of the country, and all of them are used quite widely as feeds. As a group, the cereals and their by-products are high in energy. However, they possess certain nutritive deficiencies which may prove to be quite limiting if they are not properly used. Refer to Section V—Composition of Feeds in this book for analyses, and obtain prices of available grain and by-product feeds locally. Then fill out the following table of "Available Grain and By-product Feeds":

AVAILABLE GRAIN AND BY-PRODUCT FEEDS (AS-FED)

Feed	Retail Price Per Cwt	TDN Per 100 Lb	Cost Per 100 Lb TDN	Crude Protein Per 100 Lb	Cost Per 100 Lb Crude Protein	Calcium Content	Phos-phorus Content	Carotene Mg Per Lb
						(%)	(%)	

Also, in studying the by-product feeds, be sure to understand the source of the feed and just what part of the original grain or seed goes into the by-product.

10. Why should consideration be given to the moisture content of ingredients when buying feeds and formulating rations? What are the advantages in formulating a ration on a moisture-free basis?

11. List five methods of balancing rations.

12. Using the square method, and corn (8.9% protein) and a protein supplement (44% protein), balance out a ration for 6-week-old broilers from the standpoint of protein content (refer to Chapter 24, Feeding Poultry, for requirements).

13. A farmer is feeding some hay and silage to cattle but calculates that a corn and soybean meal supplement must also be supplied. The TDN that must be supplied by the supplement is 5.8 lb/cow/day. The protein that must be supplied is 1.7 lb/cow/day. The TDN content of the No. 2 corn (as-fed) is 80% and the protein content is 8.9%. The TDN content of the soybean meal (as-fed) has been estimated at 76% and the protein content 44.4%. How much corn and soybean meal should be given to each cow? To answer the question, use one of the 5 methods of balancing rations presented in this chapter.

14. Select a cow of a certain body weight and milk production, then balance a ration for her from the standpoint of TDN and protein, using available feeds and the trial-and-error method. See Chapter 20, Feeding Dairy Cattle, Table 20-5 for requirements; and see Section V, Composition of Feeds, for the content of the feeds selected.

15. Select a specific class of sheep and prepare a balanced ration, using those feeds that are available at the lowest cost. See Chapter 21, Feeding Sheep, for requirements; and see Section V, Composition of Feeds, for the content of the feeds selected.

16. Select a specific class of swine and formulate a balanced ration using those feeds that are available at the lowest cost. Present the ration in terms of both as-fed and moisture-free bases. See Chapter 23, Feeding Swine, for requirements; and see Section V, Composition of Feeds, for the content of the feeds selected.

17. Will the *least-cost* ration always make for the greatest net returns?

18. Why are the nutritionist and the producer necessary when rations are formulated by computers?

19. What is a *spreadsheet?*

20. List the various steps involved in formulating rations by the computer.

21. What is "computer software and hardware"? How would you go about selecting computer software and hardware?

22. What is a flexible formula? How does it work?

23. Design a worksheet for formulating a mineral and vitamin premix formula. Why is such a worksheet necessary?

24. Using the net energy method, formulate a ration for a 500–lb medium-frame steer calf, which is gaining 2 lb/day.

SELECTED REFERENCES

Title of Publication	Author(s)	Publisher
Animal Nutrition, 7th Edition	L. A. Maynard, *et al.*	McGraw-Hill Book Co., New York, N.Y., 1979
Evaluation of Feeds Through Digestibility Experiments, The	B. H. Schneider W. L. Flatt	The University of Georgia Press, Athens, Ga., 1975
Feed Formulations, 3rd Edition	T. W. Perry	The Interstate Printers & Publishers, Inc., Danville, Ill., 1982
Feeds and Feeding	A. Cullison	Reston Publishing Co., Inc., Reston, Va., 1979
Feeds and Feeding, 22nd Edition	F. B. Morrison	Morrison Publishing Co., Ithaca, N.Y., 1956
Feeds and Feeding, Abridged	F. B. Morrison	Morrison Publishing Co., Ithaca, N.Y., 1961
Linear Programming Applications to Agriculture	R. R. Beneke R. Winterboer	Iowa State University Press, Ames, Iowa, 1973
Livestock Feeds and Feeding, 2nd Edition	D. C. Church	O & B Books, Inc., Corvallis, Ore., 1984
Microcomputing in Agriculture	J. Legacy T. Stilt F. Reneau	Reston Publishing Co., Inc., Reston, Va., 1984
Nutrient Requirements of Beef Cattle, 6th Revised Edition	National Research Council	National Academy Press, Washington, D.C., 1984
Nutrient Requirements of Coldwater Fishes	National Research Council	National Academy Press, Washington, D.C., 1981
Nutrient Requirements of Dairy Cattle, 6th Revised Edition, Update 1989	National Research Council	National Academy Press, Washington, D.C., 1989
Nutrient Requirements of Goats	National Research Council	National Academy Press, Washington, D.C., 1981
Nutrient Requirements of Horses, 4th Revised Edition	National Research Council	National Academy Press, Washington, D.C., 1978
Nutrient Requirements of Mink and Foxes, 2nd Revised Edition	National Research Council	National Academy Press, Washington, D.C., 1982
Nutrient Requirements of Poultry, 8th Revised Edition	National Research Council	National Academy Press, Washington, D.C., 1984
Nutrient Requirements of Rabbits, 2nd Revised Edition	National Research Council	National Academy Press, Washington, D.C., 1977
Nutrient Requirements of Sheep, 6th Revised Edition	National Research Council	National Academy Press, Washington, D.C., 1985
Nutrient Requirements of Swine, 9th Revised Edition	National Research Council	National Academy Press, Washington, D.C., 1988
Nutrient Requirements of Warmwater Fishes and Shellfishes, Revised Edition	National Research Council	National Academy Press, Washington, D.C., 1983
Nutrition of Humans & Selected Animal Species	M. L. Scott	John Wiley & Sons, New York, N.Y., 1985
Stockman's Handbook, The, 6th Edition	M. E. Ensminger	The Interstate Printers & Publishers, Inc., Danville, Ill., 1983
Spreadsheet Applications for Animal Nutrition and Feeding	R. J. Lane and T. L. Cross	Reston Publishing Co., Inc., Reston, Va., 1985

Chapter 19

FEEDING BEEF CATTLE[1]

Original painting by Tom Phillips

Contents — Page

Part I—Nutritive Needs of, and Feeds for, Beef Cattle

Part II—Feeding Breeding Beef Cattle

(Continued)

[1]Key sections, tables, and charts in this chapter were authoritatively reviewed by the following beef cattle specialists: D. R. Gill, Ph.D., Department of Animal Science, Oklahoma State University, Stillwater; W. H. Hale, Ph.D., Department of Animal Sciences, The University of Arizona, Tucson; H. W. Newland, Ph.D., Department of Animal Science, The Ohio State University, Columbus; T. W. Perry, Ph.D., Department of Animal Sciences, Purdue University, West Lafayette, Ind.; and L. M. Schake, Ph.D., Department of Animal Science, The University of Connecticut, Storrs.

Contents (Continued) Page

Part III—Feeding Stocker (Feeder) Cattle

Part IV—Feeding Finishing (Fattening) Cattle

Contents Page

Fig. 19-1. Pastures and other roughages, along with water, are the very foundation of beef production. (Courtesy, American Polled Hereford Association, Kansas City, Mo.)

Pastures and range forages, along with other roughages, are the very foundation of successful beef cattle production. In fact, it may be said that the principal function of beef cattle is to harvest vast acreages of forages, and, with or without supplementation, to convert these feeds into more nutritious and palatable products for human consumption. It is estimated (1) that 85.7% of the total feed of beef cattle is derived from roughages, and (2) that 31% of the land area of continental United States is used for grassland pasture and range, with much of this area utilized by beef cattle. If produced on well-fertilized soils, green pasture and well-cured, green, leafy hay can supply all of the nutrient requirements of beef cattle, except the need for common salt and whatever energy-rich feeds may be necessary for additional conditioning or drylot finishing.

In order to cover *Feeding Beef Cattle* with a minimum of repetition and a maximum of cohesiveness, and to enhance readership, the authors opted to present this important subject in one large chapter (rather than four separate chapters), with four parts as follows:

Part I—Nutritive Needs of, and Feeds for, Beef Cattle

Part II—Feeding Breeding Beef Cattle

Part III—Feeding Stocker (Feeder) Cattle

Part IV—Feeding Finishing (Fattening) Cattle

PART I—**Nutritive Needs of, and Feeds for, Beef Cattle**

ECONOMIC IMPORTANCE OF FEED FOR BEEF CATTLE

The feeding of beef cattle constitutes the greatest single cost item of their production. It is important, therefore, that feeding practices be as satisfactory and economical as possible.

Feed affects total profit and cow productivity. It accounts for 65 to 75% of the total cost of keeping cows, and it exerts a powerful influence on cow fertility and calf weaning weight—the two biggest success factors in the cattle business. Without doubt, faulty feeding is a major factor in the nutritional reproductive failure of cows—resulting (1) in only an 88% calf crop (calves born alive or dead), and (2) an average calf death loss of 6% from birth to weaning.

Also, feed is a major item of expense in finishing cattle. It accounts for 70 to 80% of the cost of feedlot finishing, exclusive of the purchase price of the animals.

NUTRITIVE NEEDS OF BEEF CATTLE

The nutritive requirements of beef cattle have become more critical with the shift in beef production practices. Steers were formerly permitted to make their growth primarily on roughages—pastures in the summertime and hay and other forages in the winter. After making moderate and unforced growth for 2 to 4 years, usually the animals were either turned into the feedlot or placed on more lush pastures for a reasonable degree of finishing. With this system, the growth and finishing requirements of cattle came largely at two separate periods in the life of the animal.

Under the old system of moderate growth rate, reasonably good pastures and good-quality hay fully met the nutritive needs. Such older cattle finished quickly when it was desired to ready them for market. Because of their mature size, there was little need for supplemental protein or minerals; and sufficient vitamin A was stored in the liver for the finishing period.

But fashions and sizes of beef cattle shifted radically during the present century—and with them the nutritive requirements changed. Cattle producers moved from the unhurried production and marketing of 2- to 4-year-old steers to the crowding and marketing of calves and yearlings.

Fig. 19-2. The nutritive needs of beef cattle have changed. Larger and heavier milking cows, faster gaining calves, and larger framed growing/finishing cattle make for more critical nutritive requirements. (Courtesy, American Gelbvieh Assn., Denver, Colo.)

In recent years, the introduction of crossbreeding and of the exotic breeds has resulted in heavier milking cows and faster gaining calves. Also, more and more heifers are being bred to calve as 2-year-olds. Hand in hand with these developments, scarce and high-priced grains in the early 1970s provided a preview of world food shortages in the years ahead, punctuated by some years of plenty and other years of scarcity; but, increasingly and relentlessly, with the scarce years dominating the world situation. During periods when grains are in short supply and high in price, feeder cattle will be carried to heavier weights on milk and grass before going into feedlots, then grain fed for a shorter period of time. Conversely, when grains are abundant and a better buy than forages, cattle will be fed more grain and less forage. Also, in the future, (1) the beef industry will breed and feed cattle for the most desirable combination of muscling, marbling, and external finish, and (2) biotechnology will increase production per animal dramatically. In this chapter, provision has been made for the nutritive needs created by these changes.

As feeds represent by far the greatest cost item in beef production, it is important that there be a basic understanding of the nutritive requirements. For convenience, these needs will be discussed under the following categories: (1) National Research Council Requirements, (2) energy, (3) protein, (4) minerals, (5) vitamins, and (6) water.

National Research Council (NRC) Requirements

Efficient beef production cannot be achieved unless nutrient requirements are met, and these are influenced by a number of factors: size (body weight) and reproduction/lactation are especially important in breeding cattle; and body weight and frame size are especially important in growing and finishing cattle.

Although the basic biology of all beef cattle remains the same, differences in rate of maturity and in mature size have a marked influence on the application of the basic nutrition principles to the wide range of environmental and management conditions to which beef cattle are subjected. In recent years, types of beef cattle have changed in response to economic pressures and consumer demand for leaner cuts—larger and faster gaining cattle have evolved. For this reason, frame size has been considered in calculating the nutritive requirements presented in the tables in this section. The medium-frame steer is projected to have 990 to 1,144 lb liveweight at usual market finish, and the medium-frame heifer 800 to 1,045 lb. The finished weight is projected to be over 1,144 and 1,045 lb for large-frame steers and heifers, respectively.

Tables 19-1 to 19-6 were adapted by the authors from *Nutrient Requirements of Beef Cattle,* sixth revised edition, 1984. In using these tables, note the following:

1. **Protein requirements**. These do not provide safety margins.
2. **Nutrient requirements of breeding beef cattle**. Table 19-1 presents the daily nutrient requirements; and Table 19-2 presents the nutrient requirements as a percentage of the ration. Each of these tables presents the nutrient requirements for—
   a. Pregnant yearling heifers—Last third of pregnancy.
   b. Dry pregnant mature cows—Middle third of pregnancy.
   c. Dry pregnant mature cows—Last third of pregnancy.
   d. Two-year-old heifers nursing calves—First 3-4 months postpartum—10 lb milk/day.
   e. Cows nursing calves—Average milking ability—First 3-4 months postpartum—10 lb milk/day.
   f. Cows nursing calves—Superior milking ability—First 3-4 months postpartum—20 lb milk/day.
   g. Bulls, maintenance and slow rate of growth (regain body condition).
3. **Nutrient requirements of growing and finishing cattle.** These are presented in both daily requirements (Table 19-3) and nutrient requirements in the ration (Table 19-4). Each of these tables presents the nutrient requirements for—
   a. Medium-frame steer calves.
   b. Large-frame steer calves, compensating medium-frame yearling steers, and medium-frame bulls. (Medium-frame bulls are listed separately in Table 19-4.)
   c. Large-frame bull calves and compensating large-frame yearling steers.
   d. Medium-frame heifer calves.
   e. Large-frame heifer calves and compensating medium-frame yearling heifers.
4. **Mineral requirements and maximum tolerable levels for beef cattle**. These are presented in Table 19-5.
5. **Maximum tolerable levels of certain toxic elements.** These are presented in Table 19-6.

TABLE 19-1
DAILY NUTRIENT REQUIREMENTS OF BREEDING CATTLE[1] (See footnotes at end of table.)

Weight[2]		Daily Gain[3]		Daily Consumption[4] As-Fed[5]		Daily Consumption[4] Moisture-Free Dry Matter		Energy TDN		Energy ME	Energy NE_m	Energy NE_g	Total Protein		Cal-cium	Phos-phorus	Vitamin A[6]
(lb)	*(kg)*	(lb)	*(kg)*	(lb)	*(kg)*	(lb)	*(kg)*	(lb)	*(kg)*	(Mcal)	(Mcal)	(Mcal)	(lb)	*(kg)*	(g)	(g)	(1,000 IU)
Pregnant yearling heifers—Last third of pregnancy																	
700	*318*	0.9	*0.4*	17.0	*7.7*	15.3	*7.0*	8.5	*3.9*	13.9	7.95	NA[7]	1.3	*0.6*	19	14	19
700	*318*	1.4	*0.6*	17.6	*8.0*	15.8	*7.2*	9.6	*4.4*	15.7	7.95	0.87	1.4	*0.6*	24	15	20
700	*318*	1.9	*0.9*	17.6	*8.0*	15.8	*7.2*	10.6	*4.8*	17.4	7.95	1.89	1.5	*0.7*	27	16	20
750	*341*	0.9	*0.4*	17.9	*8.1*	16.1	*7.3*	8.9	*4.0*	14.6	8.25	NA	1.3	*0.6*	20	14	20
750	*341*	1.4	*0.6*	18.4	*8.4*	16.6	*7.5*	10.0	*4.5*	16.4	8.25	0.92	1.5	*0.7*	24	16	21
750	*341*	1.9	*0.9*	18.4	*8.4*	16.6	*7.5*	11.1	*5.0*	18.2	8.25	1.99	1.6	*0.7*	28	17	21
800	*364*	0.9	*0.4*	18.7	*8.5*	16.8	*7.6*	9.2	*4.2*	15.2	8.56	NA	1.4	*0.6*	21	15	21
800	*364*	1.4	*0.6*	19.3	*8.8*	17.4	*7.9*	10.4	*4.7*	17.1	8.56	0.96	1.5	*0.7*	25	16	22
800	*364*	1.9	*0.9*	19.4	*8.8*	17.5	*8.0*	11.6	*5.3*	19.0	8.56	2.09	1.6	*0.7*	28	17	22
850	*386*	0.9	*0.4*	19.6	*8.9*	17.6	*8.0*	9.6	*4.4*	15.7	8.85	NA	1.4	*0.6*	21	16	22
850	*386*	1.4	*0.6*	20.2	*9.2*	18.2	*8.3*	10.8	*4.9*	17.8	8.85	1.01	1.6	*0.7*	25	17	23
850	*386*	1.9	*0.9*	20.3	*9.2*	18.3	*8.3*	12.1	*5.5*	19.8	8.85	2.19	1.7	*0.8*	28	18	23
900	*409*	0.9	*0.4*	20.3	*9.2*	18.3	*8.3*	9.9	*4.5*	16.3	9.15	NA	1.5	*0.7*	22	17	23
900	*409*	1.4	*0.6*	21.1	*9.6*	19.0	*8.6*	11.3	*5.1*	18.5	9.15	1.05	1.6	*0.7*	26	18	24
900	*409*	1.9	*0.9*	21.3	*9.7*	19.2	*8.7*	12.5	*5.7*	20.6	9.15	2.28	1.7	*0.8*	28	19	24
950	*432*	0.9	*0.4*	21.1	*9.6*	19.0	*8.6*	10.3	*4.7*	16.9	9.44	NA	1.5	*0.7*	23	17	24
950	*432*	1.4	*0.6*	22.0	*10.0*	19.8	*9.0*	11.7	*5.3*	19.1	9.44	1.09	1.7	*0.8*	26	19	25
950	*432*	1.9	*0.9*	22.2	*10.1*	20.0	*9.0*	13.0	*5.9*	21.3	9.44	2.38	1.8	*0.8*	29	19	25
Dry pregnant mature cows—Middle third of pregnancy																	
800	*364*	0.0	*0.0*	17.0	*7.7*	15.3	*7.0*	7.5	*3.4*	12.3	6.41	NA	1.1	*0.5*	12	12	19
900	*409*	0.0	*0.0*	18.6	*8.5*	16.7	*7.6*	8.2	*3.7*	13.4	7.00	NA	1.2	*0.5*	14	14	21
1,000	*454*	0.0	*0.0*	20.1	*9.1*	18.1	*8.2*	8.8	*4.0*	14.5	7.57	NA	1.3	*0.6*	15	15	23
1,100	*500*	0.0	*0.0*	21.7	*9.9*	19.5	*8.9*	9.5	*4.3*	15.6	8.13	NA	1.4	*0.6*	17	17	25
1,200	*545*	0.0	*0.0*	23.1	*10.5*	20.8	*9.5*	10.1	*4.6*	16.6	8.68	NA	1.4	*0.6*	18	18	26
1,300	*591*	0.0	*0.0*	24.4	*11.1*	22.0	*10.0*	10.8	*4.9*	17.7	9.22	NA	1.5	*0.7*	20	20	28
1,400	*636*	0.0	*0.0*	25.9	*11.8*	23.3	*10.6*	11.4	*5.2*	18.7	9.75	NA	1.6	*0.7*	21	21	30
Dry pregnant mature cows—Last third of pregnancy																	
800	*364*	0.9	*0.4*	18.7	*8.5*	16.8	*7.6*	9.2	*4.2*	15.0	8.56	NA	1.4	*0.6*	20	15	21
900	*409*	0.9	*0.4*	20.2	*9.2*	18.2	*8.3*	9.8	*4.5*	16.2	9.15	NA	1.5	*0.7*	22	17	23
1,000	*454*	0.9	*0.4*	21.8	*9.9*	19.6	*8.9*	10.5	*4.8*	17.3	9.72	NA	1.6	*0.7*	23	18	25
1,100	*500*	0.9	*0.4*	23.3	*10.6*	21.0	*9.5*	11.2	*5.1*	18.3	10.28	NA	1.6	*0.7*	25	20	26
1,200	*545*	0.9	*0.4*	24.8	*11.3*	22.3	*10.1*	11.8	*5.4*	19.4	10.83	NA	1.7	*0.8*	26	21	28
1,300	*591*	0.9	*0.4*	26.2	*11.9*	23.6	*10.7*	12.5	*5.7*	20.4	11.37	NA	1.8	*0.8*	28	23	30
1,400	*636*	0.9	*0.4*	27.7	*12.6*	24.9	*11.3*	13.1	*6.0*	21.5	11.90	NA	1.9	*0.9*	29	24	32
Two-year-old heifers nursing calves—First 3–4 months postpartum—10 lb (*4.5 kg*) milk/day																	
700	*318*	0.5	*0.2*	17.7	*8.0*	15.9	*7.2*	10.3	*4.7*	17.0	9.20[8]	0.87	1.8[9]	*0.8*	26	17	28
750	*341*	0.5	*0.2*	18.6	*8.5*	16.7	*7.6*	10.8	*4.9*	17.7	9.51	0.92	1.8	*0.8*	26	18	30
800	*364*	0.5	*0.2*	19.6	*8.9*	17.6	*8.0*	11.2	*5.1*	18.4	9.81	0.96	1.9	*0.9*	27	19	31
850	*386*	0.5	*0.2*	20.4	*9.3*	18.4	*8.4*	11.6	*5.3*	19.1	10.11	1.01	1.9	*0.9*	27	19	33
900	*409*	0.5	*0.2*	21.3	*9.7*	19.2	*8.7*	12.0	*5.5*	19.8	10.40	1.05	2.0	*0.9*	28	20	34
950	*432*	0.5	*0.2*	22.2	*10.1*	20.0	*9.0*	12.5	*5.7*	20.5	10.69	1.09	2.0	*0.9*	28	21	35
1,000	*454*	0.5	*0.2*	23.1	*10.5*	20.8	*9.5*	12.9	*5.9*	21.1	10.98	1.14	2.1	*1.0*	29	22	37
Cows nursing calves—Average milking ability—First 3–4 months postpartum—10 lb (*4.5 kg*) milk/day																	
800	*364*	0.0	*0.0*	19.2	*8.7*	17.3	*7.9*	10.1	*4.6*	16.6	9.81	NA	1.8	*0.8*	23	17	31
900	*409*	0.0	*0.0*	20.1	*9.1*	18.8	*8.5*	10.8	*4.9*	17.7	10.40	NA	1.9	*0.9*	24	19	33
1,000	*454*	0.0	*0.0*	22.4	*10.2*	20.2	*9.2*	11.5	*5.2*	18.8	10.98	NA	2.0	*0.9*	25	20	36
1,100	*500*	0.0	*0.0*	24.0	*10.9*	21.6	*9.8*	12.1	*5.5*	19.9	11.54	NA	2.0	*0.9*	27	22	38
1,200	*545*	0.0	*0.0*	25.6	*11.6*	23.0	*10.5*	12.8	*5.8*	21.0	12.09	NA	2.1	*1.0*	28	23	41
1,300	*591*	0.0	*0.0*	27.0	*12.3*	24.3	*11.0*	13.4	*6.1*	22.0	12.63	NA	2.2	*1.0*	30	25	43
1,400	*636*	0.0	*0.0*	28.4	*12.9*	25.6	*11.6*	14.0	*6.4*	23.0	13.15	NA	2.3	*1.0*	31	26	46
Cows nursing calves—Superior milking ability—First 3–4 months postpartum—20 lb (*9.1 kg*) milk/day																	
800	*364*	0.0	*0.0*	17.4	*7.9*	15.7	*7.1*	12.1	*5.5*	19.9	13.22	NA	2.2	*1.0*	34	22	28
900	*409*	0.0	*0.0*	20.8	*9.5*	18.7	*8.5*	13.1	*6.0*	21.5	13.81	NA	2.4	*1.1*	35	24	33
1,000	*454*	0.0	*0.0*	22.9	*10.4*	20.6	*9.4*	13.8	*6.3*	22.7	14.38	NA	2.5	*1.1*	36	25	37
1,100	*500*	0.0	*0.0*	24.8	*11.3*	22.3	*10.1*	14.5	*6.6*	23.8	14.94	NA	2.6	*1.2*	38	27	40
1,200	*545*	0.0	*0.0*	26.4	*12.0*	23.8	*10.8*	15.2	*6.9*	24.9	15.49	NA	2.7	*1.2*	39	28	42
1,300	*591*	0.0	*0.0*	28.1	*12.8*	25.3	*11.5*	15.9	*7.2*	26.0	16.03	NA	2.8	*1.3*	41	30	45
1,400	*636*	0.0	*0.0*	29.7	*13.5*	26.7	*12.1*	16.5	*7.5*	27.1	16.56	NA	2.9	*1.3*	42	31	47

(Continued)

TABLE 19–1 *(Continued)*

Weight[2] (lb)	*(kg)*	Daily Gain[3] (lb)	*(kg)*	Daily Consumption[4] As-Fed[5] (lb)	*(kg)*	Daily Consumption[4] Moisture-Free Dry Matter (lb)	*(kg)*	Energy TDN (lb)	*(kg)*	Energy ME (Mcal)	Energy NE_m (Mcal)	Energy NE_g (Mcal)	Total Protein (lb)	*(kg)*	Calcium (g)	Phosphorus (g)	Vitamin A[6] (1,000 IU)
Bulls, maintenance and slow rate of growth (regain body condition)																	
<1,300 For growth and development use requirements for bulls in Table 19–3																	
1,300	*591*	1.0	*0.5*	28.2	*12.8*	25.4	*11.5*	14.2	*6.5*	23.3	9.22	2.20	1.9	*0.9*	25	22	45
1,300	*591*	1.5	*0.7*	29.0	*13.2*	26.1	*11.9*	15.6	*7.1*	25.5	9.22	3.43	2.0	*0.9*	28	23	46
1,300	*591*	2.0	*0.9*	29.1	*13.2*	26.2	*11.9*	16.8	*7.6*	27.6	9.22	4.71	2.2	*1.0*	31	24	46
1,400	*636*	1.0	*0.5*	29.8	*13.5*	26.8	*12.2*	15.0	*6.8*	24.6	9.75	2.33	2.0	*0.9*	26	23	48
1,400	*636*	1.5	*0.7*	30.7	*14.0*	27.6	*12.5*	16.5	*7.5*	27.0	9.75	3.63	2.1	*1.0*	29	24	49
1,400	*636*	2.0	*0.9*	30.8	*14.0*	27.7	*12.6*	17.8	*8.1*	29.1	9.75	4.98	2.2	*1.0*	31	25	49
1,500	*682*	0.0	*0.0*	28.0	*12.7*	25.2	*11.5*	12.2	*5.5*	20.0	10.26	NA	1.7	*0.8*	23	23	45
1,500	*682*	1.0	*0.5*	31.4	*14.3*	28.3	*12.9*	15.8	*7.2*	25.9	10.26	2.45	2.1	*1.0*	27	24	50
1,500	*682*	1.5	*0.7*	32.2	*14.6*	29.0	*13.2*	17.3	*7.9*	28.4	10.26	3.82	2.2	*1.0*	29	25	51
1,600	*727*	0.0	*0.0*	29.4	*13.4*	26.5	*12.0*	12.8	*5.8*	21.0	10.77	NA	1.8	*0.8*	23	24	47
1,600	*727*	1.0	*0.5*	33.0	*15.0*	29.7	*13.5*	16.6	*7.5*	27.2	10.77	2.57	2.2	*1.0*	29	26	53
1,600	*727*	1.5	*0.7*	33.8	*15.4*	30.4	*13.8*	18.2	*8.3*	29.8	10.77	4.01	2.3	*1.0*	31	27	54
1,700	*773*	0.0	*0.0*	30.8	*14.0*	27.7	*12.6*	13.4	*6.1*	22.0	11.28	NA	1.9	*0.9*	26	26	49
1,700	*773*	0.5	*0.2*	32.9	*15.0*	29.6	*13.5*	15.4	*7.0*	25.3	11.28	1.26	2.1	*1.0*	27	26	52
1,800	*818*	0.0	*0.0*	32.1	*14.6*	28.9	*13.1*	14.0	*6.4*	23.0	11.77	NA	2.0	*0.9*	27	27	51
1,800	*818*	0.5	*0.2*	34.3	*15.6*	30.9	*14.0*	16.1	*7.3*	26.4	11.77	1.31	2.2	*1.0*	28	28	55
1,900	*864*	0.0	*0.0*	33.4	*15.2*	30.1	*13.7*	14.6	*6.6*	23.9	12.26	NA	2.0	*0.9*	29	29	53
1,900	*864*	0.5	*0.2*	35.8	*16.3*	32.2	*14.6*	16.8	*7.6*	27.5	12.26	1.37	2.2	*1.0*	29	29	57
2,000	*909*	0.0	*0.0*	34.8	*15.8*	31.3	*14.2*	15.2	*6.9*	24.9	12.74	NA	2.1	*1.0*	30	30	55
2,100	*955*	0.0	*0.0*	36.1	*16.4*	32.5	*14.8*	15.7	*7.1*	25.8	13.21	NA	2.2	*1.0*	32	32	58
2,200	*1,000*	0.0	*0.0*	37.3	*17.0*	33.6	*15.3*	16.3	*7.4*	26.7	13.68	NA	2.3	*1.0*	33	33	60

[1]Adapted from *Nutrient Requirements of Beef Cattle,* 6th revised edition, National Research Council—National Academy of Sciences, 1984, p. 85, Table 11.

[2]Average weight for a feeding period.

[3]Approximately 0.9 ± 0.2 lb of weight gain/day over the last third of pregnancy is accounted for by the products of conception. Daily 2.15 Mcal of NE_m and 0.1 lb of protein are provided for this requirement for a calf with a birth weight of 80 lb.

[4]Consumption should vary depending on the energy concentration of the ration and environmental conditions. These intakes are based on the energy concentration shown in the table and assuming a thermoneutral environment without snow or mud conditions. If the energy concentrations of the ration to be fed exceed the tabular value, limit feeding may be required.

[5]As-Fed was calculated using an average figure of 90% dry matter. When using silages, roots, and other wet feeds, these feeds should be converted to a moisture-free basis and the ration calculated using the moisture-free data.

[6]Vitamin A requirements per pound of ration are 1,273 IU for pregnant heifers and cows, and 1,773 IU for lactating cows and breeding bulls.

[7]Not applicable.

[8]Includes 0.34 Mcal NE_m/lb of milk produced for all heifers and cows nursing calves.

[9]Includes 0.03 lb protein/lb of milk produced for all heifers and cows nursing calves.

TABLE 19–2

NUTRIENT REQUIREMENTS IN RATION FOR BREEDING CATTLE[1] (See footnotes at end of table.)

Weight[2] (lb)	*(kg)*	Daily Gain[3] (lb)	*(kg)*	Daily Consumption[4] As-Fed (lb)	*(kg)*	Daily Consumption[4] Moisture-Free Dry Matter (lb)	*(kg)*	Moisture Basis[5] A-F (as-fed) M-F (moisture-free)	Energy TDN (%)	Energy ME (Mcal per) (lb)	*(kg)*	Energy NE_m (Mcal per) (lb)	*(kg)*	Energy NE_g (Mcal per) (lb)	*(kg)*	Total Protein (%)	Calcium (%)	Phosphorus (%)	Vitamin A (IU per) (lb)	*(kg)*
Pregnant yearling heifers—Last third of pregnancy																				
700	*318*	0.9	*0.4*	17.0	*7.7*	15.3	*7.0*	**A-F**	**49.9**	**0.82**	***1.80***	**0.47**	***1.03***	**NA**[6]	***NA***	**7.6**	**0.24**	**0.18**	**1,146**	***2,521***
								M-F	55.4	0.91	*2.00*	0.52	*1.14*	NA	*NA*	8.4	0.27	0.20	1,273	*2,801*
700	*318*	1.4	*0.6*	17.6	*8.0*	15.8	*7.2*	**A-F**	**54.3**	**0.89**	***1.96***	**0.54**	***1.19***	**0.31**	***0.63***	**8.1**	**0.30**	**0.19**	**1,146**	***2,521***
								M-F	60.3	0.99	*2.18*	0.60	*1.32*	0.34	*0.75*	9.0	0.33	0.21	1,273	*2,801*
700	*318*	1.9	*0.9*	17.6	*8.0*	15.8	*7.2*	**A-F**	**60.3**	**0.99**	***2.18***	**0.63**	***1.39***	**0.39**	***0.86***	**8.8**	**0.30**	**0.19**	**1,146**	***2,521***
								M-F	67.0	1.10	*2.42*	0.70	*1.54*	0.43	*0.95*	9.8	0.33	0.21	1,273	*2,801*
750	*341*	0.9	*0.4*	17.9	*8.1*	16.1	*7.3*	**A-F**	**49.6**	**0.84**	***1.78***	**0.47**	***1.03***	**NA**	***NA***	**7.5**	**0.24**	**0.17**	**1,146**	***2,521***
								M-F	55.1	0.90	*1.98*	0.52	*1.14*	NA	*NA*	8.3	0.27	0.19	1,273	*2,801*
750	*341*	1.4	*0.6*	18.4	*8.4*	16.6	*7.5*	**A-F**	**53.9**	**0.88**	***1.94***	**0.54**	***1.19***	**0.30**	***0.66***	**8.0**	**0.29**	**0.19**	**1,146**	***2,521***
								M-F	59.9	0.98	*2.16*	0.60	*1.32*	0.33	*0.73*	8.9	0.32	0.21	1,273	*2,801*
750	*341*	1.9	*0.9*	18.4	*8.4*	16.6	*7.5*	**A-F**	**59.9**	**0.98**	***2.16***	**0.62**	***1.37***	**0.38**	***0.83***	**8.6**	**0.33**	**0.21**	**1,146**	***2,521***
								M-F	66.5	1.09	*2.40*	0.69	*1.52*	0.42	*0.92*	9.5	0.37	0.23	1,273	*2,801*

(Continued)

TABLE 19–2 *(Continued)*

Weight[2]		Daily Gain[3]		Daily Consumption[4] As-Fed		Daily Consumption[4] Moisture-Free Dry Matter		Moisture Basis[5] A-F (as-fed) M-F (moisture-free)	Energy TDN	ME (Mcal per)		NE$_m$ (Mcal per)		NE$_g$ (Mcal per)		Total Protein	Calcium	Phosphorus	Vitamin A (IU per)	
(lb)	*(kg)*	(lb)	*(kg)*	(lb)	*(kg)*	(lb)	*(kg)*		(%)	(lb)	*(kg)*	(lb)	*(kg)*	(lb)	*(kg)*	(%)	(%)	(%)	(lb)	*(kg)*
Pregnant yearling heifers—Last third of pregnancy (Continued)																				
800	*364*	0.9	*0.4*	18.7	*8.5*	16.8	*7.6*	**A-F**	**49.3**	**0.81**	***1.78***	**0.46**	***1.01***	**NA**	***NA***	**7.4**	**0.25**	**0.18**	**1,146**	***2,521***
								M-F	54.8	0.90	*1.98*	0.51	*1.12*	NA	*NA*	8.2	0.28	0.20	1,273	*2,801*
800	*364*	1.4	*0.6*	19.3	*8.8*	17.4	*7.9*	**A-F**	**53.6**	**0.88**	***1.94***	**0.53**	***1.17***	**0.30**	***0.66***	**7.9**	**0.30**	**0.19**	**1,146**	***2,521***
								M-F	59.6	0.98	*2.16*	0.59	*1.30*	0.33	*0.73*	8.8	0.33	0.21	1,273	*2,801*
800	*364*	1.9	*0.9*	19.4	*8.8*	17.5	*8.0*	**A-F**	**59.5**	**0.97**	***2.14***	**0.62**	***1.37***	**0.38**	***0.83***	**8.4**	**0.32**	**0.19**	**1,146**	***2,521***
								M-F	66.1	1.08	*2.38*	0.69	*1.52*	0.42	*0.92*	9.3	0.35	0.21	1,273	*2,801*
850	*386*	0.9	*0.4*	19.6	*8.9*	17.6	*8.0*	**A-F**	**49.1**	**0.80**	***1.76***	**0.46**	***1.01***	**NA**	***NA***	**7.4**	**0.23**	**0.18**	**1,146**	***2,521***
								M-F	54.5	0.89	*1.96*	0.51	*1.12*	NA	*NA*	8.2	0.26	0.20	1,273	*2,801*
850	*386*	1.4	*0.6*	20.2	*9.2*	18.2	*8.3*	**A-F**	**53.4**	**0.87**	***1.92***	**0.53**	***1.17***	**0.29**	***0.63***	**7.7**	**0.27**	**0.19**	**1,146**	***2,521***
								M-F	59.3	0.97	*2.13*	0.59	*1.30*	0.32	*0.70*	8.6	0.30	0.21	1,273	*2,801*
850	*386*	1.9	*0.9*	20.3	*9.3*	18.3	*8.3*	**A-F**	**59.1**	**0.97**	***2.14***	**0.61**	***1.35***	**0.37**	***0.81***	**8.2**	**0.31**	**0.20**	**1,146**	***2,521***
								M-F	65.7	1.08	*2.38*	0.68	*1.50*	0.41	*0.90*	9.1	0.34	0.22	1,273	*2,801*
900	*409*	0.9	*0.4*	20.3	*9.3*	18.3	*8.3*	**A-F**	**48.9**	**0.80**	***1.76***	**0.46**	***1.01***	**NA**	***NA***	**7.3**	**0.23**	**0.18**	**1,146**	***2,521***
								M-F	54.3	0.89	*1.96*	0.51	*1.12*	NA	*NA*	8.1	0.26	0.20	1,273	*2,801*
900	*409*	1.4	*0.6*	21.1	*9.6*	19.0	*8.6*	**A-F**	**53.2**	**0.87**	***1.92***	**0.52**	***1.15***	**0.29**	***0.63***	**7.7**	**0.27**	**0.19**	**1,146**	***2,521***
								M-F	59.1	0.97	*2.13*	0.58	*1.28*	0.32	*0.70*	8.5	0.30	0.21	1,273	*2,801*
900	*409*	1.9	*0.9*	21.3	*9.7*	19.2	*8.7*	**A-F**	**58.9**	**0.96**	***2.12***	**0.61**	***1.35***	**0.37**	***0.81***	**8.1**	**0.29**	**0.19**	**1,146**	***2,521***
								M-F	65.4	1.07	*2.35*	0.68	*1.50*	0.41	*0.90*	9.0	0.32	0.21	1,273	*3,801*
950	*432*	0.9	*0.4*	21.1	*9.6*	19.0	*8.6*	**A-F**	**48.7**	**0.80**	***1.76***	**0.45**	***0.99***	**NA**	***NA***	**7.2**	**0.24**	**0.18**	**1,146**	***2,521***
								M-F	54.1	0.89	*1.96*	0.50	*1.10*	NA	*NA*	8.0	0.27	0.20	1,273	*2,801*
950	*432*	1.4	*0.6*	22.0	*10.0*	19.8	*9.0*	**A-F**	**53.0**	**0.87**	***1.92***	**0.52**	***1.15***	**0.29**	***0.63***	**7.6**	**0.26**	**0.19**	**1,146**	***2,521***
								M-F	58.9	0.97	*2.13*	0.58	*1.28*	0.32	*0.70*	8.4	0.29	0.21	1,273	*2,801*
950	*432*	1.9	*0.9*	22.2	*10.1*	20.0	*9.0*	**A-F**	**58.6**	**0.96**	***2.12***	**0.60**	***1.32***	**0.36**	***0.79***	**7.9**	**0.29**	**0.19**	**1,146**	***2,521***
								M-F	65.1	1.07	*2.35*	0.67	*1.47*	0.40	*0.88*	8.8	0.32	0.21	1,273	*2,801*
Dry pregnant mature cows—Middle third of pregnancy																				
800	*364*	0.0	*0.0*	17.0	*7.7*	15.3	*7.0*	**A-F**	**43.9**	**0.72**	***1.58***	**0.38**	***0.83***	**NA**	***NA***	**6.4**	**0.15**	**0.15**	**1,146**	***2,521***
								M-F	48.8	0.80	*1.76*	0.42	*0.92*	NA	*NA*	7.1	0.17	0.17	1,273	*2,801*
900	*409*	0.0	*0.0*	18.6	*8.5*	16.7	*7.6*	**A-F**	**43.9**	**0.72**	***1.58***	**0.38**	***0.83***	**NA**	***NA***	**6.3**	**0.16**	**0.16**	**1,146**	***2,251***
								M-F	48.8	0.80	*1.76*	0.42	*0.92*	NA	*NA*	7.0	0.18	0.18	1,273	*2,801*
1,000	*454*	0.0	*0.0*	20.1	*9.1*	18.1	*8.2*	**A-F**	**43.9**	**0.72**	***1.58***	**0.38**	***0.83***	**NA**	***NA***	**6.3**	**0.16**	**0.16**	**1,146**	***2,521***
								M-F	43.8	0.80	*1.76*	0.42	*0.92*	NA	*NA*	7.0	0.18	0.18	1,273	*2,801*
1,100	*500*	0.0	*0.0*	21.7	*9.9*	19.5	*8.9*	**A-F**	**43.9**	**0.72**	***1.58***	**0.38**	***0.83***	**NA**	***NA***	**6.3**	**0.17**	**0.17**	**1,146**	***2,521***
								M-F	48.8	0.80	*1.76*	0.42	*0.92*	NA	*NA*	7.0	0.19	0.19	1,273	*2,801*
1,200	*545*	0.0	*0.0*	23.1	*10.5*	20.8	*9.5*	**A-F**	**43.9**	**0.72**	***1.58***	**0.38**	***0.83***	**NA**	***NA***	**6.2**	**0.17**	**0.17**	**1,146**	***2,521***
								M-F	48.8	0.80	*1.76*	0.42	*0.92*	NA	*NA*	6.9	0.19	0.19	1,273	*2,801*
1,300	*591*	0.0	*0.0*	24.4	*11.1*	22.0	*10.0*	**A-F**	**43.9**	**0.72**	***1.58***	**0.38**	***0.83***	**NA**	***NA***	**6.2**	**0.18**	**0.18**	**1,146**	***2,521***
								M-F	48.8	0.80	*1.76*	0.42	*0.92*	NA	*NA*	6.9	0.20	0.20	1,273	*2,801*
1,400	*636*	0.0	*0.0*	25.9	*11.8*	23.3	*10.6*	**A-F**	**43.9**	**0.72**	***1.58***	**0.38**	***0.83***	**NA**	***NA***	**6.2**	**0.18**	**0.18**	**1,146**	***2,521***
								M-F	48.8	0.80	*1.76*	0.42	*0.92*	NA	*NA*	6.9	0.20	0.20	1,273	*2,801*
Dry pregnant mature cows—Last third of pregnancy																				
800	*364*	0.9	*0.4*	18.7	*8.5*	16.8	*7.6*	**A-F**	**49.1**	**0.80**	***1.76***	**0.46**	***1.01***	**NA**	***NA***	**7.4**	**0.23**	**0.18**	**1,146**	***2,521***
								M-F	54.5	0.89	*1.96*	0.51	*1.12*	NA	*NA*	8.2	0.26	0.20	1,273	*2,801*
900	*409*	0.9	*0.4*	20.2	*9.2*	18.2	*8.3*	**A-F**	**48.6**	**0.80**	***1.76***	**0.45**	***0.99***	**NA**	***NA***	**7.2**	**0.24**	**0.19**	**1,146**	***2,521***
								M-F	54.0	0.89	*1.96*	0.50	*1.10*	NA	*NA*	8.0	0.27	0.21	1,273	*2,801*
1,000	*454*	0.9	*0.4*	21.8	*9.9*	19.6	*8.9*	**A-F**	**48.2**	**0.79**	***1.75***	**0.45**	***0.99***	**NA**	***NA***	**7.1**	**0.23**	**0.18**	**1,146**	***2,521***
								M-F	53.6	0.88	*1.94*	0.50	*1.10*	NA	*NA*	7.9	0.26	0.20	1,273	*2,801*
1,100	*500*	0.9	*0.4*	23.3	*10.6*	21.0	*9.5*	**A-F**	**47.9**	**0.78**	***1.72***	**0.44**	***0.97***	**NA**	***NA***	**7.0**	**0.23**	**0.19**	**1,146**	***2,521***
								M-F	53.2	0.87	*1.91*	0.49	*1.08*	NA	*NA*	7.8	0.26	0.21	1,273	*2,801*
1,200	*545*	0.9	*0.4*	24.8	*11.3*	22.3	*10.1*	**A-F**	**47.6**	**0.78**	***1.72***	**0.44**	***0.97***	**NA**	***NA***	**7.0**	**0.23**	**0.19**	**1,146**	***2,521***
								M-F	52.9	0.87	*1.91*	0.49	*1.08*	NA	*NA*	7.8	0.26	0.21	1,273	*2,801*
1,300	*591*	0.9	*0.4*	26.2	*11.9*	23.6	*10.7*	**A-F**	**47.4**	**0.78**	***1.72***	**0.43**	***0.95***	**NA**	***NA***	**6.9**	**0.23**	**0.19**	**1,146**	***2,521***
								M-F	52.7	0.87	*1.91*	0.48	*1.06*	NA	*NA*	7.7	0.26	0.21	1,273	*2,801*
1,400	*636*	0.9	*0.4*	27.7	*12.6*	24.9	*11.3*	**A-F**	**47.3**	**0.77**	***1.70***	**0.43**	***0.95***	**NA**	***NA***	**6.8**	**0.23**	**0.19**	**1,146**	***2,521***
								M-F	52.5	0.86	*1.89*	0.48	*1.06*	NA	*NA*	7.6	0.26	0.21	1,273	*2,801*

(Continued)

TABLE 19–2 *(Continued)*

Weight[2]		Daily Gain[3]		Daily Consumption[4] As-Fed		Daily Consumption[4] Moisture-Free Dry Matter		Moisture Basis[5] A-F (as-fed) M-F (moisture-free)	Energy TDN	Energy ME (Mcal per)		Energy NE_m (Mcal per)		Energy NE_g (Mcal per)		Total Protein	Cal-cium	Phos-phorus	Vitamin A (IU per)	
(lb)	*(kg)*	(lb)	*(kg)*	(lb)	*(kg)*	(lb)	*(kg)*		(%)	(lb)	*(kg)*	(lb)	*(kg)*	(lb)	*(kg)*	(%)	(%)	(%)	(lb)	*(kg)*
Two-year-old heifers nursing calves—First 3–4 months postpartum—10 lb (*4.5 kg*) milk/day																				
700	*318*	0.5	*0.2*	17.7	*8.0*	15.9	*7.2*	**A-F**	**58.6**	**0.96**	***2.12***	**0.60**	***1.32***	**0.36**	***0.79***	**10.2**	**0.32**	**0.22**	**1,596**	***2,521***
								M-F	65.1	1.07	*2.35*	0.67	*1.47*	0.40	*0.88*	11.3	0.36	0.24	1,773	*2,801*
750	*341*	0.5	*0.2*	18.6	*8.5*	16.7	*7.6*	**A-F**	**58.0**	**0.95**	***2.10***	**0.59**	***1.31***	**0.36**	***0.79***	**9.9**	**0.31**	**0.22**	**1,596**	***2,521***
								M-F	64.4	1.06	*2.33*	0.66	*1.45*	0.40	*0.88*	11.0	0.34	0.24	1,773	*2,801*
800	*364*	0.5	*0.2*	19.6	*8.9*	17.6	*8.0*	**A-F**	**57.4**	**0.95**	***2.08***	**0.59**	***1.31***	**0.35**	***0.77***	**9.7**	**0.31**	**0.22**	**1,596**	***2,521***
								M-F	63.8	10.5	*2.31*	0.66	*1.45*	0.39	*0.86*	10.8	0.34	0.24	1,773	*2,801*
850	*386*	0.5	*0.2*	20.4	*9.3*	18.4	*8.4*	**A-F**	**56.9**	**0.94**	***2.06***	**0.59**	***1.29***	**0.34**	***0.76***	**9.5**	**0.30**	**0.21**	**1,596**	***2,521***
								M-F	63.2	1.04	*2.29*	0.65	*1.43*	0.38	*0.84*	10.6	0.33	0.23	1,773	*2,801*
900	*409*	0.5	*0.2*	21.3	*9.7*	19.2	*8.7*	**A-F**	**56.4**	**0.93**	***2.04***	**0.58**	***1.27***	**0.33**	***0.73***	**9.4**	**0.29**	**0.21**	**1,596**	***2,521***
								M-F	62.7	1.03	*2.27*	0.64	*1.41*	0.37	*0.81*	10.4	0.32	0.23	1,773	*2,801*
950	*432*	0.5	*0.2*	22.2	*10.1*	20.0	*9.0*	**A-F**	**56.1**	**0.92**	***2.02***	**0.57**	***1.25***	**0.33**	***0.73***	**9.2**	**0.28**	**0.21**	**1,596**	***2,521***
								M-F	62.3	1.02	*2.24*	0.63	*1.39*	0.37	*0.81*	10.2	0.31	0.23	1,773	*2,801*
1,000	*454*	0.5	*0.2*	23.1	*10.5*	20.8	*9.5*	**A-F**	**55.7**	**0.92**	***2.02***	**0.56**	***1.22***	**0.32**	***0.71***	**9.0**	**0.28**	**0.21**	**1,596**	***2,521***
								M-F	61.9	1.02	*2.24*	0.62	*1.36*	0.36	*0.79*	10.0	0.31	0.23	1,773	*2,801*
Cows nursing calves—Average milking ability—First 3–4 months postpartum—10 lb (*4.5 kg*) milk/day																				
800	*364*	0.0	*0.0*	19.2	*8.7*	17.3	*7.9*	**A-F**	**52.4**	**0.86**	***1.90***	**0.51**	***1.13***	**NA**	***NA***	**9.2**	**0.27**	**0.20**	**1,596**	***3,511***
								M-F	58.2	0.96	*2.11*	0.57	*1.25*	NA	*NA*	10.2	0.30	0.22	1,773	*3,901*
900	*409*	0.0	*0.0*	20.1	*9.1*	18.8	*8.5*	**A-F**	**51.6**	**0.85**	***1.86***	**0.50**	***1.09***	**NA**	***NA***	**8.9**	**0.25**	**0.20**	**1,596**	***3,511***
								M-F	57.3	0.94	*2.07*	0.55	*1.21*	NA	*NA*	9.9	0.28	0.22	1,773	*3,901*
1,000	*454*	0.0	*0.0*	22.4	*10.2*	20.2	*9.2*	**A-F**	**50.9**	**0.84**	***1.85***	**0.50**	***1.09***	**NA**	***NA***	**8.6**	**0.25**	**0.20**	**1,596**	***3,511***
								M-F	56.6	0.93	*2.05*	0.55	*1.21*	NA	*NA*	9.6	0.28	0.22	1,773	*3,901*
1,100	*500*	0.0	*0.0*	24.0	*10.9*	21.6	*9.8*	**A-F**	**50.4**	**0.83**	***1.82***	**0.49**	***1.07***	**NA**	***NA***	**8.5**	**0.24**	**0.20**	**1,596**	***3,511***
								M-F	56.0	0.92	*2.02*	0.54	*1.19*	NA	*NA*	9.4	0.27	0.22	1,773	*3,901*
1,200	*545*	0.0	*0.0*	25.6	*11.6*	23.0	*10.5*	**A-F**	**50.0**	**0.82**	***1.80***	**0.48**	***1.05***	**NA**	***NA***	**8.4**	**0.24**	**0.20**	**1,596**	***3,511***
								M-F	55.5	0.91	*2.00*	0.53	*1.17*	NA	*NA*	9.3	0.27	0.22	1,773	*3,901*
1,300	*591*	0.0	*0.0*	27.0	*12.3*	24.3	*11.0*	**A-F**	**49.6**	**0.81**	***1.78***	**0.47**	***1.03***	**NA**	***NA***	**8.2**	**0.24**	**0.20**	**1,596**	***3,511***
								M-F	55.1	0.90	*1.98*	0.52	*1.14*	NA	*NA*	9.1	0.27	0.22	1,773	*3,901*
1,400	*636*	0.0	*0.0*	28.4	*12.9*	25.6	*11.6*	**A-F**	**49.2**	**0.81**	***1.78***	**0.46**	***1.01***	**NA**	***NA***	**8.1**	**0.24**	**0.20**	**1,596**	***3,511***
								M-F	54.7	0.90	*1.98*	0.51	*1.12*	NA	*NA*	9.0	0.27	0.22	1,773	*3,901*
Cows nursing calves—Superior milking ability—First 3–4 months postpartum—20 lb (*9.1 kg*) milk/day																				
800	*364*	0.0	*0.0*	17.4	*7.9*	15.7	*7.1*	**A-F**	**70.0**	**1.14**	***2.51***	**0.77**	***1.68***	**NA**	***NA***	**12.8**	**0.43**	**0.28**	**1,596**	***3,511***
								M-F	77.3	1.27	*2.79*	0.85	*1.87*	NA	*NA*	14.2	0.48	0.31	1,773	*3,901*
900	*409*	0.0	*0.0*	20.8	*9.5*	18.7	*8.5*	**A-F**	**62.8**	**1.04**	***2.28***	**0.67**	***1.47***	**NA**	***NA***	**11.6**	**0.37**	**0.25**	**1,596**	***3,511***
								M-F	69.8	1.15	*2.53*	0.74	*1.63*	NA	*NA*	12.9	0.41	0.28	1,773	*3,901*
1,000	*454*	0.0	*0.0*	22.9	*10.4*	20.6	*9.4*	**A-F**	**60.3**	**0.99**	***2.18***	**0.63**	***1.39***	**NA**	***NA***	**11.1**	**0.35**	**0.24**	**1,596**	***3,511***
								M-F	67.0	1.10	*2.42*	0.70	*1.54*	NA	*NA*	12.3	0.39	0.27	1,773	*3,901*
1,100	*500*	0.0	*0.0*	24.8	*11.3*	22.3	*10.1*	**A-F**	**58.7**	**0.96**	***2.12***	**0.60**	***1.32***	**NA**	***NA***	**10.7**	**0.34**	**0.24**	**1,596**	***3,511***
								M-F	65.2	1.07	*2.35*	0.67	*1.47*	NA	*NA*	11.9	0.38	0.27	1,773	*3,901*
1,200	*545*	0.0	*0.0*	26.4	*12.0*	23.8	*10.8*	**A-F**	**57.3**	**0.95**	***2.08***	**0.59**	***1.29***	**NA**	***NA***	**10.4**	**0.32**	**0.23**	**1,596**	***3,511***
								M-F	63.7	1.05	*2.31*	0.65	*1.43*	NA	*NA*	11.5	0.36	0.26	1,773	*3,901*
1,300	*591*	0.0	*0.0*	28.1	*12.8*	25.3	*11.5*	**A-F**	**56.3**	**0.93**	***2.04***	**0.58**	***1.27***	**NA**	***NA***	**10.1**	**0.32**	**0.23**	**1,596**	***3,511***
								M-F	62.6	1.03	*2.27*	0.64	*1.41*	NA	*NA*	11.2	0.36	0.26	1,773	*3,901*
1,400	*636*	0.0	*0.0*	29.7	*13.5*	26.7	*12.1*	**A-F**	**55.5**	**0.91**	***2.00***	**0.56**	***1.22***	**NA**	***NA***	**9.9**	**0.32**	**0.23**	**1,596**	***3,511***
								M-F	61.7	1.01	*2.22*	0.62	*1.36*	NA	*NA*	11.0	0.35	0.25	1,773	*3,901*

(Continued)

TABLE 19–2 *(Continued)*

Weight[2]		Daily Gain[3]		Daily Consumption[4] As-Fed		Daily Consumption[4] Moisture-Free Dry Matter		Moisture Basis[5] A-F (as-fed) M-F (moisture-free)	Energy TDN	Energy ME (Mcal per)		Energy NE_m (Mcal per)		Energy NE_g (Mcal per)		Total Protein	Cal-cium	Phos-phorus	Vitamin A (IU per)	
(lb)	(kg)	(lb)	(kg)	(lb)	(kg)	(lb)	(kg)		(%)	(lb)	(kg)	(lb)	(kg)	(lb)	(kg)	(%)	(%)	(%)	(lb)	(kg)
Bulls, maintenance and slow rate of growth (regain body condition)																				
<1,300	For growth and development, use requirements for bulls in Tables 19–3 and 19–4																			
1,300	591	1.0	0.5	28.2	12.8	25.4	11.5	A-F	50.2	0.83	1.82	0.48	1.05	0.25	0.56	6.8	0.20	0.17	1,596	3,511
								M-F	55.8	0.92	2.02	0.53	1.17	0.28	0.62	7.6	0.22	0.19	1,773	3,901
1,300	591	1.5	0.7	29.0	13.2	26.1	11.9	A-F	53.7	0.88	1.94	0.53	1.17	0.30	0.66	7.1	0.22	0.17	1,596	3,511
								M-F	59.7	0.98	2.16	0.59	1.30	0.33	0.73	7.9	0.24	0.19	1,773	3,901
1,300	591	2.0	0.9	29.1	13.2	26.2	11.9	A-F	57.6	0.95	2.08	0.59	1.29	0.35	0.77	7.4	0.23	0.18	1,596	3,511
								M-F	64.0	1.05	2.31	0.65	1.43	0.39	0.86	8.2	0.26	0.20	1,773	3,901
1,400	636	1.0	0.5	29.8	13.5	26.8	12.2	A-F	50.2	0.83	1.82	0.48	1.05	0.25	0.56	6.8	0.19	0.17	1,596	3,511
								M-F	55.8	0.92	2.02	0.53	1.17	0.28	0.62	7.5	0.21	0.19	1,773	3,901
1,400	636	1.5	0.7	30.7	14.0	27.6	12.5	A-F	53.7	0.88	1.94	0.53	1.17	0.30	0.66	6.9	0.21	0.17	1,596	3,511
								M-F	59.7	0.98	2.16	0.59	1.30	0.33	0.73	7.7	0.23	0.19	1,773	3,901
1,400	636	2.0	0.9	30.8	14.0	27.7	12.6	A-F	57.6	0.95	2.08	0.59	1.29	0.35	0.77	7.2	0.23	0.18	1,596	3,511
								M-F	64.0	1.05	2.31	0.65	1.43	0.39	0.86	8.0	0.25	0.20	1,773	3,901
1,500	682	0.0	0.0	28.0	12.7	25.2	11.5	A-F	43.6	0.71	1.57	0.37	0.81	NA	NA	6.2	0.18	0.18	1,596	3,511
								M-F	48.4	0.79	1.74	0.41	0.90	NA	NA	6.9	0.20	0.20	1,773	3,901
1,500	682	1.0	0.5	31.4	14.3	28.3	12.9	A-F	50.2	0.83	1.82	0.48	1.05	0.25	0.56	6.7	0.19	0.17	1,596	3,511
								M-F	55.8	0.92	2.02	0.53	1.17	0.28	0.62	7.4	0.21	0.19	1,773	3,901
1,500	682	1.5	0.7	32.2	14.6	29.0	13.2	A-F	53.7	0.88	1.94	0.53	1.17	0.30	0.66	6.8	0.20	0.17	1,596	3,511
								M-F	59.7	0.98	2.16	0.59	1.30	0.33	0.73	7.6	0.22	0.19	1,773	3,901
1,600	727	0.0	0.0	29.4	13.4	26.5	12.0	A-F	43.6	0.71	1.57	0.37	0.81	NA	NA	6.2	0.17	0.18	1,596	3,511
								M-F	48.4	0.79	1.74	0.41	0.90	NA	NA	6.9	0.19	0.20	1,773	3,901
1,600	727	1.0	0.5	33.0	15.0	29.7	13.5	A-F	50.2	0.83	1.82	0.48	1.05	0.25	0.56	6.6	0.20	0.17	1,596	3,511
								M-F	55.8	0.92	2.02	0.53	1.17	0.28	0.62	7.3	0.22	0.19	1,773	3,901
1,600	727	1.5	0.7	33.8	15.4	30.4	13.8	A-F	53.7	0.88	1.94	0.53	1.17	0.30	0.66	6.7	0.20	0.18	1,596	3,511
								M-F	59.7	0.98	2.16	0.59	1.30	0.33	0.73	7.4	0.22	0.20	1,773	3,901
1,700	773	0.0	0.0	30.8	14.0	27.7	12.6	A-F	43.6	0.71	1.57	0.37	0.81	NA	NA	6.1	0.19	0.19	1,596	3,511
								M-F	48.4	0.79	1.74	0.41	0.90	NA	NA	6.8	0.21	0.21	1,773	3,901
1,700	773	0.5	0.2	32.9	15.0	29.6	13.5	A-F	46.8	0.77	1.63	0.42	0.93	0.20	0.43	6.3	0.18	0.17	1,596	3,511
								M-F	52.0	0.85	1.87	0.47	1.03	0.22	0.48	7.0	0.20	0.19	1,773	3,901
1,800	818	0.0	0.0	32.1	14.6	28.9	13.1	A-F	43.6	0.71	1.57	0.37	0.81	NA	NA	6.1	0.19	0.19	1,596	3,511
								M-F	48.4	0.79	1.74	0.41	0.90	NA	NA	6.8	0.21	0.21	1,773	3,901
1,800	818	0.5	0.2	34.3	15.6	30.9	14.0	A-F	46.8	0.77	1.63	0.42	0.93	0.20	0.43	6.3	0.18	0.18	1,596	3,511
								M-F	52.0	0.85	1.87	0.47	1.03	0.22	0.48	7.0	0.20	0.20	1,773	3,901
1,900	864	0.0	0.0	33.4	15.2	30.1	13.7	A-F	43.6	0.71	1.57	0.37	0.81	NA	NA	6.1	0.19	0.19	1,596	3,511
								M-F	48.4	0.79	1.74	0.41	0.90	NA	NA	6.8	0.21	0.21	1,773	3,901
1,900	864	0.5	0.2	35.8	16.3	32.2	14.6	A-F	46.8	0.77	1.63	0.42	0.93	0.20	0.43	6.2	0.18	0.18	1,596	3,511
								M-F	52.0	0.85	1.87	0.47	1.03	0.22	0.48	6.9	0.20	0.20	1,773	3,901
2,000	909	0.0	0.0	34.8	15.8	31.3	14.2	A-F	43.6	0.71	1.57	0.37	0.81	NA	NA	6.1	0.19	0.19	1,596	3,511
								M-F	48.4	0.79	1.74	0.41	0.90	NA	NA	6.8	0.21	0.21	1,773	3,901
2,100	955	0.0	0.0	36.1	16.4	32.5	14.8	A-F	43.6	0.71	1.57	0.37	0.81	NA	NA	6.1	0.20	0.20	1,596	3,511
								M-F	48.4	0.79	1.74	0.41	0.90	NA	NA	6.8	0.22	0.22	1,773	3,901
2,200	1,000	0.0	0.0	37.3	17.0	33.6	15.3	A-F	43.6	0.71	1.57	0.37	0.81	NA	NA	6.1	0.20	0.20	1,596	3,511
								M-F	48.4	0.79	1.74	0.41	0.90	NA	NA	6.8	0.22	0.22	1,773	3,901

[1]Adapted from *Nutrient Requirements of Beef Cattle,* 6th revised edition, National Research Council—National Academy of Sciences, 1984, p. 85, Table 11.

[2]Average weight for a feeding period.

[3]Approximately 0.9 ± 0.2 lb of weight gain/day over the last third of pregnancy is accounted for by the products of conception. Daily 2.15 Mcal of NE_m and 0.1 lb of protein are provided for this requirement for a calf with a birth weight of 80 lb.

[4]Consumption should vary depending on the energy concentration of the ration and environmental conditions. These intakes are based on the energy concentration shown in the table and assuming a thermoneutral environment without snow or mud conditions. If the energy concentrations of the ration to be fed exceed the tabular value, limit feeding may be required.

[5]As-fed was calculated using an average figure of 90% dry matter. When using silages, roots, and other wet feeds, these feeds should be converted to a moisture-free basis and the ration calculated using the moisture-free data.

[6]Not applicable.

TABLE 19–3
DAILY NUTRIENT REQUIREMENTS OF GROWING AND FINISHING CATTLE[1] [2](See footnotes at end of table.)

Weight		Daily Gain		Energy		Total Protein	Calcium	Phosphorus
				NE_m	NE_g			
(lb)	*(kg)*	(lb)	*(kg)*	(Mcal/day)	(Mcal/day)	(g/day)	(g/day)	(g/day)
				Medium-frame steer calves				
330	*150*	0.4	*0.2*	3.30	0.41	343	11	7
		0.9	*0.4*	3.30	0.87	428	16	9
		1.3	*0.6*	3.30	1.36	503	21	11
		1.8	*0.8*	3.30	1.87	575	27	12
		2.2	*1.0*	3.30	2.39	642	32	14
		2.6	*1.2*	3.30	2.91	702	37	16
440	*200*	0.4	*0.2*	4.10	0.50	399	12	9
		0.9	*0.4*	4.10	1.08	482	17	10
		1.3	*0.6*	4.10	1.69	554	21	12
		1.8	*0.8*	4.10	2.32	621	26	13
		2.2	*1.0*	4.10	2.96	682	31	15
		2.6	*1.2*	4.10	3.62	735	35	16
550	*250*	0.4	*0.2*	4.84	0.60	450	13	10
		0.9	*0.4*	4.84	1.28	532	17	12
		1.3	*0.6*	4.84	2.00	601	21	13
		1.8	*0.8*	4.84	2.74	664	25	14
		2.2	*1.0*	4.84	3.50	720	29	16
		2.6	*1.2*	4.84	4.28	766	33	17
660	*300*	0.4	*0.2*	5.55	0.69	499	14	12
		0.9	*0.4*	5.55	1.47	580	18	13
		1.3	*0.6*	5.55	2.29	646	22	14
		1.8	*0.8*	5.55	3.14	704	25	15
		2.2	*1.0*	5.55	4.02	755	29	16
		2.6	*1.2*	5.55	4.90	794	32	17
770	*350*	0.4	*0.2*	6.24	0.77	545	15	13
		0.9	*0.4*	6.24	1.65	625	19	14
		1.3	*0.6*	6.24	2.57	688	22	15
		1.8	*0.8*	6.24	3.53	743	25	16
		2.2	*1.0*	6.24	4.51	789	28	17
		2.6	*1.2*	6.24	5.50	822	31	18
880	*400*	0.4	*0.2*	6.89	0.85	590	16	15
		0.9	*0.4*	6.89	1.82	668	19	16
		1.3	*0.6*	6.89	2.84	728	22	17
		1.8	*0.8*	6.89	3.90	780	25	17
		2.2	*1.0*	6.89	4.98	821	27	18
		2.6	*1.2*	6.89	6.69	848	29	19
990	*450*	0.4	*0.2*	7.52	0.93	633	17	16
		0.9	*0.4*	7.52	1.99	710	20	17
		1.3	*0.6*	7.52	3.11	767	22	18
		1.8	*0.8*	7.52	4.26	815	24	19
		2.2	*1.0*	7.52	5.44	852	26	19
		2.6	*1.2*	7.52	6.65	873	28	20
1,100	*500*	0.4	*0.2*	8.14	1.01	675	19	18
		0.9	*0.4*	8.14	2.16	751	21	18
		1.3	*0.6*	8.14	3.36	805	23	19
		1.8	*0.8*	8.14	4.61	849	24	20
		2.2	*1.0*	8.14	5.89	882	26	20
		2.6	*1.2*	8.14	7.19	897	27	21
1,210	*550*	0.4	*0.2*	8.75	1.08	715	20	19
		0.9	*0.4*	8.75	2.32	790	22	20
		1.3	*0.6*	8.75	3.61	842	23	20
		1.8	*0.8*	8.75	4.95	883	24	21
		2.2	*1.0*	8.75	6.23	911	25	21
		2.6	*1.2*	8.75	7.73	921	26	21

(Continued)

TABLE 19–3 *(Continued)*

Weight		Daily Gain		Energy		Total Protein	Calcium	Phosphorus
				NE_m	NE_g			
(lb)	*(kg)*	(lb)	*(kg)*	(Mcal/day)	(Mcal/day)	(g/day)	(g/day)	(g/day)
Large-frame steer calves, compensating medium-frame yearling steers, and medium-frame bulls[3]								
330	*150*	0.4	*0.2*	3.30	0.36	361	11	7
		0.9	*0.4*	3.30	0.77	441	17	9
		1.3	*0.6*	3.30	1.21	522	22	11
		1.8	*0.8*	3.30	1.65	598	28	13
		2.2	*1.0*	3.30	2.11	671	33	14
		2.6	*1.2*	3.30	2.58	740	38	16
		3.1	*1.4*	3.30	3.06	806	44	18
		3.5	*1.6*	3.30	3.53	863	49	20
440	*200*	0.4	*0.2*	4.10	0.45	421	12	9
		0.9	*0.4*	4.10	0.96	499	17	10
		1.3	*0.6*	4.10	1.50	576	22	12
		1.8	*0.8*	4.10	2.06	650	27	14
		2.2	*1.0*	4.10	2.62	718	32	15
		2.6	*1.2*	4.10	3.20	782	37	17
		3.1	*1.4*	4.10	3.79	842	42	18
		3.5	*1.6*	4.10	4.39	892	47	20
550	*250*	0.4	*0.2*	4.84	0.53	476	13	10
		0.9	*0.4*	4.84	1.13	552	18	12
		1.3	*0.6*	4.84	1.77	628	23	13
		1.8	*0.8*	4.84	2.43	698	27	15
		2.2	*1.0*	4.84	3.10	762	31	16
		2.6	*1.2*	4.84	3.78	822	36	18
		3.1	*1.4*	4.84	4.48	877	40	19
		3.5	*1.6*	4.84	5.19	919	44	20
660	*300*	0.4	*0.2*	5.55	0.61	529	14	12
		0.9	*0.4*	5.55	1.30	603	19	13
		1.3	*0.6*	5.55	2.03	676	23	15
		1.8	*0.8*	5.55	2.78	743	27	16
		2.2	*1.0*	5.55	3.55	804	31	17
		2.6	*1.2*	5.55	4.34	859	35	18
		3.1	*1.4*	5.55	5.14	908	38	20
		3.5	*1.6*	5.55	5.95	943	42	21
770	*350*	0.4	*0.2*	6.24	0.68	579	16	13
		0.9	*0.4*	6.24	1.46	651	19	15
		1.3	*0.6*	6.24	2.28	722	23	16
		1.8	*0.8*	6.24	3.12	786	27	17
		2.2	*1.0*	6.24	3.99	843	30	18
		2.6	*1.2*	6.24	4.87	895	34	19
		3.1	*1.4*	6.24	5.77	938	37	20
		3.5	*1.6*	6.24	6.68	967	40	21
880	*400*	0.4	*0.2*	6.89	0.75	627	17	15
		0.9	*0.4*	6.89	1.61	697	20	16
		1.3	*0.6*	6.89	2.52	766	24	17
		1.8	*0.8*	6.89	3.45	828	27	18
		2.2	*1.0*	6.89	4.41	881	30	19
		2.6	*1.2*	6.89	5.38	929	33	20
		3.1	*1.4*	6.89	6.38	967	36	21
		3.5	*1.6*	6.89	7.38	989	38	22
990	*450*	0.4	*0.2*	7.52	0.82	673	18	16
		0.9	*0.4*	7.52	1.76	742	21	17
		1.3	*0.6*	7.52	2.75	809	24	18
		1.8	*0.8*	7.52	3.77	867	27	19
		2.2	*1.0*	7.52	4.81	918	29	20
		2.6	*1.2*	7.52	5.88	961	32	21
		3.1	*1.4*	7.52	6.97	995	34	22
		3.5	*1.6*	7.52	8.07	1,011	37	22
1,100	*500*	0.4	*0.2*	8.14	0.89	719	19	18
		0.9	*0.4*	8.14	1.01	785	22	19
		1.3	*0.6*	8.14	2.98	850	24	20
		1.8	*0.8*	8.14	4.08	906	27	20
		2.2	*1.0*	8.14	5.21	953	29	21
		2.6	*1.2*	8.14	6.37	993	31	22
		3.1	*1.4*	8.14	7.54	1,022	33	22
		3.5	*1.6*	8.14	8.73	1,031	35	23

(Continued)

TABLE 19–3 *(Continued)*

Weight		Daily Gain		Energy NE$_m$	Energy NE$_g$	Total Protein	Calcium	Phosphorus
(lb)	*(kg)*	(lb)	*(kg)*	(Mcal/day)	(Mcal/day)	(g/day)	(g/day)	(g/day)
Large-frame steer calves, compensating medium-frame yearling steers, and medium-frame bulls[3] (Continued)								
1,210	*550*	0.4	*0.2*	8.75	0.96	762	20	20
		0.9	*0.4*	8.75	2.05	827	23	20
		1.3	*0.6*	8.75	3.20	890	25	21
		1.8	*0.8*	8.75	4.38	944	27	22
		2.2	*1.0*	8.75	5.60	988	29	22
		2.6	*1.2*	8.75	6.84	1,023	30	23
		3.1	*1.4*	8.75	8.10	1,048	32	23
		3.5	*1.6*	8.75	9.38	1,052	34	24
1,320	*600*	0.4	*0.2*	9.33	1.02	805	22	21
		0.9	*0.4*	9.33	2.19	867	24	22
		1.3	*0.6*	9.33	3.41	930	25	22
		1.8	*0.8*	9.33	4.68	980	27	23
		2.2	*1.0*	9.33	5.98	1,021	28	23
		2.6	*1.2*	9.33	7.30	1,053	30	24
		3.1	*1.4*	9.33	8.64	1,073	31	24
		3.5	*1.6*	9.33	10.01	1,071	32	24
Large-frame bull calves and compensating large-frame yearling steers								
330	*150*	0.4	*0.2*	3.30	0.32	355	11	7
		0.9	*0.4*	3.30	0.69	438	17	9
		1.3	*0.6*	3.30	1.07	519	23	11
		1.8	*0.8*	3.30	1.47	597	28	13
		2.2	*1.0*	3.30	1.87	673	34	15
		2.6	*1.2*	3.30	2.29	745	40	17
		3.1	*1.4*	3.30	2.71	815	45	18
		3.5	*1.6*	3.30	3.14	880	51	20
		4.0	*1.8*	3.30	3.56	922	56	22
440	*200*	0.4	*0.2*	4.10	0.40	414	12	9
		0.9	*0.4*	4.10	0.85	494	18	11
		1.3	*0.6*	4.10	1.33	574	23	12
		1.8	*0.8*	4.10	1.82	649	28	14
		2.2	*1.0*	4.10	2.32	721	34	16
		2.6	*1.2*	4.10	2.84	789	39	17
		3.1	*1.4*	4.10	3.36	854	44	19
		3.5	*1.6*	4.10	3.89	912	49	21
		4.0	*1.8*	4.10	4.43	942	54	22
550	*250*	0.4	*0.2*	4.84	0.47	468	13	10
		0.9	*0.4*	4.84	1.01	547	19	12
		1.3	*0.6*	4.84	1.57	624	23	14
		1.8	*0.8*	4.84	2.15	697	28	15
		2.2	*1.0*	4.84	2.75	765	33	17
		2.6	*1.2*	4.84	3.36	830	38	18
		3.1	*1.4*	4.84	3.97	890	42	20
		3.5	*1.6*	4.84	4.60	943	47	21
		4.0	*1.8*	4.84	5.23	962	51	22
660	*300*	0.4	*0.2*	5.55	0.54	519	15	12
		0.9	*0.4*	5.55	1.15	597	19	13
		1.3	*0.6*	5.55	1.80	672	24	15
		1.8	*0.8*	5.55	2.47	741	28	16
		2.2	*1.0*	5.55	3.15	807	33	18
		2.6	*1.2*	5.55	3.85	868	37	19
		3.1	*1.4*	5.55	4.56	924	41	20
		3.5	*1.6*	5.55	5.28	971	45	22
		4.0	*1.8*	5.55	6.00	980	49	23
770	*350*	0.4	*0.2*	6.24	0.60	568	16	13
		0.9	*0.4*	6.24	1.29	644	20	15
		1.3	*0.6*	6.24	2.02	718	24	16
		1.8	*0.8*	6.24	2.77	795	28	18
		2.2	*1.0*	6.24	3.54	847	32	19
		2.6	*1.2*	6.24	4.32	904	36	20
		3.1	*1.4*	6.24	5.11	956	40	21
		3.5	*1.6*	6.24	5.92	998	44	23
		4.0	*1.8*	6.24	6.74	997	47	23

(Continued)

TABLE 19–3 *(Continued)*

Weight		Daily Gain		Energy		Total Protein	Calcium	Phosphorus
				NE_m	NE_g			
(lb)	*(kg)*	(lb)	*(kg)*	(Mcal/day)	(Mcal/day)	(g/day)	(g/day)	(g/day)
Large-frame bull calves and compensating large-frame yearling steers (Continued)								
880	*400*	0.4	*0.2*	6.89	0.67	615	17	15
		0.9	*0.4*	6.89	1.43	689	21	16
		1.3	*0.6*	6.89	2.23	761	25	18
		1.8	*0.8*	6.89	3.06	826	29	19
		2.2	*1.0*	6.89	3.91	885	32	20
		2.6	*1.2*	6.89	4.77	939	36	21
		3.1	*1.4*	6.89	5.65	986	39	22
		3.5	*1.6*	6.89	6.55	1,024	42	23
		4.0	*1.8*	6.89	7.45	1,013	45	24
990	*450*	0.4	*0.2*	7.52	0.73	661	18	17
		0.9	*0.4*	7.52	1.56	733	22	18
		1.3	*0.6*	7.52	2.44	803	25	19
		1.8	*0.8*	7.52	3.34	866	29	20
		2.2	*1.0*	7.52	4.27	922	32	21
		2.6	*1.2*	7.52	5.21	973	35	22
		3.1	*1.4*	7.52	6.18	1,016	38	23
		3.5	*1.6*	7.52	7.15	1,048	41	24
		4.0	*1.8*	7.52	8.13	1,028	44	25
1,100	*500*	0.4	*0.2*	8.14	0.79	705	20	18
		0.9	*0.4*	8.14	1.69	776	23	19
		1.3	*0.6*	8.14	2.64	844	26	20
		1.8	*0.8*	8.14	3.62	905	29	21
		2.2	*1.0*	8.14	4.62	958	32	22
		2.6	*1.2*	8.14	5.64	1,005	35	23
		3.1	*1.4*	8.14	6.68	1,045	37	24
		3.5	*1.6*	8.14	7.74	1,072	40	25
		4.0	*1.8*	8.14	8.80	1,043	42	25
1,210	*550*	0.4	*0.2*	8.75	0.85	747	21	20
		0.9	*0.4*	8.75	1.82	817	24	21
		1.3	*0.6*	8.75	2.83	884	27	22
		1.8	*0.8*	8.75	3.88	942	29	22
		2.2	*1.0*	8.75	4.96	994	32	23
		2.6	*1.2*	8.75	6.06	1,037	34	24
		3.1	*1.4*	8.75	7.18	1,072	36	25
		3.5	*1.6*	8.75	8.31	1,095	39	25
		4.0	*1.8*	8.75	9.46	1,057	41	26
1,320	*600*	0.4	*0.2*	9.33	0.91	789	22	21
		0.9	*0.4*	9.33	1.94	857	25	22
		1.3	*0.6*	9.33	3.02	923	27	23
		1.8	*0.8*	9.33	4.15	979	30	24
		2.2	*1.0*	9.33	5.30	1,027	32	24
		2.6	*1.2*	9.33	6.47	1,067	34	25
		3.1	*1.4*	9.33	7.66	1,099	36	26
		3.5	*1.6*	9.33	8.87	1,117	38	26
		4.0	*1.8*	9.33	10.10	1,071	39	26
Medium-frame heifer calves								
330	*150*	0.4	*0.2*	3.30	0.49	323	10	7
		0.9	*0.4*	3.30	1.05	409	15	9
		1.3	*0.6*	3.30	1.66	477	20	10
		1.8	*0.8*	3.30	2.29	537	25	12
		2.2	*1.0*	3.30	2.94	562	29	13
440	*200*	0.4	*0.2*	4.10	0.60	374	11	9
		0.9	*0.4*	4.10	1.31	459	16	10
		1.3	*0.6*	4.10	2.06	522	20	11
		1.8	*0.8*	4.10	2.84	574	23	12
		2.2	*1.0*	4.10	3.65	583	27	14
550	*250*	0.4	*0.2*	4.84	0.71	421	12	10
		0.9	*0.4*	4.84	1.55	505	16	11
		1.3	*0.6*	4.84	2.44	563	19	12
		1.8	*0.8*	4.84	3.36	608	23	13
		2.2	*1.0*	4.84	4.31	603	26	14

(Continued)

TABLE 19–3 *(Continued)*

Weight		Daily Gain		Energy		Total Protein	Calcium	Phosphorus
				NE_m	NE_g			
(lb)	*(kg)*	(lb)	*(kg)*	(Mcal/day)	(Mcal/day)	(g/day)	(g/day)	(g/day)
				Medium-frame heifer calves (Continued)				
660	*300*	0.4	*0.2*	5.55	0.82	465	13	11
		0.9	*0.4*	5.55	1.77	549	16	12
		1.3	*0.6*	5.55	2.79	602	19	13
		1.8	*0.8*	5.55	3.85	640	22	14
		2.2	*1.0*	5.55	4.94	621	24	15
770	*350*	0.4	*0.2*	6.24	0.92	508	14	13
		0.9	*0.4*	6.24	1.99	591	17	14
		1.3	*0.6*	6.24	3.13	638	19	14
		1.8	*0.8*	6.24	4.32	670	21	15
		2.2	*1.0*	6.24	5.55	638	23	16
880	*400*	0.4	*0.2*	6.89	1.01	549	16	14
		0.9	*0.4*	6.89	2.20	630	17	15
		1.3	*0.6*	6.89	3.46	674	19	16
		1.8	*0.8*	6.89	4.78	700	20	16
		2.2	*1.0*	6.89	6.14	654	22	16
990	*450*	0.4	*0.2*	7.52	1.11	588	17	16
		0.9	*0.4*	7.52	2.40	669	18	16
		1.3	*0.6*	7.52	3.78	708	19	17
		1.8	*0.8*	7.52	5.22	728	20	17
		2.2	*1.0*	7.52	6.70	670	20	17
1,100	*500*	0.4	*0.2*	8.14	1.20	626	18	17
		0.9	*0.4*	8.14	2.60	706	19	18
		1.3	*0.6*	8.14	4.10	741	19	18
		1.8	*0.8*	8.14	5.65	755	19	18
		2.2	*1.0*	8.14	7.25	685	19	18
1,210	*550*	0.4	*0.2*	8.75	1.29	662	19	19
		0.9	*0.4*	8.75	2.79	742	19	19
		1.3	*0.6*	8.75	4.40	773	19	19
		1.8	*0.8*	8.75	6.07	781	19	19
		2.2	*1.0*	8.75	7.79	700	19	19
		Large-frame heifer calves and compensating medium-frame yearling heifers						
330	*150*	0.4	*0.2*	3.30	0.43	342	11	7
		0.9	*0.4*	3.30	0.93	426	16	9
		1.3	*0.6*	3.30	1.47	500	21	10
		1.8	*0.8*	3.30	2.03	568	26	12
		2.2	*1.0*	3.30	2.61	630	31	14
		2.6	*1.2*	3.30	3.19	680	35	15
440	*200*	0.4	*0.2*	4.10	0.53	397	12	9
		0.9	*0.4*	4.10	1.16	480	16	10
		1.3	*0.6*	4.10	1.83	549	21	12
		1.8	*0.8*	4.10	2.62	613	25	13
		2.2	*1.0*	4.10	3.23	668	29	14
		2.6	*1.2*	4.10	3.97	708	33	16
550	*250*	0.4	*0.2*	4.84	0.63	449	13	10
		0.9	*0.4*	4.84	1.37	530	17	11
		1.3	*0.6*	4.84	2.16	596	21	13
		1.8	*0.8*	4.84	2.98	654	24	14
		2.2	*1.0*	4.84	3.82	703	28	15
		2.6	*1.2*	4.84	4.69	734	31	16
660	*300*	0.4	*0.2*	5.55	0.72	497	14	12
		0.9	*0.4*	5.55	1.57	577	17	13
		1.3	*0.6*	5.55	2.47	639	21	14
		1.8	*0.8*	5.55	3.41	693	24	15
		2.2	*1.0*	5.55	4.38	735	27	16
		2.6	*1.2*	5.55	5.37	758	30	17
770	*350*	0.4	*0.2*	6.24	0.81	543	15	13
		0.9	*0.4*	6.24	1.76	622	18	14
		1.3	*0.6*	6.24	2.78	681	21	15
		1.8	*0.8*	6.24	3.83	730	23	16
		2.2	*1.0*	6.24	4.92	767	26	17
		2.6	*1.2*	6.24	5.03	781	28	17

(Continued)

TABLE 19–3 *(Continued)*

Weight		Daily Gain		Energy		Total Protein	Calcium	Phosphorus
				NE_m	NE_g			
(lb)	*(kg)*	(lb)	*(kg)*	(Mcal/day)	(Mcal/day)	(g/day)	(g/day)	(g/day)
		Large-frame heifer calves and compensating medium-frame yearling heifers (Continued)						
880	*400*	0.4	*0.2*	6.89	0.90	588	16	15
		0.9	*0.4*	6.89	1.95	665	19	15
		1.3	*0.6*	6.89	3.07	721	21	16
		1.8	*0.8*	6.89	4.24	765	23	17
		2.2	*1.0*	6.89	5.44	797	25	18
		2.6	*1.2*	6.89	6.67	803	27	18
990	*450*	0.4	*0.2*	7.52	0.98	631	17	16
		0.9	*0.4*	7.52	2.13	707	19	17
		1.3	*0.6*	7.52	3.35	759	21	17
		1.8	*0.8*	7.52	4.63	799	23	18
		2.2	*1.0*	7.52	5.94	826	24	18
		2.6	*1.2*	7.52	7.28	824	25	19
1,100	*500*	0.4	*0.2*	8.14	1.06	672	18	18
		0.9	*0.4*	8.14	2.31	747	20	18
		1.3	*0.6*	8.14	3.63	796	21	19
		1.8	*0.8*	8.14	5.01	833	22	19
		2.2	*1.0*	8.14	6.43	854	23	19
		2.6	*1.2*	8.14	7.88	844	24	20
1,210	*550*	0.4	*0.2*	8.75	1.14	712	20	19
		0.9	*0.4*	8.75	2.47	787	21	20
		1.3	*0.6*	8.75	3.90	832	22	20
		1.8	*0.8*	8.75	5.38	865	22	20
		2.2	*1.0*	8.75	6.91	881	23	20
		2.6	*1.2*	8.75	8.47	864	23	20
1,320	*600*	0.4	*0.2*	9.33	1.21	751	21	21
		0.9	*0.4*	9.33	2.64	825	22	21
		1.3	*0.6*	9.33	4.16	867	22	21
		1.8	*0.8*	9.33	5.74	896	22	21
		2.2	*1.0*	9.33	7.37	907	22	21
		2.6	*1.2*	9.33	9.03	883	22	21

[1]Adapted from *Nutrient Requirements of Beef Cattle,* 6th revised edition, National Research Council—National Academy of Sciences, 1984, pp. 40–43, Tables 1, 2, and 3.

[2]Shrunk liveweight basis.

[3]In *Nutrient Requirements of Beef Cattle,* 1984, the energy requirements for large-frame steers, compensating medium-frame yearling steers, and medium-frame bulls are listed together. However, the protein requirements for medium-frame bulls are listed separately from the protein requirements for large-frame steers and compensating medium-frame yearling steers, because the protein requirements for medium-frame bulls are somewhat lower than for large-frame steer calves and compensating medium-frame yearling steers. In Table 19–3, however, the protein requirements of all three groups are listed together in order to facilitate use and save space.

TABLE 19–4
NUTRIENT REQUIREMENTS IN RATION FOR GROWING AND FINISHING CATTLE[1 2 3 4] (See footnotes at end of table.)

Weight		Daily Gain		Daily Consumption As-Fed		Daily Consumption Moisture-Free Dry Matter		Moisture Basis[5] A-F (as-fed) M-F (moisture-free)	Energy TDN	Energy ME (Mcal per)		Energy NE_m (Mcal per)		Energy NE_g (Mcal per)		Total Protein	Calcium	Phosphorus
(lb)	*(kg)*	(lb)	*(kg)*	(lb)	*(kg)*	(lb)	*(kg)*		(%)	(lb)	*(kg)*	(lb)	*(kg)*	(lb)	*(kg)*	(%)	(%)	(%)
								Medium-frame steer calves										
300	*136*	0.5	*0.2*	8.7	*4.0*	7.8	*3.5*	**A-F**	**48.6**	**0.80**	***1.76***	**0.45**	***1.00***	**0.23**	***0.50***	**8.6**	**0.28**	**0.18**
								M-F	54.0	0.89	*1.96*	0.50	*1.10*	0.25	*0.55*	9.6	0.31	0.20
		1.0	*0.5*	9.3	*4.2*	8.4	*3.8*	**A-F**	**52.7**	**0.90**	***1.90***	**0.51**	***1.13***	**0.28**	***0.61***	**10.3**	**0.41**	**0.22**
								M-F	58.5	0.96	*2.11*	0.57	*1.25*	0.31	*0.68*	11.4	0.45	0.24
		1.5	*0.7*	9.7	*4.4*	8.7	*4.0*	**A-F**	**56.7**	**0.94**	***2.06***	**0.58**	***1.26***	**0.34**	***0.76***	**11.9**	**0.52**	**0.25**
								M-F	63.0	1.04	*2.29*	0.64	*1.40*	0.38	*0.84*	13.2	0.58	0.28
		2.0	*0.9*	9.9	*4.5*	8.9	*4.0*	**A-F**	**60.8**	**1.00**	***2.20***	**0.63**	***1.39***	**0.40**	***0.87***	**13.3**	**0.65**	**0.29**
								M-F	67.5	1.11	*2.44*	0.70	*1.54*	0.44	*0.97*	14.8	0.72	0.32
		2.5	*1.1*	9.9	*4.5*	8.9	*4.0*	**A-F**	**66.2**	**1.10**	***2.42***	**0.71**	***1.57***	**0.46**	***1.01***	**15.0**	**0.78**	**0.33**
								M-F	73.5	1.21	*2.66*	0.79	*1.74*	0.51	*1.12*	16.7	0.87	0.37
		3.0	*1.4*	8.8	*4.0*	8.0	*3.6*	**A-F**	**76.5**	**1.30**	***2.86***	**0.86**	***1.88***	**0.58**	***1.27***	**18.0**	**1.02**	**0.42**
								M-F	85.0	1.39	*3.06*	0.95	*2.09*	0.64	*1.41*	19.9	1.13	0.47
400	*182*	0.5	*0.2*	10.8	*4.9*	9.7	*4.4*	**A-F**	**48.6**	**0.80**	***1.76***	**0.45**	***1.00***	**0.23**	***0.50***	**8.0**	**0.24**	**0.16**
								M-F	54.0	0.89	*1.96*	0.50	*1.10*	0.25	*0.55*	8.9	0.27	0.18
		1.0	*0.5*	11.6	*5.3*	10.4	*4.7*	**A-F**	**52.7**	**0.90**	***1.90***	**0.51**	***1.13***	**0.28**	***0.61***	**9.3**	**0.34**	**0.19**
								M-F	58.5	0.96	*2.11*	0.57	*1.25*	0.31	*0.68*	10.3	0.38	0.21
		1.5	*0.7*	9.7	*4.4*	10.8	*4.9*	**A-F**	**56.7**	**0.94**	***2.06***	**0.58**	***1.26***	**0.34**	***0.76***	**10.4**	**0.42**	**0.22**
								M-F	63.0	1.04	*2.29*	0.64	*1.40*	0.38	*0.84*	11.5	0.47	0.25
		2.0	*0.9*	12.2	*5.5*	11.0	*5.0*	**A-F**	**60.8**	**1.00**	***2.20***	**0.63**	***1.39***	**0.40**	***0.87***	**11.4**	**0.50**	**0.23**
								M-F	67.5	1.11	*2.44*	0.70	*1.54*	0.44	*0.97*	12.7	0.56	0.26
		2.5	*1.1*	12.2	*5.5*	11.0	*5.0*	**A-F**	**66.2**	**1.10**	***2.42***	**0.71**	***1.57***	**0.46**	***1.01***	**12.8**	**0.61**	**0.27**
								M-F	73.5	1.21	*2.66*	0.79	*1.74*	0.51	*1.12*	14.2	0.68	0.30
		3.0	*1.4*	11.1	*5.0*	10.0	*4.5*	**A-F**	**76.5**	**1.30**	***2.86***	**0.86**	***1.88***	**0.58**	***1.27***	**14.9**	**0.77**	**0.33**
								M-F	85.0	1.39	*3.06*	0.95	*2.09*	0.64	*1.41*	16.6	0.86	0.37
500	*227*	0.5	*0.2*	12.8	*5.8*	11.5	*5.2*	**A-F**	**48.6**	**0.80**	***1.76***	**0.45**	***1.00***	**0.23**	***0.50***	**7.7**	**0.23**	**0.15**
								M-F	54.0	0.89	*1.96*	0.50	*1.10*	0.25	*0.55*	8.5	0.25	0.17
		1.0	*0.5*	13.7	*6.2*	12.3	*6.0*	**A-F**	**52.7**	**0.90**	***1.90***	**0.51**	***1.13***	**0.28**	***0.61***	**8.6**	**0.29**	**0.18**
								M-F	58.5	0.96	*2.11*	0.57	*1.25*	0.31	*0.68*	9.5	0.32	0.20
		1.5	*0.7*	14.2	*6.5*	12.8	*5.8*	**A-F**	**56.7**	**0.94**	***2.06***	**0.58**	***1.26***	**0.34**	***0.76***	**9.5**	**0.36**	**0.20**
								M-F	63.0	1.04	*2.29*	0.64	*1.40*	0.38	*0.84*	10.5	0.40	0.22
		2.0	*0.9*	14.6	*6.6*	13.1	*6.0*	**A-F**	**60.8**	**1.00**	***2.20***	**0.63**	***1.39***	**0.40**	***0.87***	**10.3**	**0.42**	**0.22**
								M-F	67.5	1.11	*2.44*	0.70	*1.54*	0.44	*0.97*	11.4	0.47	0.24
		2.5	*1.1*	14.4	*6.5*	13.0	*6.0*	**A-F**	**66.2**	**1.10**	***2.42***	**0.71**	***1.57***	**0.46**	***1.01***	**11.3**	**0.50**	**0.24**
								M-F	73.5	1.21	*2.66*	0.79	*1.74*	0.51	*1.12*	12.5	0.56	0.27
		3.0	*1.4*	13.1	*6.0*	11.8	*5.4*	**A-F**	**76.5**	**1.30**	***2.86***	**0.86**	***1.88***	**0.58**	***1.27***	**13.0**	**0.62**	**0.29**
								M-F	85.0	1.39	*3.06*	0.95	*2.09*	0.64	*1.41*	14.4	0.69	0.32
600	*273*	0.5	*0.2*	14.7	*6.7*	13.2	*6.0*	**A-F**	**48.6**	**0.80**	***1.76***	**0.45**	***1.00***	**0.23**	***0.50***	**7.4**	**0.20**	**0.16**
								M-F	54.0	0.89	*1.96*	0.50	*1.10*	0.25	*0.55*	8.2	0.23	0.18
		1.0	*0.5*	15.7	*7.1*	14.1	*6.4*	**A-F**	**52.7**	**0.90**	***1.90***	**0.51**	***1.13***	**0.28**	***0.61***	**8.1**	**0.25**	**0.17**
								M-F	58.5	0.96	*2.11*	0.57	*1.25*	0.31	*0.68*	9.0	0.28	0.19
		1.5	*0.7*	16.3	*7.4*	14.7	*6.7*	**A-F**	**56.7**	**0.94**	***2.06***	**0.58**	***1.26***	**0.34**	***0.76***	**8.8**	**0.31**	**0.19**
								M-F	63.0	1.04	*2.29*	0.64	*1.40*	0.38	*0.84*	9.8	0.35	0.21
		2.0	*0.9*	16.7	*7.6*	15.0	*6.8*	**A-F**	**60.8**	**1.00**	***2.20***	**0.63**	***1.39***	**0.40**	***0.87***	**9.5**	**0.36**	**0.20**
								M-F	67.5	1.11	*2.44*	0.70	*1.54*	0.44	*0.97*	10.5	0.40	0.22
		2.5	*1.1*	16.6	*7.5*	14.9	*6.8*	**A-F**	**66.2**	**1.10**	***2.42***	**0.71**	***1.57***	**0.46**	***1.01***	**10.3**	**0.41**	**0.21**
								M-F	73.5	1.21	*2.66*	0.79	*1.74*	0.51	*1.12*	11.4	0.46	0.24
		3.0	*1.4*	15.0	*6.8*	13.5	*6.1*	**A-F**	**76.5**	**1.30**	***2.86***	**0.86**	***1.88***	**0.58**	***1.27***	**11.6**	**0.51**	**0.26**
								M-F	85.0	1.39	*3.06*	0.95	*2.09*	0.64	*1.41*	12.9	0.57	0.29

(Continued)

TABLE 19–4 *(Continued)*

Weight		Daily Gain		Daily Consumption As-Fed		Daily Consumption Moisture-Free Dry Matter		Moisture Basis[5] A-F (as-fed) M-F (moisture-free)	Energy TDN	Energy ME (Mcal per)		Energy NE$_m$ (Mcal per)		Energy NE$_g$ (Mcal per)		Total Protein	Calcium	Phosphorus
(lb)	*(kg)*	(lb)	*(kg)*	(lb)	*(kg)*	(lb)	*(kg)*		(%)	(lb)	*(kg)*	(lb)	*(kg)*	(lb)	*(kg)*	(%)	(%)	(%)
								Medium-frame steer calves (Continued)										
700	*318*	0.5	*0.2*	16.4	*7.5*	14.8	*6.7*	**A-F**	**48.6**	**0.80**	***1.76***	**0.45**	***1.00***	**0.23**	***0.50***	**7.1**	**0.20**	**0.16**
								M-F	54.0	0.89	*1.96*	0.50	*1.10*	0.25	*0.55*	7.9	0.22	0.18
		1.0	*0.5*	17.6	*8.0*	15.8	*7.2*	**A-F**	**52.7**	**0.90**	***1.90***	**0.51**	***1.13***	**0.28**	***0.61***	**7.7**	**0.24**	**0.16**
								M-F	58.5	0.96	*2.11*	0.57	*1.25*	0.31	*0.68*	8.6	0.27	0.18
		1.5	*0.7*	18.3	*8.3*	16.5	*7.5*	**A-F**	**56.7**	**0.94**	***2.06***	**0.58**	***1.26***	**0.34**	***0.76***	**8.3**	**0.28**	**0.18**
								M-F	63.0	1.04	*2.29*	0.64	*1.40*	0.38	*0.84*	9.2	0.31	0.20
		2.0	*0.9*	18.7	*8.5*	16.8	*7.6*	**A-F**	**60.8**	**1.00**	***2.20***	**0.63**	***1.39***	**0.40**	***0.87***	**8.8**	**0.31**	**0.19**
								M-F	67.5	1.11	*2.44*	0.70	*1.54*	0.44	*0.97*	9.8	0.34	0.21
		2.5	*1.1*	18.6	*8.5*	16.7	*7.6*	**A-F**	**66.2**	**1.10**	***2.42***	**0.71**	***1.57***	**0.46**	***1.01***	**9.5**	**0.36**	**0.20**
								M-F	73.5	1.21	*2.66*	0.79	*1.74*	0.51	*1.12*	10.5	0.40	0.22
		3.0	*1.4*	16.9	*7.7*	15.2	*6.9*	**A-F**	**76.5**	**1.30**	***2.86***	**0.86**	***1.88***	**0.58**	***1.27***	**10.5**	**0.44**	**0.23**
								M-F	85.0	1.39	*3.06*	0.95	*2.09*	0.64	*1.41*	11.7	0.49	0.26
800	*364*	0.5	*0.2*	18.2	*8.3*	16.4	*7.5*	**A-F**	**48.6**	**0.80**	***1.76***	**0.45**	***1.00***	**0.23**	***0.50***	**6.9**	**0.20**	**0.15**
								M-F	54.0	0.89	*1.96*	0.50	*1.10*	0.25	*0.55*	7.7	0.22	0.17
		1.0	*0.5*	19.4	*8.8*	17.5	*8.0*	**A-F**	**52.7**	**0.86**	***1.90***	**0.51**	***1.13***	**0.28**	***0.61***	**7.5**	**0.22**	**0.17**
								M-F	58.5	0.96	*2.11*	0.57	*1.25*	0.31	*0.68*	8.3	0.24	0.19
		1.5	*0.7*	20.2	*9.2*	18.2	*8.3*	**A-F**	**56.7**	**0.94**	***2.06***	**0.58**	***1.27***	**0.34**	***0.76***	**7.9**	**0.25**	**0.17**
								M-F	63.0	1.04	*2.29*	0.64	*1.41*	0.38	*0.84*	8.8	0.28	0.19
		2.0	*0.9*	20.7	*9.4*	18.6	*8.5*	**A-F**	**60.8**	**1.10**	***2.20***	**0.63**	***1.39***	**0.40**	***0.87***	**8.3**	**0.28**	**0.18**
								M-F	67.5	1.11	*2.44*	0.70	*1.54*	0.44	*0.97*	9.2	0.31	0.20
		2.5	*1.1*	20.6	*9.4*	18.5	*8.4*	**A-F**	**66.2**	**1.09**	***2.40***	**0.71**	***1.57***	**0.45**	***1.01***	**8.8**	**0.32**	**0.19**
								M-F	73.5	1.21	*2.66*	0.79	*1.74*	0.51	*1.12*	9.8	0.35	0.21
		3.0	*1.4*	18.7	*8.5*	16.8	*7.6*	**A-F**	**76.5**	**1.25**	***2.75***	**0.86**	***1.88***	**0.58**	***1.26***	**9.7**	**0.38**	**0.23**
								M-F	85.0	1.39	*3.06*	0.95	*2.09*	0.64	*1.40*	10.8	0.42	0.25
900	*409*	0.5	*0.2*	19.9	*9.0*	17.9	*8.1*	**A-F**	**48.6**	**0.80**	***1.76***	**0.45**	***0.90***	**0.23**	***0.50***	**6.8**	**0.19**	**0.16**
								M-F	54.0	0.89	*1.96*	0.50	*1.10*	0.25	*0.55*	7.6	0.21	0.18
		1.0	*0.5*	21.2	*9.6*	19.1	*8.7*	**A-F**	**52.7**	**0.86**	***1.90***	**0.51**	***1.13***	**0.28**	***0.61***	**7.2**	**0.21**	**0.16**
								M-F	58.5	0.96	*2.11*	0.57	*1.25*	0.31	*0.68*	8.0	0.23	0.18
		1.5	*0.7*	22.1	*10.0*	19.9	*9.0*	**A-F**	**56.7**	**0.94**	***2.06***	**0.58**	***1.26***	**0.34**	***0.76***	**7.6**	**0.23**	**0.17**
								M-F	63.0	1.04	*2.29*	0.64	*1.40*	0.38	*0.84*	8.4	0.25	0.19
		2.0	*0.9*	22.6	*10.3*	20.3	*9.2*	**A-F**	**60.8**	**1.10**	***2.20***	**0.63**	***1.39***	**0.40**	***0.87***	**7.9**	**0.25**	**0.18**
								M-F	67.5	1.11	*2.44*	0.70	*1.54*	0.44	*0.97*	8.8	0.28	0.20
		2.5	*1.1*	22.4	*10.2*	20.2	*9.2*	**A-F**	**66.2**	**1.09**	***2.40***	**0.71**	***1.57***	**0.46**	***1.01***	**8.4**	**0.28**	**0.18**
								M-F	73.5	1.21	*2.66*	0.79	*1.74*	0.51	*1.12*	9.3	0.31	0.20
		3.0	*1.4*	20.3	*9.2*	18.3	*8.3*	**A-F**	**76.5**	**1.25**	***2.75***	**0.86**	***1.88***	**0.58**	***1.26***	**9.0**	**0.33**	**0.21**
								M-F	85.0	1.39	*3.06*	0.95	*2.09*	0.64	*1.40*	10.1	0.37	0.23
1,000	*454*	0.5	*0.2*	21.4	*9.7*	19.3	*8.8*	**A-F**	**48.6**	**0.80**	***1.76***	**0.45**	***0.90***	**0.23**	***0.50***	**6.8**	**0.19**	**0.16**
								M-F	54.0	0.89	*1.96*	0.50	*1.10*	0.25	*0.55*	7.5	0.21	0.18
		1.0	*0.5*	23.0	*10.5*	20.7	*9.4*	**A-F**	**32.7**	**0.86**	***1.90***	**0.51**	***1.13***	**0.28**	***0.61***	**7.0**	**0.19**	**0.16**
								M-F	58.5	0.96	*2.11*	0.57	*1.25*	0.31	*0.68*	7.8	0.21	0.18
		1.5	*0.7*	23.9	*10.9*	21.5	*9.8*	**A-F**	**56.7**	**0.94**	***2.06***	**0.58**	***1.27***	**0.34**	***0.76***	**7.3**	**0.22**	**0.16**
								M-F	63.0	1.04	*2.29*	0.64	*1.41*	0.38	*0.84*	8.1	0.24	0.18
		2.0	*0.9*	24.4	*11.1*	22.0	*10.0*	**A-F**	**60.8**	**1.10**	***2.20***	**0.63**	***1.39***	**0.40**	***0.87***	**7.6**	**0.23**	**0.17**
								M-F	67.5	1.11	*2.44*	0.70	*1.54*	0.44	*0.97*	8.4	0.25	0.19
		2.5	*1.1*	24.3	*11.0*	21.9	*10.0*	**A-F**	**66.2**	**1.09**	***2.40***	**0.71**	***1.57***	**0.46**	***1.01***	**7.9**	**0.24**	**0.17**
								M-F	73.5	1.21	*2.66*	0.79	*1.74*	0.51	*1.12*	8.8	0.27	0.19
		3.0	*1.4*	22.0	*10.0*	19.8	*9.0*	**A-F**	**76.5**	**1.25**	***2.75***	**0.86**	***1.88***	**0.58**	***1.26***	**8.6**	**0.29**	**0.20**
								M-F	85.0	1.39	*3.06*	0.95	*2.09*	0.64	*1.40*	9.5	0.32	0.22

(Continued)

TABLE 19–4 *(Continued)*

Weight		Daily Gain		Daily Consumption				Moisture Basis[5] A-F (as-fed) M-F (moisture-free)	Energy							Total Protein	Calcium	Phosphorus
				As-Fed		Moisture-Free Dry Matter			TDN	ME (Mcal per)		NE$_m$ (Mcal per)		NE$_g$ (Mcal per)				
(lb)	(kg)	(lb)	(kg)	(lb)	(kg)	(lb)	(kg)		(%)	(lb)	(kg)	(lb)	(kg)	(lb)	(kg)	(%)	(%)	(%)
								Large-frame steer calves and compensating medium-frame yearling steers										
300	136	0.5	0.2	9.1	4.1	8.2	3.7	A-F	47.3	0.77	1.70	0.43	0.95	0.21	0.50	8.6	0.27	0.17
								M-F	52.5	0.86	1.89	0.48	1.06	0.23	0.51	9.5	0.30	0.19
		1.0	0.5	9.7	4.4	8.7	3.9	A-F	50.4	0.83	1.82	0.49	1.07	0.25	0.56	10.2	0.41	0.21
								M-F	56.0	0.92	2.02	0.54	1.19	0.28	0.62	11.3	0.46	0.23
		1.5	0.7	10.1	4.6	9.1	4.1	A-F	53.6	0.88	1.94	0.53	1.17	0.30	0.66	11.6	0.52	0.24
								M-F	59.5	0.98	2.16	0.59	1.30	0.33	0.73	12.9	0.58	0.27
		2.0	0.9	10.4	4.7	9.4	4.3	A-F	57.2	0.94	2.06	0.58	1.26	0.34	0.76	13.1	0.63	0.27
								M-F	63.5	1.04	2.29	0.64	1.40	0.38	0.84	14.6	0.70	0.30
		2.5	1.1	10.7	4.9	9.6	4.4	A-F	60.8	1.10	2.20	0.63	1.39	0.40	0.87	14.7	0.77	0.31
								M-F	67.5	1.11	2.44	0.70	1.54	0.44	0.97	16.3	0.85	0.34
		3.0	1.4	10.7	4.9	9.6	4.4	A-F	64.8	1.06	2.34	0.69	1.52	0.44	0.97	16.2	0.89	0.35
								M-F	72.0	1.18	2.60	0.77	1.69	0.49	1.08	18.0	0.99	0.39
		3.5	1.6	10.3	4.7	9.3	4.2	A-F	70.7	1.16	2.56	0.77	1.70	0.51	1.13	18.3	1.04	0.41
								M-F	78.5	1.29	2.84	0.86	1.89	0.57	1.25	20.3	1.16	0.45
400	182	0.5	0.2	11.2	5.1	10.1	4.6	A-F	47.3	0.77	1.70	0.43	0.95	0.21	0.46	8.0	0.23	0.15
								M-F	52.5	0.86	1.89	0.48	1.06	0.23	0.51	8.9	0.26	0.17
		1.0	0.5	9.7	4.4	10.8	4.9	A-F	50.4	0.83	1.82	0.49	1.07	0.25	0.56	9.2	0.33	0.18
								M-F	56.0	0.92	2.02	0.54	1.19	0.28	0.62	10.2	0.37	0.20
		1.5	0.7	12.6	5.7	11.3	5.1	A-F	53.6	0.88	1.94	0.53	1.17	0.30	0.66	10.3	0.42	0.21
								M-F	59.5	0.98	2.16	0.59	1.30	0.33	0.73	11.4	0.47	0.23
		2.0	0.9	13.0	5.9	11.7	5.3	A-F	57.2	0.94	2.06	0.58	1.26	0.34	0.76	11.4	0.51	0.23
								M-F	63.5	1.04	2.29	0.64	1.40	0.38	0.84	12.7	0.57	0.26
		2.5	1.1	13.2	6.0	11.9	5.4	A-F	60.8	1.10	2.20	0.63	1.39	0.40	0.87	12.5	0.59	0.27
								M-F	67.5	1.11	2.44	0.70	1.54	0.44	0.97	13.9	0.65	0.30
		3.0	1.4	13.2	6.0	11.9	5.4	A-F	64.8	1.06	2.34	0.59	1.52	0.44	0.97	13.7	0.68	0.30
								M-F	72.0	1.18	2.60	0.77	1.69	0.49	1.08	15.2	0.76	0.33
		3.5	1.6	12.8	5.8	11.5	5.2	A-F	70.7	1.16	2.56	0.77	1.70	0.51	1.13	15.2	0.81	0.32
								M-F	78.5	1.29	2.84	0.86	1.89	0.57	1.25	16.9	0.90	0.36
500	227	0.5	0.2	13.3	6.0	12.0	5.5	A-F	47.3	0.77	1.70	0.43	0.95	0.21	0.46	7.7	0.22	0.15
								M-F	52.5	0.86	1.89	0.48	1.06	0.23	0.51	8.5	0.24	0.17
		1.0	0.5	14.2	6.5	12.8	5.8	A-F	50.4	0.83	1.82	0.49	1.07	0.25	0.56	8.6	0.30	0.17
								M-F	56.0	0.92	2.02	0.54	1.19	0.28	0.62	9.5	0.33	0.19
		1.5	0.7	14.9	6.8	13.4	6.1	A-F	53.6	0.88	1.94	0.53	1.17	0.30	0.66	9.4	0.35	0.19
								M-F	59.5	0.98	2.16	0.59	1.30	0.33	0.73	10.4	0.39	0.21
		2.0	0.9	15.3	7.0	13.8	6.3	A-F	57.2	0.94	2.06	0.58	1.27	0.34	0.76	10.3	0.41	0.22
								M-F	63.5	1.04	2.29	0.64	1.41	0.38	0.84	11.4	0.46	0.24
		2.5	1.1	15.6	7.1	14.0	6.4	A-F	60.8	1.00	2.20	0.63	1.39	0.40	0.87	11.2	0.50	0.23
								M-F	67.5	1.11	2.44	0.70	1.54	0.44	0.97	12.4	0.55	0.25
		3.0	1.4	15.6	7.1	14.0	6.4	A-F	64.8	1.06	2.34	0.69	1.52	0.44	0.97	12.1	0.57	0.25
								M-F	72.0	1.18	2.60	0.77	1.69	0.49	1.08	13.4	0.63	0.28
		3.5	1.6	15.1	6.9	13.6	6.2	A-F	70.7	1.16	2.56	0.77	1.70	0.51	1.13	13.2	0.66	0.29
								M-F	78.5	1.29	2.84	0.86	1.89	0.57	1.25	14.7	0.73	0.32
600	273	0.5	0.2	15.3	7.0	13.8	6.3	A-F	47.3	0.77	1.70	0.43	0.95	0.21	0.46	7.4	0.20	0.16
								M-F	52.5	0.86	1.89	0.48	1.06	0.23	0.51	8.2	0.22	0.18
		1.0	0.5	16.2	7.4	14.6	6.6	A-F	50.4	0.83	1.82	0.49	1.07	0.25	0.56	8.1	0.26	0.16
								M-F	56.0	0.92	2.02	0.54	1.19	0.28	0.62	9.0	0.29	0.18
		1.5	0.7	17.0	7.7	15.3	7.0	A-F	53.6	0.88	1.94	0.53	1.17	0.30	0.66	8.7	0.32	0.18
								M-F	59.5	0.98	2.16	0.59	1.30	0.33	0.73	9.7	0.35	0.20
		2.0	0.9	17.6	8.0	15.8	7.2	A-F	57.2	0.94	2.06	0.58	1.27	0.34	0.76	9.5	0.36	0.20
								M-F	63.5	1.04	2.29	0.64	1.41	0.38	0.84	10.5	0.40	0.22
		2.5	1.1	17.9	8.1	16.1	7.3	A-F	60.8	1.00	2.20	0.63	1.39	0.40	0.87	10.2	0.42	0.21
								M-F	67.5	1.11	2.44	0.70	1.54	0.44	0.97	11.3	0.47	0.23

(Continued)

TABLE 19–4 *(Continued)*

Weight		Daily Gain		Daily Consumption				Moisture Basis[5] A-F (as-fed) M-F (moisture-free)	Energy							Total Protein	Calcium	Phosphorus
				As-Fed		Moisture-Free Dry Matter			TDN	ME (Mcal per)		NE_m (Mcal per)		NE_g (Mcal per)				
(lb)	*(kg)*	(lb)	*(kg)*	(lb)	*(kg)*	(lb)	*(kg)*		(%)	(lb)	*(kg)*	(lb)	*(kg)*	(lb)	*(kg)*	(%)	(%)	(%)
Large-frame steer calves and compensating medium-frame yearling steers (Continued)																		
600	*273*	3.0	*1.4*	17.9	*8.1*	16.1	*7.3*	**A-F**	**64.8**	**1.06**	***2.34***	**0.69**	***1.52***	**0.44**	***0.97***	**10.9**	**0.47**	**0.23**
(Continued)								M-F	72.0	1.18	*2.60*	0.77	*1.69*	0.49	*1.08*	12.1	0.52	0.26
		3.5	*1.6*	17.3	*7.9*	15.6	*7.1*	**A-F**	**70.7**	**1.16**	***2.56***	**0.77**	***1.70***	**0.51**	***1.13***	**11.9**	**0.55**	**0.25**
								M-F	78.5	1.29	*2.84*	0.86	*1.89*	0.57	*1.25*	13.2	0.61	0.28
700	*318*	0.5	*0.2*	17.1	*7.8*	15.4	*7.0*	**A-F**	**47.3**	**0.77**	***1.70***	**0.43**	***0.95***	**0.21**	***0.46***	**7.1**	**0.19**	**0.15**
								M-F	52.5	0.86	*1.89*	0.48	*1.06*	0.23	*0.51*	7.9	0.21	0.17
		1.0	*0.5*	18.2	*8.3*	16.4	*7.5*	**A-F**	**50.4**	**0.83**	***1.82***	**0.49**	***1.07***	**0.25**	***0.56***	**7.7**	**0.24**	**0.17**
								M-F	56.0	0.92	*2.02*	0.54	*1.19*	0.28	*0.62*	8.6	0.27	0.19
		1.5	*0.7*	19.1	*8.7*	17.2	*7.8*	**A-F**	**53.6**	**0.88**	***1.94***	**0.53**	***1.17***	**0.30**	***0.66***	**8.3**	**0.28**	**0.17**
								M-F	59.5	0.98	*2.16*	0.59	*1.30*	0.33	*0.73*	9.2	0.31	0.19
		2.0	*0.9*	19.8	*9.0*	17.8	*8.1*	**A-F**	**57.2**	**0.94**	***2.06***	**0.58**	***1.27***	**0.34**	***0.76***	**8.8**	**0.32**	**0.19**
								M-F	63.5	1.04	*2.29*	0.64	*1.41*	0.38	*0.84*	9.8	0.36	0.21
		2.5	*1.1*	20.0	*9.1*	18.0	*8.2*	**A-F**	**60.8**	**1.00**	***2.20***	**0.63**	***1.39***	**0.40**	***0.87***	**9.5**	**0.36**	**0.20**
								M-F	67.5	1.11	*2.44*	0.70	*1.54*	0.44	*0.97*	10.5	0.40	0.22
		3.0	*1.4*	20.0	*9.1*	18.0	*8.2*	**A-F**	**64.8**	**1.06**	***2.34***	**0.69**	***1.52***	**0.44**	***0.97***	**10.0**	**0.41**	**0.21**
								M-F	72.0	1.18	*2.60*	0.77	*1.69*	0.49	*1.08*	11.1	0.45	0.23
		3.5	*1.6*	19.4	*8.8*	17.5	*8.0*	**A-F**	**70.7**	**1.16**	***2.56***	**0.77**	***1.70***	**0.51**	***1.13***	**10.8**	**0.47**	**0.23**
								M-F	78.5	1.29	*2.84*	0.86	*1.89*	0.57	*1.25*	12.0	0.52	0.26
800	*364*	0.5	*0.2*	19.0	*8.6*	17.1	*7.8*	**A-F**	**47.3**	**0.77**	***1.70***	**0.43**	***0.95***	**0.21**	***0.46***	**6.9**	**0.19**	**0.16**
								M-F	52.5	0.86	*1.89*	0.48	*1.06*	0.23	*0.51*	7.7	0.21	0.18
		1.0	*0.5*	20.2	*9.2*	18.2	*8.3*	**A-F**	**50.4**	**0.83**	***1.82***	**0.49**	***1.07***	**0.25**	***0.56***	**7.5**	**0.22**	**0.16**
								M-F	56.0	0.92	*2.02*	0.54	*1.19*	0.28	*0.62*	8.3	0.24	0.18
		1.5	*0.7*	21.1	*9.6*	19.0	*8.6*	**A-F**	**53.6**	**0.88**	***1.94***	**0.53**	***1.17***	**0.30**	***0.66***	**7.9**	**0.25**	**0.17**
								M-F	59.5	0.98	*2.16*	0.59	*1.30*	0.33	*0.73*	8.8	0.28	0.19
		2.0	*0.9*	21.8	*9.9*	19.6	*8.9*	**A-F**	**57.2**	**0.94**	***2.06***	**0.58**	***1.27***	**0.34**	***0.76***	**8.4**	**0.29**	**0.18**
								M-F	63.5	1.04	*2.29*	0.64	*1.41*	0.38	*0.84*	9.3	0.32	0.20
		2.5	*1.1*	22.1	*10.0*	19.9	*9.0*	**A-F**	**60.8**	**1.00**	***2.20***	**0.63**	***1.39***	**0.40**	***0.87***	**8.8**	**0.32**	**0.19**
								M-F	67.5	1.11	*2.44*	0.70	*1.54*	0.44	*0.97*	9.8	0.35	0.21
		3.0	*1.4*	22.1	*10.0*	19.9	*9.0*	**A-F**	**64.8**	**1.06**	***2.34***	**0.69**	***1.52***	**0.44**	***0.97***	**9.4**	**0.36**	**0.20**
								M-F	72.0	1.18	*2.60*	0.77	*1.69*	0.49	*1.08*	10.4	0.40	0.22
		3.5	*1.6*	21.4	*9.7*	19.3	*8.8*	**A-F**	**70.7**	**1.16**	***2.56***	**0.77**	***1.70***	**0.51**	***1.13***	**10.0**	**0.41**	**0.22**
								M-F	78.5	1.29	*2.84*	0.86	*1.89*	0.57	*1.25*	11.1	0.45	0.24
900	*409*	0.5	*0.2*	20.7	*9.4*	18.6	*8.5*	**A-F**	**47.3**	**0.77**	***1.70***	**0.43**	***0.86***	**0.21**	***0.46***	**6.8**	**0.18**	**0.16**
								M-F	52.5	0.86	*1.89*	0.48	*0.96*	0.23	*0.51*	7.6	0.20	0.18
		1.0	*0.5*	22.0	*10.0*	19.8	*9.0*	**A-F**	**50.4**	**0.83**	***1.82***	**0.49**	***1.07***	**0.25**	***0.56***	**7.2**	**0.20**	**0.16**
								M-F	56.0	0.92	*2.02*	0.54	*1.19*	0.28	*0.62*	8.0	0.23	0.18
		1.5	*0.7*	23.1	*10.5*	20.8	*9.5*	**A-F**	**53.6**	**0.88**	***1.94***	**0.53**	***1.17***	**0.30**	***0.66***	**7.7**	**0.24**	**0.16**
								M-F	59.5	0.98	*2.16*	0.59	*1.30*	0.33	*0.73*	8.5	0.27	0.18
		2.0	*0.9*	23.8	*10.8*	21.4	*9.7*	**A-F**	**57.2**	**0.94**	***2.06***	**0.58**	***1.26***	**0.34**	***0.76***	**8.0**	**0.26**	**0.18**
								M-F	63.5	1.04	*2.29*	0.64	*1.40*	0.38	*0.84*	8.9	0.29	0.20
		2.5	*1.1*	24.2	*11.0*	21.8	*9.9*	**A-F**	**60.8**	**1.00**	***2.20***	**0.63**	***1.39***	**0.40**	***0.87***	**8.4**	**0.28**	**0.18**
								M-F	67.5	1.11	*2.44*	0.70	*1.54*	0.44	*0.97*	9.3	0.31	0.20
		3.0	*1.4*	24.1	*11.0*	21.7	*9.9*	**A-F**	**64.8**	**1.06**	***2.34***	**0.69**	***1.52***	**0.44**	***0.97***	**8.8**	**0.32**	**0.19**
								M-F	72.0	1.18	*2.60*	0.77	*1.69*	0.49	*1.08*	9.8	0.36	0.21
		3.5	*1.6*	23.4	*10.6*	21.1	*9.6*	**A-F**	**70.7**	**1.16**	***2.56***	**0.77**	***1.70***	**0.51**	***1.13***	**9.4**	**0.36**	**0.21**
								M-F	78.5	1.29	*2.84*	0.86	*1.89*	0.57	*1.25*	10.4	0.40	0.23
1,000	*454*	0.5	*0.2*	22.4	*10.2*	20.2	*9.2*	**A-F**	**47.3**	**0.77**	***1.70***	**0.43**	***0.86***	**0.21**	***0.46***	**6.8**	**0.18**	**0.15**
								M-F	52.5	0.86	*1.89*	0.48	*0.96*	0.23	*0.51*	7.5	0.20	0.17
		1.0	*0.5*	23.9	*10.9*	21.5	*9.8*	**A-F**	**50.4**	**0.83**	***1.82***	**0.49**	***1.07***	**0.25**	***0.56***	**7.0**	**0.20**	**0.15**
								M-F	56.0	0.92	*2.02*	0.54	*1.19*	0.28	*0.62*	7.8	0.23	0.17
		1.5	*0.7*	25.0	*11.4*	22.5	*10.2*	**A-F**	**53.6**	**0.88**	***1.94***	**0.53**	***1.17***	**0.30**	***0.66***	**7.4**	**0.23**	**0.16**
								M-F	59.5	0.98	*2.16*	0.59	*1.30*	0.33	*0.73*	8.2	0.25	0.18

(Continued)

TABLE 19–4 *(Continued)*

Weight		Daily Gain		Daily Consumption				Moisture Basis[5] A-F (as-fed) M-F (moisture-free)	Energy							Total Protein	Calcium	Phosphorus
				As-Fed		Moisture-Free Dry Matter			TDN	ME (Mcal per)		NE$_m$ (Mcal per)		NE$_g$ (Mcal per)				
(lb)	*(kg)*	(lb)	*(kg)*	(lb)	*(kg)*	(lb)	*(kg)*		(%)	(lb)	*(kg)*	(lb)	*(kg)*	(lb)	*(kg)*	(%)	(%)	(%)
Large-frame steer calves and compensating medium-frame yearling steers (Continued)																		
1,000 (Continued)	*454*	2.0	*0.9*	25.8	*11.7*	23.2	*10.5*	**A-F**	**57.2**	**0.94**	***2.06***	**0.58**	***1.26***	**0.34**	***0.76***	**7.7**	**0.24**	**0.16**
								M-F	63.5	1.04	*2.29*	0.64	*1.40*	0.38	*0.84*	8.6	0.27	0.18
		2.5	*1.1*	26.2	*11.9*	23.6	*10.7*	**A-F**	**60.8**	**1.00**	***2.20***	**0.63**	***1.39***	**0.40**	***0.87***	**8.0**	**0.26**	**0.17**
								M-F	67.5	1.11	*2.44*	0.70	*1.54*	0.44	*0.97*	8.9	0.29	0.19
		3.0	*1.4*	26.2	*11.9*	23.6	*10.7*	**A-F**	**64.8**	**1.06**	***2.34***	**0.69**	***1.52***	**0.44**	***0.97***	**8.4**	**0.29**	**0.18**
								M-F	72.0	1.18	*2.60*	0.77	*1.69*	0.49	*1.08*	9.3	0.32	0.20
		3.5	*1.6*	25.3	*11.5*	22.8	*10.4*	**A-F**	**70.7**	**1.16**	***2.56***	**0.77**	***1.70***	**0.51**	***1.13***	**8.8**	**0.32**	**0.19**
								M-F	78.5	1.29	*2.84*	0.86	*1.89*	0.57	*1.25*	9.8	0.35	0.21
1,100	*500*	0.5	*0.2*	24.1	*11.0*	21.7	*9.9*	**A-F**	**49.3**	**0.77**	***1.70***	**0.43**	***0.86***	**0.21**	***0.46***	**6.7**	**0.17**	**0.16**
								M-F	52.5	0.86	*1.89*	0.48	*0.96*	0.23	*0.51*	7.4	0.19	0.18
		1.0	*0.5*	25.7	*11.7*	23.1	*10.5*	**A-F**	**50.4**	**0.83**	***1.82***	**0.49**	***1.07***	**0.25**	***0.56***	**6.9**	**0.19**	**0.16**
								M-F	56.0	0.92	*2.02*	0.54	*1.19*	0.28	*0.62*	7.7	0.21	0.18
		1.5	*0.7*	26.8	*12.2*	24.1	*11.0*	**A-F**	**53.6**	**0.88**	***1.94***	**0.53**	***1.17***	**0.30**	***0.66***	**7.2**	**0.21**	**0.16**
								M-F	59.5	0.98	*2.16*	0.59	*1.30*	0.33	*0.73*	8.0	0.23	0.18
		2.0	*0.9*	27.7	*12.6*	24.9	*11.3*	**A-F**	**57.2**	**0.94**	***2.06***	**0.58**	***1.26***	**0.34**	***0.76***	**7.5**	**0.23**	**0.16**
								M-F	63.5	1.04	*2.29*	0.64	*1.40*	0.38	*0.84*	8.3	0.25	0.18
		2.5	*1.1*	28.1	*12.8*	25.3	*11.5*	**A-F**	**60.8**	**1.00**	***2.20***	**0.63**	***1.39***	**0.40**	***0.87***	**7.7**	**0.23**	**0.16**
								M-F	67.5	1.11	*2.44*	0.70	*1.54*	0.44	*0.97*	8.5	0.26	0.18
		3.0	*1.4*	28.1	*12.8*	25.3	*11.5*	**A-F**	**64.8**	**1.06**	***2.34***	**0.69**	***1.52***	**0.44**	***0.97***	**8.0**	**0.26**	**0.17**
								M-F	72.0	1.18	*2.60*	0.77	*1.69*	0.49	*1.08*	8.9	0.29	0.19
		3.5	*1.6*	27.2	*12.4*	24.5	*11.1*	**A-F**	**70.7**	**1.16**	***2.56***	**0.77**	***1.70***	**0.51**	***1.13***	**8.4**	**0.29**	**0.19**
								M-F	78.5	1.29	*2.84*	0.86	*1.89*	0.57	*1.25*	9.3	0.32	0.21
Medium-frame bulls																		
300	*136*	0.5	*0.2*	8.7	*4.0*	7.8	*3.5*	**A-F**	**48.2**	**0.79**	***1.75***	**0.44**	***0.97***	**0.22**	***0.48***	**8.7**	**0.28**	**0.18**
								M-F	53.5	0.88	*1.94*	0.49	*1.08*	0.24	*0.53*	9.7	0.31	0.20
		1.0	*0.5*	9.2	*4.2*	8.3	*3.8*	**A-F**	**51.8**	**0.85**	***1.86***	**0.50**	***1.11***	**0.27**	***0.59***	**10.4**	**0.43**	**0.22**
								M-F	57.5	0.94	*2.07*	0.56	*1.23*	0.30	*0.66*	11.6	0.48	0.24
		1.5	*0.7*	9.6	*4.4*	8.6	*3.9*	**A-F**	**55.4**	**0.91**	***2.00***	**0.56**	***1.22***	**0.32**	***0.69***	**12.1**	**0.56**	**0.25**
								M-F	61.5	1.01	*2.22*	0.62	*1.36*	0.35	*0.77*	13.4	0.62	0.28
		2.0	*0.9*	9.8	*4.5*	8.8	*4.0*	**A-F**	**59.0**	**0.97**	***2.14***	**0.61**	***1.35***	**0.40**	***0.81***	**13.7**	**0.68**	**0.30**
								M-F	65.5	1.08	*2.38*	0.68	*1.50*	0.41	*0.90*	15.2	0.75	0.33
		2.5	*1.1*	9.9	*4.5*	8.9	*4.0*	**A-F**	**63.0**	**1.04**	***2.28***	**0.67**	***1.47***	**0.42**	***0.93***	**15.3**	**0.83**	**0.33**
								M-F	70.0	1.15	*2.53*	0.74	*1.63*	0.47	*1.03*	17.0	0.92	0.37
		3.0	*1.4*	9.7	*4.4*	8.7	*4.0*	**A-F**	**68.9**	**1.13**	***2.49***	**0.76**	***1.67***	**0.49**	***1.07***	**17.4**	**0.98**	**0.39**
								M-F	76.5	1.26	*2.77*	0.84	*1.85*	0.54	*1.19*	19.3	1.09	0.43
400	*182*	0.5	*0.2*	10.7	*4.9*	9.6	*4.4*	**A-F**	**48.2**	**0.79**	***1.75***	**0.44**	***0.97***	**0.22**	***0.48***	**8.1**	**0.25**	**0.16**
								M-F	53.5	0.88	*1.94*	0.49	*1.08*	0.24	*0.53*	9.0	0.28	0.18
		1.0	*0.5*	11.4	*5.2*	10.3	*4.7*	**A-F**	**51.8**	**0.85**	***1.86***	**0.50**	***1.11***	**0.27**	***0.59***	**9.4**	**0.35**	**0.19**
								M-F	57.5	0.94	*2.07*	0.56	*1.23*	0.30	*0.66*	10.4	0.39	0.21
		1.5	*0.7*	11.9	*5.4*	10.7	*4.9*	**A-F**	**55.4**	**0.91**	***2.00***	**0.56**	***1.22***	**0.32**	***0.69***	**10.6**	**0.44**	**0.23**
								M-F	61.5	1.01	*2.22*	0.62	*1.36*	0.35	*0.77*	11.8	0.49	0.25
		2.0	*0.9*	12.2	*5.5*	11.0	*5.0*	**A-F**	**59.0**	**0.97**	***2.14***	**0.61**	***1.35***	**0.37**	***0.81***	**11.8**	**0.54**	**0.25**
								M-F	65.5	1.08	*2.38*	0.68	*1.50*	0.41	*0.90*	13.1	0.60	0.28
		2.5	*1.1*	12.3	*5.6*	11.1	*5.0*	**A-F**	**63.0**	**1.04**	***2.28***	**0.67**	***1.47***	**0.42**	***0.93***	**13.0**	**0.63**	**0.29**
								M-F	70.0	1.15	*2.53*	0.74	*1.63*	0.47	*1.03*	14.4	0.70	0.32
		3.0	*1.4*	9.7	*4.4*	10.8	*4.9*	**A-F**	**68.9**	**1.13**	***2.49***	**0.76**	***1.67***	**0.49**	***1.07***	**14.5**	**0.76**	**0.33**
								M-F	76.5	1.26	*2.77*	0.84	*1.85*	0.54	*1.19*	16.1	0.84	0.37

(Continued)

TABLE 19–4 *(Continued)*

Weight		Daily Gain		Daily Consumption				Moisture Basis[5] A-F (as-fed) M-F (moisture-free)	Energy							Total Protein	Calcium	Phosphorus
				As-Fed		Moisture-Free Dry Matter			TDN	ME (Mcal per)		NE_m (Mcal per)		NE_g (Mcal per)				
(lb)	*(kg)*	(lb)	*(kg)*	(lb)	*(kg)*	(lb)	*(kg)*		(%)	(lb)	*(kg)*	(lb)	*(kg)*	(lb)	*(kg)*	(%)	(%)	(%)
								Medium-frame bulls (Continued)										
500	*227*	0.5	*0.2*	12.7	*5.8*	11.4	*5.2*	**A-F**	**48.2**	**0.79**	***1.75***	**0.44**	***0.97***	**0.22**	***0.48***	**7.7**	**0.23**	**0.15**
								M-F	53.5	0.88	*1.94*	0.49	*1.08*	0.24	*0.53*	8.6	0.25	0.17
		1.0	*0.5*	13.4	*6.1*	12.1	*5.5*	**A-F**	**51.8**	**0.85**	***1.86***	**0.50**	***1.11***	**0.27**	***0.59***	**8.7**	**0.32**	**0.18**
								M-F	57.5	0.94	*2.07*	0.56	*1.23*	0.30	*0.66*	9.7	0.35	0.20
		1.5	*0.7*	14.1	*6.4*	12.7	*5.8*	**A-F**	**55.4**	**0.91**	***2.00***	**0.56**	***1.22***	**0.32**	***0.69***	**9.6**	**0.38**	**0.21**
								M-F	61.5	1.01	*2.22*	0.62	*1.36*	0.35	*0.77*	10.7	0.42	0.23
		2.0	*0.9*	14.4	*6.5*	13.0	*5.9*	**A-F**	**59.0**	**0.97**	***2.14***	**0.61**	***1.35***	**0.37**	***0.81***	**10.5**	**0.44**	**0.23**
								M-F	65.5	1.08	*2.38*	0.68	*1.50*	0.41	*0.90*	11.7	0.49	0.25
		2.5	*1.1*	14.5	*6.6*	13.1	*6.0*	**A-F**	**63.0**	**1.04**	***2.28***	**0.67**	***1.47***	**0.42**	***0.93***	**11.5**	**0.53**	**0.24**
								M-F	70.0	1.15	*2.53*	0.74	*1.63*	0.47	*1.03*	12.8	0.59	0.27
		3.0	*1.4*	14.2	*6.5*	12.8	*5.8*	**A-F**	**68.9**	**1.13**	***2.49***	**0.76**	***1.67***	**0.49**	***1.07***	**12.7**	**0.62**	**0.28**
								M-F	76.5	1.26	*2.77*	0.84	*1.85*	0.54	*1.19*	14.1	0.69	0.31
600	*273*	0.5	*0.2*	14.6	*6.6*	13.1	*6.0*	**A-F**	**48.2**	**0.79**	***1.75***	**0.44**	***0.97***	**0.22**	***0.48***	**7.5**	**0.22**	**0.17**
								M-F	53.5	0.88	*1.94*	0.49	*1.08*	0.24	*0.53*	8.3	0.24	0.19
		1.0	*0.5*	15.4	*7.0*	13.9	*6.3*	**A-F**	**51.8**	**0.85**	***1.86***	**0.50**	***1.11***	**0.27**	***0.59***	**8.3**	**0.27**	**0.17**
								M-F	57.5	0.94	*2.07*	0.56	*1.23*	0.30	*0.66*	9.2	0.30	0.19
		1.5	*0.7*	16.1	*7.3*	14.5	*6.6*	**A-F**	**55.4**	**0.91**	***2.00***	**0.56**	***1.22***	**0.32**	***0.69***	**9.0**	**0.32**	**0.19**
								M-F	61.5	1.01	*2.22*	0.62	*1.36*	0.35	*0.77*	10.0	0.36	0.21
		2.0	*0.9*	16.6	*7.5*	14.9	*6.8*	**A-F**	**59.0**	**0.97**	***2.14***	**0.61**	***1.35***	**0.37**	***0.81***	**9.7**	**0.39**	**0.22**
								M-F	65.5	1.08	*2.38*	0.68	*1.50*	0.41	*0.90*	10.8	0.43	0.24
		2.5	*1.1*	16.7	*7.6*	15.0	*6.8*	**A-F**	**63.0**	**1.04**	***2.28***	**0.67**	***1.47***	**0.42**	***0.93***	**10.4**	**0.45**	**0.23**
								M-F	70.0	1.15	*2.53*	0.74	*1.63*	0.47	*1.03*	11.6	0.50	0.25
		3.0	*1.4*	16.3	*7.4*	14.7	*6.7*	**A-F**	**68.9**	**1.13**	***2.49***	**0.76**	***1.67***	**0.49**	***1.07***	**11.4**	**0.51**	**0.26**
								M-F	76.5	1.26	*2.77*	0.84	*1.85*	0.54	*1.19*	12.7	0.57	0.29
700	*318*	0.5	*0.2*	16.3	*7.4*	14.7	*6.7*	**A-F**	**48.2**	**0.79**	***1.75***	**0.44**	***0.97***	**0.22**	***0.48***	**7.2**	**0.21**	**0.16**
								M-F	53.5	0.88	*1.94*	0.49	*1.08*	0.24	*0.53*	8.0	0.23	0.18
		1.0	*0.5*	17.3	*7.9*	15.6	*7.0*	**A-F**	**51.8**	**0.85**	***1.86***	**0.50**	***1.11***	**0.27**	***0.59***	**7.9**	**0.25**	**0.18**
								M-F	57.5	0.94	*2.07*	0.56	*1.23*	0.30	*0.66*	8.8	0.28	0.20
		1.5	*0.7*	18.1	*8.2*	16.3	*7.4*	**A-F**	**55.4**	**0.91**	***2.00***	**0.56**	***1.22***	**0.32**	***0.69***	**8.5**	**0.29**	**0.18**
								M-F	61.5	1.01	*2.22*	0.62	*1.36*	0.35	*0.77*	9.4	0.32	0.20
		2.0	*0.9*	18.6	*8.5*	16.7	*7.6*	**A-F**	**59.0**	**0.97**	***2.14***	**0.61**	***1.35***	**0.37**	***0.81***	**10.0**	**0.34**	**0.20**
								M-F	65.5	1.08	*2.38*	0.68	*1.50*	0.41	*0.90*	10.1	0.38	0.22
		2.5	*1.1*	18.7	*8.5*	16.8	*7.6*	**A-F**	**63.0**	**1.04**	***2.28***	**0.67**	***1.47***	**0.42**	***0.93***	**9.7**	**0.39**	**0.22**
								M-F	70.0	1.15	*2.53*	0.74	*1.63*	0.47	*1.03*	10.8	0.43	0.24
		3.0	*1.4*	18.3	*8.3*	16.5	*7.5*	**A-F**	**68.9**	**1.13**	***2.49***	**0.76**	***1.67***	**0.49**	***1.07***	**10.5**	**0.44**	**0.23**
								M-F	76.5	1.26	*2.77*	0.84	*1.85*	0.54	*1.19*	11.7	0.49	0.25
800	*364*	0.5	*0.2*	18.0	*8.2*	16.2	*7.4*	**A-F**	**48.2**	**0.79**	***1.75***	**0.44**	***0.97***	**0.22**	***0.48***	**7.0**	**0.20**	**0.17**
								M-F	53.5	0.88	*1.94*	0.49	*1.08*	0.24	*0.53*	7.8	0.22	0.19
		1.0	*0.5*	19.2	*8.7*	17.3	*7.9*	**A-F**	**51.8**	**0.85**	***1.86***	**0.50**	***1.11***	**0.27**	***0.59***	**7.6**	**0.23**	**0.17**
								M-F	57.5	0.94	*2.07*	0.56	*1.23*	0.30	*0.66*	8.4	0.25	0.19
		1.5	*0.7*	20.0	*9.1*	18.0	*8.2*	**A-F**	**55.4**	**0.91**	***2.00***	**0.56**	***1.22***	**0.32**	***0.69***	**8.1**	**0.26**	**0.18**
								M-F	61.5	1.01	*2.22*	0.62	*1.36*	0.35	*0.77*	9.0	0.29	0.20
		2.0	*0.9*	20.6	*9.4*	18.5	*8.4*	**A-F**	**59.0**	**0.97**	***2.14***	**0.61**	***1.35***	**0.37**	***0.81***	**8.6**	**0.30**	**0.19**
								M-F	65.5	1.08	*2.38*	0.68	*1.50*	0.41	*0.90*	9.5	0.33	0.21
		2.5	*1.1*	20.7	*9.4*	18.6	*8.5*	**A-F**	**63.0**	**1.04**	***2.28***	**0.67**	***1.47***	**0.42**	***0.93***	**10.0**	**0.34**	**0.21**
								M-F	70.0	1.15	*2.53*	0.74	*1.63*	0.47	*1.03*	10.1	0.38	0.23
		3.0	*1.4*	20.2	*9.2*	18.2	*8.3*	**A-F**	**68.9**	**1.13**	***2.49***	**0.76**	***1.67***	**0.49**	***1.07***	**9.7**	**0.40**	**0.22**
								M-F	76.5	1.26	*2.77*	0.84	*1.85*	0.54	*1.19*	10.8	0.44	0.24
900	*409*	0.5	*0.2*	19.7	*9.0*	17.7	*8.0*	**A-F**	**48.2**	**0.79**	***1.75***	**0.44**	***0.97***	**0.22**	***0.48***	**6.9**	**0.19**	**0.17**
								M-F	53.5	0.88	*1.94*	0.49	*1.08*	0.24	*0.53*	7.7	0.21	0.19
		1.0	*0.5*	21.0	*9.5*	18.9	*8.6*	**A-F**	**51.8**	**0.85**	***1.86***	**0.50**	***1.11***	**0.27**	***0.59***	**7.4**	**0.23**	**0.17**
								M-F	57.5	0.94	*2.07*	0.56	*1.23*	0.30	*0.66*	8.2	0.25	0.19

(Continued)

TABLE 19–4 *(Continued)*

Weight		Daily Gain		Daily Consumption: As-Fed		Daily Consumption: Moisture-Free Dry Matter		Moisture Basis[5] A-F (as-fed) M-F (moisture-free)	Energy: TDN	Energy: ME (Mcal per)		Energy: NE$_m$ (Mcal per)		Energy: NE$_g$ (Mcal per)		Total Protein	Calcium	Phosphorus
(lb)	*(kg)*	(lb)	*(kg)*	(lb)	*(kg)*	(lb)	*(kg)*		(%)	(lb)	*(kg)*	(lb)	*(kg)*	(lb)	*(kg)*	(%)	(%)	(%)
Medium-frame bulls (Continued)																		
900	*409*	1.5	*0.7*	21.9	*10.0*	19.7	*9.0*	**A-F**	**55.4**	**0.91**	***2.00***	**0.56**	***1.23***	**0.32**	***0.69***	**7.7**	**0.25**	**0.17**
(Continued)								M-F	61.5	1.01	*2.22*	0.62	*1.36*	0.35	*0.77*	8.6	0.28	0.19
		2.0	*0.9*	22.4	*10.2*	20.2	*9.2*	**A-F**	**59.0**	**0.97**	***2.14***	**0.61**	***1.35***	**0.37**	***0.81***	**8.2**	**0.28**	**0.19**
								M-F	65.5	1.08	*2.38*	0.68	*1.50*	0.41	*0.90*	9.1	0.31	0.21
		2.5	*1.1*	22.6	*10.3*	20.3	*9.2*	**A-F**	**63.0**	**1.04**	***2.28***	**0.67**	***1.47***	**0.42**	***0.93***	**8.6**	**0.31**	**0.20**
								M-F	70.0	1.15	*2.53*	0.74	*1.63*	0.47	*1.03*	9.6	0.34	0.22
		3.0	*1.4*	22.1	*10.0*	19.9	*9.0*	**A-F**	**68.9**	**1.13**	***2.49***	**0.76**	***1.67***	**0.49**	***1.07***	**9.2**	**0.35**	**0.21**
								M-F	76.5	1.26	*2.77*	0.84	*1.85*	0.54	*1.19*	10.2	0.39	0.23
1,000	*454*	0.5	*0.2*	21.3	*9.7*	19.2	*8.7*	**A-F**	**48.2**	**0.79**	***1.75***	**0.44**	***0.97***	**0.22**	***0.48***	**6.8**	**0.19**	**0.16**
								M-F	53.5	0.88	*1.94*	0.49	*1.08*	0.24	*0.53*	7.5	0.21	0.18
		1.0	*0.5*	22.7	*1.03*	2.04	*9.3*	**A-F**	**51.8**	**0.85**	***1.86***	**0.50**	***1.11***	**0.27**	***0.59***	**7.2**	**0.22**	**0.16**
								M-F	57.5	0.94	*2.07*	0.56	*1.23*	0.30	*0.66*	8.0	0.24	0.18
		1.5	*0.7*	23.7	*10.8*	21.3	*9.7*	**A-F**	**55.4**	**0.91**	***2.00***	**0.56**	***1.23***	**0.32**	***0.69***	**7.6**	**0.23**	**0.17**
								M-F	61.5	1.01	*2.22*	0.62	*1.36*	0.35	*0.77*	8.4	0.26	0.19
		2.0	*0.9*	24.2	*11.0*	21.8	*9.9*	**A-F**	**59.0**	**0.97**	***2.14***	**0.61**	***1.35***	**0.37**	***0.81***	**7.8**	**0.25**	**0.17**
								M-F	65.5	1.08	*2.38*	0.68	*1.50*	0.41	*0.90*	8.7	0.28	0.19
		2.5	*1.1*	24.4	*11.1*	22.0	*10.0*	**A-F**	**63.0**	**1.04**	***2.28***	**0.67**	***1.47***	**0.42**	***0.93***	**8.2**	**0.28**	**0.18**
								M-F	70.0	1.15	*2.53*	0.74	*1.63*	0.47	*1.03*	9.1	0.31	0.20
		3.0	*1.4*	23.9	*10.9*	21.5	*9.8*	**A-F**	**68.9**	**1.13**	***2.49***	**0.76**	***1.67***	**0.49**	***1.07***	**8.6**	**0.32**	**0.20**
								M-F	76.5	1.26	*2.77*	0.84	*1.85*	0.54	*1.19*	9.6	0.35	0.22
1,100	*500*	0.5	*0.2*	22.9	*10.4*	20.6	*9.4*	**A-F**	**48.2**	**0.79**	***1.75***	**0.44**	***0.97***	**0.22**	***0.48***	**6.7**	**0.18**	**0.17**
								M-F	53.5	0.88	*1.94*	0.49	*1.08*	0.24	*0.53*	7.4	0.20	0.19
		1.0	*0.5*	24.3	*11.0*	21.9	*10.0*	**A-F**	**51.8**	**0.85**	***1.86***	**0.50**	***1.11***	**0.27**	***0.59***	**7.0**	**0.20**	**0.17**
								M-F	57.5	0.94	*2.07*	0.56	*1.23*	0.30	*0.66*	7.8	0.22	0.19
		1.5	*0.7*	25.4	*11.5*	22.9	*10.4*	**A-F**	**55.4**	**0.91**	***2.00***	**0.56**	***1.23***	**0.32**	***0.69***	**7.3**	**0.22**	**0.17**
								M-F	61.5	1.01	*2.22*	0.62	*1.36*	0.35	*0.77*	8.1	0.24	0.19
		2.0	*0.9*	26.0	*11.8*	23.4	*10.6*	**A-F**	**59.0**	**0.97**	***2.14***	**0.61**	***1.35***	**0.37**	***0.81***	**7.6**	**0.23**	**0.17**
								M-F	65.5	1.08	*2.38*	0.68	*1.50*	0.41	*0.90*	8.4	0.26	0.19
		2.5	*1.1*	26.2	*11.9*	23.6	*10.7*	**A-F**	**63.0**	**1.04**	***2.28***	**0.67**	***1.47***	**0.42**	***0.93***	**7.8**	**0.25**	**0.18**
								M-F	70.0	1.15	*2.53*	0.74	*1.63*	0.47	*1.03*	8.7	0.28	0.20
		3.0	*1.4*	25.7	*11.7*	23.1	*10.5*	**A-F**	**68.9**	**1.13**	***2.49***	**0.76**	***1.67***	**0.49**	***1.07***	**8.3**	**0.29**	**0.19**
								M-F	76.5	1.26	*2.77*	0.84	*1.85*	0.54	*1.19*	9.2	0.32	0.21
Large-frame bull calves and compensating large-frame yearling steers																		
300	*136*	0.5	*0.2*	8.8	*4.0*	7.9	*3.6*	**A-F**	**48.2**	**0.77**	***1.70***	**0.43**	***0.95***	**0.21**	***0.46***	**8.7**	**0.28**	**0.18**
								M-F	52.5	0.86	*1.89*	0.48	*1.06*	0.23	*0.51*	9.7	0.31	0.20
		1.0	*0.5*	9.3	*4.2*	8.4	*3.8*	**A-F**	**50.4**	**0.83**	***1.82***	**0.49**	***1.07***	**0.25**	***0.56***	**10.5**	**0.42**	**0.22**
								M-F	56.0	0.92	*2.02*	0.54	*1.19*	0.28	*0.62*	11.7	0.47	0.24
		1.5	*0.7*	9.8	*4.5*	8.8	*4.0*	**A-F**	**53.6**	**0.88**	***1.95***	**0.53**	***1.17***	**0.30**	***0.66***	**12.2**	**0.57**	**0.25**
								M-F	59.5	0.98	*2.16*	0.59	*1.30*	0.33	*0.73*	13.5	0.63	0.28
		2.0	*0.9*	10.0	*4.5*	9.0	*4.1*	**A-F**	**56.3**	**0.93**	***2.04***	**0.57**	***1.25***	**0.33**	***0.73***	**13.6**	**0.68**	**0.29**
								M-F	62.5	1.03	*2.27*	0.63	*1.39*	0.37	*0.81*	15.1	0.76	0.32
		2.5	*1.1*	10.2	*4.6*	9.2	*4.2*	**A-F**	**59.9**	**0.98**	***2.16***	**0.62**	***1.37***	**0.38**	***0.83***	**15.3**	**0.82**	**0.32**
								M-F	66.5	1.09	*2.40*	0.69	*1.52*	0.42	*0.92*	17.0	0.91	0.36
		3.0	*1.4*	10.2	*4.6*	9.2	*4.2*	**A-F**	**63.5**	**1.04**	***2.30***	**0.68**	***1.49***	**0.42**	***0.93***	**16.9**	**0.97**	**0.39**
								M-F	70.5	1.16	*2.55*	0.75	*1.65*	0.47	*1.03*	18.8	1.08	0.43
		3.5	*1.6*	10.1	*4.6*	9.1	*4.1*	**A-F**	**68.0**	**1.12**	***2.46***	**0.74**	***1.62***	**0.48**	***1.05***	**18.8**	**1.12**	**0.43**
								M-F	75.5	1.24	*2.73*	0.82	*1.80*	0.53	*1.17*	20.9	1.24	0.48
		4.0	*1.8*	9.1	*4.1*	8.2	*3.7*	**A-F**	**77.4**	**1.27**	***2.79***	**0.86**	***1.90***	**0.59**	***1.31***	**22.2**	**1.38**	**0.53**
								M-F	86.0	1.41	*3.10*	0.96	*2.11*	0.66	*1.45*	24.7	1.53	0.59

(Continued)

TABLE 19–4 *(Continued)*

Weight		Daily Gain		Daily Consumption: As-Fed		Daily Consumption: Moisture-Free Dry Matter		Moisture Basis[5] A-F (as-fed) M-F (moisture-free)	Energy: TDN	Energy: ME (Mcal per)		Energy: NE$_m$ (Mcal per)		Energy: NE$_g$ (Mcal per)		Total Protein	Calcium	Phosphorus
(lb)	*(kg)*	(lb)	*(kg)*	(lb)	*(kg)*	(lb)	*(kg)*		(%)	(lb)	*(kg)*	(lb)	*(kg)*	(lb)	*(kg)*	(%)	(%)	(%)
Large-frame bull calves and compensating large-frame yearling steers (Continued)																		
400	*182*	0.5	*0.2*	10.9	*5.0*	9.8	*4.5*	**A-F**	**47.3**	**0.77**	***1.70***	**0.43**	***0.95***	**0.21**	***0.46***	**8.1**	**0.24**	**0.16**
								M-F	52.5	0.86	*1.89*	0.48	*1.06*	0.23	*0.51*	9.0	0.27	0.18
		1.0	*0.5*	11.6	*5.3*	10.4	*4.7*	**A-F**	**50.4**	**0.83**	***1.82***	**0.49**	***1.07***	**0.25**	***0.56***	**9.5**	**0.36**	**0.19**
								M-F	56.0	0.92	*2.02*	0.54	*1.19*	0.28	*0.62*	10.5	0.40	0.21
		1.5	*0.7*	12.1	*5.5*	10.9	*5.0*	**A-F**	**53.6**	**0.88**	***1.94***	**0.53**	***1.17***	**0.30**	***0.66***	**10.7**	**0.46**	**0.22**
								M-F	59.5	0.98	*2.16*	0.59	*1.30*	0.33	*0.73*	11.9	0.51	0.24
		2.0	*0.9*	12.4	*5.6*	11.2	*5.1*	**A-F**	**56.3**	**0.93**	***2.04***	**0.57**	***1.25***	**0.33**	***0.73***	**11.8**	**0.55**	**0.25**
								M-F	62.5	1.03	*2.27*	0.63	*1.39*	0.37	*0.81*	13.1	0.61	0.28
		2.5	*1.1*	12.7	*5.8*	11.4	*5.2*	**A-F**	**59.9**	**0.98**	***2.16***	**0.62**	***1.37***	**0.38**	***0.83***	**13.1**	**0.65**	**0.28**
								M-F	66.5	1.09	*2.40*	0.69	*1.52*	0.42	*0.92*	14.5	0.72	0.31
		3.0	*1.4*	12.8	*5.8*	11.5	*5.2*	**A-F**	**63.5**	**1.04**	***2.30***	**0.68**	***1.49***	**0.42**	***0.93***	**14.3**	**0.74**	**0.32**
								M-F	70.5	1.16	*2.55*	0.75	*1.65*	0.47	*1.03*	15.9	0.82	0.35
		3.5	*1.6*	12.6	*5.7*	11.3	*5.1*	**A-F**	**68.0**	**1.12**	***2.46***	**0.74**	***1.62***	**0.48**	***1.05***	**15.8**	**0.86**	**0.35**
								M-F	75.5	1.24	*2.73*	0.82	*1.80*	0.53	*1.17*	17.5	0.96	0.39
		4.0	*1.8*	11.3	*5.1*	10.2	*4.6*	**A-F**	**77.4**	**1.27**	***2.79***	**0.86**	***1.90***	**0.59**	***1.31***	**18.3**	**1.07**	**0.43**
								M-F	86.0	1.41	*3.10*	0.96	*2.11*	0.66	*1.45*	20.3	1.19	0.48
500	*227*	0.5	*0.2*	12.9	*5.9*	11.6	*5.3*	**A-F**	**47.3**	**0.77**	***1.70***	**0.43**	***0.95***	**0.21**	***0.46***	**7.7**	**0.23**	**0.17**
								M-F	52.5	0.86	*1.89*	0.48	*1.06*	0.23	*0.51*	8.6	0.25	0.19
		1.0	*0.5*	13.7	*6.2*	12.3	*5.6*	**A-F**	**50.4**	**0.83**	***1.82***	**0.49**	***1.07***	**0.25**	***0.56***	**8.8**	**0.32**	**0.19**
								M-F	56.0	0.92	*2.02*	0.54	*1.19*	0.28	*0.62*	9.8	0.36	0.21
		1.5	*0.7*	14.3	*6.5*	12.9	*5.9*	**A-F**	**53.6**	**0.88**	***1.94***	**0.53**	***1.17***	**0.30**	***0.66***	**9.8**	**0.39**	**0.20**
								M-F	59.5	0.98	*2.16*	0.59	*1.30*	0.33	*0.73*	10.9	0.43	0.22
		2.0	*0.9*	14.7	*6.7*	13.2	*6.0*	**A-F**	**56.3**	**0.93**	***2.04***	**0.57**	***1.25***	**0.33**	***0.73***	**10.6**	**0.47**	**0.23**
								M-F	62.5	1.03	*2.27*	0.63	*1.39*	0.37	*0.81*	11.8	0.52	0.25
		2.5	*1.1*	15.0	*6.8*	13.5	*6.1*	**A-F**	**59.9**	**0.98**	***2.16***	**0.62**	***1.37***	**0.38**	***0.83***	**11.6**	**0.53**	**0.25**
								M-F	66.5	1.09	*2.40*	0.69	*1.52*	0.42	*0.92*	12.9	0.59	0.28
		3.0	*1.4*	15.1	*6.9*	13.6	*6.2*	**A-F**	**63.5**	**1.04**	***2.30***	**0.68**	***1.49***	**0.42**	***0.93***	**12.6**	**0.61**	**0.28**
								M-F	70.5	1.16	*2.55*	0.75	*1.65*	0.47	*1.03*	14.0	0.68	0.31
		3.5	*1.6*	14.9	*6.8*	13.4	*6.1*	**A-F**	**68.0**	**1.12**	***2.46***	**0.74**	***1.62***	**0.48**	***1.05***	**13.8**	**0.69**	**0.32**
								M-F	75.5	1.24	*2.73*	0.82	*1.80*	0.53	*1.17*	15.3	0.77	0.35
		4.0	*1.8*	13.3	*6.0*	12.0	*5.5*	**A-F**	**77.4**	**1.27**	***2.79***	**0.86**	***1.90***	**0.59**	***1.31***	**15.8**	**0.87**	**0.36**
								M-F	86.0	1.41	*3.10*	0.96	*2.11*	0.66	*1.45*	17.5	0.97	0.40
600	*273*	0.5	*0.2*	14.8	*6.7*	13.3	*6.0*	**A-F**	**47.3**	**0.77**	***1.70***	**0.43**	***0.95***	**0.21**	***0.46***	**7.5**	**0.21**	**0.16**
								M-F	52.5	0.86	*1.89*	0.48	*1.06*	0.23	*0.51*	8.3	0.23	0.18
		1.0	*0.5*	15.7	*7.1*	14.1	*6.4*	**A-F**	**50.4**	**0.83**	***1.82***	**0.49**	***1.07***	**0.25**	***0.56***	**8.3**	**0.28**	**0.18**
								M-F	56.0	0.92	*2.02*	0.54	*1.19*	0.28	*0.62*	9.2	0.31	0.20
		1.5	*0.7*	16.4	*7.5*	14.8	*6.7*	**A-F**	**53.6**	**0.88**	***1.94***	**0.53**	***1.17***	**0.30**	***0.66***	**9.1**	**0.33**	**0.19**
								M-F	59.5	0.98	*2.16*	0.59	*1.30*	0.33	*0.73*	10.1	0.37	0.21
		2.0	*0.9*	16.9	*7.7*	15.2	*6.9*	**A-F**	**56.3**	**0.93**	***2.04***	**0.57**	***1.25***	**0.33**	***0.73***	**9.8**	**0.40**	**0.21**
								M-F	62.5	1.03	*2.27*	0.63	*1.39*	0.37	*0.81*	10.9	0.44	0.23
		2.5	*1.1*	17.2	*7.8*	15.5	*7.0*	**A-F**	**59.9**	**0.98**	***2.16***	**0.62**	***1.37***	**0.38**	***0.83***	**10.6**	**0.46**	**0.23**
								M-F	66.5	1.09	*2.40*	0.69	*1.52*	0.42	*0.92*	11.8	0.51	0.26
		3.0	*1.4*	17.2	*7.8*	15.5	*7.0*	**A-F**	**63.5**	**1.05**	***2.30***	**0.68**	***1.49***	**0.42**	***0.93***	**11.4**	**0.52**	**0.24**
								M-F	70.5	1.16	*2.55*	0.75	*1.65*	0.47	*1.03*	12.7	0.58	0.27
		3.5	*1.6*	17.0	*7.7*	15.3	*7.0*	**A-F**	**68.0**	**1.12**	***2.46***	**0.74**	***1.62***	**0.48**	***1.05***	**12.3**	**0.59**	**0.27**
								M-F	75.5	1.24	*2.73*	0.82	*1.80*	0.53	*1.17*	13.7	0.66	0.30
		4.0	*1.8*	15.3	*7.0*	13.8	*6.3*	**A-F**	**77.4**	**1.27**	***2.79***	**0.86**	***1.90***	**0.59**	***1.31***	**14.0**	**0.73**	**0.33**
								M-F	86.0	1.41	*3.10*	0.96	*2.11*	0.66	*1.45*	15.6	0.81	0.37
700	*318*	0.5	*0.2*	16.6	*7.5*	14.9	*6.8*	**A-F**	**47.3**	**0.77**	***1.70***	**0.43**	***0.95***	**0.21**	***0.46***	**7.2**	**0.20**	**0.16**
								M-F	52.5	0.86	*1.89*	0.48	*1.06*	0.23	*0.51*	8.0	0.22	0.18
		1.0	*0.5*	17.7	*8.0*	15.9	*7.2*	**A-F**	**50.4**	**0.83**	***1.82***	**0.49**	***1.07***	**0.25**	***0.56***	**7.9**	**0.26**	**0.17**
								M-F	56.0	0.92	*2.02*	0.54	*1.19*	0.28	*0.62*	8.8	0.29	0.19

(Continued)

TABLE 19–4 *(Continued)*

Weight		Daily Gain		Daily Consumption				Moisture Basis[5] A-F (as-fed) M-F (moisture-free)	Energy							Total Protein	Calcium	Phosphorus
				As-Fed		Moisture-Free Dry Matter			TDN	ME (Mcal per)		NEm (Mcal per)		NEg (Mcal per)				
(lb)	*(kg)*	(lb)	*(kg)*	(lb)	*(kg)*	(lb)	*(kg)*		(%)	(lb)	*(kg)*	(lb)	*(kg)*	(lb)	*(kg)*	(%)	(%)	(%)
Large-frame bull calves and compensating large-frame yearling steers (Continued)																		
700 (Continued)	*318*	1.5	*0.7*	18.4	*8.4*	16.6	*7.5*	**A-F**	**53.6**	**0.88**	***1.94***	**0.53**	***1.17***	0.30	***0.66***	**8.6**	**0.32**	**0.19**
								M-F	59.5	0.98	*2.16*	0.59	*1.30*	0.33	*0.73*	9.6	0.35	0.21
		2.0	*0.9*	18.9	*8.6*	17.0	*7.7*	**A-F**	**56.3**	**0.93**	***2.04***	**0.57**	***1.25***	**0.33**	***0.73***	**9.2**	**0.35**	**0.20**
								M-F	62.5	1.03	*2.27*	0.63	*1.39*	0.37	*0.81*	10.2	0.39	0.22
		2.5	*1.1*	19.3	*8.8*	17.4	*7.9*	**A-F**	**59.9**	**0.98**	***2.16***	**0.62**	***1.37***	**0.38**	***0.83***	**10.0**	**0.40**	**0.22**
								M-F	66.5	1.09	*2.40*	0.69	*1.52*	0.42	*0.92*	11.0	0.44	0.24
		3.0	*1.4*	19.4	*8.8*	17.5	*8.0*	**A-F**	**63.5**	**1.05**	***2.30***	**0.68**	***1.49***	**0.42**	***0.93***	**10.5**	**0.45**	**0.23**
								M-F	70.5	1.16	*2.55*	0.75	*1.65*	0.47	*1.03*	11.7	0.50	0.25
		3.5	*1.6*	19.1	*8.7*	17.2	*7.8*	**A-F**	**68.0**	**1.12**	***2.46***	**0.74**	***1.62***	**0.48**	***1.05***	**11.3**	**0.50**	**0.25**
								M-F	75.5	1.24	*2.73*	0.82	*1.80*	0.53	*1.17*	12.5	0.56	0.28
		4.0	*1.8*	17.2	*7.8*	15.5	*7.0*	**A-F**	**77.4**	**1.27**	***2.79***	**0.86**	***1.90***	**0.59**	***1.31***	**12.7**	**0.63**	**0.30**
								M-F	86.0	1.41	*3.10*	0.96	*2.11*	0.66	*1.45*	14.1	0.70	0.33
800	*364*	0.5	*0.2*	18.3	*8.3*	16.5	*7.5*	**A-F**	**47.3**	**0.77**	***1.70***	**0.43**	***0.95***	**0.21**	***0.46***	**7.1**	**0.19**	**0.17**
								M-F	52.5	0.86	*1.89*	0.48	*1.06*	0.23	*0.51*	7.9	0.21	0.19
		1.0	*0.5*	19.4	*8.8*	17.5	*8.0*	**A-F**	**50.4**	**0.83**	***1.82***	**0.49**	***1.07***	**0.25**	***0.56***	**7.7**	**0.23**	**0.17**
								M-F	56.0	0.92	*2.02*	0.54	*1.19*	0.28	*0.62*	8.5	0.26	0.19
		1.5	*0.7*	20.3	*9.2*	18.3	*8.3*	**A-F**	**53.6**	**0.88**	***1.94***	**0.53**	***1.17***	**0.30**	***0.66***	**8.2**	**0.28**	**0.18**
								M-F	59.5	0.98	*2.16*	0.59	*1.30*	0.33	*0.73*	9.1	0.31	0.20
		2.0	*0.9*	20.9	*9.5*	18.8	*8.5*	**A-F**	**56.3**	**0.93**	***2.04***	**0.57**	***1.25***	**0.33**	***0.73***	**8.7**	**0.32**	**0.19**
								M-F	62.5	1.03	*2.27*	0.63	*1.39*	0.37	*0.81*	9.7	0.35	0.21
		2.5	*1.1*	21.3	*9.7*	19.2	*8.7*	**A-F**	**59.9**	**0.98**	***2.16***	**0.62**	***1.37***	**0.38**	***0.83***	**9.3**	**0.36**	**0.21**
								M-F	66.5	1.09	*2.40*	0.69	*1.52*	0.42	*0.92*	10.3	0.40	0.23
		3.0	*1.4*	21.4	*9.7*	19.3	*8.8*	**A-F**	**63.5**	**1.04**	***2.30***	**0.68**	***1.49***	**0.42**	***0.93***	**9.8**	**0.41**	**0.22**
								M-F	70.5	1.16	*2.55*	0.75	*1.65*	0.47	*1.03*	10.9	0.45	0.24
		3.5	*1.6*	21.1	*9.6*	19.0	*8.6*	**A-F**	**68.0**	**1.12**	***2.46***	**0.74**	***1.62***	**0.48**	***1.05***	**10.4**	**0.45**	**0.23**
								M-F	75.5	1.24	*2.73*	0.82	*1.80*	0.53	*1.17*	11.6	0.50	0.26
		4.0	*1.8*	19.0	*8.6*	17.1	*7.8*	**A-F**	**77.4**	**1.27**	***2.79***	**0.86**	***1.90***	**0.59**	***1.31***	**11.7**	**0.55**	**0.28**
								M-F	86.0	1.41	*3.10*	0.96	*2.11*	0.66	*1.45*	13.0	0.61	0.31
900	*409*	0.5	*0.2*	20.0	*9.1*	18.0	*8.2*	**A-F**	**47.3**	**0.77**	***1.70***	**0.43**	***0.95***	**0.21**	***0.46***	**6.9**	**0.20**	**0.16**
								M-F	52.5	0.86	*1.89*	0.48	*1.06*	0.23	*0.51*	7.7	0.22	0.18
		1.0	*0.5*	21.3	*9.7*	19.2	*8.7*	**A-F**	**50.4**	**0.83**	***1.82***	**0.49**	***1.07***	**0.25**	***0.56***	**7.5**	**0.23**	**0.16**
								M-F	56.0	0.92	*2.02*	0.54	*1.19*	0.28	*0.62*	8.3	0.25	0.18
		1.5	*0.7*	22.2	*10.1*	20.0	*9.1*	**A-F**	**53.6**	**0.88**	***1.94***	**0.53**	***1.17***	**0.30**	***0.66***	**7.9**	**0.26**	**0.18**
								M-F	59.5	0.98	*2.16*	0.59	*1.30*	0.33	*0.73*	8.8	0.29	0.20
		2.0	*0.9*	22.9	*10.4*	20.6	*9.4*	**A-F**	**56.3**	**0.93**	***2.04***	**0.57**	***1.25***	**0.33**	***0.73***	**8.9**	**0.30**	**0.18**
								M-F	62.5	1.03	*2.27*	0.63	*1.39*	0.37	*0.81*	9.2	0.32	0.20
		2.5	*1.1*	23.3	*10.6*	21.0	*9.5*	**A-F**	**59.9**	**0.98**	***2.16***	**0.62**	***1.37***	**0.38**	***0.83***	**8.8**	**0.32**	**0.19**
								M-F	66.5	1.09	*2.40*	0.69	*1.52*	0.42	*0.92*	9.8	0.36	0.21
		3.0	*1.4*	23.4	*10.6*	21.1	*9.6*	**A-F**	**63.5**	**1.04**	***2.30***	**0.68**	***1.49***	**0.42**	***0.93***	**9.3**	**0.36**	**0.21**
								M-F	70.5	1.16	*2.55*	0.75	*1.65*	0.47	*1.03*	10.3	0.40	0.23
		3.5	*1.6*	23.1	*10.5*	20.8	*9.5*	**A-F**	**68.0**	**1.12**	***2.46***	**0.74**	***1.62***	**0.48**	***1.05***	**9.8**	**0.41**	**0.22**
								M-F	75.5	1.24	*2.73*	0.82	*1.80*	0.53	*1.17*	10.9	0.45	0.24
		4.0	*1.8*	20.8	*9.5*	18.7	*8.5*	**A-F**	**77.4**	**1.27**	***2.79***	**0.86**	***1.90***	**0.59**	***1.31***	**10.9**	**0.48**	**0.25**
								M-F	86.0	1.41	*3.10*	0.96	*2.11*	0.66	*1.45*	12.1	0.53	0.28
1,000	*454*	0.5	*0.2*	21.7	*9.9*	19.5	*8.9*	**A-F**	**47.3**	**0.77**	***1.70***	**0.43**	***0.95***	**0.21**	***0.46***	**6.8**	**0.19**	**0.16**
								M-F	52.5	0.86	*1.89*	0.48	*1.06*	0.23	*0.51*	7.6	0.21	0.18
		1.0	*0.5*	23.0	*10.5*	20.7	*9.4*	**A-F**	**40.4**	**0.83**	***1.82***	**0.49**	***1.07***	**0.25**	***0.56***	**7.3**	**0.23**	**0.17**
								M-F	56.0	0.92	*2.02*	0.54	*1.19*	0.28	*0.62*	8.1	0.25	0.19
		1.5	*0.7*	24.1	*11.0*	21.7	*9.9*	**A-F**	**53.6**	**0.88**	***1.94***	**0.53**	***1.17***	**0.30**	***0.66***	**7.7**	**0.24**	**0.17**
								M-F	59.5	0.98	*2.16*	0.59	*1.30*	0.33	*0.73*	8.5	0.27	0.19
		2.0	*0.9*	24.8	*11.3*	22.3	*10.1*	**A-F**	**56.3**	**0.93**	***2.04***	**0.57**	***1.25***	**0.33**	***0.73***	**7.9**	**0.27**	**0.18**
								M-F	62.5	1.03	*2.27*	0.63	*1.39*	0.37	*0.81*	8.9	0.30	0.20

(Continued)

TABLE 19–4 *(Continued)*

Weight		Daily Gain		Daily Consumption: As-Fed		Daily Consumption: Moisture-Free Dry Matter		Moisture Basis[5] A-F (as-fed) M-F (moisture-free)	Energy: TDN	Energy: ME (Mcal per)		Energy: NE$_m$ (Mcal per)		Energy: NE$_g$ (Mcal per)		Total Protein	Calcium	Phosphorus
(lb)	*(kg)*	(lb)	*(kg)*	(lb)	*(kg)*	(lb)	*(kg)*		(%)	(lb)	*(kg)*	(lb)	*(kg)*	(lb)	*(kg)*	(%)	(%)	(%)
Large-frame bull calves and compensating large-frame yearling steers (Continued)																		
1,000 (Continued)	*454*	2.5	*1.1*	25.2	*11.5*	22.7	*10.3*	A-F	59.9	0.98	2.16	0.62	1.37	0.38	0.83	8.4	0.30	0.18
								M-F	66.5	1.09	2.40	0.69	1.52	0.42	0.92	9.3	0.33	0.20
		3.0	*1.4*	25.3	*11.5*	22.8	*10.4*	A-F	63.5	1.04	2.30	0.68	1.49	0.42	0.93	8.7	0.32	0.19
								M-F	70.5	1.16	2.55	0.75	1.65	0.47	1.03	9.7	0.36	0.21
		3.5	*1.6*	25.0	*11.4*	22.5	*10.2*	A-F	68.0	1.12	2.46	0.74	1.62	0.48	1.05	9.3	0.36	0.22
								M-F	75.5	1.24	2.73	0.82	1.80	0.53	1.17	10.3	0.40	0.24
		4.0	*1.8*	22.4	*10.2*	20.2	*9.2*	A-F	77.4	1.27	2.79	0.86	1.90	0.59	0.31	10.2	0.43	0.24
								M-F	86.0	1.41	3.10	0.96	2.11	0.66	1.45	11.3	0.48	0.27
1,100	*500*	0.5	*0.2*	23.2	*10.5*	20.9	*9.5*	A-F	47.3	0.77	1.70	0.43	0.95	0.21	0.46	6.8	0.19	0.17
								M-F	52.5	0.86	1.89	0.48	1.06	0.23	0.51	7.4	0.21	0.19
		1.0	*0.5*	24.8	*11.3*	22.3	*10.1*	A-F	50.4	0.83	1.82	0.49	1.07	0.25	0.56	7.1	0.21	0.17
								M-F	56.0	0.92	2.02	0.54	1.19	0.28	0.62	7.9	0.23	0.19
		1.5	*0.7*	25.9	*11.8*	23.3	*10.6*	A-F	53.6	0.88	1.94	0.53	1.17	0.30	0.66	7.5	0.23	0.17
								M-F	59.5	0.98	2.16	0.59	1.30	0.33	0.73	8.3	0.26	0.19
		2.0	*0.9*	26.6	*12.1*	23.9	*10.9*	A-F	56.3	0.93	2.04	0.57	1.25	0.33	0.73	7.7	0.25	0.17
								M-F	62.5	1.03	2.27	0.63	1.39	0.37	0.81	8.6	0.28	0.19
		2.5	*1.1*	26.9	*12.2*	24.2	*11.0*	A-F	59.9	0.98	2.16	0.62	1.37	0.38	0.83	8.1	0.27	0.18
								M-F	66.5	1.09	2.40	0.69	1.52	0.42	0.92	9.0	0.30	0.20
		3.0	*1.4*	27.2	*12.4*	24.5	*11.1*	A-F	63.5	1.04	2.30	0.68	1.49	0.42	0.93	8.4	0.29	0.19
								M-F	70.5	1.16	2.55	0.75	1.65	0.47	1.03	9.3	0.32	0.21
		3.5	*1.6*	26.8	*12.2*	24.1	*11.0*	A-F	68.0	1.12	2.46	0.74	1.62	0.48	1.05	8.8	0.32	0.20
								M-F	75.5	1.24	2.73	0.82	1.80	0.53	1.17	9.8	0.36	0.22
		4.0	*1.8*	24.1	*11.0*	21.7	*9.9*	A-F	77.4	1.27	2.79	0.86	1.90	0.59	1.31	9.6	0.39	0.23
								M-F	86.0	1.41	3.10	0.96	2.11	0.66	1.45	10.7	0.43	0.25
Medium-frame heifer calves																		
300	*136*	0.5	*0.2*	8.3	*3.8*	7.5	*3.4*	A-F	50.4	0.83	1.82	0.49	1.07	0.25	0.56	8.6	0.26	0.19
								M-F	56.0	0.92	2.02	0.54	1.19	0.28	0.62	9.6	0.29	0.21
		1.0	*0.5*	8.9	*4.0*	8.0	*3.6*	A-F	55.8	0.92	2.03	0.57	1.25	0.32	0.71	10.3	0.40	0.20
								M-F	62.0	1.02	2.25	0.63	1.39	0.36	0.79	11.4	0.44	0.22
		1.5	*0.7*	9.1	*4.1*	8.2	*3.7*	A-F	61.7	1.02	2.24	0.65	1.42	0.40	0.87	11.8	0.53	0.24
								M-F	68.5	1.13	2.49	0.72	1.58	0.44	0.97	13.1	0.59	0.27
		2.0	*0.9*	8.9	*4.0*	8.0	*3.6*	A-F	69.3	1.13	2.50	0.76	1.67	0.50	1.09	13.6	0.67	0.30
								M-F	77.0	1.26	2.77	0.84	1.85	0.55	1.21	15.1	0.74	0.33
400	*182*	0.5	*0.2*	10.3	*4.7*	9.3	*4.2*	A-F	50.4	0.83	1.82	0.49	1.07	0.25	0.56	8.0	0.23	0.17
								M-F	56.0	0.92	2.02	0.54	1.19	0.28	0.62	8.9	0.26	0.19
		1.0	*0.5*	11.0	*5.0*	9.9	*4.5*	A-F	55.8	0.92	2.03	0.57	1.25	0.32	0.71	9.2	0.32	0.18
								M-F	62.0	1.02	2.25	0.63	1.39	0.36	0.79	10.2	0.36	0.20
		1.5	*0.7*	11.3	*5.1*	20.2	*4.6*	A-F	61.7	1.02	2.24	0.65	1.42	0.40	0.87	10.3	0.41	0.22
								M-F	68.5	1.13	2.49	0.72	1.58	0.44	0.97	11.4	0.45	0.24
		2.0	*0.9*	11.1	*5.0*	10.0	*4.5*	A-F	69.3	1.13	2.50	0.76	1.67	0.50	1.09	11.6	0.51	0.26
								M-F	77.0	1.26	2.77	0.84	1.85	0.55	1.21	12.9	0.57	0.29
500	*227*	0.5	*0.2*	12.2	*5.5*	11.0	*5.0*	A-F	50.4	0.83	1.82	0.49	1.07	0.25	0.56	7.7	0.22	0.16
								M-F	56.0	0.92	2.02	0.54	1.19	0.28	0.62	8.5	0.24	0.18
		1.0	*0.5*	13.1	*6.0*	11.8	*5.4*	A-F	55.8	0.92	2.03	0.57	1.25	0.32	0.71	8.5	0.27	0.19
								M-F	62.0	1.02	2.25	0.63	1.39	0.36	0.79	9.4	0.30	0.21
		1.5	*0.7*	13.4	*6.1*	12.1	*5.5*	A-F	61.7	1.02	2.24	0.65	1.42	0.40	0.87	9.3	0.34	0.20
								M-F	68.5	1.13	2.49	0.72	1.58	0.44	0.97	10.3	0.38	0.22
		2.0	*0.9*	13.1	*6.0*	11.8	*5.4*	A-F	69.3	1.13	2.50	0.76	1.67	0.50	1.09	10.3	0.41	0.22
								M-F	77.0	1.26	2.77	0.84	1.85	0.55	1.21	11.4	0.45	0.24
600	*273*	0.5	*0.2*	14.0	*6.4*	12.6	*5.7*	A-F	50.4	0.83	1.82	0.49	1.07	0.25	0.56	7.3	0.21	0.16
								M-F	56.0	0.92	2.02	0.54	1.19	0.28	0.62	8.1	0.23	0.18

(Continued)

TABLE 19–4 *(Continued)*

Weight		Daily Gain		Daily Consumption: As-Fed		Daily Consumption: Moisture-Free Dry Matter		Moisture Basis[5] A-F (as-fed) M-F (moisture-free)	Energy: TDN	Energy: ME (Mcal per)		Energy: NE_m (Mcal per)		Energy: NE_g (Mcal per)		Total Protein	Calcium	Phosphorus
(lb)	(kg)	(lb)	(kg)	(lb)	(kg)	(lb)	(kg)		(%)	(lb)	(kg)	(lb)	(kg)	(lb)	(kg)	(%)	(%)	(%)
Medium-frame heifer calves (Continued)																		
600 (Continued)	273	1.0	0.5	15.0	6.8	13.5	6.1	A-F	55.8	0.92	2.03	0.57	1.25	0.32	0.71	7.9	0.25	0.18
								M-F	62.0	1.02	2.25	0.63	1.39	0.36	0.79	8.8	0.28	0.20
		1.5	0.7	15.3	7.0	13.8	6.3	A-F	61.7	1.26	2.24	0.65	1.42	0.40	0.87	8.6	0.29	0.19
								M-F	68.5	1.13	2.49	0.72	1.58	0.44	0.97	9.5	0.32	0.21
		2.0	0.9	15.0	6.8	13.5	6.1	A-F	69.3	1.13	2.50	0.76	1.67	0.50	1.09	9.4	0.34	0.21
								M-F	77.0	1.26	2.77	0.84	1.85	0.55	1.21	10.4	0.38	0.23
700	318	0.5	0.2	15.7	7.1	14.1	6.4	A-F	50.4	0.83	1.82	0.49	1.07	0.25	0.56	7.1	0.20	0.17
								M-F	56.0	0.92	2.02	0.54	1.19	0.28	0.62	7.9	0.22	0.19
		1.0	0.5	16.8	7.6	15.1	6.9	A-F	55.8	0.92	2.03	0.57	1.25	0.32	0.71	7.6	0.23	0.17
								M-F	62.0	1.02	2.25	0.63	1.39	0.36	0.79	8.4	0.25	0.19
		1.5	0.7	16.9	7.7	15.5	7.0	A-F	61.7	1.26	2.24	0.65	1.42	0.40	0.87	8.1	0.25	0.18
								M-F	68.5	1.13	2.49	0.72	1.58	0.44	0.97	9.0	0.28	0.20
		2.0	0.9	16.9	7.7	15.2	7.0	A-F	69.3	1.13	2.50	0.76	1.67	0.50	1.09	8.6	0.29	0.20
								M-F	77.0	1.26	2.77	0.84	1.85	0.55	1.21	9.6	0.32	0.22
800	364	0.5	0.2	17.3	7.9	15.6	7.0	A-F	50.4	0.83	1.82	0.49	1.07	0.25	0.56	6.9	0.19	0.16
								M-F	56.0	0.92	2.02	0.54	1.19	0.28	0.62	7.7	0.21	0.18
		1.0	0.5	18.6	8.5	16.7	7.6	A-F	55.8	0.92	2.03	0.57	1.25	0.32	0.71	7.3	0.20	0.16
								M-F	62.0	1.02	2.25	0.63	1.39	0.36	0.79	8.1	0.22	0.18
		1.5	0.7	19.1	8.7	17.2	7.8	A-F	61.7	1.26	2.24	0.65	1.42	0.40	0.87	7.7	0.22	0.17
								M-F	68.5	1.13	2.49	0.72	1.58	0.44	0.97	8.5	0.24	0.19
		2.0	0.9	18.7	8.5	16.8	7.6	A-F	69.3	1.13	2.50	0.76	1.67	0.50	1.09	8.1	0.25	0.18
								M-F	77.0	1.26	2.77	0.84	1.85	0.55	1.21	9.0	0.28	0.20
900	409	0.5	0.2	19.0	8.6	17.1	7.8	A-F	50.4	0.83	1.82	0.49	1.07	0.25	0.56	6.8	0.19	0.16
								M-F	56.0	0.92	2.02	0.54	1.19	0.28	0.62	7.5	0.21	0.18
		1.0	0.5	20.3	9,2	18.3	8.3	A-F	55.8	0.92	2.03	0.57	1.25	0.32	0.71	7.0	0.20	0.16
								M-F	62.0	1.02	2.25	0.63	1.39	0.36	0.79	7.8	0.22	0.18
		1.5	0.7	20.9	9.5	18.8	8.5	A-F	61.7	1.26	2.24	0.65	1.42	0.40	0.87	7.3	0.20	0.17
								M-F	68.5	1.13	2.49	0.72	1.58	0.44	0.97	8.1	0.22	0.19
		2.0	0.9	20.3	9.2	18.3	8.3	A-F	69.3	1.13	2.50	0.76	1.67	0.50	1.09	7.7	0.23	0.17
								M-F	77.0	1.26	2.77	0.84	1.85	0.55	1.21	8.5	0.25	0.19
1,000	454	0.5	0.2	20.6	9.4	18.4	8.4	A-F	50.4	0.83	1.82	0.49	1.07	0.25	0.56	6.7	0.18	0.17
								M-F	56.0	0.92	2.02	0.54	1.19	0.28	0.62	7.4	0.20	0.19
		1.0	0.5	22.0	10.0	19.8	9.0	A-F	55.8	0.92	2.03	0.57	1.25	0.32	0.71	6.8	0.18	0.16
								M-F	62.0	1.02	2.25	0.63	1.39	0.36	0.79	7.6	0.20	0.18
		1.5	0.7	22.6	10.3	20.3	9.2	A-F	61.7	1.26	2.24	0.65	1.42	0.40	0.87	7.0	0.19	0.16
								M-F	68.5	1.13	2.49	0.72	1.58	0.44	0.97	7.8	0.21	0.18
		2.0	0.9	22.0	10.0	19.8	9.0	A-F	69.3	1.13	2.50	0.76	1.67	0.50	1.09	7.3	0.20	0.17
								M-F	77.0	1.26	2.77	0.84	1.85	0.55	1.21	8.1	0.22	0.19
Large-frame heifer calves and compensating medium-frame yearling heifers																		
300	136	0.5	0.2	8.7	4.0	7.8	3.5	A-F	48.6	0.80	1.76	0.45	1.00	0.23	0.50	8.6	0.28	0.18
								M-F	54.0	0.89	1.96	0.50	1.10	0.25	0.55	9.5	0.31	0.20
		1.0	0.5	9.3	4.2	8.4	3.8	A-F	53.1	0.88	1.94	0.52	1.15	0.29	0.63	10.2	0.41	0.22
								M-F	59.0	0.98	2.16	0.58	1.28	0.32	0.70	11.3	0.45	0.24
		1.5	0.7	9.8	4.5	8.8	4.0	A-F	57.6	0.95	2.08	0.59	1.29	0.35	0.77	11.7	0.52	0.23
								M-F	64.0	1.05	2.31	0.65	1.43	0.39	0.86	13.0	0.58	0.25
		2.0	0.9	9.9	4.5	8.9	4.0	A-F	62.6	1.03	2.26	0.67	1.47	0.41	1.00	13.1	0.62	0.27
								M-F	69.5	1.14	2.51	0.74	1.63	0.46	1.01	14.6	0.69	0.30
		2.5	1.1	9.7	4.4	8.7	4.0	A-F	69.3	1.13	2.49	0.76	1.67	0.50	1.09	15.0	0.77	0.32
								M-F	77.0	1.26	2.77	0.84	1.85	0.55	1.21	16.7	0.86	0.35
400	182	0.5	0.2	10.8	4.9	9.7	4.4	A-F	48.6	0.80	1.76	0.45	1.00	0.23	0.50	8.0	0.24	0.16
								M-F	54.0	0.89	1.96	0.50	1.10	0.25	0.55	8.9	0.27	0.18

(Continued)

TABLE 19–4 *(Continued)*

Weight		Daily Gain		Daily Consumption: As-Fed		Daily Consumption: Moisture-Free Dry Matter		Moisture Basis[5] A-F (as-fed) M-F (moisture-free)	Energy: TDN	Energy: ME (Mcal per)		Energy: NE_m (Mcal per)		Energy: NE_g (Mcal per)		Total Protein	Calcium	Phosphorus
(lb)	*(kg)*	(lb)	*(kg)*	(lb)	*(kg)*	(lb)	*(kg)*		(%)	(lb)	*(kg)*	(lb)	*(kg)*	(lb)	*(kg)*	(%)	(%)	(%)
Large-frame heifer calves and compensating medium-frame yearling heifers (Continued)																		
400	*182*	1.0	*0.5*	11.7	*5.3*	10.5	*4.8*	**A-F**	**53.1**	**0.88**	***1.94***	**0.52**	***1.15***	**0.29**	***0.63***	**10.0**	**0.32**	**0.19**
(Continued)								M-F	59.0	0.98	*2.16*	0.58	*1.28*	0.32	*0.70*	10.1	0.36	0.21
		1.5	*0.7*	12.1	*5.5*	10.9	*5.0*	**A-F**	**57.6**	**0.95**	***2.08***	**0.59**	***1.29***	**0.35**	***0.77***	**10.2**	**0.41**	**0.20**
								M-F	64.0	1.05	*2.31*	0.65	*1.43*	0.39	*0.86*	11.3	0.45	0.22
		2.0	*0.9*	12.3	*5.6*	11.1	*5.0*	**A-F**	**62.6**	**1.03**	***2.26***	**0.67**	***1.47***	**0.41**	***0.91***	**11.3**	**0.49**	**0.23**
								M-F	69.5	1.14	*2.51*	0.74	*1.63*	0.46	*1.01*	12.6	0.54	0.26
		2.5	*1.1*	9.7	*4.4*	10.8	*4.9*	**A-F**	**69.3**	**1.13**	***2.49***	**0.76**	***1.67***	**0.50**	***1.09***	**12.7**	**0.59**	**0.28**
								M-F	77.0	1.26	*2.77*	0.84	*1.85*	0.55	*1.21*	14.1	0.65	0.31
500	*227*	0.5	*0.2*	12.8	*5.8*	11.5	*5.2*	**A-F**	**48.6**	**0.80**	***1.76***	**0.45**	***1.00***	**0.23**	***0.50***	**7.6**	**0.21**	**0.15**
								M-F	54.0	0.89	*1.96*	0.50	*1.10*	0.25	*0.55*	8.4	0.23	0.17
		1.0	*0.5*	13.8	*6.3*	12.4	*5.6*	**A-F**	**53.1**	**0.88**	***1.94***	**0.52**	***1.15***	**0.29**	***0.63***	**8.5**	**0.27**	**0.18**
								M-F	59.0	0.98	*2.16*	0.58	*1.28*	0.32	*0.70*	9.4	0.30	0.20
		1.5	*0.7*	14.3	*6.5*	12.9	*5.9*	**A-F**	**57.6**	**0.95**	***2.08***	**0.59**	***1.29***	**0.35**	***0.77***	**9.3**	**0.34**	**0.18**
								M-F	64.0	1.05	*2.31*	0.65	*1.43*	0.39	*0.86*	10.3	0.38	0.20
		2.0	*0.9*	14.6	*6.6*	13.1	*6.0*	**A-F**	**62.6**	**1.03**	***2.26***	**0.67**	***1.47***	**0.41**	***0.91***	**10.1**	**0.40**	**0.22**
								M-F	69.5	1.14	*2.51*	0.74	*1.63*	0.46	*1.01*	11.2	0.44	0.24
		2.5	*1.1*	14.2	*6.5*	12.8	*5.8*	**A-F**	**69.3**	**1.13**	***2.49***	**0.76**	***1.67***	**0.50**	***1.09***	**11.2**	**0.48**	**0.23**
								M-F	77.0	1.26	*2.77*	0.84	*1.85*	0.55	*1.21*	12.4	0.53	0.26
600	*273*	0.5	*0.2*	14.7	*6.7*	13.2	*6.0*	**A-F**	**48.6**	**0.80**	***1.76***	**0.45**	***1.00***	**0.23**	***0.50***	**7.3**	**0.20**	**0.16**
								M-F	54.0	0.89	*1.96*	0.50	*1.10*	0.25	*0.55*	8.1	0.22	0.18
		1.0	*0.5*	15.7	*7.1*	14.1	*6.4*	**A-F**	**53.1**	**0.88**	***1.94***	**0.52**	***1.15***	**0.29**	***0.63***	**8.0**	**0.25**	**0.17**
								M-F	59.0	0.98	*2.16*	0.58	*1.28*	0.32	*0.70*	8.9	0.28	0.19
		1.5	*0.7*	16.4	*7.5*	14.8	*6.7*	**A-F**	**57.6**	**0.95**	***2.08***	**0.59**	***1.29***	**0.35**	***0.77***	**8.6**	**0.30**	**0.17**
								M-F	64.0	1.05	*2.31*	0.65	*1.43*	0.39	*0.86*	9.6	0.33	0.19
		2.0	*0.9*	16.7	*7.6*	15.0	*6.8*	**A-F**	**62.6**	**1.03**	***2.26***	**0.67**	***1.47***	**0.41**	***0.91***	**9.3**	**0.34**	**0.20**
								M-F	69.5	1.14	*2.51*	0.74	*1.63*	0.46	*1.01*	10.3	0.38	0.22
		2.5	*1.1*	16.2	*7.4*	14.6	*6.6*	**A-F**	**69.3**	**1.13**	***2.49***	**0.76**	***1.67***	**0.50**	***1.09***	**10.1**	**0.40**	**0.22**
								M-F	77.0	1.26	*2.77*	0.84	*1.85*	0.55	*1.21*	11.2	0.44	0.24
700	*318*	0.5	*0.2*	16.4	*7.5*	14.8	*6.7*	**A-F**	**48.6**	**0.80**	***1.76***	**0.45**	***1.00***	**0.23**	***0.50***	**7.1**	**0.19**	**0.16**
								M-F	54.0	0.89	*1.96*	0.50	*1.10*	0.25	*0.55*	7.9	0.21	0.18
		1.0	*0.5*	17.7	*8.0*	15.9	*7.2*	**A-F**	**53.1**	**0.88**	***1.94***	**0.52**	***1.15***	**0.29**	***0.63***	**7.7**	**0.23**	**0.16**
								M-F	59.0	0.98	*2.16*	0.58	*1.28*	0.32	*0.70*	8.5	0.25	0.18
		1.5	*0.7*	18.4	*8.4*	16.6	*7.5*	**A-F**	**57.6**	**0.95**	***2.08***	**0.59**	***1.29***	**0.35**	***0.77***	**8.1**	**0.26**	**0.17**
								M-F	64.0	1.05	*2.31*	0.65	*1.43*	0.39	*0.86*	9.0	0.29	0.19
		2.0	*0.9*	18.7	*8.5*	16.8	*7.6*	**A-F**	**62.6**	**1.03**	***2.26***	**0.67**	***1.47***	**0.41**	***0.91***	**8.6**	**0.30**	**0.18**
								M-F	69.5	1.14	*2.51*	0.74	*1.63*	0.46	*1.01*	9.6	0.33	0.20
		2.5	*1.1*	18.2	*8.3*	16.4	*7.5*	**A-F**	**69.3**	**1.13**	***2.49***	**0.76**	***1.67***	**0.50**	***1.09***	**9.3**	**0.34**	**0.20**
								M-F	77.0	1.26	*2.77*	0.84	*1.85*	0.55	*1.21*	10.3	0.38	0.22
800	*364*	0.5	*0.2*	18.2	*8.3*	16.4	*7.5*	**A-F**	**48.6**	**0.80**	***1.76***	**0.45**	***1.00***	**0.23**	***0.50***	**6.9**	**0.18**	**0.15**
								M-F	54.0	0.89	*1.96*	0.50	*1.10*	0.25	*0.55*	7.7	0.20	0.17
		1.0	*0.5*	20.0	*9.1*	17.6	*8.0*	**A-F**	**53.1**	**0.88**	***1.94***	**0.52**	***1.15***	**0.29**	***0.63***	**7.4**	**0.22**	**0.16**
								M-F	59.0	0.98	*2.16*	0.58	*1.28*	0.32	*0.70*	8.2	0.24	0.18
		1.5	*0.7*	20.3	*9.2*	18.3	*8.3*	**A-F**	**57.6**	**0.95**	***2.08***	**0.59**	***1.29***	**0.35**	***0.77***	**7.7**	**0.23**	**0.16**
								M-F	64.0	1.05	*2.31*	0.65	*1.43*	0.39	*0.86*	8.6	0.25	0.18
		2.0	*0.9*	20.7	*9.4*	18.6	*8.5*	**A-F**	**62.6**	**1.03**	***2.26***	**0.67**	***1.47***	**0.41**	***0.91***	**8.1**	**0.25**	**0.17**
								M-F	69.5	1.14	*2.51*	0.74	*1.63*	0.46	*1.01*	9.0	0.28	0.19
		2.5	*1.1*	20.1	*9.1*	18.1	*8.2*	**A-F**	**69.3**	**1.13**	***2.49***	**0.76**	***1.67***	**0.50**	***1.09***	**8.6**	**0.30**	**0.19**
								M-F	77.0	1.26	*2.77*	0.84	*1.85*	0.55	*1.21*	9.6	0.33	0.21
900	*409*	0.5	*0.2*	19.8	*9.0*	17.8	*8.1*	**A-F**	**48.6**	**0.80**	***1.76***	**0.45**	***1.00***	**0.23**	***0.50***	**6.8**	**0.18**	**0.16**
								M-F	54.0	0.89	*1.96*	0.50	*1.10*	0.25	*0.55*	7.5	0.20	0.18
		1.0	*0.5*	21.3	*9.7*	19.2	*8.7*	**A-F**	**53.1**	**0.88**	***1.94***	**0.52**	***1.15***	**0.29**	***0.63***	**7.1**	**0.20**	**0.16**
								M-F	59.0	0.98	*2.16*	0.58	*1.28*	0.32	*0.70*	7.9	0.22	0.18

(Continued)

TABLE 19-4 *(Continued)*

Weight		Daily Gain		Daily Consumption As-Fed		Daily Consumption Moisture-Free Dry Matter		Moisture Basis[5] A-F (as-fed) M-F (moisture-free)	Energy TDN	Energy ME (Mcal per)		Energy NE_m (Mcal per)		Energy NE_g (Mcal per)		Total Protein	Calcium	Phosphorus
(lb)	*(kg)*	(lb)	*(kg)*	(lb)	*(kg)*	(lb)	*(kg)*		(%)	(lb)	*(kg)*	(lb)	*(kg)*	(lb)	*(kg)*	(%)	(%)	(%)
Large-frame heifer calves and compensating medium-frame yearling heifers (Continued)																		
900	*409*	1.5	*0.7*	22.2	*10.1*	20.0	*9.1*	**A-F**	**57.6**	**0.95**	***2.08***	**0.59**	***1.29***	**0.35**	***0.77***	**7.4**	**0.21**	**0.16**
(Continued)								M-F	64.0	1.05	*2.31*	0.65	*1.43*	0.39	*0.86*	8.2	0.23	0.18
		2.0	*0.9*	22.6	*10.3*	20.3	*9.2*	**A-F**	**62.6**	**1.03**	***2.26***	**0.67**	***1.47***	**0.41**	***0.91***	**7.7**	**0.23**	**0.16**
								M-F	69.5	1.14	*2.51*	0.74	*1.63*	0.46	*1.01*	8.6	0.26	0.18
		2.5	*1.1*	22.0	*10.0*	19.8	*9.0*	**A-F**	**69.3**	**1.13**	***2.49***	**0.76**	***1.67***	**0.50**	***1.09***	**8.1**	**0.26**	**0.18**
								M-F	77.0	1.26	*2.77*	0.84	*1.85*	0.55	*1.21*	9.0	0.29	0.20
1,000	*454*	0.5	*0.2*	21.4	*9.7*	19.3	*8.8*	**A-F**	**48.6**	**0.80**	***1.76***	**0.45**	***1.00***	**0.23**	***0.50***	**6.7**	**0.17**	**0.16**
								M-F	54.0	0.89	*1.96*	0.50	*1.10*	0.25	*0.55*	7.4	0.19	0.18
		1.0	*0.5*	23.1	*10.5*	20.8	*9.5*	**A-F**	**53.1**	**0.88**	***1.94***	**0.52**	***1.15***	**0.29**	***0.63***	**6.9**	**0.19**	**0.16**
								M-F	59.0	0.98	*2.16*	0.58	*1.28*	0.32	*0.70*	7.7	0.21	0.18
		1.5	*0.7*	24.1	*11.0*	21.7	*9.9*	**A-F**	**57.7**	**0.95**	***2.08***	**0.59**	***1.29***	**0.35**	***0.77***	**7.2**	**0.19**	**0.16**
								M-F	64.0	1.05	*2.31*	0.65	*1.43*	0.39	*0.86*	8.0	0.21	0.18
		2.0	*0.9*	24.4	*11.1*	22.0	*10.0*	**A-F**	**62.6**	**1.03**	***2.26***	**0.67**	***1.47***	**0.41**	***0.91***	**7.4**	**0.21**	**0.16**
								M-F	69.5	1.14	*2.51*	0.74	*1.63*	0.46	*1.01*	8.2	0.23	0.18
		2.5	*1.1*	23.9	*10.9*	21.5	*9.8*	**A-F**	**69.3**	**1.13**	***2.49***	**0.76**	***1.67***	**0.50**	***1.09***	**7.7**	**0.23**	**0.16**
								M-F	77.0	1.26	*2.77*	0.84	*1.85*	0.55	*1.21*	8.6	0.25	0.18
1,100	*500*	0.5	*0.2*	23.1	*10.5*	20.8	*9.5*	**A-F**	**48.6**	**0.80**	***1.76***	**0.45**	***1.00***	**0.23**	***0.50***	**6.6**	**0.17**	**0.16**
								M-F	54.0	0.89	*1.96*	0.50	*1.10*	0.25	*0.55*	7.3	0.19	0.18
		1.0	*0.5*	24.8	*11.3*	22.3	*10.1*	**A-F**	**53.1**	**0.88**	***1.94***	**0.52**	***1.15***	**0.29**	***0.63***	**6.8**	**0.18**	**0.16**
								M-F	59.0	0.98	*2.16*	0.58	*1.28*	0.32	*0.70*	7.5	0.20	0.18
		1.5	*0.7*	25.9	*11.8*	23.3	*10.6*	**A-F**	**57.6**	**0.95**	***2.08***	**0.59**	***1.29***	**0.35**	***0.77***	**6.9**	**0.18**	**0.16**
								M-F	64.0	1.05	*2.31*	0.65	*1.43*	0.39	*0.86*	7.7	0.20	0.18
		2.0	*0.9*	26.2	*11.9*	23.6	*10.7*	**A-F**	**62.6**	**1.03**	***2.26***	**0.67**	***1.47***	**0.41**	***0.91***	**7.1**	**0.19**	**0.16**
								M-F	69.5	1.14	*2.51*	0.74	*1.63*	0.46	*1.01*	7.9	0.21	0.18
		2.5	*1.1*	25.7	*11.7*	23.1	*10.5*	**A-F**	**69.3**	**1.13**	***2.49***	**0.76**	***1.67***	**0.50**	***1.09***	**7.4**	**0.20**	**0.16**
								M-F	77.0	1.26	*2.77*	0.84	*1.85*	0.55	*1.21*	8.2	0.22	0.18

[1]Adapted from *Nutrient Requirements of Beef Cattle,* 6th revised edition, National Research Council—National Academy of Sciences, 1984, pp. 77–83, Table 10.

[2]Shrunk liveweight basis. This refers to weight after an overnight shrink without feed and water (generally equivalent to 96% of unshrunk weights taken in early morning).

[3]Vitamin A requirements are 1,000 IU per lb *(2,200 IU per kg)* of ration.

[4]This table gives reasonable examples of nutrient concentrations that should be suitable to formulate rations for specific management goals. It does not imply that rations with other nutrient concentrations when consumed in sufficient amounts would be inadequate to meet nutrient requirements.

[5]As-fed was calculated using an average figure of 90% dry matter. When feeding silages, roots, and other wet feeds, these feeds should be converted to a moisture-free basis and the ration calculated using the moisture-free data.

TABLE 19-5
MINERAL REQUIREMENTS AND MAXIMUM TOLERABLE LEVELS FOR BEEF CATTLE[1]

Mineral	Requirements Suggested Value	Requirements Range[2]	Maximum Tolerable Level[3]	Mineral	Requirements Suggested Value	Requirements Range[2]	Maximum Tolerable Level[3]
Calcium .... (%)	—	See Tables 19-1, 19-2, 19-3, 19-4	2	**Phosphorus** .... (%)	—	See Tables 19-1, 19-2, 19-3, 19-4	1
Cobalt .... (ppm)	0.10	0.07–0.11	5	**Potassium** .... (%)	0.65	0.5–0.7	3
Copper .... (ppm)	8	4–10	115	**Selenium** .... (ppm)	0.20	0.05–0.30	2
Iodine .... (ppm)	0.5	0.20–2.0	50	**Sodium** .... (%)	0.08	0.06–0.10	10[4]
Iron .... (ppm)	50	50–100	1,000	**Chlorine** .... (%)	—	—	—
Magnesium .... (%)	0.10	0.05–0.25	0.40	**Sulfur** .... (%)	0.10	0.08–0.15	0.40
Manganese .... (ppm)	40	20–50	1,000	**Zinc** .... (ppm)	30	20–40	500
Molybdenum ...... (ppm)	—	6					

[1]Adapted from *Nutrient Requirements of Beef Cattle,* 6th revised edition, National Research Council—National Academy of Sciences, 1984, p. 43.

[2]The listing of a range in which requirements are likely to be met recognizes that requirements for most minerals are affected by a variety of dietary and animal factors (body weight, sex, rate of gain). Thus, it may be better to evaluate rations based on a range of mineral requirements and for content of interfering substances than to meet a specific dietary value.

[3]From National Research Council (1980). Maximum tolerable levels are given on the basis of the ration dry matter.

[4]10% sodium chloride.

TABLE 19-6
MAXIMUM TOLERABLE LEVELS OF CERTAIN TOXIC ELEMENTS[1,2]

Element	Maximum Tolerable Level, ppm
Aluminum	1,000
Arsenic	50 (100 for organic forms)
Bromine	200
Cadmium	0.5
Fluorine	20–100
Lead	30
Mercury	2
Strontium	2,000

[1]Adapted form *Nutrient Requirements of Beef Cattle*, 6th revised edition, National Research Council—National Academy of Sciences, 1984, p. 43.

[2]National Research Council (1980), Table 4, Mineral Requirements and Maximum Tolerable Levels for Beef Cattle.

Energy

Carbohydrates, which constitute about 75% of all the dry matter of plants, are the chief sources of energy in cattle feed. Next to carbohydrates, fats are important as energy sources. In addition to supplying nitrogen, natural plant protein compounds also supply a certain amount of energy.

A relatively large portion of the feeds consumed by beef cattle is used in meeting the energy needs, regardless of whether the animals are merely being maintained (as in wintering) or fed for growth, finishing, or reproduction.

The first and most important function of feeds is that of meeting the maintenance needs. If there is not sufficient feed, as is frequently true during periods of drought or when winter rations are skimpy, the energy needs of the body are met by the breakdown of body tissue. This results in loss of condition and body weight.

Fig. 19-3. High-energy feeds being consumed by growing/finishing cattle. (Courtesy, American Angus Assn., St. Joseph, Mo.)

After the energy needs for body maintenance have been met, any surplus energy may be used for reproduction, lactation, growth, or finishing. When cattle are finished at early ages, growth and finishing are concurrent in most instances and, therefore, not easily separated.

In the finishing process, the percentage of protein, ash, and water steadily decreases as the animal matures and fattens, whereas the percentage of fat increases. Thus, the body of a calf at birth may contain about 70% water and 4% fat; whereas the body of a fat 2-year-old steer may contain only 45 to 50% water but from 30 to 35% fat. This storage of fat requires a liberal allowance of energy feeds.

Through bacterial action in the rumen, cattle are able to utilize a considerable portion of roughages as sources of energy. Yet it must be realized that with extremely bulky rations, the animal cannot consume sufficient quantities to produce the desired amount of gain. For this reason, finishing rations generally contain a considerable proportion of concentrated feeds, mostly cereal grains. On the other hand, when the energy requirements are primarily for maintenance, roughages are usually the most economical sources of energy for beef cattle.

At times, fats may be sufficiently economical to merit consideration as partial substitutes for standard energy feeds. Also, very small amounts of fatty acids are essential for beef cattle, as is true in certain other species, but no exact requirements have thus far been established. Normal cattle rations probably meet such fatty acid requirements.

Research workers of the U.S. Department of Agriculture conducted an experiment to determine some of the economic effects of limited rations, and of the possible harm to animals caused by them.[2] Identical twin calves were used and the following planes of nutrition were studied: (1) full feed—gains of more than 1.5 lb daily; (2) 75% of full feed—gains of 1.0 lb per day; (3) 62% of full feed—gains of 0.5 lb a day; and (4) a maintenance ration of about 50% of full feed—the animals neither gained nor lost weight.

All animals—including those on the low-energy rations—received ample protein, vitamins, minerals, and other nutrients. At the end of the period of retarded feeding, the steers were fed liberally until they reached a slaughter weight of 1,000 lb.

Although the low-plane-of-nutrition animals reached slaughter weight from 10 to 20 weeks later than did their twins, the former attained their weight on approximately the same total feed intake as the latter; which means that, after limited feeding ended, the retarded animals made more economical gains than did their twins. Carcass quality, amount of lean meat, and grade were not affected.

This experiment showed that, under conditions of feed scarcity, beef cattle between the ages of 6 and 12 months can be carried on a maintenance ration—so they will neither gain nor lose weight—provided the nutrient needs other than energy are supplied—without subsequent loss in feed efficiency, carcass quality, or quantity of lean meat. Also, it shows that compensatory gains occur following a low plane

[2]Winchester, C. F., and P. E. Howe, *USDA Tech. Bull. No. 1108.*

of nutrition; it shows why feedlot finishers prefer feeder cattle that have not been backgrounded at a high rate of gain.

• **Symptoms of energy deficiency (underfeeding)**—Many cattle throughout the world are underfed all or part of the year. In fact, lack of sufficient total feed is probably the most common deficiency suffered by beef cattle, although it is recognized that underfeeding is frequently complicated by a concomitant shortage of protein and other nutrients. Restricted rations often occur during periods of drought, when pastures or ranges are overstocked, or when winter rations are skimpy. Also, many range producers regularly plan that cows in good flesh should lose some condition during the winter months; they feel that it is uneconomical to feed enough to retain the fleshy condition. Fortunately, during such times of restricted feed intake, animals have nutritive reserves upon which they can draw. Although they may survive for a considerable period of time under these conditions, there is an inevitable loss in body weight and condition; and, varying with the degree of underfeeding, there may be a slowing or cessation of growth (including skeletal growth), failure to conceive, and increased mortality. Low feed intake also commonly results in increased deaths from toxic plants and from lowered resistance to parasites and diseases.

Protein

The protein allowance for beef cattle, regardless of age or system of production, should be ample to replace the daily breakdown of the tissues of the body and to provide for the growth of hair, horns, and hooves. In general, the protein needs are greatest for the growth of the young calf and for the gestating-lactating cow.

The protein requirements listed in Tables 19–1 to 19–4 are estimated needs for optimal production. They can be exceeded without toxicity or reduced animal performance. As noted, the requirements are expressed on the basis of both total and digestible protein. Nitrogen values were converted to digestible protein values by multiplying by the factor 6.25 and using an average biological value of 77.5. Cattle fed these levels of protein have gained and reproduced at optimum rates. Methods of feeding, feed preparation, and various feed additives do not appear to alter protein requirements. Feed consumption is reduced when all-concentrate rations are fed. As consumption declines, the percentage of protein in such rations should be increased proportionally.

As protein supplements ordinarily cost more per ton than grains, normally beef cattle should not be fed larger quantities of these supplements than actually are needed to balance the ration.

With stocker cattle, or in the maintenance of the beef breeding herd, it usually does not pay to add a protein supplement when a legume hay is fed. With feedlot cattle fed high-concentrate rations, or when the breeding herd is being wintered on a nonlegume roughage, sufficient protein supplement—usually 1 to 2 lb daily—should be added to the ration.

Because of rumen synthesis of essential amino acids by microorganisms, the quality of protein (or balance of essential amino acids) is of less importance in the feeding of beef cattle than in feeding some other classes of stock. Protein from plant sources, therefore, is quite satisfactory. Also, these microorganisms are able to use inorganic compounds such as ammonia, just as plants utilize chemical fertilizers—build body protein of high quality in their cells from sources of inorganic nitrogen that nonruminants cannot use. Since the life-span of these bacteria is short, further on in the digestive tract, the ruminant digests the bacteria and obtains good protein therefrom. In ruminant nutrition, therefore, even such nonprotein sources of combined nitrogen as urea and ammonia have a protein replacement value. An exception is the very young ruminant in which the rumen and its ability to synthesize are not yet well developed. For such an animal, high-quality protein in the ration is requisite to normal development. In the suckling calf, milk provides such protein.

• **Symptoms of protein deficiencies and toxicities**—Depressed appetite is the primary symptom of protein deficiency in beef cattle rations. Depressed appetite may, in turn, lead to an inadequate intake of energy; hence, protein deficiency and energy deficiency often occur together.

Other symptoms of protein deficiency are loss of weight, poor growth, irregular or delayed estrus, and reduced milk production.

Rations containing up to 40% protein have been fed to steers. Feed intake was reduced for several days when protein was added, but no signs of ammonia toxicity were evident.[3] However, excesses of nonprotein nitrogen or soluble protein may precipitate ammonia toxicity.

Minerals

Beef cattle are susceptible to the usual inefficiencies and ailments when exposed to (1) prolonged and severe mineral deficiencies, or (2) excesses of fluorine, selenium, or molybdenum (see Chapter 5, Nutritional Disorders/Toxins).

Needed minerals may be incorporated in beef cattle rations or in the water. In addition, it is recommended that all classes and ages of cattle be allowed free access to a two-compartment mineral box, with (1) salt (iodized salt in iodine-deficient areas) in one side, and (2) a suitable mineral mixture in the other side. Free-choice feeding is in the nature of cheap insurance, with the animals consuming the minerals if they are needed.

The calcium and phosphorus requirements of cattle are presented in Tables 19–1, 19–2, 19–3, and 19–4. The requirements and maximum tolerable levels of other minerals are presented in Table 19–5. Maximum tolerable levels of several elements that are known to be toxic to cattle are presented in Table 19–6. (Also see Chapter 5, Nutritional Disorders/Toxins, Table 5–1, for additional information relative to mineral toxicities.)

BEEF CATTLE MINERAL CHART

Table 19–7, Beef Cattle Mineral Chart (see pp. 720–725), presents in summary form pertinent information pertaining to the mineral needs of beef cattle.

[3]Fenderson, C. L. and W. G. Bergen, "Effect of Excess Dietary Protein on Feed Intake and Nitrogen Metabolism in Steers," *Journal of Animal Science*, Vol. 42, 1976, p. 1323.

MAJOR OR MACROMINERALS

A discussion of the major or macromineral needs of beef cattle follows.

Salt (NaCl)

Salt should be available at all times. It may be fed in the form of granulated, half ground, or block salt, but because of weathering losses, flake salt is usually not satisfactory for feeding in the open. If block salt is used, the softer types should be selected.

Most ranchers compute the yearly salt requirements on the basis of about 25 lb for each cow. Mature animals will consume 3 to 5 lb of salt per month when pastures are lush and succulent, and 1 to 1½ lb per month during the balance of the season.

The careful location of the salt supply is recognized as an important adjunct in proper range management. Through judicious scattering of the salt supply and the moving of it at proper intervals, the animals can be distributed more properly; and overgrazing of certain areas can be minimized.

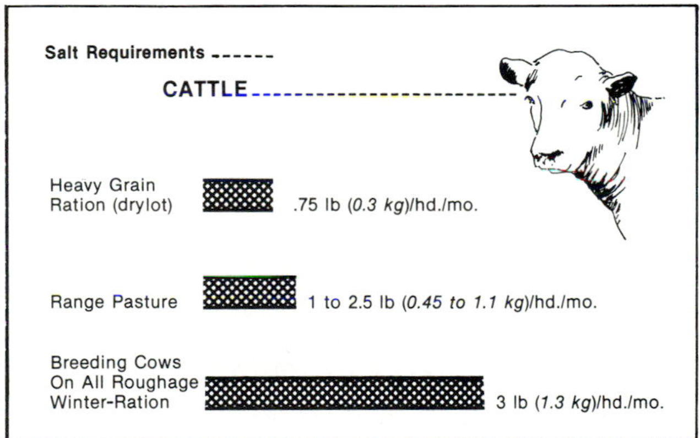

Fig. 19-4. The average salt requirements of cattle.

Calcium (Ca)

In contrast to phosphorus deficiency, calcium deficiency in beef cattle is relatively rare. In general, when the forage consists of at least one-third legume (legume hay, pasture, or silage), ample calcium will be provided. But even non-legume forages contain more calcium than cereal grains. This indicates that a mineral source of calcium is less necessary when large quantities of roughage are being consumed. Also, plants grown on calcium-rich soils may contain a higher content of this element.

As finishing cattle consume a high proportion of grains to roughages—and the grains are low in calcium—they have a greater need for a calcium supplement than do beef cattle that are being fed largely on roughages. This is especially true of cattle of the younger ages and when a long feeding period is involved.

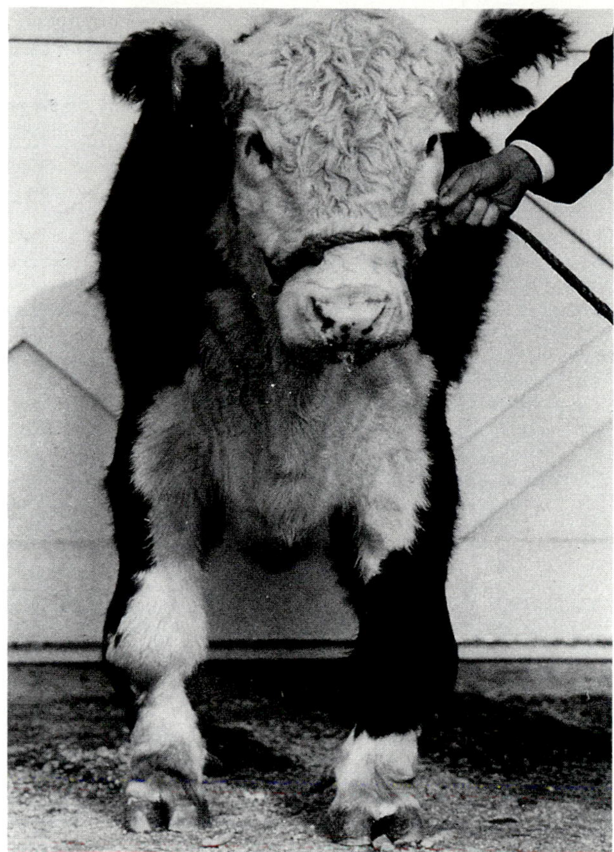

Fig. 19-5. This heifer developed rickets early in life due to a deficiency of calcium and phosphorus. Note the bowed front legs and enlarged joints. (Courtesy, USDA)

When the ration of beef cattle is suspected of being low in calcium, the animals should be given free access to a calcium supplement, with salt provided separately; or a calcium supplement may be added to the ration in keeping with nutrient requirements. (See Tables 19-1, 19-2, 19-3, 19-4, and 19-5.)

Appendix Table V-1F, Mineral Supplements, gives the composition of mineral supplements for beef cattle.

Phosphorus (P)

Phosphorus deficiencies in cattle are widepread. In some sections of the United States and other countries, the soils are so deficient in phosphorus that the feeds produced thereon do not provide enough of this mineral for cattle or other classes of stock. As a result, the cattle produced in these areas may have depraved appetites, may fail to breed regularly, and may produce markedly less milk. Growth and development are slow, and the animals become emaciated and fail to reach normal adult size. Death losses are abnormally high.

In range areas where the soils are either known or suspected to be deficient in phosphorus, cattle should always be given free access to a suitable phosphorus supplement.

TABLE
BEEF CATTLE MINERAL

Mineral; Absorption; Excretion	Conditions Usually Prevailing Where Deficiencies Are Reported	Functions of Mineral	Deficiency Symptoms; Toxicity*
Major or Macrominerals:			
SALT (NaCl, sodium and chloride)—The requirements for sodium and chlorine are commonly expressed as salt requirements because salt is an effective, economical way of supplementing rations with these elements. **Absorption**—Sodium and chlorine are mainly absorbed from the proximal portion of the small intestine, but they may also be absorbed from the distal section of the small intestine and from the large intestine. Also, some absorption of sodium and chlorine may occur from the rumen. **Excretion**—Excess salt is excreted in the urine.	Negligence; for salt is inexpensive. Deficiencies of sodium and chlorine may occur because plants have low sodium contents, because sodium losses caused by perspiration may occur in animals maintained in warm environments or used for hard work, and because sodium needs increase during lactation and during periods of rapid growth.	Sodium (Na) functions in maintaining osmotic pressure, acid-base balance, and body-fluid balance; is involved in nerve transmission and active transport of amino acids; is required for cellular uptake of glucose through activation of the glucose carrier protein; and is a major cation of extracellular fluid and provides a majority of the alkaline reserve in plasma. Chlorine (Cl) is necessary for the activation of amylase; is essential for the formation of gastric hydrochloric acid; and is involved in respiration and regulation of blood pH, through the chloride shift.	**Deficiency symptoms**—Intensive craving of salt, manifested by the animals chewing and licking various objects, and by muscle cramps. Prolonged deficiency results in lack of appetite, unthrifty appearance, and decreased production. High-producing milk cows may collapse and die when salt deficiency has been of long duration. It is noteworthy that when salt is omitted, sodium expresses its deficiency first. **Toxicity**—*The NRC gives the maximum tolerable level of salt (NaCl) as 10% of ration dry matter. As much as 3 lb (*1.4 kg*) can be consumed per cow daily without harm provided animals have access to plenty of water.
CALCIUM (Ca)—Calcium is the most abundant mineral in the body. Most of the calcium in the body is found in the bones and teeth. It constitutes 2% of the body weight. In blood, calcium is found mostly in the plasma, with a controlled concentration of 10 mg/100 ml. **Absorption**—Calcium is absorbed actively from both the duodenum and the jejunum; but most calcium is absorbed in the proximal portion of the duodenum. **Excretion**—Calcium is excreted mainly in feces with only small quantities appearing in urine.	A calcium deficiency may occur when finishing cattle are fed heavily on concentrates and limited quantities of nonlegume roughage, especially young cattle on a long feed. Adding calcium to such a ration increases the rate of gain, improves feed utilization, results in heavier, stronger bones, and enhances market grades. Also, a calcium deficiency may occur when the ration consists chiefly of dried mature grasses or cereal straws, and when cows are in heavy lactation. Osteomalacia may occur when there are high metabolic demands on calcium and phosphorus stores, such as occur during pregnancy and lactation.	Essential for bone formation, development of teeth, production of milk, transmission of nerve impulses, maintenance of normal muscle excitability (along with sodium and potassium), regulation of heart beat, movement of muscles, blood clotting (conversion of prothrombin to thrombin), and activation and stabilization of enzymes (*i.e.,* pancreatic amylase).	**Deficiency symptoms**—A deficiency of calcium results in rickets in young animals and osteomalacia in older animals. Rickets may be caused by a deficiency of calcium, phosphorus, or vitamin D. It is characterized by improper calcification of the organic matrix of bones of young, growing animals. Thus, the bones are weak, soft, and lack density. Signs include swollen tender joints; enlargement of the ends of bones; and arched back; stiffness of the legs; and development of beads on the ribs. If the cause is not corrected, calves develop bowed and deformed legs. Also, rachitic bones are highly susceptible to fracture. Osteomalacia is the result of demineralization of the bones of adult animals. This condition is characterized by weak, brittle bones that may break when stressed. **Toxicity**—Ruminants tolerate high levels of caclium. *The NRC gives the maximum calcium level as 2% of ration dry matter. However, when high levels of calcium are fed, there may be reduced feed consumption and daily gains; reduced protein and energy digestibilities; reduced absorption of tetracyclines, manganese, and zinc; and stimulation of the production of calcitonin of the thyroids. Calcitonin inhibits bone resorption, with the result that the bones may thicken (osteopetrosis) because of continued deposition but limited resorption.
PHOSPHORUS (P)—Phosphorus has varied, but extremely important, biochemical and physiological roles **Absorption**—Phosphorus absorption is dependent on source, intestinal pH, age of animal, and ration levels of sodium, calcium, iron, aluminum, manganese, potassium, magnesium, and fat. **Excretion**—Excess phosphorus is excreted primarily in the feces.	Semiarid regions are commonly associated with soils deficient in phosphorus. The phosphorus content of plants generally decreases markedly with maturity, with the result that deficiencies often occur in cattle subsisting for long periods on mature dried forage. High iron levels result in the formation of insoluble iron phosphate. Also, aluminum forms insoluble, unavailable phosphates.	Phosphorus is deposited in bones. It is also found in high concentrations in brain, muscle, liver, spleen, and kidneys. Phosphorus, as a component of phospholipids, influences cell permeability and is a component of myelin sheathing of nerves. Also, many energy transfers in cells involve the high-energy phosphate bonds in ATP. Phosphorus plays an important role in blood buffer systems. Activation of several B-vitamins (thiamin, niacin, pyridoxine, riboflavin, biotin, and pantothenic acid) to form coenzymes requires their initial phosphorylation. Phosphorus is also a part of the genetic materials DNA and RNA.	**Deficiency symptoms**—Phosphorus deficiencies in cattle are widespread. A deficiency of phosphorus results in decreased growth rates, in inefficient feed utilization, and in a depraved appetite (chewing of wood, soil, and bones—called pica); anestrus, low conception rate, and reduced milk production; low plasma phosphorus levels, and weak, fragile bones and stiffness of joints. **Toxicity**—*The NRC gives the maximum phosphorus level as 1% of ration dry matter. High phosphorus intakes may cause bone resorption, elevated plasma phosphorus levels, and urinary calculi.
MAGNESIUM (Mg)—Magnesium is the fourth most abundant cation in the body. **Absorption**—Absorption of magnesium occurs prior to the intestines, from the small intestine, and some from the large intestine. **Excretion**—Excretion of endogenous magnesium is primarily via feces. However, excess magnesium is disposed of primarily via urine.	When milk feeding of calves is prolonged without grain or hay. (Milk is rather low in magnesium.) When there is grass tetany, which is most likely to occur when beef cows in early lactation graze early spring pastures containing less than 0.2% magnesium.	Magnesium is required for skeletal development as a constituent of bone; plays an important role in neuromuscular transmission and activity; is required to activate many enzyme systems, including those involving ATP; and is required as a cofactor in decarboxylation and an activator of many peptidases. Approximately 65% of total body magnesium is contained in bone; the other 35% is distributed among various tissues and organs.	**Deficiency symptoms**—Grass tetany or grass staggers, characterized by anorexia, hyperemia, hyperirritability, convulsions, and death. Magnesium-deficient cattle exhibit loss of appetite and reduced dry matter digestibilities. Deficiencies in young cattle may result in defective bones and teeth. **Toxicity**—Normal rations will not cause toxicity. The maximum tolerable level of magnesium is considered to be 0.4% of the ration. Toxicity is characterized by loss of appetite, reduced performance, and occasional diarrhea. Also, cattle experiencing toxicity may exhibit lack of reflexes and respiration depression.

19–7
CHART (See footnotes at end of table.)

Nutrient Requirements[1]				
Daily Nutrients/ Animal	**Percentage of Ration**	**Recommended Allowances[1]**	**Practical Sources of the Mineral**	**Comments**
For young, growing animals: 2–3 g of sodium, and less than 5 g of chlorine. For lactating cows: 11 g of sodium, and 15 g of chlorine.	*Sodium concentrations of 0.06–0.10% of ration dry matter for nonlactating yearlings and calves, and no more than 0.1% dry matter for lactating beef cows. (See Table 19–5.)	Cows on pasture or fed high-roughage winter rations will consume from 1–3 lb (*0.45–1.36 kg*) salt per head per month; finishing steers fed heavy grain rations in drylot will consume 1–3.5 lb (*0.45–1.59 kg*) per head per month; a wide range due to differences in age, rations, form of salt (rock, coarse ground, or block). Most ranchers compute the yearly salt requirements on the basis of 25 lb (*10 kg*) per cow. The careful location of the salt supply is an important adjunct in range management.	Salt should be available at all times. It should be both (1) self-fed, free-choice, and (2) mixed with other ration ingredients. Free access to salt in the form of loose rock, coarse ground, or block salt. Cattle prefer loose salt to block salt, because it can be eaten more rapidly and with less effort. However, experiments with lactating cows have shown fully as good results with block salt as with loose salt even though smaller quantities were consumed. This means that the additional intake of loose salt over block salt does not appear to benefit cattle. Commercial mineral mixes (in block, or loose form) may contain ⅓ or more salt.	The salt requirements of cattle differ (1) between individuals, (2) according to whether milk is produced (being higher for lactating cows than for dry cows, because of the salt in the milk), (3) from season to season, (4) between block and loose salt (animals often consuming twice as much easy-to-get loose salt as block salt), and (5) according to the salt content of the soil, feed, and water (being higher when vegetable proteins are fed than when animal proteins are fed, higher on predominantly forage rations than on predominantly concentrate rations, and higher on lush early pasture than on more mature grasses). These are some of the reasons why free-choice feeding of salt is advocated.
*Variable, according to age weight, and type and level of production of cattle. (See Tables 19–1 and 19–3.) Because true digestibilities of calcium in feedstuffs vary, the dietary calcium requirements shown in the tables may in some instances need to be adjusted.	*Variable, according to age, weight, and type and level of production of cattle (See Tables 19–2 and 19–4.) Because true digestibilities of calcium in feedstuffs vary, the dietary calcium requirements shown in the tables may in some instances need to be adjusted.	Free access to a calcium supplement, or a calcium supplement incorporated in the ration.	Legumes are high in calcium. Also, several of the oilseed meals are good sources of calcium. Sources of supplemental calcium include calcium carbonate, ground limestone, bone meal, dicalcium phosphate, defluorinated phosphate, monocalcium phosphate, and calcium sulfate. Where both calcium and phosphorus need to be supplemented, they should be provided in a readily available and palatable form such as dicalcium phosphate, defluorinated phosphate, or bone meal.	In addition to an adequate supply of calcium, proper utilization is dependent upon (1) a highly available source of the mineral, (2) a suitable ratio between calcium and phosphorus (somewhere between 1 and 2 parts of calcium to 1 part of phosphorus). Calcium-phosphorus ratios of 2:1 have been shown to be beneficial in reducing urinary calculi. When calculi problems are encountered, even higher levels of calcium may be advisable. Ratios between calcium and phosphorus of 7:1 have been reported to be satisfactory for cattle. Generally, when cattle receive at least ⅓ of a legume forage, ample calcium will be provided. But even nonlegume forages contain more calcium than cereal grains. Plants grown on calcium-rich soils are high in calcium. Calcium availability of 70% is generally assumed for all feedstuffs.
*Variable, according to age, weight, and type and level of production. (See Tables 19–1 and 19–3.)	*Variable according to age, weight, and type and level of production. (See Tables 19–2 and 19–4.)	Free access to a phosphorus supplement, or a phosphorus supplement added to the daily ration in keeping with the nutrient requirements. Where phosphorus is added to water, either of the following methods may be employed: 1. Added by hand at rate of ¼ oz (*7 g*) of monosodium phosphate/8 gal (*30 liter*) water, or ¼ oz/head/day. 2. Added by dispenser, using stock solution of 2½ lb (*1.13 kg*) of monosodium phosphate/gal (*3.8 liter*) water (or 100 lb/40 gal [*45 kg/151 liter*] water).	Common sources of phosphorus are: dicalcium phosphate, defluorinated phosphate, bone meal, soft phosphate, sodium phosphate, ammonium polyphosphate, orthophosphates, metaphosphates, pyrophosphates, and tripolyphosphate. Oilseed meals and animal and fish products contain large amounts of phosphorus. Phytate phosphorus is not well utilized by nonruminants, but ruminants appear to use considerable quantities of this form of phosphorus.	Grains, grain by-products and high-protein supplements are fairly high in phosphorus; hence, rations high in such ingredients require little or no phosphorus supplementation. Calcium-phosphorus ratios of 2:1 are beneficial in reducing urinary calculi; and even higher levels of calcium may be necessary when urinary calculi is encountered. Ratios between calcium and phosphorus of 7:1 have been reported to be satisfactory for cattle.
*Young calves and growing-finishing cattle, 12–30 mg/kg body weight. *Beef cows, 7–9 g/day during gestation and 21, 22, and 18 g/day during early, mid, and late lactation, respectively. Magnesium requirements are increased by feeding high levels of aluminum, potassium, phosphorus, or calcium; by younger cattle and magnesium-deficient cattle; and by high levels of milk production.	*0.10% of dry matter, with a range of 0.05–0.25.		Commonly used feedstuffs vary widely in magnesium content and availability. Magnesium carbonate, oxide, and sulfate are good sources of supplemental magnesium.	Supplemental feeding of magnesium (20 g/day) reduces the incidence of grass tetany in many outbreaks.

(Continued)

TABLE 19-7

Mineral; Absorption; Excretion	Conditions Usually Prevailing Where Deficiencies Are Reported	Functions of Mineral	Deficiency Symptoms; Toxicity*
Major or Macrominerals (Continued)			
POTASSIUM (K)—Potassium is the third most abundant mineral element in the body. **Absorption**—Potassium is primarily absorbed in the small intestine. **Excretion**—Excretion is mainly via the kidneys.	When drylot finishing cattle receive high- or all-concentrate rations.	Essential for proper enzyme, muscle, and nerve function, rumen microorganism activity, and appetite.	**Deficiency symptoms**—Poor appetite and feed conversion, slow growth, stiffness, and emaciation. **Toxicity**—*The NRC gives the maximum tolerable level of potassium as 3% of ration dry matter. Toxicity from excessive intake is unlikely except (1) when water intake is restricted or water is saline, or (2) when the kidneys are not functioning properly.
SULFUR (S)—Sulfur is a component of protein, some vitamins, and several important hormones.	Cattle fed high-grain rations supplemented with nonprotein nitrogen.	Body functions that involve sulfur include protein synthesis and metabolism, fat and carbohydrate metabolism, blood clotting, endocrine function, and intra- and extracellular fluid acid-base balance. Sulfur has both structural and metabolic functions; it is found in virtually every tissue and organ of the body. Muscle has a fairly constant nitrogen to sulfur ratio of 15.3:1. The total body content of sulfur is approximately 0.15%.	**Deficiency symptoms**—Depressed appetite, loss of weight, weakness, excessive salivation, watery eyes, dullness, emaciation, and death. A lack of sulfur also results in a microbial population that does not utilize lactate. **Toxicity**—*The NRC gives the toxic level of sulfur as 0.40% of the ration dry matter. Sulfur toxicity is characterized by restlessness, diarrhea, muscular twitching, dyspnea, and in prolonged cases of inactivity followed by death.
Trace or Microminerals:			
COBALT (Co)—The cobalt requirement of cattle is actually a cobalt requirement of rumen microorganisms. The microbes incorporate cobalt into vitamin B-12, which is utilized by both microorganisms and animal tissues. **Absorption**—Vitamin B-12 is absorbed in the lower part of the small intestine. **Excretion**—Cobalt and vitamin B-12 are mainly excreted in the feces, although variable amounts are excreted in urine.	In cobalt-deficient soils where this element is not provided. Cobalt-deficient soils occur in many parts of the world, with large deficient areas in Australia, New Zealand, and along the southeast Atlantic Coast of the U.S.	The main function of cobalt is to serve as an integral part of vitamin B-12 (cobalamin). Vitamin B-12 is of importance in the metabolism of propionic acid, needed for the activity of the enzyme methylamalonyl-CoA isomerase. Vitamin B-12 is also a part of the enzyme that catalyzes the recycling of methionine from homocysteine after the loss of its labile methyl group. Vitamin B-12 is also needed for normal liver folate metabolism.	**Deficiency symptoms**—Loss of appetite and body weight, muscular wasting, severe anemia, followed by death. In severe deficiency, the mucous membranes become blanched, the skin turns pale, a fatty liver develops, and the body becomes almost totally devoid of fat. **Toxicity**—Cobalt toxicity is rare because toxic levels are about 300 times requirement levels. *The NRC gives the maximum tolerable level of cobalt as 5% of the ration dry matter.
COPPER (Cu) **Absorption**—Copper is absorbed from the upper portion of the duodenum. Zinc and silver are antagonistic to copper absorption. **Excretion**—Copper is released into bile, thence into feces. Trace amounts of copper are excreted in urine, perspiration, and milk.	In copper-deficient areas (soils), as in Florida and the Coastal Plain region. On peat and muck soils, or where soil molybdenum levels are high. Deficiencies have occurred in calves kept on an exclusive milk diet for long periods.	Copper is necessary in hemoglobin formation, iron absorption from the small intestine, iron mobilization from tissue stores, and for the oxidation of iron, permitting it to bind with the iron transport—transferrin. Copper is essential in enzyme systems, hair development and pigmentation, bone development, reproduction, and lactation.	**Deficiency symptoms**—Emaciation, depigmentation (cattle turn yellowish) and loss of hair, stunted growth, anemia, and brittle and malformed bones. Also, heat periods are suppressed, and there may be depraved appetite and diarrhea. Young calves may have straight pasterns and stand forward on their toes. Low copper intake reduces the synthesis and activity of the copper-containing enzyme, tyrosinase, which is required for pigmentation of hair, wool, and feathers. **Toxicity**—*Maximum tolerable levels for cattle are 115 ppm of dry matter. Acute toxicity may cause nausea, vomiting, salivation, abdominal pain, convulsions, paralysis, collapse, and death. Also, high copper levels may predispose animals to anemia, muscular dystrophy, decreased growth, and impaired reproduction.
IODINE (I) **Absorption**—In ruminants, the rumen is the primary absorption site. **Excretion**—Two-thirds of ingested inorganic iodine is excreted by the kidneys.	In iodine-deficient areas (soils) where iodized salt is not fed (in northwestern U.S. and in the Great Lakes Region). Where feeds come from iodine-deficient areas Substances that interfere with iodine metabolism. Rapeseed meal, soybean meal, and cottonseed meal have goitrogenic effects.	Inorganic iodine is taken up by the thyriod gland for the synthesis of thyroid hormones. Thyroid hormones have an active role in thermoregulation, intermediary metabolism, reproduction, growth and development, circulation, and muscle function.	**Deficiency symptoms**—Goiter, hairlessness in the young; retarded growth and maturity, lowered metabolic rate, and increased water retention. Occasional borderline cases may survive; in these, the moderate thyroid enlargement disappears in a few weeks. **Toxicity**—*50 ppm is the maximum tolerable level for calves. Symptoms of iodine toxicity include loss of appetite, coma, and death.
IRON (Fe) **Absorption**—Iron may be absorbed from all sections of the small intestine, but the principal site of absorption is the duodenum. Ferrous iron is absorbed to a much greater extent than ferric iron. **Excretion**—Excretion of iron occurs in urine, feces, sweat, dermis, and blood.	Calves on an exclusive milk ration (milk contains less than 10 ppm iron). Animals with excessive blood loss.	Iron has important biochemical functions in animals since it is a component of hemoglobin, myoglobin, cytochrome, and the enzymes catylase and peroxidase. Iron in these materials exists in porphyrin rings. Iron is involved in the transport of oxygen to cells and in cellular respiration.	**Deficiency symptoms**—Signs of lack of iron include anemia, reduced saturation of transferrin, listlessness, pale mucous membrane, reduced appetite and weight gain, and atrophy of the papillae of the tongue. **Toxicity**—*An iron level of 1,000 ppm is considered as the maximum tolerable level for cattle. Iron toxicity is characterized by reduced feed intake, reduced daily gain, diarrhea, hypothermia, and metabolic acidosis.

(Continued)

Nutrient Requirements[1]				
Daily Nutrients/ Animal	**Percentage of Ration**	**Recommended Allowances[1]**	**Practical Sources of the Mineral**	**Comments**
	*0.65% of the total ration dry matter, with a range of 0.5–0.7%. The needs for potassium vary with amounts of protein, phosphorus, calcium, and sodium consumed.	*0.7–1.0% of the total ration dry matter.	Roughages usually contain ample potassium. Potassium chloride is the supplement of choice.	Grains often contain less than 0.5% potassium. Excessive levels of potassium have been found to interfere with magnesium absorption. Also, excessive levels of potassium, along with high levels of phosphorus, increase the incidence of phosphatic urinary calculi.
	*0.10% of ration dry matter, with a range of 0.08–0.15%.	*The NRC suggested maximum level of sulfur in the ration is 0.4%.	Feeds high in protein are usually high in sulfur. The microbial population of the rumen has the ability to convert inorganic sulfur into organic sulfur compounds that can be used by the animal. So, either organic or inorganic sulfur can be utilized by cattle. Most feedstuffs provided to beef cattle contain sufficient sulfur to meet their needs.	Copper requirements are increased by both sulfur and molybdenum. Selenium can replace sulfur in some organic compounds.
	*0.10 ppm of ration dry matter, with a range of 0.07–0.11 ppm of dry matter.	Free access to a cobaltized mineral mixture in cobalt-deficient areas; or administering a cobalt pellet.	A cobaltized mineral mixture may be prepared by adding cobalt at the rate of 0.2 oz/100 lb (*1.25 mg/kg*) of salt as cobalt chloride or cobalt sulfate, cobalt carbonate, cobalt oxide, or a good commercial mineral mixture or salt product may be used. Also, cobalt sulfate and cobalt oxide are effective as a drench; and a cobalt pellet (composed of cobalt oxide and finely divided iron) that lodges in the reticulum is an effective preventive.	Several good commercial cobalt-containing minerals are on the market. A vitamin B–12 injection will relieve a cobalt deficiency.
	*8 ppm of dry matter, with a range of 4–10 ppm of dry matter. For presence of high levels of molybdenum and inorganic sulfate, increase the copper requirements.	*Copper deficiency can be prevented by adding 0.25–0.5% copper sulfate to salt fed free-choice. *Copper (Cu) added to total feed (dry basis) 4 ppm. Copper may also be injected as glycinate to meet the nutritional needs for the mineral.	*Salt containing 0.25–0.5% copper sulfate.	Copper deficient cattle can be returned to normal by feeding 3 g of copper sulfate or blue vitriol every 10 days. An interesting interrelation exists between copper and molybdenum. An excess of molybdenum (in the presence of sulfate) causes a condition which can be cured only by administering copper. Excess copper is toxic; it accumulates in the liver, and death may result.
*1 mg/day for a 1,100-lb (*500-kg*) cow.	*Iodized salt at rate of 0.10% of dry ration. *0.5 ppm iodine in dry matter, with a range of 0.20–2.0 ppm iodine in dry matter.	Free access to stabilized iodized salt containing 0.01% potassium iodide (0.0076% iodine).	Stabilized iodized salt containing 0.01% potassium iodide. Feed additives that supply iodine are: ethylenediamine dihydroiodide (EDDI), calcium iodate, cuprous iodide, potassium iodate, sodium iodate, potassium iodide, sodium iodide, and pentacalcium periodate.	The enlargement of the thyroid gland (goiter) is nature's way of trying to make enough thyroxin, when there is insufficient iodine in the feed. Eighty percent of hormonal iodine stored in the thyroid is thyroxin. The amount of iodine in milk is influenced by iodine intake, season, level of milk production, and use of iodine disinfectants.
	*100 ppm for calves; 50 ppm for older cattle.		Levels of iron in common feed believed to be ample. Sources of supplemental iron in decreasing order of availability are: ferrous sulfate, ferrous carbonate, ferric chloride, and ferric oxide.	After calves are past 20 weeks of age, iron does not seem to be beneficial. About 30% of all calves are affected by prenatal iron deficiency. In cattle, a majority of body iron is in the form of hemoglobin, with lesser amounts existing as protein-bound stored iron, myoglobin, and cytochrome.

(Continued)

TABLE 19–7

Mineral; Absorption; Excretion	Conditions Usually Prevailing Where Deficiencies Are Reported	Functions of Mineral	Deficiency Symptoms; Toxicity*
Trace or Microminerals (Continued) **MANGANESE (Mn)** **Absorption**—Ruminants regulate manganese levels in blood and tissue via intestinal absorption. **Excretion**—Manganese is excreted via feces, with little in the urine.	In northwestern U.S. All-concentrate rations based on corn supplemented with nonprotein nitrogen.	Manganese is essential for normal reproduction in both males and females, for bone formation, and for the functioning of the central nervous system. Also, manganese is a preferred metal cofactor for many enzymes involved in carbohydrate metabolism and in mucopolysaccharide synthesis.	**Deficiency symptoms**— In males: impaired spermatogenesis, testicular and epididymal degeneration, sex hormone inadequacy, and sterility. In females: irregular and absent estrus, delayed conception, abortion, and deformed young at birth—crooked calves. **Toxicity**—For ruminants, manganese is among the least toxic of required minerals. *With balanced rations, about 1,000 ppm is the maximum tolerable level on a short-term basis for cattle.
MOLYBDENUM (Mo)—Molybdenum is found in nearly all body cells and fluids. **Absorption**—Molybdenum is well absorbed by cattle, chiefly from the small intestine. **Excretion**—Excretion of molybdenum is primarily via urine, with small amounts excreted in bile and milk.	Molybdenum toxicity occurs only occasionally in cattle and appears to be an area problem.	Molybdenum is a constituent of the enzymes xathine oxidase, aldehyde oxidase, and sulfide oxidase; enzymes involved in the oxidation of purines and reduction of cytochrome C.	**Deficiency symptoms**—Molybdenum deficiencies have not been demonstrated in cattle. **Toxicity**—*The NRC gives the maximum tolerable level as 6 ppm. Clinical signs of molybdenum toxicity in cattle are diarrhea, loss of appetite, anemia, ataxia, and bone malformation.
SELENIUM (Se)—Initially, interest in selenium was confined to the problem of toxicity in animals. **Absorption**—Most selenium is absorbed in the duodenum. **Excretion**—Selenium is excreted in feces and urine; fecal excretion is greater than urinary excretion in ruminants.	Low selenium forage and low vitamin E. It is an area problem, but it occurs in many parts of the U.S.	Selenium functions (1) as a component of glutathione peroxidase, an enzyme that destroys peroxides in tissues, and (2) intertwined with vitamin E in a mutual sparing effect.	**Deficiency symptoms**—White muscle disease; characterized by white muscle, heart failure, and paralysis evidenced by lameness or inability to stand. Depression of glutathione peroxidase in tissues of selenium-deficient animals may account for many of the manifestations of selenium deficiency. **Toxicity**—The NRC suggests that 2 mg/kg (*2 ppm*) dry weight ration is the maximum tolerable levels for all species. Signs of toxicity include loss of appetite, loss of tail hair, sloughing of hoofs, and eventual death. Two types of selenium poisoning have been observed: (1) acute, blind staggers; and (2) chronic, alkali disease. Selenium toxicity can be counteracted by feeding some forms of sulfur. Toxic levels reported in South Dakota, North Dakota, Montana, Wyoming, Utah, Nebraska, Kansas, and Colorado.
ZINC (Zn) **Absorption**—Absorption of zinc occurs primarily from the abomasum and lower small intestine. **Excretion**—The primary route of excretion is the feces.	Zinc deficiencies have been reported in ruminants grazing forages low in zinc or high in compounds interfering with zinc utilization.	Zinc functions as both an activator and a constituent of several dehydrogenases, peptidases, and phosphates that are involved in nucleic acid metabolism, protein synthesis, and carbohydrate metabolism.	**Deficiency symptoms**—Deficiencies are characterized by decreased performance and listlessness, followed by development of swollen feet and a dermatitis that is most severe on the neck, head, and legs. Deficiencies may also result in vision impairment, excessive salivation, decreased rumen volatile fatty acid production, failure of wounds to heal normally, and impaired reproductive performance in both bulls and cows. **Toxicity**—The maximum zinc tolerance level is dependent on the ration, particularly concentrations of minerals that affect zinc absorption and utilization. *The NRC lists the maximum tolerable level of zinc as 500 ppm, but the NRC also reports that steers have been fed rations containing 1,000 ppm zinc for 13–18 months without marked reduction of performance.

[1]As used herein, the distinction between *nutrient requirements* and *recommended allowances* is as follows: In nutrient requirements, no margins of safety are included intentionally; whereas in recommended allowances, margins of safety are provided to compensate for variations in feed composition, environment, and possible losses during storage or processing.

*Where preceded by an asterisk, the toxicity levels, nutrient requirements, and recommended allowances listed herein were taken from *Nutrient Requirements of Beef Cattle,* sixth revised edition, National Research Council–National Academy of Sciences, Washington, D.C., 1984.

To be on the safe side, the general recommendation for beef cattle on both the range and in the finishing lot is to allow free choice of a suitable phosphorus supplement in a mineral box, or to add a phosphorus supplement to the ration in keeping with nutrient requirements. (See Tables 19–1, 19–2, 19–3, 19–4, and 19–5.)

When phosphorus is added to the water, either of the following methods may be employed:

1. If added by hand, add ¼ oz of monosodium phosphate per 8 gal of water, or ¼ oz per head daily.

2. If added by automatic dispenser, a stock solution of 2½ lb of monosodium phosphate per gal of water (or 100 lb to 40 gal of water) is recommended. The machine automatically proportions the stock solution to the water.

Appendix Table V–1F, Mineral Supplements, gives the composition of mineral supplements for beef cattle.

Magnesium (Mg)

Certain pastures in early spring are inadequate in magnesium, with the result that grass tetany may occur in cattle grazing on them. (See Chapter 5, Table 5–1, Nutritional Diseases and Ailments—Grass Tetany.) Lactating cows are most commonly affected. In problem areas, as high as 0.7 oz of supplemental magnesium per head daily may be required to prevent this malady.

(Continued)

Nutrient Requirements[1]				
Daily Nutrients/ Animal	**Percentage of Ration**	**Recommended Allowances[1]**	**Practical Sources of the Mineral**	**Comments**
	•40 ppm for mature cows and bulls and 20 ppm for growing-finishing cattle, with a range of 20–50 ppm. **Note well**: Requirements for manganese are increased by elevated dietary levels of calcium and phosphorus.	An intake of 40 ppm for mature breeding cattle and 20 ppm for growing-finishing cattle.	Most forages contain high levels of manganese. Manganous oxide, sulfate, and carbonate are good sources of supplemental manganese.	The manganese levels in pastures, grains, and forages are variable because of variations in plant species, soil types, soil pH, and fertilization practices. A deficiency of manganese exists in northwestern U.S., where it has been shown to cause *crooked calves.*
Requirements for molybdenum are not established. Because copper and sulfate alter molybdenum metabolism, arriving at the molybdenum requirement is impossible.	Requirements for molybdenum are not established. Because copper and sulfate alter molybdenum metabolism, arriving at the molybdenum requirement is impossible.	As a feed additive, molybdenum is not cleared by the Food and Drug Administration.	Many feeds contain 6.8–13.6 mg/lb of ration dry matter.	Excess molybdenum may cause a copper deficiency. Sulfur, in the absence of molybdenum, also may cause a copper deficiency. Increasing copper level in ration to 1 g/head daily is effective in overcoming molybdenum toxicity in beef cattle.
	The selenium requirement of beef cattle depends on the amount of vitamin E in the ration, but ranges are suggested as follows by the NRC: •Growing-finishing steers and heifers, 0.10 mg/kg (*0.1 ppm*) dry weight ration. •Breeding bulls and pregnant and lactating cows, 0.05–0.10 mg/kg (*0.05–0.10 ppm*) dry weight ration.	•0.05–0.30 mg/kg (*0.1–0.3 ppm*) dry weight of ration.	In 1979, the FDA approved the addition of selenium as either sodium selenite or sodium selenate at the rate of 0.1 ppm complete feed for beef cattle, dairy cattle, and sheep. In 1987, FDA increased the allowance of selenium in complete feeds for cattle (beef and dairy), sheep, swine, chickens, turkeys, and ducks from 0.1 ppm to 0.3 ppm.	Selenium toxicity may occur when cattle consume feeds containing 10–30 ppm of selenium on a dry matter basis for an extended period. In Israel, in a series of experiments extending over 3 years, low doses of selenium injected intramuscularly reduced the incidence of retained placenta to half that of the controls. (Eger, S., *et al.,* "Effect of Selenium and Vitamin E on the Incidence of Retained Placenta," *Journal of Dairy Science,* Vol. 68, No. 8, Aug. 1985, p. 219.)
	•30 ppm of ration dry matter, with a range of 20–40 ppm of dry matter. Beef cows with high levels of milk production have higher requirements, because milk contains 300–500 mg (*300–500 ppm*) of zinc per liter. Requirements vary according to age and growth rate, since zinc absorption decreases with age and as growth rate decreases. Requirements may be altered by dietary levels of cadmium, calcium, iron, magnesium, manganese, molybdenum, and selenium, since these minerals affect zinc absorption and/or utilization.	•20–40 ppm zinc in the total feed (air-dry basis).	Feedstuffs vary widely in zinc concentrations, with legumes usually having higher concentrations than grasses, and with protein supplements of animal origin being higher than other protein supplements.	Mild zinc deficiency in feedlot cattle results in lowered weight gains without the development of a specific syndrome.

Mineral recommendations for all classes and ages of cattle: Provide free access to a two-compartment mineral box, with (1) salt (iodized salt in iodine-deficient areas) in one side and (2) dicalcium phosphate, defluorinated phosphate, or a mixture of ⅓ salt (salt added for purposes of palatability) and ⅔ steamed bone meal in the other side. Also, the mineral requirements may be met by using a good commercial mineral, in either block or loose form. If desired, the mineral supplement may be incorporated in the ration in keeping with the recommended allowances given in this table.

Potassium (K)

Potassium deficiencies are rare, but they occasionally occur in drylot finishing cattle fed high-concentrate rations. Forages are extremely good sources of potassium. For this reason, potassium is not generally added to feeds for cattle.

Sulfur (S)

Sulfur is a component of protein, some vitamins, and several important hormones. The common sulfur-containing amino acids are methionine, cysteine, and cystine. Also, the following amino acid derivatives contain sulfur: cystathionine, taurine, and cysteic acid. Methionine is a key amino acid, because all other sulfur compounds, except the B-vitamins thiamin and biotin, can be synthesized from methionine.

All feeds contain some sulfur, but the amount usually depends on the protein content of the feed—generally speaking, the higher the protein content, the higher the sulfur content. Availability of the sulfur in the feed to microbial reduction in the rumen may be of as much concern as the actual amount that is present.

TRACE OR MICROMINERALS

A discussion of the trace or micromineral needs of beef cattle follows.

Cobalt (Co)

Deficiencies of cobalt in cattle are costly, for the affected animals become weak and emaciated and eventually die. Florida is without doubt the most serious cobalt-deficient area in the United States, but similar deficiencies of a lesser order have been observed in Michigan, Wisconsin, Massachusetts, New Hampshire, Pennsylvania, and New York. Cattle in these affected areas should have access to a cobaltized mineral mixture, made by mixing 0.2 oz of cobalt chloride, cobalt sulfate, cobalt oxide, or cobalt carbonate per 100 lb of either (1) salt, or (2) mineral mix.

In other areas of the world, cobalt deficiency is known as *Denmark disease, coastal disease, enzootic marasmus, bush sickness, salt sickness, nakuritis,* and *pining disease.*

Copper (Cu)

Copper is sometimes deficient in the soils of certain areas, notably in the state of Florida. In such areas, 0.25 to 0.5% of copper sulfate or copper oxide should be incorporated in the salt or mineral mixture. In addition to being an area disease, copper deficiencies have occurred in beef calves kept on nurse cows for periods extending beyond normal weaning age. Also, high levels of molybdenum in forage may interfere with metabolism of copper and cause copper deficiency.

Fig. 19-6. Copper deficiency in calf. Note rough coat and bleaching of hair. (Courtesy, University of Florida, Gainesville)

Iodine (I)

Iodized salt should always be fed to cattle in iodine-deficient areas (such as the northwestern United States and the Great Lakes region). This can be easily and cheaply accomplished by providing stabilized iodized salt containing 0.01% potassium iodide (0.0076% iodine). Under some conditions, organic iodine appears to be an effective aid in the prevention and treatment of foot rot and lumpy jaw (soft tissues) in cattle.

Iron (Fe)

Iron has important functions in the transport of oxygen to cells and in cellular respiration. With the exception of milk, the level of iron in common feeds is believed to be ample. Milk contains less than 10 ppm of iron, with the result that calves fed an exclusive milk diet are apt to show deficiencies. Signs of lack of iron include anemia, loss of appetite, reduced weight gains, listlessness, pale mucous membranes, and atrophy of the papillae of the tongue. Several supplemental sources of iron are available, including ferrous sulfate, ferrous carbonate, and ferric chloride.

Manganese (Mn)

A deficiency of manganese exists in some areas of the northwestern United States, where it has been shown to be one cause of *crooked calves*—calves born with enlarged joints, stiffness, twisted legs, over-knuckling, and weak and shortened bones. (See Chapter 5, Table 5-1, Nutritional Diseases and Ailments—Crooked Calves.)

Molybdenum (Mo)

Molybdenum deficiencies have not been demonstrated in mammals. The greatest concern about molybdenum is its toxicity, which has been observed in areas where pastures are grown on high-molybdenum soils—known as *teart* pastures in England, Canada, and the United States. As noted earlier, excess molybdenum interferes with copper metabolism.

Selenium (Se)

Cows grazing low-selenium pastures may be affected as follows, in comparison with cows grazing similar pastures supplemented with selenium: (1) have a higher incidence of retained placenta, (2) have higher calf death losses, (3) produce more calves with nutritional muscular dystrophy (white muscle disease), and (4) wean off lighter weight calves. Also, it has been shown that the performance of feedlot cattle fed a selenium-deficient ration is improved by providing selenium.

Zinc (Zn)

Added zinc intake has been shown to increase the rate and efficiency of gains by feedlot cattle in certain areas. This may be due to correcting a deficiency of zinc, or it may be due to the relationship between (1) phytic acid and zinc, and (2) calcium and zinc, improper ratios of which may create a need for supplemental zinc.

Vitamins

The absence of one or more vitamins in the ration may lead to a failure in growth or reproduction, or to characteristic disorders known as vitamin deficiency diseases. In severe cases, death itself may follow. Although the occasional deficiency symptoms are the most striking result of vitamin deficiencies, it must be emphasized that in practice, mild deficiencies probably cause higher total economic losses than do severe deficiencies. It is relatively uncommon for a ration, or diet, to contain so little of a vitamin that obvious symptoms of a deficiency occur. When one such case does appear, it is reasonable to suppose that there must be several cases that are too mild to produce characteristic symptoms but which are sufficiently severe to lower the state of health and the efficiency of production.

Cattle have physiological requirements for most vitamins needed by other mammals. Synthesis by microorganisms in the rumen, supplies in natural feedstuffs, and synthesis in tissues meet most of the usual requirements. Although colostrum is rich in vitamins, providing immediate protection to the newborn calf, calves have minimal stores of vitamins at birth. The ability of the calf to synthesize B vitamins and vitamin K in the rumen develops rapidly when solid feed is introduced into the ration. Vitamin D is synthesized by animals exposed to direct sunlight and is found in large amounts in sun-cured forages. High-quality forages contain large amounts of vitamin A precursors and vitamin E.

Table 19-8 lists the vitamin requirements of beef cattle.

TABLE 19-8
VITAMIN REQUIREMENTS OF BEEF CATTLE
(in Percentage or Amount per Kilogram of Dry Ration)[1]

Nutrient	Growing and Finishing Steers and Heifers	Dry Pregnant Cows	Breeding Bulls and Lactating Cows
	◄------- %	per kg dry	ration -------►
Vitamin A activity .. (IU)[2]	2,200	2,800	3,900
Vitamin D .... (IU)	275	275	275
Vitamin E .... (IU)	15-60	—	15-60

[1]From *Nutrient Requirements of Beef Cattle,* 6th rev. ed., National Research Council—National Academy of Sciences, Washington, D.C., 1984.

[2]May be vitamin A or provitamin A equivalent.

BEEF CATTLE VITAMIN CHART

Table 19-9, Beef Cattle Vitamin Chart (see next page), presents in summary form pertinent information pertaining to the vitamin needs of beef cattle.

FAT-SOLUBLE VITAMINS

A discussion of the fat soluble vitamin needs of beef cattle follows.

Vitamin A

The vitamin most likely to be deficient in beef cattle rations is vitamin A. True vitamin A is a chemically formed compound, which does not occur in plants. It is furnished in most beef cattle rations in the form of its precursor, carotene. However, plants are a variable, and sometimes undependable, source of carotene due to oxidation. Also, cattle are relatively inefficient converters of carotene to vitamin A. The latter fact was taken into consideration in the development of international standards for vitamin A, which are based on the rate at which the rat converts beta-carotene to vitamin A. The conversion rate for the rat is 1 mg of beta-carotene to 1,667 IU of vitamin A, whereas it is estimated that 1 mg of beta-carotene is equal to only 400 IU of vitamin A in cattle. Moreover, the conversion rate for cattle varies under different conditions; it is influenced by type of carotenoid, breed, individual differences in animals, and level of carotene intake. Stress conditions—such as high temperature and elevated nitrogen intake have also been suggested as causes for reduced conversion.

Under practical feeding conditions, cattle producers should consider (1) previous feeding as it influences body stores of vitamin A; (2) vitamin A destruction during feed processing or when mixed with oxidizing materials; and (3) carotene destruction in feeds during storage.

The possibility that beta-carotene may have a role in reproduction independent of its role as a vitamin A precursor has received considerable recent attention. However, further studies and evaluation are necessary.

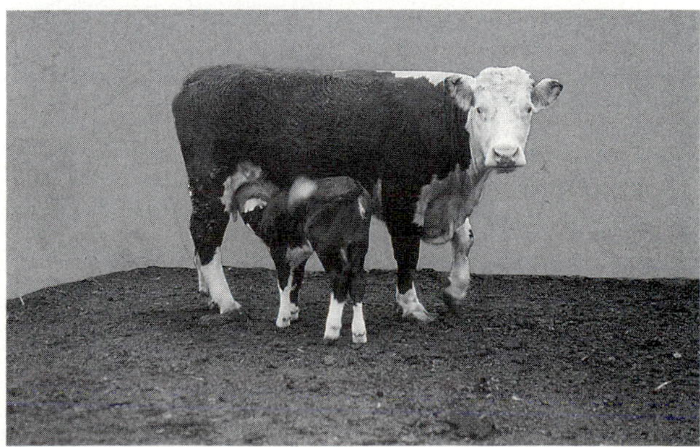

Fig. 19-7. Effect of vitamin A deficiency on reproduction. The heifer in the upper picture received a ration deficient in vitamin A, but otherwise complete. She became night blind and aborted during the last month of pregnancy; also, note the retained placenta. The heifer in the lower picture received the same ration, but during the latter part of the gestation period, a daily supplement of 1 lb of dehydrated alfalfa meal containing 50 mg of carotene was added. She produced a normal vigorous calf. (Courtesy, Calif. Ag. Exp. Sta., Davis, Calif.)

TABLE
BEEF CATTLE

Vitamins Which May Be Deficient Under Normal Conditions	Conditions Usually Prevailing Where Deficiencies Are Reported	Functions of Vitamin	Deficiency Symptoms
Fat-Soluble Vitamins:			
A—Vitamin A is found only in animals; plants contain the precursor—carotene. Vitamin A is the vitamin most likely to be of practical importance in feeding cattle.	Vitamin A deficiency is most likely to occur when cattle are fed (1) high-concentrate rations; (2) bleached pasture or hay grown under drought conditions; (3) feeds that have had excess exposure to sunlight, air, and high temperature; (4) feeds that have been heavily processed or mixed with oxidizing materials such as minerals; and (5) feeds that have been stored for long periods of time. Cattle particularly susceptible to vitamin A deficiency are: newborn calves deprived of colostrum; cattle that have been prevented from establishing or maintaining good liver stores through exposure to drought; cattle wintered without high-quality forage, and cattle exposed to stresses such as high temperatures or elevated nitrate intake.	Vitamin A functions as a component of the visual purple required for dim-light vision, and is essential for normal growth, reproduction, and maintenance of healthy epithelial tissue.	Signs of vitamin A deficiency include reduced feed intake, rough hair coat, edema of the joints and brisket, lacrimation, xerophthalmia, night blindness, slow growth, diarrhea, convulsive seizures, improper bone growth, blindness, low conception rates, abortion, stillbirths, blind calves, abnormal semen, reduced libido, and susceptibility to respiratory and other infections. Of these symptoms, only night blindness has proved unique to vitamin A deficiency. Clinical verification may include ophthalmoscopic examination, liver biopsy and assay, blood assay, testing spinal fluid pressure, conjunctival smears, and response to vitamin A therapy.
D	Young calves kept indoors, especially in the wintertime. Finishing cattle in northern U.S. on high silage and grain rations and a minimum of sun-cured hay.	Vitamin D is required for calcium and phosphorus absorption, normal mineralization of bone, and mobilization of calcium from bone.	Rickets in young calves, the symptoms of which are: decreased appetite, lowered growth rate, digestive disturbances, stiffness in gait, labored breathing, irritability, weakness, and occasionally, tetany and convulsions. Later, enlargement of the joints, slight arching of the back, bowing of the legs, and the erosion of the joint surfaces cause difficulty in locomotion. Posterior paralysis may follow fracture of vertebrae. In older animals with vitamin D deficiency, bones become weak and easily fractured, and posterior paralysis may accompany vertebral fractures. Vitamin D deficiency in the pregnant animal may result in dead, weak, or deformed calves at birth.
E	Where soils are very low in selenium. When unsaturated fats are fed.	Vitamin E is an antioxidant. It has been widely used to protect and to facilitate the uptake and storage of vitamin A. In metabolism, it is linked closely with selenium. Some deficiency signs, particularly in white muscle disease, may respond to either selenium or vitamin E, or may require both.	Muscular dystrophy (commonly called white muscle disease) in calves 2 to 12 weeks of age; characterized by heart failure and paralysis varying in severity from slight lameness to inability to stand. Also, a dystrophic tongue is often seen in affected animals. A deficiency of vitamin E may be precipitated or accentuated by feeding unsaturated fats.
K	When moldy sweet clover hay high in dicoumarol content is fed. Vitamin K deficiency results from the antagonistic action of dicoumarol that is formed in moldy sweet clover hay.		Sweet clover disease, characterized by prolonged blood clotting. Mild cases can be treated effectively with vitamin K.
Water-Soluble Vitamins:			
B Vitamins	When an antagonist is present. When ruminal synthesis is limited by lack of precursors or other problems. Vitamin B-12 is of special interest because of its role in propionate metabolism, and the practical incidence of vitamin B-12 deficiency as a secondary result of cobalt deficiency. Niacin has been reported to enhance protein synthesis by ruminal microorganisms.	Most of the established metabolic functions of B vitamins are important to cattle, as well as to other animals. Consequently, a physiological need for most B vitamins can be assumed for cattle of all ages.	Deficiency signs in young calves have been clearly demonstrated for thiamin, riboflavin, pyridoxine, pantothenic acid, biotin, nicotinic acid, vitamin B-12, and choline. Polioencephalomalacia in grain-fed cattle has been linked to thiaminase activity or production of a thiamin antimetabolite in the rumen. Affected animals have responded to intravenous administration of thiamin (2.2 mg/kg body weight).

[1]As used herein, the distinction between *nutrient requirements* and *recommended allowances* is as follows: in nutrient requirements, no margins of safety are included intentionally; whereas in recommended allowances, margins of safety are provided in order to compensate for variations in feed composition, environment, and possible losses during storage or processing.

19-9
VITAMIN CHART

Nutrient Requirements[1]: Daily Nutrients/ Animal (or Injection)	Nutrient Requirements[1]: Amount/Lb (or/kg) of Feed	Recommended Allowances[1]	Practical Sources of the Vitamin	Comments
*Variable according to class, age, and weight of cattle. (See Table 19-1.) Injection of 1 million IU of vitamin A intramuscularly will prevent deficiency symptoms for 2–4 months in growing or breeding cattle.	*Variable according to class, age, and weight of cattle. (See Tables 19-2 and 19-8.) On a dry ration basis, the vitamin A requirements are about as follows: *1. Growing-finishing steers and heifers, 1,000 IU/lb (*2,200 IU/kg*). *2. Pregnant heifers and cows, 1,270 IU/lb (*2,800 IU/kg*). *3. Lactating cows and breeding bulls, 1,770 IU/lb (*3,900 IU/kg*).	Inject newborn calves (at birth) with 250,000–1,000,000 IU of vitamin A (use the higher level under confinement production or when scours may be a problem).	Stablized vitamin A. Green pasture. Grass or legume silages. Yellow corn. Green hay not over 1 year old. The average carotene content of some common feeds is as follows: **mg Carotene per** (lb) / (*kg*) Legume hays (including alfalfa), av. quality: 9–14 / 20–31 Nonlegume hays, av. quality: 4–8 / 9–18 Dehydrated alfalfa meal, av. quality: 50–70 / 110–154 Yellow corn: 0.8–1.0 / 1.8–2.2 Silages, corn, or sorghum: 2–10 / 4–22	Carotene is rapidly destroyed by exposure to sunlight and air, especially at high temperatures. Hay over 1 year old, regardless of green color, is usually not an adequate source of carotene or vitamin A activity. Ensiling effectively preserves carotene, but the availability of carotene from corn silage may be low. The younger the animal, the quicker vitamin A deficiencies will occur. Mature animals may store sufficient vitamin A to last 6 months. When deficiency symptoms appear, they can be corrected (1) by increasing carotene intake through the introduction of high-quality forage, or (2) by supplying vitamin A in the feed or by injection.
	*125 IU/lb (*275 IU/kg*) of dry ration.	Normally, beef cattle receive sufficient vitamin D from exposure to direct sunlight or from sun-cured hay.	Exposure to direct sunlight. Sun-cured hay. Irradiated yeast.	Sun-cured alfalfa hay contains 300–1,000 IU/lb (*661–2,204 IU/kg*).
	*dl-alpha-tocopherol acetate added to dry ration at level of 6.8 to 27.3 IU/lb (*15 to 60 IU/kg*).	Generally natural feeds supply adequate quantities of alpha-tocopherol for mature cattle, although muscular dystrophy in calves occurs in certain areas.	Alpha-tocopherol, added to the ration or injected intramuscularly. Commercial vitamin E supplements. Grains contain 6–15 mg vitamin E/lb (*13–33 mg/kg*).	The incidence of white muscle disease appears to be lower when the cows receive 2–3 lb (*.91–1.36 kg*) of grain during last 60 days of pregnancy. Where supplemental vitamin E is needed, it may be added to the ration or injected intramuscularly.
			Vitamin K_1 is abundant in pasture and green roughage. Vitamin K_2 is synthesized in large amounts in the rumen. Either K_1 or K_2 effectively fulfill the vitamin K role in blood clotting mechanism.	Except when the dicoumarol content of hay is excessively high (as in moldy sweet clover hay), sufficient vitamin K is synthesized in the rumen of cattle.
		Usually, no dietary B vitamins need be supplied to cattle.	B-vitamins are abundant in milk and many other feeds, and synthesis of B vitamins by ruminal microorganisms is extensive. Calves begin microbial synthesis of B vitamins very soon after the introduction of dry feed in the ration.	

*Where preceded by an asterisk, the nutrient requirements listed herein were taken from *Nutrient Requirements of Beef Cattle,* sixth revised edition, National Research Council—National Academy of Sciences, Washington, D.C., 1984.

Vitamin D

When exposed to enough direct sunlight, beef cattle normally acquire their vitamin D needs, for the ultraviolet rays in sunlight penetrate the skin and produce vitamin D from traces of sterols in the tissues. Also, cattle obtain vitamin D from sun-cured roughages. However, the addition of vitamin D to the ration is important when cattle, especially calves, are kept in a barn most of the day, when there is limited sunshine, when the calcium:phosphorus ratio is not correct, and/or when little or no sun-cured hay is fed. Vitamin D helps build strong bones and sturdy frames.

Vitamin E

Added vitamin E may be necessary under certain conditions because of its relationship to vitamin A utilization and the prevention of white muscle disease.

Vitamin K

Under normal conditions, adequate vitamin K is synthesized in the rumen of cattle. However, symptoms of inadequacy (a bleeding syndrome known as *sweet clover disease*) occur when moldy sweet clover hay, high in dicoumarol content, is fed, since dicoumarol is a metabolic antagonist that interferes with the normal action of vitamin K.

WATER-SOLUBLE VITAMINS

A discussion of the water soluble vitamin needs of beef cattle follows. Note that this discussion is limited to the B vitamins, which are needed by the young calf, and possibly by some feedlot cattle.

B Vitamins

Dietary requirements for the B vitamins (thiamin, biotin, niacin, pyridoxine, pantothenic acid, riboflavin, and vitamin B-12) have been demonstrated experimentally for the young calf during the first 8 weeks of life, prior to the development of the functioning rumen. At this stage in life, these requirements are usually met by the milk of the dam. Later, the B vitamins appear to be synthesized in sufficient quantities by rumen bacterial fermentation, with the possible exception of feedlot cattle in which thiamin deficiencies have been reported. However, inadequacy of protein or other nutrients in the ration may impair rumen fermentation, with the result that sufficient quantities of the B vitamins will not be synthesized.

(See Appendix Tables V–1A, V–1B, V–1C, V–1D, and V–1E for the vitamin content of feeds commonly used in beef cattle rations.)

Water

Water is needed for all the essential processes of the body, such as the digestion and absorption of food nutrients, the removal of waste, and in regulating body temperature. Animals can survive for a longer period without feed than they can without water. Yet, under ordinary conditions, it can be readily provided in abundance and at little cost. Beef cattle should have an abundant supply of water before them at all times.

Fig. 19–8. Windmill, an energy-conserving watering facility. (Courtesy, *Livestock Weekly*, San Angelo, Tex.)

The water requirement is influenced by several factors, including weight, environmental temperature, rate and composition of gain, pregnancy, lactation, activity, type of ration, and feed intake. Because feeds contain some water and the oxidation of certain nutrients in feeds produces water, not all water needs must be provided by drinking. Such feeds as silages, green chop, or lush pastures are high in moisture content, while grains, hays, and dormant pasture forage are low. High-energy feeds produce much metabolic water, while low-energy feeds produce little metabolic water. These are among the complications in the matter of assessing water requirements.

Thirst in cattle is caused by changing concentrations of the body electrolytes, and this response works quite well to induce them to drink to meet their body requirements.

Saline water containing 1% soluble salts may be toxic. Excessive nitrates or alkalinity may make water unsatisfactory for cattle.

Water need is influenced by environmental temperature. Intake of water is fairly consistent up to about 40°F. Above that temperature, water intake varies as water becomes increasingly used for cooling. Where sub-zero winter temperatures occur, it may be necessary to provide heaters to make water available—*i.e.* to keep it from freezing. No further warming of the water is necessary, however.

Table 19–10 may be used as a guide to the water requirements of beef cattle.

TABLE 19–10
DAILY WATER INTAKE OF BEEF CATTLE[1]

Weight		Temperature in °F *(°C)*[2]											
		40 *(4.4)*		50 *(10.0)*		60 *(14.4)*		70 *(21.1)*		80 *(26.6)*		90 *(32.2)*	
(lb)	*(kg)*	(gal)	*(liter)*	(gal)	*(liter)*	(gal)	*(liter)*	(gal)	*(liter)*	(gal)	*(liter)*	(gal)	*(liter)*
Growing heifers, steers, and bulls													
400	*182*	4.0	*15.1*	4.3	*16.3*	5.0	*18.9*	5.8	*22.0*	6.7	*25.4*	9.5	*36.0*
600	*273*	5.3	*20.1*	5.8	*22.0*	6.6	*25.0*	7.8	*29.5*	8.9	*33.7*	12.7	*48.1*
800	*364*	6.3	*23.8*	6.8	*25.7*	7.9	*29.9*	9.2	*34.8*	10.6	*40.1*	15.0	*56.8*
Finishing cattle													
600	*273*	6.0	*22.7*	6.5	*24.6*	7.4	*28.0*	8.7	*32.9*	10.0	*37.9*	14.3	*54.1*
800	*364*	7.3	*27.6*	7.9	*29.9*	9.1	*34.4*	10.7	*40.5*	12.3	*46.6*	17.4	*65.9*
1,000	*454*	8.7	*32.9*	9.4	*35.6*	10.8	*40.9*	12.6	*47.7*	14.5	*54.9*	20.6	*78.0*
Wintering pregnant cows[3]													
900	*409*	6.7	*25.4*	7.2	*27.3*	8.3	*31.4*	9.7	*36.7*	—	—	—	—
1,100	*500*	6.0	*22.7*	6.5	*24.6*	7.4	*28.0*	8.7	*32.9*	—	—	—	—
Lactating cows													
900+	*409+*	11.4	*43.1*	12.6	*47.7*	14.5	*54.9*	16.9	*64.0*	17.9	*67.8*	16.2	*61.3*
Mature bulls													
1,400	*636*	8.0	*30.3*	8.6	*32.6*	9.9	*37.5*	11.7	*44.3*	13.4	*50.7*	19.0	*71.9*
1,600+	*727+*	8.7	*32.9*	9.4	*35.6*	10.8	*40.9*	12.6	*47.7*	14.5	*54.9*	20.6	*78.0*

[1]From *Nutrient Requirements of Beef Cattle,* sixth revised edition, National Research Council-National Academy of Sciences, 1984.

[2]Water intake of a given class of cattle in a specific management regime is a function of dry matter intake and ambient temperature. Water intake is quite constant up to 40°F *(4.4°C)*.

[3]Dry matter intake has a major influence on water intake. Heavier cows are assumed to be higher in body condition and to require less dry matter and, thus, less water intake.

Growth Stimulants and Implants

Feed additives first made headlines in 1952 when Iowa State University researchers announced the results of cattle feeding trials indicating a major breakthrough in lowering feed usage and increasing weight gains by feeding the compound diethylstilbestrol (DES).

For the next 20 years, cattle feeders and consumers greatly benefited from the increased rate and efficiency of gains accruing from the use of this product. Feeding and implanting DES increased growth rate 10 to 15%; improved feed efficiency 10%; put feedlot cattle on the market 20 to 30 days sooner; decreased feedlot costs by 3 to 3.5¢ per pound; and lowered the retail price of beef by 5 to 6¢ per pound. But the Food and Drug Administration banned the use of both oral (January 1, 1973) and implanted (April 25, 1973) DES, for the reason that stilbestrol is a carcinogen (in large enough quantities it can produce cancer in humans and animals); and the Delaney Amendment clearly states that foods can contain no residues (zero tolerance) of carcinogens.

Then, on January 24, 1974, the United States Court of Appeals invalidated the ban on DES and reinstated its use as an implant or feed additive. But the ruling of the court did nothing to change the Delaney Amendment. In the meantime, newer and more sophisticated assay techniques evolved which made it possible to detect the presence of DES residues in meat at lower levels than formerly. So, in 1979, the use of DES as feed implants for cattle and sheep was banned. But several other additives were developed subsequently and are now on the market.

Table 19–11 (see next page) summarizes the growth stimulants that are presently available and can be used. All these products have been shown to improve gain and feed efficiency of feedlot cattle significantly.

In considering the additives listed in Table 19–11, it should be noted that there is no evidence to indicate that the use of these products can or will alleviate the need for vigilant sanitation, improved nutrition, and superior management. Also, the benefits of each one must be weighed against its cost.

(Also see Chapter 13, sections on "Hormone and Hormonelike Compounds" and "Implants.")

FEEDS FOR BEEF CATTLE

Beef cattle feeding practices vary according to the relative availability of grasses, dry roughages, and grains. Where roughages are abundant and grain is limited, as in the western range states, cattle are primarily grown out on roughages. On the other hand, where grain is relatively more abundant, as in the Corn Belt and in the High Plains area of Texas and Oklahoma, finishing with more concentrates is common.

Pastures

Pastures include all crops that are harvested directly by animals.

Good pasture is the cornerstone of successful beef cattle production. It has been said that good cattle producers can

TABLE 19-11
GROWTH STIMULANTS AND IMPLANTS FOR CATTLE[1]

Additive	Method of Adminis-tering	Dosage	Increase in Daily Rate of Gain	Increase in Feed Effi-ciency	Effect on Carcass Quality	Other Comments	Withdrawal Period Prior to Slaughter
			Finishing Steers				
Antibiotic	Oral.	10 mg/100 lb body wt. daily; or 70 to 75 mg/head daily.	6%	4%	Improves carcass quality slightly; more fat deposition and marbling. Decreases liver and rumen condemnations.	Antibiotics will also reduce the disease level. More effective with high-roughage rations than with high-concentrate rations.	No withdrawal required.
Bovatec (lasalocid)	Oral.	150 to 360 mg/day. 250 to 360 mg/day.	 8%	8% 5%	No effect.	Alters rumen fermentation similar to monensin.	No withdrawal required.
Rumensin (monensin)	Oral.	50–360 mg/head/day, drylot. 50–200 mg/head/day, pasture.		10%	No effect.	Not a hormone. It results in more propionic acid and less butyric and acetic acids; hence, more energy.	No withdrawal required.
Compudose	Implant.	24 mg estradiol.	10–15%	5–10%	No effect.	Only one implant. Effective for 200 days.	No withdrawal required.
Finaplex	Implant.	140 mg.	7%	7%	The active ingredient is trenbolene acetate (TBA). It is a synthetic analog of testosterone. It gives an additional response when used in combination with an estrogen.		
Ralgro (Zeranol)	Implant.	36 mg resorcyclic acid lactone.	10%	5–10%	No effect.	Nonestrogenic.	65 days.
Steer-oid	Implant.	200 mg progesterone, 20 mg estradiol.	10–15%	5–10%	No effect.	Effective period of 140 days.	No withdrawal.
Synovex-S (for steers)	Implant.	200 mg progesterone, 20 mg estradiol benzoate.	10–15%	5–10%	No effect.	Effective period of 90 to 120 days.	No withdrawal required.
			Finishing Heifers				
Antibiotic	Oral.	10 mg/100 lb body wt. daily; or 70 to 75 mg/head daily.	6%	4%	Improves carcass quality slightly; more fat deposition and marbling.	Antibiotics will also reduce the disease level. More effective with high-roughage than with high-concentrate rations.	No withdrawal required.
Bovatec (lasalocid)	Oral.	150 to 360 mg/day.	5%	8%	No effect.	Alters rumen fermentation similar to monensin.	No withdrawal.
MGA	Oral.	0.25 to 0.50 mg daily melengestrol acetate.	10%	6%	MGA will lower the incidence of estrus in heifers and increase rate and efficiency of gain. It is not effective with pregnant heifers.	MGA is effective for heifers, but not for steers.	48 hours.
Rumensin	Oral.	50–360 mg/head/day, drylot. 50–200 mg/head/day, pasture.		15%	No effect.	Not a hormone. It results in more propionic acid and less butyric and acetic acids; hence, more energy.	No withdrawal required.
Heifer-oid	Implant.	Follow label directions.	10%	5–10%	No effect.	Effective period of 140 days. For heifers over 400 lb.	No withdrawal.
Ralgro (zeranol)	Implant.	36 mg resorcyclic acid lactone.	10%	5–10%	No effect.	Nonestrogenic.	65 days.
Synovex-H (for heifers)	Implant.	200 mg testosterone propionate. 20 mg estradiol benzoate.	10%	5–10%	No effect.	Recommended for use in heifers during last 60 to 150 days of the finishing period.	No withdrawal required.
			Suckling Calves				
Antibiotic	Oral (in creep feed).	15 to 20 mg/100 lb body wt. daily.	6%	4%		Antibiotics will also reduce the disease level. Administer in creep feed.	No withdrawal required. **Note Well:** If fed at level of 350 mg or over/day, 48-hour withdrawal required.
Ralgro (zeranol)	Implant.	36 mg resorcyclic acid lactone.	10%	5–10%	No effect.	Nonestrogenic.	65 days.
Synovex-C	Implant.	110 mg.	5–10%		No effect.	For calves of either sex.	No withdrawal required.

[1]*CAUTIONS:* FDA regulations are subject to change. Always follow the manufacturer's directions on the use of these products.

(Also see Chapter 13, Table 13-5, Implant Summary.)

be recognized by the character of their pastures, and that good cattle graze good pastures. Thus, the three go hand in hand—good producers, good pastures, and good cattle. The historic relationship and importance of cattle and pastures was extolled in an old Flemish proverb: "No grass, no cattle; no cattle, no manure; no manure, no crops."

Fig. 19-9. Good grass and plenty of milk make for heavy calves. (Courtesy, International Braford Assn., Ft. Pierce, Fla.)

Approximately 31% of the land area of continental United States is devoted to grassland pasture and range. Much of this area, especially in the far West, can be utilized only by beef cattle, sheep, or game animals. Although the term *pasture* usually suggests growing plants, it is correct to speak of pasturing stalk and stubble fields. Also, on many farms and ranches there are either low, wet areas, or rough broken areas, which cannot be used profitably for crop production. Such areas can be fenced and made available to cattle.

The type of pasture, as well as its carrying capacity and seasonable use, varies according to topography, soil, and climate. Because of the hundreds of species of grasses and legumes that are used as beef cattle pastures, each with its own best adaptation, no attempt is made to discuss the respective virtues of each variety. Instead, it is recommended that the farmer or rancher seek the advice of the local county agricultural agent, or write to the crop science department of the state university.

No method of harvesting has yet been devised that is as economical as that which can be accomplished through grazing by animals. Accordingly, successful beef cattle management necessitates as nearly year-round grazing as possible. In the northern latitudes of the United States, the grazing season is usually of about 6 months' duration, whereas in the deep South, yearlong grazing is approached. In many range areas of the West, the breeding herds obtain practically all their forage the year-round from the range, being given supplemental roughage only if the grass or browse is buried deep in snow.

During the winter months, and in periods of drought, the pasture utilized by beef cattle may consist of dried grass cured on the stalk. On a dry basis, the crude protein content of some mature, weathered grasses may be 3% or less. To supplement such feed, cattle producers commonly feed cake or cubes. The use of cake or cubes instead of meal reduces losses from wind, an especially important factor on the range. Also, dried grass may be supplemented by means of "lick tanks" or molasses, along with urea, minerals, and vitamin A.

In some instances, cattle on pasture fail to make the proper growth or gain in condition because the soil is seriously deficient in fertility or the pasture has not been well managed. In such instances, striking improvement will result from proper fertilization and management.

Hays and Other Dry Roughages

Hay is the most important harvested roughage fed to beef cattle, although many other dry roughages can be and are utilized.

The dry roughages are all high in fiber and therefore lower than concentrates in total digestible nutrients. Hay averages about 28% fiber and straw approximately 38%, whereas such concentrates as corn and wheat contain only 2 to 3% fiber. Fortunately, cattle are equipped to handle large quantities of roughages. In the first place, the paunch of a mature cow has a capacity of about 53 gal, thus providing ample storage for large quantities of less concentrated feeds. Secondly, the billions of microorganisms in the rumen attack the cellulose and pentosans of the fibrous roughages, such as hay, breaking them down into available and useful nutrients. In addition to providing nutrients at low cost, the roughages add needed bulk to cattle rations.

Fig. 19-10. Hay—the most important harvested forage fed to cattle. (Courtesy, USDA)

Roughages, as with concentrates, may be classified as carbonaceous or nitrogenous, depending on their protein content. The principal dry carbonaceous roughages used by cattle include hay from the grasses, the straws and hays from cereal grains, corncobs, the stalks and leaves of corn and the grain sorghums, and cottonseed hulls. Cured nitrogenous roughages include the various legume hays such as alfalfa, the clovers, peanut, soybean, cowpea, and velvet bean.

Although leguminous roughages are preferable, weather conditions and soils often make it more practical to produce the nonlegumes. Also, in many areas, such feeds as dry grass cured on the stalk, cereal straws, corncobs, and cottonseed hulls are abundantly available and cheap. Under such circumstances, these feeds should be used as part of the ration for wintering beef cows, for wintering stockers that are more than 1 year of age, or for supplying the limited roughage needed for finishing beef cattle.

In comparison with good-quality legume hays, the carbonaceous roughages are lower in protein content and in quality of protein, lower in calcium, and generally deficient in carotene (provitamin A). Thus, where nonlegume roughages are used for extended periods, these nutritive deficiencies should be corrected; this is especially true with the gestating-lactating cow or the young, growing calf. To the end that the feeding value of some of the common nonlegumes may be enhanced for beef cattle, the following facts are pertinent:

1. The feeding value of nonlegume hays can be increased by cutting them at an early stage and curing so as to retain as much of the protein and carotene content as possible.

2. Where dry and bleached pastures are grazed for an extended period of time, or where there is an unusually long winter, it is important that at least part of the roughage be a legume, either silage or hay, or that vitamin A either be added to the ration or injected intramuscularly; and that suitable energy, protein, and mineral supplements be provided.

3. Potentially, corncobs—which were formerly considered a waste product and of little worth—have a feeding value approaching that of hay. However, their energy cannot be utilized unless they are fortified with certain nutrients which help the microorganisms of the rumen break them down into a form which can be digested. Also, corncobs are low in palatability.

4. Cereal straws and cottonseed hulls may be incorporated into the wintering ration of pregnant cows or in the ration of finishing cattle, provided their fundamental characteristics and nutritional limitations are recognized and corrected.

Silages and Root Crops

Silage is a valuable adjunct to pastures in beef cattle production, since it is possible to use a combination of the two forages to furnish green, succulent feeds on a year-round basis. Extensive use of silage for beef cattle dates back only to about 1910. Prior to that time it was generally thought of as a feed for dairy cows. Even today, only a relatively small percentage of the beef cattle in the United States is fed silage.

Corn was the first and still remains the principal crop used in the making of silage, but many other crops are ensiled in various sections of the country. The sorghums are the leading ensilage crops in the Southwest, and grasses and legumes are the leading ensilage crops in the Northeast. Also, in different sections of the country to which they are adapted, the following feeds are ensiled: cereal grains, field peas, cowpeas, soybeans, potatoes, and numerous fruit and vegetable refuse products. A rule of thumb is that crops that are palatable and nutritious to animals as pasture, as freshly harvested feed, or as dry forage, also make palatable and nutritious silage. Conversely, crops that are unpalatable and nonnutritious as pasture, as green feed, or as dry forage, also make unpalatable and nonnutritious silage.

Fig. 19-11. Upright (tower) silos used for storing feed for beef cattle. (Courtesy, A. O. Smith Harvestore Products, Inc., Arlington Heights, Ill.)

Grass silage can be produced in those areas where the climate is too cool and the growing season too short for corn or sorghum silage. It is generally higher in protein and carotene, but lower in total digestible nutrients (TDN) and vitamin D than corn or sorghum silage. Generally, grass silage contains about 80% as much TDN as corn silage, but it is equal in TDN when 150 lb of grain per ton have been added as a preservative. Thus, grass silage usually requires the addition to the ration of less protein supplement but more total concentrates than corn or sorghum silage. This would indicate that corn or sorghum silage would be slightly preferable to grass silage in high-roughage finishing rations for beef cattle, whereas grass silage could best be used in high-roughage rations for young, growing beef cattle. However, weaner calves do not grow well on silage at low temperatures; they must have some hay and/or grain for best results.

When silage is fed to cattle, it must be remembered that, because of its high-moisture content, about 3 lb of silage are generally considered equivalent to 1 lb of dry roughage of comparable quality. A ration of 55 to 60 lb of corn silage plus ½ to ¾ lb of a protein concentrate daily will carry a dry cow through the winter. The ration may be improved, however, by replacing ⅓ to ½ of the silage with an equivalent amount of dry roughage, adding 1 lb of dry roughage for each 3 lb of silage replaced.

Silage may be successfully used for finishing steers. Long-yearling steers will eat 50 to 55 lb of a 37% dry matter corn silage plus 2 lb of a 32% protein supplement per day at the beginning of the feeding period. The amount of silage is gradually decreased as the concentrates are increased. At the end of the feeding period, the cattle should be getting around 10 to 12 lb of silage. Because of their more limited digestive capacity, the allowance of silage fed to calves should be correspondingly less.

Usually, silage provides a much cheaper succulent feed for beef cattle than roots. For this reason, the use of roots for beef cattle is very limited, and confined almost entirely to the northern areas.

Concentrates

The concentrates include those feeds which are low in fiber and high in energy. For purposes of convenience, concentrates are often further classified as (1) carbonaceous feeds, and (2) nitrogenous feeds.

In general, the use of concentrates for beef cattle is limited to (1) the finishing of cattle, (2) the development of young stock, and (3) use as limited supplements in the winter ration. Over most of the United States, the cereal grains are the chief concentrates fed to beef cattle—these grains being combined, if necessary, with protein supplements to balance the ration.

The chief carbonaceous concentrates used for beef cattle are the grains, chiefly corn, and such processed feeds as hominy feed, beet pulp, and molasses. The choice of the particular feeds is usually determined primarily by price and availability.

The nitrogenous feeds are usually subdivided into (1) natural proteins, including the oilseed meals and animal proteins, and (2) nonprotein nitrogen, consisting primarily of urea and ammonia.

Certain nonprotein nitrogen sources may be substituted for all or much of the supplemental protein required in most beef cattle rations, provided such rations are adequate in minerals and readily available carbohydrates. Among such products are urea, ammoniated molasses, ammoniated beet pulp, ammoniated cottonseed meal, ammoniated citrus pulp, and ammoniated rice hulls. The possibility exists that other products will be forthcoming. Each such product should be evaluated by controlled feeding trials.

For best results, the feeder should correct the nutritive deficiencies of the cereal grains. Most of them are low in protein, low in calcium, and lacking in vitamin D. All except yellow corn are also deficient in carotene. Regardless of whether the cereal grains are fed to growing, breeding, or finishing animals, their nutritive deficiencies can be corrected in a very effective and practical way by adding either (1) a good-quality legume hay to the ration, or (2) a protein concentrate plus suitable minerals.

UREA

Urea is not a protein. Rather, it is a simple nitrogen compound, $H_2N{-}\overset{\overset{\displaystyle O}{\|}}{C}{-}NH_2$.

From urea, microorganisms can obtain ammonia, which they synthesize into amino acids and finally bacterial protein—provided that all the other nutrients essential for protein synthesis are present.

Urea may constitute up to one-third of the total protein of the ration of cattle, provided additional fermentable energy is added in the form of molasses or grain to compensate for the lack of energy in the urea, in order to feed the rumen bacteria properly. Producers can tell the nonprotein nitrogen content (NPN) of commercial feed by looking at the feed tag, which must list the contents. A typical tag will read:

Crude protein, 40% (not more than 30%
crude protein equivalent [CPE] from NPN).

This means that the feed contains 40% crude protein, of which 10% is natural protein and 30% is NPN

Common guidelines relative to the use of urea for beef cattle are given in Table 19-12.

There is no difference in the nutritional value of liquid and dry supplements built around urea if the supplements contain the same basic nutrients. Thus, it is a matter of personal choice, convenience, and ingredient costs as to which is used by feeders.

TABLE 19-12
COMMON GUIDELINES TO THE USE OF UREA FOR CATTLE

	For Finishing Cattle	For Grower (stocker) Cattle	For Wintering Pregnant and Lactating Cows
Percent of total protein in ration from urea .... (%)	33⅓	25.0	25.0
Maximum urea/animal/day .... (lb)	0.22 *(100 g)*	0.15 *(68 g)*	–
Percent of urea, by weight of total air-dry feed consumed .... (%)	1.0	1.0	1.0
Percent of urea, by weight, of concentrate mix (grain plus protein supplement)[1] .... (%)	2.0-3.0	3.0	3.0
Percent of urea, by weight, of the protein supplement .... (%)	20-30[2]	10.0[3]	10.0
Percent of supplemental nitrogen in high-protein supplement from urea[4] ... (%)	60-90[5]	30.0	30.0
Pounds of urea added/ton of corn silage at ensiling time[6] .... (lb)	10.0 *(4.5 kg)*	10.0 *(4.5 kg)*	10.0 *(4.5 kg)*

[1]Feed intake may be depressed if over 1% is used. Yet, many beef producers are successfully using 2%.

[2]This means that as much as 60-90% of the protein value of the supplement may come from nonprotein sources. However, because such a supplement will constitute only 2-5% of the total ration fed, the first rule of thumb given in Table 19-12 still applies; namely, only ¼-⅓ of the total protein in the ration will be supplied from a nonprotein source.

[3]A protein supplement containing 10% urea provides 28.1% of the protein equivalent (281% × .10) from nonprotein nitrogen.

[4]High-urea supplements are best fed in complete mixed rations, which are *thoroughly* mixed. *Supplements containing 20-30% urea require extreme caution when being hand-fed.*

[5]In a feedlot ration, this may be equivalent to 25-40% of the total nitrogen from all sources.

[6]On a dry matter basis, corn silage ensiled at the well-dented stage contains about 8% protein. The addition of 10 lb of urea per ton (or *5 kg/1,000 kg*) of silage increases the protein content from 8-13%. However, there is loss of flexibility in feeding such a ration, and the rate of gain will be less than can be secured from higher energy, more dense rations. Also, it is extremely important that the urea be well mixed in the silage; otherwise, there is hazard of toxicity.

(Also see Chapter 11, Protein Supplements, section on "Urea.")

By-product Feeds

Innumerable by-products—both roughages and concentrates—from plant and animal processing, and from industrial manufacturing, are available and used as cattle feeds in different areas. Also, small grain stubble fields and cornstalks are utilized, and cotton fields may be grazed following harvest. Then, there are such additional by-product feeds as cull potatoes, cottonseed hulls, corncobs, cull citrus, cannery refuse, beet tops, and a host of other similar products.

(Also see Chapter 12, By-product Feeds/Crop Residues.)

Mineral Supplements

When buying and home-mixing minerals, or when buying commercial mineral mixes, the cattle raiser should first determine needs, based on (1) available feeds, (2) area (for example, the Northern Great Plains and the Southwest tend to be phosphorus-deficient areas), and (3) the age and reproduction status (pregnancy and lactation make a difference) of the animals for which the mineral mix is intended. Excesses and mineral imbalances should be avoided.

The mineral recommendations for all classes and ages of cattle, especially those fed unmixed rations or on pasture, are:

1. **When animals are on liberal grain feeding.** Provide free access to a 2-compartment mineral box, with (a) trace mineralized salt in one side, and (b) in the other side, a mixture of ⅓ trace mineralized salt (salt included to improve palatability), ⅓ defluorinated phosphate, dicalcium phosphate, or steamed bone meal, and ⅓ ground limestone or oystershell flour.

2. **When animals are primarily on roughage (pasture, hay, and/or silage).** Provide free access to a 2-compartment mineral box, with (a) trace mineralized salt in one side, and (b) in the other side a mixture of ⅓ trace mineralized salt (salt included for purposes of palatability), ⅓ defluorinated phosphate, dicalcium phosphate, or steamed bone meal.

Salt should always be available on a free-choice basis in addition to whatever mineral mix is provided.

Vitamin Supplements

The vitamin A requirements of beef cattle may be met either by (1) provitamin A (carotene) in feedstuffs, (2) supplementary vitamin A either by intramuscular injection or orally, or (3) a combination of (1) and (2). Most commercial cattle feeders add synthetic vitamin A to the ration at a sufficient level to meet the total requirements, and do not rely on the rather undependable carotene content of feeds. Also, it is noteworthy that Purdue workers demonstrated that finishing beef cattle fed rations containing "enough" carotene (2 mg/lb of dry matter) developed typical vitamin A deficient symptoms, suggesting that finishing beef cattle need some preformed vitamin A.

Table 19–13 gives the estimated carotene content of feeds in relation to appearance and methods of conservation.

Beef cattle usually receive sufficient vitamin D from exposure to direct sunlight or from sun-cured roughages.

When supplemental vitamin E is needed, it may be added to the ration or injected intramuscularly.

TABLE 19–13
ESTIMATED CAROTENE CONTENT OF FEEDS IN RELATION TO APPEARANCE AND METHODS OF CONSERVATION[1]

Feedstuff	Carotene	
	(mg/lb)	*(mg/kg)*
Dehydrated alfalfa meal, fresh, dehydrated without field curing, very bright green color[2]	110–135	*242–298*
Alfalfa leaf meal, bright green color	60–80	*132–176*
Dehydrated alfalfa meal after considerable time in storage, bright green color	50–70	*110–154*
Legume hays, including alfalfa, very quickly cured, with minimum sun exposure, bright green color, leafy	35–40	*77–88*
Legume hays, including alfalfa, good green color, leafy	18–27	*40–60*
Fresh green legumes and grasses, immature	15–40	*33–88*
Legume hays, including alfalfa, partly bleached, moderate amount of green color	9–14	*20–31*
Nonlegume hays, including timothy, cereal, and prairie hays, well cured, good green color	9–14	*20–31*
Legume silage	5–20	*11–44*
Legume hays, including alfalfa, badly bleached or discolored, traces of green color	4–8	*9–18*
Nonlegume hays, average quality, bleached some green color	4–8	*9–18*
Corn and sorghum silages, medium to good green color	2–10	*4.4–22*
Grains, mill feeds, protein concentrates, and by-product concentrates, except yellow corn and its by-products	.01–0.2	*0.2–0.4*

[1]From *Nutrient Requirements of Beef Cattle,* NRC-National Academy of Sciences, No. 579, with metric system added by the authors.

[2]Green color is not uniformly indicative of high-carotene content.

Vitamin K is synthesized in the rumen of cattle in adequate amounts under most feeding conditions.

The dietary requirements of the young calf for the B vitamins (thiamin, biotin, niacin, pyridoxine, pantothenic acid, riboflavin, and vitamin B–12) during the first 8 weeks of life, prior to the development of the functioning rumen, are usually met by milk supplied by the cow during early lactation. Later, B vitamins are synthesized by rumen bacteria in sufficient quantities in most feeding regimens.

FEED SUBSTITUTION TABLE

The successful cattle producer recognizes that feeds of similar nutritive properties can and should be interchanged in the ration as price relationships warrant, thereby making it possible at all times to obtain a balanced ration at the lowest cost. Thus, (1) the cereal grains may consist of corn, barley, wheat, oats, and/or sorghum; (2) the protein supplements may consist of soybean, cottonseed, peanut, sunlower, and/or linseed meal; (3) the roughage may include many varieties of hays and silages; and (4) a vast array of by-product feeds may be utilized. The selection of alternative feeds has been made vastly easier and more accurate through the application of modern computer techniques.

Table 19–14, Feed Substitution Table for Beef and Dairy Cattle, is a summary of the comparative values of the most common U.S. feeds. In arriving at these values, two primary factors besides chemical composition and feeding value have been considered—namely, palatability and product quality.

In using this feed substitution table, the following facts should be recognized:

1. That, for best results, different ages and groups of animals within classes should be fed differently.

2. That individual feeds differ widely in feeding value. Barley and oats, for example, vary widely in feeding value according to the hull content and the test weight per bushel, and forages vary widely according to the stage of maturity at which they are cut and how well they are cured and stored.

3. That nonlegume forages may have a higher relative value to legumes than herein indicated provided the chief need of the animal is for additional energy rather than for supplemented protein. Thus, the nonlegume forages of low value can be used to better advantage for wintering mature, dry beef cows than for young calves.

On the other hand, legumes may have a higher value relative to nonlegumes than herein indicated provided the chief need is for additional protein rather than for added energy. Thus, no protein supplement is necessary for breeding beef cows provided a good-quality legume forage is fed.

4. That, based primarily on available supply and price, certain feeds—especially those of medium protein content, such as brewers' dried grains, corn gluten feed (gluten feed), distillers' dried grains, distillers' dried solubles, peanuts, and peas (dried)—may be used interchangeably as (a) grains and by-product feeds, and/or (b) protein supplements.

5. That the feeding value of certain feeds is materially affected by preparation. Thus, wheat must be coarsely ground or rolled for cattle. The values herein reported are based on proper feed preparation in each case.

For these reasons, the comparative values of feeds shown in the feed substitution table are not absolute. Rather, they are reasonably accurate approximations based on average-quality feeds, together with experiences and experiments.

TABLE 19–14
FEED SUBSTITUTION TABLE FOR BEEF AND DAIRY CATTLE (AS-FED BASIS) (See footnotes at end of table.)

Feedstuff	Relative Feeding Value (lb for lb) In Comparison With The Designated (underlined) Base Feed Which = 100	Maximum Percentage of Base Feed (or comparable feed or feeds) Which It Can Replace For Best Results	Remarks
GRAINS, BY-PRODUCT FEEDS, ROOTS AND TUBERS:[1] (Low and Medium Protein Feeds)			
Corn, No. 2	***100***	***100***	The most important concentrate for cattle in the U.S. Grind coarsely or flake.
Almond hulls, dried, no shells	70–75	15–30	
Almond hulls and shell meal	35	15–20	
Apple pomace, air-dry	78	33⅓	Values given are for apple pomace with paper or rice hulls as press aids.
Bakery products, dried	110	15–30	
Bakery waste, not dried (30% water)	75	15–30	
Barley	90	25–100	The heavier the barley and the smaller the proportion of hulls, the higher the feeding value. Grind coarsely or roll for cattle. In Canada, where considerable barley is fed, it is often used as the only basal feed in the ration once animals are accustomed to it.
Beans (cull)	80	10	Best when cooked, but can also be fed raw. Beans should be ground. When cooked, 3–4 lb (*1.4–1.8 kg*)/head daily; when raw, 1–2 lb (*0.45–0.91 kg*). Scouring may occur if they constitute more than 15% of total ration.
Beet pulp, dried	90	50	
Beet pulp, molasses, dried	90–95	50	
Beet pulp, wet	25	40	50% the value of corn silage. May compose 40% of ration on dry matter basis.
Brewers' dried grains	80	33⅓	Not very palatable. Fed chiefly to dairy cattle. Too bulky and usually too costly to be used in finishing rations.
Brewers' grains (wet)	13–15	33⅓	Grains usually come from barley. Best to haul and feed directly. Can be stored in silo if salt is added at rate of 25 lb (*11.4 kg*) per ton of grains.
Buckwheat	55–75	33⅓	Should be ground and mixed with other grains.
Carrots (cull)	10–15	20–25	Store 3–4 weeks before using; fresh carrots cause scouring. Feed whole or sliced.
Citrus pulp, dried	80–88	25–50	
Corn-and-cob meal	85–90	100	
Corn gluten feed (gluten feed)	85–90	50	
Distillers' dried grains	73–90	33⅓	Rye distillers' dried grains are of lower value than similar products made from corn or wheat. Distillers' dried grains are used chiefly for dairy cattle.
Distillers' dried solubles	73–90	33⅓	The chief difference between distillers' dried grains and distillers' dried solubles is the higher B vitamin content of the latter. Normally this is not important for cattle.

(Continued)

TABLE 19-14 *(Continued)*

Feedstuff	Relative Feeding Value (lb for lb) In Comparison With The Designated (underlined) Base Feed Which = 100	Maximum Percentage of Base Feed (or comparable feed or feeds) Which It Can Replace For Best Results	Remarks
GRAINS, BY-PRODUCT FEEDS, ROOTS AND TUBERS:[1] (Continued)			
Fat (animal or vegetable)	225	5	Fat has 203 megacalories energy/100 lb (*45.4 kg*) for maintenance and 127 megacalories for weight gain, as compared to 92 and 60, respectively, for corn.
Hominy feed	100	50	
Manure, cattle, without bedding	75	50	Approximately 80% of the total nutrients of feeds is excreted as animal manure. However, the feeding value of manure will vary according to (1) the nutritive value of the feeds initially fed, (2) the class, age, and individuality of the animal to which the feeds were initially fed, and (3) the handling and processing of the manure.
Manure, poultry (see poultry house litter)			
Molasses, beet	75	10–40	Value is highest when used as an appetizer. May be laxative if fed at levels above 6 lb (*2.7 kg*) daily.
Molasses, cane	75	10–40	Value is highest when used as an appetizer.
Molasses, citrus	65–75	10–40	
Molasses, wood	26–30	10–20	Unpalatable.
Oats	70–90	10–100	Valuable for young stock, for breeding stock and for getting animals on feed. Oats have lowest value for finishing cattle and should be limited to ⅓ of such rations. Also, the feeding value of oats varies according to the test weight per bushel. Grind or roll for cattle.
Paunch, dried (also see "paunch-blood" under Protein Supplements of this table)	90	5–10	Dried paunch is not palatable, with the result that it depresses appetite. Rate of gain is not affected, but feed efficiency is slightly lowered.
Peas (cull), dried	88	40	Because of lack of palatability, peas will lower feed intake if they constitute more than 20% of the total ration. Also, there is bloat hazard if they exceed 40% of the ration.
Pear waste, air-dry	75	40	
Potatoes (Irish), wet	20–25	85	When fed with alfalfa hay, they are worth about 80% as much per ton as corn silage. Do not feed frozen. Sunburned, decomposed, or sprouted potatoes should not make up more than 10% of potatoes fed. Keep steers' heads down while eating to prevent choking.
Potatoes (Irish), dehydrated	88	50	Excellent source of energy, but deficient in protein, minerals, and vitamins.
Potatoes (sweet)	25	85	
Potatoes (sweet), dehydrated	95–100	50	Dehydrated sweet potatoes are more palatable than dehydrated Irish potatoes.
Poultry house litter	10–40	15–25	Poultry house litter may also be used as a protein source (see Protein Supplements, this table).
Prunes	62	15	Because of the laxative quality of prunes, they should be limited to 7% of the total ration.
Raisins (cull)	70	33⅓	
Raisin pulp	53	25	
Rice (rough rice)	80	100	
Rice bran	66⅔–75	33⅓	
Rice polishings	88	25	
Rye	96	33⅓	Not palatable when fed in large amounts.
Screenings, refuse	62–70	25–35	Should be finely ground in order to kill noxious weed seeds. Quality varies; good-quality screenings are equal to oats whereas poor-quality screenings resemble straw.
Sorghum (milo, kafir), grain	90–95	100	Varieties vary in protein content. Grind or roll for cattle.
Spelt and emmer	70–90	30–100	Similar to oats.
Wheat	100–105	50	Grind coarsely, or roll.
Wheat bran	70–90	25–33⅓	Because of its bulk and fiber, bran is not desirable for finishing rations. Bran is valuable for young animals, for breeding animals, and for starting animals on feed.
Wheat-mixed feed (mill run)	95	33⅓	Sometimes fed to the breeding herd, to young calves, and to finishing cattle being started on feed.
Wheat screenings	85	50	
Wood (cooked)	75–80	70	Wood products, which are largely cellulose and lignin, must be cooked before animals can digest them.

(Continued)

TABLE 19–14 *(Continued)*

Feedstuff	Relative Feeding Value (lb for lb) In Comparison With The Designated (underlined) Base Feed Which = 100	Maximum Percentage of Base Feed (or comparable feed or feeds) Which It Can Replace For Best Results	Remarks
PROTEIN SUPPLEMENTS:			
Soybean meal (41%)	***100***	***100***	Slightly laxative effect.
Alfalfa or clover screenings	70–75	50	Grind finely to destroy weed seeds.
Brewers' dried grains	55–65	50	Not very palatable. Fed chiefly to dairy cattle.
Copra meal (coconut oil meal), 21%	90–100	50	
Corn gluten feed (gluten feed)	65–75	50–100	
Corn gluten meal (gluten meal)	90–100	50	Somewhat unpalatable.
Cottonseed meal (41%)	100	100	
Distillers' dried grains	65–70	100	Rye distillers' grains are about 10% lower in protein than similar products made from corn or wheat. Low in palatability.
Distillers' dried solubles	70	100	
Feather meal (hydrolyzed; 84% protein)	175	50	Feather meal is unpalatable; hence, cattle must be accustomed to it gradually and it must be limited in quantity. It is best used for wintering brood cows and stocker cattle.
Legume screenings	75	75	Satisfactory, but less palatable than soybean or cottonseed meal.
Linseed meal (35%)	95	100	Linseed meal has laxative effect. Some cattle will not tolerate more than 5–8% linseed meal in the ration.
Paunch-blood feed (also see "paunch, dried" under Grains section of this table)	100	100	At slaughter, each bovine yields about 20 lb (*9.1 kg*) of paunch and 20 lb (*9.1 kg*) of blood. Dried paunch runs around 10% protein, dried blood around 80%, and a 50–50 mixture of the 2 products, around 45%.
Peanut meal (45%)	100	100	Peanut meal may become rancid if stored too long, especially in warm, moist climates.
Peas (cull), dried	65–75	50	
Poultry house litter	50–55	25	Poultry house litter may also be used as an energy source (see Grains section of this table).
Rapeseed meal (37%)	88	75	Rapeseed meal should be limited to not more than 2 lb (*0.91 kg*) per cow.
Safflower meal, well hulled (42%)	92	100	
Safflower meal, with hulls (20%)	40–45	100	Safflower meal with hulls is unpalatable. Thus, it should be mixed with more palatable feeds.
Sesame meal	90–95	25	
Soybeans, whole	95–100	95	Not satisfactory for finishing calves. Soybean allowance should be limited to amount necessary to balance the ration. Larger amounts may be unduly laxative and cause cattle to go off feed.
Sunflower meal (39%)	95–100	100	If poorly hulled and lower protein content than 39%, feeding value will be lowered accordingly. It is well liked by cattle and keeps well in storage.
DRY FORAGES AND SILAGES:[2]			All the dry nonlegume forages listed herein are satsifactory when needed minerals and either a limited amount of legume hay or a protein supplement are supplied to balance the ration.
Alfalfa hay, all analyses	***100***	***100***	Does away with or lessens protein supplement requirements.
Alfalfa silage	33⅓–50	50–85	When alfalfa silage replaces corn silage, more energy feed must be provided but less protein.
Alfalfa straw	37	50	Feed with good hay.
Apple pomace silage	17–25	50–85	Usually fed as a substitute for corn or grass silage. 50% the value of corn silage. Sometimes fed out of a stack or trench silo.
Apples	17–25	50–85	Do not feed more than 25 lb (*11.4 kg*)/mature bovine. Not recommended for finishing cattle. Danger of choking when fed whole. Relatively high handling cost.
Bagasse, dried; sugarcane or sorghum	10–20	5–10	
Barley hay	70	100	Avoid bearded varieties.
Barley silage	25–40	50–80	In silage, there is no problem with bearded varieties, which usually outyield beardless.
Barley straw	40	70	Of the cereal straws, barley ranks next to oat straw in feeding value. Feed to dry pregnant cows. Supplement daily with 5–6 lb (*2.3–2.7 kg*) alfalfa hay or 1–2 lb (*0.45–0.91 kg*) of 30–40% protein supplement.
Bean straw	34	50	Feed with good hay.

(Continued)

TABLE 19–14 *(Continued)*

Feedstuff	Relative Feeding Value (lb for lb) In Comparison With The Designated (underlined) Base Feed Which = 100	Maximum Percentage of Base Feed (or comparable feed or feeds) Which It Can Replace For Best Results	Remarks
DRY FORAGES AND SILAGES:[2] (Continued)			
Beet tops, fresh	20	33⅓–50	In the West, large acreages of fresh beet tops are grazed by cattle and sheep. Bloat may be a problem when tops are frozen. Tops are laxative. Add 2½ lb (*1.1 kg*) of ground limestone/ton of feed.
Beet top silage, sugar	17–25	33⅓–50	Feed 2 oz (*56.7 g*) of finely ground limestone or chalk with each 100 lb (*45.4 kg*) of tops, as calcium changes the oxalic acid to insoluble calcium oxalate.
Clover hay, crimson	90–100	100	Crimson clover hay has a considerably lower value if not cut at an early stage.
Clover hay, red	90–100	100	If the rest of the ration is adequate in protein, clover hay will be equal to alfalfa in feeding value; otherwise, it will be lower.
Clover straw	37	50	Feed with good hay.
Clover-timothy hay	80–90	100	Value of clover-timothy mixed hay depends on the proportion of clover present and the stage of maturity at which it is cut.
Corncobs, ground	70	90	Ground corncobs can be used as the only roughage for beef cattle if properly supplemented with proteins, minerals, and vitamins.
Corn fodder	75	80–90	
Corn husklage (shucklage)	50	80–90	Highest and best use is for dry pregnant cows. It is slightly higher in energy and more palatable than corn stover.
Corn silage	33⅓–50	50–85	
Corn (sweet) silage, cannery waste	26–40	50–85	
Corn stover	45	70–90	Corn stover will meet the energy needs of dry pregnant cows, but is deficient in protein and low in phosphorus and vitamin A. Two acres of cornstalks will carry a cow 100–120 days.
Corn (sweet) stover	50	80–90	
Cottonseed hulls	66⅔	75	Use for dry pregnant cows. Supplement daily with 4–6 (*1.8–2.7 kg*) lb of good legume hay or 1–2 lb (*0.45–0.91 kg*) of a 30–40% protein supplement.
Cowpea hay	90–100	100	
Gin trash, cotton	75	75	
Grape pomace or meal	5–15	10–15	Pomace including stems is of little value as a feed.
Grass-legume mixed hay	80–90	100	Value depends on the proportion of legume present and the stage of maturity at which it is cut.
Grass-legume silage	32–47	50–85	Unless grain is added as a preservative, grass silage requires more energy feed, but less protein supplement than corn silage when fed to finishing cattle.
Grass silage	30–45	50–85	For finishing cattle, grass silage must be supplemented with additional energy feeds, such as cereal grain or molasses, to be of the same value as corn silage.
Hop vine silage	20	50–75	It should be chopped when placed in the silo.
Hops, spent, dehydrated	80	50–65	Devoid of carotene; feed with legume hay.
Johnsongrass hay	70	100	
Lespedeza hay	80–100	100	Feeding value of lespedeza hay varies considerably with stage of maturity at which it is cut.
Mint hay	70–80	75	Cattle tire of mint hay when it is fed as the only roughage for extended periods.
Oat hay	75	100	
Oat silage	32–47	50–85	Must be chopped finely to exclude air from silo.
Oat straw	50	75	Oat straw is the best of the cereal straws. Use for dry pregnant cows. Supplement daily with 4–6 lb (*1.8–2.7 kg*) of good legume hay or 1–2 lb (*0.45–0.91 kg*) of 30–40% protein supplement.
Paper (newspaper; waste paper)	66⅔	50	Paper varies in feeding value in proportion to the cellulose (most paper is 60–90% cellulose) and lignin content. Magazine and bookstock papers are higher in cellulose and lower in lignin than newspapers; hence, of higher feeding value. Pelleting or cubing may increase the value of paper. *Caution:* Some newspapers contain heavy metals (boron, lead, barium, and antimony), sometimes used as a dye carrier in printer's ink, which may be toxic to animals. This is especially true of "funny" papers because of the quantity of heavy metals carried on the colored ink of the comics.
Pea straw	45–75	60–75	
Pea-vine hay	100–110	75–90	Can constitute the only roughage for finishing cattle.
Pea-vine silage	33⅓–50	50–85	Unless grain is added as a preservative, pea-vine silage requires more energy feed, but less protein supplement than corn silage when fed to finishing cattle.
Potato silage	25–30	50–75	About 75% the value of corn silage.
Prairie hay	65–70	100	

(Continued)

TABLE 19–14 *(Continued)*

Feedstuff	Relative Feeding Value (lb for lb) In Comparison With The Designated (underlined) Base Feed Which = 100	Maximum Percentage of Base Feed (or comparable feed or feeds) Which It Can Replace For Best Results	Remarks
DRY FORAGES AND SILAGES:[2] (Continued)			
Reed canarygrass hay	70	100	
Rice straw	47	70	High levels of rice straw can be used for wintering cattle if the straw is properly fortified.
Sawdust	75–80	70	Feeding value varies among species of trees. Digestibility is increased by cooking and other treatments.
Sorghum fodder	70	100	
Sorghum silage (grain varieties)	32–47	50–85	For finishing cattle, 85–90% as valuable as corn silage and must be supplemented in the same manner as corn silage.
Sorghum silage (sweet varieties)	25–30	50–85	Nearly equal to grain varieties in value per acre because of greater yield.
Sorghum (milo) stover	35	70–90	Can be grazed or harvested and stored either as dry feed or silage. About 2% higher in protein, but less palatable, than corn stover.
Soybean hay	85–90	50–75	Lower value than alfalfa hay, largely due to greater wastage in feeding. It may cause scouring when fed alone.
Sudangrass hay	70	100	
Sunflower silage	25–35	50–85	65–75% value of corn silage. Somewhat unpalatable and may cause constipation. Harvest for silage when ½–⅔ of heads are in bloom.
Sweet clover hay	100	100	Value of sweet clover hay varies widely. Moldy or spoiled sweet clover hay may cause sweet clover disease.
Timothy hay	70	100	
Vetch-oat hay	80–90	100	The higher the proportion of vetch, the higher the value.
Wheat hay	70	100	
Wheat straw	35	65	Of the cereal straws, wheat ranks third in nutritive value, behind oat straw and barley straw. Highest and best use is for dry pregnant cows. Supplement daily with 6 lb (*2.7 kg*) of alfalfa or 2 lb (*0.91 kg*) of a 30–40% protein supplement.

[1]Roots and tubers are of lower value than the grain and by-product feeds due to their higher moisture content.

[2]Silages are of lower value than dry forages due to their higher moisture content.

FEED PREPARATION

The physical preparation of cereal grains for cattle by soaking and cooking has been practiced by cattle exhibitors for a very long time. In recent years, many sophisticated techniques for the processing of grains have been developed, especially for feedlot cattle. Basically, however, grain is either soaked, cooked, ground, or rolled (wet or dry), and hay is either cut, shredded, ground, pelleted, or cubed.

The subject of feed preparation for all classes of livestock is fully covered in Chapter 14, Feed Processing.

FEED ALLOWANCE AND SOME SUGGESTED RATIONS[4]

Some general rules of feeding may be given, but it must be remembered that *"the eye of the master fattens his cattle."* Nevertheless, the beginner may well profit from the experience of successful feeders. It is with this hope that the suggested rations are presented.

Table 19–15, Daily Rations For Beef Cattle (see next page), contains suggested rations for different classes and ages of cattle. These are merely intended as general guides. Variations can and should be made in the rations used. The feeder should give consideration to (1) the supply of home-grown feeds, (2) the availability and price of purchased feeds, (3) the class and age of cattle, (4) the health and condition of the animals, and (5) the length of the grazing season.

[4]Insofar as possible, these rations were computed from the requirements reported by the National Research Council and applied by the authors.

TABLE
DAILY RATIONS
(As-Fed

Suggested Rations With all rations and for all classes and ages of cattle, provide free access in separate containers to (1) salt (iodized salt in iodine-deficient areas), and (2) a suitable mineral mixture.	Wintering Mature Pregnant Beef Breeding Cows (av. wt. 1,100 lb or *499 kg*)		Wintering Mature Lactating Beef Breeding Cows (av. wt. 1,100 lb or *499 kg*)		Wintering Replacement Heifers (weighing 400–500 lb or *181–227 kg* start of wintering)	
	Per Day		Per Day		Per Day	
	(lb)	*(kg)*	(lb)	*(kg)*	(lb)	*(kg)*
1. Legume hay or grass-legume mixed hay, good quality	18–20	*8.2–9.1*	30	*13.6*	13–15[3]	*5.9–6.8*[3]
Grain	—	—	—	—	2–3	*0.91–1.36*
Protein supplement	—	—	—	—	—	—
2. Grass hay or other nonlegume dry roughage	18–20	*8.2–9.1*	24–26	*10.9–11.8*	12–18[3]	*5.4–8.2*[3]
Grain	—	—	2	*0.91*	2½–4½	*1.13–2.04*
Protein Supplement	½–1	*0.23–0.45*	3	*1.36*	1¼–1½	*0.57–0.68*
3. Legume hay or grass-legume mixed hay, good quality	7–11	*3.2–5.0*	26–28	*11.8–12.7*	8–12[3]	*3.6–5.4*[3]
Grass hay or other nonlegume dry roughage	9–11	*4.1–5.0*	—	—	4–6	*1.8–2.7*
Grain	—	—	1	*0.45*	2½–4	*1.13–1.81*
Protein supplement	—	—	1	*0.45*	½–1	*0.23–0.54*
4. Corn or sorghum silage	50–55	*22.7–25*	55	*25*	25–40	*11.3–18.2*
Grain	—	—	2	*0.91*	—	—
Protein supplement	0–½	*0–0.23*	3	*1.36*	1½–1¾	*0.68–0.79*
5. Grass silage, half or more legume	50	*22.7*	50	*22.7*	25–40	*11.3–18.2*
Grain	—	—	4	*1.81*	3–4	*1.36–1.81*
Protein supplement	—	—	—	—	½	*0.23*
6. Silage (corn or sorghum silage fed with legume hay or legume silage fed with grass hay)	35	*15.9*	40	*18.1*	15–30	*6.8–13.6*
Hay	5–6	*2.3–2.7*	10	*4.5*	3–4	*1.4–1.8*
Grain	—	—	—	—	1–2	*0.45–0.91*
Protein supplement	0–½	*0–0.23*	—	—	½–1	*0.22–0.45*

[1]If stocker calves are late or the roughage is fair to poor quality, it may be desirable to add 2–4 lb (*0.91–1.81 kg*) of grain per head daily. If farm scales are available, monthly weights may be used as the criterion for grain feeding. Keep in mind that calves should gain ¾–1 lb (*0.34–0.45 kg*) daily.

[2]In general, the experienced feeder plans that cattle on full feed shall consume (1) feeds in amounts (daily: air-dry basis) equal to about 2.5–3.0% of their liveweight, (2) 70–90% concentrates, and (3) a minimum of 2–4 lb (*0.9–1.8 kg*) roughage for each 100 lb (*45 kg*) liveweight. In areas where roughage is more abundant and comparatively cheaper than grain, the proportions of roughage to grain should be somewhat higher than indicated. In computing roughage consumption, 3 lb (*1.36 kg*) of silage are considered equivalent to 1 lb (*0.45 kg*) of hay.

FITTING FOR SHOW AND SALE

All animals intended for show purposes, including both breeding animals and steers, should be placed in the proper state of condition—they should be neither too fat nor too thin. Steers should be fed to a degree of finish that will help ensure their ability to grade at least low Choice on the rail. Requirements differ slightly among breeds and lines of cattle. For example, the larger, leaner continental European breeds do not mature as quickly nor fatten as readily as the British breeds. The essentials in feeding cattle for the show might be described as similar to those in feeding steers for market, except that more attention must be given to the smallest details. A suitable ration must be selected and the animal or animals must be fed with care over a sufficiently long period.

At the beginning of show preparation, check for internal and external parasites. If any are present, apply the recommended treatment.

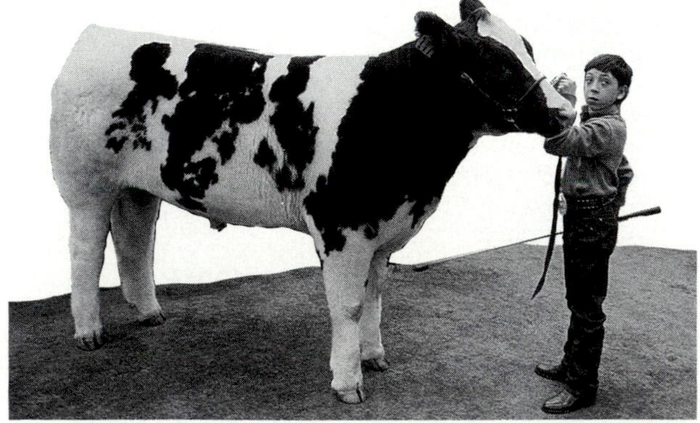

Fig. 19-12. Spot, grand champion steer at the National Western Stock Show, Denver, 1986. The 1,261-lb Shorthorn steer was shown by Brandon Horn, Lookeba, Okla. (Courtesy, National Western Stock Show, Denver, Colo.)

19-15
FOR BEEF CATTLE
Basis)

Wintering Stocker Calves Roughed Through Winter and Grazed the Following Summer. Fed for winter gain of ¾-1 lb (*0.34-0.45 kg*) per head daily (weighing 400-500 lb or *181-227 kg* start of wintering)[1]		Finishing Calves in Drylot, Generally in Winter (weighing 400-500 lb or *181-227 kg* start of feeding and 750-850 lb or *340-386 kg* at marketing)[2]		Wintering Yearlings; Roughed Through the Winter, and Generally Pasture Finished the Following Summer. Fed for winter gains of 1-1¼ lb or *0.45-0.57 kg* per head daily (weighing about 600 lb or *227 kg* start of wintering)		Finishing Yearlings in Drylot, Generally in Winter (weighing about 600 lb or *272 kg* start of feeding, and 900-1,050 lb or *409-477 kg* at marketing)[2]		Finishing Long-yearling Steers in Drylot Generally in Winter (weighing about 850 lb or *386 kg* start of feeding and 1,000-1,100 lb or *454-499 kg* at marketing)[2]	
Per Day		Per Day		Per Day		Per Day		Per Day	
(lb)	*(kg)*	(lb)	*(kg)*	(lb)	*(kg)*	(lb)	*(kg)*	(lb)	*(kg)*
12-18[3]	*5.4-8.2*	4-6	*1.8-2.7*	16-24	*7.2-10.9*	4-8	*1.8-3.6*	6-12	*2.7-5.4*
—	—	12-15	*5.4-6.8*	—	—	15-19½	*6.80-8.8*	16-22	*7.2-10.0*
—	—	1-1½	*0.45-0.68*	—	—	1-1½	*0.45-0.68*	—	—
12-18[3]	*5.4-8.2*	4-5	*1.8-2.3*	16-24	*7.2-10.9*	4-8	*1.8-3.6*	6-12	*2.7-5.4*
—	—	12-15	*5.4-6.8*	—	—	15-20	*6.8-9.1*	16½-22¾	*7.5-10.3*
¼-1½	*0.57-0.68*	1¾-2	*0.79-0.91*	1½-1¾	*0.68-0.79*	1½-2½	*0.68-1.1*	1½-1¾	*0.68-0.79*
8-12[3]	*5.4-8.2*	2-3	*0.91-1.36*	6-8	*2.7-3.6*	2-4	*0.91-1.81*	3-6	*1.4-2.7*
4-6	*1.8-2.7*	2-3	*0.91-1.36*	10-16	*4.5-7.2*	2-4	*0.91-1.81*	3-6	*1.4-2.7*
—	—	12-15	*5.4-6.8*	—	—	15-19¾	*6.8-9.0*	16-22	*7.2-10.0*
¼-1	*0.11-0.45*	1½-1¾	*0.68-0.79*	1-1½	*0.45-0.68*	1¼-1¾	*0.57-0.79*	½-¾	*0.23-0.34*
25-40	*11.3-18.1*	6-16	*2.7-7.3*	40-55	*18.2-24.9*	6-25	*2.7-11.3*	6-35	*2.7-5.9*
—	—	8-12	*3.6-5.4*	—	—	11-16	*5.0-7.3*	15-21	*6.8-9.5*
1-1¼	*0.45-0.57*	2	*0.91*	1¼-1½	*0.57-0.68*	2	*0.91*	1¼-1½	*0.57-0.68*
25-40	*11.3-18.1*	6-16	*2.7-7.3*	40-55	*18.1-24.9*	6-25	*2.7-11.3*	6-35	*2.7-15.9*
2-3	*0.91-1.36*	8-12	*3.6-5.4*	4-5	*1.8-2.3*	11-16	*5.0-7.3*	15-21	*6.8-9.5*
½	*0.23*	1-2	*0.45-0.91*	½	*0.23*	1-1½	*0.45-0.68*	1	*0.45*
15-30	*6.8-13.6*	3-8	*1.4-3.6*	20-35	*9.1-15.9*	3-15	*1.4-6.8*	3-15	*1.4-6.8*
3-4	*1.4-1.8*	1-3	*0.45-1.4*	7	*3.2*	1-4	*0.45-1.8*	1-7	*0.45-3.2*
1-2	*0.45-0.91*	8-12	*2.6-5.4*	—	—	11-16	*5.0-7.2*	15-21	*6.8-9.5*
½	*0.23*	1-2	*0.45-0.91*	½-¾	*0.23-0.34*	1-1¾	*0.45-0.79*	1-1¼	*0.45-0.57*

[3]With calves (both replacement heifers and stockers) an extra 2 lb (*0.91 kg*) of hay daily, over and above requirements, are herewith indicated to allow for wastage. Practical operators generally feed stemmy or other hay left over by calves to the cow herd.

Rules of Feeding for Show

Some general rules of feeding may be given, but most successful cattle fitters have worked out systems of their own through years of practical experience and close observation—they do not follow set rules. Nevertheless, the beginner may well profit by the experience of successful fitters, and it is with this hope that the following general rules of feeding show cattle are presented:

1. Practice economy, but avoid false economy.
2. Use care in getting animals on feed, and avoid sudden changes.
3. Provide a variety of feeds.
4. Feed a balanced ration.
5. Do not overfeed or underfeed.
6. Keep the feed box clean.
7. Supply palatable and succulent feeds.
8. Use milk replacers for young animals.
9. Provide the correct amount of bulk.
10. Do not feed damaged feeds.
11. Prepare grains.
12. Feed regularly.
13. Provide minerals.
14. Keep the animal(s) quiet and contented.
15. Provide exercise.
16. Avoid scouring.
17. Avoid sudden water changes.
18. Provide breeding stock an adequate, but lighter, ration after the fair is over.

Fitting Rations

Variations can and should be made in fitting rations, depending upon the individual animal, the relative prices of feeds, and the supply of homegrown feeds. To attain the correct state of condition, a suitable ration must be selected and the animal or animals must be fed with care over a sufficiently long period. The rations listed in Table 19–16 have been used by successful fitters. They are higher in protein content than rations normally used in commercial-finishing operations, but most experienced caretakers feel that by such means they get more bloom. In general, when show animals are being force-fed on any one of these concentrate mixtures, experienced caretakers prefer to feed a grass hay or a grass-legume mixed hay to a straight legume, because of the laxative effect and possible bloat hazard of the latter.

Ration 11 is the one which the senior author has used in fitting show steers. The cooked barley is prepared by (1) adding water in the proportion of 2 to 2½ gal to each gallon of dry barley, and (2) cooking until the kernels are thoroughly swollen and can be crushed easily between the thumb and forefinger. Each young steer also receives 4 lb daily of a supplement high in milk by-products. As the animal approaches show finish, the ration is changed by decreasing the rolled barley by 7 lb and increasing the rolled oats by 5 lb and the wheat bran by 2 lb.

TABLE 19–16
FITTING RATIONS FOR SHOW AND SALE CATTLE
(As-Fed Basis)

Rations 1 to 5 are bulky. They are recommended for use (1) by the inexperienced feeder, and (2) in starting prospective show animals on feed.

Rations 6 to 11 are less bulky and higher in energy. They are recommended for use (1) by the experienced feeder, and (2) during the latter part of the fitting period.

Ration No. 1	(lb)	(*kg*)
Rolled barley	50	*22.7*
Crushed oats	20	*9.1*
Wheat bran	20	*9.1*
Protein supplement[1]	10	*4.5*

Ration No. 2	(lb)	(*kg*)
Rolled barley	30	*13.6*
Flaked corn	20	*9.1*
Crushed oats	20	*9.1*
Wheat bran	20	*9.1*
Protein supplement[1]	10	*4.5*

Ration No. 3	(lb)	(*kg*)
Flaked corn	40	*18.1*
Crushed oats	30	*13.6*
Wheat bran	20	*9.1*
Protein supplement[1]	10	*4.5*

Ration No. 4	(lb)	(*kg*)
Crushed oats	30	*13.6*
Rolled barley	30	*13.6*
Wheat bran	20	*9.1*
Flaked corn	10	*4.5*
Protein supplement[1]	10	*4.5*

Ration No. 5	(lb)	(*kg*)
Flaked corn	55	*25.0*
Crushed oats	30	*13.6*
Protein supplement[1]	15	*6.8*

Ration No. 6	(lb)	(*kg*)
Flaked corn or sorghum	50	22.7
Rolled barley	40	*18.1*
Protein supplement[1]	10	*4.5*

Ration No. 7	(lb)	(*kg*)
Flaked corn	55	*25.0*
Crushed oats	20	*9.1*
Dried beet pulp	10	*4.5*
Protein supplement[1]	15	*6.8*

Ration No. 8	(lb)	(*kg*)
Flaked corn	40	*18.1*
Rolled barley	20	*9.1*
Crushed oats	10	*4.5*
Dried beet pulp	10	*4.5*
Wheat bran	10	*4.5*
Protein supplement[1]	10	*4.5*

Ration No. 9	(lb)	(*kg*)
Crushed oats	25	*11.3*
Rolled barley	20	*9.1*
Rolled wheat	20	*9.1*
Flaked corn	20	*9.1*
Wheat bran	10	*4.5*
Protein supplement[1]	5	*2.3*

Ration No. 10	(lb)	(*kg*)
Rolled barley	35	*15.9*
Crushed oats	20	*9.1*
Rolled wheat	20	*9.1*
Dry beet pulp	15	*6.8*
Protein supplement[1]	10	*4.5*

Ration No. 11	(lb)	(*kg*)
Rolled barley	20	*9.1*
Flaked corn	20	*9.1*
Crushed oats	20	*9.1*
Whole barley (dry wt. basis but cooked before feeding)	13	*5.9*
Commercial supplement	8	*3.6*
Linseed meal	8	*3.6*
Wheat bran	6	*2.7*
Beet pulp, dried molasses	4	*1.8*
Salt	1	*.5*

[1]The protein supplement may consist of linseed, soybean, cottonseed, or peanut meal. With most caretakers, linseed meal is the preferred protein supplement. It gives the animal a sleek hair coat and a pliable hide. Because it is a laxative feed, however, caution should be used in feeding it. Although it is true that an animal getting good clover or alfalfa hay needs less protein supplement than does one eating nonleguminous roughage, it is not possible to supply all the needed protein with hay and still get enough grain into young animals to finish them quickly.

NUTRITIONAL DISEASES AND AILMENTS

Nutritional deficiencies may be brought about by either (1) too little feed, or (2) rations that are too low in one or more nutrients

Also, forced production (such as finishing animals at early ages) and the feeding of forages and grains which are produced on leached or depleted soils have created problems in nutrition. This condition has been further aggravated through the increased confinement of cattle, many animals being confined to lots or a building all or a large part of the year. Under these unnatural conditions, nutritional diseases and ailments have become increasingly common.

Although the cause, prevention, and treatment of most of these nutritional diseases and ailments are known, they continue to reduce profits in the livestock industry simply because the available knowledge is not put into practice. Moreover, those widespread nutritional deficiencies which are not of sufficient proportions to produce clear-cut deficiency symptoms cause even greater economic losses because they go unnoticed and unrectified. Chapter 5, Nutritional Disorders/Toxins, of this book contains a summary of the important nutritional diseases and ailments affecting animals; hence, the reader is referred thereto. Some nutritional diseases common to feedlot cattle are discussed in Part IV of this chapter.

PART II—**Feeding Breeding Beef Cattle**

The beef breeding herd must be properly fed if a good calf crop is to be produced. The size of the calf crop, the vigor and size the calves attain by market time, and the feeding efficiency of the herd largely determine the profit realized.

FEEDING BROOD COWS

Feed affects total profit and cow productivity. It accounts for 65 to 75% of the total cost of keeping cows, and it exerts a powerful influence on cow fertility and calf weaning weight—the two biggest success factors in the cattle business.

Fig. 19-13. Proper feeding of brood cows is requisite to a calf being born and born alive. (Courtesy, American Polled Hereford Assn., Kansas City, Mo.)

Nutritional Requirements of Brood Cows

The nutrient allowances should be adequate to provide for maintenance, growth (if animals are immature), and reproduction and lactation. Fortunately, these needs can be met largely through feeding roughages—pasture in season, and dry forages and silages during the winter months.

The nutritional requirements of beef cows are influenced by weight and size of the female, milk production, age, and climate. Size of the cow has more influence on feed needs than any other item. Also, bigger cows produce bigger calves. Research has shown that 5 to 15 lb extra weaning weight is obtained with each 100 lb increase in cow weight. U.S. Department of Agriculture research at Clay Center, Nebraska showed that large frame brood cows produced 9% more calf weight than medium frame cows. But the large frame cows required 28% more feed than the medium frame cows. So, large cows eat more feed and wean heavier calves. The two-pronged question that each producer must determine is: Will added calf weight be sufficient to offset the added cost of maintaining a larger cow; and will the calf be so large that it causes difficulty at calving time?

Fig. 19-14. Producing calves is the first and most important requisite of brood cows. Cows should be fed and managed so that they approach a 100% calf crop and wean heavy calves. (Courtesy, American Breeders Service, DeForest, Wisc.)

Experiments and practical observations reveal that the period during which calf crop percentage is affected most by nutrition extends from 30 days before calving until 70 days after calving—until after rebreeding; a period of approximately 100 days. This, then, is the most critical period in the cow-calf business. It's when life begins—that period within which one calf is born and another is conceived. The needs for the cow during this more critical production period are approximately equal to her needs for the remainder of the year. This fact is pointed up in Fig. 19-15, showing the estimated energy requirements of a 1,000-lb beef cow during her 12-month reproductive cycle.

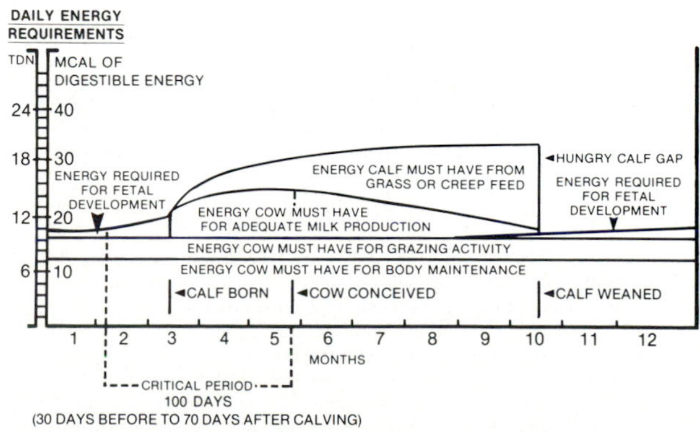

Fig. 19-15. Estimated energy requirements of a mature 1,000-lb beef cow during her 12-month reproductive cycle; based on 90-day calving season and 500-lb calf at 7 months of age. (Adapted by the authors from *Nutrient Requirements of the Cow and Calf,* Texas A&M University, B 1044, p. 7, Fig. 2)

The average daily energy requirement is about 14.5 lb of TDN. However, as shown in Fig. 19–15, the requirements are above the average for nearly 6½ months of the year. This means that, for reasons of economy, the calving season should be timed so that much of the feed can be supplied by pasture and other economical sources of homegrown energy and protein.

A second important requisite of a sound beef cattle nutrition program is to feed animals according to their requirements. It is impossible to feed the herd properly when calving occurs year-around, or when dry pregnant cows, replacement heifers, and cows nursing calves are run together. This point becomes apparent in Figs. 19–16 to 19–18, which show certain nutritive requirements for the following classes of cattle: (1) dry mature cows, last 2 to 3 months of pregnancy, weighing 1,100 lb; (2) yearling heifers, last 3 to 4 months of pregnancy, weighing 900 lb; and (3) cows nursing calves, superior milking ability, first 3 to 4 months after calving, weighing 1,100 lb.

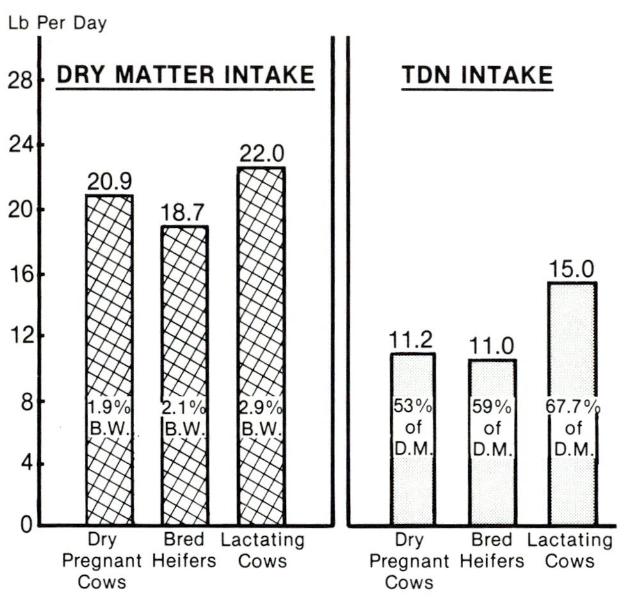

Dry Pregnant Cows—Dry pregnant cows, last third pregnancy, 1,100 lb (*500 kg*) wt. No gain.
Bred Heifers —Pregnant yearling heifers, last third pregnancy, 900 lb (*409 kg*) wt. Daily gain 1.3 lb (*0.6 kg*).
Lactating Cows —Cows nursing calves, superior milking, first 3-4 months after calving, 1,100 lb (*500 kg*) wt. No gain.

Fig. 19–16. Dry matter (in lb daily and in % of body weight) and TDN requirements (in lb daily and in % of the ration) of dry pregnant cows, bred heifers, and lactating cows. As shown, lactating cows require slightly more feed (22.0 vs 20.9 lb/day), along with a much higher energy ration (14.7% more), than dry pregnant cows. Replacement heifers, which must provide for the growth of their own bodies as well as for the growth of the fetus, consume more feed in relation to body weight (B.W.) than either dry or lactating cows. (Based on data from *Nutrient Requirements of Beef Cattle,* sixth revised edition, National Research Council-National Academy of Sciences, Washington, D.C., 1984)

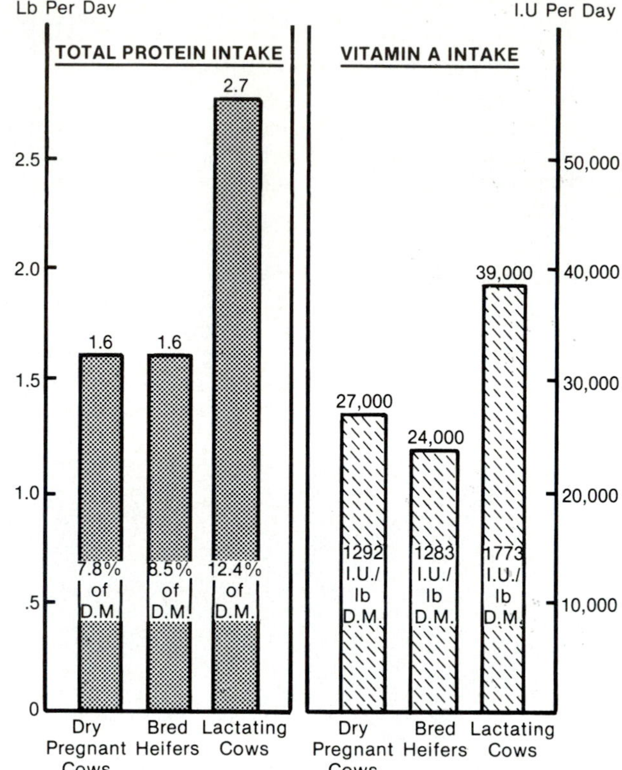

Dry Pregnant Cows—Dry pregnant cows, last third pregnancy, 1,100 lb (*500 kg*) wt. No gain.
Bred Heifers —Pregnant yearling heifers, last third pregnancy, 900 lb (*409 kg*) wt. Daily gain 1.3 lb (*0.6 kg*).
Lactating Cows —Cows nursing calves, superior milking, first 3-4 months after calving, 1,100 lb (*500 kg*) wt. No gain.

Fig. 19–17. Total protein and vitamin A requirements of dry pregnant cows, bred heifers, and lactating cows. Milk is high in protein; hence, it follows that for superior milk production the protein requirements of lactating cows are high, in terms of both crude protein intake and percent of the ration. Also, in order to meet the simultaneous protein requirements for body growth and fetal development, heifers require a higher protein ration than dry cows.

Lactating cows have the highest vitamin A requirement of the three groups, followed, in order, by dry pregnant cows and bred heifers. (Based on data from *Nutrient Requirements of Beef Cattle,* sixth revised edition, National Research Council-National Academy of Sciences, Washington, D.C., 1984)

Figs. 19–16 to 19–18 show that the requirements for lactation are the highest and most critical of all; in total feed consumed, in TDN and crude protein of the ration, and in the major minerals—calcium and phosphorus. After a cow calves, her energy needs jump about 50%, her protein needs increase sharply, her calcium and phosphorus needs nearly double, and her vitamin A requirement is 30% higher. These figures also show that the requirements of growing heifers are higher than those of mature cows.

Weight also makes a difference, as shown in Table 19–17, which gives the daily nutrient requirements at various weights of (1) dry pregnant cows, and (2) cows nursing calves.

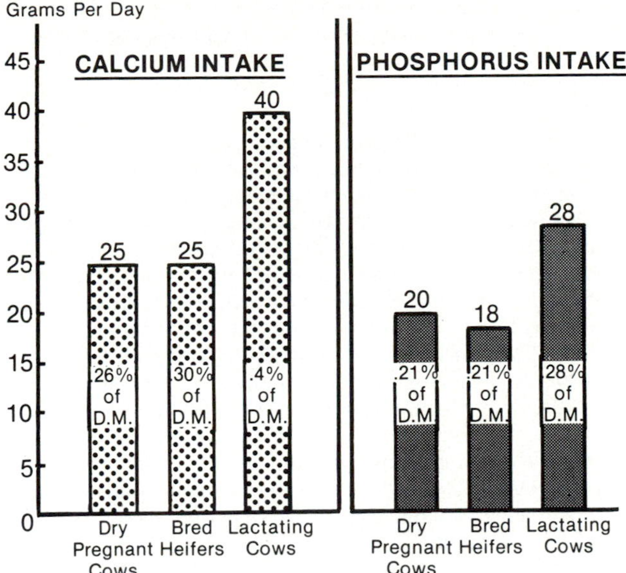

Dry Pregnant Cows—Dry pregnant cows, last third pregnancy, 1,100 lb (*500 kg*) wt. No gain.

Bred Heifers —Pregnant yearling heifers, last third pregnancy, 900 lb (*409 kg*) wt. Daily gain 1.3 lb (*0.6 kg*).

Lactating Cows —Cows nursing calves, superior milking, first 3-4 months after calving, 1,100 lb (*500 kg*) wt. No gain.

Fig. 19-18. Calcium and phosphorus requirements of dry pregnant cows, bred heifers, and lactating cows. Milk is a rich source of calcium and phosphorus; hence, it is no surprise to find that the requirements for each of these minerals are markedly higher for lactating cows than for either of the other groups. (Based on data from *Nutrient Requirements of Beef Cattle,* sixth revised edition, National Research Council-National Academy of Sciences, Washington, D.C., 1984)

TABLE 19-17
DAILY NUTRIENT REQUIREMENTS OF BEEF COWS[1]

Body Weight		TDN		Total Protein		Cal-cium	Phos-phorus
(lb)	(*kg*)	(lb)	(*kg*)	(lb)	(*kg*)	(g)	(g)
Dry pregnant mature cows (middle third of pregnancy)							
770	*350*	9.02	*4.1*	1.34	*.61*	20	15
880	*400*	9.68	*4.4*	1.45	*.66*	22	16
990	*450*	10.56	*4.8*	1.55	*.70*	23	18
1,100	*500*	11.22	*5.1*	1.64	*.74*	25	20
1,210	*550*	11.88	*5.4*	1.74	*.79*	26	21
1,320	*600*	12.54	*5.7*	1.83	*.83*	28	23
1,430	*650*	13.20	*6.0*	1.92	*.87*	30	25
Cows nursing calves, first 3 to 4 months after calving (superior milking ability)							
770	*350*	11.2	*5.1*	2.22	*1.01*	36	24
880	*400*	13.0	*5.9*	2.42	*1.10*	37	25
990	*450*	14.1	*6.4*	2.61	*1.19*	39	26
1,100	*500*	15.0	*6.8*	2.74	*1.24*	40	28
1,210	*550*	15.6	*7.1*	2.86	*1.30*	42	30
1,320	*600*	16.5	*7.5*	2.97	*1.35*	43	31
1,430	*650*	17.2	*7.8*	3.07	*1.39*	45	33

[1]Adapted by the authors from *Nutrient Requirements of Beef Cattle,* sixth revised edition, National Research Council-National Academy of Sciences, Washington, D.C., 1984, pp. 45 and 46; with U.S. Customary added by the authors.

Of course, these are minimum requirements. Hence, it would be well to add 1 percentage unit to the crude protein requirement and 3.0 percentage units to the TDN requirement, and to self-feed the minerals. This would take care of variations in feedstuffs and differences in requirements among individual cows within a herd.

Nutritional Reproductive Failure in Cows

Since cattle producers largely determine their own destiny when it comes to feeding, it is important that they know the causes of nutritional reproductive failures and how to rectify them.

A review of the literature clearly points to 3 important reproductive difficulties: (1) the small number of cows in heat and bred the first 21 days of the breeding season, (2) the low conception rate at first service, and (3) the excessive calf losses at birth or within the first 2 weeks of age. Also, it is noteworthy that each of the causes is more marked in young cows (first-calf heifers) than in mature cows.

Research throughout the country gives ample evidence that the real cause of most beef cow reproductive failure is a deficiency of one or more essential nutrients just before and immediately following calving—nutritive deficiencies during that critical 100-day period when life begins—a deficiency of energy, protein, minerals, and/or vitamins.

Based on an exhaustive review of experiments on nutritional reproductive failure in cows, the authors concluded as follows:

1. Energy is more likely to be limiting than protein in reproduction.
2. Beef cows receiving inadequate energy reproduce at a low level.
3. Phosphorus supplementation of cows on range areas deficient in phosphorus increases the calf crop.
4. Administering additional vitamin A to heifers grazing dry forage increases the calf crop.
5. The level and kind of feed before and after calving will determine how many cows will show heat—and conceive. After calving, feed requirements increase tremendously because of milk production; hence, when a cow is suckling a calf, she needs approximately 50% greater feed allowance than during the pregnancy period (see Fig. 19-15). Otherwise, she will suffer a serious loss in weight and fail to come in heat and conceive.
6. Cows in average condition should gain a minimum of 100 lb during the pregnancy period, followed by a gain of ½ to ¾ lb daily after calving and extending through the breeding season. If they are on the thin side at calving time, they should gain 1½ to 2 lb daily after they drop calves. This calls for 7 to 12 lb of TDN daily before calving (which can be provided by feeding 14 to 22 lb of average-quality hay), and 10 to 17 lb of TDN after calving (which can be provided by feeding 14 to 28 lb of hay plus 4 lb grain), with the lactating requirement dependent on both cow weight and milking ability. Additionally, there must be adequate protein, minerals, and vitamins.

In the northwestern United States, a disorder known as *weak calf syndrome* has resulted in reproductive failure of cows and in calf mortality running as high as 40% in some herds. Weak calf syndrome is caused by nutritional deficiencies in the pregnant cow, along with weather stress. (See Chapter 5, Table 5-1, Nutritional Diseases and Ailments—Weak Calf Syndrome.)

Winter Feeding

In a country as large and diverse as the United States, wide variations exist in both the length of the winter season and the available feeds. But the same principles are applicable to all areas and enterprises, and the chief objective remains the same—economically to produce high percentage calf crops with heavy birth and weaning weights.

Winter feeding is the most expensive time in cow-calf operations because the feed must be processed and brought to the animals. From an economic standpoint, therefore, it is important that wintering practices be both knowledgeable and wise. Cheaper home-grown roughages should constitute the bulk of the winter ration for dry pregnant cows. Most of the grain and the higher class roughages may be used for other classes of livestock. A practical ration may consist of silage and/or dry roughages (legume or grass hays) combined with a small quantity of protein-rich concentrates (such as soybean meal or cottonseed meal). With the use of a legume roughage, the protein-rich concentrate may be omitted. Dusty or moldy feed and frozen silage should be avoided in feeding all cattle—especially in the case of the pregnant cow, for such feed may produce complications and possible abortion.

Except during the winter months, pastures constitute most of the feed of beef cattle. By fall, however, grass is usually in short supply and relatively poor as a source of protein, certain minerals (especially phosphorus), and carotene (provitamin A). To overcome these deficiencies, the cattle raiser must resort either to (1) supplemental feeding on pasture, or (2) drylot feeding. At no other time in the operations is a possible profit so likely to be dissipated and replaced by a loss.

Fall feeding should not be delayed so long that animals begin to lose weight. The reason cattle often eat and get poor on dry, weathered grass is that it is low in energy, protein, carotene, and phosphorus and perhaps certain other minerals. These deficiencies become more acute and increase in severity as winter advances. Cattle simply cannot consume sufficient quantities of such bulky, low-quality roughage to meet their needs; and the younger the animal, the more acute the problem. Under such circumstances, the maintenance needs are met by the breakdown of body tissues, accompanied by the observed loss in weight and condition. Young animals fail to grow; it makes for lightweight calves. Also, reproduction is affected adversely; serious underfeeding results in lowered calf crops. Supplementing fall grass with a concentrated type of supplement is the practical and ideal way in which to alleviate such nutritive deficiencies.

Likewise, spring feeding should be continued until grass has attained sufficient growth and nutritive value. The calf crop will be affected adversely in pregnant cows, and milk production will be lowered in lactating cows if they subsist for prolonged periods on insufficient amounts of forage or on frosted pastures that are low in protein, vitamin A, and minerals.

When roughage is scarce and high in price, feed less of it and more concentrate; conversely, feed more concentrate and less hay when grain is plentiful. Generally speaking, 1 lb of grain can replace 2 lb of dry roughage, if they are of comparable quality.

The best calf crop is produced by cows that are kept in vigorous breeding condition—that are neither overfat nor thin and run-down. Generally speaking, this calls for winter feeding, with the maximum use of roughage. The kind and amount of concentrate needed will depend upon (1) the amount and kind of roughage given, (2) the age and condition of the cattle, and (3) whether the cows are dry or suckling calves. In total, it is important that the ration provides the kinds and amounts of nutrients needed, along with sufficient bulk to satisfy the appetite reasonably well. On a dry-feed basis, the daily requirement of dry pregnant cows is about as follows: thin cows, 2¼% of their liveweight; cows in average flesh, 2% of liveweight, and cows in good condition, 1¾% of liveweight. Cows suckling calves should receive approximately 50% more feed than dry cows of comparable weight and condition. From this it should be concluded that, unless the herd is so small as to make it impractical, dry cows should be wintered separately from those that are suckling calves. This makes it possible to limit the feed of dry cows and to effect certain other economies in their handling.

Noteworthy, too, is the composition of a calf at birth. An average 70-lb calf at birth is about 75% water, 20% protein, and 5% ash. The calf's 70-lb weight represents about 17.5 lb dry matter. From this, it is apparent that the dry gestation period does not create a heavy nutritional drain. Thus, this is a period when a cattle raiser may economize by utilizing crop residues and winter pasture.

Purebred breeders recognize that good condition and attractive appearance of breeding animals are important assets from the standpoint of favorably impressing prospective purchasers. Thus, a certain amount of the feed consumed by purebred cows—that which is above the amount needed for good health and vigor—can be rightfully charged to advertising costs. This is so because no more effective means of advertising exists than well-conditioned animals, regardless of the season of the year. Also, supplemental feeding of purebred cows that are suckling is favorably reflected in their calves—in greater growth, development, and bloom. In short, purebred breeders usually find it advantageous to feed a concentrate to cows during the winter months. Any purebred breeder who is content with less than the maximum potential should not be in the purebred business in the first place.

RATIONS FOR DRY PREGNANT COWS

Dry pregnant cows in average condition should gain in weight sufficient to account for the growth of the fetus (60 to 90 lb) plus sufficient increase in weight and condition to carry them through the suckling period. In total, they should gain 100 to 150 lb during the pregnancy period, or at the rate of approximately ½ lb daily. Of course, the size and condition of the cow is the best gauge as to the feed allowance and desired gain. As previously noted, dry cows require less supplementation than cows suckling calves. Nevertheless, they should not be permitted to lose too much flesh, unless, of course they are overfat. Also, it is recognized that it requires less feed to keep cattle from losing flesh than it does to restore them to proper condition after they have become thin. Thus, it is good economy to start feeding before cows show any signs of malnutrition. Unless a good-quality legume roughage is fed, the concentrate should provide protein, energy, and needed minerals and vitamins. Supplements for dry pregnant cows fed wheat straw should contain approximately 32% protein, 2.5% calcium, 2.25% phosphorus, 0.7% magnesium, and 0.34% sulfur.

When winter grazing is not possible, the rations in Table 19–18 may be used to meet the daily needs for energy and protein of a 1,100-lb dry pregnant cow. A combination of legume roughage with lower quality roughage (such as stalklage, straw, corncobs, or cottonseed hulls) will meet both the energy and protein requirements without the use of a supplement.

TABLE 19–18
WINTERING RATIONS FOR A 1,100-LB *(500-KG)* DRY PREGNANT COW

	Rations				
	1	2	3	4	5
	(lb */kg/*/day)				
Legume-grass hay	18 *(8)*				10 *(4.5)*
Legume-grass haylage[1]		30 *(12)*			
Corn or grain sorghum silage			35 *(15)*		
Stalklage or husklage				45 *(20)*	
Straw, cobs, or cottonseed hulls					10 *(4.5)*
Supplement[2]			.5 *(0.2)*	1 *(0.45)*	

[1]Haylage figured at 55% dry matter, corn or grain sorghum silage at 35% dry matter, stalklage or husklage at 45% dry matter.

[2]Supplement figured at 48% crude protein. Quantity to be adjusted in keeping with the protein content of the supplement. For example, if a 24% crude protein supplement is fed, the quantity of supplement should be doubled.

CALVING CONTROL (DAYTIME CALVING)

The following benefits would accrue if the majority of cows would calve during the daylight hours:

1. Improved calf survival due to birth during warmer daylight hours and readily available assistance.
2. Reduced nighttime labor from fewer cows calving.

Limited research supports the theory of altering calving time by late feeding (feeding 5 p.m. to 9 p.m., starting about 2 weeks before calving), with the bulk of calves born during daylight. Scientists at the Fort Keogh Livestock and Range Research Laboratory, Miles City, Montana, report that heifers fed between 8 and 9 p.m. had 17% more daytime births than heifers fed between 8 and 9 a.m. Other researchers report a 10 to 15% increase in daytime calving from nighttime feeding. The reason late feeding causes daytime calving has not been established.

Daytime calving has many advocates in the cow-calf industry. As a management practice, it does not require capital outlay, and, most important, it can make for increased profits from more live calves. But further experiments and experiences on the effect of feeding time on calving time are needed. In the meantime, producers may try nighttime feeding, but they shouldn't eliminate their night calving crews, because calving at night has not been totally eliminated.

RATIONS FOR COWS NURSING CALVES

Cows with calves at side should be fed for the production of milk, which requirements are more rigorous than those during pregnancy. This is important because, until weaning time, the growth of the calf is detemined chiefly by the nourishment available through the milk of its dam. The principal part of the calf's ration, therefore, may be cheaply and safely provided by giving its mother the proper feed for the production of milk. To stimulate milk flow, most beef cows need a concentrate during the winter months, and the poorer or the more limited the roughage, the greater the amount of supplement required. On the average, cows that calve in the fall should be fed a minimum of 4 to 6 lb concentrate daily throughout the winter.

The energy requirement of a cow nursing a calf is about 50% higher than that of a dry pregnant cow; and the protein, calcium, and phosphorus requirements are nearly double.

The vast majority of the nation's cows with calves at side are on pasture most, if not all, of the lactation period. So, it is important that producers recognize that the quantity and quality of pastures used for lactating cows and their calves affect milk production, daily gains and weaning weights of calves, and rebreeding of cows. Fig. 19–19 shows the interrelationship between pasture and lactating cow and

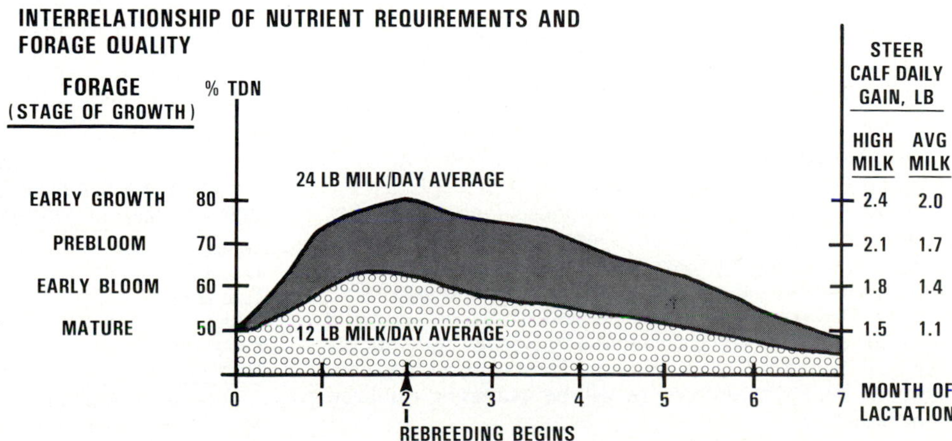

Fig. 19–19. Relationship of pasture quality to lactating beef cow requirements and nursing calf performance, assuming a 1,200-lb cow. Note that calf gains on high-producing cows will continue at 2.4 lb/day at the seventh month of lactation if pasture stays at 80% TDN, but that it will drop to 1.5 lb/day if pasture is only 50% TDN at that time.

nursing calf requirements and performance. Beef cows with a high potential for milk production that are nursing calves with a high growth potential can take advantage of a pasture management program that results in a continuous supply of nutritious forage in the early growth stage. Note that the curves shown in Fig. 19-19 include requirements for rebreeding at 60 days after calving; thus, if the pasture quality is not adequate, cows will not cycle in time to be rebred and stay on a 12-month calving interval. Of course, lower producing cows can meet their requirements on lower quality pasture.

The rations in Table 19-19 may be used for drylot feeding of beef cows nursing calves. Of course, the daily levels shown in Table 19-19 should be approached gradually so that nutritional scours will not develop in baby calves.

TABLE 19-19
WINTERING RATIONS FOR A 1,100-LB *(500-KG)* COW NURSING A CALF

	Rations				
	1	2	3	4	5
			(lb/*kg*/day)		
Legume-grass hay	30 *(13)*			20 *(9)*	10 *(4.5)*
Legume-grass haylage[1]		50 *(22)*			
Corn or grain sorghum silage			60 *(27)*		40 *(18)*
Grain				5 *(2)*	
Supplement[2]			1.5 *(0.6)*		

[1]Haylage figured at 55% dry matter; corn or grain sorghum silage figured at 35% dry matter.

[2]Supplement figured at 48% crude protein. Quantity to be adjusted in keeping with the content of the supplement. For example, if a 24% crude protein supplement is fed, the quantity of the supplement should be doubled.

CROP RESIDUES AND WINTER PASTURES

Two requisites are important in wintering the cow herd: (1) bringing them through the winter in proper condition for calving, and (2) keeping feed costs to the minimum consistent with nutritional demands. Meeting these requirements has prompted increased use of crop residues and winter pastures for brood cows. As the ever-increasing human population of the world consumes a higher proportion of grains and seeds, and their by-products, directly, cattle will utilize increasing amounts of crop residues and pastures and a minimum of products suitable for human consumption. Thus, more and more farmers with crops will include a beef herd in their operations and realize a fair return from feeds which would otherwise be wasted.

Crop Residues

Crop residues are the parts of forages that remain after harvesting a grain or seed crop. Among such crop residues are cornstalks and husklage, sorghum stalks, soybean refuse, small grain straws and chaff, and legume and grass seed straws. Crop residues left in the field, above or below the soil surface, may well constitute 4 to 5 times more energy than is harvested. This potential source of added feed, organic fertilizer, and energy will be increasingly utilized in the future.

CORN RESIDUES

Of all crop residues, that from corn is produced in greatest abundance and offers the greatest potential for expansion in cow numbers. Of course, knowing the feeding value and proper supplementation of corn residues are pertinent to their profitable use.

Fig. 19-20. Fox equipment with flail attachment harvesting corn stalklage. (Courtesy, Koehring Farm Division, Appleton, Wisc.)

Fig. 19-21. Cows grazing cornstalks. (Courtesy, Iowa State University, Ames)

Table 19-20 lists the daily nutritive requirements of a dry pregnant cow, middle third of pregnancy, weighing 1,100 lb. Table 19-21 gives the nutritive composition of air-dry corn stover and husklage.

TABLE 19-20
NUTRITIVE REQUIREMENTS OF A DRY PREGNANT COW (MIDDLE THIRD OF PREGNANCY) WEIGHING 1,100 LB *(500 KG)*[1]

Dry matter, daily	20.9 lb *(9.5 kg)*
TDN, daily	11.2 lb *(5.1 kg)*
Total protein	7.8%
Calcium	0.26%
Phosphorus	0.21%
Vitamin A	27,000 IU

[1]*Nutrient Requirements of Beef Cattle,* sixth revised edition, National Research Council-National Academy of Sciences, Washington, D.C., 1984.

TABLE 19-21
ANALYSIS OF AIR-DRY CORN STOVER AND HUSKLAGE[1]

	Corn Stover	Husklage
	(%)	(%)
TDN	48	57
Crude Protein	4.5	3.4
Calcium	0.4	0.02
Phosphorus	0.07	0.05
Vitamin A	—	—

[1]*Cow-Calf Information Roundup,* University of Illinois, 1971, p. 10, Table 2.

Not all corn refuse will be of the same composition as Table 19-21; some will be better, some poorer. The quality declines with the passing of time following grain harvest; and the more severe the weather, the greater the decline. Also, cultural practices during the growing season may alter corn residue quality.

Studies show that a 1,100-lb cow will eat approximately 22-24 lb per day of palatable, air-dry stover, or about 2 lb or more of air-dry stover per hundred pounds body weight per day. She will eat slightly larger amounts of husklage. This consumption, along with the information presented in Tables 19-20 and 19-21, suggests that stover and/or husklage rations will meet the daily energy (TDN) needs of dry pregnant cows, but such rations will be slightly deficient in protein, and low in phosphorus and vitamin A. Nevertheless, the highest and best use for corn residue is for dry pregnant cows for the period following conception to about 30 days before calving.

For corn refuse feeding, mature cows should be in medium to good condition at the start of the winter feeding period; and they should not be permitted to lose over 10 to 15% of their weight from fall through calving. Heifer weight losses should be under 5%. When weight loss approaches this limit, it's time to feed some grain or silage.

Corn residues provide adequate energy to maintain dry pregnant cows, but they must be supplemented with additional energy when fed to cows nursing calves or to young, growing animals.

The crude protein content of corn stover is on the low side for dry pregnant cows. It averages about 4.5% (Table 19-21), whereas a dry pregnant cow requires 7.8% crude protein (Table 19-20). Thus, it is recommended that ½ lb per head per day of a 30 to 40% crude protein equivalent (CPE) supplement be provided.

It follows that the protein content of corn refuse is much too low to support either productivity or growth. For example, a 1,100-lb lactating cow requires a daily allowance of 2.7 lb of crude protein. However, a daily consumption of 29 lb of stover will provide only about 1.1 lb—less than half that needed.

For nursing cows, the protein deficiency of stover and/or husklage may be corrected by supplementation, on a per head per day basis, with 2 lb of a 40% protein supplement, or 6 lb of a good legume hay. If desired, the protein supplement may be provided in the form of protein blocks, with one block provided for each 15 cows. Where hay is fed, it should be taken to the field, rather than fed in a feedlot, as this will encourage the cows to stay in the field and graze the cornstalks.

Phosphorus should be provided to all cattle fed corn residue. Calcium may be deficient, especially for lactating cows. Also, some of the trace elements may be deficient. Hence, it is recommended that all cattle on high corn refuse have free access to a complete mineral supplement. A mineral mixture with a Ca:P ratio of 1:2 is recommended for gestating cows, and a 1:1 ratio for lactating cows.

Corn residue is deficient in vitamin A, which should be supplemented. The precalving and postcalving (heavy milking) needs of approximately 27,000 and 39,000 IU per head per day, respectively (NRC-1984), may be met by feeding vitamin A supplement, by intramuscular injection of vitamin A solution, or by feeding adequate levels of green, leafy hay.

It is important that corn residue be tailored to match the cow's nutritional needs. This is relatively simple with dry pregnant cows, where supplementation with a high-phosphorus mineral and vitamin A will usually suffice. Beginning 4 to 6 weeks before calving and continuing through the lactation period, much heavier supplementation is necessary; in addition to phosphorus and vitamin A, protein must be added, and preferably some energy and calcium for nursing cows.

(Also see Chapter 12, By-product Feeds/Crop Residues.)

Winter Pasture

Where feasible, winter pasture offers cattle producers a means of reducing costs. By accumulating the feed in the field, rather than harvesting, storing, and handling the forage, the cost and labor of winter feeding can be substantially reduced. Also, costs of bedding and manure hauling can be eliminated.

Tall fescue is used as a winter pasture in the area to which it is adapted—Missouri, Illinois, Indiana, and Ohio. Usually, the new regrowth is baled in late June into round bales and left in the field. The round bales shed rain and snow and, together with the regrowth, make excellent late fall and winter grazing. Experience shows that field-stored forage has adequate quality to maintain beef cows in good condition.

Fig. 19-22. Cows on winter fescue pasture, supplemented with round bales of fescue harvested the previous June and left in the field. (Courtesy, University of Illinois, Urbana)

Fescue is a cool season grass which keeps growing during the winter in the areas to which it is adapted; and it is more palatable during the fall and winter than any other season because of the high concentration of soluble sugars. Trampling during the fall, winter, and spring months does not injure the turf.

The Ohio Station researchers reported that tall fescue winter pasture—including both standing growth and baled hay—carried 2 cows per acre for a 4-month period. The use of electric fence to strip graze the bales and regrowth increases carrying capacity by 50 to 60% over permitting the herd access to the entire field.

Range and Pasture Feeding

Good pasture or range is the cornerstone of successful beef cattle production as is attested by the following facts:

Fig. 19-23. Good pastures and good cattle go together. (Courtesy, American Red Brangus Assn., Austin, Tex.)

1. A total of 26% of the total land area of the United States (50 states) is used solely for grassland.
2. A total of 85.7% of the total feed supply of all U.S. beef cattle is derived from forage; in season this means pasture.
3. Good pasture alone will produce 200 to 400 lb of beef per acre annually (in weight of calves weaned, or in added weight of older cattle); superior pastures will do much better.
4. No method of harvesting has yet been devised which is as economical as grazing by animals.
5. Pasture gains are generally cheaper than drylot gains because (a) less labor is required, (b) grass is the cheapest of all roughages, (c) less protein supplement is required, (d) the animals scatter their own droppings, thus alleviating hauling manure and maintaining fertility, and (e) fewer buildings and less equipment are necessary.

Western pastures that receive less than 20 in. of rainfall annually are classified as the western range. The carrying capacity of much of the range is low, and little of it provides yearlong grazing. Moreover, variation in vegetative types, climate, and topography in the range country is accompanied by great diversity in the seasonal use made of it. As a result, rangelands are usually grazed during different times of the year, and the herds migrate with the season, moving to the mountains and higher elevations in summer and returning to the lower ranges in winter.

From the standpoint of vegetation and utilization by livestock, ranges differ from cultivated pastures as follows:

1. They are less productive.
2. They are more likely to progress to less palatable plants.
3. They are more difficult to restore when depleted.
4. They often serve multiple use—for wildlife production, recreation, timber production, and mineral production, as well as use by cattle and other domestic animals.

Improved ranges and pastures should be the first goal of cattle raisers, without using supplemental feeding as a substitute for good grass or as a crutch for poor range. Instead, the two—good range and proper supplemental feeding—go hand in hand.

RANGE NUTRIENT DEFICIENCIES

Growing grasses provide adequate nutrients for beef cattle in unforced production when (1) produced on fertile soils, (2) available in sufficient quantities, (3) not washy, and (4) not weathered, leached, or bleached. However, the simultaneous fulfillment of all these conditions is the exception, rather than the rule. Every cattle producer worthy of the name, forces young stock for an early market; most soils

Fig. 19-24. Drought on the range, causing a short forage supply along with limited energy and other nutrients. (Courtesy, Santa Gertrudis Breeders International, Kingsville, Tex.)

are deficient in certain nutrients, which, in turn, affect the plants and the animals feeding thereon; during droughts and early and late in the season, feed may be in short supply (thereby limiting energy and other nutrients); early spring pastures are washy and lacking in energy; and during droughts and late in the season, grasses become mature, leached, and bleached—they increase in fiber and decrease in protein, phosphorus, and carotene. To meet these conditions, a supplemental source of energy, protein, phosphorus, and vitamin A is necessary.

Energy

Hunger, due to lack of feed, is the most common deficiency on the western range. In particular, there may be a shortage of energy during droughts, late in the season, or early in the spring when grass is washy. Thus, the first and most important range pasture need is that there be sufficient feed for the animal—to provide the necessary energy required to maintain the body. Over and above these needs, any surplus energy may be used for growth, reproduction, and conditioning.

With bulky, low-quality roughages—such as dry grass cured on the stalk, common to drought periods and late in the season—animals may not be able to consume sufficient quantities to meet their energy needs; and the younger the animal, the more acute this problem. Also, very early spring grass is washy and lacking in energy. Under such energy-deficient circumstances, the maintenance needs of animals are met by the breakdown of body tissues, accompanied by the observed loss in weight and condition and the failure of young animals to grow. Also, reproduction is adversely affected; serious underfeeding results in the failure of some cows to show heat, more services per conception, lowered calf crops, and lightweight calves. Supplemental feeding is the practical way in which to alleviate such energy deficiencies.

Protein

Research indicates that a deficiency of protein results in depressed appetite, poor growth, loss of weight, reduced milk production, irregular estrus, and lowered calf crops.

The protein allowance for beef cattle, regardless of age, should be ample to replace the daily breakdown of tissues of the body, including the growth of hair, horns, and hoofs. This need is most critical for the growth of the young calf and for the gestating-lactating cow.

Mature, weathered native range grass is almost always deficient in protein—being as low as 3%, or less. Protein leaching losses due to fall and winter rains may range from 37 to 73%.

Because protein supplements ordinarily cost more per ton than grains, beef cattle should not be fed larger quantities of them than are actually needed to balance the ration. But the temptation is to feed too little of them. When on mature, weathered grass, cows should recieve about 2 lb of concentrate supplement daily—the exact amount depending upon the nutrient content of the supplement and other factors.

The protein and energy requirements are closely interdependent; hence, it follows that energy rather than feed intake should be the dietary component relative to which the nutrient needs are adjusted.

Minerals

Growth, reproduction, and lactation require adequate minerals. Although the mineral requirements of dry pregnant cows and of lactating cows are much the same everywhere, it is recognized that age and individuality make a difference. Additionally, there are area and feed differences. Thus, the informed cattle raiser will supply the specific minerals that are deficient in the ration, and in the quantities necessary, and avoid excesses and mineral imbalances.

Salt should be available at all times, on the basis of about 25 lb per range cow annually.

Fig. 19-25. Self-feeding minerals on the range. (Courtesy, Martin Jorgensen, Jr., Ideal, S.D.)

Phosphorus deficiencies are common among range beef cattle. A severe phosphorus deficiency will result in depraved appetite, emaciation, retarded growth and development, failure to reach normal adult size, failure to breed regularly, lowered calf crop, lowered milk production, and high death losses.

The New Mexico station reported (1) phosphorus losses in grasses of 49 to 83% during the winter period; (2) increased average annual calf production per range cow of 53 lb through proper mineral supplementation; and (3) that the phosphorus supply should be continuous throughout the year, and not limited to the winter months.

Iodine, copper, cobalt, and selenium supplements should be added in those areas where deficiencies of these minerals are known to exist.

The supplementation of beef cows and calves on selenium-deficient ranges can produce dramatic results. In a Central California herd in which the blood selenium concentration was less than 0.04 ppm, Farm Advisors A. O. Nelson and R. F. Miller reported (*California Agriculture*, March-April, 1987) that cows that received 2 selenium boluses in the reticulum weaned off calves weighing 37 and 44 lb more in 1985 and 1986, respectively, than the untreated controls. Injecting the calves with selenium-vitamin E at 2 to 3 months of age produced even more marked results; the selenium-treated calves were 74 lb heavier at weaning than the untreated calves.

Other mineral elements are thought to be essential, but the picture is somewhat confused; and new findings clearly indicate that we have reason to be less certain of our mineral recommendations than heretofore. For the latter reason, the judicious use of certain trace minerals may be good insurance, even though not all the requirements are known.

Vitamins

Under normal conditions, vitamin A is the vitamin most likely to be deficient in cattle rations, because dry, bleached range grass is very low in carotene (the precursor of vitamin A).

Inadequate amounts of vitamin A (carotene) during pregnancy may cause cows to abort or give birth to dead or weak calves. Extreme deficiencies may also impair the ability of cows to conceive. Bulls receiving insufficient vitamin A show a decline in sexual activity and semen quality.

In low sunshine areas, especially during the winter months, it is recommended that vitamin D be added to the ration, also.

Indirect Deficiency Losses

In addition to the usual deficiency symptoms, nutritional deficiencies on the range are usually accompanied by lowered resistance to parasites and diseases, and increased mortality of both breeding stock and calves. Also, where feed is scarce there are usually increased deaths from toxic plants.

SUPPLEMENTING EARLY SPRING RANGE

Turning to the range when the first blades of green grass appear will usually result in a temporary deficiency of energy, due to (1) washy (high water content) grasses, and (2) inadequate forage for the animal to consume. As a result, stock growers are often disappointed by the poor gains made during this period.

If there is good reason why grazing cannot be delayed until there is adequate spring growth, it is recommended that early pastures be supplemented with grass hay or straw (a legume hay will accentuate looseness, which usually exists under such circumstances), preferably placed in a rack; perhaps with an added high-energy concentrate.

SUPPLEMENTING DRY RANGE

Dry, mature, weathered, bleached grass characterizes (1) drought periods, and (2) fall-winter range. Such cured-on-the-stalk grasses are low in energy, protein, carotene, and phosphorus, and perhaps certain other minerals. These deficiencies in range plants become more acute following frost and increase in severity as winter advances. This explains the often severe shrinkage encountered on the range following the first fall freeze.

In addition to the deficiencies which normally characterize whatever plants are available, dry ranges may be plagued by a short supply of feed.

Generally speaking, a concentrated type of supplement is best used during droughts or on fall-winter ranges. However, when there is an acute shortage of forage, hay or other roughage may be used with or without a concentrate.

SUPPLEMENTING THE PUREBRED HERD ON PASTURE

Fig. 19-26. Eye-appealing purebred Hereford herd on pasture. (Courtesy, American Hereford Assn., Kansas City, Mo.)

Purebred cattle breeders and commercial cattle producers, alike, recognize the following as important profit factors: (1) percent calf crop, (2) weaning weight, and (3) number of calves a cow produces in her lifetime. But, to be successful, purebred breeders must go further. They must also give attention to (1) salability—sleek, bloomy, well-conditioned animals attract buyers and sell better; and (2) maximum development of genetic potential in characteristics of economic importance (rate of gain, feed efficiency, etc.); otherwise, they cannot make intelligent selections and breed progress. Further, the acid test of the competence of the purebred breeder is that these objectives shall be achieved without jeopardizing the breeding performance of the animal—and that's not easy.

The nutritive requirements of cattle—especially purebreds—have become more critical in recent years. Today, more and more purebreds are being production tested. Among other traits, they are being selected for more rapid and efficient gains—they are in forced production. Consequently, their nutritive requirements are more critical—especially from the standpoints of energy, protein, minerals and vitamins.

First and foremost, it must be recognized that, no matter how excellent pastures may be, they are roughages and not concentrates. Therefore, for the purebred herd, judicious supplemental feeding on grass may be warranted and profitable. In purebreds, the end result is all-important, even at somewhat added expense. In addition to supplying proper balance of nutrients, the supplement should provide variety and palatability. Also, it should not cause bloat, scours, or other digestive disturbances.

PASTURE AND RANGE SUPPLEMENTATION

Where dried grass cured on the stalk is grazed, or where insufficient pasture is available—perhaps due to drought or overstocking—supplemental feeding is necessary. Also, supplemental feeding is a way in which to extend the grazing season, both early and late.

Sorting Pasture and Range Cattle

When supplemental feeding is planned, it is strongly recommended that cattle first be sorted by age and condition groups.

Heifers should not be supplemented at the same levels as older cattle. Because they are growing, they must be fed more liberally (see Fig. 19–16). Also, heifers have need for more protein (see Fig. 19–17), and they must be fed for a longer period. But heifers should not be overfed to the point that reproduction is adversely affected.

Thin cows should be placed where they can be given extra feed and special care. Most of the ration of pregnant cows should consist of pasture plus such supplements as required —with emphasis on proteins, minerals, and vitamin A; and the kind and level of supplementation should be varied according to the quality and quantity of grass available.

Immediately before and after calving, cows that are not on pasture or range should be fed lightly and with laxative feeds. At this time, the amount of supplementation should be governed by the milk flow, the condition of the udder, the demands of the calf, and the appetite and condition of the cow. More energy, proteins, and minerals, are required for a cow suckling a calf than for a pregnant or dry cow. The nutritional requirement of cows nursing calves is approximately 50% higher than for pregnant cows.

Until weaning time, the growth of the calf is determined chiefly by the amount of milk available from its dam, plus whatever assist is given through grass or creep feeding.

Choosing a Pasture or Range Supplement

Every cattle producer faces the question of what supplement to use, when to feed, and how much to feed under existing conditions.

In supplying a supplement to range cattle, the following requisites should be observed:

1. It should balance the ration of the animal(s) to which it is fed, which means that it should supply all the nutrients needed by the animal(s) which are missing in the forage.
2. It should be fed in such a way that each animal gets its proper portion.
3. It should be fed in a form that is convenient and practical from the standpoint of the feeder, and that will least disturb the animal.

The formulations for two range cubes or pellets (one with urea, the other without) are given in Chapter 11, Protein Supplements, Tables 11–13 and 11–14.

Types and Systems of Pasture and Range Supplementation

Many different feeds may be, and are, used; among them, (1) ranch- or locally-produced hay, (2) alfalfa pellets or cubes, with or without fortification, and (3) supplements of various kinds. Likewise, many different systems of range supplementation are used, including (1) range cubes or pellets; (2) hand-feeding at intervals, rather than daily; and (3) protein blocks, liquid protein supplements, and self-feeding salt-feed mixtures. Where these feeding systems do not result in the neglect of the herd, there is no adverse effect upon the health and weight of the cows, percent calf crop, or weaning weight of calves.

Fig. 19–27. Range cubes fed on pasture or range. Many producers prefer this method of supplementation, primarily for reasons of convenience and reducing losses from wind blowing. (Courtesy, Ralston Purina Co., St. Louis, Mo.)

Fig. 19–28. Protein block in use on pasture—a means of lessening the labor attendant to the daily feeding of a protein supplement on pasture or range. (Courtesy, Moorman Mfg. Co. Inc., Quincy, Ill.)

(Also see Chapter 11, Protein Supplements, section headed, "Types and Methods of Feeding Protein Supplements.")

Confinement (Drylot) Beef Cows

Confining (drylotting) beef cows refers to the practice of confining beef cows to small quarters—to drylots, all or part of the year.

Fig. 19-29. Part of a herd of 80 brood cows in confinement in a deep, open shed on Circle S Ranch, Rockwood, Ontario, Canada. (Courtesy, *Country Guide*, Winnipeg, Manitoba, Canada)

From a feeding standpoint, the following points are pertinent in drylot beef cow operations:

1. All feed must be mechanically harvested and moved to the feedlot, rather than being harvested directly by the cows.
2. An assured, adequate, and economic feed supply must be available. The capital tied up in harvesting equipment and stored feeds may be quite large.
3. More knowledge of beef cow nutrition and ration formulation is needed.

Today, on an increasing commercial basis, beef cows are being confined to small quarters—to drylots, all or part of the year. This is a viable alternative management system for marketing forage, feeds, and crop residues through the cow herd. Low feed prices and by-product feeds, available breeding cattle, labor, facilities, and equipment may lead beef producers to consider a drylot or partial drylot beef cow/calf enterprise. Granted, the drylot herd will not replace the grazing cow, but, under some circumstances, such as high land values, it can complement the more traditional beef cow operation.

RATIONS FOR DRYLOT COWS

Rations for drylot cows generally consist of low-cost roughages—such as crop refuse, straw, cottonseed hulls, and gin trash—supplemented with protein, grain, vitamins, and minerals as required. Where available, higher quality roughages —such as silages, hays, and haylages—may be used, especially (1) during the critical 100 days, beginning 30 days before calving and extending 70 days after calving, and (2) for heifers calving as 2-year-olds. Also, during the summer and fall, green chop is frequently fed. Cows in partial confinement may, in season, graze such forages as cornstalks, grain stubble, or irrigated or native pastures.

Phase feeding according to stage of production and age of animals is recommended.

It is relatively easy to meet the nutritive requirements of a dry pregnant cow. Generallly speaking, low-quality roughages, properly supplemented, or a combination of low-quality roughages and high-quality roughages, will suffice. If only high-quality roughages are fed, they should be limited; otherwise, the cows will get too fat. However, such limited feeding does not meet the maximum fill requirements of the rumen, with the result that the cows nibble at the manure and pick up small feed particles scattered about the corral, which may make for disease and parasite problems. A simple solution to the weight and scavenger problems is to use a combination of low-quality and high-quality roughages.

The lactation requirements are much more rigorous than the dry pregnancy requirements; and the higher the milk production, the higher the nutritive requirements. Thus, requirements for two levels of milk production for lactating cows—(1) average milk production, and (2) superior milk production—are presented in Tables 19-1 and 19-2.

Under a drylot system, heifers are commonly calved as 2-year-olds, when they are still growing. Thus, during lactation they have a nutritional requirement for both growth and lactation. The nutritional requirements for these animals will be large, particularly if they are crossbreds and bred for rapid growth, considerable size, and high milk production.

The mineral needs of confinement cows may be met either by incorporating the needed minerals in the supplement which is fed, or by feeding the required minerals free-choice.

Vitamin A supplementation is extemely important for drylot cows. The carotene content of the dry forage should be disregarded and the total vitamin A requirement met by supplementation. This can be done by feeding a supplement, such as 2 lb of mill waste, containing 1 million IU of vitamin A per animal—feeding this vitamin A supplement once a month to heifers, and every other month to older cows. With older cows receiving high levels of dry forages containing normal amounts of carotene, it is probable that the vitamin A requirements are being met. However, it has been demonstrated under range conditions that (1) percent calf crop is markedly increased by supplementing with vitamin A during drought years, and (2) calves respond to vitamin A treatments given their dams 90 days prior to calving.

SEMICONFINEMENT (OR PARTIAL CONFINEMENT) COW HERDS

A semiconfinement (or partial confinement) operation takes advantage of grazing during part of the year, such as winter grazing of corn or sorghum stalks or seasonal grazing of pastures. In addition to providing low-cost feed and allowing the animals to do their own harvesting, breeding may be timed so that the calves will be dropped on clean pasture as a means of (1) preventing calf scours, and (2) stimulating milk flow.

Fattening Cull Cows for Market

There is a new profit potential in the cow business—in fattening cull cows for market. Where herd numbers are being held fairly constant, about 20% of the cows are culled each year, because of (1) poor calves, (2) being barren, (3) spoiled udders, (4) disease, (5) old age, and (6) miscellaneous reasons. Traditionally, these culls were sent to market—and the sooner the better. Today, good money can be made by holding and fattening these culls prior to slaughter.

Cutter and canner cows are in demand; and the price spread between them and Choice steers has narrowed, both on foot and on the rail—especially the latter. The reasons: (1) the increased demand for hamburger and other ground and prepared meats, for which cow beef is admirably suited; (2) fewer cows being slaughtered, because decreases in dairy cows have slowed; (3) imports of manufacturing-type beef having held at rather stable levels; and (4) better quality cow carcasses available than formerly.

All this suggests the following to cow-calf producers:

1. They cannot afford to keep marginal or barren cows. They should keep records, perform pregnancy tests, and remove loafers from the herd and fatten them for slaughter.
2. They should fatten cull cows on cheap, high-roughage rations, using a maximum of such feeds as silage, haylage, green chop, wheat pasture, and irrigated pasture. If they have good teeth and are healthy, they will make remarkable gains.
3. They should compare selling on foot vs selling on the rail. The latter may be more profitable.

Also, buying cull cows and finishing them commercially offers good opportunity. But buying such cows is a problem. Numbers are limited and scattered; and it is important that they have good teeth, and be healthy. Also, cows are usually culled because of feed shortages, breeding problems, disease, internal parasites, or age. In short, somebody had a problem with them.

FEEDING BULLS[5]

Frequently, little thought is given to the management and feeding of bulls except during the breeding season. Instead, the feeding program for herd bulls should be such as to keep them in a thrifty, vigorous condition at all times. They should neither be overfitted nor in thin, run-down condition. Also, exercise is necessary for the normal well-being of the bull.

The feeding and management of bulls differ according to age and condition. For this reason, sale bulls, young bulls, and mature bulls are treated separately in the sections that follow.

[5]The nutritive requirements of bulls for growth and maintenance at different weights, are given in Tables 19-1, 19-2, 19-3, and 19-4.

Feeding Sale Bulls

Most bull sales are held in late winter and early spring, at which time mostly yearling and 2-year-old bulls are sold.

Fig. 19-30. Brahman bull. (Courtesy, American Brahman Breeders Assn., Houston, Tex.)

In order to attract buyers, they have usually been grain fed since calfhood. Most bull buyers—especially commercial cattle producers in rougher range areas—would rather have their new bulls in less than fitted sale condition. They find that such bulls are more fertile and more apt to range with the cows when turned to pasture during the breeding season.

Sale and show bulls should be acquired 2 to 3 months ahead of the breeding season, so that they may be conditioned, or let down. Also, bear in mind that it takes about 40 days from the time a sperm cell is formed until it is ready to be ejaculated. Since the stress of handling and hauling a bull can reduce his fertility for about 40 days (it may be even longer where there is a change in elevation and climate), the rest period lets his body overcome these problems.

Handling highly conditioned sale bulls during the critical period—after the sale is over, and just ahead of the breeding season—is all-important. Experienced caretakers "let them down" and yet retain strong vigorous animals. They do this successfully by (1) providing plenty of exercise; (2) increasing the amount of bulky feeds, such as oats, in the ration; (3) cutting down gradually on the grain allowance; and (4) retaining the succulent feeds and increasing the pasture and hay.

Exercise is most important during this period, and the more excessive the finish, the more vital the exercise. Heavily fitted bulls can best be exercised by leading as much as 2 miles daily. Moderately fitted bulls can be force-exercised by placing the feed and water on opposite sides of the field.

In summary, therefore, conditioning highly fitted sale bulls for breeding consists of the gradual elimination of high-energy rations along with forced exercise.

Feeding Young Bulls

Lack of fertility in a bull may often be traced back to his early care and feeding. From weaning to three years of age, bulls should be kept separate by age groups. Young bulls should be fed more liberally than mature bulls because their growth requirements must be met before any improvement in condition can take place.

Following weaning, bulls should be fed and developed sufficiently to show their inherited characteristics, but without excessive finishing. Simultaneously, they should be given plenty of exercise. Overfeeding and lack of exercise are apt to result in infertility, low-quality sperm, and unsound feet and legs.

To achieve proper development, young bulls should gain at least 2½ lb daily from weaning to 12 to 15 months of age. This will necessitate a daily feed allowance equal to about 2½% of their body weight, with a ration comprised of 50% or more concentrate. From 15 months to 3 years old, they should make a daily gain of 2 to 2¼ lb and receive a daily feed allowance equal to 2 to 2¼% of their body weight, with the proportion of roughage increased after the first year.

If desired, the roughage may be chopped and mixed with the concentrate; and the ration may be hand-fed or self-fed. If self-fed, the feed consumption may be held at the desired level by using salt as a regulator. Other producers prefer to feed the roughage and concentrate separately; they usually free-choice the roughage and either self-feed a salt-concentrate mix or hand-feed the concentrate alone. When grain is fed separately from the roughage, about 1½ lb/100 lb of bodyweight can be fed at the beginning, gradually decreasing the grain and increasing the roughage as the animal grows older.

The least laborious and most convenient management arrangement in handling young bulls consists in allowing a group of uniform size and age the run of a pasture or enclosure of ample size, thereby providing (1) exercise, and (2) pasture in season. Of course, wherever possible, bulls should be performance tested while being developed. Ideally, this calls for individual feed and body weight records, although group feeding plus individual weight records will suffice.

Fig. 19-31. Yearling Hereford bulls on pasture. (Courtesy, American Hereford Assn., Kansas City, Mo.)

Bulls handled as recommended above will generally attain half their mature weight by the time they are 14 to 15 months of age and may be used in limited service.

During the breeding season, young bulls should be fed a grain ration consistent with pasture quality and number of cows to be bred in order to promote proper growth and development. Drought, overgrazing, and poor-quality pastures are situations in which grain supplementation is particularly needed. Heavy service and poor pasture with no supplemental feeding may shorten the breeding career of a young bull.

After the breeding season, yearling bulls generally need 5 to 6 lb of grain along with good roughage.

Feeding Mature Bulls

Winter is the proper time to condition bulls for the next breeding season. Bulls that have been running on pasture with the cows are likely to be thin; thus, they require sufficient concentrate to put them in proper flesh. Mature bulls will consume daily amounts of feeds equal to 1½ to 3% of their liveweight, depending upon condition and individuality.

The importance of having bulls in proper condition at the opening of the breeding season cannot be overemphasized. Nothing is quite so disheartening or costly as a small calf crop, with many of the calves coming late. Lack of fertility in the bull often may be traced back to his care and feeding.

Feed mature bulls all the legume hay they will eat plus 3 to 5 lb of ground or rolled grain and 1 lb of a 32% protein supplement (or equivalent) per head per day. Also, provide free access to a suitable mineral mixture. About 60 days before the bulls are turned with the cows, increase the concentrate allowance by 25 to 50%, with the amount of the increase determined by the condition of the bulls.

The mature herd bull needs no additional feed when running with the cow herd on good summer pasture.

FEEDING CALVES

Beef producers, as a whole, have lagged in applying much of what we know about feeding and managing calves. They're inclined to let mother cows and mother nature fend for the calves. More good proven practices, based on both successful experience and research, need to be put to use in feeding and handling calves.

Fig. 19-32. A good start in life! (Courtesy, American Angus Assn., St. Joseph, Mo.)

Feeding at Birth

Losing a calf means losing the profit on the cow for a whole year. Proper feeding and managing at birth can make the difference.

Recommended calving-time feeding practices follow:

1. See that the newborn calf nurses within 2 hours after birth. It is essential that it receive colostrum. The caretaker may have to assist a calf to nurse a dam that has very large teats or an udder that hangs very low. Also, weak calves should be helped to nurse.

2. Keep a close watch for signs of mastitis or injury to udders. It may be necessary to milk out a few cows for the first 2 or 3 days after calving.

3. Be sure cows have access to plenty of clean, fresh water.

Feeding Orphan and Multiple Birth Calves

Occasionally a cow dies during or immediately after parturition, leaving an orphan calf to be raised. Also, there are times when cows fail to give a sufficient quantity of milk for the newborn calf. Sometimes, there are multiple births.

If there are only a few orphans, usually they can be grafted onto other cows (or adopted)—either cows that have lost their calves or that give sufficient milk to raise two calves. When such calves cannot be grafted, they must be raised by artificial methods—without a cow.

Regardless of whether orphans are grafted or raised artificially, the problem will be simplified if the calf receives colostrum, the first milk produced by a cow after giving birth to a calf, during the first 24 hours, and preferably for the first 3 days, of its life—from its mother, from another fresh cow, or from frozen-stored colostrum. Colostrum is higher than normal milk in dry matter, protein, vitamins, and minerals. Also, it contains antibodies (a modified type of serum globulin) that give newborn calves a passive immunity against common calfhood diseases.

Because colostrum is so important for the newborn calf, producers should store a surplus of it from time to time. It can be frozen and stored for a period of 1 year or longer, then, as needed, thawed and warmed to 100° to 105°F, and fed. Also, colostrum may be fermented and stored.

Orphan calves can now be raised successfully on a milk replacer and calf starter ration, using them as directed. The milk replacer may be fed by using a bottle or pail equipped with a rubber nipple, or the calf may be taught to drink from a pail. It is important that all receptacles be kept absolutely clean and sanitary (clean and scald each time) and that feeding be at regular intervals. Dry feed should be started at the earliest possible time; not later than 1 week of age. With proper management, healthy calves may be switched entirely to a suitable dry feed at 4 to 5 weeks of age.

Basically, calves are fed according to one of 3 systems: (1) the whole milk system, (2) the combination whole milk-milk replacer system, or (3) the combination whole milk-calf starter system. Also, various combinations of these 3 systems are used, also. Further information on the subject of raising calves is presented in Chapter 20, Feeding Dairy Cattle.

Creep Feeding

Creep feeding is the supplementation of calves while they are nursing their dams. It increases weaning weight. The basis for this response is related to the lactation curve of beef cows, the increasing nutrient requirements of the calf during the nursing period, and the decline in feed quality and quantity typical of most pastures or ranges which support the cows and calves during lactation. Studies reveal that milk production of dairy cows increases up to the fourth or sixth month following freshening, then declines gradually. By contrast, maximum milk production of beef cows occurs during the first 2 months after calving, then declines.

Fig. 19-33 shows why creep feeding is important. From birth to weaning, the protein and energy requirements of a growing calf increase well beyond the ability of most beef cows to meet those needs. For example, to meet the protein and energy requirements for growth, a 100-lb calf needs 10 lb of milk, whereas a 500-lb calf needs 50 lb of milk. Since the average beef cow gives only 13 lb of milk per day throughout a 7-month suckling period, a 500-lb calf lacks 37 lb of getting enough milk from its dam at this stage of lactation to meet its needs—that's the *hungry calf gap.*

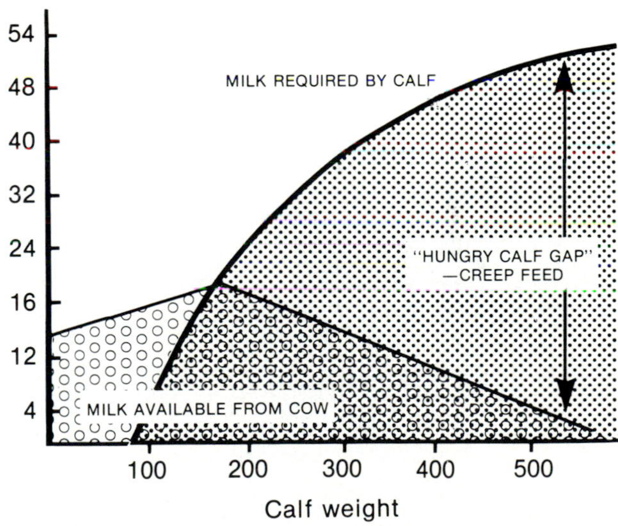

Fig. 19-33. Milk yield of a typical beef cow vs nutrient requirements of a nursing calf. This points up the need for creep feeding.

To fill the *hungry calf gap*—the nutrient requirements over and above those provided by 13 lb of milk—would require the consumption of 50 lb of green grass daily. Of course, that's a physical impossibility, because a 500-lb calf simply cannot hold that much bulk. So, the best way to fill the hungry calf gap is to creep feed with concentrate mixes.

Creep feeding is no longer primarily an emergency program to supplement drought-stricken grasses and other conditions resulting in poor pastures. Rather, continuous feeding of calves from birth to weaning is on the increase because milk in quantity, and pasture in quality, are not normally available season-long to supply the necessary nutrition (1) to produce calves that meet today's market demands, and (2) to realize the maximum genetic potential from improved breeding.

Most calves will continue to be raised on their mother's milk plus whatever pasture or other feed they share with their dams. However, more and more of them wil be creep fed in addition.

CREEP

A creep is an enclosure or feeder for feeding purposes which is accessible to the calves but through which the cows cannot pass. It allows for the feeding of the calves but not their dams. For best results, the creep should be built at a spot where the herd is inclined to loiter; on high ground, in the shade, and near watering and salting grounds. The enclosing fence may be of board, pole, or metal construction, with an entrance 16 to 20 in. wide and 3 to 3½ ft high. Self-feeders, troughs, or racks may serve as feed containers; allowing 4 to 5 in. of space per calf for self-feeding and 8 to 12 in. for hand-feeding. Also, there are on the market portable creep feeders consisting of a self-feeder to which the enclosing fence is firmly attached. These are especially suited for use on large range pastures, where frequent moving is necessary.

Fig. 19-34. Movable calf creep, with openings that will permit the calves to enter and keep the cows out. (Courtesy, National Cottonseed Products Assn., Inc., Memphis, Tenn.)

CREEP RATIONS: FEEDING DIRECTIONS

Creep-fed calves need special rations, since they are both in forced production and finishing. They are expected simultaneously to lay on fat and grow in protein tissues and skeleton. Consequently, their ration requirements are for feed high in protein, rich in readily available energy, fortified with minerals and vitamins, and with all the nutrients in proper balance. Also, the ration must be very palatable and digestible, which calls for an exacting formula. To meet these needs, more and more producers are finding it practical to buy a commercially prepared complete creep feed, or a well fortified and highly concentrated supplement to add to locally available feeds, rather than purchase individual ingredients and mix from the ground up.

Tables 19-22 and 19-23 show two creep rations, formulated by the authors, that have been widely and successfully used. A simple, yet very satisfactory, creep ration may be made by grinding and pelleting 75% alfalfa and 25% cereal grain.

TABLE 19-22
CALF CREEP RATION # 1[1] (AS-FED BASIS)

Ingredient	Precent	Per Ton	
	(%)	(lb)	*(kg)*
Oats	39.60	800.0	363.2
Corn # 2	14.80	300.0	136.2
Barley	8.90	177.5	80.7
Wheat bran	9.90	200.0	90.8
Dried molasses beet pulp	9.90	200.0	90.8
Soybean meal, 44%	9.90	200.0	90.8
Molasses	4.90	100.0	45.4
Salt	.50	10.0	4.5
Dicalcium phosphate	.50	10.0	4.5
Trace minerals[2]	.04	1.0	0.45
Vitamin A (30,000 IU/g)	.06	1.5	0.68
Total	100.00	2,000.0	907.2
Proximate analysis:	(%)		
Crude protein	14.30		
Fat	3.20		
Fiber	8.30		
Calcium	.32		
Phosphorus	.50		
TDN	69.60		

[1]*Feed preparation:* Preferably ⅛- or 3/16-in. pellets. Otherwise, steam roll and flake grains, or grind grains coarsely.

[2]See Table 19-7 for recommended trace mineral levels. Follow manufacturer's directions.

TABLE 19-23
CALF CREEP RATION # 2 (AS-FED BASIS)

Ingredient	Percent	Per Ton	
	(%)	(lb)	*(kg)*
Corn # 2	24.25	485	220.2
Alfalfa meal, 15%	22.50	450	204.3
Oats	20.00	400	181.6
Alfalfa hay (all analyses)	10.00	200	90.8
Soybean meal, 44%	6.20	124	56.3
Bran	5.00	100	45.4
Linseed meal, 35%	5.00	100	45.4
Molasses	5.00	100	45.4
Dicalcium phosphate	2.00	40	18.2
Trace minerals[1]	.05	1	0.45
Vitamin A (325,000 IU/g)[2]	—	63 g	
Total	100.00	2,000	908.0
Proximate analysis:	(%)		
Crude protein	15.10		
Fat	3.00		
Fiber	12.70		
Calcium	1.04		
Phosphorus	.73		
TDN	64.90		

[1]See Table 19-7 for recommended trace mineral levels. Follow manufacturer's directions.

[2]When 4 lb/head/day of the calf creep ration is consumed, 40,950 IU of vitamin A will be obtained in the feed.

It takes considerable effort and patience to start calves on feed. Also, a little calf psychology helps; remember that calves do not go for the privilege of eating, they need to be persuaded. One or more of the following techniques will usually prove helpful: Shut a gentle cow or a few calves in the creep, to serve as a decoy(s); scatter a little feed near the creep so that the cows will loiter nearby; and/or spread a little feed near and extending through the creep opening. It is also recognized that fall and early spring calves take to creep feeding better than late spring calves, simply because they have less grass and milk available. (Grass stimulates milk flow.)

When 3 to 4 weeks of age, calves should be started on feed very gradually. For the first 3 to 5 days, only about ¼ lb of feed per calf should be placed in the container(s) each day, and any leftover feed should be removed and given to the cows. In this manner, the feed will be kept clean and fresh. When calves are on lush pasture and their mothers are milking well, difficulty may be experienced in getting them to eat; but time and patience will pay off, and the results will become evident in 2 to 3 months.

After 5 to 7 days of hand-feeding, the creep ration can be left before the calves safely. Once they are on full feed, never let the feeder become empty; and avoid sudden changes. During the first 30 days, they will consume about 1 lb/head/day. By the end of the fifth month, they should be up to 8 lb daily. Of course, with good pastures and plenty of milk, these consumption figures may be halved. Calves will consume approximately 500 lb of creep feed per head from 1 month of age to weaning. In years of lush pasture, it will be less; in dry years, more.

NOTE WELL: Instead of allowing creep-fed calves to consume all they will eat, limited creep feeding, described in the section that follows, is gaining in popularity.

• **Salt-limited creep feeding**—Generally, limiting creep feed to about 3 lb/head/day is recommended. This will supply enough supplemental energy and protein to the dam's milk and forage to meet the requirements for normal growth of young calves. Feeding more than 3 lb/head/day may result in excessive fat deposition instead of skeletal and muscle growth.

Limited creep feeding can be accomplished by hand-feeding. But, the disadvantages of this practice are: (1) the high labor cost of feeding the calves each day; and (2) the larger calves tend to overeat and the smaller ones not to get enough, unless adequate bunk space is provided.

A common salt-limited creep feed consists of a mixture of 5 to 10% salt and 90–95% cottonseed meal. Also, either of the creep rations listed in Tables 19–22 and 19–23 may be limit-fed by adding approximately 10% salt.

CREEP GRAZING

Creep grazing refers to the practice of grazing nursing calves on separate pastures from their dams. They may either graze before the cows do, getting first choice of the more succulent, highly nutritious pastures; or they may have access to special pastures. The calves enter the special pastures through gates with openings large enough for calves, but too small for the cows to get through. In an alternative method, electric fences are positioned high enough (36 to 42 in.) for calves to pass under, but low enough to keep cows out. Limited studies indicate that as much as ½ lb extra weight gain per day may be obtained by creep grazing.

BENEFITS FROM CREEP FEEDING

The best yardstick for measuring performance in a beef breeding herd is pounds of calf weaned per cow bred. This fact, along with the demand for healthy, gain-ready feeder calves and the prices being paid, is causing cow-calf operators to consider the immediate and residual benefits of creep feeding.

Among the reasons for and the benefits from creep feeding are the following:

1. It provides a way in which to compensate for insufficient milk.
2. It makes for heavier weaning weights.
3. It facilitates fall calving.
4. It makes for more uniform calves.
5. It makes it possible for calves to achieve their full genetic growth potential.
6. It makes for economical gains, for young gains are cheap gains.
7. It is economically more efficient to feed calves directly than to feed cows too liberally.
8. It makes for attractive purebred calves.
9. It makes it easy to reinforce and improve milk.
10. It helps to control parasites.
11. It simplifies weaning.
12. It facilitates early weaning.
13. It makes for marketing flexibility.
14. It narrows the price between heifers and steers.
15. It makes for better lifetime reproductive performance of heifers.
16. It provides calves that are "bunk broke"—calves that are accustomed to feed—with the result that they will continue on feed in the feedlot with a minimum of stress, shrink, digestive disturbances, and death losses, and without the normal period of 3 weeks to get them on full feed.
17. It makes for earlier cycling and conception of the dams.
18. It leaves the dams less suckled down and in better condition.
19. It gives first-calf heifers a needed assist.
20. It usually pays.

The following rule of thumb may be used to determine whether it will pay to creep feed: It pays to creep feed when the selling price per 100 lb of calf is greater than the cost of ¼ ton (500 lb) of feed.

LIMITATIONS OF CREEP FEEDING

As with many good things, creep feeding does have its limitations; among them, the following:

1. **It isn't always profitable.** Creep feeding may not be profitable because of excessive cost of the creep ration, low response, and/or price discrimination against fleshy weaner calves.
2. **It lowers feedlot gains and efficiency.** Fleshy creep-fed calves make slightly less rapid and efficient gains than calves not creep fed when they are (a) moved directly into the feedlot following weaning; and (b) long-fed, because

creep feeding alleviates compensatory gains. Also, it may cause small type cattle to get too fleshy—to "stall" or stop growing before weaning. However, if the latter occurs, it's a sure sign that the wrong kind of cattle are being bred. However, these disadvantages may be compensated for, in part at least, by the shorter feedlot feeding period and more desirable market weights of creep-fed calves.

3. **It makes for less desirable stockers.** Creep-fed calves do not make as desirable stockers as calves that have not been creep fed, simply because the latter are normally placed on less nutritious growing rations consisting predominantly of roughages. This may be further explained in this way: One of the basic rules in feeding slaughter cattle is always to proceed to a higher plane of nutrition; never go down.

4. **It complicates selecting cows for milk production.** Creep feeding makes it difficult to put selection pressure on cows for good milk production, because creep feeding cancels differences in weaning weight due to lactation differences in dams—thereby making it impossible to determine the genetic potential for milk production. This may be an important consideration in production testing programs.

5. **It may result in heifers getting too fat.** Creep-fed heifers may get too fat, resulting in fat being deposited in their mammary system and the tendency for them to be poor milkers as cows.

6. **It is difficult in remote areas.** Creep feeding is difficult on less accessible ranges, because the very nature of creep feeding calls for close attention.

7. **It cannot be done where there are hogs, sheep, or goats running with cattle.** These animals can enter any creep opening that is big enough for a calf.

Feeding Early Weaned Calves

Early weaning refers to the practice of weaning calves earlier than the usual weaning age of about 7 months, usually within the range of 35 days to 5 months of age. Although it is not common practice among U.S. beef producers, dairy producers have been weaning 3-day-old calves for years. Also, early weaning has long been an integral part of many of the beef programs of Europe.

Currently, there is much interest in early weaning because (1) it fits into a drylot cow-calf management system; and (2) it can give a big assist in getting females, especially 2-year-old heifers, to rebreed in a short period of time.

The current interest in increasing the number of cows in the Corn Belt is largely predicated on more efficient use of crop residues, especially corn and sorghum residues. With heavy cropping, there is little or no pasture as such; hence, drylot management systems are evolving. When using crop residues as a basic feed source, the lactating cow is likely to need supplemental feed. Considering the low efficiency involved in converting supplemental energy to milk and in converting milk to meat, it is apparent that a more efficient use of feed could be achieved by giving the supplemental feed directly to the calf. A lactating cow requires about 50% more feed than a dry cow. So, rather than give her that additional feed, it is more efficient to give feed directly to the calf.

Weaning calves early from 2-year-old, first-calf heifers reduces the stress of milking and raising a calf. As a result, they recycle and rebreed earlier and grow out more rapidly. As heifers are bred for higher milk production, this reason for early weaning takes on greater importance, for the more milk they give, the slower they are to cycle. It appears doubtful that any level of nutrition will have the same effect on reproduction efficiency of the heifer as early weaning.

In addition to fitting into a drylot cow-calf system and facilitating a program of calving 2-year-old heifers, early weaning may be desirable for the following reasons:

- It may be the answer to getting one calf per cow every 12 months in intensive management systems.
- It may be the key to the most efficient feed utilization during times of droughts and other periods of feed shortages. Under such conditions, it might be advantageous to wean the calves early and to provide them with the highest quality feed available, while restricting the quality and quantity of feed fed to the cows.
- It fits in with fall calving where heavy winter feeding is required. As soon as the calves are weaned, cows may be turned to stalk or stubble fields to winter on cheap feed.
- It may make it possible to keep a particularly valuable old purebred cow in production longer.
- Young gains are cheap gains, due to (1) the higher water and lower fat content of young animals in comparison to older animals, and (2) the higher feed consumption per unit weight of young animals. Thus, the feed efficiency of early weaned calves is excellent, ranging from 3 to 4 lb of TDN per pound of gain.
- Lactating cows decline in milk production after about 1 to 2 months following parturition.
- Parasite problems are minimized in an early weaning program.

Successful early weaning of beef calves calls for the following:

1. Weaning calves that are at least 35 days of age if supplemental milk is not going to be used.
2. Using a highly palatable ration that is high in protein, available energy, minerals, and vitamins.
3. Providing a starter ration to the calves during the 2- to 3-week adjustment period before calves are actually weaned.
4. Vaccinating for blackleg, malignant edema, hemoglobinuria, pasturellosis, and enterotoxemia, and injecting with vitamins A and D, at the beginning of the adjustment period.
5. Checking calves regularly for respiratory problems and controlling flies.

Where early weaning is successful, the only responsibility of the beef cow is to produce a calf and give it a good start in life for a brief period, then go on a maintenance ration the rest of the year.

As with many good things in life, early weaning does have some disadvantages. To be successful, superior nutrition and management are essential; and the earlier the weaning age, the more exacting these requirements.

EARLY WEANING AGE

Experiments and experiences indicate that it is practical to wean calves as early as 35 days of age. This is long enough to stress the cow a bit, but short enough to get her recycling and rebred so that a calf will be produced each

12 months. Also, it allows the cow to function as a lactating animal. Weaning at 3 to 5 months of age doesn't make for early recycling; hence, it doesn't contribute to getting one calf per cow every 12 months.

RATIONS FOR EARLY WEANED CALVES

From 35 days of age on, early weaned calves can be fed any good starter rations, most of which contain dry skim milk. Two such rations are given in Chapter 20, Feeding Dairy Cattle. Most commercial feed companies manufacture a starter ration. The starter ration should be made available to the calves well ahead of weaning in order that they will be accustomed to it, thereby avoiding any setback.

MILK REPLACERS

Several reputable commercial companies now produce and sell milk replacers, which are composed of sizable amounts of milk by-products, such as dry skim milk, buttermilk, and/or whey, along with additives. Milk replacers can be fed as the only feed immediately following the colostrum period; or they may replace whole milk beginning about the seventh day.

From the standpoint of the beef cattle producer, a milk replacer is of primary interest for two uses; namely, (1) for raising orphaned and early weaned calves, and (2) for replacing nurse cows. Also, it is a valuable adjunct in certain disease control programs, especially those diseases that may be transmitted from dam to offspring.

(Also see Chapter 6, Types and Roles of Feedstuffs, section on "Milk Replacers.")

Growth Stimulants and Implants for Calves

Experiments indicate that some implants will produce a 15- to 30-lb weight advantage in calves.

Growth-promoting stimulants and implants generally increase the rate of gain and weaning weight of creep-fed calves more than noncreep-fed calves, since creep feeding makes for high-energy intake and growth stimulants improve the utilization of energy.

The growth stimulants and implants that are presently available and approved for suckling calves are listed in Table 19-11, Growth Stimulants and Implants For Cattle.

Weaning

Weaning is a traumatic experience for a calf. It represents environmental, nutritional, psychological, and, altogether too often, vaccination-castration-dehorning changes—all of which make for great stress. Generally, at weaning time the calf is moved into a strange environment to which it must adjust—a new pasture, corral, or shelter; its feed supply—milk—is suddenly removed; and its association with and protection by its mother is lost. Under such circumstances, it is small wonder that calves lose weight and become more susceptible to disease. The marvel is that they survive such mistreatment so well.

Calves should be weaned when they are 7 to 8 months old. Weaning at this age fits in well with the weight record-keeping requirements of most performance testing programs. Also, calves will be about the right age and weight for fall feeder calf sales.

Weaning earlier than 7 or 8 months may be necessary in years when pastures are short or when calves are from first-calf heifers.

The best way to wean is to remove the calves from their dams and keep them out of sight of each other. Cows and calves should never be turned together once the separation has been made. Such a practice will only prolong the weaning process, and it may also cause digestive disorders in the calf. Provide calves with plenty of water, free-choice hay, and 3 to 4 lb of grain per head per day. If calves were creep fed, continue their rations during the weaning period.

During the weaning process, calves should be confined to a small area to cut down on walking and shrinkage. In bad weather, protection should be provided from cold wind and rain; they should have access to a shed, wooded area, gorge, or other protection.

• **How to minimize weaning stress and weight loss**—Weaning calves is more a matter of preparation than of absolute separation from the dam. Minimizing stress and weight loss depends largely upon the thoroughness of the preparation—the preconditioning.

• **Drying up the cow**—With higher milking strains of beef cattle, when drying up cows, beef producers will have the same concerns as dairy producers—that of avoiding "spoiled udders." To alleviate this problem, the following procedure is recommended:

1. Do not feed milk-stimulating feeds at weaning time. Either put the cows on poorer pastures or feed a nonlegume forage.
2. Let "back pressure" in the udder build up. Examine the udder at intervals, *but do not milk it out.* If the udder fills and gets tight, rub spirits of camphor on it, *but do not milk it out.* At the end of 5 to 7 days, when the udder is soft and flabby, what little secretion remains (perhaps not more than half a cup) may be milked out if so desired.

Preconditioning

Preconditioning is a way of preparing the calf to withstand the stress and rigors of leaving its mother, learning to eat new kinds of feed, and shipping from the farm or ranch to the feedlot. To the cow-calf producer, it is a program of management, nutrition, and immunization. It, along with improved breeding based on production testing, is the trademark of the producer of feeder calves. To the feedlot operator, preconditioning is a way in which to prepare calves to fit into the program and to minimize costly and unnecessary procedures.

Changed environment; excitement of sorting, loading, and shipping; long periods without feed and/or water; movement through one or more assembly points; change of feed; and exposure to disease—all add up to *fatigue, stress, shrink,* and *lowered disease resistance.*

The term *preconditioning* generally consists of the following practices being conducted on the farm or ranch of origin and certified to by a licensed veterinarian: weaned; bunk broke and water tank or fountain trained; castrated and dehorned; vaccinated for IBR, PI-3, BVD, and *Haemophilus somnus;* and, depending on local conditions, additional

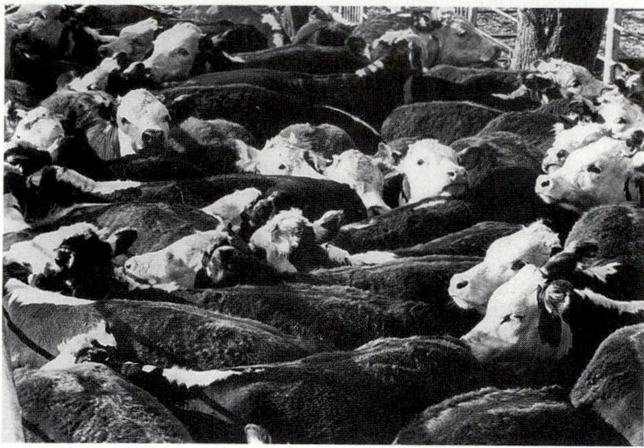

Fig. 19-35. Preconditioning, along with production testing, will be the trademark of the producer of reputation feeder calves of the future. (Courtesy, Ralston Purina Co., St. Louis, Mo.)

vaccinations may be required for blackleg and malignant edema, and for brucellosis of heifers.

It is important that the program be written down, adhered to rigidly, then certified to by both the owner and the veterinarian. The producer should take the lead in developing such a program, but the counsel of the veterinarian and potential buyers should be sought.

Preconditioning is often confused with handling newly arrived feedlot cattle, and backgrounding. This is understandable because all three of them are important phases between weaning and finishing. Yet, each of them is distinct and separate.

• **Some causes of stress**—Stress takes place when a calf is weaned, at which time the social structure is disrupted. Also, there is stress when an animal is placed in a different environment, whether it be in a pen on the home ranch or on a loaded truck on the way to market. If stress is to be lessened, therefore, it is obvious that measures must be taken before any of these steps happen.

• **Losses attributed to prefeedlot stress**—On the average, 25% of feedlot calves sicken, and 5% of the calves die. Losses from sickness are even greater—they run 2 to 5 times as great as the actual death loss. Sickness losses accrue from the expense and treatment plus the resulting inefficiency. Thus, the combined losses—death losses, and losses due to sickness—add $55 per head onto the cost of every feedlot-finished animal.

• **Cost of preconditioning**—The cost of preconditioning will run about $15 to $18 per head, the amount depending primarily upon the number and kind of vaccinations administered and the cost of feed. Some or all of this cost will be recovered by gains made during the preconditioning period.

• **Marketing agencies can lessen stress and exposure**—The above discussion pertains to preconditioning at the farm or ranch level. It is recognized, however, that not all cattle are moved directly from the cow-calf producer to the feedlot. Many of them pass through auction markets, terminal markets, and other similar intermediate locations. Hence, the marketing agencies handling feeder cattle can do much to cut down on stress and exposure to disease. Recommendations to this end are:

1. Refuse to accept sick animals, or at least isolate them. This is important because one sick animal may cause a disease break in hundreds of other cattle.
2. Isolate cattle from individual producers whenever possible. Here again, this is recommended in order to cut down the spread of infectious diseases.
3. Keep dust to a minimum.
4. Feed animals well-balanced rations similar to the rations to which they have been accustomed.
5. Move cattle to their final destination as expeditiously as possible.
6. Prevent bruises.
7. Improve sanitation in all handling facilities.
8. Reduce weather stress.
9. Keep records which will help pinpoint problem areas and unscrupulous individuals.

• **A preconditioning program**—Opinions differ widely on what constitutes properly preconditioned cattle. However, the following preconditioning program is presented with the hope that beef producers will use it (1) as a yardstick with which to compare their existing program, or (2) as a guide so that they and their local veterinarian and other advisers, may develop similar and specific programs for their respective enterprises:

1. **Handle quietly.** Calves should be handled quietly, with a minimum of excitement.

2. **Dehorn and castrate.** All calves that will eventually go into feedlots should be dehorned (although tipping of horns is acceptable), and castrated unless they are to be fed out as bulls. There is far less stress if calves are dehorned and castrated well ahead of weaning—about 2 months of age is best.

3. **Wean.** Calves should be weaned 30 days ahead of shipment.

4. **Start on feed.** Adjust calves to feed bunks and water troughs and start them on a ration similar to that which they will get in the feedlot. For the first 3 days following weaning, calves should have access to loose grass hay. Additionally, they should be started on a ration of about the following nutrient composition:

Net energy for production (NE_p) . . . (Mcal)	38.0
Crude protein . . . . (minimum %)	12.0
Calcium . . . . (%)	.5
Phosphorus . . . . (%)	.3
Vitamin A . . . . (IU/lb)	5,000.0
Roughage:concentrate ratio, approx.	40:60

If weaning calves 30 days ahead of separation from their dams is totally impractical, they should be started on a creep feed similar in composition to the above. This type of ration will be very similar to the starting ration that calves will receive when they arrive in the feedlot.

Use medicated feed only on the recommendation of your veterinarian.

5. **Vaccinate.** The preconditioning vaccination program will vary from area to area, relative to required and optional vaccinations, and in stated period of time. Most Certified Preconditioning Programs call for the following minimum requirements: Immunization at least 3 weeks prior to selling

(to be certified by a veterinarian) to include Clostridial (7 way), IBR, PI_3, BVD, *Haemophilus somnus,* and Brucellosis Calfhood Vaccination of heifers. Follow your veterinarian's advice for vaccination procedures. If a direct sale to a feedlot is involved, the calves should be vaccinated in keeping with the regular program of the feedlot.

6. **Treat for parasites.** At the time of weaning, and prior to shipment, calves should be checked for both internal and external parasites, and treated as necessary. Usually this involves (a) treating for grubs, through either spray, pour-on, or feed; (b) spraying for lice; and (c) checking for worm eggs, and worming if necessary.

7. **Reduce time from farm or ranch to feedlot.** Every effort should be made to reduce the total time between the moment calves leave the farm or ranch and when they arrive at the feedlot.

Where either truck or rail shipments are longer than 36 hours (the 28-hour law governing rail shipments may be extended to 36 hours upon written request of the owner), unload en route for the purpose of giving feed, water, and rest for a period of at least 5 consecutive hours before resuming transportation.

8. **Reduce stress and exposure to infection.** The stress and exposure to infection during the marketing and transportation periods should be reduced to a minimum.

9. **Provide preconditioning certificate.** It is extremely important that records be kept of all husbandry, nutritional, and medical histories, and that the party who sells the feeder cattle should provide the party receiving them with a written record of all of them. This will help the feedlot operator to fit the cattle to the existing program and minimize costly and unnecessary procedures. A suggested preconditioning certificate is shown in Fig. 19-36.

PRECONDITIONING CERTIFICATE

Date ________

Number of cattle ________

Steers ______ Heifers ______ Bulls ______ Breed ______ Age ______ Brand ______

PRACTICES, TREATMENT, AND VACCINATION

	Date	Product Brand	Signature of Responsible Person
Castrated			
Dehorned			
Blackleg & malignant edema vaccine			
Bovine respiratory disease complex vaccine			
Lepto vaccine			
IBR vaccine			
BVD vaccine			
Other vaccines			
Weaned			
Wormed			
Grub treated			
Lice (treatment) spray or dip			
Vitamin A.D.E. inj.			
Medication Antibiotics, sulfa, electrolytes			

Ration during preconditioning period ________

Loading point ________

The undersigned hereby declares and certifies that the practices, treatments, and vaccinations indicated above have been carried out and administered to all the cattle described and identified by this certificate.

Seller:

(Signature of owner or authorized representative) (Date)

Cattle Producer:

(Signature) (Date)

Seller's Veterinarian:

(Signature) (Date)

Fig. 19-36. Preconditioning certificate.

• **Preconditioning will increase**—On an industry-wide basis, preconditioning feeder cattle could save millions of dollars now lost in sickness, shrink, and death.

Feeder buyers will determine how quickly preconditioned cattle become generally available. They will speed their availability when they ask the seller:

1. When were they weaned (if calves are involved)?
2. Are they accustomed to bunk feeding and trough watering?
3. Have they been treated for grubs?
4. Have they been examined for internal and external parasites, and treated if necessary?
5. Have they been vaccinated for IBR, BVD, shipping fever, blackleg?
6. Is certification for the above procedures available?

Buyers will increasingly favor producers who follow such a program. Preconditioning, along with improved breeding based on production testing, will be the trademark of the producer of reputation feeder calves of the future.

• **Summary**—Studies show that preconditioning (1) increases the weight of calves when they leave the farm or ranch, and (2) improves the performance of calves during the first month in the feedlot. But the premium price needed to pay for preconditioning may be larger than the buyer is willing to pay. Nevertheless, the surgical component of preconditioning—castrating and/or dehorning—should be performed prior to weaning and shipment.

FEEDING REPLACEMENT HEIFERS

The feed and management program of replacement heifers will have a lifelong effect on their productivity. It will determine how young they may be bred, whether they calve early or late, whether they are good milkers or poor milkers, the weaning weight of their calves, and how long they remain in the herd. Also, feed accounts for 40 to 70% of the cost of raising replacement heifers; hence, it is important to know whether it is possible to effect savings on feed during the growing period without affecting reproduction adversely. It is even more important to know whether their performance as adult animals can be enhanced by proper nutrition and management.

Fig. 19-37. Well-grown Charolais heifers on pasture. (Courtesy, *Charolais Journal*, Kansas City, Mo.)

Fig. 19-38 shows optimum growth rates and weights at different ages, based on a summary of research. Heifers should not be bred before reaching $\frac{2}{3}$ of their mature weight, $\frac{9}{10}$ of their mature height, and $\frac{4}{5}$ of their mature heart girth and width at hips. Naturally, these measurements vary with different frame sizes and breeds. But once the needed growth rate is established, rations can be formulated that will allow these growth rates to be achieved.

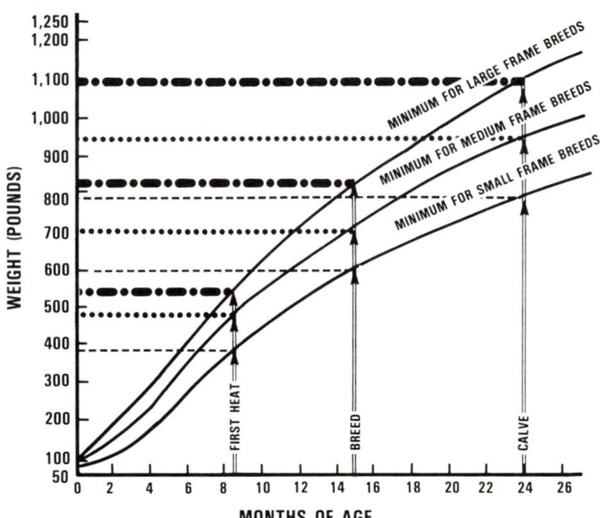

Fig. 19-38. Minimum weights in the reproductive cycle of replacement heifers of different frame sizes.

The requirements for pregnancy are really not too great. The body of an 80-lb, newborn calf contains only about 12 lb of protein, 3 lb of fat, and 3.6 lb of mineral matter. But the lactating requirements are much more rigorous. If a 2-year-old heifer gives her calf an average of $1\frac{3}{4}$ gal of milk per day over a 7-month suckling period, she will produce in that milk a total of 93 lb of protein, 107 lb of fat, 133 lb of sugar, and 20 lb of minerals.

Hence, the comparison: 12 lb of protein in the fetus vs 93 lb in the milk during the suckling period. This means that nearly 8 times more protein is required for 7 months of lactation than for 9 months of pregnancy.

Also, when breeding yearlings to calve as 2-year-olds, producers should be aware that nature has ordained that the growth of the fetus, and the lactation which follows, will take priority over the maternal requirements. Hence, when there is a nutritive deficiency, the young mother's body will be deprived, or even stunted, before the developing fetus or milk production will be materially affected.

Nutrient Requirements of Replacement Heifers

Meeting the nutrient requirements of heifers from weaning to first calving is of great importance. The requirements of heifers of different body weights and growth rates are given in Table 19–24.

TABLE 19–24
DAILY NUTRIENT REQUIREMENTS OF MEDIUM-FRAME GROWING HEIFERS[1]

Body Weight		Daily Gain		TDN[2]	Protein[2]	Calcium	Phosphorus
(lb)	*(kg)*	(lb)	*(kg)*	(lb)	(lb)	(g)	(g)
400	*182*	2.00	*0.9*	7.7	1.29	26	13
500	*227*	2.00	*0.9*	9.1	1.34	24	13
600	*273*	2.00	*0.9*	10.4	1.40	23	14
700	*318*	2.00	*0.9*	11.7	1.45	22	15
800	*364*	1.40	*0.6*	10.4	1.60	25	16
900	*409*	1.40	*0.6*	11.3	1.60	26	18

[1]The above requirements for 800– and 900–lb (*364– and 409–kg*) weights are for pregnant yearling heifers last third of pregnancy. Adapted by the authors from *Nutrient Requirements of Beef Cattle*, sixth revised edition, National Research Council-National Academy of Sciences, Washington, D.C., 1984.

[2]Pounds protein and TDN can be converted to kg by dividing by 2.2.

Rations for Replacement Heifers

In season, good pasture plus mineral supplements fed free-choice will meet the nutrient requirements for proper growth and development of heifers.

On winter range, when dry forage is of low quality, and sometimes not too abundant, 1 to 2 lb of a protein supplement should be provided in the form of cubes, blocks, meal-salt, or liquid. When consumed at the intended level, the supplement should contain sufficient vitamin A to meet the requirements. Mineral supplements should also be provided, preferably free-choice.

Where winter grazing is not available, heifers must be drylotted and fed a complete ration. Sufficient nutrients should be provided to meet the requirements and to keep heifers in a thrifty condition, neither too fat nor too thin.

Fig. 19–39. Yearling Polled Hereford replacement heifers bunk fed a winter ration in a corral. (Courtesy, *Polled Hereford World*, Kansas City, Mo.)

The wintering rations in Table 19–25 for 500–lb heifer calves should result in a rate of gain of 1 to 1.5 lb per day.

The wintering rations in Table 19–26 for 800– to 900–lb bred yearling heifers should allow a gain of 0.75 to 1 lb per day during the wintering period prior to calving.

TABLE 19–25
DAILY RATIONS FOR HEIFER CALVES (500 LB [*227 KG*]) (AS-FED BASIS)

	Rations									
	1		2		3		4		5	
	(lb)	*(kg)*	(lb)	*(kg)*	(lb)	*(kg)*	(lb)	*(kg)*	(lb)	*(kg)*
Legume-grass haylage	25	*11*								
Legume-grass hay			10	*4.5*	10	*4.5*			5	*2.3*
Corn or sorghum silage ..							30	*13.6*	20	*9.1*
Ground ear corn			4	*1.8*						
Corn, grain sorghum, or barley					3	*1.4*				
Supplement[1]							1	*.45*		

[1]Supplement contains 48% crude protein. Quantity to be adjusted in keeping with the protein content of the supplement.

TABLE 19–26
RATIONS FOR BRED YEARLING HEIFERS (800–900 LB [*364–409 KG*]) (AS-FED BASIS)

	Rations									
	1		2		3		4		5	
	(lb)	*(kg)*	(lb)	*(kg)*	(lb)	*(kg)*	(lb)	*(kg)*	(lb)	*(kg)*
Corn or sorghum silage ..	45	*20.5*	25	*11*						
Legume-grass hay			10	*4.5*	20	*9.1*			15	*2.3*
Legume-grass haylage							35	*15.9*		
Corn, grain sorghum, or barley									3	*1.4*
Supplement[1]	1.5	*0.7*								

[1]Supplement contains 48% crude protein. Quantity to be adjusted in keeping with the protein content of the supplement.

Replacement heifers should be fed rather liberally—more so than stocker cattle which are being grown for the feedlot, to the end that they will acquire most of their growth and development before calving. With limited feeding, they will not have enough weight for age to breed when they are 15 months old; and it is best not to have them calve until they are 30 months of age. Cost of the ration is important, but too limited a ration may actually be costly, since it is usually cheaper in the long run to grow heifers out well rather than to delay their development and run into management problems, higher death losses, greater pasture costs, higher interest on investment, and increased labor costs. Also, where breeding animals are sold, it is recognized that prospective purchasers are impressed by well-grown and well-conditioned young stock, whereas they are seldom interested in stunted animals at any price. The latter situation is accentuated in purebred heifers; hence, it is important that they be more liberally fed than replacement heifers for the commercial herd.

Occasionally, a replacement animal is injured by overfeeding or by fitting for show, but such losses are insignificant compared with those resulting from the thousands of undersized, poorly developed animals that are grossly underfed.

During the winter months, the feeding of a ration containing adequate protein and energy, plus the required vitamins and minerals, is necessary to keep young stock growing and gaining weight.

During their second winter, heifers bred to calve as 2-year-olds should be fed more liberally than mature cows of comparable condition. The added feed is necessary because, in addition to maintenance and developmment of the fetus, provision must be made for body growth. Even then, these heifers should have close supervision at calving time and the calves should be weaned at an early age in order to alleviate the strain of lactation.

Separate Heifers by Ages

The nutritive requirements of heifers differ according to body weight and expected daily gain (Table 19-24). Consequently, the recommended ration for a 500-lb heifer calf (Table 19-25) differs from that of an 800- to 900-lb bred heifer (Table 19-26). It is important, therefore, that replacement heifers be separated by ages for wintering, with coming yearlings in one group and coming twos in another.

Calving Two-Year-Olds

From the above, it may be concluded that more yearling heifers can and should be bred to calve as 2-year-olds. But, in doing so, the following practices should be observed in order to lessen calving difficulties:

1. Select the heaviest and highest scoring individual heifers at weaning. Weight at weaning is a means of evaluating the dam's milking ability, provided the calves have not been creep fed.
2. Keep heifers separate from older cows.
3. Start with 50% more weaner replacement heifers than needed if it is the intent to maintain the same size herd—with no provision for expansion whatsoever. This means that for every 100 cows in the herd, 20 replacement heifers are actually needed each year in order to maintain the same size herd. (There is about a 20% replacement in each herd each year.) However, 30 weaner replacement prospects (50% more than actually needed) should be held, simply because, based on averages, 10 of them will either die or have to be culled before they replace older cows.
4. Give consideration to the increased nutritional requirements of calves with increased growth rates.
5. Replacement heifers should be fed for gains of approximately 1 lb/head/day from weaning to first breeding. Following the breeding season, heifers should be managed to assure continued growth and achieve 80 to 85% of expected mature weight at the time of first calving. From breeding until calving, 1¼ lb gain/day is about right.
6. Select yearlings and coming 2-year-old heifers on the basis of individuality and rate of gain. Also, cull heifers with small pelvic openings; those with large pelvic openings (about 34 sq in.) have less calving difficulty. Avoid excessively fat heifers.
7. Breed only well-developed heifers, weighing 700 to 750 lb (depending on breed) at 13 to 14 months of age. Size at breeding is more important than age. Also, some breeds come in heat and mature a little earlier than others.
8. Breed heifers 20 days earlier than the cow herd and restrict the breeding season to 45 days. This gives a short concentrated calving period; therefore, proper attention and help can be given heifers at calving time.
9. "Flush" feed heifers to gain approximately 2.0 lb per head daily beginning 20 days before the start of and continuing through the breeding season.
10. Breed heifers to a bull known to sire small calves at birth.
11. Feed a well-balanced ration, and feed for continuous gain of 1.25 lb during the pregnancy period; but don't get them too fat.
12. Feed heifers to weigh at least 800 lb by 120 days before calving.
13. Feed heifers to gain 100 to 120 lb from 120 days prior to calving. Heifers should weigh at least 875 lb just before calving and approximately 775 lb shortly after calving.
14. Give heifers special care at calving time, including—
    a. Providing adequate facilities, including (1) a pull stall, and (2) small pens each suitable for confining a heifer and her calf for approximately 24 hours of "mothering up."
    b. Moving each heifer into the calving area approximately 2 weeks before the expected calving date.
    c. Checking heifers for calving at 2-hour intervals.
    d. Rendering assistance quickly and expertly when it is needed.
    e. Removing heifer and calf from calving area within 24 hours after birth and putting them into a clean, dry pasture or other similar area.
15. Provide superior nutrition—well balanced, and rather liberal—during the lactation period, because a heifer's nutritional requirements double after calving. This requires a good ration—one containing adequate energy and proteins, and fortified with the necessary vitamins and minerals. In season, usually this can be accomplished by keeping these heifers on good pastures, with or without supplemental feeding, both during pregnancy and lactation. When good grass is not available—in the winter, or during droughts—proper feeding must be relied upon.
16. If practical, wean early; at 2 to 6 months of age, rather than the normal 7 months. Otherwise, creep feed the calves.
17. Run heifers that calved as 2-year-olds in a separate herd until after they have had their second calf.
18. Try it (calving 2-year-olds) out on half of your replacement heifers to start with; make sure that you know what is involved before going all out.

Some breeders may wish to take another year and stick to calving out 3-year-olds. But more and more progressive, commercial cattle raisers will calve out 2-year-olds from the standpoint of cutting production costs and increasing profits.

PART III—Feeding Stocker (Feeder) Cattle

Fig. 19-40. Thin, yearling Santa Gertrudis stocker cattle shipped from Texas and turned to lush pastures in Pennsylvania, followed by a short grain feed. (Courtesy, Pennsylvania Millers and Feed Dealers Assn., Ephrata, Pa.)

Stocker cattle and stocker cattle programs have changed with the passing of time. Until the early 1900s, stocker programs involved growing purchased or homegrown calves or yearlings on grass and hay until they were 3 to 4 years of age. As calf weaning weights increased and finished slaughter cattle weights decreased, the amount of time and gain required to grow calves from weaning until the beginning of the finishing period was substantially shortened. The stocker cattle industry became a calf-yearling industry, usually starting with 300- to 500-lb calves and ending with a yearling sold to a feeder at 600 to 700 lb. Today, most stockers are calves under 500 lb weight.

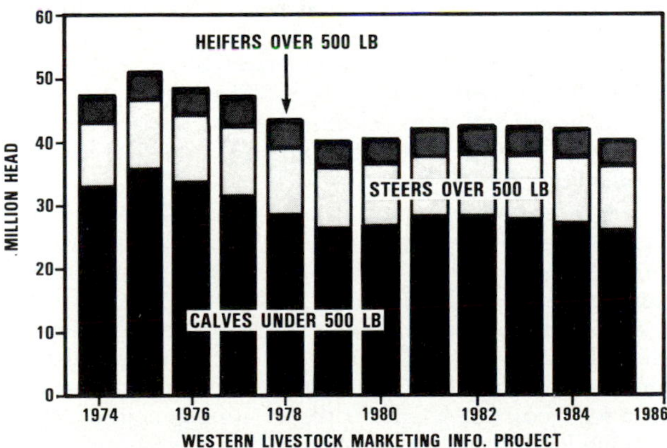

Fig. 19-41. January 1 calf and yearling feeder supply, 50 states, 1974–1985. The "over 500 lb" shows steers and heifers separately. The "under 500 lb" includes all calves, except those in feedlots. Note that calves (under 500 lb) far outnumber yearling steers and heifers. (Courtesy, *Western Livestock Round-Up,* March 1985)

The development of large feedlots and year-round feeding increased the demand for feeders ready to go on high-concentrate rations. The main reason that the larger feedlot operators like to purchase feeders ready to go on high-energy rations is that roughage is used more efficiently in growing cattle, whereas it is usually an expensive item to use in large feedlots.

Today, the stocker stage is changing again, as a result of two forces working in opposite directions, with one force favoring lengthening of the stocker stage and the other favoring shortening it:

1. Scarce and high-priced grains favor more roughage feeding and less grain feeding, resulting in carrying stockers to older ages and heavier weights, followed by a shorter feedlot period.
2. Heavier milking cows and heavier weaning weights, coupled with high-priced land, favor shortening the stocker stage, or even eliminating it, as 550-lb, or heavier, weaning weights are achieved.

In the future, both types of stocker operations will prevail, with the choice determined primarily by the price of grain and the weaning weight of the calves. Heavy-weaned calves will likely go directly into the feedlot or for slaughter. Calves with light to average weaning weights will likely be carried as stockers to 700- to 800-lb weights, thereby shortening the feedlot period and lessening grain feeding.

Today, stockers are grown according to two systems: (1) calves or light yearlings are either roughed through the winter, followed by grazing, or grazed only, then sold as feeders in late summer and fall; or (2) calves or yearlings are fed harvested roughage and grain in drylot, and then transferred to another location for finishing. Also, some new terms have evolved. Definitions of currently used stocker (feeder) terms follow.

- ***Stockers*** *are calves and yearlings, both steers and heifers, that are intended for eventual finishing and slaughtering and which are being fed and cared for in such manner that growth rather than finishing will be realized. They are generally younger and thinner than feeder cattle.*
- ***Feeders*** *are calves and yearlings, both steers and heifers, carrying more weight and/or finish than stockers, which are ready to be placed on high-energy rations for finishing and slaughtering.*

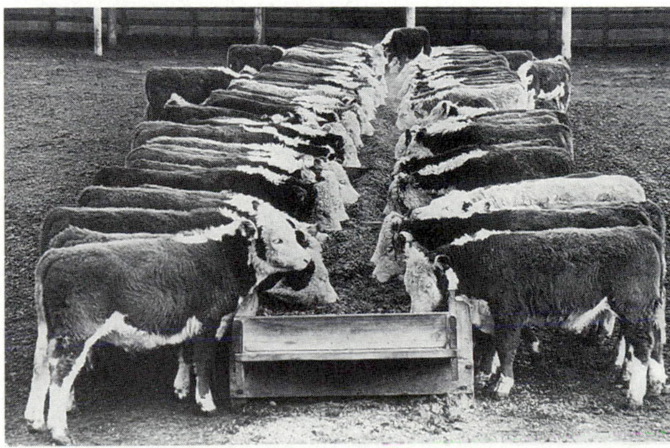

Fig. 19-42. Feeder calves, in good condition and ready to be placed on a high-energy ration for finishing and slaughtering. (Courtesy, Ralston Purina Co., St. Louis, Mo.)

• ***Replacement heifers*** *are the top end of the heifer calves selected to replace the older cows that are culled from the herd.*

• ***Preconditioning*** *refers to preparing the calf to withstand the stress and rigors of leaving its mother, learning to eat new kinds of feeds, and shipping from the farm or ranch to the feedlot or stocker grower.*

• ***Backgrounding*** *is an old practice with a new emphasis and a new name. Actually, backgrounding and the stocker stage are one and the same. Both refer to that period in the life of a calf from weaning to around an 800-lb weight, when it is ready to go on a high-energy finishing ration.* However, the term *backgrounding,* which was ushered in with the development of large feedlots, indicates a shift in emphasis. The term *stocker stage* connotes emphasis on marketing roughages through thin cattle, whereas *backgrounding* connotes emphasis on growing out feeder calves ready to go on a high-energy finishing ration. Backgrounding may be done on pasture or in the drylot, or some combination of both. At its best, the animals should be in good health, bunk broke, and ready to go on full feed.

From the above, it may be concluded that in the variable period of a calf's life between weaning and finishing, it is usually classed as either a stocker, a feeder, or a replacement heifer. Prior to weaning, calves may or may not be preconditioned.

The dividing line between stockers and feeders is not always as clear-cut as the above definitions would indicate. That is, not all thin cattle are suitable for stockers. For example, very large yearlings and most heifers are usually sold as feeders, to be placed on high-energy feeds. Also, "Okie-" type cattle are usually backgrounded for 50 to 60 days, then placed on a finishing ration.

TYPES OF STOCKER PROGRAMS

Sometimes the stocker operation is the only cattle enterprise on a farm or ranch, but more frequently it is conducted in conjunction with a cow-calf operation or it precedes the finishing program.

When the stocker enterprise is the only cattle enterprise on a farm or ranch, it is usually conducted using one of the following plans:

1. Calves or light yearlings are bought in the fall to be wintered on high-roughage rations in dry lot and sold in the spring to buyers either (a) to go on grass for the summer, or (b) to go on a drylot finishing program.

2. Lightweight calves are bought in the fall to be wintered on roughage rations, then, under the same ownership, grazed throughout the following pasture season and sold in the fall. Under this plan, usually lighter weight calves are acquired and they are wintered at a lower rate of gain than in plan 1.

3. In Kansas, Oklahoma, and Texas, calves or light yearlings are bought in the fall and grazed on small winter grains, chiefly wheat. Good wheat pastures will produce very acceptable stocker gains. The main disadvantage to the program is that, due to weather conditions, winter wheat pasture cannot always be counted upon. When it fails, the stockers must either be sold or fed a higher cost roughage.

4. In southeastern United States, which is primarily a cow-calf area, winter oats and fescue are used extensively in stocker programs. This area is turning to stocker programs in order to utilize winter pastures profitably, and to satisfy the demand for 600- to 800-lb feeder steers as a result of the expansion of feedlots.

Fig. 19-43. Stocker calves being wintered on a high-roughage ration. (Courtesy, *Livestock Breeder Journal,* Macon, Ga.)

There is a trend for more and more calves (not yearlings) to be handled according to plan 1—that is, bought in the fall, wintered on roughage, and sold directly into a finishing program. This trend will be accelerated because of heavier calves being weaned in the fall, and because it is more profitable either to use presently available pasture areas (1) for brood cows to produce more calves, and (2) for crop production.

The most common type of operation is a combination stocker-feeder program, typical of the Corn Belt and the irrigated sections of the West, where high-yielding corn and sorghum crops are produced for silage. In these areas, cattle feeders usually purchase steer calves or light yearlings in the fall or late winter; fall-graze stalk fields and small grain stubble where available; move into the drylot for the winter and feed corn or sorghum silage, supplemented with a legume hay or protein supplement; then finish on a high-energy ration either in the drylot or on pasture and sell for slaughter in the summer or fall.

An increasing number of feeders are grown under contract for, and delivered to, a feedlot for finishing. This trend has been prompted by the competition between feedlots. It is their way of assuring a continuous supply of feeders of the desired weights and quality. As a further inducement, many of the feedlots finance the grower (backgrounding) operation.

RATIONS FOR STOCKERS

For a stocker operation to be profitable, the grower must be ever aware of the following reasons back of it and feed stockers accordingly: (1) to provide a supply of the kind of cattle desired by finishing lots at the time needed; (2) to utilize roughages and other low-cost feeds, and (3) to "cheapen" the cattle.

Because of the very nature of the operation, the successful feeding of stockers requires the maximum of economy consistent with normal growth and development. This necessitates cheap feed—either pasture or range grazing or such cheap harvested roughage as hay, straw, fodder, and silage. In general, the winter feeds for stockers consist of the less desirable and less marketable roughages. It is important, therefore, that the high-roughage rations of young stockers be properly supplemented from the standpoints of proteins, minerals, and vitamins.

The feed consumption of stockers will vary somewhat with the quality of the roughage available, the age of the cattle, and the rate of gain desired. As far as practical, the stocker ration should prepare the cattle for making maximum use of the feed which follows—either the finishing ration or grass. The rate and efficiency of gain in the feedlot or on pasture varies inversely with the amount of gain made during the stocker stage—that is, the smaller the stocker gains, the higher the finishing gains; and the higher the stocker gains, the lower the finishing gains. This phenomenon is known as *compensatory gain.*

Of course, insufficient weight gains may be unprofitable to the grower. Besides, young animals can be stunted. To make maximum growth without fattening—just to maintain condition—calves of the British breeds and crossbreds should gain 1.25 lb daily, and yearlings should gain 0.9 lb daily.

The amount of gain desired in a grower program depends largely on the way the cattle are to be handled in the next stage. For example, stockers that are to be grain fed on grass or go directly to the finishing lot can be wintered more liberally, and make a higher rate of gain, than cattle that are to be turned to pasture only. Also, winter and summer gains are not exactly inversely proportional. For example, if one lot of steers gains twice as much during the winter as a second lot, its summer gains won't be limited to half those of the second lot. Rather, they will likely be 70 to 90% as much. Thus, when the stocker grower retains ownership of the cattle through the finishing stage, it may well be that the cattle that make the largest stocker gains also make the most economical and largest total gain for the entire period—from weaning to slaughter.

Tables 19-27 and 19-28 (see next page) contain some recommended rations for stocker cattle. Variations can and should be made in the rations used. The grower should give consideration to (1) the supply of homegrown feeds, (2) the availability and price of purchased feeds, (3) the class and age of cattle, (4) the health and condition of animals, and (5) the kind of feeder cattle in demand by feedlots.

In using Tables 19-28 and 19-28 as guides, it should be recognized that feeds of similar nutritive properties may be interchanged as price relationships warrant. Thus, (1) the cereal grains may consist of corn, barley, wheat, oats, and/or sorghum; (2) the protein supplement may consist of soybean, cottonseed, peanut, linseed, safflower, and/or sunflower meal; (3) the roughage may include many varieties of hays and silages; and (4) a vast array of by-product feeds may be utilized.

The following points are pertinent to the success of a stocker operation and should be kept in mind:

• **Recommended nutrient allowances**—When grower rations are formulated on the basis of percentage of nutrients in the ration, the following allowances are recommended:

Protein:	
For up to 1.5 lb daily gain	10.5%
For 1.5 lb daily gain or more	11.0%
Calcium and Phosphorus:	
For up to 500-lb liveweight	0.3-0.5%
For over 500-lb liveweight	0.25%
Vitamin A:	
Air-dry feed (10% moisture)	1,200 to 1,500 IU per lb
	15,000 IU daily per head
Implant:	
Gains of more than 1.5 lb per head daily	Include growth stimulant

• **Energy**—Satisfactory and efficient winter gains of weanling calves depend upon sufficient energy, along with a proper balance of protein and energy in the ration.

There is much information that indicates stocker cattle should not be winter fed over 1 lb of grain per 100 lb of body weight, daily, if they are to make best use of pasture the following summer. Also, delayed grain feeding on pasture until after peak pasture growth is recommended.

• **Protein**—Calves have a higher protein requirement per 100 lb of liveweight than older cattle and are more apt to be deficient on low-quality roughage. Extra energy will not be efficiently used unless protein intake is adequate.

Calves wintered on range will require .5 to .7 lb of crude protein from supplements to gain .5 lb daily. Calves wintered on good grass hay or meadow hay will require less supplemental protein and phosphorus than when wintered on range.

One pound of 41% supplement will supply about half the total protein requirement of a 500-lb calf. Three pounds of alfalfa hay or 2½ lb of dehydrated alfalfa will furnish about the same amount of protein but more energy than 1 lb of a 41% protein concentrate.

Urea is not well utilized as a protein supplement when fed to cattle on high-roughage rations. Because of this, usually it is best to use a plant protein supplement or a slow-release urea product in growing rations.

Work at Missouri and other agricultural experiment stations indicates no advantage to feeding protein supplements on pasture which contains legumes or grasses in lush condition.

• **Minerals and vitamins**—All rations should provide adequate calcium, phosphorus, and carotene or vitamin A.

Most stockers are bought in the fall to be wintered on low-quality roughage or on winter wheat or other cool-season pasture. Some are bought in the spring to be placed on bluestem, Bermudagrass, or other native pastures. Such calves usually do not have a high store of vitamin A; hence, an intraruminal or intramuscular injection of 500,000 to 1,000,000 IU/head of vitamin A upon their arrival at the place where they will go on the stocker program could be helpful, especially if a nitrate problem is anticipated on wheat pastures.

TABLE
DAILY RATION FOR STOCKER
(AS-FED

	Rations (fed for gains of							
	1		2		3		4	
	(lb)	(kg)	(lb)	(kg)	(lb)	(kg)	(lb)	(kg)
Legume hay or grass-legume mixed hay	12-18	5.4-8.2	—	—	8-12	3.6-5.4	—	—
Grass hay	—	—	12-18	5.4-8.2	4-6	1.8-2.7	—	—
Straw, corncobs, cornstalks, stalklage, cottonseed hulls	—	—	—	—	—	—	—	—
Corn or sorghum silage	—	—	—	—	—	—	25-40	11.4-18.1
Legume-grass silage, or oat silage	—	—	—	—	—	—	—	—
Legume-grass haylage, or oat haylage	—	—	—	—	—	—	—	—
Grain (corn, sorghum, barley, or oats)	—	—	—	—	—	—	—	—
Protein supplement (41% or equivalent)	—	—	1¼-1½	0.6-0.7	¼-1	0.1-0.5	1-1¼	0.5-0.6

[1]With all rations, provide suitable minerals (see Tables 19-5 and 19-7).

TABLE
DAILY RATION FOR STOCKER YEAR-
(AS-FED

	Rations (fed for gains of							
	1		2		3		4	
	(lb)	(kg)	(lb)	(kg)	(lb)	(kg)	(lb)	(kg)
Legume hay or grass-legume mixed hay	16-24	7.3-10.9	—	—	6-8	2.7-3.6	—	—
Grass hay	—	—	16-24	7.3-10.9	10-16	4.5-7.3	—	—
Straw, corncobs, cornstalks, stalklage, cottonseed hulls	—	—	—	—	—	—	—	—
Corn or sorghum silage	—	—	—	—	—	—	45-55	20.4-25.0
Legume-grass silage, or oat silage	—	—	—	—	—	—	—	—
Legume-grass haylage, or oat haylage	—	—	—	—	—	—	—	—
Grain (corn, sorghum, barley, or oats)	—	—	—	—	—	—	—	—
Protein supplement (41% or equivalent)	—	—	1½-1¾	0.7-0.8	1-1½	0.5-0.7	1¼-1½	0.6-0.7

[1]With all rations, provide suitable minerals (see Tables 19-5 and 19-7).

• **Roughage**—Feeding good-quality roughage in a drylot may be more desirable for weanling calves than wintering on the range. Winter range is most efficiently used by yearling steers and mature cattle. Meadow hay, a mixture of alfalfa and grass hay, good-quality upland hay, or silage provide a good basal ration for wintering calves.

Hay need not be processed. But it should be fed so that there will be a minimum of waste.

• **Corn or sorghum silage** does not need to be supplemented with additional grain or dry roughage unless it was put up too wet or contains very little grain as the result of drought. Sorghum silage or grass-legume silage that has more than 65 to 70% moisture will require more grain supplementation than drier silage of comparable quality.

• **Grass-legume silage** cut at the proper stage of maturity and carefully ensiled is an excellent feed for stocker cattle. Such silage contains adequate protein and need not be supplemented with a protein concentrate. However, it is much lower in energy than corn or sorghum silage; hence, the gains will be smaller unless (1) 150 to 200 lb of grain per ton are added as a preservative, or (2) it is supplemented at feeding time with some grain or other energy source.

• **Grass-legume hay** of good quality is excellent for stocker cattle. It may or may not be supplemented with grain.

• **Haylage,** made from grass-legume or straight legume forage, wilted to about 50% moisture content, and commonly stored in an oxygen-free silo, is increasing as a stocker feed. Except for the difference in moisture content, haylage has a feeding value comparable to silage or hay made from a similar crop.

• **Grass hay,** such as prairie hay, Bermudagrass, timothy, Sudangrass, Johnsongrass, etc., may make up most of the ration of stocker cattle. Energy supplementation is needed if improvement in the condition of stockers is desired. Also, such hays must be properly supplemented with protein, minerals, and vitamins.

• **Stubble and stalk fields** furnish much feed for stocker cattle, especially yearlings, in the late fall and early winter. Unless there is access to a good winter pasture, cattle grazing stalk fields should be fed 4 to 6 lb of legume hay or 1½ lb of protein concentrate daily. In addition, minerals should be provided.

19–27
CALVES (400–500 LB [*181–227 KG*])[1]
BASIS)

1.25 lb [*0.6 kg*]/head/day)											
5		**6**		**7**		**8**		**9**		**10**	
(lb)	**(*kg*)**	**(lb)**	**(*kg*)**	**(lb)**	**(*kg*)**	**(lb)**	**(*kg*)**	**(lb)**	**(*kg*)**	**(lb)**	**(*kg*)**
2–4	*0.9–1.8*	—	—	8–10	*3.6–4.5*	—	—	—	—	—	—
—	—	2–4	*0.9–1.8*	—	—	—	—	10–12	*4.5–5.4*	—	—
—	—	—	—	2–4	*0.9–1.8*	2–3	*0.9–1.4*	—	—	2	*0.9*
20–30	*9.1–13.6*	20–30	*9.1–13.6*	—	—	—	—	—	—	—	—
—	—	—	—	—	—	20–25	*9.1–11.4*	—	—	—	—
—	—	—	—	—	—	—	—	—	—	20–25	*9.1–11.4*
—	—	—	—	4–5	*1.8–2.3*	—	—	4–5	*1.8–2.3*	4–5	*1.8–2.3*
¾–1	*0.3–0.5*	1¼–1½	*0.6–0.7*	—	—	1–1½	*0.5–0.7*	1–1½	*0.5–0.7*	—	—

19–28
LINGS (600–700 LB [*273–318 KG*])[1]
BASIS)

0.9 lb [*0.4 kg*]/head/day)											
5		**6**		**7**		**8**		**9**		**10**	
(lb)	**(*kg*)**	**(lb)**	**(*kg*)**	**(lb)**	**(*kg*)**	**(lb)**	**(*kg*)**	**(lb)**	**(*kg*)**	**(lb)**	**(*kg*)**
2–4	*0.9–1.8*	—	—	6–8	*2.7–3.6*	—	—	—	—	—	—
—	—	2–4	*0.9–1.8*	—	—	—	—	16–20	*7.3–9.1*	—	—
—	—	—	—	12–15	*5.4–6.8*	10–12	*4.5–5.4*	—	—	2	*0.9*
40–50	*18.2–22.7*	40–50	*18.2–22.7*	—	—	—	—	—	—	—	—
—	—	—	—	—	—	20	*9.1*	—	—	—	—
—	—	—	—	—	—	—	—	—	—	35–40	*15.9–18.2*
—	—	—	—	5–6	*2.3–2.7*	—	—	5–6	*2.3–2.7*	5–6	*2.3–2.7*
¾–1	*0.3–0.5*	1¼–1½	*0.6–0.7*	—	—	1	*0.5*	1–1½	*0.5–0.7*	—	—

Occasionally, roughages are high in nitrates. Actually, calves can consume very high levels of nitrate provided (1) they are fed with regularity, (2) changes in feed are gradual, (3) feeds are mixed uniformly, and (4) ample water is provided at all times. When nitrate is a problem, cattle may die of oxygen insufficiency. Others in the lot will show discoloration of nonpigmented areas of the epithelium and chocolate-colored blood.

• **Turnips for grazing**—A turnip crop yielding 30 tons per acre will carry 3.5 head of stockers, averaging 700 lb weight and gaining 1.5 to 2.0 lb per day, for 90 days.

• **Grain**—Calves are unable to consume enough dry roughage to gain more than a pound a day. Thus, grain should be added in the quantity necessary to achieve the desired gains. Bear in mind that with calves of the British breeds and crossbreds it takes a gain of about 1.25 lb daily to maintain condition; yearlings of the British breeds or crossbreds will maintain condition on a gain of about 0.9 lb daily.

Some grain should be included in the ration of stocker cattle when (1) they are to be finished immediately after the wintering period, (2) they weigh less than 350 lb when started on winter feeding, and (3) heifers are to be bred when they are 13 to 15 months old.

Calves that are full-fed corn or sorghum silage high in grain content, plus 1 lb of protein concentrate or 4 to 5 lb of legume hay, need not be fed grain.

• **Pasture supplement**—Following wintering, many stocker cattle graze throughout the pasture season. Supplementing grass with a high protein feed at the rate of l lb/head/day when summer pastures drop off in quantity and protein content (generally beginning about mid-July) will usually boost average daily gain by 0.4 lb. For convenience, pasture supplements are usually fed in cube form. Usually they run 38 to 41% protein and consist primarily of cottonseed or soybean meal.

• **Growth stimulants or implants**—An approved growth stimulant or implant may increase growth in stocker cattle, on either winter rations or summer pastures. In cattle that are gaining 0.75 to 2.00 lb/head/day, a growth stimulant or implant will boost average daily gains by an additional 0.2 lb. (See Table 19–11 for FDA approved stimulants and implants.)

LEVEL OF WINTERING

The level of wintering stockers affects the gains in the next stage. Thus, calves gaining the most during the winter make the least gains on pasture the following summer. This is clearly shown in Table 19-29.

TABLE 19-29
CALVES TO YEARLINGS—EFFECT OF WINTER GAINS ON SUBSEQUENT GAINS THE FOLLOWING SUMMER AS YEARLINGS[1]

Lot No.	Winter		Summer		Total	
	(lb)	*(kg)*	(lb)	*(kg)*	(lb)	*(kg)*
Valentine, Nebraska:						
1	115	*52*	205	*93*	320	*145*
2	120	*55*	181	*82*	301	*137*
3	127	*58*	183	*83*	310	*141*
4	129	*59*	183	*83*	312	*142*
5	150	*68*	170	*77*	321	*146*
6	157	*71*	143	*65*	300	*136*
7	164	*75*	170	*77*	304	*138*
8	179	*81*	160	*73*	339	*154*
9	184	*84*	150	*68*	324	*147*
10	186	*85*	152	*69*	338	*154*
11	186	*85*	162	*74*	348	*158*
Fort Robinson, Nebraska:						
1	67	*30*	226	*103*	309	*140*
2	83	*38*	218	*99*	308	*140*
3	87	*40*	231	*105*	298	*135*
4	90	*41*	221	*100*	308	*140*
5	104	*47*	202	*92*	306	*139*
6	136	*62*	183	*83*	319	*145*

[1]*Beef-Forage Notebook,* Cattle Management, 5-C-1, University of Nebraska Extension Service.

Calves wintered to gain 1.0 lb daily make satisfactory summer pasture gains. This level is recommended for calves that are to graze season-long the following summer, provided the same ownership is retained all the way through. Somewhere in the range of 1.0 to 2.0 lb daily gain during the winter is usually desirable if calves (1) are to be sold in the spring, (2) will be on full feed 2 to 3 months after going to grass, (3) will be receiving a limited feed of grain on grass, or (4) are replacement heifers that are to be bred at 13 to 15 months of age.

Since yearlings are not growing as rapidly as calves, they may be fed for smaller gains than calves, and yet show comparable condition. Thus, for maximum growth without fattening (for just holding their condition) calves should gain approximately 1.25 lb daily, whereas yearlings need to gain only about 0.9 lb daily.

Compensatory Growth

Compensatory growth is increased growth rate in one time period as a result of growth restriction imposed during an earlier time period.

It is common practice for stocker cattle to be *roughed through* the winter as cheaply as possible, with limited daily gains. Then, in the spring, the animals are turned to lush pasture or put in a feedlot and fed a high-energy ration. Animals so managed exhibit the phenomenon of *compensatory growth;* that is, on the high-energy ration they gain faster and more efficiently than similar cattle which were fed more liberally during the wintering period. Feedlot operators were quick to sense this situation, and to take advantage of it. This is the chief reason for the popularity of Okie-type cattle. Usually, they are animals whose growth has been held back to less than their genetic potential. When fed more liberally, they exhibit a surge in growth rate and feed efficiency. Large compensatory growth usually indicates that someone (the stocker operator) has lost money while someone else (the feeder) has made money. It is noteworthy that Holsteins and the larger exotics should never be handled so as to exhibit compensatory gains. If they're held back in the winter, they're too heavy when they finish.

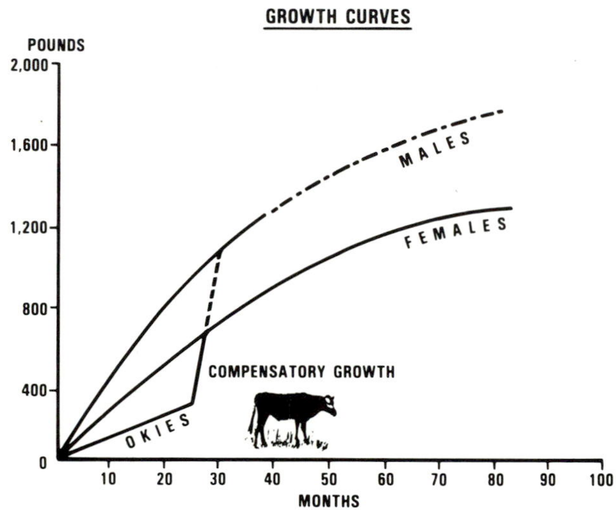

Fig. 19-44. The two curved lines show the normal growth of beef cattle under proper environment. Note that females grow most rapidly from birth to 10 months and males from birth to 15 months. Okie-type cattle grow more slowly during the first part of their lives when their environment is poor and they are under stress. When put on feedlot rations, their growth curve is steep. This rapid, economical gain is called *compensatory growth.*

WINTER PASTURES

Wherever possible, stocker calf operations are planned around a winter pasture program. Weanling calves or lightweight, thin yearlings are purchased in the fall. In some cases, homegrown calves are retained and developed under this system for sale as yearlings. As would be expected, winter pasturing of stockers is largely limited to the southern part of the United States, with the kind of pasture varying from area to area.

- **Winter wheat pastures**—Winter wheat pastures are widely used for stocker cattle in Kansas, Oklahoma, and Texas. When such pastures are good, cattle make very acceptable gains on them. However, wet weather or droughts make winter wheat pastures unreliable, with the result that it is important that there be flexibility in the stocker program, both in numbers and season of use.

• **Other cool-season pastures**—In the southern states, extensive use is made of oats, rye, ryegrass, vetch, and fescue—a perennial grass that remains green throughout the winter. This area is turning more and more to winter grazing, as a means of making profitable year-round use of their land and labor and providing 600- to 800-lb feeder cattle in greatest demand by feedlots.

Grass Tetany

Grass tetany (grass staggers or wheat pasture poisoning), a highly fatal nutritional ailment, is one of the hazards of winter pastures. Although it is more common among lactating cows, stocker cattle are affected. It generally occurs during the first 2 weeks of the pasture season, particularly in cattle grazing wheat or other cereal crops. The characteristic symptoms are nervousness, flickering of the third eyelid, twitching of the muscles (usually of the head and neck), head held high, accelerated respiration, high temperature, gnashing of the teeth, and abundant salivation. A slight stimulus may precipitate a crash to the ground and, finally, death. The condition follows a rapid course, with usually a lapse of only 2 to 6 hours between onset and death.

The subject of grass tetany is fully covered in Chapter 5, Table 5-1, Nutritional Diseases and Ailments—Grass Tetany; hence, the reader is referred thereto.

FACTS PERTINENT TO STOCKER PROGRAMS

The following points are pertinent to stocker programs:

• **Stocker programs are used by large cattle feedlots to ensure a continuous supply of feeder replacements**—Many large feedlots, which feed on a year-round basis, are effectively using stocker programs to ensure a continuous supply of feeder replacements. They usually accomplish this by buying both calves and yearlings when they are available at favorable prices, then putting them on different stocker programs, often on a contract basis, designed to stagger their readiness to move to the feedlot at the weight and time desired.

• **Stocker stage will be either lengthened or shortened, depending on the circumstances**—High-priced grains and cheap roughages, comparatively speaking, favor older and heavier stockers. So yearling stockers will increase where and when pastures and dry roughages are relatively cheap. However, the following factors favor younger and lighter stockers—calves: (1) heavier weaning weight; (2) greater efficiency of feed utilization by younger cattle; (3) the need to place larger strains of cattle (the exotics) on finishing rations at younger ages—otherwise, their carcasses become too heavy; (4) the demand for leaner beef; (5) higher remuneration from cow-calf operations; and (6) high land, labor, and interest costs mitigating against the small gains common to stocker operations.

• **Diseases are a problem**—Diseases can take a tremendous toll in a stocker program. The weanling calf is at the most susceptible stage of life to contagious diseases. This, coupled with the limited weight gains in growing operations, leaves little opportunity to recover severe losses. Plans should be made for routine immunization, starting rations, and veterinary treatment. Also, the operator will be well paid for providing TLC (tender loving care) and spotting sick animals early.

Cattle producers recognize that losses due to diseases and accidents are higher in stocker calves than in yearlings. Studies show that death losses of calves usually run about 5.0%; yearlings, about 1.0%; and 2-year-old and older cattle, about 0.5%. Most of these losses are due to stress and to viruses to which the animals are subjected between the farm or ranch and the final destination of the stockers. Shipping fever takes the heaviest toll.

• **Lightweight calves gain more efficiently**—The feed requirement for calf gain is directly related to the animal's weight. Because of low maintenance requirements, lightweight calves gain more efficiently than heavyweight calves. This is illustrated in Fig. 19-45, which shows the amount of TDN required for each 100-lb gain when calves are gaining at the rate of 1½ lb per day. All things being equal (health, genetics, management, environment), calves will gain 100 lb on less feed at the lighter weights, which is an advantage for which most stocker buyers are looking.

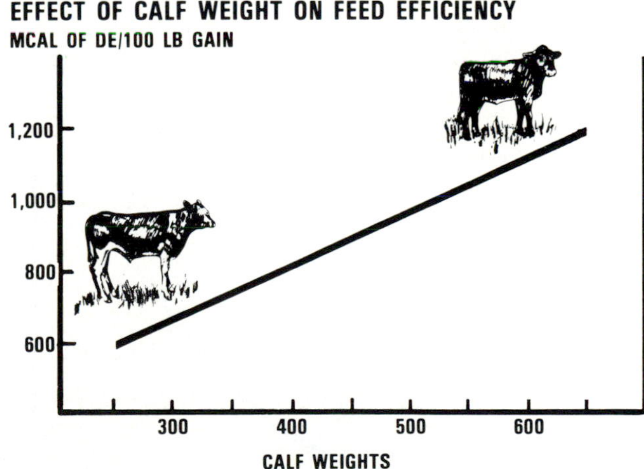

Fig. 19-45. Feed required per 100-lb gain for calves at different weights gaining 1½ lb per day. (Source: *Keys to Profitable Stocker Calf Operations,* MP-964, Texas A&M Univ. Agr. Ext. Service)

• **Lightweight calves require higher quality feed**—Although lightweight calves gain more efficiently than heavier calves, they require better quality feed for the same gain. For this reason, forage (pasture, hay, or silage) which will merely furnish a 300-lb calf sufficient feed to maintain body weight will allow a mature animal to gain weight. Fig. 19-46 illustrates this point. Note that the 250-lb calf gaining 1½ lb per day has a higher net energy requirement relative to its size than does the 600-lb calf. This emphasizes that stocker operators should buy calves to match their feed.

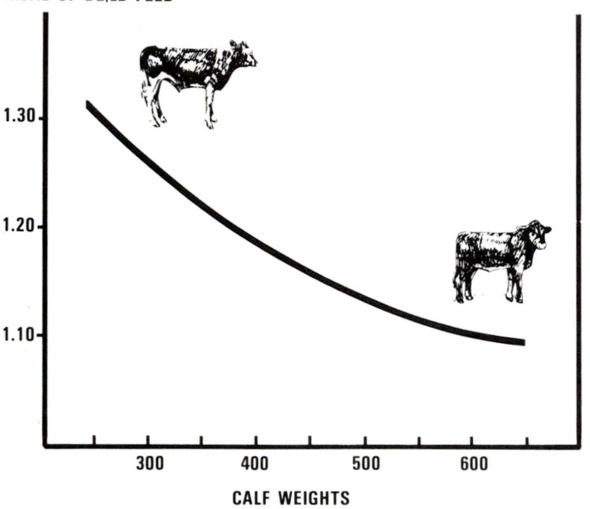

Fig. 19-46. Quality of feed required for calves of different weights to gain 1½ lb per day. (Source: *Keys to Profitable Stocker Calf Operations,* MP-964, Texas A&M Univ. Agr. Ext. Service)

• **Faster gains are cheaper**—Faster gains are cheaper because the maintenance requirement is the same regardless of the daily gain; hence, the maintenance requirement is a smaller part of the total feed consumed as gains increase. Fig. 19-47 illustrates this point. A calf gaining 2.0 lb per day requires approximately 800 Mcal of digestible energy per 100 lb gain, whereas the energy requirements per 100 lb of gain for a calf gaining 0.6 lb per day are doubled—to 1,600 Mcal. With interest rates, taxes, labor costs, and other costs rising as they have, growing calves at rates of less than 1 lb per head daily becomes increasingly difficult to justify.

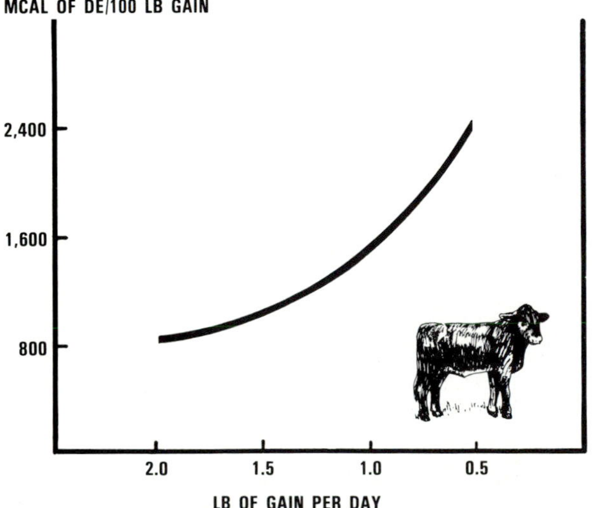

Fig. 19-47. Feed required per 100-lb gain for a 450-lb calf gaining at different rates per day. (Source: *Keys to Profitable Stocker Calf Operations,* MP-964, Texas A&M Univ. Agr. Ext. Service)

• **Compensatory growth**—Compensatory growth is making up for a bad start in life (see earlier section devoted to this subject).

• **Avoid excess feeder condition**—If cattle get too fleshy as feeders, (1) they may reach market finish before they attain desirable market weight, and (2) they will tend to gain more slowly than desirable during the feedlot finishing period. The planned rate of gain should be determined by feeder finish, growth potential, sex, and the beginning weight of the feeder calves. It will also depend upon how much condition is acceptable to the buyer. British breed or crossbred steer calves gaining at the rate of 1.25 lb daily will about maintain their condition; at 1.5 lb daily they will add some condition; at 2.0 lb daily they may be too fleshy by the time they reach 700- to 750-lb feeder weight. British breed or crossbred yearlings making daily gains of 0.9 lb will maintain growth without fattening. Larger cattle of some of the exotic breeds may make larger gains without fattening.

• **Buy and sell carefully**—Because the original weight purchased in a stocker program is a high percentage of the weight sold by the grower, any mistake made in buying or selling has a greater impact on profits or losses than it would in a finishing program. If you pay too much, buy with too much fill, pay for quality that the feeder is not willing to pay for, or sell too low, profits will be eroded away rapidly.

• **Steers are heavier**—At weaning, steer calves normally weigh about 5% more than heifer calves.

STOCKER AND GROWER CONTRACTS

Hand in hand with the development of big feedlots and year-round feeding came the need for an assured supply of feeder cattle of the desired kind on a continuous basis. To meet this need, more and more feedlots have turned to contractual arrangements with stocker growers, with numerous kinds of contracts. Usually, the cattle are owned by the feedlot, most of which are large and in a stronger financial position than the majority of stocker growers. The two most common kinds of contracts are based on either (1) a fixed cost for the gain, or (2) an agreed feed cost plus an extra charge for labor and lot rental. Usually, there is provision of adjusting for death loss. Such contracts should always be in writing, with all provisions, including weighing conditions, spelled out.

Although the use of stocker and grower contracts has increased in recent years, the concept is not new. Many Kansas bluestem pasture owners have long grown out yearlings owned by Iowa and other Corn Belt feeders.

Today, many corn farmers in the fertile irrigated area around Greeley, Colorado, make corn silage and feed stocker cattle on a contract basis, with the stockers owned by one of several large feedlots in the vicinity. Stocker cattle are also being grown under contract on the wheat pastures of Kansas, Oklahoma, and Texas; on hay and other roughages in the irrigated valleys of the West; and on sorghum silage and stalk fields throughout the Southwest.

(Also see Chapter 7, section on "Pastures for Cattle, Sheep, and Horses," • Custom cattle grazing [pasture leasing].)

PART IV—Feeding Finishing (Fattening) Cattle

The finishing of cattle is what the name implies, the laying on of fat. Additionally, there is an increase in the total muscle (red meat) mass. The ultimate aim of the finishing process is to produce beef that will best answer the requirements and desires of the consumer. This is accomplished through an improvement in the flavor, tenderness, and quality of the lean beef which results from marbling (intramuscular fat).

Fig. 19-48. Open pen cattle feedlot, with shades—the cheapest and most common type of feedlot. This is the Alta Verde Industries feedlot near Eagle Pass, Texas. (Courtesy, *West Texas Livestock Weekly,* San Angelo, Tex.)

Cattle feeders are commonly classed as either commercial feeders, or farmer-feeders, based largely on numbers. From the standpoint of statistical reporting, the U.S. Department of Agriculture commonly draws the line at 1,000 head. A commercial cattle-feeding operation is defined as one having a capacity of 1,000 head or more, at any one time.

Traditional farmer-feeders evolved with Corn Belt farming, in the north central region of the United States. Generally speaking, they market their crop, usually corn, through cattle (or hogs, or lambs), and spread the manure on the land. The purchase of feeder cattle for these enterprises is generally in the fall, with the actual feeding done during the winter months when labor is available due to limited field work. This type of operation has persisted to the present time, although it has been modernized through the years.

In addition to being larger, commercial cattle feeders generally differ from farmer-feeders in the following respects: (1) They usually feed cattle on a year-round basis, rather than during the winter months only; (2) they may grow little, or none, of their feed; (3) they are highly mechanized; (4) they are knowledgeable of costs and returns, skillful buyers and sellers, and aware of market trends; and (5) they usually do some custom feeding. Today, commercial feedlots with more than 1,000-head capacity dominate cattle feeding.

There are 2 methods of finishing cattle for market: (1) cattle feedlots, including confinement (sheltered) finishing; and (2) pasture finishing. Prior to 1900, the majority of fat cattle sent to market were 4- to 6-year-old steers that had been finished primarily on grass. Even today, the utilization of pastures continues to play an important role in all types of cattle-finishing operations; and grass finishing will increase in the future.

Fig. 19-49. Cattle being finished in a modern feedlot. (Courtesy, American Polled Hereford Assn., Kansas City, Mo.)

The principles of beef cattle nutrition are covered in Part I of this chapter; hence, they will not be repeated at this point. Instead, the application of nutrition to cattle finishing (fattening) will be discussed.

The major nutritional requirements of finishing cattle are energy, protein, minerals, vitamins, and water. The greatest need is for energy. Of course, net profit depends on how much of that energy can be converted to pounds of gain—and how efficiently.

About 75% of the cost of finishing cattle, exclusive of the purchase price of the feeders, is feedstuffs—grain, hay, silage, and miscellaneous wastes and by-products.

CATTLE FEEDING IN TRANSITION

Since the late 1940s, cattle finishing has been characterized by many changes; among them—

1. A dramatic increase in grain-fed cattle. In 1947, only 6.9 million head of market cattle were grain fed, representing 30.1% of the slaughter cattle that year. (See Fig. 19-50.)
2. Few nonfed steers and heifers being marketed. (See Fig. 19-51.)
3. A trend to larger, but fewer, feedlots. In the 1980s, feedlots with over 1,000-head capacity marketed 80% of the nation's fed cattle; and feedlots exceeding 100,000 capacity were in operation in Arizona, California, Colorado, and the northern Texas Panhandle. (See Fig. 19-52.)
4. Shifts in the geography of cattle feeding, with Texas taking the lead, Nebraska and Kansas showing big increases, and Iowa and California showing big decreases.

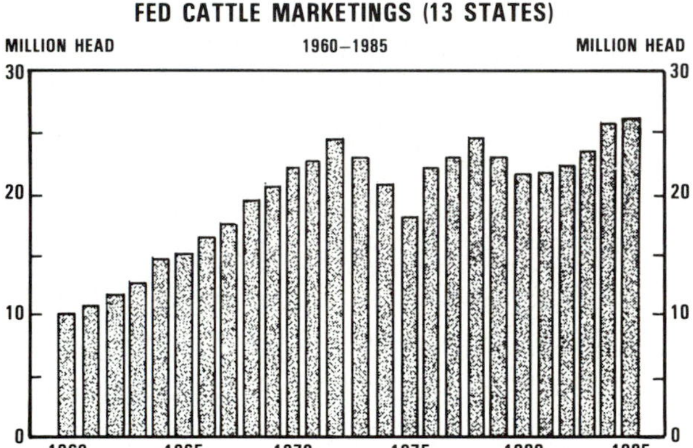

Fig. 19-50. Growth of cattle feeding in 1985, 26.1 million head of fed cattle were marketed, representing 68.5% of all the cattle slaughtered that year. (USDA sources).

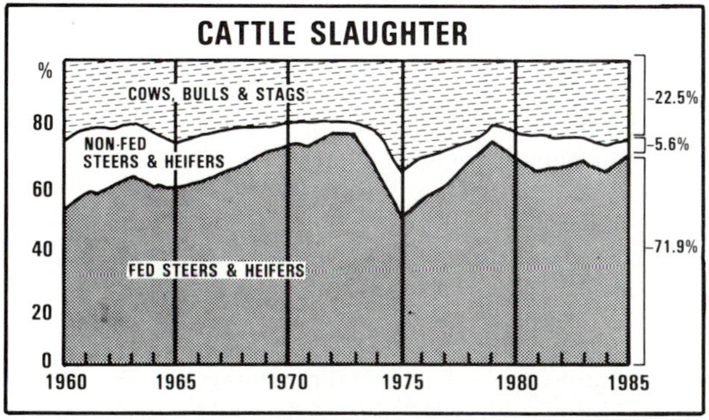

Fig. 19-51. Composition of cattle slaughter during the 25-year period, 1960-1985. In 1985, 36,288,000 head of cattle were slaughtered in the U.S., of which 71.9% were fed steers and heifers, 5.6% were nonfed steers and heifers, and 22.5% were cows, bulls, and stags. The composition of cattle slaughter varies in keeping with grain prices. (From USDA sources)

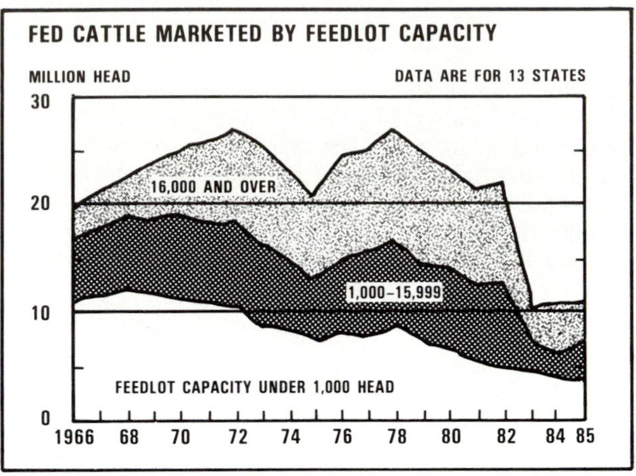

Fig. 19-52. Feedlots have increased in size. In 1985, feedlots with over 1,000 head marketed 82% of the nation's fed cattle. (Source: 1966 to 1982 data from *1983 Handbook of Agriculture Charts,* USDA, Agricultural Handbook No. 619, p. 69, chart 192; 1983 to 1985 data provided with a personal communication to the senior author from the USDA)

But bigger cattle feeding transitions are in progress! In the future, we shall see—

1. The economic viability of cattle feeding influenced by consumer demand for beef.
2. Shorter finishing periods, less grain being fed, and leaner beef. Cattle will be bred and fed for the best combination of muscling, marbling, and external finish.
3. The merger of more cattle feedlots, with fewer owners controlling more feeding.
4. More custom feeding accompanied by a trend away from feedlot ownership of cattle.
5. More integration, with cow-calf producers retaining ownership through more channels.
6. More cattle feedlot operators slaughtering their cattle and marketing their own branded beef.

KINDS OF CATTLE TO FEED[6]

All kinds of cattle may be, and are, fed. But, for maximum success, it is imperative that the right kind of cattle be selected for a particular feedlot. The cattle should match the operator's available feed, labor, shelter, and credit. Also, it is imperative that there be a suitable market outlet following finishing; for example, it would be unwise to feed lightweight heifers in an area where the strongest slaughter market is for heavy steers; nor should one finish heavy Holstein steers where the primary interest of packers is for Choice grade beef. But, assuming that a satisfactory slaughter outlet exists for different kinds of cattle, the general guides that follow will be helpful in determining what kind of cattle to feed in a given lot.

[6]Tables 19-30 to 19-34 are a consensus (or judgment) of several knowledgeable feedlot consultants and university animal scientists, based on a survey made by the authors. A consensus was resorted to for the reasons that (1) no extensive, nationwide, scientific study of this sort has ever been made, and (2) this information is much needed by cattle feeders and those who counsel with them. No claim is made relative to the scientific accuracy of Tables 19-30 to 19-34; rather, they are presented (1) because they are the best, if not the only, information of the kind presently available, and (2) with the hope that they will stimulate needed research along these lines.

Fig. 19-53. All kinds of cattle are fed. This shows some Mexican-type steers in a U.S. Feedlot. (Courtesy, Merck & Co., Inc., Rahway, N.J.)

Age and Weight of Cattle

In the early 1900s the term *feeder steer* signified to both the rancher and the Corn Belt feeder a 2½- to 3-year-old animal weighing approximately 1,000 lb. Today, cattle are referred to by ages as calves, yearlings, and 2-year-olds. This shift to younger cattle has been brought about primarily by consumer demand for smaller and lighter cuts of meat and improved feeding and management practices.

The age of cattle to feed is one of the most important questions to be decided upon by every practical cattle producer. The following factors should be considered in reaching an intelligent decision on this point.

• **Rate of gain**—When cattle are fed liberally from the time they are calves, the daily gains will reach their maximum the first year and decline with each succeeding year thereafter. On the other hand, when in comparable condition, thin but healthy 2-year-old steers will make more rapid gains in the feedlot than yearlings; likewise, yearlings will make more rapid gains than calves. Table 19-30 illustrates this situation.

TABLE 19-30
EFFECT OF AGE ON DAILY GAINS OF MEDIUM-FRAME CATTLE

Age	Daily Gains: Average of U.S. Feedlots		Daily Gains: Top 5% of U.S. Feedlots	
	(lb)	(*kg*)	(lb)	(*kg*)
Calves	2.4	*1.09*	2.7	*1.23*
Yearlings	2.8	*1..27*	3.2	*1.45*
2-year-olds	2.9	*1.32*	3.4	*1.54*

• **Economy of gain**—Calves require less feed to produce 100 lb of beef than do older cattle. This may be explained as follows:

1. The increase in body weight of older cattle is largely due to the deposition of high-energy fat, whereas the increase in body weight of young animals is due mostly to the growth of muscles, bones, and organs. Thus, the body of a calf at birth usually consists of more than 70% water, whereas the body of a fat 2-year-old steer will contain only 45% water. In the latter case, a considerable part of the water has been replaced by fat.

For feedlot efficiency figures to be very meaningful, it is necessary to know what kind of ration was fed; otherwise, on a poundage basis, it is not unlike comparing steaks and carrots in the human diet. The more concentrated and the higher the energy value of the ration, the fewer the pounds of feed required to produce 100 lb of gain. Yet, many times it is in the nature of good business to feed rations that necessitate more pounds of feed to produce 100 lb of gain simply because lower cost gains can be produced thereby. In certain areas, this applies to feeding corn silage, potato waste and other by-product feeds.

2. Calves consume a larger proportion of feed in proportion to their body weight than do older cattle.

3. Calves masticate and digest their feed more thoroughly than older cattle. Despite the fact that calves require less feed per 100 lb gain—because of the high-energy value of fat—older cattle store as much energy in their bodies for each 100 lb of total digestible nutrients consumed as do younger animals.

From the above, it is apparent that age of cattle affects the pounds of feed required to produce 100 lb of gain—that the younger the cattle, the greater the feed efficiency.

Table 19-31 shows the effect of age of cattle on feed efficiency.

TABLE 19-31
EFFECT OF AGE ON FEED EFFICIENCY OF MEDIUM-FRAME STEERS[1]

	Average of U.S. Feedlots			Top 5% of U.S. Feedlots		
	Feed/ Lb Gain	TDN/ Lb Gain	Mcal/ Lb Gain[2]	Feed/ Lb Gain	TDN/ Lb Gain	Mcal/ Lb Gain[2]
	(lb)	(lb)	(lb)	(lb)	(lb)	(lb)
Calves	7.5	6.0	9.86	7.0	5.6	9.20
Yearlings	8.0	6.4	10.52	7.5	6.0	9.86
2-year-olds and over	8.5	6.8	11.17	8.0	6.4	10.52

[1]Air dry basis (approximately 90% dry matter).
[2]Mcal metabolizable energy (ME) was calculated by assuming 1.6434 Mcal ME = 1 lb TDN.

A more accurate measure of feed efficiency than pounds of feed per pound of gain can be obtained through the use of energy conversion–the TDN or Mcal required to produce a pound of beef. It alleviates much of the inevitable disadvantage to which a relatively low-energy, bulky ration (such as a high-silage ration) is put when it is compared on a poundage basis to a more highly concentrated feed (such as an all-concentrate ration). Thus, energy requirements (in both TDN and Mcal per pound of gain) are given in Table 19-31.

• **Flexibility in marketing**—Unless they have been crowded early in life, calves will continue to make satisfactory gains at the end of the ordinary feeding period, whereas the efficiency of feed utilization decreases very sharply when mature steers are held past the time that they are finished. Therefore, under unfavorable market conditions, calves can be successfully held for a reasonable length of time, whereas prolonging the finishing period of older cattle is usually unprofitable.

- **Length of feeding period**—Calves require a somewhat longer feeding period than older cattle to reach comparable finish. To reach Choice condition, steer calves are usually full fed about 7 to 8 months; yearlings, 4 to 5 months; and 2-year-olds only about 3 to 4 months. Table 19–32 points up this situation. The longer finishing period of calves is due to the fact that they (1) make smaller daily gains, and (2) are growing as well as finishing; hence, they require more time to reach market weight and finish.

TABLE 19–32
EFFECT OF AGE ON LENGTH OF FEEDING PERIOD OF MEDIUM-FRAME CATTLE

Age	Average for U.S. Feedlots	Top 5% of U.S. Feedlots
	(days)	(days)
Steer Calves	230	210
Yearlings	140	130
2-year-olds	110	100

- **Total gain required to finish**—Calves must put on more total gain in the feedlot than older animals to attain the same degree of finish. In terms of initial weight, calves practically double their weight in the feedlot. On the average, yearlings increase in weight about 400 lb, and 2-year-olds increase their initial feedlot weight about 320 lb. Table 19–33 illustrates this situation.

TABLE 19–33
EFFECT OF AGE ON TOTAL GAIN REQUIRED TO FINISH MEDIUM-FRAME CATTLE

Age	Average for U.S. Feedlots		Top 5% of U.S. Feedlots	
	(lb)	*(kg)*	(lb)	*(kg)*
Calves	550	*250.00*	525	*238.63*
Yearlings	400	*181.81*	380	*172.72*
2-year-olds	320	*145.45*	300	*136.36*

- **Total feed consumed**—Because of their smaller size, the daily feed consumption of calves is considerably less than for older cattle. However, as calves must be fed for a longer feeding period, the total feed requirement for the entire finishing period is approximately the same for cattle of different ages.
- **Experience of the feeder**—Young cattle must be fed more expertly than older animals. Thus, the inexperienced feeder had best feed older cattle.
- **Kind and quality of feed**—Because calves are growing, it is necessary that they have more protein in the ration. Since protein supplements are higher in price than carbonaceous feeds, the younger the cattle, the more expensive the ration. Also, because of smaller digestive capacity, calves cannot utilize as much coarse roughage, pasture, or cheap by-product feeds as older cattle.

Calves also are more likely to develop peculiar eating habits than older cattle. They may reject coarse, stemmy roughages or moldy or damaged feeds that would be eaten readily by older cattle. Calves also require more elaborate preparation of the ration and attention to other small details designed to increase their appetite.

- **Comparative costs**—Calves generally cost more per 100 lb as feeders than do older cattle.
- **Dressing percentage and quality of beef**—Older cattle have a slightly higher dressing percentage than calves or baby beef. Moreover, many consumers have a decided preference for the greater flavor of beef obtained from older animals.

From the above discussion, it should be obvious that there is no best age of cattle to feed under any and all conditions. Rather, each situation requires individual study and all factors must be weighed and balanced.

Sex of Cattle

More steers than heifers are fed, because more of them are available. A portion of the heifers is held back for replacement purposes. In the future, more young bulls may be fed, because they make more rapid and efficient gains than steers or heifers. Thus, the feedlot operator must give consideration to the sex of cattle fed. First, and foremost, consideration must be given to market outlets.

Table 19–34 shows the effect of sex on rate of gain and feed efficiency. It is noteworthy that bulls gain more rapidly on less feed than steers, and that steers gain more rapidly on less feed than heifers.

TABLE 19–34
EFFECT OF SEX ON GAIN OF MEDIUM-FRAME CATTLE[1]

Sex	Average for U.S. Feedlots		Top 5% of U.S. Feedlots	
	Daily Gain	Feed/Lb Gain[2]	Daily Gain	Feed/Lb Gain[2]
	(lb)	(lb)	(lb)	(lb)
Heifers	2.4	7.5	2.7	6.9
Steers	2.8	6.9	3.2	6.6
Bulls	3.0	6.7	3.4	6.4

[1]To convert lb to kg, divide by 2.2.

[2]Air dry basis (approximately 90% dry matter).

STEERS VS HEIFERS

On the market, cattle are divided into five sex classes: steers, heifers, cows, bullocks, and bulls. The sex of feeder cattle is important to the producer from the standpoint of cost and selling price (or margin), the contemplated length of feeding period, quality of feeds available, and ease of handling. The consumer is conscious of sex differences in cattle and is of the impression that it affects the quality, finish, and conformation of the carcass.

Steers are by far the most important sex class on the market, both from the standpoint of numbers and their availability throughout the year, whereas heifers are second.

Fig. 19-54. Heifers on feed in an Oklahoma feedlot. (Courtesy, Oklahoma State University, Stillwater)

The relative merits of steers vs heifers, both from the standpoint of feedlot performance and the quality of carcass produced, has long been a controversial issue. Based on experiments and practical observations, the following conclusions and deductions seem to be warranted.

• **Length of feeding periods—**Heifers mature earlier than steers and finish sooner, thus making for a shorter feeding period when fed to the same degree of finish. In general, heifers may be ready for the market 30 to 40 days earlier than steers of the same age started on feed at the same time.

• **Market weight—**The most attractive heifer carcasses are obtained from animals weighing 650 to 950 lb on foot, showing good condition and finish but not patchy and wasty.

• **Rate and economy of gain—**Because of their slower daily gains and lower feed efficiency, the feedlot gains made by heifers are usually somewhat more costly than those made by steers of the same age.

• **Price—**Because of existing prejudices, feeder heifers can be purchased at a lower price per pound than steers, but they also bring a lower price when marketed. Thus, the net return per head may or may not be greater with heifers.

• **Carcass quality—**Carefully controlled experiments have shown conclusively that when heifers are marketed at the proper weight and degree of finish, sex makes no appreciable difference in the dressing percentage, in the retail value of the carcasses, or in the color, tenderness, and palatability of the meat.

• **Ease of handling in the feedlot—**Because of disturbances at heat periods, many feeders do not like to handle heifers in the feedlot. However, the incidence of estrus can be lowered by feeding the additive MGA. (See Chapter 13, section headed "Melengestrol Acetate [MGA].")

• **Flexibility in marketing—**If the market is unfavorable, it is usually less advisable to carry heifers on feed for a longer period than planned because (1) of possible pregnancies, and (2) they become too patchy and wasty.

• **Effect of pregnancy—**Packer buyers have long insisted that they are justified in buying finished heifers at a lower price than steers of comparable quality and finish because (1) many heifers are pregnant and have a lower dressing percentage; and (2) pregnant heifers yield less desirable carcasses.

The effect of pregnancy on feedlot heifers is discussed in Chapter 13, Feed Supplements/Additives/Implants, in the section on "Abortifacients"; hence, the reader is referred thereto.

SPAYED HEIFERS

In females, the operation corresponding to castration is known as spaying. Under most conditions, desexing heifers by the traditional surgical method is not recommended because (1) the operation is complicated and difficult, requiring a very experienced technician; (2) it is attended with more danger than castration; (3) it lowers both rate and efficiency of gains; (4) it eliminates the heifers for possible replacement purposes or sale as breeding stock; and (5) experiments and practical operations with spayed heifers have generally shown that the selling price obtained is not sufficiently higher to compensate for the lower and less efficient gains plus the attendant risk of the operation.

A summary of 11 experiments in which spayed heifers and intact open heifers were compared revealed that spaying made for 9.9% slower rate of gain and increased the feed required per 100 lb gain by 8.5%.[7]

However, spaying by one of the new methods may be justified because of the following advantages of spayed heifers: (1) They may move freely across state lines without being tested for brucellosis when destined for a feedlot, (2) they are easier to manage in the presence of bulls or steers, (3) it avoids the losses generally associated with pregnancy, and (4) they generally bring a premium price.

Three new methods of sterilizing heifers without lowering performance appear promising: (1) University of Nevada researchers have designed a tool that closes the cervix with a rubber band; (2) Washington State University is testing a chemical method that will sterilize heifers; and (3) Colorado State University veterinarians have developed a spay instrument in which the trocar point is used to penetrate the vaginal wall, the ovary is excised by rotating the inner cutting tube, and the plunger is used to store the first excised ovary in the inner tube. These techniques await further study.

(Also see Chapter 13, Feed Supplements/Additives/Implants, section on "Abortifacients.")

BULLS

The feeding of bulls (uncastrated males) instead of steers has been standard practice throughout Europe for many years. For example, since about 1954 Germany has fed and slaughtered bulls as yearlings, instead of steers, because they obtain 10 to 15% greater rate of gain and feed efficiency thereby. The practice will increase gradually in the United States, now that carcasses from young bulls are federally graded as *bullock beef* rather than *bull beef*, thereby removing the connotation that the meat is inferior to or different from steer or heifer beef.

The carcasses from older bulls are still labeled *bull beef,* to differentiate them from the carcasses of younger bulls.

[7]*Montana Farmer-Stockman*, summary prepared by R. A. Bellows, U.S. Range Livestock Experiment Station, Miles City, Mont., Oct. 2, 1975, p. 38.

Bullock beef from young bulls is graded according to the same quality standards as beef from steers and heifers.

Also, the economics of the situation favors the feeding of bulls instead of steers. The male hormones secreted by the testicles are excellent growth stimulants and will improve gain and feed efficiency by 10 to 15%. Also, bulls will produce more healthful lean meat than steers, and research has shown that bull meat is equal in value, quality, and palatability to steer meat.

Now that the carcasses from young bulls are differentiated from older bulls, only consumer acceptance remains.

The following guidelines are recommended in the feeding of bulls:

1. Start young bulls on full feed at weaning age (6 to 7 months) and finish as rapidly as possible to a market weight of about 1,100 lb.
2. Use high-energy rations because bulls tend to grow rapidly and deposit less fat than steers.
3. Feed bulls so that they are finished for market before 16 months of age.
4. Avoid exciting or stressing bulls.
5. Do not add new bulls to the pen after weaning bulls are started on feed, because this tends to encourage fighting and riding, and results in reduced gains. Even under the best management, fighting and riding may be a problem.
6. Keep bulls separate from other cattle when marketing them. If possible, do not permit the bulls to stand in the pen overnight before slaughter.
7. Bulls of beef breeding are less nervous and ride less than bulls of dairy breeding.
8. Establish a market and a slaughter endpoint before feeding bulls.

Grade of Cattle

The most profitable grade of cattle to feed will generally be that kind of cattle in which there is the greatest spread or margin between their purchase price as feeders and their selling price as finished cattle. This cannot be determined by merely comparing the existing price between the various grades at the time of purchase. Rather, it is necessary to project the differences that will probably exist, based on past records, when the animals are finished and ready for market.

As fewer grain-fed cattle are marketed in the summer and fall, the spread in price between Select and Choice fed cattle and those of the lower grades is usually the greatest during this season. On the other hand, the spread between these grades is likely to be least in late winter and early spring, when a large number of well-finished cattle are coming to market from the feedlots. However, such seasonal effects are minimal today, due to large feedlots and year-round feeding.

The length of the feeding period and the type of feed available should also receive consideration in determining the grade of cattle to feed. Thus, for a long feed and when a liberal allowance of grain is to be fed, only the better grades of feeders should be purchased. On the other hand, when a maximum quantity of coarse roughage is to be used and a short feed is planned, cattle of the medium or lower grades are most suitable. Thus, successful cattle feeders match the quality of the cattle selected with the quality of the available feed; the better the feed, the higher the grade of cattle.

Fig. 19-55. Top grade Santa Gertrudis steers on feed. (Courtesy, Santa Gertrudis Breeders International, Kingsville, Tex.)

Cattle of the lower grades should be selected with very special care to make certain that only thrifty animals are bought. Ordinarily, death losses are much higher among low-grade feeder cattle, especially when the low-grade animals are calves. The death loss in handling average or high-grade feeders seldom exceeds 1 to 2%; whereas with cull or "dogie" cattle, it frequently is 2 to 3 times this amount. Many low-grade cattle are horned, and dehorning further increases the death risk—in addition to the added labor and shrinkage resulting therefrom.

No given set of rules is applicable under any and all conditions in arriving at the particular grade of cattle to feed, but the following factors should receive consideration:

1. The feeding of high-grade cattle is favored when:
   a. The feeder is more experienced.
   b. A long feed with a maximum of grain in the ration is planned.
   c. Conditions point to a wide spread in price between grades at marketing time. Such conditions may prevail in the late summer or early fall.
2. The feeding of average or low-grade cattle is favored when:
   a. The feeder is less experienced.
   b. A short feed with a maximum of roughage or cheap by-products is planned.
   c. Conditions point to a narrow spread in price between grades at marketing time. Such conditions normally prevail in the spring.
3. In addition to the profit factors previously enumerated, it should be pointed out that with well-bred cattle the following conditions prevail:
   a. Well-bred cattle possess greater capacity for consuming large quantities of feed than steers of a more common grade, especially during the latter part of the feeding period.
   b. The higher the grade of the cattle, the higher the dressing percentage and the greater the proportionate development of the high-priced cuts.
   c. The higher the grade of the cattle, the greater the opportunities for both profit and loss.

Certainly producers who raise their own feeders should always strive to breed high-quality cattle, regardless of whether they finish them or sell them as feeders. On the other hand, the purchasers of feeder steers can well afford to appraise the situation fully prior to purchasing any particular grade.

Breeding and Type of Cattle

Although supporting data are limited, it is realized that there is considerable difference between individual animals insofar as rate and economy of gain is concerned. Also, it is recognized that these differences are greater within breeds than between breeds.

CROSSBREDS

Good crossbreds will likely show 2 to 4% improvement over the average of the parent breeds for rate and efficiency of feedlot gains. Additionally, even larger advantages accrue to the cow-calf producer. Thus, it is inevitable that an increasing number of crossbred feedlot cattle will be seen.

Fig. 19-56. Charolais crossbred steers. (Courtesy, American Charolais Assn., Kansas City, Mo.)

DAIRY BEEF

Dairy beef accounts for about 25% of the beef consumed in this country, with these animals marketed as veal calves, cull dairy cows and bulls, and finished dairy heifers and steers. Improvements in the science and technology of feeding and processing favor growing and finishing dairy beef, and minimum slaughter of veal calves.

Dairy beef is just what the term implies—*beef derived from cattle of dairy breeding, or from dairy X beef crossbreds.*

Modern cattle feeders are primarily concerned with rate and efficiency of gains, and net returns. So, most of them would just as soon feed steers of dairy breeding, either purebreds or crossbreds; some actually prefer them. Consumers demand beef that has a maximum amount of lean, with a minimum amount of waste fat, and which is tender

Fig. 19-57. Holstein steers on feed in Wisconsin. (Courtesy, Ralston Purina Co., St. Louis, Mo.)

and flavorful; and they couldn't care less whether it came from a "critter" that was black, white faced, roan, pink, yellow, or polka dot. As a result, more and more steers of dairy breeding are going the feedlot route, rather than as veal. As evidence of this transition, during the 40-year period, 1942–1982, U.S. per capita veal consumption declined from 8.2 to 0.9 lb, while per capita beef consumption increased from 61.2 to 115.9 lb in the same period of time. Also, the shift in consumer demand to more lean and less fat has been reflected in the changed Federal grades of beef. As a result when properly fed, Holstein steers will make Choice, Select, or Standard grades.

Some pertinent points relative to dairy beef follow:

- **High-growth thrust essential**—For a dairy beef program to be most successful, scientists and producers in both Britain and the United States agree that the animals should have a high growth potential, as evidenced by heavy birth weight and heavy weight at maturity. Since Holsteins are heavy at birth and mature at around 1,400 lb, in comparison with mature weights of 1,000 to 1,200 lb of the European beef breeds, it is obvious that Holsteins are ideal when it comes to producing dairy beef.
- **High-energy rations; light market weights**—If dairy steers are to be slaughtered at young ages and light weights, high-energy (low-roughage) rations are imperative. Under this system, usually young calves of either dairy or dairy X beef breeding are fed in confinement—in barns; and are fed milk replacers from 1 to 4 days of age to 200 to 300 lb, and full fed a high-concentrate ration from about 300 lb to market weight of 750 to 950 lb.

Crowding for market at an early age takes advantage of the fact that growth is generally most economical when most rapid, and that young gains are cheap gains. Also, experience shows that when Holstein calves are started on super-energy rations at around 350 to 450 lb weight and marketed under 1,100 lb, (1) there's excellent marbling with very little outside fat, and (2) many of these animals will grade Choice.

- **High-roughage rations; heavy market weights**—If roughages are relatively more abundant and cheaper than concentrates, then it may be more profitable to feed dairy beef more roughage and market at heavier weights—and to expect slower and less efficient gains. In any event, it's net returns that count, rather than rate of gain and pounds of feed required per pound of gain.

Under the high-roughage system, steers of dairy breeding are grown on maximum roughage to 600 to 750 lb weight, following which the ratio of concentrate to roughage is increased. Most dairy steers fed according to this system are marketed at weights of 1,150 to 1,400 lb, grading Select or Commercial. (Most of them are too old to grade Standard, and too lacking in marbling to grade Choice.)

• **Dairy beef has good potential**—There is ample evidence that male calves of the larger dairy breeds (Holsteins and Brown Swiss), along with dairy beef crosses, have the potential for producing acceptable beef with good feed efficiency, under a system of either (1) full feeding from an early age on a high-energy ration, or (2) growing and finishing on a maximum of roughage and marketing at older ages and heavier weights. In the final analysis, therefore, the system selected should be determined by net returns. Both methods necessitate the rearing of young, 1- to 4-day-old calves to weights of around 300 lb, with such early rearing done by either a calf-raising specialist or by the cattle feeder who will do the ultimate finishing. Such calves must usually be obtained from a wide area, and be of variable ages and sizes; hence, they are difficult to come by. Also, death losses are frequently high and discouraging.

Before producing beef from dairy steers, a market should be established. This calls for a steady flow of cattle of a predictable quality.

HAMBURGER BEEF

Consumer demand has always exerted a powerful influence upon the type of animals produced—and beef cattle are no exception. Currently changing lifestyles—more and more women working outside the home, more children out of the house for various activities, and more informality—are being reflected in consumer preferences for meals that are fast and easy to prepare, with consistent results and variety. Meeting this need calls for leaner beef (a steer with 12% fat, rather than 24%), known as manufacturing-type beef, suitable for hamburger. In the past, cull cows and beef imports filled much of this need. But these sources haven't kept pace with demand. In 1972, about 36% of the total beef supply was made into hamburger; by 1976, this figure had increased to 43%. To fill this need calls for more lean beef and less fat—for fewer grain-fed cattle and more cattle grown and finished on grass, crop residues, and by-product feeds. The cattle producer can, and must, develop methods for earning a profit through producing and marketing animals intended primarily for ground beef and hamburger. Despite this shift, it is not expected that the United States will become strictly a "hamburger society"; there will always be a demand for high-quality steaks, roasts, and other beef cuts from choice, grain-fed animals.

BOXED BEEF

With the advent of boxed beef, the steer must fit the box. This calls for steers produced to specification in size and quality.

More and more packers are fabricating and boxing beef in their plants, thereby freeing the back rooms of 200,000 supermarkets. After chilling, the carcass is subjected to a disassembly process, in which it is fabricated or broken into counter-ready cuts; vacuum-sealed; moved into storage by an automated system; loaded into refrigerated trailers; and shipped to retailers across the nation.

The following benefits accrue from central fabricating and boxing of beef:

1. Twenty-five percent of the weight of the carcass, in the form of bones and trim, is removed at the packing plant, thereby reducing shipping costs.
2. Twenty to thirty percent of the carcass is bone and fat, which ends up as waste in the retail store. When fabricating is done at the packing plant, trimmings, fats, and tallow are federally inspected and can be used as edible products; for example, fat and trim can be used in ground meats.
3. Processing equipment cost at the retail store level is usually high because of low volume and idle time. Hence, counter-ready cuts make for a saving in equipment cost at the retail store.
4. Central fabricating makes possible a regulated aging process.
5. Central fabricating aids in achieving uniformity of quality standards.
6. Central fabricating improves merchandising through more uniformity of cuts, new cuts, more variety of cuts, matching cuts to area preferences, creating a more tenable situation for meat retailers who lack expertise and interest in cutting and freeing the meat manager from butchering—thereby allowing time for personal selling and customer relations.

Beef that is not suitable for sale over the block is (1) boned out and disposed of as boneless cuts, (2) canned, (3) made into sausage, or (4) cured by drying and smoking. It is estimated that about ⅕ of all slaughter cattle are disposed of as processed meats.

FEEDLOT FINISHING FACILITIES

Feedlot finishing refers to feeding cattle in a restricted area, with the feed conveyed to the animals; and it may involve either an open pen feedlot, or confinement (sheltered) feeding.

Cattle feeding facilities and equipment are a manufacturing plant, wherein animate objects (cattle) convert feed into beef. Hence, they merit the same level of competence in planning and design as any other sophisticated manufacturing plant.

Open Pen Feedlot

An open pen feedlot is, as indicated by the name, a lot in which the cattle are in the open—usually it is without shelter (except for such natural protection as may be afforded by trees, hills, or wind fences, or perhaps a roof over the feed bunks or shades).

An open lot without shelter is the cheapest type of feedlot construction. In the Southern Plains area, where the weather is mild and shelters are unnecessary, investment costs range from $100 to $125 per head of capacity.

Fig. 19-58. Open pen feedlot with shade. (Courtesy, Carl Stevenson, Red Rock Feeding Co., Red Rock, AZ 85245)

Housing increases costs, and the more elaborate the housing, the greater the cost. It is estimated that, on the average, the cost per head capacity where housing is involved is about as follows:

Type of Facility	Sq Ft per Animal	Cost per Animal Capacity
		($)
Open shed	20	140
Cold confinement	17	185
Warm confinement (heated)	17	240

One cold confinement facility constructed in the West in the 1980s is reputed to have cost $300 per head capacity.

Confinement Feeding; Slotted (or Slatted) Floors

Currently, there is much interest in cattle confinement feeding and slotted floors. The main deterrent is cost; construction costs vary with type of structure and may range up to $300 per steer space.

Fig. 19-59. Confinement feeding on a slotted floor. (Courtesy, Pioneer Hi-Bred Corn Co., Des Moines, Iowa)

Confinement cattle feeding refers to feeding in limited quarters, generally 20 to 25 sq ft per yearling animal, which is about ⅛ the space normally allotted to a yearling in an unsurfaced lot and ⅓ that of a paved lot. The confinement is usually under roof on slotted floors.

Slotted floors are floors with slots (space) through which the feces and urine pass to a storage area immediately below or nearby.

Interest in confinement feeding and slotted floors was ushered in for the purpose of (1) automating and saving labor, (2) cutting down on bedding and facilitating manure handling, (3) lessening mud, dust, odor, and fly problems, (4) increasing gains and saving feed, (5) lessening land requirements, (6) lessening pollution, and (7) lessening complaints from area residents.

Research has shown conclusively that cattle fed during the winter months in cold areas gain faster and more efficiently if they are sheltered. However, as pointed out earlier in this chapter under the section headed "Open Pen Feedlot," the per head cost is much higher for confined or sheltered cattle. Thus, the decision on whether cattle confinement can be justified, even in the northern part of the United States, should be determined by economics. Will the cattle in confinement quarters gain sufficiently more rapidly and efficiently to justify the added cost? Of course, manure disposal and pollution control should also be considered.

COLD CONFINEMENT[8]

Cold confinement refers to a more or less open shed for confining cattle; hence, winter temperatures therein are within a few degrees of outdoor temperatures. Open sheds should be faced away from the direction of the prevailing winds. Additionally, doors or other openings in the closed walls should be provided for summer ventilation.

Fig. 19-60. Finishing cattle feed in an open shed—cold confinement. (Courtesy, *Feedlot Management*, Minneapolis, Minn.)

[8]The terms *cold confinement* and *warm confinement* refer to winter conditions. Without mechanical cooling, both systems are *warm* during the summer months.

WARM CONFINEMENT[9]

Warm confinement refers to a confinement building for cattle which is sufficiently insulated and ventilated to maintain inside winter conditions above 35°F in severe weather, and in the range of 50° to 60°F most of the time. Warm confinement makes for the maximum in feed efficiency.

FEEDS

The growth of the cattle-feeding industry in America has gone hand in hand with the production and feeding of more grains and by-product feeds. Such feeds are high in energy and low in fiber. Hence, their availability and price influence the extent, location, and type of feeding program. Roughages, which require relatively more energy to digest and metabolize than grains, are used at low levels in most finishing rations. However, due to world food shortages, a longtime trend to heavier feeders off milk and/or grass, and consumer demand for leaner meat, more roughage is being used in producing slaughter cattle. Also, roughages are important in growing programs and in warm-up rations.

Formulating rations for growing and finishing cattle is a matter of combining feeds to make a ration that will be eaten in the amount needed to supply the daily nutrient requirements of the animal.

For convenience, the commonly used beef cattle finishing feeds are herein classified as (1) concentrates, (2) by-product feeds, (3) roughages, (4) protein supplements, (5) minerals, (6) vitamins, and (7) water. Also, feed additives and implants are a part of modern cattle feeding.

Concentrates

Concentrate feeds are those which are high in energy and low in fiber. Many different kinds of concentrate feeds can be, and are, used in beef cattle finishing. Availability and price are the two most important factors determining the choice of concentrates. Consideration of the latter factor—price—necessitates that the cattle feeder be a keen student of values, and that the formulations of rations be changed in keeping with comparative feed prices.

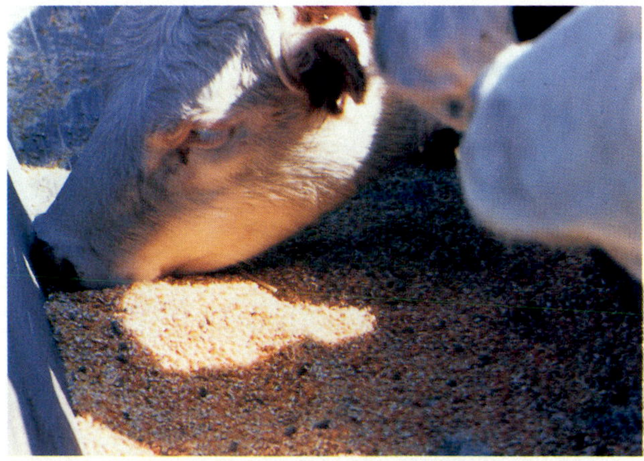

Fig. 19-61. Closeup of a steer eating concentrates from a feed bunk on a Nebraska farm. (Courtesy, USDA)

[9]*Ibid.*

Corn is the most common grain used in finishing cattle. It is palatable and rich in the energy-producing carbohydrates and fats, and low in fiber. Also, corn is easily stored, only moisture and carotene being lost over a period of time. However, corn has certain very definite limitations—it is low in protein and calcium.

The grain sorghums are assuming an increasingly important role in cattle feeding, particularly in the fringe areas of the Corn Belt, and in the South and Southwest where moisture conditions are less favorable. New and high-yielding varieties have been developed. As a result, more and more grain sorghums are being fed to cattle. The chemical composition of sorghum (milo) is similar to corn except that the protein content is generally higher and more variable. Its feeding value is greatly enhanced by steam processing and other similar methods of preparation.

Although corn and sorghum are by far the most common grains used in finishing steers, such grains as barley, rye, oats, and wheat are used in many sections of the United States and Canada. The small grains are excellent for finishing cattle when properly used. In comparison with corn feeding: (1) barley-fed cattle are more susceptible to bloat (for this reason, it is best not to use a straight legume hay along with a grain ration high in barley; a mixture of barley and dried beet pulp is commonly used in the West); (2) barley- or wheat-fed animals are more apt to tire of their ration during a long feeding period; (3) rye should not constitute more than ⅓ of the grain ration because it is unpalatable; (4) more care is necessary to prevent wheat-fed cattle from going off feed; and (5) oats should not constitute more than ½ of the ration, and preferably not more than ⅓ because of its bulk. Fortunately, these limitations can be lessened considerably by mixing these feeds together, or by mixing the grain with dried beet pulp, silage, or chopped hay. Also, it is important that the small grains be coarsely ground or properly rolled. Wheat and oats are frequently too expensive to include in cattle finishing rations.

HIGH-MOISTURE (EARLY HARVESTED) GRAIN

High-moisture grain, especially corn and sorghum, is grain that is harvested at 22 to 40% moisture content and either (1) dried, (2) stored in a silo made as airtight as possible, or (3) treated with propionic acid (or a mixture of propionic acid with either acetic acid or formic acid). Experimental tests show that the feed conversion efficiency of a high-moisture milo is improved by 8 to 15% over the same dried milo, although there is little increase in daily gain. There is less improvement from high-moisture processing of shelled corn, and the results have been most variable. High-moisture milo and corn should be ground or rolled before they are fed. It is questionable, however, if it pays to process high-moisture corn in rations that have less than 15% roughage.

(Also see Chapter 9, Silage/Haylage/High-Moisture Grain.)

MOLASSES

Molasses—cane, beet, corn, and wood—is palatable (wood molasses is the least palatable) and a good source of energy;

and cane molasses is also a good source of certain trace minerals. All molasses is low in protein. As a rule of thumb, molasses should not cost more than ¾ as much as cereal grain, pound for pound, to be economical. Generally speaking, molasses is limited to 5 to 10% of the ration, although up to 15, or even 20%, can be added if the mixing facilities will handle it (or if it is self-fed) and the price is favorable.

According to the American Feed Industry Association, nearly 25% of the nation's feedlot animals receive a liquid feed supplement.

FATS/COTTONSEED

It is recommended that 2 to 5% fat be added to high-concentrate cattle finishing rations in which milo, barley, and/or wheat are the chief grain sources. Higher levels of fat usually result in drastically lowered feed consumption. When fed at a 2 to 5% level, the energy value of fat is approximately 2¼ times that of carbohydrates. When corn is the major grain, fat additions can be expected to be less useful than with the small grains. This is understandable when it is realized that corn contains approximately 4% fat as compared to 1 to 1½% for the other feed grains.

When fat is added, the calcium and phosphorus levels of the ration should be 0.55% and 0.33%, respectively.

In recent years, the price of whole cottonseed has been favorable enough to include in cattle feeds.

(Also see Chapter 10, section on "Cottonseed" and "Fats and Oils.")

By-product Feeds

Innumerable by-products—both roughages and concentrates—from plant and animal processing, and from industrial manufacturing, are available and used as cattle feeds, in different areas, including: potatoes and potato process residue, pea vines and corn refuse silage from the canning industry, and by-products from numerous fruits and nuts.

Fig. 19-62. Drylot steers eating field-cured dry beet tops. (Courtesy, The Great Western Sugar Co., Denver, Colo.)

As is true of any ration ingredient, the requisites to effective and profitable use of each by-product feed in cattle feeding are (1) that its proximate composition be known; (2) that it be bought at a favorable price, nutritive composition considered; (3) that it be incorporated in a balanced ration; (4) that it be palatable and consumed in adequate quantity; and (5) that it not adversely affect carcass quality, particularly from the standpoint of harmful chemical residues from pesticides applied to crops. Generally speaking, the use of by-product feeds calls for ingenuity and experience in handling them, special knowledge relative to their nutritive qualities and use in balanced rations, and higher labor costs in storing and feeding them. As a result, many cattle feeders are not interested in using them, whereas others find their use a lucrative business.

The feeding value and the maximum amount that can be fed to cattle of several by-product feeds are given in Table 19-14, Feed Substitution Table For Beef and Dairy Cattle.

Roughages

The use of roughages is maximized in the stocker and backgrounding stages. However, consumer acceptance of leaner, "natural" beef is creating interest in feeding more forages and by-product feeds to finishing cattle.

Roughages are used in feedlot rations to supply bulk, energy, protein, minerals, and vitamins. They contain considerable fiber (cellulose, hemicellulose, and lignin); consequently, they have a lower available energy content than concentrates. For this reason, only limited amounts of roughages (not over 5 to 10% crude fiber for optimum gain) are normally incorporated in finishing rations, particularly toward the end of the feeding period. They are, however, used extensively in growing programs and in warm-up rations.

The amount of roughage in feedlot rations varies over a wide range—from roughage alone in some grower rations to all-concentrate rations in some finishing rations, and many roughage proportions between these two extremes. Each of these roughage-to-concentrate ratios may be highly successful and very practical under certain conditions. In the final analysis, therefore, the roughage-to-concentrate ratio for a given feedlot should be determined by (1) the available feeds and comparative prices, (2) feed processing facilities, (3) the feed handling charges, (4) the age and quality of cattle, (5) the stage in the feeding period (i.e., starting vs finishing period), (6) temperature (decrease roughage and increase concentrate in hot summer months, because high-roughage rations produce more body heat), (7) the feeder, (8) the troubles encountered (off feed, founder, scours, bloat), and (9) the results obtained by previous experience.

Also, step-wise reductions in the amount of roughage are usually made at least 3 times during the finishing period; for example—

Starter ration—70% roughage, 30% concentrate
Intermediate ration—30% roughage, 70% concentrate
Final ration—10% roughage, 90% concentrate

Such step changes in roughage to concentrate ratios should be made gradually, by blending the two mixes for 2 to 4 days.

In drylot finishing, the kind of roughage fed varies from area to area, since it is not practical to move roughages great distances. Thus, generally speaking, cattle feeders utilize those roughages that are most readily available and lowest in price.

HAY

Hay is the most important harvested roughage for feedlot cattle, although many other roughages can be, and are, utilized.

High-quality legume hays are superior from the standpoint of cattle performance—and, when they are fed, supplemental feed requirements are lower. However, lower quality hays give satisfactory results when properly supplemented.

Alfalfa is the roughage of choice in commercial feedlots. It averages 17.1% protein on a dry (moisture-free) basis and contains 58% TDN. On a net energy basis for production (NE_p), it is estimated to contain 27 Mcal per 100 lb, in comparison with a value of 56 Mcal per 100 lb for milo.

SILAGE

Increasing quantities of corn, milo, and grass silage are being fed to finishing cattle. At the present time, it is estimated that 90% of the nation's silage is made from corn and sorghum and the other 10% from hay and pasture crops, small grains, by-products and other feeds. The following silage pointers, based on experiments and practical observation, are generally observed by successful cattle feeders:

1. When cattle are given a liberal amount of grain, any silage that is fed is considered a part of the roughage ration; hence, the silage should be fed in accordance with the well-recognized rules for feeding roughages. Also, it can be assumed that about 2.5 to 3 lb of good corn silage may be substituted for 1 lb of hay in cattle rations.
2. The higher the grain content of corn or cereal grain silage, the higher the energy content and the greater the feeding value.
3. The maximum use of silage is best obtained early in the finishing period or with more mature steers that possess a larger digestive capacity.
4. Low-grade cattle may be fed either entirely on silage, or liberally on silage throughout the finishing period, but cattle that grade Select or better may return more profit if they are full fed grain during the last half of the feeding period.
5. Cattle can be placed on full feed of silage from the beginning of the feeding period without any detrimental effects.
6. When fed silage alone, cattle on full feed will consume 6 to 7 lb per 100 lb weight.
7. Sorghum silage has 60 to 90% of the value of corn silage and should be supplemented in the same manner as corn silage.
8. Grass silage should be supplemented with additional energy feeds, such as cereal grain or molasses, to be of the same value as corn silage.

HAYLAGE

Tests to date have shown that haylage, which generally contains about 40% moisture, is an excellent feed for finishing cattle. More research is needed, but indications are that costs of gains may be a little higher than for corn silage.

A fairly accurate rule of thumb is that it takes 1.5 lb of 40% moisture haylage to equal 1 lb of hay.

OTHER ROUGHAGES

Among the other roughages (other than hays and silages) used for feedlot cattle are cottonseed hulls, corncobs, cereal straws, milo stover, Bermudagrass straw, Sudangrass hay, sawdust and other wood products, oat hulls, beet tops, peanut hay, newspapers, and a host of others. When properly (1) combined with a high-quality legume roughage, and/or (2) supplemented with the necessary protein, minerals, and vitamins, all of them are excellent feeds. Availability, costs, and results should be the determining factors in their use, just as the economics of the situation should determine the use of any other feed ingredient.

(Also see Table 19–14, Feed Substitution Table for Beef and Dairy Cattle, for the feeding value and the maximum amount that can be fed to cattle of each of several roughages.)

ROUGHAGE SUBSTITUTES

The ruminant animal requires a minimum amount of roughage factor for normal rumen function. All-concentrate rations show a response to small amounts of roughage factor. As little as 1 lb per day of some low-quality roughage such as alfalfa stems, rice hulls, or sorghum straw have given improvements in efficiency ranging up to 15% of all-concentrate rations.

To meet the *roughage factor* need, different roughage substitutes have been developed and used with varying degrees of success.

Protein Supplements

The daily protein requirements of growing-finishing cattle are given in Tables 19–1 and 19–3, and the percentage of protein needed in the ration is given in Tables 19–2 and 19–4. It should be noted, however, that a larger percentage of protein is needed in rations with higher energy density. This is because fewer pounds of the high-energy ration are needed daily to meet the animal's energy requirement but the protein needs stay the same. Other factors that must be considered in formulating feedlot rations are the variations in protein content of feedstuffs, the digestibility of the protein, metabolic efficiency of protein utilization, and the previous nutritional treatment of the cattle. Age, genetic background, and health may influence efficiency of protein utilization. Because of the many factors that affect protein requirements, it is advisable to include a safety factor when balancing for protien.

Although excess protein in the ration can be partly utilized for energy, each 1% increase in protein above the required level may increase the cost of gain ¼ to ½ cent per pound. However, underfeeding protein can cost much more than over-feeding protein due to the slow gains and poor feed efficiency. There appears to be a trend among research workers and nutritionists to recommend higher protein levels than were believed necessary in recent years, especially in high-concentrate rations. Of course, price levels of protein supplements are a factor.

KINDS OF PROTEIN

Quality of protein, or balance of essential amino acids, is not a critical factor in most beef cattle finishing rations, because bacteria in the rumen "manufacture" protein that is used by cattle. However, researchers are now beginning to think that quality of protein does play a role in ruminant nutrition—that a deficiency of certain amino acids may limit microbial synthesis. Thus, the present system of establishing protein values of feeds and protein requirements of feedlot cattle on the basis of nitrogen × 6.25 is being challenged. Burroughs, of Iowa State University, has proposed a new system of evaluating protein nutrition of feedlot cattle involving two new measurements: (1) the metabolizable protein and amino acid requirements of cattle and corresponding values for feedstuffs, and (2) the urea fermentation potential of feed consumed by cattle.[10]

Since the system proposed by Iowa State workers is recent, it will need to be understood and critically appraised and evaluated. In the meantime, it is recommended that practical cattle feeders continue to choose a protein supplement(s) largely on the basis of the comparative price of a pound of protein. Also, it should be recognized that, in comparison with a single ingredient, protein mixtures may be more palatable and cause animals to eat a little more feed, and result in slightly greater gains and feed efficiency.

The leading protein supplements for finishing cattle are soybean meal, cottonseed meal, linseed meal, urea, and slow-released nonprotein nitrogen products.

Urea

Each year, urea is replacing a larger percent of the supplementary protein in cattle feedlot rations. Because of the high price of oilseed proteins, due to their increasing use for human consumption and for monogastric animals, urea and/or other nonprotein nitrogen compounds will be used increasingly as major sources of supplementary protein for feedlot cattle. A rule of thumb is that if the price of 1 lb of urea plus 6 lb of grain is less than the price of 7 lb of an oilseed meal, it will pay to use urea.

Urea or other nonprotein nitrogen compounds can furnish about 33% of the total protein requirement of feedlot cattle; higher levels cause a depression in gain and feed efficiency. It is noteworthy, however, that, even in supplements in which 90% of the protein equivalent is from urea, only about ⅓ of the total protein in the ration is supplied from nonprotein nitrogen—the remainder is supplied from grain and roughages. Guidelines relative to the use of urea for cattle are given in Table 19–12.

The following conditions are essential for proper urea utilization by feedlot cattle:

1. High level of bacteria population in the rumen. Cattle off feed or sick do not utilize urea effectively.
2. Slow release of ammonia from urea.
3. Two to four weeks' adjustment period.
4. Low level of natural protein in the ration. This suggests that certain amino acids are essential for microbial synthesis.
5. Feeds that provide a readily fermentable energy source are required in high-urea supplements. For this reason, high-fiber filler feeds (such as ground corncobs, oat hulls, rice hulls, cottonseed hulls, cellulose, paper, and sawdust) should be used with discretion in urea-containing rations.
6. Response to urea is greater on high energy-limited roughage rations than on low energy-limited roughage rations.
7. Urea supplements should be well mixed (homogeneous).
8. High-urea dry supplements should be protected from rain and kept as dry as possible, because urea is hygroscopic. (It picks up moisture.)
9. Essential nutritional factors including (a) readily available source of energy (such as grain or molasses); (b) adequate levels of calcium and phosphorus; (c) required level of trace mineral elements; (d) a nitrogen-sulfur ratio which is not wider than 15:1; (e) unidentified urea-protein synthesis factors (alfalfa meal in dry urea mixes, and distillers' solubles in liquid supplements); (f) iodized salt, to improve the palatability and mask the taste of urea; (g) fortified with proper levels of synthetic vitamin A, to furnish a minimum of 20,000 IU of vitamin A daily for growing and finishing cattle; (h) fortified with vitamin D if cattle are confined; and (i) fortified with vitamin E if natural feedstuffs are low.
10. Do not feed urea or supplements containing urea to newly arrived or shipped-in cattle for a period of 21 to 28 days.
11. Do not feed urea to cattle that have been starved or off feed for 36 hours until they have had a chance to fill the rumen with feed.
12. For growing cattle, feed a maximum of 0.15 lb of urea daily.
13. For finishing steers or heifers on grain and roughage, do not feed more than 0.22 lb of urea per head daily.
14. Formulate complete cattle rations so that no more than 33% of the crude protein or nitrogen is derived from urea. Protein supplements may contain 85 to 90% of the protein from urea; but when blended with natural feedstuffs such as grain and roughage, the total contribution of protein from urea is usually less than 33%.
15. Do not feed urea over and above the protein requirement; add only enough properly balanced urea supplement to meet the protein needs.
16. Urea should be either thoroughly mixed in a properly balanced supplement or incorporated in a complete ration.
17. If supplements containing urea are self-fed, the intake should be controlled by using a "lick wheel" for liquid supplements, or by incorporating high levels of salt and dry supplements.

AMOUNT OF PROTEIN SUPPLEMENT TO FEED

The percent protein supplement to add to the ration will depend upon the age of the cattle, the kind and amount of roughage, and the protein content of the grain(s), or other carbonaceous concentrate being fed. Also, more protein is needed in rations with higher energy density. Thus, the amount of supplement should be determined for each lot. On a percentage of total ration basis, it decreases as the cattle grow older. Here are the recommendations for crude

[10]Burroughs, W., A. H. Trenkle, and R. L. Vetter, *Proposed New System of Evaluating Protein Nutrition of Feedlot Cattle,* Iowa State University, A. S. Leaflet R161.

protein in the total ration, on an air-dry basis (which is about 10% moisture):

	Stage in Feedlot	Crude Protein
		(%)
Calves	First 60–90 days	11.0–12.0
	Next 100–200 days	10.5–11.0
	200 days to market	10.0–10.5
Yearlings	First 60–90 days	10.5–11.0
	100 days to market	10.0–10.5

For cattle fed relatively high-concentrate rations in the drylot, the following rules of thumb may be used:

1. When no legume hay is fed, add 2 lb of oilseed protein supplement per head daily.
2. When half the roughage consists of legume hay, add 1 lb of oilseed protein supplement per head daily.

When cattle are getting a full feed of good-quality legume hay, haylage, or legume silage, and limited grain, it is not necessary to add a protein supplement.

Because protein supplements are usually expensive, normally one should not add more than is required to balance the ration. Neither should they be shorted, for digestion of roughage is lowered if there is a lack of protein in the ration.

PROTEIN POINTERS

The following points should be taken into consideration to assure proper protein utilization by finishing beef cattle:

1. The protein content of the major feed ingredient(s) should be known so that the ration formulation can be precise.
2. Consideration should be given to the protein content and digestibility of the grain, because grain is the most economical source of protein and supplies the largest percent of the ration protein.
3. Higher levels of protein are needed in all-concentrate rations.
4. When possible, use a roughage high in protein so as to lessen the amount of supplemental protein necessary.
5. Urea can be successfully used at a level up to 1% of the total finishing ration to replace natural protein, provided the ration formulation permits optimum utilization of the urea by the rumen microorganisms.
6. When excessive protein levels are fed, the overage is wasted and performance may actually be reduced with high-performing animals.
7. Absolute protein requirements depend upon the age of the animal, and probably the energy content of the ration.
8. When per head consumption of corn reaches a level of 18 lb per day, no supplemental protein is required, although other additives such as minerals and vitamins are necessary. Thus, protein supplements are not needed in the late finishing stages of yearling cattle.
9. The production of single-cell protein (SCP) from solid waste as a protein supplement for feedlot cattle may increase in importance, but first many problems concerning SCP production technology must be overcome.

Minerals

Minerals play an important role in the nutrition of feedlot cattle. The amount of each mineral in the ration is important; both deficiencies and excesses are to be avoided. Therefore, an analysis should be made, particularly when new feeds and feeds from new areas are involved. The following minerals have been established as dietary essentials for feedlot cattle and other ruminants:

- **Major minerals**—Sodium, chlorine, calcium, phosphorus, magnesium, potassium, and sulfur.
- **Trace minerals**—Cobalt, copper, iodine, iron, manganese, selenium, and zinc.

When a complete mixed feed (roughage and concentrate combined) is fed to finishing cattle, it should contain 0.25 to 0.5% salt. Also, in the larger feedlots, the other needed minerals are usually incorporated in the ration as a special mineral supplement or in the protein supplement. For recommended kinds and allowances of minerals, see Tables 19–5 and 19–7. Special attention needs to be given to trace minerals in areas where there is a deficiency of one or more of them, when poor-quality roughage is fed, or when high- or all-concentrate rations are fed.

Even when minerals are added to the ration of finishing cattle, the authors favor self-feeding them in addition, as directed in Part I of this chapter under the heading, "Mineral Supplements." In particular, incoming feedlot cattle may crave minerals—especially phosphorus, due to deficiencies in their previous feeding. Such cattle, however, should be given only limited quantities of minerals until the danger of overeating has passed.

(Also see general section on "Minerals" in Part I of this chapter.)

Vitamins

Vitamins A, E, and in some cases vitamin D, should be added to feedlot rations.

The rumen organisms synthesize adequate B vitamins and vitamin K, and nothing is gained by adding these to feedlot rations of healthy cattle. Likewise, no benefit has been reported from supplemental vitamin C. Vitamin D is produced in the skin of animals in direct sunlight, but during cloudy, winter weather, or when cattle are confined, it should be added to the ration.

Feeders should watch for the following symptoms of vitamin A deficiency in feedlot cattle: rough hair coat, watery eyes, loose and watery droppings, edema (stovepipe legs), and night blindness.

Table 19–8 gives the vitamin requirements of beef cattle, whereas Table 19–35 gives the recommended levels of vitamin A for feedlot cattle, with overage for safety.

A common guideline on the level of vitamin A for feedlot cattle is to use 3,000 IU per 100 lb body weight, or 1,000 IU for each pound of total air-dry feed.

When vitamin D is needed and added to the ration, it is recommended that 4,000 to 6,000 IU of it be given per head per day. This is approximately ⅙ to ⅐ the recommended level of vitamin A.

TABLE 19-35
RECOMMENDED LEVELS OF VITAMIN A FOR FEEDLOT CATTLE

	Vitamin A/ Head/Day	Vitamin A/Ton of Supplement When it is Fed at Level of—	
		1 lb/ Head/Day[1]	2 lb/ Head/Day[1]
	(IU)	(IU)	(IU)
Cattle on growing ration in winter	10,000	20,000,000	10,000,000
Cattle on full-fed finishing ration in winter	20,000	40,000,000	20,000,000
Cattle on full-fed grain on pasture	20,000	40,000,000	20,000,000
Cattle fed in drylot in summer	30,000	60,000,000	30,000,000

[1]Air-dry basis (approximately 90% dry matter).

When grains are heat processed for feedlot cattle, some research shows that it may be advisable to provide supplemental vitamin E. The National Research Council indicates that the requirement for vitamin E is about 7 to 27 IU per pound of ration dry matter for growing and finishing cattle.

Water

Feedlot cattle should have access to plenty of clean, fresh water at all times. They will consume 7 to 12 gal per head per day. In cold climates, waterers should be equipped with heaters. Where the water supply is not limited by cost or volume, continuous-flow waterers are excellent. In order to keep the pathogen and algae content at a minimum, water tanks should be cleaned at least once a week in the winter and twice a week in the summer. In sick pens and pens of new cattle, the water tanks should be cleaned daily.

GROWTH STIMULANTS AND IMPLANTS

Table 19–11, Implants and Growth Stimulants For Cattle, which appears earlier in this chapter, summarizes the growth stimulants that are presently available and can be used; hence, the reader is referred thereto. All of these products have been shown to improve gain and feed efficiency of feedlot cattle significantly.

Other Methods of Improving Rate and Efficiency of Gain

Several other methods, in addition to additives, can be used to increase the rate and efficiency of gain of feedlot cattle. Among them are the following:

1. Feed young bulls (uncastrated males) instead of steers. The male hormones secreted by the testicles are excellent growth stimulants and will improve gain and feed efficiency by 10 to 15%. Alternatives to bulls that merit consideration are short-scrotum bulls (induced cryptorchidism), and Russian castrates. With these methods, testosterone is produced; yet, in comparison with bulls, the animals are easier to handle and may be carried to advanced ages without being labeled *bull beef*.
2. Take advantage of the genetic improvement of beef cattle by crossbreeding and the introduction of genes from the exotic breeds. This offers one of the most permanent ways of increasing the weaning weight of calves and improving performance by feedlot cattle. The selection of fast gaining and efficient cattle within the straightbreds, and in crossbreeding, may improve the efficiency of performance by feedlot cattle by 10% or more.
3. Reduce the cost of producing beef by improving the quality of cattle rations through grain processing and nutritionally balanced protein supplements.
4. Eliminate internal and external parasites, and protect cattle against the common diseases. This will save millions of dollars for both cattle feeders and consumers.
5. Keep abreast of new developments, including the discovery of new growth stimulants.

FEED PREPARATION

Prior to 1960, very little attention was given to feed processing for commercial cattle production, other than grinding or crushing grain and chopping forage. But in recent years great progress has been made and many new techniques have been developed.

Feed preparation is fully covered in Chapter 14, Feed Processing.

Mixed Rations Vs Feeding Roughages and Concentrates Separately

Fig. 19-63. Feeding mixed rations to cattle with a self-unloading truck. (Courtesy, Butler Manufacturing Co., Oswalt Division, Garden City, Kan.)

Most experiments and experiences have not shown any difference between mixed rations and the feeding of roughage and concentrates separately insofar as rate and efficiency of gain are concerned. However, a mixed ration has the following **advantages**:

1. It makes for greater efficiency in feeding and lessens the sorting at the feed bunk.
2. When the roughage is relatively unpalatable, a mixed ration forces consumption.

3. When it is desired to limit concentrate consumption, mixing with the roughage is desirable.

4. After cattle have become adjusted to the feedlot, a mixed ration makes it easier to get them on full feed.

Thus, each feeder must make a decision on the matter of mixed vs feeding roughage and concentrate separately, with relative costs and other factors considered. Most large feedlots use completely mixed rations.

All-Concentrate and High-Concentrate Rations

Based on experiments and experiences, the following conclusions relative to all-concentrate and high-concentrate rations appear to be justified:

1. Ruminants need some *roughness factor* to stimulate the rumen papillae for normal functioning. In high- or all-concentrate rations, this can be achieved partially by rolling or coarse grinding.
2. With the possible exception of whole shelled corn, and such high-fiber feeds as oats and barley, a high level of management is needed to make an all-concentrate system work under feedlot conditions. Problems associated with high-concentrate rations include acidosis, founder, and liver abscesses.
3. Some research indicates that continued high levels of performance can be better maintained by including 10 to 15% roughage in high-concentrate rations.
4. Rations having a concentrate content of 90% or more should be self-fed, and a liberal amount of feed should be available at all times.
5. Feed efficiency is improved in high- and all-concentrate rations, due to their high energy; but rate of gain is not materially affected.
6. The energy of a ration may be increased without eliminating much of the roughage by adding 4 to 5% fat.
7. Feed formulation and balance of nutrients become more critical on high- and all-concentrate rations; specifically—
    a. Higher levels of vitamin A must be added— 50,000 to 100,000 IU/head/day.
    b. The ration must be fortified with the minerals, especially the calcium and trace elements, that are normally provided through the roughage.
    c. The unidentified growth factors are usually reduced with a reduction of the roughage. To compensate therefor, 5% alfalfa meal may be added to the ration.
8. Cattle fed high- or all-concentrate rations stall and go off feed more frequently.
9. Pelleting high- or all-concentrate rations lowers daily gains.
10. In the final analysis, the comparative price of concentrates and roughages—the economics of the situation—along with management practices, will be the major determining factors.

HOW TO USE THE NET ENERGY METHOD

Use of the net energy method necessitates that the following net energy values be available:

1. The net energy requirements for growing and finishing cattle (in Mcal per animal per day), with a breakdown for steers and heifers (see Table 19–3).
2. The nutrient composition of feeds, with the net energy of each feed partitioned into energy used for body maintenance and for gain; thus, the net energy values in Mcal per unit (lb or kg) are needed for each feed for maintenance (NE_m) and for gain (NE_g)(see Part V, Composition of Feeds).

With such information available, the net energy system may be used two ways: (1) to calculate the quantity of feed necessary to meet the animal's energy needs and to compound a ration to supply the needed concentration of energy per unit of dry matter, and (2) to predict weight gains and to determine whether cattle have gained in accordance with expectations. Instructions on how to use the net energy method for these purposes, along with examples, are given in Chapter 18, Feeding Standards/Ration Formulation.

MANAGING FEEDLOT CATTLE

Although it is not possible to arrive at any overall, certain formula for success in operating a cattle feedlot, those operators who have made money have paid close attention to the details of management.

There are many facets of cattle management. Only those that are unique to cattle feedlots, and that have not been covered elsewhere in this book, will be discussed in the sections that follow.

Backgrounding

Backgrounding is the preparation of cattle from weaning until placing on finishing rations. It involves maximum roughage consumption and moderate gains.

The growing of calves from weaning until placing on finishing rations is not new. Only the term *backgrounding* is new. Likewise, some new "wrinkles" have been added to the method of conducting it.

Growing calves to the yearling stage for placement in feedlots was, and still is, known as growing stockers. Farmers have, for many years, fed high-roughage rations to calves prior to marketing. Some ranch operators have, historically, retained calves for a second grazing season. Also, wintering cattle on small grain pasture is a well-established practice in the South and Lower Plains.

The term *backgrounding* came in with the development of large commercial cattle feedlots—outfits that usually had limited amounts of available roughage and other cheap feeds, and that had need for, on a year-round basis, growthy, but unfinished, cattle of a certain weight, usually at beginning weights of 400 to 500 lb and ending weights of 600 to 700 lb.

It isn't economical for large feedlots to tie up capital for feeding cattle where limited gains are involved. It follows that high grain prices and low roughage prices make for more backgrounding. Conversely, low grain prices and high roughage prices result in less backgrounding.

KINDS OF CATTLE TO BACKGROUND

Generally speaking, the English beef breeds are best suited for backgrounding purposes. This is because they should be grown to approximately 600 to 750 lb before placing on finishing, or high-energy rations. Holsteins and some of the larger, growthier exotics are not well suited to backgrounding, unless heavy finishing weights are planned. They need to be placed on high-energy rations at weaning time; otherwise, they will not finish at desirable weights of 1,050 to 1,100 lb—they will be too heavy at market time.

RATE OF GAIN OF BACKGROUNDED CATTLE

Properly backgrounded cattle should gain from 0.75 to 1.50 lb/head/day. Cattle finishers object to cattle that have made higher gains, because it lessens, or eliminates, compensatory growth. That is, when fed high-energy rations, animals that have been backgrounded so as to make minimal daily gains usually gain better than similar cattle that have been fed more liberally during the backgrounding period. For the latter reason, when contracting for backgrounding calves, feedlots commonly specify the kind of ration and the range in gains.

Handling Newly Arrived Cattle

The most critical period for feeder cattle is the first 21 to 28 days in the feedlot. The following recommendations pertaining to incoming cattle will minimize death losses and maximize performance:

- **Provide clean, dry, comfortable quarters**—Whether it be an open lot or a building, incoming cattle should be provided with clean, dry, comfortable quarters. A dry and comfortable bed for resting is very essential because cattle are tired and have a low resistance to respiratory diseases.
- **Process upon arrival or delay it for about 28 days**—The relative merits of processing calves (1) at point of origin, (2) upon arrival at destination, or (3) 2 to 3 weeks after arrival are often debated. A common view is that stressed calves will not develop antibodies efficiently until recovered from stress; thus, vaccination should be delayed until the stress has been overcome.

In a well-designed experiment, involving 358 calves, the University of California provided the answer to this question.

The calves originated in Texas, and were in transit 32 to 38 hours with no rest stops. In all loads, ⅓ of the calves were processed at origin, ⅓ upon arrival at destination, and ⅓ were delayed 2 to 3 weeks after arrival. Processing consisted of branding, castration, ear tagging, use of a pour-on grubicide, vaccination (IBR, PI3, blackleg, and malignant edema), intramuscular injection of vitamins A, D, and E and 1 g of oxytetracycline.

Recognizing that the effect of time and place of processing on the entire feeding period from purchase to slaughter is the important thing, rather than the effect on the first month, the California workers carried part of these cattle from arrival through slaughter.

The California study showed the following relative to time of processing:

1. For the entire feeding period—from purchase to slaughter—calves processed upon arrival gained at a slightly faster rate and had a higher feed efficiency than calves which were processed immediately prior to shipment, or calves which were delay-processed 13 days after arrival.

2. At the end of the 27-day receiving period (the first 27 days after arrival), the calves processed on arrival had a weight advantage of 10 lb over those processed at origin and 11 lb over those delay-processed. At slaughter, these advantages had been increased to 46 and 26 lb, respectively. It appears, therefore, that if calves attain a weight gain advantage during the first month after arrival in the yard, they will retain that advantage throughout the entire feeding period.

3. Based on (a) rate of gain, (b) disease resistance, and (c) cost per pound of gain for feed, processing, and medication, these studies indicate that processing at point of arrival is best, and that processing at point of origin is preferable to delayed processing.

NOTE WELL: More recent studies by researchers at both the University of California and Colorado State University showed that delaying processing for 28 days causes less stress to animals than processing upon arrival or 14 days later—that delaying processing 14 days is not long enough.

- **Provide clean, fresh water**—Give the cattle easy access to clean, fresh water because they are usually dehydrated and thirsty upon arrival and will drink water before they eat feed. Open water tanks are preferable to automatic water bowls because most farm and ranch cattle are accustomed to drinking from tanks or ponds.
- **Provide a palatable ration**—Feeding a palatable ration—one that cattle will start eating soon after they are unloaded in the feedlot—will reduce the incidence of shipping fever and make the cattle recover their weight loss more rapidly.

1. **Roughage.** The best roughage for newly arrived feedlot cattle is *long grass* hay, because it is very similar in composition and taste to the grass to which most range cattle have been accustomed. Thus, cattle will usually eat long grass hay more quickly than any other roughage. In areas where grass hays are not available, or are too expensive to feed, any other nonlegume roughage can be fed, such as corn silage, sorghum silage, cottonseed hulls, corncobs, or grass-legume hay that contains more grass than legumes. Above all, do not feed high-quality alfalfa hay because it is too laxative and it will cause scouring which will trigger shipping fever. The same may be said relative to alfalfa haylage or alfalfa silage.

Corn silage of approximately 65% moisture content is an excellent feed for new cattle. If cattle do not eat the corn silage too well at the outset, the feeder should sprinkle a little grass hay on the top of it to encourage them to start eating. Also, buffering corn silage to a pH of 7.0 will increase feed intake.

2. **Concentrate.** Incoming cattle may be fed approximately 4 lb of concentrate per head daily, with a breakdown between protein supplement and grains as follows:

   a. Two pounds of a high bypass natural protein supplement, such as blood meal, corn gluten meal, or protected soybean meal, or a mixture of high bypass ingredients, preferably with a little cane molasses added to improve palatability. The protein supplement should be fortified so as to provide 50,000 IU of vitamin A daily. For heavily stressed cattle, the protein supplement should also contain a high level of antibiotic, or a combination of antibiotic and a bactericidal agent such as sulfamethazine. The following level of antibiotic-sulfamethazine is recommended:

   Feed 350 mg of antibiotic plus 350 mg of sulfamethazine per head daily to newly arrived cattle for a period of 28 days. With the antibiotic-sulfamethazine treatment, shipping fever is practically alleviated.

Do not feed urea for the first 28 days after the cattle arrive. Starvation destroys the ability of the rumen to utilize urea or other nonprotein nitrogen and makes cattle more sensitive to urea toxicity. Therefore, it is not wise to put extra stress on cattle by feeding urea during this adjustment period.

b. Two pounds of cereal grain or beet pulp per head daily, with the grain processed in the usual manner. The grain level can be raised at the rate of 1.0 lb/head/day if it seems desirable.

Fig. 19-64. Incoming cattle may be started on about 4 lb of concentrate/head/day, consisting of 2 lb of grain and 2 lb of protein supplement. (Courtesy, Ralston Purina Co., St. Louis, Mo.)

It has been, and still is, common practice to start cattle on a high-roughage ration, then gradually change them to a high-concentrate ration as they progress through the feeding program. However, based on California and New Mexico studies, it appears that for calves (not older cattle), a starting ration consisting of 75% concentrate and 25% hay (roughage) is best.

• **Satisfy mineral hunger**—Incoming cattle are usually hungry for minerals, especially if they have been on dry range forage. Thus, they should have access either to a mineral mixture consisting of two parts of dicalcium phosphate and one part of salt, or to a good commercial mineral.

• **Consider vitamin needs**—Unless incoming cattle come off green feed, vitamin A (50,000 IU per head per day) and vitamin E (100 IU per head per day) may improve gains and feed efficiency.

• **Observe, isolate, and treat sick animals**—Newly arrived cattle should be observed at least twice daily. Sick animals should be removed and treated. Treating sick animals promptly, rather than waiting until tomorrow, may mean the difference between life and death. Animals that show clinical signs of shipping fever—sunken eyes, runny nose, drooling at the mouth, labored breathing, and/or weaving (unsteady gait)—should be isolated in a separate *sick pen* or *hospital*.

Rest, fresh water, good feed, proper medication, and TLC (tender, loving care) are the cardinal essentials for preventing shipping fever and death losses.

Schedule for Getting Cattle on Feed

When new cattle arrive at the feedlot, the objective is to get them on full feed as rapidly as possible, without throwing them off feed. This is not easily accomplished because many factors influence the difficulties experienced in starting new cattle on feed, among them: (1) the length of time that the cattle have been without feed; (2) the kind of feed to which the cattle were accustomed prior to shipment; (3) the age of the cattle—young cattle adapt to a change in feed more easily than old cattle; (4) whether the cattle have been fed and watered out of troughs before; (5) the weather conditions; and (6) existing nutritional deficiencies.

TRADITIONAL PROCEDURES OF GETTING OLDER CATTLE (NOT CALVES) ON FEED

When first brought into the feedlot, cattle that are not accustomed to grain may be started on feed by either of the following procedures:

1. Self-fed long grass hay (and/or corn or sorghum silage), and hand-fed concentrate according to the following schedule (with the cattle automatically lessening their self-fed hay consumption as the grain is increased):

First day—Feed 4 lb of concentrate/head/day, consisting of 2 lb of grain and 2 lb of protein supplement.

Daily increase—Step up the grain by 1 lb/head/day until cattle are receiving 1 lb/100 lb body weight.

Increase every third day—After a level of 1 lb daily/100 lb body weight is reached, make increases every third day as follows:

Calves	¼ lb
Yearlings	½ lb
2-year-olds	1 lb

2. Hand-fed a mixed ration of chopped grass hay (and/or corn or sorghum silage) and concentrate, with the proportion of roughage decreased and the grain increased according to the following schedule:

Day	Kind of Feed	Percent of Roughage
1	Grass hay and/or nonlegume silage	100
2–4	Grass hay plus starter	60–90
5–14	Starter ration	40–60
15–21	Transition ration	15–40
22 to market	Finisher ration	5–15

Although one of the above procedures may serve as a useful guide, it is recognized that no set of instructions can replace the intuition and good judgment of an experienced feeder.

After cattle are on full feed, they may either be self-fed or hand-fed. Most large feedlots feed twice daily, barely letting the cattle clean up the previous feed before the next feeding.

SCHEDULE AND RATION FOR GETTING CALVES ON FEED

Most feeders start cattle on a high-roughage ration, then work them over to a high-concentrate ration as they progress through the feeding program. However, based on work done by California and New Mexico researchers, it appears that for calves (not older cattle) a starting ration consisting of 25% hay (roughage) and 75% concentrate is best, and that calves may be started immediately on a high-energy ration.

Among the problems encountered in new cattle are feed and water refusal, acidosis, bloat, and diarrhea. Refusal of feed and water is generally due to the fact that the animals are not used to conventional troughs and/or the feed is so different.

Acidosis generally results from feeding hungry cattle excessive levels of rapidly fermentable feeds, such as grains. The condition is characterized by an accumulation of lactic acid in the rumen and a lowering of the pH in the blood and urine. The problem can be minimized by starting cattle on a high-roughage ration and shifting them gradually to a high-concentrate ration.

Bloat occasionally occurs in new cattle, although it is more frequent during the later stages of feeding. Bloat and diarrhea in new cattle can generally be prevented by feeding generous quantities of such roughages as straw, grass hay, cottonseed hulls, or corncobs.

Amount to Feed; Rate of Gain

There are two schools of thought relative to the amount to feed cattle after they are placed in the finishing lot. The traditional method calls for getting them on full feed as quickly as possible, then keeping them on full feed until marketing. The newer concept calls for limit (controlled) feeding in the early part of the finishing period, followed by full feeding the latter part. The wise manager will choose between the two programs, recognizing that neither is adapted to every operation.

• **Full feeding (*ad lib*)**—Those favoring full feeding in the finishing lot from beginning to end point out that, once a sufficient amount of ration is consumed to meet the maintenance needs of a finishing animal, the remainder is converted to gain with remarkable efficiency. Thus, as shown in Fig. 19–65, by adding 4 lb to the daily feed intake of a 600–lb steer, rate of gain may be increased by 1$\frac{1}{10}$ lb per day. Conversely, low feed intake results in too high a percentage of the total nutrients being expended for maintenance.

Although full feeding does result in animals gaining rapidly early in the finishing period, the rate of gain tapers off as

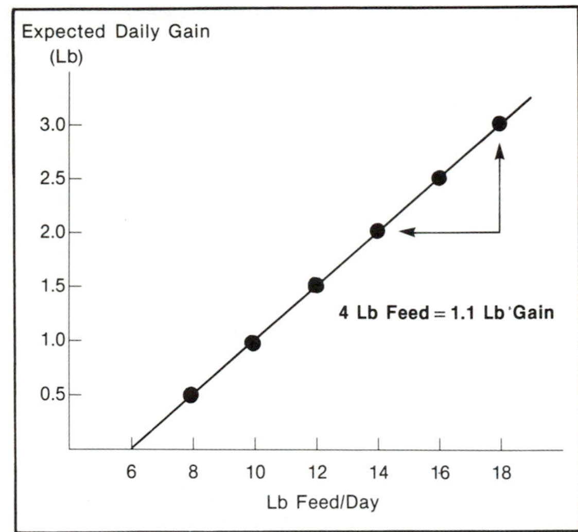

Fig. 19–65. Relationship of daily feed intake to rate of gain; 600-lb steers fed 85% concentrate ration.

the finishing period progresses. It is noteworthy, too, that high ownership costs (labor, interest on investment, etc.) favor pushing cattle for maximum gains and shortening the finishing period.

• **Limit (controlled) feeding**—Instead of giving cattle all the feed that they will consume, limit feeding calls for restricted feed intake during the early part of the finishing period—for feeding only enough to produce the desired gains, followed by full feeding. This concept evolved in the early 1980s as the result of grain prices being comparatively more favorable than forage prices. During periods of favorable grain prices, limit feeding makes it possible to feed high energy grain rations in the early part of the finishing period without getting the cattle too fleshy. Generally, limit feeding results in fewer digestive upsets, lessens feed storage and labor, minimizes feed wastage, and improves feed efficiency for the entire finishing period by about 5% because of compensatory gain.

In general, cattle will consume daily an amount (on an air-dry basis) equal to 2.5 to 3.0% of their liveweight. Feed intake will vary according to the condition of the cattle, the palatability of the feeds, the energy of the ration (in general, animals eat to meet their energy needs), the weather conditions, and the management practices. For example, older and more fleshy cattle consume less feed per 100 lb than do younger animals carrying less condition; thus, mature, overfinished steers will consume feeds in amounts equal to about 1.5% of their liveweight, whereas thin steers under 2 years of age will consume fully twice as much feed per unit liveweight.

Overfeeding is undesirable, being wasteful of feeds and creating a health hazard. When overfeeding exists, there is usually considerable leftover feed and wastage, and there is a high incidence of bloat, founder, scours, and even death. Animals that suffer from mild digestive disturbances are commonly referred to as *off feed*.

FACTORS AFFECTING FEED INTAKE

A number of factors play major roles in governing feed intake; among them, the following:

1. **Size and weight.** Large feedlot cattle consume more feed, animal for animal, than small cattle. But they may or may not be more efficient than small cattle when consideration is given to the production efficiency of their dams, and carrying them to the same degree of finish. Table 19-36, which gives the average expected dry matter intakes per head daily for cattle, illustrates this situation. Note that the weight of cattle when first placed on feed has a very significant effect on the maximum amount of feed dry matter that they will consume.

TABLE 19-36
FEED INTAKE FOR MEDIUM FRAME STEERS
(daily dry matter intake)
ON HIGH CONCENTRATE FEEDLOT RATIONS[1]

Weight When Placed on Feed, Lb[2]	Current Weight of Cattle, Lb[2]									
	300	400	500	600	700	800	900	1,000	1,100	1,200
	←——— feed intake, lb[2] ———→									
300	8.80	10.92	12.91	14.57	15.71	16.34	16.46	16.08	15.22	
400		11.59	13.70	15.48	16.73	17.46	17.69	17.41	16.64	15.39
500			14.49	16.39	17.75	18.58	18.91	18.74	18.07	16.91
600				17.29	18.76	19.71	20.14	20.07	19.50	18.43
700					19.78	20.83	21.37	21.40	20.92	19.96
800						21.96	22.60	22.73	22.35	21.48
900							23.83	24.06	23.78	23.00

[1]Oklahoma State University feed intake predictions.

[2]To convert lb to kg, divide by 2.2.

2. **Age and condition.** Older and more fleshy feedlot cattle consume less feed per unit of liveweight than do younger, leaner animals. Mature animals in good condition may be expected to consume amounts of dry matter equal to 2%, or more, of their liveweight, whereas thin animals eating high-quality roughage should eat amounts equal to 3% of their liveweight per day.

3. **Digestible nutrient content (energy density).** As digestible nutrient content (energy density) increases, consumption of feed dry matter is usually reduced. Most feeders think of energy density in terms of roughage:concentrate ratio. When density is very low, as in a high-roughage ration, animals simply cannot hold enough to make gains. As energy density increases, gain increases. It follows that feed efficiency is improved in high- and all-concentrate rations, due to their high energy. In the final analysis, however, the comparative price of concentrates and roughages—the economics of the situation—will be a major determining factor.

4. **Rumen *fill*.** The *fill* in the rumen places a ceiling on feed intake. Since low-quality roughages pass through the rumen at a slower rate than high-quality roughages, they can limit the amount of total feed that the animal can consume in a 24-hour period. Of course, this is of little consequence with fattening-type rations, which normally contain limited amounts of roughages and roughages of good quality. The amount of concentrate in the ration has a marked effect on the total pounds of feed consumed daily. Thus, when fed all-concentrate rations, the total energy intake of cattle over a 24-hour period is not much greater than with a bulky, low-concentrate ration. Apparently, this is due to the regulation of total calorie intake by the ruminant.

5. **Environmental stress.** Cattle feeders have long known that environmental stress caused by high and low temperatures, mud, and other adverse environmental factors can affect the voluntary consumption of feed. For example, feedlot cattle consume less during very hot weather than in cold weather.

The lowering of feed consumption due to heat stress may be lessened by lowering the roughage in the ration, cooling the drinking water, providing shade, and adding high levels of vitamin A.

6. **Physical makeup of the ration.** Coarsely processed grains are more palatable to cattle than finely processed grains. Thus, they will eat more of such feed.

7. **Protein or phosphorus deficiency.** A protein deficiency can markedly reduce feed intake by depressing the rumen bacterial count and consequently the rate of breakdown of feeds. A phosphorus deficiency can cause a reduction in feed intake, and even a depraved appetite.

8. **Propionic acid in the blood curbs appetite.** All mammals, including humans, have an "appetite center" at the base of the brain in the lateral region of the hypothalmus. Certain nerve cells actually regulate energy intake, causing the sensation of hunger, or preventing the animal from consuming too much. But there is a great species difference as to how much food is enough—or too much. A growing boy "wolfs down" 8 to 9% of his body weight daily. He is outdone by a baby chick, which eats 10% of its body weight. A hog eats 5 to 6% of its body weight. But a fattening steer consumes only 2 to 3% of its body weight. Why the difference? It appears that propionic acid is the triggering mechanism which tells a steer when to stop eating. It acts on the nervous center of ruminants. Since increased grain levels tend to increase propionic acid, it is conjectured that grain rations, which increase propionic acid levels of the blood, are self-limiting when it comes to feed intake. For this reason, cattle eat fewer pounds per head daily of a high-concentrate ration than of a high-roughage ration, but both groups tend to take in about the same level of energy.

9. **Heritability.** Some cattle are better "eaters" than others—they will consume more feed. This may be due either to a difference in rumen capacity and/or a difference in *threshold* for circulating metabolites in the bloodstream which affect the appetite center of the hypothalmus.

10. **Other factors affecting feed intake.** The feed intake by cattle is also affected by (a) moisture level—very high-moisture feeds reduce total dry matter intake; (b) dustiness—

dusty feeds lower total feed intake; (c) lack of water—depriving cattle of water will markedly reduce feed intake; (d) frequency of feeding—more frequent feeding results in higher feed consumption; (e) freshness of feed—cattle will consume more clean, fresh feed than stale feed; and (f) previous ration.

FACTORS AFFECTING RATE OF GAIN

In addition to feed intake, the rate of gain by feedlot cattle is influenced by the following factors:

1. **Sex.** Under feedlot conditions, at comparable weight and finish, bulls can be expected to make 10% greater gains than steers, and steers can be expected to make about 10% greater gains than heifers.

2. **Implants and growth stimulants.** The use of certain implants and growth stimulants in finishing steers and heifers usually increases gains by 8 to 12%.

Progressive Changes in Feedlot Cattle

It is important that cattle feeders be cognizant of the progressive changes in (1) rate of gain, (2) feed consumption, (3) feed efficiency, and (4) cost of gain that normally occur in feedlot cattle, from start to finish of the feeding period. Of course, many factors influence the degree of these changes.

Researchers at the Arizona Agricultural Experiment Station recorded the changes at 28-day intervals in 41 lots of feedlot cattle, with all lots taken from start to finish. Their findings are given in Table 19–37.

TABLE 19–37
AVERAGE DAILY GAIN AND FEED PER POUND OF GAIN OF CATTLE AT 28-DAY INTERVALS[1] [2]

28-Day Feeding Period	No. of Days on Feed	Pay Weight of Cattle on Feed			Average Daily Gain	Average Daily Air Dry Feed	Lb of Air Dry Feed per Lb of Gain
		Beginning	Ending	Average			
	(days)	(lb)	(lb)	(lb)	(lb)	(lb)	(lb)
1st	1–28	597	697	647	3.56	22.75	6.39
2nd	29–56	697	783	740	3.06	24.24	7.92
3rd	57–84	783	858	821	2.69	23.43	8.71
4th	85–112	858	923	890	2.30	23.83	10.36
5th	113–139	923	979	951	2.07	24.43	11.80

[1]Data collected by H. A. Meier and W. H. Hale, Ariz. Ag. Exp. Sta., unpublished. Data include 9 experiments, 41 lots of yearling steers, and 640 animals. All cattle were started on relatively high-roughage rations, with the roughage decreased as the period progressed.

[2]To convert lb to kg, divide by 2.2.

The Michigan Station researchers used a slightly different approach to obtain changes at 28-day intervals. They fed 7 lots of cattle, with each lot carrying a different length feeding period in increments of 28 days, ranging from 115 to 283 days on feed; and each group of cattle was closed out and slaughtered at the end of its feeding period. Their findings are given in Table 19–38.

TABLE 19–38
EFFECT OF LENGTH OF FEEDING PERIOD ON (1) RATE OF GAIN, (2) CONSUMPTION, AND (3) FEED EFFICIENCY[1]

	Days on Feed						
	115	143	171	199	227	255	283
No. steers	8	8	8	8	8	8	8
Av. initial weight . (lb)	655	657	685	685	672	679	673
Av. final weight . . (lb)	928	990	1,037	1,084	1,133	1,203	1,224
Av. total gain . . . . (lb)	273	333	352	399	461	524	551
Av. daily gain . . . . (lb)	2.73	2.33	2.06	2.01	2.03	2.05	1.95
Daily feed/100-lb body weight[2] . . (lb)	2.60	2.44	2.28	2.19	2.20	2.11	2.05
Total feed/100-lb gain[2] . . . . (lb)	869	861	952	962	975	968	996
Concentrate:roughage ratio	77:23	77:23	77:23	77:23	77:23	77:23	77:23

[1]Merkel, R. A., H. E. Henderson, and H. W. Newland, "Effect of Length of Feeding Period on Rate, Composition and Cost of Gain," *Michigan Beef Cattle Day Report*, Michigan State University, 1966, p. 10.

[2]To convert lb to kg, divide by 2.2.

Based on these two studies, the following deductions may be made relative to the progressive changes in feedlot cattle, from start to finish: (1) rate of gain decreases; (2) daily feed consumed per 100 lb of body weight decreases; (3) feed per 100 lb of gain increases; and (4) feed cost per 100 lb of gain increases.

Cull Out; Top Out

Obvious poor doers should be taken out early and marketed at Standard grade. Where individual weighing can be made, consideration should be given to the practicality of individually tagging (with duplicate tags, one in each ear) and weighing incoming calves; weighing them again at the end of the grower-ration period and prior to going on finishing rations; then culling out the bottom 10%.

Also, cattle should be sold when they make their grade, thereby avoiding loss in efficiency, excess finish, and too heavy weights. Usually, it is unwise to challenge a sagging market by holding and feeding for a higher market. There is no need to put feed and labor into heavy cattle at a cash discount when younger cattle will use these resources more efficiently.

Overfinishing

Experienced cattle feeders are fully aware of the fact that to carry finishing cattle to an unnecessarily high finish is usually prohibitive from a profit standpoint. This is true because the gains in weight then consist chiefly of fat but little lean tissue and water. In addition, a very fat animal eats less heartily, with the result that a small proportion of the nutrients, over and above the maintenance requirement, is available for making body tissue.

Table 19-39 and Fig. 19-66 show that the heavier the cattle, the more expensive the gains. Also, these figures point up (1) the importance of topping out finished cattle, rather than waiting until the entire lot is ready; and (2) the reason why it is generally wise to sell cattle when they are ready to go, rather than to hold for a higher market.

TABLE 19-39
COST OF ADDING ONE POUND OF WEIGHT ON BEEF STEERS AT VARIOUS WEIGHTS[1 2]

Cattle Weight, Lb	If Cost of Corn Per Bushel is—				
	$1.50	$2.00	$2.50	$3.00	$3.50
722	0.38	0.44	0.50	0.56	0.62
758	0.33	0.39	0.44	0.50	0.56
794	0.34	0.40	0.46	0.52	0.58
829	0.32	0.38	0.44	0.51	0.57
864	0.34	0.40	0.46	0.53	0.59
898	0.35	0.41	0.48	0.54	0.61
931	0.36	0.43	0.50	0.56	0.63
963	0.37	0.44	0.51	0.59	0.66
994	0.39	0.46	0.53	0.61	0.68
1,024	0.40	0.48	0.55	0.63	0.71
1,053	0.42	0.49	0.57	0.65	0.73
1,080	0.44	0.53	0.61	0.69	0.78
1,106	0.46	0.54	0.63	0.72	0.80
1,130	0.46	0.55	0.64	0.73	0.82

[1]Cost of gain at indicated weight based on corn price shown in column cost includes a gross feedlot margin of about $.25 per day, but does not cover nonfeedlot costs. Oklahoma State University data.

[2]To convert cost per lb to cost per kg, multiply cost figures by 2.2. To convert lb to kg, divide by 2.2.

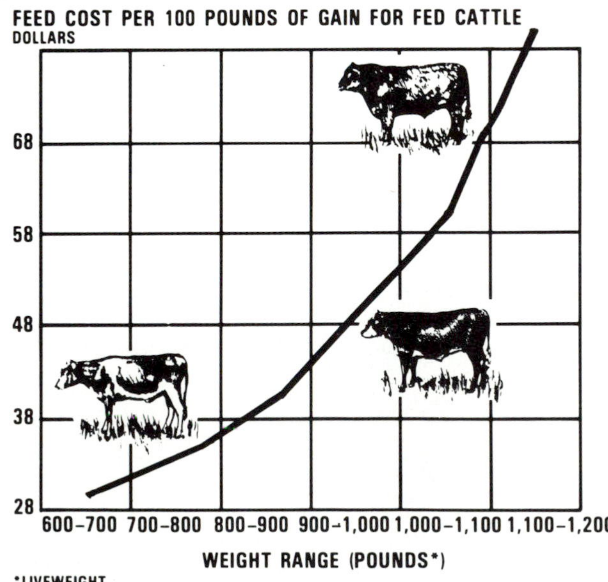

Fig. 19-66. This graph illustrates changes in feed conversion efficiency for cattle from normal feeder weights to slaughter weights. Note that feed costs per 100 lb gain more than double from 600-700 lb to 1,000-1,100 lb, and that the conversion efficiency ratio changes even more sharply when cattle pass 1,100 lb.

Feed Bunk Management

Feed bunk management is a combination of management factors involved in obtaining maximum performance, minimum digestive disorders, and keeping cattle on feed. Feed bunk management and quality control are directly involved with obtaining maximum and economical performance from cattle. It should be every feeder's goal to obtain maximum feed intake of a consistently high-quality ration, since both rate and efficiency of gain are directly related to nutrient intake.

WEATHER AFFECTS EATING AND DRINKING HABITS

During hot weather, feedlot cattle *peak* their eating during early morning and again during the evening hours—when it is cool. With heat, night drinking increases. In cool weather, they eat more during midday. Feeders should sense these changes in cattle eating habits and program their feeding accordingly.

Cattle eat more following a bad storm or a hot spell. Thus, at such times the bunks may be *slick* for 2 to 3 hours and the cattle may line up waiting to be fed. When this happens, the ration should be increased. By going to a higher roughage ration at these times, the problems from acidosis and laminitis can be minimized.

MUD PROBLEM

University of California studies show that mud can reduce cattle gains by as much as 25 to 35%. Thus, it is important that the problem be minimized, especially in high rainfall areas. Good drainage is the first essential. This should be assured at the time the feedlot is located and constructed. Mounds, preferably perpendicular to the feed bunk, will provide cattle a dry place on which to lie down. Concrete aprons along the bunk will provide them with solid footing on which to stand and feed. Also, lessening of cattle density during the winter months—fewer animals per lot—is an effective method of controlling the mud problem. Thus, many feedlots plan to feed fewer cattle during the muddy season.

RECORDS

Complete and well-kept records are a must in the operation of a cattle feedyard, even though they require a lot of time and expense. Deficient records and deficient managers generally go hand in hand. Modern computer programs make record processing easier and more efficient.

Record Forms

There is no limit to the number of different kinds of record forms that can be, and are, kept in a given feedlot. Also, there is little similarity in record forms between lots, due to differences between people, primarily managers and bookkeepers. The important things are that (1) record forms be so designed as to facilitate record keeping, with as much ease, efficiency, and accuracy as possible; and (2) records be kept.

Figs. 19-67 and 19-68 show two basic record forms; Fig. 19-67 is a Daily Record, whereas Fig. 19-68 is a Monthly, Cumulative, and Final Feed Summary. Many variations of these can be made.

DAILY RECORD

Feedlot: ____________ Pen No. ________ Date Started ____________
Month Day Year

Day of Month	Head In						Death Losses	Head Out			Daily Feed			Sold		Comments
	No.	Origin	Pur. Price	Total Pur. Wt.	Total Wt. at Lot	Av. Wt. at Lot	(cause)	No.	Total Wt.	Av. Wt.	Total Lb	Lb/ Head/ Day	To	Price/ Cwt	Carcass Grade	
1																
2																
3																
4																
5																
6																
7																
8																
9																
10																
11																
12																
13																
14																
15																
16																
17																
18																
19																
20																
21																
22																
23																
24																
25																
26																
27																
28																
29																
30																
31																

Fig. 19-67. Form for Daily Record.

MONTHLY, CUMULATIVE, AND FINAL FEED SUMMARY

Feedlot: ____________ Period: ____________
Pen No. ______ No. Head ______ Date Started ______ Date Closed ______
RATION:

Ration No.	Total Pounds	Price/Ton	Total Cost
	(lb)	$	$
#1			
#2			
#3			
#4			
#5			
#6			

FEED ANALYSIS:

Total feed fed ____________ ______ lb
Total cost of feed to date ____________ $ ______
Feed days (no. head x days) ____________ ______ no.
Net weight out ____________ ______ lb
Net weight in ____________ ______ lb
Net gain ____________ ______ lb
Av. weight out ____________ ______ lb
Av. weight in ____________ ______ lb
Feed per head per day ____________ ______ lb
Cost per head per day ____________ ______ ¢
Gain per head per day ____________ ______ lb
Cost per lb gain ____________ ______ ¢
Feed conversion (lb feed/lb gain) ____________ ______ lb

OTHER COSTS:

Milling charges ____________ ______
Mineral charges ____________ ______
Medication ____________ ______
Management ____________ ______
Labor ____________ ______
Physical plant (other than milling) ____________ ______

Fig. 19-68. Form for Monthly, Cumulative, and Final Feed Summary.

Among other necessary records are the following:

1. Feed costs, with this record kept by individual pens.
2. Grain inventory.
3. Roughage inventory.
4. Feed projections ahead.
5. Cattle receiving and movement records.
6. Sick pen and movement records.
7. Sick pen costs.
8. Mortality slips, proofs of deaths, and post-mortem reports.
9. Maintenance and repair costs.
10. Routine office bookkeeping.
11. Customer billing for feed.
12. Closeout records.

Make 28-Day Test Weights

Taking 28-day test weights will not adversely affect the performance of feedlot cattle, provided the cattle are handled properly. Check weights should include a representative cross section of the cattle in the yard, including age, weight, type, background, and sex. Where it is not convenient, or it is not desired, to weigh an entire lot of cattle, "markers"—cattle of certain odd colors, animals with tail switches clipped, etc.—may be weighed. Also, it is important that weighing be done at the same time of day, and that the lots be weighed in the same order, due to the effect of rumen fill.

CUSTOM (CONTRACT) FEEDING

Custom cattle feeding is the feeding of cattle for a fee, without taking ownership of the animals.

Capital requirements, periods of severe economic conditions (like scarce money and high interest), times of depressed feeder cattle prices, and adverse pasture conditions caused custom feeding to grow following World War II. These same forces, along with the need for high occupancy (full feedlots) and increased integration, have resulted in further expansion of custom feeding.

Most custom feeders have developed large, highly mechanized, and very efficient plants. Usually, they have on their staffs highly trained nutritionists and veterinarians who are charged with the responsibility of formulating rations and of obtaining maximum gains and feed efficiency at the lowest possible cost. Through custom feeding, they sell the use of their facilities, services, and know-how to cattle owners, usually with profit to each party.

The proportion of custom-fed cattle to cattle owned by the feedlot varies (1) in period of time—it increases in times of financial stress (when cattle feeding is not profitable, money is scarce, and interest is high); (2) according to area—for example, there is more custom feeding in Texas than in any other state; and (3) according to size of feedlot—generally speaking, the larger the feedlot, the greater the percentage of custom feeding. Some feedlots do not do any custom feeding whatsoever; others are almost wholly on a custom basis; but most lots have part of each. Feedlots that do both—those in the dual role of custom feeding and owning cattle—vary in the proportion of cattle in each category, but most of them seem to prefer about ⅔ custom-fed cattle and ⅓ ownership. It's a good bread-and-butter division; in times when fed cattle lose money, such a feedlot has sufficient assured income to pay its bills.

The ownership of custom-fed cattle is diverse. It includes (1) cow-calf producers (farmers and ranchers) who wish to retain ownership of the cattle that they produce through the feedlot phase, (2) stocker operators, (3) packers, and (4) investors, including limited partnerships, corporations, cattle buyers, cattle dealers, and others.

Custom feeding contracts should always be detailed and in writing, for a good understanding is the best way to avoid a misunderstanding. Also, contracts should be fair to both parties—to both the feedlot owner and the cattle owner.

Types of Custom Feeding Contracts

The services rendered vary from feedlot to feedlot and according to the type of contract. In some instances, the services may be so complete that the customer never sees the cattle. The feedlot operator may buy the feeder cattle, feed them, market them, and send the customer (the client) a check for the balance, after deducting input costs, interest charges, and custom feeding charges. Less complete services are usually available to suit the customer.

Both the feedlot owner and the cattle owner should analyze different types of contracts and determine which best fits their respective circumstances. Some feedlots offer several types of contracts, thereby according the cattle owner a choice.

Competition may dictate the type of contract and the changes made. But by knowing the variables and managing them correctly, the feedlot owner can write and carry out a contract that will be fair to both parties.

Generally speaking, contracts with fixed charges are the most satisfactory and the most common, primarily because there is less room for misunderstanding.

Although there are many types of custom cattle-feeding contracts, and many variations of each kind exist, most of them can be classified under one of the following types:

1. Feed cost plus daily yardage fee per head.
2. Feed cost plus markup.
3. Feed cost plus (a) daily yardage per head, and (b) markup per ton of feed.
4. Agreement to purchase contract.
5. Payment for weight gained.
6. The incentive basis contract—the higher the gain, the higher the charges.

HOGS FOLLOWING CATTLE

There was a time when hogs following feedlot cattle was commonplace. But the practice declined with the advent of large, specialized commercial feedlots of 1,000 head, or larger, capacity. Today, only a few farmer-feeders of the Corn Belt have hogs following cattle, and the practice is almost nonexistent in the large commercial cattle feedlots of the nation. The primary reasons given by cattle feeders for the decline in this practice are (1) feeds are being processed in a more sophisticated manner than formerly, with the result that few grains pass through cattle whole; (2) hogs tend to get cattle up and to get into troughs; (3) an increase in fenceline feeders, which won't keep hogs in; and (4) few hog-tight commercial feedlots.

Nevertheless, cattle feeders who have a convenient source of feeder pigs, who are not "allergic" to keeping hogs, and whose cattle lots are fenced hog-tight, can add to their net income by having hogs follow cattle. The following hog:cattle ratio is recommended, using 75- to 150-lb pigs:

	If Whole Shelled Corn is Fed	If Ground or Rolled Grain is Fed
	(pig:steer ratio)	(pig:steer ratio)
Calves	1:3	1:5
Yearlings	1:2	1:4
2-year-olds	1:1½	1:3

Pigs following cattle can be very profitable. Assuming an average gain of 100 lb per pig and a selling price of 50¢/lb, that's $50. Subtracting $10 for protein supplements and other costs leaves a net of $40/pig.

Pigs sometimes inflict injury on heifers (injuring the vulva when they are lying down); therefore, their use is generally limited to steers.

Sows may be used, but because of their size they may create problems from getting into the feed and water facilities.

CONDUCTING APPLIED FEEDLOT TESTS

When carefully conducted, and properly interpreted and used, feedlot trials can be a valuable adjunct in the operation of a large feedlot. Among their virtues, the feedlot operator can study area and feed differences. Among their limitations, usually fewer controls can be afforded than in most university-conducted experiments, thereby resulting in less accuracy. For the latter reason, most of them should be looked upon as applied tests or demonstrations *per se,* rather than carefully controlled, basic experiments; terminology which doesn't detract from their value, but which does place them in proper perspective.

The procedure for conducting tests in a commercial cattle feedlot is outlined in Chapter 15, Feed Analysis/Feed Evaluation, under the heading, "Conducting Applied Feedlot Tests"; hence, the reader is referred thereto.

FEEDLOT POLLUTION CONTROL

Pollution control is a most critical factor in site selection and operation of a cattle feedlot. Remoteness from urban development is recommended because of dust and odor. Also, before constructing a cattle feedlot, the owner should become familiar with both state and federal regulations. The state regulations can be secured from the state Department of Environmental Quality. They differ from state to state, but most states require a catch basin (detention pond) sufficient to contain the runoff from a storm of the magnitude of the largest rainfall during a 48-hour period of the most recent 10 years. A feedlot may minimize runoff by locating near the top of the slope and, if necessary, by using diversion embankments to divert runoff from other areas.

Cattle feedlots located near centers of populations are having an increasing number of complaints lodged against them because of manure, dust, flies, and odor. Lawsuits, based on the nuisance law, are being filed against them.

Pollution Laws and Regulations

Both open lot and confinement cattle systems must comply with the pollution control regulations of the U.S. Environmental Protection Agency, and of the state in which they are located.

Cattle feedlot compliance with the pollution control regulations of the U.S. Environmental Protection Agency involves the following steps: (1) obtaining a copy of the Regulatory Provisions Affecting Concentrated Animal Feeding Operations; (2) completing the Permit Application Form 1, and submitting to the appropriate authority; and (3) obtaining a copy of, and compliance with, the Feedlot Point Service Category, which sets forth applicable effluent limitations for animal feedlot operations subject to this regulation.

(Also see Chapter 17, Animal Behavior/Environment, for further information relative to pollution laws and regulations.)

PASTURE FINISHING CATTLE

Fig. 19-69. Hereford steers on bromegrass pasture near Lincoln, Nebr. (Courtesy, C. B. & Q. Railroad Co., Chicago, Ill.)

When grains are scarce and high in price, more cattle are grass finished. But, because young cattle grow and do not reach market finish under usual pasture conditions, it is impossible to finish them at early ages and light weights without either supplemental feeding on pasture and/or lot finishing at the end of the grazing season.

Advantages and Disadvantages of Grain Feeding on Pasture

The **advantages** of grain feeding cattle on pasture, compared to strictly feedlot finishing, are:

1. Pasture gains are cheaper because (a) less grain is required per 100 lb gain; (b) grass is a cheaper roughage than hay or silage; and (c) less protein supplement is required.

Generally speaking, self-feeding on pasture vs drylot indicates that pasture saves about 100 lb of dry feed per 100 lb of gain. Thus, if we assume that it requires 500 lb of gain to finish a steer, then each steer would require 500 lb less feed on pasture than in drylot. If feed costs 6¢ per pound, that's a saving of $30 per steer on feed cost.

2. Less labor is required because the cattle gather their own roughage and the labor required for feeding roughage is eliminated. In brief, grass-finished cattle do their own harvesting. Furthermore, it may be possible to get satisfactory results with but one grain feeding each day in finishing on pasture, or the animals may be self-fed, with the caretaker merely filling the feeder at intervals.

3. Handling of manure is eliminated, the maximum fertility value of the manure is conserved, and there is no pollution problem. When pastures are utilized by livestock, approximately 80% of the plant nutrients of the crop is returned to the soil.

4. Pasture finishing eliminates any requirement for buildings.

5. Finishing cattle on pasture is especially adapted to the small feeder and to areas where some of the land should be kept in permanent pasture.

The **disadvantages** of finishing cattle on pasture, compared to strictly feedlot finishing are:

1. Most feeder cattle are marketed in the fall rather than in the spring. Therefore, feeder steers purchased in the spring and intended for pasture finishing are usually scarce and high in price.
2. Though less labor is required, less labor is available. The cropping season is a rush season.
3. During midsummer, the combination of heat, and flies may cause much discomfort to the animals and reduce the gains made.
4. Pastures may become dry and parched, reducing the gains made during dry seasons.
5. The manure is usually dropped on permanent pastures year after year, which may result in the neglect of the other fields.
6. In many pastures, availability of shade and water do not present a problem. However, some areas are less fortunate in this regard.

After both the advantages and disadvantages of pasture finishing are considered, the availability of cheap, rough pasture land and the price of concentrates will usually be the determining factors in deciding upon the system to follow.

Systems of Pasture Finishing

When cattle are finished on pasture, any one of the following systems may be employed:

1. Finishing on pastures alone—no concentrates being fed.
2. Limited grain allowance during the entire pasture period, usually by adding 15% salt or 10% fat to a concentrated mixture fed in self-feeders.
3. Full feeding during the entire pasture period.
4. Full or limited grain feeding on pasture following the period of peak pasture growth.
5. Short feeding (60 to 120 days) in the feedlot at the end of the pasture period.

The system of pasture finishing that will be decided upon will depend upon the age of the cattle, the quality of the pasture, the price of concentrates, the rapidity of gains desired, and the market conditions.

Basic Considerations in Utilizing Pastures for Finishing Cattle

The following points are basic in utilizing pastures for finishing cattle:

• **Moderate winter feeding makes for most effective pasture utilization**—The more liberally beef cattle are fed during the winter, the less will be their effective utilization of pasture the following summer—the less the compensatory gains. Generally speaking, for maximum utilization of pasture, stocker calves should be fed for winter gains not in excess of 1.25 lb/head/day, and yearlings not in excess of 0.9 lb.

• **Early pastures are "washy" but high in protein**—Cattle should not be turned to pasture too early. The first growth is extremely "washy," possessing little energy. However, the crude protein content of the forages is high during the early stages of growth and rapidly decreases as the forages mature. This would indicate the importance of pasturing rather heavily during the period of maximum growth in the spring and early summer.

• **Sudden changes are to be avoided**—Changes from drylot to pastures or from less succulent to more succulent pastures should be made with care; for grass is a laxative, and the cattle may shrink severely. Also, bloat may occur in pastures that contain considerable legumes.

• **Time of starting grain feeding on pastures is determined by condition of cattle and quality of pastures**—Cattle that have been fed grain rather liberally through the winter and are in good condition should usually be fed grain from the beginning of the grazing period. On the other hand, if they have been roughed through the winter, it may be just as well to feed the grain only during the last 80 to 120 days of the grazing season, after the season of peak pasture growth. The latter recommendation is made because it is sometimes difficult to get animals to consume grain when an abundance of palatable forage is available. At peak pasture growth, the animals should be started on feed and brought to full feed as rapidly as possible.

• **Grain supplements on pastures usually make for larger daily gains and earlier marketing**—Young cattle (calves and yearlings) on summer pasture usually do not grow at their maximum potential due to energy and protein deficiencies in the feed at various times of the season. Thus, the addition of a grain supplement for cattle on pasture makes for larger daily gains and earlier marketing—either directly off grass or with a shorter drylot finishing period. The owner thus avoids late fall competition and lower prices of strictly grass cattle. Also, because cattle that are grain fed on pasture can be marketed over a longer period of time, there is greater flexibility in the operations. However, cattle that are supplemented on summer pasture often sell for less to go into feedlots because feedlot operators fear that they may not gain as rapidly as cattle that are not supplemented on pasture; they feel that more rapid gains on pasture make for less rapid and efficient gains in the feedlot—for loss of compensatory growth. To answer this question, Nevada workers studied the subsequent feedlot performance of yearling steers receiving a feed supplement while on summer pasture vs nonsupplemented cattle. The pasture-supplemented cattle were fed pellets on the ground 3 times daily, at an average daily rate of 2.18 lb/head/day, or slightly less than ½% of their body weight. The supplemented cattle outgained the nonsupplemented cattle by 0.38 lb daily, or a total of 41 lb; and each pound of supplement produced 5.7 lb of gain. At the end of the pasture season, both groups of steers were placed in the feedlot. Subsequent feedlot gains of the 2 groups were essentially the same—407 and 406 lb for the supplemented and controls, respectively; an average daily gain of 3.3 lb by each group. These results indicate that yearling steers may be profitably supplemented on summer pasture at the above rates without adversely affecting their subsequent feedlot performance.

• **Whole corn preferred to rolled corn**—When self-feeding steers on pasture, whole corn is preferred to rolled corn for the following reasons: (1) Slightly less feed is required per 100 lb gain; (2) it alleviates processing cost; and (3) it results in less incidence of founder and rumen parakeratosis because whole corn supplies some *roughness factor* in the ration to stimulate the rumen.

• **Protein supplement not needed on good pastures**—So long as pasture is green and growing, no supplemental protein is required. During drought periods and late fall when the grass matures, extra protein is needed. At such times, it is good business to add 1 lb of protein supplement to each 8 to 12 lb of grain. Usually this will increase the rate and efficiency of gain.

• **Carrying capacity of pastures will vary**—The carrying capacity of pastures will vary with the amount of grain supplement, the quality of pasture, and the age and condition of the cattle. Because of these factors, the acreage per steer will vary from ⅓ to 10.

• **Age is a factor**—Young cattle (yearlings) tend to grow as well as to fatten. Thus, older cattle will reach a high degree of finish on pastures alone. As good as the pastures are, it must be remembered that grass is still a roughage.

• **Minerals for cattle on pasture**—Salt is especially necessary when grass is being utilized. Finishing steers consume from ¾ to 1½ oz of salt per head daily. Also, cattle on pasture should have free access to a 2-compartment mineral box, with (1) trace mineralized salt in one side (salt included for purposes of palatability), and (2) in the other side, a mixture of ⅓ trace mineralized salt and ⅔ defluorinated phosphate or steamed bone meal.

• **Species of grasses or legumes will vary**—The best species of grasses or legumes or grass-legume mixtures to be seeded will vary according to the area, especially according to the soil and climatic conditions. Pasture yields vary greatly from area to area and season to season (see Chapter 7, Pasture and Range Forages, for recommended grass and/or legume species).

Temporary or supplemental pastures, such as Sudangrass or millet, are used for short periods and are usually more productive and palatable than permanent pastures. They are seeded for the purpose of providing supplemental grazing during the season when the regular permanent or rotation pastures are relatively unproductive.

Fig. 19-70. Yearling Hereford steers on Sudangrass pasture near Weeping Water, Nebr. (Courtesy, Soil Conservation Service, USDA)

• **Grass vs grass-legume mixtures should be considered**—In general, where adapted legumes can be successfully grown—either alone or with grass mixtures—the results are superior to yields obtained from pure stands of the grasses. At Washington State University, in a study of pure species of smooth brome and crested wheatgrass vs grass-alfalfa mixtures, it was found that (1) the grasses produced an average of 87 lb of beef per acre, whereas the grass-alfalfa mixtures averaged 233 lb of beef per acre; (2) when based on forage yields at monthly intervals, the same mixtures produced 3 times as much oven-dry forage per acre as the pastures seeded to grasses alone; (3) the grass-legume mixtures provided a slightly longer grazing season; (4) the grass-legume mixtures provided a higher carrying capacity in terms of animals per acre; (5) the erosion-resisting characteristics of the soil were improved by the fibrous grass roots, both while the crop was growing and after the seeding had been plowed under; and (6) the addition of grasses to legumes tended to keep out cheatgrass and other undesirable plants.[11] The two latter points are based merely on careful observation, whereas the rest of the points were proven experimentally.

• **Grain feeding will lengthen the grazing season**—At the Washington Agricultural Experiment Station, researchers found that grain feeding cattle on pasture lengthened the grazing season by an average of 57 days.

• **Self-feeding vs hand-feeding on pasture**—Self-feeding grain on pasture has generally proved superior to hand-feeding, as the animals consume more feed, make more rapid gains, and return more profit. Adding salt (usually about 15%) or fat (usually about 10%) to a concentrate

[11]Ensminger, M. E., *et al.*, *Wash. Ag. Exp. Sta. Bull.* No. 444.

mixture fed in self-feeders has proved to be a satisfactory method of controlling grain consumption at the desired level. Fresh water should be readily accessible.

• **Economy of grain feeding on pasture**—Whether it will be profitable to feed grain on pasture will depend primarily upon the price of grain, the premium paid for cattle of higher finish and grade, the season in which it is desired to market, and the area and quality of pasture.

• **Pasture bloat can be controlled**—The following preventive measures will reduce pasture bloat:

1. Avoid straight legume pasture (except for trefoil) and immature legumes.
2. Feed a coarse grass hay prior to turning onto lush pasture.
3. Feed dry forage along with pasture.
4. Avoid a rapid fill from an empty start.
5. Keep animals continuously on pasture after they are once turned out.
6. Keep salt and water conveniently accessible at all times.
7. Avoid frosted pasture.
8. Use poloxalene (Bloat Guard), oxytetracycline (Terramycin or Neo-Terramycin), or Laureth-23 (Enproal Bloat Blox) according to manufacturers' directions.

(Also see Chapter 5, Table 5-1, Nutritional Diseases and Ailments—Bloat, pasture.)

NUTRITIONAL FEEDLOT DISEASES

Several metabolic disorders, or diseases, in feedlot cattle are attributable wholly or in part to the feeding regimen. Among the more prevalent ones are acidosis, bloat, liver abscesses, and urinary calculi (water belly; urolithiasis). These are fully covered in Chapter 5, Nutritional Disorders/Toxins.

QUESTIONS FOR STUDY AND DISCUSSION

1. Do you agree or disagree with the following statement: "Pastures and other roughages are the very foundation of successful beef cattle production." Justify your answer.

2. Discuss the economic importance of feed for beef cattle.

3. In recent years, crossbreeding and the exotic breeds have produced larger and heavier milking cows, faster-gaining calves, and larger-framed growing/finishing cattle. How has this changed nutritive needs? Do the National Research Council (NRC) Requirements reflect these changes and needs?

4. Describe the symptoms of (a) energy deficiency and (b) protein deficiency in cattle.

5. For beef cattle, discuss the functions and the deficiency symptoms of each of the following minerals: salt, calcium, phosphorus, cobalt, iodine, and selenium.

6. For beef cattle, list the vitamins most apt to be deficient; then give (a) some of the deficiency symptoms, and (b) practical sources of each vitamin for use on a farm or ranch.

7. It has been said that feed additives and implants lowered the retail price of beef by 5 to 6¢ per pound. Do you agree with this statement? Justify your stand.

8. Should a growth stimulant or implant be used on calves that are not creep fed? Justify your answer.

9. Will the proportion of concentrates and roughages used in beef cattle rations change in the years ahead? If so, why?

10. Why is hay the most important harvested roughage fed to beef cattle?

11. Discuss the impact of the expanding world human population on the use of urea and other nonprotein sources for cattle.

12. Give free-choice mineral feeding recommendations (a) for cattle that are on liberal grain feeding, and (b) for cattle that are primarily on roughage (pasture, hay, and/or silage).

13. How may the vitamin A, D, and E requirements of beef cattle be met?

14. Table 19-14, Feed Substitution Table for Beef and Dairy Cattle, is a summary of the comparative values of common feeds, based upon chemical composition, palatability, and carcass quality. Why include the last two bases?

15. Some suggested rations and general rules of feeding are given. Yet, the following statement is added: "The eye of the master fattens his cattle." Why this statement?

16. Wherein does feeding cattle for show differ from feeding cattle in a commercial cattle feedlot?

17. How may feed costs, which account for 65 to 75% of the total cost of keeping cows, be lowered?

18. In comparison with medium-frame cows, large-frame cows produce 9% more calf weaning weight, but they require 28% more feed. How can a producer determine which size cow is most profitable?

19. How may a practical cattle producer meet the added energy requirements of a brood cow during the critical 100 days, extending from 30 days before calving until 70 days after calving?

20. Compare and discuss the nutritive requirements of (a) dry pregnant cows, (b) bred heifers, and (c) lactating cows. From a practical standpoint does this mean that a producer should separate different classes and ages of cattle, then feed them according to needs?

21. What is meant by the terms "calving control"/"daytime calving"? How can this phenomenon be brought about by time of feeding?

22. Discuss the interrelationship between pasture quality and lactating cow/nursing calf requirements and performance.

23. It is estimated that the annual production of corn residue in the United States is sufficient to winter more than 100 million pregnant cows. Yet, much of this potential cow feed is not utilized. Why is not more corn residue fed to cattle?

24. List and discuss the feeding value of crop residues other than corn residue that can be used for feeding cows.

25. Discuss nutrient deficiencies of cattle that are frequently encountered on U.S. ranges. How would you rectify each one?

26. Wherein does the supplementation of purebreds on pastures differ from the supplementation of commercial cattle?

27. What type and system of pasture and range supplementation would you recommend? Justify your answer.

28. What factors favor increased confinement production?

29. Discuss phase-feeding confinement cows according to stage of production and age of animals.

30. Discuss the advantages and disadvantages of (a) a semiconfinement vs (b) a year-round (total) confinement cow-calf operation.

31. How would you handle a heavily fitted sale bull from auction time to breeding season, which we shall assume to be a period of 60 days?

32. Wherein does the feed and management of a young bull 13 to 14 months old differ from the feed and management of a mature bull?

33. Discuss the feeding of a newborn calf.

34. How would you raise orphan calves or multiple birth calves?

35. What are the advantages and disadvantages of creep feeding? Under what conditions would you recommend creep feeding; under what conditions would you recommend against creep feeding?

36. Why and how may creep feeding be limited?

37. What are the advantages and disadvantages of early weaning? Outline a program for early weaning, giving the age of weaning, the feed, and the feeding schedule.

38. How would you go about weaning calves, from the standpoint of both the cows and the calves?

39. Who benefits the most from a preconditioning program, the cow-calf operator or the cattle feeder?

40. Detail a preconditioning program.

41. Discuss the weight and age of small-frame, of medium-frame, and of large-frame heifers at first heat, breeding, and calving.

42. Why should heifers be separated according to age, body weight, and expected daily gain?

43. Define the following terms: (a) stockers; (b) feeders; (c) replacement heifers; (d) preconditioning; and (e) backgrounding.

44. What are the common types of stocker programs, and what are the characteristics of each?

45. Discuss pasture supplement for stocker cattle.

46. Discuss growth stimulants or implants for stocker cattle.

47. Tell how the level of wintering stockers affects the next stage of feeding.

48. For the farmer/rancher who produces stocker calves, and who finishes them on a custom basis, is compensatory growth good or bad? Justify your answer.

49. Why do lightweight calves (a) gain more efficiently, (b) require higher quality feed, and (c) make faster and cheaper gains than heavyweight calves?

50. What provisions should be incorporated in a stocker and grower contract?

51. What factors should determine the following alternative choices of a cattle feeder: (a) cattle feedlot (including confinement feeding) vs pasture finishing; and (b) farmer-feeder vs commercial cattle feeder?

52. Discuss the transitions in cattle feeding, and the forces back of these changes.

53. Discuss how each of the following enter into the choice of the kind of cattle to feed in a given lot: (a) age and weight of cattle; (b) sex of cattle; (c) grade of cattle; and (d) breeding and type of cattle.

54. Discuss the impact of each of the following on cattle feeding: (a) dairy beef, (b) hamburger beef, and (c) boxed beef.

55. When feeding dairy beef, what will determine the choice between (a) high-energy rations and light market weight cattle; vs (b) high-roughage rations and heavy market weight?

56. With the advent of boxed beef, the steer must fit the box. How does this affect the cattle feeding industry?

57. Select a certain area for feeding cattle. Then, give for that particular area the pros and cons for each (a) an open feedlot, (b) cold confinement, and (c) warm confinement. Finally, give your recommendation.

58. What is a slotted floor? Why and how is it sometimes used in cattle feeding?

59. Why is moisture content of grains important (a) when buying feed, and (b) when balancing rations?

60. Discuss one by-product feed that is being used by a cattle feeder of your acquaintance (or by a feeder with whom you will get acquainted). Among other things, determine the following relative to it: (a) price per ton; (b) chemical composition; (c) quantity fed; and (d) replacement value (for corn or barley).

61. Discuss the present, and probable future, importance of urea in cattle finishing.

62. What minerals would you provide feedlot cattle, in what quantities would you provide them, and how would you provide them?

63. What vitamins would you provide feedlot cattle, in what quantities would you provide them, and how would you provide them?

64. Based on net returns, what implants and growth stimulants, if any, would you use for each (a) finishing steers, (b) finishing heifers, and (c) suckling calves?

65. List and discuss each of the methods, in addition to additives, that can be used to increase the rate and efficiency of gains of feedlot cattle.

66. What are all-concentrate and high-concentrate rations?

67. What is the difference between (a) preconditioning, (b) backgrounding, and (c) handling newly arrived feedlot cattle?

68. Outline, step by step, a recommended program for handling newly arrived feedlot cattle that will minimize losses and maximize performance.

69. Outline, step by step, a program for getting (a) yearling cattle, and (b) calves on full feed.

70. Why is overfinishing undesirable?

71. How may a feedlot mud problem be alleviated?

72. Discuss each of the types of custom feeding contracts.

73. Would you recommend that hogs follow feedlot cattle? Justify your answer.

74. How far should a nation go in pollution control laws and regulations, bearing in mind the following: (a) that pollution control measures cost money—hence, they will increase product prices to consumers; (b) that the ultimate in pollution control will lower food production potential in some cases; and (c) that many of the good things in life which contribute to our high standard of living, such as electricity, make for pollution?

75. Under what conditions would you recommend pasture finishing of cattle rather than feedlot finishing?

76. What are (a) the advantages, and (b) the disadvantages of grain feeding on pasture?

77. Discuss the alternate systems of pasture finishing.

78. As grain becomes scarcer and higher in price, is it likely that more cattle will be grass finished, perhaps by supplemental grain feeding?

79. Discuss each of the following basic points as they apply to pasture finishing: (a) moderate winter gains; (b) early, "washy" pasture; (c) when to grain feed on pasture; (d) effect of supplementing young cattle on pasture on subsequent feedlot performance; (e) self-feeding whole corn vs rolled corn on pasture; (f) the use of a protein supplement; (g) age of cattle; (h) grass vs grass-legume pastures; (i) self-feeding vs hand-feeding; and (j) bloat control.

80. Using current feed prices, compute the value of grass on a per steer basis if it effects a saving of 100 lb of dry feed per 100 lb of gain. What additional advantages accrue from grain feeding cattle on pasture, compared to drylot finishing?

SELECTED REFERENCES

Title of Publication	Author(s)	Publisher
Beef Cattle, 7th Edition	A. L. Neumann	John Wiley & Sons, New York, N.Y., 1977
Beef Cattle Production	J. F. Lasley	Prentice Hall, Englewood Cliff, N.J., 1981
Beef Cattle Production	K. A. Wagnon R. Albaugh G. H. Hart	The MacMillan Company, New York, N.Y., 1960
Beef Cattle Production	N.T.M. Yeates P. J. Schmidt	Butterworth Pty Limited, Brisbane, Australia, 1974
Beef Cattle Science Handbook	Ed. by M. E. Ensminger	Agriservices Foundation, Clovis, Calif., pub. annually since 1964
Beef Production and Management	G. L. Minish D. G. Fox	Reston Publishing Co., Inc., Reston, Va., 1979
Beef Production and the Beef Industry	R. E. Taylor	Burgess Publishing Co., Minneapolis, Minn., 1984
Beef Production in the South, Modified Edition	S. H. Fowler	The Interstate Printers & Publishers, Inc., Danville, Ill., 1979
Commercial Beef Cattle Production, 2nd Edition	Ed. by C. C. O'Mary I. A. Dyer	Lea & Febiger, Philadelphia, Pa., 1978
Feedlot, The, 3rd Edition	Ed. by G. B. Thompson C. C. O'Mary	Lea & Febiger, Philadelphia, Pa., 1983
Feeds and Feeding, 22nd Edition	F. B. Morrison	The Morrison Publishing Co., Ithaca, N.Y., 1956
Feeds and Feeding, Abridged	F. B. Morrison	The Morrison Publishing Co., Ithaca, N.Y., 1956
Intensive Beef Production, 2nd Edition	T. R. Preston M. B. Willis	Pergamon Press, Oxford, England, 1974
Nutrient Requirements of Beef Cattle, 6th Revised Edition	National Research Council	National Academy of Sciences, Washington, D.C., 1984
Stockman's Handbook, The, 6th Edition	M. E. Ensminger	The Interstate Printers & Publishers, Inc., Danville, Ill., 1983
World's Beef Business, The, 1st Edition	J. R. Simpson D. E. Farris	Iowa State University Press, Ames, Iowa, 1982

Original painting by Tom Phillips

Chapter 20

FEEDING DAIRY CATTLE[1]

Contents	Page

(Continued)

[1]The authors gratefully acknowledge the helpful suggestions of the following dairy specialists who reviewed this chapter: D. L. Bath, Ph.D., Extension Dairy Nutritionist, University of California, Davis; L. R. Brown, Ph.D., Extension Dairyman, The University of Connecticut, Storrs; W. H. Brown, Ph.D., Professor, Department of Animal Sciences, The University of Arizona, Tucson; R. R. Grummer, Ph.D., Department of Dairy Science, University of Wisconsin, Madison; L. M. Larsen, Ph.D., Consultant, Nutri-Systems, 426 E. Shields Ave., Fresno, CA 93704; J. W. Thomas, Ph.D., Consultant and Professor Emeritus, Department of Animal Science, Michigan State University, East Lansing; and W. H. Van Horn, Ph.D., Professor, Institute of Food and Agricultural Sciences, University of Florida, Gainesville.

Also, the following dairy specialists responded liberally to the call of the authors for counsel and advice, literature, and/or pictures: L. E. Chase, Ph.D., Associate Professor, Dairy Cattle Nutrition, Department of Animal Science, Cornell University, Ithaca, N.Y.; D. A. Hartman, Ph.D., Department of Dairy Science, Virginia Tech, Blacksburg; A. J. Heinrichs, Ph.D., Dairy Specialist, Pennsylvania State University, University Park; T. W. Howard, Ph.D., Extension Dairy Specialist, University of Wisconsin, Madison; M. F. Hutjens, Ph.D., Extension Dairyman, University of Illinois, Urbana-Champaign; J. G. Linn, Ph.D., Extension Animal Scientist, Dairy Nutrition, University of Minnesota, St. Paul; M. E. McCullough, Ph.D., Consulting Nutritionist/ Professor Emeritus, University of Georgia, Experiment; O. E. Otterby, Professor, Department of Animal Science, University of Minnesota, St. Paul; and J. Tappan, Arizona Dairy Company, Higley, Ariz.

Contents *Page*

Contents *Page*

Fig. 20-1. The attractive Maddox Dairy Headquarters, Riverdale, Calif. The herd consists of more than 3,600 lactating cows with a rolling herd average of 20,850 lb milk and 3.72% fat, on 3X daily milking. (Photo by A. H. Ensminger)

Feeding lactating dairy cows differs from feeding other classes of farm animals. There is limited time between calvings in which to get whatever milk the cow will produce. Milk production reaches a peak 2 to 6 weeks after a cow freshens, then declines during the remaining portion of the lactation period. If feeding is limited, the rate of decline is more rapid than normal and total production is lowered. Thus, reducing the ration and concomitantly lowering production is almost certain to be economically unsound.

As labor became more scarce, and more expensive, dairy producers automated—they replaced part of the labor force by machines. Eventually, it became advantageous to divide the herd into groups of cows producing at about the same level, with each group of 50 to 100 cows managed as a unit. The group concept led to changes in feeding practices, including group feeding based on the needs of a corral of cows, rather than the needs of an individual cow, and complete rations (grain and forage combined), with a different combination of feeds for each group. Next came confinement and environmental control. Other innovations will follow.

ECONOMIC IMPORTANCE OF FEED

Feed, more than any other one factor, determines the productivity and profitability of dairy cows. Within a herd, approximately 25% of the difference in milk production between cows is due to heredity; the remaining 75% is determined by environmental factors, with feed making up the largest portion. Feed accounts for about 55% (with a range from 45 to 65%) of the cost of milk production. Therefore, a good feeding program is necessary for profitable milk production.

Although it costs more to feed high producers than low producers, milk income is much higher and the net return above the cost of feed is also increased. Fig. 20-2 shows how income over feed cost improves as production per cow increases.

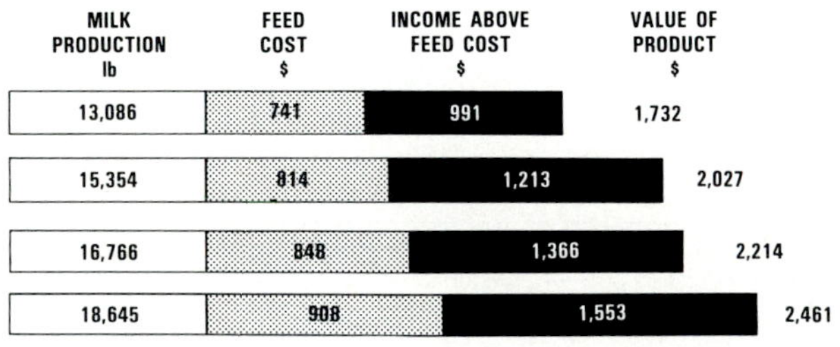

Fig. 20-2. It costs more to feed high-producing cows, but the practice pays handsome dividends. The reason: feed and overhead costs for maintenance are practically the same, regardless of the level of production. (Based on data from New York Dairy Records Laboratory; courtesy, L. R. Brown, Extension Dairyman, The University of Connecticut, Storrs)

With feed costs representing such a high percentage of the total cost of producing milk, successful managers must keep abreast of changes in feed prices from month to month in order to make needed adjustments and to maximize profits. In the days when a family farm had 10 to 20 cows, a $5 per ton increase in the price of hay made for only a $300 to $600 per year higher total feed cost. With herds of 1,000 cows, such a rise in the price of hay can mean $25,000 or more. A decrease in hay quality (and price) will also lower costs, but it may also lower income due to decrease in milk production or to refusal of hay, or both. The same story is repeated with grain. Since the price of milk and the price of feed frequently move independently of each other, the good manager must be aware of price changes and plan optimum feeding programs for current conditions.

Table 20–7, Composition of Feeds Commonly Used in Dairy Cattle Rations, consists of selected feeds from *Nutrient Requirements of Dairy Cattle,* Sixth Revised Edition, 1988, Table 7–1. In their feed compositions, the NRC committee assumed an average decrease of 4% per unit of dry matter intake above maintenance in calculating NE_{lc} values for feed ingredients, or an average discount of 8% based on their assumption that lactating cows are fed at 3 X maintenance. For the convenience of those dairy producers and dairy nutritionists who wish to use these values, the authors selected from *Nutrient Requirements of Dairy Cattle,* Sixth Revised Edition, 1988, Table 7–1, the feeds most commonly used in dairy cattle rations, and herein reproduced them in Table 20–7. **NOTE WELL:** Table 20–7 is a 3-page spread; see pages 818, 819, and 820.

NUTRITIVE NEEDS OF DAIRY CATTLE

The first consideration in any dairy feeding program is to determine the nutritive needs for body maintenance, growth, pregnancy or reproduction, and milk production.

The body uses nutrients on a priority basis, with maintenance, growth, and pregnancy generally higher in priority than milk production. There are exceptions to this rule, as there are with most rules. Thus, some cows will produce milk at the expense of their maintenance and growth.

When cows are underfed, they cannot produce at their most efficient and profitable level. It is more profitable to feed on the generous side than to feed so little that production drops to an uneconomical level. Of course, it is also unprofitable to waste nutrients. So, cows that put on flesh and fail to return it later in increased milk production should be fed somewhat different amounts and types of feeds than cows which have a higher priority for milk production.

National Research Council (NRC) Requirements

The current recommended nutritive requirements for dairy cattle are contained in *Nutrient Requirements of Dairy Cattle,* Sixth Revised Edition, 1988, prepared by the Subcommittee on Dairy Nutrition, National Research Council, and published by the National Academy Press, Washington, D.C., 1988. These are listed in Tables 20–1 to 20–5. It should be noted that the requirements listed in these tables do not allow for any margin of safety; that is, they do not provide for animal differences, feed differences, losses of certain nutrients in storage, and stresses. Accordingly, in the formulation of rations, margins of safety should be provided.

Table 20–6, shows NRC's maximum tolerable dietary levels of certain elements.

TABLE 20–1
DRY MATTER INTAKE REQUIREMENTS TO FULFILL NUTRIENT ALLOWANCES FOR MAINTENANCE, MILK PRODUCTION, AND NORMAL LIVE WEIGHT GAIN DURING MID- AND LATE LACTATION[1]

Live Wt.: (lb)		882	1,103	1,323	1,544	1,764
Live Wt.: *(kg)*		*400*	*500*	*600*	*700*	*800*
FCM (4%)[2]		Percent Live Weight[3 4]				
(lb)	*(kg)*	(%)	(%)	(%)	(%)	(%)
22	*10*	2.7	2.4	2.2	2.0	1.9
33	*15*	3.2	2.8	2.6	2.3	2.2
44	*20*	3.6	3.2	2.9	2.6	2.4
55	*25*	4.0	3.5	3.2	2.9	2.7
66	*30*	4.4	3.9	3.5	3.2	2.9
77	*35*	5.0	4.2	3.7	3.4	3.1
88	*40*	5.5	4.6	4.0	3.6	3.3
99	*45*	—	5.0	4.3	3.8	3.5
110	*50*	—	5.4	4.7	4.1	3.7
121	*55*	—	—	5.0	4.4	4.0
132	*60*	—	—	5.4	4.8	4.3

[1]Adapted by the authors from *Nutrient Requirements of Dairy Cattle,* 6th rev. ed., update 1989, NRC, National Academy Press, p. 78, Table 6–1.

[2]4% fat-corrected milk (kg) = (0.4)(kg of milk) + (15)(kg of milk fat).

[3]The probable DMI may be up to 18% less in early lactation.

[4]DMI as a percentage of live weight may be 0.02% less per 1% increase in ration moisture content above 50% if fermented feeds constitute a major portion of the ration.

NOTE: The following assumptions were made in calculating the DMI requirements shown in this table:

1. The basic or reference cow used for the calculations weighed 1,323 lb *(600 kg)* and produced milk with 4% milk fat. Other live weights in the table and corresponding fat percentages were 882 lb *(400 kg)* and 5% fat; 1,103 lb *(500 kg)* and 4.5% fat; and 1,544 and 1,764 lb *(700 and 800 kg)* and 3.5% fat.

2. The concentration of energy in the ration for the reference cow was 1.42 Mcal of NE_{lc}/kg of DM for milk yields equal to or less than 22 lb *(10 kg)*/day. It increased linearly to 1.72 Mcal of NE_{lc}/kg for milk yields equal to or greater than 88 lb *(40 kg)*/day.

3. The energy concentrations of the rations for all other cows were assumed to change linearly as their energy requirements for milk production, relative to maintenance, changed in a manner identical to that of the 1,323-lb *(600-kg)* cow as she increased in milk yield from 22 to 88 lb *(10 to 40 kg)*/day.

4. Enough DM to provide sufficient energy for cows to gain 0.055% of their body weight daily was also included in the total. If cows do not consume as much DM as they require, as calculated from this table, their energy intake will be less than their requirements. The result will be a loss of body weight, reduced milk yields, or both. If cows consume more DM than what is projected as required from this table, the energy concentration of their ration should be reduced or they may become overly fat.

TABLE 20–2
DAILY NUTRIENT REQUIREMENTS OF GROWING DAIRY CATTLE AND MATURE BULLS[1] (See footnotes at end of table.)

Live Weight		Gain		Dry Matter Intake[2]		Energy						Protein			Minerals		Vitamins	
						TDN		DE	ME	NE_m	NE_g	CP	DIP[3]	UIP[4]	Ca	P	A	D
(lb)	*(kg)*	(lb)	*(kg)*	(lb)	*(kg)*	(lb)	*(kg)*	(Mcal)	(Mcal)	(Mcal)	(Mcal)	(g)	(g)	(g)	(g)	(g)	(1,000 IU)	(1,000 IU)
Growing Large-Breed Calves Fed Only Milk or Milk Replacer																		
88	*40*	0.4	*0.2*	1.06	*0.48*	1.37	*0.62*	2.73	2.54	1.37	0.41	105	—	—	7	4	1.70	0.26
99	*45*	0.7	*0.3*	1.19	*0.54*	1.54	*0.70*	3.07	2.86	1.49	0.56	120	—	—	8	5	1.94	0.30
Growing Large-Breed Calves Fed Milk Plus Starter Mix																		
110	*50*	1.1	*0.5*	2.87	*1.30*	3.22	*1.46*	6.42	5.90	1.62	0.72	290	—	—	9	6	2.10	0.33
165	*75*	1.8	*0.8*	4.37	*1.98*	4.90	*2.22*	9.78	8.98	2.19	1.30	435	—	—	16	8	3.20	0.50
Growing Small-Breed Calves Fed Only Milk or Milk Replacer																		
55	*25*	0.4	*0.2*	0.84	*0.38*	1.08	*0.49*	2.16	2.01	0.96	0.37	84	—	—	6	4	1.10	0.16
66	*30*	0.7	*0.3*	1.12	*0.51*	1.46	*0.66*	2.90	2.70	1.10	0.52	112	—	—	7	4	1.30	0.20
Growing Small-Breed Calves Fed Milk Plus Starter Mix																		
110	*50*	1.1	*0.5*	3.15	*1.43*	3.53	*1.60*	7.06	6.49	1.62	0.72	315	—	—	10	6	2.10	0.33
165	*75*	1.3	*0.6*	3.88	*1.76*	4.34	*1.97*	8.69	7.98	2.19	0.96	387	—	—	14	8	3.20	0.50
Growing Veal Calves Fed Only Milk or Milk Replacer																		
88	*40*	0.4	*0.2*	0.99	*0.45*	1.04	*0.47*	2.07	1.89	1.37	0.55	100	—	—	7	4	1.70	0.26
110	*50*	0.9	*0.4*	1.26	*0.57*	1.30	*0.59*	2.63	2.39	1.62	0.57	125	—	—	9	5	2.10	0.33
132	*60*	1.2	*0.5*	1.76	*0.80*	1.57	*0.71*	3.17	2.84	1.85	0.81	176	—	—	13	8	2.60	0.40
165	*75*	2.0	*0.9*	3.00	*1.36*	2.67	*1.21*	5.39	4.82	2.19	1.47	300	—	—	16	9	3.20	0.50
221	*100*	2.8	*1.3*	4.41	*2.00*	3.48	*1.58*	7.06	6.22	2.72	2.26	440	—	—	20	11	4.20	0.66
276	*125*	2.8	*1.3*	5.25	*2.38*	4.15	*1.88*	8.40	7.40	3.21	2.44	524	—	—	22	13	5.30	0.82
331	*150*	2.4	*1.1*	6.00	*2.72*	4.74	*2.15*	9.60	8.46	3.69	2.29	598	—	—	24	15	6.40	0.99
Large-Breed Growing Females																		
221	*100*	1.3	*0.6*	5.80	*2.63*	4.02	*1.84*	8.13	7.03	2.72	1.22	421	57	317	17	9	4.24	0.66
221	*100*	1.5	*0.7*	6.22	*2.82*	4.37	*1.98*	8.72	7.54	2.72	1.44	452	75	346	18	9	4.24	0.66
221	*100*	1.8	*0.8*	6.66	*3.02*	4.65	*2.11*	9.32	8.06	2.72	1.66	483	92	374	18	10	4.24	0.66
331	*150*	1.3	*0.6*	7.74	*3.51*	5.31	*2.41*	10.61	9.14	3.69	1.45	562	150	283	19	11	6.36	0.99
331	*150*	1.5	*0.7*	8.27	*3.75*	5.67	*2.57*	11.33	9.76	3.69	1.71	600	173	307	19	12	6.36	0.99
331	*150*	1.8	*0.8*	8.80	*3.99*	6.04	*2.74*	12.07	10.39	3.69	1.97	639	196	331	20	12	6.36	0.99
441	*200*	1.3	*0.6*	9.68	*4.39*	6.50	*2.95*	12.99	11.14	4.57	1.65	699	239	254	20	14	8.48	1.32
441	*200*	1.5	*0.7*	10.32	*4.68*	6.92	*3.14*	13.84	11.87	4.57	1.95	749	267	274	21	14	8.48	1.32
441	*200*	1.8	*0.8*	10.96	*4.97*	7.36	*3.34*	14.71	12.62	4.57	2.25	796	295	294	22	15	8.48	1.32
551	*250*	1.3	*0.6*	11.71	*5.31*	7.67	*3.48*	15.33	13.10	5.41	1.84	718	326	229	22	16	10.60	1.65
551	*250*	1.5	*0.7*	12.46	*5.65*	8.16	*3.70*	16.32	13.94	5.41	2.18	787	359	246	23	17	10.60	1.65
551	*250*	1.8	*0.8*	13.21	*5.99*	8.67	*3.93*	17.32	14.79	5.41	2.51	857	393	263	24	17	10.60	1.65
662	*300*	1.3	*0.6*	13.80	*6.26*	8.84	*4.01*	17.69	15.05	6.20	2.02	752	413	209	23	17	12.72	1.98
662	*300*	1.5	*0.7*	14.69	*6.66*	9.42	*4.27*	18.81	16.00	6.20	2.39	814	452	223	24	18	12.72	1.98
662	*300*	1.8	*0.8*	15.57	*7.06*	9.97	*4.52*	19.95	16.97	6.20	2.77	884	490	236	25	19	12.72	1.98
772	*350*	1.3	*0.6*	16.07	*7.29*	10.05	*4.56*	20.09	17.01	6.96	2.20	874	501	193	24	18	14.84	2.31
772	*350*	1.5	*0.7*	17.09	*7.75*	10.67	*4.84*	21.36	18.09	6.96	2.60	930	545	204	25	19	14.84	2.31
772	*350*	1.8	*0.8*	18.10	*8.21*	11.33	*5.14*	22.64	19.18	6.96	3.01	985	590	214	26	20	14.84	2.31
882	*400*	1.3	*0.6*	18.50	*8.39*	11.29	*5.12*	22.58	19.03	7.69	2.37	1,007	592	182	25	19	16.96	2.64
882	*400*	1.5	*0.7*	19.67	*8.92*	12.00	*5.44*	24.00	20.23	7.69	2.80	1,070	641	190	26	20	16.96	2.64
882	*400*	1.8	*0.8*	20.86	*9.46*	12.72	*5.77*	25.44	21.44	7.69	3.24	1,135	692	198	26	21	16.96	2.64
992	*450*	1.3	*0.6*	21.15	*9.59*	12.59	*5.71*	25.18	21.12	8.40	2.53	1,151	686	176	28	19	19.08	2.97
992	*450*	1.5	*0.7*	22.49	*10.20*	13.38	*6.07*	26.78	22.46	8.40	2.99	1,224	742	182	28	20	19.08	2.97
992	*450*	1.8	*0.8*	23.86	*10.82*	14.20	*6.44*	28.40	23.81	8.40	3.46	1,298	799	187	29	21	19.08	2.97
1,103	*500*	1.3	*0.6*	24.10	*10.93*	13.98	*6.34*	27.96	23.32	9.09	2.69	1,311	785	175	28	20	21.20	3.30
1,103	*500*	1.5	*0.7*	25.64	*11.63*	14.88	*6.75*	29.74	24.81	9.09	3.18	1,395	848	179	28	20	21.20	3.30
1,103	*500*	1.8	*0.8*	27.18	*12.33*	15.79	*7.16*	31.55	26.32	9.09	3.68	1,480	913	182	29	21	21.20	3.30
1,213	*550*	1.3	*0.6*	27.39	*12.42*	15.48	*7.02*	30.95	25.67	9.77	2.84	1,490	891	180	28	20	23.32	3.63
1,213	*550*	1.5	*0.7*	29.15	*13.22*	16.47	*7.47*	32.95	27.33	9.77	3.37	1,587	963	183	28	20	23.32	3.63
1,213	*550*	1.8	*0.8*	30.96	*14.04*	17.51	*7.94*	34.99	29.02	9.77	3.90	1,685	1,035	185	29	21	23.32	3.63
1,323	*600*	1.3	*0.6*	31.11	*14.11*	17.13	*7.77*	34.24	28.23	10.43	3.00	1,694	1,007	193	28	20	25.44	3.96
1,323	*600*	1.5	*0.7*	33.19	*15.05*	18.26	*8.28*	36.50	30.09	10.43	3.55	1,805	1,088	194	28	21	25.44	3.96
1,323	*600*	1.8	*0.8*	35.26	*15.99*	19.40	*8.80*	38.79	31.98	10.43	4.11	1,919	1,170	195	29	21	25.44	3.96
Small-Breed Growing Females																		
221	*100*	0.9	*0.4*	5.31	*2.41*	3.68	*1.67*	7.35	6.34	2.72	0.91	386	38	249	15	8	4.24	0.66
221	*100*	1.1	*0.5*	5.82	*2.64*	4.01	*1.82*	8.03	6.92	2.72	1.16	422	59	275	16	8	4.24	0.66
221	*100*	1.3	*0.6*	6.31	*2.86*	4.37	*1.98*	8.71	7.51	2.72	1.40	458	80	300	17	9	4.24	0.66

(Continued)

TABLE 20–2 *(Continued)*

Live Weight		Gain		Dry Matter Intake[2]		Energy							Protein			Minerals		Vitamins	
						TDN		DE	ME	NE_m	NE_g	CP	DIP[3]	UIP[4]	Ca	P	A	D
(lb)	*(kg)*	(lb)	*(kg)*	(lb)	*(kg)*	(lb)	*(kg)*	(Mcal)	(Mcal)	(Mcal)	(Mcal)	(g)	(g)	(g)	(g)	(g)	(1,000 IU)	(1,000 IU)
								Small-Breed Growing Females (Continued)										
331	*150*	0.9	*0.4*	7.30	*3.31*	4.90	*2.22*	9.78	8.39	3.69	1.09	529	129	222	17	10	6.36	0.99
331	*150*	1.1	*0.5*	7.94	*3.60*	5.31	*2.41*	10.63	9.12	3.69	1.39	575	156	243	18	11	6.36	0.99
331	*150*	1.3	*0.6*	8.58	*3.89*	5.76	*2.61*	11.50	9.86	3.69	1.69	622	185	263	19	11	6.36	0.99
441	*200*	0.9	*0.4*	9.35	*4.24*	6.09	*2.76*	12.16	10.38	4.57	1.26	578	217	201	19	13	8.48	1.32
441	*200*	1.1	*0.5*	10.14	*4.60*	6.59	*2.99*	13.19	11.25	4.57	1.60	648	251	217	20	13	8.48	1.32
441	*200*	1.3	*0.6*	10.94	*4.96*	7.12	*3.23*	14.23	12.14	4.57	1.95	718	286	232	20	14	8.48	1.32
551	*250*	0.9	*0.4*	11.55	*5.24*	7.28	*3.30*	14.57	12.36	5.41	1.41	629	305	185	21	15	10.60	1.65
551	*250*	1.1	*0.5*	12.52	*5.68*	7.89	*3.58*	15.78	13.38	5.41	1.80	682	346	197	21	16	10.60	1.65
551	*250*	1.3	*0.6*	13.49	*6.12*	8.51	*3.86*	17.01	14.43	5.41	2.20	753	389	209	22	16	10.60	1.65
662	*300*	0.9	*0.4*	13.98	*6.34*	8.53	*3.87*	17.06	14.38	6.20	1.56	761	395	176	22	16	12.72	1.98
662	*300*	1.1	*0.5*	15.15	*6.87*	9.24	*4.19*	18.48	15.57	6.20	1.99	824	445	184	23	17	12.72	1.98
662	*300*	1.3	*0.6*	16.32	*7.40*	9.97	*4.52*	19.92	16.79	6.20	2.43	888	495	192	23	17	12.72	1.98
772	*350*	0.9	*0.4*	16.69	*7.57*	9.86	*4.47*	19.71	16.50	6.96	1.71	909	490	173	23	17	14.84	2.31
772	*350*	1.1	*0.5*	18.08	*8.20*	10.67	*4.84*	21.35	17.87	6.96	2.18	985	548	178	23	18	14.84	2.31
772	*350*	1.3	*0.6*	19.51	*8.85*	11.51	*5.22*	23.03	19.28	6.96	2.66	1,062	608	183	24	18	14.84	2.31
882	*400*	0.9	*0.4*	19.80	*8.98*	11.29	*5.12*	22.58	18.77	7.69	1.84	1,078	592	177	24	18	16.96	2.64
882	*400*	1.1	*0.5*	21.48	*9.74*	12.26	*5.56*	24.50	20.36	7.69	2.35	1,169	661	181	24	19	16.96	2.64
882	*400*	1.3	*0.6*	23.20	*10.52*	13.23	*6.00*	26.45	21.98	7.69	2.87	1,263	730	183	25	19	16.96	2.64
992	*450*	0.9	*0.4*	23.46	*10.64*	12.90	*5.85*	25.80	21.27	8.40	1.98	1,276	706	191	27	18	19.08	2.97
992	*450*	1.1	*0.5*	25.49	*11.56*	14.02	*6.36*	28.04	23.12	8.40	2.52	1,387	786	193	28	19	19.08	2.97
992	*450*	1.3	*0.6*	27.56	*12.50*	15.17	*6.88*	30.33	25.01	8.40	3.08	1,500	867	194	28	19	19.08	2.97
								Large-Breed Growing Males										
221	*100*	1.8	*0.8*	6.17	*2.80*	4.32	*1.96*	8.66	7.48	2.72	1.42	448	65	401	18	10	4.24	0.66
221	*100*	2.0	*0.9*	6.55	*2.97*	4.59	*2.08*	9.16	7.92	2.72	1.60	475	79	433	19	10	4.24	0.66
221	*100*	2.2	*1.0*	6.90	*3.13*	4.83	*2.19*	9.67	8.36	2.72	1.79	501	93	465	20	11	4.24	0.66
331	*150*	1.8	*0.8*	7.94	*3.60*	5.51	*2.50*	11.03	9.52	3.69	1.64	576	155	364	20	12	6.36	0.99
331	*150*	2.0	*0.9*	8.38	*3.80*	5.82	*2.64*	11.63	10.03	3.69	1.85	607	172	393	21	13	6.36	0.99
331	*150*	2.2	*1.0*	8.80	*3.99*	6.11	*2.77*	12.22	10.55	3.69	2.07	639	190	422	22	13	6.36	0.99
441	*200*	1.8	*0.8*	9.77	*4.43*	6.68	*3.03*	13.34	11.48	4.57	1.84	709	241	333	22	15	8.48	1.32
441	*200*	2.0	*0.9*	10.28	*4.66*	7.01	*3.18*	14.02	12.06	4.57	2.08	745	262	359	23	15	8.48	1.32
441	*200*	2.2	*1.0*	10.78	*4.89*	7.36	*3.34*	14.71	12.66	4.57	2.33	782	284	385	24	16	8.48	1.32
551	*250*	1.8	*0.8*	11.62	*5.27*	7.78	*3.53*	15.58	13.37	5.41	2.03	843	325	305	24	17	10.60	1.65
551	*250*	2.0	*0.9*	12.19	*5.53*	8.18	*3.71*	16.35	14.03	5.41	2.30	885	350	329	25	18	10.60	1.65
551	*250*	2.2	*1.0*	12.79	*5.80*	8.58	*3.89*	17.13	14.70	5.41	2.57	927	375	352	26	18	10.60	1.65
662	*300*	1.8	*0.8*	13.52	*6.13*	8.91	*4.04*	17.80	15.22	6.20	2.21	863	408	281	25	19	12.72	1.98
662	*300*	2.0	*0.9*	14.18	*6.43*	9.33	*4.23*	18.66	15.96	6.20	2.51	934	436	302	25	19	12.72	1.98
662	*300*	2.2	*1.0*	14.84	*6.73*	9.77	*4.43*	19.53	16.70	6.20	2.80	1,004	464	323	26	20	12.72	1.98
772	*350*	1.8	*0.8*	15.48	*7.02*	10.01	*4.54*	20.02	17.06	6.96	2.38	885	490	261	26	20	14.84	2.31
772	*350*	2.0	*0.9*	16.23	*7.36*	10.50	*4.76*	20.98	17.88	6.96	2.70	956	522	280	26	20	14.84	2.31
772	*350*	2.2	*1.0*	16.98	*7.70*	10.98	*4.98*	21.94	18.70	6.96	3.02	1,027	554	298	27	21	14.84	2.31
882	*400*	1.8	*0.8*	17.55	*7.96*	11.14	*5.05*	22.27	18.91	7.69	2.55	955	572	244	26	21	16.96	2.64
882	*400*	2.0	*0.9*	18.39	*8.34*	11.66	*5.29*	23.32	19.80	7.69	2.89	1,001	608	260	27	21	16.96	2.64
882	*400*	2.2	*1.0*	19.23	*8.72*	12.19	*5.53*	24.39	20.71	7.69	3.24	1,056	644	277	28	22	16.96	2.64
992	*450*	1.8	*0.8*	19.73	*8.95*	12.28	*5.57*	24.56	20.78	8.40	2.71	1,074	656	230	29	21	19.08	2.97
992	*450*	2.0	*0.9*	20.66	*9.37*	12.86	*5.83*	25.72	21.76	8.40	3.08	1,125	696	245	29	22	19.08	2.97
992	*450*	2.2	*1.0*	21.61	*9.80*	13.45	*6.10*	26.89	22.75	8.40	3.44	1,176	736	259	29	23	19.08	2.97
1,103	*500*	1.8	*0.8*	22.05	*10.00*	13.47	*6.11*	26.92	22.69	9.09	2.87	1,201	742	220	29	21	21.20	3.30
1,103	*500*	2.0	*0.9*	23.11	*10.48*	14.09	*6.39*	28.19	23.76	9.09	3.25	1,257	786	233	29	22	21.20	3.30
1,103	*500*	2.2	*1.0*	24.14	*10.95*	14.73	*6.68*	29.47	24.84	9.09	3.64	1,314	830	346	29	23	21.20	3.30
1,213	*550*	1.8	*0.8*	24.56	*11.14*	14.69	*6.66*	29.38	24.66	9.77	3.02	1,336	831	213	29	21	23.32	3.63
1,213	*550*	2.0	*0.9*	25.71	*11.66*	15.39	*6.98*	30.76	25.82	9.77	3.43	1,399	879	225	29	22	23.32	3.63
1,213	*550*	2.2	*1.0*	26.88	*12.19*	16.07	*7.29*	32.16	27.00	9.77	3.84	1,463	927	236	30	23	23.32	3.63
1,323	*600*	1.8	*0.8*	27.25	*12.36*	15.99	*7.25*	31.95	26.71	10.43	3.17	1,483	923	211	29	21	25.44	3.96
1,323	*600*	2.0	*0.9*	28.55	*12.95*	16.74	*7.59*	33.47	27.97	10.43	3.60	1,554	976	221	29	22	25.44	3.96
1,323	*600*	2.2	*1.0*	29.86	*13.54*	17.51	*7.94*	34.99	29.25	10.43	4.03	1,624	1,029	231	30	23	25.44	3.96
1,433	*650*	1.8	*0.8*	30.19	*13.69*	17.33	*7.86*	34.67	28.86	11.07	3.32	1,643	1,020	212	29	21	27.56	4.29
1,433	*650*	2.0	*0.9*	31.64	*14.35*	18.17	*8.24*	36.33	30.24	11.07	3.77	1,722	1,078	222	29	22	27.56	4.29
1,433	*650*	2.2	*1.0*	33.10	*15.01*	19.01	*8.62*	38.00	31.63	11.07	4.22	1,801	1,137	230	30	23	27.56	4.29

(Continued)

TABLE 20–2 *(Continued)*

Live Weight		Gain		Dry Matter Intake[2]		Energy						Protein			Minerals		Vitamins	
						TDN		DE	ME	NE_m	NE_g	CP	DIP[3]	UIP[4]	Ca	P	A	D
(lb)	*(kg)*	(lb)	*(kg)*	(lb)	*(kg)*	(lb)	*(kg)*	(Mcal)	(Mcal)	(Mcal)	(Mcal)	(g)	(g)	(g)	(g)	(g)	(1,000 IU)	(1,000 IU)
Large-Breed Growing Males (Continued)																		
1,544	*700*	1.8	*0.8*	33.43	*15.16*	18.79	*8.52*	37.59	31.14	11.70	3.46	1,820	1,124	219	29	22	29.68	4.62
1,544	*700*	2.0	*0.9*	35.06	*15.90*	19.71	*8.94*	39.40	32.64	11.70	3.93	1,907	1,187	227	29	22	29.68	4.62
1,544	*700*	2.2	*1.0*	36.67	*16.63*	20.62	*9.35*	41.23	34.16	11.70	4.40	1,996	1,252	235	30	23	29.68	4.62
1,654	*750*	1.8	*0.8*	37.02	*16.79*	20.37	*9.24*	40.73	33.59	12.33	3.60	2,015	1,235	232	29	22	31.80	4.95
1,654	*750*	2.0	*0.9*	38.85	*17.62*	21.37	*9.69*	42.73	35.23	12.33	4.09	2,114	1,305	239	29	23	31.80	4.95
1,654	*750*	2.2	*1.0*	40.68	*18.45*	22.38	*10.15*	44.74	36.89	12.33	4.58	2,213	1,376	246	30	23	31.80	4.95
1,764	*800*	1.8	*0.8*	38.72	*17.56*	21.30	*9.66*	42.59	35.12	12.94	3.74	2,107	1,303	216	29	22	33.92	5.28
1,764	*800*	2.0	*0.9*	40.59	*18.41*	22.34	*10.13*	44.67	36.83	12.94	4.25	2,210	1,377	221	29	23	33.92	5.28
1,764	*800*	2.2	*1.0*	42.51	*19.28*	23.40	*10.61*	46.76	38.55	12.94	4.76	2,313	1,451	227	30	23	33.92	5.28
Small-Breed Growing Males																		
221	*100*	1.1	*0.5*	5.40	*2.45*	3.79	*1.72*	7.56	6.54	2.72	1.02	392	41	287	16	8	4.24	0.66
221	*100*	1.3	*0.6*	5.82	*2.64*	4.08	*1.85*	8.15	7.04	2.72	1.23	422	58	316	17	9	4.24	0.66
221	*100*	1.5	*0.7*	6.24	*2.83*	4.37	*1.98*	8.74	7.55	2.72	1.45	453	75	345	18	9	4.24	0.66
331	*150*	1.1	*0.5*	7.23	*3.28*	4.96	*2.25*	9.92	8.55	3.69	1.20	525	129	257	18	11	6.36	0.99
331	*150*	1.3	*0.6*	7.76	*3.52*	5.31	*2.41*	10.64	9.16	3.69	1.46	563	151	282	19	11	6.36	0.99
331	*150*	1.5	*0.7*	8.29	*3.76*	5.69	*2.58*	11.36	9.78	3.69	1.71	601	174	306	19	12	6.36	0.99
441	*200*	1.1	*0.5*	9.11	*4.12*	6.09	*2.76*	12.18	10.45	4.57	1.37	630	213	232	20	13	8.48	1.32
441	*200*	1.3	*0.6*	9.70	*4.40*	6.50	*2.95*	13.02	11.17	4.57	1.66	699	241	252	20	14	8.48	1.32
441	*200*	1.5	*0.7*	10.34	*4.69*	6.95	*3.15*	13.87	11.90	4.57	1.96	751	268	273	21	14	8.48	1.32
551	*250*	1.1	*0.5*	11.00	*4.99*	7.21	*3.27*	14.41	12.31	5.41	1.53	648	296	210	21	16	10.60	1.65
551	*250*	1.3	*0.6*	11.73	*5.32*	7.70	*3.49*	15.38	13.14	5.41	1.86	718	328	228	22	16	10.60	1.65
551	*250*	1.5	*0.7*	12.48	*5.66*	8.18	*3.71*	16.35	13.97	5.41	2.19	787	361	245	23	17	10.60	1.65
662	*300*	1.1	*0.5*	12.99	*5.89*	8.31	*3.77*	16.64	14.15	6.20	1.68	707	378	193	23	17	12.72	1.98
662	*300*	1.3	*0.6*	13.85	*6.28*	8.86	*4.02*	17.74	15.09	6.20	2.04	754	415	207	23	17	12.72	1.98
662	*300*	1.5	*0.7*	14.73	*6.68*	9.44	*4.28*	18.85	16.04	6.20	2.41	814	453	221	24	18	12.72	1.98
772	*350*	1.1	*0.5*	15.13	*6.86*	9.46	*4.29*	18.91	16.01	6.96	1.82	823	461	180	23	18	14.84	2.31
772	*350*	1.3	*0.6*	16.12	*7.31*	10.08	*4.57*	20.15	17.06	6.96	2.22	877	503	191	24	18	14.84	2.31
772	*350*	1.5	*0.7*	17.11	*7.76*	10.72	*4.86*	21.41	18.13	6.96	2.62	932	547	203	25	19	14.84	2.31
882	*400*	1.1	*0.5*	17.42	*7.90*	10.63	*4.82*	21.25	17.91	7.69	1.96	947	545	171	24	19	16.96	2.64
882	*400*	1.3	*0.6*	18.54	*8.41*	11.33	*5.14*	22.64	19.08	7.69	2.39	1,010	594	180	25	19	16.96	2.64
882	*400*	1.5	*0.7*	19.71	*8.94*	12.04	*5.46*	24.06	20.27	7.69	2.82	1,073	644	189	26	20	16.96	2.64
992	*450*	1.1	*0.5*	19.91	*9.03*	11.84	*5.37*	23.70	19.87	8.40	2.10	1,083	634	166	28	19	19.08	2.97
992	*450*	1.3	*0.6*	21.21	*9.62*	12.63	*5.73*	25.26	21.18	8.40	2.55	1,155	689	174	28	19	19.08	2.97
992	*450*	1.5	*0.7*	22.56	*10.23*	13.43	*6.09*	26.84	22.51	8.40	3.01	1,227	744	180	28	20	19.08	2.97
1,103	*500*	1.1	*0.5*	22.67	*10.28*	13.14	*5.96*	26.29	21.93	9.09	2.23	1,233	726	167	28	19	21.20	3.30
1,103	*500*	1.3	*0.6*	24.17	*10.96*	14.02	*6.36*	28.04	23.39	9.09	2.71	1,315	788	173	28	20	21.20	3.30
1,103	*500*	1.5	*0.7*	25.69	*11.65*	14.91	*6.76*	29.81	24.87	9.09	3.20	1,398	851	177	28	20	21.20	3.30
1,213	*550*	1.1	*0.5*	25.73	*11.67*	14.55	*6.60*	29.08	24.12	9.77	2.36	1,400	825	174	28	19	23.32	3.63
1,213	*550*	1.3	*0.6*	27.47	*12.46*	15.52	*7.04*	31.05	25.75	9.77	2.87	1,495	895	178	28	20	23.32	3.63
1,213	*550*	1.5	*0.7*	29.24	*13.26*	16.52	*7.49*	33.03	27.40	9.77	3.39	1,591	966	181	28	20	23.32	3.63
1,323	*600*	1.1	*0.5*	29.22	*13.25*	16.07	*7.29*	32.14	26.50	10.43	2.48	1,590	933	187	28	19	25.44	3.96
1,323	*600*	1.3	*0.6*	31.22	*14.16*	17.18	*7.79*	34.35	28.32	10.43	3.02	1,699	1,012	190	28	20	25.44	3.96
1,323	*600*	1.5	*0.7*	33.25	*15.08*	18.30	*8.30*	36.59	30.17	10.43	3.57	1,810	1,091	192	28	21	25.44	3.96
Maintenance of Mature Breeding Bulls																		
1,103	*500*	—	—	17.40	*7.89*	9.57	*4.34*	19.15	15.79	9.09	—	789	472	161	20	12	21.20	3.30
1,323	*600*	—	—	19.96	*9.05*	10.98	*4.98*	21.95	18.10	10.43	—	905	573	155	24	15	25.44	3.96
1,544	*700*	—	—	22.40	*10.16*	12.33	*5.59*	24.64	20.32	11.70	—	1,016	670	148	28	18	29.68	4.62
1,764	*800*	—	—	24.76	*11.23*	13.63	*6.18*	27.24	22.46	12.94	—	1,123	764	142	32	20	33.92	5.28
1,985	*900*	—	—	27.06	*12.27*	14.88	*6.75*	29.76	24.53	14.13	—	1,227	854	135	36	22	38.16	5.94
2,205	*1,000*	—	—	29.28	*13.28*	16.10	*7.30*	32.20	26.55	15.29	—	1,328	943	129	41	25	42.40	6.60
2,426	*1,100*	—	—	31.44	*14.26*	17.31	*7.85*	34.59	28.52	16.43	—	1,426	1,029	122	45	28	46.64	7.26
2,646	*1,200*	—	—	33.56	*15.22*	18.46	*8.37*	36.92	30.44	17.53	—	1,522	1,113	115	49	30	50.88	7.92
2,867	*1,300*	—	—	35.63	*16.16*	19.60	*8.89*	39.21	32.32	18.62	—	1,616	1,196	108	53	32	55.12	8.58
3,087	*1,400*	—	—	37.68	*17.09*	20.73	*9.40*	41.45	34.17	19.68	—	1,709	1,277	102	57	35	59.36	9.24

[1]Adapted by the authors from *Nutrient Requirements of Dairy Cattle,* 6th rev. ed., update 1989, NRC, National Academy Press, pp. 81–84, Table 6-2.

[2]The data for DMI are not requirements *per se,* unlike the requirements for net energy maintenance, net energy gain, and absorbed protein. They are not intended to be estimates of voluntary intake but are consistent with the specified dietary energy concentrations. The use of rations with decreased energy concentrations will increase dry matter intake needs; metabolizable energy, digestible energy, and total digestible nutrient needs; and crude protein needs. The use of rations with increased energy concentrations will have opposite effects on these needs.

[3]DIP = degraded intake protein.

[4]UIP = undegraded intake protein.

TABLE 20-3
DAILY NUTRIENT REQUIREMENTS OF LACTATING AND PREGNANT COWS[1]

Live Weight		Energy					Total Crude Protein	Minerals		Vitamins	
		TDN		DE	ME	NE_{lc}		Ca	P	A	D
(lb)	*(kg)*	(lb)	*(kg)*	(Mcal)	(Mcal)	(Mcal)	(g)	(g)	(g)	(1,000 IU)	(1,000 IU)
Maintenance of Mature Lactating Cows[2]											
882	*400*	6.90	*3.13*	13.80	12.01	7.16	318	16	11	30	12
992	*450*	7.54	*3.42*	15.08	13.12	7.82	341	18	13	34	14
1,103	*500*	8.16	*3.70*	16.32	14.20	8.46	364	20	14	38	15
1,213	*550*	8.75	*3.97*	17.53	15.25	9.09	386	22	16	42	17
1,323	*600*	9.35	*4.24*	18.71	16.28	9.70	406	24	17	46	18
1,433	*650*	9.94	*4.51*	19.86	17.29	10.30	428	26	19	49	20
1,544	*700*	10.50	*4.76*	21.00	18.28	10.89	449	28	20	53	21
1,654	*750*	11.07	*5.02*	22.12	19.25	11.47	468	30	21	57	23
1,764	*800*	11.60	*5.26*	23.21	20.20	12.03	486	32	23	61	24
Maintenance Plus Last 2 Months of Gestation of Mature Dry Cows[3]											
882	*400*	9.15	*4.15*	18.23	15.26	9.30	875	26	16	30	12
992	*450*	9.99	*4.53*	19.91	16.66	10.16	928	30	18	34	14
1,103	*500*	10.80	*4.90*	21.55	18.04	11.00	978	33	20	38	15
1,213	*550*	11.62	*5.27*	23.14	19.37	11.81	1,027	36	22	42	17
1,323	*600*	12.39	*5.62*	24.71	20.68	12.61	1,074	39	24	46	18
1,433	*650*	13.16	*5.97*	26.23	21.96	13.39	1,120	43	26	49	20
1,544	*700*	13.91	*6.31*	27.73	23.21	14.15	1,165	46	28	53	21
1,654	*750*	14.66	*6.65*	29.21	24.44	14.90	1,209	49	30	57	23
1,764	*800*	15.39	*6.98*	30.65	25.66	15.64	1,254	53	32	61	24

Fat	Energy								Total Crude Protein		Minerals				Vitamins	
	TDN		DE		ME		NE_{lc}				Ca		P		A	D
(%)	(lb)	*(kg)*	(Mcal/lb)	*(Mcal/kg)*	(Mcal/lb)	*(Mcal/kg)*	(Mcal/lb)	*(Mcal/kg)*	(g/lb)	*(g/kg)*	(g/lb)	*(g/kg)*	(g/lb)	*(g/kg)*		
Milk Production—Nutrients/2.2 lb or/kg of Milk of Different Fat Percentages																
3.0	0.616	*0.280*	0.56	*1.23*	0.49	*1.07*	0.29	*0.64*	35	*78*	1.24	*2.73*	0.76	*1.68*	—	—
3.5	0.662	*0.301*	0.60	*1.33*	0.52	*1.15*	0.31	*0.69*	38	*84*	1.35	*2.97*	0.83	*1.83*	—	—
4.0	0.708	*0.322*	0.64	*1.42*	0.56	*1.24*	0.34	*0.74*	41	*90*	1.46	*3.21*	0.90	*1.98*	—	—
4.5	0.755	*0.343*	0.69	*1.51*	0.60	*1.32*	0.35	*0.78*	44	*96*	1.57	*3.45*	0.98	*2.13*	—	—
5.0	0.801	*0.364*	0.73	*1.61*	0.64	*1.40*	0.38	*0.83*	46	*101*	1.68	*3.69*	1.04	*2.28*	—	—
5.5	0.847	*0.385*	0.77	*1.70*	0.69	*1.48*	0.40	*0.88*	49	*107*	1.78	*3.93*	1.10	*2.43*	—	—
Live Weight Change During Lactation—Nutrients/kg of Weight Change[4]																
Weight loss	−0.99	*−2.17*	−4.34	*−9.55*	−3.75	*−8.25*	−2.25	*−4.92*	−145	*−320*	—	—	—	—	—	—
Weight gain	1.03	*2.26*	4.52	*9.96*	3.88	*8.55*	2.32	*5.12*	145	*320*	—	—	—	—	—	—

[1]Adapted by the authors from *Nutrient Requirements of Dairy Cattle,* 6th rev. ed., update 1989, NRC, National Academy Press, p. 84, Table 6-3.

[2]To allow for growth of young lactating cows, increase the maintenance allowances for all nutrients except vitamins A and D by 20% during the first lactation and 10% during the second lactation.

[3]Values for calcium assume that the cow is in calcium balance at the beginning of the last 2 months of gestation. If the cow is not in balance, then the calcium requirement can be increased from 25 to 33%.

[4]No allowance is made for mobilized calcium and phosphorus associated with live weight loss or with live weight gain. The maximum daily nitrogen available from weight loss is assumed to be 30 g or 234 g of crude protein.

TABLE 20-4
DAILY NUTRIENT REQUIREMENTS OF LACTATING COWS USING ABSORBABLE PROTEIN[1] (See footnotes at end of table.)

Live Weight		Fat	Milk		Live Weight Change		Dry Matter Intake		Energy					Protein		Minerals	
									TDN		NE_{lc}	NE_{lcdm}[2]		DIP[3]	UIP[4]	Ca	P
(lb)	*(kg)*	(%)	(lb)	*(kg)*	(lb)	*(kg)*	(lb)	*(kg)*	(lb)	*(kg)*	(Mcal)	(Mcal/lb)	*(Mcal/kg)*	(g)	(g)	(g)	(g)
Intake at 100% of the Requirement for Maintenance, Lactation, and Weight Gain																	
882	*400*	4.5	17.6	*8.0*	0.485	*0.220*	22.36	*10.14*	14.20	*6.44*	14.55	0.65	*1.43*	753	511	44	28
882	*400*	4.5	30.9	*14.0*	0.485	*0.220*	27.92	*12.66*	18.70	*8.48*	19.26	0.69	*1.52*	1,052	710	65	41
882	*400*	4.5	44.1	*20.0*	0.485	*0.220*	32.88	*14.91*	23.17	*10.51*	23.96	0.73	*1.61*	1,355	880	85	54
882	*400*	4.5	57.3	*26.0*	0.485	*0.220*	37.35	*16.94*	27.65	*12.54*	28.67	0.77	*1.69*	1,662	1,026	106	67
882	*400*	4.5	70.6	*32.0*	0.485	*0.220*	42.80	*19.41*	32.15	*14.58*	33.37	0.78	*1.72*	1,962	1,220	127	80
882	*400*	5.0	17.6	*8.0*	0.485	*0.220*	22.84	*10.36*	14.55	*6.60*	14.94	0.65	*1.44*	778	525	46	30
882	*400*	5.0	30.9	*14.0*	0.485	*0.220*	28.67	*13.00*	19.34	*8.77*	19.93	0.69	*1.53*	1,096	730	68	43
882	*400*	5.0	44.1	*20.0*	0.485	*0.220*	33.85	*15.35*	24.10	*10.93*	24.93	0.74	*1.62*	1,419	902	90	57
882	*400*	5.0	57.3	*26.0*	0.485	*0.220*	38.46	*17.44*	28.82	*13.07*	29.92	0.78	*1.72*	1,745	1,048	112	71
882	*400*	5.0	70.6	*32.0*	0.485	*0.220*	44.76	*20.30*	33.63	*15.25*	34.91	0.78	*1.72*	2,061	1,277	134	84
882	*400*	5.5	17.6	*8.0*	0.485	*0.220*	23.31	*10.57*	14.93	*6.77*	15.32	0.66	*1.45*	803	538	48	31
882	*400*	5.5	30.9	*14.0*	0.485	*0.220*	29.39	*13.33*	20.00	*9.07*	20.61	0.70	*1.55*	1,140	748	71	45

(Continued)

TABLE 20-4 *(Continued)*

Live Weight		Fat	Milk		Live Weight Change		Dry Matter Intake		Energy					Protein		Minerals	
									TDN		NE_{lc}	NE_{lcdm}[2]		DIP[3]	UIP[4]	Ca	P
(lb)	*(kg)*	(%)	(lb)	*(kg)*	(lb)	*(kg)*	(lb)	*(kg)*	(lb)	*(kg)*	(Mcal)	(Mcal/lb)	*(Mcal/kg)*	(g)	(g)	(g)	(g)
Intake at 100% of the Requirement for Maintenance, Lactation, and Weight Gain (Continued)																	
882	*400*	5.5	44.1	*20.0*	0.485	*0.220*	34.77	*15.77*	25.00	*11.34*	25.89	0.74	*1.64*	1,483	923	95	60
882	*400*	5.5	57.3	*26.0*	0.485	*0.220*	39.98	*18.13*	30.03	*13.62*	31.17	0.78	*1.72*	1,826	1,091	118	75
882	*400*	5.5	70.6	*32.0*	0.485	*0.220*	46.75	*21.20*	35.10	*15.92*	36.45	0.78	*1.72*	2,160	1,334	142	89
1,103	*500*	4.0	19.8	*9.0*	0.606	*0.275*	25.56	*11.59*	16.10	*7.30*	16.49	0.64	*1.42*	883	540	49	32
1,103	*500*	4.0	37.5	*17.0*	0.606	*0.275*	32.59	*14.78*	21.74	*9.86*	22.38	0.69	*1.51*	1,257	797	75	48
1,103	*500*	4.0	55.1	*25.0*	0.606	*0.275*	38.85	*17.62*	27.34	*12.40*	28.27	0.73	*1.61*	1,635	1,015	101	64
1,103	*500*	4.0	72.8	*33.0*	0.606	*0.275*	44.41	*20.14*	32.92	*14.93*	34.15	0.77	*1.70*	2,018	1,201	126	80
1,103	*500*	4.0	90.4	*41.0*	0.606	*0.275*	51.35	*23.29*	38.57	*17.49*	40.04	0.78	*1.72*	2,392	1,453	152	95
1,103	*500*	4.5	19.8	*9.0*	0.606	*0.275*	26.11	*11.84*	16.52	*7.49*	16.92	0.65	*1.43*	911	556	51	33
1,103	*500*	4.5	37.5	*17.0*	0.606	*0.275*	33.52	*15.20*	22.51	*10.21*	23.20	0.69	*1.53*	1,310	821	79	50
1,103	*500*	4.5	55.1	*25.0*	0.606	*0.275*	40.04	*18.16*	28.49	*12.92*	29.47	0.74	*1.62*	1,715	1,043	107	68
1,103	*500*	4.5	72.8	*33.0*	0.606	*0.275*	45.84	*20.79*	34.42	*15.61*	35.74	0.78	*1.72*	2,124	1,230	134	85
1,103	*500*	4.5	90.4	*41.0*	0.606	*0.275*	53.89	*24.44*	40.46	*18.35*	42.02	0.78	*1.72*	2,519	1,526	162	102
1,103	*500*	5.0	19.8	*9.0*	0.606	*0.275*	26.64	*12.08*	16.93	*7.68*	17.36	0.65	*1.44*	939	571	53	35
1,103	*500*	5.0	37.5	*17.0*	0.606	*0.275*	34.40	*15.60*	23.31	*10.57*	24.01	0.70	*1.54*	1,364	844	83	53
1,103	*500*	5.0	55.1	*25.0*	0.606	*0.275*	41.19	*18.68*	29.64	*13.44*	30.67	0.74	*1.64*	1,795	1,069	113	71
1,103	*500*	5.0	72.8	*33.0*	0.606	*0.275*	47.87	*21.71*	35.96	*16.31*	37.33	0.78	*1.72*	2,226	1,289	142	89
1,103	*500*	5.0	90.4	*41.0*	0.606	*0.275*	56.40	*25.58*	42.36	*19.21*	43.99	0.78	*1.72*	2,646	1,599	172	108
1,323	*600*	3.0	22.1	*10.0*	0.728	*0.330*	27.61	*12.52*	17.35	*7.87*	17.79	0.64	*1.42*	974	533	52	34
1,323	*600*	3.0	44.1	*20.0*	0.728	*0.330*	35.72	*16.20*	23.53	*10.67*	24.18	0.68	*1.49*	1,375	845	79	51
1,323	*600*	3.0	66.2	*30.0*	0.728	*0.330*	42.71	*19.37*	29.61	*13.43*	30.58	0.72	*1.58*	1,784	1,102	106	68
1,323	*600*	3.0	88.2	*40.0*	0.728	*0.330*	48.97	*22.21*	35.70	*16.19*	36.98	0.76	*1.67*	2,198	1,323	133	84
1,323	*600*	3.0	110.3	*50.0*	0.728	*0.330*	55.63	*25.23*	41.78	*18.95*	43.38	0.78	*1.72*	2,608	1,565	161	101
1,323	*600*	3.5	22.1	*10.0*	0.728	*0.330*	28.36	*12.86*	17.82	*8.08*	18.27	0.64	*1.42*	1,004	557	54	35
1,323	*600*	3.5	44.1	*20.0*	0.728	*0.330*	36.82	*16.70*	24.43	*11.08*	25.15	0.69	*1.51*	1,438	874	84	54
1,323	*600*	3.5	66.2	*30.0*	0.728	*0.330*	44.19	*20.04*	31.00	*14.06*	32.03	0.73	*1.60*	1,879	1,137	113	72
1,323	*600*	3.5	88.2	*40.0*	0.728	*0.330*	50.72	*23.00*	37.51	*17.01*	38.90	0.77	*1.69*	2,326	1,360	143	90
1,323	*600*	3.5	110.3	*50.0*	0.728	*0.330*	58.72	*26.63*	44.10	*20.00*	45.78	0.78	*1.72*	2,763	1,654	173	109
1,323	*600*	4.0	22.1	*10.0*	0.728	*0.330*	29.11	*13.20*	18.30	*8.30*	18.75	0.64	*1.42*	1,034	581	56	37
1,323	*600*	4.0	44.1	*20.0*	0.728	*0.330*	37.90	*17.19*	25.36	*11.50*	26.11	0.69	*1.52*	1,501	902	89	57
1,323	*600*	4.0	66.2	*30.0*	0.728	*0.330*	45.62	*20.69*	32.37	*14.68*	33.47	0.74	*1.62*	1,975	1,170	121	77
1,323	*600*	4.0	88.2	*40.0*	0.728	*0.330*	52.43	*23.78*	39.34	*17.84*	40.83	0.78	*1.72*	2,454	1,395	153	96
1,323	*600*	4.0	110.3	*50.0*	0.728	*0.330*	61.81	*28.03*	46.42	*21.05*	48.19	0.78	*1.72*	2,918	1,744	185	116
1,544	*700*	3.0	26.5	*12.0*	0.849	*0.385*	31.88	*14.46*	20.04	*9.09*	20.54	0.64	*1.42*	1,154	607	61	40
1,544	*700*	3.0	52.9	*24.0*	0.849	*0.385*	41.34	*18.75*	27.43	*12.44*	28.21	0.68	*1.50*	1,638	968	94	60
1,544	*700*	3.0	79.4	*36.0*	0.849	*0.385*	49.57	*22.48*	34.75	*15.76*	35.89	0.73	*1.60*	2,129	1,269	127	81
1,544	*700*	3.0	105.8	*48.0*	0.849	*0.385*	56.89	*25.80*	42.01	*19.05*	43.57	0.77	*1.69*	2,627	1,525	159	101
1,544	*700*	3.0	132.3	*60.0*	0.849	*0.385*	65.73	*29.81*	49.37	*22.39*	51.25	0.78	*1.72*	3,114	1,857	192	121
1,544	*700*	3.5	26.5	*12.0*	0.849	*0.385*	32.77	*14.86*	20.59	*9.34*	21.11	0.64	*1.42*	1,190	636	64	42
1,544	*700*	3.5	52.9	*24.0*	0.849	*0.385*	42.64	*19.34*	28.53	*12.94*	29.37	0.69	*1.52*	1,713	1,002	100	64
1,544	*700*	3.5	79.4	*36.0*	0.849	*0.385*	51.29	*23.26*	36.38	*16.50*	37.62	0.74	*1.62*	2,244	1,309	135	86
1,544	*700*	3.5	105.8	*48.0*	0.849	*0.385*	58.92	*26.72*	44.19	*20.04*	45.88	0.78	*1.72*	2,781	1,567	171	108
1,544	*700*	3.5	132.3	*60.0*	0.849	*0.385*	69.41	*31.48*	52.15	*23.65*	54.13	0.78	*1.72*	3,300	1,964	207	130
1,544	*700*	4.0	26.5	*12.0*	0.849	*0.385*	33.52	*15.20*	21.17	*9.60*	21.69	0.65	*1.43*	1,227	658	67	44
1,544	*700*	4.0	52.9	*24.0*	0.849	*0.385*	43.92	*19.92*	29.64	*13.44*	30.52	0.69	*1.53*	1,789	1,035	105	68
1,544	*700*	4.0	79.4	*36.0*	0.849	*0.385*	52.96	*24.02*	38.04	*17.25*	39.35	0.74	*1.64*	2,359	1,347	144	91
1,544	*700*	4.0	105.8	*48.0*	0.849	*0.385*	61.81	*28.03*	46.42	*21.05*	48.19	0.78	*1.72*	2,930	1,648	182	115
1,544	*700*	4.0	132.3	*60.0*	0.849	*0.385*	73.12	*33.16*	54.93	*24.91*	57.02	0.78	*1.72*	3,485	2,071	221	139
1,764	*800*	3.0	30.9	*14.0*	0.970	*0.440*	36.07	*16.36*	22.69	*10.29*	23.24	0.64	*1.42*	1,331	682	71	46
1,764	*800*	3.0	59.5	*27.0*	0.970	*0.440*	46.15	*20.93*	30.67	*13.91*	31.56	0.69	*1.51*	1,857	1,064	106	68
1,764	*800*	3.0	88.2	*40.0*	0.970	*0.440*	55.01	*24.95*	38.59	*17.50*	39.88	0.73	*1.60*	2,390	1,388	142	90
1,764	*800*	3.0	116.9	*53.0*	0.970	*0.440*	62.93	*28.54*	46.48	*21.08*	48.20	0.77	*1.69*	2,928	1,665	177	112
1,764	*800*	3.0	145.5	*66.0*	0.970	*0.440*	72.48	*32.87*	54.44	*24.69*	56.51	0.78	*1.72*	3,457	2,022	213	134
1,764	*800*	3.5	30.9	*14.0*	0.970	*0.440*	37.00	*16.78*	23.33	*10.58*	23.92	0.64	*1.42*	1,374	710	74	49
1,764	*800*	3.5	59.5	*27.0*	0.970	*0.440*	47.61	*21.59*	31.91	*14.47*	32.86	0.69	*1.52*	1,942	1,102	113	72
1,764	*800*	3.5	88.2	*40.0*	0.970	*0.440*	56.93	*25.82*	40.42	*18.33*	41.80	0.74	*1.62*	2,517	1,432	151	96
1,764	*800*	3.5	116.9	*53.0*	0.970	*0.440*	65.20	*29.57*	48.88	*22.17*	50.75	0.78	*1.72*	3,099	1,711	190	120
1,764	*800*	3.5	145.5	*66.0*	0.970	*0.440*	76.56	*34.72*	57.48	*26.07*	59.69	0.78	*1.72*	3,661	2,140	228	144
1,764	*800*	4.0	30.9	*14.0*	0.970	*0.440*	37.86	*17.17*	23.99	*10.88*	24.59	0.65	*1.43*	1,418	734	77	51

(Continued)

TABLE 20-4 *(Continued)*

Live Weight		Fat	Milk		Live Weight Change		Dry Matter Intake		Energy					Protein		Minerals	
									TDN		NE_{lc}	NE_{lcdm}[2]		DIP[3]	UIP[4]	Ca	P
(lb)	*(kg)*	(%)	(lb)	*(kg)*	(lb)	*(kg)*	(lb)	*(kg)*	(lb)	*(kg)*	(Mcal)	(Mcal/lb)	*(Mcal/kg)*	(g)	(g)	(g)	(g)
Intake at 100% of the Requirement for Maintenance, Lactation, and Weight Gain (Continued)																	
1,764	*800*	4.0	59.5	*27.0*	0.970	*0.440*	49.04	*22.24*	33.14	*15.03*	34.16	0.70	*1.54*	2,027	1,139	119	76
1,764	*800*	4.0	88.2	*40.0*	0.970	*0.440*	58.78	*26.66*	42.25	*19.16*	43.73	0.74	*1.64*	2,644	1,474	161	102
1,764	*800*	4.0	116.9	*53.0*	0.970	*0.440*	68.36	*31.00*	51.33	*23.28*	53.29	0.78	*1.72*	3,263	1,800	203	128
1,764	*800*	4.0	145.5	*66.0*	0.970	*0.440*	80.61	*36.56*	60.55	*27.46*	62.86	0.78	*1.72*	3,865	2,259	244	154
Intake at 85% of the Requirement for Maintenance and Lactation																	
882	*400*	4.5	44.1	*20.0*	−1.535	*−0.696*	25.62	*11.62*	18.72	*8.49*	19.41	0.76	*1.67*	1,066	687	85	54
882	*400*	4.5	57.3	*26.0*	−1.852	*−0.840*	30.91	*14.02*	22.58	*10.24*	23.41	0.76	*1.67*	1,310	931	106	67
882	*400*	4.5	70.6	*32.0*	−2.168	*−0.983*	36.18	*16.41*	26.44	*11.99*	27.41	0.76	*1.67*	1,554	1,187	127	80
882	*400*	5.0	44.1	*20.0*	−1.601	*−0.726*	26.70	*12.11*	19.51	*8.85*	20.23	0.76	*1.67*	1,118	720	90	57
882	*400*	5.0	57.3	*26.0*	−1.936	*−0.878*	32.30	*14.65*	23.62	*10.71*	24.47	0.76	*1.67*	1,377	987	112	71
882	*400*	5.0	70.6	*32.0*	−2.271	*−1.030*	37.93	*17.20*	27.69	*12.56*	28.72	0.76	*1.67*	1,635	1,255	134	84
882	*400*	5.5	44.1	*20.0*	−1.665	*−0.755*	27.78	*12.60*	20.31	*9.21*	21.05	0.76	*1.67*	1,169	761	95	60
882	*400*	5.5	57.3	*26.0*	−2.020	*−0.916*	33.71	*15.29*	24.63	*11.17*	25.54	0.76	*1.67*	1,443	1,042	118	75
882	*400*	5.5	70.6	*32.0*	−2.375	*−1.077*	39.65	*17.98*	28.97	*13.14*	30.03	0.76	*1.67*	1,717	1,323	142	89
1,103	*500*	4.0	55.1	*25.0*	−1.806	*−0.819*	30.14	*13.67*	22.03	*9.99*	22.83	0.76	*1.67*	1,286	810	101	64
1,103	*500*	4.0	72.8	*33.0*	−2.201	*−0.998*	36.76	*16.67*	26.86	*12.18*	27.83	0.76	*1.67*	1,590	1,134	126	80
1,103	*500*	4.0	90.4	*41.0*	−2.597	*−1.178*	43.35	*19.66*	31.69	*14.37*	32.84	0.76	*1.67*	1,894	1,458	152	95
1,103	*500*	4.5	55.1	*25.0*	−1.887	*−0.856*	31.49	*14.28*	23.02	*10.44*	23.85	0.76	*1.67*	1,350	864	107	68
1,103	*500*	4.5	72.8	*33.0*	−2.309	*−1.047*	38.54	*17.48*	28.16	*12.77*	29.18	0.76	*1.67*	1,674	1,205	134	85
1,103	*500*	4.5	90.4	*41.0*	−2.730	*−1.238*	45.58	*20.67*	33.30	*15.10*	34.52	0.76	*1.67*	1,998	1,546	162	102
1,103	*500*	5.0	55.1	*25.0*	−1.967	*−0.892*	32.83	*14.89*	23.99	*10.88*	24.87	0.76	*1.67*	1,414	917	113	71
1,103	*500*	5.0	72.8	*33.0*	−2.414	*−1.095*	40.31	*18.28*	29.46	*13.36*	30.53	0.76	*1.67*	1,758	1,275	142	89
1,103	*500*	5.0	90.4	*41.0*	−2.862	*−1.298*	47.78	*21.67*	34.91	*15.83*	36.19	0.76	*1.67*	2,103	1,633	172	108
1,323	*600*	3.0	66.2	*30.0*	−1.943	*−0.881*	32.44	*14.71*	23.68	*10.74*	24.56	0.76	*1.67*	1,399	860	106	68
1,323	*600*	3.0	88.2	*40.0*	−2.373	*−1.076*	39.60	*17.96*	28.93	*13.12*	30.00	0.76	*1.67*	1,728	1,223	133	84
1,323	*600*	3.0	110.3	*50.0*	−2.803	*−1.271*	46.79	*21.22*	34.18	*15.50*	35.44	0.76	*1.67*	2,057	1,585	161	101
1,323	*600*	3.5	66.2	*30.0*	−2.396	*−0.925*	34.05	*15.44*	24.87	*11.28*	25.79	0.76	*1.67*	1,476	924	113	72
1,323	*600*	3.5	88.2	*40.0*	−2.503	*−1.135*	41.76	*18.94*	30.52	*13.84*	31.63	0.76	*1.67*	1,830	1,308	143	90
1,323	*600*	3.5	110.3	*50.0*	−2.964	*−1.344*	49.48	*22.44*	36.16	*16.40*	37.48	0.76	*1.67*	2,184	1,692	173	109
1,323	*600*	4.0	66.2	*30.0*	−2.137	*−0.969*	35.65	*16.17*	26.06	*11.82*	27.01	0.76	*1.67*	1,552	988	121	77
1,323	*600*	4.0	88.2	*40.0*	−2.631	*−1.193*	43.92	*19.92*	32.08	*14.55*	33.27	0.76	*1.67*	1,932	1,393	153	96
1,323	*600*	4.0	110.3	*50.0*	−3.127	*−1.418*	52.19	*23.67*	38.12	*17.29*	39.52	0.76	*1.67*	2,311	1,798	185	116
1,544	*700*	3.0	79.4	*36.0*	−2.280	*−1.034*	38.06	*17.26*	27.81	*12.61*	28.83	0.76	*1.67*	1,669	1,054	127	81
1,544	*700*	3.0	105.8	*48.0*	−2.796	*−1.268*	46.68	*21.17*	34.11	*15.47*	35.36	0.76	*1.67*	2,064	1,489	159	101
1,544	*700*	3.0	132.3	*60.0*	−3.312	*−1.502*	55.30	*25.08*	40.40	*18.32*	41.88	0.76	*1.67*	2,458	1,924	192	121
1,544	*700*	3.5	79.4	*36.0*	−2.397	*−1.087*	40.02	*18.15*	29.24	*13.26*	30.30	0.76	*1.67*	1,761	1,131	135	86
1,544	*700*	3.5	105.8	*48.0*	−2.952	*−1.339*	49.28	*22.35*	36.01	*16.33*	37.32	0.76	*1.67*	2,186	1,591	171	108
1,544	*700*	3.5	132.3	*60.0*	−3.506	*−1.590*	58.54	*26.55*	42.78	*19.40*	44.34	0.76	*1.67*	2,611	2,052	207	130
1,544	*700*	4.0	79.4	*36.0*	−2.434	*−1.140*	41.96	*19.03*	30.65	*13.90*	31.78	0.76	*1.67*	1,853	1,208	144	91
1,544	*700*	4.0	105.8	*48.0*	−3.107	*−1.409*	51.86	*23.52*	37.90	*17.19*	39.28	0.76	*1.67*	2,308	1,694	182	115
1,544	*700*	4.0	132.3	*60.0*	−3.700	*−1.678*	61.78	*28.02*	45.14	*20.47*	46.79	0.76	*1.67*	2,764	2,180	221	139
1,764	*800*	3.0	88.2	*40.0*	−2.529	*−1.147*	42.23	*19.15*	30.85	*13.99*	31.98	0.76	*1.67*	1,871	1,176	142	90
1,764	*800*	3.0	110.3	*50.0*	−2.959	*−1.342*	49.41	*22.41*	36.10	*16.37*	37.42	0.76	*1.67*	2,200	1,538	169	107
1,764	*800*	3.0	132.3	*60.0*	−3.389	*−1.537*	56.58	*25.66*	41.34	*18.75*	42.86	0.76	*1.67*	2,529	1,900	196	124
1,764	*800*	3.5	88.2	*40.0*	−2.659	*−1.206*	44.39	*20.13*	32.44	*14.71*	33.62	0.76	*1.67*	1,973	1,261	151	96
1,764	*800*	3.5	110.3	*50.0*	−3.122	*−1.416*	52.10	*23.63*	38.08	*17.27*	39.46	0.76	*1.67*	2,327	1,645	181	114
1,764	*800*	3.5	132.3	*60.0*	−3.583	*−1.625*	59.82	*27.13*	43.70	*19.82*	45.31	0.76	*1.67*	2,682	2,028	211	133
1,764	*800*	4.0	88.2	*40.0*	−2.787	*−1.264*	46.55	*21.11*	34.00	*15.42*	35.25	0.76	*1.67*	2,075	1,346	161	102
1,764	*800*	4.0	110.3	*50.0*	−3.283	*−1.489*	54.82	*24.86*	40.04	*18.16*	41.51	0.76	*1.67*	2,455	1,751	193	122
1,764	*800*	4.0	132.3	*60.0*	−3.777	*−1.713*	63.06	*28.60*	46.08	*20.90*	47.76	0.76	*1.67*	2,835	2,156	225	142

[1]Adapted by the authors from *Nutrient Requirements of Dairy Cattle,* 6th rev. ed., update 1989, NRC, National Academy Press, pp. 85–86, Table 6-4.

[2]NE_{lcdm} = net energy for lactation/kg of dry matter.

[3]DIP = degraded intake protein.

[4]UIP = undegraded intake protein.

TABLE 20-5
RECOMMENDED NUTRIENT CONTENT OF RATIONS FOR DAIRY CATTLE[1] (See footnotes at end of table.)

Cow Weight (lb)	Cow Weight (*kg*)	Fat (%)	Weight Gain (lb/day)	Weight Gain (*kg/day*)	Milk Yield (lb/day)	Milk Yield (*kg/day*)	Milk Yield (lb/day)	Milk Yield (*kg/day*)	Milk Yield (lb/day)	Milk Yield (*kg/day*)	Milk Yield (lb/day)	Milk Yield (*kg/day*)	Milk Yield (lb/day)	Milk Yield (*kg/day*)
882	*400*	5.0	0.485	*0.220*	15.4	*7*	28.7	*13*	44.1	*20*	57.3	*26*	72.8	*33*
1,103	*500*	4.5	0.606	*0.275*	17.6	*8*	37.5	*17*	55.1	*25*	72.8	*33*	90.4	*41*
1,323	*600*	4.0	0.728	*0.330*	22.1	*10*	44.1	*20*	66.2	*30*	88.2	*40*	110.3	*50*
1,544	*700*	3.5	0.849	*0.385*	26.5	*12*	52.9	*24*	79.4	*36*	105.8	*48*	132.3	*60*
1,764	*800*	3.5	0.970	*0.440*	28.7	*13*	59.5	*27*	88.2	*40*	116.9	*53*	147.7	*67*

Nutrient	Lactating Cow Rations: Milk Yield	Milk Yield	Milk Yield	Milk Yield	Milk Yield	Early Lactation (Weeks 0–3)	Dry, Pregnant Cows	Calf Milk Replacer	Calf Starter Mix	Growing Heifers and Bulls[2] 3–6 Months	6–12 Months	Over 12 Mos	Mature Bulls	Maximum Tolerable Levels[3,4]
Energy:														
NE_{lc} (Mcal/lb)	0.64	0.69	0.74	0.78	0.78	0.76	0.57	—	—	—	—	—	—	—
NE_{lc} *(Mcal/kg)*	*1.42*	*1.52*	*1.62*	*1.72*	*1.72*	*1.67*	*1.25*	—	—	—	—	—	—	—
NE_m (Mcal/lb)	—	—	—	—	—	—	—	1.09	0.86	0.77	0.72	0.64	0.52	—
NE_m *(Mcal/kg)*	—	—	—	—	—	—	—	*2.40*	*1.90*	*1.70*	*1.58*	*1.40*	*1.15*	—
NE_g (Mcal/lb)	—	—	—	—	—	—	—	0.70	0.54	0.49	0.44	0.37	—	—
NE_g *(Mcal/kg)*	—	—	—	—	—	—	—	*1.55*	*1.20*	*1.08*	*0.98*	*0.82*	—	—
ME (Mcal/lb)	1.07	1.15	1.23	1.31	1.31	1.27	0.93	1.72	1.41	1.18	1.12	1.03	0.91	—
ME *(Mcal/kg)*	*2.35*	*2.53*	*2.71*	*2.89*	*2.89*	*2.80*	*2.04*	*3.78*	*3.11*	*2.60*	*2.47*	*2.27*	*2.00*	—
DE (Mcal/lb)	1.26	1.34	1.42	1.50	1.50	1.46	1.12	1.90	1.60	1.37	1.31	1.22	1.10	—
DE *(Mcal/kg)*	*2.77*	*2.95*	*3.13*	*3.31*	*3.31*	*3.22*	*2.47*	*4.19*	*3.53*	*3.02*	*2.89*	*2.69*	*2.43*	—
TDN (% of DM)	63	67	71	75	75	73	56	95	80	69	66	61	55	—
Protein equivalent:														
Crude protein (%)	12	15	16	17	18	19	12	22	18	16	12	12	10	—
DIP[5] (%)	7.8	8.7	9.6	10.3	10.4	9.7	—	—	—	4.6	6.4	7.2	—	—
UIP[6] (%)	4.4	5.2	5.7	5.9	6.2	7.0	—	—	—	8.2	4.4	2.1	—	—
Fiber content (minimum):[7]														
Crude fiber (%)	17	17	17	15	15	17	22	—	—	13	15	15	15	—
Neutral detergent fiber (NDF) (%)	28	28	28	25	25	28	35	—	—	23	25	25	25	—
Acid detergent fiber (ADF) (%)	21	21	21	19	19	21	27	—	—	16	19	19	19	—
Ether extract (minimum) (%)	3	3	3	3	3	3	3	10	3	3	3	3	3	—
Major or Macrominerals:														
Calcium (Ca) (%)	0.43	0.51	0.58	0.64	0.66	0.77	0.39[8]	0.70	0.60	0.52	0.41	0.29	0.30	2.00
Chlorine (Cl) (%)	0.25	0.25	0.25	0.25	0.25	0.25	0.20	0.20	0.20	0.20	0.20	0.20	0.20	—
Magnesium (Mg)[9] (%)	0.20	0.20	0.20	0.25	0.25	0.25	0.16	0.07	0.10	0.16	0.16	0.16	0.16	0.50
Phosphorus (P) (%)	0.28	0.33	0.37	0.41	0.41	0.48	0.24	0.60	0.40	0.31	0.30	0.23	0.19	1.00
Potassium (K)[10] (%)	0.90	0.90	0.90	1.00	1.00	1.00	0.65	0.65	0.65	0.65	0.65	0.65	0.65	3.00
Sodium (Na) (%)	0.18	0.18	0.18	0.18	0.18	0.18	0.10	0.10	0.10	0.10	0.10	0.10	0.10	—
Sulfur (S) (%)	0.20	0.20	0.20	0.20	0.20	0.25	0.16	0.29	0.20	0.16	0.16	0.16	0.16	0.40
Trace or Microminerals:														
Cobalt (Co) (ppm)	0.10	0.10	0.10	0.10	0.10	0.10	0.10	0.10	0.10	0.10	0.10	0.10	0.10	10.00
Copper (Cu)[11] (ppm)	10	10	10	10	10	10	10	10	10	10	10	10	10	100
Iodine (I)[12] (ppm)	0.60	0.60	0.60	0.60	0.60	0.60	0.25	0.25	0.25	0.25	0.25	0.25	0.25	50.00[13]
Iron (Fe) (ppm)	50	50	50	50	50	50	50	100	50	50	50	50	50	1,000
Manganese (Mn) (ppm)	40	40	40	40	40	40	40	40	40	40	40	40	40	1,000
Selenium (Se) (ppm)	0.30	0.30	0.30	0.30	0.30	0.30	0.30	0.30	0.30	0.30	0.30	0.30	0.30	2.00
Zinc (Zn) (ppm)	40	40	40	40	40	40	40	40	40	40	40	40	40	500

(Continued)

TABLE 20–5 *(Continued)*

Cow Weight		Fat	Weight Gain		Lactating Cow Rations										Early Lactation (Weeks 0–3)	Dry, Pregnant Cows	Calf Milk Replacer	Calf Starter Mix	Growing Heifers and Bulls[2]			Mature Bulls	Maximum Tolerable Levels[3,4]
					Milk Yield		Milk Yield		Milk Yield		Milk Yield		Milk Yield						3–6 Months	6–12 Months	Over 12 Mos		
(lb)	*(kg)*	(%)	(lb/day)	*(kg/day)*	(lb/day)	*(kg/day)*	(lb/day)	*(kg/day)*	(lb/day)	*(kg/day)*	(lb/day)	*(kg/day)*	(lb/day)	*(kg/day)*									
882	*400*	5.0	0.485	*0.220*	15.4	*7*	28.7	*13*	44.1	*20*	57.3	*26*	72.8	*33*									
1,103	*500*	4.5	0.606	*0.275*	17.6	*8*	37.5	*17*	55.1	*25*	72.8	*33*	90.4	*41*									
1,323	*600*	4.0	0.728	*0.330*	22.1	*10*	44.1	*20*	66.2	*30*	88.2	*40*	110.3	*50*									
1,544	*700*	3.5	0.849	*0.385*	26.5	*12*	52.9	*24*	79.4	*36*	105.8	*48*	132.3	*60*									
1,764	*800*	3.5	0.970	*0.440*	28.7	*13*	59.5	*27*	88.2	*40*	116.9	*53*	147.7	*67*									
Vitamins:[14]																							
Vitamin A				(IU/lb)	1,453		1,453		1,453		1,453		1,453		1,816	1,816	1,725	999	999	999	999	1,453	29,964
Vitamin A				*(IU/kg)*	*3,200*		*3,200*		*3,200*		*3,200*		*3,200*		*4,000*	*4,000*	*3,800*	*2,200*	*2,200*	*2,200*	*2,200*	*3,200*	*66,000*
Vitamin D				(IU/lb)	454		454		454		454		454		454	545	272	136	136	136	136	136	4,540
Vitamin D				*(IU/kg)*	*1,000*		*1,000*		*1,000*		*1,000*		*1,000*		*1,000*	*1,200*	*600*	*300*	*300*	*300*	*300*	*300*	*10,000*
Vitamin E				(IU/lb)	7		7		7		7		7		7	7	18	11	11	11	11	7	908
Vitamin E				*(IU/kg)*	*15*		*15*		*15*		*15*		*15*		*15*	*15*	*40*	*25*	*25*	*25*	*25*	*15*	*2,000*

[1]Adapted by the authors from *Nutrient Requirements of Dairy Cattle,* 6th rev. ed., update 1989, NRC, National Academy Press, p. 87, Table 6–5.

[2]The approximate weight for growing heifers and bulls at 3–6 months is 331 lb *(150 kg);* at 6–12 months, it is 551 lb *(250 kg);* and at more than 12 months, it is 882 lb *(400 kg).* The approximate average daily gain is 25 oz/day *(700 g/day).*

[3]The maximum safe levels for many of the mineral elements are not well defined and may be substantially affected by specific feeding conditions. Additional information is available in Table 20–6 and in *Mineral Tolerance of Domestic Animals* (NRC, 1980).

[4]Vitamin tolerances are discussed in detail in *Vitamin Tolerance of Animals* (NRC, 1987b).

[5]DIP = degraded intake protein.

[6]UIP = undegraded intake protein.

[7]It is recommended that 75% of the NDF in lactating cow rations be provided as forage. If this recommendation is not followed, a depression in milk fat may occur.

[8]The value for calcium assumes that the cow is in calcium balance at the beginning of the dry period. If the cow is not in balance, then the dietary calcium requirement should be increased by 25 to 33%.

[9]Under conditions conducive to grass tetany, magnesium should be increased to 0.25 or 0.30%.

[10]Under conditions of heat stress, potassium should be increased to 1.2%.

[11]The cow's copper requirement is influenced by molybdenum and sulfur in the ration.

[12]If the ration contains as much as 25% strongly goitrogenic feed on a dry basis, the iodine provided should be increased 2 times or more.

[13]Although cattle can tolerate this level of iodine, lower levels may be desirable to reduce the iodine content of milk.

[14]The following minimum quantities of B-complex vitamins are suggested per unit of milk replacer: Niacin (Nicotinic Acid, Nicotinamide), 2.6 ppm; Pantothenic Acid (Vitamin B–1), 13 ppm; Riboflavin (Vitamin B–2), 6.5 ppm; Vitamin B–6 (Pyridoxine, Pyridoxal, Pyridoxamine), 6.5 ppm; Folacin (Folic Acid), 0.5 ppm; Biotin, 0.1 ppm; Vitamin B–12 (Cobalamins), 0.07 ppm; and Choline, 0.26%. It appears that adequate amounts of these vitamins are furnished when calves have functional rumens (usually at 6 weeks of age) by a combination of rumen synthesis and natural feedstuffs.

TABLE 20–6
MAXIMUM TOLERABLE DIETARY LEVELS OF CERTAIN ELEMENTS[1]

Element	Maximum Tolerable Level
	(ppm)
Aluminum	1,000[2]
Arsenic:	
Inorganic	50
Organic	100
Bromine	200
Cadmium	0.5[3]
Fluorine	40[4]
Lead	30[3]
Mercury	2[3]
Molybdenum	10[5]
Nickel	50
Vanadium	50

[1]Adapted by the authors from *Nutrient Requirements of Dairy Cattle,* 6th rev. ed., update 1989, NRC, National Academy Press, p. 88, Table 6–6.

[2]As soluble salts of high bioavailability. Higher levels of less soluble forms found in natural substances can be tolerated.

[3]Levels are based on human food residue considerations.

[4]As sodium fluoride or fluorides of similar toxicity. The maximum safe level of fluorine for growing heifers and bulls is lower than for other dairy cattle. Somewhat higher levels are tolerated when fluorine is from less available sources such as phosphates. Morphological lesions in cattle teeth may be seen when dietary fluoride for the young exceeds 20 ppm, but a relationship between the lesions caused by fluoride levels below the maximum tolerable levels and animal performance has not been established.

[5]Toxicity related to the dietary level of copper.

TABLE

COMPOSITION OF FEEDS COMMONLY USED IN DAIRY CATTLE RATIONS

							Production			Production			
				Values as Determined at Maintenance Intake			Growing Dairy Cattle		Lactating Cows	Growing Dairy Cattle		Lactating Cows	
Entry No.	Feed Name Description	International Feed Number[2]	Dry Matter (%)	TDN (%)	DE *(Mcal/kg)*	ME *(Mcal/kg)*	NEM *(Mcal/kg)*	NEG *(Mcal/kg)*	NEL *(Mcal/kg)*	NEM (Mcal/lb)	NEG (Mcal/lb)	NEL (Mcal/lb)	Crude Protein (%)
	ALFALFA *Medicago sativa*												
1	—hay, sun-cured, early vegetative	1-00-050	90	66	*2.91*	*2.49*	*1.51*	*0.92*	*1.50*	0.69	0.42	0.68	23.0
2	—hay, sun-cured, late vegetative	1-00-054	90	63	*2.78*	*2.36*	*1.41*	*0.83*	*1.42*	0.64	0.38	0.65	20.0
3	—hay, sun-cured, early bloom	1-00-059	90	60	*2.65*	*2.22*	*1.31*	*0.74*	*1.35*	0.60	0.34	0.61	18.0
4	—hay, sun-cured, midbloom	1-00-063	90	58	*2.56*	*2.13*	*1.24*	*0.68*	*1.30*	0.57	0.31	0.59	17.0
5	—hay, sun-cured, full bloom	1-00-068	90	55	*2.43*	*2.00*	*1.14*	*0.58*	*1.23*	0.52	0.26	0.56	15.0
6	—silage, wilted, 25–45% dry matter (see similar maturity descriptions of hays)		—	—	—	—	—	—	—	—	—	—	—
	ALMOND *Prunus amygdalus*												
7	—hulls	4-00-	90	59	*2.60*	*2.18*	*1.27*	*0.70*	*1.33*	0.58	0.32	0.60	2.7
	BARLEY *Hordeum vulgare*												
8	—grain	4-00-549	88	84	*3.70*	*3.29*	*2.06*	*1.40*	*1.94*	0.94	0.64	0.88	13.5
9	—grain, Pacific Coast	4-07-939	89	86	*3.79*	*3.38*	*2.12*	*1.45*	*1.99*	0.96	0.66	0.90	10.8
	BEET, SUGAR *Beta vulgaris altissima*												
10	—pulp, dehydrated	4-00-669	91	78	*3.44*	*3.02*	*1.88*	*1.24*	*1.79*	0.86	0.57	0.81	9.7
11	—pulp w/molasses, dehydrated	4-00-672	92	78	*3.44*	*3.02*	*1.88*	*1.24*	*1.79*	0.86	0.57	0.81	10.1
	BLOOD												
12	—meal	5-00-380	92	66	*2.91*	*2.49*	*1.51*	*0.92*	*1.50*	0.69	0.42	0.68	87.2
	BREWERS' GRAINS												
13	—dehydrated	5-02-141	92	66	*2.91*	*2.49*	*1.51*	*0.91*	*1.50*	0.69	0.41	0.68	25.4
	BROME *Bromus* spp												
14	—fresh, early vegetative	2-00-892	34	74	*3.26*	*2.85*	*1.75*	*1.13*	*1.69*	0.80	0.51	0.77	18.0
	CASEIN												
15	—dehydrated (cattle)	5-01-162	91	89	*3.92*	*3.51*	*2.20*	*1.52*	*2.06*	1.00	0.69	0.94	92.7
	CITRUS *Citrus* spp												
16	—pulp w/o fines, dehydrated (dried citrus pulp)	4-01-237	91	77	*3.40*	*2.98*	*1.86*	*1.22*	*1.77*	0.85	0.55	0.80	6.7
	CORN, DENT YELLOW *Zea mays indentata*												
17	—distillers' grains, dehydrated	5-28-235	94	86	*3.79*	*3.38*	*2.12*	*1.45*	*1.99*	0.96	0.66	0.90	23.0
18	—ears, ground (corn and cob meal)	4-28-238	87	83	*3.66*	*3.25*	*2.03*	*1.37*	*1.91*	0.92	0.62	0.87	9.0
19	—gluten, meal	5-28-241	91	86	*3.79*	*3.38*	*2.12*	*1.45*	*1.99*	0.96	0.66	0.90	46.8
20	—gluten, meal, 60% protein	5-28-242	90	89	*3.92*	*3.51*	*2.20*	*1.52*	*2.06*	1.00	0.69	0.94	67.2
21	—gluten w/bran (corn gluten feed)	5-28-243	90	83	*3.66*	*3.25*	*2.03*	*1.37*	*1.91*	0.92	0.62	0.87	25.6
22	—grain, flaked	4-28-244	89	88	*3.88*	*3.47*	*2.18*	*1.50*	*2.04*	0.99	0.68	0.93	10.0
23	—grain, high-moisture	4-20-770	77	88	*3.88*	*3.47*	*2.18*	*1.50*	*2.04*	0.99	0.68	0.93	10.0
24	—grits by-product (hominy feed)	4-03-011	90	87	*3.84*	*3.42*	*2.16*	*1.48*	*2.01*	0.98	0.67	0.91	11.5
25	—silage, few ears	3-28-245	29	62	*2.73*	*2.31*	*1.38*	*0.80*	*1.40*	0.63	0.36	0.64	8.4
26	—silage, well-eared	3-28-250	33	70	*3.09*	*2.67*	*1.63*	*1.03*	*1.60*	0.74	0.47	0.73	8.1
	COTTON *Gossypium* spp												
27	—hulls	1-01-599	91	45	*1.98*	*1.55*	*0.78*	*0.25*	*0.98*	0.35	0.11	0.45	4.1
28	—seeds, w/lint	5-01-614	92	96	*4.23*	*3.83*	*2.41*	*1.69*	*2.23*	1.10	0.77	1.01	23.0
29	—seeds, w/o lint	5-01-	90	96	*4.23*	*3.82*	*2.41*	*1.69*	*2.23*	1.10	0.77	1.01	25.0
30	—seeds, meal prepressed, solv-extd, 41% protein	5-07-872	91	76	*3.35*	*2.93*	*1.82*	*1.19*	*1.74*	0.83	0.55	0.79	45.6
31	—seeds, meal prepressed, solv-extd, 44% protein	5-07-873	91	75	*3.31*	*2.89*	*1.79*	*1.16*	*1.72*	0.81	0.53	0.78	48.9
	FATS AND OILS (not exceeding 3% of diet)												
32	—fat, animal, hydrolyzed	4-00-376	99	177	*7.30*	*7.30*	*5.84*	*5.84*	*5.84*	2.65	2.65	2.65	—
	FESCUE, KENTUCKY 31 *Festuca arundinacea*												
33	—fresh, vegetative	2-01-902	29	67	*2.91*	*2.49*	*1.51*	*0.92*	*1.50*	0.69	0.42	0.68	14.5
	FISH, MENHADEN *Brevoortia tyrannus*												
34	—meal mech-extd	5-02-009	92	73	*3.22*	*2.80*	*1.73*	*1.11*	*1.67*	0.79	0.50	0.76	66.7
	FLAX *Linum usitatissimum*												
35	—seeds, meal solv-extd (linseed meal)	5-02-048	90	78	*3.44*	*3.02*	*1.88*	*1.24*	*1.79*	0.85	0.56	0.81	38.3
	MILK												
36	—skimmed dehydrated (cattle)	5-01-175	94	85	*3.75*	*3.34*	*2.10*	*1.43*	*1.96*	0.95	0.65	0.89	35.8
	MOLASSES AND SYRUP												
37	—beet, sugar, molasses, more than 48% invert sugar, more than 79.5 degrees brix	4-00-668	78	75	*3.31*	*2.89*	*1.79*	*1.16*	*1.72*	0.81	0.53	0.78	8.5
38	—citrus, syrup (citrus molasses)	4-01-241	68	75	*3.31*	*2.89*	*1.79*	*1.16*	*1.72*	0.81	0.53	0.78	8.2
39	—sugarcane, molasses, dehydrated	4-04-695	94	70	*3.09*	*2.67*	*1.63*	*1.03*	*1.60*	0.74	0.47	0.73	10.3
40	—sugarcane, molasses, more than 46% invert sugar, more than 79.5 degrees brix (Black strap)	4-04-696	75	72	*3.17*	*2.76*	*1.69*	*1.08*	*1.64*	0.77	0.49	0.75	5.8
	OATS *Avena sativa*												
41	—grain	4-03-309	89	77	*3.40*	*2.98*	*1.86*	*1.22*	*1.77*	0.85	0.55	0.80	13.3
	PEANUT *Arachis hypogaea*												
42	—kernels, meal solv-extd (peanut meal)	5-03-650	92	77	*3.40*	*2.98*	*1.86*	*1.22*	*1.77*	0.85	0.55	0.80	52.3
	PINEAPPLE *Ananas comosus*												
43	—process residue, dehydrated (pineapple bran)	4-03-722	87	68	*3.00*	*2.58*	*1.57*	*0.97*	*1.55*	0.71	0.44	0.70	4.6
	RAPE *Brassica* spp												
44	—seeds, meal solv-extd	5-03-871	91	69	*3.04*	*2.62*	*1.60*	*1.00*	*1.57*	0.73	0.45	0.71	40.6
	SORGHUM *Sorghum bicolor*												
45	—grain, 8–10% protein	4-20-893	87	80	*3.53*	*3.12*	*1.94*	*1.30*	*1.84*	0.88	0.59	0.84	9.7
46	—silage	3-04-323	30	60	*2.65*	*2.22*	*1.31*	*0.74*	*1.35*	0.60	0.34	0.61	7.5
47	—silage, dough stage	3-04-321	28	55	*2.43*	*2.00*	*1.14*	*0.58*	*1.23*	0.52	0.26	0.56	6.0
	SOYBEAN *Glycine max*												
48	—hulls	1-04-560	91	77	*3.40*	*2.98*	*1.86*	*1.22*	*1.77*	0.85	0.55	0.80	12.1
49	—seeds, heat-processed	5-04-597	90	94	*4.14*	*3.74*	*2.35*	*1.64*	*2.18*	1.07	0.75	0.99	42.2
50	—seeds, meal solv-extd, 44% protein	5-20-637	89	84	*3.70*	*3.29*	*2.06*	*1.40*	*1.94*	0.94	0.64	0.88	49.9
	SUNFLOWER, COMMON *Helianthus annuus*												
51	—seeds w/o hulls, meal solv-extd	5-04-739	93	65	*2.87*	*2.45*	*1.47*	*0.88*	*1.47*	0.67	0.40	0.67	49.8
	TIMOTHY *Phleum pratense*												
52	—hay, sun-cured, early bloom	1-04-882	90	61	*2.69*	*2.27*	*1.35*	*0.77*	*1.38*	0.61	0.35	0.63	15.0
	TRITICALE *Triticale hexaploide*												
53	—grain	4-20-362	90	84	*3.70*	*3.29*	*2.06*	*1.40*	*1.94*	0.94	0.64	0.88	17.6
	UREA												
54	—45% nitrogen, 281% protein equivalent	5-05-070	99	0	*0.0*	*0.0*	*0.0*	*0.0*	*0.0*	0.0	0.0	0.0	281.0
	WHEAT *Triticum aestivum*												
55	—bran	4-05-190	89	70	*3.09*	*2.67*	*1.63*	*1.03*	*1.60*	0.74	0.47	0.73	17.1
56	—flour by-product, less than 7% fiber (wheat shorts)	4-05-201	88	73	*3.22*	*2.80*	*1.73*	*1.11*	*1.67*	0.79	0.50	0.76	18.6
57	—flour by-product, less than 9.5% fiber (wheat middlings)	4-05-205	89	69	*3.04*	*2.62*	*1.60*	*1.00*	*1.57*	0.73	0.45	0.71	18.4
58	—grain	4-05-211	89	88	*3.88*	*3.47*	*2.18*	*1.50*	*2.04*	0.99	0.68	0.93	16.0
	WHEY												
59	—dehydrated (cattle)	4-01-182	93	81	*3.57*	*3.16*	*1.97*	*1.32*	*1.87*	0.90	0.60	0.85	14.2

[1]Selected feeds from *Nutrient Requirements of Dairy Cattle*, 6th rev. ed., update 1989, NRC, National Academy Press, p.90, Table 7–1.

[2]Some specific numbers have not been assigned by the USDA Feed Composition Data Bank.

(ON A 100% DRY MATTER BASIS)[1] NOTE WELL: This is a 3-page spread. So, see pp. 818, 819, and 820.

Entry No.	Ether Extract (%)	Total Ash (%)	Crude Fiber (%)	Neutral Detergent Fiber (%)	Acid Detergent Fiber (%)	Cellulose (%)	Lignin (%)	Macrominerals (%)						
								Calcium	Chlorine	Magnesium	Phosphorus	Potassium	Sodium	Sulfur
1	4.0	10.2	20.5	38	28	22	5	1.80	0.34	0.26	0.35	2.21	0.22	0.33
2	3.8	9.2	22.0	40	29	23	7	1.54	0.34	0.24	0.29	2.56	0.15	0.31
3	3.0	9.6	23.0	42	31	24	8	1.41	0.38	0.33	0.22	2.52	0.14	0.28
4	2.6	9.1	26.0	46	35	26	9	1.41	0.38	0.31	0.24	1.71	0.12	0.28
5	2.0	8.9	29.0	50	37	28	10	1.25	0.35	0.31	0.22	1.53	0.11	0.27
6	–	–	–	–	–	–	–	–	–	–	–	–	–	–
7	3.6	7.6	11.0	25	20	14	6	0.23	–	0.13	0.11	0.53	0.02	0.11
8	2.1	2.6	5.7	19	7	5	2	0.05	0.18	0.15	0.38	0.47	0.03	0.17
9	2.0	3.1	7.1	21	9	–	–	0.06	0.17	0.14	0.39	0.58	0.02	0.16
10	0.6	4.4	19.8	54	33	31	2	0.69	0.04	0.27	0.10	0.20	0.21	0.22
11	0.6	6.1	16.5	44	25	22	3	0.61	–	0.16	0.10	1.78	0.53	0.42
12	1.4	5.8	1.1	–	–	–	–	0.32	0.30	0.24	0.26	0.10	0.35	0.37
13	6.5	4.8	14.9	46	24	18	6	0.33	0.17	0.16	0.55	0.09	0.23	0.32
14	3.7	10.7	24.0	56	31	27	3	0.50	–	0.18	0.30	2.30	0.02	0.20
15	0.7	2.4	0.2	0	0	0	0	0.67	–	0.01	0.90	0.01	0.01	–
16	3.7	6.6	12.7	23	22	18	3	1.84	–	0.17	0.12	0.79	0.09	0.08
17	9.8	2.4	12.1	43	17	12	5	0.11	0.08	0.07	0.43	0.18	0.10	0.46
18	3.7	1.9	9.4	28	11	9	2	0.07	0.05	0.14	0.27	0.53	0.02	0.16
19	2.4	3.4	4.8	37	9	8	1	0.16	0.07	0.06	0.50	0.03	0.10	0.39
20	2.4	1.8	2.2	14	5	4	1	0.08	0.10	0.09	0.54	0.21	0.06	0.72
21	2.4	7.5	9.7	45	12	–	–	0.36	0.25	0.36	0.82	0.64	1.05	0.23
22	4.3	1.6	2.6	9	3	2	1	0.03	0.05	0.14	0.29	0.37	0.03	0.12
23	4.3	1.6	2.6	9	3	2	1	0.02	0.05	0.14	0.32	0.35	0.01	0.14
24	7.7	3.1	6.7	55	13	10	2	0.05	0.06	0.26	0.57	0.65	0.09	0.03
25	3.0	7.2	32.3	53	30	23	5	0.34	–	0.23	0.19	1.41	–	0.08
26	3.1	4.5	23.7	51	28	24	4	0.23	–	0.19	0.22	0.96	0.01	0.15
27	1.7	2.8	47.8	90	73	59	24	0.15	0.02	0.14	0.09	0.87	0.02	0.09
28	20.0	4.8	24.0	44	34	24	10	0.21	–	0.46	0.64	1.00	0.01	0.26
29	23.8	4.5	17.2	37	26	12	14	0.12	–	0.41	0.54	1.18	0.01	–
30	1.3	7.0	14.1	26	19	12	6	0.22	0.04	0.55	1.21	1.39	0.04	0.34
31	1.7	6.7	12.1	28	21	13	7	0.17	0.04	0.55	1.00	1.39	0.04	0.34
32	99.5	–	–	–	–	–	–	–	–	–	–	–	–	–
33	5.5	9.9	24.6	–	–	–	–	0.51	–	–	0.37	–	–	–
34	10.5	20.8	1.0	–	–	–	–	5.65	0.60	0.16	3.16	0.76	0.43	0.49
35	1.5	6.5	10.1	25	19	13	6	0.43	0.04	0.66	0.89	1.53	0.15	0.43
36	0.9	8.4	0.2	–	–	–	–	1.36	0.96	0.13	1.09	1.70	0.49	0.34
37	0.2	11.3	–	–	–	–	–	0.17	1.64	0.29	0.03	6.07	1.48	0.60
38	0.3	7.9	–	–	–	–	–	1.72	0.11	0.21	0.13	0.14	0.41	0.23
39	0.9	13.3	6.7	–	–	–	–	1.10	–	0.47	0.15	3.60	0.20	0.46
40	0.1	13.1	–	–	–	–	–	1.00	3.10	0.43	0.11	3.84	0.22	0.47
41	5.4	3.4	12.1	32	16	11	3	0.07	0.11	0.14	0.38	0.44	0.08	0.23
42	1.4	6.3	10.8	–	–	–	–	0.29	0.03	0.17	0.68	1.23	0.08	0.33
43	1.5	3.5	20.9	73	37	–	7	0.23	–	–	0.13	–	–	–
44	1.8	7.5	13.2	–	–	–	–	0.67	0.11	0.60	1.04	1.36	0.10	1.25
45	3.4	2.1	2.0	18	9	8	1	0.04	0.10	0.18	0.34	0.40	0.01	0.09
46	3.0	8.7	27.9	–	38	–	6	0.35	0.13	0.29	0.21	1.37	0.02	0.11
47	3.3	9.3	28.5	–	–	–	–	0.29	0.11	0.27	0.26	1.02	0.03	0.14
48	2.1	5.1	40.1	67	50	46	2	0.49	–	–	0.21	1.27	0.01	0.09
49	20.0	5.1	5.6	–	11	–	–	0.28	–	0.23	0.66	1.89	0.03	0.24
50	1.5	7.3	7.0	–	10	–	–	0.30	0.08	0.30	0.68	1.98	0.03	0.37
51	3.1	8.1	12.2	–	–	–	–	0.44	0.11	0.77	0.98	1.14	0.24	–
52	2.9	5.7	28.0	61	32	31	4	0.53	–	0.14	0.25	1.62	0.18	–
53	1.7	2.0	4.4	–	8	–	–	0.06	–	–	0.33	0.40	–	0.17
54	0.0	–	0.0	0	0	0	0	–	–	–	–	–	–	–
55	4.4	6.9	11.3	51	15	11	3	0.13	0.05	0.60	1.38	1.56	0.04	0.25
56	5.2	4.9	7.7	–	–	–	–	0.10	0.08	0.28	0.91	1.06	0.03	0.22
57	4.9	5.2	8.2	37	10	–	–	0.13	0.04	0.40	0.99	1.13	0.19	0.20
58	2.0	1.9	2.9	–	8	8	–	0.04	0.08	0.16	0.42	0.42	0.05	0.18
59	0.7	9.8	0.2	0	0	0	0	0.92	0.08	0.14	0.82	1.23	0.70	1.12

(Continued)

TABLE 20-7 *(Continued)*

Entry No.	Feed Name Description	International Feed Number[2]	Microminerals (mg/kg)							Vitamins		
			Cobalt	Copper	Iodine	Iron	Manganese	Selenium	Zinc	A Activity (1,000 IU/kg)	D (1,000 IU/kg)	E (IU/kg)
	ALFALFA *Medicago sativa*											
1	—hay, sun-cured, early vegetative	1-00-050	0.10	11	0.19	253	45	0.37	24	80	1.9	—
2	—hay, sun-cured, late vegetative	1-00-054	0.09	9	0.18	227	34	0.35	27	81	—	—
3	—hay, sun-cured, early bloom	1-00-059	0.16	11	0.17	192	31	0.34	25	56	2.0	26
4	—hay, sun-cured, midbloom	1-00-063	0.36	14	0.16	134	28	0.32	23	46	2.0	11
5	—hay, sun-cured, full bloom	1-00-068	0.33	14	0.13	150	37	0.29	25	26	2.0	11
6	—silage, wilted, 25–45% dry matter (see similar maturity descriptions of hays)		—	—	—	—	—	—	—	—	—	—
	ALMOND *Prunus amygdalus*											
7	—hulls	4-00-	0.30	11	—	301	21	—	24	—	—	—
	BARLEY *Hordeum vulgare*											
8	—grain	4-00-549	0.10	9	0.05	85	18	0.22	19	1	—	25
9	—grain, Pacific Coast	4-07-939	0.10	9	—	97	18	0.11	17	—	—	30
	BEET, SUGAR *Beta vulgaris altissima*											
10	—pulp, dehydrated	4-00-669	0.08	14	—	329	38	—	10	—	0.6	—
11	—pulp w/molasses, dehydrated	4-00-672	0.23	16	—	207	27	—	10	—	—	—
	BLOOD											
12	—meal	5-00-380	0.10	11	—	4,064	6	0.80	5	—	—	—
	BREWERS' GRAINS											
13	—dehydrated	5-02-141	0.08	23	0.07	266	40	0.76	30	0	—	29
	BROME *Bromus* spp											
14	—fresh, early vegetative	2-00-892	0.08	11	—	200	142	—	27	184	—	—
	CASEIN											
15	—dehydrated (cattle)	5-01-162	—	4	—	15	5	—	30	—	—	—
	CITRUS *Citrus* spp											
16	—pulp w/o fines, dehydrated (dried citrus pulp)	4-01-237	0.16	6	—	378	7	—	15	—	—	—
	CORN, DENT YELLOW *Zea mays indentata*											
17	—distillers' grains, dehydrated	5-28-235	0.09	48	0.05	223	23	0.48	35	1	—	—
18	—ears, ground (corn and cob meal)	4-28-238	0.31	8	0.03	91	14	0.09	14	2	—	20
19	—gluten, meal	5-28-241	0.08	30	—	423	8	1.11	29	7	—	34
20	—gluten, meal, 60% protein	5-28-242	0.05	29	0.02	313	7	0.92	35	14	—	26
21	—gluten w/bran (corn gluten feed)	5-28-243	0.10	52	0.07	471	26	0.30	72	3	—	14
22	—grain, flaked	4-28-244	0.05	4	—	30	5	0.08	14	1	—	25
23	—grain, high-moisture	4-20-770	0.05	4	—	30	6	0.08	18	1	—	25
24	—grits by-product (hominy feed)	4-03-011	0.06	15	—	75	16	0.11	3	—	—	—
25	—silage, few ears	3-28-245	—	—	—	—	—	—	—	5	—	—
26	—silage, well-eared	3-28-250	0.06	10	—	260	30	—	21	18	0.1	—
	COTTON *Gossypium* spp											
27	—hulls	1-01-599	0.02	13	—	131	119	—	22	—	—	—
28	—seeds, w/lint	5-01-614	—	9	—	151	19	—	33	—	—	—
29	—seeds, w/o lint	5-01-	—	11	—	108	14	—	36	—	—	—
30	—seeds, meal prepressed, solv-extd, 41% protein	5-07-872	0.82	20	—	223	23	—	69	—	—	—
31	—seeds, meal prepressed, solv-extd, 44% protein	5-07-873	0.82	20	—	223	23	10	69	—	—	—
	FATS AND OILS (not exceeding 3% of diet)											
32	—fat, animal, hydrolyzed	4-00-376	—	—	—	—	—	—	—	—	—	—
	FESCUE, KENTUCKY 31 *Festuca arundinacea*											
33	—fresh, vegetative	2-01-902	—	—	—	—	—	—	—	—	—	—
	FISH, MENHADEN *Brevoortia tyrannus*											
34	—meal mech-extd	5-02-009	0.17	12	1.19	524	37	2.40	1.62	—	—	13
	FLAX *Linum usitatissimum*											
35	—seeds, meal solv-extd (linseed meal)	5-02-048	0.21	29	—	354	42	0.91	—	—	—	15
	MILK											
36	—skimmed dehydrated (cattle)	5-01-175	0.12	1	—	10	2	0.13	41	—	0.4	—
	MOLASSES AND SYRUP											
37	—beet, sugar, molasses, more than 48% invert sugar, more than 79.5 degrees brix	4-00-668	0.46	22	—	87	6	—	18	—	—	5
38	—citrus, syrup (citrus molasses)	4-01-241	0.16	108	—	508	38	—	137	—	—	—
39	—sugarcane, molasses, dehydrated	4-04-695	1.21	79	2.10	250	57	—	33	—	—	—
40	—sugarcane, molasses, more than 46% invert sugar, more than 79.5 degrees brix (Black strap)	4-04-696	1.21	79	2.10	250	56	—	30	—	—	7
	OATS *Avena sativa*											
41	—grain	4-03-309	0.06	7	0.11	85	42	0.26	41	—	—	15
	PEANUT *Arachis hypogaea*											
42	—kernels, meal solv-extd (peanut meal)	5-03-650	0.12	17	0.07	154	29	—	22	—	—	—
	PINEAPPLE *Ananas comosus*											
43	—process residue, dehydrated (pineapple bran)	4-03-722	—	—	—	561	—	—	—	22	—	—
	RAPE *Brassica* spp											
44	—seeds, meal solv-extd	5-03-871	—	—	—	—	—	1.07	—	—	—	—
	SORGHUM *Sorghum bicolor*											
45	—grain, 8–10% protein	4-20-893	0.29	11	—	50	17	—	16	—	—	12
46	—silage	3-04-323	0.30	35	—	285	73	0.22	32	6	0.7	—
47	—silage, dough stage	3-04-321	0.29	27	—	187	49	0.19	27	5	0.7	—
	SOYBEAN *Glycine max*											
48	—hulls	1-04-560	0.12	18	—	324	11	—	24	—	—	—
49	—seeds, heat-processed	5-04-597	—	18	—	89	33	0.12	60	—	—	—
50	—seeds, meal solv-extd, 44% protein	5-20-637	0.20	24	—	175	35	0.11	66	—	—	—
	SUNFLOWER, COMMON *Helianthus annuus*											
51	—seeds w/o hulls, meal solv-extd	5-04-739	—	4	—	33	20	—	—	—	—	12
	TIMOTHY *Phleum pratense*											
52	—hay, sun-cured, early bloom	1-04-882	—	11	—	200	103	—	62	21	—	13
	TRITICALE *Triticale hexaploide*											
53	—grain	4-20-362	—	7	—	44	45	—	25	—	—	—
	UREA											
54	—45% nitrogen, 281% protein equivalent	5-05-070	—	—	—	—	—	—	—	—	—	—
	WHEAT *Triticum aestivum*											
55	—bran	4-05-190	0.11	14	0.07	128	125	0.43	128	1	—	21
56	—flour by-product, less than 7% fiber (wheat shorts)	4-05-201	0.12	13	—	82	132	0.49	124	—	—	61
57	—flour by-product, less than 9.5% fiber (wheat middlings)	4-05-205	0.10	22	0.12	93	126	0.83	116	—	—	—
58	—grain	4-05-211	0.14	7	0.10	61	42	0.30	50	—	—	17
	WHEY											
59	—dehydrated (cattle)	4-01-182	0.12	50	—	181	6	—	3	—	—	—

Dry Matter Intake (DMI)

Dry matter intake is of the utmost importance, especially for high-yielding lactating cows. High-producing cows must consume very large amounts of highly nutritious, digestible dry matter if they are to maximize their genetic potential for production; otherwise, they will lose weight and, subsequently, produce less milk. Conversely, middle- and late-lactation cows should be fed rations with reduced energy; otherwise, they may become too fat—a condition which can adversely affect reproduction and health. So, it is important that the caretaker monitor dry matter and energy intake closely and adjust amounts fed to fit the stage and intensity of production.

Fig. 20-3. Lactating cows at Arizona Dairy Co., Higley, Ariz., part of a herd of 4,100 lactating cows with a rolling herd average of 20,000 lb milk and 3.72% fat, on 3X milking. In order to maximize their genetic potential, high-producing cows must consume large amounts of highly nutritious, digestible dry matter. (Courtesy, James Tappan, Co-owner and Manager, Arizona Dairy Co., Higley, Ariz.)

The amount of dry matter (DM) consumed by a cow depends on a number of factors including liveweight, milk yield, stage of lactation, animal behavior and environment, previous feeding history, body condition, type and quality of feed ingredients—particularly forages, and management.

Milk production usually peaks between 4 and 8 weeks postpartum, but maximum dry matter intake usually occurs 10 to 14 weeks postpartum. However, the rise in feed intake may not occur if the cow is fed a ration containing suboptimal amounts of protein of suboptimal degradability. The lag in maximum DMI behind peak milk yield causes a negative energy balance in early lactation. To meet this energy deficit, the cow mobilizes body tissue reserves, particularly fat deposits, which results in weight loss. AS DMI increases and milk production plateaus or begins to decline, weight change stabilizes and subsequently increases in mid- and late-lactation. The cow's weight continues to increase during the dry period, much of which is due to the development of the fetus and fetal membranes during the last trimester of gestation.

Dry matter intake increases as dry matter digestibility increases from about 52 to 68%. Above approximately 68% digestibility, DMI is related to the energy requirement of the cow. The actual point at which digestibility ceases to be the limiting factor for DMI varies with the energy requirement of the cow, which is determined primarily by milk yield.

When the concentrates in a ration exceed 60 to 70%, forage intake is inadequate for normal ruminal fermentation, often a marked reduction in milk fat percentage occurs, and frequently this reduction is accompanied by lower milk production and the *fat cow* syndrome. So, for lactating cows, the grain or other concentrates in the ration should not exceed 60 to 70%.

Recent research indicates that the amount of neutral detergent fiber (NDF) in a ration is negatively correlated with DMI. Also, cows consume up to 20% more DM in the form of legumes than as grasses. Further, DMI is depressed when a major portion of the ration is composed of fermented feeds, likely as a result of organic acids, amines, and ammoniated nitrogen in such feeds. Thus, cows eat at a level of about 2.2 to 2.5% of their liveweight when they are fed only corn silage, whereas intakes above 3% of liveweight are common when cows are fed only legume hay of good quality. However, corn silage consumption can be increased by the addition of protein supplements, or by the addition of urea or other ammonium compounds.

The moisture content of the ration also affects DMI; limited data indicate that the total DMI decreases as the moisture content of the ration exceeds about 50%. However, the effect of moisture on DMI is less when it is present in the form of pasture or green chop (soilage) than when it is in the form of silage or other fermented feeds.

The many variables that affect feed intake make it difficult to predict maximum DMI with accuracy. So, NRC's Subcommittee on Dairy Cattle Nutrition prepared Table 20-1, which gives the DMI requirements for cows weighing between 880 and 1,760 lb and yielding from 22 to 132 lb of 4% fat corrected milk (FCM) daily. If cows do not consume as much DM as they require (as projected in Table 20-1), and energy concentration is not increased, energy intake will be less than required; as a result, there will be loss of liveweight or reduced milk yield, or both. If cows consume more DM than they need (as projected in Table 20-1), they may become too fat unless the energy concentration of the ration is reduced; but this is likely to happen only at lower milk yields.

The amount of DM calculated from Table 20-1 should be reduced about 18% during the first 3 weeks of lactation to reflect the fresh cow's low appetite in this period. Within reason, a cow will meet her additional energy needs from body reserves. Also, if fermented feeds, like silage, constitute a major portion of the ration, the amount of DM should be reduced by 0.02 lb/100 lb of liveweight for each 1%

increase in ration moisture content above 50%. In both cases, reduced DMI is likely to result in loss of body weight or reduced milk yield, or both, unless the cow's energy requirement can be fulfilled by increasing the energy concentration of the ration.

Energy

Cows use energy for a variety of functions. A certain amount is used for body maintenance; heifers need energy for growth; pregnant cows need additional energy for development of the fetus; and lactating cows require energy to produce milk. For optimal milk production, minimum health disorders, and optimum reproductive efficiency, cows must not be too fat or too thin.

Lack of energy is the most common deficiency of dairy rations. In young animals, an insufficient supply of energy results in retarded growth and a delay in the onset of puberty. In lactating cows, it results in a decline in milk yields and a loss in liveweight; and severe and prolonged energy deficiency also depresses reproductive performance.

Most of the energy required is supplied by carbohydrates, although fats and protein are also used as energy. All cows, except low-producing ones—those producing less than 25 to 35 lb of milk per day, need some grain if they are to produce at top levels.

• **Carbohydrates**—Carbohydrates are the major source of energy for dairy cattle. They constitute 50 to 80% of the dry matter of forages and grains. Three major categories of carbohydrates exist in feeds: (1) simple sugars (glucose or sucrose), (2) stored carbohydrates (starch), and (3) structural carbohydrate or fiber (cellulose and hemicellulose). Sugars are found in the cells of growing plants and in such feeds as molasses. Starch is the main component of grain. Cellulose and hemicellulose, which are classified as fiber, are made up of sugar molecules, as is starch, but they are bound together differently. Adult ruminants can digest fiber because the microbial population in the rumen breaks it down into usable products. However, lignin, which is not a true carbohydrate, is virtually indigestible.

• **Fat**—Rations for baby calves, with large quantities of milk or milk replacer, may contain 15 to 35% fat in the dry matter consumed. Fat is mainly used in the rations of young calves, but it may be added to the ration of lactating cows to increase energy and reduce feed dustiness. In addition to the fat present in natural feedstuffs (generally less than 2 to 3%), dairy cows are able to utilize 1 to 1½ lb of fat per day; which translates into about 5 to 6% added fat to the concentrate (grain) ration, or 3% added fat to the total mixed ration (grain and forage combined). Fat sources high in unprotected polyunsaturated fatty acids (soybean oil or corn oil) exert a greater negative effect on rumen microbes and depress fiber digestibility more than saturated fatty acid sources (tallow). Whole oilseeds (soybeans, sunflowers, or cottonseed) are good fat sources because they are slowly digested in the rumen, with the result that the oil is slowly released. Also, feeding rumen-protected fats that are resistant to microbial action in the rumen may increase milk fat percentage.

The NRC energy requirements for dairy cattle, which are presented in this chapter as Tables 20-2, 20-3, and 20-4, are expressed as digestible energy (DE), metabolizable energy (ME), net energy for maintenance (NE_m), net energy for body gain (NE_g), net energy for lactation (NE_{lc}), and total digestible nutrients (TDN). Separate energy values for each maintenance (NE_m) and gain (NE_g) are given because animals use energy for maintenance more efficiently than for growth. However, the efficiency of energy use by lactating cows for maintenance, pregnancy, and milk production is similar; so, only one energy value, net energy for lactation (NE_{lc}), is used for these functions.

The energy value of a feed may be separated into: (1) the losses that occur in digestion and metabolism, and (2) the net energy (NE) that is available to the animal for maintenance and production. The total energy in feed, which is determined by complete oxidation (burning) of the feedstuff and measurement of the heat produced, is known as *gross energy* and is expressed as calories. Common feedstuffs are similar in gross energy content, but differ in feeding value because of variations in digestibility. About 60% of the total energy in grain and 80% of the total energy in roughage is lost in feces, urine, gases, and heat. In the feed composition tables in Section V of this book, separate net energy values for each maintenance (NE_m) and gain (NE_g) are given.

In both the NRC requirement tables which are reproduced in this chapter (as Tables 20-2, 20-3, and 20-4), and in the feed composition tables in Section V of this book, energy is also expressed as total digestible nutrients (TDN). TDN is comparable to digestible energy. It has been in use longer than the net energy system and more values are available for feedstuffs. TDN is computed as follows:

TDN = Digestible nitrogen-free extract (carbohydrate) + digestible crude fiber + digestible protein + (digestible ether extract × 2.25)

NE of lactation can be calculated from TDN as follows:

$$NE_{lc}\ (\text{Mcal/lb DM}) = (\text{TDN, \% of DM} \times .01114) - .054$$

The energy requirement for maintaining a lactating cow is affected by a number of factors, especially the following: (1) *body size*—the larger the animal, the higher the maintenance energy requirement; (2) *activity*—to support grazing activity, the maintenance allowance may be increased by 10% on good pasture and up to 20% on poor pasture; and (3) *cold temperature*—under severe winter conditions without access to dry shelter, the maintenance feed allowance may be increased up to 8%. Also, during the first lactation, when a heifer is still growing, her energy needs are about 20% greater than a mature cow; and during the second lactation, her energy needs are 10% greater than a mature cow. The energy requirement for gestation is 30% of that required for maintenance alone, with most of the increase during the last 8 weeks of pregnancy.

Fig. 20–4. In comparison with confinement feeding, grazing activity may increase the maintenance allowance by 10% on good pasture and by up to 20% on poor pasture. (Courtesy, Milk Marketing, Inc., Strongsville, Ohio)

Table 20–3 includes allowances for liveweight changes during lactation. These values will aid the user in identifying the extent of dietary energy insufficiency during weight loss in early lactation and in estimating feed required to regain body weight in later lactation. The desired rate of liveweight gain will depend on the animal's body condition and stage of pregnancy.

When feed is restricted, a lactating cow will use the available energy for maintenance and reproduction at the expense of growth and milk production. Therefore, it is important to feed a ration that will fulfill the requirements for maintenance, growth, reproduction, and lactation.

For dry cows in good condition, medium quality forages will usually suffice as their primary feed. During the last 2 to 3 weeks prior to calving, concentrate feeding should be increased gradually to about 0.5% of their liveweight to allow the cow and the ruminal microorganisms to become adapted to the larger amounts of concentrate required in early lactation. Body weight losses after calving can be minimized by feeding as much of a properly balanced ration as the cow can safely use during the first 10 to 12 weeks after calving. Proper feeding immediately after calving also helps to prevent ketosis.

The NRC maintenance requirements for growing replacement heifers are 12% higher than for beef cattle. Also, NRC (1) recommends that milk or milk replacer be fed to replacement calves for at least the first month of life, (2) states that longer periods (up to 2 months) of liquid feeding may be beneficial under some conditions because of decreased disease and death losses, and (3) recommends that veal calves be fed maximum *ad libitum* amounts of milk (or milk replacer).

The most common high energy dairy feeds are: barley, beet or sugar cane molasses, beet pulp, citrus pulp, corn, corn silage, fats, high-moisture corn, high quality legume forage (hay or silage), lush pasture, oats, sorghum grain, wheat, whole cottonseeds, and whole soybeans.

- **NRC energy values of feeds vs *Feeds & Nutrition* energy value of feeds**—In *Nutrient Requirements of Dairy Cattle,* Sixth Revised Edition, energy values of feeds for TDN, digestible energy (DE), and metabolizable energy (ME) have been determined at the maintenance level of intake, whereas values for net energy for lactation are adjusted to three times (3X) the maintenance level. The NRC dairy cattle requirements for TDN, DE, and ME for animals fed above maintenance (lactating animals) have been increased to allow for the depression of digestibility of energy at these higher levels of intake. As long as formulators use both NRC energy values for feeds and NRC nutrient requirements, satisfactory rations will result.

In *Feeds & Nutrition,* Section V, the energy values of feeds are not identical to the energy values in *Nutrient Requirements of Dairy Cattle,* Sixth Revised Edition, due to varying conclusions from interpreting analytical information. However, the differences are minor and should not materially affect ration formulation. If, however, formulators are primarily interested in rations for dairy cattle, and if TDN is used as the energy measure, they may wish to refer to the NRC publication for the appropriate TDN values. As a convenience, Table 20–7, which contains TDN and NE_{lc} values from *Nutrient Requirements of Dairy Cattle,* Sixth Revised Edition, for the most common feeds used in dairy rations, is presented in this chapter. (Table 20–7 is a 3-page spread; so, see pages 818, 819, and 820.

> **NOTE WELL:** Net energy for lactation values (NE_{lc}) that conform to those used in *Nutrient Requirements of Dairy Cattle* may be calculated from TDN values of *Feeds & Nutrition,* Section V, by using the following equation:
>
> $$NE_{lc}\ (\text{Mcal/kg of DM}) = 0.0245 \times \text{TDN}\ (\%\ \text{of DM}) - 0.12$$
>
> This equation is based on an average 4% reduction in digestibility for each multiple increase in intake over maintenance intake and assumes the intake to be 3X that of maintenance.

(For a more complete and in-depth discussion of energy, also see Chapter 4, section headed "Energy [Carbohydrates and Lipids]" and the subsections under it.)

Protein

Proteins are complex chemical structures which are made up of amino acids that are linked together in many different ways. Amino acids contain carbon, hydrogen, oxygen, nitrogen, and, in some cases, sulfur. There are 22 naturally occurring amino acids. The amino acids are supplied by the digestion of microbial protein and by feed protein that escapes microbial breakdown in the rumen. The proteins in practical energy sources supply some dietary protein that escapes rumen fermentation, and this protein, plus the microbial protein produced from supplemental NPN, may be enough to produce about 44 lb of milk per day. As milk production increases, a substantial amount of additional dietary protein from protein supplements must escape rumen fermentation to meet the cow's requirement for protein.

Protein is essential for dairy cattle for maintenance, growth, milk production, and the development of the fetus. Also, it is required for the formulation of enzymes and certain hormones that control or regulate chemical reactions in the body. The protein requirement is really a requirement for amino acids.

The protein composition of feeds, and the protein requirements of dairy cattle, may be expressed as crude protein, digestible protein, degraded intake protein, undegraded intake protein, and/or nonprotein nitrogen (NPN).

1. **Crude protein.** Chemically, most proteins contain 16% nitrogen; so, crude protein is determined by finding the nitrogen content, then multiplying the result by 6.25 (100 ÷ 16 = 6.25). It is called crude protein because not all of the nitrogen in feeds is in the form of protein; rather, it is a combination of true protein and nonprotein nitrogen. Some feeds, particularly green roughages, contain ⅓ or more of their nitrogen as nonprotein nitrogenous substances such as amides, ammonium salts, amino acids, alkaloids, and other nitrogenous compounds. However, ruminal microorganisms make use of the various nitrogen sources for synthesis of microbial proteins which, in turn, are used by the cow. Consequently, for dairy cows and other ruminants, the amount of crude protein is about as good a measure for protein allowances as is the amount of true protein.

2. **Digestible protein.** This is the amount of crude protein consumed less the crude protein excreted in the feces. However, the term *apparent digestibility* is more accurate since it is recognized that a portion of the fecal nitrogen is derived from the animal and is not a feed residue.

3. **Degraded intake protein (DIP).** This refers to the intake crude protein that is broken down (degraded) by microorganisms in the rumen.

4. **Undegraded intake protein (UIP).** This is the crude protein that is not broken down in the rumen; instead, it is swept out of the rumen to the abomasum and small intestine for breakdown there and absorption as peptides and amino acids.

5. **Nonprotein nitrogen (NPN).** Feedstuffs which contain nitrogen in a form other than proteins or peptides are termed nonprotein nitrogen (NPN). Nonprotein nitrogen compounds, such as urea and ammonium salts, have a crude protein value, but they do not supply any amino acids directly. The billions of microorganisms in the rumen convert nitrogen from NPN sources into amino acids for their growth and use. Then, the microbes pass into the small intestine where they are digested and release amino acids for absorption and utilization the same as amino acids released from the digestion of true proteins (composed of amino acids) in feeds.

• **Amount of protein needed**—The amount of protein needed in the total ration of lactating cows is determined primarily by the amount of milk produced. Milk is a rich source of high-quality protein; so, as milk production increases, a substantial amount of dietary protein is necessary. Thus, a high-producing 1,320-lb cow yielding 88 lb of 3.5% protein milk daily secretes 3.08 lb of milk protein. A deficiency of protein results in lowered milk production and may depress the protein content of milk. Excess protein usually results in high cost rations.

Fig. 20-5 depicts the increase in milk yield as the protein percent of the total ration DM increases. At some point, the value of increased milk will not exceed the cost of the additional protein, with this point determined by the relative milk price and the cost of the additional protein.

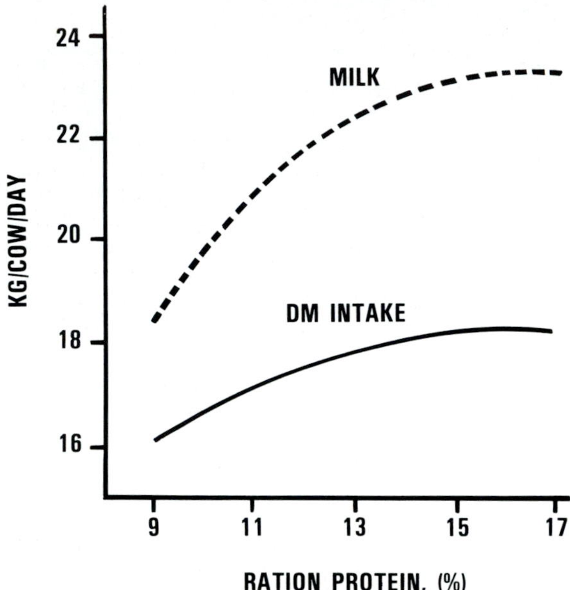

Fig. 20-5. This depicts the increase in milk yield as the protein present in the total ration DM increases. (Courtesy, J. W. Thomas, Ph.D., Professor Emeritus and Consultant, Department of Animal Science, Michigan State University, East Lansing)

The amount of protein needed in the concentrate mix depends on the kind and quality of forage fed. As the amount of legume increases, the percentage of protein in the concentrate can be lowered. For most lactating cows, the total ration (forage plus grains and protein and energy supplements) should have 19% crude protein during the first ⅓ of lactation, lowered to 14% in midlactation and 12% during the dry period. The interaction of the protein and energy supply in the rumen of the high-producing cow is very important; so, an adequate supply of degradable intake protein (DIP) is essential to maximize both feed intake and ruminal digestibility. Moe, of the USDA, Agricultural Research Service, found that high-producing cows receiving 17% protein digested feed better than those fed 14% protein.[2]

• **Excess protein**—When more protein is fed than needed, the excess is used as a source of energy. Because protein feeds are generally more expensive than carbohydrate feeds, it usually is more economical to feed only the amount needed. Besides, a large excess of dietary protein may decrease the energy supply because excess protein must be deaminated to ammonia and, for the most part transformed back into urea for excretion. Most cows fed good-quality alfalfa hay (fed free-choice), along with grain fed according to production by one of the recommended systems, will not need any supplemental protein until they

[2]*Research News,* ARS-USDA, p. 20, May 1977.

produce more than 50 to 60 lb of milk daily. As production increases above this amount, protein intake must be increased gradually, usually by decreasing the hay intake and replacing it with grain and protein concentrates. So long as the hay consumption does not fall below 12 to 15 lb daily, the supplemental concentrate need not contain more than 15 to 16% protein.

DEGRADED INTAKE PROTEIN (DIP) AND UNDEGRADED INTAKE PROTEIN (UIP)

Approximately 60% of the crude protein in the typical dairy cow ration is broken down (degraded) by microbial digestion to ammonia. The rumen microbes must convert the ammonia to microbial protein in their own cells if the dairy animal is to receive any benefit. Fermentable energy must be available for the microorganisms to grow and synthesize the necessary amino acids. If rumen ammonia levels are excessively high, the ammonia is absorbed into the blood and either recycled or excreted in the urine as urea.

All feed protein sources are not degraded in the rumen to the same extent. The optimal ration will meet both the nitgrogen requirement of rumen microorganisms for maximum synthesis of microorganism protein and allow for maximum escape or *bypass* of high quality feed protein for digestion in the small intestine. Protein synthesis by rumen microbes depends on feed intake, organic matter digestibility, feed type, protein level, and feeding system. Since 3.5 lb of microbial protein synthesis per day is near the maximum, the remainder of the protein must be derived from nondegraded (escape) protein sources. Young, fast-growing heifers and high-producing cows generally require additional nondegraded protein sources beyond their normal ration to meet total protein requirements; and the more rapid the growth and the higher the milk production, the greater the quantities of undegradable protein needed. Brewers' grain, distillers' grain, corn gluten meal, fish meal, meat meal, and heat-treated soybeans are examples of feeds with reduced rumen degradability that may be substituted in rations in which excess rumen ammonia exists and less than optimal amounts of quality protein (undegraded) pass into the small intestine.

(For an in-depth discussion of protein, also see Chapter 4, section headed "Proteins"; and for an in-depth discussion of degraded intake protein [DIP] and undegraded intake protein [UIP], also see Chapter 11, section headed "Protein Bypass [Protected Protein, Escaped Protein]," including Fig. 11-2, showing protein digestion by a cow, and Table 11-1, showing the percent of undegraded protein in common feeds.)

UREA AND OTHER NPN PRODUCTS

Urea is a nonprotein nitrogen (NPN) compound, containing about 45% N, with a protein equivalent of 281% (45% N × 6.25), and with the chemical formula—

$$O = C \begin{matrix} \diagup NH_2 \\ \diagdown NH_2 \end{matrix}$$

Using urea in the ration is similar to using degradable intake protein. It and other nonprotein nitrogen (NPN) compounds, such as ammonium salts, can be used to replace part of the protein required in dairy cattle rations after rumen function has become established. Studies have shown that when cattle are fed purified rations with only nonprotein nitrogen as a nitrogen (N) source, there is adequate microbial protein production for growing ruminants with a functional rumen to gain at about 65% of the level at which they gain when fed practical energy ingredients and protein supplements.[3] Also, studies show that there is adequate microbial protein production for lactating cows to produce 8,800 lb of milk per lactation.[4]

The following guidelines should be observed for the successful use of urea in dairy rations:

1. All rations should be assessed for protein content before either supplemental NPN or natural protein is added to the ration. Protein may not be needed.
2. Feeds most successfully supplemented with NPN are high in energy, low in protein, and low in natural NPN (such as grains and corn silage).
3. Maximum amounts of urea to feed are:
   a. 1% urea in the grain mix; 0.5% urea in the total ration.
   b. 0.5% urea in corn silage (10 lb/ton). If 0.5% is added to corn silage, the amount in the grain should be no more than 0.5%. The addition of 10 lb of urea per ton of corn silage will increase the protein content from 8 to 12% on a dry matter basis (depending on losses incurred).
   c. 0.4 lb urea per head per day, with cows in early lactation limited to 0.2 lb of urea per head per day.
4. Urea is not palatable. So, it should be mixed thoroughly with the grain mix or silage. Molasses can improve acceptability.
5. If cattle have not been fed urea previously, a 7- to 10-day adjustment period in which the urea is gradually increased will help to maintain feed intake and production.
6. Total mixed rations and/or frequent feeding of feeds containing urea maximizes utilization.
7. High levels of urea can be toxic; so, excessive intakes should be avoided. Urea should not be top-dressed; it should be mixed in the feed.

Urea can be used in making a high-protein concentrate. A mixture of 87 lb of ground shelled corn and 13 lb of urea is equivalent in energy and crude protein to 100 lb of soybean meal.

A mixture of 56 lb of shelled corn, 7 lb of urea, and 37 lb of soybean meal also equals 100 lb of soybean meal in total energy and protein equivalent, and can be used as a substitute for soybean meal.

• **Other NPN products**—Several ammoniated products or ammonium salts are used successfully as sources of nitrogen. Monoammonium phosphate, which contains about 11% nitrogen (crude protein equivalent of 68.25%), is also used for phosphorus supplementation. Ammonia (cooled to form a liquid or in a water solution) may be added to corn silage at the rate of 7 lb (5 lb of nitrogen) per ton. Urea should not be fed in the concentrate when ammonia or other NPN has been added to the corn silage.

---

[3]Oltjen, R. R., "Effects of Feeding Ruminants Nonprotein Nitrogen as the Only Nitrogen Source," *Journal of Animal Science*, 28:673, 1969.

[4]Virtanen, A. I., "On Nitrogen Metabolism in Milking Cows," *Federation Proceedings* 28:232, 1969.

Since NPN products do not provide any energy, minerals, or vitamins, these nutrients must be provided through other sources.

(For an in-depth discussion of urea and NPN, also see Chapter 4, section headed "Nonprotein Nitrogen"; and Chapter 11, section headed "Nonprotein Nitrogen [NPN] Feedstuffs.")

Minerals

Dairy cattle require at least 15 mineral elements. They are needed for both structural and regulatory functions. Minerals are needed for bone and teeth formation, and to maintain acid-base balance, water balance, and enzyme and hormone systems. They are also components of certain substances within the body, such as iron in hemoglobin. Additionally, a lactating cow needs minerals for the developing fetus and for milk production. Milk contains about 0.7% minerals; thus, a cow producing 20,000 lb of milk in a lactation secretes 140 lb of minerals per year.

The mineral elements that have been shown to be essential for dairy animals and for which signs of deficiency have been described are generally classified as (1) major or macrominerals (those required in greatest quantities and present in animal tissues at higher levels), and (2) trace or microminerals (those required in smaller amounts and generally present in tissues at lower levels). Under practical feeding conditions, it is usually necessary to provide supplemental sources of several of these elements to meet the nutritional requirements of dairy cattle.

Mineral excesses should be avoided because of interaction with other minerals and possible toxicities and undesirable interactions. *The maximum tolerable level for a mineral element has been defined as that dietary level that, when fed for a limited period, will not impair animal performance and should not produce unsafe residues in human food derived from the animal.* It is important not to exceed safe tolerances for dietary mineral elements in feeds.

When dairy cattle are fed mixed feed, in part or totally, the needed minerals are usually incorporated in the ration in keeping with the known requirements. When animals are fed an unmixed ration or are on pasture, *ad libitum* supplementation of minerals is commonly practiced.

Tables 20-2, 20-3, and 20-4 show the daily calcium and phosphorus requirements for different classes of dairy animals; Table 20-5 shows the recommended content of major and trace minerals of rations for different classes of dairy animals; and Table 20-6 shows the maximum tolerable levels of certain elements. Chapter 19, Table 19-7, presents in summary form the mineral requirements of all cattle; the mineral requirements of dairy cattle and beef cattle are similar except for the higher milk production of lactating dairy cows. Section V, Composition of Feeds, presents the macro- and micromineral content of a great array of feeds.

In addition to the minerals listed and discussed in the narrative that follows, several other elements have been shown to be essential for one or more animal species; among them, arsenic, chromium, fluorine, lead, nickel, silicon, tin, and vanadium. Currently, however, these elements are not considered to be of practical importance in the feeding of dairy cattle.

(Also see Chapter 4, section headed "Minerals," including Table 4-6, Animal Mineral Chart, for an in-depth discussion of each of the 18 mineral elements that are known to be required by at least some animal species; and see Chapter 5, Table 5-3, for an in-depth discussion of certain mineral elements that are sometimes toxic to animals.)

MAJOR OR MACROMINERALS

The major or macrominerals of importance in dairy cattle nutrition are: salt (sodium chloride), calcium, phosphorus, magnesium, potassium, and sulfur.

Salt (sodium chloride [NaCl]). Sodium and chlorine are usually provided in the form of common salt (NaCl). However, potassium chloride may be used as a source of chlorine, also. The current NRC recommendation is for a minimum of 0.43% sodium chloride in the total dairy ration, including that contributed by the feeds (Table 20-5). Excessive levels of chlorine without sodium or potassium can contribute to an acidosis condition in dairy cattle.

(Also see Chapter 4, Table 4-6—Chlorine and Sodium.)

Calcium (Ca). Whole milk contains 0.12% calcium. The NRC subcommittee based the dietary calcium requirements for dry pregnant and lactating cows (Table 20-3) on 38%

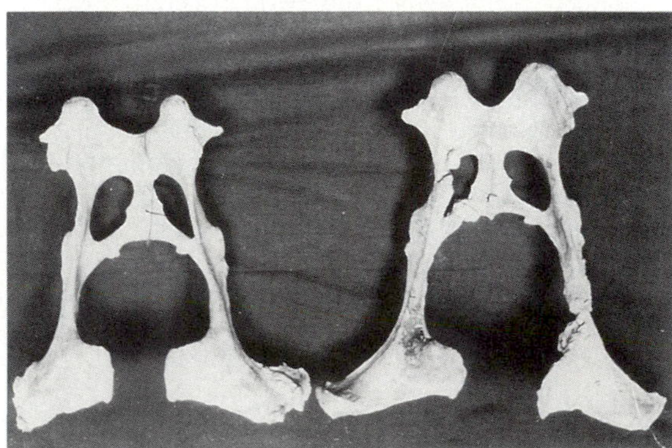

Fig. 20-6. Calcium deficiency. Lactating cows need calcium. Both hips of the cow shown above have been broken (knocked down) as a result of feeding a low-calcium ration. At lower left, the pelvis of a cow which had three breaks while the cow received a low-calcium ration. At lower right, the pelvis of the cow pictured above, showing the breaks involving both hipbones. (From *Fla. Ag. Exp. Sta. Tech. Bull. 262,* through the courtesy of R. B. Becker)

availability. Minimum calcium percentage for the complete ration (dry matter basis) recommended by the NRC for lactating cows varies from 0.43 to 0.66%, depending on level of milk production (Table 20-5). A deficiency of calcium may cause rickets, slow growth and poor bone development, easily fractured bones, reduced milk yield and increased incidence of milk fever. Feeding calcium at more than 0.95 to 1.00% (DM basis) in mixed rations may reduce dry matter intake and lower performance. The effects of variations in the calcium-to-phosphorus ratios have been overemphasized, as evidenced by studies showing that dietary calcium-to-phosphorus ratios of between 1:1 and 7:1 result in nearly equal performance, provided the animal's phosphorus intake meets its requirement.

(Also see Chapter 4, Table 4-6—Calcium.)

Phosphorus (P). Whole milk contains 0.09% phosphorus. The NRC subcommittee assumed a phosphorus availability from mixed rations fed to lactating cows of 45 to 50%. A deficiency of phosphorus may result in fragile bones, stiff joints, poor growth, low blood P (less than 4-6 mg/100 ml), depraved appetite (chewing wood, hair, and bones), and poor reproductive performance. Excessive phosphorus intakes may cause bone resorption, elevated plasma phosphorus levels, and urinary calculi.

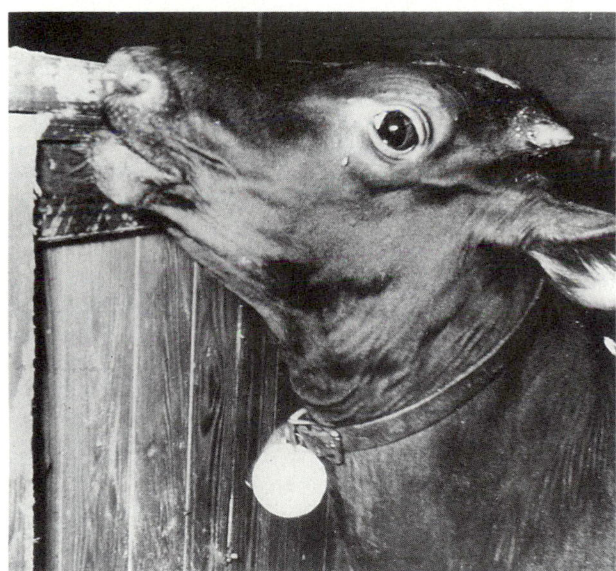

Fig. 20-7. Phosphorus-deficient calf chewing wood, a manifestation of depraved appetite. (Courtesy, Dr. S. E. Smith, Department of Animal Science, Cornell University, Ithaca, N.Y.)

(Also see Chapter 4, Table 4-6—Phosphorus.)

Magnesium (Mg). Milk contains a substantial amount of magnesium (about 0.015%). Thus, when expressed as a percentage of the ration, the magnesium requirement increases with the cow's level of milk production. Under practical conditions, magnesium deficiencies may occur (1) when calves are fed an all-milk diet for extended periods, during which their body reserves of magnesium are depleted; or (2) when dairy cattle, especially older and lactating cows, are grazing lush, rapidly growing pastures that have been highly fertilized with nitrogen or potassium, or both, during cool seasons. Under conditions conducive to grass tetany and for high-producing cows in early lactation, the suggested requirement is 0.25 to 0.30% dietary magnesium, with the supplemental magnesium provided in a readily available form such as magnesium oxide. Magnesium toxicity is not known to be a practical problem in dairy cattle.

(Also see Chapter 4, Table 4-6—Magnesium.)

Potassium (K). Milk contains about 0.15% potassium. The NRC minimum dietary potassium requirement for lactating cows is 0.90%, increased to 1.00% for high-yielding and early-lactation cows; for dry cows and young stock, it is 0.65%. Stress, especially heat stress, appears to increase the need for potassium, perhaps due to greater loss of potassium through sweat. The signs of relatively severe potassium deficiencies in lactating cows include a marked decrease in feed intake, loss in weight, decreased milk yield, pica, loss of hair glossiness, decreased pliability of hide, lower plasma and milk potassium, and higher hematocrit readings. Generally, forages contain considerably more potassium than is required by dairy cattle.

High levels of potassium (3% or above) in very lush forages grown on high potassium soils in cool weather appear to interfere with magnesium metabolism and utilization and are considered to be a factor in causing grass tetany of lactating cows.

(Also see Chapter 4, Table 4-6—Potassium.)

Sulfur (S). Milk contains 0.03% sulfur, much of which is in the form of the amino acids methionine and cystine. Sulfur is needed for microbial protein synthesis, especially when nonprotein nitrogen is fed. The NRC estimated minimum sulfur requirement for lactating cows is 0.20% of the ration; with the sulfur needs of other dairy cattle calculated from the minimum protein requirement for these animals based on a nitrogen-to-sulfur ratio of 12:1.

(Also see Chapter 4, Table 4-6—Sulfur.)

TRACE OR MICROMINERALS

The trace or microminerals of importance in dairy cattle nutrition are: cobalt, copper, iodine, iron, manganese, molybdenum, selenium, and zinc.

Cobalt (Co). Normal cow's milk averages 0.38 to 1.04 mcg of cobalt/qt. Colostrum contains 4 to 10 times more cobalt than milk. Since cobalt is a component of vitamin B-12, ruminal microorganisms are able to synthesize this vitamin only when adequate cobalt is in the ration of the cow. NRC recommends a minimum of 0.1 ppm of cobalt in the total ration. Supplements of 30 to 45 g of cobalt sulfate or 20 to 25 g of cobalt carbonate with 100 lb of salt have prevented any cobalt deficiency problems. Also, a heavy pellet containing cobalt oxide and finely divided iron, which is administered orally and remains in the reticulo-rumen, will prevent cobalt deficiency for extended periods in cattle that graze cobalt-deficient pastures.

(Also see Chapter 4, Table 4-6—Cobalt.)

Copper (Cu). Colostrum contains more copper than milk. The amount of copper in milk decreases with the length of lactation. Copper is needed for hemoglobin formation, although it is not actually contained in it. A deficiency of copper will result in anemia and bleaching of the hair. Black hair turns gray and red hair becomes yellow. NRC recommends a minimum of 10 ppm copper in the ration, with the caution that higher levels may be required for cattle grazing pastures or consuming feedstuffs that contain high levels of molybdenum or other interfering substances. Copper supplementation may be advisable under certain conditions, but it should be done with discretion. Excess copper is toxic and is a primary cause of oxidized flavor in milk.

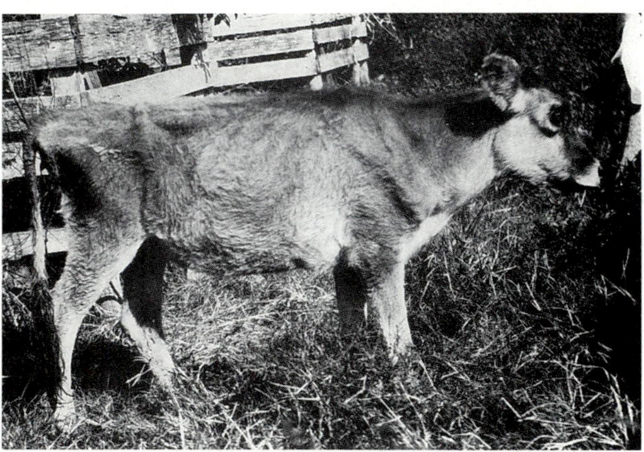

Fig. 20-8. Copper deficiency. *Top:* Registered Jersey heifer showing copper deficiency. *Bottom:* Same heifer after receiving copper. (Courtesy, R. B. Becker, University of Florida, Gainesville)

(Also see Chapter 4, Table 4-6—Copper.)

Iodine (I). About 10% of the iodine intake of lactating cows is normally excreted in milk. Iodine deficiency can be detected by analyzing milk or blood serum. Iodine concentration of less than 9.5 to 19 mcg/qt of milk or 37.8 mcg/qt of serum indicate iodine deficiency. Goiter (an enlargement of the thyroid gland) occurs in newborn calves if their mothers were fed iodine-deficient rations; necks of the calves are swollen and they are weak at birth or born dead. Much of the small amount of iodine in the body is contained in the thyroid gland as thyroxin and diiodotyrosine, both of which are contained in the protein thyroglobulin, a part of the thyroid hormone. The principal function of the thyroid gland is to regulate the metabolic rate. Many protein supplements (including soybean meal and cottonseed) are mildly goitrogenic because they reduce the availability of dietary iodine, and *Brassica* forages (cabbage, kale, rape) are highly goitrogenic. The NRC recommends that cows in lactation receive a dietary iodine concentration of 0.6 ppm, and that cows in the last 2 months of gestation be fed 0.6 ppm of iodine. When stabilized iodine is used, a level of 0.0076% in salt is adequate. The Northwest and Great Lakes regions are the most iodine-deficient areas of the United States. Lactating cows should not receive excessive dietary iodine because the resulting high iodine milk content is considered undesirable for humans. The use of iodine disinfectants as teat dips or udder washes can increase the iodine content of milk, but the main cause of high iodine levels in milk is dietary iodine.

(Also see Chapter 4, Table 4-6—Iodine.)

Iron (Fe). Iron is essential because it is a constituent in hemoglobin, the oxygen carrier in the blood. Cow's milk is low in iron—about 10 ppm. The iron requirements of a young calf are higher than those of a mature cow and are thought to be about 100 ppm until 3 months of age, and 50 ppm thereafter; as recommended by the NRC. The iron reserves of a newborn calf, which are primarily in the liver, are generally adequate to prevent serious anemia if calves are fed dry feeds at a few weeks of age. However, when calves are fed a milk diet exclusively for several weeks, they may develop iron deficiency anemia. Light colored veal is associated with low levels of muscle myoglobin and restricted iron intake. When veal calves are fed a dry ration, 40 ppm iron is sufficient to prevent severe anemia. The NRC gives an iron concentration of 1,000 ppm as the maximum tolerable level for cattle.

(Also see Chapter 4, Table 4-6—Iron.)

Manganese (Mn). Manganese deficiency in dairy cattle is seldom a problem. In general, forages contain higher levels of manganese than grains. The manganese requirement for cattle is higher for reproduction than for growth. Little experimental work has been done on the manganese requirements of dairy cattle, but rations containing 40 ppm are recommended by the NRC. Trace mineralized salt and commercial mineral supplements usually contain manganese. Manganese toxicity in cattle is unlikely.

(Also see Chapter 4, Table 4-6—Manganese.)

Molybdenum (Mo). Molybdenum is an indispensable component of the enzyme xanthine oxidase, which is found in milk and distributed widely in animal tissue. Yet a deficiency of molybdenum has never been developed or observed in cattle. Molybdenum is known largely for its toxic characteristics; molybdenum toxicosis is a practical problem in grazing cattle in several areas of the world.

There is an antagonistic relationship between molybdenum and copper. Elevated dietary molybdenum increases both the animal's requirements for copper and the amount of copper that will cause toxicosis; increased dietary copper can reduce the toxic effect of molybdenum. Thus, the relative amounts of copper and molybdenum in the diet are important in determining the occurrence of molybdenum toxicosis. If the level of copper in the body is low, a lesser amount of molybdenum is toxic; as dietary copper increases, so does tolerance to molybdenum. High levels of both molybdenum and sulfur interfere with copper absorption. Molybdenum and sulfur also influence the metabolism of copper; added dietary molybdenum decreases metabolism of copper, whereas added dietary sulfur enhances it. The signs of molybdenosis have appeared in cattle that were fed about 6 ppm of molybdenum for several months. The NRC has set the maximum tolerable level of molybdenum for cattle for relatively short feeding periods at 10 ppm. *CAUTION:* As a feed additive, molybdenum is not approved by the Food and Drug Administration.

(Also see Chapter 4, Table 4-6—Molybdenum.)

Selenium (Se). Selenium, like molybdenum, was known for its toxic characteristics long before it was discovered to be an essential nutrient. However, research has firmly established the essentiality of selenium for ruminants; it is needed in trace amounts to prevent retarded growth, reproductive problems, retained placenta, white muscle disease—a condition that occurs in calves and lambs in selenium-deficient areas, and some mastitis problems. Also, it is closely associated with vitamin E; both selenium and vitamin E protect cells from the detrimental effects of peroxidation, but each takes a different approach. Vitamin E is present in the membrane components of the cell and prevents free-radical formation, whereas selenium functions throughout the cytoplasm to destroy peroxides. This explains why selenium will correct some deficiency symptoms of vitamin E, but not others. The current NRC recommended requirements for all cows and heifers are 0.3 ppm, which is the maximum level permitted by FDA. Deficient or toxic selenium areas are widely scattered throughout the United States and the world.

(Also see Chapter 4, Table 4-6—Selenium; and Chapter 5, Table 5-1—White Muscle Disease.)

Zinc (Zn). Milk generally contains about 4 ppm of zinc, but this level has been doubled by increasing the intake of zinc in the ration. Zinc is involved in several enzyme systems and is affected adversely when excess quantities of calcium are present. The NRC recommendation of 40 ppm of zinc for all classes and ages of dairy cattle is based upon limited data. Moderate excesses of zinc are not toxic to dairy cattle. Galvanized pipes and galvanized buckets, which are commonly used to provide water to cattle, contribute zinc along the way. Thus, it is unlikely that a zinc deficiency would occur under normal circumstances; so, zinc supplementation of dairy cattle may be considered as precautionary only.

(Also see Chapter 4, Table 4-6—Zinc.)

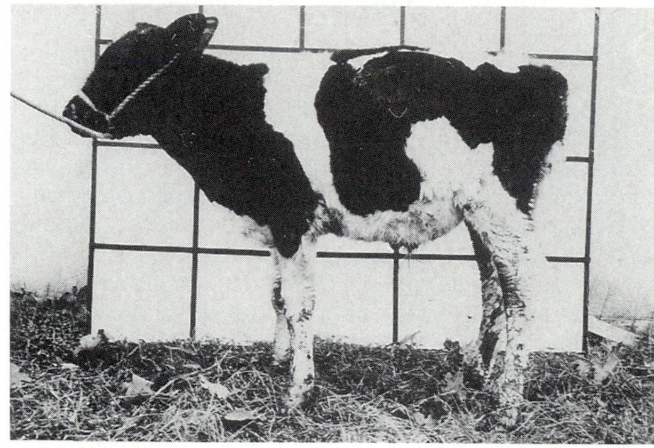

Fig. 20-9. Zinc deficiency. *Top:* Calf showing loss of hair on legs and severe scaliness, cracking, and thickening of the skin as a result of zinc deficiency. *Bottom:* The same calf after receiving supplemental zinc. (Courtesy, W. J. Miller, The University of Georgia, Athens)

Vitamins

Dairy cattle, like other animals, require vitamins for optimum performance and health.

Vitamins are classified as fat-soluble or water-soluble. The fat-soluble vitamins include vitamins A, D, E, and K; the water-soluble vitamins include the B vitamins and vitamin C.

Tables 20-2 and 20-3 present the daily vitamin A and vitamin D requirements for different classes of dairy animals; and Table 20-5 shows the recommended vitamin content of rations for different classes of dairy animals. Chapter 19, Table 19-9, presents in summary form the vitamin requirements of all cattle; the vitamin requirements of dairy cattle and beef cattle are similar except for the higher milk production of lactating cows. Section V, Composition of Feeds, shows the fat-soluble vitamin and water-soluble vitamin content of feeds.

(Also see Chapter 4, section headed "Vitamins," including Table 4-8, Animal Vitamin Chart, for an in-depth discussion of each of the vitamins.)

FAT-SOLUBLE VITAMINS

Dairy cattle require fat-soluble vitamins A, D, E, and K. Generally, all classes of dairy cattle require a dietary source of vitamins A and E. Vitamin D must either be synthesized in the skin by the action of ultraviolet radiation or be included in the ration. Rumen microbes synthesize adequate amounts of vitamin K to meet the needs of most dairy cattle with the exception of young calves, whose rumen has not begun all of its functions.

Fortunately, under normal conditions, natural feeds furnish most fat-soluble vitamins or their precursors in adequate amounts. High-quality forages contain large amounts of vitamin A precursors, and vitamin E is abundant in most feeds. Vitamin D is found in large quantities in sun-cured forages. Additionally, cattle can store adequate reserves of the fat-soluble vitamins to meet their needs for several months. Yet, when dairy producers feed limited or low-quality forage, use high levels of ensiled forage, expose cattle to little sunlight, or use milk replacers for young calves, additional vitamins will probably be needed for optimum health and high performance.

Vitamin A. Vitamin A supplementation may be desirable when (1) poor-quality or limited amounts of forage are fed, (2) forage that has been stored for a long period loses its carotene through oxidation, or (3) high levels of corn silage and low-carotene concentrates are fed.

A deficiency of vitamin A causes many problems. Some or all of the following symptoms may occur, depending on the length and severity of the deficiency: (1) night blindness—a condition that is readily detected when animals are driven among obstacles in dim light, (2) watery eyes, (3) nasal discharge, (4) coughing, (5) diarrhea, (6) pneumonia, (7) lack of coordination, (8) staggering gait, (9) convulsive seizures, (10) complete blindness, (11) stratified keratinized epithelium, (12) increased susceptibility to infection, (13) loss of appetite, (14) emaciation, (15) rough hair coat, (16) scaly skin, (17) abortion, (18) shortened gestation period, (19) birth of dead, weak, or blind calves, and (20) retained placenta.

There are a number of indicators of vitamin A deficiency which may be used before clinical signs of deficiency become evident, one of the most sensitive of which in growing calves is the elevation of cerebrospinal fluid pressure.

The vitamin A requirements of cattle can be met by carotene in feeds, supplements of vitamin A in a stabilized form, or a combination of both. For cattle, 1 mg of carotene is considered to be equivalent to 400 IU (International Units) of vitamin A. The daily vitamin A requirements for different classes of dairy cattle are shown in Tables 20–2 and 20–3; and Table 20–5 shows the recommended vitamin A content of rations for different classes of dairy animals. There are no allowances of vitamin A for milk production as intakes above those required for normal reproduction do not increase milk yield. **NOTE WELL:** The vitamin A requirements given in Tables 20–2, 20–3, and 20–5 are adequate under most practical conditions, but may be increased when animals are under certain stressful conditions such as low environmental temperature or exposure to infective bacteria.

Under most practical conditions, moderate excesses of vitamin A are not harmful. The NRC reports that the presumed safe limit of vitamin A is 30,000 IU/pound of ration for both lactating and nonlactating cows.

(Also see Chapter 4, Table 4–8—Vitamin A.)

Vitamin D. Cows fed sun-cured forage or exposed to sunlight do not need supplemental vitamin D. Even green forage, barn-cured hay, and silage have some vitamin D activity due to the irradiation of dead tissue of stems and leaves of growing plants. When animals are exposed to sunlight, vitamin D is synthesized by the skin in sufficient amounts for maintenance, growth, reproduction, and lactation. However, calves housed indoors need vitamin D supplementation due to lack of exposure to sunlight. Animal sources of vitamin D (called D_3) and plant sources (called D_2) are biologically equivalent in dairy cattle.

Fig. 20–10. Calf with severe rickets. Note the bowed legs and swollen joints. Rickets may be caused by a lack of vitamin D, calcium, or phosphorus; or by an incorrect ratio of the two minerals. (Courtesy, Michigan State University, East Lansing)

A vitamin D deficiency leads to a failure of bones to calcify normally, resulting in rickets in calves and osteomalacia in adults. Vitamin D deficiencies in calves kept indoors do occur, but deficiencies in mature cattle under normal conditions are extremely unlikely because exposure to sunlight provides adequate vitamin D. Some of the first signs of vitamin D deficiency rickets are decreases in the blood plasma concentrations of calcium or inorganic phosphorus, or both, and increases in serum phosphates.

(Also see Chapter 4, Table 4–8—Vitamin D.)

Vitamin E. Vitamin E is an antioxidant associated with selenium. It stimulates the immune system and reduces the incidence of oxidized flavor when fed at high levels (400 to 1,000 mg/cow/day); and it may aid in protection against white muscle disease, caused by deficiency of selenium.

White muscle disease is characterized by (1) a weakening of the leg muscles, resulting in calves walking with a typical crossing of the hind legs; (2) relaxation of the pasterns and splaying of the toes; (3) impaired ability to suckle, due to the musculature of the tongue being affected; and (4) in advanced cases, the calf may be unable to hold up its head and to stand.

All green feeds are good sources of vitamin E. Cows on pasture or being fed green chop, receive adequate vitamin E. However, the vitamin E content of dry feedstuffs decreases during storage.

One IU of vitamin E is defined as 1 mg of dl-alpha-tocopherol acetate. For young calves, the vitamin E requirements are estimated to range from 11 to 18 IU of vitamin E/pound of total feed (Table 20-5).

Vitamin E and selenium play a synergistic role in the nutrition of calves. Some deficiency signs, such as white muscle disease, may respond to either vitamin E or selenium; some deficiencies may require both. Also, the requirements for these nutrients may be influenced by the type of liquid in the diet.

(Also see Chapter 4, Table 4-8—Vitamin E; and Chapter 5, Table 5-1—White Muscle Disease.)

Vitamin K. Vitamin K functions as a stumulant to blood coagulation. Either vitamin K_1 (phylloquinone) or vitamin K_2 (menaquinone) will meet the needs of cattle. Green, leafy materials of any kind, both fresh and dry, are good sources of vitamin K_1. Normally, vitamin K_2 is synthesized in large amounts in the rumen; so, dietary supplementation is not recommended.

When cows consume moldy sweetclover hay, which is high in dicoumarol, blood coagulation may be impaired, followed by generalized hemorrhaging. This syndrome, commonly called *sweetclover disease* or *sweetclover poisoning,* responds to treatment with vitamin K.

(Also see Chapter 4, Table 4-8—Vitamin K.)

WATER-SOLUBLE VITAMINS

The water-soluble vitamins include biotin, choline, folacin (folic acid), inositol, niacin (nicotinic acid, nicotinamide), pantothenic acid (vitamin B-3), para-aminobenzoic acid (PABA), riboflavin (vitamin B-2), thiamin (vitamin B-1), vitamin B-6 (pyridoxine, pyridoxal, pyridoxamine), vitamin B-12 (cobalamins), and vitamin C (ascorbic acid, dehydroascorbic acid). However, a physiological need for cattle of all of these vitamins has not been demonstrated.

Until recently, it was assumed that dairy cattle with a functional rumen did not require supplemental B vitamins. The rumen microflora were believed to synthesize adequate amounts of these nutrients for the host's requirements. Besides, the B vitamins are relatively abundant in dairy feeds. But recent evidence suggests a need for supplemental niacin under certain conditions and possibly supplemental choline and thiamin in the case of mature cattle, for which microbial synthesis and quantities in feeds may be inadequate, especially during diseased conditions or periods of stress. It is assumed that dairy cattle of all ages have a physiological need for most of the B vitamins, especially biotin, choline, niacin, pantothenic acid, riboflavin, thiamin, vitamin B-6, and vitamin B-12. In young calves, deficiency signs have been demonstrated when there is inadequate intake of these vitamins, but, even without a functioning rumen, their needs for these B vitamins appear to be met when they are fed whole milk. When young calves are fed milk replacers, however, it is advisable to ascertain the adequacy of vitamin intakes until their rumens are functional. Table 20-5, footnote 14, shows the minimum quantities of B vitamins that should be provided in milk replacers.

Biotin. A biotin deficiency in calves, characterized by paralysis of the hindquarters, has been produced. Signs of deficiency did not develop when synthetic milk was supplemented with 4.5 mcg of biotin/pound of feed and fed at 10% of liveweight.

(Also see Chapter 4, Table 4-8—Biotin.)

Choline. Researchers have produced choline deficiency in calves by using a synthetic ration containing 15% casein. Within 6-8 days, the calves developed extreme weakness and labored breathing and were unable to stand. Supplementation of the ration with 236 mg of choline/quart of synthetic milk prevented the development of these signs. Adding choline to the ration may increase the percentage of milk fat in lactating cows.

(Also see Chapter 4, Table 4-8—Choline.)

Niacin (Nicotinic Acid, Nicotinamide). Niacin is required by the young preruminant calf. In order to prevent a niacin deficiency, it is recommended that niacin be added to milk replacers at a level of 2.6 ppm.

Some recent evidence suggests that rumen microorganisms may not synthesize adequate amounts of niacin to meet the needs of high-producing cows in early lactation. A summary of several studies of cows in early lactation shows a small increase in milk production and butterfat, along with a decrease in the incidence of ketosis, to the feeding of 6 g of niacin with all-natural protein. Although research results are not consistent, there appears to be significant benefits from feeding niacin to dairy cows with above average incidence of ketosis and that are overconditioned. Research studies indicate that supplementation of 6 to 12 g of niacin/day should begin 2 weeks before calving, then continue for 8 to 12 weeks after calving.

It is conjectured that the major reason for improvement in milk production that occurs with added niacin may be related to the role of niacin in carbohydrate and lipid metabolism and the resultant decrease in ketosis. Niacin may also influence rumen fermentation, as evidenced by greater microbial protein synthesis and increased levels of rumen propionate with niacin supplementation. When cows are fed heated soybean meal, rumen response to niacin is greater than it is when cows are fed unheated soybean meal.

(Also see Chapter 4, Table 4-8—Niacin [Nicotinic Acid, Nicotinamide].)

Pantothenic Acid (Vitamin B-3). Pantothenic acid deficiency in the calf is characterized by a scaly determatitis around the eyes and muzzle, loss of appetite, diarrhea, weakness (unable to stand), and convulsions.

Pantothenic acid deficiency in animals with functioning rumens is unlikely due to microbial production of pantothenic acid.

(Also see Chapter 4, Table 4-8—Pantothenic Acid [Vitamin B-3].)

Riboflavin (Vitamin V-2). Riboflavin deficiency in the calf is characterized by hyperemia (presence of blood) of the mucosa of the mouth, lesions in the corners of the mouth and along the edges of the lips, loss of hair—especially on the belly, and excess salivation.

A riboflavin deficiency in lactating cattle is unlikely because of the amounts of riboflavin that are present in feedstuffs and synthesized in the rumen.

(Also see Chapter 4, Table 4-8—Riboflavin [Vitamin B-2].)

Thiamin (Vitamin B-1). Thiamin deficiency in the calf may cause polioencephalomalacia, characterized by listlessness, muscular incoordination, progressive blindness, convulsions, and sudden death; which may be accompanied by diarrhea and dehydration. The condition is found primarily in cattle, sheep, and goats fed high-concentrate rations; and it has been linked to increased microbial thiaminase activity and the production of thiamin analogs in the rumen. Treatment consists of the IV or IM administration of thiamin at a rate of 1 mg/pound liveweight.

(Also see Chapter 4, Table 4-8—Thiamin [Vitamin B-1]; and Chapter 5, Table 5-1—Polioencephalomalacia.)

Vitamin B-6 (Pyridoxine, Pyridoxal, Pyridoxamine). Vitamin B-6 deficiency has been produced in calves fed a synthetic diet. It is characterized by loss of appetite, cessation of growth; and after about 3 months, epileptic fits in some, but not all, calves. Calves respond to vitamin B-6 therapy if it is initiated in the early stages of the disease.

(Also see Chapter 4, Table 4-8—Vitamin B-6 [Pyridoxine, Pyridoxal, Pyridoxamine].)

Vitamin B-12 (Cobalamins). Vitamin B-12 deficiency has been produced in calves under 6 weeks of age by feeding them a diet containing no animal protein. Deficiency signs include poor appetite and growth, muscular weakness, and poor general condition. It has been suggested that the vitamin B-12 requirement for dairy cattle is between 0.15 and 0.30 g/pound liveweight.

Vitamin B-12 is of special interest in the mature ruminant because of its role in propionate metabolism and because of the incidence of B-12 deficiency as a secondary result of cobalt deficiency. Certain soils have insufficient cobalt to produce levels of the element in plants that are adequate to support optimum vitamin B-12 synthesis in the rumen.

(Also see Chapter 4, Table 4-8—Vitamin B-12 [Cobalamins].)

Water

Large amounts of water are essential if a cow is to produce to her maximum capacity. It is necessary for maintaining body fluids and proper ion balance; for digesting, absorbing, and metabolizing nutrients; for eliminating waste material and excess heat from the body; for providing a fluid environment for the fetus; and for transporting nutrients to and from body tissues.

The water that dairy cattle need is supplied by drinking, by water in the feed that they consume, and by metabolic water produced by the oxidation of organic nutrients.

Fig. 20-11. Clear, clean water on a Missouri dairy farm. (Courtesy, *Holstein World*, Sandy Creek, N.Y.)

Cows drink an average of 100 to 200 lb of water per day, with heavy producers drinking up to 300 lb per day (1 gal water = 8.33 lb). Cows need 4 to 5 lb of water for each pound of milk produced. The amount of water a cow will drink depends on her size and milk yield, the quantity of dry matter consumed, the temperature and relative humidity of the air, the temperature of the water, the quality of the water, and the amount of moisture in her feed.

Good-quality, clean water is of the utmost importance in a dairy cattle feeding program—a factor often neglected by many producers. Water troughs should be cleaned routinely to ensure that the water is free from dirt and pathogenic bacteria. A common drinking trough provides an excellent means of spreading parasites and disease.

It has long been known that cattle are sometimes poisoned when they drink lake water containing blue-green algae; thus, cattle should be prevented from drinking water with heavy algae growth.

In extremely cold weather, it is a good idea to have a tank heater to keep the water from freezing. Also, frequency of watering is important. Cows stabled in a stanchion-type barn produce 3½ to 4% more milk if they have drinking cups available than if they are watered twice daily. Contrary to some opinions, cows do not produce more milk from softened than from hard water.

Dairy cattle lose water from the body in saliva, urine, feces, and milk; through sweating; and by evaporation from body surfaces and the respiratory tract. The amount of water lost from the body of cattle is influenced by the activity of the animal, ambient temperature, humidity, respiratory rate, water intake, feed consumption, and other factors.

(Also see Chapter 4, section on "Water.")

FEEDS FOR DAIRY CATTLE

For convenience, in the sections that follow the commonly used dairy cattle feeds are grouped in eight general categories: (1) forages, (2) root crops, (3) concentrates, (4) by-products, (5) special feeds and additives for dairy cattle, (6) commercial feeds, (7) feed considerations, and (8) feed substitutions. Additional information pertaining to feed ingredients and feed preparation is contained in Section II of this book.

Forages

Legumes and grasses are the major sources of forages for dairy animals. They may be harvested and fed as pasture, green chop, hay, silage, or haylage. When properly grown, harvested, and stored, they are excellent sources of nutrients. High-quality forage can make up to ⅔ of the ration dry matter with cows consuming 2½ to 3% of their body weight in forage dry matter. High-quality forages fed in balanced rations will supply much of the energy, protein, minerals, and vitamins needed for milk production.

Fig. 20–12. High-quality alfalfa hay being consumed by Holsteins at Arizona Dairy Co., Higley, Ariz. (Courtesy, James Tappan)

Average cows can produce up to 70% of their potential milk yield when fed excellent-quality forage without grain, whereas poor cows may produce their total potential without supplemental feed. The higher the level of production, the higher the percentage of the total ration which should come from grains and other concentrate feeds. Some cows have been fed rations comprised entirely of concentrate feeds. Generally speaking, all-concentrate rations are not advisable from the standpoints of either the health of the cow or the quality of the milk produced (lowered milk fat test usually accompanies such a ration).

In using any kind of forage, three important points should be kept in mind: (1) To obtain the most nutrients from forage, it must be of high quality; (2) the better the forage, the smaller the requirement for grains; and (3) the cow is, by nature, a good consumer of forage.

A high-quality forage is one that possesses the physical and chemical characteristics commonly associated with palatability, along with an abundance of feed nutrients. The most important physical indicators of forage quality that may be used in a practical way are (1) stage of maturity when cut, (2) percentage of leaves, (3) green color, (4) amount of heating, spoilage, (5) pliability of stems, (6) aroma, and (7) freedom from foreign material.

A number of attempts have been made to establish standards by which forages may be graded with some degree of uniformity, but no one system has been widely accepted, even for one kind of forage—for hay or for silage, for example. Generally speaking, forage evaluations are based on a combination of chemical analyses and personal observation of color, odor, amount of grain, and freedom from weeds and weed seeds. Additional effort is needed in this whole area so that both forage growers and dairy producers may benefit from the production of high-quality forage.

(For methods of evaluating hay, see Chapter 8, Hay; and for methods of evaluating silage, see Chapter 9, Silage/Haylage/High-Moisture Grain.)

PASTURE

Of all the feedstuffs for dairy cows, pasture is the oldest, and today the most controversial. Since the 1960s, pastures have played a decreasing role in the total feed program of U.S. dairy farms. Nevertheless, pastures are effectively and efficiently used on some dairy farms. Grazing may supply 10 to 50% of the total dry matter, and at times an even higher percentage of the protein, of the ration of lactating cows. Also, good pastures are excellent for replacement heifers.

Balancing rations of lactating cows is more difficult when utilizing pastures than when feeding stored forages. However, modern computer ration balancing techniques allow adequate adjustments to account for grazing and prevent dramatic changes in milk production. Research studies indicate that the optimum rate of grain feeding of lactating cows on pasture is 1:3 to 1:3.5 grain to milk ratio.

Good pastures provide a highly nutritious feed when fresh growth is maintained. Besides, the cows do their own harvesting and spread their manure. However, grazing results in considerable waste of the crop through trampling (unless managed very carefully), with lower yield of nutrients and milk per acre than from harvested crops. Since pasture plants vary in their growth patterns and nutritive value as weather conditions change, fluctuations in milk production can occur. Additionally, internal parasite problems—in particular, flukes and worms—are often associated with pastures, especially irrigated and swampy native pastures.

Fig. 20-13. Ayrshire cows on pasture at Laneway Farm, Rochelle, Va. (Courtesy, Ayrshire Breeders Assn., Brandon, Vt.)

In recent years, producers have largely abandoned the old system of continuous grazing on one field all season. Under continuous grazing, the more palatable plants are constantly cropped close to the ground, and the roots are depleted of their nutrient reserves as they try to get shoots up high enough for the leaves to produce additional nutrients for storage (in the roots). As a result, over a period of time, pastures produce both smaller yields and lower quality feed. New systems of pasture management, including rotation grazing, strip grazing, and zero grazing, are now being used extensively to protect the quality of the pasture. When pasture forage is grazed rapidly (within about 3 days), then allowed to recover while the animals are grazing other fields, the better plants have a chance to compete with those that are less palatable.

Fig. 20-14. A two-strand electric fence facilitates rotation grazing, strip grazing, or zero grazing. (Courtesy, Milk Marketing, Inc., Strongsville, Ohio)

Large herds are not easily handled in a pasture system. Where land prices are high, it may not be economically feasible to maintain pastures in close proximity to the corrals. Even with high-yielding pastures in irrigated areas of the country, it takes anywhere from 0.8 to 2 acres of pasture to sustain 1 cow during the growing season. In many areas, it may take several times this amount of land. When a herd reaches 100 cows or more, land may be unavailable within a radius of ½ mile of the corrals, which is considered the maximum practical distance that cows can be trailed to and from pasture. With cured feed, transportation is much more practical than with green feed, either grazed or green chopped. For these reasons, most commercial dairy producers are shifting from pasture to stored feed, usually produced on land located some distance from the corrals.

(Also see Chapter 7, Pasture and Range Forages.)

GREEN CHOP (SOILAGE OR ZERO GRAZING)

Many producers harvest and feed green chop daily. It reduces wastage from trampling and manure fouling which are inherent in grazing systems. With tall-growing crops, it can be used in place of pastures. However, cutting green chop every day can be a major problem during wet weather or during peak work periods.

It has been said that the main reason that green chopping can compete with pasture grazing is that anyone can see when a manger is empty, but it is much more difficult to see when a pasture is "empty." It requires more astute management to graze a pasture properly than it does to cut and feed it green or to harvest and store it as hay or silage.

(Also see Chapter 7, Pasture and Range Forages, Section headed "Green Chop.")

HAYS

Legume hays are best for dairy cattle. Alfalfa is by far the most popular hay crop. Some of the grass-legume combinations (such as clover and timothy) are more popular than straight grass hay. To make the best hay, alfalfa and grasses should be cut at the proper stage of maturity. With advancing maturity, plants decrease in energy, protein, calcium, phosphorus, and digestible dry matter, and increase in fiber. As crude fiber, neutral detergent fiber (NDF), and acid detergent fiber (ADF) increase, the lignin content of the plant also increases. Lignin is indigestible and makes other nutrients less digestible.

Producers can afford to pay a premium for hay of good quality. Studies have shown that if hay cut at full bloom is worth $80 per ton, hay cut at $\frac{1}{10}$ bloom is worth $140 per ton. From this, it is understandable how a producer might establish a bonus payment plan for hay that meets certain minimum standards of quality, to the financial benefit of both the hay seller and the buyer.

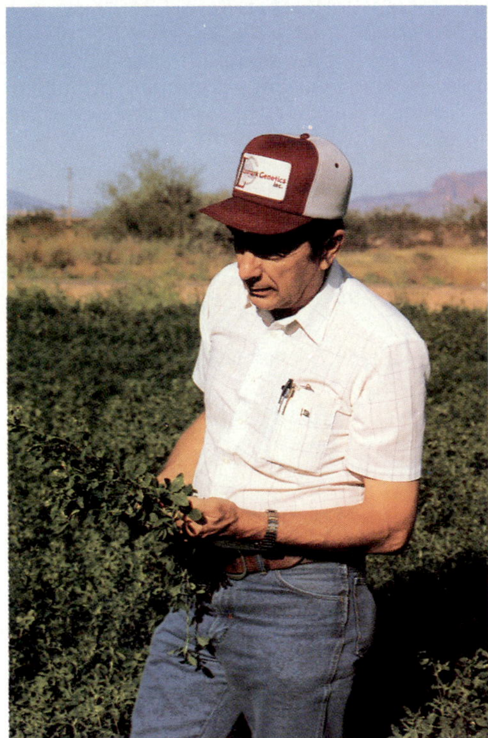

Fig. 20-15. James Tappan, successful Co-owner and Manager, Arizona Dairy Co., Higley, Ariz., examining alfalfa in the field. At Arizona Dairy, the 4,100 lactating cows, 700 dry cows, and young stock consume approximately 35,000 tons of forage (hay and silage), dry matter basis, each year. The quality of forage can make or break a dairy! (Courtesy, James Tappan)

Quality of hay and level of production go hand in hand. Since, in recent years, good-quality hay has been sold at a premium price, it is mandatory that producers utilize available hay as efficiently as possible. The best quality hay should be fed to the best cows only. To feed a low-producing cow or a dry cow top quality alfalfa in times of hay shortages is wasteful and inefficient. Therefore, the feeder needs to separate the hay and feed according to quality if at all possible.

Recently the relationship of fiber to milk and butter fat production has received considerable attention. Excessive fiber levels limit intake and energy concentration, while shortage of fiber reduces rumen digestibility and milk fat test. This is due primarily to the fact that these feeds alter the microflora of the rumen in such manner that propionic acid levels are increased, thereby lowering the amount of butterfat produced.

Because fiber is important, and because all fibers are not the same, the new NRC requirements presented in Table 20-5 give fiber minimums for crude fiber, acid detergent fiber, and neutral detergent fiber.

(Also see Chapter 8, Hay.)

SILAGES

It is often possible to produce more nutrients per acre—and thereby more milk per acre—from silage than from hay crops. This is especially true for corn, where yields are usually high. Merely putting the crop into silage does not ensure this, however. It takes special care and a proper storage facility to make this possible.

The feeding value of silage is no better than the forage that goes into the silo. Good silage is easy to make if the crop is harvested at the proper stage of maturity, cut fine, and ensiled as quickly as possible at 55 to 70% moisture. In the silo, it must be evenly distributed, firmly packed, and protected from the air. If the crop does not contain sufficient readily fermentable carbohydrate, it is important that there should be added either (1) a carbohydrate, or (2) an acid, to preserve it without fermentation.

Corn is the most widely used silage crop in the United States. In recent years, the development of hybrid varieties adapted to a shorter growing season has resulted in corn silage expanding substantially. The qualities that make corn such an excellent silage crop are its high yield, its pithy stem, and its abundance of grain. Sorghum has been used for silage, both alone and in combination with corn or soybeans. But it usually does not produce as high yields nor as high-quality feed as corn. Sunflowers have been used in some localities, particularly where corn is not well adapted. Where weather during harvest makes hay curing uncertain, silage is often produced from grasses and/or legumes. The latter type of silage is commonly referred to as hay crop silage, or grass silage. It can be made entirely from grasses or legumes, or from a mixture of both grass and legume. Grasses are easier to preserve as silage than legumes, but they are not as nutritious. Legumes are higher in protein than corn silage, but lower in energy on a comparable moisture basis.

Fig. 20-16. Four upright silos on a dairy farm. (Courtesy, Milk Marketing, Inc., Strongsville, Ohio)

Generally speaking, a higher percentage of good-quality silage is made from corn and sorghum than is made from hay-crop, but this need not be so. Consistently good hay-crop silage can be made, provided these procedures are followed: (1) it is cut early, (2) it is wilted to 55 to 70% moisture content, and (3) a suitable preservative is added. The chief reason why these special methods are necessary in ensiling hay-crop silages is that they contain a much smaller percentage of sugars than corn or sorghum silage.

Direct-cut grass-legume silage (forage that is cut and immediately put into the silo without wilting) is not usually recommended, because of excess seepage losses which carry with them many of the more soluble nutrients. When it is necessary to cut direct rather than wilt the silage, it is recommended that some dry material be added to absorb the excess moisture and trap it in the silage. Dry hay, dry grain, dried beet pulp, or straw may be used for this purpose.

Fig. 20-17. Milk cows eating round bale silage. (Courtesy, Sperry New Holland, New Holland, Pa.)

Small cereal grain forage crops (barley, oats, or wheat) are sometimes made into silage. For this purpose, they should be harvested when the grain is in the milk stage. Usually, cereal silage is not as high in nutrients as either corn or grass-legume silage. But cutting cereal grain for silage may permit double cropping and will likely produce more total nutrients per acre than could be obtained by harvesting the crop for grain—and wasting the straw.

If carefully mixed, urea can be added to corn silage, at the rate of 10 lb of urea per ton of silage. The addition of 10 lb of urea per ton of corn silage will increase the protein content from 8 to 12% on a dry matter basis. If more urea is added, the silage becomes less palatable and of less value as dairy cow feed. If urea is not properly mixed, it is possible that a cow getting a large concentration of urea might be killed from its toxicity.

(Also see Chapter 9, Silage/Haylage/High-Moisture Grain.)

HAYLAGE (LOW-MOISTURE SILAGE)

Haylage—grass and/or legumes that are wilted to 40 to 50% moisture before ensiling—is popular. Its feeding value depends on the stage of maturity when the crop is harvested and the percentage of dry matter in the haylage. Unlike wilted or direct-cut silage, there need be no limitation on how much haylage is fed. It may be fed free-choice if it is good quality. Producers should avoid overestimating the value of haylage in a feeding program. It is a good feed if properly made and stored, but it is no better than the same forage cured and stored equally well as hay or wilted silage.

(Also see Chapter 9, Silage/Haylage/High-Moisture Grain.)

Root Crops

The per acre yield of nutrients from roots is higher than for most other feed crops, but labor costs in the United States make it impractical to grow, process, and feed root crops to dairy cattle. However, in Europe mangels (fodder beets) and stock carrots have been among the more popular root crops used in dairy rations. These crops should be sliced, or in some other way reduced in size to pieces which can be consumed safely by dairy cattle. They are very high in moisture; hence, they may need to be limited for cows producing at very high levels so that there will be sufficient room in the digestive tract for nutrients other than water.

Concentrates

Concentrate feeds are those which are high in energy and low in fiber. Many different kinds of concentrate feeds are used in dairy cattle feeding. They are usually classed according to total (crude) protein content as (1) low-protein, (2) medium-protein, or (3) high-protein feeds. The chemical analysis of various feeds is shown in Section V of this book.

Three factors besides chemical composition are important in evaluating concentrates for milk cows—palatability, quality of milk produced, and cost. The most infallible way in which to appraise the first two factors is through actual feeding trials. Consideration of the third factor—cost—necessitates that dairy producers be keen students of values. They must change the formulations of their ration(s) in keeping with comparative feed prices, and do so without causing the animals to go off feed.

Corn and barley are the two chief grains used in dairy rations, although oats, sorghum grains, and wheat are also used when there is a price advantage. Corn is often used as ground ears (corn-and-cob meal), rather than as grain. In high-concentrate rations particularly, the fiber from the cobs may help to keep the digestive tract functioning normally. Grains make it feasible for dairy animals to produce at high levels. Even if forage is of high quality, it is difficult to get a production of more than 9,000 to 10,000 lb of milk during a 300-day lactation without feeding some grain. Today, many cows produce more than 20,000 lb of milk, and some herds now average more than this amount. Grains make it possible for superior cows to consume sufficient energy to sustain the high level of production of which they are genetically capable.

Mill by-products and sugar industry by-products are also extensively used in dairy rations.

Oil meals are the major source of protein, with soybean meal and cottonseed meal most commonly used, along with some safflower meal, sunflower meal, copra meal, peanut meal, canola meal, and others. Urea is also used as a source of protein.

(Also see Chapter 10, Grains/High-Energy Feeds.)

HIGH-MOISTURE GRAIN

Feeding high-moisture corn or other wet grains to dairy cattle offers the following **advantages:**

1. Grain can be harvested 2 to 3 weeks earlier than normal, thereby reducing field losses and harvest problems associated with adverse weather.
2. Storage and handling losses are reduced.
3. It lends itself to automated feeding programs.
4. The expense of drying grain is eliminated.
5. It makes for a highly palatable feed.
6. It lessens the labor of processing/grinding grain.

High moisture shelled corn should be stored at a moisture content of 25 to 30%. Ground ear corn should contain 28 to 32% moisture for proper preservation. High moisture shelled corn and ear corn should be ground before storing in conventional silos. In airtight silos, shelled corn can be stored whole, then rolled upon removal from the silo. Propionic acid can be used effectively, according to label directions, to treat and preserve high-moisture corn or barley for dairy cattle.

(Also see Chapter 9, Silage/Haylage/High-Moisture Grain.)

By-products

By-product feeds are important in dairy rations. The milling, sugar, vegetable oil, and fermentation industries provide by-products of special significance to most commercial dairy rations, and for many home-mixed feeds. However, if producers plan to incorporate by-product feeds in their rations, they should first determine the moisture content of the product, its relative feeding value and price, and the appropriate amount to feed. By-products are extremely variable in feeding value. For example, almond hulls—a widely used by-product feed in California—can range from excellent to poor as a feedstuff. Additionally, some by-product feeds may contain pesticide residues that can be excreted in the milk. For the latter reason, the producers should make sure that their feeds are free from environmental contaminants.

Many of the residues and by-products from the vegetable industry can be used effectively in dairy rations. Such feeds as citrus pulp, cull potatoes, cannery wastes, pea vines and pods, and cottonseed hulls can be used as economic alternatives to the more traditional feeds.

Fig. 20–18. Citrus residue waiting to be processed into citrus pulp, a valuable by-product. (Courtesy, Marshall E. McCullough, Professor Emeritus and Consultant, University of Georgia, Experiment)

Wheat bran and wheat middlings (or mill run) are the most widely used by-products of the milling industry. Wheat bran has always been considered to be an excellent dairy feed when the quality is high and the price is competitive. It is especially good as a conditioning feed and is widely used in fitting dairy animals for shows and sales. Wheat mill run (middlings) is usually more price competitive than wheat bran. When of high quality, it is an excellent feed, although it is not as high in digestibility or energy as the whole grains.

Beet pulp and beet molasses are used very extensively in dairy rations, in which they add to the texture and palatability as well as serve as good sources of energy. Molasses reduces the dustiness of feeds and is an excellent carrier for minerals and vitamins, as well as other ingredients where uniform mixing is important. It is usually restricted to about 5% of the grain-concentrate mix. At higher levels, it may make the feed sticky and cause mechanical mixing and feeding difficulties. Also, excess molasses may be too laxative because of its high mineral and sugar content; hence, it may reduce the utilization of other less digestible ingredients.

When available at competitive prices, the dried or wet mash from the brewing and distilling industries, as well as some of the pharmaceutical fermentations, are used extensively in dairy rations. Usually they contain some of the B vitamins and other nutrients which are supplied by rumen fermentation, and for this reason they may be more desirable in swine and poultry rations than in cattle rations.

The vegetable oil industry extracts the oil from certain seeds, leaving a residue which is high in protein and beneficial in dairy rations. The oil meals are incorporated in dairy rations primarily as a source of protein. Soybeans are a very popular oil crop; and the extracted meal is one of the most common protein supplements in dairy rations. In cotton-growing areas, cottonseed meal is widely used in dairy rations. Other oil meals used in dairy rations include peanut meal, copra (coconut) meal, canola meal, safflower meal, and several others of lesser importance. Some heating is desirable for most oil meals to destroy toxic or inhibitory materials which they contain and to decrease their degradability.

(Also see Chapter 12, By-product Feeds/Crop Residues.)

Special Feeds and Feed Additives for Dairy Cattle

Certain feeds and additives are especially adapted to dairy cattle, primarily to increase milk production and/or to affect milk composition; among such products are: (1) fats and oils, (2) fiber, and (3) the following additives: antibiotics, bovine somatotropin (BST), buffers, ionophores, isoacid (branched-chain fatty acids and valeric acid), and thyroprotein.

FATS AND OILS

Fat serves the following functions when added to dairy rations: it (1) increases the caloric density of the ration without lowering the forage (fiber) content; (2) controls dust; (3) lessens the wear and tear on feed mixing equipment; (4) facilitates pelleting of feeds; (5) increases palatability; (6) helps to homogenize and stabilize certain feed additives, especially those of a very fine particle size; and (7) increases the total amount of milk, butterfat, and SNF, but results in a slight decrease in the percentage of both butterfat and SNF.

Most forages and grains are low in lipids—they contain less than 2 to 3% fat. In general, dairy cows should be able to utilize 1 to 1½ lb of fat per day in addition to the fat present in natural feedstuffs. This means that about 3% more fat can be added to the total ration (forage plus concentrate), or that 5 to 6% fat can be added to the grain ration.

Added fat is especially effective in early lactation. Because of the increased caloric concentration provided by dietary fat and because high-producing cows are usually in negative energy balance during early lactation, fat is frequently added to the ration to increase the cow's energy intake and provide fatty acids to the udder. It may also be beneficial to provide supplemental fat when the capacity of the gastrointestinal tract limits energy intake.

An important consideration in the successful feeding of fats is that they may be used to provide increased energy without lowering the forage (fiber) intake. In early lactation, substituting fat for a portion of the starch obtained in cereal grains in the ration of cows is a way in which to maintain high energy concentrations and high fiber intakes. The substitution of fat for grain alleviates the low milk fat syndrome caused by inadequate fiber and excessive grain. A ration containing a high proportion of forage helps maintain normal rumen function and provides an environment in which fat is less inhibitory to rumen fermentation and nutrient digestion.

The type of fat (saturated or unsaturated) added to the ration greatly influences the animal's nutrient utilization, milk production, feeding behavior, ration acceptability, the amount of fat that can be fed, and milk composition. Unsaturated fats are less desirable for dairy cows because of their inhibitory effects on rumen fermentation and digestion. Animal fats (which are more saturated) and blended animal-vegetable fats have generally given the most positive responses in animal performance.

When unsaturated fats are added to the ration, the calcium and magnesium contents of the cow's total ration DM may need to be increased by 20 to 30%, because added fat increases calcium and magnesium soap formation in the rumen and the excretion of these elements in the feces.

Because vegetable oils are high in unsaturated fat, they are less satisfactory than saturated fats as ration supplements. Whole seeds, such as cottonseed, soybeans, and sunflower seeds, have been used successfully, but the added fat derived from them should not exceed 1.0 lb/cow/day. Unsaturated fats that contain high levels of oleic acid apparently exceed the hydrogenation ability of rumen microorganisms and result in the greatest milk fat depression. An increase in long-chain fatty acids in the ration (1) increases the secretion of milk and (2) inhibits the synthesis of short- and medium-chain fatty acids in mammary tissue.

Added ration fat, including feeding whole cottonseed and soybeans, decreases the protein content of milk by about 0.1%, primarily because of the lower casein content. Whole cottonseed also increases the proportion of long-chain fatty acids in milk.

Fig. 20-19. A complete (all-in-one) lactating ration containing whole cottonseed. Note the white cottonseeds. Whole cottonseed is a popular dairy feed in the West and Southwest. (Courtesy, Arizona Dairy Co., Higley, Ariz.)

In some cases, feeding rumen-protected fats that are resistant to microbial action in the rumen has increased milk fat percentage and the efficiency of milk production. However, if polyunsaturated fats are fed, this practice may result in milk with an increased proportion of polyunsaturated fatty acids and an increased susceptibility to oxidative rancidity. Limited data, indicate that calcium salts of fatty acids (which are 82% fat) and prilled forms of saturated fats are effective sources of protected fat for dairy cows and that their use may allow the total fat in the ration to reach 7 to 8% of the dietary DM of the ration.

Until the rumen becomes functional, young dairy calves require some fat in the diet. A level of 10% fat in milk replacers appears to be sufficient to supply essential fatty acids, carry fat-soluble vitamins, but insufficient to supply adequate energy for normal gains under optimum environmental temperatures. For veal production, a higher fat milk replacer (15 to 20% or more), will increase fat deposition in the carcass and is desirable. Also, 15 to 20% fat in milk replacers is needed for normal gains when calves are exposed to cold environmental temperatures.

(Also see Chapter 10, sections on "Cottonseed" and "Fats and Oils.")

FIBER

Fiber is important in dairy rations. Excessive fiber levels limit intake and energy concentration, while a shortage of fiber reduces rumen digestibility and milk fat test. Suggested minimums for crude fiber, acid detergent fiber, and neutral detergent fiber are given in Table 20–5.

The following three methods are available to measure the fiber content of feeds:

1. **Crude fiber (CF).** This is the oldest method. It is the residue of the feed that is resistant to degradation in successive boiling with dilute acid and alkali treatments. Crude fiber is not an accurate measure of total fiber or cell walls because much of the lignin and hemicellulose is lost during the analysis. Even cellulose is not totally recovered in the crude fiber fraction.

2. **Acid detergent fiber (ADF).** This consists of cellulose, lignin, lignified nitrogen compounds (heat damaged protein), and insoluble ash. Acid detergent fiber does not represent the total fiber content in feeds as it does not account for hemicellulose. Equations used to predict the digestibility or energy content of feedstuffs are usually based on ADF or include ADF as a major component.

3. **Neutral detergent fiber (NDF).** This consists of ADF plus hemicellulose. It is the best estimate of total fiber in a feed. NDF, often called cell walls, is highly correlated to feed intake, rumination, and total chewing time. With physical form considered, NDF provides the best measurement of effective fiber for formulating dairy rations.

The amount of fiber to include in the ration of dairy cattle is influenced by the body condition and level of production of the animal, the type of fiber, the particle size, the amount of total DM consumed and its bulk density, the buffering capacity of the forage, the frequency of feeding, and the economics. Lactating cows that are fed to produce large amounts of milk, or young animals that are fed to achieve rapid growth, should receive more energy and less fiber than lower producing animals. Forages that are finely ground (processed to small particle size) are rapidly consumed and fermented in the rumen, which reduces (1) the animal's chewing time, (2) ruminal fluid, and (3) the acetate-to-propionate ratio in ruminal fluid. The result of these effects is a depression in milk fat percentage. Chopped alfalfa should average about ¼ in. in length to maintain a normal milk fat percentage. Feed factors, such as small particle size of forages, that reduce the pH of ruminal fluid, decrease the number and activity of fiber-degrading bacteria and cause a depression in fiber degradation. Feeding an insufficient amount of fiber or feeding forages that have a poor buffering capacity in the rumen may have undesirable effects on rumen fermentation, fiber degradation, and milk fat percentage that are similar to those caused by reducing the particle size of the forage.

So, the general recommendation is that lactating dairy cows should receive at least one-third of the total ration dry matter as long hay or as its DM equivalent in medium-to-coarse chopped silage or other forage. A minimum of 5 lb of forage dry matter measuring 1 to 2 in. in length will meet the fiber need of most lactating cows.

The values for NDF and ADF are more accurate measures of the fiber component of feeds than are values for crude fiber. Yet, because both chemical and physical properties of feeds are involved in determining fiber quality and the energy value of feeds, there is currently no one fiber analysis that can accurately predict fiber quality and energy values for all feeds. NDF content is negatively correlated with dry matter intake and apparent digestibility of forages, but it is positively correlated with chewing time. ADF is more negatively correlated with apparent digestibility than is NDF. NDF and bulk density are positively related, which may explain the negative relationship between dry matter intake and the NDF content of the ration. According to University of Wisconsin researchers, NDF is a better predictor than ADF of dairy cow feed intake and milk production.

The optimum amount of NDF and ADF to include in the ration varies with the level of milk production and the type of forage that is fed to dairy cattle. A minimum of 21% ADF and 28% NDF is recommended for cows during the first 3 weeks of lactation (Table 20–5). During times of high milk production, however, ADF and NDF contents of the ration are usually reduced to 19 and 25%, respectively, so that adequate dietary energy can be included to meet the cow's requirement. The ADF and NDF contents of the ration should be increased in later lactation to help prevent milk fat depression and because less energy is required for milk production. Seventy-five precent of the NDF in the ration should be supplied as forage.

CAUTION: The NRC committee that prepared *Nutrient Requirements of Dairy Cattle,* Sixth Revised Edition, cautions that their NDF and ADF recommendations, which are reproduced in Table 20–5 in this chapter, should be used with caution because (1) there is little available information about the interaction of milk production, milk composition, and ration NDF when NDF is supplied by a variety of feeds at various stages of lactation and levels of milk production; (2) there are problems in assaying NDF; (3) there is lack of information when by-product feeds are included in the ration; and (4) a better understanding of the effects of NDF, ADF, lignin, feed particle size and buffering capacity, and frequency of feeding on animal performance (ruminal fermentation, milk production, milk composition, and growth) will improve the ability of producers to determine the optimal amount and kind of fiber to feed to dairy cattle.

(Also see Chapter 8, section on "NDF, ADF, and NIRS Analyses"; and Chapter 15, sections on "Neutral Detergent Fiber (NDF) and Acid Detergent Fiber (ADF)," and "Near Infrared Reflectance Spectroscopy [NIRS].")

RUMEN-PROTECTED AMINO ACIDS

Because of improved lactation when high-quality protein or mixtures of amino acids are infused postruminally, researchers have studied ways of determining the limiting amino acids for lactating dairy cows and methods of protecting these amino acids from ruminal degradation. Considerable research for protecting methionine from ruminal degradation has been undertaken because methionine is often cited as being the most limiting amino acid for lactating cows. Similar, but less, research has been conducted with lysine.

Feeding or infusing mixtures of methionine and lysine that escaped ruminal degradation have not consistently improved milk production, but these methods have increased the percentage of protein in milk and milk protein yields.

(Also see Chapter 11, section on "Protein Bypass [Protected Protein, Escaped Protein].")

FEED ADDITIVES

Many additives are used by dairy producers to increase milk production, affect milk composition, and/or improve feed efficiency; and new products are constantly evolving. Among such additives are the following: antibiotics, bovine somatotropin (BST), buffers, ionophores, isoacids (branched-chain fatty acids and valeric acid), and thyroprotein.

Antibiotics

Antibiotics, which are widely used in the diet of young dairy calves, are especially beneficial for calves exposed to adverse conditions of housing, sanitation, and disease. However, they should not be used as a substitute for good management and a clean, sanitary environment. The greatest benefits from feeding antibiotics accrue when calves are started on the antibiotic as soon as possible after birth, then continued on it during the milk or milk replacer feeding period. Antibiotics are mainly effective in increasing feed intake and growth, along with preventing diarrhea. Generally, antibiotics are fed at the following concentrations: In the milk replacer (dry basis), or in an equivalent amount of whole milk, 20 to 40 ppm; and in the starter ration, 10 to 20 ppm. For the prevention and control of disease, higher concentrations may be necessary: 50 to 100 ppm in the milk replacer, and 25 to 50 ppm in the starter.

Some studies have shown a slight increase in milk production when low levels of antibiotics are fed to lactating cows. However, this practice is not recommended because (1) of the presence of residual antibiotic in milk, and (2) the possibility of drug resistance in the animal.

Those using antibiotics should always read and follow the label directions on any antibiotic container before slaughtering animals or selling milk from cows treated with antibiotics.

(Also, for an in-depth discussion, see Chapter 13, section on "Antibiotics.")

Bovine Somatotropin (BST)

Experimentally, milk, lactose, milk fat, and protein yields have been increased significantly when exogenous bovine somatotropin (BST) has been injected into lactating cows. To meet this higher production, cows consume more total feed when BST is administered. However, the efficiency of milk production by cows is improved because a smaller proportion of the ingested nutrients in feed is needed to take care of maintenance requirements.

(Also, for an in-depth discussion, see Chapter 13, section on "Bovine Somatotropin [BST] For Dairy Cattle.")

Buffers (Mineral Salts)

Buffers are used primarily to improve the feed intake, rumen function, milk production, milk composition, and health of lactating cows. When used for young calves, buffers have given inconsistent results; they will likely be most beneficial when the calf diet being fed results in higher than normal acidity.

The common buffers are: sodium bicarbonate ($NaHCO_3$), magnesium oxide (MgO), sodium bentonite, sodium sesquicarbonate, and calcium carbonate or limestone ($CaCO_3$). Buffers function to maintain the hydrogen ion concentration in the rumen, intestines, tissues, and body fluids, or to increase the rate of passage of liquids from the rumen, or both.

Buffers are of greatest benefit to cows in the following situations: (1) during early lactation; (2) when large amounts of rapidly fermentable carbohydrates are fed, especially when they are fed at infrequent intervals; (3) when fermented forage, primarily corn silage, is the major or only forage in the ration; (4) when the concentrate and forage are fed separately; (5) when the particle size of the total ration dry matter has been reduced by chopping, grinding, or pelleting to the extent that it increases the rate of ruminal fermentation and depresses salivary secretion and buffering capacity; (6) when cows are abruptly switched from high-forage to high-concentrate rations, especially during early lactation; (7) when the animal's milk fat content is low; or (8) when off-feed problems resulting from feeding rapidly fermentable feeds are encountered.

(Also, for an in-depth discussion of buffers, see Chapter 13, section on "Buffers.")

Ionophores

Ionophores are feed additives that change the metabolism within the rumen by altering the rumen microflora to favor propionic acid production. Currently, two ionophores—Bovatec (lasalocid) and Rumensin (monensin)—are FDA-approved for replacement heifers. Both are antibiotics. Feeding Bovatec or Rumensin to replacement heifers improves liveweight gains and the efficiency of feed utilization.

(Also, for an in-depth discussion of ionophores, including approved use levels for replacement heifers, see Chapter 13, section on "Ionophores.")

Isoacids (Branched-Chain Fatty Acids and Valeric Acid)

The isoacids provide three branched-chain fatty acids (isobutyric acid, isovaleric acid, and 2-methyl butyric acid)—the same fatty acids that are made by ruminant bacteria and are present naturally in the rumen of cattle. Isoacids are essential for the growth of some rumen organisms that digest fiber. The use of isoacids may boost milk production by 8 to

10%, or 4 to 6 lb per day, with little or no increase in feed consumption. The mode of action of isoacids is not entirely clear, but it appears to be due to enhancing fiber digestion and acetate production without stimulating insulin secretion. Not all dairy cattle benefit from the use of isoacids. Moreover, the benefits of a profitable response are delayed 30 to 60 days following the initiation of isoacids.

(Also, for an in-depth discussion of isoacids, see Chapter 13, section on "Isoacids [IsoPlus].")

Thyroprotein (Iodinated Casein/Hormone)

Feeding materials with hormonal activity to lactating cows for the purpose of increasing milk production has fascinated scientists for many years, beginning with thyroprotein.

Cows that are given thyroprotein must receive additional feed to produce extra milk. Not all cows will respond to thyroprotein. Because of the problems involved in feeding the product, it is not widely used. Also, since thyroprotein is classified as a drug, milk and butterfat production records of cows fed thyroprotein are not accepted under DHIA (Dairy Herd Improvement Association) rules.

(Also see Chapter 13, section on "Thyroprotein and Goitrogens.")

Commercial Feeds

Commercial dairy feeds are just what the name implies—feeds mixed by commercial manufacturers who specialize in the business.

Several different types of commercial feeds are available for dairy cattle; among them, (1) complete dairy concentrates, (2) dry cow rations, (3) fitting rations, (4) growing or young stock rations, (5) calf starters, (6) milk replacer feeds, (7) protein supplements, and (8) mineral and vitamin premixes.

Dairy producers are major users of U.S. commercial feeds; 22% of the primary feed (mixed feed, to which a premix is sometimes added) is fed to dairy cattle.

Enlightened dairy producers should know how to determine what constitutes the best in commercial feeds for their specific needs. They will not rely solely on how the feed looks and smells.

One of the important roles of commercial feed manufacturers is that of assuring a uniform mix of all ingredients, including those added in very minute amounts. Usually, commercial feed mills offer supplements which can be combined with farm grains, thereby coupling the advantages of both commercial and home mixing.

(Also see Chapter 16, section on "Commercial Feeds.")

Feed Considerations

In addition to being nutritionally complete, the following factors should receive consideration in dairy rations:

1. **Palatability.** Palatability is important. The feeder should avoid mature, moldy, coarse, or weedy hay; finely ground hay or grain; and silages which are moldy, slimy, or too mature. It may be necessary to let animals become accustomed to new feeds before judging their palatability. Some feeds which are unpalatable when first presented to animals become quite palatable after they become accustomed to them.

2. **Variety.** Some variety in the ration is desirable, but palatability and nutritive content of individual feeds are more important than number of ingredients.

3. **Bulk.** Some bulk in the ration is desirable. Cows crave some dry forage and will eat bedding and other coarse material if sufficient fiber is not included in the ration. The proportion of grain to forage should be determined largely by comparative price. But the amount of bulk should not be decreased to so low a level that it will cause the animal to go off feed and/or lower the butterfat content of the milk. This can be helped by processing the concentrate portion of the ration as coarsely as possible.

4. **Laxativeness.** Cows that receive average amounts of legume hay and/or silage seldom become constipated. But grass hay or straw may cause some trouble. Constipation can be corrected by feeding such feeds as alfalfa, wheat bran, linseed meal, or molasses. Lush green forage also has a laxative effect.

5. **Cost.** Cost is important, but net return is even more important; hence, it may well be said that it is net return rather than cost per ton, or per bag, that is most important.

6. **Feeds affecting milk flavor.** Consumers want milk to taste like milk—not like silage, grass, or weeds.

Although feeds are not the only cause of milk flavors, they are major contributors. Feed flavors enter the milk through the digestive system, through the respiratory system, and by direct absorption. Research indicates that most feed flavors are detectable in the milk 20 minutes after the feed is consumed, and that they are usually most pronounced at the end of 2 hours.

Feed flavors that enter the milk through the respiratory system can usually be detected much sooner than those entering through the digestive system. For example, if a cow breathes air reeking with silage odors, these flavors can be detected in the milk almost immediately. Flavors that are directly absorbed by milk are less common, but they appear if the milk is left exposed for a long enough period.

The following control measures are recommended to alleviate feed flavors:

a. **Avoid sudden change to fresh, lush pasture.** Cows should be shifted from winter feeding, or old pasture, to new and lush pastures on a gradual basis. Also, cows should be taken out of such pastures 2 to 3 hours before milking. For the same reasons, freshly cut grass should not be fed immediately before milking.

b. **Control and avoid undesirable weeds.** Many weeds when eaten by cows will impart a strong flavor to milk; among them, are wild onions, skunk cabbage, some members of the mustard family, bitterweed, carrot weed, ragweed, and others. It is easier to get rid of these weeds today than formerly, so they should be eliminated from pasture and hayfields utililzed by milk cows.

c. **Feed silage after milking.** Silage flavor is both common and objectionable. It can be avoided by feeding all silages after milking, never immediately before or during milking. Usually one will be safe if silage is not fed within 2 to 4 hours of milking time, but it's safer to feed it shortly after milking. This permits the flavor-causing material to pass through the cow's digestive system before the next milking.

If cows breathe the odor of silage, it will appear as flavor in the milk. Thus, silage should never be left in the mangers or feed alleys. In fact, it is preferable that it be fed in the corral, and not in the area where the cows are being milked.

Feed Substitutions

Feed substitutions for dairy cattle and beef cattle are similar. These are presented in Chapter 19, Table 19–14, Feed Substitution Table for Cattle.

FEED PREPARATION

Fig. 20–20. Commodity shed with a bin for each feed ingredient for home-mixing. (Courtesy, Maddox Dairy, Riverdale, Calif.)

Most grains for dairy cattle should be processed before feeding, although calves under 6 months of age can be fed whole corn or oats. Fine grinding seems to be the least desirable method of processing, because of the poor palatability of the powdery texture. Coarse grinding is much better, but many producers prefer rolling to grinding. Even rolling can be undesirable if it leaves fine material. Flaking or wet rolling with the use of steam alleviates dust problems, especially when molasses is incorporated. Feed mill operators should make certain that steam rolled grains have been dried to a normal moisture level of 10 to 15% before being weighed, regardless of whether the grain is purchased or of their own processing.

Pelleted concentrates are more compact and less dusty than ground grains. Also, cows will consume pelleted feeds faster than ground or flaked grain. Cows fed pelleted grain may produce slightly more milk, with a slightly lower butterfat content, than those fed unpelleted grain.

Cows produce as well on chopped or ground hay as on long hay. However, finely ground, pelleted hay affects the amount and proportion of volatile fatty acids in the rumen, with the result that the percentage fat content of the milk is lowered. Wafering or cubing, on the other hand, has less of a depressing effect on the fat content of the milk, and will increase intake and maintain or increase the milk production slightly. Both pelleting and wafering lessen the transportation, storage, and handling charges for hay compared to long or baled hay.

(Also see Chapter 14, Feed Processing.)

RATIONS

Dairy producers must put together the available feeds so as to achieve the most profitable production. At its best, developing a dairy ration involves combining the art and the science of feeding. For small herds, individual animal response may be satisfactory. With large commercial herds, the formulating of rations must be more precise, because small costs per cow become large costs when multiplied by many cows. Yet, the most sophisticated computer must be augmented by the good judgment of the manager if the rations are to be successful in meeting the nutrient needs of individual cows and of the herd as a whole. Producers must always keep in mind that the best formula on paper is not always the best feed. A feed is of no value if it is not actually consumed.

Fig. 20–21. Lactating cows eating a complete or total mixed ration, with the forage and concentrate combined. In the late 1980s, 2 in 5 American herds were fed this way. (Courtesy, Arizona Dairy Co., Higley, Ariz.)

Also, there should be a specific ration for every need—for lactating cows, dry cows, calves, replacement heifers, dairy beef, and show and sale animals.

FEEDING LACTATING COWS

Few animal stresses are as great as those involved in the production of a large volume of milk. For each gallon of milk produced, 400 to 500 gal of blood must pass through the udder. Thus, if a cow is producing 10 gal (86 lb) of milk daily, 15 to 20 tons of blood course the udder each 24 hours. This 10 gal of milk contains more than 3 lb of fat, more than 3 lb of protein, more than 4 lb of lactose (milk sugar), and more than ½ lb of minerals. All these must be supplied in the ration over and above the nutrients needed for the body processes, wastes, and energy to sustain all of the operation.

Also, producers realize greatest profits from feeding when cows convert the maximum proportion of their feed into milk. The nutrient requirements for production depend primarily on the amount and composition of the milk. These needs for cows of all sizes and levels of production are shown in Tables 20-1 to 20-5. Rations that fulfill these requirements, plus a margin of safety, can be formulated based on composition of feeds listed in Section V. The primary concern in feeding lactating cows is to provide a ration adequate in energy, protein, fiber and roughage factor, salt, calcium, phosphorus, and vitamin A (or carotene). When allowances for these nutrients are met, other minerals and vitamins usually are present in sufficient amounts, also.

Additional considerations in feeding lactating cows include palatability of the ration; physical form; protein and mineral content of concentrates; proportion of concentrate to roughage; relative prices of ingredients; voluntary feed intake; and frequency and regularity of feeding. Thus, the proper feeding of lactating cows necessitates that producers have sufficient knowledge relative to basic nutrient requirements and principles to plan an efficient feeding program, and the experience and management ability to apply it.

Dry matter consumption is very important in feeding dairy cows. The best ration formulation on paper will not make for profitable production if the cows either fail to eat it or are given insufficient amounts of it. Also, high-producing cows must consume very large amounts of a balanced ration if they are to produce to their maximum. For this reason, producers should adjust the feeding recommendations given in Table 20-1 to fit the needs of their cows, with consideration given to body weight and condition of cows, milk production, stage of lactation, weather and other environmental factors, and type and quality of feed.

Thumb Rules for Feeding Lactating Cows

The feed requirements of lactating cows are significantly influenced by the volume and composition of the milk that they produce. Although knowledge of the nutrient requirements of the animals and of the composition of feeds is essential in order to feed properly, the ability of the cows to consume sufficient volume of feed complicates adequate feeding. Table 20-8 may be used as a guide for dry matter intake; and the two sections that follow give some thumb rules relative to the amount and kind of forage and the amount and kind of concentrate to feed.

TABLE 20-8
DAILY DRY MATTER INTAKE GUIDELINES[1]

Live Wt.: (lb) *(kg)*		900 *409*	1,100 *499*	1,200 *545*	1,300 *590*	1,500 *681*
Milk[2]		Percent of Body Weight[3]				
(lb/day)	*(kg/day)*	(%)	(%)	(%)	(%)	(%)
20	*9.1*	2.6	2.3	2.2	2.1	2.0
30	*13.6*	3.0	2.7	2.6	2.5	2.3
40	*18.2*	3.4	3.1	2.9	2.8	2.5
50	*22.7*	3.8	3.4	3.2	3.1	2.8
60	*27.2*	4.1	3.7	3.5	3.4	3.1
70	*31.8*	4.6	4.0	3.8	3.6	3.3
80	*36.3*	5.1	4.3	4.1	3.8	3.5
90	*40.9*		4.7	4.4	4.1	3.7
100	*45.4*		5.0	4.7	4.4	3.9

[1]Adapted by the authors from: Linn, J. G., M. F. Hutjens, W. T. Howard, L. H. Kilmer, and D. E. Otterby, *Feeding the Dairy Herd,* Cooperative Extension Services, Universities of Illinois, Iowa State, Minnesota, and Wisconsin, 1988, p.27, Table 19.

[2]Fat-corrected milk = (milk lb × .4) + (fat lb × 15).

[3]Intakes may be up to 18% less for cows in early lactation.

AMOUNT AND KIND OF FORAGE TO FEED

The common thumb rules for forage feeding of lactating cows follow.

1. **Forage dry matter and intake.** The forage should constitute a minimum of 40% of the total dry matter of the ration and account for an intake of approximately 1.5% of the body weight daily.
2. **Acid detergent fiber (ADF).** The ADF should constitute 19% of the ration dry matter, increased to 21% during the first 3 weeks of lactation.
3. **Neutral detergent fiber (NDF).** The NDF should constitute 25% of the ration dry matter, increased to 28% during the first 3 weeks of lactation.
4. **Hay consumption.** If good quality hay only is fed, a cow will eat about 3 lb per 100 lb of body weight.
5. **Silage.** Depending on the moisture content, 2.5 to 4.5 lb of silage are equal to (and may replace) 1 lb of hay; the lower feeding value of silage is due to its high moisture content—hay runs 10 to 15% moisture, whereas silage runs 65 to 75% moisture.
6. **Hay/grain equivalent.** It takes about 3 lb of good hay to supply the same amount of usable energy as 2 lb of grain.
7. **Pasture (grass) consumption.** Cows will consume 100 to 200 lb of pasture per day; since pasture normally contains 70 to 85% moisture, that's 15 to 60 lb of dry matter per day.
8. **Yearly hay consumption.** Except for cows fed high grain rations, it takes 5 to 6 tons of hay (or an equivalent amount in dry matter from pasture or silage) to feed 1 cow for 1 year.
9. **Forage:concentrate ratio.** If forage is very high quality, cows will eat more of it, with the result that the grain requirement will be lessened. However, over and above meeting the minimum forage requirement, the proportion of forage to concentrate should be determined primarily by the economics of the situation—that is, it should be decided on the basis of the relative price of available forage and concentrate, the milk production, and the net returns.

Fig. 20-24. Barn feeding a complete ration to lactating cows. (Courtesy, USDA)

Fig. 20-25. Corral feeding a complete ration to lactating cows. (Courtesy, James Tappan, Higley, Ariz.)

Since proteins are always the most expensive part of a ration, for practical reasons the dairy producer should not feed more of them than necessary. Nevertheless, the protein level should be in keeping with the production—the higher the yield, the higher the protein requirement. Generally speaking, this is best accomplished by varying the protein content of the grain mix according to the protein content of the roughage being fed.

The concentrate ration needed to supplement the available roughage on the dairy farm may be either home mixed or commercially manufactured. Home-mixing involves the mixing of homegrown grains and such purchased feeds as necessary to balance the ration. Commercially manufactured feeds are generally nutritionally complete, concentrate feeds. On small dairies where grain is home-grown or abundantly available locally, home mixing of concentrates is a widely used practice. However, there has been an increase in the use of commercial feeds on small dairies. Large specialized dairies purchase individual feed ingredients in bulk, then mix and feed total mixed rations.

Not all dairy producers, in the United States or in other countries, who home-mix feeds balance rations (1) on the basis of the chemical analyses of their feed ingredients, (2) with the use of computers, or (3) by combining the concentrates and forage into a complete ration. For these producers, Table 20-12, Feeding Guide For Lactating Cows, may serve as a useful guide. It shows how ingredients partitioned into four approximate protein levels (columns 2, 3, 4, and 5) may be combined to make concentrates suitable for feeding with three different qualities of roughages—excellent, medium, and poor.

Variations can and should be made in the rations listed in Table 20-12. Producers should give consideration to the supply of homegrown feeds, and to the availability and price of ingredients. Feeds of similar nutritive properties can and should be interchanged as price relationships warrant. Thus, the cereal grains may consist of corn, barley, wheat, oats, and/or sorghum; the protein supplements may consist of soybean, cottonseed, peanut and/or linseed meal; and a vast array of by-product feeds may be utilized.

Here is how to use Table 20-12: Let's assume that a producer has (1) medium-quality forage, and (2) both low- and medium-high protein (columns 2 and 4) ingredients from which to choose. How many pounds each of the low- and medium-high protein ingredients will be required in a 1,000-lb concentrate mix? Step by step, here is the answer:

1. Look under "Medium roughage—medium protein forage" 15-17% (column to the left).

2. Mix No. 6, containing 650 lb of low-protein ingredients (under column 2, under 12% ingredients) and 350 lb of medium-high protein ingredients (column 4, 18 to 28% protein), will meet the needs. The concentrates may be chosen from among those listed at the top of the respective columns of Table 20-12—the low-protein concentrates from column 2 (under 12%) and the medium-high protein concentrates from column 4 (18 to 28%).

TABLE 20-12
FEEDING GUIDE FOR LACTATING COWS (AS-FED BASIS)[1]

Note: *This shows how ingredients of 4 protein levels may be combined to make different concentrate mixes of approximate protein content to match 3 different qualities of roughages.*

(1) Suggested Grain Mix, Based on Kind of Roughage Available	(2) Low-Protein (Under 12%) Ingredients		(3) Low-Medium Protein (12–18%) Ingredients		(4) Medium-High Protein (18–28%) Ingredients		(5) High-Protein (Over 32%) Ingredients	
	(% protein)		(% protein)		(% protein)		(%protein)	
Feeds	Barley, all analyses ...... 11.7 Beet pulp w/molasses, dried . 9.3 Corn-and-cob meal .... 7.8 Corn #2 .... 8.9 Dairy feed, 12% .... 12.0 Hominy feed .... 10.3 Molasses, cane* .... 4.3 Oats, all analyses .... 11.9 Rye, all analyses* .... 12.0 Sorghum (milo) .... 10.1 Wheat, all analyses 13.1		Dairy feed, 16% .... 16.0 Wheat bran .... 15.5 Wheat middlings 16.4		Brewers' dried grains* .... 27.3 Copra (coconut) meal ..... 21.3 Corn gluten feed .... 23.0 Dairy feed, 18–24% .... 18–24 Distillers' dried grains ..... 27.3 Malt sprouts .... 22.9 Peas, field* 23.2		Dairy feed, 32–34% .. 32–24 Corn gluten meal ..... 60.8 Cottonseed meal* ..... 41.2 Linseed meal .... 35.7 Peanut meal .... 49.0 Soybean meal 44.4	
	(lb)	*(kg)*	(lb)	*(kg)*	(lb)	*(kg)*	(lb)	*(kg)*
Excellent roughage—High-protein forage, 18%: (1) legume, or (2) legume and nonlegume mixed forages of *high quality;* consisting of dry forages and/or silage.								
Mix No. 1	1,000	*454*						
Mix No. 2	900	*409*					100	*45*
Mix No. 3	800	*363*			200	*91*		
Mix No. 4	850	*386*	100	*45*			50	*23*
Medium roughage—Medium-protein forage, 15–17%: (1) legume, or (2) legume and nonlegume mixed forages of *medium quality;* consisting of dry forages and/or silage.								
Mix No. 5	800	363					200	*91*
Mix No. 6	650	*295*			350	*159*		
Mix No. 7	700	*318*	100	*45*	100	*45*	100	*45*
Mix No. 8	Straight 16% dairy feed, or ½ Mix No. 9 and ½ 16% dairy feed							
Poor roughage—Low-protein forage, under 14%: nonlegume forage; consisting of dry forages and/or silage.								
Mix No. 9	700	*318*	300	*136*				
Mix No. 10	600	*272*			200	*91*	200	*91*
Mix No. 11	600	*272*	100	*45*	100	*45*	200	*91*
Mix No. 12	500	*227*	and 500 lb *(227 kg)* 32% dairy feed					

[1]The protein compositions in columns 2 to 5 were obtained from Section V, Composition of Feeds.

Comments:

Add—To all rations (1) 1% iodized or trace-mineralized salt; (2) 1% steamed bone meal, dicalcium phosphate, or the equivalent (use monosodium phosphate or a high-phosphorus commercial mineral where alfalfa is fed liberally); (3) 1,000 IU of vitamin A/lb *(2,205 IU of vitamin A/kg)* of concentrate and, unless cows are in sunlight, add 150 IU of vitamin D/lb *(331 IU of vitamin D/kg)* of concentrate.

***Limitations**—Wheat, not more than 50% of the ration; dried molasses beet pulp, 20%; molasses, 15%; peas and brewers' dried grains, 30%; rye, 10%; and cottonseed meal, 20% of the mix for calves, but as needed for mature cows.

HOW TO BALANCE A DAIRY RATION

Good dairy producers have nutrient analyses made of all major ration ingredients and use computers to balance their rations for at least 20 nutrients.

It is recognized that rations should vary with conditions, and that many times they should be formulated to meet the conditions of a specific dairy farm. Also, good producers should know how to balance rations. Complete instructions on how to balance rations (including examples) are given in Chapter 18, Feeding Standards/Ration Formulation. By (1) following these instructions, and (2) using the nutrient requirement tables, Tables 20-1 to 20-5, it is possible to balance rations for specific weights of animals and levels of production.

COMPUTER-FORMULATED RATIONS/DISKETTE[6]

Until the late 1970s, only those dairy producers with access to a large main-frame computer could formulate a ration using the computer. Usually this was limited to those associated with a university (Extension Service) or subscribing to a time-sharing system. Many rations were formulated by using a pencil, an eraser, paper, and a calculator. Balancing rations was time-consuming, and options were limited. Then came the microcomputer! By the mid–1980s, most dairy producers owned or had access to a microcomputer. Rations could be, and were, adjusted with changes in availability, price, composition, and moisture content of ingredients.

In the past, computer programs were written by software companies and universities to convert the tabular NRC requirements into equations that computed the animal requirements. With the publication of *Nutrient Requirements of Dairy Cattle,* sixth revised edition, a new step was taken by the NRC committee. In addition to providing nutrient requirements for dairy cattle, as reproduced in this chapter in Tables 20–1 to 20–5, the committee published prediction equations. These equations may easily be translated into computer programs by programmers. Nutrient requirements for growing, lactating, or dry cows are generated by these equations.

Also, the NRC committee commissioned a software company, Microsoft® corporation, to develop a FORTRAN program that is supplied with the dairy cattle publication in the form of a 5¼ in. diskette. This program will calculate nutrient requirements from the prediction equations and from specific information supplied by the producer or nutritionist. The results of the calculations may either (1) be displayed on the computer screen, or (2) be printed. The printout is in two parts: The first part lists the requirements for the major nutrients and for vitamins A and D and is useful to the producer or nutritionist formulating rations. The second part is a very involved breakdown predicting utilization of dietary protein, information of value primarily to those who conduct research. Fig. 20–26 is a sample printout of the first page generated by this program.

```
NRC DAIRY (1988) REQUIREMENTS CALCULATED ON  3- 8-1989 AT 12:58
PREGNANT OR LACTATING CATTLE
ENERGY CONCENTRATION FED/NRC ASSUMED IS      1.000
LIVE WEIGHT IN LB IS                      1400.
MILK PRODUCTION IN KG IS                    75.
MILK FAT TEST % IS                           3.65
NUMBER OF DAYS PREGNANT IS                   0.
LACTATION NUMBER IS                          3
PROPORTIONAL FEED NEL/REQUIRED NEL IS        1.00
WEIGHT CHANGE IN LACTATION IS                 .750 LB
DRY MATTER INTAKE IS                        48.39  LB    OR     3.46   % LW
NEL NEEDED IS                               35.76  MCAL  OR      .74   MCAL/LB
ME NEEDED IS                                59.89  MCAL  OR     1.24   MCAL/LB
DE NEEDED IS                                69.11  MCAL  OR     1.43   MCAL/LB
BASELINE TDN NEEDED IS                      34.56  LB    OR    71.41   % DM
CRUDE PROTEIN INTAKE NEEDED IS               7.902 LB    OR    16.33   % DM
UNDEGRADED INTAKE PROTEIN NEEDED IS          2.761 LB    OR     5.70   % DM
DEGRADED INTAKE PROTEIN NEEDED IS            4.678 LB    OR     9.67   % DM
INTAKE PROTEIN (IP) NEEDED IS                7.439 LB    OR    15.37   % DM
CALCIUM NEEDED IS                             .285 LB    OR      .589  % DM
PHOSPHORUS NEEDED IS                          .181 LB    OR      .373  % DM
VITAMIN A NEEDED IS                       48263.     IU    OR   997.      IU/LB
VITAMIN D NEEDED IS                       19051.     IU    OR   394.      IU/LB
UNDEGRADED INTAKE PROTEIN IN IP IS                              37.11   % IP
```

Fig. 20–26. NRC dairy requirements calculated by using the data diskette supplied with *Nutrient Requirements of Dairy Cattle,* sixth revised edition, 1988.

The program runs on IBM-PC™ or PC-compatible computers. Various computer software companies have written the equations into their dairy ration formulation software, making the procedure of ration formulation more rapid and convenient. Fig. 20–27 is a computer printout of a ration for lactating cows formulated with the use of the NRC prediction equations.

[6]This section on "Computer-Formulated Rations/Diskette" was prepared specially for this book by L. M. Larsen, Ph.D., Consultant, Nutri-Systems, 426 E. Shields, Fresno, CA 93704.

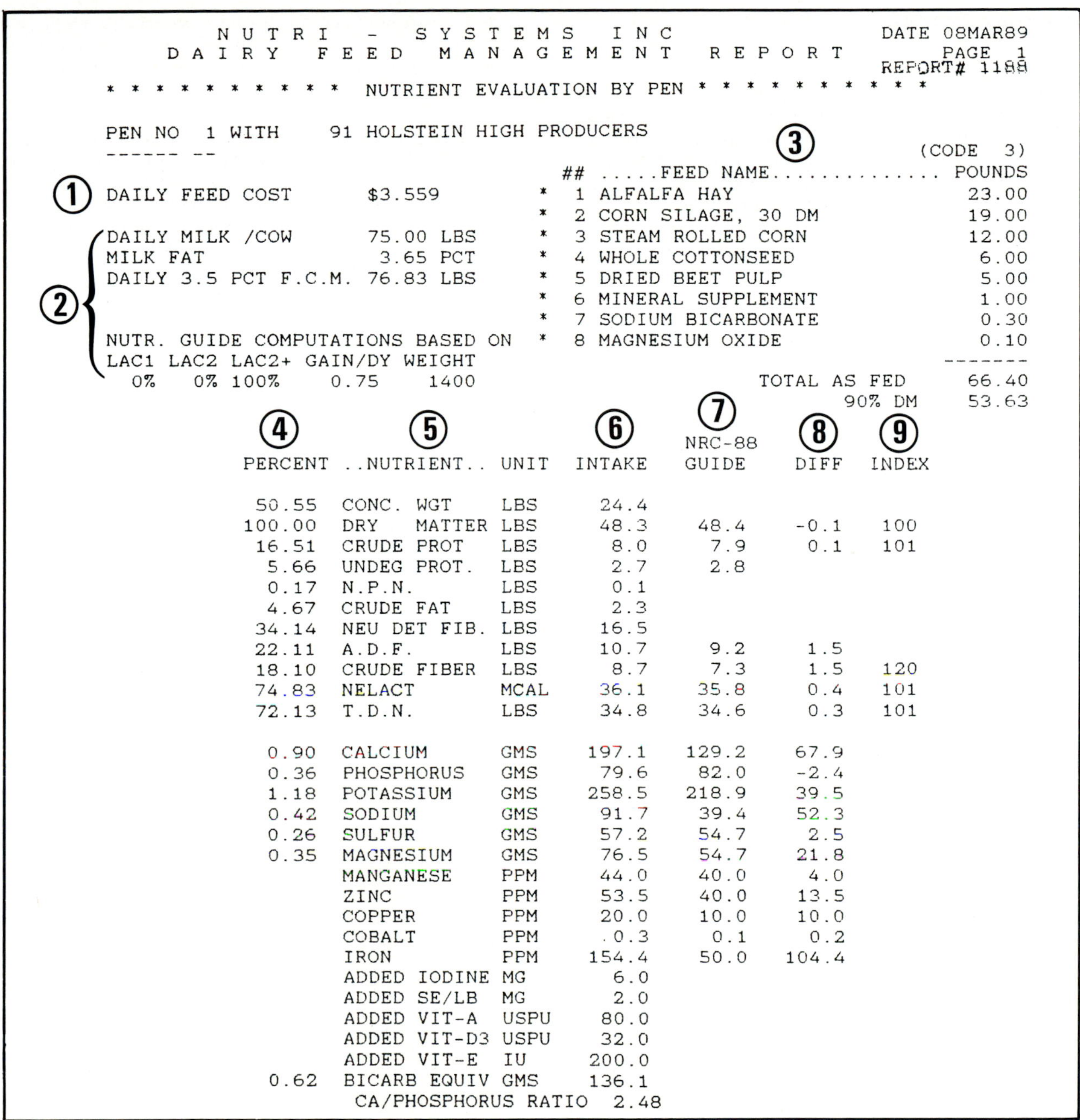

N U T R I - S Y S T E M S I N C

D A I R Y F E E D M A N A G E M E N T R E P O R T

DATE 08MAR89

PAGE 1

REPORT# 1188

* * * * * * * * * * NUTRIENT EVALUATION BY PEN * * * * * * * * * * *

PEN NO 1 WITH 91 HOLSTEIN HIGH PRODUCERS

------ --

(CODE 3)

① DAILY FEED COST $3.559

② DAILY MILK /COW 75.00 LBS

MILK FAT 3.65 PCT

DAILY 3.5 PCT F.C.M. 76.83 LBS

NUTR. GUIDE COMPUTATIONS BASED ON

LAC1	LAC2	LAC2+	GAIN/DY	WEIGHT
0%	0%	100%	0.75	1400

③

	##	.....FEED NAME.............	POUNDS
*	1	ALFALFA HAY	23.00
*	2	CORN SILAGE, 30 DM	19.00
*	3	STEAM ROLLED CORN	12.00
*	4	WHOLE COTTONSEED	6.00
*	5	DRIED BEET PULP	5.00
*	6	MINERAL SUPPLEMENT	1.00
*	7	SODIUM BICARBONATE	0.30
*	8	MAGNESIUM OXIDE	0.10

		TOTAL AS FED	66.40
		90% DM	53.63

④ PERCENT	⑤ ..NUTRIENT..	UNIT	⑥ INTAKE	⑦ NRC-88 GUIDE	⑧ DIFF	⑨ INDEX
50.55	CONC. WGT	LBS	24.4			
100.00	DRY MATTER	LBS	48.3	48.4	-0.1	100
16.51	CRUDE PROT	LBS	8.0	7.9	0.1	101
5.66	UNDEG PROT.	LBS	2.7	2.8		
0.17	N.P.N.	LBS	0.1			
4.67	CRUDE FAT	LBS	2.3			
34.14	NEU DET FIB.	LBS	16.5			
22.11	A.D.F.	LBS	10.7	9.2	1.5	
18.10	CRUDE FIBER	LBS	8.7	7.3	1.5	120
74.83	NELACT	MCAL	36.1	35.8	0.4	101
72.13	T.D.N.	LBS	34.8	34.6	0.3	101
0.90	CALCIUM	GMS	197.1	129.2	67.9	
0.36	PHOSPHORUS	GMS	79.6	82.0	-2.4	
1.18	POTASSIUM	GMS	258.5	218.9	39.5	
0.42	SODIUM	GMS	91.7	39.4	52.3	
0.26	SULFUR	GMS	57.2	54.7	2.5	
0.35	MAGNESIUM	GMS	76.5	54.7	21.8	
	MANGANESE	PPM	44.0	40.0	4.0	
	ZINC	PPM	53.5	40.0	13.5	
	COPPER	PPM	20.0	10.0	10.0	
	COBALT	PPM	0.3	0.1	0.2	
	IRON	PPM	154.4	50.0	104.4	
	ADDED IODINE	MG	6.0			
	ADDED SE/LB	MG	2.0			
	ADDED VIT-A	USPU	80.0			
	ADDED VIT-D3	USPU	32.0			
	ADDED VIT-E	IU	200.0			
0.62	BICARB EQUIV	GMS	136.1			

CA/PHOSPHORUS RATIO 2.48

Fig. 20-27. Lactating dairy ration comparing nutrient intake (column 6) with the NRC requirement (column 7).

An explanation of the numbered sections (columns) of Fig. 20-27 follows:

1. Feed cost per cow per day.
2. Input data used for computing the requirements.
3. Feedstuffs amounts given in pounds per cow per day.
4. Nutrient composition of the ration on a 100% dry matter basis.
5. Nutrient names and units.
6. Daily intake of nutrients per cow per day.
7. The NRC requirements computed from the prediction equations.
8. Deficiencies (negative values) or excesses (positive values). These values are computed by subtracting the requirements from the daily intake figures.
9. A computed index for dry matter, crude protein, NE_{lc}, and TDN. This value is computed as the percentage that nutrient intake is of the nutrient requirement.

(For a more detailed discussion of computer formulation of rations, see Chapter 18, section on "Computer Methods.")

Feeding Systems

Traditional individual feeding of lactating cows in stanchioned barns or milking parlors is giving way to new feeding systems. Although the newer methods are not as effective as feeding cows individually, they are much more economical than feeding all cows in the herd the same amount of grain, regardless of production. Additionally, they make for considerable saving in labor and facilities.

PHASE FEEDING

Phase feeding is a feeding program that is divided into periods based on milk production, milk fat percentage, feed intake, and body weight. Fig. 20–28 illustrates the shape and relationship of curves for milk production, fat percentage, dry matter intake, and body weight. Based on these curves, four distinct feeding phases of lactating cows can be identified.

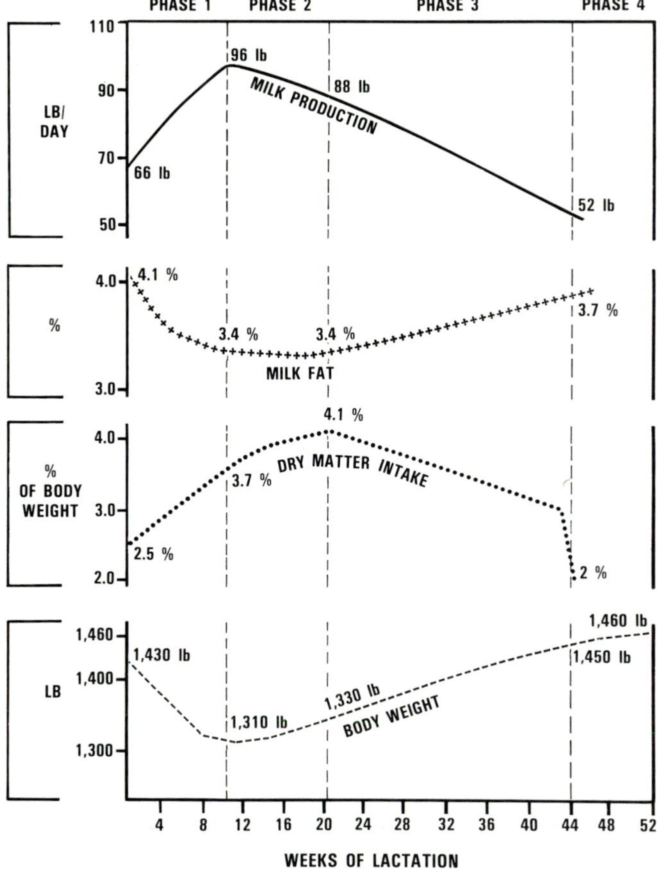

Fig. 20–28. Lactation cycle phases with corresponding changes in milk production, milk fat percentage, dry matter intake, and body weight. (*Source:* Linn, J. G., M. F. Hutjens, W. T. Howard, L. H. Kilmer, and D. E. Otterby, *Feeding the Dairy Herd,* Cooperative Ext. Services, Ill., Iowa State, Minn., and Wisc., 1988, p. 15, Fig. 6)

Producers should formulate rations to match each of these phases in order to optimize milk yield, minimize metabolic disorders, increase longevity, and increase profits. The four phases are:

1. **Phase 1, early lactation, 0 to 70 days postpartum.** During this period, milk production increases rapidly, peaking at 4 to 6 weeks after calving. But feed intake does not keep pace with nutrient needs (especially energy needs) for milk production, so body tissues are mobilized to meet these needs. During this phase, adjusting the cow to the milking ration is an important management practice. After calving, the grain should be increased by 1 to 1.5 lb daily to meet increased nutrient demands and minimize off-feed problems and acidosis. Excessive proportions of grain (over 60% of the total dry matter) can cause acidosis and low milk fat percentage. The fiber level in the total ration should not be less than 18% ADF, 28% NDF; and the forage should provide at least 21 percentage units of the NDF in the total ration. The physical form of the fiber is also important; normal rumination and digestion will be maintained if more than 50% of the forage is 1 in. in length or longer.

Protein content is critical during early lactation. Meeting or exceeding the crude protein requirements during this period helps to stimulate feed intake and permits efficient use of mobilized body tissue for milk production. Rations may need to contain 19% or more crude protein to meet the requirements during this phase. The type of protein (degradable and undegradable) and the amount of protein to feed will depend on ration ingredients, method of feeding, and milk production potential of the cow. A guideline followed by many producers is to feed 1 lb of soybean meal or equivalent protein supplement per 10 lb of milk above 50 lb. If urea is fed, it is best fed well mixed with corn silage or as part of the grain mix, and it should be limited to a maximum of 0.2 lb per cow daily.

When early lactation nutrient needs are not met, low peak production and ketosis may result. Low peak production presages low lactation production. If the grain intake is increased too rapidly or to too high a level, it may result in off-feed, acidosis, and displaced abomasum. To increase nutrient intake, (1) feed high quality forage, (2) include adequate protein in the ration, (3) increase grain intake at a constant rate after calving, (4) add 1 to 1.5 lb of fat/cow/day to the ration, (5) allow constant access to feed, and (6) minimize stresses.

2. **Phase 2, peak dry matter intake, second 10 weeks postpartum.** During this phase, cows should be fed to maintain peak milk production as long as possible. Feed intake is near maximum and can supply nutrient needs. Cows should be maintaining weight or making slight gains (see Fig. 20–28).

Grain intake may reach, but should not exceed, 2.3% of the cow's body weight (dry matter basis). High quality forage should be provided, with a minimum intake of 1.5% of the cow's body weight (dry matter basis) in order to maintain rumen function and normal fat test.

Potential problems during phase 2 include rapid drop in milk production, low fat test, silent heat periods (estrus not observed), and ketosis.

To maximize nutrient intake (1) feed forage and grain 3 or more times daily, (2) feed high quality feeds, (3) limit urea to 0.2 lb/cow/day, (4) minimize stress, and (5) use a total mixed ration.

3. **Phase 3, mid- to late-lactation, 140 to 305 days postpartum.** This is the easiest phase to manage. During this period, milk production is declining, the cow is pregnant, and nutrient intake will easily meet or exceed requirements. The level of grain feeding should be adequate to meet production requirements, and to begin to replace body weight lost during early lactation. Lactating cows require less feed to replace a pound of body tissue than dry cows; hence, it is more efficient to have cows gain body weight near the end of lactation than during the dry period. Young cows should receive additional nutrients for growth; 2-year-old heifers should receive 20% more than for maintenance, and 3-year-olds should receive 10% more.

• **Phases 1, 2, and 3 guidelines/rations**—The following guidelines and example rations, which meet the nutrient requirements given in Table 20–5, may be used by dairy producers in developing a feeding program for lactating cows suitable for phases 1, 2, and 3:

1. **Net energy-lactation (NE_l).** A minimum of 0.76 Mcal per pound of DM in early lactation, decreasing to 0.69 Mcal per pound in late lactation and 0.57 Mcal per pound during the dry period.

2. **Crude protein.** A minimum of 19% (DM basis) in early lactation, decreasing to 13% in late lactation and 12% during the dry period. Urea to be limited to a maximum of 0.4 lb per day or 1% of the grain mix.

3. **Forage dry matter.** A minimum of 1.5 lb per 100 lb body weight. High quality legume forage to be fed during early lactation.

4. **Fiber.** A minimum of 17% ADF in the DM during early lactation, increasing to 22% during the dry period. A minimum of 28% NDF during early lactation, increasing to 35% during the dry period.

5. **Minerals.** Include 0.5 to 1.0% trace mineralized salt in the grain mix; and approximately 1% calcium-phosphorus in the grain mix.

6. **Vitamins.** Incorporate in the ration sufficient vitamins A, D, and E to meet the requirements (see Table 20–5).

7. **Ration processing.** Forages and grains should not be chopped or ground too fine.

8. **Ration formulating.** Balance the ration to meet all requirements and feed it as a total mixed ration.

The example rations presented in Table 20–13 are suitable for the three phases.

TABLE 20–13
EXAMPLE RATIONS
FOR VARIOUS MILK PRODUCTION PHASES,
1,350 LB *(613 KG)* COW, 3.8% FAT TEST[1]

Item	Phase 1		Phase 2		Phase 3	
Milk (lb/day)	90		80		50	
Milk *(kg/day)*	*40.9*		*36.3*		*22.7*	
DM intake[2] (lb/day)	49		51		38	
DM intake[2] *(kg/day)*	*22.2*		*23.2*		*17.3*	
	As-Fed					
	(lb/day)	***(kg/day)***	**(lb/day)**	***(kg/day)***	**(lb/day)**	***(kg/day)***
Ration 1						
Alfalfa hay (88% DM), 140 RFV, 20% crude protein	28	*12.71*	34	*15.44*	27	*12.26*
Corn-oats[3]	21	*9.53*	24	*10.90*	16	*7.26*
Soybean meal, 44%	5.0	*2.27*				
Dical, 18% phosphorus	0.5	*0.23*	0.45	*0.20*	0.30	*0.14*
Salt, vitamins, trace mineralized	0.30	*0.14*	0.25	*0.11*	0.25	*0.11*
Weight change	−1.5	*−0.68*	—	—	+ .5	*+.23*
Ration 2 (corn silage limit fed)						
Alfalfa hay, 140 RFV, 20% CP	19	*8.63*	34	*15.44*	23	*10.44*
Corn silage (35% DM)	25	*11.35*	25	*11.35*	25	*11.35*
Corn-oats	18	*8.17*	12	*5.45*	10	*4.54*
Soybean meal, 44%	7.5	*3.41*	0.3	*0.14*	—	—
Dical, 18% phosphrous	0.45	*0.20*	0.50	*0.23*	0.3	*0.14*
Salt, vitamins, trace mineralized	0.30	*0.14*	0.25	*0.11*	0.25	*0.11*
Weight change	−1.2	*−0.54*	—	—	+ .5	*+.23*
Ration 3 (hay limit fed)[4]						
Alf-grass hay, 113 RFV, 16% CP	10	*4.54*	10	*4.54*	10	*4.54*
Corn silage	41	*18.61*	70	*31.78*	57	*25.88*
Corn-oats	16	*7.26*	11	*4.99*	6	*2.72*
Soybean meal, 44%	11.5	*5.22*	8.2	*3.72*	4.5	*2.04*
Dical, 18% phosphorus	0.40	*0.18*	0.30	*0.14*	0.25	*0.11*
Limestone	0.40	*0.18*	0.30	*0.14*	0.15	*0.07*
Salt, vitamins, trace mineralized	0.30	*0.14*	0.25	*0.11*	0.25	*0.11*
Weight change	−1.4	*−0.64*	+ .7	*+.32*	+ .5	*+.23*
Ration 4						
Alf-grass hay, 113 RFV, 16% CP	23	*10.44*	32	*14.53*	24	*10.90*
Corn-oats	22	*9.99*	22	*9.99*	19	*8.63*
Soybean meal, 44%	8.5	*3.86*	3.5	*1.59*	1.1	*0.50*
Dical, 18% phosphorus	0.45	*0.20*	0.40	*0.18*	0.25	*0.11*
Limestone	0.20	*0.09*				
Salt, vitamins, trace mineralized	0.30	*0.14*	0.25	*0.11*	0.25	*0.11*
Weight change	−1.9	*−0.86*	—	—	+ .5	*+.23*

[1]*Source:* Linn, J. G., M. F. Hutjens, W. T. Howard, L. H. Kilmer, and D. E. Otterby, *Feeding the Dairy Herd,* Cooperative Ext. Services, Ill., Iowa State, Minn., and Wisc., 1988, p. 16, Table 6.

[2]Estimated average intake during the phase.

[3]85% corn–15% oats mix.

[4]Feed amounts may have to be limited during phase 2 and 3 to avoid over-conditioning.

4. **Phase 4, dry period, 45 to 60 days before parturition.** The dry phase is important. A good dry cow program can minimize metabolic problems at or immediately following calving and increase milk yield during the subsequent lactation.

Dry cows should be fed separately from lactating cows. Their ration should be formulated to meet their specific needs: body maintenance, fetal growth, and any additional body weight not replaced during phase 3. Daily dry matter intake should be near 2% of the cow's bodyweight; with forage intake a minimum of 1% of the body weight, and grain intake according to needs, but not to exceed 1% of body weight. One-half of 1% of the body weight in grain per day is usually sufficient in dry cow feeding programs. Dry cows should not be too fat. Feeding low quality forage, such as corn stalks or grass hay, is preferred to limit feeding. A protein level of 12% is sufficient during the dry period.

Some grain should be included in the dry cow ration starting 2 weeks before calving in order (1) to change the rumen bacteria from an all-forage digestion population to a mixed population of forage and grain digesters; and (2) to minimize the stress of ration changes following calving.

The calcium and phosphorus needs of dry cows should be met, but excesses should be avoided; sometimes, rations containing more than 0.6% calcium and 0.4% phosphorus have increased milk fever problems substantially. Trace minerals, including selenium, should be supplied in most dry cow rations. Also, adequate amounts of vitamins A, D, and E should be provided in the ration to lower milk fever, lessen retained placenta, and increase calf survival.

Potential problems during phase 4 include milk fever, displaced abomasum, retained placenta, fatty liver syndrome, and poor appetite, along with other metabolic disorders and diseases associated with the fat cow syndrome.

Key management precautions during the dry period include: (1) observing body condition of dry cows and adjusting energy feeding as necessary, (2) meeting nutrient requirements but avoiding excessive feeding, (3) changing ration 2 weeks before calving to include grain and other small quantities of ingredients used in the lactating ration, (4) avoiding excess calcium and phosphorus intakes, and (5) limiting salt and other sodium-based minerals in the dry cow ration to reduce udder edema problems.

Table 20–14 lists examples of dry cow rations.

TABLE 20–14
EXAMPLE DRY COW RATIONS, 1,400 LB *(636 KG)* DRY COW[1]

Forage	As-Fed	
	(lb/day)	*(kg/day)*
Grass forage		
Orchard grass hay, 12% crude protein	25.0	*11.35*
Corn	3.0	*1.36*
Soybean meal	0.5	*0.23*
Limestone	0.15	*0.07*
Trace mineralized salt and vitamins	0.1	*0.05*
Limited legume forage[2]		
Alfalfa hay, RFV 140, 20% crude protein	12.0	*5.45*
Corn silage	43.0	*19.52*
Monosodium phosphate	0.1	*0.05*
Trace mineralized salt and vitamins	0.1	*0.05*
Limited corn silage		
Alfalfa-grass hay, RFV 113, 16% crude protein	21	*9.53*
Corn silage	20	*9.08*
Dicalcium phosphate	0.1	*0.05*
Trace mineralized salt and vitamins	0.1	*0.05*

[1]*Source:* Linn, J. G., M. F. Hutjens, W. T. Howard, L. H. Kilmer, and D. E. Otterby, *Feeding the Dairy Herd,* Cooperative Ext. Services, Ill., Iowa State, Minn., and Wisc., 1988, p. 17, Table 7.

[2]Ration contains excess energy as formulated and may over-condition cows in some situations.

CHALLENGE FEEDING (LEAD FEEDING)

Challenge feeding, or lead feeding, refers to feeding the lactating cow so that she is challenged to reach her peak (summit) production level early in lactation.

Because of the strong relationship between peak (summit) milk yield and the total milk production for the entire lactation period, emphasis should be placed on attaining maximum yield between weeks 3 and 8. According to figures provided by Iowa State University extension dairy specialist Ron Orth, for every 5 lb increase in the summit peak, rolling herd average will increase 1,000 lb.

Preparation for challenge feeding begins during the dry cow period; (1) by having the dry cow in proper condition; and (2) by making the transition from the dry cow ration to the lactating ration, thereby preparing the rumen bacteria.

After calving, challenge feeding calls for increasing the grain allowance several pounds per day above the cow's exact requirements at the time. The objective is to allow each cow to reach peak production at or near her genetic potential. The advantage of achieving maximum peak production will be maintained throughout the remaining months of lactation.

Calving time is a very traumatic experience for a high-producing cow. As a result, many cows have a depressed appetite for several days after calving. However, a skilled caretaker can get most of these cows on full feed very quickly without changing the ration markedly. But a heavy producer cannot consume sufficient energy to balance her energy output. Consequently, she will be forced to rely on body stores of fat and protein to supplement her ration. The objective in feeding the fresh cow is to keep her dependence on stored energy and protein as small and short in duration as possible. *Off-feed* is the greatest hazard and should be rigorously avoided.

During the period in which the concentrate is increased, a point will be reached when forage intake will be depressed. As this point is exceeded, each additional pound of concentrate will reduce forage intake by about ½ lb of hay equivalent. This reduces the fiber content of the ration and may increase the incidence of milk fat depression, and rumen malfunction, but this is a temporary problem and should be viewed on the basis of its effect on the total lactation period. After peak milk flow is attained and challenging the cow ceases, forage intake will gradually increase and fat test will return to normal.

Challenge feeding helps a cow reach peak production earlier than she otherwise might, thus taking advantage of the fact that her system, at the time, is physiologically adapted to heavy production.

After peak production is reached, the amount of concentrate fed should be determined by a concentrate feeding guide based on body weight, milk production, and fat test (see Table 20–9).

GROUP FEEDING

Individual feeding of lactating cows has largely given way to mechanized group feeding. The latter was developed for convenience and saving of labor, rather than for improved animal well-being or feed efficiency. Today, lactating herds with several hundred cows are common; and some herds number several thousand. In order to design a nutritional program for such large numbers that can be adapted to the specific needs of the cows, they are separated into groups according to production (and, therefore, nutritional needs).

Fig. 20-29. Corral system of group feeding. (Courtesy, The Brown Swiss Cattle Breeders Assn., Beloit, Wisc.)

When producers decide to go to group feeding, they must decide on the number of groups into which to divide the herd. To answer this question, consideration should be given to the following: (1)herd size; (2) types and costs of available feeds; (3) current type of housing, feeding, and milking system; and (4) overall economic integration of the operation—for example, labor, machinery, etc.

In large herds (more than 250 milking cows), a commonly used system is one in which a minimum of 5 groups are established: (1) high-production cows (about 90 lb of milk/head/day), (2) medium-production cows (about 65 lb of milk/head/day), (3) low-production cows (about 45 lb of milk/head/day), (4) dry cows, and (5) first calf heifers. More groups are desirable in very large herds if corrals and facilities are available. Because of feeding and social considerations, a maximum of 100 cows per group is advisable. With this program, there can be a maximum of two moves during the lactation cycle. In many cases, only one move is necessary; and in a few cases, no moves are required. This system allows each group to be fed according to need. The high-producing groups should be fed the highest quality ingredients at maximum levels. The middle-producing cows should be fed in such a way as to reduce feed cost, increase butterfat test, improve rumen function, and promote lactation persistency. The same holds true for the low-producing cows as for the medium producers except that considerable care must be exercised to avoid excessive fattening.

One of the problems inherent in group feeding concerns the behavioral adaptation of a newly introduced cow to a group. Group acceptance—*pecking order*—can pose occasional problems with a new cow, but the magnitude of the problem is usually not very great. One means of reducing this is to move several cows into a new group at the same time and just before feeding, rather than individually.

When group feeding programs are followed, grain is seldom fed in the milking parlor. This is commonly referred to as corral or bunk feeding since feeding generally takes place in bunks along the fenceline of the corrals or pens. Studies have demonstrated that cows fed their grain as a group in a common manger do as well as those fed individually in the milking parlor, but some cows may not always come into the parlor as easily when there is no grain to attract them. Some producers offer a minimum amount of feed in the parlor and the remaining amount in the corral with good success. The high producers seem to be more aggressive than the low producers; hence, they usually eat more when group fed.

Group feeding can be easily adapted to the use of complete feeds when the concentrates, roughages, and supplements are mixed into one feed rather than being fed separately. Some producers who use complete feeds prefer to feed dried roughages—especially long-stemmed hay—separately in order to enhance stimulation of the rumen and to facilitate mixing, because long hay does not lend itself to mixing in a mixer. The **advantages** of group feeding and complete feed are:

1. It allows the producer to use special formulations which are particularly important to certain animals.
2. It eliminates the necessity of supplying minerals *ad libitum*.
3. It permits precise definition of the ration that is consumed.
4. It facilitates mechanized feeding; hence, less labor is required.
5. It eliminates the problems associated with the uncontrolled preferential consumption of a specific feedstuff.
6. It results in fewer digestive disturbances, such as displaced abomasums.
7. It eliminates the practice of feeding in the parlor.
8. It allows for maximum utilization of least-cost formulation.
9. It facilitates masking of certain unpalatable feeds, such as urea.
10. It is adaptable to conventional barn systems.
11. It allows the producer to fix a ratio of fiber to concentrate proportions in the ration.
12. It reduces the hazard of micronutrient deficiencies.
13. It provides the operator with daily feed consumption figures of the group, which can then be used to improve management decisions.

Among the **disadvantages** of group feeding and complete feed are:

1. It necessitates specialized blending equipment to ensure thorough mixing.
2. It is uneconomical to divide small herds into groups to feed separate rations.
3. It is not applicable to grazing herds.
4. It is difficult to group animals in some barn designs.

5. Mismanagement can result in the so-called *fat cow syndrome* and related health problems, such as calving difficulties, poor reproduction, low production, low dry matter consumption, and metabolic disorders. In many cases, these problems do not become immediately obvious. Rather, they may take many months to develop.

• **Automatic grain feeders**—The advent of electric and computerized grain feeders was heralded as a way in which producers could keep their cows in group housing but feed them individually. However, some producers report (1) that such feeders may result in overfeeding grain, accompanied by health problems and lower profits; and (2) that such feeders may not function properly when the grain ration contains adequate fiber content.

The following general types of mechanized grain-feeding systems are available:

1. **Free-choice, electronic grain feeders.** These units allow cows equipped with an identification unit (magnet, key, or chain) access to a feeding station. This system does not restrict access time or amount of grain consumed per feeder visit. Careful management is necessary to avoid digestive problems. The major advantages of this system are low initial investment and a simple design.

2. **Preset or computerized grain feeders.** This system controls the maximum amount of grain that individual cows receive during a set period of time. Initial cost of this system varies widely depending on herd size and complexity of the system.

Regardless of which feeder system is selected, successful adoption requires superior management. The entire daily grain allocation for individual cows and/or the herd can be fed through computer-operated feeders. Maximum daily grain capacity is generally 400 to 500 lb. It is recommended that a feeder accommodate 20 to 25 cows. Stall length, protection of the unit, and location of the unit relative to cow traffic patterns can affect the success of the unit.

• **Example rations for group feeding**—Example rations for group feeding mature lactating cows at each of 3 milk production levels (90, 65, and 45 lb) are presented in Tables 20-15, 20-16, and 20-17; and example rations suitable for feeding dry cows are presented in Table 20-18.

When group feeding, first-calf heifers should be handled in a separate group and fed for both milk production and growth. Their nutrient requirements for milk production are similar to the requirements of their older counterparts producing milk at the same level, but, because of their growth, they should receive about 20% more nutrients than are required for maintenance.

Although variations can and should be made in Tables 20-15, 20-16, 20-17, and 20-18 rations, they are excellent guides. When milk yields and/or body weights differ from those used in these tables, a suitable computer program or hand calculations should be employed to obtain amounts to be fed, and to give consideration to costs of alternative feeds.

TABLE 20-15
COMPLETE RATIONS FOR 1,300 TO 1,400 LB *(591 TO 636 KG)* COWS, IN EARLY LACTATION—HIGH-PRODUCTION GROUP (90-LB MILK, 3.6% FAT AVERAGE)[1]

Feeds	All Alfalfa		¾ Alfalfa ¼ Corn Silage		⅔ Alfalfa ⅓ Corn Silage		½ Alfalfa ½ Corn Silage		⅓ Alfalfa ⅔ Corn Silage		¼ Alfalfa ¾ Corn Silage	
	All Amounts on Dry Matter Basis											
	(lb)	*(kg)*	(lb)	*(kg)*	(lb)	*(kg)*	(lb)	*(kg)*	(lb)	*(kg)*	(lb)	*(kg)*
Alfalfa, 17% crude protein, 59% TDN	24.2	*10.99*	19.2	*8.72*	17.28	*7.85*	13.59	*6.17*	9.24	*4.19*	7.03	*3.19*
Corn silage, 8% crude protein, 68.7% TDN	—	—	6.4	*2.91*	8.63	*3.92*	13.59	*6.17*	18.51	*8.40*	21.10	*9.58*
Corn, 10% crude protein, 88% TDN	18.9	*8.58*	16.1	*7.31*	15.32	*6.96*	13.11	*5.95*	11.22	*5.09*	10.16	*4.61*
44% crude protein supplement, 49% crude protein, 86% TDN	5.5	*2.50*	6.6	*3.00*	7.00	*3.18*	7.83	*3.55*	8.82	*4.00*	9.34	*4.24*
Fat, 182% TDN	1.2	*0.54*	1.2	*0.54*	1.19	*0.54*	1.19	*0.54*	1.18	*0.54*	1.19	*0.54*
Dicalcium phosphate, 22% calcium, 19% phosphorus	0.42	*0.19*	0.52	*0.24*	0.51	*0.23*	0.49	*0.22*	0.46	*0.21*	0.44	*0.20*
Limestone, 34% calcium	—	—	0.07	*0.03*	0.14	*0.06*	0.25	*0.11*	0.41	*0.19*	0.48	*0.22*
Trace mineralized salt	0.25	*0.11*	0.25	*0.11*	0.25	*0.11*	0.25	*0.11*	0.25	*0.11*	0.25	*0.11*
Mineral-vitamin mix	0.08	*0.04*	0.08	*0.04*	0.08	*0.04*	0.09	*0.04*	0.10	*0.05*	0.11	*0.05*
	Ration Nutrient Information											
Dry matter	50.5	*22.93*	50.4	*22.88*	50.4	*22.88*	50.4	*22.88*	50.2	*22.79*	50.1	*22.75*
Crude protein	8.72	*3.96*	8.71	*3.95*	8.72	*3.96*	8.72	*3.96*	8.72	*3.96*	8.72	*3.96*
Crude protein (%)	17.26		17.29		17.29		17.30		17.36		17.40	
TDN	37.52	*17.03*	37.39	*16.98*	37.43	*16.99*	37.39	*16.98*	37.33	*16.95*	37.30	*16.93*
TDN (%)	74.29		74.19		74.26		74.18		74.36		74.446	
NE_{lc} (Mcal)	38.72		38.66		38.72		38.72		38.72		38.72	
NE_{lc} (Mcal/lb or kg)	0.767	*1.69*	0.767	*1.69*	0.768	*1.69*	0.768	*1.69*	0.771	*1.70*	0.773	*1.70*
Calcium (g)	208		206		206		206		205		205	
Calcium (%)	0.91		0.90		0.90		0.90		0.90		0.90	
Phosphorus (g)	103		114		114		114		114		114	
Phosphorus (%)	0.45		0.50		0.50		0.50		0.50		0.50	
Acid detergent fiber (%)	18.04		18.12		18.07		18.24		18.09		18.05	
Ether extract (%)	5.66		5.54		5.50		5.41		5.30		5.28	
Forage:grain ratio	48:52		51:49		51:49		54:46		55:45		56:44	

[1]Ration calculations based on 1988 NRC recommendations, and on the use of either alfalfa or alfalfa and corn silage (CS) forage. Table 20-15 was prepared specially for this book by D. E. Otterby, Professor, and J. G. Linn, Extension Animal Scientist, Department of Animal Science, University of Minnesota, St. Paul; with metric added by the authors.

TABLE 20-16
COMPLETE RATIONS FOR 1,300 TO 1,400 LB *(591 TO 636 KG)* COWS, IN MID-LACTATION—MEDIUM-PRODUCTION GROUP (65-LB MILK, 3.6% FAT AVERAGE)[1]

Feeds	All Alfalfa		¾ Alfalfa ¼ Corn Silage		⅔ Alfalfa ⅓ Corn Silage		½ Alfalfa ½ Corn Silage		⅓ Alfalfa ⅔ Corn Silage		¼ Alfalfa ¾ Corn Silage		All Corn Silage	
	All Amounts on Dry Matter Basis													
	(lb)	*(kg)*	(lb)	*(kg)*	(lb)	*(kg)*	(lb)	*(kg)*	(lb)	*(kg)*	(lb)	*(kg)*	(lb)	*(kg)*
Alfalfa, 17% crude protein, 59% TDN	21.93	*9.96*	17.93	*8.14*	16.43	*7.46*	11.98	*5.44*	8.07	*3.66*	6.15	*2.79*	—	—
Corn silage, 8% crude protein, 68.7% TDN	—	—	5.98	*2.71*	8.20	*3.72*	—	—	—	—	—	—	—	—
Urea-corn silage, 12% crude protein, 70% TDN	—	—	—	—	—	—	11.98	*5.44*	16.16	*7.34*	18.46	*8.38*	26.39	*11.98*
Corn, 10% crude protein, 91% TDN	17.12	*7.77*	14.29	*6.49*	13.23	*6.01*	13.25	*6.02*	12.48	*5.67*	11.80	*5.36*	6.28	*2.85*
44% CP supplement, 49% CP, 86% TDN	2.33	*1.06*	3.21	*1.46*	3.54	*1.61*	3.62	*1.64*	4.10	*1.86*	4.33	*1.97*	7.34	*3.33*
Limestone	0.34	*0.15*	0.32	*0.15*	0.31	*0.14*	—	—	0.01	*0.01*	0.07	*0.03*	0.28	*0.13*
Trace mineralized salt	0.21	*0.10*	0.21	*0.10*	0.21	*0.10*	0.21	*0.10*	0.21	*0.10*	0.21	*0.10*	0.20	*0.09*
Mineral-vitamin mix	0.08	*0.04*	0.07	*0.03*	0.08	*0.04*	0.08	*0.04*	0.07	*0.03*	0.10	*0.05*	0.10	*0.05*
	Ration Nutrient Information													
Dry matter	42.0	*19.07*	42.0	*19.07*	42.0	*19.07*	41.5	*18.84*	41.5	*18.84*	41.5	*18.84*	40.9	*18.57*
Crude protein	6.60	*3.00*	6.60	*3.00*	6.60	*3.00*	6.60	*3.00*	6.60	*3.00*	6.60	*3.00*	6.60	*3.00*
Crude protein (%)	15.72		15.72		15.72		15.91		15.91		15.91		16.14	
TDN	29.90	*13.57*	29.85	*13.55*	29.84	*13.55*	29.56	*13.42*	29.73	*13.50*	29.70	*13.48*	29.60	*13.44*
TDN (%)	71.18		71.08		71.05		71.23		71.63		71.58		72.37	
NE_{lc} (Mcal)	30.73		30.73		30.73		30.73		30.73		30.73		30.73	
NE_{lc} (Mcal/lb or kg)	0.73	*1.61*	0.73	*1.61*	0.73	*1.61*	0.74	*1.63*	0.74	*1.63*	0.74	*1.63*	0.75	*1.65*
Calcium (g)	146		129		122		139		122		122		122	
Calcium (%)	0.77		0.68		0.64		0.74		0.65		0.65		0.66	
Phosphorus (g)	86		86		86		85		85		85		80	
Phosphorus (%)	0.45		0.45		0.45		0.45		0.45		0.45		0.43	
Acid detergent fiber (%)	19.00		19.57		19.78		19.73		19.59		19.69		19.03	
Ether extract (%)	3.51		3.36		3.34		3.00		2.84		2.74		2.95	
Forage:grain ratio	52:48		57:43		59:41		58:42		58:42		59:41		65:35	

[1]Ration calculations based on 1988 NRC recommendations, and on the use of either alfalfa or alfalfa and corn silage (CS) forage. Table 20–16 was prepared specially for this book by D. E. Otterby, Professor, and J. G. Linn, Extension Animal Scientist, Department of Animal Science, University of Minnesota, St. Paul; with metric added by the authors.

TABLE 20-17
COMPLETE RATIONS FOR 1,300 TO 1,400 LB *(591 TO 636 KG)* COWS, IN LATE LACTATION—LOW-PRODUCTION GROUP (45-LB MILK, 3.8% FAT AVERAGE)[1]

Feeds	All Alfalfa		¾ Alfalfa ¼ Corn Silage		⅔ Alfalfa ⅓ Corn Silage		½ Alfalfa ½ Corn Silage		⅓ Alfalfa ⅔ Corn Silage		¼ Alfalfa ¾ Corn Silage		All Corn Silage	
	All Amounts on Dry Matter Basis													
	(lb)	*(kg)*	(lb)	*(kg)*	(lb)	*(kg)*	(lb)	*(kg)*	(lb)	*(kg)*	(lb)	*(kg)*	(lb)	*(kg)*
Alfalfa, 17% crude protein, 59% TDN	26.14	*11.87*	21.53	*9.77*	19.74	*8.96*	14.99	*6.81*	10.46	*4.75*	7.98	*3.62*	—	—
Corn silage, 8% crude protein, 68.7% TDN	—	—	7.18	*3.26*	9.86	*4.48*	—	—	—	—	—	—	—	—
Urea-corn silage, 12% crude protein, 66% TDN	—	—	—	—	—	—	14.99	*6.81*	20.95	*9.51*	23.94	*10.87*	31.10	*14.12*
Corn, 10% crude protein, 91% TDN	10.76	*4.89*	8.21	*3.73*	6.97	*3.16*	6.74	*3.06*	4.79	*2.17*	3.97	*1.80*	2.60	*1.18*
44% CP supplement, 49% CP, 81% TDN	—	—	—	—	0.34	*0.15*	0.13	*0.06*	0.63	*0.29*	0.94	*0.43*	2.20	*1.00*
Dicalcium phosphate, 22% Ca, 19% P	0.27	*0.12*	0.26	*0.12*	0.25	*0.11*	0.30	*0.14*	0.31	*0.14*	0.31	*0.14*	0.30	*0.14*
Limestone, 34% calcium	—	—	—	—	—	—	—	—	—	—	—	—	0.12	*0.05*
Trace mineralized salt	0.19	*0.09*	0.19	*0.09*	0.19	*0.09*	0.19	*0.09*	0.19	*0.09*	0.19	*0.09*	0.18	*0.08*
Mineral-vitamin mix	0.06	*0.03*	0.06	*0.03*	0.06	*0.03*	0.07	*0.03*	0.08	*0.04*	0.09	*0.04*	0.20	*0.09*
	Ration Nutrient Information													
Dry matter	37.42	*16.99*	37.42	*16.99*	37.42	*16.99*	37.42	*16.99*	37.42	*16.99*	37.42	*16.99*	36.7	*16.66*
Crude protein	5.52	*2.51*	5.11	*2.32*	5.09	*2.31*	5.09	*2.31*	5.09	*2.31*	5.09	*2.31*	5.09	*2.31*
Crude protein (%)	14.75		13.65		13.60		13.60		13.60		13.60		13.87	
TDN	24.91	*11.31*	24.85	*11.28*	24.83	*11.27*	24.78	*11.25*	24.73	*11.23*	24.70	*11.21*	24.58	*11.16*
TDN (%)	66.52		66.40		66.35		66.22		66.10		66.02		66.98	
NE_{lc} (Mcal)	25.37		25.37		25.37		25.37		25.37		25.37		25.37	
NE_{lc} (Mcal/lb or kg)	0.677	*1.49*	0.678	*1.50*	0.678	*1.50*	0.678	*1.50*	0.678	*1.50*	0.678	*1.50*	0.691	*1.52*
Calcium (g)	195		173		165		148		129		117		96	
Calcium (%)	1.15		1.02		0.97		0.87		0.76		0.69		0.58	
Phosphorus (g)	64		64		64		64		64		64		64	
Phosphorus (%)	0.38		0.38		0.38		0.38		0.38		0.38		0.38	
Acid detergent fiber (%)	23.92		24.63		24.91		25.42		26.02		26.16		25.4	
Ether extract (%)	3.47		3.38		3.32		2.86		2.59		2.45		2.09	
Forage:grain ratio	70:30		77:23		79:21		80:20		84:16		85:15		85:15	

[1]Ration calculations based on 1988 NRC recommendations, and on the use of either alfalfa or alfalfa and corn silage (CS) forage. Table 20–17 was prepared specially for this book by D. E. Otterby, Professor, and J. G. Linn, Extension Animal Scientist, Department of Animal Science, University of Minnesota, St. Paul; with metric added by the authors.

TABLE 20-18
COMPLETE RATIONS FOR 1,300 TO 1,400 LB *(591 TO 636 KG)* DRY COWS[1]

Feeds	Alfalfa Corn Silage		Alfalfa-Grass Hay, Corn Stover		Alfalfa-Grass Hay		Oatlage		Urea-Corn Silage, Grass Hay		Grass Hay	
	All Amounts on Dry Matter Basis											
	(lb)	*(kg)*	(lb)	*(kg)*	(lb)	*(kg)*	(lb)	*(kg)*	(lb)	*(kg)*	(lb)	*(kg)*
Alfalfa, 17% CP, 59% TDN	11.09	*5.03*	—	—	—	—	—	—	—	—	—	—
Alfalfa-grass hay, 16.5% CP, 58% TDN	—	—	14.59	*6.62*	22.52	*10.25*	—	—	—	—	—	—
Grass hay, 12% CP, 60% TDN	—	—	—	—	—	—	—	—	13.50	*6.13*	17.93	*8.14*
Oatlage, 12.8% CP, 59% TDN	—	—	—	—	—	—	25.74	*11.69*	—	—	—	—
Corn silage, 8.7% CP, 68.7% TDN	12.66	*5.75*	—	—	—	—	—	—	—	—	—	—
Urea-corn silage, 12% CP, 66% TDN	—	—	—	—	—	—	—	—	12.28	*5.58*	—	—
Corn stover, 6.7% CP, 66% TDN	—	—	11.59	*5.26*	—	—	—	—	—	—	—	—
Corn, 10% CP, 88% TDN	—	—	—	—	3.47	*1.58*	1.09	*0.49*	—	—	5.46	*2.48*
44% CP supp., 49.9% CP, 81% TDN	—	—	—	—	—	—	—	—	—	—	0.53	*0.24*
Dicalcium phosphate, 22% Ca, 19% P	—	—	—	—	—	—	0.01	*0.01*	—	—	—	—
Limestone, 34% Ca	—	—	—	—	—	—	0.05	*0.02*	0.05	*0.02*	0.09	*0.04*
Monosodium phosphate, 22.5% P	0.05	*0.02*	0.10	*0.05*	0.03	*0.01*	—	—	—	—	—	—
Trace mineralized salt	0.07	*0.03*	0.07	*0.03*	0.07	*0.03*	0.07	*0.03*	0.07	*0.03*	0.07	*0.03*
Mineral-vitamin mix	0.03	*0.01*	0.02	*0.01*	0.02	*0.01*	0.06	*0.03*	0.06	*0.03*	0.06	*0.03*
	Ration Nutrient Information											
Dry matter	25.07	*11.38*	26.37	*11.97*	26.11	*11.85*	27.00	*12.26*	25.96	*11.79*	24.14	*10.96*
Crude protein	3.05	*1.38*	3.24	*1.47*	4.07	*1.85*	3.40	*1.54*	3.09	*1.40*	2.97	*1.35*
Crude protein (%)	12.18		12.29		15.58		12.60		11.92		12.31	
TDN	16.17	*7.34*	16.10	*7.31*	16.09	*7.30*	16.14	*7.33*	16.08	*7.30*	15.99	*7.26*
TDN (%)	64.49		61.08		61.64		59.78		61.98		66.27	
Calcium (g)	88		91		87		48		48		48	
Calcium (%)	0.78		0.77		0.73		0.39		0.41		0.44	
Phosphorus (g)	30		30		30		30		30		37	
Phosphorus (%)	0.27		0.25		0.25		0.25		0.27		0.34	
Acid detergent fiber (%)	30.16		37.31		28.96		33.48		35.05		31.32	
Forage:grain ratio	99:1		99:1		87:13		95:5		99:1		74:26	

[1]Ration calculations based on 1988 NRC recommendations. Table 20-18 was prepared specially for this book by D. E. Otterby, Professor, and J. G. Linn, Extension Animal Scientist, Department of Animal Science, University of Minnesota, St. Paul; with metric added by the authors.

FEEDING ON PASTURE

On the average, a lactating cow on pasture spends 8 hours daily in harvesting, or grazing. In addition, she spends another 8 hours each day in rumination—the act of chewing the cud. This means that a cow spends two-thirds of her time eating. If she is a high producer, meeting her nutritive needs within this period of time calls for good pastures plus supplemental feeding.

Problems of milk production are at a minimum during the early pasture season, when plant growth is lush. However, when the weather gets hot—the period known as the "summer slump"—it is a different story. High temperatures actually affect pasture growth more than the well-being of the cows. Many dairy producers have discontinued pasture grazing for two reasons: (1) It is difficult to keep milk production uniform when cows are on pasture, because of changing temperatures and pasture growth; and (2) with larger herds, it is not possible to have sufficient pastures in close proximity to headquarters.

Pasture of high quality has a value that is intermediate between concentrate and hay. Thus, on good pasture, it is possible to sustain a high level of production with less grain than is needed for conventional winter forage supplementation.

Fig. 20-30. Lactating cows on pasture, with natural shade. (Courtesy, *Holstein World,* Sandy Creek, N.Y.)

The following summer feeding program is recommended for most lactating cows on pasture:

1. Have good pastures and follow good pasture management. Practice a rotational system of grazing with cows kept within one area until forage is used, and then place them on another area. In this way pasture yields will be higher and cows will have a better quality feed during the entire pasture season. The length of rotation may vary from 1 to 3 days. Beyond 3 days, there will likely be some damage to the plants.

2. Consider ensiling the early pasture growth which is in excess of the amount the cows will consume, thereby avoiding overmature and/or wasted forage in the early season.

3. Feed hay and grain as a supplement to pastures when needed to obtain desired milk yield. If the pasture is of comparable quality to the forage that was fed in the winter season, the grain mix may be the same as was fed in the winter. On the other hand, if summer pastures are much better or much poorer than the quality of the winter roughage, the grain mix should be changed accordingly. Thus, the grain mix used when cows are on pasture should be formulated so as to balance the deficiencies of the grass, just the same as the winter concentrate should balance the deficiencies of the winter forage. Also, cows should be fed hay before turning them to pasture during periods when bloat is particularly hazardous, such as early and late in the season.

4. When pastures are poor and/or the weather is very hot, feed more grain and limit the hay, thereby providing needed added energy and avoiding the excess heat generated during the digestion of high-fiber rations. But do not reduce hay feeding so much as to cause a reduction in the fat content of the milk.

5. Consider supplementing summer pastures with silage. Succulence is of benefit in dairy rations; hence, if cows are accustomed to having silage during the winter, it may be continued to advantage in the summer when pastures are not doing well.

6. Provide adequate shade and water for cows on summer pasture.

FEEDING DRY COWS

Dry cows have three important needs: (1) regression of old milk producing cells in the udder and regeneration of new milk producing cells, (2) developing the unborn calf, and (3) storing body reserves for the next milking period. This necessitates that they be properly fed in late lactation and during the dry period.

Proper dry cow feeding is one of the most critical practices for successful dairy operations. It involves management practices (1) to have the cow dry 50 to 65 days, and (2) to have the cow in proper condition when beginning and finishing the dry period. Proper dry cow feeding begins during the last half to one-third of the lactation period when body reserves should be replaced. At drying off time, the cow's body condition should be classed as 2 or 3 on an arbitrary scale of 1 as too thin and 5 as too fat. At calving, the body condition should be in the 3.5 to 4.0 class. During the 60-day dry period, cows should be fed to gain 120–200 lb.

The following routine is recommended for dry cows:

1. Turn high-producing cows and cows milked 3 times daily dry 60 to 65 days before expected calving date. Turn cows more than 4 years of age dry 50 to 60 days before freshening. In many cases, dairy producers erroneously prolong the lactation of their high producers, subsequently lessening the period in which the cow can recuperate from the heavy demands of lactation. Records show that the most total milk during a lactation is produced when cows are dry 50 to 60 days.

2. For cows producing low levels of milk at the end of lactation, the caretaker should stop grain feeding, limit water intake, and stop milking abruptly to hasten drying off. However, cows producing high levels of milk at the end of lactation may require continued milking for a short period during which they are fed a low-energy ration. This should reduce stress and facilitate drying off.

3. The dry period should be a time when the digestive tract has a chance to recondition itself following the rigors of high-concentrate feeds used during lactation. Concentrate feeds put considerable stress on the rumen. Thus, roughage—especially hay—is extremely important during the dry period to promote healthy rumens. Ideally, the condition of the dry cow should be such that she will not have to gain much weight since weight can be put on much more efficiently during the latter stages of lactation. Under proper dry cow management, small amounts of grain should only be required during the last 2 weeks of the dry period for rumen acclimation.

4. The 2 weeks prior to calving should be devoted to preparing the cow for parturition and the onset of lactation and for the adaptation of the rumen microflora to a radical change in the ration. If milk fever is a problem, as it is in most herds, it is desirable that the dry cow ingest less than 100 g of calcium and 40 g of phosphorus per day but still ingest enough to satisfy the requirements for these minerals. Additionally, supplemental vitamins A, D, and E, along with added selenium, should be provided. One commonly used procedure to adapt the ruminal microflora to the lactation ration is to feed the same ration, or percentage of the same ingredients, that the cows will receive during lactation. Grain levels should not exceed 8 lb per day. It is imperative that dry cows not be in a fat condition at calving, however.

On the first day after calving, cows should be fed the same amount of grain that they were getting before calving, followed by an increase of 2 to 3 lb per day, according to the cow's appetite. The experienced caretaker is in the best position to determine how much, and what, to feed each individual cow at, and after, calving time.

FEEDING CALVES

One of the most important phases of dairy production is that of feeding and managing dairy calves. Statistics reveal that more than 20% of the dairy calves die of sickness or disease before reaching maturity. With good management, these losses may be reduced to 3 to 5%. Many calf deaths are caused by faulty nutrition or poor housing and management.

A carefully planned and executed feeding program is necessary to produce growthy, vigorous, and healthy calves. The following feeding program is recommended:

Day 1	Dam's colostrum
Day 2	Dam's colostrum
Day 3	Dam's colostrum
Day 4	Liquid feed of choice, introduce starter and water
Day 5 to weaning ...	Continue feeding program
Weaning to 12 weeks ..	Starter (up to 5 lb daily), introduce forage

Fig. 20-31. A bucket or bottle equipped with a plastic nipple is easy to clean and provides a convenient method of measuring the correct amount to feed the calf. (Courtesy, Milk Marketing, Inc., Strongsville, Ohio)

Dairy Calf Production

Physiologically, the newborn calf is not a functioning ruminant. The abomasum, which represents the largest portion of the stomach of the newborn calf, is the primary functional unit of the gastric region. The sucking reflex allows colostrum or milk to bypass the rudimentary rumen and reticulum via the esophageal groove directly into the abomasum where digestion is initiated. As the calf grows older, it consumes solid feedstuffs which serve as a mechanical stimulator on the other sections of the gastric region, thereby hastening their development. Additionally, the rumen becomes inoculated with microorganisms from the immediate environment.

Since milk is the primary product of dairy production, it is necessary to switch the young calf to cheaper feeds as expeditiously as possible. At the same time, it is important that the diet promote good health, growth, and development. Four feeds that are routinely fed to calves are (1) colostrum, (2) milk, (3) milk replacers, and (4) calf starters.

COLOSTRUM

Colostrum is the milk which is high in antibodies, and which is secreted by cows, and other mammalian females, for the first few days following parturition.

Colostrum (either dam's colostrum or mixed colostrum from first milking of older cows) should be fed to calves as soon after birth as possible (ideally within 15 minutes and certainly within 4 hours) to protect against disease. At the 4,800-cow Arizona Dairy, Higley, Arizona, every effort is made to remove calves from their mothers *wet* and without nursing, so as to minimize infection from nursing. Then, as soon as possible, they are offered colostrum from a nippled bottle. If they fail to nurse naturally, the colostrum is hand-fed, using a specially designed tube to force it into the abomasum. This system, along with the superior feeding, management, and sanitation at Arizona Dairy, has resulted in average calf losses of only 2.7% (1.4% at birth and during the first 24 hours, and 1.3% after 24 hours).

Early feeding of colostrum is necessary because:

1. Newborn calves have no antibodies to provide natural protection against disease until colostrum is received.
2. The calves' ability to absorb immunoglobulins (the disease protecting component) is substantially reduced after 24 to 36 hours.
3. Calves may become infected with highly pathogenic bacteria immediately after birth.

Calves should receive a total of 10 to 12% of their birth weight as first milking colostrum, with half of this amount received 4 to 6 hours after birth.

Surplus colostrum can be frozen and stored for a period of 1 year or longer without losing its antibody value. It may then be thawed, warmed to about 100°F, and fed as needed.

Excess colostrum is a very nutritious feed, but it has little or no immune benefit (antibody protection) to a calf beyond the first 24 hours of life. Colostrum contains about a third more solids than milk or reconstituted milk replacer, and is highly digestible. So, storage and subsequent feeding of excess colostrum is highly desirable. It may be fed fresh; frozen, then thawed prior to feeding; or stored as sour (fermented) colostrum. Since it is higher in solids than normal milk, it should be diluted 25 to 50% when fed to other than newborn calves, in order to avoid overfeeding and scours.

Naturally fermented, sour colostrum sometimes becomes putrid and unfit for consumption, especially during the summer months. The following organic acids have been used separately, and successfully, to acidify colostrum to a pH of 4.6:

Name of Acid	Concentration	Dilution
Formic acid	0.3% by weight	¼ cup acid/5 gal colostrum
Acetic acid	0.7% by weight	½ cup acid/4¼ gal colostrum
Propionic acid	1.0% by weight	1 cup per 6 gal colostrum

COLOSTRUM, MILK, AND MILK REPLACERS COMPARED

The composition of colostrum changes rapidly after calving. The first six milkings are higher in nutrients than normal milk or most reconstituted milk replacers. Table 20-19 shows the comparative composition and characteristics of colostrum, whole milk, and a reconstituted milk replacer.

TABLE 20-19
COMPOSITION AND CHARACTERISTICS OF COLOSTRUM, WHOLE MILK, AND RECONSTITUTED MILK REPLACER (1 LB *[0.45 KG]* POWDER + 7 LB *[3.18 KG]* WATER)[1]

Item	First Milking	Second Milking	Second Day	Third Day	Whole Milk	Reconstituted Milk Replacer
Specific gravity (g/ml)	1.056	1.040	1.034	1.033	1.032	
Total solids (%)	23.9	17.9	14.0	13.6	12.9	12.5
Fat (%)	6.7	5.4	4.1	4.3	4.0	2.50
Nonfat solids (%)	16.7	12.2	9.6	9.5	8.8	11.25
Protein (%)	14.0	8.4	4.6	4.1	3.1	2.8
Lactose (%)	2.7	3.9	4.5	4.7	5.0	variable
Ash (%)	1.1	1.0	0.8	0.8	0.7	variable
Vitamin A (g/100 ml)	295.0	190.0	95.0	74.0	34.0	variable
Immunoglobulins (%)	6.0	4.2	1.0	–	–	0

[1] *Source:* Linn, J. G., M. F. Hutjens, W. T. Howard, L. H. Kilmer, and D. E. Otterby, *Feeding the Dairy Herd,* Cooperative Ext. Services, Ill., Iowa State, Minn., and Wisc., 1988, p. 18, Table 8.

Milk replacers vary in quality, so the buyer/user should study the feed tag. The best milk replacers contain 22% protein, all derived from milk products—skim milk powder, buttermilk powder, dried whole whey, de-lactosed whey, casein, and/or milk albumen. Chemically modified soy protein, soy isolates, and soy concentrates are good, but as plant proteins they are less digestible than milk protein. Meat solubles, fish protein concentrate, distillers' dried solubles, brewers' dried yeast, oat flour, and wheat flour are inferior as protein sources in milk replacers.

A good milk replacer powder should contain a minimum of 15% fat, and it may contain more than 20%. The higher fat level tends to reduce the severity of diarrhea and produce additional energy for growth. Good quality animal fats are preferable to most vegetable fats. However, soy lecithin, especially when homogenized, is an acceptable fat source and improves mixing qualities of the replacer.

The calf can use two carbohydrate sources in milk replacers: lactose (milk sugar) and dextrose. However, two other carbohydrate sources, starch and sucrose (table sugar), are not satisfactory and should be excluded from milk replacers.

Mastitic milk and discard milk (unmarketable milk from cows that were treated for mastitis, metritis, or other health problems) may be fed fresh in the same manner as whole milk, or they may be fermented or preserved with an organic acid. Milk collected for three to six milkings after antibiotic treatment will ferment normally. Extremely abnormal milk (bloody or watery) should not be fed.

Properly managed feeding of mastitic milk to calves will not increase mortality or cause mastitis in these animals when they freshen. Neither will it cause diarrhea. Mastitic milk should not be fed to calves less than 2 days old, as the intestine is permeable to large protein molecules. Calves fed milk containing antibiotics should not be marketed for meat unless the required withdrawal period is observed prior to slaughter.

CALF STARTERS

A high quality, palatable calf starter should be offered when the calf is 4 days old, and not later than 10 to 12 days of age. The best starters are high in energy, contain 16 to 18% protein (20% if calves are weaned before 4 weeks of age), and are free of excessive fines. To encourage consumption, starters should consist of whole, coarsely ground, cracked, or rolled grains. Up to 5% molasses improves palatability and minimizes fines and dust. Whole grains, especially oats, can be fed with starter rations to calves up to 3 months of age. Calf starters should be fed until calves are about 12 weeks of age, with intake limited to 5 to 7 lb per calf daily.

Fig. 20-32. Calf starter and water (note two pails) provided to a calf in an individual pen. (Courtesy, Maddox Dairy, Riverdale, Calif.)

TABLE 20-22
SUGGESTED RATES FOR FEEDING WHOLE MILK, FEEDING AND DILUTION RATES FOR COLOSTRUM,[1] AND RECONSTITUTED SCHEDULE FOR MILK REPLACERS; DAILY[2]

Body Weight		Whole Milk		Solids Content		Colostrum			Water		Replacer			Water	
(lb)	*(kg)*	(lb)	*(kg)*	(lb)	*(kg)*	(lb)	*(kg)*		(lb)	*(kg)*	(lb)	*(kg)*		(lb)	*(kg)*
50–60	*22–27*	5.0	*2.3*	0.6	*0.27*	3.6	*1.6*	+	1.2	*0.54*	0.6	*0.27*	+	4	*1.8*
61–70	*28–31*	5.5	*2.5*	0.7	*0.32*	4.2	*1.9*	+	1.4	*0.65*	0.7	*0.32*	+	5	*2.3*
71–80	*32–36*	6.5	*2.9*	0.8	*0.36*	4.8	*2.2*	+	1.6	*0.73*	0.8	*0.36*	+	5.5	*2.5*
81–90	*37–40*	7.0	*3.2*	0.9	*0.41*	5.4	*2.4*	+	1.8	*0.82*	0.9	*0.41*	+	6	*2.7*
91–100	*41–45*	8.0	*3.6*	1.0	*0.45*	6.0	*2.7*	+	2.0	*0.91*	1.0	*0.45*	+	7	*3.2*
101–110	*46–50*	9.0	*4.1*	1.1	*0.50*	6.6	*3.0*	+	2.2	*1.00*	1.1	*0.50*	+	8	*3.6*

[1]Solids content is assumed to be 16%.

[2]*Source:* Linn, J. G., M. F. Hutjens, W. T. Howard, L. H. Kilmer, and D. E. Otterby, *Feeding the Dairy Herd,* Cooperative Ext. Services, Ill., Iowa State, Minn., and Wisc., 1988, p. 18, Table 10.

Milk or milk replacer may be fed by open pail, by nipple feeding from a pail or bottle, or by automated feeding equipment. Each method of feeding is satisfactory, provided it is accompanied by cleanliness and sanitation.

Most calf raisers feed twice daily, with half the amounts listed in Table 20-22 offered at each feeding. Weak or unthrifty calves may benefit from more frequent feedings. Calves need the same amount of dry matter daily, but liquid amounts may have to be reduced to avoid digestive upsets. Dry milk replacer can be added to whole or mastitic milk to increase solid content without increasing the volume of liquid fed. Once-a-day feeding of milk-fed calves has proven successful except when calves are housed in a very cold environment. If once-a-day feeding is practiced, calves should be checked for health and well being at least once in addition to feeding time.

When calves are housed in hutches and the weather is extremely cold, they should be fed a 20% fat replacer 3 times daily and the daily feed allowance should be increased by 1¼ to 1½ times in order to meet their increased energy requirements. Young calves that are doing poorly should be moved to warmer quarters.

Most producers wean calves between 4 and 8 weeks of age. When weaned at 3 weeks of age, calves may have slightly depressed growth rates, temporarily; however, by 12 weeks of age, their weights will be about the same as their later weaned mates. Weaning later than 8 weeks of age may lead to fat calves. A good practice is to wean according to starter intake—wean when the starter intake is 1 to 1½ lb/day. Starter intake can be encouraged by placing a little dry feed in the pail immediately after the liquid has been consumed. In general, early weaning (at 21 to 35 days of age) will reduce feed and labor costs. Calves eating less than 1 lb of starter per day or calves doing poorly should be fed liquid until performance improves and dry feed is consumed in satisfactory amounts.

Fig. 20-33. Group of four weaned Guernsey calves in a pen. Note the hay rack. (Courtesy, American Guernsey Cattle Club, Columbus, Ohio)

Veal Calf Production

A veal calf is defined as a young bovine animal, usually not over 4 months of age, that has subsisted largely on milk or milk replacers, thus making the color of the lean meat light, grayish pink. The majority of veal calves are of dairy breeding, consisting of bull calves and heifer calves not retained as replacements. Producers receive a premium if the lean meat of veal is light grayish pink in color (due to reduced muscle myoglobin), characteristic of feeding milk, which is naturally low in iron. Some producers attempt to enhance the desired grayish pink color by restricting iron intake and exercise. Research has shown that a dietary iron concentration of 11.4 to 13.6 mg per pound of dry matter in milk replacers is sufficient for the well being of veal calves and produces desirable grayish pink carcasses, with and without exercise.

A conversion rate of 10 lb of whole milk for 1 lb of body weight gain is normal. If milk replacer is used, the conversion rate is generally about 1.3 to 1.5 lb of dry replacer per pound of gain.

Profitable veal production depends on: (1) a low mortality rate, (2) economical housing, (3) plenty of inexpensive labor, and (4) an established market.

Two types of veal markets exist: (1) fancy veal, and (2) conventional veal. The choice of the type of veal production should be determined primarily by two factors: (1) the market price outlook, and (2) the availability of an abundance of grain and some high quality forage needed for raising conventional veal.

• **Fancy veal**—Calves for fancy veal are marketed at 300 to 400 lb weight and 3 to 4 months of age. The meat must be very light colored, indicating that the calves have been fed milk and/or milk replacers, and that they have not been fed grain or hay.

• **Conventional veal**—The production of conventional veal involves the same management and feeding practices used in raising heifer replacements. A good *starter* and milk or milk replacer normally are used from birth to 4 to 6 weeks of age. Then, a calf grower ration and a good quality forage may be fed until the calves are about 12 weeks old and weigh about 200 lb.

Preventing Calf Scours

To prevent calf scours, the producer must prevent primary infection of the newborn. This rests on strict sanitary measures and isolation, along with other preventive measures. Observance of the following practices will lessen the incidence of calf scours:

1. **Make certain that the calf gets first-milk colostrum.** Never assume that the calf has nursed. Many newborn calves never receive sufficient colostrum to protect them from calfhood diseases. So, colostrum should be fed, preferably by hand, soon after birth; large breed calves should receive a minimum of 2 to 3 qt, small breed calves should receive about 3 pt.
2. **Augment natural resistance with vitamins.** Supplementation with vitamins A, D, and E (oral or injectable) immediately after birth is helpful in increasing the calf's natural resistance to scours, especially if colostrum is low in vitamin A content.
3. **Avoid overfeeding and irregularity of feeding.** Overfeeding and irregularity are common causes of calf scours.
4. **Keep feeding utensils clean and sanitary.** Clean the feeding utensils thoroughly after each feeding, then store them upside down to drain and dry.
5. **Don't overcrowd.** In bedded areas, provide 24 to 28 sq ft per calf. In confined, elevated stalls, provide about 20 sq ft per calf.
6. **Provide adequate ventilation.** Provide a minimum of 4 air exchanges per hour in winter and 15 air exchanges per hour in the summer.
7. **Avoid damp, wet calves.** Provide plenty of dry bedding in maternity stalls; and provide adequate bedding and ventilation in the calf quarters.

• **Use of electrolytes**—Feeding an oral electrolyte solution usually is beneficial when a calf has a mild case of scours (not off feed, not depressed, and no fever). Recommended treatment follows:

1. Delete or drastically reduce the amount of milk or milk replacer fed.
2. Feed only water containing an electrolyte for 3 to 6 feedings, depending on how soon the feces become firm. Frequent feeding of small volumes is advantageous. A 100-lb calf should consume about 5 qt (10% of body weight) daily.

Oral electrolyte solutions can be purchased commercially. If not readily available, a suitable electrolyte mixture can be made by combining the following ingredients:

4 tsp of table salt
3 tsp of baking soda
½ cup of "light" corn syrup
1 gal of water

FEEDING REPLACEMENT HEIFERS/ NORMAL GROWTH FOR DAIRY HEIFERS

Fig. 20-34. Replacement heifers—group fed. (Courtesy, Milk Marketing, Inc., Strongsville, Ohio)

Between weaning and calving (12 weeks to 2-year-olds), the nutrition of heifer replacements is often neglected. At its best, the feeding and management program during this period involves 3 distinct phases: (1) weaning (about 12 weeks of age) to 1 year; (2) 1 year to 2 months before calving at 2 years; and (3) 2 months before calving to calving.

• **Replacement heifers, weaning (about 12 weeks of age) to 1 year**—During this period, replacement heifers may be fed forage free-choice and limited grain. The amount, and the protein content, of the grain mix needed will be determined by the quality of the forage being fed. Pasture can be used successfully in the feeding program of replacement heifers, provided it is supplemented with a grain mix and some dried forage, along with suitable minerals (incorporated in the grain mix or offered free-choice). Also, there should be access to clean, fresh water.

During the yearling stage, replacement heifers should not be overfed and become too fat. Overconditioning has an inhibitory effect on mammary secretory tissue development during the critical period of its maximum development between 3 and 9 months of age and results in lower milk production later in life. Overconditioning of heifers after 15 months of age does not affect mammary secretory tissue.

Table 20–23 lists some grower rations for 400-lb calves. If the protein content of the forage is good, little protein supplement will be required in the grain mix. Monensin or Lasalocid can be fed as directed on the label to heifers to improve growth and rate of gain.

TABLE 20–23
RATIONS FOR LARGE BREED DAIRY HEIFERS OF DIFFERENT WEIGHTS[1]

Weight		Rate of Gain		Ration (As-fed)			Percent
(lb)	*(kg)*	(lb/day)	*(kg/day)*		(lb)	*(kg)*	(%)
400	*182*	1.7	*0.8*	Alfalfa hay, 140 RFV, 20% CP	8.5	*3.9*	
				Grain mix, 12.0% CP	3.0	*1.4*	
				coarse ground barley			71.0
				rolled or ground oats			23.0
				molasses			5.0
				trace mineral salt			0.5
				dicalcium phosphate			0.4
				vitamin premix			0.1
		1.7	*0.8*	Alfalfa-grass hay, 113 RFV, 16% CP	7.0	*3.2*	
				Grain mix, 14.5% CP	4.5	*2.0*	
				coarse ground corn			83.1
				soybean meal			15.5
				trace mineral salt			0.5
				dicalcium phosphate			0.8
				vitamin premix			0.1
		1.7	*0.8*	Orchardgrass hay	7.0	*3.2*	
				Grain mix, 16.5% CP	5.0	*2.3*	
				rolled or coarse ground barley			65.8
				molasses			4.0
				soybean meal			27.5
				trace mineral salt			0.5
				limestone			2.1
				vitamin premix			0.1
		1.7	*0.8*	Orchardgrass hay	4.5	*2.0*	
				Corn silage	12.0	*5.4*	
				Grain mix, 19.7% CP	2.5	*1.1*	
				coarse ground corn			13.5
				rolled or coarse ground oats			6.5
				soybean meal			74.8
				trace mineral salt			1.0
				limestone			4.0
				vitamin premix			0.2
700	*318*	1.7	*0.8*	Alfalfa hay, 140 RFV, 20% CP	18.0	*8.1*	
				Grain mix	2.0	*0.9*	
				coarse ground corn			95.3
				trace mineral salt			1.5
				dicalcium phosphate			3.0
				vitamin premix			0.2
		1.7	*0.8*	Alfalfa hay, 140 RFV, 20% CP	10.0	*4.5*	
				Corn silage	15.0	*6.8*	
				Grain mix	1.0	*0.5*	
				ground ear corn			88.5
				trace mineral salt			4.0
				dicalcium phosphate			7.0
				vitamin premix			0.5
700	*318*	*1.7*	*0.8*	Alfalfa-grass hay, 113 RFV, 16% CP	12.5	*5.7*	
				Grain mix	6.0	*2.7*	
				ground ear corn			95.9
				soybean meal			2.5
				trace mineral salt			0.6
				dicalcium phosphate			0.9
				vitamin premix			0.1
		1.7	*0.8*	Orchardgrass hay	4.0	*1.9*	
				Corn silage	30.0	*13.6*	
				Grain mix	2.5	*1.1*	
				soybean meal			91.8
				trace mineral salt			2.0
				dicalcium phosphate			1.0
				limestone			5.0
				vitamin premix			0.2
1,000	*454*	1.7	*0.8*	Alfalfa hay, 113 RFV, 16% CP	13.0	*5.9*	
				Corn silage	35.0	*15.9*	
				Mineral-vitamin supplement	0.5	*0.2*	
				soybean meal			80.0
				trace mineral salt			12.0
				dicalcium phosphate			6.0
				vitamin premix			2.0
		1.7	*0.8*	Orchardgrass hay	10.0	*4.5*	
				Corn silage	38.0	*17.3*	
				Supplement	2.0	*0.9*	
				soybean meal			89.0
				trace mineral salt			3.0
				limestone			7.5
				vitamin premix			0.5
		1.7	*0.8*	Oatlage, 13% CP	50.0	*22.7*	
				Grain mix	4.0	*1.8*	
				coarse ground corn			93.9
				trace mineral salt			1.5
				limestone			3.0
				dicalcium phosphate			1.0
				vitamin premix			0.6
		1.7	*0.8*	Urea-corn silage, 0.5% urea	41.0	*18.6*	
				Grass hay	10.0	*4.5*	
				Supplement	0.5	*0.2*	
				soybean meal			64.0
				trace mineral salt			10.0
				limestone			24.0
				vitamin premix			2.0

[1]*Source:* Linn, J. G., M. F. Hutjens, W. T. Howard, L. H. Kilmer, and D. E. Otterby, *Feeding the Dairy Herd,* Cooperative Ext. Services, Ill., Iowa State, Minn., and Wisc., 1988, pp. 20–21, Table 12.

• **Replacement heifers, 1 year to 2 months before calving at 2 years**—If good quality forage is available, it may be the only feed required for heifers over 1 year of age. A suitable mineral mix should be provided on a free-choice basis. Heifers should gain 1.6 to 1.8 lb per day. If growth is not satisfactory, some grain should be provided. The rations shown in Table 20-23 indicate the amounts to feed when various forage and grain combinations are offered to 700- to 1,000-lb heifers. Heifers on good pasture require no grain or other forage. As pastures mature, dry out, or are heavily grazed, supplemental feed should be provided. Heifers that are seriously deficient in energy, phosphorus, or vitamin A may not exhibit estrus.

First estrus in heifers is dependent on size and weight, primarily weight. A general guide is that heifers will show their first estrus at 40% of their mature weight, which should be before 12 months of age. Heifers fed high planes of nutrition will show estrus at an earlier age than heifers grown at recommended rates, but underfeeding of heifers will delay estrus. Underfed or very slow growing heifers may ovulate, but estrus signs are often suppressed. Heifers in good condition and gaining weight at breeding time generally show more definite signs of estrus and have improved conception rates over heifers in poor condition and/or losing weight. Overconditioned heifers require more services per conception than heifers of normal size and weight. Table 20-24 shows desirable weights for first breeding at 15 months of age along with weights for other age categories.

TABLE 20-24
DESIRABLE WEIGHTS FOR DAIRY HEIFERS TO BE BRED AT 15 MONTHS OF AGE[1]

Age in Months	Brown Swiss or Holstein		Ayshire, Shorthorn or Guernsey		Jersey	
	(lb)	*(kg)*	(lb)	*(kg)*	(lb)	*(kg)*
Birth	90-100	*41-45*	65-75	*30-34*	55-60	*25-27*
1	120	*54*	90-100	*41-45*	70-80	*32-36*
2	170	*77*	135-145	*61-66*	110-120	*50-54*
4	270	*123*	225-235	*102-107*	190-200	*86-91*
6	370	*168*	315-325	*143-148*	270-280	*123-127*
12	670-700	*304-318*	585-600	*266-272*	510-520	*232-236*
15[2]	800-875	*363-397*	720-750	*327-341*	630-650	*286-295*
18	970-1,000	*440-454*	850-875	*363-397*	750-775	*341-352*
22	1,150-1,200	*527-545*	1,025-1,075	*465-488*	900-950	*409-431*

[1]*Source:* Linn, J. G., M. F. Hutjens, W. T. Howard, L. H. Kilmer, and D. E. Otterby, *Feeding the Dairy Herd,* Cooperative Ext. Services, Ill., Iowa State, Minn., and Wisc., 1988, p. 21, Table 13.

[2]Breed heifers in this weight range. Heifers should weigh about 60% of their mature weight when bred. With proper feeding, heifers should reach these weights and have good skeletal growth at 14 to 16 months of age.

• **Two months before calving to calving**—The feeding of heifers during this period can affect milk production during the first lactation. During the last 2 months of gestation, heifers should make daily gains of about 2.0 lb per day, in comparison with 1.7 lb during early pregnancy. Heifers that are growing rapidly at calving time, and continuing to grow during the first lactation, are more persistent milkers than full-sized heifers at calving.

The amount of grain to feed before calving will depend on forage quality, size, and condition of the heifer. A good thumb rule is to feed grain at 1% of body weight starting about 6 weeks before calving. The ration should have adequate protein, minerals, and vitamins. Excess salt intake can contribute to udder edema and should be avoided the last 2 weeks before calving.

Well-grown heifers will have a minimum of problems at calving time. But plane of nutrition can affect ease of calving in two ways: (1) calf size, and (2) fatness of the dam. Fat heifers have higher incidents of dystocia because of small pelvic openings and usually a larger than normal sized calf at birth. Underfed or poorly grown heifers will require more assistance at calving and have a higher death rate at calving than normal sized heifers.

FEEDING DAIRY BULLS

Bull calves raised for breeding purposes should be fed and handled much the same as heifers. But, since they grow slightly faster than heifers, they should receive somewhat more feed than heifers of the same age.

Older bulls should be kept in thrifty, vigorous condition, but they should not be permitted to become too fat. Mature bulls can be fed the same grain ration as the lactating cows. Depending on the quality of the roughage, usually about ½ lb of grain per 100 lb of body weight will suffice for the mature bull. Also, individual differences must be considered, for some bulls are easier keepers than others.

Fig. 20-35. A mature Jersey bull (left) in thrifty, vigorous condition, along with a Jersey cow. (Courtesy, American Jersey Cattle Club, Columbus, Ohio)

FEEDING DAIRY BEEF

Dairy beef is beef derived from cattle of dairy breeding. Dairy heifers and steers finished at market weights comparable to the finished market weights of heifers and steers of beef breeding are commonly referred to as dairy beef. Generally speaking, there are two different finishing programs and market weights for dairy beef:

1. **High-energy rations; light market weights.** These animals are full fed a high-concentrate ration from about 300 lb to market weights of 750 to 950 lb.

2. **High-roughage; heavy market weights.** These animals are grown on a maximum of roughage to 600- to 750-lb weight, following which the proportion of concentrate is increased; and they are marketed at weights of 1,150 to 1,400 lb.

(Also see Chapter 19, section on "Dairy Beef.")

FEEDING SHOW AND SALE ANIMALS

Fig. 20-36. Spunky Blevins, age 9, Mesa, Arizona, showing a registered Holstein senior heifer calf at the Arizona State Fair. (Courtesy, Arizona Dairy Co., Higley, Ariz.)

Dairy animals intended for show or sale should be fed so as to achieve a certain amount of finish or bloom, but they should not be too fat. Linseed meal, beet pulp, oats, barley, molasses, and wheat bran are popular feeds in a fitting and showing ration. Linseed meal, in particular, is used to impart a healthy bloom or shine to the hair. Likewise, good quality forages are always very important.

In fitting show and sale animals, it is important that the grain mixture be palatable and light, and that they not go off feed. Feeds are sometimes soaked before feeding to make them more palatable.

FEED AND MANAGEMENT ASPECTS OF DAIRY PRODUCTION

Higher levels of milk production per cow, larger dairy herds, and increased facility and labor costs have focused attention on the need for more efficient management.

Basically, the primary objectives of any dairy enterprise are (1) to provide each class of dairy animals—lactating cows, dry cows, replacement heifers—with the proper rations at the right time, and (2) to hold labor and facility-equipment costs to a minimum.

The enormity of the feeding operation and the need for efficiency become obvious when it is realized that a herd of 100 lactating cows with an average daily consumption of 20 lb of hay, 50 lb of silage, and 14 lb of concentrates per cow will require annually that 1,533 tons of feed be grown, purchased, and/or stored; and fed. It follows that a herd of 1,000 lactating cows will require the handling of 10 times this amount of feed, or approximately 15,330 tons.

There are innumerable management aspects of great importance to dairy production. When disregarded, many of them will materially lessen production and make the enterprise unprofitable, no matter how good the genetic capability of the animals or the feed being used. Still others are important tools from the standpoint of enhancing good management. Among the management aspects of importance in dairy production are time and frequency of feeding and watering, and use of milk and butterfat records.

Fig. 20-37. Guernsey cows fed and cared for in a stall barn on Deep Lake Farm, Lakeville, Conn. (Courtesy, American Guernsey Cattle Club, Columbus, Ohio)

Feed and Management Factors Affecting Milk Yield and Composition

Research has shown that many feed and management factors may affect milk yield and composition.

Table 20-25 summarizes many of these factors. It shows that the amount of feed consumed, changes in the energy content of the ration, protein intake, feeding practices, and other feed and management factors may affect milk yield, fat, and solids-not-fat (SNF).

TABLE 20-25
FEED AND MANAGEMENT FACTORS AFFECTING MILK YIELD AND COMPOSITION

	Yield	Fat %	SNF %
Amount of feed consumed:			
Underfeeding energy	Decrease	Increase	Decrease
Overfeeding energy	Increase	Decrease	Increase
Changes in energy content of ration:			
Increase fiber percentage	Decrease	Increase	Decrease
High grain and low forage	Increase	Decrease	Increase
Substitute shelled corn for corn-and-cob meal	Increase	Decrease	Increase
Increase starch content of ration	Increase(?)	Decrease	Increase
Rumen protected fat	Increase	Increase	Decrease
Protein intake:			
Low protein intake (% or amount)	Decrease	0	Decrease
High NPN	0	0	+ Milk NPN
Proper ratio degradable to slower degradable protein	0 or +	0	Increase?
Feeding practices:			
Grain fed only twice per day	Decrease	Decrease	0
Cooking grain	Increase	Decrease	0
Forage fed in morning before grain	?	Increase	?
Pelleted feeds or finely chopped haylage or silage	0 or +	Decrease	0
More frequent feeding	Increase	Increase	?
Addition of fat to ration	Increase	Decrease	Decrease
Feeding soybeans or cottonseeds	0 or +	0 or +	Decrease
Fish oils	0	Decrease	0
Feeding buffers if ration is less than 20% ADF	0 or +	Increase	0 or –
Feeding buffers if ration is more than 21% ADF	0	0	0
Thyroprotein	Increase	Increase	0
Other:			
Good body condition of cow	Increase	Increase	Increase
High proportion of cows freshening in summer	Decrease	Decrease	Decrease
High proportion of cows 5–20 weeks in lactation	Increase	Decrease	Decrease
Advancing age, past 6 years	0 or +	Decrease	Decrease
Temperature of more than 85°F	Decrease	Decrease	Decrease
Temperature of less than 5°F	0 or –	Increase	Increase
Mastitis	Decrease	Decrease	Decrease
More frequent milking	Increase	Decrease	0
Lush pasture	Increase?	Decrease	0 or +
Excitement and stress	Decrease	Decrease	0
Fall season	Decrease	Increase	0
Spring season	Increase	Decrease	0

Fig. 20-38. Heat stress (temperature above 85°F) of lactating cows may decrease milk production by as much as 30%, and lower the fat percentage and the SNF percentage. Evaporative coolers, like those suspended from the roof in the above picture, minimize heat stress in a hot, dry climate. (Courtesy, Arizona Dairy Co., Higley, Ariz.)

Research has shown that rations which produce rumen volatile fatty acids in the approximate range of 65% acetic, 20% propionic, and 15% butyric will usually produce normal milk fat tests. Rations that produce a higher proportion of acetic acid to propionic acid may increase fat tests. Acetic acid is the primary forerunner of butterfat, whereas propionic acid is a forerunner of sugar production and is conducive to fattening. Therefore, rations that produce a high proportion of ruminal acetic acid are best for dairy cows and those that produce more propionic acid are more suited for finishing beef cattle.

Researchers have also found that feed fat affects the fatty acid composition of milk. For example, scientists at South Dakota State University report that feeding cows sunflower seed, which is naturally high in unsaturated fatty acids and low in saturated fatty acids, will result in cows producing milk containing 50 to 60% more unsaturated fatty acids than normal milk.

Also, the vitamin A potency of milk reflects the body stores of vitamin A of the cow and the amount of carotene or vitamin A in the ration. When grazing lush spring pasture, the cow may secrete milk containing as much as 2,500 IU of vitamin A per quart, whereas when dry feeds predominate, the vitamin A may fall to 400 to 800 IU per quart. Vitamin D in milk may vary from 5 IU per quart in winter to 50 IU in summer sunshine, and to 160 IU per quart when cows are fed 200 g of irradiated yeast daily.

Time and Frequency of Feeding and Watering

Where complete rations are fed, cows are generally fed twice or three times daily. Where the forage and grain are fed separately, most dairy farms provide hay free-choice at all times, feed silage once or twice daily, and feed concentrates twice daily.

Hay may be put out in large amounts once daily, or in small amounts several times a day, depending on the type and preparation of the hay and the labor situation. Cows will eat more if moderate amounts are fed at frequent intervals, but the labor requirement may be prohibitive. At least twice-a-day feeding of hay is desirable to reduce wastage. If wastage exceeds 10%, there is need for improvement in the manner of feeding or the quality of the hay, or both.

Silage is usually fed once daily, because of the added cost of more frequent feeding. However, some of the automated systems do not have the added labor cost factor, with the result that they may be used to feed twice daily, or more frequently. Silage should be fed soon after milking, so that any residual feed flavors will disappear before the next milking. Cows should not have access to silage, or other feeds that cause off-flavored milk, for at least 2 to 3 hours prior to milking.

Grain should be fed twice, or more frequently, daily. When high-producing cows are fed in the milking parlor, they are not in the milking parlor long enough to eat all the grain that they need; hence, part of the concentrate of high producers should be fed in the manger along with the hay and/or silage. Group or corral feeding of grain has evolved, as a means of lowering labor costs. Under this system, lactating cows are grouped according to level of production. Grain is fed twice daily in the manger, right after milking, either (1) as a top dressing on the silage and/or hay, or (2) mixed with the silage and/or hay, with the grain allocation for each group or corral determined by the average level of production of each cow in the group. The advent of electronic/computerized grain feeders gives the producer the opportunity to feed concentrates frequently and to control the amount of feed that cows consume.

Calves should be fed milk or milk replacer once or twice daily, at regular intervals. It is more harmful to overfeed than to underfeed a young calf.

Water should be clean, fresh, and available at all times. If there is too little water or if the cows must stand in line to get it, milk production will decrease. In cold weather, it may be necessary to protect water from freezing, as cows cannot get sufficient water by licking ice.

Frequency of Milking

Most cows are milked 2X per day. Increasing milking frequency to 3X per day increases milk production by 10 to 25% and milking 4X per day will stimulate milk yield another 5 to 15%. Whether these increases in milk production are worth the extra expense in labor, feed, utilities, and milking supplies depends upon economic conditions on each particular dairy farm. Also, with increasing genetic capability of cows, more frequent milking (3X and 4X) appears to improve udder health and lower the incidence of mastitis.

Milk and Fat Records; Testing Programs

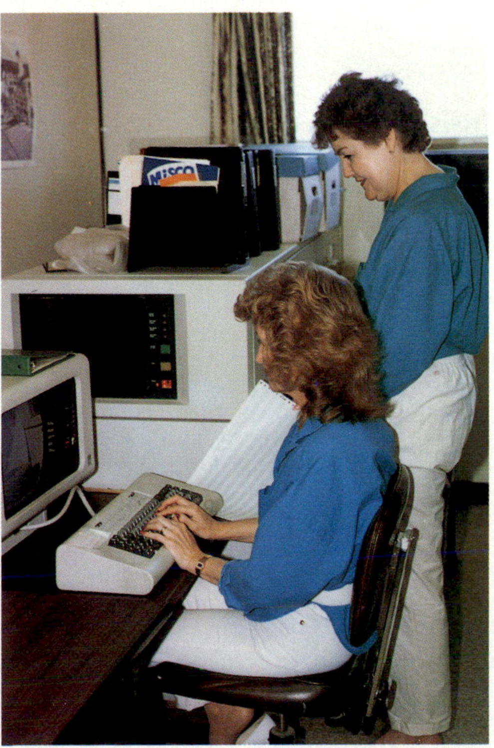

Fig. 20-39. Operating a computer in the office of Arizona Dairy Co., Higley, Ariz. (Courtesy, James Tappan)

Individual cow production and reproduction records are a must in any progressive dairy operation. They necessitate that each cow be identified. In addition to milk weight and fat test, many programs test for protein and somatic cell count. Also, most states and/or processing centers can store and provide information via computer relative to individual cow and herd records, reproduction problems, breeding dates, calving dates, genetics, heifer management, feeding, culling, health, and sales. With good and complete records, producers can identify the strengths and weaknesses of their operations, and the areas upon which improvements will reap the greatest financial rewards.

The three testing programs sponsored by Federal and state research extension services are Dairy Herd Improvement Association (DHIA), Owner-Sampler Records (OS), and Weigh-A-Day-A-Month (WADAM). Also, there is a wide range of sampling, such as sampling only one milking, alternate A.M. and P.M. weighing and sampling, etc. Each system has advantages and disadvantages. More important than the type of system used is the producer's total understanding and use of the system to obtain maximum benefits and willingness to keep records.

HEALTH DISORDERS OF DAIRY COWS

Milk production is a very intensive and demanding form of production, accompanied by much stress. Therefore, certain health problems are associated with it; among them, acidosis, bloat, displaced abomasum, fat cow syndrome, grass tetany, hardware disease, ketosis, milk fever, retained placenta, and udder edema.

Acidosis (Lactic Acidosis)

Acidosis is a metabolic disease of cattle and sheep.

When ruminants are switched from a high-roughage ration to a high-grain ration too rapidly, ruminal acidosis and atony may occur. The microflora of the rumen do not have time to adapt to the new ration, resulting in serious digestive disturbances. The pH of the rumen contents falls, and lactic acid production increases dramatically. Many of the microorganisms cannot live in this environment resulting in radical changes in the bacterial and protozoal populations. Rumen motility, then, becomes static. Afflicted animals show signs of weakness, diarrhea, and abdominal discomfort, and in many cases die.

Prevention consists in starting animals on a high-roughage ration and gradually reducing the roughage and increasing the grain; avoiding erratic feeding; and avoiding abrupt ration changes. Also, the addition of buffers to a high-grain ration aid in the prevention of acidosis.

Various treatments have been used with different degrees of success. Perhaps the most successful treatment consists in decreasing both the amount and kinds of feeds fed (lessening the total amount of the ration), then returning to a higher forage mix.

(Also see Chapter 5, Table 5-1—Acidosis [Lactic Acidosis].)

Bloat

Bloat affects all ruminants, dairy cattle included. When dairy animals are first introduced to concentrates or lush, legume pastures, bloat problems arise. Normally, the eructation process allows for the expulsion of gases that are produced in the rumen. In cases of bloat, frothing (the trapping of gas in the ingesta) prevents this process and intraruminal pressure builds, thereupon distending the left side of the abdomen until the animal is thrown off feed and goes down due to pain and the buildup of toxic metabolites.

Puncturing of the rumen of bloated cattle should be a last resort. **NOTE WELL:** Treatment, control, and prevention of bloat are detailed in the two bloat sections referred to in the parentheses that follow.

(Also see Chapter 5, Table 5-1—Bloat, Feedlot; and Bloat, Pasture.)

Displaced Abomasum

In recent years, dairy producers have encountered increasing numbers of cows with displaced abomasums. The practice of feeding dry cows liberal amounts of grain has been pointed to as one contributing factor. It has been suggested that the lack of bulk and rumen fill promotes flabby muscle tone of the rumen, thereby permitting the abomasum to become displaced. The feeding of some effective roughage is recommended.

(Also see Chapter 5, section on "Displaced Abomasum.")

Fat Cow Syndrome (Fatty Liver Syndrome)

Cows afflicted with fat cow syndrome have lowered level of liver function due to enlarged liver infiltrated with fat. General herd signs include very fat cows in the dry cow group, decreased resistance to infection, increased incidence of metabolic diseases such as ketosis, reduced feed intake, and reduced milk production and body weight.

Normally, the amount of fat in the liver is quite low (*i.e.,* 1 to 2%), although it may increase to 4 to 10% precalving. However, cows with the fat cow syndrome may have more than 20% fat accumulation in the liver.

The incidence of the fat cow syndrome may be reduced by avoiding overconditioning of cows during late lactation and the dry period and by formulating rations that maximize feed intake after calving.

Grass Tetany (Hypomagnesemia)

Grass tetany is a metabolic condition that affects cows (especially lactating cows) grazing lush pasture high in nitrogen, resulting in low absorption of magnesium, and is most common in the spring. Afflicted animals develop tetany, walk with a stiff gait, go into convulsions, and may die. During the danger period, cows grazing lush pasture should be supplemented with magnesium oxide.

(Also see Chapter 5, Table 5-1—Grass Tetany.)

Hardware Disease

The condition in which the collection of foreign objects irritates or punctures the reticulum is called reticulitis, or *hardware disease.* Ruminants are grazers by nature and are sometimes indiscriminant in their selection of feed. Often they will consume nails, pieces of wire, and other foreign objects. Due to the motility patterns of the gastric region, these objects tend to accumulate in the reticulum; and the presence of these sharp objects can pose serious problems, especially if the reticulum should be punctured.

(Also see Chapter 5, p. 211, section on "Hardware Disease.")

Ketosis

Ketosis is a metabolic disease characterized by a drop in milk production, hypoglycemia, ketonuria, and a rapid loss of weight. In general, the disorder develops within the first 30 days of lactation. While no preventative measures have proven to be 100% effective, the feeding of propylene glycol or sodium propionate has been successful in some cases; and the inclusion of niacin in the ration at a level of 6 g per head per day in the last 2 weeks of the dry period and in the fresh cow ration may aid in reducing the incidence of ketosis. Additionally, starting limited grain feeding during the latter part of the dry period is helpful in preventing ketosis.

If a cow comes down with ketosis, a glucose solution can be administered intavenously to promote rapid recovery.

(Also see Chapter 5, Table 5-1—Ketosis.)

Mastitis

Mastitis is an infection of the mammary gland caused by any one of several bacterial organisms, most frequently *staphylococcus* or *streptococcus.* Symptoms vary with degree of inflammation. Acute cases show a swollen and painful udder, and frequently cause the cow to go off feed. Chronic cases result in slightly swollen udders and small flakes in the milk.

No feed is known to cause or cure mastitis. However, the sudden addition of nutrients may result in a marked increase in milk production and cause more stress; in turn, this may cause subclinical cases. Also, feeding recommended levels of selenium and vitamin E may be helpful in preventing mastitis.

(Also see Chapter 17, section on "Mastitis.")

Milk Fever

At or soon after calving (generally within 48 to 72 hours), a sharp decrease in blood calcium (hypocalcemia) occurs in some cows, resulting in loss of appetite, subnormal temperature, and an unsteady gait. This is followed by nervousness, and, finally, collapse or complete loss of consciousness. The head is usually turned back. The name *milk fever* is a misnomer, because the body temperature is below normal.

The triggering mechanism for this drop in blood calcium is the onset of lactation—an intensive mobilization of calcium.

Feeding practices involving dry cows can markedly reduce the incidence of milk fever. When certain cows are known to have a history of milk fever, excessive calcium intake during the dry period should be avoided. If the problem persists, some nutritionists recommend the limited feeding of a high-energy, low-calcium (less than 15 g of calcium per day) ration. After calving, the calcium levels should be raised rapidly to meet the high requirements of lactation.

Recent studies have indicated that the addition of certain anions (negatively charged ions) to the ration may reduce the incidence of milk fever by aiding calcium absorption and mobilization. But more experimental work is needed relative to this method.

(Also see Chapter 5, Table 5-1—Milk Fever.)

Retained Placenta

Normally, the placenta is expelled within 3 to 6 hours after parturition. If it is retained as long as 12 hours after calving, competent assistance should be obtained.

Retained placenta occurs in about 10% of dairy cattle. It is more common following abnormally short or abnormally long pregnancies, among older cows, and following twinning. Experimentally, it has been found that a high incidence of retained afterbirth occurs when premature calving is induced by the administration of glucocorticoid drugs.

While infections such as brucellosis, vibriosis, and others have been associated with abortion and retained afterbirth, these are by no means the only causes. Its incidence increases with parturient hypocalcemia and appears to be related to the fat cow syndrome. Nutritionally, deficiencies of vitamin A, selenium, copper, and iodine have been incriminated. The prepartum injection of selenium at low doses has been shown to reduce the incidence of retained placenta. Also, it appears that fewer cases of retained placenta occur (1) when calves stay with their dams and nurse for 12 to 24 hours, and (2) when cows are kept on pasture year-round. Among cows which have previously retained the placenta, 20% are likely to do so again.

Calves born when the placenta is retained are likely to be weak. A retained placenta may cause pathological conditions resulting in uterine tissue destruction. This condition may or may not affect milk production, but it very likely will result in 5 to 10% lower fertility than for normal cows.

When a retained placenta is encountered, appropriate treatment should be administered by the veterinarian. The usual treatment consists of either antibiotics or sulfonamides, by direct infusion into the uterus or by other routes (or both).

It is seldom advisable to attempt removal of retained placenta. If the membranes are dragging in the ground, they should be cut off at the hocks. But never, never tie bricks or other objects to it. In most instances, the membranes will fall out by themselves in 1 to 2 weeks.

It is desirable to have all cows which have had retained afterbirth examined at about 30 days after calving. If pus is present, they may be treated with estrogenic hormones to induce heat and then the uterus can be infused with an antibiotic solution or perhaps with a dilute Lugol's (iodine) solution. Such examination and treatment may save considerable time with regard to the onset of normal cycles and may result in a higher conception rate.

Udder Edema

Udder edema, characterized by excessive accumulation of fluid in the intercellular spaces of the udder and forward of it, is sometimes of serious magnitude before calving. The cause is not well understood, but a reduction of blood proteins at calving time and increased blood flow without compensatory lymph removal have been suggested. It appears that high intakes of sodium chloride or potassium chloride increase the severity of udder edema, and that restriction of the salt intake will reduce the severity.

Severe edema may reduce milk production and may be one of the causes of pendulous udder.

Drugs and Pesticides

Many drugs used in the treatment of cattle diseases, along with many pesticides, are excreted in milk. Such milk should be discarded to prevent the drugs from entering the human food supply. The presence of antibiotics, sulfas, and pesticides in milk is illegal. Dairy producers should follow a residue avoidance program.

(Also see Chapter 13, section on "Food and Drug Administration [FDA].")

QUESTIONS FOR STUDY AND DISCUSSION

1. How and why does the feeding of lactating cows differ from the feeding of other classes of farm animals?

2. What motivating forces were back of group feeding? How is group feeding accomplished?

3. Discuss the economic importance of dairy cattle feeding.

4. Because it costs more to feed high producers than low producers, how can high producers return the most net profit?

5. Of what value are the National Research Council (NRC) Requirements? Why should one provide added margins of safety when using them?

6. Why is dry matter intake of such practical importance for high-producing cows?

7. Discuss dairy ration energy functions, deficiency symptoms, and sources.

8. What is the main source of energy for dairy cattle? What are the main sources of feed fats?

9. What is meant by the *net energy of lactation* (NE_{lc})? Why is this measure of feed energy so important in lactating cow rations?

10. List and discuss the factors that affect the digestion and absorption of the energy of feeds for dairy cows.

11. Discuss proteins for lactating cows from the standpoints of amount needed, excess, degraded and undegraded, rumen protected amino acids, and replacement by urea and other NPN products.

12. For dairy cattle, discuss (a) the conditions usually prevailing when a deficiency is reported, (b) the function, (c) the deficiency symptoms, and (d) the practical sources of each of the following minerals: salt, calcium, phosphorus, iodine, selenium, and zinc.

13. For dairy cattle, discuss (a) the conditions usually prevailing when a deficiency is reported, (b) the function, (c) the deficiency symptoms, and (d) the practical sources of each of the following vitamins: vitamin A, vitamin D, vitamin E, niacin, riboflavin, thiamin, and vitamin B–12.

14. How much water will a lactating dairy cow drink in a day? Why is the water requirement for a lactating cow so much higher than it is for other classes of farm animals?

15. Discuss the importance, nutrient value, and utilization of each of the following forages for dairy cattle: (a) pasture, (b) green chop, (c) hays, (d) silages, and (e) haylage.

16. List five common concentrate and five common by-product feeds for dairy cattle, and evaluate each of them for milk production.

17. Discuss the functions, amount, and type of fat to add to lactating cow rations.

18. Why is fiber so important in the rations of lactating cows? Discuss the optimum amounts of crude fiber, ADF, and NDF in the ration of dairy cows.

19. Why and how are each of the following feed additives used, practically or experimentally, in dairy rations: antibiotics, bovine somatotropin (BST), buffers, ionophores, isoacid, and thyroprotein?

20. What types of commercial dairy feeds are available? Why are dairy producers major users of U.S. commercial feeds?

21. Discuss the importance of each of the following feed factors in dairy nutrition: palatability, variety, bulk, laxativeness, cost, and effect on milk flavor.

22. How should grain be prepared for dairy cattle?

23. Under what sort of stress is the lactating cow placed? Discuss the importance of monitoring dry matter intake in dairy cows.

24. What general thumb rules should be followed when feeding (1) forages, and (2) concentrates to lactating cows?

25. Why should a dairy producer be well informed relative to the milk-feed price relationship?

26. What factors determine the amount of grain that it will pay to feed milking cows?

27. Discuss the economics of an all-forage ration versus a forage and concentrate ration for lactating cows.

28. Is there need for, or a better alternative to, Table 20–12 (or a similar table) for use in balancing lactating cow rations by producers who (a) home-mix their dairy rations, (b) are without chemical analyses of feed ingredients, (c) are without access to a computer, and (d) do not have processing and mixing equipment with which to combine their roughage(s) with their concentrates to make a complete ration?

29. Are all computer-formulated rations better than hand (pencil)-formulated rations? What are the advantages of computer-formulated rations compared to pencil-formulated rations? What is *linear programming?* List and describe the necessary data files which are generally created in steps. What is the computer spread sheet? How would you go about selecting computer software and hardware for ration formulation?

30. Define phase feeding; and discuss the four lactation cycle phases with corresponding changes in milk production, milk fat percentage, dry matter intake, and body weight.

31. Describe challenge feeding. Discuss its advantages and disadvantages.

32. Study the group feeding program presented in this chapter under the section headed "Group Feeding," and analyze it from the following standpoints: (a) the number and types of groups; (b) minimizing the behavioral adaptation in moving animals to a new production group; (c) milking parlor versus corral feeding, or a combination of both; (d) the use of a complete feed versus feeding the concentrates and roughages separately; (e) the disadvantages of group feeding; (f) the use of automatic grain feeders; (g) the group feeding nutrient specification guidelines and example rations presented in Tables 20–15 to 20–18.

33. When on pasture, how does a lactating cow divide her time between harvesting and ruminating?

34. Outline the routine for feeding and managing dry cows.

35. Outline and discuss a feeding program for young dairy calves, including colostrum, a starter ration, hay or silage, method of feeding, frequency of feeding, and age of weaning.

36. Why is it important that young calves get a good feeding of colostrum as soon after birth as possible?

37. What are veal calves? How are they fed? Why do consumers prefer grayish pink veal?

38. List recommended practices to lessen the incidence of calf scours.

39. Discuss the feeding of replacement heifers. What is the relationship of growth of dairy heifers to time of breeding?

40. Discuss the feeding of dairy bulls, dairy beef, and show and sale animals.

41. List 10 feed and management factors that affect milk yield and composition, and tell how each affects yield, fat, and SNF (see Table 20–25).

42. What factors determine whether 3X or 4X per day milking will be profitable?

43. Why, and how, should dairy producers keep individual milk, butter fat, and other records?

44. Discuss the cause, prevention, and treatment of each of the following health disorders of dairy cows: acidosis, bloat, displaced abomasum, fat cow syndrome, grass tetany, hardware disease, ketosis, mastitis, milk fever, retained placenta, and udder edema.

45. Why should dairy producers follow a drug and pesticide residue avoidance program?

SELECTED REFERENCES

Title of Publication	Author(s)	Publisher
Animal Science, 8th Edition	M. E. Ensminger	The Interstate Printers & Publishers, Inc., Danville, Ill., 1983
Dairy Cattle: Principles, Practices, Problems, Profits, 3rd Edition	D. L. Bath, *et al.*	Lea & Febiger, Philadelphia, Pa., 1985
Dairy Cattle Science, 2nd Edition	M. E. Ensminger	The Interstate Printers & Publishers, Inc., Danville, Ill., 1980
Dairy Farm Management	T. Quinn	Delmar Publishers, Inc., Albany, N.Y., 1980
Energy Metabolism of Ruminants, The	K. L. Blaxter	Hutchinson & Co., Ltd., London, England, 1962
Feed Formulations, 3rd Edition	T. W. Perry	The Interstate Printers & Publishers, Inc., Danville, Ill., 1982
Feeds and Feeding, 22nd Edition	F. B. Morrison	Morrison Publishing Co., Ithaca, N.Y., 1956
Illinois-Iowa Dairy Handbook	Cooperative Ext. Service	University of Illinois, Urbana, and Iowa State University, Ames, 1983
Large Dairy Herd Management	C. J. Wilcox, *et al.*	University Presses of Florida, Gainesville, 1978

(Continued)

(Continued)

Title of Publication	Author(s)	Publisher
Livestock Feeds and Feeding, 2nd Edition	D. C. Church	O & B Books, Inc., Corvallis, Ore., 1984
Nutrient Requirements of Dairy Cattle, 6th Revised Edition	NRC, R. W. Hemken, Chairman, Subcommittee, Dairy Cattle Nutrition	National Academy Press, Washington, D.C., 1988
Principles and Practices of Feeding Dairy Cows	W. H. Broster, *et al.*	College of Estate Management, Reading University, 1986
Science of Providing Milk for Man, The	J. R. Campbell R. T. Marshall	McGraw-Hill Book Company, New York, N.Y., 1975
Spreadsheet Applications for Animal Nutrition and Feeding	R. J. Lane T. L. Cross	Reston Publishing Co., Inc., Reston, Va., 1985
Stockman's Handbook, The, 6th Edition	M. E. Ensminger	The Interstate Printers & Publishers, Inc., Danville, Ill., 1983

Fig. 20-40. Jersey cows on pasture at Cub Hill Farm, Urbana, Ohio. (Courtesy, American Jersey Cattle Club, Columbus, Ohio)

Chapter 21

Original painting by Tom Phillips

FEEDING SHEEP[1 2]

Contents *Page*

Contents *Page*

[1]The authors gratefully acknowledge the helpful suggestions of the following authorities who reviewed this chapter: R. R. Jordan, Ph.D., Extension Animal Scientist, Department of Animal Science, University of Minnesota, St. Paul; J. B. Outhouse, Ph.D., Professor Emeritus, Department of Animal Science, Purdue University, West Lafayette, Ind.; and C. E. Terrill, Ph.D., Collaborator, USDA, Beltsville, Md.

[2]The statistics presented in the introductory section were for the mid-1980s and were obtained from: *FAO Production Yearbook*, published by FAO, Vol. 39, 1985; and *Agricultural Statistics*, USDA, 1985.

Fig. 21-1. A range band of sheep. ("Earning His Keep," an original painting by artist Tom Phillips, 3333 17th St., San Francisco, CA 94110)

The world sheep population is 1.1 billion head. Australia leads in numbers, with 150 million head; and the U.S.S.R. ranks second, with 143 million head.

Sheep produce 4.2% of the world's meat and 1.7% of the world's milk. Additionally, they produce more than 6.6 billion lb of grease wool (4 billion lb scoured wool), annually. The annual world per capita consumption is about 2.8 lb of lamb and mutton, 3.9 lb of sheep milk, and 0.8 lb of scoured wool.

In the mid-1980s, there were about 10.4 million sheep in the United States, with a total value of $638 million; and these sheep produced 380 million lb of meat and 93 million lb of grease wool. About 78% of the gross income from sheep is from meat and 22% from wool.

Although sheep are produced in every state, the 17 western states account for 82% of the nation's total production. Texas is the leading state in sheep production, with more than 18% of the nation's breeding sheep. Following Texas, the other top ranking states in numbers, in order, are: California, Wyoming, South Dakota, Utah, New Mexico, Montana, and Colorado.

United States sheep production and management systems vary greatly. Western range sheep bands number from 1,000 to 5,000 head and are handled in extensive operations in terms of land area. The intensive sheep operations of the rest of the U.S. vary from flocks of 50 ewes to 3,000 ewes. Lamb feeding operations vary from 1,000 to 50,000 head on feed at the same time.

Sheep consume a higher proportion of forages than any other class of livestock, it being estimated that 94% of the total feed supply of the U.S. sheep production is derived from forages. They are naturally adapted to grazing pastures and ranges which supply a variety of forage plants, and they thrive best on forage that is short and fine rather than high and course. Although sheep will eat considerable quantities of weeds and brush, they prefer choice grasses and legumes.

Except at lambing season, breeding sheep seldom need much grain, although when grain is less costly than forage, on the basis of energy and/or protein, it may be fed to advantage. In the northern latitudes, farm-flock ewes are frequently given ½ to 1 lb daily of a grain ration, in addition to the forage allowance, beginning about 6 weeks before lambing; and the grain allowance is usually doubled for the first 8 weeks after lambing.

Also, ewes suckling twins or on an accelerated lambing schedule are generally fed relatively high grain levels. However, many of the farm flocks of the South and the range bands of the Southwest are kept in good thrifty condition, and the lambs are raised to the marketing stage, without the feeding of any grain. In still other areas, the ewes are fed only during periods of deep snows or extended droughts. The range bands in the colder regions of the West are normally fed alfalfa hay and grain during the period of about 3 to 4 weeks that they are confined to the lambing camp.

In general, for practical reasons, the ration of ewes should consist of pastures as nearly year-round as possible, with well-cured hay and other forages available the balance of the year, plus a limited grain allowance under certain conditions. Good-quality sun-cured hay and lush pastures will not only provide sufficient protein, but they are excellent sources of most of the minerals and vitamins, also.

Fig. 21-2. Sheep and cattle rotated on superior pastures on a beautiful farm. (Courtesy, J. C. Allen and Son, Inc., West Lafayette, Ind.)

ECONOMIC IMPORTANCE OF FEED FOR SHEEP

Feed is economically important in sheep production; it accounts for 50 to 65% of the total cost of producing market lambs, and it accounts for about two-thirds of the cost of finishing feedlot lambs. This leads to the deduction that there are two ways to make money in the sheep business: (1) increase the value of the products (lamb and wool), and/or (2) reduce the cost of production. Of the two, the latter is more feasible for the vast majority of producers.

Feed is the major item of expense whether consumed as range vegetation, as permanent or improved pasture, or as rations in confinement; and whether the sheep producer be a western rancher, a purebred breeder, or a farm flock operator. It is noteworthy, too, that 5 to 10% of all ewes bred fail to lamb, and that 15 to 20% of all lambs die between birth and weaning. Although there are many causes of reproductive failure, it is recognized that faulty nutrition is a major contributing factor.

Nutritional deficiencies and diseases in sheep are of special concern because they have such widely-differing uncontrolled diets as a consequence of the great variety of conditions under which they are produced.

Fig. 21-3. Feed accounts for 50 to 65% of the total cost of producing market lambs. This shows Suffolk ewes eating concentrate feed. (Courtesy, Heupel Ranch, Santa Maria, Calif.)

NUTRITIVE NEEDS OF SHEEP

As with other classes of livestock, the nutritive requirements of sheep may be classified as (1) energy, (2) protein, (3) minerals, (4) vitamins, and (5) water.

The nutritive requirements are the values considered necessary for maintenance, optimum production, and prevention of all signs of nutritional deficiency.

National Research Council (NRC) Requirements

The National Research Council (NRC) nutritive requirements of sheep are given in the following tables: Table 21-1, 21-2, 21-3, 21-4, 21-5, 21-6, 21-7, and 21-10.

The NRC requirements are adequate for average, or below average, animals. In practical rations, margins of safety should be added to provide for below-average feeds, deterioration of feeds during transportation and storage, conditions of stress (bad weather, shipment, disease, or parasitism), and above-average animals in size, stage of production, and level of production.

TABLE 21-1
DAILY NUTRIENT REQUIREMENTS OF SHEEP (PER ANIMAL)[1] (See footnotes at end of table.)

Body Weight		Weight Gain/Loss Per Day		Daily Consumption: As-Fed[2]		Daily Consumption: Moisture-Free Dry Matter[3]		Daily Consumption: % Body Weight	Nutrients Per Animal: Energy[4]: TDN		Energy[4]: DE	Energy[4]: ME	Crude Protein		Ca	P	Vitamin A Activity	Vitamin E Activity
(lb)	*(kg)*	(lb)	*(g)*	(lb)	*(kg)*	(lb)	*(kg)*	(%)	(lb)	*(kg)*	(Mcal)	(Mcal)	(lb)	*(g)*	(g)	(g)	(IU)	(IU)
EWES[5]																		
Maintenance																		
110	*50*	0.02	*10*	2.4	*1.1*	2.2	*1.0*	2.0	1.2	*0.55*	2.4	2.0	0.21	*95*	2.0	1.8	2,350	15
132	*60*	0.02	*10*	2.7	*1.2*	2.4	*1.1*	1.8	1.3	*0.61*	2.7	2.2	0.23	*104*	2.3	2.1	2,820	16
154	*70*	0.02	*10*	2.9	*1.3*	2.6	*1.2*	1.7	1.5	*0.66*	2.9	2.4	0.25	*113*	2.5	2.4	3,290	18
176	*80*	0.02	*10*	3.2	*1.4*	2.9	*1.3*	1.6	1.6	*0.72*	3.2	2.6	0.27	*122*	2.7	2.8	3,760	20
198	*90*	0.02	*10*	3.4	*1.6*	3.1	*1.4*	1.5	1.7	*0.78*	3.4	2.8	0.29	*131*	2.9	3.1	4,230	21
Flushing—2 Weeks prebreeding and first 3 weeks of breeding																		
110	*50*	0.22	*100*	3.9	*1.8*	3.5	*1.6*	3.2	2.1	*0.94*	4.1	3.4	0.33	*150*	5.3	2.6	2,350	24
132	*60*	0.22	*100*	4.1	*1.9*	3.7	*1.7*	2.8	2.2	*1.00*	4.4	3.6	0.34	*157*	5.5	2.9	2,820	26
154	*70*	0.22	*100*	4.4	*2.0*	4.0	*1.8*	2.6	2.3	*1.06*	4.7	3.8	0.36	*164*	5.7	3.2	3,290	27
176	*80*	0.22	*100*	4.7	*2.1*	4.2	*1.9*	2.4	2.5	*1.12*	4.9	4.0	0.38	*171*	5.9	3.6	3,760	28
198	*90*	0.22	*100*	4.9	*2.2*	4.4	*2.0*	2.2	2.6	*1.18*	5.1	4.2	0.39	*177*	6.1	3.9	4,230	30
Nonlactating—First 15 weeks gestation																		
110	*50*	0.07	*30*	2.9	*1.3*	2.6	*1.2*	2.4	1.5	*0.67*	3.0	2.4	0.25	*112*	2.9	2.1	2,350	18
132	*60*	0.07	*30*	3.2	*1.4*	2.9	*1.3*	2.2	1.6	*0.72*	3.2	2.6	0.27	*121*	3.2	2.5	2,820	20
154	*70*	0.07	*30*	3.4	*1.6*	3.1	*1.4*	2.0	1.7	*0.77*	3.4	2.8	0.29	*130*	3.5	2.9	3,290	21
176	*80*	0.07	*30*	3.7	*1.7*	3.3	*1.5*	1.9	1.8	*0.82*	3.6	3.0	0.31	*139*	3.8	3.3	3,760	22
198	*90*	0.07	*30*	3.9	*1.8*	3.5	*1.6*	1.8	1.9	*0.87*	3.8	3.2	0.33	*148*	4.1	3.6	4,230	24
Last 4 weeks gestation (130–150% lambing rate expected) or last 4–6 weeks lactation suckling singles[6]																		
110	*50*	0.40 (0.10)	*180 (45)*	3.9	*1.8*	3.5	*1.6*	3.2	2.1	*0.94*	4.1	3.4	0.38	*175*	5.9	4.8	4,250	24
132	*60*	0.40 (0.10)	*180 (45)*	4.1	*1.9*	3.7	*1.7*	2.8	2.2	*1.00*	4.4	3.6	0.40	*184*	6.0	5.2	5,100	26
154	*70*	0.40 (0.10)	*180 (45)*	4.4	*2.0*	4.0	*1.8*	2.6	2.3	*1.06*	4.7	3.8	0.42	*193*	6.2	5.6	5,950	27
176	*80*	0.40 (0.10)	*180 (45)*	4.7	*2.1*	4.2	*1.9*	2.4	2.4	*1.12*	4.9	4.0	0.44	*202*	6.3	6.1	6,800	28
198	*90*	0.40 (0.10)	*180 (45)*	4.9	*2.2*	4.4	*2.0*	2.2	2.5	*1.18*	5.1	4.2	0.47	*212*	6.4	6.5	7,650	30
Last 4 weeks gestation (180–225% lambing rate expected)																		
110	*50*	0.50	*225*	4.1	*1.9*	3.7	*1.7*	3.4	2.4	*1.10*	4.8	4.0	0.43	*196*	6.2	3.4	4,250	26
132	*60*	0.50	*225*	4.4	*2.0*	4.0	*1.8*	3.0	2.6	*1.17*	5.1	4.2	0.45	*205*	6.9	4.0	5,100	27
154	*70*	0.50	*225*	4.7	*2.1*	4.2	*1.9*	2.7	2.8	*1.24*	5.4	4.4	0.47	*214*	7.6	4.5	5,950	28
176	*80*	0.50	*225*	4.9	*2.2*	4.4	*2.0*	2.5	2.9	*1.30*	5.7	4.7	0.49	*223*	8.3	5.1	6,800	30
198	*90*	0.50	*225*	5.1	*2.3*	4.6	*2.1*	2.3	3.0	*1.37*	6.0	5.0	0.51	*232*	8.9	5.7	7,650	32
First 6–8 weeks lactation suckling singles or last 4–6 weeks lactation suckling twins[6]																		
110	*50*	−0.06 (0.20)	*−25 (90)*	5.1	*2.3*	4.6	*2.1*	4.2	3.0	*1.36*	6.0	4.9	0.67	*304*	8.9	6.1	4,250	32
132	*60*	−0.06 (0.20)	*−25 (90)*	5.7	*2.6*	5.1	*2.3*	3.8	3.3	*1.50*	6.6	5.4	0.70	*319*	9.1	6.6	5,100	34
154	*70*	−0.06 (0.20)	*−25 (90)*	6.1	*2.8*	5.5	*2.5*	3.6	3.6	*1.63*	7.2	5.9	0.73	*334*	9.3	7.0	5,950	38
176	*80*	−0.06 (0.20)	*−25 (90)*	6.3	*2.9*	5.7	*2.6*	3.2	3.7	*1.69*	7.4	6.1	0.76	*344*	9.5	7.4	6,800	39
198	*90*	−0.06 (0.20)	*−25 (90)*	6.6	*3.0*	5.9	*2.7*	3.0	3.8	*1.75*	7.6	6.3	0.78	*353*	9.6	7.8	7,650	40
First 6–8 weeks lactation suckling twins																		
110	*50*	−0.13	*−60*	5.9	*2.7*	5.3	*2.4*	4.8	3.4	*1.56*	6.9	5.6	0.86	*389*	10.5	7.3	5,000	36
132	*60*	−0.13	*−60*	6.3	*2.9*	5.7	*2.6*	4.3	3.7	*1.69*	7.4	6.1	0.89	*405*	10.7	7.7	6,000	39
154	*70*	−0.13	*−60*	6.9	*3.1*	6.2	*2.8*	4.0	4.0	*1.82*	8.0	6.6	0.92	*420*	11.0	8.1	7,000	42
176	*80*	−0.13	*−60*	7.3	*3.3*	6.6	*3.0*	3.8	4.3	*1.95*	8.6	7.0	0.96	*435*	11.2	8.6	8,000	45
198	*90*	−0.13	*−60*	7.8	*3.6*	7.0	*3.2*	3.6	4.6	*2.08*	9.2	7.5	0.99	*450*	11.4	9.0	9,000	48
EWE LAMBS																		
Nonlactating—First 15 weeks gestation																		
88	*40*	0.35	*160*	3.4	*1.6*	3.1	*1.4*	3.5	1.8	*0.83*	3.6	3.0	0.34	*156*	5.5	3.0	1,880	21
110	*50*	0.30	*135*	3.7	*1.7*	3.3	*1.5*	3.0	1.9	*0.88*	3.9	3.2	0.35	*159*	5.2	3.1	2,350	22
132	*60*	0.30	*135*	3.9	*1.8*	3.5	*1.6*	2.7	2.0	*0.94*	4.1	3.4	0.35	*161*	5.5	3.4	2,820	24
154	*70*	0.28	*125*	4.1	*1.9*	3.7	*1.7*	2.4	2.2	*1.00*	4.4	3.6	0.36	*164*	5.5	3.7	3,290	26

(Continued)

TABLE 21-1 *(Continued)*

Body Weight		Weight Gain/Loss Per Day		Daily Consumption: As-Fed[2]		Daily Consumption: Moisture-Free Dry Matter[3]		Daily Consumption: % Body Weight	Nutrients Per Animal: Energy[4]: TDN		Energy[4]: DE	Energy[4]: ME	Crude Protein		Ca	P	Vitamin A Activity	Vitamin E Activity
(lb)	*(kg)*	(lb)	*(g)*	(lb)	*(kg)*	(lb)	*(kg)*	(%)	(lb)	*(kg)*	(Mcal)	(Mcal)	(lb)	*(g)*	(g)	(g)	(IU)	(IU)
EWE LAMBS (Continued)																		
Last 4 weeks gestation (100–120% lambing rate expected)																		
88	*40*	0.40	*180*	3.7	*1.7*	3.3	*1.5*	3.8	2.1	*0.94*	4.1	3.4	0.41	*187*	6.4	3.1	3,400	22
110	*50*	0.35	*160*	3.9	*1.8*	3.5	*1.6*	3.2	2.2	*1.00*	4.4	3.6	0.42	*189*	6.3	3.4	4,250	24
132	*60*	0.35	*160*	4.1	*1.9*	3.7	*1.7*	2.8	2.4	*1.07*	4.7	3.9	0.42	*192*	6.6	3.8	5,100	26
154	*70*	0.33	*150*	4.4	*2.0*	4.0	*1.8*	2.6	2.5	*1.14*	5.0	4.1	0.43	*194*	6.8	4.2	5,950	27
Last 4 weeks gestation (130–175% lambing rate expected)																		
88	*40*	0.50	*225*	3.7	*1.7*	3.3	*1.5*	3.8	2.2	*0.99*	4.4	3.6	0.44	*202*	7.4	3.5	3,400	22
110	*50*	0.50	*225*	3.9	*1.8*	3.5	*1.6*	3.2	2.3	*1.06*	4.7	3.8	0.45	*204*	7.8	3.9	4,250	24
132	*60*	0.50	*225*	4.1	*1.9*	3.7	*1.7*	2.8	2.5	*1.12*	4.9	4.0	0.46	*207*	8.1	4.3	5,100	26
154	*70*	0.47	*215*	4.4	*2.0*	4.0	*1.8*	2.6	2.5	*1.14*	5.0	4.1	0.46	*210*	8.2	4.7	5,950	27
First 6–8 weeks suckling singles (wean by 8 weeks)																		
88	*40*	–0.22	*–100*	5.1	*2.3*	4.6	*2.1*	5.2	3.2	*1.45*	6.4	5.2	0.67	*306*	8.4	5.6	4,000	32
110	*50*	–0.22	*–100*	5.7	*2.6*	5.1	*2.3*	4.6	3.5	*1.59*	7.0	5.7	0.71	*321*	8.7	6.0	5,000	34
132	*60*	–0.22	*–100*	6.1	*2.8*	5.5	*2.5*	4.2	3.8	*1.72*	7.6	6.2	0.74	*336*	9.0	6.4	6,000	38
154	*70*	–0.22	*–100*	6.7	*3.0*	6.0	*2.7*	3.9	4.1	*1.85*	8.1	6.6	0.77	*351*	9.3	6.9	7,000	40
REPLACEMENT EWE LAMBS[7]																		
66	*30*	0.50	*227*	2.9	*1.3*	2.6	*1.2*	4.0	1.7	*0.78*	3.4	2.8	0.41	*185*	6.4	2.6	1,410	18
88	*40*	0.40	*182*	3.4	*1.6*	3.1	*1.4*	3.5	2.0	*0.91*	4.0	3.3	0.39	*176*	5.9	2.6	1,880	21
110	*50*	0.26	*120*	3.7	*1.7*	3.3	*1.5*	3.0	1.9	*0.88*	3.9	3.2	0.30	*136*	4.8	2.4	2,350	22
132	*60*	0.22	*100*	3.7	*1.7*	3.3	*1.5*	2.5	1.9	*0.88*	3.9	3.2	0.30	*134*	4.5	2.5	2,820	22
154	*70*	0.22	*100*	3.7	*1.7*	3.3	*1.5*	2.1	1.9	*0.88*	3.9	3.2	0.29	*132*	4.6	2.8	3,290	22
REPLACEMENT RAM LAMBS[7]																		
88	*40*	0.73	*330*	4.4	*2.0*	4.0	*1.8*	4.5	2.5	*1.1*	5.0	4.1	0.54	*243*	7.8	3.7	1,880	24
132	*60*	0.70	*320*	5.9	*2.7*	5.3	*2.4*	4.0	3.4	*1.5*	6.7	5.5	0.58	*263*	8.4	4.2	2,820	26
176	80	0.64	*290*	6.9	*3.1*	6.2	*2.8*	3.5	3.9	*1.8*	7.8	6.4	0.59	*268*	8.5	4.6	3,760	28
220	*100*	0.55	*250*	7.3	*3.3*	6.6	*3.0*	3.0	4.2	*1.9*	8.4	6.9	0.58	*264*	8.2	4.8	4,700	30
LAMBS FINISHING—4 TO 7 MONTHS OLD[8]																		
66	*30*	0.65	*295*	3.2	*1.4*	2.9	*1.3*	4.3	2.1	*0.94*	4.1	3.4	0.42	*191*	6.6	3.2	1,410	20
88	*40*	0.60	*275*	3.9	*1.7*	3.5	*1.6*	4.0	2.7	*1.22*	5.4	4.4	0.41	*185*	6.6	3.3	1,880	24
110	*50*	0.45	*205*	3.9	*1.7*	3.5	*1.6*	3.2	2.7	*1.23*	5.4	4.4	0.35	*160*	5.6	3.0	2,350	24
EARLY WEANED LAMBS—MODERATE GROWTH POTENTIAL[8]																		
22	*10*	0.44	*200*	1.2	*0.6*	1.1	*0.5*	5.0	0.9	*0.40*	1.8	1.4	0.38	*127*	4.0	1.9	470	10
44	*20*	0.55	*250*	2.4	*1.1*	2.2	*1.0*	5.0	1.8	*0.80*	3.5	2.9	0.37	*167*	5.4	2.5	940	20
66	*30*	0.66	*300*	3.2	*1.4*	2.9	*1.3*	4.3	2.2	*1.00*	4.4	3.6	0.42	*191*	6.7	3.2	1,410	20
88	*40*	0.76	*345*	3.7	*1.7*	3.3	*1.5*	3.8	2.6	*1.16*	5.1	4.2	0.44	*202*	7.7	3.9	1,880	22
110	*50*	0.66	*300*	3.7	*1.7*	3.3	*1.5*	3.0	2.6	*1.16*	5.1	4.2	0.40	*181*	7.0	3.8	2,350	22
EARLY WEANED LAMBS—RAPID GROWTH POTENTIAL[8]																		
22	*10*	0.55	*250*	1.4	*0.7*	1.3	*0.6*	6.0	1.1	*0.48*	2.1	1.7	0.35	*157*	4.9	2.2	470	12
44	*20*	0.66	*300*	2.9	*1.3*	2.6	*1.2*	6.0	2.0	*0.92*	4.0	3.3	0.45	*205*	6.5	2.9	940	24
66	*30*	0.72	*325*	3.4	*1.6*	3.1	*1.4*	4.7	2.4	*1.10*	4.8	4.0	0.48	*216*	7.2	3.4	1,410	21
88	*40*	0.88	*400*	3.7	*1.7*	3.3	*1.5*	3.8	2.5	*1.14*	5.0	4.1	0.51	*234*	8.6	4.3	1,880	22
110	*50*	0.94	*425*	4.1	*1.9*	3.7	*1.7*	3.4	2.8	*1.29*	5.7	4.7	0.53	*240*	9.4	4.8	2,350	25
132	*60*	0.77	*350*	4.1	*1.9*	3.7	*1.7*	2.8	2.8	*1.29*	5.7	4.7	0.53	*240*	8.2	4.5	2,820	25

[1]Adapted by the authors from *Nutrient Requirements of Sheep,* sixth revised edition, NRC-National Academy of Sciences, 1985, pp. 45–47.

[2]As-fed was calculated using an average figure of 90% dry matter. When using silages, roots, and other wet feeds, these feeds should be converted to a moisture-free basis and the ration calculated using the moisture-free data.

[3]To convert dry matter to an as-fed basis, divide dry matter values by the percentage of dry matter in the particular feed.

[4]One kilogram TDN (total digestible nutrients) = 4.4 Mcal DE (digestible energy); ME (metabolizable energy) = 82% of DE. Because of rounding numbers, values in Table 1 and Table 2 may differ.

[5]Values are applicable for ewes in moderate condition. Fat ewes should be fed according to the next lower category and thin ewes at the next higher weight category.

[6]Values in parentheses are for ewes suckling lambs the last 4–6 weeks of lactation.

[7]Lambs intended for breeding; thus, maximum weight gains and finish are of secondary importance.

[8]Maximum weight gains expected.

TABLE 21-2
NUTRIENT CONCENTRATION IN RATIONS FOR SHEEP[1] [2](See footnotes at end of table.)

Body Weight		Weight Gain/Loss Per Day		Moisture Basis[3] A-F (as-fed) M-F (moisture-free)	Energy[4]					Example Diet Proportions		Crude Protein	Cal-cium	Phos-phorus	Vitamin A Activity (IU/)		Vitamin E Activity (IU/)	
					TDN[5]	DE (Mcal/)		ME (Mcal/)		Con-centrate	Forage							
(lb)	(kg)	(lb)	(g)		(%)	(lb)	(kg)	(lb)	(kg)	(%)	(%)	(%)	(%)	(%)	(lb)	(kg)	(lb)	(kg)
EWES[6]																		
Maintenance																		
154	70	0.02	10	A-F	50	4.8	2.2	4.0	1.8	0	100	8.5	0.18	0.18	5,442	2,468	31	14
				M-F	55	5.3	2.4	4.4	2.0	0	100	9.4	0.20	0.20	6,046	2,742	33	15
Flushing—2 weeks prebreeding and first 3 weeks of breeding																		
154	70	0.22	100	A-F	53	5.1	2.3	4.1	1.9	15	85	8.2	0.29	0.16	3,627	1,645	31	14
				M-F	59	5.7	2.6	4.6	2.1	15	85	9.1	0.32	0.18	4,031	1,828	33	15
Nonlactating—First 15 weeks gestation																		
154	70	0.07	30	A-F	50	4.8	2.2	4.0	1.8	0	100	8.4	0.23	0.18	4,664	2,115	31	14
				M-F	55	5.3	2.4	4.4	2.0	0	100	9.3	0.25	0.20	5,182	2,350	33	15
Last 4 weeks gestation (130–150% lambing rate expected) or last 4–6 weeks lactation suckling singles[7]																		
154	70	0.40	180	A-F	53	5.1	2.3	4.1	1.9	15	85	9.6	0.32	0.21	6,561	2,975	31	14
		(0.10)	(0.45)	M-F	59	5.7	2.6	4.6	2.1	15	85	10.7	0.35	0.23	7,290	3,306	33	15
Last 4 weeks gestation (180–225% lambing rate expected)																		
154	70	0.50	225	A-F	59	5.8	2.6	4.6	2.1	35	65	10.2	0.36	0.22	6,215	2,819	31	14
				M-F	65	6.4	2.9	5.1	2.3	35	65	11.3	0.40	0.24	6,906	3,132	33	15
First 6–8 weeks lactation suckling singles or last 4–6 weeks lactation suckling twins[7]																		
154	70	-0.06	-25	A-F	59	5.8	2.6	4.8	2.2	35	65	12.1	0.29	0.23	4,723	2,142	31	14
		(0.20)	(90)	M-F	65	6.4	2.9	5.3	2.4	35	65	13.4	0.32	0.26	5,248	2,380	33	15
First 6–8 weeks lactation suckling twins																		
154	70	-0.13	-60	A-F	59	5.8	2.6	4.8	2.2	35	65	13.5	0.35	0.26	4,962	2,250	31	14
				M-F	65	6.4	2.9	5.3	2.4	35	65	15.0	0.39	0.29	5,513	2,500	33	15
EWE LAMBS																		
Nonlactating—First 15 weeks gestation																		
121	55	0.30	135	A-F	53	5.1	2.3	4.1	1.9	15	85	9.5	0.32	0.20	3,310	1,501	31	14
				M-F	59	5.7	2.6	4.6	2.1	15	85	10.6	0.35	0.22	3,678	1,668	33	15
Last 4 weeks gestation (100–120% lambing rate expected)																		
121	55	0.35	160	A-F	57	5.6	2.5	4.6	2.1	30	70	10.6	0.35	0.20	2,550	5,622	31	14
				M-F	63	6.2	2.8	5.1	2.3	30	70	11.8	0.39	0.22	6,247	2,833	33	15
Last 4 weeks gestation (130–175% lambing rate expected)																		
121	55	0.50	225	A-F	59	5.8	2.6	4.8	2.2	40	60	11.5	0.43	0.23	5,622	2,550	31	14
				M-F	66	6.4	2.9	5.3	2.4	40	60	12.8	0.48	0.25	6,247	2,833	33	15
First 6–8 weeks lactation suckling singles (wean by 8 weeks)																		
121	55	0.22	-50	A-F	59	5.8	2.6	4.8	2.2	40	60	11.8	0.27	0.20	4,217	1,913	31	14
				M-F	66	6.4	2.9	5.3	2.4	40	60	13.1	0.30	0.22	4,686	2,125	33	15
First 6–8 weeks lactation suckling twins (wean by 8 weeks)																		
121	55	-0.22	-100	A-F	62	5.9	2.7	5.0	2.3	50	50	12.3	0.33	0.23	4,549	2,063	31	14
				M-F	69	6.6	3.0	5.5	2.5	50	50	13.7	0.37	0.26	5.054	2,292	33	15
REPLACEMENT EWE LAMBS[8]																		
66	30	0.50	227	A-F	59	5.8	2.6	4.8	2.2	35	65	11.5	0.48	0.20	2,332	1,058	31	14
				M-F	65	6.4	2.9	5.3	2.4	35	65	12.8	0.53	0.22	2,591	1,175	33	15
88	40	0.40	182	A-F	59	5.8	2.6	4.8	2.2	35	65	9.2	0.38	0.16	2,665	1,209	31	14
				M-F	65	6.4	2.9	5.3	2.4	35	65	10.2	0.42	0.18	2,961	1.343	33	15
110–154	50–70	0.25	115	A-F	53	5.1	2.3	4.1	1.9	15	85	8.2	0.28	0.15	3,110	1,410	31	14
				M-F	59	5.7	2.6	4.6	2.1	15	85	9.1	0.31	0.17	3,455	1,567	33	15

(Continued)

TABLE 21-2 *(Continued)*

Body Weight		Weight Gain/Loss Per Day		Moisture Basis[3] A-F (as-fed) M-F (moisture-free)	Energy[4]					Example Diet Proportions		Crude Protein	Cal-cium	Phos-phorus	Vitamin A Activity (IU/)		Vitamin E Activity (IU/)	
					TDN[5]	DE (Mcal/)		ME (Mcal/)		Con-centrate	Forage							
(lb)	(kg)	(lb)	(g)		(%)	(lb)	(kg)	(lb)	(kg)	(%)	(%)	(%)	(%)	(%)	(lb)	(kg)	(lb)	(kg)
REPLACEMENT RAM LAMBS[8]																		
88	40	0.73	330	A-F	57	5.6	2.5	4.6	2.1	30	70	12.2	0.39	0.19	2,332	1,058	31	14
				M-F	63	6.2	2.8	5.1	2.3	30	70	13.5	0.43	0.21	2,591	1,175	33	15
132	60	0.70	320	A-F	57	5.6	2.5	4.6	2.1	30	70	9.9	0.32	0.16	3,292	1,493	31	14
				M-F	63	6.2	2.8	5.1	2.3	30	70	11.0	0.35	0.18	3,658	1,659	33	15
176–220	80–100	0.60	270	A-F	57	5.6	2.5	4.6	2.1	30	70	8.6	0.27	0.14	3,928	1,781	31	14
				M-F	63	6.2	2.8	5.1	2.3	30	70	9.6	0.30	0.16	4,364	1,979	33	15
LAMBS FINISHING—4 TO 7 MONTHS OLD[9]																		
66	30	0.65	295	A-F	65	6.4	2.9	5.0	2.3	60	40	13.2	0.46	0.22	2.153	977	31	14
				M-F	72	7.1	3.2	5.5	2.5	60	40	14.7	0.51	0.24	2,392	1,085	33	15
88	40	0.60	275	A-F	68	6.6	3.0	5.4	2.4	75	25	10.4	0.38	0.19	2,332	1,058	31	14
				M-F	76	7.3	3.3	6.0	2.7	75	25	11.6	0.42	0.21	2,591	1,175	33	15
110	50	0.45	205	A-F	69	6.8	3.1	5.6	2.5	80	20	9.0	0.32	0.17	2,915	1,322	31	14
				M-F	77	7.5	3.4	6.2	2.8	80	20	10.0	0.35	0.19	3,239	1,469	33	15
EARLY WEANED LAMBS—MODERATE AND RAPID GROWTH POTENTIAL[9]																		
22	10	0.55	250	A-F	72	6.9	3.2	5.8	2.6	90	10	23.4	0.74	0.34	1,866	846	40	18
				M-F	80	7.7	3.5	6.4	2.9	90	10	26.2	0.82	0.38	2,073	940	44	20
44	20	0.66	300	A-F	70	6.8	3.1	5.6	2.5	85	15	15.2	0.49	0.22	1,866	846	40	18
				M-F	78	7.5	3.4	6.2	2.8	85	15	16.9	0.54	0.24	2,073	940	44	20
66	30	0.72	325	A-F	70	6.6	3.0	5.4	2.4	85	15	13.6	0.46	0.22	2,153	977	31	14
				M-F	78	7.3	3.3	6.0	2.7	85	15	15.1	0.51	0.24	2,392	1,085	33	15
88–132	40–60	0.88	400	A-F	70	6.6	3.0	5.4	2.4	85	15	13.1	0.50	0.25	2,487	1,128	31	14
				M-F	78	7.3	3.3	6.0	2.7	85	15	14.5	0.55	0.28	2,763	1,253	33	15

[1]Adapted by the authors from *Nutrient Requirements of Sheep,* sixth revised edition, NRC-National Academy of Sciences, 1985, p. 48.

[2]Values in Table 2 are calculated from daily requirements in Table 1 divided by DM intake. The exception, vitamin E daily requirements/head, are calculated from vitamin E/kg diet × DM intake.

[3]As-fed was calculated using an average figure of 90% dry matter. When using silages, roots, and other wet feeds, these feeds should be converted to a moisture-free basis and the ration calculated using the moisture-free data.

[4]One kilogram TDN = 4.4 Mcal DE (digestible energy); ME (metabolizable energy) = 82% of DE. Because of rounding numbers, values in Table 1 and Table 2 may differ.

[5]TDN calculated on following basis: hay DM, 55% TDN and on as-fed basis 50% TDN; grain DM, 83% TDN and on as-fed basis 75% TDN.

[6]Values are for ewes in moderate condition. Fat ewes should be fed according to the next lower weight category and thin ewes at the next higher weight category. Once desired or moderate weight condition is attained, use that weight category through all production stages.

[7]Values in parentheses are for ewes suckling lambs the last 4–6 weeks of lactation.

[8]Lambs intended for breeding; thus, maximum weight gains and finish are of secondary importance.

[9]Maximum weight gains expected.

Energy

Lack of energy—hunger—is probably the most common nutritional deficiency of sheep. It may result from lack of feed or from the consumption of poor-quality feed.

Inadequate amounts of feed may result from (1) overgrazing, (2) drought, (3) snow covering the feed, (4) a low dry matter content of lush, washy feeds, or (5) a low level of range feed, with the result that sheep have to walk too far to obtain adequate intake. Also, poorly digested low-quality forage leads to reduced feed intake.

The energy needs of sheep are largely met through the consumption and digestion of forages—pasture, hay, and silage. Grains, such as corn, barley, milo, wheat, and oats, are used to raise the energy level of the ration during periods when supplementation is necessary. In general, sheep subsist on an even higher proportion of forages to concentrates than do beef cattle, and this applies to finishing lambs. The bacterial action in the rumen of sheep efficiently converts

Fig. 21-4. The energy needs of sheep are largely met through consumption of forages. (Courtesy, Department of Animal Science, Texas A&M University, College Station)

forages into suitable sources of energy.

Energy intake can be controlled by limiting the amount of feed offered, by adding fiber or bulk to the ration, by feeding every other day, or by limiting the time of eating.

In addition to size, age, pregnancy, lactation, and growth, covered in the nutrient requirement tables—Tables 21-1 and 21-2, and their relationship to such nutrients as protein, which must be supplied in adequate amounts, the following factors can affect energy requirements and diet concentration:

1. **Mature size of the breed (large mature genotype).** Lambs of the larger breeds (of the larger mature genotypes) grow more rapidly, have a higher energy (feed) requirement, and utilize energy (feed) more efficiently than lambs of the smaller breeds. Table 21-3 points up the comparative efficiency in the net energy for gain (NE_g) requirements of ram lambs of small, medium, and large genotypes with projected mature weights of 209, 254, and 298 lb, respectively. Note that when weighing 110 lb and making daily gains of 0.66 lb, the NE_g requirements of small, medium, and large mature weight rams are 1,788, 1,557, and 1,320 kcal/day, respectively.

Fig. 21-5. Big sheep must be fed liberally as lambs, because lambs of the larger breeds (larger mature genotypes) grow more rapidly. (Courtesy, *The Suffolk Journal*)

TABLE 21-3
NET ENERGY REQUIREMENTS FOR LAMBS OF SMALL, MEDIUM, AND LARGE MATURE WEIGHT GENOTYPES[1 2]

Body Weight[3]	Lb	22	44	55	66	77	88	99	110
	Kg	*10*	*20*	*25*	*30*	*35*	*40*	*45*	*50*
Daily Gain[3]		kcal/d	kcal/d	kcal/d	kcal/d	kcal/d	kcal/d	kcal/d	kcal/d
(lb)	(g)								
		NE_m REQUIREMENTS[4]							
		315	530	626	718	806	891	973	1,053
		NE_g REQUIREMENTS							
		Small mature weight lambs[5]							
.22	*100*	178	300	354	406	456	504	551	596
.33	*150*	267	450	532	610	684	756	826	894
.44	*200*	357	600	708	812	912	1,008	1,102	1,192
.55	*250*	446	750	886	1,016	1,140	1,261	1,377	1,490
.66	*300*	535	900	1,064	1,219	1.368	1,513	1,652	1,788
		Medium mature weight lambs[6]							
.22	*100*	155	261	309	354	397	439	480	519
.33	*150*	233	392	463	531	596	658	719	778
.44	*200*	310	522	618	708	794	878	960	1,038
.55	*250*	388	653	771	884	993	1,097	1,199	1,297
.66	*300*	466	784	926	1,062	1,191	1,316	1,438	1,557
.77	*350*	543	914	1,080	1,238	1,390	1,536	1,678	1,816
.88	*400*	621	1,044	1,234	1,415	1,589	1,756	1,918	2,076
		Large mature weight lambs[7]							
.22	*100*	132	221	262	300	337	372	407	439
.33	*150*	197	332	392	450	505	558	610	660
.44	*200*	263	442	524	600	674	744	813	880
.55	*250*	329	553	654	750	842	930	1,016	1,099
.66	*300*	394	663	785	900	1,010	1,016	1,220	1,320
.77	*350*	461	775	916	1,050	1,179	1,303	1,423	1,540
.88	*400*	526	885	1,046	1,200	1,347	1,489	1,626	1,760
.99	*450*	592	996	1,177	1,350	1,515	1,675	1,830	1,980

[1]Adapted by the authors from *Nutrient Requirements of Sheep*, sixth revised edition, NRC-National Academy of Sciences, 1985, p. 49.

[2]Approximate mature ram weights of 209 lb (*95 kg*), 254 lb (*115 kg*), and 298 lb (*135 kg*), respectively.

[3]Weights and gains include fill.

[4]NE_m = 56 kcal • $W^{0.75}$ • d^{-1}.

[5]NE_g = 317 kcal • $W^{0.75}$ • LWG, kg • d^{-1}.

[6]NE_g = 276 kcal • $W^{0.75}$ • LWG, kg • d^{-1}.

[7]NE_g = 234 kcal • $W^{0.75}$ • LWG, kg • d^{-1}.

2. **Lambs for breeding.** The energy requirements of lambs are affected by the following:

a. **Sex.** Ram lambs gain more rapidly and have higher feed requirements than ewe lambs. Also, intact (uncastrated) male lambs use feed more efficiently for body weight gains than ewe lambs, because of the higher protein and water and lower fat content of the increased body weight.

b. **Ewe lambs bred to lamb as yearlings.** Ewe lambs bred to lamb as yearlings should be fed more liberally from weaning to breeding, during pregnancy, and during lactation.

3. **Early weaned lambs.** Early weaned lambs (lambs weaned at 5 to 8 weeks of age, or even up to 20 weeks under some range conditions) lack the rumen development and capacity to utilize bulky feeds. Hence, they should be fed palatable, high-energy, adequate-protein rations; rations containing 80 to 90% concentrate and 14 to 20% protein.

4. **Finishing lambs shifted gradually to higher energy.** Care should be exercised in changing finishing lambs to high-energy rations. They should be shifted gradually from roughage rations to more concentrated, higher-energy rations in order to avoid digestive upsets.

5. **Previous nutrient intake and condition.** It takes more energy to get thin ewes in proper condition for reproduction and lactation than for ewes that are in good condition to begin with.

6. **Last 6 weeks of gestation.** Ewes need more energy during the last 6 weeks of gestation to meet increased requirements for fetal growth and the development of the potential for high milk production. Too much energy during gestation may lead to fattening and birth difficulties; too little energy may result in low birth weights, weak lambs, and pregnancy disease in ewes.

7. **Multiple births.** At various stages of gestation, ewes carrying twins require more net energy for pregnancy (NE_{preg}) than ewes carrying singles, and ewes carrying triplets require still more NE_{preg} than ewes carrying twins. Moreover, the spread in the energy required between ewes carrying singles, twins, and triplets increases with advancing gestation. This is pointed up in Table 21-4. Note (1) that at 140 days of gestation, the NE_{preg} of ewes carrying singles, twins, and triplets is 260, 440, and 570 kcal/day, respectively; and (2) that at 100 days of gestation, the spread between ewes carrying singles versus triplets is 100 kcal/day, whereas at 140 days of gestation the spread is 119 kcal/day—16% more.

8. **Lactation.** The lactation requirements are higher than the maintenance or gestation requirements. At the peak of lactation, it is estimated that the net energy requirements of ewes suckling twins are 1.7 to 1.9 times the maintenance requirement. Also, ewes nursing twin lambs produce 20 to 40% more milk than ewes nursing singles. However, milk production during the 3rd and 4th months of lactation is only about ½ of the production during the first 2 months; hence, following peak milk production, the feed allowance should be reduced in order to maximize profits.

Fig. 21-6. Ewes suckling twins produce 20 to 40% more milk than ewes nursing singles; hence, they need more feed, especially in early lactation. (Courtesy, The American Hampshire Sheep Assn., Ashland, Mo.)

9. **Shearing.** The energy requirements may increase at shearing in cold weather due to decreased insulation.

10. **Stress.** Stress of any kind, including internal and external parasites, increases energy requirements.

11. **Environment.** The energy requirements increase as temperature, humidity, and wind depart from the comfort zone.

• **Symptoms of energy deficiency**—An energy deficiency is characterized by slowing and cessation of growth, loss of weight, reduced fertility or reproductive failure, lowered milk production and shortened lactation period, reduced quantity and quality of wool (including breaks in the fiber), lowered resistance to infection with internal parasites, and increased mortality.

TABLE 21-4

NE_{PREG} (NE_Y) REQUIREMENTS OF EWES CARRYING DIFFERENT NUMBERS OF FETUSES AT VARIOUS STAGES OF GESTATION[1]

Number of Fetuses Being Carried	Stage of Gestation (days)[2]					
	100	%[3]	120	%[3]	140	%[3]
	NE_{preg} Required (kcal/day)					
1	70	100	145	100	260	100
2	125	178	265	183	440	169
3	170	243	345	238	570	219

[1]Adapted by the authors from *Nutrient Requirements of Sheep,* sixth revised edition, NRC-National Academy of Sciences, 1985, p. 49. The (NE_Y) refers to reproductive process.

[2]For gravid uterus (plus contents) and mammary gland development only.

[3]As a percentage of a single fetus's requirement.

Protein

Sheep need protein, as do other classes of animals, for maintenance, growth, reproduction, and finishing. Additionally, sheep need protein for the production of wool—a protein product. Wool is especially rich in the sulfur-containing amino acids, cystine and methionine, which are derived from rumen synthesis. Methionine is usually the most limiting amino acid for wool production.

Green pastures and legume hays (alfalfa, clover, soybeans, lespedeza, and others) are excellent practical sources of proteins for sheep in most areas. When the ranges are bleached and dry for an extended period, or legume hays cannot be produced for winter feeding, however, it may be desirable to provide sheep with such protein-rich supplements as soybean meal, cottonseed meal, linseed meal, peanut meal, canola (rapeseed) meal, sunflower meal, or a commercial protein supplement, at the rate of about ¼ to ⅓ lb per ewe per day.

Fig. 21-7. A green grass-legume pasture, an excellent source of protein for these Hampshire ewes. (Courtesy, *The Hampshire Journal*)

The protein requirements of sheep are affected by age, growth, pregnancy, lactation, mature size, weight for age, body condition, rate of gain, and protein-energy ratio. Though correspondingly less because of their smaller body size and lower milk production, the protein requirements of ewes nursing lambs are similar to those of lactating cows.

The lamb is born with a nonfunctional rumen; hence, dietary protein must be provided through milk or a milk replacer until the rumen becomes functional. The rumen develops some degree of functionality by 2 weeks of age, but during early rumen development creep feed should be provided to supplement milk or milk replacer. By 6 to 8 weeks of age, the functioning rumen develops into a culture system for anaerobic bacteria, protozoa, and fungi. These microbes digest feedstuffs and synthesize sufficient protein to allow efficient production without milk. Ruminal microorganisms can utilize either protein or nonprotein nitrogen to synthesize microbial protein. The microbial protein, along with undigested feed protein, passes from the rumen-reticulum through the omasum to the abomasum, then to the small intestine where it is subjected to digestive processes similar to those of the nonruminant and broken down to amino acids which are absorbed and utilized by sheep. Thus, sheep can utilize protein and nonprotein nitrogen in their diets. So, the protein available for digestion in the small intestine of sheep consists of microbial protein and feed protein that has escaped microbial breakdown in the rumen.

Additional factors that can affect the protein requirements and utilization by sheep follow:

1. **Protein bypass (protected protein, escaped protein).** Quality and degradability of protein fed to sheep is more important than formerly thought. It has been shown that the protein produced by ruminal synthesis does not supply all the amino acids in quality or quantity needed for maximum growth of lambs or milk production of ewes. Moreover, it has been found that protein efficiency can be increased by protecting protein from the degradation by the microbes in the rumen and increasing the escape of protein from the rumen to the intestines where it is digested and absorbed. Because protein is a costly ingredient, it follows that increasing its efficiency reduces cost because less of it is necessary. This technology of manipulating the quantity of dietary protein rumen fermentation, thereby increasing the supply of protein (amino acids) to the small intestine, which is known as protein bypass, is detailed and illustrated in Chapter 11, Protein Supplements, under the heading "Protein Bypass (Protected Protein, Escaped Protein)."

2. **Nonprotein nitrogen (NPN).** Urea or other nonprotein nitrogen, in either liquid or dry supplements, can be used to provide all the supplemental nitrogen that may be needed in high-energy, grain-based rations, provided the diets are properly formulated and fed continuously. Among the factors which should be observed for the optimum utilization of urea are:

   a. Provide a readily available energy source, such as molasses or grain.

   b. Supply adequate and balanced levels of minerals and other nutrients.

   c. Achieve a nitrogen-sulfur ratio not wider than 10:1.

   d. When the addition of nonprotein nitrogen to high-roughage rations is planned, either (1) provide a supplement containing a readily available source of energy (molasses and/or grain) and fortified with minerals and vitamins, fed at frequent intervals, or (2) use a slow-released nonprotein nitrogen product, such as biuret.

   e. Accustom animals gradually to nonprotein nitrogen-containing feeds, which may take as long as 3 to 5 weeks before maximum use of the nitrogen is obtained.

   f. Limit urea to not more than 1.0% of the dry matter in the ration (or ⅓ of the total nitrogen in the ration, or 3% of the concentrate portion of the ration).

   g. Reduce the hazard of nonprotein nitrogen toxicity by (1) preventing sheep from consuming large amounts of NPN from an empty start and in a short time, (2) gradually accustoming animals to it, and feeding regularly, (3) providing available energy in the ration, and (4) using a slow-release product.

3. **Mature size of genotype (breed).** Table 21-5 gives the crude protein requirements of ram lambs of small, medium, and large mature weight genotypes, with projected mature weights of 209, 254, and 298 lb, respectively. It shows that the daily crude protein requirements of lambs increase with the size of the mature genotype (from small to medium, to large), and with the daily weight gains; but that the daily crude protein requirement per pound of body weight decreases as lambs become heavier because of the decrease in the protein content and the concomitant increase in the fat content of the body tissues.

TABLE 21-5
CRUDE PROTEIN REQUIREMENTS FOR LAMBS OF SMALL, MEDIUM, AND LARGE MATURE WEIGHT GENOTYPES[1 2]

Body Weight[3]	Lb	22	44	55	66	77	88	99	110
	Kg	*10*	*20*	*25*	*30*	*35*	*40*	*45*	*50*
Daily Gain[3]		g/d	g/d	g/d	g/d	g/d	g/d	g/d	g/d
(lb)	*(g)*								
					Small mature weight lambs				
.22	*100*	84	112	122	127	131	136	135	134
.33	*150*	103	121	137	140	144	147	145	143
.44	*200*	123	145	152	154	156	158	154	151
.55	*250*	142	162	167	168	168	169	164	159
.66	*300*	162	178	182	181	180	180	174	168
					Medium mature weight lambs				
.22	*100*	85	114	125	130	135	140	139	139
.33	*150*	106	132	141	145	149	153	151	149
.44	*200*	127	150	158	160	163	166	163	160
.55	*250*	147	167	174	175	177	179	175	171
.66	*300*	168	185	191	191	191	191	186	181
.77	*350*	188	203	207	206	205	204	198	192
.88	*400*	209	221	224	221	219	217	210	202
					Large mature weight lambs				
.22	*100*	94	128	134	139	145	144	150	156
.33	*150*	115	147	152	156	160	159	164	169
.44	*200*	136	166	170	173	176	174	178	182
.55	*250*	157	186	188	190	192	189	192	195
.66	*300*	179	205	206	207	208	204	206	208
.77	*350*	200	224	224	224	224	219	220	221
.88	*400*	221	243	242	241	240	234	234	234
.99	*450*	242	262	260	256	256	249	248	248

[1]Adapted by the authors from *Nutrient Requirements of Sheep,* sixth revised edition, NRC-National Academy of Sciences, 1985, p. 50.

[2]Approximate mature ram weights of 209 lb (*95 kg*), 254 lb (*115 kg*), and 298 lb (*135 kg*), respectively.

[3]Weights and gains include fill.

The relationship of size of breed (size of mature genotype) to lean-fat composition was shown in a classic study conducted by Reid *et al.*, of Cornell, involving four breeds of sheep and two crosses. These researchers found that Suffolks (a breed of large mature genotype) had the highest carcass protein and the lowest fat content among the breeds and crosses studied, especially at weights greater than 66 lb, whereas Shropshires (a breed of small genotype) had next to the lowest carcass protein and next to the highest fat contents. The statistics follow:[3]

Carcass Composition of Suffolk and Shropshire Lambs Slaughtered at 110 lb Ingesta-Free Body Weights

	Protein	Fat
	(%)	(%)
Suffolk	17	29
Shropshire	15	39

[3]Reid, J. T., *et al., Body Composition in Animals and Man,* National Academy of Sciences, Washington, D.C., 1958, p. 30, Table 9.

4. **Condition and rate of gain.** Ewes beginning pregnancy in a very thin condition have higher protein requirements than ewes in good condition. Also, the protein requirements of finishing lambs increase with rate of gain.

5. **Protein-energy ratio.** A ratio of about 30 g of protein per Mcal of DE has been shown to be adequate for mature, nonlactating ewes. Higher levels of protein are required per Mcal of DE for lactation, growth, and finishing. Also, the higher the energy concentration of the ration of finishing lambs, the greater the protein requirement.

• **Symptoms of protein deficiency**—A protein deficiency is characterized by reduced appetite, lowered feed intake, and poor feed efficiency. In turn, this makes for poor growth, poor muscular development, loss of weight, reduced reproductive efficiency, and reduced wool production. Under extreme conditions, there are severe digestive disturbances, nutritional anemia, and edema.

Minerals

Although the body contains approximately 40 mineral elements, only 16 have been demonstrated to be essential for sheep—7 major mineral constituents, and 9 trace elements. Four of the 16 essential minerals are toxic when consumed in excessive amounts, so they must be fed carefully. Tables 21-6 and 21-7 list the essential minerals, present the requirements, and give the toxic levels if known.

TABLE 21-6
MACROMINERAL REQUIREMENTS OF SHEEP (PERCENTAGE OF RATION)[1]

Nutrient	Requirement	
	As-fed[2]	Moisture-free
	(%)	(%)
Sodium	0.08–0.16	0.09–0.18
Chlorine	—	—
Calcium	0.18–0.74	0.20–0.82
Phosphorus	0.14–0.34	0.16–0.38
Magnesium	0.11–0.16	0.12–0.18
Potassium	0.45–0.72	0.50–0.80
Sulfur	0.13–0.23	0.14–0.26

[1]Adapted by the authors from *Nutrient Requirements of Sheep,* sixth revised edition, NRC-National Academy of Sciences, 1985, p. 48.

[2]As-fed was calculated using 90% dry matter (moisture-free).

TABLE
SHEEP MINERAL

Minerals Which May Be Deficient Under Normal Conditions	Conditions Usually Prevailing Where Deficiencies Are Reported	Function of Mineral	Deficiency Symptoms; Toxicity
Major or Macrominerals:			
Salt (sodium and chlorine—NaCl)	Negligence; for salt is inexpensive.	Sodium and chlorine are known to have regulatory functions in the body. They maintain osmotic pressure in cells, regulate the acid-base balance, and control water metabolism in tissues.	**Deficiency symptoms**—A deficiency of salt may result in an abnormal appetite, with the sheep trying to satisfy their craving by licking dirt, or eating toxic amounts of poisonous plants; decreased feed consumption; and decreased efficiency in the utilization of nutrients. ***Toxicity**—The maximum tolerable level of salt is 9.0% of the ration.
Calcium (Ca)	Lack of vitamin D. When finishing lambs are fed heavily on concentrates and limited quantities of legume roughage. When the feed consists largely of dried mature grasses or corn silage. Calcium-deficient areas (where pasture and range forages are deficient in Ca) are Fla., La., Neb., Va. Chronic internal parasite infections. Where there is magnesium deficiency.	Essential for development and maintenance of normal bones and teeth. Important in blood coagulation and lactation. Enables heart, nerves, and muscles to function. Regulates permeability of tissue cells. Affects availability of phosphorus and zinc.	**Deficiency symptoms**—Subnormal development of bone; rickets in young animals, and osteomalacia in adults. A high incidence of urinary calculi when there is a low calcium:high phosphorus ratio. To lessen the incidence of urinary calculi, the Ca:P ratio should be about 2:1. ***Toxicity**—If there is adequate phosphorus, sheep can tolerate a calcium-to-phosphorus ratio of 7:1 and as much as 2% calcium in the ration.
Phosphorus (P)	Lack of vitamin D. When sheep subsist for long periods on mature forages (such as dry range or grass or cereal hays). When the ration consists of a high proportion of beet by-products. When sheep subsist on pastures in phosphorus-deficient areas. Chronic internal parasite infections.	Essential for sound bones and teeth, and for the assimilation of carbohydrates and fats. A vital ingredient of the proteins in all body cells. Necessary for enzyme activation. Acts as a buffer in blood and tissue. Occupies a key position in biologic oxidation and reactions requiring energy.	**Deficiency symptoms**—Slow growth, depraved appetite, unthrifty appearance, listlessness, low level of phosphorus in the blood (less than 4 mg/100 ml of plasma), and development of knock-knees. *Caution:* A high level of phosphorus in the blood is not always an indication of adequacy in the diet; it may result from loss of weight. ***Toxicity**—Phosphorus at levels of 2 to 3 times the requirement can cause increased bone resorption in mature sheep.

TABLE 21–7
MICROMINERAL REQUIREMENTS OF SHEEP AND MAXIMUM TOLERABLE LEVELS (PPM OR MG/KG OF RATION)[1]

Nutrient	Requirement		Maximum Tolerable Level	
	As-fed[2]	Moisture-free	As-fed	Moisture-free
	(ppm or mg/kg)	(ppm or mg/kg)	(ppm or mg/kg)	(ppm or mg/kg)
Cobalt	0.09–0.18	0.1–0.2	9	10
Copper	6–10	7–11[3]	23	25[4]
Fluorine	—	—	54–135	60–150
Iodine	0.09–0.72	0.10–0.80[5]	45	50
Iron	27–45	30–50	450	500
Manganese	18–36	20–40	900	1,000
Molybdenum	0.45	0.5	9	10[4]
Selenium	0.09–0.18	0.1–0.2	1.8	2
Zinc	18–30	20–33	675	750

[1]Adapted by the authors from *Nutrient Requirements of Sheep,* sixth revised edition, NRC-National Academy of Sciences, 1985, p. 50.

[2]As-fed was calculated using 90% dry matter (moisture-free).

[3]Requirement when dietary Mo concentrations are <1 mg/kg DM.

[4]Lower levels may be toxic under some circumstances.

[5]High level for pregnancy and lactation in rations not containing goitrogens; should be increased if rations contain goitrogens.

Table 21–8, Sheep Mineral Chart, gives, in summary form, the following pertinent information relative to each mineral listed: (1) conditions usually prevailing where deficiencies are reported, (2) function, (3) deficiency symptoms/toxicity, (4) nutrient requirements, (5) recommended allowances, and (6) practical sources. Note, too, that Table 21–8 groups minerals as (1) major or macrominerals, and (2) trace or microminerals. Further elucidation on certain minerals is contained in the accompanying narrative. Fluorine is discussed because of its toxicity to sheep. There is growing evidence that fluorine, along with chromium, silicon, and vanadium, is essential for the rat and/or chick, but deficiencies in sheep under practical conditions have not been reported to date.

21–8
CHART (See footnotes at end of table.)

Mineral Requirements[1]: Minerals/Animal/Day	Mineral Requirements[1]: Mineral Content of Ration, in % or ppm	Recommended Allowances[1]	Practical Sources of the Mineral	Comments
	As-fed[2] M-F			
•Lambs in drylot consume about 9 g of salt daily. Mature sheep in drylot may consume more.	•Salt for growing lambs, %: 0.38 0.42 •Na requirement of sheep, %: 0.08–0.16 0.09–0.18 (See Table 21–6.)	•Salt for mature sheep: 0.5% of the complete feed, or 1.0% to the concentrate portion. •Range operators commonly provide ½–¾ lb (¼–⅓ *kg*) salt/ewe/month. Mature sheep in drylot may consume more.	Free access to salt. Loose salt, rather than block salt, should be provided, for the reason that sheep bite at salt blocks, rather than lick, with the result that their teeth may be broken. In iodine-deficient areas, stabilized iodized salt should always be provided.	Sheep consume about 5 times more salt/100 lb body weight than cattle, which is attributed to their high forage consumption. Sheep can consume high quantities of salt without apparent harm provided water is freely available. In alkaline areas, the water may contain enough salt to meet the requirements, and supplemental salt may not be needed.
Variable, according to class, age, and weight of sheep (see Table 21–1).	•0.18–0.74% •0.20–0.82% (See Tables 21–2 and 21–6.)	Self-feed suitable mineral, or add calcium to the ration as required to bring level of total ration slightly above requirements.	Ground limestone, or oystershell flour. Where both calcium and phosphorus are needed, use bone meal, dicalcium phosphate, or defluorinated phosphate.	Most pasture and range forage contains adequate amounts of calcium. Forage containing from 0.24–0.32% calcium is considered adequate. Calcium requirements are usually met when sheep receive at least ⅓ of a legume forage. •Blood calcium levels below 9 mg/100 ml of plasma suggest chronic low calcium intake.
Variable, according to class, age, and weight of sheep (see Table 21–1).	•0.14–0.34% •0.16–0.38% (See Tables 21–2 and 21–6.)	Self-feed suitable mineral, or add phosphorus to the ration as required to bring level of total ration slightly above requirements.	Monosodium phosphate or diammonium phosphate. Where both calcium and phosphorus are needed, use bone meal, dicalcium phosphate, or defluorinated phosphate.	The proper calcium-phosphorus ratio should be maintained. Forage containing below 0.16% phosphorus is usually considered deficient for ewes during gestation, and 0.20% borderline during lactation. •A phosphorus deficiency may be manifested when the blood phosphorus level falls below 4 mg/100 ml of plasma.

(Continued)

TABLE 21–8

Minerals Which May Be Deficient Under Normal Conditions	Conditions Usually Prevailing Where Deficiencies Are Reported	Function of Mineral	Deficiency Symptoms; Toxicity
Major or Macrominerals (Continued):			
Magnesium (Mg)	Tetany most frequently occurs in nursing ewes shortly after they are turned to pasture in the spring (grass tetany), when the magnesium requirements for lactation are high and grass is low in magnesium.	It is a constituent of bone. Also, it is necessary for many enzyme systems and for proper functioning of the nervous system. Closely associated with the metabolism of calcium and phosphorus.	**Deficiency symptoms**—Hypomagnesemic tetany, a hyperirritability of the neuromuscular system. Sometimes this condition is accompanied by hypocalcemia. Acute tetany may occur as a result of insufficient dietary magnesium or inability to mobilize skeletal magnesium. ***Toxicity**—Oral administration of 0.8% magnesium in the ration will produce toxicosis.
Potassium (K)	When finishing lambs are fed high-concentrate and urea rations and limited amounts of dry roughage. When sheep are grazing mature range forage during winter or drought periods. The potassium level of such forage may decrease to less than 0.2%.	It affects osmotic pressure and acid-base balance within the cell. It also aids in activating several enzyme systems involved in energy transfer and utilization, protein synthesis, and carbohydrate metabolism.	**Deficiency symptoms**—Poor appetite and feed conversion, progressive stiffness from front to rear, and dry wool. ***Toxicity**—The maximum tolerable level of potassium for sheep is about 3% of the ration DM.
Sulfur (S)	When finishing lambs are fed high-concentrate and urea rations and limited amounts of roughage.	Functions in synthesis of sulfur-containing amino acids (methionine and cystine) in the rumen and various compounds of the body. Wool is high in sulfur; hence, sulfur is closely related to wool production.	**Deficiency symptoms**—Loss of appetite, reduced weight gains and feed efficiency, and reduced wool growth. Also, excessive salivation, lacrimation, and shedding of wool. ***Toxicity**—It appears that 0.4% is the maximum tolerable level of dietary sulfur as sodium sulfate.
Trace or Microminerals:			
Cobalt (Co)	Cobalt-deficient areas or soils in the U.S. and Canada. The most severely deficient U.S. areas include portions of New England and the lower Atlantic Coastal Plain. Moderately deficient areas include the rest of New England, northern N.Y., northern Mich., and parts of the Central Plains.	Promote synthesis of vitamin B–12 in the rumen.	**Deficiency symptoms**—Cobalt deficiency signs are actually signs of vitamin B–12 deficiency. They are: Lack of appetite, lack of thrift, severe emaciation, weakness, anemia, decreased fertility, and decreased milk and wool production. ***Toxicity**—Approximately 204.5 mg/100 lb live weight.
Copper (Cu)	In copper-deficient areas (soils), as in Fla. and in the coastal plains region of the Southeast. Also, in several of the western states, there are areas where an excess of molybdenum induces copper deficiency.	Anemia is associated with copper deficiency. Animals suffering from inadequate copper intake appear unable to absorb iron at a normal rate, and a deficiency in hemoglobin synthesis results. Steely wool and depigmentation of black sheep.	**Deficiency symptoms**—Signs in suckling lambs include "swayback," muscular incoordination, partial paralysis of the hindquarters, and degeneration of the myelin sheath of the nerve fibers. Lambs may be born weak and may die because of their inability to nurse. Sheep suffering from a copper deficiency may produce "steely" or "stringy" wool, which is lacking in crimp, tensile strength, affinity for dyes, and elasticity. Depigmentation of the wool of black sheep has been noted as a sign of severe deficiency. ***Toxicity**—23 ppm As-fed or 25 ppm M-F (see Table 21-7), but Mo level of the ration is a factor.
Fluorine (F)	*Conditions which may result in fluorine toxicity:* High fluorine in the water supply. Use of rock phosphate that contains 3–4% fluorine.		Fluorine deficiency not reported. Rather, the hazard is fluorine toxicity. ***Toxicity**—Acute toxicity can occur at 200 ppm.
Iodine (I)	Iodine-deficient areas or soils (in northwestern U.S. and in the Great Lakes and Rocky Mountain regions) where iodized salt is not fed. Feeds from iodine-deficient areas.	Formation of thyroxin, a hormone of the thyroid gland.	**Deficiency symptoms**—Lambs born with goiter; usually stillborn or die soon after birth. Usually, such lambs have very little wool. In mature sheep an iodine deficiency may result in reduced wool yield and reduced rate of conception. ***Toxicity**—Maximum tolerable level for sheep is 45 ppm As-fed basis or 50 ppm M-F (see Table 21–7). However, much higher tolerable levels have been reported.
Iron (Fe)	Iron-deficiency anemia sometimes occurs in lambs raised on slotted floors. Loss of blood from parasite infestation can produce a secondary iron-deficiency anemia.	Hemoglobin formation.	**Deficiency symptoms**—Anemia, poor growth, lethargy, increased respiration rate, decreased resistance to infection, and in severe cases high mortality. ***Toxicity**—Signs of chronic toxicity are reduction in feed intake, growth rate, and feed efficiency. In acute toxicosis, animals exhibit loss of appetite, scanty urination, diarrhea, below normal temperature, shock, acidosis, and death.

(Continued)

Mineral Requirements[1]					
Minerals/ Animal/Day	**Mineral Content of Ration, in % or ppm**		**Recommended Allowances[1]**	**Practical Sources of the Mineral**	**Comments**
	As-Fed	**M-F**			
	*0.11, 0.14, and 0.16% for growing lambs, ewes in late pregnancy, and ewes in early lactation, respectively. *Where ewes in early lactation are grazing forage with high nitrogen and potassium content, the minimum level of magnesium in the ration is 0.2%.	*0.12, 0.15, and 0.18% for growing lambs, ewes in late pregnancy, and ewes in early lactation, respectively.		Plant protein supplements are excellent sources of Mg. Likewise, by-product feedstuffs derived from plants tend to be good sources. The common magnesium supplements are magnesium carbonate, magnesium oxide, and magnesium sulfate.	*Blood serum normally contains about 2.5 mg/100 ml.
	*0.45% for growth of lambs. 0.63–0.72 for lactation and stress.	*0.5% for growth of lambs. 0.7–0.8 for lactation and stress.	0.7 to 1.0% of total air-dry ration.	Roughages usually contain adequate potassium, with the possible exception of nonlegume silage. Potassium chloride and potassium sulfate are the supplements of choice.	The feeding of potassium chloride appears to reduce the incidence of urinary calculi in feedlot lambs. This is especially true with high-milo rations.
	*Mature ewes: 0.13–0.16% *Young lambs: 0.16–0.23%	 0.14–0.18% 0.18–0.26%	*It is recommended that a dietary nitrogen-sulfur ratio of 10:1 be maintained.	Sulfate sulfur, elemental sulfur, or sulfur-containing proteins or amino acids. Inorganic compounds are generally more convenient and economical for supplemental feeding.	*Practically all common feedstuffs contain more than 0.1% sulfur. However, mature grass and grass hays are sometimes low in sulfur. Where forages are low in sulfur or high in urea, increased weight gains and wool growth can be obtained by feeding sulfur.
	*0.09–0.18 ppm However, young, rapidly growing lambs may have a slightly higher requirement.	*0.1–0.2 ppm	*Feed cobalt at the rate of 1.4 g/100 lb (*2.5 g/100 kg*) of salt as cobalt chloride or cobalt sulfate.	A cobalt mineral mixture. Other effective methods of providing cobalt are (1) to add cobalt to the soil, or (2) to place cobalt pellets into the rumen.	Several good commercial cobalt-containing minerals are on the market in either block or loose form. Cobalt is much more effective when given by mouth than when given intravenously.
	As-fed[2] *6.3–20.7 ppm The Cu requirement varies with (1) the Mo content of the feed, and (2) the growth, pregnancy, lactation, and breed involved. (See the narrative section on "Copper" for details. Also, see Table 21-7.)	**M-F** *7–23 ppm	*Add copper sulfate to the salt at rate of 0.5%.	Salt containing 0.5% of copper sulfate.	Copper deficiencies may exist alone or along with deficiencies of cobalt and iron. An interesting interrelation exists between copper, molybdenum and sulfur. An excess of molybdenum causes a pathological condition which can be cured only by administering copper. Stores of copper in the liver, kidney, heart, lungs, pancreas, and spleen serve as a reserve for as long as 4–6 months when animals are grazing copper-deficient forage. Sheep are much more susceptible to copper toxicity than cattle. As much as 25 mg of copper in the daily ration of sheep is considered toxic; and about 9 mg/day is considered the safe tolerance level. Copper toxicity may result from feeding poultry wastes or mineral supplements designed for other species.
	*Breeding sheep should not be fed a ration containing more than 55 ppm (As-fed) or 60 ppm of fluorine on a moisture-free basis. *Finishing lambs can tolerate up to 135 ppm (As-fed) or 150 ppm of fluorine in the ration on a moisture-free basis.				Symptoms of fluorine toxicity are loss of appetite; the normal ivory color of bones changes to chalky white; bones thicken, and the teeth, especially the incisors, may become pitted and eroded to such an extent that the nerves are exposed.
	*0.09–0.72 ppm The higher levels are indicated for pregnancy and lactation. When goitrogens are present increase the iodine.	*0.1–0.8 ppm	*Free access to stabilized iodized salt containing 0.0078% iodine.	Stabilized iodized salt containing 0.0078% iodine. Calcium iodate.	Do not use iodized salt in a mixture with a concentrate to limit feed intake, as the animals may consume an excessive amount of iodine.
	*27–45 ppm	*30–50 ppm	*Intramuscular injections of iron-dextran; 2 injections, 150 mg of iron in each, given 2 to 3 weeks apart.	Ferrous gluconate, ferrous succinate, or ferrous sulfate given orally. Iron dextran injection.	A primary iron deficiency in grazing sheep is very unlikely.

(Continued)

TABLE 21–8

Minerals Which May Be Deficient Under Normal Conditions	Conditions Usually Prevailing Where Deficiencies Are Reported	Function of Mineral	Deficiency Symptoms; Toxicity
Trace or Microminerals (Continued)			
Manganese (Mn)	Lambs on a purified diet containing less than 1 ppm of manganese over a 5-month period. High calcium and iron may increase manganese requirements.	Skeletal development and reproduction.	**Deficiency symptoms**—Bone abnormalities, lack of coordination in newborn lambs, impaired growth, and depressed or disturbed reproduction. ***Toxicity**—It appears that 1,000 ppm of dietary Mn is the maximum tolerable level.
Molybdenum (Mo)	The major concern about molybdenum is that in excess it may induce a copper deficiency. Excess molybdenum in the soil such as is found in areas of Calif., Nev., and England.	It is believed that the molybdenum binds and inactivates the copper in the intestine.	**Deficiency symptoms**—A low intake of molybdenum causes excess copper to accumulate in tissues, especially the liver, even when the copper intake is moderate, thus producing fatal jaundice (easily detected in the eyes). This disease can be prevented by increasing the molybdenum intake. ***Toxicity**—High levels of molybdenum (10 to 20 ppm in forage plants) will induce copper deficiency characterized by stringy wool, lack of pigmentation in black wool, anemia, bone disorders, and infertility. Also, sheep start to scour after being turned to high molybdenum pasture (5 to 20 ppm M-F basis). The scouring can be controlled by increasing the copper level in the diet to 5 ppm.
Selenium (Se)	Areas where selenium content of crops is below 0.1 ppm, such as northwestern, northeastern, and southeastern U.S. Parts of S.D., Wyo., and Utah produce forage containing excess selenium which causes toxicity in farm animals.	Component of the enzyme glutathione peroxidase, the metabolic role of which is to protect against oxidation of polyunsaturated fatty acids and resultant tissue damage. Interrelation with vitamin E—they spare each other.	**Deficiency symptoms**—The most commonly noticed lesion from a deficiency of selenium is white muscle disease, which affects lambs 0–8 weeks of age, along with reduced growth of lambs. Additional signs of inadequate selenium are unthriftiness, infertility, early embryonic death, and periodontal disease. ***Toxicity**—Chronic selenium toxicity occurs when sheep consume feeds containing more than 3 ppm of selenium on a dry basis over a prolonged period. Toxicity signs include loss of wool, soreness and sloughing of the hooves, and marked reduction in reproductive performance.
Zinc (Zn)	Diets high in calcium adversely affecting zinc utilization.		**Deficiency symptoms**—Loss of appetite, reduction in growth rate, excessive salivation, parakeratosis, wool loss, and delayed wound healing. Ram lambs show reduced testicular development and defective spermatogenesis. In females, all phases of the reproductive process from estrus to parturition and lactation may be adversely affected. ***Toxicity**—There is a wide margin of safety between zinc requirements and zinc toxicity. However, 0.1% zinc in the diet reduced feed consumption and gain in lambs; and 0.075% induced severe copper deficiency in pregnant ewes and caused a high incidence of abortions and still births.

[1]As used herein, the distinction between "mineral requirements" and "recommended allowances" is as follows: In mineral requirements, no margins of safety are included intentionally; whereas in recommended allowances, margins of safety are provided to compensate for variations in feed composition, environment, and possible losses during storage or processing.

Where preceded by an asterisk, the requirements, recommended allowances, and other facts presented herein were taken from *Nutrient Requirements of Sheep,* 6th rev. ed., NRC-National Academy of Sciences, 1985.

[2]Estimated 90% dry matter.

MAJOR OR MACROMINERALS

The major or macrominerals involved in sheep nutrition are salt (sodium chloride), calcium, phosphorus, magnesium, potassium, and sulfur.

Salt (NaCl)

Sheep are particularly fond of salt and consume considerably more of it per 100 lb body weight than do cattle.

The total salt requirement of growing lambs approximates 0.40% of the dry matter of the ration. The sodium requirements are given in Table 21–6, but the chlorine requirements are unknown.

Range operators commonly provide 0.5 to 0.75 lb salt per ewe per month. Mature sheep in drylot may consume more. Finishing lambs consume about 0.6 lb per head per month. Loose salt, rather than block salt, should be provided, for the reason that sheep bite at salt blocks, rather than lick, with the result that their teeth may be broken. In iodine-deficient areas, stabilized iodized salt should always be provided.

When salt is added to mixed feeds, it is customary to add 0.5% to the complete ration or 1.0% to the concentrate portion. In the alkaline districts of the West, the water may contain enough salt to meet the requirements, and supplemental salt may not be necessary.

Salt may be used to limit feed intake provided adequate water is available. Such mixtures range from 10 to 50% salt, depending on the consumption desired. Trace-mineralized salt should not be used to govern feed consumption.

(Continued)

Mineral Requirements[1]				
Minerals/ Animal/Day	Mineral Content of Ration, in % or ppm	Recommended Allowances[1]	Practical Sources of the Mineral	Comments
	As-Fed *18–36 ppm **M-F** *20–40 ppm	18 ppm in As-fed ration, or 20 ppm in M-F ration.	Manganese gluconate.	
	*The minimum dietary requirement of molybdenum is not known. *The Food and Drug Administration does not recognize molybdenum as safe; hence, the law prohibits adding it to feed for sheep.	The two contrasting situations—(1) high molybdenum and copper deficiency, or (2) low molybdenum and excess copper accumulation—make it very difficult to define nutrient requirements of molybdenum and copper.	A high-molybdenum intake induces a copper deficiency even when the copper content of pasture is quite high; the scouring effect can be prevented by providing an increased copper intake. Sheep are less affected than cattle by high-molybdenum intakes. *In treating copper toxicity, both molybdenum and sulfate should be administered. Drench each lamb with 100 mg of ammonium molybdate and 1 g of sodium sulfate in 20 ml of water.	
	*0.09–0.18 ppm 0.1–0.2 ppm	Selenium as either sodium selenite or sodium selenate at the rate of 0.3 ppm of complete feed. Selenium added to salt-mineral mixture fed free-choice at rate of 90 ppm. Selenium in the limit feeding (feed supplements and salt-mineral mixtures) consumption rate for sheep of 0.7 mg per head per day.		Plants grown on the same seleniferous soils vary greatly in their uptake of selenium, with a range of 1,000 ppm to only 10–25 ppm. The most practical way to prevent livestock losses from selenium poisoning is to manage the grazing so that animals alternate between high-selenium and low-selenium areas. Selenium is a cumulative poison, but mild chronic signs can be overcome readily by feeding low-selenium forage.
	*Growth: 18 ppm 20 ppm *Reproduction: 30 ppm 33 ppm			

NOTE: Mineral recommendations for all classes and ages of sheep, especially those fed unmixed rations or on pasture, are—

1. *When sheep are on liberal grain feeding*—Provide free access to a 2-compartment mineral box, with (a) trace mineralized salt in one side, and (b) in the other side, a mixture of ⅓ trace mineralized salt (salt included for purposes of palatability), ⅓ defluorinated phosphate or steamed bone meal, and ⅓ ground limestone or oystershell flour.

2. *When sheep are primarily on roughage (pasture, hay, and/or silage)*—Provide free access to a 2-compartment mineral box, with (a) trace mineralized salt in one side, and (b) in the other side, a mixture of ⅓ trace mineralized salt and ⅔ defluorinated phosphate or steamed bone meal.

Additionally, in those areas where cobalt and/or copper deficiencies exist in the soil (and plants), add cobalt and/or copper sulfate to either the salt or salt-phosphorus mixture in the proportions indicated. If desired, the mineral supplement may be incorporated in the ration in keeping with the recommended allowances given in this table.

- **Symptoms of salt deficiency**—A deficiency of salt may result in an abnormal appetite, with the sheep trying to satisfy their craving by licking dirt, or eating toxic amounts of poisonous plants; decreased feed consumption; and decreased efficiency in the utilization of nutrients.

Calcium (Ca) and Phosphorus (P)

Calcium and phosphorus utilization depend on the presence of vitamin D and magnesium.

Most pasture and range forage contains adequate amounts of calcium, although calcium supplementation of pastures may be required in Florida, Louisiana, Nebraska, Virginia, and West Virginia.

Legumes are an excellent source of calcium; hence, the requirements for this mineral are usually met when the winter forage for breeding ewes or the roughage for finishing lambs consists of one-third or more of a good-quality legume hay. Corn silage is a poor source of calcium. Finishing lamb rations based on low-quality roughage, or high in concentrates, may require calcium supplementation.

Mature pasture and range forage in North America is almost always deficient in phosphorus. Because of the high-phosphorus content of the cereal grains, finishing lambs usually obtain an adequate allowance of this mineral unless a high proportion of beet by-products or other low-phosphorus feeds are fed.

Chronic internal parasite infections can cause a deficiency in calcium and phosphorus. Magnesium deficiency interferes with calcium absorption. Low levels of dietary phosphorus decrease the rate of calcium absorption. High levels of aluminum and iron will increase the need for phosphorus.

Fig. 21-8. Lamb fed a ration deficient in phosphorus. Note the knock-kneed conformation. (Courtesy, University of Idaho, Moscow)

• **Symptoms of calcium and phosphorus deficiency**— Rations that are decidedly lacking in both calcium and phosphorus cause rickets in young animals and osteomalacia in adults.

Signs of calcium deficiency due to low dietary intake of calcium develop slowly because the body draws on the store of calcium in the bones until it is greatly reduced. In extreme cases, which may occur in lambs on high-grain rations, low levels of calcium may result in tetany. Blood levels of calcium below 9 mg per 100 ml of serum indicate a calcium deficiency (hypocalcemia).

Phosphorus deficiency symptoms are depraved appetite, slow growth, unthrifty appearance, listlessness, knock-knees, and low level of phosphorus in the blood (less than 4 mg/ 100 ml of plasma).

Magnesium (Mg)

Magnesium is a constituent of bone. Also, it is necessary for many enzyme systems and for proper functioning of the nervous system. A deficiency of magnesium may result in grass tetany in sheep, particularly on wheat pasture.

The requirement for magnesium on a moisture-free (M-F) basis is 0.12, 0.15, and 0.18% for growing lambs, ewes in late pregnancy, and ewes in early lactation, respectively. Where ewes in early lactation are grazing forage with high nitrogen and potassium content, the minimum level of magnesium in the ration is 0.2%.

The use of intraruminal magnesium alloy pellets (bullets) weighing 30 g has effectively prevented hypomagnesemic tetany in lactating ewes.

• **Symptoms of magnesium deficiency**—Acute tetany, characterized by stiff legs and head retraction, may occur as a result of insufficient dietary magnesium.

Potassium (K)

Potassium is the third most abundant mineral in the body, accounting for approximately 0.3% of the body dry matter. It is primarily present in intracellular fluids (in skin and muscle), where it affects osmotic pressure and acid-base balance within the cells. It also aids in activating several enzyme systems involved in energy transfer and utilization, protein synthesis, and carbohydrate metabolism.

The potassium requirement for growth in lambs appears to be about 0.5% (M-F basis) of the ration. During periods of stress and during lactation, 0.7 to 0.8% may be required.

The possibility of potassium deficiency is very slight under most feeding conditions, because most forages contain adequate potassium. Nevertheless, attention must be given to the potassium supply when lambs are on high-grain rations and when sheep are grazing mature range forage during winter or drought periods. Potassium levels in mature, weathered range forage may decrease to less than 0.2%.

• **Symptoms of potassium deficiency**—A deficiency of potassium results in decreased feed intake and decreased weight gains. Listlessness, stiffness, impaired response to sudden disturbances, convulsions, and death have also been reported.

Sulfur (S)

Sulfur is used in the synthesis of the sulfur-containing amino acids, cystine and methionine, in the rumen and various compounds of the body. Also, wool is high in sulfur; hence, this element is closely related to wool production.

It is recommended that a dietary nitrogen-sulfur ratio of 10:1 be maintained. In percentage of dry matter, the sulfur requirements are as follows: mature ewes, 0.14 to 0.18%; young lambs, 0.18 to 0.26%.

Most feedstuffs contain more than 0.1% sulfur. However, mature grass and grass hays are sometimes low in sulfur and may not furnish enough of this element for optimum performance.

Where forages are low in sulfur, or where high-urea rations are fed, weight gains and growth of wool can be increased by feeding a sulfur supplement, such as sulfate sulfur, elemental sulfur, or sulfur-containing proteins or amino acids.

• **Symptoms of sulfur deficiency**— Loss of appetite, reduced weight gains and feed conversion efficiency, and reduced wool growth. Also, excessive salivation and lacrimation, and shedding of wool.

TRACE OR MICROMINERALS

The trace or microminerals involved in sheep nutrition are cobalt, copper, fluorine (because of toxicity to sheep), iodine, iron, manganese, molybdenum, selenium, and zinc.

Cobalt (Co)

Cobalt is essential for the synthesis of vitamin B-12 in the rumen. Indicators of the cobalt status of sheep are the levels of vitamin B-12 in the rumen contents, in the blood and liver, and in the feces.

Cobalt should be ingested frequently, preferably daily. This may be accomplished by adding cobalt to the salt, by adding cobalt to the soil, by placing cobalt pellets into the rumen, or by daily doses of cobalt.

The recommended amount of cobalt in the ration DM is 0.1 to 0.2 ppm. However, young, rapidly growing lambs may have a slightly higher requirement.

Cobalt-deficient areas have been widely reported in the United States and Canada. In known deficient areas, it is recommended that cobalt be added to the salt at the rate of 1.4 g per 100 lb of salt as cobalt chloride or cobalt sulfate.

Fig. 21-9. Cobalt deficiency. *Top:* Lamb fed cobalt-deficient ration containing 0.05 ppm cobalt. Lamb weighed 48.5 lb at the end of the experimental period. *Bottom:* Control lamb that received the same ration as the lamb in the top picture plus a daily allowance of 0.1 mg of cobalt as the sulfate. This lamb weighed 92.5 lb at the end of the experiment. (Courtesy, S. E. Smith, Cornell University, Ithaca, N.Y.)

• **Symptoms of cobalt deficiency**—Affected sheep show loss of appetite, lack of thrift, severe emaciation, weakness, anemia, decreased fertility, and decreased milk production.

Copper (Cu)

A copper deficiency may exist alone or in combination with deficiencies of other trace minerals. In practice, copper deficiency is frequently induced by excess molybdenum in forages.

Copper is found in adequate amounts in most feeds throughout the United States. The NRC copper requirements vary depending on the molybdenum content of the feed as follow:

	Recommended Cu Allowance	
	As-Fed	**Moisture-Free**
	(ppm)	**(ppm)**
Mo content of diet ppm $<$1.0:		
Growth	8.9–9.0	8–10
Pregnancy	8.1–9.9	9–11
Lactation	6.3–7.2	7–8
Mo content of diet ppm $>$3.0:		
Growth	15.3–18.9	17–21
Pregnancy	17.1–20.7	19–23
Lactation	12.6–15.3	14–17

Merino sheep are less efficient in absorbing copper from feedstuffs than British breeds; so, they need an additional 1 to 2 ppm in their ration.

Copper-deficient areas have been reported in Florida, in the coastal plains region of the Southeast, and in Nevada, Oregon, and other western states. In such areas, it is recommended that copper sulfate be added to the salt at the rate of 0.5%. Copper is stored in the body; reserves may last as long as 4 to 6 months when animals are grazing copper-deficient forage.

• **Symptoms of copper deficiency**—Lambs may be born weak and may die because of their inability to nurse. Suckling lambs show muscular incoordination and partial paralysis of the hindquarters.

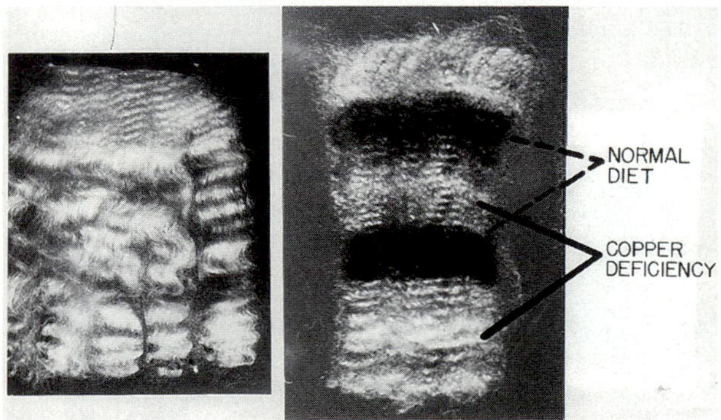

Fig. 21-10. Two samples of Australian wool, both of which show what may happen when sheep are on a copper-deficient ration. *Left:* The outer (bottom) ⅔ of this sample was produced by a sheep on a copper-deficient ration, resulting in hairlike or "steely" wool. Then copper was added to the sheep's ration, and normal, well-crimped wool was produced. *Right:* Wool sample from a normally black sheep. The white bands appeared at intervals when copper was deficient in the ration, because copper is essential for melanin or pigment production. Where such deficiencies occur under natural conditions, it is recognized that copper deficiencies result in the production of wool of lowered elasticity, tensile strength, and affinity for dyes.

Copper-deficient sheep produce "steely" wool, lacking in crimp, tensile strength, affinity for dyes, and elasticity. With a severe deficiency, the wool of black sheep is depigmented.

- **Copper toxicity**—Sheep are extremely intolerant to copper excess. Copper toxicity—characterized by hemolysis (dissolution of red corpuscles with liberation of their hemoglobin), jaundice (easily detected in the eyes), hemoglobinuria, and very dark-colored liver and kidneys—may result when sheep are fed rations high in copper and low in molybdenum. Copper toxicity may be prevented by lowering the copper level in the ration (normal is 8 to 11 ppm), or by a high-zinc ration (100 ppm on a dry matter basis). Recommended treatments for copper toxicity are (1) administering molybdenum and sulfate, or (2) drenching each lamb daily with 100 mg of ammonium molybdate and 1 g of sodium sulfate in 20 ml of water, for about 3 weeks. The Food and Drug Administration does not recognize molybdenum as safe; hence, it is not legal to add it to the feed of sheep.

Fluorine (F)

Fluorine is a cumulative poison. It occurs in some parts of the world as a result of consuming water high in fluoride or using rock phosphate that contains fluorine in amounts sufficient to be toxic—3 to 4%. Finishing lambs can tolerate up to 150 ppm of fluorine in the ration on a dry matter basis. Acute toxicity occurs at 200 ppm.

- **Symptoms of fluorine toxicity**—Affected animals exhibit loss of appetite and weight, change in bone color from ivory to chalky white, thickened bones, and pitted and eroded teeth—especially the incisors.

Iodine (I)

Iodine is necessary for the formation of thyroxin, the iodine-containing hormone of the thyroid gland. Northwestern United States and the Great Lakes region are well-known iodine-deficient areas, but other deficient areas are widely scattered throughout the United States.

The iodine requirement is 0.09 to 0.72 ppm in the ration on an as-fed basis, or 0.1 to 0.8 ppm moisture-free basis, with the higher levels indicated for pregnancy and lactation.

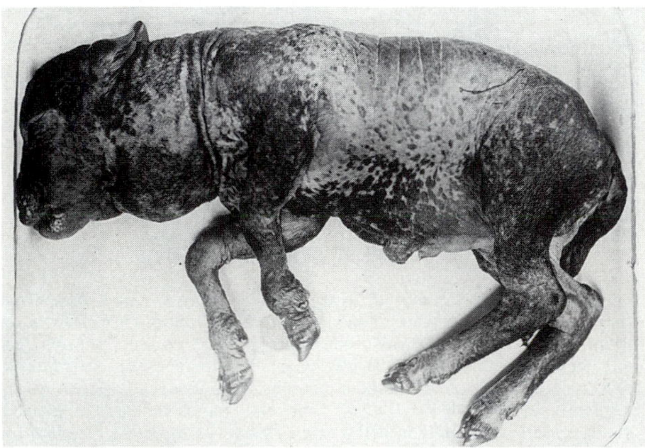

Fig. 21-11. Goiter. Woolless, goitered (big-necked) lamb stillborn due to iodine deficiency. (Courtesy, Montana State University, Bozeman)

When goitrogens such as kale or rape are fed, the dietary iodine should be increased.

Lamb losses can be prevented by feeding gestating ewes iodized salt containing 0.0078% iodine. The iodine in the salt should be stabilized to prevent losses from exposure to sunlight and moisture.

Iodized salt should not be used in a feed mixture to govern feed intake, as the animals may consume too much iodine.

- **Symptoms of iodine deficiency**—Lambs born with goiter (enlarged thyroid gland) is the most common deficiency symptom. If the condition is not too advanced, afflicted lambs may survive. Other signs are lambs born without wool, weak, or dead. An iodine deficiency in mature sheep may result in reduced wool yield and conception.

Iron (Fe)

Iron-deficiency anemia sometimes occurs (1) in lambs raised in confinement, on slotted floors, or (2) as a result of loss of blood from parasite infestation. It can be prevented (1) by giving the lambs two intramuscular injections of iron-dextran, each of 150 mg of iron, 2 to 3 weeks apart, or (2) by allowing free access to a commercial iron compound in the creep area.

- **Symptoms of iron deficiency**—Iron deficiency is characterized by anemia, poor growth, lethargy, increased respiration rate, decreased resisitance to infection, and in severe cases high mortality.

Manganese (Mn)

Although the exact requirements of sheep for manganese are unknown, it appears that 18 ppm in as-fed rations, or 20 ppm in moisture-free rations, will be adequate. It is needed for skeletal development and reproduction, as it is for various other species. A deficiency of manganese was produced in early-weaned lambs fed for a 5-month period on a ration containing less than 1 ppm of manganese. When a ration containing 8 ppm of manganese was fed to 2-year-old ewes for a 5-month period prior to breeding and throughout gestation, more services per conception (2.5 vs 1.5) were required than for ewes fed a ration containing 60 ppm manganese.

The manganese content of wool or mohair appears to be a good indicator of the manganese status of sheep.

- **Symptoms of manganese deficiency**—A manganese deficiency results in impaired growth, skeletal abnormalities and lack of coordination of the new born, and depressed or disturbed reproductive function.

Molybdenum (Mo)

A molybdenum deficiency in sheep, unrelated to copper, has not been reported. The major concern about molybdenum is its interaction with copper.

A high-molybdenum intake can induce copper deficiency in sheep even when the copper content of pasture is quite high, the effect of which can be prevented by an increased copper intake. On the other hand, when pastures provide a low intake of molybdenum, excess copper may accumulate in the body and result in a fatal jaundice, which can be prevented by increasing the dietary molybdenum.

• **Molybdenum toxicity**—Molybdenum toxicity (excess molybdenum) has been reported in California, Nevada, Oregon, and England, where it causes a scouring disease. Sheep start to scour withing a few days after being turned to high-molybdenum pasture (5 to 20 ppm on a dry matter basis; or on forage containing as little as 1 or 2 ppm molybdenum provided the dietary copper level is low and the sulfate level is high); the feces become soft, the fleece becomes stained, and the animals lose weight rapidly. Molybdenum toxicity can be controlled by increasing the copper level in the ration by 5 ppm.

Selenium (Se)

A minimum level of 0.1 ppm of selenium in feeds is considered adequate for preventing a deficiency in sheep. There are extensive areas in northwestern, northeastern, and southeastern United States where the selenium content of crops is below this level. On the other hand, parts of South Dakota, Wyoming, and Utah produce forage so high in selenium that it causes selenium toxicity.

White muscle disease (stiff lamb disease) in lambs, the main manifestation of selenium deficiency, can be prevented by (1) adding 0.3 ppm of selenium to the complete feed of ewes during gestation through weaning, or (2) adding 90 ppm of selenium to a salt-mineral mixture fed free-choice.

• **Symptoms of selenium deficiency**—Selenium deficiency has serious effects on lamb production. The main signs of such deficiency are white muscle disease, sometimes called *stiff lamb disease,* and reduced growth. If the muscle lesions are in the heart, lambs may die suddenly if subjected to exercise. Additional signs of inadequate selenium are unthriftiness, infertility, early embryonic death, and periodontal disease.

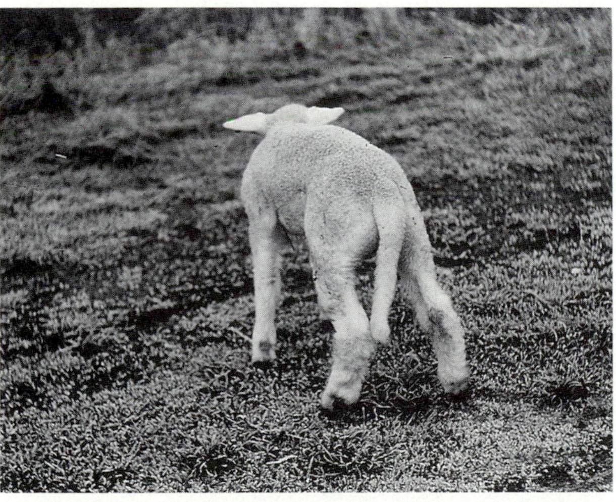

Fig. 21-12. Lamb afflicted with white muscle disease (stiff lamb disease). Note the stiff hind legs and the humped or "roached" back. Such lambs have a stilted way of moving and are usually stunted. (Courtesy, James E. Oldfield, Oregon State University, Corvallis)

• **Selenium toxicity**—Chronic selenium toxicity occurs when sheep consume plants containing more than 3 ppm selenium over a prolonged period. The extent to which plants take up selenium varies greatly by species. The most practical way in which to prevent livestock losses in high-selenium areas is to rotate the grazing between high-selenium and low-selenium areas. Although selenium is a cumulative poison, mild chronic signs can be readily overcome by feeding selenium-low forage. Also, small amounts of arsanilic acid are effective in reducing the toxicity of selenium.

Zinc (Zn)

Zinc is essential for sheep. An as-fed dietary level of 18 ppm of zinc appears to be adequate for growth, but a level of 30 ppm is necessary for normal reproduction.

• **Symptoms of zinc deficiency**—Deficiency signs include impaired growth, excessive salivation, parakeratosis, wool loss, and delayed wound healing. Ram lambs show reduced testicular development and defective spermatogenesis. In females, all phases of the reproductive process from estrus to parturition and lactation may be adversely affected.

Fig. 21-13. Zinc deficiency. *Left:* Zinc-deficient lamb after 15 weeks on the basal ration (2.7 ppm of zinc). *Right:* Lamb of the same age, which was fed the basal ration plus 100 ppm of zinc. Note the stunted growth, loss of wool, and dermatitis of skin on the lower part of the legs of the zinc-deficient lamb. (Courtesy, Purdue University, West Lafayette, Ind.)

Vitamins

Mature sheep require the fat-soluble vitamins A, D, E, and K, but they do not need added sources of the B vitamins, since the latter are normally synthesized in adequate amounts by rumen microorganisms.

Normal ewe rations are adequate in all the fat-soluble vitamins with the exception of the low carotene and/or vitamin A content of dry winter ranges. However, ewes can build up liver stores of vitamin A adequate to maintain production for 3 to 4 months. Vitamin D deficiency is not a problem unless ewes are maintained on vitamin D-deficient rations in environments devoid of sunlight. In some coastal areas of the United States, the latter condition may prevail. Unlike vitamin A, vitamin E does not appear to be stored in the body in appreciable concentrations.

Table 21-9, Sheep Vitamin Chart (see next page), presents in summary form pertinent information relative to the vitamin needs of sheep. Further information pertaining to certain vitamins is contained in the accompanying narrative.

TABLE
SHEEP VITA

Vitamin Which May Be Deficient Under Normal Conditions	Conditions Usually Prevailing Where Deficiencies Are Reported	Function of Vitamin	Deficiency Symptoms/Toxicity
Fat-Soluble Vitamins:			
A	Vitamin A deficiencies may occur when—(1) extended drought results in dry, bleached pastures; (2) winter feeding on bleached hays (especially over-ripe cereal hays or straws) with little or no green hay or silage; (3) drylot finishing on rations with little or no green forage or yellow corn, especially for feeding periods longer than 2–3 months; and (4) there is high nitrate intake, in either water or feed.	Necessary for maintaining normal epithelial tissue.	**Deficiency symptoms**—Keratinization of the respiratory, alimentary, reproductive, urinary, and ocular epithelia; lowered resisitance to infection; abnormal development of bone; birth of lambs that are weak, malformed, or dead; and night blindness. **Toxicity**—Changes in bone composition.
D	When ration consists predominantly of dehydrated hays, green feeds and seeds and their by-products. Prolonged cloudy weather or when kept inside, especially in fast-growing young lambs. When the vitamin D in feed is lost by oxidation.	Prevention of rickets in young lambs and osteomalacia in older sheep.	**Deficiency symptoms**—Rickets in young lambs and osteomalacia in older sheep. Congenital malformations in newborn lambs from extreme deficiencies. **Toxicity**—Abnormal deposition of calcium in soft tissues and brittle bones subject to deformation and fractures.
E	When lambs are making rapid growth, although this isn't always the case. When old hay is fed, as oxidation destroys vitamin E.	Prevention of white muscle disease (stiff lamb disease). As an antioxidant. Closely associated with selenium in metabolism.	**Deficiency symptoms**—Stiff lamb disease, or white muscle disease; characterized by a stiff, stilted way of moving and a "roached" back. Sometimes a paralysis of hind legs.
K	Vitamin K deficiency may occur when the dicumarol content of hay is excessively high, as when moldy sweet clover hay is fed.	Vitamin K_1 or K_2 is necessary in the blood clotting mechanism.	
Water-Soluble Vitamins:	**B** vitamin deficiencies may be evident in poorly fed and unhealthy animals.		

[1]As used herein, the distinction between "vitamin requirements" and "recommended allowances" is as follows: In vitamin requirements, no margins of safety are included intentionally; whereas in recommended allowances, margins of safety are provided in order to compensate for variations in feed composition, environment, and possible losses during storage or processing.

Where preceded by an asterisk, the vitamin requirements, recommended allowances, and other facts presented herein were taken from *Nutrient Requirements of Sheep,* 6th rev. ed., NRC-National Academy of Sciences, 1985.

21–9
MIN CHART

<table>
<tr><th colspan="2">Vitamin Requirements[1]</th><th rowspan="2">Recommended Allowances[1]</th><th rowspan="2">Practical Sources of the Vitamin</th><th rowspan="2">Comments</th></tr>
<tr><th>Vitamins/Animal/Day</th><th>Vitamin Content of Ration</th></tr>
<tr><td>Variable, according to class, age, and weight of sheep, as indicated by the following:</td><td>Variable, according to class, age, and weight of sheep (see Table 21–2).</td><td rowspan="2">Sheep that are deficient in vitamin A and weigh 70 lb (32 kg) or more should receive 100,000 IU of vitamin A by injection, and their rations should be adjusted to provide recommended levels of vitamin A or carotene. Ewes deficient in vitamin A should be given vitamin A either orally or by injection prior to breeding.</td><td rowspan="2">Stabilized vitamin A.
Green pasture. Grass or legume silages.
Yellow corn.
Green hay not over 1 year old.
*The vitamin A value of carotene from two common feeds is as follows:<table><tr><th></th><th>IU/mg</th></tr><tr><td>Dehydrated alfalfa meal</td><td>254–520</td></tr><tr><td>Silage, corn</td><td>436</td></tr></table>Where sheep are grazing forage low in carotene for extended periods, vitamin A deficiency can be prevented by (1) intramuscular injection of vitamin A, or (2) the addition of vitamin A to the ration as a pasture supplement or part of a salt mixture.</td><td rowspan="2">*Sheep do not convert carotene to vitamin A as efficiently as rats. For sheep, 1 mg of feed carotene is equivalent to 400–500 IU of vitamin A.
*It requires 200 days to deplete entirely liver storage of ewe lambs previously grazing on green feed. Because of this storage, animals that normally graze on green forage during the growing season are able to do reasonably well on a low-carotene ration of dry feed for periods of 4–6 months.</td></tr>
<tr><td colspan="2"><table><tr><th></th><th colspan="2">Mcg of B-carotene/day</th></tr><tr><th>Categories</th><th>/lb body wt.</th><th>/kg body wt.</th></tr><tr><td>All categories</td><td>31</td><td>69</td></tr><tr><td>Late gestation and in lactation</td><td>57</td><td>125</td></tr><tr><td>First 6–8 weeks of lactation of ewes suckling twins</td><td>67</td><td>147</td></tr></table><table><tr><th></th><th colspan="2">IU of Vitamin A/day</th></tr><tr><th>Categories</th><th>/lb body wt.</th><th>/kg body wt.</th></tr><tr><td>All categories</td><td>21</td><td>47</td></tr><tr><td>Late gestation and in lactation</td><td>39</td><td>85</td></tr><tr><td>First 6–8 weeks of lactation of ewes suckling twins</td><td>45</td><td>100</td></tr></table>(See Table 21–1.)</td></tr>
<tr><td>*For all sheep except early-weaned lambs: 555 IU per 220 lb (100 kg) body weight.
*For early-weaned lambs: 666 IU per 220 lb (100 kg) body weight.</td><td>Variable, according to class, age, and weight of sheep.</td><td>Breeding sheep, 500 to 800 IU/head/day.
Feeder lambs, 500 IU/head/day.</td><td>Exposure to sunlight, through irradiation.
Sun-cured hays.
Irradiated yeast.
Vitamin D_2 or vitamin D_3, which sheep use equally well.</td><td>Newborn lambs are provided with enough vitamin D from their dams to prevent rickets for 4 to 6 weeks if the ewes have adequate storage.
Sheep with white skin and short wool receive more vitamin D activity from irradiation by sunlight than do animals with dark skin or long wool.</td></tr>
<tr><td colspan="2"><table><tr><th></th><th colspan="2">*IU Vitamin E As-Fed basis</th></tr><tr><th>Categories</th><th>/lb</th><th>/kg</th></tr><tr><td>Lambs under 44 lb (20 kg) live weight</td><td>8</td><td>18</td></tr><tr><td>Lambs over 44 lb (20 kg) live weight and pregnant ewes</td><td>6</td><td>13.5</td></tr></table><table><tr><th></th><th colspan="2">*IU Vitamin E M-F basis</th></tr><tr><th>Categories</th><th>/lb</th><th>/kg</th></tr><tr><td>Lambs under 44 lb (20 kg) live weight</td><td>9</td><td>20</td></tr><tr><td>Lambs over 44 lb (20 kg) live weight and pregnant ewes</td><td>7</td><td>15</td></tr></table></td><td>Meet the requirements given in the column headed "Vitamin Content of Ration."</td><td>Wheat germ meal, dehydrated alfalfa, some green feeds, and vegetable fats are good sources of vitamin E.
Alpha-tocopherol (either in dl or d forms).</td><td>Experiments have failed to relate vitamin E deficiency with reproductive failure in sheep.
The need for vitamin E in the ration of young nursing lambs is related to the selenium level in the ration. Selenium has a sparing effect on the vitamin E requirement; the higher the selenium level in the diet, the lower the vitamin E requirement, and vice versa.</td></tr>
<tr><td></td><td></td><td>Vitamin K_2 is normally synthesized in large amounts in the rumen; no need for dietary supplementation has been established.</td><td>Green leafy materials of any kind, fresh or dry, are good sources of K_1.</td><td></td></tr>
<tr><td></td><td></td><td>*The B vitamins are not required in the diet of sheep with functioning rumens, because the microorganisms synthesize these vitamins in adequate amounts.</td><td></td><td>Addition of B vitamins has not been shown to be beneficial to mature sheep. However, young lambs (to about 2 months of age) with undeveloped rumens have been shown to have a dietary need for vitamin B–12, thiamin, pyridoxine, riboflavin, niacin, folic acid, and possibly some of the other B vitamins, since they will not be receiving these in the milk from their dams.
Cobalt is necessary for the synthesis of vitamin B–12 in the rumen.
*No supplementary dietary need for vitamin C has been shown.</td></tr>
</table>

FAT-SOLUBLE VITAMINS

The fat-soluble vitamins, which may be stored in varying quantities in the body, of importance in sheep nutrition are vitamin A, vitamin D, vitamin E, and vitamin K.

Vitamin A

Dietary vitamin A or its precursor, carotene, is necessary for maintaining normal epithelial tissues.

Sheep do not convert carotene to vitamin A as efficiently as laboratory rats. Thus, for sheep it is suggested that 1 mg of feed carotenes be considered as the equivalent of 400 to 500 IU of vitamin A. (For the rat, 1mg of beta-carotene = 1,667 IU of vitamin A).

Vitamin A is fat-soluble and is stored in the body. It takes about 200 days to deplete entirely the liver storage of vitamin A of lambs previously grazing on green feed. This explains why animals which graze green forage during the normal growing season are able to do reasonably well on a low-carotene ration for periods of 4 to 6 months.

Fig. 21-14. Green grass and/or legume forage is an excellent source of carotene, the precursor of vitamin A. This shows Suffolk ewes on green pasture. (Courtesy, National Suffolk Sheep Assn., Columbia, Mo.)

Both vitamin A and carotene may be destroyed by oxidation. Sun-cured hay is usually lower in carotene than dehydrated hay. Feedstuffs that contain little green or yellow material or that have been badly weathered, heated, or stored for long periods are low in carotene. Stabilized vitamin A, which is resistant to oxidation, may be added to rations low in carotene.

Sheep that are deficient in vitamin A and weigh 70 lb or more should recieve 100,000 IU of vitamin A by injection, followed by adjusting their rations to provide recommended levels of vitamin A or carotene. Ewes deficient in vitamin A should be given vitamin A either orally or by injection prior to breeding.

• **Symptoms of vitamin A deficiency**—Vitamin A-deficient sheep develop keratinization of the respiratory, alimentary, reproductive, urinary, and ocular epithelia, accompanied by lowered resistance to infections. Also, a deficiency of vitamin A interferes with normal development of bone; may result in the birth of weak, malformed, or dead lambs; and causes night blindness, the appearance or nonappearance of which is the most common means of determining the vitamin A status of animals.

Vitamin D

Vitamin D is required, in addition to calcium and phosphorus, for the prevention of rickets in young lambs and osteomalacia in older sheep. The addition of vitamin D to lamb rations low in this nutrient has resulted in increased growth even when there were no signs of rickets.

Fig. 21-15. Bilateral bent leg in yearling Rambouillet ram due to a deficiency of vitamin D. (Courtesy, Utah Ag. Exp. Sta.)

Since vitamin D is fat-soluble and stored, it is less important in mature animals, except during pregnancy, when the demands are greater. If pregnant ewes have adequate storage of vitamin D, they provide their newborn lambs with sufficient amounts to prevent rickets for 4 to 6 weeks.

Sheep exposed to sunlight obtain some vitamin D through irradiation, which may be sufficient to meet their requirements. Animals with white skin and/or short wool receive more vitamin D activity through irradiation than do animals with black skin and/or long wool.

Sheep on pasture seldom need additional vitamin D. But the question of adequacy arises during extended cloudy periods or when sheep are kept indoors. Under these circumstances, it is especially important that adequate vitamin D be included in the ration of fast-gaining lambs.

Early weaned lambs require 300 IU of vitamin D per 100 lb body weight, daily. All other sheep require 250 IU/100 lb body weight, daily. Sheep use either D_2 or D_3 equally well.

Sun-cured hays are fairly good sources of vitamin D. Dehydrated hays, green feeds, seeds, and by-products of seeds are poor sources. Vitamin D is oxidized, but with greater difficulty than vitamin A. However, when mixed with minerals, especially calcium carbonate, its stability is poor.

• **Symptoms of vitamin D deficiency**—A deficiency of vitamin D results in rickets in young lambs and osteomalacia in older sheep. Extreme deficiencies may cause congenital malformations in new born lambs. *CAUTION:* Vitamin D deficiency should not be confused with "spider leg syndrome," a genetic abnormality found in some sheep.

Vitamin E

Vitamin E is essential for all sheep, especially for young lambs. Unlike vitamin A, it is not stored in the body in appreciable quantities.

Vitamin E functions as an important biological antioxidant and helps to prevent white muscle disease (stiff-lamb disease) in lambs, through its association with dietary selenium in metabolism. Some signs of deficiency, such as white muscle disease or nutritional muscular dystrophy, may respond to either selenium or vitamin E, or may require both. Although vitamin E is a dietary requirement for young nursing lambs, experiments have failed to relate a deficiency of it to reproductive failure in sheep.

Stiff-lamb disease in nursing lambs is characterized by a stiffness (especially in the rear quarters), tucked-up rear flanks, and arched back. On autopsy, the disease is shown as white striations in the heart and other muscles, characterized by bilateral lesions. Affected lambs often die of pneumonia and starvation.

The need for vitamin E in the diet of young nursing lambs is related to the selenium level in the diet. Selenium has a sparing effect on the vitamin E requirement; the higher the selenium level in the diet, the lower the vitamin E requirement, and vice versa. White muscle disease in lambs is prevented by adding alpha-tocopherol and selenium to the diet. The suggested dietary levels of vitamin E are as follows: Lambs under 44 lb in weight should receive 8 IU/lb of as-fed ration; lambs over 44 lb in weight and pregnant ewes should receive 6 IU/lb of as-fed ration. (The IU is defined as 1 mg of dl-alpha-tocopherol acetate; 1 mg dl-alpha-tocopherol has the biological potency of 1.5 IU of vitamin E activity.) The above recommendations assume that dietary selenium levels are <0.05 ppm.

Fig. 21-16. Lamb afflicted with stiff-lamb disease, which in this case was associated with a lack of vitamin E in the ration. (Courtesy, Cornell University, Ithaca, N.Y.)

Values for the vitamin E requirements of sheep are presented in Tables 21-1, 21-2, and 21-10. The values presented in Table 21-1 were calculated from values per kilogram of dry feed consumed given in Table 21-2. Table 21-10 presents daily vitamin E requirements for lambs and the suggested amounts of alpha-tocopherol acetate to add to rations to provide 100% of the requirements.

TABLE 21-10
VITAMIN E REQUIREMENTS OF GROWING-FINISHING LAMBS AND SUGGESTED LEVELS OF FEED FORTIFICATION TO PROVIDE 100% OF REQUIREMENTS[1]

Body Weight		Alpha-Tocopherol Acetate			Feed Intake Per Lamb		Amount of Vitamin E Added to Concentrate			Amount of Vitamin E Added to Protein Supplement[2]		
(lb)	(*kg*)	(mg/lamb/day)[3]	(mg/lb ration)	(*mg/kg ration*)	(lb)	(*kg*)	(mg/lb)	(*mg/kg*)	(mg/ton)	(mg/lb)	(*mg/kg*)	(mg/ton)
22	*10*	5.0	44	*20*	0.50	*0.23*	9.1	*20*	18,200	133	*60*	120,000
44	*20*	10.0	44	*20*	1.00	*0.45*	9.1	*20*	18,200	60	*133*	120,000
66	*30*	15.0	33	*15*	2.10	*0.96*	6.8	*15*	13,600	45	*100*	90,000
88	*40*	20.0	33	*15*	2.86	*1.30*	6.8	*15*	13,600	45	*100*	90,000
110	*50*	25.0	33	*15*	3.50	*1.60*	6.8	*15*	13,600	45	*100*	90,000

[1]Adapted by the authors from *Nutrient Requirements of Sheep,* sixth revised edition, NRC-National Academy of Sciences, 1985, p. 51.

[2]Assumes the concentrate diet contains 15% protein supplement.

[3]Rounded values based on approximate diet intake containing recommended vitamin E levels.

Based on the average alpha-tocopherol content of feedstuffs generally used in lamb growing-finishing rations (corn, soybean meal, and alfalfa hay), the typical ration may contain less than 6.8 mg of alpha-tocopherol/lb, which could result in inadequate intake of vitamin E. In addition, preintestinal destruction of vitamin E of an orally administered dose increases from 8 to 42% as the corn content of the ration increases from 20 to 80%. So, many sheep rations heretofore believed to be adequate in vitamin E may be inadequate, thus explaining the sporadic outbreaks of white muscle disease in areas considered adequate in selenium.

Vitamin K

These are fat-soluble vitamins, one or the other of which is necessary in the blood-clotting mechanism. All green leafy materials, fresh or dry, are good sources of vitamin K_1. Vitamin K_2 is usually synthesized in large amounts in the rumen, so no need for dietary supplementation with it has been established.

WATER-SOLUBLE VITAMINS

The water-soluble vitamins which are not stored, include members of the vitamin B complex and vitamin C.

Vitamin B Complex

Digestion in the young lamb is more like that of a nonruminant animal than that of a ruminant. Thus, up to about 2 months of age, early-weaned lambs have a dietary need for biotin, choline, folacin, niacin, pantothenic acid, riboflavin, thiamin, vitamin B-6 (pyridoxine), and vitamin B-12, which are usually supplied in the ewe's milk.

Normally, the B vitamins are not required in the ration of sheep with functioning rumens (older than about 2 months or age), because the microorganisms synthesize them in adequate amounts. It is noteworthy, however, that polioencephalomalacia (PEM), a noninfectious disease of sheep, the highest incidence of which occurs in lambs, responds to a parenteral injection of 0.5 g of thiamin hydrochloride, repeated at 2-day intervals if necessary. Apparently, PEM is caused by an anti-thiamin enzyme in the rumen.

It is necessary that cobalt be present for the synthesis of vitamin B-12 in the rumen.

Vitamin C (Ascorbic Acid, Dehydroascorbic Acid)

Vitamin C is synthesized rapidly enough by the tissues to meet the animal's needs; hence, it is not a required dietary constituent for sheep.

Water

Sheep get water by drinking, and from snow, dew, and feed. The amount of water that sheep voluntarily consume is affected by temperature, rainfall, snow and dew covering, age, breed, stage of production, number of lambs carried, wool covering, respiratory rate, frequency of watering, kind and amount of feed, and exercise. On the average, mature animals consume approximately a gallon of water per day, whereas feeder lambs require about half this amount. However, sheep may go for weeks without drinking water when foraging on grasses and other feeds of high-moisture content. This condition often prevails on desert ranges in the early spring and on many of the mountain ranges during the summer months.

Fig. 21-17. Water from a mountain stream. (Courtesy, USDA)

Additives

Table 21-11 summarizes the growth stimulants that are presently available and can be used. All of these products have been shown to improve gain and feed efficiency of sheep. The information presented in Table 21-11 is the most recent available. But feed additives and implants do change from time to time; new products are developed, and sometimes old products are banned by the Food and Drug Administration. So, those using additives should always confer with local authorities and read and follow manufacturer's label directions for more comlete details on the use of a specific drug or combination of drugs.

Antibiotics may improve performance when added to creep and lamb-finishing rations. Chlortetracycline and oxytetracycline are especially effective. The response to antibiotics seems to be affected by differences in management and the amount of stress to which the lambs are subjected. There is some evidence that antibiotics reduce the incidence of enterotoxemia.

In addition to the additives listed in Table 21-11, a number of feed additives are approved for treatment of specific diseases.

TABLE 21-11
SHEEP FEED ADDITIVES AND IMPLANTS

Type of Additive	Method of Administering	Dosage	Effect On: Daily Rate of Gain (% increase)	Effect On: Feed Efficiency (% increase)	Effect On: Carcass Quality	Comments
Antibiotics (chlortetracycline and oxytetracycline)	Feeding (oral)	Aureomycin (chlortetracycline) 10 to 25 mg/lb of feed. Terramycin (oxytetracycline) 5 to 10 mg/lb of feed.	Range: 0–31 Average: 11	Range: 4–27 Average: 10	No effect to slight improvement.	Antibiotics (especially chlortetracycline and oxytetracycline) may improve performance when added to creep and lamb finishing rations. Response to antibiotics varies markedly according to differences in management and degree of stress to which lambs are subjected. There is some evidence that antibiotics reduce the incidence of enterotoxemia.
Bovatec (lasalocid)	Feeding (oral)	10–15 mg/lb complete feed, fed at rate of 15–70 mg lasalocid /day.	Range: 0–20 Average: 6–8	Range: 5–15 Average: 8–10	No effect.	Bovatec is an ionophore. In addition to increasing rate of gain and feed efficiency, Bovatec reduces rumen protein degradation and increases the amount of bypass protein. Greatest response is obtained where coccidiosis is a problem, for which purpose Bovatec was initially approved by FDA.
Ralgro (zeranol)	Implant	12 mg/head	Range: 0–25 Average: 10	Average: 6		Do not implant animals within 40 days of slaughter. Do not implant breeding animals.

FEEDS FOR SHEEP

Sheep are adapted to the consumption of a great variety of feeds, most of which are of plant origin and bulky in nature. The feeding of concentrates is usually limited to the finishing of lambs and for use by the breeding flock at such special periods as the lambing season or just before and during the breeding season. Forages constitute 100% of the ration of the vast majority of sheep during most seasons of the year.

Fig. 21-18. Sheep on a well managed grass-legume pasture. (Courtesy, Purdue Ag. Exp. Sta., West Lafayette, Ind.)

Pasture

No other class of farm animals is so well adapted to the utilization of maximum quantities of pasture as sheep. Although cattle compete with sheep for many of the same grazing areas and are also ruminants, sheep are unique in that the vast majority of the young are marketed as milk-fat animals directly off pastures. Also, in their grazing habits, sheep differ from cattle in that (1) they show a decided preference for short, fine forages, and (2) they have the gregarious or flocking instinct.

Both pastures and sheep must be well managed to maximize pasture returns. Experiments indicate that well managed seeded pastures will produce 400–500 lb, or more, of lamb per acre and average daily gains of ⅓ to ½ lb per day. Conversely, poorly managed, parasite-infested pastures may result in no gains, or even loss in weight.

Although there are great differences in plants, sheep are able to utilize the various grasses, legumes, weeds, forbs (broadleafed herbaceous plants, commonly called weeds by producers), and browse (broadleafed, woody plants, or shrubs, bushes, or small trees) that grow in millions of acres of cultivated and uncultivated land in this and other countries. This characteristic, plus the imperishable nature of wool from the standpoint of storage and transportation, has made sheep raising a frontier industry throughout the world.

Sheep are adapted to the grazing of both fenced and unfenced holdings. In this country, most farm flocks and a limited number of range bands of the Southwest are confined to fenced areas with no herder being necessary. On the other hand, most of the range pastures of western United States, Spain, the Balkans, Africa, and Asia are utilized by migratory bands under the supervision of a herder.

Regardless of the location of the area, year-round grazing is desired. In order to obtain succulent and palatable pastures, the range bands of the West are frequently moved to different altitudes at different seasons of the year. In the mountainous sections, the summer ranges are usually at high altitudes, the spring-fall grazing at intermediate altitudes, and much of the winter range is on the desert or lowland areas. On some ranges, such traveling is not possible, the ranges being used on a year-long basis.

During the winter months and following periods of extended drought, pastures usually become leached and bleached. Although sheep can and do utilize these forages, it must be realized that, in comparison with green growing grasses, they are lacking in nutrients, being especially low in protein and carotene (provitamin A) content. Consideration of these facts should be given when providing supplemental feeds.

Abundant and succulent pastures are ideal for stimulating milk production in ewes. Moreover, pastures of this type are desirable from the standpoint of the limited digestive capacity of the young lamb. Accordingly, the degree of finish carried by lambs at market time is an accurate reflection of the amount and quality of the forage available on the pasture or range.

Sheep may successfully utilize either permanent or cultivated (improved) pastures. In the range area, the vast majority of pastures are of a permanent type; whereas both types of pastures are used by native sheep. Where a choice is possible, improved grass-legume pastures are preferred. Where adapted, birdsfoot trefoil is the legume of choice because it is bloat-free and it will remain productive for a long period, provided it is adequately fertilized and managed.

Range sheep graze nearly all the year and receive most of their sustenance from native range plants. Pronounced deficiencies (singly or in combination) of range forages in energy, protein, phosphorus, and carotene (provitamin A) sometimes occur on ranges. These deficiencies are most apt to happen when the forage is mature or dormant, during overgrazing, or in periods of drought. Also, they are most marked in ewes during gestation or lactation, when the nutrient needs are highest.

Hays and Other Dry Forages

Inclement weather, extreme droughts, overstocked pastures and ranges, and leached and bleached pastures make it necessary that dry roughages be provided for sheep. They are fond of good roughage and make good use of it. In general, however, they cannot use effectively as coarse roughage as cattle.

Hays are the standard winter feed for sheep when they cannot be on the pasture or range or when the condition of the pastures is such as to require supplemental feeding. The choicest hay for sheep is a legume which has been produced on fertile soil, cut at the proper stage, and well cured. Such hay is palatable and rich in protein (and the quality of protein is good), calcium, and vitamins A and D. If legume hay cannot be secured, a high-quality grass-legume mixed hay will be entirely satisfactory and much superior to a straight grass hay. Sheep may do very well for a considerable period of time when fed only a good-quality legume hay, salt, and water.

Although legume hays are preferable for sheep, nonlegume hays are fed extensively in the sheep-raising and lamb-feeding areas. If straight grass hays must be fed, they should be cut at an early stage of maturity. Even then, they will be lower in protein, calcium, and vitamins than the legumes; hence, for best results, protein and mineral supplements should be provided.

Bright, early cut corn or sorghum fodder, or stover, and early cut, green cereal hays, and straws of many kinds are fed to sheep in different areas. The feeding value of these coarser roughages varies considerably according to the stage of maturity at which they are cut, the amount of leaves, and the green coloration. Although these forages are not satisfactory as the sole roughage, especially during the latter part of gestation and in the suckling period, they may be successfully used when mixed with liberal quantities of good-quality legume hays. Where nonlegume roughages are fed, special attention should be given to providing a suitable protein concentrate and minerals, especially calcium.

Silages and Root Crops

Silage for sheep may be made from a great variety of plants, including corn, sorghums, cereal grains, legumes, grasses, cannery refuse, pea vines, potatoes, beets, beet tops, sunflowers, and other materials. When properly preserved and fed, silages made from any of these materials are quite satisfactory.

Most practical sheep producers prefer to limit the silage allowance to about 4 to 6 lb per head per day, with the

Fig. 21-19. Ewes eating silage from a fenceline bunk. Note gravel back of fence where sheep are standing. (Courtesy, *The Sheepman's Production Handbook*, Denver, Colo.)

balance of the roughage ration consisting of hay. If a nonlegume silage is fed, it is important that the hay be a legume. If a legume hay cannot be provided when a nonlegume silage is fed, it is very necessary that a suitable protein concentrate and minerals be provided.

Roots include all plants whose roots, tubers, bulbs, or other underground vegetative parts are used for feed. The important root crops for sheep are mangels (stock beets), rutabagas (swedes), turnips, and carrots. These feeds are very succulent in nature, containing from 85 to 90% or more of water. They are highly relished by sheep and have a peculiarly beneficial effect upon the digestion and general thrift of animals. The only objection to their general use is the cost and difficulty of growing, harvesting, and storing them. For the latter reason, roots are usually (1) a by-product from food produced for human consumption, (2) grown as a second crop and harvestd by grazing sheep on them, or (3) grown and harvested mainly by producers of purebred animals and by shepherds in Europe. Presently, some sheep producers are using turnips as a fall and early winter grazing crop.

Stored roots are generally fed sliced, although some are fed whole. Where the teeth are good, little is gained from slicing roots for sheep. In this country, roots are usually limited to 5 to 6 lb per head daily, but in other countries up to 12 to 14 lb are often allowed daily per head.

Quality of Forage

The quality of forage—pasture, hay or other dry roughage, or silage—greatly affects its consumption. High-quality forage is more digestible and passes through the digestive tract more rapidly than low-quality forage; hence, sheep will consume more of it.

The most favorable nutritional response is usually obtained by feeding forage harvested before the protein content has decreased and before lignification of the fiber content has increased. Loss of leaves, weather damage, fermentation, and leaching losses reduce the value of harvested forage.

When it is necessary to feed sheep low-protein forage, intake and utilization can be increased by adding a suitable protein supplement fortified with needed minerals and vitamins.

Concentrates

The concentrates include those feeds which are low in fiber and high in digestible nutrient value. For purposes of convenience, concentrates are often further classified as (1) carbonaceous feeds, and (2) nitrogenous feeds.

Except when they are abundant and cheap, few concentrates are fed to sheep, except immediately before and after lambing, in conditioning ewes and rams for breeding, or when finishing lambs. During these periods, the most frequently used concentrates consist of the common farm grains —oats, corn, barley, wheat, rye, and the grain sorghums. Numerous by-product feeds are also utilized for sheep, including those from the flour- and corn-milling industries, beet by-products, and oil meals or cake made from soybeans, cottonseed, and flaxseed.

Feed Preparation

Sheep masticate grain more thoroughly than cattle, with the result that feed preparation is of less value for them than for cattle. Processing of grains for sheep is not necessary unless seeds are hard (such as sorghum or millet) or the sheep have "broken mouths." Pellets are increasingly being used by lamb feeders, especially when lower quality feed ingredients are used.

(Also see Chapter 14, Feed Processing, including Table 14-4 therein.)

Feed Substitution Table For Sheep

Successful sheep raisers are keen students of feed values. They recognize that feeds of similar nutritive properties can and should be interchanged in the ration as price relationships warrant, thus making it possible at all times to obtain a balanced ration at the lowest cost.

Table 21-12, Feed Substitution Table for Sheep and Goats, is a summary of the comparative values of the most common U.S. feeds. In arriving at these values, two primary factors besides chemical composition and feeding value have been considered; namely, palatability and carcass quality.

In using this feed substitution table, the following facts should be recognized:

1. That individual feeds differ widely in feeding value. Barley and oats, for example, vary widely in feeding value according to the hull content and the test weight per bushel, and forages vary widely according to the stage of maturity at which they are cut and how well they are cured and stored.

2. That nonlegume forages may have a higher relative value to legumes than herein indicated provided the chief need of the animal is for additional energy rather than for supplemental protein. Thus, the nonlegume forages of low value can be used to better advantage for wintering mature ewes than for young lambs.

On the other hand, legumes may actually have a higher value relative to nonlegumes than herein indicated provided the chief need is for additional protein rather than for added energy. Thus, no protein supplement is necessary for breeding ewes provided a good-quality legume forage is fed.

3. That, based primarily on available supply and price, certain feeds—especially those of medium protein content, such as brewers' dried grains, corn gluten feed (gluten feed), distillers' dried grains, distillers' dried solubles, and peas (dried)—are used interchangeably as (a) grains and by-products feeds, and/or (b) protein supplements.

4. That the feeding value of certain feeds is materially affected by preparation. The values herein reported are based on proper preparation in each case.

For the reasons noted above, the comparative values of feeds shown in the feed substitution table (Table 21-12) are not absolute. Rather, they are reasonably accurate approximations based on average-quality feeds.

TABLE 21–12
FEED SUBSTITUTION TABLE FOR SHEEP AND GOATS (AS-FED BASIS) (See footnotes at end of table.)

Feedstuff	Relative Feeding Value (lb for lb) in Comparison with the Designated Base Feed *(in bold italic)* Which = 100	Maximum Percentage of Base Feed (or comparable feed or feeds) Which it Can Replace for Best Results	Remarks
GRAINS, BY–PRODUCT FEEDS, ROOTS AND TUBERS:[1] **(Low and Medium Protein Feeds)**			**NOTE WELL:** Although sheep terms are used in this column, the replacement feeding value (column 2), maximum percentage replacement (column 3), and remarks (column 4) of each feed listed pertain to goats, also.
Corn, No. 2	***100***	***100***	Grinding not necessary unless (1) for old ewes with poor teeth, (2) for lambs under 5–6 weeks, (3) for incorporation in a mixed ration.
Apple pomace, dehydrated	82–86	33⅓	
Barley	90	100	It does not pay to grind barley for sheep.
Beans (cull)	80	15	
Beet pulp, dried	90	33⅓–50	Value of about 80% when used as the only concentrate for finishing lambs.
Beet pulp, molasses, dried	95	33⅓–50	Value of about 80% when used as the only concentrate for finishing lambs.
Beet pulp, wet	25	33⅓–50	
Brewers' dried grains	80–95	33⅓	Not very palatable. Fed chiefly to dairy cattle.
Citrus pulp, dried	95	25–50	
Corn gluten feed (gluten feed)	85–90	50	
Distillers' dried grains	95–100	33⅓–50	
Distillers' dried solubles	95–100	33⅓–50	
Fat	225	5	
Hominy feed	100	100	
Molasses, beet	75	10	Actual value may be higher as an appetizer.
Molasses, cane	75	10	Actual value may be higher as an appetizer.
Molasses, citrus	75	10	
Oats	80	10–100	Lower value when used as the only grain for finishing lambs. Highest value for young lambs, for breeding animals, and for starting lambs on feed. Need not be ground for sheep. Should not constitute more than ⅓ of finishing rations. Feeding value varies according to the test weight per bushel.
Peas, dried	100	40	
Rice (rough rice)	55–75	100	
Rice bran	66⅔–75	33⅓	
Rice polishings	85–90	25	
Potatoes (Irish)	25	85	Contrary to popular belief, potatoes can be fed successfully through the pregnancy and lactation periods.
Roots (chiefly mangles or stock beets, rutabagas or swedes, turnips, and carrots)	15	50	Some sheep producers believe that the feeding of high levels of roots over a long period will produce urinary calculi. Therefore, caution should be exercised in feeding them to rams and wethers (females not affected). Keep the Ca level higher than the P level. Many shepherds add roots to the ration of show sheep, for conditioning purposes.
Rye	83–87	50–100	Apparently rye is more palatable to sheep than to other classes of animals. Rye may be fed whole to sheep.
Sorghum, milo	85–100	100	All varieties have about the same feeding value. There is no advantage in grinding sorghum for sheep.
Wheat	100–110	50	May be fed as the only grain, but it is improved by mixing with another grain. Wheat may be fed whole. Wheat-fed sheep appear to be especially susecptible to founder.
Wheat bran	90	10–33⅓	Because of its bulk and fiber, wheat bran should not constitute more than 10–15% of a finishing ration. Bran is valuable for young animals, for breeding animals, and for starting animals on feed.
Wheat mixed feed (mill-run)	90–95	10–33⅓	Can be used in about the same way and in the same quantities as wheat bran for sheep.
PROTEIN SUPPLEMENTS:			
Soybean meal (41%)	***100***	***100***	
Alfalfa or clover screenings	70–75	50	Grind finely to destroy weed seeds.
Brewers' dried grains	75	100	
Copra meal (coconut meal)(21%)	90–100	50	
Corn gluten feed (gluten feed)	60–65	50–100	
Corn gluten meal (gluten meal)	100	50	
Cottonseed meal (41%)	100	100	Unlike the situation with finishing cattle, cottonseed meal is about equal to linseed meal for finishing lambs.
Distillers' dried grains	90	100	Rye distillers' dried grains are about 10% lower in protein than similar products made from corn or wheat.

(Continued)

TABLE 21–12 *(Continued)*

Feedstuff	Relative Feeding Value (lb for lb) in Comparison with the Designated Base Feed *(in bold italic)* Which = 100	Maximum Percentage of Base Feed (or comparable feed or feeds) Which it Can Replace for Best Results	Remarks
Distillers' dried solubles	90	100	
Linseed meal (35%)	90	100	
Peanut meal (45%)	100	100	
Peas, dried	65–75	50	
Safflower meal, with hulls (42%)	40–45	100	
Soybeans	95–100	100	It does not pay to grind soybeans for sheep.
Sunflower meal (35%)	85	100	
DRY FORAGES AND SILAGES:[2]			
Alfalfa hay, all analyses	***100***	***100***	
Alfalfa silage	33⅓–50	50–85	When alfalfa silage replaces corn silage, more energy feed must be provided but less protein, unless grain is used as a preservative.
Barley hay	70	50	The beards may be harmful, especially to woolly faced sheep.
Beet tops, fresh	16–25	33⅓–50	In the West, large acreages of fresh beet tops were formerly grazed by sheep and cattle.
Beet tops, dry	70	50	
Beet top silage, sugar	17–25	33⅓–50	Either provide some dry forage or feed 2 oz (*56.7 g*) of finely ground limestone to each 100 lb (*45.4 kg*) of silage.
Clover, alsike	90–100	100	
Clover hay, crimson	90–100	100	Crimson clover hay has a considerably lower value if not cut at an early stage.
Clover hay, red	90–100	100	If the rest of the ration is adequate in protein, clover hay will be equal to alfalfa in feeding value; otherwise, it will be lower.
Clover-timothy hay	80–90	100	
Corn fodder	75	100	Should be chopped.
Corn silage	33⅓–50	50–85	Although a ration in which corn silage is the only forage is sometimes fed to sheep, most feeders prefer to limit the silage and use some hay. Avoid silage that is contaminated with listeriosis.
Corn stover	35	50	Unsatisfactory for finishing lambs, but cut or shredded stover may be used as a part of the roughage for breeding ewes if fed along with a good legume.
Cowpea hay	95–100	100	
Grass hay (bluegrass, bromegrass, canarygrass, orchardgrass, timothy, and a variety of native grasses)	75	50–75	In comparison with legume hays, grass hays contain almost as much energy, have 50 to 75% less protein (they run 6–10% crude protein), and have only about 25% as much calcium (about 0.2% calcium).
Grass-legume mixed hay	80–90	100	Value depends on the proportion of legumes present and the stage of maturity at which it is cut.
Grass-legume silage	32–45	50–85	Although a ration in which grass silage is the only forage is sometimes fed to sheep, most feeders prefer to limit the silage and use some hay.
Grass silage	30–45	50–85	
Haylage (alfalfa-grass)	50	50–85	
Johnsongrass hay	70	50	
Lespedeza hay	80–100	100	Feeding value varies considerably with stage of maturity at which it is cut.
Mint hay	80–95	75	
Oat hay	75	50	Oat hay is equal to alfalfa hay in energy, but substantially lower in protein.
Pea-vine silage	33⅓–50	50–85	Unless grain is added as a preservative, pea-vine silage requires more energy feed, but less protein supplement than corn silage when fed to finishing lambs.
Pea-vine hay	100–110	75	
Prairie hay	65–70	100	
Reed canarygrass	70	100	
Sorghum fodder	70	100	
Sorghum silage (grain varieties)	32–47	50–85	Although a ration in which sorghum silage is the only forage is sometimes fed to sheep, most feeders prefer to limit the silage and use some hay.
Sorghum silage (sweet varieties)	25–30	50–85	Nearly equal to grain varieties in value per acre because of greater yields.
Sorghum stover	35	50	Unsatisfactory for finishing lambs, but cut or shredded stover may be used as part of the roughage for breeding ewes if fed along with a good legume.
Soybean hay	85–100	100	The lower value is for finishing lambs. For other classes of sheep, it is equal to alfalfa hay.
Soybean straw	68	50–75	
Sudangrass hay	50–60	50	
Sunflower hulls	40	50	
Sweet clover hay	100	100	Value of sweet clover hay varies widely. Second year sweet clover hay is less desirable than first year clover hay and is more apt to cause sweet clover disease.
Vetch-oat hay	80–90	100	The higher the proportion of vetch, the higher the value.
Wheat hay	70	50	

[1]Roots and tubers are of lower value than the grain and by-product feeds due to their higher moisture content.

[2]Silages are of lower value than dry forages due to their higher moisture content.

RATIONS FOR SHEEP

Sheep rations vary with the section of the country, depending chiefly on availability and price of local feeds. Numerous feedstuffs can be used as ration alternatives to the conventional legume hay-grain generally used by producers. These include many grain, vegetable, fruit, and food industry byproducts.

Also, sheep rations are closely associated with the age and stage of production of the animals. Except at lambing time or when emergencies occur as a result of drought or inclement weather, western bands receive little supplemental feed. Even with farm flocks, a minimum of grain is fed to breeding animals. Grain feeding usually is limited to the latter part of gestation and to the lactation period prior to turning to pasture. However, when grain is less costly than forage, on the basis of energy and/or protein, it may be fed to advantage to animals of any age or stage of production.

The following points are also pertinent in feeding sheep of all classes and ages:

1. They should be provided needed minerals; see Table 21-6, 21-7, and 21-8 for recommendations.

2. Unless grains are unusually hard, they need not be ground for sheep. The animals prefer to do their own grinding, and the feeds are not more effectively utilized when ground.

FEEDING RAMS

Rams have two functions: (1) produce high-quality semen, and (2) deposit semen in ewes. Good nutrition will optimize these two functions.

In general, rams are fed the same kind of feeds as ewes, but in slightly larger quantities. (See Table 21-13.)

Rams should be in moderate condition at the beginning of the breeding season, neither too fat nor too thin. Additional energy supplementation in the form of grain may be needed to maintain the ram's body weight and performance throughout the breeding season. During the balance of the year, pasture is usually adequate when available; otherwise, the ration may be comparable to that of the ewes.

Fig. 21-20. A Suffolk stud ram, in top form. (Painting by Tom Phillips. Courtesy, National Suffolk Sheep Assn., Columbia, Mo.)

FEEDING BREEDING EWES

Success in the sheep business is largely measured by the percentage lamb crop raised and the pounds of lambs marketed per ewe. The most important factor affecting these criteria is the feed of the ewe. Also, the yearly feed of the ewe represents about 50% of all production costs. For purposes of convenience, the feeding of ewes will be discussed under the following headings: (1) drylot (confinement) feeding, (2) flushing ewes, (3) feeding pregnant ewes, (4) feeding at lambing time, (5) feeding lactating ewes, and (6) feeding ewes in accelerated lambing.

Fig. 21-21. Suffolk breeding ewes in excellent condition, giving evidence of proper feeding. (Courtesy, Heupel Ranch, Santa Maria, Calif.)

The ideal ewe feed is low-cost, palatable, nontoxic, and nutrient rich; it satisfies the ewe, minimizes handling and feed bunk problems, and maximizes lamb and wool production.

Some producers are self-feeding complete ground mixed rations to pregnant and lactating ewes. The two main advantages are: (1) reduced labor in feeding, and (2) more efficient use of low-quality roughages. In order to prevent self-fed ewes from getting too fat, a high roughage ratio should be used, and the consumption should be controlled by limiting the time the ewes have access to self-feeders.

Table 21-13 contains suggested rations for ewes during various stages of production—maintenance, early gestation, late gestation, and lactation.

TABLE 21-13
DAILY RATIONS FOR BREEDING EWES AT VARIOUS STAGES OF PRODUCTION

Ration No.	Moisture Basis[1] A-F (as-fed) M-F (moisture-free)	Hay[2]		Corn Silage		Haylage		Corn Straw		Stover (Stalks)		Grain[3]		Protein Supplement[4]	
		(lb)	(kg)	(lb)	(kg)	(lb)	(kg)	(lb)	(kg)	(lb)	(kg)	(lb)	(kg)	(lb)	(kg)
Maintenance															
1	A-F	3.0	1.4												
	M-F	3.3	1.5												
2	A-F			6.0	2.7									0.20	0.09
	M-F			6.7	3.0									0.22	0.10
3	A-F					6.0	2.7								
	M-F					6.7	3.0								
4	A-F							3.0	1.4					0.40	0.18
	M-F							3.3	1.5					0.44	0.20
Gestation, early (first 15 weeks)															
1	A-F	3.5	1.6												
	M-F	3.9	1.8												
2	A-F	2.0	0.9									1.0	0.45		
	M-F	2.2	1.1									1.1	0.49		
3	A-F	1.8	0.8									0.6	0.27	0.20	0.09
	M-F	2.0	0.9									0.7	0.31	0.22	0.10
4	A-F			8.0	3.6									0.20	0.09
	M-F			8.9	4.0									0.22	0.10
5	A-F					7.0	3.2					0.20	0.09		
	M-F					7.8	3.5					0.22	0.10		
6	A-F	2.0	0.9							2.0	0.9	0.5	0.23		
	M-F	2.2	1.0							2.2	1.0	0.6	0.27		
7	A-F	1.0	0.45							2.0	0.9	0.5	0.23	0.30	0.14
	M-F	1.1	0.49							2.2	1.0	0.6	0.27	0.33	0.15
Gestation, late (last 4 weeks): Add 0.5–1.0 lb (0.23–0.45 kg) grain per ewe daily to any of the above rations.															
Lactation															
1	A-F	4.0	1.8									2.0–3.0	0.9–1.4	—	—
	M-F	4.4	2.0									2.2–3.3	1.1–1.5	—	—
2	A-F			10.0	4.5							1.5	0.7	0.25	0.11
	M-F			11.1	5.0							1.7	0.8	0.28	0.13
3	A-F	1.0	0.45	8.0	3.6							1.5	0.7	0.20	0.09
	M-F	1.1	0.49	8.9	4.0							1.7	0.8	0.22	0.10
4	A-F					8.9	3.6					2.0–3.0	0.9–1.4		
	M-F					8.9	4.0					2.2–3.3	1.1–1.4		

[1]As-fed was calculated using an average figure of 90% dry matter. When using silages, roots, and other wet feeds, these feeds should be converted to a moisture-free basis and the ration calculated using the moisture-free data.

[2]Alfalfa hay, midbloom, preferred.

[3]Grain may consist of corn, barley, wheat, oats, and/or grain sorghum.

[4]Protein supplement may consist of soybean, cottonseed, linseed, sunflower, safflower, or rapeseed meal.

Feeding Directions:

1. These rations are formulated to meet the requirements of a 154 lb (*70 kg*) ewe in average condition, and are designed for hand-feeding. The daily feed allowance can be increased or decreased, depending on the actual size of the ewe and the body condition.

2. Some of these rations are deficient in calcium and/or phosphorus; therefore, a supplement containing 50% trace mineral salt (for sheep) and 50% dicalcium phosphate should be fed free choice. The consumption of 0.05 lb (*0.02 kg*) per sheep per day of this mixture will provide the amounts of calcium and phosphorus needed for maintenance and the first 15 weeks of gestation; and 0.10 lb (*0.05 kg*)/day will provide the needed Ca and P for late gestation and lactation. Vitamins A and E should be added to the salt-mineral mix when sheep are fed the wheat straw and corn stover rations.

3. Ewes should gain 15 to 25 lb (*6.8 to 11.4 kg*) during gestation. During early gestation (first 15 weeks), they should gain 0.05 lb (*0.02 kg*)/day. During late gestation (last 4 weeks), they should gain 0.5 lb (*0.23 kg*)/day. If, during the last 4 weeks of pregnancy, ewes are fed 0.5 to 1.0 lb (*0.23 to 0.45 kg*) grain per head daily and gain 8 to 15 lb (*3.6 to 6.8 kg*), ketosis (lambing paralysis) can be prevented almost entirely.

4. During maintenance and early gestation, each ewe should have 14 in. (*36 cm*) of bunk feed space. In late gestation and during lactation, bunk space should be increased to 15 to 18 in. (*38 to 46 cm*).

Drylot (Confinement) Feeding

The vast majority of the nation's sheep utilize pasture in season. However, some ewe-lamb producers are drylotting all or part of the year. So, now there are two alternatives, and the producer may choose between the two.

The **advantages** of drylot (confinement) production are:
1. The virtual elimination of losses from predators.
2. Freedom from the most harmful internal parasites.
3. Lowering of the energy requirement due to limited activity.
4. The opportunity to feed ewes according to their productivity and nutrient requirements rather than their appetites.
5. It results in more rapid gains, in lambs reaching market weight at an earlier age, and in improved carcass grade.

The **disadvantages** are:
1. A higher initial capital investment, especially in buildings and equipment.
2. It requires superior management.
3. All nutritive requirements must be met.
4. External parasites and contagious diseases may be increased.
5. Animal manure disposal and bedding costs will be greater.

The Table 21-13 rations are satisfactory for ewes in confinement, and the Table 21-15 rations are excellent for creep feeding lambs raised in confinement.

The following steps are necessary for successful drylot feeding:

1. Use a big purebred ram so that the lambs inherit rapid growth potential.
2. Have the lambs born so that they will be ready for the intended market in 4 to 4½ months.
3. After 2 months lactation, cut the ration of the ewes to the same amount that they were receiving 1 month before lambing.
4. Self-feed lambs; and wean them at 90 days of age. Do not turn lambs to pasture.
5. Market the lambs at 100 to 110 lb.
6. Vaccinate all lambs against enterotoxemia (overeating disease).

Without doubt, many sheep producers can advantageously combine pasture and confinement production—using pastures for the breeding flock and confinement production for young lambs. So, sheep confinement will not eliminate good pasture systems; rather, confinement production, all or in part, will increase and will replace more and more of the conventional practice of grazing all sheep throughout the pasture season.

Flushing Ewes

Flushing is the practice of conditioning or having thin ewes gain in weight just prior to breeding. Its purpose is to increase the ovulation rate and, consequently, the lambing rate.

Fig. 21-22. Turning ewes to a fresh, luxuriant pasture 2 to 3 weeks before breeding time is a practical way to flush them. This shows Dorset ewes on pasture in Mississippi. (Courtesy, *The Dorset Journal*)

This special feeding usually begins 2 to 3 weeks prior to breeding and continues into the breeding season. It may be accomplished by turning the ewes to a fresh, luxuriant pasture 2 to 3 weeks before breeding time; or if such a pasture is not available, satisfactory results may be brought about by feeding a grain allowance of ½ to ¾ of a pound daily over a like period of time. Oats alone are excellent, or a mixture of equal parts of oats and corn is very satisfactory.

Although it is not likely that all the benefits ascribed to flushing will be fully realized under all conditions, the general feeling persists that the practice will result in a 15 to 20% increase in the lamb crop, and that the ewes will breed both earlier and more nearly at the same time. Hence, it follows that the lamb crop will be earlier and more uniform in age and size.

Mature ewes appear to respond better than yearling ewes. Also, flushing may be more beneficial early and late in the breeding season than during the peak, when the ovulation rate is highest.

Fat ewes will not benefit from flushing. Instead, they should be conditioned for breeding by stepping up the exercise.

Feeding Pregnant Ewes

If a strong, healthy crop of lambs is to be expected, the ewes must be properly fed and cared for throughout the period of pregnancy. In general, this means the feeding of a suitable and well-balanced ration, together with the necessary minerals and vitamins as required for maintenance (and growth, if the ewe is not fully mature), growth of the fleece, and development of the fetus. In addition, plenty of exercise must be provided. Suitable shelter should be made available during inclement weather, and the animals should be given access to an abundance of fresh air and sunshine at all times. Ewes should gain in weight during the entire period of pregnancy, making a total gain of 20 to 30 lb for the period. They should enter the nursing period with some reserve flesh, because the lactation requirements are much more rigorous than those of the gestation period.

After the ewes are bred, they should have access to pastures as long as they are available. When the ground is firm, winter pasture or range, stalk, or stubble fields may be grazed to advantage. Green rye or wheat pastures furnish a very succulent feed and valuable exercise for the flock. Where winter pastures are either unavailable or inadequate, supplemental feeds must be provided. The most satisfactory forage is a good-quality legume hay—alfalfa, clover, lespedeza, or soybean. The sheep producer, however, often seems to find it difficult to grow legumes on the farm or ranch and rather hard to buy such roughage at satisfactory prices. Where grass hay, such as native hays or timothy, is used, every effort should be made to cut it at an early stage of maturity and to have it properly cured. Even then, a protein supplement should be provided, together with suitable minerals. Because of the known value of legumes from the standpoint of quality of proteins, minerals, and vitamins, and the fact that grass hays are not recognized as too desirable for sheep, every effort should be made to supply at least a third of a good-quality legume roughage to pregnant ewes. A 150-lb pregnant ewe will eat 4 to 5 lb of hay daily. In order to prevent waste and protect the wool from chaff and hay seeds, suitable racks should be provided.

Fig. 21-23. Dorset ewes ready for lambing, evidencing proper feed and care. (Courtesy, *The Dorset Journal*)

Such succulent feeds as roots and silage are desirable in keeping the ewes healthy and doing well. Of the root crops, turnips seem to be preferable for pregnant ewes. Silage made from corn, milo, legumes, or grasses, and which is not frozen, spoiled, nor moldy, may be fed quite safely and is excellent feed. Ordinarily, the daily ration of roots or silage should not exceed 5 lb, which means that hay is usually fed in addition to the succulent feed.

During the last 4 to 5 weeks of pregnancy, the fetus develops very rapidly and the demands on the ewe are rather heavy. Also, ewes carrying twins or triplets, especially if they are a bit fat, are very prone to ketosis (lambing paralysis), which can be prevented by feeding a high-energy ration. So, during the last 4 to 5 weeks of pregnancy, ewes should be fed 0.5 to 1.0 lb of grain per head daily and gain 8 to 15 lb. Besides, ewes so fed milk better. The concentrate given to the farm flock usually consists of homegrown grains, whereas range bands are often given pelleted or cubed protein supplements.

Feeding at Lambing Time

At lambing, or immediately after lambing, each ewe should be placed in an individual holding or lambing pen. At this time, the grain allowance should be materially reduced, but dry roughage may be fed free-choice, when it is certain that it is of good quality and palatable. Usually, about 5 to 7 days should elapse before ewes are placed on full feed following parturition. In general, feeds of a bulky and laxative nature should be provided during the first few days. A mixture of equal parts of oats and wheat bran is excellent. Soon after lambing, the ewe should be given water with the chill removed but should not be allowed to gorge.

Feeding Lactating Ewes

Following lambing, the feed allowance of the ewe should be increased according to her capacity and needs. Although there is great individual and breed variation, ewes will yield from 1 to 4 qt of milk per day. In comparison with cow's milk, ewe's milk is richer in protein and fat and higher in ash. It must also be borne in mind that, in addition to producing milk and maintaining her body, the ewe is growing wool, which is protein in character. Immature ewes are also growing. Under these circumstances, it is but natural and normal to expect ewes to lose in condition during the suckling period. The loss in weight is primarily determined by the inherent milking qualities of the individual and by the kind and amount of feed.

In general, it is considered good practice to feed lactating ewes rather liberally for the first month or two after lambing, for lambs make the most economical gains when suckling. It is a good plan to separate the ewes with twins from those with singles, giving the former more liberal rations or the benefit of the better pastures or ranges. In fact, some large sheep operators find this practice so advisable that they regularly separate out the twin bands.

Fig. 21-24. Hampshire ewes in proper condition at lambing time. (Courtesy, American Hampshire Sheep Assn., Ashland, Mo.)

Milk production can be greatly stimulated through the proper selection of feeds. If there is not sufficient high-quality roughage for the entire winter, the most palatable and succulent portion should be reserved for use during the suckling period. Pastures should be provided as soon as possible, but in the meantime a high-quality legume hay or, better yet, a combination of hay and silage, will take care of the roughage needs. Though varying somewhat with the size and condition of the ewe and whether there are twins or a single, an adequate ration for a lactating ewe may consist of approximately 4 lb of high-quality alfalfa hay plus 1 to 2 lb of grain daily. If neither a legume hay nor legume silage is available, a protein supplement should be included in the grain ration.

As soon as the spring pasture season has arrived, the use of harvested feeds should be discontinued, being both uneconomical and unnecessary.

Lactating ewes normally reach their peak milk production about 3 to 4 weeks after lambing and produce 75% of their total milk yield during the first 8 weeks of lactation.

Weaning age varies greatly in the sheep industry. Lambs may be weaned as early as 3 to 4 weeks of age and as late as 5 to 6 months. When pasture is plentiful and lamb growth is satisfactory, there is little reason to wean lambs before they reach market weight and condition at 5 to 6 months of age.

• **Dairy sheep**—Dairy sheep, for the production of milk and cheese, are important in some countries. Annually, the United States imports more than 22 million lb of cheese made from sheep milk, a large part of which is the world-famous Roquefort cheese from France. Roquefort cheese is made from ewes' milk only in the Roquefort area of France, where it was originally ripened in caves. Similar cheese made in France from cows' milk is labeled *bleu* cheese.

Feeding Ewes in Accelerated Lambing

Accelerated lambing involves ewe lambs dropping their first lambs at 1 year of age, and lambing at intervals of 6 to 8 months thereafter.

Experimental studies and practical observation indicate (1) that it is feasible and profitable to have ewe lambs drop their first lambs at 1 year of age, *provided* they are well fed, well grown, and early dropped; and (2) that it is possible to achieve a lambing interval of 6 to 8 months, *provided* breeds with long breeding seasons are used (Dorsets, Rambouillets, or Merinos), there is superior nutrition, and hormones are used when a 6-month interval is planned.

Ewe lambs that are to be bred so that they lamb at 12 months of age should be liberally fed (1) from birth, using one of the creep rations given in Tables 21-15 or 21-16; and (2) during pregnancy and lactation, using one of the suggested rations in Table 21-13.

Accelerated lambing can (1) make for lower cost of production, (2) contribute to a more even supply of lamb throughout the year, and (3) allow for greater flexibility in production.

FEEDING RANGE SHEEP

Various geographical divisions are assumed in referring to the western range area—the native pasture area. Sometimes reference is made to the 17 range states, embracing a land area of 1.16 billion acres. At other times this area is broken down, chiefly on the basis of topography, into (1) the Great Plains area, and (2) the 11 western states. Almost half (47.7%) of the latter area is federally owned, and administered primarily by the Bureau of Land Management and the U.S. Forest Service.

In the early days of the range sheep industry, the animals were usually moved to lower winter ranges and expected to get their feed as best they could. There was precious little supplemental feeding. If the winter happened to be mild, and if a reasonable amount of grass was cured on the stalk, the band came through in pretty good shape. During an exceedingly cold winter, particularly when there was much snow, losses were severe and often diastrous. Today, the practical and successful range producer winter feeds. The progressive rancher is also equipped to meet emergency feeding periods, of which droughts are the most common.

Fig. 21-25. Range sheep being moved to summer grazing area. (Courtesy, American Sheep Producers Council, Inc., Denver, Colo.)

Ewes are normally maintained on winter grazing areas, with or without supplemental feeds, as long as possible. Usually these ranges are located at the lower altitudes and the vegetation consists of rather mature and bleached grasses or brush and browse. When the vegetation is sparse or covered by deep snow, supplemental feeds of hays, preferably alfalfa, some other legume, or concentrates are provided. Often protein supplements in the form of pellets or cubes are used, for these may be scattered about the feeding grounds, neither being blown away nor difficult for the sheep to find. Usually such expensive protein supplements are fed only when native grass hays are being utilized, high-quality alfalfa not requiring a protein supplement.

Because of the magnitude of the range sheep industry and the fact that it is a highly specialized type of operation, in the sections that follow special discussion is devoted to the feeding and management of sheep on the range.

Nutrient Deficiencies of Range Forage

Hunger, due to lack of feed, is the most common deficiency on the western range. In particular, there may be a shortage of energy during droughts, late in the season, or early in the spring when grass is washy. Under such energy-deficient conditions, sheep lose weight and condition and lambs fail to grow. Also, reproduction is adversely affected.

Mature, weathered native range grass is almost always deficient in protein—being as low as 3%, or less. Protein-leaching losses due to fall and winter rains may range from 37 to 73%.

Phosphorus deficiencies are rather common among range sheep, but calcium deficiency is seldom encountered.

Of the vitamins, vitamin A is most likely to be deficient in range forage, because dry, bleached range grass is very low in carotene (the precursor of vitamin A).

Fig. 21-26. Suffolk sheep on the range. Good sheep and good range go together. (Painting by artist Tom Phillips, 3333 17th St., San Francisco, CA 94110)

Range Supplements

Four suggested range supplements, ranging from high to low protein, are given in Table 21-14. (See next page.)

Sheep on poor or weathered range grass should be supplemented by feeding the high-protein formulation in Table 21-14, to correct both the protein and phosphorus deficiencies. Of course, the supplements in Table 21-14 may be modified in keeping with availability and cost of feeds, and yet meet known deficiencies. For example, if phosphorus is the only deficiency, it may be corrected by feeding a phosphorus supplement free-choice.

There is no one best and most practical range supplement for any and all conditions. Many different feeds may be, and are, used; among them, (1) ranch or locally produced hay, (2) alfalfa pellets or cubes, with or without fortification, and (3) supplements of various kinds.

Also, the labor attendant to the daily feeding of a pasture or range supplement can be lessened by (1) using protein blocks, or (2) self-feeding salt-feed mixtures.

Where salt is used for the purpose of governing consumption, the proportion of salt to feed may vary anywhere from 5 to 40% (with 30 and 33⅓% salt content being most common).

(Also see Chapter 11, Protein Supplements, section headed, "Self-feeding Salt-Feed Mixtures.")

TABLE 21–14
FORMULAS FOR RANGE SHEEP SUPPLEMENTS[1]

Feed[2]	Recommended Level of Protein			
	High	Medium-High	Medium-Low	Low
	(%)			
Barley, grain or corn, dent yellow, grain, grade 2 US, minimum 54 lb (*24.5 kg*)/bu	5	40	75	65
Beet, sugar, molasses, or sugar cane molasses, 48% invert sugar, minimum 79.5° Brix	5	5	5	5
Cottonseed with some hulls, solvent extracted, ground, minimum 41% protein, maximum 14% fiber, minimum 0.5% fat (cottonseed meal)	66	36	—	16
Soybean, seeds, solvent extracted, ground, maximum 7% fiber, 44% protein (soybean meal)	10	10	10	10
Urea, technical, 282% protein equivalent	—	—	5	—
Alfalfa, aerial parts, dehydrated, ground, minimum 17% protein or alfalfa, hay, sun-cured, early bloom	10	5	—	—
Vitamin A (IU/lb)	—	1,818	3,636	3,636
Vitamin A (IU/kg)	—	4,000	8,000	8,000
Calcium phosphate, monobasic, commercial	1	1	2	1
Sodium phosphate, monobasic, technical	2	2	2	2
Salt or trace mineralized salt	1	1	1	1
Total	100	100	100	100

Composition[3]	High As-Fed[4]	High *M-F*	Medium-High As-Fed[4]	Medium-High *M-F*	Medium-Low As-Fed[4]	Medium-Low *M-F*	Low As-Fed[4]	Low *M-F*
Digestible energy (Mcal/lb)	1.4	*1.5*	1.4	*1.5*	1.4	*1.5*	1.3	*1.4*
Digestible energy (Mcal/kg)	3.0	*3.3*	3.0	*3.3*	3.0	*3.3*	2.8	*3.1*
Protein (N × 6.25) (%)	30.4	*33.8*	21.9	*24.3*	23.6	*26.2*	15.9	*17.7*
Phosphorus (%)	1.8	*2.0*	1.4	*1.5*	0.8	*0.9*	1.1	*1.2*
Carotene (mg/lb)	9.0	*10.0*	4.1	*4.5*	—	—	—	—
Carotene (mg/kg)	19.8	*22.0*	9.0	*10.0*	—	—	—	—
Vitamin A (IU/lb)	—	—	1,636.0	*1,818.0*	3,273.0	*3,636.0*	3,273.0	*3,636.0*
Vitamin A (IU/kg)	—	—	3,600.0	*4,000.0*	7,200.0	*8,000.0*	7,200.0	*8,000.0*
Rate of feeding (lb/day)	0.20–0.40	*0.22–0.44*	0.20–0.40	*0.22–0.44*	0.20–0.40[5]	*0.22–0.44*[5]	0.20–0.40[5]	*0.22–0.44*[5]
Rate of feeding (kg/day)	0.09–0.18	*0.1–0.2*	0.09–0.18	*0.1–0.2*	0.09–0.18[5]	*0.1–0.2*[5]	0.09–0.18[5]	*0.1–0.2*[5]

[1]Adapted by the authors from *Nutrient Requirements of Sheep,* sixth revised edition, NRC-National Academy of Sciences, 1985, p. 52, Table 11.

[2]Feeds mixed and fed in meal or pellet form.

[3]Molasses and alfalfa hay, sun-cured, early bloom not included.

[4]Estimated 90% dry matter.

[5]In emergency situations, up to 1.1 lb (*0.5 kg*) may be fed.

HOW TO CHOOSE A RANGE SUPPLEMENT

Every range sheep producer faces the question of what supplement to use, when and how to feed it, and how much to feed under existing conditions.

In choosing a supplement for range sheep, the following requisites should be observed:

1. It should balance the ration of the animals to which it is fed. This means that it should supply all the nutrients needed by the animal that are missing in the forage. This necessitates the following three steps:

    a. Determining the approximate composition of the grazing animal's diet.

    b. Determining the sheep's requirements (refer to nutritive requirement tables, such as Tables 21–1 to 21–10).

    c. Computing the deficiencies (substract "a" from "b" above). The difference is the nutrients needed to made up the deficiencies—the nutrients that need to be provided by the supplement.

2. It should be fed in such a way that each animal gets its proper portion.

3. It should be fed in a form that is convenient and practical from the standpoint of the feeder, and that will least disturb the animals

The net profit resulting from the use of the supplement, rather than the cost per ton, should determine the choice of the supplement.

RATE OF SUPPLEMENTAL FEEDING

The time and rate of supplemental feeding is determined by the reason for feeding supplements. Supplements are fed for two primary purposes: (1) to balance rations by adding small quantities of a nutrient (such as protein, a mineral, or a vitamin) or a combination of nutrients; and (2) to provide nutrients during short-term emergencies. As an example of the latter, a supplement may be needed to prevent sheep from eating poisonous plants during periods when they are on the trail or when forage is covered with snow.

Supplemental feeding should be timed to start when it is needed. If phosphorus supplementation is required, it should be provided continuously, perhaps by free-choice feeding. When energy, protein, and/or vitamin A supplementation are involved, it takes a unique skill to recognize the nutritional state of the sheep, the range condition, and the need for supplement—both in kind and amount. The successful manager develops a grazing plan that minimizes the need for supplements, yet provides the proper supplement at the proper time and in the proper amounts.

The normal range of supplementation for sheep is ¼ to ½ lb per head per day. Rates above ½ lb approach a level that will result in reduced intake of range forage. Where range vegetation is so short as to require supplementation in excess of ½ lb per head per day, consideration should be given either to moving the sheep into a drylot or to moving them to a better grazing area.

Some managers divide their sheep according to age, condition, and twins vs single lambs. Of course, this is facilitated where there are several bands. By so doing, it is possible (1) to give the animals that require the highest level of nutrition the best pasture or range, and/or (2) to supplement according to need.

Fig. 21-27. Pregnant range ewes properly supplemented and in excellent condition. (Courtesy, USDA)

Fig. 21-28. A good start in life. (Courtesy, J. C. Allen & Son, Inc., West Lafayette, Ind.)

FEEDING GROWING-FINISHING LAMBS

The growing-finishing stage of lambs refers to that period extending from birth to weaning at 4 to 6 months of age. At no other period in the life of the sheep is the promotion of growth and the prevention of disease so important.

Proper feeding of growing-finishing lambs begins with ewes—the primary source of the diet of lambs for the first month. The ewes should be fed so that they produce plenty of milk.

Where succulent pastures are available, most practical sheep producers, including producers with both farm flocks and range bands, consider that a combination of such green forage plus the ewe's milk is ample for growing-finishing lambs. In fact, lambs are unique among farm animals, inasmuch as they may be marketed at top prices off grass. Although young cattle may be sold off grass without having any other feed, they will usually fail to attain sufficient finish to bring top prices.

Good pastures for lambs are those that are succulent and that are composed of plants that are palatable and nutritious. This means green, actively growing pastures in contrast to dormant or dried forages.

Frequently, farm-flock lambs are creep fed grain in addition to receiving their mother's milk and pastures. Usually creep feeding on the western range is too difficult. Should the range forages not be sufficiently abundant or lush to produce finished lambs, range sheep producers usually elect to sell their animals as feeders at weaning time.

"Hothouse" lambs are born out of season, in the fall or early winter, when pastures are usually unavailable. It is necessary that these animals be full fed for slaughter at 2 to 4 months of age, when they should weigh from 40 to 60 lb. In addition to the right breeding for this specialty, therefore, hothouse lambs must be carefully fed. In the first place, the ewes should be given liberal quantities of a good succulent ration in order to stimulate the milk flow. Secondly, the lambs should be creep fed with a palatable and suitable ration from the time they are 1 week of age until marketing.

Creep Feeding

The practice of supplemental feeding of nursing lambs in a separate enclosure away from their dams is known as creep feeding. Lambs will usually consume some creep feed at 7 to 10 days of age.

Creep rations can either be hand-fed or self-fed. Many sheep producers hand-feed until the lambs begin to eat regularly, then self-feed from this point on.

The amount of creep feed consumed is inversely proportional to the ewe's milk production. For this reason, (1) twin lambs usually consume more than single lambs, and (2) significant amounts of creep feed are consumed at 6 to 8 weeks of age, at which time the ewe's milk production has dropped significantly.

Until lambs are 4 to 5 weeks old, the grain should be crimped, cracked, or rolled. After 4 to 5 weeks of age, lambs show a preference for pelleted rations; and after 5 to 6 weeks, lambs may be fed whole grain unless it is extremely hard (e.g. millet).

It is important that the creep rations be very palatable. For this reason, corn, wheat bran, soybean meal, and molasses are important ingredients in a creep ration. Even then, if lambs have access to lush pasture, they may prefer it to the creep feed.

Suggested creep rations are given in Tables 21-15 and 21-16.

TABLE 21-15
SOME EXCELLENT CREEP RATIONS (AS-FED BASIS)[1]

	Unpelleted		Pelleted	
	First 2 Months	2 Months To Market	First 2 Months	2 Months To Market
	(%)	(%)	(%)	(%)
Ground corn	80	60	40	50
Ground oats		20	15	—
Soybean meal	20	10	20	10
Alfalfa hay	—	—	10	35
Bran		10	10	10
Molasses	—	—	5	5
Trace mineral salt	.5	.5	.5	.5
Limestone	1.0	1.0	1.0	1.0
Antibiotic .. (mg/lb)[2]	50	20	50	15
Vitamin A ... (IU/lb)	1,000	1,000	1,000	1,000
Vitamin D ... (IU/lb)	200	200	200	200
Vitamin E .. (mg/lb)	20	20	20	20

[1]The addition of 0.25 to 0.50% ammonium chloride will minimize urinary calculi.

[2]Chlortetracycline (Aureomycin) or oxytetracycline (Terramycin).

Feeding Directions:

1. Lambs should be started on creep feed about 10 days after birth. Although they will not consume significant amounts of feed until 3-4 weeks of age, the small amounts consumed at earlier ages are critical for establishing both rumen function and the habit of eating.
2. Feed high quality legume hay in a separate rack. Feed hay and creep ration twice daily to keep them fresh.
3. The amount of creep feed consumed by lambs 2 to 6 weeks of age is affected by the palatability of the ration (ration composition and ration form) and the location and environment of the creep area. A well-bedded, well-lighted area located close to where the ewes congregate is preferred.

TABLE 21-16
SOME SIMPLE CREEP RATIONS (AS-FED BASIS)[1]

	Unpelleted		Pelleted	
	First 2 Months	2 Months To Market	First 2 Months	2 Months To Market
	(%)	(%)	(%)	(%)
Ground corn	49	89	64	59
Crushed oats	30	—	—	—
Soybean meal	20	10	20	10
Limestone	1.0	1.0	1.0	1.0
Trace mineral salt	.5	.5	.5	.5
Alfalfa	—	—	10	25
Molasses	—	—	5	5

[1]The addition of 0.25 to 0.50% ammonium chloride will minimize urinary calculi.

Feeding Directions:

Same as presented with Table 21-15; so, see the latter.

Feeding Orphan ("Bummer") Lambs (Artificial Rearing)

Sheep producers estimate that about 10% of their lamb crop dies from starvation during the first week after birth. Some starvation results from newborn lambs sucking the scrotum and/or navel of another lamb. But most starved lambs are orphans (bummers) resulting from (1) the mother dying at lambing, (2) rejection by the mother, (3) the mother not being able to suckle the lamb because of mastitis or some similar problem, or (4) multiple births beyond the ewe's nursing capacity. Whatever the cause, the most satisfactory arrangement for the orphan is to provide a foster mother—to transfer (graft) the lamb to another ewe. The alternate to a foster mother arrangement is artificial rearing.

Observance of the following principles and practices will increase the chances of raising orphan lambs artificially:

- **Select the strong rather than the weak**—Where a multiple birth is involved and 1 or 2 lambs may be left with the mother, there are different opinions as to whether the strong or the weak should be selected for artificial rearing. Some producers select the weakest lamb(s) to raise artificially, theorizing that the stronger lamb(s) is better able to fend for itself. Others favor weaning the strongest, while leaving the weakest lamb(s) with the ewe. The authors recommend choosing the strongest and most aggressive lamb for artificial rearing for the reason that weak orphan lambs will not do well in a self-feeding system, although they can be raised satisfactorily if they are bottle fed by hand.
- **Give colostrum**—Colostrum makes for a good start in life. A newborn lamb needs 3.2 oz of colostrum per pound body weight during the first 18 hours after birth, according to the Modern Research Institute in Scotland (*Veterinary Record*, Vol. 118, March 29, pp. 351-353, 1986). Colostrum contains antibodies which impart immunity to infections for the first few weeks of life. This is important because the lamb's own immune system does not develop until it is 3 to 4 weeks old. Colostrum may be stored for this purpose. If a ewe either drops a stillborn lamb or loses her lamb within a day of birth, she may be milked and the colostrum frozen. Then, it can be warmed to 100°F and fed as needed. If ewe colostrum is not available, colostrum from a cow or a goat may be used, although it will not impart immunity to certain infections that are specific to sheep.

• **Inject orphans**—When orphan lambs are placed in the nursery, inject them with (1) vitamins A, D, E, (2) iron-dextran, and (3) selenium in selenium-deficient areas.

Also, enterotoxemia should be prevented. If the ewes were vaccinated with Type D toxoid prior to lambing, orphan lambs will receive colostral protection for 2 to 3 weeks; then, they should be vaccinated at 4 to 6 weeks of age. If the ewes were not vaccinated, the lambs should be vaccinated with the toxoid at 3 weeks of age and again at 5 weeks of age.

• **Use milk replacer**—A number of commercially prepared milk replacers are on the market. Best results will be obtained by using a replacer containing 25 to 30% fat, 20 to 25% protein provided by spray-dried milk products, and not to exceed 30 to 35% lactose; with the milk replacer diluted, mixed, and fed according to the manufacturer's directions. For comparative purposes, the composition of ewe's milk is presented in Table 21-17.

(Also see Chapter 6, Types and Roles of Feedstuffs, section on "Milk Replacers.")

TABLE 21-17
COMPOSITION OF EWE'S MILK (2.5 WEEKS POSTPARTUM)[1,2]

Dry matter	18.2%
Fat (5-10%)	7.1 g/100 g milk
Protein (true)	4.5 × 5.49 = 24.7% DM basis
Lactose	4.8 × 5.49 = 26.4% DM basis
Ash	0.85 g/100 g milk
Fiber	0.0 g/100 g milk
Caloric value (GE)	110 kcal/100 g × 5.49 = 6.04 Mcal/kg milk DM basis
Principal salts (g/100g)	
Sodium	0.040
Chlorine	0.075
Calcium	0.200
Phosphorus	0.150
Magnesium	0.016
Potassium	0.150
Citrate	0.170
Trace minerals (mg/liter)	
Aluminium	1.70
Copper	0.05-0.15
Iron	0.60-0.70
Manganese	0.06
Zinc	2.00-3.00
Vitamins (mg/liter, except where noted)	
Fat-soluble	
A	1,450 IU/liter
E (alpha-tocopherol)	15
Water-soluble	
Biotin	0.05-0.09
Folacin	0.05
Niacin	5.0
Pantothenic acid	4.0
Riboflavin	4.0
Thiamin	1.0
Vitamin B-6	0.7
Vitamin B-12	0.006-0.010
Vitamin C (ascorbic acid)	40-50

[1]Adapted by the authors from *Nutrient Requirements of Sheep,* sixth revised edition, NRC-National Academy of Sciences, 1985, p. 51, Table 8.

[2]Data provided to NRC courtesy of Dr. Robert Jenness, Biochemistry Department, University of Minnesota.

Where several orphans are being fed, up to 12 lambs of similar size and age may be grouped together in a small pen and self-fed from a multiple nipple container, allowing one nipple for each 2 to 4 lambs. When self-feeding, cool milk (50° to 60°F) should be fed because (1) it does not sour as quickly, and (2) the lambs are not apt to engorge on it. However, hand-fed lambs can safely be given warm milk, fed twice daily.

Do not use cow's milk or calf milk replacer for lambs. They contain too much lactose (milk sugar) for lambs and will cause scours.

Because milk replacer is expensive, the liquid-feeding period should be as short as possible. Lambs can be successfully weaned from milk replacer at 18 to 28 days of age.

• **Provide a good dry starter feed and water from day one**—From day one, orphan lambs should be provided access to a palatable dry ration to accustom them to eating dry feed and stimulate rumen development. This ration should be (1) palatable, (2) high in energy, (3) high in protein (22 to 24% crude protein on as-fed basis), (4) reinforced with minerals and vitamins, and (5) fed in finely ground (mash) form. A good starter ration follows:

Lamb Starter Ration:

Ingredients	**%**
Soybean meal (49% CP)	40.0
Ground corn	27.0
Alfalfa meal	15.0
Dextrose (corn sugar)	10.0
Fat (e.g. vegetable oil)	5.0
Limestone	2.0
Trace mineral salt	0.7
Vitamin premix	0.3
Total	100.0

Once the lambs have fully adjusted to the starter ration, they can be slowly switched onto the regular creep or grower ration. (See Tables 21-15 and 21-16.)

• **Maximize sanitation, observation, and TLC**—The successful artificial rearing of orphan lambs necessitates that the caretaker maximize sanitation, observation, and TLC—tender loving care.

Feeding Early Weaned Lambs

Early weaning refers to the practice of weaning lambs earlier than usual—at 6 to 8 weeks of age. Among practical sheep producers, the rule of thumb relative to early weaning is that lambs may be weaned at 45 to 60 days of age or 45 lb, whichever comes first. Currently, some flock owners are weaning lambs at 45 to 60 days of age with good results. There is much interest in early weaning because—

1. Of lambing out of season, multiple births, and more than one lamb crop per year.
2. Lactating ewes usually reach a peak in milk production 3 to 4 weeks after lambing, then decline thereafter. By 3 to 4 months after lambing, many ewes will be producing very little milk.
3. Fewer parasite problems accompany an early weaning program, provided the early weaned lambs are fed in drylot.

4. Increased knowledge of nutrition now makes it possible for scientists to improve upon milk (except for colostrum), chiefly by reinforcing it with certain minerals and vitamins.

5. Young gains are cheap gains, due to (a) the higher water and lower fat content of young animals in comparison with older animals, and (b) the higher feed consumption per unit of body maintenance of young animals.

6. Following weaning, ewes can be maintained on a limited feed allowance, thereby effecting a saving in cost.

For successful early weaning, superior nutrition and management are essential; and the earlier the weaning age, the more exacting these requirements.

Early weaning of lambs is, to a considerable extent, a matter of preparation, rather than the abrupt separation of lambs from their mothers. Lambs that are to be early weaned should be creep fed from the time they are old enough to eat. At weaning time, the separation should be made by removing the ewes from the lambs, rather than vice versa. By keeping the lambs in familiar surroundings, stress is minimized.

An early weaned lamb ration should meet the following specifications: contain a minimum of 16% crude protein; be fortified with supplemental iron if the lambs are raised on slotted floors; and have a calcium:phosphorus ratio of at least 1:1 (2:1 if urinary calculi has been experienced).

Milk replacers containing approximately 30% fat and 24% protein have been used successfully in feeding lambs receiving colostrum and weaned at 1 day of age. Replacers with reduced lactose content (from 27 to 42% on a dry matter basis) give improved performance. The milk is fed (1) cold at 36° to 40°F, rather than warm, to reduce overeating and bacterial contamination, and (2) free-choice. From the beginning, lambs are offered a very palatable solid feed in addition to the milk. The milk replacer is discontinued when the lambs are eating sufficient quantities of the dry feed (½ lb/head daily), usually at 21 to 35 days of age.

FEEDING FINISHING LAMBS

Fig. 21–29. A Texas feedlot, finishing lambs. (Courtesy, Texas Agricultural Extension Service, San Angelo, Tex.)

The primary objective of the sheep producer is that of producing milk-fat lambs suitable for slaughter at weaning time. Only when the pasture is inadequate are lambs sold via the feeder route. Almost all feeder lambs come from the range area. Some range areas produce only a small percentage of lambs which are classed as feeders, whereas in other areas almost all the lambs must be sold as feeders because the vegetation is not sufficient to promote rapid growth and finishing. It is estimated that, for the range area as a whole, an average of at least 50% of all lambs produced in any year receive additional feed after they are removed from the range and prior to slaughter. Because such lambs are fed and marketed out of season, they are sometimes referred to as *old crop lambs.*

Areas and Types of Lamb Feeding

Feeder lambs are generally sold to go into districts where grains and other concentrates are abundant or where fall and winter pastures are available. Such areas include (1) the irrigated districts of the West where the feeding of milo and other grains, by-products, and crop residues predominates, (2) the wheat-raising sections of Oklahoma, Kansas, Nebraska, and Texas, where fall grazing of the wheat fields is practiced, and (3) the Corn Belt where stubble fields and meadows are gleaned and the corn crop may be harvested by lambs.

Colorado is the leading lamb-feeding state of the nation, finishing more than one-fifth of the sheep and lambs fed in the United States. Here, locally grown alfalfa, sugar beet by-products, and barley are used extensively, along with considerable corn and protein supplements which are shipped in from outside areas. California ranks second in lamb feeding, followed by Texas, Wyoming, and Oregon.

The trend is toward fewer and large feedlots, with feeding increasing in areas near to the larger slaughter plants primarily in the northern plains and western producing areas.

Numerous feeding practices and a great variety of feeds are used in lamb-finishing operations. In general, however, all methods may be classified as either (1) field finishing, or (2) drylot finishing.

Field Finishing

Fig. 21–30. Lambs finishing on winter wheat in western Kansas. On a dry-matter basis, wheat pasture contains 60 to 80% TDN and 20 to 30% protein. Also, wheat pastures are generally parasite-free. (Courtesy, Kansas State University, Manhattan)

Field finishing refers to the partial or complete finishing of lambs on pastures or on such field crops as sorghum stubble and sorghum crops, corn fields, alfalfa pasture, winter cereal grain pasture (wheat, oats, barley, rye), sugar beet tops, and vegetable crops. In good years and on good feed, many of these lambs are finished enough to go to the packer. However, most of them are marketed as heavy feeders weighing 90 to 100 lb.

In Kansas, Oklahoma, Nebraska, and Texas, many lambs are finished primarily by fall grazing on wheat fields. In the West, alfalfa regrowth in the fall and winter is often pastured. In the Pacific Northwest, a limited number of lambs are finished by gleaning pea stubble.

In the Corn Belt, feeder lambs are sometimes used to harvest forage or grain, or both, in corn fields. They may be used (1) to consume the lower leaves only in the late summer, (2) to glean corn fields after harvest, or (3) to harvest the corn grain.

Fig. 21-31. "Sheeping down" corn on an Iowa farm. First, the crabgrass, weeds, and lower corn leaves are eaten. (Photo by A. M. Wettach, Mt. Pleasant, Iowa)

In all these systems, the lambs should be vaccinated for enterotoxemia and given access to minerals.

Field finishing decreased with the advent of fewer fences and larger harvest machinery. However, some sheep producers use electric fence or temporary woven wire fence to control field finishing and eliminate the need for herders.

Drylot Finishing

Drylot feeding is, as the name indicates, feeding under restricted conditions. This may either be (1) shelter or barn feeding, or (2) open-yard feeding.

SHELTER OR BARN FEEDING

Because of inclement weather in the fall and early winter, most of the lamb-feeding operations in the central and eastern states are in drylots which afford shelter. In some instances, the lambs are kept under cover without an exercising lot. These barns may consist of anything from open sheds to more costly and elaborate structures, including slotted floors. In the vast majority of instances, the feeds are locally grown, corn being the chief concentrate used in these feeding operations. Lambs are finished as a means of marketing the grain, conserving the fertility of the soil, and furnishing gainful work during the winter months. Most of these lambs are finished by farmers who feed one or more carloads, rather than by large operators who feed thousands of lambs. Practically all of the feeder lambs used in these operations come from the western ranges, either directly or via markets handling feeder lambs. After getting on full feed, these lambs are either hand-fed twice daily or self-fed.

OPEN-YARD FEEDING

Open-yard feeding is the common method of finishing lambs in the irrigated areas of the West, though a few eastern lamb feeding operations are in open yards. In this system, equipment costs are kept to a minimum—the facilities merely consisting of an enclosed and well-drained yard which may or may not have a natural or constructed windbreak, and the necessary feed bunks. Open-yard feeding is often used by large operators who feed thousands of lambs.

Large quantities of alfalfa and sugar beet by-products are utilized in these yards. The proportion of roughage and concentrate varies according to comparative costs, feeding value considered. In order to save labor, the practice of self-feeding is increasing. Lambs may be self-fed successfully, but it is recommended that the following precautions be taken in order to lessen the incidence of overeating disease: (1) good management, (2) vaccination against enterotoxemia, and (3) more roughage and less concentrate be used.

Finishing Lamb Rations

A great variety of feeds can be used in lamb feeding. Drylot feeders utilize grains, harvested forages, protein supplements, and by-product feeds. In general, successful feeders balance their rations by selecting those feeds which are most readily available at the lowest possible price. Properly supplemented forages and silages can be used when these are available and favorably priced, but the gains will be less than when higher energy rations are fed. Also, varying proportions of high-quality hay and feed grains can be used with the proportions depending on the relative costs of the forage and grain components of the ration. When grain is relatively inexpensive, rations with little or no forage may be used. When high-grain rations are fed, it is important (1) that the change from high forage to high grain be gradual, (2) that the lambs be vaccinated against enterotoxemia, (3) that urinary calculi be controlled, and (4) that there be superior management.

The rations given in Table 21-18 (see next page), are typical of those that are widely used to finish lambs. The corn/alfalfa hay/soybean meal rations are used in areas of the United States where these ingredients are raised. The milo/cottonseed hulls/cottonseed meal rations are typical of those used in the Southwest. These rations increase in energy value (grain content) as they go from 1 to 5 in each series. Generally, lambs should be started on rations 1 or 2, then switched to ration 3, thence 4, and thence 5; allowing a 4-7 day adjustment period on each ration before stepping up to the next higher energy level.

TABLE 21-18
GROWING–FINISHING RATIONS FOR LAMBS[1]

Ingredient	Moisture Basis[2] A-F (as-fed) M-F (moisture-free)	Rations Using Corn/Alfalfa Hay/Soybean Meal					Rations Using Milo/Cottonseed Hulls/Cottonseed Meal				
		1	2	3	4	5	1	2	3	4	5
		(%)	(%)	(%)	(%)	(%)	(%)	(%)	(%)	(%)	(%)
Corn grain (dent yellow)		31.0	41.5	51.7	63.0	73.3	—	—	—	—	—
Sorghum grain (milo)		—	—	—	—	—	19.5	32.7	46.2	60.7	73.7
Alfalfa hay (mature)		55.0	45.0	35.0	25.0	15.0	15.0	15.0	15.0	15.0	15.0
Cottonseed hulls		—	—	—	—	—	40.0	30.0	20.0	10.0	—
Soybean meal (solvent 44% CP)		7.0	6.5	6.0	5.5	5.0	—	—	—	—	—
Cottonseed meal (solvent 41% CP)		—	—	—	—	—	17.5	14.0	10.5	7.0	4.0
Molasses (cane)		6.0	6.0	6.0	5.0	5.0	6.0	6.0	6.0	5.0	5.0
Calcium carbonate		—	—	.3	.5	.7	1.0	1.3	1.3	1.3	1.3
Trace mineral salt (sheep)		.5	.5	.5	.5	.5	.5	.5	.5	.5	.5
Ammonium chloride		.5	.5	.5	.5	.5	.5	.5	.5	.5	.5
		Nutritional content									
Dry matter (%)	**A-F**	**87.5**	**87.5**	**87.5**	**87.6**	**87.6**	**89.0**	**88.9**	**88.7**	**88.7**	**88.5**
	M-F	100	100	100	100	100	100	100	100	100	100
TDN (%)	**A-F**	**75.3**	**80.1**	**84.6**	**89.2**	**93.1**	**67.9**	**72.3**	**77.2**	**82.0**	**86.8**
	M-F	65.9	70.1	74.0	78.1	82.0	60.4	64.3	68.5	72.7	76.8
Net energy for maintenance (Mcal/lb)	**A-F**	**.80**	**.86**	**.91**	**.98**	**1.04**	**.71**	**.76**	**.82**	**.88**	**.94**
	M-F	.70	.75	.80	.86	.91	.63	.68	.73	.78	.83
Net energy for maintenance (Mcal/kg)	**A-F**	**.36**	**.39**	**.41**	**.44**	**.47**	**.32**	**.35**	**.37**	**.40**	**.43**
	M-F	.32	.34	.36	.39	.41	.29	.32	.33	.35	.38
Net energy for gain (Mcal/lb)	**A-F**	**.40**	**.47**	**.53**	**.59**	**.66**	**.33**	**.38**	**.45**	**.52**	**.59**
	M-F	.35	.41	.46	.52	.58	.29	.34	.40	.46	.52
Net energy for gain (Mcal/kg)	**A-F**	**.18**	**.21**	**.24**	**.27**	**.30**	**.15**	**.17**	**.20**	**.24**	**.27**
	M-F	.16	.19	.21	.24	.26	.13	.15	.18	.21	.24
Crude protein (%)	**A-F**	**17.1**	**16.6**	**16.0**	**15.4**	**14.8**	**17.0**	**16.3**	**15.8**	**15.2**	**14.8**
	M-F	15.0	14.5	14.0	13.5	13.0	15.1	14.5	14.0	13.5	13.1
Protein bypass (%)	**A-F**	**42.3**	**44.1**	**45.8**	**47.9**	**49.7**	**44.6**	**47.6**	**51.0**	**54.7**	**57.9**
	M-F	37.0	38.6	40.1	42.0	43.5	39.7	42.3	45.2	48.5	51.2
Calcium (%)	**A-F**	**.86**	**.71**	**.70**	**.67**	**.66**	**.85**	**.96**	**.94**	**.91**	**.89**
	M-F	.75	.62	.61	.59	.58	.76	.85	.83	.81	.79
Phosphorus (%)	**A-F**	**.29**	**.30**	**.31**	**.31**	**.32**	**.37**	**.37**	**.36**	**.36**	**.36**
	M-F	.25	.26	.27	.27	.28	.33	.33	.32	.32	.32

[1]Adapted by the authors from *The Sheepman's Production Handbook*, published by the Sheep Industry Development Program, Inc., Denver, Colo., 1986, p. 44, Table 13.

[2]As-fed was calculated using an average figure of 90% dry matter. When using silages, roots, and other wet feeds, these feeds should be converted to a moisture-free basis and the ration calculated using the moisture-free data.

Feeding Directions:

1. These rations can be fed once daily in troughs or bunks if there is capacity for a day's feed. They can also be self-fed if the feeders are designed to handle such feed without bridging.
2. Offering lambs a good quality hay for 1–3 days along with rations 1 or 2 (provided free choice) can be used to start lambs on feed.
3. About 3 in. (*7.6 cm*) of self-feeder or trough space must be provided per lamb for self-feeding and about 12 in. (*30.5 cm*) if hand fed.
4. Gradually adapt the lambs to the higher energy rations by allowing 4–7 days on a ration before switching to the ration with the next higher energy level.
5. Complete mixing to prepare a uniform ration is important.
6. Lambs must not be allowed to be without feed even for a short period of time.
7. The mineral and vitamin mixture given in Table 21–19 may replace the trace mineral salt in all Table 21–18 rations.

These rations are nutritionally adequate and balanced with respect to Ca:P and N:S ratios. Feeds of similar nutritive properties can be substituted as price relationships warrant. For example, barley can be substituted for all or part of the corn or milo, wheat can replace up to ½ of the corn or milo, and dried beet pulp can replace up to 30% of the total grain ration. However, if much change in the formulation is made, the analysis should be recalculated and checked against the nutrient requirements.

The mineral and vitamin mixture given in Table 21–19 may replace the trace mineral salt in all the Table 21–18 rations.

TABLE 21-19
MINERAL AND VITAMIN MIXTURE FOR LAMB RATIONS[1]

Ingredient	Lb/Ton	*Kg/Ton*	Contribution To Complete Ration[2]
Salt, plain fine mixing	1,729.613	*785.244*	.43% salt
Sulfur, elemental[3]	200.00	*90.80*	.05% S
Cobalt carbonate ($CaCO_3$)	0.087	*0.039*	.1 ppm Co
Ethylenediamine dihydro-iodide (EDDI)	0.100	*0.045*	.2 ppm I
Manganese oxide (MnO)	10.300	*4.676*	20 ppm Mn
Zinc oxide (ZnO)	10.300	*4.676*	20 ppm Zn
Vitamin A[4]	17.6	*8.0*	600 IU/lb
Vitamin E[5]	32.0	*14.5*	10 IU/lb

[1]Adapted by the authors from *The Sheepman's Production Handbook*, published by the Sheep Industry Development Program, Inc., Denver, Colo., 1986, p. 45, Table 14.

[2]Contribution to the complete ration when 10 lb (*4.5 kg*) of the mineral and vitamin mixture is added to 1 ton of complete lamb ration.

[3]In complete rations containing ammonium sulfate instead of ammonium chloride for prevention of urinary calculi, sulfur should not be added to the mineral and vitamin premix (.5% NH_4SO_4 contributes .12% S to the ration).

[4]Contains 13,607,700 IU of vitamin A per pound.

[5]Contains 125,000 IU of vitamin E per pound.

Antibiotics are useful in reducing problems which usually occur when lambs are placed on feed. They are usually fed at a level to supply 25 to 30 mg per lamb per day, although considerably higher levels are reported to be effective during stress and problem periods.

Basic Considerations in Finishing Lambs

Although no rules of success are applicable to any and all conditions, the following basic considerations in finishing lambs are worth noting:

1. **Lamb feeding is seasonal in nature.** It usually extends from August to about the following May. This seasonal condition is due to the fact that (a) suitable feeder lambs are not available until the late summer and fall months and (b) following the growing and harvesting seasons, the feeders have available quantities of feeds which may be utilized by lambs. However, during droughts, lambs may be weaned early and fed in feedlots to avoid losses and lack of gain.

2. **Range feeder lambs preferred.** Range feeder lambs are more plentiful than native feeders, thus allowing for greater selection; and usually they are more uniform and have fewer parasites.

3. **Shearing feeder lambs may or may not be profitable.** Among the factors determining whether feeder lambs should be shorn are: (a) season of the year—heat or cold stress, (b) price at slaughter for shorn versus unshorn lambs, and (c) wool incentive price.

4. **Purchase price of lambs constitutes main cost.** In lamb feeding operations, the purchase price of the lambs represents 60 to 70% of all costs. This indicates the importance of keeping death losses to a minimum.

5. **Expect death loss of about 1 to 2%.** Experienced feeders normally expect to lose about 1 to 2% of lambs on feed. This is higher than occurs in commercial cattle feeding operations.

6. **Incoming feeder lambs are stressed.** Feeder lambs undergo considerable stress when moved from the range to the feedlot. Generally, they are gathered, sorted, weaned, left to stand overnight without feed or water prior to weighing, then loaded on trucks or railroad cars and shipped to the feedlot. In a 250-mile shipment, lambs will shrink about 5%; with greater distances, they will shrink more.

Shipping stress can be greatly lessened by—

a. Preconditioning before the lambs leave the producer's property. This involves starting them on feed, vaccinating, drenching, and shearing under some conditions.

b. Transporting lambs as rapidly as possible from range to feedlot.

7. **Highly stressed lambs should not be processed.** Incoming lambs should not be drenched, vaccinated, shorn, and/or implanted until they have rested and are eating and drinking. When the latter stage is reached—

a. Treat lambs for internal and external parasites.
b. Vaccinate for enterotoxemia (Type D).
c. Vaccinate for sore mouth—if indicated.
d. Shear—if desired.
e. Sort and lot lambs according to size.
f. Isolate and treat sick and weak lambs.

8. **Exercise care when starting on feed.** Special care is necessary in starting heavily stressed lambs on feed. Immediately upon arrival, let them have a good rest, in a comfortable place. Also, provide them free access to a good-quality grass or grass-legume hay (preferably, not over 50% legume), clean water (preferably running water), and minerals.

After they have rested for several days, start them on concentrate by feeding about ¼ lb per head daily. Gradually, this allowance should be increased until they are getting a full feed 2 to 4 weeks later. Commercial feeders place lambs on feed as rapidly as possible, without throwing them off feed.

Many systems have been successfully used in starting lambs on feed. Some commercial feeders start them on ground or pelleted rations containing 60 to 70% roughage. One Texas feeder uses cottonseed hulls to limit consumption of self-fed rations. By reducing hull levels at intervals, this feeder is able to self feed a 90 to 95% concentrate ration within 10 to 14 days. Other bulky feeds that are sometimes used to work lambs onto a high-concentrate ration include chopped hay, wheat bran, and dried beet pulp.

9. **Feed efficiency is lower and feed cost higher than for younger lambs.** Most incoming feeder lambs range from 5 to 8 months of age and from 65 to 90 lb weight; and they are marketed 60 to 90 days after they are placed on feed, at a weight of 105 to 110 lb. Feed efficiency is lower and cost of gains is higher than with younger lambs, as indicated by the following comparisons:

Age	Lb feed/Lb gain
Preweaned lambs	2.0 to 2.5
Early weaned lambs	2.5 to 4.0
Late weaned lambs	6.0 to 8.0

10. **Feed costs account for ⅔ of finishing cost.** Feed accounts for about 66% of the cost of finishing feedlot lambs, exclusive of the intitial purchase price of the feeder lambs.

11. **Most grains need not be ground.** Unless such extremely hard seeds as millet are included in the ration, it does not pay to grind feeds for finishing lambs.

12. **Feedlot gains are expensive.** Feedlot gains are expensive, usually costing more per pound than the selling price on the market. Thus, a reasonable margin or difference between the cost and selling price per hundredweight is necessary.

13. **Wether lambs make more rapid gains than ewe lambs.** Wether lambs appear to make slightly more rapid gains than ewe lambs, but they do not finish quite so early as ewe lambs.

14. **Contract feeding is popular.** Lambs are frequently fed on a contract basis, with many and varied agreements being used.

FEEDING SHOW SHEEP

No amount of blocking or trimming can counteract poor feeding and management. This does not imply that one should minimize the importance of proper grooming from the standpoint of creating a favorable and attractive appearance; rather, experienced judges rely upon their hands to determine what is underneath an artistic job of blocking and trimming. Thus, for success in the show-ring, the animals should be fed and managed to attain maximum development in body conformation and fleece quality.

Fig. 21-32. Fitting a sheep for show. (Courtesy, USDA)

In general, the feeding of show sheep differs from normal operations only in that greater effort and expense may be justified in order to produce a winner.

The most important pointers in feeding and handling show sheep are as follows:

1. **Keep show sheep healthy and free from parasites.** It is impossible to obtain proper growth, finish, and bloom in sheep that are unthrifty or infested with parasites. So, the caretaker should scrupulously follow a program designed to assure flock health, disease prevention, and parasite control.

2. **Provide a suitable ration.** In addition to being reasonably economical and well balanced, the ration for show sheep must be palatable. Many feed combinations meet these specifications. The ration selected is usually determined by (a) the availability and price of feed in the area, and (b) the preference and judgment of the shepherd.

Some suggested grain-fitting rations are listed in Tables 21-20 and 21-21. To each of these grain rations should be added (1) good-quality roughage, and (2) salt and other minerals on a free-choice basis.

• **Rations for lambs**—Lambs intended for show should be left with their mothers and fed the regular creep feed until they are about 90 days of age. At this stage, they should be weaned and placed on a lower energy, less-fattening ration than the creep feed. After weaning, the Table 21-20 concentrate mixture is recommended for show lambs, with judicious ingredient substitutions made as determined by availability and price.

TABLE 21-20
FITTING CONCENTRATE MIX FOR SHOW LAMBS

Ingredient	%
Cracked corn	50
Whole or rolled oats	35
Soybean meal	10
Molasses	4
Mineral (limestone/sheep salt-mineral mix)	1
Total	100

Lambs will consume a daily amount of feed equivalent to approximately 4% of their body weight. About ⅔ of the daily ration may consist of the concentrate mixture (Table 21-20 and ⅓ of alfalfa hay. Lambs should be hand-fed twice daily. They may be given as much of the concentrate mix as they will consume in 15 to 20 minutes, plus as much hay as they will eat more leisurely without wastage. Additionally, the lambs should have free access to a sheep mineral mix and clean, fresh water.

Lambs should be vaccinated against enterotoxemia, using Type D toxoid, at 3 weeks of age and again at 5 weeks of age.

• **Rations for yearlings and mature sheep**—The rations in Table 21-21 are for fitting yearlings and mature sheep. Show yearlings and mature sheep will eat 3 to 5 lb of grain per head daily. Additionally, they should receive a good legume hay. It is important that yearlings and mature sheep not be "fed off their feet," and not become soft and flabby as a result of lack of exercise.

TABLE 21-21
RATIONS FOR FITTING YEARLING AND MATURE SHEEP

Ingredient	Ration Number							
	1		2		3		4	
	(lb)	*(kg)*	(lb)	*(kg)*	(lb)	*(kg)*	(lb)	*(kg)*
Barley, rolled	—	—	40	*18.2*	—	—	10	*4.5*
Corn, cracked	—	—	—	—	40	*18.2*	—	—
Oats, rolled	50	*22.7*	40	*18.2*	40	*18.2*	60	*27.2*
Peas (split)	40	*18.2*	—	—	—	—	10	*4.5*
Protein supplement[1]	—	—	10	*4.5*	10	*4.5*	10	*4.5*
Wheat bran	10	*4.5*	10	*4.5*	10	*4.5*	10	*4.5*
Total	100	*45.4*	100	*45.4*	100	*45.4*	100	*45.4*

[1]Cottonseed, linseed, rapeseed (canola), soybean, and/or sunflower meal.

In general, the preceding rations are higher in protein content than rations used in commercial finishing operations, but most experienced shepherds feel that by such means they get more "bloom."

The rations listed in Tables 21–20 and 21–21 are intended as guides only; specific conditions may well warrant changes. For example, when wheat prices are favorable, this grain may replace part or all of the barley or corn in the ration. When either a lush pasture grass or a high-quality alfalfa is used as the roughage, it is usually advisable to reduce the protein supplement in the ration by as much as one-half. This is done for reasons of practicality and in order to avoid feeds that are too laxative.

3. **Provide suitable minerals.** All classes and ages of sheep should be fed minerals according to their requirements.

4. **Use clean, lush, temporary pastures in season or high-quality hay.** In season, clean luxuriant pastures produced on fertile soils may provide the roughage allowance for all show sheep, including both lambs and mature animals. The most desirable pastures are those that are classed as temporary and which are annuals or handled in such a manner that sheep have not grazed thereon for at least a year, for this will alleviate the hazard of parasites. Among the most common annuals used as pastures for show sheep are: rye, wheat, oats, field peas, and rape.

During hot weather, sheep will do better when kept in the barn during the heat of the day and turned to pasture in the cool of the evening or early in the morning. Also, the grain ration should be fed early and late during the summer months.

5. **Provide succulent feeds.** Sheep are very fond of such succulent feeds as cabbage, carrots, mangels (stock beets), turnips, and rutabagas (swedes). The experienced shepherd usually adds limited quantities of one of these feeds to the ration of show sheep. These succulent feeds are highly relished by sheep and appear to help their digestion and general thrift.

NUTRITIONAL DISORDERS AND TOXINS

There are numerous other feed and nutritional disorders of great importance in sheep production; among them, enterotoxemia, polioencephalomalacia, pregnancy disease (ketosis), urinary calculi, and poisonous plants. These and other similar problems affecting sheep are fully discussed in Chapter 5, Nutritional Disorders/Toxins.

QUESTIONS FOR STUDY AND DISCUSSION

1. In what ways may a sheep producer take practical advantage of the fact that sheep consume a higher proportion of forages than any other class of livestock?

2. What is the economic importance of feed for sheep?

3. Why should margins of safety be provided over and above the NRC requirements?

4. List and discuss each of the factors that can affect the energy requirements of sheep.

5. Discuss each of the following as they relate to the energy needs of sheep: (a) chief sources, (b) how intake may be controlled, and (c) symptoms of deficiency.

6. List and discuss each of the factors that can affect the protein requirements of sheep.

7. Discuss each of the following as they relate to the protein needs of sheep: (a) chief sources, (b) the nonfunctional rumen of newborn lambs, and (c) symptoms of deficiency.

8. Is quality of protein important in sheep nutrition? Justify your answer.

9. For sheep, for each of the following minerals (a) describe the deficiency symptoms and tell how you could distinguish between them, and (b) give practical sources: salt, calcium, phosphorus, magnesium, potassium, sulfur, cobalt, copper, iodine, selenium, and zinc.

10. What are the manifestations of toxicity in sheep of each of the following: copper, fluorine, molybdenum, and selenium?

11. For sheep, for each of the following vitamins (a) describe the deficiency symptoms and tell how you could distinguish between them, and (b) give practical sources: vitamin A, vitamin D, and vitamin E.

12. List, and discuss the effects of each of the presently FDA approved sheep feed additives and implants.

13. Most lambs are marketed as milk-fat lambs directly off pastures or ranges without having had any grain. What are the advantages of this, from an economic standpoint? Can this practice be applied to cattle or swine?

14. How do sheep differ from cattle in their grazing habits?

15. What is the physiological explanation of why sheep can consume more pounds of a high-quality forage than of a low-quality forage?

16. How does the preparation of feed for sheep and cattle differ?

17. Of what value is a feed substitution table, such as Table 21–12?

18. What are the advantages and disadvantages of drylot (confinement) feeding of sheep? Will drylot operations obsolete the use of pastures for sheep?

19. Discuss each of the following as they relate to feeding breeding ewes: (a) flushing, (b) feeding ewes carrying multiple fetuses during late pregnancy, and (c) feeding lactating ewes.

20. What are the most common nutrient deficiencies of range forage for sheep?

21. Step by step, how would you go about determining the kind of range supplement needed for sheep?

22. Range sheep operators generally agree that where range vegetation is so short as to require supplementation in excess of ½ lb per head per day, consideration should be given either to moving the sheep into a drylot or to moving them to a better grazing area. What prompts sheep producers to set an upper limit of ½ lb of supplement per head per day?

23. Discuss each of the following as they pertain to growing-finishing lambs: (a) creep feeding, (b) feeding orphan lambs, and (c) early weaning.

24. Discuss each of the following as they pertain to feeding finishing lambs: (a) areas and types of feeding, (b) field finishing, (c) drylot finishing, and (d) finishing lamb rations.

25. Why is lamb finishing so seasonal in nature; more so than cattle finishing?

26. It requires 6 to 8 lb of feed to produce 1 lb of on-foot feeder lamb. How does this efficiency of feed utilization compare with finishing cattle and growing-finishing hogs?

27. Under what conditions might it be preferable that a feedlot operator feed lambs instead of cattle?

28. List and discuss the most important pointers in feeding and handling show sheep.

SELECTED REFERENCES

Title of Publication	Author(s)	Publisher
Beginning Shepherd's Manual	B. Smith	Iowa State University Press, Ames, Iowa, 1983
Handbook For Woolgrowers	Ed. by G. R. Moule	Australian Wool Board, Melbourne, Australia, 1972
Management and Diseases of Sheep, The	Ed. by the British Council	The British Council, London, and The Commonwealth Agricultural Bureaux, Slough, England, 1979
Nutrient Requirements of Sheep, 5th Revised Edition	National Research Council	National Academy Press, Washington, D.C., 1985
Of Sheep & Shows	J. C. P. Kroge	Paddock Publishing Co., Boulder, Colo., 1972
Profitable Sheep Farming	M. McG Cooper R. J. Thomas	Farming Press Ltd., Fenton House, Ipswich, England, 1971
Sheep and Goat Handbook	Guest Professors	International Stockmen's School, Clovis, Calif., annually since 1980
Sheep & Goat Science	M. E. Ensminger R. O. Parker	Interstate Printers & Publishers, Inc., Danville, Ill., 1986
Sheep: Applied and Basic Research Information Pub. During 1979	Ed. by D. R. Lincicome	International Goat and Sheep Reseasrch, Scottsdale, Ariz., 1983
Sheep Book, The	R. Parker	Charles Scribner's Sons, New York, N.Y., 1983
Sheepman's Production Handbook, The	Ed. by G. E. Scott	Sheep Industry Development Program, Inc., Denver, Colo., 1986
Sheep Raisers Manual	W. K. Kruesi	Williamson Publishing Co., Charlotte, Vt., 1985
U.S. Sheep and Goat Industry, The	C. S. Menzies, Chairman, Task Force	Council for Agricultural Science and Technology (CAST), Report No. 94, Ames, Iowa, 1982

Fig. 21-33. Band of sheep on a western range. (Courtesy, USDA)

Goats—
Producers of
milk, meat, and mohair

Original painting by Tom Phillips

Chapter 

FEEDING GOATS[1 2]

Contents — Page

[1]The authors gratefully acknowledge the helpful suggestions of the following eminent authorities who reviewed this chapter: E. K. Cassel, Ph.D., Extension Dairy Specialist, the University of Maryland, College Park; L. D. Guthrie, Ph.D., Head, Extension Dairy Science Department, the University of Georgia, Athens; G. F. W. Haemlein, Ph.D., Professor, University of Delaware, Newark; J. E. Huston, Ph.D., Texas A&M University, San Angelo; C. N. Lee, Ph.D., University of Hawaii at Manoa, Honolulu; C. D. Lu, Ph.D., American Institute of Goat Research, Langston University, Langston, Okla.; S. J. Lyford, Jr., Ph.D., Professor, University of Massachusetts at Amherst; B. R. Moss, Ph.D., Auburn Univeristy, Auburn, Ala.; J. C. Porter, Ph.D., Extension Dairy Specialist, University of New Hampshire, Penacook; M. Shelton, Ph.D., the Texas Agricultural Experiment Station, the Texas A&M System, San Angelo; T. H. Teh, Ph.D., Prairie View A&M University, the Texas A&M System, Prairie View; and C. S. F. Williams, B.V.Sc., M.R.C.V.S., Professor, College of Veterinary Medicine, Michigan State University, East Lansing.

[2]The statistics presented in the introductory section were for the mid-1980s and were obtained from: *FAO Production Yearbook,* Published by FAO, Vol. 39, 1985; and *Agricultural Statistics,* USDA, 1985.

Fig. 22-1. Mother and kid. (Courtesy, Jodi Frediani, Santa Cruz, Calif.)

There are more than 450 million goats in the world, of which about ⅓ are in Africa. They contribute 1.4% of the world meat supply and 1.5% of the world milk supply.

Goats provide nearly ⅓ of the total meat produced in India and from 7 to 16% of the total meat produced in Turkey, Morocco, Indonesia, Nigeria, and Cyprus. In a number of countries, goat meat is preferred to other meats.

The dairy goat has long been a popular milk animal in the Old World, where it is often referred to as the cow of the poor. In some countries, goat milk accounts for up to 50% of the total milk production. Southeast Asia, Africa, and the Near East lead in the production of goat milk.

The goats of the world also produce 36 million lb of mohair and cashmere and 33 million skins, annually. The three leading mohair-producing countries of the world, by rank, are South Africa, Turkey, and the United States.

There are approximately 2,950,000 goats in the United States, consisting of about 1,600,000 Angoras, 850,000 dairy goats, and 500,000 Spanish goats. Ninety-five percent of the Angoras are located in Texas. About 85% of the gross income from Angoras is from mohair and 15% from meat. Most of the Spanish goats are also in Texas. California is the most important dairy goat state.

Most Angora and Spanish goat herds are large and produced under extensive range conditions, whereas most dairy goats are found in small herds and on small farms or farms operated on a part-time basis.

Nutritional deficiencies and diseases in goats are of special concern because they have such widely-differing uncontrolled diets as a consequence of the great variety of conditions under which they are produced. Moreover, their nutritional and management practices differ according to their primary end products. Thus, Angoras which produce mohair, and Spanish goats which produce meat, are raised primarily on rangelands, often without supplemental feed. But dairy goats, which produce milk, require well-balanced rations high in energy and protein, similar to those of lactating dairy cows.

So, the goat industry of America can be divided into three distinct types of production: (1) dairy, (2) mohair, and (3) meat. Also, mention should be made of the pygmy goat.

The dairy goat, which is kept primarily for milk production, is gaining in popularity in the United States. Presently, it consists of the following important breeds: Alpine, American La Mancha, Nubian, Oberhasli (Swiss Alpine), Saanen, and Toggenburg. According to the American Dairy Goat Association, 41,153 purebred dairy goats were registered in 1986. In most areas, goat's milk and cheese command premium prices as specialty foods.

The most numerous goat breed in America is the mohair-bearing Angora, the heavy-coated creatures kept for fiber production and brush control. Yet, few people outside the Angora district, characterized by rugged grazing lands, know what they look like. Angora goats are used to produce (1) a beautiful, long, lustrous fiber known as mohair and (2) meat, and to augment other brush-control methods.

The Spanish goat is kept for meat production and brush control. It is of uncertain origin, but in all probability its ancestors were brought from Spain by early explorers. Subsequently, dairy goat breeds have been infused. Colors vary

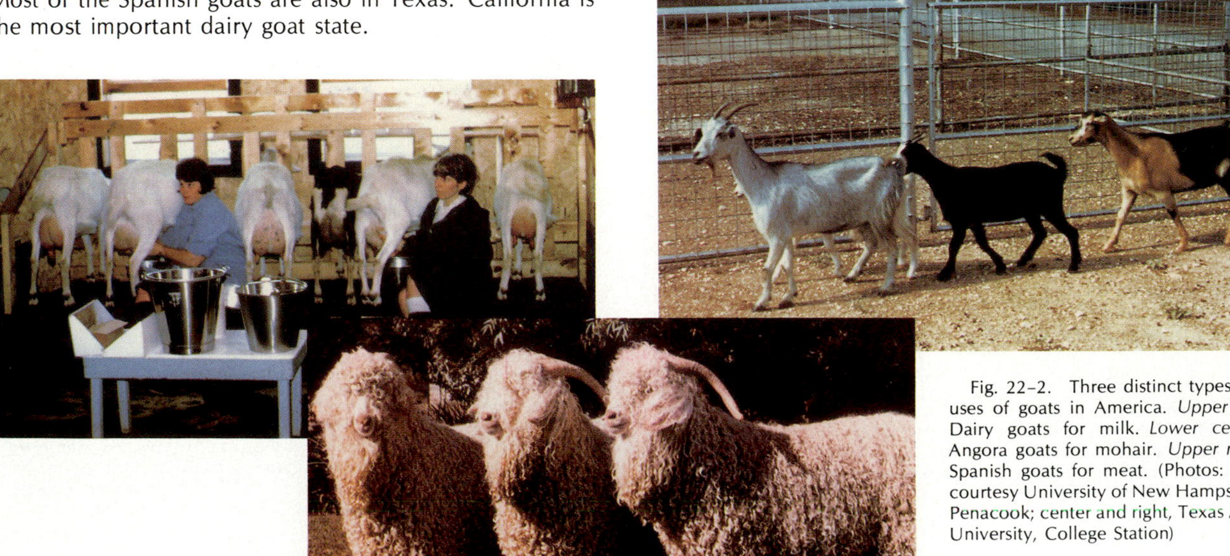

Fig. 22-2. Three distinct types and uses of goats in America. *Upper left:* Dairy goats for milk. *Lower center:* Angora goats for mohair. *Upper right:* Spanish goats for meat. (Photos: Left, courtesy University of New Hampshire, Penacook; center and right, Texas A&M University, College Station)

from solid black, brown, or white to striped, to spotted. In Mexico, the meat from young milk-fed Spanish goat kids, known as *cabrito* (Spanish for little goat), is considered a delicacy. In the United States, Spanish goats are usually slaughtered when a little older at which time the meat is known as *chevon*. Spanish goats are usually left to survive on range forages with little or no feed supplementation.

Also, small numbers of pygmy goats are found in the United States. The American pygmies are descended mostly from West African dwarf goats found in Nigeria, Ghana, and the Cameroons. They are small, adaptable animals, used for meat, milk, and research. The pygmy is smaller than the other recognized types and breeds in the United States; full grown bucks stand about 20 in. high at the withers, and does are even smaller. Pygmies may be fed the same rations as their larger counterparts, but in smaller quantities.

Because the goat industry is so diverse, the feeding methods vary accordingly. For this reason this chapter, devoted to feeding goats, is presented in three parts: Part I—Nutritive Needs and Feeds, covering the principles and practices of feeding that are applicable to all goats, regardless of type. Part II—Feeding Dairy Goats. Part III—Feeding Angora and Spanish Goats, with Angora and Spanish goats discussed together because both are produced under similar conditions.

PART I—NUTRITIVE NEEDS AND FEEDS

ECONOMIC IMPORTANCE OF FEED FOR GOATS

Due to their smaller numbers, along with producers being prone to let the animals fend for themselves, in terms of tonnage of purchased feed ingredients and feed costs, goats are unimportant in comparison to other domestic animals.

Angora and Spanish goats utilize rough, brushy range areas that are not suited to other species—many of these ranges would not otherwise be utilized and would revert to brush and wilderness. On such ranges, Angora goats are supplemental fed to a limited degree only, whereas most Spanish goats live entirely off the land and are rarely supplemented. This does not mean, however, that Angora and Spanish goats would not benefit from, and increase production, with supplemental feeding, especially during the critical periods—just before breeding (for flushing), just before and after kidding, and when feed is short.

Modern dairy goat producers generally feed well-balanced rations that are high in energy and protein and contain adequate minerals and vitamins. Many of them use commercially prepared feeds during lactation and for the young kids.

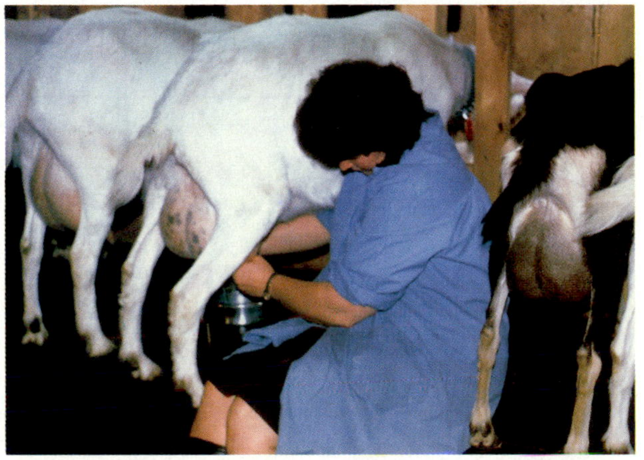

Fig. 22-3. Despite a hearty appetite and a large feed capacity, it is virtually impossible for a high-producing lactating doe to consume enough feed to meet the demands of body maintenance plus milk production during peak lactation; so, she must draw upon body reserves. (Courtesy, University of New Hampshire, Penacook)

NUTRITIVE NEEDS OF GOATS

In the past, efforts to set nutritional requirements for goats have relied heavily on the extrapolation of values derived from cattle and sheep studies. Despite their similarities as ruminants, goats exhibit significant differences from cattle and sheep in grazing habits, feed selection, water requirements, physical activities, milk composition, carcass composition, metabolic disorders, and parasites. So, the nutrient requirements of goats should be treated separately from those of other ruminants.

The hearty appetite of goats makes for a significant species difference. Lactating and growing goats will consume from 3.5 to 5.0% of their body weight (moisture-free basis) in one day, while cattle and sheep normally eat only 2.5 to 3.0%. It follows that their large feed capacity in relation to body weight makes it possible for them to consume large quantities of low quality materials. This characteristic, along with their ability to select the high-quality parts of plants, makes it possible to maintain goats successfully on poor pastures.

Since the nutritional requirements of the goat are distinctly different for milk, mohair, and meat production, specific requirements and allowances are discussed in separate feeding sections. Despite these distinctly different quantitative needs, the basic nutritional physiology of all goats is similar; hence, certain fundamentals relative to their nutritive needs—energy, protein, minerals, vitamins, and water—apply to all goats regardless of the purpose for which they are kept.

National Research Council (NRC) Requirements

The most up-to-date feeding standards for goats in the United States are those published by the National Research Council (NRC) of the National Academy of Sciences. Through the use of these standards, rations can be formulated for the different classes and categories of goats by proper use of available feedstuffs.

The nutritive requirements of goats for maintenance, various levels of activity, late pregnancy, and growth, are given in Table 22-1. Additional nutrient requirements for milk production at different fat percentages are given in Table 22-2; and additional nutrient requirements for mohair production are given in Table 22-3.

TABLE 22-1
DAILY NUTRIENT REQUIREMENTS OF GOATS[1]

Body Weight (BW)		Dry Matter per Animal[2] 0.9 Mcal ME/lb (*2 Mcal ME/kg*) Total		0.9 Mcal ME/lb % of BW	1.09 Mcal ME/lb (*2.4 Mcal ME/kg*) Total		1.09 Mcal ME/lb % of BW	Feed Energy TDN		DE	ME	NE	Protein TP	DP	Minerals Ca	P	Vitamin A	Vitamin D
(lb)	*(kg)*	(lb)	*(kg)*	(lb)	(lb)	*(kg)*	(kg)	(lb)	*(kg)*	(Mcal)	(Mcal)	(Mcal)	(g)	(g)	(g)	(g)	(1,000 IU)	(IU)
Maintenance only (includes goats under stable-fed conditions, minimal activity, and early pregnancy)																		
22	*10*	0.6	*0.28*	2.8	0.5	*0.24*	2.4	0.4	*0.16*	0.70	0.57	0.32	22	15	1	0.7	0.4	84
44	*20*	1.1	*0.48*	2.4	0.9	*0.40*	2.0	0.6	*0.27*	1.18	0.96	0.54	38	26	1	0.7	0.7	144
66	*30*	1.4	*0.65*	2.2	1.2	*0.54*	1.8	0.8	*0.36*	1.59	1.30	0.73	51	35	2	1.4	0.9	195
88	*40*	1.8	*0.81*	2.0	1.5	*0.67*	1.7	1.0	*0.45*	1.98	1.61	0.91	63	43	2	1.4	1.2	243
110	*50*	2.1	*0.95*	1.9	1.7	*0.79*	1.6	1.2	*0.53*	2.34	1.91	1.08	75	51	3	2.1	1.4	285
132	*60*	2.4	*1.09*	1.8	2.0	*0.91*	1.5	1.3	*0.61*	2.68	2.19	1.23	86	59	3	2.1	1.6	327
154	*70*	2.7	*1.23*	1.8	2.2	*1.02*	1.5	1.5	*0.68*	3.01	2.45	1.38	96	66	4	2.8	1.8	369
176	*80*	3.0	*1.36*	1.7	2.5	*1.13*	1.4	1.7	*0.75*	3.32	2.71	1.53	106	73	4	2.8	2.0	408
198	*90*	3.3	*1.48*	1.6	2.7	*1.23*	1.4	1.8	*0.82*	3.63	2.96	1.67	116	80	4	2.8	2.2	444
220	*100*	3.5	*1.60*	1.6	3.0	*1.34*	1.3	2.0	*0.89*	3.93	3.21	1.81	126	86	5	3.5	2.4	480
Maintenance plus low activity (basic plus 25% increment, includes goats under intensive management, tropical range, and early pregnancy)																		
22	*10*	0.8	*0.36*	3.6	0.7	*0.30*	3.0	0.4	*0.20*	0.87	0.71	0.40	27	19	1	0.7	0.5	108
44	*20*	1.3	*0.60*	3.0	1.1	*0.50*	2.5	0.7	*0.33*	1.47	1.20	0.68	46	32	2	1.4	0.9	180
66	*30*	1.8	*0.81*	2.7	1.5	*0.67*	2.2	1.0	*0.45*	1.99	1.62	0.92	62	43	2	1.4	1.2	243
88	*40*	2.2	*1.01*	2.5	1.9	*0.84*	2.1	1.2	*0.56*	2.47	2.02	1.14	77	54	3	2.1	1.5	303
110	*50*	2.6	*1.19*	2.4	2.2	*0.99*	2.0	1.5	*0.66*	2.92	2.38	1.34	91	63	4	2.8	1.8	357
132	*60*	3.0	*1.36*	2.3	2.5	*1.14*	1.9	1.7	*0.76*	3.35	2.73	1.54	105	73	4	2.8	2.0	408
154	*70*	3.4	*1.54*	2.2	2.8	*1.28*	1.8	1.9	*0.85*	3.76	3.07	1.73	118	82	5	3.5	2.3	462
176	*80*	3.7	*1.70*	2.1	3.1	*1.41*	1.8	2.1	*0.94*	4.16	3.39	1.91	130	90	5	3.5	2.6	510
198	*90*	4.1	*1.85*	2.1	3.4	*1.54*	1.7	2.3	*1.03*	4.54	3.70	2.09	142	99	6	4.2	2.8	555
220	*100*	4.4	*2.00*	2.0	3.7	*1.67*	1.7	2.4	*1.11*	4.91	4.01	2.26	153	107	6	4.2	3.0	600
Maintenance plus medium activity (basic plus 50% increment, includes goats on semiarid rangeland, slightly hilly pastures, and early pregnancy)																		
22	*10*	0.9	*0.43*	4.3	0.8	*0.36*	3.6	0.5	*0.24*	1.05	0.86	0.48	33	23	1	0.7	0.6	129
44	*20*	1.6	*0.72*	3.6	1.3	*0.60*	3.0	0.9	*0.40*	1.77	1.44	0.81	55	38	2	1.4	1.1	216
66	*30*	2.2	*0.98*	3.3	1.8	*0.81*	2.7	1.2	*0.54*	2.38	1.95	1.10	74	52	3	2.1	1.5	294
88	*40*	2.7	*1.21*	3.0	2.2	*1.01*	2.5	1.5	*0.67*	2.97	2.42	1.36	93	64	4	2.8	1.8	363
110	*50*	3.2	*1.43*	2.9	2.6	*1.19*	2.4	1.8	*0.80*	3.51	2.86	1.62	110	76	4	2.8	2.1	429
132	*60*	3.6	*1.64*	2.7	3.0	*1.37*	2.3	2.0	*0.91*	4.02	3.28	1.84	126	87	5	3.5	2.5	492
154	*70*	4.1	*1.84*	2.6	3.4	*1.53*	2.2	2.2	*1.02*	4.52	3.68	2.07	141	98	6	4.2	2.8	552
176	*80*	4.5	*2.03*	2.5	3.7	*1.69*	2.1	2.5	*1.13*	4.98	4.06	2.30	156	108	6	4.2	3.0	609
198	*90*	4.9	*2.22*	2.5	4.1	*1.85*	2.0	2.7	*1.24*	5.44	4.44	2.50	170	118	7	4.9	3.3	666
220	*100*	5.3	*2.41*	2.4	4.4	*2.01*	2.0	3.0	*1.34*	5.90	4.82	2.72	184	128	7	4.9	3.6	723
Maintenance plus high activity (basic plus 75% increment, includes goats on arid rangeland, sparse vegetation, mountainous pastures, and early pregnancy)																		
22	*10*	1.1	*0.50*	5.0	0.9	*0.42*	4.2	0.6	*0.28*	1.22	1.00	0.56	38	26	2	1.4	0.8	150
44	*20*	1.9	*0.84*	4.2	1.5	*0.70*	3.5	1.0	*0.47*	2.06	1.68	0.94	64	45	2	1.4	1.3	252
66	*30*	2.5	*1.14*	3.8	2.1	*0.95*	3.2	1.4	*0.63*	2.78	2.28	1.28	87	60	3	2.1	1.7	342
88	*40*	3.1	*1.41*	3.5	2.6	*1.18*	3.0	1.7	*0.78*	3.46	2.82	1.59	108	75	4	2.8	2.1	423
110	*50*	3.7	*1.67*	3.3	3.1	*1.39*	2.7	2.1	*0.93*	4.10	3.34	1.89	128	89	5	3.5	2.5	501
132	*60*	4.2	*1.92*	3.2	3.5	*1.60*	2.7	2.3	*1.06*	4.69	3.83	2.15	146	102	6	4.2	2.9	576
154	*70*	4.7	*2.14*	3.0	3.9	*1.79*	2.6	2.6	*1.19*	5.27	4.29	2.42	165	114	6	4.2	3.2	642
176	*80*	5.2	*2.37*	3.0	4.4	*1.98*	2.5	2.9	*1.32*	5.81	4.74	2.68	182	126	7	4.9	3.6	711
198	*90*	5.7	*2.59*	2.9	4.8	*2.16*	2.4	3.2	*1.44*	6.35	5.18	2.92	198	138	8	5.6	3.9	777
220	*100*	6.2	*2.81*	2.8	5.2	*2.34*	2.3	3.4	*1.56*	6.88	5.62	3.17	215	150	8	5.6	4.2	843
Additional requirements for late pregnancy (for all goat sizes)																		
		1.6	*0.71*		1.3	*0.59*		0.9	*0.40*	1.74	1.42	0.80	82	57	2	1.4	1.1	213
Additional requirements for growth—weight gain at 1.75 oz (*50 g*) per day (for all goat sizes)																		
		0.4	*0.18*		0.3	*0.15*		0.2	*0.10*	0.44	0.36	0.20	14	10	1	0.7	0.3	54
Additional requirements for growth—weight gain at 3.5 oz (*100 g*) per day (for all goat sizes)																		
		0.8	*0.36*		0.7	*0.30*		0.4	*0.20*	0.88	0.72	0.40	28	20	1	0.7	0.5	108
Additional requirements for growth—weight gain at 5.3 oz (*150 g*) per day (for all goat sizes)																		
		1.2	*0.54*		1.0	*0.45*		0.7	*0.30*	1.32	1.08	0.60	42	30	2	1.4	0.8	162

[1]Adapted by the authors from *Nutrient Requirements of Goats,* No. 15, NRC-National Academy of Sciences, 1981, pp. 10–11.

[2]Good-quality roughages furnish about 0.9 Mcal ME/lb (*2 Mcal ME/kg*) of dry matter. Roughage-concentrate mixed rations are sometimes necessary to increase the energy content of the diet to 1.09 Mcal/lb (*2.5 or 3.0 Mcal ME/kg*) of dry matter when early weaned kids or high-producing dairy goats are being fed.

TABLE 22-2
ADDITIONAL NUTRIENT REQUIREMENTS FOR MILK PRODUCTION PER POUND AT DIFFERENT FAT PERCENTAGES[1]

Fat	Feed Energy					Protein		Minerals		Vitamins	
	TDN		DE	ME	NE	TP	DP	Ca	P	A	D
(%)	(lb)	*(kg)*	(Mcal)	(Mcal)	(Mcal)	(g)	(g)	(g)	(g)	(1,000 IU)	(IU)
2.5	0.33	*0.151*	0.67	0.54	0.31	26.7	19.1	0.9	0.6	1.7	345
3.0	0.34	*0.153*	0.68	0.55	0.31	29.0	20.4	0.9	0.6	1.7	345
3.5	0.34	*0.155*	0.68	0.56	0.31	30.8	21.8	0.9	0.6	1.7	345
4.0	0.35	*0.157*	0.69	0.57	0.32	32.7	23.1	1.4	1.0	1.7	345
4.5	0.35	*0.159*	0.70	0.57	0.32	34.9	24.5	1.4	1.0	1.7	345
5.0	0.36	*0.161*	0.71	0.58	0.33	37.2	25.9	1.4	1.0	1.7	345
5.5	0.36	*0.163*	0.72	0.59	0.33	39.0	27.2	1.4	1.0	1.7	345
6.0	0.37	*0.166*	0.73	0.59	0.34	40.8	28.6	1.4	1.0	1.7	345

[1]Adapted by the authors from *Nutrient Requirements of Goats,* No. 15, NRC-National Academy of Sciences, 1981, p. 11. These requirements are in addition to those listed in Table 22-1. They include requirements for nursing single, twin, or triplet kids at the respective milk production level. To convert to requirements for milk production per kg, multiply by 2.205.

TABLE 22-3
ADDITIONAL NUTRIENT REQUIREMENTS FOR MOHAIR PRODUCTION BY ANGORA GOATS AT DIFFERENT FLEECE PRODUCTION LEVELS[1]

Annual Fleece Yield		Feed Energy					Protein	
		TDN		DE	ME	NE	TP	DP
(lb)	*(kg)*	(lb)	*(kg)*	(Mcal)	(Mcal)	(Mcal)	(g)	(g)
4.4	*2*	0.035	*0.016*	0.07	0.06	0.03	9	6
8.8	*4*	0.075	*0.034*	0.15	0.12	0.07	17	12
13.2	*6*	0.110	*0.050*	0.22	0.18	0.10	26	18
17.6	*8*	0.146	*0.066*	0.29	0.24	0.14	34	24

[1]Adapted by the authors from *Nutrient Requirements of Goats,* No. 15, NRC-National Academy of Sciences, 1981, p. 12. These requirements are in addition to those listed in Table 22-1.

Energy

Efficient utilization of nutrients depends on an adequate supply of energy, which is of paramount importance in determining the productivity of goats. Energy deficiency retards kid growth, delays puberty, reduces fertility, and depresses milk production. With continued deficiency the animals show a concurrent reduction in resistance to infectious diseases and parasites. The problem may be further complicated by deficiencies of proteins, minerals, and vitamins.

Energy limitations may result from inadequate feed intake or from a low-quality ration. Low energy intake that results from either feed restriction or low ration component digestibility prevents goats from meeting their requirements and from attaining their genetic potential. High water content of forages may also become a limiting factor.

Energy requirements are affected by age, body size, growth, pregnancy, and lactation. Energy requirements are also affected by the environment, hair growth, muscular activity, and relationships with other nutrients in the ration, which, for best results, need to be supplied in adequate amounts. Temperature, humidity, sunshine, and wind velocity may increase or decrease energy needs, depending upon the region. Stress of any kind may increase energy requirements.

Shearing mohair from Angora goats and pashmina from Cashmere goats decreases insulation and results in increased energy needs, especially during cold weather. Goats are more active and travel greater distances than sheep, which increases energy requirements. Maintenance requirements of goats on pasture, browse, and range, especially on mountainous and/or seasonal grazing, are considerably higher than those of stable-fed animals. The magnitude of this increase is presented in Table 22-1 at three levels of activity, which results from the availability of feed and water, and from the topography, elevation, and distance traveled in grazing.

Good-quality roughages furnish about 0.9 Mcal metabolizable energy (ME) per pound dry matter (DM). However, roughage-concentrate mixed rations are sometimes necessary to increase the energy content of the ration to 1.09 to 1.40 Mcal ME/lb DM when feeding early weaned kids or high producing dairy goats. So, in Table 22-1, under the heading "Dry Matter Per Animal," provision is made for two different energy levels of feeds: (1) good-quality roughage alone (0.9 Mcal ME/lb), and (2) a roughage-concentrate mix (1.09 Mcal ME/lb). The efficiency with which energy is utilized for weight gain, pregnancy, and lactation usually increases with increasing levels of ME concentration in the ration.

The energy requirement for goats as herein reported are the amounts needed (1) for maintenance, pregnancy, and growth (Table 22-1); (2) for milk production (Table 22-2); and (3) for mohair production (Table 22-3). Note that the energy requirements for the various categories are expressed as total digestible nutrients (TDN), digestible energy (DE), metabolizable energy (ME), and net energy (NE).

Fig. 22-4. Roughages furnish most of the energy required by goats. However, concentrates must be added to meet the high energy needs of rapidly growing kids and high-lactating dairy goats. (Courtesy, Rocking M Ranch, Hilmar, Calif.)

In Table 22–1, in addition to the energy requirement for maintenance only, energy values are given for three different levels of activity: (1) light, (2) medium, and (3) high. Also, additional energy allowances are made for: (1) late pregnancy (the last 60 days); and (2) three different levels of growth—50, 100, 150 g per day.

In Table 22–2, energy requirements in addition to those listed in Table 22–1 are given for milk production per pound at different fat percentages.

In Table 22–3, energy requirements in addition to those listed in Table 22–1 are given for mohair production of Angora goats of different fleece weights.

Carbohydrates and fats supply virtually all of the energy needs of the body, though a small portion may be derived from protein catabolism. Rumen microorganisms serve essential roles in the digestion of many of the complex carbohydrates consumed by goats, especially range feeds. Carbohydrates and fats, accounting for 60 to 70% of the total energy derived from the feed, are converted to volatile fatty acids (primarily acetate, propionate, and butyrate) in the rumen which are, in turn, absorbed through the rumen wall and used to supply energy.

Although fats serve as carriers of the fat-soluble vitamins and other fat-soluble substances, they are used primarily as a concentrated source of energy. A general rule of thumb is that fats supply 2.25 times the energy of carbohydrates. However, excessive amounts of fat in the ration usually reduce palatability and make the ration more susceptible to oxidation and subsequent spoilage.

Protein

Proteins are the principal constituents of the animal body and are continuously needed in the feed for growth and cell repair. The transformation of feed protein into body protein is an important process of nutrition and metabolism. Proteins consist of amino acids, which are the building blocks of all body cells. Secretions such as enzymes, hormones, mucin, and milk make for additional amino acid requirements. Proteins are, therefore, vital for animal maintenance, growth, reproduction, and milk production. However, in goats as in other ruminants, nonprotein nitrogen (NPN) can substitute for parts of the required proteins for these functions.

In Table 22–1, protein requirements are given for maintenance, activity, late pregnancy, and growth, along with the energy requirements.

In Table 22–2, protein requirements in addition to those listed in Table 22–1 are given for milk production per pound at different fat percentages.

In Table 22–3, protein requirements in addition to those listed in Table 22–1 are given for mohair production of different fleece weights of Angora goats.

Total protein (TP) is considered to be the most accurate guide for converting proteins from feed composition tables to the quantities required, but digestible protein (DP) values are also used.

Ruminal microorganisms can utilize either protein or non

TABLE
GOAT MINERAL

Mineral Which May Be Deficient Under Normal Conditions	Conditions Usually Prevalent Where Deficiencies Are Reported	Function of Mineral	Deficiency Symptoms/Toxicity
Major or Macrominerals:			
Salt (NaCl)	Negligence, for salt is inexpensive. Lactating does may require additional salt as milk contains high amounts of sodium.	Sodium chloride helps maintain osmotic pressure in body cells, upon which depends the transfer of nutrients to the cells, the removal of waste materials, and the maintenance of water balance among the tissues. Also, sodium is important in making bile, which aids in the digestion of fats and carbohydrates; and chlorine is required for the formation of hydrochloric acid in the gastric juice so vital to protein digestion. It is noteworthy that when salt is omitted, sodium expresses its deficiency first.	**Deficiency symptoms**—Loss of appetite, depraved appetite and consumption of soil and debris, emaciation, decline in milk production, a general rough appearance with poor coat and lusterless eyes. Acute deficiency symptoms include shivering, weakness, cardiac disturbances, and ultimately death. **Toxicity**—The maximum tolerable level of salt for sheep is 9.0%. For goats, a similar level of salt will likely be toxic.
Calcium (Ca)	Goats in heavy lactation. Lack of vitamin D. Calcium-deficient areas (where pasture and range forages are deficient in calcium) are Fla., La., Neb., Va., and W. Va. Feeds that contain primarily cereal grains.	Essential for the development and maintenance of good strong bones and teeth; maintains the contractability, rhythm, and tonicity of the heart muscles; antagonizes the action of the sodium and potassium on the heart; is required for normal coagulation of the blood; is necessary for proper nerve irritability; and appears to be essential for selective cellular permeability.	**Deficiency symptoms**—In young kids, retarded growth and abnormal bone development. Also, a deficiency of calcium may cause rickets in young animals and osteomalacia in adults. In lactating does, depressed milk yields and fragile bones. Milk fever can occur when calcium levels in the blood drop. **Toxicity**—If there is adequate phosphorus, sheep can tolerate a calcium:phosphorus ratio of 7:1 and as much as 2% calcium in the diet. It is postulated that goats can tolerate a similar level of calcium.

protein nitrogen to synthesize microbial protein. The microbial protein, along with the undigested feed protein, passes from the rumen-reticulum through the omasum, then to the small intestine where it is subjected to digestive processes similar to those of the nonruminant and broken down to amino acids which are absorbed and utilized by goats. Thus, goats can utilize protein and nonprotein nitrogen (such as urea) in their rations. So, the protein available for digestion in the small intestine of goats consists of microbial protein and feed protein that has escaped microbial breakdown in the rumen. But it has been shown that the protein produced by ruminal synthesis does not supply all of the amino acids in the quality or quantity needed for maximum growth of kids or milk production of does. Thus, quality and degradability of protein fed to goats is more important than formerly thought. Moreover, it has been found that protein efficiency can be increased by protecting protein from the degradation of the microbes in the rumen and increasing the escape of protein from the rumen to the intestines where it is digested and absorbed. This technology of manipulating the quantity of dietary protein rumen fermentation, thereby increasing the supply of protein (amino acids) in the small intestine, which is known as protein bypass, is detailed and illustrated in Chapter 11, Protein Supplements, under the heading "Protein Bypass (Protected Protein, Escaped Protein)."

The most commonly used protein supplements are brewers' dried grains, cottonseed meal, linseed meal, and soybean meal. One of the most economical sources of protein is good-quality alfalfa hay—fed as long hay, chopped hay, range cubes, or pellets. It can either be fed separately or mixed with the concentrate portion of the ration in a complete feed. Dehydrated alfalfa is an excellent source of protein, but it is more expensive than sun-cured hay.

Protein deficiencies in the ration deplete stores in the blood, liver, and muscles, and predispose animals to a variety of serious and even fatal ailments. Below a minimum level of 6% crude protein (CP) in the ration, feed intake will be reduced, which leads to a combined deficiency of energy and protein. This deficiency further reduces rumen function and lowers the efficiency of feed utilization. Long-term protein deficiencies retard fetal development, lead to low birth weights, affect kid growth, and depress milk production.

Minerals

If goats are fed a good concentrate, along with a good-quality hay produced on land that has been properly fertilized, few problems arising from mineral deficiencies occur.

The mineral requirements of goats are given in Tables 22-1, 22-2, and 22-4.

GOAT MINERAL CHART

Table 22-4, Goat Mineral Chart, gives a summary of the different factors involved with mineral nutrition in the goat. Further elucidation of certain minerals is contained in the accompanying narrative. Fluorine is discussed because of its toxicity.

22–4
CHART[1] (See footnotes at end of table.)

Nutrient Requirements[2]	Recommended Allowances[2]	Practical Sources	Comments
	Salt should be provided free-choice or as a component of the ration. In a complete feed, 0.5 to 1.0% salt is recommended, with proportionately higher levels in supplements.	Iodized salt in iodine-deficient areas. Can be offered free-choice or incorporated into the ration. In alkaline areas, water may contain enough salt to meet the requirements.	In range areas, salt may be added to feed to limit feed intake. If self-feeders are located near water, the level of salt in the ration should be high (25–40%). If self-feeders are some distance from water, the level of salt in the ration should be reduced. In arid regions, the salt content of some water sources can reduce intake of water and feed.
Variable according to age, sex, and class (see Tables 22-1, and 22-2).	Because milk is high in calcium, lactating does need rations with high calcium levels. In % of ration: 0.78 M-F 0.70 A-F	Ground limestone, steamed bone meal, dicalcium phosphate, and oyster shell.	The recommended ratio of calcium to phosphorus ranges from 2:1 to 4:1. If the ratio falls below 2:1, urinary calculi may develop in males. Under grazing conditions, calcium is seldom a problem with either Angora or meat-type goats.

(Continued)

TABLE 22-4

Mineral Which May Be Deficient Under Normal Conditions	Conditions Usually Prevalent Where Deficiencies Are Reported	Function of Mineral	Deficiency Symptoms/Toxicity
Major or Macrominerals (Continued):			
Phosphorus (P)	When goats subsist on pastures in phosphorus-deficient areas. When goats subsist for long periods on mature, dry forages. Lack of vitamin D.	Essential for sound bones and teeth, and for the assimilation of carbohydrate and fats. A vital ingredient of the proteins in all body cells. Necessary for enzyme activation. Acts as a buffer in blood and tissue. Occupies a key position in biologic oxidation and reactions requiring energy.	**Deficiency symptoms**—Slowed growth, depraved appetite (chewing bones, wood, hair), unthrifty appearance, rickets in young animals, osteomalacia in mature animals, and depressed milk yields in lactating does. **Toxicity**—There is no known phosphorus toxicity in goats. However, excess phosphorus consumption may decrease the absorption of calcium. Also, when phosphorus is high in relation to calcium, urinary calculi may be formed.
Magnesium (Mg)	Animals grazing lush green grass or winter cereal pastures fertilized with nitrogen and potassium.	Required for many enzyme systems and for proper functioning of the nervous system. Also, closely associated with the metabolism of calcium and phosphorus.	**Deficiency symptoms**—Loss of appetite, excitability, and calcification of soft tissues. The most noted problem associated with low magnesium is grass tetany. **Toxicity**—Magnesium toxicity of goats has not been reported under practical conditions.
Potassium (K)	When goats are grazing mature range forage during winter or drought periods. High concentrate rations.	It (1) affects osmotic pressure and acid-base balance within the cells, and (2) aids in activating several enzyme systems involved in energy transfer and utilization, protein synthesis, and carbohydrate metabolism.	**Deficiency symptoms**—Marginal deficiencies result in reduced feed intake, retarded growth, and reduced milk production. Severe deficiencies cause emaciation and poor muscular tone. **Toxicity**—The maximum tolerable level of potassium for sheep is about 3% of the ration DM. It is postulated that the toxicity level of goats is similar.
Sulfur (S)	Possibly with liberal intake of tannic acid-containing plants. This is of concern with range goats, which liberally graze and browse such plants.	Essential for synthesis of the sulfur amino acids (cystine and methionine). Sulfur is particularly high in goat hair.	**Deficiency symptoms**—Depressed appetite, loss of weight, poor growth, excessive salivation, tearing, loss of mohair, depressed milk yields. **Toxicity**—Elemental sulfur is practically devoid of toxicity.
Trace or Microminerals:			
Cobalt (Co)	In cobalt deficient areas when the cobalt level in the feed drops to 0.04 to 0.07 ppm or lower.	The only function of cobalt is that of being an integral part of vitamin B-12.	**Deficiency symptoms**—The deficiency symptoms are actually vitamin B-12 deficiencies. They are: loss of appetite, emaciation, weakness, anemia, and decreased production. **Toxicity**—In sheep, about 204.5 mg cobalt/100 lb live weight is toxic. Likely, the same applies to goats.
Copper (Cu) and Molybdenum (Mo)	Copper and molybdenum are interrelated in animal metabolism; hence, they should be considered together. The most common problem occurs when a normal or low level of copper is accompanied by a high level of molybdenum, resulting in copper being excreted and producing a copper deficiency. This condition can be corrected by adding copper.	Copper and iron are mutually involved in the formation of hemoglobin—the red pigment which carries oxygen.	Few studies on copper and molybdenum have been conducted with goats. It appears that sheep are sensitive to copper toxicity and resistant to molybdenosis, but it is not known whether this is also the case with goats.
Fluorine (F)		Necessary for sound bones and teeth.	**Deficiency symptoms**—Fluorine deficiency appears to be rare. Rather, the hazard is fluorine toxicity. **Toxicity**—With sheep, fluorine toxicity occurs at levels above 200 ppm. So, it is postulated that the toxicity level for goats is similar.
Iodine (I)	Iodine-deficient areas or soils (in northwestern U.S., and in the Great Lakes and Rocky Mountain Regions), unless iodized salt is fed.	Formation of thyroxin, a hormone of the thyroid gland.	**Deficiency symptoms**—Enlarged thyroid gland, a condition called goiter. Kids born weak or dead. **Toxicity**—The maximum tolerable level for sheep is 45 ppm A-F or 50 ppm M-F. It is postulated that the toxicity level for goats is similar.
Iron (Fe)	Iron deficiency may occur in young goat kids because of their minimal body stores at birth and the low iron content of milk.	As a component of blood hemoglobin required for oxygen transport. Iron is also required for some enzyme systems.	**Deficiency symptoms**—Anemia, poor growth, lethargy, increased respiration rate, decreased resistance to infection, and in severe cases high mortality. **Toxicity**—Free iron ions are very toxic, causing loss of appetite, diarrhea, below normal temperature, shock, acidosis, and death.

(Continued)

Nutrient Requirements[2]	Recommended Allowances[2]	Practical Sources	Comments
Variable according to age, sex, and class (see Table 22–1, and 22–2).	Can be offered free-choice or incorporated into the ration. In % of ration: 0.45 M-F 0.40 A-F	Cereal grains. Defluorinated phosphate, dicalcium phosphate, steamed bone meal, monosodium phosphate.	Phosphorus is the mineral most likely to be deficient in range forages. It is, therefore, recommended that it be supplied in range supplements. •The calcium-to-phosphorus ratio should not drop below 1.2:1.
	In % of ration: 0.25 M-F 0.22 A-F	Plant protein supplements and plant by-product feeds are excellent sources of magnesium. The common magnesium supplements are magnesium carbonate, magnesium oxide, and magnesium sulfate.	•Goats have a marginal ability to compensate for low dietary magnesium by reducing the rate of excretion.
•In growing sheep, the potassium requirement is 0.5% of the ration. In lactating dairy cattle, the requirement is 0.8% of the complete ration. These levels are also postulated as the requirements of growing and lactating goats, respectively.	In % of ration: 1.0 M-F 0.9 A-F	Roughage-based rations. Common potassium supplements are potassium chloride, potassium bicarbonate, and potassium sulfate.	
	In % of ration: 0.20 M-F 0.18 A-F A sulfur-to-nitrogen ratio of 1:10 is recommended.	Sulfates, such as sodium sulfate and ammonium sulfate, are the most available forms of sulfur for ration formulation.	Because of mohair production, Angora goats may have an elevated sulfur requirement.
•A level of 0.1 ppm in the M-F ration.	In % of ration: 0.1 to 0.2 ppm M-F 0.09 to 0.18 ppm A-F	•Cobalt sulfate or cobalt chloride added at the rate of 5.45 g per 100 lb (*12 g per 100 kg*) of salt.	
	Add copper sulfate to the salt at the rate of 0.5% Copper in total ration: 5.0 ppm M-F 4.5 ppm A-F	Salt containing 0.5% copper sulfate.	
Iodine in the ration: A-F, 0.09–0.72 ppm; M-F, 0.1–0.8 ppm. The higher levels are indicated for pregnancy and lactation.	Free access to stabilized iodized salt containing 0.0078% iodine. In total ration: 0.5 ppm M-F 0.45 ppm A-F	Iodized salt.	Iodized salt should not be used as a feed-limiter because it could lead to excessive intakes of iodine.
•0.03% ferrous iron in the ration.	•Iron-dextran (150 mg) may be injected in kids at 2 to 3 week intervals if iron deficiencies are observed. In total ration: 50 ppm M-F 45 ppm A-F	Iron-dextran is recommended as an injection; and ferrous sulfate and ferric citrate are recommended for incorporating in rations.	Iron deficiency seldom occurs in mature grazing goats.

(Continued)

TABLE 22-4

Mineral Which May Be Deficient Under Normal Conditions	Conditions Usually Prevalent Where Deficiencies Are Reported	Function of Mineral	Deficiency Symptoms/Toxicity
Trace or Microminerals (Continued):			
Manganese (Mn)	High calcium and iron may increase manganese requirements.	Skeletal development and reproduction.	**Deficiency symptoms**—Reluctance to walk, deformity of the forelegs, delayed estrus, more inseminations per conception, more abortions, and 20% reduction in birth weights. **Toxicity**—1,000 ppm appears to be the maximum tolerance level for sheep; so, it is postulated that the toxicity level for goats is similar.
Selenium (Se)			**Deficiency symptoms**—White muscle disease in young kids from birth to a few months of age, which may take one of two forms: (1) sudden unexplained death, or (2)muscular paralysis, particularly of the hind limbs, or stiffness and inability to rise. **Toxicity**—All livestock species, including goats, are susceptible to selenium toxicity. Selenium toxicity in sheep occurs from prolonged consumption of plants containing over 3 ppm Se. It is postulated that the toxicity level for goats is about the same as for sheep.
Zinc (Zn)	Rations excessively high in calcium adversely affect zinc utilization.	Needed for normal skin, bones, and hair. A component of several enzyme systems involved in digestion and respiration.	**Deficiency symptoms**—Reduced feed intake, weight loss, parakeratosis, stiffness of joints, excessive salivation, swelling of the feet and horny overgrowth, small testicles, and low libido. **Toxicity**—Levels of 1,000 ppm may be toxic.

[1]Where preceded by an asterisk, the requirements, recommended allowances, and other facts presented herein were adapted from *Nutrient Requirements of Goats,* No. 15, NRC-National Academy of Sciences, 1981.

[2]As used herein, the distinction between "nutrient requirements" and "recommended allowances" is as follows: In nutrient requirements, no margins of safety are included intentionally, whereas in recommended allowances, margins of safety are provided in order to compensate for variations in feed composition, environment, and possible losses during storage or processing.

MAJOR OR MACROMINERALS

Seven major or macrominerals are considered dietary essentials for livestock, including goats. These are: salt (sodium and chlorine), calcium, magnesium, phosphorus, potassium, and sulfur.

Salt (NaCl)

Of the various macrominerals demonstrated to be required by goats, salt is probably the most likely to be deficient in goat rations and is one of the easiest to supply. Goats require both sodium and chlorine, but sodium is the mineral element most likely to be lacking in common feeding practices.

Goats should have access to salt free-choice in loose form at all times, whether on the range, on seeded pasture, or in a corral. Where iodine is lacking, iodized salt is recommended.

Salt may be offered free-choice in a mineral mix or provided as a feed intake governor in concentrate mixes that are offered free-choice. In operations where salt is not used to control feed intake, a trace mineralized salt is usually mixed at a ratio of 1:1 with either steamed bone meal or dicalcium phosphate and offered free-choice.

If goats are deprived of salt for an extended period and are then given salt free-choice, they may consume too much initially and become sick; hence, a period of adjustment is required. Where goats have been salt-starved, they should be hand-fed salt for a period of time, with the amount increased daily until salt is left in the box from the previous day. At this stage, it is safe to provide all they will consume.

Calcium (Ca) and Phosphorus (P)

Under grazing conditions, calcium is seldom a problem for either Angora or Spanish goats, but it can be very important for high-producing dairy goats, because goat's milk is rich in calcium. A phosphorus deficiency in grazing goats is likely when they are consuming phosphorus-deficient range forages.

Production increases the demands for calcium and phosphorus. Both growth (bone development) and lactation (deposition of minerals in milk) require substantial quantities of these minerals, If there is a severe imbalance of these minerals during pregnancy and early lactation, milk fever may occur. In males, an imbalance of calcium to phosphorus often leads to the development of urinary calculi.

Most forages are good sources of dietary calcium, but are rather low in phosphorus. Legumes are excellent calcium sources, while the grasses and silages tend to be substantially lower in calcium content. Conversely, the calcium:phosphorus ratio is important. Ideally, the goat ration should have a calcium:phosphorus ratio of 1.4:1 to 4:1, but ruminants have been observed to tolerate ratios of up to 7:1.

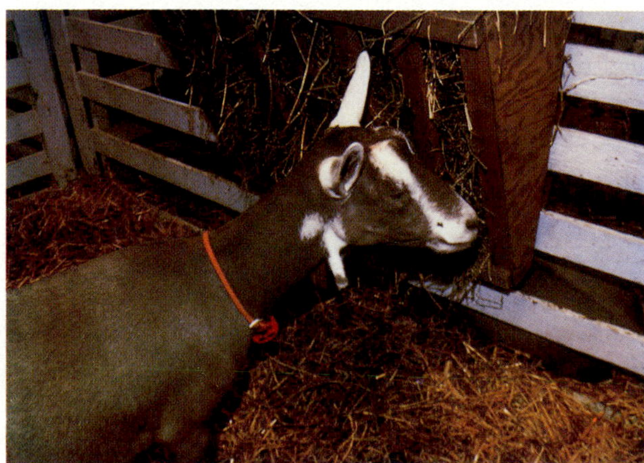

Fig. 22-5. Alfalfa hay is an excellent source of calcium for goats. (Courtesy, University of Delaware, Newark)

The source and quantity of mineral supplementation depends on the mineral composition of the total ration. Where calcium is needed, ground limestone is generally the mineral of choice. Where phosphorus is the primary need,

(Continued)

Nutrient Requirements[2]	Recommended Allowances[2]	Practical Sources	Comments
	In total ration: 40 ppm M-F 36 ppm A-F	Manganese gluconate.	
	In total ration: 0.15 ppm M-F 0.13 ppm A-F		
*Direct and indirect evidence indicates minimum requirements of 10 ppm in the ration.	In total ration: 50 ppm M-F 45 ppm A-F	Zinc carbonate. Zinc sulfate.	

it is usually provided in the form of monosodium phosphate, disodium phosphate, sodium tripolyphosphate, or feed-grade phosphoric acid. Where both calcium and phosphorus are needed, the most frequently used supplements are dicalcium phosphate, defluorinated phosphate, or steamed bone meal. The mineral supplement(s) can either be incorporated in the ration or offered free-choice. Quite often, steamed bone meal or dicalcium phosphate is mixed with equal amounts of salt and made available to goats free-choice.

Magnesium (Mg)

Magnesium is a constituent of bone. Also, it is necessary for many enzyme systems and for proper functioning of the nervous system. A deficiency of magnesium may result in grass tetany, characterized by stiff legs and head retraction. Where does graze forage with high nitrogen and potassium content, the minimum level of magnesium in the ration should be 0.22% as-fed basis.

Potassium (K)

Although potassium is required in relatively large amounts, it is usually sufficient in roughage-based rations. For growing sheep, the potassium requirement is 0.5% of the ration, whereas for lactating dairy cattle, the requirement is 0.8% of the ration. These levels are also postulated as meeting the requirements of growing and lactating goats, respectively. Ration values below these levels are infrequently encountered, and are usually caused by high-concentrate rations or severely weathered range forage.

Marginal potassium deficiencies result in reduced feed intake, retarded growth, and reduced milk production. Severe deficiencies cause emaciation and poor muscular tone.

Sulfur (S)

Since sulfur is a key constituent in two important amino acids (methionine and cystine), it should be added to goat rations when nonprotein nitrogen sources are used. This additional sulfur can then be incorporated into amino acids by the microorganisms in the rumen. Either elemental sulfur or sulfur in sulfate form can be used effectively. When urea or other NPN sources are used, the ratio of sulfur-to-nitrogen should be 1:10. More sulfur is required if tannic acid is high in the forage.

TRACE OR MICROMINERALS

Nine trace or microminerals are required by goats. These are: cobalt, copper and molybdenum, fluorine, iodine, iron, manganese, selenium, and zinc.

Goat rations are seldom deficient in trace minerals. However, for does in heavy lactation or for goats grazing on sandy soils, it may be advisable to supply a limited amount of a broad-based trace mineral mixture to guard against possible deficiencies. Care should be taken when using a trace mineral mixture because of the many interrelationships among the minerals which affect their availability for absorption.

Cobalt (Co)

Cobalt is a component of vitamin B-12, for the synthesis of which it is essential. In sheep, a cobalt intake of 0.1 ppm is considered adequate. It is postulated that the same level is adequate for goats.

Cobalt should be ingested frequently, preferably daily. This is best accomplished by adding cobalt to the salt at a level of 5.45 g/100 lb of salt, fed free-choice.

Deficiency signs include loss of appetite, emaciation, weakness, anemia, and decreased production.

Copper (Cu) and Molybdenum (Mo)

Copper and molybdenum are interrelated in animal metabolism; hence, herein they are considered together. The most common problem occurs when a normal or low level of copper is accompanied by a high level of molybdenum. In this case, copper is excreted and a deficiency occurs. This condition can be corrected by adding copper.

Few studies on copper and molybdenum have been conducted with goats. It appears that sheep are sensitive to copper toxicity and resistant to molybdenosis, but it is not known whether this is also the case with goats.

Fluorine (F)

In small amounts, fluorine helps develop strong bones and teeth, but in excessive amounts bones become porous and soft and teeth become mottled and easily worn down. Fluorine deficiency is rare; rather, the hazard is fluorine toxicity, which may be caused by high fluorine levels in ground water or in crude mineral supplements (raw rock phosphate).

Iodine (I)

Iodine is necessary for the formation of thyroxin, a hormone of the thyroid gland. A deficiency of iodine results in an enlargement of the thyroid gland, a condition called goiter. Also, kids may be born weak or dead. Iodine-deficient areas are widespread throughout the world, including parts of the United States (in northwestern United States, and in the Great Lakes and Rocky Mountain regions). Deficiencies are readily corrected by feeding iodized salt. Iodized salt should not be used for the purpose of limiting feed comsumption because it could lead to excessive intakes of iodine.

Iron (Fe)

Iron is a component of blood hemoglobin that is required for oxygen transport. It is also required for some enzyme systems. Although iron deficiency seldom occurs in mature grazing animals, it may occur in young goat kids because of their minimal body stores of iron at birth and the low iron content of milk.

If an iron deficiency is observed in young kids on a milk diet, injection of iron-dextran (150 mg) at 2 to 3 week intervals is recommended. Ferrous sulfate and ferric citrate are recommended for incorporation in rations. A level of 45 ppm of iron in an as-fed ration appears to be adequate.

Manganese (Mn)

Manganese is an essential mineral in the ration of goats, required for skeletal development and reproductive efficiency.

Deficiency symptoms are: reluctance to walk, deformity of the forelegs, delayed estrus, more inseminations per

TABLE
GOAT VITA

Vitamin Which May Be Deficient Under Normal Conditions	Conditions Usually Prevalent Where Deficiencies Are Reported	Function of Vitamin	Deficiency Symptoms
Fat-Soluble Vitamins:			
A	**D**uring extended dry periods when the supply of green forage is limited.	**R**equired for normal vision. **A**ids in reproduction and lactation. **N**eeded for maintaining normal epithelial tissue. **A**ids in resistance to infection.	**K**eratinization of the epithelia of the respiratory, alimentary, reproductive, and urinary tracts, and of the eye. Multiple infections, poor bone development, birth of abnormal offspring, and vision impairment, including night blindness.
D	**Y**oung goats kept in confinement where they have little or no access to sunlight.	**A**bsorption of calcium and phosphorus.	**B**one abnormalities, including rickets. **D**epressed growth.
E	**A**bnormally high levels of nitrates may produce vitamin E deficiencies. **W**here soils are very low in selenium.	**S**erves as a physiological antioxidant. **I**n dairy goats, the vitamin E transferred to the milk is important because of the antioxidant properties that aid in milk storage.	**E**vidence of spontaneous vitamin E deficiency signs in goats is lacking. **T**he probability of lowered productivity in goats as a result of a vitamin E deficiency is remote.
K	**V**itamin K deficiency may occur when the dicoumarol content of hay is excessively high, as when moldy sweet clover hay is fed.	**V**itamin K or K_2 is necessary in the blood clotting mechanism.	
Water-Soluble Vitamins:			
B vitamins **V**itamin C	**B** vitamin deficiencies may be evident in poorly fed and unhealthy animals. **B**–12 may be deficient if cobalt is absent or at extremely low levels, as cobalt is required for the synthesis of vitamin B–12.	**B**–1 participates as a coenzyme in the utilization of carbohydrates.	

[1]As used herein, the distinction between "nutrient requirements" and "recommended allowances" is as follows: In nutrient requirements, no margins of safety are included intentionally; whereas in recommended allowances, margins of safety are provided to compensate for variations in feed composition, environment, and possible losses during storage or processing.

conception, more abortions, and 20% reduction in birth weights.

In an experiment involving two groups of female goats for the first year of life—with one group on low manganese, 20 ppm for the first year of life and 6 ppm during the following year; and the controls receiving 100 ppm—the low manganese ration did not affect growth or bone structure, but it did affect reproduction adversely.[3]

Selenium (Se)

Selenium is essential, but only in minute amounts. It is a component of glutathione peroxidase, the metabolic role of which is to protect against oxidation of polyunsaturated fatty acids and resultant tissue damage. Also, selenium is interrelated with vitamin E—they spare each other, and with the sulfur-containing amino acids.

All livestock species, including goats, are susceptible to selenium toxicity. Selenium toxicity in sheep occurs from prolonged consumption of plants containing over 3 ppm selenium. It is postulated that the toxicity level for goats is about the same as for sheep.

[3]*Nutrient Requirements of Sheep,* sixth revised edition, NRC-National Academy of Sciences, 1985, p. 19.

Zinc (Zn)

Zinc is essential for goats. Deficiency symptoms include reduced feed intake, weight loss, parakeratosis, stiffness of joints, excessive salivation, swelling of the feet and horny overgrowth, small testicles, and low libido. Zinc must be supplied continuously because little is stored in the body in readily available form. Direct and indirect evidence indicates minimum ration requirements of 10 ppm. Levels of 1,000 ppm may be toxic.

Vitamins

Typical range or pasture diets of goats usually contain adequate levels of vitamins or vitamin precursors to maintain the normal health of goats. However, young kids, goats kept in confinement, and high-producing dairy goats may need supplemental vitamins.

GOAT VITAMIN CHART

Table 22-5, Goat Vitamin Chart, gives, in summary form, the following pertinent information relative to each vitamin listed: (1) conditions usually prevailing where deficiencies are reported, (2) function, (3) deficiency symptoms, (4) nutrient requirements, (5) recommended allowances, and (6) practical sources.

22-5
MIN CHART

Nutrient Requirements[1]	Recommended Allowances	Practical Sources	Comments
Variable according to size, sex, age, and class (see Table 22-1 and 22-2).	The recommended allowances should provide margins of safety over and above the requirements. So, add 10 to 20% to the requirements given in Tables 22-1 and 22-2.	Synthetic vitamin A. Injectable vitamin A. Yellow corn. Green forages.	Young animals, which have not built up vitamin A reserves, are more susceptible to a vitamin A deficiency than are mature animals. Goats that have had access to green feed can store sufficient vitamin A in the liver and fat to last for 3 months on a low carotene ration without showing signs of vitamin A deficiency.
Variable according to size, sex, age, and class (see Table 22-1 and 22-2).	Add 10 to 20% to the requirements given in Tables 22-1 and 22-2 to provide a margin of safety.	Sunlight action on ergosterol, a plant sterol, and on 7-dehydrocholesterol, a sterol of animal origin. Sun-cured hays. Irradiated yeast. Vitamin D_2 or vitamin D_3, which goats use equally well.	Vitamin D should be of little concern when goats are maintained on pasture or range.
		Alpha-tocopherol, added to the diet or injected intramuscularly. Grains are generally high in vitamin E.	Most goat rations contain adequate amounts of vitamin E. Hence, there is little need for vitamin E supplementation.
		Green leafy materials of any kind, fresh or dry are good sources of K_1. Vitamin K_2 is normally synthesized in large amounts in the rumen; no need for dietary supplementation has been established.	
		Only vitamin B-12 (cobalamin) is likely to be deficient in goats with functioning rumens, because the microorganisms synthesize these vitamins in adequate amounts. Adequate vitamin C is synthesized in body tissues to satisfy requirements.	The B vitamins should be included in the diets of very young kids, animals with poorly functioning rumens, sick animals, and those with radically changed diets.

FAT-SOLUBLE VITAMINS

A discussion of the fat-soluble vitamins of goats follows.

Vitamin A

Vitamin A is not contained in forages, but its precursors, the carotenes, are common in plants. Beta-carotene is the standard form of provitamin A. One milligram of beta-carotene is equivalent to approximately 400 IU of vitamin A.

Vitamin A deficiencies are likely to occur during extended dry periods when the supply of green forage is limited, or when poor-quality hays are fed. Hays that are badly weathered or have been stored for long periods generally have lost most of their carotene; hence, vitamin A should be supplemented.

Goats that are deficient in vitamin A exhibit night blindness, poor reproductive performance, a keratinization of the epithelial cells throughout the body, and bone deformities. Vitamin A supplements can be administered two ways: (1) as an additive to feed, or (2) as an intramuscular injectable in a slow-release form.

Vitamin D

Since vitamin D is abundant in sun-cured forages and can be synthesized in the body through exposure to sunlight, there is little need for dietary supplementation. It is noteworthy, however, that the physiological requirements for vitamin D increase when there is an imbalance of calcium and phosphorus. Young kids that are housed without adequate exposure to sunlight should be given supplemental vitamin D.

Vitamin E

Vitamin E is normally found in large quantities in goat rations, and supplementation is not necessary. In dairy goats, the vitamin E transferred to the milk is important because of its antioxidant properties that aid in milk storage.

Vitamin K

In adult goats, the microorganisms of the functioning rumen synthesize vitamin K.

WATER-SOLUBLE VITAMINS

Unlike monogastric animals, goats do have the ability to synthesize a number of vitamins, due primarily to the action of the ruminal microflora. In adult goats, the microorganisms of the functioning rumen synthesize the B complex vitamins. Vitamin C is synthesized in the tissues. However, when the newborn kid starts to eat, the rumen is not well developed and the microflora of the rumen are not of sufficient magnitude to synthesize adequate amounts of the B vitamins; hence, the B complex vitamins are supplied through the milk or milk replacer.

Water

Water, an essential for all metabolic processes, it important for goats. The amount of water required depends on that needed for the maintenance of normal water balance and to provide for satisfactory levels of production. The body water content of the goat varies with age and amount of fat in the body, but it may be expected to exceed 60% of the body weight.

The water requirement may be met by water consumption (drinking), but other important sources include water contained in the feed ingested and metabolic water resulting from oxidation of feed energy sources. The major avenues of water losses are those from urine, lactation, evaporation, and perspiration.

Factors affecting the water intake of goats are lactation level, environmental temperature, water content of the forage consumed, amount of exercise, and the salt and other minerals in the ration.

Goats are among the most efficient domestic animals in the use of water, approaching the camel in the low rate of water turnover per unit of body weight. They appear to be less subject to high temperature stress than wooled sheep or many breeds of cattle and require less water evaporation to control body temperature. They also have the ability to conserve water by reducing losses in urine and feces.

Lactating goats should always have a fresh, clean supply of water readily available. Nonlactating goats can get by with only a small amount of drinking water when given access to good, succulent range or pasture. However, lactating goats require from 1 to 4 gal of water per day plus about 2½ qt of water for every quart of milk produced. If there is insufficient water during lactation, milk yields will be depressed. Some water can be obtained from succulent feeds and from dew on the vegetation early in the morning.

Normally, goats consume 1.4 to 1.7 lb of water per 1 lb of dry matter, whereas cattle consume 2.1 lb of water per 1 lb of dry matter.

Wherever possible, water should be within ½ mile of the grazing area. A running stream is best. If a tank or trough is used, precautions should be taken to prevent the kids from getting into it and being unable to get out. Water troughs, bowls, tanks, or containers should be cleaned frequently to avoid a reduction in water consumption, as goats are more particular about water quality than some other animals.

Water should be no higher in salts than would be acceptable to the taste of the caretaker. Soluble salt content should be less than 3,000 ppm, but animals can become accustomed to water with salt levels as high as 6,000 ppm. Diarrhea can occur when goats are initially exposed to water with a high salt content, but they usually adjust to the water if the salt content is not extremely high. Goats tend to be more sensitive to certain salts (especially magnesium sulfate) than other animals.

Nonnutritive Factors

Nonnutritive factors are substances that cannot be classified as metabolic nutrients but can aid in the utilization of the nutrients in the feed, such as bulk and feed additives.

BULK

Forages and browse-type feeds, as well as coarse textured concentrates, contain considerable bulk. In many cases, animals can utilize bulky feed more efficiently than if the feed were finely ground because finely ground feeds pass through the digestive tract more rapidly. Finely ground feeds are not exposed to microbial fermentation and enzymatic degradation for periods sufficient to maximize utilization. Hence, they tend to be digested somewhat inefficiently.

ANTIBIOTICS AND OTHER FEED ADDITIVES

As in the case with most other farm animals, antibiotics and other compounds are often added to the rations of goats in order to improve production performance and health. While the effects of these drugs can be beneficial to the overall production scheme, certain precautions must be taken in their use. At the present time, there are over 1,000 drugs approved by the Food and Drug Administration for use in livestock production, and more than ½ of these drugs require preslaughter withdrawal from treated feed or milk-discard periods to prevent problems of contamination in the respective products.

The withdrawal and milk-discard periods of the various drugs are constantly being reviewed and revised; hence, the producer must always read and heed the label of the drug that is being used. **The label is the ultimate guide to the producer as to the proper use of the drug.** Unless it is followed, costly condemnations or seizures can result.

(Also see Chapter 13, Feed Supplements/Additives/Implants, section on "Antibiotics.")

FEEDS FOR GOATS

Goats can effectively use the same kinds of feeds as are consumed by other ruminants—grasses and legumes; hays and other dry roughages; silages, haylages, and root crops; concentrates; milk and milk replacers; and commercial feeds. Additionally, goats have a unique preference for, and succeed in feeding on, a wide assortment of browse and forbs on which other species fail.

Browse, Forbs, and Grasses/Legumes

Browse refers to the edible parts of woody vegetation, such as leaves, stems, and twigs from bushes.

Forbs refers to nongrasslike range herbs which animals eat (forbs are commonly called weeds by western ranchers).

For using browse and forbs, goats are without a peer. Mohair and meat-type goats are used extensively to graze unimproved pastures and range areas where vegetation is generally of low quality. Since goats are good browsers, they can be used effectively to control brush and undergrowth. As a result, they have been exploited *mobile pruning weapons* against encroaching browse and forbs in range areas.

Numerous types of shrubs and woody plants can be utilized as feed for goats with varying degrees of success. Table 22-6, Types of Brush Utilized by Goats, shows the relative feeding values of several types of brush that are commonly found in range areas.

TABLE 22-6
TYPES OF BRUSH UTILIZED BY GOATS[1]

Common Name	Scientific Name	Efficiency of Utilization[2]
Black persimmon	*Discaria* spp	–
Catclaw	*Acacia gregii*	++
Cedar	*Juniper ashei; J. penchoti*	+
Coral bean	*Erythrina corrallodendron*	+
Elm	*Ulmus* spp	++
Guajillo	*Acacia berlandieri*	+++
Ill-scented sumac	*Rhus* spp	++
Mesquite	*Prosopis*	–
Oak, live	*Quercus virginicus*	+++
Post oak	*Quercus stellata*	++
Shin oak	*Quercus havardii*	++
Small-leaved sumac (red)	*Rhus glabra*	++
White brush	*Lippia liguestrina*	–
Wild plum	*Prunus* spp	+
Yaupon	*Ilex vomitoria*	++

[1]Adapted by the authors from *Texas Angora Production,* Tex. Ag. Exp. Sta. Bull., B-926.

[2]Excellent utilization = +++; Good = ++; Fair = +; Poor = –.

While goats can utilize a number of types of browse and forbs that other livestock refuse, poisonous plants must be avoided. (See Chapter 5 for additional information relative to poisonous plants.) Also, goat producers should be aware that many palatable browse species are limited in value because of one or more inhibitors that bind or otherwise prevent utilization of the nutrients contained in plants. Among such inhibitors are high levels of (1) lignin in woody twigs, which is practically indigestible; (2) essential oils (terpene-based organic compounds), which inhibit growth of rumen bacteria; and (3) tannins, which depress digestion by binding and/or inhibiting enzyme activity.

Forages can provide the vast bulk of the nutrients required for maintenance. Thus, the goat rancher should have a good knowledge of the feeding value of the forages available and supplement the forages when necessary. Generally, range forages are very low in phosphorus and salt, and often

Fig. 22-6. Goats are superb browsers. (Courtesy, West Virginia University, Morgantown)

marginal in levels of vitamin A, calcium, and trace minerals. As forages mature, their nutrient value and digestibility decline, as shown in Table 22-7, which lists several range forages.

TABLE 22-7
ENERGY, PROTEIN, AND PHOSPHORUS COMPOSITION OF VARIOUS RANGE PLANTS AS INFLUENCED BY STAGE OF MATURITY[1]

Plant and Stage of Growth	Energy			Protein			Phosphorus
	TDN	DE[2]		Crude	Digestibility	Digestible	
	(%)	(kcal/lb)	*(kcal/kg)*	(%)	(%)	(%)	(%)
Curly mesquite and buffalograss:							
Green growth	60	1,200	*2,646*	12.0	65	7.8	0.11
Partly cured forage	54	1,000	*2,205*	8.8	55	4.8	0.09
Cured forage	47	900	*1,984*	5.3	28	1.5	0.06
Gramas (blue, black, and side-oats):							
New growth	56	1,100	*2,426*	11.5	76	8.7	0.14
Fruiting	52	1,000	*2,205*	7.7	49	3.8	0.12
Mostly mature	49	1,000	*2,205*	6.4	39	2.5	0.09
Mature and weathered	40	800	*1,764*	3.5	0	0.0	0.05
Bluestems:							
Very young	68	1,400	*3,087*	14.5	66	9.5	0.14
Green growth	57	1,100	*2,426*	10.4	61	6.3	0.08
Past maturity	44	900	*1,984*	3.7	0	0.0	0.04
Three-awn (purple):							
Past maturity	49	900	*1,984*	5.5	31	1.7	0.06
Mixed cured and green	48	900	*1,984*	6.1	38	2.3	0.08
Rescue grass:							
Foliage, preheading	70	1,400	*3,087*	17.0	74	12.6	0.28
Plants in head	60	1,200	*2,646*	11.7	65	7.6	0.14
Texas wintergrass:							
Luxuriant green growth	65	1,300	*2,866*	14.4	70	10.1	0.10
Green growth	56	1,100	*2,426*	10.0	59	5.9	0.12
Green growth, mature plants	50	1,000	*2,205*	7.2	46	3.3	0.08
Live oak leaves:							
New foliage	55	1,100	*2,426*	17.7	31	5.4	0.26
Foliage, mostly mature	51	1,000	*2,205*	8.9	30	2.7	0.12
Foliage	50	1,000	*2,205*	9.6	30	2.9	0.10
Shin oak leaves	50	1,000	*2,205*	9.4	31	2.9	0.15
Various forbs:							
Winter and spring	65	1,300	*2,866*	16.0	76	12.2	0.20
Summer and fall	45	900	*1,984*	18.0	79	14.2	0.15

[1]Adapted by the authors from *Nutritional Requirements of the Angora Goat* by J. E. Huston, M. Shelton, and W. C. Ellis, Tex. Ag. Exp. Sta. Bull. B-1105.

[2]Digestible energy (DE) in kilocalories per pound of dry matter was calculated on the basis that 1 lb of TDN is approximately equal to 2,000 kcal of DE.

Good-quality pasture and a supply of minerals are all that are required to feed goats at maintenance levels. For the lactating doe, pasture can replace up to one-half of the concentrate in the ration. When pastures are short or when winter limits the availability of good, fresh grass, it is advisable to provide some supplemental feed. This may consist of whole corn, range cubes, or a salt-feed mixture to limit feed intake.

Improved pasture is a necessity for lactating dairy does. In order to prevent overgrazing, grass should be allowed to get 3 to 4 in. high before animals are allowed to graze. A good management practice with goats is to divide the pasture into lots and rotate the animals from the various lots every 10 to 14 days. An electric fence provides an easy way of setting up pasture lots. Not only does this practice prevent overgrazing, it helps to break up the life cycle of internal parasites which can create health problems as well as reduce production. Some of the grasses and legumes that can be effectively used in pasture management for goats are alfalfa, alfalfa-brome mix, clover, clover and grass, timothy, and bluegrass. Since goats are ruminants, care should be exercised when fresh, lush legume pastures are first used, as bloat problems can result.

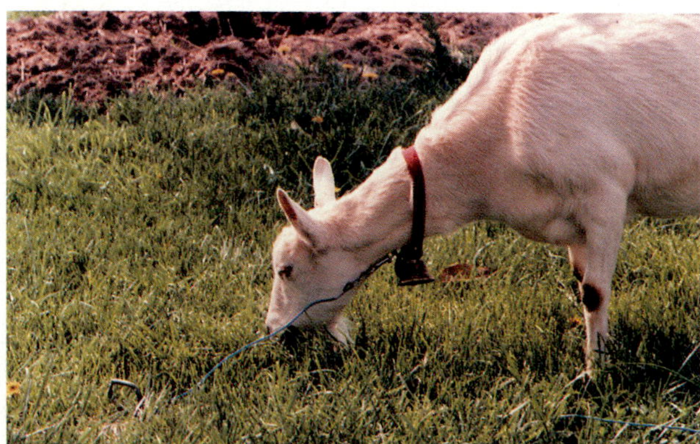

Fig. 22-7. Tethered dairy goat on lush grass. Goats tether easily. (Courtesy, University of Delaware, Newark)

Temporary pasture can provide excellent forage for lactating dairy goats. Rye, wheat, and barley are excellent for early spring or late fall grazing. Sudangrass and millet are excellent summer pastures. Rape (canola) (or a combination of rape and oats) has been used with considerable success in cooler areas.

When on pasture, high-producing dairy goats should be properly supplemented with concentrates and minerals.

Green chop (fresh herbage cut and chopped in the field), instead of pasture, is sometimes fed to lactating goats in confinement. However, this type of forage is labor intensive because the feed must be harvested daily. (Also see Chapter 7 for additional information on pasture management.)

Fig. 22–8. French Alpine does consuming green-chopped alfalfa. (Courtesy, Dr. C. Lu, The American Dairy Goat Research Institute, Langston University, Langston, Okla.)

Hays and Other Dry Roughages

With the exception of pasture and range feeds, hay and dry roughages are usually the most economical feeds for goats. A good-quality legume hay or a mixed legume and grass hay provide an excellent source of highly digestible nutrients. Mixed hays should be at least 50% legume, especially if hay is to be the primary source of feed. Grass hays require supplementation with concentrate. Except for dairy goats in lactation, it is not necessary to provide large amounts of concentrates to goats, especially if they are on maintenance or low production levels.

Fig. 22–9. High-producing dairy goats need high-quality hay, preferably alfalfa or clover. (Courtesy, University of New Hampshire, Penacook)

Hays with the highest nutritive values are those that have tender stems and are leafy. For this reason, hay from second cuttings are generally better utilized than first cuttings. Palatability of coarse, stemmy hay can be improved by crushing. The stage at which hay is cut has a direct influence on its feeding value. As the grass or legume matures, there is a steady decrease in crude protein content along with an increase in crude fiber content.

The following kinds of hays are most commonly fed to goats: alfalfa, alsike clover, red clover, ladino clover, soybean, vetch, birdsfoot trefoil, and mixed legume and grass.

In the South, cottonseed hulls are also a popular dry roughage for goats. They are bulky, containing about 43% fiber and 40% TDN.

Silages, Haylages, and Root Crops

Silages and haylages have never been used extensively for feeding goats because of the following practical reasons: (1) their high water content makes it impractical to feed them at great distances, thereby alleviating their use on most Angora and Spanish (meat) goat operations; and (2) the small number of animals in most dairy goat operations makes it impossible to feed the top 3–4 in. of silage or haylage that must be removed from the silo daily to prevent spoilage.

Dairy goats are sometimes fed silage or haylage (1) where the herd is associated with a cattle operation, thereby making it practical to share silage with the goats; or (2) where a large commercial dairy goat herd is involved. About 3 lb of silage or 2 lb of haylage may replace 1 lb of hay.

Fig. 22–10. Haylage feeding at Irish Hills Goat Farm, Tipton, Mich., owned by Gary and Nancy Abner. (Courtesy, Dr. Christine S. D. Williams, Michigan State University, East Lansing)

Goats are quite fond of root crops and garden products; and these types of feeds can be effectively incorporated in the ration for a change of routine. Carrots, beets, turnips, and cabbage are especially relished by goats. These types of feeds are high in moisture and should be fed in the same manner as silage. Roots should be chopped in order to lessen choking.

In order to prevent off-flavors in milk, it is recommended that silage, haylage, and root crops such as turnips be fed either after milking or in amounts that will be consumed 3 to 4 hours prior to milking. (For more details on silage and haylage, see Chapter 9.)

Concentrates

The concentrates used in goat rations can be classified as either energy feeds or protein supplements.

Fig. 22-11. Dairy goat eating a concentrate while on a stand ready for milking. (Courtesy, University of New Hampshire, Penacook)

ENERGY FEEDS

The common energy ingredients are corn, oats, barley, milo, and wheat, along with their by-products. The amount of energy feed to include in the ration should be determined by the production demands. A dry doe requires little or no supplementation, while a doe at the peak of lactation requires substantial amounts of energy.

Molasses, an excellent energy source, is commonly used to reduce the dustiness of feed and to increase palatability. If too much molasses is included in the ration, the feed becomes sticky and lumpy; so, it is usually limited to 5 to 8% of the mixture.

PROTEIN FEEDS

A wide variety of protein supplements can be used in rations for goats. As is the case with most other species of livestock, the oil meals are used extensively. Cottonseed meal and soybean meal are probably the most widely used sources of protein for goats, but other meals can be and are used, depending on their respective prices and availability. Among the alternative sources are copra meal, peanut meal, sunflower meal, safflower meal, rapeseed (canola) meal, corn gluten feed, brewers' dried grains, and distillers' dried grains.

Urea and other nonprotein nitrogen (NPN) sources are often used in rations for goats, but several precautions should be taken when they are incorporated in the ration. Urea can constitute up to 1% by weight of the total concentrate mix, or supply ⅓ of the protein equivalent in the total ration. Rations containing NPN should be introduced very gradually in the feeding scheme of goats, as a period of adaptation is required by the microorganisms of the rumen. In addition to urea, other NPN sources are ammoniated cottonseed meal, ammoniated rice hulls, ammoniated citrus pulp, and ammoniated beet pulp.

Milk and Milk Replacers

Unless extenuating circumstances prevail, newborn kids are seldom taken from Angora or Spanish goat mothers. However, with dairy goats, normally kids are allowed to nurse for 2 to 4 days only, or not at all. It is usually easier to train kids to nipple- or pan-feeding if they have never nursed their mothers.

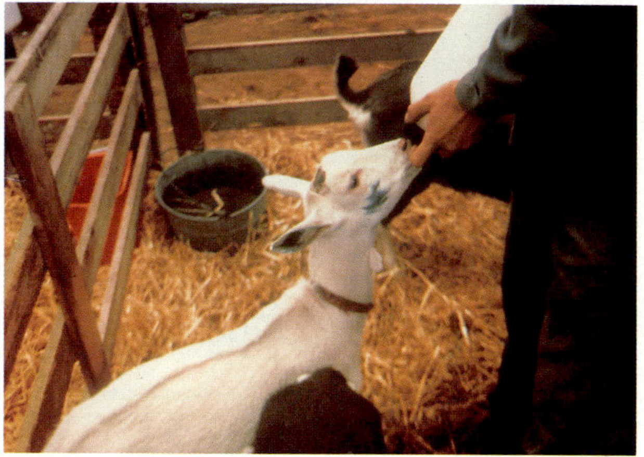

Fig. 22-12. Kid being bottle fed. (Courtesy, University of Delaware, Newark)

Recent research indicates that disease organisms, especially caprine arthritis-encephalitis (CAE), may pass from doe to kid through the milk, and that transmission may be avoided through the use of frozen colostrum from does tested and shown to be free of CAE. Colostrum from does that are known CAE carriers should be pasteurized to prevent transmission to newborn kids. (See section headed Caprine Arthritis-Encephalitis [CAE].)

When newborn kids are to be raised separately from their dams, they may be fed cow's milk or a milk replacer. The most desirable milk replacers for kids contain a minimum of 20% fat and 24% crude protein; are skimmed milk-based, rather than grain-based; and are fortified with minerals and vitamins. Because of varying formulations, care should be taken to follow the manufacturer's directions.

(Also see Chapter 6, Types and Roles of Feedstuffs, section on "Milk Replacers.")

Commercial Feeds

Commercial feeds, containing a variety of ingredients, including minerals and vitamins, are used by many goat producers. Special ingredients such as molasses and/or fat may be added to reduce dustiness and increase palatability.

Cottonseed hulls may be added to improve texture and provide fiber. In some cases, the commercial feeds are pelleted.

Commercial feeds may be available as (1) complete feeds (roughage and concentrate combined), (2) concentrates, or (3) protein supplements.

Feed Preparation

Grain can best be utilized by goats when it is processed to a limited extent. Cracking, rolling, crimping, flaking, or coarse grinding all aid in making grains more digestible. Grinding feed to a fine powder form is not desirable, especially in low-roughage rations, because it generally results in lowered palatability and digestibility. Large grains, such as corn, can be fed whole on the ground in range areas and effectively utilized with little wastage. When grains are to be mixed with hay and other ingredients or to be used on the range, pelleting is advisable.

In rangeland areas, feed supplements are sometimes offered free-choice, using salt as a governor to limit feed consumption. The proportion of salt to feed may vary from 5 to 40%. By varying the proportion of salt in the mixture, it is possible to hold the consumption of feed supplement to any level desired. The amount of salt should vary according to the level of feed consumption desired. A commonly used mixture consists of 1 part salt, 1 part cottonseed meal, and 3 parts grain. Iodized salt should never be used as a feed intake inhibitor because toxicity problems can result with the high intake of iodine. When high-salt mixes are fed, water should always be readily available.

Pelleting and cubing are used when roughages and grains are mixed together. Unless these feeds are pelleted or cubed, the various ingredients can separate out, resulting in waste and inefficient utilization of feed. Hay is sometimes chopped to reduce waste. It should be emphasized that any processing of feed creates additional feed costs; hence, feed should not be processed if the additional treatment does not increase feed efficiency or reduce waste sufficiently to offset the added cost.

When urea or other nonprotein nitrogen is to be included in the ration, the ingredients must be mixed thoroughly. It may be unwise to mix such rations in farm mixers unless (1) the urea is premixed with other ingredients to ensure its uniform distribution throughout the mix, and (2) some means is employed to prevent the urea from settling out with other fines. A molasses-urea mixture is often used to provide for these controls in mixers equipped to handle liquid molasses.

Feed Substitution Table

Sometimes, goat producers have an opportunity to obtain a feed ingredient at a favorable price, but they may not know the relative feeding value of the product with respect to the feed currently in use. In order to assist producers in making these managerial decisions, a feed substitution table giving the comparative value of feeds is needed.

Feed substitutions for goats and sheep are similar. Hence, the reader is referred to Chapter 21, Feeding Sheep, Table 21-12, Feed Substitution Table for Sheep and Goats.

NUTRITION RELATED DISORDERS

Goats are subject to several nutrition-related disorders, most of which occur in sheep, and many of which occur in other species. Many such disorders are named and discussed in Chapters 4 and 5 of this book. However, abortion, caprine arthritis-encephalitis (CAE), enterotoxemia, and posthitis of goats are discussed in the sections that follow.

Fig. 22-13. Signs of good health, evidenced by Saanen kids. (Courtesy, The American Dairy Goat Research Institute, Langston University, Langston, Okla.)

Abortion

Infectious diseases such as brucellosis are capable of causing abortions in goats. However, herein reference is made to a particular type of abortion caused by a metabolic disturbance of the functional corpus luteum, to which Angora goats are predisposed. Under normal production conditions, this malady commonly causes a low level of abortion, but catastrophic losses sometimes occur. Most abortions occur in response to stress between 90 and 110 days of gestation. Undernutrition during the critical stage of rapid fetal development and competition for nutrients between fetal and maternal organisms appear to be one explanation. It is noteworthy that the incidence of abortion is reduced in herds in which replacement does are fed for proper size and development prior to the first breeding season and during gestation.

Caprine Arthritis-Encephalitis (CAE)

It is estimated that more than 80% of the dairy goats of the United States are infected by the retrovirus that causes CAE. In kids, it causes paralysis; in adults, it causes arthritis, which, in the late stages, is similar to rheumatoid arthritis in humans.

In most herds, the expression of the disease ranges from 0 to 25%. In those animals that do show clinical symptoms, the rate of progression and the severity of the disease varies markedly. In kids, the disease may vary from a barely noticeable unsteadiness of gait to a rapid fatal paralysis. In mature

Fig. 22-14. Several small housing units for dairy kids, placed side by side. Kids should be fed and cared for separate and apart from the milking herd in order to lessen exposure to disease. (Courtesy, Rocking M Ranch, Hilmar, Calif.)

goats, the joints (front knees, hocks, and stifle joints) become swollen and disfigured, accompanied by a loss in body weight and a drop in production. The severity of the arthritis varies from years of intermittent lameness or stiffness to complete debilitation.

CAE is caused by a retrovirus, a slow acting virus that is latent, with the result that many animals may not show signs of the disease until after a long incubation period. The virus is transmitted from the doe to the kid(s) through the colostrum and milk. Does not showing symptoms of CAE may carry the virus and transmit it.

The rate of the infection in newborn goats can be reduced by more than 90% by (1) removing kids from infected does at the time they pass from the birth canal; (2) providing them colostrum that is from does identified as negative with the agar-gel immuno-diffusion (AGID) test or that has been heated to 132°F for 1 hour; (3) raising them on pasteurized goat's milk or a milk replacer; and (4) keeping them in isolation from infected goats. The AGID test can be used to monitor infection.

Enterotoxemia (Overeating Disease)

Enterotoxemia, also known as *overeating disease* or *toxic indigestion,* may be the most insidious, most often undiagnosed, and, in many herds, the most important of all goat diseases. It is a toxic reaction to *Clostridium perfringens,* Types C and D, characterized by diarrhea, depression, lack of coordination, digestive upsets, coma, and death.

In baby kids, excess feeding or sudden access to palatable feed, changes in feed, or feeding following an unusual period of starvation, may cause acute enterotoxemia and death. Causative factors for enterotoxemia in mature goats include sudden changes in concentrate feed, excessive feeding of concentrates to animals not accustomed to such feeds, sudden access to highly palatable forage, sudden change to lush pasture, sudden access to feed, and overeating by hungry goats.

Prevention consists in proper feeding along with a vaccination program, using *Clostridium perfringens* toxoid, Types C and D. Initially, all goats should be given two separate doses of the toxoid, at 2-4 week intervals. Then, all does should be given an annual booster toxoid about 1 month before kidding; all kids should be vaccinated at 3-4 weeks of age, followed by a second dose of toxoid 2 weeks later. All goats in the herd, including bucks, should receive at least two doses of toxoid annually; one when does are in late pregnancy, and another when the kids are 4 to 5 months old.

Posthitis (Sheath-rot, Pizzle-rot, Urine Scald)

Posthitis, a moderately contagious disease primarily of male goats and sheep, is characterized by ulcerative lesions with scab formation on the prepuce of males and less frequently on the vulval lips of females. It is caused by a high-protein ration in combination with the organism, *Corynebacterium renale*. The problem appears to be aggravated by confinement to areas where irritation or infection are more likely to occur. Posthitis may be serious in mature Angora wethers kept for mohair production; and it may also occur in individual breeding bucks kept in confinement. Reducing the protein intake tends to reduce the incidence and severity of the disease.

PART II—FEEDING DAIRY GOATS

A good dairy doe will average 5 lb of milk, or more, per day over a lactation period of 10 months, whereas superior animals will average 10 lb or more. Based on 200,000 official lactation records since 1968, average milk production per goat per lactation is l,643 lb with 3.8% fat. The highest official milk production on record in the United States was made by a Saanen doe that produced 6,850 lb of milk in 305 days, in 1984. The butterfat record is held by a Nubian doe that produced 384 lb in 305 days.

Fig. 22-15. With confinement housing, dairy goats need adequate feeding space, dry bedding, and good ventilation. (Courtesy, University of New Hampshire, Penacook)

Fig. 22-16. All-time, all-breed world record milk production holder. This Saanen doe produced 6,850 lb milk and 296 lb fat, in 305 days, in 1984. Bred by Gary and Sharon Swanson, Renton, Wash.; owned by Gary Lee Cox, Eagle Point, Ore. (Courtesy, T. H. Teh, Ph.D., Prairie View A&M University Research Center, Prairie View, Tex.)

In comparison with cow's milk, goat's milk has a higher percentage of small fat globules, and a sweeter flavor. Goat's milk forms a fine, soft curd during digestion, thus making it more easily digested than cow's milk for some children and for older people who cannot tolerate cow's milk. If milked in clean quarters and away from the bucks, goat's milk will be free from any unpleasant flavor or odor, but when a buck is around the milking quarters, his strong odor is quickly absorbed by warm milk. The dairy goat produces milk that is very similar in composition to that of the dairy cow (see Table 22-8).

TABLE 22-8
COMPARISON OF THE COMPOSITION OF GOAT'S MILK AND COW'S MILK

Source	Water	Protein	Fat	Sugar	Ash
	(%)	(%)	(%)	(%)	(%)
Goat	86.0	4.4	4.6	4.2	0.8
Cow	87.7	3.3	3.6	4.6	0.7

FEEDING GUIDE

Table 22-9, Dairy Goat Feeding Guide, provides a good basic guide for the nutritional program of the dairy goat.

TABLE 22-9
DAIRY GOAT FEEDING GUIDE[1]

Age	Feed	Amount Each Day
Birth to 3 days	Colostrum	All the kid wants 3 to 4 times daily.
3 days to 3 weeks	Whole milk (cow or goat, or milk replacer). Water, minerals	2 to 3 pt (*0.9 to 1.4 liter*), feed 3 times daily. All the kid wants.
3 weeks to 4 months	Whole milk. Creep/starter feed (1). Alfalfa hay (2). Water, minerals.	2 to 3 pt (*0.9 to 1.4 liter*), up to 8 weeks, feed 2 times daily. All the kid will eat, up to 1 lb (*0.45 kg*) per day. All the kid will eat. All the kid wants.
4 months to freshening	Grain mixture/grower feed (3b). Alfalfa hay or pasture (2). Water, minerals.	Up to 1 lb (*0.45 kg*) of high-protein feed. All the doe will eat. All the doe wants.
Milking doe	Grain mixture (3a). Alfalfa hay (2). Water, minerals.	Minimum of 1 lb (*0.45 kg*) up to 2 qt (*1.89 liter*) of milk per day. Add 1 lb (*0.45 kg*) grain mixture for each additional 2 qt (*1.89 liter*) of milk. All the doe will eat. All the doe wants.
Dry pregnant	Grain mixture (3b). Hay/silage or pasture (2).	Up to 1 to 2 lb (*0.45 to 0.91 kg*) mix for a dry animal. All the doe will eat.
Bucks	Roughages. Grain. Water, minerals.	All the buck wants. 1 to 2 lb (*0.45 to 0.91 kg*) during breeding season if needed. Free-choice.

(1) Creep or starter feed (see Table 22-13 for starter ration).
(2) Alfalfa hay of extremely high quality, fine stemmed, leafy, and green.
(3) Suggested grain mixtures:
(a) For a lactating doe (or select a grain mix from Table 22-11)—
55 lb (*25.0 kg*) barley or oats
15 lb (*6.8 kg*) beet pulp
20 lb (*9.1 kg*) wheat, mixed feed, or mill run
10 lb (*4.5 kg*) linseed, cottonseed, or soybean oil meal
(b) For a growing or a dry doe (or select growing ration from Table 22-13)—
15 lb (*6.8 kg*) beet pulp
50 lb (*22.7 kg*) barley or oats
15 lb (*6.8 kg*) wheat, mixed feed, or mill run
20 lb (*9.1 kg*) linseed, cottonseed, or soybean oil meal

If you use commercial dairy cow or dairy goat feed, use it according to your goat's stage of growth—growing, drying, or lactating.

[1]Adapted by the authors from *Your Dairy Goat,* Western Regional Extension Publication, WREP 47, University of California, Davis, 1981.

The dairy goat producer must combine the available feeds so as to achieve the most profitable production. At its best, developing a dairy goat feeding program involves combining the art and the science of feeding. Also, there should be a ration for each specific need—for lactating does, for dry does, for kids, for yearlings and replacements, and for bucks.

FEEDING LACTATING DAIRY DOES

Fig. 22-17. Alpine does at feed at The Coach Dairy Goat Farm, Pine Plains, N.Y. (Courtesy, The Coach Dairy Goat Farm, Pine Plains, N.Y.)

Following parturition, the feed intake should be increased gradually.

The nutritional demands upon lactating does are tremendous. It is essentially impossible for the doe to consume enough to meet the demands for body maintenance and milk production during the first few months of lactation; so, she must draw upon her body reserves to augment the nutrients consumed. Based on the National Research Council recommendations (see Tables 22-1 and 22-2), Table 22-10 shows the combined requirements for maintenance and milk production at various levels for dairy goats of three different sizes producing 4% milk. Note that provision is made for (1) five different levels of milk production, ranging from 2.5 to 20 lb daily; (2) three different body weights—130, 160, and 190 lb; (3) TDN and net energy, (4) crude protein, and (5) calcium and phosphorus. As shown in Table 22-10, a dairy goat weighing 160 lb and producing 10 lb of 4% milk daily would need 5.5 lb of TDN daily (5 Mcal of net energy), 1.0 lb of crude protein, 20 g of calcium, and 14 g of phosphorus.

TABLE 22-10
COMBINED REQUIREMENTS FOR MAINTENANCE AND MILK PRODUCTION AT VARIOUS LEVELS FOR DAIRY GOATS AT THREE DIFFERENT SIZES PRODUCING 4% MILK FAT

Daily Milk		Body Weight		TDN		Net Energy	Crude Protein		Ca	P
(lb)	*(kg)*	(lb)	*(kg)*	(lb)	*(kg)*	(Mcal)	(lb)	*(kg)*	(g)	(g)
2.5	*1.1*	130	*59.0*	2.8	*1.3*	2.4	0.42	*0.19*	8	5
		160	*72.6*	3.0	*1.4*	2.7	0.45	*0.20*	9	6
		190	*86.3*	3.5	*1.6*	2.9	0.50	*0.23*	10	6
5	*2.3*	130	*59.0*	3.4	*1.5*	3.2	0.60	*0.27*	12	8
		160	*72.6*	3.8	*1.7*	3.4	0.65	*0.30*	13	9
		190	*86.3*	4.0	*1.8*	3.7	0.70	*0.32*	14	10
10	*4.5*	130	*59.0*	5.2	*2.4*	4.8	1.95	*0.89*	19	13
		160	*72.6*	5.5	*2.5*	5.0	1.00	*0.45*	20	14
		190	*86.3*	5.8	*2.6*	5.3	1.04	*0.47*	21	15
15	*6.8*	130	*59.0*	6.9	*3.1*	6.5	1.31	*0.59*	26	18
		160	*72.6*	7.2	*3.3*	7.0	1.35	*0.61*	27	19
		190	*86.3*	7.5	*3.4*	7.4	1.40	*0.64*	28	20
20	*9.1*	130	*59.0*	8.6	*3.9*	7.9	1.67	*0.76*	33	23
		160	*72.6*	8.9	*4.0*	8.2	1.71	*0.78*	34	24
		190	*86.3*	9.2	*4.2*	8.5	1.76	*0.80*	35	25

In order to formulate a ration to meet the requirements given in Table 22-10, available feeds and feed compositions must be known (see Section V, Composition of Feeds). Also, consideration must be given to dry matter intake (DMI)—the ability of the animal to consume. Total DMI becomes a critical factor in balancing a ration because the total combination of daily forage and concentrate consumption must (1) come within this range, and (2) meet nutrient requirements. So, the lower the quality of the roughage, the higher the nutrient density of the concentrate to accomplish this objective. Experiments and experiences have shown that the daily DMI of lactating does usually ranges from 4 to 5% of their body weight. The main goal is to meet the nutrient needs at minimal cost and yet maximize production.

Some suggested rations for lactating dairy goats are given in Table 22-11. These rations meet the requirements set forth in Table 22-10. Note that when a legume hay (alfalfa or clover) is fed, a 12-14% crude protein grain mix will suffice. But, when a grass hay is fed, a 16-18% crude protein grain mix should be fed.

TABLE 22-11
RATIONS FOR LACTATING DAIRY GOATS (160-LB BODY WEIGHT)[1]

Feedstuffs	Daily Milk Production (lb)							
	2.5		5.0		10.0		15.0	
	Daily Feed							
	(lb)	*(kg)*	(lb)	*(kg)*	(lb)	*(kg)*	(lb)	*(kg)*
Ration # 1:								
Alfalfa clover hay (16% CP)	2.0	*0.9*	3.0	*1.4*	3.5	*1.6*	4.5	*2.0*
Grain mix to be selected from 4 mixes listed below (14-16% CP, 70% TDN)	3.0	*1.4*	4.0	*1.8*	6.0	*2.7*	8.0	*3.6*
Ration # 2:								
Grass hay (7% CP)	2.5	*1.1*	2.5	*1.1*	3.0	*1.4*	4.0	*1.8*
Grain mix to be selected from 4 mixes listed below (16-18% CP, 65% TDN)	3.4	*1.5*	4.6	*2.1*	7.0	*3.2*	9.0	*4.1*
Grain mixes:								
Ration	# 1		# 2		# 3		# 4	
Level of crude protein (CP) in grain mix	14%		16%		18%		20%	
Lb or kg per ton	(lb)	*(kg)*	(lb)	*(kg)*	(lb)	*(kg)*	(lb)	*(kg)*
Ingredients								
Rolled corn	900	*409.0*	800	*363.6*	720	*327.3*	656	*298.2*
Crimped oats	421	*191.0*	300	*136.0*	240	*109.1*	200	*91.0*
Beet-citrus pulp	200	*91.0*	200	*91.0*	200	*91.0*	200	*91.0*
Dried brewers' grain	—	—	150	*68.0*	200	*91.0*	200	*91.0*
40% protein supplement	300	*136.0*	—	—	516	*234.5*	—	—
Soybean meal	—	—	356	*161.8*	—	—	600	*272.7*
Molasses	150	*68.0*	150	*68.0*	100	*45.5*	100	*45.5*
Trace mineralized salt	10	*4.5*	20	*9.1*	10	*4.5*	20	*9.1*
Dicalcium phosphate	—	—	10	*4.5*	10	*4.5*	20	*9.1*
Monosodium phosphate	15	*7.0*	10	*4.5*	—	—	—	—
Magnesium oxide	4	*1.8*	4	*1.8*	4	*1.8*	4	*1.8*
Vitamins[2]								

[1]Adapted by the authors from *Extension Goat Handbook,* "Feeding," by R. S. Adams, Pennsylvania State University; B. Harris, University of Florida; M. F. Hutjens, University of Illinois; and E. T. Oleskie and F. A. Wright, Rutgers University.

[2]During winter, all mixtures should be supplemented with 6 million IU of vitamin A and 3 million IU of vitamin D per ton of grain mix.

FEEDING DRY DAIRY DOES

It is important to dry off dairy goats about 6 to 8 weeks prior to kidding. This gives the doe a brief rest from the heavy demands of lactation, enables her to meet the nutrient needs of the rapidly growing fetus, and allows her to build body reserves with which to meet the rigorous requirements of lactation which follow.

Depending on the kind and quality of the forage and the size and condition of the doe, 1 to 2 lb of 12 to 16% protein concentrate ration should be fed during the dry period. Trace-mineralized salt and water should also be available.

In the last 6 weeks of gestation, the fetus gains 70% of its birth weight, so the nutrition of the doe during this period is critical.

Some suggested rations for dry does are given in Table 22-12.

TABLE 22-12
RATIONS FOR PREGNANT OR DRY DOES

Ration	Ingredients	Amount	
		(lb)	*(kg)*
1	Pasture plus good mixed grass/legume hay	*ad libitum*	
	16% protein supplement	1.1	*0.5*
2	Mixed hay	1.1	*0.5*
	Silage	1.1	*0.5*
	Beet pulp	0.7	*0.3*
	16% protein supplement	1.1	*0.5*
3	Alfalfa hay	1.1	*0.5*
	Beets	2.2	*1.0*
	Beet pulp	1.1	*0.5*
	16% protein supplement	1.1	*0.5*

FEEDING DAIRY KIDS

Fig. 22-18. Dairy kids at a nipple bar, an efficient method of feeding. (Courtesy, University of New Hampshire, Penacook)

It is important to get 2 to 4 oz of colostrum in a kid as quickly as possible after birth.[4] Colostrum contains higher levels of total protein, milk solids, globulins, fat, and vitamin A than normal milk. It is also a laxative. Most important, it contains antibodies against disease to which the doe has immunity. Young kids are able to absorb this antibody protection effectively at birth, but by the time they are 3 days old, this ability almost disappears.

During the first 2 days of life, kids should receive at least three colostrum feedings per day. A kid will consume 1½ to 2 pt daily.

If the kid is to be raised without its mother's milk, it should be allowed to nurse for 2 or 3 days only or not at all. Most commercial dairy goat producers favor putting the newborn kid directly on bottle- or pan-feeding without allowing it to nurse its mother; after a kid nurses its mother a couple of times, it is very difficult to get it to accept a bottle or a pan. Initially, the nipple bottle is generally preferred to pan feeding as it tends to be more natural, thereby resulting in less ingested air. However, as soon as kids are strong enough and can drink milk easily, experienced dairy goat producers prefer to train them to bucket or pan feeding, which is faster and allows for easier cleaning and maintenance of utensils.

[4]Because of the hazard of caprine arthritis-encephalitis (CAE) virus, scientists recommend that kids be fed (a) colostrum only from does negative to the CAE test, (b) heat-treated colostrum (132°F for 1 hour), or (c) no colostrum at all. (See section headed Caprine Arthritis-Encephalitis [CAE].)

After the kid is removed from colostrum feeding, it can be fed cow's milk or a milk replacer, along with a starter ration, with the change made gradually. The following points are pertinent to feeding and managing kids:

1. **Be clean and sanitary.** Wash and sanitize any feeding utensils which will be used for feeding.

2. **Prepare the milk or milk replacer properly.** The milk or milk replacer should be heated to about 103°F for feeding. A young kid will consume about 2 to 3 pt of milk daily. During the first 3 days, feed should be offered 3 to 4 times daily; thereafter, twice a day feeding is adequate.

3. **Think economy.** If cow's milk is cheaper than goat's milk, use it. It is also possible to prepare a suitable kid starter on the farm by mixing 30 lb each of cornmeal, ground oats, and wheat bran, plus 10 lb of an oil meal.

4. **Get kids on dry starter as soon as possible.** At about 1 week of age, the starter can be made available to kids. Also, fine-quality hay should be fed in order to enhance the development of the rumen.

5. **Guard against overfeeding or underfeeding.** Feeding too much or too little should be avoided.

6. **Weaning.** Kids can be weaned from milk as early as 5 to 6 weeks of age, although most goat breeders delay it until 3 to 4 months of age. As young kids approach weaning age, gradually add warm water to their milk diet. This will provide them with the necessary fluids for rumen development and ease the stress of weaning them. After the kids are weaned from the milk, feed them all the bright green forage they will eat, plus ¾ to 1 lb of any good grower ration.

Suggested starter and grower rations for kids are given in Table 22-13. **NOTE WELL**: Kids should receive colostrum for the first 3 days (see Table 22-9).

TABLE 22-13
STARTER AND GROWING RATIONS FOR DAIRY KIDS[1]

	Kid Starter[2]	Growing Ration[3]
	(%)	(%)
Corn	27.6	12.9
Crimped oats	37.9	10.0
Soybean meal (44%)	10.0	8.6
Alfalfa leaf meal	18.0	10.0
Cane molasses	5.0	5.0
Cottonseed hulls	—	51.9
Trace mineralized salt	1.0	1.0
Limestone	0.3	0.4
Vitamins A, D, and E (premix)	0.2	0.2

[1]From: *Raising Goat Kids*, by T. H. Teh *et al.*, Texas A&M, International Dairy Goat Research Center, Prairie View A&M Research Center, Prairie View, Tex., Vol. A1, No. 1, Bull. March, 1985.

[2]On a dry matter basis, the kid starter should contain a minimum of 80% TDN, 16% protein, 0.6% calcium, and 0.4% phosphorus. Also see Table 22-9.

[3]The grower ration may be fed free-choice after 4 months.

FEEDING DAIRY YEARLINGS AND REPLACEMENTS

In feeding young animals, the object is to provide enough nourishment for body maintenance and growth. But too much feed, especially concentrates, causes animals to fatten, which may lead to difficulties in breeding. Beginning at 4 to 6 months of age, animals should have a good pasture (if available), high-quality hay, and a place to exercise. If the forage is good, ½ lb of grain per day should lead to ample growth. If the forage is poor, animals may require 1 to 1½ lb of grain daily. Yearlings can be fed the same grain mix that is given to the lactating does (Table 22–11). Low-quality forages should be supplemented with a 14–16% protein grain mixture. A mineral mixture of equal parts of trace mineralized salt and dicalcium phosphate is suitable for free-choice feeding.

A suggested grower ration for weaned kids and yearlings is given in Table 22–13.

FEEDING DAIRY BUCKS

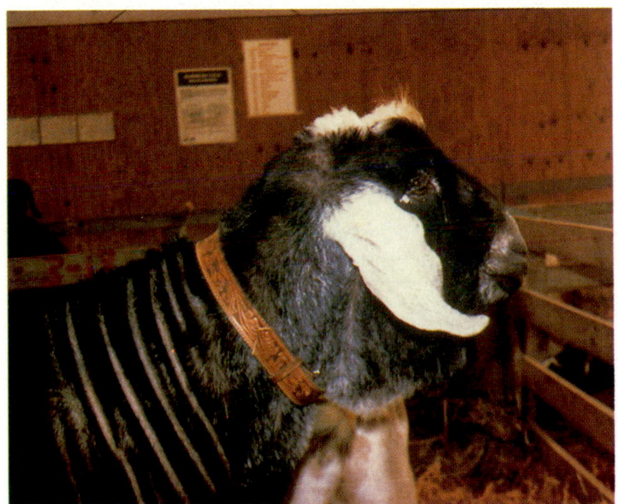

Fig. 22–19. Nubian buck. (Courtesy, University of Delaware, Newark)

Since the buck is larger than the doe, he will require more forage. Additionally, he should receive 1 to 2 lb of grain daily. Overweight bucks make inefficient breeders, so it is important to have the buck in good physical condition; but he should not be fat or have excessive fleshing. As breeding season approaches, the amount of grain should be increased to help the buck prepare for this heavy production period. Exercise is necessary to keep the herd buck healthy and sexually active. A 1:1 mixture of salt and dicalcium phosphate should be available free-choice, as well as an abundant supply of water.

CAUTION: In bucks, the incidence of urinary calculi is increased by a high concentrate ration in which there is a mineral imbalance involving excessive phosphorus intake.

A suggested ration for bucks is given in Table 22–14.

TABLE 22–14
SUGGESTED RATIONS FOR BREEDING BUCKS

	Ingredients	Amount	
		(lb)	***(kg)***
Bucks: breeding			
Out of breeding season	Good hays and pasture	*ad libitum*	
In breeding season ...	Good hays and pasture	*ad libitum*	
	plus		
	14% protein supplement ...	1.1–2.2	*0.5–1.0*
	plus		
	mineral supplements and salt		

FEEDING FOR SHOW

For success in the show-ring, an animal should be fed and managed to attain maximum development in body conformation. In general, the feeding of dairy show goats differs from normal operations only in that greater effort and expense and more liberal feed allowance may be justified in order to produce a winner. In addition to being well balanced, the ration for show goats must be palatable. No definite set of rules relative to feed allowances can be followed satisfactorily. Rather, the judgment of the skillful feeder must prevail.

Fig. 22–20. Dairy goat properly fitted and shown by a 4–H Club member. (Courtesy, University of New Hampshire, Penacook)

PART III—FEEDING ANGORA AND SPANISH (MEAT) GOATS

The basic nutrient requirements established by the National Research Council (NRC) are suited for both Angora and Spanish goats (see Table 22-1). These requirements are based upon the weight of the animals being fed, the level of activity, the stage of pregnancy, and the growth rate. The type of range generally governs the activity of goats. For example, goats that range on sparsely vegetated grassland and on seasonal mountainous ranges must travel long distances daily for grazing and watering; whereas, slightly less activity is required by goats that graze on semiarid rangeland or on slightly hilly land. For mohair production, nutrient requirements in excess of those in Table 22-1 are needed. These are provided in Table 22-3 according to the level of mohair production. Furthermore, the nutrient requirements of lactating does as given in Table 22-2 must also be considered for goats nursing kid(s). Therefore, based on Table 22-1 and 22-3, a 44-lb Angora goat kid gaining 0.22 lb per day, having medium activity, and producing 4.4 lb of mohair per year has the following requirements for energy and protein:

	Energy	Total Protein
	(Mcal DE/day)	(g/day)
Maintenance (medium activity)	1.77	55
Growth	0.88	28
Mohair	0.07	9
Total	2.72	92

Fig. 22-21. Angoras on typical range on a west Texas ranch. (Courtesy, *Sheep & Goat Raiser*)

The feed recommendations listed herein should be considered to be minimum levels. Thus, goat producers should adapt the recommendations to fit the particular operation; and, in many cases, it is advisable to use higher levels than are listed.

Since Angora and Spanish goats characteristically travel across large areas, eating the available browse (shrubs and woody plants), they have been routinely used to control brush on ranges. Brush, such as live oak, can be cut to leave a stump about 3 ft high; and the sprouts will provide additional feed. In order to maximize the production potential of range areas, goats are often placed in areas where other types of livestock are being grazed. Goats preferentially consume a certain amount of browse. Cattle prefer grasses, whereas sheep readily consume forbs (weeds) in rangelands. By combining several types of livestock, the producer can utilize these three types of plants and increase the gain per unit of land as well as manage the vegetative makeup of the land.

Available browse and forages will satisfy many of the nutritive needs of goats that are raised on ranges, but, for maximum performance, it is advisable to provide supplemental feed when range conditions become adverse. Twenty percent protein range cubes are a popular supplement. Also, shelled corn can be fed on the ground to goats, with very little waste. Usually, about ¼ to 1 lb of supplement per head per day is adequate in the winter or during dry periods when green feed is scarce. Some examples of concentrate supplements for range goats are found in Table 22-15.

TABLE 22-15
CONCENTRATE SUPPLEMENTS FOR RANGE GOATS[1]

Ingredients	Supplement A 20% Protein	Supplement B 30% Protein	Supplement C 40% Protein
	(%)	(%)	(%)
Corn or sorghum	82	58	25
Cottonseed meal or soybean meal	14	37	70
Urea	2	3	3
Dicalcium phosphate	2	2	2
Vitamin A supplement	—[2]	—[2]	—[2]
Approximate nutrient composition			
Energy:			
TDN (%)	75	72	70
DE (Mcal/lb)	1.50	1.45	1.40
Protein:			
Crude (%)	20	30	40
Digestible (%)	16	24	32
Phosphorus (%)	0.55	0.65	0.77

[1]Adapted by the authors from *Nutrient Requirements of the Angora Goat* by J. E. Huston, M. Shelton, and W. C. Ellis, Tex. Ag. Exp. Sta. Bull. B-1105.

[2]Sufficient supplement to provide 2,500 IU of vitamin A per lb of feed (5 million IU/ton of total concentrate supplement mix).

A summary of the recommended schedule for providing supplemental feed to Angora goats on average range is given in Table 22–16. If the range is extremely poor, the producer should refer to Table 22–17.

TABLE 22–16
SCHEDULE FOR SUPPLYING SUPPLEMENTAL FEED TO ANGORA GOATS ON AVERAGE RANGE[1]

Class & Weight	Period	Type of Supplement[2]	Lb/Day/ 100 Head
Wethers & dry does:			
50–80 lb	July 15 – November 15	A	20
	November 15 – March 15	B	40
Above 80 lb	November 15 – March 15	C	20
Breeding does:			
50–80 lb	July 15 – November 15	A	20
	November 15 – March 15	B	32–50
Above 80 lb	July 15 – November 15	A	5–10
	November 15 – March 15	C	20–30
Growing kids and yearlings:			
Below 40 lb	July 15 – November 15	A	35
	November 15 – March 15	B	40
Above 40 lb	July 15 – November 15	A	20
	November 15 – March 15	B	40
Developing billies:			
80–120 lb	July 15 – November 15	A	40
	November 15 – March 15	B	50

[1]Adapted by the authors from *Nutrient Requirements of the Angora Goat* by J. E. Huston, M. Shelton, and W. C. Ellis, Tex. Ag. Exp. Sta. Bull. B-1105.

[2]Compositions of the recommended supplements are listed in Table 22–15.

NOTE WELL: The same rations and feeding practices presented herein for Angora goats are suited for Spanish goats, also. However, supplemental feeding of meat goats is not the norm. Yet, Spanish goats do respond well to supplemental feeding during critical periods—just before breeding, just before and following kidding, and when range feed is short.

Fig. 22–22. Typical multicolored Spanish goats. (Courtesy, USDA)

TABLE 22–17
FEEDING REGIMEN FOR ANGORA GOATS ON EXTREMELY POOR RANGE[1]

Class	Hay Plus Concentrate			Concentrates Only			
	Hay	Concentrate[2]		Supplement[3]		Grain	
	(% body wt.)	(lb/head/day)	*(kg/head/day)*	(lb/head/day)	*(kg/head/day)*	(lb/head/day)	*(kg/head/day)*
Wethers and dry does	1	0.75	*0.34*	0.3	*0.14*	1.0	*0.45*
Breeding does:							
Dry	1	0.75	*0.34*	0.3	*0.14*	1.0	*0.45*
Pregnant	1	1.50	*0.68*	0.3	*0.14*	1.6	*0.73*
Lactating	1	1.50	*0.68*	0.3	*0.14*	2.0	*0.91*
Growing kids and yearlings	1	1.25	*0.57*	0.3	*0.14*	1.0	*0.45*
Developing billies	1	1.75	*0.79*	0.2	*0.09*	1.5	*0.68*

[1]Adapted by the authors from *Nutrient Requirements of the Angora Goat* by J. E. Huston, M. Shelton, and W. C. Ellis, Tex. Ag. Exp. Sta. Bull. B-1105.

[2]Type of concentrate depends on hay fed. If alfalfa (or other quality legume hay) is fed, corn is adequate. If grass hay is used, a supplement similar to Supplement A (Table 22–15) should be fed.

[3]Supplement should be similar to Supplement C (Table 22–15).

FEEDING DOES

On good range in spring, mature dry does will consume enough feed to satisfy all their nutrient demands except salt and phosphorus. During lactation, they may need ½ to ¾ lb of a supplement of type A as given in Table 22–15. (The total ration, supplement plus range grass, will usually run about 11 to 12% protein.)

During the summer and early fall, the quality of range feed is reduced and a higher protein supplement, such as B (Table 22–15), should be provided at the rate of 1 lb up to 10 does.

It may also be advisable to include some trace mineralized salt. Immature does (yearlings and those not fully developed) should be provided 1 lb of supplement for each 5 does. If the range is exceptionally poor, double the amount of supplement.

In late fall and winter, ranges tend to be at their lowest nutritive value. Poor ranges require supplemental feeding—supplement C, Table 22–15—at levels of 1 lb for each 3 to 5 mature does and 1 lb for each 1 to 3 yearling or underdeveloped does. These supplements should take care of the needs of late pregnancy and of lactation. For ranges with new growth, supplements should follow the recommendations given previously for good spring range.

Does with more than 2 kids should be given 25 to 50% more of the supplement during lactation than is recommended for does with singles or twins. Does which have a history of giving birth to more than 1 kid should be fed at a high rate of supplementation at least 3 weeks before expected parturition (kidding).

Additional daily supplementation of ¼ to ⅓ lb of grain or range cubes should be fed to does 1 to 2 weeks prior to turning the bucks in for the breeding season. This added feed (called flushing) improves conception rate by having does in a positive nutrient balance during breeding. When the practice of flushing is to be used, feed should be increased gradually. Likewise, at the end of the breeding season, the feed allowance should be reduced gradually, so as to avoid upsetting the appetite and the digestive tract.

- **Supplemental feeding with salt as governor**—In large, rough, and sometimes brushy pastures, hand feeding is impractical, so grain and meal mixtures with salt as an inhibitor are fed in self-feeders. The proportion of concentrates to salt varies with the amount of feed the goats should consume; usually it is 3:1, 4:1, or 5:1, with the lower amounts of salt allowing the higher feed intake.

- **Supplemental feeding of Spanish does**—Although Spanish goats are hardy and will survive on poor brushy range with little or no supplement, the better operators provide a supplemental feed to does of ¼ to ½ lb of cottonseed cake or ⅓ to ⅔ lb of yellow corn per head per day during the following critical periods: (1) beginning 3 weeks prior to breeding (for flushing), (2) beginning 3 weeks prior to kidding, (3)when kids are very young, and (4) when feed is short.

FEEDING KIDS (NURSING AND WEANED)

As long as kids are receiving adequate amounts of milk from their mothers, they do very well on good range. Additional supplementation, however, makes for more rapid growth and better prepares them for market or breeding. One pound of supplement for each 2½ to 3 kids should be provided. Older and larger kids may have their supplement reduced to 1 lb daily for each 5 kids if the range conditions are good. When range is poor, the grain and supplement should be increased to provide 1 to 1⅓ lb of grain daily per kid. In addition, kids should have access to good-quality hay.

Fig. 22–23. Two lamb carcasses (left) and two Spanish goat carcasses (known as chevon) in a meat packing plant in San Angelo, Tex. Note that the goat carcasses have less external fat. Supplemental feeding of kids makes for more rapid growth and better prepares them for market. (Courtesy, Texas A&M University, College Station)

A suitable ration for kids and yearlings is given in Table 22–18.

TABLE 22–18
RATION FOR ANGORA AND SPANISH KIDS AND YEARLINGS

Age	Ingredients	Amount
		(%)
Kids/yearlings	Alfalfa hay	32
	Cottonseed hulls	28
	Sorghum grain	18
	Barley grain	8
	Molasses	6
	Cottonseed oilmeal	6
	Salt/mineral mix	2
	Total	100

FEEDING YEARLINGS AND REPLACEMENTS

Yearling range goats fit into one of two categories: (1) those which are being retained for the breeding herd, or (2) those which are being prepared for market. Replacement does and bucks should be fed rations that allow for growth. Those which are being prepared for market should be fed so as to put on more flesh, because goats tend to be leaner than most meat animals.

To keep a replacement animal growing, it is advisable to provide good-quality forage with a minimum amount of grains in the supplement. They should be given adequate protein, minerals, and vitamin A. Playfulness is a sign of good goat health and vigor. When fed properly, they will be large enough to breed as yearlings, without interfering with their continued growth to mature size.

FEEDING WETHERS AND DRY DOES

Under present conditions, Angora wethers are kept for several years for mohair production. Hence, they are rarely fed for meat production. For those goats being prepared for market, supplemental grain should be increased gradually. If kids are to be marketed at weaning (4 to 5 months of age), it is unnecessary to castrate them. If they are to be held over until yearlings, they can remain uncastrated until about 6 to 9 months of age.

Both wethers (or muttons, as they are sometimes called) and dry does do well without supplemental feed unless the range is poor. Angora does and wethers are generally not allowed to run together. Wethers can graze higher on brush and more vigorously than does, thereby reducing the available browse for does that have a more critical need for nutrients. When supplementation is needed, dry does and wethers are fed about the same amount of grain or supplement—1 lb for each 5 animals during the summer and fall months and twice this amount during the winter. On poor range, the amount may be increased to ¾ lb per head daily when hay is also fed and to 1⅓ lb per head daily when no hay is fed.

FEEDING BUCKS

Young bucks being raised for breeding purposes should be grown in much the same manner as replacement does, except that they will require more feed because their growth is more rapid. They should be separated from the does soon after weaning. If young bucks are permitted to remain with the does, early sexual maturity in the bucks may result in their breeding young females not intended to be bred. Bucks should be growthy, but they should not be permitted to put on extra fat which may impede their muscular, skeletal, and sexual development. They may be used lightly as yearlings when properly grown out; but they should not be used if they have been on sparse feed and are not prepared for the stresses of heavy breeding.

Whether the bucks are young or mature, it is advisable to feed them ⅓ to ½ lb of grain or supplement daily per head for a week or so before turning them in with the does, with the supplementation continued throughout the breeding season. If the bucks get too fat or become inactive in their mating habits, grain can be withdrawn as one means of improving their effectiveness in breeding.

Bucks that are not in active service do not need supplemental grain or concentrates unless the range feed becomes too sparse or too mature. In the latter case, they may be given about 1 lb of supplement for each 2 to 2½ animals daily. This type of supplementation is expecially important for young bucks that are still growing. Some of the older bucks can get along well without supplementation unless the range is extremely poor.

Fig. 22-24. An outstanding Angora buck. (Courtesy, Texas A&M University, College Station)

QUESTIONS FOR STUDY AND DISCUSSION

1. Can experimental work and instruction relative to goats be justified on the basis of their contributions in terms of animal numbers, meat, milk, and fiber; (a) worldwide, and (b) in the United States?

2. Name and describe the three major types of goats, and tell how they are similar and how they are different from the standpoint of nutrition/feeding.

3. Discuss the economic importance of feed for goats.

4. Angora goats are supplementally fed on a limited basis whereas Spanish goats are seldom supplemented at all. Does this mean that it would not pay to supplement Spanish goats during critical periods—just before breeding, before and after kidding, and during feed scarcity?

5. Since cattle, sheep, and goats are all ruminants, why go to the trouble and expense of evolving with separate nutritional requirements for goats?

6. Why did the National Research Council (NRC) evolve with different nutrient requirements of goats for maintenance, various levels of activity, late pregnancy, growth, milk production, and mohair?

7. What are the signs of energy deficiency?

8. Discuss microbial synthesis in goats, and explain how it makes it possible for goats to utilize both protein and nonprotein nitrogen.

9. Discuss what happens to goats when they are fed protein-deficient rations.

10. Give the functions, deficiency symptoms, and practical sources of each of the following minerals for goats: salt, calcium, phosphorus, cobalt, iodine, and selenium.

11. Give the functions, deficiency symptoms, and practical sources of each of the following vitamins for goats: vitamin A, vitamin D, and vitamin E.

12. What are the important ways of meeting the water requirements of goats, and what are the major avenues of water losses from the body?

13. Define (a) browse, and (b) forbs. If goats can effectively use grasses and legumes, why provide them with browse and forbs?

14. What are the advantages to dividing goat pastures into lots and rotating their use every 10 to 14 days?

15. Discuss the effects of stage of maturity and cutting on hay for goats.

16. Why haven't silages and haylages been used extensively for goats?

17. What precautions should be taken when feeding silage and root crops to dairy goats?

18. How much molasses can safely be incorporated into a goat ration? Why is it not advisable to incorporate more?

19. What is the maximum recommended amount of nonprotein nitrogen that can be added to goat feeds?

20. What's a milk replacer? How and why may a milk replacer be used for dairy kids?

21. What determines the concentration of salt in rations that are to be offered free-choice to goats?

22. Discuss the cause and control/prevention of each of the following nutrition related disorders: abortion, caprine arthritis-encephalitis, enterotoxemia, and posthitis.

23. Outline a dairy goat feeding guide.

24. Discuss the nutritional demands upon lactating dairy does, and tell how they may be met.

25. What are the reasons for drying off dairy does 6 weeks before kidding?

26. List and discuss briefly pertinent points relative to feeding dairy kids.

27. How much supplement should be given to dairy goat bucks, and when should it be given?

28. Discuss the feeding of Angora and Spanish does.

SELECTED REFERENCES

Title of Publication	Author(s)	Publisher
Dairy Goat, The, Bull. No. 1160	W. F. Brannon	Cornell University Extension, Ithaca, N.Y., 1967
Dairy Goats, Pub. No. 439	B. E. Colby *et al.*	University of Massachusetts Cooperative Extension, Amherst, Mass., 1966
Dairy Goat Management, Bull. No. 334	G. W. VanderNoot D. M. Kniffen	Rutgers University Extension, New Brunswick, N.J.
Extension Goat Handbook	G. F. W. Haenlein D. L. Ace	Extension Service, USDA, Washington, D.C., 1984
Feeds and Feeding, 22nd Edition	F. B. Morrison	Morrison Publishing Co., Ithaca, N.Y., 1956
Feeds and Feeding, Abridged	F. B. Morrison	Morrison Publishing Co., Ithaca, N.Y., 1958
Goat Production	C. Gall	Academic Press, New York, N.Y., 1981
Goat Production in the Tropics	C. Devendra M. Burns	Unwin Brothers Limited, Old Woking, Surrey, England, 1983
Nutrient Requirements of Goats, No. 15	National Research Council	National Academy Press, Washington, D.C., 1981
Nutritional Requirements of the Angora Goat, Bull. B-1105	J. E. Huston M. Shelton W. C. Ellis	Texas Agricultural Experiment Station, San Angelo, Tex., 1971
Observations on the Goat	M. H. French	FAO of the United Nations, Rome, Italy, 1970
Sheep and Goat Handbook	The International Stockmen's School staff	Agriservices Foundation, Clovis, Calif., annual since 1980
Sheep and Goat Science, 5th Edition	M. E. Ensminger R. O. Parker	The Interstate Printers & Publishers, Inc., Danville, Ill., 1986
Texas Angora Goat Production, Bull. B-296		Texas Agricultural Extension Service, San Angelo, Tex., 1970
U.S. Sheep and Goat Industry, The	C. S. Menzies, Chairman, Task Force	Council for Agricultural Science and Technology (CAST), Report No. 94, Ames, Iowa, 1982

AMINO ACIDS . . . the building blocks of proteins of which there are 23. The pig has a specific requirement for each of the 10 essential amino acids.

Original painting by Tom Phillips

Chapter 23

FEEDING SWINE[1]

Contents — Page

(Continued)

[1]The authors gratefully acknowledge the assistance of the following authorities who contributed helpful information and suggestions for this chapter: R. A. Easter, Ph.D., Swine Nutritionist, Department of Animal Sciences, College of Agriculture, University of Illinois, Urbana; V. C. Speer, Ph.D., Swine Nutritionist, Department of Animal Science, Iowa State University of Science and Technology, Ames; and R. F. Wilson, Ph.D., Swine Nutritionist, Department of Animal Science, The Ohio State University, Columbus.

Contents (Continued) Page

Contents Page

In the natural state, the wild boar and his kind and kin roved through the forests, gleaning the feeds provided by nature. On a modern farm, the range is restricted and frequently entirely devoid of vegetation. More than half of the hogs marketed in this country are raised from farrow to finish in some type of confinement system—ranging all the way from simple shelters to environmentally controlled pig palaces. As a result of this confinement, domestic swine have less choice in their selection of feed than any other class of four-footed animals. For the most part they are able to consume only what the caretaker provides. This consists largely of concentrate feeds with only a small proportion of roughage. These conditions are made more critical because hogs grow much faster in proportion to their body weight than the larger farm animals, and they produce young at an earlier age. Thus, a knowledge of the nutritional needs of swine is especially important.

Fig. 23-1. Crossbred finishing hogs in an open-front finishing building with an outside concrete apron. (Courtesy, E. J. Stevermer, Iowa State University, Ames)

ECONOMIC IMPORTANCE OF FEED FOR SWINE

Knowledge of feeding swine is important from an economic standpoint, because feed accounts for approximately 65 to 75% of the total cost of producing pork. For this reason, every swine producer should endeavor to provide rations that are both satisfactory and inexpensive—rations that make for the maximum production of quality pork per unit of feed consumed—and at least cost.

Also, extensive surveys indicate that 15% of all sows bred fail to give birth to young, and that 25 to 30% of all pigs farrowed fail to live to weaning age. Although reproductive failure is heavy, baby pig losses are due to many and variable factors, certainly nutritional deficiencies play a major role.

NUTRITIVE NEEDS OF SWINE

The nutrient needs of swine are influenced by age, function, disease level, nutrient interaction, and environment. It has been established that the pig has a requirement for over 40 different nutrients. Fortunately, not all of them are of practical concern.

National Research Council (NRC) Requirements

The NRC nutritional requirements, or standards, are presented in Tables 23-1 to 23-7. It is not intended that these standards impart the impression that such figures are absolute, final, and unchangeable. Rather, they should be used as guides based on research. Also, these figures are, for the most part, requirements (rather than allowances); hence, they do not provide for margins of safety to compensate for variations in feed compositions, environment, and possible losses of nutrients during storage or processing.

In using Tables 23-1 to 23-7, the following additional points should be recognized:

1. Feedstuffs produced in various parts of the country vary in nutritive value.

2. The environment in which pigs are produced can modify the requirements.

3. Animals bred for high performance have nutritional needs that are quite different from those of average performers.

TABLE 23-1
DAILY NUTRIENT INTAKES AND REQUIREMENTS OF SWINE ALLOWED FEED *AD LIBITUM*[1]

	Swine Liveweight									
Intake and Performance Levels — Lb / *Kg*	**2.2–11** *1–5*		**11–22** *5–10*		**22–44** *10–20*		**44–110** *20–50*		**110–242** *50–110*	
	(lb)	*(g)*	(lb)	*(g)*	(lb)	*(g)*	(lb)	*(g)*	(lb)	*(g)*
Expected weight gain per day	0.4	*200*	0.6	*250*	1.0	*450*	1.5	*700*	1.8	*820*
Expected feed intake per day	0.6	*250*	1.0	*460*	2.1	*950*	4.2	*1,900*	6.9	*3,110*
Expected efficiency (gain/feed)	0.800		0.543		0.474		0.368		0.264	
Expected efficiency (feed/gain)	1.25		1.84		2.11		2.71		3.79	
Digestible energy intake (kcal per day)	850		1,560		3,230		6,460		10,570	
Metabolizable energy intake (kcal per day)	805		1,490		3,090		6,200		10,185	
Energy concentration (kcal ME per lb ration)	1,461		1,470		1,474		1,479		1,486	
Energy concentration *(kcal ME per kg ration)*	*3,220*		*3,240*		*3,250*		*3,260*		*3,275*	
Protein per day	0.1	*60*	0.2	*92*	0.4	*171*	0.6	*285*	0.9	*404*

Nutrient	**Requirement (Amount Per Day)**				
	2.2–11 lb (1–5 kg)	11–22 lb (5–10 kg)	22–44 lb (10–20 kg)	44–110 lb (20–50 kg)	110–242 lb (50–110 kg)
Indispensable amino acids:					
Arginine (g)	1.5	2.3	3.8	4.8	3.1
Histidine (g)	0.9	1.4	2.4	4.2	5.6
Isoleucine (g)	1.9	3.0	5.0	8.7	11.8
Leucine (g)	2.5	3.9	6.6	11.4	15.6
Lysine (g)	3.5	5.3	9.0	14.3	18.7
Methionine + cystine (g)	1.7	2.7	4.6	7.8	10.6
Phenylalanine + tyrosine (g)	2.8	4.3	7.3	12.5	17.1
Threonine (g)	2.0	3.1	5.3	9.1	12.4
Tryptophan (g)	0.5	0.8	1.3	2.3	3.1
Valine (g)	2.0	3.1	5.3	9.1	12.4
Linoleic acid (g)	0.3	0.5	1.0	1.9	3.1
Major or macrominerals:					
Calcium (g)	2.2	3.7	6.6	11.4	15.6
Chlorine (g)	0.2	0.4	0.8	1.5	2.5
Magnesium (g)	0.1	0.2	0.4	0.8	1.2
Phosphorus, total (g)	1.8	3.0	5.7	9.5	12.4
Phosphorus, available (g)	1.4	1.8	3.0	4.4	4.7
Potassium (g)	0.8	1.3	2.5	4.4	5.3
Sodium (g)	0.2	0.5	1.0	1.9	3.1
Trace or microminerals:					
Copper (mg)	1.50	2.76	4.75	7.60	9.33
Iodine (mg)	0.04	0.06	0.13	0.27	0.44
Iron (mg)	25	46	76	114	124
Manganese (mg)	1.00	1.84	2.85	3.80	6.22
Selenium (mg)	0.08	0.14	0.24	0.28	0.31
Zinc (mg)	25	46	76	114	155
Fat-soluble vitamins:					
Vitamin A (IU)	550	1,012	1,662	2,470	4,043
Vitamin D (IU)	55	101	190	285	466
Vitamin E (IU)	4	7	10	21	34
Vitamin K (menadione) (mg)	0.02	0.02	0.05	0.10	0.16
Water-soluble vitamins:					
Biotin (mg)	0.02	0.02	0.05	0.10	0.16
Choline (g)	0.15	0.23	0.38	0.57	0.93
Folacin (Folic acid) (mg)	0.08	0.14	0.28	0.57	0.93
Niacin (Nicotinic acid, Nicotinamide), available (mg)	5.00	6.90	11.88	19.00	21.77
Pantothenic acid (Vitamin B-3) (mg)	3.00	4.60	8.55	15.20	21.77
Riboflavin (Vitamin B-2) (mg)	1.00	1.61	2.85	4.75	6.22
Thiamin (Vitamin B-1) (mg)	0.38	0.46	0.95	1.90	3.11
Vitamin B-6 (Pyridoxine, Pyridoxal, Pyridoxamine) (mg)	0.50	0.69	1.42	1.90	3.11
Vitamin B-12 (Cobalamins) (mcg)	5.00	8.05	14.25	19.00	15.55

[1]Adapted by the authors from *Nutrient Requirements of Swine,* 9th rev. ed., National Research Council, National Academy Press, 1988, p. 51, Table 5-2.

TABLE 23-2
DAILY NUTRIENT INTAKES AND REQUIREMENTS OF INTERMEDIATE-WEIGHT BREEDING ANIMALS[1]

	Mean Gestation or Farrowing Weight of:			
	Bred Gilts, Sows, and Adult Boars		Lactating Gilts and Sows	
Lb	358.3		363.8	
Intake and Performance Levels *Kg*	*162.5*		*165.0*	
	(lb)	*(kg)*	(lb)	*(kg)*
Daily feed intake	4.2	*1.9*	11.7	*5.3*
Digestible energy (Mcal per day)	6.3		17.7	
Metabolizable energy (Mcal per day)	6.1		17.0	
		(g)		*(g)*
Crude Protein per day	0.5	*228*	1.5	*689*

Nutrient	Requirement (Amount Per Day)	
Indispensable amino acids:		
Arginine (g)	0.0	21.2
Histidine (g)	2.8	13.2
Isoleucine (g)	5.7	20.7
Leucine (g)	5.7	25.4
Lysine (g)	8.2	31.8
Methionine + cystine (g)	4.4	19.1
Phenylalanine + tyrosine (g)	8.6	37.1
Threonine (g)	5.7	22.8
Tryptophan (g)	1.7	6.4
Valine (g)	6.1	31.8
Linoleic acid (g)	1.9	5.3
Major or macrominerals:		
Calcium (g)	14.2	39.8
Chlorine (g)	2.3	8.5
Magnesium (g)	0.8	2.1
Phosphorus, total (g)	11.4	31.8
Phosphorus, available (g)	6.6	18.6
Potassium (g)	3.8	10.6
Sodium (g)	2.8	10.6
Trace or microminerals:		
Copper (mg)	9.5	26.5
Iodine (mg)	0.3	0.7
Iron (mg)	152	424
Manganese (mg)	19	53
Selenium (mg)	0.3	0.8
Zinc (mg)	95	265
Fat-soluble vitamins:		
Vitamin A (IU)	7,600	10,600
Vitamin D (IU)	380	1,060
Vitamin E (IU)	42	117
Vitamin K (menadione) (mg)	1.0	2.6
Water-soluble vitamins:		
Biotin (mg)	0.4	1.1
Choline (g)	2.4	5.3
Folacin (Folic acid) (mg)	0.6	1.6
Niacin (Nicotinic acid, Nicotinamide), available (mg)	19.0	53.0
Pantothenic acid (Vitamin B-3) (mg)	22.8	63.6
Riboflavin (Vitamin B-2) (mg)	7.1	19.9
Thiamin (Vitamin B-1) (mg)	1.9	5.3
Vitamin B-6 (Pyridoxine, Pyridoxal, Pyridoxamine) (mg)	1.9	5.3
Vitamin B-12 (Cobalamins) (mcg)	28.5	79.5

[1]Adapted by the authors from *Nutrient Requirements of Swine,* 9th rev. ed., National Research Council, National Academy Press, 1988, p. 52, Table 5-4.

TABLE 23-3
DAILY ENERGY AND FEED REQUIREMENTS OF PREGNANT GILTS AND SOWS[1]

	Weight of Bred Gilts and Sows at Mating[2]					
Lb	265		309		353	
Intake and Performance Levels *Kg*	*120*		*140*		*160*	
	(lb)	*(kg)*	(lb)	*(kg)*	(lb)	*(kg)*
Mean gestation weight[3]	314.2	*142.5*	358.3	*162.5*	402.3	*182.5*
Energy required:						
Maintenance[4] (Mcal DE per day)	4.53		5.00		5.47	
Gestation weight gain[5] (Mcal DE per day)	1.29		1.29		1.29	
Total (Mcal DE per day)	5.82		6.29		6.76	
Feed required per day[6]	4.0	*1.8*	4.2	*1.9*	4.4	*2.0*

[1]Adapted by the authors from *Nutrient Requirements of Swine,* 9th rev. ed., National Research Council, National Academy Press, 1988, p. 53, Table 5-6.

[2]Requirements are based on 55-lb *(25-kg)* maternal weight gain plus 44-lb *(20-kg)* increase in weight due to the products of conception; the total weight gain is 99 lb *(45 kg).*

[3]Mean gestation weight is weight at mating + (total weight gain/2).

[4]The animal's daily maintenance requirement is 110 kcal of DE/kg$^{0.75}$.

[5]The gestation weight gain is 1.10 Mcal of DE/day for maternal weight gain plus 0.19 Mcal of DE/day for conceptus gain.

[6]The feed required/day is based on a corn-soybean meal ration containing 3.34 Mcal of DE/kg.

TABLE 23-4
DAILY ENERGY AND FEED REQUIREMENTS OF LACTATING GILTS AND SOWS[1]

	Weight of Lactating Gilts and Sows at Postfarrowing					
Lb	320		364		408	
Intake and Performance Levels *Kg*	*145*		*165*		*185*	
	(lb)	*(kg)*	(lb)	*(kg)*	(lb)	*(kg)*
Milk yield	11.0	*5.0*	13.78	*6.25*	16.5	*7.5*
Energy required:						
Maintenance[2] (Mcal DE per day)	4.5		5.0		5.5	
Milk production[3] (Mcal DE per day)	10.0		12.5		15.0	
Total (Mcal DE per day)	14.5		17.5		20.5	
Feed required per day[4]	9.7	*4.4*	11.7	*5.3*	13.5	*6.1*

[1]Adapted by the authors from *Nutrient Requirements of Swine,* 9th rev. ed., National Research Council, National Academy Press, 1988, p. 53, Table 5-7.

[2]The animal's daily maintenance requirement is 110 kcal of DE/kg$^{0.75}$.

[3]Milk production requires 2.0 Mcal of DE/kg of milk.

[4]The feed required/day is based on a corn-soybean meal ration containing 3.34 Mcal of DE/kg.

Fig. 23-2. Lactating sows require a liberal allowance of a ration rich in energy, protein, minerals, and vitamins. (Courtesy, Hampshire Swine Registry, Peoria, Ill.)

TABLE 23-5
NUTRIENT REQUIREMENTS IN THE RATION OF SWINE ALLOWED FEED *AD LIBITUM* (90% DRY MATTER)[1]

	Swine Liveweight									
Intake and Performance Levels Lb *Kg*	**2.2–11** ***1–5***		**11–22** ***5–10***		**22–44** ***10–20***		**44–110** ***20–50***		**110–242** ***50–110***	
	(lb)	***(g)***	**(lb)**	***(g)***	**(lb)**	***(g)***	**(lb)**	***(g)***	**(lb)**	***(g)***
Expected weight gain per day	0.4	*200*	0.6	*250*	1.0	*450*	1.5	*700*	1.8	*820*
Expected feed intake per day	0.6	*250*	1.0	*460*	2.1	*950*	4.2	*1,900*	6.9	*3,110*
Expected efficiency (gain/feed)	0.800		0.543		0.474		0.368		0.264	
Expected efficiency (feed/gain)	1.25		1.84		2.11		2.71		3.79	
Digestible energy intake (kcal per day)	850		1,560		3,230		6,460		10,570	
Metabolizable energy intake (kcal per day)	805		1,490		3,090		6,200		10,185	
Energy concentration (kcal ME per lb ration)	1,461		1,470		1,474		1,479		1,486	
Energy concentration *(kcal ME per kg ration)*	*3,220*		*3,240*		*3,250*		*3,260*		*3,275*	
Protein (%)	24		20		18		15		13	
Nutrient	**Requirement (Percent or Amount/Lb *[Kg]* Ration)[2]**									
	(lb)	***(kg)***	**(lb)**	***(kg)***	**(lb)**	***(kg)***	**(lb)**	***(kg)***	**(lb)**	***(kg)***
Indispensable amino acids:										
Arginine (%)	0.60		0.50		0.40		0.25		0.10	
Histidine (%)	0.36		0.31		0.25		0.22		0.18	
Isoleucine (%)	0.76		0.65		0.53		0.46		0.38	
Leucine (%)	1.00		0.85		0.70		0.60		0.50	
Lysine (%)	1.40		1.15		0.95		0.75		0.60	
Methionine + cystine (%)	0.68		0.58		0.48		0.41		0.34	
Phenylalanine + tyrosine (%)	1.10		0.94		0.77		0.66		0.55	
Threonine (%)	0.80		0.68		0.56		0.48		0.40	
Tryptophan (%)	0.20		0.17		0.14		0.12		0.10	
Valine (%)	0.80		0.68		0.56		0.48		0.40	
Linoleic acid (%)	0.1		0.1		0.1		0.1		0.1	
Major or macrominerals:										
Calcium (%)	0.90		0.80		0.70		0.60		0.50	
Chlorine (%)	0.08		0.08		0.08		0.08		0.08	
Magnesium (%)	0.04		0.04		0.04		0.04		0.04	
Phosphorus, total (%)	0.70		0.65		0.60		0.50		0.40	
Phosphorus, available (%)	0.55		0.40		0.32		0.23		0.15	
Potassium (%)	0.30		0.28		0.26		0.23		0.17	
Sodium (%)	0.10		0.10		0.10		0.10		0.10	
Trace or microminerals:										
Copper (mg)	2.7	*6.0*	2.7	*6.0*	2.3	*5.0*	1.8	*4.0*	1.4	*3.0*
Iodine (mg)	0.06	*0.14*	0.06	*0.14*	0.06	*0.14*	0.06	*0.14*	0.06	*0.14*
Iron (mg)	45	*100*	45	*100*	36	*80*	27	*60*	18	*40*
Manganese (mg)	1.8	*4.0*	1.8	*4.0*	1.4	*3.0*	0.9	*2.0*	0.9	*2.0*
Selenium (mg)	0.14	*0.30*	0.14	*0.30*	0.11	*0.25*	0.07	*0.15*	0.05	*0.10*
Zinc (mg)	45	*100*	45	*100*	36	*80*	27	*60*	23	*50*
Fat-soluble vitamins:										
Vitamin A (IU)	2,200		2,200		1,750		1,300		1,300	
Vitamin D (IU)	220		220		200		150		150	
Vitamin E (IU)	16		16		11		11		11	
Vitamin K (menadione) (mg)	0.2	*0.5*	0.2	*0.5*	0.2	*0.5*	0.2	*0.5*	0.2	*0.5*
Water-soluble vitamins:										
Biotin (mg)	0.04	*0.08*	0.02	*0.05*	0.02	*0.05*	0.02	*0.05*	0.02	*0.05*
Choline (g)	0.27	*0.6*	0.23	*0.5*	0.18	*0.4*	0.14	*0.3*	0.14	*0.3*
Folacin (Folic acid) (mg)	0.1	*0.3*	0.1	*0.3*	0.1	*0.3*	0.1	*0.3*	0.1	*0.3*
Niacin (Nicotinic acid, Nicotinamide), available (mg)	9.1	*20.0*	6.8	*15.0*	5.7	*12.5*	4.5	*10.0*	3.2	*7.0*
Pantothenic acid (Vitamin B-3) (mg)	5.4	*12.0*	4.5	*10.0*	4.1	*9.0*	3.6	*8.0*	3.2	*7.0*
Riboflavin (Vitamin B-2) (mg)	1.8	*4.0*	1.6	*3.5*	1.4	*3.0*	1.1	*2.5*	0.9	*2.0*
Thiamin (Vitamin B-1) (mg)	0.7	*1.5*	0.5	*1.0*	0.5	*1.0*	0.5	*1.0*	0.5	*1.0*
Vitamin B-6 (Pyridoxine, Pyridoxal, Pyridoxamine) (mg)	0.9	*2.0*	0.7	*1.5*	0.7	*1.5*	0.5	*1.0*	0.5	*1.0*
Vitamin B-12 (Cobalamins) (mcg)	9.1	*20.0*	7.9	*17.5*	6.8	*15.0*	4.5	*10.0*	2.3	*5.0*

[1]Adapted by the authors from *Nutrient Requirements of Swine,* 9th rev. ed., National Research Council, National Academy Press, 1988, p. 50, Table 5-1.

[2]These requirements are based upon the following types of pigs and rations: 2.2- to 11-lb *(1- to 5-kg)* pigs, a ration that includes 25 to 75% milk products; 11- to 22-lb *(5- to 10-kg)* pigs, a corn-soybean meal ration that includes 5 to 25% milk products; 22- to 242-lb *(10- to 110-kg)* pigs, a corn-soybean meal ration. In the corn-soybean meal rations, the corn contains 8.5% protein; the soybean meal contains 44%.

TABLE 23-6
NUTRIENT REQUIREMENTS IN THE RATION OF BREEDING SWINE[1]

Intake Levels	Bred Gilts, Sows, and Adult Boars		Lactating Gilts and Sows	
Digestible energy ...... (kcal per lb ration)	1,515		1,515	
Digestible energy ..... *(kcal per kg ration)*	*3,340*		*3,340*	
Metabolizable energy .... (kcal per lb ration)	1,456		1,456	
Metabolizable energy .... *(kcal per kg ration)*	*3,210*		*3,210*	
Crude Protein .... (%)	12		13	
Nutrient	**Requirement (Percent or Amount/Lb *[Kg]* Ration)[2]**			
Indispensable amino acids:	**(lb)**	***(kg)***	**(lb)**	***(kg)***
Arginine .... (%)	0.00		0.40	
Histidine .... (%)	0.15		0.25	
Isoleucine .... (%)	0.30		0.39	
Leucine .... (%)	0.30		0.48	
Lysine .... (%)	0.43		0.60	
Methionine + cystine .... (%)	0.23		0.36	
Phenylalanine + tyrosine .... (%)	0.45		0.70	
Threonine .... (%)	0.30		0.43	
Tryptophan .... (%)	0.09		0.12	
Valine .... (%)	0.32		0.60	
Linoleic acid .... (%)	0.1		0.1	
Major or macrominerals:				
Calcium .... (%)	0.75		0.75	
Chlorine .... (%)	0.12		0.16	
Magnesium .... (%)	0.04		0.04	
Phosphorus, total .... (%)	0.60		0.60	
Phosphorus, available .... (%)	0.35		0.35	
Potassium .... (%)	0.20		0.20	
Sodium .... (%)	0.15		0.20	
Trace or microminerals:				
Copper .... (mg)	2.27	*5.00*	2.27	*5.00*
Iodine .... (mg)	0.06	*0.14*	0.06	*0.14*
Iron .... (mg)	36.29	*80.00*	36.29	*80.00*
Manganese .... (mg)	4.54	*10.00*	4.54	*10.00*
Selenium .... (mg)	0.07	*0.15*	0.07	*0.15*
Zinc .... (mg)	22.68	*50.00*	22.68	*50.00*
Fat-soluble vitamins:				
Vitamin A .... (IU)	4,000		2,000	
Vitamin D .... (IU)	200		200	
Vitamin E .... (IU)	22		22	
Vitamin K (menadione) .... (mg)	0.23	*0.50*	0.23	*0.50*
Water-soluble vitamins:				
Biotin .... (mg)	0.09	*0.20*	0.09	*0.20*
Choline .... (g)	0.57	*1.25*	0.45	*1.00*
Folacin (Folic acid) .... (mg)	0.14	*0.30*	0.14	*0.30*
Niacin (Nicotinic acid, Nicotinamide), available .... (mg)	4.54	*10.00*	4.54	*10.00*
Pantothenic acid (Vitamin B-3) .... (mg)	5.44	*12.00*	5.44	*12.00*
Riboflavin (Vitamin B-2) .... (mg)	1.70	*3.75*	1.70	*3.75*
Thiamin (Vitamin B-1) .... (mg)	0.45	*1.00*	0.45	*1.00*
Vitamin B-6 (Pyridoxine, Pyridoxal, Pyridoxamine) .... (mg)	0.45	*1.00*	0.45	*1.00*
Vitamin B-12 (Cobalamins) .... (mcg)	6.80	*15.00*	6.80	*15.00*

[1]Adapted by the authors from *Nutrient Requirements of Swine,* 9th rev. ed., National Research Council, National Academy Press, 1988, p. 52, Table 5-3.

[2]These requirements are based upon corn-soybean meal rations, feed intakes, and performance levels listed in Tables 23-2, 23-3, and 23-4. In the corn-soybean meal rations, the corn contains 8.5% protein; the soybean meal contains 44%.

TABLE 23-7
REQUIREMENTS FOR SEVERAL NUTRIENTS
OF BREEDING HERD REPLACEMENTS ALLOWED FEED *AD LIBITUM*[1]

	Weight Of:			
	Developing Gilts		Developing Boars	
Lb	44–110	110–242	44–110	110–242
Intake Levels *Kg*	*20–50*	*50–110*	*20–50*	*50–110*
Energy concentration ... (kcal ME per lb ration)	1,476	1,479	1,470	1,476
Energy concentration ... *(kcal ME per kg ration)*	*3,255*	*3,260*	*3,240*	*3,255*
Crude protein .... (%)	16	15	18	16
Nutrient:[2]				
Lysine .... (%)	0.80	0.70	0.90	0.75
Calcium .... (%)	0.65	0.55	0.70	0.60
Phosphorus, total .... (%)	0.55	0.45	0.60	0.50
Phosphorus, available .. (%)	0.28	0.20	0.33	0.25

[1]Adapted by the authors from *Nutrient Requirements of Swine,* 9th rev. ed., National Research Council, National Academy Press, 1988, p. 53, Table 5-5.

[2]Sufficient data are not available to indicate that requirements for other nutrients are different from those in Table 23-5 for animals of these weights.

Recommended Nutrient Allowances

No one nutrient is more important than another—and all are essential. Each nutrient has one or more particular and specific function to perform in the body. If the nutrient is not supplied by the ration in proper amounts, the functions (growth, reproduction, lactation) will be impaired. Since the modern swine producer is interested in maximum performance, it follows that the input of nutrients must be ample to bring this about. For this reason, the recommended allowances of the various important nutrients are higher than the minimum requirements established by the National Research Council (Tables 23-1 to 23-7). This is done in order to obtain maximum performance, and to reduce the risk of nutrient deficiencies that might occur because of differences in ingredient quality, environment, health, genetics, and animal performance on individual farms.

In Table 23-8, the authors present the recommended nutrient allowances of the swine specialists of Iowa State University, the Land Grant University in the state of Iowa, which is the leading swine producing state in the nation. It is recognized, however, that the margin of safety between requirements and allowances is based on experiments, experiences, and the judgment of the person or persons presenting the recommended allowances. For this reason, it is inevitable that recommended allowances will differ sightly from area to area and among nutritionists.

TABLE 23-8
RECOMMENDED NUTRIENT ALLOWANCES FOR SWINE[1 2 3]

	Sows, Gilts and Boars				Young Pigs		Grower-Finisher Pigs	
	Pregestation, Breeding and Gestation				Prestarter, Nursing	Starter, Creep	Grower, Re-placements	Finisher
	3 lb/day	4 lb/day	5 lb/day	Lactation	to 12 lb	to 40 lb	40–120 lb	120–240 lb
Protein, amino acids:								
Protein[4] (%)	13.00	12.00	11.00	13.00	20–24	18–20	15–17	13–15
Lysine[4] (%)	0.60	0.45	0.35	0.60	1.40	1.15	0.75–0.85	0.60–0.70
Threonine (%)	0.40	0.30	0.24	0.45	0.80	0.70	0.50	0.40
Tryptophan (%)	0.13	0.10	0.08	0.12	0.20	0.18	0.13	0.11
Major or Macrominerals:								
Salt (NaCl) (%)	0.50	0.40	0.30	0.50	0.25	0.25	0.25	0.25
Calcium (Ca) (%)	1.00	0.75	0.60	0.75	0.90	0.80	0.60	0.55
Phosphorus (P) (%)	0.75	0.60	0.50	0.60	0.70	0.65	0.50	0.45
Trace or Microminerals, added:[5]								
Copper (Cu) (ppm)	7	5	4	5	8	8	4	2
Iodine (I) (ppm)	0.20	0.14	0.11	0.14	0.14	0.14	0.07	0.04
Iron (Fe) (ppm)	100	80	60	80	100	100	50	25
Manganese (Mn) (ppm)	13	10	8	10	4	4	2	1
Selenium (Se) (ppm)	0.20	0.15	0.12	0.15	0.30	0.30	0.15	0.08
Zinc (Zn) (ppm)	65	50	40	50	100	100	50	25
Vitamins, added:[5]								
Vitamin A (IU/lb)	2,500	2,000	1,500	2,000	2,000	2,000	1,000	500
Vitamin D (IU/lb)	250	200	150	200	200	200	100	50
Vitamin E (IU/lb)	13	10	8	10	10	10	5	2.5
Biotin[6] (mg/lb)	0.13	0.10	0.08	0.10	0	0	0	0
Niacin (Nicotinic Acid, Nicotinamide) (mg/lb)	7	5	4	5	10	10	5	2.5
Pantothenic Acid (Vitamin B-3) (mg/lb)	8	6	5	6	8	8	4	2
Riboflavin (Vitamin B-2) (mg/lb)	3	2	1.5	2	2	2	1.5	0.8
Vitamin B-12 (Cobalamins) (mcg/lb)	10	7.5	6	7.5	10	10	5	2.5
Feed additives[7] (g/ton)	0–300	0–300	0–300	100–300	100–300	100–300	0–100	0–50

[1]Adapted by the authors from *Life Cycle Swine Nutrition*, Iowa State University, Ames, PM-489, June, 1988.

[2]The nutrient allowances are suggested for maximum performance, not as minimum requirements. They are based on research work with natural feedstuffs and have been found to give satisfactory results. Trace mineral and vitamin levels listed should be added to the ration in addition to those occurring in natural feedstuffs.

[3]To convert lb to kg, divide by 2.2. To convert IU/lb to IU/kg, multiply by 2.2. To convert mg/lb to mg/kg, multiply by 2.2. To convert mcg/lb to mcg/kg, multiply by 2.2. To convert g/ton (short) to g/ton (metric), divide by 0.907.

[4]Sow protein recommendations are based on corn-soybean meal rations. Other feedstuffs may require more protein to meet the amino acid requirement. Protein and lysine ranges for growing and finishing hogs allow for least cost formulation per unit of gain.

[5]Trace mineral and vitamin recommendations for finishing pigs are 50% of grower values.

[6]Biotin additions are not needed in corn-soybean meal based rations.

[7]The feed additives may be antibiotics, arsenicals or other chemotherapeutics or combinations. Levels and combinations used and stage of production for which they are used must comply with Food and Drug Administration regulations. A list of the common additives is shown in Table 23-11. High levels for sows may be beneficial just before and after breeding and at farrowing. They are not recommended during the entire gestation-lactation period unless specific diseases are present. The feed additive and the level used during growing and finishing phases should be primarily for growth promotion and improvement of feed efficiency.

Energy

Energy is the body's fuel supply. Every movement and activity of the pig's life involves the expenditure of fuel—energy for breathing, heart action, digestion, muscular movement, heat to keep the body warm, and energy to run the many chemical reactions involved in converting feed to animal use. If more energy is consumed than necessary to carry on vital functions, the excess is stored as body fat. In fact, this is what is done in finishing hogs. More energy is eaten than is needed for growth and body maintenance, with the result that the animal lays down fatty tissue with the excess.

The main nutrients supplying energy are carbohydrates. There are several forms of carbohydrates in plants. In proximate analysis these forms are identified as either nitrogen-free extract (NFE) or crude fiber. The NFE fraction includes the more soluble carbohydrates—sugars, starch, and some hemicellulose. All but hemicellulose are very digestible. Crude fiber, however, contains cellulose, hemicellulose, and lignin, all of which are highly indigestible to the pig.

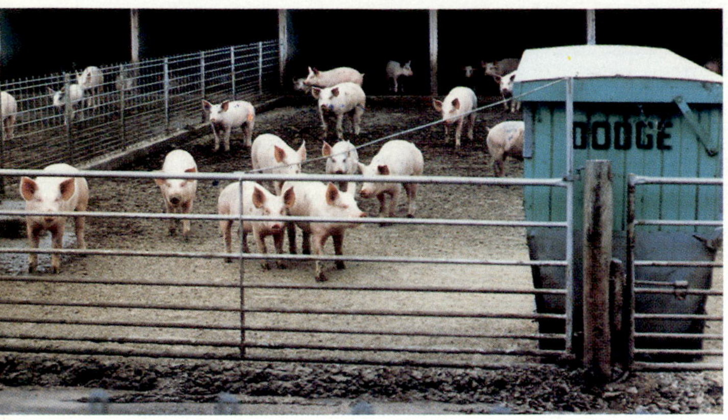

Fig. 23-3. Finishing hogs fed a complete, high-energy ration in a self-feeder. (Courtesy, E. J. Stevermer, Iowa State University, Ames)

The kind of carbohydrate a feed contains determines its value as a source of energy for the pig. Cereal grains are widely used in swine feeding because of their very high NFE (60 to 70%) and low crude fiber content.

The fats and oils constitute another group of energy nutrients. Fat is a very concentrated source of energy. It supplies approximately 2.25 times as much metabolizable energy as an equal weight of carbohydrate. Therefore, a feed high in fat, or a ration containing added fat, is much higher in energy value than a feed or ration low in fat. It follows that when high levels of fat are added to swine rations and the caloric density is increased thereby, special attention must be given to the density of the other nutrients in the ration. As caloric density is increased, pigs will reduce their voluntary intake level of the ration in order to maintain a relatively stable caloric intake. As intake is reduced, the level of nutrients other than energy will also be reduced unless changes in formulation are made. Therefore, nutrient levels in the ration must be considered in relation to the caloric density when high-energy rations are fed.

When fat is added to rations of growing-finishing hogs, daily feed intake usually decreases, daily gain increases slightly, and improvement in feed efficiency usually occurs. Response to fat is greatest in warm temperatures. When a minimum of 7.5% fat is added to the sow ration 10 to 14 days before farrowing, the fat content of colostrum and milk increases and baby pig survival improves by 2 to 3% in herds where survival rates are less than 80%. Fat supplementation during lactation reduces weight loss of sows.

Although roughages are a good source of energy for ruminants, because of their bulky nature and the restricted size of the digestive tract of hogs in comparison with ruminants, and because of the difficulty of digesting fiber by monogastrics, only limited quantities of them are contained in normal swine rations. Roughages (pastures and ground legume hays) are added to swine rations because of their protein, minerals, and vitamins, rather than for energy purposes.

Energy values are generally expressed in feed tables as total digestible nutrients (TDN), digestible energy (DE), or metabolizable energy (ME). Since energy expenditure can be measured as heat, modern nutritionists measure the energy needs of the animal in calories (a unit of heat). TDN values may be converted to DE by assuming that 1 lb of TDN is equivalent to 2,000 kilocalories of DE. ME values are calculated from the formula—

$$\mathrm{ME} = \mathrm{DE}\,\frac{(96 - 0.2 \times \%\ \text{crude protein})}{100}$$

This formula indicates that ME values are somewhat less than 96% of DE values.

ME decreases if protein is of poor quality and if there is excess protein, because the amino acids not used for protein synthesis are catabolized and used as a source of energy, and the nitrogen is excreted as urea.

Symptoms of energy deficiency are slow or interrupted growth, lowered reproduction, and offspring born dead or weak.

- **Essential fatty acids**—Pigs cannot synthesize linoleic and arachidonic acid, so they must be supplied in the ration. Because arachidonic acid can be derived *in vivo* from linoleic acid, only the latter need be a dietary constituent.

A deficiency of essential fatty acids in the young, growing pig results in a dull, dry hair coat and a scaly, dandrufflike dermatitis. In later stages, a brownish, gummy exudate and necrotic areas appear about the ears and under the flanks. Also, retarded sexual maturity, an undeveloped digestive system, and an abnormally small gallbladder have been reported.

Fig. 23-4. Fat deficiency. Littermate pigs. *Top:* Pig received fat in the ration (5.0% ether extract). *Bottom:* Pig received little fat in the ration (0.06% ether extract). Note loss of hair and scaly dandrufflike dermatitis. (Courtesy, Purdue University, West Lafayette, Ind.)

(Also see Chapter 4, section on "Energy [Carbohydrates and Lipids].")

Protein

While carbohydrates and fats are the principal sources of energy, proteins supply the building materials from which body tissue and many body regulators, such as enzymes and hormones, are made. Each protein is made up of several nitrogen compounds called amino acids.

There are 22, or more, amino acids. Of these, 10 assume unusual importance for the growing pig and are called essential amino acids. *An essential amino acid is one which the body cannot manufacture (synthesize) in sufficient quantity to permit maximum growth and performance.* Therefore, it must be supplied in the ration. The essential and nonessential amino acids for swine are identified as follows:

Essential	Nonessential
Arginine	Alanine
Histidine	Aspartic acid
Isoleucine	Asparagine
Leucine	Cysteine
Lysine	Cystine
Methionine	Glutamic acid
Phenylalanine	Glutamine
Threonine	Glycine
Tryptophan	Hydroxyproline
Valine	Proline
	Serine
	Tyrosine

The pig has a specific requirement for each of the essential amino acids. Since they are needed for the formation of every new cell, the need is most critical when growth is rapid. This makes ration formulation for the young pig very important, because the protein provided at this time must supply the amino acids for muscle growth (lean meat), internal organs, blood, bone, and all other parts associated with growth and development.

Quality of protein is a term used to describe the amino acid balance of protein. A protein is said to be of good quality when it contains all the essential amino acids in proper proportions and amounts, and to be poor quality when it is deficient in either content or balance of essential amino acids. From this it is evident that the usefulness of a protein source depends upon its amino acid composition, because the real need of the pig is for amino acids and not for protein as such.

Although it is common practice to refer to *percent protein* in a ration, this term has little significance in swine nutrition unless there is information about the amino acids present. For swine, protein quality is just as important as quantity. It is possible for pigs to perform better on a 12% protein ration, well balanced for amino acids, than on a 16% protein ration having a poor amino acid balance.

Research has shown that there are quite large differences in the availability of amino acids between feedstuffs. For example, approximately 85% of the lysine in soybean meal is digested and absorbed by the pig, compared with only 60% of the lysine in meat and bone meal. Also, the amino acid availability can be reduced substantially if a feedstuff is overheated during its processing. In the most precise type of ration formulation, adjustments are made for such differences. See Section V, Table V-3, for the Apparent Ileal Digestibility of Crude Protein and Essential Amino Acids in Feedstuffs for Swine, and Digestible Amino Acid Recommendations For Swine Feed Formulation.

Fig. 23-5. Lysine deficiency. *Top:* Pig gained 25 lb in 28 days after lysine was added to the basal ration (2.0% DL-lysine). *Bottom:* A lysine-deficient pig that received the basal ration only. This pig lost 2.0 lb in 28 days. (Courtesy, Purdue University, West Lafayette, Ind.)

From a practical standpoint, the problem of building a balanced ration for swine is centered around correcting the deficiencies in the cereal grains. Athough corn, wheat, and barley may contain from 8 to 12% protein, their protein is seriously deficient in the essential amino acid, lysine; corn is also deficient in tryptophan. Meat and bone meal is deficient in tryptophan. Moreover, the digestive tract of the pig is not adapted to extensive synthesis of proteins by microorganisms like the paunch of ruminants. Also, since protein supplements are more expensive than grain, the tendency is to feed too little of them.

Previously, it was stated that when an excess of energy is consumed by the pig it is stored in the form of fat. Protein is not stored in the body in appreciable amounts. If an excess of protein is fed, the unused nitrogen portion is discarded as urea in the urine and the carbon fraction is used as a source of energy.

Sometimes it is economical to supplement swine rations with the amino acid lysine in crystalline form. Thus, the amount of soybean meal needed in a ration can be reduced significantly if crystalline lysine is added.

For many years, the feed industry has had pure lysine and methionine available from microbiological production at prices competitive for use in swine rations in limited amounts. Crystalline tryptophan and threonine can also be purchased in feed-grade forms, but they are rather expensive.

Symptoms of protein (amino acid) deficiency are reduced feed intake accompanied by increased feed wastage, stunted growth, poor hair and skin condition, and lowered reproduction.

Minerals

Of all common farm animals, the pig is most likely to suffer from mineral deficiencies. This is due to the following peculiarities of swine husbandry:

1. Hogs are fed principally upon cereal grains and their by-products, all of which are relatively low in mineral matter, particularly in calcium.
2. The skeleton of the pig supports greater weight in proportion to its size than that of any other farm animal.
3. Hogs are fed to grow at a maximum rate for an early market, before they are mature.
4. Hogs reproduce at a younger age than other classes of livestock.

TABLE
SWINE MINERAL

Minerals Which May Be Deficient Under Normal Conditions	Conditions Usually Prevailing Where Deficiencies Are Reported	Functions of Mineral	Deficiency Symptoms/Toxicity
Major or Macrominerals: **Salt (NaCl)**	Salt deficiencies may exist when the protein supplement is all or chiefly of plant origin, although herbivorous animals require more salt than swine.	Sodium and chlorine are the principal extracellular cation and anion, respectively, in the body. Chlorine is the chief anion in gastric juice. Improves appetite, promotes growth, helps regulate body pH, and is essential for hydrochloric acid formation in the stomach.	**Deficiency symptoms**—Poor and depraved appetite, unthrifty condition, and failure to grow. **Toxicity**—Nervousness, weakness, staggering, epileptic seizures, paralysis, and death.
Calcium (Ca)	When the protein supplements are chiefly of plant origin and little forage is used. When swine are raised in confinement without vitamin D added to the ration. When feed intake is restricted during gestation. When there is a poor calcium-phosphorus ratio. Retention of calcium is affected by source of dietary protein (or phytic acid content) and the level of magnesium.	Bone and teeth formation; nerve function; muscle contraction; blood coagulation; cell permeability. Essential for milk production.	**Deficiency symptoms**—Loss of appetite and poor growth, lack of thrift, lameness and stiffness, weakened bone structure, and impaired reproduction. Severe cases may show reduced serum calcium and tetany. Rickets may develop in young pigs, or osteomalacia in older animals. Paralysis of the hind legs. **Toxicity**—An excess level of calcium tends to reduce the performance of pigs and increase the pig's zinc requirement.
Phosphorus (P)	Rations containing only plant ingredients; late gestation; lactation; high-calcium rations; swine in confinement without vitamin D added to the ration; poor calcium to phosphorus ratio. Retention of phosphorus is affected by source of dietary protein (or phytic acid content) and the level of magnesium.	Bone and teeth formation; a component of phospholipids which are important in lipid transport and metabolism and cell-membrane structure. In energy metabolism. A component of RNA and DNA, the vital cellular constituents required for protein synthesis. A constituent of several enzyme systems.	**Deficiency symptoms**—Loss of appetite and poor growth, lameness and stiffness, weakened bone structure, reduced inorganic blood phosphorus, depraved appetite, breeding difficulties, and rickets in young pigs, or osteomalacia in older animals. Paralysis of the hind legs, which is called posterior paralysis. **Toxicity**—An excess of phosphorus tends to reduce the performance of pigs, but is not toxic as such.
Magnesium (Mg)	Some research suggests that the magnesium in natural ingredients is only 50–60% available to the pig.	Essential for normal skeletal development, as a constituent of bone, cofactor in many enzyme systems, primarily in the glycolytic system.	**Deficiency symptoms**—Hyperirritability, muscular twitching, reluctance to stand, weak pasterns, loss of equilibrium, and tetany, followed by death. **Toxicity**—The toxicity level of magnesium is not known.
Potassium (K)		Major cation of intracellular fluid where it is involved in osmotic pressure and acid-base balance. Electrolyte balance and neuromuscular function. Required in enzyme reaction involving phosphorylation of creatine. Influences carbohydrate metabolism.	**Deficiency symptoms**—Loss of appetite, slow growth, poor hair and skin condition, decreased feed efficiency, inactivity, lack of coordination, and cardiac impairment. **Toxicity**—The toxic level of potassium is not well established. Pigs can tolerate up to 10 times the requirement if plenty of drinking water is provided.
Sulfur (S)		For synthesis of sulfur-containing compounds, such as glutathione, taurocholic acid, and chondroitin sulfate.	

5. The trend toward increased confinement rearing, without access to soil or forage, which would tend to balance the mineral deficiencies of the grains.

The functions of minerals are extremely diverse. They range from structural functions in some tissues to a wide variety of regulatory functions in other tissues.

Swine require at least 13 known inorganic elements, including calcium, chlorine, copper, iodine, iron, magnesium, manganese, phosphorus, potassium, selenium, sodium, sulfur, and zinc. Also, cobalt is required in the synthesis of vitamin B-12. Pigs may also require other trace elements, such as arsenic, boron, bromine, cadmium chromium, fluorine, lead, lithium, molybdenum, nickel, silicon, tin, and vanadium, which have been shown to have a physiological role in one or more species. These elements are required at such low levels, however, that their dietary essentiality for the pig has not been proven.

Several elements can be toxic to swine; among them, antimony, arsenic, cadmium, fluorine, lead, mercury, and selenium.

SWINE MINERAL CHART

Table 23-9, Swine Mineral Chart, gives in summary form the following pertinent information relative to each mineral listed: (1) conditions usually prevailing where deficiencies are reported, (2) functions, (3) deficiency symptoms/toxicity, (4) mineral requirements, (5) recommended allowances, and (6) practical sources. Further elucidation of minerals is contained in the narrative.

23-9
CHART (See footnote at end of table.)

Mineral Requirements[1]				
Minerals/ Animal/Day	**Mineral Content of Ration**	**Recommended Allowances[1]**	**Practical Sources of the Mineral**	**Comments**
•Na, variable according to class, age, and weight of swine (see Tables 23-1 and 23-2).	•Na, variable according to class, age, and weight of swine (see Table 23-5 and 23-6).	Salt variable according to class, age, and weight of swine (see Table 23-8).	Salt in loose form.	In iodine-deficient areas, stabilized iodized salt should be used. When pigs are salt starved, precaution should be taken to prevent over consumption of salt.
•Variable according to class, age, and weight of swine (see Tables 23-1 and 23-2).	•Variable according to class, age, and weight of swine (see Table 23-5, 23-6, and 23-7).	Variable according to class, age, and weight of swine (see Table 23-8).	Ground limestone, gypsum, or oyster-shell flour. Where both Ca and P are needed, use monocalcium phosphate, dicalcium phosphate, tricalcium phosphate, defluorinated phosphate, or bone meal.	Because cereal grains (which largely form the ration of swine) are low in Ca, swine are more apt to suffer from Ca deficiencies than from any of the other minerals except salt. •Most favorable Ca:P ratio is between 1:1 and 1.5:1. Sow's milk contains a Ca:P ratio of 1.3:1.
•Variable according to class, age, and weight of swine (see Tables 23-1 and 23-2).	•Variable according to class, age, and weight of swine (see Tables 23-5, 23-6, and 23-7).	Variable according to class, age, and weight of swine (see Table 23-8).	Where both Ca and P are needed, use monocalcium phosphate, dicalcium phosphate, tricalcium phosphate, defluorinated phosphate, or bone meal.	•About 60-75% of the P in cereal grains and their by-products, and in oilseed meals, is organically bound in the form of phytate and poorly available to the pig. •Most favorable Ca:P ratio is between 1:1 and 1.5:1. Sow's milk contains a Ca:P ratio of 1.3:1. Excess levels of P reduce performance of pigs.
•Variable according to class, age, and weight of swine (see Tables 23-1 and 23-2).	•0.04% of as-fed ration.	Practical rations are adequate in magnesium.	Magnesium oxide, magnesium sulfate, or magnesium carbonate. Dolomitic limestone.	Milk contains adequate magnesium for suckling pigs.
•Variable according to class, age, and weight of swine (see Tables 23-1 and 23-2).	•Variable for young pigs (see Tables 23-5 and 23-6). •No estimates available for finishing and breeding swine.	Practical rations are adequate in potassium.	Corn contains 0.33% potassium, and other cereals contain 0.42-0.49% potassium.	Potassium is the third most abundant mineral in the body of the pig, exceeded only by calcium and phosphorus.
		The addition of inorganic sulfate to low-protein rations has not been beneficial.		

(Continued)

TABLE 23-9

Minerals Which May Be Deficient Under Normal Conditions	Conditions Usually Prevailing Where Deficiencies Are Reported	Functions of Mineral	Deficiency Symptoms/Toxicity
Trace or microminerals:			
Cobalt (Co)	If vitamin B-12 is limited.	An essential component of vitamin B-12.	**Deficiency symptoms**—No deficiency symptoms reported in swine. However, supplemental cobalt prevents lesions associated with zinc deficiency. **Toxicity**—A level of 400 ppm of cobalt is toxic to the young pig and may cause loss of appetite, stiff-leggedness, humped back, incoordination, muscle tremors, and anemia.
Copper (Cu)	Suckling pigs kept off soil.	Essential element in a number of enzyme systems and necessary for synthesizing hemoglobin and preventing nutritional anemia. Hemoglobin serves as a carrier of oxygen throughout the body. When fed at 100 to 250 ppm, copper stimulates growth in pigs.	**Deficiency symptoms**—Slow growth, poor hair and skin condition, lameness and stiffness, weakened bone structure, weak and crooked legs, anemia, and cardiac and vascular disorders. **Toxicity**—Depressed hemoglobin levels and jaundice.
Iodine (I)	Iodine-deficient areas/soils (in northwestern U.S. and in the Great Lakes region) when iodized salt is not fed. Where feeds come from iodine-deficient areas.	Needed by the thyroid gland for making thyroxin, an iodine-containing hormone which controls the rate of body metabolism or heat production.	**Deficiency symptoms**—Loss of appetite, slow growth, poor hair and skin condition, impaired breeding or gestation, offspring dead or weak at birth, pigs hairless at birth, and/or enlarged thyroid. **Toxicity**—An 800-ppm iodine level in the ration depresses growth, hemoglobin level, and liver iron concentration in growing pigs. During the last 30 days of gestation and during lactation, 1,500 to 2,500 ppm of iodine was found harmful to sows.
Iron (Fe)	Suckling pigs kept off soil.	*Iron is required as a component of hemoglobin in red blood cells. Iron also is found in muscle as myoglobin, in serum as transferrin, in the placenta as interoferrin, in milk as lactoferrin, and in the liver as ferritin and hemosiderin. Iron also plays an important role in the body as a constituent of a number of metabolic enzymes.	**Deficiency symptoms**—Loss of appetite, slow growth, poor hair and skin condition, paleness of mucous membranes, high mortality in young pigs, susceptibility to disease, thumps (characterized by labored breathing), and anemia. The number of grams of hemoglobin per 100 ml of blood is a rapid, reliable indicator of the iron status of the pig. A hemoglobin level of 8 g/100 ml indicates borderline anemia; a level of 7 g or less/100 ml indicates anemia. **Toxicity**—In 3- to 10-day-old pigs, the toxic oral dose of iron from ferrous sulfate is approximately 273 mcg/lb *(601 mcg/kg)* body weight.
Manganese (Mn)		Functions as a component of several enzymes involved in carbohydrate, lipid, and protein metabolism. A component in the organic matrix of bone.	**Deficiency symptoms**—Abnormal skeletal growth, increased fat deposition, irregular or absent estrus cycles, resorbed fetuses, small and weak pigs at birth, and reduced milk production. **Toxicity**—The toxic level of manganese is not clearly defined. But high levels result in depressed feed intake, reduced growth, and limb stiffness.
Selenium (Se)	When rations consist almost exclusively of ingredients grown on selenium-deficient soils.	Functions as a part of glutathione peroxidase, an enzyme which enables the tripeptide glutathione to perform its role as a biological antioxidant in the body. The mutual sparing effect of selenium and vitamin E stems from their shared antiperoxidant roles. But high levels of vitamin E do not completely eliminate the level for selenium.	**Deficiency symptoms**—Sudden death, impaired reproduction, reduced milk production, and impaired immune response. **Toxicity**—Loss of appetite, loss of hair, fatty infiltration of the liver, degenerative changes in the liver and kidney, edema, and occasional separation of the hoof and skin at the coronary band. Dietary arsenicals help to alleviate selenium toxicity.
Zinc (Zn)	High levels of calcium in relation to zinc levels impair zinc utilization and increase the requirements.	Zinc is a component of many metalloenzymes and the hormone insulin. So, it plays an important role in protein, carbohydrate, and lipid metabolism.	**Deficiency symptoms**—Parakeratosis or swine dermatitis, pigs have a mangy appearance, reduced appetite, unthriftiness, poor growth rate, and diarrhea, and there may be vomiting. It affects swine of all ages. Zinc deficiency results in gilts producing fewer and smaller pigs; in boars with retarded testicular development; and in young pigs with retarded thymic development. **Toxicity**—Growth depression, arthritis, hemorrhage in axillary spaces, gastritis, and enteritis. High dietary calcium reduces the severity of zinc toxicity.

[1]As used herein, the distinction between "mineral requirements" and "recommended allowances" is as follows: In mineral requirements, no margins of safety are included intentionally; whereas in recommended allowances, margins of safety are provided in order to compensate for variations in feed compositions, environment, and possible losses during storage or processing.

Where preceded by an asterisk, the mineral requirements, recommended allowances, and other facts presented herein were taken from *Nutrient Requirements of Swine,* 9th rev. ed., NRC-National Academy of Sciences, 1988.

(Continued)

Mineral Requirements[1]		Recommended Allowances[1]	Practical Sources of the Mineral	Comments
Minerals/ Animal/Day	**Mineral Content of Ration**			
No requirements for cobalt have been established.		Practical rations are adequate in cobalt.	Cobalt chloride, cobalt sulfate, cobalt oxide, or cobalt carbonate. Also, several good commercial minerals containing cobalt are on the market.	
•Variable according to class, age, and weight of swine (see Tables 23–1 and 23–2).	•Variable according to class, age, and weight of swine (see Tables 23–5 and 23–6).	Variable according to class, age, and weight of swine (see Table 23–8).	Copper sulfate, copper carbonate, and copper chloride are about equally effective. The copper in copper sulfide and copper oxide is poorly available to the pig.	Beyond the suckling period, natural feedstuffs usually contain enough copper. •When fed at a level of 100 to 250 ppm, copper will increase rate and efficiency of gains of pigs to breeding age.
•Variable according to class, age, and weight of swine (see Tables 23–1 and 23–2).	•0.06 mg/lb *(0.14 mg/kg)* as-fed ration.	Variable according to class, age, and weight of swine (see Table 23–8).	Stabilized iodized salt containing 0.007% iodine. Calcium iodate, potassium iodate, and pentacalcium orthoperiodate.	The majority of the iodine in the bodies of swine is present in the thyroid.
•Variable according to class, age, and weight of swine (see Tables 23–1 and 23–2). •Newborn pigs require 7 to 16 mg of absorbed iron daily for normal growth.	•Variable according to class, age, and weight of swine (see Tables 23–5 and 23–6). •The iron requirement of young pigs fed milk or purified liquid diets is 23 to 68 mg/lb *(50 to 150 mg/kg)* of milk solids. The iron requirement of pigs fed a dry, casein-based diet is about 50% higher/unit of dry matter than for those fed a similar diet in liquid form.	Variable according to class, age, and weight of swine (see Table 23–8).	A single intramuscular injection of 100 to 200 mg of iron, in the form of iron dextran, iron dextrin, or gleptoferron given in the first 3 days of life; or Oral administration of iron from iron chelates within the first few hours of life. Ferrous sulfate, ferric chloride, ferric citrate, ferric choline citrate, and ferric ammonium citrate are effective in preventing iron deficiency anemia. The iron in ferric oxide is largely unavailable.	•Pigs are born with about 50 mg of iron. •Iron has a detoxifying effect when added to gossypol-containing diets. Add iron from soluble source to free gossypol at a weight ratio of 1:1 •Milk is deficient in iron (sow's milk contains an average of 1 mg of iron/liter). Pigs should be encouraged to eat grain ration as soon as old enough. •Natural feed ingredients usually supply enough iron to meet postweaning requirements.
•Variable according to class, age, and weight of swine (see Tables 23–1 and 23–2).	•Variable according to class, age, and weight of swine (see Tables 23–5 and 23–6).	Variable according to class, age, and weight of swine (see Table 23–8).	Manganous oxide.	Manganese is usually present in adequate amounts in most swine rations, but it may not be adequate for the optimum reproductive performance of sows.
•Variable according to class, age, and weight of swine (see Tables 23–1 and 23–2).	•The dietary requirement for selenium is between 0.1 and 0.3 ppm (see Tables 23–5 and 23–6). •In 1987, the FDA approved up to 0.3 ppm selenium in the ration of all pigs.	Variable according to class, age, and weight of swine (see Table 23–8).	Sodium selenite or sodium selenate.	Environmental stress may increase the incidence and degree of selenium deficiency. ***Caution:*** Toxic level of selenium is in range of 2.27–3.63 mg/lb *(5–8 mg/kg)* selenium in the feed.
•Variable according to class, age, and weight of swine (see Tables 23–1 and 23–2).	•Variable according to class, age, and weight of swine (see Tables 23–5 and 23–6). The zinc requirement is increased when excessive levels of calcium are fed.	Variable according to class, age, and weight of swine (see Table 23–8).	Zinc carbonate or zinc sulfate.	It has been shown that parakeratosis is caused by zinc and calcium forming an unavailable complex.

NOTE: *Mineral recommendations for all classes and ages of swine, especially those fed unmixed rations or on pastures are—*

1. *Where animals are on liberal grain feeding*—Provide free access to a 2-compartment mineral box, with (a) trace mineralized salt in one side, and (b) in the other side, a mixture of ⅓ trace mineralized salt (salt included for purposes of palatability), ⅓ defluorinated phosphate or steamed bone meal, and ⅓ ground limestone or oystershell flour.

2. *Where animals are primarily on roughage (pasture, hay, and/or silage)*—Provide free access to a 2-compartment mineral box, with (a) trace mineralized salt in one side (salt included for purposes of palatability), and (b) in the other side, a mixture of ⅓ trace mineralized salt and ⅔ defluorinated phosphate or steamed bone meal.

MAJOR OR MACROMINERALS

Salt (sodium and chlorine), calcium, and phosphorus, magnesium, potassium, and sulfur are the major or macrominerals.

Salt (NaCl)

Salt contains both sodium and chlorine, vital elements found in the fluids and soft tissues of the body. It improves the appetite, promotes growth, helps regulate body pH, and is essential for hydrochloric acid formation in the stomach.

Although swine require less salt than other classes of farm animals, it is generally advantageous to supply them with some of it, particularly if the protein supplement is not derived from animal or marine sources. A lack of salt is marked by a poor and depraved appetite, unthrifty condition, and failure to grow.

Calcium (Ca)/Phosphorus (P)

A deficiency of either calcium or phosphorus in the ration of the pig can result in poor and inefficient gains, rickets or osteomalacia, broken bones, and posterior paralysis.

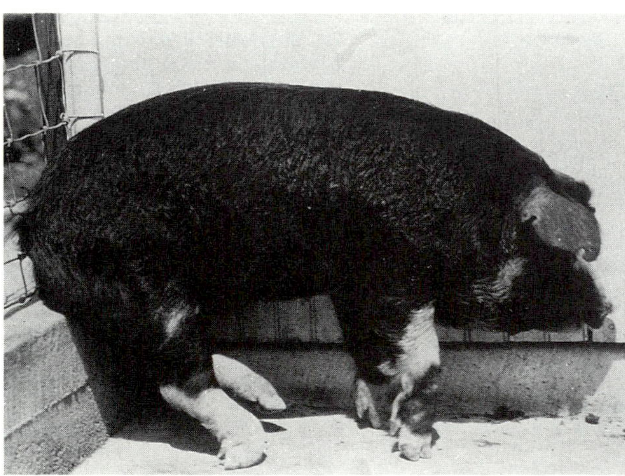

Fig. 23-6. Calcium deficiency. Note abnormal bone development and rachitic condition in advanced stage of deficiency. Lack of calcium retards normal skeletal development, but it does not usually depress total gain. (Courtesy, USDA)

A large excess of either calcium or phosphorus interferes with the absorption of the other. Thus, it is important to have a suitable ratio between the 2 minerals. The most favorable calcium to phosphorus ratio is from 1.2:1 to 1.5:1. Also, vitamin D is necessary for the proper utilization of these 2 minerals.

Fig. 23-7. Phosphorus deficiency. *Left:* Typical phosphorus-deficient pig in advanced stage of deficiency. Leg bones are weak and crooked. *Right:* This pig received the same ration as the one on the left, except that the ration was adequate in available phosphorus. (Courtesy, Purdue University, West Lafayette, Ind.)

It is important to supplement swine rations with both calcium and phosphorus. Cereal grains, which make up the bulk of swine rations, are quite low in calcium and are only fair sources of phosphorus. Moreover, much of the phosphorus in cereal grains is present in the form of phytin, a form of phosphorus which is poorly utilized. So, the availability of phosphorus in ingredients should be considered.

The following feedstuffs have a high availability of phosphourus for the pig:[2]

	Availability of the P
Alfalfa	100
Bone meal	82
Defluorinated rock phosphate	87
Dicalcium phosphate	100
Fish meal	100
Meat and bone meal	93
Monosodium phosphate	100

(For a list of the commonly used calcium and phosphorus supplements, see Chapter 13, Table 13-2.)

Magnesium (Mg)

Magnesium is a cofactor in many enzyme systems and a constituent of bone. Apparently, the magnesium requirement is met by grain-soybean meal rations, or by rations containing grain and protein supplements.

Potassium (K)

Potassium is the most abundant mineral in muscle tissue.

Grain-soybean meal rations normally contain enough potassium to meet the requirements for all classes of swine.

[2]From: *Nutrient Requirements of Swine,* 9th rev. ed., National Research Council, National Academy Press, 1988, p. 51, Table 5-2.

The dietary potassium requirement for the pig is increased by high levels of dietary chloride, sulfate, and other anions, as it is for the chick.

Sulfur (S)

Sulfur is an essential element. However, the sulfur-containing amino acids (cystine and methionine) appear adequate to meet the pig's need for synthesis of sulfur-containing compounds. The addition of inorganic sulfate to low-protein swine rations has not been beneficial.

TRACE OR MICROMINERALS

Minerals that are required in small amounts are known as trace or microminerals. These include cobalt, copper, iodine, iron, manganese, selenium, and zinc.

Fig. 23-8. Hairlessness in pigs caused by a deficiency of iodine. In iodine-deficient areas, farm animals should receive iodized salt throughout the year. (Courtesy, Department of Veterinary Pathology and Hygiene, College of Veterinary Medicine, University of Illinois, Urbana)

Cobalt (Co)

Cobalt is a component of vitamin B-12. The intestinal microflora of the pig are capable of synthesizing vitamin B-12 provided sufficient cobalt is present. But only a minimum level of dietary cobalt is necessary for this process. Intestinal synthesis is of greater importance if preformed vitamin B-12 is limiting.

There is no evidence that pigs have a requirement for cobalt, other than for vitamin B-12 synthesis.

Cobalt can partially substitute for zinc. Also, supplemental cobalt will prevent lesions associated with zinc deficiency.

A level of 400 ppm of cobalt is toxic to the young pig. Selenium and vitamin E provide some protection against toxicity from excessive levels of dietary cobalt.

Copper (Cu)

The pig requires copper for the synthesis of hemoglobin and for the synthesis and activation of several oxidative enzymes necessary for normal metabolism.

A deficiency of copper leads to poor iron mobilization. A level of 6 ppm in the ration is adequate for baby pigs.

When fed at 100 to 250 ppm, copper stimulates growth in pigs, apparently due to the antibacterial action of these high levels of copper.

Iodine (I)

The dietary iodine requirement is not well established. Moreover, it is increased by goitrogens in certain feedstuffs, including rapeseed, linseed, lentils, peanuts, and soybeans. A level of 0.14 ppm of iodine in a corn-soybean meal ration will prevent goiter in growing pigs; and a level of 0.35 ppm of added iodine will prevent iodine deficiency in sows.

The incorporation of iodized salt (0.007% iodine) at a level of 0.2% of the ration provides sufficient iodine (0.14 ppm) to meet the needs of growing pigs fed grain-soybean meal rations.

Calcium iodate, potassium iodate, and pentacalcium orthoperiodate are nutritionally available forms and more stable in salt mixtures than sodium iodide or potassium iodide.

Iron (Fe)

Iron is necessary for the formation of hemoglobin in the red blood cells and the prevention of nutritional anemia. Hemoglobin serves as a carrier of oxygen throughout the body.

As the unborn pig develops, a supply of iron is stored in its body. The amount stored varies greatly between pigs of the same litter. But in no case is the amount of iron adequate to keep the pig growing at its maximum for more than 10 days or 2 weeks after birth unless soil or some supplemental source of iron is available during the suckling period.

Sow's milk is very low in iron; and, to date, research has not uncovered any way of increasing its iron content. Thus, if suckling pigs are confined with no access to soil or feed, serious losses from anemia are likely. Once a pig begins to consume natural feedstuffs, the danger of anemia is practically nil because most feeds contain sufficient amounts of iron to meet the pig's requirement.

Anemic pigs lose their appetite and become weak and inactive. In more advanced stages of the deficiency, the pig's breathing becomes labored, a condition that is sometimes called *thumps*. In this condition they are more susceptible to other diseases and parasites. Death may occur in severe cases.

Fig. 23-9. Suckling pig with nutritional anemia, caused by a lack of iron, characterized by swollen condition about the head and paleness of the mucous membranes. (Courtesy, College of Veterinary Medicine, University of Illinois, Urbana)

For the prevention or treatment of anemia in young pigs, either (1) place a little uncontaminated sod (topsoil, from an area where hogs have not run for years) in the corner of the pen daily, (2) inject a suitable iron preparation at a level of 100 to 200 mg of iron into baby pigs at 1 to 3 days of age, (3) swab the sow's udder with iron solution, (4) give an iron-copper pill, or (5) allow access to oral iron preparations. In addition, the pigs should be encouraged to eat a grain ration as soon as they are old enough.

The postweaning dietary iron requirement is about 80 ppm.

Manganese (Mn)

Manganese functions with many enzymes in soft tissue metabolism and also in bone development. Deficiency symptoms are lameness, weakened bone structure, irregular estrus, offspring born dead or weak, and increased backfat. While manganese is usually present in adequate amounts without supplementation in most swine rations, it may not be adequate for the optimum reproductive preformance of sows.

Selenium (Se)

Selenium functions as a part of glutathione peroxidase, an enzyme which enables the tripeptide glutathione to perform its role as a biological antioxidant in the body. This explains why deficiencies of selenium and vitamin E result in similar signs. However, high levels of vitamin E do not completely eliminate the need for selenium.

Currently, the U.S. Food and Drug Administration (FDA) allows the addition of up to 0.3 ppm of selenium in the rations of all pigs.

In some cases, selenium in the ration at a level of 5 ppm will produce toxic symptoms.

Zinc (Zn)

The requirement for zinc in swine rations is very low, but when high levels of calcium are fed, zinc utilization is impaired and the requirements are increased. A zinc deficiency results in a mangelike skin condition called *parakeratosis*. Other symptoms are poor growth, inefficient feed conversion, gilts producing fewer and smaller pigs, boars with retarded testicular development, and young pigs with retarded thymic development.

Boars have a higher zinc requirement than gilts, and gilts have a higher requirement than barrows.

Fig. 23-10. Zinc deficiency. *Left:* Pig received 17 ppm of zinc and gained only 3 lb in 74 days. Note severe dermatosis ("mangy look"), or parakeratosis. *Right:* Pig received the same ration as the pig on the left, except that the ration contained 67 ppm of zinc. This pig gained 111 lb in 74 days. (Courtesy, Purdue University, West Lafayette, Ind.)

OTHER MINERALS

Other mineral elements required by the pig but present in adequate amounts in most practical rations are chromium, tin, molybdenum, nickel, vanadium, silicon, and fluorine.

FEEDS AS A SOURCE OF MINERALS

The most satisfactory source of minerals for hogs is in the feed consumed. Thus, it is important to know whether the minerals in the ration are of the right kind and sufficient in amount. Certain general characteristics of feeds in regard to calcium and phosphorus (the two predominating mineral elements of the body) are worth noting:

1. The cereal grains and their by-products and protein supplements of plant origin are low in calcium but fairly good in phosphorus. However, as mentioned earlier, the phosphorus in plants is not fully utilizable by swine.

2. The protein supplements of animal origin (skim milk, buttermilk, tankage, meat scraps, fish meal), legume forage (pasturage and hay), and rape, are all rich in calcium.

3. Most protein-rich supplements are high in phosphorus.

METHOD OF FEEDING MINERAL SUPPLEMENTS

Generally, supplementary minerals are incorporated in rations when needed (see Table 23-9). Additionally, hogs may be given free access to a suitable mineral mix. This is cheap insurance against possible needs beyond what's provided in the ration. It's a way of hedging against (1) individual pig differences in mineral requirements, and (2) feeds varying in both mineral content and availability.

Vitamins

Vitamins are complex organic compounds needed in minute amounts, which are essential for health and normal body functions. Like amino acids, each vitamin has a specific function to perform. Vitamins are classified into two groups—fat-soluble and water-soluble. The body can store reserves of the fat-soluble vitamins for a considerable period of time. But stores of the water-soluble vitamins are depleted rapidly.

Because of the greater prevalence of confinement feeding, swine are more likely to suffer from vitamin deficiencies than any other class of four-footed animals.

SWINE VITAMIN CHART

Table 23-10, Swine Vitamin Chart (see pp. 968-971), gives in summary form the following pertinent information relative to each vitamin listed: (1) conditions usually prevailing where deficiencies are reported, (2) functions, (3) deficiency symptoms/toxicity, (4) vitamin requirements, (5) recommended allowances, and (6) practical sources. Further elucidation of vitamins is contained in the narrative.

FAT-SOLUBLE VITAMINS

The primary fat-soluble vitamins of practical importance for swine are vitamins A, D, and E. Vitamin K may be of concern under some circumstances.

Vitamin A

The vitamin A needs of swine can be met by either vitamin A or carotene. Vitamin A as such does not occur in plants. However, green plants and yellow corn contain a yellow pigment called carotene which can be converted to vitamin A by the animal body. The combination of vitamin A and carotene present in the ration is refered to as its vitamin A activity.

Carotene is easily destroyed by the ultraviolet rays of the sun and by heat. The carotene content of corn and legume hay usually deteriorates quite rapidly in storage. Therefore, a synthetic concentrate is a more practical and reliable source of vitamin A than natural sources. Most commercial feed companies fortify their swine feeds with a stabilized form of vitamin A, which is active over a considerable period of time.

Vitamin A is essential for vision, reproduction, growth, and the maintenance of differentiated epithelia and mucous secretions of swine.

Swine are able to store vitamin A in the liver, and to draw from this storage during periods of low intake.

Vitamin A deficiency signs in growing pigs are incoordination of movement, loss of control of the hind legs, weakness of the back, and night blindness. Sows may fail to come into heat, they may resorb their fetuses, or they may have young born dead with various deformities and defects. Vitamin A is also needed for normal vision and growth of new cells which line the respiratory, digestive, and reproductive tracts.

Vitamin D

Vitamin D is sometimes referred to as the *sunshine* vitamin, since the action of sunlight on a compound in the skin will produce it. So long as hogs are exposed to the sun, there is no danger of a deficiency.

Living plants do not contain vitamin D. Plants that mature or are cut and cured in the sun contain some vitamin D as a result of irradiation by sunlight. Pigs can utilize equally well either vitamin D_2 (from plant products) or vitamin D_3 (from animal products). Irradiated yeast is a good source of vitamin D_2.

Vitamin D is needed for the efficient assimilation of calcium and phosphorus; hence, it is required for the growth of strong bones. A lack of vitamin D will result in stiffness and lameness, broken or deformed bones, enlargement of joints, and general unthriftiness; known as rickets in young pigs and osteomalacia in mature hogs.

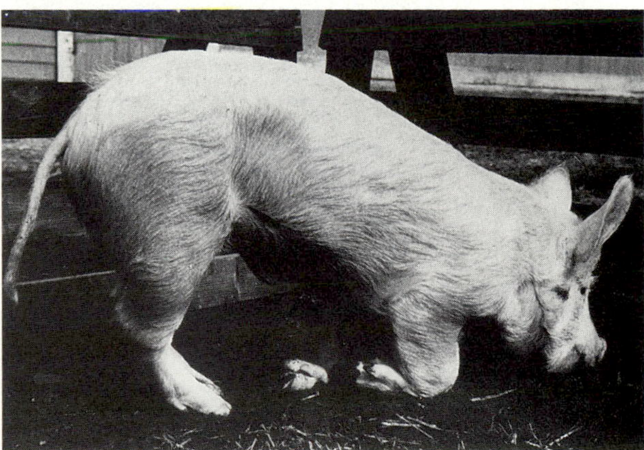

Fig. 23-11. Rickets (advanced case) caused by a deficiency of vitamin D. The pig was fed indoors, without exposure to sunlight and without adequate vitamin D. Because of leg abnormalities, it was unable to walk. Later, the pig responded to vitamin D. (Courtesy, University of Saskatchewan, Saskatoon, Canada)

It is noteworthy that the vitamin D requirement is less when a proper balance of calcium and phosphorus exists in the ration.

TABLE
SWINE VITAMIN

Vitamins Which May Be Deficient Under Normal Conditions	Conditions Usually Prevailing Where Deficiencies Are Reported	Functions of Vitamin	Deficiency Symptoms/Toxicity
Fat-soluble vitamins:			
A	Absence of green forages, either pasture or green hay—especially under drylot conditions. Where the ration consists chiefly of white corn, milo, barley, wheat, oats, or rye; or by-products of these grains; or yellow corn that has been stored more than one year.	Essential for normal maintenance and functioning of the epithelial tissues, particularly of the eye and the respiratory, digestive, reproductive, nerve, and urinary systems.	**Deficiency symptoms**—Night and day blindness, very irritable, poor appetite and slow growth, lameness, incoordination of movement, loss of control of the hind legs, and weakness of the back. Low resistance to respiratory infections. Sows may fail to come in heat, may resorb their fetuses, and may have young born dead with various deformities and defects. **Toxicity**—A roughened hair coat, scaly skin, hyperirritability and sensitivity to touch, bleeding from the cracks which appear in the skin above the hooves, blood in the urine and feces, loss of control of the legs accompanied by inability to rise, periodic tremors, and death.
D	Limited sunlight and/or limited quantities of sun-cured hay in drylot rations.	Aids in assimilation and utilization of calcium and phosphorus, and necessary in the normal bone development of animals—including the bones of the fetus.	**Deficiency symptoms**—Rickets in young pigs, or osteomalacia in mature hogs. Both conditions result in large joints and weak bones. In severe vitamin D deficiency, pigs may exhibit signs of calcium and magnesium deficiency, including tetany. **Toxicity**—Reduced feed intake and growth rate; and death. Vitamin D_3 is more toxic than D_2 in swine.
E	Rations containing excessive amounts of highly unsaturated fatty acids or oxidized fats. Swine feeds low in selenium, especially where swine are raised in confinement without access to forages.	Antioxidant. Muscle structure. Reproduction. High levels of vitamin E in the diet may increase the immune response.	**Deficiency symptoms**—Loss of appetite and slow growth. Increased embryonic mortality and muscular incoordination in suckling pigs from sows fed vitamin E-deficient rations during gestation and lactation. A wide variety of pathological conditions. **Toxicity**—Vitamin E toxicity in swine has not been demonstrated.
K	Moldy feed. High antibiotic levels, which may make for inadequate intestinal synthesis of vitamin K.	Essential for prothrombin formation and blood clotting.	**Deficiency symptoms**—Bleeding condition in young pigs, which responds to injection or oral administration of vitamin K. Slow growth and hyperirritability. **Toxicity**—Concentrations of 50 mg of menadione pyrimidinol bisulfite (MPB)/lb *(110 mg of MPB/kg)* of ration were not toxic to weanling pigs.
Water-soluble vitamins:			
Biotin	When pigs are fed dried, raw egg white or given sulfa drugs. Marginal deficiency may exist when hogs are fed cereal grain rations, housed in individual stalls or on slotted floors (which lessens coprophagy), and/or have no access to green forage.	Biotin is important metabolically as a cofactor for several enzymes.	**Deficiency symptoms**—Excessive hair loss, skin ulcerations and dermatitis, exudate around the eyes, cracking of the hooves, and cracking and bleeding of the footpads.
Choline	Baby pigs fed a synthetic milk ration containing not more than 0.8% methionine.	Involved in nerve impulses; a component of phospholipids; donor of methyl groups; and involved in the mobilization and oxidation of fatty acids in the liver.	**Deficiency symptoms**—Unthriftiness, lack of coordination, spraddled hind legs at birth, fatty infiltration of the liver, poor reproduction, poor lactation, and decreased survival of the young. **Toxicity**—No signs of choline toxicity have been reported in swine.
Folacin (Folic Acid)		Metabolic reactions involving incorporation of single carbon units into larger molecules. Folacin is involved in the conversion of serine to glycine and homocystine to methionine.	**Deficiency symptoms**—Poor growth, fading hair color, and anemia.
Niacin (Nicotinic Acid, Nicotinamide)		Niacin is a component of the coenzymes which are essential for the metabolism of carbohydrates, proteins, and lipids.	**Deficiency symptoms**—Loss of appetite and decreased gain, followed by diarrhea, occasional vomiting, dermatitis, and loss of hair.

23-10
CHART (See footnote at end of table.)

Vitamin Requirements[1]				
Vitamins/ Animal/Day	**Vitamin Content of Ration**	**Recommended Allowances[1]**	**Practical Sources of the Vitamin**	**Comments**
*Variable according to class, age, and weight of swine (see Tables 23-1 and 23-2).	*Variable according to class, age, and weight of swine (see Tables 23-5 and 23-6).	Variable according to class, age, and weight of swine (see Table 23-8).	Either vitamin A or various provitamins.	Based upon liver storage, the biopotency of 1 mg of carotene in corn fed to weanling pigs is 261 IU of vitamin A. Meals from artificially dehydrated forages are much higher in carotene than sun-cured products. Taken together, liver storage, levels of plasma vitamin A, and pressure of cerebrospinal fluid give reliable estimates of the vitamin A status of the pig.
*Variable according to class, age, and weight of swine (see Tables 23-1 and 23-2).	*Variable according to class, age, and weight of swine (see Tables 23-5 and 23-6).	Variable according to class, age, and weight of swine (see Table 23-8).	Vitamin D_2 (ergocalciferol) and vitamin D_3 (cholecalciferol) are similar in biological activity for swine. The action of ultraviolet light on the ergosterol that is present in plants forms ergocalciferol; and the photochemical conversion of 7-dehydrocholesterol in the skin of animals forms cholecalciferol. Irradiated yeast. Exposure to sunlight. Sun-cured hay (10% alfalfa in the total ration will normally supply sufficient vitamin D).	Grains, grain by-products, and high-protein feedstuffs are practically devoid of vitamin D; therefore, unless swine are exposed daily to the ultraviolet rays of the sun, the ration should be fortified with vitamin D. The vitamin D requirement is less when a proper balance of calcium and phosphorus exists in the ration. One IU vitamin D is defined as the biological activity of 0.025 mcg of cholecalciferol.
*Variable according to class, age, and weight of swine (see Tables 23-1 and 23-2). Many dietary factors affect the vitamin E requirement, including the selenium level, unsaturated fatty acids, sulfur amino acids, retinol, copper, iron, and synthetic antioxidants.	*Variable according to class, age, and weight of swine (see Tables 23-5 and 23-6).	Variable according to class, age, and weight of swine (see Table 23-8).	Alpha tocopherol. Predicting the amount of vitamin E activity in feed is difficult.	The 8 naturally occurring tocopherols differ in their biological activities, with *d*-alpha-tocopherol being the most active. One IU of vitamin E is the equivalent in biopotency of 1 mg *dl*-alpha-tocopherol acetate.
*Variable according to class, age, and weight of swine (see Tables 23-1 and 23-2).	*Supplement the as-fed ration with menadione at level of 0.2 mg/lb *(0.5 mg/kg)*.	Under practical conditions, the vitamin K requirement is met by vitamin K in feedstuffs and by intestinal synthesis.	The following water-soluble forms of menadione are commonly used to supplement swine diets: menadione sodium bisulfite complex (MSB), menadione sodium bisulfate complex (MSBC), and menadione pyrimidinol bisulfite (MPB).	Vitamin K exists in 3 forms: phylloquinone (K_1), menaquinone (K_2), and menadione (K_3).
*Variable according to class, age, and weight of swine (see Tables 23-1 and 23-2).	*Variable according to class, age, and weight of swine (see Tables 23-5 and 23-6).	Variable according to class, age, and weight of swine (see Table 23-8).	Biotin.	The protein *avidin* in raw egg white makes biotin unavailable to pigs. Heat treatment inactivates avidin and makes egg white safe for feeding to pigs.
*Variable according to class, age, and weight of swine (see Tables 23-1 and 23-2).	*Variable according to class, age, and weight of swine (see Tables 23-5 and 23-6).	Practical rations are adequate in choline.	Choline chlorides or choline dihydrogen.	Choline does not qualify as a true vitamin, because it is required at far greater levels than true vitamins and is not known to participate in any enzyme system. Studies have shown that more live pigs are born and weaned when sows receive supplemental choline throughout gestation.
*Variable according to class, age, and weight of swine (see Tables 23-1 and 23-2).	*Variable according to class, age, and weight of swine (see Tables 23-5 and 23-6).	Practical rations plus intestinal synthesis is adequate.	Synthetic folacin.	Folacin includes a group of compounds with folic acid activity.
*Variable according to class, age, and weight of swine (see Tables 23-1 and 23-2).	*Variable according to class, age, and weight of swine (see Tables 23-5 and 23-6).	Variable according to class, age, and weight of swine (see Table 23-8).	Nicotinamide. Nicotinic acid.	Niacin occurs in corn, wheat, and milo in bound form; hence, it may be unavailable to the pig. Also, the dietary tryptophan level affects the niacin requirement because of the conversion of tryptophan to niacin.

(Continued)

TABLE 23-10

Vitamins Which May Be Deficient Under Normal Conditions	Conditions Usually Prevailing Where Deficiencies Are Reported	Functions of Vitamin	Deficiency Symptoms/Toxicity
Water-soluble vitamins (Continued):			
Pantothenic Acid (Vitamin B-3)	Long periods of inadequate pantothenic acid intake.	As a component of coenzyme A, pantothenic acid is important in the catabolism and synthesis of 2-carbon units evolved during carbohydrate and fat metabolism.	**Deficiency symptoms**—A goose-stepping gait, loss of appetite, poor growth, diarrhea, loss of hair, reduced fertility, and breeding failure.
Riboflavin (Vitamin B-2)		A component of 2 coenzymes. It is important in the metabolism of proteins, fats, and carbohydrates.	**Deficiency symptoms**—Loss of appetite, poor growth, rough hair coat, diarrhea, cataracts, vomiting, reproductive failure in the sow, pigs dead or weak at birth, and crooked legs and incoordination.
Thiamin (Vitamin B-1)	Thiamin is heat labile. So, excess heat can reduce the thiamin content of the ration ingredients.	As a coenzyme in energy and protein metabolism. Promotes appetite and growth, required for normal carbohydrate metabolism, and aids reproduction.	**Deficiency symptoms**—Loss of appetite and poor growth, diarrhea, dead or weak offspring, slow pulse, low body temperature, and flabby heart.
Vitamin B-6 (Pyridoxine, Pyridoxal, Pyridoxamine)		An important cofactor for many amino acid enzyme systems. Vitamin B-6 also plays a crucial role in central nervous system function.	**Deficiency symptoms**—Loss of appetite and poor growth, unsteady gait, anemia, exudate around the eyes, and epilepticlike fits (convulsions).
Vitamin B-12 (Cobalamins)	Pigs fed ingredients of plant origin and housed on slotted floors.	Vitamin B-12 contains the trace element cobalt in its molecule. As a coenzyme, B-12 is involved in the synthesis of methyl groups derived from formate, glycine, or serene, and their transfer to homocystine to reform methionine. It is also important in the methylation of uracil to form thymine, which is converted to thymidine and used for the synthesis of DNA.	**Deficiency symptoms**—Loss of appetite, reduced weight gain, rough skin and hair coat, irritability, hypersensitivity, hind leg incoordination, and anemia.
Vitamin C (Ascorbic Acid, Dehydroascorbic Acid)		Vitamin C is a water-soluble antioxidant that is involved in the formation and maintenance of collagen, absorption and movement of iron, metabolism of fats and lipids, cholesterol control, sound teeth and bones, strong capillary walls and healthy blood vessels, metabolism of folic acid, and as a general antioxidant.	**Deficiency symptoms**—No specific symptoms noted when there is a deficiency.

[1]As used herein, the distinction between *vitamin requirements* and *recommended allowances* is as follows: In vitamin requirements, no margins of safety are included intentionally; whereas in recommended allowances, margins of safety are provided in order to compensate for variations in feed composition, environment, and possible losses during storage or processing.

Vitamin E

Vitamin E is a biological antioxidant which protects unsaturated fats against oxidation. Eight naturally occurring compounds called tocopherols have vitamin E activity, the most active of which is alpha-tocopherol. Since cell membranes in the animal body contain unsaturated fat, a vitamin E deficiency may result in oxidative damage to the cell. This is manifested in the pig by liver necrosis, pale muscle, mulberry heart, edema, and sudden death.

For many years, the primary source of vitamin E in feed was the tocopherols found in green plants and seeds. However, oxidation rapidly destroys vitamin E. For example, vitamin E losses of 50 to 70% can occur in alfalfa stored at 90°F for 12 weeks; and losses of 5 to 30% can occur during dehydratrion of alfalfa.

Also, storage of high-moisture grain or its treatment with organic acids greatly reduces its vitamin E content. Therefore, predicting the amount of vitamin E activity in feed ingredients is difficult.

The trace element selenium also functions with vitamin E in protecting the body against oxidative damage. The need for vitamin E is more acute when swine feeds are low in selenium. Thus, in areas where feed ingredients are low in selenium and where a majority of swine are raised in confinement without access to forages, supplemental vitamin E or selenium, or both, are important.

Vitamin E toxicity has not been reported in swine. Levels as high as 45 IU/lb of ration have been fed to growing pigs without toxic effects.

Vitamin K

Vitamin K exists in three forms: Phylloquinone (K_1), Menaquinone (K_2), and menadione (K_3). Menadione is the synthetic form of vitamin K which has the same cyclic structure as vitamin K_1 and K_2. Vitamin K_1 occurs naturally in green plants. Vitamin K_2 is present in microorganisms and is formed by intestinal bacteria.

(Continued)

Vitamin Requirements[1]				
Vitamins/ Animal/Day	**Vitamin Content of Ration**	**Recommended Allowances[1]**	**Practical Sources of the Vitamin**	**Comments**
*Variable according to class, age, and weight of swine (see Tables 23–1 and 23–2).	*Variable according to class, age, and weight of swine (see Tables 23–5 and 23–6).	Variable according to class, age, and weight of swine (see Table 23–8).	Calcium pantothenate (only the D isomer has vitamin activity). Dried milk products, condensed fish solubles, and alfalfa meal.	Widely distributed and occurs in practically all feedstuffs. However, the quantity present may not always be sufficient to meet the needs of the pig.
*Variable according to class, age, and weight of swine (see Tables 23–1 and 23–2).	*Variable according to class, age, and weight of swine (see Tables 23–5 and 23–6).	Variable according to class, age, and weight of swine (see Table 23–8).	Synthetic riboflavin. Yeast. Milk and milk products. Meat scraps and fish meal.	Riboflavin is apt to be lacking in swine rations that do not contain animal product sources.
*Variable according to class, age, and weight of swine (see Tables 23–1 and 23–2).	*Variable according to class, age, and weight of swine (see Tables 23–5 and 23–6).	Practical rations are adequate in thiamin.	Thiamin hydrochloride. Thiamin mononitrate. Rice polish. Wheat germ meal. Yeast. The oilseed meals. Distillers' solubles.	
*Variable according to class, age, and weight of swine (see Tables 23–1 and 23–2).	*Variable according to class, age, and weight of swine (see Tables 23–5 and 23–6).	Practical rations are adequate in vitamin B–6.	Pyridoxine hydrochloride. Rich supplemental sources include rice polish, wheat germ, and yeast.	The vitamin B–6 content of normal feed is usually sufficient.
*Variable according to class, age, and weight of swine (see Table 23–1 and 23–2).	*Variable according to class, age, and weight of swine (see Tables 23–5 and 23–6).	Variable according to class, age, and weight of swine (see Table 23–8).	Synthetic B–12, which is produced commercially by microbial fermentation. Protein supplements of animal origin. Fermentation products.	Vitamin B–12 is apt to be lacking in swine rations. Synthesis of vitamin B–12 by intestinal flora may supplement dietary sources. B–12 contains the trace element cobalt; hence, the synthesis of B–12 in the intestines is dependent on the presence of cobalt in the feed. This may be the major, if not the only, function of cobalt as an essential nutrient.
		Practical rations plus intestinal synthesis is adequate.	Vitamin C (ascorbic acid).	Normally, pigs are able to synthesize vitamin C in amounts sufficient to meet their requirements. However, there is limited evidence that dietary ascorbic acid is beneficial under some conditions.

Where preceded by an asterisk, the vitamin requirements, recommended allowances, and other facts presented herein were taken from *Nutrient Requirements of Swine,* 9th rev. ed., NRC-National Academy of Sciences, 1988.

Vitamin K is one of the essential factors necessary for proper blood clotting. Generally, sufficient amounts of vitamin K_2 are synthesized by bacteria in the digestive tract to meet the needs of swine, with the vitamin K_2 absorbed directly from the gut or obtained by coprophagy. However, this synthesis may be inadequate in situations when high antibiotic levels are used, where clotting inhibitors (dicoumarol) may be present from molds in the feed, or where there is excess calcium.

The deficiency symptoms of vitamin K are slow growth, hemorrhage, prolonged blood-clotting time, and hyperirritability.

WATER-SOLUBLE VITAMINS

Biotin, niacin, pantothenic acid, riboflavin, and vitamin B-12 are the water-soluble vitamins most likely to be deficient in swine rations. However, occasionally the other water-soluble vitamins are deficient. So, choline, folacin, thiamin, vitamin B-6, and vitamin C are also discussed in the sections that follow.

Biotin

Biotin is important metabolically as a cofactor for several enzymes.

Biotin is present in adequate amounts in most common feedstuffs, but its bioavailability varies greatly among ingredients.

In general, biotin supplementation has not improved the performance of baby pigs or of growing-finishing hogs fed a variety of feedstuffs. However, biotin supplementation of sow rations has significantly improved reproductive performance, including the number of pigs farrowed and weaned, litter weaning weight, and number of days from weaning to estrus. Also, biotin supplementation improved hoof hardness of sows.

Choline

Choline is included with the vitamins, but does not qualify as a true vitamin because it is required at far greater levels than true vitamins and is not known to participate in any enzyme system.

Choline functions as a *methyl donor* in metabolism and can lower the requirement for methionine to the extent that methionine is used for this function. It is also a constituent of some important phospholipids in the body, and it is involved in the mobilization and oxidation of the fatty acids in the liver. While choline is probably present in adequate amounts in most practical swine rations, studies have shown that more live pigs per litter are born and weaned when sows receive supplemental choline throughout gestation. Also, since proteins supply most of the choline in swine rations, a reduction in ration protein when synthetic lysine is incorporated in the ration may at the same time create a choline-deficient situation.

Choline-deficiency symptoms are lowered reproduction, pigs weak at birth, lack of coordination, and spraddled legs.

Folacin (Folic Acid)

Folacin includes a group of compounds with folic acid activity. A deficiency of folacin causes a disturbance in the metabolism of single-carbon compounds, including the synthesis of methyl groups, serine, purines, and thymine. Folacin is involved in the metabolic conversion of amino acids: of serine to glycine, and of homocysteine to methionine.

Folacin deficiency in pigs leads to slow weight gain, fading hair color, and anemia.

The folacin in feedstuffs commonly fed to swine, along with bacterial synthesis within the intestinal tract, usually adequately meets the requirement for all classes of swine.

Niacin (Nicotinic Acid, Nicotinamide)

This vitamin plays an important role in body metabolism as a constituent of two coenzymes, nicotinamide-adenine dinucleotide (NAD) and nicotinamide-adenine dinucleotide phosphate (NADP). These coenzymes in the pig are essential for the metabolism of carbohydrates, proteins, and lipids.

A niacin deficiency results in *pig pellagra,* which is characterized by diarrhea, rough skin, and retarded growth.

Recent research has shown that the niacin in cereal grains is in a bound form, and that the niacin of corn may be almost completely unavailable to swine. It should be assumed that all niacin in cereal grains and their by-products is completely unavailable. The protein source and tryptophan content of the ration can also affect the niacin requirement since tryptophan, an amino acid, can be converted to niacin.

Pantothenic Acid (Vitamin B-3)

Pantothenic acid is a constituent of coenzyme A, which plays a key role in energy metabolism.

The biological availablility of pantothenic acid is high from barley, wheat, and soybean meal, but low from corn and grain sorghum.

A lack of this vitamin may result in poor growth, diarrhea, loss of hair, and a high-stepping gait of the hind legs—often called *goose stepping.*

Dried milk products, condensed fish solubles, and alfalfa meal are good natural sources of pantothenic acid. It is also available in synthetic form as calcium pantothenate.

Fig. 23-12. Pig showing pantothenic acid-deficiency symptoms. Note high, goose-stepping gait. (Courtesy, University of California, Davis)

Riboflavin (Vitamin B-2)

Riboflavin is sometimes referred to as vitamin B-2. It functions in the body as a constituent of two coenzymes, flavin mononucleotide and flavin adenine dinucleotide. Riboflavin is important in the metabolism of proteins, fats, and carbohydrates.

In growing swine, a deficiency may cause loss of appetite, stiffness, dermatitis, and eye problems. Poor conception and reproduction have been noted in gilts fed riboflavin-deficient rations. Pigs may be born prematurely, dead, or too weak to survive.

Milk products and other animal proteins, alfalfa meal, and distillers' solubles are good natural sources of riboflavin. Also, synthetic riboflavin is available.

Thiamin (Vitamin B-1)

Thiamin is essential for carbohydrate and protein metabolism. The coenzyme thiamin pyrophosphate is essential for the oxidative decarboxylation of alpha-keto acids.

Pigs must have a dietary source of thiamin, because unlike ruminants, they cannot synthesize sufficient of it in the digestive tract. Normally, the thiamin content of feeds is sufficient to meet the needs of swine since cereal grains, which are major swine feeds, are good sources. However, deficiencies may result from: (1) heating feed ingredients excessively in processing, because thiamin is heat labile; (2) treating feedstuffs with sulfur dioxide, which inactivates the thiamin; or (3) feeding unprocessed fish or fish scraps of certain types of fish that contain the antithiamin factor known as thiaminase.

Fig. 23-13. Thiamin deficiency in littermate pigs. *Right:* Pig received no thiamin. *Left:* Pig received the equivalent of 2 mg thiamin/100 lb liveweight. Otherwise, their diets were the same. (Courtesy, USDA)

Vitamin B-6 (Pyridoxine, Pyridoxal, Pyridoxamine)

Vitamin B-6 occurs in feedstuffs as pyridoxine, pyridoxal, pyridoxamine, and pyridoxal phosphate. Pyridoxal phosphate is an important cofactor for many amino acid enzyme systems. Vitamin B-6 plays a key role in central nervous system function.

Supplementation of grain—soybean meal rations with vitamin B-6 is generally unnecessary, because the concentration and availability of vitamin B-6 in the feed ingredients will meet the pig's requirement.

A deficiency of vitamin B-6 will reduce appetite and growth rate. Advanced deficiency will result in exudate around the eyes, convulsions, lack of coordination, coma, and death.

Vitamin B-12 (Cobalamins)

Vitamin B-12 was discovered in 1948. Originally, it was known as the *animal protein factor* because of its association with ingredients of animal origin.

Vitamin B-12 stimulates the appetite, increases rate of growth, improves feed efficiency, and is necessary for normal reproduction.

Pigs require B-12 but responses to supplementation have been variable, primarily because of the synthesis of vitamin B-12 by microorganisms in the environment and within the intestinal tract, along with the pig's inclination toward coprophagy (feces eating).

Vitamin C (Ascorbic Acid, Dehydroascorbic Acid)

Vitamin C (ascorbic acid) is an intioxidant that is involved in a variety of metabolic processes. It is also essential for hydroxylation of proline and lysine, which are integral constituents of collagen. Collagen is essential for growth of cartilage and bone. Vitamin C enhances the formation of intercellular material, the formation of bone matrix, and the formation of tooth dentin.

A dietary source of vitamin C is essential for primates and guinea pigs, but domestic swine can synthesize this vitamin. However, some claims have been made that, under certain conditions—as during periods of excessive stress, pigs may not be able to synthesize vitamin C fast enough to meet their requirement. Yet, the conditions under which supplemental vitamin C may be beneficial are not well defined, so no recommendation for vitamin C is given for the pig.

UNIDENTIFIED FACTORS

Some unidentified factor or factors may, under certain circumstances, be involved in securing optimum results during the critical periods (early growth and gestation-lactation). Sources of the unknown factor or factors are distillers' dried solubles, fish solubles, dried whey, grass juice concentrate, soil, alfalfa meal, brewers' dried yeast, liver, and pasture.

Water

Water is so common that it is seldom thought of as a nutrient. However, it is the largest single part of nearly all living things. The body of a baby pig is about three-fourths water.

Water performs many tasks in the body. It makes up most of the blood which carries nutrients to the cells and carries waste products away; it is necessary in most of the body's chemical reactions; it is the body's built-in cooling system—it regulates the temperature, and it serves as a lubricant.

Life on earth would not be possible without water. An animal can live longer without feed than without water.

In general, swine will consume ¼ to ⅓ gal of water for every pound of dry feed. The higher the temperature, the greater the water consumption. It is preferable that swine have access to automatic waterers, with cool, clean water available at all times. Otherwise, they should be hand watered at least twice daily. During winter, the drinking water should not be permitted to fall below 40°F. The estimated water needs of various classes of swine are:

Class of Swine	Water Consumption (gal/head/day)
Gestating sows	2–3
Lactating sows	4–5
Weaned pigs (15–40 lb)	0.5–1
Growing pigs (40–110 lb)	1
Finishing pigs (110–240 lb)	1.5–2

Fig. 23-14. Pigs drinking from nipple waterers. (Courtesy, National Pork Producers Council, Des Moines, Iowa)

Additives

Certain feed additives have become standard ingredients of swine rations, especially for pigs from birth to market weight. They are not nutrients as such; hence, they should not be considered as dietary essentials. Although many different additives are used in swine rations, antibiotics and sulfas are most common.

Antibiotics are widely used as feed additives to stimulate growth, improve feed efficiency, secure uniformity of performance, and control infections. The response secured from their use depends on (1) the age of the pig, (2) the sanitary conditions, (3) the level fed, (4) the health and environment of the animal, (5) the type of ration, and (6) the season of the year. When fed to young, unthrifty pigs, antibiotics have increased growth rate by over 200%. For growing-finishing hogs under good sanitary conditions, antibiotics generally result in about 10% faster gains on 5% less feed. Pigs up to 100 lb weight give the greatest response to antibiotic feeding. Experimental results have been inconsistent relative to the value of antibiotics in brood sow rations, but it appears that breeding herds with a high disease level may show a favorable response.

Regardless of the standard of sanitation practices, within each swine facility there is an *environmental disease level.* It may or may not be obvious to the casual observer, depending upon its degree of severity. Just how antibiotics counter these unfavorable conditions and bring about improved performance in growing swine is not exactly known. However, they apparently do have an influence on the intestinal bacteria and improve the health of the animal, as unthrifty pigs respond more to antibiotic feeding than do healthy ones. Antibiotics may function as follows:

1. They favor bacteria that synthesize known or unknown nutrients needed by the animal.

2. They inhibit the growth of nutrient-destroying microorganisms.

3. They improve the availability andor absorption of certain nutrients.

4. They inhibit the growth of organisms that produce excessive amounts of ammonia and other toxic waste products in the intestine.

5. They prevent or control certain diseases in the intestinal tract and other parts of the body.

Antibiotics should be used with care, following the manufacturer's recommendations explicitly.

In addition to antibiotics, certain other antimicrobial compounds can be used as feed additives, either (1) at low levels for promotion of growth and improvement in feed efficiency, or (2) for treatment and prevention of disease. Among these are nitrofurans, sulfonamides, copper, and arsenicals. Such compounds, alone or in combination with other antimicrobial compounds, should be used only at approved levels and for the specific purpose for which they are authorized.

CAUTION: Toxic residues in meat, milk, and/or eggs may result from the improper use of certain antimicrobial compounds, pesticides, tranquilizers, and other chemicals. Regardless of the material used, the following safety precautions should be observed:

1. Use only those materials specifically authorized for use in swine. Regulations change frequently, so check them often.
2. Carefully follow label directions.
3. Do not mix unauthorized combinations.
4. Use minimal effective amounts.
5. Add or apply precise quantities.
6. Observe the proper interval between use and slaughter (or use for food).

Table 23–11 is a partial list of approved additives for use in swine rations.

• **Sulfonamides (sulfas)**—*Sulfonamides are organic compounds with bactericidal and growth promotant properties similar to those of the antibiotics.* But, unlike the antibiotics, they are produced chemically rather than microbiologically. Also, the therapeutic use of sulfonamides preceded that of the antibiotics; they have been widely used in human and veterinary medicine since the mid-1930s. During World War II, the sulfas were extensively used to treat battle wounds; they saved many lives and limbs, and they were hailed as the miracle drugs of the time. By 1955, two sulfas—sulfamethazine and sulfathiazole—were found to be effective in the treatment of atrophic rhinitis in swine. Soon, they were also used to treat or prevent swine bacterial enteritis, bacterial pneumonia, salmonellosis, and dysentery. Also, it was found that, in subtherapeutic doses, they would make for more rapid and efficient swine gains. They proved to be effective disease fighters and growth promotants. All went well until the 1960s and early 1970s, when the U.S. Department of Agriculture (USDA) discovered residues of sulfa drugs in many pork carcasses. The immediate concern was that a small percentage of the human population is hypersensitive to sulfas and may develop an allergic reaction after consuming low levels of sulfas in pork (or in other foods). Then, in 1980, a preliminary Food and Drug Administration (FDA) study showed that massive doses of sulfamethazine caused thyroid tumors in laboratory mice.

As a result of finding sulfa residues in pork, followed by an experiment showing that sulfamethazine produced thyroid tumors in mice, in 1987 the USDA and FDA launched a campaign to alleviate sulfa residues in pork.

Animal products are deemed to be in violation of sulfa residue regulations if levels of 0.1 ppm, or more, sulfa are found in the muscle, liver, kidney, eggs, or milk. Research at the University of Kentucky showed that as little as 1 g of sulfa per ton of complete feed can cause 100% violative liver residues and 63% kidney contaminations. This means that as little as ¼ tsp of sulfa per ton of complete feed can result in pork carcasses that are in violation. The sulfonamides are also capable of premise contamination. In 1988, the violative residue of hogs slaughtered in the United States was about 4%, down from the alarming 15% monitored by the USDA in the 1970s.

Pertinent facts about sulfas, their uses and resulting concerns, follow:

1. **Prevalence of use.** It is estimated that 60% of all the feeds that pigs receive in starter and grower-finishing rations

TABLE 23-11
PARTIAL LIST OF FEED ADDITIVES IN SWINE RATIONS[1 2]

Additive		Withdraw Before Slaughter	Manufacturers' Claims[3]													
Chemical Name	Trade Name		A	B	C	D	E	F	G	H	I	J	K	L	M	N
Apramycin	Aparlan	28 days						x								
Arsanilic acid or sodium arsanilate	Pro-Gen	5 days	x				x			x	x					
Arsanilic acid/bacitracin		5 days	x						x							
Arsanilic acid/chlortetracycline		5 days	x				x		x							
Arsanilic acid/furazolidone/oxytetracycline	Furox-O-A 390	5 days	x			x			x				x			
Arsanilic acid/oxytetracycline	Furox/OXTC	5 days	x				x		x							
Arsanilic acid/penicillin		5 days	x			x	x		x	x						
Arsanilic acid/streptomycin/penicillin		5 days	x				x		x							
Bacitracin[4]	BMD; Baciferm; Albac	None	x			x				x						
Bambermycin	Flavomycin	None	x													
Carbadox	Mecadox	10 weeks	x					x		x						
Chlortetracycline	Aureomycin; CLTC; PhiChlor	None	x			x	x		x				x		x	x
Chlortetracycline/sulfamethazine/penicillin	Aureo SP 250; PhiChlor 250	15 days	x			x	x						x			x
Chlortetracycline/sulfathiazole/penicillin	CSP 250	7 days	x				x						x			x
Furazolidone	Furox; nf-180	5 days	x	x		x	x		x		x					
Furazolidone/oxytetracycline		5 days		x												
Lincomycin	Lincomix	6 days							x	x	x					
Neomycin base[5]	Neomix; Neomycin sulfate	Varies[5]					x				x					
Nitrofurazone	nfz; Amifur	5 days					x									
Oxytetracycline[6]	Terramycin; OXTC	Varies[6]	x			x	x		x		x		x		x	
Oxytetracycline/neomycin base	Neo-terramycin; Neo/OXTC	10 days			x	x	x		x		x					
Penicillin	Penicillin P-100	None	x													
Penicillin/streptomycin		None	x			x	x									
Roxarsone	3-Nitro	5 days	x								x					
Roxarsone/bacitracin		5 days	x													
Roxarsone/chlortetracycline		5 days	x													
Tiamulin	Denagard	2 days	x							x						
Tylosin	Tylan	None	x						x	x	x		x			
Tylosin/sulfamethazine	Tylan/sulfa	15 days							x			x	x	x		
Virginiamycin	Stafac	None	x							x	x					

[1]Adapted by the authors from *Life Cycle Swine Nutrition*, Iowa State University, Ames, PM-489, June, 1988.

[2]This is only a partial list of the feed additives and combinations available. This list does not contain the use level approved for each additive. Users are advised to read the product label and adhere to use recommendations of the additive manufacturer. The regulations governing the use of these additives are subject to change.

[3]The manufacturer claims, use levels and limitations are those for which FDA clearance was obtained. It is essential that the rules and regulations governing the use of feed additives be followed. When used properly, feed additives can greatly benefit the swine producer. The following claims are indicated for the appropriate additive:

a. Promote growth and improve feed efficiency.
b. Prevention of bacterial scours of baby pigs when fed in sow's ration.
c. Treatment of bacterial scours of baby pigs when fed in sow's ration.
d. Prevention of bacterial enteritis (scours).
e. Treatment of bacterial enteritis (scours).
f. Control of bacterial enteritis (scours).
g. Prevention of swine dysentery (bloody scours).
h. Control of swine dysentery (bloody scours).
i. Treatment of swine dysentery (bloody scours).
j. Control of swine pneumonias caused by bacterial pathogens (*P. multocida* and/or *C. pyogenes*).
k. Maintenance of weight gain in presence of atrophic rhinitis.
l. Lower incidence and severity of Bordetella bronchiseptica rhinitis.
m. Aid in treatment and reducing spreading of Leptospirosis.
n. Reduce cervical abcesses.

[4]Bacitracin is available in several forms including zinc and methylene disalicylate (MD) derivatives.

[5]Neomycin levels are expressed as neomycin base, which is equivalent to 70% of the neomycin sulfate level (*i.e.* 4.9 oz *(140 g)* neomycin base is equal to 7.1 oz *(200 g)* of neomycin sulfate). Withdraw from feed 20 days before slaughter when neomycin base level is 4.9 oz *(140 g)*/ton and 5 days before slaughter when neomycin base is below 4.9 oz *(140 g)*/ton.

[6]At 17.6 oz *(500 g)*/ton level, withdraw 5 days before slaughter.

contain a sulfa, and that 80% of all pigs receive a sulfa sometime during their lives.

2. **Approved sulfas for swine.** Of the more than 5,000 sulfa compounds that have been synthesized, only two—sulfamethazine and sulfathiazole—are approved for use in swine feeds. Two additional sulfa drugs, sulfamerazine and sulfapyridine, are approved, along with sulfamethazine and sulfathiazole, for use as water medicants for swine.

Sulfamethazine and sulfathiazole are approved for use in swine feeds on the following basis: They must be used at one level only—100 g per ton; they must be used in legal combinations approved by FDA (it is illegal to use sulfas alone for swine); they are approved for the prevention and treatment of enteritis, atrophic rhinitis, cervical abscesses, and leptospirosis (Although sulfas will improve feed efficiency and rate of gain of swine, they are not approved by FDA as growth promotants.); and there is a minimal 15-day withdrawal period for feed and water containing sulfamethazine, and a minimal 7-day withdrawal period for feed and water containing sulfathiazole.

Consumers are concerned about the safety of the foods that they eat. Drug residues undermine their confidence, and reports of food that may be cancerous cause alarm. Consumer perception can make or break a food! So, both government enforcement agencies and swine producers are anxious to alleviate the sulfa residue problem.

In 1988, the U.S. Department of Agriculture, Food Safety Inspection Service, announced that they would promulgate the following intensified program:

1. Evolve with and adopt a swine indentification program that would allow the government to trace hogs back to their farm of origin.
2. Develop regulations for and implement a rapid sulfamethazine residue test for use in slaughter plants.
3. Encourage swine producers to use the Sulfa-On-Site (SOS) test, a rapid on-site sulfamethiazine residue test, on their live hogs before marketing.
4. Develop new and improved tests for the presence of sulfamethazine.

But effective sulfa control must begin with swine producers all over America. They can avoid residues by adhering rigidly to the following program:

1. Reading and following the instructions on product labels.
2. Limiting sulfa additives to approved feed levels; and not adding sulfas to water or using sulfas to treat animals if feed contains sulfas.
3. Using only approved sulfa combinations in the feed, and not adding powdered or granulated sulfas alone, which is illegal.
4. Keeping a written record of all medications.
5. Observing sulfa withdrawal times before marketing hogs—15 days for sulfamethazine combinations, and 7 days for sulfathiazole combinations.
6. Avoiding combination products with powdered sulfa. It is difficult to clean equipment following the use of powder; so, only the granular form should be used.
7. Sequencing the milling of medicated feeds, never mixing non-medicated feed following the mixing of medicated feed; and flushing feed milling and transporting equipment following the processing/transporting of medicated feeds.
8. Cleaning feed storage areas.
9. Saving until after slaughter samples of finishing feeds, properly labeled as to which hogs received the feed.
10. Identifying hogs that are ready for market, and keeping them separate from hogs receiving a sulfa; and cleaning the pens of market hogs daily after sulfa is withdrawn, to prevent sulfas from recycling through manure and urine.

If FDA obtains additional research evidence that sulfa residues are carcinogenic, and if sulfa residues are not brought under control, FDA has the following options: (1) lowering the tolerance further (below 0.1 ppm); (2) instituting stricter laws regulating the compounding of feed additives containing sulfur; or (3) banning sulfamethazine, but allowing the use of sulfathiazole which is excreted more rapidly and is less likely to cause residue problems, or banning all sulfa drugs.

When used properly and withdrawn at the specified time before marketing, sulfa drugs have been shown to be safe.

• **Porcine somatotropin (PST) for swine**—Porcine somatotropin (PST) is the scientific name for the growth hormone in swine. It is normally produced by the anterior pituitary at the base of the brain. It stimulates protein synthesis and growth in most tissues of the body, and it causes breakdown of fat deposits in adipose tissue.

Scientists have known for sometime that providing extra somatotropin to a pig will cause it to grow more rapidly and produce leaner pork. But it was not feasible to apply this knowledge practically, because the only way to get porcine somatotropin was to isolate it from the pituitary glands of slaughtered hogs, which was very expensive because of the very low yield of hormone from each hog. Recently, it has become economically feasible to mass produce this compound by recombinant DNA procedures using *E. coli* bacteria. So, large supplies of PST are now available. The major remaining barrier to the use of PST is a delivery system—a method of administering. Presently, daily injections are required to capture the benefits of growth and carcass changes. PST is a naturally occurring protein that is broken down immediately by digestion; so, an orally active feed ingredient won't work.

Experiments have consistently shown that PST can make the following great leaps forward in the swine industry: 15–20% higher average daily gain, 20–30% improvement in feed conversion efficiency, 10–15% improvement in muscle mass, and 25–30% less backfat.

In addition to evolving with a practical method of administering PST, attention is being focused on the following aspects of its use:

1. **Nutrition.** Pigs receiving PST grow faster and more efficiently, but they require more protein and amino acids in the ration.
2. **Metabolism.** Scientists estimate that PST-treated pigs have a 25% higher metabolism than untreated animals. Thus, in confinement systems, the building heating and ventilation designs and specifications will need to be revised.

Also, the ability of PST-treated animals to handle heat and cold stress needs to be studied.

3. **Market grades.** Both live hog and carcass grades will need to be changed to reflect product improvement and provide adequate incentives for producers who use PST.
4. **Reproductive efficiency.** The reproductive efficiency of gilts treated with PST needs to be studied.
5. **Consumer acceptance.** Consumer acceptance of pork obtained from PST-treated hogs is very important. Health conscious consumers need to be made aware of the fact that PST-treated hogs produce leaner carcasses, that somatotropin is a *natural* product found in the blood of normal animals, that there are no residue problems, and that PST is easily destroyed by heat or digestion.

Before PST can be used, it must have FDA approval. Then, its acceptance and use will be determined by a practical method of administering the product, and producer and consumer acceptance.

(Also see Chapter 13, section on "Porcine Somatotropin [PST] for Swine.")

FEEDS FOR SWINE

Throughout the world, swine are raised on a great variety of feeds, including numerous by-products. Except when on pasture or when ground dry forages are incorporated in the ration, they eat relatively little roughage; only 4.3% of the total feed consumed by swine in the United States is derived from roughages.

Fig. 23-15. Complete ration piped (automated) from feed bin on right to self-feeders, for growing-finishing pigs. (Courtesy, USDA)

Although corn is the chief concentrate fed to swine, the agriculture of the 50 states is very diverse, and the diet of the pig is readily adapted to feeds produced locally. A similar adaptation in feeding practices is found in other countries. Thus, in most sections of the world, swine are fed predominantly on homegrown feeds. Ireland depends largely upon potatoes and dairy by-products in combination with barley; the swine industry of Denmark has been built up to augment the dairy industry, and milk and whey supplementing homegrown and imported cereals (mostly barley); in Germany, the pig is fed on such crops as potatoes, sugar beets, and green forage; and in China, the pig is primarily a scavenger, competing very little for grains suitable for human consumption.

Thus, a swine producer generally has a wide variety of ingredients from which to choose in formulating a ration. Each ingredient may contain several nutrients in varying amounts. Because ingredients vary in price, and in amount and quality of nutrients contained, judgment must be exercised in the choice made.

The nutrient present in the largest amount determines how an ingredient is classified. Thus, corn is classed as an energy feed because it is high in calories, and soybean meal is classified as a protein supplement because it is high in protein content.

Concentrates

Because of their simple monogastric stomach, swine consume more concentrates and less roughages than any other class of large farm animals. This characteristic gives pigs limited opportunity to consume large quantities of calcium and of vitamin-rich and better quality protein roughages, with the result that they suffer from more nutritional deficiencies than any other species except poultry.

Although most concentrate feeds are not suitable as the sole ration for hogs, it must be realized that swine can utilize a larger variety of feeds to greater advantage than other farm animals. In general, the grain crops—corn, barley, wheat, oats, rye, and the sorghums—constitute the major component of the swine ration. However, sweet potatoes and peanuts are successfully and extensively used in the South, soybeans in the central states, and peas in the Northwest. In those districts where they are grown, potatoes (cull) also are utilized in considerable quantities in feeding hogs. In addition, in almost every section of the country one or more by-product feeds are fed to hogs—including the by-products of the fishing, meat-packing, milling, and dairy industries. Human food wastes, such as garbage, are also fed extensively.

The protein and vitamin requirements of the monogastric pig differ very greatly from those of the ruminant, for the latter improves the quality of proteins and creates certain vitamins through bacterial synthesis.

Despite all this, it is possible to meet the nutritive needs of the pig on concentrated feeds by keeping in mind the following facts when balancing the ration:

1. The cereal grains and their by-products are relatively good in phosphorus, but low in calcium and the other minerals.
2. Except for the carotene content of yellow corn and green peas, the grains are very poor sources of vitamin A.
3. Most cereal grains supply proteins of poor quality and quantity.
4. Protein supplements of animal origin and soybean meal generally supply proteins of high quality, whereas proteins of plant origin, other than soybean meal, generally supply proteins of low quality.
5. Because of the inadequacies of most concentrates, it is usually necessary to rely on fortifications with minerals and vitamins.

ENERGY FEEDS

Carbohydrates and fats may be classed together as energy feeds for swine. Three essential fatty acids—linoleic, linolenic, and arachidonic acid—are required by swine, but cereal grains contain adequate quantities of fat to meet the fatty acid requirements. The ingredients commonly used as sources of energy are corn, barley, sorghum, wheat, and fats and oils. Full replacement of corn with barley, oats, sorghum, or wheat will decrease dressing percentage by approximately 1% and decrease backfat by about 0.1 in.

(Also see Table 23-28, Feed Substitution Table for Swine, in this chapter; Chapter 10, Grains/High-Energy Feeds; and Chapter 12, By-product Feeds/Crop Residues.)

Corn

In this country, corn and swine production have always gone hand in hand. Normally, about one-fourth of the U.S. corn crop is fed to hogs. Because of the dominant position of corn as a swine feed, it is herein singled out for further elucidation.

Corn is an excellent energy feed for all classes of swine. It is an ideal finishing feed because it is high in digestible carbohydrate (starch), low in fiber, and very palatable. Also, it can be fed in a variety of ways. It may be fed shelled, ground, mixed, or free-choice, or even as ear corn. It may be dry or high-moisture.

In spite of its virtues, corn alone will not keep pigs alive. It contains 7 to 9% protein, but the protein is deficient in some of the essential amino acids required by the weanling pig, especially lysine and tryptophan (see Fig. 23-16). It is also so deficient in calcium and other minerals, and so inadequate in vitamin content, that pigs will die if they are limited to a ration containing only corn. So, corn must be supplemented with a protein that makes up its amino acid deficiencies. Equally important are the needed minerals and vitamins. When properly supplemented, corn is an excellent energy feed for all classes of swine.

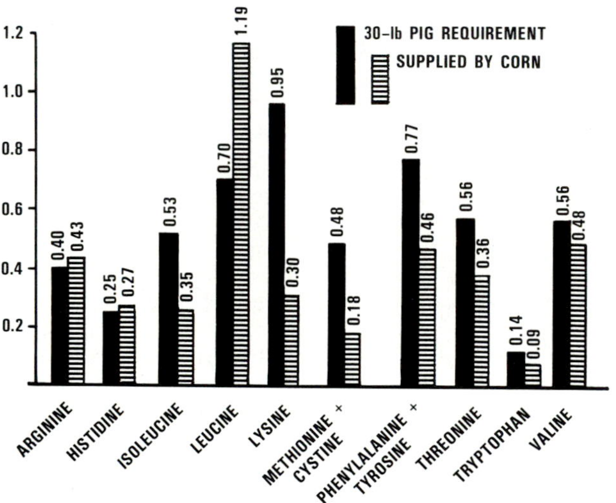

Fig. 23-16. Hogs cannot live by corn alone! This shows the amino acid content of corn (in %) in comparison with the ration requirements (in %) of a 30-lb pig. Note that corn is especially deficient in lysine and tryptophan. (Sources of data: Amino acid ration requirements [in % as-fed] of a 30-lb pig from *Nutrient Requirements of Swine*, 9th rev. ed., National Research Council, National Academy Press, 1988, p. 50, Table 5-1 [see Table 23-5 in this chapter]. Amino acid content of corn from Table V-2, Amino Acid, Composition of Feeds, in this book.)

It should be noted that corn is higher in fat than barley or wheat (4% vs less than 2%). Fat not only contributes to the high-energy content of corn, but it also improves its palatability and feeding properties in general.

• **High-moisture corn**—High-moisture corn may be substituted for dry corn on a dry matter basis with little effect on overall performance, in either feed conversion or rate of gain. It is usually fed free-choice. Protein, vitamins, and minerals can be supplemented by either hand-feeding daily or feeding free-choice in separate feeders. High-moisture corn is usually very palatable, with the result that when fed free-choice, pigs will often overeat on corn and undereat on the protein supplement. When the two are metered together, separation is likely to occur unless the corn is cracked or ground. If high-moisture corn is fed in a complete ration, the feed should be prepared frequently (every 1 or 2 days) to prevent spoilage. The feed should also be prevented from bridging in the feeders.

Thus, the choice between high moisture grain storage vs dry storage should be based largely on economics. The cost per bushel of storage capacity and the operating cost should also be compared in reaching a decision. Also, there should be an awareness that when high-moisture corn is put in high moisture storage facilities, it is suitable for feeding only and cannot be sold readily on the open market as dry corn.

(For the relative value of U.S. No. 2 corn [15.5% moisture] as affected by changes in moisture, see Chapter 16, section on "Moisture is Important," Table 16-5.)

• **High-lysine corn (Opaque-2)**—Corn is now being bred that is much higher than normal corn in lysine and tryptophan; hence, it has a better balance of the amino acids for swine. Also, the high-lysine corn (called Opaque-2) is higher in total protein than normal corn. However, Opaque-2 is lower in leucine than regular corn.

Fig. 23-17. Amino acid deficiency. Littermate pigs. *Big pig (B):* Pig fed an adequate diet containing Opaque-2 corn (high-lysine corn). *Little pig (C):* Pig fed inadequate amounts of lysine and tryptophan. (Courtesy, Cornell University, Ithaca, N.Y.)

Currently, little high-lysine corn is available commercially. However, corn breeders are working to perfect Opaque-2, for it would make for improved nutrition of both animals and people throughout the world, wherever corn (maize) is grown.

The following recommendations are made relative to using high-lysine corn:

1. Check the moisture carefully when mechanically drying high-lysine corn, because it drys more rapidly than normal corn.
2. Grind high-lysine corn more coarsely than regular corn, because it tends to powder easily during processing. A ½-in. screen works best. Some producers prefer a roller mill when processing high-lysine corn.
3. Rations formulated with high-lysine corn may contain 3% less soybean meal (a saving of 60 lb/ton) than when formulated with regular corn, because of the better balance of essential amino acids in the high-lysine corn.

If swine producers consider growing high-lysine corn, they must evaluate such economic factors as the lower yield of high-lysine corn vs the price of normal corn plus supplemental protein.

Fats and Oils

The following types of fats and oils are available and may be added to swine rations as high-energy feeds: animal fat, poultry fat, tallow, lard, corn oil, soybean oil, and other plant seed oils. When of comparable quality, there is little difference in swine performance resulting from the type of fat or oil used.

When fat is included in growing-finishing rations, it is usually added at a level of 4 to 5%. As a result of added fat, daily feed intake usually decreases, daily gain increases slightly, and feed efficiency improves. Each 1% added fat produces about 2% improvement in feed efficiency. Response to fat is greater in the summer than in the winter, because less heat is produced by pigs when they digest fat than when they digest starch in grain. High levels of fat (more than 5%) will increase backfat thickness slightly and reduce overall carcass lean. As little as 2½% added fat (50 lb/ton) reduces dust in mixing and feeding. More than 6 to 8% added fat may cause flowability problems.

When fat is added to rations, the caloric density of the ration increases. So, dietary nutrient density should increase in proportion to the increase in caloric density to maximize swine performance.

When fat is added to brood sow rations at a level of 7½% or more 10 to 14 days before farrowing, it increases the fat content of colostrum and milk and improves baby pig survival by 2 to 3% in herds where survival rates are less than 80%. However, added fat has little effect on litter size, birth weight, or weaning weight.

When oils are fed to pigs, they tend to make the body fat softer. To alleviate this problem, the level of oil must be limited and the oil must be deleted from the ration for a "hardening off" period prior to slaughter. Feeding fish oils gives a disagreeable fishy taste to the pork.

Other Energy Feeds for Swine

Yellow corn is usually the cheapest source of energy over much of the United States. But price fluctuations frequently justify consideration of other feeds; among them, the following: bakery waste, barley, beans (cull), molasses (cane or beet), oats, potatoes (cull, Irish), rye, sorghum (milo), spelt, triticale, and wheat.

Although barley, oats, wheat, milo, and rye are higher in protein than corn, it is noteworthy that their protein is generally of the same poor quality as corn. It is noteworthy, too, that all of the cereal grains have about the same vitamin and mineral deficiencies.

PROTEIN FEEDS

Protein is made up of nitrogenous compounds called amino acids. Protein feeds vary in the kind and amount of amino acids they contain. During the digestion process, the protein in feed is broken down into the various amino acids and the pig recombines them into the kind of protein needed for muscle development and repair of worn-out tissue. Thus, the real need of the pig is for amino acids, not protein as such.

The pig can synthesize some of the amino acids, with the result that they are not required in the ration. However, 10 of the amino acids are termed *essential,* because the body cannot manufacture them in sufficient quantity to permit maximum growth and performance. It is important that ingredients rich in the essential amino acids be used in formulating the ration.

Although it is a common practice to refer to *percent protein* in a ration, this term has little meaning in swine feeds unless there is knowledge concerning the amino acids present. A protein feed is considered to be of good quality when it contains all the essential amino acids in the proportions and amounts needed by the pig.

(Also see Table 23-28, Feed Substitution Table for Swine, in this chapter; Chapter 11, Protein Supplements; and Chapter 12, By-product Feeds/Crop Residues.)

Soybean Meal

Soybean meal is by far the leading high-protein supplement of the United States. In 1984, 75% of the total high-protein animal feed (including both oilseed meals and animal proteins) consisted of soybean meal.

Although soybean meal is marginal in methionine, it is otherwise very well balanced in amino acids. It must be supplemented with minerals and vitamins. Usually, it is not fed free-choice because of its high palatability, which results in pigs eating more than is needed to meet their protein needs.

Producers generally have a choice of buying soybean meal of different protein content, usualy 44 or 49%. The higher protein content meal is the more desirable to use in prestarter and starter rations because much of the hull has been removed. Thus, it is lower in fiber (not more than 3%), higher in energy, and more palatable. Growing-finishing pigs can utilize both meals about equally well; hence, the choice should be on the basis of which is the best buy as determined by the price per unit of protein.

Full-Fat Soybeans

The protein of raw soybeans is poorly utilized by young, growing pigs due to the presence of antitrypsin, a powerful growth inhibitor that affects young swine.

Cooking or roasting at the proper temperature (250°F for 2½ to 3½ minutes in a roaster) destroys this factor and makes soybeans a satisfactory feed for young pigs. However, cooking whole soybeans for brood sows is not necessary; comparable performance results when raw soybeans replace soybean meal on an equal protein basis in rations fed to gestating and/or lactating sows.

Research has shown that whole cooked soybeans can be used to replace soybean meal or other forms of protein supplement in growing-finishing rations. They will increase daily gain up to 5%, and improve feed efficiency by 5 to 10% due to the higher fat content of the whole soybeans

(17 to 18%), which makes for a higher energy ration. However, this improvement in feed efficiency may be offset by the lower protein content of whole soybeans; whole cooked beans average about 37% crude protein, whereas soybean meal usually runs 44%. Also, due to the higher energy of the beans, the protein content of the ration must be 1 to 2% higher than in a soybean meal ration in order to maintain the same protein-to-energy ratio.

The feeding of full-fat, cooked soybeans to growing-finishing pigs has little effect on grade and yield if the proper protein-to-energy ratio is maintained, but softer pork may be produced.

It is noteworthy that hogs fed cooked whole soybeans have a softer carcass than those fed a soybean meal ration. Whether this condition will influence the price packers are willing to pay for live hogs should be determined.

Other Protein Feeds for Swine

Although soybean meal is the most widely used protein supplement, many other protein feeds are suitable, and are used; among them, cottonseed meal, dried skim milk, fish meal, linseed meal, meat scraps, meat and bone scraps, peanut meal, rapeseed meal, and tankage.

Pastures

Good pasture is an excellent feed for swine, but it is not indispensable. In comparison with confinement raising, it can lower sow feed costs, help maintain high reproductive capacity for boars, and increase litter size in many cases.

Pasture was formerly thought to be an absolute essential for a successful swine operation. However, in recent years producing hogs in confinement has become a reality because of vastly improved rations, along with greater disease and parasite control. But it is still possible to utilize large amounts of forage effectively for the breeding herd.

Research reports on feed savings by swine on pasture vary considerably, depending on type of pasture, class and age of hogs, and management system. On the average, however, good pasture for growing-finishing hogs will effect a saving of 3 to 10% of the grain and of as much as 33% of the supplement, in comparison with confinement production. Bred sows and gilts on legume pasture require about half as much grain and much less supplemental protein than those in drylots. Thus, the decision on whether to raise swine on pasture or in drylot should be based primarily on (1) net returns, and (2) whether the land can be put to a more profitable alternative use.

In general, temporary pastures are preferable to permanent pastures for swine, especially from the standpoint of disease and parasite control.

Over much of the country, alfalfa, the clovers (principally ladino, red, alsike, and sweet clover), rape, and oat mixtures make excellent swine pastures. Other plants which find use as hog pastures in certain areas and under certain conditions include bluegrass, bromegrass, orchardgrass, lespedeza, carpetgrass, rye, wheat, soybeans, cowpeas, field beans, sorghum, and Sudangrass.

Fig. 23-18. Pigs in alfalfa pasture in Nebraska. (Courtesy, Agricultural Agent, Burlington Northern, Inc., St. Paul, Minn.)

Little pollution potential exists from pasture systems with low animal densities or numbers, or where pastures are rotated. However, in high-density pasture systems involving a large number of hogs, a pollution potential does exist. Frequently there is little vegetation in these pasture lots, and rainwater falling on the lot drains away, carrying some solids with it. Where the lots are steeply sloping or where a watercourse runs adjacent to or through the lots, the problem can be serious.

Where a pollution problem exists from pasture lots, the site should be abandoned or all runoff should be caught in a retention reservoir. After each rain, liquid must be removed from the reservoir by pumping and/or evaporation to ensure sufficient volume to capture all runoff from the next rain.

(Also see Chapter 7, Pasture and Range Forages, section on "Pastures for Swine.")

Dry Forages

The favorable results obtained from feeding a corn-soybean meal ration, properly fortified with minerals and vitamins, in an era of relatively cheap grain, caused many researchers and swine producers to question the wisdom of using alfalfa meal in the ration. There was never any doubt about alfalfa meal being an excellent source of quality protein, carotene (vitamin A), B vitamins (riboflavin, pantothenic acid, and niacin), vitamin D if sun-cured, calcium, and unidentified factors. But these nutrients could frequently be purchased more cheaply from other sources. Besides, its high fiber (25 to 30%) and low palatability limit the amounts of it that can be used in baby pig and in growing-finishing rations. As a result, the swing was away from the use of alfalfa meal in swine rations.

But scarce and high-priced grains caused alfalfa to return in favor as a feedstuff, particularly in rations for gestating sows. If the price of alfalfa is right—if it is a cheaper source of protein and energy than corn and other grains—it may, to advantage, be incorporated in swine rations at the following levels: up to 5% for grower-finishing hogs, up to 50% for gestating sows, and up to 10% for lactating sows. These levels serve as a safety factor to ensure the presence of certain minerals, vitamins, and possibly unidentified factors.

Alfalfa is especially well suited for winter feeding of swine. Because of its low energy and high fiber, it generates considerable heat when digested. The extra heat can be utilized to help maintain body temperature during the winter. Thus, alfalfa is more cost effective as a feed for swine during winter than during summer.

Silages

Good-quality corn or alfalfa-grass silage is an excellent feed for brood sows; hence, it may be used to advantage where it is available on dairy and beef farms. Unless the sow herd is very large, however, it will not likely pay to construct a silo especially for hogs.

Silage should be fed fresh daily in amounts the sows will clean up in 2 to 3 hours. Sows will usually eat 10 to 15 lb, and gilts 8 to 12 lb. Some wastage can be expected. It is important that the silage be supplemented with 1.0 to 1.5 lb of a good protein supplement per head daily. Additional grain should be fed if necessary to keep sows in proper condition.

When feeding silage to hogs, the following precautions should be observed:

1. Never feed silage alone, as a poor pig crop will result. Both corn and alfalfa-grass silage must be supplemented with protein and energy.
2. Silage causes digestive upsets (diarrhea) in baby pigs. Do not feed it to lactating sows.
3. Avoid feeding moldy or frozen silage. It can cause sows to abort.

Even though considerable silage may be fed to brood sows to advantage, it is important that it be of good quality and that it be properly supplemented from a nutritional standpoint.

Hogging Down Crops

Sometimes pigs are permitted to do their own harvesting. Corn is the principal crop so used, the animals being turned into the field when the grain is in the dent stage. Also, small grain crops that have been badly lodged or otherwise damaged, or that cannot be harvested because of weather, may be harvested by hogs. Soybeans and field peas may be hogged off. In the South, such crops as peanuts, sweet potatoes, chufas, and other root and tuber crops, are sometimes harvested by hogs.

Space will not permit a full discussion of this method of utilizing the various crops. As corn is the main feed hogged down, comments will be limited to this crop; but the same general principles apply to other crops, when and if they are so utilized.

Fig. 23-19. Hogging down corn. The animals are usually turned into the field when the grain is in the dent stage. (Courtesy, USDA)

Some of the **advantages** of hogging down corn are:

1. It saves labor at a busy season of the year.
2. The maximum fertility value of the manure is conserved.
3. There is less danger of infesting swine with diseases and parasites than in drylot finishing.
4. Corn that is down or badly lodged is difficult to harvest, but it may be utilized through hogging down.

Some of the **disadvantages** of hogging down corn are:

1. During wet weather, a considerable amount of corn grain is lost by being tramped into the ground.
2. During wet weather, the tramping of animals puddles the soil and lowers its tilth. This is especially noticeable in heavy clay soils.
3. It usually requires additional fencing, for it is desirable to fence off a small area, then make the pigs clean it up before moving to a new area.
4. When corn is hogged down, wheat cannot follow corn in the rotation.

Garbage

Municipal garbage has long been fed to finishing hogs, but, following World War II, the practice declined because of (1) a gradual lowering in the feeding value of garbage, and (2) other competition for garbage—notably its manufacture into lawn, greenhouse, and garden fertilizer. The recent development of garbage recycling processes, along with high-priced grains, has created renewed interest in garbage as a hog feed.

As a rule of thumb, about 4 lb of heavy garbage may be considered as equivalent to 1 lb of concentrate.

All states now have laws requiring that commercial garbage be cooked.

(Also see Chapter 12, section on "Garbage.")

FEED PREPARATION

Grinding grains improves feed efficiency; hence, it is recommended. Fine grinding will cause some bridging in self-feeders, however.

The recommended fineness for grinding various grains follows:

Grain	Fineness of Grind
Corn	Medium
Wheat	Medium to coarse
Barley	Fine
Grain sorghum	Fine to medium
Oats	Fine

As noted above, corn should be medium grind; gastric ulcers in pigs have been associated with finely ground corn. Neither should wheat be finely ground; fine grinding makes it pasty and less palatable. All other cereal grains should be finely ground.

Pelleting corn-soybean meal rations will generally improve feed utilization and increase rate of gain by at least 4 to 5%. Part of the improvement in feed efficiency is probably due to less feed wastage. Also, rations containing considerable amounts of fiber are improved by pelleting because of increased comsumption, improved carbohydrate digestibility, and reduced sorting and wastage compared to meal rations.

Other methods of feed preparation, such as cooking, soaking, or fermenting, have not been shown to be of value when swine are full fed, although there are exceptions—for example, the cooking of soybeans and potatoes does improve efficiency.

The results from liquid or paste feeding have been very inconsistent. Some research has indicated increased feed consumption and rate of gain, whereas other studies have shown a deleterious effect from liquid feeding. Therefore, it is generally recommended that liquid feeding be considered only as a mechanical means of dispensing the feed.

Although high-moisture corn does improve efficiency with some classes of livestock, there is no improvement when fed to swine. Therefore, the relative value of high-moisture corn as compared to regular corn should be computed on a dry matter basis.

(Also see Chapter 14, Table 14-4, Preparation of Feeds.)

SWINE FEEDING PROGRAM

In order to compute balanced rations, it is necessary to have available feeding standards, or allowances, and feed composition tables. The NRC requirements are given in Tables 23-1 to 23-7, and recommended allowances with margins of safety are given in Table 23-8. Feed composition tables are given in Section V.

For purposes of convenience and facilitating ration formulation, Table 23-12, Swine Feeding Program, is presented. This gives, in summary form, the recommended types of rations and feeding programs for all classes of swine—a feed for every need.

TABLE 23-12
SWINE FEEDING PROGRAM[1]

Sex/Stage	Length of Feeding Program	Season	Complete Ration[2]			Corn or Grain		Supplement[3]	
			Protein	Amount					
			(%)	(lb/day)	*(kg/day)*	(lb/day)	*(kg/day)*	(lb/day)	*(kg/day)*
Boars	From time purchased at 5–6 months of age.	Summer	12–16	4–6	*1.8–2.7*	3–5	*1.3–2.3*	0.8–1.0	*0.4–0.5*
		Winter	12–16	5–7	*2.3–3.2*	4–6	*1.8–2.7*	0.8–1.0	*0.4–0.5*
			Increase intake 1–2 lb *(0.5–0.9 kg)* during the heavy breeding season.						
Gilts									
Pregestation	From time selected at 5–6 months until breeding at 7–9 months of age.	Summer	12–16	4–6	*1.8–2.7*	3–5	*1.3–2.3*	0.8–1.0	*0.4–0.5*
		Winter	12–16	5–7	*2.3–3.2*	4–6	*1.8–2.7*	0.8–1.0	*0.4–0.5*
Flushing and breeding	For 3 weeks before breeding (do not continue after breeding).		12–14	6–9	*2.7–4.1*	5–8	*2.3–3.6*	0.8–1.0	*0.4–0.5*
Gestation		Summer	12–14	4–5	*1.8–2.3*	3–4	*1.3–1.8*	0.8–1.0	*0.4–0.5*
		Winter	11–14	5–6	*2.3–2.7*	4–5	*1.8–2.3*	0.8–1.0	*0.4–0.5*
			Increase intake 1–2 lb *(0.5–0.9 kg)* last 3 to 5 weeks of gestation if gilts seem to be too thin.						
Lactation	Wean at 3–5 weeks after farrowing.		13–16	10–14	*4.5–6.4*	Full feed complete ration.			
Sows									
Breeding and gestation	Breed back at first heat period after weaning (flushing is not beneficial with sows).	Summer	12–14	3–4	*1.3–1.8*	2–3	*0.9–1.3*	0.8–1.0	*0.4–0.5*
		Winter	11–14	4–6	*1.8–2.7*	3–4	*1.3–1.8*	0.8–1.0	*0.4–0.5*
			Increase intake 1–2 lb *(0.5–0.9 kg)* last 3 to 5 weeks of gestation if sows seem to be too thin.						
Lactation	Wean at 3–5 weeks after farrowing.		13–16	11–15	*5.0–6.8*	Full feed complete ration.			
Pigs									
Prestarter	Use only if weaned before 3 weeks and feed until pigs weigh 12 lb *(5.5 kg)*.		20–24			Full feed complete ration.			
Starter	Use as a creep feed and continue after weaning until pigs are 8 weeks of age or 40 lb *(18.2 kg)*.		18–20			Full feed complete ration.			
Grower-finisher	From 8 weeks of age or 40 lb *(18.2 kg)* until market weight.		13–17	Full feed (may limit feed after 125 lb *(56.8 kg)*).		With free-choice, corn consumption varies depending on weight of pig. Daily supplement intake should be approximately 0.75 lb *(0.34 kg)* regardless of the weight of the pig.			

[1]Adapted by the authors from *Life Cycle Swine Nutrition*, Iowa State University, Ames, PM-489, June, 1988.

[2]Assumes corn based rations.

[3]See supplements in Table 23-25.

FEED ALLOWANCES AND SUGGESTED RATIONS

In most instances, the nutrient allowances should be higher than the minimum requirements established by the National Research Council (see Tables 23-1 to 23-7). This is desirable to obtain maximum performance and reduce the risk of nutrient deficiencies that might occur because of the differences in ingredient quality, environment, health, genetics, and performance of individual animals.

From a nutritional standpoint, there is no *best* swine ration in terms of the ingredients used. So, ingredients should be selected on the basis of availability, price, and the quality of the nutrients that they contain. Corn, grain sorghum, barley, and wheat are the primary energy supplying ingredients in rations for swine. Yet, each of these feed grains is deficient in certain indispensable amino acids, minerals, and vitamins. Soybean meal, other oilseed meals or animal protein meals are commonly added as sources of supplemental amino acids to the grain. Thus, corn and soybean meal, which are the primary ingredients in the Corn Belt, may be replaced by other ingredients in other areas as determined by availability and price; then, balanced nutritionally, especially to meet the amino acid, mineral, and vitamin requirements.

Following the selection of feed ingredients, they may be fed to swine in either combined form or cafeteria style, with balanced rations achieved by any of the following mixing/feeding procedures.

• **Complete feed vs feeding grain and supplement separately**—The swine producer has two primary feeding options from which to choose: feeding complete feeds vs feeding the grain and supplement separately. In recent years, there has been an increasing trend toward the use of complete, mixed rations for all classes and ages of hogs, especially for baby pigs and growing-finishing hogs. The use of complete rations is favored because they afford a more accurate means of controlling the intake of minerals, vitamins, and feed additives. The choice between complete feeds or feeding the grain and supplement separately, or some combination of the two, should be determined by the conditions prevailing on each individual farm, including such factors as: (1) the cost of grinding and mixing; (2) the kind, amount, and palatability of available feeds; (3) facilities for handling feeds; (4) the relative cost of proteins and grains; and (5) the results obtained.

• **Commercial vs home mixed complete feeds**—Complete feeds may be commercially or home mixed. Commercial feed companies specializing in swine feeds generally manufacture and sell complete rations suitable for each class of swine. Because of the preciseness needed in formulating and mixing prestarter and starter rations, most swine producers purchase commercially prepared feeds for these needs. However, the home mixing of complete feeds for other classes of swine, and even for baby pigs, may be accomplished either (1) by "building from the ground up" (by adding each ingredient, one by one); or (2) by mixing a complete supplement, base mix, or premix with the primary ingredients as follows:

1. Purchase of a *complete supplement* which contains a substantial portion of the protein, minerals, and vitamins, along with additives, required in the complete ration and mixing it with local grain(s). Normally, the supplement is added to the grain at the rate of 200 to 300 lb per ton of complete feed.

2. Purchase of a *base mix,* which provides less of the protein requirements than a supplement, but which complements the protein quality and provides minerals, vitamins, and additives, then mixing it with grain(s) and a high protein source(s) available locally. Normally, the base mix is added to the grain and high-protein source(s) at the rate of 100 to 200 lb of complete feed.

3. Purchase of a *premix* consisting of trace minerals and/or vitamins, along with additives, and mixing it with the main ingredients of the ration. Normally, the premix is added at a level of 2 to 10 lb per ton of complete feed.

• **Recommended nutrient allowances of the swine specialists at Iowa State University**—In this chapter, the authors present the nutrient allowances, along with premixes and rations, of the swine specialists of Iowa State University, the Land Grant University in the state of Iowa, the leading swine state in the nation.

The recommended allowances and rations formulated by the swine specialists at Iowa State University consist of three premixes (Tables 23-13, 23-14, and 23-15. See pp. 984-985.) and nine finished rations (Tables 23-16 to 23-24. See pp. 986-990.). This makes for a popular and convenient method of building swine rations by blending the proper premix with the main ingredients of the ration.

Normally, premixes should be purchased from a commercial company, because they have much better quality control and mixing facilities to handle the small quantities of minerals and/or vitamins required.

Vitamin potency may decrease with extended storage and can be completely destroyed when vitamins are in contact with minerals over a prolonged period of time. If the minerals and vitamins are purchased in one premix (such as shown in Table 23-15), they should be used within 30 days of purchase for optimum vitamin potency. Also, mineral-vitamin premixes should be stored in a cool, dry, and dark place. Stabilizing agents will increase the shelf-life of mineral-vitamin premix combinations.

Fig. 23-20. Two different rations stored in outside steel bins and augered into environmentally-controlled building. (Courtesy, Iowa State University, Ames.)

Special care should be taken when blending the premix with the primary ingredients to obtain a thorough dispersion throughout the feed. A common method of blending is to mix enough premix for 1 ton of feed with 20 to 50 lb of finely ground corn or soybean meal, then add this to the mixer.

TABLE 23-13
COMPOSITION AND ANALYSIS OF TRACE MINERAL PREMIX[1]

Element	Source[2]	Amount		Percent in Premix	Parts Per Million When Added to a Complete Ration At The Following Pounds Per Ton:			
					2	3	4	5
		(lb)	*(kg)*	(%)	(ppm)	(ppm)	(ppm)	(ppm)
Copper (Cu)	Copper sulfate	1.500	*0.681*	0.38	4	6	8	10
Iodine (I)	Potassium iodide[3]	0.010	*0.005*	0.008	0.08	0.11	0.15	0.19
Iron (Fe)	Ferrous sulfate	25.000	*11.350*	5.03	50	75	101	126
Manganese (Mn)	Manganese sulfate	2.500	*1.135*	0.57	6	9	11	14
Selenium (Se)	Sodium selenite[3]	0.025	*0.011*	0.011	0.11	0.17	0.23	0.29
Zinc (Zn)	Zinc sulfate	25.000	*11.350*	5.68	57	85	114	142
	Carrier	45.965	*20.868*					
	Total	100.000	*45.400*					

[1]Adapted by the authors from *Life Cycle Swine Nutrition*, Iowa State University, Ames, PM-489, June, 1988.

[2]Other sources of trace minerals may be substituted. Iodine may be omitted if iodized salt is used.

[3]Iodine and selenium probably will be added in a separate premix form.

TABLE 23-14
COMPOSITION OF VITAMIN PREMIX[1,2]

Vitamins	Amount	Units Per Pound of Complete Ration When Added At The Following Pounds Per Ton:				
		Unit	3	5	8	10
Essential[3]						
Vitamin A (million IU)	5.0	(IU)	750.00	1,250.00	2,000.00	2,500.00
Vitamin D (million IU)	0.6	(IU)	90.00	150.00	240.00	300.00
Vitamin E (thousand IU)	26.0	(IU)	3.90	6.50	10.40	13.00
Niacin (nicotinic acid, nicotinamide) (g)	25.0	(mg)	3.75	6.25	10.00	12.50
d-Pantothenic acid (vitamin B-3) (g)	20.0	(mg)	3.00	5.00	8.00	10.00
Riboflavin (vitamin B-2) (g)	6.0	(mg)	0.90	1.50	2.40	3.00
Vitamin B-12 (cobalamins) (mg)	25.0	(mcg)	3.75	6.25	10.00	12.50
Optional[4]						
Biotin (g)	0.3	(mg)	0.05	0.08	0.12	0.15
Menadione (source of vitamin K) (g)	4.0	(mg)	0.60	1.00	1.60	2.00
Carrier	?					
Total[5] (lb)	10.0					

[1]Adapted by the authors from *Life Cycle Swine Nutrition*, Iowa State University, Ames, PM-489, June, 1988.

[2]A feed additive may be included in the vitamin premix.

[3]Most natural feedstuffs contain very little vitamin D or B-12. The amount of provitamin A (beta-carotene) in feedstuffs will depend on processing and storage, while niacin in most grains is relatively unavailable for swine. Riboflavin and pantothenic acid in natural feedstuffs can meet part of the requirement.

[4]Supplemental biotin is not necessary with corn-soybean meal based rations. It should be included in sow rations based on other grains. The vitamin K requirement is normally met by the level present in natural feedstuffs and by intestinal synthesis. A hemorrhagic or bleeding syndrome has been diagnosed which is probably due to a vitamin K antimetabolite. The antimetabolite is thought to be produced by mold occuring in one or more of the ration ingredients. When this has occurred, adding menadione has been helpful in preventing or overcoming the problem.

[5]If this premix is used in Table 23-15, dilute to 2 lb *(0.9 kg)* only.

TABLE 23-15
COMPLETE MINERAL-VITAMIN PREMIXES FOR CORN-SOYBEAN MEAL RATIONS[1 2 3 4]

Ingredients	1		2	
	(lb)	*(kg)*	(lb)	*(kg)*
Calcium carbonate	540	*245*	300	*136*
Dicalcium phosphate	1,030	*468*		
Defluorinated phosphate			1,060	*481*
Salt	250	*114*	250	*114*
Trace mineral premix (Table 23-13)	80	*36*	80	*36*
Vitamin premix (Table 23-14)[5]	40	*18*	40	*18*
Carrier (corn, middlings, or grain by-products)	60	*27*	270	*123*
Total	2,000	*908*	2,000	*908*

Calculated Analyses:

	(%)	(lb)	*(kg)*
Salt (NaCl)	12.5		
Calcium (Ca)	22.8		
Phosphorus (P)	9.5		
Copper (Cu)	0.015		
Iodine (I)	0.00031		
Iron (Fe)	0.201		
Manganese (Mn)	0.023		
Selenium (Se)	0.00046		
Zinc (Zn)	0.227		
Vitamin A (IU)		50,000	*110,000*
Vitamin D (IU)		6,000	*13,200*
Vitamin E (IU)		260	*572*
Vitamin K (optional) (mg)		40	*88*
Biotin (optional) (mg)		3	*7*
Niacin (Nicotinic Acid, Nicotinamide) (mg)		250	*550*
Pantothenic Acid (Vitamin B-3) (mg)		200	*440*
Riboflavin (Vitamin B-2) (mg)		60	*132*
Vitamin B-12 (Cobalamins) (mcg)		250	*550*

[1]Adapted by the authors from *Life Cycle Swine Nutrition,* Iowa State University, Ames, PM-489, June, 1988.

[2]Due to the instability of vitamins in the presence of trace minerals this premix should be used within 30 days of preparation.

[3]Mixing directions:

Stage	Premix		Soybean Meal, 44%		Corn	
	(lb)	*(kg)*	(lb)	*(kg)*	(lb)	*(kg)*
Gestation:						
3 lb *(1.4 kg)*/day	100	*45*	270	*123*	1,630	*740*
4 lb *(1.8 kg)*/day	80	*36*	160	*73*	1,760	*799*
5 lb *(2.3 kg)*/day	65	*30*	100	*45*	1,835	*833*
Lactation	80	*36*	270	*123*	1,650	*749*
Starter rations	65	*30*	690	*313*	1,245	*565*
Grower-finisher rations	50	*23*	300-450	*136-204*	1,700-1,500	*772-681*

[4]These premixes can be fed free-choice to sows on pasture or in other instances where free-choice minerals and vitamins are needed. Do not add or feed free-choice additional minerals or vitamins with any of the ration formulas in Tables 23-16 to 23-24, because they contain sufficient minerals and vitamins.

[5]Table 23-14 premix diluted to 2 lb *(0.9 kg)* instead of 10 lb *(4.5 kg)*. See footnote 5 under Table 23-14.

Fig. 23-21. Early-weaned pigs in double-deck nursery pens. Early weaning requires special facilities and special rations. (Courtesy, University of Illinois, Urbana)

Suggested rations are shown for each of the following classes of swine:

Table 23-16, Pregestation, Breeding, and Gestation Rations—for boars, sows, or gilts fed 3 lb per day.

Table 23-17, Pregestation, Breeding, and Gestation Rations—for boars, sows, or gilts fed 4 lb per day.

Table 23-18, Pregestation, Breeding, and Gestation Rations—for boars, sows, or gilts fed 5 lb per day.

Table 23-19, Lactation Rations.

Table 23-20, Prestarter Rations—for baby pigs before 3 weeks of age.

Table 23-21, Starter Rations.

Table 23-22, Recommended Rations for Performance Testing Boars.

Table 23-23, Swine Conditioner Rations—for newly received feeder pigs, for stress periods and convalescence.

Table 23-24, Grower-Finisher Rations—for 40 to 240 lb.

As previously indicated, the rations shown in Tables 23-16 to 23-24 call for the use of the mineral and vitamin premixes shown in Tables 23-13 to 23-15.

TABLE 23-16
PREGESTATION, BREEDING, AND GESTATION RATIONS (FOR BOARS, SOWS AND GILTS FED 3 LB *(1.4 KG)* PER DAY)[1 2 3]

Ingredient	Ration # 1		2		3		4	
	(lb)	*(kg)*	(lb)	*(kg)*	(lb)	*(kg)*	(lb)	*(kg)*
Corn, yellow (8.4% protein)[4]	1,636	*743*	1,559	*708*	1,640	*744*	1,562	*709*
Soybean meal, solvent extracted (44.0% protein)[5]	270	*123*	250	*113*	194	*88*	175	*80*
Alfalfa meal, dehydrated (17.0% protein)			100	*45*			100	*45*
Meat and bone meal (50.0% protein)					100	*45*	100	*45*
Dicalcium phosphate	50	*22*	51	*23*	31	*14*	31	*14*
Limestone	19	*8*	15	*7*	10	*5*	7	*3*
Iodized salt	10	*5*	10	*5*	10	*5*	10	*5*
Trace mineral premix (Table 23-13)	5	*2*	5	*2*	5	*2*	5	*2*
Vitamin premix (Table 23-14)	10	*5*	10	*5*	10	*5*	10	*5*
Feed additives[6]								
Total	2,000	*908*	2,000	*908*	2,000	*908*	2,000	*908*
Calculated analyses:								
Metabolizable energy (kcal/lb)	1,470		1,434		1,469		1,433	
Metabolizable energy *(kcal/kg)*	*3,234*		*3,155*		*3,232*		*3,153*	
Protein (%)	12.81		12.90		13.66		13.76	
Lysine (%)	0.60		0.60		0.60		0.60	
Threonine (%)	0.49		0.49		0.50		0.51	
Tryptophan (%)	0.14		0.15		0.13		0.14	
Calcium (Ca) (%)	1.01		1.01		1.00		1.01	
Phosphorus (P) (%)	0.75		0.75		0.75		0.75	

[1]Adapted by the authors from *Life Cycle Swine Nutrition,* Iowa State University, Ames, PM-489, June, 1988.

[2]See Table 23-12 for recommended feeding levels in drylot or confinement. These rations can be used for sows on pasture since they require only supplemental minerals or at the most 2 to 3 lb *(0.9 to 1.4 kg)* of complete feed. If less than 3 lb *(1.4 kg)* is fed per day, free-choice minerals should be available.

[3]These rations can also be used as a silage balancer. Gestating sows will eat 5 to 7 lb *(2.3 to 3.2 kg)* of corn silage daily which should be supplemented with 2 to 3 lb *(0.9 to 1.4 kg)* of one of these rations.

[4]Ground oats can replace corn up to 20% of the total ration. Ground milo, wheat, or barley can replace the corn.

[5]To replace 44% soybean meal with 47% soybean meal or whole soybeans, use the following ratios:
Each 100 lb *(45 kg)* SBM (44%) = 93 lb *(42 kg)* SBM (47%) + 7 lb *(3 kg)* corn.
Each 100 lb *(45 kg)* SBM (44%) + 35 lb *(16 kg)* corn = 135 lb *(61 kg)* whole soybeans.

[6]Feed additives are not generally recommended during gestation or for gilts during the developer period after selection unless specific disease problems exist. High levels (100-300 g/ton) may be beneficial when fed 2 weeks before and after breeding and 2 weeks before farrowing.

TABLE 23-17
PREGESTATION, BREEDING, AND GESTATION RATIONS (FOR BOARS, SOWS, OR GILTS FED 4 LB *(1.8 KG)* PER DAY)[1 2]

Ingredient	Ration # 1		2		3		4	
	(lb)	*(kg)*	(lb)	*(kg)*	(lb)	*(kg)*	(lb)	*(kg)*
Corn, yellow (8.4% protein)[3]	1,769	*803*	1,692	*768*	1,767	*802*	1,692	*768*
Soybean meal, solvent extracted (44.0% protein)[4]	160	*72*	140	*64*	90	*41*	68	*31*
Alfalfa meal, dehydrated (17.0% protein)			100	*45*			100	*45*
Meat and bone meal (50.0% protein)					100	*45*	100	*45*
Dicalcium phosphate	36	*16*	36	*16*	16	*7*	16	*7*
Limestone	15	*7*	12	*5*	7	*3*	4	*2*
Iodized salt	8	*4*	8	*4*	8	*4*	8	*4*
Trace mineral premix (Table 23-13)	4	*2*	4	*2*	4	*2*	4	*2*
Vitamin premix (Table 23-14)	8	*4*	8	*4*	8	*4*	8	*4*
Feed additives[5]								
Total	2,000	*908*	2,000	*908*	2,000	*908*	2,000	*908*
Calculated analyses:								
Metabolizable energy (kcal/lb)	1,493		1,457		1,492		1,456	
Metabolizable energy *(kcal/kg)*	*3,285*		*3,205*		*3,282*		*3,203*	
Protein (%)	10.95		11.04		11.90		11.95	
Lysine (%)	0.45		0.45		0.46		0.46	
Threonine (%)	0.41		0.41		0.43		0.43	
Tryptophan (%)	0.11		0.12		0.10		0.11	
Calcium (Ca) (%)	0.75		0.76		0.75		0.76	
Phosphorus (P) (%)	0.60		0.60		0.60		0.60	

[1]Adapted by the authors from *Life Cycle Swine Nutrition,* Iowa State University, Ames, PM-489, June, 1988.

[2]See Table 23-12 for recommended feeding levels when hand-fed in drylot or confinement. These rations can be used for gilts on pasture during gestation since they require 3 to 4 lb *(1.4 to 1.8 kg)* of feed daily. These rations can also be used for interval-fed sows or gilts if the average daily intake is approximately 4 lb *(1.8 kg).*

[3]Ground oats can replace corn up to 20% of the total ration. Ground milo, wheat, or barley can replace the corn.

[4]To replace 44% soybean meal with 47% soybean meal or whole soybeans, use the following ratios:
Each 100 lb *(45 kg)* SBM (44%) = 93 lb *(42 kg)* SBM (47%) + 7 lb *(3 kg)* corn.
Each 100 lb *(45 kg)* SBM (44%) + 35 lb *(16 kg)* corn = 135 lb *(61 kg)* whole soybeans.

[5]Feed additives are not generally recommended during gestation or for gilts during the developer period after selection unless specific disease problems exist. High levels (100-300 g/ton) may be beneficial when fed 2 weeks before and after breeding and 2 weeks before farrowing.

TABLE 23-18
PREGESTATION, BREEDING, AND GESTATION RATIONS (FOR BOARS, SOWS, OR GILTS FED 5 LB *(2.3 KG)* PER DAY)[1]

Ingredient	Ration #							
	1		2		3		4	
	(lb)	*(kg)*	(lb)	*(kg)*	(lb)	*(kg)*	(lb)	*(kg)*
Corn, yellow (8.4% protein)[2]	1,845	*838*	1,764	*801*	1,853	*841*	1,824	*828*
Soybean meal, solvent extracted (44.0% protein)[3]	100	*45*	85	*38*	20	*9*		
Alfalfa meal, dehydrated (17.0% protein)			100	*45*			50	*23*
Meat and bone meal (50.0% protein)					100	*45*	100	*45*
Dicalcium phosphate	26	*12*	26	*12*	6	*3*	7	*3*
Limestone	14	*6*	10	*5*	6	*3*	4	*2*
Iodized salt	6	*3*	6	*3*	6	*3*	6	*3*
Trace mineral premix (Table 23-13)	3	*1*	3	*1*	3	*1*	3	*1*
Vitamin premix (Table 23-14)	6	*3*	6	*3*	6	*3*	6	*3*
Feed additives[4]								
Total	2,000	*908*	2,000	*908*	2,000	*908*	2,000	*908*
Calculated analyses:								
Metabolizable energy (kcal/lb)	1,508		1,473		1,507		1,489	
Metabolizable energy *(kcal/kg)*	*3,318*		*3,241*		*3,315*		*3,276*	
Protein (%)	9.95		10.13		10.72		10.59	
Lysine (%)	0.38		0.38		0.37		0.36	
Threonine (%)	0.37		0.38		0.38		0.37	
Tryptophan (%)	0.10		0.10		0.08		0.08	
Calcium (Ca) (%)	0.61		0.60		0.61		0.61	
Phosphorus (P) (%)	0.50		0.50		0.50		0.50	

[1]Adapted by the authors from *Life Cycle Swine Nutrition,* Iowa State University, Ames, PM-489, June, 1988.

[2]Ground oats can replace corn up to 20% of the total ration. Ground milo, wheat, or barley can replace the corn.

[3]To replace 44% soybean meal with 47% soybean meal or whole soybeans, use the following ratios:
Each 100 lb *(45 kg)* SBM (44%) = 93 lb *(42 kg)* SBM (47%) + 7 lb *(3 kg)* corn.
Each 100 lb *(45 kg)* SBM (44%) + 35 lb *(16 kg)* corn = 135 lb *(61 kg)* whole soybeans.

[4]Feed additives are not generally recommended during gestation or for gilts during the developer period after selection unless specific disease problems exist. High levels (100–300 g/ton) may be beneficial when fed 2 weeks before and after breeding and 2 weeks before farrowing.

TABLE 23-19
LACTATION RATIONS[1,2]

Ingredient	Ration #							
	1		2		3		4	
	(lb)	*(kg)*	(lb)	*(kg)*	(lb)	*(kg)*	(lb)	*(kg)*
Corn, yellow (8.4% protein)	1,658	*752*	1,545	*701*	1,580	*717*	1,469	*667*
Soybean meal, solvent extracted (44.0% protein)[3]	270	*122*	283	*128*	177	*80*	260	*118*
Fat or oil source			100	*45*				
Meat and bone meal (50.0% protein)					100	*45*		
Oats (11.5% protein)							200	*90*
Beet pulp (8.0% protein)					100	*45*		
Dicalcium phosphate	35	*16*	35	*16*	15	*7*	33	*15*
Limestone	15	*7*	15	*7*	6	*3*	16	*7*
Iodized salt	10	*5*	10	*5*	10	*5*	10	*5*
Trace mineral premix (Table 23-13)	4	*2*	4	*2*	4	*2*	4	*2*
Vitamin premix (Table 23-14)	8	*4*	8	*4*	8	*4*	8	*4*
Feed additives[4]								
Total	2,000	*908*	2,000	*908*	2,000	*908*	2,000	*908*
Calculated analyses:								
Metabolizable energy (kcal/lb)	1,487		1,409		1,468		1,457	
Metabolizable energy *(kcal/kg)*	*3,271*		*3,100*		*3,230*		*3,205*	
Protein (%)	12.90		12.72		13.43		13.04	
Lysine (%)	0.60		0.60		0.60		0.60	
Threonine (%)	0.49		0.49		0.48		0.49	
Tryptophan (%)	0.14		0.14		0.13		0.15	
Calcium (Ca) (%)	0.75		0.76		0.76		0.76	
Phosphorus (P) (%)	0.61		0.60		0.60		0.60	

[1]Adapted by the authors from *Life Cycle Swine Nutrition,* Iowa State University, Ames, PM-489, June, 1988.

[2]These rations may be limit-fed from a few days before farrowing and full-fed during lactation.

[3]To replace 44% soybean meal with 47% soybean meal or whole soybeans, use the following ratios:
Each 100 lb *(45 kg)* SBM (44%) = 93 lb *(42 kg)* SBM (47%) + 7 lb *(3 kg)* corn.
Each 100 lb *(45 kg)* SBM (44%) + 35 lb *(16 kg)* corn = 135 lb *(61 kg)* whole soybeans.

[4]High levels of feed additives (100–300 g/ton) may be beneficial when fed 2 weeks before and after breeding and 2 weeks before farrowing.

TABLE 23-20
PRESTARTER RATIONS (FOR BABY PIGS BEFORE 3 WEEKS OF AGE)[1,2]

Ingredient	Ration # 1		2		3		4		5	
	(lb)	*(kg)*	(lb)	*(kg)*	(lb)	*(kg)*	(lb)	*(kg)*	(lb)	*(kg)*
Corn, yellow (8.4% protein)	917	*416*	878	*399*	866	*393*	621	*282*	774	*351*
Soybean meal, solvent extracted (44.0% protein)					635	*288*				
Soybean meal, solvent extracted, dehulled (47.0% protein)	680	*309*	580	*263*			580	*263*	725	*329*
Oat groats (dehulled oats) (16.0% protein)							200	*91*		
Skim milk, dried (33.0% protein)	100	*45*	100	*45*	200	*91*	200	*91*		
Whey, dried (12.0% protein)	200	*91*	300	*136*	200	*91*	200	*91*	400	*182*
Fish meal, menhaden (62.0% protein)			50	*23*						
Sugar							100	*45*		
Fat or oil source	40	*18*	40	*18*	40	*18*	40	*18*	40	*18*
Dicalcium phosphate	26	*12*	19	*9*	25	*12*	25	*12*	25	*12*
Limestone	19	*9*	15	*7*	16	*7*	16	*7*	18	*8*
Iodized salt	5	*2*	5	*2*	5	*2*	5	*2*	5	*2*
Trace mineral premix (Table 23-13)	5	*2*	5	*2*	5	*2*	5	*2*	5	*2*
Vitamin premix (Table 23-14)	8	*4*	8	*4*	8	*4*	8	*4*	8	*4*
Feed additives[3]	◄— 100 to 300 g per ton —►									
Total	2,000	*908*	2,000	*908*	2,000	*908*	2,000	*908*	2,000	*908*
Calculated analyses:										
Metabolizable energy (kcal/lb)	1,458		1,459		1,441		1,387		1,441	
Metabolizable energy *(kcal/kg)*	*3,208*		*3,210*		*3,170*		*3,051*		*3,170*	
Protein (%)	22.68		22.32		22.11		22.34		22.64	
Lysine (%)	1.40		1.40		1.40		1.40		1.40	
Threonine (%)	0.94		0.95		0.94		0.93		0.96	
Tryptophan (%)	0.29		0.28		0.29		0.29		0.30	
Calcium (Ca) (%)	0.91		0.91		0.91		0.90		0.91	
Phosphorus (P) (%)	0.70		0.70		0.70		0.71		0.70	

[1]Adapted by the authors from *Life Cycle Swine Nutrition,* Iowa State University, Ames, PM-489, June, 1988.

[2]The prestarter ration is normally fed in only limited amounts. It should be fed to pigs weaned before 3 weeks of age until they reach approximately 12 lb *(5.4 kg)*. They can then be switched to a starter ration. These are good rations for orphan pigs, when extreme disease outbreaks (TGE) occur, or when the sow fails to produce sufficient milk.

[3]The feed additive may be part of the vitamin premix, or if a separate premix, it should replace an equal amount of corn.

TABLE 23-21
PIG STARTER RATIONS[1,2]

Ingredient	Ration # 1		2		3		4		5	
	(lb)	*(kg)*	(lb)	*(kg)*	(lb)	*(kg)*	(lb)	*(kg)*	(lb)	*(kg)*
Corn, yellow (8.4% protein)	1,072	*487*	939	*426*	1,032	*468*	1,273	*578*	865	*393*
Soybean meal, solvent extracted (44.0% protein)					610	*277*	689	*313*		
Soybean meal, solvent extracted, dehulled (47.0% protein)	570	*259*	580	*263*						
Soybeans, full-fat, cooked (37.0% protein)									800	*363*
Oat groats, dehulled (16.0% protein)			200	*91*						
Whey, dried (12.0% protein)	300	*136*	200	*91*	300	*136*			300	*136*
Fat or oil source			20	*9*						
Dicalcium phosphate	24	*11*	25	*12*	25	*12*	3	*1*	2	*1*
Limestone	16	*7*	18	*8*	15	*7*	17	*8*	15	*7*
Iodized salt	5	*2*	5	*2*	5	*2*	5	*2*	5	*2*
Trace mineral premix (Table 23-13)	5	*2*	5	*2*	5	*2*	5	*2*	5	*2*
Vitamin premix (Table 23-14)	8	*4*	8	*4*	8	*4*	8	*4*	8	*4*
Feed additives[3]	◄— 100 to 300 g per ton —►									
Total	2,000	*908*	2,000	*908*	2,000	*908*	2,000	*908*	2,000	*908*
Calculated analyses:										
Metabolizable energy (kcal/lb)	1,483		1,472		1,460		1,473		1,525	
Metabolizable energy *(kcal/kg)*	*3,263*		*3,238*		*3,212*		*3,241*		*3,355*	
Protein (%)	19.70		20.37		19.55		20.39		20.13	
Lysine (%)	1.15		1.16		1.15		1.15		1.20	
Threonine (%)	0.82		0.82		0.82		0.81		0.85	
Tryptophan (%)	0.25		0.25		0.26		0.26		0.28	
Calcium (Ca) (%)	0.80		0.81		0.81		0.80		0.84	
Phosphorus (P) (%)	0.65		0.65		0.65		0.65		0.68	

[1]Adapted by the authors from *Life Cycle Swine Nutrition,* Iowa State University, Ames, PM-489, June, 1988.

[2]The pig starter ration can be used as a creep ration before weaning and fed after weaning until the pigs reach approximately 40 lb *(18 kg)*. They can then be switched to a grower-finisher ration.

[3]The feed additive may be part of the vitamin premix, or if a separate premix, it should replace an equal amount of corn.

TABLE 23-22
RECOMMENDED RATIONS FOR PERFORMANCE TESTING OF BOARS[1 2]

Ingredient	Conditioner Ration		Test Ration	
	(lb)	*(kg)*	(lb)	*(kg)*
Corn, yellow (8.4% protein)[3]	1,115	*506*	1,359	*617*
Soybean meal, solvent extracted (44.0% protein)[4]	500	*227*	550	*250*
Wheat middlings (15.5% protein)	200	*91*		
Whey, dried (12.0% protein)	100	*45*		
Dicalcium phosphate	50	*23*	60	*27*
Limestone	15	*7*	11	*5*
Iodized salt	5	*2*	5	*2*
Trace mineral premix (Table 23-13)	5	*2*	5	*2*
Vitamin premix (Table 23-14)	10	*5*	10	*5*
Feed additives (g/ton)	(100-300)		(0-100)	
Total	2,000	*908*	2,000	*908*
Calculated analyses:				
Metabolizable energy (kcal/lb)	1,438		1,459	
Metabolizable energy *(kcal/kg)*	*3,164*		*3,210*	
Protein (%)	17.83		17.81	
Lysine (%)	0.98		0.97	
Threonine (%)	0.71		0.70	
Tryptophan (%)	0.23		0.22	
Calcium (Ca) (%)	1.01		1.01	
Phosphorus (P) (%)	0.87		0.89	

[1]Adapted by the authors from *Life Cycle Swine Nutrition,* Iowa State University, Ames, PM-489, June, 1988.

[2]These rations normally would be used for growing boars from 40 to 250 lb *(18 to 114 kg)* body weight.

[3]If the ration is to be pelleted, 25 to 50 lb *(11 to 23 kg)* of molasses or binder can replace an equal amount of corn.

[4]To replace 44% soybean meal with 47% soybean meal or whole soybeans, use the following ratios:
Each 100 lb *(45 kg)* SBM (44%) = 93 lb *(42 kg)* SBM (47%) + 7 lb *(3 kg)* corn.
Each 100 lb *(45 kg)* SBM (44%) + 35 lb *(16 kg)* corn = 135 lb *(61 kg)* whole soybeans.

Fig. 23-22. A performance-tested Duroc boar in working condition. (Courtesy, Iowa State University, Ames)

TABLE 23-23
SWINE CONDITIONER RATIONS[1 2]

Ingredient	Ration # 1		Ration # 2		Ration # 3	
	(lb)	*(kg)*	(lb)	*(kg)*	(lb)	*(kg)*
Corn, yellow (8.4% protein)	882	*401*	840	*382*	979	*444*
Oats (11.5% protein)	300	*136*	*600*	*272*	600	*272*
Wheat middlings (15.5% protein)	300	*136*				
Soybean meal, solvent extracted (44.0% protein)[3]	170	*77*	300	*136*	350	*159*
Whey, dried (12.0% protein)	200	*91*	200	*91*		
Alfalfa meal, dehydrated (17.0% protein)	50	*23*				
Fish meal, menhaden (62.0% protein)	50	*23*				
Dicalcium phosphate	14	*6*	26	*12*	30	*14*
Limestone	14	*6*	14	*6*	16	*7*
Iodized salt	5	*2*	5	*2*	10	*5*
Trace mineral premix (Table 23-13)	5	*2*	5	*2*	5	*2*
Vitamin premix (Table 23-14)	10	*5*	10	*5*	10	*5*
Feed additives[4]	◄— 100 to 300 g/ton —►					
Total	2,000	*908*	2,000	*908*	2,000	*908*
Calculated analyses:						
Metabolizable energy (kcal/lb)	1,398		1,387		1,391	
Metabolizable energy *(kcal/kg)*	*3,076*		*3,051*		*3,060*	
Protein (%)	14.67		14.78		15.26	
Lysine (%)	0.75		0.75		0.75	
Threonine (%)	0.59		0.60		0.58	
Tryptophan (%)	0.18		0.18		0.19	
Calcium (Ca) (%)	0.73		0.71		0.71	
Phosphorus (P) (%)	0.61		0.60		0.60	

[1]Adapted by the authors from *Life Cycle Swine Nutrition,* Iowa State University, Ames, PM-489, June, 1988.

[2]These rations are recommended as the first ration fed to newly received feeder pigs, for stress periods and convalescence.

[3]To replace 44% soybean meal with 47% soybean meal use the following ratio:
Each 100 lb *(45 kg)* SBM (44%) = 93 lb *(42 kg)* SBM (47%) + 7 lb *(3 kg)* corn.

[4]Be certain that only approved feed additives and levels are used for therapy. The feed additive may be a part of the vitamin premix, or if a separate premix, it should replace an equal amount of corn.

TABLE 23-24
GROWER-FINISHER RATIONS (FOR PIGS FROM 40 TO 240 LB *(18 TO 109 KG)*)[1 2 3 4]

Ingredient	For Pigs 40–120 Lb *(18–54 Kg)*						For Pigs 121–240 Lb *(55–109 Kg)*[5]					
	1		2		3		4		5		6	
	(lb)	(lb)	(lb)	(lb)	(lb)	(lb)	(lb)	(lb)	(lb)	(lb)	(lb)	(lb)
Corn, yellow (8.4% protein)[6 7]	1,571	1,497	1,422	1,343	1,561	1,492	1,692	1,613	1,575	1,481	1,694	1,620
Soybean meal, solvent extracted (44.0% protein)[8]	380	455			310	380	265	345			190	264
Soybeans, full-fat, cooked (37.0% protein)[9]			530	610					380	475		
Meat and bone meal (50.0% protein)					110	110					100	100
Dicalcium phosphate	21	19	19	18			18	17	20	20		
Limestone	15	16	16	16	6	5	15	15	15	14	6	6
Iodized salt	5	5	5	5	5	5	5	5	5	5	5	5
Trace mineral premix (Table 23-13)	3	3	3	3	3	3	2	2	2	2	2	2
Vitamin premix (Table 23-14)	5	5	5	5	5	5	3	3	3	3	3	3
Feed additives[10]	◄——— 0 to 100 g per ton ———►						◄——— 0 to 50 g per ton ———►					
Total	2,000	2,000	2,000	2,000	2,000	2,000	2,000	2,000	2,000	2,000	2,000	2,000
Calculated analyses:												
Metabolizable energy (kcal/lb)	1,500	1,497	1,543	1,547	1,498	1,495	1,510	1,507	1,538	1,543	1,508	1,505
Metabolizable energy *(kcal/kg)*	*3,300*	*3,293*	*3,395*	*3,403*	*3,296*	*3,289*	*3,322*	*3,315*	*3,384*	*3,395*	*3,318*	*3,311*
Protein (%)	14.96	16.30	15.78	16.93	16.13	17.38	12.94	14.36	13.65	15.01	13.79	15.11
Lysine (%)	0.75	0.85	0.81	0.90	0.77	0.86	0.60	0.70	0.65	0.76	0.60	0.70
Threonine (%)	0.58	0.63	0.61	0.66	0.60	0.65	0.49	0.55	0.52	0.58	0.51	0.56
Tryptophan (%)	0.17	0.20	0.20	0.21	0.17	0.19	0.14	0.17	0.16	0.18	0.13	0.15
Calcium (Ca) (%)	0.60	0.60	0.61	0.61	0.61	0.60	0.55	0.55	0.59	0.58	0.55	0.56
Phosphorus (P) (%)	0.50	0.50	0.51	0.51	0.51	0.53	0.46	0.46	0.49	0.51	0.47	0.49

[1]Adapted by the authors from *Life Cycle Swine Nutrition,* Iowa State University, Ames, PM-489, June, 1988.

[2]Feed the ration with the higher level of soybean meal (lower level of corn) to lighter pigs in each group and decrease the soybean meal (increase the corn) until you reach the lower level as pig weights increase. If preferred, one level of protein can be fed from 40 to 240 lb *(18 to 109 kg)* with similar results as with varying the levels. To accomplish this, use the lower protein formulations from rations 1, 2, or 3 (for example, in ration No. 1 use 1571 lb *(713 kg)* of corn and 380 lb *(173 kg)* of soybean meal).

[3]If barrows and gilts are separated, use the higher range for soybean meal for the gilts and the lower range for the barrows.

[4]To convert lb to kg, divide by 2.2. To convert g/ton (short) to g/ton (metric), divide by 0.907.

[5]For potential replacement gilts, the level of dicalcium phosphate should be increased by 10 lb *(4.5 kg)* per ton. This will provide a minimum dietary level of 0.67% calcium and 0.55% phosphorus.

[6]Ground milo, wheat, or barley can replace the ground corn. Ground oats can replace corn up to 20% of the total ration.

[7]If the ration is to be pelleted, 25 to 50 lb *(11 to 23 kg)* of molasses or binder can replace an equal amount of corn.

[8]To replace 44% soybean meal with 47% soybean meal or synthetic lysine, use the following ratios:
Each 100 lb *(45 kg)* SBM (44%) = 93 lb *(42 kg)* SBM (47%) + 7 lb *(3 kg)* corn.
Each 100 lb *(45 kg)* SBM (44%) = 3 lb *(1.4 kg)* 98% lysine hydrochloride + 1 lb *(0.45 kg)* dicalcium phosphate + 96 lb *(43.6 kg)* corn.

[9]The fat content of whole soybeans increases the energy content of the ration. For maximum utilization of the ration, the protein content has been increased to maintain a similar energy-to-protein ratio.

[10]The feed additive may be part of the vitamin premix, or if it is a separate premix, it should replace an equal amount of corn.

• **Formulation and use of complete protein supplements**—In addition to the formulations given in Tables 23-13 to 23-24, Table 23-25 gives the formulas for complete protein supplements, ranging from 34.49% to 43.64% protein content, which may be used to make growing-finishing, gestation, and lactation rations; and Table 23-26 gives the mixing directions for incorporating the Table 23-25 protein supplements in complete rations.

Also, commercial supplements (usually a combined protein-mineral-vitamin-additive supplement) may be bought and mixed with the locally available grain. Table 23-27 shows the proportion of grain (of 8.4% protein) to mix with supplements ranging from 30 to 45% protein to obtain finished rations ranging from 10 to 18% protein content.

Fig. 23-23. Open front, low profile shelter with concrete apron for gestating gilts. (Courtesy, Iowa State University, Ames)

TABLE 23-25
COMPLETE PROTEIN SUPPLEMENTS[1 2 3 4]

Ingredient	1	2	3	4	5	6	7	8
	(lb)	(lb)	(lb)	(lb)	(lb)	(lb)	(lb)	(lb)
Wheat middlings (15.5% protein)[5]	192	102	59	202				
Soybean meal, solvent extracted (44.0% protein)	1,500	1,495	1,150	1,200				
Soybean meal, solvent extracted, dehulled (47.0% protein)					1,636	1,415	1,420	1,193
Alfalfa meal, dehydrated (17.0% protein)		100	200			100		
Meat and bone meal (50.0% protein)[6]			400	400			300	500
Fish meal, menhaden (62.0% protein)						150		100
Dicalcium phosphate	145	145	70	70	169	153	111	59
Limestone	75	70	33	40	90	77	64	43
Iodized salt	40	40	40	40	45	45	45	45
Trace mineral premix (Table 23-13)	16	16	16	16	20	20	20	20
Vitamin premix (Table 23-14)	32	32	32	32	40	40	40	40
Feed additives[7] (g/ton)								
Total	2,000	2,000	2,000	2,000	2,000	2,000	2,000	2,000
Calculated analyses:								
Metabolizable energy (kcal/lb)	1,228	1,203	1,167	1,222	1,260	1,243	1,249	1,254
Metabolizable energy *(kcal/kg)*	*2,702*	*2,647*	*2,567*	*2,688*	*2,772*	*2,735*	*2,748*	*2,759*
Protein (%)	34.49	34.53	37.46	37.97	38.45	38.75	40.87	43.64
Lysine (%)	2.24	2.24	2.22	2.26	2.54	2.59	2.54	2.65
Threonine (%)	1.40	1.41	1.44	1.46	1.55	1.57	1.59	1.66
Tryptophan (%)	0.49	0.50	0.45	0.45	0.52	0.52	0.49	0.48
Salt, added (%)	2.00	2.00	2.00	2.00	2.25	2.25	2.25	2.25
Calcium (%)	3.39	3.36	3.39	3.39	3.94	3.94	3.94	3.95
Phosphorus (%)	1.87	1.84	1.86	1.91	2.09	2.10	2.10	2.10

[1]Adapted by the authors from *Life Cycle Swine Nutrition,* Iowa State University, Ames, PM–489, June, 1988.

[2]These supplements can be used to make growing-finishing, gestation, or lactation rations. See Table 23–26, including the table footnote, for mixing directions.

[3]Supplements with meat and bone meal may be self-fed free-choice with shelled corn for growing-finishing pigs.

[4]To convert lb to kg, divide by 2.2. To convert g/ton (short) to g/ton (metric), divide by 0.907.

[5]The wheat middlings may be replaced with corn, corn distillers' grains with solubles, or other grain by-products.

[6]The meat and bone meal was assumed to contain 8.10% calcium and 4.10% phosphorus. If meat and bone meal with a higher concentration of calcium and phosphorus is used, the amount of dicalcium phosphate should be reduced accordingly.

[7]The concentration of feed additives will depend on the type of ration in which the supplement will be used. The concentration should be 3 to 5 times higher in supplements 1 to 4 and 4 to 6 times higher in supplements 5 to 8 than desired in the complete ration.

TABLE 23-26
COMPLETE RATIONS (USING SUPPLEMENTS IN TABLE 23-25)[1 2 3]

Ingredient	Protein Level in Complete Rations									
	13%		14%		15%		16%		17%	
	(lb)	(lb)	(lb)	(lb)	(lb)	(lb)	(lb)	(lb)	(lb)	(lb)
Corn, yellow (8.4% protein)	1,635	1,685	1,555	1,625	1,480	1,550	1,400	1,490	1,320	1,425
Supplements 1–4 (See Table 23–25.)	365		445		520		600		680	
Supplements 5–8 (See Table 23–25.)		315		375		450		510		575
Total	2,000	2,000	2,000	2,000	2,000	2,000	2,000	2,000	2,000	2,000
Calculated analyses:[4]										
Metabolizable energy (kcal/lb)	1,485	1,506	1,470	1,497	1,456	1,485	1,440	1,475	1,425	1,465
Metabolizable energy *(kcal/kg)*	*3,267*	*3,313*	*3,234*	*3,293*	*3,203*	*3,267*	*3,168*	*3,245*	*3,135*	*3,223*
Protein (%)	13.04	13.13	14.05	14.04	15.01	15.16	16.02	16.06	17.04	17.04
Lysine (%)	0.61	0.61	0.68	0.68	0.76	0.76	0.83	0.83	0.91	0.91
Threonine (%)	0.50	0.50	0.54	0.54	0.58	0.58	0.62	0.62	0.67	0.66
Tryptophan (%)	0.14	0.13	0.15	0.15	0.17	0.16	0.18	0.17	0.19	0.19
Salt, added (%)	0.37	0.35	0.45	0.42	0.52	0.51	0.60	0.57	0.68	0.65
Calcium (%)	0.62	0.63	0.76	0.75	0.88	0.89	1.02	1.01	1.15	1.14
Phosphorus (%)	0.54	0.54	0.60	0.60	0.66	0.66	0.72	0.72	0.78	0.78

[1]Adapted by the authors from *Life Cycle Swine Nutrition,* Iowa State University, Ames, PM–489, June, 1988.

[2]Suggested stages of production for using the above rations:

Grower	(——— Protein and lysine low ———)				+	+	(——— Calcium high ———)			
Finisher	+	+	+	+	+	+	(——— Calcium high ———)			
Gestation:										
3 lb *(1.4 kg)* per day	—	—	—	—	—	—	(Phosphorus marginal)		+	+
4 lb *(1.8 kg)* per day	—	—	(Phosphorus marginal)		+	+	+	+	+	+
5 lb *(2.3 kg)* per day	+	+	+	+	+	+	+	+	+	+
Lactation	—	—	(Phosphorus marginal)		+	+	+	+	+	+

[3]To convert lb to kg, divide by 2.2.

[4]Expected analysis using minimum analyses for each nutrient in the supplements from Table 23–25.

Where commercial supplements are bought to use with farm-grown grains, they may be utilized in the following ways:

1. Mixed with ground, farm-grown grain in the approximate amounts shown in Table 23–27 to make a complete ration (see the Table 23–27 footnote example for mixing directions).

2. Self-fed in separate feeders, with the ground or whole grain also being self-fed in separate self-feeders.

3. Hand-fed; with the supplement and the grain each being hand-fed in the proportions recommended in Table 23–27.

TABLE 23–27
GRAIN AND SUPPLEMENT COMBINATIONS (POUNDS)
NEEDED TO FORMULATE RATIONS OF DIFFERENT PROTEIN LEVELS (GRAIN VALUED AT 8.4% PROTEIN)[1 2 3]

Protein in Supplement		Percent Protein in Total Ration								
		10	11	12	13	14	15	16	17	18
(%)		(lb)	(lb)	(lb)	(lb)	(lb)	(lb)	(lb)	(lb)	(lb)
30	Grain	1,852	1,759	1,667	1,574	1,481	1,389	1,296	1,204	1,111
	Supplement	148	241	333	426	519	611	704	796	889
31	Grain	1,858	1,770	1,681	1,593	1,504	1,416	1,327	1,239	1,150
	Supplement	142	230	319	407	496	584	673	761	850
32	Grain	1,864	1,780	1,695	1,610	1,525	1,441	1,356	1,271	1,186
	Supplement	136	220	305	390	475	559	644	729	814
33	Grain	1,870	1,789	1,707	1,626	1,545	1,463	1,382	1,301	1,220
	Supplement	130	211	293	374	455	537	618	699	780
34	Grain	1,875	1,797	1,719	1,641	1,563	1,484	1,406	1,328	1,250
	Supplement	125	203	281	359	438	516	594	672	750
35	Grain	1,880	1,805	1,729	1,654	1,579	1,504	1,429	1,353	1,278
	Supplement	120	195	271	346	421	496	571	647	722
36	Grain	1,884	1,812	1,739	1,667	1,594	1,522	1,449	1,377	1,304
	Supplement	116	188	261	333	406	478	551	623	696
37	Grain	1,888	1,818	1,748	1,678	1,608	1,538	1,469	1,399	1,329
	Supplement	112	182	252	322	392	462	531	601	671
38	Grain	1,892	1,824	1,757	1,689	1,622	1,554	1,486	1,419	1,351
	Supplement	108	176	243	311	378	446	514	581	649
39	Grain	1,895	1,830	1,765	1,699	1,634	1,569	1,503	1,438	1,373
	Supplement	105	170	235	301	366	431	497	562	627
40	Grain	1,899	1,835	1,772	1,709	1,646	1,582	1,519	1,456	1,392
	Supplement	101	165	228	291	354	418	481	544	608
41	Grain	1,902	1,840	1,779	1,718	1,656	1,595	1,534	1,472	1,411
	Supplement	98	160	221	282	344	405	466	528	589
42	Grain	1,905	1,845	1,786	1,726	1,667	1,607	1,548	1,488	1,429
	Supplement	95	155	214	274	333	393	452	512	571
43	Grain	1,908	1,850	1,792	1,734	1,676	1,618	1,561	1,503	1,445
	Supplement	92	150	208	266	324	382	439	497	555
44	Grain	1,910	1,854	1,798	1,742	1,685	1,629	1,573	1,517	1,461
	Supplement	90	146	202	258	315	371	427	483	539
45	Grain	1,913	1,858	1,803	1,749	1,694	1,639	1,585	1,530	1,475
	Supplement	87	142	197	251	306	361	415	470	525

[1]Adapted by the authors from *Life Cycle Swine Nutrition,* Iowa State University, Ames, PM–489, June, 1988. To convert lb to kg, divide by 2.2.

[2]The grain common to the area may be substituted in Table 23–27, with the 8.4% protein content changed in keeping with the protein content of the grain used, and the proportions of grain and supplement adjusted to obtain the desired percent protein in the total ration.

[3]**Example:** In order to obtain a total ration with 15% protein, each 2,000 lb *(908 kg)* of feed should contain 1,389 lb *(631 kg)* of the 8.4% protein grain and 611 lb *(277 kg)* of the 30% supplement.

Pointers in Formulating Rations

In formulating rations and in feeding swine, the following points are noteworthy:

1. Feeds of similar nutritive properties can be interchanged in the ration as price relationships warrant.

2. If wheat, barley, oats, or grain sorghum is used instead of corn as the grain in a ration, the protein supplement may be slightly reduced.

3. When proteins of animal origin predominate, adequate mineral protection can be obtained by allowing hogs free access to a 2–compartment box or self-feeder with (a) salt (trace mineralized) in one side, and (b) a mixture of ⅓ salt (salt added for purposes of palatability) and ⅔ monosodium phosphate or other phosphorus supplement, in the other side. When supplements of plant origin constitute most of the source of proteins, add a third compartment to the mineral box and place in it a mixture of ⅓ salt (trace mineralized) and ⅔ ground limestone or oystershell flour.

4. When hogs are not exposed to sunlight or when dehydrated alfalfa meal is fed, vitamin D should be added in keeping with the recommended allowances (see Table 23–8).

5. Where the ration consists chiefly of white corn, barley, wheat, oats, rye, kafir, or by-products of these grains, there may be a deficiency of vitamin A (see Table 23–8 for recommended allowances).

6. Except for gestating sows and boars of breeding age, hogs are generally self-fed. All of the ingredients may be mixed together and placed in the same self-feeder, or the grain may be placed in one self-feeder (or compartment) and the protein supplements (including any ground alfalfa) in another. If the (a) cereal grains and (b) protein supplements (including ground alfalfa) are hand-fed, the grain and supplement should be fed separately, in the proportions indicated in the suggested rations.

7. An exception should be made to the cafeteria-style feeding when the grain ration consists of barley, oats, rye, or kafir. These feeds are higher in protein content than corn, and for this reason are generally fed as a mixed ration. Otherwise, the pigs will often eat more protein supplement than is necessary to balance the ration. Likewise, when corn is fed as the grain, sometimes such protein supplements as (a) roasted soybeans, (b) soybean meal, and (c) peanut meal are too palatable to be fed separately from the corn, especially if the corn is not of good quality.

8. Full-fed finishing hogs will consume 4 to 5 lb of feed daily per 100 lb liveweight until they weigh 100 lb. They will eat 3 to 4 lb daily per 100 lb weight from this stage until marketing.

FEED SUBSTITUTION TABLE

Successful swine producers are keen students of values. They recognize that feeds of similar nutritive properties can and should be interchanged in the ration as price relationships warrant, thus making it possible at all times to obtain a balanced ration at the lowest cost.

Table 23-28, Feed Substitution Table for Swine, is a summary of the comparative values of the most common U.S. and Canadian feeds. In arriving at these values, two primary factors besides chemical composition and feeding value have been considered; namely, palatability and carcass quality.

TABLE 23-28
FEED SUBSTITUTION TABLE FOR SWINE (AS-FED BASIS) (See footnote at end of table.)

Feedstuff	Relative Feeding Value (lb for lb) In Comparison With The Designated (underlined) Base Feed Which = 100	Maximum Percentage of Base Feed (or comparable feed or feeds) Which It Can Replace for Best Results	Remarks
ENERGY FEEDS: GRAINS, BY-PRODUCT FEEDS, ROOTS AND TUBERS:[1]			
Corn, No. 2	***100***	***100***	Corn is the leading U.S. swine feed, about 25% of the total production being fed to hogs. Corn is high in energy, but low in lysine and tryptophan. It does not pay to grind corn for growing-finishing pigs, but it should be ground for older hogs.
Alfalfa hay, early bloom, or Alfalfa meal, dehy	75–85	10–50	Low energy, good source of carotene and B vitamins, unpalatable to baby pigs. None in starter, 0–5% grower-finishing, 0–50% gestation, 0–10% lactation.
Bakery waste	95–110	20–40	Bakery wastes average about 10% protein and 13% fat. Variable salt content. May constitute up to 20% of starter rations and up to 40% of grower-finisher, gestation, and lactation rations.
Barley	90–95	100	Of variable feeding value due to wide spread in test weight per bushel. Should be ground or rolled. Low lysine.
Beans (cull)	90	33–66	Cook thoroughly; on a dry weight basis, limit to ½ the grain ration for pigs under 100 lb *(45.4 kg)* and ⅔ of the grain ration for pigs above 100 lb *(45.4 kg)*; supplement with animal protein.
Beet pulp	70–80	10	Bulky, high fiber, laxative. May constitute up to 10% of gestation and lactation rations.
Carrots (or beets, mangels, or turnips)	12–20	25	
Cassava, dried meal	85	33⅓	Low in methionine. Available as a swine feed in the tropics.
Corn and cob meal	80–90	0–70	Bulky, low energy. May constitute up to 70% of gestation rations.
Corn gluten feed (23% protein)	100	5–40	Corn gluten feed is bulky, low in lysine, and not too palatable. It may constitute up to 5% of starter rations, 10% of grower-finisher and lactation rations, and 40% of gestation rations.
Corn, high lysine	100–105	100	Superior to corn in lysine.
Corn meal	100	20	
Corn silage (25–30% D.M.)	20–30	0–90	Bulky, low energy, feed to sows only.
Emmer	80–90	80	Emmer may be used about like barley.
Fat (stabilized)	185–210	5	High energy, reduces dust.
Hominy feed	95	50	Hominy feed will produce soft pork if it constitutes more than ½ the grain ration. Hominy feed is subject to rancidity.
Millet (Proso)	85–90	50	Low lysine.
Molasses, beet	70	5	Used in pelleting. Laxative at high levels.
Molasses, cane (74% D.M.)	70	5	Used in pelleting. High levels (above 15% of the ration) cause soft, watery feces.
Molasses, citrus	70	5	It takes pigs 5–7 days to get used to the bitter taste of citrus molasses.
Oats	70–80	15–70	Grind for swine. Feeding value varies according to test weight per bushel. Oats may constitute up to 15–20% of grower-finishing rations and up to 70% of gestation rations.
Oats, groats	110–115	20	Palatable, but expensive. Primarily used in starter rations in which it may consititute up to 20%.
Peanuts	120–125	100	Peanuts are usually fed by hogging off.
Peas, dried	90–100	100	Normally peas should be fed to swine as a protein supplement. Two tons of peas equal 1 ton of grain plus 1 ton of soybean meal. Peas are low in methionine.
Potatoes, Irish (24% D.M.)	25–28	25–50	Not palatable in the raw state; must be cooked. When cooked and fed in a ratio of 3 lb *(1.4 kg)* of potatoes to l lb *(0.45 kg)* of grain, they are worth 25–28% as much as corn. May constitute 25% of grower-finisher rations and 50% of gestation rations.
Potatoes (Irish), dehy	100	33⅓	
Potatoes (sweet)	20–25	33⅓–50	Cooking also improves the feeding value of sweet potatoes.
Potatoes (sweet), dehy	90	33⅓	
Rice (rough rice)	80–85	50	Low energy, low lysine. Rice should be ground.
Rice bran	100	33⅓	If more than ⅓ of the grain consists of rice bran, soft pork will result.
Rice polishings	100–120	33⅓	Limited because feed becomes rancid in storage and soft pork will be produced.

(Continued)

TABLE 23-28 *(Continued)*

Feedstuff	Relative Feeding Value (lb for lb) In Comparison With The Designated (underlined) Base Feed Which = 100	Maximum Percentage of Base Feed (or comparable feed or feeds) Which It Can Replace for Best Results	Remarks
ENERGY FEEDS: GRAINS, BY-PRODUCT FEEDS, ROOTS AND TUBERS:[1] (Continued)			
Rice screenings	95	50	
Rye	90	20–30	Should be limited because it is unpalatable. Grind for swine. Watch for ergot.
Sorghum, grain	95	100	Check protein content and add supplement as necessary. Both very dry grain and bird-resistant varieties should be ground. Grain sorghum is low in lysine.
Spelt	65–80	25	Low energy, low lysine. Value varies according to the amount of hulls.
Sugar	70–80	0–5	High palatability, no protein. Normally, used only in starter rations.
Sunflower seed	100	50	High energy, high fiber.
Triticale	90–95	50	Higher levels not palatable. Watch for ergot.
Wheat	100–105	100	Feed whole if self-fed. Otherwise, grind coarsely; fine grinding makes it pasty and unpalatable. Low in lysine. Wheat-corn mixtures are more efficient than wheat alone.
Wheat bran	65–75	15–25	Bulky, high fiber, laxative. Bran is particularly valuable at farrowing time.
Wheat middlings	103	20	May be used as a partial substitute for grain.
Wheat standard middlings	85–100	10–30	Use as a partial grain substitute.
Wheat red dog and wheat white shorts	115–120	25	
Whey, dry	50	5–20	High lactose, very palatable. Use for baby pigs.
Whey, liquid	15	5–20	High lactose, very palatable. Use for baby pigs.
PROTEIN FEEDS:			
Soybean meal (41–50%)	***100***	***100***	Well balanced in amino acids. Best quality of all plant protein supplements. Very palatable.
Blood meal (80%)	120–130	20	High in protein (above 80%), high in lysine, but low in isoleucine. Not very palatable.
Buttermilk, dry	90–105	100	Good amino acid balance.
Buttermilk, liquid	15	100	Pound for pound, worth 1/10 as much as dried buttermilk.
Buttermilk, semisolid	33⅓–50	100	Pound for pound, worth 1/3 as much as dried buttermilk.
Canola meal (32–44%)	90	75	Low in goitrogenic compounds.
Copra meal (coconut meal)(21%)	50	25	
Corn, distillers' dried grains w/solubles	65–75	5	Used primarily as B-vitamin and unidentified sources, usually at about 5% of the ration.
Corn gluten meal (60% protein)	90	50	Bulky, low in lysine, and not too palatable.
Cottonseed meal (36–48%)	85	33⅓	Except when new glandless cottonseed meal is used, high levels may produce gossypol poisoning; hence, the level of cottonseed meal in swine rations should not exceed 8–9% of the total ration. Cottonseed meal is low in lysine.
Fish meal (60%)	115	100	Excellent balance of amino acids, and good source of calcium and phosphorus.
Linseed meal (35%)	80	25–50	Low in lysine; slightly laxative.
Malt sprouts	100	10	Malt sprouts contain a growth factor(s). They result in increased feed intake and gain.
Meat and bone meal (50%)	100	100	Low in tryptophan; good source of calcium and phosphorus.
Meat scraps (50–55%)	100	100	
Peanut meal (45%)	95	50	Becomes rancid when stored too long. Low in lysine; very palatable.
Peanuts	60–70	50	Peanuts are usually fed by hogging off. High levels will produce soft pork.
Peas, dried	50	50	
Rapeseed meal (32–44%)	85–90	33⅓	Rather unpalatable. Contains goitrogenic compounds that can be hazardous. But is usually detoxified.
Shrimp meal	90–100	50	
Skim milk, dried	90–120	100	Excellent-quality protein; very palatable; expensive. Especially good in prestarter and starter rations, of which it may constitute up to 10%.
Skim milk, liquid			Pound for pound, worth 1/10 as much as dried skim milk.
Soybeans, full fat, cooked	90–100	25–40	High energy. At high levels, will produce soft pork.
Sunflower meal (36–45%)	90–95	50	For swine, it should be combined with high-lysine supplements such as meat scraps or fish meal.
Tankage (60%)	.110	100	Good source of calcium and phosphorus. Low in tryptophan. Not palatable.
PASTURES AND DRY LEGUMES:			
Pasture, good		5–20% of grain, and 20–50% of protein supplement. It can replace all of pasture, in drylot rations.	Pasture and dry legumes are sources of good-quality proteins, of minerals, and of vitamins.
Alfalfa meal			Low energy, good source of carotene and B vitamins, unpalatable to baby pigs. For drylot rations, include 5–10% alfalfa in ration of grower-finishing pigs, up to 50% in ration of gestating sows, and up to 10% for lactating sows.

[1]Roots and tubers are of lower value than the grain and by-product feeds due to their higher moisture content.

In using this feed substitution table, the following facts should be recognized:

1. That, for best results, different ages and groups of animals within classes should be fed differently.
2. That individual feeds differ widely in feeding value. Barley and oats, for example, vary widely in feeding value according to the hull content and the test weight per bushel.
3. That, based primarily on available supply and price, certain feeds—especially those of medium protein content, such as peanuts and peas (dried)—are used interchangeably as (a) energy feeds, and/or (b) protein feeds.
4. That the feeding value of certain feeds is materially affected by preparation; thus, potatoes and beans should always be cooked for hogs. The values herein reported are based on proper feed preparation in each case.

For these reasons, the comparative values of feeds shown in Table 23-28, Feed Substitution Table for Swine, are not absolute. Rather, they are reasonably accurate approximations based on average-quality feeds.

FEEDING BABY PIGS

Baby pigs should have access to a creep feed beginning at 7 to 10 days of age. Commercial prestarters and starters are readily available, or farm-mixed rations can be used (see Table 23-20 for suggested prestarter rations, and see Table 23-21 for suggested starter rations). Pigs should receive a prestarter ration until they are about 3 weeks of age, after which they should be switched to a starter ration until they weigh about 40 lb. Early availability of quality prestarter and starter rations to young pigs will result in:

1. More uniform pigs with fewer runts.
2. Heavier weaning weights.
3. Lower mortality of baby pigs.
4. Decrease in incidence and severity of baby pig scours.
5. Less setback to young pigs when weaned from the sow. The earlier pigs are to be weaned, the more important it is that they be eating dry feed at an early age.
6. Lower weight loss by the sow.

Fig. 23-24. Farrowing crate with partially slotted floor. (Courtesy, University of Illinois, Urbana)

For successful baby pig feeding, the following pointers are pertinent:

1. Begin by giving the baby pigs a mere handful of creep feed, replenish daily. The creep feed should not be allowed to become stale or contaminated. Place feed in flat pans.
2. Once the pigs have started to eat readily, place the ration in a creep feeder so that they have access to the ration at all time. One linear foot of feeder space should be provided for each five pigs. The edge of the feeder trough should not be more than 4 in. above the floor. For maximum consumption of the creep ration, the feeder should be located near the waterer for the baby pigs.
3. Make clean, fresh water available to the young pigs in a separate waterer. It is not sufficient to rely on the sow's waterer to furnish water to the baby pigs.
4. The creep area should be light, warm, dry, and draft-free. It should be located in an area where the pigs are the least disturbed. Excitement, noise, and a change in feeding routine affect eating habits and subsequent feed consumption. Having the creep area near the sleeping area encourages more frequent eating. Arrange the creep area in such a way that it can be easily cleaned, and so that feed and water can be supplied conveniently without the producer getting into the area.
5. Individual litter creep areas are preferable. Where several litters have access to one creep area, it is advisable to limit the number of pigs to about 40 per creep.
6. If postweaning scours are encountered, the substitution of 200 to 400 lb of ground oats for a like amount of corn in rations Table 23-21 (Nos. 1, 3, 4, and 5) may be helpful.
7. By the time pigs reach a weight of 40 lb, they will have consumed about 54 lb of feed (4 lb of prestarter and 50 lb of starter).

FEEDING ORPHAN PIGS

There is no replacement for the sow's colostrum. If the newborn pig does not receive colostrum within 4 to 6 hours of birth, it has a lesser chance for survival. An orphan pig can obtain colostrum by being placed with another sow (a foster sow) that has just farrowed. If no such sow is available, the orphan can be fed a good commercial milk replacer. A homemade milk replacer can be prepared by mixing the following ingredients:

1 quart milk
1 pint half-and-half
1 raw egg
oral, water-soluble antibiotic

Portions of this mixture should be warmed to 98 to 100°F and fed about every 3 hours, with each pig receiving about ¼ cup per feeding. The use of a shallow pan for feeding is recommended. Immersing the baby pig's nose in the milk a few times will result in its drinking readily. It is extremely important that all feeding utensils be kept clean and sanitary; otherwise, scouring will occur.

The orphan pig can be fed a dry 22 to 23% crude protein prestarter from 5 to 7 days of age until about 2 to 3 weeks of age (see Table 23-20). At this time, it can be switched to a 20 to 21% crude protein pig starter (see Table 23-21).

FEEDING GROWING-FINISHING PIGS

In the practical swine enterprise, growing-finishing generally refers to that period from weaning to market weight of about 240 lb. Because hogs are finished at an early age, the process really consists of both growing and finishing. In a general way, there are 2 methods of finishing hogs for market: (1) full feeding all the time until the animals attain a market weight, and (2) limited feeding early in the period, with full feeding the last 60 to 75 days of the period before marketing.

For the production of lean (bacon) carcasses, the rate of gain should be restricted to about 1½ lb daily after a liveweight of 100 to 125 lb. This is easily accomplished by using a lighter, bulkier finishing ration (made by including 10 to 20% bran, oats, alfalfa, or other suitable bulky feed in the ration). Level of protein has no direct effect on carcass excellence, though it does affect the growth of the pig.

Neither system, full feeding nor limited feeding, can be recommended as being best for all conditions. The plan to follow should be determined by (1) market conditions, (2) type and breeding of the pigs, (3) price of feeds, (4) feeds available on the farm, (5) kind and extent of pastures available, (6) available labor, and (7) capital invested in facilities. Self-feeders are well adapted to a system of full feeding, but hand-feeding or interval feeding are necessary in any plan for limiting the ration.

Research indicates that there is little difference in the feed efficiency of group-fed pigs between (1) limited feeding of 5 lb per pig per day from 125 lb to market, and (2) self-feeding. However, limited feeding results in slower gains, increased labor or mechanization, variable performance, and increased supervision. For these reasons, the practice can be justified only when sufficient premium is paid for the modest increase achieved in the lean-to-fat ratio. It should be added that the selection of hogs with bred-in meat-type carcasses has largely alleviated the need for restricting rate of gain and using bulky rations as a means of getting leaner carcasses.

When on full feed, finishing pigs will consume 5 to 6 lb of feed daily per 100 lb liveweight up to 120 lb in weight.

Fig. 23-25. Completely slotted floor finishing pens with automated feed distribution system. (Courtesy, University of Illinois, Urbana)

From 120 lb to a finished weight of 240 lb, pigs on full feed consume about 4 lb of feed daily for each 100 lb of liveweight. With good feed and management, about 700 lb of feed are required to produce 200 lb of gain during the growing-finishing period from 40 lb to 240 lb weight. Of course, the feed consumption of the breeding herd and that of the pigs during the suckling period must be added to the growing-finishing feed consumption figures in order to arrive at the total feed requirements. Records show that about 24% of the total feed required to produce market hogs is consumed by the breeding herd (including the pigs to weaning) and 76% by the pigs after weaning. Hence, in producing a 240-lb market hog, the total feed requirements are:

	Pounds Feed	Percent of Total Feed
Breeding herd, including pigs to weaning	217	24
Pigs from 40 to 240 lb	700	76
Total	917	100

This means that 3.5 lb of feed are required to produce 1 lb of gain during the grower-finishing stage, from 40 lb weight to 240 lb weight (700 ÷ 200 = 3.5 lb). However, when the feed consumed by the breeding herd, including pigs to weaning, is included, 3.82 lb of feed are required to produce 1 lb of market hog (917 ÷ 240 = 3.82).

The protein requirements of the pig are greatest early in life. For this reason, decreasing percentages of protein supplement should be incorporated in mixed rations as the finishing process progessses.

However, ample protein should always be provided in the ration; otherwise, growth will be retarded. It is also important that the mineral and vitamin needs of growing-finishing pigs be met.

As previously indicated, pigs can utilize a great variety of concentrates. The chief ingredients of a growing-finishing ration, therefore, are usually, for practical reasons, those most readily available at the lowest possible price.

Suggested growing-finishing rations are given in Table 23-24. Note that provision is made for meeting the nutritional needs at two different stages of growth—40 to 120 lb, and 121 to 240 lb—with three different rations suggested for each stage.

FEEDING PROSPECTIVE BREEDING GILTS

Prospective breeding gilts should be kept from getting too fat. Meat-type animals can usually be left on a high-energy ration until they reach 150 to 200 lb without becoming too fat. It is neither necessary nor desirable that females intended for breeding purposes carry the same degree of finish as market animals. After selecting replacement gilts, they should be fed as follows:

1. Give about 5 lb per head per day through their second heat period.
2. Flush—full feed—after the second heat period until breeding on the third heat period.
3. After breeding, limit the feed intake to 3 to 5 lb per day. Overfeeding during gestation can cause embryonic death and thus decrease litter size.

Fig. 23-26. Duroc gilt in excellent breeding condition. (Courtesy, United Duroc Swine Registry, Peoria, Ill.)

FEEDING BOARS

The feed allowance of young boars should vary according to condition of the animals, the climatic condition, and the individuality. If the animals are inclined to get too fat, which is likely to happen when self-feeding, the ration may well contain a considerable amount of bulky feeds; otherwise, limited feeding may be necessary.

The feed requirements of the herd boar are about the same as those of a female of equal weight. He should always be kept in thrifty, vigorous condition and virile. In no case should boars be overfat, nor should they be in a thin rundown condition. Normally, the following feed allowances will suffice: for boars weighing 120 to 150 lb, 6 to 9 lb of feed daily; for mature boars, 5 to 7 lb of feed daily. A more liberal ration must be provided in the wintertime and when the sire is in heavy service. The feed allowance should be varied with the age, development, temperament, breeding demands, and roughage consumed.

Boars and pregestating/gestating sows may be fed the same rations. Rations, formulated for feeding 3, 4, or 5 lb per head daily, are presented in Tables 23-16, 23-17, and 23-18 (see pp. 986-987).

Fig. 23-27. Spotted boar in thrifty, vigorous condition. (Painting by Tom Phillips. Courtesy, National Spotted Swine Record, Inc., Bainbridge, Ind.)

FEEDING BROOD SOWS

The nutrition of brood sows is critical, for it may materially affect conception, reproduction, and lactation. Proper feeding of sows should begin with replacement gilts and continue through each stage of the breeding cycle—flushing, gestation, farrowing, and lactation.

Fig. 23-28. Crossbred white sows in open-front shelter with large concrete apron. (Courtesy, E. J. Stevermer, Department of Animal Science, Iowa State University, Ames)

Flushing Sows

The practice of conditioning or having the sows gain in weight just prior to breeding is known as flushing. The purpose of flushing is to increase the number of ova shed during estrus. About 10 to 14 days prior to expected breeding, the female should be fed a ration that will produce gains of 1 to 1¼ lb per day. Generally 6 to 8 lb per head per day of a high-energy feed that is well balanced in minerals and vitamins, is adequate. Immediately after breeding, the females should be put back on limited feeding. Continuation of a high level of feeding after breeding will result in a higher embryo mortality.

Gestation Period

The nutrients fed the pregnant gilt or sow must first take care of the usual maintenance needs. If the gilt is not fully mature, nutrients are required for both maternal growth and growth of the fetus. Quality and quantity of proteins, minerals, and vitamins become particularly important in the ration of young pregnant gilts, for their requirements are much greater and more exacting than those of the mature sow.

Approximately two-thirds of the growth of the fetus is made during the last month of the gestation period. It may be said, therefore, that the demands resulting from pregnancy are particularly accelerated during the latter third of the gestation period. Again, the increased needs are primarily for proteins, minerals, and vitamins.

During gestation, it is also necessary that body reserves be stored for subsequent use during lactation. With a large litter and a sow that is a heavy milker, the demands for milk production are generally greater than can be supplied by the ration fed at the time of lactation. Although desired gains will vary somewhat according to the initial condition, mature sows are generally fed to gain about 70 lb during the pregnancy period, and first litter pregnant gilts are fed to gain about 90 lb. This means that from the day of mating until farrowing time gilts should be fed to gain about 0.9 lb per day and mature sows about 0.7 per day. This calls for a daily feed allowance of approximately 4 to 4.5 lb per head, with variation according to environmental conditions.

It is important that the condition of dry sows be regulated so that they are neither too fat nor too thin at farrowing time. Overly fat sows may have difficulty in farrowing and give birth to weak or dead pigs. Sows that are too thin at farrowing tend to become suckled down during lactation. Thus, one way or another, limited feeding is a must for gestating gilts and sows. This may be accomplished by any one of the following feeding systems (these feeding systems are further detailed later in this chapter in the section headed "Feeding Systems"):

1. By adding sufficient bulk.
2. By interval feeding.
3. By group hand-feeding.
4. By individual feeding.

In addition to the above limited feeding systems, the use of pasture should be considered. Where available, a leguminous pasture is the ideal way in which to limit-feed gestating gilts and sows. Dry sows on good legume pasture are usually fed ½ lb less supplement and 2 lb less corn per day. In addition to limiting the feed intake, the pasture system provides valuable quality protein, minerals, vitamins, and exercise.

Suggested gestation rations, formulated for feeding 3, 4, or 5 lb per head daily, are given in Tables 23-16, 23-17, and 23-18.

Farrowing Time

It is considered good practice to feed lightly and with bulky laxative feeds from 4 to 5 days before and after farrowing. Wheat bran or oats may constitute half of the limited ration, and a small amount of linseed meal may be added.

The sow may be watered at frequent intervals before or after farrowing, but in no event should she be allowed to overeat. It is also a good plan to take the chill off the drinking water in the wintertime.

Lactation Period

The nutritive requirements of a lactating sow are more rigorous than those during gestation. They are very similar to those of a milk cow, except they are more exacting relative to quality proteins and the B vitamins because of the absence of rumen synthesis in the pig. A good lactating sow will produce an average of about 1 gal of milk daily during the suckling period. A sow's milk is also richer than cow's milk in all nutrients, especially in fat. Thus, sows suckling litters need a liberal allowance of concentrates rich in protein, calcium, phosphorus, and vitamins.

Fig. 23-29. The nutritive requirements of the lactating sow and litter are rigorous. (Courtesy, Watt Publishing Co., Mt. Morris, Ill.)

It is essential that suckling pigs receive a generous supply of milk, for at no other stage in life will they make such economical gains. The gains made by pigs from birth to weaning are largely determined by the milk production of the sows; and this in turn is dependent upon the ration fed and the sow's inherent ability to produce milk. The lactating sow should be provided with a liberal feed allowance—ranging from 2½ to 4½ lb daily for each 100 lb weight. Generous feeding during lactation, with a small shrinkage in weight, is more economical than a stingy allowance of feed, for the nutrients in milk must come either from the feed or from the sow's body. Lactating sows are commonly self-fed, because even when hand-fed they are practically on full feed.

Suggested lactation rations are given in Table 23-19.

FITTING RATIONS FOR SHOW AND SALE SWINE

Any of the rations listed in Tables 23-16, 23-17, 23-18, 23-22, or 23-24 for the respective classes and ages of swine are suitable for use in fitting show animals of similar classification. Because of the high cost of labor, the recent trend has been toward self-feeding both young breeding animals and market barrows and gilts that are being fitted for show. Many of them are left on self-feeders right up to show time, others are hand-fed only during the last month or two of the fitting period. However, most experienced exhibitors feel that they can get superior bloom and condition by either (1) hand-feeding, or (2) using a combination of hand-feeding and self-feeding (hand-feeding twice daily and allowing free access to a self-feeder). When hand-feeding, they also prefer mixing the ration with skim milk, buttermilk, or condensed buttermilk, and feeding the entire ration in the form of a slop.

Adding milk to a ration that is already properly balanced makes for a higher protein content than necessary. On the other hand, most experienced caretakers prefer using rations of higher protein content for fitting purposes. They feel they get more bloom that way. In general, however, when skim milk or buttermilk is used in slop feeding, the protein feeds of the ration may be reduced by one-half without harm to the animal.

In fitting show barrows, it may be necessary to decrease or discontinue slop feeding 2 to 4 weeks before the show to avoid paunchiness and lowering of the dressing percentage.

When oatmeal (oat groats, rolled hulled oats) is not too high priced, many successful hog caretakers replace up to 50% of the grain (corn, wheat, barley, oats, and/or sorghum) in the ration with oatmeal. They do this especially when fitting hogs—both breeding animals and barrows—in the younger age groups. Oatmeal is highly palatable, lighter, and less fattening than corn.

Suitable minerals and vitamins should always be provided.

FEEDING SYSTEMS

The choice of the feeding system(s) and the choice of the ration(s) must go hand in hand. For example, if the grain and the protein supplements are to be self-fed in separate feeders or compartments, it is important that they be of equal palatability; otherwise, pigs will consume too much of one and too little of the other. A listing and discussion of each of the common feeding systems follows.

• **Complete self-fed rations**—The trend is toward the use of complete self-fed rations for baby pigs and growing-finishing hogs, because, in comparison with free-choice feeding, they (1) lend themselves better to automation, (2) provide better control of nutrient intake, and (3) result in faster gains.

Complete rations may be formulated either by "building from the ground up" (by adding each ingredient, one by one), or by mixing a complete supplement, a base mix, or a premix with ground grain.

A survey made by the University of Illinois revealed that the most-used rations by swine producers were prepared by mixing ground grain and a commercial formula supplement (a protein, mineral, and vitamin supplement).

Where producers do their own mixing, they favor simplified rations. Fortunately, a simple ration of corn and soybean meal, fortified with minerals and vitamins, will generally give as good results as a more complex ration consisting of many ingredients. Of course, with large volume buying and computerized formulations, commercial feed companies can use more complex rations (with many ingredients) advantageously, especially from the standpoints of enhancing palatability and balancing amino acids, minerals, and vitamins.

• **Floor or drop feeding**—Floor or drop feeding is particularly suited to the controlled feeding of growing-finishing swine or the breeding herd. Feeding in the sleeping area encourages cleanliness, since pigs are less inclined to defecate where they eat. Feed wastage is reduced to a minimum when the animals do not have more feed available than they will consume at one eating. Even though automated, restricted feeding requires close attention, because the daily feed intake of pigs is affected by weather.

• **Free-choice**—Grain and protein supplements may be fed separately and free-choice. Generally, pigs fed free-choice rations in separate feeders or compartments will not make as uniform or as fast gains as pigs fed a complete mixed ration. The free-choice system requires more supervision, as the palatability of the grain or the protein supplement may vary and the pigs will then overeat or undereat the supplement or the grain. There is very little, if any, difference in economy of gain between feeding a free-choice or a complete ground mixed ration. Free-choice feeding may be the best feeding system for the small producer who does not have mixing equipment.

• **Liquid feeding**—Liquid feeding usually involves mixing predetermined amounts of feed and water prior to, or at the time of, feeding. When properly used, this method can practically eliminate feed dust in the feeding area and minimize wastage. Ratios of feed and water can be varied to produce a free-flowing liquid or a thick paste. In some cases, feed is automatically dropped into the water in the feed trough. Research has shown no difference in rate of gain of pigs full-fed on liquid or dry feeds. Neither does liquid feeding have any effect on dressing percentage, carcass measurements, or carcass quality. However, pigs full-fed liquid rations generally require more feed per pound of gain than pigs full-fed dry rations.

• **Limit feeding**—With gestating sows, limit feeding to 4 to 6 lb per head daily is a must in order to keep them from getting too fat. Overly fat sows have difficulty in farrowing and give birth to weak or dead pigs. With growing-finishing pigs, it is a way in which to increase slightly the proportion of lean to fat in the carcass. A discussion of limit feeding of (1) gilts and sows, and (2) growing-finishing pigs follows.

1. **Gilts and sows.** Replacement gilts should be started on a limited feeding program at 180 to 200 lb; and all gestating sows and gilts should be limit fed. Limit feeding may be accomplished by any one of the following methods:

a. **By feeding bulky, fibrous feeds,** such as silage, haylage, or alfalfa, with such feed constituting at least one-third of the ration. Actually, this is a way in which to lower the energy content of the ration. Although bulky feeds will hold the weight down, they usually do not lower feed cost.

b. **By interval feeding,** in which gilts or sows are turned to self-feeders for 2 to 8 hours every second or third day. Under this system, gilts will usually eat around 12 lb of feed at a time (or an average of 4 lb per day) and older sows will consume around 15 lb (or an average of 5 lb per day). The amount of feed consumed in interval feeding may be controlled either (1) by varying the interval, from every other day to twice a week, (2) by varying the length of time that the gilts and sows are left on the self-feeders (from 2 to 8 hours), or (3) by hand-feeding.

Experiments and experiences show that reproductive performance is the same with either interval feeding or daily hand-feeding a limited amount. Turning sows on to self-feeders at intervals requires less labor than daily hand-feeding, but under some conditions it results in greater stress on fences and equipment.

c. **By group hand-feeding** a limited ration to several sows. This is apt to result in the "bossy" sows getting too much and the "timid" sows getting too little. This problem can be partially alleviated by feeding over a large area.

d. **By individual feeding** in either individual stalls or in tie stalls, tethered by a neck collar or belt.

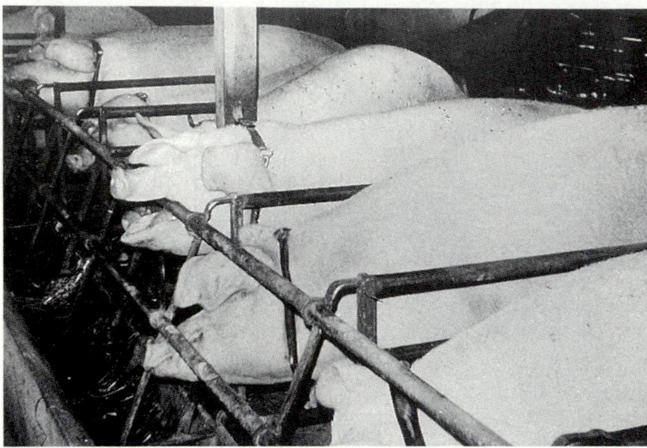

Fig. 23-30. Landrace sows tethered by neck collar and individually fed. (Photo by A. H. Ensminger)

2. **Growing-finishing pigs.** Sometimes growing-finishing pigs are limit-fed in order to produce leaner carcasses. Usually, it is started when pigs weigh around 100 lb and feed is limited to about 85 to 95% of what pigs of comparable age consume when self-fed. Limit feeding of market hogs results in slower gains, increased labor, and more mechanization. Thus, unless sufficient premium is paid for the modestly leaner carcasses, it cannot be justified.

• **Pelleted complete rations**—The use of pelleted complete rations for growing-finishing hogs will increase the average daily gain by 2 to 5% and improve the feed efficiency by approximately 5 to 10%. Thus, when a complete ration is purchased, buying a pelleted feed may be more economical than buying a meal. But the advantage of pelleting will usually not be sufficient to offset the cost of hauling grain to the mill and having a pelleted ration made. Also, pellet machines are costly; hence, the purchase of such equipment cannot be justified with the volume of hogs handled by most swine producers.

FEEDER PIG PRODUCTION

Feeder pig production refers to the production and sale of immature pigs weighing 30 to 60 lb, usually throughout the year, for growing and finishing on other farms. It makes for a two-phase system in swine production, similar to the two-phase system so well known in the cattle industry where some operators specialize in the cow-and-calf system (the production of feeder cattle) and others in finishing cattle. Until recent years, it was generally assumed that a two-phase system lent itself more logically to beef cattle than to swine because of the western range and of fewer disease problems. But several important scientific and technological developments which occurred in the swine industry in the 1950s and early 1960s ushered in considerable two-phase production of hogs. Among such developments were: (1) specific pathogen-free (SPF) herds and other improved disease control measures; (2) confined and continuous production—which increased specialization in breeding, in farrowing, and in finishing; and (3) increased mechanization.

Among the **advantages** of raising feeder pigs as compared to growing-finishing pigs are:

1. It provides an opportunity to use efficiently a maximum amount of labor, for about two-thirds of the labor of raising hogs occurs by the time pigs reach weaning age.
2. It requires less grain per dollar of product sold. To maintain a brood sow and raise a litter of pigs to 40 lb requires only 25 to 30% as much feed as is needed to feed them to market weight of 240 lb.
3. It requires less feed and manure handling.
4. It allows a rapid turnover in the volume of pigs that can be handled each year. The farrowing schedule can be planned to provide frequent sales for consistent income.
5. It can be started with relatively small capital inputs.

The main **disadvantage** to feeder pig production is that producers must depend on those engaged in growing-finishing operations for a market; hence, they are somewhat limited as to time of marketing and volume of sales.

Knowledge of the following is pertinent to successful feeder pig production:

1. **Basic requirements.** The two most important requirements of a good feeder pig program are:

   a. **High-level of management.** Since about one-fourth of the pigs born never reach weaning age, the importance of good management during this critical period is obvious.

   b. **Dependable market.** Most feeder pig producers are not equipped to feed hogs from weaning to market. Also, modern systems of multiple farrowing make it necessary to market pigs on schedule to make room for younger pigs to be farrowed and raised to weaning weights. Thus, if dependable markets are not available, feeder pig producers can find themselves in the unenviable position of having unsold pigs on their hands and more of them on the way.

2. **Variation in feeder pig prices.** Feeder pig prices are more erratic than slaughter hog prices—they move upward and downward more than slaughter hog prices, and they do so more rapidly. This is attributed to the fact that there is no well organized nationwide system of marketing feeder pigs where dependable price information is available.

3. **Methods of marketing feeder pigs.** There are three methods of marketing feeder pigs:

   a. **Contract arrangements.** This refers to the sale of feeder pigs directly from producers to finishers on a contract basis.

   b. **Competitive organized markets.** The auction method of selling feeder pigs is good for smaller feeder pig producers. Well-conducted auctions provide the advantages of (1) uniform lots in type, color, and size; (2) selling a large number of pigs quickly and at reasonable cost; (3) buyers who are interested in various sizes, qualities, and numbers; and (4) open competition.

   c. **Private treaty.** This refers to the direct negotiation between producer and buyer. Often, newspaper ads bring the two parties together and then a per head price is established after visual inspection. The major disadvantage of this system is its uncertainty.

OTHER FEED/MANAGEMENT RELATED ASPECTS

In addition to the subject matter covered earlier in this chapter, there are other feed/management related aspects of importance in swine production. Some of them will be discussed in the sections that follow.

Docking Tails

Tail biting accompanies close confinement. It results when pigs are prevented from rooting, nibbling, and chewing—from disturbing the pig's normal behavior pattern. Tail docking has become a common management practice to prevent subsequent tail biting of pigs in confinement. It should be done on all market hogs. Tails should be cut ¼ to ½ in. from the body with side-cutting pliers or another blunt instrument; the crushing action stops bleeding. The tail stump should be disinfected with a good antiseptic, and the instrument should be disinfected after docking each pig. (See Chapter 17, Animal Behavior/Environment, for additional information on tail biting and docking.)

Nutritional Anemia

Anemia is a blood condition in which there is a deficiency of hemoglobin which transports oxygen to various parts of the body. It is caused by a deficiency of iron and/or copper in the diet, and it is most likely to occur in nursing pigs that do not have access to soil.

The four basic reasons for anemia in nursing pigs raised in confinement are:

1. **Low body storage of iron in the newborn pig.** The accompanying table shows that the pig has a comparatively low tissue level of iron at birth, but a high level of iron at maturity. Thus, the iron requirement of the pig is greater than for other species.

	Iron Concentration	
Species	**Newborn**	**Adult**
	(ppm)	**(ppm)**
Pig	29	90
Human	94	74
Rabbit	135	60
Rat	59	60

2. **Low iron content of sow's colostrum and milk.** Only approximately 10% of the pig's iron requirements are met by colostrum and this drops to about 5% by the third day.
3. **Elimination of contact with soil iron.** Pigs farrowed and raised in confinement have no contact with the soil.
4. **Rapid growth rate of the nursing pig.** During the first weeks of life, the pig makes much greater increase from birth weight than other species. As a result, anemia develops more rapidly in pigs than in other species.

Anemic pigs show listlessness, rough hair coat, wrinkled skins, drooping ears and tails, pale membranes around the mouth and eyes, and labored breathing.

Any of the following anemia prevention measures may be used:

1. Inject intramuscularly in the neck or ham muscle 100 to 200 mg of iron, in the form of iron dextran, iron dextrin, or gleptoferron into baby pigs at 2 to 3 days of age. If pigs remain in confinement and do not have access to creep feed at an early age, a second injection at 2 to 3 weeks of age is desirable. Injection is the method of choice, for it assures that every pig receives its requirement.
2. Orally administer iron from iron chelates within the first few hours of life; early administration before gut closure to large molecules is crucial. To ensure daily intake by all pigs, it is important to have a preparation that is palatable and readily consumed. Also, placement of the oral preparation at the right location in the creep area is most important.
3. Give the pigs iron tablets or paste at 2 to 3 days of age. Repeat the treatment every 7 to 10 days until the pigs are eating their creep ration adequately. If pills are given, it is important to see that the pigs swallow them and not spit them out.
4. Place clean soil in the farrowing pen daily. Soil should not be contaminated with parasite eggs and other disease organisms. Iron sulfate can be sprinkled over the soil.
5. Swab sow's udder daily with a solution of 1 lb ferrous sulfate dissolved in 1 gal of warm water.

Scouring (Diarrhea, Enteritis)

Scouring is one of the major problems facing swine producers. It is estimated that about 40% of U.S. swine herds are affected with scouring, and that about 20% of pig losses between farrowing and weaning are caused by scouring. Many different etiological agents cause scouring, including nutrition, management, environment, stress, bacteria, viruses, and parasites. Although the nutritional aspects of scouring are frequently discussed, they are poorly understood. However, it is generally recognized that good nutrition, along with good management and minimal stress, will lessen the incidence of scouring and be effective treatment when an outbreak of the disease occurs.

Weaning Pigs

The optimum age to wean pigs varies considerably, depending on nutritional programs, facilities, environment, health, and management. The average age at weaning for pigs in the United States is about 4 weeks, with a range of 2 to 7 weeks. The earlier pigs are weaned, the better the required feeding and management practices. Advantage of early weaning (3 to 4 weeks of age) are:

1. Heavier pigs at 8 weeks of age, with fewer runts.
2. Lower sow feed costs.
3. More litters per year.
4. Less weight loss of the sow.
5. Greater flexibility in rebreeding or selling sows.
6. Greater turnover of sows through the farrowing unit; hence, less total farrowing area space required and lower facility cost per sow.

Successful early weaning encompasses the following:

1. A sound breeding and feeding program of gestating sows to ensure large, healthy pigs at birth; there is a high positive relationship between birth weight and weight at weaning.
2. Good milking sows that supply plenty of nutrients to get the pigs off to a good start.
3. Good baby pig and sow management during lactation to ensure strong, uniform, and healthy pigs at weaning.
4. A good creep feeding program, beginning with a good quality prestarter ration (see Table 23-20, p. 988) when suckling pigs are 7 to 10 days of age, and switching to a good quality starter ration (see Table 23-21, p. 988) after weaning.

For best results, the guidelines given in Table 23-29 should be observed when planning for early weaning.

TABLE 23-29
GUIDELINES TO SUCCESSFUL EARLY WEANING[1]

Guideline	Age in Weeks				
	1	2	3	4	5
Minimum pig weight (lb)	5	9	12	15	21
Nursery temperature at pig level (°F)	85	85	83	81	79
Minimum floor space per pig (sq ft)[2]	3	3	3	3	3
Maximum number of pigs per linear foot of feeding space	5	5	5	5	5
Maximum number of pigs per nipple waterer[3]	8	8	8	8	8
Maximum number of pigs per group	10	10	10	15	25

[1]See Appendix for conversion of U.S. customary to metric.

[2]The figures given herein are for solid floors. On slotted floors, this may be lowered 2 sq ft per pig from 1 to 5 weeks of age.

[3]Where bowls are used instead of nipples, there should be one bowl for each 12 pigs.

The age of pigs at weaning may vary from herd to herd, according to the facilities available, intensity of operation, and managerial skills of the producer. Generally, pigs can be weaned over a wide age range; however, the younger the pigs, the more demanding the management required to do it successfully. Observance of the following guides will reduce the stress at weaning:

1. Wean only pigs weighing more than 12 lb.
2. Wean over a 2- to 3-day period, weaning the larger pigs in the litter first.
3. For 3-week-old pigs provide an environmental temperature of 80–85°F.
4. Group pigs according to size.
5. Limit numbers in a pen to 30.
6. Limit feed intake for 48 hours if post-weaning scours are a problem.
7. Provide 1 feeder hole for 4 to 5 pigs and 1 waterer for each 20 to 25 pigs.
8. Medicate drinking water if scours develop.

Feed Required to Produce a Pound of Market Hog

Nationally, it has been estimated that it requires 4.0 lb of feed to produce 1 lb of on-foot hog (live) from birth to market weight (see Chapter 1, Table 1-2), exclusive of the feed required by sows and boars to produce pigs. This is high. But remember that 25 to 30% of all pigs farrowed die before weaning. Remember, too, that many swine producers are inefficient.

Table 23-30 shows realistic goals for well-managed swine operations.

TABLE 23-30
ESTIMATED FEED REQUIRED TO PRODUCE 240-LB MARKET PIG[1]

Stage of Production	Feed Required per 240-Lb Market Pig
	(lb)
Sow gestation ration (includes pregestation and breeding)	110
Boar ration	8
Lactation ration	45
Starter ration (creep to 40 lb)	54
Grower-finisher ration (40 to 240 lb)	700
Total, lb	917

Per 100 lb of pork produced $\frac{917}{240} \times 100 = 382$ **lb**

[1]See Appendix for conversion of U.S. customary to metric.

The values given in Table 23-30 are estimates based on realistic standards for apportioning the quantities of sow and boar feed to each pig and the feed conversion normally attained during the starter and grower-finisher periods. Although these data do not provide for pig deaths after weaning, normal milling losses, and feed wastage, it is assumed that these losses would likely be offset by sow weight gains which are not considered in the pounds of hog produced. Data obtained in commercial herds where accurate records have been kept indicate that 382 lb of feed per 200 lb of live hog produced is a realistic goal for a practical swine operation. However, to achieve this level of efficiency, a sound feeding and management program must be followed, including limit feeding of pregnant sows, high conception rates, large litters weaned, early weaning and rebreeding, low death losses, minimal disease problems, balanced rations, and minimal feed wastage.

Effect of Sex on Performance of Growing-Finishing Hogs

When full fed, boars consume 10 to 15% less feed daily than barrows or gilts and are 10 to 15% more efficient in feed conversion. Also, boars gain faster than barrows and gilts. Barrows gain approximately 1/10 lb faster per day than gilts, which reduced their age at slaughter by 10 days. Feed per pound of gain is similar for barrows and gilts. Gilts yield carcasses having .11 in. less backfat, .52 sq in. larger loin eye area, and 1.8% more lean cuts than barrows. Dressing percentage usually favors barrows, which is consistent with their greater depth of backfat.

Soft Pork

Feed fats are laid down in the body of pigs without undergoing much change. Thus, when finishing hogs are liberally fed high fat content feeds in which the fat is liquid at ordinary temperatures, soft pork results. This condition prevails when hogs are liberally fed such feeds as soybeans, peanuts, mast, or garbage. The oil of the cereal grains is also liquid at ordinary temperatures, but fortunately the fat content in these feeds is relatively low. When such feeds are liberally fed to swine, most of the pork fat is actually formed from the more abundant carbohydrates.

Soft pork is undesirable from the standpoint of both the processor and the consumer. It remains flabby and oily even under refrigeration. In soft pork, there is a higher shrinkage in processing; the cuts do not stand up and are unattractive in the showcase; it is difficult to slice the bacon; and the cooking losses are higher through loss of fat. For these reasons, hogs that are liberally fed those feeds known to produce soft pork are heavily discounted on the market.

The firmness of pork carcasses may be judged by (1) grasping the flank below the ham, (2) lifting one end of the cut while permitting the other end to rest on the table (a firm pork cut will not bend readily), or (3) applying a slight pressure of the thumb (not gouging) on a cut surface. Experimentally, either the iodine number or the refractive index is used in determining the degree of softness; this is a measure of the degree of unsaturation.

Unless the producer is willing to take the normal reduction in price (about $1 per cwt), it is recommended that feeds which normally produce soft pork be fed liberally only to pigs under 85 lb in weight and to the breeding herd. For growing-finishing pigs over 85 lb in weight, soybeans and peanuts should not constitute more than 10% of the ration if a serious soft pork problem is to be averted.

Experimental evidence and practical observation have shown, however, that when a ration producing hard fat is given following a period of feeds rich in unsaturated fats, the body fat gradually becomes harder. It has also been found that this process takes place more rapidly if the animals are first fasted for a period before the change in ration is made. This practice is called *hardening off.* Thus, many hogs that are, for practical reasons, finished primarily on such feeds as soybeans, peanuts, or garbage, are hardened off with a ration of corn or some other suitable grain.

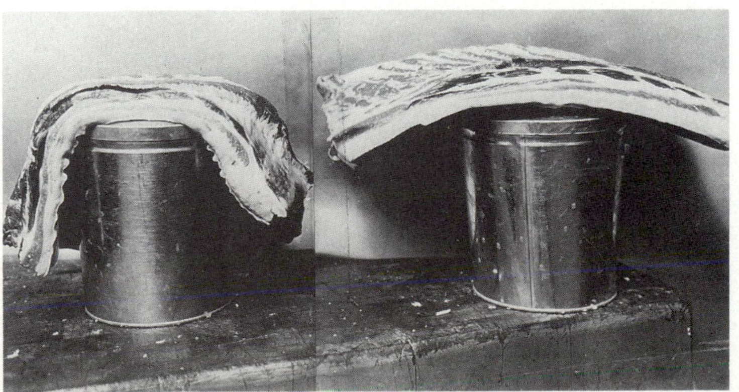

Fig. 23-31. Soft pork. Feed fats do affect body fats. The bacon belly on the left came from a hog liberally fed soybeans. (Courtesy, University of Illinois, Urbana)

Corn-Hog Ratio

The corn-hog ratio refers to the number of bushels of corn required to be equivalent in value to 100 lb of live hogs at local markets, based on average prices received by farmers for corn and hogs. During the 15-year period, 1972 to 1986, the hog-corn ratio averaged 18.6. This means that the price relationship was such that 18.6 bushels of corn equalled in value 100 lb of hogs.

A high corn-hog ratio—one above 18.6 in recent years—indicates cheap corn and high-priced hogs and likely profit to the producer—conditions that stimulate more breeding and more feeding to heavier weights. On the other hand, a low ratio, one which is below 18.6 means high-priced corn and low-priced hogs—conditions that result in less breeding and feeding of swine.

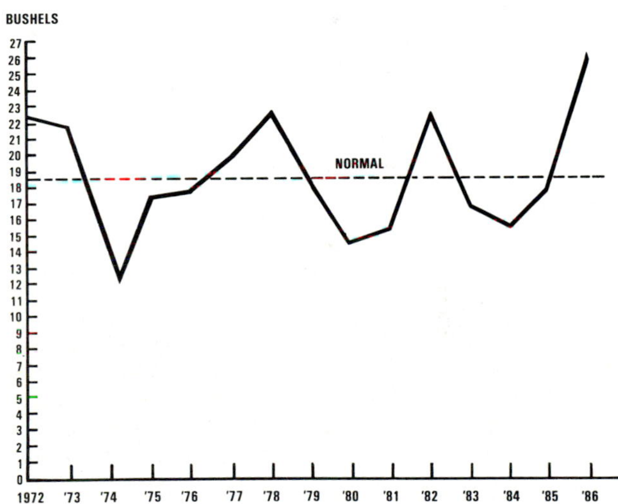

Fig. 23-32. The corn-hog ratio, 1972 to 1986. As shown, it averaged 18.6. (Based on data from *Agricultural Statistics,* 1987, USDA, p. 278, Table 412.)

Contract Hog Production (Custom Feeding/ Leasing)

A contract is an agreement between two or more persons to do or refrain from doing certain things. In recent years, swine producers have shown increasing interest in contract hog production due to (1) the high cost of capital, (2) the difficulty of many producers in obtaining adequate financing, and (3) the desire to forego the possibility of large profits for the assurance of more reliable returns. In the late 1980s, an estimated 8 to 10% of U.S. hogs were under some kind of production contract, with a much smaller number under a marketing contract.

An overview of the most common types of contract used in the swine industry follows.

PRODUCTION CONTRACTS

Investors, feed dealers, farmers and others are often interested in producing hogs, but are unwilling or unable to provide the necessary labor, facilities, and equipment. Some producers have also found contract production to be an effective method of expanding their operations. So, these entrepreneurs find producers who are willing to furnish the labor and equipment in exchange for a fixed wage or share of the profits. The resulting contracts vary considerably in form and responsibility of each party involved.

The more popular contracts provide for fixed payment, direct feeding, or profit sharing. These contracts are most commonly used for feeder pig production and hog finishing.

• **Fixed (guaranteed) payments**—These contracts guarantee the producer a fixed payment per head, usually along with provision for bonuses and penalties.

Under the fixed (guaranteed) payment, the contractor generally supplies the pigs, feed, veterinary services and medication, transportation, management assistance, and marketing. The producer normally provides the buildings and equipment, labor, utilities, and the necessary insurance. Most fixed payment contracts also provide for bonuses for keeping death losses low and feed efficiency high, and for penalties for high death losses and unmarketable animals.

• **Direct feeding**—In this type of contract, a feed dealer or cooperative contracts with producers to finish hogs. The contractor's objective is to increase feed sales and secure a reliable feed outlet.

Typically, the contractor provides the feed, directs the feeding program, and provides some management assistance. The producer agrees to purchase all feed and related services from the contractor and is responsible for all production costs. Upon sale of the hogs, the producer receives all proceeds less any outstanding balance owed to the contractor.

• **Profit sharing**—In a profit-sharing contract, the contractor and producer divide the profit on an agreed basis, such as 50–50, 60–40, depending upon who provides the majority of inputs and their value.

Normally, the contractor purchases the pigs and is responsible for all feed, veterinary expenses, transportation, and marketing costs. Over the duration of the contract, the contractor's costs are charged to the account. Then, this account balance is subtracted from the sale proceeds to determine the profit. The contractor's return depends upon the profit made on the sale of hogs and the gain received from the markup on feed, pigs, and supplies. Typically, the producer provides the facilities, labor, utilities, and insurance for his portion of the profit. The producer is usually guaranteed a minimum amount per head as long as death losses are below a set percentage.

FEEDER PIG PRODUCTION CONTRACTS

The following feeder pig production options are available:

1. **Option 1.** The contractor provides the breeding stock and contracts with the feeder pig producer to provide feeder pigs based on production criteria, such as pigs weaned per litter, with bonuses and docks based on target level. Most of the risk is retained by the producer.
2. **Option 2.** The contractor with a finishing operation provides the breeding stock, feed, and management assistance, and pays the feeder pig producer a flat fee for each pig. The fee will vary according to pig weight and current production costs. Most of the risk is assumed by the contractor.
3. **Option 3.** The contractor provides breeding stock, feed, facilities, and veterinary costs. The producer provides labor, utilities, maintenance, and manure handling. The producer is paid a fee for each pig produced and a monthly fee for each sow and boar maintained. This option is suited for owners who no longer want to be actively involved, but who have a good manager with limited cash willing to take over.
4. **Option 4.** This is a shared revenue program, with the percentage of gross sales based on inputs and services provided, and risks borne by each participant. For example, the feed dealer would receive a percentage of gross sales based on feed dealer inputs; the breeding stock supplier would receive a percentage; the management firm that supplied consultants and computerized records would receive a percentage; and the producer who supplied facilities, veterinary care, utilities, labor, and insurance would receive a percentage.

FARROW-TO-FINISH CONTRACTS

Most farrow-to-finish programs are set up on a percentage basis to reflect the relative amount of input supplied by each person or firm, with the following options available:

1. **Option 1.** Based on input costs, the following participants may share in a percentage of gross sales: The breeding stock supplier; the feed and medications supplier; the management consultant and computerized record services; and the producer-supplier of facilities, labor, veterinary care, utilities, and insurance.
2. **Option 2.** The current hog inventory is purchased by a limited partnership, which supplies sow replacements. Each of the following contract participants receives a percentage of the proceeds when hogs are marketed: The feed and medications supplier; the management agency-supplier of production and marketing guidance; and the producer-supplier of facilities, labor, utilities, veterinary costs, repairs, and manure disposal. The remaining percentage is split between the limited partnership and the general partner who manages the partnership.

BREEDING STOCK LEASING

Under a breeding stock lease, the contractor furnishes the producer with breeding age gilts and/or boars. The rent paid by the producer for the breeding stock may be either a specified number of pigs or an equivalent amount of money at designated times. The popularity and use of breeding stock leases has declined in recent years.

MARKETING CONTRACTS

Marketing contracts are of two kinds: (1) market hog contracts, and (2) feeder pig contracts.

• **Market hog contract**—A market hog contract is a forward sale contract between a buyer (normally a meat packer or a marketing agent) and a seller (normally a producer), in which the producer agrees to sell, at a specified date, a specified number of hogs to a buyer at a certain price. Normally, the following terms are detailed in a forward marketing contract: (1) the quantity, with the minimum ranging from 5,000 lb to 30,000 lb; (2) the date and location of delivery; (3) acceptable weights and grades, including premiums and discounts; (4) a description of the pricing mechanism (some contracts now price hogs on a grade and yield basis); (5) provisions for non-deliverable hogs and unacceptable carcasses; (6) provisions outlining the credit requirements of the seller and inspection of the hogs by the buyer; and (7) provisions for breach of contract.

Under a forward sale contract, the producer retains all risks of production, other than selling price. The producer uses the forward sale contract to reduce the risk of price fluctuations and to lock in an acceptable profit. But a forward sale contract may also cause the producer to miss out on greater profits if prices rise. Sometimes, a minimum price (a floor price) is used for hogs, in which the buyer guarantees the seller a minimum price (a floor price), with the seller receiving whichever price is higher at market time—the floor price, or the market price.

• **Feeder pig contracts**—Typically, feeder pig marketing contracts are between a marketing agency, often a cooperative, and a pig producer, in which the marketing agency agrees to market the pigs of a producer for a fee. A feeder pig marketing contract may contain the following provisions: (1) the producer agrees to market exclusively through the agency; and (2) the marketing agency specifies management practices and weight of feeder pigs at marketing. Essentially, producers are hiring market expertise through feeder pig contracts.

CHARACTERISTICS OF A GOOD CONTRACT

A contract is no better than the parties back of it. Moreover, it should be fair to both parties. Additionally, a good contract meets the following requisites:

1. It is in writing, and it is clear and concise.

2. It clearly defines the rights and responsibilities of both parties.

3. It contains the following: the names of both parties, the number of pigs involved, the duration of the contract, the time and method of payment, and who should supply certain inputs.

Poisons and Toxins

Swine are susceptible to a number of poisons, any one of which may be disastrous in a herd. Among them are the following: moldy feed, including three species of mycotoxins—aflatoxins, ergot poisoning, and estrogenic syndrome; pitch poisoning; lead poisoning; mercury poisoning; pesticides; plant poisoning, involving several plants that are toxic to swine; and blue-green algae.

(Also see Chapter 5, Nutritional Disorders/Toxins.)

Environmental Effects of Swine

Pollution potential, affecting the environment of both people and swine, increased as the U.S. swine industry moved toward specialization, mechanization, high animal density, and confinement.

When manure and urine are stored and undergo anaerobic digestion, dangerous and disagreeable gases are produced. The ones of primary concern are: hydrogen sulfide, ammonia, carbon dioxide, and methane.

High levels of dust particles resulting from automated dry feed handling systems, dander and hair from hogs, and dried manure particles can occur inside swine buildings. Manure gases can cling to those dust particles in such a way that inhaling these gas-laden particles is uncomfortable and objectionable. Particulate matter also includes viral, bacterial, and fungal agents from the building environment and carries them into the respiratory system of people and hogs.

To operate compatably within the community, to provide maximum self-protection, and to avoid neighbor complaints and legal actions seeking either monetary damages or court injunctions, swine producers must be aware of some basic information and strategy concerning pollution and apply pollution control measures appropriate to the location.

(Also see Chapter 12, section on "Animal Wastes [Manure and Litter]"; and Chapter 17, section on "Animal Pollution Control.")

Fig. 23-33. Confinement swine finishing building with environmental control ventilation system. (Courtesy, University of Illinois, Urbana)

QUESTIONS FOR STUDY AND DISCUSSION

1. How has the shift to confinement rearing affected the nutritional well-being of pigs?

2. What is the economic importance of feed for swine?

3. Why aren't nutritional requirements, or standards, such as NRC Tables 23-1 to 23-7, absolute, final, and unchangeable?

4. What is the difference between nutrient requirements and nutrient allowances? Why are the allowances generally higher than the requirements?

5. Discuss carbohydrates and fats/oils as energy supplying nutrients. Describe the symptoms in swine of (a) energy deficiency, and (b) essential fatty acid deficiency.

6. Define (a) essential amino acid, and (b) quality of protein. List the essential amino acid deficiencies of small grains (wheat, barley, oats), corn, and meat and bone meal. What are the symptoms of protein (amino acid) deficiency?

7. Why are protein quantity and quality most frequently the limiting factor in the rations of swine?

8. What peculiarities of swine husbandry are conducive to swine suffering from mineral deficiencies?

9. Give the (a) function, (b) deficiency symptoms, (c) practical sources, and (d) prevention of deficiency of each of the following minerals in the pig: salt, calcium, phosphorus, iodine, and iron.

10. Discuss the relationship of calcium, phosphorus, and vitamin D in swine nutrition.

11. What vitamins are most apt to be deficient in ordinary swine rations?

12. Give the (a) function, (b) deficiency symptoms, (c) practical sources, and (d) prevention of deficiency of each of the following vitamins in the pig: A, D, E, choline, pantothenic acid, riboflavin, and thiamin.

13. What tasks does water perform in the body of the pig?

14. Why are antibiotics used as feed additives? How do they function? What cautions should be observed when feeding antibiotics to swine?

15. Discuss the history, prevalence of use, benefits, and concerns stemming from the use of sulfonamides (sulfas) for swine.

16. What is porcine somatotropin (PST)? What great leaps forward do experiments show from the use of PST?

17. Why is so much corn fed to hogs in the United States, despite the fact that it is deficient in both quantity and quality of protein?

18. In what amino acids is corn particularly deficient? In what ways would Opaque-2 corn contribute to improved nutrition of animals and people throughout the world?

19. Discuss the effect of feeding fats/oils (a) to hogs in the summer vs the winter, and (b) to gestating sows beginning 10 to 14 days before farrowing.

20. How do you account for the fact that soybean meal is the leading high-protein supplement in the United States?

21. What are "full-fat soybeans"? How, and under what circumstances, may they be used in swine rations?

22. Is the use of pastures for swine outmoded in the United States? If not, how, and under what circumstances, may they be used in a practical way?

23. Will world food shortages cause a trend back to the use of more alfalfa meal for hogs? Why is alfalfa especially well suited for winter feeding of swine?

24. On your home farm (or a farm with which you are familiar), how, and under what circumstances, would you (a) utilize silage for swine, and (b) "hog off" certain crops?

25. Why has garbage declined in feeding value and importance as a swine feed in recent years? Why must commercial garbage be cooked?

26. Following the selection of feed ingredients, they may be fed to swine in either combined form or cafeteria style. How may balanced rations be achieved by each method through mixing/feeding procedures?

27. Why should there be a swine ration for every need? What classes of swine need rations suited to their particular needs?

28. Under what circumstances would you (a) buy a commercial hog feed, or (b) home mix a hog feed?

29. Under what circumstances should feed substitutions be made?

30. How would you recommend that a 4-H club or FFA member feed (a) baby pigs, and (b) orphan pigs?

31. Discuss what is unique and different about feeding each of the following classes of swine: (a) growing-finishing pigs, (b) prospective breeding gilts, (c) boars, (d) gestating sows, and (e) lactating sows.

32. List and discuss each of the common hog feeding systems. Which system would you recommend for (a) brood sows, and (b) growing-finishing hogs?

33. Why did tail biting become a problem with the advent of confinement swine production?

34. Discuss the cause, symptoms, and prevention of nutritional anemia in pigs.

35. List and discuss the advantages and disadvantages of early weaning. Give specific guidelines for early weaning.

36. Estimate the total feed required to produce a 240 lb market pig, then break it down into the amount required for each of the following stages of production: (a) sow gestation ration (including pregestation and breeding), (b) boar ration, (c) lactation ration, (d) starter ration (creep to 40 lb), and (e) grower-finisher ration (40 to 240 lb).

37. Since boars gain faster and are more efficient in feed conversion than barrows or gilts, why don't we feed more boars for market slaughter?

38. Smithfield Hams are derived from peanut-fed hogs; they are soft pork. Yet, they are highly advertised and sold at a premium. So, why be concerned about soft pork?

39. Of what significance is the corn-hog ratio?

40. What is contract hog production? Why has it increased in recent years? What are the common types of contracts and the characteristics of each?

41. Discuss poisons and toxins of swine.

42. Discuss the increased pollution potential, affecting both people and swine, which accompanied the swine industry as it moved toward specialization, mechanization, high animal density, and confinement.

SELECTED REFERENCES

Title of Publication	Author(s)	Publisher
Animal Science, 8th Edition	M. E. Ensminger	The Interstate Printers & Publishers, Inc., Danville, Ill., 1983
Digestion in the Pig	D. E. Kidder M. J. Manners	Scientechnica Bristol Kingston Press, Oldfield Park, Bath, England, 1978
Life Cycle Nutrition	P. Holden, *et al.*	Iowa State University, Ames, Iowa, 1988
Nutrient Requirements of Swine, 9th Revised Edition	National Research Council	National Research Council, National Academy Press, Washington, D.C., 1988
Pig Husbandry, 4th Edition	J. R. Luscombe	Farming Press, Ltd., Fenton House, Warfedale Road, Ipswich, England, 1972
Pig Nutrition—recent developments in	D. J. A. Cole W. Haresign	Butterworths, London, England, 1985
Pork Industry Handbook	Staff	University of Illinois, Cooperative Extension Service, 1988
Practical Pig Nutrition	C. T. Whittemore F. W. H. Elsley	Farming Press Limited, Fenton House, Ipswich, Suffolk, England, 1976
Stockman's Handbook, The, 6th Edition	M. E. Ensminger	The Interstate Printers & Publishers, Inc., Danville, Ill., 1983
Swine Feeding and Nutrition	T. J. Cunha	Academic Press, Inc., New York, N.Y., 1977
Swine Production, 4th Edition	J. L. Krider W. E. Carroll	McGraw-Hill Book Company, New York, N.Y., 1971
Swine Production and Nutrition	W. G. Pond J. H. Maner	AVI Publishing Co., Westport, Conn., 1984
Swine Production in Temperate and Tropical Environment	W. G. Pond J. H. Maner	W. H. Freeman, San Francisco, Calif., 1974
Swine Science, 5th Edition	M. E. Ensminger R. O. Parker	The Interstate Printers & Publishers, Inc., Danville, Ill., 1984

Fig. 23-34. Confinement swine breeding system using slotted floors, sow stalls, and free-stall for the boar. (Courtesy, University of Illinois, Urbana)

Fig. 23-35. Pigs eating from a circular self-feeder. (Courtesy, USDA)

Fig. 23-36. Open-sided, naturally-ventilated swine building. (Courtesy, University of Illinois, Urbana)

Fig. 23-37. Closed, environmentally-controlled swine building. (Courtesy, University of Illinois, Urbana)

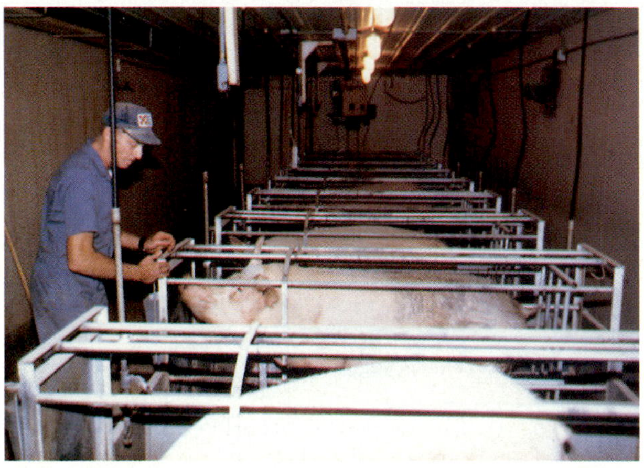

Fig. 23-38. Sows in a modular farrowing room. (Courtesy, Iowa State University, Ames)

Fig. 23-39. Sow nursing her crossbred litter. (Courtesy, USDA)

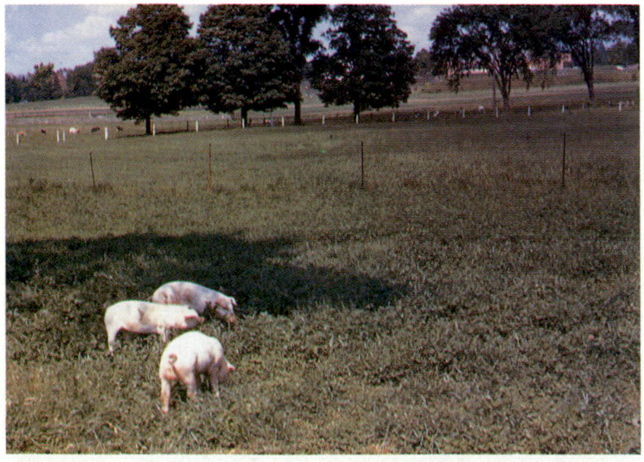

Fig. 23-40. Pigs on pasture. (Courtesy, USDA)

Fig. 23-41. Finishing pigs in pens with partially slotted floors. (Courtesy, University of Illinois, Urbana)

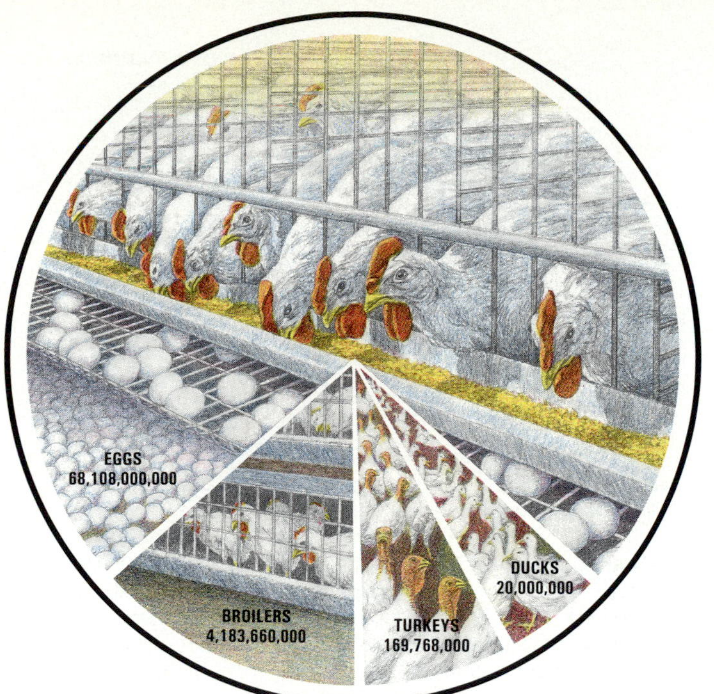

Original painting by Tom Phillips

Chapter **24**

FEEDING POULTRY[1]

Contents | Page

[1]The authors gratefully acknowledge the helpful suggestions of the following authorities who reviewed this chapter: J. D. Garlich, Ph.D., Professor, Department of Poultry Science, North Carolina State University, Raleigh; D. M. Hooge, Ph.D., Hooge Consulting Service, 1117 Sycamore Street, Turlock, CA 95380; L. S. Jensen, Ph.D., Professor, Department of Poultry Science, The University of Georgia, Athens; L. M. Larsen, Ph.D., Consultant, Nutri-Systems, 426 E. Shields, Fresno, CA 93704; P. L. Potts, Sr., Ph.D., Poultry Science Department, California Polytechnic State University, San Luis Obispo; and F. L. Stephenson, Ph.D., Professor, Department of Animal Sciences, University of Arkansas, Fayetteville.

Fig. 24-1. Eggs and feed! Egg production is an all-or-none phenomeon—birds must have enough nutrients to produce an egg; otherwise, egg size will be reduced, and eventually no egg will be produced. Also, poultry grow at a rapid rate and mature at an early age; so, proper nutrition is critical. (Courtesy, H & N, Inc., Redmond, Wash.)

Poultry feeding has changed more than the feeding of any other species—it has paced the entire livestock field. Origially, it was strictly a backyard enterprise; the mother hen did her own incubating and reared her young, and the farmer's wife fed them on table scraps and the unaccounted-for grain from the crib. Reproduction was confined to the spring months when green feeds, insects, and sunshine were available to contribute to the nutrition of the baby chicks. Feeding was largely an art rather than a science, and such commercial feeds as were sold were largely "secret formulas and patented potions." But all this has changed. Today, the vast majority of commercial poultry is produced in large units wherein the maximum of science and technology exist. Confinement production is rather commonplace, and well-balanced rations containing adequate sources of all known nutrient materials are fed for maximum production. The current trend in poultry production is toward controlled environment, which usually results in lowered feed consumption. Under such conditions, the daily feed consumption is often taken into consideration and the nutrient content of the feed (energy, amino acids, minerals, and vitamins) varied so as to compensate for the reduced feed intake and meet the requirements.

The changes in the poultry industry have resulted in greater efficiency of production, along with favorable product prices and increased human consumption. As shown in Table 24-1, during the 30-year period 1956-1986, the consumption of all red meats (beef, veal, lamb and mutton, and pork) decreased, while the consumption of chicken and turkey increased dramatically—by 141 and 156%, respectively.

TABLE 24-1
CHANGES IN PER CAPITA CONSUMPTION OF SELECTED MEAT PRODUCTS[1]

Food	1956 Consumption		1986 Consumption		Change
	(lb)	*(kg)*	(lb)	*(kg)*	(%)
Meats:					
Beef	85.4	*38.8*	78.4	*35.6*	− 8.0
Veal	9.5	*4.3*	1.9	*0.9*	− 80.0
Lamb and mutton	4.5	*2.0*	1.4	*0.6*	− 68.0
Pork	67.3	*30.6*	58.6	*26.6*	− 13.0
Poultry products:					
Chicken	24.4	*11.1*	58.8	*26.7*	+141.0
Turkey	5.2	*2.4*	13.3	*6.0*	+156.0

[1]U.S. Department of Agriculture sources.

Nutrition of poultry is more critical than that of other farm animals with regard to a number of factors. This is so because birds are quite different from four-footed animals; their digestion is more rapid, their respiration and circulation are faster, their body temperature is 8 to 10°F higher (about 106°F), they are more active, they are more sensitive to environmental influences, they grow at a more rapid rate, and they mature at an earlier age. Also, egg production is an all-or-none phenomenon—that is, birds must have enough nutrients to produce an egg; otherwise, no egg is produced.

ECONOMIC IMPORTANCE OF FEED FOR POULTRY

The economic importance of poultry feeding becomes apparent when it is realized that 55 to 75% of the total production cost of poultry is from feed, with the production of eggs toward the lower side of this range and the production of broilers and turkeys toward the upper side. For this reason, the efficient use of feed is extremely important to poultry producers.

CHANGES IN FEED EFFICIENCY

Table 24-2 shows the marked lowering of feed required to produce a unit of eggs, turkeys, and broilers since 1940. In 1940, it required 4.7 lb of feed to produce 1 lb of weight gain in broilers; in 1990, it took only 1.9 lb.

TABLE 24-2
FEED UNITS REQUIRED PER UNIT OF EGGS AND POULTRY PRODUCED, SELECTED YEARS, 1940-90

Year Ending October	Per Dozen Eggs[1,2]	Per Pound Liveweight: Turkey[3]	Per Pound Liveweight: Broiler[1,2]
	(feed units)	(feed units)	(feed units)
1940	7.4	4.50	4.7
1950	7.2	3.56	3.7
1960	6.4	3.37	3.0
1970	4.6	3.21	2.6
1980	4.1[4]	2.90[4]	2.1[4]
1990	3.75[4]	2.70[4]	1.9[4]

[1]Feed units used per dozen eggs or per pound of liveweight broiler produced. A feed unit is the economic equivalent of 1 lb of corn. 1940-1960 from *Handbook of Agricultural Charts 1965,* Ag. Hdbk. No. 300, USDA, Oct. 1965, p. 58.

[2]1970 from *Agricultural Statistics 1974,* USDA, p.358, Table 518.

[3]The turkey data are based on estimates presented in *Efficiency in Animal Feeding with Particular Reference to Nonnutritive Feed Additives,* Council for Agricultural Science and Technology, Report No. 22, Jan. 18, 1974.

[4]Estimates by the authors.

NUTRIENT NEEDS OF POULTRY

The nutrient composition of chickens and eggs shown in Table 24–3 is indicative of the relative importance of these nutrients as body and egg constituents.

TABLE 24-3
NUTRIENT COMPOSITION OF BROILERS AND EGGS

Nutrient	Broilers[1]	Egg
	(%)	(%)
Water	65.7	66
Protein	18.4	13
Fat	12.2	10[2]
Minerals	3.7	11[3]

[1]From Table 2-3 of this book.

[2]Chiefly in the yolk.

[3]Nearly all is calcium in shell.

The nutritive requirements of poultry presented in this chapter do not provide for margins of safety. Rather, the values reported represent adequacy, using as criteria growth, health, reproduction, feed efficiency, and quality of products produced.

It is recognized that a number of forces may affect the nutritive requirements of poultry; among them, the following:

1. **Temperature and humidity.** When temperature and humidity conditions deviate much from 60 to 75°F and 40 to 60% relative humidity, adjustments in nutrient levels should be made to compensate for changes in feed intake.
2. **Genetic differences.** Genetic differences among strains affect nutrient requirements; hence, in this chapter consideration has been given to differences in the requirements between broiler-type and egg-type strains of chickens.

Also, the nutrient composition of feedstuffs is variable. In order to compensate for these conditions, the nutritionist usually adds a margin of safety to the stated requirements in arriving at *nutritive allowances* to be used in ration formulation.

Fig. 24–2. Newly placed chicks in a house. The nutritive requirements of poultry are more critical than those of other farm animals because their digestion is more rapid, their respiration and circulation are faster, their body temperature is 8 to 10°F higher, they are more active, they are more sensitive to environmental influences, they grow at a more rapid rate, and they mature at an earlier age. (Courtesy, The University of Georgia, Athens)

National Research Council (NRC) Requirements

The requirements for most of the nutrients needed by poultry have been established. These differ according to the kind and age of bird and the purpose for which it is being fed. A deficiency of a nutrient can be, and often is, a limiting factor in egg production or growth.

The National Research Council (NRC) nutritive requirements are given in the following tables:

For Chickens:

Table 24–4, Nutrient Requirements of Leghorn-Type Chickens as Percentages or as Milligrams or Units per Kilogram of Diet.

Table 24–5, Body Weight and Feed Requirements of Leghorn-Type Pullets and Hens.

Table 24–6, Nutrient Requirements of Broilers as Percentages or as Milligrams or Units per Kilogram of Diet.

Table 24–7, Body Weights and Feed Requirements of Broilers.

Table 24–8, Nutrient Requirements of Meat-Type Hens for Breeding Purposes.

Table 24–9, Typical body Weights and Feed Allowances for Male and Female Meat-Type Chickens (Replacement Stock).

Table 24–10, Metabolizable Energy Required Daily by Chickens in Relation to Body Weight and Egg Production.

For Turkeys:

Table 24–11, Nutrient Requirements of Turkeys as Percentages or as Milligrams or Units per Kilogram of Feed.

Table 24–12, Growth Rate, Feed and Energy Consumption of Large-Type Turkeys.

Table 24–13, Body Weights and Feed Consumption of Large-Type Turkeys During Holding and Breeding Periods.

For Ducks:

Table 24–14, Nutrient Requirements of Pekin Ducks as Percentages or as Milligrams or Units per Kilogram of Diet.

Table 24–15, Typical Body Weights and Feed Consumption of Pekin Ducks to 8 Weeks of Age.

For Geese:

Table 24–16, Nutrient Requirements of Geese as Percentages or as Milligrams or Units per Kilogram of Diet.

For Pheasants and Bobwhite Quail:

Table 24–17, Nutrient Requirements of Pheasants and Bobwhite Quail as Percentages or as Milligrams or Units per Kilogram of Diet.

For Japanese Quail:

Table 24–18, Nutrient Requirements of Japanese Quail (Coturnix) as Percentages or as Milligrams or Units Per Kilogram of Diet.

In establishing the NRC requirements for poultry, it was further assumed that the environmental temperature in which poultry of various species and ages are grown is ideal or as near optimum as possible for efficient growth and reproduction. Therefore, the energy level of the diet was first established for each species and age of poultry, then the other nutrients were determined based upon the established level of energy. If a higher level of energy than the NRC requirement is used in the diet, feed consumption will decrease; hence, the minimum level of the other nutrients should be increased in proportion to the energy content. Similarly, if a lower dietary energy level is used, then proportionately lower levels of other nutrients should be used in the diet.

TABLE 24-4
NUTRIENT REQUIREMENTS OF LEGHORN-TYPE CHICKENS AS PERCENTAGES OR AS MILLIGRAMS OR UNITS PER KILOGRAM OF DIET[1]

	Growing			Laying		Breeding
Energy Base:	**0–6 Weeks**	**6–14 Weeks**	**14–20 Weeks**			
kcal ME/lb Diet[2]	**1,315**	**1,315**	**1,315**	**1,315**	**Daily Intake**	**1,315**
kcal ME/kg Diet[2]	***2,900***	***2,900***	***2,900***	***2,900***	**Per Hen (mg)[3]**	***2,900***
Protein (%)	18	15	12	14.5	16,000	14.5
Amino acids:						
Arginine (%)	1.00	0.83	0.67	0.68	750	0.68
Glycine and serine (%)	0.70	0.58	0.47	0.50	550	0.50
Histidine (%)	0.26	0.22	0.17	0.16	180	0.16
Isoleucine (%)	0.60	0.50	0.40	0.50	550	0.50
Leucine (%)	1.00	0.83	0.67	0.73	800	0.73
Lysine (%)	0.85	0.60	0.45	0.64	700	0.64
Methionine + cystine (%)	0.60	0.50	0.40	0.55	600	0.55
Methionine (%)	0.30	0.25	0.20	0.32	350	0.32
Phenylalanine + tyrosine (%)	1.00	0.83	0.67	0.80	880	0.80
Phenylalanine (%)	0.54	0.45	0.36	0.40	440	0.40
Threonine (%)	0.68	0.57	0.37	0.45	500	0.45
Tryptophan (%)	0.17	0.14	0.11	0.14	150	0.14
Valine (%)	0.62	0.52	0.41	0.55	600	0.55
Linoleic acid (%)	1.00	1.00	1.00	1.00	1,100	1.00
Major or macrominerals:						
Calcium (Ca) (%)	0.80	0.70	0.60	3.40	3,750	3.40
Chlorine (Cl) (%)	0.15	0.12	0.12	0.15	165	0.15
Magnesium (Mg) (mg)	600	500	400	500	55	500
Phosphorus (P), available (%)	0.40	0.35	0.30	0.32	350	0.32
Potassium (K) (%)	0.40	0.30	0.25	0.15	165	0.15
Sodium (Na) (%)	0.15	0.15	0.15	0.15	165	0.15
Trace or microminerals:						
Copper (Cu) (mg)	8	6	6	6	0.88	8
Iodine (I) (mg)	0.35	0.35	0.35	0.30	0.03	0.30
Iron (Fe) (mg)	80	60	60	50	5.50	60
Manganese (Mn) (mg)	60	30	30	30	3.30	60
Selenium (Se) (mg)	0.15	0.10	0.10	0.10	0.01	0.10
Zinc (Zn) (mg)	40	35	35	50	5.50	65
Fat-soluble vitamins:						
Vitamin A (IU)	1,500	1,500	1,500	4,000	440	4,000
Vitamin D (ICU)	200	200	200	500	55	500
Vitamin E (IU)	10	5	5	5	0.55	10
Vitamin K (mg)	0.50	0.50	0.50	0.50	0.055	0.50
Water-soluble vitamins:						
Biotin (mg)	0.15	0.10	0.10	0.10	0.011	0.15
Choline (mg)	1,300	900	500	?	?	?
Folacin (Folic Acid) (mg)	0.55	0.25	0.25	0.25	0.0275	0.35
Niacin (Nicotinic Acid, Nicotinamide) (mg)	27.0	11.0	11.0	10.0	1.10	10.0
Pantothenic Acid (Vitamin B-3) (mg)	10.0	10.0	10.0	2.20	0.242	10.0
Riboflavin (Vitamin B-2) (mg)	3.60	1.80	1.80	2.20	0.242	3.80
Thiamin (Vitamin B-1) (mg)	1.8	1.3	1.3	0.80	0.088	0.80
Vitamin B-6 (Pyridoxine, Pyridoxal, Pyridoxamine) (mg)	3.0	3.0	3.0	3.0	0.33	4.50
Vitamin B-12 (Cobalamins) (mg)	0.009	0.003	0.003	0.004	0.00044	0.004

[1]Adapted by the authors from *Nutrient Requirements of Poultry,* 8th rev. ed., NRC, National Academy Press, Washington, D.C., 1984, p. 12, Table 4.

[2]These are typical dietary energy concentrations.

[3]Assumes an average daily intake of 110 g of feed/hen daily.

TABLE 24-5
BODY WEIGHTS AND FEED REQUIREMENTS OF LEGHORN-TYPE PULLETS AND HENS[1]

Age	Body Weight[2]		Feed Consumption[3]		Typical Egg Production (Hen/Day)
(weeks)	(lb)	*(kg)*	(lb/week)	*(kg/week)*	(%)
0	0.07	*0.03*	0.11	*0.05*	—
2	0.31	*0.14*	0.20	*0.09*	—
4	0.60	*0.27*	0.40	*0.18*	—
6	0.99	*0.45*	0.57	*0.26*	—
8	1.37	*0.62*	0.73	*0.33*	—
10	1.74	*0.79*	0.86	*0.39*	—
12	2.09	*0.95*	0.95	*0.43*	—
14	2.34	*1.06*	1.01	*0.46*	—
16	2.56	*1.16*	1.01	*0.46*	—
18	2.78	*1.26*	1.01	*0.46*	—
20	3.00	*1.36*	1.01	*0.46*	—
22	3.13	*1.42*	1.17	*0.53*	10
24	3.31	*1.50*	1.30	*0.59*	38
26	3.48	*1.58*	1.48	*0.67*	64
30	3.81	*1.73*	1.70	*0.77*	88
40	4.01	*1.82*	1.70	*0.77*	80
50	4.12	*1.87*	1.70	*0.77*	74
60	4.19	*1.90*	1.65	*0.75*	68
70	4.19	*1.90*	1.63	*0.74*	62

[1]Adapted by the authors from *Nutrient Requirements of Poultry,* 8th rev. ed., NRC, National Academy Press, Washington, D.C., 1984, p. 13, Table 5.

[2]Pullets and hens of Leghorn-type strains are generally fed *ad libitum* but are occasionally control-fed to limit body weights. Values shown are typical but will vary with strain differences, season, and lighting. Specific breeder guidelines should be consulted for desired schedules of weights and feed consumption.

[3]Based on rations containing 1,315 ME kcal/lb *(2,900 ME kcal/kg).* Consumption will vary depending upon the caloric density of the ration, environmental temperature, and rate of production (see Table 24-9).

Fig. 24-3. Leghorn-type layers. (Courtesy, University of Georgia, Athens)

TABLE 24-6
NUTRIENT REQUIREMENTS OF BROILERS AS PERCENTAGES OR AS MILLIGRAMS OR UNITS PER KILOGRAM OF DIET[1]

Energy Base:		Weeks 0-3	Weeks 3-6	Weeks 6-8
kcal ME/lb Diet[2]		1,452	1,452	1,452
kcal ME/kg Diet[2]		*3,200*	*3,200*	*3,200*
Protein	(%)	23.0	20.0	18.0
Amino acids:				
Arginine	(%)	1.44	1.20	1.00
Glycine and serine	(%)	1.50	1.00	0.70
Histidine	(%)	0.35	0.30	0.26
Isoleucine	(%)	0.80	0.70	0.60
Leucine	(%)	1.35	1.18	1.00
Lysine	(%)	1.20	1.00	0.85
Methionine + cystine	(%)	0.93	0.72	0.60
Methionine	(%)	0.50	0.38	0.32
Phenylalanine + tyrosine	(%)	1.34	1.17	1.00
Phenylalanine	(%)	0.72	0.63	0.54
Threonine	(%)	0.80	0.74	0.68
Tryptophan	(%)	0.23	0.18	0.17
Valine	(%)	0.82	0.72	0.62
Linoleic acid	(%)	1.00	1.00	1.00
Major or macrominerals:				
Calcium (Ca)	(%)	1.00	0.90	0.80
Chlorine (Cl)	(%)	0.15	0.15	0.15
Magnesium (Mg)	(mg)	600	600	600
Phosphorus (P), available	(%)	0.45	0.40	0.35
Potassium (K)	(%)	0.40	0.35	0.30
Sodium (Na)	(%)	0.15	0.15	0.15
Trace or microminerals:				
Copper (Cu)	(mg)	8.0	8.0	8.0
Iodine (I)	(mg)	0.35	0.35	0.35
Iron (Fe)	(mg)	80.0	80.0	80.0
Manganese (Mn)	(mg)	60.0	60.0	60.0
Selenium (Se)	(mg)	0.15	0.15	0.15
Zinc (Zn)	(mg)	40.0	40.0	40.0
Fat-soluble vitamins:				
Vitamin A	(IU)	1,500	1,500	1,500
Vitamin D	(ICU)	200	200	200
Vitamin E	(IU)	10	10	10
Vitamin K	(mg)	0.50	0.50	0.50
Water-soluble vitamins:				
Biotin	(mg)	0.15	0.15	0.10
Choline	(mg)	1,300	850	500
Folacin (Folic Acid)	(mg)	0.55	0.55	0.25
Niacin (Nicotinic Acid, Nicotinamide)	(mg)	27.0	27.0	11.0
Pantothenic Acid (Vitamin B-3)	(mg)	10.0	10.0	10.0
Riboflavin (Vitamin B-2)	(mg)	3.60	3.60	3.60
Thiamin (Vitamin B-1)	(mg)	1.80	1.80	1.80
Vitamin B-6 (Pyridoxine, Pyridoxal, Pyridoxamine)	(mg)	3.0	3.0	2.5
Vitamin B-12 (Cobalamins)	(mg)	0.009	0.009	0.003

[1]Adapted by the authors from *Nutrient Requirements of Poultry,* 8th rev. ed., NRC, National Academy Press, Washington, D.C., 1984, p. 13, Table 6.

[2]These are typical dietary energy concentrations.

TABLE 24-7
BODY WEIGHTS AND FEED REQUIREMENTS OF BROILERS[1,2]

Age	Body Weights				Weekly Feed Consumption				Cumulative Feed Consumption				Weekly Energy Consumption		Cumulative Energy Consumption	
	Male		Female		Male		Female		Male		Female		Male	Female	Male	Female
(weeks)	(lb)	*(g)*	(lb)	*(g)*	(lb)	*(g)*	(lb)	*(g)*	(lb)	*(g)*	(lb)	*(g)*	(ME kcal/ bird)	(ME kcal/ bird)	(ME kcal/ bird)	(ME kcal/ bird)
1	0.29	*130*	0.26	*120*	0.26	*120*	0.24	*110*	0.26	*120*	0.24	*110*	385	350	385	350
2	0.71	*320*	0.66	*300*	0.57	*260*	0.53	*240*	0.84	*380*	0.77	*350*	830	770	1,215	1,120
3	1.23	*560*	1.14	*515*	0.86	*390*	0.78	*355*	1.70	*770*	1.55	*705*	1,250	1,135	2,465	2,255
4	1.90	*860*	1.74	*790*	1.18	*535*	1.10	*500*	2.88	*1,305*	2.66	*1,205*	1,710	1,600	4,175	3,855
5	2.76	*1,250*	2.45	*1,110*	1.63	*740*	1.42	*645*	4.51	*2,045*	4.08	*1,850*	2,370	2,065	6,545	5,920
6	3.73	*1,690*	3.15	*1,430*	2.16	*980*	1.76	*800*	6.67	*3,025*	5.84	*2,650*	3,135	2,560	9,680	8,480
7	4.63	*2,100*	3.85	*1,745*	2.41	*1,095*	2.01	*910*	9.08	*4,120*	7.85	*3,560*	3,505	2,910	13,185	11,390
8	5.56	*2,520*	4.54	*2,060*	2.67	*1,210*	2.14	*970*	11.75	*5,330*	9.99	*4,530*	3,870	3,105	17,055	14,495
9	6.45	*2,925*	5.18	*2,350*	2.91	*1,320*	2.23	*1,010*	14.66	*6,650*	12.21	*5,540*	4,225	3,230	21,280	17,725

[1]Adapted by the authors from *Nutrient Requirements of Poultry*, 8th rev. ed., NRC, National Academy Press, Washington, D.C., 1984, p. 14, Table 7.

[2]Typical for broilers fed well-balanced rations containing 1,452 ME kcal/lb *(3,200 ME kcal/kg)*.

Fig. 24-4. Broilers with automatic feed and water facilities. (Photo by J. C. Allen & Son, Inc., West Lafayette, Ind.)

TABLE 24-8
NUTRIENT REQUIREMENTS
OF MEAT-TYPE HENS FOR BREEDING PURPOSES[1,2]

Energy Base: kcal ME/lb Diet .... *kcal ME/kg Diet*	1,293[3] *2,850*[3]	Daily Intake Per Hen (mg)
Protein .... (%)	14.5	22,000
Amino acids:		
Arginine .... (%)	0.74	1,110
Glycine + serine .... (%)	0.62	932
Histidine .... (%)	0.14	205
Isoleucine .... (%)	0.57	850
Leucine .... (%)	0.83	1,250
Lysine .... (%)	0.51	765
Methionine + cystine .... (%)	0.55	820
Methionine .... (%)	0.35	520
Phenylalanine + tyrosine .... (%)	0.75	1,112
Phenylalanine .... (%)	0.41	610
Threonine .... (%)	0.48	720
Tryptophan .... (%)	0.13	190
Valine .... (%)	0.63	950
Major or macrominerals:		
Calcium (Ca) .... (%)	2.75	4,125
Phosphorus (P), available .... (%)	0.25	375
Sodium (Na) .... (%)	0.10	150

[1]Adapted by the authors from *Nutrient Requirements of Poultry*, 8th rev. ed., NRC, National Academy Press, Washington, D.C., 1984, p. 14, Table 8.

[2]Rations are generally fed on a limited intake basis to control body weight gains. Adjust quantity of feed offered based on desired body weights and egg production levels for specific breed or strain.

[3]Rations for laying hens generally are fed to provide daily energy intakes of 375 to 450 ME kcal/day based on body weight, environmental temperature, and rate of egg production. Percentage of nutrients shown is typical of hens given 425 ME kcal/day.

TABLE 24-9
TYPICAL BODY WEIGHTS AND FEED ALLOWANCES FOR MALE AND FEMALE MEAT-TYPE CHICKENS (REPLACEMENT STOCK)[1,2]

Age	Male Body Weight[3]		Male Feed Consumption[4]		Female Body Weight[3]		Female Feed Consumption[4]		Typical Egg Production
(weeks)	(lb)	*(g)*	(lb/week)	*(g/week)*	(lb)	*(g)*	(lb/week)	*(g/week)*	(hen/day %)
0	0.09	*40*	0.22	*100*	0.09	*40*	0.17	*75*	–
2	0.55	*250*	0.55	*250*	0.50	*225*	0.56	*225*	–
4	1.20	*545*	0.77–0.85	*350–385*	1.00	*455*	0.69–0.73	*315–330*	–
6	1.75	*795*	0.86–0.94	*390–425*	1.46	*660*	0.73–0.77	*330–350*	–
8	2.25	*1,020*	0.89–1.03	*405–475*	1.85	*840*	0.77–0.88	*350–400*	–
10	2.98	*1,250*	1.03–1.21	*475–550*	2.20	*1,000*	0.85–0.98	*385–445*	–
12	3.26	*1,480*	1.19–1.38	*540–625*	2.60	*1,180*	0.94–1.06	*425–480*	–
14	3.75	*1,700*	1.27–1.54	*575–700*	3.00	*1,360*	1.01–1.21	*460–550*	–
16	4.25	*1,930*	1.38–1.69	*625–765*	3.42	*1,550*	1.09–1.32	*495–600*	–
18	4.74	*2,150*	1.47–1.82	*665–825*	3.81	*1,730*	1.16–1.48	*525–670*	–
20	5.29	*2,400*	–[5]	–[5]	4.25	*1,930*	1.26–1.61	*570–730*	–
22	5.82	*2,640*	–	–	4.65	*2,110*	1.40–1.75	*635–795*	10
24	7.05	*3,200*	–	–	5.40	*2,450*	1.76–2.04	*800–925*	15
26	7.80	*3,540*	–	–	6.02	*2,730*	2.09–2.31	*950–1,050*	30
28	8.27	*3,750*	–	–	6.35	*2,880*	2.38–2.52	*1,078–1,141*	56
30	8.60	*3,900*	–	–	6.61	*3,000*	2.38–2.52	*1,078–1,141*	75
32	9.02	*4,090*	–	–	6.81	*3,090*	2.38–2.52	*1,078–1,141*	80
34	9.30	*4,220*	–	–	6.90	*3,130*	2.38–2.52	*1,078–1,141*	78
36	9.57	*4,340*	–	–	6.97	*3,160*	2.38–2.52	*1,078–1,141*	76
38	9.81	*4,450*	–	–	7.01	*3,180*	2.36–2.50	*1,071–1,134*	73
40	10.01	*4,540*	–	–	7.01	*3,180*	2.35–2.48	*1,064–1,127*	72

[1]Adapted by the authors from *Nutrient Requirements of Poultry,* 8th rev. ed., NRC, National Academy Press, Washington, D.C., 1984, p. 15, Table 9.

[2]Broiler-breeder strains must be grown on a controlled feeding program to limit weight. Values shown are typical but will vary according to strain. Specific breeder guidelines should be consulted for desired schedule of weights and feed allotments.

[3]Values are typical for fall-hatched chicks. Spring-hatched chicks will have decreasing natural daylight during the time of sexual maturity and usually need to be heavier to attain sexual maturity at the desired age.

[4]Adjust as required to maintain desired body weight.

[5]Males and females intermingled.

Fig. 24-5. Meat-type hens being pen-mated to one male. (Courtesy, J. C. Allen & Son, Inc., West Lafayette, Ind.)

TABLE 24-10
METABOLIZABLE ENERGY REQUIRED DAILY BY CHICKENS IN RELATION TO BODY WEIGHT AND EGG PRODUCTION[1,2]

Body Weight		Rate of Egg Production (%) 0	50	60	70	80	90
(lb)	*(kg)*	◄---- Metabolizable Energy/Hen Daily (kcal)[3] ----►					
2.2	*1.0*	130	192	205	217	229	242
3.3	*1.5*	177	239	251	264	276	289
4.4	*2.0*	218	280	292	305	317	330
5.5	*2.5*	259	321	333	346	358	371
6.6	*3.0*	296	358	370	383	395	408
7.7	*3.5*	333	395	408	420	432	445

[1]Adapted by the authors from *Nutrient Requirements of Poultry,* 8th rev. ed., NRC, National Academy Press, Washington, D.C., 1984, p. 15, Table 10.

[2]A number of formulas have been suggested for prediction of the daily energy requirements of chickens. The formula used here was derived from that in *Effect of Environment on Nutrient Requirements of Domestic Animals* (NRC, 1981).

$$\text{ME/hen daily} = W^{0.75}(173 - 1.95T) + 5.5\Delta W + 2.07EE$$

where: W = body weight (kg),
T = ambient temperature (°C),
ΔW = change in body weight in g/day, and
EE = daily egg mass (g).

[3]Temperature of 22°, egg weight of 60 g, and no change in body weight were used in calculations.

TABLE 24-11
NUTRIENT REQUIREMENTS OF TURKEYS AS PERCENTAGES OR AS MILLIGRAMS OR UNITS PER KILOGRAM OF FEED[1]

	Age (Weeks)							
Male	**0–4**	**4–8**	**8–12**	**12–16**	**16–20**	**20–24**		**Breeding**
Female	**0–4**	**4–8**	**8–11**	**11–14**	**14–17**	**17–20**	**Holding**	**Hens**
Energy Base:								
kcal ME/lb Diet[2]	**1,270**	**1,315**	**1,361**	**1,406**	**1,452**	**1,497**	**1,315**	**1,315**
kcal ME/kg Diet[2]	***2,800***	***2,900***	***3,000***	***3,100***	***3,200***	***3,300***	***2,900***	***2,900***
Protein (%)	28	26	22	19	16.5	14	12	14
Amino acids:								
Arginine (%)	1.60	1.5	1.25	1.1	0.95	0.8	0.6	0.6
Glycine and serine (%)	1.0	0.9	0.8	0.7	0.6	0.5	0.4	0.5
Histidine (%)	0.58	0.54	0.46	0.39	0.35	0.29	0.25	0.3
Isoleucine (%)	1.1	1.0	0.85	0.75	0.65	0.55	0.45	0.5
Leucine (%)	1.9	1.75	1.5	1.3	1.1	0.95	0.5	0.5
Lysine (%)	1.6	1.5	1.3	1.0	0.8	0.65	0.5	0.6
Methionine + cystine (%)	1.05	0.9	0.75	0.65	0.55	0.45	0.4	0.4
Methionine (%)	0.53	0.45	0.38	0.33	0.28	0.23	0.2	0.2
Phenylalanine + tyrosine (%)	1.8	1.65	1.4	1.2	1.05	0.9	0.8	1.0
Phenylalanine (%)	1.0	0.9	0.8	0.7	0.6	0.5	0.4	0.55
Threonine (%)	1.0	0.93	0.79	0.68	0.59	0.5	0.4	0.45
Tryptophan (%)	0.26	0.24	0.2	0.18	0.15	0.13	0.1	0.13
Valine (%)	1.2	1.1	0.94	0.8	0.7	0.6	0.5	0.58
Linoleic acid (%)	1.0	1.0	0.8	0.8	0.8	0.8	0.8	1.0
Major or macrominerals:								
Calcium (Ca) (%)	1.2	1.0	0.85	0.75	0.65	0.55	0.5	2.25
Chlorine (Cl) (%)	0.15	0.14	0.14	0.12	0.12	0.12	0.12	0.12
Magnesium (Mg) (mg)	600	600	600	600	600	600	600	600
Phosphorus (P), available (%)	0.6	0.5	0.42	0.38	0.32	0.28	0.25	0.35
Potassium (K) (%)	0.7	0.6	0.5	0.5	0.4	0.4	0.4	0.6
Sodium (Na) (%)	0.17	0.15	0.12	0.12	0.12	0.12	0.12	0.15
Trace or microminerals:								
Copper (Cu) (mg)	8	8	6	6	6	6	6	8
Iodine (I) (mg)	0.4	0.4	0.4	0.4	0.4	0.4	0.4	0.4
Iron (Fe) (mg)	80	60	60	60	50	50	50	60
Manganese (Mn) (mg)	60	60	60	60	60	60	60	60
Selenium (Se) (mg)	0.2	0.2	0.2	0.2	0.2	0.2	0.2	0.2
Zinc (Zn) (mg)	75	65	50	40	40	40	40	65
Fat-soluble vitamins:								
Vitamin A (IU)	4,000	4,000	4,000	4,000	4,000	4,000	4,000	4,000
Vitamin D[3] (ICU)	900	900	900	900	900	900	900	900
Vitamin E (IU)	12	12	10	10	10	10	10	25
Vitamin K (mg)	1.0	1.0	0.8	0.8	0.8	0.8	0.8	1.0
Water-soluble vitamins:								
Biotin (mg)	0.2	0.2	0.15	0.125	0.100	0.100	0.100	0.15
Choline (mg)	1,900	1,600	1,300	1,100	950	800	800	1,000
Folacin (Folic Acid) (mg)	1.0	1.0	0.8	0.8	0.7	0.7	0.7	1.0
Niacin (Nicotinic Acid, Nicotinamide) (mg)	70.0	70.0	50.0	50.0	40.0	40.0	40.0	30.0
Pantothenic Acid (Vitamin B-3) (mg)	11.0	11.0	9.0	9.0	9.0	9.0	9.0	16.0
Riboflavin (Vitamin B-2) (mg)	3.6	3.6	3.0	3.0	2.5	2.5	2.5	4.0
Thiamin (Vitamin B-1) (mg)	2.0	2.0	2.0	2.0	2.0	2.0	2.0	2.0
Vitamin B-6 (Pyridoxine, Pyridoxal, Pyridoxamine) (mg)	4.5	4.5	3.5	3.5	3.0	3.0	3.0	4.0
Vitamin B-12 (Cobalamins) (mg)	0.003	0.003	0.003	0.003	0.003	0.003	0.003	0.003

[1]Adapted by the authors from *Nutrient Requirements of Poultry,* 8th rev. ed., NRC, National Academy Press, Washington, D.C., 1984, p. 17, Table 11.

[2]These are typical ME concentrations for corn-soya rations. Different ME values may be appropriate if other ingredients predominate.

[3]These concentrations of vitamin D are satisfactory when the dietary concentrations of calcium and available phosphorus conform with those in this table.

TABLE 24–12
GROWTH RATE, FEED AND ENERGY CONSUMPTION OF LARGE-TYPE TURKEYS[1]

	Body Weight				Feed Consumption				Cumulative Feed Consumption				ME Consumption	
Age	Male		Female		Male		Female		Male		Female		Male	Female
(wks)	(lb)	*(kg)*	(lb)	*(kg)*	(lb/week)	*(kg/week)*	(lb/week)	*(kg/week)*	(lb)	*(kg)*	(lb)	*(kg)*	(Mcal/wk)	(Mcal/wk)
1	0.24	*0.11*	0.24	*0.11*	0.22	*0.10*	0.22	*0.10*	0.22	*0.10*	0.22	*0.10*	0.30	0.30
2	0.60	*0.27*	0.53	*0.24*	0.44	*0.20*	0.37	*0.17*	0.66	*0.30*	0.60	*0.27*	0.60	0.50
3	1.28	*0.58*	1.04	*0.47*	0.99	*0.45*	0.86	*0.39*	1.65	*0.75*	1.46	*0.66*	1.1	0.80
4	2.21	*1.0*	1.54	*0.70*	1.35	*0.61*	1.01	*0.46*	3.00	*1.36*	2.47	*1.12*	1.7	1.2
5	3.31	*1.5*	2.43	*1.1*	1.54	*0.70*	1.32	*0.60*	4.54	*2.06*	3.79	*1.72*	2.3	1.6
6	4.41	*2.0*	3.53	*1.6*	1.90	*0.86*	1.68	*0.76*	6.44	*2.92*	5.47	*2.48*	2.9	2.1
7	5.73	*2.6*	4.63	*2.1*	2.38	*1.08*	1.96	*0.89*	8.82	*4.00*	7.43	*3.37*	3.5	2.6
8	7.28	*3.3*	5.73	*2.6*	2.87	*1.30*	2.29	*1.04*	11.69	*5.30*	9.72	*4.41*	4.1	3.1
9	8.82	*4.0*	6.84	*3.1*	3.33	*1.51*	2.60	*1.18*	15.02	*6.81*	12.33	*5.59*	4.8	3.6
10	10.36	*4.7*	8.16	*3.7*	3.92	*1.78*	2.95	*1.34*	18.94	*8.59*	15.28	*6.93*	5.2	4.1
11	12.13	*5.5*	9.48	*4.3*	4.39	*1.99*	3.24	*1.47*	23.33	*10.58*	18.52	*8.40*	5.7	4.6
12	13.89	*6.3*	10.58	*4.8*	4.96	*2.25*	3.51	*1.59*	28.29	*12.83*	22.03	*9.99*	6.3	5.1
13	15.66	*7.1*	11.69	*5.3*	5.53	*2.51*	3.75	*1.70*	33.82	*15.34*	25.78	*11.69*	7.1	5.5
14	17.64	*8.0*	12.79	*5.8*	5.87	*2.66*	3.86	*1.75*	39.69	*18.00*	29.64	*13.44*	7.8	5.8
15	19.40	*8.8*	13.89	*6.3*	6.37	*2.89*	4.01	*1.82*	46.06	*20.89*	33.65	*15.26*	8.4	6.1
16	21.39	*9.7*	14.77	*6.7*	6.73	*3.05*	4.23	*1.92*	52.79	*23.94*	37.88	*17.18*	8.8	6.4
17	23.15	*10.5*	15.66	*7.1*	6.90	*3.13*	4.48	*2.03*	59.60	*27.03*	42.36	*19.21*	9.6	6.7
18	24.92	*11.3*	16.54	*7.5*	7.21	*3.27*	4.56	*2.07*	66.90	*30.34*	46.92	*21.28*	10.2	6.9
19	26.68	*12.1*	17.20	*7.8*	7.56	*3.43*	4.74	*2.15*	74.46	*33.77*	51.66	*23.43*	10.9	7.1
20	28.22	*12.8*	17.86	*8.1*	7.94	*3.60*	4.92	*2.23*	82.40	*37.37*	56.58	*25.66*	11.6	7.3
21	29.77	*13.5*	—	—	8.18	*3.71*	—	—	90.58	*41.08*	—	—	12.5	—
22	31.31	*14.2*	—	—	8.42	*3.82*	—	—	99.00	*44.90*	—	—	12.9	—
23	32.63	*14.8*	—	—	8.69	*3.94*	—	—	107.69	*48.84*	—	—	13.2	—
24	33.96	*15.4*	—	—	8.93	*4.05*	—	—	116.62	*52.89*	—	—	13.5	—

[1]Adapted by the authors from *Nutrient Requirements of Poultry,* 8th rev. ed., NRC, National Academy Press, Washington, D.C., 1984, p. 18, Table 12.

TABLE 24–13
BODY WEIGHTS AND FEED CONSUMPTION OF LARGE-TYPE TURKEYS DURING HOLDING AND BREEDING PERIODS[1,2]

	Hens					Toms			
Age	Weight		Egg Production	Feed		Weight		Feed	
(weeks)	(lb)	*(kg)*	(%)	(lb/day)	*(g/day)*	(lb)	*(kg)*	(lb/day)	*(g/day)*
20	15.4	*7.0*	—	0.44	*200*	26.5	*12.0*	0.88	*400*
25	17.6	*8.0*	—	0.47	*215*	29.8	*13.5*	0.93	*420*
30	19.8	*9.0*	Start light Stimulation	0.51	*230*	35.3	*16.0*	0.97	*440*
35	20.9	*9.5*	66	0.57	*260*	37.5	*17.0*	0.99	*450*
40	20.5	*9.3*	63	0.56	*255*	39.7	*18.0*	1.01	*460*
45	20.1	*9.1*	60	0.55	*250*	40.1	*18.2*	1.06	*480*
50	19.8	*9.0*	50	0.53	*240*	40.8	*18.5*	1.10	*500*
55	19.8	*9.0*	40	0.51	*230*	41.5	*18.8*	1.12	*510*
60	19.8	*9.0*	35	0.49	*220*	41.9	*19.0*	1.15	*520*

[1]Adapted by the authors from *Nutrient Requirements of Poultry,* 8th rev. ed., NRC, National Academy Press, Washington, D.C., 1984, p. 18, Table 13.

[2]These values are based on experimental data involving "in season" egg production (*i.e.,* November through July) of commercial stock. It is estimated that summer breeders would produce 70–90% as many eggs and consume 60–80% as much feed, respectively, as "in season" breeders.

TABLE 24-14
NUTRIENT REQUIREMENTS OF PEKIN DUCKS AS PERCENTAGES OR AS MILLIGRAMS OR UNITS PER KILOGRAM OF DIET[1,2]

	Starting (0-2 Weeks)	Growing (2-7 Weeks)	Breeding
Energy Base:			
kcal ME/lb Diet[3]	**1,315**	**1,315**	**1,315**
kcal ME/kg Diet[3]	***2,900***	***2,900***	***2,900***
Protein .... (%)	22.0	16.0	15.0
Amino acids:			
Arginine .... (%)	1.1	1.0	–
Lysine .... (%)	1.1	0.9	0.7
Methionine + cystine .... (%)	0.8	0.6	0.55
Major or macrominerals:			
Calcium (Ca) .... (%)	0.65	0.6	2.75
Chlorine (Cl) .... (%)	0.12	0.12	0.12
Magnesium (Mg) .... (mg)	500	500	500
Phosphorus (P), available .... (%)	0.40	0.35	0.35
Sodium (Na) .... (%)	0.15	0.15	0.15
Trace or microminerals:			
Manganese (Mn) .... (mg)	40.0	40.0	25.0
Selenium (Se) .... (mg)	0.14	0.14	0.14
Zinc (Zn) .... (mg)	60.0	60.0	60.0
Fat-soluble vitamins:			
Vitamin A .... (IU)	4,000	4,000	4,000
Vitamin D .... (ICU)	220	220	500
Vitamin K .... (mg)	0.4	0.4	0.4
Water-soluble vitamins:			
Niacin (Nicotinic Acid, Nicotinamide) .. (mg)	55.0	55.0	40.0
Pantothenic Acid (Vitamin B-3) ... (mg)	11.0	11.0	10.0
Riboflavin (Vitamin B-2) .... (mg)	4.0	4.0	4.0
Vitamin B-6 (Pyridoxine, Pyridoxal, Pyridoxamine) .... (mg)	2.6	2.6	3.0

[1]Adapted by the authors from *Nutrient Requirements of Poultry,* 8th rev. ed., NRC, National Academy Press, Washington, D.C., 1984, p. 20, Table 15.

[2]For nutrients not listed, see requirements for chickens as a guide.

[3]These are typical dietary energy concentrations.

Fig. 24-6. White Pekin drake. (Courtesy, USDA)

TABLE 24-15
TYPICAL BODY WEIGHTS AND FEED CONSUMPTION OF PEKIN DUCKS TO 8 WEEKS OF AGE[1]

	Body Weight				Feed Consumption By 1-Week Periods				Cumulative Feed Consumption			
Age	**Male**		**Female**		**Male**		**Female**		**Male**		**Female**	
(weeks)	**(lb)**	***(kg)***	**(lb)**	***(kg)***	**(lb)**	***(kg)***	**(lg)**	***(kg)***	**(lb)**	***(kg)***	**(lb)**	***(kg)***
0	0.11	*0.05*	0.11	*0.05*	–	–	–	–	–	–	–	–
1	0.60	*0.27*	0.60	*0.27*	0.49	*0.22*	0.49	*0.22*	0.49	*0.22*	0.49	*0.22*
2	1.72	*0.78*	1.63	*0.74*	1.70	*0.77*	1.61	*0.73*	2.18	*0.99*	2.09	*0.95*
3	3.04	*1.38*	2.82	*1.28*	2.47	*1.12*	2.45	*1.11*	4.65	*2.11*	4.52	*2.05*
4	4.32	*1.96*	4.01	*1.82*	2.82	*1.28*	2.82	*1.28*	7.50	*3.40*	7.34	*3.33*
5	5.49	*2.49*	5.07	*2.30*	3.26	*1.48*	3.15	*1.43*	10.74	*4.87*	10.50	*4.76*
6	6.53	*2.96*	6.02	*2.73*	3.59	*1.63*	3.51	*1.59*	14.33	*6.50*	14.00	*6.35*
7	7.36	*3.34*	6.75	*3.06*	3.70	*1.68*	3.59	*1.63*	18.04	*8.18*	17.60	*7.98*
8	7.96	*3.61*	7.25	*3.29*	3.70	*1.68*	3.59	*1.63*	21.74	*9.86*	21.19	*9.61*

[1]Adapted by the authors from *Nutrient Requirements of Poultry,* 8th rev. ed., NRC, National Academy Press, Washington, D.C., 1984, p. 20, Table 16.

TABLE 24-16
NUTRIENT REQUIREMENTS OF GEESE AS PERCENTAGES OR AS MILLIGRAMS OR UNITS PER KILOGRAM OF DIET[1 2]

	Starting (0-6 Weeks)	Growing (After 6 Weeks)	Breeding
Energy Base:			
kcal ME/lb Diet[3]	**1,315**	**1,315**	**1,315**
kcal ME/kg Diet[3]	***2,900***	***2,900***	***2,900***
Protein (%)	22.0	15.0	15.0
Amino acids:			
Lysine (%)	0.9	0.6	0.6
Methionine + cystine (%)	0.75	—	—
Major or macrominerals:			
Calcium (Ca) (%)	0.8	0.6	2.25
Phosphorus (P), available (%)	0.4	0.3	0.3
Fat-soluble vitamins:			
Vitamin A (IU)	1,500	1,500	4,000
Vitamin D (ICU)	200	200	200
Water-soluble vitamins:			
Niacin (Nicotinic Acid, Nicotinamide) (mg)	55.0	35.0	20.0
Pantothenic Acid (Vitamin B-3) (mg)	15.0	—	—
Riboflavin (Vitamin B-2) (mg)	4.0	2.5	4.0

[1]Adapted by the authors from *Nutrient Requirements of Poultry,* 8th rev. ed., NRC, National Academy Press, Washington, D.C., 1984, p. 19, Table 14.

[2]For nutrients not listed, see requirements for chickens as a guide.

[3]These are typical dietary energy concentrations.

Fig. 24-7. Toulouse gander. (Photo by J. C. Allen & Son, Inc., West Lafayette, Ind.)

TABLE 24-17
NUTRIENT REQUIREMENTS OF PHEASANTS AND BOBWHITE QUAIL AS PERCENTAGES OR AS MILLIGRAMS OR UNITS PER KILOGRAM OF DIET[1 2 3]

	Pheasant			Bobwhite Quail		
Energy Base:	Starting	Growing	Breeding	Starting	Growing	Breeding
kcal ME/lb Diet[4]	**1,270**	**1,225**	**1,270**	**1,270**	**1,270**	**1,270**
kcal ME/kg Diet[4]	***2,800***	***2,700***	***2,800***	***2,800***	***2,800***	***2,800***
Protein (%)	30.0	16.0	18.0	28.0	20.0	24.0
Amino acids:						
Glycine + serine (%)	1.8	1.0	—	—	—	—
Lysine (%)	1.5	0.8	—	—	—	—
Methionine + cystine (%)	1.1	0.6	0.6	—	—	—
Linoleic acid (%)	1.0	1.0	1.0	1.0	1.0	1.0
Major or macrominerals:						
Calcium (Ca) (%)	1.0	0.7	2.5	0.65	0.65	2.3
Chlorine (Cl) (%)	0.11	0.11	0.11	0.11	0.11	0.11
Phosphorus (P), available (%)	0.55	0.45	0.40	0.55	0.45	0.50
Sodium (Na) (%)	0.15	0.15	0.15	0.15	0.15	0.15
Trace or microminerals:						
Iodine (I) (mg)	0.30	0.30	0.30	0.30	0.30	0.30
Water-soluble vitamins:						
Choline (mg)	1,500.0	1,000.0	—	1,500.0	—	1,000.0
Niacin (Nicotinic Acid, Nicotinamide) (mg)	60.0	40.0	—	30.0	—	20.0
Pantothenic Acid (Vitamin B-3) (mg)	10.0	10.0	—	13.0	—	15.0
Riboflavin (Vitamin B-2) (mg)	3.5	3.0	—	3.8	—	4.0

[1]Adapted by the authors from *Nutrient Requirements of Poultry,* 8th rev. ed., NRC, National Academy Press, Washington, D.C., 1984, p. 21, Table 17.

[2]For Pheasant values not listed see requirements for turkeys as a guide.

[3]For Bobwhite Quail values not listed see requirements for Leghorn-type chickens as a guide.

[4]These are typical dietary energy concentrations.

TABLE 24-18
NUTRIENT REQUIREMENTS OF JAPANESE QUAIL (COTURNIX) AS PERCENTAGES OR AS MILLIGRAMS OR UNITS PER KILOGRAM OF DIET[1]

	Starting and Growing	Breeding
Energy Base:		
kcal ME/lb Diet[2]	**1,361**	**1,361**
kcal ME/kg Diet[2]	***3,000***	***3,000***
Protein (%)	24.0	20.0
Amino acids:		
Arginine (%)	1.25	1.26
Glycine and serine (%)	1.20	1.17
Histidine (%)	0.36	0.42
Isoleucine (%)	0.98	0.90
Leucine (%)	1.69	1.42
Lysine (%)	1.30	1.15
Methionine + cystine (%)	0.75	0.76
Methionine (%)	0.50	0.45
Phenylalanine + tyrosine (%)	1.80	1.40
Phenylalanine (%)	0.96	0.78
Threonine (%)	1.02	0.74
Tryptophan (%)	0.22	0.19
Valine (%)	0.95	0.92
Linoleic acid (%)	1.0	1.0
Major or macrominerals:		
Calcium (Ca) (%)	0.8	2.5
Chlorine (Cl) (%)	0.20	0.15
Magnesium (Mg) (mg)	300	500
Phosphorus (P), available (%)	0.45	0.55
Potassium (K) (%)	0.4	0.4
Sodium (Na) (%)	0.15	0.15
Trace or microminerals:		
Copper (Cu) (mg)	6	6
Iodine (I) (mg)	0.3	0.3
Iron (Fe) (mg)	100	60
Manganese (Mn) (mg)	90	70
Selenium (Se) (mg)	0.2	0.2
Zinc (Zn) (mg)	25	50
Fat-soluble vitamins:		
Vitamin A (IU)	5,000	5,000
Vitamin D (ICU)	1,200	1,200
Vitamin E (IU)	12	25
Vitamin K (mg)	1	1
Water-soluble vitamins:		
Biotin (mg)	0.3	0.15
Choline (mg)	2,000	1,500
Folacin (Folic Acid) (mg)	1	1
Niacin (Nicotinic Acid, Nicotinamide) (mg)	40	20
Pantothenic Acid (Vitamin B-3) (mg)	10	15
Riboflavin (Vitamin B-2) (mg)	4	4
Thiamin (Vitamin B-1) (mg)	2	2
Vitamin B-6 (Pyridoxine, Pyridoxal, Pyridoxamine) (mg)	3	3
Vitamin B-12 (Cobalamins) (mg)	0.003	0.003

[1]Adapted by the authors from *Nutrient Requirements of Poultry*, 8th rev. ed., NRC, National Academy Press, Washington, D.C., 1984, p. 22, Table 18.

[2]These are typical dietary energy concentrations.

Energy

The energy requirement may be defined as the amount of available energy that will provide for growth or egg production at a high enough level to permit maximal economic return for the production unit.

Although each primary energy source—carbohydrates, fats, proteins—has specific functions, all of them can be used to provide energy for maintenance and production of poultry. From the standpoint of providing the normal energy needs, however, the carbohydrates are by far the most important, whereas the fats rank next as an energy source.

Of the various systems of expressing energy values of feeds and nutrient requirements—gross energy (GE), digestible energy (DE), metabolizable energy (ME), and net energy (NE)—the poultry industry has found metabolizable energy to be the most reliable expression of energy needs. In general, metabolizable energy represents 25 to 90% of gross energy. Metabolizable energy, as a portion of gross energy, is usually lowest for fibrous feedstuffs and highest for fats and oils, with grain and protein supplements intermediate.

Carbohydrates, which constitute about 75% of the dry weight of plants and grain, make up a large part of poultry rations. They are composed of carbon, hydrogen, and oxygen. The group includes sugars, starch, cellulose, gums, and related substances. Carbohydrates serve as a source of heat and energy in the bird's body. A surplus taken into the body may be transformed into fat and stored as a reserve supply of heat and energy.

Although fats are used primarily to supply energy in poultry diets, they also improve the physical consistency of rations and the dispersion of microingredients in feed mixtures. The fats used for feeding poultry are derived from three sources: animal or poultry fats obtained from the rendering industry, restaurant greases, acidulated soapstocks from the vegetable oil industry, and/or mixtures thereof. The nutritional value of fats for poultry feed is determined by moisture, impurities, unsaponifiables, free fatty acids, total fatty acids, and fatty acid composition. The polyunsaturated linoleic and arachidonic acids are considered to be *essential fatty acids*. They have specific functions in the body that are not related to energy production. Birds exhibit poor growth, fatty livers, reduced egg size, and poor hatchability without these essential fatty acids. Fats for poultry feed should be stabilized against oxidation.

The metabolizable energy (ME) contribution of fats may be influenced by their fatty acid composition, free fatty acid content, level of fat inclusion in the ration, ingredient composition of the ration, and age of poultry. Fats often increase the utilization of dietary energy by poultry in excess of the increase expected when the ME of the fat is added to the ME values of the other ration constituents. Supplemental fats may increase energy utilization in adult chickens due to (1) a decreased rate of food passage through the gastrointestinal tract, and (2) the heat increment of fat being less than that of carbohydrates.

The fatty acid composition of body fat and egg fat may be altered by dietary fat. This is especially true when substantial levels of unsaturated fats are used, such as corn oil or sunflower oil. In these cases, the fatty acid composition of the body fat and/or the egg fat tends to reflect that of the dietary unsaturated fat. Feeding saturated fats following the feeding of unsaturated fats will cause the body or egg fat to become more saturated.

Because the primary function of both carbohydrates and fats is to serve as a source of energy for the body, an insufficient supply of these nutrients results in reduced growth rate or egg production in poultry.

Feed intake is governed by the energy concentration of the feed. While the bulkiness of feed can alter feed intake, the bird, for the most part, will eat to satisy its energy needs. Because of this, special attention must be given to nutrient ratios, especially the ratio of energy to various nutrients such as amino acids and minerals. So, the ME values given in the NRC tables that are reproduced in this chapter are not intended as *requirements.* Rather, they are provided to give perspective to the other nutrient requirement levels. Using the energy level for reference will enable the nutritionist to form a ratio of the amount of the nutrient per unit of energy, thereby keeping the nutrients in balance with available energy. In eating to satisfy its energy need, a bird will eat less of a high-energy diet and more of a low-energy diet. Having the nutrients in relation to dietary energy will ensure proper intake on a daily basis.

The metabolic rate of poultry is an indication of the energy needs of the bird. Several factors can affect this rate, including the following:

1. **Breed and strain.** The development of breeds and strains for specific purposes has resulted in genetic differences in the efficiency of energy utilization. For example, the University of Guelph found that Leghorn chickens obtained 3% more metabolizable energy from their diet than broiler strain chickens. In turkeys, separate lines have been developed for males and females.
2. **Activity.** Birds that have access to large areas have a higher metabolic rate than their counterparts which are confined to small cages which restrict movement.
3. **Diurnal rhythm.** Within individual birds, the metabolic rate will vary according to the time of day.
4. **Environmental temperature.** Poultry are extremely sensitive to environmental temperature, especially hot weather. Since there is little heat dissipated through sweating, much of the body heat must be lost through panting. Feed consumption will decrease about 1.5% for each rise of 1.8°F above the thermoneutral zone.
5. **Diet.** The metabolic rate of birds will be affected by the type of diet. A low energy-high protein diet will necessitate different digestive and metabolic processes than a high energy-low protein diet.
6. **Rate of growth or egg production.** A bird that is in heavy production will have different energy demands than one that is not in production.
7. **Other factors.** Other factors affecting dietary requirements for energy, in addition to those already listed, are: stress, body size, and feather coverage.

• **True metabolizable energy (TME)**—*The TME for poultry is the gross energy of the feed minus the gross energy of the excreta of food origin.* A correction for nitrogen retention may be applied to give a TME_n value. In 1975, Dr. I. R. Sibbald of the Animal Research Center, Ottawa, Canada, suggested a new method of measuring the energy content of feed ingredients based on the precision feeding of a measured amount of a single ingredient to a mature rooster. The excretra was collected so that the amount of energy that had been digested could be measured. Because the excreta (feces and urine) contain metabolic and endogenous fractions from the body, Sibbald corrected for them (he deleted them) and called the resulting energy value true metabolizable energy (TME). To clearly differentiate between the TME values and ME values, Sibblad suggested using the term apparent metabolizable energy (AME) for the traditionally determined energy values which include metabolic and endogenous fractions.

In simple formulas AME and TME are: **AME = feed energy − (fecal + urinary + gaseous energy),** whereas **TME = AME + (metabolic + endogenous energy).** In poultry, the gaseous losses are negligible and usually ignored.

TME values are easier and much less expensive to determine than the traditional ME values. In recent years, the TME values have been corrected for the energy contained in nitrogen excretion; the objective being to provide an energy measure where the energy in the ingredient is used for only one purpose—energy in the animal. So, true metabolizable energy values are currently expressed either as (1) TME, or (2) TME_n when corrected for nitrogen excretion. Following the nitrogen correction, the values obtained by use of TME assay are virtually identical to those obtained by using the more traditional apparent metabolizable energy; hence, nitrogen corrected TME(TME_n) and apparent metabolizable energy (AME) values are interchangeable.

Some scientists question the Sibbald procedure for determining TME and AME because the assays are made with starved birds. Others are reluctant to adopt the TME system of feed evaluation because of the difficulty and cost in changing from current dietary standards. Still others feel that further research studies are needed. Nevertheless, the TME system is now in use worldwide, serving as the basis for feed formulation and specifications for purchasing. Initially, the energy part of the system attracted industrial attention. Now, more and more companies are using the amino acid part of the system.

(Also see Chapter 3, section on "Hunger and Appetite"; and Chapter 4, section on "Energy Systems.")

Protein

Typical broiler starter rations contain from 21 to 24% protein, and typical laying rations from about 15 to 17% protein. Grain and millfeeds supply approximately ½ of the protein needs for most poultry rations. Additional protein is supplied from the high-protein concentrates of either animal or vegetable origin.

From the standpoint of poultry nutrition, the amino acids that make up proteins are really the essential nutrients, rather than the protein molecule itself. Hence, protein content as a measure of the nutritional value of a feed is becoming less important, and each amino acid is being considered individually. The essential amino acid requirements of each class of poultry are given in the NRC tables.

The energy content of the diet must be considered in formulating to meet the desired intake of all essential nutrients other than energy itself, including the intake of the essential amino acids. For example, if the producer uses a high-energy feed, the protein content of the feed must be high if the bird is to ingest adequate amounts of protein. Conversely, if the energy content of the feed is low, the protein content should be low, also; otherwise, the bird will consume excessive amounts of expensive protein.

Currently stated amino acid requirements have no reference to environmental conditions. Percentage requirements should probably be raised in warmer environments and lowered in colder environments in accordance with expected differences in feed or energy intakes.

The amino acid concentrations presented in the NRC tables in this chapter are intended to promote maximum growth and production. Maximum economic returns may not, however, always be assured by maximum growth and production, particularly when protein prices are high. So, at times the dietary concentrations may be somewhat reduced, lowering growth rate to some degree, but maintaining economic returns.

It has been determined that the chick requires dietary sources of protein to furnish 13 different amino acids. These amino acids are referred to as *essential,* since the chicken cannot produce them in sufficient amounts for maximum growth or egg production, and because a dietary deficiency of any one of them interferes with body protein formation and affects growth or egg production. The primary object of protein feeding, therefore, is to furnish the bird with protein which, upon digestion, will yield sufficient quantities of the 13 essential amino acids needed for top performance.

When formulating poultry rations, they must be so designed as to supply all the essential amino acids in ample amounts. Special attention needs to be given to supplying the amino acids lysine, methionine and cystine, and tryptophan, which are sometimes referred to as the "critical amino acids" in poultry nutrition. Additionally, there must be sufficient total nitrogen for the chicken to synthesize the other amino acids needed.

In practice, the amino acid requirements of poultry are met by proteins from plant and animal sources. Protein supplements that most nearly supply the essential amino acids of the bird are known as *high-quality* supplements. Usually it is necessary to choose more than one source of dietary protein, then combine them in such a way that the amino acid composition of the mixture meets the requirements of the bird.

Some high-protein feedstuffs may contain toxic compounds. For example, cottonseed meal may contain gossypol, which may impair growth and feed conversion, and discolor the yolks in stored eggs. Certain strains of rapeseed meal contain high enough levels of goitrogenic compounds to be toxic. Even soybean meal, the most used protein supplement, contains harmful substances, such as a trypsin inhibitor, but these are destroyed by proper heating.

Any excess protein consumed by the bird can be burned in the body to yield energy in somewhat the same manner as carbohydrates and fats. However, in practical feeding of poultry, it is seldom wise to use excessive protein because carbohydrates and fats are generally more economical sources of energy.

In addition to dietary energy concentration and ambient temperature, the following factors affect the amino acid requirements:

1. **The rate of growth or intensity of egg production.** The more rapid the growth and the higher the egg production, the higher the amino acid requirements.
2. **The breeds and strains.** Even within species of like body size, growth rate, or egg production, there may be differences in requirements among breeds and strains because of differences in the efficiency of digestion, nutrient absorption, and metabolism of absorbed nutrients.
3. **The protein level.** The amino acid requirements tend to increase with dietary protein.
4. **The amino acid relationships,** specifically—

   a. **Methionine-cystine.** The requirement for methionine can be met only by methionine, while the requirement for cystine may be met by cystine or methionine. This is because methionine is readily converted to cystine metabolically, while the reverse is not possible. If sulfate is deficient in the diet, a portion of the cystine that would normally be converted to sulfate may be spared by the addition of sulfates to the diet (e.g., sodium or potassium sulfate).

   b. **Phenylalanine-tyrosine.** The requirement for phenylalanine may be met only by phenylalanine, while the requirement for tyrosine may be met by tyrosine or phenylalanine.

   c. **Glycine-serine.** Glycine and serine can be used interchangeably in poultry diets.

5. **Antagonisms.** There are specific antagonisms among amino acids that may be structurally related, e.g., valine-leucine-isoleucine and arginine-lysine. Increasing one or two of such a group may raise the need for another of the same group.
6. **Imbalances.** In supplementing diets with limited amino acids, it is important to supplement first with the most limiting one, followed by the second most limiting one. Oversupplementation with only the second most limiting one may create an imbalance and accentuate the primary deficiency.
7. **Conversion of certain amino acids to vitamins.** High levels of methionine may partly compensate for a deficiency of choline or vitamin B–12 by providing needed methyl groups, and high levels of tryptophan may alleviate a niacin deficiency through metabolic conversion to niacin. Reliance on these conversions is, however, unwise, both nutritionally and economically.
8. **Amino acid availability.** The usual assumption that amino acids are 80 to 90% available is not necessarily valid. For example, the digestibility of feather protein (keratin) is improved in the processing of feather meal. Heat treatment of soybean meal inactivates compounds that interfere with trypsin digestion of protein in the intestine, thereby improving the availability of amino acids to the bird. Overheating during the drying of bloodmeal, meat scrap, and fish or fishwaste can seriously lower digestibility and the availability of specific amino acids (of which lysine is probably the most critical under practical conditions of feed formulation).

The consequences of a protein or amino acid deficiency vary with the degree of the deficiency. A *borderline deficiency* is characterized by poor growth, deformed primary wing feathers, reduced egg size, poor egg production (but hatchability is not affected), tendency toward greater deposition of carcass and liver fat, poor feed conversion into eggs or meat, and lack of melanin pigment in black- or reddish-colored feathers with low lysine. A *severe protein deficiency* is marked by stopping of feed intake, stopping of egg production, loss of body weight, resorption of ova, a tongue deformity with leucine, isoleucine, and phenylalanine deficiency, stasis of the digestive tract, and death.

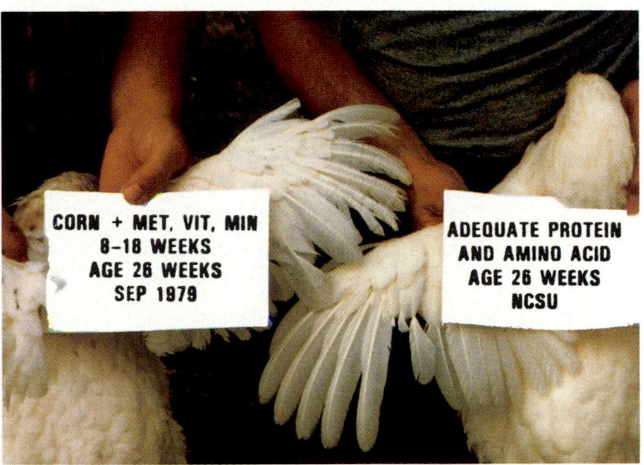

Fig. 24-8. *Left:* Deformed primary wing feathers caused by protein deficiency in early life. *Right:* Normal. (Courtesy, Department of Poultry Science, J. D. Garlich, Ph.D., North Carolina State University, Raleigh)

• **True digestibility of amino acids**[2]—Recently, there has been much interest in determining amino acid availability, using some of the same techniques that were originally developed for determination of true metabolizable energy. This interest stemmed from the fact that, frequently, amino acids cannot be completely utilized due to inherent characteristics of the feedstuff. Details follow.

The available amino acids are the amino acids in the diet that are not combined with compounds interfering with their digestion, absorption, or utilization by the bird; they are actually supplied to the sites of protein synthesis.

The digestible amino acids are those that are absorbed from the gut lumen; they are calculated as the difference between the amount of amino acids in the feed and the excreta.

Differentiation must be made between apparent and true digestibility. In addition to nonabsorbed feed, the excreta also contain materials originating in the tissue of the bird, e.g., cells sloughed from the gut wall, mucus, bile, unabsorbed digestive juices, etc. Apparent digestibility makes no correction for these endogenous components. In simple formulas, *apparent digestibility coefficient* and *true digestibility coefficient* follow:

$$\text{Apparent Digestibility Coefficient} = \frac{\text{amino acid consumed} - \text{amino acid excreted}}{\text{amino acid consumed}} \times 100$$

$$\text{True Digestibility Coefficient} = \frac{\text{amino acid consumed} - \text{amino acid excreted} + \text{endogenous amino acid}}{\text{amino acid consumed}} \times 100$$

The availability of amino acids in feedstuffs is estimated by three main methods: (1) *in vitro* tests, (2) growth tests, and (3) digestibility tests. Most of the recent research has focused on digestibility determinations.

Digestibility trials measure the digestibility of all amino acids in a feedstuff in one run. Various experimental techniques may be applied to determine the digestibility of amino acids. Dr. I. R. Sibbald of the Animal Research Center, Ottawa, Canada, developed the cockerel precision feeding technique, which estimates the true digestibility of amino acids. It consists of force feeding a precise quantity of feed (1 to 2% of body weight) to a starved adult rooster, followed by a 48-hour collection period of the excreta. Other birds are fasted throughout the experiment to allow for collection of the endogenous excreta. It appears that true amino acid digestibilities established in the chicken are valid for other bird species also.

Ration formulation on the basis of digestible amino acids has the following **advantages:**

1. It improves performance prediction under practical conditions.
2. It allows more by-product and alternate protein sources to be included in poultry feeds, along with the possibility of reducing costs while minimizing the risk of deterioration in performance.
3. It reduces the variability of poultry performance.
4. It evaluates feedstuff and rations more precisely.
5. It makes possible a more accurate supply of amino acids required for optimal performance.

The **disadvantages** or **reservations** relative (1) to the *rooster bioassay* technique for determining the true digestibility of amino acids in feedstuffs, and (2) to the use of such values in feed formulations are:

1. It does not consider the effects of microbes on dietary protein in the lower gut. Microbes both utilize and synthesize amino acids. Research has indicated that, under certain conditions, 20 to 25% of the amino acids found in the excreta are of microbial origin.
2. When conventional feed ingredients are used in poultry rations, consideration of amino acid digestibility is not particularly advantageous.

Crystalline amino acids, not being protein bound or enclosed in a feedstuff impairing their digestion, are 100% digestible and 100% available.

[2]In the preparation of this section, the authors adapted selected material from the publication: *True Faecal Digestibility of Essential Amino Acids in Feedstuffs for Poultry,* published by Eurolysine, 16 Rue Ballu 75009 Paris, France, with the permission of Eurolysine's "sister company," Heartland Lysine, Inc., 8430 West Bryn Mawr, Suite 650, Chicago, IL 60631.

Summary: When and where net returns indicate that it is profitable to employ true digestibility amino acid values, they will be used. Even though much research has been done in this area, much more is needed and many more questions await answers.

• **Table V-4, True Digestibility of Essential Amino Acids for Poultry, and True Digestible Amino Acid Recommendations for Poultry Feed Formulation**—Along with the total content of amino acids in feedstuffs, knowledge of their digestibility is essential to enable a formulator to satisfy the requirements for the projected performance. Section V, Table V-4, is presented for this purpose.

Minerals

Minerals are required for the formation of the skeleton, as components of various compounds with particular functions within the body, as activators of enzymes, and for the

TABLE
POULTRY MIN-

Minerals Which May Be Deficient Under Normal Conditions	Conditions Usually Prevailing Where Deficiencies Are Reported	Function of Mineral	Some Deficiency Symptoms	Types of Poultry Rations Usually Requiring Supplementation			
				Starting	Growing	Laying	Breeding
Major or macrominerals:							
Salt (sodium and chlorine—NaCl)	Omitted at the mill.	Improves appetite, promotes growth, helps regulate body pH, and is essential for hydrochloric acid formation in the stomach.	Chloride-deficient chicks show poor growth, high mortality, nervous symptoms, and reduced blood chloride level. Sodium deficiency in layers results in decreased egg production, poor growth, and cannibalism.	Yes	Yes	Yes	Yes
Calcium (Ca)	Imbalance of Ca:P ratio. Presence of interfering elements.	Bone formation; eggshell formation; blood clotting; and neuromuscular function.	Anorexia, thin eggshells, rickets, or osteoporosis, tetany, abnormal walk.	Yes	Yes	Yes	Yes
Phosphorus (P)		Bone formation; metabolism of carbohydrates and fats; a component of all living cells; maintenance of the acid-base balance of the body; and calcium transport in egg formation.	Anorexia, weakness, rickets. Cage-layer fatigue, characterized by birds being paralized and unable to rise from a recumbent position. But there is evidence that this condition is not due to this factor alone.	Yes	Yes	Yes	Yes
Magnesium (Mg)	Diets containing high levels of Ca and P.	Essential for normal skeletal development, as a constituent of bone, enzyme activator, primarily in glycolytic system.	Decreased egg production, depressed growth and lethargy, convulsions.	No	No	No	No
Potassium (K)	Low plant protein level/high animal protein level.	Major cation of intracellular fluid where it is involved in osmotic pressure and acid-base balance. Muscle activity. Required in enzyme reaction involving phosphorylation of creatine. Influences carbohydrate metabolism.	*Chicks:* Retarded growth and high mortality.	No	No	No	No

proper maintenance of necessary osmotic relationships within the body of the bird.

The minerals which have been shown to be essential for chickens and turkeys are calcium, chlorine, copper, iodine, iron, magnesium, manganese, molybdenum, phosphorus, potassium, selenium, sodium, sulfur, and zinc. Of these, calcium, chlorine, iodine, manganese, phosphorus, selenium, sodium, and zinc are considered to be of most practical importance since outside sources of them must be added to practical feed formulation for chickens and turkeys. Most of the pertinent facts relative to poultry minerals are summarized in Table 24-19.

Some minerals are required by poultry in relatively large amounts. They are referred to as *major* or *macrominerals*. Others are needed in very small amounts. They are referred to as *trace* or *microminerals*.

24-19
ERAL CHART (See footnote at end of table.)

Mineral Requirements[1] Mineral Content (%) of Ration (Variable According to Class, Age, and Weight of Poultry) For	See Table	Recommended Allowances[1]	Practical Sources of the Mineral	Comments
Layers Broilers Breeding hens Turkeys Ducks Geese Pheasants/Bob White Japanese Quail	24-4 24-6 24-8 24-11 24-14 24-16 24-17 24-18	0.2-0.5% of the diet.	Common table salt.	Sodium level is sometimes reduced to minimal to control the moisture level of the feces.
Layers Broilers Breeding hens Turkeys Ducks Geese Pheasants/Bob White Japanese Quail	24-4 24-6 24-8 24-11 24-14 24-16 24-17 24-18	The calcium allowance should vary with level of production and temperature. A minimum of 3.4% is believed to be optimum for layers in moderate climates. Growing rations should contain 0.8-0.9% of Ca and 0.4-0.7% of P.	Dicalcium phosphate. Limestone. Oystershell.	For young poultry, the Ca:P ratio should be about 1.2:1. However, ratios of 1:1 to 1.5:1 are well tolerated. An excess of calcium interferes with the utilization of magnesium, manganese, and zinc.
Layers Broilers Breeding hens Turkeys Ducks Geese Pheasants/Bob White Japanese Quail	24-4 24-6 24-8 24-11 24-14 24-16 24-17 24-18	Dependent on production. Laying hens require diets containing at least 3% P. Growing rations should contain 0.8-0.9% of Ca and 0.4-0.7% of P.	Defluorinated phosphate. Dicalcium phosphate. Monosodium phosphate. Phosphoric acid. Steamed bone meal.	Organic phosphorus (present in plants) is poorly utilized by growing birds, but is satisfactory for adult birds. Only about 30 to 40% of the phosphorus in plant products is available to the young chick, poult, or duckling.
Layers Broilers Turkeys Ducks Japanese Quail	24-4 24-6 24-11 24-14 24-18	600 ppm in the ration of broilers should suffice.	Magnesium oxide or magnesium sulfate.	Requirements are affected by Ca and P levels in the diet. Not normally deficient in poultry rations.
Layers Broilers Turkeys Japanese Quail	24-4 24-6 24-11 24-18		Corn contains 0.33% potassium (as-fed), and other cereals contain 0.42-0.49% potassium.	Potassium is not deficient in normal rations, due to large amounts of plant products in poultry feeds.

(Continued)

TABLE 24-19

Minerals Which May Be Deficient Under Normal Conditions	Conditions Usually Prevailing Where Deficiencies Are Reported	Function of Mineral	Some Deficiency Symptoms	Types of Poultry Rations Usually Requiring Supplementation			
				Starting	Growing	Laying	Breeding
Trace or microminerals:							
Copper (Cu)		Essential element in a number of enzyme systems and necessary for synthesizing hemoglobin and preventing nutritional anemia.	Anemia, depigmentation of feathers, digestive disorders. *Poults:* Marked cardiac hypertrophy.	Yes	Yes	Yes	Yes
Iodine (I)	Feeds produced on iodine-deficient soils.	Needed by the thyroid gland for making thyroxin, an iodine-containing hormone which controls the rate of body metabolism or heat production.	Enlarged thyroid. Eggs produced from deficient breeder hens have a lowered hatchability, prolonged hatching time, and a subsequent retardation of yolk-sac absorption.	Yes	Yes	Yes	Yes
Iron (Fe)		Necessary for formation of hemoglobin, an iron-containing compound which enables the blood to carry oxygen. Iron is also important to certain enzyme systems.	*Chicks and poults:* Microcytic, hypochromic anemia. In red-feathered chickens, complete depigmentation of the feathers occurs.	Yes	Yes	Yes	Yes
Manganese (Mn)		Necessary for growth, bone structure, and reproduction.	*Chicks and poults:* Perosis, shortened leg bones, skull deformation, parrot beak. Poor egg production, shell quality, and hatchability.	Yes	Yes	Yes	Yes
Selenium (Se)	Feeds that are grown in selenium-deficient areas.	Involved in the destruction of peroxides within the cell as a constituent of glutathione peroxidase. Useful in preventing exudative diathesis.	Exudative diathesis. Pancreatic fibrosis. Steatitis. Muscular dystrophy. With a severe selenium deficiency, growth rate is reduced and mortality is increased. *Turkeys:* Myopathies of the gizzard and heart.	Yes	Yes	Yes	Yes
Zinc (Zn)		Zinc is a component of several enzyme systems, including peptidases and carbonic anhydrase. Also, zinc is required for normal protein synthesis and metabolism and is a component of insulin.	Bone problems, poor feathering (feather fraying occurs near the ends of the feathers), retarded growth, and loss of appetite. A zinc deficiency in breeder diets reduces egg production and hatchability.	Yes	Yes	Yes	Yes

[1]As used herein, the distinction between *mineral requirements* and *recommended allowances* is as follows: In *mineral requirements,* no margins of safety are included intentionally; whereas in *recommended allowances,* margins of safety are provided in order to compensate for variations in feed composition, environment, and possible losses during storage or processing.

• **Trace minerals for a purified poultry diet**—For certain poultry experiments, it may be desirable to formulate a purified diet consisting of chemically defined ingredients, such as sugar, lard, casein, etc. When this is done, Table 24-20 may be used as a guide relative to the trace mineral elements and dietary concentrations to include in the ration.

TABLE 24-20
SUGGESTED TRACE MINERAL SUPPLEMENTS TO (CHEMICALLY DEFINED) RATIONS[1]

Element	Amount of Ration	
	(mg/lb)	***(mg/kg)***
Boron (B)	0.91	*2*
Chromium (Cr)	1.36	*3*
Fluorine (F)	9.08	*20*
Inorganic sulfate	[2]	[2]
Molybdenum (Mo)	0.45	*1*
Nickel (Ni)	0.05	*0.1*
Silicon (Si)	113.50	*250*
Tin (Sn)	1.36	*3*
Vanadium (V)	0.09	*0.2*

[1]Adapted by the authors from *Nutrient Requirements of Poultry,* 8th rev. ed., NRC, National Academy Press, Washington, D.C., 1984, p. 7, Table 1.

[2]Sulfur is supplied to the ration by methionine and cystine. There may be a response to inorganic sulfate if the ration is low in cystine.

(Continued)

Mineral Requirements[1] Mineral Content (%) of Ration (Variable According to Class, Age, and Weight of Poultry) For ...	See Table	Recommended Allowances[1]	Practical Sources of the Mineral	Comments
Layers Broilers Turkeys Ducks Geese Japanese Quail	24-4 24-6 24-11 24-14 24-16 24-18	Rations containing 6 to 8 ppm of copper should be adequate.	Copper sulfate and copper carbonate are about equally effective.	
Layers Broilers Turkeys Pheasants/Bob White Japanese Quail	24-4 24-6 24-11 24-17 24-18	Laying hens require feed containing 300 ppb. Growing chicks require feed containing 350 ppb.	Stabilized iodized salt containing 0.007% iodine. Trace mineral mixes.	
Layers Broilers Turkeys Japanese Quail	24-4 24-6 24-11 24-18	Rations should contain about 80 ppm of iron.	Alfalfa meal. Meat, liver, and fish meals.	Iron salts are used as a means of detoxifying gossypol from cottonseed meal.
Layers Broilers Turkeys Ducks Japanese Quail	24-4 24-6 24-11 24-14 24-18	Rations should contain at lease 30 ppm for layers, and 60 ppm for broilers.	Alfalfa meal, distillers' solubles, or grain by-products. Manganese sulfate, manganous chloride, manganous carbonate, and manganous dioxide.	
Layers Broilers Turkeys Ducks Japanese Quail	24-4 24-6 24-11 24-14 24-18	0.2 to 0.3 ppm added selenium.	Fish meal and brewers' dried yeast. Sodium selenite or sodium selenate.	Selenium can pose toxicity problems. Hence, care should be taken when adding it to poultry rations. In 1987, FDA provided for an increase in the maximum allowance of selenium in complete feeds for chickens, turkeys, and ducks to 0.3 ppm. Added selenium is often put into the vitamin premix to avoid separation.
Layers Broilers Turkeys Ducks Japanese Quail	24-4 24-6 24-11 24-14 24-18		Zinc carbonate or zinc sulfate.	

The common mineral supplements used by all animals are listed in Section V, Table V–1F, Mineral Supplements. (Also see Chapter 4, section on "Minerals.")

MAJOR OR MACROMINERALS

The major or macrominerals of importance in poultry are salt (sodium chloride), calcium, phosphorus, magnesium, and potassium.

Salt (Sodium Chloride/NaCl)

Sodium and chlorine are essential for all animals, including poultry. Dietary proportions of sodium, potassium, and chlorine are important determinants of acid-base balance. The proper dietary balance of sodium, potassium, and chlorine is necessary for growth, bone development, eggshell quality, and amino acid utilization. It is generally recommended that .2 to .5% of the diet consist of salt.

As in most nutrient deficiencies, a lack of salt will reduce reproductive performance and retard growth. Also, cannibalism occurs in flocks when diets are deficient in salt.

Calcium (Ca) and Phosphorus (P)

The common calcium and phosphorus supplements used by all animals are listed in Chapter 13, Table 13–2, of this book. The biological availability of calcium is high for most commonly used supplements, but the biological availability of phosphorus in supplements varies.

The calcium requirement of the laying hen is difficult to define. In Table 24–4, the listed requirement of 3.4% of the ration is believed to represent the mean dietary concentration for the quantities of feed likely to be consumed (110 g per hen per day) over a considerable range of environmental temperature. Most of the calcium in the diet of the mature laying fowl is used for eggshell formation, whereas most of the calcium in the diet of the growing bird is used for bone formation. Other functions of calcium include roles in blood clotting and neuromuscular function. High concentrations of calcium carbonate (limestone) and calcium phosphate make the ration unpalatable. Also, excess dietary calcium interferes with the availability of other minerals such as magnesium, manganese, and zinc.

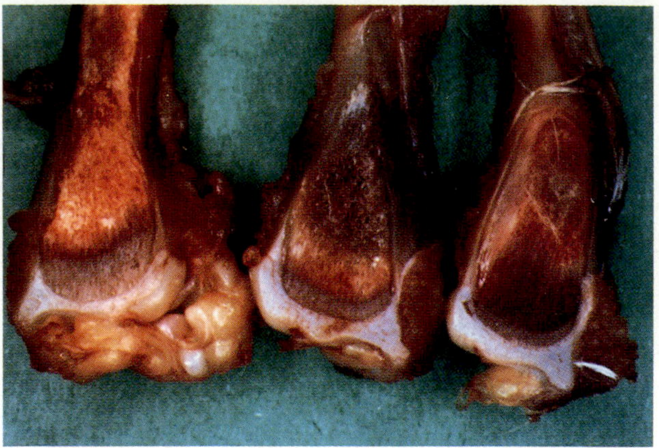

Fig. 24–10. Deficiencies of phosphorus, calcium, and/or vitamin D affect the normal calcification and development of the tibia in young chicks. The bone on the left was caused by a deficiency of phosphorus. The bone on the right was caused by a deficiency of calcium or vitamin D. The bone in the middle is from a chick that was fed a well balanced diet. (Courtesy, L. S. Jensen, Ph.D., Department of Poultry Sciences, The University of Georgia, Athens)

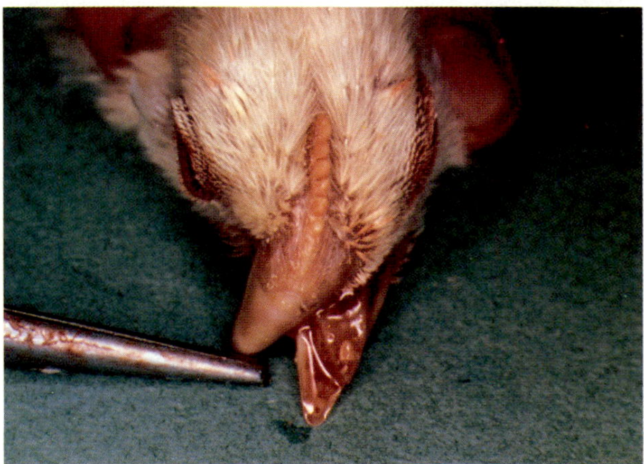

Fig. 24–9. "Rubbery beak" of a broiler, caused by feeding a high calcium layer ration which resulted in a severe phosphorus deficiency. (Courtesy, L. S. Jensen, Ph.D., Department of Poultry Science, The University of Georgia, Athens)

In addition to its function in bone formation, phosphorus is required in the metabolism of carbohydrates and fats and is a component of all living cells.

It is important that the minimum level of inorganic phosphorus be provided. The stated requirement is based on the generally greater availability of inorganic phosphorus than that of phytin phosphorus. Only about 30% of the phosphorus in plant products is available to the young chick, poult, or duckling, whereas the older bird has the ability to use most, if not all, of the phytin or organic phosphorus in plant products.

For growing chickens and turkeys, a calcium to phosphorus ratio of 1.2:1 is considered ideal. However, ratios of 1:1 to 1.5:1 are well tolerated.

Leg problems and thin eggshells occur when calcium and/or phosphorus are deficient. It is also essential, especially in poultry housed in confinement, that adequate amounts of vitamin D_3 be supplied in the ration to ensure proper absorption of these elements.

Magnesium (Mg)

Most poultry feeds contain adequate amounts of magnesium, and supplementation generally is not necessary. A dietary level of 600 ppm of magnesium is required by broilers.

Excessive magnesium levels can interfere with the absorption of other elements—notably calcium and phosphorus. If it is known that a diet contains a high level of magnesium, levels of other elements should be increased.

Chicks that are fed magnesium-deficient rations, exhibit depressed growth rates and lethargy. When disturbed, magnesium-deficient chicks become hyperirritable and go into convulsions. If the condition is not corrected, chicks become comatose and eventually die.

Potassium (K)

Potassium is found in large amounts in most plant feedstuffs, and generally supplementation is not required. It plays an essential role of cellular homeostasis, along with sodium and chlorine. In cases of potassium deficiency, generalized muscle weakness can be observed.

TRACE OR MICROMINERALS

Trace minerals are minerals that are required in small amounts. The trace minerals of special concern in poultry are copper, iron, iodine, manganese, selenium, and zinc.

As soils become leached, their content of trace minerals and the feedstuffs grown on them become borderline or deficient, thereby necessitating that more of certain minerals be added to poultry rations. Interactions between various trace minerals—such as copper and molybdenum, selenium and mercury, calcium and zinc, or calcium and manganese—are also important in poultry nutrition. Selenium is also metabolically involved with vitamin E and arsenic.

Copper (Cu) and Iron (Fe)

Both of these elements are necessary in the prevention of anemia. If a deficiency of either element exists, there is a reduction in the size of the red blood cells as well as a decreased ozygen-carrying capacity of the cells. Poultry rations should contain about 80 ppm of iron and 6 to 8 ppm of copper.

Iron salts may be used in poultry rations containing cottonseed meal to tie up gossypol—a compound that causes discoloration of yolks. However, iron salts do not prevent the pink discoloration of egg whites caused by cyclopropenoid fatty acids in cottonseed meal.

Iodine (I)

In certain areas of the United States (Northwest and Great Lakes regions), soils are deficient in iodine. In these deficient areas, it may be advisable to add iodine to the ration to prevent goiter. Feeds for laying hens should contain 300 ppb of iodine. For growing chicks, feed should contain 350 ppb of iodine.

Manganese (Mn)

This element is routinely added to poultry diets in such forms as manganous chloride, manganous carbonate, manganese dioxide, and manganese sulfate. Dietary levels of 30 ppm for layers and 60 ppm for broilers should be sufficient.

Chicks that are deficient in manganese characteristically display slipped tendons (perosis). A deficiency of either choline or biotin will produce similar symptoms. Hens suffering from a manganese deficiency will exhibit a reduction in egg production and hatchability. Many of the eggs that are laid are either thin-shelled or without shells.

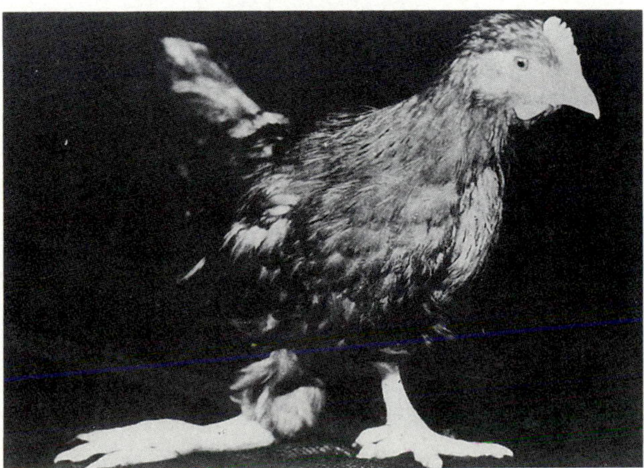

Fig. 24–11. Perosis, or slipped tendon, due to manganese deficiency. (Courtesy, Department of Poultry Science, Cornell University, Ithaca, N.Y.)

Selenium (Se)

Chicks deficient in selenium develop exudative diathesis, steatitis, and pancreatic fibrosis. Selenium-deficient turkeys develop a condition commonly referred to as white gizzard disease—a form of muscular dystrophy. Care should be taken in the incorporation of selenium in poultry rations as it is toxic at a level of 10 ppm.

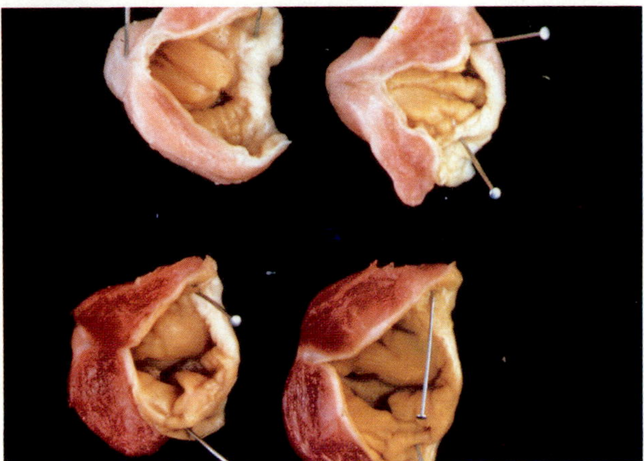

Fig. 24–12. White gizzard disease (muscular dystrophy) in turkey poults due to selenium deficiency. Note the whitish, underdeveloped gizzard at the top from a selenium-deficient poult compared to a normal gizzard below. (Courtesy, L. S. Jensen, Ph.D., Department of Poultry Science, The University of Georgia, Athens)

Zinc (Zn)

Chicks that are deficient in zinc exhibit bone problems, poor feathering, anorexia, and retarded growth. Zinc and calcium absorption are interrelated; hence, as the calcium level is increased, the zinc level should be increased, also.

Vitamins

The vitamins required by poultry, along with their deficiency symptoms and dietary sources, are shown in Table 24–21.

The column of Table 24–21 headed, "Types of Poultry Rations Usually Requiring Supplementation" indicates the types of poultry rations in which special attention must be paid to the inclusion of dietary sources of the vitamins. As shown, vitamin A, vitamin D, riboflavin, and vitamin B–12 are commonly low in most poultry rations. It is also to be emphasized that vitamin D_3, the animal form of vitamin D (made by the irradiation of 7–dehydrocholesterol), is more active for poultry, and should, therefore, be used instead of vitamin D_2, the plant form of the vitamin.

The fat-soluble vitamins (A, D, E, and K) can be stored and accumulated in the liver and other parts of the body, while only very limited amounts of the water-soluble vitamins (biotin, choline, folacin, niacin, pantothenic acid, riboflavin, thiamin, vitamin B–6, and vitamin B–12) are stored. For this reason, it is important that the water-soluble vitamins be fed regularly in the ration in adequate amounts.

Vitamin C is synthesized by poultry; hence, it is not considered as a required dietary nutrient. There is some evidence, nevertheless, of a favorable response to vitamin C by birds under stress.

Requirements for some of the vitamins may be met by the amounts occurring in natural feedstuffs. However, formulators of poultry feeds should be alert to the need for dietary

TABLE
POULTRY VITA-

Vitamins Which May Be Deficient Under Normal Conditions	Conditions Usually Prevailing Where Deficiencies Are Reported	Function of Vitamin	Some Deficiency Symptoms	Types of Poultry Rations Usually Requiring Supplementation			
				Starting	Growing	Laying	Breeding
Fat-soluble vitamins:							
Vitamin A	Old vitamin premix.	Essential for normal maintenance and functioning of the epithelial tissues, particularly of the eye and the respiratory, digestive, reproductive, nerve, and urinary systems.	***Chicks:*** Depressed growth, weakness; loss of coordination; xerophthalmia; anorexia, lowered resistance to infection; alterations in mucous membranes. ***Adults:*** Depressed production; low hatchability; discharge from nose and eyes; lowered resistance to infection; alterations in mucous membranes.	Yes	Yes	Yes	Yes
Vitamin D_3	Birds that are in confinement.	Aids in assimilation and utilization of calcium and phosphorus, and necessary in normal bone development.	***Chicks:*** Rickets, poor feathering, reduced growth. ***Adults:*** Weak bones, poor eggshell formation, reduced production and hatchability.	Yes	Yes	Yes	Yes
Vitamin E	Destruction by oxidation of the diet.	Antioxidant. Muscle structure. Reproduction.	***Chicks:*** Encephalomalacia, exudative diathesis, muscular dystrophy. ***Adults:*** Poor reproductive performance; prolonged vitamin E deficiency results in permanent sterility in the male and reproductive failure in the female. ***Poults:*** Myopathy of the gizzard.	Yes	Yes	Yes	Yes
Vitamin K	Coccidiosis. When high levels of antibodies or sulfa drugs are fed. Newly hatched chicks from deficient females.	Essential for prothrombin formation and blood clotting.	Hemorrhaging. Increased clotting time.	Yes	Yes	Yes	Yes
Water-soluble vitamins:							
Biotin	Broilers fed a milo, wheat, or wheat-barley based diet. Feeding avidin, a protein in uncooked egg white, which binds biotin and renders it unavailable nutritionally.	Involved in carbohydrate, lipid, and protein metabolism.	***Chicks:*** Cracking and degeneration of skin on feet, around beak, and perosis (slipped tendon). ***Adults:*** Reduced hatchability. ***Poults:*** Broken flight feathers, bending of the metatarsus, and dermatitis of the footpads and toes, base of beak, eye ring and vent.	Yes	No	No	Yes
Choline		Involved in nerve pulses. A component of phospholipids. Donor of methyl groups.	***Chicks, poults, ducklings:*** Retarded growth and perosis (slipped tendon). ***Adults:*** Increased mortality, lowered egg production, and increased abortion of egg yolks from ovaries.	Yes	Yes	No	No
Folacin (Folic Acid)		Related to B–12 metabolism. Metabolic reactions involving incorporation of single carbon units into larger molecules.	***Chicks:*** Poor growth, poor feathering, perosis, and anemia. ***Adults:*** Reduced hatchability and egg production. ***Turkey poults:*** Nervousness, droopy wings, and a stiff extended neck. ***Turkey breeder hens:*** Normal egg production, but reduced hatchability.	Yes	Yes	Yes	Yes

supplementation with vitamins usually assumed to be supplied by the feedstuffs.

(Also see Chapter 4, section on "Vitamins.")

24–21
MIN CHART (See footnote at end of table.)

Vitamin Requirements[1] — Vitamin Content (%) of Ration (Variable According to Class, Age, and Weight of Poultry) — For	See Table	Recommended Allowances[1]	Practical Sources of the Vitamin	Comments
Layers Broilers Turkeys Ducks Geese Japanese Quail	24–4 24–6 24–11 24–14 24–16 24–18	Variable according to class, age, and weight (see the suggested rations under each respective class).	Green forage, alfalfa meal, corn gluten meal, yellow corn, fish oils, synthetic vitamin A.	Toxicities can occur. The toxic level is on the order of 500 times the requirement. Symptoms of a vitamin A toxicity are weight loss, depressed feed intake, inflammation of epithelial tissue, and bone abnormalities. Requirements for vitamin A are expressed in either International Units (IU) or U.S. Pharmacopeia units (USP) per kilogram of diet.
Layers Broilers Turkeys Ducks Geese Japanese Quail	24–4 24–6 24–11 24–14 24–16 24–18	Variable according to class, age, and weight (see the suggested rations under each respective class).	Irradiated animal sterols, fish liver oils, vitamin A and D feeding oils, synthetic vitamin D_3.	Vitamin D_3 is more than 30 times as efficient for preventing rickets in chickens as vitamin D_2. Hypervitaminosis can occur.
Layers Broilers Turkeys Japanese Quail	24–4 24–6 24–11 24–18	Variable according to class, age, and weight (see the suggested rations under each respective class).	Alfalfa meal, vegetable oils, wheat germ, and pure vitamin concentrates such as alpha-tocopherol.	Vitamin E and selenium have a close interrelationship. In many cases, selenium can reduce the dietary requirement of vitamin E.
Layers Broilers Turkeys Ducks Japanese Quail	24–4 24–6 24–11 24–14 24–18	Variable according to class, age, and weight (see the suggested rations under each respective class).	Green pasture, alfalfa meal, synthetic vitamin K (menadione sodium bisulfite).	
Layers Broilers Turkeys Japanese Quail	24–4 24–6 24–11 24–18	Variable according to class, age, and weight (see the suggested rations under each respective class).	Grains, soybean meal, alfalfa meal, dried yeast, milk products, green pasture.	Availability in wheat and barley is extremely low.
Layers Broilers Turkeys Pheasants/Bob White Japanese Quail	24–4 24–6 24–11 24–17 24–18	Variable according to class, age, and weight (see the suggested rations under each respective class).	Fish products and pure vitamin.	Recent evidence indicates that choline is synthesized by mature chickens in quantities adequate for egg production. Choline may be a factor in egg size of quail. Dietary requirements of growing quail appear to be higher than those for chicks or poults.
Layers Broilers Turkeys Japanese Quail	24–4 24–6 24–11 24–18	Variable according to class, age, and weight (see the suggested rations under each respective class).	Alfalfa meal, wheat, soybean meal, and liver preparations.	

(Continued)

TABLE 24-21

Vitamins Which May Be Deficient Under Normal Conditions	Conditions Usually Prevailing Where Deficiencies Are Reported	Function of Vitamin	Some Deficiency Symptoms	Types of Poultry Rations Usually Requiring Supplementation			
				Starting	Growing	Laying	Breeding
Water-soluble vitamins (Continued):							
Niacin (Nicotinic Acid, Nicotinamide)	A predominately corn-soybean ration.	Required by all living cells, and an essential component of important metabolic enzyme systems involved in glycolysis and tissue respiration.	***Chicks:*** Enlargement of hock joints and perosis, retarded growth, and inflammation of mouth and tongue ("black tongue"). ***Adults:*** No symptoms observed in hen except on protein-deficient diet. ***Turkey poults:*** A hock disorder similar to perosis.	Yes	Yes	Yes	Yes
Pantothenic Acid (Vitamin B-3)	Use of artificially dried corn (heating destroys the pantothenic acid) and the omission of milk byproducts from the diet.	Part of coenzyme A, a necessary factor for intermediary metabolism.	***Chicks:*** Poor growth, ragged feather development, degeneration of skin around beak, eyes, and vent, and liver damage. ***Adults:*** Reduced hatchability. Mortality is high in newly hatched chicks from pantothenic acid-deficient hens.	Yes	Yes	Yes	Yes
Riboflavin (Vitamin B-2)		A component of enzyme systems essential to normal metabolic processes.	***Chicks:*** Curled toe paralysis, reduced growth, and diarrhea. ***Adults:*** Poor hatchability with many dying during 2nd week of incubation.	Yes	Yes	Yes	Yes
Thiamin (Vitamin B-1)		As a coenzyme in energy metabolism. Promotes appetite and growth, and required for normal carbohydrate metabolism.	***Chicks:*** Anorexia, loss of coordination, poor feathering, polyneuritis. ***Adults:*** Blue comb, paralysis.	No	No	No	No
Vitamin B-6 (Pyridoxine, Pyridoxal, Pyridoxamine)		As coenzyme in protein and nitrogen metabolism. Involved in red blood cell formation. Important in endocrine systems.	***Chicks:*** Poor growth, lack of coordination, and convulsions. ***Adults:*** Reduced body weight, egg production, and hatchability.	No	No	No	No
Vitamin B-12 (Cobalamins)		Numerous metabolic functions, and essential for normal growth and reproduction in poultry.	***Chicks:*** Poor growth, perosis, mortality. ***Adults:*** Reduced hatchability, fatty heart, liver, and kidneys.	Yes	Yes	Yes	Yes

[1]As used herein, the distinction between *vitamin requirements* and *recommended allowances* is as follows: In *vitamin requirements,* no margins of safety are included intentionally; whereas in *recommended allowances,* margins of safety are provided in order to compensate for variations in feed compositions, environment, and possible losses during storage or processing.

FAT-SOLUBLE VITAMINS

Vitamin A and vitamin D have long been routinely added to poultry feed. Today, vitamin E and vitamin K are also included in vitamin supplements for most poultry feeds.

Vitamin A

The requirement of vitamin A in poultry is dependent upon the following factors:

1. **Individual bird variations.** Metabolic rates and subsequent nutrient requirements vary from bird to bird, even though they may be of the same strain.
2. **Type of bird.** There are genetic differences between different strains of birds with regards to their respective abilities to utilize vitamin A.
3. **Production and stress.** Vitamin A requirements vary according to the type and stage of production. The layer needs more vitamin A than the growing chicken, since vitamin A is passed into the egg. Likewise, birds that are subjected to environmental stress probably have different vitamin A requirements than birds under minimal stress.

(Continued)

Vitamin Requirements[1] — Vitamin Content (%) of Ration (Variable According to Class, Age, and Weight of Poultry) — For ...	See Table	Recommended Allowances[1]	Practical Sources of the Vitamin	Comments
Layers Broilers Turkeys Ducks Geese Pheasants/Bob White Japanese Quail	24–4 24–6 24–11 24–14 24–16 24–17 24–18	Variable according to class, age, and weight (see the suggested rations under each respective class).	Chemically synthesized nicotinic acid, liver, yeast, and fermentation products, and most grasses.	Some niacin can be synthesized in the body through the conversion of tryptophan. The niacin in cereal grains and by-products is virtually unavailable and should not be included in the available niacin calculation.
Layers Broilers Turkeys Ducks Geese Pheasants/Bob White Japanese Quail	24–4 24–6 24–11 24–14 24–16 24–17 24–18	Variable according to class, age, and weight (see the suggested rations under each respective class).	Pure calcium pantothenate, alfalfa meal, dried milk products, and fermentation residues.	
Layers Broilers Turkeys Ducks Geese Pheasants/Bob White Japanese Quail	24–4 24–6 24–11 24–14 24–16 24–17 24–18	Variable according to class, age, and weight (see the suggested rations under each respective class).	Alfalfa meal, milk products, distillers' solubles, fermentation products, and pure vitamin.	
Layers Broilers Turkeys Japanese Quail	24–4 24–6 24–11 24–18	Variable according to class, age, and weight (see the suggested rations under each respective class).	Cereal grains and their by-products. Synthetic thiamin.	
Layers Broilers Turkeys Ducks Japanese Quail	24–4 24–6 24–11 24–14 24–18	Variable according to class, age, and weight (see the suggested rations under each respective class).	Milk products, meat and fish by-products, soybean meal.	
Layers Broilers Turkeys Japanese Quail	24–4 24–6 24–11 24–18	Variable according to class, age, and weight (see the suggested rations under each respective class).	Fish meal, fish solubles, meat meal, liver preparations, fermentation products, and commercial vitamin B–12 concentrate.	Body reserves are rapidly depleted in hens that are fed high-protein diets.

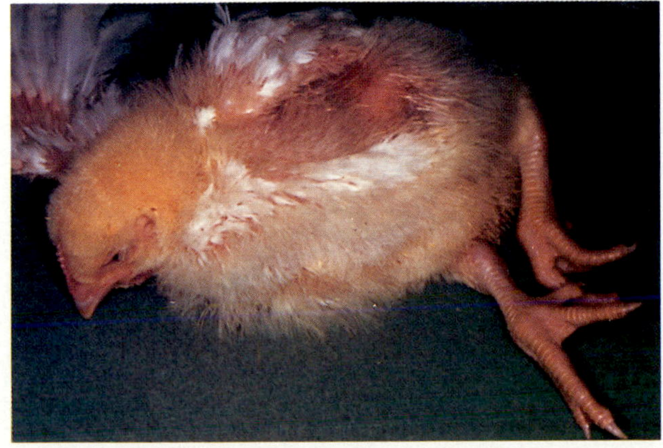

Fig. 24–13. Acute vitamin A deficiency in a young broiler chick. Note the incoordination. (Courtesy, L. S. Jensen, Ph.D., Department of Poultry Science, The University of Georgia, Athens)

4. **Destruction of the vitamin.** Vitamin A in feed can be destroyed when fats in the feed become rancid. Likewise, certain processing methods can reduce the activity of the vitamin in the feed. Parasites and bacteria in the gut can also destroy vitamin A before it is absorbed.

5. **Absorbability.** Since vitamin A absorption is dependent on a lipoprotein in the blood, deficiencies of protein and/or fat can reduce absorption.

Young chicks are more susceptible to vitamin A deficiencies than adults as it takes a relatively long period for adult birds to deplete their body stores. Deficiency symptoms of vitamin A in chicks are characterized by retarded growth, emaciation, general weakness, staggered gait, ruffled plummage, lowered resistance to infection, xeropthalmia and disruption of mucosal membranes. In adults, eye problems are prevalent along with decreased egg production and hatchability.

Excessive amounts of vitamin A can be toxic, but these levels must be on the order of 500 times the recommended allowances. Symptoms of vitamin A hypervitaminosis are anorexia, emaciation, inflammation of epithelial tissues, abnormalities of bone, swelling of the eyelids, and mortality.

Vitamin D

Cholecalciferol (D_3) is the form of vitamin D that has the highest activity in poultry feed. Today, most birds are reared in confinement where exposure to sunlight is insufficient for the conversion of 7-dehydrocholesterol—the precursor of vitamin D in the skin—to vitamin D_3. Thus, vitamin D_3 is routinely added to poultry feed. The cost of supplementation is small—especially when one considers the consequences of such deficiencies.

Fig. 24-14. A chick deficient in vitamin D, showing ungainly manner of balancing body. The beak is also soft and rubbery. (Courtesy, Department of Poultry Science, Cornell University, Ithaca, N.Y.)

The dietary requirements of vitamin D depend on four factors: (1) exposure to sunlight; (2) Ca:P ratio; (3) levels of calcium and phosphorus in the feed; and (4) intensity of production.

Leg problems, poor growth, and poor feathering are the common deficiency symptoms in growing birds. Egg production, hatchability, and eggshell quality will be decreased in deficient hens.

Vitamin E

As with vitamins A and D, vitamin E is extremely susceptible to destruction from the oxidation of fats in the feed. To prevent this, antioxidants are commonly added to poultry feeds. Also, vitamin E is often added to feed in an esterified form to protect it from destruction.

The three classical symptoms of vitamin E deficiency in chicks are encephalomalacia, exudative diathesis, and nutritional muscular dystrophy. Encephalomalacia is a condition whereby there is a necrosis in the brain. Chicks exhibit an outstretching of the legs with the toes curled; and the head is often in a retracted position. Prior to these symptoms of acute toxicity, chicks display a generalized lack of coordination. Exudative diathesis is a condition in which the walls of capillaries become highly permeable. Nutritional muscular dystrophy in chicks is analogous to stiff lamb disease in sheep and white muscle disease in calves.

Fig. 24-15. A chick with nutritional encephalomalacia, due to a lack of vitamin E. Note head retraction and loss of control of legs. (Courtesy, Department of Poultry Science, Cornell University, Ithaca, N.Y.)

Reproduction is impaired in vitamin E-deficient adult birds. Degeneration of the testes is observed in deficient males—a condition that can lead to permanent sterility if not corrected in time. Layers suffering from a vitamin E deficiency do not show a dramatic drop in egg production, but hatchability is severely reduced.

In vitamin E-deficient turkey poults, a myopathy of the gizzard can be observed.

Selenium and vitamin E are closely related in physiological functions, but they cannot replace each other in the diet.

Vitamin K

Vitamin K occurs as a number of naturally occurring and synthetic compounds with varying solubilities in fat and water. Menadione is a fat-soluble, synthetic compound that can be considered as the reference standard for vitamin K activity. Two naturally occurring forms are K_1 or phylloquinone and K_2 or minaquinone. Water-soluble forms include menadione sodium bisulfite (MSB), menadione sodium bisulfite complex (MSBC), and menadione dimethylpyrimidol (MPB). The theoretical activity of these compounds can be calculated on the basis of the proportion of menadione present in the molecule.

Vitamin K should be added to starter rations and rations that incorporate drugs, such as sulfaquinoxaline, which reduce the microbial population of the gut. When heavy parasitic infections occur, such as in coccidiosis, the vitamin K requirement is increased. Birds that are deficient in vitamin K have a greatly increased susceptibility to hemorrhaging and an increased clotting time. Newly hatched chicks from vitamin K-deficient females will have low storage of vitamin K and are generally susceptible to injury.

WATER-SOLUBLE VITAMINS

Of the water-soluble vitamins, biotin, niacin, pantothenic acid, riboflavin, and vitamin B-12 may be low in poultry feeds. Young chicks are most susceptible to vitamin deficiencies.

The requirements for the water-soluble vitamins are interrelated in some instances. They are also dependent upon the nature of the diet. The type of carbohydrate, protein concentration, and amino acid balance are major factors determining the dietary requirements for several vitamins.

Biotin

The availability of biotin in many of the cereal grains is very low. Hence, biotin levels and availability should be carefully monitored in poultry rations. Eggs are high in biotin. Therefore, careful consideration should be given to biotin in rations for layers. The symptoms of biotin deficiency in chicks are cracking and degeneration of skin on the feet and around the beak, and perosis. Reduced hatchability is observed in biotin-deficient hens.

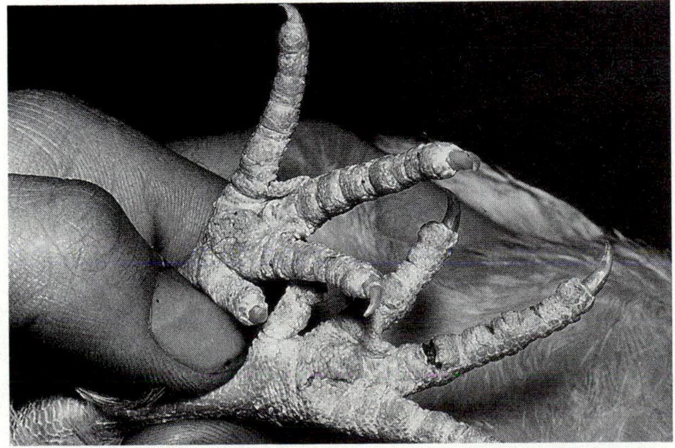

Fig. 24-16. Biotin deficiency. Note the severe lesions on the bottom of the feet. (Courtesy, Department of Poultry Science, University of Wisconsin, Madison)

Choline

Since choline levels can be marginal in starter and grower rations, choline is generally added to these types of feeds. Much like the deficiency symptoms of several of the other water-soluble vitamins, perosis (slipped tendon) can be observed in choline-deficient chicks. Since older birds can synthesize choline on the cellular level, it is difficult to produce a choline deficiency in laying hens.

Folacin (Folic Acid)

Because feeds rich in folacin —for example, alfalfa meal and liver meal—are used less frequently today in practical poultry rations, most poultry feeds are supplemented with folacin.

Deficiency symptoms of folacin in chicks are poor growth, abnormal coloration of feathers, perosis, and anemia. Hens that are deficient in folacin are observed to have reduced egg production and hatchability. Poults fed folacin-deficient rations may develop a paralysis of the neck.

Niacin (Nicotinic Acid/Nicotinamide)

Synthetic niacin is routinely added to starter and breeder rations. Some niacin can be synthesized in the body from tryptophan; but since corn, which is notably low in tryptophan, is the most widely used feedstuff in poultry feed, this means of fulfilling the niacin requirement should be viewed with skepticism.

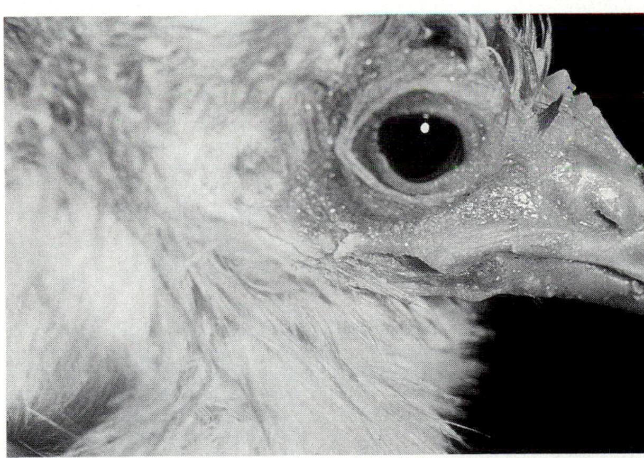

Fig. 24-17. "Spectacled eye" in a niacin-deficient chick. Also, note the loss of feathers around the eye. (Courtesy, University of Wisconsin, Madison)

Chicks deficient in niacin show an enlargement of the hock joints, perosis, dermatitis, retarded growth, and an inflammation of the mouth and tongue. In addition to these symptoms, poor feathering and hyperirritability may be observed.

Pantothenic Acid (Vitamin B-3)

Pantothenic acid is widely distributed in nature. However, it is routinely added to starter and breeder rations—often in the form of calcium pantothenate—to ensure against any possibility of deficiency. Deficiency symptoms in chicks are rather nonspecific—poor growth, poor feathering, liver damage, and lesions around the beak, eyes, and vent. Lowered hatchability and a high mortality rate of newly hatched chicks can result from feeding hens a diet deficient in pantothenic acid.

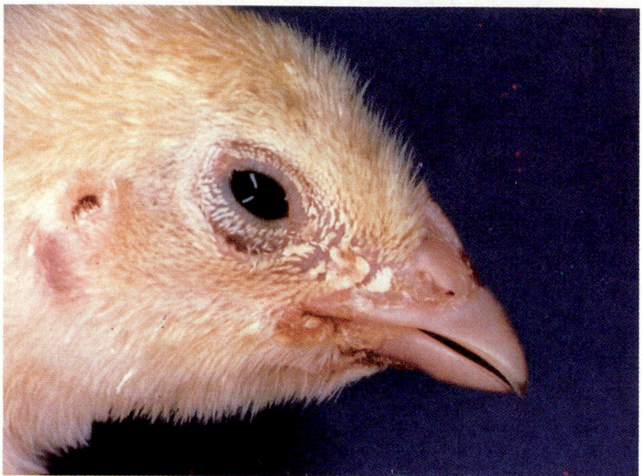

Fig. 24-18. Severe dermatitis caused by pantothenic acid deficiency in chick. (Courtesy, L. S. Jensen, Ph.D., Department of Poultry Science, The University of Georgia, Athens)

Riboflavin (Vitamin B-2)

All types of poultry rations should be supplemented with riboflavin. Chicks suffering from a riboflavin deficiency display a characteristic curled toe paralysis as well as depressed growth and diarrhea. In curled-toe paralysis, the brachial and sciatic nerves become greatly enlarged. Poor egg production and hatchability are observed in riboflavin-deficient hens.

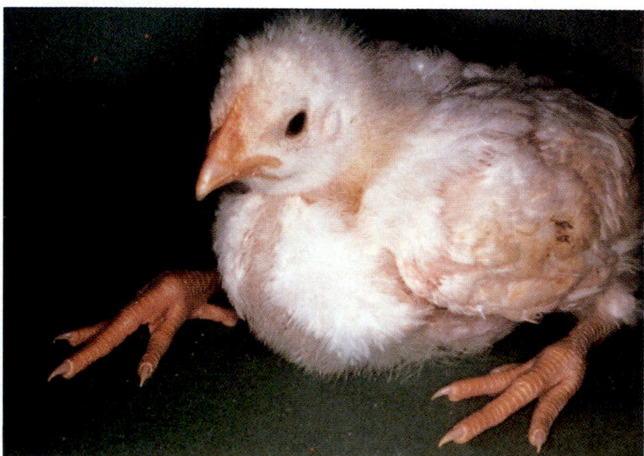

Fig. 24-19. Riboflavin deficiency in chick. Note the curled-toe paralysis and squatting on the hocks. (Courtesy, L. S. Jensen, Ph.D., Department of Poultry Science, The University of Georgia, Athens)

Thiamin (Vitamin B-1)

Although the requirement for thiamin is high in poultry, it is rarely added as a supplement to the ration because most feed ingredients contain high levels. Polyneuritis—a type of paralysis—is common in thiamin-deficient birds. Prior to this acute deficiency condition, anorexia, emaciation, ruffled feathers, and incoordination are observed in thiamin-deficient chicks. When polyneuritis sets in, a progressive paralysis is observed—beginning first in the toes and ultimately reaching the head, whereupon the head is retracted so that it lies on the back. Deficient adults frequently have a blue comb.

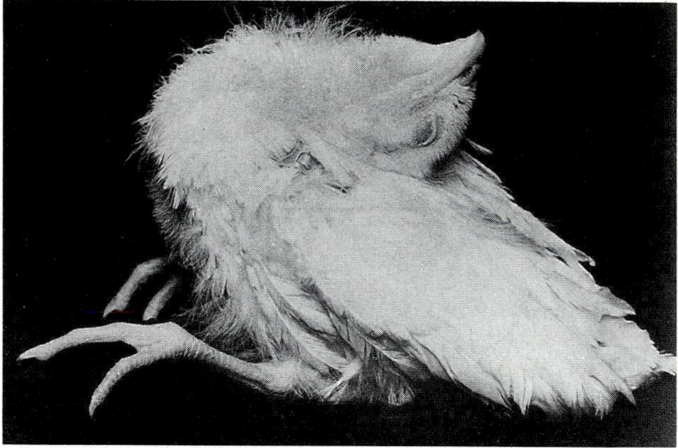

Fig. 24-20. Thiamin (B-1) deficiency, resulting in acute stage of polyneuritis. Note characteristic head retraction. (Courtesy, Department of Poultry Science, University of Wisconsin, Madison)

Vitamin B-6 (Pyridoxine, Pyridoxal, Pyridoxamine)

Deficiencies of vitamin B-6 are rare in poultry, since most feeds used in poultry rations contain relatively high levels of the vitamin. Deficiency symptoms in chicks are characterized by depressed growth and neurological problems such as poor coordination and convulsions. Mature birds deficient in vitamin B-6 exhibit anorexia, rapid loss of weight, and lowered egg production and hatchability.

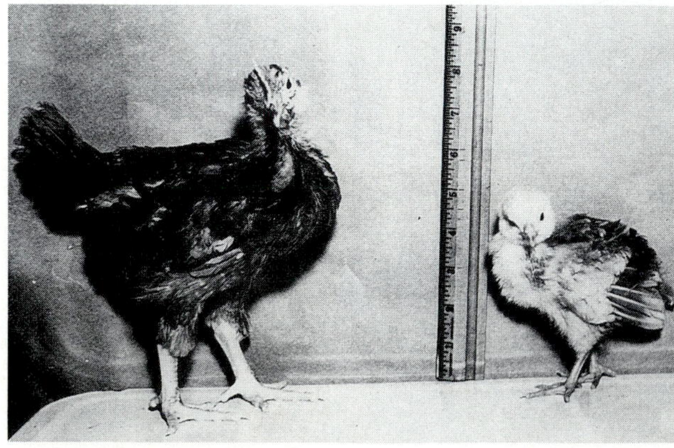

Fig. 24-21. Pyridoxine (B-6) deficiency. *Left:* Normal, control chick. *Right:* Chick shows retarded growth and abnormal feathering due to vitamin B-6 deficiency. (Courtesy, University of Georgia, Athens)

Vitamin B–12 (Cobalamins)

Vitamin B–12 is found only in animal products and bacterial fermentation products. Consequently, poultry feeds—which consist primarily of plant products—are supplemented with vitamin B–12.

Liver stores of vitamin B–12 may be high enough to sustain adults for several months, but high-protein diets can accelerate this depletion process.

Birds deficient in vitamin B–12 exhibit poor feed conversions, depressed growth, reduced hatchability, and in some cases perosis (slipped tendon). Fatty livers, kidneys, and hearts can be observed in some deficient birds.

Fig. 24–22. Vitamin B–12 deficiency. The perky chick at right and his smaller companion are both 3½ weeks old. *Left:* The small chick, fed a ration deficient in vitamin B–12, weighed 157 g. *Right:* The larger chick, fed the same ration plus vitamin B–12, weighed 280 g. (Courtesy, Merck Chemical Division, Rahway, N.J.)

Unidentified Growth Factors (UGF)

In addition to the vitamins listed in Table 24–21, certain unidentified growth factors (UGF) are important in poultry nutrition. They are referred to as *unidentified* or *unknown* because they have not yet been isolated or synthesized in the laboratory. Nevertheless, numerous reports attest to the favorable responses to the dietary inclusion of these products in poultry rations, including stimulation of growth, increased egg production and hatchability, reduced liver fat, improved product quality, and reduced toxicity of minerals. A diet that supplies the specific levels of all the known nutrients but which does not supply the unidentified factors is inadequate for best performance. Rich sources of unidentified factors are egg yolk, whey, yeast, fish and meat by-products, soybeans, green forages, and fermentation by-products. Most nutritionists recognize the possible benefits of adding some UGF supplementation to the diets of broilers and breeding hens. Commonly, the UGF source is added to the diet at a level of 1 to 3%, although antibiotic fermentation residue products may be used at levels ranging from 2 to 8 lb per ton. This practice may have a twofold advantage: (1) providing possible unidentified growth factor responses, and (2) supplying additional amounts of some of the known vitamins.

(Also see Chapter 13, section on "Unidentified Factors.")

Water

Poultry should have free access to clean, fresh water at all times. It is needed as a solvent, a lubricant, and a temperature control device.

A general rule is that chickens drink approximately twice as much water by weight as the feed they consume.

The daily water consumption of chickens and turkeys is given in Table 24–22.

Of course, water consumption varies according to the kind of feed, temperature, humidity, activity of the bird, and rate of egg production. During hot weather, chickens will consume about twice as much water as they do under conditions of average temperature.

TABLE 24-22
DAILY WATER CONSUMPTION
BY CHICKENS AND TURKEYS OF DIFFERENT AGES[1 2]

	Per 1,000 Birds					
Age	Leghorn-Type Pullets		Chicken Broilers[3]		Turkeys[3]	
(week)	(U.S. gal)	*(liter)*	(U.S. gal)	*(liter)*	(U.S. gal)	*(liter)*
1	5	*19*	5	*19*	10	*38*
2	10	*38*	13	*50*	20	*76*
3	12	*45*	24	*90*	30	*114*
4	17	*64*	37	*140*	40	*151*
5	22	*83*	53	*200*	50	*189*
6	25	*95*	69	*260*	60	*227*
7	28	*106*	85	*320*	75	*284*
8	30	*114*	100	*380*	95	*360*
9	35	*132*			115	*435*
10	38	*144*			125	*473*
12	40	*151*			150	*568*
15	42	*158*			160	*606*
20	45	*170*			200	*757*
			Laying or Breeding			
35	50	*189*			M 240 F 130	*908* *492*

[1]Adapted by the authors from *Nutrient Requirements of Poultry,* 8th rev. ed., NRC, National Academy Press, Washington, D.C., 1984, p. 8, Table 2.

[2]Will vary considerably depending on temperature and ration composition.

[3]Mixed sexes.

Water is the largest single constituent of poultry tissue; it constitutes 85% of the body of a baby chick, 58% of the body of an adult bird, and 66% of an egg. Yet, water is often neglected. Birds can lose 98% of their body fat or 50% of their body protein and still survive. However, a 10% loss in body water causes serious physiological disorders, and 20% loss in body water will cause death.

In addition to being readily available, water quality is important. Water should be tested to determine that salts, pesticides, and microorganisms are at acceptable levels and that the water is palatable to poultry. Water that adversely affects growth, reproduction, or productivity should not be used.

Poultry producers should be concerned about the concentration of the following components of water:

1. **Turbidity.** This may be due to silt, clay, algae, or organic matter. Muddy water is not only unpalatable to the birds, but it may cause clogging of the water system.

2. **Taste, odor, and color.** Poultry water should be tasteless, odorless, and colorless. Taste is primarily due to the presence of salts in the water; ferrous and manganese sulfates give a bitter taste to water. A rotten egg odor is due to the presence of hydrogen sulfide. Reddish or brownish color is due to the presence of iron; and bluish color is caused by copper.

3. **Total dissolved solids.** Total dissolved solids is an expression of total ions (both cations and anions) in the water.

4. **Hardness.** Hardness is an expression of the total calcium and magnesium content of water in calcium carbonate equivalent. Water that contains high concentrations of calcium and magnesium is not suitable for poultry.

5. **pH.** The pH is an expression of acidity or alkalinity. Water with a pH of 7 is neutral, below 7 is acidic, and above 7 is alkaline. A low pH makes the water unpalatable and may lower performance and egg quality.

6. **Sulfate.** A high sulfate content is laxative and causes wet litter.

7. **Nitrate and nitrite.** Nitrate may be due to contamination by chemical fertilizer or manure, or due to the water dissolving rocks that contain nitrate. Nitrate as such is not toxic, but the microorganisms in the intestinal tract convert it to nitrite, which is very toxic.

8. **Sodium and chlorine.** High concentrations of sodium and chlorine in the water increase the water consumption and lead to wet litter.

9. **Toxic elements.** Water may be contaminated with toxic elements, such as arsenic, lead, and/or selenium.

10. **Microbial contamination.** Different types of microbes may be present in water. The number of microbes is not as important as the type. Where microbial contamination of water exists, the water should be chlorinated.

11. **Industrial contamination.** Sometimes, water is contaminated with industrial residues, and is not suitable for poultry.

Water made unacceptable by one or more of the above components may produce inferior results in poultry growth, egg production, and egg quality.

Ideally, the water provided to poultry should be suitable for human consumption.

(Also see Chapter 4, section on "Water.")

Additives

Modern poultry feeds commonly contain one or more nonnutritive additives. These additives are used for a variety of reasons. They are not nutrients, but some of them improve production under certain circumstances. Others prevent rancidity in the feed. There is no evidence of a nutritional deficiency when they are omitted from a ration. Among such additives are the following:

1. **Antibiotics.** The primary reasons for using antibiotics in poultry feeds are for their growth-stimulating and improved efficiency of feed conversion effects, for which purpose they are generally used in both broiler and market turkey rations. Also, egg production is frequently improved by dietary supplementation with antibiotics. The reasons for the beneficial effects of antibiotics still remain obscure, but the best explanation for their growth-stimulating activity is the disease level theory, based on the fact that antibiotics have failed to show any measurable effect on birds maintained under germ-free conditions.

Antibiotics are generally fed to poultry at levels of 0.5 to 25 mg/pound of diet, depending upon the particular antibiotic used. Higher levels of antibiotics (100 to 400 g per ton of feed) are used for disease control purposes.

High levels of calcium in a laying mash will inhibit assimilation of certain tetracycline-type antibiotics to the bloodstream and reduce their effectiveness.

2. **Antifungal agents.** Feeds provide an excellent environment for the growth of fungi (molds), such as *Aspergillus flavus, Fusarium,* and *Candida albicans,* which are detrimental to the health of poultry. *Aspergillus flavus* produces a potent toxin which is referred to as aflatoxin. *Candida albicans* is the causative agent of a condition in poultry called thrust or moniliasis.

Several compounds have been introduced as feed additives to prevent the growth of molds in feeds. The products of choice are: propionic acid, acetic acid, and sodium propionate.

For an in-depth discussion of fungi (molds), see Chapter 5, Table 5-3—Mycotoxins [Toxin producing Fungi or Molds]; and Chapter 13, section on "Mold [Fungi] Inhibitors.")

3. **Antioxidants.** *Antioxidants are compounds that prevent oxidative rancidity in polyunsaturated fats.* They are used to prevent rancidity in poultry feeds. The antioxidants which are presently accepted for addition to fat in poultry feeds are butylated hydroxyanisole (BHA), butylated hydroxytoluene (BHT), and ethoxyquin. They are used at a level of ¼ lb per ton.

Antioxidants are added to feed fats to stabilize them against rancidity. BHT and BHA are commonly used to stabilize fat.

(Also see Chapter 13, section on "Antioxidants.")

4. **Arsenicals.** These products exert much the same effects as the antibiotics; hence, they are often added to poultry feeds to improve performance. It would appear that the action of arsenicals and antibiotics is very similar, since the effects of the two are not considered to be additive. Arsanilic acid and sodium arsanilate are the most widely used growth-promoting arsenicals. They are FDA-approved for use alone or in certain drug combinations for chickens and turkeys. When used according to directions, they increase rate of gain and improve feed efficiency of chickens and turkeys; improve pigmentation in growing chickens and turkeys; and increase egg production in layers.

(Also see Chapter 13, section on "Arsenicals.")

5. **Drugs for disease prevention and control.** Poultry rations frequently contain drugs designed to prevent specific

diseases. For example, a wide variety of chemical substances, sold under various trade names, is available for use in the prevention of coccidiosis. These drugs are known as coccidiostats.

Turkey rations are frequently formulated with drugs for the prevention of blackhead. This class of drugs, known as histomonostats, also contains a wide variety of chemical substances sold under various trade names.

(Also see section in this chapter on "Salmonella.")

6. **Flavor additives.** *Flavoring agents are feed additives that are designed to increase palatability and feed intake.*

Chickens have the ability to differentiate between sucrose solutions, for which they show a preference, and saccharin solutions which they avoid. Other studies indicate that chickens possess a sense of taste, but little or no ability to smell.

Flavor additives may be useful in preventing early "starve-outs" and in keeping birds on feed during periods of disease and stress. Also, there is some indication that feed flavors may improve feed intake and production when such unpalatable feeds as blood meal, fish solubles, and fermentation by-products are fed, or when dusty and finely ground wheat or milo are fed. Some poultry producers in the South feel that flavors improve the feed intake and production of layers and broilers in hot weather. But additional research work is needed before flavors can be established as advantageous feed additives.

(Also see Chapter 13, section on "Flavoring Agents.")

7. **Grit.** The use of grit is controversial. Some research indicates that hens fed an all-mash ration do not need grit, but there is evidence that as a component of or supplement to all-mash it may improve feed utilization and increase production under certain conditions. The primary function of grit is to help the gizzard grind food materials that pass through it. It is definitely needed when birds consume whole grains or coarse, fibrous feedstuffs. Crushed granite or other hard, insoluble material can be used for grit.

(Also see Chapter 13, section on "Grit.")

8. **Xanthophylls.** Feeds that contain large amounts of xanthophylls produce a deep yellow color in the beak, skin, shanks, feet, fat, and egg yolks of poultry. The consumer associates this pigmentation with quality and in many cases is willing to pay a premium price for a bird of this type. Also, processors of egg yolks are frequently interested in producing dark-colored yolks to maximize coloration of egg noodles and other food products. The latter can be accomplished by adding about 60 mg of xanthophyll per kilogram of diet. In recognition of these consumer preferences, many producers add ingredients that contain xanthophylls to poultry rations.

It is not necessary to incorporate high levels of xanthophyll in starter and grower rations. Low levels can be maintained through these periods of feeding, but finishing rations should contain high levels of these pigment-producing compounds. Alfalfa meal, yellow corn, and corn gluten meal are sources that are commonly used. Common dried algae and marigold petal meal are even richer sources of xanthophyll. The synthetic carotenoid canthaxanthin (red) is approved by the Food and Drug Administration and is now widely used by producers in the United States.

Table 24–23 lists xanthophyll-rich feedstuffs, along with the quantities that they contain.

TABLE 24–23
XANTHOPHYLL-RICH FEEDSTUFFS[1]

Feedstuffs	Xanthophylls	
	(mg/lb)	*(mg/kg)*
Alfalfa meal, 17% protein	118	*260*
Alfalfa meal, 20% protein	127	*280*
Alfalfa meal, 22% protein	150	*330*
Alfalfa juice protein, 40% protein	363	*800*
Algae, common, dried	908	*2,000*
Corn, yellow	8	*17*
Corn gluten meal, 41% protein	79	*175*
Corn gluten meal, 60% protein	132	*290*
Marigold petal meal	3,178	*7,000*

[1]Adapted by the authors from *Nutrient Requirements of Poultry,* 8th rev. ed., NRC, National Academy Press, Washington, D.C., 1984, p. 9, Table 3.

(Also see Chapter 13, section on "Additives That Enhance Market Value.")

FEEDS FOR POULTRY

A wide variety of feedstuffs can be, and is, used in poultry rations. Broadly speaking, these may be classed as energy feedstuffs, protein supplements, mineral supplements, and vitamin supplements.

1. **Energy feedstuffs.** The major energy sources of poultry feeds are the cereal grains and their by-products and fats. Corn is the most important grain used by poultry, supplying about one-third of the total feed which they consume.

Oats, barley, and the sorghum grains are also used extensively in poultry rations. Oats are lower in energy than corn and are generally too expensive for broiler and layer rations. But oats can be used very effectively in feeds for replacement birds. Barley is less palatable than corn and is lower in vitamin A and energy. The sorghum grains can be readily substituted in the place of corn as an energy feed; hence, in the southern states, they are being used extensively. Wheat is used when the price is right. Although millet is seldom used in poultry rations, it can be freely substituted for corn.

Molasses can be used effectively in poultry rations provided that its level of usage is closely monitored. Excessive amounts cause wet droppings.

Rye is seldom used in poultry rations because it tends to depress growth and cause sticky, pasty droppings when used at moderate to high levels.

Animal and vegetable fats are now used extensively in poultry feed. In addition to their high energy value, fats reduce the dustiness of feed mixtures, increase their palatability, and improve the texture and appearance of the feed. However, the use of fats in poultry feeds requires good mixing equipment. Also, it is necessary that the fat be properly stabilized in order to prevent rancidity.

Many other energy feedstuffs, including a great array of milling by-products (for example, the corn gluten and bran feeds and the wheat processing by-products), are used in poultry feeds.

2. **Protein supplements.** The usefulness of a protein feedstuff depends upon its ability to furnish the essential amino acids required by the bird, the digestibility of the protein, and the presence or absence of toxic substances. As a general rule, several different sources of protein produce better results than single protein sources. Both animal and plant protein supplements are used for poultry. Most of the protein supplements of animal origin contribute minerals and vitamins which significantly affect their value in poultry rations, but they are generally more variable in composition than the plant protein supplements.

Among the animal protein supplements commonly used in poultry rations are meat and fish by-products, milk by-products, and such miscellaneous animal by-products as blood meal, hydrolyzed poultry feathers, and poultry by-product meal.

Fish meals—primarily herring, menhaden, and anchovy—are used extensively in poultry rations, since they are high in protein and extremely well balanced in essential amino acid composition. However, since fish meals are high in fat, they tend to create a fishy flavor in meat and eggs when used in large amounts. This, together with the high cost of fish meal, restricts its use to one of a secondary source of protein supplementation.

The common plant protein supplements used in poultry feeding include the oilseed meals (soybean meal, cottonseed meal, peanut meal, rapeseed meal, and limited amounts of linseed meal), corn gluten meal, and alfalfa meal and other legume meals. Soybean meal is, by far, the most widely used protein supplement in poultry rations.

Care should be exercised when cottonseed meal is used in poultry rations—especially for laying birds. Gossypol—a compound found in the gland of the cottonseed—produces a mottling of egg yolks, even in extremely small amounts. Cyclopropenoid fatty acids—another class of compounds found in cottonseed meal—impart a pink color to egg whites. Because of these compounds, cottonseed meal is generally not used in laying rations, but it can be used effectively in growing and replacement rations. Linseed meal can be used effectively in limited amounts, but it may depress growth and cause diarrhea if fed at high levels. It is recommended that linseed meal not exceed 3 to 5% of the ration. Corn gluten meal and alfalfa meal are used extensively, both for their protein value, and for their high content of carotenoids—compounds that impart a deep yellow pigmentation to the skin.

3. **Mineral supplements.** Mineral supplements are required by poultry for skeletal development in growing birds, for eggshell formation in laying hens, and for certain other regulatory processes in the body.

The common calcium supplements used in poultry feeding are ground limestone, crushed oystershells or oystershell flour, bone meal, calcite, chalk, and marble.

Most of the phosphorus in plant products is in organic form and not well utilized by young chicks or turkey poults. Hence for poultry, emphasis is placed upon inorganic phosphorus sources in feed formulation. Bone meal, dicalcium phosphate, defluorinated phosphate, and colloidal phosphate are used where phosphorus is needed in the ration.

Salt is added to most poultry rations at a 0.2 to 0.5% level. Too much salt will result in increased water consumption and wet droppings.

Fig. 24-23. Broiler breeders. For high hatchability and good development of embryos, breeders require adequate amounts of the mineral manganese and of vitamins A, D, E, niacin, pantothenic acid, riboflavin, and vitamin B-12. (Courtesy, P. L. Potts, Sr., Ph.D., Poultry Department, California Polytechnic University, San Luis Obispo)

4. **Vitamin supplements.** A great many vitamins are important in present-day poultry feed formulation. Formerly, a wide variety of crude feedstuffs were added to poultry formulas primarily for their vitamin content. Today, many of these have been replaced by special vitamin supplements, which in many cases are chemically pure sources that need to be used only in very minute amounts. In modern poultry feed formulation and production, premixes often represent the common sense approach to providing both mineral and vitamin needs for poultry.

Presence of Substances Affecting Product Quality

The composition of the feed can affect the product. The color of the skin or shanks of a broiler or of the yolk of an egg is primarily due to the carotenoid pigments consumed in the feed. Corn, alfalfa meal, and corn gluten meal are the main feeds used to contribute these pigments.

Screw process cottonseed meal, which is high in gossypol, when fed to laying hens may cause egg yolk discoloration in stored eggs. Some fish products may impart off-flavors to poultry meat or eggs. Thus, certain feedstuffs may be undesirable simply because of the effect they produce on the end product.

• **Cholesterol in eggs**—Without doubt, the per capita consumption of eggs has decreased because of relating the fat and cholesterol content of eggs to heart disease, although changes in breakfast eating habits have also contributed to the decline. But scientists and producers are coming to the rescue!

Recently, low-cholesterol eggs have been produced (1) by manipulating the feed ingredients in layer rations, in some cases with added lighting changes; (2) by additives to the ration; and (3) by solvent extraction of the yolk. At this stage, most of these techniques are secretive; some of them are in the process of being patented. More experimental work is needed; and the profitability (the price that consumers are willing to pay for low-cholesterol eggs, less the cost of production) must be determined.

(Also see Chapter 13, section on "Additives That Enhance Market Value.")

FEED PREPARATION

The usual end product resulting from mixing poultry feedstuffs is a ground feed known as a mash. While this mash is usually in the form of a ground mixture, it can be processed to produce pellets or crumbles.

Pellets or crumbles cost slightly more than the same ration in mash form. Yet, they are used for broilers and turkeys because of improved feed efficiency (fewer pounds of feed to produce a pound of gain).

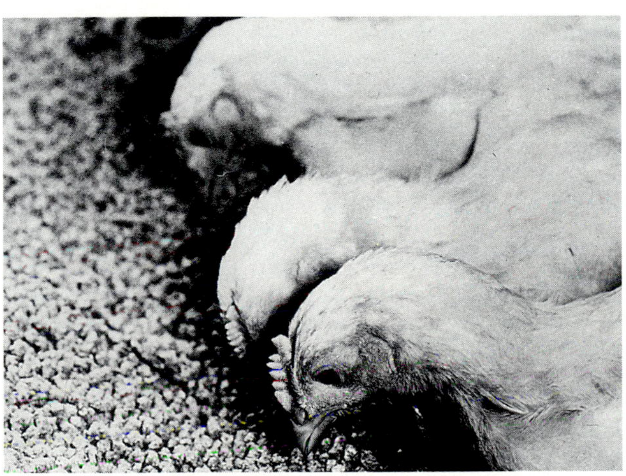

Fig. 24-24. Six-week-old White Rock broilers eating pellets. (Courtesy, Ralston Purina Company, St. Louis, Mo.)

(Also see Chapter 14, Feed Processing.)

POULTRY RATIONS

In 1984, 109.6 million tons of commercial feeds were produced in the United States, of which 87% was primary feed (mixed feed) and 13% consisted of secondary feed (feed to which a formula feed supplement is added at the rate of 100 lb per ton or more). A total of 35.5%—more than one-third—of the primary feed was fed to poultry.

The poultry producer has the following alternatives for purchasing and preparing rations:

1. Purchase of a commercially prepared complete feed.
2. Purchase of a commercially prepared protein supplement, reinforced with vitamins and minerals, which may be blended with local grain.
3. Purchase of a commercially prepared vitamin-mineral premix which may be mixed with an oil meal, and then blended with local grain.
4. Purchase of individual ingredients (including vitamins and minerals) and mixing the feed from the "ground" up.

Fig. 24-25. Bulk feed bins at a poultry operation. (Courtesy, Department of Poultry Science, The University of Georgia, Athens)

Factors Involved in Formulating Poultry Rations

Before anyone can intelligently formulate a poultry ration, it is necessary to know (1) the nutrient requirements of the particular birds to be fed; (2) the availability, nutrient content, and cost of feedstuffs; (3) the acceptability and physical condition of feedstuffs; (4) the average daily consumption of the birds to be fed; and (5) the presence of substances harmful to product quality.

Formulating Rations

The increasing complexity of poultry rations, along with larger and larger enterprises, makes it imperative that producers who choose to mix feed be absolutely sure that they will have a nutritionally balanced and adequate ration.

When fed free-choice, birds tend to eat to satisfy their energy requirements. Consequently, it is possible, within limits, to regulate the intake of all nutrients, except water, by including them in the diet in specific ratios to available energy. Thus, the energy content of the diet must be considered in formulating to meet a desired intake of all essential nutrients other than energy itself.

The larger commercial feed companies, and the larger poultry producers who do their own mixing or formulating, generally rely on the services of a nutritionist and the use of a computer in formulating their rations. **(NOTE:** See Chapter 18, section on "Computer Methods" for computer ration formulation.) Even though they are more time-consuming, and fewer factors can be considered simultaneously, a good job can be done in formulating rations by the hand method.

(Also see Chapter 18, Feeding Standards/Ration Formulation.)

FEED SUBSTITUTION TABLE

Successful poultry producers are well educated in economics. They know that feed conversion rate alone does not determine success. The cost of producing each pound of broiler, turkey, or dozen eggs is the determinant of success. For this reason, they want to obtain a balanced ration that will make for the highest net returns. To help poultry producers decide what feeds might be interchangeable with the ones they are using and to what extent certain feeds can be replaced, the authors prepared Table 24–24, Feed Substitution Table for Poultry.

TABLE 24–24
FEED SUBSTITUTION TABLE FOR POULTRY (AS-FED)

Feedstuff	Relative Feeding Value (lb for lb) in Comparison With the Designated (underlined) Base Feed Which = 100	Maximum Percentage of Base Feed (or comparable feed or feeds) Which It Can Replace for Best Results	Remarks
GRAINS, BY-PRODUCT FEEDS:			
Corn, No. 2	100	100	Corn is the most widely used grain in poultry feed.
Bakery wastes	75	50	Bakery wastes are very similar to cereal grains in composition.
Barley	80–85	50	Barley is very low in vitamin A; less palatable than corn.
Beans (cull)	90	5	
Cassava	85	20	Extremely low in protein.
Hominy	95	50	Not generally used in poultry rations; low energy value, but is good source of linoleic acid.
Millet	95–100	65	Best when used as a 50:50 mix with barley, corn, or oats; can be used as a scratch feed.
Molasses	70	5	High levels of molasses will produce wet droppings.
Oats	70–80	50	Usually too expensive for broiler and layer rations, but is used extensively in replacement rations.
Peas, dried	90–100	5–10	
Rice, rough	80–85	20–50	Rice may be deficient in vitamin A.
Rice, bran	50	5–10	Rice bran is high in fat and susceptible to oxidation.
Rice polishings	85–90	5–10	
Rye	90	25–30	Rye is not a very good poultry feed as it can depress growth and cause sticky droppings.
Sorghum	100	100	
Triticale	80–90	30	Triticale is somewhat lower in feeding value than either wheat or corn.
Wheat	90–95	100	Wheat is the most variable of the cereal grains in protein; very low in xanthophylls; should be processed to increase palatability.
Wheat bran	75	10–15	
PROTEIN SUPPLEMENTS:			
Soybean meal (48%)	100	100	Well balanced in amino acids. Best quality of all plant protein supplements. Very palatable.
Babassu oil meal	50	20	Similar to copra meal.
Blood meal, flash- or ring-dried	120	5–20	Excellent source of lysine.
Copra meal (coconut meal)	50	25	Copra meal should be supplemented with some animal protein.
Corn gluten meal	50–75	25	Animal protein must be used along with corn gluten meal.
Cottonseed meal	85	80[1]	Gossypol, a compound found in cottonseed meal, can cause discoloration of egg yolks. The addition of iron in the ration may minimize the dangers of gossypol poisoning. Cyclopropenoid fatty acids in cottonseed meal can cause discoloration of egg white. Glandless cottonseed meal is recommended.
Feather meal	50	5	Feather protein is deficient in methionine, lysine, histidine, and tryptophan.
Fish meal, anchovy and menhaden	115	50–65	Expensive. Excellent balance of amino acids and good source of calcium and phosphorus. Most poultry rations incorporate some fish meal at levels of about 2–5% of the ration. Fish meal can impart fishy flavors if fed at too high levels.
Linseed meal	80	10	Linseed meal depresses growth and can cause diarrhea. Levels should be restricted to a maximum of 5% of the ration.
Meat and bone meal	100	20–50	Tends to be deficient in tryptophan.
Meat meal (50–55% protein)	100	50–65	Meat meal is high in phosphorus and low in methionine and cystine. It is recommended that the maximum level of usage not exceed 10% of the ration.
Peanut meal (41% protein)	95	75–100	If peanut meal is to be substituted for soybean meal, lysine must be added.
Poultry by-product meal	100	20–50	
Rapeseed meal	85–90	33⅓	It may contain goitrogenic compounds that may be hazardous. But the newer varieties of rape (known as canola) are much lower in goitrogenic compounds.
Safflower seed meal (decorticated)	95	50–100	Safflower seed meal can be incorporated at levels up to 15% of the ration. Deficient in lysine and methionine.
Sesame meal	95–100	100	Extremely deficient in lysine.
Sunflower seed meal	95–100	100	
Yeast, torula	100	60	Yeast is a good source of the B-complex vitamins.

[1]When cottonseed meal is substituted for soybean meal at this level, it must be degossypolized.

SPECIAL FEEDING PROGRAMS; FEED INTAKE AND SUGGESTED RATIONS

The nutritive requirements of poultry vary according to species (between chickens, turkeys, ducks, and geese), according to age, and according to the type of production—whether the birds are kept for meat production, layers, or breeders. For this reason, many different rations are required.

To be successful, rations must meet the nutritive requirements of the birds to which they are fed. The National Research Council nutritive requirements are given in Tables 24-4 through 24-18. In using these tables to formulate practical rations, it must be remembered that they are minimum requirements; which means that they do not provide for any margins of safety. Further, the protein-energy relationships shown therein should be retained.

Birds eat primarily to satisfy their energy needs. Also, the temperature of the environment has an important influence on feed intake—the warmer the environment, the less the feed intake. Therefore, the requirement of all nutrients, expressed as a percent of the diet, is dependent upon the environmental temperature. Other factors affecting feed intake are health, genetics, form of feed, nutritional balance, stress, body size, and rate of egg production or growth.

It is believed that feed intake is, in part, controlled by the amount of glucose in the blood. It has been observed that the addition of fat to the diet results in overconsumption on the part of the bird. As a result, some variation in the protein:energy ratio may be tolerated. In general, when free-choice feeding dietary protein levels that are low in relation to energy, fat deposition is markedly increased; with higher levels of protein, less fat is deposited. Increasing the protein level above that required for maximum growth rate reduces fat deposition still further.

FEEDING CHICKENS

Today, most commercial chickens are provided complete rations with all the needed nutrients available in the quantiies necessary, and with the feed and water available on an *ad libitum* or free-choice basis. Formulations are varied according to the type of production—whether the birds are bred and kept for egg production (layers), hatchery production (breeders), or meat production (broilers). Also, consideration is given to age; sex; stage and level of production; temperature, disease level, and other stresses; management; and other factors.

Feeding Layers

Chickens kept for the production of eggs for human consumption (Leghorn-type) have small body size and are prolific layers. They are generally fed *ad libitum* during the laying period. Occasionally, layers will consume excess feed during the latter phases of egg production (following peak production) with resultant obesity, reduced feed efficiency, and higher incidence of fatty liver syndrome. When this situation is detected, limiting feed intake to 85 to 95% of full feed consumption is desirable. Data on feed consumption in a particular flock, together with information on body weight, ambient temperature, and rate of egg production, may be used to determine the degree of feed restriction.

Fig. 24-26. A modern cage facility in a commercial laying house. Note the 3 tiers of cages, with a feed trough and an automatic egg collecting belt in front of each row of cages. (Courtesy, P. L. Potts, Sr., Ph.D., Poultry Department, California Polytechnic University, San Luis Obispo)

These additional pointers are pertinent to feeding layers:

1. The largest item of cost in the production of eggs is feed. It normally constitutes 50 to 70% of the total cost; though, in exceptional cases, it can run as low as 45% or as high as 75%.

2. In the final analysis, the objective of feeding laying hens is to produce a dozen eggs of good quality at the lowest possible feed cost. Thus, the actual cost of the feed that a layer eats in producing a dozen eggs—not the price per pound of feed—determines the economy of the ration.

3. Feed consumption per bird varies primarily with egg production and body size. It is also influenced by the health of the birds and the environment, especially the temperature.

4. Normally, a mature Leghorn, or other lightweight bird, eats about 82.5 lb of feed per year and produces about 22 doz eggs in that same period of time. Hence, it requires about 3.75 lb of feed to produce 1 doz eggs. A bird of the heavier breeds eats 95 to 115 lb of feed per year; hence, they are not as efficient egg producers. With lightweight layers, the producer should aim for a feed efficiency of 3.5 to 4.0 lb of feed per dozen eggs.

5. Feed may affect egg quality. Deficiencies of calcium, phosphorus,manganese, and vitamin D_3 lead to poor shell quality. Yolk color is almost entirely dependent on the bird's diet. Low vitamin A levels may increase the incidence of blood spots.

Example rations for *layers* producing eggs for human consumption are given in Table 24-25, p. 1044.

TABLE 24-25a
EXAMPLE LAYER RATIONS[A,B,1]

Ingredient	Protein Level of Rations[C]											
	15%		16%		17%		18%		19%		20%	
	(lb)	*(kg)*	(lb)	*(kg)*	(lb)	*(kg)*	(lb)	*(kg)*	(lb)	*(kg)*	(lb)	*(kg)*
Ground yellow corn[2,3]	1,457	*662.3*	1,403	*637.7*	1,339	*608.6*	1,242	*564.5*	1,177	*535*	1,120	*509.1*
Alfalfa meal, 17%	25	*11.3*	25	*11.3*	25	*11.3*	25	*11.3*	25	*11.3*	25	*11.3*
Soybean meal, dehulled	292.2	*132.8*	340.6	*155*	393.6	*179*	451.6	*205.3*	504.6	*229.4*	554	*251.8*
Meat and bone meal, 47%[5]	50	*23.0*	50	*23.0*	50	*23.0*	50	*23.0*	50	*23.0*	50	*23.0*
DL-Methionine or equivalent	1.0	*0.5*	1.0	*0.5*	1.0	*0.5*	1.0	*0.5*	1.0	*0.5*	1.0	*0.5*
Dicalcium phosphate[6]	9	*4.1*	8	*3.6*	8	*3.6*	7	*3.1*	7	*3.1*	7	*3.1*
Ground limestone[7]	159	*72.3*	159	*72.3*	159	*72.3*	174	*79.1*	174	*79.1*	174	*79.1*
Iodized salt[4]	7	*3.1*	7	*3.1*	7	*3.1*	7	*3.1*	7	*3.1*	7	*3.1*
Stabilized yellow grease, or equivalent	—	—	7	*3.1*	18	*8.2*	43	*19.5*	55	*25*	62	*28.2*
Antioxidant	[9]	[9]	[9]	[9]	[9]	[9]	[9]	[9]	[9]	[9]	[9]	[9]
Mineral and vitamin supplements:[12]												
Calcium pantothenate (mg)	5,000		4,500		4,500		4,500		4,000		4,000	
Manganese[11] (Mn) (g)	52		52		52		52		52		52	
Selenium[25] (Se) (mg)	90.8		90.8		90.8		90.8		90.8		90.8	
Zinc[17] (Zn) (g)	16		16		16		16		16		16	
Vitamin A (IU)	6,000,000		6,000,000		6,000,000		6,000,000		6,000,000		6,000,000	
Vitamin D_3 (IU)	2,000,000		2,000,000		2,000,000		2,000,000		2,000,000		2,000,000	
Vitamin K[20]	—		—		—		—		—		—	
Choline (mg)	274,000		231,000		184,000		140,000		94,000		50,000	
Niacin (Nicotinic Acid, Nicotinamide) (mg)	12,000		12,000		12,000		12,000		12,000		12,000	
Riboflavin (Vitamin B-2) (mg)	2,000		2,000		2,000		2,000		2,000		2,000	
Vitamin B-12 (Cobalamins) (mg)	6		6		6		6		6		6	
Totals[21]	2,000	*909.4*	2,000	*909.3*	2,000	*909.3*	2,000	*909.4*	2,000	*909.4*	2,000	*909.1*
Calculated analysis:[27]												
Metabolizable energy (kcal)	1,306.2	*2,873.6*	1,303.9	*2,868.6*	1,303.4	*2,867.5*	1,304.1	*2,869*	1,304.5	*2,870*	1,301	*2,862.2*
Protein (%)	15.08		16.02		17.03		18.01		19.03		20.00	
Lysine (%)	0.68		0.75		0.83		0.91		0.98		1.06	
Methionine (%)	0.31		0.32		0.33		0.34		0.35		0.36	
Methionine + cystine (%)	0.54		0.57		0.59		0.62		0.64		0.67	
Fat (%)	3.29		3.54		3.98		5.05		5.54		5.76	
Fiber (%)	2.20		2.20		2.21		2.18		2.18		2.19	
Calcium (Ca) (%)	3.25		3.24		3.24		3.50		3.50		3.51	
Total phosphorus (P) (%)	0.52		0.52		0.53		0.52		0.53		0.54	
Available phosphorus[13] (P) (%)	0.45		0.45		0.45		0.45		0.45		0.45	
Vitamins (units or mg/lb or kg):												
Vitamin A activity (IU)	5,904	*12,988.8*	5,842	*12,852.4*	5,770	*12,694*	5,660	*12,452*	5,586	*12,239.2*	5,522	*12,148.4*
Vitamin D_3, added (IU)	1,000	*2,200*	1,000	*2,200*	1,000	*2,200*	1,000	*2,200*	1,000	*2,200*	1,000	*2,200*
Choline (mg)	500.13	*1,100.3*	500.34	*1,100.7*	500.05	*1,100.1*	500.39	*1,100.9*	500.48	*1,101.1*	500.66	*1,101.5*
Niacin (Nicotinic Acid, Nicotinamide) (mg)	15.40	*33.9*	15.48	*34.1*	15.55	*34.2*	15.50	*34.1*	15.56	*34.2*	15.64	*34.4*
Pantothenic Acid (Vitamin B-3) (mg)	5.01	*11*	4.88	*10.7*	4.99	*10.9*	5.07	*11.2*	4.95	*10.8*	5.01	*11*
Riboflavin (Vitamin B-2) (mg)	1.84	*4.0*	1.85	*4.1*	1.86	*4.1*	1.86	*4.1*	1.87	*4.1*	1.88	*4.1*

[A]Adapted by the authors from *NECC Chicken and Turkey Rations,* prepared by the New England College Poultry Conference Board, by poultry specialists from, and distribution by, the New England Land-Grant Universities: University of Connecticut, Storrs; University of Maine, Orono; University of Massachusetts, Amherst; University of New Hampshire, Durham; University of Rhode Island, Kingston; and University of Vermont, Burlington.

[B]See footnotes following Table 24-31, p. 1054.

[C]Six rations varying in protein levels are presented. The ration which best meets the needs of a particular flock may be determined from the factors presented below.

See "Factors to Consider in Determining Which Layer Ration to Feed," which follows:

FACTORS TO CONSIDER IN DETERMINING WHICH LAYER RATION TO FEED

A. Feed consumption of the hens must be known to select the appropriate layer ration. Feed consumption may vary depending on type of bird, bird weight, pen temperature, rate of production, energy content of ration, disease problems and many other factors.

B. **Table 24-25b.** SUGGESTED MINIMUM DAILY PROTEIN INTAKE.

Production Status	Daily Protein Required*
	(g)
Coming into production	19–20
90% hen day egg production	18–19
80% hen day egg production	17–18
70% hen day egg production	16–17
60% and under egg production	15–16

*Under conditions of higher production, disease, or stress, add 2 g protein to the above.

D. When feed consumption is not known, we suggest using not less than 17% protein in the diet.

WEIGHT CONVERSION TABLE

1 pound	= 453.57 grams or	0.4536 kilogram	
1 ounce	= 28.349 grams	1 kilogram = 2.2046 pounds	
1 kilogram (kg)	= 1,000 grams	1 gram (g) = 1,000 milligrams	
1 milligram (mg)	= 100 micrograms (gammas)		
1 part per million (ppm)	= 1 milligram/kilogram or 1 microgram/gram		
1 part per million (ppm)	= 0.454 mg/lb or 0.907 g/ton		

C. **Table 24-25c.** A GUIDE FOR ESTIMATING DIETARY PROTEIN LEVEL BASED ON FEED INTAKE.

Lbs Feed Consumed Per/100 Birds/Day	% Protein In Laying Ration					
	15	16	17	18	19	20
	Protein Intake					
	(g)	(g)	(g)	(g)	(g)	(g)
18–19	12	13	14	15	16	17
20–21	14	15	15	16	17	18
22–23	15	16	17	18	19	20
24–25	16	17	18	20	21	22
26–27	18	19	20	21	22	24
28–29	19	20	22	23	24	26
30–31	20	22	23	25	26	27

Example—Your birds are consuming 24 lb per 100 birds per day and laying at a rate of 80% hen day egg production. Their need for protein is 17–18 g (Table 24-25b). Therefore to supply 18 g of protein you should use the 17% protein ration (Table 24-25c).

PHASE FEEDING

The trend is to phase feeding laying hens. Phase feeding refers to changes in the laying hen's diet (1) to adjust to age and state of production of the hen, (2) to adjust for season of the year and for temperature and climatic changes, (3) to account for differences in body weight and nutrient requirements of different strains of birds, and (4) to adjust one or more nutrients as other nutrients are changed for economic or availability reasons. Research has shown, for example, that a hen laying at the rate of 60% has different nutritional requirements than one laying at the rate of 80%; hens have different requirements in summer and in winter; a 24-week-old layer has different needs than one 54 weeks old. The main objective, therefore, of phase feeding is to reduce the waste of nutrients caused by feeding more than a bird actually needs under different sets of conditions. In this way, feed efficiency can be improved and the cost of producing a dozen eggs reduced.

A phase feeding program for laying hens generally calls for use of a rather high-protein feed (usually 17 to 18%) from the onset of egg production through the peak production period. Thereafter, a lower level of protein (about 16%) is fed for the next 5 or 6 months, followed by still lower levels (usually 15%) until the laying period is completed. This general plan takes age into consideration, but for greatest benefits other factors will also need to be considered.

Although phase feeding has its advantages, it does present some problems: it is a complicated procedure, it necessitates a knowledgeable poultry producer, and it requires more bulk bins, closer check on feed deliveries, etc.

Phase feeding is practiced widely in commercial operations. Although it does not promise to bring about large increases in egg production, it can help production reach a higher peak and sustain it longer if other conditions are right. Most of all, phase feeding offers a good potential for lowering costs and increasing income. Like many other developments, it favors the larger operator.

MOLTING

Molting is the process of shedding and renewing of feathers. It is a normal process of chickens and other feathered species; and it occurs in both sexes. In the wild state, birds usually shed and renew old, worn plumage before the beginning of the cold weather and their migratory flights. Since undomesticated birds lay only a few eggs, molting and reproduction are not usually associated.

Chickens kept for commercial egg production have a different molting pattern. They have been bred for high performance, and their environment, with respect to temperature and light, is usually modified to remove major seasonal influences. A natural molt does not normally occur until after a period of 8 to 12 months of egg production. If nothing is done to alter the normal molting cycle, it requires about 4 months for a hen to drop her feathers and grow a new set. It is possible, however, to speed the process through a program of forced molting, thereby recycling the hens for another period of egg production and improved egg quality.

Molting is controlled by the gonads and the thyroid gland and is associated with a drop in estrogen levels and a decreasing rate of egg production. Egg production is not greatly affected by the process of natural molting, but molting is prolonged when the birds are kept in production. High producers tend to molt late; but once production ceases, molting is rapid.

Several factors affect the onset and length of molting; among them, (1) weight and physical condition of the bird, (2) length of light exposure, (3) nutrition of the bird, and (4) environmental influences, such as temperature and humidity. Thus, if one drastically reduces the amount of light or starves the birds in such a way as to knock them out of production, molting can be induced or speeded up.

In a natural molt, a chicken loses feathers from various sections of its body in the following order: head; neck; feather tracts of the breast, thighs, and back; and wing and tail feathers. Some birds molt earlier than others; and some molt more slowly than others. A high producing flock generally molts late and rapidly.

Decreasing day length is the normal trigger for molting. Therefore, lighting programs for layers should provide either constant or increasing day length. Minor stresses, such as temporary feed or water shortage, disease, cold temperature, or sudden changes in the lighting program, can also initiate a partial or premature molt, evidenced by a considerable number of feathers on the floor of the poultry house. In these cases, chickens lose some head and neck feathers. If the molt continues beyond this point, a more severe drop in performance can be expected.

• **Force (induced) molting**—Under force molting, a layer flock is induced to shed and replace its feathers at a time selected by the flock manager. This may come near the end of a normal laying cycle, or a flock may be force molted earlier as part of a multiple molting program.

An induced molt causes all of the hens in a flock to go out of production for a period of time. During this period, regression and rejuvenation of the reproductive tract occur, accompanied by the loss and replacement of feathers. After a molt, the hen's reproductive rate usually peaks slightly below the previous peak rate, and egg quality is improved.

Forced molting has been practiced in a limited way since the turn of the century, but always as economic conditions dictated. Today, it is estimated that approximately 60% of the nation's laying flocks are recycled, and that 90% or more of the California flocks are recycled. The first and most important reason for induced molting is that it usually improves profits and, therefore, is part of a planned replacement policy. The question of whether it pays to recycle depends primarily upon the relative performance of all-pullet versus recycled flocks.

• **Types of recycling**—Egg producing hens may be molted one or more times, giving rise to the following types of recycling programs:

1. **Two-cycle molting program.** This involves one molt and two cycles of egg production. The hens are force molted after about 10 to 12 months of egg production, brought back into egg production for about 6 months, then sold. United States commercial egg producers commonly peak their first-cycle (all-pullet) flocks at 85 to 95% production. Second-cycle flocks commonly peak between 75 and 85%—about 10% lower than first-cycle flocks, but a sizeable number of

flocks now peak in excess of 85%. If molting is done properly, it will result in a return of all egg quality characteristics equivalent to a 10- to 12-months-old pullet, thereby alleviating the deterioration of quality traits with extended laying periods; many plants will not carton eggs from layers beyond 70 weeks of age in the first cycle or from layers beyond 30 weeks in the second cycle.

The two-cycle molting program generally extends laying life by an additional 6 to 8 months beyond the standard age at which all-pullet flocks are sold; and each hen produces an extra 100 or more eggs, which reduces the overall hen replacement cost per dozen.

2. **Three-cycle molting program.** This involves two molts and three cycles of egg production. The hens are first molted after about 9 months of production, then held through another shorter production period, molted again, followed by an even shorter period of lay, then sold. This periodic molting program totals about 24 months; longer programs are seldom profitable. Commercial egg producers commonly peak three-cycle flocks at 75%, which is 10 to 20% below first-cycle (all-pullet) flocks, and 7% below two-cycle flocks.

• **Methods of molting**—Many satisfactory molting programs are in use, but most of them are simply modifications of two basic concepts—feed withdrawal, followed by a low nutrient feed intake period. Modifications involve the number of days without feed, the composition of the feed, and the duration of the low-nutrient intake period. Other minor factors include the choice of lighting programs during the molt and whether water should be restricted. Of the many satisfactory molting programs, three that feature different basic procedures follow:

1. **Conventional force molting program.** This procedure, which is sometimes referred to as an on-again/off-again program, is outlined in Table 24–26.

TABLE 24–26
CONVENTIONAL FORCE MOLTING PROGRAM
(ON-AGAIN/OFF-AGAIN PROGRAM)

Day	Feed	Water	Light
1	None	None	8 hours
2	None	None	8 hours
3	Egg-type layers: 10 lb/100 hens	Water	8 hours
4	None	None	8 hours
5	Egg-type layers: 10 lb/100 hens	Water	8 hours
6	None	None	8 hours
7	Egg-type layers: 10 lb/100 hens	Water	8 hours
8	None	None	8 hours
9	Egg-type layers: 10 lb/100 hens	Water	8 hours
10 through 55–60	Restricted feeding—about 75% of full feed intake	Water	8 hours
61	Full-feed layer ration	Water	14–16 hours

Additional pertinent feeding instructions relative to the conventional molting program follow:

a. Self-feed oystershell from the start of molting until 2 weeks after egg production is reestablished, then return to controlled shell feeding.

b. Do not use skip-a-day feeding programs until after 10 days following the start of the molting procedure.

2. **California force molting program.** California has a higher percentage of force molting than any other state. For this reason, the California method is presented. It is characterized by simplicity, low cost and high subsequent performance, and with all birds getting equal treatment and having uniform recovery. It is presented in Table 24–27.

TABLE 24–27
CALIFORNIA FORCE MOLTING PROGRAM

Day	Feed	Water	Light
1 through 10 to 14	None	Water	Discontinue artificial light or limit to 8 hours
11 to 15 through 35	Full feed cracked grain or low-protein, low-calcium molt mash	Water	Discontinue artificial light or limit to 8 hours
36 through 68	Full feed laying mash	Water	14–16 hours

Additional pertinent information about the California molting program follows:

a. The 10 to 14 days without feed will usually result in 25 to 30% loss in body weight, with less than 1% mortality.

b. The feed withholding period (from day 1 through up to day 10 to 14) should end when the accumulated mortality reaches 1.25%, when body weight reduction for a 3.6-lb hen reaches 30%, or when it has gone 14 days—whichever comes first.

3. **"Fast-molting" and "fast-fast molting."** The *fast-molting* involves elimination of the resting period or low nutrient period and a return to 50% rate of lay in 4 to 6 weeks instead of the typical 8 weeks of the conventional system. The *fast-fast molt* goes one step further; it involves reducing the initial feed withholding period to 4 to 6 days (instead of 10 to 14) and elimination of the resting period. The fast-fast molt is extremely fast; flocks can be back to 50% production by the third week, instead of the typical 8 weeks of the conventional system. Preliminary studies indicate that the early production achieved by the fast-molt and by the fast-fast molt is at the expense of later production and egg quality. But more experimental work is needed.

• **Will force molting pay?**—There is no simple answer to this question, since it depends on a variety of economic factors. However, the following general guidelines should be taken into consideration:

1. Recycling becomes less profitable as egg prices increase and as price differential between egg sizes decreases.

2. As the cost of replacement pullets increases, it becomes more advantageous to recycle.

3. The profitability of recycling increases as the price paid for *spent* hens decreases.

4. Using replacement pullets instead of recycling ties up additional capital.

Feeding Breeders

Fig. 24-27. Broiler breeder birds for the production of hatching eggs. Note the plastic, slat-raised floors. Separate feeders for males and females are now common in breeder houses better to control the body weight of the two sexes. (Courtesy, L. S. Jensen, Ph.D., Department of Poultry Science, The University of Georgia, Athens)

Fig. 24-28. Single sire pedigree pens. (Courtesy, Indian River Company, Nacogdoches, Tex.)

Fig. 24-29. Broiler breeders, showing nesting. (Courtesy, California Polytechnic University, San Luis Obispo)

The following pointers are pertinent to feeding breeders:

1. The nutritive requirements for breeding flocks are more rigorous than those for commercial laying flocks. Breeders require greater amounts of vitamins A, D, E, and B-1, and of riboflavin, pantothenic acid, niacin, and manganese than do laying flocks. Rations with these added ingredients in the right proportions give high hatchability and good development of chicks. Such rations cost more than normal layer rations.

2. Broiler breeder replacement pullets should receive low energy diets, in the range of 1090 to 1135 kcal/pound and/or the feed intake should be restricted, to avoid excess fat accumulation at the time they reach sexual maturity.

3. Broiler breeder hens, which are heavy and have a high energy requirement for maintenance, require approximately 400 to 450 kcal ME per hen per day, for maximum egg production. Since these hens tend to become over-fat when fed high energy diets, it appears best to limit the energy content of their diets to approximately 1200 to 1250 kcal/pound and/or to restrict their feed intake in some way so that they do not obtain much more than about 420 kcal ME per hen per day.

4. Male breeders require slightly less energy than females during growth. The lower fat deposition in the male compared with the female is offset by the energy needs for more rapid growth. Being larger, the adult male cock requires considerably more energy than the hen for maintenance, but this is largely offset by the hen's need for energy for egg production.

Example rations for chicken breeders producing eggs for hatching are given in Table 24–28.

TABLE 24–28
EXAMPLE CHICKEN BREEDER RATIONS[A,B,1,14]

Ingredient	Body Weight			
	Egg-Type Breeders 3½–5 Lb		Meat-Type Breeders 5–8 Lb	
	(lb)	*(kg)*	(lb)	*(kg)*
Ground yellow corn[2,3]	1,305	*593.2*	1,379	627
Wheat middlings	—	—	70	*32*
Alfalfa meal, 17%	50	*23*	50	23
Soybean meal, dehulled	324	*147.3*	248	*113*
Fish meal, herring, 65%[4,5]	50	*23*	50	*23*
Meat and bone meal, 47%[5]	50	*23*	50	*23*
Dicalcium phosphate[6]	6	*2.7*	4	*1.8*
Ground limestone[7]	157	*71.4*	142	*64.5*
DL-Methionine, or equivalent	0.5	*0.2*	0.4	*0.2*
Stabilized yellow grease, or equivalent	51	*23*	[16]	[16]
Iodized salt[4]	7	*3.1*	7	*3.1*
Antioxidant	[9]	[9]	[9]	[9]
Mineral and vitamin supplements:[12]				
Calcium pantothenate (mg)	6,000		6,000	
Manganese[11] (Mn) (g)	52		52	
Selenium[25] (Se) (mg)	90.8		90.8	
Zinc[17] (Zn) (g)	16		16	
Vitamin A (IU)	4,000,000		4,000,000	
Vitamin D_3 (IU)	2,000,000		2,000,000	
Vitamin E (IU)	2,000		2,000	
Vitamin K[20]	—		—	
Choline (mg)	172,000		208,000	
Niacin (Nicotinic Acid, Nicotinamide) (mg)	10,000		10,000	
Riboflavin (Vitamin B-2) (mg)	3,000		3,000	
Vitamin B-12 (Cobalamins) (mg)	6		6	
Totals[21]	2,000.5	*909.9*	2,000.4	*910.6*
Calculated analysis:[27]				
Metabolizable energy (kcal)	1,337	*2,941*	1,295	*2,849*
Protein (%)	17.01		16.03	
Lysine (%)	0.87		0.78	
Methionine (%)	0.33		0.31	
Methionine + cystine (%)	0.59		0.56	
Fat (%)	5.81		3.57	
Fiber (%)	2.42		2.67	
Calcium (%)	3.27		2.98	
Total phosphorus (P) (%)	0.54		0.54	
Available phosphorus[13] (P) (%)	0.46		0.46	
Vitamins (units or mg/lb or kg):				
Vitamin A activity (IU)	5,983	*13,163*	6,067	*13,347*
Vitamin D_3, added (IU)	1,000	*2,200*	1,000	*2,200*
Choline (mg)	500.40	*1,101*	500.57	*1,101*
Niacin (Nicotinic Acid, Nicotinamide) (mg)	15.21	*33.5*	17.00	*37.4*
Pantothenic Acid (Vitamin B-3) (mg)	5.67	*12.5*	5.74	*12.6*
Riboflavin (Vitamin B-2) (mg)	2.49	*5.5*	2.51	*5.5*

[A]Adapted by the authors from *NECC Chicken and Turkey Rations*, prepared by the New England College Poultry Conference Board, by poultry specialists from, and distribution by, the New England Land-Grant Universities: University of Connecticut, Storrs; University of Maine, Orono; University of Massachusetts, Amherst; University of New Hampshire, Durham; University of Rhode Island, Kingston; and University of Vermont, Burlington.

[B]See footnotes following Table 24–31, p. 1054.

Feeding Replacement Pullets

Pullets generally perform well during their laying year when their nutrient requirements have been met during the growing period.

Fig. 24-30. Newly hatched baby chicks—future replacement pullets.

Leghorn-type pullets are seldom restricted-fed during the growing period because varying the lighting during growth from 6 to 20 weeks of age can be used to control feed consumption and sexual development.

However, pullets of heavy breeds tend to accumulate excessive amounts of body fat; so, it is common practice to restrict the feed intake of these birds to produce pullets with leaner bodies at the time of sexual maturity. This is beneficial because: (1) it produces healthier pullets, and (2) it reduces feed costs during the pullet rearing period.

Also, some producers restrict the feed intake of light breeds, based on research which shows that, even among the light breeds, restricted feed intake results (1) in slightly higher mortality during the rearing period, but (2) in lower mortality and higher egg production during the laying year.

Most feed restriction programs are started at 9 to 12 weeks of age. The methods commonly employed follow:

1. **Skip-a-day method.** This involves feeding pullets on alternate days only, from 9 weeks to sexual maturity. When fed every-other-day, pullets consume more feed on the days that feed is available than they would normally consume on a daily basis. They are unable, however, to consume enough feed in one day to satisfy their total energy requirements for 2 days. Thus, growth and body fat content are reduced.

2. **Daily restriction of feed.** Under this system, the producer determines the amount of feed that would normally be consumed by the pullets each day, then provides them with a fraction of that amount on a daily basis. Often this fraction is 75 to 85% of the amount of feed that would be consumed on a free-choice basis. Typical body weight increases and feed requirements of pullets are shown in Table 24-9.

3. **Bulky, low-energy or low-protein and/or amino acid imbalanced rations.** Another form of restriction involves (a) the feeding of bulky, low-energy rations, or (b) the feeding of low-protein and/or amino acid imbalanced rations on a free-choice basis during the period 12 to 20 weeks of age. The rations are formulated to be adequate in all nutrients, but are sufficiently low in energy or low in protein and/or imbalanced in amino acids that pullets cannot consume enough to satisfy their energy or protein needs for maximum growth. Under such a program, it is possible to restrict the growth of young pullets by 10 to 15%, an amount comparable to the growth depression with the skip-a-day and daily restriction methods.

These further pointers are pertinent to feeding replacement chicks (pullets):

1. Feed accounts for approximately 60% of the cost of raising replacement pullets.

2. Replacement chicks are usually fed a diet lower in energy than broiler chicks. Also, feed and daily light periods may be restricted, so as to permit the pullets to reach larger body size before they start to lay than would be the case were they full fed, and fully lighted.

3. Always use complete starter feeds for chicks, and give chicks starter feeds without grain supplement until they are 5 weeks old.

4. When chicks are 5 weeks old, change to the growing ration.

The NRC requirements for replacement pullets are presented in tables in the earlier section of this chapter headed, "National Research Council (NRC) Requirements."

Example *starter* and *grower* rations (1) for egg-type and meat-type strains, from days 1 to 35, and (2) for egg- and meat-type strains, from day 36 until egg production begins are given in Table 24-29.

TABLE 24-29
EXAMPLE REPLACEMENT CHICKEN RATIONS[A,B,C,1,14,26]

Ingredient	Starter (For egg- and meat-type strains, from days 1 to 35)				Grower (For egg- and meat-type strains, from day 36 until egg production begins)					
	20% Protein		18% Protein		15% Protein		14% Protein		12% Protein	
	(lb)	***(kg)***	**(lb)**	***(kg)***	**(lb)**	***(kg)***	**(lb)**	***(kg)***	**(lb)**	***(kg)***
Ground yellow corn[2,3]	1,267	*576*	1,310	*595.4*	1,412	*641.8*	1,438	*653.6*	1,481	*673.2*
Wheat middlings	130	*59.1*	200	*90.9*	254	*115.5*	254	*115.5*	323	*146.8*
Alfalfa meal, 17%	25	*11.3*	25	*11.3*	25	*11.3*	25	*11.3*	25	*11.3*
Soybean meal, dehulled	422	*192*	309	*140.4*	217	*98.6*	217	*98.6*	104.8	*47.6*
Fish meal, herring, 65%[4,5]	50	*23*	50	*23*	—	—	—	—	—	—
Meat and bone meal, 47%[5]	50	*23*	50	*23*	50	*23*	—	—	—	—
Lysine	—	—	—	—	—	—	1	*0.5*	1.2	*0.5*
Dicalcium phosphate[6]	10	*4.5*	9	*4.1*	14	*6.4*	30	*13.6*	29	*13.2*
Ground limestone[7]	19	*8.6*	20	*9.1*	21	*9.5*	28	*12.7*	29	*13.2*
Stabilized yellow grease, or equivalent	20	*9.1*	20	*9.1*	[16]	[16]	[16]	[16]	[16]	[16]
Iodized salt[4]	7	*3.1*	7	*3.1*	7	*3.1*	7	*3.1*	7	*3.1*
Antibiotic supplement	[8]	[8]	—	—	—	—	—	—	—	—
Antioxidant	[9]	[9]	[9]	[9]	[9]	[9]	[9]	[9]	[9]	[9]
Coccidiostat	[10]	[10]	[10]	[10]	[10]	[10]	[10]	[10]	[10]	[10]
Mineral and vitamin supplements:[12]										
Calcium pantothenate (mg)	4,000		4,000		3,000		3,000		3,000	
Manganese[11] (Mn) (g)	52		52		52		52		52	
Selenium[25] (Se) (mg)	90.8		90.8		90.8		90.8		90.8	
Vitamin A (IU)	3,000,000		3,000,000		3,000,000		3,000,000		3,000,000	
Vitamin D_3 (IU)	1,000,000		1,000,000		1,000,000		1,000,000		1,000,000	
Vitamin K[20]	—		—		—		—		—	
Choline (mg)	213,000		298,000		84,000		125,000		209,000	
Niacin (Nicotinic Acid, Nicotinamide) (mg)	10,000		10,000		10,000		10,000		10,000	
Riboflavin (Vitamin B-2) (mg)	1,500		1,500		1,500		1,500		1,500	
Vitamin B-12 (Cobalamins) (mg)	6		6		6		6		6	
Totals[21]	2,000	*909.7*	2,000	*909.4*	2,000	*909.2*	2,000	*908.9*	2,000	*908.9*
Calculated analysis:[27]										
Metabolizable energy (kcal)	1,361	*2,994*	1,362	*2,996*	1,343	*2,955*	1,341	*2,950*	1,342	*2,952*
Protein (%)	20.03		18.01		15.03		14.01		12.01	
Lysine (%)	1.04		0.89		0.63		0.63		0.49	
Methionine (%)	0.34		0.32		0.25		0.24		0.21	
Methionine + cystine (%)	0.64		0.59		0.48		0.46		0.41	
Fat (%)	4.48		4.70		3.76		3.54		3.74	
Fiber (%)	2.67		2.83		3.01		3.00		3.15	
Calcium (%)	0.90		0.90		0.89		0.90		0.90	
Total phosphorus (%)	0.66		0.66		0.66		0.66		0.65	
Available phosphorus[13] (%)	0.41		0.41		0.40		0.40		0.40	
Vitamins (units or mg/lb or kg):										
Vitamin A activity (IU)	4,188	*9,214*	4,237	*9,321*	4,354	*9,579*	4,381	*9,638*	4,430	*9,746*
Vitamin D_3 added (IU)	500	*1,100*	500	*1,100*	500	*1,100*	500	*1,100*	500	*1,100*
Choline (mg)	600.20	*1,320*	600.11	*1,320*	420.48	*925*	420.20	*924*	419.76	*923*
Niacin (Nicotinic Acid, Nicotinamide) (mg)	19.08	*42*	20.49	*45*	20.87	*45.9*	20.53	*45.2*	21.84	*48*
Pantothenic Acid (Vitamin B-3) (mg)	5.34	*11.7*	5.27	*11.6*	4.69	*10.3*	4.72	*10.4*	4.64	*10.2*
Riboflavin (Vitamin B-2) (mg)	1.78	*3.9*	1.76	*3.9*	1.68	*3.7*	1.64	*3.6*	1.62	*3.6*

[A]Adapted by the authors from *NECC Chicken and Turkey Rations*, prepared by the New England College Poultry Conference Board, by poultry specialists from, and distribution by, the New England Land-Grant Universities: University of Connecticut, Storrs; University of Maine, Orono; University of Massachusetts, Amherst; University of New Hampshire, Durham; University of Rhode Island, Kingston; and University of Vermont, Burlington.

[B]See footnotes following Table 24-31, p. 1054.

[C]Equivalent to 14% protein + 1 lb lysine.

Feeding Broilers

Fig. 24-31. Typical broiler farm, showing two houses and two bulk feed bins. (Courtesy, The University of Georgia, Athens)

The following pointers pertaining to feeding broilers, roasters, and capons are pertinent:

1. Feed is the largest cost item in broiler production, representing 60 to 75% of the total cost.

2. Producers aim for broilers with an average weight of over 4.0 lb at 6 to 7 weeks of age, feed conversion of less than 2.0, and mortality under 3.0%. Many good producers are achieving feed conversion of about 1.9.

3. Some operations use a 2-stage ration program (starter and finisher) for broilers, but most are using at least 3 stages in their feeding programs (starter, grower, and finisher) to reduce costs and make more efficient use of the nutrients. In a 3-stage program the starter feed should be used for 3 to 4 weeks, the grower for about 2 weeks, and the finisher for the remainder of the feeding period.

Fig. 24-32. Six-week-old broilers. Note 2 continuous pan feeders and 4 rows of waterers. (Courtesy, P. L. Potts, Sr., Ph.D., Poultry Department, California Polytechnic University, San Luis Obispo)

Fig. 24-33. Nutrition research has played a major role in the development of the modern broiler industry. Battery brooders such as these are extensively used in research to evaluate feeds and additives. (Courtesy, University of Georgia, Athens)

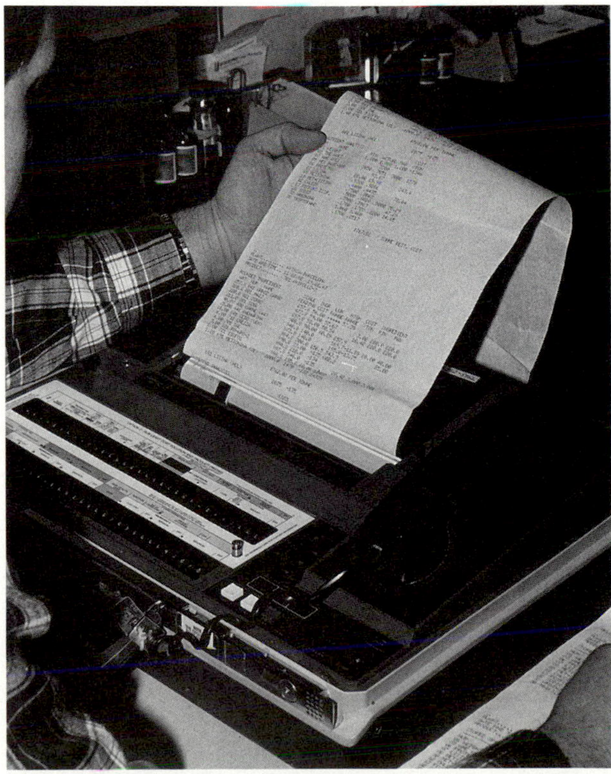

Fig. 24-34. Broiler rations are routinely computed by least cost formulation. The computer determines the combination of feedstuffs that will meet the nutrient specifications set by the nutritionist at the lowest cost per unit of feed. (Courtesy, University of Georgia, Athens)

The NRC requirements for broilers are presented in the tables in the earlier section of this chapter headed, "National Research Council (NRC) Requirements."

Example broiler *starter* and *finisher* rations are given in Table 24–30.

TABLE 24–30
EXAMPLE BROILER RATIONS[A,B,1]

Ingredient	Starter[18] (For Broiler Chicks Until 24 Days of Age)		Finisher[C] (For Broilers From Day 25 to Market)	
	(lb)	*(kg)*	(lb)	*(kg)*
Ground yellow corn[3]	1,106	*503*	1,235	*561*
Alfalfa meal, 17%	—	—	25	*11*
Soybean meal, dehulled	605	*275*	420	*191*
Corn gluten meal, 60%	50	*23*	75	*34*
Fish meal, herring, 65%[4,5]	50	*23*	50	*23*
Meat and bone meal, 47%[5]	50	*23*	50	*23*
Dicalcium phosphate[6]	10	*4*	9	*4*
Ground limestone[7]	16	*7*	14	*6.3*
DL-methionine, or equivalent	0.8	*0.3*	—	—
Stabilized yellow grease, or equivalent	106	*48*	115	*52.2*
Iodized salt[4]	7	*3*	7	*3*
Antibiotic supplement	[8]	[8]	[8]	[8]
Antioxidant	[9]	[9]	[9]	[9]
Coccidiostat	[10]	[10]	[10]	[10]
Mineral and vitamin supplement:[12]				
Calcium pantothenate (mg)	5,000		5,000	
Manganese[15] (Mn) (g)	75		75	
Organic arsenical supplement[19]	0.1		0.1	
Selenium[25] (Se) (mg)	90.8		90.8	
Zinc[24] (Zn) (g)	30		30	
Vitamin A (IU)	4,000,000		4,000,000	
Vitamin D_3 (IU)	1,000,000		1,000,000	
Vitamin E (IU)	2,000		2,000	
Vitamin K[20] (mg)	2,000		1,000	
Choline (mg)	503,000		672,000	
Niacin (Nicotinic Acid, Nicotinamide) (mg)	20,000		20,000	
Riboflavin (Vitamin B–2) (mg)	3,000		3,000	
Vitamin B-12 (Cobalamins) (mg)	12		12	
Totals[21]	2,000.9	*909.3*	2,000.1	*908.5*
Calculated Analysis:[27]				
Metabolizable energy (kcal)	1,399	*3,078*	*1,494*	*3,287*
Protein (%)	24.08		21.00	
Lysine (%)	1.30		1.05	
Methionine (%)	0.45		0.38	
Methionine + cystine (%)	0.81		0.71	
Fat (%)	8.20		8.92	
Fiber (%)	1.97		2.22	
Calcium (%)	0.84		0.80	
Total phosphorus (%)	0.64		0.60	
Available phosphorus[13] (%)	0.40		0.38	
Vitamins (units or mg/lb or kg):				
Vitamin A activity (IU)	3,769	*8,292*	5,424	*11,933*
Vitamin D_3 (added) (IU)	500	*1,100*	500	*1,100*
Choline (mg)	800.03	*1,760*	800.48	*1,761*
Niacin (Nicotinic Acid, Nicotinamide) (mg)	21.36	*47*	21.33	*46.9*
Pantothenic Acid (Vitamin B–3) (mg)	5.69	*12.5*	5.51	*12.1*
Riboflavin (Vitamin B–2) (mg)	2.44	*5.4*	2.49	*5.5*
Xanthophyll[28] (mg)	9.50	*20.9*	14.05	*30.9*

[A]Adapted by the authors from *NECC Chicken and Turkey Rations,* prepared by the New England College Poultry Conference Board, by poultry specialists from, and distribution by, the New England Land-Grant Universities: University of Connecticut, Storrs; University of Maine, Orono; University of Massachusetts, Amherst; University of New Hampshire, Durham; University of Rhode Island, Kingston; and University of Vermont, Burlington.

[B]See footnotes following Table 24–31, p. 1054.

[C]For preslaughter drug withdrawal times see section headed "Regulations Regarding Feed Additives," following "Footnotes."

FEEDING TURKEYS

Fig. 24-35. Turkeys grown in confinement. (Courtesy, P. L. Potts, Sr., Ph.D., Poultry Department, California Polytechnic University, San Luis Obispo)

Feeding turkeys involves two distinct areas of emphasis: feeding market turkeys, and feeding turkey breeders.

• **Feeding market turkeys**—Most market turkeys are of the large type. Males (toms) are usually marketed at 19 to 25 weeks of age and at liveweights of 23 to 35 lb. Younger toms are often sold as oven-ready dressed birds; older toms are generally further processed or used in the restaurant trade. Females (hens) are commonly marketed at 16 to 17 weeks of age and at about 14 lb weight. Medium- and small-type turkeys (roasters/fryers) are often sold at younger ages and lighter liveweights—normally, about 10 lb.

Fig. 24-36. Broad breasted white market turkeys being grown on the range. (Courtesy, The University of Georgia, Athens)

The formulations of the rations fed to market turkeys should be changed as the birds grow. Thus, the nutrient requirements shown in Table 24-11 provide for such changes at 3- to 4-week intervals; in practice, however, the changes may occur more or less frequently than indicated in Table 24-11. Nutritional adjustments are often made for expected ambient temperature variations in order to assure that the birds consume the necessary amount of protein, minerals, and vitamins, regardless of changes in feed consumption.

• **Feeding turkey breeders**—The feeding programs for breeder stock are usually divided into prebreeding (holding) and breeding periods. The prebreeding, or holding, rations may be fed from the time the breeders are selected, at about 16 weeks of age. Holding rations are usually formulated at medium energy levels in order to stabilize development and weight gains after market age. Hens are fed the holding ration until the time of light stimulation, at about 30 weeks of age; thereafter, breeder rations are fed. Toms may be fed a nutritionally balanced holding ration from the time of breeder selection throughout the breeding season. In some programs, the body weight of toms is controlled by limited feeding. The light stimulation of toms is normally initiated at about 26 weeks of age.

Fig. 24-37. Turkey eggs and newly hatched poult. Approximately 10% of all incubated turkey eggs are infertile. The B vitamins, vitmin E, and selenium have been shown to affect the quality of the semen. In the diet of the turkey hen, these same nutrients, along with magnesium, have been shown to affect egg fertility. (Courtesy, Nicholas Turkey Breeding Farms, Sonoma, Calif.)

It is not necessary to feed low-energy rations or to restrict feed intake of turkey hens in the prebreeding, or holding, period. Corn-soybean meal type rations may be fed *ad libitum*. Growth restriction does not result in any consistent improvement in reproductive performance. Nevertheless, the use of holding rations for turkey breeders is common practice. These rations usually contain medium energy concentrations so as to stabilize development and weight gains after mature body weight is attained. Care should be taken to provide turkey breeder hens adequate intake of minerals and vitamins during the holding period so that they are not depleted of these nutrients prior to the onset of lay.

These further pointers are pertinent to feeding turkeys:

1. Prevent poult "starve out." Upon arrival, poults should be encouraged to consume feed and water as soon as possible. It may be necessary to dip the beaks of some of them in feed and water to start them eating and drinking.
2. Turkeys grow faster than chickens; hence, they have relatively higher feed and protein requirements.
3. Young turkeys use feed efficiently. Small White turkeys raised to a liveweight of 6 to 8 lb at 14 to 16 weeks of age require 3 lb of feed per pound of live turkey produced.

Large White turkeys require about 3.5 to 3.75 lb of feed to produce 1 lb of liveweight, when grown to a market weight of about 30 lb and 24 weeks of age.

4. A high-fiber, low-energy holding ration retards sexual maturity and may result in some desirable effects upon later reproductive performance. The holding ration limits energy intake, but should not limit protein, vitamins, and minerals. When a holding ration is used, the birds should be switched to the breeder ration 2 weeks prior to egg production.

5. Good range provides green feed and tends to reduce feed costs. However, it may make for higher losses from blackhead and other diseases, and predators; and range turkey operations may make the neighbors unhappy because of dust, odors, and noise.

6. As they approach maturity, turkeys fed for market purposes should be fed rations that are quite different from those that are fed to turkey breeders.

The NRC requirements for turkeys are presented in Tables 24–11, 24–12, and 24–13 in the earlier section of this chapter headed, "National Research Council (NRC) Requirements."

Example rations for market turkeys and breeders are given in Table 24–31, Example Turkey Rations.

TABLE 24–31
EXAMPLE TURKEY RATIONS[A,B,C,D,1]

Ingredient	Starter (0–8 Weeks)		Grower (8–16 Weeks)		Finisher (16–Market)[C,22]		Breeder	
	(lb)	*(kg)*	(lb)	*(kg)*	(lb)	*(kg)*	(lb)	*(kg)*
Ground yellow corn[2,3]	929	*422.3*	1,199	*545*	1,490	*677.2*	1,218	*554*
Wheat middlings	100	*45.4*	50	*23*	—	—	250	*114*
Alfalfa meal, 17%	25	*11.3*	25	*11.3*	25	*11.3*	60	*27.3*
Soybean meal, dehulled	675	*307*	570	*259*	335	*152.3*	190	*86.4*
Fish meal, herring, 65%[4,5]	100	*45.4*	—	—	—	—	100	*45.4*
Meat and bone meal, 47%[5]	100	*45.4*	50	*23*	50	*23*	50	*23*
Dicalcium phosphate[6]	13	*6*	32	*14.5*	23	*10.5*	10	*4.5*
Ground limestone[7]	10	*4.5*	14	*6*	17	*8*	92	*42*
DL-Methionine, or equivalent	0.6	*0.3*	—	—	—	—	—	—
Stabilized yellow grease, or equivalent	40	*18.2*	50	*23*	50	*23*	20	*9*
Iodized salt[4]	8	*3.6*	10	*4.5*	10	*4.5*	10	*4.5*
Antibiotic supplement	[8]	[8]	—	—	—	—	—	—
Coccidiostat or antihistomonal	[10]	[10]	[10]	[10]	[10]	[10]	—	—
Antioxidant	[9]	[9]	[9]	[9]	[9]	[9]	[9]	[9]
Mineral and vitamin supplements:[D,12]								
Calcium pantothenate (mg)	4,500		4,500		6,000		10,000	
Manganese[23] (Mn) (g)	30		30		30		30	
Selenium[25] (Se) (mg)	181.6		181.6		181.6		181.6	
Zinc[24] (Zn) (g)	30		30		30		30	
Vitamin A (IU)	7,500,000		7,500,000		7,500,000		4,000,000	
Vitamin D_3 (IU)	1,700,000		1,700,000		1,700,000		1,700,000	
Vitamin E (IU)	10,000		5,000		—		30,000	
Biotin (mg)	100		100		100		100	
Choline (mg)	674,000		388,000		417,000		427,000	
Niacin (Nicotinic Acid, Nicotinamide) (mg)	42,000		46,000		48,000		50,000	
Riboflavin (Vitamin B-2) (mg)	4,000		5,000		5,000		5,000	
Vitamin B-12 (Cobalamins) (mg)	6		6		6		6	
Totals[21]	2,000.6	*909.4*	2,000	*909.3*	2,000	*909.8*	2,000	*910.1*
Calculated analysis:[27]								
Metabolizable energy (kcal)	1,322	*2,908*	1,371	*3,016*	1,440	*3,168*	1,291	*2,840*
Protein (%)	27.31		21.09		16.22		17.01	
Lysine (%)	1.60		1.10		0.75		0.87	
Methionine (%)	0.48		0.33		0.27		0.32	
Methionine + cystine (%)	0.88		0.65		0.53		0.57	
Fat (%)	5.31		5.43		5.84		4.88	
Fiber (%)	2.60		2.47		2.28		3.25	
Calcium (%)	1.17		1.00		0.93		2.26	
Total phosphorus (P) (%)	0.91		0.80		0.69		0.70	
Available phosphorus[13] (P) (%)	0.65		0.55		0.46		0.61	
Vitamins (units or mg/lb or kg):								
Vitamin A activity (IU)	6,054	*13,319*	6,361	*13,994*	6,691	*14,720*	6,382	*14,040*
Vitamin D_3, added (IU)	850	*1,870*	850	*1,870*	850	*1,870*	850	*1,870*
Choline (mg)	1,000.07	*2,200*	700.08	*1,540*	600.00	*1,320*	650.24	*1,430*
Niacin (Nicotinic Acid, Nicotinamide) (mg)	35.94	*79.1*	34.51	*75.9*	33.84	*74.4*	42.07	*92.6*
Pantothenic Acid (Vitamin B-3) (mg)	6.07	*13.4*	5.59	*12.3*	5.63	*12.4*	8.11	*17.8*
Riboflavin (Vitamin B-2) (mg)	3.19	*7*	3.44	*7.6*	3.38	*7.4*	3.65	*8*

[A]Adapted by the authors from *NECC Chicken and Turkey Rations,* prepared by the New England College Poultry Conference Board, by poultry specialists from, and distribution by, the New England Land-Grant Universities: University of Connecticut, Storrs; University of Maine, Orono; University of Massachusetts, Amherst; University of New Hampshire, Durham; University of Rhode Island, Kingston; and University of Vermont, Burlington.

[B]See footnotes following this table.

[C]For preslaughter drug withdrawal times, see section following footnotes entitled "Regulations Regarding Feed Additives."

[D]Folacin may be required under certain conditions. It may be added at the rate of 1,000 mg per ton of feed.

(Continued)

FOOTNOTES FOR TABLES 24-25, 24-28, 24-29, 24-30, AND 24-31

[1]Wherever substitutions are made in the rations, the total nutrient content should be adjusted to meet established requirements.

[2]Two to four hundred pounds of coarsely ground wheat or yellow hominy may be used to replace an equal amount of corn. If wheat is used, add 200,000 IU of vitamin A for each 100 lb of corn removed.

[3]There is usually some loss of provitamin A activity in corn and alfalfa meal during storage. If stored ingredients are used, it may be advisable to increase the added vitamin A level of the ration by 1,000 or 2,000 IU/lb. This can be accomplished by increasing the recommended supplement by 2,000,000 or 4,000,000 IU/ton of feed.

[4]The added salt level should be reduced by the amount supplied by the fish meal and other by-product ingredients.

[5]Poultry by-product meal may be substituted for all of the meat and bone scrap and up to 50% of the fish meal. Correct for calcium and phosphorus loss due to substitutions of poultry by-product meal.

[6]Based on an 18.5% phosphorus product, steamed bone meal or defluorinated rock phosphate may replace the dicalcium phosphate on a phosphorus basis.

[7]Based on 35% calcium, low magnesium limestone.

[8]An antibiotic may be used in these rations at the level recommended by the manufacturer (see section following footnotes entitled "Regulations Regarding Feed Additives").

[9]1,2-dihydro-6-ethoxy-2,2,4-trimethylquinoline (ethoxyquin) is recommended in the chick starter, broiler and breeder rations at the 0.0125% level to help prevent the appearance of encephalomalacia (crazy chick disease). If desired, it, or an equivalent antioxidant, may be added to help prevent the oxidation of dietary components. Total ethoxyquin from all sources must not exceed 0.25 lb per ton.

[10]A coccidiostat or antihistomonal drug may be used in these rations, as required, at levels recommended by the manufacturer (see section following footnotes entitled "Regulations Regarding Feed Additives").

[11]This amount of manganese will be furnished by 0.5 lb of manganese sulfate or 0.21 lb manganous oxide (70% feeding grades). An equivalent amount of manganese may be added from other acceptable sources.

[12]Caution should be used when high potency vitamin mixes are involved. It is recommended that 10 lb be the minimum amount of any item added to a ton of feed to insure proper mixing. Thus, high potency vitamin, mineral, or drug mixes should be premixed with a carrier (such as corn meal) to such a dilution that 10 lb of the final mix will be added for each ton of feed mixed. Minerals and vitamins should not be premixed together.

[13]Available phosphorus has been taken as 30% of total phosphorus from plant sources for chicks, and 75% of total phosphorus from plant sources for adult birds. Phosphorus from other than plant sources is considered to be 100% utilized.

[14]For those persons wanting a specific restricted feeding program, specific programs are available from individual breeders or Extension specialists.

[15]This amount of manganese will be furnished by 0.7 lb manganese sulfate or 0.3 lb manganous oxide (70% feeding grades). An equivalent amount of manganese may be added from other acceptable sources.

[16]Stabilized fats may replace an equal amount of cereal grains to provide a higher energy level, control dust, and aid pelleting. Where maintaining body weight in layers is a problem, increase fat by 1 or 2% during the winter by replacing an equal amount of cereal grains.

[17]Approximately this amount of zinc will be furnished by 29 g of zinc carbonate or 20 g of zinc oxide. An equivalent amount of zinc may be used from other acceptable sources.

[18]Feed starting ration until birds are 35 days old.

[19]Based on 3-nitro-4 hydroxyphenylarsonic acid at a level of 45 g (0.1 lb) per ton. Other compounds that may be used at a level recommended by the manufacturer are sodium arsanilate or arsanilic acid (see section following footnotes entitled "Regulations Regarding Feed Additives").

[20]In the absence of alfalfa or if the birds are raised on wire, 2 g of vitamin K activity should be added. Values in the broiler rations are based on menadione. Other compounds supplying equivalent levels of vitamin K may be used.

[21]If an even 2,000 lb is desired, adjust by removing or adding ground yellow corn.

[22]May be fed with grain after 20 weeks.

[23]This amount of manganese will be furnished by approximately 0.3 lb of manganese sulfate or 0.13 lb of manganous oxide (70% feeding grades). An equivalent amount of manganese may be added from other acceptable sources.

[24]This amount of zinc will be furnished by approximately 53 g of zinc carbonate or 37 g of zinc oxide. An equivalent amount of zinc may be added from other acceptable sources.

[25]Federal law, which strictly regulates the addition of selenium to poultry rations, should be consulted. Selenium, as sodium selenite or sodium selenate, may be added to complete feed for chickens at a level not to exceed 0.1 ppm, and to complete feed for turkeys at a level not to exceed 0.2 ppm. It shall be incorporated into each ton of complete feed for chickens by a premix containing not more than 90.8 mg of added selenium and weighing not less than 1 lb. It shall be incorporated into each ton of complete feed for turkeys by a premix containing no more than 181.6 mg of added selenium and weighing not less than 2 lb.

[26]For heavy caged layer pullets we suggest feeding 18% protein 0-6 weeks, 14% protein 7-12 weeks, and 12% protein 13-20 weeks of age.

[27]Any discrepancies in calculated analysis that occur in the decimal part of the figures are due to rounding errors.

[28]These are not highly pigmented diets. The xanthophyll of natural ingredients is variable, so if more pigment is desired use a high potency source of xanthophyll.

REGULATIONS REGARDING FEED ADDITIVES

The Food and Drug Administration of the U.S. Department of Health and Human Services has published a series of regulations concerning the use of additives, such as arsenicals, antibiotics, coccidiostats and other drugs, in animal feeds. For information concerning the use of any additive, consult the feed tag or label. If you still have questions about proper use of the additive, especially in conjunction with other additives, see your veterinarian, feed dealer or drug supplier.

FEEDING DUCKS

Until 1975, the annual production of ducks in the United States was approximately 10 million. Then, between 1975 and 1985, annual production doubled. In 1985, 21.6 million ducks were marketed in the United States. In 1989, U.S. per capita consumption of duck was estimated to be 0.66 lb.

Ducks are grown successfully in two different types of environments: (1) in an open rearing system in which the growing house opens onto an exercise yard with water for wading or swimming; and (2) in a confinement system in which they are raised in environmentally controlled houses, with litter or with a combination of litter and wire floors.

Typically, ducks are provided with two or three feeds during the growing period: when only two feeds are provided during the growing period, a 22% protein starter ration is usually fed the first 2 weeks, followed by a grower-finisher

Fig. 24–38. White Pekin ducks in an open rearing system—on the range, at Ward Duck Co., La Puente, Calif. (Courtesy, P. L. Potts, Sr., Ph.D., California Polytechnic University, San Luis Obispo)

Fig. 24–39. Ducks in an environmentally controlled house. (Courtesy, Cherry Valley Farms Limited, Rothwell, Lincoln, England)

ration. When 3 feeds are provided during the growing period, they consist of a 22% protein starter ration, an 18% protein grower ration, and a 16% protein finisher ration.

The following pointers are pertinent to feeding ducks:

1. Ducks should be fed pellets rather than mash. Use ⅛ in. pellets for starter rations, and 3/16 in. pellets for older ducks. Pellets will make for a saving of 15 to 20% in the feed required to produce a market duck.
2. Ducks are very susceptible to aflatoxicosis; so, monitoring feeds, for aflatoxin is important.
3. Ducks are nearly as good foragers as geese.
4. Ducks should be ready for market between 7½ and 8 weeks of age.
5. When used, holding rations are designed to maintain breeding ducks from about 8 weeks of age until the breeding season commences, without their getting too fat. It is recommended that birds fed holding rations be limited to about ½ lb per bird per day.
6. When a holding ration is used, breeder diet should be substituted for it about 4 weeks before eggs are desired for hatching purposes.
7. When feeding ducks, pellet quality, proper feather development, and limiting carcass fat disposition are concerns, in addition to proper growth, and satisfactory feed conversion.
8. Commercial ducks grow as rapidly and efficiently as commercial broilers.

The NRC requirements for ducks are presented in Tables 24–14 and 24–15, in the earlier section of this chapter headed, "National Research Council (NRC) Requirements."

Examples of three-phase grower rations, and of a breeder ration, for ducks are given in Table 24–32.

Fig. 24–40. White Pekin ducks swimming in a creek. (Photo by J. C. Allen & Son, West Lafayette, Ind.)

TABLE 24-32a
EXAMPLE MARKET DUCK AND BREEDER DUCK RATIONS [1,2]

Ingredients and Analysis	Starter (0–2 weeks)		Grower (2–4 weeks)		Finisher (4–8 weeks)		Breeder	
	(lb)	*(kg)*	(lb)	*(kg)*	(lb)	*(kg)*	(lb)	*(kg)*
Ingredients:								
Yellow corn	1,209	*548.9*	1,420	*644.8*	1,489	*676.0*	1,309.5	*594.5*
Soybean meal, 48.5%	510	*231.5*	320	*145.3*	260	*118.1*	318	*144.4*
Meat and bone meal, 50%	80	*36.3*	80	*36.3*	80	*36.3*	76	*34.5*
Fish meal, 60%	56	*25.4*	65	*29.5*	50	*22.7*	60	*27.3*
Dried whey, delactosed	—	—	—	—	—	—	45	*20.4*
Animal-vegetable fat	50	*22.7*	30	*13.6*	40	*18.2*	—	—
Dicalcium phosphate	—	—	—	—	3	*1.4*	—	—
Limestone	13	*5.9*	10	*4.5*	8	*3.6*	112	*50.8*
Salt	7	*3.2*	6	*2.7*	6	*2.7*	6	*2.7*
Trace mineral mix (Table 24-32a)	2	*0.9*	2	*0.9*	2	*0.9*	2	*0.9*
Vitamin mix (Table 24-32b)	20	*9.1*	15	*6.8*	10	*4.5*	20	*9.1*
Methionine, hydroxy analogue	3	*1.4*	2	*0.9*	2	*0.9*	1.5	*0.7*
Pellet binder	50	*22.7*	50	*22.7*	50	*22.7*	50	*22.7*
Total	2,000.0	*908.0*	2,000.0	*908.0*	2,000.0	*908.0*	2,000.0	*908.0*
Calculated analysis:								
Metabolizable energy (cal./lb)	1,400		1,425		1,450		1,300	
Metabolizable energy *(cal./kg)*	*3,087*		*3,142*		*3,197*		*2,867*	
Protein (%)	21.4		18.0		16.4		17.6	
Fat (%)	5.6		5.0		5.5		3.3	
Fiber (%)	2.6		2.6		2.6		2.4	
Calcium (%)	0.85		0.80		0.75		2.75	
Available phosphorus (%)	0.4		0.4		0.4		0.4	
Total phosphorus (%)	0.60		0.58		0.57		0.57	
Ash (%)	5.3		4.7		4.4		10.1	

Footnotes Table 24-32c.

TABLE 24-32b
THE TRACE MINERAL MIX [1]

Mineral	Percent per Kg of Mix
	(%)
Trace or microminerals:	
Copper (Cu)	0.30
Iodine (I)	0.06
Iron (Fe)	3.00
Manganese (Mn)	6.50
Zinc (Zn)	6.50

TABLE 24-32c
THE VITAMIN MIX [1]

Vitamin	Amount per Lb or Kg of Mix (lb)	*(kg)*
Fat-soluble vitamins:		
Vitamin A (IU)	400,000	*880,000*
Vitamin D_3 (ICU)	60,000	*132,000*
Vitamin E (IU)	500	*1,100*
Vitamin K (msb)[3] (mg)	200	*440*
Water-soluble vitamins:		
Choline (mg)	13,018	*28,639.6*
Niacin (Nicotinic Acid, Nicotinamide) (mg)	2,500	*5,500*
Pantothenic Acid (Vitamin B-3) (mg)	300	*660*
Riboflavin (Vitamin B-2) (mg)	300	*660*
Vitamin B-12 (Cobalamins) (mg)	0.4	*0.88*

Footnotes for Tables 24-32a, 24-32b, and 24-32c:

[1]*From: Complete Duck Grower and Breeder Rations,* Purdue University, West Lafayette, Ind.)

[2]For best results, all rations should be pelleted.

[3]Menadione sodium bisulfite.

FEEDING GEESE

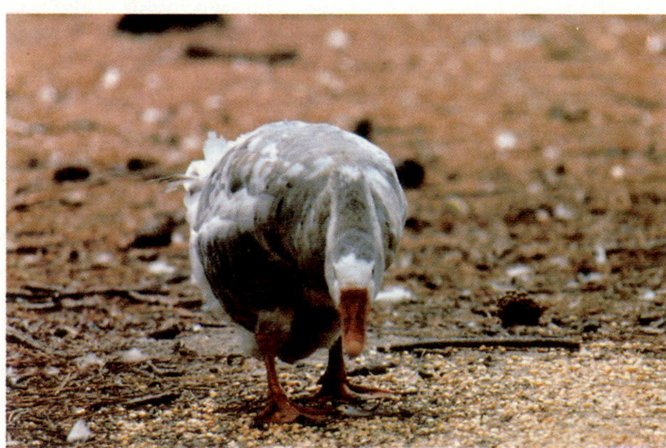

Fig. 24-41. Goose. (Courtesy, USDA)

Geese are very hardy, highly disease resistant, are the closest grazers known, and can live almost entirely on good pasture. Yet, the production of geese for meat purposes has never enjoyed the popularity in the United States that it has in some European countries.

Geese are raised under the following variety of feeding programs:

1. The production of *farm geese,* with the goslings given a starter feed for about 2 weeks, followed by foraging the farm for a variety of pasture and grain feedstuffs; then, marketed at about 18 weeks of age after liberal grain feeding for the last 2 or 3 weeks.

2. The goslings are limit-fed prepared feed throughout the growing period, but are allowed considerable foraging in addition; then, marketed at about 14 weeks of age following liberal feeding of a high-energy finishing ration.

3. The goslings are full-fed in confinement and marketed as *junior* or *green geese* at about 10 weeks of age.

4. The raising and use of geese for weeding purposes. Weeder geese are used with great success to control and eradicate troublesome grass and certain weeds in a great variety of crops and plantings, including cotton, hops, mint, onions, garlic, strawberries, nurseries, corn, orchards, groves, and vineyards. The geese eat grass and young weeds as quickly as they appear, but they do not touch certain cultivated plants. They will work continuously from daylight to dark, 7 days a week (even on bright moonlight nights) nipping off the grass and weeds as promptly as new growth appears.

At the end of the weeding season, geese are generally brought from the field and placed in pens for fattening for 3 or 4 weeks, until they weigh 10 to 12 lb or more. Markets are highest during the 4 to 6 weeks prior to Thanksgiving and Christmas.

The carrying of geese over from one season to the next for weeding purposes is not recommended, because older geese are less active in hot weather than young birds.

5. The production in some European countries of goose livers for *pate de foie gras.* In this program, the geese are grown to about 12 weeks of age, following which they are force-fed a high-grain ration for the production of livers of high-fat content.

For breeding purposes, geese are fed a prebreeding (holding) ration beginning 6 to 8 weeks before the breeding season, followed by a breeding ration formulated for the intensive production of fertile eggs.

The following additional pointers are pertinent to feeding geese:

1. Rations for geese should be pelleted, with $\frac{3}{32}$- or $\frac{3}{16}$-in. pellets preferred. Mash and crumbles cause too much feed wastage and should not be used.

2. Although all rations may be home-mixed, a commercially prepared ration is recommended for young goslings and breeders during the laying season.

3. Succulent green feed should provide the bulk of the ration for young growing geese.

The NRC requirements for geese are presented in Table 24–16, in the earlier section of this chapter headed, "National Research Council (NRC) Requirements."

Example rations for geese are given in Table 24–33.

TABLE 24–33
EXAMPLE GEESE RATIONS[1]

Ingredient	Starter 0–3 Weeks		Grower 3 Weeks to Market		Breeder Layer	
	(lb)	*(kg)*	(lb)	*(kg)*	(lb)	*(kg)*
Ground yellow corn	600	*272.3*	600	*272.3*	300	*136.2*
Ground oats	400	*181.5*	500	*227.0*	700	*317.8*
Ground barley	375	*170.2*	375	*170.2*	210	*95.3*
Dehydrated alfalfa	72	*32.6*	72	*32.6*	250	*113.5*
Soybean oil meal, 44% protein	500	*227.0*	400	*181.5*	450	*204.3*
Limestone	20	*9.1*	20	*9.1*	50	*22.7*
Dicalcium phosphate	20	*9.1*	20	*9.1*	25	*11.4*
Iodized salt	10	*4.5*	10	*4.5*	10	*4.5*
Vitamin premix	3	*1.7*	3	*1.7*	5	*2.3*
Total	2,000	*908.0*	2,000	*908.0*	2,000	*908.0*

[1]*From: Goose Production in North Dakota,* North Dakota State University, Fargo.

FEEDING PIGEONS

Pigeons are raised primarily for sport and hobby—being widely used in racing, shows, and training to perform tricks. But many people consider pigeon (squab) to be an epicurean delight, and a limited demand for pigeon meat exists. Squabs are marketable as early as 28 days of age, at which time their dressed weight is about 1 lb. At this age, squabs are tender and self-basting due to the fat under the skin.

Pigeons grow more rapidly than most birds during the first 20 days of life. They receive their first nourishment from "pigeon milk" regurgitated from the parent pigeon's crop. Pigeon milk is a thick, creamy, semi-digested substance high in protein and fat, but low in carbohydrate. When 20 to 40 days of age, squabs may be fed a pigeon feed. Unlike other forms of poultry, pigeons will not eat mash, so pigeon feed either consists of whole or cracked grains or commercially prepared pellets.

Fig. 24–42. Pigeons at feed. (Courtesy, Ralston Purina Company, St. Louis, Mo.)

Most pigeon producers feed (1) a complete pelleted ration, or (2) a complete pelleted feed plus whole or cracked grain. The most common grains are corn, wheat, sorghum, and peas. Grains can be offered in an open trough or cafeteria style where the self-feeder has individual compartments for each type of grain. If an open trough is used, it is recommended that pigeons be fed twice daily. Only enough feed should be offered at each feeding period as will be eaten in 1 hour.

Commercial pigeon feeds are available. But the fancier may prepare a suitable ration of grains, plus a free-choice mineral mixture, similar to the ration shown in Table 24–34.

Fig. 24–43. Pheasant.

TABLE 24–34
EXAMPLE RATION FOR PIGEONS[1]

Grain Mix (Whole Grains)	Amount	
	(lb)	(kg)
Yellow corn	35	15.8
Grain sorghum	20	9.1
Cowpeas or field peas	20	9.1
Wheat	15	6.8
Oat groats	5	2.3
Hempseed	5	2.3
Total	100	45.4
Mineral mix (fed in separate hopper free-choice):		
Medium-sized ground oyster shells	50	22.7
Grit (appropriate size)	25	11.3
Bone meal or dicalcium phosphate	20	9.1
Salt	5	2.3
Total	100	45.4

[1]Adopted by the authors from *Managing Game Birds,* p. 6, Table 2, published by Michigan State University, East Lansing.

Fig. 24–44. Bobwhite quail.

FEEDING GAME BIRDS: PHEASANTS, BOBWHITE QUAIL, JAPANESE QUAIL, CHUKARS, PARTRIDGES, GROUSE, AND DOVES

Game birds may be propagated for many different purposes. Fanciers may keep game birds as pets. Some growers produce dressed birds, especially pheasants and Bobwhite quail, for specialty markets in stores and restaurants. Other growers produce pheasants, Bobwhite quail, chukars and various other types of partridges and/or grouse for game release farms (shooting preserves). Still others produce birds for release to the wild.

Japanese quail are fed for egg production. They mature at 5 to 6 weeks of age and lay up to 300 eggs per year. As is true of other game birds, Japanese quail must be fed higher protein rations than chickens so as to achieve fast early growth.

Fig. 24–45. Japanese quail.

The NRC requirements for pheasants and Bobwhite quail are presented in Table 24–17, and the NRC requirements for Japanese quail are given in Table 24–18; in an earlier section of this chapter headed, "National Research Council (NRC) Requirements."

Commercial game bird feeds are available in most areas. Generally, these are of the following types: *starter ration* containing 28% protein, for the first 6 weeks; *grower ration* containing 20% protein, for 7 to 14 weeks of age; and, depending on whether they are being marketed for game-release for dressed game birds, *finisher ration* or *flight conditioner ration* containing 15% protein, from 15 weeks until market. Commercial game bird feeds may be in the form of mash, crumbles, or pellets. **NOTE:** Game birds require a higher level of protein in early life than chickens.

Example rations for home-mixing of game bird rations are given in Table 24–35. Grit should be available in a separate feeder, with the size grit determined by the size of the birds.

TABLE 24-35
EXAMPLE RATIONS FOR GAME BIRDS

Item	Starter	Grower	Breeder
	(%)	(%)	(%)
Alfalfa meal, sun-cured	7.5	5.5	5.0
Corn, yellow, ground	14.0	44.0	56.7
Meat and bone meal	7.0	0.0	0.0
Fish meal	7.0	0.0	0.0
Sorghum, ground	8.0	0.0	0.0
Soybean meal, 44% protein	41.0	35.0	14.7
Wheat, ground	12.0	0.0	0.0
Wheat middlings	2.0	0.0	0.0
Wheat bran	0.0	12.0	16.8
Limestone, ground	0.0	1.0	4.1
$CaHPO_4 \cdot 2H_2O$	0.0	1.5	1.5
Salt, iodized	0.7	0.4	0.5
DL-methionine	0.3	0.1	0.2
Premix[1]	0.5	0.5	0.5
Calculated Analysis:			
Protein (%)	28.1	20.8	15.1
Metabolizable energy (kcal/kg)(or kcal/2.2 lb)	2,720	2,660	2,570
Calcium (%)	1.0	0.94	2.15
Total phosphorus (%)	0.76	0.76	0.74
Available phosphorus (%)	0.52	0.45	0.44

[1]Premix should contain: in mg per kg (or per 2.2 lb) diet—$MnSO_4 \cdot H_2O$, 40; ZnO, 60; vitamin B-12, (cobalimins) 0.005; menadione sodium bisulfite, 2; riboflavin (vitamin B-2), 6; niacin (nicotinic acid, nicotinamide), 40; calcium pantothenate, 20; folacin (folic acid), 0.5; antioxidant, 100; antibiotic, 10; in IU—vitamin A, 5,000; vitamin D_3, 1,500; vitamin E, 20. An equivalent commercial premix can be used, but follow the directions of the supplier.

Specific information relative to feeding pheasants, Bobwhite quail, and Japanese quail follows:

• **Feeding pheasants**—following the starter phase, a grain supplement consisting of equal portions of (1) corn or grain sorghum, and (2) oats, barley, or wheat can be effectively used in limited amounts; but the grain supplement should not consititute more than one-fourth of the entire ration. Pheasants can be fed and managed to produce fertile eggs at any time of the year. Egg production requires proper feeds and light stimulation.

• **Feeding Bobwhite quail**—Wild quail survive on native grass seeds and insects. Confinement-reared quail require nutritionally balanced rations to promote growth and health. After 6 weeks of age, Bobwhites should be fed according to whether they will be utilized as breeders, flight conditioned for shooting preserves, or processed for meat purposes.

• **Feeding Japanese quail**—Laying rations for Japanese quail should contain about 25% protein; additionally, a free-choice supply of calcium (limestone or oystershell) should be available. Some growers may wish to supplement commercial feeds with small seeds or cracked grain. When Japanese quail are fed whole seeds, fine grit should be provided.

FEEDING GUINEA FOWL

In recent years, there has been a growing interest, worldwide, in the production of guinea fowl. In France and Italy, the commercial production of guinea fowl is a highly profitable industry.

Fig. 24-46. Guinea fowl. (courtesy, USDA)

Hens lay for 35 weeks and produce 175 eggs. Baby guineas are called *guinea poults* or *keets.* At 86 days of age, the keets weigh about 3.3 lb, made with a feed conversion of 2.7 to 3.2.

A 3-phase feeding program is normally followed, consisting of a 24% protein starter, a 19–21% grower, and a 17–18% finisher.

FEEDING OSTRICHES

Feeding ostriches is different! The ostrich is the largest bird in the world. At maturity, it may stand 10 ft tall and weigh more than 330 lb. Young ostriches grow very rapidly; they reach full size in about 6 months, but they do not attain sexual maturity until 3 to 4 years of age. They may live to 70 years of age. The ostrich is the only bird that eliminates its urine and feces separately.

Fig. 24-47. Ostrich.

Ostriches are valued primarily for their skins, which are made into fine quality leather. Spasmodically, the plumes are popular for decorations and accessories. The eggs may be used for human food, but the meat is seldom consumed because it is tough and has an unpleasant taste.

In their native habitat, ostriches feed on succulent plants, fruit, and leaves, as well as on occasional insects, lizards, birds, mice, and turtles. They also eat much sand and gravel to aid in grinding food for digestion. Ostriches drink water when they can find it, but they can survive for long periods without drinking if the plants they eat are green and moist.

Ostrich farms in the United States are relatively new; so, it remains to be seen whether ostriches in this country will be a passing fad or an infant industry.

Good nutrition is essential to the success of ostrich production. Because of the rapid growth of ostrich *chicks* (they reach full size at about 6 months of age), they must receive a diet that is high in protein, and that contains the essential amino acids, along with adequate energy, minerals, and vitamins. Also, the nutrition of breeders affects egg production and hatchability.

The nutritive requirements of ostriches and turkeys appear to be very similar. So, the rations presented in Table 24-31, Example Turkey Rations, may be used for ostriches. Because of the very rapid growth of ostrich chicks, however, it is suggested that they be continued on the high-protein turkey *starter ration* given in 24-31 longer than the 8 weeks suggested for turkey poults.

FEEDING PEAFOWL

The peafowl belongs to the same family as pheasants and chickens, differing primarily in plumage. Although they are prized as ornamental birds (blue, white, black, or green—with blue most common), peafowl are edible and are regarded as a delicacy on special occasions, perhaps more for rareness than for taste.

The care and management of peafowl is similar to that of turkeys. Peachicks may be fed a high-protein (28 to 30%) turkey or game-bird starter feed, preferably crumbles. At about 6 to 8 weeks of age, a game bird grower diet may be substituted for the starter, and small amounts of cracked corn, wild bird seed, or chopped green grass (lawn clippings) may be added to the ration.

When roaming free, adult peafowl eat a variety of seeds, insects, and plants. Additionally, they should be provided some turkey or game-bird feed, bird seed, or grain, with the allowance increased in cold weather.

Fig. 24-48. Peafowl.

SALMONELLA/TOXIC INORGANIC ELEMENTS

Salmonella and/or toxic inorganic elements may be problems in poultry.

• ***Salmonella***—The bacteria that cause the disease Salmonellosis in animals and humans have been around for a very long time. The name *Salmonella* is after the American bacteriologist and veterinarian, Daniel E. Salmon, who first isolated the bacteria in 1885. In recent years, *Salmonella* has been much in the news, primarily because new and sophisticated equipment has enabled scientists more minutely to screen the microbe field, and more intense effort has been made to look for microbes.

Salmonella is a family of bacteria that consists of more than 2,000 different strains, which may be found in feeds—especially animal by-products (meat meal, poultry by-product meal, and fish meal), and in human foods—including broilers, turkeys, and eggs. The disease, known as salmonellosis, may occur if foods contaminated with *Salmonella* are eaten raw, not properly cooked, or mishandled after cooking. The symptoms of the illness are diarrhea, nausea, vomiting, and sometimes fever. The illness may occur within 6 to 72 hours after eating the contaminated food and may last 2 to 6 days.

Salmonella in human foods can be destroyed by proper cooking at a temperature of 160°F, or more, at the centermost part of the thickest item being cooked. The bacteria may also be inactivated by treating dressed poultry with 1%

lactic acid, but it will color the meat slightly; and the bacteria may be completely destroyed by irradiation, but such treatment is not approved by FDA, presently. So, consumers should be urged (1) to refrigerate all animal products, (2) to fully cook all foods of animal origin, and (3) to avoid the recontamination of all foods after cooking. Additionally, producers should use all animal drugs and medications in compliance with FDA regulations and in accord with the directions on the label; and all producers and processors should continue their vigilance in reducing *Salmonella* in all feeds and facilities by rigid sanitation and heat treatment, and by preventing humanly edible animal products from becoming contaminated by *Salmonella*.

Despite all the scare stories, however, American consumers have the blessed assurance that they enjoy the safest and most abundant high quality animal products in the world.

(For an in-depth discussion of *Salmonella* and salmonellosis, see Chapter 5, Table 5-3—Salmonellosis; Chapter 13, sections on "Antibiotics" and "Chemotherapeutics"; and Chapter 23, section on "Additives.")

• **Toxic inorganic elements**—Poultry are susceptible to a number of toxins, any one of which may prove disastrous in a flock. Among them are a number of inorganic elements such as arsenic, lead, and selenium. It is important, therefore, that the poultry producer guard against toxic levels of inorganic elements.

For an in-depth discussion of toxic inorganic elements, see Chapter 5, section on "Potential Poisons," including Table 5-3, Potential Poisons.

QUESTIONS FOR STUDY AND DISCUSSION

1. How do you account for the fact that poultry feeding has changed more than the feeding of any other species?

2. During the last half of the 20th century, per capita red meat consumption declined, while chicken and turkey consumption increased. What forces caused this difference?

3. What is the economic importance of feed for poultry?

4. In 1940, it required 4.7 lb of feed to produce 1 lb of weight gain of broilers vs 1.9 lb of feed in 1990; whereas, during this same period of time, feed required to produce 1 lb of beef has been lowered by about a pound only. Why have broiler producers made more marked progress in feed efficiency than beef cattle producers?

5. Explain how temperature/humidity and genetics may affect the nutritive requirements of poultry.

6. Why do nutritionists add a margin of safety to *requirements,* known as *allowances,* in ration formulation?

7. What are the NRC requirements? How are they used?

8. Define *energy requirement.* What are the primary energy sources? Discuss the relative importance of each energy source.

9. How does the nutritionist use *metabolizable energy (ME)* in ration formulation?

10. What is *true metabolizable energy (TME)*? How does it differ from ME?

11. From the standpoint of poultry nutrition, the amino acids that make up proteins are really the essential nutrients, whereas in cattle nutrition the protein molecule suffices. Why the difference?

12. What factors affect the amino acid requirements of poultry?

13. What is meant by *true digestibility of amino acids?* Explain the difference between *apparent digestibility coefficient* and *true digestibility coefficient.*

14. What minerals are of most practical importance for chickens and turkeys because outside sources of them must be added to most feed formulations?

15. Discuss the function, deficiency symptoms, and practical sources of each of the following major minerals in the nutrition of poultry: salt/(NaCl), calcium, and phosphorus.

16. Discuss the function, deficiency symptoms, and practical sources of each of the following trace minerals in the nutrition of poultry: iodine, manganese, selenium, and zinc.

17. What vitamins are commonly low in poultry rations? Discuss the relative effectiveness of vitamin D_2 (the plant form) and vitamin D_3 (the animal form) in meeting the vitamin D needs of poultry.

18. Discuss the function, deficiency symptoms, and practical sources of each of the following fat-soluble vitamins in the nutrition of poultry: vitamins A, D, and E.

19. Discuss the function, deficiency symptoms, and practical sources of each of the following water-soluble vitamins in the nutrition of poultry: biotin, riboflavin, thiamin, vitamin B-6, and vitamin B-12.

20. What components of water should poultry producers be concerned about when they are in high concentration?

21. What are the primary reasons for using antibiotics in poultry feeds? What is the best explanation for the growth-stimulating activity of antibiotics?

22. Why do poultry producers use each of the following additives: antifungal agents, antioxidants, arsenicals, grit, and marigold petal meal?

23. What major energy feedstuffs are fed to poultry? What major protein feedstuffs are fed to poultry?

24. Why are many consumers concerned about the cholesterol content of eggs? Should egg producers evolve with low-cholesterol eggs?

25. In 1984, more than one-third of the primary commercial feed used in the United States was fed to poultry. Why is so much U.S. commercially-prepared feed fed to poultry?

26. Should a poultry producer use a protein supplement of plant origin, animal origin, or a combination of the two?

27. Why do poultry producers use more feed in the form of crumbles than producers of other classes of livestock?

28. How may a poultry producer use a *feed substitution table* advantageously?

29. What are the main controls of feed intake of poultry?

30. Describe *phase feeding* and *molting* of layers. Why do so many poultry producers apply these practices?

31. How do the nutritive requirements for breeder chickens differ from those for commercial layers?

32. List and discuss three methods of restricted (limited) feeding commonly employed in feeding pullets of the heavy breeds.

33. In modern broiler production, what is considered average (a) market weight, (b) market age, (c) feed conversion, and (d) mortality?

34. What are the primary differences between feeding market turkeys and feeding turkey breeders?

35. Why do the protein requirements of turkeys from 1 day of age to marketing differ so widely?

36. What forces caused the production of ducks in the United States to double from 1975 to 1985?

37. Discuss the two major types of environments in which ducks are grown.

38. List and discuss the variety of programs under which geese are raised in the United States.

39. What is pigeon milk?

40. Outline a feeding program for a game bird farm raising pheasants, Bobwhite quail, Japanese quail, and other game birds.

41. Discuss the feeding of guinea fowl, ostriches, and peafowl.

42. Discuss the importance of *toxic inorganic elements* and *Salmonella*, from the standpoints of (a) poultry and (b) people consuming poultry products.

43. Is the farm poultry flock a thing of the past? If so, why?

SELECTED REFERENCES

Title of Publication	Author(s)	Publisher
Animal Science, 8th Edition	M. E. Ensminger	The Interstate Printers & Publishers, Inc., Danville, Ill., 1983
Calcium in Broiler, Layer, and Turkey Nutrition	R. H. Harms, *et al.*	National Feed Ingredients Assoc., West Des Moines, Iowa, 1976
Commercial Chicken Production, 3rd Edition	M. O. North	The AVI Publishing Co., Inc., Westport, Conn., 1984
Feeds for Livestock, Poultry and Pets	M. H. Gutcho	Noyes Data Corporation, Park Ridge, N.J., 1973
Manual of Poultry Production in the Tropics	R. R. Say	CAB International, Wallingford, Oxon, U.K., 1987
Modern Waterfowl Management and Breeding Guide	O. Grow	American Bantam Association, 1972
Nutrient Requirements of Poultry and Nutritional Research	C. Fisher K. N. Boorman	Butterworths & Co., Ltd., London, England, 1986
Nutrient Requirements of Poultry, No. 1, 8th Revised Edition	National Research Council, Subcommittee on Poultry Nutrition	National Academy of Sciences, Washington, D.C., 1984
Nutrition of the Chicken, 3rd Edition	M. L. Scott M. C. Nesheim R. J. Young	M. L. Scott & Associates, Ithaca, N.Y., 1982
Nutrition of the Turkey	M. L. Scott	M. L. Scott of Ithaca, Ithaca, N.Y., 1987
Phosphorus in Poultry and Game Bird Nutrition	R. H. Harms B. L. Damron	National Feed Ingredients Assoc., West Des Moines, Iowa, 1977
Physiology and Biochemistry of the Domestic Fowl, Vol. 1	Ed. by D. J. Bell, B. M. Freeman	Academic Press, New York, N.Y., 1971
Poultry: Feeds and Nutrition	H. Patrick P. J. Schiable	The AVI Publishing Co., Inc., Westport, Conn., 1980
Poultry Meat and Egg Production	C. R. Parkhurst G. J. Mountney	Van Nostrand Reinhold Co., New York, N.Y., 1988
Poultry Production, 12th Edition	M. C. Nesheim R. E. Austic L. E. Card	Lea & Febiger, Philadelphia, Pa., 1979
Poultry Science, 2nd Edition	M. E. Ensminger	The Interstate Printers & Publishers, Inc., Danville, Ill., 1980
Raising Poultry Successfully	W. Graves	Williamson Publishing, Charlotte, Vt., 1985

Fig. 24–49. Buff Cochins

Fig. 24–50. Buff-Laced Polish

Fig. 24–51. Mottled Houdans

Fig. 24–52. Narragansett Turkeys

Fig. 24–53. Rouen Ducks

Fig. 24–54. African Geese

Figs. 24–49 to 54. Paintings by famous artists of different varieties of chickens, turkeys, and waterfowl. (Courtesy, Watt Publishing Co., Mt. Morris, Ill.)

Original painting by Tom Phillips

Chapter **25**

FEEDING HORSES[1]

Contents — Page

[1]The authors gratefully acknowledge the helpful suggestions of the following eminent authorities who reviewed this chapter: H. F. Hintz, Ph.D., Professor of Animal Nutrition, Cornell University, Ithaca, New York; R. D. Scoggins, D.V.M., Equine Extension Veterinarian, University of Illinois, Urbana-Champaign; W. J. Tyznik, Ph.D., Animal Science Department, The Ohio State University, Columbus; David McGlothlin, Manager, Horse Division, Harris Farms, Coalinga, Calif.; and Charles Pollard, Pollard Ranch, Clovis, Calif.

Contents — Page

(Continued)

Contents (Continued)

Contents

Fig. 25-1. Northern Dancer, the most commercially successful Thoroughbred sire that the world has ever known, shown at age 25. Age 25 in a horse is comparable to a 75-year-old person. (Courtesy, Windfields Farm, Chesapeake City, Md.)

From time immemorial, feed has been the most important influence in the environment of the horse. Through fossil remains, it is evident that the horse's evolution has always paralleled the soil and vegetation. In the beginning, little *Eohippus,* which was about the size of a Fox Terrier dog, had four toes on the front foot and three on the hind, had soft teeth, and was adapted to feeding on the herbage of the swamp. Gradually, it grew taller, its teeth grew stronger and harder, its legs grew longer, and all but one toe disappeared, thereby enabling it to feed farther from water and adapting it to the prairies.

It is only natural, in a world so big, that some equines should fare better than others. Thus, the ponderous horse of Flanders, progenitor of the modern draft horse, was the product of fertile soils, a mild climate, and abundant vegetation; whereas the diminutive, hardy Shetland Pony evolved on the scanty vegetation native to the long, cold winters of the Shetland Isles.

But the effect of feed and nutrition as creative forces on horses did not end with their domestication, about 5,000 years ago. At that time, humans replaced nature and assumed responsibility as their keepers. When one considers that among wild bands 95% foal crops were common and unsoundnesses were relatively unknown, it is apparent that horses haven't fared so well with humans serving as their providers.

Unless horses are fed properly, their maximum potential in reproduction, growth, body form, speed, endurance, style, and attractiveness cannot be achieved. Additionally, the following conditions make it imperative that the nutrition of horses be the best that science and technology can devise:

1. **Confinement.** Many horses are kept in stables or corrals.
2. **Fitting yearlings.** When forcing young equines, it is important to their development and soundness that the ration be nutritionally balanced.
3. **Racing 2-year-olds.** In the United States, we race more 2-year-olds than any other nation in the world; our richest races are for them. If the nutrient content of the ration is

not adequate, there is bound to be more breakdown on the track than with older horses. This is costly.

Fig. 25-2. Some horses fared better than others in their development. The diminutive Shetland Pony (left) evolved on the scanty vegetation of the Shetland Isles, whereas the giant Shire (right) was the product of fertile soils and abundant vegetation of Flanders.

4. **Stress.** Stress may be caused by excitement, temperament, fatigue, number of horses together, previous nutrition, breed, age, and management. Race and show horses are always under stress; and the more tired they become and the greater the speed, the greater the stress. Thus, the ration for race and show horses should be scientifically formulated, rather than based on fads, foibles, and trade secrets. The greater the stress, the more exacting the nutritive requirements.

5. **Horses are unique.** They differ from other farm animals in that they have greater individual value; are kept for recreation, sport, and work; are fed for a longer life of usefulness; have a smaller digestive tract; should not carry surplus weight; and are fed for nerve, mettle, animation, character of muscle, and athletic performance.

ECONOMIC IMPORTANCE OF FEED FOR HORSES

Although feed constitutes the greatest single cost item in the horse business, its relative economic importance varies more widely than with any other class of livestock. This is so because many horses are kept for recreation, sport, or hobby purposes, whereas the vast majority of other species are kept for strictly business reasons—for making money. It follows that the cost of land, buildings and equipment, and labor for horses is usually much higher in relation to the cost of feed than for other animals.

Where no pasture is available, a 1,000–lb horse will, on the average, consume about 25 lb of feed (hay and grain) daily, or 4.6 tons per year. The cost of horse feed will vary in keeping with prevailing feed prices, the quality of the ration, and the proportion of concentrate and roughage fed. Also, caretakers are prone to feed a good many additives, all of which cost money. In total, U.S. horse owners spend $9 billion each year on feed, medication, and services for their horses.

In addition to the cost of horse feed as such, faulty nutrition makes for hidden costs of great economic importance. Horse owners are losing millions of dollars through inefficiency. On the average, they are producing a 50% foal crop —which means that they are keeping two mares a whole year to produce one foal. They are retiring an appalling number of horses from tracks, shows, and other uses due to unsoundnesses. Many of the bone ailments that plague breeders and trainers—the sprains, spavins, splints, and ringbones—are the tragic result of improper skeletal development during the fetal and early growth stages.

NUTRITIVE NEEDS OF HORSES

Meeting the nutrient needs of horses is a major factor in determining their efficiency and years of service. As with other animals, the horse needs nutrients for maintenance, growth, fattening, reproduction, and production. With horses, the production need is for work—mostly for recreation and sport, and as cow ponies. Unlike most animals, however, the work is usually irregular and often very strenuous—characteristics which create a particular stress on the animal and make the job of feeding according to the nutritive needs very difficult.

• **The digestive tract of horses makes for differences**—The digestive tract of equines makes for difficulties in meeting their nutritive needs. It differs anatomically and physiologically from that of the cow (ruminant) as follows:

1. It is smaller, with the result that the horse cannot eat as much roughage as cattle. Also, it functions best at two-thirds capacity. Because of its small size, if a horse is fed too much roughage, labored breathing and quick tiring may result. Actually, the horse's stomach is designed for almost constant intake of small quantities of feed (such as happens when a horse is grazing on pasture), rather than large amounts at one time.

2. Without feed, the horse's stomach will empty completely in 24 hours, whereas it takes about 72 hours (3 times as long) for the cow's stomach to empty. At the time of eating, feed passes through the horse's stomach very rapidly —so much so that the feed eaten at the beginning of the meal passes into the intestine before the last part of the meal is completed.

3. The cow has four compartments (rumen, reticulum, omasum, and the abomasum or true stomach), whereas the horse has one.

4. There is comparatively little microbial action in the stomach of the horse, but much such action in the first compartment (rumen) of the cow.

5. The primary seats of microbial activity in ruminants and horses occupy different locations in the digestive system in relation to the small intestine. In cattle (and sheep), the rumen precedes the small intestine; in horses, the cecum follows it. As a result, the efficiency of absorption of nutrients synthesized by the microorganisms is likely to be lower in a horse than in a ruminant.

The limited protein synthesis in the horse (limited when compared with ruminants), and the lack of efficiency of absorption due to the cecum being on the lower end of the gut (thereby not giving the small intestine a chance at the ingesta after it leaves the cecum), clearly indicate that horse rations should contain high-quality proteins, adequate in amino acids.

In comparison to a cow, therefore, a horse should be fed less roughage, more and higher quality protein, and perhaps added B vitamins. Actually, the nutrient requirements of a horse more nearly parallel those of a pig than a cow.

• **Horses have special nutritive needs for bone formation and maintenance**—In addition to meeting the nutritive needs common to all species—the needs for maintenance, growth, fattening, reproduction, and production—the formation and maintenance of sound bones in horses is extremely important and very complicated. The 205 bones of the horse's skeleton contain 25% ash or minerals, 20% protein, 10% fat, and 45% water.

Since bone contains 20% protein, proper protein nutrition is very important; for example, the weanling requires 0.6 to 0.7% of the amino acid lysine in the ration.

The ash of bone contains about 36% calcium, 17% phosphorus, 0.8% magnesium, and small, but important amounts of iron and copper—some 16 to 20 minerals are present and essential.

But bone formation and maintenance require certain vitamins, too. Vitamin A is essential for sound bones and teeth; and vitamin D, along with calcium and phosphorus, is vital. A deficiency of vitamin D in the horse's ration results in reduced bone calcification, swollen joints, stiffness of gait, softness of bones, bone deformities, and frequent cases of fracture.

Vitamin C is involved in the formation and maintenance of intracellular material, including collagen and related substances in the bones and soft tissues. Also, vitamin C supplementation may be beneficial for horses under stress such as during rapid growth and high-level performance, in hot weather, or when something interferes with vitamin synthesis.

Other minerals and vitamins may also be beneficial for the proper formation and maintenance of the bones of the horse, as well as for a healthy horse. So, many owners with high-level performance animals supplement with additional minerals and vitamins to ensure adequate nutrition.

• **Horses under stress have special nutritive needs**—Some scientists and horse owners feel that higher nutritive levels are helpful under high stress conditions, such as when horses are subjected to training, transporting, racing, performing, crowds, and various environmental conditions.

A high-level performance horse is the counterpart of a human athlete. However, the equine athlete is disadvantaged by earlier training than the human athlete. For example, if humans have a life span of 77 years and horses have a life span of 22 years, a 2-year-old horse is comparable to a 7-year-old human. How many human athletes are subjected to training at age 7?

• **Horses should be fed as individuals**—Most high-performing horses are *prima donnas;* hence, they require individual attention. They must eat their feed; otherwise, no matter how nutritionally complete it may be, it won't do them any good.

Meeting the nutritive needs of horses is further complicated because these needs do not necessarily remain the same from day to day or from period to period. The age and size of the animal; the stage of gestation or lactation of a mare; the kind and degree of activity; climatic conditions; the kind, quality, and amount of feed; the system of management; and the health, condition, and temperament of the animal are all continually exerting a powerful influence in determining its nutritive needs. How well the caretaker understands, anticipates, interprets, and meets these requirements usually determines the success or failure of the ration. For these reasons, no set of instructions, calculator, or book of knowledge can substitute for experience and born horse intuition. Skill and good judgment are essential.

National Research Council (NRC) Requirements

The nutrient requirements of horses are given in Tables 25-1, 25-2, 25-3, 25-4, and 25-5, which follow.

In addition to the NRC tables presented in this section, *Nutrient Requirements of Horses,* 5th rev. ed., 1989, also presents tables giving the daily nutrition requirements of horses with mature weights of 1540 lb *(700 kg),* 1760 lb *(800 kg),* and 1980 lb *(900 kg).* Also, the new NRC computer program calculates energy requirements based on the foal's body weight, age, and average daily gain; then, it calculates the requirements for most of the other nutrients based on this energy requirement.

TABLE 25-1
DAILY NUTRIENT REQUIREMENTS PER PONY, 440 LB *(200 KG)* MATURE WEIGHT[1] (See footnotes at end of table.)

Animal	Body Weight		Daily Gain	Energy		Protein		Calcium	Phosphorus	Magnesium	Potassium	Vitamin A Activity
				TDN	Digestible Energy	Crude Protein	Lysine					
	(lb)	*(kg)*	(lb)	(lb)	(Mcal)	(lb)	(g)	(g)	(g)	(g)	(g)	(1,000 IU)
Mature horses												
Maintenance	440	*200*		3.7	7.4	0.65	10	8	6	3.0	10.0	6
Stallions (breeding season)	440	*200*		4.7	9.3	0.81	13	11	8	4.3	14.1	9
Pregnant mares[2]												
9 months	440	*200*		4.1	8.2	0.79	13	16	12	3.9	13.1	12
10 months	440	*200*		4.2	8.4	0.81	13	16	12	4.0	13.4	12
11 months	440	*200*		4.5	8.9	0.86	14	17	13	4.3	14.2	12
Lactating mares												
Foaling to 3 months	440	*200*		6.9	13.7	1.51	24	27	18	4.8	21.2	12
3 months to weaning	440	*200*		6.1	12.2	1.16	18	18	11	3.7	14.8	12
Working horses												
Light work[3]	440	*200*		4.6	9.3	0.81	13	11	8	4.3	14.1	9
Moderate work[4]	440	*200*		5.5	11.1	0.97	16	14	10	5.1	16.9	9
Intense work[5]	440	*200*		7.4	14.8	1.30	21	18	13	6.8	22.5	9

(Continued)

TABLE 25-1 *(Continued)*

Animal	Body Weight		Daily Gain	Energy		Protein		Cal- cium	Phos- phorus	Magne- sium	Potas- sium	Vitamin A Activity
				TDN	Digestible Energy	Crude Protein	Lysine					
	(lb)	*(kg)*	(lb)	(lb)	(Mcal)	(lb)	(g)	(g)	(g)	(g)	(g)	(1,000 IU)
Growing horses												
Weanling, 4 months	165	*75*	0.88	3.7	7.3	0.80	15	16	9	1.6	5.0	3
Weanling, 6 months												
Moderate growth	209	*95*	0.66	3.8	7.6	0.83	16	13	7	1.8	5.7	4
Rapid growth	209	*95*	0.88	4.4	8.7	0.95	18	17	9	1.9	6.0	4
Yearling, 12 months												
Moderate growth	309	*140*	0.44	4.4	8.7	0.86	17	12	7	2.4	7.6	6
Rapid growth	309	*140*	0.66	5.2	10.3	1.01	19	15	8	2.5	7.9	6
Long yearling, 18 months												
Not in training	375	*170*	0.22	4.2	8.3	0.82	16	10	6	2.7	8.8	8
In training	375	*170*	0.22	5.8	11.6	1.14	22	14	8	3.7	12.2	8
2-year-old, 24 months												
Not in training	408	*185*	0.11	4.0	7.9	0.74	13	9	5	2.8	9.4	8
In training	408	*185*	0.11	5.7	11.4	1.06	19	13	7	4.1	13.5	8

[1] Adapted by the authors from *Nutrient Requirements of Horses,* 5th rev. ed., NRC-National Academy of Sciences, 1989, p. 42. To convert lb to kg, divide by 2.2.

[2] Mares should gain weight during late gestation to compensate for tissue deposition. However, nutrient requirements are based on maintenance body weight.

[3] Examples are horses used in Western and English pleasure, bridle path hack, equitation, etc.

[4] Examples are horses used in ranch work, roping, cutting, barrel racing, jumping, etc.

[5] Examples are horses in race training, polo, etc.

TABLE 25-2
DAILY NUTRIENT REQUIREMENTS PER HORSE, 880 LB *(400 KG)* MATURE WEIGHT[1]

Animal	Body Weight		Daily Gain	Energy		Protein		Cal- cium	Phos- phorus	Magne- sium	Potas- sium	Vitamin A Activity
				TDN	Digestible Energy	Crude Protein	Lysine					
	(lb)	*(kg)*	(lb)	(lb)	(Mcal)	(lb)	(g)	(g)	(g)	(g)	(g)	(1,000 IU)
Mature horses												
Maintenance	880	*400*		6.7	13.4	1.17	19	16	11	6.0	20.0	12
Stallions (breeding season)	880	*400*		8.4	16.8	1.47	23	20	15	7.7	25.5	18
Pregnant mares[2]												
9 months	880	*400*		7.5	14.9	1.43	23	28	21	7.1	23.8	24
10 months	880	*400*		7.6	15.1	1.46	23	29	22	7.3	24.2	24
11 months	880	*400*		8.1	16.1	1.55	25	31	23	7.7	25.7	24
Lactating mares												
Foaling to 3 months	880	*400*		11.5	22.9	2.50	40	45	29	8.7	36.8	24
3 months to weaning	880	*400*		9.9	19.7	1.84	29	29	18	6.9	26.4	24
Working horses												
Light work[3]	880	*400*		8.4	16.8	1.47	23	20	15	7.7	25.5	18
Moderate work[4]	880	*400*		10.5	20.1	1.76	28	25	17	9.2	30.6	18
Intense work[5]	880	*400*		13.4	26.8	2.35	38	33	23	12.3	40.7	18
Growing horses												
Weanling, 4 months	320	*145*	1.87	6.8	13.5	1.48	28	33	18	3.2	9.8	7
Weanling, 6 months												
Moderate growth	397	*180*	1.21	6.5	12.9	1.41	27	25	14	3.4	10.7	8
Rapid growth	397	*180*	1.54	7.3	14.5	1.53	30	30	16	3.6	11.1	8
Yearling, 12 months												
Moderate growth	584	*265*	1.29	7.8	15.6	1.57	30	23	13	4.5	14.5	12
Rapid growth	584	*265*	1.29	8.6	17.1	2.12	33	27	15	4.6	14.8	12
Long yearling, 18 months												
Not in training	728	*330*	1.60	7.9	15.9	1.57	30	21	12	5.3	17.3	15
In training	728	*330*	1.60	10.3	21.6	2.12	41	29	16	7.1	23.4	15
2-year-old, 24 months												
Not in training	805	*365*	1.77	7.7	15.3	1.42	26	19	11	5.7	18.7	16
In training	805	*365*	1.77	10.8	21.5	2.00	37	27	15	7.9	26.2	16

[1] Adapted by the authors from *Nutrient Requirements of Horses,* 5th rev. ed., NRC-National Academy of Sciences, 1989, p. 43. To convert lb to kg, divide by 2.2.

[2] Mares should gain weight during late gestation to compensate for tissue deposition. However, nutrient requirements are based on maintenance body weight.

[3] Examples are horses used in Western and English pleasure, bridle path hack, equitation, etc.

[4] Examples are horses used in ranch work, roping, cutting, barrel racing, jumping, etc.

[5] Examples are horses in race training, polo, etc.

TABLE 25-3
DAILY NUTRIENT REQUIREMENTS PER HORSE, 1,100 LB *(500 KG)* MATURE WEIGHT[1]

Animal	Body Weight		Daily Gain	Energy		Protein		Cal-cium	Phos-phorus	Magne-sium	Potas-sium	Vitamin A Activity
				TDN	Digestible Energy	Crude Protein	Lysine					
	(lb)	*(kg)*	(lb)	(lb)	(Mcal)	(lb)	(g)	(g)	(g)	(g)	(g)	(1,000 IU)
Mature horses												
Maintenance	1,100	*500*		8.2	16.4	1.44	23	20	14	7.5	25.0	15
Stallions (breeding season)	1,100	*500*		10.3	20.5	1.79	29	25	18	9.4	31.2	22
Pregnant mares[2]												
9 months	1,100	*500*		9.1	18.2	1.75	28	35	26	8.7	29.1	30
10 months	1,100	*500*		9.3	18.5	1.78	29	35	27	8.9	29.7	30
11 months	1,100	*500*		9.9	19.7	1.89	30	37	28	9.4	31.5	30
Lactating mares												
Foaling to 3 months	1,100	*500*		14.2	28.3	3.12	50	56	36	10.9	46.0	30
3 months to weaning	1,100	*500*		14.2	24.3	2.29	37	36	22	8.6	33.0	30
Working horses												
Light work[3]	1,100	*500*		10.3	20.5	1.79	29	25	18	9.4	31.2	22
Moderate work[4]	1,100	*500*		12.3	24.6	2.15	34	30	21	11.3	37.4	22
Intense work[5]	1,100	*500*		16.4	32.8	2.87	46	40	29	15.1	49.9	22
Growing horses												
Weanling, 4 months	385	*175*	1.87	7.2	14.4	1.58	30	34	19	3.7	11.3	8
Weanling, 6 months												
Moderate growth	473	*215*	1.43	7.5	15.0	1.64	32	29	16	4.0	12.7	10
Rapid growth	473	*215*	1.87	8.6	17.2	1.88	36	36	20	4.3	13.3	10
Yearling, 12 months												
Moderate growth	715	*325*	1.10	9.5	18.9	1.86	36	29	16	5.5	17.8	15
Rapid growth	715	*325*	1.43	10.7	21.3	2.09	40	34	19	5.7	18.2	15
Long yearling, 18 months												
Not in training	880	*400*	0.77	9.9	19.8	1.95	38	27	15	6.4	21.1	18
In training	880	*400*	0.77	13.6	26.5	2.61	50	36	20	8.6	28.2	18
2-year-old, 24 months												
Not in training	990	*450*	0.44	9.4	18.8	1.75	32	24	13	7.0	23.1	20
In training	990	*450*	0.44	13.2	26.3	2.44	45	34	19	9.8	32.2	20

[1]Adapted by the authors from *Nutrient Requirements of Horses,* 5th rev. ed., NRC-National Academy of Sciences, 1989, p. 43. To convert lb to kg, divide by 2.2.

[2]Mares should gain weight during late gestation to compensate for tissue deposition. However, nutrient requirements are based on maintenance body weight.

[3]Examples are horses used in Western and English pleasure, bridle path hack, equitation, etc.

[4]Examples are horses used in ranch work, roping, cutting, barrel racing, jumping, etc.

[5]Examples are horses in race training, polo, etc.

TABLE 25-4
DAILY NUTRIENT REQUIREMENTS PER HORSE, 1,320 LB *(600 KG)* MATURE WEIGHT[1]

Animal	Body Weight		Daily Gain	Energy		Protein		Cal-cium	Phos-phorus	Magne-sium	Potas-sium	Vitamin A Activity
				TDN	Digestible Energy	Crude Protein	Lysine					
	(lb)	*(kg)*	(lb)	(lb)	(Mcal)	(lb)	(g)	(g)	(g)	(g)	(g)	(1,000 IU)
Mature horses												
Maintenance	1,320	*600*		9.7	19.4	1.70	27	24	17	9.0	30.0	18
Stallions (breeding season)	1,320	*600*		12.5	24.3	2.12	34	30	21	11.2	36.9	27
Pregnant mares[2]												
9 months	1,320	*600*		10.8	21.5	2.07	33	41	31	10.3	34.5	36
10 months	1,320	*600*		10.9	21.9	2.11	34	42	32	10.5	35.1	36
11 months	1,320	*600*		11.7	23.3	2.24	36	44	34	11.2	37.2	36
Lactating mares												
Foaling to 3 months	1,320	*600*		16.9	33.7	3.74	60	67	43	13.1	55.2	36
3 months to weaning	1,320	*600*		14.5	28.9	2.75	44	43	27	10.4	39.6	36
Working horses												
Light work[3]	1,320	*600*		12.6	24.3	2.12	34	30	21	11.2	36.9	27
Moderate work[4]	1,320	*600*		14.6	29.1	2.55	41	36	25	13.4	44.2	27
Intense work[5]	1,320	*600*		19.4	38.8	3.40	54	47	34	17.8	59.0	27
Growing horses												
Weanling, 4 months	440	*200*	2.20	8.3	16.5	1.80	35	40	22	4.3	13.0	9
Weanling, 6 months												
Moderate growth	539	*245*	1.65	8.5	17.0	1.86	36	34	19	4.6	14.5	11
Rapid growth	539	*245*	2.09	9.6	19.2	2.10	40	40	22	4.9	15.1	11
Yearling, 12 months												
Moderate growth	825	*375*	1.43	11.4	22.7	2.24	43	36	20	6.4	20.7	17
Rapid growth	825	*375*	1.76	12.6	25.1	2.47	48	41	22	6.6	21.2	17
Long yearling, 18 months												
Not in training	1,045	*475*	0.99	12.0	23.9	2.36	45	33	18	7.7	25.1	21
In training	1,045	*475*	0.99	16.0	32.0	3.13	60	44	24	10.2	33.3	21

(Continued)

TABLE 25–4 *(Continued)*

Animal	Body Weight		Daily Gain	Energy: TDN	Energy: Digestible Energy	Protein: Crude Protein	Protein: Lysine	Calcium	Phosphorus	Magnesium	Potassium	Vitamin A Activity
Growing horses (Continued)	(lb)	*(kg)*	(lb)	(lb)	(Mcal)	(lb)	(g)	(g)	(g)	(g)	(g)	(1,000 IU)
2-year-old, 24 months												
Not in training	1,188	*540*	0.66	11.8	23.5	2.18	40	31	17	8.5	27.9	24
In training	1,188	*540*	0.66	16.7	32.3	3.00	55	43	24	11.6	38.4	24

[1]Adapted by the authors from *Nutrient Requirements of Horses,* 5th rev. ed., NRC-National Academy of Sciences, 1989, p. 44. To convert lb to kg, divide by 2.2.
[2]Mares should gain weight during late gestation to compensate for tissue deposition. However, nutrient requirements are based on maintenance body weight.
[3]Examples are horses used in Western and English pleasure, bridle path hack, equitation, etc.
[4]Examples are horses used in ranch work, roping, cutting, barrel racing, jumping, etc.
[5]Examples are horses in race training, polo, etc.

TABLE 25–5
NUTRIENT CONCENTRATIONS IN RATIONS FOR HORSES AND PONIES[1]

Animal	Ration Proportions[2]: Conc.	Ration Proportions[2]: Hay	Moisture Basis A-F (As-fed) M-F (Moisture-free)	TDN	Digestible Energy		Crude Protein	Lysine	Calcium	Phosphorus	Magnesium	Potassium	Vitamin A Activity	
	(%)	(%)		(lb)	(Mcal/lb)	*(Mcal/kg)*	(%)	(%)	(%)	(%)	(%)	(%)	(IU/lb)	*(IU/kg)*
Mature horses														
Maintenance	0	100	**A-F**	**0.40**	**0.80**	***1.80***	**7.2**	**0.25**	**0.21**	**0.15**	**0.08**	**0.27**	**750**	***1,650***
			M-F	0.45	0.90	*2.00*	8.0	0.28	0.24	0.17	0.09	0.30	830	*1,830*
Stallions	30	70	**A-F**	**0.50**	**1.00**	***2.15***	**8.6**	**0.30**	**0.26**	**0.19**	**0.10**	**0.33**	**1,080**	***2,370***
			M-F	0.55	1.10	*2.40*	9.6	0.34	0.29	0.21	0.11	0.36	1,200	*2,640*
Pregnant mares														
9 months	20	80	**A-F**	**0.45**	**0.90**	***2.00***	**8.9**	**0.31**	**0.39**	**0.29**	**0.10**	**0.32**	**1,510**	***3,330***
			M-F	0.50	1.00	*2.25*	10.0	0.35	0.43	0.32	0.10	0.35	1,680	*3,710*
10 months	20	80	**A-F**	**0.45**	**0.90**	***2.00***	**9.0**	**0.32**	**0.39**	**0.30**	**0.10**	**0.33**	**1,490**	***3,280***
			M-F	0.50	1.00	*2.25*	10.0	0.35	0.43	0.32	0.10	0.36	1,660	*3,650*
11 months	30	70	**A-F**	**0.50**	**1.00**	***2.15***	**9.5**	**0.33**	**0.41**	**0.31**	**0.10**	**0.35**	**1,490**	***3,280***
			M-F	0.55	1.10	*2.40*	10.6	0.37	0.45	0.34	0.11	0.38	1,660	*3,650*
Lactating mares														
Foaling to 3 months	50	50	**A-F**	**0.55**	**1.10**	***2.35***	**12.0**	**0.41**	**0.47**	**0.30**	**0.09**	**0.38**	**1,130**	***2,480***
			M-F	0.60	1.20	*2.60*	13.2	0.46	0.52	0.34	0.10	0.42	1,250	*2,750*
3 months to weaning	35	65	**A-F**	**0.53**	**1.05**	***2.20***	**10.0**	**0.34**	**0.33**	**0.20**	**0.08**	**0.30**	**1,240**	***2,720***
			M-F	0.58	1.15	*2.45*	11.0	0.37	0.36	0.22	0.09	0.33	1,370	*3,020*
Working horses														
Light work[3]	35	65	**A-F**	**0.53**	**1.05**	***2.20***	**8.8**	**0.32**	**0.27**	**0.19**	**0.10**	**0.34**	**1,100**	***2,420***
			M-F	0.58	1.15	*2.45*	9.8	0.35	0.30	0.22	0.11	0.37	1,220	*2,690*
Moderate work[4]	50	50	**A-F**	**0.55**	**1.10**	***2.40***	**9.4**	**0.35**	**0.28**	**0.22**	**0.11**	**0.36**	**970**	***2,140***
			M-F	0.60	1.20	*2.65*	10.4	0.37	0.31	0.23	0.12	0.39	1,100	*2,420*
Intense work[5]	65	35	**A-F**	**0.60**	**1.20**	***2.55***	**10.3**	**0.36**	**0.31**	**0.23**	**0.12**	**0.39**	**800**	***1,760***
			M-F	0.65	1.30	*2.85*	11.4	0.40	0.35	0.25	0.13	0.43	890	*1,950*
Growing horses														
Weanling, 4 months	70	30	**A-F**	**0.63**	**1.25**	***2.60***	**13.1**	**0.54**	**0.62**	**0.34**	**0.07**	**0.27**	**650**	***1,420***
			M-F	0.70	1.40	*2.90*	14.5	0.60	0.68	0.38	0.08	0.30	720	*1,580*
Weanling, 6 months														
Moderate growth	70	30	**A-F**	**0.63**	**1.25**	***2.60***	**13.0**	**0.55**	**0.50**	**0.28**	**0.07**	**0.27**	**760**	***1,680***
			M-F	0.70	1.40	*2.90*	14.5	0.61	0.56	0.31	0.08	0.30	850	*1,870*
Rapid growth	70	30	**A-F**	**0.63**	**1.25**	***2.60***	**13.1**	**0.55**	**0.55**	**0.30**	**0.07**	**0.27**	**670**	***1,470***
			M-F	0.70	1.40	*2.90*	14.5	0.61	0.61	0.34	0.08	0.30	740	*1,630*
Yearling, 12 months														
Moderate growth	60	40	**A-F**	**0.58**	**1.15**	***2.50***	**11.3**	**0.48**	**0.39**	**0.21**	**0.07**	**0.27**	**890**	***1,950***
			M-F	0.65	1.30	*2.80*	12.6	0.53	0.43	0.24	0.08	0.30	980	*2,160*
Rapid growth	60	40	**A-F**	**0.58**	**1.15**	***2.50***	**11.3**	**0.48**	**0.40**	**0.22**	**0.07**	**0.27**	**790**	***1,730***
			M-F	0.65	1.30	*2.80*	12.6	0.53	0.45	0.25	0.08	0.30	870	*1,920*
Long yearling, 18 months														
Not in training	45	55	**A-F**	**0.53**	**1.05**	***2.30***	**10.1**	**0.43**	**0.31**	**0.17**	**0.07**	**0.27**	**930**	***2,050***
			M-F	0.58	1.15	*2.50*	11.3	0.48	0.34	0.19	0.08	0.30	1,030	*2,270*
In training	50	50	**A-F**	**0.55**	**1.10**	***2.40***	**10.8**	**0.45**	**0.32**	**0.18**	**0.08**	**0.27**	**740**	***1,620***
			M-F	0.60	1.20	*2.65*	12.0	0.50	0.36	0.20	0.09	0.30	820	*1,800*
2-year-old, 24 months														
Not in training	35	65	**A-F**	**0.50**	**1.00**	***2.20***	**9.4**	**0.38**	**0.28**	**0.15**	**0.08**	**0.27**	**1,080**	***2,380***
			M-F	0.58	1.15	*2.45*	10.4	0.42	0.31	0.17	0.09	0.30	1,200	*2,640*
In training	50	50	**A-F**	**0.55**	**1.10**	***2.40***	**10.1**	**0.41**	**0.31**	**0.17**	**0.09**	**0.29**	**840**	***1,840***
			M-F	0.60	1.20	*2.65*	11.3	0.45	0.34	0.20	0.10	0.32	930	*2,040*

[1]Adapted by the authors from *Nutrient Requirements of Horses,* 5th rev. ed., NRC-National Academy of Sciences, 1989, pp. 46 and 47.
[2]Values assume a concentrate feed containing 1.5 Mcal/lb *(3.3 Mcal/kg)* and hay containing 0.91 Mcal/lb *(2.00 Mcal/kg)* of dry matter.
[3]Examples are horses used in Western and English pleasure, bridle path hack, equitation, etc.
[4]Examples are horses used in ranch work, roping, cutting, barrel racing, jumping, etc.
[5]Examples are horses in race training, polo, etc.

Nutrient Requirements Vs Allowances

In ration formulation, two words are commonly used—*requirements* and *allowances.* Requirements do not provide for margins of safety. Thus, to feed a horse on the basis of meeting the bare requirements would not be unlike building a bridge without providing margins of safety for heavier than average loads or for floods. No competent engineer would be so foolish as to design such a bridge. Likewise, knowledgeable horse nutritionists provide for margins of safety—they provide for the necessary nutritive allowances. They allow for variations in feed composition; possible losses during storage and processing; day-to-day, and period-to-period, differences in needs of animals; age and size of animal; stage of gestation and lactation; the kind and degree of activity; the amount of stress; the system of management; the health, condition, and temperament of the animal; and the kind, quality, and amount of feed—all of which exert a powerful influence in determining nutritive needs.

Scientists normally list the mineral and vitamin requirements of horses on the basis of body weight or units per day. To facilitate application by horse owners, the authors of this book give recommended nutrient allowances in Tables 25-8 and 25-9 three ways: (1) per horse daily, (2) nutrient concentration in ration, and (3) nutrient concentration per ton of feed (as-fed basis).

It is recognized that nutrient allowances provided in the ration (in either a complete ration [with the grain and hay mixed together], or just in the concentrate) are subject to some variation because they vary with the consumption of the ration. For example, if the horse eats half as much feed as recommended, the percent of nutrients (protein, minerals, vitamins) must be doubled in order that the same amount of each nutrient will be consumed. The latter problem can be alleviated by providing (1) the protein/mineral/vitamin/unidentified factor mix, or (2) the mineral/vitamin/ unidentified factor mix on a per horse per day basis, as a separate additive top-dressed on the feed of each horse. The hazards of providing the additive on a per horse per day basis are: caretakers giving too much, too little, or none at all of the additive; and/or the additive segregating out and being left uneaten in the bottom of the feed container.

Recommended Nutrient Allowances

Presently available information indicates that the recommended nutrient allowances given in Table 25-6 will meet the minimum requirements for horses and provide reasonable margins of safety. Additional recommended allowance figures for minerals are given in Table 25-8, Horse Mineral Chart, pp. 1078–1081, under the heading "Nutrient Allowances"; and additional recommended allowance figures for vitamins are given in Table 25-9, pp. 1086–1091, Horse Vitamin Chart, under the heading "Nutrient Allowances."

TABLE 25-6
RECOMMENDED NUTRIENT ALLOWANCES FOR HORSES (TOTAL RATION/AS-FED BASIS)[1] (See footnotes at end of table.)

	Mature Horses (Consuming 25 lb feed/horse/day. Idle horses require less feed and/or consume more roughage than heavily worked horses or lactating mares.)					Young Horses, Based on Mature Weight 1,000 lb				
	Idle Horses/ Light Work/ Moderate Work (1,000 lb Wt.)	Heavy Training/ Heavy Work (1,000 lb Wt.)	Stallions in Breeding Season (1,000 lb Wt.)	Mares, Last 90 Days Gestation (1,000 lb Wt.)	Mares, Peak of Lactation (1,000 lb Wt.)	Creep Feed (250 lb Body Wt/11 lb Feed Daily)	Weanlings (450 lb Body Wt/12 lb Feed Daily)	Yearlings (650 lb Body Wt/13 lb Feed Daily)	2-Yr-Olds & 3-Yr-Olds (800 lb Body Wt/14 lb Feed Daily)	2-Yr-Olds in Light Training (800 lb Body Wt/15 lb Feed Daily)
Digestible Energy:										
TDN[2] (%)	55	62.50	75	62.50	75	75	75	70	60	65
Mcal per (lb)	0.8	1.2	1.0	0.90	1.10	1.25	1.25	1.15	1.00	1.10
Mcal per *(kg)*[3]	*1.80*	*2.55*	*2.15*	*2.0*	*2.35*	*2.60*	*2.60*	*2.50*	*2.20*	*2.40*
Crude Protein (%)	9.0	11.0	14.0	13.0	14.0	18.0	16.0	14.0	13.0	13.0
Lysine (%)	0.25	0.36	0.30	0.32	0.41	0.54	0.55	0.48	0.38	0.41
Major or Macrominerals:										
Salt (%)	0.75	0.75	0.75	0.75	0.75	0.75	0.75	0.75	0.75	0.75
Calcium (%)	0.21	0.31	0.26	0.29	0.47	0.62	0.55	0.40	0.28	0.31
Phosphorus (%)	0.15	0.23	0.19	0.30	0.30	0.34	0.30	0.22	0.15	0.17
Magnesium (%)	0.08	0.12	0.10	0.10	0.09	0.07	0.07	0.07	0.08	0.09
Potassium (%)	0.27	0.39	0.33	0.33	0.38	0.27	0.27	0.27	0.27	0.29
Sulfur (%)	0.15	0.15	0.15	0.15	0.15	0.15	0.15	0.15	0.15	0.15
Trace or Microminerals:										
Cobalt (ppm)[4]	0.11	0.11	0.11	0.11	0.11	0.11	0.11	0.11	0.11	0.11
Copper (ppm)	25	25	25	25	30	40	40	30	25	25
Iodine (ppm)	0.11	0.11	0.11	0.11	0.11	0.11	0.11	0.11	0.11	0.11
Iron (ppm)	40	60	90	90	90	90	80	60	60	60
Manganese (ppm)	46	46	46	46	46	46	46	46	46	46
Selenium (ppm)	0.11	0.11	0.11	0.11	0.11	0.11	0.11	0.11	0.11	0.11
Zinc (ppm)	80	90	90	100	100	100	100	100	90	90
	(/lb)	**(/lb)**	**(/lb)**	**(/lb)**	**(/lb)**	**(/lb)**	**(/lb)**	**(/lb)**	**(/lb)**	**(/lb)**
Fat-soluble Vitamins in Feed:										
Vitamin A (IU)	1,045	1,045	1,045	1,569	1,569	1,045	1,045	1,045	1,045	1,045
Vitamin D (IU)	156	156	156	314	314	419	419	419	419	419
Vitamin E (IU)	26	41	41	41	41	41	41	41	41	41
Vitamin K (mg)	0.32	0.32	0.32	0.32	0.32	0.30	0.30	0.30	0.30	0.30

(Continued)

TABLE 25-6 *(Continued)*

	Mature Horses (Consuming 25 lb feed/horse/day. Idle horses require less feed and/or consume more roughage than heavily worked horses or lactating mares.)					Young Horses, Based on Mature Weight 1,000 lb				
	Idle Horses/ Light Work/ Moderate Work (1,000 lb Wt.)	Heavy Training/ Heavy Work (1,000 lb Wt.)	Stallions in Breeding Season (1,000 lb Wt.)	Mares, Last 90 Days Gestation (1,000 lb Wt.)	Mares, Peak of Lactation (1,000 lb Wt.)	Creep Feed (250 lb Body Wt/11 lb Feed Daily)	Weanlings (450 lb Body Wt/12 lb Feed Daily)	Yearlings (650 lb Body Wt/13 lb Feed Daily)	2-Yr-Olds & 3-Yr-Olds (800 lb Body Wt/14 lb Feed Daily)	2-Yr-Olds in Light Training (800 lb Body Wt/15 lb Feed Daily)
	(/lb)	(/lb)	(/lb)	(/lb)	(/lb)	(/lb)	(/lb)	(/lb)	(/lb)	(/lb)
Water-soluble Vitamins in Feed:										
Biotin (mg)	0.1	0.1	0.1	0.1	0.1	0.1	0.1	0.1	0.1	0.1
Choline (mg)	20	30	30	30	30	62.5	62.5	62.5	62.5	62.5
Folacin (mg)	0.8	1.2	1.2	1.2	1.2	3.0	3.0	3.0	3.0	3.0
Niacin (mg)	10	20.8	10	10	10	10	10	10	10	10
Pantothenic acid (mg)	10	20.8	10	10	10	10	10	10	10	10
Riboflavin (mg)	1.6	1.6	1.6	1.6	1.6	1.6	1.6	1.6	1.6	1.6
Thiamin (B-1) (mg)	1.57	2.61	1.57	1.57	1.57	1.57	1.57	1.57	1.57	1.57
Vitamin B-6 (mg)	1.0	1.0	1.0	1.0	1.0	0.5	0.5	0.5	0.5	0.5
Vitamin B-12 (mg)	0.005	0.006	0.006	0.006	0.006	0.007	0.007	0.007	0.007	0.007
Vitamin C (Ascorbic acid) (mg)	2.4	4.0	4.0	4.0	4.0	3.75	3.75	3.75	3.75	3.75

[1]Where hay is fed separately, double the amounts shown in this table should be added to the concentrate.

[2]1 lb TDN = 2 Mcal or 2,000 Kcal.

[3]1 kg = 2.2 lb or 1,000 g.

[4]1 ppm (parts per million) = 1 mg/kg.

In the discussion that follows, the nutrient requirements and recommended allowances of the horse are discussed under these headings: (1) energy (carbohydrates and fats), (2) protein, (3) minerals, (4) vitamins, and (5) water.

Energy

The energy requirements of horses for various activities are hard to develop because it is difficult to express quantitatively the type of exercise, the intensity and duration of work, the condition and training of the animals, the ability and weight of the rider and driver, the degree of fatigue, and the environmental temperature—all of which influence energy requirements. Based on Cornell studies by Pagan, Hintz reported the energy requirements given in Table 25-7. The Cornell researchers found that the amount of energy expended was proportional to the body weight of the riderless horse or the combined weight of the horse plus the rider, and that the amount of energy expended was exponentially related to speed. Additional studies are needed to determine the energy expenditures at speeds faster than the 13 miles per hour reported in Table 25-7.

TABLE 25-7
DIGESTIBLE ENERGY REQUIREMENTS FOR VARIOUS ACTIVITIES OF LIGHT HORSES[1]

Gait	Speed (Miles/Hour)[2]	DE/Hour (Kcal/kg of Wt.)[3]
Slow walk	2.2	1.7
Fast walk	3.6	2.5
Slow trot	7.5	6.5
Medium trot	9.3	9.5
Fast trot/slow canter	11.2	13.7
Medium canter	13.0	19.5

[1]Hintz, H. F., Energy Requirements of Horses, *Feed Management,* Vol. 37, No. 2, Feb. 1986, p. 15.

[2]To convert to metric, see Appendix, Weights and Measures.

[3]Body weight of horse plus weight of rider and tack.

Fig. 25-3. When racing, horses may require up to 100 times more energy when running than at rest. (Courtesy, *The Backstretch,* Detroit, Mich.)

A lack of energy intake may cause slow and stunted growth in foals and loss of weight, poor condition, and excessive fatigue in mature horses. Excess energy may result in obese horses, which are more susceptible to stress and founder and have lowered reproductive efficiency and decreased longevity.

The caretaker may base individual horse energy requirements on observation. If the horse is too thin, increase the energy intake; if too fat, decrease energy intake.

Although proteins, fats, and carbohydrates can be used to provide energy for maintenance, work, or fattening of horses, the carbohydrates (primarily grains and roughages) are by far the most important.

CARBOHYDRATES

Increased energy for horses is generally met by increasing the grain and decreasing the roughage. But, since the horse naturally evolved as a grazing animal, its digestive system often has difficulty in handling large quantities of cereal grains (starchy material). Evidence of the horses' difficulty with high grain rations is manifested clinically as founder, colic, and/or loss of appetite. So, to promote normal physiological activity of the gastrointestinal tract, some coarse roughage is necessary; finely ground roughage (or pelleted forage) will not suffice. But the precise amount of fiber needed has not been determined. It is generally recommended that horses be fed 1.0 to 2.0 lb of roughage per 100 lb of body weight, daily. Young horses and working (or running) horses must have rations in which a large part of the carbohydrate content is low in fiber, and in the form of nitrogen-free extract.

FATS

Some fat in the ration is desirable, because fat is the carrier of the fat-soluble vitamins (vitamins A, D, E, and K), and because the horse needs some linoleic acid (an essential fatty acid) in the ration. A ration devoid of fats results in reduced growth, scaly skin, and a rough, thin hair coat. Although the fatty acid requirements of horses have not been determined, it is thought that most ordinary farm rations contain ample quantities of these nutrients.

In the past, most horse owners and scientists were of the opinion that horses could not tolerate high-fat rations. But, studies indicate that they will readily consume 10 to 20% added fat in the ration, without difficulty—and even with benefit.[2] In an endurance trial conducted at the Colorado Station, horses fed fat supplemented rations (9% added fat) outperformed their counterparts that were fed either (1) starch supplemented rations, or (2) protein supplemented rations.[3] Subsequent experiments and experiences indicate that the amount of the energy and the energy density needed in the ration increase dramatically as work intensity increases.

The horse's utilization of feed energy for muscular work involves (1) the production of energy-rich adenosine triphosphate (ATP) from carbohydrates, fats, and proteins by means of oxidative processes, and (2) the contraction of the muscle fibers induced by ATP. When exercising lightly or moderately—at a gait no faster than cantering—the cardiovascular system of the horse can supply enough oxygen to the muscle tissue to permit the complete oxidation of energy sources (ATP) needed for muscle contraction, with little or no formation of lactic acid. This is called *aerobic exercise*. For such exercise, normal rations consisting of grain and hay suffice.

When the exercise of the equine athlete is very intense—as in training, racing, cutting, reining, and polo—and the horse is unable to transport enough oxygen to the muscles to oxidize completely the energy sources (ATP), aerobic capacity is augmented in part by nonoxidative processes. This is called *anaerobic exercise*. Glycogen serves as the primary energy source during anaerobic exercise. But if a sufficient amount of oxygen is not present, large amounts of lactic acid build up, resulting in heavy breathing, fatigue, soreness, and a prolonged recovery period; and the elimination of the lactic acid buildup requires extra oxygen. So, it is theorized (1) that vegetable or animal fats, which provide about 2.25 times as much energy as the carbohydrates, will conserve muscle glycogen for that all important anaerobic period of high energy demand—such as the stretch run; and (2) that a high fat ration will allow the horse to store and maintain higher levels of muscle glycogen, thereby imparting more staying power or endurance.

Also, when a substantial increase in energy (calories) is needed for the high performance horse, energy-dense fats, which contain more than twice as much energy per unit weight as carbohydrate feeds, are safer, lessening founder, colic, and other digestive disturbances. Also, when fed 5 to 20% high-quality fat in a properly processed ration, performance horses usually clean up their feed better. As with any ration change, the conversion to a high fat ration should be gradual.

Research at Texas A&M University also shows that a high fat diet will increase the reproductive and lactation performance of broodmares.

Unsaturated fats—vegetable oils, particularly corn oil or safflower oil—also have the added virtue of imparting gloss, or sheen, to the hair when as little as 2 oz (4 Tbsp) are fed twice daily. Except for vegetable oils (unsaturated fats) producing an attractive coat, there is no difference between animal and vegetable fats; hence, the choice may be determined by economics—which is the best buy.

Feeds high in fat content are likely to become rancid, and rancid feeds are unpalatable, if not actually injurious in some instances. Thus, when fat is added to the ration, it is important that it be stabilized by the use of an antioxidant. Also, there should be added protein (to maintain the protein-calorie ratio), minerals, and vitamins.

(Also see Chapter 10, section on "Fats and Oils.")

Protein

Except for the proteins built by bacterial action in the cecum, horses must have amino acids or more complex protein compounds in the ration.

[2]Tyznik, W. J., "Energy for Horses," paper presented at the 1975 California Livestock Symposium.

[3]Slade, L. M., and P. L. Hambleton, "Feeding the Horse for Endurance," *Stud Managers Handbook*, Vol. 12, 1976, p. 140, published by Agriservices Foundation, edited by M. E. Ensminger.

Horses of all ages and kinds require adequate amounts of protein of suitable quality for maintenance, growth, finishing, reproduction, and work. Of course, the protein requirements for growth, reproduction, and lactation are the greatest and most critical. (See Table 25-6.) The protein requirements for work are minimal and are not increased by work load; they're the same for maintenance, medium exercise, and intense exercise, according to researchers at Washington State University.[4] Therefore, the protein requirements for work are essentially the same as the maintenance requirements. Of course, when total feed intake increases to meet the added energy requirements of working horses, total protein intake also increases, even though the percent protein in the ration remains the same.

If more protein is consumed than is needed, the extra protein is broken down by the body and utilized for energy. However, protein is not as effectively and efficiently utilized for energy as are carbohydrates and fats. Furthermore, the by-products of protein metabolism, such as urea, are excreted in the urine.

A deficiency of protein in the ration of the horse may result in the following deficiency symptoms: depressed appetite, poor growth, loss of weight, reduced milk production, irregular estrus, lowered foal crops, loss of condition, and lack of stamina.

Since the vast majority of protein requirements given in feeding standards are on a minimum basis, the allowances for young animals should be higher. *CAUTION:* Increasing the protein in foal rations to stimulate growth without providing adequate levels of minerals may cause skeletal deformities and leg weaknesses which cannot be reversed.

In the case of ruminants (cattle and sheep), there is tremendous bacterial action in the paunch. These bacteria build body proteins of high quality from sources of inorganic nitrogen that nonruminants (humans, rats, swine, poultry, and dogs) cannot. Farther on in the digestive tract, the ruminant digests the bacteria and obtains good proteins therefrom. Although the horse is not a ruminant, apparently the same bacterial process occurs to a more limited extent in the cecum, that greatly enlarged blind pouch of the large intestine of the horse. However, it is much more limited than in ruminants, and the cecum is located beyond the small intestine, the main area for digestion and absorption of nutrients. This points up the fallacy of relying entirely on cecal synthesis in the horse; above all, it must be remembered that little cecal synthesis exists in young equines.

The limited protein synthesis in the horse (as compared with ruminants) and the lack of efficiency of absorption due to the cecum being on the lower end of the gut (thereby not giving the small intestine a chance at the digesta after it leaves the cecum), clearly indicate that horse rations should contain high-quality proteins, adequate in essential amino acids. Most cereal grains, such as oats, corn, or barley, are deficient in lysine, tryptophan, and methionine for optimum growth. Some protein supplements, such as linseed and cottonseed meal, do not contain adequate lysine. Proteins from animal sources (such as dried skimmed milk) and from alfalfa and soybeans are high quality. High-quality protein is especially important for young equines, because cecum synthesis is very limited early in life. So, in practical horse feeding, foals should be provided with some protein feeds of animal origin in order to supplement the protein found in grains and forages. In feeding mature horses, a safe plan to follow is to provide plant protein from several sources, unless a high-quality protein source such as dried skimmed milk or soybean meal is used.

In recognition that lysine is the first limiting amino acid of horses and is thus an indicator of the quality of protein which horses require, the lysine requirements of horses are given in Tables 25-1 through 25-6.

There is evidence that nonprotein nitrogen (urea) can be substituted for protein in the ration of the horse, but the conversion to protein is inefficient as compared with ruminants. Up to 5% of urea in the total ration does not appear to be harmful. Nevertheless, in recognition of the more limited bacterial action in the horse and the hazard of toxicity, most state laws forbid the use of such nonprotein nitrogen sources as urea in horse rations.

The extent to which the horse's ration is supplemented with proteins depends on the age of the horse and on the quality of the forage fed. Growing or lactating animals require somewhat more protein than horses that are idle, gestating, or working. Also, grass hays are generally low in quality and quantity of proteins and require more supplementation than legumes.

PROTEIN POISONING

Some opinions to the contrary, protein poisoning as such has never been documented. There is no proof that heavy feeding of high-protein feeds to horses is harmful, provided (1) the ration is balanced in all other respects, (2) the animal's kidneys are normal and healthy (a large excess of protein in terms of body needs increases the work of the kidneys for the excretion of the urea), (3) any ration change to high-protein feed is made gradually, as is recommended in any change in feed, and (4) there is adequate exercise and normal metabolism.

Some horses do appear to be allergic to certain proteins or to excesses of specific amino acids, as a result of which they may develop *protein bumps.*

Minerals

When we think of minerals for the horse, we instinctively think of bones and unsoundnesses. This is so because (1) a horse's skeleton is very large, weighing 100 lb or more in a full-grown horse, of which more than half consists of organic matter and minerals, and (2) experienced trainers estimate that ⅓ of the horses in training require treatments for unsoundnesses, in one form or another. But in addition to furnishing structural material for the growth of bones, teeth, and tissues, minerals regulate many of the life processes.

In an amazingly short time after birth, a healthy foal can run almost as fast as its mother—and on legs almost as long. In fact, the cannon bones (the lower leg bones extending from the knees and hocks to the fetlocks) are as long at the time of birth as they will ever be. This indicates that important

[4]Patterson, P. H., C. N. Coon, and I. M. Hughes, "Protein Requirements for Mature Working Horses," *Journal of Animal Science,* Vol. 61, No. 1, July 1985, p. 187.

development of the skeleton takes place in the fetus, before the foal is born. It is evident, therefore, that adequate minerals must be provided the broodmare if the bones of her offspring are to be sound.

The mineral requirements of mares in lactation are even more rigorous than those during gestation. Mares weighing about 1,000 lb will produce an average of 2 gal, or more, of milk per day throughout the 7-month suckling period. That's a total of 3,612 lb of milk. Since fresh mare's milk contains 0.7% ash, this amount of mare's milk contains 25.28 lb of mineral (3,612 × 0.7% = 25.28). Here's how this phenomenon works: When properly fed before breeding, in early pregnancy, and when barren, mineral deposits are made in the mare's skeleton. Then at those times when the mineral demands are greater than can be obtained from the feed—the last of pregnancy, and during lactation—the mare draws from the stored reserves in her skeleton. Of course, if there hasn't been proper storage in the mare's skeleton, something must "give"—and that something is the mother. Nature has ordained that growth of the fetus, and the lactation that follows, shall take priority over the maternal requirements. Hence, when there is a mineral deficiency, the mare's body will be deprived, or even stunted if she is young, before the developing fetus or milk production will be materially affected.

The proper development of the bone is particularly important in the horse, as evidenced by the stress and strain on the skeletal structure of the racehorse, especially when racing the 2-year-old. Since the greatest development of the skeleton takes place in the young, growing animal, it is evident that adequate minerals must be provided at an early age if the bone is to remain sound.

The typical horse ration of grass hay and farm grains is usually deficient in calcium, but adequate in phosphorus. Also, salt is almost always deficient; and many horse rations do not contain sufficient iodine and certain other trace elements. Thus, horses usually need special mineral supplements. But they should not be fed either more or less minerals than needed. Also, it is recognized that mineral allowances given with the ration or in a mineral box should vary according to the mineral content of the soil on which feeds are grown.

Although acute mineral deficiency diseases and actual death losses are relatively rare, inadequate supplies of any one of the essential mineral elements may result in lack of thrift, poor gains, inefficient feed utilization, lowered reproduction, and decreased performance in racing, showing, riding, or whatnot. This does not mean that all mineral elements known to be required by at least one animal species must always be included in horse mineral supplements. Rather, only the specific minerals that are deficient in the ration—and in the quantities necessary—should be supplied. *Excesses and mineral imbalances are to be avoided.*

Table 25-8, Horse Mineral Chart, pp. 1078–1081, lists the minerals required by horses and gives pertinent information pertaining to each. Minerals may be incorporated in the ration in keeping with the recommended allowances given in this table and in Table 25-6, pp. 1072–1073. Additionally, horses should have free access to salt.

The daily mineral requirements and recommended allowances vary with the mature weight, age, and type and level of productivity of the horse. Likewise, the mineral requirements and recommended allowances of the total ration vary with the percent dry matter in the ration, and with the age and the type and level of productivity of the horse.

• **Metabolic bone disease (MBD)**—In recent years, there has been a great increase in metabolic bone disease in growing horses, especially epiphysitis, contracted tendons, and osteochondritis dissecans (OCD). A brief description of each of these conditions follows:

1. **Epiphysitis.** This is an inflammation of the growth plate of the long bones, primarily found at the lower end of the radius above the knee, but it may be noticeable at the distal tibial and the distal metacarpal and metatarsal bones. (See Fig. 25-4.) Epiphysitis results in a firm and painful swelling.

Fig. 25-4. Severe epiphysitis which was copper responsive. This colt was fed an alfalfa-grass hay and 3 lb of grain per day. But it was found that the hay was copper deficient (3 ppm). The clinical signs of epiphysitis disappeared in about 2 months after switching to a hay that contained a higher level of copper (10–15 ppm). (Courtesy, Colorado State University, Ft. Collins)

2. **Contracted tendons.** This involves a shortening of the flexor tendons, causing the heels to be raised and the pasterns to be straight or, in severe cases, to knuckle forward with the horse walking on its toe. Contracted tendons may be present at birth, or they may be acquired during growth.

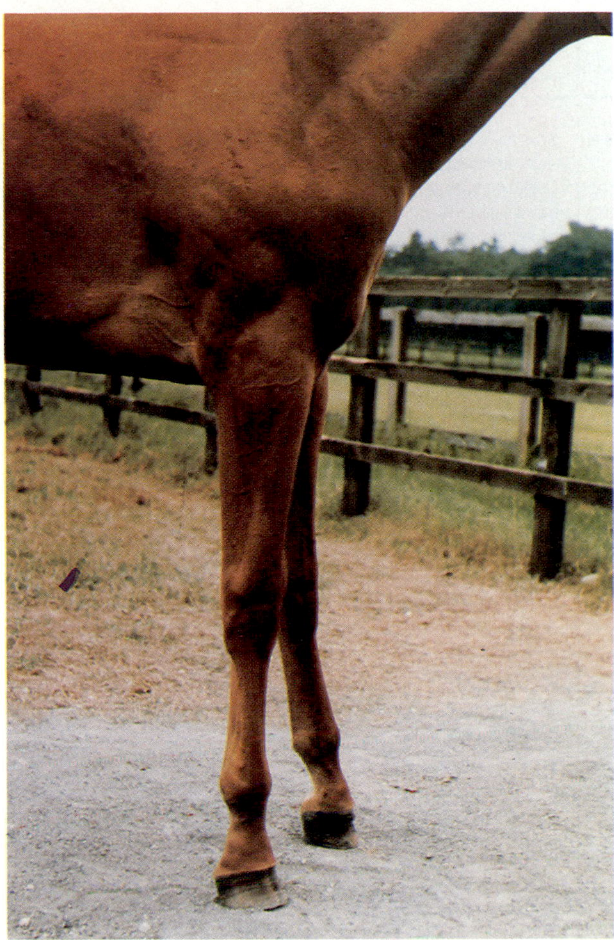

Fig. 25-5. Contracted flexor tendon of the left forelimb, showing characteristic raised heel and straight pastern. (Courtesy, Beaufort Cottage Stables, Newmarket, Great Britain)

3. **Osteochondritis dissecans (OCD).** This is a condition in which the cartilage in a growing foal does not properly convert into bone. It may appear in either of two forms: (a) The form in which it is localized in one or a few joints (most commonly the stifle and hock joints, although any joint may be involved), usually without any clinical signs; and (b) the second and less common form, which most commonly affects the more distal limb joints such as the pastern and fetlock, although it may affect any joint, including those of the back.

At this time, the cause of the increase in the incidence of the above bone diseases is not entirely clear. However, it appears that the major factors are: (1) rapid growth and excess weight, (2) injury to the epiphysis, (3) nutritional imbalances, (4) genetic predisposition, (5) limited forced exercise, (6) exercise on hard ground, and (7) faulty conformation.

Based on field observations and a study conducted by Ohio State University, involving 384 yearlings raised on 19 breeding farms in Ohio and Kentucky, and including the Thoroughbred, Standardbred, Arabian, and Quarter Horse breeds, there is strong evidence that calcium, phosphorus, copper, and zinc deficiencies/imbalances and/or masking are involved. They reported that the average calcium content of the rations on farms with the fewest skeletal problems was 1.16% ± 0.09 and the phosphorus content was 0.72% ± 0.08.[5]

On the other hand, Krook and Maylin of Cornell University theorize that overfeeding of calcium to the growing horse or the pregnant mare is the primary cause of osteochondrosis in the foal. Their explanation: A dietary calcium overload causes an excessive secretion of the hormone calcitonin which acts directly on the bone of growing horses; calcitonin inhibits (1) the conversion of cartilage to bone, and (2) the resorption of calcium from bone. In the pregnant mare, calcitonin can be transferred through the placenta to the fetus; which might explain how a foal could be born with osteochondrosis. The Cornell scientists also theorize that osteochondrosis would predispose racehorses to fractures. They recommend that alfalfa hay not be fed to pregnant mares and growing horses because of its high calcium content, and that calcium be limited to 34 g per horse per day.[6]

Both the Ohio State and Cornell scientists recognize the seriousness of metabolic bone disease. But they differ markedly as to the cause. Ohio State researchers submit strong evidence that deficiencies of calcium, phosphorus, copper, and zinc are the primary causes, whereas the Cornell scientists theorize that high calcium is the main cause.

Further experimental studies are needed. In the meantime, horse owners and caretakers are admonished to practice the old adage: "Use moderation in all things."

Based on experiments (including unpublished work at both Ohio State and Cornell) and experiences, the authors of *Feeds & Nutrition* recommend (1) that breeders continue to feed alfalfa hay to pregnant mares and growing horses, and (2) that the levels of calcium, phosphorus, copper, iron, manganese, and zinc be in keeping with the recommendations given in Table 25-6 of this chapter.

NOTE WELL: The more rapid the growth of young equines, the more important it is that the nutrients in the ration be fine-tuned to meet their needs.

---

[5]Knight, Debra A., et al, *Correlation of Dietary Minerals to Incidence and Severity of Metabolic Bone Disease in Ohio and Kentucky,* College of Veterinary Medicine, The Ohio State University; paper presented at the 1985 American Association of Equine Practitioners meeting in Toronto, Canada.

[6]Sellnow, L., "Linking Breakdowns and Diet," *The Blood-Horse,* May 3, 1986, p. 3162.

TABLE
HORSE MINERAL

Minerals Which May Be Deficient Under Normal Conditions	Conditions Usually Prevailing Where Deficiencies Are Reported	Function of Mineral	Some Deficiency Symptoms	Practical Sources of the Mineral
Major or macrominerals:				
Salt (NaCl)	Negligence, for salt is cheap. The salt requirement is greatly increased under conditions which cause heavy sweating, thereby resulting in large losses of this mineral from the body. Unless it is replaced, fatigue will result. For this reason, when engaged in hard work and perspiring profusely, horses should receive liberal allowances of salt.	Salt serves as both a condiment and a nutrient. Sodium and chlorine help maintain osmotic pressure in body cells, upon which depends the transfer of nutrients to the cells and the removal of waste materials. Sodium is associated with muscle contraction and is important in making bile, which aids in the digestion of fats and carbohydrates. Chlorine is required for the formation of hydrochloric acid in the gastric juice so vital to protein digestion.	In warm or hot weather, workhorses show heat stress. Long-term symptoms of sodium deficiency are depraved appetite, rough hair coat, reduced growth of young animals, and decreased milk production.	Salt provided free-choice, preferably in loose form, or 0.5–1.0% salt added to the ration. It is very difficult for horses to eat very hard block or rock salt. This often results in inadequate consumption. Also, if there is much competition for a salt block, the more timid animals may not get their requirements. Iodized salt should be used in iodine-deficient areas.
Calcium (Ca)	The typical horse ration of grass hay and farm grains—usually deficient in calcium.	Builds strong bones and sound teeth. Very important during lactation. Affects availability of phosphorus. Calcium and phosphorus comprise ¾ of the ash of the skeleton and from ⅓–½ of the minerals of milk.	A deficiency of calcium in young animals is generally characterized by poorly formed, soft bone, which may bend or bow; and a severe deficiency may cause rickets. A deficiency of calcium in older animals results in porous, fragile bones. Because deficiency conditions may not be completely reversible, prevention is imperative.	Ground limestone or oystershell flour. When both calcium and phosphorus are needed, use steamed bone meal or dicalcium phosphate (see Table 13-2). Horses absorb 55 to 75% of the calcium in a typical ration.
Phosphorus (P)	Horses grazed on phosphorus-deficient areas or fed for a long period on mature, weathered forage.	Important in the development of bones and teeth. Essential to metabolism of carbohydrates and fats, and enzyme activation.	Rickets in young horses; osteomalacia in mature horses.	Monosodium phosphate, disodium phosphate, or sodium tripolyphosphate. Where both calcium and phosphorus are needed, use steamed bone meal or dicalcium phosphate (see Table 13-2). Horses absorb 35–55% of the phosphorus in a typical ration.
Magnesium (Mg)	Horses fed high grain-low forage ration, which characterizes most horses at hard work (as in racing and showing). Lactating mares grazing on lush spring pastures low in magnesium or in which Mg is unavailable.	Reduces stress and irritability. Magnesium is important in enzyme systems, bone formation, and calcium and phosphorus metabolism.	Horses under stress are keyed up, high-strung, and jumpy. Foals fed a purified ration deficient in magnesium develop nervousness, muscular tremors, convulsive paddling of the legs and, in some cases, die. Grass tetany.	Magnesium sulfate. Magnesium oxide.
Potassium (K)	When stabled horses are fed high-concentrate rations. Excessive sweating.	Major cation of intracellular fluid where it is involved in osmotic pressure and acid-base balance. Muscle activity. Required in enzyme reaction involving phosphorylation of creatine. Influences carbohydrate metabolism.	Reduced appetite, growth retardation, unsteady gait, general muscle weakness, pica, diarrhea, distended abdomen, emaciation, followed by death. Fatigue. Abnormal electrocardiograms.	Potassium chloride. Roughages usually contain ample potassium.
Sulfur (S)		Sulfur is an integral part of the amino acids methionine and cystine.		

25–8
CHART (See footnotes at end of table.)

Classes/Function	Nutrient Requirements[1 2] Per Horse Daily	In Ration A-F	Per Ton Ration A-F		Nutrient Allowances[1 2] Per Horse Daily	In Ration A-F	Per Ton Ration A-F		Comments
	(g)	(%)	(lb)	*(kg)*	(g)	(%)	(lb)	*(kg)*	
Maintenance: 1,000-lb *(454-kg)* horse	85	0.5–1.0	10–20	*4.5–9.1*	85	0.75	15	*6.8*	**H**orses require both sodium and chlorine, but the requirement for chlorine is approximately half that of sodium. Generally, the chlorine requirements will be met if the sodium needs are adequate. **S**odium and chlorine are low in feeds of plant origin. **T**here is little danger of overfeeding salt unless a salt-starved animal is suddenly exposed to too much salt, or if liberal amounts of water are not available. **E**xcessive salt intake may result in high water intake, excessive urine excretion, digestive disturbances, or death from salt cramps.
Gestation/Lactation: 1,000-lb *(454-kg)* mare	85	0.5–1.0	10–20	*4.5–9.1*	85	0.75	15	*6.8*	
Growth: 450-lb *(204.5-kg)* weanling	41	0.5–1.0	10–20	*4.5–9.1*	41	0.75	15	*6.8*	
Working: 1,000-lb *(454-kg)* horse	85	0.5–1.0	10–20	*4.5–9.1*	85	0.75	15	*6.8*	
Maintenance: 1,000-lb *(454-kg)* horse	20	0.175	3.5	*1.6*	23	0.21	4.1	*1.8*	**T**he calcium-phosphourus ratio should be maintained close to 1.1:1 although 2:1 is acceptable. Narrower ratios may cause osteomalacia in mature horses. When there is a shortage of calcium in the ration, it is withdrawn from the bones. **F**eeding excess calcium interferes with the utilization of magnesium, manganese, and iron—and perhaps in the utilization of zinc.
Gestation/Lactation: 1,000-lb *(454-kg)* mare	56	0.495	9.9	*4.5*	64	0.57	11.3	*5.1*	
Growth: 450-lb *(204.5-kg)* weanling	36	0.66	13.2	*6.0*	41	0.76	15.1	*6.8*	
Working: 1,000-lb *(454-kg)* horse	40	0.35	7.0	*3.2*	46	0.41	8.1	*3.7*	
Maintenance: 1,000-lb *(454-kg)* horse	14	0.125	2.5	*1.1*	16.1	0.14	2.8	*1.3*	**F**or the growing horse, the calcium-phosphorus ratio should be maintained close to 1.1:1, although 2:1 is acceptable. **T**he mature horse can tolerate a Ca:P ratio as wide as 4:1 or 5:1 provided adequate levels of phosphorus are available. **E**xcess phosphorus can cause bighead. **I**f plenty of vitamin D is present, the ratio of calcium to phosphorus becomes less important.
Gestation/Lactation: 1,000-lb *(454-kg)* mare	36	0.315	6.3	*2.9*	41.4	0.37	7.3	*3.3*	
Growth: 450-lb *(204.5-kg)* weanling	19	0.35	7.0	*3.2*	21.9	0.4	8.0	*3.7*	
Working: 1,000-lb *(454-kg)* horse	29	0.255	5.1	*2.3*	33.4	0.3	5.9	*2.7*	
Maintenance: 1,000-lb *(454-kg)* horse	7.5	0.065	1.3	*0.6*	8.6	0.08	1.5	*0.7*	**E**xcess of magnesium upsets calcium and phosphorus metabolism. **R**ations containing 50% forage will likely contain sufficient magnesium for unstressed horses.
Gestation/Lactation: 1,000-lb *(454-kg)* mare	10.9	0.096	1.92	*0.9*	12.5	0.11	2.2	*1.0*	
Growth: 450-lb *(204.5-kg)* weanling	5.7	0.105	2.1	*1.0*	6.6	0.12	2.4	*1.1*	
Working: 1,000-lb *(454-kg)* horse	15.1	0.135	2.7	*1.2*	17.4	0.16	3.1	*1.4*	
Maintenance: 1,000-lb *(454-kg)* horse	25.0	0.22	4.4	*2.0*	28.8	0.26	5.1	*2.3*	**A** ration that contains at least 50% forage can be expected to meet potassium requirements.
Gestation/Lactation: 1,000-lb *(454-kg)* mare	46.0	0.405	8.1	*3.7*	52.9	0.47	9.3	*4.2*	
Growth: 450-lb *(204.5-kg)* weanling	18.2	0.335	6.7	*3.0*	20.9	0.39	7.7	*3.5*	
Working: 1,000-lb *(454-kg)* horse	49.9	0.44	8.8	*4.0*	57.4	0.51	10.1	*4.6*	
Maintenance: 1,000-lb *(454-kg)* horse					17.0	0.15	3.0	*1.36*	**T**he precise sulfur requirement is not known, but an allowance of 0.15% of the total ration appears to be adequate. **I**f the protein requirement of the ration is met, the sulfur intake will usually be at least 0.15%, which appears to be adequate.
Gestation/Lactation: 1,000-lb *(454-kg)* mare					17.0	0.15	3.0	*1.36*	
Growth: 450-lb *(204.5-kg)* weanling					8.2	0.15	3.0	*1.36*	
Working: 1,000-lb *(454-kg)* horse					17.0	0.15	3.0	*1.36*	

(Continued)

TABLE 25-8

Minerals Which May Be Deficient Under Normal Conditions	Conditions Usually Prevailing Where Deficiencies Are Reported	Function of Mineral	Some Deficiency Symptoms	Practical Sources of the Mineral
Trace or micro-minerals:				
Cobalt (Co)	Animals grazed in cobalt-deficient areas, such as Australia, Western Canada, and the following states of U.S.: Fla., Mich., Wisc., N.H., Penn., and N.Y.	Cobalt is required for the synthesis of vitamin B-12 in the intestinal tract of the horse.	Anemia. Severe weight loss.	Cobaltized mineral mix made by adding cobalt at the rate of 0.2 oz/100 lb (*5.7 g/45.4 kg*) of salt as cobalt chloride, cobalt sulfate, cobalt oxide, or cobalt carbonate. Also, several good commercial cobalt-containing minerals are on the market.
Copper (Cu)	Suckling foals. Mare's milk, along with milk from other species, is low in copper. Deficiency occurs in regions where soils contain too little copper or where horses are getting an excess of molybdenum, sulfur, or zinc.	Copper, along with iron and vitamin B-12 is necessary for hemoglobin formation, although it forms no part of the hemoglobin molecule (or red blood cells). Closely associated with normal bone development in young growing animals.	Anemia, characterized by fewer than normal red cells and less than normal amount of hemoglobin. Abnormal bone development in young equines, including an increased incidence of epiphysitis, contracted tendons, and osteochondritis dissecans (OCD).	Trace mineralized salt containing copper sulfate or copper carbonate.
Iodine (I)	Iodine-deficient areas or soils (in Northwestern U.S. and in the Great Lakes region) when iodized salt is not fed. Use of feeds that come from iodine-deficient areas.	Iodine is needed by the thyroid gland in making thyroxin, an iodine-containing compound which controls the rate of body metabolism or heat production.	Foals born dead, or very weak with enlarged thyroid glands (goiter) and unable to stand or nurse. Higher than normal incidence of navel ill.	Stabilized iodized salt containg 0.01% potassium iodide (0.0076% iodine). Calcium iodate.
Iron (Fe)	Suckling foals kept away from soil and feed other than milk. Horses subjected to pressure from racing, showing, or other heavy use. Such animals require added iron in their daily ration. Excessive blood loss from a wound or heavy parasite infestation.	Necessary for formation of hemoglobin, an iron-containing compound which enables the blood to carry oxygen. Also, important to certain enzyme systems.	Iron-deficiency anemia, characterized by fewer than normal red cells and less than normal amount of hemoglobin. Anemic horses tire easily. **NOTE WELL**: Iron deficiency anemia may also result from heavy parasitization.	Ferrous sulfate administered orally. Trace mineralized salt. Cane molasses. Iron oxide should not be used as a source of iron for horses because it is poorly absorbed.
Manganese (Mn)	Excess calcium and phosphorus which decreases absorption of manganese.	Essential for normal bone formation (as a component of the organic matrix). Thought to be an activator of enzyme systems. Growth and reproduction.	Poor growth. Lameness, shortening and bowing of legs, and enlarged joints. Impaired reproduction (testicular degeneration of males; defective ovulation of females).	Trace mineralized salt containing 0.25% manganese (or more).
Selenium (Se)	Muscle disorders and lowered serum selenium.		Infertility. Myositis (muscular discomfort or pain).	Forages or grains grown on soils known to have adequate selenium. Sodium selenate. Sodium selenite.
Zinc (Zn)	Feeds low in zinc. Excess calcium may reduce the absorption and utilization of zinc.	Important in many enzyme systems. Required for normal protein synthesis and metabolism. Imparts gloss or *bloom* to the hair coat.	Rough, dull hair coat. Loss of appetite.	Zinc carbonate. Zinc sulfate.

[1]All "nutrient requirements" given in this table were adapted by the authors from *Nutrient Requirements of Horses,* 5th rev. ed., NRC—National Academy of Sciences, 1989. The "nutrient allowances" given in this table represent the authors' best judgment based on current research; it is intended that they meet the nutrient requirements, and provide adequate margins of safety in addition.

[2]Feed consumption of a mature 1,000-lb (*454-kg*) horse estimated at 25 lb (*11.36 kg*) per day. Feed consumption of a 450-lb (*204.5-kg*) weanling estimated at 12 lb (*5.45 kg*) per day.

(Continued)

Classes/Function	Nutrient Requirements[1 2]			Nutrient Allowances[1 2]			Comments
	Per Horse Daily	In Ration A-F	Per Ton Ration A-F	Per Horse Daily	In Ration A-F	Per Ton Ration A-F	
	(mg)	(ppm, or mg/kg)	(g/ton)	(mg)	(ppm, or mg/kg)	(g/ton)	
Maintenance: 1,000–lb *(454–kg)* horse	1.13	0.1	0.091	1.3	0.11	0.104	The disease called *salt sick* in Florida is due to a cobalt deficiency associated with a copper deficiency. The cobalt requirement for horses is very low, for horses have remained in good health while grazing pastures so low in cobalt that ruminants confined to them have died.
Gestation/Lactation: 1,000–lb *(454–kg)* mare	1.13	0.1	0.091	1.3	0.11	0.104	
Growth: 450–lb *(204.5–kg)* weanling	0.54	0.1	0.091	0.6	0.11	0.100	
Working: 1,000–lb *(454–kg)* horse	1.13	0.1	0.091	1.3	0.11	0.104	
Maintenance: 1,000–lb *(454–kg)* horse	113.4	10.0	9.070	283.4	25	22.675	A copper deficiency in horses has been reported in Australia. In high-molybdenum areas, more copper may be added to horse rations; but excesses and toxicity should be avoided.
Gestation/Lactation: 1,000–lb *(454–kg)* mare	113.4	10.0	9.070	340.1	30	27.210	
Growth: 450–lb *(204.5–kg)* weanling	54.4	10.0	9.070	217.7	40	36.280	
Working: 1,000–lb *(454–kg)* horse	113.4	10.0	9.070	283.4	25	22.675	
Maintenance: 1,000–lb *(454–kg)* horse	1.13	0.1	0.091	1.3	0.11	0.104	Enlargement of the thyroid gland (goiter) is nature's way of trying to make enough thyroxin (an iodine-containing hormone) when there is insufficient iodine in the feed. Feeding excess iodine continuously will also produce goiter in foals. Iodine deficiency seldom occurs in coastal areas because of the abundance of iodine from spray drift from ocean or sea water.
Gestation/Lactation: 1,000–lb *(454–kg)* mare	1.13	0.1	0.091	1.3	0.11	0.104	
Growth: 450–lb *(204.5–kg)* weanling	0.54	0.1	0.091	0.6	0.11	0.100	
Working: 1,000–lb *(454–kg)* horse	1.13	0.1	0.091	1.3	0.11	0.104	
Maintenance: 1,000–lb *(454–kg)* horse	453.5	40	36.280	453.5	40	36.280	The horse's body contains about 0.004% iron. Milk is deficient in iron, and the iron content of the mother cannot be increased through feeding iron. Thus, foals should be individually or creep fed as soon as they are old enough. A variable store of both iron and copper is located in the liver and spleen, and some iron is found in the kidneys. Too much iron may be harmful.
Gestation/Lactation: 1,000–lb *(454–kg)* mare	566.9	50	45.350	1,020.4	90	81.630	
Growth: 450–lb *(204.5–kg)* weanling	272.1	50	45.350	489.8	90	81.630	
Working: 1,000–lb *(454–kg)* horse	453.5	40	36.280	680.3	60	54.420	
Maintenance: 1,000–lb *(454–kg)* horse	453.5	40	36.280	521.5	46	41.720	Most natural feedstuffs are rich in manganese.
Gestation/Lactation: 1,000–lb *(454–kg)* mare	453.5	40	36.280	521.5	46	41.720	
Growth: 450–lb *(204.5–kg)* weanling	217.7	40	36.280	250.4	46	41.734	
Working: 1,000–lb *(454–kg)* horse	453.5	40	36.280	521.5	46	41.720	
Maintenance: 1,000–lb *(454–kg)* horse	1.13	0.1	0.091	1.3	0.11	0.104	Excess selenium results in selenium poisoning, or alkali disease (see Chapter 5, Nutritional Disorders/Toxins).
Gestation/Lactation: 1,000–lb *(454–kg)* mare	1.13	0.1	0.091	1.3	0.11	0.104	
Growth: 450–lb *(204.5–kg)* weanling	0.54	0.1	0.091	0.6	0.11	0.100	
Working: 1,000–lb *(454–kg)* horse	1.13	0.1	0.091	1.3	0.11	0.104	
Maintenance: 1,000–lb *(454–kg)* horse	453.5	40	36.280	907.0	80	72.560	If zinc in the feed is on the low side, the addition of zinc should improve the hair coat. Excess zinc prevents calcium utilization and produces signs of calcium deficiency. The toxicity level exceeds 1,000 ppm.
Gestation/Lactation: 1,000–lb *(454–kg)* mare	453.5	40	36.280	1,133.8	100	90.700	
Growth: 450–lb *(204.5–kg)* weanling	217.7	40	36.280	544.2	100	90.700	
Working: 1,000–lb *(454–kg)* horse	453.5	40	36.280	1,020.4	90	81.630	

MAJOR OR MACROMINERALS

The major or macrominerals required by the horse are salt (sodium chloride), calcium, phosphorus, magnesium, potassium, and sulfur.

Salt (NaCl)

Salt, which serves as both a condiment and a nutrient, is needed by all classes of animals, but more especially by herbivora (grass-eating animals, such as the horse). It may be provided in the form of granulated, rock, or block salt. In general, the form selected is determined by price and availability. It should be pointed out, however, that it is difficult for horses to eat very hard block and rock salt. This often results in inadequate consumption. Also, if there is much competition for the salt block, the more timid animals may not get their requirements.

Iodized salt should be provided in iodine-deficient areas. Trace-mineralized salt is recommended, because it is a simple, safe means of providing iodine and other trace minerals at only slightly greater cost than common salt.

The horse requires both sodium and chlorine. Generally, the chlorine requirement will be met if the sodium needs are adequate.

A deficiency of sodium over a long period of time results in depraved appetite, rough coat, reduced growth, and lowered milk production.

The salt requirement is greatly increased under conditions which cause heavy sweating, thereby resulting in large losses of this mineral from the body. Unless it is replaced, fatigue will result. For this reason, when engaged in hard work and perspiring profusely, horses should receive liberal allowances of salt.

On the average, a horse needs about 3 oz of salt daily or 1⅓ lb per week, although salt requirements vary with work and temperature.

Salt can be fed free-choice to horses, provided they have not been salt starved. If the animals have not previously been fed salt for a considerable length of time, they may overeat, resulting in digestive disturbances and even death from salt cramps. Salt-starved animals should first be hand-fed salt, and the daily allowance should be increased gradually until they start leaving a little in the mineral box. When this point is reached, self-feeding may be followed.

When added to the concentrate ration, salt should be added at a level of 0.5 to 1.0%

Calcium (Ca) and Phosphorus (P)

Horses are more apt to suffer from a lack of calcium and phosphorus than from any of the other minerals except salt. These 2 minerals comprise about 75% of the ash of the skeleton and from 33 to 50% of the minerals of milk.

One of the major problems in the formulation of rations is assessment of bioavailability of calcium and phosphorus. In a typical ration, horses absorb only 55 to 75% of the calcium and 35 to 55% of the phosphorus. Several factors account for this low absorption, including the Ca:P ratio, level of intake, source of calcium and phosphorus, and the presence of organic inhibitors such as oxalate and phytate. Also, due to poor utilization, the calcium and phosphorus requirements of aged animals (animals over 20 years of age) are higher than for younger animals.

In considering the calcium and phosphorus requirements of horses, it is important to realize that the proper utilization of these minerals by the body is dependent upon three factors: (1) an adequate supply of calcium and phosphorus in an available form; (2) a suitable ratio between them; and (3) sufficient vitamin D to make possible the assimilation and utilization of the calcium and phosphorus. If plenty of vitamin D is present (as provided either by the action of sunlight or through the ration), the ratio of calcium to phosphorus becomes less important. Also, less vitamin D is needed when there is a desirable calcium-phosphorus ratio.

Normally, the calcium to phosphorus ratio should be about 1.1:1. However, the ratio varies according to age. For example, older horses can have a calcium to phosphorus ratio of 2:1. Provided adequate phosphorus is fed, weanling foals will *tolerate* a 3:1 ratio and mature horses a 5:1 ratio. It is important, however, to have more calcium than phosphorus; but not too much calcium. Feeding excessive calcium interferes with the utilization of magnesium, manganese, and iron—and perhaps with the utilization of zinc.

The pregnant mare deposits an amount equivalent to 10–12% of her body weight in products of conception; and these products contain approximately 1.2% calcium and 0.6% phosphorus. Since most of these minerals are deposited in the bones, and approximately 90% of the bone development occurs during the last 90 days of gestation, about 6 g of calcium and 3 g of phosphorus per day will be deposited during this period in a 1,100-lb pregnant mare.

The calcium and phosphorus requirements for lactation depend on level of production. Milk production varies among mares and with stage of lactation. During peak production, mares will give up to 4½ gal (38.7 lb/*17.6 kg*) of milk per day, with each pound containing 0.45 g of calcium and 0.2 g of phosphorus. Thus, a mare producing 4½ gal of milk per day will deposit 17.4 g of calcium and 7.7 g of phosphorus per day.

Lack of either or both calcium and phosphorus can result in bone disorders, with the type and severity of the disorder dependent upon the age of the animal and the degree and duration of the deficiency. Deficiency in young horses is generally characterized by poorly formed, soft bones, which may bend or bow; and deficiency in older animals, by porous, fragile bones. Because these conditions are not completely reversible, prevention is imperative.

A deficiency of either calcium or phosphorus will cause rickets in foals. Also, there is substantial evidence that lack of calcium and phosphorus, along with deficiencies of copper and zinc, cause epiphysitis, contracted tendons, and osteochondritis dissecans (OCD) in young horses.

Bone disturbances (called osteodystrophia febrosa, nutritional secondary hyperparathyroidism, osteomalacia, osteoporosis, and Miller's disease) develop in adult horses fed rations containing limited calcium and high prosphorus. The disease develops when rations with a calcium-phosphorus ratio of 0.8:1 are fed for 6 to 12 months, and it progresses rapidly when the ratio is 0.6:1.

After giving consideration to all the above mentioned factors, and after considering the experiments, the authors recommend that calcium and phosphorus be provided to lactating mares and young horses at the levels in the total ration given in Tables 25-6 and 25-8.

Generally speaking, legume forages, such as alfalfa hay or pasture, are rich in calcium; cereal grains and their by-products—oats, corn, barley, and wheat bran—are fair to good sources of phosphorus; and the protein supplements—linseed meal, soybean meal, and dried skim milk—are good sources of both calcium and phosphorus. So, by selecting and combining the common horse feeds properly, the maintenance needs of most horses can be met. (See Fig. 25-6.)

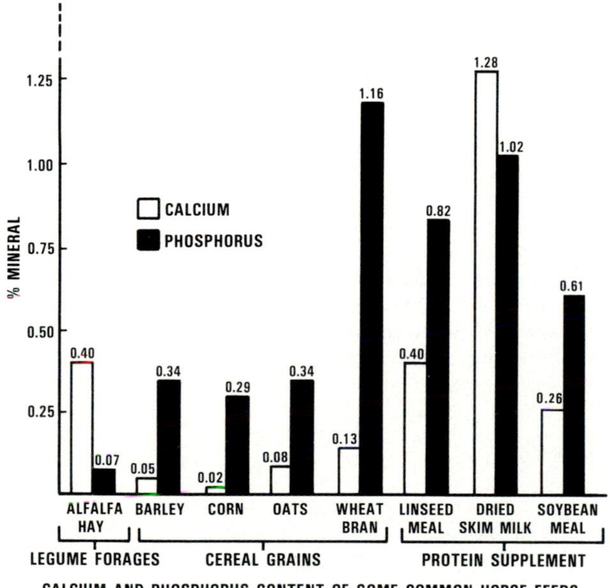

Fig. 25-6. Calcium and phosphorus content of some common horse feeds (as-fed basis).

When both calcium and phosphorus supplementation are needed, the authors favor the use of high quality steamed bone meal for horses, because bone meal contains many ingredients in addition to calcium and phosphorus. It is a good source of iron, manganese, and zinc, and it contains such trace minerals as copper and cobalt. However, it is increasingly difficult to get good bone meal. Some of the imported products are high in fat, rancid, and/or odorous and unpalatable. Where good bone meal is not available, dicalcium phosphate is generally recommended.

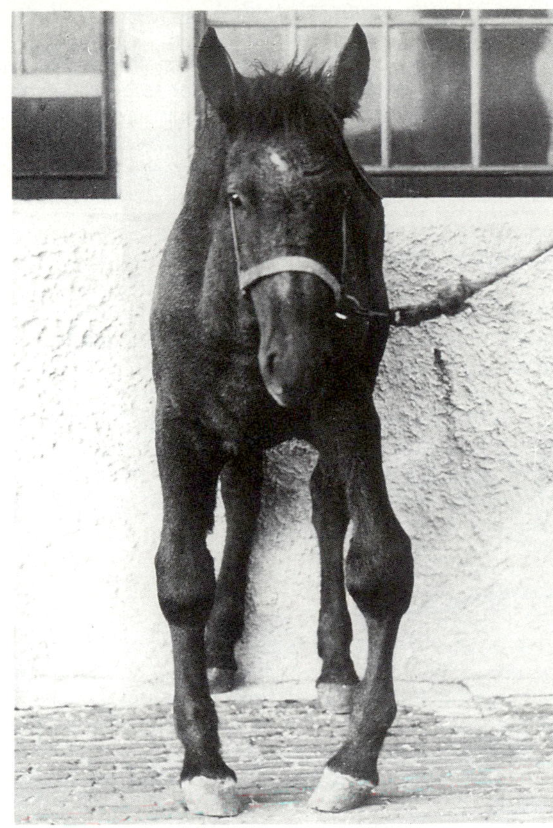

Fig. 25-7. Foal with severe rickets. Note the enlarged joints and crooked legs. (Courtesy, Department of Veterinary Pathology and Hygiene, College of Veterinary Medicine, University of Illinois, Urbana)

Fig. 25-8. Calcium-phosphorus imbalance. Horse with *big head disease* (nutritional secondary hyperparathyroidism) resulting from feeding a ration low in calcium and high in phosphorus, commonly associated with a ration excessively high in wheat bran. Note that the upper jaw is enlarged because calcium is replaced by fibrous connective tissue. (Courtesy, National Academy of Sciences, and College of Veterinary Medicine, Texas A&M University, College Station)

When calcium alone is needed, ground limestone or oystershell flour are commonly used.

When phosphorus alone is needed, defluorinated rock phosphate, sodium monophosphate, or sodium polyphosphate are the minerals of choice. Sodium monophosphate and sodium polyphosphate are not palatable; hence, it is important that they be combined with more palatable products.

Magnesium (Mg)

Horse rations containing 50% forage will likely contain sufficient magnesium for unstressed horses, unless the forage is known to be deficient in magnesium. But horses at hard work (as in racing and showing) consume more grain (which is low in magnesium) and less forage. Also, horses being raced or shown, or otherwise stressed, are frequently keyed up, high-strung, and jumpy, similar to the nervousness that characterizes animals and humans known to be suffering from a magnesium deficiency.

Potassium (K)

Forage-consuming animals generally require about 0.3% of potassium in their rations. A ration that contains at least 50% forage can be expected to meet potassium requirements. However, a horse ration that does not contain roughage, molasses, or oil meals may be deficient in potassium.

Significant amounts of potassium are lost during heavy sweating.

A reduced appetite is an early sign of a potassium deficiency. A severe deficiency may cause muscle tremors and erratic heart beat.

Sulfur (S)

Inorganic sulfur is not known to be an essential dietary constituent of the horse. If the protein requirement of the ration is met, the sulfur intake will usually be at least 0.15%, which appears to be adequate.

TRACE OR MICROMINERALS

The need for trace minerals may be inferred from the many reports on the value of blackstrap molasses (a good source of trace minerals) for horses fed low-quality hays.

Although the horse's requirements for most of the trace minerals are not known, there is evidence that the need is similar to that of other species.

Trace minerals may be supplied (1) as part of either the concentrate or complete ration, and/or (2) in a trace mineralized (TM) salt. When incorporated in the concentrate mix, TM salt should be added at the rate of 1%. When TM salt is fed free-choice, it should be placed in a conveniently located covered mineral box in amounts that will be consumed in not more than 1 to 2 weeks. When remaining in a mineral box longer than this period of time, it may become unpalatable and there may be losses of some elements.

A discussion of each of the trace minerals follows. They are listed alphabetically, and not necessarily in order of importance.

Cobalt (Co)

Cobalt is required for the synthesis of vitamin B-12 in the intestinal tract of the horse. A lack of cobalt and/or B-12 will result in anemia. However, the cobalt requirement of the horse is very low, for horses have remained in good health while grazing pastures so low in cobalt that ruminants confined to them have died.

Copper (Cu)

A copper deficiency has been reported in Australia in horses grazing pastures low in copper. Also, mare's milk (along with milk from all species) is low in copper; and its copper concentration decreases greatly during the first weeks of lactation. The feeling persists that the presence of 5 to 25 ppm of molybdenum in forages causes disturbances in copper utilization in horses. Cornell researchers found that the addition of high levels of molybdenum (27.4 and 107.3 ppm) to the ration decreased copper absorption and retention as a consequence of increased excretion of dietary copper in feces and increased excretion of absorbed copper in bile.[7]

Copper is of special interest to horse breeders because, in addition to its effect on iron metabolism, it is closely associated with normal bone development in young growing animals of all species. Abnormal bone development has been reported in foals on low copper rations.

There are wide species differences in tolerance to copper. Horses are very tolerant to copper, whereas sheep are very sensitive to it. The maximum tolerable levels of copper for growing animals in ppm, according to the National Academy of Sciences, are: horse, 800 ppm; chicken, 300 ppm; swine, 250 ppm; cattle, 100 ppm; and sheep, 25 ppm.

The authors' recommended copper allowances are: 30 ppm of the total ration for lactating mares, and 40 ppm for young horses. (See Tables 25-6 and 25-8.)

Iodine (I)

Pregnant mares are very susceptible to iodine deficiency. Where such a deficiency exists, the foals are usually stillborn or so weak that they cannot stand and suck. There is also some evidence to indicate that navel ill in foals may be lessened by feeding iodine to broodmares. Iodine deficiency is more prevalent in inland regions of the United States, but it seldom occurs in the coastal areas because of the abundance of iodine from ocean and sea water.

If iodized salt (0.0076% iodine) is fed, additional iodine supplementation is unnecessary. Organic iodine supplements, such as kelp, are not any better than iodized salt.

Excessive and continued iodine supplementation, such as feeding kelp meal in addition to iodized salt or trace mineralized salt, may cause goiter.

---

[7]Cymbalak, N. F., *et al*, "Influence of Dietary Molybdenum on Copper Metabolism in Ponies," *The Journal of Nutrition*, Vol. III, No. 1, January 1981, p. 96.

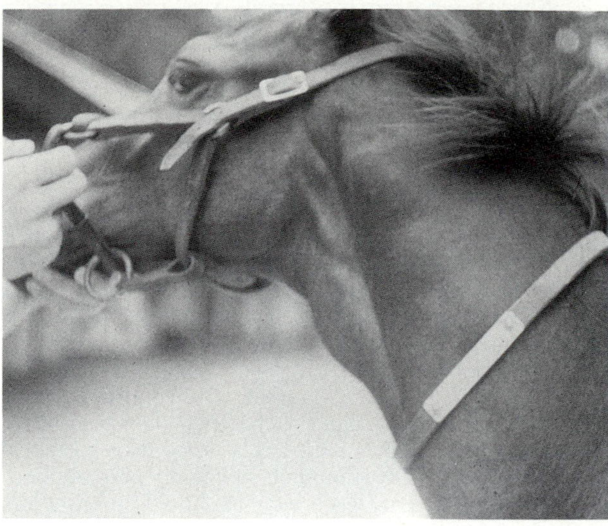

Fig. 25-9. Goiter caused by feeding excess iodine. Goiter is usually caused by an iodine deficiency, but it may result from feeding too much iodine over a long period of time. (Courtesy, Dr. D. E. Cooperrider, Chief, Diagnostic Laboratories, Division of Animal Industry, Kissimmee, Fla.)

Iron (Fe)

The National Research Council estimates the maintenance requirements of the horse for iron at 40 ppm, and the requirements of foals at 50 ppm. However, it has been reported that horses which are subjected to pressure from racing, showing, or other heavy use, require higher levels of iron. Also, it is important that the iron be in biologically available form (iron oxide should not be used as a source of iron for horses because it is poorly absorbed).

Manganese (Mn)

Feeds containing 60 to 70 ppm of manganese are recommended, with the higher levels fed to foals, stressed horses, and breeding animals.

Since most natural feedstuffs are rich in manganese, it can be assumed that part of the requirement for this element will be met by the normal ration.

Selenium (Se)

Selenium is an essential mineral for horses. Deficient animals have muscle disorders and lowered serum selenium.

It is recommended that horse rations contain 0.1 ppm of selenium in the complete feed.

Excess selenium (above 5 ppm) may result in selenium poisoning, or alkali disease (see Chapter 5, Nutritional Disorders/Toxins).

Zinc (Zn)

A level of 80 to 100 ppm of zinc is recommended, with young and highly stressed animals receiving the upper level.

Zinc is necessary for the maintenance and development of skin and hair. Since beautiful hair coats are important for horses, fortifying the daily ration with zinc will prevent any possibility of a zinc deficiency; and if the zinc in the feed is on the low side, it should improve the hair coat.

CHELATED TRACE MINERALS

The word chelate is derived from the Greek *chele,* meaning a claw or pincerlike organ. Those selling chelated minerals generally recommend a smaller quantity of them (but at a higher price per pound) and extoll their "fenced-in" properties.

When it comes to synthetic chelating agents, much needs to be learned about their selectivity toward minerals, the kind and quantity most effective, their mode of action, and their behavior with different species of animals and with varying rations.

It is possible that their use may actually create a mineral imbalance. These answers, and more, should be forthcoming through carefully controlled experiments.

MINERAL IMBALANCES

Having the right balance and forms of minerals can be very important. The more calcium you feed, the more phosphorus you need. The more copper you feed, the more manganese you need.

Also, minerals can be fed in several different forms. For example, iron can be fed as an oxide, sulfite, sulfate, or as a proteinate. Oxides may be absorbed at about 2 to 5%, while sulfites may be absorbed at up to 10%, and sulfates at 25%.

Thus, the requirements of any mineral may be modified (1) by another mineral which enhances or interferes with its utilization, or (2) by the form of the mineral.

From the above, it is apparent that excess fortification of the horse's ration with one or more mineral elements may prove more detrimental than helpful. Thus, caretakers who know and care will avoid harmful imbalances; they will provide minerals on the basis of *recommended allowances* (see Table 25-6). Also, when fortifying rations with minerals, consideration should be given to the minerals provided by the ingredients of the normal ration, for it is the total composition of the feed that counts.

FEEDING MINERALS

With the exception of sodium, the self-feeding of the major minerals cannot be relied upon to meet the needs of horses. This is so because horses consume such supplements on the basis of palatability, rather than because of dietary need. As a result, the free-choice intake of minerals among individual horses will vary from too little to too much. Sometimes minerals are incorporated in a salt mix, but salt consumption is erratic and variable according to the sodium content of the feedstuffs being fed. So, the only way to ensure that each horse receives the needed major minerals is to incorporate the proper amounts in the animal's feed and/or water.

Trace minerals may be added to the ration and/or incorporated in the salt. In either case, the amounts and proportions of trace minerals should be selected with care because the improper use of trace minerals can lead to induced deficiencies. Theoretically, the total ration (grain plus forage) should be balanced in trace mineral content, with the trace mineral mix providing only the minerals needed and with each one in the right amount. Of course, this isn't practical. Therefore, a trace mineral mix must contain an array of minerals in adequate levels to meet a wide variety of conditions. Fortunately, the horse is tolerant to most trace mineral excesses.

When horses are on pasture and no grain or protein supplement is being fed, minerals may be self-fed, usually as either a commercially manufactured mineral block or as a mineral mixture. A suitable home-mixed mineral for self-feeding on pasture may be prepared as follows:

1. **Where the pasture is primarily grass.** Prepare a mixture containing two parts of calcium to one part of phosphorus.

2. **Where the pasture is primarily a legume.** Prepare a mixture containing one part of calcium to one part of phosphorus.

To each of the above mixes, add one-third trace-mineralized salt to provide the microminerals and improve the palatability.

Vitamins

Unfortunately, there are no warning signals to tell a caretaker when a horse is not getting enough of certain vitamins. But a continuing inadequate supply of any one of several vitamins can produce illness which is very difficult to diagnose until it becomes severe, at which time it is difficult and expensive—if not too late—to treat. The important thing, therefore, is to ensure against such deficiencies occurring. But caretakers should not shower a horse with mistaken kindness through using shotgun-type vitamin preparations. Instead, the quantity of each vitamin should be based on available scientific knowledge.

TABLE
HORSE VITAMIN

Vitamins Which May Be Deficient Under Normal Conditions	Conditions Usually Prevailing Where Deficiencies Are Reported	Function of Vitamin	Some Deficiency Symptoms	Practical Sources of the Vitamin
Fat-soluble vitamins:				
A	Extended drought, bleached hays. Stall feeding where there is little or no green forage or yellow corn. Following great stress, as when race or show horses are put in training. The younger the animal, the quicker vitamin A deficiencies will show up. Mature animals may store sufficient vitamin A to last 6 months.	Promotes growth and stimulates appetite. Assists in reproduction and lactation. Keeps the mucous membranes of respiratory and other tracts in healthy condition. Makes for normal vision. Prevents night blindness.	Loss of appetite, poor growth, reproductive problems, nerve degeneration, night blindness, lachrymation (tears), keratinization of the cornea and skin, uneven and poor hoof development, a predisposition to respiratory infection, incoordination, progressive weakness, convulsive seizures, certain bone disorders, and finicky appetite.	Stabilized vitamin A. Green grass. Green grass or legume hay not over 1 year old. Carrots, yellow corn.
D	Limited sunlight and/or limited sun-cured hay, especially when horse is kept inside most of the time.	Assimilation and utilization of calcium and phosphorus, necessary in normal bone development—including the bones of the fetus.	Rickets in foals, osteomalacia in mature horses. Both conditions result in large joints and weak bones. Rickets is characterized by reduced bone calcification, stiff and swollen joints, stiffness of gait, irritability, and reduction in serum calcium and phosphorus. Osteomalacia results in bones which soften, become distorted, and fracture easily.	Either vitamin D_2 (the plant form) or D_3 (the animal form) is equally effective for the horse. Exposure to sunlight. Sun-cured hays.
E	More vitamin E may be destroyed or used by horses during times of stress or strain than can be obtained through normal feeds.	As an antioxidant. As an occasional replacement for selenium. Improves reproduction. Prevents anhidrosis.	Lowered breeding preformance in both mares and stallions. Anhidrosis—a dry, dull hair coat; elevated temperature; and high blood pressure. Anhidrosis has been successfully treated by the oral administration of 1,000 to 3,000 IU of vitamin E daily.	Alpha-tocopherol acetate, a stable form of vitamin E. Wheat germ meal and wheat germ oil. Green plants. Green hays.

It has long been known that the vitamin content of feeds varies considerably according to soil, climatic conditions, and curing and storing.

Deficiencies may occur during periods (1) of extended drought or in other conditions of restriction in diet, (2) when production is being forced, or during stress, (3) when large quantities of highly refined feeds are being fed, or (4) when low-quality forages are utilized.

Although the occasional deficiency symptoms are the most striking result in vitamin deficiencies, it must be emphasized that in practice mild deficiencies probably cause higher total economic losses than do severe deficiencies. It is relatively uncommon for a ration to contain so little of a vitamin that obvious symptoms of a deficiency occur. When such a case does appear, it is reasonable to suppose that there must be several cases that are too mild to produce characteristic symptoms but that are sufficiently severe to lower the state of health and the efficiency of production.

Certain vitamins are necessary for the growth, development, health, and reproduction of horses. Deficiencies of vitamins A and D are sometimes encountered. Also, indications are that vitamin E and some of the B vitamins are required by horses. Further, it is recognized that single, uncomplicated vitamin deficiencies are the exception rather than the rule.

High-quality, leafy, green forages plus plenty of sunshine generally give horses most of the vitamins they need. Horses get carotene (which they can convert to vitamin A) and riboflavin from green pasture and green hay not over a year old, and they get vitamin D from sunlight and sun-cured hay. If plenty of green forage and sunlight are not available, the caretaker should get the advice of a nutritionist or veterinarian on the use of vitamin additives to the feed.

Tables 25-6, p. 1072, and 25-9, pp. 1086–1091, list the vitamins most commonly involved in horse nutrition; and Table 25-9, Horse Vitamin Chart gives pertinent information pertaining to each. Although there is no evidence of deficiencies of certain vitamins, it is possible that more of them may be destroyed or used by horses during stress or strain than can be obtained through normal feeds or synthesized by the intestinal microflora of the horse; hence, adding them to the ration may assure maximum performance.

25-9
CHART (See footnotes at end of table.)

Classes/Function	Nutrient Requirements[1 2]			Nutrient Allowances[1 2]			Comments
	Per Horse Daily	In Ration A-F	Per Ton Ration A-F	Per Horse Daily	In Ration A-F	Per Ton Ration A-F	
	(IU)	(IU/lb)	(IU/ton)	(IU)	(IU/lb)	(IU/ton)	
Maintenance:							Vitamin A is not synthesized in the cecum.
1,000-lb *(454-kg)* horse	22,725	909	1,818,000	26,134	1,045	2,090,700	Hay over 1 year old, regardless of green color, is usually not an adequate source of carotene or vitamin A activity.
Gestation/Lactation:							
1,000-lb *(454-kg)* mare	34,100	1,364	2,728,000	39,215	1,569	3,137,200	When deficiency symptoms appear, add stabilized vitamin A to the ration.
Growth:							
450-lb *(204.5-kg)* weanling	10,908	909	1,818,000	12,544	1,045	2,090,700	It is wasteful to feed more vitamin A than is needed. Also exceedingly high levels over an extended period of time may cause bone fragility, hyperostosis, and exfoliated epithelium.
Working:							
1,000-lb *(454-kg)* horse	22,725	909	1,818,000	26,134	1,045	2,090,700	
Maintenance:							The vitamin D requirement is less when a proper balance of calcium and phosphorus exists in the ration.
1,000-lb *(454-kg)* horse	3,400	136	272,000	3,910	156	312,800	When animals are exposed to direct sunlight, the ultraviolet light produces vitamin D from traces of cholesterol in the skin. Stabled horses, exercised in the early morning, will not get sufficient vitamin D in this manner.
Gestation/Lactation:							
1,000-lb *(454-kg)* mare	6,825	273	546,000	7,849	314	627,900	Too much vitamin D may harm a horse. Vitamin D toxicity is characterized by calcification of the blood vessels, heart, and other soft tissues, and by bone abnormalities. Toxic level of vitamin D has not been established in the horse, but a level 50 times the requirement may be harmful.
Growth:							
450-lb *(204.5-kg)* weanling	4,368	364	728,000	5,023	419	837,200	
Working:							
1,000-lb *(454-kg)* horse	3,400	136	272,000	3,910	156	312,800	
Maintenance:							Utilization of vitamin E is dependent on adequate selenium.
1,000-lb *(454-kg)* horse	575	23	46,000	661	26	52,900	
Gestation/Lactation:							
1,000-lb *(454-kg)* mare	900	36	72,000	1,035	41	82,800	
Growth:							
450-lb *(204.5-kg)* weanling	432	36	72,000	497	41	82,800	
Working:							
1,000-lb *(454-kg)* horse	900	36	72,000	1,035	41	82,800	

(Continued)

TABLE 25-9

Vitamins Which May Be Deficient Under Normal Conditions	Conditions Usually Prevailing Where Deficiencies Are Reported	Function of Vitamin	Some Deficiency Symptoms	Practical Sources of the Vitamin
K	Following intestinal disorders.	Concerned with blood coagulation. It converts precursor proteins to the active blood clotting factors.	Increased clotting time of the blood and lowered level of prothrombin.	Green pasture. Well-cured hays. Cereal grains. Milk. Menadione (vitamin K_3).
Water-soluble vitamins: Biotin	Sulfa drugs kill intestinal organisms; hence, when they are used an extended period, there may be a deficiency of biotin.	Biotin plays an important role in the metabolism of carbohydrates, fats, and proteins.	In all animals, a deficiency of biotin will depress growth and cause a loss of hair and/or a dermatitis.	Alfalfa hay, blackstrap molasses, cottonseed meal, soybean meal, peanut meal, milk, wheat bran, synthetic biotin, and yeast (brewers', torula).
Choline	Ration low in methionine, an amino acid.	Prevention of fatty livers, the transmitting of nerve impulses, and the metabolism of fat.	Slow growth and fatty livers are the deficiency symptoms.	Feed sources, such as alfalfa hay, blackstrap molasses, and cereal grains. Body manufacture of choline from excess of the amino acid methionine. Choline chloride. Choline dihydrogen.
Folacin (Folic Acid)		In all vertebrates, folacin is essential for normal growth and reproduction, for the prevention of blood disorders, and for important biochemical mechanisms in each cell.	Poor growth. Anemia.	Alfalfa hay, the oil meals (soybean, cottonseed, and linseed), skimmed milk, and wheat and wheat by-products. Synthetic folacin, wheat germ, and yeast (brewers', torula).
Niacin (Nicotinic Acid, Nicotinamide)		Constituent of two important coenzymes. They are involved in the release of energy from carbohydrates, fats, and proteins, and in the synthesis of fatty acids, protein, and DNA.	Reduced growth and appetite. Skin rashes, diarrhea, nerve disorders.	Green alfalfa. Niacin is widely distributed in feeds; fermentation solubles and certain oil meals are especially good sources. Synthetic niacin.
Pantothenic Acid (Vitamin B-3)		Part of coenzyme A, which plays a key role in body metabolism.	Poor growth, skin rashes, poor appetite, nervous disorders.	Safflower meal, blackstrap molasses, wheat bran, and milk. Calcium pantothenate.
Riboflavin (Vitamin B-2)	When green feeds (pasture, hay, or silage) are not available.	Riboflavin has an essential role in the oxidative mechanisms of the cells.	Periodic ophthalmia (or moon blindness), characterized by catarrhal conjunctivitis in one or both eyes, accompanied by photophobia, and lachrymation. Decreased rate of growth and feed efficiency. Porous and weak bones; ligaments and joints impaired.	Green pasture. Green hay. Milk and milk products. Synthetic riboflavin. Yeast.

(Continued)

Classes/Function	Nutrient Requirements[1 2] Per Horse Daily	In Ration A-F	Per Ton Ration A-F	Nutrient Allowances[1 2] Per Horse Daily	In Ration A-F	Per Ton Ration A-F	Comments
	(mg)	(mg/lb)	(mg/ton)	(mg)	(mg/lb)	(mg/ton)	
Maintenance:							High levels of vitamin K will overcome bleeding due to dicoumarol.
1,000-lb *(454-kg)* horse				8.0	0.32	640	
Gestation/Lactation:							Vitamin K is generally (1) widely distributed in normal feeds, and/or (2) synthesized in adequate amounts by the intestinal microflora of the horse.
1,000-lb *(454-kg)* mare				8.0	0.32	640	
Growth:							
450-lb *(204.5-kg)* weanling				3.6	0.30	600	
Working:							
1,000-lb *(454-kg)* horse				8.0	0.32	640	
Maintenance:							Biotin is closely related metabolically to folacin, pantothenic acid, and vitamin B-12.
1,000-lb *(454-kg)* horse				2.5	0.1	200	
Gestation/Lactation:							
1,000-lb *(454-kg)* mare				2.5	0.1	200	
Growth:							
450-lb *(204.5-kg)* weanling				1.2	0.1	200	
Working:							
1,000-lb *(454-kg)* horse				2.5	0.1	200	
Maintenance:							Choline content of normal feeds is usually sufficient.
1,000-lb *(454-kg)* horse				500	20.0	40,000	
Gestation/Lactation:							
1,000-lb *(454-kg)* mare				750	30.0	60,000	
Growth:							
450-lb *(204.5-kg)* weanling				750	62.5	125,000	
Working:							
1,000-lb *(454-kg)* horse				750	30.0	60,000	
Maintenance:							Folacin is widely distributed in horse feeds.
1,000-lb *(454-kg)* horse				20	0.8	1,600	Also, folacin is synthesized in the lower gut.
Gestation/Lactation:							
1,000-lb *(454-kg)* mare				30	1.2	2,400	
Growth:							
450-lb *(204.5-kg)* weanling				36	3.0	6,000	
Working:							
1,000-lb *(454-kg)* horse				30	1.2	2,400	
Maintenance:							There is some evidence that niacin is synthesized by the horse.
1,000-lb *(454-kg)* horse				250	10.0	20,000	
Gestation/Lactation:							The horse can convert the essential amino acid tryptophan into niacin. Hence, it is important to make certain that the ration is adequate in niacin; otherwise, the horse will use tryptophan to supply niacin needs.
1,000-lb *(454-kg)* mare				250	10.0	20,000	
Growth:							
450-lb *(204.5-kg)* weanling				250	20.8	41,600	
Working:							
1,000-lb *(454-kg)* horse				250	10.0	20,000	
Maintenance:							Grain is very deficient in pantothenic acid.
1,000-lb *(454-kg)* horse				250	10.0	20,000	Of all the B vitamins, pantothenic acid is most likely to be deficient under stable (confinement) conditions.
Gestation/Lactation:							
1,000-lb *(454-kg)* mare				250	10.0	20,000	
Growth:							
450-lb *(204.5-kg)* weanling				250	20.8	41,600	
Working:							
1,000-lb *(454-kg)* horse				250	10.0	20,000	
Maintenance:							Lack of vitamin B-2 is not the only cause of moon blindness. Sometimes, moon blindness follows leptospirosis, and it may be caused by an allergic reaction.
1,000-lb *(454-kg)* horse	22.8	0.91	1,820	40.0	1.6	3,200	
Gestation/Lactation:							
1,000-lb *(454-kg)* mare	22.8	0.91	1,820	40.0	1.6	3,200	
Growth:							
450-lb *(204.5-kg)* weanling	10.9	0.91	1,820	19.2	1.6	3,200	
Working:							
1,000-lb *(454-kg)* horse	22.8	0.91	1,820	40.0	1.6	3,200	

(Continued)

TABLE 25-9

Vitamins Which May Be Deficient Under Normal Conditions	Conditions Usually Prevailing Where Deficiencies Are Reported	Function of Vitamin	Some Deficiency Symptoms	Practical Sources of the Vitamin
Thiamin (Vitamin B-1)	Poor-quality hay and grain. When sulfa drugs or antibiotics are given to the horse, the synthesis of B vitamins is impaired. Consumption of bracken fern (*Pteris aquilina*) and horsetail (*Equisetum* spp) will cause thiamin deficiency due to the antithiamin compounds that they contain.	In energy metabolism. Without thiamin, there would be no energy. In the working of the peripheral nerves. Promotes appetite and growth.	A thiamin deficiency has been produced experimentally. Decreased feed consumption (loss of weight), anemia, incoordination (especially in the hindquarters), lowered blood thiamin, elevated blood pyruvic acid, enlarged heart, and nervous symptoms.	Wheat and wheat by-products. Oilseed meals. Oat grain and groats. Thiamin hydrochloride. Yeast (brewers', torula).
Vitamin B-6 (Pyridoxine, Pyridoxal, Pyridoxamine)		In its coenzyme forms, it is involved in a large number of physiologic functions, particularly protein, carbohydrate, and fat metabolism.	No deficiency symptoms of vitamin B-6 have been reported in the horse. So, it is thought to be synthesized in the cecum.	Green pasture, alfalfa hay, wheat bran, wheat germ, and yeast (brewers', torula).
Vitamin B-12 (Cobalamins)	When few, or no feeds of animal origin are fed. Where cobalt is not present in the feed, thereby precluding the synthesis of vitamin B-12 in the gastrointestinal tract.	Coenzyme in several enzyme systems. Closely linked with choline, folacin, and pantothenic acid.	Loss of appetite and poor growth.	Protein supplements of animal origin. Fermentation products. Cobalamins, yeast.
Vitamin C (Ascorbic Acid, Dehydroascorbic Acid)	The vitamin C requirements of fish and humans have been observed to increase in periods of stress. So, it is conjectured that heavily stressed horses may require more vitamin C than they can synthesize.	Formation and maintenance of collagen. More rapid healing of wounds. Sound bones.	No deficiency symptoms in horses noted. In humans and monkeys, scurvy is the main deficiency symptom. Also, in humans sudden death from severe internal hemorrhage and heart failure are always a danger.	Ordinary rations and body synthesis provide adequate vitamin C for horses. Well-cured hays and green pastures are good sources of vitamin C.
Unidentified factors	Since the U.S. foal crop is only around 50%, it is obvious that there is room for improvement somewhere along the line; and perhaps unidentified factors are involved. Also, optimal results with horses during the critical periods (growth, gestation-lactation, and when under stress as in racing or showing) appear to be dependent upon providing unidentified factors through such ingredients as distillers' dried solubles, dehydrated alfalfa meal, condensed fish solubles, brewers' dried yeast, antibiotic fermentation residues, dried whey, and corn fermentation solubles.			

[1]As used herein, the distinction between "nutrient requirements" and "nutrient allowances" is as follows: In nutrient requirements, no margins of safety are included intentionally; whereas in nutrient allowances, margins of safety are provided in order to compensate for variations in feed composition, environment, and possible losses during storage or processing. The "nutrient requirements" in Table 25-9 were adapted by the authors from *Nutrient Requirements of Horses*, 5th rev.ed., NRC-National Academy of Sciences, 1989. The "nutrient allowances" were developed by the authors, based on experiments and experiences; it is intended that they meet the nutrient requirements, and provide adequate margins of safety in addition.

FAT-SOLUBLE VITAMINS

The fat-soluble vitamins, which are stored in the body in appreciable quantities, are vitamin A (carotene), vitamin D, vitamin E, and vitamin K.

Vitamin A

Vitamin A is not synthesized in the cecum; thus, it must be provided in the feed, either (1) as vitamin A, or (2) as carotene, the precursor of vitamin A. For horses, 1 mg of beta-carotene is equivalent to 400 IU of vitamin A.

Horses that have been consuming green forage for 4 to 6 weeks usually store sufficient vitamin A in the liver to maintain adequate levels of plasma vitamin A for 3 to 6 months.

It is noteworthy (1) that the absorption of vitamin A is adversely affected by the presence of parasites in the intestinal tract, and (2) that the presence of enough protein of good quality enhances the conversion of carotene to vitamin A.

Severe deficiency of vitamin A may cause night blindness (impaired adaptation to darkness), lacrimation (tears), keratinization of the cornea and skin, reproductive difficulties, poor or uneven hoof development, difficulty in breathing, incoordination, convulsive seizures, progressive weakness, and poor appetite. There is also some evidence that deficiency of this vitamin may cause or contribute to certain leg bone weaknesses. When vitamin A deficiency symptoms appear, the caretaker should add a stabilized vitamin A product to the ration.

A considerable margin of safety in vitamin A and carotene is provided in the recommended allowances due to the oxidative destruction of these materials in feeds during storage. But, it is wasteful to feed more vitamin A than is needed. Also, feeding exceedingly high levels of vitamin A over an extended period of time may cause bone fragility, hyperostosis, and exfoliated epithelium. When fed as directed, the vast majority of horse feeds won't provide excesses of vitamin A.

(Continued)

Classes/Function	Nutrient Requirements[1,2] Per Horse Daily	In Ration A-F	Per Ton Ration A-F	Nutrient Allowances[1,2] Per Horse Daily	In Ration A-F	Per Ton Ration A-F	Comments
	(mg)	(mg/lb)	(mg/ton)	(mg)	(mg/lb)	(mg/ton)	
Maintenance:							Thiamin is synthesized in the lower gut of the horse by bacterial action, but there is some doubt as to its sufficiency. When neither green pasture nor high-quality roughage is available, thiamin hydrochloride should be added to the ration. Since carbohydrate metabolism is increased during physical exertion, it is important that B-1 be available in quantity at such times.
1,000-lb *(454-kg)* horse	34.1	1.36	2,720	39.2	1.57	3,140	
Gestation/Lactation:							
1,000-lb *(454-kg)* mare	34.1	1.36	2,720	39.2	1.57	3,140	
Growth:							
450-lb *(204.5-kg)* weanling	16.4	1.36	2,720	18.9	1.57	3,140	
Working:							
1,000-lb *(454-kg)* horse	56.8	2.27	4,540	65.3	2.61	5,220	
Maintenance:							Normally, horse rations contain adequate vitamin B-6. Also, it appears to be synthesized in the cecum. Yet, these sources may not be adequate for the maximum performance of the horse.
1,000-lb *(454-kg)* horse				25.0	1.0	2,000	
Gestation/Lactation:							
1,000-lb *(454-kg)* mare				25.0	1.0	2,000	
Growth:							
450-lb *(204.5-kg)* weanling				6.0	0.5	1,000	
Working:							
1,000-lb *(454-kg)* horse				25.0	1.0	2,000	
Maintenance:							It is reported that horses in poor nutritional condition showing anemia respond to the administration of vitamin B-12.
1,000-lb *(454-kg)* horse				0.125	0.005	10	
Gestation/Lactation:							
1,000-lb *(454-kg)* mare				0.150	0.006	12	
Growth:							
450-lb *(204.5-kg)* weanling				0.084	0.007	14	
Working:							
1,000-lb *(454-kg)* horse				0.150	0.006	12	
Maintenance:							Dietary need is clearly evident for humans, monkeys, guinea pigs, fruit-eating bats, and bulbul birds. However, vitamin C is probably required by other species, but synthesized in the body; the only question is whether the horse can synthesize enough vitamin C when under stress.
1,000-lb *(454-kg)* horse				60	2.4	4,800	
Gestation/Lactation:							
1,000-lb *(454-kg)* mare				100	4.0	8,000	
Growth:							
450-lb *(204.5-kg)* weanling				45	3.75	7,500	
Working:							
1,000-lb *(454-kg)* horse				100	4.0	8,000	

[2]Feed consumption of mature 1,000-lb *(454-kg)* horse estimated at 25 lb *(11.36 kg)* per day. Feed consumption of 450-lb *(204.5-kg)* weanling estimated at 12 lb *(5.45 kg)* per day.

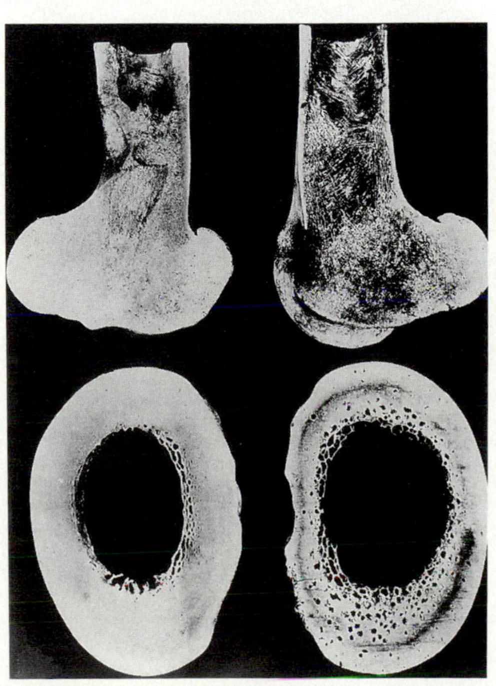

Fig. 25-10. (Left) Vitamin A made the difference! *Upper:* On the right is shown the sagittal section of the distal end of the femur of a vitamin-deficient horse compared to normal bone (left). *Lower:* On the right is shown the cross section of the cannon bone from a vitamin-deficient horse compared to normal bone (left). (Courtesy, Calif. Ag. Exp. Sta.)

Vitamin D

For horses, both D_2 (the plant form) and D_3 (the animal form) are equally effective, so there is no need to use some of each.

Foals sometimes develop rickets because of insufficient vitamin D, calcium, or phosphorus. Rickets is characterized by reduced bone calcification, stiff and swollen joints, stiffness of gait, irritability, and reduction in serum calcium and phosphorus. It can be prevented by exposing the animal to direct sunlight as much as possible, by allowing free access to a suitable mineral mixture, or by providing good-quality, sun-cured hay or luxuriant pasture grown on well-fertilized soil. In northern areas that do not have adequate sunshine or where foals are kept indoors for extended periods, many caretakers provide the foal with a vitamin D supplement.

With vitamin D, as with vitamin A, there is need for adequacy without harmful excesses. Too much vitamin D may harm a horse. Vitamin D toxicity is characterized by calcification of the blood vessels, heart, and other soft tissues, and by bone abnormalities. Also, there is general weakness and loss of body weight. Although the toxic level of vitamin D in the horse has not been established, a level 50 times the requirement may be harmful.

The vitamin D requirement is less when a proper balance of calcium and phosphorus exists in the ration.

Vitamin D occurs naturally in only a few common feeds, primarily sun-cured forages.

Vitamin D_2, vitamin D_3, cod and certain other fish liver oils, and irradiated yeast are the main supplementary sources.

Vitamin E

Vitamin E, or tocopherol, is associated with reproduction. Also, it prevents and corrects anhidrosis (dry, dull hair coat).

Most practical rations contain liberal quantities of vitamin E, perhaps enough except under conditions of work, stress, or reproduction, or when there is interference with its utilization. Green forages, especially alfalfa, are good sources.

Anhidrosis in horses—a condition showing dull hair coat, elevated temperature, high blood pressure, and labored breathing—has been successfully treated by the oral administration of 1,000 to 3,000 IU of vitamin E daily for 1 month.

It is now preferred to use milligrams of alpha-tocopherol equivalents as a summation term for all vitamin E activity. However, feed composition tables generally give values in IU, and IU is still used for labeling most feed products.

The requirements for vitamin E are influenced by interrelationships with other essential nutrients—increased by the presence of interfering substances, and spared by the presence of other substances that may be protective or that may assume part of its functions. The recommended allowances of vitamin E are given in Table 25-6.

Vitamin K

When vitamin K is deficient, the coagulation time of the blood is increased and the prothrombin level is decreased. This is the main justification in adding this vitamin to the ration of the horse. Also, vitamin K has value in veterinary medicine as an aid in controlling hemorrhage.

WATER-SOLUBLE VITAMINS

The large amounts of water which pass through the horse's body daily tend to carry out the water-soluble vitamins, thereby depleting the supply. Thus, they must be supplied in the horse's ration on a day-to-day basis. All of the water-soluble vitamins except vitamin C are known as B vitamins.

Vitamins of the B-complex, particularly biotin, choline, folacin (folic acid), niacin, pantothenic acid, riboflavin, thiamin (B-1), vitamin B-6 (pyridoxine), and vitamin B-12, may be essential, especially for (1) young horses before the synthesis of the B-complex vitamins by the microflora begins, and (2) horses that are under stress, as in racing and showing. However, it is not clear which ones are needed, in what quantities they are needed, and their status from the standpoints of synthesis and absorption in the horse. Healthy horses usually get enough of them either in natural rations or by synthesis in the intestinal tract. However, when neither green pasture nor high-quality dry forage is available, it may be in the nature of good insurance to provide them, especially for horses that are under stress.

Although some of the B vitamins and unidentified factors are synthesized in the cecum of the horse, it is doubtful that microbial activity is sufficient to meet the needs during the critical periods—growth, reproduction, and when animals are subjected to great stress as in showing or racing. Also, there is reason to question the efficacy of absorption this far down the digestive tract; for, in comparison with that of humans and other animals, the cecum is on the wrong end of the digestive tract. Moreover, it is known that horses fed thiamin-deficient rations lose weight, become nervous, and show incoordination in the hindquarters; then, when thiamin is added to the ration, this condition is cured. For these reasons, in valuable horses it is not wise to rely solely on bacterial synthesis. The B vitamins, along with unidentified factors, may be provided by adding to the ration such ingredients as distillers' dried solubles, dried brewers' yeast, dried fish solubles, or animal liver meal; usually through a reputable commercial feed.

Biotin

Ordinary equine rations probably contain ample biotin, or horses synthesize all they need. But, in recognition that biotin is required of all species, and that it plays an important role in the metabolism of carbohydrates, fats, and proteins, it is possible that adding biotin to the ration of the horse may assure maximum performance.

Fig. 25-11 indicates that biotin is essential for sound hooves.

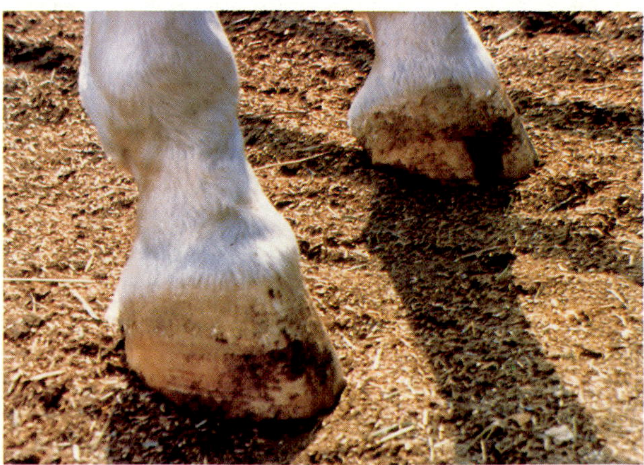

Fig. 25-11. Prior to treatment, the hooves of this horse had become weak, cracked, and crumbly, and were unable to hold shoes; and the animal had to be retired from competition. Following supplementation with 20 mg of biotin per day for 5 months, the hooves regained strength and were able to hold shoes; and the horse was returned to competition. (Courtesy, Colorado State University, Ft. Collins)

From this it should not be concluded that all hoof problems are due to biotin deficiency. They are not. Actually, several nutrients are known to influence hoof growth—biotin among them. Studies indicate that a complete balanced ration is essential for proper hoof growth. Without doubt, heritability is also a factor.

Choline

Choline is a metabolic essential for building and maintaining cell structure and in the transmission of nerve impulses. Choline deficiency has been produced in rats, dogs, chickens, pigs, and other species. Slow growth is a nonspecific symptom.

The dietary requirement for choline depends on the level of methionine (an amino acid) in the ration. Also, it is noteworthy that all naturally occurring fats contain some choline; however, normal horse feeds are low in fat. Hence, the addition to the ration of 500 mg of choline per day is the recommended allowance for the maintenance of a 1,000-lb horse.

Folacin (Folic Acid)

There is no single compound vitamin with the name folacin; rather, the term *folacin* is used to designate folic acid and a group of closely related substances which are essential for vertebrates.

Folacin is widely distributed in horse feeds. Also, it is synthesized in the lower digestive tract of the horse. Hence, it is unlikely that a dietary source is required, although a small amount may be in the nature of cheap insurance.

Niacin (Nicotinic Acid, Nicotinamide)

Niacin is a collective term which includes nicotinic acid and nicotinamide, both natural forms of the vitamin with equal niacin activity.

Some evidence indicates that niacin is synthesized by the horse. Also, the horse can convert the essential amino acid tryptophan into niacin. Hence, it is important to make certain that the ration is adequate in niacin; otherwise, the horse will use tryptophan to supply niacin needs. Niacin is widely distributed in feeds; fermentation solubles, and certain oil meals are especially good sources. Only a modest addition of niacin to the ration is indicated.

Pantothenic Acid (Vitamin B-3)

Intestinal synthesis of pantothenic acid has been found to occur in all species studied. In the case of the horse, such synthesis appears to be sufficiently extensive to meet body needs, at least in part. However, of all the B vitamins, pantothenic acid is most likely to be deficient under stable (confinement) conditions. As indicated in Table 25-9, a daily allowance of 250 mg of pantothenic acid is recommended for a 1,000-lb gestating or lactating mare.

Riboflavin (Vitamin B-2)

A deficiency of riboflavin may cause periodic ophthalmia (moon blindness), characterized by catarrhal conjunctivitis in one or both eyes, accompanied by photophobia, and lacrimation. Repeated attacks affect the retina, lens, and ocular fluids and cause impaired vision or blindness. But it is known that lack of this vitamin is not the only factor causing this condition. Sometimes moon blindness follows leptospirosis in horses, and it may be caused by a localized hypersensitivity or allergic reaction. Periodic opthalmia caused by lack of riboflavin may be prevented by feeding green hay and green pasture, supplying feeds high in riboflavin, or by adding crystalline riboflavin to the ration.

Two properties of riboflavin lend support to riboflavin supplementation for the horse: (1) it is destroyed by light, and it is destroyed by heat in an alkaline solution; and (2) body storage is very limited, so day-to-day needs must be provided in the ration.

Thiamin (Vitamin B-1)

Vitamin B-1 is synthesized in the lower gut of the horse by bacterial action, but there is some doubt as to its sufficiency and as to the amount absorbed always meeting the full requirements.

A thiamin deficiency has been produced experimentally. It is characterized by loss of appetite, loss of weight, anemia, incoordination (especially of the hind legs), lower blood thiamin, elevated blood pyruvic acid, and dilated and hypertrophied heart.

Vitamin B-1 is required for normal carbohydrate metabolism. Since carbohydrate metabolism is increased during physical exertion, it is important that B-1 be available in quantity at such times.

Vitamin B-6 (Pyridoxine, Pyridoxal, Pyridoxamine)

There is no evidence that deficiencies of vitamin B-6 occur in horses on commonly fed rations; and it is not expected that deficiencies should occur in view of the widespread distribution of vitamin B-6 in feedstuffs and the probable synthesis of B-6 in the cecum. Yet, these sources may not be adequate to assure maximum performance of the horse. So, the daily supplementation of 25 mg of vitamin B-6 per horse may be in the nature of cheap insurance, especially because of the important role of the vitamin.

Vitamin B-6, in its coenzyme forms, is involved in a large number of physiologic functions, particularly the metabolism of protein, carbohydrate, and fat. Also, it is involved in clinical problems, including (1) anemia that is not iron responsive, (2) kidney stones, and (3) the physiologic demands of pregnancy.

Vitamin B-12 (Cobalamins)

It has been reported that horses in poor nutritional condition showing anemia respond to the administration of vitamin

B-12. An allowance of 0.084 mg of B-12 per day is recommended for weanlings.

Vitamin B-12 injections are frequently given to horses to improve performance and to prevent or cure anemia. There is no experimental evidence that such shots are either helpful or harmful.

Vitamin C (Ascorbic Acid, Dehydroascorbic Acid)

A dietary need for ascorbic acid is limited to humans, monkeys, guinea pigs, fruit-eating bats, and bulbul birds. But, the vitamin is probably required by all other species, including the horse, but is synthesized adequately in the body. It is noteworthy, however, that catfish and trout require dietary sources of vitamin C when they are stressed by being raised in extremely crowded conditions. This theory of stress has been carried over to human nutrition. So, it is conjectured that heavily stressed horses may not be able to synthesize sufficient vitamin C for maximum performance; hence, adding the vitamin to the ration may make for added assurance.

VITAMIN IMBALANCES

Experiments have shown that the amounts needed of certain vitamins may be affected by the supply of another vitamin or of some other nutritive essential. Also, it is known that excess fortification of the horse's ration with certain vitamins may prove more detrimental than helpful. Thus, caretakers should avoid harmful imbalances; they should provide vitamins on the basis of recommended allowances. Also, when fortifying with vitamins, consideration should be given to the vitamins provided by the ingredients of the normal ration, for it is the total composition of the feed that counts.

UNIDENTIFIED FACTORS

Since the U.S. foal crop is only around 50%, and since horses under stress (racing, showing, etc.) frequently become temperamental in their eating habits, it is obvious that there is room for improvement in the ration somewhere along the line. Perhaps unidentified factors are involved.

Unidentified factors include those vitamins which the chemist has not yet isolated and identified. For this reason, they are sometimes referred to as the vitamins of the future. There is mounting evidence of the importance of unidentified factors for animals, including humans. Among other things, they lower the incidence of ulcers in humans and swine. For horses, they appear to increase growth and improve feed efficiency and breeding performance when added to rations thought to be complete with regard to known nutrients. The anatomical and physiological mechanism of the digestive system of the horse, plus the stresses and strains to which modern horses are subjected, would indicate the wisdom of adding unidentified factor sources to the ration of the horse. Unidentified factors appear to be of special importance during breeding, gestation, lactation, and growth.

Three highly regarded unidentified factor sources are dried whey product, corn fermentation solubles, and dehydrated alfalfa meal.

Water

Fig. 25-12. Free access to clean, fresh water is desirable. (Courtesy, American Quarter Horse Assn., Amarillo, Tex.)

Horses can survive for a longer period without feed than they can without water. The loss of 10% body water will result in disorders; the loss of 20% body water will cause death.

Water is essential for the various physiological processes of the horse, such as the production of saliva and sweating.

Normally, horses will drink about 1 gal of water per 100-lb body weight per day; so, on an average day, a 1,000-lb horse will drink about 10 gal of water. But, in addition to size, the amount of water consumed depends on the following:

1. **Age.** The younger the horse, the more water it will drink per unit of body weight. This is because the bodies of younger animals contain more water than older animals. Thus, the body of a newborn foal contains 73% water, whereas the body of a mature horse contains only 62% water (See Chapter 2, Table 2-2).
2. **Temperature and humidity.** A rise in the environmental temperature from 55° to 75°F will increase water requirements by 15 to 20%, Also, humidity is a factor.
3. **Work.** Depending on the kind and severity, work may increase water consumption by 20 to 300%.
4. **Ration.** Water consumption increases with the amount of feed consumed daily, the dry matter of the ration, and the kind of feed—it's higher on a high-fiber hay ration than on a high-grain ration. Impaction colic may occur with limited water intake and a dry, high-fiber ration.
5. **Lactation.** Lactation may increase water consumption by 50 to 100% above maintenance.

Free access to ample quantities of clean, fresh water is desirable. When this is not possible, horses should be watered at approximately the same times daily. Opinions vary among caretakers as to the proper time and method of watering horses. All agree, however, that regularity and frequency are desirable. Most caretakers agree that water may be given before, during, or after feeding.

Frequent, small waterings between feedings are desirable during warm weather or when animals are being put to hard use. Do not allow horses to drink heavily when they are hot, because they may founder; and do not allow horses to drink heavily just before being put to work.

Water should be available in both stalls and corrals. All waterers should have drains for easy cleaning, and should be heated to 40° to 45°F during the winter months in cold regions.

Fig. 25-13. Automatic waterer of good design positioned in a stall. (Courtesy, *Equus Magazine*, Gaithersburg, Md.)

Antibiotics

Antibiotics are not nutrients; they are drugs. They are chemical substances, produced by molds or bacteria, which have the ability to inhibit the growth of or to destroy other microorganisms.

The senior author was a member of the research team that conducted the first U.S. study on feeding antibiotics to foals, which study was subsequently used in obtaining Food and Drug Administration (FDA) approval for feeding Aureomycin to foals. This experiment revealed that an 85 mg level of Aureomycin, fed to foals, from 5 days of age to 5 months, produced 22 lb more weight.[8]

Certain antibiotics, at stipulated levels, are approved by the FDA for growth promotion and for the improvement of feed efficiency of young equines up to 1 year of age. Unless there is a disease, however, there is no evidence to warrant the continuous feeding of antibiotics to mature horses. Such practice may even be harmful. Hence, when antibiotics are needed for therapeutic purposes, it is best to seek the advice of a veterinarian.

It appears that antibiotics may be especially helpful for young foals which suffer setbacks from infections, digestive disturbances, inclement weather, and other stress factors. Also, horses may benefit from antibiotics (1) when being transported from one location to another—for example, when being moved to a new show or track; (2) when there is a low disease level in the herd; or (3) when mares are foaling.

The poorer the sanitation and management, the greater the response from antibiotics. It follows, therefore, that there is a temptation to use antibiotics as a *crutch,* rather than improve the regimen.

When added to feed, the level of antibiotics should be in keeping with the directions of the manufacturer and with the Food and Drug Administration regulations.

(Also see Chapter 13, section on "Antibiotics.")

FEEDS FOR HORSES

Individual feeds vary widely in feeding value. Oats and barley, for example, differ in feeding value according to the hull content and weight per bushel, and forages vary according to the stage of maturity at which they are cut and how well they are cured and stored. Also, the feeding value of certain feeds is materially affected by preparation.

Regardless of the feeds selected, they should be of sound quality, and not moldy, spoiled, or dusty. This applies to both hay and grain. The careful selection of feeds is more important for horses than for any other class of livestock.

More than one kind of hay makes for appetite appeal. In season, any good pasture can replace part or all of the hay unless work or training conditions make substitution impractical.

Good-quality oats and timothy hay always have been considered standard feeds for light horses. However, feeds of similar nutritive properties can be interchanged in the ration as price relationships warrant; among them, the grains—oats, corn, barley, wheat, and sorghum; the protein supplements—linseed meal, soybean meal, cottonseed meal, and canola meal; and hays of many varieties. Feed substitution makes it possible to obtain a balanced ration at lowest cost.

During the winter months, it is well to add a few sliced carrots to the ration, an occasional bran mash, or a small amount of linseed meal. Also, a bran mash or linseed meal may be used to regulate the bowels.

The proportion of concentrates must be increased and the roughages decreased as energy needs rise with the greater amount, severity, or speed of work. A horse that works at a trot needs considerably more feed than one that works at a walk. For this reason, riding horses in medium to light use require somewhat less grain and more hay in proportion to body weight than light horses that are racing. Also, from an esthetic standpoint, large, paunchy stomachs are objectionable on horses that are used for recreation and sport.

In addition to making for a nutritionally complete ration, the following factors should be considered when choosing horse feeds: cost, palatability, preparation, variety, bulk, and laxativeness.

For purposes of convenience in the discussion that follows, the authors have classed feeds as (1) forages, (2) concentrates, (3) protein supplements, (4) special feeds, and (5), treats.

---

[8]*Wash. Ag. Exp. Sta. Circ. 263,* April 1955.

Forages

Horses are herbivores, ordained by nature to feed largely on forages. In unstressed-surroundings, and when not forced to exert excessively, they maintain themselves quite well on pasture and hay. As horses' nutrient requirements increase with the amount and severity of work, however, their limited stomach capacity prohibits them from consuming sufficient bulky feeds to meet their needs, necessitating that more concentrated feeds in easily digestible form be provided. Nevertheless, some bulk in the ration is necessary in order to keep the digestive tract functioning normally.

Forages are the vegetative portions of plants without the seed or fruit. The term *roughage* is often used to designate these feeds because they are generally higher in fiber and lower in energy than other feeds. The quality of such feeds is usually determined by their fiber and nutrient content, digestibility, and palatability. Their color, texture, and freedom from weeds and other foreign materials are indicative of their value, but these characteristics should be augmented with chemical analyses whenever posssible.

Forages may be classified as (1) pasture, (2) hay, or (3) silage.

PASTURES

The great horse breeding centers of the world are characterized by good pastures. Thus, the bluegrass area of Kentucky is known for its lush pastures produced on residual limestone soils. In short, good caretakers, good pastures, and good horses go hand in hand—good pasture is the cornerstone of successful horse production. Yet, it is becoming increasingly difficult to provide good pastures for many horses, especially those in suburban areas. Also, it is recognized that many folks are prone to overrate the quality of their grass.

In season, there is no finer forage for horses than superior pastures—pastures that are much more than gymnasiums. This is especially true of idle horses, broodmares, and young stock. In fact, pastures have a very definite place for all horses, with the possible exception of animals at heavy work or in training. Even with the latter groups, pastures may be used with discretion. Horses in use may be turned to pasture at nights or over the weekend. Certainly, the total benefits derived from pasture are to the good, although pasturing may have some laxative effects and produce a greater tendency to sweat.

Fig. 25-14. Mares and foals congregating for shade in a pasture. (Courtesy, Dr. Craig H. Wood, University of Kentucky, Lexington)

In addition to the nutritive value of the grass, pasture provides invaluable exercise on natural footing—with plenty of sunshine, fresh air, and lowered feed costs as added benefits. Feeding on pasture is the ideal existence for young stock and breeding animals.

The use of a temporary pasture (grown in a regular crop rotation), instead of a parasite-infested permanent pasture, is recommended. Legume pastures are excellent for horses, as equines are less subject to bloat than cattle or sheep. The specific grass or grass-legume mixture will vary from area to area, according to differences in soil, temperature, and rainfall (see Chapter 7, Pasture and Range Forages).

Fig. 25-15. Supplemental feeding on pasture, using a portable feeder of the type shown in the insert. (Pasture scene: Courtesy, Dr. Craig H. Wood, University of Kentucky, Lexington. Insert: Courtesy, *Equus Magazine*, Gaithersburg, Md.)

The parasites in horse pastures most responsible for equine colic can be reduced dramatically by picking up the manure twice a week. An Ohio State University study showed that a pasture routinely cleaned in this manner had 18 times fewer parasites than an uncleaned pasture. Manure can be removed manually or mechanically. In England, a power sweeper is available, consisting of a small trailer fitted with a vacuum pump powered by a tractor or small engine.

Horse pastures should be well drained and not rough or stony. All dangerous places—such as pits, stumps, poles, and tanks—should be guarded. Shade, water, and suitable minerals should be available in all pastures.

Most horse pastures can be improved by seeding new and better varieties of grasses and legumes, and by fertilizing and management. Also, caretakers need to give attention to balancing pastures nutritionally. Early-in-the-season pastures are of high-water content and lack energy. Mature, weathered grass is almost always deficient in protein (being as low as 3% or less) and low in carotene (the precursor of vitamin A). But these deficiencies can be corrected by proper supplemental feeding.

Five grass species—orchardgrass, reed canarygrass, fescue, smooth bromegrass, and Bermudagrass—account for the major portion of seeded grasses in the United States. The leading legumes are alfalfa, trefoil, and clover.

Sorghums or Sudangrasses in the growing stage should never be grazed by horses, because of the hazard of sorghum poisoning or cystitis, a cyanide-related toxicity. This disease, which occurs more frequently in mares than in stallions or geldings, is characterized by continuous urination, mares appearing to be constantly in heat, and incoordination in the gait. Animals seldom recover after either the incoordination or the dribbling of urine becomes evident. Apparently hay from Sudangrass or from hybrid Sudangrasses will not produce the same malady.

HAYS

Through mistaken kindness or carelessness, horses are often fed too much hay or other roughage, with the result that they have hay bellies (distended digestive tracts), breathe laboriously, and tire quickly. With cattle and sheep, on the other hand, it is usually advisable to feed considerable roughage. This difference between horses and ruminants is due primarily to the relatively small size of the simple stomach of the horse in comparison with the fourfold stomach of the ruminant.

Under most conditions, the roughage requirement of horses ranges from 0.5 to 1.0% of body weight, or from 5 to 10 lb of roughage daily for a 1,000-lb horse.

Usually, young horses and idle horses can be provided with an unlimited allowance of hay. In fact, much good will result from feeding young and idle horses more roughage and less grain. But one should gradually increase the grain and decrease the hay as work or training begins.

Racehorses should receive a minimum of roughage, since they need a maximum of energy. When limiting the allowance of roughage, it is sometimes necessary to muzzle greedy horses (gluttons) to prevent them from eating the bedding.

Fig. 25-16. Horses are fond of good-quality hay. (Courtesy, Dr. Craig H. Wood, University of Kentucky, Lexington)

Hay native to the locality is usually fed. However, horseowners everywhere prefer good-quality timothy. With young stock and breeding animals especially, it is desirable that a sweet grass-legume mixture of alfalfa or clover hay be fed. The legume provides a source of high-quality proteins and certain minerals and vitamins (see Table 25-10).

TABLE 25-10
FEEDING VALUE OF GRASS-LEGUMES VS TIMOTHY

Mixture	Crude Protein	Calcium	Phosphrous
	(%)	(%)	(%)
Grass-alfalfa	21	1.2	.4
Grass-clover	15	.8	.3
Timothy	8	.6	.2

Horses like variety. Therefore, if at all possible, it is wise to have more than one kind of hay in the stable. For example, timothy may be provided at one feeding and a grass-legume mixed hay at the other feeding. Good caretakers often vary the amount of alfalfa fed, for increased amounts of alfalfa in the ration will increase urination and give a softer consistency to the bowel movements. This means that elimination from kidneys and bowels can be carefully regulated by the amount and frequency of alfalfa feedings. Naturally, such regulation is more important with irregular use and idleness. On the other hand, in some areas alfalfa is fed as the sole roughage with good results.

The most important factor affecting hay quality is the stage of maturity at which it is cut. As the plant matures, the stem-to-leaf ratio increases, the percentage of digestible nutrients (such as protein and calcium) decreases, and the digestibility and voluntary intake decrease. Workers at the Pennsylvania Station cut two hay crops, alfalfa and orchardgrass, on three different dates—June 3, June 13, and June 23—then determined the nutritive value of the hays for horses. Their results are given in Table 25-11.

TABLE 25-11
EFFECT OF DATE OF CUTTING ON NUTRITIVE VALUE OF HAY FED TO HORSES[1]

Hay	Date	Crude Protein	Crude Fiber	Digestibility of Energy
	(%)	(%)	(%)	(%)
Alfalfa	June 3	15.1	28.3	65
	June 13	14.4	32.7	56
	June 23	9.1	38.7	52
Orchardgrass	June 3	14.0	30.9	59
	June 13	10.9	32.3	55
	June 23	6.5	38.0	49

[1]Adapted from Darlington, J. M., and T. V. Herschberger, *Journal of Animal Science,* Vol. 27, No. 6, p. 1573.

As shown in Table 25-11, each 10-day delay in the cutting date of both alfalfa and orchardgrass resulted in a decrease in crude protein, an increase in crude fiber, and a lowering of digestibility of energy.

(Also see Chapter 8, Hay.)

Alfalfa (Lucerne)

Alfalfa is an important, perennial, leguminous forage plant with trifoliate leaves and bluish-purple flowers. It is grown widely, principally for hay. Alfalfa is capable of surviving dry periods because of its extraordinarily long root system, and it is adapted to widely varying conditions of climate and soil. It yields the highest tonnage per acre and has the highest protein content of the legume hays.

Good-quality alfalfa hay is excellent for horses. It averages about 15.3% protein, which is of high quality; and it is a good source of most minerals and vitamins. In addition to being used as a hay, alfalfa is an ingredient of most all-pelleted feeds.

When fed alfalfa hay, horses urinate more than when fed grass hay; and there may be a strong smell of ammonia in the barn because alfalfa is high in nitrogen.

Oat Hay

Oat hay is an excellent feed for horses. It's easy to cure, and horses like it. Early cutting (in the soft dough stage) greatly increases its feeding value, due to the higher protein content. Even though considerable energy is stored in the kernels at maturity, shattering of the grain during harvesting of mature oats results in energy losses and decreased feeding value compared with early cut hay.

Oat hay is low in protein; hence, its feeding value is greatly enhanced when it is fed with alfalfa or some other legume.

Timothy

Timothy is the preferred hay by most horse owners, although it has become scarce and expensive.

Timothy is easy to harvest and cure. However, in comparison with hay made from the legumes, it is low in crude protein and minerals, particularly calcium.

As with all other forages, the feeding value of timothy is affected by the stage of growth of the plants at the time of cutting. With increasing maturity, (1) the percentage of crude protein decreases, (2) the percentage of crude fiber increases, (3) the hay becomes less palatable, and (4) the digestibility decreases. However, delaying cutting until timothy has reached the full bloom stage, or later, usually results in the highest yields. When both yield and quality are considered, the best results are obtained when timothy is cut for hay at the early bloom stage.

SILAGES

Well-preserved silage of good quality, free from mold and not frozen, affords a highly nutritious succulent forage for horses during the winter months. *Because horses are more susceptible than cattle or sheep to botulism or other digestive disturbances resulting from the feeding of poor quality silage, nothing but choice, fresh silage, free from mold or spoilage, should ever be fed.*

Various types of silages may be fed successfully to horses, but corn silage and grass-legume silage are most common. If the silage contains much grain, the concentrate allowance should be reduced accordingly.

Silage should not be used as the only roughage for horses. Usually it should be fed in such quantity as to replace not more than ⅓ to ½ of the roughage allowance, considering that ordinarily 1 lb of hay is equivalent to approximately 3 lb of wet silage. This means that the silage allowance usually does not exceed 10 to 15 lb daily per head for mature animals, although much larger amounts have been used satisfactorily in some instances. Silage is expecially suited for the winter feeding of idle horses, broodmares, and growing foals.

Concentrates

Horses cannot handle as large quantities of roughages as ruminants. When used for heavy work, for pleasure, or for racing, they must be even more restricted in their roughage allowance and should receive a higher proportion of concentrates.

Because of less bulk and lower shipping and handling costs, the concentrates used for horse feeding are less likely to be locally grown than the roughages. Even so, the vast majority of grains fed to horses are homegrown, thus varying from area to area, according to the grain crops best adapted.

Of all the concentrates, heavy oats most nearly meet the needs of horses; and, because of the uniformly good results obtained from their use, they have always been recognized as the preferred grain for horses. Corn is also widely used as a horse feed, particularly in the central states. Despite occasional prejudice to the contrary, barley is a good horse feed. As proof of the latter assertion, it is noteworthy that the Arab—who was a good horseman—fed barley almost exclusively. Also, wheat, wheat bran, molasses, and commercial-mixed feeds are extensively used. It is to be emphasized, therefore, that careful attention should be given to the prevailing price of feeds available locally, for many feeds are well suited to horses. Often substitutions can be made that will result in a marked saving without affecting the nutritive value of the ration. So, the primary consideration in selecting the cereal grain(s) is the cost per unit of energy.

(Also see Chapter 10, Grains/High-Energy Feeds.)

BARLEY

Barley is the leading horse grain in western United States.

Compared with corn, barley contains somewhat more protein (crude protein: barley, 13%; corn, 10%) and fiber (due to the hulls) and somewhat less carbohydrate and fat. Like oats, the feeding value of barley is quite variable, due to the wide spread in test weight per bushel. Most caretakers feel that it is preferable to feed barley along with more bulky feeds; for example, 25% oats or 15% wheat bran.

When fed to horses, barley should always be steam rolled or coarsely ground.

BRAN MASH

Feeding a bran mash is the traditional way of regulating the bowels of horses on idle days and at such other times as required.

The mash is prepared by filling a 2- to 2½-gal bucket with wheat bran, pouring enough boiling hot water over it to make it the consistency of breakfast oatmeal, covering the bucket with a blanket and allowing it to steam until cool, then feeding it to the horse.

Occasionally, when horses are offered a bran mash for the first time, they may refuse to eat it. When this occurs, they may be enticed to eat the mash by either (1) introducing them to a little of it by hand, or (2) sprinkling some sugar, or some other well-liked feed, over it.

CORN (MAIZE)

Corn ranks second to oats as a horse feed. It is palatable, nutritious, and rich in energy-producing carbohydrate and fat. Additionally, it is the only cereal grain that contains carotene, the precursor of vitamin A. But, it has certain very definite limitations. It lacks quality (being especially low in the amino acids lysine and tryptophan) and quantity of proteins (it runs about 10%), and it is deficient in minerals, particularly calcium.

Corn may be fed to horses on the cob, shelled, cracked, as corn-and-cob meal, or flaked.

DRIED BREWERS' GRAINS

Dried brewers' grains are a by-product of beer production. They are what is left after the sugar and other solubles have been fermented. Dried brewers' grains are lower in energy and higher in protein and fiber than some grains; they are a wholesome, nutritious, and palatable feed for horses.

MOLASSES (CANE OR BEET)

Molasses is a by-product of sugar factories, with cane molasses coming from sugarcane and beet molasses coming from sugar beets. Cane molasses is slightly preferred to beet molasses for horses, although either is satisfactory.

For horses, molasses is 80 to 95% as valuable as oats, pound for pound. However, molasses is used primarily as an appetizer.

In hot, humid areas, molasses should be limited to 5% of the ration; otherwise, mold may develop. Where mustiness is a problem, add calcium propionate to the feed according to the manufacturer's directions.

OATS

Oats are the leading U.S. horse feed. They normally weigh 32 lb per bushel, but the best oats for horses are heavier. The feeding value varies according to the hull content and test weight per bushel.

Because of their bulky nature, oats form a desirable loose mass in the stomach, which prevents impaction.

Oats may be rolled, crimped, or fed whole, but they do not have to be processed for mature horses with sound teeth.

Oat groats, oats with the hulls removed, are excellent for foal rations, but they are relatively high in price.

SORGHUM (MILO)

The production of grain sorghum has increased in the United States in recent years; and with more production, increased feeding to horses has followed. Properly used, it is a good horse feed. In comparison with the other commonly used cereal grains fed to horses, sorghum grain is more variable in protein content, less palatable (because of the presence of tannic acid), and has a harder seed. The small, round grain should be steam rolled, coarsely crimped, or coarsely ground.

WHEAT

When the price is favorable, wheat may be fed to horses. In comparison with the other cereal grains, it is higher in protein. Wheat should be rolled or coarsely cracked. In order to lessen the possibility of colic, wheat should be limited to 20% of the concentrate ration and fed in combination with more bulky feeds.

WHEAT BRAN

Wheat bran is the coarse outer covering of the wheat kernel. It contains a fair amount of protein (averaging about 16%) and a good amount of phosphorus. Bran is valuable for horses because of its bulky nature and laxative properties. Also, it is very palatable. However, high levels of wheat bran have been associated with nutritional secondary hyperparathyroidism (big head disease); attributed to the high phosphorus content of wheat bran, resulting in a calcium/phosphorus imbalance. (See Fig. 25-6.) So, wheat bran should not constitute more than 10 to 15% of the concentrate portion of the ration.

Protein Supplements

The extent to which the horse's ration is supplemented with proteins depends primarily on the age of the horse and on the quality of the forage fed. Growing or lactating animals require somewhat more protein than horses that are idle, gestating, or working. Also, grass hays and farm grains are generally low in quality and quantity of proteins and require more supplementation than legumes.

In practical horse feeding, foals should be provided with some protein feeds of animal origin in order to supplement the proteins found in grains and forages. In feeding mature horses, a safe plan to follow is to provide plant protein from several sources.

In general, feeds of high-protein content are more expensive than those high in carbohydrates or fats. Accordingly, there is a temptation to feed too little protein. On the other hand, when protein feeds are the cheapest—as is often true of cull peas in certain sections of the West—excess quantities of them may be fed as energy feeds without harm, provided the ration is balanced in all other respects. Any amino acids that are left over, after the protein requirements have been met, are deaminated or broken down in the body. In this process, a part of each amino acid is converted to energy, and the remainder is excreted via the kidneys.

The following oil meals are most commonly used as protein supplements for horses: linseed meal, soybean meal, cottonseed meal, sunflower meal, and rapeseed meal (canola meal).

(Also see Chapter 11, Protein Supplements.)

COTTONSEED MEAL

Among the oilseed meals, cottonseed meal ranks second in tonnage to soybean meal.

The protein content of cottonseed meal can vary from about 22% in meal made from undecorticated (unhulled) seed to 60% in flour made from seed from which the hulls have been removed completely. Thus, in screening out the residual hulls, which are low in protein and high in fiber, the processor is able to make a cottonseed meal of the protein content desired—usually 41, 44, or 50%.

Cottonseed meal is low in lysine and tryptophan and deficient in vitamin D, carotene (vitamin A value), and calcium. But, it is rich in phosphorus. Also, unless glandless seed is used, it contains a toxic substance known as gossypol, varying in amounts with the seed and the processing.

Some prejudices to the contrary, good grade cottonseed meal is satisfactory for mature horses. But, because of its deficiency in lysine and tryptophan, young, growing horses do not gain as rapidly or as efficiently when fed cottonseed meal as when fed soybean meal.

LINSEED MEAL

Linseed meal is a by-product of flaxseed following oil extraction by either of two processes: (1) the mechanical process (which is known as the *old process*); or (2) the solvent process *(new process)*. If solvent extracted, it must be so designated. Horse owners prefer the mechanical process, for the remaining meal is more palatable, has a higher fat content, and imparts more gloss to the hair coat.

Linseed meal averages about 35% protein content. For horses, the proteins of linseed meal do not effectively make good the deficiencies of the cereal grains—linseed meal being low in the amino acids lysine and tryptophan. Also, linseed meal is lacking in carotene and vitamin D, and is only fair in calcium and the B vitamins. Because of its deficiencies, linseed meal should not be fed to horses as the sole protein supplement.

Because of its laxative nature, linseed meal in limited quantities is a valuable addition to the ration of horses. Also, if prepared by the old process, it imparts a desirable *bloom* to the hair of show and sale animals.

SOYBEAN MEAL

Soybean meal is the most widely used protein supplement in the United States. It is the ground residue (soybean oil cake or soybean oil chips) remaining after the removal of most of the oil from soybeans. The oil is extracted by either of three processes: (1) the expeller process; (2) the hydraulic process; or (3) the solvent process. Although a name descriptive of the extraction process must be used in the brand name, well-cooked soybean meal produced by each of the extraction processes is of approximately the same feeding value.

Soybean meal normally contains 41, 44, or 50% protein, according to the amount of hull removed. The protein is of better quality than the other protein-rich supplements of plant origin; it contains a higher level of lysine than the other oilseed meals. However, it is low in calcium, phosphorus, carotene, and vitamin D.

Soybean meal is satisfactory as the only protein supplement to grain for mature horses, providing a high-quality ground legume is incorporated in the ration and adequate sources of calcium and phosphorus are provided. For foals, it is best that a dried milk by-product be included.

SUNFLOWER MEAL

The development of high oil-yielding varieties by Russian scientists stirred worldwide interest in the use of sunflowers as an oilseed crop. Some of these varieties yield over 50% oil.

Sunflower meal (41% protein or better) can be used as a protein supplement for horses provided (1) it is good quality, and (2) care is taken to supply adequate lysine, for sunflower meal is low in this amino acid. When incorporated in well-balanced rations, properly processed sunflower meal of good quality may supply up to one-third of the protein supplement of horses.

UREA

It is recognized that horses frequently consume urea-containing cubes and blocks intended for cattle and sheep, particularly on the western range. Moreover, it appears that mature horses are able to do so without untoward effects. The latter observation was confirmed in one limited experiment[9] in which 4 horses consumed an average of 4.57 lb per day of a urea-containing supplement, or 0.55 lb/head/day of feed urea (262%), for 5 months. Also, the Louisiana Station[10] did not find urea detrimental or toxic to horses when it constituted up to 5% of the grain ration, with up to 0.5 lb per day of urea consumed. There are reports, however, of urea toxicity in foals, in which bacterial action is more limited than in older horses. Also, most state feed control laws limit the sale of urea-containing feeds to ruminants.

Special Feeds

Special feeds may be needed from time to time for imparting bloom or gloss to the hair, or for promoting growth of young stock.

BLOOM IMPARTING FEEDS

Bloom or gloss is important in horses. But sometimes they lack this desired quality—their hair is dull and dry. Feeding a well-balanced ration will usually rectify this situation. Also,

[9]*Veterinary Medicine,* Vol. 58, No. 12, Dec. 1963, pp. 945–46.

[10]"Non-Toxicity of Urea Feeding to Horses," *Veterinary Medicine/Small Animal Clinician,* Nov. 1965.

feeding either of the following products will make for an attractive, shiny coat:

1. **Corn oil.** Feed at the rate of 2 oz (4 Tbsp) per horse per day.
2. **Whole flaxseed soaked.** Put a handful of whole flaxseed in a teacup, cover it with water, let it stand overnight, then pour it over the morning feed. Repeat twice each week.

Unless the horse is afflicted with lice, mange, or some other ailment, either of the above treatments will impart bloom or gloss to the coat.

LYSINE

Protein quality is important for horses. Because of more limited amino acid synthesis in the horse than in ruminants, plus the fact that the cecum is located beyond the small intestine—the main area for digestion and absorption of nutrients, it is generally recommended that high-quality protein rations, adequate in amino acids, be fed to equines. This is especially important for young equines, because cecal synthesis is very limited in early life.

Fortunately, the amino acid content of proteins from different sources varies. Thus, the deficiencies of one protein may be improved by combining it with another, and the mixture of the two proteins often will have a higher feeding value than either one alone. It is for this reason, along with added palatability, that a considerable variety of feeds in the horse ration is desirable.

Cornell University reported that the addition of lysine to the ration of growing horses increased weight gains, feed consumption, and feed efficiency. But, this experimental ration contained linseed meal as the major source of protein. Normally, it is much more practical to supply a source of good-quality protein, such as milk protein or soybean meal (perhaps along with some linseed meal), rather than add lysine to linseed meal.

It is not recommended that owners spend money on lysine for horses. Instead, a well-balanced ration should be fed to horses of all ages, and especially high-quality proteins should be incorporated in the ration of young equines. There is no experimental evidence that the addition of lysine will improve a good ration.

MILK BY-PRODUCTS

The superior nutritive values of milk by-products are due to their high-quality proteins, vitamins, a good mineral balance, and the beneficial effect of the milk sugar, lactose. In addition, these products are palatable and highly digestible. They are an ideal feed for young equines and for balancing the deficiencies of the cereal grains. Most foal rations contain one or more milk by-products, primarily dried skim milk, with some dried whey and dried buttermilk included at times. The chief limitation to their wider use is price.

MILK REPLACER

As indicated by the name, *a milk replacer is a replacement for milk.*

Foals suckling their dams generally develop very satisfactorily up to weaning time. But the most critical period in the entire life of a horse is that interval from weaning time (about 6 months of age, or earlier) until 1 year of age. This is especially so in the case of young horses being fitted for shows or sales, where condition is so important. Thus, where valuable weanlings or yearlings are to be shown or sold, the use of a milk replacer may be practical.

(Also see Chapter 6, Types and Roles of Feedstuffs, section on "Milk Replacers.")

Treats

Horses are fed a great variety of treats. On a government horse breeding establishment in Brazil, the senior author saw a large, well-manicured vegetable garden growing everything from carrots to melons, just for horses. Also, trainers recognize that most racehorses, which are the *prima donnas* of the equine world, don't "eat like a horse"; they eat like people—and sometimes they're just as finicky. Their menus may include a choice of carrots or other roots, fruit, pumpkins, squashes, or melons, sugar or honey, and innumerable other goodies.

Ask caretakers why they feed treats to their horses and you'll get a variety of answers. However, high on the list of reasons will be (1) appetizers; (2) a source of nutrients and as conditioners; (3) rewards; (4) alleviating obesity (dieting the horse); or (5) folklore.

Sometimes, even nutritionists overlook the fact that, when evaluated on a dry matter basis, high-water content tubers, fruits, and melons have almost the same nutrient value as the cereal grains. This becomes apparent in Table 25-12 which gives the energy value on a moisture-free basis of several horse treats compared with barley, corn, oats, and timothy.

TABLE 25-12
ENERGY VALUE OF SEVERAL FEEDS ON A MOISTURE-FREE BASIS

Feed	Water	Dry Matter	Energy Value (TDN) As-Fed	Energy Value (TDN) Dry Matter Basis
	(%)	(%)	(%)	(%)
Barley	10	90	77	85
Corn	10	90	80	90
Oats	11	89	68	76
Timothy hay, mature	14	86	41	48
Apples	82	18	13	74
Carrots	88	12	10	82
Melons	94	6	5	80
Potatoes	79	21	18	85
Sugar beets	87	13	10	77

Palatability

Palatability is important, for horses must eat their feed if it's to do them any good. But many horses are finicky simply because they're spoiled. For the latter, stepping up the exercise and halving the ration will usually effect a miraculous cure.

Also, it seems possible that well-liked feeds are digested somewhat better than those which are equally nutritious, but less palatable.

Palatability is particularly important when feeding horses that are being used hard, as in racing or showing. Unless the ration is consumed, such horses will not obtain sufficient nutrients to permit maximum performance. For this reason, lower quality feeds, such as straw or stemmy hay, should be fed to mature, idle horses.

Familiarity and habit are important factors concerned with the palatability of horse feeds. For example, horses have to learn to eat pellets, and very frequently they will back away from feeds with new and unfamiliar odors. For this reason, any change in feeds should be made gradually.

Occasionally, the failure of horses to eat a normal amount of feed is due to a serious nutritive deficiency. For example, if horses are fed a ration made up of palatable feeds, but deficient in one or more required vitamins or minerals, they may eat normal amounts for a time. Then when the body reserves of the lacking nutrient(s) are exhausted, they will usually consume much less feed, due to an impairment of their health and a consequent lack of appetite. If the deficiency is not continued until the horses are injured permanently, they will usually recover their appetites if some feed is added which supplies the nutritive lack and makes the ration complete.

FEED PREPARATION

The physical preparation of cereal grains for horses has been practiced by caretakers for a very long time. Basically, grain is either soaked, cooked, ground, or rolled (wet or dry), and hay is either fed long, pelleted, or cubed.

For horses, flaking or extruding are the preferred methods of grain preparation; they make for light rations and fewer digestive disturbances. For animals with good teeth, the value of oats is increased only 5% by processing.

Hay for horses is usually fed long or incorporated in an all-pelleted ration (with the grain and hay combined).

Further elucidation of feed preparation for horses is contained in Chapter 14, Feed Processing.

FEED ALLOWANCES AND SUGGESTED RATIONS

Proper feeding of horses calls for giving them (1) the right allowance, and (2) balanced rations. The ration may be either a home-mixed feed or a commercial feed.

In formulating rations, the following points are pertinent:

1. The effects of the rations on horses should be observed. Calculations alone will not ensure success.
2. Rations should be palatable, economical, and practical. They need not be complicated mixtures.
3. The roughage should be selected first. Then, the concentrates (grains) may be selected which, when combined with the roughage, will provide a balanced ration.

Feed Allowances

When given all the feed that they will consume, mature horses will generally eat an amount equivalent to about 2.5% of their body weight, daily. Growing foals and lactating mares eat more heartily—they will consume up to 3% of their body weight.

Because the horse has a limited digestive capacity, the amount of concentrates must be increased and the roughages decreased when the energy needs rise with more work. The following general guides may be used for the daily ration of horses under usual conditions.

- **Horses at light work (1 to 3 hours per day of riding or driving)**—Allow ⅖ to ½ lb of grain and 1¼ to 1½ lb of hay per day per 100 lb of body weight.
- **Horses at medium work (3 to 5 hours per day of riding or driving)**—Allow about ¾ lb of grain and 1 to 1¼ lb of hay per 100 lb of body weight.
- **Horses at hard work (5 to 8 hours per day of riding or driving)**—Allow about 1¼ to 1⅓ lb of grain and 1 to 1¼ lb of hay per 100 lb of body weight.

As will be noted from these recommendations, the total allowance of both concentrates and hay should be about 2 to 2½ lb daily per 100 lb of body weight.

The recommended feed allowances on the basis of animal weight are equally applicable to equines of all sizes, including ponies and donkeys; simply vary as necessary according to the work performed and the individuality of the animal.

About 6 to 12 lb of grain daily is an average grain ration for a light horse at medium or light work. Racehorses in training usually consume 10 to 16 lb of grain per day; the exact amount varies with the individual requirements and the amount of work. The hay allowance averages about 1 to 1¼ lb daily per 100 lb of body weight, but it is restricted as the grain allowance is increased. Light feeders should not be overworked.

Horses differ just as people do, in feed required and tendency to put on weight. Moreover, the age and degree of activity of horses are very important factors; the amount of feed should be increased in keeping with the amount, severity, and speed of work; and feed requirements are influenced by weather—for example, under ideal fall weather conditions, a horse may require 14 lb of 60% TDN feed daily, whereas in the same area the same horse may require 16 lb daily of the same feed in July and August, and 20 lb in the winter.

The quantities of feeds recommended are intended as guides only. The allowance, especially of the concentrates, should be increased when the horse is too thin and decreased when the horse is too fat.

Suggested Rations

A ration is the amount of feed given to a horse in a day, or a 24-hour period. To most caretakers, however, the word implies the feeds fed to an animal without limitation of the time in which they are consumed.

To supply all the needs—maintenance, growth, reproduction and lactation, and work (running)—horses must receive feeds in quality and quantity to furnish the necessary energy (carbohydrates and fats), proteins, minerals, vitamins, and perhaps unknown factors and additives. Such rations are said to be balanced. Moreover, the feed must be palatable—horses must like it. Also, liberal margins of safety have been provided to compensate for variations in feed composition, environment, possible losses of nutrients during storage, and differences among individual animals.

In feeding horses, as with other classes of livestock, it is recognized that nutritional deficiencies (especially deficiencies of certain vitamins and minerals) may not be of sufficient proportions to cause clear-cut deficiency symptoms. Yet, such deficiencies without outward signs may cause great economic losses because they go unnoticed and unrectified. Accordingly, sufficient minerals and vitamins should always be present, but care should be taken to avoid imbalances.

Several suggested rations are given in Table 25-13.

TABLE 25-13
RATIONS FOR HORSES[1] (AS-FED BASIS)

Age, Sex, and Use	Daily Allowance	Kind of Hay (More than one kind of hay makes for variety and appetite appeal. In season, any good pasture can replace part or all of the hay except for horses at work or in training.)	Suggested Grain Rations (With all rations, and for all classes and ages of horses, provide suitable minerals; see Table 25-8, Horse Mineral Chart, for recommendations.)		
			Ration No. 1	Ration No. 2	Ration No. 3
Mature idle horses; Stallions, mares, and geldings (weighing 900–1,400 lb; or *409–636 kg*)	1½–1¾ lb (*0.7–0.8 kg*) of hay per 100 lb (*45 kg*) body weight.	Pasture in season; or grass-legume mixed hay.	(With grass hay, add ¾ lb (*0.34 kg*) daily of a high-protein supplement.)		
			(lb) (*kg*)	(lb) (*kg*)	(lb) (*kg*)
Light horses at work, in riding, driving, and racing (weighing 900–1,400 lb; or *409–636 kg*)	**Light use**—⅖–½ lb (*0.18–0.2 kg*) grain and 1¼–1½ lb (*0.6–0.7 kg*) hay per 100 lb (*45 kg*) body weight. **Medium use**—¾–1 lb (*0.3–0.5 kg*) grain and 1–1¼ lb (*0.5–0.6 kg*) hay per 100 lb (*45 kg*) body weight. **Hard use**—1¼–1⅓ lb (*0.5–0.6 kg*) grain and 1–1¼ lb (*0.5–0.57 kg*) hay per 100 lb (*45 kg*) body weight.	Grass hay.	Oats 100 *45*	Oats .... 70 *32* Corn 30 *14*	Oats .... 70 *32* Barley 30 *14*
Stallions in breeding season (weighing 900–1,400 lb; or *409–636 kg*)	¾–1½ lb (*0.3–0.7 kg*) grain per 100 lb (*45 kg*) body weight, together with a quantity of hay within same range.	Grass-legume mixed (or ⅓–½ legume hay, with balance grass hay).	Oats .... 55 *25* Wheat .... 20 *9* Wheat bran ... 20 *9* Linseed meal .. 5 *2*	Corn .... 35 *16* Oats .... 35 *16* Wheat .... 15 *7* Wheat bran ... 15 *7*	Oats 100 *45*
Pregnant/lactating mares (weighing 900–1,400 lb; or *409–636 kg*)	¾–1½ lb (*0.3–0.7 kg*) grain per 100 lb (*45 kg*) body weight, together with a quantity of hay within same range.	Grass-legume mixed; or ⅓–½ legume hay, with balance grass hay (straight grass hay may be used first half of pregnancy).	Oats .... 80 *36* Wheat bran ... 20 *9*	Barley .... 45 *20* Oats .... 45 *20* Wheat bran ... 10 *5*	Oats .... 95 *43* Linseed meal .. 5 *2*
Foals before weaning (weighing 100–350 lb, or *45–159 kg;* with projected mature weights of 900–1,400 lb; or *409–636 kg*)	½–¾ lb (*0.2–0.3 kg*) grain per 100 lb (*45 kg*) body weight, together with a quantity of hay within same range.	Legume hay.	Oats .... 50 *23* Wheat bran ... 40 *18* Linseed meal .. 10 *5*	Oats .... 30 *14* Barley .... 30 *14* Wheat bran ... 30 *14* Linseed meal .. 10 *5*	Oats .... 80 *36* Wheat bran ... 20 *9*
			(Rations balanced basis of following assumptions: Mares of mature weights of 600, 800, 1,000, and 1,200 lb [or *273, 364, 455,* and *545 kg*] may produce 36, 42, 44, and 49 lb [or *16, 19, 20,* and *22 kg*] of milk daily.)		
Weanlings (weighing 350–450 lb; or *159–204 kg*)	1–1½ lb (*0.5–0.7 kg*) grain and 1½–2 lb (*0.7–0.9 kg*) hay per 100 lb (*45 kg*) body weight.	Grass-legume mixed (or ½ legume hay, with balance grass hay).	Oats .... 30 *14* Barley .... 30 *14* Wheat bran ..... 30 *14* Linseed meal 10 *5*	Oats .... 70 *32* Wheat bran ... 15 *7* Linseed meal .. 15 *7*	Oats .... 80 *36* Linseed meal .. 20 *9*
Yearlings: 2nd summer (weighing 450–700 lb; or *204–317 kg*)	Good luxuriant pastures. (If in training or for other reasons without access to pastures, the ration should be intermediate between the adjacent upper and lower groups.)				
Yearling or rising 2-year-old; 2nd winter (weighing 700–1,000 lb; or *317–454 kg*)	½–1 lb (*0.2–0.5 kg*) of grain and 1–1½ lb (*0.5–0.7 kg*) hay per 100 lb (*45 kg*) body weight.	Grass-legume mixed; or ⅓ to ½ legume hay, with remainder grass hay.	Oats .... 80 *36* Wheat bran 20 *9*	Barley .... 35 *16* Oats .... 35 *16* Wheat bran ... 15 *7* Linseed meal .. 15 *7*	Oats 100 *45*

[1]Mineral recommendations for all classes and ages of horses, especially those fed unmixed rations or on pasture:

a. *When animals are on liberal grain feeding*—Provide free access to a 2-compartment mineral box, with (1) trace mineralized salt in one side; and (2) in the other side, a mixture of ⅓ trace mineralized salt (salt included for purposes of palatability), ⅓ defluorinated phosphate or steamed bone meal, and ⅓ ground limestone or oystershell flour.

b. *When animals are primarily on roughage (pasture, hay, and/or silage)*—Provide free access to a 2-compartment mineral box, with (1) trace mineralized salt in one side (salt included for purposes of palatability); and (2) in the other side, a mixture of ⅓ trace mineralized salt and ⅔ defluorinated phosphate or steamed bone meal.

Overfeeding

Overfeeding may result in the following bad consequences: If done suddenly, it may cause founder (laminitis), colic, or enterotoxemia; if prolonged, it will likely result in obesity (too fat).

The main qualities desired in horses are trimness, action, spirit, and endurance. These qualities cannot be obtained in horses that are overfed and fat. The latter is especially true with horses used for racing, where the carrying of any surplus body weight must be avoided.

Home-Mixed Feeds

When selecting rations, one should compare the cost of home-mixed vs commercial feeds. If only small quantities are required or little storage space is available, it may be more satisfactory to buy ready-mixed feeds.

When home-mixed feeds are used, feeds of similar nutritive properties can be interchanged in the ration as price relationships warrant. This makes it possible to obtain a balanced ration at lowest cost.

Any recommended quantities of feeds should be used as guides only. The caretaker should increase the feed, especially the concentrates, when the horse is too thin and decrease the feed if it gets too fat.

Sudden changes in the ration should be avoided, especially when changing from a less concentrated ration to a more concentrated one. When this rule of feeding is ignored, digestive disturbances result and the horse goes *off feed.* In either adding or omitting one or more ingredients, the change should be made gradually. Likewise, caution should be exercised in turning horses to pasture or in transferring them to more lush grazing.

In general, horses may be given as much nonlegume roughage as they will eat. But they must be accustomed gradually to legumes because legumes may be laxative.

Commercial Horse Feeds

Commercial horse feeds are feeds mixed by manufacturers who specialize in the feed business.

Commercial feed manufacturers are able to purchase feed in quantity lots, making possible price advantages and the scientific control of quality. Many caretakers have found that because of the small quantities of feed usually involved, and the complexities of horse rations, they have more reason to rely on good commercial feeds than do owners of other classes of farm animals.

The nutritive requirements of horses vary according to age, weight, use or demands, growth, stage of gestation or lactation, and environment. Also, part of the horse ration may be homegrown. It would appear, therefore, that the commercial feeds shown in Table 25-14 are necessary if one is to meet most horse needs.

TABLE 25-14
COMMERCIAL HORSE FEEDS AND NEEDS

Needed Horse Feeds	Prevailing Conditions	Crude Protein	Used For
		(%)	
Complete (hay and grain combined in a pellet).	For the owner who must buy all feeds.	13	All horses 10 months or older.
Concentrate.	For the owner who has satisfactory hay and/or pasture.	14	All horses 10 months or older.
Protein supplement.	For supplementing available hay and grain.	25	All horses 10 months or older.
Foal ration.	For creep feeding.	18	Two weeks to 10 months of age.
Protein-salt block.	For free-choice feeding in corral or on pasture.	20	All horses 10 months of age or older.
Enriched vitamin-trace mineral-unidentified factor supplement.	For the owner who has hay and grain that meet all needs except vitamin-trace mineral-unidentified factors.		All horses not receiving any of the above feeds.

SWEET FEED

A sweet feed is a feed to which has been added one or more ingredients that are sweet. Most commonly, it is molasses (approximately 10%); although brown sugar (about 5%) is sometimes used, and occasionally honey.

The horse has a *sweet tooth;* hence, it's not easy to switch an animal from a sweet feed to what may be a more nutritious ration.

In addition to enhancing palatability, sweet feed controls dust and keeps the minerals and vitamins blended.

FEED SUBSTITUTION TABLE

Successful horse owners are keen students of values. They recognize that feeds of similar nutritive properties can and should be interchanged in the ration as price relationships warrant, thereby making it possible at all times to obtain a balanced ration at the lowest cost.

Table 25-15, Feed Substitution Table for Horses, is a summary of the comparative values of the most common U.S. horse feeds. In arriving at these values, chemical composition, feeding value, and palatability have been considered. The comparative values shown are not absolute. Rather, they are reasonably accurate approximations based on average-quality feeds.

TABLE 25-15
FEED SUBSTITUTION TABLE FOR HORSES (AS-FED BASIS) (See footnotes at end of table.)

Feedstuffs	Relative Feeding Value (lb for lb) in Comparison with the Designated Base Feed (in bold italic) Which = 100	Maximum Percentage of Base Feed (or comparable feed or feeds) Which it Can Replace for Best Results	Remarks
GRAINS, BY-PRODUCT FEEDS, ROOTS AND TUBERS:[1] **(Low and Medium Protein Feeds)**			
Oats	***100***	***100***	The leading grain for horses . The feeding value of oats varies according to the hull content and test weight per bushel. Need not be ground.
Barley	110	100	Most caretakers feel that it is preferable to feed barley along with more bulky feeds; for example, 25% oats or 15% wheat bran. Crush for horses.
Beet pulp, dried	100	33⅓	Not palatable to horses.
Beet pulp, molasses, dried	100	33⅓	Not palatable to horses.
Brewers' dried grains	100	50	
Carrots	15–25	10	Horses are very fond of carrots.
Corn, No. 2	115	100	Ranks second to oats as a light horse feed.
Corn gluten feed (gluten feed)	100	50	It has a lower value than indicated when forage is of low-protein content.
Distillers' dried grains	90–100	25	
Distillers' dried solubles	90–100	25	
Hominy feed	115	100	
Molasses, beet	80–95	10	In hot, humid areas, molasses should be limited to 5%; otherwise, mold may develop unless an inhibitor is used. Cane molasses is slightly preferred to beet molasses.
Molasses, cane	80–95	10	In hot, humid areas, molasses should be limited to 5%; otherwise, mold may develop unless an inhibitor is used.
Peas, dried	100	40	
Rice (rough rice)	115	50	Grind for horses.
Rye	115	33⅓	Higher levels or abrupt changes to rye may cause digestive disturbances. Not palatable.
Sorghum, grain	110–115	85	All varieties have about the same feeding value. Crush for horses.
Wheat	115	50	Wheat should be mixed with a more bulky feed in order to prevent colic.
Wheat bran	100	20	Valuable for horses because of its bulky nature and laxative properties.
Wheat-mixed feed (mill run)	105	20	Excessive quantities will cause colic or other digestive upsets.
PROTEIN SUPPLEMENTS:			
Linseed meal (35%)	***100***	***100***	Linseed meal (old process) is the preferred protein supplement for horses. It is valued because of its laxative properties and because of the sleek hair coat which it imparts.
Brewers' dried grains	65–70	50	
Buttermilk, dried	100	100	May be used in place of dried skimmed milk for foals.
Copra meal (coconut meal); (21%)	90–100	50	
Corn gluten feed (gluten feed)	70	100	
Corn gluten meal (gluten meal)	100	50	Somewhat unpalatable to horses.
Cottonseed meal (41%)	100	100	Satisfactory if limited to amount necessary to balance ordinary rations.
Peanut meal (45%)	100	100	
Peas, dried	75	50	
Skimmed milk, dried	100	100	Especially valuable for young equines; for creep feeding until past weaning.
Soybean meal (41%)	100	100	
Soybeans	100	100	Soybeans should be limited to ⅓ of the concentrate ration.
Sunflower meal (41%)	100	33⅓	Sunflower meal should not constitute more than ⅓ of the protein supplement for palatability reasons.
Whey, dried	50	50	Whey may be laxative.
DRY FORAGES AND SILAGES:[2]			
Timothy hay	***100***	***100***	The preferred hay of caretakers.
Alfalfa hay, all analyses	133⅓	100	Good quality alfalfa is excellent for horses. Alfalfa may be ground and pelleted. It provides high-quality proteins, and certain minerals and vitamins. It is somewhat laxative. Contrary to some "old wives' tales," it will not damage the kidneys.
Barley hay	100	100	Lower value if not cut at the early dough stage.
Bromegrass hay	100	100	
Clover hay, crimson	125	100	Crimson clover hay has considerably lower value if not cut at an early stage.
Clover hay, red	125	100	Clover hay should be well cured and free from dust and mold.
Clover-timothy hay	110–115	100	Value of clover-timothy mixed hay depends on the proportion of clover present and the stage of maturity at which it is cut.
Corn fodder	100	50	Preferably fed along with a good legume hay. It is best to shred the fodder.
Corn silage	45–55	33⅓–50	
Corn stover	60	50	Preferably fed along with a good legume hay. It is best to shred the stover.
Cowpea hay	110	100	

(Continued)

TABLE 25-15 *(Continued)*

Feedstuffs	Relative Feeding Value (lb for lb) in Comparison with the Designated Base Feed (in bold italic) Which = 100	Maximum Percentage of Base Feed (or comparable feed or feeds) Which it Can Replace for Best Results	Remarks
DRY FORAGES AND SILAGES (Continued):			
Grass-legume mixed hay	110–115	100	
Grass-legume silage	45–50	33⅓–50	
Grass silage	40–45	33⅓–50	
Johnsongrass hay	90–95	100	
Lespedeza hay	115	100	
Oat hay	100	100	Lower value if not cut at the early dough stage.
Orchardgrass	100	100	Should be cut before maturity.
Prairie hay	100	100	
Reed canarygrass	90–95	100	
Sorghum fodder	100	50	Preferably fed along with a good legume hay. It is best to shred the fodder.
Sorghum silage	40–45	33⅓–50	
Sorghum stover	60	50	Preferably fed along with a good legume hay. It is best to shred the stover.
Soybean hay	110	100	
Sudangrass hay	90–95	100	
Vetch-oat hay	110–115	100	The higher the proportion of vetch, the higher the value.
Wheat hay	100	100	

[1]Roots and tubers are of lower value than the grain and by-product feeds due to their higher moisture content.

[2]Well preserved silage of good quality, free from mold and not frozen, affords a highly nutritious succulent forage for horses during the winter months—especially for idle horses, broodmares, and growing colts. Silages are of lower value than dry forages due to their higher moisture content.

FEEDING PLEASURE HORSES

Keeping pleasure horses—horses used for recreation and sport—in peak condition makes for greater satisfaction when they are used.

It is difficult to feed pleasure horses because they are used irregularly. Also, most pleasure horses are worked lightly, perhaps 1 to 3 hr of riding per day. Others are worked medium hard, as when ridden 3 to 5 hr per day. Still others are worked very hard, as when raced or when ridden 5 to 8 hr per day. The recommended daily feed allowance per 100 lb body weight of pleasure horses in light, medium, and hard use follows:

	Lb Daily/100 Lb Weight of Horse		
	Light Use	Medium Use	Hard Use
Hay	1¼–1½	1–1¼	1–1¼
Grain	⅖–½	¾–1	1¼–1⅓

As shown, the roughage content of the ration decreases and the concentrate content increases as the amount of work increases. This is because the digestibility and the efficiency of conversion are greater for high-energy concentrates than for roughages.

Of course, horses differ in temperament and in ease of keeping. Also, no two horses will perform the same amount of work with an equal expenditure of energy, and no two persons will get the same amount of work out of the same horse. So, the feed allowance should be increased if the horse fails to maintain condition, and it should be decreased if the animal becomes too fat.

In season, pasture may replace hay, all or in part, according to the quality of the pasture. But the concentrate allowance of the working horse should remain about the same on pasture as in the stable or dry corral. There is a tendency of the pastured working horse to sweat and tire more easily (be *soft*), probably due to the high water content of green forage.

In addition to forage and grain, pleasure horses should be provided with suitable minerals.

The mineral requirement of the working horse differs from the idle horse mainly in the salt requirements, due to the loss of salt in perspiration.

The vitamin requirements of working horses are approximately the same as those of idle horses, except for the increase in the B-complex requirements due to the greater carbohydrate metabolism of the working horse.

FEEDING HORSES IN TRAINING (EQUINE ATHLETES)

Strenuous exercise of equine athletes in heavy training—such as training for racing, endurance trials, cutting, roping, jumping, or hunting—does not appear to alter significantly the requirement for any specific nutrient except calories. Exercise creates little or no increase in demand by the horse for protein, calcium, phosphorus, most trace minerals, and most fat soluble vitamins. So, if the ration contains adequate nutrients for maintenance, any increase in the requirements for heavy training may be met by the increased feed intake required to meet the increased energy requirements for exercise. Of course, if the ration is deficient for maintenance,

Fig. 25-17. Pleasure horses must be properly fed if they are to exhibit maximum style and animation. (Courtesy, American Morgan Horse Assn., Westmoreland, N.Y.)

exercise could worsen the situation. Likewise, if the ration is marginal for young animals in training, heavy exercise could make for unsoundnesses.

Horses in training will eat about 1½ to 1¾ lb of grain and ¾ to 1 lb of hay per 100 lb liveweight.

- **Glycogen overload**—Perhaps horse trainers can take a page out of the book of human athletes—maybe they already have. Experiments and experiences, some of them a century old, with human athletes engaged in grueling and extended endurance events exceeding 30 to 60 minutes—such as long-distance and marathon runners, cross-country skiers, and conquering Mt. Everest—show that diets win races.[11] Studies show a strong relationship between carbohydrate intake and stored muscle glycogen (popularly known as animal starch), and between glycogen and stamina. The higher the glycogen content, the better the performance. The key to extraordinarily high levels of muscle glycogen is the manipulation of the diet as follows:

1. **Step 1: Reduce muscle glycogen.** Beginning about 13 days in advance of the event, and continuing for 7 days, reduce the glycogen stores in the muscle to very low levels by decreasing the carbohydrates in the diet and exercising to exhaustion the same muscles that will be used in the event.

2. **Step 2: Keep glycogen content of the exercising muscles low.** This is accomplished by adhering to a diet consisting amost exclusively of fat and protein for about 3 days.

3. **Step 3: Provide a carbohydrate-rich diet to produce glycogen overload.** This is accomplished by abruptly adding large quantities of carbohydrate (bread, corn, beans, potatoes) to the diet, while continuing the intake of fat and protein, for about 3 days, thereby storing above normal amounts of muscle glycogen. This change from a predominantly fat and protein diet to one high in carbohydrates has produced as much as an 8-fold increase in muscle glycogen and a 3-fold increase in the length of time that strenuous exercise can be performed continuously. Food made the difference!

The discovery that glycogen levels in the muscle can be markedly enhanced by dietary manipulation, a phenomenon known as *glycogen loading,* is being increasingly used by human athletes subjected to endurance at high speed. It is improving performance, which is especially marked in the final stages of the event. The rare, juicy steaks of the training table, and the compulsory bacon and eggs a few hours before the game, are giving way to glycogen loading.

[11]Van Itallie, T. B., "If Only We Knew," *Nutrition Today,* Vol. 3, No. 2, 1968, p. 3; Astrand, P., "Something Old and Something New—Very New," *Nutrition Today,* Vol. 3, June 1968, p. 9; and Slovic, P., "What Helps The Long Distance Runner Run?," *Nutrition Today,* Vol. 10, No. 3, 1975, p. 18.

Fig. 25-18. Cutting horse in action. Horses in heavy training have a higher nutritional requirement than most pleasure horses. (Courtesy, Vim Mathews, Orland, Calif.)

It is reasonable to assume that a similar manipulation of the ration would have a like effect on equine athletes—that the speed and endurance of horses would be markedly improved by glycogen depletion followed by feeding large quantities of high energy and easily digested carbohydrates, such as corn, barley, oats, and sugar. Indeed, glycogen overload could make for a winner in racing, cutting, reining —horses that work at anaerobic levels. In fact, it is reputed that glycogen loading is one of the best kept secrets of professional horse trainers. However, the application of glycogen overload in horses should await controlled experimental studies.

FEEDING RACEHORSES

It is recognized that some unsoundnesses may be inherited, others may be due to accident and injury, and still others may be due to subjecting horses to stress and strain far beyond the capability of even the best structure and tissue. However, nutritional deficiencies appear to be the major cause of unsoundnesses in racehorses.

Racehorses are equine athletes whose nutritive requirements are the most exacting, but the most poorly met, of all animals. This shocking statement is true because racehorses are commonly handled as follows:

1. Started in training at 16 to 18 months of age, which is comparable to an adolescent boy or girl doing sweatshop labor.
2. Moved from track to track under all sorts of conditions.
3. Trained the year around, raced innumerable times each year, and forced to run when fatigued.
4. Outdoors only a short time each day—usually before sunup, with the result that the sun's rays have little chance to produce vitamin D from the cholesterol in the skin.
5. Without opportunity for even a few mouthfuls of grass —a rich, natural source of the B vitamins and unidentified factors.
6. Fed oats, grass hay, and possibly bran—produced in unknown areas, and on soils of unknown composition. Such an oats-grass hay-bran ration is almost always deficient in vitamins A and D and the B vitamins, and lopsided and low in calcium and phosphorus.
7. Given a potion of some concoction of questionable value—if not downright harmful.

By contrast, human athletes—college football teams and participants in the Olympics, for example—are usually required to eat at a special training table, supervised by nutrition experts. They are fed the best diet that science can formulate and technology can prepare. It's adequate in protein, rich in readily available energy, and fortified and balanced in vitamins and minerals.

It's small wonder, therefore, that so many equine athletes go unsound, whereas most human athletes compete year after year until overtaken by age.

Indeed, highly strung and highly stressed, racehorses need special rations just as human athletes do—and for the same reasons; and, the younger the age, the more acute the need. This calls for rations adequate in protein, rich in readily available energy, fortified with vitamins, minerals, and unidentified factors—and with all nutrients in proper balance.

A racehorse is asked to develop a large amount of horsepower in a period of 1 to 3 minutes. The oxidations that occur in a racehorse's body are at a higher pitch than in an idle or little-worked horse, and, therefore, more vitamins are required.

Also, racehorses are the prima donnas of the equine world; most of them are temperamental, and no two of them can be fed alike. They vary in rapidity of eating, in the quantity of feed that they will consume, in the proportion of concentrate to roughage that they will take, and in response to different caretakers. Thus, for best results, they must be fed as individuals.

During the racing season, the hay fed to a racehorse should be limited to 7 or 8 lb, whereas the concentrate allowance may range up to 16 lb. Heavy roughage eaters may have to be muzzled, to keep them from eating their bedding. A bran mash is commonly fed once a week.

NOTE WELL: Because the affliction known as *bleeders,* which is so common among racehorses, appears to be more stress-induced than nutritional, it is fully covered in Chapter 17, Animal Behavior/Environment, in the section headed "Bleeders—a stress/environmental interaction of equines."

FEEDING BROODMARES

Broodmares require good quality, balanced rations. The nutrients of primary concern are energy, protein, minerals, and vitamins. The nutrient requirements of a broodmare change considerably as she advances from being open (not pregnant), through pregnancy, and into lactation. In general, during the first 7 months of pregnancy the requirements are very similar to those for maintenance at the time of breeding. However, from the eighth month on, the requirements of the pregnant mare increase by 20 to 50%. During lactation, the mare requires even more nutrients, up to twice the amounts that she was receiving at breeding time.

Fig. 25-19. Broodmares and foals on lush pastures. (Courtesy, The American Morgan Horse Assn., Inc., Westmoreland, N.Y.)

Proper feeding of broodmares begins with conditioning them prior to the breeding season by providing adequate and proper feed and the right amount of exercise.

On horse breeding establishments, pregnancy is commonly determined 45 to 90 days after service by using one of the following methods: rectal palpation, a commercially prepared kit to test for pregnant mare serum gonadotropin (PMSG), or an ultrasound scanner. Early pregnancy diagnosis makes it possible (1) to cull mares more intelligently, and (2) to separate pregnant mares from barren ones and feed each group differently.

Broodmares need a ration that will meet their body needs plus (1) the needs of the fetus, or (2) furnishing the nutrients required for milk production. If work is also being performed, additional energy feeds must be provided. Moreover, for the young, growing mare additional proteins, minerals, and vitamins, above the ordinary requirements, must be provided; otherwise, the fetus will not develop normally, or milk will be produced at the expense of the tissues of the dam. Also, protein deficiency may affect undesirably the fertility of the mare.

As with the females of all species, the nutritive requirements for milk production in the mare are much more rigorous than the pregnancy requirements. It is estimated that, 2 months following foaling, mares of mature weights of 600, 800, 1,000, and 1,200 lb may produce 36, 42, 44, and 49 lb of milk daily. Thus, it can be appreciated that a mare's feed requirements during the suckling period are not far different from those of a high-producing dairy cow. In general, it is important that the ration of the gestating-lactating mare supply sufficient energy, protein, calcium, and phosphorus; and vitamins A and D (the D being provided through the feed if the animal is not exposed to sunlight), and riboflavin.

The correct feeding of a broodmare that is worked is often simpler than the feeding of an idle one, for the condition of the animal can be regulated more carefully under working conditions. In addition to a ration that will meet the maintenance and work requirements largely through high-energy feeds, the working broodmare needs ample protein, calcium, and phosphorus with which to take care of the growth of the fetus and/or milk production.

The broodmare should be fed and watered with care immediately before and after foaling. For the first 24 hours after parturition, she may have a little hay and a limited amount of water from which the chill has been taken. A light feed of bran or a wet bran mash is suitable for the first feed and the following meal may consist of oats or a mixture of oats and bran. A reasonably generous allowance of good-quality hay is permissible after the first day. If confined to the stable, as may be necessary in inclement weather, the mare should be kept on a limited and light grain and hay ration for about 10 days after foaling. Feeding too much grain at this time is likely to produce digestive disturbances in the mare, and even more hazardous, it may produce too much milk, which may cause indigestion in the foal. If weather conditions are favorable and it is possible to allow the mare to foal on a clean, lush pasture, she will regulate her own feed needs most admirably.

In comparison with geldings or unbred mares, the following differences in feeding gestating-lactating broodmares should be observed:

1. A greater quantity of feed is necessary—from 20 to 100% more—the highest requirement being during lactation.
2. Dusty or moldy feed and frozen silage should be avoided in feeding all horses, but especially in feeding the broodmare, for such feed may produce complications and possible abortion.
3. More proteins are necessary for the broodmare.
4. More attention must be given to supplying the necessary minerals and vitamins.
5. The bowels should be carefuly regulated through providing regular exercise and feeding such laxative feeds as bran, linseed meal, and alfalfa hay.
6. A few days before and after foaling, the ration should be (a) decreased, and (b) lightened by using wheat bran.
7. Regular and ample exercise is a necessary adjunct to proper feeding of the broodmare.

FEEDING STALLIONS

Horse breeders have long recognized the importance of proper feeding of the stallion. The Bible implies that well-fed stallions have greater sex drive than poorly fed stallions (Jeremiah 5:8).

The program throughout the entire year should be such as to keep the stallion in a vigorous, thrifty condition at all times. Immediately before the breeding season, the feed might very well be increased in quantity so that the stallion will gain in weight. The quantity of grain fed will vary with the individual temperament and feeding ability of the stallion, the work and exercise provided, services allowed, available pastures, and quality of roughage. Usually this will be between ¾ and 1½ lb daily of the grain mixture per 100 lb body weight, together with a quantity of hay within the same range.

During the breeding season, the stallion's ration should contain more protein and additional minerals and vitamins

Fig. 25-20. Seattle Slew, Triple Crown winner and great breeding stallion, grazing in his Kentucky paddock. (Courtesy, Dr. Craig H. Wood, University of Kentucky, Lexington)

than are given in rations fed work horses or stallions not in service. During the balance of the year (when not in service), the stallion may be provided a ration like that of other horses similarly handled. In season, pastures are an excellent source of both nutrients and exercise.

In addition to the grain and roughage, needed minerals and vitamins should be provided. During the winter months or when little work or exercise is provided, the stallion should recieve a succulent feed, such as carrots. Also, laxative feeds, such as wheat bran or linseed meal, should be supplied at these times. Plenty of fresh, clean water should be provided at all times.

Overfitted, heavy stallions should be regarded with suspicion, for they may be uncertain breeders. On the other hand, a poor, thin, run-down condition is also to be avoided.

FEEDING FOALS

Growth is the very foundation of horse production. This is so because horses cannot perform properly or possess the necessary speed and endurance if their growth has been stunted or their skeletons have been injured by inadequate rations during early age. Naturally, these requirements become increasingly acute when horses are forced for early use, such as the training and racing of the 2-year-old. Also, unless foals are rather liberally fed when young, they never attain the much desired body form, which is so important where young stock is sold or shown.

As with all young mammals, milk from the dam gives the foal a good start in life. Within 30 minutes to 2 hours after birth, the foal should be up on its feet and getting its first feed of colostrum.

But milk is not the perfect food, as once claimed. It is deficient in iron and copper, with the result that suckling young may suffer from anemia. This nutritional deficiency may be prevented, and increased growth, durability, and soundness may be obtained, by feeding foals separately from their dams; either (1) by tying the mare while the foal eats, or (2) by providing a creep for the foals.

Fig. 25-22. Foal creep. With this arrangement, foals can be fed separately from their dams. (Courtesy, Beaufort Cottage Stables, Newmarket, Great Britain)

At birth, foals have little, if any, natural immunity to disease and infection. Because of this, it is important that they nurse soon after birth so that they obtain colostrum (the first milk secreted for 12 to 24 hours after birth) from their dams. Consumption of colostrum is critical because it is a rich source of antibodies which will provide the foal with temporary immunity. Additionally, colostrum is very nutritious. It is much higher in energy, protein, mineral, and vitamin content than later milk.

Since mares have not been selected for milk production, they are not heavy milkers. The mare will produce 22 to 50 lb of milk per day, depending on the stage of lactation. However, mare's milk is a good feed for the young foal in terms of energy, protein, minerals, and vitamins. It contains

Fig. 25-21. (Left) In an amazingly short time after birth, a healthy foal can run almost as fast as its mother—and on legs almost as long. (Courtesy, American Quarter Horse Assn., Amarillo, Tex.)

about 20% protein on a dry basis; and it contains 0.45 g of calcium and 0.2 g of phosphorus per lb of milk. Peak lactation occurs about 6 to 8 weeks after birth; and usually there is a drop in milk production after 3 months. It is at this time that consideration needs to be given to individual feeding or creep feeding—to filling the gap between what the mare produces and what the foal needs.

When the foal is between 10 days and 3 weeks of age, it will begin to nibble on grain and hay. In order to promote thrift and early development and to avoid setback at weaning time, it is important to encourage the foal to eat supplementary feed as early as possible. For this purpose, a low-built grain box should be provided especially for the foal. If on pasture, the supplemental feeding of foals can be accomplished in either of the following two ways: (1) by bringing the mares and foals to a central area where the mares may be tied while the foals eat, or (2) by feeding the foals in a creep. The choice between individual feeding and creep feeding may be left to the caretaker; the important thing is that foals receive supplemental feed.

A creep is an enclosure for feeding purposes, made accessible to the foal(s), but through which the dam cannot pass. For best results, the creep should be built at a spot where the mares are inclined to loiter. The ideal location is on high ground, well drained, in the shade, and near the place of watering. Keeping the salt supply nearby will be helpful in holding mares near the creep.

It is important that foals be started on feed carefully, and at an early age. At first, only a small amount of feed should be placed in the trough each day, any surplus being removed and given to other horses. In this manner, the feed will be kept clean and fresh, and the foals will not be consuming any moldy or sour feed.

Rolled oats and wheat bran, to which a little brown sugar has been added, is especially palatable as a starting ration.

Table 25-16 gives the formulation of an excellent foal ration, which may be either individually fed or creep fed.

TABLE 25-16
FOAL RATION

Ingredients	Percent	Amount in 500-lb Mix	
	(%)	(lb)	*(kg)*
Corn (flaked)	37.4	187.0	*87.9*
Soybean meal (41%)	33.0	165.0	*74.9*
Oats (rolled)	23.0	115.0	*52.2*
Brewers' yeast	0.5	2.5	*1.1*
Molasses	3.0	15.0	*6.8*
Steamed bone meal or dicalcium phosphate	2.0	10.0	*4.5*
Salt (trace mineralized)	1.0	5.0	*2.3*
Vitamins A and D premix	0.1	0.5	*0.2*
Total	100.0	500.0	*227.0*

Because of the difficulty in formulating and home mixing a foal ration, the purchase of a good commercial feed usually represents a wise investment.

In addition to its grain ration, the foal should be given good-quality hay (preferably a legume), unless it is on good pasture.

Free access to salt should be provided, plenty of fresh water should be available at all times.

When foals are on luxuriant pasture and their mothers are milking well, difficulty may be experienced in getting them to eat. Thus, patience on the part of the caretaker is extremely important. However, foals are curious. Usually, they'll examine a creep. But it may be necessary to start them on the creep ration by first letting them nibble a little feed from the hand.

Giving creep-fed foals all they want to eat is not recommended. Although they will not founder or overeat when fed in this manner, a creep feed may increase body weight to the point at which it will cause bone disorders. The heavy body weight may lead to problems with epiphysitis and other bone abnormalities. So, creep feeding should be limited as follows: Feed 1 lb of concentrate per month of age per day, with a maximum allowance of 6 lb per day.

Under such a system of care and management, the foal will become less dependent upon its dam, and the weaning process will be facilitated. If properly cared for, at 12 months of age foals will normally attain 75% of their mature body weight and 90% of their mature wither height. Most Thoroughbred and Standardbred breeders plan to have the animals attain full height by the time they are 2 years of age. However, such results require liberal feeding from the beginning.

It is well recognized that the forced development of race, show, and sale horses must be done expertly if the animals are to remain durable and sound. This calls for particular emphasis on the kind of ration, feed allowance, and exercise.

• **Orphaned foal**—Occasionally a mare dies during or immediately after parturition, leaving an orphan foal to be raised. Also, there are times when a mare rejects her foal or fails to give sufficient milk for the newborn foal. Sometimes there are twins. In such cases, it is necessary to resort to other milk supplies. The problem will be simplified if the foal has at least received the colostrum from the dam, for it does play a very important part in the well-being of the newborn young.

If at all possible, the foal should be shifted to another mare.

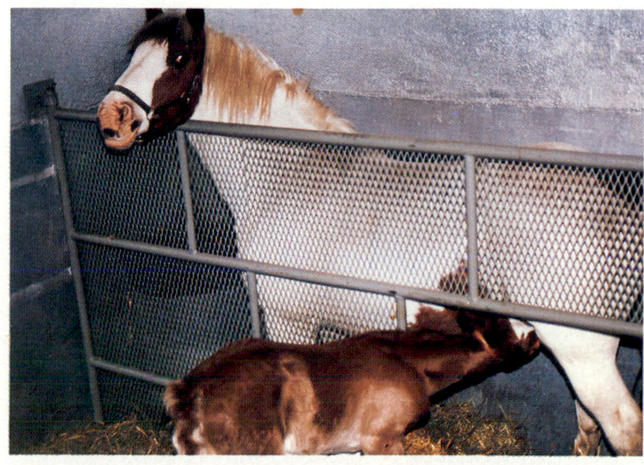

Fig. 25-23. Foster gate protecting orphan foal from foster mother. (Courtesy, Kildangan Stud, Monasterevin, Co. Kildare, Ireland)

Some breeding establishments regularly follow the plan of breeding a mare that is a good milk producer but whose foal is expected to be of little value. Her own foal is either destroyed or raised on a bottle, and the mare is used as a foster mother or nurse mare.

Some nurseries keep a supply of colostrum on hand. They remove colostrum from mares (1) whose foals were stillborn, or (2) that produce excess milk, then store it in a freezer for future use for foals that do not receive colostrum from their dams. When needed, it can be removed from the freezer, heated, and fed. This is an excellent practice.

If no colostrum is available, the foal should be placed on either (1) cow's milk made as nearly as possible to the same composition as mare's milk; or (2) a synthetic milk replacer.

A comparison of cow's and mare's milk, is given in Table 25-17.

TABLE 25-17
COMPOSITION OF MILK FROM COWS AND MARES

Source	Water	Protein	Fat	Sugar	Ca	P
	(%)	(%)	(%)	(%)	(%)	(%)
Cow	87.17	3.55	3.69	4.88	0.12	0.10
Mare, 2 months after foaling	90.78	1.99	1.50	5.67	0.09	0.03

As can be observed, in comparison with cow s milk, mare's milk is higher in percentage of water and sugar and lower in other components.

For best results in raising the orphan foal, milk from a fresh cow, low in butterfat, should be used. To about a pint of milk, add a tablespoonful of sugar and from 3 to 5 tablespoonfuls of lime water. Warm to body temperature and for the first few days feed about one-fourth of a pint every hour. After 3 or 4 weeks the sugar can be stopped, and at 5 to 6 weeks skimmed milk can be used entirely.

Orphan foals may also be raised on a synthetic milk replacer made especially for foals, fed according to the directions of the manufacturer. Here again the situation is simplified if the foal has first received colostrum.

For the first few days, the milk (either cow's milk or milk replacer) may be fed by using a bottle and rubber nipple. Later, the foal should be taught to drink from a pail. It is important that all receptacles be kept absolutely clean and sanitary (cleaned and scalded each time), and that feeding be at regular intervals. Grain feeding should be started at the earliest possible time with the orphan foal.

• **Shaker foal (equine botulism)**—This is the name given to the disease in foals in which *Cl botulism* is present in the tissues, where it produces toxins. Most commonly, horses contract botulism by ingestion of a preformed toxin, which can be present in contaminated feed. One well known clinical sign of the disease is quivering or shaking knees, from which the name is derived. Other symptoms include depression, loss of appetite, and muscle paralysis. The disease is fatal in 80% to 90% of the cases; and afflicted foals may die within 24 hours. Equine botulism strikes adult horses as well as foals. The disease may be effectively prevented by vaccinating broodmares with 3 doses of a toxoid, given one month apart, during the eighth, ninth, and tenth months of pregnancy, followed by annual boosters. An annual single-dose booster 2 to 4 weeks before parturition is all that is required to provide immunity to the mare and subsequent foals. Adult horses should also receive an initial 3-dose vaccination and a yearly booster.

(Also see Chapter 5, Table 5-3—Botulism.)

FEEDING WEANLINGS

The most critical period in the entire life of a horse is that interval from weaning time (about 6 months of age) until 1 year of age. Foals suckling their dams and receiving no grain may develop very satisfactorily up to weaning time. However, lack of preparation prior to weaning and neglect following the separation from the dam may prevent the animal from gaining proper size and conformation. The primary objective in the breeding of horses is the economical production of a well-developed, sound individual at maturity. To achieve this result requires good care and management of weanlings.

Fig. 25-24. Weanlings on pasture. (Courtesy, USDA)

No great setback or disturbances will be encountered at weaning time provided the foals have developed a certain independence as a result of proper grain feeding during the suckling period. Generally, weanlings should receive 1 to 1½ lb of grain and 1½ to 2 lb of hay daily per each 100 lb of liveweight. The amount of feed will vary somewhat with the individuality of the animal, the quality of roughage, available pastures, the price of feeds, and whether the weanling is being developed for show, race, or sale. Naturally, animals being developed for early use or sale should be fed more liberally, although it is equally important to retain clean, sound joints, legs, and feet—a condition which cannot be obtained so easily in heavily fitted animals.

Based on studies involving the creep feeding of foals, followed by assigning weanlings to five different energy/protein levels, the Kentucky Station reported as follows: If a system of creep feeding suckling foals is followed by feeding weanlings a high-energy, protein, and calcium ration, the animals will be taller and heavier at 12 months of age, but there is risk of developing bone disorders.[12] This study would indicate that weanlings should be limit-fed rather than full-fed.

Because of the rapid development of bone and muscle in weanlings, it is important that, in addition to ample quantity of feed, the ration also provide quality of proteins, and adequate minerals and vitamins.

FEEDING YEARLINGS

If young animals have been fed and cared for so that they are well grown and thrifty as yearlings, usually little difficulty will be experienced at any later date.

When on pasture, yearlings that are being grown for show or sale should receive grain in addition to grass. They should be confined to their stalls in the daytime during the hot days and turned out at night (because of not being exposed to sunshine, adequate vitamin D must be provided). This point needs to be emphasized when forced development is desired; for, fine as pastures may be, they are roughages rather than concentrates.

The winter feeding program for the rising 2-year-olds should be such as to produce plenty of bone and muscle rather than fat. From ½ to 1 lb of grain and 1 to 1½ lb of hay should be fed for each 100 lb of liveweight. The quantity will vary with the quality of the roughage, the individuality of the animal, and the use for which the animal is produced. In producing for sale, more liberal feeding may be economical. Minerals should be incorporated in the ration; and an abundance of fresh, pure water should be available.

Fig. 25-25. Yearlings wintering on a pasture with access to an open shed, shown eating hay from a rack along the wall. (Courtesy, Dr. Craig H. Wood, University of Kentucky, Lexington)

[12]Thompson, K. N., S. G. Jackson, and J. P. Baker, "Effect of Nutrition on Growth in Horses—Part II," *Equine Data Line*, University of Kentucky, March, 1986, p. 2.

FEEDING TWO- AND THREE-YEAR-OLDS

Except for the fact that the 2- and 3-year-olds will be larger, and, therefore, will require more feed, a description of their proper care and management would be merely a repitition of the principles that have already been discussed for the yearling.

Fig. 25-26. Three-year-olds being grain fed on pasture. (Courtesy, American Quarter Horse Assn., Amarillo, Tex.)

With the 2-year-old that is to be raced, however, the care and feeding at this time become matters of extreme importance. Once the young horse is placed in training, the ration should be adequate enough to allow for continued development and to provide necessary maintenance and additional energy for work. This means that special attention must be given to providing adequate proteins, minerals, and vitamins in the ration. Overexertion must be avoided, the animal must be well groomed, and the feet must be cared for properly. In brief, every precaution must be taken if the animal is to remain sound—a most difficult task when animals are raced at an early age, even though the right genetic makeup and the proper environment are present.

FITTING FOR SHOW OR SALE

Each year, many horses are fitted for shows or sales. In both cases, a fattening process is involved, but exercise is doubly essential.

For horses that are being fitted for shows, the conditioning process is also a matter of hardening, and the horses are used daily in harness or under saddle. Regardless of whether a sale or a show is the major objective, fleshing should be obtained without sacrificing action or soundness or without causing filling of the legs and hocks.

In fattening horses, the animals should be brought to full feed rather gradually, until the ration reaches a maximum of about 2 lb of grain daily for each 100 lb of liveweight. When on full feed, horses make surprising gains. Daily weight gains of 4 to 5 lb are not uncommon. Such animals soon become fat, sleek, and attractive. This is probably the basis for the statement that "fat will cover up a multitude of sins in a horse."

Fig. 25–27. A beautifully fitted horse—bloomy and attractive. (Courtesy, American Quarter Horse Assn., Amarillo, Tex.)

Although exercise is desirable from the standpoint of keeping the animals sound, it is estimated that such activity decreases the daily rate of gains by as much as 20%. Because of the greater cost of gains and the expense involved in bringing about forced exercise, most feeders of sale horses limit the exercise to that obtained naturally from running in a paddock.

In comparison with finishing cattle or sheep, there is more risk in fattening horses. Heavily fed horses kept in idleness are likely to become blemished and injured through playfulness, and there are more digestive disorders among liberally fed horses than in other classes of stock handled in a similar manner.

In fitting show horses, the finish must remain firm and hard, the action superb, and the soundness unquestioned. Thus, they must be carefully fed, groomed, and exercised to bring them to proper bloom.

Breeders who fit and sell yearlings or younger animals may feed a palatable milk replacer or commercial feed to advantage.

FEEDING SYSTEMS

Most horses are hand-fed. The grain ration usually is divided into 3 equal feeds given morning, noon, and night. Because a digestive tract distended with hay is a hindrance in hard work, most of the hay should be fed at night. The common practice is to feed ¼ of the daily hay allowance at each of the morning and noon feedings and the remaining ½ at night when the animals have plenty of time to eat leisurely.

Usually the grain ration is fed first and then the roughage. This way, the animals usually eat bulky forages more slowly.

A few caretakers do self-feed high-energy rations, but, sooner or later, those who do usually founder a valuable horse. Except for the use of reasonably hard salt-protein blocks, salt-feed mixes in meal form (never in pellet form), or high-roughage rations, the self-feeding of horses is not recommended.

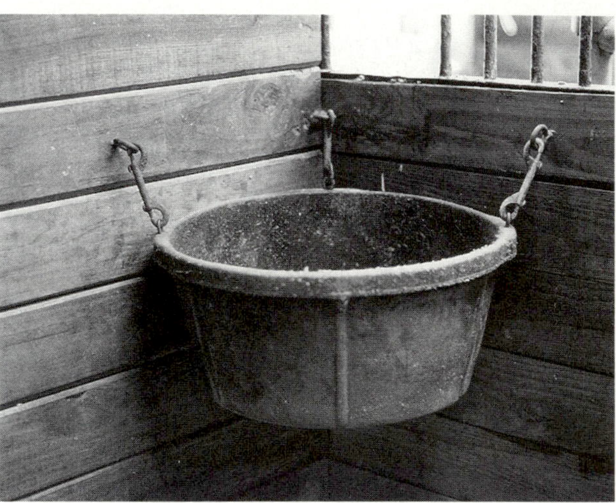

Fig. 25–28. A round feed container suspended by eye hooks and located in one corner of the stall, at shoulder level or lower, will keep the horse from accidentally cutting or puncturing itself while it eats. (Courtesy, *Equus* Magazine, Gaithersburg, Md.)

Fig. 25–29. A good hayrack. Note the rack's nearly vertical bars and curved form. These features, along with positioning the rack in the corner of the stall, are designed to alleviate injury to the horse. (Courtesy, *Equus* Magazine, Gaithersburg, Md.)

ART OF FEEDING

Feeding horses is both an art and a science. The art is knowing how to feed and how to take care of each horse's individual requirements. The science is meeting the nutritive requirements with the right combination of ingredients.

Starting Horses on Feed

Horses must be accustomed to changes in feed gradually. In general, they may be given as much nonlegume roughage as they will consume. But they must be accustomed gradually to high-quality legumes, which may be very laxative. This can be done by slowly replacing the nonlegume roughage with greater quantities of legumes. Also, as the grain ration is increased, the roughage is decreased.

Starting horses on grain requires care and good judgment. Usually it is advisable first to accustom them to a bulky type of ration; a starting ration with considerable rolled oats is excellent for this purpose.

The keenness of the appetite and the consistency of the feces are an excellent index of a horse's capacity to take more feed. In all instances, scouring should be avoided.

General Feeding Rules

Observance of the following rules will help avoid some of the common difficulties that result from poor feeding practices:

1. Know the approximate weight and age of each animal.
2. Feed by weight of feed, not by volume (volume as determined by a coffee can or marked bucket). Horses do not require a certain volume of feed; rather, they require a certain weight of nutrients based on their body weight.
3. Avoid sudden changes in the ration.
4. Never feed moldy, musty, dusty, or frozen feed.
5. Feed regularly. Horses anticipate their feed.
6. Look for problems at feeding time; don't just dump the feed and run. Look for injuries and abnormalities.
7. Check the feces. Any change in quantity, odor, color, or composition may presage trouble.
8. Inspect the feedbox frequently to see if the horse goes off feed. Feed refusal means (1) the horse was over-fed, (2) something is wrong with the feed, or (3) the horse is sick.
9. Keep the feed and water containers clean. Scrub them periodically to insure proper sanitation.
10. Do not overfeed. Some horses suffer from obesity, while others suffer from deficiency. Fat horses not receiving adequate exercise are predisposed to colic and founder. An old Arab proverb cautions: "The two greatest enemies of horses are fat and rest."
11. Force aggressive eaters to slow down. Some horses may bolt their feed when fed in deep narrow feed boxes. Their eating may be slowed by scattering the feed in a larger box, or by placing large round stones, bricks, or salt blocks in the feed container.
12. Accord timid eaters solitude to eat. Feed them where it is quiet and they will not be disturbed.
13. Do not feed from the hand; this can lead to *nibbling.*
14. Exercise stalled horses daily. It improves their appetite, digestion, and overall well being. This may be accomplished by riding, longeing, walking, ponying, swimming, treadmilling.
15. Avoid excessive exercise (to the point of fatigue and stress), rough treatment, noise, and excitement.
16. Do not feed concentrates 1 hour before or within 1 hour after hard work.
17. Feed horses as individuals; consider their likes and temperaments. Learn the peculiarities and desires of each animal because each one is different.
18. Gradually decrease the condition of horses that have been fitted for show or sale. Many caretakers accomplish this difficult task, and yet retain strong vigorous animals, by cutting down gradually on the feed and increasing the exercise.
19. Prevent wood chewing. This habit usually results from boredom, lack of exercise, lack of adequate roughage, or lack of phosphorus; so, alleviate the causes.
20. Make certain that the horse's teeth are sound.
21. Know the signs of a well-fed, healthy horse, any departure from which constitutes a warning signal.

Signs of a Well Fed, Healthy Horse

The signs of a well-fed, healthy horse are given in Chapter 17, Animal Behavior/Environment.

Fig. 25-30. Signs of a well-fed and healthy horse—it is bright-eyed and "bushy-tailed." (Courtesy, Ruth White, White Horse Ranch, Naper, Nebr.)

OTHER FEED AND MANAGEMENT ASPECTS

Several other feed and management aspects are pertinent to feeding horses; among them, (1) bolting feed, (2) eating bedding, (3) pica—wood chewing, and (4) soil analyses. For further information on bolting feed, eating bedding, and pica—wood chewing, see Chapter 17, Animal Behavior/Environment; and for further information on soil analysis, see Chapter 5, Nutritional Disorders/Toxins.

NUTRITIONAL DISEASES

Chapter 5, Nutritional Disorders/Toxins, contains a summary of the important nutritional diseases and ailments and nutrition related disorders affecting horses, as well as other classes of farm animals.

QUESTIONS FOR STUDY AND DISCUSSION

1. What conditions pertaining to the care and use of horses make it imperative that their nutrition be the best that science and technology can devise?

2. Discuss the economic importance of feed for horses.

3. Explain why the nutritive needs of horses differ from other species primarily because of their (a) digestive tract, and (b) special needs for bone formation and maintenance.

4. What is the difference between "nutritive requirements" and "nutritive allowances"?

5. Note that Table 25-6 recommends much higher allowances of several nutrients for lactating mares and young horses than for mature horses that are idle or performing light or moderate work. Why is this so?

6. How do you explain (a) that the energy requirements of a horse increase with weight (with or without rider) and speed, and (b) that the energy requirements in racing may be 100 times the energy requirements at rest?

7. Recent experimental work indicates that horses will readily consume 10 to 20% added fat to the ration, with benefit. When, and for what reasons, should fat be added to a horse ration; and what type of fat should be used?

8. Are the protein requirements of the horse increased by work load? Explain.

9. What is meant by "protein poisoning"? Is there such a thing?

10. How can a lactating mare produce milk that contains more minerals than the feed that she is consuming at the time?

11. Describe the following bone diseases and the factors thought to be responsible for each of them: epiphysitis, contracted tendons, and osteochondritis dissecans (OCD). What preventive measures would you recommend?

12. List the minerals that are most apt to be deficient in horse rations; then, for each of them give (a) the deficiency symptoms, and (b) practical sources.

13. Upon what three factors does the proper utilization of calcium and phosphorus depend?

14. Why is incorporating needed minerals in the feed or water preferable to free-choice feeding?

15. List the vitamins that are most apt to be deficient in horse rations; then, for each of them give (a) the deficiency symptoms, and (b) practical sources.

16. If there is no evidence of deficiencies of certain vitamins, can adding them to horse rations be justified?

17. What is meant by "unidentified factors"? List three rich sources of unidentified factors.

18. What factors influence the amount of water consumed by a horse?

19. What roles may antibiotics play as an additive?

20. Is there a need and a place for pastures in modern horse production?

21. Why do caretakers prefer timothy hay and oats to all other feeds?

22. Why do caretakers use each of the following: old process linseed meal, bran mash, sweet feed, corn oil, and treats?

23. Why is feed palatability so important in horse rations?

24. What are the preferred methods of preparing feeds for horses?

25. Give recommended allowances for grain and hay for horses at (a) light work, (b) medium work, and (c) heavy work.

26. Under what circumstances would you recommend that a caretaker use (a) a home-mixed feed, or (b) a commercial feed?

27. Of what value is a feed substitution table?

28. Discuss the proper feeding of each of the following classes of horses: (a) pleasure horses, (b) horses in training, (c) racehorses, (d) broodmares, (e) stallions, (f) foals, (g) weanlings, (h) yearlings, (i) 2- and 3-year-olds, and (j) horses being fitted for show or sale.

29. Discuss (a) free-choice feeding of foals, and (b) feeding the orphaned foal.

30. Discuss the advantages and disadvantages of (a) hand-feeding, and (b) self-feeding of horses.

31. List what you consider to be the six most important horse feeding rules.

32. Write down the ration that is fed to a certain horse (your horse or a friend's horse). Then, evaluate it from the standpoints of (a) rate of feeding; (b) protein content; (c) content of salt, iodine, calcium, and phosphorus; and (d) content of vitamins A, D, and riboflavin.

33. List all the ways in which the nutritive requirements and the feeding of horses are different from those for cattle, sheep, swine, and poultry.

SELECTED REFERENCES

Title of Publication	Author(s)	Publisher
Animal Science, 8th Edition	M. E. Ensminger	The Interstate Printers & Publishers, Inc., Danville, Ill., 1983
Breeding and Raising Horses, Ag. Hdbk. No. 394	M. E. Ensminger	Agricultural Research Service, USDA, Washington, D.C., 1972
Complete Encyclopedia of Horses, The	M. E. Ensminger	A. S. Barnes & Co., Inc., Cranbury, N.J., 1977. (Now, Oak Tree Publications, San Diego, Calif.)
Feeding and Care of the Horse	L. D. Lewis	Lea & Febiger, Philadelphia, Penn., 1982
Feeding Ponies	W. C. Miller	J. A. Allen & Co., London, England, 1968
Horse Feeding and Nutrition	T. J. Cunha	Academic Press, New York, N.Y., 1980
Horse, The	P. D. Rossdale	The California Thoroughbred Breeders Assn., Arcadia, Calif., 1972
Horse, The	J. W. Evans, *et al*	W. H. Freeman and Co., San Francisco, 1977
Horse Science Handbook, Vols. 1, 2, and 3	Ed. by M. E. Ensminger	Agriservices Foundation, Clovis, Calif., 1963, 1964, and 1966
Horsemanship and Horse Care, Ag. Info. Bull. No. 353	M. E. Ensminger	Agricultural Research Service, USDA, Washington, D.C., 1972
Horses and Horsemanship	M. E. Ensminger	The Interstate Printers & Publishers, Inc., Danville, Ill., 1977
Horses and Tack	M. E. Ensminger	Houghton Mifflin Company, Boston, Mass., 1977
Light Horse Management	R. C. Barbalace	Caballus Publishers, Fort Collins, Colo., 1974
Light Horses, Farmers' Bull. No. 2127	M. E. Ensminger	U.S. Department of Agriculture, Washington, D.C., 1965
Nutrient Requirements of Horses, No. 6, 4th Revised Edition	National Research Council	National Academy of Sciences, Washington, D.C., 1978
Shetland Pony	L. F. Bedell	Iowa State University Press, Ames, Iowa, 1959
Stud Managers' Handbook	Ed. by M. E. Ensminger	Agriservices Foundation, Clovis, Calif., annually since 1965

Fig. 25-31. Signs of good nutrition and health. (Courtesy, American Quarter Horse Assn., Amarillo, Tex.)

Fig. 25-32. A matched pair of Morgans, successfully shown in many open carriage competitions; driven by the owner, Hope Jenkins Jones. Show horses are always under stress; and the greater the stress, the more exacting the nutritive requirements. (Courtesy, American Morgan Horse Assn., Westmoreland, N.Y.)

Chapter

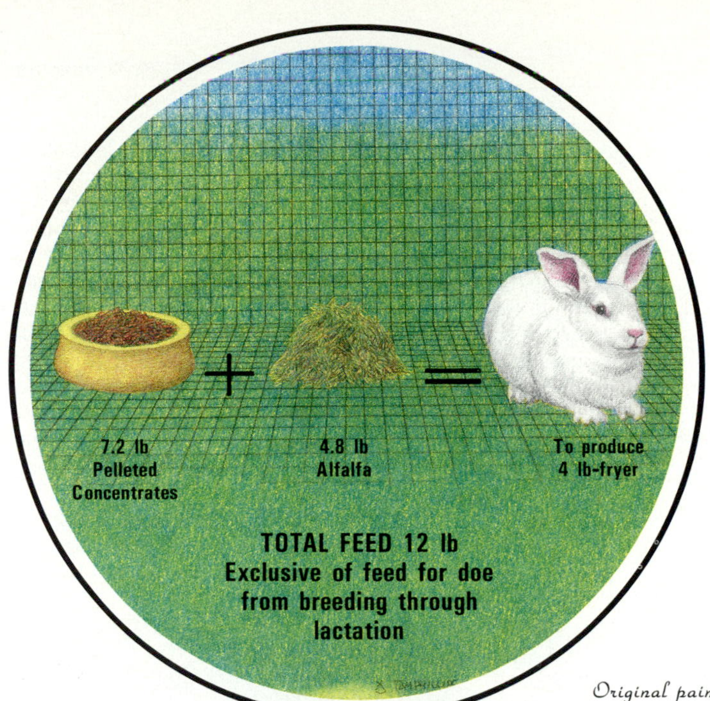

Original painting by Tom Phillips

FEEDING RABBITS[1]

Contents *Page*

Contents *Page*

[1]The authors gratefully acknowledge the helpful suggestions of the following authoritative reviewers of this chapter: P. R. Cheeke, Ph.D., Professor of Comparative Nutrition, Department of Animal Science, Oregon State University, Corvallis; L. B. Daniels, Ph.D., Department of Animal Sciences, University of Arkansas Division of Agriculture, Fayetteville; D. J. Harris, Ph.D., General Manager, Farms Division, Pel-Freez Rabbit Meat, Inc., Rogers, Ark.; L. J. Heppler, D.D.S., Heppler Farms, Oregon City, Ore.; S. D. Lukefahr, Ph.D., Assistant Professor, Department of Food Science and Animal Industries, Alabama Agricultural and Mechanical University, Normal; J. I. McNitt, Ph.D., Rabbit Production Specialist, Southern University, Baton Rouge, La.; and T. E. Reed, D.V.M., President, The American Rabbit Breeders Assn., Inc., Markle, Ind.

Rabbits were first domesticated in Africa. There is evidence that they were used for food as early as 3,000 years ago in Asia and 2,000 years ago in Europe.

Rabbit meat is a staple food in European diets. Estimated annual per capita consumption is: Hungary, 8.8 lb; France, 7.9 lb; Spain, 7.9 lb; Italy, 6.2 lb; and Portugal, 4.4 lb. By comparison, the estimated annual per capita rabbit consumption in the United States is a mere 2 oz.

Fig. 26-1. A 10,000-doe rabbitry in Hungary, which produces about 300,000 fryers annually. Hungary has the world's largest rabbitries. (Photo by David J. Harris; print courtesy of Pel-Freez Rabbit Meat, Inc., Rogers, Ark.)

Commercial rabbit production in the United States began about 1900, when Belgian Hares were imported from England. During this period, rabbitries sprang up at a tremendous rate, centering primarily in southern California. The demand for quality breeding stock during this time was so great that rabbits were sold at highly inflated prices—in the thousands of dollars. As with most economic booms, the bubble burst, and rabbits dwindled to their present status as a minor segment of the animal industry.

Fig. 26-2. Angora rabbits in China, used for wool production. The better types of Angoras will produce from 2 to 2½ in. of wool in approximately 11 weeks, and an annual growth of 8 to 10 in. weighing 12 to 16 oz. When fed properly balanced rations, 100 lb of feed will, on the average, produce 1 lb of wool. (Photo made in China by Audrey H. Ensminger)

Through the years, rabbits have been accorded little attention in American agriculture; yet, large numbers of them are being used in medicinal and biological research, and for meat, fur, and pets, and as a hobby. Very little is known about their nutrient requirements, and little research is presently being conducted concerning their nutritional needs.

While rabbit production in the United States is rather limited in comparison with other animal industries, it warrants serious consideration because about 200,000 people are either directly or indirectly engaged in it, and because of its potential.

Commercial operations produce rabbits primarily for meat, with much of the demand centered around the ethnic eating habits of immigrants in urban areas. Prices for fur are currently so low that it should be considered only as a supplemental income. The vast majority of rabbits are raised by part-time backyard operators as a sideline or hobby.

Fig. 26-3. An outdoor rabbitry suitable for backyard production. (Courtesy, Peter R. Cheeke, Oregon State University, Corvallis)

Commercial rabbit raising is an extremely intensive form of animal production. Californian and New Zealand Whites are the most widely used breeds for meat production because of their large litters, good mothering ability, large size, and the preference of many processors for white rabbits. New Zealand Whites and Florida Whites are the most widely used rabbits for research. The American Rabbit Breeders Association recognizes more than 50 different breeds.

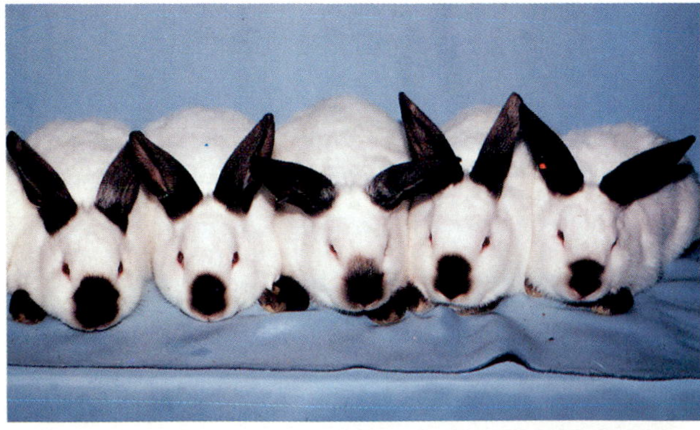

Fig. 26-4. Californian rabbits, a major meat breed. Compared to the New Zealand Whites, the Californian is somewhat smaller and is finished out at a lighter weight. (Courtesy, Pel-Freez Rabbit Meat, Inc., Rogers, Ark.)

Most full-time commercial operations involving a family with a minimum of hired help require 500 to 1,000 does in order to be an economic unit. In the past, rabbit producers routinely produced 4 litters per doe annually. Today, with the high costs of labor, feed, and facilities, most commercial operations must produce 8 litters per doe annually to make a profit. To attain the 8 litters per doe level of production, young rabbits are weaned at 4 to 5 weeks of age and raised separately from their mother until they reach market weight, generally at 8 to 9 weeks of age. These intensive breeding programs require that does be rebred anywhere from 7 to 14 days after kindle (after giving birth). Since 1 doe can produce 8 litters per year, she has the ability to produce from 130 to 250 lb of live weight per year through her reproductive ability. In large operations, 1 buck is kept for as many as 20 to 30 does and produces more than 1,000 offspring annually. From this, it may be concluded that 4 does can produce 700 lb live weight, or 400 lb of dressed carcass weight, annually—as much as an average beef cow can produce in 1½ years—and produce it on less feed.

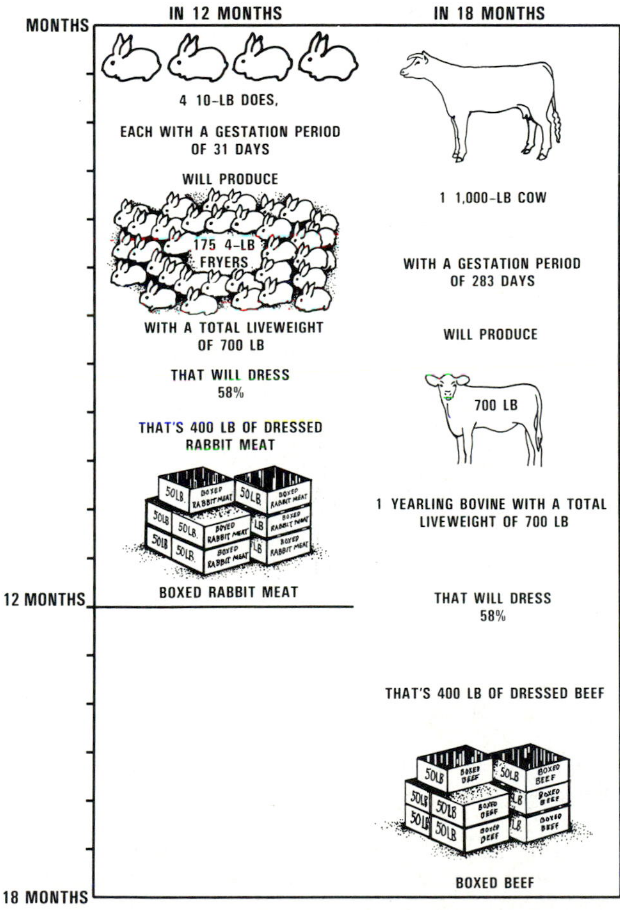

Fig. 26-5. Four 10-lb does will produce 400 lb of dressed meat in 12 months, whereas it takes one 1,000-lb cow 18 months to produce the same amount of dressed (carcass) meat. And that's not all! Rabbits produce the same amount of meat on less feed; and the ready-to-eat yield of rabbit meat after deboning and cooking is 80% of the initial dressed (carcass) weight vs 49% for the beef animal.

Also, it is noteworthy that rabbits have the potential for a constant state of reproduction. Does can be rebred on the day they kindle, and the young can be weaned at 28 days. Since the gestation period is 31 days, a doe can have a litter 3 days after weaning. Thus, it is theoretically possible to have a doe with three litters simultaneously: one weaned litter (4 weeks), one litter in the nest box, and one litter in utero. **NOTE WELL**: Such intensive breeding is still experimental and is not recommended until further studies are completed.

The meat of the domestic rabbit is white, high in protein (25%), low in fat (4%), and low in cholesterol and sodium—facts that should please the discriminating American consumer. When the public is made aware of the high quality of rabbit meat, consumption of rabbit should increase; but this will happen only after vigorous educational promotion by the industry, extolling the virtues of rabbit meat and telling how to prepare and serve it.

Rabbits are also used extensively in biomedical teaching and research. Over 600,000 rabbits are used annually for these purposes, helping researchers investigate a host of maladies, including cardiac diseases, hypertension, antibody production, endocrinology, venereal disease, and virology. Instructors find rabbits invaluable for laboratory demonstrations. Since large numbers of adult rabbits are needed for these purposes, a market exists. But producing for such an outlet is risky as the demand is unpredictable. The producer must have on hand enough rabbits to meet the sporadic demands, yet, at the same time, keep the number of adult breeder rabbits at as low a level as possible in order to minimize feed costs.

Despite their prolificacy, rapid growth rate, good feed conversion, use of humanly inedible foods, and nutritious meat, there are many failures in the rabbit business; primarily, because of the intensive labor requirements, disease problems, and unstable markets.

ECONOMIC IMPORTANCE OF FEED FOR RABBITS

Rabbit feed is very important nutritionally and economically.

Confined rabbit production precludes the choice of feeds by rabbits. It follows that big differences in their performance can be expected as a result of different diet compositions and feeding methods.

Annually, an estimated 250,000 tons of commercial feeds are fed to the more than 10 million rabbits produced in the United States. Most rabbitries use commercially pelleted rabbit feeds.

As a meat-type animal, the rabbit compares very favorably in feed conversion with the more traditional animals. With a balanced ration, feed conversions of 3:1 can be obtained in fryer rabbits. While this conversion rate is not as favorable as the 2.1:1 achieved in broilers, it does compare very favorably with the steer (9:1). Protein efficiency in rabbits is approximately 6:1, compared to 1.9:1 in broilers and

10.6:1 in steers. (See Chapter 1, Food and Animals—a global perspective, Table 1-2, for the feed to food efficiency rating of rabbits in comparison with other animal species.)

Since the rabbit possesses an enlarged cecum, similar to that of the horse, approximately 40% of commercial rabbit rations consist of alfalfa or other roughage material. So, it follows that rabbits are not in direct competition with humans for foodstuffs. In the future, if competition for human food becomes more acute, the use of roughages will assume added importance in the livestock industry; and the rabbit will out-compete many other animals in the ability to utilize roughages.

NUTRITIVE NEEDS OF RABBITS

Fig. 26-6. This shows the growth rate of rabbits from 14 to 56 days of age. Growth is influenced primarily by nutrient intake; and the nutritive requirements become increasingly acute when young are in forced production. (Photo by D. J. Harris. Print courtesy Wm. K. Sanchez, Manna Pro Corp., Portland, Ore.)

Rabbits are nonruminant herbivores. The digestive anatomy and physiology of the rabbit closely resemble those of the horse, and in many ways the nutritive requirements are similar. Several differences, such as the habit of coprophagy (the ingestion of fecal material) and decreased fiber utilization in rabbits, alter the requirements somewhat. Nevertheless, the types of feeds used for rabbits and horses are very similar.

Rabbits excrete two types of feces—hard and soft. The hard feces (or day feces, containing about 40% water), which are produced in the large intestine, are the fecal pellets which are most commonly seen. The soft feces (or night feces, containing about 70% water), are produced in the cecum, excreted in grape-like clusters, and consumed directly from the anus in a peculiar type of coprophagic behavior displayed by rabbits as early as 3 weeks of age.

NOTE WELL: Night feces is really a misnomer, because soft feces are often produced during the day.

Coprophagy by rabbits plays an important role in the modification of their nutritive requirements. It has been

Fig. 26-7. Two types of feces. *Left:* regular hard pellets. *Right:* soft feces in a grapelike cluster. (Courtesy, Pel-Freez Rabbit Meat, Inc., Rogers, Ark.)

speculated that the act of coprophagy may aid in the absorption of some of the essential amino acids and certain vitamins—including vitamin K and the B complex vitamins. By recycling the digesta, certain feeds are digested and absorbed that were not utilized the first time. The small intestine is the main site of absorption for these nutrients, but a good deal of this synthesis takes place in the cecum. Since the cecum lies behind the small intestine, much of that which is digested in the cecum is not absorbed but is passed in the feces. Coprophagy enables the rabbits to recycle these nutrients which may not have been absorbed in the cecum or the large intestine. In addition to the microbial synthesis of many of these nutrients, some fiber digestion takes place in the cecum.

National Research Council (NRC) Requirements

Although rabbits have been used extensively in biological and medical research, very limited research data have been published concerning their nutritive needs. However, the National Research Council (NRC) has established requirements for energy, crude protein, crude fiber, essential amino acids for growth, and some minerals and vitamins for rabbits. These are given in Table 26-1. The NRC requirements should be used only as guides; they are subject to further refinement as the results of additional research become available.

A prominent French nutritionist, F. Lebas, published his updated version of the nutrient requirements for rabbits herewith presented in Table 26-2. Note that Lebas lists a requirement for *indigestible fiber,* based on his and other studies indicating that indigestible fiber aids in preventing enteritis.

The nutritional requirements of rabbits differ according to type and level of production; hence, both Tables 26-1 and 26-2 make provisions for such differences. Additionally, the nutritional requirements are affected by (1) breeds, largely as a matter of size and feed capacity, (2) environmental temperature, and (3) stress.

TABLE 26-1
NUTRIENT REQUIREMENTS OF RABBITS FED AD LIBITUM (PERCENTAGE OR AMOUNT PER LB OR KG OF DIET)[1]

Nutrients[2]	Unit	Growth		Maintenance		Gestation		Lactation	
		(lb)	*(kg)*	(lb)	*(kg)*	(lb)	*(kg)*	(lb)	*(kg)*
Energy and protein:									
TDN	(%)	65	*65*	55	*55*	58	*58*	70	*70*
Digestible energy	(kcal)	1,134	*2,500*	952	*2,100*	1,134	*2,500*	1,134	*2,500*
Fat[3]	(%)	2	*2*	2	*2*	2	*2*	2	*2*
Crude protein	(%)	16	*16*	12	*12*	15	*15*	17	*17*
Crude fiber[3]	(%)	10-12	*10-12*	14	*14*	10-12	*10-12*	10-12	*10-12*
Minerals:									
Calcium	(%)	0.4	*0.4*	—	—[4]	0.45	*0.45*[3]	0.75	*0.75*[3]
Phosphorus	(%)	0.22	*0.22*	—	—[4]	0.37	*0.37*[3]	0.5	*0.5*
Magnesium	(mg)	136-181	*300-400*	136-181	*300-400*	136-181	*300-400*	136-181	*300-400*
Potassium	(%)	0.6	*0.6*	0.6	*0.6*	0.6	*0.6*	0.6	*0.6*
Sodium[3,5]	(%)	0.2	*0.2*	0.2	*0.2*	0.2	*0.2*	0.2	*0.2*
Chlorine[3,5]	(%)	0.3	*0.3*	0.3	*0.3*	0.3	*0.3*	0.3	*0.3*
Copper	(mg)	1.4	*3.0*	1.4	*3.0*	1.4	*3.0*	1.4	*3.0*
Iodine[3]	(mg)	0.1	*0.2*	0.1	*0.2*	0.1	*0.2*	0.1	*0.2*
Iron[4]		—	—	—	—	—	—	—	—
Manganese	(mg)	3.9	*8.5*	1.1	*2.5*	1.1	*2.5*	1.1	*2.5*
Zinc[4]		—	—	—	—	—	—	—	—
Vitamins:									
Vitamin A	(IU)	263	*580*	—	—[4]	4,545[6]	*10,000*[6]	—	—[4]
Vitamin A as carotene	(mg)	0.38	*0.83*[3,7]	—	—[8]	0.38	*0.83*[3,7]	—	—[8]
Vitamin D[9]		—	—	—	—	—	—	—	—
Vitamin E	(mg)	18	*40*[10]	—	—[4]	18	*40*[10]	18	*40*[10]
Vitamin K	(mg)	—	—[11]	—	—[11]	0.1	*0.2*[3]	—	—[11]
Choline	(g)	0.5	*1.2*[3]	—	—[12]	—	—[12]	—	—[12]
Niacin	(mg)	82	*180*	—	—[12]	—	—[12]	—	—[12]
Vitamin B-6	(mg)	18	*39*	—	—[12]	—	—[12]	—	—[12]
Amino acids:									
Arginine	(%)	0.6	*0.6*	—	—[9]	—	—[9]	—	—[9]
Glycine	(%)	—	—[4]	—	—[9]	—	—[9]	—	—[9]
Histidine	(%)	0.3	*0.3*	—	—[9]	—	—[9]	—	—[9]
Isoleucine	(%)	0.6	*0.6*	—	—[9]	—	—[9]	—	—[9]
Leucine	(%)	1.1	*1.1*	—	—[9]	—	—[9]	—	—[9]
Lysine	(%)	0.65	*0.65*	—	—[9]	—	—[9]	—	—[9]
Methionine + cystine	(%)	0.6	*0.6*	—	—[9]	—	—[9]	—	—[9]
Phenylalanine + tyrosine	(%)	1.1	*1.1*	—	—[9]	—	—[9]	—	—[9]
Threonine	(%)	0.6	*0.6*	—	—[9]	—	—[9]	—	—[9]
Tryptophan	(%)	0.2	*0.2*	—	—[9]	—	—[9]	—	—[9]
Valine	(%)	0.7	*0.7*	—	—[9]	—	—[9]	—	—[9]

[1]Adapted by the authors from *Nutrient Requirements of Rabbits,* 2nd rev. ed., NRC-National Academy of Sciences, 1977, p. 14, Table 1.

[2]Nutrients not listed indicate dietary need unknown or not demonstrated.

[3]May not be minimum but known to be adequate.

[4]Quantitative requirement not determined, but dietary need demonstrated.

[5]May be met with 0.5% NaCl.

[6]Based on studies at the Oregon State University Rabbit Research Center.

[7]Converted from amount per rabbit per day using an air-dry feed intake of 60 g per day for a 1-kg rabbit.

[8]Quantitative requirement not determined.

[9]Probably required, amount unknown.

[10]Estimated.

[11]Intestinal synthesis probably adequate.

[12]Dietary need unknown.

TABLE 26–2
NUTRIENT REQUIREMENTS OF RABBITS[1]
(IN PERCENTAGE, PPM, OR KCAL/KG OF DIET)

Nutrients	Unit	Class of Rabbit				
		Growing 4–12 Weeks	Maintenance	Gestation	Lactation	Does and Litters Fed One Diet
Energy and protein:						
Digestible energy	kcal/kg[2]	2,500	2,200	2,500	2,700	2,500
Metabolizable energy	kcal/kg[2]	2,400	2,120	2,400	2,600	2,410
Fat	%	3	3	3	5	3
Crude protein	%	15	13	18	18	17
Crude fiber	%	14	15–16	14	12	14
Indigestible fiber	%	12	13	12	10	12
Minerals:						
Calcium	%	0.5	0.6	0.8	1.1	1.1
Phosphorus	%	0.3	0.4	0.5	0.8	0.8
Magnesium	%	0.03	—	0.04	0.04	0.04
Potassium	%	0.8	—	0.9	0.9	0.9
Sodium	%	0.4	—	0.4	0.4	0.4
Chlorine	%	0.4	—	0.4	0.4	0.4
Sulfur	%	0.04	—	—	—	0.04
Cobalt	ppm	1	—	—	1	1
Copper	ppm	5	—	—	5	5
Iodine	ppm	0.2	0.2	0.2	0.2	0.2
Iron	ppm	50	50	50	50	50
Manganese	ppm	8.5	2.5	2.5	2.5	8.5
Zinc	ppm	50	—	70	70	70
Vitamins:						
Vitamin A	IU/kg[2]	6,000	—	12,000	12,000	10,000
Carotene	ppm	0.83	—	0.83	0.83	0.83
Vitamin D	IU/kg[2]	900	—	900	900	900
Vitamin E	ppm	50	50	50	50	50
Vitamin K	ppm	0	0	2	2	2
Folacin (Folic Acid)	ppm	1	0	0	—	—
Pantothenic Acid (Vitamin B-3)	ppm	20	0	0	—	—
Riboflavin (Vitamin B-2)	ppm	6	0	0	—	4
Thiamin (Vitamin B-1)	ppm	2	0	0	—	2
Vitamin B-6 (Pyridoxine, Pyridoxal, Pyridoxamine)	ppm	40	0	0	—	2
Vitamin B-12 (Cobalamins)	ppm	0.01	0	0	0	—
Vitamin C (Ascorbic Acid, Dehydro-ascorbic Acid)	ppm	0	0	0	0	0
Amino acids:						
Arginine	%	0.9	—	—	0.8	0.9
Histidine	%	0.35	—	—	0.43	0.4
Isoleucine	%	0.60	—	—	0.70	0.65[3]
Leucine	%	1.05	—	—	1.25	1.2
Lysine	%	0.6	—	—	0.75	0.7
Phenylalanine[3]	%	1.20	—	—	1.40	1.25
Sulfur amino acids	%	0.5	—	—	0.6	0.55
Threonine	%	0.55	—	—	0.7	0.6
Tryptophan	%	0.18	—	—	0.22	0.2
Valine	%	0.70	—	—	0.85	0.8

[1]Adapted by the authors from Lebas, F., *The Journal of Applied Rabbit Research,* vol. 3, no. 2, 1980, p. 15.

[2]To convert per lb basis, divide by 2.2.

[3]Values from Cheeke, *et al., Rabbit Production,* 1987, p. 168.

Energy

Rabbit production is intensive in nature, and the energy demands on the doe are high if she is to produce up to eight litters a year. Likewise, growth of young juniors creates a high-energy requirement.

Carbohydrates (starch and cellulose) and fats are the primary sources of energy for rabbits. Starch is found in cereal grains and tubers; cellulose is the structural component (fiber) of plants.

Good-quality legume hays and supplemental concentrates or commercial pellets are routinely used to supply a high level of energy during peak production periods. The energy requirements for maintaining dry does or young bucks not in service are low, with the result that good-quality hay should be sufficient.

While no requirement for dietary fat has been established, rations routinely contain 2 to 5% fat. It has been suggested that higher fat levels might be feasible, but caution should be exercised in order to prevent digestive disturbances, such as scours. Also, high levels of fat may reduce feed pellet quality, resulting in pellets that break apart easily.

Protein

All production parameters—fur, growth, reproduction, and lactation—require high levels of good-quality protein. Workers at the Arkansas Station recently completed research showing that gestating/lactating does have a very high dietary protein requirement—much higher than the NRC requirement shown in Table 26-1; does fed a diet that contained 26% crude protein the last 2 weeks of gestation and during early lactation produced kits that were 30% heavier and had higher quality fur than does fed an 18% crude protein ration during late gestation and early lactation.[2]

Rabbits require certain amino acids in the diet. The essential amino acid profile is very similar to that of the chick and the pig. Table 26-1 shows the essential amino acid requirements for the growing rabbit. Of the essential amino acids, lysine is most likely to be deficient in rabbit feeds.

Fig. 26-8. Protein quality, affected by processing, made the difference! *Left:* rabbit fed raw soybeans. *Right:* rabbit fed solvent extracted soybean meal. The raw beans contained more fat and fat soluble vitamins than the processed meal. (Photo by D. J. Harris, Pel-Freez Rabbit Meat, Inc., Rogers, Ark. Print courtesy Wm. K. Sanchez, Manna Pro Corp., Portland, Ore.)

[2]Unpublished research. Reported in a personal communication from Professor L. B. Daniels, The University of Arkansas, to the senior author, 1986.

Studies have shown that nonprotein nitrogen (NPN), such as urea, is of little value in rabbit rations. This is attributed to the fact that NPN sources are degraded and absorbed in the small intestine and subsequently eliminated as waste products before the NPN ever reaches the cecum where it might be transformed into bacterial protein.

Coprophagy has been shown to increase the biological value of certain low-quality proteins. There are indications that some increase in value is observed with the coprophagy of high-quality protein, but the magnitude of the increase is not large. Confirmationn of the benefits of coprophagy in protein utilization in rabbits is presented in Table 26-3. Casein was compared to gelatin in two feeding systems; in one system the rabbits were allowed to consume their feces, in the other they were not. Rabbits on the casein diet ate more than those on the gelatin diet. Rabbits that were allowed to consume their feces demonstrated better protein utilization in both diets, but the resulting increase in apparent biological value of the gelatin diet as a result of coprophagy was quite dramatic.

TABLE 26-3
EFFECT OF COPROPHAGY ON VOLUNTARY INTAKE AND PROTEIN PARAMETERS OF MATURE RABBITS[1]

Source of Protein[2]	Dry Matter Consumption	Digestion Coefficient of Nitrogen	Nitrogen Balance	Apparent Biological Value
	(g/day/$W_{kg}^{.75}$)	(%)	(mg/day/$W_{kg}^{.75}$)	
Casein—coprophagy allowed	34.2	96.1	458.3	37.8
Casein—coprophagy prevented	27.2	88.3	236.1	26.8
Gelatin—coprophagy allowed	19.8	87.9	89.4	12.8
Gelatin—coprophagy prevented	17.1	70.7	−186.2	−48.6

[1]Adapted by the authors from Kennedy, L. G., and T. V. Hershberger, "Protein Quality for the Nonruminant Herbivore," *Journal of Animal Science,* 1974, 39:510.

[2]*Casein diet:* casein, 22.25%; corn starch, 18.62%; glucose monohydrate, 25.62%; alphacel, 15.51%; corn oil, 10.00%; vitamin and mineral mix, 8%.

Gelatin diet: gelatin, 19.22%; corn starch, 18.62%;glucose monohydrate, 25.62%; alphacel, 18.54%; corn oil, 10.00%; vitamin and mineral mix, 8%.

Legumes are excellent sources of protein, and alfalfa is used extensively in rabbit rations. Oilseed meals are widely used as protein supplements when high-protein levels are required. Animal and fish products are seldom included in rabbit diets because of their high costs.

Minerals

Little research has been reported concerning the mineral requirements of rabbits. It is generally thought that the requirements involve the same elements as are required by other animals; but for many elements, quantitative requirements have not yet been established.

RABBIT MINERAL CHART

A summary of the various minerals, along with their respective requirements and nutritional considerations, is presented in Table 26–4.

(Also see Chapter 4, Nutrients/ Metabolism, Table 4–6, Animal Mineral Chart.)

TABLE
RABBIT MIN

Minerals Which May Be Deficient Under Normal Conditions	Conditions Usually Prevailing Where Deficiencies Are Reported	Functions of Mineral	Some Deficiency Symptoms
Major or macrominerals:			
Salt (NaCl)	Negligence, for salt is cheap.	Sodium and chlorine help maintain osmotic pressure in body cells, upon which depends the transfer of nutrients to the cells and the removal of waste materials. Also, sodium is important in making bile, which aids in the digestion of fats and carbohydrates; and chlorine is required for the formation of hydrochloric acid in the gastric juice so vital to protein digestion.	Depressed growth.
Calcium (Ca)	Rations of grass hay and farm grains.	Essential for development and maintenance of normal bones and teeth. Important in blood coagulation and lactation. Enables heart, nerves, and muscles to function. Regulates permeability of tissue cells. Affects availability of phosphorus and zinc.	Rickets; tetany; brittle bones.
Phosphorus (P)	High legume forage ration without supplemental phosphorous.	Essential for sound bones and teeth, and for the assimilation of carbohydrates and fats. A vital ingredient of the proteins in all body cells. Necessary for enzyme activation. Acts as a buffer in blood and tissue. Occupies a key position in biologic oxidation, and reactions requiring energy.	Rickets; tetany; brittle bones.
Magnesium (Mg)		Necessary for many enzyme systems and for proper functioning of the nervous system. Closely associated with the metabolism of calcium and phosphorus.	Poor fur growth; fur chewing; hyperirritability.
Potassium (K)		Essential for proper enzyme, muscle and nerve function, cecal mircroorganism activity, and appetite.	Muscular dystrophy.
Trace or microminerals:			
Cobalt (Co)	Feeds from cobalt-deficient areas.	Constituent of vitamin B–12.	Anemia.
Copper (Cu)		Copper, along with iron, is necessary for hemoglobin formation, although it forms no part of the hemoglobin molecule of red blood cells.	Anemia; graying of hair.
Iodine (I)	Feeds grown in iodine-deficient areas.	Needed for the production of thyroxin, an iodine-containing hormone that regulates metabolic rate.	
Iron (Fe)		Necessary for formation of hemoglobin, an iron-containing compound which enables the blood to carry oxygen. Also, important to certain enzyme systems.	Microcytic and hypochromic anemia.
Manganese (Mn)	Excess calcium and phosphorus decreases absorption of manganese.	Considered essential in utilization of calcium and phosphorus, for proper functioning of mammary glands and normal reproduction.	Abnormalities of the skeletal system.
Zinc (Zn)		Component of several enzyme systems and also required for normal protein synthesis.	Weight loss, alopecia, graying of hair, dermatitis, low hematocrit, reproductive problems.

[1]As used herein, the distinction between "mineral requirements" and "recommended allowances" is as follows: In mineral requirements, no margins of safety are included intentionally, whereas in recommended allowances, margins of safety are provided in order to compensate for variations in feed composition, environment, and possible losses during storage or processing. Where preceded by an asterisk, the mineral requirements, allowances, and other facts presented herein were taken from *Nutrient Requirements of Rabbits,* 2nd rev. ed., NRC-National Academy of Sciences, 1977.

MAJOR OR MACROMINERALS

The major or macrominerals needed with regard to rabbit nutrition are salt (soldium chloride), calcium, phosphorus, magnesium, and potassium.

Salt (NaCl)

In rabbit operations, salt is generally included in the ration at the rate of .5 to 1%. In areas that are known to be iodine deficient, iodized salt is routinely used. Rabbits enjoy salt,

26–4
ERAL CHART

Mineral Requirements[1] Mineral Content of Ration	Recommended Allowances[1]		Practical Sources of the Mineral	Comments
	Daily Nutrients/ Animal	Percent of Total Ration		
*0.5%.		0.5–1.0	Salt spools. Can be added to feed.	Sodium and chlorine are low in feeds of plant origin. There is little danger of overfeeding salt unless a salt-starved animal is suddenly given access to too much salt or if liberal amounts of water are not available.
Variable according to age and production (see Table 26–1).		0.4–1.0	Ground limestone or oystershell flour. When both Ca and P are needed, use bone meal, dicalcium phosphate, or defluorinated phosphate.	The Ca:P ratio should be maintained close to 1:1, although 2:1 is acceptable when the higher calcium content is due to the presence of legume. Where there is a shortage of calcium in the ration, it is withdrawn from the bones.
Variable according to age and production (see Table 26–1).		0.22–0.50	Monosodium phosphate. Monoammonium phosphate. When both Ca and P are needed, use bone meal, dicalcium phosphate, or defluorinated phosphate.	(Same as stated for Ca under "Comments" above.) If plenty of vitamin D is present, the ratio of Ca:P becomes less important. Apparently phosphorus cannot be withdrawn from the bone.
*300–400 ppm.		0.03–0.04	Magnesium sulfate. Magnesium oxide.	In high Ca diets, Mg may become deficient due to interference from the Ca in absorption.
*0.6%.		0.6	Potassium chloride. Roughages.	
			Cobalt chloride, cobalt sulfate, cobalt oxide, or cobalt carbonate.	Cobalt-deficient areas are: Fla., Mass., Mich., N.H., N.Y., N.C., Pa., S.C., and Wisc. Also, Australia and western Canada.
*3 ppm.		0.00003	Copper carbonate. Copper sulfate.	Trace mineralized salt containing copper is satisfactory.
*0.2 ppm.		0.2 ppm	Iodized salt.	
		0.0050	Trace mineralized salt.	
Variable according to age and production (see Table 26–1).	1 mg for growing animals; 0.3 mg for mature animals.	0.000025–0.000085	Trace mineralized salt.	Manganese is needed for growth and reproduction of most animals.
			Zinc carbonate. Zinc sulfate.	

and if a salt spool is made available, they will satisfy their salt requirements. One major disadvantage to using salt spools—especially in all-wire hutches—is that the salt is corrosive to the metal. Commercially prepared feed probably contains adequate salt, but the producer should check the feed label to make sure.

Calcium (Ca) and Phosphorus (P)

Calcium absorption in the rabbit is extremely efficient; and an excess of calcium in the diet may affect the requirements of other minerals—notably magnesium and phosphorus. The suggested dietary calcium levels have been set at 0.4% for growth, 0.45% for gestation, and 0.75% for lactation. The recommended levels of phosphorus are 0.22% for growth, 0.37% for gestation, and 0.5% for lactation. When formulating rations, the producer should make sure that the calcium to phosphorus ratio falls in the range of 1:1 to 1.5:1. When the ratio falls below 1:1 or rises above 1.5:1, imbalances can occur.

At Washington State University, in a study with rabbits, the effect of soil phosphorus on plants, and, in turn, the effects of these plants on animals, was established.[3] Generation after generation, rabbits were fed alfalfa, with one group receiving hay produced on low-phosphorus soils and the other group eating alfalfa grown on high-phosphorus soils. The rabbits in the low-phorphorus, soil-alfalfa group (1) were retarded in growth—with 9.8% lower weaning weights, (2) required 12% more matings per conception, and (3) had a 47% lower breaking strength of bones than the rabbits on the high-phosphorus soil-alfalfa group.

Fig. 26-9. Rabbit with bowed legs and enlarged joints resulting from eating alfalfa produced on low-phosphorus soils. (Courtesy, Dr. Wilton W. Heinemann, Washington State University, Prosser)

The following general characteristics of feeds in regard to calcium and phosphorus are important in rabbit rations:

1. The cereal grains and their by-products and straws, dried mature grasses, and protein supplements of plant origin are low in calcium.
2. The protein supplements of animal origin and legume forage are rich in calcium.
3. The cereal grains and their by-products are fairly high or even rich in phosphorus.
4. Almost all protein-rich supplements are high in phosphorus.
5. Beet by-products and dried, mature nonleguminous forages (such as grass hays and fodders) are likely to be low in phosphorus.
6. The calcium and phosphorus content of plants can be increased through fertilizing the soil upon which they are grown.

Fig. 26-10. Milk supplement top-dressed on rabbit feed. Milk is ideal for balancing out the calcium and other deficiencies of cereal grains. (Courtesy, Pel-Freez Rabbit Meat, Inc., Rogers, Ark.)

It is noteworthy that from one-half to two-thirds of the phosphorus in seeds and grains, and their by-products, is present as phytate, the salt of myoinositol hexaphosphate, which is unavailable to monogastric animals. But phytate can be hydrolyzed by ruminants because of the presence of microbial phytase in the rumen. Arkansas researchers reported that phytate phosphorus is hydrolyzed in the digestive tract of, and made available to, rabbits (pseudo-ruminants) much like it is in ruminants; thus, formulas for rabbit rations can be based on total phosphorus, rather than on available phosphorus.[4]

(Also see Chapter 4, Nutrients/Metabolism, Table 4-6, Animal Mineral Chart—Phosphorus.)

In considering the calcium and phosphorus requirements of rabbits, it is important to realize that the proper utilization of these minerals by the body is dependent upon 3 factors: (1) an adequate supply of calcium and phosphorus, (2) a suitable ratio between them (somewhere between 1 to 2 parts of calcium to 1 to 2 parts of phosphorus), and (3) sufficient vitamin D to make possible the assimilation and utilization of the calcium and phosphorus. Many "cure-alls" and commercial mineral mixtures fail to take all these factors into consideration.

If plenty of vitamin D is present (as provided either by sunlight or through the ration), the ratio of calcium to phosphorus becomes less important. Also, less vitamin D is needed when there is a desirable calcium-phosphorus ratio.

Toxicities in rabbits due to excess calcium or phosphorus are rare.

[3]Heinemann, W. W., *et al.*, *Wash. Ag. Exp. Sta., Tech. Bull. 24,* June 1957.

[4]Nelson, T. S., L. B. Daniels, L. A. Schriver, and L. K. Kirby, "Hydrolysis of Phytate Phosphorus by Young Rabbits," *Arkansas Farm Research,* July-August, 1985, p.8.

Magnesium (Mg)

Rabbits suffering from the lack of dietary magnesium exhibit poor growth, fur-chewing habits, and hyperirritability. A number of minerals are interrelated; so what might seem to be an adequate level of magnesium may actually be a deficiency due to an excessive amount of another mineral with which it is competing for absorption and utilization—for example, calcium. Under normal conditions, .03 to .04% magnesium in the ration should be adequate.

Excess magnesium in the form of magnesium sulfate (Epsom Salts) will cause severe diarrhea.

Potassium (K)

A form of muscular dystrophy which resembles a deficiency of vitamin E occurs in rabbits suffering from a deficiency of potassium. It has been suggested that the ration should contain about .6% potassium. Generally, a diet consisting of 50% roughage is adequate in fulfilling the potassium requirement of rabbits.

TRACE OR MICROMINERALS

Work concerning the role of trace minerals in the nutrition of the rabbit has been somewhat limited, centering primarily on cobalt, iodine, iron, copper, manganese, and zinc.

Cobalt (Co)

Vitamin B-12 contains cobalt as the central element of a complex porphyrin structure. Since the microorganisms in the cecum of the rabbit have the ability to synthesize vitamin B-12, all of the precursors—including cobalt—must be supplied. At present, no quantitative requirement for cobalt by rabbits has been established.

In the United States, cobalt deficiency has been reported in Florida, Massachusetts, Michigan, New Hampshire, New York, North Carolina, Pennsylvania, South Carolina, and Wisconsin. Also, deficiencies occur in Australia and in western Canada (see Chapter 5, Fig. 5-30).

In cobalt-deficient areas, most commercial rabbit pellets contain approximately 1.0 ppm cobalt.

Iodine (I)

The nutritive needs for iodine have not been determined for rabbits. However, it is recommended that iodized salt be used routinely, especially if the operation is in an area known to be deficient in iodine. Rations should contain at least 0.1 mg of iodine per pound.

Iron (Fe) and Copper (Cu)

If there are dietary deficiencies of either of these elements, anemia will result. Requirements for iron have yet to be established, but an iron deficiency in rabbits is unlikely.

The NRC has established a copper requirement of 3 ppm for rabbit diets. A copper deficiency will cause a graying of black fur, due to the role of copper in the synthesis of melanin, a pigment in hair and fur.

Manganese (Mn)

The manganese requirement for rabbit rations is 3.9 mg/lb in growing rations, and 1.1 mg/lb in maintenance, gestation, and lactation rations.

Deficiencies of manganese in rabbits have been well documented. Bone deformities such as crooked legs, brittleness, decreased bone weight, decreased ash content, and smaller size bone have been observed in manganese deficient rabbits.

Selenium (Se)

No selenium deficiency symptoms have ever been demonstrated in rabbits. Instead, the rabbit relies on vitamin E to prevent peroxide formation.

Zinc (Zn)

The requirements for zinc in rabbits have not been quantitated, but the need for dietary zinc has been clearly established. Zinc functions as a component of several enzyme systems and is involved in the transfer of carbon dioxide in red blood cells, in calcification of bones, and in the synthesis and metabolism of proteins and nucleic acids. Symptoms of a zinc deficiency in rabbits include lowered feed consumption, weight loss, loss of hair, dermatitis, lowered hematocrit, and reproductive problems.

Fig. 26-11. Weighing experimental rabbits in a feeding trial. (Courtesy, Pel-Freez Rabbit Meat, Inc., Rogers, Ark.)

Vitamins

As with mineral nutrition in rabbits, the amount of information concerning the vitamin requirements is very limited.

RABBIT VITAMIN CHART

Table 26–5, Rabbit Vitamin Chart, lists the vitamins that have been studied in rabbits, and gives pertinent information pertaining to each.

(Also see Chapter 4, Nutrients/Metabolism, Table 4–8, Animal Vitamin Chart.)

TABLE
RABBIT VITA

Vitamins Which May Be Deficient Under Normal Conditions	Conditions Usually Prevailing Where Deficiencies Are Reported	Functions of Vitamins	Some Deficiency Symptoms
Fat-soluble vitamins:			
A	When on high-forage rations that have been stored a long time and have lost their carotene value.	Promotes growth and stimulates appetite. Assists in reproduction and lactation. Keeps the mucous membranes of respiratory and other tracts in healthy condition. Makes for normal vision. Prevents night blindness.	Retarded growth, incoordination and paralysis, blindness, drooping ears, and hydrocephalus (enlarged head) of fetuses from vitamin A-deficient does.
D	In confinement rearing where the rabbits do not have access to sunlight or sun-cured hay.	Assimilation and utilization of calcium and phosphorus, necessary in normal bone development—including the bones of the fetus.	Rickets.
E		Serves as insurance against destruction of vitamin A. Makes for improved reproduction. Protection of cellular lipids from oxidation.	Nutritional muscular dystrophy of skeletal and cardiac muscle, paralysis, and fatty livers.
K	Intestinal disorders or when antibiotics are used.	Concerned with blood coagulation.	Prolonged bleeding following a minor injury and abortion and placental hemorrhage in does.
Water-soluble vitamins:			
Biotin	The presence in feeds of avidin, a biotin antagonist.	Important in the metabolism of carbohydrates, fats, proteins.	Loss of hair and dermatitis.
Choline	Rations low in methionine, an amino acid.	Prevention of fatty livers, transmitting nerve impulses, and the metabolism of fat.	Depressed growth, fatty and cirrhotic liver, and necrotic kidneys.
Folacin (Folic Acid)		Essential for normal growth and reproduction, for the prevention of blood disorders, and for important biochemical mechanisms within each cell.	
Niacin (Nicotinic Acid, Nicotinamide)		Constituent of coenzymes which produce energy within the cells.	Loss of appetite, diarrhea, and emaciation.
Pantothenic Acid (Vitamin B-3)	No deficiency ever produced in rabbits.	Constituent of coenzyme A (CoA). It plays a key role in energy metabolism.	
Riboflavin (Vitamin B-2)	Destruction of riboflavin by light or by heat in an alkaline solution.	In the oxidative mechanisms in the cells.	Retarded growth and lowered feed efficiency.
Thiamin (Vitamin B-1)	Prolonged feeding of a thiamin-deficient diet.	A cofactor of certain enzymes involved in carbohydrate and fat metabolism.	Loss of appetite, muscle paralysis, and accumulation of pyruvic acid in the blood.
Vitamin B-6 (Pyridoxine, Pyridoxal, Pyridoxamine)		Key constituent of cofactors involving amino acid and energy metabolism.	Depressed growth, acrodynia (dermatitis), convulsions, and paralysis.
Vitamin B-12 (Cobalamins)	Lack of cobalt.	Two coenzyme forms—coenzyme B-12, and methyl B-12.	Retarded growth.

[1]As used herein, the distinction between "vitamin requirements" and "recommended allowances" is as follows: In vitamin requirements, no margins of safety are included intentionally, whereas in recommended allowances, margins of safety are provided in order to compensate for variations in feed composition, environment, and possible losses during storage or processing. Where preceded by an asterisk, the vitamin requirements, allowances, and other facts presented herein were taken from *Nutrient Requirements of Rabbits,* 2nd rev. ed., NRC-National Academy of Sciences, 1977.

26–5
MIN CHART

Vitamin Requirements[1] Vitamin Content of Ration	Recommended Allowances[1] Daily Nutrients Per Rabbit	Per Lb of Ration	Practical Sources of the Vitamin	Comments
Variable according to age and production (see Table 26–1).	23 mcg of carotene per pound of ration has been shown to prevent symptoms of vitamin A deficiency.	For does in production, 4,545 IU/lb feed.	Stabilized vitamin A. Green grass or legume.	Vitamin A is not synthesized in the cecum. A vitamin A level of 86,355 IU/lb of diet, which is only 19 times the vitamin A requirement, is toxic to rabbits.
	No allowances have been recommended for vitamin D supplementation.		Sun-cured hays. Exposure to sunlight. Commercially available supplements of either D_2 or D_3.	The vitamin D requirement is directly related to the Ca:P ratio and their respective levels. Vitamin D toxicity is of greater concern than deficiency in rabbits. Toxicity symptoms are loss of appetite, impaired movement, and calcification of soft tissues such as the kidneys and the arteries.
*18 mg/lb for rabbits in production.	0.5 mg/lb of body weight.	25 IU	Stabilized vitamin E. Germ or gum oils of plants. Green plants and hays. Cereals.	Selenium does not exert a vitamin E-sparing effect in rabbits; instead, rabbits depend entirely on vitamin E for protection against peroxides.
*0.1 mg/lb for gestating does.	No allowances have been recommended for vitamin K supplementation.		Synthetic vitamin K. Green grass. Well-cured hays.	Studies indicate that dietary vitamin K is required for reproduction but not growth.
			Widely distributed in nature.	Deficiencies are not associated with normal rations.
*0.5 g/lb for growing rabbit.		0.12%	Alfalfa hay and meal. Choline chloride. Rice polishings.	
			Alfalfa hay/meal and oil seed meals.	
*82 mg/lb for growing rabbit.	5 mg/lb of body weight.		Nicotinamide (synthetic). Nicotinic acid (synthetic). Rice polishings. Yeast (brewers', torula).	Limited amounts of niacin can be synthesized from the amino acid tryptophan. Supplemental niacin in some cases can increase growth rate.
			Calcium pantothenate (synthetic), rice polishings, yeast (brewers', torula), alfalfa hay and meal.	
Not known.			Synthetic riboflavin, alfalfa hay, whey, yeast.	
Not known.			Thiamin hydrochloride, rice polishings.	
*18 mg/lb for growing rabbit.		0.5 mg	Pyridoxine hydrochloride. Rice polishings. Alfalfa hay/meal.	Normally, rabbit rations contain adequate vitamin B-6.
			Meat animal by-products. Marine by-products.	

FAT-SOLUBLE VITAMINS

Vitamin A and Vitamin E are probably the only two vitamins for which there are serious needs for dietary supplementation.

Vitamin A

The vitamin A requirement of does in production is 4,545 IU/lb of feed. Generally, either a deficiency or a toxicity of vitamin A will affect the reproductive performance of the female before other symptoms can be recognized. Additional symptoms of vitamin A deficiency are: retarded growth, incoordination and paralysis, blindness, drooping ears, and hydrocephalus (enlarged head) of fetuses born to vitamin A-deficient does.

Research indicates that 23 mcg of carotene per pound of body weight per day are adequate to prevent the symptoms of vitamin A deficiency.

A vitamin A level of 86,355 IU/lb of diet, which is only 19 times the vitamin A requirement, is toxic to rabbits. Toxicity symptoms in does are: reproductive problems, including abortion; fetal resorption; small, weak litters with a high mortality the first week; and kits with hydrocephalus.

Fig. 26-12. Hydrocephalus in newborn rabbits caused by vitamin A toxicity. (Courtesy, Peter R. Cheeke, Oregon State University, Corvallis)

Vitamin D

If rabbits are exposed to sunlight for sufficient periods, this will supply their needs for vitamin D. Also, sun-cured alfalfa meal, a common feed, is an excellent source of vitamin D activity. Hence, vitamin D is rarely added to the ration. If rabbits are reared in confinement where access to sunlight is limited or are fed a ration containing a serious imbalance of calcium and phosphorus, there may be a need to supply vitamin D to the ration in either the D_2 or D_3 form. Both forms are equivalent in activity in rabbits. No quantitative requirement for vitamin D has been established for rabbits, but deficiency symptoms—rickets—have been produced experimentally.

Vitamin D toxicity is of greater concern than deficiency in rabbits. Toxicity symptoms are loss of appetite, impaired movement, and calcification of soft tissues such as the kidneys and the arteries.

Vitamin E

The amount of data pertaining to the requirement for vitamin E in rabbits is limited. However, the NRC suggests a level of 18 mg per pound of diet. Unlike other animal species, selenium does not exert a vitamin E-sparing effect in rabbits; instead, rabbits depend entirely on vitamin E for protection against peroxides.

Symptoms of a vitamin E deficiency in rabbits are muscular dystrophy, reproductive failure, and fatty livers.

Vitamin K

Bacterial synthesis within the cecum and subsequent coprophagy should, for the most part, supply the needs for vitamin K. However, research has indicated that vitamin K is required for reproduction. Hence, a level of 0.1 mg per pound of ration is suggested for pregnant does. If drugs are being used that could reduce the microflora of the gut or if rabbits are prevented from coprophagizing, supplemental vitamin K may be indicated.

Symptoms of vitamin K deficiency are prolonged bleeding following a minor injury, and abortion and placental hemorrhage in does.

WATER-SOLUBLE VITAMINS

Through bacterial synthesis in the cecum, along with coprophagy, the rabbit is able to fulfill many of the vitamin B complex requirements. The requirements for pantothenic acid, riboflavin, biotin, folic acid, and B-12 are met through this route. Research has indicated that some supplementation of niacin (11 mg/kg body weight), pyridoxine (39 ppm of the diet), and choline (.12% of the diet) will aid in production. Cobalt must be supplied in the diet as it is a precursor of vitamin B-12.

Biotin

A biotin deficiency in rabbits can be experimentally induced by the feeding of raw egg whites. Symptoms of this type of deficiency are a loss of hair and dermatitis. Since biotin is widely distributed in feed, deficiencies of this vitamin are rare.

Choline

The classification of choline as a vitamin is questionable, because (1) it is utilized in relatively large quantities, and (2) the body can synthesize considerable quantities of it.

Rabbits suffering from a deficiency of choline exhibit depressed growth, necrosis of the kidneys, and fatty, cirrhotic livers. These symptoms are prevented when the diet contains at least 0.12% choline.

Folacin (Folic Acid)

Folacin is the term used to designate folic acid and a group of closely related substances which are essential for all vertebrates. It is essential for normal growth and reproduction, for the prevention of blood disorders, and for important biochemical mechanisms within each cell. Normal rabbit feeds, including alfalfa hay/meal and the oil seed meals, are high in folacin. Deficiencies of folacin in rabbits are unknown.

Niacin (Nicotinic Acid, Nicotinamide)

Niacin serves in coenzymes to produce energy within the cells.

While substantial synthesis of niacin occurs in rabbits, niacin fed to rabbits at the level of 5 mg per pound of body weight can improve growth in some instances. Symptoms of a niacin deficiency in rabbits are diarrhea, loss of appetite, and emaciation.

Pantothenic Acid (Vitamin B-3)

Pantothenic acid is an important constituent of coenzyme A (CoA). It plays a key role in body metabolism.

No deficiency of pantothenic acid in rabbits has ever been produced.

Riboflavin (Vitamin B-2)

Riboflavin has an essential role in the oxidative mechanisms in the cells.

A deficiency of riboflavin causes retarded growth and lowered feed efficiency in rabbits.

Thiamin (Vitamin B-1)

Thiamin is a cofactor for certain enzymes involved in carbohydrate and fat metabolism.

A deficiency of thiamin in rabbits causes loss of appetite, muscle paralysis, and accumulation of pyruvic acid in the blood.

Vitamin B-6 (Pyridoxine, Pyridoxal, Pyridoxamine)

A deficiency of pyridoxine in rabbits will produce acrodynia (dermatitis), convulsions, and paralysis. This can be avoided by ensuring that rabbits receive at least 40 mcg of the vitamin per gram of diet.

Vitamin B-12 (Cobalamins)

Vitamin B-12 is not a single substance; rather, it consists of several closely related compounds with similar activity. The term cobalamins is applied to this group of substances because all of them contain cobalt. In the body, vitamin B-12 functions in two coenzyme forms—coenzyme B-12, and methyl B-12. Meat animal by-products and marine by-products are high in vitamin B-12. Vitamin B-12 is synthesized in the gut of rabbits if cobalt is present.

Water

The water requirement of rabbits is influenced by a number of factors; among them, the following:

1. **Temperature and humidity.** When the temperature and humidity increase to levels above the zone of thermoneutrality, water within the body becomes an important means of heat dissipation.

2. **Stage of production.** A doe with a litter of seven can drink up to 1 gal of water per day.

3. **Composition of feed.** Feeds that are high in protein and fiber increase the need for water because of the increased need to excrete end products produced in the digestion and metabolism of these feed components. The inclusion of succulent feeds into the ration provides an additional source of water. But care should be taken to restrict the use of this type of feed as the high water content may restrict the level of nutrient intake.

Plenty of clean, fresh water should be available to rabbits at all times. Because of the possible presence of parasites, water should not come from ponds to which dogs and wild animals have access. Also, well water should be tested for impurities.

Because water provides a medium for cross-contamination of bacteria and parasites between rabbits, water bowls should be disinfected routinely. If a drip system of watering is being used, the producer should clean the system periodically, especially when new stock is being introduced to the herd.

Nonnutritive Factors

Two nonnutritive factors—fiber and feed additives—are discussed with regards to their respective influences on rabbit nutrition.

FIBER

Roughages make up a large percentage of rabbit rations, and the digestibility of fiber plays an important role in the utilization of feed by rabbits. Experiments have shown that rabbits do not digest fiber efficiently. In a study designed to compare the digestibility of an all-alfalfa ration and a ration containing both alfalfa and grain, rabbits were observed to digest the same amount of crude protein on a percentage basis as the horse and the pony; but fiber digestibility in the rabbit was less than one-half that of either the horse or the pony in both rations (see Table 26-6).

TABLE 26-6
COMPARISON OF DIGESTIBILITY OF ALFALFA AND ALFALFA-GRAIN RATION IN HORSES, PONIES, AND RABBITS[1]

	Organic Matter	Crude Protein	Crude Fiber
	(%)	(%)	(%)
Alfalfa composition[2]	90.1	19.7	25.2
Digestion coefficients:			
Horse	60.4	74.0	34.7
Pony	62.5	76.2	38.1
Rabbit	54.3	73.7	16.2
Alfalfa-grain composition[2,3]	91.6	16.7	17.8
Digestion coefficients:			
Horse	71.1	77.3	38.6
Pony	72.4	79.6	40.9
Rabbit	65.2	73.2	18.1

[1]Adapted by the authors from Slade, L. M., and H. F. Hintz, "Comparison of Digestion in Horses, Ponies, Rabbits, and Guinea Pigs," *Journal of Animal Science*, 1969, 28:842-843.

[2]Composition based on a dry matter basis.

[3]Ration consisted of 50% alfalfa, 30% rolled barley, 9% wheat mill run, 10% molasses, and 1% salt.

Although fiber is not a useful energy source for rabbits, it is a very important component of rabbit feeds. Numerous studies have shown that low fiber diets cause increased enteritis. Also, when fiber is lacking in the ration, rabbits will sometimes resort to eating their own fur. Fur then accumulates in the digestive tract, resulting in obstruction. The inclusion of some roughage in the ration will remedy the problem in many cases.

Nonproducing animals can be fed rations that contain up to 16 to 22% fiber while animals in production should be limited to 12 to 14% fiber because of the greater need for digestible nutrients.

The form of fiber can affect its value to the rabbit. There is some evidence indicating that finely ground fiber can cause diarrhea. Thus, for proper utilization, fiber should be fed in a coarse form.

FEED ADDITIVES

Feed additives in rabbit rations can be divided into three groups: coccidiostats, antibiotics, and antioxidants.

Anticoccidial Drugs

Coccidiosis is the most prevalent parasitic disease in rabbits. Four species of this protozoan live in the intestine of the rabbit, while an additional species infiltrates the liver. The intestinal species cause diarrhea, loss of appetite, weight loss, and sometimes death. The liver species—the most pathogenic form of coccidiosis in rabbits—enters via the intestine and travels up the bile ducts to the liver. While this form may not be lethal, infected livers must be condemned. Treatment for either the intestinal or liver forms of coccidiosis in rabbits is as follows: sulfaquinoxaline (1) administered in the drinking water at 0.04% for 14 to 30 days; or (2) given in the feed at 0.025% for 20 days, or for 2 days out of every 8. Rabbits should not be treated for coccidiosis within 10 days of slaughter if they are to be used for food. Since most rabbit producers feed only one feed or at the most two, the addition of anticoccidial drugs to the feed becomes impractical. The best method to administer these drugs is through the drinking water. Withdrawing these drugs can then be facilitated by merely changing the water.

Antibiotics

In some cases, the addition of low levels of antibiotics in the rations of various livestock has been shown to exert growth-promoting effects. Oxytetracycline is approved by the Food and Drug Administration as a feed additive for rabbits at a use level of 10 g/ton, as an aid in stimulating growth and improving feed efficiency. (See Chapter 13, Feed Supplements & Additives/Implants.)

Antioxidants

Several antioxidants can be added to rabbit rations in order to prevent spoilage due to the autoxidation of fats within the feed. Also, vitamin E is an effective, natural antioxidant. Compared to the composition of rations for other livestock, rabbit rations are relatively low in fat; but the addition of antioxidants to the ration can ensure against feed spoilage and increase the storage life of the feed. (See Chapter 13, Feed Supplements/Additives/Implants.)

FEEDS FOR RABBITS

Generally speaking, a good-quality legume hay is sufficient to meet most of the maintenance needs of the rabbit; but as production pressures intensify, there arises a need for additional feed sources that are higher in energy. A variety of feedstuffs has been used successfully in feeding rabbits.

Table 26–7 lists many of the feeds that are used in rabbit rations, along with their respective energy values. Chemical composition of these, and other feeds can be found in Section V—Composition of Feeds.

Energy Feeds

Production demands necessitate high-energy levels in rations. Traditionally, cereal grains and their by-products have filled this need. Whole grains from barley, buckwheat, corn, oats, rye, wheat, and the grain sorghums have been widely used in rabbit rations.

The digestibility of certain grains can be increased with minimal processing. For example, corn is more easily digested when it is cracked.

Cereal by-products, such as red dog flour, wheat bran, mill run, and middlings, offer excellent alternatives to the feeding of whole grains; but since these by-products tend to have laxative properties, the level of usage in the ration should be carefully monitored, and sufficient fiber should be present. Economic considerations should be the determining factor as to which feed should be used.

TABLE 26–7
ENERGY VALUES (AS-FED) OF FEEDS COMMONLY FED TO RABBITS[1]

Feed	Dry Matter	TDN	Digestible Energy	
	(%)	(%)	(kcal/lb)	*(kcal/kg)*
Alfalfa:				
Fresh	24*	14*	281*	*620**
Hay, early bloom	91	50*	900	*1,990*
Hay, full bloom	91	44	830	*1,830*
Meal, dehy, 17% protein	92*	53*	890	*1,960*
Meal, dehy, 20% protein	92*	59*	940	*2,080*
Bakery waste	91	101*	1,901*	*4,190**
Barley:				
Grain	89*	75*	1,310	*2,880*
Grain, Pacific Coast	89*	75*	1,200	*2,640*
Beet pulp, dehy	91	70*	1,090	*2,400*
Bermudagrass hay	91*	43*	860	*1,900*
Bermudagrass, coastal hay	91*	40*	860	*1,900*
Clover, red, hay	87*	49*	810	*1,780*
Corn grain	88	83*	1,170	*2,570*
Cottonseed meal, solvent-extracted 41% protein	91	67*	1,402*	*3,090**
Linseed meal, solvent-extracted	90*	68*	1,556*	*3,430**
Milk, dehy	95	118*	2,350*	*5,180**
Oat grain	89*	65*	1,190	*2,630*
Peanut meal without hulls, solvent-extracted	93	90*	1,869*	*4,120**
Rye grain	87	77*	1,350	*2,960*
Sorghum grain	90	—	1,280	*2,820*
Soybean meal, solvent-extracted	89*	82*	1,710*	*3,770**
Timothy hay, mid-bloom	89	32*	940	*2,080*
Wheat:				
Grain hard red spring	88	79*	1,360	*2,990*
Grain hard red winter	89*	84*	1,360	*3,000*
Grain soft white winter	90	79*	1,360	*2,990*
Bran	89*	57*	840	*1,840*

[1]Values obtained from two sources: (1) Those followed by an asterisk came from *Nutrient Requirements of Rabbits,* 2nd rev. ed., NRC-National Academy of Sciences, 1977, p. 18, Table 14; (2) those without an asterisk came from Part V—Composition of Feeds of this book.

Protein Supplements

Fish and animal by-products are seldom used as protein supplements in rabbit rations. But various oilseed meals are routinely added. Peanut meal, the protein supplement most readily eaten by rabbits, provides an excellent source of both protein and energy. Soybean meal, brewers' dried grains, sunflower meal, rapeseed meal (canola meal), safflower meal, sesame meal, linseed meal, and cottonseed meal can all be used interchangeably to different extents.

Amino acid balance in dietary protein is important in the growing rabbit. The Rowett Research Institute, Bucksburn, Scotland, found that the addition of both lysine and methionine to a cereal-peanut meal diet improved the growth rate of weaned rabbits. The minimum requirements for normal growth were found to be 6.2 g methionine plus cystine and 9.4 g lysine/kg diet.[5]

Dry Forages

Rabbits have the ability to utilize dry forages as a large portion of their diet. A good-quality hay will alleviate much of the expense of feeding more costly grains. When using hays, the producer must make certain that they are clean and free from dirt and mold, since it does not take much nutritionally related stress to cause digestive disturbances in rabbits. Once a young rabbit develops scours, precious weight is lost, lowering the efficiency of feed utilization, and the animal becomes highly susceptible to disease and other stresses.

LEGUME HAYS

Legumes are probably the best roughage feed for rabbits, with alfalfa preferred. These hays are high in protein and are extremely palatable to rabbits. A good-quality legume hay can be the sole feed source for bucks not in service, dry does, and young, growing replacement stock. Up to 40% of the ration of pregnant and lactating does can consist of legume hay. Besides being excellent sources of nutrients, legume hays also supply much of the bulk and fiber that are needed to maintain healthy rabbits. Some of the most commonly used legumes in rabbit rations are alfalfa, clover (except sweet clover), vetch, lespedeza, kudzu, cowpea, and peanut.

GRASS HAYS

Typical commercial rabbit feeds contain about 40% alfalfa meal, or occasionally some other legume. However, when the price is right, and when incorporated in nutritionally balanced rations, high quality grass hays can be used successfully in rabbit rations, often at a saving in feed costs. Among such grass hays are Bermudagrass, bromegrass, fescue, Johnsongrass, orchardgrass, timothy, ryegrass, and Sudangrass. In comparison with legume hays, grass hays are generally lower in protein and calcium, lower in vitamin A activity, higher in fiber, and less palatable.

Research workers at the Arkansas Station reported (1) that Coastal Bermudagrass is equal to alfalfa for growing rabbits, and (2) that the Bermudagrass cost 40% less than alfalfa at the time and place of the study, thereby greatly reducing the cost of the ration.[6]

Grasses (Green)

A wide variety of grasses can be fed to rabbits—from high-quality alfalfa to lawn clippings. Since fresh grasses are high in water and bulk, it is wise to use a supplemental feed to ensure against any nutritional deficiencies. A number of plants should not be used in rabbit rations because they are either of no nutritional value or they contain compounds that are toxic to rabbits. (Also see the section on "Poisonous Plants" in Chapter 5, Nutritional Disorders/Toxins.)

Miscellaneous Feeds

A wide variety of feeds can be used in rabbit rations. Often many feedstuffs can come from the garden or table scraps—especially in backyard production. Fats, meats, and spoiled foods should not be used, but garden trimmings and vegetables are low cost and may be efficient.

MISCELLANEOUS ROUGHAGES

Scraps and by-products from various friuts can be fed to rabbits. Potato peelings are readily eaten. The pulp and peelings of citrus fruits as well as apples are often good sources of cheap, supplemental feed. Caution should be exercised when using fruit pulp, lest the rabbits consume too much fiber.

ROOTS AND TUBERS

The addition of roots and tubers to the rations of rabbits can be beneficial, especially in the winter when fresh greens are not available. They are highly palatable and are good sources of vitamins and minerals. However, the water content of roots and tubers tends to be extremely high (about 90%), and the protein level is quite low (1 to 4%). Thus, the producer should not incorporate too high a level of them in the ration. A deficiency in some of the nutrients may result when feeding roots and tubers, because rabbits preferentially eat this type of feed first, subsequently neglecting the higher quality feeds. For this reason, the daily allowance of roots and tubers in a maintenance ration should be limited to 1.5% of body weight. Rabbits in production should not be fed any roots or tubers.

SHRUBS AND TREES

Occasionally, twigs from woody plants can be given to rabbits. While the nutritional value of such feeds may be doubtful, they provide the rabbit with something on which to chew, and some additional fiber.

---

[5]Spreadbury, D., "A study of the protein and amino acid requirements of the growing New Zealand White rabbit with emphasis on lysine and the sulphur-containing amino acids," *British Journal of Nutrition*, 1978, 39, 601-6.

[6]Daniels, L. B., L. A. Shriver, and T. S. Nelson, Evaluation of Bermudagrass in Diets of Domestic Rabbits, *Arkansas Farm Research*, May-June, 1985, p. 2.

FEED PREPARATION

Fig. 26-13. *Left:* pelleted feed. *Right:* unpelleted feed. (Courtesy, Pel-Freez Rabbit Meat, Inc., Rogers, Ark.)

The size, type, and intensity of operation will, in most cases, determine the type of feed preparation. Since most backyard operations are small and do not have access to feed mixing facilities, they generally use commercially available complete pelleted rations, along with some additional hay.

Fig. 26-14. Supplementary hay being fed from the top of the cage. (Courtesy, Pel-Freez Rabbit Meat, Inc., Rogers, Ark.)

If rabbit production is a sideline to a farm operation, homegrown grains and hays are fed with very little processing involved. Commercial operations must obtain feed in large quantities; hence, the processing and preparation of feed for them warrants considerable attention.

Nutritionally complete pelleted rations are available commercially and are being used extensively and successfully. They generally come in 2 types: (1) production rations, which are high in protein and energy; and (2) maintenance rations, which are somewhat lower in protein and energy. The complete pelleted rations generally contain 50 to 60% concentrate and 40 to 50% roughage, with micronutrients supplied at nutritionally adequate levels. The size of the pellet is important. It should be ⅛ to $\frac{5}{32}$ in. in diameter and ⅛ to ¼ in. long. If the pellets are larger than these specifications, the rabbits will bite off pieces of them, with the result that there will be fines and considerable waste. Also, smaller pellets result in kits getting on dry feed more quickly.

Grains can be fed whole to rabbits, but oats and barley should be rolled and corn cracked in order to maximize digestibility. Rabbits are extremely sensitive to moldy and dusty feeds, so care should be taken to provide only clean, quality grains.

Protein supplements should be in cake, pelleted, or crumble form. If the supplement is fed in mash form, there is the possibility that it will settle out from the rest of the feed and be wasted. Since protein supplements are probably the most costly components of the ration, it is imperative that they be used as efficiently as possible. Cottonseed meal must be degossypolized before feeding it to rabbits, and soybean seeds must be heat-treated in order to make them palatable.

Forages should be clean, leafy, and free from mold. In order to maximize the utilization of forages, they should be cut in lengths of about 3 in. and fed in a hay rack. If forage is fed in a form longer than 3 in., rabbits will pull the feed from the hay rack, drag it out on the floor of the hutch, chew a piece from the end, then leave the remainder.

Fig. 26-15. Long pellets (left) vs short pellets (right). Long pellets result in considerable fines and wastage. (Courtesy, Pel-Freez Rabbit Meat, Inc., Rogers, Ark.)

FEED ALLOWANCES AND SUGGESTED RATIONS

Generally, the term nutrient requirements implies a rigid, inflexible level of a certain nutrient for a particular nutritional demand. In reality, the actual nutrient requirement for a particular animal in a fixed set of environmental influences is seldom as precise as has been stated by a group of scientists agreeing on a hypothetical figure. Rather, nutrient allowances are given in order that the producer might have a general idea as to the range in which the real nutrient requirement should fall.

When formulating rations for rabbits, one should consider the following:

- **Fat**—Maintenance rations should contain 2 to 4% fat. Production rations should have 3 to 6% fat.

• **Nitrogen-free extract**—This indicator of the carbohydrate portion of the ration should constitute 42 to 50% of maintenance rations and 44 to 52% of production rations.

• **Crude protein**—For adult rabbits not in production, a ration containing 12 to 15% crude protein will suffice. Rabbits 6 months or older that are growing and/or being fattened, should recieve a 16 to 18% crude protein ration. During the last 2 weeks of gestation and early lactation, does should be fed a ration containing 24 to 26% crude protein.

• **Ash**—Ash may run from 5 to 6.5% in all rabbit rations.

• **Fiber**—Rabbits on maintenance rations, should be fed more fiber than rabbits on production regimens since fiber feeds tend to be cheaper than either energy or protein feeds. Most maintenance rations contain 16 to 22% crude fiber, whereas production rations contain 12 to 16%.

A listing of the nutrient allowances for rabbits is given in Table 26-8.

TABLE 26-8
NUTRIENT ALLOWANCES FOR RABBITS[1]

	Crude Protein[2]	Fat	Fiber	Nitrogen-Free Extract	Ash
	(%)	(%)	(%)	(%)	(%)
Pregnant does	15–17 (16)	3–6	12–16	44–52	5–6.5
Does with litters	24–26 (25)	3–6	12–16	44–52	5–6.5
Dry does	12–15 (13)	2–4	16–22	42–50	5–6.5
Herd bucks	12–15 (13)	2–4	16–22	42–50	5–6.5
Developing young (after weaning)	16–18 (17)	3–6	12–16	44–52	5–6.5

[1]Allowances are on the basis of air-dried feed (as-fed basis).

[2]Values in parentheses represent what the authors feel is optimal for most rations.

Feeding Guide and Rations

A rabbit feeding guide, along with suggested rations, is presented in Table 26-9.

TABLE 26-9
RABBIT FEEDING GUIDE AND SUGGESTED RATIONS[1]

Age, Sex, Production	Total Daily Feed Per Day As-fed		Total Daily Feed As % Live Weight As-fed	Suggested Rations[2] Ingredient	Total Diet A	B
	(oz)	*(g)*	(%)		(%)	(%)
Normal growth, does or bucks,						
average 6.5 lb (*2.95 kg*)	5.1	*145*	5.8	Alfalfa hay	60	
				Corn, grain	21.5	
				Barley, grain	15	
				Soybean meal	3	
				Salt	0.5	
Normal growth and fattening, does or bucks,[3] at body weight of:						
4 lb (*1.8 kg*)	4.0	*113*	6.2	Alfalfa hay	40	50
5 lb (*2.3 kg*)	4.8	*136*	6.0	Corn	—	23.5
6 lb (*2.7 kg*)	5.4	*154*	5.7	Wheat bran	5	5
7 lb (*3.2 kg*)	6.1	*172*	5.4	Barley, grain	31.5	11
				Oats, grain	18	—
				Soybean meal	5	10
				Salt	0.5	0.5
Maintenance, does or bucks, at body weight of:						
5 lb (*2.3 kg*)	3.2	*91*	4.0	Alfalfa hay	70	—
10 lb (*4.5 kg*)	5.3	*150*	3.3	Clover	—	70
15 lb (*6.8 kg*)	7.2	*204*	3.0	Oats, grain	19.5	29.5
				Wheat, grain	10	—
				Salt	0.5	0.5
Pregnant does, at body weight of:						
5 lb (*2.3 kg*)	4.0	*113*	5.0	Alfalfa	—	50
10 lb (*4.5 kg*)	6.6	*186*	4.1	Clover hay	50	—
15 lb (*6.8 kg*)	9.0	*254*	3.7	Oats, grain	43.5	45.5
				Soybean meal	6	4
				Salt	0.5	0.5
Lactating doe, at body weight of 10 lb (*4.5 kg*) with a litter of 7	18.3	*520*	3.4	Alfalfa hay	40	40
				Wheat, grain	25	25
				Sorghum, grain	24.5	22.5
				Soybean meal	10	12
				Salt	0.5	0.5

[1]Adapted by the authors from *Nutrient Requirements of Rabbits,* No. 9, 1st and 2nd rev. ed., NRC-National Academy of Sciences, 1966 and 1977.

[2]In iodine-deficient areas, use iodized salt.

[3]Ration can be used for pregnant and lactating does, also.

FEED SUBSTITUTION TABLE

In order to utilize effectively feeds which might become available to the producer as alternatives to traditional feeds, a feed substitution table for rabbits has been developed. (See Table 26-10.)

TABLE 26-10
RABBIT FEED SUBSTITUTION TABLE (See footnote at end of table.)

Feedstuff	Relative Feeding Value (lb for lb) in Comparison with the Designated Base Feed (in bold italic) Which = 100	Maximum Percentage of Base Feed (or comparable feed or feeds) Which it Can Replace for Best Results	Remarks
GRAINS, BY-PRODUCT FEEDS, ROOTS AND TUBERS:[1] **(Low and Medium Protein Feeds)**			
Oats	***100***	***100***	Most preferred concentrate by rabbits; should be rolled.
Barley	110	100	Should be rolled.
Beets:			Do not feed to rabbits in production or young less than 3 months old. Do not feed to other rabbits in excess of 1½% of body weight.
Garden	10	10	
Mangel	10	10	
Sugar	10	25	
Buckwheat	95	100	
Cabbage, aerial	10	10	Same precautions as with beets; can cause goiter if fed in high amounts.
Carrots, roots	10	10	Same precautions as with beets.
Chicory	65	10	Same precautions as with beets.
Corn	125	100	Should be cracked in order to maximize digestibility.
Kale	15	10	In excessive amounts, kale produces a strong odor in the urine.
Kohlrabi	15	10	
Potato, roots	25	10	Cut out green buds before use.
Potato, peelings	25	10	
Rutabagas, roots	10	10	Same precautions as with beets.
Rye	100	35	May have palatability problems.
Sorghum	125	100	
Sunflower seeds	115	100	Excellent feed, but it is commonly used for other purposes than rabbit feed.
Sweet potato, roots	25	10	Same precautions as with beets.
Turnips, roots	10	10	Same precautions as with beets.
Wheat, grain	120	100	
Wheat, bran	120	100	Has laxative properties; make sure the ration has adequate fiber.
Wheat, middlings	130	100	Has laxative properties; make sure the ration has adequate fiber.
Wheat, mill run	115	100	Has laxative properties; make sure the ration has adequate fiber.
Wheat, red dog	125	100	Has laxative properties; make sure the ration has adequate fiber.
PROTEIN SUPPLEMENTS:			
Soybean meal	***100***	***100***	
Brewers' dried grain	65	50	
Cottonseed meal	100	100	Should be degossypolized. Can replace soybean meal up to 7% of the ration.
Linseed meal	80	100	Has laxative properties.
Peanut meal	100	100	Most preferred protein supplement by rabbits.
Sesame seed meal	100	100	
Soybean seeds	85	100	Unpalatable if fed raw.
Sunflower meal	70		
DRY FORAGES:			
Alfalfa, hay	***100***	***100***	Eliminates or lessens protein supplement requirement.
Bluegrass hay	50	33⅓	
Clover hay	60	100	If the rest of the ration is adequate in protein, clover is equal to alfalfa in feed value; otherwise, it will be lower.
Cowpea hay	70	100	
Johnsongrass hay	30	25	
Kudzu hay	65	100	
Lespedeza hay	60	100	Feed value varies with the stage of maturity at which it is cut.
Oat hay	40	33⅓	Lower feed value if not cut at early dough stage.
Peanut hay	50	100	
Prairie grass hay	15	10	
Sudangrass hay	55	33⅓	
Timothy hay	35	33⅓	
Vetch hay	70	100	

(Continued)

TABLE 26-10 *(Continued)*

Feedstuff	Relative Feeding Value (lb for lb) in Comparison with the Designated Base Feed (in bold italic) Which = 100	Maximum Percentage of Base Feed (or comparable feed or feeds) Which it Can Replace for Best Results	Remarks
GRASSES:			
Alfalfa	***100***	***100***	
Bermudagrass	90	33⅓	
Canadian bluegrass	100	50	
Carpetgrass	90	33⅓	
Clover, red	90	100	
Colonial bentgrass	115	100	
Crotolaria	65	33⅓	
Dallisgrass	75	33⅓	
Dandelion	70	33⅓	
Foxtail millet	80	33⅓	
Kentucky bluegrass	95	50	
Kudzu	95	100	
Lespedeza	95	100	
Meadow fescue	90	50	
Napiergrass	90	50	Low crude protein but high TDN values.
Oatgrass	70	33⅓	Low crude protein content.
Orchardgrass	90	50	
Pigeon pea	100	100	
Red fescue	100	50	
Redtop	90	50	
Rhodesgrass	65	33⅓	Low crude protein.
Sudangrass	50	33⅓	

[1]Roots and tubers are of lower value than grain and by-product feeds due to their higher moisture content.

FEEDING RABBITS

When good nutrition and proper management procedures are followed, growing rabbits should obtain feed conversion rates of 2.8 to 4.0. Most producers estimate that it takes about 100 lb of feed to grow one litter to market weight of fryers. This figure includes the feed consumed by the doe from the time of mating to the weaning of the litter.

When including working does, their young, bucks, and replacement does, the average feed efficiency for rabbits ranges from 4.5:1 to 5:1.

Rabbits should be fed according to their level of production. Thus, lactating does have a much greater feed requirement than dry or pregnant does.

Feeding Dry Does and Herd Bucks

Since these animals are not involved in an intensive type of production, they can be fed maintenance rations—rations that will supply just enough nutrients to keep them healthy and sound. A good-quality leafy, fine-stemmed legume hay plus trace mineral salt can maintain dry does and bucks not in service if the feed is available free-choice. If the legume is not of good quality or if a carbonaceous hay is to be offered free-choice, supplementation with a grain-protein mixture or an all-grain pellet is recommended. A general rule of thumb is to feed 2 oz of supplement for every 8 lb of liveweight. A buck that is in active service requires more feed. Four to 6 oz of a complete pelleted ration, or a good-quality legume hay plus 2 oz of supplement per 8 lb of body weight, should suffice. If the sexually active buck begins to put on weight, the amount of feed should be reduced.

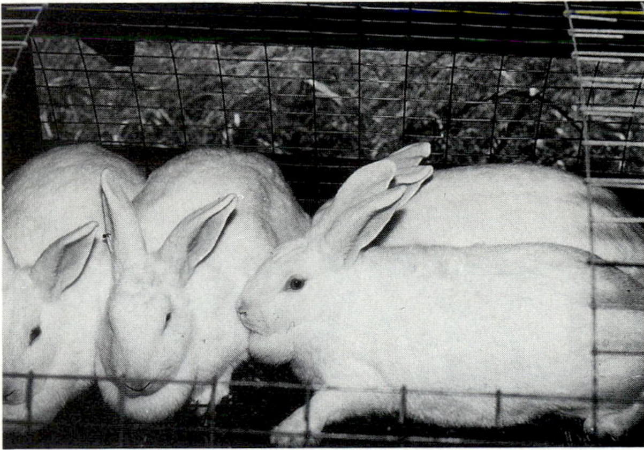

Fig. 26-16. Breeding age (20 weeks) New Zealand White Does. (Courtesy, Pel-Freez Rabbit Meat, Inc., Rogers, Ark.)

Feeding Pregnant Does

Immediately after mating, but before pregnancy is confirmed, the does should be kept on a maintenance level ration. Pregnancy should be confirmed before the plane of nutrition is increased, because if the doe does not conceive, she may become overweight.

Fig. 26-17. Confirming pregnancy by palpation. (Photo by D. J. Harris. Print courtesy of Wm. K. Sanchez, Manna Pro Corp., Portand, Ore.)

The greatest nutritional demands do not occur until the latter stages of pregnancy. Many rabbit producers feed only commercial pellets to pregnant does to minimize disease and nutritional problems inherent in the feeding of hay.

A recommended practice for feeding pregnant does is as follows:

1. If a doe is not nursing, she should be fed 4 to 6 oz per day of a commercial feed from the time she is bred until she is ready to kindle, with the level of feeding varying according to the level of energy in the ration; generally, more of a low-energy ration must be fed in order to obtain performance comparable to a high-energy ration.
2. A day or two before kindling, the doe may eat less; her feed allowance should be reduced accordingly.

Feeding Lactating Does

Lactation creates extremely high nutrient demands on the doe. Either a complete pelleted ration or a combination of good-quality hay and supplement will provide the necessary energy and protein.

Fig. 26-18. A lactating doe and her litter of seven. The lactation requirements are rigorous. (Courtesy, Pel-Freez Rabbit Meat, Inc., Rogers, Ark.)

After kindling, gradually increase the allowance over the next few days to full feed. During lactation, the doe should continue to receive feed *ad libitum*.

When the young are weaned, the doe can be placed on a maintenance ration until she is rebred and diagnosed as pregnant.

Feeding Juniors

In the medium-size breeds, 2 to 4 oz of a pelleted supplement daily and free-choice of good-quality hay will be adequate. If a complete pelleted ration is to be used, the juniors should be fed 4 to 6 oz per day. When juniors are being grown for breeding stock, a ration of 99% alfalfa pellet (15 to 16% protein) and 1% salt may be used. Young does and bucks should not be fed to the point that they become obese, because obese rabbits often have reproductive problems. Besides, the practice is economically wasteful.

Fig. 26-19. New Zealand White fryers at 8 weeks of age. (Courtesy, Pel-Freez Rabbit Meat, Inc., Rogers, Ark.)

Feeding Orphan Litters

When a doe dies at or after kindling, the producer must make provisions for feeding the orphan litter. If a recently kindled doe with a small litter is available, some of the young may be transferred (fostered) to her. This method is far more practical than feeding each kit by hand. However, since this practice is not always possible, a procedure like the following may be used:

1. For the first two weeks of life, feed cow's or goat's milk, or a commercial milk replacer. Heat the milk to body temperature, then feed it by using an eyedropper or a doll's nursing bottle. The eyes of the young rabbits will open at about day 10.
2. After the initial nursing period, solid food, such as fresh grass and rolled oats, can be offered in addition to milk. This will stimulate development of the gut.
3. When the young are about 17 days of age, they can be taught to drink from a pan and offered small quantities of a good growing ration.
4. Gradually, the quantity of solid feed can be increased.

FEEDING METHODS

A standard rule is that rabbits should never be exposed to radical changes in their diets. If feed changes are to be made, they should be made gradually, taking as long as 5 to 10 days. If, for example, the ration is to be lowered from 100% alfalfa to 40% alfalfa, a gradual daily reduction in the alfalfa content of the feed is recommended until the desired feeding level has been attained. Occasionally, when making a slow transition in the ration, some rabbits will waste a large amount of feed by digging and searching for the old, familiar feed. When this happens, an abrupt switch to the new ration may be made.

Hand-feeding

While this method of feeding involves considerable time and labor, it enables producers to keep a close watch relative to the general condition and feeding habits of their animals.

Fig. 26–20. Hand-feeding rabbits. (Courtesy, Pel-Freez Rabbit Meat, Inc., Rogers, Ark.)

Rabbits can be fed once, twice, or three times daily. However, in most operations, a once-a-day feeding practice is the best. Less labor is involved; and the rabbits are more likely to maintain an active appetite if they are allowed to clean up their feed before the next feeding. Since rabbits eat about 2½ times as much at night as during the day, a once-a-day feeding should be offered in the evening. If more than one daily feeding is used, the last feeding of the day should provide the largest quantity of feed. The number of feedings is not critical, but it is important that the time of feeding be regular from day to day. Rabbits are creatures of habit, and any break of routine can cause digestive problems.

If rabbits are hand-fed only the amount of feed that they will clean up in a day, it is easy to detect an animal that is off feed; hence, the daily allowance may serve as an important health check.

Self-feeding

Growth tends to be more rapid and efficient in self-fed rabbits than those that are fed by hand. This improved growth rate is due to the fact that self-fed rabbits have access to feed at all times; consequently, they eat more frequently and chew their food more thoroughly.

Animals that are on maintenance rations may consume more feed than necessary if feed is provided *ad libitum*. Thus, only lactating does and market rabbits 4 to 8 weeks of age should be fed free-choice.

Filling self-feeders with enough pelleted feed to last several days may cause problems in areas with high humidity, resulting in pellets absorbing moisture, expanding, lodging in the feeders, and molding.

OTHER FEED AND MANAGEMENT ASPECTS

In any discussion of rabbit nutrition, certain managerial and nutritionally related health aspects should be considered.

Much of the equipment used in rabbit husbandry—especially in backyard operations—can be made from scrap materials or materials adapted from other types of livestock operations. Feed and watering equipment are good examples.

Feed Equipment

Several types of feeders are available commercially, or feeders can be made from various materials. The type of feeder used will depend on the cost, method of feeding, and physical form of the feed.

Fig. 26–21. Two common rabbit feeders. *Left:* A solid-bottom feeder. *Right:* A self-cleaning feeder with a screened bottom which allows the fines to filter out without clogging the feeder. (Courtesy, Pel-Freez Rabbit Meat, Inc., Rogers, Ark.)

Water Equipment

Numerous types of water equipment may be used. For a small operation, cans, crocks, or bottles are satisfactory, provided they are kept clean. For modest to large rabbitries, an automatic watering system should be installed, so as to (1) provide clean, fresh water at all times, and (2) minimize labor and equipment costs. Currently, the following two types of watering systems are being used in commercial rabbitries: (1) the water cup system, which is identical to that used by poultry producers; and (2) the dew drop system, which operates on low water pressure and utilizes a pressure-reducing valve or an overhead tank or reservoir. Both systems are relatively inexpensive and easy to maintain.

Fur Eating

Quite often, if there is a deficiency of fiber, protein, or some minerals in the feed, rabbits will resort to eating their fur. If too much fur is eaten, there is a danger of gastrointestinal obstruction which, if unnoticed and unrectified, can result in death. When fur eating is a problem, the ration should be reviewed and corrective measures should be taken.

Rabbits suffering from a blockage in the gastrointestinal tract should be given 10 cc of fresh pineapple juice for three consecutive days, along with free-choice hay or straw. The pineapple juice can be administered with an eyedropper, spoon, or stomach tube—preferably, the latter.

Fig. 26-22. Fur-chewed rabbits. (Courtesy, Pel-Freez Rabbit Meat, Inc., Rogers, Ark.)

Enteric Diseases

Enteric diseases are a major cause of death in young rabbits worldwide. Previously, most diarrheal diseases were called enteritis complex or mucoid enteritis. Currently, three specific enteric diseases have emerged—enterotoxemia, mucoid enteropathy, and Tyzzer's disease.

- **Enterotoxemia**—This is an explosive diarrheal disease, primarily of rabbits 4 to 8 weeks of age, although it may affect rabbits of any age. The clinical signs are profuse diarrhea, loss of appetite, rough hair coat, and death within 48 hours. Several bacteria have been implicated as causing this disease—especially *Clostridium spiroforme, C. perfringens,* and *Escherichia coli.* Because of the rapidity of death, treatment is seldom attempted. The diet is a big factor in the development of the disease; much less enterotoxemia is seen when high fiber diets are fed. The feeding of hay, straw, and whole oats appears to lessen the incidence of the disease.
- **Mucoid enteropathy**—This is a diarrheal disease of rabbits of any age. While the cause of the disease is unknown, it has been shown to be the result of constipation—an impaction in the digestive tract. The clinical signs of the disease are gelatinous or mucous-covered feces, loss of appetite, loss of weight, drinking of large quantities of water, and often a bloated abdomen due to excess water in the stomach. There is no effective treatment. Changing to a new batch of feed will usually check the disease, for unknown reason.
- **Tyzzer's disease—**Tyzzer's disease, named after the man who discovered it in Japanese Waltzing mice in 1917, is caused by *Bacillus piliformis* and is associated with poor sanitation and stress. It results in severe diarrhea and death in young rabbits 6–12 weeks of age. The disease also affects rodents, cats, and foals. It is characterized by profuse diarrhea, loss of appetite, dehydration, and death within 1 to 3 days. Post-mortem signs are very similar to those of enterotoxemia, except for white spots of salt-grain size on the liver. No treatment is effective, but control is accomplished by not allowing dogs access to the area where feed and nesting materials are stored.

Fig. 26-23. Dutch breed, which originated in Holland. Rabbits are nonruminant herbivores. Their digestive anatomy and physiology closely resemble those of the horse. (Courtesy, Pel-Freez Rabbit Meat, Inc., Rogers, Ark.)

QUESTIONS FOR STUDY AND DISCUSSION

1. Why is per capita rabbit consumption so much higher in Europe than in the United States?

2. Discuss the efficiency of rabbits versus beef cattle in the production of meat.

3. What characteristics of rabbit meat should appeal to health conscious consumers?

4. Discuss the economic importance of feed for rabbits.

5. Of what importance is the cecum of the rabbit from the standpoint of the kind of feed eaten?

6. What is coprophagy and what role does it play in rabbit nutrition?

7. Of what value are the NRC requirements for rabbits?

8. Why did the French nutritionist Lebas list a requirement for "indegestible fiber" for rabbits?

9. Do rabbits have an essential amino acid requirement? If so, what amino acids are most likely to be deficient?

10. Why doesn't the rabbit utilize nonprotein nitrogen effectively

11. Discuss the experiment confirming the benefits of coprophagy in protein utilization in rabbits.

12. What considerations should be reviewed when supplying calcium and phosphorus to rabbits?

13. Why can rabbit ration formulas be based on total phosphorus rather than on available phosphorus?

14. Analyze the following situation: A rabbit producer is certain that the ration being fed contains adequate magnesium. Yet, the animals are showing all the symptoms of a magnesium deficiency. What is likely the problem, and how can it be remedied?

15. Cobalt is not required in most nonruminant rations. Why is it of concern in the nutrition of rabbits?

16. Describe vitamin A deficiencies in rabbits. Should there be concern about vitamin A toxicity?

17. If a rabbit shows symptoms of muscular dystrophy, what nutrient(s) might be deficient in the ration?

18. What factors influence the water requirements of rabbits?

19. How does the rabbit compare with the horse in the digestibility of fiber? Why is fiber a very important component of rabbit feed?

20. What is coccidiosis, and how is it treated in rabbits?

21. List and discuss the rabbit feeds commonly used as sources of (a) energy, (b) protein, and (c) dry forages.

22. Discuss the preparation of feeds for rabbits.

23. What's the difference between nutrient requirements and nutrient allowances?

24. Suggest a ration, including the daily amount, for a 10 lb pregnant doe.

25. Outline a feeding program for each of the following: (a) dry does, (b) pregnant does, (c) lactating does, and (d) juniors.

26. A producer has an orphan litter. How may they be fed?

27. What are the advantages and disadvantages of handfeeding in rabbit operations?

28. If fur eating is encountered, what would you do about it?

29. Discuss the feed-related aspects of enteric diseases.

SELECTED REFERENCES

Title of Publication	Author(s)	Publisher
Commercial Rabbit Raising, Ag. Hdbk. No. 309	R. B. Casady P. B. Sawin J. VanDam	Agricultural Research Service, USDA, Washington, D.C., 1971
Domestic Rabbit, The	J. C. Sandford	Granada Publishing, Ltd., London, England, 1957
Nutrient Requirements of Rabbits, 2nd Revised Edition	National Research Council	National Academy of Science, Washington, D.C., 1977
Rabbit Feeding and Nutrition	P. R. Cheeke	Academic Press, Orlando, Fla., 1987
Rabbit Production	P. R. Cheeke, *et al.*	The Interstate Printers and Publishers, Inc., Danville, Ill., 1987
Raising Small Meat Animals	V. M. Giammattei	The Interstate Printers and Publishers, Inc., Danville, Ill., 1976

Also, information on rabbits may be obtained from the American Rabbit Breeders Association, 1925 S. Main, P.O. Box 426, Dept. H7, Bloomington, IL 61701.

Fig. 26–24. Two-week-old kits in a nest box. Confined production precludes the choice of feeds by rabbits. (Courtesy, Manna Pro Corp., Portland, Ore.)

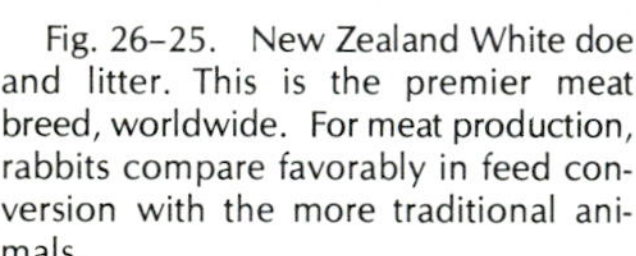

Fig. 26–25. New Zealand White doe and litter. This is the premier meat breed, worldwide. For meat production, rabbits compare favorably in feed conversion with the more traditional animals.

Fig. 26–26. California doe and litter. Rabbit feed is very important nutritionally. (Courtesy, Pel-Freeze Rabbit Meat, Inc., Rogers, Ark.)

Chapter 27

Original painting by Tom Phillips

FEEDING MINK[1]

Contents | Page

[1]The authors gratefully acknowledge the helpful suggestions of the following eminent authorities who reviewed this chapter: J. Adair, Senior Instructor—Mink, Department of Animal Science, Oregon State University, Corvallis, Ore.; E. Alden, Funbo-Lovsta Research Station, Department of Animal Nutrition and Management, Swedish University of Agricultural Sciences, Uppsala, Sweden; Dr. W. L. Leoschke, Ph.D., Fur Animal Nutrition Research, Chemistry Department, Valparaiso University, Valparaiso, Ind.; B. Smith, Editor, Fur Rancher, Eden Prairie, Minn.; and Dr. W. Wustenberg, DVM, Bay City, Ore.

Foxes were raised commercially for fur earlier than mink (*Mustela vison*), but, in the 1940s U.S. fashions for women changed to slimmer, less bulky styles, for which fox fur was not suited but mink fur was ideal. Thus, the decline of the fox industry and the rise of the mink industry in the 1940s was primarily related to fashions. Hand in hand with this development, the sciences of genetics and nutrition teamed up to make for tremendous growth in commercial mink production. More than 20 different fur colors were created, ranging from white (a Scandinavian specialty) to demibuff (a rich mahogany) to *standard* (formerly brown, now black); and new diets were formulated and fed to maximize production, fur color, and quality.

Fig. 27-1. Mink in native habitat. An excellent swimmer, it feeds on fish and amphibians in addition to birds and small rodents. (Courtesy, John Adair, Department of Animal Science, Oregon State University, Corvallis)

Today, U.S. fur sales are booming with retail sales of approximately $1.5 billion annually and mink the most popular fur by far. About 2,000 U.S. mink ranches produce more than 4 million pelts each year, with half of the production centered in three states—by rank: Wisconsin, Minnesota, and Utah.

A large part of the world mink output comes from Scandinavia. In 1984, Denmark, Norway, Finland, and Sweden together produced 13.8 million mink pelts, nearly half of the world's production of 28 million.

ECONOMIC IMPORTANCE OF FEED FOR MINK

Feed costs represent the largest single cost in mink production.

Mink are carnivorous animals. While limited amounts of grains and cereal by-products are used in mink diets, meat and fish and their by-products compose the greatest portion (80–85%) of the traditional standard mink feed. However, the supplies of fresh meat (including horsemeat) and fish are becoming both scarce and expensive; additionally, storage, handling, and feeding of these high-moisture feeds makes for additional costs. The high cost of feed has prompted pursuit of two alternatives: (1) greater dependence on meat and fish wastes and by-products (scraps, trimmings, viscera), and (2) formulation of diets totally from dry ingredients because of the ease with which they can be stored and fed.

In the mid-1980s, most ranchers with a four kit average produced a pelt from 110–120 lb of wet feed (cereal + fresh frozen ingredients, containing 65–67% water); and this included the feed consumed by the parents of the kits. This wet feed cost 10–15¢/lb delivered to the ranch. So, the feed cost of each pelt was $11.50 to $17.25. During this same period, in the mid-1980s, pelleted mink feed cost 30–40¢/lb (*i.e.,* 33–44¢/lb on a dehydrated, moisture-free basis), but it required little more than ⅓ as much poundage of the pellets as of the wet feed to produce each pelt; which, along with the saving in storage, handling, and feeding costs, made the dry feed competitive.

In the mid-1980s, more than 10 to 15% of the North American mink were raised on 100% pellets and another 10% received some pellets along with the wet ranch mix. Some of the latter were on a "5 + 2 program," *i.e.,* 7 days of pellets in the hoppers with concurrent redi-mix, or ranch mix, on the wire, and with no feeding of the mink on weekends *except* pellets in the hoppers.

As we learn more about the nutritional requirements and feeding behavior of mink, more and more pellets will be used.

Fig. 27-2. An adult male mink will consume 200 lb of wet feed per year, and an adult female, 130 lb. (Courtesy, Ralston Purina Company, St. Louis, Mo.)

NUTRITIVE NEEDS OF MINK

Mink require the same nutrients as all other animals. The main difference lies in the amounts of each nutrient and the types of feeds which best supply them.

Mink are flesh eaters and have a simple digestive system much like that found in the pig, dog, fish, monkey, and humans. They have the shortest digestive tract of any animal other than fish, along with limited digestive capacity. Also, feed passage through the digestive tract is rapid, averaging 2⅓ hours vs 72 hours for the ruminant cow. So, mink need diets composed of highly digestible ingredients.

CARBOHYDRATES

Mink, in common with other species, do not have a carbohydrate requirement as such. The function of carbohydrates in mink diets is to supply energy.

Cereal grains and their by-products are the chief sources of carbohydrates, primarily starch. Carbohydrates may supply 15–30% of the ME for growth and for fur development, and 10–20% of the ME for pregnancy and lactation.

The digestibility of cereal grains can be significantly increased either by (1) heat treatment (cooking, popping, steam rolling, toasting), or (2) fine grinding (almost to flour consistency).

If a high level (around 20%) of finely ground cereal is included in the ration, some fiber such as wheat or rice bran may be needed in order to prevent soft stools which tend to stick to the pen wire. In the winter months, this is particularly important as fur matting may result.

FATS

Fat (lipid), which is the most concentrated source of energy, is most commonly supplied in mink diets by meat and fish and their by-products. For high-energy diets, rendered fats and oils (tallow, lard, fish oils, and/or vegetable oils) are usually added. The cereals and their by-products are generally low in fat.

In addition to furnishing energy, fat provides essential fatty acids; notably, linoleic, linolenic, and arachidonic acids, small quantities of which need to be supplied in the feed. Linoleic and linolenic acids occur mainly in vegetable oil, while arachidonic acid is found in animal fat. Danish researchers report that, for optimal growth from birth to weaning, kits have a linoleic acid requirement of 5% of the ME. Since mink diets traditionally include high levels of fat, there have been no reported cases of deficiencies of essential fatty acids in practical rations.

The digestibility of fat depends largely upon the composition of the fatty acids. But it generally ranges from 80–90%, with an average of about 85%.

High-energy diets are, of necessity, relatively high in fats. The level of fat in the diet depends on the life-stage of the mink. Generally, low-fat diets are fed prior to and during breeding—in February, March, and early April. Then, fat levels are raised to meet the high energy needs of lactating mothers and fast-growing kits. Dr. Wm. L. Leoschke, Valparaiso University, Valparaiso, Indiana, noted mink authority, recommends that fat supply the following percentages of the total ME of the diet: for growth, 44–53%; for fur development (including late growth), 42–47%; for pregnancy, 34–37%; and for lactation, 47–50%.[2] In the Scandinavian countries, an average of 35–45% of the metabolizable energy in mink diets is of fat origin (Eva Alden, Funbo-Lovsta Research Station, Department of Animal Nutrition and Management, Swedish University of Agricultural Sciences, Uppsala, Sweden, personal communication to the senior author, 1986).

There is clear evidence that yellow fat disease (steatitis), a disease in which mortality in affected mink generally runs about 50%, is caused by feeding highly unsaturated fatty acids that have become rancid. Care in the selection of ingredients, along with the use of antioxidants and cold storage (0° to −10°F), in fats placed in storage or incorporated in mixed diets will alleviate yellow fat disease.

The role of high fat levels in causing wet-belly is controversial. Leoschke (in a personal communication to the senior author, 1986) states: "Excessive fat levels in the fall ranch diet do contribute to wet-belly disease of the mink." However, others have not encountered wet-belly in certain strains of mink receiving high levels of fat in the diet, indicating a strain difference.

Researchers generally agree that fur color is not related to the *quantity* of fat fed. However, there is evidence that poor color in dark pelts is related to the *quality* of fat fed to mink. In a personal communication to the senior author (1986), Leoschke stated: "Loss of fur quality in dark mink and other color phases is primarily related to the use of high levels of unstabilized polyunsaturated fatty acids present in stored horse meat or fatty fish, which can bring about 'fire engine red' dark mink pelts via oxidation of the fatty acids present in the fur of the mink during the final weeks of fur development in late November and early December. Thus, the use of salmon waste or brown rockfish in the furring diet can bring about 'fire engine red' darks."

Protein

The protein requirement of mink is actually the requirement for essential amino acids. It follows that the protein requirement is related to the protein quality in a given feedstuff.

The protein of mink feed should be of high biological value, especially during the critical periods of gestation, lactation, and immediately following weaning. That is, the protein should contain relatively balanced and desirable levels of the essential amino acids which are available to the animal. Muscle tissue contains large amounts of lysine and methionine, while fur contains considerable amounts of methionine, arginine, and cystine. Requirements for essential amino acids in mink have not been established, but it can safely be assumed that mink have an essential amino acid requirement somewhat similar to those of other nonruminant, mammalian livestock—for example, swine.

Meat is a high quality protein feedstuff for mink, as it (1) contains an amino acid pattern similar to the actual amino acid requirements of the animal, and (2) is highly digestible. Likewise, properly processed fish meal is an excellent source of high-quality protein for mink. However, excessive heating of fish products during dehydration can result in impairment of protein quality by the well known Maillard or browning reaction, resulting in the destruction of the amino acids lysine, cystine/methionine, and tryptophan, and the bonding of the amino acid arginine in an indigestible form.

Feed consumption of mink is determined primarily by the taste appeal and caloric density of the diet. Because of the critical role of dietary energy density in the determination of mink feed intake, it is logical to relate the protein requirements of the mink to the energy content of the diet rather than merely to list them as a percentage of protein in the diet.

---

[2]*Nutrient Requirements of Mink and Foxes*, No. 7, 2nd rev. ed., NRC-National Academy of Sciences, 1982, p. 9.

It is important to emphasize that the protein recommendations given in Table 27–1 are minimum requirements of mink during different phases of the life cycle. So, producers may wish to provide higher protein allowances than these so as to have a margin of safety.

The protein quality of mink feed may be changed by amino acid supplementation or by heat processing.

• **Amino acid supplementation**—A number of experiments have been conducted on the value of supplementing practical ranch diets with specific amino acids. These studies indicate: (1) that supplementing high-fish diets with 0.05% methionine (dry basis) will improve growth and fur quality of mink; (2) that dry diet formulations may be improved by lysine and methionine supplementation; and (3) that the dehydration of fish meals and poultry by-product meals may destroy lysine, with the result that dry diets and pellet formulations may be improved by lysine supplementation.

• **Heat processing of protein feedstuffs**—Heat processing of mink feedstuffs increases the value of some products but decreases the value of others.

Heating eggs a minimum of 5 minutes at 196°F denatures avidin, a protein that binds the vitamin biotin in an unavailable linkage. Heating eggs also denatures egg proteins that bind iron in a structure unavailable to the mink. Heat processing of raw soybean oil meal is essential for the denaturation of a trypsin inhibitor.

TABLE
MINK MIN-

Minerals Which May Be Deficient Under Normal Conditions	Conditions Usually Prevailing Where Deficiencies Are Reported	Function of Mineral	Deficiency/Excess/Symptoms
Major or macrominerals:			
Salt (NaCl)	Lactation, when high amounts of sodium are secreted in milk.	Sodium chloride helps maintain osmotic pressure in body cells, upon which depends the transfer of nutrients to the cells, the removal of waste materials, and the maintenance of water balance among the tissues. Also, sodium is important in making bile, which aids in the digestion of fats and carbohydrates; and chlorine is required for the formation of hydrochloric acid in the gastric juice so vital to protein digestion. It is noteworthy that when salt is omitted, sodium expresses its deficiency first.	**Deficiency**: Nursing sickness in lactating females, characterized by loss of appetite, emaciation, weakness, incoordination, and coma. **Excess**: 1.5% salt (dry basis) in diet of kits results in reduced reproduction during the following breeding period.
Calcium (Ca)	Diets that have a serious imbalance of calcium and phosphorus or are deficient in vitamin D. When sunlight is restricted.	Essential for the development and maintenance of good strong bones and teeth; maintains the contractability, rhythm, and tonicity of the heart muscles; antagonizes the action of the sodium and potassium on the heart; is required for normal coagulation of the blood; is necessary for proper nerve irritability; and appears to be essential for selective cellular permeability.	**Deficiency**: Abnormal bone growth. **Excess**: On a high Ca and low vitamin D diet, mink kits show the following signs, progressively: difficulty in walking, they crawl, unable to stand, knobs on the ribs, concave spinal column (lordosis), and leg bones bend and enlarge at the ends.
Phosphorus (P)	Diets that have a serious imbalance of calcium and phosphorus.	Essential for sound bones and teeth, and for the assimilation of carbohydrate and fats. A vital ingredient of the proteins in all body cells. Necessary for enzyme activation. Acts as a buffer in blood and tissue. Occupies a key position in biologic oxidation and reactions requiring energy.	**Deficiency**: Abnormal bone growth.
Magnesium (Mg)			**Deficiency**: Magnesium deficiency is not a problem. **Excess**: High magnesium may decrease the absorption of Ca and P.
Potassium (K)			
Trace or microminerals:			
Copper (Cu)			
Iodine (I)			
Iron (Fe)	Some ocean fish (Pacific hake, Atlantic whiting, and coalfish) can induce an iron deficiency.	Iron is a part of the hemoglobin molecule, essential for oxygen transport.	**Deficiency**: Cotton fur—an almost complete lack of pigmentation in the under fur; depressed growth; emaciation; anemia.
Manganese (Mn)			**Deficiency**: Pastel mink are prone to manganese deficiency, where it results in "screw neck" or head tilting—a birth defect. The condition can be prevented by supplementing pregnant mink with 1,000 ppm of manganese.
Selenium (Se)			
Zinc (Zn)			**Deficiency**: There is no evidence of zinc deficiency.

[1]As used herein, the distinction between "mineral requirements" and "recommended allowances" is as follows: In mineral requirements, no margins of safety are included intentionally, whereas in recommended allowances, margins of safety are provided in order to compensate for variations in feed composition, environment, and possible losses during storage or processing. Where preceded by an asterisk, the mineral requirements, recommended allowances, and other facts presented herein were taken from *Nutrient Requirements of Mink and Foxes,* No. 7, 2nd rev. ed., NRC-National Academy of Sciences, 1982.

[2]These requirements apply provided (1) the vitamin D concentration is 820 IU/kg dry feed, and (2) the calcium-to-phosphorus ratio is between 0.75:1.0 and 1.7:1.0.

Conversely, heat processing fish and poultry by-products may lower their nutritional value.

Because so many of the common diet ingredients are protein-rich, protein deficiencies are rarely encountered in mink; and the symptoms are rather nonspecific—depressed growth; poor fur development; and poor reproduction-lactation performance.

Minerals[3]

Most of the research on the mineral needs of mink has been limited to salt, calcium, phosphorus, and iron, which is indicative of the main mineral problem areas. The proper combination of common mink feed ingredients will reduce or eliminate the need for supplementation with other minerals.

There are indications that a high-ash diet decreases the energy value and digestibility of the feed. For this reason, it is recommended that mink diets not exceed 8 to 10% ash. A summary of pertinent information concerning mineral nutrition in mink is presented in Table 27-4, Mink Mineral Chart; and additional information is presented in the narrative that follows.

[3]All mineral guidelines in this section are on a dry matter basis unless otherwise stated.

27–4
ERAL CHART

Mineral Requirements[1]		Recommended Allowances in Dry (M-F) Diet[1]	Practical Sources of the Mineral	Comments
Minerals/ Animal/day	Mineral Content of Dry (M-F) Diet			
	•1.3 to 1.5% in the dry diet (0.5% salt in wet feed) will prevent nursing sickness.	•0.5% salt is rountinely added to wet feeds, especially for mink in lactation.	Salt (sodium chloride).	Fresh meat contains very little salt. Higher levels of salt will increase water intake. Salt is inexpensive and is routinely added to mink diets.
	•Requirements are variable according to age, sex, and production (see Table 27-1). •For growing mink, 0.4 to 1.0% Ca.[2]	0.5 to 1.0%	Bone meal Defluorinated rock phosphate. Dicalcium phosphate. Ground limestone. Oystershell. Fish meals.	For growing mink, the Ca:P ratio should be 1:1 to 1.2:1. Meat and meat by-products are extremely low in calcium. Vitamin D is necessary for proper calcium absorption.
	•Requirements are variable according to age, sex, and production (see Table 27-1). •For growing mink, 0.4 to 0.8 P.[2]	0.4 to 0.9%	Monosodium phosphate. Dicalcium phosphate. Steamed bone meal.	Meat products are generally good sources of phosphorus. On the average, only 30% availability of P should be assumed.
	•440 mg/kg of diet.			Excess of calcium or phosphorus in the diet may decrease the absorption of magnesium, and vice versa.
		•Add 0.3% potassium to breeder and grower diets.		Potassium is adequate in diets containing 10 to 30% cereal, because cereals are rich in potassium.
	•4.5-6.0 ppm.			The copper requirement is adequately met by diets containing fish.
	•0.2 ppm for breeder and growth diets.			Fish containing diets provide adequate iodine.
	•The minimum requirement is not known.	•44 and 40 ppm for breeder and grower diets, respectively.		
	•The minimum requirement is not known.	•0.1 ppm.		
	•66 and 59 ppm on a dry matter basis for breeder and grower diets, respectively.			There is no evidence of zinc deficiency in mink.

SALT (NaCl)

In mink diets, the threat of salt deficiencies arising in breeder, grower, and maintenance diets is negligible, since (1) on a dry weight basis, animal products, which are used extensively in mink diets, contain about 1% salt, and (2) fish contain about 2% salt.

During lactation, females sometime deplete their salt reserves because milk is relatively high in sodium. When this occurs, the female becomes thin and emaciated, and her fur has an unthrifty appearance. This condition is commonly called *nursing sickness.* Although most mink diets are fortified with salt, a lactating female that becomes extremely thin should be immediately separated from her kits if they are old enough to eat solid food.

CALCIUM (Ca) AND PHOSPHORUS (P)

The ratio of calcium to phosphorus should be 1:1 to 1.2:1, but mink can get along satisfactorily on levels up to 2:1. Generally, meats and their by-products are extremely low in calcium but high in phosphorus. For example, on an as-fed basis, beef kidney contains 0.01% calcium and 0.22% phosphorus—a ratio of 1 part of calcium to 22 parts of phosphorus. On the other hand, fish meals and whole fish or animal products containing bone provide a good balance of calcium and phosphorus.

It is generally recommended that calcium constitute 0.5 to 1.0% of the dry diet, and that phosphorus make up 0.4 to 0.9% on a dry basis, depending on the age and type of production of the animal.

On a rachitogenic diet high in calcium and low in phosphorus and vitamin D, mink kits develop rickets, characterized by difficulty in walking, knobs on the ribs, lordosis (the

TABLE
MINK VITAMIN

Vitamin Which May Be Deficient Under Normal Conditions	Conditions Usually Prevailing Where Deficiencies Are Reported	Function of Vitamin	Deficiency/Excess/Symptoms
Fat-soluble vitamins:			
Vitamin A		Essential for (1) normal maintenance and functioning of the epithelial tissues, particularly of the eye and the respiratory, digestive, reproductive, nerve, and urinary systems; (2) the production of visual pigments in the eye, which are necessary for vision in dim light; and (3) growth of bony structures.	**Deficiency**: Night blindness, lack of coordination, xerophthalmia, metaplasia of epithelial tissues, depressed growth, fatty livers, and abnormal development of the skull. **Excess**: Loss of appetite; bone change with bony outgrowth (spurs) from bone, decalcification, and fractures; loss of fur; protruding eyes; very sensitive skin; and lowered reproduction.
Vitamin D	Confinement, where animals do not have access to sunlight.	Vitamin D is associated with calcium absorption, transportation, and deposition. Vitamin D, phosphorus, and calcium all play a role in the prevention of rickets, and their effectiveness depends upon proper amounts of each.	**Deficiency**: Rickets and abnormal bone development. **Excess**: Loss of appetite, nausea, loss of weight, and digestive disorders.
Vitamin E (alpha-tocopherol)	Feeds containing large amounts of fish scrap, horsemeat, or poultry offal.	Necessary for reproduction. Protection of vitamin A and carotene from oxidation. Hence, it *stretches* the supply of vitamin A. As an antioxidant in feed.	**Deficiency**: Yellow fat disease: abnormal eating behavior, blood in urine, and anemia. Cotton fur may accompany yellow fat disease if rancid fat is fed. On a synthetic diet (without consumption of rancid fat), the signs are: sudden death due to minor stress, and erythrocyte (red blood corpuscle) fragility.
Vitamin K	Little experimental work, but a deficiency of vitamin K in practical diets is unlikely.		
Water-soluble vitamins:			
Biotin	Diets containing raw egg or turkey offal, due to the presence of avidin. Avidin is a protein found in egg white and oviduct tissue, which binds biotin and prevents its absorption. Including as little as 5% spray-dried eggs in the diet without supplementing with biotin.	As an essential component of a coenzyme concerned in carbon dioxide fixation. Biotin enters into several reactions of intermediary metabolism.	**Deficiency**: Graying or banding of underfur in dark mink and sometimes loss of hair. On biotin-free purified diets, deficient animals show "spectacle eyes," crusty feet, yellow or bloody exudate, and a dermatitis of the foot pads, in addition to the gray underfur.
Folacin (Folic Acid)		Enters into certain enzyme systems that are concerned with nucleic acid metabolism. Essential for the formation of blood cells.	**Deficiency**: Loss of appetite and weight, anemia, and bloody diarrhea.
Niacin (Nicotinic Acid, Nicotinamide)		As a constituent of coenzymes I and II. These coenzymes are essential in biologic oxidation, especially in the oxidation of carbohydrates.	**Deficiency**: Loss of appetite and weight and diarrhea.

spinal column in the thoracic region becomes concave), and the leg bones bending and enlarging at the ends.

IRON (Fe)

The precise iron requirement of mink is not known. It appears, however, that 20 to 30 ppm will meet the minimum requirement providing no interfering factors are present. In order to provide a margin of safety, an allowance of 60 to 80 ppm of iron in the feed is recommended.

When certain raw ocean fish (Pacific hake, Atlantic whiting, or coalfish) are fed to mink, severe anemia and *cotton fur* may result. In cotton fur, there is a depigmentation of the underfur which severely reduces the value of the pelt.

The malfunction of iron is caused by very high levels of trimethylamine oxide (TMAO) present in the species of fish named, which is broken down by enzymes present in the fish digestive track to yield several products, including formaldehyde (FA). Both TMAO and FA interfere with iron absorption and may cause anemia and cotton fur. Freezing raw fish accentuates the problem while cooking the fish at 200°F destroys or inactivates the causative factor.

Cotton fur may also be caused by feeding rancid fat, which interferes with iron absorption. The problem can be corrected by administering iron parenterally. However, feeding iron supplements has met with mixed success.

Vitamins

The experimental basis for specifying the vitamin requirements and allowances of mink is rather limited. However, based on experiments and experiences, some guidelines have been established. These are presented in Table 27–5 and in the narrative that follows.

27–5
CHART (See footnote at end of table.)

Vitamin Requirements[1]		Recommended Allowances of Dry (M-F) Diet[1]	Practical Sources of the Vitamin	Comments
Vitamins/Animal/Day	**Vitamin Content of Dry (M-F) Diet**			
•For growing mink, 91 IU/lb (*200 IU/kg*) of body weight, or 145 IU/100 kcal ME.	•Young mink require 2,695 IU/lb (*5,930 IU/kg*) dry matter.	For young mink, 2,965 IU/lb (*6,523 IU/kg*) dry matter.	Synthetic vitamin A. Liver. Milk fat.	Mink are poor converters of carotene; hence, the carotene content of the diet should be disregarded in supplying the vitamin A requirements of mink.
		Where mink are not exposed to sunlight, add 227 IU of vitamin D/lb feed (*500 IU/kg feed*).	Synthetic vitamin D. Sunlight. Fish-liver oils.	A deficiency of vitamin D is unlikely when mink are allowed exposure to the sun. Supplementation of vitamin D will not overcome the effects of an acute calcium-phosphorus imbalance.
•Young mink require 0.66 mg/100 kcal ME.	•Young mink require 12.3 mg/lb (*27 mg/kg*) dry matter.	For young mink, 13.5 mg/lb (*29.7 mg/kg*) dry matter.	Alpha-tocopherol. Wheat germ oil. Liver. Cereal grains.	Selenium can spare vitamin E to a limited degree. Iron and copper destroy vitamin E. Synthetic antioxidants added to the diet can aid in the prevention of yellow fat disease.
•Young mink require 3.0 mcg/100 kcal ME.	•Young mink require 0.05 mg/lb (*0.12 mg/kg*) dry matter.	For young mink, 0.05 mg/lb (*0.12 mg/kg*) dry matter.	Synthetic biotin. Yeast. Liver. Milk. Molasses.	Cooking raw eggs and turkey offal at 196°F (*91°C*) for 5 minutes destroys the antivitamin avidin. Deficiencies of biotin are not normally encountered on conventional mink diets.
•Young mink require 13.0 mcg/100 kcal ME.	•For young mink, 0.23 mg/lb (*0.5 mg/kg*) of dry feed, or 0.135 mg/100 kcal ME, may not be minimum, but is adequate.	For young mink, 0.23 mg/lb (*0.51 mg/kg*) dry feed.	Synthetic folic acid. Yeast. Liver. Meat.	Typical ranch feeds contain adequate folic acid.
•Young mink require 0.5 mg/100 kcal ME.	•Young mink require 9.1 mg/lb (*20 mg/kg*) dry matter.	For young mink, 10.01 mg/lb (*22.02 mg/kg*) dry matter.	Synthetic niacin. Meat. Yeast. Cereals. Fish. Liver.	Not likely to be deficient because fish, meat, liver, and mink milk are good sources. Mink are unable to convert sufficient tryptophan into niacin.

TABLE 27–5

Vitamin Which May Be Deficient Under Normal Conditions	Conditions Usually Prevailing Where Deficiencies Are Reported	Function of Vitamin	Deficiency/Excess/Symptoms
Water-soluble vitamins (Continued):			
Pantothenic Acid (Vitamin B–3)		Part of coenzyme A, a necessary factor for intermediary metabolism. Maintenance of hair and skin.	**Deficiency**: Loss of appetite, diarrhea, weakness, emaciation, and dehydration.
Riboflavin (Vitamin B–2)		Primarily in protein metabolism. Essential for normal growth, maintaining a healthy condition of the skin, and reproduction.	**Deficiency**: Loss of appetite, loss of weight, and extreme weakness. Subsequently, riboflavin-deficient kits reproduce poorly even when fed adequate riboflavin thereafter.
Thiamin (Vitamin B–1)	Diet containing raw fish that have thiaminase.	Carbohydrate metabolism of all living cells. Promotes growth, appetite, and digestion.	**Deficiency**: Chastek paralysis. Loss of appetite and weight, diarrhea, and paralysis.
Vitamin B–6 (Pyridoxine, Pyridoxal, Pyridoxamine)		Functions in the metabolism of protein. Also, necessary for nerves, proper heart function, blood regeneration, and prevention of anemia.	**Deficiency**: In growing kits, there is reduced feed intake, loss of weight, diarrhea, brown exudate around the nose, excessive tears, swelling and puffiness around the nose and face region, listlessness, muscular incoordination, convulsions, and finally death. In breeding females, there is lowered conception and number of kits per litter. In males, there is degeneration of the testes.
Vitamin B–12 (Cobalamins)		Necessary for growth. Red blood formation. It is essential to prevent and cure anemia, and to facilitate the development of lots of erythrocytes capable of carrying ample oxygen from the lungs to the muscle. Increased food utilization.	**Deficiency**: Loss of appetite, loss of weight and fatty degeneration of liver.
Vitamin C (Ascorbid Acid, Dehydroascorbic Acid)	No requirement for vitamin C for mink has been demonstrated.		

[1]As used herein, the distinction between "vitamin requirements" and "recommended allowances" is as follows: In vitamin requirements, no margins of safety are included intentionally; whereas in recommended allowances, margins of safety are provided in order to compensate for variations in feed composition, environment, and possible losses during storage or processing. In this table, most allowances are 10% higher than the requirements.

Where preceded by an asterisk, the vitamin requirements, recommended allowances, and other facts presented herein were taken from *Nutrient Requirements of Mink and Foxes,* No. 7, 2nd rev. ed., NRC-National Academy of Sciences, 1982.

FAT-SOLUBLE VITAMINS

Vitamin A and vitamin E are the fat-soluble vitamins which have received the most attention in mink nutrition. Little information is available concerning the needs of mink for either vitamin D or vitamin K.

Vitamin A

For growing kits, the NRC recommends a daily intake of vitamin A at a level of 90 IU/lb body weight, or 145 IU of vitamin A/100 kcal of ME. Little vitamin A is stored in the liver when 45 IU of vitamin A per pound of body weight daily is fed; but at the 182 IU/lb level, liver storage is significant.

Mink are inefficient in converting carotene to vitamin A. As a result, alfalfa meal and other plant sources of carotene are rarely used. However, liver from fish or other animals, milk fat, and synthetic vitamin A are good sources of vitamin A.

Symptoms of a vitamin A deficiency include poor appetite, retarded growth, night blindness, opaque lenses and encrusted conjunctivas, fatty liver, abnormal enlargement of the skull, and a loss of coordination particularly in the hindquarters.

Excess (hypervitaminosis) vitamin A is characterized by loss of appetite; bone outgrowths (spurs), decalcification, and fractures; loss of fur, bulging eyes, sensitive skin, and lowered reproduction. This toxic condition can be produced by feeding 5 to 10% of whale liver, an extremely potent source of vitamin A.

Vitamin D

If mink have access to sunlight, there is generally little need for vitamin D supplementation in the diet. A calcium-phosphorus imbalance, together with marginal levels of

(Continued)

Vitamin Requirements[1]		Recommended Allowances of Dry (M-F) Diet[1]	Practical Sources of the Vitamin	Comments
Vitamins/Animal/Day	Vitamin Content of Dry (M-F) Diet			
•Young mink require 0.2 mg/100 kcal ME.	•Young mink require 3.64 mg/lb (*8.0 mg/kg*) dry matter.	For young mink, 4.0 mg/lb (*8.8 mg/kg*) dry matter.	Calcium pantothenate. Organ meats. Eggs. Cereals. Fish solubles. Certain vegetables.	
•Young mink require 0.04 mg/100 kcal ME.	•Young mink require 0.73 mg/lb (*1.6 mg/kg*) dry matter.	For young mink, 0.80 mg/lb (*1.76 mg/kg*) dry matter.	Synthetic riboflavin. Liver. Milk products. Kidney. Heart. Eggs. Soybean.	
•Young mink require 0.033 mg/100 kcal ME.	•Young mink require 0.59 mg/lb (*1.3 mg/kg*) dry matter.	For young mink, 0.65 mg/lb (*1.43 mg/kg*) dry matter.	Synthetic thiamin. Pork. Cereal grains. Oilseed meals. Milk products. Brewers' yeast.	Fish known to contain thiaminase should be cooked before feeding.
•Young mink require 0.04 mg/100 kcal ME.	•Young mink require 0.73 mg/lb (*1.6 mg/kg*) dry matter.	For young mink, 0.8 mg/lb (*1.76 mg/kg*) dry matter.	Vitamin B–6. Yeast. Liver. Cereal grains. Meat. Egg yolk. Milk.	
•Young mink require 0.8 mcg/100 kcal ME.	•Young mink require 14.8 mcg/lb (*32.6 mcg/kg*) dry matter.	For young mink, 16.3 mcg/lb (*35.86 mcg/kg*) dry matter.	Vitamin B–12. Fish meal. Meat scraps. Liver. Dried skim milk.	Generally not needed as a supplement since animal protein is high in B–12.

dietary vitamin D, will produce rickets and bone problems. Since careful attention is given to the Ca:P ratio in the formulation of mink diets, and since mink generally receive adequate amounts of sunlight, vitamin D is generally not added to mink diets. However, young kits grow extremely fast—so fast that they sometimes get rickets in spite of what seems to be adequate light and dietary vitamin D. A simple postmortem test can tell mink producers whether they are running into vitamin D problems. If the legs of a dead kit bend double without breaking, good-quality cod-liver oil (8 oz per 100 lb of feed, excluding added water) and fresh, ground, green bone should be added to the diet. In many cases, an improvement can be observed within a few days.

Vitamin E

Vitamin E acts both as a vitamin and as an antioxidant.

Because of the high fat levels incorporated in mink diets, the role of nutritional antioxidants is paramount. For young mink, vitamin E should be fed at a minimum level of 12.3 mg/lb of dry feed. Studies with selenium have indicated that this element can spare the vitamin E requirement in mink to a limited extent. Conversely, iron and copper can destroy vitamin E.

A deficiency of vitamin E can cause *yellow fat disease,* or *steatitis.* Kits suffering from this disease are anemic and may have bloody urine. Necropsy reveals a yellow discoloration of fat. If rancid fat is fed, cotton fur may also be a problem, and the frequency of "red hips" (poor quality fur on the hips) increases. Sometimes, vitamin E deficiencies can be corrected through the use of selenium or sulfur amino acids; however, vitamin E is less toxic than selenium and less expensive than the sulfur amino acids—cystine and methionine.

It is noteworthy that an uncomplicated (without consumption of rancid fat) vitamin E deficiency can be produced by feeding mink a purified diet. The signs: sudden death due to minor stress and erythrocyte (red blood corpuscle) fragility.

Vitamin K

Very little research has been conducted concerning the requirement of vitamin K—an important factor in blood clotting—in mink. However, a deficiency of vitamin K in practical diets appears unlikely.

WATER-SOLUBLE VITAMINS

Many of the meat and animal products are excellent sources of the B-complex vitamins.

Biotin

Young mink require 0.05 mg of biotin per pound of feed on a dry feed basis.

Biotin deficiencies can result in mink that are fed raw egg, due to the presence of the antivitamin avidin. Adequate cooking of the eggs thoroughly denatures avidin and renders the feed safe from biotin deficiency problems. If high levels of offal from turkey breeder hens (40% or more on a dry basis), are included in the diet, biotin deficiencies can occur; but, as with eggs, cooking at 196°F for 5 minutes will render the offal safe for feeding. If the ingredients in question are not cooked, supplementation with biotin is mandatory; moreover, the level of supplementation must be high enough to permit binding by avidin, plus enough surplus biotin to meet the requirements of the animal.

Biotin deficiency symptoms include achromotrichia (absence of pigment in the hair, or graying), poor-quality fur, hair loss, scaling of the skin, conjunctivitis, and fatty liver, followed by death.

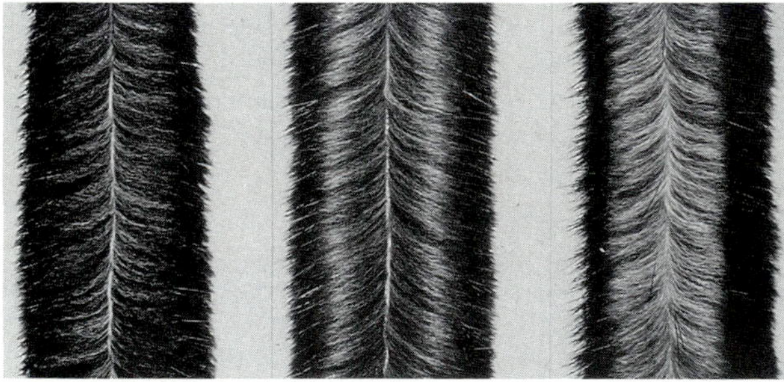

Fig. 27–3. Biotin deficiency in dark mink. *Left to right:* normal; gray-banded; and gray underfur. (Courtesy, John Adair, Department of Animal Science, Oregon State University, Corvallis)

As little as 5% spray-dried chicken eggs in mink diets unsupplemented with biotin may cause fur graying and total reproductive failure. *CAUTION:* Spray drying eggs does not denature avidin; moreover, it makes the avidin much more heat resistant, with the result that cooking may not destroy it.

Folacin (Folic Acid)

The required level of folic acid in feed based on purified diet studies is 0.213 mg/lb of dry feed. Liver, muscle meat, and yeast are excellent sources of folic acid.

While practical rations normally contain adequate amounts of folic acid, mink fed folic acid-deficient purified diets display loss of appetite and weight, hemorrhaging of the gastrointestinal tract, and fatty degeneration of the liver.

Niacin (Nicotinic Acid, Nicotinamide)

Mink cannot synthesize sufficient amounts of niacin from tryptophan to meet their physiological requirements for this vitamin. Based on the use of purified diets, the NRC has tentatively established the niacin requirement for growing mink at 9.1 mg/lb of dry feed. Niacin, however, is unlikely to be deficient in mink diets as meat, liver, fish, cereals, and mink milk are good sources of the vitamin.

Deficiency symptoms include generalized weakness, appetite and weight loss, and diarrhea followed by coma.

Pantothenic Acid (Vitamin B-3)

Pantothenic acid, an integral constituent of coenzyme A, is required at a level of 3.64 mg/lb of dry feed. Eggs, organ meats, fish solubles, cereals, and certain vegetables are high in pantothenic acid. In addition, calcium pantothenate, a commercial compound, is relatively cheap and easy to use.

Mink fed purified diets deficient in pantothenic acid exhibit loss of appetite, diarrhea and bloody feces, weakness, emaciation, and dehydration.

Riboflavin (Vitamin B-2)

The riboflavin requirements for growth and fur production in mink have been established at 0.73 mg/lb of dry feed. Most mink diets contain sufficient riboflavin, so supplementation is unnecessary.

Mink fed riboflavin-deficient diets display loss of appetite and weight, and extreme weakness. When riboflavin-deficient kits are retained for breeding, they reproduce poorly even when fed adequate riboflavin thereafter.

Thiamin (Vitamin B-1)

Young mink fed a purified diet require 0.59 mg of thiamin per pound of dry feed.

Freshwater fish and some species of saltwater fish may contain the enzyme thiaminase, which inactivates thiamin.[4] The enzyme is mainly concentrated in the innards of fish. When these fish are fed raw to mink, thiamin deficiencies will result. These deficiencies are characterized by *Chastek paralysis.* Symptoms of this disease are loss of appetite and loss of weight, weakness, incoordination, diarrhea, and finally, paralysis followed by death.

If still eating, supplemental feeding of thiamin will restore the animals to good health. Milk showing severe thiamin deficiency may recover following a single intraperitoneal injection of thiamin hydrochloride.

[4]For a list of fish containing thiaminase, see *Nutrient Requirements of Mink and Foxes,* No. 7, 2nd rev. ed., NRC-National Academy of Sciences, 1982, pp. 64–65, Table 15.

Cooking fish at 181°F for at least 5 minutes denatures thiaminase. As a result, fish containing thiaminase should be cooked before being fed to mink to prevent a possible thiamin deficiency. **NOTE WELL**: Cooling of fish after cooking is very important in order to minimize bacterial population build-up.

In addition to cooking, other practical approaches to the use of thiaminase-containing fish are: (1) feeding such fish on alternate days and giving a thiamin supplement, and (2) monitoring the urinary excretion of thiamin and not allowing it to go below 150 mcg per animal per 24 hours.

(Also see Chapter 4, Nutrients/Metabolism, Table 4-8, Animal Vitamin Chart—Thiamin.)

Vitamin B-6 (Pyridoxine, Pyridoxal, Pyridoxamine)

For young mink, the NRC requirement is 0.73 mg/lb of dry grower diet. Supplemental vitamin B-6 is not generally necessary because many of the feeds used in mink diets—for example, liver, meat, fish, and milk products—contain high quantities of pyridoxine.

Deficiency symptoms in growing kits are: reduced feed intake, loss of weight, diarrhea, brown exudate around the nose, excessive tears, swelling and puffiness around the nose and face region, listlessness, muscular incoordination, convulsions, and finally death. In breeding females, there is lowered conception and number of kits. In males, there is degeneration of the testes.

Vitamin B-12 (Cobalamins)

Vitamin B-12 is normally found in large amounts in meat and animal by-products; hence, it is seldom necessary to provide supplemental sources in mink diets. The dietary requirement for vitamin B-12 has been established at 14.8 mg per pound of dry feed. Mink suffering from a vitamin B-12 deficiency exhibit a loss of appetite, emaciation, and severe fatty degeneration of the liver.

Vitamin C (Ascorbic Acid, Dehydroascorbic Acid)

On diets adequate in other nutrients, no requirement for vitamin C for growth or reproduction has been demonstrated.

Other Vitamin Factors

Currently, there are no definitive requirements for inositol or choline. A level of 250 mg of inositol per kilogram of dry feed is adequate for mink on a purified diet. Finnish and Russian researchers recommend a supplementation of 20- 40 mg choline per animal per day.

Water

Clean, fresh water should be available free-choice to mink at all times.

The water requirements of mink depend on various factors, including the water content of the feed, the stage of the life cycle of the mink, and the weather.

On dry feed containing about 10% water, mink drink more water than when on wet feed containing 65–75% water. Experiments show that mink need 2.8 oz of water per 1 oz of dry feedstuff, and that it does not matter whether the water is mixed with the feed or given separately. This points up the critical need for water when mink are on dry feed.

In the last part of the growth period, a male mink weighing 4.5 lb and a female mink weighing 2.25 lb have a daily water requirement of approximately 0.6 lb and 0.4 lb, respectively. The most critical stages are during lactation and right after suckling kits start eating solid feed.

In the winter, it may be necessary to heat the water to prevent freezing. If the watering system is not automatic, the waterers should be checked often during the cold weather to make sure that fresh, unfrozen water is always available to mink. In the summer, heat can place heavy stresses on mink, which can be partially alleviated by clean, cool water. The water systems should be cleaned routinely to prevent the growth of pathogenic organisms.

Nonnutritive Additives

Most mink feeds contain one or more nonnutritive factors, which are not nutrients as such. Nevertheless, some of them may improve production and prevent or control diseases under certain circumstances, whereas others prevent rancidity in feed. Among such factors are antibiotics and antioxidants.

1. **Antibiotics.** Antibiotics may be fed to mink (1) at low levels to modify bacterial populations, and (2) at high levels to suppress disease-producing organisms.

The response of mink to dietary antibiotics has been variable and may depend upon (1) the quality of the diet, (2) the health of the animal, and (3) the resistance of the bacteria harbored by the mink.

Although the results have not been consistent, most experiments and experiences attribute the following to the feeding of antibiotics: (1) increased growth rate of weaned kits, (2) improved pelt quality, and (3) increased weaning weights and reduced kit mortality when fed to female breeders.

Currently, chlortetracycline and oxytetracycline are approved as feed additives in mink diets at levels of 20 to 50 g/ton of dry feed as an aid in increasing growth rate, feed efficiency, and pelt size. Oxytetracycline is also approved at higher levels (50 to 100 g/ton) as an aid in the prevention and treatment of diarrhea.

When adding antibiotics (or any drug) to a feed, the producer should read the drug label and follow the directions carefully. Never exceed the recommended level.

NOTE: In some countries (for example, in Sweden and Denmark), the use of antibiotics is limited to treatment of diseases.

(Also see Chapter 13, Feed Supplements/Additives/Implants, section on "Antibiotics.")

2. **Antioxidants.** Antioxidants are commonly used in mink feeds to prevent rancidity due to the autoxidation of fats. When improperly stored, or when stored for prolonged periods, horsemeat and fish are especially prone to oxidation; and when fed to mink, may cause yellow fat disease (steatitis). The compounds which are currently being used for addition to fat in mink rations are Santoquin (ethoxyquin), BHT (butylated hydroxytoluene), and BHA (butylated hydroxyanisole). When levels of 55 mg per pound of wet feed of these compounds are used, yellow fat disease (steatitis) can be prevented. These compounds are capable of inhibiting

the detrimental effects of oxygen on unsaturated fats, fat-soluble vitamins, and other related components of the diet, but are not capable of reducing the physiological vitamin E requirement.

(Also see Chapter 13, Feed Supplements/Additives/Implants, section on "Antioxidants.")

NUTRITIONAL DISORDERS AND TOXINS

All animals are subject to certain nutritional disorders and toxins, a subject that is fully covered in this book in Chapter 5, Nutritional Disorders/Toxins. The nutritional disorders and toxins covered in the sections that follow are peculiar to mink, although they are not always limited to mink.

Disorders Related to Nutrition

Since mink are by nature carnivorous, their feeding has involved different types of ingredients and different storage and handling practices and problems than those of most other species of animals. Mink are prone to the nutrition-related disorders described in the sections that follow.

CHASTEK PARALYSIS (THIAMIN DEFICIENCY)

Freshwater fish and some species of saltwater fish contain the enzyme thiaminase, which inactivates thiamin (vitamin B-1). Among such fish are: whitefish, freshwater smelt, carp, goldfish, creek chub, fathead minnow, buckeye shiner, sucker, channel catfish, bullhead and minnow, white bass, sugar pike, burbot, and saltwater herring.

If mink are fed raw fish containing thiaminase, thiamin is destroyed and thiamin deficiency known as Chastek paralysis, may occur.

Thiamin deficiency is ushered in by loss of appetite and extreme weakness, which is followed by a lurching, rolling gait; convulsions; and paralysis (Chastek paralysis), either in the hindquarters or in the whole animal. In adult mink, diarrhea is usually present in the last stage of the disease. In young kits, death occurs very soon after the first symptoms are observed.

Sick animals should be given an IP injection of thiamin (1 mg/lb), and the thiaminase-containing fish should either be cooked (at 181 °F for at least 5 minutes) or deleted from the ration. If mink are still eating, supplementing feed with thiamin will restore the animals to good health.

(Also see (1) earlier section in this chapter on "Thiamin"; and (2) Chapter 4, Nutrients/Metabolism, Table 4-8, Animal Vitamin Chart—Thiamin.)

COTTON FUR (IRON DEFICIENCY)

Cotton fur, which is caused by a deficiency of iron, is characterized by an almost complete lack of pigmentation of the underfur. This condition generally indicates anemia in mink. It is usually caused by feeding certain raw fish (Pacific hake, coalfish, whiting) containing high levels of trimethylamine oxide (TMAO) which interferes with iron absorption. The condition can be prevented by cooking the offending fish, or by feeding it less frequently.

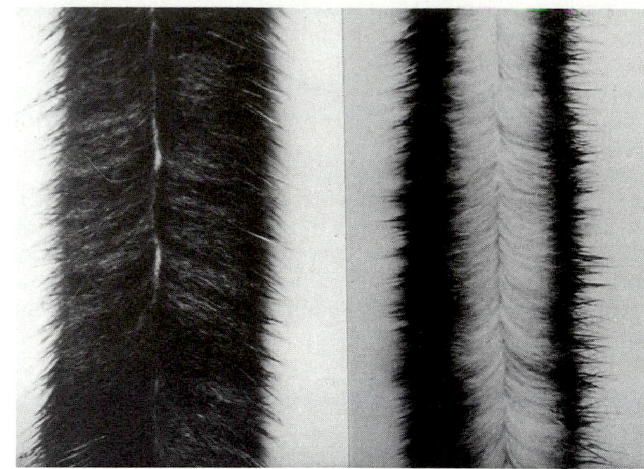

Fig. 27-4. Cotton fur in mink. Pelts parted to show underfur. *Left:* normal fur. *Right:* cotton fur. (Courtesy, John Adair, Department of Animal Science, Oregon State University, Corvallis)

(Also see earlier section in this chapter on "Iron.")

GRAY UNDERFUR (BIOTIN DEFICIENCY)

Gray underfur generally indicates a biotin deficiency. It is usually caused by feeding high levels of uncooked eggs or turkey offal to young mink. These feed ingredients contain avidin, which inactivates biotin, a vitamin required for pigmentation. Affected mink may be injected with 1 mg of biotin twice weekly for 4 weeks. Prevention consists of cooking eggs and turkey offal and/or adding biotin to the ration.

(Also see earlier section in this chapter on "Biotin.")

NURSING SICKNESS

Nursing sickness is a disease of lactating female mink. It is most frequently seen in females with large litters—more than seven kits. Females nursing large litters produce about 2.5 qt of milk a day throughout the lactation period. If a female loses a lot of weight during lactation, it indicates (1) that she has not been able to consume enough feed to prevent loss of weight, and (2) that she has had to convert her own body fat and muscle tissue to make up for the deficit not provided by the feed.

Nursing sickness occurs near the end of, or just after, the lactation period. It is characterized by loss of appetite—partially or totally, loss of weight, fatigue, dehydration, and kidney failure. Death occurs within a few days after the onset of clinical signs. Nursing kits of affected females should be weaned or fostered as soon as possible. To minimize the psychological stress due to loss of her kits, a male kit should be left with the ailing mother. Also, affected females should be tempted to eat by providing them with fresh liver, freshly killed sparrows, or similar feed. Treatment by the veterinarian may include injecting an electrolyte or saline solution.

The following preventive measures will lessen the incidence of nursing sickness: (1) encouraging kits to start eating solid food as early as possible by placing trays containing soft food in the pen; (2) including 0.5% salt (NaCl) in wet feed, or 1.3 to 1.5% in dry feed; (3) feeding lactating females a high-energy ration; and (4) providing plenty of fresh water and feed to nursing females at all times.

URINARY CALCULI (GRAVEL, STONES, WATER BELLY, UROLITHIASIS)

Urinary calculi are sometimes a problem, with losses exceeding 10%.

The disorder is most frequently seen in male kits during the summer, and in pregnant and lactating females during the spring. Males are more prone to urinary calculi than females, due to the greater difficulty in passing stones caused by the inelasticity of the penis bone.

Mink urinary calculi consist primarily of magnesium ammonium phosphate hexahydrate ($MgNH_4PO_4 \bullet 6H_2O$, struvite stones). The cause is unknown. However, the formation of the bladder stones appears to be favored by the presence of alkaline urine—above a pH of 6.0.

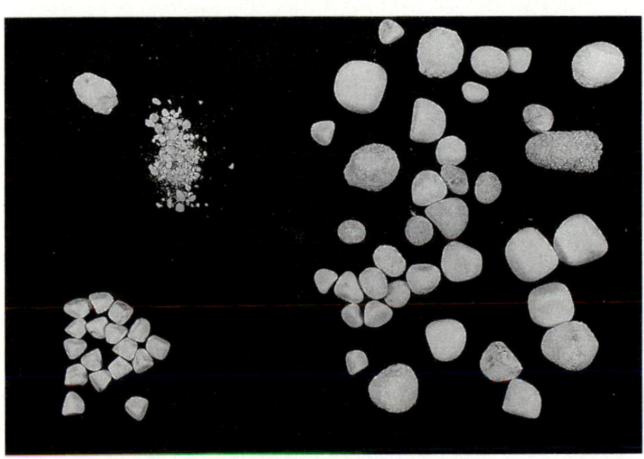

Fig. 27-5. Mink bladder stones (urinary calculi). (*Photo:* University of Wisconsin. *Print:* Courtesy, W. L. Leoschke, Fur Animal Nutrition Research, Valparaiso University, Valparaiso, Ind.)

The disease is characterized by dribbling of the urine and staining of the pelt around the urinary orifice. However, there may be sudden death, without prior symptoms, caused by uremia (urea in the blood).

A urine pH level below 6.0 appears to prevent urinary calculi. So, the following preventive measures may be used: (1) the addition to the diet of 1 g of ammonium chloride per mink per day; (2) the addition to the diet of 2.0% phosphoric acid (75% acid concentration) on a dry matter basis, or the addition of 0.8% phosphoric acid (75% feed grade) to diets containing 15 to 20% fortified cereal and 80 to 85% fresh/frozen feedstuffs; or (3) the acid preservation (with sulfuric, acetic, or formic acid) of fish silage, with the level of the feeding of the silage limited to (a) 10% during reproduction and early growth and (b) 30% during the late growth and furring periods.

(Also see Chapter 5, Nutritional Disorders/Toxins, Table 5-1, Nutritional Diseases and Ailments—Urinary Calculi.)

WET-BELLY DISEASE

This is a condition, primarily affecting male mink in the late summer and autumn, characterized by dribbling of urine and staining of the pelt around the urinary orifice. In some cases, the pelt is only damp and the urine can be washed out; in other cases, affected areas of the pelt must be discarded. Hence, the condition is of economic importance.

Although the cause of this condition is not known, some researchers have reported little or no problem with certain strains of animals, indicating a genetic difference. Other researchers have implicated the following causes: (1) high calorie density diets; (2) high calcium diets with wide calcium-to-phosphorus ratios; and (3) high levels of raw poultry waste, whole fish, and tripe, due to the presence of *Proteus* organisms in these feedstuffs and triggered by the stress of cold weather.

The incidence of wet-belly may be reduced by restricting feed intake from mid-October to pelting, and by cooking poultry waste, whole fish and tripe. Males with wet-belly should not be used for breeding.

YELLOW FAT DISEASE (STEATITIS)

This nutritional disease is common in young, rapidly growing kits. Affected mink may exhibit slight locomotor disturbances followed by death, or they may be found dead without prior symptoms. Postmortem examination reveals yellow internal or subcutaneous fat.

Yellow fat disease is caused by feeding rancid, unsaturated fatty acids, in which all the vitamin E is used as an antioxidant.

Treatment consists in discontinuing the feeding of rancid fat, adding stabilized vitamin E to the feed, and injecting affected kits with 10 to 20 mg of vitamin E for several days. Because of adding antioxidants to raw products, the disease has become relatively rare.

Disorders Caused by Toxic Substances

Toxicities may result from a host of substances in mink feeds. Some of the more common nonnutritional toxic substances are listed and discussed in the sections that follow.

BOTULISM

Botulism is a food poisoning caused by ingestion of food containing *Clostridium botulinum*. The toxins produced by these bacteria are among the most potent poisons known. There are several types of botulism toxin; mink are especially susceptible to Type C.

Occasionally, botulism causes heavy losses in unvaccinated mink. When an animal consumes feed that contains the toxin, the onset of the disease is rapid; usually many mink are found dead within 24 hours. The toxin affects the nerve centers, causing muscular incoordination and stiffness, followed by paralysis and death.

Treatment is usually ineffective. The only successful control is annual vaccination of all kits with botulism toxoid (generally Type C vaccine) soon after weaning.

(Also see Chapter 5, Nutritional Disorders/Toxins, Table 5-3, Potential Poisons—Botulism.)

CLOSTRIDIUM PERFRINGENS

This is a food poisoning caused by the bacterium *Clostridium perfringens,* a spore-forming anaerobe widely distributed in soil, sewage, and unsanitary food processing plants. Meat is the main source of infection of mink.

The most common symptoms are loss of appetite and diarrhea. Treatment consists in changing the diet, improving sanitation in feed handling, and adding an antibiotic to the feed.

ESTROGENS, SYNTHETIC

In the past, the residues of diethylstilbestrol (DES), a synthetic estrogen, in meat, caused sterility, decreased litter size, and poor kit survival in mink. This problem was ushered in with the use of DES implants for steers beginning in 1956. But, in 1979, the use of DES as feed or implants for cattle and sheep was banned by the FDA; so, this problem no longer exists.

HISTAMINE

Histamine is formed by decarboxylation (removal of the COOH groups) of histidine, one of the essential amino acids, by the enzymatic activity of certain bacteria (*Clostridium, Proteus, Salmonella,* and *Escherichia coli*). Histamine is poisonous to mink. The symptoms: decreased feed consumption, diarrhea, reduced gains, vomiting, and bloated stomachs. Diets that contain high levels of acid-preserved feedstuffs are especially prone to histamine formation.

LEAD

In the past, lead poisoning (plumbism) in mink was usually associated with the use of paint containing lead on cages and on feed and water equipment. Afflicted mink exhibit loss of appetite, muscular incoordination, stiffness, trembling, dehydration, convulsions, and a discharge around the eyes.

Lead paint is now banned by law; so, lead poisoning is now a minor problem.

(Also see Chapter 5, Nutritional Disorders/Toxins, Table 5-3, Potential Poisons—Lead.)

MERCURY

Mercury may produce toxicity in mink. Methyl mercury, an organic compound of mercury, is much more toxic to mink than inorganic compounds.

Mercury from industrial plants may be discharged into water, then accumulated and concentrated in fish and shellfish as methyl mercury. To date, however, methyl mercury poisoning of mink on high-fish diets has not been a problem.

Methyl mercury is also used to treat seed grains to control fungicides. However, the Food and Drug Administration prohibits the use of mercury-treated grain for food or feed. But accidents do happen!

The clinical signs of mercury poisoning are loss of appetite and weight, incoordination, tremors, paralysis, convulsions, and high-pitched vocalization.

Although known cases of mercury poisoning of mink have been limited, there is widespread concern over environmental pollution by this element.

(Also see Chapter 5, Nutritional Disorders/Toxins, Table 5-3, Potential Poisons—Mercury.)

MYCOTOXINS

Mycotoxins are poisons produced by fungi or molds that grow on grains, forages, and other feedstuffs. Although there are many mycotoxin-producing fungi, *Aspergillus flavus* which produces aflatoxin, is the most important from the standpoint of mink production. Kits are more susceptible than mature mink. Aflatoxin is fatal in doses of more than 227 mcg/lb.

Aflatoxin can cause serious economic loss and mortality when consumed by mink. It is carcinogenic and has a toxic effect on the liver. The symptoms: loss of appetite, loss of weight, liver damage, followed by death.

(Also see Chapter 5, Nutritional Disorders/Toxins, Table 5-3, Potential Poisons—Mycotoxins.)

NITROSAMINES (DMNA)

When used as a preservative in fish meal, sodium nitrite, formaldehyde, or sodium benzoate may react with the di- and tri-methylamines present in some fish (and fish meal) to form N-nitrosodimethylamine (dimethylnitrosamine), an extremely toxic substance to mink, as well as to cattle, sheep, and swine. The symptoms of nitrosamine poisoning in mink are: loss of appetite, ascites (accumulation of fluid in the abdominal cavity), liver damage, and extensive internal hemorrhaging.

In mink, 3.2 mg/lb body weight of N-nitrosodimethylamine administered by subcutaneous injection constitutes a lethal dose.

PESTICIDES (CHLORINATED HYDROCARBONS)

Pesticides are chemicals used to destroy, prevent, or control pests. Among them, the chlorinated hydrocarbon pesticides are of special concern to mink producers. The most commonly used (or formerly used; some are now banned) chlorinated hydrocarbon on animals and in their environment are: aldrin, benzene hexzachloride (BHC), chlordane, DDT, dieldrin, heptachlor, lindane, methoxychlor, TDE (Rothane), and toxaphene. Because of their wide agricultural use and persistent nature, sometimes they have become troublesome pollutants in mink feed and water, where they have caused reproductive failure.

(Also see Chapter 5, Nutritional Disorders/Toxins, Table 5-3, Potential Poisons—Pesticide Poisoning.)

POLYCHLORINATED BIPHENYLS (CHLORINATED HYDROCARBONS)

The polychlorinated biphenyls (PCBs) are chlorinated hydrocarbons, which, like the chlorinated hydrocarbon pesticides, persist in the environment and may cause cancer. The PCBs were first used in the United States in 1930. Subsequently, they were produced in huge quantities and widely used in everything from electrical equipment, to paints, to adhesives, to caulking compounds. Although the production of PCBs was halted in 1977, they had already become a major pollutant of many lakes and rivers across the nation. Their wide use prior to 1977, along with their persistence, indicate that they will be of considerable concern for many years.

Directly or indirectly, PCBs have found their way into mink feeds where they have caused reproductive problems. Experimentally, as little as 2 parts per million (ppm) PCB fed to mink for 8 months caused reproductive failure; and higher levels in PCB-contaminated fish or beef were lethal to adult mink. The symptoms of PCB poisoning in mink are: loss of appetite and weight, bleeding gastric ulcers and bloody stools, kidney damage, and fatty liver.

(Also see Chapter 5, Nutritional Disorders/Toxins, Table 5-3, Potential Poisons—Polychlorinated Biphenyls.)

SALMONELLA

Several species of salmonella may cause the toxic disease of animals known as salmonellosis. In mink, the infection is commonly caused by *Salmonella dublin* which may be found in raw meat, or *Salmonella typhimurium,* which may be found in raw eggs and poultry.

Salmonella infection is particularly dangerous to mink during gestation because it causes abortion. At other times, salmonella do not produce any specific symptoms in mink, although they may contribute to unthrifty animals and deaths.

Salmonella infection can be treated with antibiotics, given in the feed. In addition, it is important to avoid raw slaughterhouse and poultry offal during gestation. Heat treatment kills salmonella bacteria.

(Also see Chapter 5, Nutritional Disorders/Toxins, Table 5-3, Potential Poisons—Salmonellosis.)

THYROID GLANDS

Slaughterhouse products that contain thyroid glands, and thus the hormone thyroxin, may cause reproductive problems when fed to mink. Diets containing 15% gullet trimmings (including the thyroid/parathyroid tissue) cause a marked decrease in the number of females that whelp, the number of kits whelped, and the kit birth weight and survival.

NOTE: Because of strong indication of a connection between cases of thyrotoxicosis and the consumption of certain ground beef products made from trimmings that contained cattle thyroid glands, stemming from an outbreak of thyrotoxicosis that occurred in 1985 in Minnesota, South Dakota, and Iowa, the USDA ordered an immediate ban on the use of the larynx muscles and thyroid glands in human food. Under the proposal, published December 15, 1986, the products may be used in pet food.

FEEDS FOR MINK

In the early days of the mink industry, there was always a good supply of horsemeat available to provide a relatively cheap feed for mink. As the industry grew, and the role of the horse in American agriculture gradually declined, horsemeat became more expensive and more difficult to obtain. Mink producers then turned to fish and meat animal by-products as sources of feed. More recently, with the great expansion of the U.S. poultry industry, poultry offal has become abundant. Many of these products are economical and of good feeding value, but the problems inherent with wet, bulky feeds remain. Additionally, the nutritional value of the feed varies from batch to batch.

To avoid problems inherent in wet feeds, the mink industry developed new dry feeds which are acceptable to fussy eaters. Today, mink can be raised on 100% dry ingredients, which generally consist of dehydrated combinations of fish meal, poultry meal, liver, cereals, molasses, and other ingredients, fed in pellet form. In addition to the convenience and ease of storage, dry feed makes it possible to use special mixes for different life cycles; for example, breeding animals and kits can be fed mixes designed for their special needs, and energy-deficient feed mixes can be designed for use in periods when the breeder wants mink to lose weight. Because of these several advantages of dry feed mixes, the shift from predominantly wet feed to partly or all dry feed will likely continue.

Fig. 27-6. At home on land or water; in his native habitat, the mink's varied diet includes fish and mice.

Animal Products

Since mink are carnivorous, a diet based primarily on meat and animal by-products is traditional. Depending on the animal source and type of by-product, these feeds can be highly variable in digestibility, moisture, and nutrient content. The main factors to be considered in evaluating a meat source are (1) freshness; (2) freedom from oxidation or rancidity, toxic residues, enzymes or metabolic inhibitors that may interfere with the utilization of nutrients and infectious agents that can be present in raw meats; (3) lack of control of product composition and quality at the packing or processing plant; (4) difficulty of freezing and thawing without loss of nutrients; (5) storage and bulk handling problems; (6) amount of available information for feed evaluation; and finally, (7) price.

Fig. 27-7. Mink raised in captivity have a far better life than their counterparts in the wild. Feeds are formulated to meet all of their nutritional requirements, and the environment is altered so as to minimize stress. (Courtesy, Michigan State University, East Lansing)

The most valuable meat products are those which are composed of muscle tissue, fat, and certain glands. Inclusion of connective tissue, cartilage, bone, and trimmings tends to reduce feeding value.

MEAT AND MEAT BY-PRODUCTS

Almost everything from the meat packing industry that is not sold for human consumption can be utilized in some way as mink feed. Many of these by-products represent a heterogenous collection of parts deemed unsuitable for human consumption and, therefore, tend to be highly variable in their nutritional qualities.

Availability, cost, type of mink operation, and quality of feed should be the determining factors in arriving at a choice of feeds.

Today, the pet food industry is using large quantities of the meat and meat by-products which are also used by the mink industry. This competition plays an important role in the availability and costs of these feeds.

Skeletal muscle from horses has been the traditional feed for mink, but skeletal muscle from other animals is now widely used. Rabbits, slunks (unborn calves), and baby chicks are also excellent mink feeds.

The feeding value of organ meats, such as kidneys, liver, and hearts, is high. Kidneys can be fed at levels up to 25% of the diet. Like muscle meats, organ meats tend to be low in calcium; hence, the overall Ca:P ratio of the diet should be carefully scrutinized when considerable quantities of organ meats are fed. Trash organs, such as lungs and gullets, can be used in mink feeding programs to provide cheap sources of relatively high-quality protein. However, gullets should always have the thyroid gland trimmed off prior to feeding because the high-iodine content of the gland may cause toxicity problems, such as impaired reproduction. Liver is one of the most important feed ingredients for mink. Since it is an abundant source of protein, vitamins, and minerals as well as a source of several unidentified factors which are related to good reproduction, most mink nutritionists recommend the incorporation of liver in gestation and lactation diets. Since pork liver sometimes contains salmonella organisms which induce abortion in pregnant mink, some producers do not feed it to mink during gestation.

Poultry products, which tend to be high in digestible protein and fat, are extensively used in mink diets; and they are becoming more and more abundant with the expansion of the poultry industry. Eggs and entrails have high feeding values, but eggs should be cooked to denature avidin—the biotin antagonist. Offal from hen turkeys, fed at high levels, can cause graying of the undercoat, but cooking destroys this causative factor; alternatively, the offal can be supplemented with biotin. Poultry feet and legs are of relatively low feeding value for mink; however, the younger the birds, the higher the digestibility and nutritive value of their feet and legs. Cull (spent) hens may be killed, ground whole (without the feathers), and fed to mink. They should be chilled rapidly after killing to prevent bacterial contamination.

When favorably priced, milk and dairy products can be used effectively in mink diets. Cottage cheese has been shown to be an excellent quality protein feed and can be included in mink diets, up to a level of 15% wet weight. Dried whey and skim milk are economical feeds, but care should be taken not to feed tham at too high levels, as they have laxative properties. Since mink cannot utilize the high lactose and mineral content of these products, about 3% is the maximum recommended level of dried whey and dried milk, generally speaking. Milk albumin has a high biological value, being especially rich in methionine, cystine, and lysine, and having a digestibility of over 90%. It is available in two processed forms: (1) a frozen form that can be initially added at the level of 5% of the diet and gradually increased to 10%, and (2) a dry form which can be initially incorporated as 3% of the diet and raised to 4% during lactation.

FISH AND FISH BY-PRODUCTS

Fish and fish by-products provide high-quality protein in mink diets. A wide variety of fish can be fed to mink as scraps, trimmings, or whole fish with the relative feeding value of each product depending on the species of fish and the nature of the product. Menhaden and mackerel are very oily fish. Since fish oils are extremely unstable, these oily fish are generally excluded from mink diets. Additionally, fish that are high in moisture are difficult to preserve and are discriminated against for reasons of practicality. Also, storage problems occur with fish that are large and round. Because fish oil is generally high in unsaturated fatty acids, rancidity problems from feeding fish are a constant worry to mink producers. Antioxidants should be added, and all fish products should be frozen promptly for storage and used within a relatively short time.

The producer should know which types of fish contain thiaminase, an enzyme that breaks down the thiamin in the feed, or trimethylamine oxide (TMAO), which can bind the iron in the feed and cause anemia. Proper cooking of thiaminase- and TMAO-containing fish, or feeding them less frequently, will alleviate the problem.

(Also see earlier sections on "Iron" and "Thiamin.")

Cereal By-products

Large amounts of processed grains and cereal by-products from the breakfast food industry are now used in mink diets. Often such items as skim milk, wheat germ meal, alfalfa meal, yeast, and meat and fish meals are added, thus giving rise to the term *supplemented cereals.*

Cereals can be used in breeder diets at levels of 10 to 15% and in production diets at 10 to 25%. When supplemented cereals are used, the cereal portion of the diet can be increased to 15 to 20% in breeding diets and 15 to 35% in production diets.

Cooking or fine grinding increase the digestibility of cereals.

Dried bakery products can be used at levels up to 15 to 20% of the diet since they are already cooked, and some contain high levels of fat.

Dry Feeds

Experiments and experiences have shown that mink producers can achieve as good results with dry pellets as with conventional wet feeds, and in some cases even better results. However, a dry feed must (1) be carefully formulated, (2) consist of high-quality ingredients, and (3) be properly pelleted. Moreover, mink must have easy access to a good and plentiful supply of drinking water when on dry feed.

Some mink do not adjust well to dry diets, especially during lactation and early growth. When this problem is encountered, the dry feed should be softened with water.

Miscellaneous Feeds

Mink diets can include ingredients obtained from the garden or from various other feed sources. Carrots, lettuce, tomatoes, turnips, and other vegetable greens can compose up to 8% of mink rations. Beet pulp, apple pomace, molasses, potatoes, and kelp are among the great array of feed ingredients that can be, and are, used to a limited extent in mink diets. The main value of such materials is their vitamin content. They should be evaluated against other vitamin sources, taking into consideration such things as palatability, cost, and availability.

Vitamin Supplements

In order to ensure against vitamin deficiencies, such products as yeasts (torula or brewers'), distillers' dried solubles, and wheat germ meal, are routinely added to mink diets. These sources provide large amounts of the B complex vitamins and contribute unidentified growth factors and high-quality protein. Natural or synthetic sources of the fat-soluble vitamins are also added, often by means of supplemented cereals.

FEED PREPARATION

The digestive tract of the mink, a carnivorous animal, is very different from that of other animals. The gastrointestinal tract is extremely short, and the rate of passage of feed is very rapid—2 to 4 hours, as compared to 1 to 2 days in humans. When the length of the digestive tract is related to the length of the body of the animal, the gastrointestinal tract of the mink is only ¼ as long as that of the cow, or ½ as long as that of the rabbit. The stomach is relatively small, having a capacity of about 2 oz. These facts underscore the importance of formulating and preparing mink rations carefully in order to get maximum utilization from the feed.

Because of the relatively small digestive and absorptive area in mink, coarse fibrous feeds cannot be utilized effectively. Fiber should be minimal, and feed should be ground and well mixed.

Unground feeds do not maintain a uniform consistency. Besides, fine grinding of cereal grains increases their digestibility. So, cereals and other nonmeat and nonfish ingredients should be finely ground.

Cooking is essential in the preparation of many mink feeds. Fish that contain either thiaminase, which destroys thiamin, or trimethylamine oxide (TMAO), which interferes with iron metabolism, causing cotton fur, should be cooked for 20 minutes at low pressure (less than 15 lb per square inch). If pressure cooking is not used, fish should be simmered for at least 2 hours. If cooking is impractical, alternate feeding of thiaminase- and nonthiaminase-containing fish, and of TMAO-containing and non-TMAO-containing fish, may be practiced. Also, eggs and turkey hen offal should be cooked and/or supplemented with biotin in order to destroy avidin, which inactivates biotin.

In order to improve digestibility, cereals should be cooked if they are not finely ground.

Proper freezing of wet feeds is the key to good mink nutrition. Freezing inhibits bacterial growth in the feed and slows the enzymatic processes which can destroy the product. Fresh meat and fish should be chilled immediately, then quick frozen at −10°F, or colder, then stored at 0°F. The reason for rapid freezing is that smaller ice crystals are formed than if the process is extended over a long period of time; and these smaller crystals help maintain the quality of the product during thawing. Thawing is critical to the quality of frozen feeds. If the package is too large, outer portions of the feed that have thawed can be undergoing bacterial and enzymatic breakdown, while the interior of the block remains frozen. Feed packages should be thin (4 to 5 in.) in order to ensure rapid freezing and thawing.

Freezer burn—a drying out of the food during freezing—can be alleviated by the use of moistureproof bags. Feeds that have been thawed should be used immediately and never refrozen. Since spoilage starts immediately after thawing, any unused feed should probably be discarded. Even though freezing provides an excellent means of storing mink feeds, frozen products should not be stored for more than 6 months. Also, frozen products that have been stored for the longest period should be used first.

Fish and fish offal may be ensiled, usually with the addition of an acid preservative (sulfuric, acetic, or formic acid) or of an enzyme-impeding or enzyme-destroying substance. When good quality fish products are processed rapidly, properly preserved with an acid or an enzyme inhibitor, and stored in a good tank or silo, they will keep for several months.

Dry diets are now being fed successfully to mink. Because of their several advantages, they will be used more widely in ranch practices in the future. Dry feeds are easier to transport, store, and feed than fresh meat and fish products, and they can be standardized more effectively with regard to nutrient composition.

FEED ALLOWANCES AND SUGGESTED RATIONS

The various stages of mink production can be broken down on a calendar basis (see Table 27-6). From December 1 to March 1, mink are fed winter maintenance diets. This type of diet is the most inexpensive type of mink diet; traditionally, consisting of meats and fish having low nutritive value, along with fortified cereals. The breeding season begins the first part of March; and the nutritional needs for the breeding stock are subsequently increased. To fulfill these needs mink are fed a breeder-grower diet containing liver and red meats. Following the breeding season, males are placed on a lower quality diet or are given smaller quantities of their breeder-grower diet. The gestation period of production lasts from mid-March to the first of May. Throughout this period and lactation, females are fed the breeder-grower type of diet as the demands of production are high. Breeder-grower diets are fed to kits from weaning until the middle of September to ensure adequate nutrition throughout this rapid period of growth. In mid-September, young kits are then placed on a grower-pelter regimen—a diet of lower nutritive value than the breeder-grower diet, but higher than the maintenance diet.

TABLE 27-6
TYPES OF RATIONS FED TO MINK THROUGHOUT THE YEAR[1]

Stage of Production	Type of Ration: Maintenance	Breeder-Grower	Grower-Pelter
Breeder females	Dec. 1–March 1	March 1–July 15	July 15–Dec. 1
Breeder males	Dec. 1–March 1	March 1–April 1	April 1–Dec. 1
Kits		June 1–Sept. 15	Sept. 15–Dec. 1
Nonpregnant females			May 1–Dec. 1

[1]Adapted by the authors from *Fundamentals of Mink Ranching,* H. F. Travis and P. J. Schaible, Circ. Bull. 229, Michigan State University, p. 73.

The producer must always consider costs in relation to the nutritional efficiency of each ingredient. The ideal mink diet is one which provides the proper nutrients for the particular stage of production, with reasonable margins of safety and no deficiencies. A high-energy, high-protein diet fed to an animal which is not in production is not a good diet; it is wasteful and costly. On the other hand, a maintenance level diet fed to a high-producing female will limit her production potential.

The National Research Council (NRC) has developed some basic diet formulations, which are presented in Table 27-7. These formulas do not give specific recommendations with regard to the various levels of production.

Sometimes producers need to convert rations or ingredients from wet to dry basis, or from dry to as-fed. Formulations for accomplishing these conversions are presented in Chapter 18, Feeding Standards/Ration Formulation.

TABLE 27-7
SUGGESTED RANGES OF COMPOSITION OF PRACTICAL DIETS FOR MINK ON A "DRY MATTER" (MOISTURE-FREE) BASIS[1]

Ingredients	Percent
Fortified cereal[2]	15–30
Liver	0–10[3]
Quality protein feedstuffs (cooked eggs, whole poultry, whole fish, horsemeat, rabbits, nutria, etc.)	0–30[4]
Beef by-products (tripe, lungs, lips, udders, spleen, etc.)	10–30
Poultry by-products (heads, entrails, feet)	10–70
Fish scrap	10–50
Fat supplementation (rendered animal fat or vegetable oils)	0–6[5]
Proximate analysis[6] of diet	**Percent**
Protein	25–40
Fat	18–30
Carbohydrate	20–50
Ash	6–12

[1]Adapted by the authors from *Nutrient Requirements of Mink and Foxes,* No.7, 2nd rev. ed., NRC-National Academy of Sciences, 1982, p. 38.

[2]May consist of single-cooked grains such as oat groats or wheat in combination with vitamin and trace mineral supplementation or commercially prepared fortified cereal mixtures.

[3]Reproduction-lactation diets (March-May) often contain 5–10% beef liver, although necessity for this has not been accepted universally.

[4]Level of quality protein feedstuffs is often increased during the critical fur development and reproduction-lactation phases—a practice consistent with the higher protein requirements of the mink during these critical periods.

[5]That level of fat supplementation that provides proper protein/energy balance for each phase of the life cycle.

[6]That proximate analysis consistent with the optimum nutritional balance for each phase of the life cycle.

FEEDING BREEDING STOCK

All successful operations require that animals should be fed according to their nutritional demands. If production is to be maximized, each animal should be fed according to needs—no more, no less. For breeding stock, this calls for rations designed to meet the special needs for maintenance, breeding, gestation, and lactation.

Maintenance

Throughout the autumn months, mink should be allowed to gain a little weight to prepare for winter. In the winter, the production status of mink is at a minimum, and a maintenance ration is in order. Only enough nutrients should be supplied to meet the demands of keeping the animal warm and maintaining normal metabolic functions. During this period, diets that are cheaper and less nutritious than those used in the months of production can be used.

Breeding

Fig. 27-8. As the breeding season approaches, the animals should be kept a little hungry to stir activity. They should be alert and inquisitive, as shown. (Courtesy, Michigan State University, East Lansing)

As the breeding season approaches, males should be in such condition that they look like a good athlete—neither too fat nor too thin. Females should be kept on a maintenance diet and should be kept a little hungry to stir activity. They should neither be underfed nor overfed, either of which will affect reproduction adversely. Following the mating season, males can be returned to a maintenance diet. If certain males are thin or in poor condition, they should be liberally fed a high-quality diet until they return to optimum physical condition.

Gestation

After breeding, females should be fed at a level slightly above maintenance requirements until the latter stages of gestation. This means that the diet can have as little as 15% fat on a dry basis. In the first week of May, increase the fat level to produce a highly digestible, high-energy feed. This will prepare the female for the rigorous demands of lactation.

Lactation

The greatest nutritional stress on the breeding female occurs during lactation. Newborn kits weigh about ⅔ oz, but, by 3 weeks of age, they weigh about ¼ lb. During this period of production, the female can consume in excess of 1 lb of wet feed per day.

Care must be taken to provide vitamins and minerals at proper levels. Nursing sickness can occur in lactating females when the diet is deficient in sodium or lacking in energy density. Considerable quantities of sodium are excreted in the milk. Thus, if it is not available in the diet to replenish body losses, the female may lose weight rapidly and die. Salt is routinely added to lactation diets at a level of 0.5% on a wet feed basis.

With the rigorous demands of lactation, high-quality, albeit expensive, feed is essential. High energy density of the ration is essential; otherwise, the lactating female may not be able to consume sufficient feed to meet her energy needs. Good-quality protein, such as whey, skim milk, meat, fish, and/or cooked eggs, should be provided.

Fig. 27-9. The mother of a fast-growing litter should be fed to meet the rigorous nutritional demands of lactation. (Courtesy, Maple Leaf Mills, Ltd., Ontario, Canada)

FEEDING KITS

The early growth of kits is very rapid (see Fig. 27–10). It follows that this is one of the most critical periods in the life of a mink and requires the best possible nutrition.

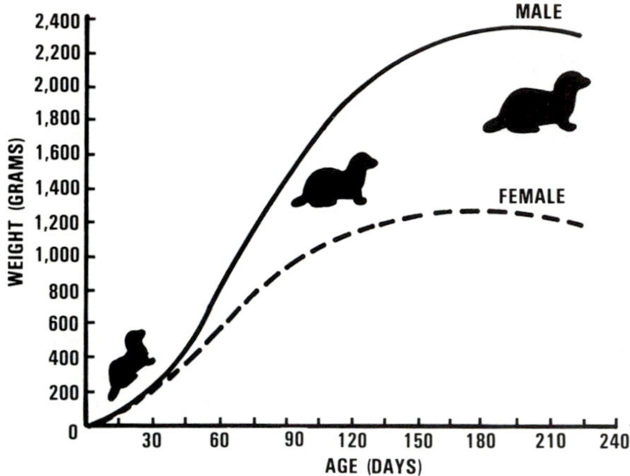

Fig. 27–10. Standard dark mink growth curve. In this figure, the various color types of mink are not differentiated; rather, it is assumed that they will grow similarly if given adequate nutrition in satisfactory environments. (*Source:* Adapted by the authors from *Nutrient Requirements of Mink and Foxes,* No. 7, 2nd rev. ed., NRC-National Academy of Sciences, 1982, p. 4. *Original source:* Wehr, N. B., J. E. Oldfield, and J. Adair, Oregon State University, Corvallis)

Kits will begin to eat solid feed at 4 weeks of age. They should be weaned at 6 to 8 weeks of age. At weaning, they are generally separated from the mother but are housed as a group for an additional couple of weeks. When kits are grouped together, they compete for feed and eat extremely fast, sometimes resulting in digestive upsets. It is important that their feed be thoroughly thawed before feeding, as partially frozen feed can create digestive disorders in young kits. Demands for high-quality feeds are high in kits up to 16 weeks of age. The protein content of the feed can be reduced after 16 weeks of age, but the kits should remain on a full feed regimen.

Fig. 27–11. Kits are normally weaned from their mother at 6 to 8 weeks of age but are usually housed together for an additional couple of weeks. (Courtesy, Ralston Purina Company, St. Louis, Mo.)

In order to correct for overfeeding and to sharpen their appetites, kits can be allowed to go unfed one day a week following separation. This should only be done after they are housed singly; otherwise, fighting, sometimes resulting in death, may occur.

METHODS OF FEEDING AND WATERING

Since mink have a short digestive tract and cannot accommodate large quantities of feed at one time, it is necessary to feed them more frequently than many other species. It is recommended that mink be fed once or twice daily during breeding and gestation. Lactating females should be fed two or three times a day, while adult mink on maintenance rations can be fed once daily if the weather isn't too cold. Young, growing kits need to be fed four times a day.

In addition to feeding conventional wet feed, some mink ranchers are now providing pellets (dry feed) free-choice in hoppers. This allows the feeder to be less precise in the amount of wet feed given at each feeding. Also, under this system, the wet feed may be omitted on weekends without throwing the mink off feed during the 2 days that they are provided free-choice pellets only. Other ranchers are self-feeding pellets only, without any wet feed whatsoever.

Many mink ranchers feed by hand. While this method involves more time and labor, the producer is able to keep close watch over the animals. Overfeeding and underfeeding can be carefully monitored by hand-feeding. In addition, any unhealthy animals can be spotted and treated quickly before the disease has a chance to spread throughout the operation. Automatic feeding is much more timesaving, but considerable wastage of feed can occur, and the personal touch in management is diminished.

Several types of watering systems can be adapted to mink ranching. One of the most important factors to be considered is the prevention of freezing in the cold, winter months. This can be accomplished by frequent draining of the water vessels and pipes or by the use of electric heating cables. The simplest, but the most laborious, method of watering is the use of the sprinkling can. In large operations, the time involved with this system is too great to be economical. Manually operated mechanical systems, whereby pipes are laid out so that there is a hole over the water containers, can be used effectively. Periodically a water valve is opened, and water flows to the drinking cups. Fully automatic watering systems—for example, the nipple valve—are excellent in warm weather but tend to freeze up in cold weather.

Fig. 27-12. Violet variety of mink begging for food. (Courtesy, USDA)

FEED CONTAMINANTS

Many animal and fish by-products contain compounds that can be detrimental to the production of mink. For this reason, it is important that the producer obtain feed from reliable sources that routinely check the quality of their ingredients.

Microorganisms from animal and fish products can be passed on to mink if the feed has been improperly stored or prepared. Meat from diseased animals should never be fed to mink. Cold storage of animal and fish products will minimize bacterial growth, and cooking will further reduce the threat of enteritis and other diseases.

In recent years, the effects of industrial chemicals and wastes on the environment have been carefully studied and scrutinized. Many of these products present potential health hazards to mink; hence, care should be exercised to prevent contaminated feed from being incorporated in mink diets.

Currently, the Scandinavian countries are using chemical and microbiological methods for evaluating the quality of feedstuffs and mixtures. This concept is also being used in some of the larger U.S. units.

(Also see earlier section on "Disorders Caused by Toxic Substances.")

QUESTIONS FOR STUDY AND DISCUSSION

1. What caused the rise of the U.S. mink industry in the 1940s?

2. It takes an average of 70 skins to make one full length mink coat; it requires an average of 115 lb of wet feed to produce one mink ready for pelting; and wet feed costs an average of about 12½¢/lb. What is the feed cost alone to produce one mink coat?

3. Why are mink feeds generally more expensive than the feeds normally fed to other animals?

4. Define (a) metabolizable energy, and (b) energy density.

5. The quantity of energy consumed by mink is directly related to the need to satisfy their energy requirement. What is the practical significance of this fact?

6. In what ways may the digestibility of cereal grains be increased for mink?

7. Discuss the nutritive importance of fat in mink diets.

8. Why should the protein of mink feed be of high biological value?

9. Present pertinent information relative to the needs of mink for each of the following minerals: salt, calcium, phosphorus, and iron.

10. Present pertinent information relative to the needs of mink for each of the following vitamins: vitamin A, vitamin E, biotin, and thiamin.

11. For what purposes are antibiotics and antioxidants used in mink production?

12. Discuss the cause, prevention, and treatment of each of the following disorders related to nutrition: Chastek paralysis, cotton fur, gray underfur, nursing sickness, urinary calculi, wet-belly disease, and yellow fat disease.

13. Discuss the symptoms, prevention, and treatment of each of the disorders caused by the following toxic substances: botulism, histamine, mercury, mycotoxins, salmonella, and thyroid glands.

14. In order and in period of time, there has been a trend shift in the wet feed of mink from horsemeat, to fish, to poultry offal. What forces were back of these shifts?

15. Discuss the advantages of dry feeds. Discuss the proportion of dry feed that may be fed to mink, and the frequency and method of feeding dry feed.

16. What are the special nutritive needs of mink for lactation and growth?

17. What are the advantages of each of the following methods of feeding mink: hand-feeding, self-feeding, and a combination of hand-feeding and self-feeding?

SELECTED REFERENCES

Title of Publication	Author(s)	Publisher
Bibliography of Mustelids, A, Part II	A. V. Shump, *et al.*	Mink Farmers' Research Foundation, Pittsville, Wisc., 1976
Blue Book of Mink Farming, The	P. J. Schaible	Editorial Service Co., Inc., Milwaukee, Wisc., 1966
Fundamentals of Mink Ranching, Circ. Bull. 229	H. F. Travis P. J. Schaible	Michigan State University, East Lansing, Mich.
Fur Farming, 2nd Edition.	R. G. Hodgson	Fur Trade Journal of Canada, Bewdley, Ontario, Canada, 1967
Make Mine Mink	E. R. Bowness	Canada Mink Breeders Assn., Mississauga, Ontario, Canada, 1974
Mink Production	G. Joergensen, Editor	Danish Fur Breeders Ass'n. (Danish Edition) and Scientifur (English Edition), 1985
Modern Mink Management	A. H. Leonard	A. H. Leonard, Webster Groves, Mo., 1966
Nutrient Requirements of Mink and Foxes, No. 7, 1st Revised Edition	National Research Council	National Academy of Sciences, Washington, D.C., 1968

Fig. 27-13. Pastel mink in native habitat. (Courtesy, Department of Animal Science, Oregon State University, Corvallis)

Chapter 28

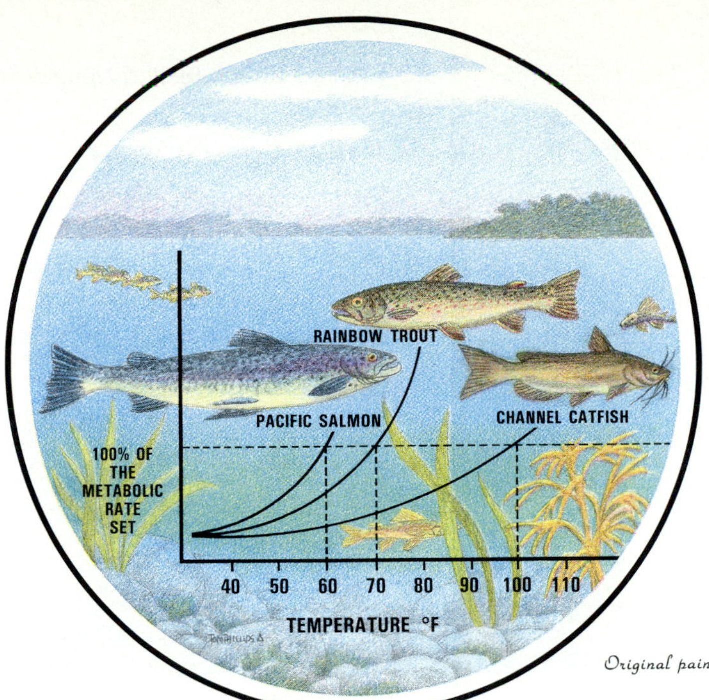

Original painting by Tom Phillips

FEEDING FISH[1]

Contents — Page

[1]The authors gratefully acknowledge the helpful suggestions of the following authorities who reviewed this chapter: H. K. Dupree, Ph.D., Scientific Director, Fish Farming Experiment Station, U.S. Department of the Interior, Fish and Wildlife Service, Stuttgart, Ark.; L. G. Fowler, Ph.D., Assistant Director, Abernathy Salmon Cultural Development Center, U.S. Department of the Interior, Fish and Wildlife Service, Longview, Wash.; D. M. Gatlin III, Ph.D., Assistant Professor, Department of Agriculture, University of Arkansas at Pine Bluff; B. P. Grant, Ph.D., Director, International Aquaculture Research Center, Hagerman, Idaho; J. E. Halver, Ph.D., School of Fisheries, University of Washington, Seattle; J. D. Hendricks, Ph.D., Department of Food Science and Technology, Oregon State University, Corvallis; and D. Wenger, Wenger Manufacturing, Sabetta, Kans.

Although recent research has shown the advantage of scientific fish feeding, aquaculture is an old term with a new look. It is derived from the latin word *aqua,* which means water, and from the work culture, which means to till, to cultivate, and to grow. So, *aquaculture is the controlled cultivation and harvest of aquatic animals and plants.* It involves the production of a marketable crop of fish or other commodity in water.

Aquaculture dates back more than 4,000 years to China, Japan, and Egypt. The practice of fish farming can also be found in the societies of India and Java 3,000 years ago, and in Europe 2,500 years ago. Until recent years, most fish feeds were the result of trial and error. Today, the nutritional requirements of fish are based on research, although some gaps in our knowledge still exist.

Fig. 28–1. Fish farming. (Courtesy, USDA)

Aquaculture can be divided into three types of production —freshwater, brackish water, and marine. Within these three types of production, three systems of management can be listed as follows:

1. **Hatchery propagation.** Young fish are hatched and raised to a size where they can be released into natural populations to grow and reproduce.
2. **Capture of young fish.** In this system, wild fish are captured, transferred to managed water, and grown to market weight by supplemental feeding or on natural foods.
3. **Management of entire life cycle.** Catfish and trout production in the United States are examples of this system. Young fish are hatched and grown either to be marketed or kept as replacement breeder stock.

Within the classification of vertebrates, there are more species of fish than of any other animal. Estimates of the number of fish species range from 15,000 to 17,000, as compared to 8,600 avian species and 4,500 mammalian species. Among this vast number of species of fish, representatives of almost every type of feeding behavior can be found.

Presently, fish contribute less than 1% of the world food supplies in terms of dietary energy, less than 5% of total protein, and less than 14% of animal protein. But as health concerns and dieting increase throughout the world, more fish will be included in human diets.

ECONOMIC IMPORTANCE OF FISH

Fig. 28–2. The manufacture/delivery of fish feeds is big business. This shows a bulk feed truck delivering feed from the manufacturer to bulk feed bins on a fish farm. (Courtesy, Department of Fisheries and Allied Aquaculture, Auburn University, Auburn, Ala.)

Today, U.S. fish farming is a big and growing business. In 1984, U.S. Aquaculture production had a value of $503,030,000, with catfish leading, baitfish ranking second, and trout ranking third. Fig. 28–3 shows the comparative value of all U.S. Aquaculture production in 1980 and in 1984.

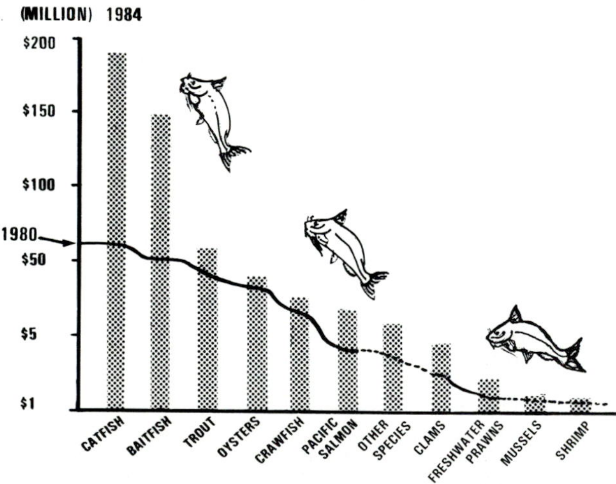

Fig. 28–3. In 1980, U.S. Aquaculture production had a total value of $203,178,000, with a breakdown by species as shown by the curve. In 1984, the total value was $503,030,000, representing a 2½ fold increase, with a breakdown by species as shown by the bars. **NOTE**: "Other species" include abalone, buffalo, carp, mullet, sturgeon, tilapia, and lesser species. (*From:* Annual series of *Current Fisheries Statistics, Fisheries of the United States,* National Marine Fisheries Service, U.S. Department of Commerce)

Fish are very efficient feed converters. It takes only 1.5 to 1.7 lb of feed to produce 1 lb of fish; this compares with 2.1 lb of feed to produce 1 lb of broiler. Also, protein efficiency (pounds of protein in feed required to produce 1 lb of fish protein) is very excellent, 2.1:1, although it is exceeded by broilers (1.9:1). Besides, fish consume proteins which are not generally used for human consumption; for example, trash fish and animal by-products—they recycle protein. For these reasons, plus the fact that fish do not compete with crop production, it is expected that both freshwater and brackish-water fisheries, including fish production in small ponds and paddy fields, will receive increasing attention in the future.

Fish are poikilotherms (cold-blooded animals); that is, their body temperature changes to that of the environment. From a production standpoint, this offers both advantages and disadvantages when comparing fish farming with the production of homothermal (warm-blooded animal) species. Since the body temperature of fish is the same as the temperature of their environment, little or no energy is required to maintain their body temperature. By contrast, considerable energy is required to maintain the body temperature of warm-blooded animals. However, fish are very susceptible to stresses which result from environmental changes, especially rapid fluctuations in water temperature. It follows that the production of freshwater fish is more extensive than that of marine fish because it is easier to monitor and control the environmental stresses which affect their growth rate.

Freshwater fish production can be divided into two classifications—coldwater culture (40 to 60°F); and warmwater culture (70 to 100°F). In the United States, channel catfish is the predominant warmwater species, whereas trout and Pacific salmon represent the bulk of coldwater fish production. In other parts of the world, the carps are the major warmwater fish.

Fig. 28-4. Coldwater species, trout and salmon, being produced in Scandinavia in sea pens equipped with automated, pneumatic feeding systems. (Courtesy, International Aquaculture Research Center, Hagerman, Idaho)

NUTRITIVE NEEDS OF FISH

Fish in the wild seldom show signs of nutrient deficiency: natural foods are relatively well-balanced nutritionally and growth rate is proportional to quantity, not quality, of food available. In confinement, where natural food is limited or absent, nutritional requirements become critical. Although fish require essentially the same nutrients for the same metabolic functions as land animals, there are the following differences:

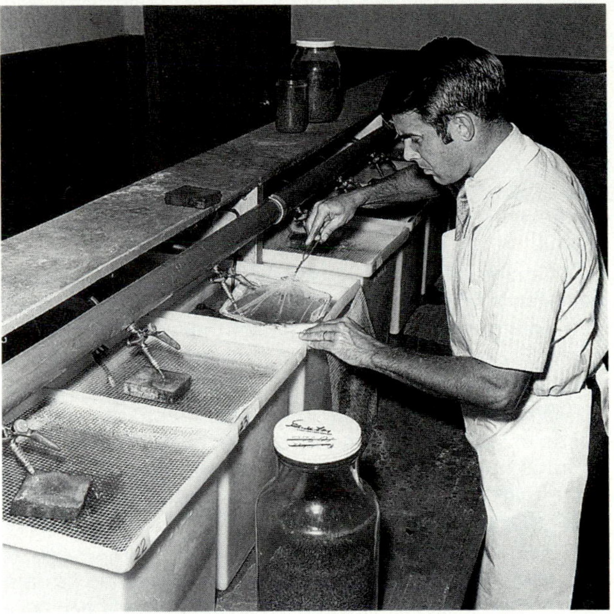

Fig. 28-5. Nutritional requirements of fish must be determined through feeding experiments conducted in highly controlled environments, such as tanks, aquariums, or raceways, although practical feeding may be done in ponds. (Courtesy, Auburn University, Auburn, Ala.)

1. The energy requirements of fish are lower, resulting in superior feed efficiency.
2. Vitamin C (ascorbic acid) is synthesized poorly, or not at all, by most fish.
3. Most fish require dietary omega-3 fatty acids, whereas warm-blooded animals do not.
4. Fish can absorb some soluble minerals from the water, thereby reducing the need for those minerals in the diet.

The nutritional requirements of fish do not vary greatly among species. There are exceptions, however, usually associated with coldwater or warmwater fish, finfish or crustaceans, and freshwater or marine species. These differences generally relate to the requirement of essential fatty acids, and to carbohydrate utilization.

Fish can be divided into three types of eaters—carnivores, herbivores, and omnivores.

1. **Carnivores.** Consume primarily animal material. Foods consumed by this type of fish may be as small as a microscopic crustacean or insect or as large as an amphibian or a small mammal.
2. **Herbivores.** Consume primarily available vegetation and decayed organic materials in the environment.
3. **Omnivores.** Consume both animal organisms and plant materials.

Some anatomic adaptations in the mouth and digestive system, relating to their foods and feeding habits, are observed in some fishes.

Based on well-defined differences in the organization and structure of the mouth, it is possible to classify fish according to their feeding habits into the following categories:

1. **Predators.** These fish, of which trout and salmon are examples, feed on organisms that are generally large enough to be seen with the naked eye. Teeth are well developed and act as a means of grasping and holding the prey. Some predators rely primarily on sight to hunt while others rely on the senses of taste and touch or on lateral-line sense organs.

2. **Grazers.** These fish, of which the mullet is an example, graze in the same sense as grazing animals. Generally, they feed continuously at the bottom of the water habitat on either plants or small animal organisms. Ingested food is taken by well-defined bites.

3. **Strainers.** These fish, of which menhaden are an example, select food primarily by size, rather than by type. An adult menhaden can strain in excess of 6 gal of water per minute through its gill rakers. Through this process of rapid straining, the menhaden is able to concentrate a relatively large mass of plankton and other organisms.

4. **Suckers.** This group of fish, of which buffalo fish are an example, feeds primarily on the bottom of their habitat—sucking in mud, filtering, and extracting disgestible material.

5. **Parasites.** Some fish, notably the sea lamprey, attach themselves to other animals and exist on the body fluids of the host.

In addition to the anatomical adaptations to eating, fish have developed behavioral feeding patterns which are sensitive to environmental stimuli. The fish farmer should recognize the influences of environment on feeding behavior if the efficiency of production is to be maximized. By knowing the behavioral patterns of the particular species of fish, the producer can adapt a system of feeding which will best utilize labor and feed. Some environmental influences on feeding behavior follow:

1. **Time of day.** Some fish depend largely on their sense of sight in locating food, while others rely primarily on their senses of taste, touch, and smell. One would expect the fish using sight for feed to be active feeders during the daylight hours. Night feeders rely on the other senses to locate food.

2. **Season of the year.** Some fish, such as largemouth bass, cease feeding activity during their spawning season. Most fish increase feed intake in the spring in the temperate climatic regions when the water temperature starts to rise. As a result, the peak growth period for most fish occurs in the spring and early summer.

3. **Rapid changes of light intensity.** Some fish, such as the yellow perch, show peak feed activity at dawn and dusk.

4. **Physical contact with the food.** Quite often the texture of a potential food source is felt before the fish will consume it.

5. **Water temperature.** Apart from feed and environmental qualities (dissolved oxygen, etc.), water temperature is probably the most important factor affecting appetite and amount of feed consumed. Sharply reduced feed consumption in trout occurs when the temperature decreases to 38°F. The lower limit for feed consumption by catfish is about 50°F.

Size and age of fish are also important factors in determining the nutrient requirements of fish.

Fig. 28-6. The nutrient requirements of catfish are affected by many factors, including size and age. Three sizes/ages of catfish are shown above. *Top,* fingerlings; *Center,* 2 lb market fish; *Bottom,* brood fish. (Top and bottom pictures courtesy Department of Fisheries and Allied Aquaculture, Auburn University, Auburn, Ala. Center picture, courtesy USDA)

Feeding fish is complex. Fish nutrition is truly "an art and a science" because of the complications imposed by the responsiveness of fish to their environment. As a result, fish farmers cannot merely turn to a reference source to find out how much, when, or what to feed. They must be able to recognize the numerous factors affecting feed utilization and adapt their management programs accordingly.

Even though the digestive anatomy, physiology, and feeding habits are different in fish than in the warm-blooded domestic animal, the ultimate nutritional requirements can be expressed in the same terms—maintenance, growth, and reproduction. Since the water environment can supply a number of nutrients, and since such environments vary widely, it is not easy to identify the limiting nutrients to be included in a formulated feed. Most of the research relative to the nutrient requirements of fish has centered around salmon, trout, catfish, and carp. However, experiments and experiences indicate that the nutritional requirements of all coldwater fishes as a group, and of all warmwater fishes as a group, are similar. So, the information presented in this chapter for fish of each group can be used for formulating diets for other species within the group.

One very important factor in establishing the nutrient requirements of fish is the effect of water temperature. To deal with this problem, the National Research Council (NRC) reports Standard Environmental Temperatures (SET) for various species of fish in order that fish producers will have an idea of the applicability of the requirements to their particular systems. The Standard Environmental Temperatures (SET's) used by the National Research Council are: 59°F for chinook salmon, 50°F for rainbow trout, and 86°F for channel catfish. As water temperature deviates from these standards, nutrient requirements increase. If the temperature of the water is lower or higher than the SET for a particular species of fish, feed intake is usually reduced.

A discussion of the nutrient requirements of fish follows, with separate sections devoted to energy, protein, minerals, vitamins, water, and nonnutritional factors.

Energy

The energy requirements of fish are lower than the energy requirements of warm-blooded animals due to the following differences in energy metabolism:

1. Fish have a low basal energy need, because, being cold-blooded, they do not expend energy to maintain body temperature. By not having to regulate body temperature, more energy is available for growth, fattening, and reproduction. It has been estimated that only 70% of the dietary calories are used for maintenance, thus leaving a sizeable number of calories for growth, fattening, and reproduction.

2. Fish have a low energy need for locomotion and voluntary activity; they have no need for large antigravitation muscles because (1) of the buoyancy of their water environment, and (2) a streamlined fish moving through water represents one of nature's most efficient modes of locomotion.

3. Fish use little energy requirement for protein catabolism and waste nitrogen excretion, since ammonia is the principal end product of their protein catabolism as contrasted with the more complex urea and uric acid formed by terrestrial homeotherms. Fishes excrete ammonia principally through the gills and expend little energy in the excretion process.

4. Fish use proteins and fats for energy, preferentially, and carbohydrates sparingly; however, carbohydrates are used more efficiently by warmwater fish than by coldwater fish. By contrast, warm-blooded farm animals and poultry use carbohydrates for energy, preferentially.

The amount of energy required by fish is affected by species, diet, size, age, sex, reproductive stage, water temperature, water quality, light exposure (darkness and the rest period that accompanies it, decreases the energy requirement in some species), and activity.

• **Energy value of feedstuffs for fish**—The standard methods of measuring and expressing energy value of feedstuffs for farm animals and poultry is fully covered (1) in Chapter 4, Nutrients/Metabolism, and (2) in the preliminary section of Section 5, Composition of Feeds. Additionally, four standard energy values for warm-blooded animals —total digestible nutrients (TDN), digestible energy (DE), metabolizable energy (ME), and net energy (NE)—for a great array of feeds are reported in the Composition of Feeds tables in this book. So, only the differences between fish and warm-blooded animals relative to the energy value of feedstuffs will be discussed in this section.

Digestible energy (DE) and metabolizable energy (ME) are used to express feed values and energy requirements of fishes. Theoretically, ME should be superior to DE, since, in fish, it accounts for energy losses from protein via urine and gills. Practically, however, ME offers little advantage over DE in evaluating useful energy in feeds for fish because there are no urinary or gill losses of fats and carbohydrates; hence, the ME and DE of the fat and carbohydrate fractions of feed ingredients or mixed feeds are identical. Gill losses do affect the protein portion of the diet; fish excrete ammonia through the gills directly into the water with little energy expenditure, with the result that protein has a higher energy value for fish than for mammals. Because of their efficiency in eliminating nitrogenous wastes through the gill tissue directly into the water, fish are able to utilize more protein in their diet than is required for maximum growth. Nevertheless—protein is more expensive than fats and carbohydrates; hence, only the optimal amount of protein required for maintenance, growth, and reproduction should be included in the diet, and the less expensive fats and carbohydrates should supply the energy and spare the protein for growth.

From the above, it may be deduced that the energy value of feedstuffs for fish and warm-blooded animals is not far different; it is slightly higher in fish due to their greater efficiency in digesting protein.

• **DE versus ME values**—Insufficient data are available to ascertain whether DE values alone are sufficient or if the extra work and expense associated with determining ME is justified. Moreover, there is lack of agreement relative to the method to determine energy values of feedstuffs for fish; *i.e.*, direct method or indirect method. The direct method involves confining fish in metabolism chambers, along with the difficulties encountered in simultaneously collecting fecal, gill, and urinary excretions. The procedures followed are very similar to the digestion trials with other organisms. Indirect methods commonly involve the use of chromic oxide as a marker and the partial collection of wastes.

Digestibility values are easier to determine; besides, the fish are not stressed when allowed to feed voluntarily. In order to determine whether DE or ME values should be used in evaluating fish feeds, we need to know the effect of feeds on stress, species of fish, temperature, feeding level, activity, water oxygen content, and metabolite accumulation. These and many other questions need to be answered through research.

• **Energy values of fish feeds**—Suitable feeds for fish are included in this book in Section V—Composition of Feeds. Since only limited energy values for fish foods are available, it is suggested that the metabolizable energy values for poultry be used. As a result of a comparison of efficiencies of energy and protein retention between rainbow trout and broiler chickens, Cowley *et al.* concluded that the retention of dietary energy and protein by the rainbow trout is only moderately superior to that of broiler chickens.[2]

Table 28-1 lists the limited, presently available feed ingredient ME values for trout and the DE values for channel catfish and tilapia.

TABLE 28-1
METABOLIZABLE ENERGY VALUES FOR TROUT AND DIGESTIBLE ENERGY VALUES FOR CHANNEL CATFISH AND TILAPIA OF SELECTED FEED INGREDIENTS[1 2]

Feed Name	Metabolizable Energy–Trout	Digestible Energy	
		Channel Catfish	Tilapia
	(kcal/lb)	(kcal/lb)	(kcal/lb)
Alfalfa meal, 17% protein	577	304	459
Blood meal, spray dehy	1,182		
Brewers' grains, dehy	1,086		
Casein	1,777		
Corn, dent yellow grain	682		
Nonheated		500	1,118
Heated		1,150	1,372
Corn, distillers' solubles	954		
Corn gluten meal	1,364		
Cottonseed meal, solv extd 41%	945	1,159	
Cottonseed meal, w/o hulls 50%	1,123		
Crab meal, process residue	1,345		
Fishmeal, anchovy	1,682		
Fishmeal, herring	1,727		
Fishmeal, menhaden		1,773	1,836
Fishmeal, white fish	1,245		
Fish solubles, dehy	1,523		
Gelatin	1,841		
Meat meal w/bone	1,209	1,577	1,336
Poultry by-product meal	1,164		
Poultry, feathers, hydrolyzed	1,309	1,550	
Rapeseed meal (Canola meal)	1,123		
Soybean meal, solv extd		1,173	1,518
Soybean meal, w/o hulls	1,314		
Soybean seeds, heat processed, 175°C	1,654		
Wheat, flour by-product	559		
Wheat, hard red winter, grain		1,159	1,314
Wheat, middlings	559		
Whey, dehy, low lactose	964		
Yeast, brewers', dehy	1,232		
Yeast, torula	1,404		
Oil, menhaden		4,014	
Oil, soybean		4,059	

[1]Adapted by the authors from *Nutrient Requirements of Warmwater Fishes and Shellfishes*, NRC-National Academy Press, 1983, pp. 41–42, Table 12.

[2]To convert kcal/lb to kcal/kg, multiply by 2.2.

[2]Cowley, C. B., A. M. Mackie, and J. G. Bell, *Nutrition and Feeding in Fish*, Academic Press, 1985, p. 113.

As with traditional livestock, fish derive their energy from three sources—carbohydrates, fats, and proteins.

CARBOHYDRATES

Carbohydrates are a major source of energy for humans and domestic animals, but not for fish. The primary sources of carbohydrates in fish feeds are plant feedstuffs, including cereal grains, wheat by-products, soybean meal, and cottonseed meal.

No carbohydrate requirement has been established for fish because carbohydrates do not supply any essential nutrients that cannot be obtained from other nutrients in the feed; *i.e.*, the energy requirement of fish may be satisfied by fat or protein, as well as by carbohydrate. If sufficient energy nutrients are not available in the feed, the organism will catabolize protein for energy at the expense of growth and tissue repair. The use of carbohydrate for energy to save protein for other purposes is known as the *protein-sparing effect* of carbohydrate.

Carbohydrate in excess of the immediate energy need is converted into fat and deposited in various tissues as reserve energy for use during periods of less abundant feed.

In addition to being a source of energy, carbohydrates may also serve as precursors of nonessential amino acids. Starches also improve the pelleting quality of fish feeds.

Dietary fiber is not utilized by fish. Levels over 10% in salmonid feeds and over 20% in catfish feeds reduce nutrient intake and impair the digestibility of feeds. Most of the fiber in the feed ultimately becomes a pollutant in the water.

• **Carbohydrates in salmonid feeds**—Nutritionists have placed the maximum carbohydrate level of salmonid diets at 12 to 20%. The gross energy from carbohydrates available to mammals is 4.15 kcal/g, whereas the value for trout is only 1.6 kcal/g, a 40% relative efficiency. It is noteworthy, however, that recent research has shown that the processing method affects the availability of starch; extrusion processing (as used in making floating pellets) increases the availability of carbohydrate dramatically in salmonid diets.

• **Carbohydrates in feeds for warmwater fish**—Studies have shown that channel catfish and carp can utilize higher levels of carbohydrates than trout—they can use up to 25% carbohydrates as effectively as fats as an energy source. Starches are more readily utilized than sugars.

FATS (LIPIDS)

In nature, fat is the major source of energy for fish. In addition to providing energy, fats serve several other functions for fish such as reserve energy storage, insulation of the body, cushion for vital organs, lubrication, transport of fat-soluble vitamins, and maintenance of neutral buoyancy.

The average gross energy value of fat is 9.45 kcal/g, and the fish has the ability to use 8 kcal for every gram of fat, representing about 84% efficiency of energy utilization.

In nature, fish tend to deposit fat peculiar to the species. However, the diet of fish will alter the type of deposited fat; so, the fat of cultured fish tends to be similar to the fat ingested.

Polyunsaturated (soft) fats in fish feeds are digested easily in both warm and cold water, but saturated (hard) fats are digested efficiently only in warm water.

The following factors should be considered in evaluating fats for fish feeds:

1. **Digestibility.** Fish and vegetable oils, which are polyunsaturated, are more easily digested by fish than saturated fats such as beef tallow, especially at colder temperatures.

2. **Optimal level in the feed.** The optimal level of dietary fat in fish feeds will vary according to the protein content of the feed and the type of fat. Fats are a primary source of energy for fish and have a protein-sparing effect; therefore, high levels in the feed are beneficial. However, excessive fat levels in the feed can hamper the pelleting of feed and cause spoilage during storage if not adequately protected with antioxidants. Also, trout and carp that are fed excessive levels of fat can develop liver lipoid disease (LLD)—ceroidosis. This disease is characterized by a yellow-brown discoloration of the liver due to the accumulation of pigmented insoluble fat which results in the damage or destruction of the parenchymal cells.

Little research has been reported on the effect of fat deficiency in fish. However, it is known (1) that a deficiency of fat in the diet will result in fish utilizing protein for energy; and (2) that a deficiency of the omega-3 fatty acid will result in poor growth, elevated tissue levels of some fatty acids, necrosis of the caudal fin, fatty/pale liver, dermal depigmentation, increased muscle water content, syncope accentuated by stress, increased mitochondrial swelling, increased respiration rate of liver homogenates, heart myopathy, and lowered hemoglobin level.

3. **Fatty acids essential for fish.** Fish require the omega-3 linolenic family of fatty acids. However, the requirement of omega-3 fatty acids does not seem to be as large for warmwater fishes as for coldwater fishes. It is recommended the omega-3 fatty acids be incorporated in salmonoid fish diets at a level of 1%. This amount may be supplied by the addition of fish oil such as herring, menhaden, or salmon at 3 to 5% of the dry diet.

4. **Presence of toxic substances.** Since many pesticides and environmental contaminants are fat-soluble, the producer should make sure that the fat being used has been analyzed and found to be free from harmful contaminants.

5. **Quality of the fat.** Rancidity of fats, especially of polyunsaturated oils, due to oxidation can be a problem in fish feeds. Oxidation of fats often results in the destruction of vitamins, especially vitamin E, and favors mold growth and breakdown of other nutrients. Also, the feeding of rancid fats can impart off-flavor to processed fish. So, only fresh oils protected with antioxidants should be used; and feeds should be stored in a cool, dry area to minimize oxidation of the lipids in the feed.

• **Fat requirements for coldwater fishes**—When there is little or no fat in the feed, trout form their own fat from carbohydrates and proteins. Feed manufacturers use fish oil and vegetable oil in fish feeds as the primary energy source.

Linolenic fatty acids (omega-3 type) are essential for trout and salmon and should be incorporated at a level of at least 1% of the ration for maximum growth response.

The level of dietary fat required for trout and salmon depends on the size of the fish, the protein level in the feed, and the kind of supplemental fat. The recommended percent of fat and protein in the ration dry matter for different ages of trout and salmon follows:

Feed/Size of Fish	Protein	Fat
	(%)	(%)
Starter feeds (fry)	50	19
Grower feeds (fingerlings)	40	15
Production feeds (older fish)	35	12

Some fatty acid deficiency signs reported for trout include poor growth, necrosis of the caudal fin, pale and fatty livers, dermal pigmentation, increased muscle water content, and reduced hemoglobin.

• **Fat requirements for warmwater fishes**—The available data suggest that warmwater fish feeds should contain 10 to 15% fat, and that more than 15% fat in the feed will not improve growth or increase protein deposition.

The level of the omega-3 type fatty acids required by warmwater fishes is unknown, but it is believed to be between 0.3 and 0.5% of the dry diet.

Feed fats affect the taste and storage quality of catfish products. Rancid or off-flavor fish oil has an adverse effect on the flavor of fresh and frozen fish. Fish reared on soybean oil, safflower oil, or corn oil have a better flavor than those fed beef tallow or fish oil.

The principal EFA deficiency signs of warmwater fishes are reduced growth rate, reduced feed efficiency, and in some cases increased mortality.

PROTEIN AS AN ENERGY SOURCE

In fish feeds, fats and carbohydrates are the primary sources of energy, but excess protein, or protein that is deficient in one or more essential amino acids, is also utilized for energy. In order to metabolize protein for energy, it must be deaminated and the carbon skeleton altered in such a manner as to permit it to enter the energy metabolic pathway. However, fish are relatively efficient in using protein, deriving 3.9 of the 4.65 gross kilocalories per gram from protein, for an 84% efficiency.

Protein

The primary objective of fish husbandry is to produce fish flesh, which contains more than 50% protein on a dry weight basis.

Weight gain of fish is essentially proportional to the protein content of the diet at a range of 20 to 40%, but above 50% does not improve growth. Excess protein can be detrimental to fish if not properly balanced with adequate lipid levels; besides, excess protein makes for added feed cost. Fish digest the protein in feeds into amino acids, which are then absorbed into the blood and carried to the cells. Amino acids serve a threefold role in the nutrition of fish: (1) to meet the requirements for formation of the functional body proteins (hormones, enzymes, and products of respiration); (2) to provide tissue repair and growth; and (3) to provide energy.

Fish can synthesize some amino acids, but usually not in sufficient quantity to satisfy their total requirement. So, certain amino acids must be supplied in the feed due to the inability of fish to synthesize them. Fish require the same ten essential amino acids as higher mammals, and they require them in about the same amounts as do broilers. When fish are fed feeds lacking in one or more of the essential amino acids, they become inactive, and lose both appetite and weight. The National Research Council amino acid requirements for fish are given in Table 28-2.

TABLE 28-2
COMPARATIVE AMINO ACID REQUIREMENTS OF GROWING ANIMALS, FOR SELECTED SPECIES[1]

Amino Acid	Japanese Eel	Common Carp	Channel Catfish	Chinook Salmon	Chick	Swine	Rat
Arginine	4.5 (1.7/37.7)	4.2 (1.6/38.5)	4.3 (1.03/24)	6.0 (2.4/40)	5.6 (1.00/18)	1.2 (0.16/13)	5.0 (0.60/12)
Histidine	2.1 (0.8/37.7)	2.1 (0.8/38.5)	1.5 (0.37/24)	1.8 (0.7/40)	1.4 (0.26/18)	1.2 (0.15/13)	2.5 (0.30/12)
Isoleucine	4.0 (1.5/37.7)	2.3 (0.9/38.5)	2.6 (0.62/24)	2.2 (0.9/41)	3.3 (0.60/18)	3.4 (0.41/13)	4.2 (0.50/12)
Leucine	5.3 (2.0/37.7)	3.4 (1.3/38.5)	3.5 (0.84/24)	3.9 (1.6/41)	5.6 (1.00/18)	3.7 (0.48/13)	6.3 (0.75/12)
Lysine	5.3 (2.0/37.7)	5.7 (2.2/38.5)	5.0 (1.5/30)	5.0 (2.0/40)	4.7 (0.85/18)	4.4 (0.57/13)	5.8 (0.70/12)
Methionine[2]	5.0 (1.9/37.7)[3]	3.1 (1.2/38.5)	2.3 (0.56/24)	4.0 (1.6/40)[3]	3.3 (0.60/18)[3]	2.3 (0.30/13)	5.0 (0.60/12)
Phenylalanine[4]	5.8 (2.2/37.7)	6.5 (2.5/38.5)	5.0 (1.2/24)	5.1 (2.1/41)	5.6 (1.00/18)	4.4 (0.57/13)	6.7 (0.80/12)
Threonine	4.0 (1.5/37.7)	3.9 (1.5/38.5)	2.0 (0.53/24)	2.2 (0.9/40)	3.1 (0.56/18)	2.8 (0.37/13)	4.2 (0.50/12)
Tryptophan	1.1 (0.4/37.7)	0.8 (0.3/38.5)	0.5 (0.12/24)	0.5 (0.2/40)	0.9 (0.17/18)	0.8 (0.10/13)	1.3 (0.15/12)
Valine	4.0 (1.5/37.7)	3.6 (1.4/38.5)	3.0 (0.71/24)	3.2 (1.3/40)	3.4 (0.62/18)	3.2 (0.41/13)	5.0 (0.60/12)

[1]From: *Nutrient Requirements of Warmwater Fishes and Shellfishes,* NRC-National Academy Press, 1983, p. 4, Table 3. **NOTE:** Requirements are expressed as percent of protein. In parentheses the numerators are requirements as percent of diet dry matter and the denominators are percent total protein in the diet DM. Data for chinook salmon, chick, swine, and rat are from NRC (1981), NRC (1977a), NRC (1979), and NRC (1978), respectively; data for eel and carp are from information of Nose and Arai (Cowey and Sargent, 1979); data for catfish are from Wilson and coworkers (Harding *et al.,* 1977; Wilson *et al.,* 1977, 1978, 1980; Robinson *et al.,* 1980a,b and 1981).

[2]In the absence of cystine.

[3]Methionine plus cystine.

[4]In the absence of tyrosine.

Several factors determine the requirement for protein in fish feeds; among them—

1. **Temperature.** The protein requirement increases with a rise in temperature.

2. **Age.** Older fish have a lower protein requirement than young fish.

3. **Species.** There are species differences; for example, young catfish require less crude protein than salmonids (see Table 28-3).

4. **Energy content of the feed.** For optimal growth and feed efficiency, there should be a balance between the protein and energy content of the feed.

5. **Feeding rate.** The feeding rate determines the daily amount of protein received by the fish. When the feed allowance is increased, the protein level in the feed can be reduced; when the feed allowance is decreased, the protein level in the feed should be increased to assure that fish receive sufficient dietary protein.

A number of protein deficiency symptoms can be documented. Eye problems are evident—resulting in a cloudy condition in the lens, as well as cataract formation. Osteoporosis and a dystrophy of the epiphyseal cartilages can often be seen. On occasion, tryptophan—an essential amino acid—is deficient in a particular ration, with the result that scoliosis and lordosis occur (reported in salmon and trout).

6. **Size.** The protein requirements for fish are size-related (see Table 28-3).

Table 28-3 gives the recommended protein levels for different species and sizes of fish.

TABLE 28-3
PROTEIN REQUIREMENTS (PERCENT OF TOTAL RATION, BY WEIGHT) FOR RAPID GROWTH IN DIFFERENT KINDS OF FISH[1]

Species	Fry To Fingerlings	Fingerlings To Subadults	Adults and Brood Fish
	(%)	(%)	(%)
Trout and salmon	50	35–40	30–32
Channel catfish	35–40	25–36	28–32
Common carp	43–47	37–42	28–32
Largemouth bass	40	40	35
Striped bass	40	36	35
Eel	50–60	45–50	

[1]Adapted by the authors from: Dupree, H. K. and J. V. Huner, Editors, *Third Report to the Fish Farmers,* U.S. Department of the Interior, Fish and Wildlife Service, 1984, p. 142, Table 11.1.

The quality of protein, reflecting the amino acid content, is most important in optimizing the utilization of dietary proteins. If a ration is deficient in any of the ten essential amino acids, poor growth and decreased efficiency of feed conversion will result, despite a high total protein level in the feed. Animal protein sources are generally of higher quality than plant sources. Synthetic amino acids can be added to fish feed, but more research needs to be done to establish benefits and costs therefrom. However, amino acid balance at reasonable cost is best achieved by using a combination of animal proteins, particularly fish meal, and vegetable proteins.

Fish meal appears to be a highly desirable feed ingredient in fish formulas. Substitutions can be made for most of the ingredients in standard fish formulations, but whenever fish meal is left out poorer growth and feed conversion usually result. Starter diets usually contain at least 15% fish meal (starter trout and salmon diets generally contain at least 50% fish meal), and production and brood stock diets normally contain more than 5% fish meal. Other commonly used protein ingredients for fish diets are: soybean meal, cottonseed meal, corn gluten meal, meat meal, poultry by-product meal, hydrolyzed feather meal, dried blood meal, and dried skimmilk.

Fish cannot utilize nonprotein nitrogen sources such as urea and diammonium citrate, which even nonruminant animals can utilize to a limited extent; thus, these have no value as a feed source for fish. They can even be toxic at high levels.

The chemical composition of fish tissue can be altered by the levels and ingredients in feeds. Within limits, there is a general increase in the percentage of protein in the carcass in relation to the amount in the feed. Also, there is a direct relation between the percentage of protein in the feed and the water content of the carcass. Fish fed low-protein feeds tend to have more fat, less water, and less protein.

• **Protein in salmonid feeds**—The protein requirements for salmon and trout are similar (see Table 28-3).

The level of protein required in feed varies with the quality and proportion of natural proteins that make up the feed. Between 0.5 and 0.7 lb of dietary protein, in a balanced hatchery feed, is required to produce a pound of trout. The requirement for protein is also temperature-dependent. The optimal protein level in the feed for chinook salmon is 40% at 47°F and 55% at 58°F.

• **Protein in catfish feeds**—Protein utilization of catfish is affected by the protein source and water temperature.

Channel catfish convert the best protein source, fish meal, better than they do the best plant source, soybean meal. In catfish feeds, generally 20% of the dietary protein requirement consists of animal protein.

Better efficiency is obtained for all proteins when they are fed to catfish at temperatures between 75°F and 88°F than at 65°F or below.

Minerals

Fish, like domesticated animals, require minerals. However, across the gills and skin, fish have the ability to absorb from, and excrete into, the water a number of minerals, thereby reducing the mineral requirement in the diet. For this reason, research concerning the dietary mineral requirements of fish has been difficult to conduct, and the results have been inconclusive. Most researchers agree that fish require all of the macro- and micro-elements required by other animals for enzymes and cofactors. Calcium and cobalt are readily absorbed and excreted through the gills. While phosphate, chloride, and sulfate ions may be absorbed from the water, they are more readily absorbed in the digestive tract.

FISH MINERAL CHART

More research is necessary on the requirements, functions, and interactions of minerals in the diet and water of fish. In the meantime, based on presently available information, Table 28-4 summarizes the mineral deficiency symptoms and gives (1) the *requirements* for coldwater fish, and (2) the recommended *allowances* for warmwater fish. **NOTE WELL**: The figures for coldwater fish are *requirements,* whereas the figures for warmwater fish are *allowances.*

TABLE 28-4
FISH MINERAL CHART

	Deficiency Symptoms	Coldwater Fish Requirement in Feed[1]	Warmwater Fish Recommended Allowance in Feed	
		(%)	(%)	(mg/kg)[2]
Major or Macrominerals:				
Calcium	*Trout:* Poor growth, appetite, and feed efficiency. *Catfish:* Reduced growth and lower carcass ash, calcium, and phosphorus.	0.2–1.0	0.35–0.45[3]	3,500–4,500[3]
Phosphorus	*Coldwater and warmwater fish:* Reduced growth, appetite, feed conversion, and bone ash. Additionally, salmonids show skeletal abnormalities and bone deformities.	0.7–0.8 (inorganic P)	0.45[3] (available P)	4,500[3] (available P)
Magnesium	*Coldwater and warmwater fish:* Poor growth, loss of appetite, sluggishness, muscle flacidity, high mortality, and depressed magnesium levels in the body. Additionally, trout show renal calculi, vertebral curvature, degeneration of the muscles, and high mortality.	>.006	0.04[3]	400[3]
Trace or Microminerals:				
Cobalt	*Coldwater and warmwater fish:* Required for the synthesis of vitamin B-12 by gut bacteria.		0.000005[4]	0.05[4]
Copper	*Warmwater fish:* Depressed growth.		0.0005[4]	5.0[4]
Iodine	*Salmonids:* Thyroid hyperplasia (goiter).	0.00006–0.00011	0.0005[4]	5.0[4]
Iron	*Coldwater and warmwater fish:* Hypochromic, microcytic anemia.		0.003[4]	30.0[4]
Manganese	*Coldwater and warmwater fish:* Depressed growth and loss of appetite. Additionally, trout show abnormal tail growth and shortening of the body.		0.001[4]	10.0[4]
Selenium	*Salmonids:* Muscular dystrophy and exudative diathesis.	0.1–0.35	0.1[4]	1,000[4]
Zinc	*Salmonids:* Cataracts, caudal fin erosion, and depressed growth. *Channel catfish:* Poor growth and appetite.	0.0015–0.003	0.015[4]	150[4]

[1]From: Hilton, J. W. and S. J. Slinger, *Nutrition and Feeding of Rainbow Trout,* Department of Nutrition, College of Biological Science, University of Guelph, Guelph, Ontario; published by Department of Fisheries and Oceans, Ottawa, 1981, p. 6, Table 3.

[2]1 mg is the same as 1 ppm.

[3]From: Lovell, R. T., *Nutrition and Feeding of Channel Catfish* (Revised), Southern Cooperative Series Bul. No. 296, 1984, p. 28. These are dietary mineral requirements for catfish.

[4]From: Gatlin, D. M. III, University of Arkansas at Pine Bluff, Pine Bluff, Ark., in a personal communication to the senior author, 1986. These are recommendations for channel catfish.

Fig. 28–7. Thyroid hyperplasia (goiter) induced by iodine deficiency. When thyroid tumors in fish were first described in 1891, it was believed that the condition represented a form of throat cancer. After considerable debate and research, it was found that the tumors resulted from a dietary iodine deficiency—goiter. Today, iodine is routinely added to fish diets, and thyroid tumors are rarely seen in commercial fish production. (From Gaylord & Marsh; print provided by J. E. Halver, School of Fisheries, University of Washington, Seattle)

Because of the difficulties in determining the mineral requirements of fish, along with the lack of sufficient fish mineral research, some nutritionists suggest that a trace mineral mixture suitable for poultry be included in fish feeds.

Vitamins

As the digestive system of the fish is monogastric in structure and function, there is definite need for vitamins in fish diets. In general, the vitamin requirements of fish resemble those of monogastric animals, with a few exceptions. Fish represent one of the few types of higher animals that have been shown to have a requirement for vitamin C. There is not enough bacterial activity in the gut of the fish to satisfy either the B complex or the vitamin K requirements.

Many of the presently available requirement levels for vitamins involved the use of semipurified diets and were based to a large extent on liver storage of a particular vitamin. Additional studies are needed further to validate earlier requirements, using improved purified diets, and particularly, practical diets. In assessing the requirement for a particular vitamin, as many as possible of the following parameters should be employed: weight gain; feed efficiency; mortality; absence of deficiency symptoms; vitamin tissue storage levels; blood analysis—hemoglobin level, hematocrit value, and red blood cell count; carcass moisture, protein, and lipid content; and vitamin-specific enzyme activities.

FISH VITAMIN CHART

The presently recommended vitamin requirements and allowances for fish are given in Table 28–5. Vitamin losses in fish feeds occur due to oxidation and reaction with other feed components. For this reason, amounts in excess of the requirements should be added—the latter are known as recommended allowances.

Prior to 1970, only "complete" fish feeds contained added vitamins. Today, almost all feeds for fish contain vitamin supplements. It is noteworthy that the vitamin requirements of fish differ between species and increase as the fish grows.

TABLE 28–5
FISH VITAMIN CHART

	Coldwater Fishes[1]				Warmwater Fishes[2]			
	Requirements[3]		Recommended Allowances[4]		Supplemental Allowances[5]		Complete Diet Allowance[5]	
	(per lb body wt./day)	*(per kg body wt./day)*	(per lb dry feed)	*(per kg dry feed)*	(per lb dry feed)	*(per kg dry feed)*	(per lb dry feed)	*(per kg dry feed)*
Fat-Soluble Vitamins:								
A .... (IU)	34	*75*	1,136	*2,500*	1,000	*2,200*	2,500	*5,500*
D .... (IU)	33	*72*	1,091	*2,400*	100	*220*	454	*1,000*
E .... (IU)	0.45	*1*	13.6	*30*	5	*11*	23	*50*
K .... (mg)	0.05	*0.1*	4.5	*10*	2.3	*5*	4.5	*10*
Water-Soluble Vitamins:								
Biotin .... (mg)	0.02	*0.05*	0.45	*1*	0	*0*	0.05	*0.1*
Choline .... (mg)	14–23	*30–50*	1,364	*3,000*	200	*440*	250	*550*
Folacin (Folic Acid) .... (mg)	0.09	*0.20*	2.3	*5*	0	*0*	2.3	*5*
Inositol .... (mg)	8–9	*18–20*	182	*400*	0	*0*	45	*100*
Niacin (Nicotinic Acid, Nicotinamide) .... (mg)	1.4–2.7	*3–6*	68	*150*	7.7–12.7[6]	*17–28*[6]	45	*100*
Pantothenic Acid (Vitamin B-3) .... (mg)	0.45	*1*	18	*40*	3.2–5[6]	*7–11*[6]	23	*50*
Riboflavin (Vitamin B-2) .... (mg)	0.2–0.5	*0.5–1.0*	9	*20*	0.9–3.2[6]	*2–7*[6]	9.1	*20*
Thiamin (Vitamin B-1) .... (mg)	0.07–0.09	*0.15–0.20*	4.5	*10*	0	*0*	9.1	*20*
Vitamin B-6 (Pyridoxine, Pyridoxal, Pyridoxamine) ... (mg)	0.09–0.18	*0.2–0.4*	4.5	*10*	5	*11*	9.1	*20*
Vitamin B-12 (Cobalamins) .... (mg)	0.0003	*0.0006*	0.009	*0.02*	0.9–4.5	*2–10*	9.1	*20*
Vitamin C (Ascorbic Acid, Dehydroascorbic Acid) ..... (mg)	1.4–2.7	*3–6*	45	*100*	0–45[6]	*0–100*[6]	14–45[6]	*30–100*[6]

[1]From: *Nutrient Requirements of Coldwater Fishes,* NRC-National Academy of Sciences, 1981, p. 41, Table 7.

[2]From: *Nutrient Requirements of Warmwater Fishes,* NRC-National Academy of Sciences, 1977, p. 18; and *Third Report to the Fish Farmers,* edited by Dupree, H. K. and J. V. Huner, published by the U.S. Department of the Interior, Fish and Wildlife Service, 1984, p. 147, Table 11.5.

[3]Based on young fish.

[4]Total vitamin contribution from all sources. Other amounts may be more appropriate to offset losses resulting from the effects of formulation and storage, or when feeding other than small fish at SET.

[5]These amounts do not allow for processing or storage losses. Other amounts may be more appropriate for various species and under various environmental conditions.

[6]Highest amounts probably appropriate when "standing crop" of fish exceeds 500 kg/hectare of water surface.

VITAMIN DEFICIENCY SYMPTOMS

The symptoms of vitamin deficiencies are given in Table 28-6.

TABLE 28-6
SYMPTOMS OF VITAMIN DEFICIENCIES IN FISHES[1]

	Vitamin Deficiency (Hypovitaminosis) Symptoms
Fat-Soluble Vitamins:	
A	*Salmonids:* Cataracts, intolerance to light, anemia. *Catfish:* Exophthalmia, edema, hemorrhaged kidneys. *Carp:* Faded color, exophthalmia, warped operculum, fin and skin hemorrhages.
D	*Salmonids:* Lethargy, tetany-like contractions, fatty liver, muscles, and carcass, droopy tails. *Catfish:* Low bone ash. *Carp:* None tested.
E	*Salmonids:* Anemia, depigmentation, heart abnormality. *Catfish:* Depigmentation, fatty liver, anemia, mortality. *Carp:* Muscular dystrophy, mortality.
K	*Salmonids:* Hemorrhages, increased prothrombin time, pale gills. *Catfish:* Skin hemorrhages. *Carp:* Not tested.
Water-Soluble Vitamins:	
Biotin	*Salmonids:* Loss of appetite, pale gills, high glycogen in liver. *Catfish:* Depigmentation, anemia. *Carp:* Poor growth.
Choline	*Salmonids:* Hemorrhages, fatty livers, poor growth. *Catfish:* Enlarged liver, hemorrhages in kidneys and intestines, reduced weight gain. *Carp:* Fatty liver.
Folacin (Folic Acid)	*Salmonids:* Anemia, fragility of caudal fin, lethargy, pale gills. *Catfish:* Loss of appetite and lethargy. *Carp:* None detected.
Inositol	*Salmonids:* Loss of appetite, poor growth, poor feed efficiency. *Catfish:* None detected. *Carp:* Skin lesions.
Niacin (Nicotinic Acid, Nicotinamide)	*Salmonids:* Swollen gills, intestinal lesions, poor coordination, anemia. *Catfish:* Skin and fin lesions, hemorrhages, prominate eyeball, tetany, lethargy, mortality. *Carp:* Skin hemorrhages, mortality.
Pantothenic acid (Vitamin B-3)	*Salmonids:* Clubbed gills, anemia, sluggish, prostration. *Catfish:* Clubbed gills, anemia, erosion of skin, barbels mortality. *Carp:* Poor growth, anemia, skin hemorrhages, exophthalmia.
Riboflavin (Vitamin B-2)	*Salmonids:* Cataracts, anemia, dark coloration. *Catfish:* Short body, dwarfism, opaque eye lens. *Carp:* Skin and fin hemorrhages, mortality.
Thiamin (Vitamin B-1)	*Salmonids:* Convulsions, neuritis. *Catfish:* Dark coloration, reduced gains, lethargy, mortality. *Carp:* Fin congestion, nervousness, fading body color.
Vitamin B-6 (Pyridoxine, Pyridoxal, Pyridoxamine)	*Salmonids:* Anemia, nervousness, fits, erratic swimming. *Catfish:* Nervousness, tetany, erratic swimming, greenish blue color, mortality. *Carp:* Nervous disorders.
Vitamin B-12 (Cobalimins)	*Salmonids:* Anemia, fragmented and immature erythrocytes. *Catfish:* Reduced hematocrit, reduced weight gains. *Carp:* None detected.
Vitamin C (Ascorbic acid, Dehydro ascorbic Acid)	*Salmonids:* Spinal deformities, anemia, lethargy, prostration. *Catfish:* Scoliosis, lordosis, reduced growth, dislocated vertebrae, popeye, bloated belly, reduced bone collagen. *Carp:* None tested.

[1]In the preparation of this table, the authors adapted material from two primary sources: Salmonids from *Nutrition and Feeding of Rainbow Trout,* Hilton, J. W. and S. J. Slinger, Department of Nutrition, College of Biological Science, University of Guelph, Guelph, Ontario, published by Department of Fisheries and Oceans, Ottawa, 1981, p. 5, Table 2; and catfish and carp from *Nutrient Requirements of Warmwater Fishes and Shellfishes,* NRC-National Academy of Sciences, 1983, p. 16, Table 8. **NOTE WELL**: In addition to catfish and carp, other warmwater species are listed in the NRC publication listed above.

FAT-SOLUBLE VITAMINS

The fat-soluble vitamins, A, D, E, and K, are all of practical importance in fish feeds.

Vitamin A

Vitamin A is essential for the normal structure and functions of eye and gill and for general maintenance of differentiated epithilia in various physiological systems.

Fish are susceptible both to deficiencies and excesses of vitamin A. At certain temperatures and levels some species of fish are capable of using pro-vitamin, beta-carotene, to fulfill their requirements, while other species must have vitamin A *per se* added to the diet.

Vitamin A can be supplied by rich feed sources, synthetic vitamin A, or fish oils.

• **Toxicity of vitamin A**—The producer should be aware of the possibility of toxic amounts of vitamin A when feeding certain fish by-products, such as shark or tuna viscera or whale oil. Whale oil contains large amounts of kitol, a compound that is an inactive form of vitamin A. When heated to temperatures in excess of 392°F, kitol becomes as biologically active as pure vitamin A. Kitol is thought to be a defense mechanism against hypervitaminosis-A in the whale.

Vitamin D

Dietary vitamin D is needed by fish for facilitating mobilization, transport, absorption, and utilization of calcium and phosphorus. Rainbow trout fingerlings require a minimum of 727 to 1,090 IU of D_3 per pound of semipurified diet. A minimum level of 450 IU of D_3 per pound of diet will prevent deficiency symptoms in warmwater fish. Vitamin D_3 has greater activity for fish than vitamin D_2.

Channel catfish on a deficient diet show reduced weight gain, along with lower body ash, calcium, and phosphorus.

Vitamin E

Vitamin E has two functions for fish: (1) as an intracellular antioxidant and (2) metabolic functions unrelated to cellular antioxidants.

The requirement for vitamin E in fish diets is dependent on several factors—the amount and type of dietary polyunsaturated fatty acids, storage facilities, vitamer used, and diet preparations. Antioxidants, such as butylated hydroxyanisole (BHA), ethoxyquin, and butylated hydroxytoluene (BHT), can be added to the diet to prevent spoilage, but vitamin E is still required as an intracellular antioxidant. Cases of both hypovitaminosis and hypervitaminosis of vitamin E have been reported in fish.

Vitamin E is generally provided in the form of synthetic dl-alpha-tocopherol.

Vitamin K

Research relative to the dietary requirements of vitamin K in fish has been extremely limited. However, qualitatively, vitamin K has been shown to be a dietary essential, involved in electron transport and oxidative phosphorylation reaction of cellular metabolism. The normal supplemental source is menadione (vitamin K_3).

WATER-SOLUBLE VITAMINS

All of the B vitamins and vitamin C are soluble in water. Water-soluble vitamins must be replenished almost daily, for there is little tissue storage. Fish require biotin, choline, folacin, inositol, niacin, pantothenic acid, riboflavin, thiamin, vitamin B-6, vitamin B-12, and vitamin C.

Biotin

Biotin plays an important role in the metabolism of carbohydrates, fats, and proteins. Most animal products are good sources of biotin, but diets high in animal products are susceptible to oxidation, and protective measures are necessary.

Biotin is synthesized in some fish by intestinal flora, but supplemental biotin in the diet may be required for maximum growth. Most diets containing fish meal will probably contain sufficient biotin for normal growth of young fish.

Choline

Choline, which is a key part of lecithin, is vital for the prevention of fatty livers, the transmission of nerve impulses, and the metabolism of fat. Choline hydrochloride is routinely added as a supplement to fish diets. In order to prevent this additive from reacting with vitamin K and alpha-tocopherol, it is recommended that it be added to the feed in a water carrier and that vitamin E and K be added in a fat carrier. Since choline also reacts with vitamin C, another water-soluble vitamin, they should be added to fish feeds separately.

Folacin (Folic Acid)

Folacin is essential for the synthesis of nucleic acids, DNA, and RNA, and thus necessary for normal erythrocyte formation.

In the past, folacin levels were low in dry diets for trout, with the result that feeding these diets often resulted in anemia. Today, the problem is recognized and the supplementation with folacin is routine. In pond culture, fish have access to natural sources of folic acid—insects and algae—and subsequent supplementation of folic acid in the diets has not been as critical. Folacin is extremely susceptible to degradation in processing and storage. In anticipation of this destruction, levels in excess of metabolic needs are routinely added to fish foods.

Inositol

The classification of inositol as a vitamin is disputed; more properly perhaps, it should be classified as an essential nutrient.

Research on inositol requirements in fish is rather limited. The compound is stable throughout feed preparation and storage. Since inositol is abundant in most biological tissues, the risks of dietary deficiencies occurring are minimal in most conventional diets.

Niacin (Nicotinic Acid, Nicotinamide)

Niacin is a collective term for nicotinic acid and nicotinamide. In the body, they serve as coenzymes, to produce energy within the cells precisely when needed and in the amount necessary.

Mammals have the ability to use limited amounts of tryptophan to synthesize niacin. Fish, to a very limited degree, have this ability; but supplementary niacin is necessary in the diet to maximize growth.

Nicotinic acid and nicotinamide provide about the same biological activity for fish, and both forms are stable in multi-vitamin premixes. However, losses in nicotinic acid activity due to processing of extruded feeds may be 20% or more.

Pantothenic Acid (Vitamin B-3)

Pantothenic acid is a dietary essential for fish. It is an important constituent of coenzyme A, and it plays a key role in body metabolism and release of energy from all three energy-yielding nutrients—carbohydrates, lipids, and proteins.

Nutritional gill disease (caused by pantothenic acid deficiency) is one of the more common nutritional disorders found in trout hatcheries. The free form of pantothenic acid is heat and pH sensitive, and processing and storage can cause substantial losses of the vitamin. Pantothenic acid as salts of Na^{+} or Ca^{++} should be added to the diet as these forms are relatively stable.

Riboflavin (Vitamin B-2)

Riboflavin is a component of two flavoprotein coenzymes, needed for the breakdown of pyruvate, fatty acids, and amino acids.

In most fish diets, the producer does not have to be concerned with the loss of riboflavin in storage if the feed is protected from light. Storage of fish feed in light-proof bags is recommended.

Thiamin (Vitamin B-1)

Thiamin is essential for the metabolism of carbohydrates and lipids, and for the direct oxidative cellular metabolism of glucose.

Several dietary factors can be instrumental in determining the thiamin requirement of fish. Many fish contain relatively high levels of thiaminase—an enzyme that breaks thiamin into two inactive molecules. Since many fish diets contain fish products, the producer should be aware of the potential risks of feeding raw fish (see Chapter 27, Feeding Mink, sections on "Thiamin" and "Chastek Paralysis"). Cooking destroys the enzyme. The dietary level of fat can also affect the thiamin requirement. A high-fat diet can have a thiamin-sparing effect, but it will not totally alleviate the requirement for thiamin. Rather, a high-fat diet will merely delay the onset of deficiency symptoms if there is insufficient thiamin.

Thiamin hydrochloride and thiamin mononitrate are the usual synthetic sources of thiamin used in diet supplementation.

Vitamin B-6 (Pyridoxine, Pyridoxal, Pyridoxamine)

Vitamin B-6 is a key coenzyme in protein metabolism. It follows that fish consuming high-protein diets—fish that are in the growing stage—have high requirements for pyridoxine. In carnivores, the pyridoxine requirement is especially high, and the body stores are rapidly depleted.

Phosphorylated pyridoxal compounds are relatively stable in feed, unless the premix also contains trace minerals. In general, when needed, it is recommended that the vitamin be added to feed in the form of pyridoxine hydrochloride.

Vitamin B-12 (Cobalamins)

Vitamin B-12, also known as cyanocobalamin, is a large molecule that contains cobalt.

Animal products have traditionally been considered excellent sources of vitamin B-12 but care should be taken in processing and storage since B-12 is easily destroyed. Cool and dry storage, as well as rapid turnover of stored feeds, is recommended in order to ensure adequate dietary levels of vitamin B-12.

Vitamin C (Ascorbic Acid, Dehydroascorbic Acid)

Vitamin C is primarily involved in the formation of procollagen and is necessary for the biosynthesis of collagen and cartilage, tissue formation and repair, and calcification of bone. Also, it functions in the conversion of hydroxyproline to proline and metabolism of tyrosine.

Most birds and mammals can synthesize vitamin C, but most fish cannot. The amount of vitamin C required by fish is dependent on several factors—degree of stress, size of fish, and dietary nutrients. In the past, citrus fruits, cabbage, and some animal products were used for vitamin C supplementation, but synthetic ascorbic acid is now widely used. Vitamin C is very unstable to heat, moisture, and oxidation.

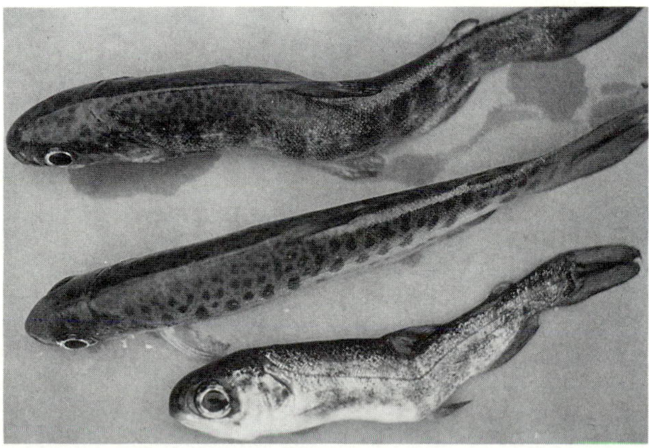

Fig. 28-8. Vitamin C deficiency symptoms in coho salmon (top and bottom). *Top:* SCOLIOSIS—a lateral curvature of the spine. *Bottom:* LORDOSIS—a forward curvature of the spine. *Center:* Normal (control) fish that received adequate vitamin C. (Courtesy, Dr. J. E. Halver, School of Fisheries, University of Washington, Seattle)

Water

In addition to its role as a nutrient carrier, water serves additional functions in fish. In freshwater fish, the concentration of ions in the blood is stronger than in the water of the environment. Since water always diffuses from the area of weakest ionic concentration to the strongest, fresh water readily diffuses through the gills and digestive tract into the fish. In saltwater fish, the blood ion concentration is weaker than that of marine water, consequently forcing the fish to absorb nutrients from the environment.

Water also acts as a medium for carrying oxygen. The amount of dissolved oxygen in water depends upon the movement of air over the water's surface, the movement of the water itself, the population of aquatic plants, the amount of sunshine, the population and activity of aquatic animals, and water temperature. Water temperature is critical because less oxygen is dissolved as the temperature increases and, equally important, as the temperature increases, the metabolic rate of aquatic animals accelerates, increasing the oxygen demand. On a physiological basis, if the water temperature is increased and there is less oxygen in the water, the fish has to increase its respiratory rate; and the demand for energy in the fish to sustain this activity is increased.

Nonnutritive Factors

While nonnutritive factors do not directly contribute to the maintenance, growth, or reproduction of fish, they should be considered in the formulation of rations as they can affect feed efficiency or the quality of the final marketable product. Three nonnutritive factors—antioxidants, fiber, and pigment-producing factors—warrant discussion concerning fish nutrition.

ANTIOXIDANTS

Due to the high-fat content of fish diets, along with the highly unsaturated nature of the fats, the dangers of oxidation and subsequent spoilage of feeds can present major problems. One means of controlling these problems is the addition of antioxidants to the feed, such as ethoxyquin, butylated hydroxyanisole (BHA), butylated hydroxytoluene (BHT), or vitamin E. The main disadvantage in the use of these chemicals is that the levels allowable in feed are adequate only for the prevention of oxidation of animal fats. Vegetable oils and fish oils are highly unsaturated, with the result that high levels of antioxidants are required—far higher than the regulatory guidelines will permit. Vitamin E functions in the fish's physiological system as well as in feed preservation. But the level of vitamin E in fish feeds must be adequate to prevent oxidation of oils and still meet the nutritional requirements of the fish.

FIBER

Due to the simplicity in the structure of the gastrointestinal tract of fish, the digestibility of fiber is extremely low—about 10%. Very little microbial breakdown of fiber has been noted. Herbivore fish have the ability to tolerate higher amounts of fiber than carnivores; nevertheless, it is still recommended that crude fiber not exceed 10% of fish diets—preferably not more than 5 or 6%. Fiber does serve as a source of bulk, which facilitates the passage of ingesta through the digestive tract.

PIGMENT-PRODUCING FACTORS

Quite often, producers wish to enhance the skin and tissue color in order to make the product more attractive to the consumer. This can be achieved through the addition of certain ingredients to the diet. Paprika fed at a level of 2% of the diet will improve the coloration of brook trout. Xanthophylls from alfalfa meal, corn gluten meal, and dried egg products will increase yellow pigmentation in brown trout skin. The use of shrimp or prawn wastes, which contain carotinoids, will produce a healthy "rosy" color when fed to trout. Also, commercial sources of coloring agents are now available for addition to fish feeds. Species differences with regard to the amount of color change have been observed, so it is possible that what works with one type of fish will not be effective with another.

FEEDS FOR FISH

Feed for commercial fish production accounts for 30 to 50% of the total expense. So, if fish farmers wish to maximize production at the lowest cost, they must know the nutritional requirements of the fish species which they produce. Also, they must recognize and, to the extent practical, maintain the environment to which fish are exposed. Producers have the good fortune of being able to select a type of fish which can adapt to their environment.

Fig. 28-9. A feed truck delivering bulk feed to feed bins on a fish farm. (Courtesy, Agricultural Experiment Station, University of Arkansas at Pine Bluff)

Good quality feeds, along with proper processing and feeding techniques, are essential in the intensive production of fish. Feeds that are not eaten do not produce fish; worse yet, they often reduce production and fish quality by contributing to oxygen demand and water quality deterioration. Several types of fish feeds are available, with the choice determined by the formulation and nutrient content, the cost of the feed, the behavioral characteristics of the species, and the size of fish. Also, the feed preferences of different species often change during their life span, and are often influenced by the type of culture system being used and the level of production. Suitable feed ingredients are included in this book in Section V—Composition of Feeds.

The processing methods used in preparing fish feeds are similar to those used in preparing feeds for domestic animals (see Chapter 14, Feed Processing).

Natural Foods and Feeds

As the term implies, natural foods are obtained from the immediate environment. Small fish feed upon algae and zooplankton. As fish grow, they devour progressively larger natural foods—insects, worms, mollusks, crustaceans, small fish, tadpoles, frogs, and plants.

Pondfish culturalists take advantage of the natural foods present. The insects, worms, and forage fish which pond fish consume are high in water—containing 75 to 80%. The remaining components are: protein, 12–15%; fat, 3–7%; ash, 1–4%; and carbohydrate, less than 1%. During warm weather when insects hatch and bottom organisms are abundant, a pond can provide a considerable amount of food for fish. The food production can be increased by pond fertilization with chemical fertilizers, organic materials, and animal manures. Because the environment tends to be highly variable in its production of biomass, this method of providing food is inefficient unless the producer is utilizing large bodies of water. However, natural food organisms are relied upon to provide nutrients lacking in supplemental feeds used in pond culture.

• **Fertilizing farm ponds**—A fertilized pond will usually produce three to four times as many pounds of fish each year as a nonfertilized pond. The well-fertilized, well-managed pond will usually produce 200 to 400 lb of fish per acre per year.

Fertilizer may be applied to ponds to produce phytoplankton (tiny green plants or "bloom"). The increased phytoplankton provide more food for small fish and insects, which, in turn, are eaten by larger fish.

Fertilizer recommendations vary from area to area, depending primarily on soil type, water quality, and temperature. Also, the use of organic matter such as barnyard manure, hay, or oilseed protein meals in conjunction with the chemical fertilizer may be desirable. The fish farmer should obtain the fertilizer recommendation (kind of fertilizer, and rate and schedule of application) from the local County Extension Agent, or from the fish specialist(s) of the U.S. Department of the Interior Fish and Wildlife Service, the U.S. Department of Agriculture, or the university of the state in which the fish farm is located.

Artificial Wet Feeds

Artificial wet feeds contain various organs, meats, and byproducts from animals, poultry, and fish. The most common ingredients in wet feeds are liver, spleen, ovaries, intestines, blood, testicles, condemned meat, trash fish, kidneys, fish scraps, mollusks, brain, meat trimmings, heart, poultry offal and by-products, and milk by-products. When feeding artificial wet diets, it is important to ensure that sufficient amounts of the omega-3 fatty acid (linolenic acid family) is present. Also, fish products contain thiaminase, necessitating heat treatment of these fish ingredients or supplemental thiamin.

Formulated Feeds

Most commercial fish producers are now using dry formulated feeds, that contain a combination of both plant and animal ingredients. The most commonly used grains are wheat and corn. The most commonly used grain by-products are brewers' grains, corn gluten meal, cottonseed meal, peanut meal, soybean meal, rice bran, and various wheat by-products. The most commonly used animal by-products are blood meal, feather meal (hydrolyzed), fish meal, poultry by-product meal, shrimp meal, and whey.

Formulated feeds may be either supplemental or complete rations.

- **Supplemental fish feeds**—These feeds are formulated to provide adequate energy and protein, but they may be deficient in minerals and vitamins which the fish are expected to obtain from natural foods. Such feeds are fed to fish reared in low densities in ponds.

- **Complete fish feeds**—These feeds are formulated to provide all the essential nutrients required by fish for optimal growth. If high densities of fish are being reared, a complete feed must be provided, as natural feeds will be limited or absent. Complete feeds must be (1) of a physical consistency that will allow them to be fed in the water with minimum leaching, yet ingested and digested by the fish; (2) properly sized for different sizes of fish; (3) palatable to fish so that it will be readily consumed and not left to dissipate into the water; and (4) relatively free from dust and fine particles, which are not well consumed, and which, in excess, may cause water pollution.

Fig. 28-10. Concrete raceways for trout. Fish grown in raceways require feed that is more complete than when fish are in ponds where they have access to natural feeds. (Courtesy, USDA)

Formulated feeds are manufactured in compressed (sinking) pellets, expanded (floating) pellets, moist or semi-moist pellets, crumbles (granules), meals, or flakes.

Fig. 28-11. Pellets for fish. *Left:* compressed or sinking pellets. *Right:* expanded or floating pellets. (Courtesy, Department of Fisheries and Allied Aquaculture, Auburn University, Auburn, Ala.)

Dry pelleted feeds have several advantages over other feeds; among them, (1) availability at all times of the year in any quantity, (2) the size of pellets can be altered so as to be suitable for the size of the fish consuming it, (3) they give improved feed conversions and lower feed cost per unit of weight gain, and result in less waste and contamination of the water than meals or wet feeds, and (4) they lend themselves to lower cost bulk handling, storage, and automatic feeding.

- **Compressed or sinking pellets**—These pellets are made by adding steam to the feed as it is pelleted. The steam increases the moisture content by 5 to 6% and raises the temperature to 150-180°F during processing. The feed mixture is forced through a die (dies are available in different sizes) to extrude a compressed, dense pellet. The pellets are cooled and air dried to no more than 10% moisture immediately after pelleting.

- **Expanded or floating pellets**—These pellets require higher temperatures and pressures than compressed or sinking pellets. Under these conditions, raw starch is gelatinized; and bonds are formed within the gelatinized starch to give a durable, water-stable pellet. The sudden release of pressure following extrusion allows water vapor to expand and the ensuing entrapment of gas creates a buoyant, floating food particle. The major disadvantages of floating pellets are the higher cost (8 to 15% higher than sinking pellets), greater bulk (which reduces feed intake), and possibly some vitamin destruction due to the high temperature required in processing. Many fish producers prefer floating feeds because they can observe the fish feeding, which aids in management and reduces feed wastage due to overfeeding. Also, extruded pellets are very durable and breakage of pellets with the production of fines and consequent wastage is considerably reduced. Recent studies with catfish have shown that feeding 15% of the ration as floating feed and 85% as sinking feed gives better feed utilization and is more economical than feeding either alone.

• **Moist and semi-moist pellets**—Moist pellets, which contain 30–50% moisture are made from variable amounts of either fresh or frozen pasteurized fish, together with some dry ingredients. No heat is required for pelleting moist feeds. Refrigeration must be used to protect moist feeds against spoilage. After extrusion, moist pellets should be quick-frozen and stored at −14°F. Moist pelleted feed spoils rapidly when thawed. Also, moist feeds cost more to manufacture, ship, and store than dry pelleted feeds because they must be kept frozen. Salmon producers are the major users of moist feeds.

Semi-moist pellets contain 20–25% moisture, which is intermediate in moisture between moist and dry pellets. They do not require refrigeration, but they must be protected by adequate mold inhibitors and preservatives.

• **Crumbles (granules)**—These are made by crushing pellets, followed by screening out the granules to the desired sizes. The finished feed should be sized and contain not more than 15% oversize or undersize granules. When adding oil or fat, not more than 3% fat should be included at the time of mixing; added fat may be sprayed on crumbles after manufacture.

• **Meals and flakes**—Meals are often coated with vegetable oils or animal fats to increase energy level and improve flotation. They are usually scattered over the surface of the water. Meals are fed to bait minnows, goldfish, and fry of striped bass, grass carp, and sunfish.

Flake feed is prepared for aquarium fishes. It is usually sprinkled on the water surface.

FEED HANDLING AND STORAGE

Two rules should be followed in the handling and storage of pelleted dry feed for fish. They are:

1. **Handle with care.** Pellets are fragile and easily broken. When breakage occurs, fines from the pellet represent lost feed and can cause water pollution.

In order to reduce the fines, avoid rough treatment of feed. If the feed is bagged, do not walk or stand on the bags. Use machinery that will not break the pellets. Screw-type augers are known to be extremely hard on pelleted feeds.

2. **Store feed in a cool, dry place and use it within 90 days.** Due to the high protein and fat levels in fish diets, the potential of ingredient spoilage is extremely great. The storage area should be kept clean and adequately ventilated. Protection from insects and rodents and from chemical contaminants should receive high priority in feed storage. When storing bagged feed, use multiwalled paper and/or plastic bags instead of burlap. Besides affording extra protection from breakage, plastic-lined paper bags protect the feed from moisture and oxidation. Storage bins should be clean and protected from the heat. Gravity flow bins are probably best because the feed that has been stored for the longest time is the first to be used. Moist feeds should be refrigerated at temperatures of −10°F or lower.

(Also see Chapter 14, Feed Processing.)

RATIONS FOR FISH

Because the cost of feed is the largest single item of expense in fish production, the fish farmer must utilize feed as efficiently as possible. An excess of a certain ingredient could well be as economically detrimental to overall production expenses as a deficiency. If producers are to obtain efficient feed conversions, they must adapt the feed to the particular needs of the fish. Water temperature, size and species of fish, and stocking density are probably the most critical factors to consider when formulating a ration.

Trout and Salmon Diets

In trout and salmon diets, protein levels should not be less than 45% in starter diets, not less than 40% in production diets, and not less than 35% in brood stock diets. Fat levels should be 15 to 20% in starter diets, 10–15% in production diets, and 10 to 15% in brood stock diets. Crude fiber should not exceed 4% in starter diets, and not exceed 5% in production and brood stock diets. The starter diet should not be fed longer than necessary to get the fish off to a good start; prolonged feeding of the starter can cause gill irritation, so it is important to switch to larger size particles as soon as possible. A list of commonly used feed ingredients for trout, along with their recommended feeding levels, is found in Table 28–7.

TABLE 28–7
SUGGESTED INGREDIENTS FOR TROUT FEEDS[1]

Ingredient	Quality	Percent of Diet; Recommended Level in Parentheses
Soybean oil meal	Solvent extracted; dehulled; minimum 47.5% protein	0–25 (20)
Cottonseed meal	Prime quality; solvent extracted; dehulled; minimum 48% protein; less than 0.4% free gossypol	0–20 (10)
Corn gluten meal	Maximum 3% fiber; minimum 60% protein	0–20 (10)
Meat meal	Maximum 7% fat; maximum 2.5% fiber; minimum 50% protein	0–15 (10)
Blood meal	Maximum 3.5% fiber; minimum 0.5% fat; minimum 80% protein	0–10 (5)
Hydrolyzed feather meal	Maximum 3% fiber; minimum 1.0% fat; minimum 85% protein	0–15 (10)
Poultry by-product meal	Maximum 2% fiber; minimum 12.5% fat; minimum 60% protein	0–15 (10)
Crab meal	Maximum 11% fiber; minimum 31% protein	0–10 (10)
Shrimp meal	Minimum 40% protein	0–10 (5)
Fish solubles (condensed)	Maximum 0.5% fiber; minimum 7% fat; minimum 30% protein	0–10 (5)
Whey, dried	Maximum 6% water; maximum 10% ash; maximum 3% salt; minimum 12% protein	0–10 (10)
Yeast, brewers' dried or torula	Maximum 3% fiber; minimum 35% protein	0–15 (10)
Fish meal:	Maximum 12% fat; maximum 10% moisture; maximum 5% salt	10–50 (25–40) —diet should contain minimum of 7% fish meal protein
Herring	Minimum 67.5% protein	
Anchovy	Minimum 65% protein	
Menhaden	Minimum 60% protein	

[1]Prepared especially for this book by L. Orme, Director, U.S. Department of the Interior, Fish and Wildlife Service, Diet Development Center, Spearfish, S.D.

Formulation specifications for trout are presented in Table 28–8; and diets for Pacific salmon are given in Table 28–9.

TABLE 28–8
FORMULATION SPECIFICATIONS FOR TROUT DIETS[1]

	Percent of Diet		
	Starter	Production (Grower)	Brood Stock
	(%)	(%)	(%)
Fish meal (herring, anchovy, mackerel, capelin): —minimum crude protein 65% —stabilized with ethoxyquin —maximum fat level 12% —maximum moisture 10% —maximum salt level not to exceed 3% —maximum ash 15%	45–50	25–35	30–35
Wheat middlings: —minimum crude protein 16% —maximum crude fiber 9.5%	0–15	10–30	15–35
Wheat gluten meal: —minimum crude protein 80%	0–3	0–2	0–1
Soybean meal[2]: —minimum crude protein 48%	5–10	5–15	5–20
Corn gluten meal: —minimum crude protein 60%	0–10	0–10	0–10
Dehydrated alfalfa meal: —minimum crude protein 17% —maximum crude fiber 27%	—	0–3	0–5
Dried whey: —minimum crude protein 13% —partially delactosed	0–5	0–3	0–5
Yeast, dried brewers': —45% crude protein	0–5	0–5	0–5
Corn distillers' dried solubles: —minimum crude protein 27%	0–10	0–10	0–10
Animal by-product meals:			
—hydrolyzed feather meal, minimum crude protein 85%	0–5	0–7	0–7
—Poultry by-product meal, minimum crude protein 60%	0–5	0–7	0–7
—blood meal, ring dried or spray dried, minimum crude protein 80%	0–5	0–7	0–7
—meat meal, minimum crude protein 50%	0–5	0–7	0–7
Fat supplement:			
—marine oil (salmon, capelin, herring, mackerel, etc.)	5–15	5–15	5–15
—vegetable oil (soybean oil, canola oil)	0–5	0–8	0–5
—animal fat and grease, all oils and fats to be stabilized	0–5	0–5	0–5
Vitamin premix	2–4	2–4	2–4
Mineral premix	2–4	2–4	2–4

[1]Adapted by the authors from *Nutrition and Feeding of Rainbow Trout,* by Hilton, J. W., and S. J. Slinger, Department of Nutrition, College of Biological Science, University of Guelph, Guelph, Ontario; published by Department of Fisheries and Oceans, Ottawa, 1981, p. 8.

[2]Soybean meal is not commonly used in Pacific salmon diets due to poor growth, probably because of unpalatability.

TABLE 28–9
DIETS FOR PACIFIC SALMON—
OREGON MOIST AND ABERNATHY DRY FORMULATIONS[1]

Ingredients	Starter		Small Pellets or Crumbles[2]		Large Pellets	
	Moist	Dry	Moist	Dry	Moist	Dry
	(%)	(%)	(%)	(%)	(%)	(%)
Herring meal (70% crude protein)[3]	49.9	58.0	47.5	55.0	28.0	50.0
Wheat germ meal (25% crude protein)	10.0	—	Remainder	5.0	Remainder	5.0
Dried whey (12% crude protein)	8.0	5.0	4.0	5.0	5.0	5.0
Trace mineral premix	0.1	0.1	0.1	0.1	0.1	0.1
Vitamin premix	1.5	1.5	1.5	1.5	1.5	1.5
Wet fish (pasteurized)	20.0	—	30.0	—	30.0	—
Fish oil (stabilized)	10.0	12.0	7.0	9.0	7.75	9.0
Vitamin C	0.15	0.1	0.15	0.1	0.15	0.1
Choline chloride (70% product)	0.5	0.5	0.5	0.5	0.5	0.5
Binder	—	2.0	3.0	2.0	—	2.0
Cottonseed meal (47% crude protein)	—	—	—	—	15.0	—
Corn distillers' solubles (25% crude protein)	—	—	—	—	4.0	—
Dried blood meal (spray or flash, 80% crude protein)	—	10.0	—	10.0	—	10.0
Wheat middlings (15% crude protein)	—	Remainder	—	Remainder	—	Remainder

[1]Formulations provided for this book by L. G. Fowler, U.S. Fish and Wildlife Service, Longview, WA 98632

[2]Small pellets are moist feeds, and crumbles are dry feeds.

[3]Anchovy meal (65% crude protein) allowed in dry crumbles and large pellets.

Fig. 28–12. *Right:* Sedimentation zones upstream from trout raceway outflows allow solids to be removed to side-stream stabilization lagoons, reducing eutrophication (the process of becoming high in dissolved nutrients) of streams receiving farm effluent by 75% or greater. (Courtesy, Blake Grant, Director, International Aquaculture Research Center, Hagerman, Idaho)

Catfish Diets

Catfish feeds are generally lower in protein than those of coldwater fish. Fingerlings require 35 to 40% protein, production (grower) fish 25–36%, and brood stock 28–32% protein. Examples of various catfish feeds are found in Tables 28–10 and 28–11.

TABLE 28–10
FEED INGREDIENTS AND PERCENT COMPOSITION OF SOME PRACTICAL FEED FORMULATIONS FOR CHANNEL CATFISH[1]

Ingredient	Percent Protein	Fish Farming Experimental Station[2]	Texas A&M University		Auburn University	
			1	2	Fingerlings (36% protein)	Production (32% protein)
Alfalfa meal	17.5	3.5				
Blood meal	75.3	5.0				
Bone meal	11.2					
Brewers' grains	25.0					
Corn, distillers' grains	27.1					
Corn, distillers' solubles	27.3	8.0			5.0	
Corn grain, yellow	9.6		30.0	30.5	23.5	29.1
Corn grain, flint	9.9					
Cottonseed meal	40.8	10.0				
Cottonseed meal (without hulls)	50.0					
Dicalcium phosphate			0.25	1.5	1.5	1.0
Fats, animal				2.0	2.5	1.5
Fish meal, catfish	55.3					
Fish meal, menhaden	61.1	12.0		9.0	10.0	8.0
Fish meal, tuna	59.4					
Fish meal, white	61.9					
Grains, distillers'	27.4					
Liver meal, animal	66.5					
Meat meal, whole animal	54.3					
Meat meal (with bone)	50.5		15.0			
Oats, cereal by-product	14.6					
Peanut meal	44.0				18.0	
Poultry by-product meal	57.8					
Poultry feathers, hydrolyzed	85.4	5.0				
Rice bran	12.7	25.0				10.0[3]
Rice polishings (dust)	12.1	10.0				
Soybean meal	44.0		47.5			
Soybean meal (without hulls)	48.8	20.0		54.4	37.0	48.3
Wheat bran	15.1					
Wheat middlings	16.7		0.85			
Wheat grain	14.9					
Wheat shorts	16.4					
Whey (low lactose)	16.5		2.5			
Yeast (brewers')	45.1					
Vitamin premix			[4]	[4]	[5]	[5]
Mineral premix			0.50	0.50	[6]	[6]
Limestone			0.90			

[1]Adapted by the authors from *Third Report to the Fish Farmers,* by Dupree, H. K., and J. V. Huner, U.S. Department of the Interior, Fish and Wildlife Service, 1984, p. 151, Table 11.6.

[2]Herring fish meal (70% protein) may be substituted at 10% of the formulation. Wheat shorts, wheat middlings, or cereal grains may be substituted for the rice bran.

[3]Wheat shorts may be substituted for the rice bran.

[4]Vitamin premixes as published by the National Research Council (footnote 1, above).

[5]Quantities of vitamins per ton (grams, unless otherwise indicated): vitamin A, 4 million IU; vitamin D, 2 million IU; vitamin B-12, 8 mg; vitamin E, 50; menadione, 10; choline chloride (70%), 500; niacin, 80; riboflavin, 12; pyridoxine, 10; thiamin, 10; pantothenic acid, 32; folic acid, 2; and ethoxyquin, 125. Ascorbic acid (335 g) is added to the feed during pelleting, or top-dressed on extruded feed.

[6]Quantities of minerals per ton (milligrams): manganese, 110; zinc, 105; iron, 36; copper, 4.5; iodide, 2.3; and cobalt, 0.5.

TABLE 28–11
EXTRUDED (FLOATING) AND HARD (SINKING) PELLET FORMULA FOR CATFISH IN PONDS, RACEWAYS, AND CAGES[1] (ALSO MINNOWS, GOLDFISH, CARP, AND BUFFALO)

Ingredient	Percent
Fish meal, menhaden, minimum 61% protein	10.0
Soybean meal, solv extd, w/o hulls, minimum 49% protein	35.0
Cottonseed meal, solv extd, minimum 41% protein	12.0
Wheat, whole grain (ground)[2]	31.7
Rice bran, with germs, solv extd	3.5
Fat, animal or plant[3]	5.0
Dicalcium phosphate	1.0
Vitamin premix[4]	0.8
Mineral premix[5]	1.0
Analysis:	
Crude protein, more than	32.0
Crude fiber, less than	3.5
Crude fat, more than	7.0

[1]Feed formula prepared for this book by H. Dupree, Fish Farming Experimental Station, U.S. Fish and Wildlife Service, Stuttgart, Ark.

[2]Wheat may be replaced up to 25% with corn.

[3]Sprayed on after manufacture.

[4]A. If the feed is to be used as a supplemental pond feed (*i.e.,* low intensity pond culture), the premix should be as follows (per ton of feed basis): vitamin A, 900,000–1,800,000 IU; vitamin D_3, 450,000–900,000 IU; vitamin E, 27 g; thiamin, 0.9 g; riboflavin, 8.1 g; pyridoxine, 2.7 g; pantothenic acid, 9–18 g; nicotinic acid, 12 g; and ascorbic acid, 55 g.

B. If the feed is to be used as a complete feed (*i.e.,* raceways, cages, and high intensity pond culture), the premix should be as follows (per ton of feed basis): vitamin A, 4,000,000 IU; vitamin D_3, 2,000,000 IU; vitamin E, 50 g; vitamin K, 10 g; thiamin, 10 g; riboflavin, 12 g; pyridoxine, 10 g; pantothenic acid, 32 g; nicotinic acid, 80 g; folic acid, 2.0 g; choline chloride, 500 g; ascorbic acid, 350 g; and vitamin B-12, 8 mg.

[5]Mineral premix should provide the following (per ton of feed basis): manganese, 100 g; iodide, 2.5 g; copper, 3.9 g; zinc, 80 g; iron, 40 g; and cobalt, 45 mg.

FEEDING COLDWATER FISH

In the United States, trout and salmon are the two most common species of coldwater fish grown commercially.

The art and science of feeding coldwater fish has progressed dramatically over the last 100 years. Originally, producers relied on natural feeds to grow trout, followed years later by the addition of wet diet supplements to natural feeds, and finally to the recent development of complete diets.

Variations in feeding habits occur among the different types of coldwater fish. For example, rainbow trout are surface feeders whereas brown trout are bottom feeders. Therefore, the type of feed pellet to be used must be given careful consideration. Since trout consume their feed in about 5 to 10 minutes, the producer does not have to be concerned with the pellets falling apart unless too much feed is being used and remains uneaten.

The producer should feed according to the stocking rate, the size of the fish, the type of pond or culture facility, the water temperature, and the energy content of the feed. Feed consumption is markedly affected by water temperature, decreasing in cold weather. Also feed intake is affected by the energy content of the feed, since fish, like terrestrial animals, eat to satisfy their energy needs. Feed consumption is reduced in polluted water.

A feeding guide for salmonid fish is given in Table 28–12.

TABLE 28–12
FISH FEED GUIDE FOR SALMONIDS[1] [2]

Number Fish		Granule/Pellet	Water Temperature °F/°*C*									
			43/*6*	45/*7*	46/*8*	48/*9*	50/*10*	52/*11*	54/*12*	55/*13*	57/*14*	59/*15*
(per lb)	(*per kg*)	(Granule No.)	◄—— % body weight per day ——►									
1,182	*2,600*	1	2.9	3.4	3.7	3.9	4.6	4.8	5.2	5.8	6.0	6.4
591	*1,300*	1	2.8	3.3	3.6	3.8	4.4	4.7	4.9	5.6	5.9	6.1
318	*700*	2	2.7	3.0	3.3	3.6	4.1	4.5	4.8	5.1	5.6	5.8
182	*400*	2	2.6	2.8	3.0	3.2	3.9	4.0	4.6	4.9	5.0	5.1
91	*200*	3	2.3	2.6	2.8	3.0	3.6	3.8	4.3	4.5	4.6	4.7
59	*130*	3–4	2.1	2.3	2.5	2.8	3.3	3.6	3.7	3.9	4.0	4.1
41	*90*	4	1.9	2.0	2.1	2.4	2.7	2.9	3.0	3.2	3.6	3.8
		(Pellet No.)										
18	*40*	3/32	1.6	1.7	1.8	1.9	2.0	2.1	2.4	2.6	3.0	3.2
14	*30*	3/32	1.5	1.6	1.7	1.8	1.8	1.9	2.0	2.2	2.8	2.9
9	*20*	1/8	1.3	1.4	1.5	1.6	1.7	1.8	1.9	2.1	2.4	2.5
7	*15*	1/8	1.2	1.3	1.4	1.5	1.6	1.7	1.8	2.0	2.3	2.4
4	*10*	3/16	1.1	1.2	1.3	1.4	1.5	1.6	1.7	1.8	1.9	2.0
2	*5*	3/16	1.0	1.1	1.2	1.3	1.4	1.5	1.6	1.7	1.8	1.9
1	*2*	1/4	0.8	0.9	1.0	1.0	1.1	1.1	1.2	1.3	1.5	1.6

[1]Adapted by the authors from *Nutrition and Feeding of Rainbow Trout,* by Hilton, J. W., and S. J. Slinger, Department of Nutrition, College of Biological Science, University of Guelph, Guelph, Ontario; published by the Department of Fisheries and Oceans, Ottawa, 1981.

[2]Feeding rates based on a single strain of rainbow trout fed dry diets containing about 3,000 kcal digestible energy/kg.

The feeding rate for salmonids is commonly expressed as a percentage of body weight fed per day. As shown in Table 28–12, (1) smaller fish require feed at a greater percentage of their body weight per day than larger fish, and (2) water temperature has a marked effect on feed requirement, being very low at 43°F. Very low and very high temperatures constitute a considerable stress to salmonids. But feeding guides are just that—guides. Thus, feed as a percentage of body weight for growing fish can vary between 0.5 and 10%, depending on numerous factors. For example, in the spring when the water begins to warm and when the photoperiod is increasing, it is possible to feed up to twice the amounts shown in most feeding guide tables.

In addition to feeding rates, the following feeding practices are important:

1. Frequency of feeding, with swim-up fry being fed small amounts of feed 20–24 times per day, gradually reduced to one to three times per day.

2. Feed particle size, hardness and texture, palatability, and placement of feed in relation to fish size are important. Very small fish will not travel far for feed.

3. Changes in feed intake and feed size should be made gradually, extending over a few days.

4. Most salmonids should be selected for brood stock at 2 to 3 years of age. At this time, they should be switched from the production or grower diet to a brood stock diet and fed only once a day. Depending on the temperature of the water, breeders should be taken off feed 3 to 6 weeks before spawning, then gradually brought back to full feed over a period of 2 to 4 weeks after spawning. Overfeeding before spawning reduces reproductive performance.

FEEDING WARMWATER FISH

Warmwater fish production involves numerous species of fish, but the major industries are catfish and carp farming. There have been attempts to produce predacious fish commercially, but due to the food requirements and behavioral characteristics of the fish, success has been very limited.

Carp production is the oldest form of aquaculture—dating back 4,000 years to China. Today, different types of carp culture can be found throughout the world, quite often in conjunction with another type of agriculture production—for example, polyculture in rice fields.

The catfish industry in the United States has experienced phenomenal growth in recent years. In 1960, only 400 acres were under catfish culture, and the production was a mere 300,000 lb. By 1985, catfish farming was more than a $200 million-a-year industry producing over 290 million lb of catfish on about 100,000 acres. Indications are that the industry will continue to expand as consumers become more familiar with the product.

Feeding Catfish

Fig. 28–13. A basketful of farm-raised channel catfish ready for market, weighing 2 to 2½ lb. (Courtesy, Mississippi State University, Mississippi State)

Catfish raised for food and for fee-fishing in the Southwest are fed to liveweights of 2 to 2½ lb. Most catfish marketed in the Southeast—the area of greatest production—are processed at liveweights of ¾ to 2 lb. If good management practices are followed, fingerlings will have feed conversion rates of 0.9 to 1.0, and marketed fish will have conversion rates of 1.5 to 1.7.

Catfish, being omnivores, have well-defined stomachs and can effectively digest meat; but in commercial production the use of fresh meat diets is extremely limited. Most feeds are in dry pellet form, which make for economical storage and handling. Catfish diets generally contain the oil meals, distillers' solubles, and fish meal, in addition to vitamin and mineral premixes.

Since a sizable portion of the diet of catfish can come from the natural feed chain, ponds containing fingerlings should be fertilized in order to establish an optimum level of plankton growth. But fingerling producers must avoid over fertilizing.

It is generally recommended that 50 lb of 16–20–4 or 16–20–0 inorganic fertilizer per acre be applied every 10 days until an adequate plankton bloom is present. The producer can tell if the pond needs more fertilizer by sticking an arm into the water up to the elbow. If the hand can be seen, more fertilizer is needed.

In general, catfish are bottom feeders, but they can be taught to feed at the surface. While floating pellets are more expensive, some producers justify the added cost because they can routinely observe the condition of the fish at feeding.

The water temperature, fish weight, and water quality are the primary factors affecting the feed consumption of warmwater fish.

Table 28–13 presents a feeding guide for channel catfish in ponds. Table 28–14 gives a feeding guide for catfish in raceways. The pond feed is a supplemental, 36% protein diet; the raceway feed is a complete formulation.

TABLE 28–13
TYPICAL SPRING-SUMMER-FALL FEEDING SCHEDULE FOR CHANNEL CATFISH IN PONDS STOCKED AT 2,000–3,000 FINGERLINGS PER ACRE AS 5-IN. FISH AND HARVESTED AS 1.1-LB FOOD FISH, IN SOUTHEASTERN UNITED STATES[1,2]

Date	Water Temperature		Fish Weight		Feed Allowance Per Day
	(°F)	*(°C)*	(lb)	*(kg)*	(% of fish wt.)
April 15	68	*20.0*	0.04	*0.02*	2.0
April 30	72	*22.2*	0.06	*0.03*	2.5
May 15	78	*25.5*	0.11	*0.05*	2.8
May 30	80	*26.6*	0.16	*0.07*	3.0
June 15	83	*28.3*	0.21	*0.10*	3.0
June 30	84	*28.8*	0.28	*0.13*	3.0
July 15	85	*29.4*	0.35	*0.16*	2.8
July 30	85	*29.4*	0.42	*0.19*	2.5
August 15	86	*30.0*	0.60	*0.27*	2.2
August 30	86	*30.0*	0.75	*0.34*	1.8
September 15	83	*28.3*	0.89	*0.40*	1.6
September 30	79	*26.1*	1.01	*0.46*	1.4
October 15	73	*22.7*	1.10	*0.50*	1.1

[1]Adapted by the authors from *Third Report to the Fish Farmers,* by Dupree, H. K., and J. V. Huner, U.S. Department of the Interior, Fish and Wildlife Service, 1984, p. 156, Table 11.8.

[2]Feed allowances are based on data obtained with rations containing 36% protein and about 2.88 kcal of digestible energy per gram of protein. If feeds of lower protein and energy concentrations are used, daily allowances should be increased proportionally. Data adapted from R. T. Lovell, 1977. Feeding practices. Pages 50–55 in Nutrition and feeding of channel catfish. *Southern Cooperative Series Bul. 218.* Alabama Ag. Exp. Sta., Auburn University, Auburn, Ala.

TABLE 28–14
FEEDING RATES (PERCENT BODY WEIGHT PER DAY) FOR CHANNEL CATFISH FED A COMPLETE FEED (25% FLOATING, 75% SINKING FEED) IN RACEWAYS[1]

Water Temperature	Size and Weight		
	1–2 in. 0.001–0.004 lb	2–5 in. 0.004–0.04 lb	over 5 in. over 0.04 lb
(°F)	(% body weight)	(% body weight)	(% body weight)
Below 55°	1	1	1
At 55°	3	2	1.5
Above 55°	5	3	2

[1]Adapted by the authors from *Fish Hatchery Management,* by Piper, R. G., I. B. McElwain, L. E. Orme, J. P. McCraren, L. G. Fowler, and J. R. Leonard, U.S. Department of the Interior, Fish and Wildlife Service, Washington, D.C., 1982, p. 252, Table 29.

Multiple daily feeding can increase growth rate. In such a production system, feed is offered in the maximum amounts that can be metabolized in the pond. Feeding frequency varies with water temperature (see Table 28–15). A rule of thumb is that 90% of the feed should be eaten in 15 minutes or less.

TABLE 28–15
SUGGESTED MAXIMUM FEEDING RATES AND FREQUENCIES FOR CHANNEL CATFISH[1]

Water Temperature		Feeding Frequency	Feeding Rates
(°F)	(°C)	(times per day)	(% of total fish wt.)
90	*32.2*	1	1
80–86	*26.6–30.0*	2	3
68–80	*20.0–26.6*	1	2½
58–68	*14.4–20.0*	1	1½
50–58	*10.0–14.4*	0.5[2]	¾–1
50	*10.0*	0.3[3]	½–1

[1]Adapted by the authors from *Third Report to the Fish Farmers,* by Dupree, H. K., and J. V. Huner, U.S. Department of the Interior, Fish and Wildlife Service, 1984, p. 156, Table 11.9.

[2]Feed once on alternate days.

[3]Feed once every 3 to 4 days.

• **Rules of thumb for feeding catfish**—The following rules of thumb for feeding catfish are generally followed:

1. **Be aware of water temperature.** Catfish make their most efficient gains in water temperatures of around 84°F. When the water temperature falls below 55°F, daily supplemental feeding should be reduced to about 1%, or less, of body weight. In the winter months, catfish will require about ¼ the amount of feed that would be fed in the summer. If the water temperature drops below 45°F, it is only necessary to feed at the rate of 0.5% of body weight every 4 to 5 days.

2. **Do not full-feed on rainy days or when it has been overcast for extended periods (more than 4 days).** It has been found that catfish will show a dramatic drop in feed consumption on rainy days. If the sky is overcast for several days, feeding activity decreases but will return to normal once the sky clears. This is due to the fact that on overcast days, plants and algae take up more oxygen than they give off, thereby decreasing the level of oxygen in the water.

3. **Until catfish reach a weight of 1 lb, the maximum feeding rate should be 3% of body weight per day.**

4. **The feeding rate of catfish weighing more than 1 lb should not exceed 2% of body weight per day.**

• **Feeding catfish fry and fingerlings**—Catfish feeding begins with the fry and fingerling stages. Newly hatched catfish, which are called fry, live on nutrients from the yolk sac for 3 to 10 days (depending upon the water temperature), following which they accept food from a variety of sources. Fry can either be stocked in a rearing trough or moved directly to a specially prepared rearing pond. Feed for trough-feeding of fry should be small in particle size, high in animal protein, and high in fat. In troughs, fry are held until they have absorbed their yolk sacs and have developed a grayish color. At the latter stage, they are called swim-up fry because they rise to the surface to seek feed. As soon as possible, the fry should be transferred to ponds with high zooplankton densities to utilize the natural food source. At this time, supplemental feeding should begin, using a 36% high quality protein catfish feed for fry and fingerlings.

Catfish fingerlings have been shown to utilize feeds extremely efficiently—demonstrating feed conversions of 0.9 to 1.0. Of course, as the fish increase in size, the total efficiency will decrease because the natural foods will constitute smaller and smaller amounts of the total food intake.

Feeding Carp

Carp are the most extensively cultured fish in the world. They grow well under a variety of cultural conditions, use natural foods efficiently, and respond well to supplemental feeding.

The feeding principles of carp production can be generally considered to be applicable to the less commonly produced warmwater food and bait fish—for example, minnows, buffalo fish, and barbs.

Carp grow fastest in water which is 77° to 86°F. When the temperature drops to 60°F, growth is inhibited; and if the temperature falls below 55°F, feeding activity is greatly decreased. However, carp still require supplemental feed in cold water to avoid excessive weight loss and debilitation. For every 18°F above 55°F, there is a two- to threefold increase in feed consumption by carp.

Some researchers recommend that at least 50% of the feed for carp consist of natural feeds. Carp eat plankton primarily, along with small animals close to the shore and the bottom. But they will also eat some of the natural vegetation. With emphasis on the utilization of natural feeds, pond fertilization becomes an important production technique.

Artificial feeding—supplementary feeding is probably a more appropriate term—is commonly practiced. Soybean, corn, and wheat are the most widely used feedstuffs; but barley, oats, rye, beans, potatoes, millet, rice bran, vetch, and grass seeds may be used.

Carp are slow eaters. It generally takes them 30 minutes to an hour to finish eating a dry feed as compared to 5 minutes for trout. The pellets should be water stable in order to prevent leaching of nutrients, wastage of feed, and possible reduction in water quality.

Feeds for fry carp and related species are generally in a fine, flourlike consistency. It is best to feed on all sides of the pond in order to ensure that all the young fish have adequate access to feed.

In the warm regions of the world, where rice is commonly grown, carp are sometimes raised in rice fields. The warm water provides an excellent environment for rapid growth of both the rice and carp. The waste products from the carp fertilize the rice, and the carp consume some of the aquatic organisms that can interfere with the growth of the rice. Carp grow most efficiently during the summer, which is the time that rice will have its greatest growth.

In most countries, fertilization of ponds with manure is a common practice (see Fig. 28–14). Manure and organic

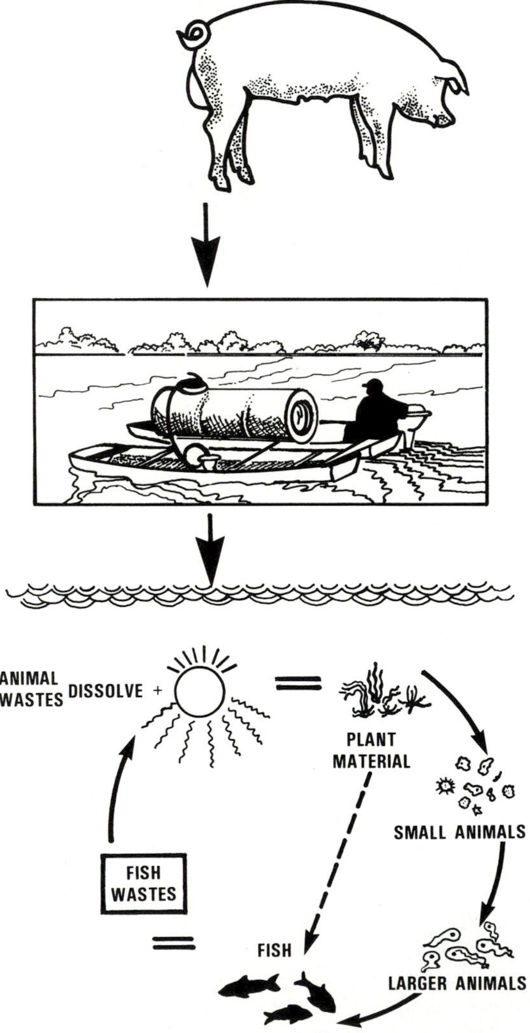

Fig. 28–14. Manure can often be used to fertilize ponds populated by carp. The animal wastes are collected and dispersed in the pond where they are dissolved. When the minerals and organic material of the manure that has been dissolved in the pond combine with sunlight and plant photosynthetic material, stimulation of vegetative growth occurs. Small microscopic organisms consume the plant material and are in turn consumed by larger animals; *i.e.*, snails, tadpoles, etc. From the population of plants and animals, a source of natural food is available to the fish that are being produced. Wastes from the fish are then released in the water, and the cycle begins again. Fertilization of ponds with manure has been successful in carp production, but there is some skepticism as to the merits of fertilizing ponds with manure for other types of fish.

wastes can be used to fertilize ponds, but the producer should be aware of possible nitrate-nitrite toxicity and oxygen depletion in the ponds. By using manure as a pond fertilizer, two purposes are served. First, it provides a convenient method of dealing with the potential pollution problem resulting from the feeding of livestock. Second, nutrients from the manure can be utilized to support the natural food system of the pond.

The Chinese have fed manure to fish since the days of the Ming Dynasty; hence, the use of manure in fish culture is of long standing in that country. In Kwang Tung Province, in the heart of the land of fish and rice, the senior author was fascinated with an integrated hog-fish production operation on a commune where, in 1 year, they produced 49,700 pigs and had a fish catch of 863,000 lb. The fish were fed solely on natural feeds produced in the ponds from manure flushed from the nearby hog barns, plus grass clippings from the pond banks. This unique operation provided a method of handling manure plus pollution control and, at the same time, recycled and used the feed twice—first through the hogs, and second through the fish. It is noteworthy, too, that the Chinese were feeding manure to fish without the hazard of nitrate-nitrite toxicity and/or of oxygen starvation.

The Chinese Academy of Agricultural and Forestry Sciences recommends the following practices when feeding manure to fish: (1) mixing the manure with plant materials, then composting it before feeding (although manure may be, and is, fed to fish without composting, pollution may be lessened by first composting); (2) fertilizing and fermenting the composted manure; and (3) feeding manure to the little fingerlings, rather than to the big fish. But, according to the fish experts of China, the feed of fish should vary according to species (carp are especially well adapted to the use of animal waste) and age. They also report that, in addition to manure, freshwater fish of China are fed a great variety of foods, including silage, the leaves of such crops as sweet potatoes and turnips, soybeans, bean cake and curd, distillers' grains, and wheat. It's noteworthy, too, that, in addition to supplying needed food, the freshwater fish of China lessen the mosquito menace.

Fig. 28–15. A hog-fish combination on Kwang Li People's Commune, Kaw Yao County, Kwang Tung Province, where, in 1971, they produced 49,700 pigs and had a fish catch of 863,000 lb. (Photo by A. Ensminger)

FEEDING SYSTEMS

In all feeding systems, the producer must compare the costs of labor to the costs of automation. In general, larger operations can justify more automation. The producer must also take into consideration the frequency of feeding to be followed. Small fish must be fed frequently—8 to 20 times daily for trout—in order to prevent cannibalism and irregular growth. As the fish become large, the frequency of feeding can eventually be reduced to once or twice daily.

There are three methods of feeding fish—hand-feeding, semiautomatic feeding, and automatic feeding.

Hand-feeding

Fig. 28-16. While hand-feeding requires considerable labor, it enables the farmer to check the health and general condition of the fish. (Courtesy, Mississippi State Univeristy, Mississippi State)

This is the oldest form of feeding. It involves the fish culturist walking along the water banks and spreading the feed from a ladle or by hand. It is best to feed along all sides of the pond. This ensures that all the fish have access to feed with a minimum of competition; hence, more pounds per acre are yielded. The farmer must be attentive to the amount of feed that will be consumed. Overfeeding will reduce profits as well as cause pollution. On small operations, the expense of this additional labor is far less than the expense of converting to more automatic methods.

Semiautomatic Feeding

In operations that involve intense production, a certain amount of automation is necessary. Fish can be fed from boats or by blowing feed from mechanical equipment traveling along the edge of the water. In extremely large operations, feed can be released from airplanes flying close to the water.

Fig. 28-17. Feeding fish by blowing feed into the pond by a feed blower. (Courtesy, Agricultural Experiment Station, University of Arkansas at Pine Bluff.)

Automatic Feeding (Mechanical Feeders)

Automatic feeding can be divided into two classes of feeder systems: (1) the demand type, which is activated by the fish; and (2) the automatic type, which is activated by a time clock. In the demand feeder, fish can obtain the food mechanically tripping a feed release trigger. This type

Fig. 28-18. An automatic, demand type feeder. With this type of feeder, fish obtain the food by tripping a feed release trigger in the water. (Courtesy, Ralston Purina Co., St. Louis, Mo.)

of feeder can create some wastage of feed, but the producer must consider the cost of labor vs the cost of wasted feed. An important advantage of the trigger-type self-feeder is that it can be adapted to almost any size of operation. The second type of automatic feeder is designed to dispense fixed amounts of feed at preset time intervals; it may offer more feed than is eaten or less than is required. Since these feeders are rather sophisticated, the cost cannot be justified unless the operation is large and intensive.

Many large operators find that the time and labor involved in filling self-feeders can be considerable; hence, most large operations use a truck or tractor equipped with feed blowers.

OTHER FEED AND MANAGEMENT ASPECTS

When discussing the various nutritional aspects of fish culture, two additional areas should be considered—feeds affecting the quality and flavor of fish, and types of specialized aquatic production.

Feed Ingredients Affecting Fish Quality and Flavor

In many cases, fish farmers can alter the texture and flavor of their product merely by altering the composition of the ration. With this in mind, they should be aware of the demands of the consumer and attempt to produce a uniform product.

Trout that are fed wet or moist diets tend to be high in fat content and have a soft meat texture, whereas those fed dry diets have a more desirable flavor with a firmer texture. If the wet diet contains fresh sea fish, there is a possibility that an off-flavor ("sardine" taste) will result. Also, off-flavor may be caused by spoiled feed, by feed containing oxidized (rancid) oils, or by algae in the water.

Carp fed bread and potatoes develop a wet, soft meat consistency which is considered undesirable. Supplemental feeding of cereal grains will produce carp that have a desirable flavor as well as a firm meat texture. If no supplemental feeding is provided—that is, if carp get all their required nutrients from natural sources—an undesirable wild flavor may result.

Catfish flavor is affected by several factors. Excessive plankton bloom, feeds high in fish oil, presence of muskgrass, overfeeding, chemicals, and organic debris can all produce an off-flavor in catfish. When the water temperature is high—as in late summer—there is a greater chance of off-flavors occurring in the meat.

Specialized Aquatic Production

In the United States, the consumer is extremely sensitive to both the physical texture and organoleptic qualities of meat. The average American consumer is relatively unaware of the possibilities for variety in fish foods which result from such specialized productions as aquarium fish, crayfish, eel, lobster, prawn, predacious fish, shrimp, sturgeon, and tilapia.

In countries outside the North American continent, specialized fish culture is common. For example, work in the development of efficient eel farming has been done in Europe and Japan; and the Soviet Union has been noted for its production of caviar from sturgeon.

AQUARIUM FISH (HOBBY FISH/TROPICAL FISH) CULTURE

This group includes representatives of several families, and over 100 species of small, colorful, and unique fishes.

Although aquarium fish are not cultured for food, they do represent a major segment of the pet industry. Also, they are used extensively in research; for example, in water quality testing. There are many commercial feeds for aquarium fish, in a variety of shapes and formulations. Most of these feeds are nutritionally adequate; but they are closed formulas and are subject to periodic change in composition, thereby limiting their use for experiments. Expensive carotenoids and other pigments are commonly included in the feed to promote bright coloration of the fish.

Fig. 28-19. Aquarium fish. (Courtesy, USDA)

CRAYFISH (CRAWFISH) CULTURE

Louisiana is the leading U.S. producer of crayfish, with approximately 125,000 acres devoted thereto, which is 90% of the nation's crawfish acreage. About 8% of the U.S. crayfish acreage is in Texas, and the remaining 2% is in the states of Mississippi, South Carolina, and Arkansas. Annual harvests range from 500 to 2,000 lb per acre. Some crayfish are farmed on rice land as a secondary crop.

Feeds for crayfish include natural food and agricultural by-products. The feeding of specific formulations manufactured for crayfish is neither necessary nor economically sound.

The simplest type of culture involves the trapping of crayfish from marginal wetlands, usually adjacent to crop or timber lands. However, higher levels of production are achieved by the construction and management of ponds specifically for crayfish.

EEL (ELVERS) CULTURE

Eel is considered a gourmet food in Japan, Taiwan, and in most European countries; and it commands high prices.

In various countries around the world, extensive production of eel in brackish-water ponds and inlets is being conducted.

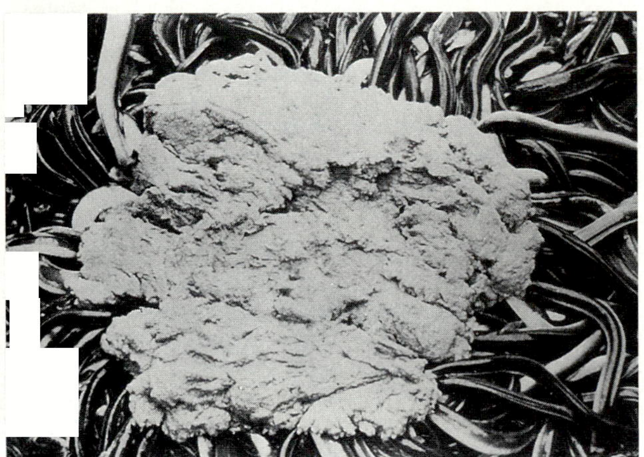

Fig. 28-20. Young Japanese eel feeding on paste food. (Courtesy, Dr. P. Ghittino, Torino, Italy)

Almost all eels produced in Japan are from pond units, and the techniques of production are becomming more and more refined. In the past, the natural feeds were supplemented only by fresh fish diets. As a result, feed conversion rates were extremely poor (5 to 15:1). Today, modern Japanese diets make for conversion rates of 1.2 to 1.6:1 using a paste feed, and of 1.0 to 1.6:1 using a dry pellet. The diets generally range from 20 to 30% protein.

American eel culturists have been left to the mercy of unpredictable overseas markets because there are virtually no domestic sales in North America.

LOBSTER (SHELLFISH) CULTURE

Lobsters are large hard-shelled marine crustaceans that are closely related to crabs, crayfish, and shrimp. The American lobster, *Homarus americanus,* which is perhaps the largest lobster, lives along the Atlantic coast from Labrador to Virginia; it may reach a length of 3 ft and a weight of 44 lb. Most European lobsters are smaller.

Lobsters live on the bottom of the ocean near shore and hide in holes or under rocks. A lobster is a predator that sits in its burrow all day, waving its feelers outside the entrance and holding its claws ready to capture any prey that comes near. Lobsters eat crabs, snails, small fish, and even other lobsters. At night, the lobster walks along the ocean bottom seeking food.

Compared to farm animal and poultry nutrition, and even fish nutrition, lobster nutrition is in its infancy. Until recently, lobsters required natural food diets to grow and survive—such foods as brine shrimp, crab, shrimp, or fish were used.

Attempts to substitute cereal-based or purified diets for these natural foods have usually resulted in poor growth and low survival. However, with the development of a successful purified diet in 1980 by researchers at the Bodega Marine Laboratory of the University of California, nutritional requirements can now be determined and thereby provide knowledge of diets from readily available feedstuffs. Although the commercial production of lobsters is not yet feasible, research continues.

PRAWN (LARGE SHRIMP) CULTURE

Prawns and other crustaceans are in heavy demand in the United States and throughout much of the world.

Because larvae must spend some time in brackish water, seed stocks must be especially handled before being made available to the freshwater culturist. The requirement of water temperatures of 75° to 95°F for prawn culture greatly limits the length of the growing season in the United States.

Feeding need not begin until the small prawns reach a length of 1 to 1½ in.—usually in 30 to 40 days. Special shrimp rations or the cheaper sinking catfish feeds, offered at about 3% of the body weight of the prawns, are suitable.

Fig. 28-21. A giant male Malaysian (freshwater prawn), *Macrobrachiam rosenbergii.* Note the long claws; the male uses these to protect the female until her shell hardens. (Courtesy, Mississippi State University, Mississippi State)

PREDACIOUS FISH CULTURE

There is little profit in the commercial production of predatory fish such as bass, perch, and pike. Feed conversion rates are highly variable (2:1 to 10:1), and high rates of cannibalism have been reported. In order to reduce cannibalism effectively, stocking densities must be extremely low, and the ponds must have heavy vegetation accompanied by a large population of forage fish.

Recent research dealing with the training of bass and other predatory fish to accept formulated fish feeds has yielded promising results for the future.

SHRIMP (SHELLFISH) CULTURE

Shrimp is the common name applied to about 2,000 species of crustaceans, whereas the term prawn may refer to any large shrimp. Shrimp, which have a very high market value, are shellfish and are closely related to crabs, crayfish, and lobsters.

Shrimps are predominantly marine, but several families live in fresh or brackish waters, and one species that lives in swamps in South America is semiterrestrial. Most are bottom dwellers, some even burrow into the sea bottom; others swim in the open sea. Shrimps are filter feeders and scavengers. They commonly feed on small animal or plant organisms. Several genera form the basis of a major fishery, and some shrimps are commercially raised, with Japan leading in number of shrimp farms and volume of commercial shrimp production.

Because shrimp are slow feeders, feeds that remain stable in water for several hours must be used. The pellets should be small (0.08 to 0.16 in.). Shrimp farmers who culture shrimp in large tanks or small earthen ponds generally feed a mixture containing 40% dry pellets (formulations similar to broiler rations), 40% shrimp waste, and 20% clam waste.

STURGEON CULTURE

The Russians have done considerable research on the production of sturgeon. Both wild and cultured fish are used in caviar production. Dry diets using fish and animal meals, distillers' solubles, yeast, and vitamins have been developed and are being used extensively.

TILAPIA CULTURE

Tilapia are fast-growing, rapidly-reproducing, and easy-to-manage fish that are raised throughout the world. Since these omnivorous fish are extremely sensitive to cold temperature, tilapia culture is limited to tropical climates. They were originally introduced in the southern region of the United States as a means of controlling aquatic vegetation in lakes and ponds. They are still useful in this aspect of pond management. Today, they are also being produced in limited numbers as commercial food fish and as sport fish.

Research has indicated that tilapia utilize many plant ingredients to about the same degree as trout utilize animal products. Thus, a program combining pond fertilization with the feeding of a dry diet and chopped leaves and grasses should provide a suitable ration for culturing tilapia.

NUTRITIONAL TOXICANTS OF FISH

In addition to being susceptible to excesses and deficiencies of certain nutrients, fish are highly susceptible to a number of chemicals that often are found in water—either naturally or as a result of soil pollution due to past agricultural or industrial practices. Metal toxicity can result from the earth surrounding the pond. In the past, pesticides and industrial chemicals have been routinely dumped in waterways. Organisms, such as red tide, can grow in the water and poison fish.

Heavy Metal Toxicities

Several metals have been shown to be toxic to fish. In some cases, the toxicity may result from natural deposits of the metal in the environment while often it can be traced to needless pollution by humans. The amount of literature dealing with heavy metal toxicity in fish is limited. Mercury poisoning in fish has received the most attention, but cases of toxicities from arsenic, chromium, copper, and selenium have been documented. A list of heavy metal toxicities, along with symptoms, is given in Table 28-16.

Of all the forms of mercury found in the environment, methylmercury appears to be the most toxic. This is the most prevalent form found in mercury-contaminated fish. Fish absorb mercury at an extremely rapid rate while they metabolize and excrete it very slowly; thus, there is a cumulative effect of the element when fish are exposed to a contaminated environment. The element is widely distributed through the body and the effects can be devastating. Cataracts result when mercury accumulates in the eye. Reproduction is impaired when the gonads are affected. In general, there is a decreased ability of the fish afflicted by heavy metal toxicants to absorb sodium; hence, the osmotic regulatory mechanism of the fish is adversely affected.

Organic Compound Toxicities

Numerous naturally occurring and synthetic organic compounds elicit toxic responses in fish. Aflatoxin, nitrosamines, cyclopropenoid fatty acids, and tannic acid have all been shown to induce liver cancer in fish. Gossypol—a toxin present in glanded cottonseed—causes loss of appetite and ceroid accumulation in the livers. Phytic acid, which ties up zinc in feed, and growth inhibitors found in soybean meal can be destroyed by proper heating during processing. Chlorinated hydrocarbons occur as contaminants of fish meal and can cause mortality when present in fry feeds; and brood fish transfer these compounds from the feed to their eggs, resulting in low hatchability and high mortality of swim-up fry. Toxaphene affects the utilization of vitamin C in catfish and can cause the *broken back syndrome.* A list of symptoms of organic toxicants is given in Table 28-16.

TABLE 28-16
COMPOUNDS TOXIC TO FISH

Compound	Symptoms of Toxicity
Heavy metals:	
Arsenic	Weight loss; liver and kidney damage; hepatoma.
Chromium	Coughing; sloughing of mucus from anal areas; pathologic changes in the intestines.
Copper	Pigmentation alterations; growth retardation; congestion in gills, liver, and lower digestive tract; lateral tail flexion; loss of response to touch; damage to the kidney and liver.
Mercury	Cataract formation; alteration of sodium absorption; lowered egg production.
Selenium	Ascites; exophthalmia; reduced red blood cell count; edema of spleen, kidney, skeletal muscle, stomach, and ovary.
Organic compounds:	
Aflatoxins	Liver tumors; necrotic livers; edematous gills; internal hemorrhaging.
Antibiotics	Depends on antibiotic used; antimycin A—changes in coloration; erythromycin—fatigue; inflamed ovaries; pustules on liver and kidneys.
Cyclopropenoid fatty acids	Synergistic with aflatoxins in producing tumors.
Detergents	Damage to taste buds; clubbed gills.
Gossypol	Anorexia; cocarcinogenic with aflatoxin B_1.
Pesticides	Sterility; weakness; nerve disorders; gastrointestinal dysfunctions.
Sulfonamides	Kidney disorders; growth retardation; edema.
Tannic acid	Liver tumors.

QUESTIONS FOR STUDY AND DISCUSSION

1. Rank in order the five leading species groups in monetary value of aquaculture production; and explain why they rank as they do. What accounts for the tremendous increases since 1980?

2. Why is freshwater fish production more prevalent than marine fish production?

3. List the primary differences in the nutrient requirements of fish and land animals.

4. List and discuss five environmental influences on the feeding behavior of fish.

5. What is meant by Standard Environmental Temperature (SET)? What are the SET values for salmon, trout, and catfish?

6. How do the energy requirements of fish differ from the energy requirements of warm-blooded animals?

7. Discuss the relative merits of digestible energy (DE) and metabolizable energy (ME) of feedstuffs as measures of the energy value of fish feeds.

8. Since only limited energy values of fish feeds are available, what metabolizable energy values should a fish producer use?

9. Discuss the relative importance of carbohydrates, fats, and proteins as energy sources for fish.

10. Discuss the comparative maximum levels of carbohydrates that may be used as energy for each salmonids and warmwater fish.

11. List and discuss the factors that should be considered in evaluating fats for fish feeds.

12. Compare the amino acid requirements of fish and broilers.

13. List and discuss the factors that determine the requirement for protein in fish feeds.

14. Explain how fish and warm-blooded animals differ in their ability to absorb and excrete a number of minerals.

15. What causes thyroid hyperplasia (goiter) in fish? How can it be prevented?

16. What are the major differences between the vitamin requirements of fish and higher animals? Describe the symptoms of vitamin C deficiency in fish.

17. Physiologically, how do freshwater and saltwater fishes differ in their use of water?

18. Discuss the role and importance of each of the following nonnutritive factors in fish rations: antioxidants, fiber, and pigment-producers.

19. Define each of the following terms as they apply to fish: natural foods and feeds, artificial wet feeds, formulated feeds, supplemental feeds, complete feeds, compressed or sinking pellets, expanded or floating pellets, moist pellets, crumbles, and meals/flakes.

20. Why and how are farm ponds fertilized?

21. What are the primary differences between the rations of salmonids, catfish, and carp?

22. Discuss the effect of size and water temperature on the feeding rate of salmonids.

23. Discuss the effect of water temperature, fish weight, and water quality on the feed consumption of warmwater fish.

24. How and why do the Chinese use manure to fertilize ponds populated by carp?

25. List three methods of feeding fish, and give the advantages and the disadvantages of each system.

26. What's unique about each of the following specialized types of aquatic production: aquarium fish, crayfish, eel, lobster, prawn, predatory fish, shrimp, sturgeon, and tilapia?

27. List and discuss three heavy metals and three organic compounds that may be toxic to fish.

Fig. 28–22. Harvesting market-size channel catfish. (Courtesy, Auburn University, Auburn, Ala.)

SELECTED REFERENCES

Title of Publication	Author(s)	Publisher
Artificial Propagation of Marine Fish	J. E. Shelbourne	T. F. H. Publications, Jersey City, N.J.
Commercial Catfish Farming	J. S. Lee	The Interstate Printers & Publishers, Inc., Danville, Ill., 1973
Finfish Nutrition and Fishfeed Technology, Vols. 1 and 2	J. E. Halver K. Tiews	Heenemann, Berlin, Germany, 1979
Fish Farming Handbook	E. E. Brown J. B. Gratzek	AVI Publishing Co., Inc., Wesport, Conn., 1980
Fish Hatchery Management	R. G. Piper, *et al.*	U.S. Department of the Interior, Fish and Wildlife Service, Washington, D.C., 1982
Fish Nutrition, 2nd Edition	Edited by J. E. Halver	Academic Press, New York, N.Y., 1989
Fish Physiology	Edited by: W. S. Hoar D. J. Randall	Academic Press, New York, N.Y., 1969
Ichthyology	K. F. Lagler J. E. Bardach R. R. Miller	John Wiley & Sons, Inc., New York, N.Y., 1962
Manual of Fish Culture, Part 3, Section B	A. M. Phillips	Bureau of Sport Fisheries and Wildlife, 1970
Nutrient Requirements of Coldwater Fishes	G. L. Rumsey, *et al.*	National Research Council, National Academy Press, Washington, D.C., 1981
Nutrient Requirements of Warmwater Fishes and Shellfishes	R. R. Stickney, *et al.*	National Research Council, National Academy Press, Washington, D.C., 1983
Nutrition and Feeding in Fish	Edited by: C. B. Crowley A. M. Mackie J. G. Bell	Academic Press, New York, N.Y., 1985
Nutrition and Feeding of Channel Catfish, SCSB-296	Edited by: E. H. Robinson R. T. Lovell	Texas Agricultural Experiment Station, Dept. of Agric. Communications, Texas A&M Univ., College Station, Tex., 1984
Nutrition and Feeding of Rainbow Trout	J. W. Hilton S. J. Slinger	Department of Fisheries and Oceans, Ottawa, Canada, 1981
Physiology of Fishes, The	M. E. Brown	Academic Press, New York, N.Y., 1957
Third Report To The Fish Farmers	Edited by: H. K. Dupree J. V. Huner	U.S. Fish and Wildlife Service, Washington, D.C., 1984
World Fish Farming—cultivation & economics	E. E. Brown	AVI Publishing Co., Inc., Wesport, Conn., 1977

Fig. 28-23. Fish. (Courtesy, USDA)

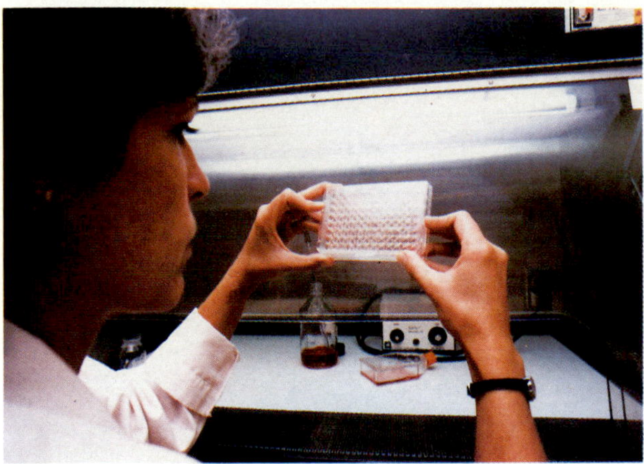

Fig. 28-24. Fish micronutrient requirements being set for demand feeding. (Courtesy, International Aquaculture Research Center, Hagerman, Idaho)

Fig. 28-25. Bulk feed with storage bins and mechanized distribution on farms save labor and bag costs. (Courtesy, International Aquaculture Research Center, Hagerman, Idaho)

Fig. 28-26. Catfish pond in which the fish have just been fed by a mechanical blower mounted on a truck. Note the floating type feed shown on the water. (Courtesy, Auburn University, Auburn, Ala.)

Fig. 28-27. Removing channel catfish egg mass from spawning container. (Courtesy, Auburn University, Auburn, Ala.)

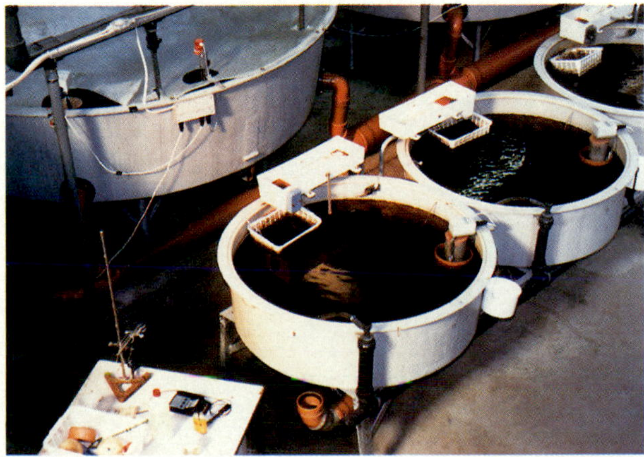

Fig. 28-28. Eel farming in Denmark. (Courtesy, International Aquaculture Research Center, Hagerman, Idaho)

Fig. 28-29. Aquarium fish. (Courtesy, USDA)

TA BASE, DEFICIENCY DISEASE, DICOUMAROL,
BLE NUTRIENT, EMBDEN-MEYERHOF PATHW
SIFIER, ESTROGEN, EXTRUDED, FAT SOLU
OT FINISHING, FERRITIN, FOOD AND DI
MINISTRATION, GASTRIC JUICE, GOS
OL, HEMICELLULOSE, HOMOGENIZ
POCALCEMIA, INTRADERMAL, L
IN, LIGNIFICATION, METABOL
RBIDITY, NEUTRAL DETERG
BER, OSTEOMALACIA, OV
NG, OXYTOCIN, PEAR
ORAL, PHOSPHOLIP
ROMBIN, PROXIM
SIS, RADIOACT
ATION, RETI
RUMINAT
SELENI
SPEC
TO

4 Glossary

Tom Phillips

SECTION IV

GLOSSARY

It is important that nutritionists and keepers of herds and flocks use the correct feed and nutrition terms and know what they mean. They need to know old terms, many of which have been redefined by the relentless wheels of progress; and they need to know a host of new terms. They need to know about new grains and forages; and about new by-products, recycled feeds, and crop residues. They need to know about new technology in feed processing; and about new supplements, additives, and implants.

To meet these needs, two glossary chapters are presented: Chapter 29, Glossary of Nutrition Terms; and Chapter 30, Glossary of Feedstuffs. These glossaries cover a wide range of terms. Each term has been defined as simply as possible, yet every effort has been made to retain technical correctness. Because of space limitations, many terms that are covered elsewhere in the text are not repeated in Chapters 29 and 30. Instead, the reader may refer to them in the index. It is the authors' fond hope that this compilation of terms will provide a common understanding in communication in feeds and nutrition.

ABOUT THE WORLD OF ANIMALS, the preceding two-page spread, is from an original painting by the noted artist, Tom Phillips, prepared specially for this book. It portrays the artist's conception of what is in Chapter 29, Glossary of Nutrition Terms, and Chapter 30, Glossary of Feedstuffs.

Chapter **29**

The Lascaux cave paintings in southern France, dating from about 15,000 B.C., seem to indicate the importance of the OX or BULL in early western culture.
Around 1400 B.C., the Phoenicians used it as shown (left) and called it *ALEPH,* which meant *OX.* Then, they let it represent the beginning sound of the word, which was "A."
The above symbol for OX, found in cuneiform picture writing of the fertile crescent (about 3,300 B.C.), is actually a bull's head with horns and ears.
A
After 900 B.C., the Greeks got hold of it, turned it around (above left) and called it *ALPHA.* So, the *OX* stands at the head of the alphabet.
Original painting by Tom Phillips

GLOSSARY OF NUTRITION TERMS

A glossary of terms frequently used in discussing matters related to feeds and nutrition is presented in this chapter. (See Chapter 30 for a glossary of feedstuffs.)

Because of space limitations, most terms that are defined or explained elsewhere in this book are not repeated in this chapter. Thus, if a particular term is not listed herein, the reader should look in the index or in the particular chapter where it is discussed.

A

ABLACTATION. The act of weaning.

ACCLIMATIZE. The process of becoming adjusted to a new environment.

ACETONE. A by-product of the breakdown of fats for energy. It builds up when the body's glycogen stores are depleted, which happens when carbohydrate is not available for fuel.

ACETYL-CoA. The chief precursor of lipids, and important intermediary in the Kreb's cycle; formed by an acetyl group attaching itself to coenzyme A during the oxidation of amino acids, fatty acids, or pyruvate.

ACHROMOTRICIA. Loss of pigment in hair.

ACID. A substance that neutralizes base substances by donating H ions. Essentially, acids are hydrogen donors—in solution, they provide H ions.

ACID DETERGENT FIBER (ADF). This indicates the amount of highly indigestible plant material in a feed or forage. It is composed mainly of cellulose, lignin, and silica. The lower the ADF, the higher the digestibility or available energy.

ACIDIFICATION. The act of reducing the pH of a substance or solution; increasing the acidity.

ACTIN. Muscle fiber fraction that complexes with myosin to bring about muscle contraction.

ACTIVATE. To make an inactive substance capable of exerting its proper effect.

ADDITIVE. An ingredient or combination of ingredients added to a basic feed mix to help fulfill a specific need.

ADIPOSE TISSUE. Fatty tissue.

AD LIBITUM. Free-choice access to feed.

ADRENAL GLAND. One of the endocrine (ductless) glands of the body, located near the kidney, which secretes hormones needed to utilize nutrients.

AERIAL PART. The aboveground part of a plant.

AEROBE. The term usually applied to microorganisms that require oxygen to live and reproduce.

AFTERMATH. The regrowth of range or seeded pasture forage after grazing or mowing.

AGAR. An extract of seaweed that forms a firm gel at temperatures of 104°F and lower. It is used as a gelling and stabilizing agent in foods and as a microbiological plating medium.

AGGLOMERATED FEED. A mixture of feeds in compacted or extruded form.

AGROFORESTRY. A production system in which trees for timber are grown in forage crops.

AIR ASHED. Reduced by combustion in air to a mineral residue.

AIR DRY (approximately 90% dry matter). This refers to feed that is dried by means of natural air movement, usually in the open. It may be either an actual or an assumed dry matter content; the latter is approximately 90%. Most feeds are fed in the air dry state.

AIV PRESERVATIVE. A combination of hydrochloric and sulfuric acids used to increase the acidity of silage and to improve its keeping quality and nutritive value, developed by Dr. A. I. Virtanin of Finland.

ALBUMEN. One of the important proteins of blood, milk, eggs, and other substances.

ALGAE. Single-celled plants which contain about 50% protein on a dry basis and offer an attractive possibility as a protein source. They synthesize proteins by the use of solar energy. Cultivated freshwater algae will produce about 10 times as much protein per unit of land area as soybeans.

ALKALI. It usually refers to a soluble salt or a mixture of soluble salts present in some soils of arid or semiarid regions in quantity detrimental to ordinary agriculture.

ALKALOSIS. Disturbance in the acid-base balance in which there is a reduction of the acid partner in the buffer system or an increase in the base. In either case, the necessary 20:1 ratio between base and acid is upset by an increase in the relative amount of base.

ALLERGY. A severe reaction, or sensitivity, which occurs in some individuals following the introduction of certain antigens into their bodies.

ALOPECIA. Loss of hair.

ALPHA-TOCOPHEROL EQUIVALENT. A standard unit of measurement (in milligrams) for designating vitamin E requirements, since potencies of the other members of the vitamin E group vary.

AMBIENT TEMPERATURE. The prevailing or surrounding temperature.

AMERICAN FEED INDUSTRY ASSOCIATION (AFIA). The address is 1701 N. Ft. Myer Drive, Arlington, VA 22209. The American Feed Industry Association is a nationwide organization of feed manufacturers banded together (1) to improve the quality and promote the use of commercial feeds, (2) to encourage high standards on the part of its members, and (3) to protect the best interests of the feed manufacturer and the livestock producer in legislative programs.

AMINO ACIDS. Nitrogen-containing compounds that constitute the "building blocks" or units from which more complex proteins are formed. They contain both an amino (NH_2) group and a carboxyl (COOH) group.

AMMONIA. A colorless gaseous alkaline compound of nitrogen and hydrogen (NH_3) that is lighter than air, of extremely pungent smell and taste, and very soluble in water.

AMMONIATED. Combined or impregnated with ammonia or an ammonium compound.

AMPHOTERIC. Having properties of both an acid and a base and therefore able to function as either. Amino acids have this dual chemical nature because of their structure—they contain both an acid (carboxyl, COOH) and a base (amino, NH_2) group.

AMYLOPECTIN. A polysaccharide, the insoluble part of starch, which forms a paste in hot water and thickens during cooking.

AMYLOSE. A simple single sugar; a carbohydrate containing a single saccharide (sugar) unit.

ANABOLISM. The conversion of simple substances into more complex substances by living cells (constructive metabolism).

ANAEROBE. A microorganism that does not require air or free oxygen to live and reproduce.

ANALOGUE (ANALOG). Anything that is analogous or similar to something else.

ANAPHYLAXIS. An acute allergic reaction.

ANASARCA (DROPSY). Generalized edema.

ANEURISM. The dilation of an artery wall.

ANIMAL NUTRITIONIST. A professional who possesses an earned bachelor's, or higher, degree with a major in animal science, chemistry, biochemistry, or related sciences from an accredited college or university, and who is able to apply the knowledge of feeds and their nutrients to the growth, maintenance, production, and health of animals.

ANIMAL PROTEIN. Protein derived from meat-packing or rendering plants, surplus milk or milk products, and marine sources. It includes proteins from meat, milk, poultry, eggs, fish, and their products.

ANIMAL PROTEIN FACTOR (APF). The term formerly used to refer to an unidentified growth factor essential for poultry and swine and present in protein feeds of animal origin. It is now known to be the same as Vitamin B-12.

ANION. An ion carrying a negative (-) electrical charge.

ANOREXIA. A lack or loss of appetite for food.

ANOXIA. Lack of oxygen in the blood or tissues. This condition may result from various types of anemia, reduction in the flow of blood to tissues, or lack of oxygen in the air at high altitudes.

ANTACID. Any substance which counteracts or neutralizes acidity.

ANTAGONIST. A substance that counteracts the action of another substance. The antagonist prevents the normal action because its molecular structure is so like that of the first substance that it almost fits into its position in a metabolic process; it gets in the way and prevents the reaction from taking place.

ANTHELMINTIC (VERMIFUGE). A product which expels or destroys internal parasites.

ANTIBODY. A protein substance (modified type of blood-serum globulin) developed or synthesized by lymphoid tissue of the body in response to an antigenic stimulus. Each antigen elicits production of a specific antibody. In disease defense, the animal must have an encounter with the pathogen (antigen) before a specific antibody is developed in its blood.

ANTIGEN. A high molecular weight substance (usually protein) which, when foreign to the bloodstream of an animal, stimulates formation of a specific antibody and reacts specifically *in vivo* or *in vitro* with its homologous antibody.

ANTIHISTAMINES. Any of various compounds used to treat certain allergic reactions in the body.

ANTIMETABOLITE. A substance bearing a close structural resemblance to one required for normal physiological functioning, which exerts its effect by replacing or interfering with the utilization of the essential metabolite.

ANTIOXIDANT. A compound that prevents oxidative rancidity of polyunsaturated fats. Antioxidants are used to prevent rancidity in feeds and foods.

ANTIVITAMIN. Any substance which inhibits the normal function of a vitamin.

APPETITE. The immediate desire to eat when food is present. Loss of appetite in an animal is usually caused by illness or stress.

APPETIZER. A substance that stimulates the appetite.

ARIBOFLAVINOSIS. A group of clinical manifestations of riboflavin deficiency.

ARTIFICIALLY DRIED. Dried, or dehydrated, by other than natural means, usually by heat from fossil fuel sources.

AS-FED. This refers to feed as normally fed to animals. It may range from 0 to nearly 100% dry matter. Moisture-free feed consists of 100% dry matter.

ASH. The mineral matter of a feed. The residue that remains after complete incineration of the organic matter.

ASPIRATING. The process of removing chaff, dust, or other light materials by use of air.

ASSAY. Determination of (1) the purity or potency of a substance, or (2) the amount of any particular constituent of a mixture.

ASSIMILATION. A physiological term referring to a group of processes by which the nutrients in feed are made available to and used by the body; includes digestion, absorption, distribution, and metabolism.

ATAXIA. Lack of coordination.

ATOM. A particle of matter indivisible by chemical means. It is the fundamental building block of the chemical elements. The elements, such as iron and sulfur, differ from each other because they contain different kinds of atoms.

ATROPHY. A wasting away of a part of the body, usually muscular, induced by injury or disease.

AVAILABLE CRUDE PROTEIN (ACP). The crude protein that the animal can use, calculated by substracting ICP from CP.

AVAILABLE NUTRIENT. A nutrient which can be digested, absorbed, and/or used in the body.

AVERAGE DAILY GAIN (ADG). The average daily liveweight increase of an animal.

AVIDIN. A protein in egg white (albumen) which combines with biotin and makes it unavailable; an antivitamin.

AVOIRDUPOIS WEIGHTS AND MEASURES. Avoirdupois is a French word, meaning *to weigh*. The old English system of weights and measures is referred to as the Avoirdupois System, or U.S. Customary Weights and Measures, to differentiate it from the Metric System.

B

BACTERIA. Microscopic, single-cell plants, found in most environments, often referred to as microbes; some are beneficial, others are capable of causing disease.

BACTERICIDE. A product that destroys bacteria.

BALANCED RATION. One which provides an animal the proper amounts and proportions of all the required nutrients.

BASAL DIET. A diet common to all groups of experimental animals to which the experimental substance(s) is added.

BASAL METABOLIC RATE (BMR). The heat produced by an animal during complete rest (but not sleeping) following fasting, when using just enough energy to maintain vital cellular activity, respiration, and circulation, the measured value of which is called the basal metabolic rate (BMR). Basal conditions include thermoneutral environment, resting, postabsorptive state (digestive processes are quiescent), consciousness quiescence, and sexual repose. It is determined in humans 14 to 18 hours after eating and when at absolute rest. It is measured by means of a calorimeter and is expressed in calories per square meter of body surface.

BASE. A chemical substance that is capable of neutralizing acid by accepting hydrogen ions from the acid. *Alkali* is a synonymous term.

BATTERY. A series of pens or cages used to house animals in concentrated confinement rearing systems.

BILE. A yellowish-green fluid produced by the liver and stored in the gallbladder (except for the horse and other animals that do not have a gallbladder), which empties through the bile duct into the duodenum of the small intestine, where it participates in the digestion of fats.

BIOASSAY. Determination of the relative effective strength of a substance (as a vitamin, hormone, or drug) by comparing its effect on a living test organism with that of a standard preparation.

BIOCHEMISTRY. The chemistry of living things—plants and animals.

BIOFLAVONOID. Compounds that are widely distributed in nature as pigments in flowers, fruits, tree barks, vegetables, and grains. They are not considered essential nutrients.

BIOLOGIC ACTIVITY. Degree of effect in an organism of a specific vitamin; means of measuring required amount of vitamin to prevent a deficiency.

BIOLOGICAL. Of or pertaining to life and life processes, including the study of structure, functioning, growth, origin, evolution, and distribution of living organisms.

BIOLOGICAL VALUE OF A PROTEIN. The percentage of the protein of a feed or feed mixture which is usable as a protein by the animal. Thus, the biological value of a protein is a reflection of the kinds and amounts of amino acids available to the animal after digestion. A protein which has a high biological value is said to be of *good quality*.

BIOSYNTHESIS. The production of a new material in living cells or tissues.

BLENDED. Combined or mixed so as to render the constituent parts indistinguishable from one another, such as when two or more feed ingredients are mixed.

BLOAT. A disorder of ruminants caused by excess gas in the rumen.

BLOCK. Feed or feeds compressed into a solid mass and cohesive enough to hold its form, usually weighing from 15 to 500 lb.

BLOOM. Said of an animal that has beauty and freshness. An animal in bloom has a glossy hair coat and presents an attractive appearance.

BOLTING FEED. Animals that ingest or eat too rapidly or greedily are said to be bolting their feed.

BOLUS.

- Regurgitated feed that has been chewed and is ready to be swallowed, known as the cud.
- A large pill for dosing animals.

BOMB CALORIMETER. An instrument used to measure the gross energy content of any material, in which the feed (or other substance) tested is placed in an enclosed chamber and burned in the presence of oxygen.

BRAND NAME. Any word, name, symbol, or device, or any combination of these, often registered as a trademark or name, which identifies a product and distinguishes it from others.

BRICK. A feed or feeds, other than pellets, compressed into a solid mass and weighing less than 2 lb.

BRITISH THERMAL UNIT (BTU). The amount of energy required to raise 1 lb of water 1 °F; equivalent to 252 calories.

BRIX. A term commonly used to indicate the sugar (sucrose) content of molasses. It is expressed in degrees and was originally used to indicate the percentage by weight of sugar in sucrose solutions, with each degree Brix being equal to 1% sucrose.

BROILER. A young chicken.

BUFFER. A substance in a solution that makes the degree of acidity (hydrogen-ion concentration) resistant to change when an acid or base is added.

BUSHEL. A unit of capacity equal to 2,150.42 cubic in. (approximately 1.25 cu ft).

BY-PRODUCT FEEDS. The innumerable roughages and concentrates obtained as secondary products from plant and animal processing, and from industrial manufacturing.

C

CACHEXIA. A general lack of nutrition and wasting of tissues due to malnutrition, a chronic disease, or a terminal illness.

CAKE (PRESSCAKE). The mass resulting from the pressing of seeds, meat, or fish in order to remove oils, fats, or other liquids.

CALCIFEROL. The same as vitamin D_2.

CALCIFICATION. The process by which organic tissue becomes hardened by a deposit of calcium salts.

CALCITONIN. One of two hormones responsible for fine regulation of blood calcium level; the other is parathyroid hormone. Calcitonin, secreted by the thyroid gland, inhibits bone resorption and release of calcium to the blood, thereby lowering blood calcium.

CALCULI. Mineral deposits that occur in the kidney or urinary tract.

CALORIC. Pertaining to heat or energy.

CALORIMETER. An instrument for measuring the amount of energy in a substance.

CANNED. A term applied to a feed which has been processed, packaged, sealed, and sterilized for preservation in cans or similar containers.

CANNIBALISM.

- The habit of one animal pecking at or eating on another animal, such as a fowl pecking at or eating on another fowl or one pig biting the tail of another.
- The eating of young, such as a sow may do after farrowing or a doe rabbit may do if disturbed soon after kindling.

CANNULA (CANULA). A tubular device used to connect an internal structure with the outside of the animal. A metal-, rubber-, or glass-tube cannula may be inserted into the body cavity to allow the escape of fluids or gas. Liquids and other materials may be introduced into the body through a cannula.

CARBON DIOXIDE. One of the two waste gases produced by rumen fermentation. (The other is methane.) The production of carbon dioxide represents an energy loss. The chemical formula is CO_2.

CARCASS WEIGHT. Weight of the carcass of an animal following slaughter, as it hangs on the rail, expressed either as warm (hot) or chilled (cold) carcass weight.

CARCINOGEN. Any cancer-producing substance or agent.

CARCINOGENIC. Cancer-producing.

CARNIVORE. A flesh-eating animal.

CARRIER. An edible material to which ingredients are added to facilitate their uniform incorporation into feeds. The active particles are absorbed, impregnated, or coated into or onto the edible material in such a way as to carry the active ingredient physically.

CARRYING CAPACITY. The number of animal units (one cow, plus suckling calf—if there is a calf, or one heifer two years old or over; or equivalent) a property or area will carry on a year-round basis. This includes the land grazed plus the land necessary to produce the winter feed.

CATABOLISM. The conversion or breaking down of complex substances into more simple compounds by living cells (destructive metabolism).

CATALYST. Any substance which speeds up the rate of a chemical reaction without being destroyed or inactivated in the process. Enzymes are organic catalysts.

CATION. Ion that carries a positive (+) electric charge.

CEREAL. A plant in the grass family (*Gramineae*), the seeds of which are used for human and animal food; *e.g.,* maize and wheat.

CHAFF. Glumes, husks, or other seed covering, together with other plant parts, separated from seed in threshing or processing.

CHELATE. The word *chelate* is derived from the Greek word meaning *claw.* It refers to a cyclic compound which is formed between an organic molecule and a metalic ion, the latter being held within the organic molecule as if by a claw. Examples of naturally occurring chelates are the chlorophylls, cytochromes, hemoglobin, and vitamin B-12.

CHEMOSTATIC CONTROL OF APPETITE. Certain cells located at the base of the brain in the lateral region of the hypothalamus regulate feed intake by sensing mechanisms. Factors shown to influence these cells and thereby alter feed intake include the volatile fatty acids. (Also see THERMOSTATIC CONTROL OF APPETITE.)

CHIPPED. Cut or broken into fragments.

CHLOROPHYLL. The green coloring material in plants which is essential for one phase of the conversion of solar energy into organic material, the process known as photosynthesis.

CHOLESTEROL. A white, fat-soluble substance found in animal fats and oils, bile, blood, brain tissue, nervous tissue, the liver, kidneys, and adrenal glands. It is important in metabolism and is a precursor of certain hormones. Some have implicated cholesterol as a factor in arteriosclerosis.

CHOPPED. Reduced in particle size by cutting with knives or other sharp instruments.

CHYME. Semifluid feed mass in the gastrointestinal tract following gastric digestion.

CIRRHOSIS OF THE LIVER. Progressive destruction of the liver cells and an abnormal increase of connective tissue.

CLEANED. Removal of material by such methods as scalping, aspirating, magnetic separation, or by any other method.

CLEANINGS. Chaff, weed seeds, dust, and other foreign matter removed from cereal grains.

CLIPPING. Removal of the end of whole grain.

COAGULATED. Curdled, clotted, or congealed, usually brought about by the action of a coagulant.

COEFFICIENT OF DIGESTIBILITY. The percentage value of a food nutrient that is absorbed. For example, if a food contains 10 g of nitrogen and it is found that 9.5 g are absorbed, the digestibility is 95%.

COENZYME. A substance, usually containing a vitamin, which works with an enzyme (protein mainly) to perform a certain function.

COLLAGEN. A white, papery transparent type of connective tissue which is of protein composition. It forms gelatin when heated with water.

COLLOID. Glutinous, gluelike; a dispersion of matter throughout a medium.

COLOSTRUM. The milk secreted by mammalian females for the first few days following parturition, which is both high in antibodies and laxative.

COMBUSTION. The combination of substances with oxygen accompanied by the liberation of heat.

COMFORT ZONE. The temperature range within which no demand is made on temperature regulating mechanism.

COMMERCIAL FEEDS. Feeds mixed by manufacturers who specialize in the feed business.

COMMINUTE. Synonym of chopped.

COMPACTION. Closely packed feed in the stomach and intestines of an animal causing constipation and/or digestive disturbances.

COMPENSATORY GROWTH. Accelerated growth following a period of limited feed intake.

COMPLETE PROTEIN. A protein that contains the essential amino acids in quantities sufficient for maintenance of the body and for a normal rate of growth. Such proteins are said to have a high biologic value. Animal protein sources are usually more complete than plant protein sources.

COMPLETE RATION. All feedstuffs (forages, grains, and processed feeds) combined in one feed. A complete ration fits well into mechanized feeding and the use of computers to formulate least-cost rations.

COMPOSITE SAMPLE. A combination of individual samples taken at selected intervals to minimize the effect of the variability of an individual sample.

CONCENTRATE. A broad classification of feedstuffs which are high in energy and low in crude fiber (under 18%). For convenience, concentrates are often broken down into (1) carbonaceous feeds, and (2) nitrogenous feeds.

CONDENSED. Reduced to denser form by removal of moisture.

CONDITION.

- The state of health, as evidenced by the hair-coat, and general appearance.
- The amount of flesh or finish (fat covering).

CONDITIONING. Achieving predetermined moisture characteristics and/or temperature of ingredients or a mixture of ingredients prior to further processsing.

CONDOMINIUMS. The condominium concept in animals refers to an operation in which there is separate ownership and management of animals and lots, but joint ownership of part of the facilities (perhaps the feed mill, feed trucks, and other equipment).

CONGENITAL. Malformation existing at birth; acquired during development in the uterus and not through heredity.

CONTENTMENT. A stress-free condition exhibited by healthy animals; the cow will stretch on rising, the sheep will stand or lie quietly, the pig will curl its tail, and the horse will look completely unworried when resting.

COOKED. Heated to alter chemical or physical characteristics or to sterilize.

COPROPHAGY. The ingestion of fecal material.

COW POOLS. A business organization or cooperative which cares for and milks cows in a centralized location.

CRACKED. Particle size reduced by combined breaking and crushing action.

CREATININE. A nitrogenous compound arising from protein metabolism and excreted in the urine.

CREEP. An enclosure or feeder used for supplemental feeding of nursing young, which excludes their dams.

CRIBBER (WIND SUCKER, or STUMP SUCKER). A horse that has the vice of biting or setting the teeth against some object, such as the manger, while sucking air.

CRIMPED. Rolled between corrugated rollers. The grain to which this term refers may be tempered or conditioned before crimping, and may be cooled afterward.

CROP RESIDUES. The plant growth that remains after harvesting a grain or seed crop, such as cornstalks and husklage, sorghum stalks, soybean refuse, small grain straws and chaff, and legume and grass seed straws.

CRUDE FAT. Material that is extracted from moisture-free feeds by ether. It consists largely of fats and oils with small amounts of waxes, resins, and pigments. In calculating the energy value of a feed, the fat is considered to have 2.25 times as much energy as either nitrogen-free extract or protein.

CRUDE FIBER (CF). Crude fiber contains cellulose, hemicellulose, and lignin. It is determined by treating a sample of feed with ether to remove fats, then the residue is boiled alternately in a weak acid and a weak base. The remaining residue contains crude fiber and ash. Then, the ash is subtracted out and the crude fiber is what remains.

CRUDE PROTEIN (CP). This is a mixture of true protein and nonprotein nitrogen. It indicates the capacity of a feed to meet an animal's protein needs, although it is of little value in predicting energy availability.

CRUMBLES. Pelleted feed reduced to granular form.

CUBING. Refers to the practice of compressing long or coarsely cut hay into cubes about 1¼ in. square and 2 in. long, with a bulk density of 30 to 32 lb per cu ft.

CUD. A bolus of previously eaten feed which is regurgitated by a ruminant animal for further chewing.

CUDDING. The act of chewing the cud. In ruminants, cudding is a sign of good health and is one of the first things to disappear in sickness.

CURD. The coagulated or thickened part of milk. Curd from whole mild consists of casein, fat, and whey, whereas curd from skim milk contains casein and whey, but only traces of fat.

CUSTOM CATTLE FEEDING. The feeding of cattle for a fee, without taking ownership of the animals.

CYANOCOBALAMIN. Same as Vitamin B-12.

CYSTITIS. Inflammation of the urinary bladder.

D

D-ACTIVATED. Plant or animal sterol fractions which have been Vitamin D activated by ultraviolet light or by other means.

DATA BASE. An organized body of information placed in computers for systematic storage and retrieval; usually deals with a specific topic or project.

DEBEAKING. The removal of part of the beak of chickens and poults to prevent cannibalism. Many broiler growers and some egg producers have chicks debeaked at the hatchery, with about ⅔ of the beak removed. Turkeys are usually debeaked at 2 to 3 weeks of age, with at least ½ of the upper beak removed.

DECORTIFICATION.

- Removal of the bark, hull, husk, or shell from a plant, seed, or root.
- Removal of portions of the cortical substance of a structure or organ, as in the brain, kidney, and lung.

DEFECATION. The evacuation of fecal material from the rectum.

DEFICIENCY DISEASE. A disease caused by a lack of one or more basic nutrients, such as a vitamin, a mineral, or an amino acid.

DEFLUORINATED. Processed in such manner that the fluorine content is reduced to a level which is nontoxic under normal use.

DEHULLED. Grains or other seeds with the outer covering removed.

DEHYDRATE. To remove most or all moisture from a substance for the purpose of preservation, primarily through artificial drying.

DENATURING. The term denaturing applies to a wide variety of structural changes and involves different effects on individual proteins. Mild heat treatment of protein is advantageous. But extensive heat treatment can result in lowering of protein quality by the well known Millard or browning reaction, in which the free amino groups of the peptide chain react with the aldehyde group of reducing sugars such as glucose or lactose to yield an amino-sugar complex that is no longer available to the animal.

DEPRAVED APPETITE (PICA). A craving for and eating of unnatural substances, such as dirt, hair, dung, wood, etc.

DESICCATE. To dry completely.

DIALYSIS. Separating substances in solution by taking advantage of the different rates at which they pass through a semipermeable membrane.

DIARRHEA. Abnormal frequency and liquidity of fecal discharge.

DICOUMAROL. A chemical compound found in spoiled sweet clover hay or made synthetically. It is an anticoagulant and can cause internal hemorrhages when eaten by animals. Medically, it is used as an anticoagulant. The trade name is Dicumarol.

DIET. Feed (food) ingredient or mixture of ingredients, including water, which is consumed.

DIGESTIBLE DRY MATTER (DDM). This refers to digestiblity as predicted by ADF or measured by feeding trials with animals.

DIGESTIBLE DRY MATTER INTAKE (DDMI). This is an estimate of how much DDM an animal will consume.

DIGESTIBLE NUTRIENT. The part of each feed nutrient that is digested or absorbed by the animal.

DIGESTIBLE PROTEIN. That protein of the ingested food protein which is absorbed.

DIGESTION COEFFICIENT (coefficient of digestibility). The difference between the nutrients consumed and the nutrients excreted expressed as a percentage.

DILUENT. An edible substance used to mix with and reduce the concentration of nutrients and/or additives to make them more acceptable to animals, safer to use, and more capable of being mixed uniformly in a feed. (It may also be a carrier.)

DISACCHARIDE. Compound sugars composed of two molecules of a monosaccharide. The three common members are: sucrose (table sugar), lactose (milk sugar), and maltose (grain sugar).

DISTENSION OF RUMEN. The limitation of feed intake by the physical fill of the rumen. When bulky rations high in fiber are consumed, the animal's intake is limited by the amount of feedstuff which can be packed into the rumen.

DIURESIS. An increased excretion of urine.

DIURETIC. An agent that increases the production of urine.

DORMANT. Plants cured on the stem; applied to most nongrowing range plants after the seeds have formed.

DOUGH STAGE. Stage at which grain seeds are soft and immature.

DRESSED. Made uniform in texture by breading or screening of lumps from feed and/or the application of liquid(s).

DRIED. Materials from which water or other liquids have been removed.

DRUGS. Substances of mineral, vegetable, or animal origin used in the relief of pain, for the cure of disease, or for the enhancement of production.

DRY. Nonlactating female. The dry period is the time between lactations (when a female is not secreting milk).

DRY FORAGES AND ROUGHAGES. Feeds in the dry state that are bulky and low in weight per unit volume; they contain more than 18% crude fiber and are relatively low in energy.

DRYLOT. A relatively small enclosure without vegetation, either (1) with shelter, or (2) an open yard, in which animals may be confined.

DRYLOT FEEDING. Feeding in a drylot.

DRY MATTER (DM). That part of a feed which is not water. It is computed by determining the percentage of water and subtracting the water content from 100%.

DRY MATTER BASIS. A method of expressing the level of a nutrient contained in a feed on the basis that the material contains no moisture.

DRY MATTER INTAKE (DMI). This is an estimate of the amount of forage an animal will eat with only forage fed.

DRY-RENDERED. Residues of animal tissues cooked in open steam-jacketed vessels until the water has evaporated. Fat is removed by draining and pressing the solid residue.

DRY ROLLING. Refers to processing grains without added steam or other moisture.

DUNG. The feces (manure) or excrement of animals and birds.

DYSTROPHY. Degeneration.

E

EARLY BLOOM. Period between initiation of bloom to stage at which one-tenth of the plants are in bloom.

EARLY LEAF. Stage at which the plant reaches one-third of its growth before blooming.

EARLY WEANING. The practice of weaning young animals earlier than usual; weaning beef calves at 35 days to 5 months of age, weaning lambs at 3 to 4 weeks of age, weaning pigs at 3 to 4 weeks of age, and weaning foals under 5 months of age.

EASY KEEPER. An animal that grows or fattens rapidly on limited feed.

EATING WITH RELISH. In healthy animals, the appetite is good and the feed is attacked with relish (as indicated by eagerness to get to the trough, wagging the tail, etc.).

ECZEMA. A condition involving the inflammation of the skin with lesions of either a dry or weeping nature. It is most commonly caused by allergies or exposure to chemicals or other irritants, but it may be due to deficiencies of such nutrients as essential fatty acids and vitamins.

EDEMA. Swelling of a part or all of the body due to the accumulation of excess water.

EFFICIENCY OF FEED CONVERSION. This is expressed as units of feed per unit of product—meat, milk, or eggs.

ELECTROLYTE. A chemical compound which in solution dissociates by releasing ions. (An ion is an atomic particle that carries a positive (+) or a negative (-) charge.

ELEMENT. One of the 103 known chemical substances that cannot be divided into simpler substances by chemical means.

EMACIATED. An excessively thin condition of the body.

EMBDEN-MEYERHOF PATHWAY. The final pathway within cells by which glucose is metabolized to pyruvic acid. Pyruvic acid is metabolized by the rumen microorganisms to form one of the volatile fatty acids. Animal cells oxidize pyruvic acid to carbon dioxide, energy, water, and heat.

EMULSIFIER. An agent that breaks down large fat globules to smaller, uniformly-distributed particles. This action is accomplished in the intestine chiefly by the bile acids, which lower surface tension of the fat particles. Emulsification greatly increases the surface area of fat, facilitating contact with fat-digesting enzymes.

ENDOCRINE. Pertaining to glands and their secretions that pass directly into the blood or lymph instead of into a duct (secreting internally). Hormones are secreted by endocrine glands.

ENDOGENOUS. Originating within the body; e.g., hormones and enzymes.

ENDOSPERM. The carbohydrate portion of seed.

ENEMA. The introduction of a liquid into the intestines by way of the anus.

ENERGY. The capacity to perform work; available power. Energy is manifested in various forms—motion, position, light, heat, and sound.

ENERGY FEEDS. Feeds that are high in energy and low in fiber (under 18%), and that generally contain less than 20% protein.

ENSILED. Materials that have been subjected to anaerobic fermentation to form silage.

ENTERITIS. Inflammation of the intestines.

ENZYMATIC. Related to an enzyme.

ERGOSTEROL. A plant sterol which, when activated by ultraviolet rays, becomes vitamin D_2. It is also called provitamin D_2 and ergosterin.

ERGOT.

- A fungus disease of plants.
- The horny growth at the back of the fetlock joint; the spurs of a horse's hoofs.

ERUCTATION. The act of belching gas from the stomach.

ESSENTIAL AMINO ACIDS. Those amino acids which cannot be made in the body from other substances or which cannot be made in sufficient quantity to supply the animal's needs; hence, they must be supplied in the ration.

ESSENTIAL FATTY ACID. A fatty acid that cannot be synthesized in the body or that cannot be made in sufficient quantities for the body's needs; hence, they must be supplied in the ration.

ESTER. A compound produced by the reaction between an acid and an alcohol with elimination of a molecule of water. This process is called esterification. For example, a triglyceride is a glycerol ester of fatty acids.

ESTROGEN. A general term for the principal female sex hormones.

ETHER EXTRACT (EE). Fatty substances of feeds and foods that are soluble in ether.

EVAPORATED. Reduced to a denser form; concentrated as by evaporation or distillation.

EXCRETA. The products of excretion—primarily feces and urine.

EXOGENOUS. Provided from outside of the organism.

EXPANDED. As applied to feed, it refers to an increase in volume as the result of a sudden reduction in the surrounding pressure.

EXPELLER PROCESS. A process for the mechanical extraction of oil from seeds, involving the use of a screw press.

EXPERIMENT. The word *experiment* is derived from the Latin *experimentum,* meaning proof from experience. It is a procedure used to discover or to demonstrate a fact or general truth.

EXTRACELLULAR FLUID. Fluid outside the cell. It comprises about one-third of the total body fluid and includes tissue fluid, blood plasma, cerebrospinal fluid, fluid in the eye, and fluid of the gastrointestinal tract.

EXTRACTED. Having removed fat or oil from materials by heat, mechanical pressure, or solvent. Similar terms: expeller extracted, hydraulic extracted, solvent extracted.

EXTRINSIC FACTOR. A dietary substance which was formerly thought to interact with the intrinsic factor of the gastric secretion to produce the antianemic factor, now known to be vitamin B–12. (Also see INTRINSIC FACTOR.)

EXTRUDED. A type of feed preparation in which the feed is forced through a die under pressure.

EXUDATE. Material that escapes from blood vessels and is deposited in tissues or tissue surfaces; characterized by a high content of protein, cells, or other cellular solid matter.

F

FACTOR. In nutrition, any specific chemical substance found in feed.

FACULTATIVE ANAEROBES. Microorganisms capable of living either in the presence or absence of oxygen.

FARMER-FEEDER. A cattle feeder who feeds fewer than 1,000 head at one time.

FAT. The term fat is frequently used in a general sense to include both fats and oils, or a mixture of the two. Both fats and oils have the same general structure and chemical properties, but they have different physical characteristics. The melting points of most fats are such that they are solid at ordinary room temperatures, while oils have lower melting points and are liquids at these temperatures.

FAT SOLUBLE. Soluble in fats and fat solvents but generally not soluble in water.

FATTENING. The deposition of energy in the form of fat within the body tissues.

FDA. Food and Drug Administration, a federal regulatory agency.

FEBRILE. Condition characterized by fever.

FECES. The excreta discharged from the digestive tract through the anus.

FEED (or FEEDSTUFF). Any naturally occurring ingredient, or material, fed to animals for the purpose of sustaining them.

FEED ADDITIVES AND IMPLANTS. Nonnutritive products that improve the rate and/or efficiency of gain of animals, prevent certain diseases, or preserve feeds.

FEED EFFICIENCY. The ratio expressing the number of units of feed required for one unit of production (meat, milk, eggs) by an animal. Also known as feed conversion.

FEED GRADE. Feedstuffs suitable for animals, but not for human consumption.

FEED GRAIN. Any of several grains most commonly used for livestock or poultry feed, such as corn, sorghum, oats, and barley.

FEEDER'S MARGIN. The difference between the cost per hundredweight of feeder animals and the selling price per hundredweight of the same animals when finished.

FEEDLOT. A lot or plot of land on which animals are fed or finished for market.

FEEDLOT FINISHING. Refers to finishing cattle or lambs in a restricted area, with the feed conveyed to them; and it may involve feeding (1) in open pen feedlot, or (2) under shelter.

FEEDSTUFF. Any product, of natural or artificial origin, that has nutritional value in the ration when properly prepared.

FERMENTATION. Chemical changes brought about by enzymes produced by various microorganisms.

FERMENTATION PRODUCT. Product formed as a result of an enzymatic transformation of organic substrates.

FERMENTED. Acted upon by yeasts, molds, or bacteria in a controlled aerobic or anaerobic process in the manufacture of such products as alcohols, acids, vitamins of the B complex group, or antibiotics.

FERRITIN. Protein-iron compound in which iron is stored in tissues; the storage form of iron in the body.

FIBER CONTENT OF A FEED. The amount of hard-to-digest carbohydrates. Most fiber is made up of cellulose and lignin.

FIBROUS. High in cellulose and/or lignin content.

FILL.

- A term designating the fullness of the digestive tract of an animal.
- With market animals, the fill refers to the amount of feed and water consumed upon their arrival at the market and prior to selling.

FINES. Any material which will pass through a screen whose openings are smaller than the specified minimum size.

FINISH. To fatten a slaughter animal. Also, the degree of fatness of such an animal.

FINISHING ANIMALS. The laying on of fat.

FISTULA. An abnormal tubelike passage from some part of the body to another part or to the exterior—sometimes surgically inserted. Rumen fistulas consist of a capped opening to the rumen which permits concentrates to be removed. (Also see CANNULA.)

FITTING. The conditioning of an animal for show or sale, which usually involves a combination of special feeding plus exercise and grooming.

FLAKES. An ingredient rolled or cut into flat pieces with or without prior steam conditioning.

FLATULENCE. Excessive formation of gas in the stomach or intestines.

FLAVORING AGENTS. Feed additives that are supposed to increase palatability and feed intake.

FLORA. Plant life. In nutrition it generally refers to the bacteria present in the digestive tract (microflora).

FLUSHING. The practice of feeding females more generously 2 to 3 weeks before breeding. The beneficial effects attributed to this practice are (1) more eggs (ova) are shed, and this results in more offspring; (2) the females come in heat more promptly; and (3) conception is more certain.

FODDER. Coarse feeds, such as corn or sorghum stalks.

FOOD. When used in reference to animals, it is synonymous with feed.

FOOD AND DRUG ADMINISTRATION (FDA). The federal agency in the Department of Health, Education, and Welfare that is charged with the responsibility of safeguarding American consumers against injury, unsanitary food, and fraud. It protects industry against unscrupulous competition, and it inspects and analyzes samples and conducts independent research on such things as toxicity (using laboratory animals), disappearance curves for pesticides, and long-range effects of drugs.

FOOD GRAIN. Grains most commonly used for human food—chiefly wheat, rice, and rye.

FORAGE. The vegetative portion of plants in a fresh, dried, or ensiled state, which is fed to livestock (as pasture, hay, or silage). Although most forages are roughages, a finely ground forage is not. Conversely, many roughages (such as corncobs, straws, and plastic tabs) are not forages.

FORMULA FEED. A feed consisting of two or more ingredients mixed in specified proportions.

FORTIFY. Nutritionally, to add one or more nutrients or feedstuffs.

FOWLER'S SOLUTION. A tonic containing arsenic which is sometimes used in conditioning show animals, particularly those breeds that have a very heavy hair coat. Poisoning can result from its excessive or continued use.

FREE-CHOICE. Free to eat two or more feeds at will.

FREEZE DRYING. See LYOPHILIZATION.

FRESH.

- A cow that has recently given birth to a calf and is lactating.
- Newly produced or gathered feed material; not stored, cured, or preserved.

FULL BLOOM. When two-thirds or more of the plants are in bloom.

FULL-FEED. The term indicating that animals are being provided as much feed as they will consume safely without going off feed.

FUNGI. Plants that contain no chlorophyll, flowers, or leaves, such as molds, mushrooms, toadstools, and yeasts. They may get their nourishment from either dead or living organic matter.

FUNGICIDE. Any substance used to kill fungi.

FUTURES TRADING. The futures market is a way in which to provide (1) an insurance medium in the marketing field, and (2) the facilities and machinery for underwriting price risks.

G

GAIN RATIO. See RATIOS.

GALACTOSE. A hexose sugar (monosaccharide) obtained along with glucose from lactose (milk sugar) hydrolysis.

GASTRIC. Pertaining to the stomach.

GASTRIC JUICE. A clear liquid secreted by the wall of the stomach. It contains hydrochloric acid and the enzymes rennin, pepsin, and gastric lipase.

GASTRITIS. An inflammation of the stomach, especially of the lining, or mucous membrane.

GASTROENTERITIS. Inflammation of the membranes of the stomach and the intestine.

GASTROINTESTINAL. Pertaining to the stomach and intestines.

GAVAGE. Introduction of material (as nutrients) into the stomach by means of a stomach tube.

GELATINIZED. Having had the starch granules completely ruptured by a combination of moisture, heat, and pressure, and in some instances, by mechanical shear.

GERM. Embryo of a seed.

GESTATION. The period of embryonic and fetal development from fertilization to birth; pregnancy.

GLOSSITIS. A swollen, reddened tongue; a symptom of riboflavin deficiency.

GLUCONEOGENESIS. Formation of glucose from protein or fat.

GLUCOSE. A hexose monosaccharide obtained upon the hydrolysis of starch and certain other carbohydrates. Also called dextrose.

GLUTEN. The tough, viscid, nitrogenous substance remaining when the flour of wheat or other grain is washed to remove the starch.

GLYCERIDE. An ester of glycerol and fatty acids.

GLYCEROL. An alcohol containing three carbons and three hydroxy groups. It is most commonly found in chemical combinations with fats in compounds called triglycerides.

GLYCOGENESIS. Conversion of glucose to glycogen.

GLYCOGENOLYSIS. Conversion of glycogen to glucose.

GLYCOLYSIS. Conversion of carbohydrate to lactic acid in animals or pyruvic acid in enzymatic reactions.

GOITROGENIC. Producing or tending to produce goiter.

GOSSYPOL. A toxic yellow pigment found in cottonseed, which is toxic to swine and certain other nonruminants, and which may cause discoloration of egg yolks during cold storage.

GRAB SAMPLE. A single sample, not taken at a set time or a specific flow.

GRAIN. Seed from cereal plants.

GRAIN SCREENINGS. The small imperfect grains, weed seeds, and other foreign material that is separated in cleaning grain by a screen.

GRAS (GENERALLY RECOGNIZED AS SAFE). A designation of food additives that have been judged as safe for human consumption by a panel of expert pharmocologists and toxicologists who consider available data, including experience of common use in food. The common use factor is often the major criterion on which judgment is based.

GRAZE. To consume standing vegetation, as by livestock or wild animals.

GRAZING FEE. A charge, usually on a monthly basis, for grazing of livestock.

GREEN REVOLUTION. Refers to the adoption of high-yielding varieties of wheat and rice. It all started with a short-strawed wheat developed by Dr. Norman Borlaug, a Rockefeller Foundation scientist stationed in Mexico, who evolved new spring wheats that helped Mexico and Southeast Asia close the food gap, a development that brought him fame and a Nobel Prize. The first of these improved wheat varieties was released in 1948.

GRIND. To reduce to small segments by impact, shearing, or attrition (as in a mill).

GRITS. Coarsely ground grain from which the bran and germ have been removed, usually screened to uniform particle size.

GROATS. Grain from which the hulls have been removed.

GROUND. Reduced in particle size by impact, shearing, or attrition.

GROW OUT. To feed animals so that they attain a certain desired amount of growth with little or no fattening.

GROWTH. May be defined as the increase in size of the muscles, bones, internal organs, and other parts of the body.

GROWTHY. Describes an animal that is large and well developed for its age.

GRUEL. A feed prepared by mixing ground ingredients with hot or cold water.

H

HARD KEEPER. An animal that is unthrifty and grows or fattens slowly regardless of the quantity or quality of feed.

HAY BELLY. Said of a horse with a distended barrel due to excessive feeding of bulky rations, such as hay, straw, or grass. Also called *grass belly.*

HAY QUALITY. Refers to the physical and chemical characteristics of hay commonly associated with palatability and abundance of feed nutrients.

HEALTH. A state of complete physical, mental, and social well being; not merely the absence of disease and infirmity.

HEAT INCREMENT (HI). The increase in heat production following consumption of feed when the animal is in a thermoneutral environment.

HEAT LABILE. Unstable to heat.

HEAT OF FERMENTATION (HF). The heat produced in the digestive tract as a result of microbial action.

HEAT-PROCESSED. Subjected to a method of preparation involving the use of elevated temperatures, with or without pressure.

HEAT-RENDERED. Melted, extracted, or clarified through use of heat. Usually water and fat are removed.

HEME. Iron-containing, nonprotein portion of the hemoglobin.

HEMICELLULOSE. A carbohydrate classified in the crude fiber fraction of feedstuffs that is similar to cellulose, except that it contains pentoses (5-carbon sugars) and uronic acid in addition to hexoses.

HEMOGLOBIN. The oxygen-carrying, red-pigmented protein of the red blood cells.

HEMORRHAGE. Escape of blood from the vessels; bleeding.

HERB. A flowering plant with one or more stems that die back to the ground each year without persistent stems aboveground; ferns, grasses, and forbs as distinct from shrubs and trees.

HERBIVORE. A plant-eating animal.

HIGH-LYSINE CORN (OPAQUE-2). Corn that is much higher than normal corn in lysine and tryptophan; hence, it has a better balance of the amino acids for monogastric animals. Also, high-lysine corn is higher in total protein, but lower in leucine than regular corn.

HIGH-NUTRIENT DENSITY. A high level of a specific nutrient in a given amount (weight) of food.

HIVES. Small swellings under or within the skin. They appear suddenly over large portions of the body and can be caused by feed allergies.

HOME RANGE. The area around an animal's established home which is traversed in its normal activities.

HOMOGENIZED. Particles broken down into evenly distributed globules small enough to remain emulsified for long periods of time.

HORMONE. A body-regulating chemical secreted by an endocrine gland into the bloodstream, thence transported to another region within the animal where it elicits a physiological response.

HORSEPOWER. A unit of power equal to 746 watts.

HULLS. Outer covering of grain or other seed, especially when dry.

HYDRAULIC PROCESS. A process for the mechanical extraction of oil from seeds, involving the use of a hydraulic press. Sometimes referred to as *old process.*

HYDROGENATION. The chemical addition of hydrogen to any unsaturated compound.

HYDROLYSIS. The splitting of a substance into smaller units by chemically adding water to the material.

HYGROSCOPIC. The degree of tendency to absorb and retain moisture. For example, nonfat dry milk is hygroscopic in that it tends to absorb moisture.

HYPERTHYROIDISM. Overactivity of the thyroid gland.

HYPERTROPHIED. Having increased in size beyond the normal growth.

HYPERVITAMINOSIS. An abnormal condition resulting from the intake of an excess of one or more vitamins.

HYPOCALCEMIA. Below normal concentration of ionic calcium in blood resulting in convulsions, as in tetany or parturient paresis (milk fever).

HYPOGLYCEMIA. A reduction in concentration of blood glucose below normal.

HYPOKALEMIA. Low potassium levels in the blood.

HYPOMAGNESEMIA. An abnormally low level of magnesium in the blood.

HYPONATREMIA. Decreased level of sodium in the blood.

I

ICU (INTERNATIONAL CHICK UNIT). The vitamin D requirements of poultry are generally expressed as ICU, based on using the chick as the assay animal because of the unequal activity of vitamin D_2 and vitamin D_3 for this species.

IMMUNOGLOBULINS. A family of proteins found in body fluids which have the property of combining with antigens, and when the antigen is pathogenic, sometimes inactivating it and producing a state of immunity. Also called antibodies.

IMPACTION. See COMPACTION.

IMPLANT. A substance that is implanted into the body for the purpose of growth promotion or controlling some physiological function.

INCOMPLETE PROTEIN. A feed protein which lacks sufficient amount of one or more of the essential amino acids.

INERT. Relatively inactive.

INGEST. To eat or take in through the mouth.

INGESTA. Food or drink taken into the stomach.

INGESTION. The taking in of food and drink.

INGREDIENT. A constituent feed material.

INOCULATION. The addition of the proper bacteria to legume seed to enable the plant to use nitrogen from the air.

INSOLUBLE CRUDE PROTEIN (ICP). This is the protein that is not available for digestion by rumen microbes due to lignin tie-up or heat damage.

INSOLUBLE CRUDE PROTEIN/CRUDE PROTEIN RATIO (ICP/CP). This indicates the percent of unavailable protein. In high-quality forages, the ICP fraction should be under 10% CP. If over 10%, use ACP value instead of CP. High ratio values indicate heat damage, mature late-cut forage, and/or grass-type forage.

INSULIN. A hormone secreted by the pancreas into the blood, which regulates sugar metabolism.

INTESTINAL JUICE. A clear liquid secreted by glands in the wall of the small intestine. It contains the enzymes lactase, maltase, and sucrase, and several peptidases.

INTOLERANCE. Sensitivity or allergy to certain foods.

INTRACELLULAR. Inside the cells.

INTRADERMAL. Into, or between, the layers of the skin.

INTRAMUSCULAR. Within the muscle.

INTRAPERITONEAL. Within the peritoneal cavity.

INTRAVENOUS. Within the vein or veins.

INTRINSIC FACTOR. A chemical substance secreted by the stomach which is necessary for the absorption of vitamin B-12. The exact chemical nature of intrinsic factor is not known, but it is thought to be a mucoprotein or mucopolysaccharide. A deficiency of this factor may lead to a deficiency of vitamin B-12, and, ultimately, to pernicious anemia.

IN VITRO. Occuring in an artificial environment, as in a test tube.

IN VIVO. Occuring in the living body.

IODINE NUMBER. A number which denotes the degree of unsaturation of a fat or fatty acid. It is the amount of iodine in grams which can be taken up by 100 g of fat.

ION. Molecular constituent of one or more atoms that is a free-wandering particle in solution. An ion carries a positive or negative electric charge. Ions carrying negative charges are called anions; those carrying positive charges are called cations.

IRRADIATED YEAST. Yeast that has been treated with ultraviolet light. Yeast contains considerable ergosterol, which, when exposed to ultraviolet light, produces vitamin D_2.

IRRADIATION. Exposing to ultraviolet light.

ISOLATED SOYBEAN PROTEIN. Protein obtained from soybeans that can be spun into fibers, flavored, colored, and fabricated into meatlike products, including beef steaks, chicken, pork chops, ham, bacon, lamb chops, or sausage—all difficult to distinguish from the real products.

ISOMER. The possession of two or more distinct compounds of the same molecular formula, each molecule possessing an identical number of atoms of each element but in different arrangement.

ISOTOPE. Element that has the same number of protons (atomic number) as another element but a different number of neutrons (atomic mass).

IU (INTERNATIONAL UNIT). A standard unit of potency of a biologic agent (e.g., a vitamin, hormone, antibiotic, antitoxin) as defined by the International Conference for Unification of Formulae. Potency is based on the bioassay that produces a particular effect agreed on internationally. Also called a *USP unit.*

J

JOULE. A proposed international unit (4,184 j = 1 calorie) for expressing mechanical, chemical, or electrical energy, as well as the concept of heat. In the future, energy requirements and feed values will likely be expressed by this unit.

K

KERATIN. A sulfur-containing protein which is the primary component of epidermis, hair, wool, hoof, horn, and the organic matrix of the teeth.

KERNEL. The whole grain of a cereal. The meats of nuts and drupes (single-stoned fruits).

KETOGENIC. Conducive to the production of ketones, products of fatty acid oxidation that eventually are broken down to carbon dioxide and water.

KETOSIS. A condition characterized by an abnormally high concentration of keytone (acetone) bodies in fluids and tissues—acetonemia.

KIBBLED. Cracked or crushed baked dough, or extruded feed that has been cooked prior to or during the extrusion process.

KILOGRAM. A unit of weight; 1 kg = 2.2046 lb, 1 lb = 0.4536 kg.

KJELDAHL METHOD. A method of determining the amount of nitrogen in an organic compound. The quantity of nitrogen measured is then multiplied by 6.25 to calculate the protein content of the feed or compound analyzed. The method was developed by the Danish chemist, J. G. C. Kjeldahl, in 1883.

KREBS CITRIC ACID CYCLE. Also called TCA (tricarboxylic acid) cycle and the Krebs cycle. The pathway by which animal cells use oxygen to oxidize carbohydrates to carbon dioxide, energy, water, and heat.

KWASHIORKOR. A syndrome produced by a severe protein deficiency, with characteristic changes in pigmentation of the skin and hair, a bulging belly, edema, skin lesions, anemia, and apathy.

L

LABILE. Unstable. Easily destroyed.

LACHRYMATION. Tearing; secreting and conveying tears.

LACTIC ACID. A compound formed in the body in the metabolism of carbohydrates.

LACTOSE (MILK SUGAR). A disaccharide found in milk having the formula $C_{12}H_{22}O_{11}$. It hydrolyzes to glucose and galactose. Commonly known as milk sugar.

LAMINITIS (or FOUNDER). An inflamation of the sensitive laminae under the horny wall of the hoof, characterized by ridges running around the hoof, often associated with overfeeding. All feet may be affected, but the front feet are most susceptible.

LATE BLOOM. When blossoms begin to dry and fall.

LAXATIVE. A feed or drug that will induce bowel movements and relieve constipation.

LAYER. An adult chicken which produces eggs.

LEAVES. Lateral outgrowths of stems that constitute part of the foliage of a plant, typically a flattened green blade, which function in photosynthesis.

LECITHIN. A phospholipid found in many animal tissues.

LEGUME. A plant that has the ability to work symbiotically with bacteria to fix nitrogen from the air.

LIGNIFICATION. The process of impregnating cell walls of a plant with lignin.

LIGNIN. A practically indigestible compound which along with cellulose is a major component of the cell wall of certain plant materials, such as wood, hulls, straws, and overripe hays.

LIMITED FEEDING. Feeding animals less than they would like to eat. Giving sufficient feed to maintain weight and growth, but not enough for their potential production or finishing.

LIMITING AMINO ACID. The essential amino acid of a protein which shows the greatest percentage deficit in comparison with the amino acids contained in the same quantity of another protein selected as a standard.

LINOLEIC ACID. An essential fatty acid.

LIPASE. A fat-splitting enzyme, present in gastric juice and pancreatic juice. It acts on fats to produce fatty acids and glycerol.

LIPOLYSIS. The hydrolysis of fats by enzymes, acids, alkalis, or other means to yield glycerol and fatty acids.

LIQUID PROTEIN SUPPLEMENTS. Protein products which usually contain molasses and urea, with added vitamins and trace minerals.

LIVER ABSCESSES. Single or multiple abscesses on the liver, observed at slaughter. Usually the abscess consists of a central mass of necrotic liver surrounded by pus and a wall of connective tissue. At slaughter, those livers affected with abscesses are condemned for human food.

LIVEWEIGHT. Weight of an animal on foot.

LYMPH. The slightly yellow, transparent fluid occupying the lymphatic channels of the body.

LYOPHILIZATION. The evaporation of water from a frozen product with the aid of high vacuum. Also called freeze drying.

M

MACRO (or MAJOR) MINERALS. The major minerals—calcium, phosphorus, sodium, chlorine, potassium, magnesium, and sulfur.

MAINTENANCE REQUIREMENT. A ration which is adequate to prevent any loss or gain of tissue in the body when there is no production.

MAIZE. A cereal grain, commonly called corn in the United States.

MALIGNANT. A cancerous growth, as distinguished from a benign growth.

MALNUTRITION. Any disorder of nutrition. Commonly used to indicate a state of inadequate nutrition.

MALTOSE. A disaccharide, also known as malt sugar, having the formula $C_{12}H_{22}O_{11}$. Obtained from the partial hydrolysis of starch. It hydrolyzes to glucose.

MANURE. A mixture of animal excrements (consisting of undigested feeds plus certain body wastes) and bedding.

MARASMUS. From the Greek *marasmos,* meaning *a dying away.* A progressive wasting and emaciation. *Enzootic* marasmus is a condition of malnutrition in domestic animals due to a deficiency of one or more trace elements, especially cobalt and copper. In human infants, the condition results primarily from lack of calories (energy foods).

MASH. A mixture of ingredients in meal form.

MASTICATION. The chewing of feed.

MEAL.

- A feed ingredient having a particle size somewhat larger than flour.
- Mixtures of concentrate feeds, usually in which all of the ingredients are ground.

MEAT ANALOGS. Food material usually prepared from vegetable protein to resemble specific meats in texture, color, and flavor.

MEATS.

- Animal tissues used as food.
- The edible parts of nuts or fruits.

MECHANICALLY EXTRACTED. A method of extracting the fat content from oilseeds by the application of heat and mechanical pressure. The hydraulic and expeller processes are both methods of mechanical extraction.

MECONIUM. Excrement accumulated in the bowels during fetal development.

MEDICATED FEED. Any feed which contains drug ingredients intended or represented for the cure, mitigation, treatment, or prevention of diseases of animals (other than humans).

METABOLISM. Refers to all the changes that take place in the nutrients after they are absorbed from the digestive tract, including (1) the building-up processes in which the absorbed nutrients are used in the formation or repair of body tissues, and (2) the breaking-down processes in which nutrients are oxidized for the production of heat and work.

METABOLITE. Any substance produced by metabolism or by a metabolic process.

METHANE. One of two primary waste gases produced by rumen fermentation. (The other is carbon dioxide.) The production of methane results in a loss of energy. The chemical formula is CH_4. Methane is produced by the reduction of carbon dioxide by hydrogen.

METHIONINE. An essential sulfur-containing amino acid that can be readily made synthetically.

MICROBE. Same as microorganism.

MICROFLORA. Microbial life characteristric of a region, such as the bacteria and protozoa populating the rumen.

MICROINGREDIENT. Any ration component, such as minerals, vitamins, antibiotics, and drugs, normally measured in milligrams or micrograms per kilogram or in parts per million.

MICROORGANISM. Any organism of microscopic size, applied especially to bacteria and protozoa.

MID-BLOOM. Period during which $\frac{1}{10}$ to $\frac{2}{3}$ of the plants are in bloom.

MILK ALBUMIN. The coagulated protein fraction from whey.

MILK EJECTION OR *LET-DOWN*. The process, controlled by oxytocin, in which milk is forced from the alveoli, where it is stored, into the larger ducts and cisterns, where it is available to sucklings or the milker.

MILK FAT. The lipid components of milk.

MILK STAGE. The plant period after bloom when the seeds begin to form.

MILL BY-PRODUCT. A secondary product obtained in addition to the principal product in milling practice.

MILL RUN. The state in which a by-product material comes from the mill, ungraded and usually uninspected.

MINERALS (ASH). The inorganic elements of animals and plants, determined by burning off the organic matter and weighing the residue, which is called ash.

MINERAL SUPPLEMENT. A rich source of one or more of the inorganic elements needed to perform certain essential body functions.

MIXING. To combine by agitation two or more materials to a specific degree of dispersion.

MOISTURE. A term used to indicate the water contained in feeds—expressed as a percentage.

MOISTURE-FREE (M-F, OVEN-DRY, 100% DRY MATTER). This refers to any substance that has been dried in an oven at 221°F until all the moisture has been removed.

MOLDS (FUNGI). Fungi which are distinguised by the formation of mycelium (a network of filaments, or threads), or by spore masses.

MORBIDITY. A state of sickness.

MUCUS. Viscid fluid secreted by mucous membranes and glands, consisting mainly of mucin (a glycoprotein), inorganic salts, and water. Mucus serves to lubricate and protect the gastrointestinal mucosa and helps to move the feed mass along the digestive tract.

MYCOTOXINS. Toxic metabolites produced by molds during growth. Sometimes present in feed materials.

MYOSIN. Muscle fiber fraction that complexes with actin to bring about muscle contraction.

N

NATIONAL RESEARCH COUNCIL (NRC). A division of the National Academy of Sciences established in 1916 to promote the effective utilization of scientific and technical resources. Periodically, this private, nonprofit organization of scientists publishes bulletins giving nutrient requirements and allowances of domestic animals, copies of which are available on a charge basis through the National Academy of Sciences, National Research Council, 2101 Constitution Avenue, N.W., Washington, DC 20418.

NATIVE PASTURES. Unseeded pastures native to the area.

NECROSIS. Cell death.

NEPHRITIS. Inflammation of the nephrons of the kidneys.

NEUTRAL DETERGENT FIBER (NDF). This indicates the amount of cell wall material or plant structural fiber in a feed. NDF contains ADF plus hemicellulose. The lower the NDF the more forage the animal will eat; hence, low NDF is desired.

NEW PROCESS. Pertains to the extraction of oil from seeds. Same as solvent process (solvent extracted).

NITROGEN. A chemical element essential to life. Animals get it from protein feeds; plants get it from the soil; and some bacteria get it directly from the air.

NITROGEN BALANCE. The nitrogen in the feed intake minus the nitrogen in the feces, minus the nitrogen in the urine.

NITROGEN FIXATION. Conversion of free nitrogen of the atmosphere to organic nitrogen compounds by symbiotic or nonsymbiotic microbial activity.

NITROGEN-FREE EXTRACT (NFE). It consists principally of sugars, starches, pentoses, and nonnitrogenous organic acids. The percentage is determined by subtracting the sum of the percentages of moisture, crude protein, crude fat, crude fiber, and ash from 100.

NONPROTEIN NITROGEN (NPN). Nitrogen which comes from other than a protein source but may be used by a ruminant in the building of protein. NPN sources include compounds like urea and anhydrous ammonia, which are used in feed formulations for ruminants only.

NOURISH. To provide food or other substances necessary for life and growth.

NOXIOUS. Harmful, not wholesome.

NUTRIENT ALLOWANCES. Nutrient recommendations that allow for variations in feed composition; possible losses during storage and processing; day-to-day and period-to-period differences in needs of animals; age and size of animal; stage of gestation and lactation; the kind and degree of activity; the amount of stress; the system of management; the health, condition, and temperament of the animal; and the kind, quality, and amount of feed—all of which exert a powerful influence in determining nutritive needs.

NUTRIENT REQUIREMENTS. This refers to meeting the animal's minimum needs, without margins of safety, for maintenance, growth, fitting, reproduction, lactation, and work. To meet these nutritive requirements, the different classes of animals must receive sufficient feed to furnish the necessary quantity of energy (carbohydrates and fats), protein, minerals, and vitamins.

NUTRIENT TO CALORIE RATIO. The energy needs of animals and their requirements of the several nutrients are quantitatively correlated. For those nutrients that are needed to metabolize energy, it is logical to consider that the amount of energy metabolized determines their requirements. Hence, it is logical to express nutrients in weight per unit of energy needed. For example, it is suggested that the protein to calorie ratio should be expressed as grams of protein per 1,000 kcal metabolizable energy (g protein/1,000 kcal ME). If the ME is corrected for nitrogen retained or lost from the body, then the abbreviation should be g protein/1,000 kcal ME_n. This same dimension may easily be extended to other nutrients, such as g calcium/1,000 kcal, or mg riboflavin/1,000 kcal, etc.

NUTRIENTS. The chemical substances found in feed materials that can be used, and are necessary, for the growth, maintenance, production, and health of animals. The chief classes of nutrients are carbohydrates, fats, proteins, minerals, vitamins, and water.

NUTRITION. This is the science of the use of feed (food) as it relates to health. It includes all the processes by which the living organism ingests, digests, absorbs, and uses the nutrients in feeds (foods) for maintenance, growth, work, and reproduction.

NUTRITIVE RATIO (NR). The ratio of digestible protein to other digestible nutrients in a feedstuff or ration. (The NR of shelled corn is about 1:10.)

NUTRITURE. Nourishment.

O

OBESE. Overweight due to a surplus of body fat.

OFFAL. It usually refers to that part of the animal carcass not sold for meat, consisting principally of the digestive tract.

OFF FEED. Not eating with a normal, healthy appetite.

OIL. Althouth fats and oils have the same general structure and chemical properties, they have different physical characteristics. The melting points of oils are such that they are liquid at ordinary room temperatures.

OIL CROPS. Crops grown primarily for oil, including soybeans, cottonseed, rape (canola), peanuts, flaxseed, sunflower, safflower, and castor bean.

OLD PROCESS. Pertains to the extraction of oil from seeds. Same as hydraulic process.

OLFACTORY. Pertaining to the sense of smell.

OPEN PEN FEEDLOT. A lot in which cattle or sheep are in the open—usually it is without shelter (except for such natural protection as may be afforded by trees, hills, or wind fences, or perhaps a roof over the feed bunks or shades).

ORAL. Pertaining to the mouth.

ORTS. Leftover feed which an animal refuses to eat.

OSMOTIC PRESSURE. The pressure that causes water or another solvent to move from a solution with a low concentration of solid (solute) to one having a high concentration of solute.

OSSIFICATION. The process of bone formation; the calcification of bone with advancing maturity.

OSTEITIS. Inflammation of a bone.

OSTEOMALACIA. A bone disease of adult animals caused by lack of vitamin D, inadequate intake of calcium or phosphorus, or an incorrect dietary ratio of calcium and phosphorus.

OSTEOPOROSIS. Abnormal porosity and fragility of bone as the result of (1) a calcium, phosphorus, and/or vitamin D deficiency, or (2) an incorrect ratio between the two minerals.

OVERFEEDING. Fed to excess.

OVERFINISHING. Excess finishing or fatness—a wasteful practice.

OVERRIPE. Stage after the plant is mature and the seeds are ripe (applies mostly to range plants).

OXIDATION. The combination with oxygen, or the loss of a hydrogen, or the loss of an electron, all of which render an ion more electropositive. The animal combines carbon from feedstuffs with inhaled oxygen to produce carbon dioxide, energy (as ATP), water, and heat. (Also see KREBS CITRIC ACID CYCLE.)

OXYTOCIN. The hormone that controls milk let-down.

OYSTER SHELL. Consists mainly of calcium salts; it is a useful and attractive source of calcium for layers to aid egg shell strength.

P

PALATABILITY. The result of the following factors sensed by the animal in locating and consuming feed: appearance, odor, taste, texture, temperature, and, in some cases, auditory properties of the feed (like the sound of pigs eating corn). These factors are affected by the physical and chemical nature of the feed.

PALATABLE FEEDS. Feeds that are well liked and eaten with relish.

PANTOTHENIC ACID. One of the B vitamins. It is a constituent of coenzyme A, which plays an essential role in fat and cholesterol synthesis.

PAPILLAE. Small nipple-shaped projections located on the interior of the rumen wall.

PARATHYROID GLANDS. Four small endocrine glands situated beside the thyroid gland, concerned chiefly with maintaining normal calcium in the body.

PARATHYROID HORMONE. The parathyroid hormone, secreted by the parathyroid glands, stimulates the release of calcium from bone and the elevation of the blood calcium level.

PARTS PER BILLION (PPB). It equals micrograms per kilogram or microliters per liter.

PARTS PER MILLION (PPM). It equals milligrams per kilogram or milliliters per liter.

PEARLED. Dehulled grains reduced into smaller smooth particles by machine brushing, or abrasion.

PECTIN. A nondigestible polysaccharide found in the soft tissue of fruits and vegetables. One of the richest sources is lemon or orange rind. Pectin possesses the ability to gel, and it is often used as a base for fruit jellies.

PEELINGS. Outer layers or coverings which have been removed by stripping or tearing.

PELLET BINDERS. Products that enhance the firmness of pellets.

PELLETS. Ground feed compacted by steaming and forcing the material through die openings.

PEPTIDE LINKAGE. The characteristic joining of amino acids to form proteins. Such a chain of amino acids is called a peptide.

PER ORAL. Administration through the mouth.

PER OS. Oral administration (by the mouth).

pH. A measure of the acidity or alkalinity of a solution. Values range from 0 (most acid) to 14 (most alkaline), with neutrality at pH 7.

PHASE FEEDING. Refers to changes in the animal's diet (1) to adjust for age and stage of production, (2) to adjust for season of the year and for temperature and climatic changes, (3) to account for differences in body weight and nutrient requirements of different strains of animals, or (4) to adjust one or more nutrients as other nutrients are changed for economic or availability reasons.

PHOSPHOLIPIDS. Fatlike substances containing phosphorus and nitrogen, along with fatty acids and cholesterol.

PHOTOSYNTHESIS. The process whereby green plants utilize the energy of the sun to build complex organic molecules containing energy.

PHYSIOLOGICAL. Pertaining to the science which deals with the functions of living organisms or their parts.

PHYSIOLOGICAL FUEL VALUES. Units, expressed in calories, used in the United States to measure food energy in human nutrition. It is similar to metabolizable energy.

PHYSIOLOGICAL SALINE. A salt solution (0.9% NaCl) having the same osmotic pressure as the blood plasma.

PHYTATE. A salt of phytic acid (a phosphorus compound occurring in the outer layers of cereal grains. It binds phosphorus and other minerals [e.g. zinc] making them unavailable to monogastric animals. The calcium of the insoluble calcium phytate is unabsorbable from the intestine).

PICA. Depraved appetite characterized by eating dirt, sand, hair, feces, etc. This condition is usually caused by nutritional deficiencies and is particularly prevalent among animals in confinement.

PLANT PROTEINS. This group includes the common oilseed by-products—soybean meal, cottonseed meal, linseed meal, peanut meal, safflower meal, sunflower meal, rapeseed meal, and coconut (or copra) meal. They vary in protein content and nutrient value, depending on the seed from which they are produced, the amount of hull and/or seed coat included, and the method of oil extraction used.

PLASMA. The colorless fluid portion of the blood in which the corpuscles are suspended. It is often used as a basis for measurement of bloodborne nutrients and their metabolites.

POLAR. Miscible in water.

POLY. Many.

POLYDIPSIA. Excessive or abnormal thirst.

POLYNEURITIS. Neuritis of several peripheral nerves at the same time, caused by metallic and other poisons, infectious disease, or vitamin deficiency. In humans, alcoholism is also a major cause of polyneuritis.

POLYUNSATURATED FATTY ACIDS. Fatty acids having more than one double bond. Linoleic and linolenic acids, which contain 2 and 3 double bonds, respectively, are essential in the diet of humans.

POPPED. The expansion of whole or cracked grain by heat. It may entail pressure as well.

POT-BELLIED. Designating any individual that has developed an abnormally large abdomen.

POULT. A young turkey of either sex from a day old to a few weeks.

PREBLOOM. The last third of plant growth before blooming.

PRECONDITIONING. A way of preparing an animal (usually referring to calves) to withstand the stress and rigors of leaving its mother, learning to eat new kinds of feed, and shipping from the farm or ranch to the feedlot.

PRECURSOR. A compound that can be used by the body to form another compound; for example, carotene is a precursor of vitamin A.

PREHENSION. The seizing (grasping) and conveying of feed to the mouth.

PREMIX. A uniform mixture of one or more microingredients and a carrier, used in the introduction of microingredients into a larger mixture.

PRESERVATIVES. A number of materials are available to incorporate into silage and hay, with claims made that they will improve the preservation of nutrients, nutritive value, and/or palatability of the feed.

PRESSED. Compacted or molded by pressure; having fat, oil, or juice extracted under pressure.

PRESSURE COOKER. An airtight container for the cooking of feed at high temperature under steam pressure.

PRODUCT. A substance produced from one or more other substances as a result of chemical or physical change.

PROLACTIN. The lactating-stimulating hormone of the anterior lobe of the pituitary gland.

PROSTAGLANDIN. A group of naturally occurring long-chain fatty acid derivatives having local hormonelike actions of widely diverse forms.

PROTEIN. From the Greek meaning *of first rank, importance.* Complex organic compounds made up chiefly of amino acids present in characteristic proportions for each specific protein. At least 24 amino acids have been identified and may occur in combinations to form an almost limitless number of proteins. Proteins always contain carbon, hydrogen, oxygen, and nitrogen; and, in addition, they usually contain sulfur and frequently phosphorus. Crude protein is determined by finding the nitrogen content and multiplying the result by 6.25. The nitrogen content of proteins averages about 16% ($100 \div 16 = 6.25$). Proteins are essential in all plant and animal life as components of the active protoplasm of each living cell.

PROTEIN-BOUND IODINE (PBI) TEST. Test used to measure thyroid activity by determining the amount of iodine bound to thyroxin and in transit in the plasma.

PROTEIN BUMPS. Some horses appear to be allergic to certain proteins or to excesses of specific amino acids, as a result of which they may develop *protein bumps.*

PROTEIN EQUIVALENT. A term indicating the total nitrogenous contribution of a substance in comparison with the nitrogen content of protein (usually plant protein). For example, the nonprotein nitrogen (NPN) compound urea contains approximately 45% nitrogen and has a protein equivalent of 281% ($6.25 \times 45\%$).

PROTEIN-SPARING. An effect in which less protein is used by the animal to meet the animal's glucose needs in times of glucose shortage. Propionic acid is protein-sparing in that it can be converted to glucose. Acetic and butyric acids cannot be converted to glucose. Likewise, fat cannot be converted. The glucogenic amino acids may be converted to glucose.

PROTEIN SUPPLEMENTS. Products that contain more than 20% protein or protein equivalent.

PROTHROMBIN. Plasma fraction that is converted to thrombin during the clotting of blood.

PROTOPLASM. The essential protein sustance of living cells.

PROVITAMIN. The material from which an animal may produce vitamins; e.g., carotene (provitamin A) in plants is converted to vitamin A in animals.

PROVITAMIN A. Carotene.

PROXIMATE ANALYSIS. A chemical scheme for evaluating feedstuffs, in which a feedstuff is partitioned into the six fractions: (1) moisture (water) or dry matter (DM); (2) total (crude) protein (CP or TP—N × 6.25); (3) ether extract (EE) or fat; (4) ash (mineral salts); (5) crude fiber (CF)—the incompletely digested carbohydrates; and (6) nitrogen-free extract (NFE)—the more readily digested carbohydrates (calculated rather than measured chemically).

PUFFED. Whole, cracked, or processed grain expended by pressure and heating.

PULLET. A female chicken from a day old up to 12 months of lay when it becomes a hen.

PULP. The solid residue remaining after extraction of juices from fruits, roots, or stems.

PURGATIVE. A laxative.

PURIFIED DIET. A mixture of the known essential dietary nutrients in a pure form that is fed to experimental (test) animals in nutrition studies.

PUTREFACTION. The decomposition of proteins by microorganisms under anaerobic conditions.

PYRUVIC ACID. An organic 3-carbon acid that is a key intermediate in carbohydrate metabolism. It is the end result of the Embden-Myerhoff pathway (anaerobic glycolysis). Pyruvic acid may be further metabolized to any of the three volatile fatty acids—acetic, propionic, or butyric acids.

Q

QUALITY. A term used to denote the desirability and/or acceptance of an animal or feed product.

QUALITY OF PROTEIN. A term used to describe the amino acid balance of protein. A protein is said to be of good quality when it contains all of the essential amino acids in proper proportions and amounts needed by a specific animal; and it is said to be poor quality when it is deficient in either content or balance of essential amino acids.

R

RACHITIC. Adjective used to denote the condition of rickets.

RADIANT HEAT. Heat transmitted by radiation (as from the sun) in contrast to that transmitted by conduction or convection.

RADIATION. Electromagnetic phenomena that have properties combining both wave and particle functions; spans the entire spectrum from low frequency radio waves through white light to high frequency gamma rays.

RADIOACTIVE. Giving off atomic energy in the form of alpha, beta, or gamma rays.

RADIO ACTIVE ^{131}I UPTAKE TESTS. Tests of thyroid function using a radioactive isotope of iodine, ^{131}I.

RADIOACTIVE ISOTOPE. A chemical element that changes into another with the emission of rays (alpha, beta, or gamma). Among the naturally radioactive elements are uranium and radium; some other elements can be made radioactive by bombardment with neutrons or deutrons, or by other means.

RADIOACTIVE TRACER. A small quantity of radioactive isotope used to follow biological, chemical, or other processes by detection, determination, or localization of the radioactivity.

RANCID. A term used to describe fats that have undergone hydrolytic or oxidative decomposition.

RANGE CUBES. Large pellets produced for feeding on the ground on pasture or range.

RANGE FORAGES. Native forages suitable for grazing by livestock that are produced in the western range area.

RANGELAND. Land that, for the most part, produces native forage suitable for grazing by livestock. It embraces rather extensive areas of land suitable for grazing, but not suitable for cultivation—especially in arid, semiarid, or forested regions.

RATE OF PASSAGE. The time taken by undigested residues from a given meal to reach the feces. (A stained indigestible material is commonly used to estimate rate of passage.)

RATION(S). The amount of feed supplied to an animal for a definite period, usually for a 24-hour period. However, by practical usage, the word ration implies the feed fed to an animal without limitation to the time in which it is consumed.

RATIOS. *Weight ratio, gain ratio,* and *conformation score ratio* are used to indicate the performance of an individual in relation to the average of all animals of the same group. It is calculated as follows:

$$\frac{\textbf{Individual record}}{\textbf{Average of animals in group}} \times 100$$

It is a record or index of individual deviation from the group average expressed in terms of percentage. Thus, if the average bull test station gain was 3.00 lb per day, the gain ratio of a bull gaining 3.30 lb per day would be 110. A ratio of 100 is average for a particular group. Thus, ratios above 100 indicate animals above average, whereas ratios below 100 indicate animals below average.

RED MEAT. Meat that is red when raw, due to the red coloration of myoglobin, the pigment of muscle. Red meats include beef, veal, pork, mutton, and lamb muscle tissue with attendant fat and bone.

REGURGITATION. The casting up (backward flow) of undigested food from the stomach to the mouth, as by ruminants.

RELATIVE FEED VALUE (RFV). This is a measure of forage intake and energy value that takes into account CP, ADF, and NDF. RFV is expressed as a percent compared to full bloom alfalfa at 100% RFV. High RFV reflects high quality, greater intake, higher digestibility, and improved performance.

RENNIN. Milk-curdling enzyme of the gastric juice found in young animals such as calves.

RESIDUE. That which remains of any particular substance.

RESPIRATORY QUOTIENT (RQ). A ratio indicating the relation of the volume of carbon dioxide given off in respiration to that of the oxygen consumed. The RQ is used to indicate the *type* of feed being metabolized. This is possible because carbohydrates, fats, and proteins differ in the relative amounts of oxygen and carbon contained in their molecules. Thus, the RQ is near 1 when the body is burning chiefly carbohydrates; near 0.7 when chiefly fats; near 0.8 when chiefly proteins; and sometimes exceeding 1 when carbohydrates are being changed to fats for storage.

RESTRICTED FEEDING. Reduction of the *ad libitum* intake to avoid over-fatness and/or to improve productive efficiency.

RETARDED GROWTH. Slower than normal rate of growth.

RETINOL EQUIVALENT (RE). Measure of vitamin A activity currently adopted by FAO/WHO and U.S. National Research Council's recommendations of vitamin A, replacing the term IU (international unit). The measure accounts for dietary variances in preformed vitamin A (retinol) and its precursor, carotene. One RE (retinol equivalent) equals 3.33 IU or 1 mcg retinol.

RETORT-CHARRED. Material partially burned in a closed retort, as is done in the manufacture of bone black.

ROLLED. Compressed into flat particles by passing between rollers.

ROOTS. Subterranean parts of plants.

ROUGHAGE. A coarse, bulky feed that is high in fiber content and coarse textured. If a high-fiber feed is finely ground, it is no longer a roughage even though the fiber concentration is unchanged.

RUMEN CULTURE. A liquid or dried preparation consisting of microbes from the rumen.

RUMEN FILL. The *fill* in the rumen places a ceiling on feed intake. Since low-quality roughages pass through the rumen at a slower rate than high-quality roughages, they can limit the amount of total feed that the animal can consume in a 24-hour period.

RUMEN FLORA. The microorganisms of the rumen.

RUMINATION. The act of regurgitating previously eaten feed and chewing a soft mass of coarse feed particles, called a bolus or cud. Each bolus is chewed for about a minute, then swallowed again. Ruminants may spend 8 hours or more per day in rumination, the amount of time varying according to the nature of the ration. Coarse, fibrous rations result in more time ruminating. Rechewing does not improve digestibility. Rather, rumination has an important bearing on the amount of feed the animal can eat and utilize. Feed particle size must be reduced to allow passage of the material from the rumen. It follows that high-quality forages require much less rechewing and pass out of the rumen at a faster rate; hence, they allow the ruminant to eat more.

S

SACCHARIDES. Referring to sugars. The prefixes mono-, di-, tri-, and poly- denote the number of sugars contained in the saccharide.

SALINE. Consisting of or containing salt.

SALIVA. A clear, somewhat viscid solution secreted by glands within the mouth. It may contain the enzymes salivary amylase and salivary maltase.

***SALMONELLA*.** A pathogenic, diarrhea-producing organism, of which there are over 100 known strains, sometimes present in contaminated feeds.

SALMONELLOSIS. Infection caused by *Salmonella* (a genus of microbes), characterized by violent diarrhea, abdominal cramps, painful straining or defecation, and fever.

SAPONIFICATION. The formation of soap and glycerol from the reaction of fat with alkali.

SATIETY. Full satisfaction of desire; may refer to satisfaction of appetite.

SATURATED FAT. A completely hydrogenated fat—each carbon atom is associated with the maximum number of hydrogens; there are no double bonds.

SCALPING. The removal of larger material by screening.

SCOURING. One of the major problems facing livestock producers is scouring (diarrhea) in young animals. It may be due to feeding practices, management practices, environment, or disease.

SCRATCH. A poultry feed term referring to whole, cracked, or coarsely cut grain.

SCREENED. A feedstuff that has been separated into various sized particles by passing over or through screens.

SCURVY. A hemorrhagic disease caused by a lack of vitamin C.

SELENIUM. An element that functions as a part of glutathione peroxidase, an enzyme which enables the tripeptide glutathione to perform its role as a biological antioxidant in the body. This explains why deficiencies of selenium and vitamin E result in similar signs—loss of appetite and slow growth.

SELF-FED. Provided with a part or all of the ration on a continuous basis, thereby permitting the animal to eat at will.

SELF-FEEDER. A feed container by means of which animals can eat at will. See *AD LIBITUM*.

SEMIPERMEABLE MEMBRANE. A membrane that allows the passage of only certain solids but is freely permeable to water.

SEPARATING. Classification of particle by size, shape, and/or density.

SERUM. The colorless fluid portion of blood remaining after clotting and removal of corpuscles. It differs from plasma in that the fibrinogen has been removed.

SHORT FEED. To feed for a short period, usually less than 120 days for cattle.

SHREDDED. Cut into long, narrow pieces.

SHRINKAGE.

- A term indicating the amount of loss in body weight when animals are exposed to adverse conditions, such as being transported, severe weather, or shortage of feed.
- The loss in carcass weight during the aging process.

SIFTED. Materials that have been passed through wire sieves to separate particles of different sizes. The common connotation is the separation of finer material than would be done by screening.

SLEEK COAT AND PLIABLE AND ELASTIC SKIN. A sleek, oily coat and a pliable and elastic skin characterize healthy animals. When the hair coat loses its luster and the skin becomes dry, scurfy, and hidebound, there is usually trouble.

SLOTTED FLOORS. Floors with slots through which the feces and urine pass to a storage area below or nearby.

SMALL GRAIN REFUSE. This refers to (1) straw, and (2) tailings—the chaff and grain behind the combine.

SOAP. A compound formed along with glycerol from the reaction of fat with alkali.

SODIUM BENTONITE (CLAY). Used as a pellet binder. Also shows promise for improving the nitrogen utilization of ruminants.

SODIUM BICARBONATE ($NaHCO_3$). A chemical compound which is known to function as a buffer and pH agent, maintaining sufficient alkaline reserves (buffering capacity) in the animal to ensure normal physiological and metabolic functions.

SOFT PORK. Feed fats are laid down in the body without undergoing much change. Thus, when finishing hogs are liberally fed on high fat content feeds in which the fat is liquid at ordinary temperatures, soft pork results. This condition prevails when hogs are liberally fed such feeds as soybeans, peanuts, mast, or garbage.

SOLUBLES. Liquid containing dissolved substances obtained from processing animal or plant materials. It may contain some fine suspended solids.

SOLUTION. A uniform liquid mixture of two or more substances molecularly dispersed within one another.

SOLVENT-EXTRACTED. Fat or oil removed from materials (such as oilseeds) by organic solvents. Also called *new process.*

SORGHUM. A cereal grass used mainly for feed grain or silage. Often grown in the drier areas of the United States.

SPECIFIC DYNAMIC ACTION (SDA). The increased production of heat by the body as a result of a stimulus to metabolic activity caused by ingesting food.

SPECIFIC GRAVITY. The ratio of the weight of a body to the weight of an equal volume of water.

$$\textbf{Specific gravity} = \frac{\textbf{Weight of body in air}}{\textbf{Wt. of body in air minus wt. in } \mathbf{H_2O}}$$

SPECIFIC HEAT.

- The heat-absorbing capacity of a substance in relation to that of water.
- The heat expressed in calories required to raise the temperature of 1 g of a substance 1°C.

SPF. Specific pathogen-free.

SPRAY-DEHYDRATED. Material which has been dried by spraying onto the surface of a heated drum. It is recovered by scraping it from the drum.

STABILIZED. Made more resistant to chemical change by the addition of a particular substance.

STALKS. The main stem of a herbaceous plant; often with its dependent parts, as leaves, twigs, and fruit.

STEAMED. Treatment of ingredients with steam to alter physical and/or chemical properties.

STEARIN. The fat formed from the reaction of stearic acid with glycerol.

STEMS. The coarse, aerial parts of plants which serve as supporting structures for leaves, buds, flowers, and fruit.

STEROL. An alcohol of high molecular weight, such as cholesterol and ergosterol.

STOCKING RATE. This is a range management term pertaining to animal numbers in relation to carrying capacity of a unit. Stocking too lightly wastes forage, while stocking too heavily results in reduced plant vigor and less forage produced per plant, as well as a change of forage plant cover from an abundance of valuable forage plants to an abundance of worthless plants.

STOMATITIS. Inflammation of the oral mucosa.

STOOL (FECES). Fecal material; evacuation from the digestive tract.

STOVER. Fodder; mature cured stalks from which seeds have been removed, such as stalks of corn without ears or stalks of sorghum without heads.

STRAW. The plant residue remaining after separation of the seeds in threshing.

STRESS. Any physical or emotional factor to which an animal fails to make a satisfactory adaptation. Stress may be caused by excitement, temperament, fatigue, shipping, disease, heat or cold, nervous strain, number of animals together, previous nutrition, breed, age, or management. The greater the stress, the more exacting the nutritive requirements.

STUBBLE. The basal portion of herbaceous plants remaining after the top portion has been harvested either mechanically or by grazing animals.

STUMP SUCKER. See CRIBBER.

SUBCUTANEOUS. Situated or occurring beneath the skin.

SUCCULENCE. A condition of plants characterized by juiciness, freshness, and tenderness, making them appetizing to animals. Succulence is provided in such feeds as silage, root crops, and grasses.

SUCKLE. To nurse at the breast or mammary glands.

SUCRASE. An enzyme present in intestinal juice which acts on sucrose to produce glucose and fructose.

SUCROSE. A disaccharide having the formula $C_{12}H_{22}O_{11}$. It is hydrolyzed to glucose and fructose. Commonly known as cane, beet, or table sugar.

SUGAR. A sweet, crystallizable substance that consists essentially of sucrose, and that occurs naturally in the most readily available amounts in sugarcane, sugar beet, sugar maple, sorghum, and sugar palm.

SUN-CURED. Material dried by exposure in open air to the direct rays of the sun.

SUPPLEMENT. A feed or feed mixture used to improve the nutritional value of a basal feed (e.g., protein supplement—soybean meal). Supplements are usually rich in protein, minerals, vitamins, antibiotics, or a combination of part or all of these; and they are usually combined with basal feeds to produce a complete feed.

SWEET FEED. Refers to a commercial horse feed which is characterized by its sweetness due to the addition of molasses.

SYMBIOSIS. The living together in intimate association of two dissimilar organisms, with a resulting mutual benefit. The ruminant animal and its microorganisms in the paunch are a well-known example of symbiosis.

SYNDROME. A combination of symptoms resulting from a single cause.

SYNTHESIS. The bringing together of two or more substances to form a new material.

SYNTHETICS. Artificially produced products that may be similar to natural products.

T

TAIL BITING. An abnormal behavior, characterized by one pig biting the tail of another.

TDN. See TOTAL DIGESTIBLE NUTRIENT.

TDN TO MCAL. One pound of TDN = 2.0 Mcal or 2,000 kcal. It is recognized, however, that the roughage component in a ration affects its energy value. Thus, when converting all-roughage rations from TDN to calories, some scientists figure that 1 lb of TDN = 1,500 kcal, instead of 2,000.

TEART. Molybdenosis of farm animals caused by feeding on vegetation grown on soil that contains high levels of molybdenum.

TEMPERING. See CONDITIONING.

TETANY. A condition in an animal in which there are localized, spasmodic muscular contractions.

TETHER. To tie an animal with a rope or chain to allow grazing but prevent straying.

THERM. The amount of heat required to raise the temperature of 1,000 kg of water 1°C, or 1,000 lb of water 4°F. One therm is 1,000 large Calories, or 1 megacalorie (Mcal).

THERMAL. Refers to heat.

THERMOSTATIC CONTROL OF APPETITE. The theory that food intake is altered by special receptors in the hypothalamus that sense heat and cold.

THIAMINASE. An enzyme that splits the thiamin molecule into two biologically inactive molecules, found especially in raw freshwater fish and in bracken fern.

THIAMIN-STIMULATING HORMONE (TSH). Hormone secreted by the anterior pituitary gland that regulates uptake of iodine and synthesis of thyroxin by the thyroid gland.

THRIFTY. Healthy and vigorous in appearance.

THROMBOSIS. The obstruction of a blood vessel by the formation of a blood clot.

THYROXIN. The iodine-containing hormone produced by the thyroid gland.

TOASTED. Browned, dried, or parched by exposure to a fire, or to gas or electric heat.

TOCOPHEROL. Any of various forms of an alcohol having the properties of vitamin E.

TONIC. A drug, medicine, or feed designed to stimulate the appetite.

TOTAL DIGESTIBLE NUTRIENTS (TDN). A term which indicates the energy value of a feedstuff. The TDN is computed by use of the following formula:

$$\% \text{ TDN} = \% \text{ DCP} + \% \text{ DCF} + \% \text{ DNFE} + (\% \text{ DEE} \times 2.25)$$

where DCP = digestible crude protein; DCF = digestible crude fiber; DNFE = digestible nitrogen-free extract; and DEE = digestible ether extract.

TOTAL MILK SOLIDS. Primarily milk fat, proteins, lactose, and minerals.

TOXEMIA. A condition produced by the presence of poisons (toxins) in the blood.

TOXIC. Of a poisonous nature.

TRACE ELEMENT. A chemical element used in minute amounts by organisms and held essential to their physiology. The essential trace elements are cobalt, copper, iodine, iron, manganese, selenium, and zinc.

TRACE MINERAL. A mineral nutrient required by animals in micro amounts only (measurable in milligrams per pound or smaller units).

TRACER ISOTOPE. An isotope of an element, a small amount of which may be incorporated into a sample of material (the carrier) to follow (trace) the sites of its deposition in the tissues and its paths of excretion. The tracer may be radioactive, in which case observations are made by means of a geiger counter. If the tracer is stable, mass spectrometers, density measurements, or neutron activation analysis may be employed to determine isotopic composition. Tracers are also called labels, or tags, and materials are said to be labeled, or tagged, when radioactive tracers are incorporated in them.

TRANSFERRIN. An iron-bearing protein complex, a serum beta-globulin; the transport form of iron in the body.

TREATS. This refers to such well-liked things as carrots or other roots, apples and other fruits, pumpkins, squashes, melons, molasses, sugar, honey, or innumerable other goodies.

TRIGLYCERIDE. A compound of 3 fatty acids esterified to glycerol. A neutral fat, synthesized from carbohydrate, stored in adipose tissue. It releases free fatty acids into the blood on being hydrolyzed by enzymes.

TRUE PROTEIN. A nitrogenous compound which will hydrolyze completely to amino acids.

TRYPSIN. A protein-splitting (proteolytic) enzyme secreted by the pancreas that acts in the small intestine to reduce proteins to shorter chain polypeptides and dipeptides.

TRYPTOPHAN. One of the essential amino acids.

TUBER. A short, thickened, fleshy stem or terminal portion of a stem or rhizome that is usually formed underground, bears minute scale leaves each with a bud capable under suitable conditions of developing into a new plant, and constitutes the resting stage of various plants such as the potato or the Jerusalem artichoke.

TWENTY-EIGHT-HOUR LAW IN RAIL SHIPMENTS. This law prohibits transporting livestock by rail for a longer period than 28 consecutive hours without unloading, feeding, watering, and resting 5 consecutive hours before resuming transportation. On request of the owner, the period can be extended to 36 hours.

U

ULTRA-VIOLET. Rays in sunlight which enable vitamin D to be synthesized under the skin.

UNDERFEEDING. Usually, this refers to providing insufficient energy. The degree of lowered production therefrom is related to the extent of underfeeding and the length of time it exists.

UNIDENTIFIED FACTORS. These are referred to as *unidentified* or *unknown* factors because they have not yet been isolated or synthesized in the laboratory. Nevertheless, rich sources of these factors and their effects have been well established. There is evidence that growth factors exist in dried whey, marine and packinghouse by-products, distillers' solubles, antibiotic fermentation residues, alfalfa meal, and certain green forages. There is also evidence that at least one unknown hatchability factor is in fish solubles and green forage. Most of the unidentified factor sources are added to the ration at a level of 1 to 3%.

UNSATURATED FAT. A fat having one or more double bonds; not completely hydrogenated.

UNSATURATED FATTY ACID. Any one of several fatty acids containing one or more double bonds, such as oleic, linoleic, linolenic, and arachidonic acids.

UNTHRIFTINESS. Lack of vigor, poor growth or development; the quality or state of being unthrifty in animals.

UREASE. An enzyme which acts on urea to produce carbon dioxide and ammonia. It is found in the jackbean and the soybean, and is produced by certain microorganisms in the rumen.

UREMIA. A toxic accumulation of urinary constituents in the blood.

URINE. Liquid or semisolid matter produced in the kidneys and discharged through the ureters to the urinary bladder, thence voided via the urethra. Normally, it is a clear, transparent, amber-colored, slightly acid fluid.

USP (UNITED STATES PHARMACOPOEIA). A unit of measurement or potency of biologicals that usually coincides with an international unit. (Also see IU.)

V

VACUUM-DEHYDRATED. Freed of moisture after removal of surrounding air while in an airtight enclosure.

VALENCE. Power of an element or a radical to combine with or to replace other elements or radicals. Atoms of various elements combine in definite proportions. The valence number of an element is the number of atoms of hydrogen with which one atom of the element can combine.

VEALER. Calves fed for early slaughter, usually under 3 months of age.

VEGETABLE PROTEIN. The form of protein in all feed sources of vegetable (plant) origin. These are not as well balanced in essential amino acids as animal protein sources; hence, supplementation is required.

VFA. See VOLATILE FATTY ACIDS.

VILLI. Small threadlike projections attached to the interior side of the wall of the small intestine, which increase its absorptive capacity.

VITAMIN PRODUCT LABELS. When a product is marketed as a vitamin supplement *per se,* the quantitative guarantees (unit/lb) of vitamins A and D are expressed in USP units; of E in IU; and of other vitamins in milligrams per pound.

VITAMINS. Complex organic compounds that function as parts of enzyme systems essential for the transformation of energy and the regulation of metabolism of the body, and required in minute amounts by one or more animal species for normal growth, production, reproduction, and/or health. All vitamins must be present in the ration for normal functioning, except for B vitamins in the ruminants (cattle and sheep) and vitamin C.

VITAMIN SUPPLEMENTS. Rich synthetic or natural feed sources of one or more of the complex organic compounds, called vitamins, that are required in minute amounts by animals for normal growth, production, reproduction, and/ or health.

VOID. To evacuate feces and/or urine.

VOLATILE FATTY ACIDS (VFA). Commonly used in reference to acetic, propionic, and butyric acids found especially in rumen contents and/or silage.

VOMITING. The forcible expulsion of the contents of the stomach through the mouth.

W

WAFERS. A form of compressed feed based on fibrous ingredients in which the finished form usually has a diameter or cross-section measurement greater than its length.

WARM CONFINEMENT. Refers to a confinement building which is sufficiently insulated and ventilated to maintain inside winter conditions above 35°F in severe weather, and in the range of 50° to 60°F most of the time.

WASTE GASES. As microbial metabolism proceeds within the rumen, the waste gases, methane (CH_4), and carbon dioxide (CO_2), are formed. These gases represent a waste of dietary energy.

WEANING. The stopping of young animals from suckling their mothers.

WEIGHT RATIO. See RATIOS.

WET MILLED. Subjected to a milling process while containing moisture.

WET RENDERED. Cooked with steam under pressure in closed tanks.

WILTED. To become limp because of water loss.

WIND SUCKER. See CRIBBER.

WOOD CHEWING. Abnormal behavior in horses characterized by chewing on wood.

X

XANTHOPHYLL. Pigment in feed that imparts color to chicken egg yolks, and to the beak, skin, and shanks of yellow-skinned breeds of chickens. Synthetic pigments are also used.

XEROSIS. Abnormal dryness.

Y

YEAST. A source of protein.

Z

ZERO GRAZING. Feeding of green forage as green chop in a lot or stall.

(Above) Fig. 29-1. Branding time on famed Tequesquite Ranch, Albert, New Mexico. (Courtesy, Sherrie Mitchell, Tequesquite Ranch)

(Below) Fig. 29-2. "Stalkin'," watercolor by noted artist Tom Phillips, 3333 17th Street, San Francisco, CA 94110.

Chapter 30

Original painting by Tom Phillips

GLOSSARY OF FEEDSTUFFS

Table 30 (1) lists feedstuffs by the common names most generally used, followed by the genus and species—the latin, or scientific, names; and (2) gives, in summary form, pertinent information relative to each of them, including a brief description, place of origin, adaptation, cultural characteristics, importance, and use. Because of space limitations, many of the feeds described elsewhere in this book are not repeated in this chapter. For example, pertinent information about grasses and legumes is presented in Chapter 7; information about the cereal grains is presented in Chapter 10; and pertinent information about the pulse proteins is presented in Chapter 11. Thus, if a particular feedstuff is not listed in Chapter 30, the reader should look in the index or in the particular chapter and section where it is discussed. But no claim is made to listing all feedstuffs, or to listing the most important feedstuffs of a given area.

The authors recognize that a number of range plants which are generally considered to be poisonous may, in some cases, be consumed in limited amounts, especially by some species and at certain seasons. However, those poisonous plants that are responsible for the most extensive U.S. livestock losses are not listed in this chapter in order that the reader will not be misled relative to their feeding value.

The chemical composition of many of the feeds listed in this chapter can be found in Section V—Composition of Feeds.

TABLE 30-1
GLOSSARY OF FEEDSTUFFS

Feedstuff	Description	Place of Origin Geographical Adaptation Cultural Characteristics	Importance/Use	Comments
		A		
ACACIA *Acacia* spp	Leguminous trees or shrubs. Orange-yellow flowers are found in dense heads, spikes, or auxiliary racemes.	Most of the 400 or more species are native to Africa and Australia. Widely distributed throughout the tropics and subtropics, occasionally in temperate regions.	Many species with low tannin content are considered to be useful browse. Leaves and young shoots are readily eaten by livestock. Almost all of the species in the U.S. provide palatable browse.	Pods are too tough and/or too bitter to be used for browse. CATCLAW is the most important species in the U.S.
ACORNS *Quercus* spp	Nuts of the oak species.	Acorns of the white oak group are the most widely used feed.	Very nutritious; high in fats and oils. Can be an important feed for domestic and wild animals.	Acorns are high in tannin; hence, excess levels may produce toxicities which result in the following abnormalities: cows drying up; deformed calves, known as *acorn calves;* and hens producing eggs with olive-colored yolks and lowered hatchability. Acorns have been used throughout history as human food.
ADLAY (JOB'S TEARS) *Coix lachryma-jobi*	Robust, broad-leaved, loose-growing, branched annual grass. Panicles yield attractive shiny grains resembling tears.	Native to southeastern Asia. Widely cultivated throughout the tropics of the world.	Cultivated as a forage crop but grain from the cereal varieties can be ground and fed to poultry.	Grains are sometimes used as ornamental beads.
AFRICAN LOCUST BEAN (NITTA TREE) *Parkia filicoides*	Spreading, leguminous tree. Pods grow in bunches, measure up to 10 in. (*25 cm*) in length, and contain a yellow, dry pulp embedded in dark brown seeds.	Native to Africa.	Pods can be harvested during the dry season when feed is scarce. All parts of the pods are suitable for feed.	Sometimes used to improve the soil because it extracts nutrients from the deep layers of the soil.
AFRICAN MARIGOLD (AZTEC MARIGOLD) *Tagetes erecta*	Stout, branching annual herb. Yellow to orange flower heads are 2 to 4 in. (*5 to 10 cm*) across.	Introduced to the U.S. from Mexico.	Used to produce zempa meal—a meal composed of dried flower petals.	Zempa meal enhances skin pigmentation and egg yolk color.
AFRICAN OIL PALM *Elaeis guineensis*	Palm of 65 to 80 ft (*20 to 24 m*) in height. Fruit consists of a soft outer skin and a fibrous layer which is rich in palm oil. Inside the fibrous layer is a nut with a shell and a kernel.	Native to tropical West Africa.	Solvent-extracted meal is rather unpalatable and should be well mixed in feed to mask the flavor. Seldom fed to poultry, but can be used as a substitute for wheat middlings in poultry rations.	Produces firm butter in dairy cattle and firm pork in swine.
ALDER *Alnus* spp	A genus of trees and shrubs. Male and female flower clusters are produced separately on different parts of the same branch. Leaves are shed in autumn while still green. Propagates by underground rhizomes, suckers, and seeds.	Widely distributed in the Northern Hemisphere and the Andes of South America.	Palatibility ranges from poor to fair, but due to its wide distribution, it often provides good secondary browse. Less palatable to cattle than to sheep or goats.	Roots often contain N-fixing nodules like those found in legumes. Alder chips and shavings have been fed, experimentally, to animals.
ALFILERIA (FILAREE, HERONSBILL, PINCLOVER, PINGRASS) *Erodium cicutarium*	Annual or biennial herb of the geranium family. Hairy weed of the rosette type, but may sometimes form clumps. Low, spreading red stems.	Native to Europe. Introduced in the U.S. from the Mediterranean region. Common in all of the western range states. Thrives in desert and foothill areas. Grows well on sandy, barren soils. Spreads extremely rapidly.	Considered to be one of the best winter forages in Arizona. Provides excellent spring forage for all classes of range livestock.	Troublesome weed in lawns, gardens, and cultivated fields.
ALGAROBA *Prosopsis chilensis*	Leguminous tree; up to 33 ft (*10 m*) tall.	Native to South America. Introduced to dry areas throughout world. Drought-resistant.	Cultivated as a shade tree and for forage. Seeds can be ground and used as a concentrate feed. Ripe pods are highly palatable, green pods are bitter and unpalatable.	Kiln-dried pods are better than air-dried pods. Tree can become a weed.
ALKALI SACATON (FINE-TOP SALT GRASS) *Sporobolus airoides*	Hardy bunchgrass. Grows to 3 ft (*91 cm*) in height. Deep, coarse roots.	Found on alkali flats, rocky soils, bottom lands, and desert soils.	Highly palatable in early stages of growth. Tough and unpalatable upon maturing.	Mineral content is extremely high. Good forage grass in alkali regions.

(Continued)

TABLE 30-1 *(Continued)*

Feedstuff	Description	Place of Origin Geographical Adaptation Cultural Characteristics	Importance/Use	Comments
ALYCE CLOVER *Alysicarpus vaginalis*	Summer annual legume. Not a true clover. Low-spreading, coarse-stemmed plant.	Native to Asia. Widely used in the Gulf Coast states. Adapted to moderately fertile soil. Intolerant of wet soils.	Used to a limited extent for pasture, hay, and soil conservation in southern U.S.	Very susceptible to root-knot nematode.
ANISE *Pimpinella anisum*	Annual herb related to the carrot. Grows up to 2 ft (*61 cm*) in height. Leaves resemble parsley. Flowers and seeds are produced in large, loose clusters.	Native to the Mediterranean region.	Leaves are used for flavoring and garnishing in human foods. Seeds are used to flavor curry powder. Oil extracted from the seeds is used in beverages, drugs, and cosmetics. By-products can be fed to livestock.	Anise oil is sometimes used to increase the palatability of feeds.
ANNATTO *Bixa orellana*	Shrub or small tree. Fruits are heart-shaped and covered with short stiff hairs. At maturity the fruits split open.	Native to Central America.	The seeds are seldom used as a protein feed, but are used as pigmentation enhancers in poultry feeds.	Bixin—a coloring compound used for foods and dyes—is produced from the seeds.
ANTELOPE GRASS *Echinochloa pyramidalis*	Tall perennial grass. Grows up to 15 ft (*4.6 m*) tall.	Adapted to wet clay soils alongside water and in wet meadows. Drought-resistant.	Valuable dry season forage. High yields can be obtained after the coarse, rainy season growth is burned off.	
APACHE-PLUME *Fallugia paradoxa*	Multibranched, often evergreen, shrub. May grow to 7 ft (*2 m*) in height. Feathery-plumed seed clusters resemble an Apache war bonnet.	Found in the southwestern U.S. and Mexico. Commonly found at elevations of 5,000 to 8,500 ft (*1,500 to 2,592 m*). Most commonly found on sandy or clay loams. Thrives on deep, moist, rich soils.	Of considerable importance in some range areas due to its local abundance. Palatability ranges from fair to good for cattle, sheep, and goats. Its chief value is as a winter forage.	Withstands close grazing. Can be used for erosion control.
ARROWROOT (YUQUILLA) *Maranta arundinacea*	Perennial herb. Multibranched stems may reach 6 ft (*1.8 m*) in height. Roots are rich in starch.	Native to South America and the West Indies.	Starch is extracted from the roots to form tapioca. The residue, called bittie, is sometimes fed to cattle and swine.	Fine bittie is fed to swine, coarse bittie to cattle.
ARUM, WHITESPOT GIANT *Amorphallus campanalatus*	Herbaceous plant that grows up to 4 ft (*1.2 m*) in height. Produces edible, large, roundish bulbs.		Bulbs can be fed to swine fresh or ensiled.	
ASPEN *Populus grandidentata*	Deciduous tree.	Temperate and cool subtropical regions.	Leaves and treated wood scraps can be fed to livestock.	Used primarily for pulpwood.
B				
BABUL (GUM ARABIC TREE) *Acacia arabica*	An acacia tree; a legume.	Native to the dry, sandy areas of northern Africa, Arabia, and India.	Leaves and pods provide excellent browse.	Pods have been reported to contain tannic acid.
BAHIAGRASS *Paspalum notatum*	Low growing, sod-forming, deep-rooted perennial grass. Stout, short rhizomes form dense sods on sandy soils.	Native to tropical America. Drought-resistant. Adapted to sandy regions having heavy to moderate rainfalls with short, dry seasons. In the U.S., it is grown primarily in the southeastern coastal plains states.	Produces high yields of palatable, nutritious forage for pasture and hay when managed properly. Excellent forage for cow-calf operations and good for growing animals. Rated as poor for finishing. Used for erosion control.	Forage quality is highest in the spring, but feeding value remains satisfactory throughout the growing season.
BALSAM APPLE *Momordica balsamina*	High climbing annual vine. Fruit splits into three sections when ripe.	Dry regions of the tropics throughout the world.	Immature fruit is boiled as a vegetable. Plant is well liked by camels.	
BALSAMROOT *Balsamorhiza* spp	Weeds having thickened, resinous taproots with thin corky bark and a fibrous yellow center. Leaves are roughly arrow-shaped but are cleft to the midrib.	Native to the western U.S. Withstands trampling and close grazing.	Arrowleaf balsamroot is an important forage plant on spring ranges. Leaves and flowers are eaten by all livestock.	Once used by Indians for food.
BAMBARRA GROUNDNUT (CONGO GOOBER, EARTH PEA, KAFFIR PEA, MADAGASCAR GROUNDNUT, STONE GROUNDNUT) *Voandzeia subterranea*	Short-lived, creeping herbaceous legume. Branched stems. Hairless leaves. Pods mature underground like the peanut but yield only one nut per pod.	Native to the district of Bambarra on the upper Niger in Africa. Well adapted to poor soils.	Nuts fed to swine and poultry. Tops fed to cattle.	Nuts contain only about 6% oil. Hence, they are not used for oil extraction.

(Continued)

TABLE 30-1 *(Continued)*

Feedstuff	Description	Place of Origin Geographical Adaptation Cultural Characteristics	Importance/Use	Comments
BAMBOO *Arundinaria* spp, *Chusquea* spp, *Bambusa* spp	Diverse and multispecied type of plant. Ranges in height from 2 ft (*61 cm*) to over 100 ft (*30.5 m*).	Found throughout the world, especially in the tropics and subtropics.	Sprouts used as human food. Whole plants and shoots used as forage.	Bamboo is the most useful and versatile plant in the Orient.
BAOBAB (MONKEY-BREAD TREE, DEAD RAT TREE, BOTTLE TREE) *Adansonia digitata*	One of the largest and longest-lived trees in the world. Trunk is often hollow, growing to 100 ft (*30.5 m*) in circumference.	Tropics.	Leaves and fruit used for feed and food.	
BAUHINIA *Piliostigma thonningii*	Small, leguminous tree yielding heavy crops of large brown pods.	Sandy soils of Africa.	Pods are highly palatable to cattle.	
BEARDGRASS (See BLUESTEM.)				
BEECHNUTS *Fagus* spp	Seeds (covered by a spiny husk) from the beech tree.	Found throughout the eastern U.S.; usually in association with sugar maple.	Of little importance as a feed. Used primarily as forage for hogs and wild animals.	
BENTGRASS (REDTOP) *Agrostis* spp	Moderately tall fine-stemmed, creeping annual or perennial grass.	Grown over much of the northern half of the U.S. Thrives on soils too low in lime or fertility for bluegrass.	Most species cultivated for lawn cover and golf putting greens. Redtop is grown to a limited extent for pasture and hay. It is among the lowest in palatability of the northern grasses.	To all but one species, the common name *bent* is applied, the one exception being REDTOP.
BENTONITE (CLAY)	Naturally occuring mineral consisting of tri-layered aluminum silicate, montmorillonite, and possibly calcium or sodium.	Volcanic origin. Swells greatly upon wetting. Characterized by dispersing and absorbing qualities.	Used as a binding agent in feed pellets and as an anticaking agent in nonmedicated feed.	May tie up medication in medicated feed.
BERMUDAGRASS (BAHAMA GRASS, DEVIL'S GRASS, STAR GRASS, WIREGRASS, JOINT GRASS) *Cynodon dactylon*	Long lived, warm season, perennial grass; forms a dense turf.	Probably originated in Africa, but widespread throughout tropical, subtropical, and mild temperate areas of the world. Prefers heavy soils, but when well fertilized, will grow on deep sands. Hardy grass capable of withstanding drought or waterlogging.	Used primarily for pasture and lawn cover. Also used for hay. Produces excellent yields of palatable, nutritious forage when managed properly. Rated as an excellent forage for cow-calf operations, good for growing, but poor for finishing.	COASTAL BERMUDAGRASS, a sterile hybrid, is now widely grown in the southern U.S. COASTCROSS-1, another hybrid, is less winter-hardy, but more digestible and produces higher gains than Coastal Bermudagrass. Hybrids are established vegetatively by planting sprigs of rhizomes and stolons or clippings from mature forage.
BIRCH *Betula* spp	Tree with slender branches and hard, colored wood.	Native to eastern North America.	Leaves and twigs used for browse.	
BIRDWOOD GRASS *Cenchrus setigerus*	Perennial grass growing to 4 ft (*1.2 m*) in height.	Areas receiving 100 to 200 in. of rainfall but with long, dry seasons. Used as a pasture grass in Africa, India, and Pakistan.	Pasture. Withstands heavy grazing. Palatable, but not very productive.	
BISCUITROOT *Cogswellia* spp	Perennial herbs of the carrot family. Two types of root systems: (1) deep-set, elongated, woody taproots, and (2) fleshy tuberlike roots. Short stems.	Dry regions of western North America. Early growing and maturing.	Valuable as forage on range in the spring or early summer. Palatability ranges from poor to good for cattle but is poor for horses.	It was once an important food of the Indians.
BITTIE (BITTY)	Residue following the extraction of starch from arrowroot.		Fine bittie is fed to swine; coarse bittie to cattle. Low in protein, fiber, and fat.	
BLACKBERRY, TRAILING *Rubus macropetalus*	Aggressive trailing shrub. Produces a fruit commonly relished by humans.	Widely distributed throughout the Pacific Northwest of North America.	Palatability of the leafy foliage ranges from fair for cattle to good for sheep.	Deer and game birds are very fond of the berries.
BLACKBRUSH (BURROWBRUSH) *Coleogyne ramosissima*	Intricately branched shrub; 1 to 6 ft (*2.5 cm to 1.8 m*) tall. Foliage is scant but persists throughout the year.	Southwestern and Rocky Mountain areas of the U.S.	Provides fair forage for cattle, sheep, and goats during the winter. Sheep and goats utilize it better than cattle.	
BLACK CUTCH (CATECHU) *Acacia catechu*	Moderate-sized deciduous, leguminous trees.	Tropical regions.	Good source of browse for cattle.	Cattle eat the leaves and the lower small branches.
BLACK MEDIC *Medicago lupulina*	Annual legume. Procumbent stems rise in thick stands. Stems are small and leafy.	Native to Europe and Asia. Prefers reasonably fertile soils that are not distinctly acid and a good supply of moisture.	Greatest value is in pasture. Low forage yields.	No commercial cultivars are recognized. A troublesome weed in lawns.

(Continued)

TABLE 30-1 *(Continued)*

Feedstuff	Description	Place of Origin Geographical Adaptation Cultural Characteristics	Importance/Use	Comments
BLACK SEED GRASS *Chloris virgata*	Annual grass; leafy; grows to 3 ft *(91 cm)* in height.	Tropical and subtropical regions. Has a short growing season.	Palatable grazing when young. Fair-quality hay if cut at flowering stage.	
BLADDER SENNA *Colutea* spp	Yellow-flowered, leguminous shrub.	Cultivated in Europe for its bladdery pod as a wildlife feed.	Browse.	
BLUEBELL (LUNGWORTS) *Mertensia* spp	Well-known perennials of the borage family. Stems are 4 to 60 in. *(10 cm to 1.5 m)* high. Thick tuberlike roots. Blue or purple, nodding, mostly bractless flowers.	Native to the Northern Hemisphere. Found throughout the U.S., especially in the Far West.	Larger species are choice sheep feed, especially valuable for growing lambs and flushing ewes. Good forage for cattle in some areas. Horses rarely consume bluebells.	
BLUEBERRY *Vaccinium* spp	Woody shrubs; 2 to 10 ft *(61 cm to 3 m)* tall; clustered smooth-skinned, waxy berries.	Found in the wilds of the U.S. and are also grown commercially. Inhabit acid soils, generally associated with coniferous forests.	As browse, rated fair for sheep, poor for cattle.	Deer, elk, caribou, bears, birds, and rodents browse these plants extensively.
BLUEGRASS *Poa* spp	Grass distinguished by small awnless spikelets. Lemmas with a heavy midnerve.	Approximately 200 species are distributed throughout the world, primarily in cool temperate regions. Sixty-nine species are found in North America.	Among the most palatable of the grasses. Widely used for pasture, hay, and lawns.	KENTUCKY BLUEGRASS and CANADA BLUEGRASS are the most important species of bluegrass. Other well-known species are ANNUAL BLUEGRASS, BIG BLUEGRASS, MUTTON BLUEGRASS, ROUGHSTALK BLUEGRASS, SANDBERG BLUEGRASS, and TEXAS BLUEGRASS.
BLUESTEM (BEARDGRASS, TURKEY-FOOT, BROOMSEDGE) *Andropogon* spp	Large genus of grasses. Stems are either solid or pithy. Two spikelets are produced at each node of the rachis—one is fertile, one sterile.	Well-represented throughout the warmer regions of the world.	Several species are considered to be good forage grasses. They are palatable when young and tender, but become coarse and tough with maturity. Bluestem cures well and makes a good dry forage for cattle and horses.	BIG BLUESTEM and LITTLE BLUESTEM are probably the most prevalent constituent of wild hay in the prairies of the U.S. Although rather unpalatable, BROOMSEDGE produces vegetative cover on soils of very low fertility.
BORNEO TALLOW NUT *Shorea stenoptera*	Tree, 18 to 50 ft *(5.5 to 15.3 m)* tall. Kernels are enclosed in a thin, brittle case which sits in an acornlike cup.	Tropical areas. Nuts are collected from the ground, dried in the sun, and the shells separated from the kernels by pounding in mortars and hand picking.	Seed meal (usually solvent extracted) is sometimes fed to livestock. Should not be used in chick diets and should be limited to 10% in layer diets due to the presence of tannic acid.	High levels will produce greenish-brown yolks in eggs.
BOUNDARY MARSH (LAND MARSH, DRAGON'S BLOOD) *Cordyline terminalis*	Leaves vary from green to red-striped, depending on variety.	Found primarily in South America, Hawaii, and the Caribbean.	Used to mark boundary lines. Also used for livestock feed, medicine, perfume, and fiber.	In Hawaii, leaves from the green varieties are used to make hula skirts.
BREADFRUIT *Artocarpus altilis*	Broad tree with large, heavy leathery lobed leaves. Green seedless fruits often weigh more than 10 lb *(4.5 kg)* apiece.	Native to the East Indies. Widespread throughout tropical lowlands of the world.	Fruits often prepared for food in a manner similar to potatoes. Commonly fed to pigs but can be fed to all types of livestock.	Staple food in the diet of Pacific Island peoples.
BREADNUT TREE (RAMON) *Brosimum alicastrum*	Evergreen tree. Yields up to 90 lb *(40.9 kg)* of seeds.	Native to tropical America.	Seeds, leaves, and fruits are used for feed. Commonly used for forage in the dry season.	
BRISTLEGRASS *Setaria* spp	Annual or perennial grasses. Numerous bristles on the inflorescence.	Bristlegrass encompasses a wide variety of grasses adapted to a wide range of conditions.	In general, bristlegrasses are palatable and nutritious. Plains bristlegrass is used extensively as a forage in the southwestern U.S.	Foxtail millet is the most widely used bristlegrass (see MILLET, FOXTAIL). Some species are serious weeds in cultivated crops.
BROMEGRASS *Bromus* spp	Annual and perennial grasses; flat leaf blades; spreading panicles; open seed heads.	Large genus; about 36 species native to the U.S. Most species are found in north temperate zone. Vary greatly in adaptability.	Some species are important forages; others are considered troublesome weeds.	The species of major agricultural importance are CALIFORNIA BROMEGRASS, RESCUEGRASS, SMOOTH BROME, and MOUNTAIN BROME.
BROOMSEDGE (See BLUESTEM.)				
BUFFALO BEAN GRASS (CORN GRASS) *Rottboellia exaltata*	Annual grass. Grows up to 10 ft *(3.1 m)* tall.	Adapted to wet, shady areas on disturbed soils.	Excellent forage if not allowed to grow too tall and develop irritating stiff hairs.	
BUFFALO BUNCHGRASS *Festuca scabrella*	Erect, tufted perennial bunchgrass, 1 to 4 ft *(30.5 cm to 1.2 m)* high. Large tufts of prominently ridged leaves that have a bluish hue. Reproduction is by seed only.	Widely distributed throughout Canada and the northern U.S. One of the principal grasses of Montana and Idaho. Thrives on prairies, rocky cliffs, hills, and mountain slopes up to 10,000 ft *(3,050 m)*, and on dry, open woods. Prefers dry, deep, sandy loam soils.	High palatability. Relished by cattle and horses. Somewhat too hard a grass for sheep.	Because of the large tussock habit of growth, it is hard to mow. Unable to withstand heavy trampling.

(Continued)

TABLE 30-1 *(Continued)*

Feedstuff	Description	Place of Origin Geographical Adaptation Cultural Characteristics	Importance/Use	Comments
BUFFALOGRASS *Buchloe dactyloides*	Low-growing perennial grass. Spreads by stolons. Seeds are enclosed in a hard bur. Grayish green foliage turns to a light straw color at maturity. Plants are unisexual.	Grows primarily in the central and southern Great Plains of the U.S. Usually found on hard clay soils. Tolerates alkaline soils but not sandy soils. Very drought-resistant.	Widely regarded as a range pasture grass. It is highly palatable and nutritious in the green summer stages. Also provides valuable winter grazing in its dry cured stage.	Tolerant of heavy grazing.
BUFFALOGRASS (COW GRASS) *Paspalum conjugatum*	Creeping perennial grass.	Widely distributed throughout tropical and subtropical areas, especially the humid tropics. Prefers moist, heavy soils.	Pasture at immature stages.	At maturity, seeds are produced that can lodge in the throats of livestock, resulting in choking.
BUFFELGRASS (ANJAN, BLUE BUFFALOGRASS, AFRICAN FOXTAIL, RHODESIAN FOXTAIL) *Cenchrus ciliaris*	Tufted perennial grass; up to 2 ft (*61 cm*) tall. Large, strong root system. Occurs in bunch or spreading types.	Native to South Africa. In the U.S., it is grown primarily in South Texas for pasture. Thrives on light, sandy soil. Drought-resistant, but sensitive to cold and waterlogging.	Good pasture grass in dry areas. High protein content and highly digestible when young, but quality decreases with maturity.	Can withstand close grazing and fire once established.
BULRUSH *Scirpus mucronatus*	Fast-spreading weed; tuberlike roots.	Associated with rice production.	Browse or hay.	
BUR-CLOVER *Medicago* spp	Annual legume resembling clover except that the pods are spiny and coiled.	Adapted to warm temperate regions. Prefers moist, well-drained soils. Very tolerant of alkali. Does poorly on soils of low fertility.	Used mostly for pasture. Animals must acquire a taste for bur-clover. Sheep are fond of ripe pods. In the southern U.S., it is commonly combined with Bermudagrass.	Burs are eaten when softened by rain. Burs may get tangled in the wool of sheep and reduce its value.
BURRO GRASS *Scleropogon brevifolius*	Tufted grass. Wiry stolons. Flat leaves. Spikelets with twisted awns.	Semiarid plains and valleys of the southwestern U.S.	Range grass.	Commonly associated with TOBOSA.
		C		
CACAO (COCOA) *Theobroma cacao*	Evergreen tree. Grows to 25 ft (*7.6 m*) in height. Large seeds are found in large red or yellow pods.	Indigenous to South America, but primarily cultivated in West Africa. Adapted to tropical rain-forest regions within 20° of the equator.	Seeds are used for the manufacture of cocoa butter, cocoa, and chocolate. Dried, ground pods are rich in potassium and are similar to corn-and-cob meal. Cocoa shells, beans, and oil cake are highly nutritious, but their use is limited because of the presence of the drug theobromine. Pods contain very little theobromine.	Cocoa products can be detoxified by cooking in water for 1½ hours, filtering, and drying.
CACTI *Family Cactaceae*	Succulent perennials. Range in size from small to treelike growth.	Widely distributed throughout semi-arid and desert regions. Can withstand long droughts.	Furnish emergency forage in drought.	When fed in large amounts, cactus can cause scours.
CACTUS, PRICKLY PEAR *Opuntia engelmanni*	Semierect, spiny, branched perennial. Grows in dense patches. Flat stems. Showy flowers are bright yellow.	Found from western Texas to California and throughout Mexico. Drought- and heat-resistant.	Of greatest value as an emergency feed in drought when the spines are burned off and the plant fed to livestock. Valuable source of water in arid regions. High mineral content. Will not maintain livestock as a sole source of feed.	Fruit is sweet and sometimes used for human consumption.
CALCITE	Hexagonal crystals of calcium carbonate.	Commonly found in limestone, chalk, dogtoothed spar, stalactites, and stalagmites.	Source of calcium.	
CAMEL THORN *Acacia giraffae*	Leguminus tree which grows up to 50 ft (*15 m*) tall. Dark grey bark. Straight leaves have a white thorn at their base, Gold, spherical flower heads. Pods are straight or crescent-shaped.	Found in savannas and along riverbeds.	Fallen pods are readily eaten by livestock.	Green pods are poisonous.
CANARYGRASS *Phalaris* spp	Annuals or perennial grasses. In the southern U.S., most species are summer annuals. Some species in the northern U.S. are winter annuals.	Widely distributed throughout the world.	Most species are used for forage. The name, canarygrass, was derived from *P. canariensis,* which is commonly used for bird seed.	REED CANARYGRASS and HARDINGGRASS are the only species cultivated for forage in the U.S. Both are perennials and must be harvested or grazed in the early stages of growth to obtain palatable, high quality forage.

(Continued)

TABLE 30-1 *(Continued)*

Feedstuff	Description	Place of Origin Geographical Adaptation Cultural Characteristics	Importance/Use	Comments
CANIHUA *Chenopodium pallidicaule*		Important crop in the Andes. Adapted to high elevations.	Small, bitter seeds are used primarily for human consumption.	
CANNA (GRUYA, AUSTRALIAN ARROWROOT, QUEENSLAND ARROWROOT, TOUSLES-MOIS) *Canna edulis*	Perennial reed or canelike herb. Up to 10 ft *(3 m)* tall. Rhizomes bear tubers.	Tropical regions.	Leaves and tubers are used for animal feed.	Tubers are used primarily for human consumption.
CAPER BUSH *Capparis spinosa*	Straggly, spiny shrub. Grows to 3 ft *(0.9 m)* in height.	Widely cultivated in warm regions, especially in the Mediterranean countries.	Young, flower buds are picked and used as feed.	
CARDOON *Cynara cardunculus*	Robust plant with thick stalks and very large spiny leaves.	Not cultivated in the U.S.	Aerial part is sometimes used as cattle feed.	Cultivated as food for its roots and celerylike stalks.
CARIB GRASS (ALEMAN GRASS, MALOJILLA) *Eriochloa polystachya*	Branching, grass with trailing stems that root at the lower nodes.	Grown in the Gulf Coast states. Adapted to humid regions with evenly distributed rainfall. Intolerant of either cold or drought. Withstands waterlogging.	Well-suited for grazing and hay.	
CAROB (ST. JOHN'S BREAD, LOCUST BEAN) *Ceratonia siliqua*	Medium-size leguminous evergreen tree. Glossy, dark brown leaves. Straight, brown pods, up to 8 in. *(20 cm)* in length. Seeds are imbedded in the fleshy pods.	Native to the Mediterranean region. Widespread in many arid and subtropical areas.	Cultivated for its sugar-rich pods. Beans, pods, and carob germ are fed to livestock. Pulp is highly indigestible and unpalatable. Tannic acid in carob beans reduces the digestibility of feed if used in large amounts.	In eastern Mediterranean countries, the seeds are used for weight because of their uniformity of size; hence, the derivation of the word "carat."
CARPETGRASS *Axonopus affinis*	Warm season, low growing, sod forming perennial. Abundant rooting of creeping surface runners.	Indigenous to Central Americas and the West Indies. Introduced into the U.S. during the colonial period. Adapted to the warmer areas of the southeastern U.S. where the ground seldom freezes. Thrives on soils with a high water table. However, it does not grow well in swamps or waterlogged soils.	Chief component in unimproved pasture of the southeastern U.S. Low nutritional value and dry matter yield. Extremely low mineral content, even when fertilized.	Used for lawns, turfs, and firebreaks.
CASSAVA (MANIOC, TAPIOCA, BRAZILIAN ARROWROOT, YUCA) *Manihot esculenta*	Herbaceous shrub or tree. Grows up to 13 ft *(4 m)* in height. Fingerlike leaves. Tuberous, starchy roots form clusters at the base of the stem. Two types of cassava exist: (1) bitter varieties—contain high amounts of prussic acid and must be processed before use, and (2) sweet varieties—contain low amounts of prussic acid and can be fed raw.	Cultivated throughout the tropics and subtropics. Very little cultivated in the U.S.	Cassava roots, either cooked or raw, are fed to pigs, cattle, sheep, and goats in many areas. Molasses or water must be added to cassava-based poultry and swine rations in mash form to make them palatable. However, pelleting alleviates this need. Leaves are richer in protein and minerals than any other part of the plant.	Root meal is extremely deficient in methionine. Young leaves of the sweet varieties should be boiled for 15 minutes or dried for 3 weeks to inactivate the prussic acid.
CASTOR BEAN *Ricinus communis*	Seeds (not really beans) encased in a spiny outer shell.	Grown as an annual herb in temperate climates, a perennial tree in tropical climates.	Oil is commonly extracted from the beans for use in industry and medicine. Seeds contain a toxin called ricin but most animals are resistant to it, especially birds. Detoxified meal can constitute up to 40% of poultry rations. Ruminants should not be fed rations containing more than 5% castor pomace or 10% castor hulls.	Sometimes the plant is cultivated as an ornamental.
CATJANG (HINDU COWPEA) *Vigna sinensis* (cylindrica group)	Erect, semibushy legume with uptilted cylindrical pods 3 to 4 in. *(8 to 12 cm)* long. Bluish-purple to white flowers. Hoe-shaped leaves.	Matures very rapidly (55 to 90 days).	Seeds and aerial parts are used for feed.	
CENTRO (BUTTERFLY PEA) *Centrosema pubescens*	Creeping, twining perennial leguminous herb. Trifoliate leaves. Large flowers. Very leafy and does not produce any woody growth.	Native to South America. Adapted to dry conditions.	Fairly palatable once animals become used to it.	Can be used with many grasses.

(Continued)

TABLE 30-1 *(Continued)*

Feedstuff	Description	Place of Origin Geographical Adaptation Cultural Characteristics	Importance/Use	Comments
CENTURY PLANT *Agave americana*	Has a large candelabralike stalk that can reach 30 ft (*9 m*) in height. Leaves grow next to the ground in a large rosette.	Native to Mexico. Found in hot, dry areas in the Mediterranean region and in Africa.	The skin is removed from the leaves, and the remaining parts are chopped and mixed into hay.	Saponin may cause illness if excessive amounts are fed. Sugar-rich sap is used for mescal and pulque—Mexican liquors.
CHESTNUTS *Castanea dentata*	Sweet, edible nuts of trees belonging to the genus *Castanea*. Fruits have spiny bur and become fibrous upon maturity.	Used to be abundant throughout the hardwood belt of the U.S., especially in the Appalachian region. Blight has killed most of the chestnuts in the U.S.	Shells and meats can be fed to livestock. Meats are low in fiber, medium in energy.	
CHICKWEED *Alsine* spp	Annual or perennial herbs. Weak, spreading stems. White flowers.	Most species are found in moist or wet areas.	Palatability is considered fair in cattle and fairly good in sheep.	One of the most common types of weeds in gardens and cultivated fields.
CHINESE LANTERN TREE *Dichrostachys glomerata*	Small tree or shrub with twisted branches and stout, straight, or slightly curved thorns. Forms dense and impenetrable thickets. Pods are twisted and in large bunches.		Pods are well-liked by cattle.	Difficult to eradicate.
CHOUMOELLIER	Hybrid of cabbage, kohlrabi, and kale.	Grown in New Zealand and Australia.	Forage.	It may be grazed by sheep.
CHUFA *Cyperus esculentus*	Tuber-producing plant. The tubers are small (about ¾ in. [*19 mm*] long), cylindrical, and hard. Grasslike top has simple leaves and a flower stalk that may grow to 3 ft (*0.9 m*) in height.	Cultivated primarily in southern Europe and Africa. Minor crop of the southern U.S. Grows best on sandy soils.	When properly supplemented with protein feeds and corn, chufas can be used to fatten swine.	Tubers may produce soft pork if improperly used.
CLAM SHELLS	Rich in calcium carbonate.		Calcium supplement.	Ground finely for feed.
CLEMATIS *Clematis* spp	Perennials with enlarged stem joints. Large compound opposite leaves. Showy flowers lack petals.	About 24 species are native to the western U.S.	When young and tender, clematis is rated as fair forage for sheep. Cattle occasionally eat the leaves of climbing species.	Indians used to chew this plant as a remedy for sore throats and colds.
CLIFFROSE (QUININE BUSH) *Cowania stansburiana*	Leafy, evergreen shrub. Fragrant wildrose-shaped flowers. Twigs have a bitter (quininelike) taste.	Found in the southwestern U.S. and Mexico. Adapted to dry, rocky foothills and mesas up through elevations of 8,000 ft (*2,440 m*). Characteristic of limestone regions.	Where abundant, cliffrose is an important browse plant for sheep and cattle. Especially valuable as winter browse.	Branches are brittle and easily broken by cattle.
CLOVER *Trifolium* spp	Perennial or annual legumes. Flowers of all species are borne in heads. Number of florets or individual flowers per head ranges from 5 to 200. Seeds per pod range from 1 to 8 depending on species. Leaves are trifoliate in most cases.	About 250 species of clover are recognized throughout the world. They are believed to have originated in southwestern Asia Minor and southeastern Europe. Wild species are found on all continents except Australia. In general, clovers thrive in cool, moist climates on soils having an available supply of calcium, phosphorus, and potassium. Photoperiodism (length of daylight) is critical in many species and varieties. Most are long-day plants.	Second only to alfalfa in importance as a legume forage crop.	The major species in the U.S. are ALSIKE CLOVER, ARROWLEAF CLOVER, CRIMSON CLOVER, HOP CLOVER, LADINO CLOVER (WHITE CLOVER), PERSIAN CLOVER, RED CLOVER, ROSE CLOVER, SEASIDE CLOVER, STRAWBERRY CLOVER, SUBTERRANEAN CLOVER, and WHITETIP CLOVER. Species of local importance are BALL CLOVER, BIGFLOWERED CLOVER, CLUSTER CLOVER, LAPPA CLOVER, STRIATE CLOVER, and ZIGZAG CLOVER.
CLUBHEAD CUT GRASS (RICE CUT GRASS) *Leersia hexandra*	Perennial grass. Grows to 4 ft (*1.2 m*) tall.	Grows along water and in swamps in the humid tropics and subtropics. Prefers fertile soils. Tolerates waterlogging and overgrazing.	Palatable forage when young.	Burned off to provide dry season grazing.
CLUSTER FIG *Ficus glomerata*	Large tree that provides dense shade.		Fruits and leaves are used for livestock feed.	
COCOA (See CACAO.)				
COFFEE *Coffea arabica*	Shrub that grows up to 16 ft (*5 m*). White, fragrant starlike flowers. Dark green, glossy leaves. Fruit is two-seeded and deep crimson in color.	Tropical rain-forest plant. Prefers cool temperatures of highlands but is susceptible to freezing.	The leaves are sometimes used for forage. Coffee pulp—a by-product of coffee processing—is used in some areas as a roughage for cattle. Oil cake is unpalatable.	

(Continued)

TABLE 30-1 *(Continued)*

Feedstuff	Description	Place of Origin Geographical Adaptation Cultural Characteristics	Importance/Use	Comments
COLUMBINE *Aquilegia* spp	Perennial herbs. Slender to stout, mostly perpendicular, often branched taproots. Attractive ornate flowers.	Native to the Old World and North America. In the U.S., most species are found in the West. Grows in moist areas.	Generally considered to be of secondary importance as browse. Palatability is fair for sheep, poor for cattle, zero for horses.	
CORAL	The exoskeleton of small marine organisms. Almost pure calcium carbonate. About 37% calcium and 0% phosphorus.		Calcium supplement.	
COROZO PALM *Orbignya cohune*	Oil-bearing palms. Large fruit with thick, hard nutshells and one or more kernels. Fruits grow in bunches measuring about 3 ft (*0.9 m*) in length and weighing up to 55 lb (*25 kg*).	Tropical areas.	Highly palatable oil meal can be fed effectively to livestock—up to 30% in swine diets and up to 10% in poultry.	Palm nuts are used primarily for the manufacture of soap.
COTTONTOP (ARIZONA COTTONGRASS) *Trichachne californica*	Fast-growing, coarse, leafy perennial grass. Slender, erect stems grow to 40 in. (*102 cm*) in height. Strong, wooly, knotted rootstocks. Silky-cottony panicle with paired lance-shaped spikelets covered with long white hairs.	Common on the deserts and foothills of southern New Mexico and Arizona and in the woodlands and semidesert areas of the Southwest, except California. Responds quickly to spring and summer rains.	Highly palatable forage when immature. Upon maturity, it becomes tough. Valuable grass when grown in combination with slower-growing grasses (*e.g.,* grama). The animals will graze the cottontop first, allowing the other grasses time to become established.	Cures well on the ground and makes good winter forage.
COTTONWOOD *Populus* spp	Relative of the aspen. Seeds are covered with tufts of cottony hairs.	Widely distributed in the Northern Hemisphere outside of the tropics. Typical of moist sites, such as along stream and pond banks and in depressions where moisture accumulates.	Palatable forage, but of lower forage value than aspen.	NARROWLEAF COTTONWOOD, next to Aspen, is probably the most widely distributed species of *Populus* in the western U.S.
CRABGRASS *Digitaria* spp	Erect or prostrate annuals and perennials.	Adapted to a wide variety of conditions.	Good forage grasses.	CRABGRASS is a common weed in cultivated fields.
CRAMBE (ABYSSINIAN KALE, ABYSSINIAN CABBAGE) *Crambe abyssinica*	Erect annual herb. Grows to 3 ft (*0.9 m*) in height. Upon maturity, the leaves drop off and the pods and stems turn tan in color.		Oil meal must be treated with ammonia or sodium carbonate or heated to neutralize sulfur compounds that make the feed unpalatable. Treated oil meal can be fed to ruminants at levels up to 50% of the supplement.	Oil is extracted from the seeds. The oil meal is high in protein.
CROW FOOT GRASS *Dactyloctenium aegyptium*	Annual grass with soft, slightly succulent leaves. Grows to 30 in. (*76 cm*) in height.	Grows either on denuded land in semiarid areas or as a weed in cultivated fields.	Forage.	Reported to contain cyanogenetic glucosides. Hence, at certain times, it may be poisonous.
CROWN VETCH *Coronilla varia*	Perennial legume. Vetchlike leaves. Arrangement of florets resemble a crown. Deep taproot with numerous lateral roots. Flowers are variegated white to purple. Upon drying, the pods break in segments, each of which contains a rod-shaped seed.	Found predominantly in the Mediterranean region, central and southern Europe, southwestern Asia, and northern Africa. Grown in northern ⅔ of the U.S. Prefers fertile, well-drained soils of pH 6 or above. However, once established, it is relatively tolerant of soil acidity and infertility.	Palatable forage, especially well-suited to the improvement of permanent bluegrass pasture. Used extensively as a ground cover for the stabilization of roadbanks and along waterways.	The three most widely used cultivars in the U.S. are CHEMUNG, PENNGIFT, and EMERALD. Crown vetch is nonbloating.
CURLY MESQUITE *Hilaria belangeri*	Stolon-forming grass. Grows in tufts. Slender stems up to 1 ft (*30.5 cm*) tall. Short, narrow leaves.	Native to central Texas, Arizona, and throughout Mexico. Grows on dry soils that vary from clay to gravelly in texture. Highly drought-resistant. Responds readily to summer rains.	Highly regarded for its forage value. Very palatable to all classes of range livestock.	Produces a fair amount of forage for its size.
		D		
DAL GRASS (BAMBOO GRASS) *Hymenachne amplexicaulis*	Perennial grass. Stout culms. Rooting at lower nodes.	Found in tropics and subtropics of the Old and the New Worlds. Semiaquatic type of grass.	Used for pasture, hay, and silage.	Relished by buffaloes.

(Continued)

TABLE 30-1 *(Continued)*

Feedstuff	Description	Place of Origin Geographical Adaptation Cultural Characteristics	Importance/Use	Comments
DALLISGRASS *Paspalum dilatatum*	Fast-growing, stout perennial. Smooth leaves; deep root system. Grows in clumps. Slender stems usually droop with weight of the seed.	Native to Argentina, Brazil, and Uruguay. Widely used throughout the Cotton Belt of the U.S. Prefers moist, fertile, clay, and loam bottomland. Drought- and moisture-tolerant.	Highly palatable. Dallisgrass and legume combination used primarily for pasture. Produces good yields of hay.	In southeastern U.S., growth occurs from early spring to late fall; but it is dormant during the coldest winter months.
DATE PALM *Phoenix dactylifera*	Strong palm tree of up to 100 ft (*30.5 m*) in height. Long, stiff leaves. Cylindrical fruits of 1 to 2 in. (*2.5 to 5 cm*) grow profusely on long strands. Fruits lose most of their moisture before harvest.	Found throughout tropical regions of the world. Most dates produced in the U.S. are grown in California and Arizona.	Whole fruits, ground seeds, and dried pulp can be fed to livestock. Ground seeds are good ruminant feeds when supplemented with a protein feed. Dried pulp can be fed to all classes of livestock.	
DEERBRUSH (BLUEBRUSH, MOUNTAIN LILAC, SWEETBIRCH) *Ceanothus integerrimus*	Deciduous shrub, 4 to 12 ft (*1.2 to 3.7 m*) high. Loose and slender branches.	Mountainous and hilly regions of the western U.S. Commonly found in ponderosa pine and mixed conifer areas. Prefers well-drained, moderately fertile soils.	One of the most valuable browse plants of the west. Excellent browse for cattle, sheep, goats, and deer. Fair to good for horses.	In California, this plant provides more forage than any other browse species.
DEERVETCH *Lotus* spp	Perennial or annual legumes. Numerous species can be divided into two groups: (1) low plants with small, whitish, wooly leaves, and (2) upright, dark green tall plants.	Worldwide distribution. Low-growing species found on dry, well-drained soils. Tall species prefer moist soils.	Forage value of the various species ranges from low to high. The tall species are generally more palatable.	Some common types are SPANISH CLOVER, DOUGLAS DEERVETCH, BIG DEERVETCH, STREAM DEERVETCH, and DESERT DEERVETCH.
DOUM PALM *Hyphaene* spp	One of the few types of branching palms. Grows to 50 ft (*15.3 m*) in height. Produces extremely hard, fibrous kernels.	Desert regions of Africa.	Kernel meal is comparable to corn in feeding value. Produces less back fat than corn.	Kernels formerly used for buttons.
DRAGON'S HEAD *Lallomentia iberica*	Annual herb with short vegetative cycle.	Adapted to dry climates.	Grown primarily for its seeds from which a drying oil is extracted. The oil cake can be fed to horses, rabbits, and ruminants.	High in protein (35–45%).
DROPSEED *Sporobolus* spp	Most species are bunchgrasses, but a few have creeping rhizomes. Most species are perennials. All species are characterized by their high yields of seeds which shatter at maturity.	Commonly found in the southwestern U.S. Grows primarily at low elevations in desert, semidesert, and plains areas.	Most dropseeds are of poor quality. They are generally palatable, but their coarseness reduces feeding value. However, sand dropseed is considered to be a high-quality forage.	Presence on ranges usually indicates overgrazing, drought, or unfavorable soil conditions.
DUBI GRASS *Urochloa bolbodes*	Tufted perennial grass. Grows to 40 in. (*102 cm*) in height.	Open grasslands and bush regions.	Highly palatable forage.	
DWARF KOA (BUNDLE FLOWER) *Desmanthus viragatus*	Resembles koa haole, but has slender, pithy stems, small leaflets, and narrow pods.	Native of the tropical and subtropical regions of the New World.	Palatable forage.	Makes excellent growth. Withstands cutting and grazing.
		E		
EGGSHELL MEAL	Dried mixture of eggshells, shell membranes, and egg contents.	Product of egg-breaking plants.	Mineral supplement. About 94% of the shell is calcium carbonate. Other 6% contains calcium and magnesium phosphates, boron, copper, chromium, and iodine.	
ELEPHANT'S EAR (EARPOD, MULLATO EAR, MONKEY EAR) *Enterolobium cyclocarpum*	Spreading, deciduous leguminous tree. Produces brown ear-shaped pods.	Tropical feed.	Browse.	Seeds are sometimes used for human food.
ELEPHANT'S FOOD *Portulacaria afra*	Shrub. Grows to 10 ft (*3 m*) in height. Fleshy leaves. Terminal spikes bear many small pink flowers.	Native to Africa.	One type is heavily browsed by goats and sheep.	
EPHEDRA (JOINTFIR) *Ephedra* spp	Shrubs or small trees. Jointed branches.	Widely distributed throughout the world. Grows on dry, open sites in valleys and on slopes, mesas, and foothills.	Not very palatable, but used for forage in emergency periods such as drought or scarcity of quality forage.	VINE, CALIFORNIA, NEVADA, TORREY, and LONGLEAF JOINTFIRS are commonly found in the U.S.

(Continued)

TABLE 30-1 *(Continued)*

Feedstuff	Description	Place of Origin Geographical Adaptation Cultural Characteristics	Importance/Use	Comments
ERIGERON (WILD DAISY) *Erigeron* spp	Large genus of annual, biennial, and perennial herbs. Heads contain numerous small flowers.	Widely distributed throughout the temperate and mountainous regions of the world. Numerous species are found in western U.S.	Palatability is highly variable. More palatable to goats and sheep than to cattle. Unpalatable to horses. Low feeding value.	
EUROPEAN HACKBERRY *Celtis australis*	Heavy, hardwood tree.		Young leaves are palatable to cattle. Mature leaves are unpalatable.	Wood is used for ship-rods and walking sticks.
		F		
FALSE-MESQUITE (BASTARD-MESQUITE, FALSE-CATCLAW) *Calliandra eriophylla*	Shrub; up to 2 ft (*61 cm*) tall. Branches are stiff and bluish gray. Young twigs are soft and hairy.	Distributed from western Texas to central Arizona and south to Mexico. Found on warm, open, sunny areas of dry mesas and foothills, especially common between the elevations of 3,000 to 5,000 ft (*915 to 1,525 m*). Grows on a wide variety of soils, but particularly characteristic of sandy gravel and gravelly clay soils.	Valuable forage on many ranges. Palatability varies from fairly good to very good. Withstands heavy grazing.	Has about half the crude fiber and about the same lipid content as alfalfa.
FENUGREEK *Trigonella foenumgraecum*	Annual herbaceous legume.	Native to southern Europe. Adapted to conditions where moisture is insufficient for berseem.	The whole plant has been used for forage. The seeds are sometimes used as condiments.	May impart off-flavors to milk. Sometimes used after cotton or sugarcane as a short rotation crop.
FERNLEAF *Pedicularis* spp	Annual or perennial weeds of the figwort family. Distinguished by a spurless, strongly, two-lipped corolla.	Genus of over 250 species, most of which are native to the Northern Hemisphere, especially in the Old World. About 30 species are found in the western U.S., primarily in the mountains.	Vary in palatability from zero to fair as forage. More palatable to sheep than cattle.	
FESCUE *Festuca* spp	Annual and perennial grasses. Classified by leaf types: (1) broad-leaved, or (2) fine-leaved.	About 100 species are found throughout temperate or cool zones. Vary widely in growth habitat.	Perennial species provide excellent forage and turf. Annual species may be troublesome weeds.	TALL FESCUE and MEADOW FESCUE, both broad-leaved species, are the most widely cultivated fescues. Some of the important fine-leaved species are IDAHO FESCUE, RED FESCUE, SHEEP FESCUE, and CHEWING'S FESCUE.
FINE NEEDLEGRASS *Aristida mutabilis*	Small annual grass of up to 15 in. (*38 cm*) tall.	Adapted to dry conditions. One of the first grasses to colonize bare ground in arid and semiarid areas.	Range grazing.	
FINGERGRASS *Chloris* spp	Tufted perennials or annuals with flat or scabrous blades. Several species have showy and feathery spikes.	Found on the ranges of the southwestern U.S. and in Hawaii.	Range grazing.	
FIREWEED *Chamaenerion angustifolium*	Perennial herb of the primrose family. Grows up to 9 ft (*2.7 m*) tall. Rapidly growing rootstock. Produces large numbers of seeds.	Widely distributed throughout North America, Europe, and Asia. Adapted to a wide variety of soils and climates. Moderately drought-resistant.	One of the most important range forage weeds. Palatability ranges from fair to good for sheep and zero to fair for cattle. Produces forage throughout the summer.	Name is derived from the fact that the plant flourishes on newly burned-off forest lands.
FLAMBOYANT (ROYAL POINCIANA, FLAME TREE) *Delonix regia*	Leguminous tree of up to 65 ft (*20 m*) in height. Red flowers. Its dark brown pods may reach 3.5 ft (*1 m*) in length.	Native to Madagascar. Grows well only in frost-free regions.	Pods are used for feed.	
FLAT PEA *Lathyrus sylvestris*	Perennial legume.	Native to Europe. Used on logged-off land in the Pacific Coast states.	Used for forage and green manure. Contains high amount of highly digestible protein.	Poisonous when fed in large amounts after the seed has formed.
FLUTED PUMPKIN (TELFAIRIA) *Telfairia occidentalis*	Vines having very long shoots (up to 45 ft [*14 m*] in length), purple flowers, and large gourds.	Native to Africa. Tropical plant.	Leaves are relished by sheep and goats.	Leaves are used primarily for human consumption.
FOXTAIL *Alopecurus* spp	Low or moderately tall perennial grasses or some annuals with flat blades and soft, dense, spikelike panicles.	Native to Europe and Asia. Has been cultivated in the U.S. since about 1750. Adapted to temperate climates, especially to low, overflow areas.	All species are highly palatable and nutritious, but are not found in abundance in the U.S. Meadow foxtail is used as a meadow grass in the eastern U.S. and the Pacific Northwest.	CREEPING FOXTAIL and MEADOW FOXTAIL are the most important species found in the U.S. The awns of ripe foxtail may pierce animal tissue.

(Continued)

TABLE 30-1 *(Continued)*

Feedstuff	Description	Place of Origin Geographical Adaptation Cultural Characteristics	Importance/Use	Comments
		G		
GALLETA (CURLY GRASS) *Hilaria jamesii*	Erect perennial grass. Strong scaly rootstocks. Numerous wiry leaves. Fine, hairy flower turns from purple to white upon maturity.	Important forage of the southwestern U.S. Probably the second most valuable range grass of New Mexico. Grows on valley slopes. Fast growth during summer rains. Extremely drought-resistant.	During the growing stages, it is of good to very good palatability and high feeding value. After growth ceases, it becomes dry and unpalatable. Also used as a soil binder.	Not easily killed by overstocking.
GAMAGRASS *Tripsacum* spp	Robust perennial grass. Broad, flat blades.	Eastern U.S., Mexico, and Central America.	Good forage grasses but not of importance in the U.S.	Related to corn; some hybrids with corn have been successful.
GAMBA GRASS *Andropogon gayanus*	Tall perennial grass; up to 6.5 ft (*2 m*) tall.	Found in areas receiving 24 to 48 in. (*61 to 122 cm*) of rainfall with dry seasons of 5 to 6 months. Does not tolerate waterlogging or overgrazing.	Highly palatable and nutritious in immature stages, however, quality decreases rapidly as the grass matures.	Frequent burning suppresses growth of this grass.
GENTIAN *Gentiana* spp	Large genus of annual, biennial, and perennial herbs.	Native to the cooler regions of the world, primarily in mountainous regions and north temperate zones. Grows primarily in meadows and moist open sites on mountains.	Low palatability; poor for cattle and poor to fair for sheep. Tonic and stomachic properties of gentian may have a beneficial effect for range animals.	Long used for medicinal purposes —primarily as tonics. Also fermented and distilled as a liquor which is popular in the Alps.
GERANIUMS *Geranium* spp	Perennial, annual, and occasionally biennial herbs. Peculiar fruiting structure resembles a Maypole. Leaves emit a distinctive odor when crushed. Stems, leaves, stalks, and portions of the inflorescense have simple, sticky hairs.	Diverse genus. Occurs up to 10,000 ft (*3,050 m*) in elevation in dry mountain meadows, in open damp timberland, and in the grasslands of plains and foothills. Prefers a rich loam soil with partial shade.	Moderate forage value. More valuable for sheep than for cattle. Sheep eat most of the herbage in the spring but later eat only the flowers. Cattle eat only the tender herbage.	Certain geraniums have been used for medicinal purposes, primarily astringents.
GOLDENWEEDS *Haplopappus* spp	Large and variable genus. Largely herbaceous but many species are half-shrubs with woody roots and crowns.	Widely distributed throughout the western ranges of the U.S., primarily in semidesert areas.	A few species provide fairly good winter forage for sheep but most species are very low in palatability.	Several woody species have been used for medicinal purposes by Indians and Mexicans.
GORSE (FURZE) *Ulex europaeus*	Shrubbery perennial with yellow flowers. Densely branched, spiny shrub that has spread throughout agricultural land and left it almost worthless.	Introduced from Europe as an ornamental. In the U.S., it is commonly found in the Pacific Coast and northeastern states.	Forage.	In Europe, it is used for forage and fuel.
GRAMA *Bouteloua* spp	Vary greatly according to species.	Some 18 species are native to the U.S.; found primarily throughout the Great Plains and West. Summer growers. Most species cure naturally; thus, stands from previous year's growth are palatable and nutritious.	Valuable forage for range and pasture.	SIDE-OATS GRAMA and BLUE GRAMA are the two most widely grown species in the U.S. BLACK GRAMA is of local importance.
GRASSPEA (CHICKLING VETCH) *Lathyrus sativus*	Annual herb with many varieties.	Tropical areas.	Foliage and seed are used for forage. Whole plant is a good source of carotene.	Seeds are sometimes used for human consumption.
GUANO (MANURE)	Manure.	Generally refers to the partially decomposed excrement of sea fowl, found on some coasts or islands.		In the tropics, the term *guano* sometimes refers to any kind of manure used as livestock feed.
GUAR (CLUSTER BEAN) *Cyamopsis tetragonoloba*	Erect, annual, leguminous herb.	Native to India and Pakistan. In the U.S., it is grown primarily in Texas and Oklahoma.	Poor-quality hay, but the straw and stubble can provide adequate pasturage. Untreated guar meal can constitute up to 25% of cattle rations. Detoxified meal can be used as the sole protein source for cattle.	Grown primarily for its seeds containing mannogalactan—a product used as a stabilizer for ice cream and frozen dessert.
GUATEMALA GRASS (HONDURAS GRASS) *Tripsacum laxum*	Tall, broad-leaved perennial; up to 11.5 ft (*3.5 m*) tall.	Grows in humid areas. Prefers rich soils and tolerates acidity.	Not suited for grazing, but can be used as green forage or hay.	More persistent, but less productive and nutritious than elephantgrass.
GUAVA *Psidium guajava*	Small evergreen tree cultivated for its fruit. Fruits vary in shape from spherical to pyriform; sweet to slightly acid pulp is soft when ripe.	Found throughout the tropics. Some is grown in the Gulf Coast states.	Livestock use the tree for browse.	Fruit is mainly used for jellies and jams.

(Continued)

TABLE 30-1 *(Continued)*

Feedstuff	Description	Place of Origin Geographical Adaptation Cultural Characteristics	Importance/Use	Comments
		H		
HAIRGRASS *Deschampsia* spp	Low or moderately tall annuals or usually perennials. Shiny pale or purple spikelets in narrow or open panicles.	Distributed throughout the U.S. Tufted hairgrass is often the dominant grass in mountain meadows.	Considered to be an excellent forage.	
HAIRY INDIGO *Indigofera hirsuta*	Climbing annual legume with heavy foliage on fine stems. Leaves are covered with short hairs.	Adapted to the Coastal Plains of the U.S. Prefers sandy loam soils.	Produce large amounts of high-quality forage. Used for hay, pasture, cover crops, and green manure.	
HARDINGGRASS (See CANARYGRASS.)				
HAWKSBEARD *Crepis* spp	Annual and biennial weeds.	Found in foothill and mountain areas. Prefers well-drained soils.	Sheep are fond of all species.	
HAWKWEED *Hieracium* spp	Hairy, glandular, or smooth perennial herbs.	Distributed throughout the northern temperate regions of the world and in the Andes of South America. About 25 species are found in the western U.S. Abundant in woodlands, moderately dry, open habitats, and, in a few cases, moist meadows.	Several species are valuable range plants, being closely grazed by sheep and goats and fair forage for cattle and horses. Among these species are WOOLYWEED, HOUNDS-TONGUE, HAWKWEED, and SLENDER HAWKWEED.	Several species are considered to be serious pests in the eastern U.S.; among them, ORANGE HAWKWEED and SHAGGY HAWKWEED.
HEMP (MARIJUANA) *Cannabis sativa*	Erect dioecious annual; only the female plant yields seeds. May grow as high as 16 ft (*5 m*).	Adapted to most tropical and temperate regions. Requires rich bottomland.	In tropical countries, hemp is cultivated for fiber and its oil-rich seed. The resultant oil cake is of high fiber and low digestibility. Hence, it is used primarily as a ruminant feed. Whole seeds are used for poultry feeds.	Illegal to grow in the U.S. because of its narcotic properties upon smoking or ingestion, but can be found growing wild in pastures and ditches and along roadsides. Was grown extensively in the U.S. for fiber during World War II.
HERRINGBONE GRASS *Tetrapogon* spp	Annual grasses.	Found in arid and semiarid areas.	Forage.	
HOCK GRASS *Latipes senegalensis*	Short-lived perennial grass.	Semiarid regions.	Low in productivity but very palatable and provides good pasture in dry areas.	
HONEY LOCUST (SWEET LOCUST) *Gledetsia triacanthos*	Tall, spiny tree bearing straight or crescent-shaped pods 16 to 20 in. (*41 to 51 cm*) long.	Dry regions.	Pods provide excellent source of browse.	
HOOK GRASS *Latipes senegalensis*	Small, short-lived perennial with numerous short leaves.	Semiarid regions.	Palatable and well-grazed.	Low productivity.
HORSE CHESTNUT *Aesculus hippocastanum*	Tree; up to 65 ft (*20 m*) tall. Yields spherical or elongated spiny fruit.	Native to northern Greece. Thrives in moist temperate areas.	Fruit is sometimes used for livestock feed.	
HUCKLEBERRY (See BLUEBERRY.)				
HURRICANE GRASS (SEYMOUR GRASS, BARBADOS SOUR GRASS, PITTED BLUESTEM) *Bothriochloa pertusa*	Tufted perennial.	Dry areas. Withstands heavy grazing and short dry spells. Low-yielding.	Used for pasture, hay, and mulching.	Considered by some to be a weed.
		I		
ILLIPI *Bassia longifolia*	Large deciduous tree with spreading branches and a large, round crown. Bears seeds from which the oil is extracted.	Native to South America.	Browse.	Flowers are used as a food and a source of alcohol. Oil meal contains saponins and is toxic.
IMPERIAL GRASS *Axoponus scoparius*	Densely tufted perennial grass. Large, blunt-ended, hairy leaves.	Prefers well-drained soils in areas of high rainfall. Tolerates drought on deep soils. Tolerant of high and low temperatures.	Used primarily as a silage grass.	Persistent only when cut.
INDIANGRASS *Sorghastrum nutans*	Tall, erect perennial, warm season grass. Short, scaly rhizomes.	Found east of the Rocky Mountains from Canada to the Gulf of Mexico. Thrives on fertile bottom soils but will grow on sandy soils and dry slopes.	Fresh Indiangrass is highly palatable. As it dries, palatability decreases.	Common constituent of prairie hay in the Great Plains.

(Continued)

TABLE 30-1 *(Continued)*

Feedstuff	Description	Place of Origin Geographical Adaptation Cultural Characteristics	Importance/Use	Comments
INDIAN RICEGRASS *Oryzopsis hymenoides*	Densely tufted perennial grass. Grows to 1 to 2 ft *(30 to 61 cm)* in height. Long, slender leaves.	Widely distributed throughout the western U.S. Tolerant of drought and alkali soils. Found primarily on dry, sandy soils.	Considered to be an outstanding winter forage.	It was once commonly found throughout the western ranges, but was almost wiped out by overgrazing. At one time, ricegrass seed was harvested by Indians for use as meal and flour when the corn crop failed.
INDIAN SANDBAR *Cenchrus cacharticus*	Erect annual growing to 2 ft *(61 cm)*.	Adapted to hot, dry tropical areas with short growing seasons. Generally found on sandy soils.	When young, it serves as a good pasture grass and hay.	
INDIAN SHOT *Canna indica*	Slender tuberous plant that can grow to 40 in. *(102 cm)*.	Tropical areas of the Americas.	Tubers are used as a source of energy.	Grown in gardens in South America for its decorative foliage.
INDIAN WHEAT, WOOLY (WOOLY PLANTAIN) *Plantago purshii*	Small, silvery annual weed. Small flowers are found in a dense cylindrical spike, resembling a spike of wheat.	Distributed throughout western U.S., except California. Grows in pastures, waste areas, and along roadsides. Prefers dry, open loamy soils.	Considered to be fair- to good-quality forage. Outstanding feed on desert lambing grounds.	
INDIGO (ANIL) *Ingigofera* spp	Tall, leguminous herb. Narrow, pinnate leaves. Slender pods. Small flowers.	Found throughout the tropical regions of the world.	Foliage is sometimes used for hay, grazing, or silage. Some species may be toxic. HAIRY INDIGO is one species commonly fed to livestock.	Indigo has long been used as a source of dye (brilliant blue).
INGA *Inga* spp	A genus of leguminous trees and shrubs.	Tropical regions.	Pods contain an edible pulp.	
IVORY PALM (COROZO NUT, TAGUS PALM) *Phytelephas macrocarpa*	Slow-growing palm. Produces large, white seeds in large, round clusters of spiked fruits at the base of the palm.	Native to South America.	Nut meal provides an excellent source of energy for all types of livestock. However, it is very low in protein.	Seeds formerly used for buttons.
		J		
JACK BEAN (HORSE BEAN) *Canavalia ensiformis*	Fast-growing, erect annual legume. Deep-rooted.	Grown primarily in the southern U.S. Drought-resistant.	Forage. Palatable only when dried. Can be toxic in large amounts. Maximum safe levels in cattle feed is 30% of total feed.	Heat treatment renders the seeds and pods harmless. Sometimes used for food or green manure.
JACKFRUIT *Artocarpus integrafolia*	Tree up to 50 ft *(15 m)* in height. The sweetly acid fruits weigh up to 100 lb *(45.4 kg)* apiece and grow along the trunk of the tree.	Native to Malaysia. Commonly grown in the tropical regions, especially in Brazil.	Leaves and fruit rinds are used as forage for cattle.	The fruit, containing very little protein or fat, is an important food in the eastern tropics.
JAMBOLUM (JAVA PLUM) *Syzygium cuminii*	Glabrous tree with white flowers and purplish red edible berries.	Popular feed in India and Malaysia.	Leaves and seeds are fed to livestock.	
JAPANESE CANE *Saccharum sinense*	Leafy, thin, hard cane.	Grows in tropical regions. Adapted to poor soils and dry climates.	Cultivated for forage in the same manner as Napiergrass, but yields and feeding value are lower.	Cultivated primarily for syrup.
JARAGUA GRASS (THATCHING GRASS) *Hyparrhenia rufa*	Grass that can reach up to 10 ft *(3 m)* in height.	Native to the Old World, but is grown to a limited extent in the tropical Americas. Adapted to soils of low fertility. Does not withstand close, continuous grazing.	Palatable forage before tussocks are formed.	Flowering stands must be mowed or burned.
JERUSALEM ARTICHOKE *Helianthus tuberosus*	Herbaceous perennial of the sunflower family. Produces fleshy rootstocks with tubers.	Native to North America. Cultivated in tropical and subtropical countries.	Fresh immature tops can be used for forage. Tubers are often fed to swine. Also used for human food.	Tubers contain inulin rather than starch; do not store well.
JOB'S TEARS (See ADLAY.)				
JOJOBA (COFFEEBUSH, GOATBERRY, PIGNUT, BUSHNUT, BUCKNUT) *Simmondsia californica*	Bushy, blue-green shrub. Produces nutlike fruit.	Native to Arizona and southern California. Found primarily on sandy or gravelly soils on dry foothills and mesas.	Important browse species of the Southwest. Good to very good winter forage; fair summer forage.	Withstands heavy browsing.
JUJUBE (CHINESE DATE) *Ziziphus jujuba*	Medium size deciduous tree. Single stoned fruit turns dark brown when ripe.	Native to China. Thrives in warm, dry climates.	Both the leaves and fruits can be fed to livestock.	

(Continued)

TABLE 30-1 *(Continued)*

Feedstuff	Description	Place of Origin Geographical Adaptation Cultural Characteristics	Importance/Use	Comments
JUNEBERRY (SERVICEBERRY) *Amelanchier* spp	Shrubs or small trees. Showy flowers have 5 white petals. The succulent fruit is usually purplish or bluish-black.	Native to North America. Found primarily in northern temperate zones.	Several species are used for browse, ranging from poor to good in palatability.	
JUNEGRASS *Koeleria cristata*	Tufted, perennial bunchgrass.	Native to western North America. It also occurs in Europe and Asia. It is one of the most common of the western grasses, thriving on a wide variety of soils.	Fair to good forage. Relished early in the growing season by all types of livestock, but sheep do not graze the stalks after seed maturity. Not a heavy yielding grass because its leaves are short and basal.	Closely resembles bluegrass. Tends to grow in scattered clumps rather than in solid stands.
		K		
KALE *Brassica oleracea*	Hardy biennial plant. Cultivated as an annual for the production of food.	Native to the eastern Mediterranean region. In the northern U.S., it is planted in the spring for summer production. In the southern U.S., it is planted in the fall and harvested throughout the winter.	Green stems and leaves rich in carotene. Dehydrated leaf meal is used for protein and vitamin supplementation—carotene and riboflavin. In poultry rations, it can be substituted for alfalfa meal.	Grown in the northern Pacific Coast region for forage; particularly for sheep.
KANGAROO GRASS (RED OAT-GRASS, BLUEGRASS) *Themeda triandra*	Tufted perennial; blue or green, hairy or smooth leaves.	Extremely variable in habits and size. Grows in clay soils in areas with 24 to 36 in. *(61 to 91 cm)* of rain annually. Easily overgrazed when young. Low carrying capacity.	Poor nutritive value. Becomes unpalatable upon maturity.	Maintained as dominant grass by annual burnings. Does not recover from fire until rain has fallen.
KANGKONG (SWAMP CABBAGE) *Ipomoea reptans*	Variable water and marsh plant; creeping, hollow, water-filled stems; shiny green leaves; funnel-shaped flowers.	Grows in ponds that are fertilized with feces.	Pig and cattle feed in Southeast Asia. Sometimes cultivated for food. Palatable and high-yielding.	Heavy feeding may cause watery feces.
KAPOK *Ceiba pentandra*	Tree with large seed pods with cottonlike down. The seeds are black and about the size of small peas.	Native to tropical America. However, most of it is grown in Southeast Asia.	Seeds yield an oil that is used in edible products and soaps. Ground seeds can constitute 70% of cattle rations. Fiber of the oil cake is used much like hemp meal.	Cultivated primarily for the production of its down which is impervious to water, making it suitable for life jackets and rafts.
KELP (SEAWEED) *Fucaceae* and *Laminariaceae*	Generally fed in dry meal form. High in certain vitamins and minerals, notably iodine.	Digestibility of protein is low.	Used primarily as a source of vitamins and minerals. Can constitute up to 10% of cattle feeds. Sheep can be fed up to 35 g/head/day.	Feeding large amounts may result in iodine toxicity. Can produce a fishy odor in pork. Has a positive effect on egg yolk color in poultry.
KENAF (WILD STOCKROSE, DECCAN HEMP, ABARI HEMP) *Hibiscus cannabinus*	Erect shrub; 3 to 5 ft *(0.9 to 1.5 m)* tall. Hairy leaves and conical fruit with hairy seeds. Stems are primarily cellulose but are also rich in lipid.	Cultivated primarily in India, China, and Thailand. Adapted to a wide variety of conditions.	Good source of energy and protein (leaves may contain up to 30% protein). Young plants are ensiled easily.	Sometimes cultivated for fiber.
KIKUYU GRASS *Pennisetum clandestium*	Vigorous, aggressive perennial grass. Spreads by rhizomes.	Tropical regions. Prefers well-drained soils. Drought-resistant. Withstands heavy grazing. Used for permanent pasture at high elevations with evenly distributed rainfall.	Excellent pasture. Contains high levels of digestible protein through maturity.	Stocking rates should be light until the grass is well established.
KLEINGRASS *Panicum coloratum*	Bunch- and sod-forming panicgrass. Slender stems can grow to 4 ft *(1.2 m)*. Abundant dark green leaves. Fibrous root system. Spreads by tillers.	Introduced from Africa, primarily to South Texas. Adapted to moist, heavy soils.	Used for pasture, hay, and silage. Suitable for cattle only.	Causes *photosensitization* in lambs and some adult sheep. Cattle are not affected.
KOA HAOLE (IPIL-IPIL, WHITE POPINAC, LEAD TREE, WILD TAMARIND, SHACK SHACK) *Leucaena glauca*	Deep-rooted leguminous tree or arborescent shrub. Yellow-white flowers in long-stalked heads.	Native to Mexico. Widely cultivated throughout the tropics, especially on dry wastelands. Intolerant of heavy grazing. For prepared fields, it is best to mix koa haole with a grass cover—commonly guinea grass.	Valuable browse plant; proven excellent for fattening. Young foliage is extremely palatable, rich in protein. Pods and seeds are sometimes used as a concentrate for cattle. Inclusion of 5% dried koa haole may increase egg hatchability.	May cause loss of hair primarily in the tails of horses and asses due to the presence of glucoside mimosine.
KOHLRABI (STEM TURNIP, COLINABO) *Brassica oleracea*	Cultivated for turniplike enlargement of the stem above the ground. Leaves arise from the enlarged stem.	Related to cabbage.	Leaves and *bulbs* (enlarged stems) are relished by sheep. Bulbs are tender and succulent when young but become tough upon maturity.	

(Continued)

TABLE 30-1 *(Continued)*

Feedstuff	Description	Place of Origin Geographical Adaptation Cultural Characteristics	Importance/Use	Comments
KOKKO TREE (LEBBECK TREE) *Albizzia lebbek*	Leguminous tree having a large, spreading crown.	Tropical regions.	Leaves and twigs are commonly used for browse.	Good shade tree.
KORDOFAN PEA *Clitoria ternatea*	Tall, slender, climbing legume with pubescent stems.	Widely distributed throughout the tropical regions of the world.	Commonly used as forage, especially with Sudangrass, sweet sorghum, and sunn hemp.	
KUDZU *Pueraria lobata*	Coarse, woody, perennial, leguminous vine. Forms long runners that root and form crowns at the nodes. Long, broad leaves.	Introduced into the U.S. from Japan. In the U.S., it is commonly found in the Southeast. Adapted to subtropical and humid, warm temperate climates. Drought-resistant. Aboveground parts killed by frost. Cannot withstand trampling.	Equal to alfalfa in nutritive value and palatability. Valuable cover for rough land, providing pasture and erosion control. Makes excellent silage. Hay is rather coarse.	Difficult to establish, but once established, it is long-lived and hardy. A troublesome weed in woodlands. Close grazing will generally kill the stand. Difficult to harvest mechanically because of the viny growth of the plant.
KYASUWA GRASS (CHINA GRASS) *Pennisetum polystachyon*	Annual or perennial grass; up to 6.5 ft (*2 m*) tall.	Grows in areas with long, dry seasons and up to 24 to 36 in. (*61 to 91 cm*) of rain. Resistant to drought and waterlogging.	Valuable as a short season, leafy pasture or hay.	
		L		
LABLAB (EGYPTIAN BEAN, HYACINTH BEAN, BONAVIST BEAN) *Dolichos lablab*	Robust annual or perennial twining legume.	Native to India. Adapted to the same climatic conditions as cowpea.	Cultivated primarily for its seeds, but is also grown as a hay or silage crop. Crop residues can be fed to livestock. Both the pods and the seeds can be used as concentrate feeds.	
LAMB'S QUARTER *Chenopodium album*	Annual weed of 1 to 10 ft (*30 cm to 3 m*) in height.	Introduced from Eurasia. Range plant.	Range forage.	Young leaves are sometimes used in salads or as a cooked vegetable.
LAND GRASS (SWEET GRASS) *Panicum laevifolium*	Annual summer grass. Fairly thick, fibrous stems.	Found on old lands; usually growing rapidly on unploughed land.	Pasture and hay.	Does not dry readily. Hence, it generally makes poor quality hay.
LENTIL (RED DAHL, SPLIT PEA) *Lens culinaris*	Multibranched annual; 1 to 2 ft (*30 to 61 cm*) tall. Produces edible rounded lens-shaped seeds in short pods—2 seeds per pod.	Native to Asia. Extremely important legume of the Near East and India. Widely cultivated in temperate, subtropical, and tropical regions.	Seeds are used primarily for food. The vines are used for forage.	
LIGNIN SULFONATE	A salt of the extract of spent sulfite liquor obtained from the sulfite digestion of wood or abaca.	Produced in two forms: (1) liquid (minimum 50% dry matter by weight); (2) dried (minimum 96% dry matter by weight).	Pelleting aid; binding agent; surfactant; source of metabolizable energy.	
LOCUSTS *Schistocerea gregaria*	Migratory grasshoppers that travel in swarms.	Used in Africa for food and feed.	Dried locust meal is readily consumed by swine and poultry. However, it may impart an off-flavor to meat.	Freshly killed locusts have an objectionable odor; sun-curing alleviates this problem.
LOVEGRASS *Eragrostis* spp	Annual or perennial grasses of various habits. Many lovegrasses produce an abundance of seed.	Over 250 species of lovegrass are known—about 40 are native to the U.S. Produces abundant growth on soils of low fertility.	In India and Australia, it is cultivated as a forage crop. Minimal use as a forage in the U.S. One species, Teff, is grown as a cereal in Ethiopia. Widely used for erosion control.	The most important species in the U.S. are BOER LOVEGRASS, WEEPING LOVEGRASS, LEHMANN LOVEGRASS, and SAND LOVEGRASS.
LOVEROOT (WILD CELERY, LOVAGE, OSHA, WILD PARSLEY, LIGUSTICUM) *Ligusticum* spp	Grow from aromatic, deep-set, stout, woody, and sometimes branched taproots. Roots have a celerylike flavor. Stems are rather stout. Flowers are white and pink. Leaves are large and divided into many segments.	Found primarily in the Northern Hemisphere. Typically, plants of higher elevations, being found throughout the mountains of the western U.S. Prefer moist and fertile soils.	Most species are highly palatable. Sheep are especially fond of loveroots.	Commonly associated with BLUEGRASSES, BROMEGRASSES, COLUMBINE, COW-PARSNIP, FALSE HELLOBORE, SEDGES, LARKSPURS, and WILLOWS.
LUCUNTU GRASS *Ischaemum timorense*	Perennial grass.	Found in areas receiving more than 50 in. (*127 cm*) of rain with short, dry seasons. Prefers heavy soils with a high water table. Dormant during the dry season.	Used as a pasture grass in humid areas.	Once established, it is difficult to eradicate.

(Continued)

TABLE 30-1 *(Continued)*

Feedstuff	Description	Place of Origin Geographical Adaptation Cultural Characteristics	Importance/Use	Comments
		M		
MANGO *Mangifera indica*	Large, spreading evergreen tree. Some varieties produce leaves that smell like turpentine when they are crushed. Produces round or oval fruits weighing up to l lb (*454 g*). In the center of the fruit, there is a large, fibrous, flat seed containing a kernel.	Native to India but cultivated throughout the tropics of the world.	Fruit and kernels can be fed to livestock. Kernels can constitute up to 20% of poultry rations.	Fruits are commonly used for human consumption and are seldom fed to animals.
MANILA TAMARIND *Pithecolobium dulce*	Large, rapidly growing, leguminous tree. Produces spirally twisted red pods with black seeds.	Found in tropical areas.	Pods are very palatable. Can be used for lopping or browse.	Cultivated as a hedge or shade tree.
MANNAGRASS *Glyceria* spp	Usually tall aquatic or marsh perennials.	Confined to marshes and wet lands.	Forage.	TALL MANNAGRASS and FOWL MANNAGRASS are the most valued species found in the U.S.
MANZANITA *Arctostaphylos* spp	Most species are upright shrubs growing up to 7 ft (*2 m*) in height.	Commonly found on dry sites, typically in full sunlight on well drained soils in coniferous forests—especially on the arid chaparrals of the California and Oregon foothills. Can thrive on poor soils.	In the Southwest, manzanitas are considered to be of low value as forage for cattle and as good forage for sheep. Goats eat both the leaves and the bark.	Excellent cover for watersheds. Valuable forage for wildlife.
MARGOSA TREE (NEEM TREE) *Azadirachta indica*	Trunk exudes a tenacious gum.	Found throughout southern and southeastern Asia.	Fruit and leaves can be fed to livestock.	Oil cake is used for fertilizer.
MARIPOSA *Calochortus* spp	Perennials of the lily family. Branched, leafy stems with husk-coated bulblike roots bearing grass-like leaves.	Found in the western U.S., particularly California and Oregon. Grows in dry, open prairies and foothills up to the higher, moist, and shady alpine regions.	When fresh and succulent, palatability is good for sheep and fair for cattle.	
MARVEL (DELHI GRASS) *Dichanthium annulatum*	Coarse, stemmy, tufted perennial grass.	Found occasionally in grasslands on rocky ground. Drought-resistant.	Pasture and hay.	
MEADOWRUE *Thalictrum* spp	Erect, perennial herbs of the buttercup family. Flowers lack petals; male flowers are often very attractive with their numerous colorful stamens.	Most abundant in the temperate regions of the Northern Hemisphere, but a few species are found in the Andes and South Africa. About 13 species are found in the western U.S., mostly in the Rocky Mountains. Prefer rich, moist soils and some shade.	Abundant locally, thereby providing considerable forage. Palatability ranges from zero to poor in cattle and poor to fair in sheep.	Some species, not found in the U.S., have active chemical properties which render them unsuitable for forage.
MELICGRASS *Melica* spp	Perennial grasses. Grow in dense or loose clumps.	About 60 species occur throughout the temperate zones of the world. Seventeen are found in the U.S., primarily in California. Most species prefer the moist, fertile soils of slopes and timber areas.	Because they tend to be widely scattered on rangelands, their forage value is considered to be of secondary importance. Most species are readily consumed by livestock.	ONIONGRASS and SHOWY ONIONGRASS are the two most important species in the U.S.
MESQUITE *Prosopis* spp	Leguminous shrub.	Characteristic of warm, dry subtropical and tropical climates. Abundant in the southern U.S.	Pods provide good forage. Leaves are rarely eaten. Also used for fence posts and fuel.	HONEY MESQUITE, VELVET MESQUITE, and COMMON MESQUITE are the three primary species in the U.S.
MEXICAN GRASS *Ixophorus unisetus*	Tufted, broad-leaved grass. Grows to 6.5 ft (*2 m*) tall.	Grows in wet areas on moist, fertile soils. Growth stops in dry weather.	Palatable, quality forage.	Does not persist when grazed.
MILLET, FOXTAIL *Setaria italica*	Rapidly growing, fine-stemmed, and leafy annual cereal grass. Flowering stems are leafy throughout their length.	First cultivated in China over 4,000 years ago. Grown throughout the Plains and central states. Warm weather crop.	Used for hay, pasture, and green fodder. Seed is used for bird feed.	Feeding value is greatest during the period from flowering to milk stage of the seeds. Many varieties are used for grain only.
MILLET, JAPANESE (BARNYARD GRASS, CHIWAGA) *Echinochloa crusgalli*	Tall, robust, coarse annual grass. Grows to 4 ft (*1.2 m*) or more. Large, broad leaves. Dense, drooping seedheads.	Native to Asia. Grown to a limited extent in the northeastern U.S. Requires moist, well-drained fertile soils. Tolerant of temporary flooding.	High yields of coarse forage. Fair feeding value. Should be strip-grazed.	Superior in cool summers to SUDANGRASS or FOXTAIL MILLET. Troublesome weed in corn and soybeans, and sometimes in alfalfa and clover fields.

(Continued)

TABLE 30-1 *(Continued)*

Feedstuff	Description	Place of Origin Geographical Adaptation Cultural Characteristics	Importance/Use	Comments
MILLET, PEARL (CATTAIL MILLET, BULRUSH MILLET, INDIAN MILLET, HORSE MILLET) *Pennisetum typhoides*	Tall, erect annual grass. Grows to 15 ft (*4.6 m*). Coarse, pithy stems grow in dense clumps. Long, pointed leaves. Plants tiller profusely.	Native to India. Adapted to the southern U.S. Cultivated extensively in India and Africa for grain. Rapid rate of growth and maturity.	Used for pasture and silage for livestock.	May cause a depression in milk fat test in dairy cows.
MIMOSA BUSH *Acacia farnesiana*	Extremely thorny, leguminous bush.	Native to tropical regions.	Leaves and pods provide excellent browse.	Sometimes used for hedges.
MINT *Mentha* spp	Aromatic perennial herbs.	About 15 species can be found in the U.S. Typically, they grow in wet or moist places, most commonly at medium elevations in the West.	Palatability ranges from fair to fairly good for cattle and fair to good for sheep.	SPEARMINT and PEPPERMINT are grown for their aromatic oils. After steam distillation, plant residues are fed to animals.
MOUNTAIN-DANDELIONS *Agoseris* spp	Annual and perennial herbs. Strong, deep taproots. Short-branched crowns.	Over 30 species can be found in the western U.S. Adaptable to a wide variety of conditions. Most abundant on moderately dry flats and meadows.	Moderately grazed by cattle and horses. Sheep relish mountain-dandelions.	
MOUNTAIN-MAHOGANY *Cercocarpus* spp	Evergreen shrubs or small trees of the rose family. Flowers lack petals.	Found in the western U.S. Adapted to dry, interior, and mountainous regions.	Important browse plant providing yearlong forage for cattle, sheep, and goats.	Broad-leaf species are more palatable than the narrow-leaf species. Many species are used for firewood.
MOWRA (MOWRAH, MAHUA) *Bassia latifolia*	Large, deciduous tree with a short trunk, spreading branches, and a large rounded crown.	Native to South Asia.	Cattle consume the leaves, flowers, and fruit. Flowers are used as a vegetable and for the production of alcohol. Fleshy seeds yield edible fat.	Oil cake contains saponin and is therefore toxic.
MUHLY GRASS *Muhlenbergia* spp	Numerous species of perennials and some annual grasses. Small, one-flowered spikelets.	Widely distributed throughout the Americas and Asia. Commonly found in Mexico and the southwestern U.S.	Annual species are poor or worthless forage. Perennial species are highly variable in palatability, ranging from fairly to highly palatable. Palatability declines with maturity.	BUSH MUHLY, SPIKE MUHLY, and MOUNTAIN MUHLY are the most important species in the U.S. Several perennial species are serious weeds.
MULBERRY *Morus* spp	Small trees. Rather large leaves. Fruits resemble blackberries in size and shape.		Browse.	Often planted as ornamentals.
MULE'S-EARS *Wyethia amplexicaulis*	Smooth, coarse, tufted perennial herb. Grows up to 2 ft (*61 cm*). Thick, woody taproot. Leaves resemble the ears of a mule.	Widely distributed throughout the western areas of the U.S. and Canada. Grows on open flats, gentle slopes, and parks up to elevations of 9,000 ft (*2,745 m*). Prefers heavy, compact clay and gumbo soils and also very fine, sandy black topsoil. Resists trampling well.	Flower heads are eaten by all types of browsing livestock. More palatable for sheep than for cattle. Forage becomes tough and unpalatable upon maturity.	Once used by Indians for food.
MULGA *Acacia aneura*	Gregarious trees.	Native to Australia. Adapted to zero rainfall areas.	Important forage in extremely arid regions of Australia.	
MUSTARD *Brassica* spp	Lobed leaves. Yellow flowers. Linear beaked pods.	Most species are native to Europe and western Asia.	Cultivated primarily for its oil and as a condiment. Detoxified mustard meal has been fed successfully to all classes of livestock.	Name applies to numerous species, the most common of which are WHITE MUSTARD, BLACK MUSTARD, INDIAN MUSTARD, and WILD MUSTARD. May cause off-flavors in milk.
		N		
NADI BLUEGRASS (ANTIGUA HAY GRASS, INDIA BLUESTEM) *Dichanthium caricosum*	Vigorous, leafy, tufted perennial grass.	Found in tropical areas. Can withstand drought and heavy grazing.	In some areas, it is used for pasture and hay, but it is generally considered to be rather unpalatable.	
NAL GRASS (GIANT REED) *Arundo donax*	Quick-growing perennial. Woody stems.	Native to the warm regions of the Old World. Found throughout southern U.S. and tropical America.	Only the leaves of this grass are browsed because the stems become highly unpalatable at maturity.	
NAPIERGRASS (ELEPHANTGRASS, UGANDA GRASS) *Pennisetum purpureum*	High-yielding, tall, erect perennial grass with thick stems.	Native to Africa. Grown in the warmest regions of the U.S. Prefers deep soils of moderate to fairly heavy texture. Tolerant of short drought but not waterlogging.	Commonly used for hay, silage, and rotational grazing.	The strain, MERKER GRASS, is fine-stemmed and is more resistant to eyespot disease and drought than the other varieties.

(Continued)

TABLE 30-1 *(Continued)*

Feedstuff	Description	Place of Origin Geographical Adaptation Cultural Characteristics	Importance/Use	Comments
NATALGRASS *Rhynchelytrum roseum*	Short-lived perennial, sometimes considered to be an annual. Slender culms and flat blades.	Adapted to Florida, southern Texas, and southern California. Does not tolerate freezing temperatures. Thrives on well-drained, poor, dry, sandy soils.	As a pasture grass, it is nutritious and palatable. As a hay, it is similar to timothy hay in nutritive value.	Used in areas that are heavily infested with nematodes.
NEEDLEGRASS *Stipa* spp	Feathery-awned grass. Each spikelet has one flower terminating in a needlelike awn.	About 30 species are found in the western U.S. Adapted to temperate climates.	Good source of forage because of its abundance, long-growing period and ability to cure well on the ground. Provides good early spring and winter grazing. However, the needlelike awns can cause sore mouths in livestock.	Some of the most important species are NEEDLE-AND-THREAD GRASS, PORCUPINEGRASS, GREEN NEEDLEGRASS, and CALIFORNIA NEEDLEGRASS.
NIGER PLANT *Guizotia abyssinica*	Herbaceous plant	Native to the tropical regions of Africa.	At the flowering stage, it is used as a forage for sheep. It may also be ensiled. Whole seeds are used for pet bird feed; and the palatable oil cake is similar in feeding value to peanut meal and hulls.	Seeds are used for the production of oil.
NILE GRASS *Acroceras macrum*	Perennial grass. Rhizomes and stolons form a dense sward.	Commonly found on flood plains, river banks, and swamp edges in eastern and southern Africa. Withstands waterlogging but not drought.	Provides very palatable and nutritious dry season grazing.	Gives large yields of hay.
		O		
OAK *Quercus* spp	Very large genus of trees (about 500 species). LIVE OAKS have thick evergreen, persistent leaves. CHESTNUT OAKS have chestnutlike leaves. The third class of oaks is dwarf, shrubby, shinnery oaks (SHIN OAK).	Restricted to the northern hemisphere; primarily in the temperate regions and tropical mountain areas.	Leaves of deciduous oaks are generally more tender and nutritious than those of live oaks. Sheep and goats browse oaks more heavily than do cattle or horses. See ACORNS.	Eastern oaks have too much tannin to be used as browse. Several species of oak are considered to be poisonous, and care should be exercised when allowing livestock to graze in areas where these species abound.
OATGRASS *Dathonia* spp	Perennial bunchgrass.	Widely distributed in warm and temperate regions of the world, especially in South Africa and Australia. Seven species occur in the U.S.—six of them in the western U.S.	All of the western species, except for poverty oatgrass, are fairly palatable. Poverty oatgrass is of no value to livestock.	Some of the important species are CALIFORNIA OATGRASS, PARRY OATGRASS, TIMBER OATGRASS, ONE-SPIKE OATGRASS, and FLAT-STEM OATGRASS.
OATGRASS, TALL *Arrhenatherum elatius*	Hardy, upright perennial bunchgrass with many leaves. Seed heads resemble oats.	Native to Europe. Widely distributed in northern and central U.S. and also in the Pacific Northwest. Well adapted to light textured soils.	Suitable for pasture and hay.	Shorter lived than most bunchgrasses. Frequently seeded with other grasses and legumes.
OCEANSPRAY *Sericotheca discolor*	Attractive symmetrical shrub of the rose family. Spreading branches. Long-lived flowers.	Found in areas from British Columbia to western Montana and California.	Palatability ranges from zero to fair for cattle and poor to fair for sheep.	Leafage and young twigs are more readily eaten in the fall than in the summer or spring.
OKRA (GUMBO, LADY'S FINGER) *Hibiscus esculentus*	Annual. Fruit is a long pod, generally ribbed and spineless in cultivated varieties.	Cultivated as a vegetable in warm climates.	Seeds of mature pods are sometimes used for chicken feed. In some areas, the seeds are extracted for oil and the resulting oil meal can be fed to livestock. The leaves can be used as forage.	Generally similar in culture and exposure to SUMMER SQUASH.
ORCHARDGRASS *Dactylis glomerata*	Long-lived, bunch-type perennial grass. Folded leaf blades and compressed sheaths.	Native to Europe. Cultivated in the U.S. since the colonial period. In the U.S., it is found primarily in the Northeast, eastern Great Plains, western mountain regions, and throughout the irrigated areas of the West. Grows rapidly in cool temperatures. Shade and drought-tolerant. Does not tolerate wet soils or waterlogging.	Well-suited for spring pasture and rotational grazing, often in combination with a legume. Also used extensively for hay and silage. Palatability and nutritive value decreases rapidly as the grass approaches maturity. One of the earliest growing grasses in the spring.	Grown alone, it will yield 1 to 2 tons of hay per acre (*2,240 to 4,480 kg/ha*). With clover or alfalfa, yields can be increased to 2 to 3 tons per acre (*4,480 to 6,720 kg/ha*).

(Continued)

TABLE 30-1 *(Continued)*

Feedstuff	Description	Place of Origin Geographical Adaptation Cultural Characteristics	Importance/Use	Comments
ORCHID-TREE (MOUNTAIN EBONY, NAPOLEON'S HAT) *Bauhinia variegata*	Small, deciduous, leguminous tree with variegated lavender orchid-like flowers and cleft leaves that resemble the hoof of an animal.	Grows readily in all tropical countries.	Important browse of the tropics.	
OWLCLOVER *Orthocarpus* spp	Genus of annual and perennial weeds. Considerable variation in shape, size, color of flowers, and margins of leaves among the various species.	Found primarily in western North America. Most species occur in California.	Low in palatability. Rated as worthless to poor forage for cattle, fair for sheep.	YELLOW OWLCLOVER is the most common and widespread species. Other common species include BUTTER-AND-EGGS, PURPLE OWLCLOVER, and PURPLE-WHITE OWLCLOVER.
		P		
PAINTBRUSH (PAINTEDCUP) *Castilleja* spp	Annual or perennial herbs. Most species are partially parasitic on the roots of other plants. Striking flowers appear in terminal leafy spikes. Stems are usually tufted and are either branched or unbranched.	Found primarily in western North America. Grow in a wide range of elevations, from low elevations to above the timberline.	Variable in palatability, depending on species. Some species are worthless while others provide good forage.	SNAPDRAGONS are the most widely known example of paintbrushes.
PALMO MIDDLINGS	Wheat middlings and 7 to 10% palm oil.	Of the same or slightly less feeding value as standard wheat middlings.	May be fed to swine or other livestock.	
PALMYRA PALM *Borassus flabellifer*	Tall, erect palm; up to 98 ft *(30 m)*. Fan-shaped leaves. Black stem.	Native to Africa but widely cultivated in India.	Fruit and leaves can be fed to livestock.	Cultivated for human use for its fruit and sugar-rich sap that is used for sugar and wine.
PAMPASGRASS *Cortaderia selloana*	Dioecious perennial reed.	Warm regions of the New World.	Can be used for supplementary dry-land pasture for cattle.	Cultivated primarily as an ornamental.
PANGOLAGRASS *Digitaria decumbens*	Creeping perennial grass. Grows to a height of 4 ft *(1.2 m)*. Produces many semidecumbent surface runners that form roots at the joints or nodes.	Native to South Africa. Adapted to subtropical and tropical areas. In the U.S., it is grown in the Gulf Coast states and California. Adapted to fertile, moist, well-drained soils. Somewhat drought-resistant. Withstands heavy grazing.	One of the most important forage grasses in regions to which it is adapted. Highly palatable. Superior in dry matter and protein yields to Coastal Bermudagrass, bahiagrass, and guineagrass.	Should be grazed rotationally.
PANICGRASS, BLUE *Panicum antidotale*	Deep rooted, erect glabrous perennial grass. Forms tough crowns via short, thick rhizomes.	Native to India. Used in the southwestern U.S. for dry-land and irrigated pastures. Adapted to heavy loam or dark, clay soils.	Useful for either hay or pasture. Palatability decreases upon maturing.	Intolerant of close, intensive grazing or cutting.
PAPAYA (PAW PAW, PAPAY, PAPAW, LECHOSA, TREE MELON) *Carica papaya*	Small, fast-growing branchless tree; up to 25 ft *(8 m)* tall. The large stem is hollow between the nodes. Very large compound leaves. Very large melonlike fruits, averaging 3 lb *(1.4 kg)* but weighing up to 20 lb *(9 kg)*. Fruit rind is thin, smooth, and tender. The pulp is mild flavored.	Native to Central America. Widely distributed throughout the tropical regions of the world.	Fruits, when properly supplemented, make a good feed for swine. Leaves have also been fed to livestock.	Fruit contains the enzyme papain—sometimes used as a meat tenderizer. Where grown, the fruit is a favorite human food.
PAPER-BARK THORN *Acacia sieberiana*	Light-colored leguminous tree up to 49 ft *(15 m)* tall. Flat-topped crown. Long, hairy leaves have straight white thorns at their base. Straight, often hairy, pods. Hairy branches and leaves.	Tropical areas.	Pods and young shoots are readily eaten.	
PARAGRASS (MALOJILLO) *Panicum purpurascens*	Coarse, creeping, sod-forming perennial grass. Roots readily at the nodes. The nodes, leaves, and leaf sheaths are hairy. Sparse seed producer.	Native to Africa. Commonly cultivated in most tropical countries. Grown along the Gulf Coast region of the U.S. Adapted to moist soils and can withstand waterlogging.	Commonly used for hay and pasture. Although coarse in texture, paragrass hay is of good quality if cut before the stems become woody. Should be used for pasture only after the grass is well-sodded and is at least 1½ ft *(46 cm)* high.	Should be grazed rotationally. Occasional disking may stimulate growth.

(Continued)

TABLE 30-1 *(Continued)*

Feedstuff	Description	Place of Origin Geographical Adaptation Cultural Characteristics	Importance/Use	Comments
PARSNIP *Pastinaca sativa*	Biennial, commonly grown as an annual. Root is large (up to 3 in. [*7.6 cm*] in diameter) and tapering from 6 to 15 in. (*15 to 38 cm*) long, depending on the variety.	Believed to be native to the Mediterranean region. Cultivated in the same manner as carrots, but the growing season is longer.	In the U.S., the roots are cultivated primarily for human consumption, but on rare occasions they are fed to livestock.	Grows wild in wastelands and pastures.
PASPALUM (See BAHIAGRASS, DALLISGRASS, VASEY GRASS.)				
PEAVINE *Lathyrus* spp	Smooth, weak stemmed, trailing or climbing plants.	Peavines are well represented in the western U.S. Found on moist, rich soils and on dry scablands, in open exposures, and in the shade of coniferous and broadleaf timber—being most typical of open aspen areas.	Palatability varies considerably. Trailing or climbing species with tendrils range from fair to good forage for cattle, sheep, and goats. The erect species range from poor to fair forage. All species are used for forage, primarily in the summer and fall. After the first heavy frosts, most species dry up and disappear. Can also be used for silage.	The best known peavine is the SWEET PEA. Can withstand heavy grazing.
PECAN *Carya illinoinensis*	Large tree; up to 100 ft (*30 m*) in height. Large compound leaves have a dozen or more long, oval leaflets. Oval fruits have thin, hard, woody shells.	Native to the lower Mississippi Valley westward through Texas and northern Mexico.	Shells, a by-product of pecan production, can be fed to ruminants as a source of fiber.	Improved varieties are widely cultivated.
PEEPUL (BO-TREE) *Ficus religiosa*	Large glabrous tree with shiny, broad-based, pointed leaves.	Found in tropical regions.	The leaves and branches are commonly topped for forage.	Commonly used as an avenue tree.
PENNYCRESS *Thlaspialpestre* spp	Annual and perennial herbs. Alternate, usually clasping stem leaves broadened at the base into ear-shaped appendages. Two-celled pods contain 2 to 8 seeds in each cell.	Widely distributed in temperate, Arctic, and alpine regions, primarily in the Northern Hemisphere, but can also be found in South America, South Africa, and Australia. Grows chiefly on well-drained soils in mountainous regions.	The species native to the western U.S. are fair to fairly good sheep feed and poor to fair cattle feed.	May impart off-flavors to milk. Considered to be a pest in many cultivated fields.
PENTSTEMON *Pentstemon* spp	Perennial plants having opposite leaves and showy, two lipped flowers. Strong root systems.	Most species are native to North America, most of which are found in the western U.S.	Highly variable in palatability—from worthless to very good. Large, shrubby species vary from poor to fair forage for sheep. Low, shrubby species are of negligible forage value. Herbaceous species with thick, glaucous foliage is worthless as forage. Herbaceous with smooth, green foliage rates a fair forage for cattle.	Many species are cultivated as ornamentals. Abundant in overgrazed areas.
PEONY, BROWNS (WILD PEONY) *Paeonia brownii*	Robust, somewhat succulent perennial herb of the buttercup family. Large, thick, leathery flowers. Stems are erect when young but droop upon maturing.	Native to North America. Grows primarily in the West Coast states. Extremely varied in habit adaptation.	The palatability rating is fair for sheep and zero for cattle.	Early maturing, thereby limiting the time when it is palatable.
PERILLA *Perilla* spp	Genus of Asiatic mints. Branched annual up to 3 ft (*0.9 m*) tall.	Most perilla is produced in Asia.	Oil meal is used for feeding sheep and cattle. Fiber is poorly digested.	
PHILLIPESARA (INDIAN MOTH BEAN) *Phaseolus aconitifolius*	Creeping, herbaceous, annual legume.	Native to India and Pakistan. Drought-resistant.	Used as a hot season forage. Produces palatable pasture and hay.	Useful in mixtures of lablab, pigeonpea and Sudangrass.
PHLOX *Phlox* spp	Very showy plants, often growing in masses. Reproduces by seed or creeping rootstocks. Clustered flowers.	Confined to North America. Most of the western phloxes grow on open sites, often on dry and gravelly or rocky soils.	Flowers are relished by sheep. Species having small and prickly leaves are often avoided. Palatability of species with larger and more tender leaves is fair for sheep, cattle, and horses.	
PIGTAIL GRASS *Ctenium concinnum*	Perennial grass with dense, broad tufts. Hard, narrow leaves. Single spikes twist spirally at maturity.	Grows on seasonally wet grasslands.	Forage.	

(Continued)

TABLE 30-1 *(Continued)*

Feedstuff	Description	Place of Origin Geographical Adaptation Cultural Characteristics	Importance/Use	Comments
PINE *Pinus sylvestris*	Tree having needlelike, long leaves and dark bark.	Widely grown in temperate climates and the cooler areas of subtropical countries for timber and pulpwood.	Excessive browsing of yellow pine (*P. ponderosa*) can cause abortion. In times of famine, leaves can be fed to ruminants, if intake is carefully monitored. One lb (*454 g*) of leaves per day can be fed safely to sheep.	
PINEGRASS *Calamagrostis rubescens*	Tufted reedgrass; up to 40 in. (*102 cm*) tall. Spreads by creeping rhizomes.	Widely distributed from British Columbia to Manitoba southward to northern California and central Colorado. Ranges from sea level to about 10,000 ft (*3,050 m*), but is commonly found at medium levels beneath stands of ponderosa pine. Drought-resistant.	Due to its abundance in some localities, it can be considered to be an important range forage. Depending on the particular locality, it ranges from worthless to fair forage for sheep and from poor to good forage for cattle and horses.	It is most readily consumed in the spring when it is young and tender. It becomes extremely tough upon maturing. Can withstand heavy trampling.
PINHOLE GRASS (SWEET PITTED GRASS) *Brothriochloa insculpta*	Sweet-smelling grass. Grows to 43 in. (*109 cm*) in height. Develops stolons when well-grazed.	Well-adapted to the tropical black soils of Africa and Asia.	Valuable in natural pasture.	
PLANTAIN *Plantago* spp	Weeds characterized by small, bracted, greenish flowers on naked stems. The seed pods of most species are two-celled.	Widely distributed at low elevations throughout the U.S. Found primarily in waste places, maritime, or alpine localities throughout the world, especially in temperate regions.	Some species are fairly abundant in the southwestern U.S. and are fairly palatable. Consequently, they provide important spring forage. However, plantain is very low yielding.	Some of the annual species are referred to as INDIANWHEAT.
POKE (POKEWEED, SCOKE, GARGET) *Phytolacca americana*	Large, taprooted perennial. Strong-growing top may reach 10 ft (*3 m*) or more in height.	Native throughout eastern North America. Found in rich lowlands.	Tender shoots can be fed to livestock.	Both the roots and berries are poisonous, but the roots are the more toxic of the two.
POLEMONIUM, WHITE (WHITE SKUNKLEAF, SKUNKWEED) *Polemonium albiflorum*	Leafy, relatively robust, perennial herb of the phlox family. May grow to 40 in. (*102 cm*) tall. Herbage is glandular-hairy. Faint skunklike odor.	Largely confined to the aspen and spruce belts of Idaho, Wyoming, Utah, and Nevada. Prefers moist to moderately dry, sometimes gravelly soils.	Palatability ranges from worthless to fair in cattle and fair to good in sheep.	
POLOXALINE	Nonionic polymer of polyoxypropylene-polyoxyethylene glycol.		Surfactant used in the manufacture of flakes from feed grains and also as a bloat preventative.	
POLYETHYLENE ROUGHAGE REPLACEMENT	Pellets formed by the catalytic polymerization of ethylene.		Feed additive. Source of roughage to promote healthy ruminal activity.	
POPLAR (See ASPEN.)				
POPPY (OPIUM) *Papaver somniferum*	Rigid, spiny-leaved annual. Urn-shaped fruits. Very fine seeds. Large four-part white to purple flowers.	Indigenous to Asia and cultivated in many areas throughout the world.	The oil meal is sometimes fed to livestock as a protein feed; not very palatable and low in energy.	Rarely used in American markets. Oil meal is generally not fed to young or breeding stock because of its mildly narcotic properties.
PORCUPINEGRASS *Stipa spartea*	One of the needlegrasses. Grows to a height of 3½ ft (*1 m*). Awns become sharp and hard at maturity.	Grows on the prairies of the Midwest.	Fairly palatable forage that is high in protein.	Animals may develop sore mouths from eating mature porcupinegrass. Seeds caught in the wool of sheep work downward into the flesh, thereupon decreasing the quality of the carcass.
PRICKLY PEAR (See CACTUS, PRICKLY PEAR.)				
PUERO (TROPICAL KUDZU) *Pueraria phaseoloides*	Vigorous, densely growing leguminous vine. Deep root system.	Cultivated in tropical regions. Prefers high rainfall and fertile clay soils. Tolerant of short dry spells and shade.	Palatable, high-yielding forage. Combines well with molassesgrass.	Young plants cannot withstand trampling or heavy grazing.
PUMPKIN (CALABAZA) *Cucurbita* spp	Edible fruits of trailing annual plants having large 5-pointed leaves. Fruits are highly variable in size, color, shape and weight. They have a moderately hard rind with a thick, edible flesh surrounding a central seed cavity.	Native to tropical and subtropical climates.	Seeds are very high in oil and can be fed to livestock, especially swine. Hulls and oil meal can be fed to livestock.	Livestock fed a ration too high in pumpkin seeds will develop indigestion.
PUPAE, SILKWORM *Bombyx mori*	By-product of the silk industry.	Best utilized when the highly unsaturated fatty acids are removed.	Can be used to replace fish meal in rations for nonruminants. Also used as a fish food in China.	True protein is actually only 75% of the crude protein content.

(Continued)

TABLE 30-1 *(Continued)*

Feedstuff	Description	Place of Origin Geographical Adaptation Cultural Characteristics	Importance/Use	Comments
		Q		
QUACKGRASS *Agropyron repens*	Troublesome weed grass. Spreads by seeds and creeping rhizomes. Stems grow up to 3 ft *(0.9 m)*. Thin, flat leaf blades.	Native to Europe. Widely distributed throughout most temperate regions of the world.	Good forage qualities. Used for pasture, hay, and silage.	Good soil binder for embankments.
QUAKING GRASS *Briza* spp	Low annual or perennial grasses. Erect culms. Flat blades. Open, showy panicles.	Introduced into the U.S. from Europe.	Of little agricultural importance in the U.S. LITTLE QUAKING GRASS is sometimes used for spring forage in some parts of California.	Three species are found in the U.S.: (1) BIG QUAKING GRASS, (2) LITTLE QUAKING GRASS, and (3) *Briza media*.
		R		
RABBITBRUSH (RABBITSAGE, YELLOWBRUSH, RAYLESS-GOLDENROD) *Chrysothamnus* spp	Shrubby plants having narrow leaves and an abundance of showy yellow flower clusters.	All species are confined to western North America. Most species grow at comparatively low elevations on fairly deep, heavy soils.	Most species are of little forage value. However, LANCELEAF RABBITBRUSH and TWISTLEAF RABBITBRUSH rate as fair to good forage. SMALL RABBITBRUSH is abundant in desert areas and supplies considerable winter forage for sheep.	Indians used to chew the bark as a form of crude chewing gum.
RABBITFOOT GRASS *Polypogon monspeliensis*	Annual grass. Can grow to 3 ft *(0.9 m)*.	Introduced from Europe. In the U.S., it is confined primarily to the western coastal states.	Palatable to livestock. Sometimes sufficiently abundant in meadows to be of importance.	A common weed in the western U.S.
RAMIE (CHINA GRASS, RHEA) *Boehmeria nivea*	Nettlelike, shrubby plant. Slender stems. Cordate leaves.	Native to China. In the U.S., it is grown in Florida, Texas, Louisiana, and California.	Sometimes used as a source of forage. Leaves are low in fiber and high in protein, carotene, and minerals and have proven to be of particular value to poultry and swine. Can be fed fresh, dried (leaf meal), or as silage (with molasses). Five percent leaf meal will fulfill the requirements for vitamins A and D in poultry. Half of the soybean meal in swine rations can be replaced by young ramie.	Cultivated primarily as a fiber crop.
RATANY, RANGE (PURPLE HEATHER) *Krameria glandulosa*	Low, bushy, diffusely branched shrub, up to 2 ft *(61 cm)* high. Bluish-green twigs and bluish leaves. Sweet-scented flowers. Spiny burlike fruit.	Found from western Texas to southern Utah and southern California. Common at elevations between 2,000 and 4,500 ft *(610 and 1,373 m)*. Grows on dry, hot foothills.	Ranks as fair to fairly good forage.	Good emergency forage. Goats browse it freely.
RAVENNA GRASS *Erianthus ravennae*	Coarse, tall, perennial, reedlike grass with dense silky panicles.	Native to Europe.	Forage.	Often cultivated as an ornamental.
RED-AND-YELLOW PEA *Lotus wrightii*	Small perennial leguminous herb. Dark green, bushy, fine-leaved herbage. Bright yellow and red flowers. Deep taproot.	Scattered throughout the southwestern U.S. Grows on sandy, gravelly, or clay soils in dry parks, open ridges, or open stands of timber. Can tolerate drought and abuse.	Where plentiful, it is considered valuable as a range plant. Palatability is high for all livestock, especially sheep.	Particularly appetizing in the latter part of the growing season.
RED BUSH WILLOW *Combretum apiculatum*	Shrub or tree up to 20 ft *(6 m)* tall. Yellow, oval leaves are broad and leathery. Flowers occur in short spikes. Reddish-brown fruit has 4 wings.	Found in tropical areas.	Forage.	
RED OAT (ALGERIAN OAT) *Avena byzantina*	Annual grass of the oat genus.	Adapted to warmer climates than common oats and more resistant to drought.	Sometimes used for forage and/or grain.	
RED SANDALWOOD TREE *Adenanthera pavonia*	Medium-sized tree of up to 50 ft *(15.3 m)* that bears bright red seeds.	Widely distributed throughout the tropics of the world.	Browse.	The red seeds are sometimes used for food and ornaments.

(Continued)

TABLE 30-1 *(Continued)*

Feedstuff	Description	Place of Origin Geographical Adaptation Cultural Characteristics	Importance/Use	Comments
REDTOP *Agrostis alba*	Perennial grass. Stems are both upright and creeping. Flat, sharp-pointed leaves. Forms a less dense sod than Kentucky bluegrass.	Believed to be native to Europe, but some forms of it are native to North America. Distribution in the U.S. is very similar to that of Kentucky bluegrass in the East. However, it can be grown in most areas of the U.S., except in the drier regions and the exteme South. Thrives on very acid soils, poor clay soils of low fertility, and on poorly drained land. Matures from mid-June to mid-August, depending on the latitude.	Slightly less palatable than Kentucky bluegrass. Prior to 1940, redtop was the second most important pasture grass in the U.S. It has now declined to minor significance. If grown on good soil, it is of satisfactory quality as a forage. However, of the common northern pasture grasses, redtop is among the lowest in palatability.	Used mainly on poor or wet land for hay or pasture, usually mixed with alsike clover or annual lespedeza.
REEDGRASS *Calamagrostis* spp	Difficult to identify due to the great variation among the various species. Most species are tall, robust perennials with open or narrow panicles.	Distributed throughout the cool and temperate regions of the world. Of the 26 species found in the U.S., most are found in the western mountains. Many species grow in wet meadows, bogs, marshy areas, stream banks, and moist, open woods. Others are typically dry-land plants.	Palatability varies according to species. Several are highly palatable. Others, especially the dry-land species, are considered to be of inferior quality. As a group, however, reedgrasses are rated as fair to good forage.	BLUEJOINT and PINEGRASS are the most important reedgrasses. The other species are regarded as secondary forage plants.
REINDEER MOSS *Cladonia rangiferna*	Gray, erect, multibranched moss.	Found in arctic and northern temperate regions.	Major source of feed for caribou and reindeer.	Sometimes consumed by humans.
RESCUEGRASS *Bromus catharticus*	Short-lived perennial bromegrass.	Native to Argentina. Adapted to humid areas with mild winters.	On good soils, rescuegrass provides a good amount of palatable forage.	Furnishes good winter pasture in the southern U.S.
RHODESGRASS *Chloris gayana*	Fine-stemmed, leafy, perennial grass. Grows up to 3 ft (*0.9 m*). Produces abundant seeds and can also spread by running stolons that can reach 6 ft (*2 m*) in length.	Native to South Africa. Adapted in the U.S. to the Gulf Coast and southwestern states. Relatively drought-resistant, but ample moisture is required for good production. Grows well on sandy soils, well-drained peat soils, and alkaline soils.	In tropical countries, it is noted as one of the best hay grasses. If grown under favorable conditions, high yields of palatable forage can be obtained.	With irrigation, it has succeeded on soil too alkaline for other crops. Withstands heavy trampling.
RIB GRASS *Plantago lanceolata*	Perennial herb.	Grows in temperate climates.	Leaves are used for food and feed.	
ROSE *Rosa* spp	Erect, trailing, or climbing shrubs, usually with prickly stems. Leaflets are saw-toothed. Colored, fragrant flowers occur singly or in clusters at the ends of the branchlets. Petals are rounded or heart-shaped.	Originally native to the cooler and temperate climates of the Northern Hemisphere. Today, roses are cultivated throughout the world. Wild roses are commonly found on the western ranges of the U.S. growing in a wide variety of conditions—from hot, dry areas to cool, moist places.	As a group, roses are rated as poor to fair forage for cattle and fairly good for sheep.	CHEROKEE ROSE is a weedy pest in pastures in southeastern U.S.
ROSE-OF-CHINA (CHINESE HIBISCUS) *Hibiscus rosa-sinensis*	Shrub, but can sometimes grow as a tree up to 35 ft (*10.7 m*). Profuse, large, ornate flowers.	Commonly found in the tropics and subtropics.	Sometimes browsed by cattle.	Generally grown for hedges.
ROUGH PEA (SINGLETARY PEA, CALEY PEA, WILD WINTER PEA) *Lathyrus hirsutus*	Winter annual legume. It generally reseeds well by means of a hard seed.	Introduced from the Mediterranean region to the southern third of the U.S.	Used as cover and pasture crops.	Should be used for hay and/or pasture before the ripening of the seeds.
ROYAL PALM *Roystonea regia*	Tall palm.	Grows in Central America and Cuba.	Fruits are sometimes fed to swine. When fresh fruits are fed to swine, they will eat the outer fleshy part and leave the kernel. Once the kernel has dried, swine will also eat it.	

(Continued)

TABLE 30-1 *(Continued)*

Feedstuff	Description	Place of Origin Geographical Adaptation Cultural Characteristics	Importance/Use	Comments
RUSH *Juncus* spp	Grasslike, perennial plants. Unbranched, hairless stems. Stiff leaf blades. Small, homely, clustered flowers. One- to three-celled fruit capsule contains small, cinnamon-colored seeds. Often with tailed appendages.	Generally found in swamps or other wet places.	Palatability of the more tender species ranks as fair to fairly good for cattle and sheep.	Fine-leaved meadow types of rushes generally provide the best forage.
RUSSIAN-THISTLE *Salsola kali*	Annual weed that can grow to 4 ft (*1.2 m*) in height and form a dense bushlike plant ranging from 2 to 6 ft (*61 cm to 2 m*) in diameter.	Native to the U.S.S.R. Widely distributed over the western U.S. and Canada. Salt-resistant and drought-resistant.	On early spring ranges, Russian-thistle rates as fair forage for all range livestock. However, once the plant matures and forms thorns, it is useless. If harvested before maturity, it can be ensiled.	Russian-thistle is a host of the sugar beet leafhopper which carries a virus that causes curly-top of sugar beets and "blight" of beans, tomatoes, and other plants.
RUTABAGA (SWEDE, SWEDISH TURNIP, TURNIP-ROOTED CABBAGE, RUSSIAN TURNIP, LAURENTIAN TURNIP) *Brassica napus*	Biennial plant cultivated as an annual. Glabrous leaves. Yellow-fleshed roots.	Grown primarily in northern Eurasia and northern North America. Extensively grown in Europe. Prefers cool climate.	Used as a human food and livestock feed. Sheep prefer it to all other roots.	More nutritious than turnips. Can produce off-flavors in milk.
RYEGRASS *Lolium* spp	Bunchgrasses. No creeping habit of growth. Perennial ryegrasses grow to 3 ft (*0.9 m*) in height with erect culms. Awns are usually absent. Annual ryegrasses behave like short-lived perennials. Taller than perennial ryegrasses, sometimes growing to 4¼ ft (*1.3 m*). Typically, awns are present.	Commonly used for forage in Australia, New Zealand, the British Isles, and the temperate areas of western Europe and the U.S. Can be grown on a wide variety of soils. Tolerant of short periods of flooding. Less winter-hardy than other grasses; *e.g.,* timothy and orchardgrass.	Commonly used for hay, pasture, silage, soil conservation, and turf. Can be seeded alone or with other grasses or legumes. Hay is of excellent quality for horses. In the southeastern states, ryegrass is used for fall, winter, and spring pasture. Used for overseeding lawns in warm areas to produce green growth in the winter.	The common name ryegrass is generally applied to the species ITALIAN RYEGRASS and PERENNIAL RYEGRASS.
		S		
SABI GRASS *Urochloa mosambicensis*	Perennial grass. Grows to 4 ft (*1.2 m*) in height.	Tropical grass. Drought-resistant.	Provides good palatable grazing, even when dry.	
SACATON *Sporobolus wrightii*	Extremely robust perennial bunchgrass. Leaf stalks can grow to 6 (*2 m*) and sometimes 8 ft (*2.4 m*) tall. Should not be confused with its relative, alkali sacaton, which is smaller and less coarse.	Occurs from Arizona to western Texas and southward into Mexico. Found primarily on low alluvial flats, bottomlands, and channels subject to flooding. Will not tolerate alkali soils. Drought-resistant.	Young shoots are relished by cattle and horses. Much of the herbage cures well and provides fairly good winter forage. When properly prepared, sacaton makes good, nutritious hay, especially for horses.	Ranchers sometimes burn off the dead growth in late winter to promote new growth.
SAGEBRUSH *Artemisia* spp	Woody shrubs of the Mayweed tribe.	Sagebrushes occur in a wide variety of climates on a diverse range of soils. They compete with grasses for moisture and nutrients and, consequently, lower the forage value of range areas.	In the U.S., sagebrushes are of the highest forage value in the southwest. Palatability ranges from worthless to very good. In tall stands, sagebrush is worthless as a forage.	Abundant sagebrush growth is often indicative of overgrazing. FRINGED SAGEBRUSH (western North America, northern Asia, and Europe) and BIG SAGEBRUSH (western North America) are species most commonly used for browse.
SAGO PALM *Metroxylon sagu*	Palm tree up to 50 ft (*15.3 m*) high. Large pinnate leaves. Creeping or ascending stout stems.	Grows well in freshwater swamps.	When the palm is 12 years old, the trunk is cut up and the soft material in the center is harvested to extract the starch, called sago meal. This meal is very digestible and can be fed to all classes of livestock.	High levels of sago meal—over 50% of the ration in swine, 25% in poultry—may decrease production.
SAINFOIN *Onobrychio viriaefolia*	Erect, deep-rooted, perennial legume. Stout stems originate from a crown. Leaves are oddly pinnate with 13 to 21 leaflets per leaf. Flowers are generally pink, though sometimes white. A single kidney-shaped bean is produced in a bean-shaped, bilaterally compressed pod.	Found in most of southern Europe eastward to Lake Baykal in Siberia. Very little is grown in the U.S. However, it is well adapted to the dry calcareous soils of the northern Rocky Mountain region. Long-lived on dry land but short-lived on irrigated land. Grows well on low phosphorus soils.	Shows promise as a pasture and hay crop in the northern plains of the U.S. About equal to alfalfa in feeding value.	Sainfoin is nonbloating.

(Continued)

TABLE 30-1 *(Continued)*

Feedstuff	Description	Place of Origin Geographical Adaptation Cultural Characteristics	Importance/Use	Comments
SAINT AUGUSTINE GRASS *Stenotaphrum secundatum*	Broad-leaved, creeping, perennial, sod-forming grass. Stolons with long intervals. Short, leafy, and flat branches.	Native to the West Indies, Australia, and southern Mexico. In the U.S., it is found along the southern Atlantic and Gulf Coast regions from South Carolina to Texas. Adapted to almost all soil types, especially muck soils. Thrives in partial shade and can withstand salt spray.	Generally not considered to be an important pasture grass but it can be of some local importance. Considered to be the most dependable pasture grass for the organic soils of southern Florida.	Well suited and primarily used for lawns. Forms a dense sod that can withstand heavy trampling.
SALAL *Gaultheria shallon*	Low, spreading shrub. Grows up to 6 ft (*2 m*) or more. Can form dense impenetrable thickets.	Very common undershrub of the forests of the northwestern U.S. Occurs on the Pacific slope west of the Cascade and Sierra Nevada mountains from Alaska to central California. Thrives on acid soils in forested areas.	Generally considered as worthless to low value as forage for livestock. However, in some areas of local abundance, it may be grazed extensively.	The WINTERGREENS, closely related to salal, are of limited forage value.
SALMONBERRY *Rubus spectabilis*	Erect, vigorous shrub closely related to raspberries and blackberries. Large, juicy salmon-colored fruit.	Found in the northwestern regions of the continental U.S., western Canada, and Alaska. Grows along drainage lines in moist canyon bottoms, streams, and moist flats and in swamps. Prefers rich, moist soils of alluvial bottomlands.	Important browse for game animals. However, it is seldom used as browse for domestic animals.	Fruit is highly prized for food by the Alaskan Indians.
SALTBUSH *Atriplex* spp	Some species are shrubs, others are herbs.	Well adapted to the arid regions of the western U.S. Generally tolerant of drought and alkali conditions.	Some species are unsurpassed as browse plants with regards to volume of browse and usefulness as winter pasture.	FOURWING SALTBUSH and SHADSCALE SALTBUSH are the two most important species in the western U.S. Several species are considered to be weeds. Plants of this genus serve as a host for the sugar beet leafhopper.
SALTGRASS *Distichlis* spp	Dioecious, low perennial grass. Extensive creeping, scaly rhizomes.	Four species are found in the U.S. Occupy alkaline or saline areas.	Generally considered to be of little forage value, except where better grasses are not available.	High in salt content.
SAL TREE *Shorea robusta*	Tree with light brown, closed grain, hard wood.	Native to India.	Leaves can be used for forage.	One of the most important commercial timber trees of India. Seeds are used for fat extraction.
SAMAN (RAIN TREE, MONKEY POD, COW TAMARIND) *Samanea saman*	Large, lofty canopied leguminous tree. Large symmetrical crown. Fernlike leaves close at night permitting rain to fall to the grass beneath the tree.	Found in tropical regions.	Leaves and pods are relished by livestock.	
SANDBUR *Cenchrus* spp	Low, branching annuals, or sometimes perennials. Deciduous burs.	Widespread from the northeastern U.S. to Central America. Prefers sandy soils.	Excellent forage before burs are formed. Spiny burs can get in the wool of sheep, and may injure both humans and animals.	Troublesome weed in lawns, gardens, and cultivated fields.
SANDWORT *Arenaria* spp	Annual or perennial weeds. Small white flowers are borne in open or contracted clusters or solitary in the leaf axils.	Widely distributed throughout the western U.S. Commonly found in dry, sandy or gravelly soils and on moderately moist, rich loams.	On the average, palatability ranges from poor to fair but in many areas sandwort is considered to be worthless as forage.	
SAUSAGE TREE *Kigelia pinnata*	Lofty, wide-spreading tree. Grows up to 50 ft (*15.3 m*). Cylindrical, sausagelike fruits can weigh up to 13 lb (*6 kg*).	Native to Africa.	Fruits can be used for browse.	
SAVANNAH GRASS (FLAT JOINT GRASS, TROPICAL CARPET GRASS) *Axoponus compressus*	Shallow rooted perennial. Grows to 32 in. (*81 cm*).	Found in tropical regions. Withstands poor conditioning and seasonal flooding in the dry season. Susceptible to insect attack.	Palatable pasture grass.	Can be heavily grazed, but is outyielded by some other grasses.

(Continued)

TABLE 30-1 *(Continued)*

Feedstuff	Description	Place of Origin Geographical Adaptation Cultural Characteristics	Importance/Use	Comments
SEDGE *Carex* spp	One of the world's largest genuses of flowering plants. Perennials propagated by rootstocks. Solid, unjointed, usually three-angled stems. Mostly basal, closed sheaths. Flat, long, thin grasslike leaves. Most species range from 8 to 20 in. *(20 to 51 cm)* in height, but some species can grow to 3 ft *(0.9 m)*.	Adapted to a wide variety of soils that are moderately fertile and that receive favorable moisture. Prefer neutral or acid soils. Most sedges are adapted to full sunlight.	The small group of slope- and timber-inhabiting species are usually most valuable for forage in the spring and fall. Small-leaved, low, dry-land species are often an important range forage, being most palatable in the spring. Moist-meadow species are the most palatable sedges providing fine, green, tender foliage. Wet-site species are good forage for cattle but grow on land too wet for sheep. Robust, large-leaved, wet-site sedges are of low palatablility for sheep, fair for cattle.	OVALHEAD SEDGE (western U.S.) and THREADLEAF SEDGE (Great Plains and western U.S.) are the two most important sedges.
SEMAPHORE-GRASS *Pleuropogon* spp	Soft annual or perennial grasses. Single culms. Flat blades. Loose racemes.	Found primarily in the western U.S.	Palatable forage. However, there is too little semaphore-grass to be of real economic value.	
SERRADELLA *Ornithopus sativus*	Small, semiviny, annual leguminous herb.	Adapted to moist soils in cooler climates.	Used for forage.	
SERVICEBERRY (JUNEBERRY, SHADBLOW, SHADBUSH, SARVISBERRY) *Amelanchier* spp	Shrubs or trees, up to 20 ft *(6 m)*, but generally smaller. Fruits produced in small, rather open clusters or racemes, which are sweet when ripe. Simple, toothed or entire leaves. Showy white flowers.	Distributed throughout the U.S.	Palatability ranges from poor to good, depending on the particular species.	UTAH SERVICEBERRY is a good browse species.
SESBANIA *Sesbania grandiflora*	Small, fast-growing, leguminous tree. Few branches. Long pods.	Adapted to humid tropical regions. Cultivated on low dikes between rice fields.	Leaves and young shoots are well liked by cattle, goats, and poultry. If cut back to a suitable height, sesbania can provide abundant fresh forage throughout the dry season.	Large white flowers are used for human food.
SICKLEBUSH (CHRISTMAS BUSH) *Dichrostachys cinerea*	Leguminous shrub. Grows up to 10 ft *(3 m)*. Gray bark. Solitary thorns, often with one or two leaves. Pendulous flower heads are pink in the upper part, yellow in the lower. Dark brown twisted pods are borne in clusters.	Adapted to bush country.	Pods are readily eaten and provide browse in the dry season.	Can become a problem in overgrazed uncleared areas, forming impenetrable thickets.
SIGNAL GRASS (PALISADE GRASS, BREAD GRASS, CEYLON SHEEP GRASS) *Brachiaria brizantha*	Coarse, broad-leaved perennial grass. Rhizomatous or stoloniferous. Grows up to 6.5 ft *(2 m)*. Variable in hairiness, leafiness, and yield.	Adapted to most soils in sheltered areas in regions receiving over 30 in. *(76 cm)* of rainfall. Can tolerate moderate droughts. Binds loose soils.	Provides valuable pasture. Prostrate types have a high leaf-to-stem ratio and are high in protein.	
SILICON DIOXIDE			Additive used as an anticaking agent and a grinding aid.	
SISAL HEMP (HENEQUEN) *Agave sisalana*	Perennial stemless plant. Thick succulent leaves have smooth edges and a sharp, terminal spine.	Grown chiefly in Mexico and Cuba for the fiber extracted from the leaves for use in rope and twine.	Leaf waste has been successfully fed to cattle and rabbits.	Ferments rapidly and should be used within 48 hours, sun-cured, or ensiled with molasses.
SISSOO (ROSEWOOD) *Dalbergia sissoo*	Large deciduous hardwood tree.	Important timber tree of Mexico and Central America.	Small leaves provide good forage and can be fed fresh or ensiled. Cattle relish the leaves and young shoots.	
SKUNKBUSH (SQUAWBERRY, SKUNKBRUSH, SQUAW-BRUSH) *Rhus trilobata*	Multibranched shrub. Grows up to 7 ft *(2 m)*. Disagreeable odor.	Widely distributed from Alberta to Illinois, northern Mexico, California, and southern Oregon.	Generally considered to be of low palatability, but in the Southwest, it is rated as fair to good browse forage for cattle and sheep.	Indians once prized the tough pliable shoots for basketmaking.

(Continued)

TABLE 30-1 *(Continued)*

Feedstuff	Description	Place of Origin Geographical Adaptation Cultural Characteristics	Importance/Use	Comments
SKUNKLEAF, BLUE *Polemonium pulcherrimum*	Well-known perennial of the phlox family. Numerous slender stems branch out from the base. Root starts as a slender taproot and then becomes branched. Compound leaves can be hairy or smooth. Flowers droop in the bud, but become erect at blooming.	Widespread throughout the 11 western states. Typically inhabits the lodgepole pine, aspen, spruce-fir, and alpine belts. Usually found in scattered, small patches in open meadows, canyon bottoms, parks and grassy slopes, primarily in moist, relatively coarse-textured soils.	Considered to be of low forage value. Palatability ranges from zero to poor for cattle and poor to fairly good for sheep.	It is best for grazing in midsummer.
SLOUGHGRASS, AMERICAN *Beckmannia syzigache*	Erect, stout annual grass.	Widely distributed throughout the U.S.	Palatable hay and pasture.	Frequently cut for hay.
SMILO GRASS *Oryzopsis miliacea*	Perennial ricegrass.	Native to the Mediterranean region.	Introduced into California as a forage crop.	
SNAILS *Helix* spp	Freshwater, marine, or terrestial gastropod mollusk.		In many areas of the world, snails are being cooked, dried, and ground as a high-protein feedstuff for poultry and swine.	Snails have about the same feeding value as fish or meat meal.
SNOWBRUSH *Ceanothus velutinus*	Shiny, smooth evergreen shrub. Many crooked stems branch from the base.	Well known in the Rocky Mountains, Cascades, and Sierras. Found in a wide variety of sites, occurring on about any well-drained soil.	Forage value is very low. Worthless for cattle and horses. Sheep and goats will browse new shoots and blossoms.	
SODOM APPLE *Solanum incanum*	Small shrub or perennial herb.	Found in tropical areas.	Sheep and cattle avoid this plant. Goats will eat the leaves when no other browse is available.	
SORREL (INDIAN SORREL, JAMAICA SORREL, ROSELLE) *Hibiscus sabdariffa*	Bushy annual up to 6 ft (*2 m*) in height.	Extensively grown in tropical countries for the dark red calices surrounding the fruit.	A red, nonalcoholic drink is made from the calices. Leaves and seeds are fed to livestock.	
SOTOL *Dasylirion*	Very similar to yucca.	Native to Mexico and the southwestern U.S.	Can be used as emergency feed. Compact heads may be ensiled.	
SOUR CLOVER *Melilotus indica*	Winter annual legume. Sour clover in thick stands branches very little, but when growing alone, it branches profusely. Stems are like those of alfalfa, but are a little harder. Leaves are three-parted and abundant.	Native to the Mediterranean region. Adapted to the soils of the southwestern U.S. and the delta lands of the Mississippi River. Prefers mild climates.	In some places, sour clover can be used for pasture, but it tends to be rather unpalatable.	Tends to cause scours in cattle.
SOURGRASS *Trichachne insularis*	Perennial grass. Slender, erect or ascending racemes. White to brownish, silky panicle.	Found in South America, Mexico, and the southwestern U.S.	Generally considered to be of low palatability because of its sour taste.	
SPANISH CLOVER *Desmodium unicatum*	Spreading perennial legume. Its upright stems spread when cut. Large root system.	Native to tropical America. Grows best in wet lowlands. Grows well with grasses.	Forage.	
SPANISH LIME (MAMONCILLO) *Melicocca bijuga*	Medium-sized tree. Leaves have two pairs of elliptic leaflets. Greenish-white, fragrant flowers. Fruits have yellow translucent pulp and large round seeds.	Found in tropical climates.	Fruit and seeds are edible.	Seeds are sometimes roasted and used for human food.
SPEAR GRASS (PILI, TANGLEHEAD GRASS) *Heteropogon contortius*	Quick-growing grass. Stems up to 4 ft (*1.2 m*) high. Long awns twist together upon maturity.	Widely distributed. Abundant in areas receiving less than 30 in. (*76 cm*) of rainfall.	Valuable forage before flowering.	Awns can injure the mouths and skins of livestock.
SQUIRRELTAIL *Sitanion* spp	Perennial bunchgrasses.	Native to western North America.	Good forage value where young.	Mature squirreltail can injur the mouths of livestock.
STARGRASS *Cynodon plectostachyus*	Spreading perennial grass. Stout, fast-growing stolons form a dense turf.	Adapted to a more tropical climate than Bermudagrass and is more productive than Bermudagrass in dry areas. Commonly found on dry lake beds.	Valuable pasture for dry areas.	

(Continued)

TABLE 30-1 *(Continued)*

Feedstuff	Description	Place of Origin Geographical Adaptation Cultural Characteristics	Importance/Use	Comments
STICKSEED (BURSEED, STICK-TIGHT, BEGGARTICK) *Lappula* spp	Consists of about 40 species of annuals, biennials, and perennials of the borage family. Plants are usually covered with rough hairs. Ovary develops into 4-barbed nutlets.	Native to the northern temperate zone. Widely distributed in the western U.S. Commonly found on rangelands, especially where the land is overgrazed.	Rated as worthless to poor forage for cattle and horses, poor to fair for sheep and goats.	Burs often become matted in the hair and wool of livestock.
STYLO (BRAZILIAN LUCERNE) *Stylosanths gracilis*	Perennial legume.	Native to South America. Adapted to a wide range of conditions.	Used in place of alfalfa. Mixes well with grasses. Palatability is low during the rains but is readily eaten in the dry season.	Becomes woody and unpalatable if allowed to get too tall.
SULLA (FRENCH HONEY SUCKLE) *Hedysarum coronarium*	Deep-rooted, perennial legume. Yellow thorny pods turn brown at maturity.	Adapted to warm climates with mild winters. Drought resistant.	Of great importance as a forage crop in North Africa and South Europe.	
SUMMER CYPRESS *Kochia scoparia*	Fast growing, densely branched shrub. Dense foliage.	Native to Europe.	Browse.	
SURINAM GRASS (KENYA SHEEP GRASS) *Brachiaria decumbens*	Trailing grass. Stems up to 2 ft *(61 cm)*.	Adapted to well-drained soils in humid climates. Withstands drought but not waterlogging.	Valuable grass of the East African savannas.	
SWEET CLOVER *Melilotus* spp	Most cultivated species of the legume are biennial but a few are annual. The first season's growth consists of one multibranched stem. Deep-set taproot. Trifoliate, toothed leaves. White or yellow flowers. Pods usually contain one seed, sometimes two, and are loosely attached so that seeds shatter as they mature.	Native to temperate Europe and Asia. Grown primarily in the Corn Belt and central third of the U.S. Adapted to a wide range of soils and climatic conditions, but will not tolerate acid soils.	Commonly used for pasture and silage. First-year growth provides substantial pasture in the southern areas. Fairly good hay can be made if cut in the bud stage. However, it is hard to cure and is seldom harvested as hay. Best silage is made when sweet clover is harvested in the prebloom stage.	Rated as one of the best soil-improving crops. Contains the antimetabolite coumarin, a substance that inhibits proper blood clotting and consequently causes hemorrhaging, commonly called *sweet clover disease.* Low-coumarin cultivars are available.
SWEETROOT *Osmorhiza* spp	Perennial herbs. Seeds are long, narrow ribbed, club shaped, and tipped with a short beak. Flowers are usually yellow or white. Plant has a characteristic licoricelike odor and taste, especially in the thick clustered roots.	Found exclusively in North America. Most species are found in the western range areas. Occurs in mountains, open woods, moist meadows, and in rich, coniferous woods.	All species are at least fairly palatable to livestock, but their abundance is limited, thereby limiting their importance as range forages. Sheep and goats graze them closely.	SWEET ANISE is one of the choice range weeds. Most species remain green throughout the summer grazing season.
SWEET THORN *Acacia karoo*	Leguminous shrubs or trees. Grow up to 65 ft *(20 m)*. Long leaves have white thorns at their base. Golden yellow flower heads. Slender, straight pods.	Occurs along rivers in tropical regions.	Although the leaves and thorns are free from prussic acid, they are seldom browsed, probably because of the thorns.	
SWITCHGRASS *Panicum virgatum*	Erect, often purplish perennial panicgrass. Numerous scaly, creeping rootstocks.	Found in all states east of the Mississippi River, in the southwestern U.S., south through Mexico into Central America.	Occasionally cultivated for pasture and hay. It is more valuable as a hay crop.	Excellent soil-binder. Becomes woody and unpalatable at maturity.
SWORD BEAN *Canavalia gladiata*	Twining or climbing legume. Broad pods with a rough surface. White to pink flowers. Flat seeds.		Seeds contain saponin. Therefore, raw, sword beans should not be fed in large amounts. Heat treatment renders the seeds and pods safe.	
T				
TAGASASTE (WHITE-FLOWERED LUCERNE TREE, ESCALON) *Cytisus proliferus*	Ornamental leguminous shrub.	Native to the Mediterranean region. Grows in sandy soils. Drought resistant.	High-yielding, highly valued forage bush. Green leaves can be lopped for poultry.	Animals used to grazing this tree will eat even the thick stems.
TAMARIND *Tamarindus indica*	Medium sized leguminous tree up to 38 ft *(12 m)* tall. Dense rounded crown of feathery leaves. Pale yellow flowers with dark red buds. Pods are about 1 in. *(2.5 cm)* wide and 6 in. *(15 cm)* long.	Widely cultivated in tropical regions.	Pods contain a dark brown edible, but sour, pulp. Leaves and seeds can also be used as feed.	

(Continued)

TABLE 30-1 *(Continued)*

Feedstuff	Description	Place of Origin Geographical Adaptation Cultural Characteristics	Importance/Use	Comments
TAMARUGO *Prosopsis tamarugo*	Dense, thorny, leguminous bush. Grows to 40 ft (*12 m*) tall. Fine, leathery leaves. Edible fruits. Slightly curved pods fall to the ground when ripe.	Cultivated on a large scale in Norte Grande in Chile. Exceptionally drought-resistant. Has the ability to absorb atmospheric moisture through its leaves and transport the water down to the roots.	Although the digestibility of the leaves and seeds is low (45 to 50%), it can maintain sheep in arid areas.	
TANNIA (NEW COCOYAM) *Xanthosoma sagittifolium*	Coarse herb. Trunk can grow to 5 ft (*1.5 m*) in height. Large, dark leaves. Produces solid bulbs in the form of an underground stem.	Native to the New World.	When cooked, roots provide excellent feed, especially for swine. Leaves are well liked by goats, sheep, and cattle.	
TANSY MUSTARD *Sophia* spp	Species in the western U.S. are annuals or biennials. Sharp, pungent taste. Two-celled pods split open at maturity, thereupon releasing minute seeds. Most species have hairy leaves.	In the western U.S., tansy mustard is found in a wide variety of soils, being abundant in the open, in fields, along roadsides and ditches, and in other places where the mineral soil is recently exposed.	Considered to be of rather low feeding value. Palatability rates as fair for sheep and goats and poor for cattle.	Horses rarely graze these herbs.
TARO (EDDOE, OLD COCOYAM, KALO, CHINESE POTATO, MALANGA) *Colocasia esculenta*	Large, coarse herb. Crowns have large, oblong-oval leaves. Large spherical underground tubers.	Important crop of the tropics.	Cooked tubers can be fed to livestock. Leaves are relished by cattle and sheep.	Uncooked taro roots can irritate the digestive tract if fed in large quantities.
TEFF *Eragrostis tef*	Leafy, quick-maturing annual. Stems can grow to 4 ft (*1.2 m*) in height.	Grows in dry areas with a short rainy season. Prefers heavy soils. Highly resistant to disease and pests but not to weed competition.	Cultivated primarily for its grain. Used also as an annual hay grass for arid areas. Elsewhere, it is too stemmy and unproductive for cultivation.	An economically important cereal grass in Africa.
TELFAIRIA (FLUTED PUMPKIN) *Telfairia occidentalis*	Viny plants that produce very long shoots (greater than 45 ft [*14 m*]), purplish flowers, and immense gourds.	Native to Africa.	Leaves are used primarily for human consumption, but are also relished by sheep and goats.	
TEOSINTE (BUFFALOGRASS) *Euchlaena mexicana*	Coarse, cornlike annual.	Cultivated in tropical areas. Requires rich, moist soils and tropical climates. Hence, very little is grown in the U.S.	In some areas, teosinte is grown as a high-yielding forage.	
THREE-AWN (NEEDLEGRASS, WIREGRASS, POVERTY GRASS) *Aristida* spp	Annual and perennial grasses. Depend on seeds for reproduction. Seeds characterized by 3-branched beard at the tip.	Widely distributed throughout the western U.S., especially the Southwest. Found primarily on dry, sandy soils. Common grasses on semidesert areas, plains, and lower elevations in the mountains.	In the Southwest, three-awns are considered to be good spring and summer forage before the seeds mature. The small annuals and a few perennials produce little leafage and are considered to be of little forage value. A large number of perennials are leafy and provide considerable forage. Once mature, three-awn is unpalatable.	Barbed seeds may cause trouble in grazing animals, resulting in sore mouths and inflaming the eyes, ears, and nostrils of livestock. Generally classified as an undesirable range species which appears when overgrazing is practiced.
TOBOSA *Hilaria mutica*	Erect, perennial very much resembling galleta. Spikelets are bearded at the base. Culms form a tough rhizomatous base.	Grows from western Texas to Arizona and into Mexico. Prefers fine, somewhat compact soils on open flats, swales, depressions, and, to a certain extent, foothills. Commonly found in areas subject to flooding in the rainy season.	When green and succulent, tobosa rates as good palatable forage, especially for cattle and horses.	Withstands heavy grazing. Of little value as winter forage.
TOCO GRASS (BATIKI BLUEGRASS) *Ischaemum aristatum*	Perennial grass growing up to 2 ft (*61 cm*). Forms a dense mat.	Grows in seasonally wet or water-logged areas. Withstands poor management.	Provides good pasture if well grazed. Recovers quickly from over-grazing.	Can cause off-flavors in milk.
TODDY PALM (WILD DATE PALM) *Phoenix sylvestris*	Fast-growing palm. Grows to 50 ft (*15.3 m*). Arched leaves divide into numerous leaflets.	Found in tropical areas.	Leaves can be used for forage.	Sap is extracted and distilled to form a potent spirit.
TORPEDOGRASS (PANIC RAMPANT) *Panicum repens*	Tall, erect perennial. Spreads by creeping rhizomes.	Found along tropical and subtropical coasts of both hemispheres. Widespread in marshy areas near fresh or salt water and in sandy soils. Tolerates flooding and drought.	Palatable pasture over a long growing season. Resistant to heavy trampling.	Can become a troublesome weed in ditches and drains.

(Continued)

TABLE 30-1 *(Continued)*

Feedstuff	Description	Place of Origin Geographical Adaptation Cultural Characteristics	Importance/Use	Comments
TREFOIL *Lotus* spp	Deep rooted, fine stemmed leafy perennial legumes. Somewhat decumbent as single plants but erect in thick stands. Leaves contain 5 leaflets. Yellow, showy flowers. Spreading seed pods resemble a bird's foot.	Found in a variety of conditions.	High feeding value as a forage, especially as a late summer feed. Primarily used for pasture, either alone or in mixtures. Good hay can also be obtained.	BIG TREFOIL and BIRDSFOOT TREFOIL are the most important species.
TRIDENS *Tridens* spp	Erect, tufted perennial grasses. Rarely rhizomatous or stoloniferous. Three-toothed lemma.	Most species are found in the southern half of the U.S.	In general, tridens are of little economic importance. However, three species—*T. grandiflorus, T. elongatus,* and *T. pilosus*—have been found to be of some value on range.	
TRISETUM *Trisetum* spp	Perennial or, rarely, annual grasses. Flat leaves. Narrow, spikelike, somewhat open flower heads with numerous small, 3- to 5-flowered individual flower groups. Lemmas are generally short and hairy at the base, 2-toothed at the apex and bear short, straight awns or longer, bent and twisted awns from the back below the tip and between the teeth.	Widely distributed in the temperate and cooler or mountain regions of both hemispheres.	The species found in the western U.S. are fairly good to good forage for cattle and horses. Their value to sheep is somewhat less.	
TUCUM PALMS *Astrocaryum tucuma*		Native to the Amazon valley.	Tucum nut oil meal is occasionally used for cattle rations in place of coconut meal.	
TUMBLEGRASS *Schedonnardus paniculatus*	Low, tufted perennial. Stiff, slender, divergent spikes. Leaves crowded at the base. At maturity, the plant breaks away and rolls in the wind as a tumbleweed.	Found primarily in the prairies and plains of the U.S.	Represents a small portion of the forage of the Great Plains.	
TUMBLE MUSTARD (HEDGE MUSTARD, JIM HILL MUSTARD, TALL MUSTARD, TALL SISYMBRIUM, TUMBLING MUSTARD) *Sisymbrium altissimum*	Rank, ungainly biennial weed. Erect, multibranched stems. Stiff, narrow, multiseeded fruit. Small, yellowish white flowers.	Commonly found in most of the western U.S., generally below the ponderosa pine belt. Grows along railways and highways and on waste grounds, range, and cultivated fields.	When young and tender, it rates as fair in palatability, but as it matures, feeding value declines markedly.	Considered to be a troublesome weed. Commonly associated with RUSSIAN-THISTLE.
TUNG TREE *Aleurites fordii*	Small trees. Dark green, ovate leaves. Clusters of white flowers produced in the spring. Spherical, angled drape contains 3 to 5 seeds.	Native to China. Cultivated in subtropical regions including the Gulf Coast States.	Foliage and undetoxified oil meal are unpalatable and toxic. Experimental methods of detoxifying the oil meal have been proven relatively effective, but detoxification is not currently commercially feasible. When supplemented with lysine, the detoxified oil meal produces good growth in chickens.	Tung oil is largely used for paints, varnishes, and finishes.
		U		
UCUHUBA *Virola surinamensis*	Nut-bearing tree.	Native to Brazil. Grows in tropical climates.	Nuts are used for the extraction of fat. Oil meal is sometimes fed to dairy cattle.	
UMBRELLA THORN *Acacia tortilis*	Leguminous tree. Dark grey bark. Flat-topped crown. The leaves are usually hairy and have pairs of white thorns. Flower heads are cream-colored. Pods are irregularly twisted.	Found in South Africa.	Leaves and fruits are often eaten by livestock.	
		V		
VALERIAN *Valeriana* spp	Perennial herbs. Fleshy, strong-scented roots and rootstocks.	Found in all of the western states. Thrive in moist sites at moderate to high elevations.	Generally of secondary importance as forage on most ranges. Valerians rate as poor to fair in palatability for cattle, fairly good to good for sheep.	Plants can be recognized by their characteristic odor, described as smelling like "dirty feet." Roots are processed for a stimulatory drug.

(Continued)

TABLE 30-1 *(Continued)*

Feedstuff	Description	Place of Origin Geographical Adaptation Cultural Characteristics	Importance/Use	Comments
VASEY GRASS *Paspalum urvillei*	Erect perennial grass that grows in tufts or bunches about 1 ft (*30.5 cm*) in diameter.	Native to Argentina and Uruguay. Commonly found in the lower southern U.S. Adapted to low, fertile wetlands.	Seldom cultivated, but utilized where found growing. Makes high yields of quality hay. Excellent pasture forage.	Cannot withstand heavy grazing.
VELDTGRASS, PERENNIAL *Ehrharta calycina*	Perennial bunchgrass. Leafy stems grow to 3 ft (*0.9 m*) in height.	Native to South Africa. Adapted to light soils. Drought-resistant.	Highly palatable. Proved valuable for range reseeding of the sandy, coastal soils of California.	
VELVET BEAN *Stizolobium* spp	Vigorous growing, summer annual legume. Vines can grow more than 25 ft (*7.6 m*) in length. Trifoliate leaves. Two types of pods: (1) dense, black, velvety pubescence, and (2) white or gray hairs. Pods generally contain 3 to 6 seeds per pod, depending on species. Numerous fleshy surface roots.	Probably originated in India. Adapted to warm climates. Most velvet beans are grown in the South Atlantic and Gulf Coast states. Intolerant of cold, wet soils.	Excellent source of winter pasture. Many dairymen prefer corn-velvetbean silage to corn silage alone. Meal can be fed to livestock effectively. Seldom used for hay because of handling difficulties.	Excellent green manure crop.
VETCH *Vicia* spp	Most species are viny annual legumes. Large flowers, spherical seeds, and elongated, somewhat compressed pods.	Of the approximately 150 species of vetch, about 15 are native to the U.S. Cultivated vetches are native to Europe and adjacent parts of Asia and Africa. Widely distributed throughout the temperate regions of the world. 75% of the vetch acreage in the U.S. is in Oklahoma, Arkansas, Texas, and Louisiana. Adapted to most types of soils. More tolerant of acid soils than the other legumes. Require cool temperature but mild winters. No species are drought-resistant.	All commercial species make good pasturage, hay, silage, and green manure. Can also be used for green feed and as cover crops.	The most important cultivated vetches of the U.S. are BIGFLOWER VETCH, COMMON VETCH, HAIRY VETCH, HUNGARIAN VETCH, and PURPLE VETCH. Reseeding (hard-seeded) common vetches have been developed through interspecific hybridization.
VINE-MESQUITE *Panicum obtusum*	Perennial range grass that produces creeping stems up to 10 ft (*3 m*) in length.	Native from Missouri west to Colorado and south into Mexico. Adapted to areas of low rainfall.	Used for grazing and erosion control.	
		W		
WATER HYACINTH (MILLION DOLLAR WEED) *Eichhornia crassipas*	Unattached, free floating water plant. Leaves above the water surface. Roots below the water surface.	Found in tropical regions where it multiplies rapidly and clogs lakes, rivers, and ponds. Difficult to eradicate.	Fresh water hyacinth is unpalatable due to prickly crystals. Boiled water hyacinth is commonly fed to swine in Asia. Hay is unpalatable to cattle unless 20% molasses is added (urea may also be added). For ensilage, it should be allowed to wilt in the shade for 48 hours before it is chopped and ensiled. Molasses should be added, and it is advisable to add urea and salt. Also, a combination of 4 parts water hyacinth to 1 part straw makes good silage. Plant juice can be used for beef protein-concentrate.	Easily harvested by nets.
WATERLEAF (CATS-BREECHES, RAGGED-BREECHES, BEAR-CABBAGE, BALLHEAD WATERLEAF) *Hydrophyllum capitatum*	Low, perennial herb. Ball-shaped clusters of violet-blue flowers. Deep lobed, long stalked hairy leaves. Short indistinct stems. Numerous fleshy roots clustered on a short, underground rootstock.	Ranges from British Columbia and Montana to Colorado and central California. Most commonly found at medium elevations on fertile, semishaded sites in or bordering on woodlands, in canyon bottoms, and on bushy hillsides.	Represents an important part of the early forage crop of rangelands. Readily grazed. Palatability is fairly good to very good for sheep and fair for cattle.	Young, tender shoots are sometimes eaten by humans.

(Continued)

TABLE 30-1 *(Continued)*

Feedstuff	Description	Place of Origin Geographical Adaptation Cultural Characteristics	Importance/Use	Comments
WATERLEAF *Talinum triangulare*	Fleshy herb up to 1 ft (*30.5 cm*).	Found in tropical regions.	Used primarily for human consumption but is sometimes used as a palatability improver in pasture.	
WESTERN CONEFLOWER *Rudbeckia occidentalis*	Coarse perennial herb of the sunflower family. The flower heads are cone-shaped.	Found throughout the western U.S., especially along open or shaded stream banks, hillsides, well drained mountain swales, open parks, and open aspen stands.	Although it is low in palatability, it often provides considerable forage, because of its commonness, large size, and abundant foliage.	In some areas, WESTERN CONEFLOWER provides limited protection against erosion.
WHEATGRASS *Agropyron* spp	The important species are hardy, erect, cool season perennial grasses. Either sod-forming or have a bunch type of growth. Some species have rhizomes. Seed heads resemble wheat heads.	About 150 species are widely distributed throughout the temperate regions of the world. Over 30 species are native to North America. In the U.S., native species are found in vast areas of the central and northern Great Plains, the intermountain region, and higher elevations of the Rocky Mountains. Thirteen species are found in Alaska. QUACKGRASS, a weed, is the predominant wheatgrass in the eastern U.S.	Most important of the grasses in the western U.S. for the production of highly nutritious early season forage. Also used extensively to control wind and water erosion.	The most important native wheatgrasses of the U.S. are WESTERN WHEAT GRASS (central and northern Great Plains), BLUEBUNCH WHEATGRASS (northern intermountain region), and SLENDER WHEATGRASS (widespread over mountains, foothills, and high plains). CRESTED WHEATGRASS is the most important species introduced into the U.S.
WHYNNE GRASS (MOLASSES GRASS) *Melinis minutiflora*	Spreading perennial grass. Forms large tussocks with hairy stems of up to 6 ft (*1.8 m*) tall. Sticky red leaves have a molasseslike odor.	Introduced into the U.S. from Africa by way of Brazil. Grows in areas with 31 to 71 in. (*79 to 180 cm*) of rainfall. Requires well-drained soils. Drought-resistant. Susceptible to overgrazing, fire, and waterlogging.	Once cattle become used to the smell, whynne grass provides palatable grazing.	Said to repel snakes and insects and to be useful for controlling ticks.
WILD-RYE GRASS *Elymus* spp	Most species are perennial bunch growers, but some form sods. Foliage is tall, coarse, and harsh. Many species have twisted or bearded flower heads.	Fairly large genus which is found in the north temperate zone. The western U.S. is the area containing the most species. In the western U.S., wild-rye grasses occur from the lower semidesert areas to the aspen and spruce belts. Some typically grow in bottomlands and meadows while others are found in brush and woodland areas.	Wild-ryes are coarse at maturity, but are of moderate to high palatability if harvested early. They are sometimes useful in mixtures and for the quick establishment of cover.	Highly susceptible to ergot fungus, a very toxic organism to livestock. The most useful wild-rye species of the U.S. are CANADA WILD-RYE (Plains and western states), GIANT WILD-RYE (western states), BLUE WILD-RYE (western states), RUSSIAN WILD-RYE (northern Great Plains and intermountain states), and BASIN WILD-RYE (western states.).
WILLOW *Salix* spp	One of the most common woody plants. Vary in height from 0.5 ft to 100 ft (*15 cm to 30.5 m*). Deciduous, short-stemmed, alternate simple leaves are generally long, narrow, or taper-pointed. Most mature leaves are green with some white hairs. Dioecious.	Composed of about 250 species, most of which are native to the north temperate and arctic zones. Most species grow in moist areas around freshwater streams and rivers. Prefer deep, fertile soils on moist banks, flats, or slopes bordering on very wet sites. Seldom grow in the shade.	One of the principle sources of browse in the western mountain regions. Palatability averages fair for cattle and good for sheep. Good, late summer feed as the foliage becomes more palatable as the season advances.	Livestock tend to browse willows very closely.
WINTERFAT (WHITE SAGE) *Eurotia lanata*	Low, silvery-white shrub. Grows up to 15 ft (*4.6 m*). New stems are produced each year from the basal part of the previous year's stems and from the woody crown. Covered with a mat of silvery-white hair. Deep taproot with numerous lateral roots.	Widely distributed throughout the western U.S. and Canada. Frequently found in lower foothills, plains, and valley on dry soils moderately impregnated with salt or white alkali material. Drought resistant.	Grazed by all types of livestock. Considered to be good forage for goats and horses.	Overgrazing has dramatically reduced its population.
WOODRUSH *Juncoides* spp	Perennial, leafy, grasslike plants. Closed leaf sheaths, hollow stems. Capsules are 1- to 3-celled.	Native to North America. Found primarily in the western U.S. in moist to wet sites, either in the open or shade.	Palatability is highly variable. As a group, woodrushes rate as poor to fair for sheep and fair to fairly good for cattle.	
WOOLLYFINGER GRASS (PANGOLA RIVER GRASS) *Digitaria pentzii*	Densely tufted perennial grass up to 4 ft (*1.2 m*) tall. Variably hairy leaves and stems.	Grows under a variety of conditions. Often found on sandy soils.	Valuable dry season grass, but unpalatable during the grazing season. Also used for soil stabilization.	Intolerant of heavy grazing.

(Continued)

TABLE 30-1 *(Continued)*

Feedstuff	Description	Place of Origin Geographical Adaptation Cultural Characteristics	Importance/Use	Comments
		X, Y, Z		
YAMPA *Carum gairdneri*	Smooth, slender, erect perennial of the carrot family. Fleshy, tuberous roots resemble tiny sweet potatoes. Solitary, smooth, slender, and sometimes branched stems 1 to 4 ft (*2.5 cm to 1.2 m*) in height.	Distributed throughout the western U.S. Grows on moderately moist soils in open, weedy parks, in stands of aspen, ponderosa pine, and Douglas fir, in moist meadows, and in open sites with sagebrush and wheatgrass.	Furnishes a good amount of forage on western ranges. Palatability ranges from fair to high.	Used for food by American Indians.
YARROW, WESTERN (MILFOIL, WILDTANSY, WOOLY YARROW) *Achillea millefolium*	Strong-scented perennial herb. Leaves often form rosettes. Dense flower clusters resemble the top of a derby hat.	One of the most widely distributed and abundant herbaceous plants of the western U.S. Occurs in a wide variety of habitats. Prefers sandy and gravelly loam soils. Relatively drought resistant.	Forage value is widely variable. Sheep are fond of its flowers. Palatability ranges from poor to good in sheep and zero to fair for cattle.	COMMON YARROW, a close relative, is widely distributed in the eastern U.S. Contains glucosides and alkaloids which, if consumed in large amounts by sheep, may cause diarrhea and enuresis.
YUCCA *Yucca* spp	Shrubs with cup-shaped flowers and stiff, sword-shaped leaves.	Confined to North and Central America, Bermuda, and the West Indies. In the U.S., most are found in the desert areas of the Southwest.	When finely chopped or shredded, yucca can be used as forage for cattle.	

Fig. 30–1. Soybean pods and seed. (Courtesy, American Soybean Assn., St. Louis, Mo.)

Fig. 30–2. Sorghum, headed out. (Courtesy, USDA)

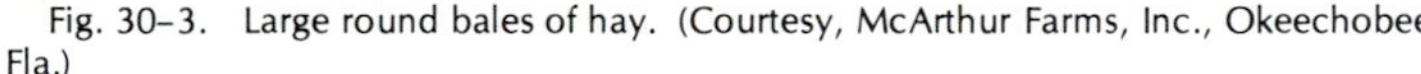

Fig. 30-3. Large round bales of hay. (Courtesy, McArthur Farms, Inc., Okeechobee, Fla.)

Fig. 30-4. Guernesy cows on pasture. (Courtesy, The American Guernsey Cattle Club, Columbus, Ohio)

Fig. 30-5. Band of sheep on a western range. (Courtesy, USDA)

Fig. 30-6. Nubian dairy goat feeding on hay. (Courtesy, University of Delaware, Newark)

1. Inspector taking sample of mixed feed.

2. Checking grain sample by compartments in probe.

3. Rack of tubes ready to be placed in the block digestor.

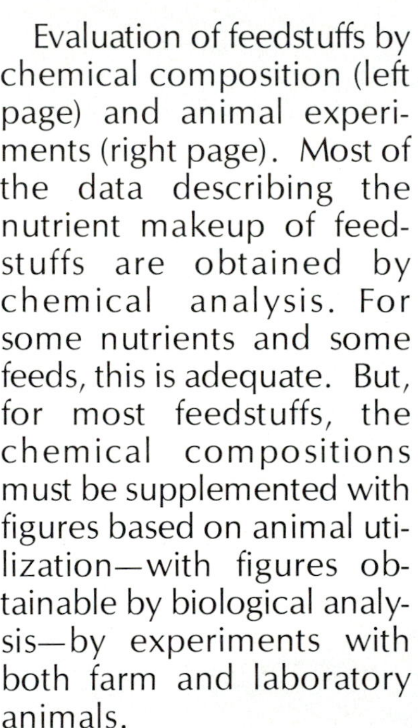

Evaluation of feedstuffs by chemical composition (left page) and animal experiments (right page). Most of the data describing the nutrient makeup of feedstuffs are obtained by chemical analysis. For some nutrients and some feeds, this is adequate. But, for most feedstuffs, the chemical compositions must be supplemented with figures based on animal utilization—with figures obtainable by biological analysis—by experiments with both farm and laboratory animals.

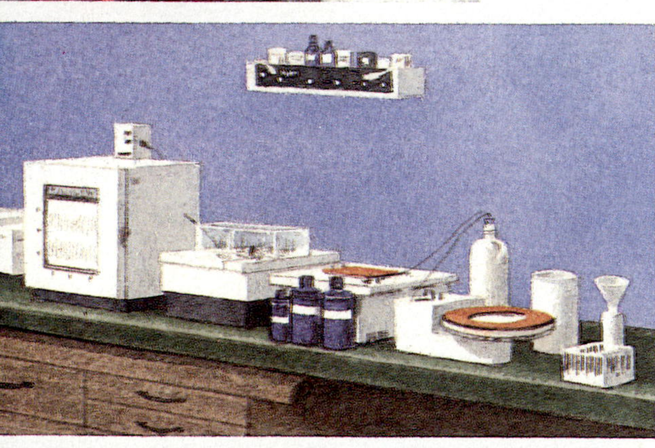

4. Automated readout of 40 sample/hour from block digester. Readings are in percent nitrogen or crude protein.

5. Automated analysis of lysine in grain or feed hydrolysates.

6. Weighing samples for auto analyzer.

7. Liquid Chromatograph with auto sampler, used to determine the levels of vitamins, amino acids, or drugs in samples of feeds, premixes, and pharmaceutical products.

8. A crude fiber analysis.

The most accurate, but the most time consuming and costly, method of determining the value of a feedstuff involves live animal trials, using laboratory and/or farm animals. Through such experiments, the digestibility, availability of nutrients, and palatability of a feedstuff can be determined.

5

Composition of Feeds

SECTION V

COMPOSITION OF FEEDS

Both livestock producers and nutritionists have need for accurate and up-to-date composition of feedstuffs in order (1) to formulate rations for maximum production and net returns, and (2) to predict the production response of animals when they are fed rations of a given composition. In recognition of this need and its importance, compositions of a great array of feeds are presented in Section V.

COMPOSITION OF FEEDS, the painting on the preceding two-page spread, is from an original by the noted artist, Tom Phillips, prepared specially for this book. It portrays the artist's conception of the evaluation of feedstuffs (1) by chemical composition (left page), and (2) by animal experiments (right page).

COMPOSITION OF FEEDS

Contents *Page*

In addition to the discussion that follows pertaining to composition of feeds, the reader is referred to Chapter 15, Feed Analysis/Feed Evaluation, and Chapter 18, Feeding Standards/Ration Formulation.

Both nutritionists and livestock producers should have access to accurate and up-to-date composition of feedstuffs in order to formulate rations for maximum production and net returns. The ultimate goal of feedstuff analysis, and the reason for feed composition tables, is to be able to predict the productive responses of animals when they are fed rations of a given composition. In recognition of this need and its importance, the authors spared no time or expense in compiling the feed composition tables presented in this section. At the outset, a survey of the industry was made in order to determine what kind of feed composition tables would be most useful, in both format and content. Secondly, it was decided to utilize, to the extent available, the feed compositions which, for many years, were compiled by Lorin Harris, Utah State University, now carried forward by the USDA, National Agricultural Library, Feed Composition Bank. These data were augmented by the authors with feed compositions from the National Academy of Sciences, NRC, and from experimental reports, industries, and other reliable sources.

To facilitate quick and easy use, the feeds in Table V are classified and separated into the following subtables:

1. **Table V-1A, Energy Feeds.** This includes feeds which are high in energy and low in fiber (under 18%), and which generally contain less than 20% protein.
2. **Table V-1B, Protein Feeds.** This includes feeds that contain more than 20% protein or protein equivalent.
3. **Table V-1C, Dry Forages.** This includes feeds which are bulky, low in weight per unit volume, and relatively low in energy, and which in the dry state contain more than 18% crude fiber.
4. **Table V-1D, Silages and Haylages.** Silage (ensilage) is fermentable high-moisture forage stored under anaerobic conditions in a silo, consisting of either green crops or crops to which moisture has been added, chopped when stored and containing 65 to 70% moisture.

   Haylages are low-moisture silages, made from grasses and/or legumes that are wilted to 40 to 55% moisture content before ensiling.

5. **Table V-1E, Pasture and Range Plants.** This includes grass, browse, and other plants that are harvested by grazing animals.
6. **Table V-1F, Mineral Supplements.** This includes rich natural and synthetic sources of inorganic elements needed to perform certain essential body functions.
7. **Table V-1G, Vitamin Supplements.** This includes rich synthetic or natural feed sources of one or more complex compounds, called vitamins, that are required by animals in minute amounts for normal growth, production, reproduction, and/or health.
8. **Table V-2, Amino Acids.** This gives the known amino acid composition of certain feeds.
9. **Table V-3, Apparent Ileal Digestibility of Crude Protein and Essential Amino Acids in Feedstuffs for swine, and Digestible Amino Acid Recommendations for Swine Feed Formulation.** The protein requirement of the pig is primarily a requirement for essential amino acids. Currently, apparent ileal digestibility data give the best practical assessment of the availability of amino acids in various feed ingredients.
10. **Table V-4, True Digestibility of Essential Amino Acids for Poultry, and True Digestibility Amino Acid Recommendations for Poultry Feed Formulation.** Like all monogastric animals, poultry do not require protein as such; instead, they need well-defined amounts of available amino acids to perform at a desired level. Currently, digestibility assays give the best assessment of the availability of amino acids in various feed ingredients.

(Also see Chapter 24, Section on "Protein . . . • True digestibility of amino acids.")

Some feeds fit the criteria of more than one of the above classes. For example, whole soybeans are used as both an energy feed and a protein feed; hence, they are listed in both Table V-1A and Table V-1B.

In Tables V-1A to V-1E, and V-1G, covering 6 of the respective feedstuff classifications indicated above, values for each feed are presented in tabular form on 4 pages (2 double-page spreads), with one page devoted to each of the following categories (see 4 pages in miniature that follow, listing the compositions called for in the table headings of the 4 full pages):

Page 1270, left-hand, proximate analysis
Page 1271, right-hand, energy
Page 1272, left-hand, minerals
Page 1273, right-hand, vitamins

p. 1270, Left-Hand Page

MOISTURE BASIS:
A-F (as-fed) or
M-F (moisture-free)

PROXIMATE ANALYSIS:

Dry Matter
Ash
Crude Fiber
Neutral Det. Fib. (NDF)
Acid Det. Fib. (ADF)
Lignin
Ether Extract (Fat)
N-Free Extract
Crude Protein

DIGESTIBLE PROTEIN:

Ruminant
Swine
Horse

p. 1271, Right-Hand Page

TDN:

Ruminant
Swine
Horse

DIGESTIBLE ENERGY:

Ruminant
Swine
Horse

METABOLIZABLE ENERGY:

Ruminant
Swine
Poultry—ME_n
Horse

NET ENERGY:

Ruminant—NE_m
Ruminant—NE_g
Lactating Cows—NE_{lc}

p. 1272, Left-Hand Page

MOISTURE BASIS
A-F (as-fed) or
M-F (moisture-free)

DRY MATTER

MACRO MINERALS:

Calcium (Ca)
Phosphorus (P)
Sodium (Na)
Chlorine (Cl)
Magnesium (Mg)
Potassium (K)
Sulfur (S)

MICRO MINERALS:

Cobalt (Co)
Copper (Cu)
Iodine (I)
Iron (Fe)
Manganese (Mn)
Selenium (Se)
Zinc (Zn)

p. 1273, Right-Hand Page

FAT-SOLUBLE VITAMINS:

A
Carotene (Provitamin A)
D
E
K

WATER-SOLUBLE VITAMINS:

B–12
Biotin
Choline
Folacin (Folic Acid)
Niacin
Pantothenic Acid (B–3)
(Pyridoxine) B–6
Riboflavin (B–2)
Thiamin (B–1)

Table V–1F, Mineral Supplements, Composition, shows, in tabular form, for each mineral supplement the values listed in the two miniature pages of a two-page spread which follows:

p. 1482, Left-Hand Page

MOISTURE BASIS
A-F (as-fed) or
M-F (moisture-free)

PROXIMATE ANALYSIS:

Dry Matter
Ash
Crude Fiber
Ether Extract (Fat)
N-Free Extract
Crude Protein

DIGESTIBLE PROTEIN:

Ruminant
Nonruminant
Horse

p. 1483, Right-Hand Page

MACRO MINERALS:

Calcium (Ca)
Phosphorus (P)
Sodium (Na)
Chlorine (Cl)
Magnesium (Mg)
Potassium (K)
Sulfur (S)

MICRO MINERALS:

Cobalt (Co)
Copper (Cu)
Fluorine (F)
Iodine (I)
Iron (Fe)
Manganese (Mn)
Selenium (Se)
Zinc (Zn)

Table V–2, Amino Acid Composition of Feeds, shows, in tabular form, on a single page, the listings on the miniature page that follows:

p. 1496

Moisture Basis
Dry Matter
Crude Protein

AMINO ACIDS

Arginine
Cystine
Glycine
Histidine
Isoleucine
Leucine
Lysine
Methionine
Phenylalanine
Serine
Threonine
Tryptophan
Tyrosine
Valine

FEED NAMES

Ideally, a feed name should conjure up the same meaning to all those who use it, and it should provide helpful information. This was the guiding philosophy of the authors when choosing the names given in the Feed Composition Tables. Genus and species—latin names—are also included. To facilitate worldwide usage, the International Feed Number of each feed is given. To the extent possible, consideration was also given to source (or parent material), variety or kind, stage of maturity, processing, part eaten, and grade.

Where feeds are known by more than one name, cross-referencing was used.

MOISTURE CONTENT OF FEEDS

It is necessary to know the moisture content of feeds in ration formulation and buying. Usually, the composition of a feed is expressed according to one or more of the following bases:

1. **As-fed; A-F (wet, fresh).** This refers to feed as normally fed to animals. As-fed may range from near 0% to 100% dry matter.

2. **Air-dry (approximately 90% dry matter).** This refers to feed that is dried by means of natural air movement, usually in the open. It may either be an actual or an assumed dry matter content; the latter is approximately 90%. Most feeds are fed in an air-dry state.

3. **Moisture-free; M-F (oven-dry, 100% dry matter).** This refers to a sample of feed that has been dried in an oven at 221°F until all the moisture has been removed.

Where available, feed compositions are presented on both **As-Fed (A-F)** and Moisture-Free (M-F) bases. Formulas for adjusting moisture content from moisture-free to as-fed, or as-fed to moisture-free, are given in Chapter 18, in the section headed "Adjusting Moisture Content."

CAROTENE

Where carotene has been converted to vitamin A, the conversion rate of the rat has been used as the standard value, with 1 mg of β-carotene equal to 1,667 IU of vitamin A. Based on this standard, comparative efficiencies of each animal species are given in Chapter 4, Table 4–8, Vitamin A.

PERTINENT INFORMATION ABOUT DATA

The information which follows is pertinent to the feed composition tables presented in this section.

- **Variations in composition**—Feeds vary in their composition. Thus, actual analysis of a feedstuff should be obtained and used whenever possible, especially where a large lot of feed from one source is involved. Many times, however, it is either impossible to determine actual compositions or there is insufficient time to obtain such analysis. Under such circumstances, tabulated data may be the only information available.

- **Feed compositions change**—Feed compositions change over a period of time, primarily due to (1) the introduction of new varieties, and (2) modifications in the manufacturing process from which by-products evolve.

- **Biological value**—The response of animals when fed a feed is termed the biological value, which is a function of its chemical composition and the ability of the animal to derive useful nutrient value from the feed. The latter relates to the digestibility, or availability, of the nutrients in the feed. Thus, soft coal and shelled corn may have the same gross energy value in a bomb calorimeter but markedly different useful energy values (TDN, digestible energy, metabolizable energy, and net energy) when consumed by an animal. Biological tests of feeds are more laborious and costly than chemical analysis, but they are much more accurate in predicting the response of animals to a feed.

- **Where information is not available**—Where information is not available or reasonable estimates could not be made, no values are shown. Hopefully, such information will become available in the future.

- **Calculated on a dry matter (DM) basis**—All data were calculated on a 100% dry matter basis (moisture-free), then converted to an as-fed basis by multiplying the decimal equivalent of the DM content times the compositional value shown in the table.

- **Fiber**—Four values relating to dietary fiber are given in the feed composition tables—crude fiber, neutral detergent fiber (NDF), acid detergent fiber (ADF), and lignin.

Crude fiber, methods for the determination of which were developed more than 100 years ago, is declining as a measure of low digestible material in the more fibrous feeds. The newer method of forage analysis, developed by Van Soest and associates of the U.S. Department of Agriculture, separates feed dry matter into two fractions: a neutral detergent fibrous fraction; and an acid detergent fibrous fraction. Also the amount of lignin in the ADF may be determined. (See Chapter 15, Fig. 15–4.)

1. **Crude fiber (CF).** This fraction is an indicator of the relative indigestibility and bulkiness of the sample. It is the residue that remains after boiling a feed in a weak acid, and then in a weak alkali, in an attempt to imitate the process that occurs in the digestive tract. This procedure is based on the supposition that carbohydrates which are readily dissolved also will be readily digested by animals, and that those not soluble under such conditions are not readily digested. Unfortunately, the treatment dissolves much of the lignin, a nondigestible component. Hence, crude fiber is only an approximation of the indigestible material in feedstuffs. Nevertheless, it is a rough indicator of the energy value of feeds. Also, the crude fiber value is needed for the computation of TDN.

(Also see Chapter 15, section on "Crude Fiber [CF].")

2. **Neutral detergent fiber (NDF).** This is the fraction of the feed which is not soluble in neutral detergent. It consists of plant cell walls, including lignin, cellulose, and hemicellulose. NDF is closely related to feed intake because it contains all the fiber components that occupy space in the rumen and are slowly digested. The lower the NDF, the more forage the animal will eat; hence, a low percentage of NDF is desirable.

(Also see Chapter 15, section on "Neutral Detergent Fiber [NDF] and Acid Detergent Fiber [ADF].")

3. **Acid detergent fiber (ADF).** This is the fraction of the feed which is not soluble in acid detergent. It consists of cellulose (digestible) and lignin (indigestible). ADF is an indicator of forage digestibility because it contains a high proportion of lignin which is the indigestible fiber fraction. The lower the ADF, the more feed an animal can digest; hence, a low percentage of ADF is desirable.

(Also see Chapter 15, section on "Neutral Detergent Fiber [NDF] and Acid Detergent Fiber [ADF].")

4. **Lignin.** This fraction is essentially indigestible by all animals and is the substance that limits the availability of cellulose carbohydrates in the plant cell wall to rumen bacteria.

The acid detergent fiber procedure is used as a preparatory step in determining the lignin content of a forage sample. Hemicellulose is solubilized during this procedure, while the lignocellulose fraction of the feed remains insoluble. Cellulose is then separated from lignin by the addition of sulfuric acid. Only lignin and acid-insoluble ash remain upon completion of this step. This residue is then ashed, and the difference of the weights before and after ashing yields the amount of lignin present in the feed.

• **Nitrogen-free extract**—The nitrogen-free extract was calculated with mean data as: mean nitrogen-free extract (%) = 100 − % ash − % crude fiber − % ether extract − % protein.

• **Protein values**—Both crude protein and digestible protein values are given. Crude protein is determined by finding the nitrogen content and multiplying the result by 6.25. The nitrogen content of proteins averages about 16% (100 ÷ 16 = 6.25).

In addition to the compositions given in the Feed Composition Tables in this section, bypass protein values of selected feeds are given in Chapter 11, Table 11–1 of this book.

• **Ruminant values**—The ruminant values represent a pooling of cattle, sheep, and goat data.

• **Energy**—Many of the energy values given in the feed composition tables were derived from complex formulas developed by L. E. Harris and other animal scientists.

The following four measures of energy are shown:

1. **Total digestible nutrients (TDN).** This value is given because there are more of them, and because it has been the standard method of expressing the energy value of feeds for many years. However, the following disadvantages are inherent in the TDN system: (a) Only digestive losses are considered—it does not take into account other important losses, such as those in the urine, gases, and increased heat production; (b) there is a poor relationship between crude fiber and NFE digestibility in certain feeds; and (c) it overestimates roughages in relation to concentrates when animals are fed for high rates of production, due to the higher heat loss per pound of TDN in high-fiber feeds.

2. **Digestible energy (DE).** Digestible energy is that portion of the gross energy in a feed that is not excreted in the feces. It is roughly comparable to TDN.

For most animals, digestible energy is relatively easy to determine. With poultry, however, true digestibility is very difficult to measure because undigested residues and urinary wastes are excreted together.

3. **Metabolizable energy (ME).** Metabolizable energy represents that portion of the gross energy that is not lost in the feces, urine, and gas (mainly methane). It does not take into account the energy lost as heat, commonly called heat increment. As a result, it overevaluates roughages compared with concentrates, as do TDN and DE.

Metabolizable energy is considered to be the most accurate evaluation of the energy of feedstuffs for the scientific formulation of poultry feeds.

4. **Net energy (NE).** Net energy represents the energy fraction in a feed that is left after the fecal, urinary, gas, and heat losses are deducted from the GE. Because of its greater accuracy, net energy is being used increasingly in ration formulations, especially in computerized formulations for large operations. However, net energy is difficult and expensive to determine.

Two systems of net energy evaluation are presently used: (a) net energy for maintenance (NE_m) and net energy for gain (NE_g), and (b) net energy for lactation (NE_{lc}).

Note to dairy producers and nutritionists: In *Nutrient Requirements of Dairy Cattle,* Sixth Revised Edition, 1988, Table 7–1, Composition of Feeds Commonly Used in Dairy Cattle Diets on a 100% Dry Matter Basis, the NRC committee assumed an average decrease of 4% per unit of dry matter intake above maintenance in calculating NE_{lc} values for feed ingredients, or an average discount of 8% based on their assumption that lactating cows are fed at 3X maintenance. For the convenience of those dairy producers and dairy nutritionists who wish to use these values, the authors selected from *Nutrient Requirements of Dairy Cattle,* Sixth Revised Edition, 1988, Table 7–1, the feeds most commonly used in dairy cattle rations and reproduced them in Chapter 20, Feeding Dairy Cattle, pages 818, 819, and 820, Table 20–7, in a three-page spread.

• **Minerals**—The level of minerals in forages is largely determined by the mineral content of the soil on which the feeds are grown. Calcium, phosphorus, iodine, and selenium are well-known examples of soil nutrient—plant nutrient relationships.

• **Vitamins**—Generally speaking, it is unwise to rely on harvested feeds as a source of carotene (vitamin A value), unless the forage being fed is fresh (pasture or green chop) or of a good green color and not over a year old.

The authors are very grateful to Lorin E. Harris, Ph.D., and Clyde R. Richards, Ph.D., Utah State University, Logan, for their interest and invaluable assistance in preparing the Feed Composition Tables for this book. Also, in the preparation of Table V–1F, Composition of Mineral Supplements, the authors acknowledge with appreciation the review and added values provided by International Minerals and Chemical Corporation, Northbrook, Illinois; and R. F. Klay, Ph.D., Research Department, Moorman Mfg. Co., Quincy, Illinois.

(Above) Fig. V–1. "Courtin' Time," a water color by artist Tom Phillips, commissioned by Dr. Roscoe E. Dean, M.D., Wessington Springs, S.D. ("Courtin' Time" is reproduced in this book through the courtesy of Dr. Dean)

(Below) Fig. V–2. "Morgans on a Misty Morning." (Courtesy, Linda H. Hopson, Homeward Farm, Herkimer, N.Y., and the American Morgan Horse Assn.)

TABLE V-1A ENERGY FEEDS, COMPOSITION OF FEEDS, DATA EXPRESSED **AS-FED** AND **MOISTURE-FREE** (See footnote at end of table.)

Entry Number	Feed Name Description	International Feed Number	Moisture Basis: A-F (as-fed) or M-F (moisture-free)	Proximate Analysis: Dry Matter	Ash	Crude Fiber	Neutral Det. Fib. (NDF)	Acid Det. Fib. (ADF)	Lignin	Ether Extract (Fat)	N-Free Extract	Crude Protein	Digestible Protein: Ruminant	Swine	Horse
				%	%	%	%	%	%	%	%	%	%	%	%
	ACORNS, OAK *Quercus* spp														
1	FRESH	4-07-755	A-F	50	1.2	9.3	–	–	–	2.3	35.0	2.2	0.4	1.0	1.3
			M-F	100	2.5	18.6	–	–	–	4.7	70.0	4.4	0.9	2.1	2.5
	ADLAY (JOB'S TEARS) *Coix lacrymajobi*														
2	SEEDS	4-08-448	A-F	89	2.6	4.5	–	–	–	5.1	62.8	13.8	9.7	10.8	10.1
			M-F	100	2.9	5.1	–	–	–	5.7	70.7	15.6	10.9	12.2	11.4
	ALFALFA (LUCERNE) *Medicago sativa*														
3	SEED SCREENINGS	5-08-326	A-F	90	5.1	11.3	–	–	–	9.7	33.1	31.0	26.0	29.6	26.7
			M-F	100	5.6	12.5	–	–	–	10.8	36.7	34.3	28.8	32.8	29.5
	ALMOND *Prunus amygdalus*														
4	HULLS[1]	4-00-359	A-F	90	6.1	13.5	28.8	25.2	–	2.9	63.8	4.1	0.0	2.0	2.4
			M-F	100	6.8	14.9	32.0	28.0	–	3.2	70.6	4.5	0.0	2.2	2.7
	ANIMAL														
5	FAT, HYDROLYZED	5-00-376	A-F	99	–	–	–	–	–	98.4	–	–	–	–	–
			M-F	100	–	–	–	–	–	99.2	–	–	–	–	–
6	RUMEN CONTENTS, DEHY (PAUNCH PRODUCT)	1-09-327	A-F	87	9.6	24.4	–	–	–	1.8	36.2	15.3	10.3	10.1	10.3
			M-F	100	11.0	28.0	–	–	–	2.1	41.4	17.6	11.8	11.6	11.8
7	TALLOW	4-08-127	A-F	97	0.1	–	–	–	–	96.8	–	1.5	-1.6	-0.4	0.3
			M-F	100	0.1	–	–	–	–	99.5	–	1.6	-1.6	-0.4	0.3
	ANIMAL–POULTRY														
8	FAT	4-00-409	A-F	99	–	–	–	–	–	99.1	–	–	–	–	–
			M-F	100	–	–	–	–	–	100.0	–	–	–	–	–
	ANISE *Pimpinella anisum*														
9	SEEDS, MEAL MECH EXTD	4-00-415	A-F	90	10.3	18.6	–	–	–	9.6	36.9	14.1	7.6	11.0	10.3
			M-F	100	11.5	20.8	–	–	–	10.7	41.2	15.7	8.5	12.3	11.5
	APPLE *Malus* spp														
10	FRUIT, FRESH	4-00-421	A-F	18	0.4	1.3	–	–	–	0.4	15.3	0.5	-0.1	0.1	0.2
			M-F	100	2.2	7.3	–	–	–	2.2	85.5	2.8	-0.5	0.7	1.3
11	POMACE, WET	4-00-424	A-F	23	1.1	4.7	–	–	–	1.1	14.6	1.3	-0.4	0.7	0.8
			M-F	100	5.0	20.7	–	–	–	4.7	64.0	5.6	-1.7	3.2	3.5
12	POMACE, DEHY	4-00-423	A-F	89	3.0	16.2	–	–	–	4.3	61.2	4.4	-0.4	2.3	2.7
			M-F	100	3.4	18.2	–	–	–	4.8	68.6	5.0	-0.4	2.6	3.0
	ARTICHOKE, JERUSALEM *Helianthus tuberosus*														
13	TUBERS, FRESH	4-20-653	A-F	20	1.7	0.8	–	–	–	0.1	15.5	1.9	–	–	–
			M-F	100	8.3	3.9	–	–	–	0.5	77.6	9.7	–	–	–
	ASPARAGUS, GARDEN *Asparagus officinalis*														
14	STEM BUTTS, MEAL	1-00-437	A-F	91	14.6	29.0	–	–	–	0.9	32.3	14.2	8.8	9.0	9.3
			M-F	100	16.0	31.9	–	–	–	1.0	35.5	15.6	9.7	9.8	10.2
	ATLAS SORGHUM *Sorghum bicolor*														
15	GRAIN	4-04-342	A-F	89	1.9	2.0	–	–	–	3.3	70.5	11.4	7.5	8.6	8.1
			M-F	100	2.1	2.3	–	–	–	3.8	79.1	12.7	8.4	9.6	9.1
	BAKERY														
16	WASTE, DEHY (DRIED BAKERY PRODUCT)	4-00-466	A-F	91	3.7	1.2	–	1.6	–	10.9	65.3	10.1	6.3	8.6	7.1
			M-F	100	4.0	1.3	–	1.8	–	12.0	71.6	11.1	6.9	9.5	7.8
	BANANA *Musa* spp														
17	FRUIT, DEHY	4-00-484	A-F	86	2.6	1.0	–	–	–	0.5	78.6	3.5	-0.4	1.6	2.0
			M-F	100	3.0	1.2	–	–	–	0.6	91.1	4.1	-0.5	1.8	2.3
18	PEELINGS, DEHY	4-00-486	A-F	91	12.8	9.5	–	–	–	4.6	55.6	8.6	2.9	6.0	5.9
			M-F	100	14.0	10.5	–	–	–	5.0	61.1	9.4	3.2	6.6	6.5
	BARLEY *Hordeum vulgare*														
19	GRAIN, ALL ANALYSES	4-00-549	A-F	88	2.4	5.0	16.8	10.7	1.5	1.7	67.7	11.7	8.8	9.6	9.6
			M-F	100	2.7	5.7	19.0	12.1	1.7	1.9	76.5	13.2	10.0	10.8	10.8
20	GRAIN, GRADE 1, 47 lb/bu (*60.5 kg/hl*)	4-00-535	A-F	88	2.3	4.8	–	–	–	2.0	67.3	11.5	7.6	8.7	8.3
			M-F	100	2.6	5.5	–	–	–	2.2	76.6	13.1	8.7	10.0	9.4
21	GRAIN, LIGHT, LESS THAN 36 lb/bu (*46.3 kg/hl*)	4-00-566	A-F	89	3.3	7.6	–	–	–	2.0	63.7	12.3	9.4	9.4	8.9
			M-F	100	3.7	8.5	–	–	–	2.2	71.7	13.9	10.5	10.6	10.0
22	GRAIN, PACIFIC COAST	4-07-939	A-F	89	2.5	6.5	–	–	–	2.0	68.2	9.5	7.1	6.9	6.7
			M-F	100	2.8	7.3	–	–	–	2.2	76.9	10.8	8.1	7.8	7.6
23	GRAIN SCREENINGS	4-00-542	A-F	89	3.1	7.5	–	9.8	–	2.3	64.5	11.5	8.9	8.7	8.3
			M-F	100	3.4	8.4	–	11.0	–	2.6	72.6	13.0	10.0	9.8	9.3
24	MALT SPROUTS, DEHY	4-00-545	A-F	93	5.6	14.2	–	–	–	1.4	48.8	22.9	19.2	21.5	17.2
			M-F	100	6.1	15.3	–	–	–	1.5	52.6	24.6	20.7	23.2	18.6
	BEAN *Phaseolus vulgaris*														
25	SEEDS, KIDNEY	5-00-600	A-F	89	3.6	4.1	–	–	–	1.2	57.8	21.8	14.6	20.6	16.5
			M-F	100	4.0	4.6	–	–	–	1.4	65.3	24.7	16.5	23.2	18.6
26	SEEDS, NAVY	5-00-623	A-F	89	4.0	4.4	–	–	–	1.4	56.7	22.9	20.2	21.6	17.6
			M-F	100	4.5	5.0	–	–	–	1.5	63.4	25.6	22.5	24.2	19.7
27	SEEDS, PINTO	5-00-624	A-F	90	4.3	4.0	–	–	–	1.3	57.9	22.7	17.2	21.4	17.3
			M-F	100	4.8	4.5	–	–	–	1.4	64.2	25.1	19.1	23.7	19.1

ENERGY FEEDS

Entry Number	TDN Ruminant	TDN Swine	TDN Horse	Digestible Energy Ruminant		Digestible Energy Swine		Digestible Energy Horse		Metabolizable Energy Ruminant		Metabolizable Energy Swine		Metabolizable Energy Poultry ME_n		Metabolizable Energy Horse		Net Energy Ruminant NE_m		Net Energy Ruminant NE_g		Net Energy Lactating Cows NE_{lc}	
	%	%	%	Mcal		kcal		Mcal		Mcal		kcal		kcal		Mcal		Mcal		Mcal		Mcal	
				lb	*kg*	lb	*kg*	lb	*kg*	lb	*kg*	lb	*kg*	lb	*kg*	lb	*kg*	lb	*kg*	lb	*kg*	lb	*kg*
1	**23**	**39**	**23**	**0.60**	***1.33***	**771**	***1700***	**0.44**	***0.98***	**0.51**	***1.12***	**727**	***1602***	–	–	**0.36**	***0.80***	**0.40**	***0.87***	**0.26**	***0.56***	**0.38**	***0.84***
	47	77	46	1.21	*2.66*	1543	*3401*	0.89	*1.95*	1.01	*2.23*	1453	*3204*	–	–	0.73	*1.60*	0.79	*1.75*	0.51	*1.13*	0.76	*1.69*
2	**67**	**73**	**59**	**1.37**	***3.03***	**1462**	***3223***	**1.08**	***2.37***	**1.21**	***2.66***	**1381**	***3044***	–	–	**0.88**	***1.95***	**0.77**	***1.71***	**0.52**	***1.14***	**0.73**	***1.62***
	75	82	66	1.55	*3.41*	1646	*3630*	1.21	*2.67*	1.36	*3.00*	1555	*3428*	–	–	0.99	*2.19*	0.87	*1.92*	0.58	*1.28*	0.83	*1.82*
3	**78**	**73**	–	**1.52**	***3.36***	**1452**	***3202***	–	–	**1.35**	***2.99***	**1264**	***2786***	–	–	–	–	**0.82**	***1.80***	**0.55**	***1.21***	**0.77**	***1.70***
	87	80	–	1.69	*3.72*	1608	*3546*	–	–	1.50	*3.31*	1400	*3086*	–	–	–	–	0.91	*2.00*	0.61	*1.34*	0.85	*1.88*
4	**66**	**73**	**45**	**1.34**	***2.95***	**1457**	***3213***	**0.86**	***1.89***	**1.15**	***2.54***	**1375**	***3032***	–	–	**0.70**	***1.55***	**0.77**	***1.70***	**0.51**	***1.12***	**0.66**	***1.46***
	73	81	50	1.48	*3.27*	1615	*3560*	0.95	*2.09*	1.28	*2.82*	1524	*3360*	–	–	0.78	*1.71*	0.85	*1.88*	0.56	*1.24*	0.73	*1.61*
5	**223**	**209**	–	**4.46**	***9.84***	**4144**	***9135***	–	–	**4.31**	***9.49***	**3834**	***8452***	**3685**	***8125***	–	–	**2.95**	***6.49***	**2.23**	***4.91***	**2.43**	***5.35***
	225	211	–	4.50	*9.92*	4177	*9210*	–	–	4.34	*9.57*	3865	*8521*	3715	*8191*	–	–	2.97	*6.55*	2.25	*4.95*	2.45	*5.39*
6	**51**	**42**	**45**	**1.00**	***2.20***	–	–	**0.85**	***1.88***	**0.83**	***1.83***	–	–	–	–	**0.70**	***1.54***	**0.47**	***1.03***	**0.25**	***0.54***	**0.50**	***1.10***
	58	48	52	1.14	*2.51*	–	–	0.98	*2.15*	0.95	*2.09*	–	–	–	–	0.80	*1.77*	0.54	*1.18*	0.28	*0.62*	0.57	*1.26*
7	**203**	–	–	**4.07**	***8.97***	–	–	–	–	**3.91**	***8.62***	–	–	**3304**	***7285***	–	–	**2.60**	***5.73***	**1.95**	***4.29***	**2.21**	***4.87***
	209	–	–	4.18	*9.22*	–	–	–	–	4.02	*8.86*	–	–	3395	*7484*	–	–	2.67	*5.88*	2.00	*4.41*	2.27	*5.00*
8	**188**	**196**	–	**3.57**	***7.87***	**3750**	***8267***	–	–	**3.40**	***7.51***	**3617**	***7973***	**3482**	***7677***	–	–	**2.65**	***5.85***	**1.99**	***4.38***	**2.03**	***4.48***
	189	198	–	3.60	*7.94*	3783	*8340*	–	–	3.43	*7.57*	3648	*8044*	3513	*7744*	–	–	2.68	*5.90*	2.01	*4.42*	2.05	*4.52*
9	**67**	**65**	–	**1.22**	***2.70***	**1304**	***2875***	–	–	**1.05**	***2.32***	**1226**	***2702***	–	–	–	–	**0.72**	***1.58***	**0.47**	***1.03***	**0.69**	***1.52***
	74	73	–	1.37	*3.01*	1457	*3211*	–	–	1.18	*2.59*	1369	*3018*	–	–	–	–	0.80	*1.77*	0.52	*1.15*	0.77	*1.70*
10	**13**	**16**	**12**	**0.27**	***0.60***	**319**	***703***	**0.22**	***0.48***	**0.24**	***0.53***	**302**	***666***	–	–	**0.18**	***0.39***	**0.15**	***0.34***	**0.10**	***0.23***	**0.15**	***0.32***
	74	89	66	1.53	*3.37*	1782	*3929*	1.21	*2.67*	1.34	*2.95*	1688	*3722*	–	–	0.99	*2.19*	0.86	*1.90*	0.57	*1.26*	0.82	*1.81*
11	**17**	**17**	**9**	**0.33**	***0.73***	**337**	***744***	**0.18**	***0.41***	**0.29**	***0.64***	**317**	***700***	–	–	**0.15**	***0.33***	**0.17**	***0.38***	**0.11**	***0.24***	**0.17**	***0.37***
	74	74	42	1.46	*3.21*	1483	*3270*	0.81	*1.78*	1.27	*2.79*	1395	*3075*	–	–	0.66	*1.46*	0.76	*1.67*	0.48	*1.06*	0.74	*1.62*
12	**60**	**69**	**40**	**1.21**	***2.66***	**1380**	***3043***	**0.77**	***1.71***	**1.04**	***2.28***	**1300**	***2867***	–	–	**0.64**	***1.40***	**0.66**	***1.46***	**0.42**	***0.91***	**0.64**	***1.42***
	68	77	45	1.35	*2.98*	1548	*3413*	0.87	*1.91*	1.16	*2.56*	1459	*3216*	–	–	0.71	*1.57*	0.74	*1.63*	0.47	*1.03*	0.72	*1.59*
13	–	–	–	–	–	–	–	–	–	–	–	–	–	–	–	–	–	–	–	–	–	–	–
	–	–	–	–	–	–	–	–	–	–	–	–	–	–	–	–	–	–	–	–	–	–	–
14	**43**	**26**	**40**	**0.87**	***1.91***	–	–	**0.76**	***1.68***	**0.69**	***1.52***	–	–	–	–	**0.63**	***1.38***	**0.37**	***0.82***	**0.15**	***0.32***	**0.44**	***0.96***
	47	29	44	0.95	*2.10*	–	–	0.84	*1.85*	0.76	*1.67*	–	–	–	–	0.69	*1.52*	0.41	*0.90*	0.16	*0.36*	0.48	*1.06*
15	**80**	**77**	**67**	**1.52**	***3.34***	**1533**	***3379***	**1.22**	***2.68***	**1.35**	***2.98***	**1450**	***3197***	–	–	**1.00**	***2.20***	**0.79**	***1.74***	**0.53**	***1.16***	**0.75**	***1.65***
	90	86	75	1.70	*3.75*	1720	*3792*	1.37	*3.01*	1.51	*3.34*	1627	*3588*	–	–	1.12	*2.47*	0.89	*1.95*	0.59	*1.31*	0.84	*1.85*
16	**82**	**89**	–	**1.62**	***3.57***	**1794**	***3954***	–	–	**1.43**	***3.16***	**1683**	***3711***	**1739**	***3834***	–	–	**1.00**	***2.20***	**0.70**	***1.54***	**0.85**	***1.88***
	89	98	–	1.78	*3.91*	1966	*4335*	–	–	1.57	*3.46*	1845	*4068*	1906	*4203*	–	–	1.09	*2.41*	0.77	*1.69*	0.94	*2.06*
17	**64**	**80**	–	**1.33**	***2.94***	**1604**	***3537***	–	–	**1.17**	***2.58***	**1522**	***3355***	–	–	–	–	**0.76**	***1.68***	**0.51**	***1.12***	**0.72**	***1.59***
	74	93	–	1.55	*3.41*	1859	*4099*	–	–	1.36	*2.99*	1764	*3889*	–	–	–	–	0.88	*1.95*	0.59	*1.30*	0.84	*1.84*
18	**58**	**78**	**37**	**1.24**	***2.74***	**1554**	***3425***	**0.72**	***1.60***	**1.07**	***2.35***	**1469**	***3240***	–	–	**0.59**	***1.31***	**0.71**	***1.56***	**0.45**	***1.00***	**0.68**	***1.51***
	64	85	41	1.36	*3.01*	1707	*3762*	0.80	*1.75*	1.17	*2.59*	1614	*3558*	–	–	0.65	*1.44*	0.78	*1.71*	0.50	*1.10*	0.75	*1.66*
19	**75**	**70**	**73**	**1.55**	***3.42***	**1396**	***3078***	**1.31**	***2.88***	**1.17**	***2.57***	**1331**	***2934***	**1180**	***2602***	**1.07**	***2.36***	**0.78**	***1.73***	**0.52**	***1.16***	**0.82**	***1.81***
	85	79	82	1.75	*3.86*	1578	*3478*	1.48	*3.26*	1.32	*2.91*	1504	*3316*	1334	*2941*	1.21	*2.67*	0.89	*1.95*	0.59	*1.31*	0.93	*2.05*
20	**71**	**70**	**69**	**1.42**	***3.13***	**1403**	***3092***	**1.25**	***2.76***	**1.25**	***2.76***	**1323**	***2917***	**1105**	***2436***	**1.03**	***2.26***	**0.74**	***1.63***	**0.49**	***1.07***	**0.71**	***1.55***
	81	80	79	1.62	*3.56*	1597	*3521*	1.43	*3.15*	1.43	*3.15*	1507	*3322*	1258	*2774*	1.17	*2.58*	0.84	*1.86*	0.55	*1.22*	0.80	*1.77*
21	**69**	**67**	**67**	**1.35**	***2.98***	**1345**	***2964***	**1.21**	***2.68***	**1.18**	***2.61***	**1266**	***2790***	–	–	**1.00**	***2.20***	**0.72**	***1.58***	**0.47**	***1.03***	**0.69**	***1.52***
	77	76	75	1.52	*3.36*	1514	*3338*	1.37	*3.02*	1.33	*2.94*	1425	*3142*	–	–	1.12	*2.47*	0.81	*1.78*	0.53	*1.16*	0.78	*1.71*
22	**75**	**70**	**66**	**1.51**	***3.32***	**1405**	***3097***	**1.20**	***2.64***	**1.34**	***2.96***	**1324**	***2918***	**1171**	***2582***	**0.98**	***2.16***	**0.81**	***1.77***	**0.54**	***1.20***	**0.76**	***1.67***
	85	80	74	1.70	*3.75*	1585	*3495*	1.35	*2.98*	1.51	*3.34*	1494	*3293*	1322	*2914*	1.11	*2.44*	0.91	*2.00*	0.61	*1.35*	0.86	*1.89*
23	**71**	**69**	**65**	**1.43**	***3.14***	**1133**	***2498***	**1.18**	***2.61***	**1.26**	***2.77***	**1057**	***2330***	**815**	***1797***	**0.97**	***2.14***	**0.78**	***1.73***	**0.52**	***1.15***	**0.74**	***1.64***
	80	77	73	1.60	*3.53*	1274	*2809*	1.33	*2.93*	1.42	*3.12*	1189	*2620*	917	*2021*	1.09	*2.41*	0.88	*1.94*	0.59	*1.30*	0.84	*1.84*
24	**66**	**61**	–	**1.31**	***2.89***	**1219**	***2688***	–	–	**1.13**	***2.50***	**1046**	***2307***	**722**	***1592***	–	–	**0.69**	***1.52***	**0.43**	***0.95***	**0.67**	***1.48***
	71	66	–	1.41	*3.11*	1313	*2894*	–	–	1.22	*2.69*	1126	*2483*	778	*1714*	–	–	0.74	*1.63*	0.47	*1.03*	0.72	*1.60*
25	**73**	**79**	–	**1.46**	***3.22***	**1584**	***3493***	–	–	**1.29**	***2.85***	**1388**	***3059***	–	–	–	–	–	–	–	–	–	–
	82	89	–	1.65	*3.64*	1789	*3945*	–	–	1.46	*3.22*	1567	*3455*	–	–	–	–	–	–	–	–	–	–
26	**76**	**78**	–	**1.53**	***3.37***	**1737**	***3830***	–	–	**1.36**	***3.00***	**1531**	***3376***	**1052**	***2320***	–	–	**0.83**	***1.83***	**0.56**	***1.24***	**0.78**	***1.72***
	85	87	–	1.71	*3.77*	1941	*4280*	–	–	1.52	*3.36*	1711	*3772*	1176	*2593*	–	–	0.93	*2.05*	0.63	*1.39*	0.87	*1.93*
27	**74**	**78**	–	**1.47**	***3.25***	**1569**	***3460***	–	–	**1.30**	***2.87***	**1372**	***3025***	–	–	–	–	**0.81**	***1.79***	**0.54**	***1.20***	**0.77**	***1.69***
	82	87	–	1.63	*3.60*	1740	*3835*	–	–	1.44	*3.18*	1521	*3354*	–	–	–	–	0.90	*1.98*	0.60	*1.33*	0.85	*1.87*

(Continued)

TABLE V-1A ENERGY FEEDS, MINERAL AND VITAMIN COMPOSITION OF FEEDS, DATA EXPRESSED **AS-FED** AND **MOISTURE-FREE**

Entry Number	Feed Name Description	Moisture Basis: A-F (as-fed) or M-F (moisture-free)	Dry Matter	Macro Minerals							Micro Minerals						
				Calcium (Ca)	Phosphorus (P)	Sodium (Na)	Chlorine (Cl)	Magnesium (Mg)	Potassium (K)	Sulfur (S)	Cobalt (Co)	Copper (Cu)	Iodine (I)	Iron (Fe)	Manganese (Mn)	Selenium (Se)	Zinc (Zn)
			%	%	%	%	%	%	%	%	ppm or mg/kg	ppm or mg/kg	ppm or mg/kg	%	ppm or mg/kg	ppm or mg/kg	ppm or mg/kg
	ACORNS, OAK *Quercus* spp																
1	FRESH	A-F	50	–	–	–	–	–	–	–	–	–	–	–	–	–	–
		M-F	100	–	–	–	–	–	–	–	–	–	–	–	–	–	–
	ADLAY (JOB'S TEARS) *Coix lacrymajobi*																
2	SEEDS	A-F	89	0.02	–	–	–	–	–	–	–	–	–	0.005	–	–	–
		M-F	100	0.02	–	–	–	–	–	–	–	–	–	0.005	–	–	–
	ALFALFA (LUCERNE) *Medicago sativa*																
3	SEED SCREENINGS	A-F	90	1.11	0.34	0.02	–	0.45	2.12	–	1.201	23.6	–	0.069	68.7	–	80.0
		M-F	100	1.23	0.37	0.02	–	0.50	2.35	–	1.330	26.2	–	0.077	76.1	–	88.6
	ALMOND *Prunus amygdalus*																
4	HULLS	A-F	90	0.19	0.09	–	–	–	0.48	0.10	–	–	–	–	–	–	–
		M-F	100	0.21	0.10	–	–	–	0.53	0.11	–	–	–	–	–	–	–
	ANIMAL																
5	FAT, HYDROLYZED	A-F	99	–	–	–	–	–	–	–	–	–	–	–	–	–	–
		M-F	100	–	–	–	–	–	–	–	–	–	–	–	–	–	–
6	RUMEN CONTENTS, DEHY (PAUNCH PRODUCT)	A-F	87	0.69	0.58	–	–	–	–	–	–	–	–	–	–	–	–
		M-F	100	0.79	0.67	–	–	–	–	–	–	–	–	–	–	–	–
7	TALLOW	A-F	97	–	–	–	–	–	–	–	–	–	–	–	–	–	–
		M-F	100	–	–	–	–	–	–	–	–	–	–	–	–	–	–
	ANIMAL-POULTRY																
8	FAT	A-F	99	–	–	–	–	–	0.23	–	–	–	–	–	–	–	–
		M-F	100	–	–	–	–	–	0.23	–	–	–	–	–	–	–	–
	ANISE *Pimpinella anisum*																
9	SEEDS, MEAL MECH EXTD	A-F	90	1.23	0.21	–	–	–	1.47	–	–	–	–	–	–	–	–
		M-F	100	1.37	0.24	–	–	–	1.64	–	–	–	–	–	–	–	–
	APPLE *Malus* spp																
10	FRUIT, FRESH	A-F	18	0.01	0.01	0.01	0.01	0.05	0.14	0.01	–	2.2	–	0.001	3.1	–	–
		M-F	100	0.06	0.06	0.06	0.06	0.28	0.78	0.06	–	12.3	–	0.006	17.2	–	–
11	POMACE, WET	A-F	23	0.02	0.02	–	–	–	0.11	–	–	–	–	–	–	–	–
		M-F	100	0.10	0.10	–	–	–	0.47	–	–	–	–	–	–	–	–
12	POMACE, DEHY	A-F	89	0.11	0.10	0.12	–	0.06	0.43	0.02	–	–	–	0.027	7.2	–	–
		M-F	100	0.13	0.12	0.14	–	0.07	0.49	0.02	–	–	–	0.030	8.1	–	–
	ARTICHOKE, JERUSALEM *Helianthus tuberosus*																
13	TUBERS, FRESH	A-F	20	–	0.06	–	–	–	0.41	–	–	–	–	–	–	–	–
		M-F	100	–	0.30	–	–	–	2.05	–	–	–	–	–	–	–	–
	ASPARAGUS, GARDEN *Asparagus officinalis*																
14	STEM BUTTS, MEAL	A-F	91	–	–	–	–	–	–	–	–	–	–	–	–	–	–
		M-F	100	–	–	–	–	–	–	–	–	–	–	–	–	–	–
	ATLAS SORGHUM *Sorghum bicolor*																
15	GRAIN	A-F	89	–	–	–	–	–	–	–	–	–	–	–	–	–	–
		M-F	100	–	–	–	–	–	–	–	–	–	–	–	–	–	–
	BAKERY																
16	WASTE, DEHY (DRIED BAKERY PRODUCT)	A-F	91	0.14	0.22	1.02	1.47	0.16	0.40	0.02	1.224	11.0	–	0.017	65.0	–	17.8
		M-F	100	0.15	0.24	1.12	1.61	0.18	0.43	0.02	1.342	12.1	–	0.018	71.2	–	19.5
	BANANA *Musa* spp																
17	FRUIT, DEHY	A-F	86	–	–	–	–	–	–	–	–	–	–	–	–	–	–
		M-F	100	–	–	–	–	–	–	–	–	–	–	–	–	–	–
18	PEELINGS, DEHY	A-F	91	–	–	–	–	–	–	–	–	–	–	–	–	–	–
		M-F	100	–	–	–	–	–	–	–	–	–	–	–	–	–	–
	BARLEY *Hordeum vulgare*																
19	GRAIN, ALL ANALYSES	A-F	88	0.05	0.34	0.03	0.12	0.13	0.46	0.15	0.171	7.6	0.044	0.008	16.0	0.158	39.3
		M-F	100	0.06	0.39	0.03	0.13	0.15	0.52	0.17	0.193	8.6	0.050	0.009	18.1	0.179	44.4
20	GRAIN, GRADE 1, 47 lb/bu *(60.5 kg/hl)*	A-F	88	0.05	0.35	–	–	–	–	–	–	–	–	–	–	–	–
		M-F	100	0.06	0.39	–	–	–	–	–	–	–	–	–	–	–	–
21	GRAIN, LIGHT, LESS THAN 36 lb/bu *(46.3 kg/hl)*	A-F	89	0.05	0.32	–	–	–	–	–	–	–	–	–	–	–	–
		M-F	100	0.06	0.36	–	–	–	–	–	–	–	–	–	–	–	–
22	GRAIN, PACIFIC COAST	A-F	89	0.05	0.34	0.02	0.15	0.12	0.51	0.14	0.087	8.1	–	0.009	16.0	0.101	15.2
		M-F	100	0.06	0.39	0.02	0.17	0.14	0.58	0.16	0.098	9.1	–	0.010	18.0	0.114	17.1
23	GRAIN SCREENINGS	A-F	89	0.32	0.29	0.02	–	0.12	0.80	0.13	–	–	–	0.006	–	–	–
		M-F	100	0.36	0.33	0.02	–	0.14	0.90	0.15	–	–	–	0.006	–	–	–
24	MALT SPROUTS, DEHY	A-F	93	0.18	0.63	0.88	0.36	0.17	0.25	0.79	–	5.9	–	0.018	29.4	0.416	56.4
		M-F	100	0.19	0.68	0.95	0.39	0.18	0.27	0.85	–	6.3	–	0.020	31.7	0.448	60.7
	BEAN *Phaseolus vulgaris*																
25	SEEDS, KIDNEY	A-F	89	0.11	0.43	0.01	–	0.10	0.93	–	0.508	6.1	–	0.007	17.0	0.355	24.3
		M-F	100	0.12	0.49	0.01	–	0.12	1.05	–	0.573	6.9	–	0.008	19.2	0.401	27.5
26	SEEDS, NAVY	A-F	89	0.17	0.54	0.04	0.06	0.13	1.31	0.23	–	9.9	–	0.010	21.1	–	–
		M-F	100	0.19	0.61	0.05	0.06	0.15	1.47	0.26	–	11.0	–	0.012	23.6	–	–
27	SEEDS, PINTO	A-F	90	0.18	0.48	0.01	–	0.24	2.59	–	0.551	13.0	–	0.014	17.3	–	40.0
		M-F	100	0.20	0.53	0.01	–	0.27	2.87	–	0.611	14.4	–	0.015	19.2	–	44.4

ENERGY FEEDS

Entry Number	Fat-Soluble Vitamins: A (1 mg Carotene = 1667 IU Vit A)	Carotene (Provitamin A)	D	E	K	Water-Soluble Vitamins: B–12	Biotin	Choline	Folacin (Folic Acid)	Niacin	Pantothenic Acid (B–3)	(Pyridoxine) B–6	Riboflavin (B–2)	Thiamin (B–1)
	IU/g	ppm or mg/kg	IU/kg	ppm or mg/kg	ppm or mg/kg	ppb or mcg/kg	ppm or mg/kg	ppm or mg/kg	ppm or mg/kg	ppm or mg/kg	ppm or mg/kg	ppm or mg/kg	ppm or mg/kg	ppm or mg/kg
1	–	–	–	–	–	–	–	–	–	–	–	–	–	–
	–	–	–	–	–	–	–	–	–	–	–	–	–	–
2	–	–	–	–	–	–	–	–	–	**30**	–	–	**2.0**	**3.0**
	–	–	–	–	–	–	–	–	–	34	–	–	2.3	3.4
3	–	–	–	–	–	–	–	–	–	–	–	–	–	–
	–	–	–	–	–	–	–	–	–	–	–	–	–	–
4	–	–	–	–	–	–	–	–	–	–	–	–	–	–
	–	–	–	–	–	–	–	–	–	–	–	–	–	–
5	–	–	–	–	–	–	–	–	–	–	–	–	–	–
	–	–	–	–	–	–	–	–	–	–	–	–	–	–
6	–	–	–	–	–	–	–	–	–	–	–	–	–	–
	–	–	–	–	–	–	–	–	–	–	–	–	–	–
7	–	–	–	–	–	–	–	–	–	–	–	–	–	–
	–	–	–	–	–	–	–	–	–	–	–	–	–	–
8	–	–	–	**7.9**	–	–	–	–	–	–	–	–	–	–
	–	–	–	7.9	–	–	–	–	–	–	–	–	–	–
9	–	–	–	–	–	–	–	–	–	–	–	–	–	–
	–	–	–	–	–	–	–	–	–	–	–	–	–	–
10	–	–	–	–	–	–	–	–	–	**6**	–	–	**0.2**	**0.4**
	–	–	–	–	–	–	–	–	–	31	–	–	1.2	2.5
11	–	–	–	–	–	–	–	–	–	–	–	–	–	–
	–	–	–	–	–	–	–	–	–	–	–	–	–	–
12	–	–	–	–	–	–	–	–	–	–	–	–	–	–
	–	–	–	–	–	–	–	–	–	–	–	–	–	–
13	–	–	–	–	–	–	–	–	–	–	–	–	–	–
	–	–	–	–	–	–	–	–	–	–	–	–	–	–
14	–	–	–	–	–	–	–	–	–	–	–	–	–	–
	–	–	–	–	–	–	–	–	–	–	–	–	–	–
15	–	–	–	–	–	–	**0.26**	–	–	**34**	**13.4**	–	**2.0**	–
	–	–	–	–	–	–	0.29	–	–	39	15.0	–	2.2	–
16	**7.0**	**4.2**	–	**40.9**	–	–	**0.07**	**917**	**0.19**	**26**	**8.2**	**4.29**	**1.4**	**2.9**
	7.7	4.6	–	44.9	–	–	0.07	1005	0.20	28	9.0	4.70	1.5	3.2
17	–	–	–	–	–	–	–	–	–	–	–	–	–	–
	–	–	–	–	–	–	–	–	–	–	–	–	–	–
18	–	–	–	–	–	–	–	–	–	–	–	–	–	–
	–	–	–	–	–	–	–	–	–	–	–	–	–	–
19	**3.4**	**2.0**	–	**23.2**	**0.22**	–	**0.15**	**1036**	**0.57**	**76**	**7.9**	**5.80**	**1.6**	**4.5**
	3.8	2.3	–	26.2	0.24	–	0.17	1171	0.64	86	9.0	6.55	1.8	5.0
20	–	–	–	–	–	–	–	–	–	–	–	–	–	**4.3**
	–	–	–	–	–	–	–	–	–	–	–	–	–	4.9
21	–	–	–	–	–	–	–	–	–	–	–	–	–	–
	–	–	–	–	–	–	–	–	–	–	–	–	–	–
22	–	–	–	**26.2**	–	–	**0.15**	**976**	**0.50**	**47**	**7.1**	**2.89**	**1.5**	**4.2**
	–	–	–	29.6	–	–	0.17	1102	0.56	53	8.0	3.26	1.7	4.7
23	–	–	–	–	–	–	–	–	–	–	**7.9**	–	**1.1**	–
	–	–	–	–	–	–	–	–	–	–	8.9	–	1.2	–
24	–	–	–	**3.7**	–	–	**4.09**	**1591**	**0.20**	**55**	**9.0**	**8.62**	**2.8**	**8.3**
	–	–	–	4.0	–	–	4.40	1713	0.22	59	9.6	9.28	3.0	8.9
25	–	–	–	–	–	–	–	–	–	**24**	–	–	**1.8**	**5.7**
	–	–	–	–	–	–	–	–	–	27	–	–	2.1	6.4
26	–	–	–	**1.0**	–	–	**0.11**	**1017**	**1.29**	**24**	**2.4**	**0.30**	**1.7**	**6.3**
	–	–	–	1.1	–	–	0.12	1136	1.44	27	2.7	0.33	1.9	7.1
27	–	–	–	–	–	–	–	–	–	**22**	**2.2**	–	**3.1**	**8.6**
	–	–	–	–	–	–	–	–	–	24	2.5	–	3.4	9.6

(Continued)

TABLE V-1A ENERGY FEEDS, COMPOSITION OF FEEDS, DATA EXPRESSED **AS-FED** AND **MOISTURE-FREE**—*(Continued)*

Entry Number	Feed Name Description	International Feed Number	Moisture Basis: A-F (as-fed) or M-F (moisture-free)	Proximate Analysis: Dry Matter	Ash	Crude Fiber	Neutral Det. Fib. (NDF)	Acid Det. Fib. (ADF)	Lignin	Ether Extract (Fat)	N-Free Extract	Crude Protein	Digestible Protein: Ruminant	Swine	Horse
				%	%	%	%	%	%	%	%	%	%	%	%
	BEAN, LIMA *Phaseolus limensis*														
28	SEEDS	**5-00-613**	**A-F**	**90**	**4.1**	**4.5**	**–**	**–**	**–**	**1.3**	**58.9**	**20.7**	**16.7**	**19.4**	**15.1**
			M-F	100	4.6	5.1	–	–	–	1.5	65.7	23.1	18.6	21.7	16.8
	BEAN, MUNG *Phaseolus aureus*														
29	SEEDS	**5-08-185**	**A-F**	**90**	**3.8**	**3.9**	**–**	**–**	**–**	**1.3**	**57.1**	**23.9**	**19.6**	**22.6**	**18.7**
			M-F	100	4.2	4.3	–	–	–	1.4	63.5	26.6	21.8	25.2	20.8
	BEAN, TEPARY *Phaseolus acutifolius, latifolius*														
30	SEEDS	**5-08-349**	**A-F**	**90**	**4.2**	**3.4**	**–**	**–**	**–**	**1.4**	**59.3**	**22.2**	**18.0**	**20.9**	**16.7**
			M-F	100	4.6	3.8	–	–	–	1.5	65.5	24.5	19.9	23.1	18.4
	BEET, COMMON RED *Beta vulgaris, crassa*														
31	ROOTS, FRESH	**4-08-350**	**A-F**	**13**	**1.3**	**0.9**	**–**	**–**	**–**	**0.1**	**9.0**	**1.6**	**1.0**	**1.2**	**1.1**
			M-F	100	10.1	6.6	–	–	–	0.8	70.1	12.5	8.1	9.4	8.9
	BEET, MANGEL *Beta vulgaris, macrorrhiza*														
32	ROOTS, FRESH	**4-00-637**	**A-F**	**11**	**1.1**	**0.8**	**–**	**–**	**–**	**0.1**	**7.7**	**1.3**	**0.9**	**1.0**	**0.9**
			M-F	100	9.6	7.4	–	–	–	0.7	70.5	11.8	8.1	9.2	8.4
	BEET, SUGAR *Beta vulgaris, altissima*														
33	ROOTS, FRESH	**4-00-677**	**A-F**	**20**	**1.1**	**1.2**	**–**	**–**	**–**	**0.1**	**16.1**	**1.4**	**0.3**	**0.5**	**1.0**
			M-F	100	5.5	5.9	–	–	–	0.6	81.2	6.8	1.5	2.5	4.8
34	MOLASSES, MORE THAN 48% INVERT SUGAR, MORE THAN 79.5 DEGREES BRIX	**4-00-668**	**A-F**	**78**	**8.9**	**–**	**–**	**–**	**–**	**0.2**	**62.2**	**6.6**	**3.6**	**4.5**	**4.5**
			M-F	100	11.4	–	–	–	–	0.2	79.9	8.5	4.6	5.8	5.8
35	MOLASSES, STEFFENS	**4-08-351**	**A-F**	**79**	**8.8**	**–**	**–**	**–**	**–**	**–**	**62.1**	**7.8**	**4.6**	**5.6**	**5.4**
			M-F	100	11.2	–	–	–	–	–	78.9	9.9	5.8	7.1	6.9
36	PULP, WET	**4-00-671**	**A-F**	**11**	**0.5**	**2.3**	**–**	**–**	**–**	**0.2**	**6.7**	**1.2**	**0.7**	**0.9**	**0.9**
			M-F	100	4.7	21.3	–	–	–	2.1	60.8	11.2	6.1	8.2	7.9
37	PULP, DEHY	**4-00-669**	**A-F**	**91**	**4.8**	**18.2**	**53.6**	**26.3**	**4.5**	**0.5**	**58.4**	**8.8**	**4.3**	**3.6**	**6.2**
			M-F	100	5.3	20.1	59.0	29.0	5.0	0.6	64.3	9.7	4.7	4.0	6.8
38	PULP WITH MOLASSES, DEHY	**4-00-672**	**A-F**	**92**	**5.7**	**15.2**	**–**	**24.5**	**2.4**	**0.6**	**61.1**	**9.3**	**6.1**	**2.3**	**6.5**
			M-F	100	6.2	16.6	–	26.6	2.6	0.6	66.5	10.1	6.7	2.5	7.1
	BREAD, WHEAT *Triticum aestivum*														
39	DEHY	**4-07-944**	**A-F**	**95**	**2.3**	**0.3**	**–**	**–**	**–**	**2.3**	**78.0**	**12.4**	**8.2**	**9.4**	**8.9**
			M-F	100	2.4	0.3	–	–	–	2.4	81.9	13.0	8.6	9.8	9.3
	BREWERS														
40	GRAINS, WET	**5-02-142**	**A-F**	**22**	**0.9**	**3.1**	**9.2**	**5.0**	**1.1**	**1.5**	**10.7**	**5.8**	**4.2**	**5.5**	**4.5**
			M-F	100	3.9	14.1	42.0	23.0	5.0	6.8	48.7	26.4	19.3	25.0	20.6
41	GRAINS, DEHY	**5-02-141**	**A-F**	**92**	**3.6**	**13.0**	**38.7**	**23.9**	**4.6**	**6.6**	**41.6**	**27.3**	**20.1**	**21.8**	**21.0**
			M-F	100	4.0	14.1	42.0	26.0	5.0	7.1	45.2	29.6	21.8	23.7	22.8
	BROCCOLI *Brassica oleracea, italica*														
42	STEMS WITH HEADS, FRESH, IMMATURE	**4-08-186**	**A-F**	**11**	**1.1**	**1.5**	**–**	**–**	**–**	**0.3**	**4.4**	**3.6**	**2.9**	**3.0**	**2.7**
			M-F	100	10.1	13.8	–	–	–	2.8	40.4	33.0	26.5	27.9	25.1
	BROMEGRASS, CHESS *Bromus secalinus*														
43	SEEDS	**4-08-360**	**A-F**	**90**	**3.6**	**8.2**	**–**	**–**	**–**	**1.7**	**66.4**	**9.7**	**6.0**	**7.1**	**6.8**
			M-F	100	4.0	9.2	–	–	–	1.9	74.1	10.8	6.7	7.9	7.6
	BROOMCORN (MILLET, PROSO; HOG MILLET) *Panicum miliaceum*														
44	GRAIN	**4-03-120**	**A-F**	**90**	**2.9**	**5.3**	**–**	**14.9**	**3.2**	**3.6**	**66.3**	**11.6**	**7.9**	**7.9**	**8.3**
			M-F	100	3.3	6.0	–	16.6	3.6	4.0	73.9	12.9	8.8	8.8	9.3
	BUCKWHEAT, COMMON *Fagopyrum sagittatum*														
45	GRAIN	**4-00-994**	**A-F**	**88**	**2.1**	**10.6**	**–**	**14.9**	**–**	**2.4**	**61.5**	**11.1**	**8.0**	**8.5**	**7.2**
			M-F	100	2.4	12.1	–	17.0	–	2.8	70.2	12.6	9.1	9.7	8.2
46	GRAIN, HULLS ADDED, LESS THAN 22% FIBER	**1-08-003**	**A-F**	**89**	**4.2**	**18.2**	**–**	**–**	**–**	**4.9**	**43.5**	**18.5**	**13.5**	**12.8**	**12.7**
			M-F	100	4.7	20.4	–	–	–	5.5	48.7	20.7	15.1	14.3	14.2
47	GRAIN, HULLS ADDED, MORE THAN 30% FIBER	**1-08-002**	**A-F**	**88**	**3.2**	**28.6**	**–**	**–**	**–**	**3.4**	**39.8**	**13.3**	**9.0**	**8.3**	**8.7**
			M-F	100	3.6	32.4	–	–	–	3.9	45.1	15.1	10.2	9.4	9.8
	CACAO (COCOA) *Theobroma cacao*														
48	SEED COATS (SHELLS)	**1-01-053**	**A-F**	**92**	**5.2**	**13.8**	**–**	**–**	**–**	**8.9**	**50.1**	**14.5**	**9.9**	**9.2**	**9.5**
			M-F	100	5.6	14.9	–	–	–	9.6	54.2	15.7	10.7	9.9	10.3
	CANNA *Canna* spp														
49	TUBERS, FRESH	**4-01-127**	**A-F**	**31**	**2.3**	**1.0**	**–**	**–**	**–**	**0.2**	**26.0**	**1.1**	**0.5**	**0.4**	**0.6**
			M-F	100	7.4	3.4	–	–	–	0.8	84.8	3.6	1.6	1.4	2.0
	CAROB BEAN *Ceratonia siliqua*														
50	PODS WITH SEEDS	**4-08-370**	**A-F**	**88**	**2.5**	**8.7**	**–**	**–**	**–**	**2.6**	**68.5**	**5.5**	**1.9**	**3.3**	**3.6**
			M-F	100	2.8	9.9	–	–	–	3.0	78.0	6.3	2.1	3.8	4.0
51	SEEDS	**4-01-133**	**A-F**	**87**	**2.7**	**7.6**	**–**	**7.8**	**–**	**2.0**	**58.2**	**16.1**	**8.4**	**12.9**	**11.9**
			M-F	100	3.1	8.8	–	9.0	–	2.3	67.2	18.6	9.7	14.9	13.7
	CARROT *Daucus* spp														
52	ROOTS, FRESH[1]	**4-01-145**	**A-F**	**11**	**1.0**	**1.1**	**1.0**	**0.9**	**–**	**0.2**	**8.1**	**1.2**	**0.8**	**0.8**	**0.8**
			M-F	100	8.4	9.5	9.0	8.0	–	1.3	70.7	10.0	7.3	7.2	7.0

Entry Number	TDN			Digestible Energy						Metabolizable Energy								Net Energy					
	Ruminant	Swine	Horse	Ruminant		Swine		Horse		Ruminant		Swine		Poultry ME_n		Horse		Ruminant NE_m		Ruminant NE_g		Lactating Cows NE_{lc}	
	%	%	%	Mcal		kcal		Mcal		Mcal		kcal		kcal		Mcal		Mcal		Mcal		Mcal	
				lb	*kg*	lb	*kg*	lb	*kg*	lb	*kg*	lb	*kg*	lb	*kg*	lb	*kg*	lb	*kg*	lb	*kg*	lb	*kg*
28	78	78	–	1.55	3.42	1556	3430	–	–	1.39	3.05	1360	2999	–	–	–	–	–	–	–	–	–	–
	87	87	–	1.73	3.82	1737	3830	–	–	1.55	3.41	1519	3348	–	–	–	–	–	–	–	–	–	–
29	79	81	–	1.53	3.37	1612	3554	–	–	1.36	3.00	1412	3113	–	–	–	–	0.82	1.80	0.55	1.21	0.77	1.70
	88	90	–	1.70	3.75	1792	3951	–	–	1.51	3.33	1570	3460	–	–	–	–	0.91	2.00	0.61	1.35	0.86	1.89
30	69	80	–	1.37	3.02	1606	3542	–	–	1.20	2.65	1406	3100	–	–	–	–	–	–	–	–	–	–
	76	89	–	1.52	3.34	1775	3914	–	–	1.33	2.93	1554	3426	–	–	–	–	–	–	–	–	–	–
31	10	10	9	0.20	0.44	206	453	0.17	0.38	0.17	0.38	194	427	–	–	0.14	0.31	0.10	0.22	0.06	0.14	0.10	0.21
	81	80	73	1.53	3.38	1593	3512	1.32	2.91	1.34	2.96	1503	3313	–	–	1.08	2.39	0.77	1.71	0.50	1.09	0.75	1.65
32	9	9	8	0.18	0.39	181	399	0.14	0.32	0.15	0.34	171	377	–	–	0.12	0.26	0.09	0.21	0.06	0.14	0.09	0.20
	80	83	72	1.59	3.51	1652	3641	1.31	2.90	1.40	3.09	1560	3439	–	–	1.08	2.38	0.85	1.88	0.56	1.24	0.81	1.79
33	17	18	17	0.33	0.73	355	783	0.30	0.66	0.30	0.65	336	742	–	–	0.25	0.54	0.17	0.38	0.12	0.26	0.17	0.36
	84	89	83	1.67	3.69	1786	3937	1.50	3.31	1.49	3.27	1692	3730	–	–	1.23	2.71	0.88	1.93	0.58	1.29	0.83	1.83
34	61	57	–	1.20	2.64	1130	2491	–	–	1.04	2.29	1061	2338	875	1929	–	–	0.70	1.54	0.47	1.04	0.65	1.43
	78	73	–	1.54	3.38	1451	3198	–	–	1.33	2.94	1362	3002	1123	2476	–	–	0.90	1.98	0.60	1.33	0.83	1.83
35	61	–	–	1.21	2.67	–	–	–	–	1.06	2.34	–	–	–	–	–	–	–	–	–	–	–	–
	77	–	–	1.54	3.40	–	–	–	–	1.35	2.98	–	–	–	–	–	–	–	–	–	–	–	–
36	8	7	7	0.16	0.35	140	308	0.12	0.27	0.14	0.30	130	287	–	–	0.10	0.22	0.08	0.17	0.05	0.11	0.08	0.17
	72	64	60	1.44	3.18	1272	2804	1.11	2.45	1.25	2.76	1187	2618	–	–	0.91	2.01	0.71	1.57	0.44	0.97	0.70	1.55
37	67	67	59	1.31	2.89	1305	2878	1.09	2.40	1.10	2.43	1225	2700	294	648	0.89	1.97	0.73	1.60	0.47	1.03	0.70	1.55
	74	73	65	1.44	3.18	1436	3166	1.20	2.64	1.21	2.68	1348	2971	323	713	0.98	2.17	0.80	1.76	0.52	1.14	0.77	1.70
38	69	70	62	1.39	3.05	1390	3064	1.13	2.50	1.21	2.67	1307	2881	299	660	0.93	2.05	0.73	1.61	0.47	1.04	0.71	1.55
	75	76	67	1.51	3.33	1514	3338	1.23	2.72	1.32	2.91	1424	3139	326	719	1.01	2.23	0.80	1.76	0.52	1.14	0.77	1.69
39	67	82	–	1.33	2.94	1646	3629	–	–	1.15	2.54	1557	3434	1482	3268	–	–	0.71	1.56	0.44	0.98	0.69	1.52
	70	86	–	1.40	3.09	1727	3808	–	–	1.21	2.67	1634	3603	1556	3429	–	–	0.74	1.63	0.47	1.03	0.72	1.60
40	15	18	–	0.32	0.70	351	774	–	–	0.27	0.60	305	673	–	–	–	–	0.18	0.40	0.12	0.26	0.17	0.38
	68	80	–	1.44	3.17	1602	3533	–	–	1.25	2.76	1394	3074	–	–	–	–	0.83	1.82	0.54	1.19	0.79	1.74
41	65	66	48	1.25	2.76	1045	2303	–	–	1.01	2.22	1038	2288	1047	2308	–	–	0.64	1.41	0.39	0.86	0.67	1.48
	71	71	52	1.36	2.99	1134	2500	–	–	1.09	2.41	1127	2484	1137	2506	–	–	0.69	1.53	0.42	0.93	0.73	1.61
42	–	–	–	–	–	–	–	–	–	–	–	–	–	–	–	–	–	–	–	–	–	–	–
	–	–	–	–	–	–	–	–	–	–	–	–	–	–	–	–	–	–	–	–	–	–	–
43	68	70	65	1.35	2.97	1398	3082	1.18	2.60	1.18	2.60	1318	2905	–	–	0.97	2.13	0.72	1.60	0.47	1.04	0.70	1.53
	76	78	72	1.50	3.32	1560	3440	1.32	2.90	1.32	2.90	1471	3242	–	–	1.08	2.38	0.81	1.78	0.53	1.16	0.78	1.71
44	74	68	62	1.47	3.25	1353	2983	1.14	2.51	1.31	2.88	1273	2807	1311	2890	0.93	2.06	0.75	1.66	0.50	1.09	0.72	1.59
	82	75	70	1.64	3.62	1508	3324	1.27	2.80	1.45	3.21	1419	3128	1461	3221	1.04	2.30	0.84	1.85	0.55	1.22	0.80	1.77
45	63	69	62	1.26	2.77	1377	3036	1.13	2.50	1.09	2.40	1297	2859	1200	2645	0.93	2.05	0.65	1.43	0.41	0.90	0.63	1.40
	72	78	71	1.43	3.15	1570	3462	1.29	2.85	1.24	2.74	1479	3261	1368	3016	1.06	2.33	0.74	1.63	0.47	1.03	0.72	1.60
46	52	75	50	1.16	2.56	–	–	0.94	2.07	0.99	2.18	–	–	–	–	0.77	1.70	0.68	1.49	0.43	0.95	0.66	1.45
	59	84	56	1.30	2.87	–	–	1.05	2.32	1.11	2.44	–	–	–	–	0.86	1.90	0.76	1.67	0.48	1.06	0.74	1.62
47	33	51	42	0.88	1.94	–	–	0.80	1.77	0.71	1.56	–	–	–	–	0.66	1.45	0.56	1.23	0.33	0.72	0.57	1.25
	37	58	48	1.00	2.20	–	–	0.91	2.01	0.80	1.77	–	–	–	–	0.75	1.65	0.63	1.40	0.37	0.82	0.64	1.41
48	–	–	–	–	–	–	–	–	–	–	–	–	–	–	–	–	–	–	–	–	–	–	–
	–	–	–	–	–	–	–	–	–	–	–	–	–	–	–	–	–	–	–	–	–	–	–
49	25	28	21	0.49	1.08	568	1253	0.39	0.86	0.43	0.95	539	1188	–	–	0.32	0.70	0.26	0.57	0.17	0.38	0.25	0.55
	82	93	69	1.59	3.51	1851	4080	1.27	2.79	1.41	3.10	1756	3870	–	–	1.04	2.29	0.84	1.86	0.56	1.23	0.81	1.77
50	69	73	56	1.37	3.01	1462	3222	1.03	2.26	1.20	2.65	1381	3045	–	–	0.84	1.86	0.73	1.62	0.48	1.06	0.70	1.54
	79	83	63	1.56	3.43	1665	3671	1.17	2.58	1.37	3.02	1573	3469	–	–	0.96	2.11	0.84	1.84	0.55	1.21	0.80	1.76
51	70	61	70	1.34	2.95	1215	2679	1.26	2.78	1.17	2.59	1140	2513	–	–	1.04	2.28	0.69	1.52	0.45	0.98	0.66	1.46
	80	70	81	1.54	3.40	1403	3093	1.46	3.21	1.35	2.99	1316	2901	–	–	1.20	2.64	0.80	1.76	0.52	1.13	0.77	1.69
52	10	10	8	0.19	0.43	208	458	0.14	0.31	0.17	0.38	195	430	208	458	0.12	0.26	0.10	0.23	0.07	0.16	0.10	0.22
	84	90	67	1.69	3.72	1805	3979	1.23	2.72	1.50	3.30	1695	3736	1805	3979	1.01	2.23	0.91	2.00	0.61	1.35	0.86	1.89

(Continued)

TABLE V-1A ENERGY FEEDS, MINERAL AND VITAMIN COMPOSITION OF FEEDS, DATA EXPRESSED **AS-FED** AND **MOISTURE-FREE**—*(Continued)*

Entry Number	Feed Name Description	Moisture Basis: A-F (as-fed) or M-F (moisture-free)	Dry Matter	Macro Minerals							Micro Minerals						
				Calcium (Ca)	Phosphorus (P)	Sodium (Na)	Chlorine (Cl)	Magnesium (Mg)	Potassium (K)	Sulfur (S)	Cobalt (Co)	Copper (Cu)	Iodine (I)	Iron (Fe)	Manganese (Mn)	Selenium (Se)	Zinc (Zn)
			%	%	%	%	%	%	%	%	ppm or mg/kg	ppm or mg/kg	ppm or mg/kg	%	ppm or mg/kg	ppm or mg/kg	ppm or mg/kg
	BEAN, LIMA *Phaseolus limensis*																
28	SEEDS	**A-F**	**90**	**0.08**	**0.38**	**0.02**	**0.03**	**0.18**	**1.61**	**0.20**	–	**8.2**	–	**0.009**	**16.1**	–	–
		M-F	100	0.09	0.42	0.02	0.03	0.20	1.80	0.22	–	9.1	–	0.010	18.0	–	–
	BEAN, MUNG *Phaseolus aureus*																
29	SEEDS	**A-F**	**90**	**0.13**	**0.35**	**0.01**	–	–	**1.04**	–	–	–	–	**0.009**	–	–	–
		M-F	100	0.15	0.38	0.01	–	–	1.15	–	–	–	–	0.009	–	–	–
	BEAN, TEPARY *Phaseolus acutifolius, latifolius*																
30	SEEDS	**A-F**	**90**	–	–	–	–	–	–	–	–	–	–	–	–	–	–
		M-F	100	–	–	–	–	–	–	–	–	–	–	–	–	–	–
	BEET, COMMON RED *Beta vulgaris, crassa*																
31	ROOTS, FRESH	**A-F**	**13**	**0.02**	**0.04**	**0.08**	**0.03**	**0.02**	**0.31**	**0.02**	–	**1.3**	–	**0.002**	**7.0**	–	–
		M-F	100	0.18	0.28	0.62	0.23	0.15	2.40	0.15	–	10.2	–	0.011	54.3	–	–
	BEET, MANGEL *Beta vulgaris, macrorrhiza*																
32	ROOTS, FRESH	**A-F**	**11**	**0.02**	**0.02**	**0.07**	**0.16**	**0.02**	**0.25**	**0.02**	–	**0.6**	–	**0.002**	–	–	–
		M-F	100	0.18	0.22	0.63	1.41	0.20	2.30	0.20	–	5.5	–	0.016	–	–	–
	BEET, SUGAR *Beta vulgaris, altissima*																
33	ROOTS, FRESH	**A-F**	**20**	**0.09**	**0.05**	**0.10**	**0.10**	**0.13**	**0.30**	**0.01**	–	**1.6**	–	**0.002**	**22.8**	–	**6.5**
		M-F	100	0.47	0.24	0.49	0.49	0.66	1.52	0.06	–	8.1	–	0.008	114.7	–	32.7
34	MOLASSES, MORE THAN 48% INVERT SUGAR, MORE THAN 79.5 DEGREES BRIX	**A-F**	**78**	**0.12**	**0.03**	**1.16**	**1.28**	**0.23**	**4.73**	**0.46**	**0.362**	**16.8**	–	**0.007**	**4.5**	–	**14.0**
		M-F	100	0.16	0.03	1.48	1.64	0.29	6.07	0.60	0.465	21.6	–	0.009	5.8	–	18.0
35	MOLASSES, STEFFENS	**A-F**	**79**	**0.11**	**0.02**	**0.92**	**1.99**	**0.01**	**4.66**	**0.44**	–	–	–	–	–	–	–
		M-F	100	0.14	0.03	1.17	2.53	0.01	5.92	0.56	–	–	–	–	–	–	–
36	PULP, WET	**A-F**	**11**	**0.10**	**0.01**	**0.02**	–	**0.02**	**0.02**	**0.02**	–	–	–	**0.004**	–	–	**1.1**
		M-F	100	0.87	0.10	0.19	–	0.22	0.19	0.22	–	–	–	0.033	–	–	10.0
37	PULP, DEHY	**A-F**	**91**	**0.63**	**0.09**	**0.19**	**0.04**	**0.26**	**0.18**	**0.20**	**0.074**	**12.5**	–	**0.027**	**34.2**	–	**0.7**
		M-F	100	0.70	0.10	0.21	0.04	0.28	0.20	0.22	0.081	13.7	–	0.030	37.7	–	0.8
38	PULP WITH MOLASSES, DEHY	**A-F**	**92**	**0.56**	**0.09**	**0.48**	–	**0.15**	**1.63**	**0.39**	**0.209**	**14.7**	–	**0.017**	**18.4**	–	**5.1**
		M-F	100	0.61	0.10	0.53	–	0.16	1.78	0.42	0.227	16.0	–	0.018	20.1	–	5.5
	BREAD, WHEAT *Triticum aestivum*																
39	DEHY	**A-F**	**95**	**0.06**	**0.11**	–	–	–	–	–	–	–	–	–	–	–	–
		M-F	100	0.07	0.11	–	–	–	–	–	–	–	–	–	–	–	–
	BREWERS																
40	GRAINS, WET	**A-F**	**22**	**0.06**	**0.12**	**0.06**	**0.03**	**0.03**	**0.02**	**0.07**	**0.022**	**4.9**	–	**0.006**	**9.0**	–	**23.2**
		M-F	100	0.29	0.54	0.28	0.13	0.15	0.09	0.34	0.101	22.2	–	0.027	40.9	–	106.0
41	GRAINS, DEHY	**A-F**	**92**	**0.30**	**0.51**	**0.21**	**0.15**	**0.15**	**0.09**	**0.30**	**0.076**	**21.7**	**0.066**	**0.024**	**37.2**	–	**27.3**
		M-F	100	0.33	0.55	0.23	0.17	0.17	0.09	0.32	0.083	23.6	0.072	0.026	40.4	–	29.6
	BROCCOLI *Brassica oleracea, italica*																
42	STEMS WITH HEADS, FRESH, IMMATURE	**A-F**	**11**	**0.10**	**0.08**	**0.02**	–	–	**0.38**	–	–	–	–	**0.001**	–	–	–
		M-F	100	0.92	0.73	0.14	–	–	3.49	–	–	–	–	0.010	–	–	–
	BROMEGRASS, CHESS *Bromus secalinus*																
43	SEEDS	**A-F**	**90**	–	–	–	–	–	–	–	–	–	–	–	–	–	–
		M-F	100	–	–	–	–	–	–	–	–	–	–	–	–	–	–
	BROOMCORN (MILLET, PROSO; HOG MILLET) *Panicum miliaceum*																
44	GRAIN	**A-F**	**90**	**0.03**	**0.30**	–	–	**0.16**	**0.43**	–	–	–	–	**0.007**	–	–	–
		M-F	100	0.03	0.34	–	–	0.18	0.48	–	–	–	–	0.008	–	–	–
	BUCKWHEAT, COMMON *Fagopyrum sagittatum*																
45	GRAIN	**A-F**	**88**	**0.10**	**0.33**	**0.05**	**0.04**	**0.10**	**0.45**	**0.14**	**0.049**	**9.5**	–	**0.005**	**33.7**	–	**8.8**
		M-F	100	0.11	0.37	0.06	0.05	0.12	0.51	0.16	0.056	10.8	–	0.005	38.4	–	10.0
46	GRAIN, HULLS ADDED, LESS THAN 22% FIBER	**A-F**	**89**	–	**0.48**	–	–	–	**0.66**	–	–	–	–	–	–	–	–
		M-F	100	–	0.54	–	–	–	0.74	–	–	–	–	–	–	–	–
47	GRAIN, HULLS ADDED, MORE THAN 30% FIBER	**A-F**	**88**	–	**0.37**	–	–	–	**0.68**	–	–	–	–	–	–	–	–
		M-F	100	–	0.42	–	–	–	0.77	–	–	–	–	–	–	–	–
	CACAO (COCOA) *Theobroma cacao*																
48	SEED COATS (SHELLS)	**A-F**	**92**	–	–	–	–	–	–	–	–	–	–	–	–	–	–
		M-F	100	–	–	–	–	–	–	–	–	–	–	–	–	–	–
	CANNA *Canna* spp																
49	TUBERS, FRESH	**A-F**	**31**	–	–	–	–	–	–	–	–	–	–	–	–	–	–
		M-F	100	–	–	–	–	–	–	–	–	–	–	–	–	–	–
	CAROB BEAN *Ceratonia siliqua*																
50	PODS WITH SEEDS	**A-F**	**88**	**0.20**	**0.30**	–	–	–	–	–	–	–	–	–	–	–	–
		M-F	100	0.23	0.34	–	–	–	–	–	–	–	–	–	–	–	–
51	SEEDS	**A-F**	**87**	**0.27**	**0.19**	–	–	–	–	–	–	–	–	–	–	–	–
		M-F	100	0.31	0.22	–	–	–	–	–	–	–	–	–	–	–	–
	CARROT *Daucus* spp																
52	ROOTS, FRESH	**A-F**	**11**	**0.05**	**0.04**	**0.06**	**0.06**	**0.02**	**0.32**	**0.02**	–	**1.2**	–	**0.002**	**3.6**	–	–
		M-F	100	0.40	0.35	0.48	0.50	0.20	2.80	0.17	–	10.4	–	0.013	31.5	–	–

Entry Number	Fat-Soluble Vitamins					Water-Soluble Vitamins								
	A (1 mg Carotene = 1667 IU Vit A)	Carotene (Provitamin A)	D	E	K	B-12	Biotin	Choline	Folacin (Folic Acid)	Niacin	Pantothenic Acid (B-3)	(Pyridoxine) B-6	Riboflavin (B-2)	Thiamin (B-1)
	IU/g	ppm or mg/kg	IU/kg	ppm or mg/kg	ppm or mg/kg	ppb or mcg/kg	ppm or mg/kg	ppm or mg/kg	ppm or mg/kg	ppm or mg/kg	ppm or mg/kg	ppm or mg/kg	ppm or mg/kg	ppm or mg/kg
28	–	–	–	–	–	–	–	–	3.31	20	8.4	–	1.7	4.6
	–	–	–	–	–	–	–	–	3.70	22	9.4	–	1.9	5.1
29	–	–	–	–	–	–	–	–	–	25	–	–	2.1	3.9
	–	–	–	–	–	–	–	–	–	27	–	–	2.4	4.3
30	–	–	–	–	–	–	–	–	–	–	–	–	–	–
	–	–	–	–	–	–	–	–	–	–	–	–	–	–
31	–	–	–	–	–	–	–	–	–	4	–	–	0.5	0.3
	–	–	–	–	–	–	–	–	–	31	–	–	3.9	2.4
32	0.2	0.1	–	–	–	–	–	–	0.17	3	1.0	0.43	0.4	0.3
	1.5	0.9	–	–	–	–	–	–	1.58	31	9.4	3.94	3.9	2.4
33	–	–	–	–	–	–	–	–	–	–	–	–	–	–
	–	–	–	–	–	–	–	–	–	–	–	–	–	–
34	–	–	–	4.0	–	–	–	827	–	41	4.5	–	2.3	–
	–	–	–	5.1	–	–	–	1062	–	53	5.8	–	2.9	–
35	–	–	–	–	–	–	–	–	–	–	–	–	–	–
	–	–	–	–	–	–	–	–	–	–	–	–	–	–
36	–	–	–	–	–	–	–	–	–	–	–	–	–	–
	–	–	–	–	–	–	–	–	–	–	–	–	–	–
37	0.4	0.2	1	–	–	–	–	820	–	17	1.4	–	0.7	0.4
	0.4	0.2	1	–	–	–	–	902	–	18	1.5	–	0.8	0.4
38	0.4	0.2	–	–	–	–	–	814	–	16	1.5	–	0.7	–
	0.4	0.2	–	–	–	–	–	887	–	18	1.7	–	0.7	–
39	–	–	–	–	–	–	–	–	–	29	–	–	2.0	2.8
	–	–	–	–	–	–	–	–	–	30	–	–	2.1	2.9
40	–	–	–	5.5	–	–	–	–	–	–	–	–	–	–
	–	–	–	25.0	–	–	–	–	–	–	–	–	–	–
41	0.8	0.5	–	26.7	–	3.6	0.44	1651	0.22	44	8.2	1.03	1.5	0.6
	0.8	0.5	–	29.0	–	3.9	0.48	1792	0.24	47	8.9	1.11	1.6	0.7
42	–	–	–	–	–	–	–	–	–	9	–	–	2.3	1.0
	–	–	–	–	–	–	–	–	–	83	–	–	21.1	9.2
43	–	–	–	–	–	–	–	–	–	–	–	–	–	–
	–	–	–	–	–	–	–	–	–	–	–	–	–	–
44	–	–	–	–	–	–	–	438	–	24	10.9	–	3.3	7.5
	–	–	–	–	–	–	–	488	–	27	12.2	–	3.7	8.3
45	–	–	–	–	–	–	–	439	–	18	11.5	–	4.7	3.7
	–	–	–	–	–	–	–	501	–	21	13.1	–	5.4	4.2
46	–	–	–	–	–	–	–	–	–	–	–	–	–	–
	–	–	–	–	–	–	–	–	–	–	–	–	–	–
47	–	–	–	–	–	–	–	–	–	–	–	–	–	–
	–	–	–	–	–	–	–	–	–	–	–	–	–	–
48	–	–	3	–	–	–	–	–	–	–	–	–	–	–
	–	–	3	–	–	–	–	–	–	–	–	–	–	–
49	–	–	–	–	–	–	–	–	–	–	–	–	–	–
	–	–	–	–	–	–	–	–	–	–	–	–	–	–
50	–	–	–	–	–	–	–	–	–	–	–	–	–	–
	–	–	–	–	–	–	–	–	–	–	–	–	–	–
51	–	–	–	–	–	–	–	–	–	–	–	–	–	–
	–	–	–	–	–	–	–	–	–	–	–	–	–	–
52	129.9	77.9	–	6.9	–	–	0.01	–	0.14	7	3.5	1.39	0.6	0.7
	1129.4	677.5	–	60.2	–	–	0.07	–	1.21	58	30.1	12.05	4.9	5.8

(Continued)

TABLE V-1A ENERGY FEEDS, COMPOSITION OF FEEDS, DATA EXPRESSED **AS-FED** AND **MOISTURE-FREE**—*(Continued)*

Entry Number	Feed Name Description	International Feed Number	Moisture Basis: A-F (as-fed) or M-F (moisture-free)	Proximate Analysis									Digestible Protein		
				Dry Matter	Ash	Crude Fiber	Neutral Det. Fib. (NDF)	Acid Det. Fib. (ADF)	Lignin	Ether Extract (Fat)	N-Free Extract	Crude Protein	Ruminant	Swine	Horse
				%	%	%	%	%	%	%	%	%	%	%	%
	CARROT (Continued)														
53	ROOTS, DEHY	4-26-070	A-F	89	8.6	8.0	–	–	–	1.2	61.9	9.2	5.8	6.5	8.1
			M-F	100	9.7	9.0	–	–	–	1.4	69.6	10.3	6.5	7.3	9.1
	CASSAVA *Manihot* spp														
54	TUBERS, FRESH	4-01-150	A-F	32	1.3	1.5	–	–	–	0.3	28.2	1.2	–0.1	0.5	0.6
			M-F	100	3.9	4.6	–	–	–	1.0	86.9	3.6	–0.2	1.4	2.0
55	TUBERS, DEHY, MEAL	4-01-152	A-F	91	2.1	4.1	–	–	–	0.6	81.5	2.2	–0.8	1.6	0.9
			M-F	100	2.3	4.5	–	–	–	0.7	90.1	2.4	–0.8	1.7	1.0
56	STARCH BY-PRODUCT (TAPIOCA MEAL; MANIHOT MEAL)	4-08-572	A-F	87	1.8	4.6	–	–	–	0.7	78.8	0.9	–1.8	–0.8	–0.1
			M-F	100	2.1	5.3	–	–	–	0.8	90.8	1.0	–2.1	–0.9	–0.1
	CATTLE *Bos taurus*														
57	MANURE, WITHOUT BEDDING, DEHY	1-01-190	A-F	93	17.9	30.7	–	43.8	–	2.5	30.1	12.0	7.7	6.9	7.6
			M-F	100	19.1	33.0	–	47.0	–	2.7	32.3	12.9	8.2	7.5	8.1
	CHESTNUT *Castanea* spp														
58	KERNELS WITH COATS, FRESH	4-01-207	A-F	48	1.0	1.1	–	–	–	1.5	41.0	2.9	1.2	1.7	1.9
			M-F	100	2.1	2.3	–	–	–	3.2	86.3	6.1	2.4	3.6	3.9
59	KERNELS WITHOUT COATS, WITHOUT STARCH	4-01-208	A-F	89	1.4	15.4	–	–	–	3.7	62.7	5.8	–1.2	3.6	3.8
			M-F	100	1.6	17.3	–	–	–	4.2	70.4	6.5	–1.3	4.0	4.2
	CHICKPEA (GARBANZO; GRAM PEA) *Cicer arietinum*														
60	SEEDS	5-01-218	A-F	89	2.9	7.0	–	–	–	4.1	56.0	19.1	14.3	17.8	13.3
			M-F	100	3.3	7.9	–	–	–	4.6	62.8	21.4	16.1	20.0	14.9
	CHUFA *Cyperus esculentus, sativus*														
61	TUBERS, FRESH	4-08-374	A-F	27	1.8	2.0	–	–	–	1.8	18.8	2.1	1.1	1.4	1.4
			M-F	100	6.8	7.5	–	–	–	6.8	70.9	7.9	4.1	5.3	5.4
	CITRUS *Citrus* spp														
62	PULP, WET	4-08-376	A-F	18	1.4	2.3	–	–	–	0.6	12.8	1.2	0.5	0.7	0.8
			M-F	100	7.7	12.6	–	–	–	3.3	69.9	6.6	2.8	4.1	4.3
63	PULP WITHOUT FINES, DEHY (DRIED CITRUS PULP)[1]	4-01-237	A-F	91	6.0	11.6	20.9	20.0	–	3.4	63.9	6.1	2.7	3.8	4.0
			M-F	100	6.6	12.8	23.0	22.0	–	3.7	70.2	6.7	3.0	4.2	4.4
64	PULP, AMMONIATED, DEHY	4-01-238	A-F	87	4.6	13.2	–	–	–	5.6	51.9	12.1	8.2	9.3	8.7
			M-F	100	5.3	15.1	–	–	–	6.4	59.4	13.8	9.3	10.6	10.0
65	SYRUP (MOLASSES)	4-01-241	A-F	67	5.1	–	–	–	–	0.2	55.7	5.8	2.0	3.9	3.9
			M-F	100	7.6	–	–	–	–	0.3	82.7	8.5	3.0	5.8	5.8
	CITRUS, GRAPEFRUIT *Citrus paradisi*														
66	PULP WITHOUT FINES, DEHY (DRIED GRAPEFRUIT PULP)	4-01-244	A-F	90	5.4	13.2	–	–	–	3.2	61.5	6.9	1.7	4.6	4.6
			M-F	100	6.0	14.6	–	–	–	3.5	68.3	7.7	1.9	5.1	5.2
	CITRUS, LEMON *Citrus limon*														
67	PULP WITHOUT FINES, DEHY (DRIED LEMON PULP)	4-01-247	A-F	93	5.3	14.7	–	–	–	1.4	65.0	6.5	3.0	4.1	4.3
			M-F	100	5.7	15.9	–	–	–	1.5	70.0	7.0	3.2	4.4	4.6
	CITRUS, ORANGE *Citrus sinensis*														
68	PULP WITHOUT FINES, DEHY (DRIED ORANGE PULP)	4-01-254	A-F	88	3.7	8.4	–	–	–	1.7	66.9	7.5	5.9	5.1	5.1
			M-F	100	4.2	9.6	–	–	–	1.9	75.9	8.5	6.7	5.8	5.8
69	PULP WITHOUT FINES, AMMONIATED, DEHY	4-01-255	A-F	89	5.6	12.4	–	–	–	3.3	53.0	14.7	10.4	11.6	10.8
			M-F	100	6.3	13.9	–	–	–	3.7	59.6	16.5	11.7	13.0	12.1
	COCOA (CACAO) *Theobroma cacao*														
70	SEED COATS (SHELLS)	1-01-053	A-F	92	5.2	13.8	–	–	–	8.9	50.1	14.5	9.9	9.2	9.5
			M-F	100	5.6	14.9	–	–	–	9.6	54.2	15.7	10.7	9.9	10.3
	CORN, DENT WHITE *Zea mays, indentata*														
71	GRAIN	4-02-928	A-F	90	1.2	2.8	–	–	–	3.9	71.7	10.8	6.9	8.0	7.7
			M-F	100	1.3	3.1	–	–	–	4.3	79.3	11.9	7.7	8.9	8.5
	CORN, DENT YELLOW *Zea mays, indentata*														
72	GRAIN, ALL ANALYSES	4-02-935	A-F	88	1.3	2.3	–	3.8	–	3.6	71.0	9.9	8.1	7.3	7.0
			M-F	100	1.5	2.6	–	4.3	–	4.1	80.7	11.2	9.2	8.2	7.9
73	GRAIN, GRADE 1, 56 lb/bu *(72.1 kg/hl)*	4-02-930	A-F	86	1.2	2.1	–	–	–	3.9	70.4	8.8	6.8	6.4	6.2
			M-F	100	1.4	2.4	–	–	–	4.5	81.4	10.2	7.9	7.3	7.2
74	GRAIN, GRADE 2, 54 lb/bu *(69.5 kg/hl)*	4-02-931	A-F	87	1.2	2.1	–	–	–	4.0	71.3	8.9	6.8	7.0	6.2
			M-F	100	1.4	2.4	–	–	–	4.5	81.5	10.2	7.8	8.0	7.1
75	GRAIN, GRADE 3, 52 lb/bu *(66.9 kg/hl)*	4-02-932	A-F	86	1.2	2.1	–	–	–	3.7	70.3	8.7	6.5	–	5.6
			M-F	100	1.4	2.4	–	–	–	4.3	81.8	10.1	7.6	–	6.5
76	GRAIN, GRADE 4, 49 lb/bu *(63.1 kg/hl)*	4-02-933	A-F	85	1.2	2.1	–	–	–	3.7	69.2	9.0	5.5	6.5	6.3
			M-F	100	1.5	2.5	–	–	–	4.3	81.2	10.6	6.4	7.7	7.4
77	GRAIN, GRADE 5, 46 lb/bu *(59.2 kg/hl)*	4-02-934	A-F	86	1.3	2.4	–	–	–	3.8	69.1	9.3	5.7	6.7	6.5
			M-F	100	1.5	2.8	–	–	–	4.4	80.5	10.8	6.6	7.9	7.6
78	GRAIN, GRADE SAMPLE (INFERIOR)	4-02-929	A-F	80	1.3	2.0	–	–	–	3.2	64.9	8.7	5.3	6.3	6.1
			M-F	100	1.6	2.5	–	–	–	4.0	81.0	10.8	6.7	7.9	7.6
79	GRAIN, HIGH MOISTURE	4-20-770	A-F	77	1.2	2.1	17.5	3.8	1.3	3.3	61.9	8.1	4.9	5.9	5.7
			M-F	100	1.6	2.7	22.9	4.9	1.7	4.3	80.8	10.6	6.4	7.7	7.4
80	GRAIN, FLAKED[1]	4-28-244	A-F	89	0.9	0.6	8.0	2.7	–	2.0	75.4	9.9	6.2	9.4	7.0
			M-F	100	1.0	0.7	9.0	3.0	–	2.2	84.9	11.2	7.0	10.6	7.9
81	BRAN	4-02-991	A-F	89	1.9	9.6	54.9	14.2	–	5.1	64.0	8.4	4.8	5.9	4.4
			M-F	100	2.2	10.8	61.7	16.0	–	5.7	71.9	9.4	5.4	6.6	4.9

ENERGY FEEDS

Entry Number	TDN Ruminant	TDN Swine	TDN Horse	Digestible Energy Ruminant		Digestible Energy Swine		Digestible Energy Horse		Metabolizable Energy Ruminant		Metabolizable Energy Swine		Poultry ME_n		Horse		Net Energy Ruminant NE_m		Ruminant NE_g		Lactating Cows NE_{lc}	
	%	%	%	Mcal		kcal		Mcal		Mcal		kcal		kcal		Mcal		Mcal		Mcal		Mcal	
				lb	kg	lb	kg	lb	kg	lb	kg	lb	kg	lb	kg	lb	kg	lb	kg	lb	kg	lb	kg
53	76	73	77	1.52	3.35	1469	3238	1.53	3.38	1.25	2.76	1379	3041	–	–	1.25	2.76	–	–	–	–	–	–
	85	82	86	1.71	3.76	1650	3638	1.72	3.80	1.41	3.10	1550	3417	–	–	1.41	3.10	–	–	–	–	–	–
54	26	26	24	0.51	1.13	511	1127	0.43	0.95	0.45	0.99	482	1063	–	–	0.35	0.78	0.28	0.61	0.18	0.41	0.26	0.58
	80	79	73	1.58	3.48	1576	3475	1.32	2.92	1.39	3.07	1486	3277	–	–	1.08	2.39	0.86	1.89	0.57	1.25	0.82	1.80
55	72	66	–	1.43	3.16	1326	2923	–	–	1.26	2.79	1246	2748	–	–	–	–	0.79	1.73	0.52	1.15	0.75	1.64
	79	73	–	1.59	3.49	1465	3230	–	–	1.40	3.08	1377	3036	–	–	–	–	0.87	1.91	0.58	1.27	0.82	1.82
56	70	79	–	1.41	3.10	1587	3499	–	–	1.25	2.74	1505	3318	–	–	–	–	–	–	–	–	–	–
	81	91	–	1.62	3.58	1829	4032	–	–	1.43	3.16	1734	3823	–	–	–	–	–	–	–	–	–	–
57	44	–	25	0.88	1.94	1281	2825	0.53	1.17	0.70	1.54	1222	2695	–	–	0.43	0.96	0.36	0.79	0.13	0.29	0.43	0.96
	47	–	27	0.94	2.08	1374	3029	0.57	1.25	0.75	1.65	1311	2890	–	–	0.47	1.03	0.39	0.85	0.14	0.31	0.47	1.03
58	41	44	33	0.81	1.79	879	1938	0.61	1.34	0.73	1.60	834	1839	–	–	0.50	1.10	0.43	0.94	0.29	0.63	0.40	0.89
	86	93	70	1.71	3.78	1851	4081	1.28	2.83	1.53	3.36	1756	3871	–	–	1.05	2.32	0.90	1.98	0.61	1.33	0.85	1.87
59	45	67	47	1.11	2.45	1342	2959	0.89	1.96	0.94	2.07	1263	2785	–	–	0.73	1.61	0.71	1.56	0.46	1.01	0.68	1.51
	51	75	53	1.25	2.75	1508	3325	1.00	2.20	1.05	2.33	1420	3130	–	–	0.82	1.81	0.80	1.76	0.52	1.14	0.77	1.69
60	78	80	–	1.56	3.45	1607	3543	–	–	1.40	3.08	1408	3104	–	–	–	–	0.89	1.97	0.62	1.36	0.83	1.84
	88	90	–	1.75	3.86	1801	3971	–	–	1.57	3.45	1578	3479	–	–	–	–	1.00	2.21	0.69	1.52	0.94	2.06
61	18	24	12	0.39	0.86	480	1059	0.23	0.50	0.34	0.75	455	1004	–	–	0.19	0.41	0.23	0.51	0.15	0.34	0.22	0.48
	68	91	44	1.47	3.24	1812	3996	0.85	1.88	1.28	2.83	1718	3787	–	–	0.70	1.54	0.87	1.91	0.58	1.27	0.82	1.81
62	15	15	10	0.29	0.63	299	659	0.18	0.40	0.25	0.55	282	622	–	–	0.15	0.33	0.14	0.32	0.09	0.21	0.14	0.31
	82	82	53	1.56	3.43	1634	3601	0.99	2.19	1.37	3.02	1542	3401	–	–	0.81	1.80	0.79	1.74	0.51	1.12	0.76	1.68
63	75	46	48	1.41	3.10	1386	3055	–	–	1.12	2.46	1097	2420	606	1336	–	–	0.74	1.63	0.48	1.06	0.77	1.71
	83	50	53	1.55	3.41	1523	3357	–	–	1.23	2.71	1206	2659	666	1468	–	–	0.81	1.79	0.53	1.16	0.85	1.87
64	76	65	42	1.40	3.09	1293	2850	0.80	1.77	1.24	2.73	1216	2680	–	–	0.66	1.45	0.69	1.53	0.45	0.99	0.67	1.47
	87	74	48	1.60	3.53	1479	3260	0.92	2.03	1.41	3.12	1391	3066	–	–	0.75	1.66	0.79	1.75	0.51	1.13	0.76	1.68
65	51	54	–	1.01	2.22	1084	2390	–	–	0.86	1.89	1022	2253	–	–	–	–	0.57	1.26	0.38	0.83	0.52	1.15
	75	80	–	1.49	3.29	1609	3547	–	–	1.27	2.81	1517	3344	–	–	–	–	0.85	1.87	0.56	1.23	0.78	1.71
66	72	70	48	1.38	3.04	1402	3091	0.91	2.00	1.21	2.66	1321	2913	–	–	0.75	1.64	0.70	1.55	0.45	1.00	0.68	1.50
	80	78	54	1.53	3.37	1556	3430	1.01	2.22	1.34	2.96	1466	3232	–	–	0.83	1.82	0.78	1.72	0.50	1.10	0.75	1.66
67	73	69	57	1.38	3.05	1378	3038	1.05	2.32	1.21	2.66	1296	2858	–	–	0.86	1.90	0.70	1.54	0.44	0.97	0.68	1.50
	78	74	61	1.49	3.29	1484	3272	1.13	2.50	1.30	2.87	1396	3078	–	–	0.93	2.05	0.75	1.66	0.48	1.05	0.73	1.61
68	79	71	61	1.45	3.19	1413	3115	1.11	2.46	1.28	2.83	1333	2939	–	–	0.91	2.01	0.71	1.58	0.47	1.02	0.69	1.51
	89	80	69	1.64	3.62	1603	3534	1.26	2.79	1.45	3.21	1513	3335	–	–	1.04	2.28	0.81	1.79	0.53	1.16	0.78	1.72
69	64	62	56	1.27	2.81	1230	2712	1.03	2.28	1.11	2.44	1153	2542	–	–	0.85	1.87	0.67	1.49	0.43	0.94	0.66	1.45
	72	69	63	1.43	3.16	1382	3047	1.16	2.56	1.24	2.74	1296	2856	–	–	0.95	2.10	0.76	1.67	0.48	1.06	0.74	1.62
70	–	–	–	–	–	–	–	–	–	–	–	–	–	–	–	–	–	–	–	–	–	–	–
	–	–	–	–	–	–	–	–	–	–	–	–	–	–	–	–	–	–	–	–	–	–	–
71	77	78	65	1.53	3.38	1563	3446	1.19	2.62	1.37	3.01	1479	3261	–	–	0.97	2.15	0.81	1.78	0.54	1.19	0.76	1.68
	85	86	72	1.70	3.74	1729	3812	1.31	2.90	1.51	3.33	1636	3607	–	–	1.08	2.38	0.89	1.97	0.60	1.32	0.84	1.86
72	80	79	64	1.63	3.59	1514	3338	1.17	2.57	1.26	2.78	1472	3246	1523	3359	0.96	2.11	0.86	1.90	0.59	1.31	0.82	1.81
	91	90	73	1.85	4.09	1721	3795	1.32	2.92	1.43	3.16	1674	3690	1732	3818	1.09	2.39	0.98	2.16	0.67	1.48	0.93	2.05
73	81	77	60	1.52	3.35	1546	3408	1.10	2.43	1.36	3.00	1464	3229	1470	3241	0.90	1.99	0.78	1.72	0.53	1.16	0.74	1.63
	94	89	70	1.76	3.88	1788	3943	1.28	2.81	1.57	3.47	1694	3736	1701	3750	1.05	2.31	0.91	2.00	0.61	1.34	0.85	1.88
74	80	81	61	1.57	3.47	1586	3498	1.11	2.45	1.45	3.19	1526	3365	1567	3456	0.91	2.01	0.89	1.96	0.61	1.35	0.84	1.85
	92	92	70	1.80	3.96	1813	3998	1.27	2.81	1.66	3.65	1745	3846	1792	3950	1.04	2.30	1.01	2.24	0.70	1.55	0.96	2.11
75	73	73	–	1.45	3.20	1469	3238	–	–	1.19	2.63	1380	3043	–	–	–	–	0.79	1.73	0.53	1.17	0.76	1.68
	84	85	–	1.69	3.72	1708	3766	–	–	1.38	3.05	1605	3538	–	–	–	–	0.91	2.01	0.62	1.36	0.88	1.95
76	80	76	60	1.50	3.30	1515	3339	1.10	2.43	1.34	2.95	1435	3163	–	–	0.90	1.99	0.77	1.70	0.52	1.14	0.73	1.60
	94	89	71	1.75	3.87	1776	3915	1.29	2.85	1.57	3.46	1682	3708	–	–	1.06	2.34	0.90	1.99	0.61	1.34	0.85	1.88
77	73	76	61	1.46	3.23	1515	3339	1.10	2.43	1.30	2.87	1434	3162	–	–	0.91	2.00	0.77	1.70	0.52	1.14	0.73	1.61
	85	88	71	1.71	3.76	1765	3891	1.29	2.84	1.52	3.35	1671	3685	–	–	1.06	2.33	0.90	1.98	0.60	1.33	0.85	1.87
78	68	71	58	1.37	3.01	1411	3111	1.05	2.33	1.22	2.68	1336	2946	–	–	0.87	1.91	0.72	1.58	0.48	1.06	0.68	1.50
	85	88	72	1.70	3.76	1762	3884	1.32	2.90	1.52	3.34	1668	3678	–	–	1.08	2.38	0.90	1.98	0.60	1.33	0.85	1.87
79	71	68	54	1.41	3.11	1354	2984	0.99	2.18	1.27	2.80	1282	2826	–	–	0.81	1.79	0.79	1.75	0.55	1.22	0.74	1.63
	92	88	71	1.85	4.07	1768	3898	1.29	2.84	1.66	3.66	1674	3692	–	–	1.06	2.33	1.04	2.28	0.72	1.59	0.96	2.12
80	–	85	–	–	–	1694	3734	–	–	–	–	1608	3546	–	–	–	–	–	–	–	–	–	–
	–	95	–	–	–	1908	4205	–	–	–	–	1811	3993	–	–	–	–	–	–	–	–	–	–
81	67	73	66	1.37	3.02	1464	3227	1.21	2.66	1.20	2.65	1382	3048	–	–	0.99	2.18	0.76	1.67	0.50	1.10	0.72	1.59
	76	82	75	1.54	3.39	1645	3626	1.35	2.99	1.35	2.97	1553	3425	–	–	1.11	2.45	0.85	1.88	0.56	1.24	0.81	1.79

(Continued)

TABLE V-1A ENERGY FEEDS, MINERAL AND VITAMIN COMPOSITION OF FEEDS, DATA EXPRESSED **AS-FED** AND **MOISTURE-FREE**—*(Continued)*

Entry Number	Feed Name Description	Moisture Basis: A-F (as-fed) or M-F (moisture-free)	Dry Matter	Macro Minerals: Calcium (Ca)	Phosphorus (P)	Sodium (Na)	Chlorine (Cl)	Magnesium (Mg)	Potassium (K)	Sulfur (S)	Micro Minerals: Cobalt (Co)	Copper (Cu)	Iodine (I)	Iron (Fe)	Manganese (Mn)	Selenium (Se)	Zinc (Zn)
			%	%	%	%	%	%	%	%	ppm or mg/kg	ppm or mg/kg	ppm or mg/kg	%	ppm or mg/kg	ppm or mg/kg	ppm or mg/kg
	CARROT (Continued)																
53	ROOTS, DEHY	**A-F**	**89**	**0.33**	**0.29**	**0.89**	–	**0.15**	**2.23**	**0.15**	–	**9.9**	–	**0.010**	**28.0**	–	–
		M-F	100	0.37	0.32	1.00	–	0.17	2.50	0.17	–	11.1	–	0.012	31.5	–	–
	CASSAVA *Manihot* spp																
54	TUBERS, FRESH	**A-F**	**32**	**0.05**	**0.05**	–	–	–	**0.33**	–	–	–	–	–	–	–	–
		M-F	100	0.15	0.15	–	–	–	1.01	–	–	–	–	–	–	–	–
55	TUBERS, DEHY, MEAL	**A-F**	**91**	**0.14**	**0.09**	–	–	–	**0.24**	–	–	–	–	–	–	–	–
		M-F	100	0.15	0.10	–	–	–	0.26	–	–	–	–	–	–	–	–
56	STARCH BY-PRODUCT (TAPIOCA MEAL; MANIHOT MEAL)	**A-F**	**87**	–	**0.03**	–	–	–	**0.23**	–	–	–	–	–	–	–	–
		M-F	100	–	0.04	–	–	–	0.27	–	–	–	–	–	–	–	–
	CATTLE *Bos taurus*																
57	MANURE, WITHOUT BEDDING, DEHY	**A-F**	**93**	**1.35**	**1.08**	–	–	–	**0.47**	–	–	–	–	–	–	–	–
		M-F	100	1.45	1.15	–	–	–	0.50	–	–	–	–	–	–	–	–
	CHESTNUT *Castanea* spp																
58	KERNELS WITH COATS, FRESH	**A-F**	**48**	**0.03**	**0.09**	**0.01**	–	–	**0.45**	–	–	–	–	**0.002**	–	–	–
		M-F	100	0.06	0.19	0.01	–	–	0.96	–	–	–	–	0.004	–	–	–
59	KERNELS WITHOUT COATS, WITHOUT STARCH	**A-F**	**89**	–	–	–	–	–	–	–	–	–	–	–	–	–	–
		M-F	100	–	–	–	–	–	–	–	–	–	–	–	–	–	–
	CHICKPEA (GARBANZO; GRAM PEA)																
60	SEEDS	**A-F**	**89**	**0.15**	**0.33**	**0.03**	–	–	**0.80**	–	–	–	–	**0.007**	–	–	–
		M-F	100	0.17	0.37	0.03	–	–	0.89	–	–	–	–	0.008	–	–	–
	CHUFA *Cyperus esculentus, sativus*																
61	TUBERS, FRESH	**A-F**	**27**	**0.01**	**0.07**	–	–	–	**0.14**	–	–	–	–	–	–	–	–
		M-F	100	0.04	0.26	–	–	–	0.53	–	–	–	–	–	–	–	–
	CITRUS *Citrus* spp																
62	PULP, WET	**A-F**	**18**	–	–	–	–	–	–	–	–	–	–	–	–	–	–
		M-F	100	–	–	–	–	–	–	–	–	–	–	–	–	–	–
63	PULP WITHOUT FINES, DEHY (DRIED CITRUS PULP)	**A-F**	**91**	**1.69**	**0.12**	**0.07**	–	**0.15**	**0.71**	**0.17**	**0.169**	**5.0**	–	**0.033**	**6.6**	–	**13.7**
		M-F	100	1.86	0.13	0.08	–	0.17	0.78	0.19	0.185	5.4	–	0.036	7.3	–	15.1
64	PULP, AMMONIATED, DEHY	**A-F**	**87**	**1.66**	**0.12**	–	–	**0.07**	–	–	–	–	–	–	–	–	–
		M-F	100	1.90	0.14	–	–	0.08	–	–	–	–	–	–	–	–	–
65	SYRUP (MOLASSES)	**A-F**	**67**	**1.18**	**0.09**	**0.28**	**0.07**	**0.14**	**0.09**	**0.16**	**0.109**	**72.8**	–	**0.035**	**40.9**	–	**92.4**
		M-F	100	1.76	0.13	0.41	0.11	0.21	0.14	0.23	0.162	108.0	–	0.051	60.7	–	137.1
	CITRUS, GRAPEFRUIT *Citrus paradisi*																
66	PULP WITHOUT FINES, DEHY (DRIED GRAPEGRUIT PULP)	**A-F**	**90**	**1.34**	**0.17**	–	–	**0.39**	–	–	–	–	–	–	–	–	–
		M-F	100	1.48	0.18	–	–	0.43	–	–	–	–	–	–	–	–	–
	CITRUS, LEMON *Citrus limon*																
67	PULP WITHOUT FINES, DEHY (DRIED LEMON PULP)	**A-F**	**93**	–	–	–	–	–	–	–	–	–	–	–	–	–	–
		M-F	100	–	–	–	–	–	–	–	–	–	–	–	–	–	–
	CITRUS, ORANGE *Citrus sinensis*																
68	PULP WITHOUT FINES, DEHY (DRIED ORANGE PULP)	**A-F**	**88**	**0.62**	**0.10**	–	–	–	–	–	–	–	–	–	–	–	–
		M-F	100	0.71	0.11	–	–	–	–	–	–	–	–	–	–	–	–
69	PULP WITHOUT FINES, AMMONIATED, DEHY	**A-F**	**89**	–	–	–	–	–	–	–	–	–	–	–	–	–	–
		M-F	100	–	–	–	–	–	–	–	–	–	–	–	–	–	–
	COCOA (CACAO) *Theubroma cacao*																
70	SEED COATS (SHELLS)	**A-F**	**92**	–	–	–	–	–	–	–	–	–	–	–	–	–	–
		M-F	100	–	–	–	–	–	–	–	–	–	–	–	–	–	–
	CORN, DENT WHITE *Zea mays, indentata*																
71	GRAIN	**A-F**	**90**	**0.04**	**0.27**	–	–	–	–	–	**0.062**	**6.0**	–	**0.003**	**8.8**	–	–
		M-F	100	0.04	0.30	–	–	–	–	–	0.069	6.6	–	0.003	9.7	–	–
	CORN, DENT YELLOW *Zea mays, indentata*																
72	GRAIN, ALL ANALYSES	**A-F**	**88**	**0.05**	**0.28**	**0.01**	**0.05**	**0.11**	**0.33**	**0.11**	**0.378**	**3.5**	–	**0.004**	**5.7**	**0.127**	**19.4**
		M-F	100	0.05	0.31	0.01	0.06	0.13	0.37	0.13	0.429	4.0	–	0.004	6.4	0.144	22.0
73	GRAIN, GRADE 1, 56 lb/bu (*72.1 kg/hl*)	**A-F**	**86**	**0.03**	**0.27**	**0.01**	**0.05**	**0.10**	**0.28**	**0.12**	–	**4.1**	–	**0.003**	**5.6**	–	–
		M-F	100	0.03	0.31	0.01	0.06	0.11	0.33	0.14	–	4.8	–	0.003	6.4	–	–
74	GRAIN, GRADE 2, 54 lb/bu (*69.5 kg/hl*)	**A-F**	**87**	**0.02**	**0.29**	**0.02**	**0.04**	**0.11**	**0.31**	**0.12**	**0.029**	**3.8**	–	**0.003**	**5.3**	–	**13.7**
		M-F	100	0.03	0.33	0.02	0.05	0.13	0.36	0.14	0.033	4.3	–	0.003	6.1	–	15.6
75	GRAIN, GRADE 3, 52 lb/bu (*66.9 kg/hl*)	**A-F**	**86**	**0.02**	**0.25**	–	–	–	–	–	–	–	–	**0.002**	**5.5**	–	–
		M-F	100	0.02	0.29	–	–	–	–	–	–	–	–	0.002	6.4	–	–
76	GRAIN, GRADE 4, 49 lb/bu (*63.1 kg/hl*)	**A-F**	**85**	**0.04**	**0.26**	**0.01**	**0.05**	**0.13**	**0.27**	**0.10**	–	**2.2**	–	**0.003**	**4.9**	–	–
		M-F	100	0.04	0.30	0.01	0.06	0.15	0.32	0.12	–	2.5	–	0.003	5.8	–	–
77	GRAIN, GRADE 5, 46 lb/bu (*59.2 kg/hl*)	**A-F**	**86**	–	–	–	–	–	–	–	–	–	–	–	–	–	–
		M-F	100	–	–	–	–	–	–	–	–	–	–	–	–	–	–
78	GRAIN, GRADE SAMPLE (INFERIOR)	**A-F**	**80**	–	**0.28**	–	–	–	–	–	–	–	–	–	–	–	–
		M-F	100	–	0.35	–	–	–	–	–	–	–	–	–	–	–	–
79	GRAIN, HIGH MOISTURE	**A-F**	**77**	**0.01**	**0.25**	**0.01**	**0.04**	**0.11**	**0.28**	**0.11**	–	**2.2**	–	**0.003**	**5.3**	–	**25.4**
		M-F	100	0.02	0.33	0.01	0.05	0.15	0.37	0.14	–	2.9	–	0.004	6.9	–	33.2
80	GRAIN, FLAKED	**A-F**	**89**	–	–	–	–	–	–	–	–	–	–	–	–	–	–
		M-F	100	–	–	–	–	–	–	–	–	–	–	–	–	–	–
81	BRAN	**A-F**	**89**	**0.03**	**0.20**	–	**0.05**	**0.26**	**0.64**	**0.07**	–	–	–	–	**15.9**	–	–
		M-F	100	0.04	0.22	–	0.06	0.29	0.72	0.08	–	–	–	–	17.9	–	–

Entry Number	Fat-Soluble Vitamins					Water-Soluble Vitamins								
	A (1 mg Carotene = 1667 IU Vit A)	Carotene (Provitamin A)	D	E	K	B–12	Biotin	Choline	Folacin (Folic Acid)	Niacin	Pantothenic Acid (B–3)B–6	(Pyri-doxine) (B–2)	Ribo-flavin (B–1)	Thiamin
	IU/g	ppm or mg/kg	IU/kg	ppm or mg/kg	ppm or mg/kg	ppb or mcg/kg	ppm or mg/kg	ppm or mg/kg	ppm or mg/kg	ppm or mg/kg	ppm or mg/kg	ppm or mg/kg	ppm or mg/kg	ppm or mg/kg
53	**829.5**	–	–	–	–	–	–	–	–	**45**	–	–	**3.7**	**4.5**
	932.0	–	–	–	–	–	–	–	–	51	–	–	4.2	5.1
54	–	–	–	–	–	–	–	–	–	–	–	–	–	–
	–	–	–	–	–	–	–	–	–	–	–	–	–	–
55	–	–	–	–	–	–	–	–	–	–	–	–	–	–
	–	–	–	–	–	–	–	–	–	–	–	–	–	–
56	–	–	–	–	–	–	–	–	–	–	–	–	–	–
	–	–	–	–	–	–	–	–	–	–	–	–	–	–
57	–	–	–	–	–	–	–	–	–	–	–	–	–	–
	–	–	–	–	–	–	–	–	–	–	–	–	–	–
58	–	–	–	–	–	–	–	–	–	**6**	–	–	**2.2**	**2.2**
	–	–	–	–	–	–	–	–	–	13	–	–	4.6	4.6
59	–	–	–	–	–	–	–	–	–	–	–	–	–	–
	–	–	–	–	–	–	–	–	–	–	–	–	–	–
60	–	–	–	–	–	–	–	–	–	**29**	–	–	**1.2**	**2.4**
	–	–	–	–	–	–	–	–	–	33	–	–	1.3	2.7
61	–	–	–	–	–	–	–	–	–	–	–	–	–	–
	–	–	–	–	–	–	–	–	–	–	–	–	–	–
62	–	–	–	–	–	–	–	–	–	–	–	–	–	–
	–	–	–	–	–	–	–	–	–	–	–	–	–	–
63	**0.4**	**0.2**	–	–	–	–	–	**789**	–	**22**	**14.0**	–	**2.1**	**1.5**
	0.4	0.2	–	–	–	–	–	867	–	24	15.4	–	2.3	1.6
64	–	–	–	–	–	–	–	–	–	–	–	–	–	–
	–	–	–	–	–	–	–	–	–	–	–	–	–	–
65	–	–	–	–	–	–	–	–	–	**27**	**17.2**	–	**6.2**	–
	–	–	–	–	–	–	–	–	–	40	25.5	–	9.2	–
66	–	–	–	–	–	–	–	–	–	–	–	–	–	–
	–	–	–	–	–	–	–	–	–	–	–	–	–	–
67	–	–	–	–	–	–	–	–	–	–	–	–	–	–
	–	–	–	–	–	–	–	–	–	–	–	–	–	–
68	–	–	–	–	–	–	–	–	–	–	–	–	–	–
	–	–	–	–	–	–	–	–	–	–	–	–	–	–
69	–	–	–	–	–	–	–	–	–	–	–	–	–	–
	–	–	–	–	–	–	–	–	–	–	–	–	–	–
70	–	–	**3**	–	–	–	–	–	–	–	–	–	–	–
	–	–	3	–	–	–	–	–	–	–	–	–	–	–
71	**0.7**	**0.4**	–	–	–	–	**0.06**	–	–	**16**	**4.0**	–	**1.4**	**4.6**
	0.7	0.4	–	–	–	–	0.07	–	–	17	4.4	–	1.5	5.1
72	**9.5**	**5.7**	–	**20.9**	**0.22**	–	**0.07**	**504**	**0.31**	**23**	**5.1**	**6.16**	**1.1**	**3.7**
	10.8	6.5	–	23.8	0.25	–	0.08	573	0.35	26	5.8	7.01	1.2	4.2
73	–	–	–	–	–	–	–	–	–	–	**5.8**	–	–	–
	–	–	–	–	–	–	–	–	–	–	6.7	–	–	–
74	**2.9**	**1.7**	–	**21.6**	–	–	**0.06**	**569**	**0.35**	**24**	**3.9**	**6.88**	**1.3**	**3.5**
	3.3	2.0	–	24.7	–	–	0.07	650	0.40	28	4.5	7.87	1.5	4.0
75	–	–	–	–	–	–	–	–	–	–	–	–	–	–
	–	–	–	–	–	–	–	–	–	–	–	–	–	–
76	–	–	–	–	–	–	–	–	–	–	–	–	–	–
	–	–	–	–	–	–	–	–	–	–	–	–	–	–
77	–	–	–	–	–	–	–	–	–	–	–	–	–	–
	–	–	–	–	–	–	–	–	–	–	–	–	–	–
78	–	–	–	–	–	–	–	–	–	–	–	–	–	–
	–	–	–	–	–	–	–	–	–	–	–	–	–	–
79	–	–	–	–	–	–	–	–	–	–	–	–	–	–
	–	–	–	–	–	–	–	–	–	–	–	–	–	–
80	–	–	–	**0.8**	–	–	–	–	–	–	–	–	–	–
	–	–	–	0.9	–	–	–	–	–	–	–	–	–	–
81	–	–	–	–	–	–	**0.09**	–	–	**41**	**5.2**	–	**1.5**	**4.6**
	–	–	–	–	–	–	0.10	–	–	47	5.9	–	1.7	5.2

(Continued)

TABLE V-1A ENERGY FEEDS, COMPOSITION OF FEEDS, DATA EXPRESSED **AS-FED** AND **MOISTURE-FREE**—*(Continued)*

Entry Number	Feed Name Description	International Feed Number	Moisture Basis: A-F (as-fed) or M-F (moisture-free)	Proximate Analysis: Dry Matter	Ash	Crude Fiber	Neutral Det. Fib. (NDF)	Acid Det. Fib. (ADF)	Lignin	Ether Extract (Fat)	N-Free Extract	Crude Protein	Digestible Protein: Ruminant	Swine	Horse
				%	%	%	%	%	%	%	%	%	%	%	%
	CORN, DENT YELLOW (Continued)														
82	DISTILLERS GRAINS, DEHY[1]	5-02-842	A-F	93	2.2	11.5	40.0	15.8	–	8.9	43.1	27.8	20.1	19.4	22.7
			M-F	100	2.4	12.3	43.0	17.0	–	9.5	46.2	29.7	21.5	20.8	24.3
83	DISTILLERS SOLUBLES, DEHY[1]	5-28-237	A-F	93	7.2	4.6	21.4	6.5	–	8.6	45.2	27.4	21.5	20.0	22.3
			M-F	100	7.7	4.9	23.0	7.0	–	9.3	48.6	29.5	23.2	21.5	24.1
84	EARS, GROUND (CORN AND COB MEAL)[1]	4-28-238	A-F	87	1.7	8.2	24.4	9.6	–	3.2	65.7	7.8	4.7	5.6	5.4
			M-F	100	1.9	9.4	28.0	11.0	–	3.7	75.9	9.0	5.5	6.5	6.2
85	GERM MEAL, WET MILLED, SOLV EXTD	5-28-240	A-F	92	3.9	12.2	–	–	–	1.5	53.5	20.7	16.6	19.4	14.9
			M-F	100	4.2	13.3	–	–	–	1.6	58.3	22.6	18.1	21.2	16.2
86	GLUTEN, MEAL, 60% PROTEIN[1]	5-28-242	A-F	90	1.7	1.8	12.6	4.5	–	2.1	23.7	60.8	54.5	–1.3	–0.4
			M-F	100	1.9	2.0	14.0	5.0	–	2.3	26.3	67.5	60.6	–1.4	–0.5
87	GLUTEN FEED[1]	5-28-243	A-F	90	6.6	8.7	40.5	10.8	–	2.1	49.4	23.0	19.8	18.2	17.7
			M-F	100	7.4	9.7	45.0	12.0	–	2.4	55.0	25.6	22.0	20.2	19.6
88	GRITS (HOMINY GRITS)[1]	4-03-011	A-F	90	2.8	4.8	49.5	11.7	–	6.5	65.8	10.3	7.3	7.6	7.3
			M-F	100	3.1	5.3	55.0	13.0	–	7.2	72.9	11.4	8.1	8.5	8.1
89	MOLASSES, MORE THAN 43% DEXTROSE EQUIVALENT, MORE THAN 50% TOTAL DEXTROSE, MORE THAN 78 DEGREES BRIX	4-12-273	A-F	73	8.0	–	–	–	–	–	64.2	0.3	–1.9	–1.1	–0.4
			M-F	100	11.0	–	–	–	–	–	88.6	0.4	–2.7	–1.5	–0.6
90	OIL	4-16-450	A-F	99	–	–	–	–	–	99.0	–	–	–	–	–
			M-F	100	–	–	–	–	–	100.0	–	–	–	–	–
91	STARCH	4-16-449	A-F	90	0.2	0.1	–	–	–	0.1	89.2	0.6	–2.2	–1.2	–0.3
			M-F	100	0.2	0.2	–	–	–	0.2	98.8	0.6	–2.5	–1.3	–0.4
	CORN, FLINT *Zea mays, indurata*														
92	GRAIN	4-02-948	A-F	89	1.5	1.9	–	–	–	4.3	70.9	9.9	6.1	7.2	7.0
			M-F	100	1.7	2.1	–	–	–	4.9	80.2	11.1	6.9	8.2	7.9
	CORN, OPAQUE 2, HIGH LYSINE *Zea mays*														
93	GRAIN	4-11-445	A-F	90	1.6	3.0	–	–	–	4.4	71.0	10.1	6.3	7.4	7.2
			M-F	100	1.7	3.3	–	–	–	4.9	78.8	11.2	7.0	8.3	8.0
	CORN, POP *Zea mays, everta*														
94	GRAIN	4-02-964	A-F	90	1.5	1.9	–	–	–	5.0	70.0	11.5	7.5	8.7	8.2
			M-F	100	1.7	2.1	–	–	–	5.6	77.9	12.8	8.4	9.7	9.2
	CORN, SWEET *Zea mays, saccharata*														
95	GRAIN	4-02-977	A-F	91	2.1	4.0	–	–	–	7.4	66.2	11.2	7.3	8.4	8.0
			M-F	100	2.3	4.4	–	–	–	8.1	72.9	12.4	8.0	9.3	8.8
96	CANNERY RESIDUE, FRESH	2-02-975	A-F	77	3.5	17.0	–	22.3	–	1.9	47.6	6.8	3.9	4.1	–
			M-F	100	4.6	22.2	–	29.0	–	2.5	62.0	8.8	5.1	5.4	–
	COTTON *Gossypium* spp														
97	SEEDS[1]	5-13-749	A-F	91	4.7	19.4	40.0	30.9	–	20.4	24.5	21.8	17.7	20.5	16.2
			M-F	100	5.2	21.4	44.0	34.0	–	22.5	27.0	24.0	19.4	22.6	17.9
	COWPEA, COMMON *Vigna sinensis*														
98	PODS WITH SEEDS, DEHY	5-01-648	A-F	92	4.7	10.8	–	–	–	1.3	53.8	21.7	17.5	20.4	16.0
			M-F	100	5.1	11.7	–	–	–	1.4	58.3	23.5	19.0	22.1	17.3
99	SEEDS	5-01-661	A-F	89	3.5	5.0	–	–	–	1.5	55.5	23.8	19.6	22.4	18.7
			M-F	100	3.9	5.6	–	–	–	1.6	62.2	26.7	21.9	25.1	20.9
	DARSO SORGHUM *Sorghum bicolor*														
100	GRAIN	4-04-357	A-F	90	1.5	2.1	–	–	–	3.2	72.9	10.1	6.9	7.4	7.1
			M-F	100	1.6	2.3	–	–	–	3.5	81.3	11.2	7.7	8.3	8.0
	DASHEEN (TARO) *Colocasia esculenta*														
101	TUBERS, FRESH	4-10-463	A-F	28	1.2	0.6	–	–	–	0.1	24.2	1.5	0.5	1.0	0.9
			M-F	100	4.3	2.2	–	–	–	0.4	87.7	5.4	1.8	3.6	3.4
	DATE PALM *Phoenix dactylifera*														
102	PITS	4-05-648	A-F	92	2.9	14.4	–	–	–	8.2	57.8	8.3	4.7	5.8	5.7
			M-F	100	3.2	15.7	–	–	–	9.0	63.0	9.1	5.1	6.3	6.3
	DISTILLERS PRODUCTS (ALSO SEE CORN; RYE; SORGHUM; WHEAT)														
103	GRAINS, DEHY	5-02-144	A-F	93	1.5	12.8	–	–	–	7.4	43.5	27.3	18.9	26.0	22.3
			M-F	100	1.6	13.8	–	–	–	8.0	47.0	29.5	20.4	28.1	24.1
104	SOLUBLES, DEHY	5-02-147	A-F	92	6.2	3.4	–	–	–	8.9	44.8	28.8	24.0	27.4	24.0
			M-F	100	6.7	3.7	–	–	–	9.7	48.6	31.3	26.0	29.8	26.1
	DURRA SORGHUM *Sorghum bicolor, durra*														
105	GRAIN	4-04-365	A-F	89	3.0	2.1	–	–	–	3.6	70.1	10.2	5.7	7.6	7.3
			M-F	100	3.4	2.4	–	–	–	4.1	78.7	11.5	6.4	8.5	8.2
	EMMER *Triticum dicoccum*														
106	GRAIN	4-01-830	A-F	91	3.5	9.6	–	–	–	2.0	63.9	11.7	9.4	8.9	8.4
			M-F	100	3.9	10.6	–	–	–	2.2	70.4	12.9	10.3	9.8	9.3
	FATS AND OILS														
107	FAT, ANIMAL, HYDROLYZED	4-00-376	A-F	99	–	–	–	–	–	98.4	–	–	–	–	–
			M-F	100	–	–	–	–	–	99.2	–	–	–	–	–
108	FAT (LARD), SWINE	4-04-790	A-F	99	–	–	–	–	–	99.3	–	–	–	–	–
			M-F	100	–	–	–	–	–	100.0	–	–	–	–	–

ENERGY FEEDS

Entry Number	TDN Ruminant	TDN Swine	TDN Horse	DE Ruminant		DE Swine		DE Horse		ME Ruminant		ME Swine		ME Poultry ME_n		ME Horse		NE Ruminant NE_m		NE Ruminant NE_g		NE Lactating Cows NE_{lc}	
	%	%	%	Mcal		kcal		Mcal		Mcal		kcal		kcal		Mcal		Mcal		Mcal		Mcal	
				lb	kg	lb	kg	lb	kg	lb	kg	lb	kg	lb	kg	lb	kg	lb	kg	lb	kg	lb	kg
82	81	65	—	1.54	3.41	1246	2746	—	—	1.28	2.82	1196	2636	894	1970	—	—	0.87	1.92	0.59	1.30	0.84	1.86
	87	69	—	1.65	3.64	1333	2938	—	—	1.37	3.02	1279	2820	956	2108	—	—	0.93	2.05	0.63	1.39	0.90	1.99
83	80	78	—	1.55	3.42	1474	3250	—	—	1.34	2.96	1419	3128	1324	2919	—	—	0.92	2.03	0.63	1.40	0.86	1.89
	87	84	—	1.67	3.68	1587	3499	—	—	1.45	3.19	1528	3369	1426	3143	—	—	0.99	2.19	0.68	1.51	0.92	2.03
84	72	69	56	1.45	3.20	1410	3109	1.03	2.28	1.26	2.78	1260	2779	1238	2730	0.85	1.87	0.86	1.90	0.60	1.31	0.76	1.68
	83	79	65	1.68	3.70	1630	3593	1.20	2.64	1.46	3.21	1457	3211	1431	3155	0.98	2.16	1.00	2.20	0.69	1.52	0.88	1.95
85	68	69	—	1.41	3.11	1568	3456	—	—	1.24	2.73	1372	3025	772	1702	—	—	0.73	1.61	0.47	1.04	0.71	1.55
	74	76	—	1.54	3.39	1708	3765	—	—	1.35	2.98	1495	3295	841	1854	—	—	0.80	1.76	0.52	1.14	0.77	1.69
86	—	—	—	—	—	—	—	—	—	—	—	—	—	3370	7431	—	—	—	—	—	—	—	—
	—	—	—	—	—	—	—	—	—	—	—	—	—	3745	8256	—	—	—	—	—	—	—	—
87	75	75	—	1.44	3.17	1375	3031	—	—	1.21	2.67	1037	2287	784	1729	—	—	0.82	1.80	0.55	1.21	0.78	1.72
	83	84	—	1.60	3.52	1529	3371	—	—	1.34	2.96	1154	2543	872	1924	—	—	0.91	2.00	0.61	1.35	0.87	1.91
88	84	82	49	1.69	3.73	1620	3571	0.91	2.01	1.29	2.85	1533	3381	1313	2894	0.75	1.65	0.88	1.95	0.61	1.34	0.91	2.00
	93	91	54	1.88	4.14	1796	3959	1.01	2.23	1.43	3.15	1700	3748	1455	3208	0.83	1.83	0.98	2.16	0.67	1.48	1.01	2.22
89	—	—	—	—	—	—	—	—	—	—	—	—	—	—	—	—	—	—	—	—	—	—	—
	—	—	—	—	—	—	—	—	—	—	—	—	—	—	—	—	—	—	—	—	—	—	—
90	203	208	—	3.57	7.86	3421	7543	—	—	3.40	7.50	3438	7579	3917	8635	—	—	2.65	5.84	1.99	4.38	2.20	4.86
	205	210	—	3.60	7.94	3456	7619	—	—	3.43	7.57	3473	7656	3956	8722	—	—	2.68	5.90	2.01	4.42	2.22	4.90
91	88	87	—	1.68	3.70	1663	3665	—	—	1.40	3.08	1671	3684	1673	3689	—	—	0.97	2.14	0.68	1.49	0.92	2.03
	98	97	—	1.86	4.09	1842	4060	—	—	1.55	3.41	1851	4080	1853	4086	—	—	1.07	2.37	0.75	1.66	1.02	2.25
92	83	79	61	1.56	3.44	1582	3487	1.12	2.46	1.39	3.07	1498	3304	—	—	0.92	2.02	0.80	1.77	0.54	1.19	0.76	1.67
	94	89	69	1.76	3.88	1787	3940	1.26	2.79	1.57	3.47	1693	3733	—	—	1.04	2.28	0.91	2.00	0.61	1.35	0.86	1.89
93	80	79	61	1.53	3.38	1643	3623	1.12	2.47	1.37	3.01	1557	3432	1527	3367	0.92	2.03	0.81	1.78	0.54	1.20	0.76	1.68
	89	88	68	1.70	3.75	1824	4022	1.25	2.75	1.52	3.34	1728	3810	1696	3738	1.02	2.25	0.90	1.98	0.60	1.33	0.85	1.87
94	85	79	61	1.59	3.50	1589	3503	1.12	2.46	1.42	3.13	1505	3317	—	—	0.92	2.02	0.82	1.80	0.55	1.21	0.77	1.70
	95	88	68	1.77	3.90	1769	3900	1.25	2.74	1.58	3.49	1675	3693	—	—	1.02	2.25	0.91	2.00	0.61	1.35	0.86	1.89
95	89	82	48	1.64	3.61	1633	3601	0.90	1.99	1.47	3.24	1548	3412	—	—	0.74	1.64	0.83	1.84	0.56	1.24	0.79	1.73
	98	90	53	1.80	3.97	1797	3962	1.00	2.19	1.62	3.56	1703	3754	—	—	0.82	1.80	0.92	2.02	0.62	1.36	0.86	1.90
96	54	—	42	1.05	2.31	—	—	0.79	1.74	0.90	1.99	—	—	—	—	0.65	1.43	0.57	1.26	0.36	0.79	0.56	1.23
	70	—	55	1.37	3.01	—	—	1.03	2.27	1.18	2.59	—	—	—	—	0.84	1.86	0.74	1.63	0.47	1.03	0.72	1.60
97	87	71	—	1.57	3.46	1430	3152	—	—	1.39	3.07	1243	2739	—	—	—	—	0.97	2.13	0.67	1.49	0.91	2.01
	95	79	—	1.72	3.80	1572	3467	—	—	1.53	3.38	1367	3013	—	—	—	—	1.06	2.34	0.74	1.63	1.00	2.21
98	71	69	—	1.41	3.11	1377	3036	—	—	1.24	2.72	1193	2630	—	—	—	—	0.72	1.59	0.46	1.02	0.70	1.53
	76	75	—	1.53	3.37	1492	3290	—	—	1.34	2.95	1292	2849	—	—	—	—	0.78	1.72	0.50	1.10	0.75	1.66
99	76	75	—	1.48	3.27	1500	3307	—	—	1.32	2.90	1309	2886	—	—	—	—	0.80	1.76	0.53	1.18	0.75	1.66
	85	84	—	1.66	3.66	1682	3708	—	—	1.47	3.25	1468	3236	—	—	—	—	0.89	1.97	0.60	1.32	0.84	1.86
100	79	78	68	1.51	3.33	1564	3447	1.23	2.70	1.34	2.96	1480	3263	—	—	1.01	2.22	0.80	1.76	0.54	1.18	0.76	1.67
	88	87	75	1.69	3.72	1743	3842	1.37	3.01	1.50	3.30	1649	3637	—	—	1.12	2.47	0.89	1.96	0.60	1.32	0.84	1.86
101	23	18	22	0.46	1.01	354	780	0.39	0.87	0.41	0.90	331	729	—	—	0.32	0.71	0.24	0.52	0.16	0.35	0.23	0.50
	83	64	79	1.66	3.66	1283	2828	1.43	3.15	1.47	3.24	1198	2641	—	—	1.17	2.58	0.86	1.90	0.57	1.26	0.82	1.80
102	76	83	—	1.52	3.36	1669	3680	—	—	1.35	2.97	1463	3225	1138	2509	—	—	0.87	1.92	0.59	1.31	0.82	1.80
	82	90	—	1.65	3.63	1804	3977	—	—	1.46	3.21	1581	3485	1230	2711	—	—	0.94	2.08	0.64	1.41	0.88	1.95
103	78	82	—	1.61	3.54	1637	3609	—	—	1.43	3.16	1433	3160	1316	2901	—	—	0.93	2.06	0.65	1.43	0.87	1.92
	84	89	—	1.74	3.85	1779	3922	—	—	1.56	3.43	1558	3434	1430	3153	—	—	1.02	2.24	0.70	1.55	0.95	2.09
104	68	75	31	1.39	3.07	1493	3291	0.62	1.37	1.22	2.69	1409	3107	—	—	0.51	1.12	0.77	1.70	0.51	1.12	0.74	1.63
	75	81	34	1.52	3.34	1627	3587	0.68	1.49	1.33	2.93	1536	3387	—	—	0.56	1.22	0.84	1.86	0.56	1.22	0.80	1.77
105	76	78	63	1.48	3.26	1569	3460	1.15	2.53	1.31	2.89	1486	3276	—	—	0.94	2.07	0.79	1.74	0.53	1.16	0.75	1.65
	86	88	70	1.66	3.66	1762	3884	1.29	2.83	1.47	3.24	1668	3677	—	—	1.05	2.32	0.89	1.95	0.59	1.30	0.84	1.85
106	72	67	66	1.39	3.07	1349	2974	1.20	2.64	1.22	2.69	1269	2797	—	—	0.98	2.16	0.72	1.59	0.47	1.03	0.69	1.53
	80	74	72	1.53	3.38	1485	3275	1.32	2.90	1.35	2.97	1397	3080	—	—	1.08	2.38	0.79	1.75	0.51	1.13	0.77	1.69
107	223	209	—	4.46	9.84	4144	9135	—	—	4.31	9.49	3834	8452	3685	8125	—	—	2.95	6.49	2.23	4.91	2.43	5.35
	225	211	—	4.50	9.92	4177	9210	—	—	4.34	9.57	3865	8521	3715	8191	—	—	2.97	6.55	2.25	4.95	2.45	5.39
108	—	—	—	—	—	3669	8089	—	—	—	—	3537	7798	3936	8677	—	—	—	—	—	—	—	—
	—	—	—	—	—	3694	8143	—	—	—	—	3561	7850	3962	8735	—	—	—	—	—	—	—	—

(Continued)

TABLE V-1A ENERGY FEEDS, MINERAL AND VITAMIN COMPOSITION OF FEEDS, DATA EXPRESSED **AS-FED** AND **MOISTURE-FREE**—*(Continued)*

		Moisture Basis:		Macro Minerals							Micro Minerals						
Entry Number	Feed Name Description	A-F (as-fed) or M-F (moisture-free)	Dry Matter	Calcium (Ca)	Phosphorus (P)	Sodium (Na)	Chlorine (Cl)	Magnesium (Mg)	Potassium (K)	Sulfur (S)	Cobalt (Co)	Copper (Cu)	Iodine (I)	Iron (Fe)	Manganese (Mn)	Selenium (Se)	Zinc (Zn)
			%	%	%	%	%	%	%	%	ppm or mg/kg	ppm or mg/kg	ppm or mg/kg	%	ppm or mg/kg	ppm or mg/kg	ppm or mg/kg
	CORN, DENT YELLOW (Continued)																
82	DISTILLERS GRAINS, DEHY	A-F	93	0.09	0.39	0.09	0.07	0.07	0.16	0.43	0.076	38.9	0.048	0.020	19.3	0.352	41.7
		M-F	100	0.10	0.42	0.09	0.08	0.07	0.18	0.46	0.082	41.7	0.051	0.021	20.7	0.377	44.7
83	DISTILLERS SOLUBLES, DEHY	A-F	93	0.30	1.30	0.23	0.26	0.60	1.70	0.37	0.167	77.9	0.079	0.052	72.0	0.371	88.0
		M-F	100	0.32	1.40	0.24	0.28	0.65	1.83	0.40	0.180	83.9	0.085	0.056	77.6	0.400	94.8
84	EARS, GROUND (CORN AND COB MEAL)	A-F	87	0.06	0.24	0.02	0.04	0.12	0.46	0.14	0.273	6.8	0.023	0.008	19.9	0.074	12.1
		M-F	100	0.07	0.27	0.02	0.05	0.14	0.53	0.16	0.315	7.9	0.026	0.010	23.0	0.086	14.0
85	GERM MEAL, WET MILLED, SOLV EXTD	A-F	92	0.04	0.51	0.04	0.04	0.16	0.35	0.31	—	4.5	—	0.034	3.8	0.340	104.8
		M-F	100	0.04	0.55	0.04	0.04	0.17	0.38	0.34	—	4.9	—	0.037	4.1	0.370	114.2
86	GLUTEN, MEAL, 60% PROTEIN	A-F	90	0.07	0.45	0.05	0.09	0.08	0.18	0.65	0.045	26.1	0.018	0.023	6.3	0.829	30.6
		M-F	100	0.08	0.50	0.05	0.10	0.09	0.20	0.72	0.051	29.0	0.020	0.026	7.0	0.921	34.0
87	GLUTEN FEED	A-F	90	0.32	0.74	0.12	0.22	0.33	0.57	0.21	0.087	47.1	0.066	0.043	23.1	0.272	64.6
		M-F	100	0.36	0.82	0.14	0.25	0.36	0.64	0.23	0.097	52.3	0.074	0.048	25.7	0.302	71.8
88	GRITS (HOMINY GRITS)	A-F	90	0.05	0.51	0.08	0.05	0.24	0.59	0.03	0.055	13.6	—	0.007	14.5	—	—
		M-F	100	0.05	0.57	0.09	0.06	0.26	0.65	0.03	0.061	15.1	—	0.008	16.1	—	—
89	MOLASSES, MORE THAN 43% DEXTROSE EQUIVALENT, MORE THAN 50% TOTAL DEXTROSE, MORE THAN 78 DEGREES BRIX	A-F	73	—	—	—	—	—	—	—	—	—	—	—	—	—	—
		M-F	100	—	—	—	—	—	—	—	—	—	—	—	—	—	—
90	OIL	A-F	99	—	—	—	—	—	—	—	—	—	—	—	—	—	—
		M-F	100	—	—	—	—	—	—	—	—	—	—	—	—	—	—
91	STARCH	A-F	90	0.00	0.03	0.06	—	0.00	0.00	—	—	2.3	1.151	0.001	—	0.021	3.0
		M-F	100	0.00	0.03	0.07	—	0.00	0.00	—	—	2.6	1.275	0.001	—	0.023	3.3
	CORN, FLINT *Zea mays, indurata*																
92	GRAIN	A-F	89	—	0.27	—	—	—	0.32	—	—	11.5	—	0.003	7.0	—	—
		M-F	100	—	0.31	—	—	—	0.36	—	—	13.0	—	0.003	7.9	—	—
	CORN, OPAQUE 2, HIGH LYSINE *Zea mays*																
93	GRAIN	A-F	90	0.03	0.20	—	—	0.13	0.35	0.10	—	—	—	—	—	—	—
		M-F	100	0.03	0.22	—	—	0.14	0.39	0.11	—	—	—	—	—	—	—
	CORN, POP *Zea mays, everta*																
94	GRAIN	A-F	90	—	0.30	—	—	—	—	—	—	2.4	—	0.003	—	—	—
		M-F	100	—	0.33	—	—	—	—	—	—	2.6	—	0.003	—	—	—
	CORN, SWEET *Zea mays, saccharata*																
95	GRAIN	A-F	91	0.03	0.32	—	—	—	—	—	—	5.4	—	0.005	—	—	—
		M-F	100	0.03	0.35	—	—	—	—	—	—	6.0	—	0.005	—	—	—
96	CANNERY RESIDUE, FRESH	A-F	77	0.25	0.54	0.02	—	0.18	0.88	0.10	—	5.4	—	0.016	—	—	—
		M-F	100	0.32	0.70	0.03	—	0.24	1.15	0.13	—	7.0	—	0.020	—	—	—
	COTTON *Gossypium* spp																
97	SEEDS	A-F	91	0.14	0.69	0.03	—	0.32	1.11	0.24	—	49.0	—	0.014	11.1	—	—
		M-F	100	0.16	0.76	0.03	—	0.35	1.22	0.26	—	53.9	—	0.016	12.2	—	—
	COWPEA, COMMON *Vigna sinensis*																
98	PODS, WITH SEEDS, DEHY	A-F	92	—	—	—	—	—	—	—	—	—	—	—	—	—	—
		M-F	100	—	—	—	—	—	—	—	—	—	—	—	—	—	—
99	SEEDS	A-F	89	0.09	0.44	0.04	0.04	0.26	1.16	0.25	—	4.4	—	0.021	40.2	—	—
		M-F	100	0.10	0.50	0.04	0.05	0.29	1.30	0.28	—	5.0	—	0.024	45.1	—	—
	DARSO SORGHUM *Sorghum bicolor*																
100	GRAIN	A-F	90	0.08	0.30	—	—	—	—	—	—	—	—	—	—	—	—
		M-F	100	0.09	0.34	—	—	—	—	—	—	—	—	—	—	—	—
	DASHEEN (TARO) *Colocasia esculenta*																
101	TUBERS, FRESH	A-F	28	0.02	0.07	—	—	—	—	—	—	—	—	—	—	—	—
		M-F	100	0.07	0.25	—	—	—	—	—	—	—	—	—	—	—	—
	DATE PALM *Phoenix dactylifera*																
102	PITS	A-F	92	0.60	0.18	0.01	—	0.14	2.20	—	—	—	—	—	—	—	—
		M-F	100	0.65	0.20	0.01	—	0.15	2.40	—	—	—	—	—	—	—	—
	DISTILLERS PRODUCTS (ALSO SEE CORN; RYE; SORGHUM; WHEAT)																
103	GRAINS, DEHY	A-F	93	0.12	0.54	0.05	0.05	0.09	0.20	0.46	0.092	47.9	—	0.027	35.0	—	—
		M-F	100	0.13	0.59	0.05	0.06	0.10	0.21	0.49	0.099	51.7	—	0.029	37.8	—	—
104	SOLUBLES, DEHY	A-F	92	0.24	1.35	0.45	—	0.53	1.97	—	0.196	71.6	—	0.031	64.1	—	138.0
		M-F	100	0.26	1.47	0.49	—	0.58	2.14	—	0.213	77.9	—	0.034	69.7	—	150.0
	DURRA SORGHUM *Sorghum bicolor, durra*																
105	GRAIN	A-F	89	—	—	—	—	—	—	—	—	—	—	—	—	—	—
		M-F	100	—	—	—	—	—	—	—	—	—	—	—	—	—	—
	EMMER *Triticum dicoccum*																
106	GRAIN	A-F	91	0.05	0.36	—	—	—	0.47	—	—	31.4	—	0.006	78.1	—	—
		M-F	100	0.06	0.40	—	—	—	0.52	—	—	34.6	—	0.006	86.0	—	—
	FATS AND OILS																
107	FAT, ANIMAL, HYDROLYZED	A-F	99	—	—	—	—	—	—	—	—	—	—	—	—	—	—
		M-F	100	—	—	—	—	—	—	—	—	—	—	—	—	—	—
108	FAT (LARD), SWINE	A-F	99	—	—	—	—	—	—	—	—	—	—	—	—	—	—
		M-F	100	—	—	—	—	—	—	—	—	—	—	—	—	—	—

Entry Number	Fat-Soluble Vitamins: A (1 mg Carotene = 1667 IU Vit A)	Carotene (Provitamin A)	D	E	K	Water-Soluble Vitamins: B-12	Biotin	Choline	Folacin (Folic Acid)	Niacin	Pantothenic Acid (B-3)	(Pyridoxine) B-6	Riboflavin (B-2)	Thiamin (B-1)
	IU/g	ppm or mg/kg	IU/kg	ppm or mg/kg	ppm or mg/kg	ppb or mcg/kg	ppm or mg/kg	ppm or mg/kg	ppm or mg/kg	ppm or mg/kg	ppm or mg/kg	ppm or mg/kg	ppm or mg/kg	ppm or mg/kg
82	**5.2**	**3.1**	–	–	–	**0.3**	**0.41**	**1113**	**1.00**	**38**	**11.3**	**4.22**	**5.0**	**1.8**
	5.6	3.3	–	–	–	0.3	0.44	1191	1.07	41	12.1	4.51	5.3	1.9
83	**1.1**	**0.7**	–	**45.9**	–	**4.2**	**1.49**	**4751**	**1.34**	**124**	**23.3**	**9.41**	**15.1**	**6.8**
	1.2	0.7	–	49.4	–	4.5	1.61	5116	1.45	133	25.0	10.14	16.3	7.3
84	**5.3**	**3.2**	–	**17.5**	–	–	**0.03**	**357**	**0.24**	**17**	**4.2**	**5.97**	**0.9**	**2.9**
	6.1	3.7	–	20.2	–	–	0.04	412	0.28	20	4.8	6.89	1.0	3.3
85	**3.4**	**2.0**	–	**85.8**	–	–	**0.22**	**1586**	**0.20**	**39**	**4.2**	–	**3.8**	**4.5**
	3.7	2.2	–	93.5	–	–	0.24	1728	0.22	42	4.6	–	4.1	4.9
86	–	–	–	**14.6**	–	–	–	–	–	–	–	**6.39**	–	–
	–	–	–	16.2	–	–	–	–	–	–	–	7.10	–	–
87	**9.8**	**5.9**	–	**12.1**	–	–	**0.33**	**1514**	**0.27**	**70**	**13.6**	**13.93**	**2.2**	**2.0**
	10.9	6.5	–	13.5	–	–	0.36	1684	0.30	78	15.1	15.49	2.5	2.2
88	**15.4**	**9.2**	–	–	–	–	**0.13**	**1154**	**0.31**	**47**	**8.2**	**10.95**	**2.1**	**8.1**
	17.0	10.2	–	–	–	–	0.15	1280	0.34	52	9.1	12.14	2.4	8.9
89	–	–	–	–	–	–	–	–	–	–	–	–	–	–
	–	–	–	–	–	–	–	–	–	–	–	–	–	–
90	–	–	–	–	–	–	–	–	–	–	–	–	–	–
	–	–	–	–	–	–	–	–	–	–	–	–	–	–
91	–	–	–	–	–	–	–	–	–	–	–	–	**0.2**	**0.9**
	–	–	–	–	–	–	–	–	–	–	–	–	0.3	1.0
92	–	–	–	–	–	–	–	–	–	**16**	–	–	–	–
	–	–	–	–	–	–	–	–	–	18	–	–	–	–
93	**7.8**	**4.7**	–	–	–	–	–	**518**	–	**19**	**4.7**	–	**1.1**	–
	8.6	5.2	–	–	–	–	–	575	–	22	5.2	–	1.2	–
94	–	–	–	–	–	–	**0.08**	–	–	**17**	**3.3**	**4.36**	**1.1**	–
	–	–	–	–	–	–	0.09	–	–	19	3.7	4.85	1.3	–
95	–	–	–	–	–	–	–	–	–	**32**	**8.2**	–	**1.8**	–
	–	–	–	–	–	–	–	–	–	35	9.0	–	1.9	–
96	**17.3**	**10.4**	–	–	–	–	–	–	–	–	–	–	–	–
	22.5	13.5	–	–	–	–	–	–	–	–	–	–	–	–
97	–	–	–	–	–	–	–	–	–	–	–	–	–	–
	–	–	–	–	–	–	–	–	–	–	–	–	–	–
98	–	–	–	–	–	–	–	–	–	–	–	–	–	–
	–	–	–	–	–	–	–	–	–	–	–	–	–	–
99	**0.4**	**0.2**	–	–	–	–	–	–	–	**24**	**15.5**	–	**2.3**	**9.3**
	0.4	0.2	–	–	–	–	–	–	–	27	17.3	–	2.5	10.4
100	**4.5**	**2.7**	–	–	–	–	–	–	–	**28**	**13.3**	–	**1.4**	–
	5.0	3.0	–	–	–	–	–	–	–	31	14.8	–	1.5	–
101	–	–	–	–	–	–	–	–	–	–	–	–	–	–
	–	–	–	–	–	–	–	–	–	–	–	–	–	–
102	–	–	–	–	–	–	–	–	–	–	–	–	–	–
	–	–	–	–	–	–	–	–	–	–	–	–	–	–
103	**13.0**	**7.8**	–	**30.5**	–	–	–	**2645**	–	**47**	**11.9**	**6.00**	**6.6**	**2.5**
	14.0	8.4	–	32.9	–	–	–	2858	–	51	12.9	6.48	7.1	2.6
104	**1.9**	**1.1**	–	–	–	**2.9**	**2.84**	**4992**	–	**143**	**25.3**	**8.66**	**11.3**	**6.9**
	2.0	1.2	–	–	–	3.1	3.09	5425	–	155	27.5	9.42	12.3	7.5
105	–	–	–	–	–	–	**0.30**	–	–	**48**	**8.6**	**1.96**	**0.8**	–
	–	–	–	–	–	–	0.33	–	–	54	9.7	2.21	0.9	–
106	–	–	–	–	–	–	–	–	–	–	–	–	–	–
	–	–	–	–	–	–	–	–	–	–	–	–	–	–
107	–	–	–	–	–	–	–	–	–	–	–	–	–	–
	–	–	–	–	–	–	–	–	–	–	–	–	–	–
108	–	–	–	**22.8**	–	–	–	–	–	–	–	–	–	–
	–	–	–	23.0	–	–	–	–	–	–	–	–	–	–

(Continued)

TABLE V-1A ENERGY FEEDS, COMPOSITION OF FEEDS, DATA EXPRESSED **AS-FED** AND **MOISTURE-FREE**—*(Continued)*

Entry Number	Feed Name Description	International Feed Number	Moisture Basis: A-F (as-fed) or M-F (moisture-free)	Proximate Analysis									Digestible Protein		
				Dry Matter	Ash	Crude Fiber	Neutral Det. Fib. (NDF)	Acid Det. Fib. (ADF)	Lignin	Ether Extract (Fat)	N-Free Extract	Crude Protein	Ruminant	Swine	Horse
				%	%	%	%	%	%	%	%	%	%	%	%
	FATS AND OILS (Continued)														
109	FAT, ANIMAL-POULTRY	4-00-409	A-F	99	–	–	–	–	–	99.1	–	–	–	–	–
			M-F	100	–	–	–	–	–	100.0	–	–	–	–	–
110	OIL, CORN	4-07-882	A-F	99	–	–	–	–	–	99.0	–	–	–	–	–
			M-F	100	–	–	–	–	–	100.0	–	–	–	–	–
111	OIL, SOYBEAN	4-07-983	A-F	100	0.3	–	–	–	–	95.0	7.3	1.4	–1.8	–0.6	0.2
			M-F	100	0.3	–	–	–	–	95.5	7.3	1.4	–1.8	–0.6	0.2
112	TALLOW, ANIMAL	4-08-127	A-F	97	0.1	–	–	–	–	96.8	–	1.5	–1.6	–0.4	0.3
			M-F	100	0.1	–	–	–	–	99.5	–	1.6	–1.6	–0.4	0.3
	FETERITA SORGHUM *Sorghum bicolor, caudatum*														
113	GRAIN	4-04-369	A-F	89	1.6	1.7	–	1.8	–	2.8	71.7	11.7	9.0	8.8	8.4
			M-F	100	1.8	1.9	–	2.0	–	3.1	80.1	13.0	10.0	9.9	9.4
114	HEADS	4-04-370	A-F	90	3.2	7.4	–	–	–	2.6	65.7	10.7	6.8	8.0	7.6
			M-F	100	3.6	8.3	–	–	–	2.9	73.3	11.9	7.6	8.9	8.5
	FLAX, COMMON *Linum usitatissimum*														
115	SEEDS	5-02-042	A-F	96	5.0	6.3	–	–	–	35.9	30.7	17.8	13.9	16.5	11.2
			M-F	100	5.2	6.6	–	–	–	37.5	32.1	18.6	14.5	17.2	11.7
116	SEED SCREENINGS	4-02-056	A-F	91	6.2	12.1	–	–	–	9.3	47.1	16.6	9.3	13.2	12.2
			M-F	100	6.8	13.2	–	–	–	10.2	51.6	18.2	10.2	14.5	13.4
	GARBAGE, HOTEL AND RESTAURANT														
117	BOILED, WET	4-07-865	A-F	23	1.3	0.7	–	–	–	5.4	11.9	3.6	2.5	2.8	2.6
			M-F	100	5.5	3.0	–	–	–	23.7	52.0	15.8	11.1	12.4	11.6
118	BOILED, DEHY, GROUND	4-07-879	A-F	92	6.3	2.2	10.4	4.6	1.8	19.6	48.0	16.1	11.6	12.8	11.8
			M-F	100	6.9	2.4	11.3	5.0	1.9	21.3	52.1	17.4	12.6	13.8	12.8
	GARBANZO (CHICKPEA; GRAM PEA) *Cicer arietinum*														
119	SEEDS	5-01-218	A-F	89	2.9	7.0	–	–	–	4.1	56.0	19.1	14.3	17.8	13.3
			M-F	100	3.3	7.9	–	–	–	4.6	62.8	21.4	16.1	20.0	14.9
	GOOSEFOOT (LAMB'S QUARTERS) *Chenopodium album*														
120	SEEDS	5-08-424	A-F	90	9.6	15.1	–	–	–	4.5	41.9	18.8	15.0	17.6	13.0
			M-F	100	10.7	16.8	–	–	–	5.0	46.6	20.9	16.6	19.5	14.4
	GRAIN SCREENINGS—SEE BARLEY; WHEAT														
	GRAM PEA (CHICKPEA; GARBANZO) *Cicer arietinum*														
121	SEEDS	5-01-218	A-F	89	2.9	7.0	–	–	–	4.1	56.0	19.1	14.3	17.8	13.3
			M-F	100	3.3	7.9	–	–	–	4.6	62.8	21.4	16.1	20.0	14.9
	GRAPES *Vitis spp*														
122	FRUIT (RAISINS), DEHY, CULL	4-08-427	A-F	87	4.1	10.1	–	–	–	4.3	62.2	6.4	1.5	4.2	4.3
			M-F	100	4.7	11.6	–	–	–	4.9	71.4	7.4	1.8	4.8	4.9
123	POMACE, WET (MARC)	2-02-206	A-F	37	2.7	9.7	–	–	–	1.8	18.0	5.2	0.8	3.5	1.1
			M-F	100	7.3	25.9	–	–	–	4.9	48.1	13.8	2.2	9.4	3.0
124	POMACE, DEHY (MARC)	1-01-208	A-F	90	7.5	27.9	48.1	49.2	31.8	7.6	35.3	12.1	1.6	7.1	7.7
			M-F	100	8.3	30.9	53.2	54.4	35.2	8.4	39.0	13.4	1.7	7.9	8.5
	GRAPEFRUIT *Citrus paradisi*														
125	PULP WITHOUT FINES, DEHY (DRIED GRAPEFRUIT PULP)	4-01-244	A-F	90	5.4	13.2	–	–	–	3.2	61.5	6.9	1.7	4.6	4.6
			M-F	100	6.0	14.6	–	–	–	3.5	68.3	7.7	1.9	5.1	5.2
	GRASS														
126	SEED SCREENINGS	4-26-071	A-F	94	–	18.7	–	29.6	5.5	–	–	10.0	6.1	7.3	7.0
			M-F	100	–	20.0	–	31.6	5.9	–	–	10.7	6.5	7.8	7.5
	HEGARI SORGHUM *Sorghum bicolor, caffrorum*														
127	GRAIN	4-04-398	A-F	89	1.4	2.3	–	–	–	2.5	72.7	10.4	6.6	7.7	7.4
			M-F	100	1.6	2.5	–	–	–	2.8	81.4	11.7	7.4	8.7	8.3
128	HEADS	4-04-399	A-F	90	6.6	14.5	–	–	–	1.9	59.1	7.7	4.2	5.3	5.3
			M-F	100	7.3	16.1	–	–	–	2.1	65.8	8.6	4.6	5.9	5.9
	HOG MILLET (BROOMCORN; MILLET, PROSO) *Panicum miliaceum*														
129	GRAIN	4-03-120	A-F	90	2.9	5.3	–	14.9	3.2	3.6	66.3	11.6	7.9	7.9	8.3
			M-F	100	3.3	6.0	–	16.6	3.6	4.0	73.9	12.9	8.8	8.8	9.3
130	**HOMINY GRITS (CORN, DENT YELLOW, GRITS)** *Zea mays, indentata*	4-03-011	A-F	90	2.8	4.8	–	–	–	6.5	65.8	10.3	7.3	7.6	7.3
			M-F	100	3.1	5.3	–	–	–	7.2	72.9	11.4	8.1	8.5	8.1
	HORSE BEAN *Vicia faba, equina*														
131	SEEDS	5-02-407	A-F	88	3.3	7.9	–	–	–	1.3	50.1	25.5	21.1	24.3	19.4
			M-F	100	3.8	8.9	–	–	–	1.5	56.8	29.0	24.0	27.5	22.0
	HORSE CHESTNUT, COMMON *Aesculus hippocastanum*														
132	KERNELS WITH COATS	4-02-409	A-F	90	2.2	2.5	–	–	–	6.5	72.1	7.1	1.8	4.7	4.7
			M-F	100	2.4	2.8	–	–	–	7.2	79.8	7.8	2.0	5.2	5.3
	IVORY PALM, COMMON *Phytelephas macrocarpa*														
133	KERNELS WITH COATS, MEAL SOLV EXTD	4-02-433	A-F	89	1.1	7.1	–	–	–	0.9	75.4	4.8	0.8	2.6	3.0
			M-F	100	1.3	8.0	–	–	–	1.0	84.4	5.3	0.9	2.9	3.3

Entry Number	TDN Ruminant	TDN Swine	TDN Horse	Digestible Energy Ruminant		Digestible Energy Swine		Digestible Energy Horse		Metabolizable Energy Ruminant		Metabolizable Energy Swine		Metabolizable Energy Poultry ME_n		Metabolizable Energy Horse		Net Energy Ruminant NE_m		Net Energy Ruminant NE_g		Net Energy Lactating Cows NE_{lc}	
	%	%	%	Mcal		kcal		Mcal		Mcal		kcal		kcal		Mcal		Mcal		Mcal		Mcal	
				lb	kg	lb	kg	lb	kg	lb	kg	lb	kg	lb	kg	lb	kg	lb	kg	lb	kg	lb	kg
109	188	196	–	3.57	7.87	3750	8267	–	–	3.40	7.51	3617	7973	3482	7677	–	–	2.65	5.85	1.99	4.38	2.03	4.48
	189	198	–	3.60	7.94	3783	8340	–	–	3.43	7.57	3648	8044	3513	7744	–	–	2.68	5.90	2.01	4.42	2.05	4.52
110	203	208	–	3.57	7.86	3421	7543	–	–	3.40	7.50	3438	7579	3917	8635	–	–	2.65	5.84	1.99	4.38	2.20	4.86
	205	210	–	3.60	7.94	3456	7619	–	–	3.43	7.57	3473	7656	3956	8722	–	–	2.68	5.90	2.01	4.42	2.22	4.90
111	193	–	–	3.86	8.51	3412	7523	–	–	3.70	8.15	3287	7247	4050	8929	–	–	2.40	5.30	1.79	3.94	2.09	4.61
	194	–	–	3.88	8.55	3430	7561	–	–	3.71	8.19	3303	7283	4070	8974	–	–	2.42	5.32	1.80	3.96	2.10	4.63
112	203	–	–	4.07	8.97	–	–	–	–	3.91	8.62	–	–	3304	7285	–	–	2.60	5.73	1.95	4.29	2.21	4.87
	209	–	–	4.18	9.22	–	–	–	–	4.02	8.86	–	–	3395	7484	–	–	2.67	5.88	2.00	4.41	2.27	5.00
113	79	76	71	1.51	3.32	1524	3361	1.28	2.82	1.34	2.95	1442	3178	–	–	1.05	2.31	0.79	1.74	0.53	1.16	0.75	1.65
	88	85	79	1.68	3.71	1704	3756	1.43	3.15	1.50	3.30	1611	3552	–	–	1.17	2.59	0.88	1.95	0.59	1.30	0.84	1.84
114	73	71	63	1.41	3.10	1417	3123	1.15	2.54	1.24	2.73	1336	2945	–	–	0.95	2.08	0.74	1.63	0.48	1.07	0.71	1.56
	81	79	71	1.57	3.47	1581	3486	1.29	2.84	1.38	3.05	1491	3287	–	–	1.06	2.33	0.82	1.82	0.54	1.19	0.79	1.74
115	–	–	–	–	–	–	–	–	–	–	–	–	–	–	–	–	–	–	–	–	–	–	–
	–	–	–	–	–	–	–	–	–	–	–	–	–	–	–	–	–	–	–	–	–	–	–
116	58	70	–	1.17	2.57	1403	3092	–	–	0.99	2.18	1321	2912	–	–	–	–	0.60	1.31	0.35	0.78	0.60	1.32
	64	77	–	1.28	2.82	1538	3392	–	–	1.09	2.30	1449	3195	–	–	–	–	0.65	1.44	0.39	0.86	0.66	1.45
117	20	25	–	0.42	0.92	501	1105	–	–	0.38	0.83	478	1053	–	–	–	–	0.24	0.53	0.17	0.37	0.22	0.49
	89	110	–	1.83	4.02	2191	4831	–	–	1.64	3.61	2090	4607	–	–	–	–	1.06	2.33	0.74	1.62	0.98	2.16
118	82	97	–	1.65	3.63	1947	4293	–	–	1.48	3.26	1855	4090	–	–	–	–	0.94	2.08	0.65	1.44	0.88	1.94
	89	106	–	1.79	3.94	2113	4659	–	–	1.60	3.53	2013	4439	–	–	–	–	1.02	2.26	0.71	1.57	0.95	2.10
119	78	80	–	1.56	3.45	1607	3543	–	–	1.40	3.08	1408	3104	–	–	–	–	0.89	1.97	0.62	1.36	0.83	1.84
	88	90	–	1.75	3.86	1801	3971	–	–	1.57	3.45	1578	3479	–	–	–	–	1.00	2.21	0.69	1.52	0.94	2.06
120	55	45	–	1.16	2.56	892	1967	–	–	0.99	2.18	747	1647	–	–	–	–	0.64	1.41	0.39	0.87	0.63	1.39
	61	50	–	1.29	2.85	992	2186	–	–	1.10	2.42	830	1830	–	–	–	–	0.71	1.56	0.44	0.96	0.70	1.54
121	78	80	–	1.56	3.45	1607	3543	–	–	1.40	3.08	1408	3104	–	–	–	–	0.89	1.97	0.62	1.36	0.83	1.84
	88	90	–	1.75	3.86	1801	3971	–	–	1.57	3.45	1578	3479	–	–	–	–	1.00	2.21	0.69	1.52	0.94	2.06
122	44	72	45	1.11	2.44	1449	3195	0.85	1.87	0.94	2.07	1369	3019	–	–	0.70	1.53	0.72	1.59	0.47	1.04	0.69	1.52
	51	83	52	1.27	2.80	1664	3668	0.97	2.15	1.08	2.38	1572	3466	–	–	0.80	1.76	0.83	1.82	0.54	1.19	0.79	1.75
123	12	–	11	0.37	0.81	–	–	0.23	0.50	0.30	0.65	–	–	–	–	0.19	0.41	0.26	0.57	0.16	0.34	0.26	0.56
	32	–	30	0.98	2.17	–	–	0.61	1.34	0.79	1.74	–	–	–	–	0.50	1.10	0.68	1.51	0.42	0.92	0.68	1.50
124	24	–	18	0.48	1.06	–	–	0.41	0.89	0.30	0.66	–	–	703	1551	0.33	0.73	–	–	–	–	–	–
	27	–	20	0.53	1.17	–	–	0.45	0.99	0.33	0.73	–	–	778	1715	0.37	0.81	–	–	–	–	–	–
125	72	70	48	1.38	3.04	1402	3091	0.91	2.00	1.21	2.66	1321	2913	–	–	0.75	1.64	0.70	1.55	0.45	1.00	0.68	1.50
	80	78	54	1.53	3.37	1556	3430	1.01	2.22	1.34	2.96	1466	3232	–	–	0.83	1.82	0.78	1.72	0.50	1.10	0.75	1.66
126	–	–	–	–	–	–	–	–	–	–	–	–	–	–	–	–	–	–	–	–	–	–	–
	–	–	–	–	–	–	–	–	–	–	–	–	–	–	–	–	–	–	–	–	–	–	–
127	80	76	70	1.52	3.35	1529	3371	1.27	2.80	1.35	2.98	1446	3189	–	–	1.04	2.30	0.79	1.74	0.53	1.16	0.75	1.65
	90	86	79	1.70	3.74	1711	3772	1.42	3.13	1.51	3.33	1618	3567	–	–	1.17	2.57	0.88	1.94	0.59	1.30	0.84	1.84
128	70	66	52	1.33	2.93	1321	2911	0.96	2.12	1.16	2.55	1242	2737	–	–	0.79	1.74	0.67	1.47	0.42	0.93	0.65	1.44
	78	74	58	1.48	3.26	1470	3241	1.07	2.36	1.29	2.84	1382	3047	–	–	0.88	1.94	0.74	1.64	0.47	1.03	0.72	1.60
129	74	68	62	1.47	3.25	1353	2983	1.14	2.51	1.31	2.88	1273	2807	1311	2890	0.93	2.06	0.75	1.66	0.50	1.09	0.72	1.59
	82	75	70	1.64	3.62	1508	3324	1.27	2.80	1.45	3.21	1419	3128	1461	3221	1.04	2.30	0.84	1.85	0.55	1.22	0.80	1.77
130	84	82	49	1.69	3.73	1620	3571	0.91	2.01	1.29	2.85	1533	3381	1313	2894	0.75	1.65	0.88	1.95	0.61	1.34	0.91	2.00
	93	91	54	1.88	4.14	1796	3959	1.01	2.23	1.43	3.15	1700	3748	1455	3208	0.83	1.83	0.98	2.16	0.67	1.48	1.01	2.22
131	74	74	71	1.42	3.13	1483	3269	–	–	1.26	2.77	1294	2853	–	–	–	–	0.74	1.64	0.49	1.08	0.71	1.56
	84	84	80	1.61	3.56	1683	3710	–	–	1.43	3.14	1469	3238	–	–	–	–	0.84	1.86	0.55	1.22	0.80	1.77
132	56	86	48	1.32	2.91	1728	3810	0.90	1.99	1.15	2.53	1641	3618	–	–	0.74	1.63	0.84	1.86	0.57	1.26	0.79	1.75
	62	96	53	1.46	3.22	1912	4216	1.00	2.20	1.27	2.80	1816	4003	–	–	0.82	1.81	0.93	2.05	0.63	1.39	0.88	1.93
133	78	75	67	1.47	3.24	1490	3286	1.22	2.69	1.30	2.87	1408	3105	–	–	1.00	2.20	0.75	1.66	0.49	1.09	0.72	1.58
	88	83	75	1.65	3.63	1670	3681	1.37	3.01	1.46	3.22	1578	3478	–	–	1.12	2.47	0.84	1.85	0.55	1.22	0.80	1.77

(Continued)

TABLE V-1A ENERGY FEEDS, MINERAL AND VITAMIN COMPOSITION OF FEEDS, DATA EXPRESSED **AS-FED** AND **MOISTURE-FREE**—*(Continued)*

Entry Number	Feed Name Description	Moisture Basis: A-F (as-fed) or M-F (moisture-free)	Dry Matter	Macro Minerals: Calcium (Ca)	Phosphorus (P)	Sodium (Na)	Chlorine (Cl)	Magnesium (Mg)	Potassium (K)	Sulfur (S)	Micro Minerals: Cobalt (Co)	Copper (Cu)	Iodine (I)	Iron (Fe)	Manganese (Mn)	Selenium (Se)	Zinc (Zn)
			%	%	%	%	%	%	%	%	ppm or mg/kg	ppm or mg/kg	ppm or mg/kg	%	ppm or mg/kg	ppm or mg/kg	ppm or mg/kg
	FATS AND OILS (Continued)																
109	FAT, ANIMAL-POULTRY	A-F	99	—	—	—	—	—	0.23	—	—	—	—	—	—	—	—
		M-F	100	—	—	—	—	—	0.23	—	—	—	—	—	—	—	—
110	OIL, CORN	A-F	99	—	—	—	—	—	—	—	—	—	—	—	—	—	—
		M-F	100	—	—	—	—	—	—	—	—	—	—	—	—	—	—
111	OIL, SOYBEAN	A-F	100	—	—	—	—	—	—	—	—	—	—	—	—	—	—
		M-F	100	—	—	—	—	—	—	—	—	—	—	—	—	—	—
112	TALLOW, ANIMAL	A-F	97	—	—	—	—	—	—	—	—	—	—	—	—	—	—
		M-F	100	—	—	—	—	—	—	—	—	—	—	—	—	—	—
	FETERITA SORGHUM *Sorghum bicolor, caudatum*																
113	GRAIN	A-F	89	0.02	0.30	—	—	—	—	—	—	—	—	—	—	—	—
		M-F	100	0.02	0.34	—	—	—	—	—	—	—	—	—	—	—	—
114	HEADS	A-F	90	—	—	—	—	—	—	—	—	—	—	—	—	—	—
		M-F	100	—	—	—	—	—	—	—	—	—	—	—	—	—	—
	FLAX, COMMON *Linum usitatissimum*																
115	SEEDS	A-F	96	0.28	0.55	—	—	—	—	—	—	—	—	—	—	—	—
		M-F	100	0.29	0.58	—	—	—	—	—	—	—	—	—	—	—	—
116	SEED SCREENINGS	A-F	91	0.34	0.43	—	—	0.39	0.77	0.23	—	—	—	0.010	—	—	—
		M-F	100	0.37	0.47	—	—	0.43	0.84	0.25	—	—	—	0.010	—	—	—
	GARBAGE, HOTEL AND RESTAURANT																
117	BOILED, WET	A-F	23	0.10	0.06	—	—	0.01	—	—	—	5.0	—	0.010	5.0	—	—
		M-F	100	0.44	0.28	—	—	0.06	—	—	—	22.0	—	0.043	22.0	—	—
118	BOILED, DEHY, GROUND	A-F	92	0.42	0.32	—	—	0.18	—	—	—	20.9	—	0.032	12.6	—	—
		M-F	100	0.46	0.35	—	—	0.19	—	—	—	22.7	—	0.035	13.7	—	—
	GARBANZO (CHICKPEA; GRAM PEA) *Cicer arietinum*																
119	SEEDS	A-F	89	0.15	0.33	0.03	—	—	0.80	—	—	—	—	0.007	—	—	—
		M-F	100	0.17	0.37	0.03	—	—	0.89	—	—	—	—	0.008	—	—	—
	GOOSEFOOT (LAMB'S QUARTERS) *Chenopodium album*																
120	SEEDS	A-F	90	—	—	—	—	—	—	—	—	—	—	—	—	—	—
		M-F	100	—	—	—	—	—	—	—	—	—	—	—	—	—	—
	GRAIN SCREENINGS—SEE BARLEY; WHEAT																
	GRAM PEA (CHICKPEA; GARBANZO) *Cicer arietinum*																
121	SEEDS	A-F	89	0.15	0.33	0.03	—	—	0.80	—	—	—	—	0.007	—	—	—
		M-F	100	0.17	0.37	0.03	—	—	0.89	—	—	—	—	0.008	—	—	—
	GRAPES *Vitis* spp																
122	FRUIT (RAISINS), DEHY, CULL	A-F	87	—	—	—	—	—	—	—	—	—	—	—	—	—	—
		M-F	100	—	—	—	—	—	—	—	—	—	—	—	—	—	—
123	POMACE, WET (MARC)	A-F	37	—	—	—	—	—	—	—	—	—	—	—	—	—	—
		M-F	100	—	—	—	—	—	—	—	—	—	—	—	—	—	—
124	POMACE, DEHY (MARC)	A-F	90	0.52	0.15	0.08	0.01	0.09	0.82	—	—	—	—	—	36.8	—	21.9
		M-F	100	0.58	0.17	0.09	0.01	0.10	0.91	—	—	—	—	—	40.7	—	24.2
	GRAPEFRUIT *Citrus paradisi*																
125	PULP WITHOUT FINES, DEHY (DRIED GRAPEFRUIT PULP)	A-F	90	1.34	0.17	—	—	0.39	—	—	—	—	—	—	—	—	—
		M-F	100	1.48	0.18	—	—	0.43	—	—	—	—	—	—	—	—	—
	GRASS																
126	SEED SCREENINGS	A-F	94	—	—	—	—	—	—	—	—	—	—	—	—	—	—
		M-F	100	—	—	—	—	—	—	—	—	—	—	—	—	—	—
	HEGARI SORGHUM *Sorghum bicolor, caffrorum*																
127	GRAIN	A-F	89	0.02	0.27	—	—	—	—	—	—	—	—	—	—	—	—
		M-F	100	0.02	0.30	—	—	—	—	—	—	—	—	—	—	—	—
128	HEADS	A-F	90	0.10	0.19	—	—	0.10	—	—	—	—	—	—	13.1	—	—
		M-F	100	0.11	0.21	—	—	0.11	—	—	—	—	—	—	14.6	—	—
	HOG MILLET (BROOMCORN; MILLET, PROSO) *Panicum miliaceum*																
129	GRAIN	A-F	90	0.03	0.30	—	—	0.16	0.43	—	—	—	—	0.007	—	—	—
		M-F	100	0.03	0.34	—	—	0.18	0.48	—	—	—	—	0.008	—	—	—
130	**HOMINY GRITS (CORN, DENT YELLOW, GRITS)** *Zea mays, indentata*	A-F	90	0.05	0.51	0.08	0.05	0.24	0.59	0.03	0.055	13.6	—	0.007	14.5	—	—
		M-F	100	0.05	0.57	0.09	0.06	0.26	0.65	0.03	0.061	15.1	—	0.008	16.1	—	—
	HORSE BEAN *Vicia faba, equina*																
131	SEEDS	A-F	88	0.13	0.54	—	—	—	1.17	—	—	—	—	—	—	—	—
		M-F	100	0.15	0.62	—	—	—	1.33	—	—	—	—	—	—	—	—
	HORSE CHESTNUT, COMMON *Aesculus hippocastanum*																
132	KERNELS WITH COATS	A-F	90	—	—	—	—	—	—	—	—	—	—	—	—	—	—
		M-F	100	—	—	—	—	—	—	—	—	—	—	—	—	—	—
	IVORY PALM, COMMON *Phytelephas macrocarpa*																
133	KERNELS WITH COATS, MEAL SOLV EXTD	A-F	89	—	—	—	—	—	—	—	—	—	—	—	—	—	—
		M-F	100	—	—	—	—	—	—	—	—	—	—	—	—	—	—

Entry Number	Fat-Soluble Vitamins: A (1 mg Carotene = 1667 IU Vit A)	Carotene (Provitamin A)	D	E	K	Water-Soluble Vitamins: B–12	Biotin	Choline	Folacin (Folic Acid)	Niacin	Pantothenic Acid (B–3)	(Pyridoxine) B–6	Riboflavin (B–2)	Thiamin (B–1)
	IU/g	ppm or mg/kg	IU/kg	ppm or mg/kg	ppm or mg/kg	ppb or mcg/kg	ppm or mg/kg	ppm or mg/kg	ppm or mg/kg	ppm or mg/kg	ppm or mg/kg	ppm or mg/kg	ppm or mg/kg	ppm or mg/kg
109	–	–	–	**7.9**	–	–	–	–	–	–	–	–	–	–
	–	–	–	7.9	–	–	–	–	–	–	–	–	–	–
110	–	–	–	–	–	–	–	–	–	–	–	–	–	–
	–	–	–	–	–	–	–	–	–	–	–	–	–	–
111	–	–	–	–	–	–	–	–	–	–	–	–	–	–
	–	–	–	–	–	–	–	–	–	–	–	–	–	–
112	–	–	–	–	–	–	–	–	–	–	–	–	–	–
	–	–	–	–	–	–	–	–	–	–	–	–	–	–
113	**1.0**	**0.6**	–	–	–	–	–	–	–	**50**	**14.0**	–	**1.6**	–
	1.1	0.7	–	–	–	–	–	–	–	56	15.7	–	1.8	–
114	–	–	–	–	–	–	–	–	–	–	–	–	–	–
	–	–	–	–	–	–	–	–	–	–	–	–	–	–
115	–	–	–	–	–	–	–	–	–	–	–	–	–	–
	–	–	–	–	–	–	–	–	–	–	–	–	–	–
116	–	–	–	–	–	–	–	–	–	–	–	–	–	–
	–	–	–	–	–	–	–	–	–	–	–	–	–	–
117	–	–	–	–	–	–	–	–	–	–	–	–	–	–
	–	–	–	–	–	–	–	–	–	–	–	–	–	–
118	–	–	–	–	–	–	–	–	–	–	–	–	–	–
	–	–	–	–	–	–	–	–	–	–	–	–	–	–
119	–	–	–	–	–	–	–	–	–	**29**	–	–	**1.2**	**2.4**
	–	–	–	–	–	–	–	–	–	33	–	–	1.3	2.7
120	–	–	–	–	–	–	–	–	–	–	–	–	–	–
	–	–	–	–	–	–	–	–	–	–	–	–	–	–
121	–	–	–	–	–	–	–	–	–	**29**	–	–	**1.2**	**2.4**
	–	–	–	–	–	–	–	–	–	33	–	–	1.3	2.7
122	–	–	–	–	–	–	–	–	–	–	–	–	–	–
	–	–	–	–	–	–	–	–	–	–	–	–	–	–
123	–	–	–	–	–	–	–	–	–	–	–	–	–	–
	–	–	–	–	–	–	–	–	–	–	–	–	–	–
124	–	–	–	–	–	–	–	**253**	–	**18**	**3.1**	–	**2.2**	–
	–	–	–	–	–	–	–	279	–	20	3.4	–	2.5	–
125	–	–	–	–	–	–	–	–	–	–	–	–	–	–
	–	–	–	–	–	–	–	–	–	–	–	–	–	–
126	–	–	–	–	–	–	–	–	–	–	–	–	–	–
	–	–	–	–	–	–	–	–	–	–	–	–	–	–
127	–	–	–	–	–	–	–	–	–	–	–	–	–	–
	–	–	–	–	–	–	–	–	–	–	–	–	–	–
128	–	–	–	–	–	–	–	–	–	–	–	–	–	–
	–	–	–	–	–	–	–	–	–	–	–	–	–	–
129	–	–	–	–	–	–	–	**438**	–	**24**	**10.9**	–	**3.3**	**7.5**
	–	–	–	–	–	–	–	488	–	27	12.2	–	3.7	8.3
130	**15.4**	**9.2**	–	–	–	–	**0.13**	**1154**	**0.31**	**47**	**8.2**	**10.95**	**2.1**	**8.1**
	17.0	10.2	–	–	–	–	0.15	1280	0.34	52	9.1	12.14	2.4	8.9
131	–	–	–	–	–	–	–	–	–	–	–	–	–	–
	–	–	–	–	–	–	–	–	–	–	–	–	–	–
132	–	–	–	–	–	–	–	–	–	–	–	–	–	–
	–	–	–	–	–	–	–	–	–	–	–	–	–	–
133	–	–	–	–	–	–	–	–	–	–	–	–	–	–
	–	–	–	–	–	–	–	–	–	–	–	–	–	–

(Continued)

TABLE V-1A ENERGY FEEDS, COMPOSITION OF FEEDS, DATA EXPRESSED **AS-FED** AND **MOISTURE-FREE**—*(Continued)*

Entry Number	Feed Name Description	International Feed Number	Moisture Basis: A-F (as-fed) or M-F (moisture-free)	Proximate Analysis: Dry Matter	Ash	Crude Fiber	Neutral Det. Fib. (NDF)	Acid Det. Fib. (ADF)	Lignin	Ether Extract (Fat)	N-Free Extract	Crude Protein	Digestible Protein: Ruminant	Swine	Horse
				%	%	%	%	%	%	%	%	%	%	%	%
	JACK BEAN, COMMON *Canavalia ensiformis*														
134	SEEDS	5-02-435	A-F	89	2.9	8.5	–	–	–	2.8	47.6	27.4	23.0	26.1	22.7
			M-F	100	3.2	9.5	–	–	–	3.1	53.4	30.7	25.8	29.3	25.5
	JOB'S TEARS (ADLAY) *Coix lacrymajobi*														
135	SEEDS	4-08-448	A-F	89	2.6	4.5	–	–	–	5.1	62.8	13.8	9.7	10.8	10.1
			M-F	100	2.9	5.1	–	–	–	5.7	70.7	15.6	10.9	12.2	11.4
	KAFIR SORGHUM *Sorghum bicolor, caffrorum*														
136	GRAIN	4-04-428	A-F	89	1.5	2.0	–	1.8	–	2.8	72.0	10.8	6.6	8.3	7.7
			M-F	100	1.7	2.2	–	2.0	–	3.1	80.8	12.1	7.4	9.4	8.7
137	HEADS	4-04-429	A-F	89	3.0	7.4	–	–	–	2.6	66.6	9.8	3.6	7.1	6.9
			M-F	100	3.4	8.3	–	–	–	2.9	74.6	10.9	4.0	8.0	7.7
	KALO SORGHUM *Sorghum bicolor*														
138	GRAIN	4-04-430	A-F	89	1.7	1.6	–	–	–	3.2	70.9	11.8	7.8	9.0	8.5
			M-F	100	1.9	1.8	–	–	–	3.6	79.5	13.2	8.8	10.0	9.5
	KELP (SEAWEED) *Laminariales* (order), *Fucales* (order)														
139	WHOLE, DEHY	1-08-073	A-F	91	35.0	6.5	–	–	–	0.5	42.4	6.5	2.9	2.2	3.4
			M-F	100	38.6	7.1	–	–	–	0.5	46.7	7.1	3.2	2.4	3.7
	KOALIANG SORGHUM *Sorghum bicolor, nervosum*														
140	GRAIN	4-04-431	A-F	90	2.1	1.9	–	2.2	–	3.9	71.1	10.7	6.8	7.9	7.6
			M-F	100	2.4	2.1	–	2.5	–	4.3	79.3	11.9	7.6	8.9	8.5
	LAMB'S QUARTER (GOOSEFOOT) *Chenopodium album*														
141	SEEDS	5-08-424	A-F	90	9.6	15.1	–	–	–	4.5	41.9	18.8	15.0	17.6	13.0
			M-F	100	10.7	16.8	–	–	–	5.0	46.6	20.9	16.6	19.5	14.4
	LARD (FAT) *Sus scrofa*														
142	SWINE	4-04-790	A-F	99	–	–	–	–	–	99.3	–	–	–	–	–
			M-F	100	–	–	–	–	–	100.0	–	–	–	–	–
	LEMON *Citrus limon*														
143	PULP WITHOUT FINES, DEHY (DRIED LEMON PULP)	4-01-247	A-F	93	5.3	14.7	–	–	–	1.4	65.0	6.5	3.0	4.1	4.3
			M-F	100	5.7	15.9	–	–	–	1.5	70.0	7.0	3.2	4.4	4.6
	LENTIL, COMMON *Lens culinaris*														
144	SEEDS	5-02-506	A-F	88	2.6	3.4	–	–	–	1.0	57.1	24.4	19.3	23.1	19.4
			M-F	100	2.9	3.8	–	–	–	1.1	64.5	27.6	21.8	26.2	21.9
	LUCERNE—SEE ALFALFA														
	MAIZE—SEE CORN														
	MALT SPROUTS (BARLEY) *Hordeum vulgare*														
145	DEHY	5-00-545	A-F	93	5.6	14.2	–	–	–	1.4	48.8	22.9	19.2	21.5	17.2
			M-F	100	6.1	15.3	–	–	–	1.5	52.6	24.6	20.7	23.2	18.6
	MANGEL, BEET *Beta vulgaris, macrorrhiza*														
146	ROOTS, FRESH	4-00-637	A-F	11	1.1	0.8	–	–	–	0.1	7.7	1.3	0.9	1.0	0.9
			M-F	100	9.6	7.4	–	–	–	0.7	70.5	11.8	8.1	9.2	8.4
	MANURE, CATTLE *Bos taurus*														
147	WITHOUT BEDDING, DEHY	1-01-190	A-F	93	17.9	30.7	–	43.8	–	2.5	30.1	12.0	7.7	6.9	7.6
			M-F	100	19.1	33.0	–	47.0	–	2.7	32.3	12.9	8.2	7.5	8.1
	MESQUITE, COMMON *Prosopis chilensis*														
148	PODS WITH SEEDS, SUN-CURED	4-10-398	A-F	89	4.4	25.0	–	–	–	2.8	44.5	12.4	11.1	–	–
			M-F	100	4.9	28.1	–	–	–	3.1	50.0	13.9	12.5	–	–
	MILLET *Setaria* spp														
149	GRAIN	4-03-098	A-F	90	2.7	5.8	–	–	–	4.0	65.2	12.1	6.8	8.8	8.7
			M-F	100	3.1	6.4	–	–	–	4.5	72.6	13.5	7.5	9.8	9.7
	MILLET, FOXTAIL *Setaria italica*														
150	GRAIN	4-03-102	A-F	89	3.4	7.4	–	–	–	4.1	63.0	11.4	7.5	8.6	8.2
			M-F	100	3.8	8.3	–	–	–	4.6	70.6	12.8	8.4	9.6	9.2
	MILLET, PROSO (BROOMCORN; HOG MILLET) *Panicum miliaceum*														
151	GRAIN	4-03-120	A-F	90	2.9	5.3	–	14.9	3.2	3.6	66.3	11.6	7.9	7.9	8.3
			M-F	100	3.3	6.0	–	16.6	3.6	4.0	73.9	12.9	8.8	8.8	9.3
	MILO SORGHUM *Sorghum bicolor, subglabrescens*														
152	GRAIN	4-04-444	A-F	89	1.6	2.2	20.6	4.7	–	2.8	71.9	10.1	6.8	7.2	7.1
			M-F	100	1.8	2.5	23.2	5.3	–	3.1	81.2	11.4	7.7	8.1	8.1
153	GLUTEN WITH BRAN (GLUTEN FEED)	5-08-089	A-F	89	6.4	7.1	–	–	–	4.1	48.3	23.2	18.9	19.7	17.9
			M-F	100	7.2	8.0	–	–	–	4.6	54.2	26.0	21.2	22.1	20.1
154	HEADS	4-04-446	A-F	90	4.9	8.9	–	–	–	2.6	64.9	8.6	6.6	6.1	6.0
			M-F	100	5.4	9.9	–	–	–	2.9	72.2	9.6	7.3	6.8	6.7

Entry Number	TDN Ruminant	TDN Swine	TDN Horse	Digestible Energy Ruminant		Digestible Energy Swine		Digestible Energy Horse		Metabolizable Energy Ruminant		Metabolizable Energy Swine		Metabolizable Energy Poultry ME_n		Metabolizable Energy Horse		Net Energy Ruminant NE_m		Net Energy Ruminant NE_g		Net Energy Lactating Cows NE_{lc}	
	%	%	%	Mcal		kcal		Mcal		Mcal		kcal		kcal		Mcal		Mcal		Mcal		Mcal	
				lb	*kg*	lb	*kg*	lb	*kg*	lb	*kg*	lb	*kg*	lb	*kg*	lb	*kg*	lb	*kg*	lb	*kg*	lb	*kg*
134	**81**	**77**	–	**1.51**	***3.32***	**1548**	***3414***	–	–	**1.34**	***2.96***	**1354**	***2985***	–	–	–	–	**0.77**	***1.69***	**0.51**	***1.12***	**0.73**	***1.61***
	91	87	–	1.69	*3.73*	1737	*3828*	–	–	1.50	*3.32*	1518	*3347*	–	–	–	–	0.86	*1.90*	0.57	*1.26*	0.82	*1.80*
135	**67**	**73**	**59**	**1.37**	***3.03***	**1462**	***3223***	**1.08**	***2.37***	**1.21**	***2.66***	**1381**	***3044***	–	–	**0.88**	***1.95***	**0.77**	***1.71***	**0.52**	***1.14***	**0.73**	***1.62***
	75	82	66	1.55	*3.41*	1646	*3630*	1.21	*2.67*	1.36	*3.00*	1555	*3428*	–	–	0.99	*2.19*	0.87	*1.92*	0.58	*1.28*	0.83	*1.82*
136	**75**	**81**	**70**	**1.47**	***3.24***	**1521**	***3354***	**1.26**	***2.77***	**1.30**	***2.87***	**1437**	***3169***	**1529**	***3372***	**1.03**	***2.27***	**0.79**	***1.74***	**0.53**	***1.16***	**0.75**	***1.64***
	85	91	78	1.65	*3.64*	1707	*3764*	1.41	*3.11*	1.46	*3.22*	1613	*3557*	1717	*3785*	1.16	*2.55*	0.88	*1.95*	0.59	*1.30*	0.84	*1.85*
137	**47**	**72**	**62**	**0.94**	***2.06***	**1430**	***3153***	**1.14**	***2.51***	**0.76**	***1.68***	**1349**	***2975***	–	–	**0.93**	***2.06***	**0.07**	***0.15***	**−0.14**	***−0.31***	**0.24**	***0.52***
	52	80	70	1.05	*2.31*	1602	*3531*	1.28	*2.81*	0.85	*1.88*	1511	*3332*	–	–	1.05	*2.31*	0.08	*0.17*	−0.16	*−0.35*	0.27	*0.58*
138	**80**	**77**	**69**	**1.52**	***3.35***	**1530**	***3373***	**1.25**	***2.75***	**1.35**	***2.99***	**1448**	***3191***	–	–	**1.02**	***2.26***	**0.79**	***1.75***	**0.53**	***1.17***	**0.75**	***1.65***
	90	86	77	1.71	*3.76*	1715	*3782*	1.40	*3.08*	1.52	*3.35*	1623	*3578*	–	–	1.15	*2.53*	0.89	*1.96*	0.59	*1.31*	0.84	*1.85*
139	**29**	–	–	**0.58**	***1.27***	–	–	–	–	**0.40**	***0.87***	–	–	–	–	–	–	–	–	–	–	–	–
	32	–	–	0.63	*1.40*	–	–	–	–	0.44	*0.96*	–	–	–	–	–	–	–	–	–	–	–	–
140	**81**	**79**	**64**	**1.54**	***3.39***	**1577**	***3477***	**1.17**	***2.58***	**1.37**	***3.03***	**1494**	***3293***	–	–	**0.96**	***2.11***	**0.80**	***1.77***	**0.54**	***1.19***	**0.76**	***1.67***
	91	88	72	1.72	*3.79*	1761	*3882*	1.30	*2.87*	1.53	*3.38*	1667	*3676*	–	–	1.07	*2.36*	0.90	*1.97*	0.60	*1.32*	0.85	*1.86*
141	**55**	**45**	–	**1.16**	***2.56***	**892**	***1967***	–	–	**0.99**	***2.18***	**747**	***1647***	–	–	–	–	**0.64**	***1.41***	**0.39**	***0.87***	**0.63**	***1.39***
	61	50	–	1.29	*2.85*	992	*2186*	–	–	1.10	*2.42*	830	*1830*	–	–	–	–	0.71	*1.56*	0.44	*0.96*	0.70	*1.54*
142	–	–	–	–	–	**3669**	***8089***	–	–	–	–	**3537**	***7798***	**3936**	***8677***	–	–	–	–	–	–	–	–
	–	–	–	–	–	3694	*8143*	–	–	–	–	3561	*7850*	3962	*8735*	–	–	–	–	–	–	–	–
143	**73**	**69**	**57**	**1.38**	***3.05***	**1378**	***3038***	**1.05**	***2.32***	**1.21**	***2.66***	**1296**	***2858***	–	–	**0.86**	***1.90***	**0.70**	***1.54***	**0.44**	***0.97***	**0.68**	***1.50***
	78	74	61	1.49	*3.29*	1484	*3272*	1.13	*2.50*	1.30	*2.87*	1396	*3078*	–	–	0.93	*2.05*	0.75	*1.66*	0.48	*1.05*	0.73	*1.61*
144	**74**	**84**	–	**1.48**	***3.26***	**1675**	***3692***	–	–	**1.31**	***2.89***	**1471**	***3244***	–	–	–	–	**0.81**	***1.79***	**0.55**	***1.21***	**0.77**	***1.69***
	84	95	–	1.67	*3.68*	1894	*4175*	–	–	1.48	*3.27*	1663	*3667*	–	–	–	–	0.92	*2.03*	0.62	*1.37*	0.87	*1.91*
145	**66**	**61**	–	**1.31**	***2.89***	**1219**	***2688***	–	–	**1.13**	***2.50***	**1046**	***2307***	**722**	***1592***	–	–	**0.69**	***1.52***	**0.43**	***0.95***	**0.67**	***1.48***
	71	66	–	1.41	*3.11*	1313	*2894*	–	–	1.22	*2.69*	1126	*2483*	778	*1714*	–	–	0.74	*1.63*	0.47	*1.03*	0.72	*1.60*
146	**9**	**9**	**8**	**0.18**	***0.39***	**181**	***399***	**0.14**	***0.32***	**0.15**	***0.34***	**171**	***377***	–	–	**0.12**	***0.26***	**0.09**	***0.21***	**0.06**	***0.14***	**0.09**	***0.20***
	80	83	72	1.59	*3.51*	1652	*3641*	1.31	*2.90*	1.40	*3.09*	1560	*3439*	–	–	1.08	*2.38*	0.85	*1.88*	0.56	*1.24*	0.81	*1.79*
147	**44**	–	**25**	**0.88**	***1.94***	**1281**	***2825***	**0.53**	***1.17***	**0.70**	***1.54***	**1222**	***2695***	–	–	**0.43**	***0.96***	**0.36**	***0.79***	**0.13**	***0.29***	**0.43**	***0.96***
	47	–	27	0.94	*2.08*	1374	*3029*	0.57	*1.25*	0.75	*1.65*	1311	*2890*	–	–	0.47	*1.03*	0.39	*0.85*	0.14	*0.31*	0.47	*1.03*
148	–	–	–	–	–	–	–	–	–	–	–	–	–	–	–	–	–	–	–	–	–	–	–
	–	–	–	–	–	–	–	–	–	–	–	–	–	–	–	–	–	–	–	–	–	–	–
149	**61**	**66**	**61**	**1.54**	***3.39***	**1315**	***2900***	**1.11**	***2.46***	**1.17**	***2.58***	**1224**	***2697***	**1441**	***3178***	**0.91**	***2.01***	**0.86**	***1.90***	**0.59**	***1.29***	**0.81**	***1.79***
	68	73	68	1.71	*3.77*	1463	*3226*	1.24	*2.73*	1.30	*2.87*	1361	*3001*	1603	*3535*	1.02	*2.24*	0.96	*2.11*	0.65	*1.44*	0.90	*1.99*
150	**76**	**72**	**57**	**1.45**	***3.19***	1436	***3167***	**1.04**	***2.30***	**1.28**	***2.82***	**1355**	***2988***	–	–	**0.86**	***1.89***	**0.75**	***1.65***	**0.49**	***1.09***	**0.71**	***1.58***
	85	80	64	1.62	*3.57*	1610	*3549*	1.17	*2.58*	1.43	*3.16*	1519	*3349*	–	–	0.96	*2.12*	0.84	*1.85*	0.55	*1.22*	0.80	*1.77*
151	**74**	**68**	**62**	**1.47**	***3.25***	**1353**	***2983***	**1.14**	***2.51***	**1.31**	***2.88***	**1273**	***2807***	**1311**	***2890***	**0.93**	***2.06***	**0.75**	***1.66***	**0.50**	***1.09***	**0.72**	***1.59***
	82	75	70	1.64	*3.62*	1508	*3324*	1.27	*2.80*	1.45	*3.21*	1419	*3128*	1461	*3221*	1.04	*2.30*	0.84	*1.85*	0.55	*1.22*	0.80	*1.77*
152	**76**	**77**	**68**	**1.41**	***3.12***	**1520**	***3350***	**1.23**	***2.71***	**1.13**	***2.49***	**1474**	***3250***	**1467**	***3234***	**1.01**	***2.23***	**0.75**	***1.66***	**0.50**	***1.09***	**0.74**	***1.62***
	86	87	77	1.60	*3.52*	1716	*3782*	1.39	*3.06*	1.27	*2.81*	1664	*3669*	1656	*3651*	1.14	*2.51*	0.85	*1.87*	0.56	*1.23*	0.83	*1.83*
153	**74**	**74**	–	**1.45**	***3.20***	**1484**	***3271***	–	–	**1.29**	***2.83***	**1223**	***2696***	**975**	***2150***	–	–	**0.78**	***1.72***	**0.52**	***1.14***	**0.74**	***1.63***
	84	83	–	1.63	*3.59*	1665	*3670*	–	–	1.44	*3.18*	1372	*3025*	1094	*2413*	–	–	0.87	*1.93*	0.58	*1.28*	0.83	*1.83*
154	**73**	**72**	**57**	**1.46**	***3.21***	**1445**	***3185***	**1.05**	***2.32***	**1.29**	***2.84***	**1363**	***3005***	–	–	**0.86**	***1.90***	**0.76**	***1.69***	**0.51**	***1.11***	**0.73**	***1.61***
	81	80	64	1.62	*3.58*	1607	*3543*	1.17	*2.58*	1.43	*3.16*	1516	*3343*	–	–	0.96	*2.12*	0.85	*1.88*	0.56	*1.24*	0.81	*1.79*

(Continued)

TABLE V-1A ENERGY FEEDS, MINERAL AND VITAMIN COMPOSITION OF FEEDS, DATA EXPRESSED AS-FED AND MOISTURE-FREE—*(Continued)*

Entry Number	Feed Name Description	Moisture Basis: A-F (as-fed) or M-F (moisture-free)	Dry Matter	Macro Minerals							Micro Minerals						
				Calcium (Ca)	Phosphorus (P)	Sodium (Na)	Chlorine (Cl)	Magnesium (Mg)	Potassium (K)	Sulfur (S)	Cobalt (Co)	Copper (Cu)	Iodine (I)	Iron (Fe)	Manganese (Mn)	Selenium (Se)	Zinc (Zn)
			%	%	%	%	%	%	%	%	ppm or mg/kg	ppm or mg/kg	ppm or mg/kg	%	ppm or mg/kg	ppm or mg/kg	ppm or mg/kg
	JACK BEAN, COMMON *Canavalia ensiformis*																
134	SEEDS	A-F	89	—	—	—	—	—	—	—	—	—	—	—	—	—	—
		M-F	100	—	—	—	—	—	—	—	—	—	—	—	—	—	—
	JOB'S TEARS (ADLAY) *Coix lacrymajobi*																
135	SEEDS	A-F	89	0.02	—	—	—	—	—	—	—	—	—	0.005	—	—	—
		M-F	100	0.02	—	—	—	—	—	—	—	—	—	0.005	—	—	—
	KAFIR SORGHUM *Sorghum bicolor, caffrorum*																
136	GRAIN	A-F	89	0.03	0.31	0.05	0.10	0.15	0.34	0.16	0.387	7.0	—	0.007	15.8	0.797	13.5
		M-F	100	0.04	0.35	0.06	0.11	0.17	0.38	0.18	0.435	7.8	—	0.008	17.8	0.894	15.2
137	HEADS	A-F	89	0.08	0.25	—	—	0.24	—	—	—	—	—	—	—	—	—
		M-F	100	0.09	0.28	—	—	0.27	—	—	—	—	—	—	—	—	—
	KALO SORGHUM *Sorghum bicolor*																
138	GRAIN	A-F	89	—	—	—	—	—	—	—	—	—	—	—	—	—	—
		M-F	100	—	—	—	—	—	—	—	—	—	—	—	—	—	—
	KELP (SEAWEED) *Laminariales* (order), *Fucales* (order)																
139	WHOLE, DEHY	A-F	91	2.47	0.28	—	—	0.85	—	—	—	—	—	—	—	—	—
		M-F	100	2.72	0.31	—	—	0.93	—	—	—	—	—	—	—	—	—
	KOALIANG SORGHUM *Sorghum bicolor, nervosum*																
140	GRAIN	A-F	90	0.03	0.29	—	—	—	—	—	—	—	—	—	—	—	—
		M-F	100	0.03	0.32	—	—	—	—	—	—	—	—	—	—	—	—
	LAMB'S QUARTER (GOOSEFOOT) *Chenopodium album*																
141	SEEDS	A-F	90	—	—	—	—	—	—	—	—	—	—	—	—	—	—
		M-F	100	—	—	—	—	—	—	—	—	—	—	—	—	—	—
	LARD (FAT) *Sus scrofa*																
142	SWINE	A-F	99	—	—	—	—	—	—	—	—	—	—	—	—	—	—
		M-F	100	—	—	—	—	—	—	—	—	—	—	—	—	—	—
	LEMON *Citrus limon*																
143	PULP WITHOUT FINES, DEHY (DRIED LEMON PULP)	A-F	93	—	—	—	—	—	—	—	—	—	—	—	—	—	—
		M-F	100	—	—	—	—	—	—	—	—	—	—	—	—	—	—
	LENTIL, COMMON *Lens culinaris*																
144	SEEDS	A-F	88	0.08	0.38	0.03	—	—	0.79	—	—	—	—	0.007	—	—	—
		M-F	100	0.09	0.42	0.03	—	—	0.89	—	—	—	—	0.008	—	—	—
	LUCERNE—SEE ALFALFA																
	MAIZE—SEE CORN																
	MALT SPROUTS (BARLEY) *Hordeum vulgare*																
145	DEHY	A-F	93	0.18	0.63	0.88	0.36	0.17	0.25	0.79	—	5.9	—	0.018	29.4	0.416	56.4
		M-F	100	0.19	0.68	0.95	0.39	0.18	0.27	0.85	—	6.3	—	0.020	31.7	0.448	60.7
	MANGEL, BEET *Beta vulgaris, macrorrhiza*																
146	ROOTS, FRESH	A-F	11	0.02	0.02	0.07	0.16	0.02	0.25	0.02	—	0.6	—	0.002	—	—	—
		M-F	100	0.18	0.22	0.63	1.41	0.20	2.30	0.20	—	5.5	—	0.016	—	—	—
	MANURE, CATTLE *Bos taurus*																
147	WITHOUT BEDDING, DEHY	A-F	93	1.35	1.08	—	—	—	0.47	—	—	—	—	—	—	—	—
		M-F	100	1.45	1.15	—	—	—	0.50	—	—	—	—	—	—	—	—
	MESQUITE, COMMON *Prosopis chilensis*																
148	PODS WITH SEEDS, SUN-CURED	A-F	89	—	—	—	—	—	—	—	—	—	—	—	—	—	—
		M-F	100	—	—	—	—	—	—	—	—	—	—	—	—	—	—
	MILLET *Setaria* spp																
149	GRAIN	A-F	90	0.05	0.29	0.04	0.14	0.16	0.43	0.13	0.044	21.8	—	0.007	29.9	—	13.9
		M-F	100	0.05	0.32	0.04	0.16	0.18	0.48	0.14	0.049	24.3	—	0.008	33.3	—	15.4
	MILLET, FOXTAIL *Setaria italica*																
150	GRAIN	A-F	89	—	0.41	—	—	—	0.31	—	—	—	—	0.010	—	—	—
		M-F	100	—	0.46	—	—	—	0.35	—	—	—	—	0.011	—	—	—
	MILLET, PROSO (BROOMCORN; HOG MILLET) *Panicum miliaceum*																
151	GRAIN	A-F	90	0.03	0.30	—	—	0.16	0.43	—	—	—	—	0.007	—	—	—
		M-F	100	0.03	0.34	—	—	0.18	0.48	—	—	—	—	0.008	—	—	—
	MILO SORGHUM *Sorghum bicolor, subglabrescens*																
152	GRAIN	A-F	89	0.04	0.30	0.04	0.08	0.13	0.31	0.11	0.471	4.3	0.061	0.005	15.8	0.201	16.9
		M-F	100	0.05	0.34	0.04	0.09	0.14	0.35	0.12	0.531	4.9	0.069	0.006	17.9	0.227	19.1
153	GLUTEN WITH BRAN (GLUTEN FEED)	A-F	89	0.20	0.80	—	—	—	1.60	—	0.089	25.0	—	0.030	48.6	—	26.5
		M-F	100	0.23	0.90	—	—	—	1.80	—	0.100	28.0	—	0.034	54.6	—	29.8
154	HEADS	A-F	90	0.12	0.20	—	—	0.15	0.50	0.12	—	—	—	—	—	—	—
		M-F	100	0.14	0.23	—	—	0.17	0.56	0.13	—	—	—	—	—	—	—

Entry Number	Fat-Soluble Vitamins					Water-Soluble Vitamins								
	A (1 mg Carotene = 1667 IU Vit A)	Carotene (Provitamin A)	D	E	K	B–12	Biotin	Choline	Folacin (Folic Acid)	Niacin	Pantothenic Acid (B–3)	(Pyri-doxine) B–6	Ribo-flavin (B–2)	Thiamin (B–1)
	IU/g	ppm or mg/kg	IU/kg	ppm or mg/kg	ppm or mg/kg	ppb or mcg/kg	ppm or mg/kg	ppm or mg/kg	ppm or mg/kg	ppm or mg/kg	ppm or mg/kg	ppm or mg/kg	ppm or mg/kg	ppm or mg/kg
134	**–**	**–**	**–**	**–**	**–**	**–**	**–**	**–**	**–**	**–**	**–**	**–**	**–**	**–**
	–	–	–	–	–	–	–	–	–	–	–	–	–	–
135	**–**	**–**	**–**	**–**	**–**	**–**	**–**	**–**	**–**	**30**	**–**	**–**	**2.0**	**3.0**
	–	–	–	–	–	–	–	–	–	34	–	–	2.3	3.4
136	**0.6**	**0.4**	**–**	**–**	**–**	**–**	**0.24**	**439**	**0.20**	**38**	**12.0**	**6.68**	**1.2**	**3.8**
	0.7	0.4	–	–	–	–	0.27	493	0.22	43	13.4	7.50	1.4	4.3
137	**–**	**–**	**–**	**–**	**–**	**–**	**–**	**–**	**–**	**–**	**–**	**–**	**–**	**–**
	–	–	–	–	–	–	–	–	–	–	–	–	–	–
138	**–**	**–**	**–**	**–**	**–**	**–**	**0.26**	**–**	**–**	**33**	**4.9**	**5.11**	**1.2**	**–**
	–	–	–	–	–	–	0.29	–	–	37	5.5	5.73	1.3	–
139	**–**	**–**	**–**	**–**	**–**	**–**	**–**	**–**	**–**	**–**	**–**	**–**	**–**	**–**
	–	–	–	–	–	–	–	–	–	–	–	–	–	–
140	**–**	**–**	**–**	**–**	**–**	**–**	**–**	**–**	**–**	**–**	**–**	**–**	**–**	**–**
	–	–	–	–	–	–	–	–	–	–	–	–	–	–
141	**–**	**–**	**–**	**–**	**–**	**–**	**–**	**–**	**–**	**–**	**–**	**–**	**–**	**–**
	–	–	–	–	–	–	–	–	–	–	–	–	–	–
142	**–**	**–**	**–**	**22.8**	**–**	**–**	**–**	**–**	**–**	**–**	**–**	**–**	**–**	**–**
	–	–	–	23.0	–	–	–	–	–	–	–	–	–	–
143	**–**	**–**	**–**	**–**	**–**	**–**	**–**	**–**	**–**	**–**	**–**	**–**	**–**	**–**
	–	–	–	–	–	–	–	–	–	–	–	–	–	–
144	**–**	**–**	**–**	**–**	**–**	**–**	**–**	**–**	**–**	**20**	**–**	**–**	**2.2**	**3.7**
	–	–	–	–	–	–	–	–	–	22	–	–	2.5	4.2
145	**–**	**–**	**–**	**3.7**	**–**	**–**	**4.09**	**1591**	**0.20**	**55**	**9.0**	**8.62**	**2.8**	**8.3**
	–	–	–	4.0	–	–	4.40	1713	0.22	59	9.6	9.28	3.0	8.9
146	**0.2**	**0.1**	**–**	**–**	**–**	**–**	**–**	**–**	**0.17**	**3**	**1.0**	**0.43**	**0.4**	**0.3**
	1.5	0.9	–	–	–	–	–	–	1.58	31	9.4	3.94	3.9	2.4
147	**–**	**–**	**–**	**–**	**–**	**–**	**–**	**–**	**–**	**–**	**–**	**–**	**–**	**–**
	–	–	–	–	–	–	–	–	–	–	–	–	–	–
148	**–**	**–**	**–**	**–**	**–**	**–**	**–**	**–**	**–**	**–**	**–**	**–**	**–**	**–**
	–	–	–	–	–	–	–	–	–	–	–	–	–	–
149	**–**	**–**	**–**	**–**	**–**	**–**	**–**	**739**	**0.22**	**48**	**9.0**	**–**	**1.5**	**6.6**
	–	–	–	–	–	–	–	822	0.25	54	10.1	–	1.6	7.3
150	**–**	**–**	**–**	**–**	**–**	**–**	**–**	**–**	**–**	**33**	**–**	**–**	**1.1**	**3.8**
	–	–	–	–	–	–	–	–	–	37	–	–	1.2	4.3
151	**–**	**–**	**–**	**–**	**–**	**–**	**–**	**438**	**–**	**24**	**10.9**	**–**	**3.3**	**7.5**
	–	–	–	–	–	–	–	488	–	27	12.2	–	3.7	8.3
152	**0.4**	**0.2**	**0**	**12.1**	**0.22**	**–**	**0.23**	**638**	**0.21**	**37**	**11.0**	**4.69**	**1.1**	**4.1**
	0.5	0.3	0	13.7	0.25	–	0.26	720	0.23	42	12.4	5.29	1.3	4.7
153	**–**	**–**	**–**	**–**	**–**	**–**	**0.22**	**1746**	**–**	**99**	**19.9**	**10.83**	**2.7**	**5.7**
	–	–	–	–	–	–	0.25	1959	–	112	22.3	12.15	3.0	6.4
154	**1.0**	**0.6**	**–**	**–**	**–**	**–**	**–**	**–**	**–**	**–**	**–**	**–**	**2.0**	**–**
	1.2	0.7	–	–	–	–	–	–	–	–	–	–	2.2	–

(Continued)

TABLE V-1A ENERGY FEEDS, COMPOSITION OF FEEDS, DATA EXPRESSED **AS-FED** AND **MOISTURE-FREE**—*(Continued)*

Entry Number	Feed Name Description	International Feed Number	Moisture Basis: A-F (as-fed) or M-F (moisture-free)	Proximate Analysis: Dry Matter	Ash	Crude Fiber	Neutral Det. Fib. (NDF)	Acid Det. Fib. (ADF)	Lignin	Ether Extract (Fat)	N-Free Extract	Crude Protein	Digestible Protein: Ruminant	Swine	Horse
				%	%	%	%	%	%	%	%	%	%	%	%
	MOLASSES AND SYRUP														
155	BEET, SUGAR, MOLASSES, MORE THAN 48% INVERT SUGAR,	4-00-668	A-F	**78**	**8.9**	**–**	**–**	**–**	**–**	**0.2**	**62.2**	**6.6**	**3.6**	**4.5**	**4.5**
	MORE THAN 79.5 DEGREES BRIX		M-F	100	11.4	–	–	–	–	0.2	79.9	8.5	4.6	5.8	5.8
156	BEET, SUGAR, MOLASSES, STEFFENS	4-08-351	A-F	**79**	**8.8**	**–**	**–**	**–**	**–**	**–**	**62.1**	**7.8**	**4.6**	**5.6**	**5.4**
			M-F	100	11.2	–	–	–	–	–	78.9	9.9	5.8	7.1	6.9
157	CITRUS, SYRUP (CITRUS MOLASSES)	4-01-241	A-F	**67**	**5.1**	**–**	**–**	**–**	**–**	**0.2**	**55.7**	**5.8**	**2.0**	**3.9**	**3.9**
			M-F	100	7.6	–	–	–	–	0.3	82.7	8.5	3.0	5.8	5.8
158	CORN, MOLASSES, MORE THAN 43% DEXTROSE EQUIVALENT,	4-02-888	A-F	**73**	**8.0**	**–**	**–**	**–**	**–**	**–**	**64.2**	**0.3**	**–1.9**	**–1.1**	**–0.4**
	MORE THAN 50% TOTAL DEXTROSE, MORE THAN 78 DEGREES BRIX		M-F	100	11.0	–	–	–	–	–	88.6	0.4	–2.7	–1.5	–0.6
159	SYRUP, PEAR	4-08-487	A-F	**76**	**5.3**	**6.4**	**–**	**–**	**–**	**2.5**	**61.0**	**1.2**	**–1.2**	**–0.3**	**0.3**
			M-F	100	6.9	8.4	–	–	–	3.3	79.8	1.6	–1.6	–0.4	0.4
160	SUGAR CANE, MOLASSES, DEHY	4-04-695	A-F	**94**	**12.5**	**6.3**	**–**	**–**	**–**	**0.9**	**65.0**	**9.7**	**5.3**	**7.0**	**6.8**
			M-F	100	13.3	6.7	–	–	–	0.9	68.8	10.3	5.7	7.4	7.2
161	SUGAR CANE, MOLASSES, MORE THAN 46% INVERT SUGAR,	4-04-696	A-F	**74**	**9.8**	**0.4**	**–**	**0.3**	**0.2**	**0.2**	**59.7**	**4.3**	**0.6**	**1.3**	**2.7**
	MORE THAN 79.5 DEGREES BRIX (BLACKSTRAP)		M-F	100	13.2	0.5	–	0.4	0.3	0.2	80.2	5.8	0.8	1.8	3.7
162	WOOD, MOLASSES	4-05-502	A-F	**62**	**4.1**	**0.5**	**–**	**1.2**	**–**	**0.3**	**56.6**	**0.6**	**–1.3**	**–0.6**	**–0.1**
			M-F	100	6.5	0.9	–	2.0	–	0.5	91.1	1.0	–2.2	–1.0	–0.1
	OAK *Quercus* spp														
163	NUTS (ACORNS), FRESH	4-07-755	A-F	**50**	**1.2**	**9.3**	**–**	**–**	**–**	**2.3**	**35.0**	**2.2**	**0.4**	**1.0**	**1.3**
			M-F	100	2.5	18.6	–	–	–	4.7	70.0	4.4	0.9	2.1	2.5
	OATS *Avena sativa**														
164	GRAIN, ALL ANALYSES	4-03-309	A-F	**89**	**3.1**	**10.7**	**26.4**	**14.2**	**2.7**	**4.7**	**58.9**	**11.9**	**9.2**	**9.7**	**9.1**
			M-F	100	3.4	11.9	29.6	15.9	3.0	5.2	66.1	13.3	10.3	10.9	10.2
165	GRAIN, GRADE 1, HEAVY, 36 lb/bu (*46.3 kg/hl*)	4-03-312	A-F	**89**	**3.0**	**10.6**	**–**	**–**	**–**	**4.8**	**59.1**	**11.0**	**9.3**	**8.3**	**7.9**
			M-F	100	3.4	12.0	–	–	–	5.4	66.7	12.4	10.5	9.3	8.9
166	GRAIN, GRADE 1, 34 lb/bu (*43.8 kg/hl*)	4-03-313	A-F	**90**	**–**	**11.1**	**–**	**–**	**–**	**4.5**	**–**	**12.1**	**9.8**	**–**	**8.6**
			M-F	100	–	12.3	–	–	–	5.0	–	13.4	10.9	–	9.5
167	GRAIN, GRADE 2, 32 lb/bu (*41.2 kg/hl*)	4-03-316	A-F	**89**	**3.3**	**10.8**	**–**	**–**	**–**	**4.1**	**59.7**	**11.3**	**7.4**	**8.5**	**8.1**
			M-F	100	3.7	12.1	–	–	–	4.7	67.0	12.6	8.3	9.5	9.1
168	GRAIN, LIGHT, LESS THAN 27 lb/bu (*34.7 kg/hl*)	4-03-318	A-F	**91**	**4.2**	**14.4**	**–**	**–**	**–**	**4.5**	**55.7**	**11.9**	**8.2**	**9.0**	**8.6**
			M-F	100	4.6	15.9	–	–	–	4.9	61.5	13.1	9.1	10.0	9.4
169	GRAIN, GRADE SAMPLE (INFERIOR)	4-03-310	A-F	**89**	**4.1**	**11.7**	**–**	**–**	**–**	**4.1**	**57.2**	**11.4**	**7.6**	**8.7**	**8.2**
			M-F	100	4.6	13.2	–	–	–	4.6	64.7	12.9	8.5	9.8	9.3
170	GRAIN, PACIFIC COAST	4-07-999	A-F	**91**	**3.8**	**11.2**	**–**	**–**	**–**	**5.0**	**61.8**	**9.1**	**6.8**	**6.5**	**6.4**
			M-F	100	4.2	12.3	–	–	–	5.5	68.0	10.0	7.5	7.2	7.0
171	CEREAL BY-PRODUCT (FEEDING OAT MEAL; OAT MIDDLINGS)	4-03-303	A-F	**91**	**2.3**	**3.6**	**–**	**–**	**–**	**6.4**	**63.7**	**14.8**	**10.9**	**11.6**	**10.8**
			M-F	100	2.5	4.0	–	–	–	7.0	70.2	16.3	12.1	12.8	11.9
172	GROATS	4-03-331	A-F	**90**	**2.1**	**2.5**	**–**	**–**	**–**	**6.2**	**63.0**	**15.8**	**11.1**	**13.7**	**11.6**
			M-F	100	2.4	2.8	–	–	–	6.9	70.3	17.6	12.3	15.3	13.0
	OLIVE *Olea europaea*														
173	FRUIT WITHOUT PITS, DEHY	4-03-415	A-F	**93**	**2.9**	**27.9**	**–**	**–**	**–**	**21.3**	**30.1**	**10.8**	**–1.2**	**8.0**	**7.7**
			M-F	100	3.1	29.9	–	–	–	22.9	32.4	11.6	–1.3	8.6	8.3
	ORANGE *Citrus sinensis*														
174	PULP WITHOUT FINES, DEHY (DRIED ORANGE PULP)[1]	4-01-254	A-F	**88**	**3.7**	**8.4**	**18.5**	**14.1**	**–**	**1.7**	**66.9**	**7.5**	**5.9**	**5.1**	**5.1**
			M-F	100	4.2	9.6	21.0	16.0	–	1.9	75.9	8.5	6.7	5.8	5.8
175	PULP WITHOUT FINES, AMMONIATED, DEHY	4-01-255	A-F	**89**	**5.6**	**12.4**	**–**	**–**	**–**	**3.3**	**53.0**	**14.7**	**10.4**	**11.6**	**10.8**
			M-F	100	6.3	13.9	–	–	–	3.7	59.6	16.5	11.7	13.0	12.1
	PALM, DATE *Phoenix dactylifera*														
176	PITS	4-05-648	A-F	**92**	**2.9**	**14.4**	**–**	**–**	**–**	**8.2**	**57.8**	**8.3**	**4.7**	**5.8**	**5.7**
			M-F	100	3.2	15.7	–	–	–	9.0	63.0	9.1	5.1	6.3	6.3
	PALM, ROYAL *Roystonea regis*														
177	SEEDS	1-08-477	A-F	**86**	**5.5**	**22.8**	**–**	**–**	**–**	**8.3**	**43.8**	**6.1**	**2.7**	**2.0**	**3.2**
			M-F	100	6.4	26.4	–	–	–	9.6	50.6	7.1	3.1	2.3	3.7
	PARSNIP, GARDEN *Pastinaca sativa*														
178	ROOTS, FRESH	4-03-536	A-F	**17**	**1.1**	**1.4**	**–**	**1.5**	**–**	**0.4**	**12.4**	**1.7**	**1.0**	**1.2**	**1.2**
			M-F	100	6.2	8.2	–	9.0	–	2.1	73.5	10.0	5.9	7.2	7.0
	PAUNCH PRODUCTS														
179	DEHY (RUMEN CONTENTS; PAUNCH MEAL)	1-09-327	A-F	**87**	**9.6**	**24.4**	**–**	**–**	**–**	**1.8**	**36.2**	**15.3**	**10.3**	**10.1**	**10.3**
			M-F	100	11.0	28.0	–	–	–	2.1	41.4	17.6	11.8	11.6	11.8
	PEA *Pisum* spp														
180	SEED COATS (PEA HULLS)	1-03-602	A-F	**90**	**3.1**	**37.5**	**–**	**–**	**–**	**0.9**	**37.5**	**10.8**	**5.5**	**9.9**	**6.8**
			M-F	100	3.5	41.7	–	–	–	1.0	41.8	12.1	6.2	11.0	7.5
181	SPLIT SEED BY-PRODUCT (PEA FEED; PEA MEAL)	1-08-478	A-F	**90**	**3.5**	**23.7**	**–**	**–**	**–**	**1.4**	**43.7**	**17.7**	**14.5**	**12.1**	**12.0**
			M-F	100	3.9	26.3	–	–	–	1.6	48.6	19.7	16.1	13.4	13.4
	PEA, FIELD *Pisum sativum, arvense*														
182	SEEDS	5-08-481	A-F	**91**	**2.9**	**5.9**	**–**	**–**	**–**	**1.3**	**57.8**	**23.2**	**19.9**	**21.8**	**17.7**
			M-F	100	3.1	6.5	–	–	–	1.4	63.5	25.4	21.9	24.0	19.5
	PEA, GARDEN *Pisum sativum, sativum*														
183	SEEDS	5-08-482	A-F	**89**	**2.9**	**5.7**	**–**	**–**	**–**	**1.7**	**55.1**	**23.8**	**19.5**	**22.5**	**18.6**
			M-F	100	3.3	6.4	–	–	–	1.9	61.8	26.7	21.9	25.2	20.9

*The chemical composition and feeding value of oats are quite variable, due to the wide spread in hull content and weight per bushel.

Entry Number	TDN Ruminant	TDN Swine	TDN Horse	Digestible Energy Ruminant		Digestible Energy Swine		Digestible Energy Horse		Metabolizable Energy Ruminant		Metabolizable Energy Swine		Metabolizable Energy Poultry ME_n		Metabolizable Energy Horse		Net Energy Ruminant NE_m		Net Energy Ruminant NE_g		Net Energy Lactating Cows NE_{lc}	
	%	%	%	Mcal		kcal		Mcal		Mcal		kcal		kcal		Mcal		Mcal		Mcal		Mcal	
				lb	kg	lb	kg	lb	kg	lb	kg	lb	kg	lb	kg	lb	kg	lb	kg	lb	kg	lb	kg
155	61	57	–	1.20	2.64	1130	2491	–	–	1.04	2.29	1061	2338	875	1929	–	–	0.70	1.54	0.47	1.04	0.65	1.43
	78	73	–	1.54	3.38	1451	3198	–	–	1.33	2.94	1362	3002	1123	2476	–	–	0.90	1.98	0.60	1.33	0.83	1.83
156	61	–	–	1.21	2.67	–	–	–	–	1.06	2.34	–	–	–	–	–	–	–	–	–	–	–	–
	77	–	–	1.54	3.40	–	–	–	–	1.35	2.98	–	–	–	–	–	–	–	–	–	–	–	–
157	51	54	–	1.01	2.22	1084	2390	–	–	0.86	1.89	1022	2253	–	–	–	–	0.57	1.26	0.38	0.83	0.52	1.15
	75	80	–	1.49	3.29	1609	3547	–	–	1.27	2.81	1517	3344	–	–	–	–	0.85	1.87	0.56	1.23	0.78	1.71
158	–	–	–	–	–	–	–	–	–	–	–	–	–	–	–	–	–	–	–	–	–	–	–
	–	–	–	–	–	–	–	–	–	–	–	–	–	–	–	–	–	–	–	–	–	–	–
159	61	70	–	1.22	2.68	1406	3100	–	–	1.07	2.36	1334	2940	–	–	–	–	–	–	–	–	–	–
	80	92	–	1.59	3.51	1841	4058	–	–	1.40	3.09	1746	3848	–	–	–	–	–	–	–	–	–	–
160	66	70	61	1.36	2.99	1208	2663	1.12	2.48	1.18	2.60	1127	2485	1227	2706	0.92	2.03	0.70	1.55	0.44	0.97	0.68	1.51
	70	74	65	1.44	3.17	1279	2820	1.19	2.62	1.25	2.75	1194	2632	1300	2866	0.98	2.15	0.74	1.64	0.47	1.03	0.73	1.60
161	60	56	–	1.22	2.68	1135	2502	–	–	1.12	2.46	995	2194	870	1918	–	–	0.77	1.70	0.53	1.18	0.64	1.41
	81	76	–	1.63	3.60	1526	3364	–	–	1.50	3.31	1338	2950	1170	2579	–	–	1.04	2.28	0.72	1.58	0.86	1.89
162	52	51	–	1.03	2.27	1033	2278	–	–	0.91	2.02	976	2151	–	–	–	–	–	–	–	–	–	–
	83	82	–	1.66	3.66	1664	3669	–	–	1.47	3.25	1571	3464	–	–	–	–	–	–	–	–	–	–
163	23	39	23	0.60	1.33	771	1700	0.44	0.98	0.51	1.12	727	1602	–	–	0.36	0.80	0.40	0.87	0.26	0.56	0.38	0.84
	47	77	46	1.21	2.66	1543	3401	0.89	1.95	1.01	2.23	1453	3204	–	–	0.73	1.60	0.79	1.75	0.51	1.13	0.76	1.69
164	69	65	65	1.36	3.00	1278	2818	1.19	2.63	1.19	2.62	1026	2263	1150	2536	0.98	2.15	0.80	1.77	0.54	1.19	0.70	1.55
	77	72	73	1.53	3.37	1433	3160	1.34	2.94	1.33	2.94	1151	2538	1290	2844	1.10	2.41	0.90	1.98	0.60	1.33	0.79	1.74
165	72	69	50	1.42	3.13	1370	3021	0.94	2.06	1.21	2.67	1291	2847	–	–	0.77	1.69	0.73	1.60	0.48	1.05	0.70	1.54
	81	77	57	1.60	3.53	1547	3410	1.06	2.33	1.37	3.02	1457	3213	–	–	0.87	1.91	0.82	1.81	0.54	1.18	0.79	1.74
166	–	–	–	–	–	–	–	–	–	–	–	–	–	1195	2635	–	–	–	–	–	–	–	–
	–	–	–	–	–	–	–	–	–	–	–	–	–	1328	2927	–	–	–	–	–	–	–	–
167	68	–	53	1.36	2.99	1293	2850	0.99	2.18	1.19	2.62	1248	2752	–	–	0.81	1.79	0.72	1.59	0.47	1.04	0.69	1.53
	76	–	60	1.52	3.36	1450	3197	1.11	2.45	1.33	2.94	1400	3086	–	–	0.91	2.01	0.81	1.79	0.53	1.16	0.78	1.72
168	59	65	49	1.25	2.76	1299	2864	0.92	2.04	1.08	2.38	1220	2689	–	–	0.76	1.67	0.70	1.55	0.45	0.99	0.68	1.50
	66	72	54	1.38	3.04	1433	3159	1.02	2.25	1.19	2.62	1345	2966	–	–	0.84	1.84	0.78	1.71	0.50	1.10	0.75	1.66
169	66	66	51	1.32	2.91	1323	2917	0.95	2.10	1.15	2.54	1245	2744	–	–	0.78	1.73	0.70	1.55	0.46	1.00	0.68	1.50
	75	75	58	1.49	3.29	1494	3295	1.08	2.38	1.30	2.87	1406	3100	–	–	0.88	1.95	0.80	1.75	0.51	1.13	0.77	1.69
170	70	73	48	1.41	3.11	1270	2799	0.89	1.97	1.24	2.73	1190	2623	1199	2644	0.73	1.62	0.76	1.68	0.50	1.11	0.73	1.61
	78	80	52	1.55	3.42	1397	3079	0.98	2.17	1.36	3.00	1309	2886	1320	2909	0.81	1.78	0.84	1.85	0.55	1.22	0.80	1.77
171	86	77	57	1.72	3.79	1641	3618	1.04	2.30	1.55	3.42	1554	3426	1432	3158	0.86	1.89	0.95	2.10	0.67	1.47	0.89	1.95
	95	85	62	1.90	4.18	1810	3991	1.15	2.54	1.71	3.78	1714	3779	1580	3483	0.95	2.08	1.05	2.32	0.73	1.62	0.98	2.15
172	87	84	59	1.72	3.80	1410	3108	1.08	2.39	1.46	3.21	1328	2928	1475	3251	0.89	1.96	1.02	2.24	0.72	1.58	0.88	1.94
	98	94	66	1.92	4.24	1574	3471	1.21	2.66	1.63	3.59	1483	3269	1647	3630	0.99	2.18	1.14	2.50	0.80	1.77	0.98	2.17
173	51	–	–	1.02	2.25	–	–	–	–	0.84	1.85	–	–	–	–	–	–	–	–	–	–	–	–
	55	–	–	1.10	2.42	–	–	–	–	0.90	1.99	–	–	–	–	–	–	–	–	–	–	–	–
174	79	71	61	1.45	3.19	1413	3115	1.11	2.46	1.28	2.83	1333	2939	–	–	0.91	2.01	0.71	1.58	0.47	1.02	0.69	1.51
	89	80	69	1.64	3.62	1603	3534	1.26	2.79	1.45	3.21	1513	3335	–	–	1.04	2.28	0.81	1.79	0.53	1.16	0.78	1.72
175	64	62	56	1.27	2.81	1230	2712	1.03	2.28	1.11	2.44	1153	2542	–	–	0.85	1.87	0.67	1.49	0.43	0.94	0.66	1.45
	72	69	63	1.43	3.16	1382	3047	1.16	2.56	1.24	2.74	1296	2856	–	–	0.95	2.10	0.76	1.67	0.48	1.06	0.74	1.62
176	68	75	31	1.39	3.07	1493	3291	0.62	1.37	1.22	2.69	1409	3107	–	–	0.51	1.12	0.77	1.70	0.51	1.12	0.74	1.63
	75	81	34	1.52	3.34	1627	3587	0.68	1.49	1.33	2.93	1536	3387	–	–	0.56	1.22	0.84	1.86	0.56	1.22	0.80	1.77
177	64	–	12	1.29	2.83	–	–	0.31	0.68	1.12	2.47	–	–	–	–	0.25	0.55	–	–	–	–	–	–
	74	–	14	1.49	3.28	–	–	0.35	0.78	1.30	2.86	–	–	–	–	0.29	0.64	–	–	–	–	–	–
178	15	14	12	0.28	0.61	274	605	0.21	0.47	0.25	0.54	259	571	–	–	0.17	0.38	0.14	0.30	0.09	0.20	0.13	0.29
	89	81	68	1.64	3.61	1619	3568	1.25	2.75	1.45	3.20	1528	3368	–	–	1.03	2.26	0.81	1.78	0.52	1.15	0.78	1.71
179	51	42	45	1.00	2.20	–	–	0.85	1.88	0.83	1.83	–	–	–	–	0.70	1.54	0.47	1.03	0.25	0.54	0.50	1.10
	58	48	52	1.14	2.51	–	–	0.98	2.15	0.95	2.09	–	–	–	–	0.80	1.77	0.54	1.18	0.28	0.62	0.57	1.26
180	69	82	44	1.17	2.58	–	–	0.84	1.85	1.00	2.20	–	–	–	–	0.69	1.52	0.44	0.97	0.21	0.47	0.48	1.06
	77	91	49	1.30	2.87	–	–	0.93	2.06	1.11	2.45	–	–	–	–	0.77	1.69	0.49	1.08	0.24	0.53	0.54	1.19
181	78	55	62	1.34	2.96	–	–	1.13	2.49	1.17	2.59	–	–	–	–	0.93	2.04	0.57	1.26	0.33	0.74	0.58	1.28
	87	61	69	1.49	3.29	–	–	1.25	2.77	1.30	2.87	–	–	–	–	1.03	2.27	0.64	1.40	0.37	0.82	0.64	1.42
182	76	82	–	1.51	3.34	1638	3611	–	–	1.34	2.96	1435	3163	1108	2442	–	–	0.80	1.77	0.54	1.18	0.76	1.68
	83	90	–	1.66	3.66	1798	3964	–	–	1.47	3.25	1575	3473	1216	2681	–	–	0.88	1.94	0.59	1.30	0.84	1.84
183	77	80	–	1.49	3.28	1609	3547	–	–	1.32	2.91	1410	3108	–	–	–	–	0.79	1.74	0.53	1.17	0.75	1.65
	86	90	–	1.67	3.67	1804	3977	–	–	1.48	3.26	1580	3484	–	–	–	–	0.89	1.96	0.59	1.31	0.84	1.85

(Continued)

TABLE V-1A ENERGY FEEDS, MINERAL AND VITAMIN COMPOSITION OF FEEDS, DATA EXPRESSED **AS-FED** AND **MOISTURE-FREE**—*(Continued)*

Entry Number	Feed Name Description	Moisture Basis: A-F (as-fed) or M-F (moisture-free)	Dry Matter	Macro Minerals							Micro Minerals						
				Calcium (Ca)	Phosphorus (P)	Sodium (Na)	Chlorine (Cl)	Magnesium (Mg)	Potassium (K)	Sulfur (S)	Cobalt (Co)	Copper (Cu)	Iodine (I)	Iron (Fe)	Manganese (Mn)	Selenium (Se)	Zinc (Zn)
			%	%	%	%	%	%	%	%	ppm or mg/kg	ppm or mg/kg	ppm or mg/kg	%	ppm or mg/kg	ppm or mg/kg	ppm or mg/kg
	MOLASSES AND SYRUP																
155	BEET, SUGAR, MOLASSES, MORE THAN 48% INVERT SUGAR,	A-F	78	0.12	0.03	1.16	1.28	0.23	4.73	0.46	0.362	16.8	–	0.007	4.5	–	14.0
	MORE THAN 79.5 DEGREES BRIX	M-F	100	0.16	0.03	1.48	1.64	0.29	6.07	0.60	0.465	21.6	–	0.009	5.8	–	18.0
156	BEET, SUGAR, MOLASSES, STEFFENS	A-F	79	0.11	0.02	0.92	1.99	0.01	4.66	0.44	–	–	–	–	–	–	–
		M-F	100	0.14	0.03	1.17	2.53	0.01	5.92	0.56	–	–	–	–	–	–	–
157	CITRUS, SYRUP (CITRUS MOLASSES)	A-F	67	1.18	0.09	0.28	0.07	0.14	0.09	0.16	0.109	72.8	–	0.035	40.9	–	92.4
		M-F	100	1.76	0.13	0.41	0.11	0.21	0.14	0.23	0.162	108.0	–	0.051	60.7	–	137.1
158	CORN, MOLASSES, MORE THAN 43% DEXTROSE EQUIVALENT,	A-F	73	–	–	–	–	–	–	–	–	–	–	–	–	–	–
	MORE THAN 50% TOTAL DEXTROSE, MORE THAN 78 DEGREES BRIX	M-F	100	–	–	–	–	–	–	–	–	–	–	–	–	–	–
159	SYRUP, PEAR	A-F	76	0.11	–	–	–	–	–	–	–	–	–	–	–	–	–
		M-F	100	0.14	–	–	–	–	–	–	–	–	–	–	–	–	–
160	SUGAR CANE, MOLASSES, DEHY	A-F	94	1.04	0.42	0.19	–	0.44	3.40	0.43	1.145	74.9	–	0.024	54.1	–	31.2
		M-F	100	1.10	0.45	0.20	–	0.47	3.60	0.46	1.213	79.4	–	0.025	57.3	–	33.0
161	SUGAR CANE, MOLASSES, MORE THAN 46% INVERT SUGAR,	A-F	74	0.74	0.08	0.16	2.26	0.31	2.98	0.35	1.180	48.9	1.564	0.020	43.7	–	15.6
	MORE THAN 79.5 DEGREES BRIX (BLACKSTRAP)	M-F	100	1.00	0.11	0.22	3.04	0.42	4.01	0.47	1.587	65.7	2.103	0.027	58.8	–	20.9
162	WOOD, MOLASSES	A-F	62	1.17	0.05	0.03	0.12	0.07	0.04	0.03	–	–	–	–	12.6	–	–
		M-F	100	1.88	0.07	0.05	0.20	0.11	0.06	0.05	–	–	–	–	20.3	–	–
	OAK *Quercus* spp																
163	NUTS (ACORNS), FRESH	A-F	50	–	–	–	–	–	–	–	–	–	–	–	–	–	–
		M-F	100	–	–	–	–	–	–	–	–	–	–	–	–	–	–
	OATS *Avena sativa**																
164	GRAIN, ALL ANALYSES	A-F	89	0.08	0.34	0.05	0.09	0.14	0.40	0.21	0.056	6.0	0.112	0.007	35.8	0.215	34.9
		M-F	100	0.09	0.38	0.06	0.10	0.16	0.45	0.23	0.063	6.7	0.125	0.008	40.1	0.241	39.2
165	GRAIN, GRADE 1, HEAVY, 36 lb/bu (*46.3 kg/hl*)	A-F	89	0.07	0.34	–	–	–	–	–	–	18.5	–	–	–	–	–
		M-F	100	0.08	0.38	–	–	–	–	–	–	20.9	–	–	–	–	–
166	GRAIN, GRADE 1, 34 lb/bu (*43.8 kg/hl*)	A-F	90	0.09	0.32	0.06	0.12	–	0.37	–	–	–	–	–	38.2	–	–
		M-F	100	0.10	0.36	0.07	0.13	–	0.41	–	–	–	–	–	42.5	–	–
167	GRAIN, GRADE 2, 32 lb/bu (*41.2 kg/hl*)	A-F	89	0.06	0.27	–	–	–	–	–	–	–	–	–	–	–	–
		M-F	100	0.07	0.30	–	–	–	–	–	–	–	–	–	–	–	–
168	GRAIN, LIGHT, LESS THAN 27 lb/bu (*34.7 kg/hl*)	A-F	91	–	–	–	–	–	–	–	–	–	–	–	–	–	–
		M-F	100	–	–	–	–	–	–	–	–	–	–	–	–	–	–
169	GRAIN, GRADE SAMPLE (INFERIOR)	A-F	89	–	–	–	–	–	–	–	–	–	–	–	–	–	–
		M-F	100	–	–	–	–	–	–	–	–	–	–	–	–	–	–
170	GRAIN, PACIFIC COAST	A-F	91	0.10	0.31	0.06	0.12	0.17	0.38	0.20	–	–	–	0.008	38.0	0.076	–
		M-F	100	0.11	0.34	0.07	0.13	0.19	0.42	0.22	–	–	–	0.008	41.8	0.084	–
171	CEREAL BY-PRODUCT (FEEDING OAT MEAL; OAT MIDDLINGS)	A-F	91	0.07	0.44	0.09	0.05	0.14	0.50	0.22	0.046	5.2	–	0.039	43.8	–	139.5
		M-F	100	0.08	0.48	0.10	0.06	0.16	0.55	0.24	0.051	5.7	–	0.043	48.3	–	153.8
172	GROATS	A-F	90	0.08	0.43	0.05	0.08	0.11	0.35	0.20	–	6.0	0.108	0.008	27.8	–	0.0
		M-F	100	0.08	0.48	0.06	0.09	0.13	0.39	0.22	–	6.7	0.120	0.009	31.0	–	0.1
	OLIVE *Olea europaea*																
173	FRUIT WITHOUT PITS, DEHY	A-F	93	–	–	–	–	–	–	–	–	–	–	–	–	–	–
		M-F	100	–	–	–	–	–	–	–	–	–	–	–	–	–	–
	ORANGE *Citrus sinensis*																
174	PULP WITHOUT FINES, DEHY (DRIED ORANGE PULP)	A-F	88	0.62	0.10	–	–	–	–	–	–	–	–	–	–	–	–
		M-F	100	0.71	0.11	–	–	–	–	–	–	–	–	–	–	–	–
175	PULP WITHOUT FINES, AMMONIATED, DEHY	A-F	89	–	–	–	–	–	–	–	–	–	–	–	–	–	–
		M-F	100	–	–	–	–	–	–	–	–	–	–	–	–	–	–
	PALM, DATE *Phoenix dactylifera*																
176	PITS	A-F	92	0.60	0.18	0.01	–	0.14	2.20	–	–	–	–	–	–	–	–
		M-F	100	0.65	0.20	0.01	–	0.15	2.40	–	–	–	–	–	–	–	–
	PALM, ROYAL *Roystonea regis*																
177	SEEDS	A-F	86	–	–	–	–	–	–	–	–	–	–	–	–	–	–
		M-F	100	–	–	–	–	–	–	–	–	–	–	–	–	–	–
	PARSNIP, GARDEN *Pastinaca sativa*																
178	ROOTS, FRESH	A-F	17	0.05	0.07	0.01	–	–	0.49	–	–	–	–	0.001	–	–	–
		M-F	100	0.30	0.43	0.06	–	–	2.86	–	–	–	–	0.004	–	–	–
	PAUNCH PRODUCTS																
179	DEHY (RUMEN CONTENTS; PAUNCH MEAL)	A-F	87	0.69	0.58	–	–	–	–	–	–	–	–	–	–	–	–
		M-F	100	0.79	0.67	–	–	–	–	–	–	–	–	–	–	–	–
	PEA *Pisum* spp																
180	SEED COATS (PEA HULLS)	A-F	90	0.67	0.13	–	–	–	–	–	–	–	–	–	–	–	–
		M-F	100	0.75	0.14	–	–	–	–	–	–	–	–	–	–	–	–
181	SPLIT SEED BY-PRODUCT (PEA FEED; PEA MEAL)	A-F	90	–	–	–	–	–	–	–	–	–	–	–	–	–	–
		M-F	100	–	–	–	–	–	–	–	–	–	–	–	–	–	–
	PEA, FIELD *Pisum sativum, arvense*																
182	SEEDS	A-F	91	0.16	0.38	0.00	–	0.14	1.36	–	1.700	11.7	–	0.020	21.2	0.393	46.8
		M-F	100	0.17	0.41	0.00	–	0.15	1.49	–	1.866	12.9	–	0.022	23.3	0.432	51.4
	PEA, GARDEN *Pisum sativum, sativum*																
183	SEEDS	A-F	89	0.08	0.40	–	–	–	0.90	–	–	–	–	–	–	–	–
		M-F	100	0.09	0.45	–	–	–	1.01	–	–	–	–	–	–	–	–

*The chemical composition and feeding value of oats are quite variable, due to the wide spread in hull content and weight per bushel.

Entry Number	Fat-Soluble Vitamins					Water-Soluble Vitamins								
	A (1 mg Carotene = 1667 IU Vit A)	Carotene (Provitamin A)	D	E	K	B-12	Biotin	Choline	Folacin (Folic Acid)	Niacin	Pantothenic Acid (B-3)	(Pyridoxine) B-6	Riboflavin (B-2)	Thiamin (B-1)
	IU/g	ppm or mg/kg	IU/kg	ppm or mg/kg	ppm or mg/kg	ppb or mcg/kg	ppm or mg/kg	ppm or mg/kg	ppm or mg/kg	ppm or mg/kg	ppm or mg/kg	ppm or mg/kg	ppm or mg/kg	ppm or mg/kg
155	–	–	–	4.0	–	–	–	827	–	41	4.5	–	2.3	–
	–	–	–	5.1	–	–	–	1062	–	53	5.8	–	2.9	–
156	–	–	–	–	–	–	–	–	–	–	–	–	–	–
	–	–	–	–	–	–	–	–	–	–	–	–	–	–
157	–	–	–	–	–	–	–	–	–	27	17.2	–	6.2	–
	–	–	–	–	–	–	–	–	–	40	25.5	–	9.2	–
158	–	–	–	–	–	–	–	–	–	–	–	–	–	–
	–	–	–	–	–	–	–	–	–	–	–	–	–	–
159	–	–	–	–	–	–	–	–	–	–	–	–	–	–
	–	–	–	–	–	–	–	–	–	–	–	–	–	–
160	–	–	–	5.2	–	–	–	–	–	–	–	–	–	–
	–	–	–	5.5	–	–	–	–	–	–	–	–	–	–
161	–	–	–	5.4	–	–	0.69	764	0.11	36	37.4	4.21	2.8	0.9
	–	–	–	7.3	–	–	0.92	1027	0.15	49	50.3	5.67	3.8	1.2
162	–	–	–	–	–	–	–	–	–	–	–	–	–	–
	–	–	–	–	–	–	–	–	–	–	–	–	–	–
163	–	–	–	–	–	–	–	–	–	–	–	–	–	–
	–	–	–	–	–	–	–	–	–	–	–	–	–	–
164	0.2	0.1	–	14.9	–	–	0.27	967	0.39	14	9.9	2.53	1.4	6.0
	0.2	0.1	–	16.8	–	–	0.30	1084	0.44	16	11.1	2.84	1.5	6.8
165	–	–	–	–	–	–	–	–	–	–	–	–	–	–
	–	–	–	–	–	–	–	–	–	–	–	–	–	–
166	–	–	–	20.1	–	–	0.11	1106	0.30	18	13.1	1.31	1.1	–
	–	–	–	22.3	–	–	0.12	1229	0.34	20	14.5	1.45	1.2	–
167	–	–	–	–	–	–	–	–	–	–	–	–	–	–
	–	–	–	–	–	–	–	–	–	–	–	–	–	–
168	–	–	–	–	–	–	–	–	–	–	–	–	–	–
	–	–	–	–	–	–	–	–	–	–	–	–	–	–
169	–	–	–	–	–	–	–	–	–	–	–	–	–	–
	–	–	–	–	–	–	–	–	–	–	–	–	–	–
170	–	–	–	20.2	–	–	–	917	–	14	11.7	–	1.2	–
	–	–	–	22.2	–	–	–	1009	–	16	12.8	–	1.3	–
171	–	–	–	23.7	–	–	0.22	1157	0.46	24	17.6	–	1.7	7.0
	–	–	–	26.1	–	–	0.24	1277	0.51	26	19.4	–	1.9	7.7
172	–	–	–	14.8	–	–	–	1132	0.51	10	13.8	1.00	1.2	6.5
	–	–	–	16.5	–	–	–	1264	0.57	11	15.4	1.12	1.3	7.2
173	–	–	–	–	–	–	–	–	–	–	–	–	–	–
	–	–	–	–	–	–	–	–	–	–	–	–	–	–
174	–	–	–	–	–	–	–	–	–	–	–	–	–	–
	–	–	–	–	–	–	–	–	–	–	–	–	–	–
175	–	–	–	–	–	–	–	–	–	–	–	–	–	–
	–	–	–	–	–	–	–	–	–	–	–	–	–	–
176	–	–	–	–	–	–	–	–	–	–	–	–	–	–
	–	–	–	–	–	–	–	–	–	–	–	–	–	–
177	–	–	–	–	–	–	–	–	–	–	–	–	–	–
	–	–	–	–	–	–	–	–	–	–	–	–	–	–
178	–	–	–	–	–	–	–	–	–	2	–	–	0.7	0.6
	–	–	–	–	–	–	–	–	–	10	–	–	4.3	3.8
179	–	–	–	–	–	–	–	–	–	–	–	–	–	–
	–	–	–	–	–	–	–	–	–	–	–	–	–	–
180	–	–	–	–	–	–	–	–	–	–	–	–	–	–
	–	–	–	–	–	–	–	–	–	–	–	–	–	–
181	–	–	–	–	–	–	–	–	–	–	–	–	–	–
	–	–	–	–	–	–	–	–	–	–	–	–	–	–
182	–	–	–	–	–	–	0.19	654	0.36	34	7.4	1.01	1.4	4.1
	–	–	–	–	–	–	0.21	718	0.40	37	8.2	1.11	1.5	4.5
183	–	–	–	–	–	–	–	–	–	–	–	–	–	–
	–	–	–	–	–	–	–	–	–	–	–	–	–	–

(Continued)

TABLE V-1A ENERGY FEEDS, COMPOSITION OF FEEDS, DATA EXPRESSED **AS-FED** AND **MOISTURE-FREE**—*(Continued)*

Entry Number	Feed Name Description	International Feed Number	Moisture Basis: A-F (as-fed) or M-F (moisture-free)	Proximate Analysis: Dry Matter	Ash	Crude Fiber	Neutral Det. Fib. (NDF)	Acid Det. Fib. (ADF)	Lignin	Ether Extract (Fat)	N-Free Extract	Crude Protein	Digestible Protein: Ruminant	Swine	Horse
				%	%	%	%	%	%	%	%	%	%	%	%
	PEAR *Pyrus* spp														
184	CANNERY RESIDUE, WET	**4-08-486**	**A-F**	**15**	**0.3**	**2.6**	–	–	–	**0.2**	**11.5**	**0.6**	**0.1**	**0.3**	**0.3**
			M-F	100	2.0	17.1	–	–	–	1.3	75.7	3.9	0.5	1.7	2.2
185	POMACE, WET	**4-26-119**	**A-F**	**17**	**0.3**	**5.6**	–	–	–	**0.4**	**10.1**	**0.7**	**0.1**	–	**0.2**
			M-F	100	1.6	32.7	–	–	–	2.2	59.3	4.2	0.7	–	1.0
186	POMACE, DEHY	**4-03-661**	**A-F**	**90**	**2.4**	**21.1**	–	–	–	**1.7**	**60.2**	**5.0**	**1.7**	**2.8**	**3.1**
			M-F	100	2.6	23.4	–	–	–	1.8	66.6	5.5	1.9	3.1	3.5
187	SYRUP	**4-08-487**	**A-F**	**76**	**5.3**	**6.4**	–	–	–	**2.5**	**61.0**	**1.2**	**–1.2**	**–0.3**	**0.3**
			M-F	100	6.9	8.4	–	–	–	3.3	79.8	1.6	–1.6	–0.4	0.4
	PEARL MILLET *Pennisetum glaucum*														
188	GRAIN	**4-03-118**	**A-F**	**90**	**2.2**	**3.7**	–	–	–	**4.3**	**67.2**	**13.0**	**8.9**	**7.9**	**9.4**
			M-F	100	2.5	4.1	–	–	–	4.7	74.4	14.3	9.8	8.8	10.4
	PIGEON-GRASS (VERBENA, EUROPEAN) *Verbena officinalis*														
189	SEEDS	**4-26-082**	**A-F**	**89**	**6.3**	**17.1**	–	–	–	**5.9**	**45.6**	**14.2**	**9.3**	–	–
			M-F	100	7.1	19.2	–	–	–	6.6	51.2	15.9	10.4	–	–
	PIGWEED *Axyris* spp														
190	SEEDS	**4-08-580**	**A-F**	**90**	**3.3**	**15.9**	–	–	–	**6.2**	**47.8**	**16.8**	**12.3**	**13.5**	**12.4**
			M-F	100	3.7	17.7	–	–	–	6.9	53.1	18.7	13.7	15.0	13.8
	PINEAPPLE *Ananas comosus*														
191	CANNERY RESIDUE, DEHY (PINEAPPLE BRAN)[1]	**4-03-722**	**A-F**	**87**	**3.0**	**18.2**	**63.5**	**32.2**	–	**1.3**	**60.5**	**4.0**	**0.7**	**2.0**	**2.4**
			M-F	100	3.5	20.9	73.0	37.0	–	1.5	69.5	4.6	0.8	2.3	2.7
192	CANNERY RESIDUE, MOLASSES ADDED, DEHY (PINEAPPLE BRAN, MOLASSES ADDED)	**4-08-489**	**A-F**	**87**	**3.2**	**15.9**	–	–	–	**1.0**	**63.4**	**3.9**	**0.8**	**1.9**	**2.3**
			M-F	100	3.7	18.2	–	–	–	1.1	72.5	4.5	0.9	2.2	2.6
	POP CORN *Zea mays, everta*														
193	GRAIN	**4-02-964**	**A-F**	**90**	**1.5**	**1.9**	–	–	–	**5.0**	**70.0**	**11.5**	**7.5**	**8.7**	**8.2**
			M-F	100	1.7	2.1	–	–	–	5.6	77.9	12.8	8.4	9.7	9.2
	POTATO *Solanum tuberosum*														
194	TUBERS, FRESH	**4-03-787**	**A-F**	**24**	**1.1**	**0.6**	–	–	–	**0.1**	**19.5**	**2.2**	**1.4**	**0.7**	**1.5**
			M-F	100	4.8	2.4	–	–	–	0.4	83.2	9.3	5.7	3.1	6.4
195	TUBERS, BOILED	**4-03-784**	**A-F**	**24**	**1.3**	**0.7**	–	–	–	**0.1**	**19.6**	**2.2**	**0.3**	**1.5**	**1.5**
			M-F	100	5.3	3.0	–	–	–	0.3	82.1	9.3	1.1	6.3	6.4
196	TUBERS, DEHY	**4-07-850**	**A-F**	**91**	**7.2**	**2.1**	–	**2.7**	–	**0.5**	**73.3**	**8.1**	**2.9**	**5.6**	**5.5**
			M-F	100	7.9	2.3	–	3.0	–	0.5	80.5	8.9	3.2	6.1	6.1
197	CANNERY RESIDUE, WET	**4-03-777**	**A-F**	**12**	**0.3**	**1.1**	–	–	–	**0.1**	**9.4**	**1.0**	**0.5**	**0.7**	**0.7**
			M-F	100	2.9	9.0	–	–	–	0.7	78.9	8.5	4.6	5.8	5.8
198	CANNERY RESIDUE, DEHY	**4-03-775**	**A-F**	**89**	**3.0**	**6.5**	–	–	–	**0.3**	**71.5**	**7.4**	**5.7**	**1.9**	**5.1**
			M-F	100	3.4	7.3	–	–	–	0.4	80.5	8.4	6.4	2.2	5.7
199	DISTILLERS RESIDUE, DEHY	**5-03-773**	**A-F**	**96**	**6.7**	**20.6**	–	–	–	**3.1**	**42.4**	**22.9**	**18.5**	**21.5**	**17.0**
			M-F	100	7.0	21.5	–	–	–	3.2	44.3	23.9	19.3	22.5	17.8
200	PEELINGS, FRESH	**4-03-774**	**A-F**	**23**	**1.2**	**0.7**	–	–	–	**0.1**	**19.1**	**2.1**	**0.8**	**1.5**	**1.4**
			M-F	100	5.0	3.1	–	–	–	0.3	82.7	8.9	3.6	6.6	6.1
	PUMPKIN *Cucurbita pepo*														
201	FRUIT, FRESH	**4-03-815**	**A-F**	**9**	**0.8**	**1.3**	–	–	–	**1.0**	**4.4**	**1.6**	**1.1**	**1.1**	**1.2**
			M-F	100	8.7	14.5	–	–	–	10.9	48.7	17.3	12.6	12.4	12.7
202	SEEDS	**5-03-817**	**A-F**	**94**	**7.6**	**13.0**	–	–	–	**17.2**	**17.7**	**38.3**	**32.2**	**36.9**	**34.6**
			M-F	100	8.1	13.9	–	–	–	18.3	18.8	40.9	34.3	39.3	36.9
	RAISINS (GRAPES) *Vitis* spp														
203	DEHY, CULL	**4-08-427**	**A-F**	**87**	**4.1**	**10.1**	–	–	–	**4.3**	**62.2**	**6.4**	**1.5**	**4.2**	**4.3**
			M-F	100	4.7	11.6	–	–	–	4.9	71.4	7.4	1.8	4.8	4.9
	RICE *Oryza sativa*														
204	GRAIN, GROUND (GROUND ROUGH RICE; GROUND PADDY RICE)	**4-03-938**	**A-F**	**89**	**5.3**	**8.6**	–	–	–	**1.6**	**65.9**	**7.5**	**4.0**	**5.1**	**5.1**
			M-F	100	6.0	9.7	–	–	–	1.8	74.1	8.4	4.5	5.7	5.7
205	BRAN WITH GERMS (RICE BRAN)	**4-03-928**	**A-F**	**91**	**11.3**	**11.9**	**28.0**	**25.7**	**3.6**	**13.5**	**41.0**	**13.0**	**8.6**	**9.5**	**9.4**
			M-F	100	12.5	13.1	30.9	28.4	4.0	14.9	45.2	14.3	9.5	10.4	10.4
206	GROATS, POLISHED (RICE, POLISHED)[1]	**4-03-942**	**A-F**	**89**	**0.5**	**0.4**	**14.2**	**0.9**	–	**0.5**	**80.3**	**7.0**	**3.6**	**5.9**	**4.7**
			M-F	100	0.6	0.4	16.0	1.0	–	0.5	90.6	7.9	4.0	6.7	5.3
207	GROATS (RICE, BROWN)	**4-03-936**	**A-F**	**88**	**1.0**	**0.8**	–	–	–	**1.8**	**77.2**	**7.4**	**3.2**	**5.1**	**5.1**
			M-F	100	1.2	0.9	–	–	–	2.0	87.6	8.4	3.6	5.7	5.7
208	HULLS, AMMONIATED	**1-05-698**	**A-F**	**92**	–	**44.7**	–	–	–	**0.9**	**16.9**	**10.4**	**6.3**	**5.6**	**6.4**
			M-F	100	–	48.6	–	–	–	1.0	18.4	11.3	6.9	6.1	6.9
209	POLISHINGS	**4-03-943**	**A-F**	**90**	**7.6**	**3.2**	–	**3.6**	–	**12.6**	**54.9**	**12.0**	**8.6**	**10.1**	**8.6**
			M-F	100	8.4	3.5	–	4.0	–	13.9	60.9	13.3	9.5	11.2	9.5
	RUMEN CONTENTS														
210	DEHY (PAUNCH PRODUCT; PAUNCH MEAL)	**1-09-327**	**A-F**	**87**	**9.6**	**24.4**	–	–	–	**1.8**	**36.2**	**15.3**	**10.3**	**10.1**	**10.3**
			M-F	100	11.0	28.0	–	–	–	2.1	41.4	17.6	11.8	11.6	11.8
	RUTABAGA *Brassica napus, napobrassica*														
211	ROOTS, FRESH	**4-04-001**	**A-F**	**12**	**0.8**	**1.2**	–	**1.7**	–	**0.2**	**8.8**	**1.1**	**0.8**	**0.9**	**0.9**
			M-F	100	7.0	10.1	–	14.0	–	1.4	72.4	9.2	6.9	7.7	7.6
	RYE *Secale cereale*														
212	GRAIN, ALL ANALYSES	**4-04-047**	**A-F**	**87**	**1.6**	**2.2**	–	–	–	**1.5**	**70.0**	**12.0**	**8.4**	**9.1**	**8.7**
			M-F	100	1.9	2.5	–	–	–	1.7	80.1	13.8	9.6	10.4	9.9

Entry Number	TDN Ruminant	TDN Swine	TDN Horse	Digestible Energy Ruminant		Digestible Energy Swine		Digestible Energy Horse		Metabolizable Energy Ruminant		Metabolizable Energy Swine		Metabolizable Energy Poultry ME_n		Metabolizable Energy Horse		Net Energy Ruminant NE_m		Net Energy Ruminant NE_g		Net Energy Lactating Cows NE_{lc}	
	%	%	%	Mcal		kcal		Mcal		Mcal		kcal		kcal		Mcal		Mcal		Mcal		Mcal	
				lb	*kg*	lb	*kg*	lb	*kg*	lb	*kg*	lb	*kg*	lb	*kg*	lb	*kg*	lb	*kg*	lb	*kg*	lb	*kg*
184	11	11	10	0.22	0.49	227	501	0.18	0.39	0.19	0.42	214	471	–	–	0.15	0.32	0.12	0.26	0.08	0.17	0.11	0.25
	72	75	63	1.45	3.19	1495	3295	1.16	2.56	1.26	2.77	1406	3100	–	–	0.95	2.10	0.77	1.70	0.49	1.09	0.75	1.65
185	–	7	–	–	–	137	301	–	–	–	–	130	286	–	–	–	–	–	–	–	–	–	–
	–	40	–	–	–	803	1771	–	–	–	–	764	1685	–	–	–	–	–	–	–	–	–	–
186	65	36	50	1.27	2.79	726	1600	0.94	2.07	1.10	2.42	657	1450	–	–	0.77	1.70	0.65	1.43	0.40	0.89	0.64	1.41
	72	40	56	1.40	3.10	804	1772	1.04	2.30	1.21	2.68	728	1606	–	–	0.85	1.88	0.72	1.59	0.45	0.98	0.71	1.56
187	61	70	–	1.22	2.68	1406	3100	–	–	1.07	2.36	1334	2940	–	–	–	–	–	–	–	–	–	–
	80	92	–	1.59	3.51	1841	4058	–	–	1.40	3.09	1746	3848	–	–	–	–	–	–	–	–	–	–
188	65	65	64	1.36	3.01	1351	2977	1.16	2.57	0.99	2.19	1269	2799	1155	2546	0.96	2.11	0.63	1.39	0.39	0.85	0.67	1.48
	72	72	71	1.51	3.33	1494	3294	1.29	2.84	1.10	2.43	1404	3096	1278	2817	1.06	2.33	0.70	1.54	0.43	0.94	0.74	1.64
189	57	–	–	–	–	–	–	–	–	–	–	–	–	–	–	–	–	–	–	–	–	–	–
	64	–	–	–	–	–	–	–	–	–	–	–	–	–	–	–	–	–	–	–	–	–	–
190	61	59	–	1.26	2.79	1182	2606	–	–	1.09	2.41	1106	2437	–	–	–	–	0.69	1.53	0.44	0.98	0.67	1.48
	68	66	–	1.40	3.10	1314	2896	–	–	1.21	2.68	1229	2709	–	–	–	–	0.77	1.70	0.49	1.09	0.75	1.65
191	64	61	50	1.27	2.81	1223	2696	0.93	2.04	1.11	2.45	1147	2530	–	–	0.76	1.67	0.70	1.55	0.46	1.00	0.67	1.49
	73	70	57	1.46	3.23	1405	3097	1.06	2.34	1.27	2.81	1318	2906	–	–	0.87	1.92	0.81	1.78	0.52	1.15	0.77	1.71
192	63	64	53	1.24	2.74	1279	2819	0.98	2.17	1.08	2.38	1202	2649	–	–	0.81	1.78	0.66	1.45	0.41	0.91	0.64	1.41
	72	73	61	1.42	3.14	1463	3226	1.13	2.48	1.23	2.72	1375	3032	–	–	0.92	2.03	0.75	1.65	0.47	1.05	0.73	1.61
193	85	79	61	1.59	3.50	1589	3503	1.12	2.46	1.42	3.13	1505	3317	–	–	0.92	2.02	0.82	1.80	0.55	1.21	0.77	1.70
	95	88	68	1.77	3.90	1769	3900	1.25	2.74	1.58	3.49	1675	3693	–	–	1.02	2.25	0.91	2.00	0.61	1.35	0.86	1.89
194	19	20	19	0.38	0.84	398	878	0.35	0.77	0.34	0.74	377	830	–	–	0.29	0.63	0.20	0.45	0.14	0.30	0.19	0.43
	81	85	82	1.62	3.58	1695	3736	1.48	3.27	1.43	3.16	1602	3533	–	–	1.22	2.68	0.87	1.92	0.58	1.28	0.83	1.82
195	17	21	19	0.33	0.73	416	918	0.35	0.77	0.29	0.63	394	869	–	–	0.29	0.63	0.16	0.35	0.10	0.21	0.16	0.35
	69	88	81	1.38	3.05	1742	3841	1.47	3.23	1.19	2.63	1649	3635	–	–	1.20	2.65	0.67	1.48	0.40	0.89	0.67	1.48
196	75	80	70	1.42	3.14	1565	3450	1.27	2.80	1.26	2.77	1479	3261	1280	2822	1.04	2.30	0.85	1.88	0.58	1.28	0.84	1.85
	83	88	77	1.56	3.45	1718	3788	1.39	3.07	1.38	3.04	1624	3581	1405	3098	1.14	2.52	0.94	2.06	0.64	1.40	0.92	2.03
197	9	9	9	0.18	0.39	187	413	0.17	0.36	0.15	0.34	177	389	–	–	0.14	0.30	0.10	0.21	0.06	0.14	0.09	0.20
	73	79	77	1.48	3.27	1577	3477	1.39	3.07	1.29	2.85	1487	3278	–	–	1.14	2.52	0.81	1.79	0.53	1.16	0.78	1.71
198	79	–	70	1.47	3.23	31	67	1.27	2.79	1.30	2.86	-24	-52	–	–	1.04	2.29	0.73	1.60	0.47	1.04	0.70	1.53
	90	–	79	1.65	3.64	34	76	1.43	3.14	1.46	3.22	-27	-59	–	–	1.17	2.58	0.82	1.80	0.53	1.18	0.78	1.73
199	63	52	–	1.26	2.77	1036	2284	–	–	1.08	2.37	875	1929	–	–	–	–	0.60	1.33	0.35	0.78	0.61	1.35
	66	54	–	1.32	2.90	1083	2387	–	–	1.12	2.48	914	2016	–	–	–	–	0.63	1.39	0.37	0.81	0.64	1.41
200	18	20	19	0.36	0.79	400	882	0.34	0.75	0.32	0.70	379	835	–	–	0.28	0.61	0.19	0.43	0.13	0.28	0.18	0.41
	79	87	81	1.56	3.44	1736	3826	1.47	3.23	1.37	3.03	1643	3621	–	–	1.20	2.65	0.84	1.85	0.55	1.22	0.80	1.77
201	8	7	–	0.16	0.35	147	325	–	–	0.14	0.31	139	307	–	–	–	–	0.09	0.20	0.06	0.13	0.08	0.18
	88	81	–	1.75	3.86	1616	3562	–	–	1.56	3.45	1525	3362	–	–	–	–	0.97	2.14	0.66	1.46	0.91	2.00
202	91	72	–	1.72	3.80	1434	3162	–	–	1.55	3.41	1244	2744	–	–	–	–	0.92	2.02	0.63	1.38	0.86	1.89
	97	76	–	1.84	4.05	1529	3372	–	–	1.65	3.64	1327	2925	–	–	–	–	0.98	2.15	0.67	1.48	0.91	2.01
203	44	72	45	1.11	2.44	1449	3195	0.85	1.87	0.94	2.07	1369	3019	–	–	0.70	1.53	0.72	1.59	0.47	1.04	0.69	1.52
	51	83	52	1.27	2.80	1664	3668	0.97	2.15	1.08	2.38	1572	3466	–	–	0.80	1.76	0.83	1.82	0.54	1.19	0.79	1.75
204	68	72	59	1.35	2.98	1492	3290	1.09	2.39	1.18	2.61	1409	3107	1210	2668	0.89	1.96	0.71	1.57	0.46	1.01	0.68	1.51
	76	81	67	1.52	3.35	1678	3699	1.22	2.69	1.33	2.93	1584	3493	1360	2999	1.00	2.21	0.80	1.76	0.52	1.14	0.77	1.70
205	64	69	–	1.10	2.42	1474	3250	–	–	0.99	2.18	1347	2971	920	2028	–	–	0.62	1.38	0.38	0.84	0.53	1.18
	71	77	–	1.21	2.67	1626	3586	–	–	1.09	2.40	1487	3278	1015	2238	–	–	0.69	1.52	0.42	0.93	0.59	1.30
206	78	86	–	1.59	3.51	1697	3741	–	–	1.44	3.17	1658	3656	1399	3085	–	–	0.90	1.98	0.62	1.37	0.85	1.87
	88	97	–	1.80	3.96	1916	4223	–	–	1.63	3.58	1872	4127	1580	3483	–	–	1.01	2.24	0.70	1.55	0.96	2.11
207	78	–	–	1.56	3.43	1666	3674	–	–	1.39	3.07	1642	3619	–	–	–	–	0.83	1.82	0.56	1.23	0.78	1.71
	88	–	–	1.76	3.89	1889	4165	–	–	1.58	3.47	1861	4103	–	–	–	–	0.94	2.06	0.64	1.40	0.88	1.94
208	–	–	–	–	–	–	–	–	–	–	–	–	–	–	–	–	–	–	–	–	–	–	–
	–	–	–	–	–	–	–	–	–	–	–	–	–	–	–	–	–	–	–	–	–	–	–
209	81	88	–	1.56	3.45	1684	3713	–	–	1.43	3.16	1555	3428	1367	3015	–	–	0.88	1.93	0.60	1.32	0.83	1.82
	89	97	–	1.73	3.82	1866	4114	–	–	1.59	3.50	1723	3798	1515	3340	–	–	0.97	2.14	0.67	1.47	0.92	2.02
210	51	42	45	1.00	2.20	–	–	0.85	1.88	0.83	1.83	–	–	–	–	0.70	1.54	0.47	1.03	0.25	0.54	0.50	1.10
	58	48	52	1.14	2.51	–	–	0.98	2.15	0.95	2.09	–	–	–	–	0.80	1.77	0.54	1.18	0.28	0.62	0.57	1.26
211	11	10	7	0.20	0.43	202	445	0.13	0.29	0.17	0.38	191	421	–	–	0.11	0.24	0.10	0.21	0.06	0.14	0.09	0.20
	87	83	58	1.60	3.53	1656	3650	1.07	2.36	1.41	3.12	1564	3448	–	–	0.88	1.94	0.78	1.73	0.50	1.11	0.76	1.67
212	73	75	75	1.42	3.12	1474	3249	1.35	2.96	1.18	2.60	1319	2909	1202	2651	1.10	2.43	0.80	1.75	0.54	1.18	0.74	1.63
	84	86	86	1.62	3.57	1685	3716	1.54	3.39	1.35	2.97	1509	3327	1375	3031	1.26	2.78	0.91	2.01	0.61	1.35	0.85	1.86

(Continued)

ENERGY FEEDS

TABLE V-1A ENERGY FEEDS, MINERAL AND VITAMIN COMPOSITION OF FEEDS, DATA EXPRESSED **AS-FED** AND **MOISTURE-FREE**—*(Continued)*

Entry Number	Feed Name Description	Moisture Basis: A-F (as-fed) or M-F (moisture-free)	Dry Matter	Macro Minerals Calcium (Ca)	Phosphorus (P)	Sodium (Na)	Chlorine (Cl)	Magnesium (Mg)	Potassium (K)	Sulfur (S)	Micro Minerals Cobalt (Co)	Copper (Cu)	Iodine (I)	Iron (Fe)	Manganese (Mn)	Selenium (Se)	Zinc (Zn)
			%	%	%	%	%	%	%	%	ppm or mg/kg	ppm or mg/kg	ppm or mg/kg	%	ppm or mg/kg	ppm or mg/kg	ppm or mg/kg
	PEAR *Pyrus spp*																
184	CANNERY RESIDUE, WET	A-F	15	–	–	–	–	–	–	–	–	–	–	–	–	–	–
		M-F	100	–	–	–	–	–	–	–	–	–	–	–	–	–	–
185	POMACE, WET	A-F	17	–	–	–	–	–	–	–	–	–	–	–	–	–	–
		M-F	100	–	–	–	–	–	–	–	–	–	–	–	–	–	–
186	POMACE, DEHY	A-F	90	–	–	–	–	–	–	–	–	–	–	–	–	–	–
		M-F	100	–	–	–	–	–	–	–	–	–	–	–	–	–	–
187	SYRUP	A-F	76	0.11	–	–	–	–	–	–	–	–	–	–	–	–	–
		M-F	100	0.14	–	–	–	–	–	–	–	–	–	–	–	–	–
	PEARL MILLET *Pennisetum glaucum*																
188	GRAIN	A-F	90	0.05	0.31	0.04	0.14	0.16	0.43	0.13	0.045	22.1	–	0.006	31.0	–	13.3
		M-F	100	0.05	0.34	0.04	0.16	0.18	0.48	0.14	0.049	24.5	–	0.007	34.3	–	14.7
	PIGEON-GRASS (VERBENA, EUROPEAN) *Verbena officinalis*																
189	SEEDS	A-F	89	–	–	–	–	–	–	–	–	–	–	–	–	–	–
		M-F	100	–	–	–	–	–	–	–	–	–	–	–	–	–	–
	PIGWEED *Axyris spp*																
190	SEEDS	A-F	90	–	–	–	–	–	–	–	–	–	–	–	–	–	–
		M-F	100	–	–	–	–	–	–	–	–	–	–	–	–	–	–
	PINEAPPLE *Ananas comosus*																
191	CANNERY RESIDUE, DEHY (PINEAPPLE BRAN)	A-F	87	0.20	0.11	–	–	–	–	–	–	–	–	0.049	–	–	–
		M-F	100	0.23	0.13	–	–	–	–	–	–	–	–	0.057	–	–	–
192	CANNERY RESIDUE, MOLASSES ADDED, DEHY (PINEAPPLE BRAN, MOLASSES ADDED)	A-F	87	–	–	–	–	–	–	–	–	–	–	–	–	–	–
		M-F	100	–	–	–	–	–	–	–	–	–	–	–	–	–	–
	POP CORN *Zea mays, everta*																
193	GRAIN	A-F	90	–	0.30	–	–	–	–	–	–	2.4	–	0.003	–	–	–
		M-F	100	–	0.33	–	–	–	–	–	–	2.6	–	0.003	–	–	–
	POTATO *Solanum tuberosum*																
194	TUBERS, FRESH	A-F	24	0.01	0.06	0.02	0.07	0.03	0.51	0.02	–	6.7	–	0.002	9.8	–	–
		M-F	100	0.04	0.24	0.09	0.28	0.14	2.17	0.09	–	28.4	–	0.008	41.7	–	–
195	TUBERS, BOILED	A-F	24	0.01	0.05	–	–	–	–	–	–	–	–	–	–	–	–
		M-F	100	0.04	0.22	–	–	–	–	–	–	–	–	–	–	–	–
196	TUBERS, DEHY	A-F	91	0.07	0.20	0.01	0.36	0.11	1.97	0.08	–	–	–	–	2.3	–	2.0
		M-F	100	0.08	0.22	0.01	0.40	0.12	2.16	0.09	–	–	–	–	2.5	–	2.2
197	CANNERY RESIDUE, WET	A-F	12	0.02	0.03	–	–	–	0.16	–	–	–	–	–	–	–	–
		M-F	100	0.18	0.23	–	–	–	1.33	–	–	–	–	–	–	–	–
198	CANNERY RESIDUE, DEHY	A-F	89	0.14	0.23	–	–	–	–	–	–	–	–	–	–	–	–
		M-F	100	0.16	0.25	–	–	–	–	–	–	–	–	–	–	–	–
199	DISTILLERS RESIDUE, DEHY	A-F	96	–	–	–	–	–	–	–	–	–	–	–	–	–	–
		M-F	100	–	–	–	–	–	–	–	–	–	–	–	–	–	–
200	PEELINGS, FRESH	A-F	23	0.03	0.04	–	–	–	–	–	–	–	–	–	–	–	–
		M-F	100	0.14	0.19	–	–	–	–	–	–	–	–	–	–	–	–
	PUMPKIN *Cucurbita pepo*																
201	FRUIT, FRESH	A-F	9	–	0.04	–	–	–	0.24	–	–	–	–	–	–	–	–
		M-F	100	–	0.39	–	–	–	2.60	–	–	–	–	–	–	–	–
202	SEEDS	A-F	94	–	–	–	–	–	–	–	–	–	–	–	–	–	–
		M-F	100	–	–	–	–	–	–	–	–	–	–	–	–	–	–
	RAISINS (GRAPES) *Vitis spp*																
203	DEHY, CULL	A-F	87	–	–	–	–	–	–	–	–	–	–	–	–	–	–
		M-F	100	–	–	–	–	–	–	–	–	–	–	–	–	–	–
	RICE *Oryza sativa*																
204	GRAIN, GROUND (GROUND ROUGH RICE; GROUND PADDY RICE)	A-F	89	0.07	0.32	0.06	0.07	0.13	0.47	0.05	–	–	–	–	18.0	–	15.0
		M-F	100	0.07	0.36	0.07	0.08	0.14	0.53	0.05	–	–	–	–	20.2	–	16.9
205	BRAN WITH GERMS (RICE BRAN)	A-F	91	0.07	1.44	0.03	0.07	0.85	1.69	0.18	1.383	11.0	–	0.019	337.6	–	37.4
		M-F	100	0.08	1.59	0.04	0.08	0.94	1.87	0.20	1.526	12.1	–	0.021	372.4	–	41.3
206	GROATS, POLISHED (RICE, POLISHED)	A-F	89	0.02	0.11	0.01	0.04	0.09	0.23	0.08	0.846	5.4	–	0.002	29.6	–	13.7
		M-F	100	0.03	0.13	0.02	0.04	0.10	0.26	0.09	0.955	6.1	–	0.002	33.4	–	15.4
207	GROATS (RICE, BROWN)	A-F	88	0.03	0.20	0.02	0.07	0.08	0.30	0.04	0.727	3.7	–	0.003	20.3	–	14.3
		M-F	100	0.04	0.23	0.03	0.07	0.09	0.34	0.05	0.824	4.2	–	0.004	23.0	–	16.2
208	HULLS, AMMONIATED	A-F	92	0.15	0.19	–	–	–	–	–	–	–	–	–	–	–	–
		M-F	100	0.16	0.21	–	–	–	–	–	–	–	–	–	–	–	–
209	POLISHINGS	A-F	90	0.05	1.34	0.04	0.11	0.60	1.28	0.17	3.890	8.0	–	0.009	126.8	–	63.2
		M-F	100	0.05	1.49	0.05	0.12	0.66	1.41	0.19	4.311	8.8	–	0.009	140.5	–	70.0
	RUMEN CONTENTS																
210	DEHY (PAUNCH PRODUCT; PAUNCH MEAL)	A-F	87	0.69	0.58	–	–	–	–	–	–	–	–	–	–	–	–
		M-F	100	0.79	0.67	–	–	–	–	–	–	–	–	–	–	–	–
	RUTABAGA *Brassica napus, napobrassica*																
211	ROOTS, FRESH	A-F	12	0.06	0.04	0.01	–	0.02	0.23	0.03	–	1.0	–	0.004	1.2	–	–
		M-F	100	0.49	0.29	0.08	–	0.18	1.86	0.27	–	7.9	–	0.028	9.9	–	–
	RYE *Secale cereale*																
212	GRAIN, ALL ANALYSES	A-F	87	0.06	0.31	0.02	0.03	0.12	0.46	0.15	–	7.5	–	0.007	72.0	–	28.1
		M-F	100	0.07	0.36	0.03	0.03	0.14	0.52	0.17	–	8.6	–	0.008	82.3	–	32.2

Entry Number	Fat-Soluble Vitamins					Water-Soluble Vitamins								
	A (1 mg Carotene = 1667 IU Vit A)	Carotene (Provitamin A)	D	E	K	B–12	Biotin	Choline	Folacin (Folic Acid)	Niacin	Pantothenic Acid (B–3)	(Pyri-doxine) B–6	Ribo-flavin (B–2)	Thiamin (B–1)
	IU/g	ppm or mg/kg	IU/kg	ppm or mg/kg	ppm or mg/kg	ppb or mcg/kg	ppm or mg/kg	ppm or mg/kg	ppm or mg/kg	ppm or mg/kg	ppm or mg/kg	ppm or mg/kg	ppm or mg/kg	ppm or mg/kg
184	–	–	–	–	–	–	–	–	–	–	–	–	–	–
	–	–	–	–	–	–	–	–	–	–	–	–	–	–
185	–	–	–	–	–	–	–	–	–	–	–	–	–	–
	–	–	–	–	–	–	–	–	–	–	–	–	–	–
186	–	–	–	–	–	–	–	–	–	–	–	–	–	–
	–	–	–	–	–	–	–	–	–	–	–	–	–	–
187	–	–	–	–	–	–	–	–	–	–	–	–	–	–
	–	–	–	–	–	–	–	–	–	–	–	–	–	–
188	4.3	2.6	–	–	–	–	–	790	–	52	8.8	–	1.8	7.1
	4.8	2.9	–	–	–	–	–	874	–	57	9.7	–	2.0	7.9
189	–	–	–	–	–	–	–	–	–	–	–	–	–	–
	–	–	–	–	–	–	–	–	–	–	–	–	–	–
190	–	–	–	–	–	–	–	–	–	–	–	–	–	–
	–	–	–	–	–	–	–	–	–	–	–	–	–	–
191	78.4	47.0	–	–	–	–	–	–	–	–	–	–	–	–
	90.0	54.0	–	–	–	–	–	–	–	–	–	–	–	–
192	–	–	–	–	–	–	–	–	–	–	–	–	–	–
	–	–	–	–	–	–	–	–	–	–	–	–	–	–
193	–	–	–	–	–	–	0.08	–	–	17	3.3	4.36	1.1	–
	–	–	–	–	–	–	0.09	–	–	19	3.7	4.85	1.3	–
194	–	–	–	–	–	–	–	–	–	17	–	–	0.5	1.2
	–	–	–	–	–	–	–	–	–	74	–	–	2.0	5.0
195	–	–	–	–	–	–	–	–	–	–	–	–	–	–
	–	–	–	–	–	–	–	–	–	–	–	–	–	–
196	–	–	–	–	–	–	0.10	2622	0.61	33	20.0	14.12	1.0	–
	–	–	–	–	–	–	0.11	2879	0.66	37	22.0	15.50	1.1	–
197	–	–	–	–	–	–	–	–	–	–	–	–	–	–
	–	–	–	–	–	–	–	–	–	–	–	–	–	–
198	–	–	–	–	–	–	–	–	–	–	–	–	–	–
	–	–	–	–	–	–	–	–	–	–	–	–	–	–
199	–	–	–	–	–	–	–	–	–	–	–	–	–	–
	–	–	–	–	–	–	–	–	–	–	–	–	–	–
200	–	–	–	–	–	–	–	–	–	–	–	–	–	–
	–	–	–	–	–	–	–	–	–	–	–	–	–	–
201	–	–	–	–	–	–	–	–	–	–	–	–	–	–
	–	–	–	–	–	–	–	–	–	–	–	–	–	–
202	–	–	–	–	–	–	–	–	–	–	–	–	–	–
	–	–	–	–	–	–	–	–	–	–	–	–	–	–
203	–	–	–	–	–	–	–	–	–	–	–	–	–	–
	–	–	–	–	–	–	–	–	–	–	–	–	–	–
204	–	–	–	14.0	–	–	–	926	0.25	40	7.1	–	0.7	–
	–	–	–	15.7	–	–	–	1041	0.28	45	8.0	–	0.8	–
205	–	–	–	60.4	–	–	0.43	1230	2.20	299	22.8	13.24	2.6	22.4
	–	–	–	66.7	–	–	0.47	1357	2.42	330	25.2	14.61	2.8	24.8
206	–	–	–	3.5	–	–	–	901	0.15	15	3.5	0.39	0.6	0.7
	–	–	–	4.0	–	–	–	1017	0.17	17	3.9	0.45	0.6	0.7
207	–	–	–	10.3	–	–	0.09	–	0.19	43	10.7	7.00	0.6	2.9
	–	–	–	11.7	–	–	0.10	–	0.21	49	12.1	7.94	0.7	3.3
208	–	–	–	–	–	–	–	–	–	–	–	–	–	–
	–	–	–	–	–	–	–	–	–	–	–	–	–	–
209	–	–	–	90.2	–	–	0.62	1248	–	506	46.4	27.89	1.8	20.0
	–	–	–	100.0	–	–	0.68	1383	–	560	51.4	30.90	2.0	22.1
210	–	–	–	–	–	–	–	–	–	–	–	–	–	–
	–	–	–	–	–	–	–	–	–	–	–	–	–	–
211	–	–	–	–	–	–	–	–	–	10	–	–	0.7	0.6
	–	–	–	–	–	–	–	–	–	85	–	–	5.4	4.8
212	0.1	0.1	–	14.5	–	–	0.06	419	0.62	14	7.5	–	1.7	4.1
	0.2	0.1	–	16.6	–	–	0.06	479	0.71	16	8.5	–	1.9	4.7

(Continued)

TABLE V-1A ENERGY FEEDS, COMPOSITION OF FEEDS, DATA EXPRESSED **AS-FED** AND **MOISTURE-FREE**—*(Continued)*

Entry Number	Feed Name Description	International Feed Number	Moisture Basis: A-F (as-fed) or M-F (moisture-free)	Proximate Analysis: Dry Matter	Ash	Crude Fiber	Neutral Det. Fib. (NDF)	Acid Det. Fib. (ADF)	Lignin	Ether Extract (Fat)	N-Free Extract	Crude Protein	Digestible Protein: Ruminant	Swine	Horse
				%	%	%	%	%	%	%	%	%	%	%	%
	RYE (Continued)														
213	BRAN	**4-04-022**	**A-F**	**91**	**4.7**	**6.9**	**–**	**–**	**–**	**3.1**	**60.2**	**15.8**	**10.6**	**11.1**	**11.7**
			M-F	100	5.2	7.6	–	–	–	3.4	66.4	17.5	11.7	12.2	12.9
214	DISTILLERS GRAINS, DEHY	**5-04-023**	**A-F**	**92**	**2.3**	**12.3**	**–**	**–**	**–**	**6.0**	**48.3**	**23.0**	**13.8**	**21.7**	**17.5**
			M-F	100	2.5	13.4	–	–	–	6.5	52.6	25.1	15.0	23.6	19,0
215	DISTILLERS GRAINS WITH SOLUBLES, DEHY	**5-04-024**	**A-F**	**90**	**6.4**	**8.1**	**–**	**–**	**–**	**4.1**	**44.7**	**27.2**	**22.6**	**25.9**	**22.3**
			M-F	100	7.1	8.9	–	–	–	4.5	49.4	30.1	24.9	28.6	24.7
216	FLOUR	**4-04-028**	**A-F**	**89**	**0.9**	**0.5**	**–**	**–**	**–**	**1.2**	**75.8**	**10.4**	**6.6**	**7.7**	**7.4**
			M-F	100	1.0	0.6	–	–	–	1.4	85.3	11.7	7.4	8.7	8.3
217	FLOUR BY-PRODUCT, LESS THAN 8.5% FIBER (RYE MIDDLINGS)	**4-04-031**	**A-F**	**90**	**3.8**	**5.2**	**–**	**6.3**	**–**	**3.4**	**60.0**	**17.1**	**13.7**	**13.7**	**12.7**
			M-F	100	4.2	5.8	–	7.0	–	3.8	67.1	19.1	15.3	15.4	14.1
218	MILL RUN, LESS THAN 9.5% FIBER (RYE FEED)	**4-04-034**	**A-F**	**88**	**4.7**	**4.9**	**–**	**–**	**–**	**3.1**	**59.9**	**15.6**	**11.9**	**11.2**	**11.5**
			M-F	100	5.4	5.5	–	–	–	3.5	67.9	17.6	13.5	12.7	13.0
	SAFFLOWER *Carthamus tinctorius*														
219	SEEDS	**4-07-958**	**A-F**	**93**	**3.0**	**23.6**	**–**	**37.2**	**–**	**30.8**	**20.9**	**14.9**	**7.2**	**11.7**	**10.9**
			M-F	100	3.2	25.3	–	40.0	–	33.1	22.4	16.0	7.7	12.6	11.7
	SAGRAIN SORGHUM *Sorghum bicolor*														
220	GRAIN	**4-08-603**	**A-F**	**90**	**1.5**	**2.1**	**–**	**–**	**–**	**3.5**	**73.4**	**9.5**	**5.8**	**6.9**	**6.7**
			M-F	100	1.7	2.3	–	–	–	3.9	81.6	10.6	6.4	7.7	7.4
	SCHROCK SORGHUM *Sorghum bicolor, schrock*														
221	GRAIN	**4-04-452**	**A-F**	**89**	**1.6**	**3.1**	**–**	**–**	**–**	**3.1**	**70.8**	**10.0**	**6.3**	**7.4**	**7.1**
			M-F	100	1.9	3.5	–	–	–	3.5	79.8	11.3	7.1	8.3	8.0
	SCREENINGS, GRAIN, CEREAL (ALSO SEE BARLEY; WHEAT)														
222	ALL ANALYSES	**4-02-156**	**A-F**	**90**	**5.4**	**12.0**	**–**	**–**	**–**	**3.7**	**56.7**	**12.1**	**8.7**	**7.8**	**8.7**
			M-F	100	6.0	13.4	–	–	–	4.1	63.2	13.4	9.7	8.7	9.7
223	REFUSE	**4-02-151**	**A-F**	**91**	**9.0**	**18.6**	**–**	**36.3**	**–**	**4.4**	**46.2**	**12.6**	**8.9**	**9.6**	**9.1**
			M-F	100	9.9	20.5	–	40.0	–	4.9	50.9	13.8	9.8	10.6	10.0
224	UNCLEANED	**4-02-153**	**A-F**	**92**	**10.2**	**17.1**	**–**	**–**	**–**	**5.8**	**45.2**	**13.7**	**9.5**	**10.6**	**10.0**
			M-F	100	11.1	18.6	–	–	–	6.3	49.1	14.9	10.3	11.6	10.8
	SEAWEED (KELP) *Laminariales* (order), *Fucales* (order)														
225	WHOLE, DEHY	**1-08-073**	**A-F**	**91**	**35.0**	**6.5**	**–**	**–**	**–**	**0.5**	**42.4**	**6.5**	**2.9**	**2.2**	**3.4**
			M-F	100	38.6	7.1	–	–	–	0.5	46.7	7.1	3.2	2.4	3.7
	SHALLU SORGHUM *Sorghum bicolor, roxburghi*														
226	GRAIN	**4-04-456**	**A-F**	**90**	**1.9**	**2.0**	**–**	**–**	**–**	**3.7**	**71.0**	**11.5**	**7.5**	**8.7**	**8.2**
			M-F	100	2.1	2.2	–	–	–	4.1	78.8	12.7	8.4	9.6	9.1
	SORGHUM *Sorghum bicolor*														
227	GRAIN, ALL ANALYSES	**4-04-383**	**A-F**	**90**	**1.8**	**2.6**	**16.2**	**8.1**	**1.2**	**2.7**	**71.6**	**11.5**	**7.1**	**8.7**	**8.2**
			M-F	100	1.9	2.8	18.0	9.0	1.3	2.9	79.5	12.8	7.9	9.6	9.2
228	GRAIN, LESS THAN 9% PROTEIN[1]	**4-08-138**	**A-F**	**89**	**2.1**	**2.2**	**16.0**	**8.0**	**–**	**2.9**	**72.4**	**8.9**	**5.3**	**5.7**	**6.2**
			M-F	100	2.4	2.5	18.0	9.0	–	3.3	81.8	10.1	6.0	6.4	7.0
229	GRAIN, 9-12% PROTEIN	**4-08-139**	**A-F**	**89**	**1.9**	**2.4**	**–**	**–**	**–**	**2.7**	**72.2**	**9.8**	**6.1**	**5.6**	**6.9**
			M-F	100	2.1	2.6	–	–	–	3.1	81.1	11.0	6.9	6.3	7.8
230	GRAIN, MORE THAN 12% PROTEIN	**4-08-140**	**A-F**	**89**	**2.3**	**1.8**	**–**	**–**	**–**	**1.5**	**71.8**	**11.6**	**7.6**	**8.8**	**8.3**
			M-F	100	2.6	2.0	–	–	–	1.7	80.7	13.0	8.6	9.9	9.3
231	DISTILLERS GRAINS, DEHY	**5-04-374**	**A-F**	**94**	**4.3**	**12.1**	**–**	**–**	**–**	**8.3**	**38.3**	**30.8**	**24.7**	**29.4**	**26.1**
			M-F	100	4.6	12.9	–	–	–	8.8	40.8	32.9	26.3	31.4	27.9
232	DISTILLERS GRAINS WITH SOLUBLES, DEHY	**5-04-375**	**A-F**	**95**	**4.2**	**10.1**	**–**	**–**	**–**	**9.4**	**38.0**	**33.1**	**27.8**	**31.7**	**28.6**
			M-F	100	4.4	10.7	–	–	–	9.9	40.1	34.9	29.3	33.4	30.2
233	ATLAS, GRAIN	**4-04-342**	**A-F**	**89**	**1.9**	**2.0**	**–**	**–**	**–**	**3.3**	**70.5**	**11.4**	**7.5**	**8.6**	**8.1**
			M-F	100	2.1	2.3	–	–	–	3.8	79.1	12.7	8.4	9.6	9.1
234	DARSO, GRAIN	**4-04-357**	**A-F**	**90**	**1.5**	**2.1**	**–**	**–**	**–**	**3.2**	**72.9**	**10.1**	**6.9**	**7.4**	**7.1**
			M-F	100	1.6	2.3	–	–	–	3.5	81.3	11.2	7.7	8.3	8.0
235	DURRA, GRAIN	**4-04-365**	**A-F**	**89**	**3.0**	**2.1**	**–**	**–**	**–**	**3.6**	**70.1**	**10.2**	**5.7**	**7.6**	**7.3**
			M-F	100	3.4	2.4	–	–	–	4.1	78.7	11.5	6.4	8.5	8.2
236	FETERITA, GRAIN	**4-04-369**	**A-F**	**89**	**1.6**	**1.7**	**–**	**1.8**	**–**	**2.8**	**71.7**	**11.7**	**9.0**	**8.8**	**8.4**
			M-F	100	1.8	1.9	–	2.0	–	3.1	80.1	13.0	10.0	9.9	9.4
237	FETERITA, HEADS	**4-04-370**	**A-F**	**90**	**3.2**	**7.4**	**–**	**–**	**–**	**2.6**	**65.7**	**10.7**	**6.8**	**8.0**	**7.6**
			M-F	100	3.6	8.3	–	–	–	2.9	73.3	11.9	7.6	8.9	8.5
238	HEGARI, GRAIN	**4-04-398**	**A-F**	**89**	**1.4**	**2.3**	**–**	**–**	**–**	**2.5**	**72.7**	**10.4**	**6.6**	**7.7**	**7.4**
			M-F	100	1.6	2.5	–	–	–	2.8	81.4	11.7	7.4	8.7	8.3
239	HEGARI, HEADS	**4-04-399**	**A-F**	**90**	**6.6**	**14.5**	**–**	**–**	**–**	**1.9**	**59.1**	**7.7**	**4.2**	**5.3**	**5.3**
			M-F	100	7.3	16.1	–	–	–	2.1	65.8	8.6	4.6	5.9	5.9
240	KALO, GRAIN	**4-04-430**	**A-F**	**89**	**1.7**	**1.6**	**–**	**–**	**–**	**3.2**	**70.9**	**11.8**	**7.8**	**9.0**	**8.5**
			M-F	100	1.9	1.8	–	–	–	3.6	79.5	13.2	8.8	10.0	9.5
241	KAFIR, GRAIN	**4-04-428**	**A-F**	**89**	**1.5**	**2.0**	**–**	**1.8**	**–**	**2.8**	**72.0**	**10.8**	**6.6**	**8.3**	**7.7**
			M-F	100	1.7	2.2	–	2.0	–	3.1	80.8	12.1	7.4	9.4	8.7
242	KAFIR, HEADS	**4-04-429**	**A-F**	**89**	**3.0**	**7.4**	**–**	**–**	**–**	**2.6**	**66.6**	**9.8**	**3.6**	**7.1**	**6.9**
			M-F	100	3.4	8.3	–	–	–	2.9	74.6	10.9	4.0	8.0	7.7
243	KOALIANG, GRAIN	**4-04-431**	**A-F**	**90**	**2.1**	**1.9**	**–**	**2.2**	**–**	**3.9**	**71.1**	**10.7**	**6.8**	**7.9**	**7.6**
			M-F	100	2.4	2.1	–	2.5	–	4.3	79.3	11.9	7.6	8.9	8.5
244	MILO, GRAIN	**4-04-444**	**A-F**	**89**	**1.6**	**2.2**	**20.6**	**4.7**	**–**	**2.8**	**71.9**	**10.1**	**6.8**	**7.2**	**7.1**
			M-F	100	1.8	2.5	23.2	5.3	–	3.1	81.2	11.4	7.7	8.1	8.1
245	MILO, HEADS	**4-04-446**	**A-F**	**90**	**4.9**	**8.9**	**–**	**–**	**–**	**2.6**	**64.9**	**8.6**	**6.6**	**6.1**	**6.0**
			M-F	100	5.4	9.9	–	–	–	2.9	72.2	9.6	7.3	6.8	6.7
246	MILO, GLUTEN WITH BRAN (GLUTEN FEED)	**5-08-089**	**A-F**	**89**	**6.4**	**7.1**	**–**	**–**	**–**	**4.1**	**48.3**	**23.2**	**18.9**	**19.7**	**17.9**
			M-F	100	7.2	8.0	–	–	–	4.6	54.2	26.0	21.2	22.1	20.1

Entry Number	TDN Ruminant	TDN Swine	TDN Horse	DE Ruminant		DE Swine		DE Horse		ME Ruminant		ME Swine		Poultry MEn		ME Horse		Ruminant NEm		Ruminant NEg		Lactating Cows NElc	
	%	%	%	Mcal		kcal		Mcal		Mcal		kcal		kcal		Mcal		Mcal		Mcal		Mcal	
				lb	kg	lb	kg	lb	kg	lb	kg	lb	kg	lb	kg	lb	kg	lb	kg	lb	kg	lb	kg
213	56	61	66	1.23	2.72	1211	2670	1.20	2.65	1.06	2.34	1134	2499	–	–	0.99	2.17	0.73	1.62	0.48	1.05	0.70	1.55
	61	67	73	1.36	3.00	1336	2945	1.33	2.92	1.17	2.58	1250	2756	–	–	1.09	2.40	0.81	1.78	0.53	1.16	0.78	1.71
214	54	79	–	1.08	2.38	1586	3497	–	–	0.90	1.99	1387	3057	–	–	–	–	0.45	1.00	0.22	0.49	0.50	1.09
	59	86	–	1.18	2.59	1727	3807	–	–	0.98	2.17	1509	3328	–	–	–	–	0.49	1.09	0.24	0.53	0.54	1.19
215	73	68	–	1.46	3.21	1354	2985	–	–	1.29	2.83	1173	2586	–	–	–	–	0.78	1.72	0.52	1.14	0.74	1.64
	80	75	–	1.61	3.55	1497	3299	–	–	1.42	3.13	1296	2858	–	–	–	–	0.86	1.90	0.57	1.26	0.82	1.81
216	75	76	–	1.49	3.30	1522	3356	–	–	1.33	2.93	1440	3175	–	–	–	–	–	–	–	–	–	–
	84	86	–	1.68	3.71	1713	3777	–	–	1.49	3.29	1621	3573	–	–	–	–	–	–	–	–	–	–
217	75	67	68	1.49	3.29	1348	2973	1.24	2.72	1.32	2.92	1269	2798	–	–	1.01	2.23	0.87	1.92	0.60	1.31	0.81	1.79
	83	75	76	1.67	3.67	1506	3321	1.38	3.04	1.48	3.26	1418	3126	–	–	1.13	2.50	0.97	2.14	0.67	1.47	0.91	2.00
218	65	67	65	1.29	2.85	1340	2955	1.19	2.61	1.12	2.48	1262	2782	–	–	0.97	2.14	0.68	1.49	0.43	0.95	0.66	1.45
	73	76	74	1.46	3.23	1520	3351	1.35	2.97	1.27	2.81	1431	3155	–	–	1.10	2.43	0.77	1.69	0.49	1.08	0.74	1.64
219	83	–	–	1.07	2.36	–	–	–	–	1.19	2.62	–	–	–	–	–	–	0.93	2.06	0.64	1.42	0.87	1.92
	89	–	–	1.15	2.53	–	–	–	–	1.27	2.81	–	–	–	–	–	–	1.00	2.21	0.69	1.52	0.94	2.06
220	82	80	65	1.55	3.41	1591	3508	1.19	2.63	1.38	3.04	1507	3322	–	–	0.98	2.15	0.81	1.78	0.54	1.19	0.76	1.68
	91	88	73	1.72	3.79	1768	3898	1.32	2.92	1.53	3.38	1675	3692	–	–	1.09	2.39	0.90	1.98	0.60	1.33	0.85	1.87
221	79	76	66	1.50	3.31	1523	3357	1.19	2.63	1.34	2.95	1440	3175	–	–	0.98	2.16	0.78	1.72	0.52	1.15	0.74	1.63
	89	86	74	1.69	3.73	1716	3783	1.34	2.96	1.51	3.32	1623	3579	–	–	1.10	2.43	0.88	1.94	0.59	1.30	0.83	1.84
222	62	55	53	1.23	2.71	1493	3291	0.98	2.16	1.06	2.34	1409	3107	831	1833	0.80	1.77	0.63	1.39	0.39	0.85	0.62	1.37
	69	61	59	1.37	3.02	1663	3666	1.09	2.41	1.18	2.60	1570	3461	926	2042	0.90	1.98	0.70	1.55	0.43	0.95	0.69	1.53
223	52	61	–	1.03	2.28	1216	2681	–	–	0.86	1.89	1138	2509	–	–	–	–	0.41	0.91	0.19	0.41	0.47	1.03
	57	67	–	1.14	2.51	1339	2952	–	–	0.95	2.08	1253	2763	–	–	–	–	0.46	1.00	0.21	0.45	0.51	1.13
224	60	65	–	1.19	2.62	1297	2858	–	–	1.01	2.23	1217	2682	–	–	–	–	0.62	1.36	0.37	0.81	0.62	1.36
	65	70	–	1.29	2.84	1408	3104	–	–	1.10	2.42	1321	2913	–	–	–	–	0.67	1.47	0.40	0.88	0.67	1.47
225	29	–	–	0.58	1.27	–	–	–	–	0.40	0.87	–	–	–	–	–	–	–	–	–	–	–	–
	32	–	–	0.63	1.40	–	–	–	–	0.44	0.96	–	–	–	–	–	–	–	–	–	–	–	–
226	81	78	66	1.54	3.39	1560	3439	1.21	2.66	1.37	3.02	1476	3254	–	–	0.99	2.18	0.80	1.77	0.54	1.18	0.76	1.67
	90	87	74	1.71	3.76	1731	3817	1.34	2.95	1.52	3.35	1638	3612	–	–	1.10	2.42	0.89	1.96	0.60	1.31	0.84	1.86
227	67	–	71	1.34	2.96	1570	3462	1.28	2.82	1.17	2.58	1426	3143	–	–	1.05	2.31	0.53	1.18	0.30	0.66	0.55	1.21
	75	–	79	1.49	3.29	1743	3843	1.42	3.13	1.30	2.87	1583	3489	–	–	1.17	2.57	0.59	1.31	0.33	0.73	0.61	1.35
228	75	76	65	1.50	3.30	1518	3347	1.19	2.62	1.33	2.93	1433	3159	1505	3318	0.97	2.14	0.78	1.73	0.53	1.16	0.74	1.64
	84	86	74	1.69	3.73	1715	3781	1.34	2.96	1.50	3.31	1619	3570	1701	3749	1.10	2.42	0.89	1.95	0.59	1.31	0.84	1.85
229	73	76	68	1.42	3.14	1517	3345	1.23	2.70	1.11	2.44	1432	3158	1517	3345	1.01	2.22	0.73	1.62	0.48	1.06	0.76	1.67
	82	85	76	1.60	3.53	1706	3760	1.38	3.04	1.24	2.74	1610	3550	1706	3760	1.13	2.49	0.82	1.82	0.54	1.19	0.86	1.88
230	69	–	–	1.39	3.06	–	–	–	–	1.22	2.69	–	–	–	–	–	–	0.76	1.67	0.50	1.11	0.72	1.59
	78	–	–	1.56	3.44	–	–	–	–	1.37	3.02	–	–	–	–	–	–	0.85	1.88	0.56	1.24	0.81	1.79
231	78	77	–	1.56	3.45	1535	3385	–	–	1.39	3.06	1338	2949	–	–	–	–	0.85	1.88	0.57	1.27	0.80	1.77
	83	82	–	1.67	3.68	1637	3609	–	–	1.48	3.26	1426	3144	–	–	–	–	0.91	2.00	0.61	1.35	0.86	1.89
232	81	82	–	1.63	3.58	1649	3635	–	–	1.45	3.19	1442	3178	–	–	–	–	0.88	1.94	0.60	1.32	0.83	1.83
	86	87	–	1.72	3.78	1739	3835	–	–	1.53	3.37	1521	3353	–	–	–	–	0.93	2.05	0.63	1.39	0.88	1.93
233	80	77	67	1.52	3.34	1533	3379	1.22	2.68	1.35	2.98	1450	3197	–	–	1.00	2.20	0.79	1.74	0.53	1.16	0.75	1.65
	90	86	75	1.70	3.75	1720	3792	1.37	3.01	1.51	3.34	1627	3588	–	–	1.12	2.47	0.89	1.95	0.59	1.31	0.84	1.85
234	79	78	68	1.51	3.33	1564	3447	1.23	2.70	1.34	2.96	1480	3263	–	–	1.01	2.22	0.80	1.76	0.54	1.18	0.76	1.67
	88	87	75	1.69	3.72	1743	3842	1.37	3.01	1.50	3.30	1649	3637	–	–	1.12	2.47	0.89	1.96	0.60	1.32	0.84	1.86
235	76	78	63	1.48	3.26	1569	3460	1.15	2.53	1.31	2.89	1486	3276	–	–	0.94	2.07	0.79	1.74	0.53	1.16	0.75	1.65
	86	88	70	1.66	3.66	1762	3884	1.29	2.83	1.47	3.24	1668	3677	–	–	1.05	2.32	0.89	1.95	0.59	1.30	0.84	1.85
236	79	76	71	1.51	3.32	1524	3361	1.28	2.82	1.34	2.95	1442	3178	–	–	1.05	2.31	0.79	1.74	0.53	1.16	0.75	1.65
	88	85	79	1.68	3.71	1704	3756	1.43	3.15	1.50	3.30	1611	3552	–	–	1.17	2.59	0.88	1.95	0.59	1.30	0.84	1.84
237	73	71	63	1.41	3.10	1417	3123	1.15	2.54	1.24	2.73	1336	2945	–	–	0.95	2.08	0.74	1.63	0.48	1.07	0.71	1.56
	81	79	71	1.57	3.47	1581	3486	1.29	2.84	1.38	3.05	1491	3287	–	–	1.06	2.33	0.82	1.82	0.54	1.19	0.79	1.74
238	80	76	70	1.52	3.35	1529	3371	1.27	2.80	1.35	2.98	1446	3189	–	–	1.04	2.30	0.79	1.74	0.53	1.16	0.75	1.65
	90	86	79	1.70	3.74	1711	3772	1.42	3.13	1.51	3.33	1618	3567	–	–	1.17	2.57	0.88	1.94	0.59	1.30	0.84	1.84
239	70	66	52	1.33	2.93	1321	2911	0.96	2.12	1.16	2.55	1242	2737	–	–	0.79	1.74	0.67	1.47	0.42	0.93	0.65	1.44
	78	74	58	1.48	3.26	1470	3241	1.07	2.36	1.29	2.84	1382	3047	–	–	0.88	1.94	0.74	1.64	0.47	1.03	0.72	1.60
240	80	77	69	1.52	3.35	1530	3373	1.25	2.75	1.35	2.99	1448	3191	–	–	1.02	2.26	0.79	1.75	0.53	1.17	0.75	1.65
	90	86	77	1.71	3.76	1715	3782	1.40	3.08	1.52	3.35	1623	3578	–	–	1.15	2.53	0.89	1.96	0.59	1.31	0.84	1.85
241	75	81	70	1.47	3.24	1521	3354	1.26	2.77	1.30	2.87	1437	3169	1529	3372	1.03	2.27	0.79	1.74	0.53	1.16	0.75	1.64
	85	91	78	1.65	3.64	1707	3764	1.41	3.11	1.46	3.22	1613	3557	1717	3785	1.16	2.55	0.88	1.95	0.59	1.30	0.84	1.85
242	47	72	62	0.94	2.06	1430	3153	1.14	2.51	0.76	1.68	1349	2975	–	–	0.93	2.06	0.07	0.15	−0.14	−0.31	0.24	0.52
	52	80	70	1.05	2.31	1602	3531	1.28	2.81	0.85	1.88	1511	3332	–	–	1.05	2.31	0.08	0.17	−0.16	−0.35	0.27	0.58
243	81	79	64	1.54	3.39	1577	3477	1.17	2.58	1.37	3.03	1494	3293	–	–	0.96	2.11	0.80	1.77	0.54	1.19	0.76	1.67
	91	88	72	1.72	3.79	1761	3882	1.30	2.87	1.53	3.38	1667	3676	–	–	1.07	2.36	0.90	1.97	0.60	1.32	0.85	1.86
244	76	77	68	1.41	3.12	1520	3350	1.23	2.71	1.13	2.49	1474	3250	1467	3234	1.01	2.23	0.75	1.66	0.50	1.09	0.74	1.62
	86	87	77	1.60	3.52	1716	3782	1.39	3.06	1.27	2.81	1664	3669	1656	3651	1.14	2.51	0.85	1.87	0.56	1.23	0.83	1.83
245	73	72	57	1.46	3.21	1445	3185	1.05	2.32	1.29	2.84	1363	3005	–	–	0.86	1.90	0.76	1.69	0.51	1.11	0.73	1.61
	81	80	64	1.62	3.58	1607	3543	1.17	2.58	1.43	3.16	1516	3343	–	–	0.96	2.12	0.85	1.88	0.56	1.24	0.81	1.79
246	74	74	–	1.45	3.20	1484	3271	–	–	1.29	2.83	1223	2696	975	2150	–	–	0.78	1.72	0.52	1.14	0.74	1.63
	84	83	–	1.63	3.59	1665	3670	–	–	1.44	3.18	1372	3025	1094	2413	–	–	0.87	1.93	0.58	1.28	0.83	1.83

(Continued)

TABLE V-1A ENERGY FEEDS, MINERAL AND VITAMIN COMPOSITION OF FEEDS, DATA EXPRESSED **AS-FED** AND **MOISTURE-FREE**—*(Continued)*

Entry Number	Feed Name Description	Moisture Basis: A-F (as-fed) or M-F (moisture-free)	Dry Matter	Macro Minerals							Micro Minerals						
				Calcium (Ca)	Phosphorus (P)	Sodium (Na)	Chlorine (Cl)	Magnesium (Mg)	Potassium (K)	Sulfur (S)	Cobalt (Co)	Copper (Cu)	Iodine (I)	Iron (Fe)	Manganese (Mn)	Selenium (Se)	Zinc (Zn)
			%	%	%	%	%	%	%	%	ppm or mg/kg	ppm or mg/kg	ppm or mg/kg	%	ppm or mg/kg	ppm or mg/kg	ppm or mg/kg
	RYE (Continued)																
213	BRAN	A-F	91	–	–	–	–	–	–	–	–	–	–	–	–	–	–
		M-F	100	–	–	–	–	–	–	–	–	–	–	–	–	–	–
214	DILTILLERS GRAINS, DEHY	A-F	92	0.15	0.48	0.17	0.05	0.17	0.07	0.44	–	–	–	–	18.4	–	–
		M-F	100	0.16	0.52	0.18	0.05	0.18	0.08	0.48	–	–	–	–	20.0	–	–
215	DISTILLERS GRAINS WITH SOLUBLES, DEHY	A-F	90	–	–	–	–	–	–	–	–	–	–	–	–	–	–
		M-F	100	–	–	–	–	–	–	–	–	–	–	–	–	–	–
216	FLOUR	A-F	89	0.02	0.29	–	–	–	0.46	–	–	–	–	–	–	–	–
		M-F	100	0.02	0.32	–	–	–	0.52	–	–	–	–	–	–	–	–
217	FLOUR BY-PRODUCT, LESS THAN 8.5% FIBER (RYE MIDDLINGS)	A-F	90	0.06	0.63	–	–	–	0.63	–	–	–	–	–	44.0	–	–
		M-F	100	0.07	0.70	–	–	–	0.70	–	–	–	–	–	49.1	–	–
218	MILL RUN, LESS THAN 9.5% FIBER (RYE FEED)	A-F	88	0.07	0.63	–	–	0.23	0.81	0.04	–	–	–	–	–	–	–
		M-F	100	0.08	0.71	–	–	0.26	0.92	0.04	–	–	–	–	–	–	–
	SAFFLOWER *Carthamus tinctorius*																
219	SEEDS	A-F	93	0.24	0.57	0.06	–	0.34	0.74	0.06	–	10.0	–	0.032	1.1	–	30.0
		M-F	100	0.26	0.61	0.06	–	0.36	0.79	0.06	–	10.7	–	0.035	1.2	–	32.2
	SAGRAIN SORGHUM *Sorghum bicolor*																
220	GRAIN	A-F	90	0.02	0.27	–	–	–	–	–	–	–	–	–	–	–	–
		M-F	100	0.02	0.30	–	–	–	–	–	–	–	–	–	–	–	–
	SCHROCK SORGHUM *Sorghum bicolor, schrock*																
221	GRAIN	A-F	89	0.02	0.30	–	–	–	–	–	–	–	–	–	–	–	–
		M-F	100	0.02	0.34	–	–	–	–	–	–	–	–	–	–	–	–
	SCREENINGS, GRAIN, CEREAL (ALSO SEE BARLEY; WHEAT)																
222	ALL ANALYSES	A-F	90	0.33	0.35	0.40	–	0.12	0.30	–	–	–	–	–	44.4	–	–
		M-F	100	0.37	0.39	0.45	–	0.14	0.34	–	–	–	–	–	49.4	–	–
223	REFUSE	A-F	91	0.31	0.33	0.25	–	0.22	0.18	0.30	–	–	–	0.025	–	0.653	–
		M-F	100	0.35	0.36	0.28	–	0.24	0.20	0.33	–	–	–	0.027	–	0.719	–
224	UNCLEANED	A-F	92	0.37	0.41	0.26	–	0.22	0.18	0.30	–	–	–	0.025	–	–	–
		M-F	100	0.40	0.45	0.28	–	0.24	0.20	0.33	–	–	–	0.027	–	–	–
	SEAWEED (KELP) *Laminariales* (order), *Fucales* (order)																
225	WHOLE, DEHY	A-F	91	2.47	0.28	–	–	0.85	–	–	–	–	–	–	–	–	–
		M-F	100	2.72	0.31	–	–	0.93	–	–	–	–	–	–	–	–	–
	SHALLU SORGHUM *Sorghum bicolor, roxburghi*																
226	GRAIN	A-F	90	–	–	–	–	–	–	–	–	–	–	–	–	–	–
		M-F	100	–	–	–	–	–	–	–	–	–	–	–	–	–	–
	SORGHUM *Sorghum bicolor*																
227	GRAIN, ALL ANALYSES	A-F	90	0.05	0.32	0.03	0.08	0.14	0.35	0.15	0.275	9.7	–	0.007	9.8	–	42.4
		M-F	100	0.06	0.35	0.03	0.09	0.16	0.38	0.17	0.305	10.8	–	0.007	10.9	–	47.1
228	GRAIN, LESS THAN 9% PROTEIN	A-F	89	0.03	0.27	0.04	–	–	0.35	–	0.067	9.7	0.023	0.002	15.4	–	13.7
		M-F	100	0.03	0.31	0.05	–	–	0.40	–	0.075	11.0	0.025	0.003	17.4	–	15.4
229	GRAIN, 9-12% PROTEIN	A-F	89	0.03	0.27	0.02	–	0.15	0.34	0.14	0.067	9.7	0.023	0.004	15.4	–	13.7
		M-F	100	0.04	0.30	0.02	–	0.17	0.38	0.16	0.075	10.9	0.025	0.004	17.3	–	15.4
230	GRAIN, MORE THAN 12% PROTEIN	A-F	89	0.03	0.29	0.04	–	0.17	0.34	0.16	–	–	–	0.005	–	–	–
		M-F	100	0.03	0.32	0.05	–	0.19	0.38	0.18	–	–	–	0.005	–	–	–
231	DISTILLERS GRAINS, DEHY	A-F	94	0.15	0.69	0.05	–	0.18	0.36	0.17	–	–	–	0.005	–	–	–
		M-F	100	0.16	0.74	0.05	–	0.19	0.38	0.18	–	–	–	0.005	–	–	–
232	DISTILLERS GRAINS WITH SOLUBLES, DEHY	A-F	95	0.17	0.92	–	–	–	–	–	–	–	–	–	104.5	–	–
		M-F	100	0.18	0.97	–	–	–	–	–	–	–	–	–	110.2	–	–
233	ATLAS, GRAIN	A-F	89	–	–	–	–	–	–	–	–	–	–	–	–	–	–
		M-F	100	–	–	–	–	–	–	–	–	–	–	–	–	–	–
234	DARSO, GRAIN	A-F	90	0.08	0.30	–	–	–	–	–	–	–	–	–	–	–	–
		M-F	100	0.09	0.34	–	–	–	–	–	–	–	–	–	–	–	–
235	DURRA, GRAIN	A-F	89	–	–	–	–	–	–	–	–	–	–	–	–	–	–
		M-F	100	–	–	–	–	–	–	–	–	–	–	–	–	–	–
236	FETERITA, GRAIN	A-F	89	0.02	0.30	–	–	–	–	–	–	–	–	–	–	–	–
		M-F	100	0.02	0.34	–	–	–	–	–	–	–	–	–	–	–	–
237	FETERITA, HEADS	A-F	90	–	–	–	–	–	–	–	–	–	–	–	–	–	–
		M-F	100	–	–	–	–	–	–	–	–	–	–	–	–	–	–
238	HEGARI, GRAIN	A-F	89	0.02	0.27	–	–	–	–	–	–	–	–	–	–	–	–
		M-F	100	0.02	0.30	–	–	–	–	–	–	–	–	–	–	–	–
239	HEGARI, HEADS	A-F	90	0.10	0.19	–	–	0.10	–	–	–	–	–	–	13.1	–	–
		M-F	100	0.11	0.21	–	–	0.11	–	–	–	–	–	–	14.6	–	–
240	KALO, GRAIN	A-F	89	–	–	–	–	–	–	–	–	–	–	–	–	–	–
		M-F	100	–	–	–	–	–	–	–	–	–	–	–	–	–	–
241	KAFIR, GRAIN	A-F	89	0.03	0.31	0.05	0.10	0.15	0.34	0.16	0.387	7.0	–	0.007	15.8	0.797	13.5
		M-F	100	0.04	0.35	0.06	0.11	0.17	0.38	0.18	0.435	7.8	–	0.008	17.8	0.894	15.2
242	KAFIR, HEADS	A-F	89	0.08	0.25	–	–	0.24	–	–	–	–	–	–	–	–	–
		M-F	100	0.09	0.28	–	–	0.27	–	–	–	–	–	–	–	–	–
243	KOALIANG, GRAIN	A-F	90	0.03	0.29	–	–	–	–	–	–	–	–	–	–	–	–
		M-F	100	0.03	0.32	–	–	–	–	–	–	–	–	–	–	–	–
244	MILO, GRAIN	A-F	89	0.04	0.30	0.04	0.08	0.13	0.31	0.11	0.471	4.3	0.061	0.005	15.8	0.201	16.9
		M-F	100	0.05	0.34	0.04	0.09	0.14	0.35	0.12	0.531	4.9	0.069	0.006	17.9	0.227	19.1
245	MILO, HEADS	A-F	90	0.12	0.20	–	–	0.15	0.50	0.12	–	–	–	–	–	–	–
		M-F	100	0.14	0.23	–	–	0.17	0.56	0.13	–	–	–	–	–	–	–
246	MILO, GLUTEN WITH BRAN (GLUTEN FEED)	A-F	89	0.20	0.80	–	–	–	1.60	–	0.089	25.0	–	0.030	48.6	–	26.5
		M-F	100	0.23	0.90	–	–	–	1.80	–	0.100	28.0	–	0.034	54.6	–	29.8

Entry Number	Fat-Soluble Vitamins: A (1 mg Carotene = 1667 IU Vit A)	Carotene (Provitamin A)	D	E	K	Water-Soluble Vitamins: B-12	Biotin	Choline	Folacin (Folic Acid)	Niacin	Pantothenic Acid (B-3)	(Pyridoxine) B-6	Riboflavin (B-2)	Thiamin (B-1)
	IU/g	ppm or mg/kg	IU/kg	ppm or mg/kg	ppm or mg/kg	ppb or mcg/kg	ppm or mg/kg	ppm or mg/kg	ppm or mg/kg	ppm or mg/kg	ppm or mg/kg	ppm or mg/kg	ppm or mg/kg	ppm or mg/kg
213	–	–	–	–	–	–	–	–	–	–	–	–	–	–
	–	–	–	–	–	–	–	–	–	–	–	–	–	–
214	–	–	–	–	–	–	–	–	–	17	5.2	–	3.3	1.3
	–	–	–	–	–	–	–	–	–	18	5.7	–	3.6	1.4
215	–	–	–	–	–	–	–	–	–	63	17.4	–	8.2	3.1
	–	–	–	–	–	–	–	–	–	69	19.2	–	9.0	3.4
216	–	–	–	–	–	–	–	–	–	8	10.3	–	0.9	2.4
	–	–	–	–	–	–	–	–	–	9	11.6	–	1.0	2.7
217	–	–	–	–	–	–	–	–	–	17	23.0	–	2.4	3.3
	–	–	–	–	–	–	–	–	–	19	25.7	–	2.7	3.7
218	–	–	–	–	–	–	–	–	–	–	–	–	–	–
	–	–	–	–	–	–	–	–	–	–	–	–	–	–
219	–	–	–	–	–	–	–	–	–	–	–	–	–	–
	–	–	–	–	–	–	–	–	–	–	–	–	–	–
220	–	–	–	–	–	–	–	–	–	–	–	–	–	–
	–	–	–	–	–	–	–	–	–	–	–	–	–	–
221	–	–	–	–	–	–	0.25	–	–	–	–	–	–	–
	–	–	–	–	–	–	0.29	–	–	–	–	–	–	–
222	–	–	–	–	–	–	–	1044	1.06	10	12.8	–	1.8	–
	–	–	–	–	–	–	–	1163	1.18	12	14.3	–	2.0	–
223	–	–	–	–	–	–	–	–	–	47	23.0	2.39	0.7	0.5
	–	–	–	–	–	–	–	–	–	52	25.3	2.63	0.7	0.5
224	–	–	–	–	–	–	–	–	–	–	–	–	–	–
	–	–	–	–	–	–	–	–	–	–	–	–	–	–
225	–	–	–	–	–	–	–	–	–	–	–	–	–	–
	–	–	–	–	–	–	–	–	–	–	–	–	–	–
226	–	–	–	–	–	–	–	–	–	–	–	–	–	–
	–	–	–	–	–	–	–	–	–	–	–	–	–	–
227	2.0	1.2	–	–	–	–	0.26	686	0.22	47	10.2	5.41	1.2	4.5
	2.2	1.3	–	–	–	–	0.29	762	0.24	52	11.3	6.00	1.4	5.0
228	–	–	–	2.2	–	–	0.29	763	0.22	48	12.8	4.63	1.3	4.4
	–	–	–	2.5	–	–	0.32	862	0.25	54	14.4	5.23	1.5	5.0
229	–	–	–	1.3	–	–	0.29	762	0.22	48	12.8	4.60	1.3	4.4
	–	–	–	1.5	–	–	0.32	857	0.25	54	14.4	5.17	1.5	5.0
230	–	–	–	–	–	–	–	–	–	–	–	–	–	–
	–	–	–	–	–	–	–	–	–	–	–	–	–	–
231	–	–	–	–	–	–	0.31	805	–	–	–	–	–	–
	–	–	–	–	–	–	0.33	858	–	–	–	–	–	–
232	–	–	–	–	–	–	–	844	–	61	12.3	–	4.2	1.3
	–	–	–	–	–	–	–	891	–	64	13.0	–	4.4	1.4
233	–	–	–	–	–	–	0.26	–	–	34	13.4	–	2.0	–
	–	–	–	–	–	–	0.29	–	–	39	15.0	–	2.2	–
234	4.5	2.7	–	–	–	–	–	–	–	28	13.3	–	1.4	–
	5.0	3.0	–	–	–	–	–	–	–	31	14.8	–	1.5	–
235	–	–	–	–	–	–	0.30	–	–	48	8.6	1.96	0.8	–
	–	–	–	–	–	–	0.33	–	–	54	9.7	2.21	0.9	–
236	1.0	0.6	–	–	–	–	–	–	–	50	14.0	–	1.6	–
	1.1	0.7	–	–	–	–	–	–	–	56	15.7	–	1.8	–
237	–	–	–	–	–	–	–	–	–	–	–	–	–	–
	–	–	–	–	–	–	–	–	–	–	–	–	–	–
238	–	–	–	–	–	–	–	–	–	–	–	–	–	–
	–	–	–	–	–	–	–	–	–	–	–	–	–	–
239	–	–	–	–	–	–	–	–	–	–	–	–	–	–
	–	–	–	–	–	–	–	–	–	–	–	–	–	–
240	–	–	–	–	–	–	0.26	–	–	33	4.9	5.11	1.2	–
	–	–	–	–	–	–	0.29	–	–	37	5.5	5.73	1.3	–
241	0.6	0.4	–	–	–	–	0.24	439	0.20	38	12.0	6.68	1.2	3.8
	0.7	0.4	–	–	–	–	0.27	493	0.22	43	13.4	7.50	1.4	4.3
242	–	–	–	–	–	–	–	–	–	–	–	–	–	–
	–	–	–	–	–	–	–	–	–	–	–	–	–	–
243	–	–	–	–	–	–	–	–	–	–	–	–	–	–
	–	–	–	–	–	–	–	–	–	–	–	–	–	–
244	0.4	0.2	0	12.1	0.22	–	0.23	638	0.21	37	11.0	4.69	1.1	4.1
	0.5	0.3	0	13.7	0.25	–	0.26	720	0.23	42	12.4	5.29	1.3	4.7
245	1.0	0.6	–	–	–	–	–	–	–	–	–	–	2.0	–
	1.2	0.7	–	–	–	–	–	–	–	–	–	–	2.2	–
246	–	–	–	–	–	–	0.22	1746	–	99	19.9	10.83	2.7	5.7
	–	–	–	–	–	–	0.25	1959	–	112	22.3	12.15	3.0	6.4

(Continued)

TABLE V-1A ENERGY FEEDS, COMPOSITION OF FEEDS, DATA EXPRESSED **AS-FED** AND **MOISTURE-FREE**—*(Continued)*

Entry Number	Feed Name Description	International Feed Number	Moisture Basis: A-F (as-fed) or M-F (moisture-free)	Proximate Analysis: Dry Matter	Ash	Crude Fiber	Neutral Det. Fib. (NDF)	Acid Det. Fib. (ADF)	Lignin	Ether Extract (Fat)	N-Free Extract	Crude Protein	Digestible Protein: Ruminant	Swine	Horse
				%	%	%	%	%	%	%	%	%	%	%	%
	SORGHUM (Continued)														
247	SAGRAIN, GRAIN	4-08-603	A-F	**90**	**1.5**	**2.1**	–	–	–	**3.5**	**73.4**	**9.5**	**5.8**	**6.9**	**6.7**
			M-F	100	1.7	2.3	–	–	–	3.9	81.6	10.6	6.4	7.7	7.4
248	SCHROCK, GRAIN	4-04-452	A-F	**89**	**1.6**	**3.1**	–	–	–	**3.1**	**70.8**	**10.0**	**6.3**	**7.4**	**7.1**
			M-F	100	1.9	3.5	–	–	–	3.5	79.8	11.3	7.1	8.3	8.0
249	SHALLU, GRAIN	4-04-456	A-F	**90**	**1.9**	**2.0**	–	–	–	**3.7**	**71.0**	**11.5**	**7.5**	**8.7**	**8.2**
			M-F	100	2.1	2.2	–	–	–	4.1	78.8	12.7	8.4	9.6	9.1
250	SUDANGRASS, GRAIN	4-08-520	A-F	**92**	**12.0**	**25.4**	–	–	–	**2.4**	**38.4**	**14.2**	**9.9**	**11.1**	**10.3**
			M-F	100	13.0	27.5	–	–	–	2.6	41.6	15.4	10.7	12.0	11.2
	SOYBEAN *Glycine max*														
251	SEEDS[1]	5-04-610	A-F	**92**	**5.1**	**5.4**	–	**10.1**	–	**17.2**	**25.9**	**38.4**	**34.5**	**31.5**	**34.9**
			M-F	100	5.6	5.8	–	11.0	–	18.7	28.1	41.7	37.5	34.2	37.9
252	MILL FEED	4-04-594	A-F	**90**	**5.0**	**33.7**	–	–	–	**1.9**	**36.4**	**12.6**	**8.1**	**6.0**	**9.1**
			M-F	100	5.5	37.6	–	–	–	2.1	40.7	14.1	9.0	6.7	10.2
253	MILL RUN	4-04-595	A-F	**89**	**4.4**	**33.8**	–	–	–	**2.3**	**36.7**	**12.3**	**8.3**	**9.4**	**8.9**
			M-F	100	5.0	37.8	–	–	–	2.6	41.0	13.7	9.2	10.5	9.9
254	OIL	4-07-983	A-F	**100**	**0.3**	–	–	–	–	**95.0**	**7.3**	**1.4**	**–1.8**	**–0.6**	**0.2**
			M-F	100	0.3	–	–	–	–	95.5	7.3	1.4	–1.8	–0.6	0.2
	SPELT *Triticum spelta*														
255	GRAIN	4-04-651	A-F	**90**	**3.5**	**9.1**	–	–	–	**1.9**	**63.4**	**12.0**	**8.0**	**9.1**	**8.6**
			M-F	100	3.9	10.2	–	–	–	2.1	70.5	13.3	8.9	10.1	9.6
	SUDANGRASS SORGHUM *Sorghum bicolor, sudanense*														
256	GRAIN	4-08-520	A-F	**92**	**12.0**	**25.4**	–	–	–	**2.4**	**38.4**	**14.2**	**9.9**	**11.1**	**10.3**
			M-F	100	13.0	27.5	–	–	–	2.6	41.6	15.4	10.7	12.0	11.2
	SUGARCANE *Saccharum officinarum*														
257	MOLASSES, DEHY	4-04-695	A-F	**94**	**12.5**	**6.3**	–	–	–	**0.9**	**65.0**	**9.7**	**5.3**	**7.0**	**6.8**
			M-F	100	13.3	6.7	–	–	–	0.9	68.8	10.3	5.7	7.4	7.2
258	MOLASSES, MORE THAN 46% INVERT SUGAR, MORE THAN 79.5 DEGREES BRIX	4-04-696	A-F	**74**	**9.8**	**0.4**	–	**0.3**	**0.2**	**0.2**	**59.7**	**4.3**	**0.6**	**1.3**	**2.7**
			M-F	100	13.2	0.5	–	0.4	0.3	0.2	80.2	5.8	0.8	1.8	3.7
	SUNFLOWER *Helianthus* spp														
259	SEEDS	5-08-530	A-F	**94**	**3.7**	**22.7**	–	–	–	**32.3**	**14.4**	**20.9**	**16.7**	**19.6**	**14.9**
			M-F	100	4.0	24.1	–	–	–	34.4	15.3	22.2	17.8	20.8	15.8
	SWEET CORN—SEE CORN, SWEET														
	SWEET POTATO *Ipomoea batatas*														
260	TUBERS, DEHY, MEAL	4-08-536	A-F	**89**	**3.8**	**3.7**	–	–	–	**1.0**	**74.5**	**6.4**	**0.9**	**4.9**	**4.3**
			M-F	100	4.3	4.1	–	–	–	1.1	83.3	7.2	1.0	5.5	4.8
261	CANNERY RESIDUE, DEHY	4-08-535	A-F	**90**	**6.0**	**9.6**	–	–	–	**0.3**	**71.8**	**2.5**	**–0.5**	**0.6**	**1.2**
			M-F	100	6.7	10.6	–	–	–	0.3	79.6	2.8	–0.5	0.6	1.3
262	TUBERS, FRESH	4-04-788	A-F	**33**	**1.1**	**1.4**	–	**2.6**	–	**0.4**	**28.5**	**1.7**	**0.5**	**0.9**	**1.0**
			M-F	100	3.2	4.2	–	8.0	–	1.2	86.4	5.0	1.5	2.7	3.1
	SWINE *Sus scrofa*														
263	FAT	4-04-790	A-F	**99**	–	–	–	–	–	**99.3**	–	–	–	–	–
			M-F	100	–	–	–	–	–	100.0	–	–	–	–	–
	TALLOW														
264	ANIMAL	4-08-127	A-F	**97**	**0.1**	–	–	–	–	**96.8**	–	**1.5**	**–1.6**	**–0.4**	**0.3**
			M-F	100	0.1	–	–	–	–	99.5	–	1.6	–1.6	–0.4	0.3
	TARO (DASHEEN) *Colocasia esculenta*														
265	TUBERS, FRESH	4-10-463	A-F	**28**	**1.2**	**0.6**	–	–	–	**0.1**	**24.2**	**1.5**	**0.5**	**1.0**	**0.9**
			M-F	100	4.3	2.2	–	–	–	0.4	87.7	5.4	1.8	3.6	3.4
	TOMATO *Lycopersicon esculentum*														
266	FRUIT, FRESH	4-05-040	A-F	**6**	**0.5**	**0.5**	–	–	–	**0.3**	**3.7**	**1.0**	**0.7**	**0.8**	**0.7**
			M-F	100	8.2	9.1	–	–	–	5.0	61.3	16.4	11.6	12.9	12.0
267	POMACE, WET	5-05-042	A-F	**25**	**1.0**	**8.5**	–	–	–	**3.6**	**6.6**	**5.4**	**4.3**	**5.0**	**3.8**
			M-F	100	4.0	33.7	–	–	–	14.4	26.3	21.5	17.2	20.1	15.1
268	POMACE, DEHY	5-05-041	A-F	**92**	**6.8**	**25.0**	**50.4**	**46.5**	**10.5**	**9.8**	**29.3**	**21.0**	**11.9**	**19.7**	**15.2**
			M-F	100	7.4	27.2	54.8	50.5	11.4	10.7	31.9	22.9	12.9	21.5	16.6
	TRITICALE *Triticale hexaploide*														
269	GRAIN	4-20-362	A-F	**89**	**1.8**	**3.0**	**11.9**	–	–	**1.5**	**67.3**	**15.4**	**11.1**	**12.2**	**11.3**
			M-F	100	2.0	3.3	13.3	–	–	1.7	75.7	17.3	12.5	13.7	12.7
	TURNIP *Brassica rapa, rapa*														
270	ROOTS, FRESH[1]	4-05-067	A-F	**9**	**0.8**	**1.1**	**4.0**	**3.1**	–	**0.2**	**5.9**	**1.2**	**0.9**	**0.4**	**0.9**
			M-F	100	8.7	11.5	44.0	34.0	–	1.9	64.8	13.1	9.8	4.4	9.4
	VELVETBEAN *Mucuna* spp														
271	PODS WITH SEEDS, SUN-CURED	4-05-087	A-F	**89**	**4.7**	**12.9**	–	–	–	**4.3**	**49.8**	**17.3**	**12.9**	**13.9**	**13.0**
			M-F	100	5.3	14.5	–	–	–	4.8	55.9	19.4	14.5	15.7	14.6
	VERBENA, EUROPEAN (PIGEON-GRASS) *Verbena officinalis*														
272	SEEDS	4-26-082	A-F	**89**	**6.3**	**17.1**	–	–	–	**5.9**	**45.6**	**14.2**	**9.3**	–	–
			M-F	100	7.1	19.2	–	–	–	6.6	51.2	15.9	10.4	–	–

ENERGY FEEDS

Entry Number	TDN			Digestible Energy						Metabolizable Energy								Net Energy					
	Ruminant	Swine	Horse	Ruminant		Swine		Horse		Ruminant		Swine		Poultry ME_n		Horse		Ruminant NE_m		Ruminant NE_g		Lactating Cows NE_{lc}	
	%	%	%	Mcal		kcal		Mcal		Mcal		kcal		kcal		Mcal		Mcal		Mcal		Mcal	
				lb	kg	lb	kg	lb	kg	lb	kg	lb	kg	lb	kg	lb	kg	lb	kg	lb	kg	lb	kg
247	82	80	65	1.55	3.41	1591	3508	1.19	2.63	1.38	3.04	1507	3322	–	–	0.98	2.15	0.81	1.78	0.54	1.19	0.76	1.68
	91	88	73	1.72	3.79	1768	3898	1.32	2.92	1.53	3.38	1675	3692	–	–	1.09	2.39	0.90	1.98	0.60	1.33	0.85	1.87
248	79	76	66	1.50	3.31	1523	3357	1.19	2.63	1.34	2.95	1440	3175	–	–	0.98	2.16	0.78	1.72	0.52	1.15	0.74	1.63
	89	86	74	1.69	3.73	1716	3783	1.34	2.96	1.51	3.32	1623	3579	–	–	1.10	2.43	0.88	1.94	0.59	1.30	0.83	1.84
249	81	78	66	1.54	3.39	1560	3439	1.21	2.66	1.37	3.02	1476	3254	–	–	0.99	2.18	0.80	1.77	0.54	1.18	0.76	1.67
	90	87	74	1.71	3.76	1731	3817	1.34	2.95	1.52	3.35	1638	3612	–	–	1.10	2.42	0.89	1.96	0.60	1.31	0.84	1.86
250	47	–	–	0.94	2.08	–	–	–	–	0.76	1.68	–	–	–	–	–	–	–	–	–	–	–	–
	51	–	–	1.02	2.25	–	–	–	–	0.83	1.82	–	–	–	–	–	–	–	–	–	–	–	–
251	84	93	–	1.69	3.72	1820	4012	–	–	1.52	3.34	1605	3539	1534	3382	–	–	0.93	2.04	0.64	1.41	0.86	1.90
	92	101	–	1.83	4.04	1977	4359	–	–	1.65	3.63	1744	3846	1667	3674	–	–	1.01	2.22	0.70	1.53	0.94	2.07
252	48	26	–	1.21	2.67	529	1167	–	–	1.04	2.29	419	924	362	798	–	–	0.63	1.39	0.39	0.86	0.61	1.34
	53	30	–	1.35	2.98	591	1303	–	–	1.16	2.56	468	1032	404	890	–	–	0.71	1.56	0.43	0.96	0.68	1.50
253	37	–	–	0.74	1.64	–	–	–	–	0.57	1.25	–	–	298	657	–	–	–	–	–	–	–	–
	42	–	–	0.83	1.83	–	–	–	–	0.64	1.40	–	–	333	735	–	–	–	–	–	–	–	–
254	193	–	–	3.86	8.51	3412	7523	–	–	3.70	8.15	3287	7247	4050	8929	–	–	2.40	5.30	1.79	3.94	2.09	4.61
	194	–	–	3.88	8.55	3430	7561	–	–	3.71	8.19	3303	7283	4070	8974	–	–	2.42	5.32	1.80	3.96	2.10	4.63
255	68	67	66	1.37	3.01	1337	2947	1.20	2.64	1.20	2.64	1257	2772	–	–	0.98	2.17	0.72	1.58	0.46	1.02	0.69	1.52
	76	74	73	1.52	3.35	1487	3278	1.33	2.94	1.33	2.94	1399	3084	–	–	1.09	2.41	0.80	1.75	0.51	1.13	0.77	1.69
256	47	–	–	0.94	2.08	–	–	–	–	0.76	1.68	–	–	–	–	–	–	–	–	–	–	–	–
	51	–	–	1.02	2.25	–	–	–	–	0.83	1.82	–	–	–	–	–	–	–	–	–	–	–	–
257	66	70	61	1.36	2.99	1208	2663	1.12	2.48	1.18	2.60	1127	2485	1227	2706	0.92	2.03	0.70	1.55	0.44	0.97	0.68	1.51
	70	74	65	1.44	3.17	1279	2820	1.19	2.62	1.25	2.75	1194	2632	1300	2866	0.98	2.15	0.74	1.64	0.47	1.03	0.73	1.60
258	60	56	–	1.22	2.68	1135	2502	–	–	1.12	2.46	995	2194	870	1918	–	–	0.77	1.70	0.53	1.18	0.64	1.41
	81	76	–	1.63	3.60	1526	3364	–	–	1.50	3.31	1338	2950	1170	2579	–	–	1.04	2.28	0.72	1.58	0.86	1.89
259	78	84	–	1.56	3.44	1685	3715	–	–	1.38	3.05	1476	3254	–	–	–	–	0.89	1.95	0.60	1.33	0.83	1.83
	83	90	–	1.66	3.66	1792	3951	–	–	1.47	3.25	1570	3461	–	–	–	–	0.94	2.08	0.64	1.41	0.88	1.95
260	72	64	68	1.42	3.12	1280	2823	1.23	2.72	1.25	2.75	1202	2651	–	–	1.01	2.23	0.76	1.67	0.50	1.11	0.72	1.60
	81	72	76	1.58	3.49	1431	3156	1.38	3.04	1.39	3.07	1344	2963	–	–	1.13	2.49	0.85	1.87	0.56	1.24	0.81	1.78
261	69	76	59	1.35	2.98	1514	3337	1.08	2.38	1.18	2.60	1431	3154	–	–	0.88	1.95	0.71	1.57	0.46	1.01	0.69	1.51
	76	84	65	1.50	3.30	1679	3701	1.20	2.63	1.31	2.89	1587	3498	–	–	0.98	2.16	0.79	1.74	0.51	1.12	0.76	1.68
262	27	27	25	0.52	1.16	530	1169	0.45	0.99	0.46	1.02	501	1104	–	–	0.37	0.81	0.28	0.63	0.19	0.42	0.27	0.60
	81	81	75	1.59	3.51	1610	3549	1.36	2.99	1.40	3.09	1519	3349	–	–	1.11	2.45	0.86	1.90	0.57	1.26	0.82	1.81
263	–	–	–	–	–	3669	8089	–	–	–	–	3537	7798	3936	8677	–	–	–	–	–	–	–	–
	–	–	–	–	–	3694	8143	–	–	–	–	3561	7850	3962	8735	–	–	–	–	–	–	–	–
264	203	–	–	4.07	8.97	–	–	–	–	3.91	8.62	–	–	3304	7285	–	–	2.60	5.73	1.95	4.29	2.21	4.87
	209	–	–	4.18	9.22	–	–	–	–	4.02	8.86	–	–	3395	7484	–	–	2.67	5.88	2.00	4.41	2.27	5.00
265	23	18	22	0.46	1.01	354	780	0.39	0.87	0.41	0.90	331	729	–	–	0.32	0.71	0.24	0.52	0.16	0.35	0.23	0.50
	83	64	79	1.66	3.66	1283	2828	1.43	3.15	1.47	3.24	1198	2641	–	–	1.17	2.58	0.86	1.90	0.57	1.26	0.82	1.80
266	5	5	3	0.10	0.21	92	204	0.07	0.14	0.08	0.19	87	192	–	–	0.05	0.12	0.05	0.11	0.03	0.07	0.05	0.10
	86	77	58	1.60	3.53	1549	3416	1.08	2.39	1.41	3.12	1460	3218	–	–	0.89	1.96	0.80	1.76	0.52	1.14	0.77	1.70
267	16	–	–	0.31	0.68	–	–	–	–	0.26	0.57	–	–	–	–	–	–	0.15	0.33	0.09	0.19	0.16	0.34
	62	–	–	1.23	2.71	–	–	–	–	1.04	2.28	–	–	–	–	–	–	0.60	1.33	0.34	0.75	0.62	1.36
268	60	47	–	1.21	2.66	934	2058	–	–	1.03	2.27	783	1727	796	1754	–	–	0.61	1.35	0.37	0.81	0.61	1.35
	66	51	–	1.31	2.89	1015	2238	–	–	1.12	2.47	852	1877	865	1907	–	–	0.67	1.47	0.40	0.88	0.67	1.47
269	75	–	79	1.44	3.17	1453	3203	1.42	3.13	1.27	2.80	1420	3130	1420	3130	1.16	2.57	0.75	1.65	0.49	1.09	0.71	1.57
	84	–	89	1.62	3.56	1634	3603	1.60	3.52	1.43	3.15	1597	3521	1597	3521	1.31	2.89	0.84	1.86	0.56	1.22	0.80	1.77
270	8	7	6	0.16	0.34	146	322	0.11	0.24	0.14	0.31	138	304	–	–	0.09	0.20	0.09	0.19	0.06	0.13	0.08	0.18
	85	80	66	1.70	3.74	1594	3513	1.21	2.66	1.51	3.33	1503	3314	–	–	0.99	2.18	0.94	2.06	0.64	1.40	0.88	1.94
271	74	59	71	1.38	3.04	1182	2607	1.28	2.82	1.21	2.67	1106	2439	–	–	1.05	2.31	0.68	1.50	0.43	0.95	0.66	1.45
	83	66	80	1.55	3.41	1328	2927	1.44	3.17	1.36	2.99	1242	2739	–	–	1.18	2.60	0.76	1.68	0.49	1.07	0.74	1.63
272	57	–	–	–	–	–	–	–	–	–	–	–	–	–	–	–	–	–	–	–	–	–	–
	64	–	–	–	–	–	–	–	–	–	–	–	–	–	–	–	–	–	–	–	–	–	–

(Continued)

TABLE V-1A ENERGY FEEDS, MINERAL AND VITAMIN COMPOSITION OF FEEDS, DATA EXPRESSED **AS-FED** AND **MOISTURE-FREE**—*(Continued)*

Entry Number	Feed Name Description	Moisture Basis: A-F (as-fed) or M-F (moisture-free)	Dry Matter	Macro Minerals							Micro Minerals						
				Calcium (Ca)	Phosphorus (P)	Sodium (Na)	Chlorine (Cl)	Magnesium (Mg)	Potassium (K)	Sulfur (S)	Cobalt (Co)	Copper (Cu)	Iodine (I)	Iron (Fe)	Manganese (Mn)	Selenium (Se)	Zinc (Zn)
			%	%	%	%	%	%	%	%	ppm or mg/kg	ppm or mg/kg	ppm or mg/kg	%	ppm or mg/kg	ppm or mg/kg	ppm or mg/kg
	SORGHUM (Continued)																
247	SAGRAIN, GRAIN	A-F	90	0.02	0.27	–	–	–	–	–	–	–	–	–	–	–	–
		M-F	100	0.02	0.30	–	–	–	–	–	–	–	–	–	–	–	–
248	SCHROCK, GRAIN	A-F	89	0.02	0.30	–	–	–	–	–	–	–	–	–	–	–	–
		M-F	100	0.02	0.34	–	–	–	–	–	–	–	–	–	–	–	–
249	SHALLU, GRAIN	A-F	90	–	–	–	–	–	–	–	–	–	–	–	–	–	–
		M-F	100	–	–	–	–	–	–	–	–	–	–	–	–	–	–
250	SUDANGRASS, GRAIN	A-F	92	–	–	–	–	–	–	–	–	–	–	–	–	–	–
		M-F	100	–	–	–	–	–	–	–	–	–	–	–	–	–	–
	SOYBEAN *Glycine max*																
251	SEEDS	A-F	92	0.25	0.60	0.00	0.03	0.27	1.66	0.22	–	18.2	–	0.009	36.4	0.111	56.9
		M-F	100	0.27	0.65	0.00	0.03	0.29	1.80	0.24	–	19.8	–	0.010	39.6	0.120	61.8
252	MILL FEED	A-F	90	0.47	0.18	–	–	0.32	1.51	–	–	–	–	–	28.5	–	–
		M-F	100	0.52	0.20	–	–	0.36	1.69	–	–	–	–	–	31.8	–	–
253	MILL RUN	A-F	89	0.49	0.18	–	–	–	–	0.06	–	–	–	–	–	–	–
		M-F	100	0.54	0.20	–	–	–	–	0.07	–	–	–	–	–	–	–
254	OIL	A-F	100	–	–	–	–	–	–	–	–	–	–	–	–	–	–
		M-F	100	–	–	–	–	–	–	–	–	–	–	–	–	–	–
	SPELT *Triticum spelta*																
255	GRAIN	A-F	90	0.12	0.38	–	–	–	–	–	–	–	–	–	–	–	–
		M-F	100	0.13	0.42	–	–	–	–	–	–	–	–	–	–	–	–
	SUDANGRASS SORGHUM *Sorghum bicolor, sudanense*																
256	GRAIN	A-F	92	–	–	–	–	–	–	–	–	–	–	–	–	–	–
		M-F	100	–	–	–	–	–	–	–	–	–	–	–	–	–	–
	SUGARCANE *Saccharum officinarum*																
257	MOLASSES, DEHY	A-F	94	1.04	0.42	0.19	–	0.44	3.40	0.43	1.145	74.9	–	0.024	54.1	–	31.2
		M-F	100	1.10	0.45	0.20	–	0.47	3.60	0.46	1.213	79.4	–	0.025	57.3	–	33.0
258	MOLASSES, MORE THAN 46% INVERT SUGAR, MORE THAN 79.5 DEGREES BRIX	A-F	74	0.74	0.08	0.16	2.26	0.31	2.98	0.35	1.180	48.9	1.564	0.020	43.7	–	15.6
		M-F	100	1.00	0.11	0.22	3.04	0.42	4.01	0.47	1.587	65.7	2.103	0.027	58.8	–	20.9
	SUNFLOWER *Helianthus* spp																
259	SEEDS	A-F	94	0.16	0.67	0.02	–	0.37	0.68	0.28	–	23.5	–	0.006	21.9	–	68.6
		M-F	100	0.17	0.71	0.02	–	0.39	0.72	0.29	–	25.0	–	0.006	23.3	–	73.0
	SWEET CORN—SEE CORN, SWEET																
	SWEET POTATO *Ipomoea batatas*																
260	TUBERS, DEHY, MEAL	A-F	89	0.12	0.15	–	–	–	–	–	–	–	–	–	–	–	–
		M-F	100	0.13	0.16	–	–	–	–	–	–	–	–	–	–	–	–
261	CANNERY RESIDUE, DEHY	A-F	90	–	–	–	–	–	–	–	–	–	–	–	–	–	–
		M-F	100	–	–	–	–	–	–	–	–	–	–	–	–	–	–
262	TUBERS, FRESH	A-F	33	0.03	0.05	0.02	0.02	0.05	0.35	0.04	–	1.4	–	0.002	3.7	–	–
		M-F	100	0.10	0.14	0.05	0.06	0.16	1.07	0.13	–	4.2	–	0.005	11.1	–	–
	SWINE *Sus scrofa*																
263	FAT	A-F	99	–	–	–	–	–	–	–	–	–	–	–	–	–	–
		M-F	100	–	–	–	–	–	–	–	–	–	–	–	–	–	–
	TALLOW																
264	ANIMAL	A-F	97	–	–	–	–	–	–	–	–	–	–	–	–	–	–
		M-F	100	–	–	–	–	–	–	–	–	–	–	–	–	–	–
	TARO (DASHEEN) *Colocasia esculenta*																
265	TUBERS, FRESH	A-F	28	0.02	0.07	–	–	–	–	–	–	–	–	–	–	–	–
		M-F	100	0.07	0.25	–	–	–	–	–	–	–	–	–	–	–	–
	TOMATO *Lycopersicon esculentum*																
266	FRUIT, FRESH	A-F	6	0.01	0.03	–	–	–	0.25	–	–	–	–	0.001	–	–	–
		M-F	100	0.17	0.49	–	–	–	4.21	–	–	–	–	0.016	–	–	–
267	POMACE, WET	A-F	25	–	0.16	–	–	–	–	–	–	–	–	–	–	–	–
		M-F	100	–	0.64	–	–	–	–	–	–	–	–	–	–	–	–
268	POMACE, DEHY	A-F	92	0.39	0.55	–	–	0.18	3.34	–	–	30.0	–	0.424	47.1		–
		M-F	100	0.43	0.60	–	–	0.20	3.63	–	–	32.6	–	0.460	51.2	–	–
	TRITICALE *Triticale hexaploide*																
269	GRAIN	A-F	89	0.04	0.30	0.01	–	0.23	0.51	–	0.078	8.3	–	0.005	42.5	–	31.2
		M-F	100	0.04	0.34	0.01	–	0.26	0.57	–	0.087	9.3	–	0.005	47.8	–	35.1
	TURNIP *Brassica rapa, rapa*																
270	ROOTS, FRESH	A-F	9	0.06	0.03	0.01	0.06	0.02	0.26	0.04	–	2.0	–	0.002	3.9	–	2.7
		M-F	100	0.64	0.32	0.10	0.65	0.20	2.82	0.43	–	21.3	–	0.012	42.7	–	29.4
	VELVETBEAN *Mucuna* spp																
271	PODS WITH SEEDS, SUN-CURED	A-F	89	0.24	0.38	0.14	0.22	0.21	1.19	0.15	–	–	–	0.013	–	–	–
		M-F	100	0.27	0.42	0.16	0.24	0.23	1.33	0.17	–	–	–	0.015	–	–	–
	VERBENA, EUROPEAN (PIGEON-GRASS) *Verbena officinalis*																
272	SEEDS	A-F	89	–	–	–	–	–	–	–	–	–	–	–	–	–	–
		M-F	100	–	–	–	–	–	–	–	–	–	–	–	–	–	–

Entry Number	Fat-Soluble Vitamins: A (1 mg Carotene = 1667 IU Vit A)	Carotene (Provitamin A)	D	E	K	Water-Soluble Vitamins: B-12	Biotin	Choline	Folacin (Folic Acid)	Niacin	Pantothenic Acid (B-3)	(Pyridoxine) B-6	Riboflavin (B-2)	Thiamin (B-1)
	IU/g	ppm or mg/kg	IU/kg	ppm or mg/kg	ppm or mg/kg	ppb or mcg/kg	ppm or mg/kg	ppm or mg/kg	ppm or mg/kg	ppm or mg/kg	ppm or mg/kg	ppm or mg/kg	ppm or mg/kg	ppm or mg/kg
247	–	–	–	–	–	–	–	–	–	–	–	–	–	–
	–	–	–	–	–	–	–	–	–	–	–	–	–	–
248	–	–	–	–	–	–	0.25	–	–	–	–	–	–	–
	–	–	–	–	–	–	0.29	–	–	–	–	–	–	–
249	–	–	–	–	–	–	–	–	–	–	–	–	–	–
	–	–	–	–	–	–	–	–	–	–	–	–	–	–
250	–	–	–	–	–	–	–	–	–	–	–	–	–	–
	–	–	–	–	–	–	–	–	–	–	–	–	–	–
251	1.5	0.9	–	33.7	–	–	0.38	2931	–	23	16.0	11.04	2.9	11.3
	1.6	1.0	–	36.6	–	–	0.42	3184	–	24	17.4	12.00	3.2	12.2
252	–	–	–	–	–	–	0.22	444	0.22	24	13.3	2.24	3.5	2.2
	–	–	–	–	–	–	0.25	495	0.25	27	14.8	2.51	3.9	2.5
253	–	–	–	–	–	–	–	–	–	24	13.1	–	3.5	–
	–	–	–	–	–	–	–	–	–	27	14.7	–	3.9	–
254	–	–	–	–	–	–	–	–	–	–	–	–	–	–
	–	–	–	–	–	–	–	–	–	–	–	–	–	–
255	–	–	–	–	–	–	–	–	–	48	–	–	–	–
	–	–	–	–	–	–	–	–	–	53	–	–	–	–
256	–	–	–	–	–	–	–	–	–	–	–	–	–	–
	–	–	–	–	–	–	–	–	–	–	–	–	–	–
257	–	–	–	5.2	–	–	–	–	–	–	–	–	–	–
	–	–	–	5.5	–	–	–	–	–	–	–	–	–	–
258	–	–	–	5.4	–	–	0.69	764	0.11	36	37.4	4.21	2.8	0.9
	–	–	–	7.3	–	–	0.92	1027	0.15	49	50.3	5.67	3.8	1.2
259	–	–	–	–	–	–	–	–	–	–	–	–	3.3	0.4
	–	–	–	–	–	–	–	–	–	–	–	–	3.5	0.5
260	117.3	70.4	–	–	–	–	–	–	–	–	–	–	–	–
	131.2	78.7	–	–	–	–	–	–	–	–	–	–	–	–
261	–	–	–	–	–	–	–	–	–	–	–	–	–	–
	–	–	–	–	–	–	–	–	–	–	–	–	–	–
262	236.8	142.1	–	–	–	–	–	–	–	7	–	–	0.7	1.1
	718.8	431.2	–	–	–	–	–	–	–	20	–	–	2.0	3.4
263	–	–	–	22.8	–	–	–	–	–	–	–	–	–	–
	–	–	–	23.0	–	–	–	–	–	–	–	–	–	–
264	–	–	–	–	–	–	–	–	–	–	–	–	–	–
	–	–	–	–	–	–	–	–	–	–	–	–	–	–
265	–	–	–	–	–	–	–	–	–	–	–	–	–	–
	–	–	–	–	–	–	–	–	–	–	–	–	–	–
266	–	–	–	–	–	–	–	–	–	6	–	–	0.4	0.6
	–	–	–	–	–	–	–	–	–	108	–	–	6.2	9.2
267	–	–	–	–	–	–	–	–	–	–	–	–	–	–
	–	–	–	–	–	–	–	–	–	–	–	–	–	–
268	–	–	–	–	–	–	–	–	–	–	–	–	6.1	11.3
	–	–	–	–	–	–	–	–	–	–	–	–	6.7	12.3
269	–	–	–	–	–	–	–	457	–	–	–	–	0.4	–
	–	–	–	–	–	–	–	514	–	–	–	–	0.5	–
270	–	–	–	–	–	–	–	92	0.26	7	1.7	–	0.6	0.7
	–	–	–	–	–	–	–	1009	2.84	72	19.0	–	6.5	7.1
271	–	–	–	–	–	–	–	–	–	–	–	–	–	–
	–	–	–	–	–	–	–	–	–	–	–	–	–	–
272	–	–	–	–	–	–	–	–	–	–	–	–	–	–
	–	–	–	–	–	–	–	–	–	–	–	–	–	–

(Continued)

TABLE V-1A ENERGY FEEDS, COMPOSITION OF FEEDS, DATA EXPRESSED **AS-FED** AND **MOISTURE-FREE**—*(Continued)*

Entry Number	Feed Name Description	International Feed Number	Moisture Basis: A-F (as-fed) or M-F (moisture-free)	Proximate Analysis: Dry Matter	Ash	Crude Fiber	Neutral Det. Fib. (NDF)	Acid Det. Fib. (ADF)	Lignin	Ether Extract (Fat)	N-Free Extract	Crude Protein	Digestible Protein: Ruminant	Swine	Horse
				%	%	%	%	%	%	%	%	%	%	%	%
	VETCH *Vicia spp*														
273	SEEDS	5-08-546	A-F	91	3.1	5.7	–	–	–	0.8	51.5	29.6	24.7	28.2	25.0
			M-F	100	3.4	6.3	–	–	–	0.9	56.8	32.6	27.3	31.1	27.6
	WHEAT *Tritricum aestivum*														
274	GRAIN, ALL ANALYSES[1]	4-05-211	A-F	89	1.8	2.6	–	7.1	–	1.8	69.7	13.1	10.5	11.1	9.5
			M-F	100	2.0	2.9	–	8.0	–	2.0	78.4	14.7	11.7	12.4	10.7
275	GRAIN, HARD RED SPRING	4-05-258	A-F	88	1.7	2.6	37.9	11.0	–	1.8	67.4	14.2	10.0	11.7	10.4
			M-F	100	1.9	2.9	43.3	12.6	–	2.1	76.9	16.2	11.4	13.3	11.8
276	GRAIN, HARD RED WINTER	4-05-268	A-F	89	1.7	2.6	24.8	3.9	0.9	1.6	69.8	12.8	8.8	9.9	9.3
			M-F	100	2.0	2.9	28.0	4.4	1.0	1.8	78.8	14.5	9.9	11.2	10.5
277	GRAIN, SOFT RED WINTER	4-05-294	A-F	88	1.9	2.3	–	–	–	1.6	71.2	11.4	8.6	8.7	8.2
			M-F	100	2.1	2.6	–	–	–	1.8	80.5	12.9	9.7	9.8	9.3
278	GRAIN, SOFT WHITE WINTER[1]	4-05-337	A-F	90	1.5	2.3	12.6	3.6	–	1.5	75.0	10.2	6.4	7.5	7.2
			M-F	100	1.6	2.6	14.0	4.0	–	1.7	82.9	11.3	7.0	8.3	8.0
279	GRAIN, SOFT WHITE WINTER, PACIFIC COAST	4-08-555	A-F	89	1.9	2.5	–	–	–	1.9	72.9	10.0	6.2	7.3	7.1
			M-F	100	2.1	2.8	–	–	–	2.2	81.7	11.2	7.0	8.2	7.9
280	BRAN	4-05-190	A-F	89	5.9	10.0	40.9	12.0	2.6	4.0	53.6	15.5	12.0	11.8	13.1
			M-F	100	6.7	11.2	45.9	13.5	3.0	4.5	60.2	17.5	13.5	13.3	14.7
281	DISTILLERS GRAINS, DEHY	5-05-193	A-F	93	3.0	11.8	–	–	–	6.7	40.2	31.6	26.5	30.2	27.1
			M-F	100	3.3	12.6	–	–	–	7.2	43.0	33.9	28.4	32.4	29.0
282	ENDOSPERM	4-05-197	A-F	88	1.2	0.3	–	–	–	1.1	74.2	11.1	7.3	8.4	8.0
			M-F	100	1.4	0.3	–	–	–	1.3	84.4	12.6	8.3	9.5	9.0
283	FLOUR, LESS THAN 1.5% FIBER	4-05-199	A-F	88	1.5	0.9	–	–	–	1.7	70.5	13.7	12.6	10.7	10.0
			M-F	100	1.7	1.0	–	–	–	1.9	79.9	15.5	14.3	12.1	11.3
284	GLUTEN	5-05-221	A-F	90	1.8	3.2	–	–	–	2.0	19.8	63.4	55.8	61.8	63.3
			M-F	100	2.0	3.6	–	–	–	2.2	21.9	70.3	61.8	68.5	70.1
285	GRAIN SCREENINGS	4-05-216	A-F	89	3.2	5.3	–	–	–	2.8	64.1	13.3	9.6	10.6	9.7
			M-F	100	3.6	6.0	–	–	–	3.2	72.3	15.0	10.8	12.0	10.9
286	MIDDLINGS, LESS THAN 9.5% FIBER[1]	4-05-205	A-F	89	4.7	7.7	32.9	8.9	–	4.3	55.7	16.4	12.4	13.9	12.1
			M-F	100	5.3	8.7	37.0	10.0	–	4.9	62.7	18.5	14.0	15.6	13.7
287	MILL RUN, LESS THAN 9.5% FIBER	4-05-206	A-F	90	5.1	8.2	–	9.9	–	4.1	57.4	15.1	11.4	12.0	11.0
			M-F	100	5.7	9.1	–	11.0	–	4.6	63.9	16.7	12.6	13.4	12.3
288	RED DOG, LESS THAN 4% FIBER	4-05-203	A-F	88	2.4	2.9	–	–	–	3.4	64.0	15.6	12.9	13.6	11.5
			M-F	100	2.7	3.3	–	–	–	3.8	72.5	17.6	14.6	15.4	13.0
289	SHORTS, LESS THAN 7% FIBER	4-05-201	A-F	88	4.4	6.4	–	–	–	4.6	56.5	16.5	12.9	13.2	12.2
			M-F	100	5.0	7.2	–	–	–	5.2	63.9	18.7	14.6	15.0	13.8
290	MIDDLINGS, PALM OIL ADDED	4-08-557	A-F	93	6.0	7.6	–	–	–	8.3	55.3	16.2	11.7	12.9	11.9
			M-F	100	6.4	8.1	–	–	–	8.9	59.2	17.3	12.5	13.8	12.8
	WHEAT, DURAM *Triticum duram*														
291	GRAIN	4-05-224	A-F	88	1.6	2.2	–	–	–	1.8	68.2	13.8	9.7	10.8	10.1
			M-F	100	1.8	2.6	–	–	–	2.0	77.8	15.7	11.0	12.3	11.5
	WHEAT, WHITE CLUB *Triticum compactum*														
292	GRAIN	4-05-318	A-F	88	1.6	–	–	–	–	1.7	–	8.6	5.0	6.1	6.0
			M-F	100	1.8	–	–	–	–	1.9	–	9.8	5.7	7.0	6.8
	WOOD														
293	MOLASSES	4-05-502	A-F	62	4.1	0.5	–	1.2	–	0.3	56.6	0.6	-1.3	-0.6	-0.1
			M-F	100	6.5	0.9	–	2.0	–	0.5	91.1	1.0	-2.2	-1.0	-0.1

[1]Neutral Detergent Fiber (NDF) and Acid Detergent Fiber (ADF) values taken from *Nutrient Requirements of Dairy Cattle,* 6th rev. ed., NRC, National Academy Press, 1988, pp. 90–110, Table 7–1.

TABLE V-1B PROTEIN FEEDS, COMPOSITION OF FEEDS, DATA EXPRESSED **AS-FED** AND **MOISTURE-FREE** (See footnote at end of table.)

Entry Number	Feed Name Description	International Feed Number	Moisture Basis: A-F (as-fed) or M-F (moisture-free)	Proximate Analysis: Dry Matter	Ash	Crude Fiber	Neutral Det. Fib. (NDF)	Acid Det. Fib. (ADF)	Lignin	Ether Extract (Fat)	N-Free Extract	Crude Protein	Digestable Protein: Ruminant	Swine	Horse
				%	%	%	%	%	%	%	%	%	%	%	%
	ACACIA, SWEET *Acacia farnesiana*														
1	SEEDS	5-09-110	A-F	87	–	–	–	–	–	–	–	47.9	41.4	46.4	46.0
			M-F	100	–	–	–	–	–	–	–	55.0	47.6	53.3	52.9
	ALFALFA (LUCERNE) *Medicago sativa*														
2	SEEDS	5-08-325	A-F	88	4.4	8.1	–	–	–	10.6	32.0	33.2	27.9	31.9	29.3
			M-F	100	5.0	9.2	–	–	–	12.0	36.2	37.6	31.6	36.1	33.2
	ALGAE, GREEN *Scenedesmus quadricauda*														
3	WHOLE, FAN AIR DRIED	5-07-749	A-F	93	14.3	6.1	–	–	–	5.3	20.7	46.7	33.6	45.2	44.2
			M-F	100	15.3	6.5	–	–	–	5.7	22.3	50.2	36.2	48.6	47.5
4	**AMMONIUM, POLYPHOSPHATE SOLUTION**	6-08-042	A-F	60	–	–	–	–	–	–	–	54.8	–	–	–
			M-F	100	–	–	–	–	–	–	–	92.0	–	–	–

ENERGY FEEDS

Entry Number	TDN Ruminant	TDN Swine	TDN Horse	Digestible Energy Ruminant		Digestible Energy Swine		Digestible Energy Horse		Metabolizable Energy Ruminant		Metabolizable Energy Swine		Poultry ME_n		Metabolizable Energy Horse		Net Energy Ruminant NE_m		Ruminant NE_g		Lactating Cows NE_{lc}	
	%	%	%	Mcal lb	Mcal kg	kcal lb	kcal kg	Mcal lb	Mcal kg	Mcal lb	Mcal kg	kcal lb	kcal kg	kcal lb	kcal kg	Mcal lb	Mcal kg	Mcal lb	Mcal kg	Mcal lb	Mcal kg	Mcal lb	Mcal kg
273	64	81	–	1.37	3.01	1616	3562	–	–	1.19	2.63	1415	3119	–	–	–	–	0.80	1.76	0.53	1.18	0.76	1.67
	71	89	–	1.51	3.32	1782	3928	–	–	1.32	2.90	1560	3439	–	–	–	–	0.88	1.94	0.59	1.30	0.84	1.84
274	77	79	76	1.54	3.40	1544	3404	1.36	2.99	1.28	2.82	1485	3274	1402	3092	1.11	2.45	0.88	1.93	0.60	1.33	0.81	1.79
	87	89	85	1.73	3.82	1735	3826	1.53	3.36	1.44	3.17	1669	3679	1576	3474	1.25	2.76	0.98	2.17	0.68	1.49	0.92	2.02
275	78	74	76	1.56	3.43	1405	3098	1.36	2.99	1.39	3.07	1339	2952	1225	2701	1.11	2.45	0.87	1.91	0.60	1.31	0.81	1.78
	89	84	86	1.78	3.91	1604	3536	1.55	3.42	1.59	3.50	1528	3370	1399	3084	1.27	2.80	0.99	2.18	0.68	1.50	0.92	2.04
276	78	–	76	1.57	3.45	1530	3373	1.36	3.00	1.40	3.09	1458	3214	1454	3205	1.12	2.46	0.88	1.94	0.61	1.34	0.82	1.81
	88	–	86	1.77	3.90	1728	3809	1.54	3.38	1.58	3.49	1646	3630	1642	3620	1.26	2.78	1.00	2.19	0.69	1.51	0.93	2.05
277	78	73	74	1.57	3.45	1464	3228	1.33	2.94	1.40	3.09	1383	3049	1403	3093	1.09	2.41	0.89	1.95	0.61	1.35	0.83	1.82
	89	83	84	1.77	3.91	1656	3652	1.51	3.32	1.59	3.50	1565	3450	1587	3499	1.24	2.73	1.00	2.21	0.69	1.52	0.94	2.06
278	80	76	76	1.60	3.54	1526	3364	1.36	2.99	1.44	3.16	1442	3180	1299	2864	1.11	2.46	0.91	2.00	0.62	1.38	0.84	1.86
	89	84	83	1.77	3.91	1687	3719	1.50	3.31	1.59	3.50	1595	3516	1436	3167	1.23	2.71	1.00	2.21	0.69	1.52	0.93	2.06
279	79	76	72	1.58	3.48	1595	3517	1.29	2.85	1.41	3.11	1510	3329	1446	3188	1.06	2.33	0.88	1.95	0.61	1.34	0.82	1.82
	88	85	80	1.77	3.89	1787	3939	1.45	3.19	1.58	3.48	1691	3729	1620	3572	1.19	2.61	0.99	2.18	0.68	1.50	0.92	2.04
280	63	57	44	1.26	2.78	1119	2466	0.84	1.84	1.09	2.40	1027	2265	556	1225	0.69	1.51	0.67	1.48	0.42	0.93	0.64	1.41
	70	64	50	1.42	3.12	1256	2768	0.94	2.07	1.22	2.70	1153	2543	624	1375	0.77	1.70	0.75	1.66	0.48	1.05	0.72	1.58
281	78	80	–	1.52	3.35	1600	3527	–	–	1.35	2.96	1398	3081	–	–	–	–	0.81	1.79	0.54	1.19	0.77	1.70
	84	86	–	1.63	3.59	1714	3778	–	–	1.44	3.18	1497	3301	–	–	–	–	0.87	1.92	0.58	1.28	0.83	1.82
282	–	75	–	–	–	1491	3288	–	–	–	–	1410	3109	–	–	–	–	–	–	–	–	–	–
	–	85	–	–	–	1697	3741	–	–	–	–	1605	3538	–	–	–	–	–	–	–	–	–	–
283	87	72	78	1.57	3.46	1635	3605	1.40	3.08	1.40	3.09	1550	3417	1354	2986	1.15	2.53	0.77	1.69	0.51	1.13	0.73	1.61
	99	82	89	1.78	3.92	1854	4087	1.59	3.50	1.59	3.51	1757	3873	1535	3385	1.30	2.87	0.87	1.92	0.58	1.28	0.83	1.82
284	81	92	–	1.61	3.55	1830	4035	–	–	1.44	3.18	1613	3556	–	–	–	–	0.90	1.99	0.62	1.37	0.84	1.85
	90	101	–	1.79	3.94	2027	4470	–	–	1.60	3.52	1787	3940	–	–	–	–	1.00	2.20	0.69	1.52	0.93	2.05
285	63	62	66	1.26	2.79	1137	2506	1.20	2.65	1.10	2.42	1061	2339	1273	2806	0.99	2.18	0.63	1.39	0.39	0.86	0.62	1.37
	71	70	75	1.43	3.14	1282	2827	1.36	2.99	1.24	2.73	1197	2638	1436	3165	1.11	2.45	0.71	1.57	0.44	0.97	0.70	1.54
286	74	68	59	1.39	3.07	1321	2912	1.08	2.39	1.16	2.57	1225	2702	940	2072	0.89	1.96	0.78	1.72	0.52	1.15	0.79	1.73
	83	77	66	1.57	3.45	1487	3279	1.22	2.69	1.31	2.89	1380	3042	1058	2333	1.00	2.20	0.88	1.94	0.59	1.30	0.88	1.95
287	71	72	58	1.46	3.21	1438	3170	1.07	2.35	1.14	2.52	1254	2765	803	1771	0.88	1.93	0.76	1.68	0.50	1.11	0.76	1.68
	79	80	65	1.62	3.58	1600	3527	1.19	2.62	1.27	2.81	1395	3076	894	1971	0.97	2.15	0.85	1.87	0.56	1.23	0.85	1.87
288	77	72	69	1.47	3.23	1429	3149	1.26	2.77	1.39	3.06	1305	2876	1164	2566	1.03	2.27	0.82	1.80	0.55	1.22	0.77	1.70
	87	82	79	1.66	3.67	1619	3570	1.42	3.14	1.57	3.47	1479	3260	1319	2908	1.17	2.57	0.92	2.04	0.63	1.38	0.87	1.92
289	76	71	59	1.47	3.24	1413	3116	1.08	2.39	1.25	2.75	1327	2926	1001	2206	0.89	1.96	0.85	1.88	0.58	1.28	0.79	1.75
	86	80	67	1.66	3.66	1598	3524	1.23	2.70	1.41	3.11	1501	3309	1132	2496	1.00	2.21	0.96	2.12	0.66	1.45	0.90	1.98
290	83	77	43	1.56	3.43	1534	3382	0.82	1.80	1.38	3.04	1449	3194	–	–	0.67	1.48	0.80	1.76	0.53	1.17	0.76	1.68
	88	82	46	1.67	3.67	1642	3621	0.88	1.93	1.48	3.26	1551	3420	–	–	0.72	1.58	0.86	1.89	0.57	1.25	0.81	1.80
291	74	70	76	1.51	3.33	1401	3089	1.36	3.00	1.35	2.97	1322	2914	1452	3200	1.12	2.46	0.83	1.83	0.57	1.25	0.78	1.72
	85	80	87	1.72	3.80	1599	3525	1.55	3.42	1.54	3.39	1508	3325	1657	3652	1.27	2.81	0.95	2.09	0.65	1.43	0.89	1.96
292	–	–	–	–	–	–	–	–	–	–	–	–	–	–	–	–	–	–	–	–	–	–	–
	–	–	–	–	–	–	–	–	–	–	–	–	–	–	–	–	–	–	–	–	–	–	–
293	52	51	–	1.03	2.27	1033	2278	–	–	0.91	2.02	976	2151	–	–	–	–	–	–	–	–	–	–
	83	82	–	1.66	3.66	1664	3669	–	–	1.47	3.25	1571	3464	–	–	–	–	–	–	–	–	–	–

PROTEIN FEEDS

Entry Number	TDN Ruminant	TDN Swine	TDN Horse	Digestible Energy Ruminant		Digestible Energy Swine		Digestible Energy Horse		Metabolizable Energy Ruminant		Metabolizable Energy Swine		Poultry ME_n		Metabolizable Energy Horse		Net Energy Ruminant NE_m		Ruminant NE_g		Lactating Cows NE_{lc}	
	%	%	%	Mcal lb	Mcal kg	kcal lb	kcal kg	Mcal lb	Mcal kg	Mcal lb	Mcal kg	kcal lb	kcal kg	kcal lb	kcal kg	Mcal lb	Mcal kg	Mcal lb	Mcal kg	Mcal lb	Mcal kg	Mcal lb	Mcal kg
1	–	–	–	–	–	–	–	–	–	–	–	–	–	–	–	–	–	–	–	–	–	–	–
	–	–	–	–	–	–	–	–	–	–	–	–	–	–	–	–	–	–	–	–	–	–	–
2	81	79	–	1.58	3.49	1577	3476	–	–	1.42	3.13	1381	3044	–	–	–	–	0.87	1.91	0.59	1.31	0.81	1.78
	92	89	–	1.79	3.96	1786	3936	–	–	1.61	3.55	1564	3447	–	–	–	–	0.98	2.16	0.67	1.48	0.92	2.02
3	52	47	–	1.11	2.44	938	2069	–	–	1.13	2.49	787	1734	–	–	–	–	0.83	1.83	0.56	1.23	0.79	1.73
	56	50	–	1.19	2.62	1009	2224	–	–	1.21	2.67	846	1865	–	–	–	–	0.89	1.97	0.60	1.32	0.84	1.86
4	–	–	–	–	–	–	–	–	–	–	–	–	–	–	–	–	–	–	–	–	–	–	–
	–	–	–	–	–	–	–	–	–	–	–	–	–	–	–	–	–	–	–	–	–	–	–

(Continued)

TABLE V-1A ENERGY FEEDS, MINERAL AND VITAMIN COMPOSITION OF FEEDS, DATA EXPRESSED **AS-FED** AND **MOISTURE-FREE**—*(Continued)*

Entry Number	Feed Name Description	Moisture Basis: A-F (as-fed) or M-F (moisture-free)	Dry Matter	Macro Minerals							Micro Minerals						
				Calcium (Ca)	Phosphorus (P)	Sodium (Na)	Chlorine (Cl)	Magnesium (Mg)	Potassium (K)	Sulfur (S)	Cobalt (Co)	Copper (Cu)	Iodine (I)	Iron (Fe)	Manganese (Mn)	Selenium (Se)	Zinc (Zn)
			%	%	%	%	%	%	%	%	ppm or mg/kg	ppm or mg/kg	ppm or mg/kg	%	ppm or mg/kg	ppm or mg/kg	ppm or mg/kg
	VETCH *Vicia spp*																
273	SEEDS	A-F	91	–	–	–	–	–	–	–	–	–	–	–	–	–	–
		M-F	100	–	–	–	–	–	–	–	–	–	–	–	–	–	–
	WHEAT *Triticum aestivum*																
274	GRAIN, ALL ANALYSES	A-F	89	0.05	0.35	0.06	0.08	0.14	0.41	0.18	0.442	5.8	0.090	0.006	41.5	0.256	31.4
		M-F	100	0.06	0.39	0.06	0.09	0.15	0.15	0.20	0.497	6.5	0.101	0.006	46.7	0.288	35.2
275	GRAIN, HARD RED SPRING	A-F	88	0.04	0.37	0.02	0.08	0.14	0.36	0.15	0.123	6.0	–	0.006	37.0	0.263	37.9
		M-F	100	0.05	0.42	0.02	0.09	0.16	0.41	0.17	0.140	6.8	–	0.007	42.2	0.300	43.3
276	GRAIN, HARD RED WINTER	A-F	89	0.04	0.37	0.02	0.05	0.12	0.43	0.14	0.145	5.1	–	0.004	30.4	0.289	35.2
		M-F	100	0.05	0.42	0.02	0.06	0.13	0.49	0.15	0.163	5.7	–	0.004	34.3	0.326	39.8
277	GRAIN, SOFT RED WINTER	A-F	88	0.05	0.36	0.01	0.07	0.10	0.41	0.11	0.103	7.0	–	0.003	33.4	0.042	42.1
		M-F	100	0.06	0.40	0.01	0.08	0.11	0.46	0.12	0.117	8.0	–	0.004	37.8	0.047	47.7
278	GRAIN, SOFT WHITE WINTER	A-F	90	–	0.40	0.02	–	–	–	0.12	0.136	7.1	–	0.004	36.2	0.046	27.1
		M-F	100	–	0.44	0.02	–	–	–	0.13	0.150	7.8	–	0.004	40.0	0.051	30.0
279	GRAIN, SOFT WHITE WINTER, PACIFIC COAST	A-F	89	0.09	0.31	0.01	–	0.13	0.45	0.16	–	–	–	0.006	–	–	–
		M-F	100	0.10	0.35	0.01	–	0.15	0.51	0.18	–	–	–	0.006	–	–	–
280	BRAN	A-F	89	0.13	1.16	0.06	0.05	0.58	1.23	0.22	0.075	11.0	0.066	0.015	114.9	0.641	94.6
		M-F	100	0.14	1.30	0.06	0.06	0.65	1.38	0.25	0.084	12.4	0.074	0.017	129.0	0.719	106.2
281	DISTILLERS GRAINS, DEHY	A-F	93	0.11	0.58	–	–	–	–	–	–	–	–	–	15.0	–	–
		M-F	100	0.12	0.63	–	–	–	–	–	–	–	–	–	16.1	–	–
282	ENDOSPERM	A-F	88	–	–	–	–	–	–	–	–	–	–	–	–	–	–
		M-F	100	–	–	–	–	–	–	–	–	–	–	–	–	–	–
283	FLOUR, LESS THAN 1.5% FIBER	A-F	88	0.04	0.29	–	–	–	0.05	–	–	–	–	–	–	–	–
		M-F	100	0.04	0.33	–	–	–	0.06	–	–	–	–	–	–	–	–
284	GLUTEN	A-F	90	0.06	0.23	0.06	–	0.04	0.02	0.95	0.049	11.6	0.058	0.006	18.1	3.753	38.5
		M-F	100	0.07	0.25	0.07	–	0.04	0.02	1.06	0.054	12.8	0.064	0.007	20.1	4.158	42.6
285	GRAIN SCREENINGS	A-F	89	0.13	0.34	0.05	–	0.21	0.81	0.20	1.252	2.3	–	0.012	28.9	0.603	38.9
		M-F	100	0.14	0.38	0.05	–	0.24	0.91	0.22	1.412	2.6	–	0.014	32.5	0.681	43.9
286	MIDDLINGS, LESS THAN 9.5% FIBER	A-F	89	0.13	0.89	0.01	0.04	0.34	0.98	0.17	0.502	15.9	0.109	0.009	114.0	0.736	96.9
		M-F	100	0.15	1.00	0.01	0.04	0.38	1.10	0.19	0.565	17.9	0.123	0.011	128.3	0.828	109.1
287	MILL RUN, LESS THAN 9.5% FIBER	A-F	90	0.10	1.02	–	–	0.48	1.20	0.30	0.209	18.5	–	0.010	104.1	–	–
		M-F	100	0.11	1.13	–	–	0.53	1.34	0.34	0.232	20.6	–	0.011	115.8	–	–
288	RED DOG, LESS THAN 4% FIBER	A-F	88	0.06	0.51	0.01	0.14	0.18	0.52	0.24	0.117	6.3	–	0.005	52.1	0.324	65.0
		M-F	100	0.07	0.58	0.02	0.16	0.21	0.59	0.27	0.133	7.1	–	0.006	59.1	0.367	73.7
289	SHORTS, LESS THAN 7% FIBER	A-F	88	0.09	0.80	0.03	0.05	0.27	0.93	0.21	0.105	11.5	–	0.008	114.1	0.476	102.4
		M-F	100	0.10	0.91	0.03	0.06	0.31	1.05	0.23	0.119	13.0	–	0.009	129.1	0.538	115.9
290	MIDDLINGS, PALM OIL ADDED	A-F	93	–	–	–	–	–	–	–	–	–	–	–	–	–	–
		M-F	100	–	–	–	–	–	–	–	–	–	–	–	–	–	–
	WHEAT, DURAM *Triticum durum*																
291	GRAIN	A-F	88	0.10	0.36	–	–	0.16	0.44	–	0.079	6.8	–	0.005	30.7	–	19.3
		M-F	100	0.11	0.41	–	–	0.18	0.50	–	0.090	7.8	–	0.005	35.0	–	22.0
	WHEAT, WHITE CLUB *Triticum compactum*																
292	GRAIN	A-F	88	–	–	–	–	–	–	–	–	–	–	–	–	–	–
		M-F	100	–	–	–	–	–	–	–	–	–	–	–	–	–	–
	WOOD																
293	MOLASSES	A-F	62	1.17	0.05	0.03	0.12	0.07	0.04	0.03	–	–	–	–	12.6	–	–
		M-F	100	1.88	0.07	0.05	0.20	0.11	0.06	0.05	–	–	–	–	20.3	–	–

TABLE V-1B PROTEIN FEEDS, MINERAL AND VITAMIN COMPOSITION OF FEEDS, DATA EXPRESSED **AS-FED** AND **MOISTURE-FREE**

Entry Number	Feed Name Description	Moisture Basis: A-F (as-fed) or M-F (moisture-free)	Dry Matter	Macro Minerals							Micro Minerals						
				Calcium (Ca)	Phosphorus (P)	Sodium (Na)	Chlorine (Cl)	Magnesium (Mg)	Potassium (K)	Sulfur (S)	Cobalt (Co)	Copper (Cu)	Iodine (I)	Iron (Fe)	Manganese (Mn)	Selenium (Se)	Zinc (Zn)
			%	%	%	%	%	%	%	%	ppm or mg/kg	ppm or mg/kg	ppm or mg/kg	%	ppm or mg/kg	ppm or mg/kg	ppm or mg/kg
	ACACIA, SWEET *Acacia farnesiana*																
1	SEEDS	A-F	87	–	–	–	–	–	–	–	–	–	–	–	–	–	–
		M-F	100	–	–	–	–	–	–	–	–	–	–	–	–	–	–
	ALFALFA (LUCERNE) *Medicago sativa*																
2	SEEDS	A-F	88	–	–	–	–	–	–	–	–	–	–	–	–	–	–
		M-F	100	–	–	–	–	–	–	–	–	–	–	–	–	–	–
	ALGAE, GREEN *Scenedesmus quadricauda*																
3	WHOLE, FAN AIR DRIED	A-F	93	1.76	2.05	0.21	–	1.49	1.36	–	–	46.5	–	0.214	139.5	–	83.8
		M-F	100	1.89	2.20	0.23	–	1.60	1.46	–	–	50.0	–	0.230	150.0	–	90.1
4	**AMMONIUM, POLYPHOSPHATE SOLUTION**	A-F	60	0.10	13.44	–	–	–	–	0.50	–	–	–	0.505	–	–	–
		M-F	100	0.17	22.58	–	–	–	–	0.85	–	–	–	0.848	–	–	–

ENERGY FEEDS

Entry Number	Fat-Soluble Vitamins: A (1 mg Carotene = 1667 IU Vit A)	Carotene (Provitamin A)	D	E	K	Water-Soluble Vitamins: B-12	Biotin	Choline	Folacin (Folic Acid)	Niacin	Pantothenic Acid (B-3)	(Pyridoxine) B-6	Riboflavin (B-2)	Thiamin (B-1)
	IU/g	ppm or mg/kg	IU/kg	ppm or mg/kg	ppm or mg/kg	ppb or mcg/kg	ppm or mg/kg	ppm or mg/kg	ppm or mg/kg	ppm or mg/kg	ppm or mg/kg	ppm or mg/kg	ppm or mg/kg	ppm or mg/kg
273	–	–	–	–	–	–	–	–	–	–	–	–	–	–
	–	–	–	–	–	–	–	–	–	–	–	–	–	–
274	–	–	–	**15.5**	–	**0.9**	**0.10**	**918**	**0.43**	**59**	**11.3**	**3.74**	**1.3**	**4.3**
	–	–	–	17.4	–	1.0	0.11	1032	0.49	66	12.7	4.20	1.4	4.8
275	–	–	–	**12.7**	–	–	**0.11**	**1010**	**0.41**	**56**	**9.6**	**5.11**	**1.4**	**4.2**
	–	–	–	14.4	–	–	0.13	1153	0.46	64	11.0	5.83	1.6	4.8
276	–	–	–	**11.1**	–	–	**0.11**	**1004**	**0.38**	**53**	**10.1**	**3.01**	**1.3**	**4.5**
	–	–	–	12.5	–	–	0.12	1133	0.43	60	11.4	3.40	1.5	5.1
277	–	–	–	**15.6**	–	–	–	**892**	**0.41**	**53**	**10.1**	**3.21**	**1.5**	**4.7**
	–	–	–	17.7	–	–	–	1009	0.46	60	11.4	3.63	1.7	5.3
278	–	–	–	**30.9**	–	–	–	–	–	**62**	**11.2**	**4.79**	–	–
	–	–	–	34.2	–	–	–	–	–	69	12.3	5.29	–	–
279	–	–	–	–	–	–	–	**973**	–	**46**	**11.1**	–	**1.1**	–
	–	–	–	–	–	–	–	1090	–	52	12.4	–	1.2	–
280	**4.4**	**2.6**	–	**14.3**	–	–	**0.38**	**1232**	**1.77**	**197**	**28.0**	**10.34**	**3.6**	**8.4**
	4.9	2.9	–	16.0	–	–	0.42	1383	1.98	221	31.4	11.61	4.0	9.4
281	**1.8**	**1.1**	–	–	–	–	–	–	–	**56**	**8.2**	–	**3.7**	**2.0**
	2.0	1.2	–	–	–	–	–	–	–	60	8.7	–	4.0	2.1
282	–	–	–	–	–	–	–	–	–	–	–	–	–	–
	–	–	–	–	–	–	–	–	–	–	–	–	–	–
283	–	–	–	–	–	–	–	**972**	–	**26**	**13.0**	–	**0.8**	**5.8**
	–	–	–	–	–	–	–	1102	–	29	14.7	–	0.9	6.6
284	–	–	**0**	**34.1**	–	**73.1**	**0.00**	**577**	**0.74**	**74**	**5.8**	**2.26**	**0.7**	**0.9**
	–	–	0	37.7	–	81.0	0.00	640	0.82	82	6.4	2.51	0.7	1.0
285	–	–	–	–	–	–	–	**869**	**0.43**	**58**	**11.3**	–	**0.9**	**6.3**
	–	–	–	–	–	–	–	980	0.49	65	12.7	–	1.0	7.2
286	**5.1**	**3.1**	–	**23.8**	–	–	**0.24**	**1246**	**1.24**	**95**	**17.8**	**9.14**	**2.0**	**14.2**
	5.8	3.5	–	26.9	–	–	0.27	1403	1.39	107	20.0	10.29	2.3	15.9
287	–	–	–	**31.9**	–	–	**0.31**	**1005**	**1.08**	**116**	**13.7**	**11.09**	**2.1**	**15.2**
	–	–	–	35.5	–	–	0.34	1118	1.20	129	15.2	12.33	2.4	17.0
288	–	–	–	**37.4**	–	–	**0.11**	**1453**	**0.82**	**46**	**13.3**	**5.40**	**2.2**	**21.8**
	–	–	–	42.4	–	–	0.12	1648	0.93	52	15.0	6.12	2.5	24.7
289	**5.1**	**3.1**	–	**36.0**	–	–	–	**1697**	**1.51**	**105**	**21.9**	–	**4.1**	**19.5**
	5.8	3.5	–	40.7	–	–	–	1920	1.71	119	24.8	–	4.6	22.1
290	–	–	–	–	–	–	–	–	–	–	–	–	–	–
	–	–	–	–	–	–	–	–	–	–	–	–	–	–
291	–	–	–	–	–	–	–	–	**0.39**	**44**	–	–	**1.2**	**6.4**
	–	–	–	–	–	–	–	–	0.44	51	–	–	1.4	7.3
292	–	–	–	–	–	–	–	–	–	**55**	**11.1**	**4.46**	–	–
	–	–	–	–	–	–	–	–	–	62	12.6	5.07	–	–
293	–	–	–	–	–	–	–	–	–	–	–	–	–	–
	–	–	–	–	–	–	–	–	–	–	–	–	–	–

PROTEIN FEEDS

Entry Number	Fat-Soluble Vitamins: A (1 mg Carotene = 1667 IU Vit A)	Carotene (Provitamin A)	D	E	K	Water-Soluble Vitamins: B-12	Biotin	Choline	Folacin (Folic Acid)	Niacin	Pantothenic Acid (B-3)	(Pyridoxine) B-6	Riboflavin (B-2)	Thiamin (B-1)
	IU/g	ppm or mg/kg	IU/kg	ppm or mg/kg	ppm or mg/kg	ppm or mg/kg	ppm or mg/kg	ppm or mg/kg	ppm or mg/kg	ppm or mg/kg	ppm or mg/kg	ppm or mg/kg	ppm or mg/kg	ppm or mg/kg
1	–	–	–	–	–	–	–	–	–	–	–	–	–	–
	–	–	–	–	–	–	–	–	–	–	–	–	–	–
2	–	–	–	–	–	–	–	–	–	–	–	–	–	–
	–	–	–	–	–	–	–	–	–	–	–	–	–	–
3	**424.9**	**254.9**	–	–	–	–	–	–	–	–	–	–	–	–
	456.7	274.0	–	–	–	–	–	–	–	–	–	–	–	–
4	–	–	–	–	–	–	–	–	–	–	–	–	–	–
	–	–	–	–	–	–	–	–	–	–	–	–	–	–

(Continued)

TABLE V-1B PROTEIN FEEDS, COMPOSITION OF FEEDS, DATA EXPRESSED **AS-FED** AND **MOISTURE-FREE**—*(Continued)*

Entry Number	Feed Name Description	International Feed Number	Moisture Basis: A-F (as-fed) or M-F (moisture-free)	Proximate Analysis									Digestible Protein		
				Dry Matter	Ash	Crude Fiber	Neutral Det. Fib. (NDF)	Acid Det. Fib. (ADF)	Lignin	Ether Extract (Fat)	N-Free Extract	Crude Protein	Ruminant	Swine	Horse
				%	%	%	%	%	%	%	%	%	%	%	%
	ANIMAL														
5	BLOOD, SPRAY DEHY (BLOOD FLOUR)	5-00-381	A-F	93	4.1	1.0	–	–	–	1.2	0.2	86.0	82.6	84.2	88.6
			M-F	100	4.4	1.1	–	–	–	1.3	0.2	93.0	89.3	91.1	95.8
6	BLOOD, MEAL	5-00-380	A-F	91	5.3	1.0	–	–	–	1.3	3.1	80.5	57.2	60.1	82.5
			M-F	100	5.8	1.1	–	–	–	1.5	3.4	88.2	62.6	65.9	90.4
7	LIVER, MEAL	5-00-389	A-F	93	6.3	1.4	–	–	–	15.7	3.2	66.1	57.9	64.4	66.1
			M-F	100	6.8	1.5	–	–	–	17.0	3.5	71.4	62.5	69.6	71.4
8	MEAT, MEAL RENDERED	5-00-385	A-F	94	28.1	2.7	–	–	–	9.1	3.3	50.7	43.8	49.1	48.5
			M-F	100	29.9	2.9	–	–	–	9.7	3.5	54.0	46.7	52.3	51.7
9	MEAT WITH BONE, MEAL RENDERED	5-00-388	A-F	93	28.0	2.4	–	–	–	10.0	2.6	50.4	45.8	43.3	48.3
			M-F	100	30.0	2.6	–	–	–	10.7	2.8	54.0	49.1	46.4	51.7
10	TANKAGE, MEAL RENDERED	5-00-386	A-F	92	21.1	1.8	–	–	–	8.7	0.1	60.5	52.8	58.9	59.8
			M-F	100	22.9	2.0	–	–	–	9.4	0.1	65.6	57.3	63.9	64.9
11	TANKAGE WITH BONE, MEAL RENDERED	5-00-387	A-F	93	28.2	2.2	–	–	–	12.7	3.1	46.6	40.2	45.1	44.1
			M-F	100	30.4	2.4	–	–	–	13.7	3.3	50.2	43.2	48.6	47.4
	AVOCADO *Persea spp*														
12	SEEDS, MEAL SOLV EXTD	5-00-451	A-F	91	18.1	17.6	–	–	–	1.1	36.0	18.6	8.2	17.3	12.6
			M-F	100	19.8	19.3	–	–	–	1.2	39.4	20.4	9.0	19.0	13.7
	BABASSU *Orbignya spp*														
13	KERNELS WITH COATS, MEAL MECH EXTD (BABASSU OIL MEAL)	5-00-454	A-F	92	5.0	13.3	–	–	–	6.4	45.1	22.3	19.2	21.0	16.7
			M-F	100	5.4	14.5	–	–	–	7.0	49.0	24.2	20.8	22.8	18.1
14	KERNELS WITH COATS, MEAL SOLV EXTD (BABASSU OIL MEAL)	5-00-455	A-F	93	6.3	17.6	–	–	–	2.3	45.3	21.2	17.1	19.9	15.4
			M-F	100	6.8	19.0	–	–	–	2.5	48.9	22.9	18.4	21.5	16.6
	BARLEY *Hordeum vulgare*														
15	MALT SPROUTS, DEHY[1]	5-00-545	A-F	93	5.6	14.2	43.7	16.7	–	1.4	48.8	22.9	19.2	21.5	17.2
			M-F	100	6.1	15.3	47.0	18.0	–	1.5	52.6	24.6	20.7	23.2	18.6
	BEAN *Phaseolus vulgaris*														
16	SEEDS, KIDNEY	5-00-600	A-F	89	3.6	4.1	–	–	–	1.2	57.8	21.8	14.6	20.6	16.5
			M-F	100	4.0	4.6	–	–	–	1.4	65.3	24.7	16.5	23.2	18.6
17	SEEDS, NAVY	5-00-623	A-F	89	4.0	4.4	–	–	–	1.4	56.7	22.9	20.2	21.6	17.6
			M-F	100	4.5	5.0	–	–	–	1.5	63.4	25.6	22.5	24.2	19.7
18	SEEDS, PINTO	5-00-624	A-F	90	4.3	4.0	–	–	–	1.3	57.9	22.7	17.2	21.4	17.3
			M-F	100	4.8	4.5	–	–	–	1.4	64.2	25.1	19.1	23.7	19.1
	BEAN, LIMA *Phaseolus limensis*														
19	SEEDS	5-00-613	A-F	90	4.1	4.5	–	–	–	1.3	58.9	20.7	16.7	19.4	15.1
			M-F	100	4.6	5.1	–	–	–	1.5	65.7	23.1	18.6	21.7	16.8
	BEAN, MUNG *Phaseolus aureus*														
20	SEEDS	5-08-185	A-F	90	3.8	3.9	–	–	–	1.3	57.1	23.9	19.6	22.6	18.7
			M-F	100	4.2	4.3	–	–	–	1.4	63.5	26.6	21.8	25.2	20.8
	BEAN, TEPARY *Phaseolus acutifolius, latifolius*														
21	SEEDS	5-08-349	A-F	90	4.2	3.4	–	–	–	1.4	59.3	22.2	18.0	20.9	16.7
			M-F	100	4.6	3.8	–	–	–	1.5	65.5	24.5	19.9	23.1	18.4
	BEECHNUT *Fagus spp*														
22	MEAL MECH EXTD	4-00-630	A-F	87	7.1	23.4	–	–	–	4.3	35.2	17.0	11.9	13.7	12.6
			M-F	100	8.2	26.9	–	–	–	4.9	40.5	19.5	13.7	15.7	14.5
	BEET, SUGAR *Beta vulgaris, altissima*														
23	PULP, AMMONIATED, DEHY	5-26-073	A-F	90	11.0	14.7	–	–	–	0.5	47.4	16.9	13.7	–	–
			M-F	100	12.2	16.3	–	–	–	0.5	52.4	18.7	15.1	–	–
	BLOOD														
24	SPRAY DEHY (BLOOD FLOUR)	5-00-381	A-F	93	4.1	1.0	–	–	–	1.2	0.2	86.0	82.6	84.2	88.6
			M-F	100	4.4	1.1	–	–	–	1.3	0.2	93.0	89.3	91.1	95.8
25	MEAL	5-00-380	A-F	91	5.3	1.0	–	–	–	1.3	3.1	80.5	57.2	60.1	82.5
			M-F	100	5.8	1.1	–	–	–	1.5	3.4	88.2	62.6	65.9	90.4
	BREWERS														
26	GRAINS, DEHY	5-02-141	A-F	92	3.6	13.0	38.7	23.9	4.6	6.6	41.6	27.3	20.1	21.8	21.0
			M-F	100	4.0	14.1	42.0	26.0	5.0	7.1	45.2	29.6	21.8	23.7	22.8
27	GRAINS, WET	5-02-142	A-F	22	0.9	3.1	9.2	5.0	1.1	1.5	10.7	5.8	4.2	5.5	4.5
			M-F	100	3.9	14.1	42.0	23.0	5.0	6.8	48.7	26.4	19.3	25.0	20.6
	BUCKWHEAT, COMMON *Fagopyrum sagittatum*														
28	MIDDLINGS	5-00-991	A-F	89	4.6	7.4	–	–	–	7.2	41.3	28.2	24.9	26.8	23.6
			M-F	100	5.2	8.3	–	–	–	8.1	46.5	31.8	28.1	30.3	26.6
	BUTTERMILK, CATTLE *Bos taurus*														
29	CONDENSED	5-01-159	A-F	29	3.6	0.1	–	–	–	2.4	12.4	10.8	9.2	10.0	9.5
			M-F	100	12.2	0.3	–	–	–	8.3	42.3	36.9	31.1	33.9	32.4
30	DEHY	5-01-160	A-F	92	9.1	0.3	–	–	–	5.2	45.9	31.7	28.7	29.5	27.3
			M-F	100	9.9	0.4	–	–	–	5.6	49.7	34.4	31.1	32.0	29.5
	CACAO (COCOA) *Theobroma cacao*														
31	SEEDS	5-01-569	A-F	95	5.8	5.3	–	–	–	16.4	43.2	24.6	9.1	23.2	18.9
			M-F	100	6.1	5.6	–	–	–	17.3	45.3	25.8	9.5	24.4	19.9

PROTEIN FEEDS

Entry Number	TDN Ruminant	TDN Swine	TDN Horse	Digestible Energy Ruminant		Digestible Energy Swine		Digestible Energy Horse		Metabolizable Energy Ruminant		Metabolizable Energy Swine		Metabolizable Energy Poultry ME_n		Metabolizable Energy Horse		Net Energy Ruminant NE_m		Net Energy Ruminant NE_g		Net Energy Lactating Cows NE_{lc}	
	%	%	%	Mcal		kcal		Mcal		Mcal		kcal		kcal		Mcal		Mcal		Mcal		Mcal	
				lb	kg	lb	kg	lb	kg	lb	kg	lb	kg	lb	kg	lb	kg	lb	kg	lb	kg	lb	kg
5	84	–	–	1.69	3.72	1224	2698	–	–	1.52	3.35	880	1940	1255	2767	–	–	0.96	2.12	0.67	1.47	0.89	1.97
	91	–	–	1.83	4.03	1323	2916	–	–	1.64	3.62	951	2097	1357	2991	–	–	1.04	2.29	0.72	1.59	0.96	2.13
6	61	61	–	1.20	2.65	1220	2690	–	–	0.99	2.19	1012	2231	1282	2826	–	–	0.63	1.38	0.38	0.84	0.60	1.33
	66	67	–	1.32	2.91	1336	2946	–	–	1.09	2.40	1109	2444	1404	3096	–	–	0.69	1.51	0.42	0.92	0.66	1.45
7	89	93	–	1.79	3.94	1867	4116	–	–	1.62	3.57	1645	3627	1306	2878	–	–	1.00	2.21	0.70	1.55	0.93	2.04
	97	101	–	1.93	4.26	2017	4446	–	–	1.75	3.85	1777	3918	1410	3109	–	–	1.08	2.38	0.76	1.67	1.00	2.21
8	67	64	–	1.20	2.64	936	2064	–	–	0.89	1.95	1009	2225	947	2088	–	–	0.52	1.15	0.28	0.62	0.69	1.53
	71	68	–	1.28	2.81	998	2200	–	–	0.94	2.08	1075	2371	1009	2225	–	–	0.56	1.23	0.30	0.66	0.74	1.63
9	66	68	–	1.32	2.91	1028	2267	–	–	1.14	2.51	981	2162	946	2086	–	–	0.71	1.57	0.45	1.00	0.69	1.52
	71	73	–	1.41	3.11	1102	2430	–	–	1.22	2.70	1051	2317	1014	2236	–	–	0.76	1.68	0.49	1.07	0.74	1.63
10	67	67	–	1.34	2.96	1112	2451	–	–	1.17	2.58	951	2096	1212	2673	–	–	0.74	1.62	0.48	1.05	0.71	1.56
	73	73	–	1.46	3.21	1207	2660	–	–	1.27	2.80	1032	2275	1316	2901	–	–	0.80	1.76	0.52	1.14	0.77	1.69
11	63	68	–	1.44	3.17	1382	3047	–	–	1.26	2.79	1199	2644	1188	2618	–	–	0.90	1.99	0.62	1.36	0.85	1.86
	68	73	–	1.55	3.42	1488	3280	–	–	1.36	3.00	1291	2846	1278	2818	–	–	0.97	2.14	0.67	1.47	0.91	2.00
12	52	8	–	1.04	2.30	162	358	–	–	0.87	1.91	71	156	–	–	–	–	0.50	1.11	0.27	0.59	0.53	1.17
	57	9	–	1.14	2.52	178	391	–	–	0.95	2.09	78	171	–	–	–	–	0.55	1.21	0.29	0.65	0.58	1.28
13	80	69	–	1.49	3.28	1372	3024	–	–	1.32	2.90	1188	2619	–	–	–	–	0.75	1.65	0.49	1.07	0.72	1.58
	87	74	–	1.62	3.56	1489	3282	–	–	1.43	3.15	1289	2843	–	–	–	–	0.81	1.79	0.53	1.16	0.78	1.72
14	63	54	–	1.26	2.78	1077	2374	–	–	1.08	2.39	915	2017	–	–	–	–	0.62	1.36	0.37	0.81	0.62	1.36
	68	58	–	1.36	3.00	1162	2561	–	–	1.17	2.58	987	2176	–	–	–	–	0.66	1.46	0.40	0.87	0.66	1.46
15	66	61	–	1.31	2.89	1219	2688	–	–	1.13	2.50	1046	2307	722	1592	–	–	0.69	1.52	0.43	0.95	0.67	1.48
	71	66	–	1.41	3.11	1313	2894	–	–	1.22	2.69	1126	2483	778	1714	–	–	0.74	1.63	0.47	1.03	0.72	1.60
16	73	79	–	1.46	3.22	1584	3493	–	–	1.29	2.85	1388	3059	–	–	–	–	–	–	–	–	–	–
	82	89	–	1.65	3.64	1789	3945	–	–	1.46	3.22	1567	3455	–	–	–	–	–	–	–	–	–	–
17	76	78	–	1.53	3.37	1737	3830	–	–	1.36	3.00	1531	3376	1052	2320	–	–	0.83	1.83	0.56	1.24	0.78	1.72
	85	87	–	1.71	3.77	1941	4280	–	–	1.52	3.36	1711	3772	1176	2593	–	–	0.93	2.05	0.63	1.39	0.87	1.93
18	74	78	–	1.47	3.25	1569	3460	–	–	1.30	2.87	1372	3025	–	–	–	–	0.81	1.79	0.54	1.20	0.77	1.69
	82	87	–	1.63	3.60	1740	3835	–	–	1.44	3.18	1521	3354	–	–	–	–	0.90	1.98	0.60	1.33	0.85	1.87
19	78	78	–	1.55	3.42	1556	3430	–	–	1.39	3.05	1360	2999	–	–	–	–	–	–	–	–	–	–
	87	87	–	1.73	3.82	1737	3830	–	–	1.55	3.41	1519	3348	–	–	–	–	–	–	–	–	–	–
20	79	81	–	1.53	3.37	1612	3554	–	–	1.36	3.00	1412	3113	–	–	–	–	0.82	1.80	0.55	1.21	0.77	1.70
	88	90	–	1.70	3.75	1792	3951	–	–	1.51	3.33	1570	3460	–	–	–	–	0.91	2.00	0.61	1.35	0.86	1.89
21	69	80	–	1.37	3.02	1606	3542	–	–	1.20	2.65	1406	3100	–	–	–	–	–	–	–	–	–	–
	76	89	–	1.52	3.34	1775	3914	–	–	1.33	2.93	1554	3426	–	–	–	–	–	–	–	–	–	–
22	37	–	–	0.74	1.63	–	–	–	–	0.57	1.25	–	–	–	–	–	–	–	–	–	–	–	–
	42	–	–	0.85	1.87	–	–	–	–	0.65	1.44	–	–	–	–	–	–	–	–	–	–	–	–
23	70	–	–	–	–	–	–	–	–	–	–	–	–	–	–	–	–	–	–	–	–	–	–
	77	–	–	–	–	–	–	–	–	–	–	–	–	–	–	–	–	–	–	–	–	–	–
24	84	–	–	1.69	3.72	1224	2698	–	–	1.52	3.35	880	1940	1255	2767	–	–	0.96	2.12	0.67	1.47	0.89	1.97
	91	–	–	1.83	4.03	1323	2916	–	–	1.64	3.62	951	2097	1357	2991	–	–	1.04	2.29	0.72	1.59	0.96	2.13
25	61	61	–	1.20	2.65	1220	2690	–	–	0.99	2.19	1012	2231	1282	2826	–	–	0.63	1.38	0.38	0.84	0.60	1.33
	66	67	–	1.32	2.91	1336	2946	–	–	1.09	2.40	1109	2444	1404	3096	–	–	0.69	1.51	0.42	0.92	0.66	1.45
26	65	66	48	1.25	2.76	1045	2303	–	–	1.01	2.22	1038	2288	1047	2308	–	–	0.64	1.41	0.39	0.86	0.67	1.48
	71	71	52	1.36	2.99	1134	2500	–	–	1.09	2.41	1127	2484	1137	2506	–	–	0.69	1.53	0.42	0.93	0.73	1.61
27	15	18	–	0.32	0.70	351	774	–	–	0.27	0.60	305	673	–	–	–	–	0.18	0.40	0.12	0.26	0.17	0.38
	68	80	–	1.44	3.17	1602	3533	–	–	1.25	2.76	1394	3074	–	–	–	–	0.83	1.82	0.54	1.19	0.79	1.74
28	74	76	–	1.48	3.26	1524	3360	–	–	1.31	2.89	1332	2936	–	–	–	–	0.82	1.82	0.56	1.23	0.78	1.71
	83	86	–	1.67	3.68	1719	3789	–	–	1.48	3.26	1502	3311	–	–	–	–	0.93	2.05	0.63	1.39	0.87	1.93
29	26	22	–	0.52	1.14	442	974	–	–	0.46	1.02	383	844	–	–	–	–	0.29	0.63	0.20	0.43	0.27	0.59
	88	76	–	1.77	3.89	1503	3314	–	–	1.58	3.48	1303	2872	–	–	–	–	0.98	2.15	0.67	1.48	0.91	2.01
30	82	77	–	1.56	3.44	1578	3479	–	–	1.29	2.84	1381	3045	1248	2752	–	–	0.88	1.94	0.60	1.32	0.85	1.87
	89	84	–	1.69	3.73	1710	3770	–	–	1.40	3.08	1497	3300	1353	2982	–	–	0.95	2.10	0.65	1.43	0.92	2.03
31	59	91	–	1.19	2.61	1812	3996	–	–	1.00	2.21	1593	3511	–	–	–	–	0.58	1.29	0.33	0.74	0.60	1.31
	62	95	–	1.24	2.74	1903	4196	–	–	1.05	2.32	1672	3687	–	–	–	–	0.61	1.35	0.35	0.77	0.63	1.38

(Continued)

TABLE V-1B PROTEIN FEEDS, MINERAL AND VITAMIN COMPOSITION OF FEEDS, DATA EXPRESSED **AS-FED** AND **MOISTURE-FREE**—*(Continued)*

Entry Number	Feed Name Description	Moisture Basis: A-F (as-fed) or M-F (moisture-free)	Dry Matter	Macro Minerals							Micro Minerals						
				Calcium (Ca)	Phosphorus (P)	Sodium (Na)	Chlorine (Cl)	Magnesium (Mg)	Potassium (K)	Sulfur (S)	Cobalt (Co)	Copper (Cu)	Iodine (I)	Iron (Fe)	Manganese (Mn)	Selenium (Se)	Zinc (Zn)
			%	%	%	%	%	%	%	%	ppm or mg/kg	ppm or mg/kg	ppm or mg/kg	%	ppm or mg/kg	ppm or mg/kg	ppm or mg/kg
	ANIMAL																
5	BLOOD, SPRAY DEHY (BLOOD FLOUR)	A-F	93	0.41	0.30	0.38	0.25	0.15	0.15	0.60	–	8.2	–	0.277	6.4	–	–
		M-F	100	0.44	0.33	0.42	0.27	0.17	0.16	0.65	–	8.8	–	0.300	6.9	–	–
6	BLOOD, MEAL	A-F	91	0.29	0.25	0.32	0.30	0.22	0.09	0.34	0.088	12.6	–	0.372	5.3	0.731	4.4
		M-F	100	0.32	0.28	0.35	0.33	0.24	0.10	0.37	0.097	13.8	–	0.407	5.8	0.801	4.8
7	LIVER, MEAL	A-F	93	0.56	1.26	–	–	0.10	–	–	0.135	89.4	–	0.064	8.8	–	61.8
		M-F	100	0.61	1.36	–	–	0.11	–	–	0.146	96.5	–	0.069	9.5	–	66.8
8	MEAT, MEAL RENDERED	A-F	94	8.61	4.58	1.05	1.11	0.25	0.55	0.46	2.250	9.6	–	0.050	11.8	0.505	74.3
		M-F	100	9.18	4.88	1.11	1.18	0.27	0.58	0.49	2.398	10.2	–	0.053	12.6	0.538	79.2
9	MEAT WITH BONE, MEAL RENDERED	A-F	93	10.00	4.94	0.72	0.75	1.02	1.33	0.25	0.181	1.5	1.317	0.066	13.3	0.263	94.3
		M-F	100	10.72	5.30	0.77	0.80	1.09	1.43	0.27	0.194	1.6	1.412	0.070	14.3	0.282	101.1
10	TANKAGE, MEAL RENDERED	A-F	92	5.87	3.09	1.67	1.73	0.36	0.55	0.70	0.153	38.8	–	0.211	19.2	–	–
		M-F	100	6.37	3.36	1.81	1.88	0.39	0.60	0.76	0.166	42.1	–	0.229	20.8	–	–
11	TANKAGE WITH BONE, MEAL RENDERED	A-F	93	11.16	5.41	–	–	–	–	0.26	–	–	–	–	–	0.261	–
		M-F	100	12.01	5.82	–	–	–	–	0.28	–	–	–	–	–	0.281	–
	AVOCADO *Persea spp*																
12	SEEDS, MEAL SOLV EXTD	A-F	91	–	–	–	–	–	–	–	–	–	–	–	–	–	–
		M-F	100	–	–	–	–	–	–	–	–	–	–	–	–	–	–
	BABASSU *Orbignya spp*																
13	KERNELS WITH COATS, MEAL MECH EXTD (BABASSU OIL MEAL)	A-F	92	0.15	0.80	–	0.02	0.96	–	–	–	41.1	–	0.035	293.8	–	–
		M-F	100	0.16	0.87	–	0.02	1.04	–	–	–	44.6	–	0.038	318.9	–	–
14	KERNELS WITH COATS, MEAL SOLV EXTD (BABASSU OIL MEAL)	A-F	93	–	–	–	–	–	–	–	–	–	–	–	–	–	–
		M-F	100	–	–	–	–	–	–	–	–	–	–	–	–	–	–
	BARLEY *Hordeum vulgare*																
15	MALT SPROUTS, DEHY	A-F	93	0.18	0.63	0.88	0.36	0.17	0.25	0.79	–	5.9	–	0.018	29.4	0.416	56.4
		M-F	100	0.19	0.68	0.95	0.39	0.18	0.27	0.85	–	6.3	–	0.020	31.7	0.448	60.7
	BEAN *Phaseolus vulgaris*																
16	SEEDS, KIDNEY	A-F	89	0.11	0.43	0.01	–	0.10	0.93	–	0.508	6.1	–	0.007	17.0	0.355	24.3
		M-F	100	0.12	0.49	0.01	–	0.12	1.05	–	0.573	6.9	–	0.008	19.2	0.401	27.5
17	SEEDS, NAVY	A-F	89	0.17	0.54	0.04	0.06	0.13	1.31	0.23	–	9.9	–	0.010	21.1	–	–
		M-F	100	0.19	0.61	0.05	0.06	0.15	1.47	0.26	–	11.0	–	0.012	23.6	–	–
18	SEEDS, PINTO	A-F	90	0.18	0.48	0.01	–	0.24	2.59	–	0.551	13.0	–	0.014	17.3	–	40.0
		M-F	100	0.20	0.53	0.01	–	0.27	2.87	–	0.611	14.4	–	0.015	19.2	–	44.4
	BEAN, LIMA *Phaseolus limensis*																
19	SEEDS	A-F	90	0.08	0.38	0.02	0.03	0.18	1.61	0.20	–	8.2	–	0.009	16.1	–	–
		M-F	100	0.09	0.42	0.02	0.03	0.20	1.80	0.22	–	9.1	–	0.010	18.0	–	–
	BEAN, MUNG *Phaseolus aureus*																
20	SEEDS	A-F	90	0.13	0.35	0.01	–	–	1.04	–	–	–	–	0.009	–	–	–
		M-F	100	0.15	0.38	0.01	–	–	1.15	–	–	–	–	0.009	–	–	–
	BEAN, TEPARY *Phaseolus acutifolius, latifolius*																
21	SEEDS	A-F	90	–	–	–	–	–	–	–	–	–	–	–	–	–	–
		M-F	100	–	–	–	–	–	–	–	–	–	–	–	–	–	–
	BEECHNUT *Fagus spp*																
22	MEAL MECH EXTD	A-F	87	–	–	–	–	–	–	–	–	–	–	–	–	–	–
		M-F	100	–	–	–	–	–	–	–	–	–	–	–	–	–	–
	BEET, SUGAR *Beta vulgaris, altissima*																
23	PULP, AMMONIATED, DEHY	A-F	90	–	–	–	–	0.10	0.60	0.40	0.272	0.8	–	0.016	22.8	–	1.6
		M-F	100	–	–	–	–	0.11	0.66	0.44	0.301	0.9	–	0.018	25.2	–	1.8
	BLOOD																
24	SPRAY DEHY (BLOOD FLOUR)	A-F	93	0.41	0.30	0.38	0.25	0.15	0.15	0.60	–	8.2	–	0.277	6.4	–	–
		M-F	100	0.44	0.33	0.42	0.27	0.17	0.16	0.65	–	8.8	–	0.300	6.9	–	–
25	MEAL	A-F	91	0.29	0.25	0.32	0.30	0.22	0.09	0.34	0.088	12.6	–	0.372	5.3	0.731	4.4
		M-F	100	0.32	0.28	0.35	0.33	0.24	0.10	0.37	0.097	13.8	–	0.407	5.8	0.801	4.8
	BREWERS																
26	GRAINS, DEHY	A-F	92	0.30	0.51	0.21	0.15	0.15	0.09	0.30	0.076	21.7	0.066	0.024	37.2	–	27.3
		M-F	100	0.33	0.55	0.23	0.17	0.17	0.09	0.32	0.083	23.6	0.072	0.026	40.4	–	29.6
27	GRAINS, WET	A-F	22	0.06	0.12	0.06	0.03	0.03	0.02	0.07	0.022	4.9	–	0.006	9.0	–	23.2
		M-F	100	0.29	0.54	0.28	0.13	0.15	0.09	0.34	0.101	22.2	–	0.027	40.9	–	106.0
	BUCKWHEAT, COMMON *Fagopyrum sagittatum*																
28	MIDDLINGS	A-F	89	–	1.02	–	–	–	0.98	–	–	–	–	–	–	–	–
		M-F	100	–	1.15	–	–	–	1.11	–	–	–	–	–	–	–	–
	BUTTERMILK, CATTLE *Bos taurus*																
29	CONDENSED	A-F	29	0.44	0.26	0.31	0.12	0.19	0.23	0.03	–	–	–	–	–	–	–
		M-F	100	1.51	0.89	1.06	0.41	0.65	0.79	0.10	–	–	–	–	–	–	–
30	DEHY	A-F	92	1.32	0.94	0.83	0.44	0.48	0.83	0.08	–	1.0	–	0.001	3.5	–	40.2
		M-F	100	1.43	1.01	0.90	0.48	0.52	0.90	0.09	–	1.1	–	0.001	3.8	–	43.6
	CACAO (COCOA) *Theobroma cacao*																
31	SEEDS	A-F	–	–	–	–	–	–	–	–	–	–	–	–	–	–	–
		M-F	100	–	–	–	–	–	–	–	–	–	–	–	–	–	–

PROTEIN FEEDS

Entry Number	Fat-Soluble Vitamins: A (1 mg Carotene = 1667 IU Vit A)	Carotene (Provitamin A)	D	E	K	Water-Soluble Vitamins: B-12	Biotin	Choline	Folacin (Folic Acid)	Niacin	Pantothenic Acid (B-3)	(Pyridoxine) B-6	Riboflavin (B-2)	Thiamin (B-1)
	IU/g	ppm or mg/kg	IU/kg	ppm or mg/kg	ppm or mg/kg	ppm or mg/kg	ppm or mg/kg	ppm or mg/kg	ppm or mg/kg	ppm or mg/kg	ppm or mg/kg	ppm or mg/kg	ppm or mg/kg	ppm or mg/kg
5	–	–	–	–	–	**12.2**	**0.28**	**597**	**0.37**	**22**	**3.2**	**4.43**	**2.9**	**0.3**
	–	–	–	–	–	13.2	0.30	645	0.40	24	3.5	4.79	3.1	0.3
6	–	–	–	–	–	**44.3**	**0.09**	**780**	**0.10**	**31**	**2.3**	**4.41**	**2.0**	**0.3**
	–	–	–	–	–	48.5	0.09	854	0.11	34	2.6	4.83	2.2	0.4
7	–	–	–	–	–	**501.3**	**0.02**	**11370**	**5.56**	**205**	**29.2**	–	**36.2**	**0.2**
	–	–	–	–	–	541.5	0.02	12281	6.01	221	31.5	–	39.1	0.2
8	–	–	–	**0.9**	–	**75.2**	**0.12**	**1980**	**0.39**	**56**	**6.0**	**4.23**	**5.2**	**0.2**
	–	–	–	1.0	–	80.1	0.13	2110	0.42	60	6.4	4.51	5.5	0.2
9	–	–	–	**0.9**	–	**118.4**	**0.10**	**2049**	**0.37**	**51**	**5.5**	**5.86**	**4.7**	**0.2**
	–	–	–	0.9	–	126.9	0.11	2195	0.40	55	5.9	6.28	5.0	0.2
10	–	–	–	–	–	**89.4**	–	**2203**	**1.54**	**38**	**3.2**	–	**2.2**	**0.3**
	–	–	–	–	–	97.1	–	2391	1.67	41	3.5	–	2.4	0.4
11	–	–	–	**0.8**	–	**104.4**	**0.07**	**2067**	**0.57**	**58**	**4.8**	–	**5.0**	**0.2**
	–	–	–	0.9	–	112.4	0.08	2225	0.62	63	5.2	–	5.4	0.2
12	–	–	–	–	–	–	–	–	–	–	–	–	–	–
	–	–	–	–	–	–	–	–	–	–	–	–	–	–
13	**0.3**	**0.2**	–	–	–	–	–	**728**	–	**14**	**6.2**	**14.75**	**2.3**	–
	0.3	0.2	–	–	–	–	–	791	–	16	6.7	16.01	2.5	–
14	–	–	–	–	–	–	–	–	–	–	–	–	–	–
	–	–	–	–	–	–	–	–	–	–	–	–	–	–
15	–	–	–	**3.7**	–	–	**4.09**	**1591**	**0.20**	**55**	**9.0**	**8.62**	**2.8**	**8.3**
	–	–	–	4.0	–	–	4.40	1713	0.22	59	9.6	9.28	3.0	8.9
16	–	–	–	–	–	–	–	–	–	**24**	–	–	**1.8**	**5.7**
	–	–	–	–	–	–	–	–	–	27	–	–	2.1	6.4
17	–	–	–	**1.0**	–	–	**0.11**	**1017**	**1.29**	**24**	**2.4**	**0.30**	**1.7**	**6.3**
	–	–	–	1.1	–	–	0.12	1136	1.44	27	2.7	0.33	1.9	7.1
18	–	–	–	–	–	–	–	–	–	**22**	**2.2**	–	**3.1**	**8.6**
	–	–	–	–	–	–	–	–	–	24	2.5	–	3.4	9.6
19	–	–	–	–	–	–	–	–	**3.31**	**20**	**8.4**	–	**1.7**	**4.6**
	–	–	–	–	–	–	–	–	3.70	22	9.4	–	1.9	5.1
20	–	–	–	–	–	–	–	–	–	**25**	–	–	**2.1**	**3.9**
	–	–	–	–	–	–	–	–	–	27	–	–	2.4	4.3
21	–	–	–	–	–	–	–	–	–	–	–	–	–	–
	–	–	–	–	–	–	–	–	–	–	–	–	–	–
22	–	–	–	–	–	–	–	–	–	–	–	–	–	–
	–	–	–	–	–	–	–	–	–	–	–	–	–	–
23	–	**1.2**	–	–	–	–	–	–	–	–	–	–	–	–
	–	1.3	–	–	–	–	–	–	–	–	–	–	–	–
24	–	–	–	–	–	**12.2**	**0.28**	**597**	**0.37**	**22**	**3.2**	**4.43**	**2.9**	**0.3**
	–	–	–	–	–	13.2	0.30	645	0.40	24	3.5	4.79	3.1	0.3
25	–	–	–	–	–	**44.3**	**0.09**	**780**	**0.10**	**31**	**2.3**	**4.41**	**2.0**	**0.3**
	–	–	–	–	–	48.5	0.09	854	0.11	34	2.6	4.83	2.2	0.4
26	**0.8**	**0.5**	–	**26.7**	–	**3.6**	**0.44**	**1651**	**0.22**	**44**	**8.2**	**1.03**	**1.5**	**0.6**
	0.8	0.5	–	29.0	–	3.9	0.48	1792	0.24	47	8.9	1.11	1.6	0.7
27	–	–	–	**5.5**	–	–	–	–	–	–	–	–	–	–
	–	–	–	25.0	–	–	–	–	–	–	–	–	–	–
28	–	–	–	–	–	–	–	–	–	–	–	–	–	–
	–	–	–	–	–	–	–	–	–	–	–	–	–	–
29	**25.6**	**15.4**	–	–	–	–	–	–	–	–	–	–	**12.6**	–
	87.2	52.3	–	–	–	–	–	–	–	–	–	–	42.8	–
30	–	–	–	**6.3**	–	**19.6**	**0.29**	**1746**	**0.39**	**9**	**37.0**	**2.47**	**30.6**	**3.4**
	–	–	–	6.8	–	21.2	0.31	1891	0.42	9	40.1	2.67	33.1	3.7
31	–	–	–	–	–	–	–	–	–	–	–	–	–	–
	–	–	–	–	–	–	–	–	–	–	–	–	–	–

(Continued)

TABLE V-1B PROTEIN FEEDS, COMPOSITION OF FEEDS, DATA EXPRESSED **AS-FED** AND **MOISTURE-FREE**—*(Continued)*

Entry Number	Feed Name Description	International Feed Number	Moisture Basis: A-F (as-fed) or M-F (moisture-free)	Proximate Analysis: Dry Matter	Ash	Crude Fiber	Neutral Det. Fib. (NDF)	Acid Det. Fib. (ADF)	Lignin	Ether Extract (Fat)	N-Free Extract	Crude Protein	Digestible Protein: Ruminant	Swine	Horse
				%	%	%	%	%	%	%	%	%	%	%	%
	CASEIN														
32	ACID PRECIPITATED, DEHY[1]	5-01-162	A-F	91	2.2	0.2	0	0	–	0.6	3.6	84.0	81.5	82.2	86.5
			M-F	100	2.4	0.2	0	0	–	0.7	3.9	92.7	89.9	90.8	95.5
	CASTOR BEAN *Ricinus communis*														
33	SEEDS WITHOUT TOXIN, MEAL	5-01-155	A-F	87	7.5	35.5	–	–	–	1.0	16.6	26.0	20.0	24.7	21.3
			M-F	100	8.6	41.0	–	–	–	1.2	19.2	30.0	23.1	28.5	24.6
	CATTLE *Bos taurus*														
34	BUTTERMILK, CONDENSED	5-01-159	A-F	29	3.6	0.1	–	–	–	2.4	12.4	10.8	9.2	10.0	9.5
			M-F	100	12.2	0.3	–	–	–	8.3	42.3	36.9	31.1	33.9	32.4
35	BUTTERMILK, DEHY	5-01-160	A-F	92	9.1	0.3	–	–	–	5.2	45.9	31.7	28.7	29.5	27.3
			M-F	100	9.9	0.4	–	–	–	5.6	49.7	34.4	31.1	32.0	29.5
36	CHEESE RIND	5-01-163	A-F	85	7.7	0.2	–	–	–	19.8	11.7	45.5	39.4	44.1	43.6
			M-F	100	9.0	0.3	–	–	–	23.3	13.7	53.7	46.4	52.0	51.4
37	HEARTS, FRESH	5-01-164	A-F	26	1.1	0.2	–	–	–	11.2	–	17.5	15.3	17.1	17.5
			M-F	100	4.3	0.7	–	–	–	43.8	–	68.7	60.1	66.9	68.3
38	KIDNEYS, FRESH	5-01-165	A-F	27	1.1	–	–	–	–	9.2	1.0	16.0	13.9	15.6	15.6
			M-F	100	4.0	–	–	–	–	34.0	3.7	59.5	51.7	57.8	57.9
39	LIPS, FRESH	5-07-940	A-F	30	–	–	–	–	–	7.0	–	18.0	15.7	17.5	17.6
			M-F	100	–	–	–	–	–	23.3	–	60.0	52.2	58.3	58.5
40	LIVERS, FRESH	5-01-166	A-F	28	1.4	0.2	–	–	–	5.1	1.9	19.5	17.0	19.0	19.4
			M-F	100	4.9	0.6	–	–	–	18.3	6.7	69.6	60.9	67.8	69.4
41	LUNGS, FRESH	5-07-941	A-F	24	0.9	0.7	–	–	–	9.2	–	15.4	13.4	15.0	15.2
			M-F	100	3.8	2.9	–	–	–	38.9	–	65.0	56.7	63.3	64.2
42	MILK, FRESH	5-01-168	A-F	12	0.8	–	–	–	–	3.6	4.7	3.3	3.2	3.2	2.6
			M-F	100	6.2	–	–	–	–	29.5	37.6	26.7	25.5	25.9	20.9
43	MILK, DEHY	5-01-167	A-F	95	5.4	0.2	–	–	–	26.3	38.1	25.3	20.7	24.0	19.8
			M-F	100	5.6	0.2	–	–	–	27.6	39.9	26.6	21.8	25.1	20.8
44	SKIM MILK, FRESH	5-01-170	A-F	10	0.7	–	–	–	–	0.1	5.8	3.0	2.5	3.0	2.5
			M-F	100	6.9	–	–	–	–	1.0	60.6	31.2	26.0	30.9	26.0
45	SKIM MILK, DEHY	5-01-175	A-F	94	8.0	0.2	0.0	–	–	1.1	51.6	33.3	30.0	31.8	28.9
			M-F	100	8.4	0.2	0.0	–	–	1.2	54.8	35.4	31.8	33.7	30.7
46	SPLEENS, FRESH	5-07-942	A-F	24	1.5	1.0	–	–	–	3.9	1.3	16.5	14.5	16.1	16.5
			M-F	100	6.0	4.0	–	–	–	16.1	5.3	68.7	60.1	66.9	68.3
47	TONGUES, FRESH	5-08-129	A-F	32	0.9	–	–	–	–	15.0	0.4	16.4	14.2	15.9	15.6
			M-F	100	2.8	–	–	–	–	46.9	1.3	51.2	44.2	49.6	48.6
48	UDDERS, FRESH	5-07-943	A-F	20	1.5	0.3	–	–	–	6.1	0.6	11.9	10.3	11.6	11.6
			M-F	100	7.4	1.2	–	–	–	30.0	2.7	58.6	50.9	56.9	57.0
49	WHEY, FRESH	4-08-134	A-F	7	0.7	–	–	–	–	0.3	5.1	0.9	0.6	0.7	0.7
			M-F	100	9.4	–	–	–	–	4.3	73.9	13.2	8.8	10.0	9.5
50	WHEY, ALBUMIN	5-01-177	A-F	92	29.1	0.9	–	–	–	1.0	12.8	48.4	41.8	46.9	46.1
			M-F	100	31.6	0.9	–	–	–	1.1	13.9	52.5	45.4	50.9	50.1
51	WHEY, CONDENSED	4-01-180	A-F	54	5.0	0.3	–	–	–	0.4	41.2	6.9	4.5	5.2	5.0
			M-F	100	9.3	0.5	–	–	–	0.8	76.6	12.8	8.4	9.7	9.2
52	WHEY, DEHY	4-01-182	A-F	93	8.8	0.2	0.3	0.2	–	0.8	70.2	13.3	8.9	13.1	9.6
			M-F	100	9.4	0.2	0.3	0.2	–	0.8	75.3	14.2	9.5	14.1	10.3
53	WHEY, LOW LACTOSE, DEHY	4-01-186	A-F	93	15.4	0.2	–	–	–	1.0	60.0	16.7	11.5	13.3	12.3
			M-F	100	16.5	0.2	–	–	–	1.1	64.3	17.9	12.4	14.3	13.2
	CHARLOCK (MUSTARD) *Brassica kaber*														
54	SEEDS	5-08-461	A-F	96	5.5	5.0	–	–	–	38.8	23.6	23.0	18.6	21.6	17.1
			M-F	100	5.7	5.2	–	–	–	40.5	24.6	24.0	19.4	22.6	17.8
	CHEESE, CATTLE *Bos taurus*														
55	RIND	5-01-163	A-F	85	7.7	0.2	–	–	–	19.8	11.7	45.5	39.4	44.1	43.6
			M-F	100	9.0	0.3	–	–	–	23.3	13.7	53.7	46.4	52.0	51.4
	CHICKEN *Gallus domesticus*														
56	FLESH, FRESH	5-28-310	A-F	30	1.0	–	–	–	–	6.9	–	21.6	–	–	–
			M-F	100	3.4	–	–	–	–	23.4	–	73.2	–	–	–
57	BROILERS, WHOLE	5-07-945	A-F	24	0.8	–	–	–	–	4.8	–	18.4	–	–	–
			M-F	100	3.3	–	–	–	–	20.2	–	76.5	–	–	–
58	BY-PRODUCT, FRESH (VISCERA WITH FEET, WITH HEADS)	5-07-951	A-F	39	4.9	–	–	0.5	–	14.7	3.0	17.6	15.1	17.0	16.2
			M-F	100	12.4	–	–	1.4	–	37.5	7.7	44.9	38.5	43.3	41.5
59	BY-PRODUCT, WITHOUT FEET, FRESH (VISCERA WITHOUT FEET, WITH HEADS)	5-07-952	A-F	34	1.1	0.2	–	0.7	–	16.1	3.0	13.8	11.7	13.2	12.4
			M-F	100	3.1	0.7	–	1.9	–	47.2	8.7	40.3	34.2	38.7	36.2
60	EGGS WITH SHELLS, FRESH	5-01-213	A-F	34	10.7	–	–	–	–	10.6	–	12.8	–	–	–
			M-F	100	31.4	–	–	–	–	31.1	–	37.5	–	–	–
61	EGGS WITHOUT SHELLS, FRESH	5-08-114	A-F	25	1.0	0.0	–	–	–	11.1	1.0	12.3	10.6	11.9	11.6
			M-F	100	4.0	0.0	–	–	–	43.6	3.9	48.5	41.7	46.9	45.5
62	FEET, FRESH	5-07-947	A-F	33	5.4	–	–	–	–	7.5	–	17.7	15.3	17.2	17.0
			M-F	100	16.6	–	–	–	–	23.1	–	54.5	47.1	52.8	52.3
63	GIZZARDS, FRESH	5-07-948	A-F	25	1.5	–	–	–	–	2.7	0.7	20.1	17.7	19.6	20.4
			M-F	100	6.0	–	–	–	–	10.8	2.8	80.4	70.8	78.6	81.6
64	HEADS, FRESH	5-07-949	A-F	33	–	–	–	–	–	6.0	–	19.0	16.5	18.4	18.4
			M-F	100	–	–	–	–	–	18.2	–	57.6	50.0	55.9	55.8
65	HENS, CARCASS	5-08-095	A-F	33	1.0	–	–	–	–	12.1	–	19.9	–	–	–
			M-F	100	3.1	–	–	–	–	36.6	–	60.3	–	–	–
66	WHOLE, FRESH, DAY OLD	5-07-946	A-F	13	0.8	–	–	1.0	–	3.4	0.2	7.4	6.4	7.2	7.2
			M-F	100	6.1	–	–	8.0	–	26.3	1.6	57.9	50.3	56.2	56.2
	CHICKPEA (GARBANZO; GRAM PEA) *Cicer arietinum*														
67	SEEDS	5-01-218	A-F	89	2.9	7.0	–	–	–	4.1	56.0	19.1	14.3	17.8	13.3
			M-F	100	3.3	7.9	–	–	–	4.6	62.8	21.4	16.1	20.0	14.9

Entry Number	TDN Ruminant	TDN Swine	TDN Horse	Digestible Energy Ruminant		Digestible Energy Swine		Digestible Energy Horse		Metabolizable Energy Ruminant		Metabolizable Energy Swine		Metabolizable Energy Poultry ME_n		Metabolizable Energy Horse		Net Energy Ruminant NE_m		Net Energy Ruminant NE_g		Net Energy Lactating Cows NE_{lc}	
	%	%	%	Mcal lb	Mcal kg	kcal lb	kcal kg	Mcal lb	Mcal kg	Mcal lb	Mcal kg	kcal lb	kcal kg	kcal lb	kcal kg	Mcal lb	Mcal kg	Mcal lb	Mcal kg	Mcal lb	Mcal kg	Mcal lb	Mcal kg
32	81	80	–	1.56	3.44	1590	3506	–	–	1.22	2.68	1391	3068	1867	4116	–	–	0.82	1.81	0.55	1.22	0.78	1.72
	89	88	–	1.72	3.80	1755	3869	–	–	1.34	2.96	1535	3385	2061	4543	–	–	0.91	2.00	0.61	1.35	0.86	1.89
33	25	–	–	0.50	1.11	–	–	–	–	0.33	0.73	–	–	–	–	–	–	0.07	0.16	-0.13	-0.29	0.23	0.51
	29	–	–	0.58	1.28	–	–	–	–	0.38	0.85	–	–	–	–	–	–	0.08	0.18	-0.15	-0.33	0.27	0.59
34	26	22	–	0.52	1.14	442	974	–	–	0.46	1.02	383	844	–	–	–	–	0.29	0.63	0.20	0.43	0.27	0.59
	88	76	–	1.77	3.89	1503	3314	–	–	1.58	3.48	1303	2872	–	–	–	–	0.98	2.15	0.67	1.48	0.91	2.01
35	82	77	–	1.56	3.44	1578	3479	–	–	1.29	2.84	1381	3045	1248	2752	–	–	0.88	1.94	0.60	1.32	0.85	1.87
	89	84	–	1.69	3.73	1710	3770	–	–	1.40	3.08	1497	3300	1353	2982	–	–	0.95	2.10	0.65	1.43	0.92	2.03
36	78	80	–	1.69	3.73	1609	3547	–	–	1.54	3.39	1371	3023	–	–	–	–	1.06	2.34	0.76	1.68	0.98	2.15
	91	95	–	2.00	4.40	1897	4183	–	–	1.81	4.00	1617	3564	–	–	–	–	1.25	2.76	0.90	1.98	1.15	2.53
37	–	–	–	–	–	–	–	–	–	–	–	–	–	–	–	–	–	–	–	–	–	–	–
	–	–	–	–	–	–	–	–	–	–	–	–	–	–	–	–	–	–	–	–	–	–	–
38	–	–	–	–	–	–	–	–	–	–	–	–	–	–	–	–	–	–	–	–	–	–	–
	–	–	–	–	–	–	–	–	–	–	–	–	–	–	–	–	–	–	–	–	–	–	–
39	–	–	–	–	–	–	–	–	–	–	–	–	–	–	–	–	–	–	–	–	–	–	–
	–	–	–	–	–	–	–	–	–	–	–	–	–	–	–	–	–	–	–	–	–	–	–
40	29	31	–	0.58	1.28	615	1356	–	–	0.53	1.17	544	1200	–	–	–	–	0.34	0.76	0.25	0.54	0.32	0.70
	104	110	–	2.08	4.59	2201	4852	–	–	1.90	4.18	1947	4293	–	–	–	–	1.23	2.71	0.88	1.93	1.13	2.49
41	–	41	–	–	–	831	1832	–	–	–	–	701	1546	770	1697	–	–	–	–	–	–	–	–
	–	175	–	–	–	3509	7737	–	–	–	–	2962	6530	3249	7164	–	–	–	–	–	–	–	–
42	16	15	–	0.32	0.71	309	681	–	–	0.30	0.65	275	606	–	–	–	–	0.19	0.42	0.14	0.30	0.18	0.40
	128	125	–	2.61	5.75	2500	5510	–	–	2.40	5.29	2223	4902	–	–	–	–	1.53	3.37	1.11	2.46	1.46	3.22
43	113	109	–	2.27	4.99	2183	4813	–	–	2.09	4.61	1935	4266	–	–	–	–	1.27	2.81	0.92	2.02	1.17	2.57
	119	115	–	2.38	5.24	2290	5049	–	–	2.20	4.84	2030	4475	–	–	–	–	1.34	2.95	0.96	2.12	1.22	2.70
44	9	9	–	0.18	0.39	188	415	–	–	0.16	0.35	166	365	–	–	–	–	0.10	0.22	0.07	0.16	0.10	0.21
	92	96	–	1.86	4.11	1968	4339	–	–	1.66	3.66	1732	3819	–	–	–	–	1.06	2.33	0.74	1.63	1.00	2.20
45	80	86	–	1.43	3.15	1758	3876	–	–	1.07	2.37	1630	3593	1152	2539	–	–	0.69	1.52	0.43	0.95	0.83	1.82
	85	92	–	1.52	3.35	1867	4115	–	–	1.14	2.51	1730	3815	1223	2696	–	–	0.73	1.62	0.46	1.01	0.88	1.94
46	23	24	–	0.47	1.04	476	1048	–	–	0.43	0.94	419	923	–	–	–	–	0.28	0.61	0.20	0.43	0.25	0.56
	97	99	–	1.95	4.30	1973	4350	–	–	1.76	3.89	1737	3829	–	–	–	–	1.14	2.52	0.81	1.78	1.05	2.32
47	–	–	–	–	–	–	–	–	–	–	–	–	–	–	–	–	–	–	–	–	–	–	–
	–	–	–	–	–	–	–	–	–	–	–	–	–	–	–	–	–	–	–	–	–	–	–
48	23	22	–	0.45	1.00	448	987	–	–	0.41	0.91	396	873	–	–	–	–	0.27	0.60	0.20	0.43	0.25	0.55
	111	110	–	2.22	4.90	2205	4860	–	–	2.04	4.50	1951	4301	–	–	–	–	1.34	2.95	0.96	2.12	1.22	2.70
49	7	–	–	0.13	0.29	–	–	–	–	0.12	0.26	–	–	–	–	–	–	–	–	–	–	–	–
	94	–	–	1.88	4.15	–	–	–	–	1.70	3.74	–	–	–	–	–	–	–	–	–	–	–	–
50	59	–	–	1.28	2.83	–	–	–	–	1.11	2.44	–	–	–	–	–	–	0.76	1.67	0.50	1.09	0.73	1.60
	64	–	–	1.39	3.07	–	–	–	–	1.20	2.65	–	–	–	–	–	–	0.82	1.81	0.54	1.18	0.79	1.74
51	47	46	–	0.94	2.07	928	2046	–	–	0.84	1.85	878	1936	–	–	–	–	0.56	1.24	0.39	0.86	0.52	1.15
	87	86	–	1.74	3.84	1726	3806	–	–	1.56	3.43	1633	3601	–	–	–	–	1.04	2.30	0.72	1.60	0.97	2.13
52	76	77	–	1.51	3.33	1444	3183	–	–	1.28	2.83	1411	3110	880	1939	–	–	0.87	1.92	0.59	1.30	0.78	1.71
	82	83	–	1.62	3.57	1549	3415	–	–	1.38	3.03	1514	3337	944	2081	–	–	0.93	2.06	0.63	1.40	0.83	1.83
53	74	75	–	1.40	3.09	1242	2737	–	–	1.14	2.52	1245	2744	931	2053	–	–	0.75	1.66	0.49	1.08	0.77	1.69
	79	80	–	1.50	3.31	1330	2932	–	–	1.22	2.70	1333	2939	997	2199	–	–	0.81	1.78	0.52	1.15	0.82	1.81
54	91	113	–	2.04	4.50	2263	4989	–	–	1.87	4.11	2008	4428	–	–	–	–	1.32	2.91	0.95	2.10	1.21	2.66
	95	118	–	2.13	4.69	2360	5204	–	–	1.95	4.29	2095	4618	–	–	–	–	1.38	3.03	0.99	2.19	1.26	2.77
55	78	80	–	1.69	3.73	1609	3547	–	–	1.54	3.39	1371	3023	–	–	–	–	1.06	2.34	0.76	1.68	0.98	2.15
	91	95	–	2.00	4.40	1897	4183	–	–	1.81	4.00	1617	3564	–	–	–	–	1.25	2.76	0.90	1.98	1.15	2.53
56	–	–	–	–	–	–	–	–	–	–	–	–	–	–	–	–	–	–	–	–	–	–	–
	–	–	–	–	–	–	–	–	–	–	–	–	–	–	–	–	–	–	–	–	–	–	–
57	–	–	–	–	–	–	–	–	–	–	–	–	–	–	–	–	–	–	–	–	–	–	–
	–	–	–	–	–	–	–	–	–	–	–	–	–	–	–	–	–	–	–	–	–	–	–
58	–	–	–	–	–	–	–	–	–	–	–	–	–	–	–	–	–	–	–	–	–	–	–
	–	–	–	–	–	–	–	–	–	–	–	–	–	–	–	–	–	–	–	–	–	–	–
59	–	–	–	–	–	–	–	–	–	–	–	–	–	–	–	–	–	–	–	–	–	–	–
	–	–	–	–	–	–	–	–	–	–	–	–	–	–	–	–	–	–	–	–	–	–	–
60	–	–	–	–	–	–	–	–	–	–	–	–	–	–	–	–	–	–	–	–	–	–	–
	–	–	–	–	–	–	–	–	–	–	–	–	–	–	–	–	–	–	–	–	–	–	–
61	32	35	–	0.65	1.43	694	1529	–	–	0.60	1.33	619	1365	–	–	–	–	0.39	0.86	0.28	0.62	0.35	0.78
	127	137	–	2.54	5.61	2731	6020	–	–	2.37	5.22	2437	5373	–	–	–	–	1.53	3.37	1.11	2.45	1.39	3.07
62	–	–	–	–	–	–	–	–	–	–	–	–	–	–	–	–	–	–	–	–	–	–	–
	–	–	–	–	–	–	–	–	–	–	–	–	–	–	–	–	–	–	–	–	–	–	–
63	–	–	–	–	–	–	–	–	–	–	–	–	–	–	–	–	–	–	–	–	–	–	–
	–	–	–	–	–	–	–	–	–	–	–	–	–	–	–	–	–	–	–	–	–	–	–
64	–	–	–	–	–	–	–	–	–	–	–	–	–	–	–	–	–	–	–	–	–	–	–
	–	–	–	–	–	–	–	–	–	–	–	–	–	–	–	–	–	–	–	–	–	–	–
65	–	–	–	–	–	–	–	–	–	–	–	–	–	–	–	–	–	–	–	–	–	–	–
	–	–	–	–	–	–	–	–	–	–	–	–	–	–	–	–	–	–	–	–	–	–	–
66	–	–	–	–	–	–	–	–	–	–	–	–	–	–	–	–	–	–	–	–	–	–	–
	–	–	–	–	–	–	–	–	–	–	–	–	–	–	–	–	–	–	–	–	–	–	–
67	78	80	–	1.56	3.45	1607	3543	–	–	1.40	3.08	1408	3104	–	–	–	–	0.89	1.97	0.62	1.36	0.83	1.84
	88	90	–	1.75	3.86	1801	3971	–	–	1.57	3.45	1578	3479	–	–	–	–	1.00	2.21	0.69	1.52	0.94	2.06

(Continued)

TABLE V-1B PROTEIN FEEDS, MINERAL AND VITAMIN COMPOSITION OF FEEDS, DATA EXPRESSED **AS-FED** AND **MOISTURE-FREE**—*(Continued)*

Entry Number	Feed Name Description	Moisture Basis: A-F (as-fed) or M-F (moisture-free)	Dry Matter	Macro Minerals							Micro Minerals						
				Calcium (Ca)	Phosphorus (P)	Sodium (Na)	Chlorine (Cl)	Magnesium (Mg)	Potassium (K)	Sulfur (S)	Cobalt (Co)	Copper (Cu)	Iodine (I)	Iron (Fe)	Manganese (Mn)	Selenium (Se)	Zinc (Zn)
			%	%	%	%	%	%	%	%	ppm or mg/kg	ppm or mg/kg	ppm or mg/kg	%	ppm or mg/kg	ppm or mg/kg	ppm or mg/kg
	CASEIN																
32	ACID PRECIPITATED, DEHY	A-F	91	0.61	0.82	0.01	–	0.01	0.01	–	–	4.1	–	0.002	3.5	–	31.8
		M-F	100	0.67	0.90	0.01	–	0.01	0.01	–	–	4.5	–	0.002	3.9	–	35.1
	CASTOR BEAN *Ricinus communis*																
33	SEEDS WITHOUT TOXIN, MEAL	A-F	87	–	–	–	–	–	–	–	–	–	–	–	–	–	–
		M-F	100	–	–	–	–	–	–	–	–	–	–	–	–	–	–
	CATTLE *Bos taurus*																
34	BUTTERMILK, CONDENSED	A-F	29	0.44	0.26	0.31	0.12	0.19	0.23	0.03	–	–	–	–	–	–	–
		M-F	100	1.51	0.89	1.06	0.41	0.65	0.79	0.10	–	–	–	–	–	–	–
35	BUTTERMILK, DEHY	A-F	92	1.32	0.94	0.83	0.44	0.48	0.83	0.08	–	1.0	–	0.001	3.5	–	40.2
		M-F	100	1.43	1.01	0.90	0.48	0.52	0.90	0.09	–	1.1	–	0.001	3.8	–	43.6
36	CHEESE RIND	A-F	85	0.98	0.56	0.81	0.60	0.02	0.28	–	–	–	–	–	–	–	–
		M-F	100	1.16	0.66	0.95	0.71	0.03	0.32	–	–	–	–	–	–	–	–
37	HEARTS, FRESH	A-F	26	0.01	0.20	0.08	–	0.01	0.19	–	0.147	1.6	–	0.004	0.2	0.081	8.4
		M-F	100	0.04	0.77	0.32	–	0.04	0.73	–	0.575	6.1	–	0.015	0.8	0.317	33.0
38	KIDNEYS, FRESH	A-F	27	0.02	0.22	0.20	–	0.01	0.18	–	–	3.2	–	0.007	0.7	–	15.7
		M-F	100	0.06	0.82	0.73	–	0.04	0.66	–	–	11.7	–	0.023	2.6	–	58.4
39	LIPS, FRESH	A-F	30	–	–	–	–	–	–	–	–	–	–	–	–	–	–
		M-F	100	–	–	–	–	–	–	–	–	–	–	–	–	–	–
40	LIVERS, FRESH	A-F	28	0.01	0.23	0.10	–	0.01	0.20	–	–	6.1	–	0.005	2.8	–	26.6
		M-F	100	0.04	0.82	0.35	–	0.04	0.72	–	–	21.9	–	0.017	9.9	–	95.0
41	LUNGS, FRESH	A-F	24	0.01	0.16	0.16	–	0.01	0.08	–	0.099	1.1	0.074	0.008	0.1	0.083	13.1
		M-F	100	0.06	0.69	0.69	–	0.03	0.33	–	0.416	4.6	0.311	0.033	0.5	0.347	55.4
42	MILK, FRESH	A-F	12	0.12	0.09	0.05	0.11	0.01	0.14	0.04	0.001	0.1	–	0.002	–	–	2.8
		M-F	100	0.93	0.75	0.38	0.92	0.10	1.13	0.32	0.005	0.8	–	0.010	–	–	23.0
43	MILK, DEHY	A-F	95	0.89	0.70	0.36	1.48	0.09	1.05	0.31	0.005	0.9	–	0.018	0.4	–	21.7
		M-F	100	0.93	0.74	0.38	1.55	0.09	1.10	0.32	0.005	0.9	–	0.019	0.5	–	22.8
44	SKIM MILK, FRESH	A-F	10	0.13	0.10	0.04	0.05	0.01	0.12	0.03	0.011	1.1	–	0.001	0.2	–	4.9
		M-F	100	1.31	1.04	0.47	0.54	0.12	1.29	0.32	0.111	11.6	–	0.009	2.3	–	51.0
45	SKIM MILK, DEHY	A-F	94	1.28	1.02	0.51	0.90	0.12	1.60	0.32	0.113	11.7	–	0.001	2.1	0.124	38.5
		M-F	100	1.36	1.09	0.54	0.96	0.13	1.70	0.34	0.120	12.4	–	0.001	2.3	0.131	40.9
46	SPLEENS, FRESH	A-F	24	0.01	0.27	0.14	–	0.01	0.22	–	0.241	0.2	0.183	0.012	–	–	19.5
		M-F	100	0.02	1.13	0.58	–	0.05	0.91	–	0.997	0.6	0.757	0.048	–	–	80.9
47	TONGUES, FRESH	A-F	32	0.01	0.18	0.07	–	–	0.20	–	–	–	–	0.003	–	–	–
		M-F	100	0.03	0.56	0.22	–	–	0.63	–	–	–	–	0.007	–	–	–
48	UDDERS, FRESH	A-F	20	0.53	0.28	0.12	–	0.02	0.16	–	0.310	0.5	–	0.003	0.6	0.110	21.2
		M-F	100	2.62	1.37	0.58	–	0.08	0.79	–	1.528	2.6	–	0.011	2.8	0.542	104.4
49	WHEY, FRESH	A-F	7	0.06	0.05	–	–	–	0.19	–	–	–	–	0.003	0.2	–	–
		M-F	100	0.81	0.71	–	–	–	2.75	–	–	–	–	0.029	3.2	–	–
50	WHEY, ALBUMIN	A-F	92	10.86	4.03	–	–	–	–	–	–	–	–	–	–	–	–
		M-F	100	11.79	4.37	–	–	–	–	–	–	–	–	–	–	–	–
51	WHEY, CONDENSED	A-F	54	0.39	0.47	–	–	–	–	–	–	–	–	–	–	–	–
		M-F	100	0.72	0.88	–	–	–	–	–	–	–	–	–	–	–	–
52	WHEY, DEHY	A-F	93	0.86	0.76	0.62	0.07	0.13	1.11	1.04	0.111	46.5	–	0.017	5.9	–	3.2
		M-F	100	0.92	0.82	0.66	0.08	0.14	1.19	1.11	0.119	49.9	–	0.019	6.3	–	3.4
53	WHEY, LOW LACTOSE, DEHY	A-F	93	1.49	1.11	1.44	1.03	0.21	2.95	1.07	–	7.0	9.854	0.025	8.0	0.052	7.9
		M-F	100	1.60	1.18	1.54	1.10	0.23	3.16	1.15	–	7.5	10.554	0.027	8.6	0.056	8.4
	CHARLOCK (MUSTARD) *Brassica kaber*																
54	SEEDS	A-F	96	–	–	–	–	–	–	–	–	–	–	–	–	–	–
		M-F	100	–	–	–	–	–	–	–	–	–	–	–	–	–	–
	CHEESE, CATTLE *Bos taurus*																
55	RIND	A-F	85	0.98	0.56	0.81	0.60	0.02	0.28	–	–	–	–	–	–	–	–
		M-F	100	1.16	0.66	0.95	0.71	0.03	0.32	–	–	–	–	–	–	–	–
	CHICKEN *Gallus domesticus*																
56	FLESH, FRESH	A-F	30	0.01	0.20	0.06	–	–	0.29	–	–	–	–	0.002	–	–	–
		M-F	100	0.04	0.69	0.20	–	–	0.97	–	–	–	–	0.005	–	–	–
57	BROILERS, WHOLE	A-F	24	0.01	0.20	–	–	–	–	–	–	–	–	0.002	–	–	–
		M-F	100	0.04	0.82	–	–	–	–	–	–	–	–	0.008	–	–	–
58	BY-PRODUCT, FRESH (VISCERA WITH FEET, WITH HEADS)	A-F	39	–	–	–	–	–	–	–	–	–	–	–	–	–	–
		M-F	100	–	–	–	–	–	–	–	–	–	–	–	–	–	–
59	BY-PRODUCT, WITHOUT FEET, FRESH (VISCERA WITHOUT FEET, WITH HEADS)	A-F	34	0.34	0.24	–	–	–	–	–	–	–	–	–	–	–	–
		M-F	100	1.00	0.70	–	–	–	–	–	–	–	–	–	–	–	–
60	EGGS WITH SHELLS, FRESH	A-F	34	–	–	–	–	–	–	–	–	–	–	–	–	–	–
		M-F	100	–	–	–	–	–	–	–	–	–	–	–	–	–	–
61	EGGS WITHOUT SHELLS, FRESH	A-F	25	0.07	0.20	0.13	–	0.01	0.12	–	0.166	0.7	0.384	0.003	0.3	0.156	13.7
		M-F	100	0.28	0.79	0.50	–	0.04	0.47	–	0.654	2.6	1.511	0.009	1.2	0.613	53.9
62	FEET, FRESH	A-F	33	2.10	0.76	0.12	–	0.03	0.08	–	2.020	0.7	0.120	0.004	0.6	–	15.8
		M-F	100	6.45	2.33	0.38	–	0.10	0.26	–	6.216	2.1	0.370	0.010	1.7	–	48.6
63	GIZZARDS, FRESH	A-F	25	0.01	0.11	0.07	–	–	0.24	–	–	–	–	0.003	–	–	–
		M-F	100	0.04	0.42	0.26	–	–	0.96	–	–	–	–	0.012	–	–	–
64	HEADS, FRESH	A-F	33	–	–	–	–	–	–	–	–	–	–	–	–	–	–
		M-F	100	–	–	–	–	–	–	–	–	–	–	–	–	–	–
65	HENS, CARCASS	A-F	33	0.01	0.20	–	–	–	–	–	–	–	–	0.002	–	–	–
		M-F	100	0.04	0.60	–	–	–	–	–	–	–	–	0.005	–	–	–
66	WHOLE, FRESH, DAY OLD	A-F	13	–	–	–	–	–	–	–	–	–	–	–	–	–	–
		M-F	100	–	–	–	–	–	–	–	–	–	–	–	–	–	–
	CHICKPEA (GARBANZO; GRAM PEA) *Cicer arietinum*																
67	SEEDS	A-F	89	0.15	0.33	0.03	–	–	0.80	–	–	–	–	0.007	–	–	–
		M-F	100	0.17	0.37	0.03	–	–	0.89	–	–	–	–	0.008	–	–	–

Entry Number	Fat-Soluble Vitamins: A (1 mg Carotene = 1667 IU Vit A)	Carotene (Provitamin A)	D	E	K	Water-Soluble Vitamins: B–12	Biotin	Choline	Folacin (Folic Acid)	Niacin	Pantothenic Acid (B–3)	(Pyridoxine) B–6	Riboflavin (B–2)	Thiamin (B–1)
	IU/g	ppm or mg/kg	IU/kg	ppm or mg/kg	ppm or mg/kg	ppm or mg/kg	ppm or mg/kg	ppm or mg/kg	ppm or mg/kg	ppm or mg/kg	ppm or mg/kg	ppm or mg/kg	ppm or mg/kg	ppm or mg/kg
32	–	–	–	–	–	–	**0.04**	**208**	**0.47**	**1**	**2.7**	**0.42**	**1.5**	**0.4**
	–	–	–	–	–	–	0.05	229	0.52	1	2.9	0.47	1.7	0.5
33	–	–	–	–	–	–	–	–	–	–	–	–	–	–
	–	–	–	–	–	–	–	–	–	–	–	–	–	–
34	**25.6**	**15.4**	–	–	–	–	–	–	–	–	–	–	**12.6**	–
	87.2	52.3	–	–	–	–	–	–	–	–	–	–	42.8	–
35	–	–	–	**6.3**	–	**19.6**	**0.29**	**1746**	**0.39**	**9**	**37.0**	**2.47**	**30.6**	**3.4**
	–	–	–	6.8	–	21.2	0.31	1891	0.42	9	40.1	2.67	33.1	3.7
36	–	–	–	–	–	–	–	–	–	–	–	–	–	–
	–	–	–	–	–	–	–	–	–	–	–	–	–	–
37	–	–	–	**0.9**	–	**27.9**	**0.01**	**623**	**0.01**	**58**	**9.5**	**0.83**	**7.9**	**3.4**
	–	–	–	3.6	–	109.2	0.03	2440	0.03	226	37.1	3.25	31.1	13.3
38	–	–	–	**7.1**	–	**258.6**	**0.81**	**524**	**1.85**	**45**	**25.5**	**0.96**	**22.5**	**2.4**
	–	–	–	26.2	–	959.6	3.02	1943	6.88	168	94.6	3.56	83.4	8.8
39	–	–	–	–	–	–	–	–	–	–	–	–	–	–
	–	–	–	–	–	–	–	–	–	–	–	–	–	–
40	–	–	–	**7.1**	–	**425.8**	**0.98**	**1424**	**2.33**	**75**	**46.1**	**5.03**	**25.8**	**1.8**
	–	–	–	25.4	–	1523.2	3.51	5092	8.35	269	164.9	18.00	92.2	6.3
41	–	–	–	**3.0**	–	**100.2**	**0.03**	**1879**	**0.22**	**12**	**0.6**	**0.43**	**2.0**	**0.7**
	–	–	–	12.7	–	423.3	0.12	7933	0.92	49	2.6	1.81	8.4	2.8
42	–	–	–	–	–	–	–	**904**	–	**1**	**8.4**	–	**1.7**	**0.3**
	–	–	–	–	–	–	–	7311	–	10	68.0	–	13.8	2.4
43	–	–	**0**	–	–	–	**0.38**	–	–	**8**	**22.7**	**4.71**	**19.6**	**3.8**
	–	–	0	–	–	–	0.40	–	–	9	23.8	4.94	20.6	3.9
44	–	–	–	–	–	–	–	–	–	**1**	**3.5**	–	**2.0**	**0.4**
	–	–	–	–	–	–	–	–	–	12	36.9	–	20.8	4.6
45	–	–	**0**	**9.1**	–	**50.9**	**0.33**	**1394**	**0.62**	**11**	**36.4**	**4.10**	**19.1**	**3.7**
	–	–	0	9.6	–	54.1	0.35	1480	0.66	12	38.6	4.35	20.3	3.9
46	–	–	–	**13.6**	–	**59.3**	**0.04**	**491**	**1.16**	**6**	**2.0**	**0.32**	**3.7**	**0.7**
	–	–	–	56.5	–	246.0	0.16	2036	4.82	25	8.2	1.32	15.3	3.1
47	–	–	–	–	–	–	–	–	–	**50**	–	–	**2.9**	**1.2**
	–	–	–	–	–	–	–	–	–	156	–	–	9.1	3.8
48	–	–	–	**10.0**	–	**114.0**	**0.06**	**877**	**0.06**	**21**	**9.5**	**1.39**	**3.0**	**6.6**
	–	–	–	49.3	–	561.5	0.30	4319	0.30	102	46.7	6.85	14.6	32.7
49	–	–	–	–	–	–	–	–	–	**1**	**5.3**	–	**1.4**	**0.3**
	–	–	–	–	–	–	–	–	–	14	76.7	–	20.3	4.3
50	–	–	–	–	–	–	–	–	–	**2**	**7.3**	–	**8.8**	**0.7**
	–	–	–	–	–	–	–	–	–	2	7.9	–	9.6	0.7
51	–	–	–	–	–	–	–	–	–	**3**	**11.8**	–	**14.2**	**2.6**
	–	–	–	–	–	–	–	–	–	5	22.0	–	26.5	4.8
52	–	–	–	**0.2**	–	**18.9**	**0.35**	**1790**	**0.85**	**11**	**46.2**	**3.21**	**27.4**	**4.0**
	–	–	–	0.2	–	20.3	0.38	1921	0.91	11	49.6	3.45	29.4	4.3
53	–	–	–	–	–	**35.9**	**0.50**	**4096**	**0.89**	**18**	**74.5**	**4.48**	**47.6**	**5.0**
	–	–	–	–	–	38.4	0.54	4387	0.96	19	79.8	4.79	50.9	5.4
54	–	–	–	–	–	–	–	–	–	–	–	–	–	–
	–	–	–	–	–	–	–	–	–	–	–	–	–	–
55	–	–	–	–	–	–	–	–	–	–	–	–	–	–
	–	–	–	–	–	–	–	–	–	–	–	–	–	–
56	**230.0**	–	–	–	–	–	–	–	–	**101**	–	–	**1.4**	**0.8**
	779.7	–	–	–	–	–	–	–	–	342	–	–	4.8	2.7
57	–	–	–	–	–	–	–	–	–	**55**	–	–	**3.8**	**0.7**
	–	–	–	–	–	–	–	–	–	230	–	–	15.6	2.9
58	–	–	–	–	–	–	–	–	–	–	–	–	–	–
	–	–	–	–	–	–	–	–	–	–	–	–	–	–
59	–	–	–	–	–	–	–	–	–	–	–	–	–	–
	–	–	–	–	–	–	–	–	–	–	–	–	–	–
60	–	–	–	–	–	–	–	–	–	–	–	–	–	–
	–	–	–	–	–	–	–	–	–	–	–	–	–	–
61	–	–	–	**21.0**	–	**9.3**	**0.24**	**5583**	**0.38**	**1**	**22.8**	**1.92**	**3.0**	**1.0**
	–	–	–	82.6	–	36.6	0.94	21983	1.48	3	89.8	7.57	12.0	4.0
62	–	–	–	**4.3**	–	**18.0**	**0.03**	**170**	**0.79**	**38**	**4.1**	**0.62**	**0.9**	**0.1**
	–	–	–	13.1	–	55.4	0.08	523	2.43	117	12.6	1.91	2.8	0.3
63	–	–	–	–	–	–	–	–	–	**45**	–	–	**2.0**	**0.3**
	–	–	–	–	–	–	–	–	–	180	–	–	8.0	1.2
64	–	–	–	–	–	–	–	–	–	–	–	–	–	–
	–	–	–	–	–	–	–	–	–	–	–	–	–	–
65	–	–	–	–	–	–	–	–	–	**74**	–	–	**2.1**	**0.8**
	–	–	–	–	–	–	–	–	–	225	–	–	6.5	2.5
66	–	–	–	–	–	–	–	–	–	–	–	–	–	–
	–	–	–	–	–	–	–	–	–	–	–	–	–	–
67	–	–	–	–	–	–	–	–	–	**29**	–	–	**1.2**	**2.4**
	–	–	–	–	–	–	–	–	–	33	–	–	1.3	2.7

(Continued)

TABLE V-1B PROTEIN FEEDS, COMPOSITION OF FEEDS, DATA EXPRESSED **AS-FED** AND **MOISTURE-FREE**—*(Continued)*

Entry Number	Feed Name Description	International Feed Number	Moisture Basis: A-F (as-fed) or M-F (moisture-free)	Proximate Analysis									Digestible Protein		
				Dry Matter	Ash	Crude Fiber	Neutral Det. Fib. (NDF)	Acid Det. Fib. (ADF)	Lignin	Ether Extract (Fat)	N-Free Extract	Crude Protein	Ruminant	Swine	Horse
				%	%	%	%	%	%	%	%	%	%	%	%
	CITRUS *Citrus* spp														
68	MOLASSES, AMMONIATED	5-01-240	A-F	61	4.7	–	–	–	–	2.2	32.5	21.4	18.0	20.4	18.5
			M-F	100	7.7	–	–	–	–	3.5	53.5	35.2	29.6	33.7	30.5
69	PULP AMMONIATED, DEHY	4-01-238	A-F	87	4.6	13.2	–	–	–	5.6	51.9	12.1	8.2	9.3	8.7
			M-F	100	5.3	15.1	–	–	–	6.4	59.4	13.8	9.3	10.6	10.0
70	SEEDS, MEAL MECH EXTD	5-01-239	A-F	88	6.0	10.2	–	–	–	10.2	30.7	31.3	26.3	29.9	27.2
			M-F	100	6.8	11.5	–	–	–	11.6	34.8	35.4	29.8	33.9	30.7
	CITRUS, ORANGE *Citrus sinensis*														
71	PULP WITHOUT FINES, AMMONIATED, DEHY	4-01-255	A-F	89	5.6	12.4	–	–	–	3.3	53.0	14.7	10.4	11.6	10.8
			M-F	100	6.3	13.9	–	–	–	3.7	59.6	16.5	11.7	13.0	12.1
	CLOVER, RED *Trifolium pratense*														
72	SEEDS	5-08-004	A-F	88	6.7	9.2	–	–	–	7.8	31.9	32.1	27.0	30.8	28.1
			M-F	100	7.7	10.5	–	–	–	8.9	36.4	36.6	30.7	35.0	32.0
73	SEED SCREENINGS	5-08-005	A-F	90	5.9	12.7	–	–	–	5.9	37.3	28.6	23.8	27.2	23.9
			M-F	100	6.5	14.1	–	–	–	6.5	41.2	31.6	26.4	30.2	26.5
	CLOVER, SWEET, YELLOW *Melilotus officinalis*														
74	SEED SCREENINGS	5-08-007	A-F	87	8.6	15.5	–	–	–	3.6	41.0	18.5	14.7	17.3	12.8
			M-F	100	9.9	17.8	–	–	–	4.1	47.0	21.2	16.9	19.8	14.7
	COCOA (CACAO) *Theobroma cacao*														
75	SEEDS	5-01-569	A-F	95	5.8	5.3	–	–	–	16.4	43.2	24.6	9.1	23.2	18.9
			M-F	100	6.1	5.6	–	–	–	17.3	45.3	25.8	9.5	24.4	19.9
	COCONUT *Cocos nucifera*														
76	KERNELS WITH COATS, MEAL MECH EXTD (COPRA MEAL)	5-01-572	A-F	92	6.4	12.1	–	18.3	–	6.8	45.0	21.2	17.6	15.5	15.4
			M-F	100	7.0	13.2	–	20.0	–	7.4	49.2	23.1	19.2	16.9	16.9
77	KERNELS WITH COATS, MEAL SOLV EXTD (COPRA MEAL)	5-01-573	A-F	91	6.0	14.4	–	21.9	–	2.1	47.3	21.3	17.2	15.5	15.6
			M-F	100	6.6	15.8	–	24.0	–	2.3	52.0	23.4	18.8	17.1	17.1
	COMMON LESPEDEZA—SEE LESPEDEZA, COMMON-KOREAN														
	CORN *Zea mays*														
78	DISTILLERS GRAINS, DEHY	5-02-842	A-F	93	2.2	11.5	–	–	–	8.9	43.1	27.8	20.1	19.4	22.7
			M-F	100	2.4	12.3	–	–	–	9.5	46.2	29.7	21.5	20.8	24.3
79	DISTILLERS SOLUBLES, DEHY	5-02-844	A-F	93	7.2	4.6	–	–	–	8.6	45.2	27.4	21.5	20.0	22.3
			M-F	100	7.7	4.9	–	–	–	9.3	48.6	29.5	23.2	21.5	24.1
80	FURFURAL RESIDUE, AMMONIATED	5-08-397	A-F	94	5.7	51.7	–	–	–	0.4	1.9	34.6	29.2	33.2	30.4
			M-F	100	6.0	54.8	–	–	–	0.4	2.0	36.8	31.0	35.2	32.3
81	GERM MEAL, WET MILLED, SOLV EXTD	5-02-898	A-F	92	3.9	12.2	–	–	–	1.5	53.5	20.7	16.6	19.4	14.9
			M-F	100	4.2	13.3	–	–	–	1.6	58.3	22.6	18.1	21.2	16.2
82	GLUTEN FEED	5-02-903	A-F	90	6.6	8.7	–	–	–	2.1	49.4	23.0	19.8	18.2	17.7
			M-F	100	7.4	9.7	–	–	–	2.4	55.0	25.6	22.0	20.2	19.6
83	GLUTEN MEAL[1]	5-02-900	A-F	91	3.1	4.5	33.7	8.2	–	2.2	38.4	43.2	36.3	41.7	40.3
			M-F	100	3.4	4.9	37.0	9.0	–	2.4	42.0	47.3	39.7	45.7	44.1
	COTTON *Gossypium* spp														
84	SEEDS WITHOUT LINT	5-13-749	A-F	91	4.7	19.4	–	–	–	20.4	24.5	21.8	17.7	20.5	16.2
			M-F	100	5.2	21.4	–	–	–	22.5	27.0	24.0	19.4	22.6	17.9
85	SEEDS, MEAL MECH EXTD, 36% PROTEIN	5-01-625	A-F	92	6.4	15.0	–	–	–	4.5	28.8	37.2	29.4	35.8	33.5
			M-F	100	7.0	16.3	–	–	–	4.9	31.3	40.5	32.0	38.9	36.5
86	SEEDS, MEAL MECH EXTD, 41% PROTEIN	5-01-617	A-F	93	6.1	11.9	25.9	18.5	5.6	4.7	28.9	41.0	35.1	33.8	37.7
			M-F	100	6.6	12.9	28.0	20.0	6.0	5.0	31.2	44.3	37.9	36.5	40.8
87	SEEDS, MEAL, PREPRESSED, SOLV EXTD, 41% PROTEIN[1]	5-07-872	A-F	90	6.4	12.9	23.4	17.1	–	1.0	28.8	41.3	35.4	39.8	38.2
			M-F	100	7.0	14.2	26.0	19.0	–	1.2	31.9	45.7	39.1	44.1	42.3
88	SEEDS, MEAL SOLV EXTD, 41% PROTEIN	5-01-621	A-F	91	6.5	12.1	23.6	18.4	5.5	1.5	29.6	41.2	31.3	35.1	38.2
			M-F	100	7.1	13.4	26.0	20.2	6.0	1.6	32.5	45.4	34.5	38.6	42.0
89	SEEDS, MEAL SOLV EXTD, 46% PROTEIN[1]	5-26-100	A-F	92	7.2	8.9	25.8	19.3	–	1.6	26.8	47.6	41.1	46.1	45.2
			M-F	100	7.8	9.6	28.0	21.0	–	1.8	29.1	51.7	44.6	50.0	49.1
90	SEEDS, MEAL SOLV EXTD, 48% PROTEIN	5-26-101	A-F	90	7.4	7.0	–	–	–	1.9	24.7	49.3	42.7	47.8	47.4
			M-F	100	8.2	7.7	–	–	–	2.1	27.4	54.7	47.3	53.0	52.5
91	SEEDS, LOW GOSSYPOL, MEAL SOLV EXTD	5-01-633	A-F	93	5.8	12.7	–	–	–	1.2	31.5	41.5	35.5	40.0	38.3
			M-F	100	6.3	13.7	–	–	–	1.3	34.0	44.8	38.3	43.2	41.3
92	SEEDS WITHOUT HULLS, MEAL, PREPRESSED, SOLV EXTD, 50% PROTEIN	5-07-874	A-F	93	6.6	8.2	–	–	–	1.3	26.8	50.3	40.7	48.7	48.2
			M-F	100	7.1	8.8	–	–	–	1.4	28.8	54.0	43.8	52.4	51.8
93	SEEDS, MEAL SOLV EXTD, AMMONIATED	5-09-352	A-F	91	6.6	14.1	–	–	–	3.8	–	44.5	–	–	–
			M-F	100	7.3	15.5	–	–	–	4.2	–	48.9	–	–	–
	COTTON, GLANDLESS *Gossypium* spp														
94	SEEDS, MEAL SOLV EXTD	5-08-979	A-F	95	7.3	2.5	–	–	–	1.9	23.8	59.8	52.1	58.2	58.7
			M-F	100	7.6	2.7	–	–	–	2.0	25.0	62.7	54.6	61.0	61.6
	COWPEA, COMMON *Vigna sinensis*														
95	PODS WITH SEEDS, DEHY	5-01-648	A-F	92	4.7	10.8	–	–	–	1.3	53.8	21.7	17.5	20.4	16.0
			M-F	100	5.1	11.7	–	–	–	1.4	58.3	23.5	19.0	22.1	17.3
96	SEEDS	5-01-661	A-F	89	3.5	5.0	–	–	–	1.5	55.5	23.8	19.6	22.4	18.7
			M-F	100	3.9	5.6	–	–	–	1.6	62.2	26.7	21.9	25.1	20.9
	CRAB *Callinectes sapidus, cancer* spp, *Paralithodes camschatica*														
97	CANNERY RESIDUE, MEAL (CRAB MEAL)	5-01-663	A-F	92	41.1	10.7	–	–	–	2.2	5.9	32.2	27.1	30.9	27.9
			M-F	100	44.6	11.6	–	–	–	2.4	6.4	35.0	29.4	33.5	30.3

Entry Number	TDN Ruminant	TDN Swine	TDN Horse	DE Ruminant		DE Swine		DE Horse		ME Ruminant		ME Swine		Poultry ME_n		ME Horse		Ruminant NE_m		Ruminant NE_g		Lactating Cows NE_{lc}	
	%	%	%	Mcal		kcal		Mcal		Mcal		kcal		kcal		Mcal		Mcal		Mcal		Mcal	
				lb	*kg*	lb	*kg*	lb	*kg*	lb	*kg*	lb	*kg*	lb	*kg*	lb	*kg*	lb	*kg*	lb	*kg*	lb	*kg*
68	**46**	–	–	**0.93**	***2.05***	–	–	–	–	**0.81**	***1.80***	–	–	–	–	–	–	–	–	–	–	–	–
	76	–	–	1.53	*3.37*	–	–	–	–	1.34	*2.96*	–	–	–	–	–	–	–	–	–	–	–	–
69	**76**	**65**	**42**	**1.40**	***3.09***	**1293**	***2850***	**0.80**	***1.77***	**1.24**	***2.73***	**1216**	***2680***	–	–	**0.66**	***1.45***	**0.69**	***1.53***	**0.45**	***0.99***	**0.67**	***1.47***
	87	74	48	1.60	*3.53*	1479	*3260*	0.92	*2.03*	1.41	*3.12*	1391	*3066*	–	–	0.75	*1.66*	0.79	*1.75*	0.51	*1.13*	0.76	*1.68*
70	**75**	**69**	–	**1.49**	***3.27***	**1390**	***3064***	–	–	**1.32**	***2.91***	**1208**	***2663***	–	–	–	–	**0.82**	***1.80***	**0.55**	***1.22***	**0.77**	***1.70***
	85	79	–	1.68	*3.71*	1574	*3469*	–	–	1.49	*3.29*	1368	*3015*	–	–	–	–	0.93	*2.04*	0.63	*1.38*	0.87	*1.92*
71	**64**	**62**	**56**	**1.27**	***2.81***	**1230**	***2712***	**1.03**	***2.28***	**1.11**	***2.44***	**1153**	***2542***	–	–	**0.85**	***1.87***	**0.67**	***1.49***	**0.43**	***0.94***	**0.66**	***1.45***
	72	69	63	1.43	*3.16*	1382	*3047*	1.16	*2.56*	1.24	*2.74*	1296	*2856*	–	–	0.95	*2.10*	0.76	*1.67*	0.48	*1.06*	0.74	*1.62*
72	**77**	**66**	–	**1.49**	***3.28***	**1310**	***2888***	–	–	**1.33**	***2.92***	**1135**	***2501***	–	–	–	–	**0.79**	***1.75***	**0.53**	***1.17***	**0.75**	***1.65***
	88	75	–	1.70	*3.74*	1491	*3288*	–	–	1.51	*3.33*	1292	*2848*	–	–	–	–	0.90	*1.99*	0.61	*1.34*	0.85	*1.88*
73	**70**	**64**	–	**1.38**	***3.04***	**1279**	***2819***	–	–	**1.21**	***2.67***	**1104**	***2433***	–	–	–	–	**0.74**	***1.62***	**0.48**	***1.06***	**0.71**	***1.56***
	78	71	–	1.53	*3.37*	1416	*3121*	–	–	1.34	*2.95*	1222	*2694*	–	–	–	–	0.81	*1.79*	0.53	*1.17*	0.78	*1.72*
74	–	–	–	–	–	–	–	–	–	–	–	–	–	–	–	–	–	–	–	–	–	–	–
	–	–	–	–	–	–	–	–	–	–	–	–	–	–	–	–	–	–	–	–	–	–	–
75	**59**	**91**	–	**1.19**	***2.61***	**1812**	***3996***	–	–	**1.00**	***2.21***	**1593**	***3511***	–	–	–	–	**0.58**	***1.29***	**0.33**	***0.74***	**0.60**	***1.31***
	62	95	–	1.24	*2.74*	1903	*4196*	–	–	1.05	*2.32*	1672	*3687*	–	–	–	–	0.61	*1.35*	0.35	*0.77*	0.63	*1.38*
76	**74**	**76**	–	**1.52**	***3.34***	**1500**	***3306***	–	–	**1.33**	***2.94***	**1538**	***3392***	**681**	***1502***	–	–	**0.84**	***1.85***	**0.57**	***1.25***	**0.80**	***1.75***
	81	83	–	1.66	*3.65*	1638	*3612*	–	–	1.46	*3.21*	1681	*3705*	745	*1641*	–	–	0.92	*2.03*	0.62	*1.37*	0.87	*1.91*
77	**69**	**73**	–	**1.37**	***3.01***	**1460**	***3218***	–	–	**1.20**	***2.64***	**1386**	***3055***	**692**	***1525***	–	–	**0.74**	***1.63***	**0.48**	***1.06***	**0.70**	***1.55***
	75	80	–	1.50	*3.31*	1602	*3532*	–	–	1.31	*2.90*	1521	*3353*	759	*1674*	–	–	0.81	*1.79*	0.53	*1.16*	0.77	*1.70*
78	**81**	**65**	–	**1.54**	***3.41***	**1246**	***2746***	–	–	**1.28**	***2.82***	**1196**	***2636***	**894**	***1970***	–	–	**0.87**	***1.92***	**0.59**	***1.30***	**0.84**	***1.86***
	87	69	–	1.65	*3.64*	1333	*2938*	–	–	1.37	*3.02*	1279	*2820*	956	*2108*	–	–	0.93	*2.05*	0.63	*1.39*	0.90	*1.99*
79	**80**	**78**	–	**1.55**	***3.42***	**1474**	***3250***	–	–	**1.34**	***2.96***	**1419**	***3128***	**1324**	***2919***	–	–	**0.92**	***2.03***	**0.63**	***1.40***	**0.86**	***1.89***
	87	84	–	1.67	*3.68*	1587	*3499*	–	–	1.45	*3.19*	1528	*3369*	1426	*3143*	–	–	0.99	*2.19*	0.68	*1.51*	0.92	*2.03*
80	**43**	–	–	**0.85**	***1.88***	–	–	–	–	**0.67**	***1.47***	–	–	–	–	–	–	–	–	–	–	–	–
	45	–	–	0.90	*1.99*	–	–	–	–	0.71	*1.56*	–	–	–	–	–	–	–	–	–	–	–	–
81	**68**	**69**	–	**1.41**	***3.11***	**1568**	***3456***	–	–	**1.24**	***2.73***	**1372**	***3025***	**772**	***1702***	–	–	**0.73**	***1.61***	**0.47**	***1.04***	**0.71**	***1.55***
	74	76	–	1.54	*3.39*	1708	*3765*	–	–	1.35	*2.98*	1495	*3295*	841	*1854*	–	–	0.80	*1.76*	0.52	*1.14*	0.77	*1.69*
82	**75**	**75**	–	**1.44**	***3.17***	**1375**	***3031***	–	–	**1.21**	***2.67***	**1037**	***2287***	**784**	***1729***	–	–	**0.82**	***1.80***	**0.55**	***1.21***	**0.78**	***1.72***
	83	84	–	1.60	*3.52*	1529	*3371*	–	–	1.34	*2.96*	1154	*2543*	872	*1924*	–	–	0.91	*2.00*	0.61	*1.35*	0.87	*1.91*
83	**78**	**80**	–	**1.50**	***3.31***	**1637**	***3609***	–	–	**1.24**	***2.74***	**1437**	***3168***	**1367**	***3015***	–	–	**0.84**	***1.85***	**0.57**	***1.25***	**0.80**	***1.75***
	85	88	–	1.64	*3.62*	1792	*3951*	–	–	1.36	*2.99*	1573	*3467*	1497	*3300*	–	–	0.92	*2.03*	0.62	*1.37*	0.87	*1.92*
84	**87**	**71**	–	**1.57**	***3.46***	**1430**	***3152***	–	–	**1.39**	***3.07***	**1243**	***2739***	–	–	–	–	**0.97**	***2.13***	**0.67**	***1.49***	**0.91**	***2.01***
	95	79	–	1.72	*3.80*	1572	*3467*	–	–	1.53	*3.38*	1367	*3013*	–	–	–	–	1.06	*2.34*	0.74	*1.63*	1.00	*2.21*
85	**66**	**67**	–	**1.32**	***2.91***	**1360**	***2999***	–	–	**1.15**	***2.52***	**1180**	***2602***	**891**	***1964***	–	–	**0.73**	***1.62***	**0.47**	***1.05***	**0.71**	***1.56***
	72	73	–	1.44	*3.16*	1480	*3264*	–	–	1.25	*2.75*	1284	*2831*	969	*2137*	–	–	0.80	*1.76*	0.52	*1.14*	0.77	*1.69*
86	**72**	**69**	–	**1.49**	***3.29***	**1305**	***2878***	–	–	**1.12**	***2.47***	**1197**	***2638***	**1025**	***2261***	–	–	**0.72**	***1.59***	**0.64**	***1.41***	**0.76**	***1.67***
	77	75	–	1.61	*3.56*	1410	*3109*	–	–	1.21	*2.67*	1293	*2851*	1108	*2443*	–	–	0.78	*1.72*	0.69	*1.52*	0.82	*1.81*
87	**72**	**61**	–	**1.51**	***3.33***	**1184**	***2610***	–	–	**1.14**	***2.51***	**1064**	***2347***	**974**	***2147***	–	–	**0.70**	***1.55***	**0.64**	***1.41***	**0.76**	***1.67***
	80	68	–	1.67	*3.69*	1311	*2889*	–	–	1.26	*2.78*	1178	*2598*	1078	*2377*	–	–	0.78	*1.71*	0.71	*1.56*	0.84	*1.85*
88	**68**	**61**	–	**1.48**	***3.27***	**1209**	***2666***	–	–	**1.17**	***2.57***	**1069**	***2356***	**889**	***1960***	–	–	**0.74**	***1.64***	**0.63**	***1.40***	**0.73**	***1.61***
	75	67	–	1.63	*3.60*	1330	*2933*	–	–	1.28	*2.83*	1176	*2592*	978	*2156*	–	–	0.82	*1.80*	0.70	*1.54*	0.80	*1.77*
89	**71**	**64**	–	**1.42**	***3.13***	**1282**	***2826***	–	–	**1.25**	***2.74***	**1105**	***2436***	**950**	***2095***	–	–	**0.77**	***1.70***	**0.51**	***1.12***	**0.74**	***1.63***
	77	70	–	1.54	*3.40*	1392	*3068*	–	–	1.35	*2.98*	1199	*2644*	1032	*2275*	–	–	0.84	*1.85*	0.55	*1.22*	0.80	*1.77*
90	**73**	**65**	–	**1.44**	***3.17***	**1292**	***2847***	–	–	**1.27**	***2.79***	**1116**	***2459***	**983**	***2166***	–	–	**0.80**	***1.77***	**0.54**	***1.18***	**0.76**	***1.67***
	81	72	–	1.59	*3.51*	1431	*3155*	–	–	1.40	*3.09*	1236	*2725*	1089	*2401*	–	–	0.89	*1.96*	0.60	*1.31*	0.84	*1.85*
91	**68**	**63**	–	**1.36**	***3.01***	**1265**	***2788***	–	–	**1.19**	***2.62***	**1089**	***2400***	–	–	–	–	**0.71**	***1.57***	**0.45**	***1.00***	**0.69**	***1.52***
	74	68	–	1.47	*3.24*	1364	*3008*	–	–	1.28	*2.83*	1174	*2589*	–	–	–	–	0.77	*1.69*	0.49	*1.08*	0.75	*1.64*
92	**70**	–	–	**1.43**	***3.15***	**1295**	***2854***	–	–	**1.25**	***2.76***	**1185**	***2613***	**971**	***2141***	–	–	**0.76**	***1.67***	**0.49**	***1.08***	**0.73**	***1.60***
	75	–	–	1.54	*3.38*	1392	*3068*	–	–	1.35	*2.97*	1274	*2809*	1044	*2301*	–	–	0.81	*1.79*	0.53	*1.16*	0.78	*1.72*
93	**66**	**66**	–	**1.30**	***2.88***	**1326**	***2923***	–	–	**1.07**	***2.36***	**1169**	***2576***	–	–	–	–	**0.66**	***1.46***	**0.43**	***0.94***	**0.70**	***1.54***
	72	73	–	1.43	*3.16*	1457	*3212*	–	–	1.18	*2.59*	1284	*2831*	–	–	–	–	0.73	*1.60*	0.47	*1.03*	0.77	*1.69*
94	**81**	**78**	–	**1.62**	***3.58***	**1560**	***3438***	–	–	**1.45**	***3.19***	**1359**	***2996***	**948**	***2089***	–	–	**0.92**	***2.03***	**0.63**	***1.39***	**0.86**	***1.90***
	85	82	–	1.70	*3.75*	1635	*3605*	–	–	1.52	*3.34*	1425	*3141*	993	*2190*	–	–	0.97	*2.13*	0.66	*1.46*	0.91	*2.00*
95	**71**	**69**	–	**1.41**	***3.11***	**1377**	***3036***	–	–	**1.24**	***2.72***	**1193**	***2630***	–	–	–	–	**0.72**	***1.59***	**0.46**	***1.02***	**0.70**	***1.53***
	76	75	–	1.53	*3.37*	1492	*3290*	–	–	1.34	*2.95*	1292	*2849*	–	–	–	–	0.78	*1.72*	0.50	*1.10*	0.75	*1.66*
96	**76**	**75**	–	**1.48**	***3.27***	**1500**	***3307***	–	–	**1.32**	***2.90***	**1309**	***2886***	–	–	–	–	**0.80**	***1.76***	**0.53**	***1.18***	**0.75**	***1.66***
	85	84	–	1.66	*3.66*	1682	*3708*	–	–	1.47	*3.25*	1468	*3236*	–	–	–	–	0.89	*1.97*	0.60	*1.32*	0.84	*1.86*
97	**27**	–	–	**0.54**	***1.18***	**686**	***1511***	–	–	**0.35**	***0.78***	**555**	***1224***	**827**	***1823***	–	–	**0.07**	***0.16***	–	–	**0.25**	***0.54***
	29	–	–	0.58	*1.29*	744	*1640*	–	–	0.39	*0.85*	602	*1328*	897	*1977*	–	–	0.08	*0.18*	–	–	0.27	*0.59*

(Continued)

TABLE V-1B PROTEIN FEEDS, MINERAL AND VITAMIN COMPOSITION OF FEEDS, DATA EXPRESSED **AS-FED** AND **MOISTURE-FREE**—*(Continued)*

Entry Number	Feed Name Description	Moisture Basis: A-F (as-fed) or M-F (moisture-free)	Dry Matter	Macro Minerals							Micro Minerals						
				Calcium (Ca)	Phosphorus (P)	Sodium (Na)	Chlorine (Cl)	Magnesium (Mg)	Potassium (K)	Sulfur (S)	Cobalt (Co)	Copper (Cu)	Iodine (I)	Iron (Fe)	Manganese (Mn)	Selenium (Se)	Zinc (Zn)
			%	%	%	%	%	%	%	%	ppm or mg/kg	ppm or mg/kg	ppm or mg/kg	%	ppm or mg/kg	ppm or mg/kg	ppm or mg/kg
	CITRUS *Citrus spp*																
68	MOLASSES, AMMONIATED	A-F	61	0.76	0.16	–	–	0.08	–	–	–	–	–	–	–	–	–
		M-F	100	1.25	0.26	–	–	0.13	–	–	–	–	–	–	–	–	–
69	PULP, AMMONIATED, DEHY	A-F	87	1.66	0.12	–	–	0.07	–	–	–	–	–	–	–	–	–
		M-F	100	1.90	0.14	–	–	0.08	–	–	–	–	–	–	–	–	–
70	SEEDS, MEAL MECH EXTD	A-F	88	1.10	0.67	–	–	0.60	1.31	–	–	6.6	–	0.030	7.5	–	7.5
		M-F	100	1.25	0.75	–	–	0.68	1.49	–	–	7.5	–	0.033	8.5	–	8.5
	CITRUS, ORANGE *Citrus sinensis*																
71	PULP WITHOUT FINES, AMMONIATED, DEHY	A-F	89	–	–	–	–	–	–	–	–	–	–	–	–	–	–
		M-F	100	–	–	–	–	–	–	–	–	–	–	–	–	–	–
	CLOVER, RED *Trifolium pratense*																
72	SEEDS	A-F	88	0.29	0.54	–	–	–	–	–	–	–	–	–	–	–	–
		M-F	100	0.33	0.61	–	–	–	–	–	–	–	–	–	–	–	–
73	SEED SCREENINGS	A-F	90	0.41	0.67	–	–	–	–	–	–	–	–	–	–	–	–
		M-F	100	0.46	0.74	–	–	–	–	–	–	–	–	–	–	–	–
	CLOVER, SWEET, YELLOW *Melilotus officinalis*																
74	SEED SCREENINGS	A-F	87	0.82	0.41	–	–	–	–	–	–	–	–	–	–	–	–
		M-F	100	0.94	0.47	–	–	–	–	–	–	–	–	–	–	–	–
	COCOA (CACAO) *Theobroma cacao*																
75	SEEDS	A-F	95	–	–	–	–	–	–	–	–	–	–	–	–	–	–
		M-F	100	–	–	–	–	–	–	–	–	–	–	–	–	–	–
	COCONUT *Cocos nucifera*																
76	KERNELS WITH COATS, MEAL MECH EXTD (COPRA MEAL)	A-F	92	0.19	0.60	0.04	–	0.30	1.65	0.34	0.127	16.7	–	0.068	70.1	–	48.5
		M-F	100	0.21	0.65	0.04	–	0.33	1.80	0.37	0.139	18.2	–	0.075	76.6	–	53.0
77	KERNELS WITH COATS, MEAL SOLV EXTD (COPRA MEAL)	A-F	91	0.17	0.60	0.04	0.03	0.31	1.41	–	–	–	–	–	54.5	–	–
		M-F	100	0.19	0.66	0.04	0.03	0.34	1.55	–	–	–	–	–	59.8	–	–
	COMMON LESPEDEZA–SEE LESPEDEZA, COMMON-KOREAN																
	CORN *Zea mays*																
78	DISTILLERS GRAINS, DEHY	A-F	93	0.09	0.39	0.09	0.07	0.07	0.16	0.43	0.076	38.9	0.048	0.020	19.3	0.352	41.7
		M-F	100	0.10	0.42	0.09	0.08	0.07	0.18	0.46	0.082	41.7	0.051	0.021	20.7	0.377	44.7
79	DISTILLERS SOLUBLES, DEHY	A-F	93	0.30	1.30	0.23	0.26	0.60	1.70	0.37	0.167	77.9	0.079	0.052	72.0	0.371	88.0
		M-F	100	0.32	1.40	0.24	0.28	0.65	1.83	0.40	0.180	83.9	0.085	0.056	77.6	0.400	94.8
80	FURFURAL RESIDUE, AMMONIATED	A-F	94	0.16	0.08	–	–	–	–	–	–	–	–	–	–	–	–
		M-F	100	0.17	0.09	–	–	–	–	–	–	–	–	–	–	–	–
81	GERM MEAL, WET MILLED, SOLV EXTD	A-F	92	0.04	0.51	0.04	0.04	0.16	0.35	0.31	–	4.5	–	0.034	3.8	0.340	104.8
		M-F	100	0.04	0.55	0.04	0.04	0.17	0.38	0.34	–	4.9	–	0.037	4.1	0.370	114.2
82	GLUTEN FEED	A-F	90	0.32	0.74	0.12	0.22	0.33	0.57	0.21	0.087	47.1	0.066	0.043	23.1	0.272	64.6
		M-F	100	0.36	0.82	0.14	0.25	0.36	0.64	0.23	0.097	52.3	0.074	0.048	25.7	0.302	71.8
83	GLUTEN MEAL	A-F	91	0.15	0.46	0.09	0.06	0.06	0.03	0.20	0.077	27.7	–	0.039	7.7	1.015	173.7
		M-F	100	0.16	0.51	0.10	0.07	0.06	0.03	0.22	0.085	30.3	–	0.043	8.5	1.111	190.2
	COTTON *Gossypium spp*																
84	SEEDS WITHOUT LINT	A-F	91	0.14	0.69	0.03	–	0.32	1.11	0.24	–	49.0	–	0.014	11.1	–	–
		M-F	100	0.16	0.76	0.03	–	0.35	1.22	0.26	–	53.9	–	0.016	12.2	–	–
85	SEEDS, MEAL MECH EXTD, 36% PROTEIN	A-F	92	0.18	0.95	0.04	–	0.68	1.34	0.26	0.150	16.5	–	0.016	22.6	–	56.7
		M-F	100	0.20	1.04	0.05	–	0.74	1.46	0.28	0.163	17.9	–	0.017	24.5	–	61.7
86	SEEDS, MEAL MECH EXTD, 41% PROTEIN	A-F	93	0.19	1.07	0.04	0.04	0.53	1.33	0.40	0.626	18.5	–	0.018	22.3	–	61.8
		M-F	100	0.21	1.16	0.05	0.05	0.57	1.44	0.43	0.676	20.0	–	0.019	24.1	–	66.8
87	SEEDS, MEAL, PREPRESSED, SOLV EXTD, 41% PROTEIN	A-F	90	0.16	1.07	0.04	0.06	0.48	1.25	0.31	0.738	18.2	–	0.018	20.4	–	62.7
		M-F	100	0.17	1.18	0.04	0.07	0.53	1.38	0.34	0.817	20.2	–	0.019	22.5	–	69.4
88	SEEDS, MEAL SOLV EXTD, 41% PROTEIN	A-F	91	0.17	1.11	0.04	0.04	0.54	1.37	0.25	0.483	19.5	–	0.019	20.6	–	60.7
		M-F	100	0.19	1.22	0.05	0.05	0.59	1.51	0.27	0.531	21.4	–	0.021	22.7	–	66.7
89	SEEDS, MEAL SOLV EXTD, 46% PROTEIN	A-F	92	–	–	–	–	–	–	–	–	–	–	–	–	–	–
		M-F	100	–	–	–	–	–	–	–	–	–	–	–	–	–	–
90	SEEDS, MEAL SOLV EXTD, 48% PROTEIN	A-F	90	0.20	1.20	–	–	–	–	–	–	–	–	–	–	–	–
		M-F	100	0.22	1.33	–	–	–	–	–	–	–	–	–	–	–	–
91	SEEDS, LOW GOSSYPOL, MEAL SOLV EXTD	A-F	93	–	–	–	–	–	–	–	–	–	–	–	–	–	–
		M-F	100	–	–	–	–	–	–	–	–	–	–	–	–	–	–
92	SEEDS WITHOUT HULLS, MEAL, PREPRESSED, SOLV EXTD, 50% PROTEIN	A-F	93	0.18	1.16	0.05	0.05	0.46	1.45	0.52	0.042	14.5	–	0.012	23.0	–	73.8
		M-F	100	0.19	1.24	0.06	0.05	0.50	1.56	0.56	0.046	15.6	–	0.012	24.8	–	79.4
93	SEEDS, MEAL SOLV EXTD, AMMONIATED	A-F	91	–	–	–	–	–	–	–	–	–	–	–	–	–	–
		M-F	100	–	–	–	–	–	–	–	–	–	–	–	–	–	–
	COTTON, GLANDLESS *Gossypium spp*																
94	SEEDS, MEAL SOLV EXTD	A-F	95	–	–	–	–	–	–	–	–	–	–	–	–	–	–
		M-F	100	–	–	–	–	–	–	–	–	–	–	–	–	–	–
	COWPEA, COMMON *Vigna sinensis*																
95	PODS WITH SEEDS, DEHY	A-F	92	–	–	–	–	–	–	–	–	–	–	–	–	–	–
		M-F	100	–	–	–	–	–	–	–	–	–	–	–	–	–	–
96	SEEDS	A-F	89	0.09	0.44	0.04	0.04	0.26	1.16	0.25	–	4.4	–	0.021	40.2	–	–
		M-F	100	0.10	0.50	0.04	0.05	0.29	1.30	0.28	–	5.0	–	0.024	45.1	–	–
	CRAB *Callinectes sapidus, cancer spp, Paralithodes camschatica*																
97	CANNERY RESIDUE, MEAL (CRAB MEAL)	A-F	92	14.46	1.58	0.88	1.51	0.94	0.45	0.25	–	32.7	0.557	0.435	132.8	–	–
		M-F	100	15.69	1.72	0.95	1.63	1.02	0.49	0.27	–	35.5	0.605	0.472	144.0	–	–

PROTEIN FEEDS

Entry Number	Fat-Soluble Vitamins: A (1 mg Carotene = 1667 IU Vit A)	Carotene (Provitamin A)	D	E	K	Water-Soluble Vitamins: B–12	Biotin	Choline	Folacin (Folic Acid)	Niacin	Pantothenic Acid (B–3)	(Pyri-doxine) B–6	Ribo-flavin (B–2)	Thiamin (B–1)
	IU/g	ppm or mg/kg	IU/kg	ppm or mg/kg	ppm or mg/kg	ppm or mg/kg	ppm or mg/kg	ppm or mg/kg	ppm or mg/kg	ppm or mg/kg	ppm or mg/kg	ppm or mg/kg	ppm or mg/kg	ppm or mg/kg
68	–	–	–	–	–	–	–	–	–	–	–	–	–	–
	–	–	–	–	–	–	–	–	–	–	–	–	–	–
69	–	–	–	–	–	–	–	–	–	–	–	–	–	–
	–	–	–	–	–	–	–	–	–	–	–	–	–	–
70	–	–	–	–	–	–	–	–	–	–	–	–	–	–
	–	–	–	–	–	–	–	–	–	–	–	–	–	–
71	–	–	–	–	–	–	–	–	–	–	–	–	–	–
	–	–	–	–	–	–	–	–	–	–	–	–	–	–
72	–	–	–	–	–	–	–	–	–	–	–	–	–	–
	–	–	–	–	–	–	–	–	–	–	–	–	–	–
73	**4.4**	**2.7**	–	–	–	–	–	–	–	–	–	–	–	–
	4.9	2.9	–	–	–	–	–	–	–	–	–	–	–	–
74	–	–	–	–	–	–	–	–	–	–	–	–	–	–
	–	–	–	–	–	–	–	–	–	–	–	–	–	–
75	–	–	–	–	–	–	–	–	–	–	–	–	–	–
	–	–	–	–	–	–	–	–	–	–	–	–	–	–
76	–	–	–	–	–	–	–	**1046**	**1.08**	**27**	**6.1**	–	**3.3**	**0.8**
	–	–	–	–	–	–	–	1143	1.18	30	6.6	–	3.6	0.9
77	–	–	–	–	–	–	–	**1089**	**0.30**	**24**	**6.5**	**4.36**	**3.5**	–
	–	–	–	–	–	–	–	1196	0.33	26	7.2	4.78	3.8	–
78	**5.2**	**3.1**	–	–	–	**0.3**	**0.41**	**1113**	**1.00**	**38**	**11.3**	**4.22**	**5.0**	**1.8**
	5.6	3.3	–	–	–	0.3	0.44	1191	1.07	41	12.1	4.51	5.3	1.9
79	**1.1**	**0.7**	–	**45.9**	–	**4.2**	**1.49**	**4751**	**1.34**	**124**	**23.3**	**9.41**	**15.1**	**6.8**
	1.2	0.7	–	49.4	–	4.5	1.61	5116	1.45	133	25.0	10.14	16.3	7.3
80	–	–	–	–	–	–	–	–	–	–	–	–	–	–
	–	–	–	–	–	–	–	–	–	–	–	–	–	–
81	**3.4**	**2.0**	–	**85.8**	–	–	**0.22**	**1586**	**0.20**	**39**	**4.2**	–	**3.8**	**4.5**
	3.7	2.2	–	93.5	–	–	0.24	1728	0.22	42	4.6	–	4.1	4.9
82	**9.8**	**5.9**	–	**12.1**	–	–	**0.33**	**1514**	**0.27**	**70**	**13.6**	**13.93**	**2.2**	**2.0**
	10.9	6.5	–	13.5	–	–	0.36	1684	0.30	78	15.1	15.49	2.5	2.2
83	**27.3**	**16.3**	–	**29.3**	–	–	**0.19**	**360**	**0.30**	**50**	**10.0**	**7.98**	**1.5**	**0.2**
	29.8	17.9	–	32.0	–	–	0.21	394	0.33	55	10.9	8.73	1.6	0.2
84	–	–	–	–	–	–	–	–	–	–	–	–	–	–
	–	–	–	–	–	–	–	–	–	–	–	–	–	–
85	–	–	–	**15.1**	–	–	**1.12**	**2739**	**3.75**	**29**	**9.7**	–	**5.3**	**4.5**
	–	–	–	16.4	–	–	1.21	2980	4.08	32	10.6	–	5.7	4.9
86	**0.4**	**0.2**	–	**32.3**	–	–	**0.91**	**2753**	**2.45**	**35**	**10.2**	**5.00**	**5.2**	**7.1**
	0.4	0.2	–	34.9	–	–	0.99	2974	2.65	38	11.0	5.41	5.6	7.6
87	–	–	–	–	–	–	**0.55**	**2861**	**2.57**	**40**	**7.3**	**4.11**	**5.3**	**3.3**
	–	–	–	–	–	–	0.61	3166	2.85	44	8.1	4.55	5.8	3.7
88	–	–	–	**14.6**	–	–	**0.55**	**2780**	**2.55**	**41**	**13.7**	**5.41**	**4.7**	**7.3**
	–	–	–	16.1	–	–	0.61	3058	2.81	45	15.1	5.95	5.2	8.0
89	–	–	–	–	–	–	–	–	–	–	–	–	–	–
	–	–	–	–	–	–	–	–	–	–	–	–	–	–
90	–	–	–	–	–	–	–	**3316**	–	**51**	**15.5**	–	**6.0**	–
	–	–	–	–	–	–	–	3674	–	56	17.1	–	6.6	–
91	–	–	–	–	–	–	–	–	–	–	–	–	–	**20.3**
	–	–	–	–	–	–	–	–	–	–	–	–	–	21.9
92	–	–	–	**11.3**	–	–	**0.44**	**2962**	**0.93**	**45**	**14.3**	**6.29**	**4.9**	**8.2**
	–	–	–	12.1	–	–	0.48	3184	1.00	48	15.4	6.76	5.3	8.8
93	–	–	–	–	–	–	–	–	–	–	–	–	–	–
	–	–	–	–	–	–	–	–	–	–	–	–	–	–
94	–	–	–	–	–	–	–	–	–	–	–	–	–	–
	–	–	–	–	–	–	–	–	–	–	–	–	–	–
95	–	–	–	–	–	–	–	–	–	–	–	–	–	–
	–	–	–	–	–	–	–	–	–	–	–	–	–	–
96	**0.4**	**0.2**	–	–	–	–	–	–	–	**24**	**15.5**	–	**2.3**	**9.3**
	0.4	0.2	–	–	–	–	–	–	–	27	17.3	–	2.5	10.4
97	–	–	–	–	–	**437.6**	**0.07**	**2008**	**0.11**	**45**	**6.5**	**6.62**	**6.1**	**0.4**
	–	–	–	–	–	474.8	0.07	2179	0.12	49	7.0	7.18	6.7	0.5

(Continued)

TABLE V-1B PROTEIN FEEDS, COMPOSITION OF FEEDS, DATA EXPRESSED **AS-FED** AND **MOISTURE-FREE**—*(Continued)*

Entry Number	Feed Name Description	International Feed Number	Moisture Basis: A-F (as-fed) or M-F (moisture-free)	Proximate Analysis: Dry Matter	Ash	Crude Fiber	Neutral Det. Fib. (NDF)	Acid Det. Fib. (ADF)	Lignin	Ether Extract (Fat)	N-Free Extract	Crude Protein	Digestible Protein: Ruminant	Swine	Horse
				%	%	%	%	%	%	%	%	%	%	%	%
	CRACKLINGS, SWINE *Sus scrofa*														
98	RENDERED	5-04-791	A-F	93	27.9	6.7	–	–	–	13.2	–	50.7	43.9	48.2	48.7
			M-F	100	30.1	7.2	–	–	–	14.2	–	54.6	47.3	51.9	52.4
	CRAMBE, ABYSSINIAN *Crambe abyssinica*														
99	SEEDS WITHOUT HULLS, MEAL MECH EXTD	5-16-453	A-F	92	5.8	4.2	–	–	–	4.9	31.4	45.8	–	–	–
			M-F	100	6.3	4.6	–	–	–	5.3	34.1	49.8	–	–	–
	DIAMMONIUM PHOSPHATE														
100	$(NH_4)_2HPO_4$	6-00-370	A-F	98	35.5	–	–	–	–	–	–	112.9	–	–	–
			M-F	100	36.3	–	–	–	–	–	–	115.5	–	–	–
	DISTILLERS PRODUCTS (ALSO SEE CORN; RYE; SORGHUM; WHEAT)														
101	GRAINS, DEHY	5-02-144	A-F	93	1.5	12.8	–	–	–	7.4	43.5	27.3	18.9	26.0	22.3
			M-F	100	1.6	13.8	–	–	–	8.0	47.0	29.5	20.4	28.1	24.1
102	SOLUBLES, DEHY	5-02-147	A-F	92	6.2	3.4	–	–	–	8.9	44.8	28.8	24.0	27.4	24.0
			M-F	100	6.7	3.7	–	–	–	9.7	48.6	31.3	26.0	29.8	26.1
	EGGS—SEE CHICKEN														
	FEATHERS, POULTRY														
103	MEAL, HYDROLYZED	5-03-795	A-F	93	3.2	1.4	–	6.1	–	5.1	–	83.8	74.1	61.2	86.1
			M-F	100	3.4	1.5	–	6.6	–	5.5	–	90.2	79.7	65.9	92.7
	FISH														
104	LIVER, MEAL MECH EXTD	5-01-968	A-F	93	6.1	1.2	–	–	–	17.3	5.4	62.8	54.9	61.2	62.3
			M-F	100	6.6	1.3	–	–	–	18.6	5.8	67.7	59.2	65.9	67.2
105	MEAL MECH EXTD	5-01-977	A-F	92	21.4	0.7	–	–	–	6.0	–	64.3	57.1	59.1	64.1
			M-F	100	23.3	0.8	–	–	–	6.6	–	70.2	62.3	64.5	70.0
106	SOLUBLES, CONDENSED	5-01-969	A-F	50	10.1	0.5	–	–	–	6.1	2.2	31.5	28.0	29.3	30.9
			M-F	100	20.0	1.0	–	–	–	12.2	4.4	62.5	55.6	58.1	61.3
107	SOLUBLES, DEHY	5-01-971	A-F	93	12.7	2.0	–	–	–	9.0	8.7	60.4	52.7	58.8	59.7
			M-F	100	13.7	2.1	–	–	–	9.7	9.4	65.1	56.8	63.4	64.3
	FISH, ANCHOVY *Engraulis ringen*														
108	MEAL MECH EXTD	5-01-985	A-F	92	14.7	1.0	–	–	–	4.1	6.7	65.4	57.3	55.0	65.4
			M-F	100	16.0	1.1	–	–	–	4.5	7.3	71.1	62.3	59.8	71.1
	FISH, MENHADEN *Brevoortia tyrannus*														
109	MEAL MECH EXTD	5-02-009	A-F	92	19.1	0.9	–	–	–	9.6	0.8	61.2	49.6	49.6	60.7
			M-F	100	20.9	1.0	–	–	–	10.5	0.8	66.8	54.1	54.1	66.2
	FISH, SARDINE *Clupea* spp														
110	FLESH, FRESH	5-07-312	A-F	29	2.4	–	–	–	–	8.5	–	19.0	–	–	–
			M-F	100	8.2	–	–	–	–	29.4	–	65.5	–	–	–
111	MEAL MECH EXTD	5-02-015	A-F	93	15.8	1.0	–	–	–	5.0	6.1	65.2	53.5	63.6	65.1
			M-F	100	17.0	1.1	–	–	–	5.4	6.5	70.0	57.4	68.2	69.8
	FISH, SHARK *Selachii* (order)														
112	MEAL MECH EXTD	5-02-018	A-F	91	13.4	0.5	–	–	–	2.6	1.8	72.8	64.1	67.0	73.8
			M-F	100	14.7	0.5	–	–	–	2.8	2.0	79.9	70.3	73.5	81.1
	FISH, TUNA *Thunnus thynnus*														
113	FLESH, FRESH	5-09-278	A-F	29	1.4	–	–	–	–	3.5	–	24.9	–	–	–
			M-F	100	4.7	–	–	–	–	12.2	–	86.0	–	–	–
114	MEAL MECH EXTD	5-02-023	A-F	93	21.9	0.8	–	–	–	6.9	4.2	59.0	43.9	54.2	58.0
			M-F	100	23.6	0.9	–	–	–	7.4	4.5	63.6	47.4	58.5	62.6
	FISH, WHITE *Gadidae* (family), *Lophiidae* (family), *Rajidae* (family)														
115	MEAL MECH EXTD	5-02-025	A-F	91	23.1	0.5	–	–	–	4.7	0.1	62.6	58.3	61.0	62.3
			M-F	100	25.4	0.6	–	–	–	5.1	0.1	68.8	64.0	67.0	68.5
	FLAX, COMMON *Linum usitatissimum*														
116	SEEDS, GROUND	5-02-042	A-F	96	5.0	6.3	–	–	–	35.9	30.7	17.8	13.9	16.5	11.2
			M-F	100	5.2	6.6	–	–	–	37.5	32.1	18.6	14.5	17.2	11.7
117	SEEDS, MEAL SOLV EXTD, 33% PROTEIN (LINSEED MEAL)	5-26-089	A-F	91	6.1	9.5	–	–	–	1.0	41.3	33.2	28.0	31.8	29.1
			M-F	100	6.7	10.4	–	–	–	1.1	45.4	36.5	30.8	34.9	31.9
118	SEEDS, MEAL SOLV EXTD, 35% PROTEIN (LINSEED MEAL)	5-26-090	A-F	90	5.8	8.9	–	–	–	1.7	38.2	35.7	30.3	34.3	32.0
			M-F	100	6.4	9.9	–	–	–	1.9	42.3	39.6	33.6	38.0	35.4
119	SEEDS, MEAL SOLV EXTD, 37% PROTEIN (LINSEED MEAL)[1]	5-26-091	A-F	90	5.8	9.3	22.5	17.1	–	1.0	37.4	36.6	31.2	35.2	33.0
			M-F	100	6.4	10.3	25.0	19.0	–	1.1	41.5	40.7	34.6	39.1	36.7
120	SEED SCREENINGS, MEAL MECH EXTD	5-02-054	A-F	91	8.6	11.2	22.9	15.5	6.4	7.3	39.6	24.8	20.4	23.5	19.6
			M-F	100	9.4	12.2	25.0	17.0	7.0	8.0	43.3	27.1	22.3	25.7	21.4
	FURFURAL, CORN *Zea mays*														
121	RESIDUE, AMMONIATED	5-08-397	A-F	94	5.7	51.7	–	–	–	0.4	1.9	34.6	29.2	33.2	30.4
			M-F	100	6.0	54.8	–	–	–	0.4	2.0	36.8	31.0	35.2	32.3
	GARBANZO (CHICKPEA; GRAM PEA) *Cicer arietinum*														
122	SEEDS	5-01-218	A-F	89	2.9	7.0	–	–	–	4.1	56.0	19.1	14.3	17.8	13.3
			M-F	100	3.3	7.9	–	–	–	4.6	62.8	21.4	16.1	20.0	14.9

PROTEIN FEEDS

Entry Number	TDN Ruminant	TDN Swine	TDN Horse	Digestible Energy Ruminant		Digestible Energy Swine		Digestible Energy Horse		Metabolizable Energy Ruminant		Metabolizable Energy Swine		Metabolizable Energy Poultry ME_n		Metabolizable Energy Horse		Net Energy Ruminant NE_m		Net Energy Ruminant NE_g		Net Energy Lactating Cows NE_{lc}	
	%	%	%	Mcal		kcal		Mcal		Mcal		kcal		kcal		Mcal		Mcal		Mcal		Mcal	
				lb	kg	lb	kg	lb	kg	lb	kg	lb	kg	lb	kg	lb	kg	lb	kg	lb	kg	lb	kg
98	–	123	–	–	–	2458	5419	–	–	–	–	2191	4831	–	–	–	–	–	–	–	–	–	–
	–	132	–	–	–	2646	5833	–	–	–	–	2359	5200	–	–	–	–	–	–	–	–	–	–
99	–	–	–	–	–	–	–	–	–	–	–	–	–	–	–	–	–	–	–	–	–	–	–
	–	–	–	–	–	–	–	–	–	–	–	–	–	–	–	–	–	–	–	–	–	–	–
100	–	–	–	–	–	–	–	–	–	–	–	–	–	–	–	–	–	–	–	–	–	–	–
	–	–	–	–	–	–	–	–	–	–	–	–	–	–	–	–	–	–	–	–	–	–	–
101	76	83	–	1.52	3.36	1669	3680	–	–	1.35	2.97	1463	3225	1138	2509	–	–	0.87	1.92	0.59	1.31	0.82	1.80
	82	90	–	1.65	3.63	1804	3977	–	–	1.46	3.21	1581	3485	1230	2711	–	–	0.94	2.08	0.64	1.41	0.88	1.95
102	78	82	–	1.61	3.54	1637	3609	–	–	1.43	3.16	1433	3160	1316	2901	–	–	0.93	2.06	0.65	1.43	0.87	1.92
	84	89	–	1.74	3.85	1779	3922	–	–	1.56	3.43	1558	3434	1430	3153	–	–	1.02	2.24	0.70	1.55	0.95	2.09
103	67	62	–	1.21	2.66	1238	2729	–	–	0.84	1.84	1004	2213	1104	2434	–	–	0.48	1.05	0.24	0.53	0.65	1.43
	72	67	–	1.30	2.87	1332	2936	–	–	0.90	1.98	1080	2382	1188	2619	–	–	0.51	1.13	0.26	0.58	0.70	1.54
104	97	96	–	1.94	4.28	1914	4219	–	–	1.77	3.90	1688	3722	–	–	–	–	1.12	2.48	0.80	1.76	1.03	2.28
	104	103	–	2.09	4.61	2062	4547	–	–	1.91	4.21	1819	4011	–	–	–	–	1.21	2.67	0.86	1.90	1.11	2.45
105	67	66	–	1.34	2.95	1317	2903	–	–	1.17	2.57	1138	2508	1174	2587	–	–	0.64	1.40	0.39	0.86	0.63	1.39
	73	72	–	1.46	3.22	1438	3169	–	–	1.27	2.81	1242	2738	1281	2825	–	–	0.69	1.53	0.42	0.93	0.69	1.51
106	41	44	–	0.85	1.87	866	1909	–	–	0.78	1.73	736	1623	755	1665	–	–	0.54	1.20	0.38	0.84	0.44	0.97
	82	87	–	1.68	3.71	1717	3784	–	–	1.55	3.42	1459	3217	1497	3300	–	–	1.08	2.37	0.75	1.66	0.87	1.92
107	77	66	–	1.50	3.30	1467	3234	–	–	1.21	2.66	1278	2818	1322	2915	–	–	0.81	1.78	0.54	1.19	0.77	1.70
	83	71	–	1.61	3.56	1581	3485	–	–	1.30	2.87	1377	3036	1424	3140	–	–	0.87	1.92	0.58	1.28	0.83	1.83
108	72	69	–	1.45	3.19	1370	3020	–	–	1.27	2.81	1124	2478	1245	2745	–	–	0.79	1.73	0.52	1.14	0.75	1.65
	79	75	–	1.57	3.47	1489	3283	–	–	1.39	3.05	1222	2695	1354	2985	–	–	0.85	1.88	0.56	1.24	0.81	1.79
109	67	61	–	1.33	2.94	1578	3479	–	–	1.16	2.55	1194	2633	1292	2848	–	–	0.75	1.65	0.49	1.08	0.72	1.58
	73	67	–	1.45	3.21	1723	3799	–	–	1.26	2.79	1304	2875	1411	3110	–	–	0.82	1.81	0.53	1.18	0.79	1.73
110	–	–	–	–	–	–	–	–	–	–	–	–	–	–	–	–	–	–	–	–	–	–	–
	–	–	–	–	–	–	–	–	–	–	–	–	–	–	–	–	–	–	–	–	–	–	–
111	70	67	–	1.40	3.09	1327	2925	–	–	1.23	2.71	1148	2531	1313	2896	–	–	0.76	1.67	0.49	1.08	0.73	1.60
	75	72	–	1.51	3.32	1425	3141	–	–	1.32	2.90	1233	2717	1410	3109	–	–	0.81	1.79	0.53	1.16	0.78	1.72
112	55	71	–	1.10	2.43	1423	3138	–	–	0.93	2.04	1237	2726	–	–	–	–	0.55	1.21	0.31	0.68	0.54	1.18
	61	78	–	1.21	2.67	1563	3445	–	–	1.02	2.24	1358	2993	–	–	–	–	0.60	1.32	0.34	0.75	0.59	1.30
113	–	–	–	–	–	–	–	–	–	–	–	–	–	–	–	–	–	–	–	–	–	–	–
	–	–	–	–	–	–	–	–	–	–	–	–	–	–	–	–	–	–	–	–	–	–	–
114	64	65	–	1.34	2.95	1468	3236	–	–	1.26	2.77	1074	2369	1276	2812	–	–	0.85	1.88	0.57	1.27	0.68	1.50
	69	70	–	1.44	3.18	1583	3489	–	–	1.36	2.99	1158	2554	1375	3032	–	–	0.92	2.02	0.62	1.37	0.74	1.62
115	70	68	–	1.40	3.09	1383	3050	–	–	1.23	2.71	1202	2650	1174	2588	–	–	0.78	1.71	0.51	1.13	0.74	1.63
	77	75	–	1.54	3.39	1520	3351	–	–	1.35	2.98	1321	2912	1289	2843	–	–	0.85	1.88	0.56	1.24	0.81	1.79
116	–	–	–	–	–	–	–	–	–	–	–	–	–	–	–	–	–	–	–	–	–	–	–
	–	–	–	–	–	–	–	–	–	–	–	–	–	–	–	–	–	–	–	–	–	–	–
117	69	65	–	1.37	3.02	1467	3235	–	–	1.20	2.64	1280	2822	644	1419	–	–	0.73	1.61	0.48	1.05	0.70	1.55
	76	71	–	1.50	3.31	1612	3555	–	–	1.31	2.90	1407	3101	707	1559	–	–	0.80	1.77	0.52	1.15	0.77	1.71
118	70	67	–	1.41	3.10	1336	2945	–	–	1.24	2.73	1156	2549	–	–	–	–	0.75	1.65	0.49	1.08	0.72	1.58
	78	74	–	1.56	3.43	1479	3260	–	–	1.37	3.02	1280	2822	–	–	–	–	0.83	1.83	0.54	1.20	0.79	1.75
119	69	65	–	1.38	3.05	1306	2879	–	–	1.21	2.67	1129	2489	–	–	–	–	0.73	1.61	0.48	1.05	0.70	1.55
	77	73	–	1.54	3.39	1451	3199	–	–	1.35	2.97	1254	2765	–	–	–	–	0.81	1.79	0.53	1.17	0.78	1.72
120	54	59	–	1.07	2.37	1177	2595	–	–	0.90	1.98	1009	2224	–	–	–	–	0.52	1.14	0.28	0.62	0.54	1.19
	59	64	–	1.17	2.59	1287	2837	–	–	0.98	2.16	1103	2431	–	–	–	–	0.56	1.24	0.31	0.68	0.59	1.30
121	43	–	–	0.85	1.88	–	–	–	–	0.67	1.47	–	–	–	–	–	–	–	–	–	–	–	–
	45	–	–	0.90	1.99	–	–	–	–	0.71	1.56	–	–	–	–	–	–	–	–	–	–	–	–
122	78	80	–	1.56	3.45	1607	3543	–	–	1.40	3.08	1408	3104	–	–	–	–	0.89	1.97	0.62	1.36	0.83	1.84
	88	90	–	1.75	3.86	1801	3971	–	–	1.57	3.45	1578	3479	–	–	–	–	1.00	2.21	0.69	1.52	0.94	2.06

(Continued)

TABLE V-1B PROTEIN FEEDS, MINERAL AND VITAMIN COMPOSITION OF FEEDS, DATA EXPRESSED **AS-FED** AND **MOISTURE-FREE**—*(Continued)*

Entry Number	Feed Name Description	Moisture Basis: A-F (as-fed) or M-F (moisture-free)	Dry Matter	Macro Minerals							Micro Minerals						
				Calcium (Ca)	Phosphorus (P)	Sodium (Na)	Chlorine (Cl)	Magnesium (Mg)	Potassium (K)	Sulfur (S)	Cobalt (Co)	Copper (Cu)	Iodine (I)	Iron (Fe)	Manganese (Mn)	Selenium (Se)	Zinc (Zn)
			%	%	%	%	%	%	%	%	ppm or mg/kg	ppm or mg/kg	ppm or mg/kg	%	ppm or mg/kg	ppm or mg/kg	ppm or mg/kg
	CRACKLINGS, SWINE *Sus scrofa*																
98	RENDERED	A-F	93	11.25	5.78	1.11	–	–	–	–	–	–	–	–	14.4	–	–
		M-F	100	12.11	6.22	1.19	–	–	–	–	–	–	–	–	15.5	–	–
	CRAMBE, ABYSSINIAN *Crambe abyssinica*																
99	SEEDS WITHOUT HULLS, MEAL MECH EXTD	A-F	92	–	–	–	–	–	–	–	–	–	–	–	–	–	–
		M-F	100	–	–	–	–	–	–	–	–	–	–	–	–	–	–
	DIAMMONIUM PHOSPHATE																
100	$(NH_4)_2HPO_4$	A-F	98	0.50	20.09	0.04	–	0.45	–	2.47	–	80.7	–	1.514	504.3	–	302.6
		M-F	100	0.52	20.54	0.04	–	0.46	–	2.53	–	82.5	–	1.548	515.7	–	309.4
	DISTILLERS PRODUCTS (ALSO SEE CORN; RYE; SORGHUM; WHEAT)																
101	GRAINS, DEHY	A-F	93	0.12	0.54	0.05	0.05	0.09	0.20	0.46	0.092	47.9	–	0.027	35.0	–	–
		M-F	100	0.13	0.59	0.05	0.06	0.10	0.21	0.49	0.099	51.7	–	0.029	37.8	–	–
102	SOLUBLES, DEHY	A-F	92	0.24	1.35	0.45	–	0.53	1.97	–	0.196	71.6	–	0.031	64.1	–	138.0
		M-F	100	0.26	1.47	0.49	–	0.58	2.14	–	0.213	77.9	–	0.034	69.7	–	150.0
	EGGS–SEE CHICKEN																
	FEATHERS, POULTRY																
103	MEAL, HYDROLYZED	A-F	93	0.30	0.62	0.63	0.28	0.18	0.27	1.50	0.116	7.3	0.044	0.023	11.9	0.913	71.9
		M-F	100	0.33	0.67	0.68	0.30	0.19	0.29	1.61	0.125	7.9	0.047	0.025	12.9	0.983	77.3
	FISH																
104	LIVER, MEAL MECH EXTD	A-F	93	–	–	–	–	–	–	–	–	–	–	–	–	–	–
		M-F	100	–	–	–	–	–	–	–	–	–	–	–	–	–	–
105	MEAL MECH EXTD	A-F	92	6.63	3.61	1.11	1.25	0.21	0.40	0.25	0.110	15.1	–	0.038	23.6	–	99.1
		M-F	100	7.24	3.94	1.21	1.37	0.23	0.44	0.27	0.120	16.5	–	0.042	25.8	–	108.2
106	SOLUBLES, CONDENSED	A-F	50	0.16	0.57	2.45	2.93	0.03	1.64	0.12	0.069	46.6	1.111	0.028	13.2	–	43.2
		M-F	100	0.32	1.14	4.86	5.81	0.06	3.24	0.25	0.137	92.4	2.202	0.055	26.2	–	85.6
107	SOLUBLES, DEHY	A-F	93	0.40	1.27	1.70	–	0.30	2.50	0.45	–	20.0	–	0.095	50.4	2.692	76.7
		M-F	100	0.43	1.37	1.83	–	0.32	2.69	0.48	–	21.5	–	0.102	54.3	2.901	82.6
	FISH, ANCHOVY *Engraulis ringen*																
108	MEAL MECH EXTD	A-F	92	3.74	2.48	0.88	1.00	0.25	0.72	0.78	0.173	9.1	3.137	0.022	11.0	1.355	105.0
		M-F	100	4.07	2.70	0.95	1.08	0.27	0.78	0.84	0.188	9.9	3.411	0.024	11.9	1.473	114.2
	FISH, MENHADEN *Brevoortia tyrannus*																
109	MEAL MECH EXTD	A-F	92	5.19	2.88	0.41	0.55	0.15	0.70	0.56	0.153	10.3	1.091	0.055	37.0	2.147	144.2
		M-F	100	5.67	3.14	0.45	0.60	0.17	0.77	0.61	0.167	11.3	1.191	0.060	40.4	2.344	157.5
	FISH, SARDINE *Clupea* spp																
110	FLESH, FRESH	A-F	29	0.03	0.21	–	–	–	–	–	–	–	–	0.002	–	–	–
		M-F	100	0.11	0.73	–	–	–	–	–	–	–	–	0.007	–	–	–
111	MEAL MECH EXTD	A-F	93	4.61	2.68	0.18	0.41	0.10	0.32	–	0.183	20.2	–	0.030	23.2	1.772	–
		M-F	100	4.95	2.88	0.19	0.44	0.11	0.35	–	0.197	21.7	–	0.033	24.9	1.903	–
	FISH, SHARK *Selachii* (order)																
112	MEAL MECH EXTD	A-F	91	3.48	1.86	0.33	–	0.17	–	–	–	51.1	–	0.016	90.1	–	112.6
		M-F	100	3.82	2.04	0.36	–	0.19	–	–	–	56.1	–	0.018	99.0	–	123.6
	FISH, TUNA *Thunnus thynnus*																
113	FLESH, FRESH	A-F	29	–	–	0.04	–	–	–	–	–	–	–	0.001	–	–	–
		M-F	100	–	–	0.14	–	–	–	–	–	–	–	0.004	–	–	–
114	MEAL MECH EXTD	A-F	93	7.86	4.21	0.74	1.01	0.23	0.72	0.68	0.178	10.3	–	0.036	8.4	4.300	210.7
		M-F	100	8.48	4.54	0.80	1.09	0.25	0.77	0.73	0.192	11.1	–	0.039	9.1	4.636	227.2
	FISH, WHITE *Gadidae* (family), *Lophiidae* (family), *Rajidae* (family)																
115	MEAL MECH EXTD	A-F	91	6.60	3.98	0.46	–	0.18	0.45	–	3.358	4.1	–	0.026	9.4	1.612	69.1
		M-F	100	7.25	4.37	0.51	–	0.20	0.49	–	3.688	4.5	–	0.029	10.3	1.771	75.9
	FLAX, COMMON *Linum usitatissimum*																
116	SEEDS, GROUND	A-F	96	0.28	0.55	–	–	–	–	–	–	–	–	–	–	–	–
		M-F	100	0.29	0.58	–	–	–	–	–	–	–	–	–	–	–	–
117	SEEDS, MEAL SOLV EXTD, 33% PROTEIN (LINSEED MEAL)	A-F	91	0.38	0.80	–	–	0.59	1.37	0.49	–	–	–	–	–	–	–
		M-F	100	0.41	0.88	–	–	0.65	1.51	0.54	–	–	–	–	–	–	–
118	SEEDS, MEAL SOLV EXTD, 35% PROTEIN (LINSEED MEAL)	A-F	90	0.40	0.82	0.14	–	0.60	1.37	0.39	–	–	–	–	–	–	–
		M-F	100	0.44	0.91	0.15	–	0.66	1.52	0.43	–	–	–	–	–	–	–
119	SEEDS, MEAL SOLV EXTD, 37% PROTEIN (LINSEED MEAL)	A-F	90	0.39	0.86	–	–	0.59	1.37	–	–	–	–	–	–	–	–
		M-F	100	0.43	0.95	–	–	0.66	1.52	–	–	–	–	–	–	–	–
120	SEED SCREENINGS, MEAL MECH EXTD	A-F	91	0.42	0.56	0.14	–	0.39	0.77	0.23	–	–	–	0.010	–	–	–
		M-F	100	0.46	0.62	0.15	–	0.43	0.84	0.25	–	–	–	0.010	–	–	–
	FURFURAL, CORN *Zea mays*																
121	RESIDUE, AMMONIATED	A-F	94	0.16	0.08	–	–	–	–	–	–	–	–	–	–	–	–
		M-F	100	0.17	0.09	–	–	–	–	–	–	–	–	–	–	–	–
	GARBANZO (CHICKPEA; GRAM PEA) *Cicer arietinum*																
122	SEEDS	A-F	89	0.15	0.33	0.03	–	–	0.80	–	–	–	–	0.007	–	–	–
		M-F	100	0.17	0.37	0.03	–	–	0.89	–	–	–	–	0.008	–	–	–

Entry Number	Fat-Soluble Vitamins					Water-Soluble Vitamins								
	A (1 mg Carotene = 1667 IU Vit A)	Carotene (Provitamin A)	D	E	K	B-12	Biotin	Choline	Folacin (Folic Acid)	Niacin	Pantothenic Acid (B-3)	(Pyri-doxine) B-6	Ribo-flavin (B-2)	Thiamin (B-1)
	IU/g	ppm or mg/kg	IU/kg	ppm or mg/kg	ppm or mg/kg	ppm or mg/kg	ppm or mg/kg	ppm or mg/kg	ppm or mg/kg	ppm or mg/kg	ppm or mg/kg	ppm or mg/kg	ppm or mg/kg	ppm or mg/kg
98	–	–	–	–	–	–	–	–	–	–	–	–	**4.0**	–
	–	–	–	–	–	–	–	–	–	–	–	–	4.3	–
99	–	–	–	–	–	–	–	–	–	–	–	–	–	–
	–	–	–	–	–	–	–	–	–	–	–	–	–	–
100	–	–	–	–	–	–	–	–	–	–	–	–	–	–
	–	–	–	–	–	–	–	–	–	–	–	–	–	–
101	**13.0**	**7.8**	–	**30.5**	–	–	–	**2645**	–	**47**	**11.9**	**6.00**	**6.6**	**2.5**
	14.0	8.4	–	32.9	–	–	–	2858	–	51	12.9	6.48	7.1	2.6
102	**1.9**	**1.1**	–	–	–	**2.9**	**2.84**	**4992**	–	**143**	**25.3**	**8.66**	**11.3**	**6.9**
	2.0	1.2	–	–	–	3.1	3.09	5425	–	155	27.5	9.42	12.3	7.5
103	–	–	–	–	–	**80.4**	**0.04**	**894**	**0.22**	**21**	**8.9**	**4.39**	**2.0**	**0.1**
	–	–	–	–	–	86.5	0.05	962	0.23	23	9.6	4.72	2.2	0.1
104	–	–	–	–	–	–	–	–	–	–	–	–	–	–
	–	–	–	–	–	–	–	–	–	–	–	–	–	–
105	–	–	–	**19.2**	–	**258.6**	–	**3644**	–	**75**	**15.0**	**14.68**	**5.6**	**0.8**
	–	–	–	20.9	–	282.3	–	3979	–	82	16.3	16.03	6.1	0.9
106	**2.2**	**1.3**	–	–	–	**506.6**	**0.14**	**3370**	**0.22**	**176**	**35.7**	**12.20**	**12.9**	**5.5**
	4.4	2.6	–	–	–	1004.2	0.28	6680	0.44	348	70.7	24.19	25.5	11.0
107	–	–	–	**6.1**	–	**485.9**	**0.40**	**5525**	**0.57**	**256**	**50.4**	**19.71**	**13.5**	**7.4**
	–	–	–	6.5	–	523.6	0.43	5953	0.62	276	54.3	21.24	14.6	8.0
108	–	–	–	**3.7**	–	**214.5**	**0.20**	**3700**	**0.16**	**81**	**10.0**	**4.71**	**7.3**	**0.5**
	–	–	–	4.0	–	233.2	0.21	4023	0.17	88	10.9	5.12	8.0	0.6
109	–	–	–	**6.8**	–	**122.0**	**0.18**	**3112**	**0.15**	**55**	**8.6**	**3.80**	**4.8**	**0.6**
	–	–	–	7.4	–	133.2	0.20	3398	0.17	60	9.4	4.15	5.3	0.6
110	–	–	–	–	–	–	–	–	–	–	–	–	–	–
	–	–	–	–	–	–	–	–	–	–	–	–	–	–
111	–	–	–	–	–	**238.0**	**0.10**	**3277**	–	**75**	**11.0**	–	**5.4**	**0.3**
	–	–	–	–	–	255.5	0.11	3518	–	81	11.8	–	5.8	0.3
112	–	–	–	–	–	–	–	**3663**	–	**64**	**9.0**	–	**6.8**	–
	–	–	–	–	–	–	–	4021	–	70	9.9	–	7.5	–
113	–	–	–	–	–	–	–	–	–	–	–	–	–	–
	–	–	–	–	–	–	–	–	–	–	–	–	–	–
114	–	–	–	**5.6**	–	**300.1**	**0.20**	**2993**	–	**144**	**7.7**	–	**6.8**	**1.5**
	–	–	–	6.0	–	323.5	0.22	3227	–	155	8.4	–	7.3	1.6
115	–	–	–	**8.9**	–	**84.3**	**0.08**	**4295**	**0.35**	**59**	**9.9**	**5.30**	**9.1**	**1.7**
	–	–	–	9.8	–	92.6	0.09	4719	0.38	65	10.9	5.83	10.0	1.8
116	–	–	–	–	–	–	–	–	–	–	–	–	–	–
	–	–	–	–	–	–	–	–	–	–	–	–	–	–
117	–	–	–	–	–	–	–	**1212**	–	**29**	**14.1**	–	**2.9**	–
	–	–	–	–	–	–	–	1332	–	31	15.5	–	3.1	–
118	–	–	–	**5.9**	–	–	–	**1216**	**2.85**	**30**	–	**9.93**	**2.9**	**9.4**
	–	–	–	6.5	–	–	–	1346	3.15	33	–	10.99	3.2	10.4
119	–	–	–	–	–	–	–	–	–	–	–	–	–	–
	–	–	–	–	–	–	–	–	–	–	–	–	–	–
120	–	–	–	–	–	–	–	–	–	–	–	–	–	–
	–	–	–	–	–	–	–	–	–	–	–	–	–	–
121	–	–	–	–	–	–	–	–	–	–	–	–	–	–
	–	–	–	–	–	–	–	–	–	–	–	–	–	–
122	–	–	–	–	–	–	–	–	–	**29**	–	–	**1.2**	**2.4**
	–	–	–	–	–	–	–	–	–	33	–	–	1.3	2.7

(Continued)

TABLE V-1B PROTEIN FEEDS, COMPOSITION OF FEEDS, DATA EXPRESSED **AS-FED** AND **MOISTURE-FREE**—*(Continued)*

Entry Number	Feed Name Description	International Feed Number	Moisture Basis: A-F (as-fed) or M-F (moisture-free)	Proximate Analysis: Dry Matter	Ash	Crude Fiber	Neutral Det. Fib. (NDF)	Acid Det. Fib. (ADF)	Lignin	Ether Extract (Fat)	N-Free Extract	Crude Protein	Digestible Protein: Ruminant	Swine	Horse
				%	%	%	%	%	%	%	%	%	%	%	%
	GELATIN														
123	PROCESS RESIDUE (GELATIN BY-PRODUCTS)	**5-14-503**	**A-F**	**90**	**–**	**–**	**–**	**–**	**–**	**0.0**	**–**	**87.5**	**77.5**	**85.8**	**90.6**
			M-F	100	–	–	–	–	–	0.1	–	97.4	86.3	95.5	100.8
	GOAT *Capra hircus*														
124	MILK, FRESH	**5-02-128**	**A-F**	**13**	**0.8**	**–**	**–**	**–**	**–**	**4.2**	**4.8**	**3.4**	**2.8**	**3.2**	**2.6**
			M-F	100	6.0	–	–	–	–	31.6	36.6	25.8	21.0	24.3	19.8
	GRAM PEA (CHICKPEA; GARBANZO) *Cicer arietinum*														
125	SEEDS	**5-01-218**	**A-F**	**89**	**2.9**	**7.0**	**–**	**–**	**–**	**4.1**	**56.0**	**19.1**	**14.3**	**17.8**	**13.3**
			M-F	100	3.3	7.9	–	–	–	4.6	62.8	21.4	16.1	20.0	14.9
	HORSE *Equus caballus*														
126	MEAT, FRESH	**5-07-980**	**A-F**	**31**	**–**	**0.3**	**–**	**–**	**–**	**10.0**	**–**	**19.5**	**17.0**	**19.0**	**19.2**
			M-F	100	–	0.9	–	–	–	32.5	–	63.6	55.4	61.8	62.6
127	MEAT WITH BONE, FRESH	**5-07-981**	**A-F**	**36**	**–**	**–**	**–**	**–**	**–**	**7.0**	**–**	**18.5**	**16.0**	**17.9**	**17.6**
			M-F	100	–	–	–	–	–	19.4	–	51.4	44.4	49.8	48.8
128	MILK, FRESH	**5-02-401**	**A-F**	**17**	**0.7**	**–**	**–**	**–**	**–**	**3.0**	**–**	**4.2**	**3.4**	**3.9**	**3.1**
			M-F	100	4.0	–	–	–	–	17.6	–	24.7	20.0	23.2	18.6
	HORSE BEAN *Vicia faba, equina*														
129	SEEDS	**5-02-407**	**A-F**	**88**	**3.3**	**7.9**	**–**	**–**	**–**	**1.3**	**50.1**	**25.5**	**21.1**	**24.3**	**19.4**
			M-F	100	3.8	8.9	–	–	–	1.5	56.8	29.0	24.0	27.5	22.0
	JACK BEAN, COMMON *Canavalia ensiformis*														
130	SEEDS	**5-02-435**	**A-F**	**89**	**2.9**	**8.5**	**–**	**–**	**–**	**2.8**	**47.6**	**27.4**	**23.0**	**26.1**	**22.7**
			M-F	100	3.2	9.5	–	–	–	3.1	53.4	30.7	25.8	29.3	25.5
	KAPOK *Ceiba pentandra*														
131	SEEDS, MEAL MECH EXTD, 30% PROTEIN	**5-02-468**	**A-F**	**87**	**6.4**	**20.0**	**–**	**–**	**–**	**7.2**	**22.7**	**30.6**	**27.3**	**29.3**	**26.5**
			M-F	100	7.4	23.0	–	–	–	8.3	26.1	35.2	31.3	33.7	30.5
132	SEEDS, MEAL MECH EXTD, 40% PROTEIN	**5-02-469**	**A-F**	**89**	**7.7**	**17.1**	**–**	**–**	**–**	**7.9**	**14.6**	**42.2**	**37.1**	**40.7**	**39.3**
			M-F	100	8.6	19.1	–	–	–	8.8	16.3	47.2	41.5	45.6	44.1
	KOREAN LESPEDEZA—SEE LESPEDEZA, COMMON-KOREAN														
	LENTIL, COMMON *Lens culinaris*														
133	SEEDS	**5-02-506**	**A-F**	**88**	**2.6**	**3.4**	**–**	**–**	**–**	**1.0**	**57.1**	**24.4**	**19.3**	**23.1**	**19.4**
			M-F	100	2.9	3.8	–	–	–	1.1	64.5	27.6	21.8	26.2	21.9
	LESPEDEZA, COMMON-KOREAN *Lespedeza striata-stipulacea*														
134	SEEDS	**5-08-456**	**A-F**	**92**	**5.1**	**9.6**	**–**	**–**	**–**	**7.6**	**32.8**	**36.6**	**29.6**	**35.2**	**32.8**
			M-F	100	5.6	10.5	–	–	–	8.3	35.8	39.9	32.3	38.4	35.8
	LESPEDEZA, SERICEA (CHINESE LESPEDEZA) *Lespedeza cuneata*														
135	SEEDS	**5-08-457**	**A-F**	**92**	**3.8**	**13.5**	**–**	**–**	**–**	**4.2**	**37.3**	**33.5**	**28.2**	**32.1**	**29.3**
			M-F	100	4.1	14.6	–	–	–	4.6	40.4	36.3	30.6	34.8	31.7
	LINSEED—SEE FLAX														
	LIVER														
136	MEAL	**5-00-389**	**A-F**	**93**	**6.3**	**1.4**	**–**	**–**	**–**	**15.7**	**3.2**	**66.1**	**57.9**	**64.4**	**66.1**
			M-F	100	6.8	1.5	–	–	–	17.0	3.5	71.4	62.5	69.6	71.4
	LOCUST, NEW MEXICO *Robinia neomexicana*														
137	SEEDS	**5-09-055**	**A-F**	**89**	**–**	**–**	**–**	**–**	**–**	**–**	**–**	**36.5**	**31.1**	**35.1**	**33.0**
			M-F	100	–	–	–	–	–	–	–	41.0	34.9	39.4	37.1
138	PODS WITH SEEDS, DEHY	**5-26-099**	**A-F**	**92**	**–**	**–**	**–**	**–**	**–**	**–**	**–**	**–**	**–**	**–**	**–**
			M-F	100	–	–	–	–	–	–	–	–	–	–	–
	LUCERNE—SEE ALFALFA														
	LUPINE, YELLOW, SWEET *Lupinus luteus*														
139	SEEDS	**5-08-458**	**A-F**	**89**	**4.5**	**14.0**	**–**	**–**	**–**	**4.9**	**25.7**	**39.8**	**35.4**	**38.4**	**36.7**
			M-F	100	5.1	15.7	–	–	–	5.5	28.9	44.8	39.8	43.2	41.3
	MAIZE—SEE CORN														
	MALT SPROUTS (BARLEY) *Hordeum vulgare*														
140	DEHY	**5-00-545**	**A-F**	**93**	**5.6**	**14.2**	**–**	**–**	**–**	**1.4**	**48.8**	**22.9**	**19.2**	**21.5**	**17.2**
			M-F	100	6.1	15.3	–	–	–	1.5	52.6	24.6	20.7	23.2	18.6
	MANURE, POULTRY														
141	WITH LITTER, DEHY	**5-05-587**	**A-F**	**86**	**17.8**	**15.0**	**–**	**–**	**8.0**	**2.6**	**25.7**	**24.6**	**14.3**	**23.3**	**19.8**
			M-F	100	20.8	17.5	–	–	9.4	3.0	30.0	28.7	16.7	27.2	23.2
142	WITHOUT LITTER, DEHY	**5-14-015**	**A-F**	**90**	**29.2**	**12.0**	**36.3**	**14.4**	**2.1**	**2.0**	**21.6**	**25.4**	**21.0**	**24.1**	**20.4**
			M-F	100	32.4	13.3	40.2	16.0	2.3	2.2	23.9	28.2	23.2	26.7	22.6
	MEAT														
143	MEAL RENDERED	**5-00-385**	**A-F**	**94**	**28.1**	**2.7**	**–**	**–**	**–**	**9.1**	**3.3**	**50.7**	**43.8**	**49.1**	**48.5**
			M-F	100	29.9	2.9	–	–	–	9.7	3.5	54.0	46.7	52.3	51.7

Entry Number	TDN Ruminant	TDN Swine	TDN Horse	Digestible Energy Ruminant		Digestible Energy Swine		Digestible Energy Horse		Metabolizable Energy Ruminant		Metabolizable Energy Swine		Poultry ME_n		Metabolizable Energy Horse		Net Energy Ruminant NE_m		Net Energy Ruminant NE_g		Lactating Cows NE_{lc}	
	%	%	%	Mcal lb	Mcal kg	kcal lb	kcal kg	Mcal lb	Mcal kg	Mcal lb	Mcal kg	kcal lb	kcal kg	kcal lb	kcal kg	Mcal lb	Mcal kg	Mcal lb	Mcal kg	Mcal lb	Mcal kg	Mcal lb	Mcal kg
123	–	–	–	–	–	1131	2494	–	–	–	–	970	2138	970	2138	–	–	–	–	–	–	–	–
	–	–	–	–	–	1259	2776	–	–	–	–	1079	2379	1079	2379	–	–	–	–	–	–	–	–
124	17	–	–	0.34	0.75	–	–	–	–	0.32	0.70	–	–	–	–	–	–	–	–	–	–	–	–
	130	–	–	2.59	5.71	–	–	–	–	2.41	5.32	–	–	–	–	–	–	–	–	–	–	–	–
125	78	80	–	1.56	3.45	1607	3543	–	–	1.40	3.08	1408	3104	–	–	–	–	0.89	1.97	0.62	1.36	0.83	1.84
	88	90	–	1.75	3.86	1801	3971	–	–	1.57	3.45	1578	3479	–	–	–	–	1.00	2.21	0.69	1.52	0.94	2.06
126	–	–	–	–	–	–	–	–	–	–	–	–	–	–	–	–	–	–	–	–	–	–	–
	–	–	–	–	–	–	–	–	–	–	–	–	–	–	–	–	–	–	–	–	–	–	–
127	–	–	–	–	–	–	–	–	–	–	–	–	–	–	–	–	–	–	–	–	–	–	–
	–	–	–	–	–	–	–	–	–	–	–	–	–	–	–	–	–	–	–	–	–	–	–
128	18	–	–	0.36	0.80	–	–	–	–	0.33	0.73	–	–	–	–	–	–	–	–	–	–	–	–
	107	–	–	2.15	4.74	–	–	–	–	1.97	4.33	–	–	–	–	–	–	–	–	–	–	–	–
129	74	74	71	1.42	3.13	1483	3269	–	–	1.26	2.77	1294	2853	–	–	–	–	0.74	1.64	0.49	1.08	0.71	1.56
	84	84	80	1.61	3.56	1683	3710	–	–	1.43	3.14	1469	3238	–	–	–	–	0.84	1.86	0.55	1.22	0.80	1.77
130	81	77	–	1.51	3.32	1548	3414	–	–	1.34	2.96	1354	2985	–	–	–	–	0.77	1.69	0.51	1.12	0.73	1.61
	91	87	–	1.69	3.73	1737	3828	–	–	1.50	3.32	1518	3347	–	–	–	–	0.86	1.90	0.57	1.26	0.82	1.80
131	57	48	–	1.15	2.52	966	2130	–	–	0.98	2.16	818	1803	–	–	–	–	0.60	1.33	0.37	0.81	0.60	1.32
	65	56	–	1.32	2.90	1111	2449	–	–	1.13	2.48	940	2072	–	–	–	–	0.69	1.53	0.42	0.93	0.69	1.51
132	62	52	–	1.26	2.79	1040	2292	–	–	1.09	2.41	883	1948	–	–	–	–	0.69	1.52	0.44	0.97	0.67	1.47
	69	58	–	1.42	3.12	1164	2567	–	–	1.23	2.70	989	2181	–	–	–	–	0.77	1.70	0.49	1.09	0.75	1.65
133	74	84	–	1.48	3.26	1675	3692	–	–	1.31	2.89	1471	3244	–	–	–	–	0.81	1.79	0.55	1.21	0.77	1.69
	84	95	–	1.67	3.68	1894	4175	–	–	1.48	3.27	1663	3667	–	–	–	–	0.92	2.03	0.62	1.37	0.87	1.91
134	65	75	–	1.41	3.10	1501	3310	–	–	1.23	2.72	1308	2884	–	–	–	–	0.84	1.84	0.56	1.24	0.79	1.74
	71	82	–	1.53	3.38	1637	3610	–	–	1.34	2.96	1427	3145	–	–	–	–	0.91	2.01	0.61	1.35	0.86	1.90
135	61	71	–	1.29	2.85	1424	3139	–	–	1.12	2.47	1236	2725	–	–	–	–	0.74	1.63	0.48	1.05	0.71	1.57
	66	77	–	1.40	3.09	1543	3401	–	–	1.21	2.67	1339	2952	–	–	–	–	0.80	1.76	0.52	1.14	0.77	1.70
136	89	93	–	1.79	3.94	1867	4116	–	–	1.62	3.57	1645	3627	1306	2878	–	–	1.00	2.21	0.70	1.55	0.93	2.04
	97	101	–	1.93	4.26	2017	4446	–	–	1.75	3.85	1777	3918	1410	3109	–	–	1.08	2.38	0.76	1.67	1.00	2.21
137	–	–	–	–	–	–	–	–	–	–	–	–	–	–	–	–	–	–	–	–	–	–	–
	–	–	–	–	–	–	–	–	–	–	–	–	–	–	–	–	–	–	–	–	–	–	–
138	–	–	–	–	–	–	–	–	–	–	–	–	–	–	–	–	–	–	–	–	–	–	–
	–	–	–	–	–	–	–	–	–	–	–	–	–	–	–	–	–	–	–	–	–	–	–
139	77	65	–	1.43	3.15	1302	2870	–	–	1.26	2.78	1126	2482	–	–	–	–	0.71	1.57	0.46	1.02	0.68	1.51
	86	73	–	1.61	3.54	1464	3228	–	–	1.42	3.13	1267	2792	–	–	–	–	0.80	1.76	0.52	1.14	0.77	1.70
140	66	61	–	1.31	2.89	1219	2688	–	–	1.13	2.50	1046	2307	722	1592	–	–	0.69	1.52	0.43	0.95	0.67	1.48
	71	66	–	1.41	3.11	1313	2894	–	–	1.22	2.69	1126	2483	778	1714	–	–	0.74	1.63	0.47	1.03	0.72	1.60
141	47	9	–	0.99	2.19	175	386	–	–	0.83	1.83	88	194	–	–	–	–	0.52	1.16	0.30	0.66	0.54	1.18
	55	10	–	1.16	2.56	205	451	–	–	0.97	2.14	103	226	–	–	–	–	0.61	1.35	0.35	0.77	0.63	1.38
142	46	–	–	0.92	2.02	–	–	–	–	0.74	1.63	–	–	–	–	–	–	0.54	1.18	0.30	0.66	0.55	1.22
	51	–	–	1.01	2.24	–	–	–	–	0.82	1.81	–	–	–	–	–	–	0.59	1.31	0.33	0.73	0.61	1.35
143	67	64	–	1.20	2.64	936	2064	–	–	0.89	1.95	1009	2225	947	2088	–	–	0.52	1.15	0.28	0.62	0.69	1.53
	71	68	–	1.28	2.81	998	2200	–	–	0.94	2.08	1075	2371	1009	2225	–	–	0.56	1.23	0.30	0.66	0.74	1.63

(Continued)

TABLE V-1B PROTEIN FEEDS, MINERAL AND VITAMIN COMPOSITION OF FEEDS, DATA EXPRESSED **AS-FED** AND **MOISTURE-FREE**—*(Continued)*

Entry Number	Feed Name Description	Moisture Basis: A-F (as-fed) or M-F (moisture-free)	Dry Matter	Macro Minerals: Calcium (Ca)	Phosphorus (P)	Sodium (Na)	Chlorine (Cl)	Magnesium (Mg)	Potassium (K)	Sulfur (S)	Micro Minerals: Cobalt (Co)	Copper (Cu)	Iodine (I)	Iron (Fe)	Manganese (Mn)	Selenium (Se)	Zinc (Zn)
			%	%	%	%	%	%	%	%	ppm or mg/kg	ppm or mg/kg	ppm or mg/kg	%	ppm or mg/kg	ppm or mg/kg	ppm or mg/kg
	GELATIN																
123	PROCESS RESIDUE (GELATIN BY-PRODUCTS)	**A-F**	**90**	**0.49**	–	–	–	**0.05**	–	–	–	–	–	–	–	–	–
		M-F	100	0.55	–	–	–	0.05	–	–	–	–	–	–	–	–	–
	GOAT *Capra hircus*																
124	MILK, FRESH	**A-F**	**13**	**0.13**	**0.11**	**0.04**	**0.18**	**0.03**	**0.19**	**0.00**	–	**0.3**	–	**0.001**	–	–	–
		M-F	100	1.02	0.80	0.30	1.38	0.22	1.43	0.03	–	2.3	–	0.001	–	–	–
	GRAM PEA (CHICKPEA; GARBANZO) *Cicer arietinum*																
125	SEEDS	**A-F**	**89**	**0.15**	**0.33**	**0.03**	–	–	**0.80**	–	–	–	–	**0.007**	–	–	–
		M-F	100	0.17	0.37	0.03	–	–	0.89	–	–	–	–	0.008	–	–	–
	HORSE *Equus caballus*																
126	MEAT, FRESH	**A-F**	**31**	**0.02**	**0.33**	**0.06**	–	**0.01**	**0.12**	–	**0.066**	**0.1**	**0.091**	**0.006**	**0.2**	–	**18.5**
		M-F	100	0.07	1.06	0.18	–	0.04	0.38	–	0.214	0.2	0.295	0.017	0.5	–	60.2
127	MEAT WITH BONE, FRESH	**A-F**	**36**	–	–	–	–	–	–	–	–	–	–	–	–	–	–
		M-F	100	–	–	–	–	–	–	–	–	–	–	–	–	–	–
128	MILK, FRESH	**A-F**	**17**	**0.15**	**0.05**	**0.03**	–	**0.01**	**0.10**	–	–	**0.0**	–	–	–	–	–
		M-F	100	0.90	0.31	0.20	–	0.08	0.60	–	–	0.3	–	–	–	–	–
	HORSE BEAN *Vicia faba, equina*																
129	SEEDS	**A-F**	**88**	**0.13**	**0.54**	–	–	–	**1.17**	–	–	–	–	–	–	–	–
		M-F	100	0.15	0.62	–	–	–	1.33	–	–	–	–	–	–	–	–
	JACK BEAN, COMMON *Canavalia ensiformis*																
130	SEEDS	**A-F**	**89**	–	–	–	–	–	–	–	–	–	–	–	–	–	–
		M-F	100	–	–	–	–	–	–	–	–	–	–	–	–	–	–
	KAPOK *Ceiba pentandra*																
131	SEEDS, MEAL MECH EXTD, 30% PROTEIN	**A-F**	**87**	–	–	–	–	–	–	–	–	–	–	–	–	–	–
		M-F	100	–	–	–	–	–	–	–	–	–	–	–	–	–	–
132	SEEDS, MEAL MECH EXTD, 40% PROTEIN	**A-F**	**89**	–	–	–	–	–	–	–	–	–	–	–	–	–	–
		M-F	100	–	–	–	–	–	–	–	–	–	–	–	–	–	–
	KOREAN LESPEDEZA–SEE LESPEDEZA, COMMON-KOREAN																
	LENTIL, COMMON *Lens culinaris*																
133	SEEDS	**A-F**	**88**	**0.08**	**0.38**	**0.03**	–	–	**0.79**	–	–	–	–	**0.007**	–	–	–
		M-F	100	0.09	0.42	0.03	–	–	0.89	–	–	–	–	0.008	–	–	–
	LESPEDEZA, COMMON-KOREAN *Lespedeza striata-stipulacea*																
134	SEEDS	**A-F**	**92**	–	–	–	–	–	–	–	–	–	–	–	–	–	–
		M-F	100	–	–	–	–	–	–	–	–	–	–	–	–	–	–
	LESPEDEZA, SERICEA (CHINESE LESPEDEZA) *Lespedeza cuneata*																
135	SEEDS	**A-F**	**92**	–	–	–	–	–	–	–	–	–	–	–	–	–	–
		M-F	100	–	–	–	–	–	–	–	–	–	–	–	–	–	–
	LINSEED–SEE FLAX																
	LIVER																
136	MEAL	**A-F**	**93**	**0.56**	**1.26**	–	–	**0.10**	–	–	**0.135**	**89.4**	–	**0.064**	**8.8**	–	**61.8**
		M-F	100	0.61	1.36	–	–	0.11	–	–	0.146	96.5	–	0.069	9.5	–	66.8
	LOCUST, NEW MEXICO *Robinia Neomexicana*																
137	SEEDS	**A-F**	**89**	–	–	–	–	–	–	–	–	–	–	–	–	–	–
		M-F	100	–	–	–	–	–	–	–	–	–	–	–	–	–	–
138	PODS WITH SEEDS, DEHY	**A-F**	**92**	–	–	–	–	–	–	–	–	–	–	–	–	–	–
		M-F	100	–	–	–	–	–	–	–	–	–	–	–	–	–	–
	LUCERNE–SEE ALFALFA																
	LUPINE, YELLOW, SWEET *Lupinus luteus*																
139	SEEDS	**A-F**	**89**	**0.23**	**0.39**	–	–	–	**0.81**	–	–	–	–	–	–	–	–
		M-F	100	0.26	0.44	–	–	–	0.91	–	–	–	–	–	–	–	–
	MAIZE–SEE CORN																
	MALT SPROUTS (BARLEY) *Hordeum vulgare*																
140	DEHY	**A-F**	**93**	**0.18**	**0.63**	**0.88**	**0.36**	**0.17**	**0.25**	**0.79**	–	**5.9**	–	**0.018**	**29.4**	**0.416**	**56.4**
		M-F	100	0.19	0.68	0.95	0.39	0.18	0.27	0.85	–	6.3	–	0.020	31.7	0.448	60.7
	MANURE, POULTRY																
141	WITH LITTER, DEHY	**A-F**	**86**	**2.67**	**1.69**	**0.41**	–	**0.43**	**1.32**	–	–	**283.3**	–	**0.046**	**281.1**	**0.559**	**359.4**
		M-F	100	3.12	1.98	0.47	–	0.50	1.55	–	–	331.0	–	0.054	328.4	0.653	419.9
142	WITHOUT LITTER, DEHY	**A-F**	**90**	**8.07**	**2.22**	**0.61**	**0.86**	**0.56**	**1.99**	**0.16**	–	**24.6**	–	–	–	–	**366.4**
		M-F	100	8.95	2.46	0.68	0.96	0.62	2.20	0.18	–	27.2	–	–	–	–	405.9
	MEAT																
143	MEAL RENDERED	**A-F**	**94**	**8.61**	**4.58**	**1.05**	**1.11**	**0.25**	**0.55**	**0.46**	**2.250**	**9.6**	–	**0.050**	**11.8**	**0.505**	**74.3**
		M-F	100	9.18	4.88	1.11	1.18	0.27	0.58	0.49	2.398	10.2	–	0.053	12.6	0.538	79.2

Entry Number	Fat-Soluble Vitamins: A (1 mg Carotene = 1667 IU Vit A)	Carotene (Provitamin A)	D	E	K	Water-Soluble Vitamins: B-12	Biotin	Choline	Folacin (Folic Acid)	Niacin	Pantothenic Acid (B-3)	(Pyri-doxine) B-6	Ribo-flavin (B-2)	Thiamin (B-1)
	IU/g	ppm or mg/kg	IU/kg	ppm or mg/kg	ppm or mg/kg	ppm or mg/kg	ppm or mg/kg	ppm or mg/kg	ppm or mg/kg	ppm or mg/kg	ppm or mg/kg	ppm or mg/kg	ppm or mg/kg	ppm or mg/kg
123	–	–	–	–	–	–	–	–	–	–	–	–	–	–
	–	–	–	–	–	–	–	–	–	–	–	–	–	–
124	–	–	–	–	–	–	–	–	–	–	–	–	–	–
	–	–	–	–	–	–	–	–	–	–	–	–	–	–
125	–	–	–	–	–	–	–	–	–	29	–	–	1.2	2.4
	–	–	–	–	–	–	–	–	–	33	–	–	1.3	2.7
126	–	–	–	7.6	–	43.5	0.03	320	0.25	5	1.5	0.21	–	0.4
	–	–	–	24.7	–	141.7	0.08	1043	0.83	16	4.8	0.70	–	1.4
127	–	–	–	–	–	–	–	–	–	–	–	–	–	–
	–	–	–	–	–	–	–	–	–	–	–	–	–	–
128	–	–	–	–	–	2.0	–	–	0.00	1	4.5	0.45	0.3	0.4
	–	–	–	–	–	12.1	–	–	0.01	4	26.7	2.65	1.8	2.6
129	–	–	–	–	–	–	–	–	–	–	–	–	–	–
	–	–	–	–	–	–	–	–	–	–	–	–	–	–
130	–	–	–	–	–	–	–	–	–	–	–	–	–	–
	–	–	–	–	–	–	–	–	–	–	–	–	–	–
131	–	–	–	–	–	–	–	–	–	–	–	–	–	–
	–	–	–	–	–	–	–	–	–	–	–	–	–	–
132	–	–	–	–	–	–	–	–	–	–	–	–	–	–
	–	–	–	–	–	–	–	–	–	–	–	–	–	–
133	–	–	–	–	–	–	–	–	–	20	–	–	2.2	3.7
	–	–	–	–	–	–	–	–	–	22	–	–	2.5	4.2
134	–	–	–	–	–	–	–	–	–	–	–	–	–	–
	–	–	–	–	–	–	–	–	–	–	–	–	–	–
135	–	–	–	–	–	–	–	–	–	–	–	–	–	–
	–	–	–	–	–	–	–	–	–	–	–	–	–	–
136	–	–	–	–	–	501.3	0.02	11370	5.56	205	29.2	–	36.2	0.2
	–	–	–	–	–	541.5	0.02	12281	6.01	221	31.5	–	39.1	0.2
137	–	–	–	–	–	–	–	–	–	–	–	–	–	–
	–	–	–	–	–	–	–	–	–	–	–	–	–	–
138	–	–	–	–	–	–	–	–	–	–	–	–	–	–
	–	–	–	–	–	–	–	–	–	–	–	–	–	–
139	–	–	–	–	–	–	–	–	–	–	–	–	–	–
	–	–	–	–	–	–	–	–	–	–	–	–	–	–
140	–	–	–	3.7	–	–	4.09	1591	0.20	55	9.0	8.62	2.8	8.3
	–	–	–	4.0	–	–	4.40	1713	0.22	59	9.6	9.28	3.0	8.9
141	–	–	–	–	–	–	–	–	–	–	–	–	–	–
	–	–	–	–	–	–	–	–	–	–	–	–	–	–
142	–	–	–	–	–	21.1	–	–	–	19	–	–	11.7	–
	–	–	–	–	–	23.4	–	–	–	21	–	–	13.0	–
143	–	–	–	0.9	–	75.2	0.12	1980	0.39	56	6.0	4.23	5.2	0.2
	–	–	–	1.0	–	80.1	0.13	2110	0.42	60	6.4	4.51	5.5	0.2

(Continued)

TABLE V-1B PROTEIN FEEDS, COMPOSITION OF FEEDS, DATA EXPRESSED **AS-FED** AND **MOISTURE-FREE**—*(Continued)*

Entry Number	Feed Name Description	International Feed Number	Moisture Basis: A-F (as-fed) or M-F (moisture-free)	Proximate Analysis: Dry Matter	Ash	Crude Fiber	Neutral Det. Fib. (NDF)	Acid Det. Fib. (ADF)	Lignin	Ether Extract (Fat)	N-Free Extract	Crude Protein	Digestible Protein: Ruminant	Swine	Horse
				%	%	%	%	%	%	%	%	%	%	%	%
	MEAT (Continued)														
144	WITH BLOOD, MEAL RENDERED (TANKAGE)	5-00-386	A-F	92	21.1	1.8	–	–	–	8.7	0.1	60.5	52.8	58.9	59.8
			M-F	100	22.9	2.0	–	–	–	9.4	0.1	65.6	57.3	63.9	64.9
145	WITH BLOOD, WITH BONE, MEAL RENDERED (TANKAGE)	5-00-387	A-F	93	28.2	2.2	–	–	–	12.7	3.1	46.6	40.2	45.1	44.1
			M-F	100	30.4	2.4	–	–	–	13.7	3.3	50.2	43.2	48.6	47.4
146	WITH BONE, MEAL RENDERED	5-00-388	A-F	93	28.0	2.4	–	–	–	10.0	2.6	50.4	45.8	43.3	48.3
			M-F	100	30.0	2.6	–	–	–	10.7	2.8	54.0	49.1	46.4	51.7
	MILK														
147	FRESH (CATTLE *Bos taurus*)	5-01-168	A-F	12	0.8	–	–	–	–	3.6	4.7	3.3	3.2	3.2	2.6
			M-F	100	6.2	–	–	–	–	29.5	37.6	26.7	25.5	25.9	20.9
148	DEHY (CATTLE *Bos taurus*)	5-01-167	A-F	95	5.4	0.2	–	–	–	26.3	38.1	25.3	20.7	24.0	19.8
			M-F	100	5.6	0.2	–	–	–	27.6	39.9	26.6	21.8	25.1	20.8
149	SKIMMED, FRESH (CATTLE *Bos taurus*)	5-01-170	A-F	10	0.7	–	–	–	–	0.1	5.8	3.0	2.5	3.0	2.5
			M-F	100	6.9	–	–	–	–	1.0	60.6	31.2	26.0	30.9	26.0
150	SKIMMED, DEHY (CATTLE *Bos taurus*)	5-01-175	A-F	94	8.0	0.2	0.0	–	–	1.1	51.6	33.3	30.0	31.8	28.9
			M-F	100	8.4	0.2	0.0	–	–	1.2	54.8	35.4	31.8	33.7	30.7
151	FRESH (GOAT *Capra hircus*)	5-02-128	A-F	13	0.8	–	–	–	–	4.2	4.8	3.4	2.8	3.2	2.6
			M-F	100	6.0	–	–	–	–	31.6	36.6	25.8	21.0	24.3	19.8
152	FRESH (HORSE *Equus caballus*)	5-02-401	A-F	17	0.7	–	–	–	–	3.0	–	4.2	3.4	3.9	3.1
			M-F	100	4.0	–	–	–	–	17.6	–	24.7	20.0	23.2	18.6
153	FRESH (SHEEP *Ovis aries*)	5-08-510	A-F	19	0.9	–	–	–	–	7.3	5.9	4.6	3.8	4.4	3.5
			M-F	100	4.7	–	–	–	–	39.0	31.6	24.7	20.1	23.3	18.7
154	FRESH (SWINE *Sus scrofa*)	5-08-537	A-F	20	1.0	–	–	–	–	6.7	5.1	7.3	–	–	–
			M-F	100	5.0	–	–	–	–	33.3	25.4	36.3	–	–	–
	MILKWEED, COMMON *Asclepias syriaca*														
155	SEEDS	5-09-137	A-F	86	–	–	–	–	–	–	–	31.8	26.9	30.5	28.0
			M-F	100	–	–	–	–	–	–	–	37.0	31.2	35.5	32.5
	MILO SORGHUM *Sorghum bicolor, subglabrescens*														
156	GLUTEN	5-08-087	A-F	90	1.5	3.3	–	–	–	4.3	38.5	42.5	36.5	41.1	39.7
			M-F	100	1.7	3.7	–	–	–	4.8	42.7	47.2	40.5	45.6	44.0
	MOLASSES AND SYRUP *Saccharum officinarum*														
157	SUGARCANE, MOLASSES, AMMONIATED	5-04-702	A-F	65	5.9	–	–	–	–	–	–	26.3	15.5	25.3	23.7
			M-F	100	9.1	–	–	–	–	–	–	40.5	23.9	38.9	36.5
	MONOAMMONIUM PHOSPHATE														
158	$NH_4H_2PO_4$	6-09-338	A-F	98	53.0	–	–	–	–	–	–	69.4	–	–	–
			M-F	100	54.2	–	–	–	–	–	–	71.0	–	–	–
	MUSTARD *Brassica* spp														
159	SEEDS, MEAL MECH EXTD	5-03-154	A-F	93	7.1	10.8	–	–	–	5.1	37.9	32.0	27.6	30.6	27.6
			M-F	100	7.6	11.6	–	–	–	5.5	40.8	34.5	29.6	33.0	29.7
	MUSTARD (CHARLOCK) *Brassica kaber*														
160	SEEDS	5-08-461	A-F	96	5.5	5.0	–	–	–	38.8	23.6	23.0	18.6	21.6	17.1
			M-F	100	5.7	5.2	–	–	–	40.5	24.6	24.0	19.4	22.6	17.8
	NAVY BEAN *Phaseolus vulgaris*														
161	SEEDS	5-00-623	A-F	89	4.0	4.4	–	–	–	1.4	56.7	22.9	20.2	21.6	17.6
			M-F	100	4.5	5.0	–	–	–	1.5	63.4	25.6	22.5	24.2	19.7
	ORANGE *Citrus sinensis*														
162	PULP WITHOUT FINES, AMMONIATED, DEHY	4-01-255	A-F	89	5.6	12.4	–	–	–	3.3	53.0	14.7	10.4	11.6	10.8
			M-F	100	6.3	13.9	–	–	–	3.7	59.6	16.5	11.7	13.0	12.1
	PALM *Elaeis* spp														
163	KERNELS WITH COATS, MEAL SOLV EXTD	5-03-486	A-F	90	3.9	14.2	–	–	–	0.4	53.1	18.2	13.8	17.0	12.3
			M-F	100	4.3	15.8	–	–	–	0.4	59.2	20.3	15.4	18.9	13.7
	PEA *Pisum* spp														
164	SEEDS	5-03-600	A-F	89	2.9	5.5	–	–	–	1.1	56.7	23.2	19.0	21.9	18.0
			M-F	100	3.2	6.1	–	–	–	1.2	63.4	26.0	21.2	24.5	20.1
165	SPLIT SEED BY-PRODUCT (PEA FEED; PEA MEAL)	1-08-478	A-F	90	3.5	23.7	–	–	–	1.4	43.7	17.7	14.5	12.1	12.0
			M-F	100	3.9	26.3	–	–	–	1.6	48.6	19.7	16.1	13.4	13.4
	PEA, FIELD *Pisum sativum, arvense*														
166	SEEDS	5-08-481	A-F	91	2.9	5.9	–	–	–	1.3	57.8	23.2	19.9	21.8	17.7
			M-F	100	3.1	6.5	–	–	–	1.4	63.5	25.4	21.9	24.0	19.5
	PEA, GARDEN *Pisum sativum, sativum*														
167	SEEDS	5-08-482	A-F	89	2.9	5.7	–	–	–	1.7	55.1	23.8	19.5	22.5	18.6
			M-F	100	3.3	6.4	–	–	–	1.9	61.8	26.7	21.9	25.2	20.9
	PEANUT *Arachis hypogaea*														
168	PODS WITH SEEDS, MEAL SOLV EXTD	5-03-656	A-F	92	4.7	13.7	–	–	–	1.2	25.4	47.4	40.9	45.9	44.9
			M-F	100	5.1	14.8	–	–	–	1.3	27.5	51.3	44.2	49.6	48.7
169	SEEDS WITHOUT HULLS, MEAL MECH EXTD (PEANUT MEAL)	5-03-649	A-F	93	5.0	6.2	13.2	5.6	1.0	5.6	26.7	49.2	45.5	46.2	47.0
			M-F	100	5.4	6.7	14.2	6.1	1.1	6.0	28.8	53.1	49.1	49.9	50.7

Entry Number	TDN Ruminant	TDN Swine	TDN Horse	Digestible Energy Ruminant		Digestible Energy Swine		Digestible Energy Horse		Metabolizable Energy Ruminant		Metabolizable Energy Swine		Metabolizable Energy Poultry ME_n		Metabolizable Energy Horse		Net Energy Ruminant NE_m		Net Energy Ruminant NE_g		Net Energy Lactating Cows NE_{lc}	
	%	%	%	Mcal lb	Mcal kg	kcal lb	kcal kg	Mcal lb	Mcal kg	Mcal lb	Mcal kg	kcal lb	kcal kg	kcal lb	kcal kg	Mcal lb	Mcal kg	Mcal lb	Mcal kg	Mcal lb	Mcal kg	Mcal lb	Mcal kg
144	67	67	–	1.34	2.96	1112	2451	–	–	1.17	2.58	951	2096	1212	2673	–	–	0.74	1.62	0.48	1.05	0.71	1.56
	73	73	–	1.46	3.21	1207	2660	–	–	1.27	2.80	1032	2275	1316	2901	–	–	0.80	1.76	0.52	1.14	0.77	1.69
145	63	68	–	1.44	3.17	1382	3047	–	–	1.26	2.79	1199	2644	1188	2618	–	–	0.90	1.99	0.62	1.36	0.85	1.86
	68	73	–	1.55	3.42	1488	3280	–	–	1.36	3.00	1291	2846	1278	2818	–	–	0.97	2.14	0.67	1.47	0.91	2.00
146	66	68	–	1.32	2.91	1028	2267	–	–	1.14	2.51	981	2162	946	2086	–	–	0.71	1.57	0.45	1.00	0.69	1.52
	71	73	–	1.41	3.11	1102	2430	–	–	1.22	2.70	1051	2317	1014	2236	–	–	0.76	1.68	0.49	1.07	0.74	1.63
147	16	15	–	0.32	0.71	309	681	–	–	0.30	0.65	275	606	–	–	–	–	0.19	0.42	0.14	0.30	0.18	0.40
	128	125	–	2.61	5.75	2500	5510	–	–	2.40	5.29	2223	4902	–	–	–	–	1.53	3.37	1.11	2.46	1.46	3.22
148	113	109	–	2.27	4.99	2183	4813	–	–	2.09	4.61	1935	4266	–	–	–	–	1.27	2.81	0.92	2.02	1.17	2.57
	119	115	–	2.38	5.24	2290	5049	–	–	2.20	4.84	2030	4475	–	–	–	–	1.34	2.95	0.96	2.12	1.22	2.70
149	9	9	–	0.18	0.39	188	415	–	–	0.16	0.35	166	365	–	–	–	–	0.10	0.22	0.07	0.16	0.10	0.21
	92	96	–	1.86	4.11	1968	4339	–	–	1.66	3.66	1732	3819	–	–	–	–	1.06	2.33	0.74	1.63	1.00	2.20
150	80	86	–	1.43	3.15	1758	3876	–	–	1.07	2.37	1630	3593	1152	2539	–	–	0.69	1.52	0.43	0.95	0.83	1.82
	85	92	–	1.52	3.35	1867	4115	–	–	1.14	2.51	1730	3815	1223	2696	–	–	0.73	1.62	0.46	1.01	0.88	1.94
151	17	–	–	0.34	0.75	–	–	–	–	0.32	0.70	–	–	–	–	–	–	–	–	–	–	–	–
	130	–	–	2.59	5.71	–	–	–	–	2.41	5.32	–	–	–	–	–	–	–	–	–	–	–	–
152	18	–	–	0.36	0.80	–	–	–	–	0.33	0.73	–	–	–	–	–	–	–	–	–	–	–	–
	107	–	–	2.15	4.74	–	–	–	–	1.97	4.33	–	–	–	–	–	–	–	–	–	–	–	–
153	26	–	–	0.51	1.13	–	–	–	–	0.48	1.05	–	–	–	–	–	–	–	–	–	–	–	–
	136	–	–	2.73	6.02	–	–	–	–	2.55	5.63	–	–	–	–	–	–	–	–	–	–	–	–
154	27	–	–	0.53	1.18	–	–	–	–	0.50	1.10	–	–	–	–	–	–	–	–	–	–	–	–
	133	–	–	2.66	5.86	–	–	–	–	2.48	5.47	–	–	–	–	–	–	–	–	–	–	–	–
155	–	–	–	–	–	–	–	–	–	–	–	–	–	–	–	–	–	–	–	–	–	–	–
	–	–	–	–	–	–	–	–	–	–	–	–	–	–	–	–	–	–	–	–	–	–	–
156	83	93	–	1.63	3.59	1870	4122	–	–	1.46	3.22	1650	3637	–	–	–	–	0.90	1.99	0.62	1.37	0.84	1.86
	92	104	–	1.81	3.99	2075	4575	–	–	1.62	3.58	1831	4038	–	–	–	–	1.00	2.21	0.69	1.52	0.94	2.06
157	47	–	–	0.95	2.08	–	–	–	–	0.82	1.81	–	–	–	–	–	–	–	–	–	–	–	–
	73	–	–	1.46	3.21	–	–	–	–	1.27	2.79	–	–	–	–	–	–	–	–	–	–	–	–
158	–	–	–	–	–	–	–	–	–	–	–	–	–	–	–	–	–	–	–	–	–	–	–
	–	–	–	–	–	–	–	–	–	–	–	–	–	–	–	–	–	–	–	–	–	–	–
159	68	65	–	1.39	3.07	1298	2862	–	–	1.22	2.68	1119	2468	–	–	–	–	0.78	1.71	0.51	1.13	0.74	1.64
	73	70	–	1.50	3.30	1396	3078	–	–	1.31	2.88	1204	2654	–	–	–	–	0.84	1.84	0.55	1.21	0.80	1.76
160	91	113	–	2.04	4.50	2263	4989	–	–	1.87	4.11	2008	4428	–	–	–	–	1.32	2.91	0.95	2.10	1.21	2.66
	95	118	–	2.13	4.69	2360	5204	–	–	1.95	4.29	2095	4618	–	–	–	–	1.38	3.03	0.99	2.19	1.26	2.77
161	76	78	–	1.53	3.37	1737	3830	–	–	1.36	3.00	1531	3376	1052	2320	–	–	0.83	1.83	0.56	1.24	0.78	1.72
	85	87	–	1.71	3.77	1941	4280	–	–	1.52	3.36	1711	3772	1176	2593	–	–	0.93	2.05	0.63	1.39	0.87	1.93
162	64	62	56	1.27	2.81	1230	2712	1.03	2.28	1.11	2.44	1153	2542	–	–	0.85	1.87	0.67	1.49	0.43	0.94	0.66	1.45
	72	69	63	1.43	3.16	1382	3047	1.16	2.56	1.24	2.74	1296	2856	–	–	0.95	2.10	0.76	1.67	0.48	1.06	0.74	1.62
163	69	63	–	1.29	2.85	1258	2773	–	–	1.12	2.48	1085	2392	–	–	–	–	0.63	1.38	0.39	0.85	0.62	1.37
	77	70	–	1.44	3.18	1402	3092	–	–	1.25	2.76	1210	2667	–	–	–	–	0.70	1.54	0.43	0.95	0.69	1.52
164	77	–	–	1.56	3.44	1483	3268	–	–	1.39	3.07	1293	2850	959	2115	–	–	0.87	1.91	0.59	1.31	0.81	1.79
	87	–	–	1.75	3.85	1660	3659	–	–	1.56	3.44	1447	3191	1074	2368	–	–	0.97	2.14	0.67	1.47	0.91	2.00
165	78	55	62	1.34	2.96	–	–	1.13	2.49	1.17	2.59	–	–	–	–	0.93	2.04	0.57	1.26	0.33	0.74	0.58	1.28
	87	61	69	1.49	3.29	–	–	1.25	2.77	1.30	2.87	–	–	–	–	1.03	2.27	0.64	1.40	0.37	0.82	0.64	1.42
166	76	82	–	1.51	3.34	1638	3611	–	–	1.34	2.96	1435	3163	1108	2442	–	–	0.80	1.77	0.54	1.18	0.76	1.68
	83	90	–	1.66	3.66	1798	3964	–	–	1.47	3.25	1575	3473	1216	2681	–	–	0.88	1.94	0.59	1.30	0.84	1.84
167	77	80	–	1.49	3.28	1609	3547	–	–	1.32	2.91	1410	3108	–	–	–	–	0.79	1.74	0.53	1.17	0.75	1.65
	86	90	–	1.67	3.67	1804	3977	–	–	1.48	3.26	1580	3484	–	–	–	–	0.89	1.96	0.59	1.31	0.84	1.85
168	68	77	–	1.34	2.96	1454	3207	–	–	1.17	2.57	1267	2793	1043	2300	–	–	0.71	1.56	0.45	0.99	0.69	1.51
	74	83	–	1.45	3.20	1574	3471	–	–	1.26	2.79	1371	3023	1129	2489	–	–	0.76	1.68	0.49	1.07	0.74	1.64
169	81	86	–	1.54	3.40	1957	4316	–	–	1.47	3.24	1823	4019	1207	2662	–	–	0.86	1.89	0.58	1.28	0.81	1.78
	87	93	–	1.66	3.67	2113	4659	–	–	1.59	3.50	1968	4339	1303	2873	–	–	0.92	2.04	0.63	1.38	0.87	1.92

(Continued)

TABLE V-1B PROTEIN FEEDS, MINERAL AND VITAMIN COMPOSITION OF FEEDS, DATA EXPRESSED **AS-FED** AND **MOISTURE-FREE**—*(Continued)*

Entry Number	Feed Name Description	Moisture Basis: A-F (as-fed) or M-F (moisture-free)	Dry Matter	Macro Minerals							Micro Minerals						
				Calcium (Ca)	Phosphorus (P)	Sodium (Na)	Chlorine (Cl)	Magnesium (Mg)	Potassium (K)	Sulfur (S)	Cobalt (Co)	Copper (Cu)	Iodine (I)	Iron (Fe)	Manganese (Mn)	Selenium (Se)	Zinc (Zn)
			%	%	%	%	%	%	%	%	ppm or mg/kg	ppm or mg/kg	ppm or mg/kg	%	ppm or mg/kg	ppm or mg/kg	ppm or mg/kg
	MEAT (Continued)																
144	WITH BLOOD, MEAL RENDERED (TANKAGE)	A-F	92	5.87	3.09	1.67	1.73	0.36	0.55	0.70	0.153	38.8	–	0.211	19.2	–	–
		M-F	100	6.37	3.36	1.81	1.88	0.39	0.60	0.76	0.166	42.1	–	0.229	20.8	–	–
145	WITH BLOOD, WITH BONE, MEAL RENDERED (TANKAGE)	A-F	93	11.16	5.41	–	–	–	–	0.26	–	–	–	–	–	0.261	–
		M-F	100	12.01	5.82	–	–	–	–	0.28	–	–	–	–	–	0.281	–
146	WITH BONE, MEAL RENDERED	A-F	93	10.00	4.94	0.72	0.75	1.02	1.33	0.25	0.181	1.5	1.317	0.066	13.3	0.263	94.3
		M-F	100	10.72	5.30	0.77	0.80	1.09	1.43	0.27	0.194	1.6	1.412	0.070	14.3	0.282	101.1
	MILK																
147	FRESH (CATTLE *Bos taurus*)	A-F	12	0.12	0.09	0.05	0.11	0.01	0.14	0.04	0.001	0.1	–	0.002	–	–	2.3
		M-F	100	0.93	0.75	0.38	0.92	0.10	1.13	0.32	0.005	0.8	–	0.010	–	–	23.0
148	DEHY (CATTLE *Bos taurus*)	A-F	95	0.89	0.70	0.36	1.48	0.09	1.05	0.31	0.005	0.9	–	0.018	0.4	–	21.7
		M-F	100	0.93	0.74	0.38	1.55	0.09	1.10	0.32	0.005	0.9	–	0.019	0.5	–	22.8
149	SKIMMED, FRESH (CATTLE *Bos taurus*)	A-F	10	0.13	0.10	0.04	0.05	0.01	0.12	0.03	0.011	1.1	–	0.001	0.2	–	4.9
		M-F	100	1.31	1.04	0.47	0.54	0.12	1.29	0.32	0.111	11.6	–	0.009	2.3	–	51.0
150	SKIMMED, DEHY (CATTLE *Bos taurus*)	A-F	94	1.28	1.02	0.51	0.90	0.12	1.60	0.32	0.113	11.7	–	0.001	2.1	0.124	38.5
		M-F	100	1.36	1.09	0.54	0.96	0.13	1.70	0.34	0.120	12.4	–	0.001	2.3	0.131	40.9
151	FRESH (GOAT *Capra hircus*)	A-F	13	0.13	0.11	0.04	0.18	0.03	0.19	0.00	–	0.3	–	0.001	–	–	–
		M-F	100	1.02	0.80	0.30	1.38	0.22	1.43	0.03	–	2.3	–	0.001	–	–	–
152	FRESH (HORSE *Equus caballus*)	A-F	17	0.15	0.05	0.03	–	0.01	0.10	–	–	0.0	–	–	–	–	–
		M-F	100	0.90	0.31	0.20	–	0.08	0.60	–	–	0.3	–	–	–	–	–
153	FRESH (SHEEP *Ovis aries*)	A-F	19	0.21	0.15	0.04	0.08	0.02	0.15	–	–	0.1	–	0.001	0.1	–	2.6
		M-F	100	1.10	0.82	0.22	0.41	0.08	0.82	–	–	0.6	–	0.001	0.3	–	13.7
154	FRESH (SWINE *Sus scrofa*)	A-F	20	–	–	–	–	–	–	–	–	–	–	–	–	–	–
		M-F	100	–	–	–	–	–	–	–	–	–	–	–	–	–	–
	MILKWEED, COMMON *Asclepias syriaca*																
155	SEEDS	A-F	86	–	–	–	–	–	–	–	–	–	–	–	–	–	–
		M-F	100	–	–	–	–	–	–	–	–	–	–	–	–	–	–
	MILO SORGHUM *Sorghum bicolor, subglabrescens*																
156	GLUTEN	A-F	90	–	–	–	–	–	–	–	–	–	–	–	–	–	–
		M-F	100	–	–	–	–	–	–	–	–	–	–	–	–	–	–
	MOLASSES AND SYRUP *Saccharum officinarum*																
157	SUGARCANE, MOLASSES, AMMONIATED	A-F	65	0.79	0.13	–	–	–	–	–	–	–	–	–	–	–	–
		M-F	100	1.22	0.20	–	–	–	–	–	–	–	–	–	–	–	–
	MONOAMMONIUM PHOSPHATE																
158	$NH_4H_2PO_4$	A-F	98	0.38	24.42	0.08	–	0.46	0.14	0.82	–	85.7	–	0.991	461.7	–	639.6
		M-F	100	0.39	24.99	0.08	–	0.47	0.14	0.84	–	87.7	–	1.014	472.6	–	654.7
	MUSTARD *Brassica* spp																
159	SEEDS, MEAL MECH EXTD	A-F	93	–	–	–	–	–	–	–	–	–	–	–	–	–	–
		M-F	100	–	–	–	–	–	–	–	–	–	–	–	–	–	–
	MUSTARD (CHARLOCK) *Brassica kaber*																
160	SEEDS	A-F	96	–	–	–	–	–	–	–	–	–	–	–	–	–	–
		M-F	100	–	–	–	–	–	–	–	–	–	–	–	–	–	–
	NAVY BEAN *Phaseolus vulgaris*																
161	SEEDS	A-F	89	0.17	0.54	0.04	0.06	0.13	1.31	0.23	–	9.9	–	0.010	21.1	–	–
		M-F	100	0.19	0.61	0.05	0.06	0.15	1.47	0.26	–	11.0	–	0.012	23.6	–	–
	ORANGE *Citrus sinensis*																
162	PULP WITHOUT FINES, AMMONIATED, DEHY	A-F	89	–	–	–	–	–	–	–	–	–	–	–	–	–	–
		M-F	100	–	–	–	–	–	–	–	–	–	–	–	–	–	–
	PALM *Elaeis* spp																
163	KERNELS WITH COATS, MEAL SOLV EXTD	A-F	90	–	–	–	–	–	–	–	–	–	–	–	–	–	–
		M-F	100	–	–	–	–	–	–	–	–	–	–	–	–	–	–
	PEA *Pisum* spp																
164	SEEDS	A-F	89	0.12	0.41	0.04	0.05	0.12	0.95	–	–	–	–	0.007	2.9	–	23.0
		M-F	100	0.14	0.46	0.05	0.06	0.14	1.06	–	–	–	–	0.008	3.2	–	25.7
165	SPLIT SEED BY-PRODUCT (PEA FEED; PEA MEAL)	A-F	90	–	–	–	–	–	–	–	–	–	–	–	–	–	–
		M-F	100	–	–	–	–	–	–	–	–	–	–	–	–	–	–
	PEA, FIELD *Pisum sativum, arvense*																
166	SEEDS	A-F	91	0.16	0.38	0.00	–	0.14	1.36	–	1.700	11.7	–	0.020	21.2	0.393	46.8
		M-F	100	0.17	0.41	0.00	–	0.15	1.49	–	1.866	12.9	–	0.022	23.2	0.432	51.4
	PEA, GARDEN *Pisum sativum, sativum*																
167	SEEDS	A-F	89	0.08	0.40	–	–	–	0.90	–	–	–	–	–	–	–	–
		M-F	100	0.09	0.45	–	–	–	1.01	–	–	–	–	–	–	–	–
	PEANUT *Arachis hypogaea*																
168	PODS WITH SEEDS, MEAL SOLV EXTD	A-F	92	0.20	0.61	0.07	–	0.04	1.23	–	–	–	–	–	29.9	–	20.4
		M-F	100	0.22	0.66	0.08	–	0.04	1.33	–	–	–	–	–	32.4	–	22.0
169	SEEDS WITHOUT HULLS, MEAL MECH EXTD (PEANUT MEAL)	A-F	93	0.20	0.56	0.12	0.03	0.26	1.16	0.22	0.111	15.4	0.067	0.030	25.5	–	33.0
		M-F	100	0.22	0.61	0.13	0.03	0.28	1.25	0.24	0.119	16.6	0.072	0.033	27.6	–	35.6

PROTEIN FEEDS

Entry Number	Fat-Soluble Vitamins: A (1 mg Carotene = 1667 IU Vit A)	Carotene (Provitamin A)	D	E	K	Water-Soluble Vitamins: B-12	Biotin	Choline	Folacin (Folic Acid)	Niacin	Pantothenic Acid (B-3)	(Pyridoxine) B-6	Riboflavin (B-2)	Thiamin (B-1)
	IU/g	ppm or mg/kg	IU/kg	ppm or mg/kg	ppm or mg/kg	ppm or mg/kg	ppm or mg/kg	ppm or mg/kg	ppm or mg/kg	ppm or mg/kg	ppm or mg/kg	ppm or mg/kg	ppm or mg/kg	ppm or mg/kg
144	–	–	–	–	–	89.4	–	2203	1.54	38	3.2	–	2.2	0.3
	–	–	–	–	–	97.1	–	2391	1.67	41	3.5	–	2.4	0.4
145	–	–	–	0.8	–	104.4	0.07	2067	0.57	58	4.8	–	5.0	0.2
	–	–	–	0.9	–	112.4	0.08	2225	0.62	63	5.2	–	5.4	0.2
146	–	–	–	0.9	–	118.4	0.10	2049	0.37	51	5.5	5.86	4.7	0.2
	–	–	–	0.9	–	126.9	0.11	2195	0.40	55	5.9	6.28	5.0	0.2
147	–	–	–	–	–	–	–	904	–	1	8.4	–	1.7	0.3
	–	–	–	–	–	–	–	7311	–	10	68.0	–	13.8	2.4
148	–	–	0	–	–	–	0.38	–	–	8	22.7	4.71	19.6	3.8
	–	–	0	–	–	–	0.40	–	–	9	23.8	4.94	20.6	3.9
149	–	–	–	–	–	–	–	–	–	1	3.5	–	2.0	0.4
	–	–	–	–	–	–	–	–	–	12	36.9	–	20.8	4.6
150	–	–	0	9.1	–	50.9	0.33	1394	0.62	11	36.4	4.10	19.1	3.7
	–	–	0	9.6	–	54.1	0.35	1480	0.66	12	38.6	4.35	20.3	3.9
151	–	–	–	–	–	–	–	–	–	–	–	–	–	–
	–	–	–	–	–	–	–	–	–	–	–	–	–	–
152	–	–	–	–	–	2.0	–	–	0.00	1	4.5	0.45	0.3	0.4
	–	–	–	–	–	12.1	–	–	0.01	4	26.7	2.65	1.8	2.6
153	–	–	–	16.1	–	4.1	0.40	–	0.05	5	4.1	0.72	4.1	1.0
	–	–	–	85.9	–	22.0	2.12	–	0.27	27	22.0	3.85	22.0	5.5
154	–	–	–	–	–	–	–	–	–	–	–	–	–	–
	–	–	–	–	–	–	–	–	–	–	–	–	–	–
155	–	–	–	–	–	–	–	–	–	–	–	–	–	–
	–	–	–	–	–	–	–	–	–	–	–	–	–	–
156	–	–	–	–	–	–	–	–	–	–	–	–	–	–
	–	–	–	–	–	–	–	–	–	–	–	–	–	–
157	–	–	–	–	–	–	–	–	–	–	–	–	–	–
	–	–	–	–	–	–	–	–	–	–	–	–	–	–
158	–	–	–	–	–	–	–	–	–	–	–	–	–	–
	–	–	–	–	–	–	–	–	–	–	–	–	–	–
159	–	–	–	–	–	–	–	–	–	–	–	–	–	–
	–	–	–	–	–	–	–	–	–	–	–	–	–	–
160	–	–	–	–	–	–	–	–	–	–	–	–	–	–
	–	–	–	–	–	–	–	–	–	–	–	–	–	–
161	–	–	–	1.0	–	–	0.11	1017	1.29	24	2.4	0.30	1.7	6.3
	–	–	–	1.1	–	–	0.12	1136	1.44	27	2.7	0.33	1.9	7.1
162	–	–	–	–	–	–	–	–	–	–	–	–	–	–
	–	–	–	–	–	–	–	–	–	–	–	–	–	–
163	–	–	–	–	–	–	–	–	–	–	–	–	–	–
	–	–	–	–	–	–	–	–	–	–	–	–	–	–
164	1.2	0.7	–	3.0	–	–	0.18	547	0.22	31	27.8	1.97	1.8	4.6
	1.3	0.8	–	3.3	–	–	0.20	612	0.25	34	31.1	2.21	2.0	5.2
165	–	–	–	–	–	–	–	–	–	–	–	–	–	–
	–	–	–	–	–	–	–	–	–	–	–	–	–	–
166	–	–	–	–	–	–	0.19	654	0.36	34	7.4	1.01	1.4	4.1
	–	–	–	–	–	–	0.21	718	0.40	37	8.2	1.11	1.5	4.5
167	–	–	–	–	–	–	–	–	–	–	–	–	–	–
	–	–	–	–	–	–	–	–	–	–	–	–	–	–
168	–	–	–	2.6	–	–	0.39	1819	–	170	53.2	–	8.2	7.3
	–	–	–	2.8	–	–	0.42	1969	–	184	57.6	–	8.8	7.9
169	–	–	–	2.4	–	–	0.33	1975	0.66	173	47.6	6.12	9.1	5.7
	–	–	–	2.6	–	–	0.36	2132	0.71	186	51.4	6.61	9.8	6.2

(Continued)

TABLE V-1B PROTEIN FEEDS, COMPOSITION OF FEEDS, DATA EXPRESSED **AS-FED** AND **MOISTURE-FREE**—*(Continued)*

Entry Number	Feed Name Description	International Feed Number	Moisture Basis: A-F (as-fed) or M-F (moisture-free)	Proximate Analysis									Digestible Protein		
				Dry Matter	Ash	Crude Fiber	Neutral Det. Fib. (NDF)	Acid Det. Fib. (ADF)	Lignin	Ether Extract (Fat)	N-Free Extract	Crude Protein	Ruminant	Swine	Horse
				%	%	%	%	%	%	%	%	%	%	%	%
	PEANUT (Continued)														
170	SEEDS WITHOUT HULLS, MEAL SOLV EXTD (PEANUT MEAL)	5-03-650	A-F	93	5.8	7.7	–	–	–	2.2	27.9	49.0	42.3	47.4	46.7
			M-F	100	6.3	8.3	–	–	–	2.4	30.1	52.9	45.7	51.3	50.5
	PERILLA, COMMON *Perilla frutescens*														
171	SEEDS, MEAL MECH EXTD	5-03-690	A-F	92	8.2	20.5	–	–	–	8.6	16.2	38.0	33.8	36.6	34.5
			M-F	100	9.0	22.4	–	–	–	9.4	17.7	41.5	37.0	40.0	37.7
	POPPY *Papaver* spp														
172	SEEDS, MEAL MECH EXTD	5-03-751	A-F	89	12.4	11.6	–	–	–	7.9	20.7	36.6	29.7	35.2	33.1
			M-F	100	13.9	13.0	–	–	–	8.8	23.2	41.0	33.3	39.5	37.1
	POULTRY														
173	BY-PRODUCT, MEAL RENDERED	5-03-798	A-F	94	14.8	2.2	–	–	–	13.1	2.6	61.2	53.4	53.3	60.5
			M-F	100	15.8	2.3	–	–	–	13.9	2.7	65.3	57.0	56.8	64.5
174	FEATHERS, MEAL, HYDROLYZED	5-03-795	A-F	93	3.2	1.4	–	6.1	–	5.1	–	83.8	74.1	61.2	86.1
			M-F	100	3.4	1.5	–	6.6	–	5.5	–	90.2	79.7	65.9	92.7
175	MANURE WITH LITTER, DEHY	5-05-587	A-F	86	17.8	15.0	–	–	8.0	2.6	25.7	24.6	14.3	23.3	19.8
			M-F	100	20.8	17.5	–	–	9.4	3.0	30.0	28.7	16.7	27.2	23.2
176	MANURE WITHOUT LITTER, DEHY	5-14-015	A-F	90	29.2	12.0	36.3	14.4	2.1	2.0	21.6	25.4	21.0	24.1	20.4
			M-F	100	32.4	13.3	40.2	16.0	2.3	2.2	23.9	28.2	23.2	26.7	22.6
	RAPE *Brassica napus*														
177	SEEDS, MEAL MECH EXTD	5-03-870	A-F	92	6.9	12.0	–	–	–	7.3	30.1	35.6	30.4	34.2	31.7
			M-F	100	7.5	13.1	–	–	–	7.9	32.7	38.7	33.0	37.2	34.5
178	SEEDS, MEAL SOLV EXTD, 34% PROTEIN	5-26-092	A-F	90	7.0	13.0	–	–	–	2.5	33.5	34.0	–	–	–
			M-F	100	7.8	14.4	–	–	–	2.8	37.2	37.8	–	–	–
	RAPE, SUMMER *Brassica napus, annua*														
179	SEEDS, MEAL, PREPRESSED, SOLV EXTD [1]	5-08-135	A-F	92	7.2	9.3	33.1	16.6	–	1.1	33.9	40.5	34.6	39.0	37.2
			M-F	100	7.8	10.1	36.0	18.0	–	1.2	36.8	44.0	37.6	42.4	40.5
	RICE *Oryza sativa*														
180	HULLS, AMMONIATED	1-05-698	A-F	92	–	44.7	–	–	–	0.9	16.9	10.4	6.3	5.6	6.4
			M-F	100	–	48.6	–	–	–	1.0	18.4	11.3	6.9	6.1	6.9
	RUBBER TREE, PARA *Hevea brasiliensis*														
181	KERNELS, MEAL MECH EXTD	5-03-959	A-F	91	5.1	6.6	–	–	–	14.7	35.3	29.3	23.8	28.0	24.7
			M-F	100	5.6	7.2	–	–	–	16.1	38.8	32.3	26.1	30.8	27.2
	RYE *Secale cereale*														
182	DISTILLERS GRAINS, DEHY	5-04-023	A-F	92	2.3	12.3	–	–	–	6.0	48.3	23.0	13.8	21.7	17.5
			M-F	100	2.5	13.4	–	–	–	6.5	52.6	25.1	15.0	23.6	19.0
183	DISTILLERS GRAINS WITH SOLUBLES, DEHY	5-04-024	A-F	90	6.4	8.1	–	–	–	4.1	44.7	27.2	22.6	25.9	22.3
			M-F	100	7.1	8.9	–	–	–	4.5	49.4	30.1	24.9	28.6	24.7
	SAFFLOWER *Carthamus tinctorius*														
184	SEEDS, MEAL SOLV EXTD, 20% PROTEIN	5-26-095	A-F	92	4.6	32.2	–	39.6	–	1.1	32.7	21.6	17.4	20.2	15.8
			M-F	100	5.0	34.9	–	43.0	–	1.2	35.5	23.4	18.9	22.0	17.2
185	SEEDS WITHOUT HULLS, MEAL MECH EXTD	5-08-499	A-F	91	6.5	12.8	–	–	–	6.0	23.9	42.0	36.0	40.6	39.0
			M-F	100	7.2	14.0	–	–	–	6.6	26.2	46.1	39.5	44.5	42.8
186	SEEDS WITHOUT HULLS, MEAL SOLV EXTD, 42% PROTEIN	5-26-094	A-F	92	6.5	14.6	–	19.2	–	1.3	26.3	42.7	36.7	41.3	39.8
			M-F	100	7.2	16.0	–	21.0	–	1.5	28.8	46.7	40.1	45.1	43.5
187	SEEDS WITHOUT HULLS, MEAL SOLV EXTD	5-07-959	A-F	91	7.7	13.1	–	–	–	1.2	26.2	42.8	36.0	41.4	39.9
			M-F	100	8.5	14.4	–	–	–	1.3	28.7	47.0	39.5	45.4	43.9
	SERICEA LESPEDEZA—SEE LESPEDEZA, SERICEA														
	SESAME *Sesamum indicum*														
188	SEEDS	5-08-509	A-F	95	5.8	10.6	–	–	–	44.1	10.5	23.7	19.2	22.3	18.0
			M-F	100	6.1	11.2	–	–	–	46.6	11.1	25.0	20.3	23.6	19.0
189	SEEDS, MEAL MECH EXTD [1]	5-04-220	A-F	93	10.3	5.6	15.8	15.8	–	8.7	23.0	45.0	38.5	43.5	42.3
			M-F	100	11.2	6.1	17.0	17.0	–	9.4	24.8	48.6	41.6	47.0	45.7
190	SEEDS, MEAL SOLV EXTD, 44% PROTEIN	5-26-096	A-F	92	13.1	6.8	–	–	–	1.4	25.8	45.0	39.4	44.4	42.3
			M-F	100	14.2	7.4	–	–	–	1.5	28.0	48.9	42.8	48.3	46.0
	SHEEP *Ovis aries*														
191	MILK, FRESH	5-08-510	A-F	19	0.9	–	–	–	–	7.3	5.9	4.6	3.8	4.4	3.5
			M-F	100	4.7	–	–	–	–	39.0	31.6	24.7	20.1	23.3	18.7
	SHRIMP *Pandalus* spp, *Penaeus* spp														
192	CANNERY RESIDUE, MEAL (SHRIMP MEAL)	5-04-226	A-F	90	26.7	14.1	–	16.6	–	3.9	6.7	38.7	33.0	37.3	35.4
			M-F	100	29.6	15.6	–	18.4	–	4.3	7.4	43.0	36.7	41.4	39.3
	SORGHUM *Sorghum bicolor*														
193	DISTILLERS GRAINS, DEHY	5-04-374	A-F	94	4.3	12.1	–	–	–	8.3	38.3	30.8	24.7	29.4	26.1
			M-F	100	4.6	12.9	–	–	–	8.8	40.8	32.9	26.3	31.4	27.9
194	DISTILLERS GRAINS WITH SOLUBLES, DEHY	5-04-375	A-F	95	4.2	10.1	–	–	–	9.4	38.0	33.1	27.8	31.7	28.6
			M-F	100	4.4	10.7	–	–	–	9.9	40.1	34.9	29.3	33.4	30.2
195	GLUTEN MEAL	5-04-388	A-F	90	1.7	4.9	–	–	–	4.4	34.9	44.4	38.2	43.0	41.8
			M-F	100	1.9	5.4	–	–	–	4.8	38.7	49.2	42.4	47.6	46.3
196	MILO, GLUTEN MEAL	5-08-087	A-F	90	1.5	3.3	–	–	–	4.3	38.5	42.5	36.5	41.1	39.7
			M-F	100	1.7	3.7	–	–	–	4.8	42.7	47.2	40.5	45.6	44.0

PROTEIN FEEDS

Entry Number	TDN			Digestible Energy						Metabolizable Energy								Net Energy					
	Ruminant	Swine	Horse	Ruminant		Swine		Horse		Ruminant		Swine		Poultry ME_n		Horse		Ruminant NE_m		Ruminant NE_g		Lactating Cows NE_{lc}	
	%	%	%	Mcal		kcal		Mcal		Mcal		kcal		kcal		Mcal		Mcal		Mcal		Mcal	
				lb	kg	lb	kg	lb	kg	lb	kg	lb	kg	lb	kg	lb	kg	lb	kg	lb	kg	lb	kg
170	73	74	–	1.43	3.15	1296	2857	–	–	1.29	2.84	1264	2787	1229	2709	–	–	0.78	1.72	0.51	1.13	0.74	1.63
	79	80	–	1.54	3.40	1401	3088	–	–	1.39	3.07	1367	3013	1328	2928	–	–	0.84	1.86	0.55	1.22	0.80	1.76
171	61	48	–	1.24	2.73	952	2099	–	–	1.06	2.35	801	1766	–	–	–	–	0.66	1.44	0.41	0.90	0.64	1.42
	67	52	–	1.35	2.98	1040	2293	–	–	1.16	2.56	875	1929	–	–	–	–	0.72	1.58	0.44	0.98	0.70	1.55
172	64	43	–	1.28	2.82	854	1882	–	–	1.11	2.45	712	1570	–	–	–	–	0.73	1.62	0.48	1.06	0.70	1.55
	72	48	–	1.43	3.16	957	2110	–	–	1.25	2.74	798	1760	–	–	–	–	0.82	1.81	0.54	1.18	0.79	1.73
173	74	76	–	1.44	3.17	1406	3101	–	–	1.21	2.66	1301	2869	1300	2865	–	–	0.81	1.78	0.53	1.18	0.76	1.68
	79	81	–	1.53	3.37	1499	3305	–	–	1.29	2.83	1387	3058	1385	3054	–	–	0.86	1.89	0.57	1.25	0.81	1.79
174	67	62	–	1.21	2.66	1238	2729	–	–	0.84	1.84	1004	2213	1104	2434	–	–	0.48	1.05	0.24	0.53	0.65	1.43
	72	67	–	1.30	2.87	1332	2936	–	–	0.90	1.98	1080	2382	1188	2619	–	–	0.51	1.13	0.26	0.58	0.70	1.54
175	47	9	–	0.99	2.19	175	386	–	–	0.83	1.83	88	194	–	–	–	–	0.52	1.16	0.30	0.66	0.54	1.18
	55	10	–	1.16	2.56	205	451	–	–	0.97	2.14	103	226	–	–	–	–	0.61	1.35	0.35	0.77	0.63	1.38
176	46	–	–	0.92	2.02	–	–	–	–	0.74	1.63	–	–	–	–	–	–	0.54	1.18	0.30	0.66	0.55	1.22
	51	–	–	1.01	2.24	–	–	–	–	0.82	1.81	–	–	–	–	–	–	0.59	1.31	0.33	0.73	0.61	1.35
177	71	67	–	1.41	3.11	1366	3011	–	–	1.24	2.73	1185	2612	908	2002	–	–	0.79	1.75	0.53	1.16	0.75	1.66
	77	72	–	1.53	3.38	1485	3274	–	–	1.35	2.97	1288	2840	988	2177	–	–	0.86	1.90	0.57	1.26	0.82	1.81
178	–	–	–	–	–	–	–	–	–	–	–	–	–	–	–	–	–	–	–	–	–	–	–
	–	–	–	–	–	–	–	–	–	–	–	–	–	–	–	–	–	–	–	–	–	–	–
179	70	62	–	1.40	3.08	1249	2753	–	–	1.22	2.69	1075	2369	–	–	–	–	0.75	1.65	0.49	1.07	0.72	1.58
	76	68	–	1.52	3.34	1357	2993	–	–	1.33	2.93	1168	2575	–	–	–	–	0.81	1.79	0.53	1.17	0.78	1.72
180	–	–	–	–	–	–	–	–	–	–	–	–	–	–	–	–	–	–	–	–	–	–	–
	–	–	–	–	–	–	–	–	–	–	–	–	–	–	–	–	–	–	–	–	–	–	–
181	85	85	–	1.69	3.73	1704	3757	–	–	1.53	3.36	1496	3298	–	–	–	–	0.96	2.11	0.67	1.47	0.89	1.96
	93	94	–	1.86	4.11	1874	4130	–	–	1.68	3.70	1645	3626	–	–	–	–	1.05	2.32	0.73	1.62	0.98	2.16
182	54	79	–	1.08	2.38	1586	3497	–	–	0.90	1.99	1387	3057	–	–	–	–	0.45	1.00	0.22	0.49	0.50	1.09
	59	86	–	1.18	2.59	1727	3807	–	–	0.98	2.17	1509	3328	–	–	–	–	0.49	1.09	0.24	0.53	0.54	1.19
183	73	68	–	1.46	3.21	1354	2985	–	–	1.29	2.83	1173	2586	–	–	–	–	0.78	1.72	0.52	1.14	0.74	1.64
	80	75	–	1.61	3.55	1497	3299	–	–	1.42	3.13	1296	2858	–	–	–	–	0.86	1.90	0.57	1.26	0.82	1.81
184	46	–	–	0.87	1.92	1089	2401	–	–	0.86	1.90	929	2047	623	1374	–	–	0.51	1.12	0.27	0.59	0.40	0.89
	50	–	–	0.95	2.09	1182	2606	–	–	0.94	2.06	1008	2222	676	1491	–	–	0.55	1.21	0.29	0.65	0.44	0.96
185	70	63	–	1.40	3.08	1070	2359	–	–	1.22	2.70	912	2010	780	1720	–	–	0.76	1.67	0.50	1.10	0.72	1.60
	77	69	–	1.53	3.37	1173	2587	–	–	1.34	2.96	1000	2204	855	1886	–	–	0.83	1.83	0.55	1.20	0.79	1.75
186	66	56	–	1.15	2.53	1508	3325	–	–	0.97	2.15	1317	2904	885	1951	–	–	0.58	1.27	0.33	0.74	0.56	1.24
	72	62	–	1.26	2.77	1647	3632	–	–	1.06	2.35	1439	3172	967	2131	–	–	0.63	1.39	0.37	0.81	0.61	1.35
187	69	54	–	1.34	2.96	1073	2365	–	–	1.17	2.57	913	2012	837	1845	–	–	0.75	1.66	0.49	1.08	0.72	1.59
	76	59	–	1.47	3.25	1179	2598	–	–	1.28	2.83	1003	2210	919	2026	–	–	0.83	1.82	0.54	1.19	0.79	1.74
188	99	–	–	1.97	4.34	–	–	–	–	1.80	3.96	–	–	–	–	–	–	–	–	–	–	–	–
	104	–	–	2.08	4.59	–	–	–	–	1.90	4.19	–	–	–	–	–	–	–	–	–	–	–	–
189	71	70	–	1.41	3.12	1554	3425	–	–	1.24	2.73	1373	3026	985	2172	–	–	0.75	1.66	0.49	1.08	0.72	1.59
	76	76	–	1.53	3.37	1677	3698	–	–	1.34	2.95	1482	3267	1064	2345	–	–	0.81	1.79	0.53	1.16	0.78	1.72
190	69	81	–	1.38	3.04	1610	3549	–	–	1.13	2.48	1389	3063	1178	2598	–	–	0.71	1.56	0.46	1.01	0.79	1.75
	75	88	–	1.50	3.30	1750	3858	–	–	1.23	2.70	1510	3329	1281	2824	–	–	0.77	1.70	0.50	1.10	0.86	1.90
191	26	–	–	0.51	1.13	–	–	–	–	0.48	1.05	–	–	–	–	–	–	–	–	–	–	–	–
	136	–	–	2.73	6.02	–	–	–	–	2.55	5.63	–	–	–	–	–	–	–	–	–	–	–	–
192	41	–	–	0.82	1.82	–	–	–	–	0.65	1.43	–	–	871	1920	–	–	0.29	0.65	0.08	0.17	0.38	0.84
	46	–	–	0.92	2.02	–	–	–	–	0.72	1.59	–	–	967	2131	–	–	0.33	0.72	0.08	0.18	0.42	0.93
193	78	77	–	1.56	3.45	1535	3385	–	–	1.39	3.06	1338	2949	–	–	–	–	0.85	1.88	0.57	1.27	0.80	1.77
	83	82	–	1.67	3.68	1637	3609	–	–	1.48	3.26	1426	3144	–	–	–	–	0.91	2.00	0.61	1.35	0.86	1.89
194	81	82	–	1.63	3.58	1649	3635	–	–	1.45	3.19	1442	3178	–	–	–	–	0.88	1.94	0.60	1.32	0.83	1.83
	86	87	–	1.72	3.78	1739	3835	–	–	1.53	3.37	1521	3353	–	–	–	–	0.93	2.05	0.63	1.39	0.88	1.93
195	80	91	–	1.60	3.52	1581	3486	–	–	1.43	3.15	1386	3056	1237	2727	–	–	0.85	1.88	0.58	1.28	0.80	1.76
	89	100	–	1.77	3.90	1752	3862	–	–	1.58	3.49	1536	3385	1370	3021	–	–	0.94	2.08	0.64	1.42	0.89	1.95
196	83	93	–	1.63	3.59	1870	4122	–	–	1.46	3.22	1650	3637	–	–	–	–	0.90	1.99	0.62	1.37	0.84	1.86
	92	104	–	1.81	3.99	2075	4575	–	–	1.62	3.58	1831	4038	–	–	–	–	1.00	2.21	0.69	1.52	0.94	2.06

(Continued)

TABLE V-1B PROTEIN FEEDS, MINERAL AND VITAMIN COMPOSITION OF FEEDS, DATA EXPRESSED **AS-FED** AND **MOISTURE-FREE**—*(Continued)*

Entry Number	Feed Name Description	Moisture Basis: A-F (as-fed) or M-F (moisture-free)	Dry Matter	Macro Minerals: Calcium (Ca)	Phosphorus (P)	Sodium (Na)	Chlorine (Cl)	Magnesium (Mg)	Potassium (K)	Sulfur (S)	Micro Minerals: Cobalt (Co)	Copper (Cu)	Iodine (I)	Iron (Fe)	Manganese (Mn)	Selenium (Se)	Zinc (Zn)
			%	%	%	%	%	%	%	%	ppm or mg/kg	ppm or mg/kg	ppm or mg/kg	%	ppm or mg/kg	ppm or mg/kg	ppm or mg/kg
	PEANUT (Continued)																
170	SEEDS WITHOUT HULLS, MEAL SOLV EXTD (PEANUT MEAL)	A-F	93	0.36	0.61	0.03	0.03	0.27	1.16	0.31	–	–	–	–	–	–	–
		M-F	100	0.39	0.66	0.03	0.03	0.30	1.25	0.33	–	–	–	–	–	–	–
	PERILLA, COMMON *Perilla frutescens*																
171	SEEDS, MEAL MECH EXTD	A-F	92	0.56	0.47	–	–	–	–	–	–	–	–	–	–	–	–
		M-F	100	0.61	0.51	–	–	–	–	–	–	–	–	–	–	–	–
	POPPY *Papaver* spp																
172	SEEDS, MEAL MECH EXTD	A-F	89	–	–	–	–	–	–	–	–	–	–	–	–	–	–
		M-F	100	–	–	–	–	–	–	–	–	–	–	–	–	–	–
	POULTRY																
173	BY-PRODUCT, MEAL RENDERED	A-F	94	3.97	2.06	0.78	0.54	0.14	0.51	0.53	4.926	19.9	3.101	0.064	16.5	0.920	193.5
		M-F	100	4.23	2.20	0.83	0.58	0.14	0.55	0.56	5.250	21.2	3.305	0.069	17.6	0.980	206.2
174	FEATHERS, MEAL, HYDROLYZED	A-F	93	0.30	0.62	0.63	0.28	0.18	0.27	1.50	0.116	7.3	0.044	0.023	11.9	0.913	71.9
		M-F	100	0.33	0.67	0.68	0.30	0.19	0.29	1.61	0.125	7.9	0.047	0.025	12.9	0.983	77.3
175	MANURE WITH LITTER, DEHY	A-F	86	2.67	1.69	0.41	–	0.43	1.32	–	–	283.3	–	0.046	281.1	0.559	359.4
		M-F	100	3.12	1.98	0.47	–	0.50	1.55	–	–	331.0	–	0.054	328.4	0.653	419.9
176	MANURE WITHOUT LITTER, DEHY	A-F	90	8.07	2.22	0.61	0.86	0.56	1.99	0.16	–	24.6	–	–	–	–	366.4
		M-F	100	8.95	2.46	0.68	0.96	0.62	2.20	0.18	–	27.2	–	–	–	–	405.9
	RAPE *Brassica napus*																
177	SEEDS, MEAL MECH EXTD	A-F	92	0.66	1.04	–	–	0.50	0.83	–	–	6.8	–	0.018	55.3	0.959	43.2
		M-F	100	0.72	1.14	–	–	0.54	0.90	–	–	7.4	–	0.019	60.2	1.043	47.0
178	SEEDS, MEAL SOLV EXTD, 34% PROTEIN	A-F	90	–	–	–	–	–	–	–	–	–	–	–	–	–	–
		M-F	100	–	–	–	–	–	–	–	–	–	–	–	–	–	–
	RAPE, SUMMER *Brassica napus, annua*																
179	SEEDS, MEAL, PREPRESSED, SOLV EXTD	A-F	92	0.66	0.93	–	–	–	–	–	–	–	–	–	–	–	–
		M-F	100	0.72	1.01	–	–	–	–	–	–	–	–	–	–	–	–
	RICE *Oryza sativa*																
180	HULLS, AMMONIATED	A-F	92	0.15	0.19	–	–	–	–	–	–	–	–	–	–	–	–
		M-F	100	0.16	0.21	–	–	–	–	–	–	–	–	–	–	–	–
	RUBBER TREE, PARA *Hevea brasiliensis*																
181	KERNELS, MEAL MECH EXTD	A-F	91	–	–	–	–	–	–	–	–	–	–	–	–	–	–
		M-F	100	–	–	–	–	–	–	–	–	–	–	–	–	–	–
	RYE *Secale cereale*																
182	DISTILLERS GRAINS, DEHY	A-F	92	0.15	0.48	0.17	0.05	0.17	0.07	0.44	–	–	–	–	18.4	–	–
		M-F	100	0.16	0.52	0.18	0.05	0.18	0.08	0.48	–	–	–	–	20.0	–	–
183	DISTILLERS GRAINS WITH SOLUBLES, DEHY	A-F	90	–	–	–	–	–	–	–	–	–	–	–	–	–	–
		M-F	100	–	–	–	–	–	–	–	–	–	–	–	–	–	–
	SAFFLOWER *Carthamus tinctorius*																
184	SEEDS, MEAL SOLV EXTD, 20% PROTEIN	A-F	92	0.31	0.61	–	–	0.32	0.74	0.20	–	9.6	–	0.043	17.7	–	39.6
		M-F	100	0.34	0.66	–	–	0.35	0.80	0.22	–	10.4	–	0.046	19.2	–	43.0
185	SEEDS WITHOUT HULLS, MEAL MECH EXTD	A-F	91	–	–	–	–	–	–	–	–	–	–	–	–	–	–
		M-F	100	–	–	–	–	–	–	–	–	–	–	–	–	–	–
186	SEEDS WITHOUT HULLS, MEAL SOLV EXTD, 42% PROTEIN	A-F	92	0.38	1.08	–	–	1.18	1.18	0.34	1.832	80.6	–	0.091	36.6	–	168.5
		M-F	100	0.41	1.18	–	–	1.29	1.29	0.38	2.000	88.0	–	0.100	40.0	–	184.0
187	SEEDS WITHOUT HULLS, MEAL SOLV EXTD	A-F	91	0.35	1.42	0.04	0.16	0.92	1.05	0.06	2.022	88.6	–	0.082	40.3	–	186.3
		M-F	100	0.38	1.56	0.05	0.18	1.01	1.15	0.06	2.221	97.3	–	0.090	44.2	–	204.6
	SERICEA LESPEDEZA—SEE LESPEDEZA, SERICEA																
	SESAME *Sesamum indicum*																
188	SEEDS	A-F	95	0.97	0.72	–	–	–	–	–	–	–	–	–	–	–	–
		M-F	100	1.02	0.76	–	–	–	–	–	–	–	–	–	–	–	–
189	SEEDS, MEAL MECH EXTD	A-F	93	2.01	1.36	0.05	0.07	0.46	1.25	0.33	–	–	–	0.010	47.7	–	99.6
		M-F	100	2.17	1.46	0.05	0.07	0.50	1.35	0.35	–	–	–	0.010	51.5	–	107.5
190	SEEDS, MEAL SOLV EXTD, 44% PROTEIN	A-F	92	2.01	1.28	–	–	–	–	–	–	–	–	–	47.5	–	–
		M-F	100	2.18	1.39	–	–	–	–	–	–	–	–	–	51.6	–	–
	SHEEP *Ovis aries*																
191	MILK, FRESH	A-F	19	0.21	0.15	0.04	0.08	0.02	0.15	–	–	0.1	–	0.001	0.1	–	2.6
		M-F	100	1.10	0.82	0.22	0.41	0.08	0.82	–	–	0.6	–	0.001	0.3	–	13.7
	SHRIMP *Pandalus* spp, *Penaeus* spp																
192	CANNERY RESIDUE, MEAL (SHRIMP MEAL)	A-F	90	10.40	1.85	1.57	1.04	0.54	0.83	–	–	–	–	0.011	29.8	–	28.4
		M-F	100	11.55	2.06	1.74	1.15	0.60	0.92	–	–	–	–	0.012	33.0	–	31.5
	SORGHUM *Sorghum bicolor*																
193	DISTILLERS GRAINS, DEHY	A-F	94	0.15	0.69	0.05	–	0.18	0.36	0.17	–	–	–	0.005	–	–	–
		M-F	100	0.16	0.74	0.05	–	0.19	0.38	0.18	–	–	–	0.005	–	–	–
194	DISTILLERS GRAINS WITH SOLUBLES, DEHY	A-F	95	0.17	0.92	–	–	–	–	–	–	–	–	–	104.5	–	–
		M-F	100	0.18	0.97	–	–	–	–	–	–	–	–	–	110.2	–	–
195	GLUTEN MEAL	A-F	90	0.03	0.27	–	–	0.16	0.48	–	–	–	–	–	15.6	–	–
		M-F	100	0.04	0.30	–	–	0.17	0.53	–	–	–	–	–	17.3	–	–
196	MILO, GLUTEN MEAL	A-F	90	–	–	–	–	–	–	–	–	–	–	–	–	–	–
		M-F	100	–	–	–	–	–	–	–	–	–	–	–	–	–	–

Entry Number	Fat-Soluble Vitamins					Water-Soluble Vitamins								
	A (1 mg Carotene = 1667 IU Vit A)	Carotene (Provitamin A)	D	E	K	B-12	Biotin	Choline	Folacin (Folic Acid)	Niacin	Pantothenic Acid (B-3)	(Pyridoxine) B-6	Riboflavin (B-2)	Thiamin (B-1)
	IU/g	ppm or mg/kg	IU/kg	ppm or mg/kg	ppm or mg/kg	ppm or mg/kg	ppm or mg/kg	ppm or mg/kg	ppm or mg/kg	ppm or mg/kg	ppm or mg/kg	ppm or mg/kg	ppm or mg/kg	ppm or mg/kg
170	–	–	–	**2.9**	–	–	–	**1896**	–	**178**	**36.8**	**5.95**	**5.3**	–
	–	–	–	3.2	–	–	–	2049	–	192	39.8	6.43	5.7	–
171	–	–	–	–	–	–	–	–	–	–	–	–	–	–
	–	–	–	–	–	–	–	–	–	–	–	–	–	–
172	–	–	–	–	–	–	–	–	–	–	–	–	–	–
	–	–	–	–	–	–	–	–	–	–	–	–	–	–
173	–	–	–	**2.2**	–	**304.1**	**0.09**	**6052**	**0.51**	**54**	**12.4**	**4.43**	**10.6**	**0.2**
	–	–	–	2.4	–	324.1	0.09	6451	0.54	57	13.2	4.72	11.3	0.2
174	–	–	–	–	–	**80.4**	**0.04**	**894**	**0.22**	**21**	**8.9**	**4.39**	**2.0**	**0.1**
	–	–	–	–	–	86.5	0.05	962	0.23	23	9.6	4.72	2.2	0.1
175	–	–	–	–	–	–	–	–	–	–	–	–	–	–
	–	–	–	–	–	–	–	–	–	–	–	–	–	–
176	–	–	–	–	–	**21.1**	–	–	–	**19**	–	–	**11.7**	–
	–	–	–	–	–	23.4	–	–	–	21	–	–	13.0	–
177	–	–	–	**18.8**	–	–	–	**6532**	–	**155**	**9.0**	–	**3.0**	**1.8**
	–	–	–	20.4	–	–	–	7103	–	168	9.8	–	3.3	1.9
178	–	–	–	–	–	–	–	–	–	–	–	–	–	–
	–	–	–	–	–	–	–	–	–	–	–	–	–	–
179	–	–	–	–	–	–	–	–	–	–	–	–	–	–
	–	–	–	–	–	–	–	–	–	–	–	–	–	–
180	–	–	–	–	–	–	–	–	–	–	–	–	–	–
	–	–	–	–	–	–	–	–	–	–	–	–	–	–
181	–	–	–	–	–	–	–	–	–	–	–	–	–	–
	–	–	–	–	–	–	–	–	–	–	–	–	–	–
182	–	–	–	–	–	–	–	–	–	**17**	**5.2**	–	**3.3**	**1.3**
	–	–	–	–	–	–	–	–	–	18	5.7	–	3.6	1.4
183	–	–	–	–	–	–	–	–	–	**63**	**17.4**	–	**8.2**	**3.1**
	–	–	–	–	–	–	–	–	–	69	19.2	–	9.0	3.4
184	–	–	–	**0.9**	–	–	–	**1541**	–	**12**	**36.2**	**474.43**	**2.2**	–
	–	–	–	1.0	–	–	–	1673	–	13	39.3	515.00	2.4	–
185	–	–	–	–	–	–	**1.42**	**2608**	**0.44**	**22**	**86.6**	–	**4.0**	**18.2**
	–	–	–	–	–	–	1.56	2860	0.49	24	95.0	–	4.4	20.0
186	–	–	–	**0.6**	–	–	**1.56**	**3156**	**1.47**	**21**	**38.2**	**10.71**	**2.3**	**4.2**
	–	–	–	0.7	–	–	1.70	3447	1.60	23	41.7	11.70	2.5	4.6
187	–	–	–	**0.7**	–	–	**1.71**	**3242**	**1.61**	**22**	**39.5**	**11.83**	**2.5**	**4.6**
	–	–	–	0.8	–	–	1.88	3562	1.77	24	43.4	13.00	2.8	5.1
188	–	–	–	–	–	–	–	–	–	–	–	–	–	–
	–	–	–	–	–	–	–	–	–	–	–	–	–	–
189	**0.7**	**0.4**	–	–	–	–	–	**1533**	–	**19**	**5.9**	**12.45**	**3.4**	**2.8**
	0.8	0.5	–	–	–	–	–	1655	–	20	6.4	13.44	3.6	3.0
190	–	–	–	–	–	–	–	**1517**	–	–	**6.3**	–	**3.7**	–
	–	–	–	–	–	–	–	1649	–	–	6.8	–	4.0	–
191	–	–	–	**16.1**	–	**4.1**	**0.40**	–	**0.05**	**5**	**4.1**	**0.72**	**4.1**	**1.0**
	–	–	–	85.9	–	22.0	2.12	–	0.27	27	22.0	3.85	22.0	5.5
192	–	–	–	–	–	–	–	**5497**	–	–	–	–	**3.9**	–
	–	–	–	–	–	–	–	6102	–	–	–	–	4.4	–
193	–	–	–	–	–	–	**0.31**	**805**	–	–	–	–	–	–
	–	–	–	–	–	–	0.33	858	–	–	–	–	–	–
194	–	–	–	–	–	–	–	**844**	–	**61**	**12.3**	–	**4.2**	**1.3**
	–	–	–	–	–	–	–	891	–	64	13.0	–	4.4	1.4
195	–	–	–	–	–	–	–	**680**	–	**37**	**9.3**	–	**1.5**	–
	–	–	–	–	–	–	–	754	–	41	10.4	–	1.7	–
196	–	–	–	–	–	–	–	–	–	–	–	–	–	–
	–	–	–	–	–	–	–	–	–	–	–	–	–	–

(Continued)

TABLE V-1B PROTEIN FEEDS, COMPOSITION OF FEEDS, DATA EXPRESSED **AS-FED** AND **MOISTURE-FREE**—*(Continued)*

Entry Number	Feed Name Description	International Feed Number	Moisture Basis: A-F (as-fed) or M-F (moisture-free)	Proximate Analysis: Dry Matter	Ash	Crude Fiber	Neutral Det. Fib. (NDF)	Acid Det. Fib. (ADF)	Lignin	Ether Extract (Fat)	N-Free Extract	Crude Protein	Digestible Protein: Ruminant	Swine	Horse
				%	%	%	%	%	%	%	%	%	%	%	%
	SOYBEAN *Glycine max*														
197	SEEDS[1]	5-04-610	A-F	92	5.1	5.4	–	9.2	–	17.2	25.9	38.4	34.5	31.5	34.9
			M-F	100	5.6	5.8	–	10.0	–	18.7	28.1	41.7	37.5	34.2	37.9
198	SEEDS, LOW PROTEIN, LOW CARBOHYDRATES, MEAL SOLV EXTD	5-04-613	A-F	–	–	–	–	–	–	–	–	–	–	–	–
			M-F	100	–	–	–	–	–	–	–	–	–	–	–
199	SEEDS, MEAL MECH EXTD, 41% PROTEIN	5-04-600	A-F	90	6.0	6.0	–	–	–	4.7	30.4	42.9	36.6	37.5	40.1
			M-F	100	6.7	6.7	–	–	–	5.2	33.8	47.7	40.7	41.7	44.6
200	SEEDS, MEAL SOLV EXTD, 44% PROTEIN	5-20-637	A-F	89	6.4	6.2	12.5	8.9	–	1.5	30.6	44.4	37.8	40.4	41.9
			M-F	100	7.2	7.0	14.0	10.0	–	1.7	34.3	49.8	42.4	45.4	47.1
201	SEEDS WITHOUT HULLS, MEAL SOLV EXTD, 49% PROTEIN	5-20-638	A-F	90	6.1	3.7	6.6	6.2	–	1.2	29.8	49.0	42.4	44.6	47.1
			M-F	100	6.8	4.1	7.4	6.9	–	1.4	33.2	54.6	47.3	49.7	52.4
202	SEED COATS	1-04-560	A-F	91	4.6	36.2	59.4	42.4	1.8	2.0	37.0	10.8	6.5	3.4	6.7
			M-F	100	5.1	40.0	65.6	46.8	2.0	2.2	40.9	11.9	7.2	3.8	7.4
203	FLOUR, SOLV EXTD	5-04-593	A-F	93	6.1	2.5	–	–	–	0.8	32.3	51.6	44.7	50.1	49.7
			M-F	100	6.5	2.7	–	–	–	0.8	34.7	55.3	47.9	53.7	53.2
204	FLOUR BY-PRODUCT (SOYBEAN MILL FEED)	4-04-594	A-F	90	5.0	33.7	–	–	–	1.9	36.4	12.6	8.1	6.0	9.1
			M-F	100	5.5	37.6	–	–	–	2.1	40.7	14.1	9.0	6.7	10.2
205	LECITHIN	4-04-562	A-F	–	–	–	–	–	–	–	–	–	–	–	–
			M-F	100	–	–	–	–	–	–	–	–	–	–	–
206	MILL RUN (GROATS BY-PRODUCT)	4-04-595	A-F	89	4.4	33.8	–	–	–	2.3	36.7	12.3	8.3	9.4	8.9
			M-F	100	5.0	37.8	–	–	–	2.6	41.0	13.7	9.2	10.5	9.9
207	PROTEIN CONCENTRATE, MORE THAN 70% PROTEIN	5-08-038	A-F	92	3.5	0.1	–	–	–	0.5	3.4	84.3	74.5	82.5	86.7
			M-F	100	3.8	0.1	–	–	–	0.6	3.7	91.8	81.2	89.9	94.5
208	PROTEIN PRODUCT, CHEMICALLY MODIFIED	5-26-010	A-F	–	–	–	–	–	–	–	–	–	–	–	–
			M-F	100	–	–	–	–	–	–	–	–	–	–	–
	STARFISH *Asteroidea* (class)														
209	WHOLE, MEAL	5-08-527	A-F	97	43.9	1.9	–	–	–	5.8	14.3	30.6	25.5	29.2	25.6
			M-F	100	45.5	2.0	–	–	–	6.0	14.8	31.7	26.4	30.2	26.6
	SUGARCANE *Saccharum officinarum*														
210	MOLASSES, AMMONIATED	5-04-702	A-F	65	5.9	–	–	–	–	–	–	26.3	15.5	25.3	23.7
			M-F	100	9.1	–	–	–	–	–	–	40.5	23.9	38.9	36.5
	SUNFLOWER, COMMON *Helianthus annuus*														
211	SEEDS WITHOUT HULLS, MEAL MECH EXTD, 41% PROTEIN	5-26-097	A-F	92	6.7	13.2	–	–	–	7.5	23.9	40.7	36.5	–	–
			M-F	100	7.3	14.3	–	–	–	8.2	26.0	44.2	39.7	–	–
212	SEEDS WITHOUT HULLS, MEAL SOLV EXTD, 44% PROTEIN	5-26-098	A-F	93	7.7	11.0	–	–	–	2.9	24.6	46.8	42.1	–	–
			M-F	100	8.3	11.8	–	–	–	3.1	26.5	50.3	45.3	–	–
	SWEET CLOVER, YELLOW *Melilotus officinalis*														
213	SEED SCREENINGS	5-08-007	A-F	87	8.6	15.5	–	–	–	3.6	41.0	18.5	14.7	17.3	12.8
			M-F	100	9.9	17.8	–	–	–	4.1	47.0	21.2	16.9	19.8	14.7
	SWINE *Sus scrofa*														
214	CRACKLINGS, RENDERED	5-04-791	A-F	93	27.9	6.7	–	–	–	13.2	–	50.7	43.9	48.2	48.7
			M-F	100	30.1	7.2	–	–	–	14.2	–	54.6	47.3	51.9	52.4
215	MILK, FRESH	5-08-537	A-F	20	1.0	–	–	–	–	6.7	5.1	7.3	–	–	–
			M-F	100	5.0	–	–	–	–	33.3	25.4	36.3	–	–	–
	TOMATO *Lycopersicon esculentum*														
216	POMACE, DEHY	5-05-041	A-F	92	6.8	25.0	50.4	46.5	10.5	9.8	29.3	21.0	11.9	19.7	15.2
			M-F	100	7.4	27.2	54.8	50.5	11.4	10.7	31.9	22.9	12.9	21.5	16.6
	TURKEY *Meleagris gallopavo*														
217	VISCERA, FRESH	5-08-616	A-F	31	2.1	–	–	–	–	13.3	–	13.4	11.4	12.9	12.3
			M-F	100	6.8	–	–	–	–	43.1	–	43.4	37.0	41.8	39.7
218	VISCERA, FRESH, POULT	5-07-985	A-F	26	2.0	0.2	–	–	–	8.4	1.2	14.2	12.3	13.8	13.6
			M-F	100	7.7	0.9	–	–	–	32.3	4.8	54.4	47.0	52.7	52.1
219	VISCERA, FRESH, MATURE BIRDS	5-07-984	A-F	33	2.4	–	–	–	–	14.7	–	14.2	–	–	–
			M-F	100	7.2	–	–	–	–	44.4	–	43.1	–	–	–
	UREA														
220	45% NITROGEN, 281% PROTEIN EQUIVALENT[1]	5-05-070	A-F	99	–	–	0	0	–	–	–	281.7	–	–	–
			M-F	100	–	–	0	0	–	–	–	285.0	–	–	–
	UREA, CONDITIONER ADDED														
221	42% NITROGEN, 262% PROTEIN EQUIVALENT	5-20-705	A-F	100	–	–	–	–	–	–	–	261.9	–	–	–
			M-F	100	–	–	–	–	–	–	–	262.2	–	–	–
	WHALE *Balaena glacialis, balaenoptera* spp, *Physeter catodon*														
222	MEAT, FRESH	5-07-986	A-F	29	1.0	–	–	–	–	7.5	–	20.6	18.0	20.1	20.6
			M-F	100	3.4	–	–	–	–	25.8	–	70.8	62.0	69.0	70.7
223	MEAT, MEAL RENDERED	5-05-160	A-F	91	4.0	0.3	–	–	–	7.6	8.0	71.4	62.8	67.8	72.2
			M-F	100	4.4	0.3	–	–	–	8.4	8.8	78.1	68.7	74.2	79.0
	WHEAT *Triticum aestivum*														
224	DISTILLERS GRAINS, DEHY	5-05-193	A-F	93	3.0	11.8	–	–	–	6.7	40.2	31.6	26.5	30.2	27.1
			M-F	100	3.3	12.6	–	–	–	7.2	43.0	33.9	28.4	32.4	29.0
225	DISTILLERS SOLUBLES, DEHY	5-05-195	A-F	94	6.8	3.4	–	–	–	2.2	50.5	31.1	–	–	–
			M-F	100	7.2	3.6	–	–	–	2.3	53.7	33.1	–	–	–

PROTEIN FEEDS

Entry Number	TDN			Digestible Energy						Metabolizable Energy								Net Energy					
	Ruminant	Swine	Horse	Ruminant		Swine		Horse		Ruminant		Swine		Poultry ME_n		Horse		Ruminant NE_m		Ruminant NE_g		Lactating Cows NE_{lc}	
	%	%	%	Mcal		kcal		Mcal		Mcal		kcal		kcal		Mcal		Mcal		Mcal		Mcal	
				lb	*kg*	lb	*kg*	lb	*kg*	lb	*kg*	lb	*kg*	lb	*kg*	lb	*kg*	lb	*kg*	lb	*kg*	lb	*kg*
197	**84**	**93**	–	**1.69**	***3.72***	**1820**	***4012***	–	–	**1.52**	***3.34***	**1605**	***3539***	**1534**	***3382***	–	–	**0.93**	***2.04***	**0.64**	***1.41***	**0.86**	***1.90***
	92	101	–	1.83	*4.04*	1977	*4359*	–	–	1.65	*3.63*	1744	*3846*	1667	*3674*	–	–	1.01	*2.22*	0.70	*1.53*	0.94	*2.07*
198	–	–	–	–	–	–	–	–	–	–	–	–	–	–	–	–	–	–	–	–	–	–	–
	–	–	–	–	–	–	–	–	–	–	–	–	–	–	–	–	–	–	–	–	–	–	–
199	**77**	**79**	–	**1.53**	***3.38***	**1579**	***3482***	–	–	**1.37**	***3.01***	**1302**	***2870***	**1102**	***2429***	–	–	**0.86**	***1.89***	**0.58**	***1.29***	**0.81**	***1.78***
	85	88	–	1.70	*3.76*	1755	*3870*	–	–	1.52	*3.35*	1447	*3190*	1224	*2699*	–	–	0.95	*2.10*	0.65	*1.43*	0.90	*1.98*
200	**76**	**75**	–	**1.45**	***3.19***	**1565**	***3450***	–	–	**1.17**	***2.59***	**1430**	***3153***	**1005**	***2216***	–	–	**0.79**	***1.74***	**0.53**	***1.16***	**0.75**	***1.66***
	85	84	–	1.62	*3.57*	1756	*3872*	–	–	1.32	*2.90*	1605	*3538*	1128	*2486*	–	–	0.89	*1.95*	0.59	*1.30*	0.85	*1.87*
201	**78**	**77**	–	**1.51**	***3.32***	**1591**	***3508***	–	–	**1.10**	***2.42***	**1433**	***3160***	**1124**	***2478***	–	–	**0.72**	***1.59***	**0.47**	***1.03***	**0.79**	***1.75***
	87	86	–	1.68	*3.70*	1773	*3909*	–	–	1.22	*2.70*	1597	*3521*	1253	*2761*	–	–	0.81	*1.78*	0.52	*1.15*	0.88	*1.95*
202	**69**	**47**	**39**	**1.20**	***2.65***	**851**	***1877***	**0.75**	***1.65***	**1.03**	***2.26***	**305**	***671***	**301**	***665***	**0.61**	***1.35***	**0.75**	***1.66***	**0.50**	***1.09***	**0.72**	***1.59***
	77	52	43	1.33	*2.92*	940	*2073*	0.83	*1.82*	1.13	*2.50*	336	*742*	333	*734*	0.68	*1.50*	0.83	*1.84*	0.55	*1.21*	0.80	*1.75*
203	**76**	–	–	**1.55**	***3.42***	**1591**	***3509***	–	–	**1.37**	***3.03***	**1558**	***3434***	–	–	–	–	**0.88**	***1.95***	**0.60**	***1.33***	**0.83**	***1.83***
	81	–	–	1.66	*3.66*	1705	*3760*	–	–	1.47	*3.25*	1669	*3680*	–	–	–	–	0.95	*2.09*	0.65	*1.42*	0.89	*1.96*
204	**48**	**26**	–	**1.21**	***2.67***	**529**	***1167***	–	–	**1.04**	***2.29***	**419**	***924***	**362**	***798***	–	–	**0.63**	***1.39***	**0.39**	***0.86***	**0.61**	***1.34***
	53	30	–	1.35	*2.98*	591	*1303*	–	–	1.16	*2.56*	468	*1032*	404	*890*	–	–	0.71	*1.56*	0.43	*0.96*	0.68	*1.50*
205	–	–	–	–	–	–	–	–	–	–	–	–	–	–	–	–	–	–	–	–	–	–	–
	–	–	–	–	–	–	–	–	–	–	–	–	–	–	–	–	–	–	–	–	–	–	–
206	**37**	–	–	**0.74**	***1.64***	–	–	–	–	**0.57**	***1.25***	–	–	**298**	***657***	–	–	–	–	–	–	–	–
	42	–	–	0.83	*1.83*	–	–	–	–	0.64	*1.40*	–	–	333	*735*	–	–	–	–	–	–	–	–
207	**76**	**91**	–	**1.47**	***3.25***	**1958**	***4316***	–	–	**1.14**	***2.52***	**1733**	***3821***	**1121**	***2472***	–	–	**0.76**	***1.67***	**0.49**	***1.09***	**0.72**	***1.60***
	82	100	–	1.61	*3.54*	2133	*4703*	–	–	1.24	*2.74*	1889	*4164*	1222	*2695*	–	–	0.82	*1.82*	0.54	*1.19*	0.79	*1.74*
208	–	–	–	–	–	–	–	–	–	–	–	–	–	–	–	–	–	–	–	–	–	–	–
	–	–	–	–	–	–	–	–	–	–	–	–	–	–	–	–	–	–	–	–	–	–	–
209	**36**	–	–	**1.05**	***2.32***	–	–	–	–	**0.87**	***1.91***	–	–	–	–	–	–	**0.73**	***1.61***	**0.46**	***1.02***	**0.71**	***1.57***
	38	–	–	1.09	*2.40*	–	–	–	–	0.90	*1.98*	–	–	–	–	–	–	0.76	*1.67*	0.48	*1.06*	0.74	*1.62*
210	**47**	–	–	**0.95**	***2.08***	–	–	–	–	**0.82**	***1.81***	–	–	–	–	–	–	–	–	–	–	–	–
	73	–	–	1.46	*3.21*	–	–	–	–	1.27	*2.79*	–	–	–	–	–	–	–	–	–	–	–	–
211	**68**	**76**	–	**1.38**	***3.04***	**1524**	***3361***	–	–	**1.13**	***2.48***	**1327**	***2927***	–	–	–	–	**0.63**	***1.38***	**0.42**	***0.92***	**0.75**	***1.66***
	74	83	–	1.50	*3.30*	1657	*3653*	–	–	1.23	*2.70*	1443	*3181*	–	–	–	–	0.68	*1.50*	0.45	*1.00*	0.82	*1.80*
212	**65**	**69**	–	**1.22**	***2.70***	**1366**	***3012***	–	–	**1.01**	***2.23***	**1173**	***2585***	**919**	***2026***	–	–	**0.59**	***1.30***	**0.34**	***0.74***	**0.63**	***1.40***
	70	74	–	1.32	*2.90*	1469	*3239*	–	–	1.09	*2.40*	1261	*2780*	988	*2178*	–	–	0.64	*1.40*	0.36	*0.80*	0.68	*1.50*
213	–	–	–	–	–	–	–	–	–	–	–	–	–	–	–	–	–	–	–	–	–	–	–
	–	–	–	–	–	–	–	–	–	–	–	–	–	–	–	–	–	–	–	–	–	–	–
214	–	**123**	–	–	–	**2458**	***5419***	–	–	–	–	**2191**	***4831***	–	–	–	–	–	–	–	–	–	–
	–	132	–	–	–	2646	*5833*	–	–	–	–	2359	*5200*	–	–	–	–	–	–	–	–	–	–
215	**27**	–	–	**0.53**	***1.18***	–	–	–	–	**0.50**	***1.10***	–	–	–	–	–	–	–	–	–	–	–	–
	133	–	–	2.66	*5.86*	–	–	–	–	2.48	*5.47*	–	–	–	–	–	–	–	–	–	–	–	–
216	**60**	**47**	–	**1.21**	***2.66***	**934**	***2058***	–	–	**1.03**	***2.27***	**783**	***1727***	**796**	***1754***	–	–	**0.61**	***1.35***	**0.37**	***0.81***	**0.61**	***1.35***
	66	51	–	1.31	*2.89*	1015	*2238*	–	–	1.12	*2.47*	852	*1877*	865	*1907*	–	–	0.67	*1.47*	0.40	*0.88*	0.67	*1.47*
217	–	–	–	–	–	–	–	–	–	–	–	–	–	–	–	–	–	–	–	–	–	–	–
	–	–	–	–	–	–	–	–	–	–	–	–	–	–	–	–	–	–	–	–	–	–	–
218	**30**	**29**	–	**0.59**	***1.30***	**583**	***1284***	–	–	**0.54**	***1.20***	**516**	***1137***	–	–	–	–	**0.36**	***0.79***	**0.26**	***0.57***	**0.33**	***0.72***
	113	111	–	2.26	*4.99*	2229	*4915*	–	–	2.08	*4.59*	1974	*4351*	–	–	–	–	1.36	*3.00*	0.98	*2.17*	1.25	*2.75*
219	–	–	–	–	–	–	–	–	–	–	–	–	–	–	–	–	–	–	–	–	–	–	–
	–	–	–	–	–	–	–	–	–	–	–	–	–	–	–	–	–	–	–	–	–	–	–
220	–	–	–	–	–	–	–	–	–	–	–	–	–	–	–	–	–	–	–	–	–	–	–
	–	–	–	–	–	–	–	–	–	–	–	–	–	–	–	–	–	–	–	–	–	–	–
221	–	–	–	–	–	–	–	–	–	–	–	–	–	–	–	–	–	–	–	–	–	–	–
	–	–	–	–	–	–	–	–	–	–	–	–	–	–	–	–	–	–	–	–	–	–	–
222	–	–	–	–	–	–	–	–	–	–	–	–	–	–	–	–	–	–	–	–	–	–	–
	–	–	–	–	–	–	–	–	–	–	–	–	–	–	–	–	–	–	–	–	–	–	–
223	**80**	**93**	–	**1.70**	***3.75***	**1861**	***4103***	–	–	**1.53**	***3.38***	**1641**	***3617***	–	–	–	–	**1.02**	***2.26***	**0.72**	***1.59***	**0.95**	***2.09***
	88	102	–	1.86	*4.10*	2036	*4488*	–	–	1.68	*3.69*	1795	*3957*	–	–	–	–	1.12	*2.47*	0.79	*1.74*	1.04	*2.28*
224	**78**	**80**	–	**1.52**	***3.35***	**1600**	***3527***	–	–	**1.35**	***2.96***	**1398**	***3081***	–	–	–	–	**0.81**	***1.79***	**0.54**	***1.19***	**0.77**	***1.70***
	84	86	–	1.63	*3.59*	1714	*3778*	–	–	1.44	*3.18*	1497	*3301*	–	–	–	–	0.87	*1.92*	0.58	*1.28*	0.83	*1.82*
225	–	–	–	–	–	–	–	–	–	–	–	–	–	–	–	–	–	–	–	–	–	–	–
	–	–	–	–	–	–	–	–	–	–	–	–	–	–	–	–	–	–	–	–	–	–	–

(Continued)

TABLE V-1B PROTEIN FEEDS, MINERAL AND VITAMIN COMPOSITION OF FEEDS, DATA EXPRESSED **AS-FED** AND **MOISTURE-FREE**—*(Continued)*

Entry Number	Feed Name Description	Moisture Basis: A-F (as-fed) or M-F (moisture-free)	Dry Matter	Macro Minerals							Micro Minerals						
				Calcium (Ca)	Phosphorus (P)	Sodium (Na)	Chlorine (Cl)	Magnesium (Mg)	Potassium (K)	Sulfur (S)	Cobalt (Co)	Copper (Cu)	Iodine (I)	Iron (Fe)	Manganese (Mn)	Selenium (Se)	Zinc (Zn)
			%	%	%	%	%	%	%	%	ppm or mg/kg	ppm or mg/kg	ppm or mg/kg	%	ppm or mg/kg	ppm or mg/kg	ppm or mg/kg
	SOYBEAN *Glycine max*																
197	SEEDS	A-F	92	0.25	0.60	0.00	0.03	0.27	1.66	0.22	–	18.2	–	0.009	36.4	0.111	56.9
		M-F	100	0.27	0.65	0.00	0.03	0.29	1.80	0.24	–	19.8	–	0.010	39.6	0.120	61.8
198	SEEDS, LOW PROTEIN, LOW CARBOHYDRATES, MEAL SOLV EXTD	A-F	–	–	–	–	–	–	–	–	–	–	–	–	–	–	–
		M-F	100	–	–	–	–	–	–	–	–	–	–	–	–	–	–
199	SEEDS, MEAL MECH EXTD, 41% PROTEIN	A-F	90	0.26	0.61	0.18	0.07	0.26	1.79	0.33	0.178	21.7	–	0.017	31.3	0.102	57.2
		M-F	100	0.29	0.68	0.20	0.08	0.29	1.98	0.37	0.198	24.1	–	0.019	34.8	0.113	63.6
200	SEEDS, MEAL SOLV EXTD, 44% PROTEIN	A-F	89	0.35	0.64	0.03	–	0.27	1.98	0.41	1.381	19.9	–	0.017	31.6	0.486	50.5
		M-F	100	0.40	0.71	0.04	–	0.31	2.22	0.47	1.550	22.3	–	0.019	35.5	0.546	56.6
201	SEEDS WITHOUT HULLS, MEAL SOLV EXTD, 49% PROTEIN	A-F	90	0.25	0.63	0.00	0.07	0.37	1.79	0.41	2.693	13.5	0.152	0.010	49.5	–	51.1
		M-F	100	0.28	0.70	0.00	0.08	0.41	1.99	0.46	3.000	15.0	0.169	0.011	55.2	–	56.9
202	SEED COATS	A-F	91	0.45	0.19	0.03	–	–	1.15	0.08	0.109	16.1	–	0.030	9.9	–	21.8
		M-F	100	0.49	0.21	0.03	–	–	1.27	0.09	0.121	17.8	–	0.033	11.0	–	24.1
203	FLOUR, SOLV EXTD	A-F	93	0.38	0.65	0.00	0.20	0.25	1.90	0.41	0.943	14.8	–	0.015	30.3	0.278	30.9
		M-F	100	0.40	0.70	0.00	0.22	0.27	2.04	0.44	1.010	15.8	–	0.016	32.4	0.298	33.1
204	FLOUR BY-PRODUCT (SOYBEAN MILL FEED)	A-F	90	0.47	0.18	–	–	0.32	1.51	–	–	–	–	–	28.5	–	–
		M-F	100	0.52	0.20	–	–	0.36	1.69	–	–	–	–	–	31.8	–	–
205	LECITHIN	A-F	–	–	–	–	–	–	–	–	–	–	–	–	–	–	–
		M-F	100	–	–	–	–	–	–	–	–	–	–	–	–	–	–
206	MILL RUN (GROATS BY-PRODUCT)	A-F	89	0.49	0.18	–	–	–	–	0.06	–	–	–	–	–	–	–
		M-F	100	0.54	0.20	–	–	–	–	0.07	–	–	–	–	–	–	–
207	PROTEIN CONCENTRATE, MORE THAN 70% PROTEIN	A-F	92	0.11	0.68	0.12	–	0.02	0.01	–	0.389	14.1	0.321	0.014	5.5	0.137	29.6
		M-F	100	0.12	0.74	0.13	–	0.02	0.01	–	0.424	15.4	0.350	0.015	6.0	0.149	32.3
208	PROTEIN PRODUCT, CHEMICALLY MODIFIED	A-F	–	–	–	–	–	–	–	–	–	–	–	–	–	–	–
		M-F	100	–	–	–	–	–	–	–	–	–	–	–	–	–	–
	STARFISH *Asteroidea* (class)																
209	WHOLE, MEAL	A-F	97	15.70	0.45	–	–	–	–	–	–	–	–	–	–	–	–
		M-F	100	16.27	0.47	–	–	–	–	–	–	–	–	–	–	–	–
	SUGARCANE *Saccharum officinarum*																
210	MOLASSES, AMMONIATED	A-F	65	0.79	0.13	–	–	–	–	–	–	–	–	–	–	–	–
		M-F	100	1.22	0.20	–	–	–	–	–	–	–	–	–	–	–	–
	SUNFLOWER, COMMON *Helianthus annuus*																
211	SEEDS WITHOUT HULLS, MEAL MECH EXTD, 41% PROTEIN	A-F	92	–	–	–	–	–	–	–	–	–	–	–	–	–	–
		M-F	100	–	–	–	–	–	–	–	–	–	–	–	–	–	–
212	SEEDS WITHOUT HULLS, MEAL SOLV EXTD, 44% PROTEIN	A-F	93	–	–	–	–	–	–	–	–	–	–	–	–	–	–
		M-F	100	–	–	–	–	–	–	–	–	–	–	–	–	–	–
	SWEET CLOVER, YELLOW *Melilotus officinalis*																
213	SEED SCREENINGS	A-F	87	0.82	0.41	–	–	–	–	–	–	–	–	–	–	–	–
		M-F	100	0.94	0.47	–	–	–	–	–	–	–	–	–	–	–	–
	SWINE *Sus scrofa*																
214	CRACKLINGS, RENDERED	A-F	93	11.25	5.78	1.11	–	–	–	–	–	–	–	–	14.4	–	–
		M-F	100	12.11	6.22	1.19	–	–	–	–	–	–	–	–	15.5	–	–
215	MILK, FRESH	A-F	20	–	–	–	–	–	–	–	–	–	–	–	–	–	–
		M-F	100	–	–	–	–	–	–	–	–	–	–	–	–	–	–
	TOMATO *Lycopersicon esculentum*																
216	POMACE, DEHY	A-F	92	0.39	0.55	–	–	0.18	3.34	–	–	30.0	–	0.424	47.1	–	–
		M-F	100	0.43	0.60	–	–	0.20	3.63	–	–	32.6	–	0.460	51.2	–	–
	TURKEY *Meleagris gallopavo*																
217	VISCERA, FRESH	A-F	31	–	–	–	–	–	–	–	–	–	–	–	–	–	–
		M-F	100	–	–	–	–	–	–	–	–	–	–	–	–	–	–
218	VISCERA, FRESH, POULT	A-F	26	–	–	–	–	–	–	–	–	–	–	–	–	–	–
		M-F	100	–	–	–	–	–	–	–	–	–	–	–	–	–	–
219	VISCERA, FRESH, MATURE BIRDS	A-F	33	–	–	–	–	–	–	–	–	–	–	–	–	–	–
		M-F	100	–	–	–	–	–	–	–	–	–	–	–	–	–	–
	UREA																
220	45% NITROGEN, 281% PROTEIN EQUIVALENT	A-F	99	–	–	–	–	0.00	–	–	–	6.9	–	0.018	–	–	6.9
		M-F	100	–	–	–	–	0.00	–	–	–	7.0	–	0.019	–	–	7.0
	UREA, CONDITIONER ADDED																
221	42% NITROGEN, 262% PROTEIN EQUIVALENT	A-F	100	–	–	–	–	–	–	–	–	–	–	–	–	–	–
		M-F	100	–	–	–	–	–	–	–	–	–	–	–	–	–	–
	WHALE *Balaena glacialis, balaenoptera* spp, *Physeter catodon*																
222	MEAT, FRESH	A-F	29	0.01	0.14	0.08	–	–	0.02	–	–	–	–	–	–	–	–
		M-F	100	0.03	0.48	0.28	–	–	0.07	–	–	–	–	–	–	–	–
223	MEAT, MEAL RENDERED	A-F	91	0.40	0.56	–	–	–	–	–	–	–	–	–	–	–	–
		M-F	100	0.44	0.61	–	–	–	–	–	–	–	–	–	–	–	–
	WHEAT *Triticum aestivum*																
224	DISTILLERS GRAINS, DEHY	A-F	93	0.11	0.58	–	–	–	–	–	–	–	–	–	15.0	–	–
		M-F	100	0.12	0.63	–	–	–	–	–	–	–	–	–	16.1	–	–
225	DISTILLERS SOLUBLES, DEHY	A-F	94	–	–	–	–	–	–	–	–	–	–	–	–	–	–
		M-F	100	–	–	–	–	–	–	–	–	–	–	–	–	–	–

PROTEIN FEEDS

Entry Number	Fat-Soluble Vitamins: A (1 mg Carotene = 1667 IU Vit A)	Carotene (Provitamin A)	D	E	K	Water-Soluble Vitamins: B–12	Biotin	Choline	Folacin (Folic Acid)	Niacin	Pantothenic Acid (B–3)	(Pyridoxine) B–6	Riboflavin (B–2)	Thiamin (B–1)
	IU/g	ppm or mg/kg	IU/kg	ppm or mg/kg	ppm or mg/kg	ppm or mg/kg	ppm or mg/kg	ppm or mg/kg	ppm or mg/kg	ppm or mg/kg	ppm or mg/kg	ppm or mg/kg	ppm or mg/kg	ppm or mg/kg
197	1.5	0.9	–	33.7	–	–	0.38	2931	–	23	16.0	11.04	2.9	11.3
	1.6	1.0	–	36.6	–	–	0.42	3184	–	24	17.4	12.00	3.2	12.2
198	–	–	–	–	–	–	–	–	–	–	–	–	–	–
	–	–	–	–	–	–	–	–	–	–	–	–	–	–
199	0.4	0.2	–	6.5	–	–	0.33	2623	6.39	31	14.3	7.22	3.4	3.9
	0.4	0.2	–	7.3	–	–	0.36	2916	7.10	34	15.8	8.02	3.8	4.3
200	–	–	–	3.0	0.22	2.0	0.36	2706	0.69	26	13.8	5.90	3.0	6.6
	–	–	–	3.4	0.25	2.2	0.41	3036	0.77	29	15.5	6.62	3.4	7.4
201	–	–	0	3.3	–	2.0	0.38	2772	0.59	24	14.1	5.59	2.9	3.5
	–	–	0	3.7	–	2.2	0.42	3089	0.66	27	15.7	6.23	3.3	3.9
202	–	–	–	6.6	–	–	–	588	–	25	13.4	1.70	3.6	1.6
	–	–	–	7.3	–	–	–	649	–	27	14.8	1.88	4.0	1.8
203	0.7	0.4	–	–	–	–	0.73	2258	–	43	14.2	–	2.9	2.7
	0.8	0.5	–	–	–	–	0.78	2420	–	46	15.2	–	3.1	2.9
204	–	–	–	–	–	–	0.22	444	0.22	24	13.3	2.24	3.5	2.2
	–	–	–	–	–	–	0.25	495	0.25	27	14.8	2.51	3.9	2.5
205	–	–	–	–	–	–	–	–	–	–	–	–	–	–
	–	–	–	–	–	–	–	–	–	–	–	–	–	–
206	–	–	–	–	–	–	–	–	–	24	13.1	–	3.5	–
	–	–	–	–	–	–	–	–	–	27	14.7	–	3.9	–
207	–	–	–	–	0.02	–	–	2	–	5	3.5	–	0.7	0.3
	–	–	–	–	0.02	–	–	2	–	5	3.8	–	0.8	0.4
208	–	–	–	–	–	–	–	–	–	–	–	–	–	–
	–	–	–	–	–	–	–	–	–	–	–	–	–	–
209	–	–	–	–	–	–	–	–	–	–	–	–	8.8	–
	–	–	–	–	–	–	–	–	–	–	–	–	9.1	–
210	–	–	–	–	–	–	–	–	–	–	–	–	–	–
	–	–	–	–	–	–	–	–	–	–	–	–	–	–
211	–	–	–	–	–	–	–	–	–	–	–	–	–	–
	–	–	–	–	–	–	–	–	–	–	–	–	–	–
212	–	–	–	–	–	–	–	–	–	–	–	–	–	–
	–	–	–	–	–	–	–	–	–	–	–	–	–	–
213	–	–	–	–	–	–	–	–	–	–	–	–	–	–
	–	–	–	–	–	–	–	–	–	–	–	–	–	–
214	–	–	–	–	–	–	–	–	–	–	–	–	4.0	–
	–	–	–	–	–	–	–	–	–	–	–	–	4.3	–
215	–	–	–	–	–	–	–	–	–	–	–	–	–	–
	–	–	–	–	–	–	–	–	–	–	–	–	–	–
216	–	–	–	–	–	–	–	–	–	–	–	–	6.1	11.3
	–	–	–	–	–	–	–	–	–	–	–	–	6.7	12.3
217	–	–	–	–	–	–	–	–	–	–	–	–	–	–
	–	–	–	–	–	–	–	–	–	–	–	–	–	–
218	–	–	–	–	–	–	–	–	–	–	–	–	–	–
	–	–	–	–	–	–	–	–	–	–	–	–	–	–
219	–	–	–	–	–	–	–	–	–	–	–	–	–	–
	–	–	–	–	–	–	–	–	–	–	–	–	–	–
220	–	–	–	–	–	–	–	–	–	–	–	–	–	–
	–	–	–	–	–	–	–	–	–	–	–	–	–	–
221	–	–	–	–	–	–	–	–	–	–	–	–	–	–
	–	–	–	–	–	–	–	–	–	–	–	–	–	–
222	–	–	–	–	–	–	–	–	–	–	–	–	0.8	0.9
	–	–	–	–	–	–	–	–	–	–	–	–	2.7	3.1
223	–	–	–	–	–	88.0	–	–	–	104	2.6	8.32	8.3	1.3
	–	–	–	–	–	96.2	–	–	–	114	2.9	9.10	9.1	1.4
224	1.8	1.1	–	–	–	–	–	–	–	56	8.2	–	3.7	2.0
	2.0	1.2	–	–	–	–	–	–	–	60	8.7	–	4.0	2.1
225	–	–	–	–	–	–	–	–	–	–	–	–	–	–
	–	–	–	–	–	–	–	–	–	–	–	–	–	–

(Continued)

TABLE V-1B PROTEIN FEEDS, COMPOSITION OF FEEDS, DATA EXPRESSED **AS-FED** AND **MOISTURE-FREE**—*(Continued)*

Entry Number	Feed Name Description	International Feed Number	Moisture Basis: A-F (as-fed) or M-F (moisture-free)	Proximate Analysis: Dry Matter	Ash	Crude Fiber	Neutral Det. Fib. (NDF)	Acid Det. Fib. (ADF)	Lignin	Ether Extract (Fat)	N-Free Extract	Crude Protein	Digestible Protein: Ruminant	Swine	Horse
				%	%	%	%	%	%	%	%	%	%	%	%
	WHEAT (Continued)														
226	GERM MEAL	5-05-218	A-F	88	4.3	3.1	–	4.4	–	8.5	48.1	24.4	22.9	23.1	19.4
			M-F	100	4.9	3.5	–	5.0	–	9.6	54.4	27.6	26.0	26.2	21.9
227	GLUTEN	5-05-221	A-F	91	0.9	0.4	–	–	–	0.8	9.8	79.0	71.1	77.3	80.9
			M-F	100	1.0	0.4	–	–	–	0.9	10.8	86.9	78.2	85.0	88.9
	WHEY, CATTLE *Bos taurus*														
228	FRESH[1]	4-08-134	A-F	7	0.7	–	0	0	–	0.3	5.1	0.9	0.6	0.7	0.7
			M-F	100	9.4	–	0	0	–	4.3	73.9	13.2	8.8	10.0	9.5
229	CONDENSED	4-01-180	A-F	54	5.0	0.3	–	–	–	0.4	41.2	6.9	4.5	5.2	5.0
			M-F	100	9.3	0.5	–	–	–	0.8	76.6	12.8	8.4	9.7	9.2
230	DEHY	4-01-182	A-F	93	8.8	0.2	0.3	0.2	–	0.8	70.2	13.3	8.9	13.1	9.6
			M-F	100	9.4	0.2	0.3	0.2	–	0.8	75.3	14.2	9.5	14.1	10.3
231	LOW LACTOSE, DEHY (DRIED WHEY PRODUCT)[1]	4-01-186	A-F	93	15.4	0.2	0	0	–	1.0	60.0	16.7	11.5	13.3	12.3
			M-F	100	16.5	0.2	0	0	–	1.1	64.3	17.9	12.4	14.3	13.2
232	ALBUMIN (DRIED MILK ALBUMIN)	5-01-177	A-F	92	29.1	0.9	–	–	–	1.0	12.8	48.4	41.8	46.9	46.1
			M-F	100	31.6	0.9	–	–	–	1.1	13.9	52.5	45.4	50.9	50.1
	YEAST, BREWERS *Saccharomyces cerevisiae*														
233	DEHY	7-05-527	A-F	93	6.5	3.0	–	3.7	–	0.9	38.8	43.8	39.0	–	–
			M-F	100	7.0	3.2	–	4.0	–	1.0	41.7	47.1	41.9	–	–
	YEAST, IRRADIATED *Saccharomyces cerevisiae*														
234	DEHY	7-05-529	A-F	94	6.2	6.2	–	–	–	1.1	32.4	48.1	–	–	–
			M-F	100	6.6	6.5	–	–	–	1.2	34.5	51.2	–	–	–
	YEAST, TORULA *Torulopsis utilis*														
235	DEHY	7-05-534	A-F	93	8.0	2.5	–	3.7	–	1.6	31.5	49.6	45.1	40.6	–
			M-F	100	8.6	2.7	–	4.0	–	1.7	33.8	53.3	48.5	43.7	–

[1]Neutral Detergent Fiber (NDF) and Acid Detergent Fiber (ADF) values taken from *Nutrient Requirements of Dairy Cattle,* 6th rev. ed., NRC, Natinal Academy Press, 1988, pp. 90–110, Table 7–1.

TABLE V-1C DRY FORAGES, COMPOSITION OF FEEDS, DATA EXPRESSED **AS-FED** AND **MOISTURE-FREE** (See footnote at end of table.)

Entry Number	Feed Name Description	International Feed Number	Moisture Basis: A-F (as-fed) or M-F (moisture-free)	Proximate Analysis: Dry Matter	Ash	Crude Fiber	Neutral Det. Fib. (NDF)	Acid Det. Fib. (ADF)	Lignin	Ether Extract (Fat)	N-Free Extract	Crude Protein	Digestable Protein: Ruminant	Swine	Horse
				%	%	%	%	%	%	%	%	%	%	%	%
	ACACIA *Acacia* spp														
1	PODS, SUN-CURED	1-00-006	A-F	91	4.5	17.8	–	–	–	1.9	56.8	10.4	5.3	5.7	6.4
			M-F	100	4.9	19.5	–	–	–	2.0	62.1	11.4	5.8	6.2	7.0
	ADAM'S NEEDLE *Yucca smalliana*														
2	HAY, SUN-CURED	1-08-560	A-F	93	6.9	38.6	–	–	–	2.2	38.3	6.6	3.0	2.2	3.4
			M-F	100	7.5	41.7	–	–	–	2.4	41.4	7.1	3.2	2.4	3.7
	ALDER *Alnus* spp														
3	LEAVES, SUN-CURED	1-00-015	A-F	85	5.0	14.5	–	–	–	5.1	41.7	18.7	8.8	13.1	12.9
			M-F	100	5.9	17.0	–	–	–	6.0	49.1	22.0	10.3	15.5	15.2
	ALFALFA (LUCERNE) *Medicago sativa*														
4	HAY, SUN-CURED, ALL ANALYSES	1-00-078	A-F	90	8.6	28.2	35.4	30.9	8.9	1.7	35.9	16.0	11.2	7.5	11.9
			M-F	100	9.5	31.2	39.2	34.2	9.8	1.9	39.7	17.7	12.4	8.3	13.2
5	HAY, PREBLOOM, SUN-CURED	1-00-054	A-F	90	8.3	20.7	38.3	29.6	5.5	4.1	36.4	20.2	15.3	14.3	14.0
			M-F	100	9.2	23.1	42.7	33.0	6.1	4.5	40.6	22.5	17.0	15.9	15.6
6	HAY, EARLY BLOOM, SUN-CURED	1-00-059	A-F	91	8.4	25.8	36.8	29.0	5.8	2.6	35.8	17.9	13.3	12.3	12.2
			M-F	100	9.2	28.5	40.7	32.0	6.4	2.9	39.6	19.8	14.7	13.5	13.5
7	HAY, MIDBLOOM, SUN-CURED	1-00-063	A-F	91	7.8	25.5	43.2	33.4	6.7	3.3	37.4	17.1	12.0	11.5	11.6
			M-F	100	8.6	28.0	47.4	36.7	7.4	3.6	41.1	18.8	13.2	12.6	12.7
8	HAY, FULL BLOOM, SUN-CURED	1-00-068	A-F	91	7.1	27.3	45.0	35.2	6.9	3.1	37.9	15.5	11.3	10.1	10.3
			M-F	100	7.8	30.1	49.5	38.7	7.6	3.4	41.7	17.0	12.5	11.1	11.3
9	HAY, MATURE, SUN-CURED	1-00-071	A-F	91	6.7	29.3	50.1	40.1	11.3	2.9	37.0	15.2	11.3	9.8	10.1
			M-F	100	7.4	32.1	55.0	44.0	12.4	3.2	40.6	16.7	12.4	10.8	11.1
10	HAY, BROWN, SUN-CURED	1-00-103	A-F	89	9.2	25.2	–	–	–	1.5	36.6	16.1	9.7	10.7	10.8
			M-F	100	10.4	28.4	–	–	–	1.7	41.3	18.2	11.0	12.1	12.2
11	HAY, STEMMY, SUN-CURED	1-00-093	A-F	91	6.7	36.2	–	–	–	1.0	34.8	12.5	8.5	–	7.0
			M-F	100	7.4	39.7	–	–	–	1.1	38.2	13.6	9.3	–	7.7
12	HAY, RAINED ON, SUN-CURED	1-00-130	A-F	89	6.6	33.5	–	–	–	0.8	30.4	17.9	12.9	12.2	12.2
			M-F	100	7.4	37.6	–	–	–	0.9	34.1	20.0	14.5	13.7	13.6
13	LEAVES, MEAL, SUN-CURED	1-00-146	A-F	88	9.2	15.0	–	–	–	2.7	40.7	20.1	15.8	14.3	13.9
			M-F	100	10.5	17.1	–	–	–	3.1	46.3	22.9	18.0	16.3	15.9
14	LEAVES, MEAL, DEHY	1-00-137	A-F	92	10.4	18.3	–	–	–	3.0	39.9	20.6	14.4	14.5	14.2
			M-F	100	11.3	19.8	–	–	–	3.3	43.3	22.3	15.6	15.7	15.4
15	STEMS, MEAL, SUN-CURED	1-00-165	A-F	91	7.6	36.5	–	–	–	1.4	34.1	11.3	7.1	6.4	7.1
			M-F	100	8.4	40.2	–	–	–	1.6	37.5	12.4	7.8	7.0	7.8

PROTEIN FEEDS

Entry Number	TDN Ruminant %	TDN Swine %	TDN Horse %	DE Ruminant Mcal lb	DE Ruminant Mcal kg	DE Swine kcal lb	DE Swine kcal kg	DE Horse Mcal lb	DE Horse Mcal kg	ME Ruminant Mcal lb	ME Ruminant Mcal kg	ME Swine kcal lb	ME Swine kcal kg	Poultry ME_n kcal lb	Poultry ME_n kcal kg	ME Horse Mcal lb	ME Horse Mcal kg	Ruminant NE_m Mcal lb	Ruminant NE_m Mcal kg	Ruminant NE_g Mcal lb	Ruminant NE_g Mcal kg	Lactating Cows NE_{lc} Mcal lb	Lactating Cows NE_{lc} Mcal kg
226	83	80	–	1.66	3.66	1727	3807	–	–	1.50	3.30	1522	3357	1223	2696	–	–	0.96	2.11	0.67	1.48	0.89	1.95
	94	91	–	1.88	4.14	1954	4308	–	–	1.69	3.73	1723	3798	1384	3051	–	–	1.08	2.38	0.76	1.67	1.00	2.21
227	84	–	–	1.69	3.73	2002	4413	–	–	1.52	3.35	1771	3905	–	–	–	–	0.96	2.11	0.67	1.47	0.89	1.96
	93	–	–	1.86	4.10	2200	4850	–	–	1.67	3.69	1946	4291	–	–	–	–	1.06	2.33	0.73	1.62	0.98	2.16
228	7	–	–	0.13	0.29	–	–	–	–	0.12	0.26	–	–	–	–	–	–	–	–	–	–	–	–
	94	–	–	1.88	4.15	–	–	–	–	1.70	3.74	–	–	–	–	–	–	–	–	–	–	–	–
229	47	46	–	0.94	2.07	928	2046	–	–	0.84	1.85	878	1936	–	–	–	–	0.56	1.24	0.39	0.86	0.52	1.15
	87	86	–	1.74	3.84	1726	3806	–	–	1.56	3.43	1633	3601	–	–	–	–	1.04	2.30	0.72	1.60	0.97	2.13
230	76	77	–	1.51	3.33	1444	3183	–	–	1.28	2.83	1411	3110	880	1939	–	–	0.87	1.92	0.59	1.30	0.78	1.71
	82	83	–	1.62	3.57	1549	3415	–	–	1.38	3.03	1514	3337	944	2081	–	–	0.93	2.06	0.63	1.40	0.83	1.83
231	74	75	–	1.40	3.09	1242	2737	–	–	1.14	2.52	1245	2744	931	2053	–	–	0.75	1.66	0.49	1.08	0.77	1.69
	79	80	–	1.50	3.31	1330	2932	–	–	1.22	2.70	1333	2939	997	2199	–	–	0.81	1.78	0.52	1.15	0.82	1.81
232	59	–	–	1.28	2.83	–	–	–	–	1.11	2.44	–	–	–	–	–	–	0.76	1.67	0.50	1.09	0.73	1.60
	64	–	–	1.39	3.07	–	–	–	–	1.20	2.65	–	–	–	–	–	–	0.82	1.81	0.54	1.18	0.79	1.74
233	73	70	–	1.46	3.21	–	–	–	–	1.28	2.83	1299	2865	928	2047	–	–	0.81	1.79	0.54	1.19	0.77	1.69
	78	76	–	1.57	3.45	–	–	–	–	1.38	3.04	1396	3078	998	2199	–	–	0.87	1.92	0.58	1.28	0.83	1.82
234	72	–	–	1.43	3.16	–	–	–	–	1.26	2.77	–	–	–	–	–	–	–	–	–	–	–	–
	76	–	–	1.53	3.37	–	–	–	–	1.34	2.95	–	–	–	–	–	–	–	–	–	–	–	–
235	72	64	–	1.49	3.29	1287	2837	–	–	1.27	2.81	1096	2416	840	1851	–	–	0.82	1.81	0.55	1.21	0.78	1.71
	78	69	–	1.60	3.53	1383	3049	–	–	1.37	3.02	1178	2597	902	1989	–	–	0.88	1.95	0.59	1.30	0.84	1.84

DRY FORAGES

Entry Number	TDN Ruminant %	TDN Swine %	TDN Horse %	DE Ruminant Mcal lb	DE Ruminant Mcal kg	DE Swine kcal lb	DE Swine kcal kg	DE Horse Mcal lb	DE Horse Mcal kg	ME Ruminant Mcal lb	ME Ruminant Mcal kg	ME Swine kcal lb	ME Swine kcal kg	Poultry ME_n kcal lb	Poultry ME_n kcal kg	ME Horse Mcal lb	ME Horse Mcal kg	Ruminant NE_m Mcal lb	Ruminant NE_m Mcal kg	Ruminant NE_g Mcal lb	Ruminant NE_g Mcal kg	Lactating Cows NE_{lc} Mcal lb	Lactating Cows NE_{lc} Mcal kg
1	55	56	–	1.11	2.45	–	–	–	–	0.93	2.06	–	–	–	–	–	–	0.50	1.11	0.27	0.59	0.53	1.17
	61	61	–	1.21	2.68	–	–	–	–	1.02	2.25	–	–	–	–	–	–	0.55	1.21	0.29	0.64	0.58	1.28
2	51	–	29	0.98	2.16	–	–	0.59	1.30	0.80	1.76	–	–	–	–	0.48	1.07	0.41	0.91	0.18	0.40	0.47	1.03
	55	–	31	1.06	2.33	–	–	0.64	1.40	0.86	1.90	–	–	–	–	0.52	1.15	0.45	0.98	0.20	0.44	0.51	1.12
3	42	–	48	0.84	1.84	–	–	0.90	1.98	0.67	1.48	–	–	–	–	0.74	1.63	–	–	–	–	–	–
	49	–	57	0.98	2.17	–	–	1.06	2.33	0.79	1.74	–	–	–	–	0.87	1.91	–	–	–	–	–	–
4	51	32	48	1.03	2.28	–	–	0.90	1.98	0.90	1.99	–	–	–	–	0.74	1.62	0.48	1.06	0.25	0.55	0.49	1.07
	56	35	53	1.14	2.52	–	–	0.99	2.19	1.00	2.20	–	–	–	–	0.81	1.79	0.53	1.18	0.28	0.61	0.54	1.18
5	54	64	48	1.25	2.75	–	–	0.90	1.98	1.02	2.24	–	–	–	–	0.74	1.63	0.57	1.26	0.34	0.74	0.58	1.27
	60	72	54	1.39	3.07	–	–	1.00	2.21	1.13	2.50	–	–	–	–	0.82	1.82	0.64	1.41	0.37	0.82	0.65	1.42
6	52	50	48	1.15	2.54	–	–	0.90	1.99	0.96	2.12	–	–	–	–	0.74	1.63	0.60	1.33	0.36	0.80	0.55	1.21
	58	55	53	1.27	2.80	–	–	1.00	2.19	1.06	2.35	–	–	–	–	0.82	1.80	0.67	1.47	0.40	0.88	0.61	1.33
7	52	54	46	1.12	2.46	–	–	0.86	1.90	0.94	2.07	–	–	–	–	0.71	1.56	0.52	1.14	0.28	0.62	0.51	1.13
	57	59	50	1.23	2.70	–	–	0.95	2.09	1.03	2.28	–	–	–	–	0.78	1.71	0.57	1.25	0.31	0.68	0.56	1.24
8	51	50	44	1.09	2.39	–	–	0.83	1.83	0.88	1.95	–	–	–	–	0.68	1.50	0.49	1.09	0.26	0.57	0.52	1.15
	56	55	48	1.19	2.63	–	–	0.91	2.01	0.97	2.14	–	–	–	–	0.75	1.65	0.54	1.20	0.29	0.63	0.58	1.27
9	55	47	43	1.22	2.68	–	–	0.82	1.81	0.93	2.06	–	–	–	–	0.67	1.48	0.47	1.04	0.24	0.53	0.51	1.12
	61	52	47	1.33	2.94	–	–	0.90	1.98	1.03	2.26	–	–	–	–	0.74	1.63	0.52	1.14	0.26	0.58	0.56	1.23
10	49	42	48	0.99	2.18	–	–	0.90	1.99	0.82	1.80	–	–	–	–	0.74	1.64	0.48	1.05	0.25	0.55	0.51	1.11
	55	47	55	1.12	2.46	–	–	1.02	2.25	0.92	2.04	–	–	–	–	0.84	1.85	0.54	1.18	0.28	0.62	0.57	1.26
11	49	–	–	0.99	2.18	–	–	–	–	0.78	1.73	–	–	–	–	–	–	–	–	–	–	–	–
	54	–	–	1.09	2.40	–	–	–	–	0.86	1.90	–	–	–	–	–	–	–	–	–	–	–	–
12	47	–	50	1.00	2.20	–	–	0.94	2.07	0.83	1.83	–	–	–	–	0.77	1.70	0.44	0.96	0.21	0.46	0.48	1.05
	53	–	56	1.12	2.47	–	–	1.05	2.32	0.93	2.05	–	–	–	–	0.86	1.90	0.49	1.07	0.24	0.52	0.54	1.18
13	56	63	56	1.13	2.49	–	–	1.03	2.27	0.96	2.12	–	–	–	–	0.85	1.86	0.58	1.28	0.35	0.77	0.58	1.28
	64	72	64	1.29	2.84	–	–	1.18	2.59	1.10	2.42	–	–	–	–	0.96	2.13	0.66	1.46	0.40	0.87	0.66	1.46
14	58	62	54	1.16	2.55	–	–	1.01	2.22	0.98	2.16	918	2024	734	1619	0.83	1.82	0.62	1.36	0.37	0.82	0.62	1.36
	63	67	59	1.26	2.77	–	–	1.09	2.41	1.06	2.35	995	2193	796	1754	0.89	1.97	0.67	1.47	0.40	0.88	0.67	1.47
15	42	–	37	0.88	1.95	–	–	0.73	1.60	0.71	1.56	–	–	–	–	0.60	1.31	0.41	0.91	0.19	0.41	0.47	1.03
	46	–	41	0.97	2.14	–	–	0.80	1.76	0.78	1.71	–	–	–	–	0.65	1.44	0.45	1.00	0.20	0.45	0.51	1.13

(Continued)

TABLE V-1B PROTEIN FEEDS, MINERAL AND VITAMIN COMPOSITION OF FEEDS, DATA EXPRESSED **AS-FED** AND **MOISTURE-FREE**—*(Continued)*

Entry Number	Feed Name Description	Moisture Basis: A-F (as-fed) or M-F (moisture-free)	Dry Matter	Macro Minerals							Micro Minerals						
				Calcium (Ca)	Phosphorus (P)	Sodium (Na)	Chlorine (Cl)	Magnesium (Mg)	Potassium (K)	Sulfur (S)	Cobalt (Co)	Copper (Cu)	Iodine (I)	Iron (Fe)	Manganese (Mn)	Selenium (Se)	Zinc (Zn)
			%	%	%	%	%	%	%	%	ppm or mg/kg	ppm or mg/kg	ppm or mg/kg	%	ppm or mg/kg	ppm or mg/kg	ppm or mg/kg
	WHEAT (Continued)																
226	GERM MEAL	A-F	88	0.06	0.95	0.02	0.06	0.25	0.94	0.27	0.120	9.2	–	0.006	132.5	0.463	119.4
		M-F	100	0.06	1.07	0.03	0.07	0.28	1.06	0.30	0.136	10.4	–	0.007	149.9	0.524	135.1
227	GLUTEN	A-F	90	0.06	0.23	0.06	–	0.04	0.02	0.95	0.049	11.6	0.058	0.006	18.1	3.753	38.5
		M-F	100	0.07	0.25	0.07	–	0.04	0.02	1.06	0.054	12.8	0.064	0.007	20.1	4.158	42.6
	WHEY, CATTLE *Bos taurus*																
228	FRESH	A-F	7	0.06	0.05	–	–	–	0.19	–	–	–	–	0.003	0.2	–	–
		M-F	100	0.81	0.71	–	–	–	2.75	–	–	–	–	0.029	3.2	–	–
229	CONDENSED	A-F	54	0.39	0.47	–	–	–	–	–	–	–	–	–	–	–	–
		M-F	100	0.72	0.88	–	–	–	–	–	–	–	–	–	–	–	–
230	DEHY	A-F	93	0.86	0.76	0.62	0.07	0.13	1.11	1.04	0.111	46.5	–	0.017	5.9	–	3.2
		M-F	100	0.92	0.82	0.66	0.08	0.14	1.19	1.11	0.119	49.9	–	0.019	6.3	–	3.4
231	LOW LACTOSE, DEHY (DRIED WHEY PRODUCT)	A-F	93	1.49	1.11	1.44	1.03	0.21	2.95	1.07	–	7.0	9.854	0.025	8.0	0.052	7.9
		M-F	100	1.60	1.18	1.54	1.10	0.23	3.16	1.15	–	7.5	10.554	0.027	8.6	0.056	8.4
232	ALBUMIN (DRIED MILK ALBUMIN)	A-F	92	10.86	4.03	–	–	–	–	–	–	–	–	–	–	–	–
		M-F	100	11.79	4.37	–	–	–	–	–	–	–	–	–	–	–	–
	YEAST, BREWERS *Saccharomyces cerevisiae*																
233	DEHY	A-F	93	0.14	1.36	0.07	0.07	0.24	1.69	0.43	0.506	38.4	0.358	0.009	6.7	0.911	39.0
		M-F	100	0.15	1.47	0.08	0.08	0.26	1.82	0.46	0.544	41.3	0.384	0.009	7.2	0.979	41.9
	YEAST, IRRADIATED *Saccharomyces cerevisiae*																
234	DEHY	A-F	94	0.78	1.42	–	–	–	2.14	–	–	–	–	–	–	–	–
		M-F	100	0.83	1.51	–	–	–	2.28	–	–	–	–	–	–	–	–
	YEAST, TORULA *Torulopsis utilis*																
235	DEHY	A-F	93	0.55	1.61	0.01	0.02	0.14	1.92	0.55	0.031	11.9	2.502	0.011	9.3	–	99.5
		M-F	100	0.59	1.73	0.01	0.02	0.15	2.06	0.59	0.033	12.8	2.689	0.012	10.0	–	107.0

TABLE V-1C DRY FORAGES, MINERAL AND VITAMIN COMPOSITION OF FEEDS, DATA EXPRESSED **AS-FED** AND **MOISTURE-FREE**

Entry Number	Feed Name Description	Moisture Basis: A-F (as-fed) or M-F (moisture-free)	Dry Matter	Macro Minerals							Micro Minerals						
				Calcium (Ca)	Phosphorus (P)	Sodium (Na)	Chlorine (Cl)	Magnesium (Mg)	Potassium (K)	Sulfur (S)	Cobalt (Co)	Copper (Cu)	Iodine (I)	Iron (Fe)	Manganese (Mn)	Selenium (Se)	Zinc (Zn)
			%	%	%	%	%	%	%	%	ppm or mg/kg	ppm or mg/kg	ppm or mg/kg	%	ppm or mg/kg	ppm or mg/kg	ppm or mg/kg
	ACACIA *Acacia* spp																
1	PODS, SUN-CURED	A-F	91	–	–	–	–	–	–	–	–	–	–	–	–	–	–
		M-F	100	–	–	–	–	–	–	–	–	–	–	–	–	–	–
	ADAMS'S NEEDLE *Yucca smaliana*																
2	HAY, SUN-CURED	A-F	93	–	–	–	–	–	–	–	–	–	–	–	–	–	–
		M-F	100	–	–	–	–	–	–	–	–	–	–	–	–	–	–
	ALDER *Alnus* spp																
3	LEAVES, SUN-CURED	A-F	85	–	–	–	–	–	–	–	–	–	–	–	–	–	–
		M-F	100	–	–	–	–	–	–	–	–	–	–	–	–	–	–
	ALFALFA (LUCERNE) *Medicago sativa*																
4	HAY, SUN-CURED, ALL ANALYSES	A-F	90	1.28	0.24	0.07	0.33	0.30	1.85	0.25	0.250	10.0	–	0.019	41.4	–	21.9
		M-F	100	1.42	0.26	0.08	0.37	0.33	2.05	0.28	0.277	11.0	–	0.020	45.8	–	24.3
5	HAY, PREBLOOM, SUN-CURED	A-F	90	1.34	0.30	0.10	0.31	0.19	2.25	0.48	0.256	10.2	–	0.021	42.2	–	33.5
		M-F	100	1.50	0.33	0.12	0.34	0.21	2.51	0.54	0.285	11.4	–	0.024	47.1	–	37.4
6	HAY, EARLY BLOOM, SUN-CURED	A-F	91	1.48	0.20	0.14	0.34	0.31	2.32	0.27	0.264	11.4	–	0.021	32.8	0.497	27.3
		M-F	100	1.63	0.22	0.15	0.38	0.34	2.56	0.30	0.292	12.6	–	0.023	36.2	0.549	30.2
7	HAY, MIDBLOOM, SUN-CURED	A-F	91	1.27	0.22	0.11	0.34	0.32	1.42	0.26	0.359	16.1	–	0.021	55.1	–	28.1
		M-F	100	1.39	0.24	0.12	0.38	0.35	1.56	0.28	0.394	17.7	–	0.023	60.5	–	30.9
8	HAY, FULL BLOOM, SUN-CURED	A-F	91	1.08	0.22	0.06	–	0.25	1.42	0.27	0.210	9.0	–	0.015	38.5	–	23.7
		M-F	100	1.19	0.24	0.07	–	0.27	1.56	0.30	0.230	9.9	–	0.016	42.3	–	26.1
9	HAY, MATURE, SUN-CURED	A-F	91	1.07	0.19	0.07	–	0.20	1.88	0.23	0.370	12.5	–	0.015	35.1	–	20.1
		M-F	100	1.18	0.21	0.08	–	0.22	2.07	0.25	0.406	13.7	–	0.017	38.5	–	22.1
10	HAY, BROWN, SUN-CURED	A-F	89	1.47	0.28	–	–	–	–	–	–	–	–	–	–	–	–
		M-F	100	1.66	0.32	–	–	–	–	–	–	–	–	–	–	–	–
11	HAY, STEMMY, SUN-CURED	A-F	91	0.89	0.20	–	–	0.29	1.80	–	–	–	–	0.014	–	–	–
		M-F	100	0.98	0.21	–	–	0.32	1.98	–	–	–	–	0.015	–	–	–
12	HAY, RAINED ON, SUN-CURED	A-F	89	2.04	0.21	0.05	–	0.24	2.16	–	–	2.5	–	0.026	22.1	–	23.8
		M-F	100	2.29	0.23	0.06	–	0.27	2.42	–	–	2.8	–	0.029	24.8	–	26.6
13	LEAVES, MEAL, SUN-CURED	A-F	88	2.32	0.24	–	–	0.36	1.47	–	–	–	–	0.031	29.8	–	–
		M-F	100	2.64	0.27	–	–	0.41	1.67	–	–	–	–	0.035	33.9	–	–
14	LEAVES, MEAL, DEHY	A-F	92	1.61	0.25	0.07	0.31	0.35	2.22	0.42	0.200	10.4	–	0.036	37.8	–	15.9
		M-F	100	1.75	0.27	0.07	0.34	0.38	2.40	0.46	0.217	11.2	–	0.039	40.9	–	17.2
15	STEMS, MEAL, SUN-CURED	A-F	91	–	–	–	–	–	–	–	–	–	–	–	–	–	–
		M-F	100	–	–	–	–	–	–	–	–	–	–	–	–	–	–

PROTEIN FEEDS

Entry Number	Fat-Soluble Vitamins: A (1 mg Carotene = 1667 IU Vit A)	Carotene (Provitamin A)	D	E	K	Water-Soluble Vitamins: B-12	Biotin	Choline	Folacin (Folic Acid)	Niacin	Pantothenic Acid (B-3)	(Pyridoxine) B-6	Riboflavin (B-2)	Thiamin (B-1)
	IU/g	ppm or mg/kg	IU/kg	ppm or mg/kg	ppm or mg/kg	ppm or mg/kg	ppm or mg/kg	ppm or mg/kg	ppm or mg/kg	ppm or mg/kg	ppm or mg/kg	ppm or mg/kg	ppm or mg/kg	ppm or mg/kg
226	**–**	**–**	**–**	**141.2**	**–**	**–**	**0.22**	**3062**	**2.12**	**68**	**18.6**	**9.97**	**6.0**	**23.1**
	–	–	–	159.7	–	–	0.24	3465	2.40	77	21.0	11.28	6.8	26.2
227	**–**	**–**	**0**	**34.1**	**–**	**73.1**	**0.00**	**577**	**0.74**	**74**	**5.8**	**2.26**	**0.7**	**0.9**
	–	–	0	37.7	–	81.0	0.00	640	0.82	82	6.4	2.51	0.7	1.0
228	**–**	**–**	**–**	**–**	**–**	**–**	**–**	**–**	**–**	**1**	**5.3**	**–**	**1.4**	**0.3**
	–	–	–	–	–	–	–	–	–	14	76.7	–	20.3	4.3
229	**–**	**–**	**–**	**–**	**–**	**–**	**–**	**–**	**–**	**3**	**11.8**	**–**	**14.2**	**2.6**
	–	–	–	–	–	–	–	–	–	5	22.0	–	26.5	4.8
230	**–**	**–**	**–**	**0.2**	**–**	**18.9**	**0.35**	**1790**	**0.85**	**11**	**46.2**	**3.21**	**27.4**	**4.0**
	–	–	–	0.2	–	20.3	0.38	1921	0.91	11	49.6	3.45	29.4	4.3
231	**–**	**–**	**–**	**–**	**–**	**35.9**	**0.50**	**4096**	**0.89**	**18**	**74.5**	**4.48**	**47.6**	**5.0**
	–	–	–	–	–	38.4	0.54	4387	0.96	19	79.8	4.79	50.9	5.4
232	**–**	**–**	**–**	**–**	**–**	**–**	**–**	**–**	**–**	**2**	**7.3**	**–**	**8.8**	**0.7**
	–	–	–	–	–	–	–	–	–	2	7.9	–	9.6	0.7
233	**–**	**–**	**–**	**2.1**	**–**	**1.1**	**1.04**	**3847**	**9.69**	**443**	**81.5**	**36.67**	**34.1**	**85.2**
	–	–	–	2.3	–	1.1	1.12	4134	10.41	476	87.6	39.40	36.6	91.6
234	**–**	**–**	**–**	**–**	**–**	**–**	**–**	**–**	**–**	**–**	**–**	**–**	**18.5**	**–**
	–	–	–	–	–	–	–	–	–	–	–	–	19.7	–
235	**–**	**–**	**–**	**–**	**–**	**4.0**	**1.19**	**2981**	**25.66**	**512**	**107.5**	**34.48**	**47.7**	**6.8**
	–	–	–	–	–	4.3	1.27	3203	27.58	550	115.6	37.06	51.3	7.3

DRY FORAGES

Entry Number	Fat-Soluble Vitamins: A (1 mg Carotene = 1667 IU Vit A)	Carotene (Provitamin A)	D	E	K	Water-Soluble Vitamins: B-12	Biotin	Choline	Folacin (Folic Acid)	Niacin	Pantothenic Acid (B-3)	(Pyridoxine) B-6	Riboflavin (B-2)	Thiamin (B-1)
	IU/g	ppm or mg/kg	IU/kg	ppm or mg/kg	ppm or mg/kg	ppm or mg/kg	ppm or mg/kg	ppm or mg/kg	ppm or mg/kg	ppm or mg/kg	ppm or mg/kg	ppm or mg/kg	ppm or mg/kg	ppm or mg/kg
1	**–**	**–**	**–**	**–**	**–**	**–**	**–**	**–**	**–**	**–**	**–**	**–**	**–**	**–**
	–	–	–	–	–	–	–	–	–	–	–	–	–	–
2	**–**	**–**	**–**	**–**	**–**	**–**	**–**	**–**	**–**	**–**	**–**	**–**	**–**	**–**
	–	–	–	–	–	–	–	–	–	–	–	–	–	–
3	**–**	**–**	**–**	**–**	**–**	**–**	**–**	**–**	**–**	**–**	**–**	**–**	**–**	**–**
	–	–	–	–	–	–	–	–	–	–	–	–	–	–
4	**45.0**	**27.0**	**2**	**55.9**	**–**	**–**	**0.18**	**892**	**3.07**	**43**	**18.1**	**–**	**9.5**	**3.1**
	49.8	29.9	2	61.9	–	–	0.20	988	3.40	48	20.1	–	10.5	3.4
5	**300.2**	**180.1**	**–**	**–**	**–**	**–**	**–**	**–**	**–**	**–**	**–**	**–**	**–**	**–**
	335.1	201.0	–	–	–	–	–	–	–	–	–	–	–	–
6	**210.9**	**126.5**	**2**	**23.5**	**–**	**–**	**–**	**–**	**–**	**–**	**–**	**–**	**–**	**–**
	233.0	139.8	2	26.0	–	–	–	–	–	–	–	–	–	–
7	**50.5**	**30.3**	**1**	**–**	**–**	**–**	**–**	**–**	**–**	**–**	**–**	**–**	**9.6**	**–**
	55.5	33.3	2	–	–	–	–	–	–	–	–	–	10.6	–
8	**98.5**	**59.1**	**–**	**–**	**–**	**–**	**–**	**–**	**–**	**–**	**–**	**–**	**–**	**–**
	108.4	65.0	–	–	–	–	–	–	–	–	–	–	–	–
9	**17.6**	**10.6**	**1**	**–**	**–**	**–**	**–**	**–**	**–**	**–**	**–**	**–**	**–**	**–**
	19.3	11.6	1	–	–	–	–	–	–	–	–	–	–	–
10	**–**	**–**	**–**	**–**	**–**	**–**	**–**	**–**	**–**	**–**	**–**	**–**	**–**	**–**
	–	–	–	–	–	–	–	–	–	–	–	–	–	–
11	**–**	**–**	**–**	**–**	**–**	**–**	**–**	**–**	**–**	**–**	**–**	**–**	**–**	**–**
	–	–	–	–	–	–	–	–	–	–	–	–	–	–
12	**–**	**–**	**1**	**–**	**–**	**–**	**–**	**–**	**–**	**–**	**–**	**–**	**–**	**–**
	–	–	1	–	–	–	–	–	–	–	–	–	–	–
13	**97.7**	**58.6**	**–**	**–**	**–**	**–**	**–**	**–**	**–**	**–**	**–**	**–**	**20.9**	**–**
	111.3	66.8	–	–	–	–	–	–	–	–	–	–	23.8	–
14	**230.1**	**138.1**	**0**	**–**	**–**	**–**	**0.33**	**1597**	**–**	**47**	**33.1**	**–**	**15.2**	**5.5**
	249.4	149.6	0	–	–	–	0.36	1730	–	51	35.9	–	16.5	5.5
15	**52.1**	**31.2**	**–**	**–**	**–**	**–**	**–**	**–**	**–**	**–**	**–**	**–**	**10.6**	**–**
	57.3	34.4	–	–	–	–	–	–	–	–	–	–	11.7	–

(Continued)

TABLE V-1C DRY FORAGES, COMPOSITION OF FEEDS, DATA EXPRESSED **AS-FED** AND **MOISTURE-FREE**—*(Continued)*

Entry Number	Feed Name Description	International Feed Number	Moisture Basis: A-F (as-fed) or M-F (moisture-free)	Proximate Analysis: Dry Matter	Ash	Crude Fiber	Neutral Det. Fib. (NDF)	Acid Det. Fib. (ADF)	Lignin	Ether Extract (Fat)	N-Free Extract	Crude Protein	Digestible Protein: Ruminant	Swine	Horse
				%	%	%	%	%	%	%	%	%	%	%	%
	ALFALFA (Continued)														
16	MEAL, DEHY, 13% PROTEIN	1-00-021	A-F	89	8.8	27.3	–	–	–	1.9	38.1	13.4	9.0	8.3	8.7
			M-F	100	9.8	30.6	–	–	–	2.2	42.6	14.9	10.0	9.3	9.7
17	MEAL, DEHY, 15% PROTEIN[1]	1-00-022	A-F	90	9.1	26.6	45.9	36.9	–	2.2	37.0	15.6	10.9	10.2	10.4
			M-F	100	10.0	29.4	51.0	41.0	–	2.5	40.9	17.3	12.1	11.3	11.5
18	MEAL, DEHY, 17% PROTEIN	1-00-023	A-F	92	9.7	24.0	41.3	31.5	9.7	2.8	37.8	17.4	12.6	11.7	11.8
			M-F	100	10.6	26.2	45.0	34.3	10.6	3.0	41.2	18.9	13.7	12.8	12.8
19	MEAL, DEHY, 20% PROTEIN[1]	1-00-024	A-F	92	10.2	20.8	38.6	28.5	–	3.3	37.1	20.2	15.1	14.4	14.0
			M-F	100	11.1	22.7	42.0	31.0	–	3.6	40.5	22.1	16.4	15.7	15.2
20	MEAL, DEHY, 22% PROTEIN[1]	1-07-851	A-F	93	10.2	18.3	36.3	26.0	–	4.1	38.1	22.2	16.4	16.4	15.5
			M-F	100	11.0	19.8	39.0	28.0	–	4.4	41.0	23.9	17.7	17.7	16.6
21	STRAW	1-08-323	A-F	93	6.8	40.6	–	–	–	1.5	34.6	9.2	5.2	4.5	5.4
			M-F	100	7.3	43.8	–	–	–	1.6	37.3	9.9	5.7	4.9	5.9
	ALFALFA-BROMEGRASS, SMOOTH *Medicago sativa-Bromus inermis*														
22	HAY, SUN-CURED	1-00-255	A-F	91	6.1	31.0	–	–	–	2.1	37.7	14.1	9.6	8.9	9.2
			M-F	100	6.7	34.1	–	–	–	2.3	41.5	15.5	10.5	9.8	10.2
	ALFALFA-GRASS *Medicago sativa-grass*														
23	HAY, SUN-CURED	1-08-331	A-F	91	6.7	30.3	–	36.6	–	2.1	37.8	14.5	9.9	9.2	9.5
			M-F	100	7.4	33.2	–	40.0	–	2.3	41.3	15.9	10.9	10.1	10.4
	ALFALFA-ORCHARDGRASS *Medicago sativa-Dactylis glomerata*														
24	HAY, SUN-CURED	1-00-322	A-F	89	7.3	28.7	–	–	–	1.9	36.8	14.8	10.2	9.5	9.8
			M-F	100	8.1	32.0	–	–	–	2.2	41.2	16.5	11.4	10.7	11.0
	ALFALFA-TIMOTHY *Medicago sativa-Phleum pratense*														
25	HAY, WEATHERED, SUN-CURED	1-00-330	A-F	93	5.0	30.6	–	–	–	1.3	46.3	9.8	5.7	5.0	5.9
			M-F	100	5.4	32.9	–	–	–	1.4	49.8	10.5	6.2	5.4	6.3
	ALFILERIA, REDSTEM (FILAREE) *Erodium cicutarium*														
26	HAY, SUN-CURED	1-08-333	A-F	89	11.8	23.4	–	–	–	2.9	40.2	10.9	6.8	6.1	6.8
			M-F	100	13.2	26.2	–	–	–	3.3	45.1	12.2	7.7	6.9	7.6
	ALMOND *Prunus amygdalus*														
27	HULLS[1]	4-00-359	A-F	90	6.1	13.5	28.8	25.2	–	2.9	63.8	4.1	0.0	2.0	2.4
			M-F	100	6.8	14.9	32.0	28.0	–	3.2	70.6	4.5	0.0	2.2	2.7
28	SHELLS, GROUND	1-07-754	A-F	92	4.0	39.6	–	–	–	3.7	40.0	4.4	1.1	0.3	1.8
			M-F	100	4.4	43.1	–	–	–	4.0	43.6	4.8	1.2	0.4	1.9
	ALSIKE CLOVER *Trifolium hybridum*														
29	HAY, SUN-CURED	1-01-313	A-F	88	7.6	26.2	–	–	–	2.4	39.1	12.4	8.3	7.5	8.0
			M-F	100	8.7	29.9	–	–	–	2.8	44.5	14.2	9.5	8.6	9.1
	ALYCE CLOVER *Alysicarpus vaginalis*														
30	HAY, SUN-CURED	1-00-361	A-F	90	5.7	36.2	–	–	–	1.6	35.3	10.9	6.7	6.1	6.8
			M-F	100	6.4	40.3	–	–	–	1.7	39.3	12.2	7.4	6.8	7.6
	ANIMAL														
31	RUMEN CONTENTS, DEHY	1-09-327	A-F	87	9.6	24.4	–	–	–	1.8	36.2	15.3	10.3	10.1	10.3
			M-F	100	11.0	28.0	–	–	–	2.1	41.4	17.6	11.8	11.6	11.8
	APPLE *Malus* spp														
32	POMACE, DEHY[1]	4-00-423	A-F	89	3.0	16.2	–	23.1	–	4.3	61.2	4.4	–0.4	2.3	2.7
			M-F	100	3.4	18.2	–	26.0	–	4.8	68.6	5.0	–0.4	2.6	3.0
33	POMACE WITHOUT PECTIN, DEHY	4-00-425	A-F	91	3.3	24.2	–	–	–	7.3	49.4	7.0	3.5	4.6	4.7
			M-F	100	3.6	26.5	–	–	–	8.0	54.2	7.7	3.8	5.1	5.2
	ASH *Fraxinus* spp														
34	LEAVES, SUN-CURED	1-00-434	A-F	85	5.7	12.7	–	–	–	4.7	49.6	12.4	7.9	7.6	8.1
			M-F	100	6.7	14.9	–	–	–	5.5	58.3	14.6	9.3	9.0	9.5
	ASPARAGUS, GARDEN *Asparagus officinalis*														
35	STEM BUTTS, MEAL	1-00-437	A-F	91	14.6	29.0	–	–	–	0.9	32.3	14.2	8.8	9.0	9.3
			M-F	100	16.0	31.9	–	–	–	1.0	35.5	15.6	9.7	9.8	10.2
	ASPEN, QUAKING *Populus tremuloides*														
36	WOOD, BOILED, GROUND	1-26-075	A-F	90	–	–	–	–	–	–	–	–	–	–	–
			M-F	100	–	–	–	–	–	–	–	–	–	–	–
37	LEAVES, SUN-CURED	1-00-439	A-F	85	7.7	20.8	–	–	–	4.6	37.2	14.6	9.2	9.6	9.8
			M-F	100	9.1	24.5	–	–	–	5.4	43.8	17.2	10.8	11.3	11.5
	ATLAS SORGHUM *Sorghum bicolor*														
38	FODDER WITH HEADS, SUN-CURED	1-04-334	A-F	91	6.6	20.2	–	–	–	2.8	55.1	6.5	–	–	3.2
			M-F	100	7.2	22.2	–	–	–	3.1	60.5	7.1	–	–	3.5
39	STOVER WITHOUT HEADS, SUN-CURED	1-04-336	A-F	90	7.1	28.3	–	–	–	2.1	48.8	3.7	0.5	–0.2	1.3
			M-F	100	7.9	31.5	–	–	–	2.3	54.2	4.1	0.6	–0.2	1.4
	BAGASSE (SUGARCANE) *Saccharum officinarum*														
40	PULP, DEHY	1-04-686	A-F	91	2.9	42.3	78.8	54.5	12.8	0.7	43.8	1.4	–1.5	–2.3	–0.5
			M-F	100	3.1	46.5	86.5	59.8	14.0	0.8	48.0	1.6	–1.7	–2.5	–0.6
	BAHIAGRASS *Paspalum notatum*														
41	HAY, SUN-CURED	1-00-462	A-F	90	5.7	28.1	66.6	33.9	5.1	1.8	46.0	8.5	4.7	4.0	5.0
			M-F	100	6.3	31.2	73.9	37.7	5.7	2.0	51.1	9.5	5.2	4.5	5.5

DRY FORAGES

Entry Number	TDN Ruminant	TDN Swine	TDN Horse	Digestible Energy Ruminant		Digestible Energy Swine		Digestible Energy Horse		Metabolizable Energy Ruminant		Metabolizable Energy Swine		Metabolizable Energy Poultry ME_n		Metabolizable Energy Horse		Net Energy Ruminant NE_m		Net Energy Ruminant NE_g		Net Energy Lactating Cows NE_{lc}	
	%	%	%	Mcal		kcal		Mcal		Mcal		kcal		kcal		Mcal		Mcal		Mcal		Mcal	
				lb	kg	lb	kg	lb	kg	lb	kg	lb	kg	lb	kg	lb	kg	lb	kg	lb	kg	lb	kg
16	49	39	43	0.99	2.17	–	–	0.82	1.80	0.81	1.79	–	–	–	–	0.67	1.48	0.47	1.03	0.24	0.53	0.50	1.11
	55	44	48	1.10	2.43	–	–	0.91	2.01	0.91	2.00	–	–	–	–	0.75	1.65	0.52	1.15	0.27	0.59	0.56	1.24
17	53	31	45	1.07	2.35	–	–	0.86	1.89	0.89	1.97	586	1293	696	1535	0.70	1.55	0.54	1.20	0.31	0.68	0.56	1.23
	59	34	50	1.18	2.60	–	–	0.95	2.09	0.99	2.18	648	1429	770	1698	0.78	1.71	0.60	1.33	0.34	0.75	0.62	1.36
18	55	44	47	1.12	2.47	643	1418	0.89	1.96	0.96	2.12	601	1326	682	1504	0.73	1.61	0.60	1.32	0.36	0.79	0.57	1.25
	60	48	51	1.22	2.69	701	1546	0.97	2.14	1.05	2.32	656	1446	744	1640	0.80	1.75	0.65	1.44	0.39	0.86	0.62	1.37
19	57	48	50	1.07	2.36	943	2079	0.94	2.08	0.93	2.06	872	1923	737	1625	0.77	1.70	0.58	1.28	0.34	0.75	0.59	1.30
	62	52	55	1.17	2.58	1030	2270	1.03	2.27	1.02	2.25	952	2099	805	1774	0.84	1.86	0.64	1.40	0.37	0.82	0.64	1.42
20	60	49	52	1.20	2.65	991	2186	0.98	2.15	1.02	2.26	841	1855	768	1692	0.80	1.76	0.63	1.39	0.38	0.84	0.63	1.38
	65	53	56	1.29	2.85	1068	2355	1.05	2.32	1.10	2.43	907	1999	827	1823	0.86	1.90	0.68	1.49	0.41	0.90	0.67	1.49
21	43	–	34	0.89	1.95	–	–	0.67	1.47	0.71	1.55	–	–	–	–	0.55	1.20	0.40	0.88	0.17	0.38	0.46	1.01
	46	–	36	0.96	2.11	–	–	0.72	1.58	0.76	1.68	–	–	–	–	0.59	1.30	0.43	0.95	0.18	0.41	0.50	1.09
22	58	41	45	1.11	2.44	–	–	0.85	1.88	0.93	2.06	–	–	–	–	0.70	1.54	0.59	1.30	0.35	0.76	0.59	1.31
	63	46	49	1.22	2.69	–	–	0.94	2.06	1.03	2.26	–	–	–	–	0.77	1.69	0.65	1.42	0.38	0.84	0.65	1.43
23	51	42	45	1.01	2.23	–	–	0.86	1.89	0.83	1.84	–	–	–	–	0.70	1.55	0.50	1.10	0.26	0.58	0.53	1.16
	55	46	50	1.11	2.44	–	–	0.94	2.07	0.91	2.01	–	–	–	–	0.77	1.70	0.54	1.20	0.29	0.63	0.58	1.27
24	52	42	47	1.03	2.27	–	–	0.88	1.93	0.86	1.89	–	–	–	–	0.72	1.58	0.49	1.08	0.26	0.57	0.52	1.14
	58	46	52	1.15	2.54	–	–	0.98	2.16	0.96	2.11	–	–	–	–	0.80	1.77	0.55	1.20	0.29	0.64	0.58	1.27
25	52	37	47	1.04	2.30	–	–	0.89	1.96	0.86	1.90	–	–	–	–	0.73	1.61	0.48	1.07	0.25	0.55	0.52	1.15
	56	40	51	1.12	2.47	–	–	0.96	2.11	0.93	2.05	–	–	–	–	0.78	1.73	0.52	1.15	0.27	0.59	0.56	1.23
26	48	41	35	0.97	2.15	–	–	0.68	1.49	0.80	1.77	–	–	–	–	0.56	1.23	0.46	1.02	0.24	0.52	0.50	1.09
	54	45	39	1.09	2.41	–	–	0.76	1.68	0.90	1.98	–	–	–	–	0.62	1.37	0.52	1.14	0.26	0.58	0.56	1.23
27	66	73	45	1.34	2.95	1457	3213	0.86	1.89	1.15	2.54	1375	3032	–	–	0.70	1.55	0.77	1.70	0.51	1.12	0.66	1.46
	73	81	50	1.48	3.27	1615	3560	0.95	2.09	1.28	2.82	1524	3360	–	–	0.78	1.71	0.85	1.88	0.56	1.24	0.73	1.61
28	49	–	22	0.98	2.16	–	–	0.48	1.06	0.80	1.76	–	–	–	–	0.39	0.87	0.46	1.02	0.23	0.51	0.50	1.11
	53	–	24	1.06	2.35	–	–	0.52	1.15	0.87	1.92	–	–	–	–	0.43	0.94	0.51	1.12	0.25	0.56	0.55	1.21
29	51	42	41	1.01	2.24	–	–	0.78	1.72	0.85	1.86	–	–	–	–	0.64	1.41	0.49	1.08	0.26	0.58	0.51	1.13
	58	48	46	1.16	2.55	–	–	0.89	1.96	0.96	2.12	–	–	–	–	0.73	1.60	0.56	1.23	0.30	0.66	0.59	1.29
30	44	–	38	0.89	1.96	–	–	0.74	1.63	0.71	1.57	–	–	–	–	0.61	1.34	0.30	0.66	0.08	0.18	0.38	0.85
	49	–	43	0.99	2.18	–	–	0.83	1.82	0.80	1.75	–	–	–	–	0.68	1.49	0.33	0.73	0.09	0.20	0.43	0.94
31	51	42	45	1.00	2.20	–	–	0.85	1.88	0.83	1.83	–	–	–	–	0.70	1.54	0.47	1.03	0.25	0.54	0.50	1.10
	58	48	52	1.14	2.51	–	–	0.98	2.15	0.95	2.09	–	–	–	–	0.80	1.77	0.54	1.18	0.28	0.62	0.57	1.26
32	60	69	40	1.21	2.66	1380	3043	0.77	1.71	1.04	2.28	1300	2867	–	–	0.64	1.40	0.66	1.46	0.42	0.91	0.64	1.42
	68	77	45	1.35	2.98	1548	3413	0.87	1.91	1.16	2.56	1459	3216	–	–	0.71	1.57	0.74	1.63	0.47	1.03	0.72	1.59
33	62	–	–	1.25	2.75	–	–	–	–	1.07	2.37	–	–	–	–	–	–	–	–	–	–	–	–
	68	–	–	1.37	3.02	–	–	–	–	1.18	2.60	–	–	–	–	–	–	–	–	–	–	–	–
34	62	–	–	1.24	2.73	–	–	–	–	1.08	2.37	–	–	–	–	–	–	–	–	–	–	–	–
	73	–	–	1.46	3.21	–	–	–	–	1.27	2.79	–	–	–	–	–	–	–	–	–	–	–	–
35	43	26	40	0.87	1.91	–	–	0.76	1.68	0.69	1.52	–	–	–	–	0.63	1.38	0.37	0.82	0.15	0.32	0.44	0.96
	47	29	44	0.95	2.10	–	–	0.84	1.85	0.76	1.67	–	–	–	–	0.69	1.52	0.41	0.90	0.16	0.36	0.48	1.06
36	–	–	–	–	–	–	–	–	–	–	–	–	–	–	–	–	–	–	–	–	–	–	–
	–	–	–	–	–	–	–	–	–	–	–	–	–	–	–	–	–	–	–	–	–	–	–
37	45	58	36	1.00	2.20	–	–	0.70	1.54	0.83	1.83	–	–	–	–	0.57	1.26	0.56	1.24	0.34	0.74	0.56	1.24
	53	68	42	1.17	2.58	–	–	0.82	1.81	0.98	2.16	–	–	–	–	0.67	1.48	0.66	1.46	0.39	0.87	0.66	1.46
38	55	–	44	1.09	2.40	–	–	0.80	1.76	0.90	1.98	–	–	–	–	0.66	1.45	0.54	1.19	0.27	0.59	0.56	1.23
	60	–	48	1.20	2.64	–	–	0.88	1.93	0.98	2.17	–	–	–	–	0.72	1.59	0.59	1.31	0.30	0.65	0.61	1.35
39	53	32	33	1.00	2.21	–	–	0.65	1.44	0.83	1.83	–	–	–	–	0.54	1.18	0.44	0.96	0.21	0.46	0.48	1.06
	59	35	37	1.12	2.46	–	–	0.73	1.60	0.92	2.03	–	–	–	–	0.60	1.31	0.48	1.07	0.23	0.51	0.53	1.17
40	43	–	–	0.88	1.93	–	–	–	–	0.69	1.53	–	–	–	–	–	–	0.37	0.82	0.15	0.32	0.40	0.88
	48	–	–	0.96	2.12	–	–	–	–	0.76	1.68	–	–	–	–	–	–	0.41	0.90	0.16	0.36	0.44	0.96
41	46	38	42	0.96	2.12	–	–	0.81	1.78	0.79	1.74	–	–	–	–	0.66	1.46	0.41	0.90	0.18	0.40	0.46	1.01
	51	42	47	1.07	2.36	–	–	0.90	1.98	0.88	1.93	–	–	–	–	0.73	1.62	0.45	1.00	0.20	0.45	0.51	1.13

(Continued)

TABLE V-1C DRY FORAGES, MINERAL AND VITAMIN COMPOSITION OF FEEDS, DATA EXPRESSED **AS-FED** AND **MOISTURE-FREE**—*(Continued)*

Entry Number	Feed Name Description	Moisture Basis: A-F (as-fed) or M-F (moisture-free)	Dry Matter	Macro Minerals							Micro Minerals						
				Calcium (Ca)	Phosphorus (P)	Sodium (Na)	Chlorine (Cl)	Magnesium (Mg)	Potassium (K)	Sulfur (S)	Cobalt (Co)	Copper (Cu)	Iodine (I)	Iron (Fe)	Manganese (Mn)	Selenium (Se)	Zinc (Zn)
			%	%	%	%	%	%	%	%	ppm or mg/kg	ppm or mg/kg	ppm or mg/kg	%	ppm or mg/kg	ppm or mg/kg	ppm or mg/kg
	ALFALFA (Continued)																
16	MEAL, DEHY, 13% PROTEIN	A-F	89	–	–	–	–	–	–	–	–	–	–	–	–	–	–
		M-F	100	–	–	–	–	–	–	–	–	–	–	–	–	–	–
17	MEAL, DEHY, 15% PROTEIN	A-F	90	1.24	0.22	0.07	0.44	0.28	2.24	0.18	0.172	9.5	0.117	0.028	27.8	0.281	19.4
		M-F	100	1.37	0.24	0.08	0.48	0.31	2.48	0.20	0.190	10.5	0.129	0.031	30.8	0.311	21.4
18	MEAL, DEHY, 17% PROTEIN	A-F	92	1.40	0.23	0.10	0.47	0.29	2.38	0.23	0.302	8.6	0.148	0.041	31.0	0.335	19.3
		M-F	100	1.52	0.25	0.11	0.52	0.32	2.60	0.25	0.329	9.3	0.162	0.045	33.8	0.365	21.1
19	MEAL, DEHY, 20% PROTEIN	A-F	92	1.59	0.28	0.11	0.47	0.33	2.41	0.50	0.259	12.2	0.135	0.036	45.2	0.285	21.8
		M-F	100	1.74	0.31	0.13	0.51	0.36	2.63	0.55	0.283	13.3	0.147	0.039	49.4	0.311	23.8
20	MEAL, DEHY, 22% PROTEIN	A-F	93	1.69	0.30	0.12	0.52	0.31	2.40	0.30	0.311	9.8	0.166	0.036	36.4	0.534	19.5
		M-F	100	1.82	0.33	0.13	0.56	0.33	2.58	0.32	0.336	10.5	0.179	0.039	39.2	0.576	21.0
21	STRAW	A-F	93	–	0.13	–	–	–	–	–	–	–	–	–	–	–	–
		M-F	100	–	0.14	–	–	–	–	–	–	–	–	–	–	–	–
	ALFALFA-BROMEGRASS, SMOOTH *Medicago sativa-Bromus inermis*																
22	HAY, SUN-CURED	A-F	91	1.02	0.25	0.21	0.43	0.52	1.77	0.21	0.079	15.5	–	0.012	36.2	–	–
		M-F	100	1.12	0.28	0.23	0.47	0.57	1.94	0.23	0.086	17.1	–	0.013	39.8	–	–
	ALFALFA-GRASS *Medicago sativa-grass*																
23	HAY, SUN-CURED	A-F	91	1.33	0.23	0.11	–	0.28	2.31	0.22	–	11.4	–	0.020	60.7	–	20.1
		M-F	100	1.45	0.25	0.12	–	0.30	2.53	0.24	–	12.4	–	0.022	66.4	–	22.0
	ALFALFA-ORCHARDGRASS *Medicago sativa-Dactylis glomerata*																
24	HAY, SUN-CURED	A-F	89	–	–	–	–	–	–	–	–	–	–	–	–	–	–
		M-F	100	–	–	–	–	–	–	–	–	–	–	–	–	–	–
	ALFALFA-TIMOTHY *Medicago sativa-Phleum pratense*																
25	HAY, WEATHERED, SUN-CURED	A-F	93	–	–	–	–	–	–	–	–	–	–	–	–	–	–
		M-F	100	–	–	–	–	–	–	–	–	–	–	–	–	–	–
	ALFILERIA, REDSTEM (FILAREE) *Erodium cicutarium*																
26	HAY, SUN-CURED	A-F	89	1.57	0.41	–	–	–	–	–	–	–	–	–	–	–	–
		M-F	100	1.76	0.46	–	–	–	–	–	–	–	–	–	–	–	–
	ALMOND *Prunus amygdalus*																
27	HULLS	A-F	90	0.19	0.09	–	–	–	0.48	0.10	–	–	–	–	–	–	–
		M-F	100	0.21	0.10	–	–	–	0.53	0.11	–	–	–	–	–	–	–
28	SHELLS, GROUND	A-F	92	–	–	–	–	–	–	–	–	–	–	–	–	–	–
		M-F	100	–	–	–	–	–	–	–	–	–	–	–	–	–	–
	ALSIKE CLOVER *Trifolium hybridum*																
29	HAY, SUN-CURED	A-F	88	1.14	0.22	0.40	0.68	0.40	1.95	0.17	–	5.3	–	0.023	60.5	–	–
		M-F	100	1.30	0.25	0.46	0.78	0.45	2.22	0.19	–	6.0	–	0.026	69.0	–	–
	ALYCE CLOVER *Alysicarpus vaginalis*																
30	HAY, SUN-CURED	A-F	90	–	–	–	–	–	–	–	–	–	–	–	–	–	–
		M-F	100	–	–	–	–	–	–	–	–	–	–	–	–	–	–
	ANIMAL																
31	RUMEN CONTENTS, DEHY	A-F	87	0.69	0.58	–	–	–	–	–	–	–	–	–	–	–	–
		M-F	100	0.79	0.67	–	–	–	–	–	–	–	–	–	–	–	–
	APPLE *Malus spp*																
32	POMACE, DEHY	A-F	89	0.11	0.10	0.12	–	0.06	0.43	0.02	–	–	–	0.027	7.2	–	–
		M-F	100	0.13	0.12	0.14	–	0.07	0.49	0.02	–	–	–	0.030	8.1	–	–
33	POMACE WITHOUT PECTIN, DEHY	A-F	91	–	–	–	–	–	–	–	–	–	–	–	–	–	–
		M-F	100	–	–	–	–	–	–	–	–	–	–	–	–	–	–
	ASH *Fraxinus spp*																
34	LEAVES, SUN-CURED	A-F	85	–	–	–	–	–	–	–	–	–	–	–	–	–	–
		M-F	100	–	–	–	–	–	–	–	–	–	–	–	–	–	–
	ASPARAGUS, GARDEN *Asparagus officinalis*																
35	STEM BUTTS, MEAL	A-F	91	–	–	–	–	–	–	–	–	–	–	–	–	–	–
		M-F	100	–	–	–	–	–	–	–	–	–	–	–	–	–	–
	ASPEN, QUAKING *Populus tremuloides*																
36	WOOD, BOILED, GROUND	A-F	90	–	–	–	–	–	–	–	–	–	–	–	–	–	–
		M-F	100	–	–	–	–	–	–	–	–	–	–	–	–	–	–
37	LEAVES, SUN-CURED	A-F	85	–	–	–	–	–	–	–	–	–	–	–	–	–	–
		M-F	100	–	–	–	–	–	–	–	–	–	–	–	–	–	–
	ATLAS SORGHUM *Sorghum bicolor*																
38	FODDER WITH HEADS, SUN-CURED	A-F	91	0.73	0.27	–	–	–	–	–	–	–	–	–	–	–	–
		M-F	100	0.80	0.30	–	–	–	–	–	–	–	–	–	–	–	–
39	STOVER WITHOUT HEADS, SUN-CURED	A-F	90	0.35	0.10	–	–	–	–	–	–	–	–	–	–	–	–
		M-F	100	0.39	0.11	–	–	–	–	–	–	–	–	–	–	–	–
	BAGASSE (SUGARCANE) *Saccharum officinarum*																
40	PULP, DEHY	A-F	91	0.47	0.26	0.04	–	0.08	0.34	0.09	–	–	–	0.019	–	–	–
		M-F	100	0.51	0.29	0.04	–	0.08	0.37	0.10	–	–	–	0.021	–	–	–
	BAHIAGRASS *Paspalum notatum*																
41	HAY, SUN-CURED	A-F	90	0.45	0.20	–	–	0.17	–	–	–	–	–	0.006	–	–	–
		M-F	100	0.50	0.22	–	–	0.19	–	–	–	–	–	0.006	–	–	–

Entry Number	Fat-Soluble Vitamins					Water-Soluble Vitamins								
	A (1 mg Carotene = 1667 IU Vit A)	Carotene (Provitamin A)	D	E	K	B-12	Biotin	Choline	Folacin (Folic Acid)	Niacin	Pantothenic Acid (B-3)	(Pyridoxine) B-6	Riboflavin (B-2)	Thiamin (B-1)
	IU/g	ppm or mg/kg	IU/kg	ppm or mg/kg	ppm or mg/kg	ppm or mg/kg	ppm or mg/kg	ppm or mg/kg	ppm or mg/kg	ppm or mg/kg	ppm or mg/kg	ppm or mg/kg	ppm or mg/kg	ppm or mg/kg
16	94.3	56.5	–	–	–	–	–	–	–	–	–	6.80	10.2	–
	105.3	63.2	–	–	–	–	–	–	–	–	–	7.60	11.4	–
17	124.1	74.5	–	81.9	9.59	–	0.25	1573	1.56	42	20.7	6.27	10.6	3.0
	137.3	82.3	–	90.6	10.61	–	0.28	1739	1.73	46	22.9	6.94	11.7	3.3
18	200.3	120.2	–	105.7	8.24	–	0.33	1369	4.37	37	29.7	7.18	12.9	3.4
	218.5	131.1	–	115.3	8.98	–	0.36	1494	4.77	40	32.4	7.83	14.1	3.7
19	265.4	159.2	–	143.3	14.19	–	0.35	1417	2.96	48	35.5	8.72	15.2	5.4
	289.7	173.8	–	156.4	15.50	–	0.39	1547	3.24	52	38.8	9.52	16.6	5.9
20	391.4	234.8	–	221.3	11.65	–	0.33	1605	5.15	50	39.0	8.28	17.6	5.9
	421.8	253.0	–	238.4	12.55	–	0.36	1729	5.55	54	42.0	8.92	19.0	6.3
21	–	–	–	–	–	–	–	–	–	–	–	–	–	–
	–	–	–	–	–	–	–	–	–	–	–	–	–	–
22	39.4	23.7	–	–	–	–	–	–	–	25	21.4	–	6.1	–
	43.4	26.0	–	–	–	–	–	–	–	27	23.5	–	6.7	–
23	28.9	17.3	–	–	–	–	–	–	–	–	–	–	–	–
	31.6	18.9	–	–	–	–	–	–	–	–	–	–	–	–
24	–	–	–	–	–	–	–	–	–	–	–	–	–	–
	–	–	–	–	–	–	–	–	–	–	–	–	–	–
25	–	–	–	–	–	–	–	–	–	–	–	–	–	–
	–	–	–	–	–	–	–	–	–	–	–	–	–	–
26	–	–	–	–	–	–	–	–	–	–	–	–	–	–
	–	–	–	–	–	–	–	–	–	–	–	–	–	–
27	–	–	–	–	–	–	–	–	–	–	–	–	–	–
	–	–	–	–	–	–	–	–	–	–	–	–	–	–
28	–	–	–	–	–	–	–	–	–	–	–	–	–	–
	–	–	–	–	–	–	–	–	–	–	–	–	–	–
29	272.1	163.2	–	–	–	–	–	–	–	–	–	–	15.1	4.2
	310.1	186.0	–	–	–	–	–	–	–	–	–	–	17.2	4.8
30	–	–	–	–	–	–	–	–	–	–	–	–	–	–
	–	–	–	–	–	–	–	–	–	–	–	–	–	–
31	–	–	–	–	–	–	–	–	–	–	–	–	–	–
	–	–	–	–	–	–	–	–	–	–	–	–	–	–
32	–	–	–	–	–	–	–	–	–	–	–	–	–	–
	–	–	–	–	–	–	–	–	–	–	–	–	–	–
33	–	–	–	–	–	–	–	–	–	–	–	–	–	–
	–	–	–	–	–	–	–	–	–	–	–	–	–	–
34	–	–	–	–	–	–	–	–	–	–	–	–	–	–
	–	–	–	–	–	–	–	–	–	–	–	–	–	–
35	–	–	–	–	–	–	–	–	–	–	–	–	–	–
	–	–	–	–	–	–	–	–	–	–	–	–	–	–
36	–	–	–	–	–	–	–	–	–	–	–	–	–	–
	–	–	–	–	–	–	–	–	–	–	–	–	–	–
37	–	–	–	–	–	–	–	–	–	–	–	–	–	–
	–	–	–	–	–	–	–	–	–	–	–	–	–	–
38	78.6	47.1	–	–	–	–	–	–	–	–	–	–	–	–
	86.4	51.8	–	–	–	–	–	–	–	–	–	–	–	–
39	34.0	20.4	–	–	–	–	–	–	–	–	–	–	–	–
	37.7	22.6	–	–	–	–	–	–	–	–	–	–	–	–
40	–	–	–	–	–	–	–	–	–	–	–	–	–	–
	–	–	–	–	–	–	–	–	–	–	–	–	–	–
41	–	–	–	–	–	–	–	–	–	–	–	–	–	–
	–	–	–	–	–	–	–	–	–	–	–	–	–	–

(Continued)

TABLE V-1C DRY FORAGES, COMPOSITION OF FEEDS, DATA EXPRESSED **AS-FED** AND **MOISTURE-FREE**—*(Continued)*

Entry Number	Feed Name Description	International Feed Number	Moisture Basis: A-F (as-fed) or M-F (moisture-free)	Proximate Analysis: Dry Matter	Ash	Crude Fiber	Neutral Det. Fib. (NDF)	Acid Det. Fib. (ADF)	Lignin	Ether Extract (Fat)	N-Free Extract	Crude Protein	Digestible Protein: Ruminant	Swine	Horse
				%	%	%	%	%	%	%	%	%	%	%	%
	BAHIAGRASS (Continued)														
42	HAY, MATURE, SUN-CURED	**1-00-461**	**A-F**	**91**	**5.4**	**30.5**	**–**	**–**	**–**	**1.5**	**49.2**	**4.3**	**1.0**	**0.2**	**1.7**
			M-F	100	5.9	33.6	–	–	–	1.6	54.2	4.7	1.1	0.3	1.8
	BANANA *Musa spp*														
43	LEAVES WITHOUT PETIOLES, WITHOUT VEINS, MEAL	**1-00-482**	**A-F**	**92**	**12.1**	**16.5**	**–**	**–**	**–**	**12.1**	**31.7**	**19.7**	**–**	**–**	**–**
			M-F	100	13.1	17.9	–	–	–	13.1	34.5	21.4	–	–	–
	BARLEY *Hordeum vulgare*														
44	HAY, SUN-CURED	**1-00-495**	**A-F**	**88**	**6.6**	**23.6**	**–**	**–**	**–**	**1.9**	**48.5**	**7.8**	**4.3**	**3.4**	**4.4**
			M-F	100	7.5	26.7	–	–	–	2.1	54.9	8.8	4.9	3.8	5.0
45	STRAW	**1-00-498**	**A-F**	**91**	**6.7**	**37.9**	**77.5**	**51.1**	**6.9**	**1.7**	**41.1**	**4.0**	**0.6**	**0.0**	**1.4**
			M-F	100	7.3	41.5	84.8	55.9	7.6	1.9	44.9	4.4	0.7	0.0	1.6
	BARNYARD GRASS (MILLET, JAPANESE) *Echinochloa crusgalli*														
46	HAY, SUN-CURED	**1-03-105**	**A-F**	**87**	**8.5**	**26.3**	**–**	**–**	**–**	**1.7**	**41.8**	**8.8**	**5.4**	**4.4**	**5.3**
			M-F	100	9.8	30.2	–	–	–	2.0	48.0	10.1	6.2	5.0	6.0
	BEAN *Phaseolus spp*														
47	HAY, SUN-CURED	**1-00-583**	**A-F**	**90**	**6.3**	**24.1**	**–**	**–**	**–**	**1.7**	**44.4**	**13.9**	**9.4**	**8.7**	**9.1**
			M-F	100	7.0	26.7	–	–	–	1.9	49.1	15.4	10.4	9.6	10.1
48	STRAW	**1-00-585**	**A-F**	**89**	**8.7**	**36.8**	**–**	**50.0**	**–**	**1.3**	**35.2**	**7.1**	**3.5**	**2.8**	**3.9**
			M-F	100	9.7	41.3	–	56.0	–	1.5	39.5	8.0	4.0	3.2	4.4
49	VINES, DEHY, GROUND	**1-00-582**	**A-F**	**90**	**8.7**	**22.3**	**–**	**–**	**–**	**3.7**	**40.8**	**14.8**	**10.4**	**9.5**	**9.8**
			M-F	100	9.6	24.7	–	–	–	4.1	45.2	16.4	11.5	10.5	10.9
	BEAN, MUNG *Phaseolus aureus*														
50	HAY, SUN-CURED	**1-00-616**	**A-F**	**90**	**7.9**	**23.8**	**–**	**–**	**–**	**2.2**	**46.5**	**9.8**	**5.9**	**5.1**	**6.0**
			M-F	100	8.8	26.4	–	–	–	2.4	51.6	10.9	6.5	5.7	6.6
	BEAN, NAVY *Phaseolus vulgaris*														
51	PODS, SUN-CURED (BEANS, FIELD)	**1-08-346**	**A-F**	**92**	**3.9**	**34.8**	**–**	**–**	**–**	**1.0**	**45.0**	**7.1**	**3.4**	**2.7**	**3.8**
			M-F	100	4.2	37.9	–	–	–	1.1	49.0	7.7	3.7	2.9	4.2
	BEAN, TEPARY *Phaseolus acutifolius, latifolius*														
52	HAY, SUN-CURED	**1-08-348**	**A-F**	**90**	**10.5**	**24.8**	**–**	**–**	**–**	**2.9**	**34.7**	**17.1**	**12.2**	**11.5**	**11.6**
			M-F	100	11.7	27.6	–	–	–	3.2	38.6	19.0	13.6	12.8	12.9
	BEET, SUGAR *Beta vulgaris, altissima*														
53	TOPS WITH CROWN, DEHY	**1-00-640**	**A-F**	**85**	**16.4**	**11.9**	**–**	**–**	**–**	**1.0**	**44.9**	**10.7**	**6.6**	**4.9**	**6.7**
			M-F	100	19.4	14.0	–	–	–	1.2	52.9	12.6	7.7	5.8	7.9
54	LEAVES, SUN-CURED	**1-00-641**	**A-F**	**91**	**29.8**	**6.1**	**–**	**–**	**–**	**6.6**	**25.3**	**23.2**	**17.6**	**16.9**	**16.3**
			M-F	100	32.8	6.7	–	–	–	7.2	27.8	25.5	19.3	18.5	17.9
55	HULLS	**1-00-643**	**A-F**	**85**	**8.4**	**28.2**	**–**	**–**	**–**	**2.1**	**35.4**	**11.1**	**6.2**	**6.5**	**7.0**
			M-F	100	9.9	33.1	–	–	–	2.5	41.5	13.0	7.3	7.6	8.2
	BEGGARWEED (TICKCLOVER, CHEROKEE) *Desmodium tortuosum*														
56	HAY, SUN-CURED	**1-08-540**	**A-F**	**91**	**7.8**	**28.4**	**–**	**–**	**–**	**2.3**	**37.2**	**15.2**	**10.6**	**9.8**	**10.1**
			M-F	100	8.6	31.2	–	–	–	2.5	40.9	16.7	11.6	10.8	11.1
	BENTGRASS, COLONIAL *Agrostis tenuis*														
57	HAY, SUN-CURED	**1-00-684**	**A-F**	**87**	**6.3**	**27.6**	**–**	**–**	**–**	**2.2**	**44.9**	**6.2**	**2.6**	**2.1**	**3.2**
			M-F	100	7.2	31.7	–	–	–	2.5	51.5	7.1	3.0	2.4	3.7
	BERMUDAGRASS *Cynodon dactylon*														
58	HAY, SUN-CURED	**1-00-703**	**A-F**	**91**	**8.0**	**28.4**	**–**	**–**	**–**	**1.8**	**43.7**	**9.2**	**5.5**	**4.5**	**5.4**
			M-F	100	8.8	31.2	–	–	–	2.0	48.0	10.0	6.1	5.0	6.0
	BERMUDAGRASS, COASTAL *Cynodon dactylon*														
59	HAY, SUN-CURED[1]	**1-00-716**	**A-F**	**91**	**6.3**	**27.0**	**69.2**	**34.6**	**–**	**2.0**	**43.9**	**11.7**	**7.5**	**6.7**	**7.4**
			M-F	100	6.9	29.7	76.0	38.0	–	2.2	48.3	12.8	8.2	7.4	8.1
60	GROUND	**1-10-609**	**A-F**	**87**	**6.1**	**24.3**	**–**	**–**	**–**	**3.3**	**38.2**	**15.1**	**10.9**	**–**	**10.7**
			M-F	100	7.0	27.9	–	–	–	3.8	43.9	17.4	12.5	–	12.3
	BERSEEM CLOVER (CLOVER, EGYPTIAN) *Trifolium alexandrinum*														
61	HAY, SUN-CURED	**1-01-340**	**A-F**	**90**	**11.6**	**21.9**	**–**	**–**	**–**	**2.2**	**40.3**	**13.7**	**9.4**	**8.5**	**8.9**
			M-F	100	13.0	24.4	–	–	–	2.5	44.9	15.2	10.4	9.5	10.0
	BIRDSFOOT TREFOIL *Lotus corniculatus*														
62	HAY, SUN-CURED	**1-05-044**	**A-F**	**91**	**6.7**	**29.3**	**–**	**–**	**–**	**1.9**	**38.9**	**13.9**	**9.6**	**8.7**	**9.1**
			M-F	100	7.4	32.3	–	–	–	2.1	42.9	15.3	10.6	9.6	10.0
	BLACK GRASS (RUSH, SALTMEADOW) *Juncus gerardii*														
63	HAY, SUN-CURED	**1-03-970**	**A-F**	**88**	**7.1**	**24.9**	**–**	**–**	**–**	**2.4**	**46.3**	**7.5**	**4.0**	**3.2**	**4.2**
			M-F	100	8.0	28.2	–	–	–	2.7	52.5	8.5	4.6	3.6	4.8
	BLUEGRASS, CANADA *Poa compressa*														
64	HAY, SUN-CURED	**1-00-762**	**A-F**	**92**	**6.5**	**27.6**	**–**	**–**	**–**	**2.4**	**45.8**	**9.5**	**4.1**	**4.8**	**5.7**
			M-F	100	7.0	30.1	–	–	–	2.6	49.9	10.3	4.5	5.2	6.2
	BLUEGRASS, FOWL (FOWL MEADOW GRASS) *Poa palustris*														
65	HAY, SUN-CURED	**1-00-767**	**A-F**	**85**	**7.0**	**32.3**	**–**	**–**	**–**	**1.6**	**37.2**	**6.9**	**3.4**	**2.8**	**3.8**
			M-F	100	8.2	38.0	–	–	–	1.9	43.8	8.1	4.1	3.3	4.5

Entry Number	TDN Ruminant	TDN Swine	TDN Horse	Digestible Energy Ruminant		Digestible Energy Swine		Digestible Energy Horse		Metabolizable Energy Ruminant		Metabolizable Energy Swine		Metabolizable Energy Poultry ME_n		Metabolizable Energy Horse		Net Energy Ruminant NE_m		Net Energy Ruminant NE_g		Net Energy Lactating Cows NE_{lc}	
	%	%	%	Mcal		kcal		Mcal		Mcal		kcal		kcal		Mcal		Mcal		Mcal		Mcal	
				lb	*kg*	lb	*kg*	lb	*kg*	lb	*kg*	lb	*kg*	lb	*kg*	lb	*kg*	lb	*kg*	lb	*kg*	lb	*kg*
42	**48**	**30**	**38**	**0.96**	***2.12***	–	–	**0.73**	***1.62***	**0.79**	***1.73***	–	–	–	–	**0.60**	***1.33***	**0.44**	***0.96***	**0.21**	***0.46***	**0.48**	***1.06***
	53	33	42	1.06	*2.33*	–	–	0.81	*1.78*	0.87	*1.91*	–	–	–	–	0.66	*1.46*	0.48	*1.06*	0.23	*0.51*	0.53	*1.17*
43	–	–	–	–	–	–	–	–	–	–	–	–	–	–	–	–	–	–	–	–	–	–	–
	–	–	–	–	–	–	–	–	–	–	–	–	–	–	–	–	–	–	–	–	–	–	–
44	**50**	**40**	**42**	**0.99**	***2.17***	–	–	**0.81**	***1.77***	**0.79**	***1.73***	–	–	–	–	**0.66**	***1.46***	**0.45**	***0.99***	**0.22**	***0.49***	**0.50**	***1.10***
	57	46	48	1.11	*2.46*	–	–	0.91	*2.01*	0.89	*1.96*	–	–	–	–	0.75	*1.65*	0.51	*1.11*	0.25	*0.56*	0.57	*1.25*
45	**43**	–	**28**	**0.84**	***1.86***	–	–	**0.57**	***1.26***	**0.64**	***1.40***	–	–	–	–	**0.47**	***1.03***	**0.29**	***0.64***	**0.07**	***0.15***	**0.42**	***0.91***
	47	–	31	0.92	*2.03*	–	–	0.63	*1.38*	0.70	*1.54*	–	–	–	–	0.51	*1.13*	0.32	*0.70*	0.08	*0.16*	0.45	*1.00*
46	**47**	**34**	**38**	**0.94**	***2.07***	–	–	**0.73**	***1.62***	**0.77**	***1.70***	–	–	–	–	**0.60**	***1.32***	**0.43**	***0.95***	**0.21**	***0.46***	**0.47**	***1.04***
	54	39	44	1.08	*2.38*	–	–	0.84	*1.85*	0.89	*1.95*	–	–	–	–	0.69	*1.52*	0.49	*1.09*	0.24	*0.53*	0.54	*1.19*
47	**59**	**48**	**52**	**1.13**	***2.48***	–	–	**0.97**	***2.13***	**0.95**	***2.10***	–	–	–	–	**0.79**	***1.74***	**0.52**	***1.15***	**0.29**	***0.64***	**0.54**	***1.20***
	66	53	57	1.25	*2.75*	–	–	1.07	*2.35*	1.05	*2.32*	–	–	–	–	0.88	*1.93*	0.58	*1.28*	0.32	*0.71*	0.60	*1.33*
48	**43**	–	**30**	**0.87**	***1.91***	–	–	**0.60**	***1.32***	**0.69**	***1.53***	–	–	–	–	**0.49**	***1.08***	**0.36**	***0.79***	**0.14**	***0.30***	**0.42**	***0.93***
	49	–	33	0.97	*2.15*	–	–	0.67	*1.47*	0.78	*1.72*	–	–	–	–	0.55	*1.21*	0.40	*0.88*	0.15	*0.34*	0.47	*1.05*
49	**60**	**55**	**42**	**1.15**	***2.54***	–	–	**0.80**	***1.77***	**0.98**	***2.16***	–	–	–	–	**0.66**	***1.45***	**0.56**	***1.23***	**0.32**	***0.71***	**0.57**	***1.25***
	66	61	46	1.28	*2.81*	–	–	0.89	*1.96*	1.08	*2.39*	–	–	–	–	0.73	*1.60*	0.62	*1.36*	0.36	*0.78*	0.63	*1.39*
50	**50**	**43**	**43**	**1.01**	***2.23***	–	–	**0.82**	***1.80***	**0.84**	***1.84***	–	–	–	–	**0.67**	***1.47***	**0.49**	***1.08***	**0.26**	***0.57***	**0.52**	***1.15***
	55	48	47	1.12	*2.47*	–	–	0.90	*1.99*	0.93	*2.04*	–	–	–	–	0.74	*1.63*	0.54	*1.20*	0.29	*0.64*	0.58	*1.27*
51	**49**	**29**	**42**	**0.97**	***2.14***	–	–	**0.81**	***1.78***	**0.79**	***1.75***	–	–	–	–	**0.66**	***1.46***	**0.44**	***0.98***	**0.21**	***0.47***	**0.49**	***1.08***
	53	32	46	1.06	*2.33*	–	–	0.88	*1.94*	0.86	*1.91*	–	–	–	–	0.72	*1.59*	0.48	*1.06*	0.23	*0.51*	0.53	*1.17*
52	**48**	**48**	**43**	**1.01**	***2.22***	–	–	**0.82**	***1.80***	**0.83**	***1.83***	–	–	–	–	**0.67**	***1.48***	**0.51**	***1.13***	**0.28**	***0.61***	**0.53**	***1.18***
	53	53	48	1.12	*2.46*	–	–	0.91	*2.00*	0.92	*2.04*	–	–	–	–	0.75	*1.64*	0.57	*1.25*	0.31	*0.68*	0.59	*1.31*
53	**51**	**45**	–	**1.02**	***2.25***	–	–	–	–	**0.86**	***1.89***	–	–	–	–	–	–	–	–	–	–	–	–
	60	54	–	1.20	*2.65*	–	–	–	–	1.01	*2.23*	–	–	–	–	–	–	–	–	–	–	–	–
54	–	–	–	–	–	–	–	–	–	–	–	–	–	–	–	–	–	–	–	–	–	–	–
	–	–	–	–	–	–	–	–	–	–	–	–	–	–	–	–	–	–	–	–	–	–	–
55	**29**	**34**	**36**	**0.75**	***1.66***	–	–	**0.69**	***1.52***	**0.58**	***1.29***	–	–	–	–	**0.57**	***1.25***	**0.43**	***0.95***	**0.21**	***0.47***	**0.47**	***1.03***
	34	40	42	0.88	*1.94*	–	–	0.81	*1.78*	0.69	*1.51*	–	–	–	–	0.66	*1.46*	0.50	*1.11*	0.25	*0.55*	0.55	*1.21*
56	**48**	**44**	**45**	**1.00**	***2.20***	–	–	**0.85**	***1.88***	**0.82**	***1.81***	–	–	–	–	**0.70**	***1.54***	**0.50**	***1.11***	**0.27**	***0.60***	**0.53**	***1.17***
	52	48	49	1.10	*2.42*	–	–	0.94	*2.07*	0.91	*2.00*	–	–	–	–	0.77	*1.69*	0.56	*1.22*	0.30	*0.66*	0.58	*1.29*
57	**48**	**35**	**35**	**0.96**	***2.12***	–	–	**0.68**	***1.51***	**0.79**	***1.75***	–	–	–	–	**0.56**	***1.24***	**0.45**	***0.99***	**0.23**	***0.50***	**0.48**	***1.07***
	56	40	40	1.10	*2.43*	–	–	0.78	*1.73*	0.91	*2.01*	–	–	–	–	0.64	*1.42*	0.51	*1.13*	0.26	*0.58*	0.55	*1.22*
58	**43**	**35**	**40**	**0.88**	***1.93***	–	–	**0.77**	***1.70***	**0.64**	***1.41***	–	–	–	–	**0.63**	***1.40***	**0.29**	***0.65***	**0.07**	***0.16***	**0.40**	***0.88***
	47	39	44	0.96	*2.12*	–	–	0.85	*1.87*	0.70	*1.55*	–	–	–	–	0.70	*1.53*	0.32	*0.71*	0.08	*0.18*	0.44	*0.96*
59	**49**	**43**	**46**	**1.04**	***2.30***	–	–	**0.86**	***1.90***	**0.66**	***1.44***	–	–	–	–	**0.71**	***1.56***	**0.31**	***0.68***	**0.09**	***0.20***	**0.53**	***1.16***
	54	48	50	1.15	*2.53*	–	–	0.95	*2.10*	0.72	*1.59*	–	–	–	–	0.78	*1.72*	0.34	*0.75*	0.10	*0.22*	0.58	*1.28*
60	**55**	–	**37**	**1.11**	***2.44***	–	–	**0.70**	***1.53***	**0.91**	***2.00***	–	–	–	–	**0.57**	***1.25***	**0.55**	***1.22***	**0.30**	***0.67***	**0.56**	***1.24***
	63	–	43	1.27	*2.80*	–	–	0.80	*1.76*	1.04	*2.30*	–	–	–	–	0.65	*1.44*	0.64	*1.40*	0.35	*0.77*	0.64	*1.42*
61	**52**	**43**	**43**	**1.04**	***2.30***	–	–	**0.81**	***1.79***	**0.87**	***1.92***	–	–	–	–	**0.67**	***1.47***	**0.52**	***1.15***	**0.29**	***0.63***	**0.54**	***1.19***
	58	48	48	1.16	*2.56*	–	–	0.91	*2.00*	0.97	*2.14*	–	–	–	–	0.74	*1.64*	0.58	*1.28*	0.32	*0.71*	0.60	*1.33*
62	**54**	**41**	**46**	**0.91**	***2.01***	–	–	**0.87**	***1.91***	**0.83**	***1.84***	–	–	–	–	**0.71**	***1.56***	**0.55**	***1.22***	**0.32**	***0.69***	**0.57**	***1.25***
	59	46	51	1.01	*2.22*	–	–	0.96	*2.11*	0.92	*2.03*	–	–	–	–	0.78	*1.73*	0.61	*1.34*	0.35	*0.77*	0.62	*1.37*
63	**49**	**40**	**38**	**0.98**	***2.17***	–	–	**0.73**	***1.61***	**0.81**	***1.79***	–	–	–	–	**0.60**	***1.32***	**0.47**	***1.04***	**0.25**	***0.55***	**0.50**	***1.11***
	55	45	43	1.11	*2.46*	–	–	0.83	*1.82*	0.92	*2.03*	–	–	–	–	0.68	*1.50*	0.54	*1.18*	0.28	*0.62*	0.57	*1.26*
64	**57**	**42**	**41**	**1.14**	***2.52***	–	–	**0.79**	***1.75***	**0.97**	***2.13***	–	–	–	–	**0.65**	***1.43***	**0.61**	***1.35***	**0.37**	***0.81***	**0.61**	***1.35***
	62	46	45	1.25	*2.75*	–	–	0.86	*1.90*	1.06	*2.33*	–	–	–	–	0.71	*1.56*	0.67	*1.47*	0.40	*0.88*	0.67	*1.47*
65	**44**	**24**	**31**	**0.87**	***1.93***	–	–	**0.62**	***1.36***	**0.71**	***1.56***	–	–	–	–	**0.51**	***1.12***	**0.38**	***0.85***	**0.17**	***0.38***	**0.43**	***0.96***
	51	29	37	1.03	*2.27*	–	–	0.73	*1.61*	0.83	*1.84*	–	–	–	–	0.60	*1.32*	0.45	*1.00*	0.20	*0.45*	0.51	*1.13*

(Continued)

TABLE V-1C DRY FORAGES, MINERAL AND VITAMIN COMPOSITION OF FEEDS, DATA EXPRESSED **AS-FED** AND **MOISTURE-FREE**—*(Continued)*

Entry Number	Feed Name Description	Moisture Basis: A-F (as-fed) or M-F (moisture-free)	Dry Matter	Macro Minerals							Micro Minerals						
				Calcium (Ca)	Phosphorus (P)	Sodium (Na)	Chlorine (Cl)	Magnesium (Mg)	Potassium (K)	Sulfur (S)	Cobalt (Co)	Copper (Cu)	Iodine (I)	Iron (Fe)	Manganese (Mn)	Selenium (Se)	Zinc (Zn)
			%	%	%	%	%	%	%	%	ppm or mg/kg	ppm or mg/kg	ppm or mg/kg	%	ppm or mg/kg	ppm or mg/kg	ppm or mg/kg
	BAHIAGRASS (Continued)																
42	HAY, MATURE, SUN-CURED	A-F	91	–	–	–	–	–	–	–	–	–	–	–	–	–	–
		M-F	100	–	–	–	–	–	–	–	–	–	–	–	–	–	–
	BANANA *Musa* spp																
43	LEAVES WITHOUT PETIOLES, WITHOUT VEINS, MEAL	A-F	92	–	–	–	–	–	–	–	–	–	–	–	–	–	–
		M-F	100	–	–	–	–	–	–	–	–	–	–	–	–	–	–
	BARLEY *Hordeum vulgare*																
44	HAY, SUN-CURED	A-F	88	0.21	0.25	0.12	–	0.14	1.30	0.15	0.059	3.9	–	0.027	34.8	–	–
		M-F	100	0.24	0.28	0.14	–	0.16	1.47	0.17	0.067	4.4	–	0.030	39.4	–	–
45	STRAW	A-F	91	0.27	0.07	0.13	0.61	0.21	2.16	0.16	0.061	4.9	–	0.019	15.1	–	6.8
		M-F	100	0.30	0.07	0.14	0.67	0.23	2.37	0.17	0.067	5.4	–	0.021	16.6	–	7.4
	BARNYARD GRASS (MILLET, JAPANESE) *Echinochloa crusgalli*																
46	HAY, SUN-CURED	A-F	87	0.20	–	–	–	–	2.11	–	–	–	–	–	–	–	–
		M-F	100	0.23	–	–	–	–	2.42	–	–	–	–	–	–	–	–
	BEAN *Phaseolus* spp																
47	HAY, SUN-CURED	A-F	90	–	–	–	–	–	–	–	–	–	–	–	–	–	–
		M-F	100	–	–	–	–	–	–	–	–	–	–	–	–	–	–
48	STRAW	A-F	89	1.65	0.13	–	–	0.12	1.02	–	–	–	–	–	–	–	–
		M-F	100	1.85	0.14	–	–	0.13	1.14	–	–	–	–	–	–	–	–
49	VINES, DEHY, GROUND	A-F	90	–	–	–	–	–	–	–	–	–	–	–	–	–	–
		M-F	100	–	–	–	–	–	–	–	–	–	–	–	–	–	–
	BEAN, MUNG *Phaseolus aureus*																
50	HAY, SUN-CURED	A-F	90	–	–	–	–	–	–	–	–	–	–	–	–	–	–
		M-F	100	–	–	–	–	–	–	–	–	–	–	–	–	–	–
	BEAN, NAVY *Phaseolus vulgaris*																
51	PODS, SUN-CURED (BEANS, FIELD)	A-F	92	0.78	0.10	–	–	–	2.02	–	–	–	–	–	–	–	–
		M-F	100	0.85	0.11	–	–	–	2.20	–	–	–	–	–	–	–	–
	BEAN, TEPARY *Phaseolus acutifolius, latifolius*																
52	HAY, SUN-CURED	A-F	90	–	–	–	–	–	–	–	–	–	–	–	–	–	–
		M-F	100	–	–	–	–	–	–	–	–	–	–	–	–	–	–
	BEET, SUGAR *Beta vulgaris, altissima*																
53	TOPS WITH CROWN, DEHY	A-F	85	0.37	0.20	–	–	0.33	3.57	–	–	5.2	–	0.051	35.9	–	16.8
		M-F	100	0.44	0.24	–	–	0.39	4.21	–	–	6.1	–	0.060	42.3	–	19.8
54	LEAVES, SUN-CURED	A-F	91	–	–	–	–	–	–	–	–	–	–	–	–	–	–
		M-F	100	–	–	–	–	–	–	–	–	–	–	–	–	–	–
55	HULLS	A-F	85	–	–	–	–	–	–	–	–	–	–	–	–	–	–
		M-F	100	–	–	–	–	–	–	–	–	–	–	–	–	–	–
	BEGGARWEED (TICKCLOVER, CHEROKEE) *Desmodium tortuosum*																
56	HAY, SUN-CURED	A-F	91	1.05	0.27	–	–	–	2.32	–	–	–	–	–	–	–	–
		M-F	100	1.16	0.30	–	–	–	2.55	–	–	–	–	–	–	–	–
	BENTGRASS, COLONIAL *Agrostis tenuis*																
57	HAY, SUN-CURED	A-F	87	–	0.18	–	–	–	1.40	–	–	–	–	–	–	–	–
		M-F	100	–	0.20	–	–	–	1.60	–	–	–	–	–	–	–	–
	BERMUDAGRASS *Cynodon dactylon*																
58	HAY, SUN-CURED	A-F	91	0.43	0.16	0.07	–	0.16	1.40	0.19	0.111	24.3	0.105	0.027	99.4	–	53.0
		M-F	100	0.47	0.17	0.08	–	0.17	1.53	0.21	0.122	26.6	0.115	0.029	109.0	–	58.1
	BERMUDAGRASS, COASTAL *Cynodon dactylon*																
59	HAY, SUN-CURED	A-F	91	0.38	0.17	–	–	0.16	1.46	0.19	–	–	–	0.028	–	–	18.2
		M-F	100	0.42	0.18	–	–	0.17	1.61	0.21	–	–	–	0.030	–	–	20.0
60	GROUND	A-F	87	0.29	0.23	–	–	–	–	–	–	–	–	–	–	–	–
		M-F	100	0.33	0.26	–	–	–	–	–	–	–	–	–	–	–	–
	BERSEEM CLOVER (CLOVER, EGYPTIAN) *Trifolium alexandrinum*																
61	HAY, SUN-CURED	A-F	90	2.28	0.26	–	–	0.36	2.03	–	–	–	–	–	–	–	–
		M-F	100	2.54	0.29	–	–	0.40	2.26	–	–	–	–	–	–	–	–
	BIRDSFOOT TREFOIL *Lotus corniculatus*																
62	HAY, SUN-CURED	A-F	91	1.54	0.21	0.06	–	0.46	1.74	0.23	0.100	8.4	–	0.021	26.0	–	69.9
		M-F	100	1.70	0.23	0.07	–	0.51	1.92	0.25	0.111	9.3	–	0.023	28.7	–	77.2
	BLACK GRASS (RUSH, SALTMEADOW) *Juncus gerardii*																
63	HAY, SUN-CURED	A-F	88	–	0.09	–	–	–	1.53	–	–	–	–	–	–	–	–
		M-F	100	–	0.10	–	–	–	1.74	–	–	–	–	–	–	–	–
	BLUEGRASS, CANADA *Poa compressa*																
64	HAY, SUN-CURED	A-F	92	0.28	0.24	0.10	–	0.30	1.73	0.12	–	–	–	0.028	84.9	–	–
		M-F	100	0.30	0.26	0.11	–	0.33	1.88	0.13	–	–	–	0.030	92.6	–	–
	BLUEGRASS, FOWL (FOWL MEADOW GRASS) *Poa palustris*																
65	HAY, SUN-CURED	A-F	85	0.22	0.22	–	–	–	–	–	–	–	–	–	–	–	–
		M-F	100	0.26	0.26	–	–	–	–	–	–	–	–	–	–	–	–

Entry Number	Fat-Soluble Vitamins					Water-Soluble Vitamins								
	A (1 mg Carotene = 1667 IU Vit A)	Carotene (Provitamin A)	D	E	K	B-12	Biotin	Choline	Folacin (Folic Acid)	Niacin	Pantothenic Acid (B-3)	(Pyridoxine) B-6	Riboflavin (B-2)	Thiamin (B-1)
	IU/g	ppm or mg/kg	IU/kg	ppm or mg/kg	ppm or mg/kg	ppm or mg/kg	ppm or mg/kg	ppm or mg/kg	ppm or mg/kg	ppm or mg/kg	ppm or mg/kg	ppm or mg/kg	ppm or mg/kg	ppm or mg/kg
42	–	–	–	–	–	–	–	–	–	–	–	–	–	–
	–	–	–	–	–	–	–	–	–	–	–	–	–	–
43	**245.8**	**147.5**	–	–	–	–	–	–	–	**53**	–	–	**43.4**	**1.2**
	267.2	160.3	–	–	–	–	–	–	–	58	–	–	47.2	1.3
44	**77.4**	**46.4**	**1**	–	–	–	–	–	–	–	–	–	–	–
	87.5	52.5	1	–	–	–	–	–	–	–	–	–	–	–
45	**3.5**	**2.1**	**1**	–	–	–	–	–	–	–	–	–	–	–
	3.9	2.3	1	–	–	–	–	–	–	–	–	–	–	–
46	–	–	–	–	–	–	–	–	–	–	–	–	–	–
	–	–	–	–	–	–	–	–	–	–	–	–	–	–
47	–	–	–	–	–	–	–	–	–	–	–	–	–	–
	–	–	–	–	–	–	–	–	–	–	–	–	–	–
48	–	–	–	–	–	–	–	–	–	–	–	–	–	–
	–	–	–	–	–	–	–	–	–	–	–	–	–	–
49	–	–	–	–	–	–	–	–	–	–	–	–	–	–
	–	–	–	–	–	–	–	–	–	–	–	–	–	–
50	**166.5**	**99.9**	–	–	–	–	–	–	–	–	–	–	–	–
	184.5	110.7	–	–	–	–	–	–	–	–	–	–	–	–
51	–	–	–	–	–	–	–	–	–	–	–	–	–	–
	–	–	–	–	–	–	–	–	–	–	–	–	–	–
52	–	–	–	–	–	–	–	–	–	–	–	–	–	–
	–	–	–	–	–	–	–	–	–	–	–	–	–	–
53	–	–	–	–	–	–	–	–	–	–	–	–	–	–
	–	–	–	–	–	–	–	–	–	–	–	–	–	–
54	–	–	–	**346.9**	–	–	–	–	–	–	–	–	–	–
	–	–	–	381.2	–	–	–	–	–	–	–	–	–	–
55	–	–	–	–	–	–	–	–	–	–	–	–	–	–
	–	–	–	–	–	–	–	–	–	–	–	–	–	–
56	–	–	–	–	–	–	–	–	–	–	–	–	–	–
	–	–	–	–	–	–	–	–	–	–	–	–	–	–
57	–	–	–	–	–	–	–	–	–	–	–	–	–	–
	–	–	–	–	–	–	–	–	–	–	–	–	–	–
58	**87.5**	**52.5**	–	–	–	–	–	–	–	–	–	–	–	–
	96.0	57.6	–	–	–	–	–	–	–	–	–	–	–	–
59	**123.7**	**74.2**	–	–	–	–	–	–	–	–	–	–	–	–
	136.2	81.7	–	–	–	–	–	–	–	–	–	–	–	–
60	–	–	–	–	–	–	–	–	–	–	–	–	–	–
	–	–	–	–	–	–	–	–	–	–	–	–	–	–
61	–	–	–	–	–	–	–	–	–	–	–	–	–	–
	–	–	–	–	–	–	–	–	–	–	–	–	–	–
62	**217.8**	**130.6**	**1**	–	–	–	–	–	–	–	–	–	**14.6**	**6.2**
	240.4	144.2	2	–	–	–	–	–	–	–	–	–	16.1	6.8
63	–	–	–	–	–	–	–	–	–	–	–	–	–	–
	–	–	–	–	–	–	–	–	–	–	–	–	–	–
64	**378.4**	**227.0**	–	–	–	–	–	–	–	–	–	–	–	–
	412.6	247.5	–	–	–	–	–	–	–	–	–	–	–	–
65	**177.1**	**106.3**	–	–	–	–	–	–	–	–	–	–	–	–
	208.4	125.0	–	–	–	–	–	–	–	–	–	–	–	–

(Continued)

TABLE V-1C DRY FORAGES, COMPOSITION OF FEEDS, DATA EXPRESSED **AS-FED** AND **MOISTURE-FREE**—*(Continued)*

Entry Number	Feed Name Description	International Feed Number	Moisture Basis: A-F (as-fed) or M-F (moisture-free)	Proximate Analysis: Dry Matter	Ash	Crude Fiber	Neutral Det. Fib. (NDF)	Acid Det. Fib. (ADF)	Lignin	Ether Extract (Fat)	N-Free Extract	Crude Protein	Digestible Protein: Ruminant	Swine	Horse
				%	%	%	%	%	%	%	%	%	%	%	%
	BLUEGRASS, KENTUCKY *Poa pratensis*														
66	HAY, SUN-CURED	1-00-776	A-F	89	5.9	26.8	–	–	–	3.0	44.3	9.1	5.3	4.6	5.4
			M-F	100	6.6	30.0	–	–	–	3.4	49.7	10.2	5.9	5.1	6.1
67	HAY, EARLY BLOOM, SUN-CURED	1-05-681	A-F	90	4.5	25.5	–	–	–	3.6	45.5	11.0	6.9	6.2	6.9
			M-F	100	5.0	28.3	–	–	–	4.0	50.5	12.2	7.6	6.9	7.6
68	HAY, MATURE, SUN-CURED	1-00-774	A-F	89	6.5	29.7	–	–	–	2.6	44.9	5.6	2.2	1.5	2.7
			M-F	100	7.3	33.3	–	–	–	2.9	50.3	6.3	2.4	1.6	3.0
	BLUEJOINT REEDGRASS *Calamagrostis canadensis*														
69	HAY, SUN-CURED	1-03-903	A-F	93	6.5	31.8	–	–	–	2.2	44.5	8.4	5.6	3.8	4.8
			M-F	100	7.0	34.0	–	–	–	2.4	47.6	9.0	6.0	4.1	5.2
	BLUESTEM *Andropogon* spp														
70	HAY, SUN-CURED	1-00-819	A-F	90	6.3	30.8	–	–	–	2.2	45.6	4.9	1.5	0.8	2.1
			M-F	100	7.0	34.3	–	–	–	2.4	50.8	5.4	1.7	0.9	2.4
	BROMEGRASS *Bromus* spp														
71	HAY, SUN-CURED	1-00-890	A-F	91	7.1	29.8	–	31.7	4.3	1.9	43.3	8.7	4.8	4.1	5.1
			M-F	100	7.8	32.9	–	34.9	4.7	2.1	47.7	9.5	5.3	4.5	5.6
72	HAY, PREBLOOM, SUN-CURED	1-00-887	A-F	88	8.3	29.2	59.7	35.1	4.1	2.3	38.9	9.2	5.2	4.7	5.5
			M-F	100	9.4	33.3	68.0	40.0	4.7	2.6	44.3	10.5	5.9	5.4	6.3
73	HAY, MATURE, SUN-CURED[1]	1-00-889	A-F	92	6.0	33.0	62.6	39.6	–	1.7	45.2	6.1	2.1	1.8	3.0
			M-F	100	6.6	35.9	68.0	43.0	–	1.8	49.1	6.6	2.3	1.9	3.3
	BROMEGRASS, RESCUE *Bromus catharticus*														
74	HAY, SUN-CURED	1-00-931	A-F	90	8.1	24.6	–	–	–	3.2	44.5	9.8	5.8	5.1	5.9
			M-F	100	9.0	27.3	–	–	–	3.5	49.3	10.9	6.5	5.7	6.6
	BROMEGRASS, SMOOTH *Bromus inermis*														
75	HAY, SUN-CURED	1-00-947	A-F	90	7.1	29.0	70.5	33.4	5.1	2.6	38.4	12.4	6.8	7.4	7.9
			M-F	100	8.0	32.4	78.7	37.3	5.7	2.9	42.9	13.9	7.6	8.3	8.9
	BROOMCORN (HOG MILLET; MILLET, PROSO) *Panicum miliaceum*														
76	HAY, SUN-CURED	1-03-119	A-F	90	7.6	24.4	–	–	–	2.1	46.8	9.4	5.5	4.8	5.6
			M-F	100	8.4	27.0	–	–	–	2.4	51.8	10.4	6.1	5.3	6.2
77	STOVER WITHOUT HEADS, SUN-CURED	1-26-104	A-F	90	5.7	36.3	–	–	–	1.8	47.9	3.9	–	–	–
			M-F	100	6.3	40.3	–	–	–	2.0	53.2	4.3	–	–	–
	BUCKWHEAT, COMMON *Fagopyrum sagittatum*														
78	HULLS	1-00-987	A-F	88	1.4	42.7	–	–	–	0.8	40.0	3.0	0.2	-0.7	0.8
			M-F	100	1.6	48.5	–	–	–	0.9	45.5	3.5	0.2	-0.8	0.9
79	STRAW	1-00-988	A-F	88	8.9	34.9	–	–	–	0.9	39.6	4.1	0.9	0.2	1.6
			M-F	100	10.0	39.5	–	–	–	1.1	44.7	4.6	1.0	0.2	1.8
	BUFFALOGRASS *Buchloe dactyloides*														
80	HAY, SUN-CURED	1-01-003	A-F	90	11.9	24.9	–	–	–	1.5	45.1	6.9	3.7	2.5	3.7
			M-F	100	13.2	27.5	–	–	–	1.7	49.9	7.6	4.1	2.8	4.1
	BUR-CLOVER, TOOTHED *Medicago hispida*														
81	HAY, SUN-CURED	1-01-030	A-F	90	8.7	22.5	–	–	–	2.5	38.9	17.0	12.2	11.5	11.5
			M-F	100	9.7	25.1	–	–	–	2.8	43.4	19.0	13.6	12.8	12.9
	CACAO (COCOA) *Theobroma cacao*														
82	PODS, SUN-CURED	1-08-387	A-F	92	7.0	31.7	–	–	–	1.3	44.2	7.5	3.8	3.1	4.2
			M-F	100	7.6	34.6	–	–	–	1.4	48.2	8.2	4.1	3.3	4.5
	CANADA BLUEGRASS–SEE BLUEGRASS, CANADA														
	CANARYGRASS *Phalaris canariensis*														
83	HAY, SUN-CURED	1-01-091	A-F	92	8.3	32.5	–	–	–	1.0	42.5	7.7	4.6	3.2	4.3
			M-F	100	9.0	35.3	–	–	–	1.1	46.2	8.4	5.0	3.5	4.7
	CANARYGRASS, REED *Phalaris arundinacea*														
84	HAY, SUN-CURED	1-01-104	A-F	89	7.3	30.2	62.9	32.7	–	2.7	40.0	9.1	5.8	4.6	5.5
			M-F	100	8.1	33.9	70.5	36.6	–	3.0	44.8	10.2	6.4	5.1	6.1
	CARPETGRASS *Axonopus* spp														
85	HAY, SUN-CURED	1-01-138	A-F	92	9.5	30.0	–	–	–	1.9	43.4	7.3	3.6	2.8	4.0
			M-F	100	10.3	32.6	–	–	–	2.0	47.2	7.9	3.9	3.1	4.3
	CATTAIL, NARROWLEAF *Typha angustifolia*														
86	HAY, SUN-CURED	1-08-372	A-F	91	8.2	30.8	–	–	–	1.7	44.3	5.8	2.3	1.6	2.9
			M-F	100	9.0	33.9	–	–	–	1.9	48.8	6.4	2.6	1.8	3.1
	CEREALS														
87	IMMATURE, DEHY	1-26-069	A-F	92	14.4	16.0	–	–	–	4.8	32.5	24.4	19.0	–	–
			M-F	100	15.6	17.4	–	–	–	5.2	35.3	26.5	20.7	–	–
	CHAMISE (GREASEWOOD) *Adenostoma fasciculatum*														
88	FAN AIR DRIED, PELLETED	1-07-860	A-F	90	3.0	18.6	–	–	–	4.9	57.1	6.4	1.9	2.1	3.3
			M-F	100	3.3	20.7	–	–	–	5.4	63.5	7.1	2.1	2.4	3.7
	CHINESE LESPEDEZA–SEE LESPEDEZA, SERICEA														

Entry Number	TDN Ruminant	TDN Swine	TDN Horse	Digestible Energy Ruminant		Digestible Energy Swine		Digestible Energy Horse		Metabolizable Energy Ruminant		Metabolizable Energy Swine		Metabolizable Energy Poultry ME_n		Metabolizable Energy Horse		Net Energy Ruminant NE_m		Net Energy Ruminant NE_g		Net Energy Lactating Cows NE_{lc}	
	%	%	%	Mcal lb	Mcal kg	kcal lb	kcal kg	Mcal lb	Mcal kg	Mcal lb	Mcal kg	kcal lb	kcal kg	kcal lb	kcal kg	Mcal lb	Mcal kg	Mcal lb	Mcal kg	Mcal lb	Mcal kg	Mcal lb	Mcal kg
66	54	44	37	1.06	2.34	–	–	0.73	1.60	0.89	1.96	–	–	–	–	0.60	1.31	0.51	1.12	0.28	0.62	0.53	1.17
	61	50	42	1.19	2.63	–	–	0.81	1.80	1.00	2.20	–	–	–	–	0.67	1.47	0.57	1.26	0.31	0.69	0.60	1.31
67	55	53	41	1.10	2.42	–	–	0.79	1.73	0.93	2.04	–	–	–	–	0.64	1.42	0.57	1.25	0.33	0.72	0.57	1.26
	61	59	46	1.22	2.69	–	–	0.87	1.93	1.03	2.27	–	–	–	–	0.72	1.58	0.63	1.38	0.36	0.80	0.64	1.41
68	41	34	32	0.89	1.97	–	–	0.64	1.40	0.72	1.59	–	–	–	–	0.52	1.15	0.45	1.00	0.23	0.50	0.49	1.08
	46	38	36	1.00	2.21	–	–	0.71	1.57	0.81	1.78	–	–	–	–	0.59	1.29	0.51	1.12	0.25	0.56	0.55	1.21
69	54	36	38	1.05	2.32	–	–	0.75	1.65	0.87	1.92	–	–	–	–	0.61	1.35	0.48	1.05	0.24	0.53	0.52	1.14
	58	39	41	1.12	2.48	–	–	0.80	1.76	0.93	2.05	–	–	–	–	0.66	1.44	0.51	1.13	0.26	0.57	0.55	1.22
70	41	31	33	0.88	1.95	–	–	0.65	1.43	0.71	1.56	–	–	–	–	0.53	1.17	0.44	0.96	0.21	0.47	0.48	1.06
	45	34	37	0.98	2.17	–	–	0.72	1.59	0.79	1.74	–	–	–	–	0.59	1.31	0.49	1.07	0.24	0.52	0.54	1.18
71	48	35	39	1.12	2.48	–	–	0.75	1.65	0.95	2.09	–	–	–	–	0.61	1.35	0.50	1.11	0.27	0.60	0.50	1.11
	53	38	43	1.24	2.73	–	–	0.83	1.82	1.05	2.30	–	–	–	–	0.68	1.49	0.55	1.22	0.30	0.66	0.55	1.22
72	50	33	34	1.01	2.23	–	–	0.67	1.48	0.84	1.85	–	–	–	–	0.55	1.21	0.51	1.13	0.28	0.63	0.53	1.17
	58	38	39	1.15	2.54	–	–	0.77	1.69	0.96	2.11	–	–	–	–	0.63	1.38	0.58	1.28	0.32	0.71	0.60	1.33
73	47	30	37	0.93	2.06	–	–	0.71	1.58	0.76	1.67	–	–	–	–	0.59	1.29	0.43	0.94	0.20	0.44	0.48	1.05
	51	32	40	1.02	2.24	–	–	0.78	1.71	0.82	1.81	–	–	–	–	0.64	1.40	0.46	1.02	0.21	0.47	0.52	1.14
74	51	46	37	1.03	2.27	–	–	0.72	1.59	0.86	1.89	–	–	–	–	0.59	1.31	0.51	1.12	0.28	0.61	0.53	1.18
	56	51	41	1.14	2.52	–	–	0.80	1.76	0.95	2.10	–	–	–	–	0.66	1.45	0.57	1.25	0.31	0.68	0.59	1.30
75	53	41	47	1.12	2.47	–	–	0.89	1.95	0.95	2.09	–	–	–	–	0.73	1.60	0.54	1.19	0.31	0.67	0.53	1.17
	60	46	53	1.25	2.76	–	–	0.99	2.18	1.06	2.33	–	–	–	–	0.81	1.79	0.60	1.33	0.34	0.75	0.59	1.30
76	51	42	42	1.02	2.24	–	–	0.81	1.79	0.84	1.86	–	–	–	–	0.66	1.46	0.49	1.08	0.26	0.57	0.52	1.14
	56	47	47	1.13	2.48	–	–	0.90	1.98	0.93	2.06	–	–	–	–	0.74	1.62	0.54	1.19	0.29	0.63	0.57	1.27
77	46	–	–	–	–	–	–	–	–	–	–	–	–	–	–	–	–	–	–	–	–	–	–
	51	–	–	–	–	–	–	–	–	–	–	–	–	–	–	–	–	–	–	–	–	–	–
78	14	–	32	0.57	1.25	–	–	0.63	1.39	0.39	0.87	–	–	–	–	0.52	1.14	0.37	0.82	0.15	0.34	0.43	0.95
	16	–	36	0.65	1.42	–	–	0.72	1.58	0.45	0.99	–	–	–	–	0.59	1.29	0.42	0.93	0.17	0.38	0.49	1.08
79	37	15	29	0.78	1.72	–	–	0.58	1.28	0.61	1.34	–	–	–	–	0.48	1.05	0.33	0.73	0.12	0.25	0.40	0.89
	42	17	33	0.88	1.95	–	–	0.66	1.45	0.69	1.52	–	–	–	–	0.54	1.19	0.38	0.83	0.13	0.29	0.46	1.01
80	48	30	35	0.93	2.05	–	–	0.69	1.53	0.76	1.66	–	–	–	–	0.57	1.25	0.40	0.88	0.18	0.39	0.45	1.00
	53	33	39	1.03	2.27	–	–	0.77	1.69	0.84	1.84	–	–	–	–	0.63	1.39	0.44	0.97	0.19	0.43	0.50	1.11
81	54	52	49	1.08	2.39	–	–	0.92	2.03	0.91	2.01	–	–	–	–	0.75	1.66	0.54	1.18	0.30	0.66	0.55	1.21
	60	58	55	1.21	2.66	–	–	1.03	2.26	1.01	2.24	–	–	–	–	0.84	1.85	0.60	1.32	0.34	0.74	0.61	1.35
82	49	29	40	0.96	2.13	–	–	0.76	1.68	0.79	1.74	–	–	–	–	0.63	1.38	0.43	0.95	0.20	0.44	0.48	1.06
	53	32	43	1.05	2.32	–	–	0.83	1.84	0.86	1.89	–	–	–	–	0.68	1.51	0.47	1.03	0.22	0.48	0.52	1.15
83	50	26	39	1.02	2.25	–	–	0.75	1.64	0.79	1.75	–	–	–	–	0.61	1.35	0.40	0.89	0.18	0.39	0.46	1.02
	55	28	42	1.11	2.45	–	–	0.81	1.79	0.86	1.90	–	–	–	–	0.67	1.47	0.44	0.97	0.19	0.42	0.50	1.11
84	44	36	43	0.93	2.06	–	–	0.82	1.80	0.76	1.68	–	–	–	–	0.67	1.47	0.46	1.02	0.24	0.52	0.50	1.10
	49	41	48	1.05	2.31	–	–	0.91	2.01	0.85	1.88	–	–	–	–	0.75	1.65	0.52	1.15	0.27	0.59	0.56	1.23
85	44	30	35	0.91	2.00	–	–	0.68	1.51	0.73	1.61	–	–	–	–	0.56	1.24	0.42	0.93	0.19	0.43	0.47	1.05
	47	33	38	0.99	2.18	–	–	0.74	1.64	0.79	1.75	–	–	–	–	0.61	1.35	0.46	1.01	0.21	0.46	0.52	1.14
86	41	28	34	0.88	1.93	–	–	0.67	1.48	0.70	1.54	–	–	–	–	0.55	1.21	0.41	0.91	0.19	0.41	0.47	1.03
	45	31	38	0.97	2.13	–	–	0.74	1.63	0.77	1.70	–	–	–	–	0.61	1.34	0.46	1.00	0.21	0.45	0.51	1.13
87	56	–	–	1.13	2.48	–	–	–	–	0.92	2.02	–	–	–	–	–	–	–	–	–	–	–	–
	61	–	–	1.23	2.70	–	–	–	–	1.00	2.20	–	–	–	–	–	–	–	–	–	–	–	–
88	43	64	–	0.87	1.91	–	–	–	–	0.69	1.53	–	–	–	–	–	–	–	–	–	–	–	–
	48	71	–	0.96	2.13	–	–	–	–	0.77	1.70	–	–	–	–	–	–	–	–	–	–	–	–

(Continued)

TABLE V-1C DRY FORAGES, MINERAL AND VITAMIN COMPOSITION OF FEEDS, DATA EXPRESSED **AS-FED** AND **MOISTURE-FREE**—*(Continued)*

Entry Number	Feed Name Description	Moisture Basis: A-F (as-fed) or M-F (moisture-free)	Dry Matter	Macro Minerals							Micro Minerals						
				Calcium (Ca)	Phosphorus (P)	Sodium (Na)	Chlorine (Cl)	Magnesium (Mg)	Potassium (K)	Sulfur (S)	Cobalt (Co)	Copper (Cu)	Iodine (I)	Iron (Fe)	Manganese (Mn)	Selenium (Se)	Zinc (Zn)
			%	%	%	%	%	%	%	%	ppm or mg/kg	ppm or mg/kg	ppm or mg/kg	%	ppm or mg/kg	ppm or mg/kg	ppm or mg/kg
	BLUEGRASS, KENTUCKY *Poa pratensis*																
66	HAY, SUN-CURED	A-F	89	0.40	0.27	0.10	0.55	0.19	1.66	0.12	–	8.8	–	0.025	76.2	–	–
		M-F	100	0.45	0.30	0.11	0.62	0.21	1.87	0.13	–	9.9	–	0.028	85.6	–	–
67	HAY, EARLY BLOOM, SUN-CURED	A-F	90	–	–	–	–	–	–	–	–	–	–	–	–	–	–
		M-F	100	–	–	–	–	–	–	–	–	–	–	–	–	–	–
68	HAY, MATURE, SUN-CURED	A-F	89	0.24	0.20	0.13	0.39	0.10	1.51	0.16	–	–	–	0.016	57.0	–	–
		M-F	100	0.26	0.23	0.15	0.44	0.12	1.70	0.18	–	–	–	0.018	63.9	–	–
	BLUEJOINT REEDGRASS *Calamagrostis canadensis*																
69	HAY, SUN-CURED	A-F	93	0.39	0.14	–	–	–	0.10	–	–	–	–	–	–	–	–
		M-F	100	0.42	0.15	–	–	–	0.11	–	–	–	–	–	–	–	–
	BLUESTEM *Andropogon* spp																
70	HAY, SUN-CURED	A-F	90	–	–	0.01	0.04	–	–	–	–	–	–	–	–	–	–
		M-F	100	–	–	0.01	0.04	–	–	–	–	–	–	–	–	–	–
	BROMEGRASS *Bromus* spp																
71	HAY, SUN-CURED	A-F	91	0.32	0.15	0.03	–	0.09	1.49	0.18	–	–	–	0.019	–	–	–
		M-F	100	0.36	0.16	0.03	–	0.10	1.64	0.20	–	–	–	0.020	–	–	–
72	HAY, PREBLOOM, SUN-CURED	A-F	88	0.28	0.33	0.02	–	0.08	2.04	0.18	–	–	–	–	–	–	–
		M-F	100	0.32	0.37	0.02	–	0.09	2.32	0.20	–	–	–	–	–	–	–
73	HAY, MATURE, SUN-CURED	A-F	92	0.40	0.09	0.02	–	0.08	1.75	–	–	–	–	–	–	–	–
		M-F	100	0.43	0.09	0.02	–	0.09	1.90	–	–	–	–	–	–	–	–
	BROMEGRASS, RESCUE *Bromus catharticus*																
74	HAY, SUN-CURED	A-F	90	–	–	–	–	–	–	–	–	–	–	–	–	–	–
		M-F	100	–	–	–	–	–	–	–	–	–	–	–	–	–	–
	BROMEGRASS, SMOOTH *Bromus inermis*																
75	HAY, SUN-CURED	A-F	90	0.34	0.24	0.57	0.48	0.21	1.88	0.23	–	10.1	–	0.015	41.8	–	–
		M-F	100	0.38	0.26	0.63	0.54	0.23	2.10	0.26	–	11.3	–	0.016	46.6	–	–
	BROOMCORN (HOG MILLET; MILLET, PROSO) *Panicum miliaceum*																
76	HAY, SUN-CURED	A-F	90	–	–	–	–	–	–	–	–	–	–	–	–	–	–
		M-F	100	–	–	–	–	–	–	–	–	–	–	–	–	–	–
77	STOVER WITHOUT HEADS, SUN-CURED	A-F	90	–	–	–	–	–	–	–	–	–	–	–	–	–	–
		M-F	100	–	–	–	–	–	–	–	–	–	–	–	–	–	–
	BUCKWHEAT, COMMON *Fagopyrum sagittatum*																
78	HULLS	A-F	88	0.26	0.02	–	–	–	0.27	–	–	–	–	–	–	–	–
		M-F	100	0.29	0.02	–	–	–	0.31	–	–	–	–	–	–	–	–
79	STRAW	A-F	88	1.25	0.11	–	–	–	2.42	–	–	–	–	–	–	–	–
		M-F	100	1.41	0.12	–	–	–	2.74	–	–	–	–	–	–	–	–
	BUFFALOGRASS *Buchloe dactyloides*																
80	HAY, SUN-CURED	A-F	90	0.56	0.12	–	–	–	1.38	–	–	–	–	–	–	–	–
		M-F	100	0.62	0.13	–	–	–	1.53	–	–	–	–	–	–	–	–
	BUR-CLOVER, TOOTHED *Medicago hispida*																
81	HAY, SUN-CURED	A-F	90	–	–	–	–	–	–	–	–	–	–	–	–	–	–
		M-F	100	–	–	–	–	–	–	–	–	–	–	–	–	–	–
	CACAO (COCOA) *Theubroma cacao*																
82	PODS, SUN-CURED	A-F	92	–	–	–	–	–	–	–	–	–	–	–	–	–	–
		M-F	100	–	–	–	–	–	–	–	–	–	–	–	–	–	–
	CANADA BLUEGRASS–SEE BLUEGRASS, CANADA																
	CANARYGRASS *Phalaris canariensis*																
83	HAY, SUN-CURED	A-F	92	0.33	0.24	–	–	0.24	2.16	–	–	11.0	–	0.014	85.0	–	–
		M-F	100	0.36	0.26	–	–	0.26	2.35	–	–	11.9	–	0.015	92.4	–	–
	CANARYGRASS, REED *Phalaris arundinacea*																
84	HAY, SUN-CURED	A-F	89	0.32	0.21	0.01	–	0.19	2.60	–	–	10.6	–	0.014	82.5	–	–
		M-F	100	0.36	0.24	0.02	–	0.22	2.91	–	–	11.9	–	0.015	92.4	–	–
	CARPETGRASS *Axonopus* spp																
85	HAY, SUN-CURED	A-F	92	0.39	0.12	–	–	–	–	–	–	–	–	–	–	–	–
		M-F	100	0.43	0.14	–	–	–	–	–	–	–	–	–	–	–	–
	CATTAIL, NARROWLEAF *Typha angustifolia*																
86	HAY, SUN-CURED	A-F	91	–	–	–	–	–	–	–	–	–	–	–	–	–	–
		M-F	100	–	–	–	–	–	–	–	–	–	–	–	–	–	–
	CEREALS																
87	IMMATURE, DEHY	A-F	92	0.65	0.46	–	–	–	–	–	–	–	–	–	–	–	–
		M-F	100	0.71	0.50	–	–	–	–	–	–	–	–	–	–	–	–
	CHAMISE (GREASEWOOD) *Adenostoma fasciculatum*																
88	FAN AIR DRIED, PELLETED	A-F	90	–	–	–	–	–	–	–	–	–	–	–	–	–	–
		M-F	100	–	–	–	–	–	–	–	–	–	–	–	–	–	–
	CHINESE LESPEDEZA–SEE LESPEDEZA, SERICEA																

Entry Number	Fat-Soluble Vitamins					Water-Soluble Vitamins								
	A (1 mg Carotene = 1667 IU Vit A)	Carotene (Provitamin A)	D	E	K	B–12	Biotin	Choline	Folacin (Folic Acid)	Niacin	Pantothenic Acid (B–3)	(Pyri-doxine) B–6	Ribo-flavin (B–2)	Thiamin (B–1)
	IU/g	ppm or mg/kg	IU/kg	ppm or mg/kg	ppm or mg/kg	ppm or mg/kg	ppm or mg/kg	ppm or mg/kg	ppm or mg/kg	ppm or mg/kg	ppm or mg/kg	ppm or mg/kg	ppm or mg/kg	ppm or mg/kg
66	–	–	–	–	–	–	–	–	–	–	–	–	**9.9**	–
	–	–	–	–	–	–	–	–	–	–	–	–	11.1	–
67	–	–	–	–	–	–	–	–	–	–	–	–	–	–
	–	–	–	–	–	–	–	–	–	–	–	–	–	–
68	–	–	–	–	–	–	–	–	–	–	–	–	–	–
	–	–	–	–	–	–	–	–	–	–	–	–	–	–
69	**9.1**	**5.5**	–	–	–	–	–	–	–	–	–	–	–	–
	9.8	5.9	–	–	–	–	–	–	–	–	–	–	–	–
70	**62.4**	**37.4**	–	–	–	–	–	–	–	–	–	–	–	–
	69.5	41.7	–	–	–	–	–	–	–	–	–	–	–	–
71	**50.3**	**30.2**	–	–	–	–	–	–	–	–	–	–	–	–
	55.4	33.2	–	–	–	–	–	–	–	–	–	–	–	–
72	**95.1**	**57.1**	**1**	–	–	–	–	–	–	–	–	–	–	–
	108.4	65.0	1	–	–	–	–	–	–	–	–	–	–	–
73	**40.0**	**24.0**	**1**	–	–	–	–	–	–	–	–	–	–	–
	43.5	26.1	1	–	–	–	–	–	–	–	–	–	–	–
74	–	–	–	–	–	–	–	–	–	–	–	–	–	–
	–	–	–	–	–	–	–	–	–	–	–	–	–	–
75	–	–	–	–	–	–	–	–	–	–	–	–	–	–
	–	–	–	–	–	–	–	–	–	–	–	–	–	–
76	–	–	–	–	–	–	–	–	–	–	–	–	–	–
	–	–	–	–	–	–	–	–	–	–	–	–	–	–
77	–	–	–	–	–	–	–	–	–	–	–	–	–	–
	–	–	–	–	–	–	–	–	–	–	–	–	–	–
78	–	–	–	–	–	–	–	–	–	–	–	–	–	–
	–	–	–	–	–	–	–	–	–	–	–	–	–	–
79	–	–	–	–	–	–	–	–	–	–	–	–	–	–
	–	–	–	–	–	–	–	–	–	–	–	–	–	–
80	–	–	–	–	–	–	–	–	–	–	–	–	–	–
	–	–	–	–	–	–	–	–	–	–	–	–	–	–
81	–	–	–	–	–	–	–	–	–	–	–	–	–	–
	–	–	–	–	–	–	–	–	–	–	–	–	–	–
82	–	–	–	–	–	–	–	–	–	–	–	–	–	–
	–	–	–	–	–	–	–	–	–	–	–	–	–	–
83	–	–	–	–	–	–	–	–	–	–	–	–	–	–
	–	–	–	–	–	–	–	–	–	–	–	–	–	–
84	**28.2**	**16.9**	–	–	–	–	–	–	–	–	–	–	**8.5**	**3.6**
	31.6	18.9	–	–	–	–	–	–	–	–	–	–	9.5	4.0
85	–	–	–	–	–	–	–	–	–	–	–	–	–	–
	–	–	–	–	–	–	–	–	–	–	–	–	–	–
86	–	–	–	–	–	–	–	–	–	–	–	–	–	–
	–	–	–	–	–	–	–	–	–	–	–	–	–	–
87	–	–	–	–	–	–	–	–	–	–	–	–	–	–
	–	–	–	–	–	–	–	–	–	–	–	–	–	–
88	–	–	–	–	–	–	–	–	–	–	–	–	–	–
	–	–	–	–	–	–	–	–	–	–	–	–	–	–

(Continued)

TABLE V-1C DRY FORAGES, COMPOSITION OF FEEDS, DATA EXPRESSED AS-FED AND MOISTURE-FREE—*(Continued)*

Entry Number	Feed Name Description	International Feed Number	Moisture Basis: A-F (as-fed) or M-F (moisture-free)	Proximate Analysis: Dry Matter	Ash	Crude Fiber	Neutral Det. Fib. (NDF)	Acid Det. Fib. (ADF)	Lignin	Ether Extract (Fat)	N-Free Extract	Crude Protein	Digestible Protein: Ruminant	Swine	Horse
				%	%	%	%	%	%	%	%	%	%	%	%
	CLOVER, ALSIKE *Trifolium hybridum*														
89	HAY, SUN-CURED	1-01-313	A-F	88	7.6	26.2	–	–	–	2.4	39.1	12.4	8.3	7.5	8.0
			M-F	100	8.7	29.9	–	–	–	2.8	44.5	14.2	9.5	8.6	9.1
	CLOVER, ALYCE *Alysicarpus vaginalis*														
90	HAY, SUN-CURED	1-00-361	A-F	90	5.7	36.2	–	–	–	1.6	35.3	10.9	6.7	6.1	6.8
			M-F	100	6.4	40.3	–	–	–	1.7	39.3	12.2	7.4	6.8	7.6
	CLOVER, CRIMSON *Trifolium incarnatum*														
91	HAY, SUN-CURED	1-01-328	A-F	88	7.8	28.1	–	–	–	2.0	35.2	14.7	10.2	9.5	9.8
			M-F	100	8.9	31.9	–	–	–	2.3	40.1	16.8	11.6	10.9	11.1
92	STRAW	1-08-379	A-F	88	7.0	38.8	–	–	–	1.5	32.9	7.5	3.9	3.2	4.2
			M-F	100	8.0	44.2	–	–	–	1.7	37.5	8.6	4.5	3.7	4.8
	CLOVER, EGYPTIAN (BERSEEM CLOVER) *Trifolium alexandrinum*														
93	HAY, SUN-CURED	1-01-340	A-F	90	11.6	21.9	–	–	–	2.2	40.3	13.7	9.4	8.5	8.9
			M-F	100	13.0	24.4	–	–	–	2.5	44.9	15.2	10.4	9.5	10.0
	CLOVER, LADINO *Trifolium repens*														
94	HAY, SUN-CURED	1-01-378	A-F	89	8.4	18.5	32.1	28.5	5.9	2.4	39.9	20.0	15.4	14.1	13.8
			M-F	100	9.4	20.8	36.0	32.0	6.6	2.7	44.7	22.4	17.3	15.9	15.5
	CLOVER, LADINO-GRASS *Trifolium repens-grass*														
95	HAY, SUN-CURED	1-01-511	A-F	89	7.3	19.9	–	–	–	2.2	43.2	16.5	11.7	11.1	11.1
			M-F	100	8.2	22.3	–	–	–	2.5	48.5	18.5	13.2	12.4	12.5
	CLOVER, PERSIAN *Trifolium resupinatum*														
96	HAY, SUN-CURED	1-01-389	A-F	90	9.3	27.7	–	–	–	1.6	36.7	14.7	10.1	9.4	9.7
			M-F	100	10.3	30.8	–	–	–	1.8	40.8	16.3	11.3	10.5	10.8
	CLOVER, RED *Trifolium pratense*														
97	HAY, SUN-CURED, ALL ANALYSES	1-01-415	A-F	88	6.7	27.1	49.5	36.2	8.8	2.5	39.2	13.0	7.7	8.0	8.4
			M-F	100	7.5	30.7	56.0	41.0	10.0	2.8	44.3	14.7	8.7	9.0	9.5
98	HAY, IMMATURE, SUN-CURED	1-01-394	A-F	87	8.5	17.8	–	–	–	3.4	38.9	18.7	13.7	13.0	12.8
			M-F	100	9.7	20.4	–	–	–	3.9	44.6	21.4	15.7	14.9	14.7
99	HAY, MATURE, SUN-CURED	1-01-405	A-F	91	6.1	31.6	–	–	–	2.2	39.7	11.8	7.5	6.8	7.4
			M-F	100	6.7	34.6	–	–	–	2.4	43.4	12.9	8.2	7.4	8.1
	CLOVER, RED-GRASS *Trifolium pratense-grass*														
100	HAY, FULL BLOOM, SUN-CURED	1-01-532	A-F	89	5.9	29.9	–	–	–	1.6	37.4	14.0	9.6	8.9	9.2
			M-F	100	6.6	33.7	–	–	–	1.8	42.2	15.8	10.8	10.0	10.4
	CLOVER, SUBTERRANEAN *Trifolium subterraneum*														
101	HAY, SUN-CURED	1-20-278	A-F	90	10.0	9.1	–	–	–	3.3	40.1	27.5	21.7	–	20.5
			M-F	100	11.1	10.1	–	–	–	3.7	44.6	30.5	24.1	–	22.8
	CLOVER, SWEET, YELLOW *Melilotus officinalis*														
102	HAY, SUN-CURED, ALL ANALYSES	1-04-754	A-F	89	7.6	28.9	–	–	–	1.9	36.4	13.7	9.5	8.6	9.0
			M-F	100	8.6	32.6	–	–	–	2.2	41.1	15.4	10.7	9.7	10.1
103	LEAVES, SUN-CURED	1-04-757	A-F	91	10.8	9.7	–	–	–	3.2	41.5	25.3	19.5	18.8	17.9
			M-F	100	12.0	10.8	–	–	–	3.5	45.8	28.0	21.5	20.7	19.8
104	STEMS, SUN-CURED	1-04-758	A-F	92	6.7	39.1	–	–	–	1.1	35.3	9.7	5.1	5.0	5.9
			M-F	100	7.3	42.5	–	–	–	1.2	38.5	10.6	5.6	5.4	6.4
	CLOVER, WHITE *Trifolium repens*														
105	HAY, SUN-CURED	1-01-464	A-F	90	9.0	21.9	–	–	–	2.4	40.1	16.9	12.4	11.4	11.5
			M-F	100	9.9	24.2	–	–	–	2.6	44.4	18.8	13.7	12.6	12.7
	CLOVER-GRASS *Trifolium spp-grass*														
106	HAY, SUN-CURED	1-01-473	A-F	88	6.3	29.5	–	–	–	2.4	39.8	10.3	5.9	5.6	6.4
			M-F	100	7.1	33.4	–	–	–	2.7	45.1	11.7	6.7	6.4	7.2
	CLOVER-TIMOTHY *Trifolium spp-Phleum pratense*														
107	HAY, FULL BLOOM, SUN-CURED	1-01-484	A-F	90	5.7	32.2	–	–	–	3.1	39.6	9.5	5.5	4.8	5.7
			M-F	100	6.3	35.8	–	–	–	3.4	44.0	10.5	6.2	5.4	6.3
	COCOA (CACAO) *Theobroma cacao*														
108	PODS, SUN-CURED	1-08-387	A-F	92	7.0	31.7	–	–	–	1.3	44.2	7.5	3.8	3.1	4.2
			M-F	100	7.6	34.6	–	–	–	1.4	48.2	8.2	4.1	3.3	4.5
	COFFEE *Coffea spp*														
109	GROUNDS, WET	1-01-576	A-F	74	1.2	21.5	–	–	–	9.3	31.5	10.2	6.7	6.1	6.6
			M-F	100	1.6	29.2	–	–	–	12.6	42.7	13.8	9.1	8.3	8.9
110	HULLS	1-01-577	A-F	90	4.9	32.6	83.7	63.5	18.1	7.4	29.5	15.6	–	–	11.0
			M-F	100	5.4	36.2	93.0	70.6	20.1	8.2	32.8	17.3	–	–	12.2
	CORN *Zea mays*														
111	FODDER WITH EARS, WITH HUSKS, SUN-CURED	1-02-775	A-F	90	4.9	32.6	–	–	–	7.4	29.5	15.6	4.0	3.6	11.0
			M-F	100	5.4	36.2	–	–	–	8.2	32.8	17.3	4.4	4.0	12.2
112	FODDER WITH EARS, WITH HUSKS, DOUGH STAGE, SUN-CURED	1-02-771	A-F	74	5.2	20.9	–	–	–	2.1	37.4	8.0	4.3	4.2	4.9
			M-F	100	7.0	28.4	–	–	–	2.9	50.8	10.9	5.9	5.7	6.6
113	FODDER WITH EARS, WITH HUSKS, MATURE, SUN-CURED	1-02-772	A-F	82	4.4	18.6	–	–	–	1.9	50.6	6.6	3.3	2.6	3.6
			M-F	100	5.4	22.6	–	–	–	2.3	61.7	8.0	4.0	3.2	4.4

DRY FORAGES

Entry Number	TDN			Digestible Energy						Metabolizable Energy								Net Energy					
	Ruminant	Swine	Horse	Ruminant		Swine		Horse		Ruminant		Swine		Poultry ME_n		Horse		Ruminant NE_m		Ruminant NE_g		Lactating Cows NE_{lc}	
	%	%	%	Mcal		kcal		Mcal		Mcal		kcal		kcal		Mcal		Mcal		Mcal		Mcal	
				lb	kg	lb	kg	lb	kg	lb	kg	lb	kg	lb	kg	lb	kg	lb	kg	lb	kg	lb	kg
89	51	42	41	1.01	2.24	–	–	0.78	1.72	0.85	1.86	–	–	–	–	0.64	1.41	0.49	1.08	0.26	0.58	0.51	1.13
	58	48	46	1.16	2.55	–	–	0.89	1.96	0.96	2.12	–	–	–	–	0.73	1.60	0.56	1.23	0.30	0.66	0.59	1.29
90	44	–	38	0.89	1.96	–	–	0.74	1.63	0.71	1.57	–	–	–	–	0.61	1.34	0.30	0.66	0.08	0.18	0.38	0.85
	49	–	43	0.99	2.18	–	–	0.83	1.82	0.80	1.75	–	–	–	–	0.68	1.49	0.33	0.73	0.09	0.20	0.43	0.94
91	50	41	43	1.00	2.21	–	–	0.82	1.81	0.84	1.84	–	–	–	–	0.67	1.49	0.54	1.19	0.31	0.68	0.55	1.21
	57	46	49	1.14	2.52	–	–	0.94	2.06	0.95	2.10	–	–	–	–	0.77	1.69	0.61	1.35	0.35	0.77	0.63	1.38
92	40	–	29	0.83	1.82	–	–	0.58	1.28	0.65	1.44	–	–	–	–	0.48	1.05	0.36	0.80	0.15	0.32	0.42	0.94
	46	–	33	0.94	2.07	–	–	0.66	1.46	0.75	1.64	–	–	–	–	0.54	1.20	0.41	0.91	0.17	0.37	0.48	1.07
93	52	43	43	1.04	2.30	–	–	0.81	1.79	0.87	1.92	–	–	–	–	0.67	1.47	0.52	1.15	0.29	0.63	0.54	1.19
	58	48	48	1.16	2.56	–	–	0.91	2.00	0.97	2.14	–	–	–	–	0.74	1.64	0.58	1.28	0.32	0.71	0.60	1.33
94	58	60	57	1.16	2.55	–	–	1.05	2.31	0.99	2.18	–	–	–	–	0.86	1.89	0.58	1.27	0.34	0.75	0.58	1.28
	65	67	64	1.30	2.87	–	–	1.17	2.59	1.11	2.44	–	–	–	–	0.96	2.12	0.65	1.43	0.38	0.84	0.65	1.44
95	58	55	54	1.13	2.49	–	–	1.00	2.20	0.96	2.11	–	–	–	–	0.82	1.80	0.55	1.22	0.32	0.70	0.56	1.24
	65	62	60	1.27	2.80	–	–	1.12	2.47	1.08	2.37	–	–	–	–	0.92	2.02	0.62	1.37	0.36	0.79	0.63	1.40
96	50	39	45	1.00	2.19	–	–	0.85	1.88	0.82	1.81	–	–	–	–	0.70	1.54	0.46	1.02	0.23	0.52	0.50	1.10
	56	43	50	1.11	2.44	–	–	0.95	2.09	0.91	2.01	–	–	–	–	0.78	1.72	0.51	1.13	0.26	0.57	0.55	1.22
97	52	44	42	1.21	2.67	–	–	0.81	1.78	0.95	2.10	–	–	–	–	0.66	1.46	0.50	1.11	0.28	0.61	0.53	1.16
	59	50	48	1.37	3.02	–	–	0.91	2.01	1.08	2.38	–	–	–	–	0.75	1.65	0.57	1.26	0.31	0.69	0.60	1.31
98	56	62	50	1.12	2.46	–	–	0.93	2.04	0.95	2.09	–	–	–	–	0.76	1.67	0.58	1.28	0.35	0.76	0.58	1.28
	64	71	57	1.28	2.82	–	–	1.06	2.34	1.09	2.40	–	–	–	–	0.87	1.92	0.66	1.46	0.40	0.88	0.66	1.46
99	52	39	41	1.03	2.27	–	–	0.79	1.75	0.85	1.88	–	–	–	–	0.65	1.43	0.49	1.07	0.25	0.56	0.52	1.14
	56	42	45	1.13	2.49	–	–	0.87	1.92	0.94	2.06	–	–	–	–	0.71	1.57	0.53	1.18	0.28	0.61	0.57	1.25
100	51	39	47	1.02	2.25	–	–	0.88	1.93	0.85	1.87	–	–	–	–	0.72	1.59	0.48	1.06	0.25	0.56	0.51	1.12
	58	44	53	1.15	2.54	–	–	0.99	2.18	0.96	2.11	–	–	–	–	0.81	1.79	0.54	1.19	0.29	0.63	0.57	1.26
101	–	–	–	–	–	–	–	–	–	–	–	–	–	–	–	–	–	–	–	–	–	–	–
	–	–	–	–	–	–	–	–	–	–	–	–	–	–	–	–	–	–	–	–	–	–	–
102	49	39	43	0.98	2.16	–	–	0.81	1.79	0.81	1.78	–	–	–	–	0.67	1.47	0.49	1.07	0.26	0.57	0.51	1.13
	55	44	48	1.11	2.44	–	–	0.92	2.03	0.91	2.01	–	–	–	–	0.75	1.66	0.55	1.21	0.29	0.64	0.58	1.28
103	56	–	–	1.12	2.48	–	–	–	–	0.95	2.10	–	–	–	–	–	–	–	–	–	–	–	–
	62	–	–	1.24	2.74	–	–	–	–	1.05	2.31	–	–	–	–	–	–	–	–	–	–	–	–
104	45	–	37	0.91	2.00	–	–	0.72	1.59	0.73	1.60	–	–	–	–	0.59	1.30	0.40	0.88	0.17	0.38	0.46	1.01
	49	–	40	0.99	2.17	–	–	0.78	1.72	0.79	1.75	–	–	–	–	0.64	1.41	0.43	0.95	0.19	0.41	0.50	1.10
105	57	52	50	1.11	2.45	–	–	0.94	2.07	0.94	2.07	–	–	–	–	0.77	1.70	0.53	1.18	0.30	0.66	0.55	1.22
	63	58	56	1.23	2.71	–	–	1.04	2.30	1.04	2.29	–	–	–	–	0.86	1.88	0.59	1.31	0.33	0.73	0.61	1.35
106	51	38	38	1.01	2.22	–	–	0.74	1.62	0.84	1.84	–	–	–	–	0.60	1.33	0.47	1.04	0.25	0.55	0.50	1.11
	58	43	43	1.14	2.51	–	–	0.83	1.84	0.95	2.09	–	–	–	–	0.68	1.51	0.54	1.18	0.28	0.62	0.57	1.26
107	51	39	34	1.01	2.23	–	–	0.67	1.47	0.84	1.85	–	–	–	–	0.55	1.21	0.49	1.08	0.26	0.57	0.52	1.14
	56	43	38	1.12	2.48	–	–	0.74	1.63	0.93	2.05	–	–	–	–	0.61	1.34	0.54	1.20	0.29	0.63	0.58	1.27
108	49	29	40	0.96	2.13	–	–	0.76	1.68	0.79	1.74	–	–	–	–	0.63	1.38	0.43	0.95	0.20	0.44	0.48	1.06
	53	32	43	1.05	2.32	–	–	0.83	1.84	0.86	1.89	–	–	–	–	0.68	1.51	0.47	1.03	0.22	0.48	0.52	1.15
109	–	–	–	–	–	–	–	–	–	–	–	–	–	–	–	–	–	–	–	–	–	–	–
	–	–	–	–	–	–	–	–	–	–	–	–	–	–	–	–	–	–	–	–	–	–	–
110	45	–	–	–	–	–	–	–	–	–	–	–	–	–	–	–	–	–	–	–	–	–	–
	50	–	–	–	–	–	–	–	–	–	–	–	–	–	–	–	–	–	–	–	–	–	–
111	45	45	44	0.91	2.00	–	–	0.84	1.85	0.74	1.64	–	–	–	–	0.69	1.52	0.44	0.97	0.12	0.26	0.46	1.01
	50	50	49	1.01	2.22	–	–	0.93	2.05	0.83	1.82	–	–	–	–	0.76	1.69	0.49	1.08	0.13	0.29	0.51	1.12
112	47	37	34	0.90	1.99	–	–	0.65	1.43	0.76	1.68	–	–	–	–	0.53	1.17	0.42	0.92	0.23	0.50	0.44	0.96
	64	50	46	1.23	2.70	–	–	0.88	1.94	1.03	2.28	–	–	–	–	0.72	1.59	0.57	1.25	0.31	0.68	0.59	1.31
113	56	44	–	1.04	2.30	–	–	–	–	0.89	1.95	–	–	–	–	–	–	0.59	1.30	0.37	0.81	0.58	1.28
	69	54	–	1.27	2.80	–	–	–	–	1.08	2.38	–	–	–	–	–	–	0.72	1.59	0.45	0.99	0.71	1.56

(Continued)

TABLE V-1C DRY FORAGES, MINERAL AND VITAMIN COMPOSITION OF FEEDS, DATA EXPRESSED **AS-FED** AND **MOISTURE-FREE**—*(Continued)*

Entry Number	Feed Name Description	Moisture Basis: A-F (as-fed) or M-F (moisture-free)	Dry Matter	Macro Minerals Calcium (Ca)	Phosphorus (P)	Sodium (Na)	Chlorine (Cl)	Magnesium (Mg)	Potassium (K)	Sulfur (S)	Micro Minerals Cobalt (Co)	Copper (Cu)	Iodine (I)	Iron (Fe)	Manganese (Mn)	Selenium (Se)	Zinc (Zn)
			%	%	%	%	%	%	%	%	ppm or mg/kg	ppm or mg/kg	ppm or mg/kg	%	ppm or mg/kg	ppm or mg/kg	ppm or mg/kg
	CLOVER, ALSIKE *Trifolium hybridum*																
89	HAY, SUN-CURED	**A-F**	**88**	**1.14**	**0.22**	**0.40**	**0.68**	**0.40**	**1.95**	**0.17**	–	**5.3**	–	**0.023**	**60.5**	–	–
		M-F	100	1.30	0.25	0.46	0.78	0.45	2.22	0.19	–	6.0	–	0.026	69.0	–	–
	CLOVER, ALYCE *Alysicarpus vaginalis*																
90	HAY, SUN-CURED	**A-F**	**90**	–	–	–	–	–	–	–	–	–	–	–	–	–	–
		M-F	100	–	–	–	–	–	–	–	–	–	–	–	–	–	–
	CLOVER, CRIMSON *Trifolium incarnatum*																
91	HAY, SUN-CURED	**A-F**	**88**	**1.23**	**0.19**	**0.34**	**0.55**	**0.25**	**2.10**	**0.25**	–	–	**0.059**	**0.062**	**183.3**	–	–
		M-F	100	1.40	0.22	0.39	0.63	0.28	2.40	0.28	–	–	0.067	0.070	208.7	–	–
92	STRAW	**A-F**	**88**	–	–	–	–	–	–	–	–	–	–	–	–	–	–
		M-F	100	–	–	–	–	–	–	–	–	–	–	–	–	–	–
	CLOVER, EGYPTIAN (BERSEEM CLOVER) *Trifolium alexandrinum*																
93	HAY, SUN-CURED	**A-F**	**90**	**2.28**	**0.26**	–	–	**0.36**	**2.03**	–	–	–	–	–	–	–	–
		M-F	100	2.54	0.29	–	–	0.40	2.26	–	–	–	–	–	–	–	–
	CLOVER, LADINO *Trifolium repens*																
94	HAY, SUN-CURED	**A-F**	**89**	**1.30**	**0.30**	**0.12**	**0.27**	**0.42**	**2.17**	**0.19**	**0.144**	**8.4**	**0.268**	**0.042**	**109.7**	–	**15.2**
		M-F	100	1.45	0.34	0.13	0.30	0.47	2.44	0.21	0.161	9.4	0.301	0.047	123.1	–	17.0
	CLOVER, LADINO-GRASS *Trifolium repens-grass*																
95	HAY, SUN-CURED	**A-F**	**89**	**0.95**	**0.22**	–	–	**0.30**	**1.78**	–	–	**14.3**	–	**0.016**	**39.6**	–	–
		M-F	100	1.07	0.24	–	–	0.34	2.00	–	–	16.1	–	0.017	44.5	–	–
	CLOVER, PERSIAN *Trifolium resupinatum*																
96	HAY, SUN-CURED	**A-F**	**90**	**1.49**	**0.21**	–	–	–	–	–	–	–	–	–	–	–	–
		M-F	100	1.66	0.23	–	–	–	–	–	–	–	–	–	–	–	–
	CLOVER, RED *Trifolium pratense*																
97	HAY, SUN-CURED, ALL ANALYSES	**A-F**	**88**	**1.22**	**0.22**	**0.16**	**0.28**	**0.34**	**1.60**	**0.15**	**0.138**	**18.8**	**0.217**	**0.022**	**95.2**	–	**32.5**
		M-F	100	1.38	0.25	0.18	0.32	0.38	1.81	0.16	0.156	21.2	0.245	0.024	107.7	–	36.7
98	HAY, IMMATURE, SUN-CURED	**A-F**	**87**	**1.35**	**0.32**	–	–	**0.34**	**2.82**	–	**0.201**	**18.4**	–	**0.063**	**75.6**	–	**45.4**
		M-F	100	1.55	0.37	–	–	0.39	3.24	–	0.230	21.1	–	0.073	86.7	–	52.0
99	HAY, MATURE, SUN-CURED	**A-F**	**91**	**1.07**	**0.19**	–	–	**0.32**	**1.55**	–	**0.110**	**8.5**	–	**0.021**	**39.3**	–	**32.0**
		M-F	100	1.18	0.21	–	–	0.35	1.70	–	0.120	9.4	–	0.023	43.0	–	35.0
	CLOVER, RED-GRASS *Trifolium pratense-grass*																
100	HAY, FULL BLOOM, SUN-CURED	**A-F**	**89**	–	–	–	–	–	–	–	**0.137**	–	–	–	–	–	–
		M-F	100	–	–	–	–	–	–	–	0.155	–	–	–	–	–	–
	CLOVER, SUBTERRANEAN *Trifolium subterraneum*																
101	HAY, SUN-CURED	**A-F**	**90**	–	–	–	–	–	–	–	–	–	–	–	–	–	–
		M-F	100	–	–	–	–	–	–	–	–	–	–	–	–	–	–
	CLOVER, SWEET, YELLOW *Melilotus officinalis*																
102	HAY, SUN-CURED, ALL ANALYSES	**A-F**	**89**	**1.44**	**0.24**	**0.08**	**0.33**	**0.39**	**1.35**	**0.42**	–	**8.8**	–	**0.015**	**95.4**	–	–
		M-F	100	1.63	0.27	0.09	0.37	0.44	1.53	0.47	–	10.0	–	0.017	107.7	–	–
103	LEAVES, SUN-CURED	**A-F**	**91**	**2.83**	**0.29**	–	–	–	–	–	–	–	–	–	–	–	–
		M-F	100	3.12	0.32	–	–	–	–	–	–	–	–	–	–	–	–
104	STEMS, SUN-CURED	**A-F**	**92**	**0.57**	**0.30**	–	–	–	–	–	–	–	–	–	–	–	–
		M-F	100	0.62	0.33	–	–	–	–	–	–	–	–	–	–	–	–
	CLOVER, WHITE *Trifolium repens*																
105	HAY, SUN-CURED	**A-F**	**90**	**1.71**	**0.29**	–	–	–	–	–	–	–	–	–	–	–	–
		M-F	100	1.90	0.32	–	–	–	–	–	–	–	–	–	–	–	–
	CLOVER-GRASS *Trifolium spp-grass*																
106	HAY, SUN-CURED	**A-F**	**88**	**0.78**	**0.22**	**0.17**	**0.63**	**0.23**	**1.58**	**0.14**	–	**7.0**	–	**0.019**	**82.2**	–	**17.7**
		M-F	100	0.88	0.25	0.19	0.71	0.26	1.79	0.16	–	8.0	–	0.021	93.2	–	20.0
	CLOVER-TIMOTHY *Trifolium spp-Phleum pratense*																
107	HAY, FULL BLOOM, SUN-CURED	**A-F**	**90**	–	–	–	–	–	–	–	–	–	–	–	–	–	–
		M-F	100	–	–	–	–	–	–	–	–	–	–	–	–	–	–
	COCOA (CACAO) *Theobroma cacao*																
108	PODS, SUN-CURED	**A-F**	**92**	–	–	–	–	–	–	–	–	–	–	–	–	–	–
		M-F	100	–	–	–	–	–	–	–	–	–	–	–	–	–	–
	COFFEE *Coffea spp*																
109	GROUNDS, WET	**A-F**	**74**	**0.09**	**0.06**	–	–	–	–	–	–	–	–	–	–	–	–
		M-F	100	0.12	0.08	–	–	–	–	–	–	–	–	–	–	–	–
110	HULLS	**A-F**	**90**	–	–	–	–	–	–	–	–	–	–	–	–	–	–
		M-F	100	–	–	–	–	–	–	–	–	–	–	–	–	–	–
	CORN *Zea mays*																
111	FODDER WITH EARS, WITH HUSKS, SUN-CURED	**A-F**	**90**	**0.45**	**0.23**	**0.03**	**0.17**	**0.26**	**0.84**	**0.13**	–	**6.9**	–	**0.009**	**61.4**	–	–
		M-F	100	0.50	0.25	0.03	0.19	0.29	0.93	0.14	–	7.7	–	0.010	68.2	–	–
112	FODDER WITH EARS, WITH HUSKS, DOUGH STAGE, SUN-CURED	**A-F**	**74**	–	–	–	–	–	–	–	–	–	–	–	–	–	–
		M-F	100	–	–	–	–	–	–	–	–	–	–	–	–	–	–
113	FODDER WITH EARS, WITH HUSKS, MATURE, SUN-CURED	**A-F**	**82**	–	–	–	–	–	–	–	–	–	–	–	–	–	–
		M-F	100	–	–	–	–	–	–	–	–	–	–	–	–	–	–

Entry Number	Fat-Soluble Vitamins: A (1 mg Carotene = 1667 IU Vit A)	Carotene (Provitamin A)	D	E	K	Water-Soluble Vitamins: B-12	Biotin	Choline	Folacin (Folic Acid)	Niacin	Pantothenic Acid (B-3)	(Pyridoxine) B-6	Riboflavin (B-2)	Thiamin (B-1)
	IU/g	ppm or mg/kg	IU/kg	ppm or mg/kg	ppm or mg/kg	ppm or mg/kg	ppm or mg/kg	ppm or mg/kg	ppm or mg/kg	ppm or mg/kg	ppm or mg/kg	ppm or mg/kg	ppm or mg/kg	ppm or mg/kg
89	272.1	163.2	–	–	–	–	–	–	–	–	–	–	15.1	4.2
	310.1	186.0	–	–	–	–	–	–	–	–	–	–	17.2	4.8
90	–	–	–	–	–	–	–	–	–	–	–	–	–	–
	–	–	–	–	–	–	–	–	–	–	–	–	–	–
91	32.9	19.8	–	–	–	–	–	–	–	–	–	–	–	–
	37.5	22.5	–	–	–	–	–	–	–	–	–	–	–	–
92	–	–	–	–	–	–	–	–	–	–	–	–	–	–
	–	–	–	–	–	–	–	–	–	–	–	–	–	–
93	–	–	–	–	–	–	–	–	–	–	–	–	–	–
	–	–	–	–	–	–	–	–	–	–	–	–	–	–
94	239.5	143.7	–	–	–	–	–	–	–	10	1.0	–	15.2	3.7
	268.7	161.2	–	–	–	–	–	–	–	11	1.1	–	17.0	4.2
95	94.6	56.7	–	–	–	–	–	–	–	–	–	–	8.2	13.1
	106.2	63.7	–	–	–	–	–	–	–	–	–	–	9.3	14.8
96	–	–	–	–	–	–	–	–	–	–	–	–	–	–
	–	–	–	–	–	–	–	–	–	–	–	–	–	–
97	40.5	24.3	–	–	–	–	0.09	–	–	38	9.9	–	15.7	2.0
	45.9	27.5	–	–	–	–	0.11	–	–	43	11.2	–	17.8	2.2
98	–	–	–	–	–	–	–	–	–	–	–	–	–	–
	–	–	–	–	–	–	–	–	–	–	–	–	–	–
99	–	–	–	–	–	–	–	–	–	–	–	–	–	–
	–	–	–	–	–	–	–	–	–	–	–	–	–	–
100	16.6	10.0	–	–	–	–	–	–	–	–	–	–	–	–
	18.7	11.2	–	–	–	–	–	–	–	–	–	–	–	–
101	–	–	–	–	–	–	–	–	–	–	–	–	–	–
	–	–	–	–	–	–	–	–	–	–	–	–	–	–
102	145.8	87.4	2	–	–	–	–	–	–	–	–	–	–	–
	164.6	98.8	2	–	–	–	–	–	–	–	–	–	–	–
103	–	–	–	–	–	–	–	–	–	–	–	–	–	–
	–	–	–	–	–	–	–	–	–	–	–	–	–	–
104	–	–	–	–	–	–	–	–	–	–	–	–	–	–
	–	–	–	–	–	–	–	–	–	–	–	–	–	–
105	92.1	55.3	–	115.5	–	–	–	–	–	–	–	–	–	–
	102.2	61.3	–	128.1	–	–	–	–	–	–	–	–	–	–
106	24.0	14.4	–	–	–	–	–	–	–	–	–	–	–	–
	27.2	16.3	–	–	–	–	–	–	–	–	–	–	–	–
107	–	–	–	–	–	–	–	–	–	–	–	–	–	–
	–	–	–	–	–	–	–	–	–	–	–	–	–	–
108	–	–	–	–	–	–	–	–	–	–	–	–	–	–
	–	–	–	–	–	–	–	–	–	–	–	–	–	–
109	–	–	–	–	–	–	–	–	–	–	–	–	–	–
	–	–	–	–	–	–	–	–	–	–	–	–	–	–
110	–	–	–	–	–	–	–	–	–	–	–	–	–	–
	–	–	–	–	–	–	–	–	–	–	–	–	–	–
111	6.6	4.0	1	–	–	–	–	–	–	–	–	–	–	–
	7.3	4.4	1	–	–	–	–	–	–	–	–	–	–	–
112	–	–	–	–	–	–	–	–	–	–	–	–	–	–
	–	–	–	–	–	–	–	–	–	–	–	–	–	–
113	–	–	–	–	–	–	–	–	–	–	–	–	–	–
	–	–	–	–	–	–	–	–	–	–	–	–	–	–

(Continued)

TABLE V-1C DRY FORAGES, COMPOSITION OF FEEDS, DATA EXPRESSED **AS-FED** AND **MOISTURE-FREE**—*(Continued)*

Entry Number	Feed Name Description	International Feed Number	Moisture Basis: A-F (as-fed) or M-F (moisture-free)	Proximate Analysis									Digestible Protein		
				Dry Matter	Ash	Crude Fiber	Neutral Det. Fib. (NDF)	Acid Det. Fib. (ADF)	Lignin	Ether Extract (Fat)	N-Free Extract	Crude Protein	Ruminant	Swine	Horse
				%	%	%	%	%	%	%	%	%	%	%	%
	CORN (Continued)														
114	FODDER WITH EARS, WITH HUSKS, WEATHER DAMAGED, SUN-CURED	1-08-390	A-F	85	6.5	25.7	–	–	–	0.8	41.2	10.8	6.9	6.2	6.8
			M-F	100	7.6	30.2	–	–	–	0.9	48.5	12.7	8.1	7.3	8.0
115	STOVER WITHOUT EARS, WITHOUT HUSKS, SUN-CURED[1]	1-02-776	A-F	85	6.1	29.3	57.0	33.2	–	1.1	43.2	5.4	2.3	1.5	2.7
			M-F	100	7.2	34.4	67.0	39.0	–	1.3	50.8	6.4	2.7	1.7	3.1
116	HUSKS, SUN-CURED	1-02-785	A-F	89	3.2	30.0	–	–	–	0.8	51.2	3.3	0.7	–0.5	0.9
			M-F	100	3.6	33.9	–	–	–	0.9	57.9	3.7	0.8	–0.6	1.1
117	LEAVES, SUN-CURED	1-02-788	A-F	87	8.5	24.0	65.3	42.1	6.2	2.6	44.4	7.6	3.0	3.3	4.3
			M-F	100	9.7	27.6	74.9	48.3	7.1	3.0	51.0	8.7	3.4	3.8	4.9
118	COBS, GROUND[1]	1-02-782	A-F	90	1.6	32.2	80.1	31.5	–	0.6	52.7	2.8	–0.4	–1.0	0.4
			M-F	100	1.8	35.8	89.0	35.0	–	0.7	58.7	3.1	–0.4	–1.2	0.5
	CORN, SWEET *Zea mays, saccharata*														
119	FODDER WITH EARS, WITH HUSKS, SUN-CURED	1-08-407	A-F	88	9.0	26.4	–	–	–	1.8	41.3	9.2	5.9	4.7	5.5
			M-F	100	10.3	30.1	–	–	–	2.1	47.1	10.5	6.7	5.4	6.3
	COTTON *Gossypium* spp														
120	BOLLS, SUN-CURED	1-01-596	A-F	91	6.2	34.5	–	–	–	2.8	38.9	9.0	2.3	4.4	5.3
			M-F	100	6.8	37.8	–	–	–	3.0	42.6	9.8	2.6	4.8	5.8
121	GIN BY-PRODUCT (GIN TRASH)	1-08-413	A-F	89	14.4	30.3	–	44.9	–	2.4	32.9	9.1	5.2	4.5	5.4
			M-F	100	16.2	34.1	–	50.4	–	2.7	36.9	10.2	5.9	5.1	6.1
122	HULLS[1]	1-01-599	A-F	90	2.6	43.2	81.0	65.7	21.0	1.5	39.3	3.8	–0.4	–0.2	1.3
			M-F	100	2.9	47.8	90.0	73.0	23.3	1.7	43.5	4.2	–0.5	–0.2	1.4
123	LEAVES, SUN-CURED	1-01-598	A-F	92	16.1	10.3	–	–	–	7.0	43.4	15.4	10.7	10.0	10.3
			M-F	100	17.4	11.2	–	–	–	7.6	47.0	16.7	11.6	10.8	11.1
124	STEMS, SUN-CURED	1-01-601	A-F	92	4.2	43.8	–	–	–	0.8	37.8	5.7	2.2	1.4	2.7
			M-F	100	4.6	47.5	–	–	–	0.9	41.0	6.2	2.4	1.6	3.0
	COWPEA, COMMON *Vigna sinensis*														
125	HAY, SUN-CURED	1-01-645	A-F	90	10.5	24.4	–	–	–	2.6	35.1	17.7	12.2	12.0	12.0
			M-F	100	11.7	27.1	–	–	–	2.9	38.8	19.6	13.4	13.3	13.3
126	STRAW	1-01-649	A-F	91	5.4	43.8	–	–	–	1.2	33.9	7.0	3.3	2.6	3.7
			M-F	100	5.9	48.0	–	–	–	1.4	37.2	7.6	3.6	2.8	4.1
	CRABGRASS *Digitaria* spp														
127	HAY, SUN-CURED	1-01-667	A-F	90	7.0	30.5	–	–	–	2.2	43.5	7.1	2.2	2.7	3.8
			M-F	100	7.8	33.8	–	–	–	2.4	48.2	7.8	2.4	3.0	4.3
	CRESTED WHEATGRASS *Agropyron desertorum*														
128	HAY, SUN-CURED	1-05-418	A-F	92	6.4	30.8	–	33.2	5.1	2.1	42.2	10.3	6.4	5.5	6.3
			M-F	100	7.0	33.6	–	36.2	5.5	2.3	46.0	11.2	6.9	6.0	6.8
	CRIMSON CLOVER *Trifolium incarnatum*														
129	HAY, SUN-CURED	1-01-328	A-F	88	7.8	28.1	–	–	–	2.0	35.2	14.7	10.2	9.5	9.8
			M-F	100	8.9	31.9	–	–	–	2.3	40.1	16.8	11.6	10.9	11.1
	DALLISGRASS *Paspalum dilatatum*														
130	HAY, SUN-CURED	1-01-737	A-F	91	7.9	30.8	–	–	–	1.9	41.1	9.2	5.3	4.6	5.5
			M-F	100	8.7	33.9	–	–	–	2.1	45.2	10.1	5.8	5.0	6.0
	DOLICHOS (HYACINTH BEAN) *Dolichos lablab*														
131	HAY, SUN-CURED	1-01-805	A-F	90	7.1	33.6	–	–	–	1.6	34.2	13.8	9.0	8.6	9.0
			M-F	100	7.9	37.2	–	–	–	1.7	37.9	15.3	9.9	9.6	10.0
	DURRA SORGHUM *Sorghum bicolor, durra*														
132	FODDER, WITH HEADS, SUN-CURED	1-04-359	A-F	90	5.2	24.1	–	–	–	2.8	51.4	6.4	2.9	2.2	3.3
			M-F	100	5.8	26.8	–	–	–	3.1	57.2	7.1	3.2	2.4	3.7
	EMMER *Triticum dicoccum*														
133	HAY, SUN-CURED	1-01-827	A-F	90	9.1	32.8	–	–	–	2.0	36.4	9.7	5.8	5.1	5.9
			M-F	100	10.1	36.4	–	–	–	2.2	40.5	10.8	6.4	5.6	6.5
	FESCUE, MEADOW *Festuca elatior*														
134	HAY, SUN-CURED	1-01-912	A-F	88	7.9	28.0	74.0	43.8	–	2.4	41.0	8.2	4.6	3.9	5.3
			M-F	100	9.0	32.0	84.6	50.0	–	2.7	46.9	9.4	5.3	4.4	6.0
	FESCUE, TALL (ALTA) *Festuca arundinacea*														
135	HAY, SUN-CURED	1-05-684	A-F	89	5.9	32.6	61.7	35.6	–	2.0	41.4	7.2	3.6	2.9	3.9
			M-F	100	6.6	36.6	69.3	40.0	–	2.2	46.5	8.1	4.0	3.2	4.4
	FILAREE (ALFILERIA, REDSTEM) *Erodium cicutarium*														
136	HAY, SUN-CURED	1-08-333	A-F	89	11.8	23.4	–	–	–	2.9	40.2	10.9	6.8	6.1	6.8
			M-F	100	13.2	26.2	–	–	–	3.3	45.1	12.2	7.7	6.9	7.6
	FLAT PEA *Lathyrus sylvestris*														
137	HAY, SUN-CURED	1-02-030	A-F	91	6.5	26.6	–	–	–	3.3	30.4	23.9	18.2	17.5	16.8
			M-F	100	7.2	29.3	–	–	–	3.7	33.5	26.3	20.0	19.3	18.5
	FLAX, COMMON *Linum usitatissimum*														
138	FIBER PROCESS RESIDUE, DEHY	1-02-036	A-F	91	6.0	35.8	–	–	–	2.7	36.7	10.0	5.6	5.3	6.1
			M-F	100	6.6	39.2	–	–	–	3.0	40.2	11.0	6.2	5.8	6.7
139	STRAW	1-02-038	A-F	93	5.5	48.1	–	–	–	2.2	32.4	5.0	2.2	0.8	2.2
			M-F	100	5.9	51.6	–	–	–	2.3	34.8	5.3	2.3	0.8	2.3

DRY FORAGES

Entry Number	TDN Ruminant	TDN Swine	TDN Horse	Digestible Energy Ruminant		Digestible Energy Swine		Digestible Energy Horse		Metabolizable Energy Ruminant		Metabolizable Energy Swine		Poultry MEn		Metabolizable Energy Horse		Net Energy Ruminant NEm		Net Energy Ruminant NEg		Net Energy Lactating Cows NElc	
	%	%	%	Mcal		kcal		Mcal		Mcal		kcal		kcal		Mcal		Mcal		Mcal		Mcal	
				lb	kg	lb	kg	lb	kg	lb	kg	lb	kg	lb	kg	lb	kg	lb	kg	lb	kg	lb	kg
114	46	34	46	0.92	2.04	–	–	0.87	1.91	0.76	1.68	–	–	–	–	0.71	1.56	0.43	0.96	0.22	0.48	0.47	1.03
	54	41	54	1.09	2.40	–	–	1.02	2.24	0.89	1.97	–	–	–	–	0.83	1.84	0.51	1.12	0.26	0.57	0.55	1.22
115	51	26	36	1.01	2.23	–	–	0.70	1.54	0.85	1.87	–	–	–	–	0.57	1.26	0.51	1.12	0.29	0.63	0.52	1.15
	59	31	42	1.19	2.62	–	–	0.82	1.81	1.00	2.19	–	–	–	–	0.67	1.48	0.60	1.32	0.34	0.74	0.61	1.35
116	55	29	41	1.11	2.44	–	–	0.79	1.74	0.94	2.06	–	–	–	–	0.65	1.43	0.69	1.51	0.44	0.97	0.66	1.46
	62	32	47	1.25	2.75	–	–	0.89	1.97	1.06	2.33	–	–	–	–	0.73	1.62	0.78	1.71	0.50	1.09	0.75	1.65
117	53	39	35	1.07	2.35	–	–	0.68	1.49	0.90	1.98	–	–	–	–	0.56	1.22	0.55	1.22	0.32	0.71	0.56	1.23
	61	45	40	1.22	2.70	–	–	0.78	1.71	1.03	2.28	–	–	–	–	0.64	1.40	0.63	1.40	0.37	0.82	0.64	1.42
118	44	28	28	0.91	2.00	–	–	0.57	1.25	0.74	1.62	–	–	–	–	0.46	1.02	0.39	0.87	0.17	0.37	0.44	0.96
	50	32	31	1.01	2.23	–	–	0.63	1.39	0.82	1.80	–	–	–	–	0.52	1.14	0.44	0.96	0.19	0.42	0.49	1.07
119	56	34	38	1.03	2.28	–	–	0.73	1.61	0.86	1.90	–	–	–	–	0.60	1.32	0.43	0.95	0.21	0.46	0.47	1.04
	64	39	43	1.18	2.60	–	–	0.83	1.83	0.99	2.17	–	–	–	–	0.68	1.50	0.49	1.08	0.24	0.53	0.54	1.19
120	40	34	33	0.80	1.77	–	–	0.65	1.44	0.62	1.37	–	–	–	–	0.54	1.18	0.28	0.62	0.06	0.14	0.38	0.83
	44	37	36	0.88	1.93	–	–	0.72	1.58	0.68	1.50	–	–	–	–	0.59	1.29	0.31	0.68	0.07	0.15	0.41	0.91
121	41	24	25	0.96	2.11	–	–	0.51	1.13	0.72	1.59	–	–	–	–	0.42	0.93	0.45	0.98	0.22	0.49	0.45	1.00
	46	26	28	1.08	2.37	–	–	0.58	1.27	0.81	1.78	–	–	–	–	0.47	1.04	0.50	1.10	0.25	0.54	0.51	1.12
122	42	–	29	0.86	1.90	–	–	0.59	1.30	0.60	1.31	–	–	–	–	0.48	1.07	0.25	0.55	0.03	0.08	0.39	0.86
	47	–	32	0.95	2.10	–	–	0.65	1.44	0.66	1.45	–	–	–	–	0.54	1.18	0.28	0.61	0.04	0.08	0.43	0.95
123	34	–	–	0.67	1.48	–	–	–	–	0.49	1.08	–	–	–	–	–	–	–	–	–	–	–	–
	36	–	–	0.73	1.61	–	–	–	–	0.53	1.17	–	–	–	–	–	–	–	–	–	–	–	–
124	40	–	33	0.84	1.85	–	–	0.66	1.46	0.66	1.46	–	–	–	–	0.54	1.19	0.37	0.82	0.15	0.32	0.44	0.97
	43	–	36	0.91	2.01	–	–	0.72	1.58	0.72	1.58	–	–	–	–	0.59	1.29	0.40	0.89	0.16	0.35	0.48	1.05
125	54	48	46	1.14	2.52	–	–	0.86	1.90	0.92	2.02	–	–	–	–	0.71	1.56	0.58	1.28	0.34	0.75	0.56	1.24
	60	53	50	1.26	2.78	–	–	0.95	2.10	1.02	2.24	–	–	–	–	0.78	1.72	0.64	1.41	0.38	0.83	0.62	1.38
126	38	–	30	0.82	1.80	–	–	0.61	1.34	0.64	1.41	–	–	–	–	0.50	1.10	0.37	0.81	0.14	0.32	0.44	0.96
	42	–	33	0.90	1.98	–	–	0.67	1.47	0.70	1.55	–	–	–	–	0.55	1.20	0.40	0.89	0.16	0.35	0.48	1.05
127	47	33	35	0.94	2.08	–	–	0.69	1.51	0.77	1.69	–	–	–	–	0.56	1.24	0.43	0.95	0.21	0.45	0.48	1.05
	52	36	39	1.05	2.30	–	–	0.76	1.68	0.85	1.88	–	–	–	–	0.62	1.37	0.48	1.05	0.23	0.50	0.53	1.17
128	49	38	41	0.99	2.19	–	–	0.78	1.73	0.82	1.80	–	–	–	–	0.64	1.42	0.48	1.06	0.25	0.55	0.52	1.14
	53	41	44	1.08	2.39	–	–	0.85	1.88	0.89	1.96	–	–	–	–	0.70	1.54	0.52	1.15	0.27	0.59	0.56	1.24
129	50	41	43	1.00	2.21	–	–	0.82	1.81	0.84	1.84	–	–	–	–	0.67	1.49	0.54	1.19	0.31	0.68	0.55	1.21
	57	46	49	1.14	2.52	–	–	0.94	2.06	0.95	2.10	–	–	–	–	0.77	1.69	0.61	1.35	0.35	0.77	0.63	1.38
130	49	33	37	0.97	2.14	–	–	0.73	1.60	0.80	1.76	–	–	–	–	0.60	1.32	0.44	0.98	0.22	0.47	0.49	1.07
	53	36	41	1.07	2.36	–	–	0.80	1.77	0.88	1.93	–	–	–	–	0.66	1.45	0.49	1.08	0.24	0.52	0.54	1.18
131	51	34	42	0.99	2.19	–	–	0.81	1.78	0.82	1.80	–	–	–	–	0.66	1.46	0.45	0.99	0.22	0.49	0.49	1.08
	56	37	47	1.10	2.42	–	–	0.90	1.98	0.91	2.00	–	–	–	–	0.74	1.62	0.50	1.10	0.25	0.55	0.54	1.20
132	53	45	40	1.06	2.33	–	–	0.76	1.68	0.89	1.95	–	–	–	–	0.62	1.38	0.52	1.14	0.29	0.63	0.54	1.19
	59	50	44	1.18	2.60	–	–	0.85	1.87	0.99	2.17	–	–	–	–	0.69	1.53	0.58	1.27	0.32	0.70	0.60	1.32
133	43	29	33	0.89	1.97	–	–	0.66	1.45	0.72	1.58	–	–	–	–	0.54	1.19	0.41	0.91	0.19	0.42	0.47	1.02
	48	32	37	0.99	2.19	–	–	0.73	1.62	0.80	1.76	–	–	–	–	0.60	1.33	0.46	1.02	0.21	0.47	0.52	1.14
134	53	35	37	1.06	2.35	–	–	0.72	1.58	0.90	1.98	–	–	–	–	0.59	1.30	0.55	1.20	0.32	0.70	0.56	1.22
	61	40	42	1.22	2.68	–	–	0.82	1.81	1.02	2.26	–	–	–	–	0.67	1.48	0.62	1.38	0.36	0.80	0.64	1.40
135	48	31	41	0.95	2.10	–	–	0.78	1.71	0.78	1.72	–	–	–	–	0.64	1.40	0.44	0.96	0.21	0.47	0.48	1.06
	54	34	46	1.07	2.36	–	–	0.87	1.92	0.88	1.93	–	–	–	–	0.72	1.58	0.49	1.08	0.24	0.53	0.54	1.19
136	48	41	35	0.97	2.15	–	–	0.68	1.49	0.80	1.77	–	–	–	–	0.56	1.23	0.46	1.02	0.24	0.52	0.50	1.09
	54	45	39	1.09	2.41	–	–	0.76	1.68	0.90	1.98	–	–	–	–	0.62	1.37	0.52	1.14	0.26	0.58	0.56	1.23
137	54	–	–	1.13	2.49	–	–	–	–	0.96	2.11	–	–	–	–	–	–	0.60	1.33	0.36	0.80	0.60	1.33
	60	–	–	1.25	2.75	–	–	–	–	1.05	2.32	–	–	–	–	–	–	0.67	1.47	0.40	0.88	0.67	1.47
138	38	34	33	0.75	1.66	–	–	0.66	1.46	0.57	1.27	–	–	–	–	0.54	1.19	0.39	0.85	0.16	0.35	0.45	0.99
	41	37	37	0.83	1.82	–	–	0.72	1.59	0.63	1.39	–	–	–	–	0.59	1.31	0.42	0.93	0.18	0.39	0.49	1.08
139	38	–	–	0.81	1.78	–	–	–	–	0.66	1.46	–	–	–	–	–	–	0.37	0.81	0.14	0.30	0.44	0.96
	41	–	–	0.87	1.91	–	–	–	–	0.71	1.56	–	–	–	–	–	–	0.39	0.87	0.15	0.33	0.47	1.04

(Continued)

TABLE V-1C DRY FORAGES, MINERAL AND VITAMIN COMPOSITION OF FEEDS, DATA EXPRESSED **AS-FED** AND **MOISTURE-FREE**—*(Continued)*

Entry Number	Feed Name Description	Moisture Basis: A-F (as-fed) or M-F (moisture-free)	Dry Matter	Macro Minerals							Micro Minerals						
				Calcium (Ca)	Phosphorus (P)	Sodium (Na)	Chlorine (Cl)	Magnesium (Mg)	Potassium (K)	Sulfur (S)	Cobalt (Co)	Copper (Cu)	Iodine (I)	Iron (Fe)	Manganese (Mn)	Selenium (Se)	Zinc (Zn)
			%	%	%	%	%	%	%	%	ppm or mg/kg	ppm or mg/kg	ppm or mg/kg	%	ppm or mg/kg	ppm or mg/kg	ppm or mg/kg
	CORN (Continued)																
114	FODDER WITH EARS, WITH HUSKS, WEATHER DAMAGED, SUN-CURED	**A-F**	**85**	–	–	–	–	–	–	–	–	–	–	–	–	–	–
		M-F	100	–	–	–	–	–	–	–	–	–	–	–	–	–	–
115	STOVER WITHOUT EARS, WITHOUT HUSKS, SUN-CURED	**A-F**	**85**	**0.49**	**0.08**	**0.06**	–	**0.34**	**1.24**	**0.15**	–	**4.3**	–	**0.018**	**115.9**	–	–
		M-F	100	0.57	0.10	0.07	–	0.40	1.45	0.17	–	5.1	–	0.021	136.0	–	–
116	HUSKS, SUN-CURED	**A-F**	**89**	**0.16**	**0.12**	–	–	–	**0.57**	–	–	–	–	–	–	–	–
		M-F	100	0.18	0.14	–	–	–	0.65	–	–	–	–	–	–	–	–
117	LEAVES, SUN-CURED	**A-F**	**87**	**0.60**	**0.22**	–	–	–	**1.56**	–	–	–	–	–	–	–	–
		M-F	100	0.69	0.25	–	–	–	1.79	–	–	–	–	–	–	–	–
118	COBS, GROUND	**A-F**	**90**	**0.11**	**0.04**	–	–	**0.06**	**0.78**	**0.42**	**0.117**	**6.6**	–	**0.021**	**5.6**	–	–
		M-F	100	0.12	0.04	–	–	0.07	0.87	0.47	0.130	7.3	–	0.023	6.2	–	–
	CORN, SWEET *Zea mays, saccharata*																
119	FODDER WITH EARS, WITH HUSKS, SUN-CURED	**A-F**	**88**	–	**0.17**	–	–	–	**0.98**	–	–	–	–	–	–	–	–
		M-F	100	–	0.19	–	–	–	1.12	–	–	–	–	–	–	–	–
	COTTON *Gossypium* spp																
120	BOLLS, SUN-CURED	**A-F**	**91**	**0.72**	**0.13**	–	–	**0.24**	**2.42**	–	–	–	–	–	–	–	–
		M-F	100	0.78	0.14	–	–	0.27	2.64	–	–	–	–	–	–	–	–
121	GIN BY-PRODUCT (GIN TRASH)	**A-F**	**89**	**0.70**	**0.20**	–	–	–	–	–	–	–	–	–	–	–	–
		M-F	100	0.78	0.23	–	–	–	–	–	–	–	–	–	–	–	–
122	HULLS	**A-F**	**90**	**0.13**	**0.09**	**0.02**	**0.02**	**0.13**	**0.78**	**0.08**	**0.018**	**12.0**	–	**0.012**	**107.8**	–	**19.8**
		M-F	100	0.15	0.09	0.02	0.02	0.14	0.87	0.09	0.020	13.3	–	0.014	119.2	–	21.9
123	LEAVES, SUN-CURED	**A-F**	**92**	**3.92**	**0.22**	–	–	**0.82**	**1.22**	–	–	–	–	–	**25.6**	–	–
		M-F	100	4.25	0.24	–	–	0.89	1.33	–	–	–	–	–	27.8	–	–
124	STEMS, SUN-CURED	**A-F**	**92**	–	–	–	–	–	–	–	–	–	–	–	–	–	–
		M-F	100	–	–	–	–	–	–	–	–	–	–	–	–	–	–
	COWPEA, COMMON *Vigna sinensis*																
125	HAY, SUN-CURED	**A-F**	**90**	**1.26**	**0.31**	**0.24**	**0.15**	**0.41**	**2.04**	**0.32**	**0.064**	–	–	**0.055**	**438.4**	–	–
		M-F	100	1.40	0.35	0.27	0.17	0.45	2.26	0.35	0.070	–	–	0.060	485.1	–	–
126	STRAW	**A-F**	**91**	–	–	–	–	–	–	–	–	–	–	–	–	–	–
		M-F	100	–	–	–	–	–	–	–	–	–	–	–	–	–	–
	CRABGRASS *Digitaria* spp																
127	HAY, SUN-CURED	**A-F**	**90**	–	–	–	–	–	–	–	–	–	–	–	–	–	–
		M-F	100	–	–	–	–	–	–	–	–	–	–	–	–	–	–
	CRESTED WHEATGRASS *Agropyron desertorum*																
128	HAY, SUN-CURED	**A-F**	**92**	**0.24**	**0.14**	–	–	–	–	–	**0.219**	–	–	–	–	–	–
		M-F	100	0.26	0.15	–	–	–	–	–	0.239	–	–	–	–	–	–
	CRIMSON CLOVER *Trifolium incarnatum*																
129	HAY, SUN-CURED	**A-F**	**88**	**1.23**	**0.19**	**0.34**	**0.55**	**0.25**	**2.10**	**0.25**	–	–	**0.059**	**0.062**	**183.3**	–	–
		M-F	100	1.40	0.22	0.39	0.63	0.28	2.40	0.28	–	–	0.067	0.070	208.7	–	–
	DALLISGRASS *Paspalum dilatatum*																
130	HAY, SUN-CURED	**A-F**	**91**	**0.46**	**0.18**	–	–	**0.67**	–	–	–	–	–	**0.011**	–	–	–
		M-F	100	0.51	0.20	–	–	0.74	–	–	–	–	–	0.012	–	–	–
	DOLICHOS (HYACINTH BEAN) *Dolichos lablab*																
131	HAY, SUN-CURED	**A-F**	**90**	–	–	–	–	–	–	–	–	–	–	–	–	–	–
		M-F	100	–	–	–	–	–	–	–	–	–	–	–	–	–	–
	DURRA SORGHUM *Sorghum bicolor, durra*																
132	FODDER, WITH HEADS, SUN-CURED	**A-F**	**90**	–	–	–	–	–	–	–	–	–	–	–	–	–	–
		M-F	100	–	–	–	–	–	–	–	–	–	–	–	–	–	–
	EMMER *Triticum dicoccum*																
133	HAY, SUN-CURED	**A-F**	**90**	–	–	–	–	–	–	–	–	–	–	–	–	–	–
		M-F	100	–	–	–	–	–	–	–	–	–	–	–	–	–	–
	FESCUE, MEADOW *Festuca elatior*																
134	HAY, SUN-CURED	**A-F**	**88**	**0.33**	**0.25**	–	–	**0.44**	**1.61**	–	**0.119**	–	–	–	**21.4**	–	–
		M-F	100	0.37	0.29	–	–	0.50	1.84	–	0.135	–	–	–	24.5	–	–
	FESCUE, TALL (ALTA) *Festuca arundinacea*																
135	HAY, SUN-CURED	**A-F**	**89**	**0.35**	**0.21**	**0.05**	–	**0.20**	**2.12**	–	–	–	–	–	–	–	–
		M-F	100	0.39	0.24	0.06	–	0.23	2.38	–	–	–	–	–	–	–	–
	FILAREE (ALFILERIA, REDSTEM) *Erodium cicutarium*																
136	HAY, SUN-CURED	**A-F**	**89**	**1.57**	**0.41**	–	–	–	–	–	–	–	–	–	–	–	–
		M-F	100	1.76	0.46	–	–	–	–	–	–	–	–	–	–	–	–
	FLAT PEA *Lathyrus sylvestris*																
137	HAY, SUN-CURED	**A-F**	**91**	–	**0.30**	–	–	–	**1.99**	–	–	–	–	–	–	–	–
		M-F	100	–	0.33	–	–	–	2.19	–	–	–	–	–	–	–	–
	FLAX, COMMON *Linum usitatissimum*																
138	FIBER PROCESS RESIDUE, DEHY	**A-F**	**91**	**1.21**	**0.22**	–	–	–	–	–	–	–	–	–	–	–	–
		M-F	100	1.33	0.24	–	–	–	–	–	–	–	–	–	–	–	–
139	STRAW	**A-F**	**93**	**0.51**	**0.05**	–	**0.25**	**0.29**	**1.62**	**0.25**	–	–	–	–	**7.6**	–	–
		M-F	100	0.55	0.06	–	0.27	0.31	1.74	0.27	–	–	–	–	8.2	–	–

Entry Number	Fat-Soluble Vitamins: A (1 mg Carotene = 1667 IU Vit A)	Carotene (Provitamin A)	D	E	K	Water-Soluble Vitamins: B-12	Biotin	Choline	Folacin (Folic Acid)	Niacin	Pantothenic Acid (B-3)	(Pyri-doxine) B-6	Ribo-flavin (B-2)	Thiamin (B-1)
	IU/g	ppm or mg/kg	IU/kg	ppm or mg/kg	ppm or mg/kg	ppm or mg/kg	ppm or mg/kg	ppm or mg/kg	ppm or mg/kg	ppm or mg/kg	ppm or mg/kg	ppm or mg/kg	ppm or mg/kg	ppm or mg/kg
114	–	–	–	–	–	–	–	–	–	–	–	–	–	–
	–	–	–	–	–	–	–	–	–	–	–	–	–	–
115	**6.3**	**3.8**	**1**	–	–	–	–	–	–	–	–	–	–	–
	7.4	4.5	1	–	–	–	–	–	–	–	–	–	–	–
116	–	–	–	–	–	–	–	–	–	–	–	–	–	–
	–	–	–	–	–	–	–	–	–	–	–	–	–	–
117	–	–	**5**	–	–	–	–	–	–	–	–	–	–	–
	–	–	5	–	–	–	–	–	–	–	–	–	–	–
118	**1.0**	**0.6**	–	–	–	–	–	–	–	**7**	**3.8**	–	**1.0**	**0.9**
	1.2	0.7	–	–	–	–	–	–	–	8	4.2	–	1.1	1.0
119	–	–	–	–	–	–	–	–	–	–	–	–	–	–
	–	–	–	–	–	–	–	–	–	–	–	–	–	–
120	–	–	–	–	–	–	–	–	–	–	–	–	–	–
	–	–	–	–	–	–	–	–	–	–	–	–	–	–
121	–	–	–	–	–	–	–	–	–	–	–	–	–	–
	–	–	–	–	–	–	–	–	–	–	–	–	–	–
122	–	–	–	–	–	–	–	–	–	–	–	–	**3.7**	–
	–	–	–	–	–	–	–	–	–	–	–	–	4.1	–
123	–	–	–	–	–	–	–	–	–	–	–	–	–	–
	–	–	–	–	–	–	–	–	–	–	–	–	–	–
124	–	–	–	–	–	–	–	–	–	–	–	–	–	–
	–	–	–	–	–	–	–	–	–	–	–	–	–	–
125	**52.7**	**31.6**	–	–	–	–	–	–	–	–	–	–	–	–
	58.3	35.0	–	–	–	–	–	–	–	–	–	–	–	–
126	–	–	–	–	–	–	–	–	–	–	–	–	–	–
	–	–	–	–	–	–	–	–	–	–	–	–	–	–
127	–	–	–	–	–	–	–	–	–	–	–	–	–	–
	–	–	–	–	–	–	–	–	–	–	–	–	–	–
128	**34.2**	**20.5**	–	–	–	–	–	–	–	–	–	–	–	–
	37.2	22.3	–	–	–	–	–	–	–	–	–	–	–	–
129	**32.9**	**19.8**	–	–	–	–	–	–	–	–	–	–	–	–
	37.5	22.5	–	–	–	–	–	–	–	–	–	–	–	–
130	–	–	–	–	–	–	–	–	–	–	–	–	–	–
	–	–	–	–	–	–	–	–	–	–	–	–	–	–
131	–	–	–	–	–	–	–	–	–	–	–	–	–	–
	–	–	–	–	–	–	–	–	–	–	–	–	–	–
132	–	–	–	–	–	–	–	–	–	–	–	–	–	–
	–	–	–	–	–	–	–	–	–	–	–	–	–	–
133	–	–	–	–	–	–	–	–	–	–	–	–	–	–
	–	–	–	–	–	–	–	–	–	–	–	–	–	–
134	**105.8**	**63.4**	–	**118.6**	–	–	–	–	–	–	–	–	–	–
	120.9	72.5	–	135.6	–	–	–	–	–	–	–	–	–	–
135	**30.8**	**18.5**	–	–	–	–	–	–	–	–	–	–	–	–
	34.6	20.7	–	–	–	–	–	–	–	–	–	–	–	–
136	–	–	–	–	–	–	–	–	–	–	–	–	–	–
	–	–	–	–	–	–	–	–	–	–	–	–	–	–
137	–	–	–	–	–	–	–	–	–	–	–	–	–	–
	–	–	–	–	–	–	–	–	–	–	–	–	–	–
138	–	–	–	–	–	–	–	–	–	–	–	–	–	–
	–	–	–	–	–	–	–	–	–	–	–	–	–	–
139	–	–	–	–	–	–	–	–	–	–	–	–	–	–
	–	–	–	–	–	–	–	–	–	–	–	–	–	–

(Continued)

TABLE V-1C DRY FORAGES, COMPOSITION OF FEEDS, DATA EXPRESSED **AS-FED** AND **MOISTURE-FREE**—*(Continued)*

Entry Number	Feed Name Description	International Feed Number	Moisture Basis: A-F (as-fed) or M-F (moisture-free)	Proximate Analysis									Digestible Protein		
				Dry Matter	Ash	Crude Fiber	Neutral Det. Fib. (NDF)	Acid Det. Fib. (ADF)	Lignin	Ether Extract (Fat)	N-Free Extract	Crude Protein	Ruminant	Swine	Horse
				%	%	%	%	%	%	%	%	%	%	%	%
	FOWL MEADOW GRASS (BLUEGRASS, FOWL) *Poa palustris*														
140	HAY, SUN-CURED	1-00-767	A-F	85	7.0	32.3	–	–	–	1.6	37.2	6.9	3.4	2.8	3.8
			M-F	100	8.2	38.0	–	–	–	1.9	43.8	8.1	4.1	3.3	4.5
	FOXTAIL, MEADOW *Alopecurus pratensis*														
141	HAY, SUN-CURED	1-02-072	A-F	90	8.4	25.5	–	–	–	2.0	41.1	12.8	8.5	7.7	8.2
			M-F	100	9.4	28.4	–	–	–	2.2	45.8	14.2	9.5	8.6	9.2
	FOXTAIL MILLET *Setaria italica*														
142	HAY, SUN-CURED	1-03-099	A-F	87	7.5	25.7	–	–	–	2.5	43.7	7.5	3.7	3.2	4.2
			M-F	100	8.6	29.6	–	–	–	2.9	50.3	8.6	4.3	3.7	4.8
	FURZE (GORSE, COMMON) *Ulex europaeus*														
143	BROWSE, SUN-CURED	1-08-425	A-F	94	7.0	38.5	–	–	–	2.0	35.4	11.6	7.3	6.5	7.3
			M-F	100	7.4	40.7	–	–	–	2.1	37.5	12.3	7.7	6.9	7.7
	GAMAGRASS *Tripsacum* spp														
144	HAY, SUN-CURED	1-02-081	A-F	89	6.2	29.2	–	–	–	1.8	44.1	7.6	3.9	3.2	4.2
			M-F	100	7.0	32.8	–	–	–	2.0	49.6	8.5	4.4	3.6	4.8
	GORSE, COMMON (FURZE) *Ulex europaeus*														
145	BROWSE, SUN-CURED	1-08-425	A-F	94	7.0	38.5	–	–	–	2.0	35.4	11.6	7.3	6.5	7.3
			M-F	100	7.4	40.7	–	–	–	2.1	37.5	12.3	7.7	6.9	7.7
	GRAMA *Bouteloua* spp														
146	HAY, SUN-CURED	1-02-162	A-F	89	8.4	29.1	–	–	–	1.5	44.5	5.6	2.2	1.5	2.7
			M-F	100	9.5	32.6	–	–	–	1.7	50.0	6.2	2.4	1.6	3.0
	GRAPE *Vitis* spp														
147	POMACE, DEHY (MARC)	1-02-208	A-F	90	7.5	27.9	48.1	49.2	31.8	7.6	35.3	12.1	1.6	7.1	7.7
			M-F	100	8.3	30.9	53.2	54.4	35.2	8.4	39.0	13.4	1.7	7.9	8.5
	GRASS														
148	HAY, SUN-CURED, ALL ANALYSES	1-02-250	A-F	89	7.3	29.7	–	34.4	–	2.3	41.2	8.9	5.2	4.4	5.7
			M-F	100	8.1	33.2	–	38.5	–	2.5	46.1	10.0	5.8	4.9	6.4
149	HAY, IMMATURE, SUN-CURED	1-02-212	A-F	87	8.7	21.3	–	–	–	2.2	40.1	14.5	9.1	9.4	9.6
			M-F	100	10.0	24.6	–	–	–	2.5	46.2	16.7	10.5	10.8	11.1
150	HAY, MATURE, SUN-CURED	1-02-246	A-F	90	5.1	31.6	–	–	–	2.1	47.8	3.9	0.7	-0.1	1.4
			M-F	100	5.6	34.9	–	–	–	2.3	52.9	4.3	0.7	-0.1	1.5
151	IMMATURE, DEHY	1-02-239	A-F	92	11.5	16.5	–	–	–	4.1	36.8	22.7	16.3	16.4	15.8
			M-F	100	12.5	18.0	–	–	–	4.5	40.2	24.7	17.8	17.9	17.3
152	STRAW	1-08-430	A-F	85	5.7	35.0	–	–	–	2.0	37.8	4.5	1.8	0.7	2.0
			M-F	100	6.7	41.2	–	–	–	2.4	44.5	5.3	2.1	0.8	2.3
	GRASS-LEGUME														
153	HAY, SUN-CURED	1-02-301	A-F	89	5.6	31.6	–	33.9	–	2.3	39.5	10.3	6.0	5.6	6.3
			M-F	100	6.2	35.4	–	38.0	–	2.5	44.3	11.5	6.7	6.3	7.1
154	HAY, IMMATURE, SUN-CURED	1-02-275	A-F	90	7.3	28.0	–	–	–	2.9	37.9	13.8	9.3	8.6	9.0
			M-F	100	8.2	31.1	–	–	–	3.3	42.2	15.3	10.4	9.6	10.0
	GREASEWOOD (CHAMISE) *Adenostoma fasciculatum*														
155	FAN AIR DRIED, PELLETED	1-07-860	A-F	90	3.0	18.6	–	–	–	4.9	57.1	6.4	1.9	2.1	3.3
			M-F	100	3.3	20.7	–	–	–	5.4	63.5	7.1	2.1	2.4	3.7
	GUAR *Cyamopsis tetragonoloba*														
156	HAY, SUN-CURED	1-02-327	A-F	92	12.3	20.5	–	–	–	1.5	42.0	16.0	12.0	10.5	10.7
			M-F	100	13.3	22.2	–	–	–	1.6	45.6	17.3	13.0	11.3	11.6
157	LEAVES, SUN-CURED	1-02-328	A-F	93	14.6	10.0	–	–	–	2.7	36.7	28.7	22.4	21.7	20.5
			M-F	100	15.7	10.8	–	–	–	2.9	39.6	31.0	24.1	23.4	22.1
	HARDINGGRASS *Phalaris tuberosa, stenoptera*														
158	HAY, SUN-CURED	1-02-353	A-F	93	6.4	32.6	–	–	–	1.7	46.7	5.8	2.2	1.5	2.8
			M-F	100	6.8	35.0	–	–	–	1.9	50.1	6.2	2.4	1.6	3.0
	HEGARI SORGHUM *Sorghum bicolor, caffrorum*														
159	FODDER WITH HEADS, DOUGH STAGE, SUN-CURED	1-04-390	A-F	88	7.5	18.5	–	–	–	2.0	52.8	7.2	3.6	2.9	4.0
			M-F	100	8.5	21.0	–	–	–	2.3	60.0	8.2	4.1	3.3	4.5
160	STOVER WITHOUT HEADS, SUN-CURED	1-08-515	A-F	87	8.9	25.3	–	34.1	–	1.9	46.0	5.2	1.9	1.2	2.5
			M-F	100	10.2	29.0	–	39.0	–	2.2	52.6	6.0	2.2	1.4	2.8
	HOG MILLET (BROOMCORN; MILLET, PROSO) *Panicum miliaceum*														
161	HAY, SUN-CURED	1-03-119	A-F	90	7.6	24.4	–	–	–	2.1	46.8	9.4	5.5	4.8	5.6
			M-F	100	8.4	27.0	–	–	–	2.4	51.8	10.4	6.1	5.3	6.2
162	STOVER WITHOUT HEADS, SUN-CURED	1-26-104	A-F	90	5.7	36.3	–	–	–	1.8	47.9	3.9	–	–	–
			M-F	100	6.3	40.3	–	–	–	2.0	53.2	4.3	–	–	–
	HOPS, COMMON *Humulus lupulus*														
163	FRUIT (HOPS), SPENT, DEHY	5-02-396	A-F	93	5.6	22.7	–	–	–	4.7	37.2	23.1	6.9	21.8	17.4
			M-F	100	6.0	24.3	–	–	–	5.1	39.9	24.8	7.4	23.3	18.7
	HORSE BEAN *Vicia faba, equina*														
164	HAY, SUN-CURED	1-02-402	A-F	91	5.5	22.0	–	–	–	0.8	49.8	13.4	8.9	8.2	8.7
			M-F	100	6.0	24.0	–	–	–	0.9	54.5	14.6	9.8	9.0	9.5

Entry Number	TDN Ruminant	TDN Swine	TDN Horse	Digestible Energy Ruminant		Digestible Energy Swine		Digestible Energy Horse		Metabolizable Energy Ruminant		Metabolizable Energy Swine		Poultry ME_n		Metabolizable Energy Horse		Net Energy Ruminant NE_m		Net Energy Ruminant NE_g		Lactating Cows NE_{lc}	
	%	%	%	Mcal		kcal		Mcal		Mcal		kcal		kcal		Mcal		Mcal		Mcal		Mcal	
				lb	kg	lb	kg	lb	kg	lb	kg	lb	kg	lb	kg	lb	kg	lb	kg	lb	kg	lb	kg
140	44	24	31	0.87	1.93	–	–	0.62	1.36	0.71	1.56	–	–	–	–	0.51	1.12	0.38	0.85	0.17	0.38	0.43	0.96
	51	29	37	1.03	2.27	–	–	0.73	1.61	0.83	1.84	–	–	–	–	0.60	1.32	0.45	1.00	0.20	0.45	0.51	1.13
141	57	42	44	1.08	2.37	–	–	0.84	1.86	0.90	1.99	–	–	–	–	0.69	1.52	0.48	1.07	0.25	0.56	0.51	1.13
	64	47	49	1.20	2.64	–	–	0.94	2.07	1.01	2.22	–	–	–	–	0.77	1.69	0.54	1.19	0.28	0.62	0.57	1.26
142	51	38	35	1.00	2.19	–	–	0.68	1.49	0.83	1.83	–	–	–	–	0.56	1.22	0.46	1.01	0.24	0.52	0.49	1.08
	59	43	40	1.15	2.52	–	–	0.78	1.72	0.95	2.10	–	–	–	–	0.64	1.41	0.53	1.16	0.27	0.60	0.56	1.24
143	33	–	37	0.83	1.83	–	–	0.73	1.60	0.64	1.42	–	–	–	–	0.60	1.31	0.45	0.99	0.21	0.47	0.50	1.10
	35	–	39	0.88	1.93	–	–	0.77	1.70	0.68	1.50	–	–	–	–	0.63	1.39	0.48	1.05	0.23	0.50	0.53	1.16
144	40	34	38	0.88	1.94	–	–	0.74	1.63	0.71	1.56	–	–	–	–	0.61	1.34	0.45	0.99	0.22	0.49	0.49	1.07
	45	38	43	0.99	2.18	–	–	0.83	1.84	0.79	1.75	–	–	–	–	0.68	1.51	0.51	1.11	0.25	0.56	0.55	1.21
145	33	–	37	0.83	1.83	–	–	0.73	1.60	0.64	1.42	–	–	–	–	0.60	1.31	0.45	0.99	0.21	0.47	0.50	1.10
	35	–	39	0.88	1.93	–	–	0.77	1.70	0.68	1.50	–	–	–	–	0.63	1.39	0.48	1.05	0.23	0.50	0.53	1.16
146	39	27	35	0.84	1.86	–	–	0.68	1.49	0.67	1.47	–	–	–	–	0.55	1.22	0.40	0.89	0.18	0.40	0.46	1.00
	44	31	39	0.94	2.08	–	–	0.76	1.67	0.75	1.65	–	–	–	–	0.62	1.37	0.45	1.00	0.20	0.45	0.51	1.13
147	24	–	18	0.48	1.06	–	–	0.41	0.89	0.30	0.66	–	–	703	1551	0.33	0.73	–	–	–	–	–	–
	27	–	20	0.53	1.17	–	–	0.45	0.99	0.33	0.73	–	–	778	1715	0.37	0.81	–	–	–	–	–	–
148	51	35	40	1.08	2.38	–	–	0.76	1.68	0.91	2.00	–	–	–	–	0.63	1.38	0.53	1.17	0.30	0.66	0.52	1.15
	57	39	44	1.21	2.66	–	–	0.85	1.88	1.01	2.24	–	–	–	–	0.70	1.54	0.59	1.31	0.33	0.74	0.58	1.29
149	52	47	46	1.03	2.28	–	–	0.87	1.92	0.87	1.91	–	–	–	–	0.71	1.57	0.50	1.10	0.27	0.61	0.52	1.14
	61	54	54	1.19	2.63	–	–	1.00	2.21	1.00	2.21	–	–	–	–	0.82	1.82	0.58	1.27	0.32	0.70	0.60	1.32
150	44	31	34	0.92	2.03	–	–	0.67	1.47	0.74	1.64	–	–	–	–	0.55	1.21	0.45	0.98	0.22	0.48	0.49	1.07
	48	34	37	1.02	2.24	–	–	0.74	1.63	0.82	1.81	–	–	–	–	0.61	1.33	0.49	1.09	0.24	0.53	0.54	1.19
151	60	69	51	1.19	2.63	–	–	0.96	2.11	1.02	2.24	–	–	–	–	0.79	1.73	0.68	1.49	0.42	0.94	0.66	1.46
	65	75	56	1.30	2.87	–	–	1.05	2.31	1.11	2.45	–	–	–	–	0.86	1.89	0.74	1.63	0.46	1.02	0.72	1.59
152	40	–	26	0.83	1.83	–	–	0.54	1.18	0.66	1.46	–	–	–	–	0.44	0.97	0.38	0.84	0.17	0.37	0.43	0.95
	47	–	31	0.98	2.15	–	–	0.63	1.39	0.78	1.72	–	–	–	–	0.52	1.14	0.45	0.99	0.20	0.44	0.51	1.12
153	52	37	38	1.04	2.30	–	–	0.74	1.64	0.87	1.92	–	–	–	–	0.61	1.34	0.47	1.04	0.24	0.54	0.50	1.11
	58	41	43	1.17	2.57	–	–	0.83	1.83	0.98	2.15	–	–	–	–	0.68	1.50	0.53	1.17	0.27	0.60	0.57	1.25
154	55	46	41	1.08	2.37	–	–	0.78	1.72	0.90	1.99	–	–	–	–	0.64	1.41	0.51	1.13	0.28	0.62	0.54	1.18
	61	51	45	1.20	2.64	–	–	0.87	1.91	1.01	2.22	–	–	–	–	0.71	1.57	0.57	1.26	0.31	0.69	0.60	1.31
155	43	64	–	0.87	1.91	–	–	–	–	0.69	1.53	–	–	–	–	–	–	–	–	–	–	–	–
	48	71	–	0.96	2.13	–	–	–	–	0.77	1.70	–	–	–	–	–	–	–	–	–	–	–	–
156	52	46	51	1.04	2.29	–	–	0.96	2.11	0.86	1.90	–	–	–	–	0.78	1.73	0.49	1.08	0.25	0.56	0.52	1.15
	57	49	55	1.13	2.48	–	–	1.04	2.28	0.93	2.06	–	–	–	–	0.85	1.87	0.53	1.17	0.27	0.61	0.57	1.25
157	–	–	–	–	–	–	–	–	–	–	–	–	–	–	–	–	–	–	–	–	–	–	–
	–	–	–	–	–	–	–	–	–	–	–	–	–	–	–	–	–	–	–	–	–	–	–
158	47	30	37	0.96	2.11	–	–	0.72	1.59	0.77	1.71	–	–	–	–	0.59	1.30	0.44	0.98	0.21	0.46	0.49	1.08
	50	33	40	1.02	2.26	–	–	0.77	1.71	0.83	1.83	–	–	–	–	0.63	1.40	0.48	1.05	0.23	0.50	0.53	1.16
159	50	45	–	1.00	2.20	–	–	–	–	0.83	1.82	–	–	–	–	–	–	–	–	–	–	–	–
	57	51	–	1.13	2.50	–	–	–	–	0.94	2.07	–	–	–	–	–	–	–	–	–	–	–	–
160	49	32	34	0.95	2.09	–	–	0.66	1.45	0.78	1.72	–	–	–	–	0.54	1.19	0.42	0.92	0.20	0.44	0.46	1.02
	56	36	38	1.08	2.39	–	–	0.75	1.66	0.89	1.96	–	–	–	–	0.62	1.36	0.48	1.05	0.23	0.50	0.53	1.16
161	51	42	42	1.02	2.24	–	–	0.81	1.79	0.84	1.86	–	–	–	–	0.66	1.46	0.49	1.08	0.26	0.57	0.52	1.14
	56	47	47	1.13	2.48	–	–	0.90	1.98	0.93	2.06	–	–	–	–	0.74	1.62	0.54	1.19	0.29	0.63	0.57	1.27
162	–	–	–	–	–	–	–	–	–	–	–	–	–	–	–	–	–	–	–	–	–	–	–
	–	–	–	–	–	–	–	–	–	–	–	–	–	–	–	–	–	–	–	–	–	–	–
163	32	52	–	0.89	1.96	1033	2277	–	–	0.71	1.56	874	1927	–	–	–	–	0.58	1.27	0.33	0.73	0.59	1.30
	34	55	–	0.96	2.11	1107	2441	–	–	0.76	1.68	937	2066	–	–	–	–	0.62	1.36	0.36	0.79	0.63	1.39
164	52	48	59	1.06	2.34	–	–	1.09	2.40	0.88	1.95	–	–	–	–	0.89	1.97	0.53	1.16	0.29	0.64	0.55	1.21
	57	53	65	1.16	2.55	–	–	1.19	2.63	0.97	2.13	–	–	–	–	0.98	2.15	0.58	1.27	0.32	0.70	0.60	1.32

(Continued)

TABLE V-1C DRY FORAGES, MINERAL AND VITAMIN COMPOSITION OF FEEDS, DATA EXPRESSED **AS-FED** AND **MOISTURE-FREE**—*(Continued)*

Entry Number	Feed Name Description	Moisture Basis: A-F (as-fed) or M-F (moisture-free)	Dry Matter	Macro Minerals							Micro Minerals						
				Calcium (Ca)	Phos-phorus (P)	Sodium (Na)	Chlo-rine (Cl)	Mag-nesium (Mg)	Potas-sium (K)	Sulfur (S)	Cobalt (Co)	Copper (Cu)	Iodine (I)	Iron (Fe)	Man-ganese (Mn)	Sele-nium (Se)	Zinc (Zn)
			%	%	%	%	%	%	%	%	ppm or mg/kg	ppm or mg/kg	ppm or mg/kg	%	ppm or mg/kg	ppm or mg/kg	ppm or mg/kg
	FOWL MEADOW GRASS (BLUEGRASS, FOWL) *Poa palustris*																
140	HAY, SUN-CURED	**A-F**	**85**	**0.22**	**0.22**	–	–	–	–	–	–	–	–	–	–	–	–
		M-F	100	0.26	0.26	–	–	–	–	–	–	–	–	–	–	–	–
	FOXTAIL, MEADOW *Alopecurus pratensis*																
141	HAY, SUN-CURED	**A-F**	**90**	**0.95**	**0.18**	**0.01**	–	**0.12**	**1.57**	–	–	–	–	–	–	–	–
		M-F	100	1.06	0.20	0.01	–	0.13	1.75	–	–	–	–	–	–	–	–
	FOXTAIL, MILLET *Setaria italica*																
142	HAY, SUN-CURED	**A-F**	**87**	**0.29**	**0.16**	**0.09**	**0.11**	**0.20**	**1.69**	**0.14**	–	–	–	–	**120.1**	–	–
		M-F	100	0.33	0.18	0.10	0.13	0.23	1.94	0.16	–	–	–	–	138.1	–	–
	FURZE (GORSE, COMMON) *Ulex europaeus*																
143	BROWSE, SUN-CURED	**A-F**	**94**	–	–	–	–	–	–	–	–	–	–	–	–	–	–
		M-F	100	–	–	–	–	–	–	–	–	–	–	–	–	–	–
	GAMAGRASS *Tripsacum* spp																
144	HAY, SUN-CURED	**A-F**	**89**	–	–	–	–	–	–	–	–	–	–	–	–	–	–
		M-F	100	–	–	–	–	–	–	–	–	–	–	–	–	–	–
	GORSE, COMMON (FURZE) *Ulex europaeus*																
145	BROWSE, SUN-CURED	**A-F**	**94**	–	–	–	–	–	–	–	–	–	–	–	–	–	–
		M-F	100	–	–	–	–	–	–	–	–	–	–	–	–	–	–
	GRAMA *Bouteloua* spp																
146	HAY, SUN-CURED	**A-F**	**89**	**0.34**	**0.18**	–	–	–	–	–	–	–	–	–	–	–	–
		M-F	100	0.38	0.20	–	–	–	–	–	–	–	–	–	–	–	–
	GRAPE *Vitis* spp																
147	POMACE, DEHY (MARC)	**A-F**	**90**	**0.52**	**0.15**	**0.08**	**0.01**	**0.09**	**0.82**	–	–	–	–	–	**36.8**	–	**21.9**
		M-F	100	0.58	0.17	0.09	0.01	0.10	0.91	–	–	–	–	–	40.7	–	24.2
	GRASS																
148	HAY, SUN-CURED, ALL ANALYSES	**A-F**	**89**	**0.44**	**0.18**	**0.01**	–	**0.20**	**1.38**	**0.19**	**0.133**	**6.5**	–	**0.013**	**75.4**	–	**15.1**
		M-F	100	0.49	0.20	0.01	–	0.22	1.55	0.21	0.149	7.3	–	0.015	84.4	–	16.9
149	HAY, IMMATURE, SUN-CURED	**A-F**	**87**	–	–	–	–	–	–	–	–	–	–	–	–	–	–
		M-F	100	–	–	–	–	–	–	–	–	–	–	–	–	–	–
150	HAY, MATURE, SUN-CURED	**A-F**	**90**	**0.34**	**0.14**	–	–	**0.26**	–	–	–	**21.0**	–	**0.011**	**44.7**	–	–
		M-F	100	0.38	0.16	–	–	0.29	–	–	–	23.3	–	0.012	49.5	–	–
151	IMMATURE, DEHY	**A-F**	**92**	**0.65**	**0.45**	–	–	–	–	–	–	–	–	–	**86.7**	–	–
		M-F	100	0.71	0.50	–	–	–	–	–	–	–	–	–	94.8	–	–
152	STRAW	**A-F**	**85**	–	–	–	–	–	–	–	–	–	–	–	–	–	–
		M-F	100	–	–	–	–	–	–	–	–	–	–	–	–	–	–
	GRASS-LEGUME																
153	HAY, SUN-CURED	**A-F**	**89**	**0.66**	**0.19**	**0.06**	–	**0.23**	**1.62**	**0.13**	**0.118**	**5.0**	–	**0.014**	**42.5**	–	**23.2**
		M-F	100	0.74	0.22	0.07	–	0.26	1.82	0.14	0.132	5.6	–	0.015	47.7	–	26.0
154	HAY, IMMATURE, SUN-CURED	**A-F**	**90**	**0.99**	**0.31**	–	–	**0.14**	**1.95**	–	**0.387**	**6.0**	–	**0.008**	**104.7**	–	–
		M-F	100	1.10	0.35	–	–	0.15	2.17	–	0.430	6.6	–	0.008	116.4	–	–
	GREASEWOOD (CHAMISE) *Adenostoma fasciculatum*																
155	FAN AIR DRIED, PELLETED	**A-F**	**90**	–	–	–	–	–	–	–	–	–	–	–	–	–	–
		M-F	100	–	–	–	–	–	–	–	–	–	–	–	–	–	–
	GUAR *Cyamopsis tetragonoloba*																
156	HAY, SUN-CURED	**A-F**	**92**	–	–	–	–	–	–	–	–	–	–	–	–	–	–
		M-F	100	–	–	–	–	–	–	–	–	–	–	–	–	–	–
157	LEAVES, SUN-CURED	**A-F**	**93**	–	–	–	–	–	–	–	–	–	–	–	–	–	–
		M-F	100	–	–	–	–	–	–	–	–	–	–	–	–	–	–
	HARDINGGRASS *Phalaris tuberosa, stenoptera*																
158	HAY, SUN-CURED	**A-F**	**93**	–	–	–	–	–	–	–	–	–	–	–	–	–	–
		M-F	100	–	–	–	–	–	–	–	–	–	–	–	–	–	–
	HEGARI SORGHUM *Sorghum bicolor, caffrorum*																
159	FODDER WITH HEADS, DOUGH STAGE, SUN-CURED	**A-F**	**88**	–	–	–	–	–	–	–	–	–	–	–	–	–	–
		M-F	100	–	–	–	–	–	–	–	–	–	–	–	–	–	–
160	STOVER WITHOUT HEADS, SUN-CURED	**A-F**	**87**	**0.33**	**0.08**	–	–	–	–	–	–	–	–	–	–	–	–
		M-F	100	0.38	0.09	–	–	–	–	–	–	–	–	–	–	–	–
	HOG MILLET (BROOMCORN; MILLET, PROSO) *Panicum miliaceum*																
161	HAY, SUN-CURED	**A-F**	**90**	–	–	–	–	–	–	–	–	–	–	–	–	–	–
		M-F	100	–	–	–	–	–	–	–	–	–	–	–	–	–	–
162	STOVER WITHOUT HEADS, SUN-CURED	**A-F**	**90**	–	–	–	–	–	–	–	–	–	–	–	–	–	–
		M-F	100	–	–	–	–	–	–	–	–	–	–	–	–	–	–
	HOPS, COMMON *Humulus lupulus*																
163	FRUIT (HOPS), SPENT, DEHY	**A-F**	**93**	–	–	–	–	–	–	–	–	–	–	–	–	–	–
		M-F	100	–	–	–	–	–	–	–	–	–	–	–	–	–	–
	HORSE BEAN *Vicia faba, equina*																
164	HAY, SUN-CURED	**A-F**	**91**	–	–	–	–	–	–	–	–	–	–	–	–	–	–
		M-F	100	–	–	–	–	–	–	–	–	–	–	–	–	–	–

Entry Number	Fat-Soluble Vitamins					Water-Soluble Vitamins								
	A (1 mg Carotene = 1667 IU Vit A)	Carotene (Provitamin A)	D	E	K	B–12	Biotin	Choline	Folacin (Folic Acid)	Niacin	Pantothenic Acid (B–3)	(Pyridoxine) B–6	Riboflavin (B–2)	Thiamin (B–1)
	IU/g	ppm or mg/kg	IU/kg	ppm or mg/kg	ppm or mg/kg	ppm or mg/kg	ppm or mg/kg	ppm or mg/kg	ppm or mg/kg	ppm or mg/kg	ppm or mg/kg	ppm or mg/kg	ppm or mg/kg	ppm or mg/kg
140	**177.1**	**106.3**	–	–	–	–	–	–	–	–	–	–	–	–
	208.4	125.0	–	–	–	–	–	–	–	–	–	–	–	–
141	–	–	–	–	–	–	–	–	–	–	–	–	–	–
	–	–	–	–	–	–	–	–	–	–	–	–	–	–
142	**86.9**	**52.1**	–	–	–	–	–	–	–	–	–	–	–	–
	100.0	60.0	–	–	–	–	–	–	–	–	–	–	–	–
143	–	–	–	–	–	–	–	–	–	–	–	–	–	–
	–	–	–	–	–	–	–	–	–	–	–	–	–	–
144	–	–	–	–	–	–	–	–	–	–	–	–	–	–
	–	–	–	–	–	–	–	–	–	–	–	–	–	–
145	–	–	–	–	–	–	–	–	–	–	–	–	–	–
	–	–	–	–	–	–	–	–	–	–	–	–	–	–
146	–	–	–	–	–	–	–	–	–	–	–	–	–	–
	–	–	–	–	–	–	–	–	–	–	–	–	–	–
147	–	–	–	–	–	–	–	**253**	–	**18**	**3.1**	–	**2.2**	–
	–	–	–	–	–	–	–	279	–	20	3.4	–	2.5	–
148	**40.8**	**24.5**	–	–	–	–	–	–	–	–	–	–	–	–
	45.7	27.4	–	–	–	–	–	–	–	–	–	–	–	–
149	–	–	–	–	–	–	–	–	–	–	–	–	–	–
	–	–	–	–	–	–	–	–	–	–	–	–	–	–
150	**7.8**	**4.7**	–	–	–	–	–	–	–	–	–	–	–	–
	8.6	5.2	–	–	–	–	–	–	–	–	–	–	–	–
151	**492.6**	**295.5**	–	–	–	–	**0.28**	–	–	**65**	**10.7**	–	**16.1**	**5.4**
	538.2	322.8	–	–	–	–	0.31	–	–	71	11.6	–	17.6	5.9
152	–	–	–	–	–	–	–	–	–	–	–	–	–	–
	–	–	–	–	–	–	–	–	–	–	–	–	–	–
153	**18.0**	**10.8**	**2**	–	–	–	–	–	–	–	–	–	–	–
	20.2	12.1	3	–	–	–	–	–	–	–	–	–	–	–
154	**203.3**	**122.0**	–	–	–	–	–	–	–	–	–	–	–	–
	226.0	135.6	–	–	–	–	–	–	–	–	–	–	–	–
155	–	–	–	–	–	–	–	–	–	–	–	–	–	–
	–	–	–	–	–	–	–	–	–	–	–	–	–	–
156	–	–	–	–	–	–	–	–	–	–	–	–	–	–
	–	–	–	–	–	–	–	–	–	–	–	–	–	–
157	–	–	–	–	–	–	–	–	–	–	–	–	–	–
	–	–	–	–	–	–	–	–	–	–	–	–	–	–
158	–	–	–	–	–	–	–	–	–	–	–	–	–	–
	–	–	–	–	–	–	–	–	–	–	–	–	–	–
159	–	–	–	–	–	–	–	–	–	–	–	–	–	–
	–	–	–	–	–	–	–	–	–	–	–	–	–	–
160	–	–	–	–	–	–	–	–	–	–	–	–	–	–
	–	–	–	–	–	–	–	–	–	–	–	–	–	–
161	–	–	–	–	–	–	–	–	–	–	–	–	–	–
	–	–	–	–	–	–	–	–	–	–	–	–	–	–
162	–	–	–	–	–	–	–	–	–	–	–	–	–	–
	–	–	–	–	–	–	–	–	–	–	–	–	–	–
163	–	–	–	–	–	–	–	–	–	–	–	–	–	–
	–	–	–	–	–	–	–	–	–	–	–	–	–	–
164	–	–	–	–	–	–	–	–	–	–	–	–	–	–
	–	–	–	–	–	–	–	–	–	–	–	–	–	–

(Continued)

TABLE V-1C DRY FORAGES, COMPOSITION OF FEEDS, DATA EXPRESSED **AS-FED** AND **MOISTURE-FREE**—*(Continued)*

Entry Number	Feed Name Description	International Feed Number	Moisture Basis: A-F (as-fed) or M-F (moisture-free)	Proximate Analysis: Dry Matter	Ash	Crude Fiber	Neutral Det. Fib. (NDF)	Acid Det. Fib. (ADF)	Lignin	Ether Extract (Fat)	N-Free Extract	Crude Protein	Digestible Protein: Ruminant	Swine	Horse
				%	%	%	%	%	%	%	%	%	%	%	%
	HORSEBEAN (Continued)														
165	STRAW	1-02-404	A-F	87	10.2	33.8	–	–	–	1.5	31.4	10.1	5.9	5.5	6.3
			M-F	100	11.7	38.9	–	–	–	1.7	36.0	11.6	6.8	6.4	7.2
	HYACINTH BEAN (DOLICHOS) *Dolichos lablab*														
166	HAY, SUN-CURED	1-01-805	A-F	90	7.1	33.6	–	–	–	1.6	34.2	13.8	9.0	8.6	9.0
			M-F	100	7.9	37.2	–	–	–	1.7	37.9	15.3	9.9	9.6	10.0
	INDIGO *Indigofera* spp														
167	HAY, CUT 1, SUN-CURED	1-02-425	A-F	89	5.1	22.2	–	–	–	2.2	48.9	10.7	6.7	5.9	6.6
			M-F	100	5.7	24.9	–	–	–	2.5	54.9	12.0	7.5	6.7	7.5
	JOHNSONGRASS SORGHUM *Sorghum halepense*														
168	HAY, SUN-CURED	1-04-407	A-F	91	7.7	30.4	–	–	–	2.0	43.7	6.7	3.0	2.4	3.6
			M-F	100	8.6	33.6	–	–	–	2.2	48.3	7.5	3.3	2.7	4.0
	JUNEGRASS *Koeleria cristata*														
169	HAY, SUN-CURED	1-02-436	A-F	89	6.9	30.3	–	–	–	2.5	40.8	8.1	4.4	3.7	4.6
			M-F	100	7.8	34.2	–	–	–	2.9	46.0	9.1	4.9	4.1	5.2
	KAFIR SORGHUM *Sorghum bicolor, caffrorum*														
170	CHAFF	1-04-420	A-F	92	7.9	19.6	–	–	–	1.0	56.8	6.7	1.3	2.3	3.5
			M-F	100	8.6	21.3	–	–	–	1.1	61.7	7.3	1.5	2.6	3.8
171	FODDER WITH HEADS, LATE DOUGH, SUN-CURED	1-04-417	A-F	91	11.2	23.8	–	–	–	2.1	44.3	10.0	6.3	5.2	6.0
			M-F	100	12.3	26.0	–	–	–	2.3	48.5	10.9	6.9	5.7	6.6
172	FODDER WITH HEADS, SUN-CURED	1-04-418	A-F	90	9.4	25.7	–	–	–	2.2	44.5	8.3	4.2	3.8	4.8
			M-F	100	10.4	28.5	–	–	–	2.5	49.4	9.2	4.7	4.2	5.3
173	STOVER WITHOUT HEADS, SUN-CURED	1-04-419	A-F	82	8.1	26.9	–	–	–	1.5	40.7	4.6	1.6	0.9	2.1
			M-F	100	9.9	32.9	–	–	–	1.9	49.7	5.7	1.9	1.1	2.6
	KELP (SEAWEED) *Laminariales* (order), *Fucales* (order)														
174	WHOLE, DEHY	1-08-073	A-F	91	35.0	6.5	–	–	–	0.5	42.4	6.5	2.9	2.2	3.4
			M-F	100	38.6	7.1	–	–	–	0.5	46.7	7.1	3.2	2.4	3.7
	KENTUCKY BLUEGRASS *Poa pratensis*														
175	HAY, SUN-CURED	1-00-776	A-F	89	5.9	26.8	–	–	–	3.0	44.3	9.1	5.3	4.6	5.4
			M-F	100	6.6	30.0	–	–	–	3.4	49.7	10.2	5.9	5.1	6.1
176	HAY, EARLY BLOOM, SUN-CURED	1-05-681	A-F	90	4.5	25.5	–	–	–	3.6	45.5	11.0	6.9	6.2	6.9
			M-F	100	5.0	28.3	–	–	–	4.0	50.5	12.2	7.6	6.9	7.6
177	HAY, MATURE, SUN-CURED	1-00-774	A-F	89	6.5	29.7	–	–	–	2.6	44.9	5.6	2.2	1.5	2.7
			M-F	100	7.3	33.3	–	–	–	2.9	50.3	6.3	2.4	1.6	3.0
	KOA HAOLE (LEAD TREE, WHITE POPINAC) *Leucaena glauca*														
178	HAY, SUN-CURED	1-02-492	A-F	89	5.1	29.8	–	–	–	1.9	39.2	12.7	8.4	7.7	8.2
			M-F	100	5.7	33.6	–	–	–	2.1	44.2	14.3	9.5	8.7	9.2
	KUDZU *Pueraria* spp														
179	HAY, SUN-CURED	1-02-478	A-F	91	5.8	35.7	–	–	–	1.9	34.8	13.1	8.7	8.0	8.5
			M-F	100	6.4	39.1	–	–	–	2.1	38.1	14.3	9.5	8.7	9.3
	LADINO CLOVER *Trifolium repens*														
180	HAY, SUN-CURED	1-01-378	A-F	89	8.4	18.5	32.1	28.5	5.9	2.4	39.9	20.0	15.4	14.1	13.8
			M-F	100	9.4	20.8	36.0	32.0	6.6	2.7	44.7	22.4	17.3	15.9	15.5
	LADINO CLOVER-GRASS *Trifolium repens-grass*														
181	HAY, SUN-CURED	1-01-511	A-F	89	7.3	19.9	–	–	–	2.2	43.2	16.5	11.7	11.1	11.1
			M-F	100	8.2	22.3	–	–	–	2.5	48.5	18.5	13.2	12.4	12.5
	LEAD TREE, WHITE POPINAC (KOA HAOLE) *Leucaena glauca*														
182	HAY, SUN-CURED	1-02-492	A-F	89	5.1	29.8	–	–	–	1.9	39.2	12.7	8.4	7.7	8.2
			M-F	100	5.7	33.6	–	–	–	2.1	44.2	14.3	9.5	8.7	9.2
	LESPEDEZA, COMMON *Lespedeza striata*														
183	HAY, SUN-CURED, ALL ANALYSES	1-08-591	A-F	89	4.7	28.4	–	–	–	2.5	39.4	13.8	5.9	8.7	9.1
			M-F	100	5.3	32.0	–	–	–	2.8	44.4	15.6	6.7	9.8	10.2
184	HAY, PREBLOOM, SUN-CURED	1-20-881	A-F	89	6.4	22.8	–	–	–	2.7	43.1	14.3	7.2	9.2	9.5
			M-F	100	7.2	25.5	–	–	–	3.0	48.3	16.0	8.0	10.2	10.6
185	HAY, EARLY BLOOM, SUN-CURED	1-20-882	A-F	91	5.5	29.1	–	–	–	3.6	37.9	14.8	10.8	–	–
			M-F	100	6.0	32.0	–	–	–	4.0	41.7	16.3	11.9	–	–
186	HAY, MIDBLOOM, SUN-CURED	1-02-554	A-F	91	4.5	26.2	–	–	–	2.3	46.5	11.4	6.1	6.5	7.2
			M-F	100	4.9	28.8	–	–	–	2.5	51.2	12.6	6.7	7.2	7.9
187	HAY, FULL BLOOM, SUN-CURED	1-20-887	A-F	89	5.0	27.4	–	–	–	1.9	42.2	12.8	8.5	7.8	8.3
			M-F	100	5.6	30.7	–	–	–	2.1	47.3	14.3	9.5	8.7	9.2
188	LEAVES, SUN-CURED	1-02-564	A-F	89	6.3	20.1	–	–	–	2.8	42.8	17.4	12.5	11.8	11.8
			M-F	100	7.1	22.5	–	–	–	3.2	47.8	19.4	14.0	13.2	13.2
189	STEMS, SUN-CURED	1-02-565	A-F	89	3.6	38.2	–	–	–	1.1	38.1	8.5	4.7	4.0	4.9
			M-F	100	4.1	42.7	–	–	–	1.2	42.6	9.5	5.2	4.4	5.5
190	STRAW	1-08-452	A-F	90	4.6	29.2	–	–	–	2.3	47.1	6.8	3.2	2.5	3.6
			M-F	100	5.1	32.4	–	–	–	2.6	52.3	7.6	3.6	2.8	4.0
	LESPEDEZA, COMMON-KOREAN *Lespedeza striata-stipulacea*														
191	HAY, PREBLOOM, SUN-CURED	1-26-024	A-F	92	5.4	27.6	–	–	–	2.9	43.6	12.5	6.2	–	8.4
			M-F	100	5.9	30.0	–	–	–	3.1	47.4	13.6	6.7	–	9.1
192	HAY, EARLY BLOOM, SUN-CURED	1-26-025	A-F	93	6.3	28.5	–	–	–	4.1	39.6	14.5	–	–	10.0
			M-F	100	6.8	30.6	–	–	–	4.4	42.6	15.6	–	–	10.8

DRY FORAGES

Entry Number	TDN Ruminant	TDN Swine	TDN Horse	DE Ruminant		DE Swine		DE Horse		ME Ruminant		ME Swine		Poultry ME_n		ME Horse		NE Ruminant NE_m		NE Ruminant NE_g		NE Lactating Cows NE_{lc}	
	%	%	%	Mcal		kcal		Mcal		Mcal		kcal		kcal		Mcal		Mcal		Mcal		Mcal	
				lb	kg	lb	kg	lb	kg	lb	kg	lb	kg	lb	kg	lb	kg	lb	kg	lb	kg	lb	kg
165	45	22	31	0.87	1.92	–	–	0.62	1.37	0.70	1.55	–	–	–	–	0.51	1.12	0.36	0.80	0.15	0.32	0.42	0.93
	52	25	36	1.00	2.21	–	–	0.71	1.57	0.81	1.78	–	–	–	–	0.58	1.29	0.42	0.91	0.17	0.37	0.48	1.07
166	51	34	42	0.99	2.19	–	–	0.81	1.78	0.82	1.80	–	–	–	–	0.66	1.46	0.45	0.99	0.22	0.49	0.49	1.08
	56	37	47	1.10	2.42	–	–	0.90	1.98	0.91	2.00	–	–	–	–	0.74	1.62	0.50	1.10	0.25	0.55	0.54	1.20
167	54	50	48	1.07	2.36	–	–	0.90	1.99	0.90	1.98	–	–	–	–	0.74	1.63	0.54	1.18	0.30	0.67	0.55	1.21
	60	56	54	1.20	2.65	–	–	1.02	2.24	1.01	2.23	–	–	–	–	0.83	1.84	0.60	1.32	0.34	0.75	0.62	1.36
168	51	31	35	1.01	2.23	–	–	0.68	1.51	0.84	1.85	–	–	–	–	0.56	1.24	0.48	1.06	0.25	0.56	0.51	1.13
	56	34	39	1.12	2.46	–	–	0.76	1.66	0.93	2.04	–	–	–	–	0.62	1.37	0.53	1.18	0.28	0.61	0.57	1.25
169	42	35	33	0.91	2.00	–	–	0.66	1.45	0.74	1.62	–	–	–	–	0.54	1.19	0.45	1.00	0.23	0.50	0.49	1.08
	48	39	38	1.02	2.26	–	–	0.74	1.64	0.83	1.83	–	–	–	–	0.61	1.34	0.51	1.13	0.26	0.57	0.55	1.22
170	42	42	–	0.84	1.85	–	–	–	–	0.66	1.45	–	–	–	–	–	–	–	–	–	–	–	–
	46	45	–	0.91	2.01	–	–	–	–	0.72	1.58	–	–	–	–	–	–	–	–	–	–	–	–
171	55	38	39	1.04	2.29	–	–	0.76	1.68	0.86	1.90	–	–	–	–	0.62	1.38	0.46	1.00	0.23	0.50	0.50	1.09
	60	42	43	1.14	2.50	–	–	0.83	1.84	0.94	2.08	–	–	–	–	0.68	1.51	0.50	1.10	0.25	0.54	0.54	1.20
172	52	37	37	1.01	2.22	–	–	0.71	1.57	0.83	1.83	–	–	–	–	0.59	1.29	0.45	1.00	0.23	0.50	0.49	1.08
	58	41	41	1.12	2.46	–	–	0.79	1.75	0.92	2.03	–	–	–	–	0.65	1.43	0.50	1.11	0.25	0.55	0.55	1.20
173	47	24	30	0.93	2.05	–	–	0.59	1.30	0.77	1.71	–	–	–	–	0.48	1.07	0.45	0.99	0.24	0.53	0.47	1.05
	57	30	36	1.14	2.51	–	–	0.72	1.59	0.95	2.08	–	–	–	–	0.59	1.30	0.55	1.21	0.29	0.64	0.58	1.28
174	29	–	–	0.58	1.27	–	–	–	–	0.40	0.87	–	–	–	–	–	–	–	–	–	–	–	–
	32	–	–	0.63	1.40	–	–	–	–	0.44	0.96	–	–	–	–	–	–	–	–	–	–	–	–
175	54	44	37	1.06	2.34	–	–	0.73	1.60	0.89	1.96	–	–	–	–	0.60	1.31	0.51	1.12	0.28	0.62	0.53	1.17
	61	50	42	1.19	2.63	–	–	0.81	1.80	1.00	2.20	–	–	–	–	0.67	1.47	0.57	1.26	0.31	0.69	0.60	1.31
176	55	53	41	1.10	2.42	–	–	0.79	1.73	0.93	2.04	–	–	–	–	0.64	1.42	0.57	1.25	0.33	0.72	0.57	1.26
	61	59	46	1.22	2.69	–	–	0.87	1.93	1.03	2.27	–	–	–	–	0.72	1.58	0.63	1.38	0.36	0.80	0.64	1.41
177	41	34	32	0.89	1.97	–	–	0.64	1.40	0.72	1.59	–	–	–	–	0.52	1.15	0.45	1.00	0.23	0.50	0.49	1.08
	46	38	36	1.00	2.21	–	–	0.71	1.57	0.81	1.78	–	–	–	–	0.59	1.29	0.51	1.12	0.25	0.56	0.55	1.21
178	53	41	45	1.03	2.28	–	–	0.85	1.87	0.86	1.90	–	–	–	–	0.70	1.53	0.49	1.08	0.26	0.58	0.52	1.14
	59	46	51	1.17	2.57	–	–	0.96	2.11	0.97	2.14	–	–	–	–	0.78	1.73	0.55	1.22	0.30	0.65	0.58	1.28
179	50	34	41	1.00	2.21	–	–	0.79	1.73	0.82	1.82	–	–	–	–	0.65	1.42	0.47	1.03	0.24	0.52	0.51	1.11
	55	38	45	1.10	2.41	–	–	0.86	1.90	0.90	1.99	–	–	–	–	0.71	1.56	0.51	1.13	0.26	0.57	0.55	1.22
180	58	60	57	1.16	2.55	–	–	1.05	2.31	0.99	2.18	–	–	–	–	0.86	1.89	0.58	1.27	0.34	0.75	0.58	1.28
	65	67	64	1.30	2.87	–	–	1.17	2.59	1.11	2.44	–	–	–	–	0.96	2.12	0.65	1.43	0.38	0.84	0.65	1.44
181	58	55	54	1.13	2.49	–	–	1.00	2.20	0.96	2.11	–	–	–	–	0.82	1.80	0.55	1.22	0.32	0.70	0.56	1.24
	65	62	60	1.27	2.80	–	–	1.12	2.47	1.08	2.37	–	–	–	–	0.92	2.02	0.62	1.37	0.36	0.79	0.63	1.40
182	53	41	45	1.03	2.28	–	–	0.85	1.87	0.86	1.90	–	–	–	–	0.70	1.53	0.49	1.08	0.26	0.58	0.52	1.14
	59	46	51	1.17	2.57	–	–	0.96	2.11	0.97	2.14	–	–	–	–	0.78	1.73	0.55	1.22	0.30	0.65	0.58	1.28
183	44	46	45	0.98	2.15	–	–	0.86	1.89	0.80	1.77	–	–	–	–	0.70	1.55	0.53	1.16	0.29	0.65	0.54	1.19
	50	52	51	1.10	2.42	–	–	0.97	2.13	0.91	2.00	–	–	–	–	0.79	1.75	0.59	1.30	0.33	0.73	0.61	1.35
184	49	53	48	1.04	2.30	–	–	0.90	1.99	0.87	1.92	–	–	–	–	0.74	1.63	0.55	1.21	0.32	0.70	0.56	1.24
	55	59	54	1.17	2.57	–	–	1.01	2.22	0.97	2.15	–	–	–	–	0.83	1.82	0.62	1.36	0.35	0.78	0.63	1.38
185	52	–	–	1.03	2.28	–	–	0.87	1.91	–	–	–	–	–	–	–	–	–	–	–	–	–	–
	57	–	–	1.13	2.50	–	–	0.95	2.10	–	–	–	–	–	–	–	–	–	–	–	–	–	–
186	52	48	48	1.06	2.34	–	–	0.90	1.97	0.89	1.95	–	–	–	–	0.73	1.62	0.54	1.18	0.30	0.66	0.55	1.22
	57	53	52	1.17	2.58	–	–	0.99	2.18	0.98	2.15	–	–	–	–	0.81	1.78	0.59	1.30	0.33	0.73	0.61	1.35
187	53	44	48	1.05	2.32	–	–	0.90	1.98	0.88	1.94	–	–	–	–	0.74	1.62	0.51	1.12	0.28	0.62	0.53	1.17
	59	49	54	1.18	2.60	–	–	1.00	2.21	0.99	2.18	–	–	–	–	0.82	1.82	0.57	1.26	0.31	0.69	0.60	1.31
188	52	60	53	1.10	2.42	–	–	0.99	2.19	0.92	2.04	–	–	–	–	0.81	1.79	0.59	1.29	0.35	0.77	0.59	1.30
	58	67	60	1.23	2.70	–	–	1.11	2.44	1.03	2.28	–	–	–	–	0.91	2.00	0.66	1.45	0.39	0.86	0.66	1.45
189	40	–	39	0.87	1.91	–	–	0.75	1.65	0.69	1.53	–	–	–	–	0.61	1.35	0.42	0.92	0.20	0.43	0.47	1.03
	45	–	43	0.97	2.14	–	–	0.84	1.84	0.78	1.71	–	–	–	–	0.69	1.51	0.47	1.03	0.22	0.48	0.52	1.15
190	44	38	39	0.95	2.10	–	–	0.75	1.64	0.78	1.72	–	–	–	–	0.61	1.35	0.49	1.07	0.26	0.56	0.52	1.14
	49	43	43	1.06	2.33	–	–	0.83	1.83	0.86	1.90	–	–	–	–	0.68	1.50	0.54	1.19	0.28	0.63	0.57	1.26
191	52	–	–	1.05	2.31	–	–	0.80	1.77	0.86	1.90	–	–	–	–	0.66	1.44	0.51	1.13	0.23	0.51	0.53	1.18
	57	–	–	1.14	2.51	–	–	0.87	1.92	0.93	2.06	–	–	–	–	0.71	1.57	0.56	1.23	0.25	0.55	0.58	1.28
192	54	–	38	1.08	2.37	–	–	0.71	1.56	0.88	1.94	–	–	–	–	0.58	1.27	0.53	1.16	0.24	0.53	0.55	1.21
	58	–	41	1.16	2.55	–	–	0.76	1.68	0.95	2.09	–	–	–	–	0.62	1.37	0.57	1.25	0.26	0.57	0.59	1.30

(Continued)

TABLE V-1C DRY FORAGES, MINERAL AND VITAMIN COMPOSITION OF FEEDS, DATA EXPRESSED **AS-FED** AND **MOISTURE-FREE**—*(Continued)*

Entry Number	Feed Name Description	Moisture Basis: A-F (as-fed) or M-F (moisture-free)	Dry Matter	Macro Minerals							Micro Minerals						
				Calcium (Ca)	Phosphorus (P)	Sodium (Na)	Chlorine (Cl)	Magnesium (Mg)	Potassium (K)	Sulfur (S)	Cobalt (Co)	Copper (Cu)	Iodine (I)	Iron (Fe)	Manganese (Mn)	Selenium (Se)	Zinc (Zn)
			%	%	%	%	%	%	%	%	ppm or mg/kg	ppm or mg/kg	ppm or mg/kg	%	ppm or mg/kg	ppm or mg/kg	ppm or mg/kg
	HORSEBEAN (Continued)																
165	STRAW	**A-F**	**87**	**–**	**–**	**–**	**–**	**–**	**–**	**–**	**–**	**–**	**–**	**–**	**–**	**–**	**–**
		M-F	100	–	–	–	–	–	–	–	–	–	–	–	–	–	–
	HYACINTH BEAN (DOLICHOS) *Dolichos lablab*																
166	HAY, SUN-CURED	**A-F**	**90**	**–**	**–**	**–**	**–**	**–**	**–**	**–**	**–**	**–**	**–**	**–**	**–**	**–**	**–**
		M-F	100	–	–	–	–	–	–	–	–	–	–	–	–	–	–
	INDIGO *Indigofera* spp																
167	HAY, CUT 1, SUN-CURED	**A-F**	**89**	**1.34**	**0.19**	**–**	**–**	**0.50**	**–**	**–**	**–**	**–**	**–**	**–**	**–**	**–**	**–**
		M-F	100	1.50	0.21	–	–	0.56	–	–	–	–	–	–	–	–	–
	JOHNSONGRASS SORGHUM *Sorghum halepense*																
168	HAY, SUN-CURED	**A-F**	**91**	**0.80**	**0.27**	**0.01**	**–**	**0.31**	**1.22**	**0.09**	**–**	**–**	**–**	**0.054**	**–**	**–**	**–**
		M-F	100	0.89	0.30	0.01	–	0.35	1.35	0.10	–	–	–	0.059	–	–	–
	JUNEGRASS *Koeleria cristata*																
169	HAY, SUN-CURED	**A-F**	**89**	**–**	**–**	**–**	**–**	**–**	**–**	**–**	**–**	**–**	**–**	**–**	**–**	**–**	**–**
		M-F	100	–	–	–	–	–	–	–	–	–	–	–	–	–	–
	KAFIR SORGHUM *Sorghum bicolor, caffrorum*																
170	CHAFF	**A-F**	**92**	**–**	**–**	**–**	**–**	**–**	**–**	**–**	**–**	**–**	**–**	**–**	**–**	**–**	**–**
		M-F	100	–	–	–	–	–	–	–	–	–	–	–	–	–	–
171	FODDER WITH HEADS, LATE DOUGH, SUN-CURED	**A-F**	**91**	**–**	**–**	**–**	**–**	**–**	**–**	**–**	**–**	**–**	**–**	**–**	**–**	**–**	**–**
		M-F	100	–	–	–	–	–	–	–	–	–	–	–	–	–	–
172	FODDER WITH HEADS, SUN-CURED	**A-F**	**90**	**0.35**	**0.18**	**–**	**–**	**0.26**	**1.53**	**–**	**–**	**–**	**–**	**–**	**–**	**–**	**–**
		M-F	100	0.39	0.20	–	–	0.29	1.70	–	–	–	–	–	–	–	–
173	STOVER WITHOUT HEADS, SUN-CURED	**A-F**	**82**	**0.50**	**0.08**	**–**	**–**	**–**	**–**	**–**	**–**	**–**	**–**	**–**	**–**	**–**	**–**
		M-F	100	0.61	0.10	–	–	–	–	–	–	–	–	–	–	–	–
	KELP (SEAWEED) *Laminariales* (order), *Fucales* (order)																
174	WHOLE, DEHY	**A-F**	**91**	**2.47**	**0.28**	**–**	**–**	**0.85**	**–**	**–**	**–**	**–**	**–**	**–**	**–**	**–**	**–**
		M-F	100	2.72	0.31	–	–	0.93	–	–	–	–	–	–	–	–	–
	KENTUCKY BLUEGRASS *Poa pratensis*																
175	HAY, SUN-CURED	**A-F**	**89**	**0.40**	**0.27**	**0.10**	**0.55**	**0.19**	**1.66**	**0.12**	**–**	**8.8**	**–**	**0.025**	**76.2**	**–**	**–**
		M-F	100	0.45	0.30	0.11	0.62	0.21	1.87	0.13	–	9.9	–	0.028	85.6	–	–
176	HAY, EARLY BLOOM, SUN-CURED	**A-F**	**90**	**–**	**–**	**–**	**–**	**–**	**–**	**–**	**–**	**–**	**–**	**–**	**–**	**–**	**–**
		M-F	100	–	–	–	–	–	–	–	–	–	–	–	–	–	–
177	HAY, MATURE, SUN-CURED	**A-F**	**89**	**0.24**	**0.20**	**0.13**	**0.39**	**0.10**	**1.51**	**0.16**	**–**	**–**	**–**	**0.016**	**57.0**	**–**	**–**
		M-F	100	0.26	0.23	0.15	0.44	0.12	1.70	0.18	–	–	–	0.018	63.9	–	–
	KOA HAOLE (LEAD TREE, WHITE POPINAC) *Leucaena glauca*																
178	HAY, SUN-CURED	**A-F**	**89**	**–**	**–**	**–**	**–**	**–**	**–**	**–**	**–**	**–**	**–**	**–**	**–**	**–**	**–**
		M-F	100	–	–	–	–	–	–	–	–	–	–	–	–	–	–
	KUDZU *Pueraria* spp																
179	HAY, SUN-CURED	**A-F**	**91**	**2.15**	**0.32**	**–**	**–**	**0.73**	**–**	**–**	**–**	**–**	**–**	**–**	**–**	**–**	**–**
		M-F	100	2.35	0.35	–	–	0.80	–	–	–	–	–	–	–	–	–
	LADINO CLOVER *Trifolium repens*																
180	HAY, SUN-CURED	**A-F**	**89**	**1.30**	**0.30**	**0.12**	**0.27**	**0.42**	**2.17**	**0.19**	**0.144**	**8.4**	**0.268**	**0.042**	**109.7**	**–**	**15.2**
		M-F	100	1.45	0.34	0.13	0.30	0.47	2.44	0.21	0.161	9.4	0.301	0.047	123.1	–	17.0
	LADINO CLOVER-GRASS *Trifolium repens-grass*																
181	HAY, SUN-CURED	**A-F**	**89**	**0.95**	**0.22**	**–**	**–**	**0.30**	**1.78**	**–**	**–**	**14.3**	**–**	**0.016**	**39.6**	**–**	**–**
		M-F	100	1.07	0.24	–	–	0.34	2.00	–	–	16.1	–	0.017	44.5	–	–
	LEAD TREE, WHITE POPINAC (KOA HAOLE) *Leucaena glauca*																
182	HAY, SUN-CURED	**A-F**	**89**	**–**	**–**	**–**	**–**	**–**	**–**	**–**	**–**	**–**	**–**	**–**	**–**	**–**	**–**
		M-F	100	–	–	–	–	–	–	–	–	–	–	–	–	–	–
	LESPEDEZA, COMMON *Lespedeza striata*																
183	HAY, SUN-CURED, ALL ANALYSES	**A-F**	**89**	**0.78**	**0.25**	**–**	**–**	**0.20**	**1.21**	**0.21**	**–**	**7.1**	**–**	**0.019**	**99.6**	**–**	**21.3**
		M-F	100	0.88	0.29	–	–	0.23	1.37	0.24	–	8.0	–	0.022	112.3	–	24.0
184	HAY, PREBLOOM, SUN-CURED	**A-F**	**89**	**1.03**	**0.20**	**–**	**–**	**0.21**	**1.07**	**–**	**–**	**–**	**–**	**0.031**	**159.5**	**–**	**–**
		M-F	100	1.16	0.22	–	–	0.24	1.20	–	–	–	–	0.034	178.4	–	–
185	HAY, EARLY BLOOM, SUN-CURED	**A-F**	**91**	**1.07**	**0.22**	**–**	**–**	**0.23**	**1.00**	**–**	**–**	**–**	**–**	**0.364**	**232.5**	**–**	**–**
		M-F	100	1.17	0.24	–	–	0.25	1.10	–	–	–	–	0.400	255.5	–	–
186	HAY, MIDBLOOM, SUN-CURED	**A-F**	**91**	**1.07**	**0.17**	**–**	**–**	**0.22**	**0.94**	**–**	**–**	**–**	**–**	**0.029**	**210.6**	**–**	**–**
		M-F	100	1.18	0.19	–	–	0.24	1.04	–	–	–	–	0.032	231.9	–	–
187	HAY, FULL BLOOM, SUN-CURED	**A-F**	**89**	**1.02**	**0.18**	**–**	**–**	**0.20**	**0.93**	**–**	**–**	**–**	**–**	**0.027**	**135.2**	**–**	**–**
		M-F	100	1.14	0.21	–	–	0.23	1.04	–	–	–	–	0.031	151.4	–	–
188	LEAVES, SUN-CURED	**A-F**	**89**	**1.30**	**0.20**	**–**	**–**	**–**	**0.92**	**–**	**–**	**–**	**–**	**–**	**–**	**–**	**–**
		M-F	100	1.46	0.22	–	–	–	1.03	–	–	–	–	–	–	–	–
189	STEMS, SUN-CURED	**A-F**	**89**	**0.64**	**0.13**	**–**	**–**	**–**	**0.89**	**–**	**–**	**–**	**–**	**–**	**–**	**–**	**–**
		M-F	100	0.72	0.15	–	–	–	1.00	–	–	–	–	–	–	–	–
190	STRAW	**A-F**	**90**	**–**	**–**	**–**	**–**	**–**	**–**	**–**	**–**	**–**	**–**	**–**	**–**	**–**	**–**
		M-F	100	–	–	–	–	–	–	–	–	–	–	–	–	–	–
	LESPEDEZA, COMMON-KOREAN *Lespedeza striata-stipulacea*																
191	HAY, PREBLOOM, SUN-CURED	**A-F**	**92**	**1.07**	**0.20**	**–**	**–**	**0.22**	**1.10**	**–**	**–**	**–**	**–**	**0.032**	**164.1**	**–**	**–**
		M-F	100	1.16	0.22	–	–	0.24	1.20	–	–	–	–	0.035	178.4	–	–
192	HAY, EARLY BLOOM, SUN-CURED	**A-F**	**93**	**1.09**	**0.22**	**–**	**–**	**0.23**	**1.02**	**–**	**0.038**	**0.2**	**–**	**0.038**	**237.6**	**–**	**–**
		M-F	100	1.17	0.24	–	–	0.25	1.10	–	0.040	0.2	–	0.041	255.5	–	–

DRY FORAGES

Entry Number	Fat-Soluble Vitamins					Water-Soluble Vitamins								
	A (1 mg Carotene = 1667 IU Vit A)	Carotene (Provitamin A)	D	E	K	B–12	Biotin	Choline	Folacin (Folic Acid)	Niacin	Pantothenic Acid (B–3)	(Pyridoxine) B–6	Riboflavin (B–2)	Thiamin (B–1)
	IU/g	ppm or mg/kg	IU/kg	ppm or mg/kg	ppm or mg/kg	ppm or mg/kg	ppm or mg/kg	ppm or mg/kg	ppm or mg/kg	ppm or mg/kg	ppm or mg/kg	ppm or mg/kg	ppm or mg/kg	ppm or mg/kg
165	**–** –	**–** –	**–** –	**–** –	**–** –	**–** –	**–** –	**–** –	**–** –	**–** –	**–** –	**–** –	**–** –	**–** –
166	**–** –	**–** –	**–** –	**–** –	**–** –	**–** –	**–** –	**–** –	**–** –	**–** –	**–** –	**–** –	**–** –	**–** –
167	**–** –	**–** –	**–** –	**–** –	**–** –	**–** –	**–** –	**–** –	**–** –	**–** –	**–** –	**–** –	**–** –	**–** –
168	**58.8** 64.9	**35.3** 38.9	**–** –	**–** –	**–** –	**–** –	**–** –	**–** –	**–** –	**–** –	**–** –	**–** –	**–** –	**–** –
169	**–** –	**–** –	**–** –	**–** –	**–** –	**–** –	**–** –	**–** –	**–** –	**–** –	**–** –	**–** –	**–** –	**–** –
170	**–** –	**–** –	**–** –	**–** –	**–** –	**–** –	**–** –	**–** –	**–** –	**–** –	**–** –	**–** –	**–** –	**–** –
171	**–** –	**–** –	**–** –	**–** –	**–** –	**–** –	**–** –	**–** –	**–** –	**–** –	**–** –	**–** –	**–** –	**–** –
172	**–** –	**–** –	**–** –	**–** –	**–** –	**–** –	**–** –	**–** –	**–** –	**–** –	**–** –	**–** –	**–** –	**–** –
173	**–** –	**–** –	**–** –	**–** –	**–** –	**–** –	**–** –	**–** –	**–** –	**–** –	**–** –	**–** –	**–** –	**–** –
174	**–** –	**–** –	**–** –	**–** –	**–** –	**–** –	**–** –	**–** –	**–** –	**–** –	**–** –	**–** –	**–** –	**–** –
175	**–** –	**–** –	**–** –	**–** –	**–** –	**–** –	**–** –	**–** –	**–** –	**–** –	**–** –	**–** –	**9.9** 11.1	**–** –
176	**–** –	**–** –	**–** –	**–** –	**–** –	**–** –	**–** –	**–** –	**–** –	**–** –	**–** –	**–** –	**–** –	**–** –
177	**–** –	**–** –	**–** –	**–** –	**–** –	**–** –	**–** –	**–** –	**–** –	**–** –	**–** –	**–** –	**–** –	**–** –
178	**–** –	**–** –	**–** –	**–** –	**–** –	**–** –	**–** –	**–** –	**–** –	**–** –	**–** –	**–** –	**–** –	**–** –
179	**66.9** 73.2	**40.1** 43.9	**–** –	**–** –	**–** –	**–** –	**–** –	**–** –	**–** –	**–** –	**–** –	**–** –	**7.6** 8.4	**–** –
180	**239.5** 268.7	**143.7** 161.2	**–** –	**–** –	**–** –	**–** –	**–** –	**–** –	**–** –	**10** 11	**1.0** 1.1	**–** –	**15.2** 17.0	**3.7** 4.2
181	**94.6** 106.2	**56.7** 63.7	**–** –	**–** –	**–** –	**–** –	**–** –	**–** –	**–** –	**–** –	**–** –	**–** –	**8.2** 9.3	**13.1** 14.8
182	**–** –	**–** –	**–** –	**–** –	**–** –	**–** –	**–** –	**–** –	**–** –	**–** –	**–** –	**–** –	**–** –	**–** –
183	**73.9** 83.3	**44.3** 50.0	**–** –	**–** –	**–** –	**–** –	**–** –	**–** –	**–** –	**–** –	**–** –	**–** –	**8.7** 9.8	**–** –
184	**–** –	**–** –	**–** –	**–** –	**–** –	**–** –	**–** –	**–** –	**–** –	**–** –	**–** –	**–** –	**–** –	**–** –
185	**–** –	**–** –	**–** –	**–** –	**–** –	**–** –	**–** –	**–** –	**–** –	**–** –	**–** –	**–** –	**–** –	**–** –
186	**–** –	**–** –	**–** –	**–** –	**–** –	**–** –	**–** –	**–** –	**–** –	**–** –	**–** –	**–** –	**–** –	**–** –
187	**–** –	**–** –	**–** –	**–** –	**–** –	**–** –	**–** –	**–** –	**–** –	**–** –	**–** –	**–** –	**–** –	**–** –
188	**–** –	**–** –	**–** –	**–** –	**–** –	**–** –	**–** –	**–** –	**–** –	**–** –	**–** –	**–** –	**–** –	**–** –
189	**–** –	**–** –	**–** –	**–** –	**–** –	**–** –	**–** –	**–** –	**–** –	**–** –	**–** –	**–** –	**–** –	**–** –
190	**–** –	**–** –	**–** –	**–** –	**–** –	**–** –	**–** –	**–** –	**–** –	**–** –	**–** –	**–** –	**–** –	**–** –
191	**220.8** 240.0	**132.5** 144.0	**–** –	**–** –	**–** –	**–** –	**–** –	**–** –	**–** –	**–** –	**–** –	**–** –	**–** –	**–** –
192	**213.2** 229.2	**127.9** 137.5	**–** –	**–** –	**–** –	**–** –	**–** –	**–** –	**–** –	**–** –	**–** –	**–** –	**–** –	**–** –

(Continued)

DRY FORAGES

TABLE V-1C DRY FORAGES, COMPOSITION OF FEEDS, DATA EXPRESSED **AS-FED** AND **MOISTURE-FREE**—*(Continued)*

Entry Number	Feed Name Description	International Feed Number	Moisture Basis: A-F (as-fed) or M-F (moisture-free)	Proximate Analysis: Dry Matter	Ash	Crude Fiber	Neutral Det. Fib. (NDF)	Acid Det. Fib. (ADF)	Lignin	Ether Extract (Fat)	N-Free Extract	Crude Protein	Digestible Protein: Ruminant	Swine	Horse
				%	%	%	%	%	%	%	%	%	%	%	%
	LESPEDEZA, COMMON-KOREAN (Continued)														
193	HAY, MIDBLOOM, SUN-CURED	1-26-026	A-F	93	4.3	27.9	–	–	–	2.5	47.2	11.1	6.0	–	7.2
			M-F	100	4.6	30.0	–	–	–	2.7	50.8	11.9	6.5	–	7.7
194	HAY, FULL BLOOM, SUN-CURED	1-26-027	A-F	93	5.3	30.2	–	–	–	3.0	41.1	13.4	–	–	9.0
			M-F	100	5.7	32.5	–	–	–	3.2	44.2	14.4	–	–	9.7
	LESPEDEZA, SERICEA (CHINESE LESPEDEZA) *Lespedeza cuneata*														
195	HAY, SUN-CURED, ALL ANALYSES	1-02-607	A-F	90	4.8	29.9	–	–	–	2.0	42.9	10.7	3.5	5.9	6.6
			M-F	100	5.3	33.1	–	–	–	2.2	47.5	11.8	3.9	6.5	7.4
196	HAY, PREBLOOM, SUN-CURED	1-09-172	A-F	93	5.2	20.6	–	–	–	5.3	44.6	17.3	9.0	11.6	11.7
			M-F	100	5.5	22.2	–	–	–	5.8	47.9	18.6	9.7	12.5	12.6
197	HAY, EARLY BLOOM, SUN-CURED	1-02-600	A-F	95	5.2	22.1	–	–	–	5.3	45.4	16.7	8.5	11.0	11.2
			M-F	100	5.5	23.3	–	–	–	5.6	48.0	17.6	9.0	11.6	11.8
198	HAY, MIDBLOOM, SUN-CURED	1-09-173	A-F	90	4.7	20.9	–	–	–	4.6	44.7	15.2	6.8	9.9	10.1
			M-F	100	5.2	23.2	–	–	–	5.1	49.7	16.9	7.5	10.9	11.2
199	HAY, FULL BLOOM, SUN-CURED	1-02-601	A-F	94	4.1	23.1	–	–	–	5.0	48.6	13.6	4.9	8.3	8.8
			M-F	100	4.3	24.4	–	–	–	5.3	51.5	14.4	5.2	8.8	9.4
	LUCERNE–SEE ALFALFA														
	MAIZE–SEE CORN														
	MARSH PLANTS														
200	HAY, SUN-CURED	1-03-182	A-F	89	6.9	28.7	–	–	–	2.1	42.4	8.8	4.7	4.3	5.2
			M-F	100	7.7	32.3	–	–	–	2.4	47.7	9.9	5.3	4.8	5.8
	MEADOW FESCUE *Festuca elatior*														
201	HAY, SUN-CURED	1-01-912	A-F	88	7.9	28.0	74.0	43.8	–	2.4	41.0	8.2	4.6	3.9	5.3
			M-F	100	9.0	32.0	84.6	50.0	–	2.7	46.9	9.4	5.3	4.4	6.0
	MEADOW FOXTAIL *Alopecurus pratensis*														
202	HAY, SUN-CURED	1-02-072	A-F	90	8.4	25.5	–	–	–	2.0	41.1	12.8	8.5	7.7	8.2
			M-F	100	9.4	28.4	–	–	–	2.2	45.8	14.2	9.5	8.6	9.2
	MEADOW PLANTS, INTERMOUNTAIN														
203	HAY, SUN-CURED	1-03-181	A-F	95	8.2	31.2	–	–	–	2.4	45.2	8.2	5.0	3.6	4.7
			M-F	100	8.6	32.7	–	–	–	2.5	47.5	8.7	5.2	3.8	4.9
	MILLET, FOXTAIL *Setaria italica*														
204	HAY, SUN-CURED	1-03-099	A-F	87	7.5	25.7	–	–	–	2.5	43.7	7.5	3.7	3.2	4.2
			M-F	100	8.6	29.6	–	–	–	2.9	50.3	8.6	4.3	3.7	4.8
205	STRAW	1-03-095	A-F	90	5.5	37.4	–	–	–	1.6	41.4	3.8	0.6	–0.1	1.3
			M-F	100	6.1	41.7	–	–	–	1.8	46.2	4.2	0.7	–0.2	1.5
	MILLET, JAPANESE (BARNYARD GRASS) *Echinochloa crusgalli*														
206	HAY, SUN-CURED	1-03-105	A-F	87	8.5	26.3	–	–	–	1.7	41.8	8.8	5.4	4.4	5.3
			M-F	100	9.8	30.2	–	–	–	2.0	48.0	10.1	6.2	5.0	6.0
	MILLET, PROSO (BROOMCORN; HOG MILLET) *Panicum miliaceum*														
207	HAY, SUN-CURED	1-03-119	A-F	90	7.6	24.4	–	–	–	2.1	46.8	9.4	5.5	4.8	5.6
			M-F	100	8.4	27.0	–	–	–	2.4	51.8	10.4	6.1	5.3	6.2
	MILO SORGHUM *Sorghum bicolor, subglabrescens*														
208	FODDER WITH HEADS, SUN-CURED	1-04-433	A-F	89	8.5	21.8	–	–	–	2.9	49.4	6.5	2.5	2.3	3.4
			M-F	100	9.6	24.4	–	–	–	3.2	55.5	7.3	2.8	2.5	3.8
209	STOVER WITHOUT HEADS, SUN-CURED	1-04-434	A-F	91	11.8	31.1	–	–	–	1.3	42.5	4.4	–0.8	0.4	1.8
			M-F	100	13.0	34.1	–	–	–	1.4	46.6	4.9	–0.9	0.4	2.0
	MINT *Mentha* spp														
210	HAY, SUN-CURED	1-03-124	A-F	88	7.7	20.5	–	–	–	2.1	44.6	12.8	8.5	7.9	8.3
			M-F	100	8.8	23.3	–	–	–	2.4	50.9	14.6	9.7	9.0	9.5
	MOSS, SPHAGNUM *Sphagnum* spp														
211	WHOLE, MOLASSES ADDED, SUN-CURED	1-03-135	A-F	82	7.5	7.4	–	–	–	0.5	57.6	9.3	3.9	5.0	5.7
			M-F	100	9.1	9.0	–	–	–	0.6	70.0	11.3	4.7	6.1	6.9
	MUHLY, BUSH *Muhlenbergia porteri*														
212	HAY, SUN-CURED	1-03-139	A-F	93	16.4	27.9	–	–	–	1.6	42.3	5.1	0.2	0.9	2.3
			M-F	100	17.6	29.9	–	–	–	1.7	45.3	5.5	0.2	1.0	2.5
	MUSTARD *Brassica* spp														
213	HAY, SUN-CURED	1-03-153	A-F	90	5.5	29.7	–	–	–	10.4	28.5	15.9	11.2	10.5	10.7
			M-F	100	6.1	33.0	–	–	–	11.5	31.7	17.7	12.5	11.7	11.9
	NAPIERGRASS *Pennisetum purpureum*														
214	HAY, SUN-CURED	1-08-462	A-F	89	10.5	34.0	–	–	–	1.8	34.6	8.2	3.8	3.8	4.7
			M-F	100	11.8	38.2	–	–	–	2.0	38.8	9.2	4.2	4.2	5.3
	NATALGRASS *Rhynchelytrum roseum*														
215	HAY, SUN-CURED	1-03-180	A-F	91	5.5	37.0	–	–	–	1.7	40.2	6.6	3.0	2.3	3.5
			M-F	100	6.0	40.6	–	–	–	1.9	44.2	7.3	3.3	2.5	3.8

DRY FORAGES

Entry Number	TDN Ruminant	TDN Swine	TDN Horse	Digestible Energy Ruminant		Digestible Energy Swine		Digestible Energy Horse		Metabolizable Energy Ruminant		Metabolizable Energy Swine		Poultry ME_n		Metabolizable Energy Horse		Net Energy Ruminant NE_m		Net Energy Ruminant NE_g		Lactating Cows NE_{lc}	
	%	%	%	Mcal lb	Mcal kg	kcal lb	kcal kg	Mcal lb	Mcal kg	Mcal lb	Mcal kg	kcal lb	kcal kg	kcal lb	kcal kg	Mcal lb	Mcal kg	Mcal lb	Mcal kg	Mcal lb	Mcal kg	Mcal lb	Mcal kg
193	55	–	44	1.09	2.41	–	–	0.80	1.77	0.89	1.97	–	–	–	–	0.66	1.45	0.54	1.19	0.26	0.57	0.56	1.23
	59	–	47	1.18	2.59	–	–	0.86	1.90	0.96	2.12	–	–	–	–	0.71	1.56	0.58	1.28	0.28	0.61	0.60	1.32
194	54	–	44	1.08	2.37	–	–	0.80	1.76	0.88	1.94	–	–	–	–	0.65	1.44	0.53	1.17	0.25	0.54	0.55	1.21
	58	–	47	1.16	2.55	–	–	0.86	1.89	0.95	2.09	–	–	–	–	0.70	1.55	0.57	1.26	0.26	0.58	0.59	1.30
195	42	40	44	0.93	2.06	–	–	0.83	1.84	0.76	1.67	–	–	–	–	0.68	1.51	0.50	1.09	0.26	0.58	0.52	1.15
	46	45	49	1.03	2.27	–	–	0.92	2.03	0.84	1.85	–	–	–	–	0.76	1.67	0.55	1.21	0.29	0.64	0.58	1.28
196	60	74	46	1.25	2.75	–	–	0.88	1.93	1.07	2.36	–	–	–	–	0.72	1.59	0.68	1.50	0.43	0.94	0.67	1.47
	65	79	50	1.34	2.96	–	–	0.94	2.08	1.15	2.54	–	–	–	–	0.77	1.70	0.73	1.62	0.46	1.01	0.72	1.58
197	59	72	46	1.24	2.72	–	–	0.87	1.92	1.06	2.33	–	–	–	–	0.71	1.57	0.68	1.50	0.42	0.93	0.67	1.47
	62	76	48	1.31	2.88	–	–	0.92	2.03	1.12	2.46	–	–	–	–	0.76	1.66	0.72	1.58	0.45	0.98	0.71	1.56
198	57	67	45	1.18	2.60	–	–	0.86	1.89	1.01	2.22	–	–	–	–	0.70	1.55	0.64	1.40	0.39	0.86	0.63	1.38
	64	74	50	1.31	2.89	–	–	0.95	2.10	1.12	2.47	–	–	–	–	0.78	1.72	0.71	1.56	0.44	0.96	0.70	1.54
199	57	68	44	1.21	2.66	–	–	0.84	1.86	1.03	2.27	–	–	–	–	0.69	1.53	0.66	1.46	0.41	0.90	0.65	1.44
	61	72	47	1.28	2.82	–	–	0.89	1.97	1.09	2.40	–	–	–	–	0.73	1.62	0.70	1.55	0.43	0.95	0.69	1.53
200	47	36	38	0.96	2.12	–	–	0.73	1.61	0.79	1.74	–	–	–	–	0.60	1.32	0.46	1.01	0.23	0.51	0.49	1.09
	53	40	43	1.08	2.38	–	–	0.82	1.82	0.89	1.96	–	–	–	–	0.68	1.49	0.52	1.14	0.26	0.58	0.56	1.22
201	53	35	37	1.06	2.35	–	–	0.72	1.58	0.90	1.98	–	–	–	–	0.59	1.30	0.55	1.20	0.32	0.70	0.56	1.22
	61	40	42	1.22	2.68	–	–	0.82	1.81	1.02	2.26	–	–	–	–	0.67	1.48	0.62	1.38	0.36	0.80	0.64	1.40
202	57	42	44	1.08	2.37	–	–	0.84	1.86	0.90	1.99	–	–	–	–	0.69	1.52	0.48	1.07	0.25	0.56	0.51	1.13
	64	47	49	1.20	2.64	–	–	0.94	2.07	1.10	2.22	–	–	–	–	0.77	1.69	0.54	1.19	0.28	0.62	0.57	1.26
203	55	36	37	1.06	2.34	–	–	0.73	1.61	0.88	1.93	–	–	–	–	0.60	1.32	0.47	1.04	0.23	0.51	0.52	1.14
	58	38	39	1.11	2.45	–	–	0.77	1.69	0.92	2.03	–	–	–	–	0.63	1.38	0.50	1.10	0.25	0.54	0.54	1.20
204	51	38	35	1.00	2.19	–	–	0.68	1.49	0.83	1.83	–	–	–	–	0.56	1.22	0.46	1.01	0.24	0.52	0.49	1.08
	59	43	40	1.15	2.52	–	–	0.78	1.72	0.95	2.10	–	–	–	–	0.64	1.41	0.53	1.16	0.27	0.60	0.56	1.24
205	42	–	29	0.87	1.91	–	–	0.59	1.30	0.69	1.53	–	–	–	–	0.48	1.07	0.39	0.85	0.16	0.36	0.44	0.98
	47	–	33	0.97	2.13	–	–	0.66	1.45	0.77	1.70	–	–	–	–	0.54	1.19	0.43	0.95	0.18	0.40	0.50	1.09
206	47	34	38	0.94	2.07	–	–	0.73	1.62	0.77	1.70	–	–	–	–	0.60	1.32	0.43	0.95	0.21	0.46	0.47	1.04
	54	39	44	1.08	2.38	–	–	0.84	1.85	0.89	1.95	–	–	–	–	0.69	1.52	0.49	1.09	0.24	0.53	0.54	1.19
207	51	42	42	1.02	2.24	–	–	0.81	1.79	0.84	1.86	–	–	–	–	0.66	1.46	0.49	1.08	0.26	0.57	0.52	1.14
	56	47	47	1.13	2.48	–	–	0.90	1.98	0.93	2.06	–	–	–	–	0.74	1.62	0.54	1.19	0.29	0.63	0.57	1.27
208	57	43	36	1.07	2.36	–	–	0.70	1.54	0.90	1.99	–	–	–	–	0.57	1.26	0.49	1.07	0.26	0.57	0.51	1.13
	63	48	40	1.20	2.65	–	–	0.78	1.73	1.01	2.23	–	–	–	–	0.64	1.42	0.55	1.20	0.29	0.64	0.58	1.27
209	48	19	29	0.96	2.11	–	–	0.59	1.29	0.78	1.72	–	–	–	–	0.48	1.06	0.44	0.98	0.21	0.47	0.49	1.07
	52	21	32	1.05	2.31	–	–	0.64	1.42	0.86	1.88	–	–	–	–	0.53	1.16	0.49	1.07	0.23	0.52	0.53	1.18
210	49	49	48	1.01	2.23	–	–	0.89	1.97	0.84	1.86	–	–	–	–	0.73	1.61	0.51	1.13	0.29	0.63	0.53	1.17
	56	55	54	1.15	2.54	–	–	1.02	2.25	0.96	2.12	–	–	–	–	0.84	1.84	0.59	1.29	0.33	0.72	0.61	1.34
211	42	–	–	0.84	1.85	–	–	–	–	0.68	1.50	–	–	–	–	–	–	–	–	–	–	–	–
	51	–	–	1.02	2.25	–	–	–	–	0.83	1.83	–	–	–	–	–	–	–	–	–	–	–	–
212	30	19	26	0.72	1.59	–	–	0.55	1.21	0.54	1.18	–	–	–	–	0.45	0.99	0.33	0.73	0.10	0.23	0.41	0.91
	32	21	28	0.77	1.70	–	–	0.59	1.29	0.58	1.27	–	–	–	–	0.48	1.06	0.36	0.78	0.11	0.25	0.44	0.98
213	–	–	–	–	–	–	–	–	–	–	–	–	–	–	–	–	–	–	–	–	–	–	–
	–	–	–	–	–	–	–	–	–	–	–	–	–	–	–	–	–	–	–	–	–	–	–
214	45	22	29	0.91	2.00	–	–	0.58	1.28	0.74	1.62	–	–	–	–	0.48	1.05	0.41	0.89	0.18	0.40	0.46	1.01
	51	25	32	1.02	2.25	–	–	0.65	1.43	0.83	1.82	–	–	–	–	0.53	1.18	0.46	1.00	0.21	0.45	0.51	1.13
215	48	–	33	0.95	2.10	–	–	0.66	1.46	0.78	1.71	–	–	–	–	0.54	1.20	0.42	0.92	0.19	0.42	0.47	1.04
	53	–	37	1.05	2.30	–	–	0.73	1.60	0.85	1.88	–	–	–	–	0.60	1.31	0.46	1.01	0.21	0.46	0.52	1.14

(Continued)

DRY FORAGES

TABLE V-1C DRY FORAGES, MINERAL AND VITAMIN COMPOSITION OF FEEDS, DATA EXPRESSED **AS-FED** AND **MOISTURE-FREE**—*(Continued)*

Entry Number	Feed Name Description	Moisture Basis: A-F (as-fed) or M-F (moisture-free)	Dry Matter	Macro Minerals							Micro Minerals						
				Calcium (Ca)	Phos-phorus (P)	Sodium (Na)	Chlo-rine (Cl)	Mag-nesium (Mg)	Potas-sium (K)	Sulfur (S)	Cobalt (Co)	Copper (Cu)	Iodine (I)	Iron (Fe)	Man-ganese (Mn)	Sele-nium (Se)	Zinc (Zn)
			%	%	%	%	%	%	%	%	ppm or mg/kg	ppm or mg/kg	ppm or mg/kg	%	ppm or mg/kg	ppm or mg/kg	ppm or mg/kg
	LESPEDEZA, COMMON-KOREAN (Continued)																
193	HAY, MIDBLOOM, SUN-CURED	**A-F**	**93**	**1.11**	**0.24**	–	–	**0.25**	**0.93**	–	–	–	–	**0.030**	**220.9**	–	–
		M-F	100	1.19	0.26	–	–	0.27	1.00	–	–	–	–	0.033	237.5	–	–
194	HAY, FULL BLOOM, SUN-CURED	**A-F**	**93**	**0.98**	**0.21**	–	–	**0.22**	**0.91**	–	–	–	–	**0.030**	**140.9**	–	–
		M-F	100	1.05	0.23	–	–	0.24	0.98	–	–	–	–	0.033	151.5	–	–
	LESPEDEZA, SERICEA (CHINESE LESPEDEZA) *Lespedeza cuneata*																
195	HAY, SUN-CURED, ALL ANALYSES	**A-F**	**90**	**0.93**	**0.22**	–	–	**0.20**	**0.99**	–	–	–	–	**0.027**	**91.1**	–	–
		M-F	100	1.03	0.25	–	–	0.22	1.10	–	–	–	–	0.030	100.8	–	–
196	HAY, PREBLOOM, SUN-CURED	**A-F**	**93**	**1.43**	**0.24**	–	–	–	**0.64**	–	–	–	–	–	–	–	–
		M-F	100	1.54	0.26	–	–	–	0.69	–	–	–	–	–	–	–	–
197	HAY, EARLY BLOOM, SUN-CURED	**A-F**	**95**	**1.46**	**0.25**	–	–	–	**0.65**	–	–	–	–	–	–	–	–
		M-F	100	1.54	0.26	–	–	–	0.69	–	–	–	–	–	–	–	–
198	HAY, MIDBLOOM, SUN-CURED	**A-F**	**90**	–	–	–	–	–	–	–	–	–	–	–	–	–	–
		M-F	100	–	–	–	–	–	–	–	–	–	–	–	–	–	–
199	HAY, FULL BLOOM, SUN-CURED	**A-F**	**94**	**0.95**	**0.21**	–	–	–	–	–	–	–	–	–	–	–	–
		M-F	100	1.01	0.22	–	–	–	–	–	–	–	–	–	–	–	–
	LUCERNE–SEE ALFALFA																
	MAIZE–SEE CORN																
	MARSH PLANTS																
200	HAY, SUN-CURED	**A-F**	**89**	**0.32**	**0.10**	–	–	**0.26**	**0.68**	–	–	–	–	**0.010**	–	–	–
		M-F	100	0.36	0.11	–	–	0.29	0.77	–	–	–	–	0.012	–	–	–
	MEADOW FESCUE *Festuca elatior*																
201	HAY, SUN-CURED	**A-F**	**88**	**0.33**	**0.25**	–	–	**0.44**	**1.61**	–	**0.119**	–	–	–	**21.4**	–	–
		M-F	100	0.37	0.29	–	–	0.50	1.84	–	0.135	–	–	–	24.5	–	–
	MEADOW FOXTAIL *Alopecurus pratensis*																
202	HAY, SUN-CURED	**A-F**	**90**	**0.95**	**0.18**	**0.01**	–	**0.12**	**1.57**	–	–	–	–	–	–	–	–
		M-F	100	1.06	0.20	0.01	–	0.13	1.75	–	–	–	–	–	–	–	–
	MEADOW PLANTS, INTERMOUNTAIN																
203	HAY, SUN-CURED	**A-F**	**95**	**0.58**	**0.17**	**0.11**	–	**0.16**	**1.50**	–	–	–	–	–	–	–	–
		M-F	100	0.61	0.18	0.12	–	0.17	1.58	–	–	–	–	–	–	–	–
	MILLET, FOXTAIL *Setaria italica*																
204	HAY, SUN-CURED	**A-F**	**87**	**0.29**	**0.16**	**0.09**	**0.11**	**0.20**	**1.69**	**0.14**	–	–	–	–	**120.1**	–	–
		M-F	100	0.33	0.18	0.10	0.13	0.23	1.94	0.16	–	–	–	–	138.1	–	–
205	STRAW	**A-F**	**90**	**0.08**	–	–	–	–	**1.43**	–	–	–	–	–	–	–	–
		M-F	100	0.09	–	–	–	–	1.60	–	–	–	–	–	–	–	–
	MILLET, JAPANESE (BARNYARD GRASS) *Echinochloa crusgalli*																
206	HAY, SUN-CURED	**A-F**	**87**	**0.20**	–	–	–	–	**2.11**	–	–	–	–	–	–	–	–
		M-F	100	0.23	–	–	–	–	2.42	–	–	–	–	–	–	–	–
	MILLET, PROSO (BROOMCORN; HOG MILLET) *Panicum miliaceum*																
207	HAY, SUN-CURED	**A-F**	**90**	–	–	–	–	–	–	–	–	–	–	–	–	–	–
		M-F	100	–	–	–	–	–	–	–	–	–	–	–	–	–	–
	MILO SORGHUM *Sorghum bicolor, subglabrescens*																
208	FODDER WITH HEADS, SUN-CURED	**A-F**	**89**	**0.35**	**0.18**	–	–	–	–	–	–	–	–	–	–	–	–
		M-F	100	0.40	0.20	–	–	–	–	–	–	–	–	–	–	–	–
209	STOVER WITHOUT HEADS, SUN-CURED	**A-F**	**91**	**0.51**	**0.15**	**0.10**	–	**0.23**	**0.41**	–	–	–	–	–	–	–	–
		M-F	100	0.56	0.16	0.11	–	0.26	0.45	–	–	–	–	–	–	–	–
	MINT *Mentha* spp																
210	HAY, SUN-CURED	**A-F**	**88**	**1.50**	**0.19**	–	–	–	–	–	–	–	–	–	–	–	–
		M-F	100	1.71	0.22	–	–	–	–	–	–	–	–	–	–	–	–
	MOSS, SPHAGNUM *Sphagnum* spp																
211	WHOLE, MOLASSES ADDED, SUN-CURED	**A-F**	**82**	–	–	–	–	–	–	–	–	–	–	–	–	–	–
		M-F	100	–	–	–	–	–	–	–	–	–	–	–	–	–	–
	MUHLY, BUSH *Muhlenbergia porteri*																
212	HAY, SUN-CURED	**A-F**	**93**	–	–	–	–	–	–	–	–	–	–	–	–	–	–
		M-F	100	–	–	–	–	–	–	–	–	–	–	–	–	–	–
	MUSTARD *Brassica* spp																
213	HAY, SUN-CURED	**A-F**	**90**	–	–	–	–	–	–	–	–	–	–	–	–	–	–
		M-F	100	–	–	–	–	–	–	–	–	–	–	–	–	–	–
	NAPIERGRASS *Pennisetum purpureum*																
214	HAY, SUN-CURED	**A-F**	**89**	–	–	–	–	–	–	–	–	–	–	–	–	–	–
		M-F	100	–	–	–	–	–	–	–	–	–	–	–	–	–	–
	NATALGRASS *Rhynchelytrum roseum*																
215	HAY, SUN-CURED	**A-F**	**91**	**0.47**	**0.31**	–	–	–	–	–	–	–	–	–	–	–	–
		M-F	100	0.51	0.34	–	–	–	–	–	–	–	–	–	–	–	–

Entry Number	Fat-Soluble Vitamins: A (1 mg Carotene = 1667 IU Vit A)	Carotene (Provitamin A)	D	E	K	Water-Soluble Vitamins: B–12	Biotin	Choline	Folacin (Folic Acid)	Niacin	Pantothenic Acid (B–3)	(Pyri-doxine) B–6	Ribo-flavin (B–2)	Thiamin (B–1)
	IU/g	ppm or mg/kg	IU/kg	ppm or mg/kg	ppm or mg/kg	ppm or mg/kg	ppm or mg/kg	ppm or mg/kg	ppm or mg/kg	ppm or mg/kg	ppm or mg/kg	ppm or mg/kg	ppm or mg/kg	ppm or mg/kg
193	**85.3**	**51.2**	–	–	–	–	–	–	–	–	–	–	–	–
	91.7	55.0	–	–	–	–	–	–	–	–	–	–	–	–
194	**19.4**	**11.6**	–	–	–	–	–	–	–	–	–	–	–	–
	20.8	12.5	–	–	–	–	–	–	–	–	–	–	–	–
195	**59.3**	**35.6**	–	–	–	–	–	–	–	–	–	–	**8.7**	–
	65.6	39.4	–	–	–	–	–	–	–	–	–	–	9.7	–
196	**62.0**	**37.2**	–	–	–	–	–	–	–	–	–	–	–	–
	66.7	40.0	–	–	–	–	–	–	–	–	–	–	–	–
197	–	–	–	–	–	–	–	–	–	–	–	–	–	–
	–	–	–	–	–	–	–	–	–	–	–	–	–	–
198	–	–	–	–	–	–	–	–	–	–	–	–	–	–
	–	–	–	–	–	–	–	–	–	–	–	–	–	–
199	–	–	–	–	–	–	–	–	–	–	–	–	–	–
	–	–	–	–	–	–	–	–	–	–	–	–	–	–
200	–	–	–	–	–	–	–	–	–	–	–	–	–	–
	–	–	–	–	–	–	–	–	–	–	–	–	–	–
201	**105.8**	**63.4**	–	**118.6**	–	–	–	–	–	–	–	–	–	–
	120.9	72.5	–	135.6	–	–	–	–	–	–	–	–	–	–
202	–	–	–	–	–	–	–	–	–	–	–	–	–	–
	–	–	–	–	–	–	–	–	–	–	–	–	–	–
203	**53.1**	**31.9**	–	–	–	–	–	–	–	–	–	–	–	–
	55.8	33.5	–	–	–	–	–	–	–	–	–	–	–	–
204	**86.9**	**52.1**	–	–	–	–	–	–	–	–	–	–	–	–
	100.0	60.0	–	–	–	–	–	–	–	–	–	–	–	–
205	–	–	–	–	–	–	–	–	–	–	–	–	–	–
	–	–	–	–	–	–	–	–	–	–	–	–	–	–
206	–	–	–	–	–	–	–	–	–	–	–	–	–	–
	–	–	–	–	–	–	–	–	–	–	–	–	–	–
207	–	–	–	–	–	–	–	–	–	–	–	–	–	–
	–	–	–	–	–	–	–	–	–	–	–	–	–	–
208	**2.9**	**1.8**	–	–	–	–	–	–	–	–	–	–	–	–
	3.3	2.0	–	–	–	–	–	–	–	–	–	–	–	–
209	–	–	–	–	–	–	–	–	–	–	–	–	–	–
	–	–	–	–	–	–	–	–	–	–	–	–	–	–
210	–	–	–	–	–	–	–	–	–	–	–	–	–	–
	–	–	–	–	–	–	–	–	–	–	–	–	–	–
211	–	–	–	–	–	–	–	–	–	–	–	–	–	–
	–	–	–	–	–	–	–	–	–	–	–	–	–	–
212	–	–	–	–	–	–	–	–	–	–	–	–	–	–
	–	–	–	–	–	–	–	–	–	–	–	–	–	–
213	–	–	–	–	–	–	–	–	–	–	–	–	–	–
	–	–	–	–	–	–	–	–	–	–	–	–	–	–
214	–	–	–	–	–	–	–	–	–	–	–	–	–	–
	–	–	–	–	–	–	–	–	–	–	–	–	–	–
215	–	–	–	–	–	–	–	–	–	–	–	–	–	–
	–	–	–	–	–	–	–	–	–	–	–	–	–	–

(Continued)

TABLE V-1C DRY FORAGES, COMPOSITION OF FEEDS, DATA EXPRESSED **AS-FED** AND **MOISTURE-FREE**—*(Continued)*

Entry Number	Feed Name Description	International Feed Number	Moisture Basis: A-F (as-fed) or M-F (moisture-free)	Proximate Analysis									Digestible Protein		
				Dry Matter	Ash	Crude Fiber	Neutral Det. Fib. (NDF)	Acid Det. Fib. (ADF)	Lignin	Ether Extract (Fat)	N-Free Extract	Crude Protein	Ruminant	Swine	Horse
				%	%	%	%	%	%	%	%	%	%	%	%
	NATIVE GRASS–SEE MEADOW, INTERMOUNTAIN; MARSH; PRAIRIE, MIDWEST														
	NEEDLEGRASS *Stipa* spp														
216	HAY, SUN-CURED	1-03-202	A-F	88	6.2	30.0	–	–	–	2.2	41.9	7.8	4.1	3.4	4.4
			M-F	100	7.1	34.0	–	–	–	2.5	47.5	8.8	4.7	3.9	5.0
	OAK, LIVE *Quercus virginiana*														
217	LEAVES, DEHY	1-03-249	A-F	93	6.0	28.4	–	–	–	2.4	47.1	9.5	5.4	4.7	5.6
			M-F	100	6.4	30.4	–	–	–	2.6	50.5	10.1	5.8	5.0	6.0
	OATGRASS, POVERTY *Danthonia spicata*														
218	HAY, SUN-CURED	1-01-751	A-F	92	3.9	29.5	–	–	–	2.9	47.6	7.8	4.5	3.3	4.4
			M-F	100	4.2	32.2	–	–	–	3.2	51.9	8.5	4.9	3.6	4.8
	OATGRASS, TALL *Arrhenatherum elatius*														
219	HAY, SUN-CURED	1-03-259	A-F	88	6.7	29.6	–	–	–	2.0	42.5	6.7	3.1	2.5	3.6
			M-F	100	7.7	33.8	–	–	–	2.3	48.6	7.7	3.5	2.9	4.1
	OATS *Avena sativa*														
220	CHAFF	1-03-305	A-F	91	10.2	25.8	–	–	–	1.9	47.0	6.5	1.0	2.2	3.4
			M-F	100	11.2	28.3	–	–	–	2.1	51.4	7.1	1.1	2.4	3.7
221	HAY, SUN-CURED, ALL ANALYSES	1-03-280	A-F	91	7.2	29.1	–	34.8	–	2.2	43.6	8.6	4.5	4.1	5.0
			M-F	100	7.9	32.0	–	38.4	–	2.4	48.1	9.5	5.0	4.5	5.5
222	HAY, IMMATURE, SUN-CURED	1-03-272	A-F	91	8.5	23.8	56.8	32.4	3.6	2.4	44.1	12.6	9.2	7.6	8.1
			M-F	100	9.3	26.0	62.1	35.4	3.9	2.6	48.3	13.8	10.1	8.3	8.9
223	HAY, DOUGH STAGE, SUN-CURED	1-03-276	A-F	92	6.8	27.5	60.7	33.1	5.5	3.0	46.1	8.6	5.1	4.0	5.0
			M-F	100	7.4	29.9	66.0	36.0	6.0	3.2	50.1	9.4	5.5	4.4	5.5
224	GRAIN, CLIPPED BY-PRODUCT	1-03-269	A-F	92	10.8	25.2	–	–	–	2.3	45.0	8.8	3.5	4.2	5.2
			M-F	100	11.8	27.4	–	–	–	2.5	48.8	9.6	3.8	4.6	5.6
225	HULLS	1-03-281	A-F	92	6.1	30.9	68.6	37.4	6.5	1.3	50.5	3.7	1.2	2.1	3.3
			M-F	100	6.6	33.4	74.3	40.5	7.0	1.4	54.7	4.0	1.2	2.3	3.6
226	STRAW	1-03-283	A-F	92	7.2	37.2	65.7	43.1	7.0	2.0	41.6	4.1	0.8	0.0	2.4
			M-F	100	7.8	40.4	71.3	46.8	7.6	2.2	45.2	4.4	0.9	0.0	2.6
227	STRAW, TREATED WITH SODIUM HYDROXIDE, DEHY	1-03-285	A-F	89	4.4	55.0	–	–	–	1.2	26.4	1.6	–2.1	–2.0	–0.4
			M-F	100	5.0	62.0	–	–	–	1.4	29.8	1.8	–2.4	–2.3	–0.4
	OATS, WILD *Avena fatua*														
228	HAY, SUN-CURED	1-03-392	A-F	92	6.9	31.9	–	–	–	2.6	44.0	6.7	3.1	2.4	3.6
			M-F	100	7.5	34.6	–	–	–	2.9	47.7	7.3	3.4	2.6	3.9
	OATS-PEA *Avena sativa-Pisum* spp														
229	HAY, SUN-CURED	1-03-398	A-F	88	7.4	26.9	–	–	–	3.0	38.7	11.7	8.3	6.9	7.4
			M-F	100	8.5	30.6	–	–	–	3.4	44.2	13.3	9.5	7.8	8.5
	ORCHARDGRASS *Dactylis glomerata*														
230	HAY, SUN-CURED	1-03-438	A-F	89	6.5	31.0	64.1	36.0	–	2.8	39.7	9.4	5.7	4.8	5.6
			M-F	100	7.3	34.7	71.8	40.3	–	3.1	44.4	10.5	6.4	5.4	6.3
231	HAY, FULL BLOOM, SUN-CURED	1-03-427	A-F	93	8.3	33.4	58.5	34.3	4.3	2.9	39.5	8.7	5.0	4.0	5.0
			M-F	100	8.9	36.0	63.1	37.0	4.6	3.1	42.6	9.4	5.4	4.4	5.4
	PANICUM *Panicum* spp														
232	HAY, SUN-CURED	1-03-496	A-F	92	6.0	30.7	–	–	–	2.0	45.4	8.3	4.3	3.7	4.8
			M-F	100	6.5	33.2	–	–	–	2.2	49.1	9.0	4.7	4.1	5.2
	PAPER														
233	WASTE	1-26-072	A-F	92	0.5	63.4	86.5	73.6	22.1	3.6	23.9	0.6	–	–	–
			M-F	100	0.5	68.9	94.0	80.0	24.0	3.9	26.0	0.7	–	–	–
	PARAGRASS *Brachiaria mutica*														
234	HAY, SUN-CURED	1-03-517	A-F	91	6.7	33.7	–	–	–	0.9	45.5	3.9	0.4	–0.1	1.4
			M-F	100	7.4	37.1	–	–	–	1.0	50.2	4.3	0.4	–0.1	1.5
	PAUNCH PRODUCT														
235	DEHY (RUMEN CONTENTS; PAUNCH MEAL)	1-09-327	A-F	87	9.6	24.4	–	–	–	1.8	36.2	15.3	10.3	10.1	10.3
			M-F	100	11.0	28.0	–	–	–	2.1	41.4	17.6	11.8	11.6	11.8
	PEA *Pisum* spp														
236	HAY, SUN-CURED	1-03-572	A-F	87	7.4	23.7	–	–	–	2.6	41.7	11.8	7.7	7.0	7.6
			M-F	100	8.5	27.2	–	–	–	3.0	47.9	13.5	8.8	8.0	8.7
237	STRAW	1-03-577	A-F	87	5.7	34.3	–	–	–	1.5	37.7	7.8	4.0	3.5	4.4
			M-F	100	6.5	39.5	–	–	–	1.8	43.4	8.9	4.6	4.0	5.1
	PEA, FIELD *Pisum sativum, arvense*														
238	HAY (VINES WITHOUT SEEDS, WITH PODS), SUN-CURED	1-03-607	A-F	88	6.8	23.4	–	–	–	2.3	42.5	12.7	8.5	7.8	8.2
			M-F	100	7.8	26.7	–	–	–	2.6	48.4	14.5	9.7	8.9	9.4
	PEANUT *Arachis hypogaea*														
239	HAY, SUN-CURED	1-03-619	A-F	91	8.2	30.3	–	37.2	–	3.3	39.1	9.9	5.7	5.2	6.0
			M-F	100	9.0	33.4	–	41.0	–	3.6	43.1	10.9	6.3	5.7	6.6
240	HULLS (PODS)	1-08-028	A-F	91	3.8	57.2	69.0	60.8	21.8	2.1	20.7	7.3	1.7	2.9	4.0
			M-F	100	4.2	62.9	75.8	66.8	23.9	2.3	22.8	8.0	1.9	3.1	4.4

Entry Number	TDN Ruminant	TDN Swine	TDN Horse	Digestible Energy Ruminant		Digestible Energy Swine		Digestible Energy Horse		Metabolizable Energy Ruminant		Metabolizable Energy Swine		Metabolizable Energy Poultry ME_n		Metabolizable Energy Horse		Net Energy Ruminant NE_m		Net Energy Ruminant NE_g		Net Energy Lactating Cows NE_{lc}	
	%	%	%	Mcal		kcal		Mcal		Mcal		kcal		kcal		Mcal		Mcal		Mcal		Mcal	
				lb	*kg*	lb	*kg*	lb	*kg*	lb	*kg*	lb	*kg*	lb	*kg*	lb	*kg*	lb	*kg*	lb	*kg*	lb	*kg*
216	**42**	**34**	**35**	**0.90**	***1.98***	–	–	**0.69**	***1.52***	**0.73**	***1.60***	–	–	–	–	**0.57**	***1.25***	**0.45**	***0.99***	**0.23**	***0.50***	**0.49**	***1.07***
	47	39	40	1.02	*2.24*	–	–	0.78	*1.72*	0.82	*1.82*	–	–	–	–	0.64	*1.41*	0.51	*1.13*	0.26	*0.57*	0.55	*1.22*
217	**17**	**43**	**43**	**0.70**	***1.55***	–	–	**0.81**	***1.80***	**0.52**	***1.14***	–	–	–	–	**0.67**	***1.47***	**0.52**	***1.14***	**0.28**	***0.61***	**0.54**	***1.20***
	18	46	46	0.75	*1.66*	–	–	0.87	*1.92*	0.56	*1.23*	–	–	–	–	0.72	*1.58*	0.55	*1.22*	0.30	*0.65*	0.58	*1.28*
218	**59**	**44**	**39**	**1.13**	***2.48***	–	–	**0.76**	***1.67***	**0.95**	***2.10***	–	–	–	–	**0.62**	***1.37***	**0.53**	***1.16***	**0.29**	***0.64***	**0.55**	***1.21***
	64	48	43	1.23	*2.71*	–	–	0.82	*1.82*	1.04	*2.29*	–	–	–	–	0.68	*1.49*	0.57	*1.27*	0.32	*0.70*	0.60	*1.32*
219	**46**	**32**	**34**	**0.93**	***2.04***	–	–	**0.67**	***1.48***	**0.76**	***1.67***	–	–	–	–	**0.55**	***1.21***	**0.43**	***0.95***	**0.21**	***0.47***	**0.47**	***1.04***
	52	36	39	1.06	*2.33*	–	–	0.76	*1.68*	0.87	*1.91*	–	–	–	–	0.63	*1.38*	0.49	*1.09*	0.24	*0.53*	0.54	*1.19*
220	**33**	**33**	**36**	**0.81**	***1.78***	–	–	**0.71**	***1.56***	**0.63**	***1.39***	–	–	–	–	**0.58**	***1.28***	**0.43**	***0.95***	**0.20**	***0.44***	**0.48**	***1.05***
	37	36	40	0.89	*1.95*	–	–	0.77	*1.70*	0.69	*1.52*	–	–	–	–	0.63	*1.40*	0.47	*1.04*	0.22	*0.49*	0.52	*1.15*
221	**52**	**36**	**38**	**1.04**	***2.29***	–	–	**0.74**	***1.62***	**0.93**	***2.04***	–	–	–	–	**0.60**	***1.33***	**0.57**	***1.26***	**0.33**	***0.73***	**0.55**	***1.22***
	57	40	42	1.15	*2.53*	–	–	0.81	*1.79*	1.02	*2.25*	–	–	–	–	0.67	*1.47*	0.63	*1.39*	0.37	*0.81*	0.61	*1.35*
222	**53**	**47**	**45**	**1.06**	***2.33***	–	–	**0.86**	***1.89***	**0.88**	***1.94***	–	–	–	–	**0.70**	***1.55***	**0.51**	***1.13***	**0.28**	***0.61***	**0.54**	***1.19***
	58	51	49	1.16	*2.55*	–	–	0.94	*2.07*	0.96	*2.12*	–	–	–	–	0.77	*1.69*	0.56	*1.24*	0.30	*0.67*	0.59	*1.30*
223	**50**	**43**	**38**	**0.99**	***2.17***	–	–	**0.73**	***1.61***	**0.82**	***1.80***	–	–	–	–	**0.60**	***1.32***	**0.46**	***1.00***	**0.22**	***0.49***	**0.46**	***1.02***
	55	47	41	1.07	*2.36*	–	–	0.80	*1.75*	0.89	*1.96*	–	–	–	–	0.65	*1.44*	0.50	*1.09*	0.24	*0.54*	0.50	*1.11*
224	**50**	**37**	**37**	**1.00**	***2.20***	–	–	**0.72**	***1.60***	**0.82**	***1.80***	–	–	–	–	**0.59**	***1.31***	**0.46**	***1.01***	**0.22**	***0.49***	**0.50**	***1.10***
	55	41	40	1.08	*2.38*	–	–	0.79	*1.73*	0.89	*1.96*	–	–	–	–	0.64	*1.42*	0.49	*1.09*	0.24	*0.53*	0.54	*1.19*
225	**32**	**27**	**23**	**0.64**	***1.41***	–	–	**0.48**	***1.06***	**0.46**	***1.01***	**394**	***868***	**162**	***357***	**0.40**	***0.87***	**0.16**	***0.36***	**−0.05**	***−0.12***	**0.30**	***0.67***
	35	29	24	0.69	*1.52*	–	–	0.52	*1.15*	0.49	*1.09*	427	*940*	175	*386*	0.43	*0.94*	0.18	*0.39*	−0.06	*−0.13*	0.33	*0.72*
226	**46**	–	**44**	**1.15**	***2.53***	–	–	**0.84**	***1.85***	**0.78**	***1.72***	–	–	–	–	**0.69**	***1.52***	**0.43**	***0.94***	**0.20**	***0.43***	**0.48**	***1.06***
	50	–	48	1.24	*2.74*	–	–	0.91	*2.01*	0.84	*1.86*	–	–	–	–	0.75	*1.65*	0.46	*1.02*	0.21	*0.47*	0.52	*1.15*
227	**61**	–	–	**1.23**	***2.71***	–	–	–	–	**1.06**	***2.33***	–	–	–	–	–	–	–	–	–	–	–	–
	69	–	–	1.38	*3.05*	–	–	–	–	1.19	*2.63*	–	–	–	–	–	–	–	–	–	–	–	–
228	**49**	**34**	**33**	**0.99**	***2.18***	–	–	**0.66**	***1.44***	**0.81**	***1.79***	–	–	–	–	**0.54**	***1.18***	**0.46**	***1.02***	**0.23**	***0.51***	**0.50**	***1.11***
	53	37	36	1.07	*2.37*	–	–	0.71	*1.57*	0.88	*1.94*	–	–	–	–	0.58	*1.29*	0.50	*1.11*	0.25	*0.55*	0.55	*1.20*
229	**52**	**43**	**37**	**1.03**	***2.26***	–	–	**0.72**	***1.58***	**0.86**	***1.89***	–	–	–	–	**0.59**	***1.30***	**0.49**	***1.08***	**0.27**	***0.59***	**0.52**	***1.14***
	59	49	42	1.17	*2.58*	–	–	0.82	*1.81*	0.98	*2.16*	–	–	–	–	0.67	*1.48*	0.56	*1.24*	0.30	*0.67*	0.59	*1.30*
230	**51**	**37**	**40**	**1.23**	***2.71***	–	–	**0.77**	***1.69***	**0.96**	***2.11***	–	–	–	–	**0.63**	***1.38***	**0.50**	***1.09***	**0.27**	***0.59***	**0.52**	***1.15***
	57	42	45	1.38	*3.03*	–	–	0.86	*1.89*	1.07	*2.36*	–	–	–	–	0.70	*1.55*	0.56	*1.22*	0.30	*0.66*	0.58	*1.29*
231	**50**	**34**	**31**	**1.02**	***2.25***	–	–	**0.62**	***1.38***	**0.83**	***1.82***	–	–	–	–	**0.51**	***1.13***	**0.46**	***1.01***	**0.22**	***0.49***	**0.50**	***1.10***
	54	36	34	1.10	*2.42*	–	–	0.67	*1.48*	0.89	*1.97*	–	–	–	–	0.55	*1.22*	0.49	*1.09*	0.24	*0.53*	0.54	*1.19*
232	**47**	**37**	**40**	**0.98**	***2.16***	–	–	**0.77**	***1.71***	**0.80**	***1.76***	–	–	–	–	**0.63**	***1.40***	**0.48**	***1.05***	**0.24**	***0.54***	**0.51**	***1.13***
	51	40	43	1.06	*2.33*	–	–	0.84	*1.84*	0.87	*1.91*	–	–	–	–	0.69	*1.51*	0.52	*1.14*	0.26	*0.58*	0.56	*1.23*
233	**25**	–	–	**0.92**	***2.02***	–	–	–	–	**0.75**	***1.66***	–	–	–	–	–	–	–	–	–	–	–	–
	27	–	–	1.00	*2.20*	–	–	–	–	0.82	*1.80*	–	–	–	–	–	–	–	–	–	–	–	–
234	**41**	**21**	**35**	**0.86**	***1.89***	–	–	**0.69**	***1.51***	**0.68**	***1.50***	–	–	–	–	**0.56**	***1.24***	**0.38**	***0.84***	**0.16**	***0.35***	**0.44**	***0.98***
	46	24	39	0.94	*2.08*	–	–	0.76	*1.67*	0.75	*1.65*	–	–	–	–	0.62	*1.37*	0.42	*0.93*	0.18	*0.39*	0.49	*1.08*
235	**51**	**42**	**45**	**1.00**	***2.20***	–	–	**0.85**	***1.88***	**0.83**	***1.83***	–	–	–	–	**0.70**	***1.54***	**0.47**	***1.03***	**0.25**	***0.54***	**0.50**	***1.10***
	58	48	52	1.14	*2.51*	–	–	0.98	*2.15*	0.95	*2.09*	–	–	–	–	0.80	*1.77*	0.54	*1.18*	0.28	*0.62*	0.57	*1.26*
236	**51**	**45**	**42**	**1.02**	***2.24***	–	–	**0.79**	***1.75***	**0.85**	***1.87***	–	–	–	–	**0.65**	***1.43***	**0.50**	***1.10***	**0.27**	***0.60***	**0.52**	***1.15***
	58	52	48	1.17	*2.57*	–	–	0.91	*2.01*	0.97	*2.14*	–	–	–	–	0.75	*1.64*	0.57	*1.26*	0.31	*0.69*	0.60	*1.32*
237	**43**	**26**	**35**	**0.85**	***1.88***	–	–	**0.67**	***1.49***	**0.68**	***1.51***	–	–	–	–	**0.55**	***1.22***	**0.48**	***1.07***	**0.26**	***0.57***	**0.51**	***1.12***
	49	30	40	0.98	*2.16*	–	–	0.77	*1.71*	0.79	*1.73*	–	–	–	–	0.64	*1.40*	0.56	*1.23*	0.30	*0.66*	0.59	*1.29*
238	**52**	**47**	**46**	**1.04**	***2.29***	–	–	**0.86**	***1.89***	**0.87**	***1.91***	–	–	–	–	**0.70**	***1.55***	**0.51**	***1.12***	**0.28**	***0.62***	**0.53**	***1.17***
	59	53	52	1.18	*2.61*	–	–	0.98	*2.15*	0.99	*2.18*	–	–	–	–	0.80	*1.77*	0.58	*1.28*	0.32	*0.71*	0.60	*1.33*
239	**48**	**39**	**32**	**0.95**	***2.10***	–	–	**0.64**	***1.41***	**0.78**	***1.71***	–	–	–	–	**0.52**	***1.15***	**0.39**	***0.85***	**0.16**	***0.36***	**0.45**	***0.99***
	53	43	35	1.05	*2.32*	–	–	0.70	*1.55*	0.86	*1.89*	–	–	–	–	0.58	*1.27*	0.43	*0.94*	0.18	*0.40*	0.49	*1.09*
240	**16**	–	–	**0.33**	***0.72***	–	–	–	–	**0.14**	***0.32***	–	–	–	–	–	–	**−0.16**	***−0.36***	**−0.38**	***−0.83***	**0.11**	***0.24***
	18	–	–	0.36	*0.79*	–	–	–	–	0.16	*0.35*	–	–	–	–	–	–	−0.18	*−0.39*	−0.41	*−0.91*	0.12	*0.26*

(Continued)

TABLE V-1C DRY FORAGES, MINERAL AND VITAMIN COMPOSITION OF FEEDS, DATA EXPRESSED **AS-FED** AND **MOISTURE-FREE**—*(Continued)*

Entry Number	Feed Name Description	Moisture Basis: A-F (as-fed) or M-F (moisture-free)	Dry Matter	Macro Minerals							Micro Minerals						
				Calcium (Ca)	Phos-phorus (P)	Sodium (Na)	Chlo-rine (Cl)	Mag-nesium (Mg)	Potas-sium (K)	Sulfur (S)	Cobalt (Co)	Copper (Cu)	Iodine (I)	Iron (Fe)	Man-ganese (Mn)	Sele-nium (Se)	Zinc (Zn)
			%	%	%	%	%	%	%	%	ppm or mg/kg	ppm or mg/kg	ppm or mg/kg	%	ppm or mg/kg	ppm or mg/kg	ppm or mg/kg
	NATIVE GRASS–SEE MEADOW, INTERMOUNTAIN; MARSH; PRAIRIE, MIDWEST																
	NEEDLEGRASS *Stipa spp*																
216	HAY, SUN-CURED	A-F	88	–	–	–	–	–	–	–	–	–	–	–	–	–	–
		M-F	100	–	–	–	–	–	–	–	–	–	–	–	–	–	–
	OAK, LIVE *Quercus virginiana*																
217	LEAVES, DEHY	A-F	93	–	–	–	–	–	–	–	–	–	–	–	–	–	–
		M-F	100	–	–	–	–	–	–	–	–	–	–	–	–	–	–
	OATGRASS, POVERTY *Danthonia spicata*																
218	HAY, SUN-CURED	A-F	92	–	–	–	–	–	–	–	–	–	–	–	–	–	–
		M-F	100	–	–	–	–	–	–	–	–	–	–	–	–	–	–
	OATGRASS, TALL *Arrhenatherum elatius*																
219	HAY, SUN-CURED	A-F	88	0.30	0.27	0.01	–	0.17	1.74	–	–	–	–	–	32.2	–	–
		M-F	100	0.34	0.31	0.01	–	0.19	1.99	–	–	–	–	–	36.8	–	–
	OATS *Avena sativa*																
220	CHAFF	A-F	91	0.49	0.13	–	–	–	0.86	–	–	–	–	–	–	–	–
		M-F	100	0.54	0.14	–	–	–	0.94	–	–	–	–	–	–	–	–
221	HAY, SUN-CURED, ALL ANALYSES	A-F	91	0.29	0.23	0.17	0.47	0.26	1.35	0.21	0.067	4.4	–	0.037	89.6	–	40.8
		M-F	100	0.32	0.25	0.18	0.52	0.29	1.49	0.23	0.073	4.8	–	0.041	98.7	–	45.0
222	HAY, IMMATURE, SUN-CURED	A-F	91	0.30	0.39	–	–	–	–	–	0.043	–	–	–	–	–	–
		M-F	100	0.33	0.43	–	–	–	–	–	0.047	–	–	–	–	–	–
223	HAY, DOUGH STAGE, SUN-CURED	A-F	92	0.22	0.20	–	–	0.16	1.41	–	–	16.3	–	0.009	48.6	0.157	36.0
		M-F	100	0.24	0.22	–	–	0.18	1.53	–	–	17.7	–	0.010	52.9	0.171	39.2
224	GRAIN, CLIPPED BY-PRODUCT	A-F	92	–	–	–	–	–	–	–	–	–	–	–	–	–	–
		M-F	100	–	–	–	–	–	–	–	–	–	–	–	–	–	–
225	HULLS	A-F	92	0.14	0.14	0.04	0.08	0.08	0.57	0.14	–	4.1	–	0.011	18.8	–	–
		M-F	100	0.15	0.15	0.04	0.08	0.09	0.62	0.15	–	4.5	–	0.012	20.4	–	–
226	STRAW	A-F	92	0.22	0.06	0.39	0.72	0.16	2.35	0.21	–	9.5	–	0.016	29.0	–	5.5
		M-F	100	0.24	0.07	0.42	0.78	0.17	2.55	0.23	–	10.3	–	0.017	31.5	–	5.9
227	STRAW, TREATED WITH SODIUM HYDROXIDE, DEHY	A-F	89	–	–	–	–	–	–	–	–	–	–	–	–	–	–
		M-F	100	–	–	–	–	–	–	–	–	–	–	–	–	–	–
	OATS, WILD *Avena fatua*																
228	HAY, SUN-CURED	A-F	92	0.22	0.25	–	–	–	–	–	–	–	–	–	–	–	–
		M-F	100	0.24	0.27	–	–	–	–	–	–	–	–	–	–	–	–
	OATS-PEA *Avena sativa-Pisum spp*																
229	HAY, SUN-CURED	A-F	88	0.71	0.22	–	–	–	1.02	–	–	–	–	–	–	–	–
		M-F	100	0.81	0.25	–	–	–	1.17	–	–	–	–	–	–	–	–
	ORCHARDGRASS *Dactylis glomerata*																
230	HAY, SUN-CURED	A-F	89	0.34	0.23	0.01	0.37	0.16	2.68	0.23	0.339	12.9	–	0.014	162.7	–	32.0
		M-F	100	0.38	0.26	0.02	0.41	0.18	3.00	0.26	0.379	14.5	–	0.015	182.3	–	35.8
231	HAY, FULL BLOOM, SUN-CURED	A-F	93	–	–	–	–	–	–	–	–	–	–	–	–	–	–
		M-F	100	–	–	–	–	–	–	–	–	–	–	–	–	–	–
	PANICUM *Panicum spp*																
232	HAY, SUN-CURED	A-F	92	–	–	–	–	–	–	–	–	–	–	–	–	–	–
		M-F	100	–	–	–	–	–	–	–	–	–	–	–	–	–	–
	PAPER																
233	WASTE	A-F	92	0.09	0.06	–	–	–	–	–	–	–	–	–	–	–	–
		M-F	100	0.10	0.07	–	–	–	–	–	–	–	–	–	–	–	–
	PARAGRASS *Brachiaria mutica*																
234	HAY, SUN-CURED	A-F	91	0.35	0.35	–	–	–	1.45	–	–	–	–	–	–	–	–
		M-F	100	0.39	0.39	–	–	–	1.60	–	–	–	–	–	–	–	–
	PAUNCH PRODUCT																
235	DEHY (RUMEN CONTENTS; PAUNCH MEAL)	A-F	87	0.69	0.58	–	–	–	–	–	–	–	–	–	–	–	–
		M-F	100	0.79	0.67	–	–	–	–	–	–	–	–	–	–	–	–
	PEA *Pisum spp*																
236	HAY, SUN-CURED	A-F	87	–	–	–	–	–	–	–	–	–	–	–	–	–	–
		M-F	100	–	–	–	–	–	–	–	–	–	–	–	–	–	–
237	STRAW	A-F	87	–	–	–	–	–	–	–	–	–	–	–	–	–	–
		M-F	100	–	–	–	–	–	–	–	–	–	–	–	–	–	–
	PEA, FIELD *Pisum sativum, arvense*																
238	HAY (VINES WITHOUT SEEDS, WITH PODS), SUN-CURED	A-F	88	–	–	–	–	–	–	–	–	–	–	–	–	–	–
		M-F	100	–	–	–	–	–	–	–	–	–	–	–	–	–	–
	PEANUT *Arachis hypogaea*																
239	HAY, SUN-CURED	A-F	91	1.12	0.14	–	–	0.44	1.25	0.21	0.072	–	–	–	–	–	–
		M-F	100	1.23	0.16	–	–	0.49	1.38	0.23	0.080	–	–	–	–	–	–
240	HULLS (PODS)	A-F	91	0.24	0.06	0.12	–	0.15	0.87	0.09	0.109	16.2	–	0.029	62.5	–	21.9
		M-F	100	0.26	0.07	0.13	–	0.17	0.95	0.10	0.119	17.8	–	0.032	68.7	–	24.1

Entry Number	Fat-Soluble Vitamins: A (1 mg Carotene = 1667 IU Vit A)	Carotene (Provitamin A)	D	E	K	Water-Soluble Vitamins: B-12	Biotin	Choline	Folacin (Folic Acid)	Niacin	Pantothenic Acid (B-3)	(Pyridoxine) B-6	Riboflavin (B-2)	Thiamin (B-1)
	IU/g	ppm or mg/kg	IU/kg	ppm or mg/kg	ppm or mg/kg	ppm or mg/kg	ppm or mg/kg	ppm or mg/kg	ppm or mg/kg	ppm or mg/kg	ppm or mg/kg	ppm or mg/kg	ppm or mg/kg	ppm or mg/kg
216	–	–	–	–	–	–	–	–	–	–	–	–	–	–
	–	–	–	–	–	–	–	–	–	–	–	–	–	–
217	**146.2**	**87.7**	–	–	–	–	–	–	–	–	–	–	–	–
	156.6	93.9	–	–	–	–	–	–	–	–	–	–	–	–
218	–	–	–	–	–	–	–	–	–	–	–	–	–	–
	–	–	–	–	–	–	–	–	–	–	–	–	–	–
219	–	–	–	–	–	–	–	–	–	–	–	–	–	–
	–	–	–	–	–	–	–	–	–	–	–	–	–	–
220	**45.6**	**27.4**	–	–	–	–	–	–	–	–	–	–	–	–
	49.9	29.9	–	–	–	–	–	–	–	–	–	–	–	–
221	**45.0**	**27.0**	**1**	–	–	–	–	–	–	–	–	–	–	–
	49.6	29.7	2	–	–	–	–	–	–	–	–	–	–	–
222	–	–	–	–	–	–	–	–	–	–	–	–	–	–
	–	–	–	–	–	–	–	–	–	–	–	–	–	–
223	**37.5**	**22.5**	–	–	–	–	–	–	–	–	–	–	–	–
	40.8	24.5	–	–	–	–	–	–	–	–	–	–	–	–
224	–	–	–	–	–	–	–	–	–	–	–	–	–	–
	–	–	–	–	–	–	–	–	–	–	–	–	–	–
225	–	–	–	–	–	–	–	**260**	**0.96**	**9**	**3.1**	**2.18**	**1.5**	**0.6**
	–	–	–	–	–	–	–	281	1.04	10	3.4	2.37	1.7	0.7
226	**5.8**	**3.5**	**1**	–	–	–	–	–	–	–	–	–	–	–
	6.3	3.8	1	–	–	–	–	–	–	–	–	–	–	–
227	–	–	–	–	–	–	–	–	–	–	–	–	–	–
	–	–	–	–	–	–	–	–	–	–	–	–	–	–
228	–	–	–	–	–	–	–	–	–	–	–	–	–	–
	–	–	–	–	–	–	–	–	–	–	–	–	–	–
229	–	–	–	–	–	–	–	–	–	–	–	–	–	–
	–	–	–	–	–	–	–	–	–	–	–	–	–	–
230	**28.9**	**17.3**	–	**170.7**	–	–	–	–	–	–	–	–	**6.1**	**2.6**
	32.4	19.4	–	191.1	–	–	–	–	–	–	–	–	6.8	2.9
231	–	–	–	–	–	–	–	–	–	–	–	–	–	–
	–	–	–	–	–	–	–	–	–	–	–	–	–	–
232	–	–	–	–	–	–	–	–	–	–	–	–	–	–
	–	–	–	–	–	–	–	–	–	–	–	–	–	–
233	–	–	–	–	–	–	–	–	–	–	–	–	–	–
	–	–	–	–	–	–	–	–	–	–	–	–	–	–
234	–	–	–	–	–	–	–	–	–	–	–	–	–	–
	–	–	–	–	–	–	–	–	–	–	–	–	–	–
235	–	–	–	–	–	–	–	–	–	–	–	–	–	–
	–	–	–	–	–	–	–	–	–	–	–	–	–	–
236	–	–	–	–	–	–	–	–	–	–	–	–	–	–
	–	–	–	–	–	–	–	–	–	–	–	–	–	–
237	–	–	–	–	–	–	–	–	–	–	–	–	–	–
	–	–	–	–	–	–	–	–	–	–	–	–	–	–
238	–	–	–	–	–	–	–	–	–	–	–	–	–	–
	–	–	–	–	–	–	–	–	–	–	–	–	–	–
239	**52.6**	**31.5**	–	–	–	–	–	–	–	–	–	–	**8.8**	–
	58.0	34.8	–	–	–	–	–	–	–	–	–	–	9.7	–
240	**1.3**	**0.8**	–	–	–	–	–	–	–	–	–	–	–	–
	1.5	0.9	–	–	–	–	–	–	–	–	–	–	–	–

(Continued)

TABLE V-1C DRY FORAGES, COMPOSITION OF FEEDS, DATA EXPRESSED **AS-FED** AND **MOISTURE-FREE**—*(Continued)*

Entry Number	Feed Name Description	International Feed Number	Moisture Basis: A-F (as-fed) or M-F (moisture-free)	Proximate Analysis: Dry Matter	Ash	Crude Fiber	Neutral Det. Fib. (NDF)	Acid Det. Fib. (ADF)	Lignin	Ether Extract (Fat)	N-Free Extract	Crude Protein	Digestible Protein: Ruminant	Swine	Horse
				%	%	%	%	%	%	%	%	%	%	%	%
	PEARL MILLET *Pennisetum glaucum*														
241	HAY, SUN-CURED	1-03-112	A-F	87	8.9	32.2	–	–	–	1.8	37.2	7.3	4.6	3.1	4.1
			M-F	100	10.2	36.9	–	–	–	2.0	42.5	8.4	5.3	3.5	4.7
	PEAVINE *Lathyrus* spp														
242	HAY, SUN-CURED	1-03-666	A-F	91	6.8	24.9	–	–	–	2.6	36.6	19.8	15.4	13.9	13.6
			M-F	100	7.5	27.5	–	–	–	2.9	40.3	21.8	17.0	15.3	15.0
	PERSIAN CLOVER *Trifolium resupinatum*														
243	HAY, SUN-CURED	1-01-389	A-F	90	9.3	27.7	–	–	–	1.6	36.7	14.7	10.1	9.4	9.7
			M-F	100	10.3	30.8	–	–	–	1.8	40.8	16.3	11.3	10.5	10.8
	POVERTY OATGRASS *Danthonia spicata*														
244	HAY, SUN-CURED	1-01-751	A-F	92	3.9	29.5	–	–	–	2.9	47.6	7.8	4.5	3.3	4.4
			M-F	100	4.2	32.2	–	–	–	3.2	51.9	8.5	4.9	3.6	4.8
	PRAIRIE GRASS, MIDWEST (PRAIRIE HAY)														
245	HAY, SUN-CURED	1-03-191	A-F	91	7.2	30.7	–	–	–	2.1	45.2	5.8	2.9	1.6	2.9
			M-F	100	8.0	33.7	–	–	–	2.3	49.6	6.4	3.2	1.7	3.1
246	HAY, IMMATURE, SUN-CURED	1-03-183	A-F	90	8.4	28.6	–	–	–	2.3	43.0	7.9	4.1	3.4	4.4
			M-F	100	9.3	31.7	–	–	–	2.6	47.7	8.7	4.6	3.8	4.9
247	HAY, MATURE, SUN-CURED	1-03-187	A-F	90	8.3	28.4	–	–	–	3.6	44.2	5.6	2.2	1.5	2.7
			M-F	100	9.2	31.5	–	–	–	3.9	49.0	6.2	2.4	1.6	3.0
	QUACKGRASS *Agropyron repens*														
248	HAY, SUN-CURED	1-03-827	A-F	91	6.7	29.8	–	–	–	2.3	43.0	9.5	5.5	4.8	5.7
			M-F	100	7.3	32.7	–	–	–	2.5	47.1	10.4	6.0	5.2	6.2
	RAMIE *Boehmeria nivea*														
249	LEAVES, SUN-CURED	1-03-861	A-F	91	17.6	13.5	–	–	–	4.1	36.1	20.0	14.7	14.0	13.8
			M-F	100	19.3	14.8	–	–	–	4.5	39.5	21.9	16.1	15.4	15.1
	RED CLOVER *Trifolium pratense*														
250	HAY, SUN-CURED, ALL ANALYSES	1-01-415	A-F	88	6.7	27.1	49.5	36.2	8.8	2.5	39.2	13.0	7.7	8.0	8.4
			M-F	100	7.5	30.7	56.0	41.0	10.0	2.8	44.3	14.7	8.7	9.0	9.5
251	HAY, IMMATURE, SUN-CURED	1-01-394	A-F	87	8.5	17.8	–	–	–	3.4	38.9	18.7	13.7	13.0	12.8
			M-F	100	9.7	20.4	–	–	–	3.9	44.6	21.4	15.7	14.9	14.7
252	HAY, MATURE, SUN-CURED	1-01-405	A-F	91	6.1	31.6	–	–	–	2.2	39.7	11.8	7.5	6.8	7.4
			M-F	100	6.7	34.6	–	–	–	2.4	43.4	12.9	8.2	7.4	8.1
	RED CLOVER-GRASS *Trifolium pratense-grass*														
253	HAY, SUN-CURED	1-01-536	A-F	89	6.4	28.4	–	–	–	1.6	39.5	12.8	8.5	7.8	8.3
			M-F	100	7.2	32.0	–	–	–	1.8	44.5	14.4	9.6	8.8	9.3
	REDTOP *Agrostis alba*														
254	HAY, SUN-CURED	1-03-885	A-F	92	6.0	28.4	–	–	–	2.8	47.4	7.4	3.7	3.0	4.1
			M-F	100	6.6	30.9	–	–	–	3.1	51.5	8.1	4.0	3.2	4.4
	REED CANARYGRASS *Phalaris arundinacea*														
255	HAY, SUN-CURED	1-01-104	A-F	89	7.3	30.2	62.9	32.7	–	2.7	40.0	9.1	5.8	4.6	5.5
			M-F	100	8.1	33.9	70.5	36.6	–	3.0	44.8	10.2	6.4	5.1	6.1
	REEDGRASS, BLUEJOINT *Calamagrostis canadensis*														
256	HAY, SUN-CURED	1-03-903	A-F	93	6.5	31.8	–	–	–	2.2	44.5	8.4	5.6	3.8	4.8
			M-F	100	7.0	34.0	–	–	–	2.4	47.6	9.0	6.0	4.1	5.2
	RESCUE BROMEGRASS *Bromus catharticus*														
257	HAY, SUN-CURED	1-00-931	A-F	90	8.1	24.6	–	–	–	3.2	44.5	9.8	5.8	5.1	5.9
			M-F	100	9.0	27.3	–	–	–	3.5	49.3	10.9	6.5	5.7	6.6
	RHODESGRASS *Chloris gayana*														
258	HAY, SUN-CURED	1-03-913	A-F	90	8.2	32.2	–	–	–	1.5	42.6	5.7	2.6	1.5	2.8
			M-F	100	9.1	35.7	–	–	–	1.7	47.2	6.3	2.8	1.7	3.1
	RICE *Oryza sativa*														
259	HULLS	1-08-075	A-F	92	19.0	38.9	71.9	62.3	9.6	1.0	30.3	3.0	0.1	–0.9	0.6
			M-F	100	20.6	42.2	78.0	67.6	10.4	1.1	32.9	3.2	0.2	–1.0	0.7
260	STRAW	1-03-925	A-F	91	15.4	31.9	64.4	50.1	4.4	1.3	38.2	3.9	0.7	–0.1	1.4
			M-F	100	17.0	35.1	71.0	55.2	4.9	1.4	42.1	4.3	0.8	–0.1	1.5
	RUMEN CONTENTS														
261	DEHY (PAUNCH PRODUCT; PAUNCH MEAL)	1-09-327	A-F	87	9.6	24.4	–	–	–	1.8	36.2	15.3	10.3	10.1	10.3
			M-F	100	11.0	28.0	–	–	–	2.1	41.4	17.6	11.8	11.6	11.8
	RUSH *Juncus* spp														
262	HAY, SUN-CURED	1-03-960	A-F	90	5.8	28.4	–	–	–	1.9	44.7	9.0	5.9	4.4	5.3
			M-F	100	6.5	31.6	–	–	–	2.2	49.8	10.0	6.6	4.9	5.9
	RUSH, SALTMEADOW (BLACK GRASS) *Juncus gerardii*														
263	HAY, SUN-CURED	1-03-970	A-F	88	7.1	24.9	–	–	–	2.4	46.3	7.5	4.0	3.2	4.2
			M-F	100	8.0	28.2	–	–	–	2.7	52.5	8.5	4.6	3.6	4.8

Entry Number	TDN Ruminant	TDN Swine	TDN Horse	Digestible Energy Ruminant		Digestible Energy Swine		Digestible Energy Horse		Metabolizable Energy Ruminant		Metabolizable Energy Swine		Metabolizable Energy Poultry ME_n		Metabolizable Energy Horse		Net Energy Ruminant NE_m		Net Energy Ruminant NE_g		Net Energy Lactating Cows NE_{lc}	
	%	%	%	Mcal		kcal		Mcal		Mcal		kcal		kcal		Mcal		Mcal		Mcal		Mcal	
				lb	*kg*	lb	*kg*	lb	*kg*	lb	*kg*	lb	*kg*	lb	*kg*	lb	*kg*	lb	*kg*	lb	*kg*	lb	*kg*
241	50	24	30	0.94	*2.06*	–	–	0.61	*1.33*	0.77	*1.69*	–	–	–	–	0.50	*1.09*	0.38	*0.84*	0.17	*0.37*	0.44	*0.96*
	57	28	35	1.07	*2.36*	–	–	0.69	*1.53*	0.88	*1.93*	–	–	–	–	0.57	*1.25*	0.44	*0.96*	0.19	*0.42*	0.50	*1.10*
242	55	56	53	1.12	*2.46*	–	–	0.99	*2.18*	0.94	*2.08*	–	–	–	–	0.81	*1.78*	0.57	*1.26*	0.33	*0.73*	0.58	*1.28*
	61	62	59	1.23	*2.71*	–	–	1.09	*2.40*	1.04	*2.29*	–	–	–	–	0.89	*1.97*	0.63	*1.39*	0.37	*0.80*	0.64	*1.41*
243	50	39	45	1.00	*2.19*	–	–	0.85	*1.88*	0.82	*1.81*	–	–	–	–	0.70	*1.54*	0.46	*1.02*	0.23	*0.52*	0.50	*1.10*
	56	43	50	1.11	*2.44*	–	–	0.95	*2.09*	0.91	*2.01*	–	–	–	–	0.78	*1.72*	0.51	*1.13*	0.26	*0.57*	0.55	*1.22*
244	59	44	39	1.13	*2.48*	–	–	0.76	*1.67*	0.95	*2.10*	–	–	–	–	0.62	*1.37*	0.53	*1.16*	0.29	*0.64*	0.55	*1.21*
	64	48	43	1.23	*2.71*	–	–	0.82	*1.82*	1.04	*2.29*	–	–	–	–	0.68	*1.49*	0.57	*1.27*	0.32	*0.70*	0.60	*1.32*
245	46	31	34	0.94	*2.06*	–	–	0.67	*1.48*	0.76	*1.68*	–	–	–	–	0.55	*1.21*	0.44	*0.97*	0.21	*0.46*	0.48	*1.07*
	50	34	37	1.03	*2.27*	–	–	0.74	*1.62*	0.84	*1.84*	–	–	–	–	0.60	*1.33*	0.48	*1.06*	0.23	*0.51*	0.53	*1.17*
246	48	35	35	0.97	*2.13*	–	–	0.68	*1.50*	0.79	*1.75*	–	–	–	–	0.56	*1.23*	0.45	*0.99*	0.22	*0.49*	0.49	*1.08*
	54	38	39	1.07	*2.36*	–	–	0.76	*1.67*	0.88	*1.93*	–	–	–	–	0.62	*1.37*	0.50	*1.10*	0.25	*0.54*	0.54	*1.20*
247	48	37	27	0.97	*2.14*	–	–	0.55	*1.22*	0.80	*1.75*	–	–	–	–	0.45	*1.00*	0.47	*1.03*	0.24	*0.53*	0.50	*1.11*
	54	41	30	1.08	*2.37*	–	–	0.61	*1.35*	0.88	*1.94*	–	–	–	–	0.50	*1.11*	0.52	*1.14*	0.27	*0.59*	0.56	*1.23*
248	41	38	39	0.92	*2.03*	–	–	0.76	*1.67*	0.74	*1.64*	–	–	–	–	0.62	*1.37*	0.48	*1.05*	0.25	*0.54*	0.51	*1.13*
	45	42	43	1.01	*2.22*	–	–	0.83	*1.83*	0.81	*1.79*	–	–	–	–	0.68	*1.50*	0.52	*1.15*	0.27	*0.59*	0.56	*1.24*
249	–	–	–	–	–	–	–	–	–	–	–	–	–	–	–	–	–	–	–	–	–	–	–
	–	–	–	–	–	–	–	–	–	–	–	–	–	–	–	–	–	–	–	–	–	–	–
250	52	44	42	1.21	*2.67*	–	–	0.81	*1.78*	0.95	*2.10*	–	–	–	–	0.66	*1.46*	0.50	*1.11*	0.28	*0.61*	0.53	*1.16*
	59	50	48	1.37	*3.02*	–	–	0.91	*2.01*	1.08	*2.38*	–	–	–	–	0.75	*1.65*	0.57	*1.26*	0.31	*0.69*	0.60	*1.31*
251	56	62	50	1.12	*2.46*	–	–	0.93	*2.04*	0.95	*2.09*	–	–	–	–	0.76	*1.67*	0.58	*1.28*	0.35	*0.76*	0.58	*1.28*
	64	71	57	1.28	*2.82*	–	–	1.06	*2.34*	1.09	*2.40*	–	–	–	–	0.87	*1.92*	0.66	*1.46*	0.40	*0.88*	0.66	*1.46*
252	52	39	41	1.03	*2.27*	–	–	0.79	*1.75*	0.85	*1.88*	–	–	–	–	0.65	*1.43*	0.49	*1.07*	0.25	*0.56*	0.52	*1.14*
	56	42	45	1.13	*2.49*	–	–	0.87	*1.92*	0.94	*2.06*	–	–	–	–	0.71	*1.57*	0.53	*1.18*	0.28	*0.61*	0.57	*1.25*
253	51	39	46	1.01	*2.23*	–	–	0.86	*1.90*	0.84	*1.85*	–	–	–	–	0.71	*1.55*	0.48	*1.05*	0.25	*0.55*	0.51	*1.11*
	57	44	51	1.14	*2.51*	–	–	0.97	*2.14*	0.95	*2.09*	–	–	–	–	0.80	*1.75*	0.54	*1.18*	0.28	*0.62*	0.57	*1.26*
254	50	41	37	1.02	*2.25*	–	–	0.73	*1.60*	0.84	*1.86*	–	–	–	–	0.60	*1.31*	0.50	*1.11*	0.27	*0.59*	0.53	*1.17*
	54	45	40	1.11	*2.45*	–	–	0.79	*1.74*	0.92	*2.02*	–	–	–	–	0.65	*1.42*	0.55	*1.21*	0.29	*0.64*	0.58	*1.27*
255	44	36	43	0.93	*2.06*	–	–	0.82	*1.80*	0.76	*1.68*	–	–	–	–	0.67	*1.47*	0.46	*1.02*	0.24	*0.52*	0.50	*1.10*
	49	41	48	1.05	*2.31*	–	–	0.91	*2.01*	0.85	*1.88*	–	–	–	–	0.75	*1.65*	0.52	*1.15*	0.27	*0.59*	0.56	*1.23*
256	54	36	38	1.05	*2.32*	–	–	0.75	*1.65*	0.87	*1.92*	–	–	–	–	0.61	*1.35*	0.48	*1.05*	0.24	*0.53*	0.52	*1.14*
	58	39	41	1.12	*2.48*	–	–	0.80	*1.76*	0.93	*2.05*	–	–	–	–	0.66	*1.44*	0.51	*1.13*	0.26	*0.57*	0.55	*1.22*
257	51	46	37	1.03	*2.27*	–	–	0.72	*1.59*	0.86	*1.89*	–	–	–	–	0.59	*1.31*	0.51	*1.12*	0.28	*0.61*	0.53	*1.18*
	56	51	41	1.14	*2.52*	–	–	0.80	*1.76*	0.95	*2.10*	–	–	–	–	0.66	*1.45*	0.57	*1.25*	0.31	*0.68*	0.59	*1.30*
258	52	25	33	0.97	*2.15*	–	–	0.65	*1.44*	0.80	*1.76*	–	–	–	–	0.54	*1.18*	0.40	*0.87*	0.17	*0.38*	0.45	*1.00*
	58	28	37	1.08	*2.38*	–	–	0.72	*1.60*	0.89	*1.96*	–	–	–	–	0.59	*1.31*	0.44	*0.97*	0.19	*0.42*	0.50	*1.10*
259	11	–	12	0.27	*0.60*	–	–	0.30	*0.67*	0.16	*0.35*	–	–	36	*79*	0.25	*0.55*	–0.26	*–0.57*	–0.47	*–1.04*	0.08	*0.17*
	12	–	13	0.30	*0.65*	–	–	0.33	*0.72*	0.17	*0.38*	–	–	39	*86*	0.27	*0.59*	–0.28	*–0.62*	–0.51	*–1.13*	0.08	*0.19*
260	40	12	22	0.80	*1.76*	–	–	0.47	*1.04*	0.62	*1.37*	–	–	–	–	0.39	*0.85*	0.34	*0.75*	0.12	*0.26*	0.41	*0.91*
	44	13	24	0.88	*1.94*	–	–	0.52	*1.15*	0.69	*1.51*	–	–	–	–	0.43	*0.94*	0.37	*0.82*	0.13	*0.29*	0.46	*1.01*
261	51	42	45	1.00	*2.20*	–	–	0.85	*1.88*	0.83	*1.83*	–	–	–	–	0.70	*1.54*	0.47	*1.03*	0.25	*0.54*	0.50	*1.10*
	58	48	52	1.14	*2.51*	–	–	0.98	*2.15*	0.95	*2.09*	–	–	–	–	0.80	*1.77*	0.54	*1.18*	0.28	*0.62*	0.57	*1.26*
262	56	38	42	1.06	*2.34*	–	–	0.80	*1.75*	0.89	*1.96*	–	–	–	–	0.65	*1.44*	0.48	*1.05*	0.25	*0.55*	0.51	*1.12*
	62	43	46	1.18	*2.61*	–	–	0.89	*1.95*	0.99	*2.19*	–	–	–	–	0.73	*1.60*	0.53	*1.17*	0.28	*0.61*	0.57	*1.25*
263	49	40	38	0.98	*2.17*	–	–	0.73	*1.61*	0.81	*1.79*	–	–	–	–	0.60	*1.32*	0.47	*1.04*	0.25	*0.55*	0.50	*1.11*
	55	45	43	1.11	*2.46*	–	–	0.83	*1.82*	0.92	*2.03*	–	–	–	–	0.68	*1.50*	0.54	*1.18*	0.28	*0.62*	0.57	*1.26*

(Continued)

TABLE V-1C DRY FORAGES, MINERAL AND VITAMIN COMPOSITION OF FEEDS, DATA EXPRESSED **AS-FED** AND **MOISTURE-FREE**—*(Continued)*

Entry Number	Feed Name Description	Moisture Basis: A-F (as-fed) or M-F (moisture-free)	Dry Matter	Macro Minerals							Micro Minerals						
				Calcium (Ca)	Phosphorus (P)	Sodium (Na)	Chlorine (Cl)	Magnesium (Mg)	Potassium (K)	Sulfur (S)	Cobalt (Co)	Copper (Cu)	Iodine (I)	Iron (Fe)	Manganese (Mn)	Selenium (Se)	Zinc (Zn)
			%	%	%	%	%	%	%	%	ppm or mg/kg	ppm or mg/kg	ppm or mg/kg	%	ppm or mg/kg	ppm or mg/kg	ppm or mg/kg
	PEARL MILLET *Pennisetum glaucum*																
241	HAY, SUN-CURED	**A-F**	**87**	–	–	–	–	–	–	–	–	–	–	–	–	–	–
		M-F	100	–	–	–	–	–	–	–	–	–	–	–	–	–	–
	PEAVINE *Lathyrus* spp																
242	HAY, SUN-CURED	**A-F**	**91**	–	–	–	–	–	–	–	–	–	–	–	–	–	–
		M-F	100	–	–	–	–	–	–	–	–	–	–	–	–	–	–
	PERSIAN CLOVER *Trifolium resupinatum*																
243	HAY, SUN-CURED	**A-F**	**90**	**1.49**	**0.21**	–	–	–	–	–	–	–	–	–	–	–	–
		M-F	100	1.66	0.23	–	–	–	–	–	–	–	–	–	–	–	–
	POVERTY OATGRASS *Danthonia spicata*																
244	HAY, SUN-CURED	**A-F**	**92**	–	–	–	–	–	–	–	–	–	–	–	–	–	–
		M-F	100	–	–	–	–	–	–	–	–	–	–	–	–	–	–
	PRAIRIE GRASS, MIDWEST (PRAIRIE HAY)																
245	HAY, SUN-CURED	**A-F**	**91**	**0.32**	**0.13**	–	–	**0.24**	**0.98**	–	–	–	–	**0.008**	–	–	–
		M-F	100	0.35	0.14	–	–	0.26	1.08	–	–	–	–	0.009	–	–	–
246	HAY, IMMATURE, SUN-CURED	**A-F**	**90**	**0.51**	**0.17**	–	–	**0.22**	**0.98**	–	–	–	–	**0.009**	–	–	–
		M-F	100	0.57	0.19	–	–	0.24	1.08	–	–	–	–	0.009	–	–	–
247	HAY, MATURE, SUN-CURED	**A-F**	**90**	**0.41**	**0.08**	–	–	–	–	–	–	–	–	–	–	–	–
		M-F	100	0.45	0.09	–	–	–	–	–	–	–	–	–	–	–	–
	QUACKGRASS *Agropyron repens*																
248	HAY, SUN-CURED	**A-F**	**91**	**0.30**	**0.11**	–	–	–	–	–	–	–	–	–	**42.1**	–	–
		M-F	100	0.33	0.12	–	–	–	–	–	–	–	–	–	46.1	–	–
	RAMIE *Boehmeria nivea*																
249	LEAVES, SUN-CURED	**A-F**	**91**	–	–	–	–	–	–	–	–	–	–	–	–	–	–
		M-F	100	–	–	–	–	–	–	–	–	–	–	–	–	–	–
	RED CLOVER *Trifolium pratense*																
250	HAY, SUN-CURED, ALL ANALYSES	**A-F**	**88**	**1.22**	**0.22**	**0.16**	**0.28**	**0.34**	**1.60**	**0.15**	**0.138**	**18.8**	**0.217**	**0.022**	**95.2**	–	**32.5**
		M-F	100	1.38	0.25	0.18	0.32	0.38	1.81	0.16	0.156	21.2	0.245	0.024	107.7	–	36.7
251	HAY, IMMATURE, SUN-CURED	**A-F**	**87**	**1.35**	**0.32**	–	–	**0.34**	**2.82**	–	**0.201**	**18.4**	–	**0.063**	**75.6**	–	**45.4**
		M-F	100	1.55	0.37	–	–	0.39	3.24	–	0.230	21.1	–	0.073	86.7	–	52.0
252	HAY, MATURE, SUN-CURED	**A-F**	**91**	**1.07**	**0.19**	–	–	**0.32**	**1.55**	–	**0.110**	**8.5**	–	**0.021**	**39.3**	–	**32.0**
		M-F	100	1.18	0.21	–	–	0.35	1.70	–	0.120	9.4	–	0.023	43.0	–	35.0
	RED CLOVER-GRASS *Trifolium pratense-grass*																
253	HAY, SUN-CURED	**A-F**	**89**	**1.04**	**0.20**	–	–	**0.40**	**1.25**	–	**0.133**	**14.2**	–	**0.015**	**47.1**	–	–
		M-F	100	1.18	0.23	–	–	0.45	1.41	–	0.150	16.0	–	0.017	53.1	–	–
	REDTOP *Agrostis alba*																
254	HAY, SUN-CURED	**A-F**	**92**	**0.39**	**0.20**	**0.06**	**0.06**	**0.20**	**1.74**	**0.23**	**0.134**	**3.6**	**0.092**	**0.015**	**207.7**	–	–
		M-F	100	0.43	0.22	0.07	0.07	0.22	1.89	0.25	0.146	3.9	0.100	0.016	225.5	–	–
	REED CANARYGRASS *Phalaris arundinacea*																
255	HAY, SUN-CURED	**A-F**	**89**	**0.32**	**0.21**	**0.01**	–	**0.19**	**2.60**	–	–	**10.6**	–	**0.014**	**82.5**	–	–
		M-F	100	0.36	0.24	0.02	–	0.22	2.91	–	–	11.9	–	0.015	92.4	–	–
	REEDGRASS, BLUEJOINT *Calamagrostis canadensis*																
256	HAY, SUN-CURED	**A-F**	**93**	**0.39**	**0.14**	–	–	–	**0.10**	–	–	–	–	–	–	–	–
		M-F	100	0.42	0.15	–	–	–	0.11	–	–	–	–	–	–	–	–
	RESCUE BROMEGRASS *Bromus catharticus*																
257	HAY, SUN-CURED	**A-F**	**90**	–	–	–	–	–	–	–	–	–	–	–	–	–	–
		M-F	100	–	–	–	–	–	–	–	–	–	–	–	–	–	–
	RHODESGRASS *Chloris gayana*																
258	HAY, SUN-CURED	**A-F**	**90**	**0.35**	**0.27**	–	–	–	**1.20**	–	–	–	–	–	–	–	–
		M-F	100	0.39	0.30	–	–	–	1.33	–	–	–	–	–	–	–	–
	RICE *Oryza sativa*																
259	HULLS	**A-F**	**92**	**0.11**	**0.10**	**0.02**	**0.07**	**0.41**	**0.64**	**0.08**	**2.046**	**3.1**	–	**0.010**	**295.0**	–	**22.0**
		M-F	100	0.12	0.10	0.02	0.08	0.45	0.69	0.09	2.220	3.4	–	0.010	320.1	–	23.9
260	STRAW	**A-F**	**91**	**0.19**	**0.07**	**0.28**	–	**0.10**	**1.20**	–	–	–	–	–	**313.9**	–	–
		M-F	100	0.21	0.08	0.31	–	0.11	1.32	–	–	–	–	–	345.8	–	–
	RUMEN CONTENTS																
261	DEHY (PAUNCH PRODUCT; PAUNCH MEAL)	**A-F**	**87**	**0.69**	**0.58**	–	–	–	–	–	–	–	–	–	–	–	–
		M-F	100	0.79	0.67	–	–	–	–	–	–	–	–	–	–	–	–
	RUSH *Juncus* spp																
262	HAY, SUN-CURED	**A-F**	**90**	**0.31**	**0.10**	–	–	**0.26**	**0.68**	–	–	–	–	**0.010**	–	–	–
		M-F	100	0.35	0.11	–	–	0.29	0.76	–	–	–	–	0.011	–	–	–
	RUSH, SALTMEADOW (BLACK GRASS) *Juncus gerardii*																
263	HAY, SUN-CURED	**A-F**	**88**	–	**0.09**	–	–	–	**1.53**	–	–	–	–	–	–	–	–
		M-F	100	–	0.10	–	–	–	1.74	–	–	–	–	–	–	–	–

Entry Number	Fat-Soluble Vitamins: A (1 mg Carotene = 1667 IU Vit A)	Carotene (Provitamin A)	D	E	K	Water-Soluble Vitamins: B–12	Biotin	Choline	Folacin (Folic Acid)	Niacin	Pantothenic Acid (B–3)	(Pyridoxine) B–6	Riboflavin (B–2)	Thiamin (B–1)
	IU/g	ppm or mg/kg	IU/kg	ppm or mg/kg	ppm or mg/kg	ppm or mg/kg	ppm or mg/kg	ppm or mg/kg	ppm or mg/kg	ppm or mg/kg	ppm or mg/kg	ppm or mg/kg	ppm or mg/kg	ppm or mg/kg
241	–	–	–	–	–	–	–	–	–	–	–	–	–	–
	–	–	–	–	–	–	–	–	–	–	–	–	–	–
242	**48.0**	**28.8**	–	–	–	–	–	–	–	–	–	–	**8.4**	–
	52.9	31.7	–	–	–	–	–	–	–	–	–	–	9.3	–
243	–	–	–	–	–	–	–	–	–	–	–	–	–	–
	–	–	–	–	–	–	–	–	–	–	–	–	–	–
244	–	–	–	–	–	–	–	–	–	–	–	–	–	–
	–	–	–	–	–	–	–	–	–	–	–	–	–	–
245	–	–	**1**	–	–	–	–	–	–	–	–	–	–	–
	–	–	1	–	–	–	–	–	–	–	–	–	–	–
246	–	–	–	–	–	–	–	–	–	–	–	–	–	–
	–	–	–	–	–	–	–	–	–	–	–	–	–	–
247	**15.6**	**9.3**	–	–	–	–	–	–	–	–	–	–	–	–
	17.3	10.4	–	–	–	–	–	–	–	–	–	–	–	–
248	**56.6**	**33.9**	–	–	–	–	–	–	–	–	–	–	–	–
	62.0	37.2	–	–	–	–	–	–	–	–	–	–	–	–
249	–	–	–	–	–	–	–	–	–	–	–	–	–	–
	–	–	–	–	–	–	–	–	–	–	–	–	–	–
250	**40.5**	**24.3**	–	–	–	–	**0.09**	–	–	**38**	**9.9**	–	**15.7**	**2.0**
	45.9	27.5	–	–	–	–	0.11	–	–	43	11.2	–	17.8	2.2
251	–	–	–	–	–	–	–	–	–	–	–	–	–	–
	–	–	–	–	–	–	–	–	–	–	–	–	–	–
252	–	–	–	–	–	–	–	–	–	–	–	–	–	–
	–	–	–	–	–	–	–	–	–	–	–	–	–	–
253	**18.7**	**11.2**	–	–	–	–	–	–	–	–	–	–	–	–
	21.1	12.7	–	–	–	–	–	–	–	–	–	–	–	–
254	**6.1**	**3.7**	–	–	–	–	–	–	–	–	–	–	–	–
	6.6	4.0	–	–	–	–	–	–	–	–	–	–	–	–
255	**28.2**	**16.9**	–	–	–	–	–	–	–	–	–	–	**8.5**	**3.6**
	31.6	18.9	–	–	–	–	–	–	–	–	–	–	9.5	4.0
256	**9.1**	**5.5**	–	–	–	–	–	–	–	–	–	–	–	–
	9.8	5.9	–	–	–	–	–	–	–	–	–	–	–	–
257	–	–	–	–	–	–	–	–	–	–	–	–	–	–
	–	–	–	–	–	–	–	–	–	–	–	–	–	–
258	–	–	–	–	–	–	–	–	–	–	–	–	–	–
	–	–	–	–	–	–	–	–	–	–	–	–	–	–
259	–	–	–	**7.5**	–	–	–	–	–	**28**	**7.9**	**0.07**	**0.5**	**2.2**
	–	–	–	8.1	–	–	–	–	–	31	8.6	0.08	0.6	2.4
260	–	–	–	–	–	–	–	–	–	–	–	–	–	–
	–	–	–	–	–	–	–	–	–	–	–	–	–	–
261	–	–	–	–	–	–	–	–	–	–	–	–	–	–
	–	–	–	–	–	–	–	–	–	–	–	–	–	–
262	–	–	–	–	–	–	–	–	–	–	–	–	–	–
	–	–	–	–	–	–	–	–	–	–	–	–	–	–
263	–	–	–	–	–	–	–	–	–	–	–	–	–	–
	–	–	–	–	–	–	–	–	–	–	–	–	–	–

(Continued)

TABLE V-1C DRY FORAGES, COMPOSITION OF FEEDS, DATA EXPRESSED **AS-FED** AND **MOISTURE-FREE**—*(Continued)*

Entry Number	Feed Name Description	International Feed Number	Moisture Basis: A-F (as-fed) or M-F (moisture-free)	Proximate Analysis: Dry Matter	Ash	Crude Fiber	Neutral Det. Fib. (NDF)	Acid Det. Fib. (ADF)	Lignin	Ether Extract (Fat)	N-Free Extract	Crude Protein	Digestible Protein: Ruminant	Swine	Horse
				%	%	%	%	%	%	%	%	%	%	%	%
	RUSSIAN THISTLE (TUMBLING TUMBLEWEED) *Salsola kali, tenuifolia*														
264	HAY, SUN-CURED	1-03-988	A-F	86	13.1	25.1	–	–	–	1.7	36.3	10.2	6.6	5.7	6.4
			M-F	100	15.2	29.0	–	–	–	2.0	42.0	11.8	7.6	6.5	7.3
	RUSSIAN WILD-RYE *Elymus junceus*														
265	HAY, SUN-CURED	1-08-171	A-F	90	7.2	33.4	–	–	–	1.7	40.5	7.2	4.0	2.9	3.9
			M-F	100	8.0	37.1	–	–	–	1.9	45.0	8.0	4.5	3.2	4.4
	RYE *Secale cereale*														
266	HAY, SUN-CURED	1-04-004	A-F	93	4.2	31.1	–	–	1.9	1.9	48.4	7.9	4.1	3.4	4.4
			M-F	100	4.5	33.3	–	–	2.0	2.0	51.8	8.5	4.4	3.6	4.8
267	STRAW	1-04-007	A-F	91	3.8	38.3	–	–	–	1.4	44.7	2.8	–0.3	–1.1	0.5
			M-F	100	4.2	42.1	–	–	–	1.5	49.2	3.0	–0.4	–1.2	0.6
	RYEGRASS *Lolium* spp														
268	HAY, SUN-CURED[1]	1-04-057	A-F	88	7.1	25.3	56.3	37.0	–	1.8	46.2	7.5	3.8	3.2	4.2
			M-F	100	8.1	28.8	64.0	42.0	–	2.1	52.5	8.5	4.3	3.6	4.8
	SAFFLOWER *Carthamus tinctorius*														
269	HAY, MATURE, SUN-CURED	1-04-104	A-F	91	1.4	53.1	–	–	–	4.7	28.3	3.8	0.6	–0.2	1.3
			M-F	100	1.5	58.2	–	–	–	5.1	31.0	4.2	0.6	–0.2	1.5
270	HULLS	1-04-105	A-F	91	1.6	52.9	–	–	–	3.4	29.7	3.3	0.1	–0.6	0.9
			M-F	100	1.8	58.2	–	–	–	3.7	32.7	3.6	0.1	–0.7	1.0
	SAINFOIN *Onobrychis* spp														
271	HAY, SUN-CURED	1-04-143	A-F	86	6.5	22.7	–	–	–	3.2	39.1	14.2	9.9	9.2	9.4
			M-F	100	7.6	26.5	–	–	–	3.7	45.6	16.6	11.5	10.7	11.0
	SALTBUSH, NUTTALL *Atriplex nuttallii*														
272	BROWSE, SUN-CURED	1-26-111	A-F	90	16.5	21.3	–	–	–	1.6	37.2	13.4	9.9	–	–
			M-F	100	18.3	23.7	–	–	–	1.8	41.3	14.9	11.0	–	–
273	BROWSE, MATURE, SUN-CURED	1-26-112	A-F	91	16.0	26.1	–	–	–	4.4	37.1	7.4	3.1	–	–
			M-F	100	17.6	28.7	–	–	–	4.8	40.8	8.1	3.4	–	–
	SALTGRASS *Distichlis* spp														
274	HAY, SUN-CURED	1-04-168	A-F	89	11.4	28.3	–	–	–	1.8	40.0	8.0	4.3	3.5	4.5
			M-F	100	12.7	31.6	–	–	–	2.1	44.7	8.9	4.8	4.0	5.1
	SAWDUST														
275	WOOD	1-07-714	A-F	90	0.7	71.5	–	–	–	0.8	16.7	0.3	–2.5	–3.2	–1.4
			M-F	100	0.8	79.4	–	–	–	0.9	18.6	0.3	–2.8	–3.6	–1.6
	SEAWEED (KELP) *Laminariales* (order), *Fucales* (order)														
276	WHOLE, DEHY	1-08-073	A-F	91	35.0	6.5	–	–	–	0.5	42.4	6.5	2.9	2.2	3.4
			M-F	100	38.6	7.1	–	–	–	0.5	46.7	7.1	3.2	2.4	3.7
	SEDGE *Carex* spp														
277	HAY, SUN-CURED	1-04-193	A-F	89	6.4	28.0	–	–	–	2.1	44.4	8.4	4.2	3.9	4.9
			M-F	100	7.2	31.3	–	–	–	2.4	49.7	9.4	4.7	4.4	5.5
	SLOUGHGRASS, AMERICAN *Beckmannia syzigachne*														
278	HAY, MATURE, SUN-CURED	1-04-237	A-F	93	9.4	29.5	–	–	–	1.3	45.6	7.5	4.6	3.0	4.1
			M-F	100	10.1	31.6	–	–	–	1.4	48.9	8.0	4.9	3.2	4.4
	SORGHUM *Sorghum bicolor*														
279	FODDER WITH HEADS, SUN-CURED	1-07-960	A-F	90	8.9	25.6	–	–	–	2.0	47.4	6.2	2.4	2.0	3.2
			M-F	100	9.9	28.4	–	–	–	2.2	52.6	6.9	2.6	2.2	3.5
280	STOVER WITHOUT HEADS, SUN-CURED	1-04-302	A-F	92	8.9	29.9	–	39.9	–	1.6	46.8	4.4	0.7	0.4	1.8
			M-F	100	9.7	32.6	–	43.6	–	1.8	51.1	4.9	0.7	0.4	2.0
281	ATLAS, FODDER WITH HEADS, SUN-CURED	1-04-334	A-F	91	6.6	20.2	–	–	–	2.8	55.1	6.5	–	–	3.2
			M-F	100	7.2	22.2	–	–	–	3.1	60.5	7.1	–	–	3.5
282	ATLAS, STOVER WITHOUT HEADS, SUN-CURED	1-04-336	A-F	90	7.1	28.3	–	–	–	2.1	48.8	3.7	0.5	–0.2	1.3
			M-F	100	7.9	31.5	–	–	–	2.3	54.2	4.1	0.6	–0.2	1.4
283	DURRA, FODDER WITH HEADS, SUN-CURED	1-04-359	A-F	90	5.2	24.1	–	–	–	2.8	51.4	6.4	2.9	2.2	3.3
			M-F	100	5.8	26.8	–	–	–	3.1	57.2	7.1	3.2	2.4	3.7
284	HEGARI, FODDER WITH HEADS, DOUGH STAGE, SUN-CURED	1-04-390	A-F	88	7.5	18.5	–	–	–	2.0	52.8	7.2	3.6	2.9	4.0
			M-F	100	8.5	21.0	–	–	–	2.3	60.0	8.2	4.1	3.3	4.5
285	HEGARI, STOVER WITHOUT HEADS, SUN-CURED	1-08-515	A-F	87	8.9	25.3	–	34.1	–	1.9	46.0	5.2	1.9	1.2	2.5
			M-F	100	10.2	29.0	–	39.0	–	2.2	52.6	6.0	2.2	1.4	2.8
286	JOHNSONGRASS, HAY, SUN-CURED	1-04-407	A-F	91	7.7	30.4	–	–	–	2.0	43.7	6.7	3.0	2.4	3.6
			M-F	100	8.6	33.6	–	–	–	2.2	48.3	7.5	3.3	2.7	4.0
287	KAFIR, CHAFF	1-04-420	A-F	92	7.9	19.6	–	–	–	1.0	56.8	6.7	1.3	2.3	3.5
			M-F	100	8.6	21.3	–	–	–	1.1	61.7	7.3	1.5	2.6	3.8
288	KAFIR, FODDER WITH HEADS, SUN-CURED	1-04-418	A-F	90	9.4	25.7	–	–	–	2.2	44.5	8.3	4.2	3.8	4.8
			M-F	100	10.4	28.5	–	–	–	2.5	49.4	9.2	4.7	4.2	5.3
289	KAFIR, FODDER WITH HEADS, LATE BLOOM, SUN-CURED	1-04-417	A-F	91	11.2	23.8	–	–	–	2.1	44.3	10.0	6.3	5.2	6.0
			M-F	100	12.3	26.0	–	–	–	2.3	48.5	10.9	6.9	5.7	6.6
290	KAFIR, STOVER WITHOUT HEADS, SUN-CURED	1-04-419	A-F	82	8.1	26.9	–	–	–	1.5	40.7	4.6	1.6	0.9	2.1
			M-F	100	9.9	32.9	–	–	–	1.9	49.7	5.7	1.9	1.1	2.6
291	MILO, FODDER WITH HEADS, SUN-CURED	1-04-433	A-F	89	8.5	21.8	–	–	–	2.9	49.4	6.5	2.5	2.3	3.4
			M-F	100	9.6	24.4	–	–	–	3.2	55.5	7.3	2.8	2.5	3.8
292	MILO, STOVER WITHOUT HEADS, SUN-CURED	1-04-434	A-F	91	11.8	31.1	–	–	–	1.3	42.5	4.4	–0.8	0.4	1.8
			M-F	100	13.0	34.1	–	–	–	1.4	46.6	4.9	–0.9	0.4	2.0

Entry Number	TDN Ruminant	TDN Swine	TDN Horse	Digestible Energy Ruminant		Digestible Energy Swine		Digestible Energy Horse		Metabolizable Energy Ruminant		Metabolizable Energy Swine		Metabolizable Energy Poultry ME_n		Metabolizable Energy Horse		Net Energy Ruminant NE_m		Net Energy Ruminant NE_g		Net Energy Lactating Cows NE_{lc}	
	%	%	%	Mcal lb	Mcal kg	kcal lb	kcal kg	Mcal lb	Mcal kg	Mcal lb	Mcal kg	kcal lb	kcal kg	kcal lb	kcal kg	Mcal lb	Mcal kg	Mcal lb	Mcal kg	Mcal lb	Mcal kg	Mcal lb	Mcal kg
264	39	29	33	0.79	1.73	–	–	0.65	1.44	0.62	1.36	–	–	–	–	0.54	1.18	0.33	0.74	0.12	0.27	0.40	0.89
	45	34	39	0.91	2.00	–	–	0.76	1.67	0.71	1.57	–	–	–	–	0.62	1.37	0.39	0.85	0.14	0.31	0.47	1.02
265	53	27	34	1.10	2.41	–	–	0.67	1.48	0.84	1.84	–	–	–	–	0.55	1.21	0.41	0.91	0.19	0.42	0.46	1.02
	59	30	38	1.22	2.68	–	–	0.75	1.64	0.93	2.05	–	–	–	–	0.61	1.35	0.46	1.01	0.21	0.46	0.52	1.14
266	43	39	44	0.86	1.90	–	–	0.83	1.84	0.68	1.50	–	–	–	–	0.68	1.51	0.35	0.77	0.12	0.27	0.43	0.94
	46	41	47	0.93	2.04	–	–	0.89	1.97	0.73	1.61	–	–	–	–	0.73	1.61	0.37	0.82	0.13	0.29	0.46	1.01
267	39	–	32	0.78	1.72	–	–	0.64	1.40	0.60	1.32	–	–	–	–	0.52	1.15	0.25	0.55	0.03	0.07	0.35	0.78
	43	–	35	0.86	1.89	–	–	0.70	1.54	0.66	1.46	–	–	–	–	0.57	1.26	0.27	0.60	0.04	0.08	0.39	0.86
268	53	37	40	1.06	2.34	–	–	0.76	1.68	0.89	1.97	–	–	–	–	0.62	1.38	0.54	1.19	0.31	0.68	0.55	1.22
	60	42	45	1.21	2.66	–	–	0.87	1.91	1.02	2.24	–	–	–	–	0.71	1.56	0.61	1.35	0.35	0.78	0.63	1.38
269	–	–	–	–	–	–	–	–	–	–	–	–	–	–	–	–	–	–	–	–	–	–	–
	–	–	–	–	–	–	–	–	–	–	–	–	–	–	–	–	–	–	–	–	–	–	–
270	15	–	–	0.30	0.65	–	–	–	–	0.11	0.25	–	–	414	914	–	–	–	–	–	–	–	–
	16	–	–	0.32	0.72	–	–	–	–	0.12	0.27	–	–	456	1004	–	–	–	–	–	–	–	–
271	53	52	43	1.05	2.32	–	–	0.81	1.78	0.89	1.95	–	–	–	–	0.66	1.46	0.53	1.18	0.31	0.68	0.54	1.20
	61	60	50	1.23	2.70	–	–	0.94	2.08	1.03	2.28	–	–	–	–	0.77	1.70	0.62	1.37	0.36	0.79	0.63	1.40
272	35	–	–	0.69	1.53	–	–	–	–	0.57	1.26	–	–	–	–	–	–	–	–	–	–	–	–
	39	–	–	0.77	1.70	–	–	–	–	0.64	1.40	–	–	–	–	–	–	–	–	–	–	–	–
273	42	–	–	0.83	1.82	–	–	–	–	0.70	1.55	–	–	–	–	–	–	–	–	–	–	–	–
	46	–	–	0.91	2.00	–	–	–	–	0.77	1.70	–	–	–	–	–	–	–	–	–	–	–	–
274	45	28	32	0.90	1.99	–	–	0.64	1.42	0.73	1.60	–	–	–	–	0.53	1.16	0.39	0.87	0.17	0.38	0.45	0.99
	51	31	36	1.01	2.22	–	–	0.72	1.58	0.81	1.79	–	–	–	–	0.59	1.30	0.44	0.97	0.19	0.42	0.50	1.11
275	32	–	–	0.65	1.43	–	–	–	–	0.47	1.04	–	–	–	–	–	–	–	–	–	–	–	–
	36	–	–	0.72	1.59	–	–	–	–	0.53	1.16	–	–	–	–	–	–	–	–	–	–	–	–
276	29	–	–	0.58	1.27	–	–	–	–	0.40	0.87	–	–	–	–	–	–	–	–	–	–	–	–
	32	–	–	0.63	1.40	–	–	–	–	0.44	0.96	–	–	–	–	–	–	–	–	–	–	–	–
277	47	37	39	0.96	2.12	–	–	0.75	1.66	0.79	1.74	–	–	–	–	0.62	1.36	0.47	1.03	0.24	0.53	0.50	1.11
	52	42	44	1.08	2.38	–	–	0.84	1.86	0.88	1.95	–	–	–	–	0.69	1.53	0.53	1.16	0.27	0.60	0.56	1.24
278	49	30	39	0.97	2.14	–	–	0.76	1.68	0.79	1.75	–	–	–	–	0.62	1.37	0.46	1.00	0.22	0.49	0.47	1.03
	52	32	42	1.04	2.30	–	–	0.81	1.80	0.85	1.87	–	–	–	–	0.67	1.47	0.49	1.08	0.24	0.52	0.50	1.10
279	51	34	36	1.02	2.24	–	–	0.71	1.56	0.84	1.86	–	–	–	–	0.58	1.28	0.51	1.12	0.28	0.61	0.53	1.17
	56	38	40	1.13	2.49	–	–	0.79	1.73	0.94	2.06	–	–	–	–	0.64	1.42	0.56	1.24	0.31	0.68	0.59	1.30
280	47	27	33	0.92	2.02	–	–	0.66	1.46	0.74	1.64	–	–	–	–	0.54	1.20	0.40	0.89	0.18	0.39	0.42	0.93
	51	29	37	1.00	2.21	–	–	0.72	1.59	0.81	1.79	–	–	–	–	0.59	1.31	0.44	0.97	0.19	0.43	0.46	1.02
281	55	–	44	1.09	2.40	–	–	0.80	1.76	0.90	1.98	–	–	–	–	0.66	1.45	0.54	1.19	0.27	0.59	0.56	1.23
	60	–	48	1.20	2.64	–	–	0.88	1.93	0.98	2.17	–	–	–	–	0.72	1.59	0.59	1.31	0.30	0.65	0.61	1.35
282	53	32	33	1.00	2.21	–	–	0.65	1.44	0.83	1.83	–	–	–	–	0.54	1.18	0.44	0.96	0.21	0.46	0.48	1.06
	59	35	37	1.12	2.46	–	–	0.73	1.60	0.92	2.03	–	–	–	–	0.60	1.31	0.48	1.07	0.23	0.51	0.53	1.17
283	53	45	40	1.06	2.33	–	–	0.76	1.68	0.89	1.95	–	–	–	–	0.62	1.38	0.52	1.14	0.29	0.63	0.54	1.19
	59	50	44	1.18	2.60	–	–	0.85	1.87	0.99	2.17	–	–	–	–	0.69	1.53	0.58	1.27	0.32	0.70	0.60	1.32
284	50	45	–	1.00	2.20	–	–	–	–	0.83	1.82	–	–	–	–	–	–	–	–	–	–	–	–
	57	51	–	1.13	2.50	–	–	–	–	0.94	2.07	–	–	–	–	–	–	–	–	–	–	–	–
285	49	32	34	0.95	2.09	–	–	0.66	1.45	0.78	1.72	–	–	–	–	0.54	1.19	0.42	0.92	0.20	0.44	0.46	1.02
	56	36	38	1.08	2.39	–	–	0.75	1.66	0.89	1.96	–	–	–	–	0.62	1.36	0.48	1.05	0.23	0.50	0.53	1.16
286	51	31	35	1.01	2.23	–	–	0.68	1.51	0.84	1.85	–	–	–	–	0.56	1.24	0.48	1.06	0.25	0.56	0.51	1.13
	56	34	39	1.12	2.46	–	–	0.76	1.66	0.93	2.04	–	–	–	–	0.62	1.37	0.53	1.18	0.28	0.61	0.57	1.25
287	42	42	–	0.84	1.85	–	–	–	–	0.66	1.45	–	–	–	–	–	–	–	–	–	–	–	–
	46	45	–	0.91	2.01	–	–	–	–	0.72	1.58	–	–	–	–	–	–	–	–	–	–	–	–
288	52	37	37	1.01	2.22	–	–	0.71	1.57	0.83	1.83	–	–	–	–	0.59	1.29	0.45	1.00	0.23	0.50	0.49	1.08
	58	41	41	1.12	2.46	–	–	0.79	1.75	0.92	2.03	–	–	–	–	0.65	1.43	0.50	1.11	0.25	0.55	0.55	1.20
289	55	38	39	1.04	2.29	–	–	0.76	1.68	0.86	1.90	–	–	–	–	0.62	1.38	0.46	1.00	0.23	0.50	0.50	1.09
	60	42	43	1.14	2.50	–	–	0.83	1.84	0.94	2.08	–	–	–	–	0.68	1.51	0.50	1.10	0.25	0.54	0.54	1.20
290	47	24	30	0.93	2.05	–	–	0.59	1.30	0.77	1.71	–	–	–	–	0.48	1.07	0.45	0.99	0.24	0.53	0.47	1.05
	57	30	36	1.14	2.51	–	–	0.72	1.59	0.95	2.08	–	–	–	–	0.59	1.30	0.55	1.21	0.29	0.64	0.58	1.28
291	57	43	36	1.07	2.36	–	–	0.70	1.54	0.90	1.99	–	–	–	–	0.57	1.26	0.49	1.07	0.26	0.57	0.51	1.13
	63	48	40	1.20	2.65	–	–	0.78	1.73	1.01	2.23	–	–	–	–	0.64	1.42	0.55	1.20	0.29	0.64	0.58	1.27
292	48	19	29	0.96	2.11	–	–	0.59	1.29	0.78	1.72	–	–	–	–	0.48	1.06	0.44	0.98	0.21	0.47	0.49	1.07
	52	21	32	1.05	2.31	–	–	0.64	1.42	0.86	1.88	–	–	–	–	0.53	1.16	0.49	1.07	0.23	0.52	0.53	1.18

(Continued)

TABLE V-1C DRY FORAGES, MINERAL AND VITAMIN COMPOSITION OF FEEDS, DATA EXPRESSED **AS-FED** AND **MOISTURE-FREE**—*(Continued)*

Entry Number	Feed Name Description	Moisture Basis: A-F (as-fed) or M-F (moisture-free)	Dry Matter	Macro Minerals: Calcium (Ca)	Phosphorus (P)	Sodium (Na)	Chlorine (Cl)	Magnesium (Mg)	Potassium (K)	Sulfur (S)	Micro Minerals: Cobalt (Co)	Copper (Cu)	Iodine (I)	Iron (Fe)	Manganese (Mn)	Selenium (Se)	Zinc (Zn)
			%	%	%	%	%	%	%	%	ppm or mg/kg	ppm or mg/kg	ppm or mg/kg	%	ppm or mg/kg	ppm or mg/kg	ppm or mg/kg
	RUSSIAN THISTLE (TUMBLING TUMBLEWEED) *Salsola kali, tenuifolia*																
264	HAY, SUN-CURED	A-F	86	1.31	0.22	–	–	0.65	5.92	–	–	–	–	–	–	–	–
		M-F	100	1.51	0.25	–	–	0.75	6.85	–	–	–	–	–	–	–	–
	RUSSIAN WILD-RYE *Elymus junceus*																
265	HAY, SUN-CURED	A-F	90	–	–	–	–	–	–	–	–	–	–	–	–	–	–
		M-F	100	–	–	–	–	–	–	–	–	–	–	–	–	–	–
	RYE *Secale cereale*																
266	HAY, SUN-CURED	A-F	93	0.31	0.18	0.03	–	0.12	1.26	–	–	–	–	–	–	–	31.7
		M-F	100	0.33	0.19	0.03	–	0.13	1.35	–	–	–	–	–	–	–	33.9
267	STRAW	A-F	91	0.22	0.08	0.12	0.22	0.07	0.88	0.10	–	3.6	–	–	6.0	–	–
		M-F	100	0.24	0.09	0.13	0.24	0.08	0.97	0.11	–	4.0	–	–	6.6	–	–
	RYEGRASS *Lolium spp*																
268	HAY, SUN-CURED	A-F	88	–	–	–	–	–	–	–	–	–	–	–	–	–	–
		M-F	100	–	–	–	–	–	–	–	–	–	–	–	–	–	–
	SAFFLOWER *Carthamus tinctorius*																
269	HAY, MATURE, SUN-CURED	A-F	91	–	–	–	–	–	–	–	–	–	–	–	–	–	–
		M-F	100	–	–	–	–	–	–	–	–	–	–	–	–	–	–
270	HULLS	A-F	91	–	–	–	–	–	–	–	–	–	–	–	–	–	–
		M-F	100	–	–	–	–	–	–	–	–	–	–	–	–	–	–
	SAINFOIN *Onobrychis spp*																
271	HAY, SUN-CURED	A-F	86	–	–	–	–	–	–	–	–	–	–	–	–	–	–
		M-F	100	–	–	–	–	–	–	–	–	–	–	–	–	–	–
	SALTBUSH, NUTTALL *Atriplex nuttallii*																
272	BROWSE, SUN-CURED	A-F	90	1.82	0.12	–	–	–	4.52	–	–	–	–	–	–	–	–
		M-F	100	2.02	0.13	–	–	–	5.02	–	–	–	–	–	–	–	–
273	BROWSE, MATURE, SUN-CURED	A-F	91	–	–	–	–	–	–	–	–	–	–	–	–	–	–
		M-F	100	–	–	–	–	–	–	–	–	–	–	–	–	–	–
	SALTGRASS *Distichlis spp*																
274	HAY, SUN-CURED	A-F	89	–	–	–	–	–	–	–	–	–	–	–	–	–	–
		M-F	100	–	–	–	–	–	–	–	–	–	–	–	–	–	–
	SAWDUST																
275	WOOD	A-F	90	–	–	–	–	–	–	–	–	–	–	–	–	–	–
		M-F	100	–	–	–	–	–	–	–	–	–	–	–	–	–	–
	SEAWEED (KELP) *Laminariales* (order), *Fucales* (order)																
276	WHOLE, DEHY	A-F	91	2.47	0.28	–	–	0.85	–	–	–	–	–	–	–	–	–
		M-F	100	2.72	0.31	–	–	0.93	–	–	–	–	–	–	–	–	–
	SEDGE *Carex spp*																
277	HAY, SUN-CURED	A-F	89	–	–	–	–	–	–	–	–	2.9	–	–	–	–	–
		M-F	100	–	–	–	–	–	–	–	–	3.3	–	–	–	–	–
	SLOUGHGRASS, AMERICAN *Beckmannia syzigachne*																
278	HAY, MATURE, SUN-CURED	A-F	93	0.38	0.13	–	–	–	–	–	–	–	–	–	–	–	–
		M-F	100	0.40	0.14	–	–	–	–	–	–	–	–	–	–	–	–
	SORGHUM *Sorghum bicolor*																
279	FODDER WITH HEADS, SUN-CURED	A-F	90	0.56	0.17	0.02	–	0.27	1.12	–	–	–	–	–	–	–	–
		M-F	100	0.62	0.19	0.02	–	0.30	1.24	–	–	–	–	–	–	–	–
280	STOVER WITHOUT HEADS, SUN-CURED	A-F	92	0.37	0.10	–	–	–	1.10	–	–	–	–	–	–	–	–
		M-F	100	0.40	0.11	–	–	–	1.20	–	–	–	–	–	–	–	–
281	ATLAS, FODDER WITH HEADS, SUN-CURED	A-F	91	0.73	0.27	–	–	–	–	–	–	–	–	–	–	–	–
		M-F	100	0.80	0.30	–	–	–	–	–	–	–	–	–	–	–	–
282	ATLAS, STOVER WITHOUT HEADS, SUN-CURED	A-F	90	0.35	0.10	–	–	–	–	–	–	–	–	–	–	–	–
		M-F	100	0.39	0.11	–	–	–	–	–	–	–	–	–	–	–	–
283	DURRA, FODDER WITH HEADS, SUN-CURED	A-F	90	–	–	–	–	–	–	–	–	–	–	–	–	–	–
		M-F	100	–	–	–	–	–	–	–	–	–	–	–	–	–	–
284	HEGARI, FODDER WITH HEADS, DOUGH STAGE, SUN-CURED	A-F	88	–	–	–	–	–	–	–	–	–	–	–	–	–	–
		M-F	100	–	–	–	–	–	–	–	–	–	–	–	–	–	–
285	HEGARI, STOVER WITHOUT HEADS, SUN-CURED	A-F	87	0.33	0.08	–	–	–	–	–	–	–	–	–	–	–	–
		M-F	100	0.38	0.09	–	–	–	–	–	–	–	–	–	–	–	–
286	JOHNSONGRASS, HAY, SUN-CURED	A-F	91	0.80	0.27	0.01	–	0.31	1.22	0.09	–	–	–	0.054	–	–	–
		M-F	100	0.89	0.30	0.01	–	0.35	1.35	0.10	–	–	–	0.059	–	–	–
287	KAFIR, CHAFF	A-F	92	–	–	–	–	–	–	–	–	–	–	–	–	–	–
		M-F	100	–	–	–	–	–	–	–	–	–	–	–	–	–	–
288	KAFIR, FODDER WITH HEADS, SUN-CURED	A-F	90	0.35	0.18	–	–	0.26	1.53	–	–	–	–	–	–	–	–
		M-F	100	0.39	0.20	–	–	0.29	1.70	–	–	–	–	–	–	–	–
289	KAFIR, FODDER WITH HEADS, LATE BLOOM, SUN-CURED	A-F	91	–	–	–	–	–	–	–	–	–	–	–	–	–	–
		M-F	100	–	–	–	–	–	–	–	–	–	–	–	–	–	–
290	KAFIR, STOVER WITHOUT HEADS, SUN-CURED	A-F	82	0.50	0.08	–	–	–	–	–	–	–	–	–	–	–	–
		M-F	100	0.61	0.10	–	–	–	–	–	–	–	–	–	–	–	–
291	MILO, FODDER WITH HEADS, SUN-CURED	A-F	89	0.35	0.18	–	–	–	–	–	–	–	–	–	–	–	–
		M-F	100	0.40	0.20	–	–	–	–	–	–	–	–	–	–	–	–
292	MILO, STOVER WITHOUT HEADS, SUN-CURED	A-F	91	0.51	0.15	0.10	–	0.23	0.41	–	–	–	–	–	–	–	–
		M-F	100	0.56	0.16	0.11	–	0.26	0.45	–	–	–	–	–	–	–	–

Entry Number	Fat-Soluble Vitamins: A (1 mg Carotene = 1667 IU Vit A)	Carotene (Provitamin A)	D	E	K	Water-Soluble Vitamins: B-12	Biotin	Choline	Folacin (Folic Acid)	Niacin	Pantothenic Acid (B-3)	(Pyridoxine) B-6	Riboflavin (B-2)	Thiamin (B-1)
	IU/g	ppm or mg/kg	IU/kg	ppm or mg/kg	ppm or mg/kg	ppm or mg/kg	ppm or mg/kg	ppm or mg/kg	ppm or mg/kg	ppm or mg/kg	ppm or mg/kg	ppm or mg/kg	ppm or mg/kg	ppm or mg/kg
264	–	–	–	–	–	–	–	–	–	–	–	–	–	–
	–	–	–	–	–	–	–	–	–	–	–	–	–	–
265	–	–	–	–	–	–	–	–	–	–	–	–	–	–
	–	–	–	–	–	–	–	–	–	–	–	–	–	–
266	**10.1**	**6.1**	–	–	–	–	–	–	–	–	–	–	–	–
	10.9	6.5	–	–	–	–	–	–	–	–	–	–	–	–
267	–	–	–	–	–	–	–	–	–	–	–	–	–	–
	–	–	–	–	–	–	–	–	–	–	–	–	–	–
268	**175.8**	**105.5**	–	–	–	–	–	–	–	–	–	–	–	–
	199.9	119.9	–	–	–	–	–	–	–	–	–	–	–	–
269	–	–	–	–	–	–	–	–	–	–	–	–	–	–
	–	–	–	–	–	–	–	–	–	–	–	–	–	–
270	–	–	–	–	–	–	–	–	–	–	–	–	–	–
	–	–	–	–	–	–	–	–	–	–	–	–	–	–
271	–	–	–	–	–	–	–	–	–	–	–	–	–	–
	–	–	–	–	–	–	–	–	–	–	–	–	–	–
272	–	–	–	–	–	–	–	–	–	–	–	–	–	–
	–	–	–	–	–	–	–	–	–	–	–	–	–	–
273	–	–	–	–	–	–	–	–	–	–	–	–	–	–
	–	–	–	–	–	–	–	–	–	–	–	–	–	–
274	–	–	–	–	–	–	–	–	–	–	–	–	–	–
	–	–	–	–	–	–	–	–	–	–	–	–	–	–
275	–	–	–	–	–	–	–	–	–	–	–	–	–	–
	–	–	–	–	–	–	–	–	–	–	–	–	–	–
276	–	–	–	–	–	–	–	–	–	–	–	–	–	–
	–	–	–	–	–	–	–	–	–	–	–	–	–	–
277	–	–	–	–	–	–	–	–	–	–	–	–	–	–
	–	–	–	–	–	–	–	–	–	–	–	–	–	–
278	**84.0**	**50.4**	–	–	–	–	–	–	–	–	–	–	–	–
	90.0	54.0	–	–	–	–	–	–	–	–	–	–	–	–
279	–	–	–	–	–	–	–	–	–	–	–	–	–	–
	–	–	–	–	–	–	–	–	–	–	–	–	–	–
280	–	–	–	–	–	–	–	–	–	–	–	–	–	–
	–	–	–	–	–	–	–	–	–	–	–	–	–	–
281	**78.6**	**47.1**	–	–	–	–	–	–	–	–	–	–	–	–
	86.4	51.8	–	–	–	–	–	–	–	–	–	–	–	–
282	**34.0**	**20.4**	–	–	–	–	–	–	–	–	–	–	–	–
	37.7	22.6	–	–	–	–	–	–	–	–	–	–	–	–
283	–	–	–	–	–	–	–	–	–	–	–	–	–	–
	–	–	–	–	–	–	–	–	–	–	–	–	–	–
284	–	–	–	–	–	–	–	–	–	–	–	–	–	–
	–	–	–	–	–	–	–	–	–	–	–	–	–	–
285	–	–	–	–	–	–	–	–	–	–	–	–	–	–
	–	–	–	–	–	–	–	–	–	–	–	–	–	–
286	**58.8**	**35.3**	–	–	–	–	–	–	–	–	–	–	–	–
	64.9	38.9	–	–	–	–	–	–	–	–	–	–	–	–
287	–	–	–	–	–	–	–	–	–	–	–	–	–	–
	–	–	–	–	–	–	–	–	–	–	–	–	–	–
288	–	–	–	–	–	–	–	–	–	–	–	–	–	–
	–	–	–	–	–	–	–	–	–	–	–	–	–	–
289	–	–	–	–	–	–	–	–	–	–	–	–	–	–
	–	–	–	–	–	–	–	–	–	–	–	–	–	–
290	–	–	–	–	–	–	–	–	–	–	–	–	–	–
	–	–	–	–	–	–	–	–	–	–	–	–	–	–
291	**2.9**	**1.8**	–	–	–	–	–	–	–	–	–	–	–	–
	3.3	2.0	–	–	–	–	–	–	–	–	–	–	–	–
292	–	–	–	–	–	–	–	–	–	–	–	–	–	–
	–	–	–	–	–	–	–	–	–	–	–	–	–	–

(Continued)

TABLE V-1C DRY FORAGES, COMPOSITION OF FEEDS, DATA EXPRESSED **AS-FED** AND **MOISTURE-FREE**—*(Continued)*

Entry Number	Feed Name Description	International Feed Number	Moisture Basis: A-F (as-fed) or M-F (moisture-free)	Proximate Analysis: Dry Matter	Ash	Crude Fiber	Neutral Det. Fib. (NDF)	Acid Det. Fib. (ADF)	Lignin	Ether Extract (Fat)	N-Free Extract	Crude Protein	Digestible Protein: Ruminant	Swine	Horse
				%	%	%	%	%	%	%	%	%	%	%	%
	SORGHUM (Continued)														
293	SORGO, FODDER WITH HEADS, SUN-CURED	1-04-460	A-F	87	6.5	26.0	–	–	–	2.0	46.4	6.0	2.8	1.9	3.1
			M-F	100	7.5	29.9	–	–	–	2.4	53.4	6.9	3.2	2.2	3.5
294	SUDANGRASS, HAY, SUN-CURED	1-04-480	A-F	91	10.7	26.2	60.2	20.4	35.5	1.6	41.7	10.9	4.7	6.1	6.8
			M-F	100	11.8	28.7	66.0	22.4	39.0	1.7	45.8	12.0	5.2	6.7	7.5
	SOYBEAN *Glycine max*														
295	HAY, SUN-CURED	1-04-558	A-F	89	7.2	30.6	–	35.7	–	2.3	35.0	14.1	9.5	8.9	9.3
			M-F	100	8.0	34.3	–	40.0	–	2.5	39.3	15.8	10.7	10.0	10.4
296	HULLS	1-04-560	A-F	91	4.6	36.2	59.4	42.4	1.8	2.0	37.0	10.8	6.5	3.4	6.7
			M-F	100	5.1	40.0	65.6	46.8	2.0	2.2	40.9	11.9	7.2	3.8	7.4
297	STRAW	1-04-567	A-F	88	5.6	38.9	–	–	–	1.3	37.4	4.6	1.3	0.6	1.9
			M-F	100	6.4	44.3	–	–	–	1.5	42.7	5.2	1.5	0.7	2.2
	SPANISHMOSS *Tillandsia usneoides*														
298	WHOLE, SUN-CURED	1-04-868	A-F	90	8.3	26.7	–	–	–	2.4	47.7	4.9	–0.3	0.8	2.2
			M-F	100	9.2	29.7	–	–	–	2.6	53.0	5.5	–0.3	0.9	2.4
	SPHAGNUM MOSS *Sphagnum* spp														
299	WHOLE, MOLASSES ADDED, SUN-CURED	1-03-135	A-F	82	7.5	7.4	–	–	–	0.5	57.6	9.3	3.9	5.0	5.7
			M-F	100	9.1	9.0	–	–	–	0.6	70.0	11.3	4.7	6.1	6.9
	STARGRASS *Cynodon plectostachyus*														
300	HAY, SUN-CURED	1-13-407	A-F	90	8.6	28.8	–	–	–	1.5	44.4	7.7	2.0	–	2.8
			M-F	100	9.6	32.0	–	–	–	1.7	49.3	8.6	2.2	–	3.1
	SUBTERRANEAN CLOVER *Trifolium subterraneum*														
301	HAY, SUN-CURED	1-20-278	A-F	90	10.0	9.1	–	–	–	3.3	40.1	27.5	21.7	–	20.5
			M-F	100	11.1	10.1	–	–	–	3.7	44.6	30.5	24.1	–	22.8
	SUDANGRASS SORGHUM *Sorghum bicolor, sudanense*														
302	HAY, SUN-CURED	1-04-480	A-F	91	10.7	26.2	60.2	20.4	35.5	1.6	41.7	10.9	4.7	6.1	6.8
			M-F	100	11.8	28.7	66.0	22.4	39.0	1.7	45.8	12.0	5.2	6.7	7.5
	SUGAR BEET *Beta vulgaris, altissima*														
303	TOPS WITH CROWNS, DEHY	1-00-640	A-F	85	16.4	11.9	–	–	–	1.0	44.9	10.7	6.6	4.9	6.7
			M-F	100	19.4	14.0	–	–	–	1.2	52.9	12.6	7.7	5.8	7.9
304	LEAVES, SUN-CURED	1-00-641	A-F	91	29.8	6.1	–	–	–	6.6	25.3	23.2	17.6	16.9	16.3
			M-F	100	32.8	6.7	–	–	–	7.2	27.8	25.5	19.3	18.5	17.9
305	HULLS	1-00-643	A-F	85	8.4	28.2	–	–	–	2.1	35.4	11.1	6.2	6.5	7.0
			M-F	100	9.9	33.1	–	–	–	2.5	41.5	13.0	7.3	7.6	8.2
	SUGARCANE (BAGASSE) *Saccharum officinarum*														
306	DEHY	1-04-686	A-F	91	2.9	42.3	78.8	54.5	12.8	0.7	43.8	1.4	–1.5	–2.3	–0.5
			M-F	100	3.1	46.5	86.5	59.8	14.0	0.8	48.0	1.6	–1.7	–2.5	–0.6
	SUMMER CYPRESS, BELVEDERE *Kochia scoparia*														
307	HAY, SUN-CURED	1-04-717	A-F	91	12.6	23.3	–	–	–	1.4	41.8	12.1	7.6	7.1	7.7
			M-F	100	13.8	25.6	–	–	–	1.6	45.8	13.2	8.3	7.8	8.4
	SUNFLOWER *Helianthus* spp														
308	HULLS	1-04-720	A-F	92	2.6	56.0	–	–	–	3.0	26.6	3.6	1.0	–0.4	1.1
			M-F	100	2.8	61.0	–	–	–	3.2	29.0	3.9	1.1	–0.4	1.2
	SWEET CLOVER, YELLOW *Melilotus officinalis*														
309	HAY, SUN-CURED	1-04-754	A-F	89	7.6	28.9	–	–	–	1.9	36.4	13.7	9.5	8.6	9.0
			M-F	100	8.6	32.6	–	–	–	2.2	41.1	15.4	10.7	9.7	10.1
310	LEAVES, SUN-CURED	1-04-757	A-F	91	10.8	9.7	–	–	–	3.2	41.5	25.3	19.5	18.8	17.9
			M-F	100	12.0	10.8	–	–	–	3.5	45.8	28.0	21.5	20.7	19.8
311	STEMS, SUN-CURED	1-04-758	A-F	92	6.7	39.1	–	–	–	1.1	35.3	9.7	5.1	5.0	5.9
			M-F	100	7.3	42.5	–	–	–	1.2	38.5	10.6	5.6	5.4	6.4
	SWEET POTATO *Ipomoea batatas*														
312	HAY, SUN-CURED	1-04-779	A-F	90	10.2	21.3	–	23.4	–	3.2	43.5	11.9	4.6	7.0	7.6
			M-F	100	11.3	23.6	–	26.0	–	3.6	48.3	13.2	5.1	7.7	8.4
	TALL FESCUE (ALTA) *Festuca arundinacea*														
313	HAY, SUN-CURED	1-05-684	A-F	89	5.9	32.6	61.7	35.6	–	2.0	41.4	7.2	3.6	2.9	3.9
			M-F	100	6.6	36.6	69.3	40.0	–	2.2	46.5	8.1	4.0	3.2	4.4
	TEOSINTE *Euchlaena* spp														
314	HAY, SUN-CURED	1-08-538	A-F	89	10.3	26.4	–	–	–	1.9	41.7	9.1	5.3	4.5	5.4
			M-F	100	11.5	29.5	–	–	–	2.1	46.6	10.2	5.9	5.1	6.1
	TICKCLOVER, CHEROKEE (BEGGARWEED) *Desmodium tortuosum*														
315	HAY, SUN-CURED	1-08-540	A-F	91	7.8	28.4	–	–	–	2.3	37.2	15.2	10.6	9.8	10.1
			M-F	100	8.6	31.2	–	–	–	2.5	40.9	16.7	11.6	10.8	11.1
	TIMOTHY *Phleum pratense*														
316	HAY, SUN-CURED, ALL ANALYSES[1]	1-04-893	A-F	91	4.6	30.3	63.7	36.4	–	2.4	47.3	6.8	2.8	2.4	3.3
			M-F	100	5.1	33.2	70.0	40.0	–	2.6	51.8	7.4	3.0	2.6	3.6
317	HAY, PREBLOOM, SUN-CURED	1-04-881	A-F	89	6.3	28.2	56.0	29.4	2.8	2.5	39.8	12.4	7.5	7.4	8.0
			M-F	100	7.1	31.6	62.7	32.9	3.1	2.8	44.6	13.9	8.4	8.3	8.9

Entry Number	TDN Ruminant	TDN Swine	TDN Horse	DE Ruminant		DE Swine		DE Horse		ME Ruminant		ME Swine		ME Poultry ME_n		ME Horse		NE Ruminant NE_m		NE Ruminant NE_g		NE Lactating Cows NE_{lc}	
	%	%	%	Mcal		kcal		Mcal		Mcal		kcal		kcal		Mcal		Mcal		Mcal		Mcal	
				lb	kg	lb	kg	lb	kg	lb	kg	lb	kg	lb	kg	lb	kg	lb	kg	lb	kg	lb	kg
293	51	35	36	0.99	2.18	–	–	0.71	1.56	0.82	1.81	–	–	–	–	0.58	1.28	0.45	0.99	0.23	0.50	0.48	1.06
	59	41	42	1.14	2.50	–	–	0.81	1.79	0.94	2.08	–	–	–	–	0.67	1.47	0.52	1.14	0.26	0.58	0.56	1.22
294	51	35	41	1.00	2.20	–	–	0.79	1.74	0.82	1.82	–	–	–	–	0.65	1.43	0.48	1.07	0.25	0.55	0.52	1.14
	56	38	45	1.10	2.42	–	–	0.87	1.91	0.90	1.99	–	–	–	–	0.71	1.57	0.53	1.17	0.28	0.61	0.57	1.25
295	49	40	41	1.11	2.45	–	–	0.79	1.74	0.86	1.90	–	–	–	–	0.65	1.43	0.52	1.14	0.29	0.63	0.55	1.21
	55	44	46	1.25	2.75	–	–	0.89	1.95	0.97	2.13	–	–	–	–	0.73	1.60	0.58	1.28	0.32	0.70	0.62	1.36
296	69	47	39	1.20	2.65	851	1877	0.75	1.65	1.03	2.26	305	671	301	665	0.61	1.35	0.75	1.66	0.50	1.09	0.72	1.59
	77	52	43	1.33	2.92	940	2073	0.83	1.82	1.13	2.50	336	742	333	734	0.68	1.50	0.83	1.84	0.55	1.21	0.80	1.75
297	36	–	29	0.71	1.57	–	–	0.58	1.27	0.54	1.19	–	–	–	–	0.47	1.04	0.21	0.46	0.00	0.00	0.32	0.71
	41	–	33	0.81	1.79	–	–	0.66	1.45	0.62	1.36	–	–	–	–	0.54	1.19	0.24	0.53	0.00	0.00	0.37	0.81
298	51	34	33	0.99	2.19	–	–	0.65	1.43	0.82	1.81	–	–	–	–	0.53	1.18	0.44	0.98	0.22	0.48	0.49	1.07
	57	38	37	1.10	2.43	–	–	0.72	1.59	0.91	2.01	–	–	–	–	0.59	1.31	0.49	1.09	0.24	0.53	0.54	1.19
299	42	–	–	0.84	1.85	–	–	–	–	0.68	1.50	–	–	–	–	–	–	–	–	–	–	–	–
	51	–	–	1.02	2.25	–	–	–	–	0.83	1.83	–	–	–	–	–	–	–	–	–	–	–	–
300	49	–	35	0.90	1.98	–	–	0.65	1.44	0.74	1.62	–	–	–	–	0.53	1.17	0.41	0.90	0.08	0.18	0.41	0.90
	54	–	39	1.00	2.20	–	–	0.73	1.60	0.82	1.80	–	–	–	–	0.59	1.30	0.45	1.00	0.09	0.20	0.45	1.00
301	–	–	–	–	–	–	–	–	–	–	–	–	–	–	–	–	–	–	–	–	–	–	–
	–	–	–	–	–	–	–	–	–	–	–	–	–	–	–	–	–	–	–	–	–	–	–
302	51	35	41	1.00	2.20	–	–	0.79	1.74	0.82	1.82	–	–	–	–	0.65	1.43	0.48	1.07	0.25	0.55	0.52	1.14
	56	38	45	1.10	2.42	–	–	0.87	1.91	0.90	1.99	–	–	–	–	0.71	1.57	0.53	1.17	0.28	0.61	0.57	1.25
303	51	45	–	1.02	2.25	–	–	–	–	0.86	1.89	–	–	–	–	–	–	–	–	–	–	–	–
	60	54	–	1.20	2.65	–	–	–	–	1.01	2.23	–	–	–	–	–	–	–	–	–	–	–	–
304	–	–	–	–	–	–	–	–	–	–	–	–	–	–	–	–	–	–	–	–	–	–	–
	–	–	–	–	–	–	–	–	–	–	–	–	–	–	–	–	–	–	–	–	–	–	–
305	29	34	36	0.75	1.66	–	–	0.69	1.52	0.58	1.29	–	–	–	–	0.57	1.25	0.43	0.95	0.21	0.47	0.47	1.03
	34	40	42	0.88	1.94	–	–	0.81	1.78	0.69	1.51	–	–	–	–	0.66	1.46	0.50	1.11	0.25	0.55	0.55	1.21
306	43	–	–	0.88	1.93	–	–	–	–	0.69	1.53	–	–	–	–	–	–	0.37	0.82	0.15	0.32	0.40	0.88
	48	–	–	0.96	2.12	–	–	–	–	0.76	1.68	–	–	–	–	–	–	0.41	0.90	0.16	0.36	0.44	0.96
307	51	36	43	0.99	2.17	–	–	0.82	1.81	0.81	1.79	–	–	–	–	0.67	1.48	0.44	0.96	0.21	0.45	0.48	1.06
	56	40	47	1.08	2.38	–	–	0.90	1.98	0.89	1.96	–	–	–	–	0.74	1.63	0.48	1.05	0.23	0.50	0.53	1.16
308	26	–	–	0.52	1.14	–	–	–	–	0.34	0.74	–	–	–	–	–	–	–	–	–	–	–	–
	28	–	–	0.57	1.25	–	–	–	–	0.37	0.81	–	–	–	–	–	–	–	–	–	–	–	–
309	49	39	43	0.98	2.16	–	–	0.81	1.79	0.81	1.78	–	–	–	–	0.67	1.47	0.49	1.07	0.26	0.57	0.51	1.13
	55	44	48	1.11	2.44	–	–	0.92	2.03	0.91	2.01	–	–	–	–	0.75	1.66	0.55	1.21	0.29	0.64	0.58	1.28
310	56	–	–	1.12	2.48	–	–	–	–	0.95	2.10	–	–	–	–	–	–	–	–	–	–	–	–
	62	–	–	1.24	2.74	–	–	–	–	1.05	2.31	–	–	–	–	–	–	–	–	–	–	–	–
311	45	–	37	0.91	2.00	–	–	0.72	1.59	0.73	1.60	–	–	–	–	0.59	1.30	0.40	0.88	0.17	0.38	0.46	1.01
	49	–	40	0.99	2.17	–	–	0.78	1.72	0.79	1.75	–	–	–	–	0.64	1.41	0.43	0.95	0.19	0.41	0.50	1.10
312	49	49	39	1.02	2.25	–	–	0.76	1.67	0.85	1.86	–	–	–	–	0.62	1.37	0.51	1.13	0.28	0.62	0.54	1.18
	55	54	44	1.13	2.50	–	–	0.84	1.85	0.94	2.07	–	–	–	–	0.69	1.52	0.57	1.26	0.31	0.69	0.60	1.31
313	48	31	41	0.95	2.10	–	–	0.78	1.71	0.78	1.72	–	–	–	–	0.64	1.40	0.44	0.96	0.21	0.47	0.48	1.06
	54	34	46	1.07	2.36	–	–	0.87	1.92	0.88	1.93	–	–	–	–	0.72	1.58	0.49	1.08	0.24	0.53	0.54	1.19
314	50	33	37	0.97	2.14	–	–	0.71	1.57	0.80	1.76	–	–	–	–	0.59	1.29	0.43	0.94	0.20	0.45	0.47	1.04
	56	37	41	1.08	2.39	–	–	0.80	1.76	0.89	1.96	–	–	–	–	0.65	1.44	0.48	1.05	0.23	0.50	0.53	1.17
315	48	44	45	1.00	2.20	–	–	0.85	1.88	0.82	1.81	–	–	–	–	0.70	1.54	0.50	1.11	0.27	0.60	0.53	1.17
	52	48	49	1.10	2.42	–	–	0.94	2.07	0.91	2.00	–	–	–	–	0.77	1.69	0.56	1.22	0.30	0.66	0.58	1.29
316	53	38	43	1.06	2.35	–	–	0.81	1.79	0.85	1.88	–	–	–	–	0.67	1.47	0.50	1.10	0.27	0.58	0.51	1.13
	58	42	47	1.17	2.57	–	–	0.89	1.96	0.93	2.06	–	–	–	–	0.73	1.61	0.55	1.20	0.29	0.64	0.56	1.24
317	57	43	56	1.15	2.53	–	–	1.04	2.29	0.97	2.15	–	–	–	–	0.85	1.88	0.57	1.25	0.33	0.73	0.58	1.27
	64	48	63	1.28	2.83	–	–	1.17	2.57	1.09	2.41	–	–	–	–	0.96	2.11	0.64	1.40	0.37	0.82	0.64	1.42

(Continued)

TABLE V-1C DRY FORAGES, MINERAL AND VITAMIN COMPOSITION OF FEEDS, DATA EXPRESSED **AS-FED** AND **MOISTURE-FREE**—*(Continued)*

Entry Number	Feed Name Description	Moisture Basis: A-F (as-fed) or M-F (moisture-free)	Dry Matter	Macro Minerals							Micro Minerals						
				Calcium (Ca)	Phosphorus (P)	Sodium (Na)	Chlorine (Cl)	Magnesium (Mg)	Potassium (K)	Sulfur (S)	Cobalt (Co)	Copper (Cu)	Iodine (I)	Iron (Fe)	Manganese (Mn)	Selenium (Se)	Zinc (Zn)
			%	%	%	%	%	%	%	%	ppm or mg/kg	ppm or mg/kg	ppm or mg/kg	%	ppm or mg/kg	ppm or mg/kg	ppm or mg/kg
	SORGHUM (Continued)																
293	SORGO, FODDER WITH HEADS, SUN-CURED	A-F	87	0.33	0.14	–	0.55	0.30	1.26	–	–	–	–	–	113.7	–	–
		M-F	100	0.38	0.16	–	0.63	0.35	1.45	–	–	–	–	–	130.7	–	–
294	SUDANGRASS, HAY, SUN-CURED	A-F	91	0.47	0.28	0.01	–	0.34	1.90	0.06	0.116	28.6	–	0.015	69.5	–	34.6
		M-F	100	0.51	0.31	0.02	–	0.37	2.08	0.06	0.127	31.4	–	0.017	76.3	–	38.0
	SOYBEAN *Glycine max*																
295	HAY, SUN-CURED	A-F	89	1.13	0.22	0.10	0.13	0.71	0.92	0.25	0.083	8.0	0.216	0.026	94.3	–	21.5
		M-F	100	1.26	0.24	0.11	0.15	0.81	1.04	0.28	0.093	9.0	0.242	0.029	105.8	–	24.1
296	HULLS	A-F	91	0.45	0.19	0.03	–	–	1.15	0.08	0.109	16.1	–	0.030	9.9	–	21.8
		M-F	100	0.49	0.21	0.03	–	–	1.27	0.09	0.121	17.8	–	0.033	11.0	–	24.1
297	STRAW	A-F	88	1.40	0.05	0.11	–	0.81	0.49	0.23	–	–	–	0.027	44.9	–	–
		M-F	100	1.59	0.06	0.12	–	0.92	0.56	0.26	–	–	–	0.030	51.1	–	–
	SPANISHMOSS *Tillandsia usneoides*																
298	WHOLE, SUN-CURED	A-F	90	–	0.04	–	–	–	0.46	–	–	–	–	–	–	–	–
		M-F	100	–	0.05	–	–	–	0.52	–	–	–	–	–	–	–	–
	SPHAGNUM MOSS *Sphagnum* spp																
299	WHOLE, MOLASSES ADDED, SUN-CURED	A-F	82	–	–	–	–	–	–	–	–	–	–	–	–	–	–
		M-F	100	–	–	–	–	–	–	–	–	–	–	–	–	–	–
	STARGRASS *Cynodon plectostachyus*																
300	HAY, SUN-CURED	A-F	90	–	–	–	–	–	–	–	–	–	–	–	–	–	–
		M-F	100	–	–	–	–	–	–	–	–	–	–	–	–	–	–
	SUBTERRANEAN CLOVER *Trifolium subterranean*																
301	HAY, SUN-CURED	A-F	90	–	–	–	–	–	–	–	–	–	–	–	–	–	–
		M-F	100	–	–	–	–	–	–	–	–	–	–	–	–	–	–
	SUDANGRASS SORGHUM *Sorghum bicolor, sudanense*																
302	HAY, SUN-CURED	A-F	91	0.47	0.28	0.01	–	0.34	1.90	0.06	0.116	28.6	–	0.015	69.5	–	34.6
		M-F	100	0.51	0.31	0.02	–	0.37	2.08	0.06	0.127	31.4	–	0.017	76.3	–	38.0
	SUGAR BEET *Beta vulgaris, altissima*																
303	TOPS WITH CROWNS, DEHY	A-F	85	0.37	0.20	–	–	0.33	3.57	–	–	5.2	–	0.051	35.9	–	16.8
		M-F	100	0.44	0.24	–	–	0.39	4.21	–	–	6.1	–	0.060	42.3	–	19.8
304	LEAVES, SUN-CURED	A-F	91	–	–	–	–	–	–	–	–	–	–	–	–	–	–
		M-F	100	–	–	–	–	–	–	–	–	–	–	–	–	–	–
305	HULLS	A-F	85	–	–	–	–	–	–	–	–	–	–	–	–	–	–
		M-F	100	–	–	–	–	–	–	–	–	–	–	–	–	–	–
	SUGARCANE (BAGASSE) *Saccharum officinarum*																
306	DEHY	A-F	91	0.47	0.26	0.04	–	0.08	0.34	0.09	–	–	–	0.019	–	–	–
		M-F	100	0.51	0.29	0.04	–	0.08	0.37	0.10	–	–	–	0.021	–	–	–
	SUMMER CYPRESS, BELVEDERE *Kochia scoparia*																
307	HAY, SUN-CURED	A-F	91	–	–	–	–	–	–	–	–	–	–	–	–	–	–
		M-F	100	–	–	–	–	–	–	–	–	–	–	–	–	–	–
	SUNFLOWER *Helianthus* spp																
308	HULLS	A-F	92	0.15	0.03	0.02	–	0.12	0.61	0.10	–	9.5	–	0.004	15.0	–	17.1
		M-F	100	0.16	0.03	0.03	–	0.13	0.67	0.10	–	10.3	–	0.004	16.3	–	18.7
	SWEET CLOVER, YELLOW *Melilotus officinalis*																
309	HAY, SUN-CURED	A-F	89	1.44	0.24	0.08	0.33	0.39	1.35	0.42	–	8.8	–	0.015	95.4	–	–
		M-F	100	1.63	0.27	0.09	0.37	0.44	1.53	0.47	–	10.0	–	0.017	107.7	–	–
310	LEAVES, SUN-CURED	A-F	91	2.83	0.29	–	–	–	–	–	–	–	–	–	–	–	–
		M-F	100	3.12	0.32	–	–	–	–	–	–	–	–	–	–	–	–
311	STEMS, SUN-CURED	A-F	92	0.57	0.30	–	–	–	–	–	–	–	–	–	–	–	–
		M-F	100	0.62	0.33	–	–	–	–	–	–	–	–	–	–	–	–
	SWEET POTATO *Ipomoea batatas*																
312	HAY, SUN-CURED	A-F	90	0.08	0.32	–	–	–	–	–	–	–	–	–	–	–	–
		M-F	100	0.09	0.36	–	–	–	–	–	–	–	–	–	–	–	–
	TALL FESCUE (ALTA) *Festuca arundinacea*																
313	HAY, SUN-CURED	A-F	89	0.35	0.21	0.05	–	0.20	2.12	–	–	–	–	–	–	–	–
		M-F	100	0.39	0.24	0.06	–	0.23	2.38	–	–	–	–	–	–	–	–
	TEOSINTE *Euchlaena* spp																
314	HAY, SUN-CURED	A-F	89	–	0.17	–	–	–	0.88	–	–	–	–	–	–	–	–
		M-F	100	–	0.19	–	–	–	0.98	–	–	–	–	–	–	–	–
	TICKCLOVER, CHEROKEE (BEGGARWEED) *Desmodium tortuosum*																
315	HAY, SUN-CURED	A-F	91	1.05	0.27	–	–	–	2.32	–	–	–	–	–	–	–	–
		M-F	100	1.16	0.30	–	–	–	2.55	–	–	–	–	–	–	–	–
	TIMOTHY *Phleum pratense*																
316	HAY, SUN-CURED, ALL ANALYSES	A-F	91	0.38	0.17	0.03	0.49	0.11	1.43	0.11	0.071	4.3	0.034	0.010	45.2	–	15.5
		M-F	100	0.41	0.19	0.03	0.53	0.12	1.57	0.12	0.077	4.7	0.037	0.011	49.5	–	17.0
317	HAY, PREBLOOM, SUN-CURED	A-F	89	0.41	0.36	0.06	–	0.10	2.72	0.12	–	23.0	–	0.021	79.5	–	59.8
		M-F	100	0.45	0.40	0.07	–	0.11	3.05	0.13	–	25.8	–	0.024	89.0	–	67.0

Entry Number	Fat-Soluble Vitamins: A (1 mg Carotene = 1667 IU Vit A)	Carotene (Provitamin A)	D	E	K	Water-Soluble Vitamins: B-12	Biotin	Choline	Folacin (Folic Acid)	Niacin	Pantothenic Acid (B-3)	(Pyridoxine) B-6	Riboflavin (B-2)	Thiamin (B-1)
	IU/g	ppm or mg/kg	IU/kg	ppm or mg/kg	ppm or mg/kg	ppm or mg/kg	ppm or mg/kg	ppm or mg/kg	ppm or mg/kg	ppm or mg/kg	ppm or mg/kg	ppm or mg/kg	ppm or mg/kg	ppm or mg/kg
293	3.9	2.3	–	–	–	–	–	–	–	–	–	–	–	–
	4.5	2.7	–	–	–	–	–	–	–	–	–	–	–	–
294	–	–	–	–	–	–	–	–	–	–	–	–	–	–
	–	–	–	–	–	–	–	–	–	–	–	–	–	–
295	53.1	31.8	1	26.3	–	–	–	–	–	–	–	–	–	–
	59.5	35.7	1	29.5	–	–	–	–	–	–	–	–	–	–
296	–	–	–	6.6	–	–	–	588	–	25	13.4	1.70	3.6	1.6
	–	–	–	7.3	–	–	–	649	–	27	14.8	1.88	4.0	1.8
297	–	–	–	–	–	–	–	–	–	–	–	–	–	–
	–	–	–	–	–	–	–	–	–	–	–	–	–	–
298	–	–	–	–	–	–	–	–	–	–	–	–	–	–
	–	–	–	–	–	–	–	–	–	–	–	–	–	–
299	–	–	–	–	–	–	–	–	–	–	–	–	–	–
	–	–	–	–	–	–	–	–	–	–	–	–	–	–
300	–	–	–	–	–	–	–	–	–	–	–	–	–	–
	–	–	–	–	–	–	–	–	–	–	–	–	–	–
301	–	–	–	–	–	–	–	–	–	–	–	–	–	–
	–	–	–	–	–	–	–	–	–	–	–	–	–	–
302	–	–	–	–	–	–	–	–	–	–	–	–	–	–
	–	–	–	–	–	–	–	–	–	–	–	–	–	–
303	–	–	–	–	–	–	–	–	–	–	–	–	–	–
	–	–	–	–	–	–	–	–	–	–	–	–	–	–
304	–	–	–	346.9	–	–	–	–	–	–	–	–	–	–
	–	–	–	381.2	–	–	–	–	–	–	–	–	–	–
305	–	–	–	–	–	–	–	–	–	–	–	–	–	–
	–	–	–	–	–	–	–	–	–	–	–	–	–	–
306	–	–	–	–	–	–	–	–	–	–	–	–	–	–
	–	–	–	–	–	–	–	–	–	–	–	–	–	–
307	43.9	26.3	–	–	–	–	–	–	–	–	–	–	–	–
	48.1	28.9	–	–	–	–	–	–	–	–	–	–	–	–
308	–	–	–	–	–	–	–	–	–	–	–	–	–	–
	–	–	–	–	–	–	–	–	–	–	–	–	–	–
309	145.8	87.4	2	–	–	–	–	–	–	–	–	–	–	–
	164.6	98.8	2	–	–	–	–	–	–	–	–	–	–	–
310	–	–	–	–	–	–	–	–	–	–	–	–	–	–
	–	–	–	–	–	–	–	–	–	–	–	–	–	–
311	–	–	–	–	–	–	–	–	–	–	–	–	–	–
	–	–	–	–	–	–	–	–	–	–	–	–	–	–
312	–	–	–	–	–	–	–	–	–	–	–	–	–	–
	–	–	–	–	–	–	–	–	–	–	–	–	–	–
313	30.8	18.5	–	–	–	–	–	–	–	–	–	–	–	–
	34.6	20.7	–	–	–	–	–	–	–	–	–	–	–	–
314	–	–	–	–	–	–	–	–	–	–	–	–	–	–
	–	–	–	–	–	–	–	–	–	–	–	–	–	–
315	–	–	–	–	–	–	–	–	–	–	–	–	–	–
	–	–	–	–	–	–	–	–	–	–	–	–	–	–
316	39.8	23.8	2	57.6	–	–	0.06	741	2.09	31	7.2	–	9.2	1.5
	43.5	26.1	2	63.1	–	–	0.07	811	2.29	34	7.9	–	10.1	1.7
317	186.1	111.6	–	–	–	–	–	–	–	–	–	–	–	–
	208.4	125.0	–	–	–	–	–	–	–	–	–	–	–	–

(Continued)

TABLE V-1C DRY FORAGES, COMPOSITION OF FEEDS, DATA EXPRESSED **AS-FED** AND **MOISTURE-FREE**—*(Continued)*

Entry Number	Feed Name Description	International Feed Number	Moisture Basis: A-F (as-fed) or M-F (moisture-free)	Proximate Analysis									Digestible Protein		
				Dry Matter	Ash	Crude Fiber	Neutral Det. Fib. (NDF)	Acid Det. Fib. (ADF)	Lignin	Ether Extract (Fat)	N-Free Extract	Crude Protein	Ruminant	Swine	Horse
				%	%	%	%	%	%	%	%	%	%	%	%
	TIMOTHY (Continued)														
318	HAY, EARLY BLOOM, SUN-CURED	**1-04-882**	**A-F**	**89**	**5.1**	**30.3**	**54.4**	**30.4**	**3.8**	**2.6**	**41.7**	**9.5**	**5.4**	**4.9**	**5.8**
			M-F	100	5.7	33.9	61.0	34.1	4.3	2.9	46.8	10.7	6.0	5.5	6.5
319	HAY, MIDBLOOM, SUN-CURED	**1-04-883**	**A-F**	**89**	**5.6**	**30.3**	**58.0**	**33.6**	**4.4**	**2.3**	**42.1**	**8.6**	**5.0**	**4.1**	**4.7**
			M-F	100	6.3	34.0	65.3	37.8	4.9	2.6	47.4	9.7	5.6	4.7	5.3
320	HAY, FULL BLOOM, SUN-CURED[1]	**1-04-884**	**A-F**	**89**	**4.6**	**31.1**	**60.5**	**33.8**	**–**	**2.7**	**43.4**	**6.8**	**3.5**	**2.6**	**1.4**
			M-F	100	5.2	35.1	68.0	38.0	–	3.0	48.9	7.7	4.0	2.9	1.6
	TIMOTHY-CLOVER *Phleum pratense-Trifolium* spp														
321	HAY, SUN-CURED	**1-04-973**	**A-F**	**89**	**5.1**	**31.7**	**–**	**–**	**–**	**2.0**	**42.7**	**7.8**	**4.1**	**3.4**	**4.4**
			M-F	100	5.7	35.5	–	–	–	2.2	47.8	8.7	4.6	3.8	4.9
	TREFOIL, BIG *Lotus uliginosus*														
322	HAY, SUN-CURED	**1-08-545**	**A-F**	**89**	**7.3**	**21.9**	**–**	**–**	**–**	**2.7**	**45.4**	**12.0**	**7.8**	**7.1**	**7.6**
			M-F	100	8.2	24.5	–	–	–	3.0	50.9	13.4	8.7	7.9	8.6
	TREFOIL, BIRDSFOOT *Lotus corniculatus*														
323	HAY, SUN-CURED[1]	**1-05-044**	**A-F**	**91**	**6.7**	**29.3**	**42.8**	**32.8**	**–**	**1.9**	**38.9**	**13.9**	**9.6**	**8.7**	**9.1**
			M-F	100	7.4	32.3	47.0	36.0	–	2.1	42.9	15.3	10.6	9.6	10.0
	TUMBLEWEED, TUMBLING (RUSSIAN THISTLE) *Salsola kali, tenuifolia*														
324	HAY, SUN-CURED	**1-03-988**	**A-F**	**86**	**13.1**	**25.1**	**–**	**–**	**–**	**1.7**	**36.3**	**10.2**	**6.6**	**5.7**	**6.4**
			M-F	100	15.2	29.0	–	–	–	2.0	42.0	11.8	7.6	6.5	7.3
	VELVETBEAN *Mucuna* spp														
325	HAY, SUN-CURED	**1-05-080**	**A-F**	**93**	**7.4**	**27.5**	**–**	**–**	**–**	**3.1**	**38.4**	**16.4**	**9.5**	**10.8**	**11.0**
			M-F	100	8.0	29.6	–	–	–	3.3	41.4	17.7	10.3	11.7	11.9
	VETCH *Vicia* spp														
326	HAY, SUN-CURED[1]	**1-05-106**	**A-F**	**89**	**7.8**	**24.8**	**42.7**	**29.4**	**–**	**2.7**	**35.1**	**18.4**	**13.4**	**12.7**	**12.6**
			M-F	100	8.8	27.9	48.0	33.0	–	3.0	39.6	20.7	15.1	14.3	14.2
	VETCH-OATS *Vicia* spp-*Avena sativa*														
327	HAY, SUN-CURED	**1-05-132**	**A-F**	**88**	**8.2**	**27.3**	**–**	**–**	**–**	**2.7**	**37.4**	**11.9**	**8.5**	**7.1**	**7.6**
			M-F	100	9.4	31.2	–	–	–	3.1	42.8	13.6	9.7	8.1	8.7
	VETCH-WHEAT *Vicia* spp-*Triticum aestivum*														
328	HAY, SUN-CURED	**1-05-136**	**A-F**	**90**	**7.2**	**28.8**	**–**	**–**	**–**	**2.2**	**36.4**	**15.4**	**11.4**	**10.1**	**10.3**
			M-F	100	8.0	32.0	–	–	–	2.4	40.5	17.1	12.7	11.2	11.4
	WATERGRASS *Hydrochloa caroliniensis*														
329	HAY, SUN-CURED	**1-05-154**	**A-F**	**94**	**18.0**	**31.2**	**–**	**–**	**–**	**1.1**	**36.3**	**7.6**	**3.8**	**3.1**	**4.2**
			M-F	100	19.1	33.1	–	–	–	1.2	38.5	8.1	4.1	3.3	4.5
	WHEAT *Triticum aestivum*														
330	HAY, SUN-CURED[1]	**1-05-172**	**A-F**	**89**	**7.0**	**25.7**	**60.5**	**36.5**	**–**	**2.0**	**46.4**	**7.7**	**3.8**	**3.3**	**4.4**
			M-F	100	7.9	29.0	68.0	41.0	–	2.2	52.3	8.7	4.3	3.8	4.9
331	CHAFF	**1-05-192**	**A-F**	**93**	**15.6**	**30.4**	**–**	**–**	**–**	**1.9**	**39.5**	**5.4**	**2.5**	**–1.1**	**2.5**
			M-F	100	16.8	32.7	–	–	–	2.0	42.6	5.9	2.7	–1.2	2.7
332	STRAW	**1-05-175**	**A-F**	**90**	**6.9**	**37.4**	**70.3**	**47.7**	**8.4**	**1.8**	**40.4**	**3.2**	**0.3**	**–0.6**	**0.6**
			M-F	100	7.7	41.7	78.4	53.2	9.4	2.0	45.0	3.6	0.3	–0.7	0.7
	WHEATGRASS, CRESTED *Agropyron desertorum*														
333	HAY, SUN-CURED	**1-05-418**	**A-F**	**92**	**6.4**	**30.8**	**–**	**33.2**	**5.1**	**2.1**	**42.2**	**10.3**	**6.4**	**5.5**	**6.3**
			M-F	100	7.0	33.6	–	36.2	5.5	2.3	46.0	11.2	6.9	6.0	6.8
	WHEATGRASS, INTERMEDIATE *Agropyron intermedium*														
334	HAY, SUN-CURED	**1-05-431**	**A-F**	**90**	**9.5**	**31.1**	**–**	**36.1**	**5.5**	**1.8**	**40.0**	**7.7**	**4.0**	**3.3**	**4.3**
			M-F	100	10.5	34.5	–	40.1	6.1	2.0	44.4	8.6	4.5	3.7	4.8
	WHEATGRASS, SLENDER *Agropyron trachycaulum*														
335	HAY, SUN-CURED	**1-05-436**	**A-F**	**91**	**7.3**	**32.8**	**–**	**–**	**–**	**2.2**	**41.0**	**8.1**	**4.8**	**3.6**	**4.6**
			M-F	100	7.9	35.9	–	–	–	2.4	44.9	8.9	5.3	4.0	5.1
	WHEATGRASS, TALL *Agropyron elongatum*														
336	HAY, SUN-CURED	**1-08-170**	**A-F**	**89**	**8.4**	**32.0**	**–**	**–**	**–**	**2.0**	**40.3**	**6.4**	**4.0**	**2.2**	**3.4**
			M-F	100	9.4	36.0	–	–	–	2.2	45.3	7.2	4.5	2.5	3.8
	WHITE CLOVER *Trifolium repens*														
337	HAY, SUN-CURED	**1-01-464**	**A-F**	**90**	**9.0**	**21.9**	**–**	**–**	**–**	**2.4**	**40.1**	**16.9**	**12.4**	**11.4**	**11.5**
			M-F	100	9.9	24.2	–	–	–	2.6	44.4	18.8	13.7	12.6	12.7
	WILD-RYE, RUSSIAN *Elymus junceus*														
338	HAY, SUN-CURED	**1-08-171**	**A-F**	**90**	**7.2**	**33.4**	**–**	**–**	**–**	**1.7**	**40.5**	**7.2**	**4.0**	**2.9**	**3.9**
			M-F	100	8.0	37.1	–	–	–	1.9	45.0	8.0	4.5	3.2	4.4
	WINTERFAT, COMMON *Eurotia lanata*														
339	BROWSE, SUN-CURED	**1-08-559**	**A-F**	**93**	**9.6**	**27.4**	**–**	**–**	**–**	**1.9**	**40.8**	**12.9**	**8.5**	**7.8**	**8.3**
			M-F	100	10.4	29.6	–	–	–	2.1	44.1	13.9	9.2	8.4	9.0
	WOOD														
340	SAWDUST	**1-07-714**	**A-F**	**90**	**0.7**	**71.5**	**–**	**–**	**–**	**0.8**	**16.7**	**0.3**	**–2.5**	**–3.2**	**–1.4**
			M-F	100	0.8	79.4	–	–	–	0.9	18.6	0.3	–2.8	–3.6	–1.6

[1]Neutral Detergent Fiber (NDF) and Acid Detergent Fiber (ADF) values taken from *Nutrient Requirements of Dairy Cattle*, 6th rev. ed., NRC, National Academy Press, 1988, pp. 90–110, Table 7–1.

Entry Number	TDN Ruminant	TDN Swine	TDN Horse	Digestible Energy Ruminant		Digestible Energy Swine		Digestible Energy Horse		Metabolizable Energy Ruminant		Metabolizable Energy Swine		Metabolizable Energy Poultry ME_n		Metabolizable Energy Horse		Net Energy Ruminant NE_m		Net Energy Ruminant NE_g		Net Energy Lactating Cows NE_{lc}	
	%	%	%	Mcal		kcal		Mcal		Mcal		kcal		kcal		Mcal		Mcal		Mcal		Mcal	
				lb	kg	lb	kg	lb	kg	lb	kg	lb	kg	lb	kg	lb	kg	lb	kg	lb	kg	lb	kg
318	**52**	**40**	**38**	**1.03**	***2.27***	**–**	**–**	**0.74**	***1.63***	**0.86**	***1.89***	**–**	**–**	**–**	**–**	**0.61**	***1.34***	**0.52**	***1.14***	**0.29**	***0.63***	**0.54**	***1.18***
	58	44	43	1.16	*2.55*	–	–	0.83	*1.83*	0.96	*2.12*	–	–	–	–	0.68	*1.50*	0.58	*1.28*	0.32	*0.71*	0.60	*1.33*
319	**53**	**37**	**51**	**1.06**	***2.33***	**–**	**–**	**0.94**	***2.08***	**0.88**	***1.95***	**–**	**–**	**–**	**–**	**0.77**	***1.70***	**0.51**	***1.13***	**0.28**	***0.63***	**0.53**	***1.18***
	59	41	57	1.19	*2.62*	–	–	1.06	*2.34*	1.00	*2.19*	–	–	–	–	0.87	*1.92*	0.58	*1.28*	0.32	*0.71*	0.60	*1.33*
320	**51**	**36**	**39**	**1.03**	***2.26***	**–**	**–**	**0.75**	***1.65***	**0.86**	***1.89***	**–**	**–**	**–**	**–**	**0.62**	***1.36***	**0.48**	***1.07***	**0.26**	***0.57***	**0.51**	***1.13***
	58	41	44	1.16	*2.55*	–	–	0.85	*1.87*	0.97	*2.13*	–	–	–	–	0.69	*1.53*	0.55	*1.20*	0.29	*0.64*	0.58	*1.27*
321	**50**	**34**	**38**	**0.99**	***2.18***	**–**	**–**	**0.73**	***1.60***	**0.82**	***1.80***	**–**	**–**	**–**	**–**	**0.60**	***1.31***	**0.46**	***1.01***	**0.23**	***0.51***	**0.49**	***1.09***
	56	38	42	1.11	*2.44*	–	–	0.81	*1.79*	0.91	*2.02*	–	–	–	–	0.67	*1.47*	0.51	*1.13*	0.26	*0.57*	0.55	*1.22*
322	**54**	**50**	**45**	**1.07**	***2.36***	**–**	**–**	**0.85**	***1.87***	**0.90**	***1.98***	**–**	**–**	**–**	**–**	**0.70**	***1.53***	**0.53**	***1.17***	**0.30**	***0.65***	**0.55**	***1.20***
	60	56	50	1.20	*2.65*	–	–	0.95	*2.09*	1.01	*2.22*	–	–	–	–	0.78	*1.72*	0.59	*1.31*	0.33	*0.73*	0.61	*1.35*
323	**54**	**41**	**46**	**0.91**	***2.01***	**–**	**–**	**0.87**	***1.91***	**0.83**	***1.84***	**–**	**–**	**–**	**–**	**0.71**	***1.56***	**0.55**	***1.22***	**0.32**	***0.69***	**0.57**	***1.25***
	59	46	51	1.01	*2.22*	–	–	0.96	*2.11*	0.92	*2.03*	–	–	–	–	0.78	*1.73*	0.61	*1.34*	0.35	*0.77*	0.62	*1.37*
324	**39**	**29**	**33**	**0.79**	***1.73***	**–**	**–**	**0.65**	***1.44***	**0.62**	***1.36***	**–**	**–**	**–**	**–**	**0.54**	***1.18***	**0.33**	***0.74***	**0.12**	***0.27***	**0.40**	***0.89***
	45	34	39	0.91	*2.00*	–	–	0.76	*1.67*	0.71	*1.57*	–	–	–	–	0.62	*1.37*	0.39	*0.85*	0.14	*0.31*	0.47	*1.02*
325	**56**	**51**	**46**	**1.12**	***2.47***	**–**	**–**	**0.87**	***1.91***	**0.94**	***2.08***	**–**	**–**	**–**	**–**	**0.71**	***1.57***	**0.55**	***1.22***	**0.31**	***0.69***	**0.57**	***1.26***
	61	55	49	1.21	*2.66*	–	–	0.94	*2.06*	1.02	*2.24*	–	–	–	–	0.77	*1.69*	0.60	*1.32*	0.34	*0.74*	0.61	*1.35*
326	**55**	**52**	**48**	**1.09**	***2.41***	**–**	**–**	**0.91**	***2.00***	**0.92**	***2.04***	**–**	**–**	**–**	**–**	**0.74**	***1.64***	**0.55**	***1.22***	**0.32**	***0.71***	**0.56**	***1.24***
	62	58	54	1.23	*2.72*	–	–	1.02	*2.25*	1.04	*2.30*	–	–	–	–	0.84	*1.85*	0.62	*1.38*	0.36	*0.80*	0.64	*1.40*
327	**51**	**40**	**37**	**1.00**	***2.21***	**–**	**–**	**0.72**	***1.58***	**0.83**	***1.84***	**–**	**–**	**–**	**–**	**0.59**	***1.29***	**0.47**	***1.04***	**0.25**	***0.55***	**0.50**	***1.11***
	58	46	42	1.14	*2.52*	–	–	0.82	*1.80*	0.95	*2.10*	–	–	–	–	0.67	*1.48*	0.54	*1.19*	0.29	*0.63*	0.57	*1.26*
328	**58**	**43**	**45**	**1.10**	***2.42***	**–**	**–**	**0.86**	***1.89***	**0.93**	***2.04***	**–**	**–**	**–**	**–**	**0.71**	***1.55***	**0.50**	***1.11***	**0.27**	***0.60***	**0.53**	***1.16***
	64	48	51	1.22	*2.69*	–	–	0.96	*2.11*	1.03	*2.27*	–	–	–	–	0.78	*1.73*	0.56	*1.23*	0.30	*0.66*	0.59	*1.29*
329	**39**	**14**	**27**	**0.79**	***1.75***	**–**	**–**	**0.55**	***1.22***	**0.61**	***1.34***	**–**	**–**	**–**	**–**	**0.45**	***1.00***	**0.30**	***0.66***	**0.07**	***0.16***	**0.39**	***0.87***
	42	15	28	0.84	*1.86*	–	–	0.59	*1.30*	0.65	*1.42*	–	–	–	–	0.48	*1.06*	0.32	*0.70*	0.08	*0.17*	0.42	*0.92*
330	**49**	**38**	**40**	**0.94**	***2.07***	**–**	**–**	**0.76**	***1.68***	**0.79**	***1.74***	**–**	**–**	**–**	**–**	**0.63**	***1.38***	**0.45**	***0.99***	**0.22**	***0.49***	**0.54**	***1.19***
	56	43	45	1.06	*2.33*	–	–	0.86	*1.90*	0.89	*1.96*	–	–	–	–	0.71	*1.56*	0.51	*1.11*	0.25	*0.56*	0.61	*1.34*
331	**35**	**19**	**24**	**0.76**	***1.68***	**–**	**–**	**0.51**	***1.12***	**0.58**	***1.28***	**–**	**–**	**–**	**–**	**0.42**	***0.92***	**0.33**	***0.73***	**0.11**	***0.24***	**0.41**	***0.91***
	37	20	26	0.82	*1.81*	–	–	0.55	*1.20*	0.63	*1.38*	–	–	–	–	0.45	*0.99*	0.36	*0.79*	0.12	*0.25*	0.45	*0.98*
332	**40**	**–**	**36**	**0.86**	***1.90***	**–**	**–**	**0.71**	***1.56***	**0.70**	***1.54***	**–**	**–**	**–**	**–**	**0.58**	***1.28***	**0.36**	***0.79***	**0.14**	***0.30***	**0.39**	***0.86***
	44	–	40	0.96	*2.12*	–	–	0.79	*1.73*	0.78	*1.72*	–	–	–	–	0.65	*1.42*	0.40	*0.88*	0.15	*0.34*	0.44	*0.96*
333	**49**	**38**	**41**	**0.99**	***2.19***	**–**	**–**	**0.78**	***1.73***	**0.82**	***1.80***	**–**	**–**	**–**	**–**	**0.64**	***1.42***	**0.48**	***1.06***	**0.25**	***0.55***	**0.52**	***1.14***
	53	41	44	1.08	*2.39*	–	–	0.85	*1.88*	0.89	*1.96*	–	–	–	–	0.70	*1.54*	0.52	*1.15*	0.27	*0.59*	0.56	*1.24*
334	**48**	**27**	**33**	**1.04**	***2.28***	**–**	**–**	**0.65**	***1.44***	**0.80**	***1.76***	**–**	**–**	**–**	**–**	**0.54**	***1.18***	**0.40**	***0.89***	**0.18**	***0.40***	**0.46**	***1.01***
	53	30	37	1.15	*2.54*	–	–	0.73	*1.60*	0.89	*1.96*	–	–	–	–	0.59	*1.31*	0.45	*0.99*	0.20	*0.44*	0.51	*1.12*
335	**52**	**32**	**35**	**1.01**	***2.22***	**–**	**–**	**0.68**	***1.50***	**0.83**	***1.83***	**–**	**–**	**–**	**–**	**0.56**	***1.23***	**0.44**	***0.98***	**0.21**	***0.47***	**0.49**	***1.08***
	57	35	38	1.10	*2.43*	–	–	0.75	*1.64*	0.91	*2.00*	–	–	–	–	0.61	*1.35*	0.49	*1.07*	0.23	*0.52*	0.53	*1.18*
336	**47**	**26**	**31**	**0.97**	***2.14***	**–**	**–**	**0.61**	***1.35***	**0.77**	***1.69***	**–**	**–**	**–**	**–**	**0.50**	***1.11***	**0.40**	***0.89***	**0.18**	***0.40***	**0.45**	***1.00***
	53	30	35	1.09	*2.40*	–	–	0.69	*1.52*	0.86	*1.89*	–	–	–	–	0.57	*1.25*	0.45	*1.00*	0.20	*0.45*	0.51	*1.12*
337	**57**	**52**	**50**	**1.11**	***2.45***	**–**	**–**	**0.94**	***2.07***	**0.94**	***2.07***	**–**	**–**	**–**	**–**	**0.77**	***1.70***	**0.53**	***1.18***	**0.30**	***0.66***	**0.55**	***1.22***
	63	58	56	1.23	*2.71*	–	–	1.04	*2.30*	1.04	*2.29*	–	–	–	–	0.86	*1.88*	0.59	*1.31*	0.33	*0.73*	0.61	*1.35*
338	**53**	**27**	**34**	**1.10**	***2.41***	**–**	**–**	**0.67**	***1.48***	**0.84**	***1.84***	**–**	**–**	**–**	**–**	**0.55**	***1.21***	**0.41**	***0.91***	**0.19**	***0.42***	**0.46**	***1.02***
	59	30	38	1.22	*2.68*	–	–	0.75	*1.64*	0.93	*2.05*	–	–	–	–	0.61	*1.35*	0.46	*1.01*	0.21	*0.46*	0.52	*1.14*
339	**46**	**40**	**44**	**0.96**	***2.12***	**–**	**–**	**0.83**	***1.84***	**0.78**	***1.73***	**–**	**–**	**–**	**–**	**0.68**	***1.51***	**0.48**	***1.05***	**0.24**	***0.53***	**0.51**	***1.13***
	49	43	47	1.04	*2.29*	–	–	0.90	*1.99*	0.85	*1.87*	–	–	–	–	0.74	*1.63*	0.51	*1.13*	0.26	*0.57*	0.55	*1.22*
340	**32**	**–**	**–**	**0.65**	***1.43***	**–**	**–**	**–**	**–**	**0.47**	***1.04***	**–**	**–**	**–**	**–**	**–**	**–**	**–**	**–**	**–**	**–**	**–**	**–**
	36	–	–	0.72	*1.59*	–	–	–	–	0.53	*1.16*	–	–	–	–	–	–	–	–	–	–	–	–

DRY FORAGES

TABLE V-1C DRY FORAGES, MINERAL AND VITAMIN COMPOSITION OF FEEDS, DATA EXPRESSED **AS-FED** AND **MOISTURE-FREE**—*(Continued)*

				Macro Minerals							Micro Minerals						
Entry Number	Feed Name Description	Moisture Basis: A-F (as-fed) or M-F (moisture-free)	Dry Matter	Calcium (Ca)	Phosphorus (P)	Sodium (Na)	Chlorine (Cl)	Magnesium (Mg)	Potassium (K)	Sulfur (S)	Cobalt (Co)	Copper (Cu)	Iodine (I)	Iron (Fe)	Manganese (Mn)	Selenium (Se)	Zinc (Zn)
			%	%	%	%	%	%	%	%	ppm or mg/kg	ppm or mg/kg	ppm or mg/kg	%	ppm or mg/kg	ppm or mg/kg	ppm or mg/kg
	TIMOTHY (Continued)																
318	HAY, EARLY BLOOM, SUN-CURED	**A-F**	**89**	**0.46**	**0.25**	**0.09**	**–**	**0.11**	**2.14**	**0.12**	**–**	**57.1**	**–**	**0.019**	**91.8**	**–**	**55.3**
		M-F	100	0.51	0.29	0.10	–	0.13	2.41	0.13	–	64.0	–	0.021	103.0	–	62.0
319	HAY, MIDBLOOM, SUN-CURED	**A-F**	**89**	**0.32**	**0.20**	**0.08**	**–**	**0.12**	**1.61**	**0.12**	**–**	**14.3**	**–**	**0.014**	**49.9**	**–**	**38.2**
		M-F	100	0.36	0.23	0.10	–	0.13	1.82	0.13	–	16.0		0.015	56.1	–	43.0
320	HAY, FULL BLOOM, SUN-CURED	**A-F**	**89**	**0.36**	**0.21**	**0.11**	**0.55**	**0.10**	**1.77**	**0.12**		**25.7**	**–**	**0.013**	**82.4**	**–**	**47.9**
		M-F	100	0.41	0.24	0.12	0.62	0.11	2.00	0.13	–	29.0	–	0.014	93.0	–	54.0
	TIMOTHY-CLOVER *Phleum pratense-Trifolium* spp																
321	HAY, SUN-CURED	**A-F**	**89**	**0.62**	**0.17**	**0.17**	**0.48**	**0.17**	**1.33**	**0.13**	**–**	**6.3**	**–**	**0.011**	**48.7**	**–**	**–**
		M-F	100	0.69	0.19	0.19	0.54	0.20	1.49	0.14	–	7.0	–	0.013	54.5	–	–
	TREFOIL, BIG *Lotus ulginosus*																
322	HAY, SUN-CURED	**A-F**	**89**	**–**	**–**	**–**	**–**	**–**	**–**	**–**	**–**	**–**	**–**	**–**	**–**	**–**	**–**
		M-F	100	–	–	–	–	–	–	–	–	–	–	–	–	–	–
	TREFOIL, BIRDSFOOT *Lotus corniculatus*																
323	HAY, SUN-CURED	**A-F**	**91**	**1.54**	**0.21**	**0.06**	**–**	**0.46**	**1.74**	**0.23**	**0.100**	**8.4**	**–**	**0.021**	**26.0**	**–**	**69.9**
		M-F	100	1.70	0.23	0.07	–	0.51	1.92	0.25	0.111	9.3	–	0.023	28.7	–	77.2
	TUMBLEWEED, TUMBLING (RUSSIAN THISTLE) *Salsola kali, tenuifolia*																
324	HAY, SUN-CURED	**A-F**	**86**	**1.31**	**0.22**	**–**	**–**	**0.65**	**5.92**	**–**	**–**	**–**	**–**	**–**	**–**	**–**	**–**
		M-F	100	1.51	0.25	–	–	0.75	6.85	–	–	–	–	–	–	–	–
	VELVETBEAN *Mucuna* spp																
325	HAY, SUN-CURED	**A-F**	**93**	**–**	**0.24**	**–**	**–**	**–**	**2.20**	**–**	**–**	**–**	**–**	**–**	**–**	**–**	**–**
		M-F	100	–	0.26	–	–	–	2.37	–	–	–	–	–	–	–	–
	VETCH *Vicia* spp																
326	HAY, SUN-CURED	**A-F**	**89**	**1.21**	**0.30**	**0.46**	**–**	**0.24**	**1.88**	**0.13**	**0.315**	**8.8**	**0.437**	**0.044**	**53.9**	**–**	**–**
		M-F	100	1.36	0.34	0.52	–	0.27	2.12	0.15	0.355	9.9	0.492	0.049	60.8	–	–
	VETCH-OATS *Vicia* spp-*Avena sativa*																
327	HAY, SUN-CURED	**A-F**	**88**	**0.76**	**0.27**	**–**	**–**	**–**	**1.51**	**–**	**–**	**–**	**–**	**–**	**–**	**–**	**–**
		M-F	100	0.87	0.31	–	–	–	1.72	–	–	–	–	–	–	–	–
	VETCH-WHEAT *Vicia* spp-*Triticum aestivum*																
328	HAY, SUN-CURED	**A-F**	**90**	**–**	**–**	**–**	**–**	**–**	**–**	**–**	**–**	**–**	**–**	**–**	**–**	**–**	**–**
		M-F	100	–	–	–	–	–	–	–	–	–	–	–	–	–	–
	WATERGRASS *Hydrochloa caroliniensis*																
329	HAY, SUN-CURED	**A-F**	**94**	**–**	**–**	**–**	**–**	**–**	**–**	**–**	**–**	**–**	**–**	**–**	**–**	**–**	**–**
		M-F	100	–	–	–	–	–	–	–	–	–	–	–	–	–	–
	WHEAT *Triticum aestivum*																
330	HAY, SUN-CURED	**A-F**	**89**	**0.13**	**0.18**	**0.19**	**–**	**0.11**	**0.88**	**0.19**	**–**	**–**	**–**	**0.018**	**–**	**–**	**–**
		M-F	100	0.15	0.20	0.21	–	0.12	1.00	0.22	–	–	–	0.020	–	–	–
331	CHAFF	**A-F**	**93**	**0.19**	**0.08**	**–**	**–**	**–**	**0.52**	**–**	**–**	**–**	**–**	**–**	**–**	**–**	**–**
		M-F	100	0.21	0.08	–	–	–	0.56	–	–	–	–	–	–	–	–
332	STRAW	**A-F**	**90**	**0.16**	**0.05**	**0.13**	**0.29**	**0.11**	**1.27**	**0.17**	**0.041**	**3.2**	**–**	**0.015**	**36.7**	**–**	**5.8**
		M-F	100	0.18	0.05	0.14	0.32	0.12	1.41	0.19	0.046	3.6	–	0.016	40.9	–	6.5
	WHEATGRASS, CRESTED *Agropyron desertorum*																
333	HAY, SUN-CURED	**A-F**	**92**	**0.24**	**0.14**	**–**	**–**	**–**	**–**	**–**	**0.219**	**–**	**–**	**–**	**–**	**–**	**–**
		M-F	100	0.26	0.15	–	–	–	–	–	0.239	–	–	–	–	–	–
	WHEATGRASS, INTERMEDIATE *Agropyron intermedium*																
334	HAY, SUN-CURED	**A-F**	**90**	**0.33**	**0.17**	**–**	**–**	**0.05**	**1.49**	**–**	**–**	**–**	**–**	**–**	**–**	**–**	**–**
		M-F	100	0.37	0.19	–	–	0.06	1.65	–	–	–	–	–	–	–	–
	WHEATGRASS, SLENDER *Agropyron trachycaulum*																
335	HAY, SUN-CURED	**A-F**	**91**	**0.32**	**0.23**	**–**	**–**	**0.22**	**2.45**	**0.11**	**–**	**–**	**–**	**–**	**–**	**–**	**–**
		M-F	100	0.35	0.25	–	–	0.24	2.68	0.12	–	–	–	–	–	–	–
	WHEATGRASS, TALL *Agropyron elongatum*																
336	HAY, SUN-CURED	**A-F**	**89**	**–**	**–**	**–**	**–**	**–**	**–**	**–**	**–**	**–**	**–**	**–**	**–**	**–**	**–**
		M-F	100	–	–	–	–	–	–	–	–	–	–	–	–	–	–
	WHITE CLOVER *Trifolium repens*																
337	HAY, SUN-CURED	**A-F**	**90**	**1.71**	**0.29**	**–**	**–**	**–**	**–**	**–**	**–**	**–**	**–**	**–**	**–**	**–**	**–**
		M-F	100	1.90	0.32	–	–	–	–	–	–	–	–	–	–	–	–
	WILD-RYE, RUSSIAN *Elymus junceus*																
338	HAY, SUN-CURED	**A-F**	**90**	**–**	**–**	**–**	**–**	**–**	**–**	**–**	**–**	**–**	**–**	**–**	**–**	**–**	**–**
		M-F	100	–	–	–	–	–	–	–	–	–	–	–	–	–	–
	WINTERFAT, COMMON *Eurotia lanata*																
339	BROWSE, SUN-CURED	**A-F**	**93**	**–**	**–**	**–**	**–**	**–**	**–**	**–**	**–**	**–**	**–**	**–**	**–**	**–**	**–**
		M-F	100	–	–	–	–	–	–	–	–	–	–	–	–	–	–
	WOOD																
340	SAWDUST	**A-F**	**90**	**–**	**–**	**–**	**–**	**–**	**–**	**–**	**–**	**–**	**–**	**–**	**–**	**–**	**–**
		M-F	100	–	–	–	–	–	–	–	–	–	–	–	–	–	–

Entry Number	Fat-Soluble Vitamins: A (1 mg Carotene = 1667 IU Vit A)	Carotene (Provitamin A)	D	E	K	Water-Soluble Vitamins: B-12	Biotin	Choline	Folacin (Folic Acid)	Niacin	Pantothenic Acid (B-3)	(Pyri-doxine) B-6	Ribo-flavin (B-2)	Thiamin (B-1)
	IU/g	ppm or mg/kg	IU/kg	ppm or mg/kg	ppm or mg/kg	ppm or mg/kg	ppm or mg/kg	ppm or mg/kg	ppm or mg/kg	ppm or mg/kg	ppm or mg/kg	ppm or mg/kg	ppm or mg/kg	ppm or mg/kg
318	**78.0**	**46.8**	–	**11.6**	–	–	–	–	–	–	–	–	–	–
	87.5	52.5	–	13.0	–	–	–	–	–	–	–	–	–	–
319	**79.0**	**47.4**	**2**	–	–	–	–	–	–	–	–	–	–	–
	88.9	53.3	2	–	–	–	–	–	–	–	–	–	–	–
320	–	–	–	–	–	–	–	–	–	–	–	–	–	–
	–	–	–	–	–	–	–	–	–	–	–	–	–	–
321	**39.7**	**23.8**	–	–	–	–	–	–	–	–	–	–	–	–
	44.5	26.7	–	–	–	–	–	–	–	–	–	–	–	–
322	–	–	–	–	–	–	–	–	–	–	–	–	–	–
	–	–	–	–	–	–	–	–	–	–	–	–	–	–
323	**217.8**	**130.6**	**1**	–	–	–	–	–	–	–	–	–	**14.6**	**6.2**
	240.4	144.2	2	–	–	–	–	–	–	–	–	–	16.1	6.8
324	–	–	–	–	–	–	–	–	–	–	–	–	–	–
	–	–	–	–	–	–	–	–	–	–	–	–	–	–
325	–	–	–	–	–	–	–	–	–	–	–	–	–	–
	–	–	–	–	–	–	–	–	–	–	–	–	–	–
326	–	–	–	–	–	–	–	–	–	–	–	–	–	–
	–	–	–	–	–	–	–	–	–	–	–	–	–	–
327	–	–	–	–	–	–	–	–	–	–	–	–	–	–
	–	–	–	–	–	–	–	–	–	–	–	–	–	–
328	–	–	–	–	–	–	–	–	–	–	–	–	–	–
	–	–	–	–	–	–	–	–	–	–	–	–	–	–
329	–	–	–	–	–	–	–	–	–	–	–	–	–	–
	–	–	–	–	–	–	–	–	–	–	–	–	–	–
330	**126.3**	**75.8**	**1**	–	–	–	–	–	–	–	–	–	**15.1**	–
	142.3	85.4	2	–	–	–	–	–	–	–	–	–	17.0	–
331	–	–	–	–	–	–	–	–	–	–	–	–	–	–
	–	–	–	–	–	–	–	–	–	–	–	–	–	–
332	**3.3**	**2.0**	**1**	–	–	–	–	–	–	–	–	–	**2.2**	–
	3.7	2.2	1	–	–	–	–	–	–	–	–	–	2.4	–
333	**34.2**	**20.5**	–	–	–	–	–	–	–	–	–	–	–	–
	37.2	22.3	–	–	–	–	–	–	–	–	–	–	–	–
334	–	–	–	–	–	–	–	–	–	–	–	–	–	–
	–	–	–	–	–	–	–	–	–	–	–	–	–	–
335	–	–	–	–	–	–	–	–	–	–	–	–	–	–
	–	–	–	–	–	–	–	–	–	–	–	–	–	–
336	–	–	–	–	–	–	–	–	–	–	–	–	–	–
	–	–	–	–	–	–	–	–	–	–	–	–	–	–
337	**92.1**	**55.3**	–	**115.5**	–	–	–	–	–	–	–	–	–	–
	102.2	61.3	–	128.1	–	–	–	–	–	–	–	–	–	–
338	–	–	–	–	–	–	–	–	–	–	–	–	–	–
	–	–	–	–	–	–	–	–	–	–	–	–	–	–
339	–	–	–	–	–	–	–	–	–	–	–	–	–	–
	–	–	–	–	–	–	–	–	–	–	–	–	–	–
340	–	–	–	–	–	–	–	–	–	–	–	–	–	–
	–	–	–	–	–	–	–	–	–	–	–	–	–	–

TABLE V-1D SILAGES AND HAYLAGES, COMPOSITION OF FEEDS, DATA EXPRESSED **AS-FED** AND **MOISTURE-FREE** (See footnote at end of table.)

Entry Number	Feed Name Description	International Feed Number	Moisture Basis: A-F (as-fed) or M-F (moisture-free)	Proximate Analysis									Digestible Protein		
				Dry Matter	Ash	Crude Fiber	Neutral Det. Fib. (NDF)	Acid Det. Fib. (ADF)	Lignin	Ether Extract (Fat)	N-Free Extract	Crude Protein	Ruminant	Swine	Horse
				%	%	%	%	%	%	%	%	%	%	%	%
	ALFALFA (LUCERNE) SILAGE *Medicago sativa*														
1	ALL ANALYSES	3-00-212	A-F	27	2.6	8.6	–	10.7	–	0.9	10.4	4.7	3.4	–	–
			M-F	100	9.4	31.7	–	39.2	–	3.2	38.2	17.4	12.3	–	–
2	LESS THAN 30% DRY MATTER	3-08-149	A-F	26	2.4	7.4	10.4	10.6	1.4	1.0	10.8	4.7	3.3	–	–
			M-F	100	9.2	28.2	39.5	40.3	5.4	3.8	41.0	17.8	12.3	–	–
3	30–50% DRY MATTER	3-08-150	A-F	43	3.8	14.8	17.3	15.1	2.3	1.1	15.9	7.9	5.5	–	–
			M-F	100	8.7	34.0	39.9	34.8	5.3	2.5	36.6	18.2	12.7	–	–
4	MORE THAN 50% DRY MATTER	3-08-151	A-F	57	5.2	19.4	30.5	23.9	5.7	1.1	21.5	9.9	6.8	–	–
			M-F	100	9.2	34.0	53.4	41.9	9.9	1.9	37.7	17.2	11.9	–	–
5	WILTED	3-00-221	A-F	41	3.7	12.7	–	10.7	–	1.4	15.9	7.4	5.1	–	–
			M-F	100	9.0	30.9	–	25.9	–	3.5	38.6	18.0	12.4	–	–
6	PREBLOOM, WILTED	3-00-215	A-F	35	3.0	9.8	–	–	–	1.4	13.3	7.2	5.0	–	–
			M-F	100	8.6	28.4	–	–	–	4.0	38.3	20.7	14.5	–	–
7	EARLY BLOOM, WILTED	3-00-216	A-F	35	2.8	11.0	–	13.7	–	1.1	13.9	5.8	3.5	–	–
			M-F	100	8.2	31.8	–	39.6	–	3.2	40.2	16.7	10.0	–	–
8	MIDBLOOM, WILTED	3-00-217	A-F	38	3.0	11.9	–	–	–	1.2	15.6	6.4	4.5	–	–
			M-F	100	7.9	31.2	–	–	–	3.1	41.1	16.8	11.8	–	–
9	FULL BLOOM, WILTED	3-00-218	A-F	36	2.8	12.1	–	–	–	1.0	15.1	5.4	3.6	–	–
			M-F	100	7.7	33.2	–	–	–	2.7	41.5	14.9	9.8	–	–
10	AIV ADDED	3-00-225	A-F	30	2.9	6.7	–	–	–	2.3	11.2	6.9	5.0	–	–
			M-F	100	9.5	22.3	–	–	–	7.8	37.5	22.9	16.7	–	–
11	CORN GRAIN ADDED	3-00-226	A-F	23	2.3	7.0	–	–	–	0.9	9.0	4.1	3.0	–	–
			M-F	100	9.7	30.0	–	–	–	4.0	38.6	17.7	12.9	–	–
12	FORMIC ACID ADDED	3-08-850	A-F	26	1.8	7.8	–	–	–	1.3	10.3	5.0	3.7	–	–
			M-F	100	7.0	29.8	–	–	–	5.0	39.2	19.0	14.1	–	–
13	MOLASSES ADDED	3-00-238	A-F	30	2.6	7.8	–	–	–	1.5	12.4	5.6	3.9	–	–
			M-F	100	8.7	26.1	–	–	–	4.9	41.6	18.7	13.0	–	–
	ALFALFA-BROMEGRASS, SMOOTH, SILAGE *Medicago sativa-Bromus inermis*														
14	ALL ANALYSES	3-00-268	A-F	26	2.0	9.4	–	–	–	1.1	10.4	3.6	2.3	–	–
			M-F	100	7.4	35.6	–	–	–	4.1	39.3	13.6	8.6	–	–
15	LESS THAN 30% DRY MATTER	3-08-146	A-F	25	–	7.7	–	–	–	1.1	–	3.8	2.5	–	–
			M-F	100	–	30.8	–	–	–	4.4	–	15.2	10.0	–	–
16	30–50% DRY MATTER	3-08-147	A-F	40	–	13.1	–	–	–	0.9	–	6.2	4.1	–	–
			M-F	100	–	33.0	–	–	–	2.3	–	15.5	10.3	–	–
17	MORE THAN 50% DRY MATTER	3-08-148	A-F	56	–	18.7	–	–	–	1.9	–	8.0	5.2	–	–
			M-F	100	–	33.7	–	–	–	3.4	–	14.4	9.3	–	–
18	WILTED	3-00-269	A-F	46	4.1	15.2	–	–	–	1.1	18.6	7.1	4.8	–	–
			M-F	100	8.8	33.0	–	–	–	2.3	40.4	15.5	10.3	–	–
	ALFALFA-ORCHARDGRASS, SILAGE *Medicago sativa-Dactylis glomerata*														
19	LESS THAN 30% DRY MATTER	3-08-145	A-F	28	–	8.8	–	–	–	1.1	–	4.8	3.3	–	–
			M-F	100	–	31.4	–	–	–	4.1	–	17.1	11.7	–	–
20	30–50% DRY MATTER	3-08-144	A-F	37	4.0	11.2	–	16.8	3.7	1.5	13.7	6.7	4.7	–	–
			M-F	100	10.9	30.2	–	45.4	10.0	4.0	36.9	18.1	12.6	–	–
21	MORE THAN 50% DRY MATTER	3-08-143	A-F	57	5.1	20.3	–	–	–	2.0	20.6	9.2	6.2	–	–
			M-F	100	9.0	35.5	–	–	–	3.5	36.0	16.1	10.8	–	–
22	**ALFALFA-TIMOTHY, SILAGE** *Medicago sativa-Phleum pratense*	3-00-345	A-F	28	2.2	9.7	–	–	–	1.0	10.9	3.8	2.4	–	–
			M-F	100	8.1	35.2	–	–	–	3.5	39.5	13.7	8.7	–	–
	APPLE, SILAGE *Malus* spp														
23	FRUIT	4-00-419	A-F	12	0.6	1.8	–	–	–	0.7	8.6	0.7	0.2	0.4	0.4
			M-F	100	4.8	14.6	–	–	–	5.8	69.3	5.5	1.9	3.1	3.4
24	POMACE	4-00-420	A-F	21	1.0	4.4	–	–	–	1.3	12.8	1.6	0.8	1.1	1.1
			M-F	100	4.9	20.6	–	–	–	6.3	60.4	7.8	3.9	5.2	5.2
25	**ARTICHOKE, JERUSALEM, SILAGE** *Helianthus tuberosus*	3-22-136	A-F	30	5.6	8.2	–	–	–	0.5	13.5	2.3	–	–	–
			M-F	100	18.5	27.2	–	–	–	1.7	45.0	7.6	–	–	–
	ATLAS SORGHUM–SEE SORGHUM, SILAGE														
26	**BARLEY, SILAGE** *Hordeum vulgare*	3-00-512	A-F	31	3.2	9.4	–	–	–	1.2	14.3	3.2	1.8	–	–
			M-F	100	10.2	30.0	–	–	–	3.9	45.6	10.3	5.7	–	–
27	**BARNYARD GRASS (MILLET, JAPANESE)-SOYBEAN, SILAGE** *Echinochloa crusgalli-Glycine max*	3-26-066	A-F	21	2.9	7.3	–	–	–	1.1	6.9	2.9	1.7	–	–
			M-F	100	13.7	34.6	–	–	–	5.2	32.8	13.7	8.1	–	–
28	**BEAN, LIMA, SILAGE** *Phaseolus limensis*	3-00-610	A-F	25	2.4	7.7	–	–	–	0.5	11.5	2.9	1.7	–	–
			M-F	100	9.6	30.8	–	–	–	2.0	46.0	11.6	6.9	–	–
29	**BEAN, MUNG, SILAGE** *Phaseolus aureus*	3-00-620	A-F	27	7.2	5.2	–	–	–	1.2	9.9	3.8	2.4	–	–
			M-F	100	26.4	19.1	–	–	–	4.5	36.1	13.9	8.9	–	–
	BEET, MANGEL, SILAGE *Beta vulgaris, macrorrhiza*														
30	ROOTS	4-00-636	A-F	10	1.9	0.8	–	–	–	0.1	6.4	1.0	0.5	0.8	0.7
			M-F	100	18.8	7.6	–	–	–	1.2	62.5	9.9	4.5	7.5	6.9
31	TOPS WITH CROWNS	3-00-635	A-F	13	2.8	1.9	–	–	–	0.6	5.7	1.8	1.2	–	–
			M-F	100	22.1	14.5	–	–	–	4.6	44.9	13.9	9.2	–	–
	BEET, SUGAR–SEE SUGAR BEET, SILAGE														

SILAGES AND HAYLAGES

Entry Number	TDN Ruminant	TDN Swine	TDN Horse	Digestible Energy Ruminant		Digestible Energy Swine		Digestible Energy Horse		Metabolizable Energy Ruminant		Metabolizable Energy Swine		Metabolizable Energy Poultry ME_n		Metabolizable Energy Horse		Net Energy Ruminant NE_m		Net Energy Ruminant NE_g		Net Energy Lactating Cows NE_{lc}	
	%	%	%	Mcal		kcal		Mcal		Mcal		kcal		kcal		Mcal		Mcal		Mcal		Mcal	
				lb	kg	lb	kg	lb	kg	lb	kg	lb	kg	lb	kg	lb	kg	lb	kg	lb	kg	lb	kg
1	16	–	–	0.28	0.61	–	–	–	–	0.27	0.59	–	–	–	–	–	–	0.16	0.36	0.09	0.20	0.16	0.35
	59	–	–	1.02	2.25	–	–	–	–	0.98	2.17	–	–	–	–	–	–	0.59	1.31	0.33	0.74	0.58	1.29
2	14	–	–	0.31	0.68	–	–	–	–	0.26	0.57	–	–	–	–	–	–	0.13	0.28	0.06	0.14	0.14	0.31
	53	–	–	1.16	2.56	–	–	–	–	0.97	2.14	–	–	–	–	–	–	0.49	1.07	0.24	0.52	0.54	1.18
3	23	–	–	0.48	1.07	–	–	–	–	0.40	0.88	–	–	–	–	–	–	0.21	0.47	0.10	0.23	0.23	0.51
	53	–	–	1.11	2.45	–	–	–	–	0.92	2.03	–	–	–	–	–	–	0.49	1.07	0.24	0.52	0.54	1.18
4	30	–	–	0.63	1.38	–	–	–	–	0.52	1.14	–	–	–	–	–	–	0.27	0.59	0.13	0.28	0.30	0.66
	52	–	–	1.10	2.41	–	–	–	–	0.90	1.99	–	–	–	–	–	–	0.47	1.04	0.22	0.49	0.52	1.15
5	24	–	–	0.49	1.07	–	–	–	–	0.41	0.90	–	–	–	–	–	–	0.23	0.52	0.13	0.28	0.25	0.54
	59	–	–	1.18	2.61	–	–	–	–	0.99	2.18	–	–	–	–	–	–	0.57	1.26	0.31	0.69	0.59	1.31
6	22	–	–	0.44	0.97	–	–	–	–	0.37	0.82	–	–	–	–	–	–	0.22	0.49	0.13	0.28	0.22	0.49
	63	–	–	1.27	2.79	–	–	–	–	1.08	2.37	–	–	–	–	–	–	0.64	1.40	0.37	0.82	0.64	1.42
7	18	–	–	0.39	0.86	–	–	–	–	0.33	0.72	–	–	–	–	–	–	0.17	0.36	0.08	0.17	0.18	0.40
	52	–	–	1.13	2.49	–	–	–	–	0.94	2.07	–	–	–	–	–	–	0.48	1.05	0.22	0.50	0.53	1.16
8	23	–	–	0.47	1.03	–	–	–	–	0.39	0.87	–	–	–	–	–	–	0.23	0.51	0.13	0.29	0.24	0.52
	61	–	–	1.22	2.70	–	–	–	–	1.03	2.28	–	–	–	–	–	–	0.61	1.34	0.35	0.76	0.62	1.37
9	21	–	–	0.43	0.94	–	–	–	–	0.36	0.79	–	–	–	–	–	–	0.20	0.45	0.11	0.24	0.21	0.47
	59	–	–	1.17	2.59	–	–	–	–	0.98	2.16	–	–	–	–	–	–	0.56	1.23	0.30	0.66	0.59	1.29
10	20	–	–	0.37	0.82	–	–	–	–	0.30	0.66	–	–	–	–	–	–	0.18	0.41	0.11	0.23	0.18	0.40
	66	–	–	1.24	2.72	–	–	–	–	1.00	2.21	–	–	–	–	–	–	0.61	1.35	0.35	0.77	0.61	1.34
11	13	–	–	0.27	0.59	–	–	–	–	0.22	0.49	–	–	–	–	–	–	0.14	0.31	0.08	0.17	0.14	0.32
	54	–	–	1.14	2.52	–	–	–	–	0.95	2.09	–	–	–	–	–	–	0.60	1.31	0.34	0.74	0.61	1.35
12	17	–	–	0.34	0.75	–	–	–	–	0.29	0.64	–	–	–	–	–	–	0.18	0.39	0.11	0.23	0.18	0.39
	65	–	–	1.30	2.87	–	–	–	–	1.11	2.45	–	–	–	–	–	–	0.67	1.47	0.40	0.88	0.67	1.47
13	18	–	–	0.40	0.88	–	–	–	–	0.33	0.72	–	–	–	–	–	–	0.21	0.46	0.13	0.28	0.20	0.44
	61	–	–	1.33	2.94	–	–	–	–	1.09	2.41	–	–	–	–	–	–	0.69	1.53	0.42	0.93	0.67	1.47
14	16	–	–	0.31	0.69	–	–	–	–	0.26	0.57	–	–	–	–	–	–	0.15	0.34	0.09	0.19	0.16	0.35
	59	–	–	1.18	2.60	–	–	–	–	0.99	2.17	–	–	–	–	–	–	0.58	1.29	0.32	0.72	0.61	1.33
15	14	–	–	0.28	0.61	–	–	–	–	0.23	0.50	–	–	–	–	–	–	0.13	0.29	0.07	0.15	0.14	0.31
	55	–	–	1.10	2.43	–	–	–	–	0.91	2.00	–	–	–	–	–	–	0.52	1.14	0.26	0.58	0.56	1.23
16	21	–	–	0.42	0.93	–	–	–	–	0.34	0.76	–	–	–	–	–	–	0.19	0.43	0.09	0.21	0.21	0.47
	53	–	–	1.06	2.34	–	–	–	–	0.87	1.91	–	–	–	–	–	–	0.49	1.07	0.24	0.52	0.54	1.18
17	30	–	–	0.60	1.32	–	–	–	–	0.49	1.09	–	–	–	–	–	–	0.28	0.61	0.14	0.31	0.30	0.67
	54	–	–	1.08	2.38	–	–	–	–	0.89	1.96	–	–	–	–	–	–	0.50	1.11	0.25	0.55	0.55	1.20
18	26	–	–	0.53	1.16	–	–	–	–	0.44	0.97	–	–	–	–	–	–	0.24	0.54	0.13	0.28	0.26	0.58
	57	–	–	1.15	2.53	–	–	–	–	0.95	2.10	–	–	–	–	–	–	0.53	1.17	0.28	0.61	0.57	1.25
19	15	–	–	0.30	0.67	–	–	–	–	0.25	0.55	–	–	–	–	–	–	0.14	0.31	0.07	0.15	0.15	0.34
	54	–	–	1.08	2.38	–	–	–	–	0.89	1.96	–	–	–	–	–	–	0.50	1.11	0.25	0.55	0.55	1.20
20	20	–	–	0.42	0.93	–	–	–	–	0.35	0.77	–	–	–	–	–	–	0.19	0.41	0.09	0.20	0.20	0.45
	54	–	–	1.14	2.52	–	–	–	–	0.95	2.09	–	–	–	–	–	–	0.50	1.11	0.25	0.55	0.55	1.20
21	31	–	–	0.64	1.40	–	–	–	–	0.52	1.16	–	–	–	–	–	–	0.29	0.63	0.14	0.31	0.31	0.69
	54	–	–	1.11	2.45	–	–	–	–	0.92	2.02	–	–	–	–	–	–	0.50	1.11	0.25	0.55	0.55	1.20
22	16	–	–	0.32	0.70	–	–	–	–	0.27	0.59	–	–	–	–	–	–	0.15	0.34	0.08	0.18	0.16	0.36
	58	–	–	1.16	2.55	–	–	–	–	0.96	2.12	–	–	–	–	–	–	0.56	1.23	0.30	0.66	0.59	1.29
23	9	10	5	0.19	0.41	206	453	0.10	0.23	0.16	0.36	194	428	–	–	0.08	0.19	0.10	0.22	0.07	0.15	0.10	0.22
	76	83	43	1.51	3.33	1658	3656	0.82	1.82	1.32	2.92	1567	3454	–	–	0.68	1.49	0.82	1.81	0.54	1.18	0.79	1.73
24	15	16	8	0.30	0.66	314	691	0.16	0.34	0.26	0.57	295	650	–	–	0.13	0.28	0.16	0.36	0.10	0.23	0.16	0.35
	68	74	37	1.40	3.10	1479	3261	0.74	1.62	1.21	2.68	1391	3066	–	–	0.60	1.33	0.77	1.70	0.49	1.08	0.75	1.64
25	–	–	–	–	–	–	–	–	–	–	–	–	–	–	–	–	–	–	–	–	–	–	–
	–	–	–	–	–	–	–	–	–	–	–	–	–	–	–	–	–	–	–	–	–	–	–
26	16	–	–	0.32	0.70	–	–	–	–	0.25	0.55	–	–	–	–	–	–	0.13	0.29	0.05	0.12	0.16	0.35
	51	–	–	1.01	2.22	–	–	–	–	0.80	1.76	–	–	–	–	–	–	0.42	0.93	0.17	0.38	0.51	1.13
27	–	–	–	–	–	–	–	–	–	–	–	–	–	–	–	–	–	–	–	–	–	–	–
	–	–	–	–	–	–	–	–	–	–	–	–	–	–	–	–	–	–	–	–	–	–	–
28	14	–	–	0.29	0.63	–	–	–	–	0.24	0.53	–	–	–	–	–	–	0.13	0.29	0.07	0.15	0.14	0.31
	57	–	–	1.15	2.53	–	–	–	–	0.96	2.11	–	–	–	–	–	–	0.52	1.15	0.27	0.59	0.56	1.23
29	14	–	–	0.29	0.63	–	–	–	–	0.24	0.52	–	–	–	–	–	–	0.12	0.27	0.05	0.12	0.14	0.30
	53	–	–	1.05	2.32	–	–	–	–	0.86	1.89	–	–	–	–	–	–	0.44	0.97	0.19	0.43	0.50	1.11
30	5	8	6	0.12	0.27	167	367	0.11	0.23	0.10	0.23	157	347	–	–	0.09	0.19	0.08	0.17	0.05	0.10	0.07	0.16
	51	81	54	1.20	2.64	1618	3567	1.02	2.24	1.00	2.21	1527	3367	–	–	0.84	1.84	0.73	1.60	0.45	1.00	0.71	1.57
31	7	–	–	0.14	0.32	–	–	–	–	0.12	0.26	–	–	–	–	–	–	0.07	0.15	0.04	0.08	0.07	0.16
	56	–	–	1.13	2.49	–	–	–	–	0.94	2.06	–	–	–	–	–	–	0.55	1.21	0.29	0.64	0.58	1.27

(Continued)

TABLE V-1D SILAGES AND HAYLAGES, MINERAL AND VITAMIN COMPOSITION OF FEEDS, DATA EXPRESSED **AS-FED** AND **MOISTURE-FREE**

Entry Number	Feed Name Description	Moisture Basis: A-F (as-fed) or M-F (moisture-free)	Dry Matter	Macro Minerals: Calcium (Ca)	Phosphorus (P)	Sodium (Na)	Chlorine (Cl)	Magnesium (Mg)	Potassium (K)	Sulfur (S)	Micro Minerals: Cobalt (Co)	Copper (Cu)	Iodine (I)	Iron (Fe)	Manganese (Mn)	Selenium (Se)	Zinc (Zn)
			%	%	%	%	%	%	%	%	ppm or mg/kg	ppm or mg/kg	ppm or mg/kg	%	ppm or mg/kg	ppm or mg/kg	ppm or mg/kg
	ALFALFA (LUCERNE), SILAGE *Medicago sativa*																
1	ALL ANALYSES	**A-F**	**27**	**0.48**	**0.07**	**0.04**	**0.11**	**0.09**	**0.64**	**0.09**	–	**3.0**	–	**0.008**	**13.5**	–	**11.1**
		M-F	100	1.75	0.27	0.16	0.41	0.33	2.35	0.31	–	11.1	–	0.028	49.7	–	40.7
2	LESS THAN 30% DRY MATTER	**A-F**	**26**	**0.39**	**0.08**	**0.04**	–	**0.10**	**0.62**	**0.10**	–	–	–	**0.008**	–	–	–
		M-F	100	1.48	0.29	0.16	–	0.36	2.36	0.36	–	–	–	0.030	–	–	–
3	30-50% DRY MATTER	**A-F**	**43**	**0.58**	**0.13**	**0.05**	–	**0.13**	**0.97**	**0.15**	–	**3.7**	–	**0.012**	**21.7**	–	**8.9**
		M-F	100	1.32	0.31	0.11	–	0.30	2.23	0.34	–	8.5	–	0.028	50.0	–	20.5
4	MORE THAN 50% DRY MATTER	**A-F**	**57**	**0.54**	**0.17**	**0.04**	–	**0.19**	**1.14**	**0.22**	–	**6.4**	–	**0.014**	**23.7**	–	**9.4**
		M-F	100	0.95	0.30	0.07	–	0.34	2.00	0.38	–	11.2	–	0.024	41.4	–	16.5
5	WILTED	**A-F**	**41**	**0.55**	**0.12**	**0.06**	**0.17**	**0.16**	**0.94**	**0.15**	–	**3.8**	–	**0.013**	**21.4**	–	–
		M-F	100	1.33	0.30	0.15	0.41	0.38	2.29	0.36	–	9.2	–	0.031	51.9	–	–
6	PREBLOOM, WILTED	**A-F**	**35**	–	–	–	–	–	–	–	–	–	–	–	–	–	–
		M-F	100	–	–	–	–	–	–	–	–	–	–	–	–	–	–
7	EARLY BLOOM, WILTED	**A-F**	**35**	–	–	–	–	–	–	–	–	–	–	–	–	–	–
		M-F	100	–	–	–	–	–	–	–	–	–	–	–	–	–	–
8	MIDBLOOM, WILTED	**A-F**	**38**	–	–	–	–	–	–	–	–	–	–	–	–	–	–
		M-F	100	–	–	–	–	–	–	–	–	–	–	–	–	–	–
9	FULL BLOOM, WILTED	**A-F**	**36**	–	–	–	–	–	–	–	–	–	–	–	–	–	–
		M-F	100	–	–	–	–	–	–	–	–	–	–	–	–	–	–
10	AIV ADDED	**A-F**	**30**	–	–	–	–	–	–	–	–	–	–	–	**6.6**	–	–
		M-F	100	–	–	–	–	–	–	–	–	–	–	–	22.0	–	–
11	CORN GRAIN ADDED	**A-F**	**23**	–	–	–	–	–	–	–	–	–	–	–	–	–	–
		M-F	100	–	–	–	–	–	–	–	–	–	–	–	–	–	–
12	FORMIC ACID ADDED	**A-F**	**26**	–	–	–	–	–	–	–	–	–	–	–	–	–	–
		M-F	100	–	–	–	–	–	–	–	–	–	–	–	–	–	–
13	MOLASSES ADDED	**A-F**	**30**	**0.49**	**0.09**	**0.05**	–	**0.10**	**0.78**	**0.11**	–	–	–	**0.009**	–	–	–
		M-F	100	1.64	0.30	0.16	–	0.34	2.61	0.36	–	–	–	0.030	–	–	–
	ALFALFA-BROMEGRASS, SMOOTH, SILAGE *Medicago sativa-Bromus inermis*																
14	ALL ANALYSES	**A-F**	**26**	**0.17**	**0.05**	–	–	**0.05**	**0.49**	–	**0.030**	**3.0**	–	**0.004**	**8.0**	–	–
		M-F	100	0.64	0.20	–	–	0.20	1.86	–	0.113	11.2	–	0.015	30.2	–	–
15	LESS THAN 30% DRY MATTER	**A-F**	**25**	**0.16**	**0.05**	**0.11**	–	**0.05**	**0.47**	**0.06**	–	–	–	**0.004**	–	–	–
		M-F	100	0.64	0.20	0.42	–	0.20	1.86	0.23	–	–	–	0.015	–	–	–
16	30-50% DRY MATTER	**A-F**	**40**	**0.25.**	**0.08**	**0.17**	–	**0.08**	**0.74**	**0.09**	–	–	–	**0.006**	–	–	–
		M-F	100	0.64	0.20	0.42	–	0.20	1.86	0.23	–	–	–	0.015	–	–	–
17	MORE THAN 50% DRY MATTER	**A-F**	**56**	**0.36**	**0.11**	**0.23**	–	**0.11**	**1.03**	**0.13**	–	–	–	**0.009**	–	–	–
		M-F	100	0.64	0.20	0.42	–	0.20	1.86	0.23	–	–	–	0.015	–	–	–
18	WILTED	**A-F**	**46**	**0.72**	**0.12**	–	–	–	–	–	–	–	–	–	–	–	–
		M-F	100	1.56	0.26	–	–	–	–	–	–	–	–	–	–	–	–
	ALFALFA-ORCHARDGRASS, SILAGE *Medicago sativa-Dactylis glomerata*																
19	LESS THAN 30% DRY MATTER	**A-F**	**28**	**0.24**	**0.08**	**0.03**	–	**0.09**	**0.81**	**0.08**	–	–	–	**0.007**	–	–	–
		M-F	100	0.86	0.29	0.10	–	0.33	2.90	0.27	–	–	–	0.025	–	–	–
20	30-50% DRY MATTER	**A-F**	**37**	**0.32**	**0.11**	**0.04**	–	**0.12**	**1.07**	**0.10**	–	–	–	**0.010**	–	–	–
		M-F	100	0.86	0.29	0.10	–	0.33	2.90	0.27	–	–	–	0.025	–	–	–
21	MORE THAN 50% DRY MATTER	**A-F**	**57**	**0.43**	**0.17**	**0.06**	–	**0.19**	**1.66**	**0.15**	–	–	–	**0.015**	–	–	–
		M-F	100	0.75	0.30	0.10	–	0.33	2.90	0.27	–	–	–	0.025	–	–	–
22	**ALFALFA-TIMOTHY, SILAGE** *Medicago sativa-Phleum pratense*	**A-F**	**28**	–	–	–	–	–	–	–	–	–	–	–	–	–	–
		M-F	100	–	–	–	–	–	–	–	–	–	–	–	–	–	–
	APPLE, SILAGE *Malus* spp																
23	FRUIT	**A-F**	**12**	–	–	–	–	–	–	–	–	–	–	–	–	–	–
		M-F	100	–	–	–	–	–	–	–	–	–	–	–	–	–	–
24	POMACE	**A-F**	**21**	**0.02**	**0.02**	–	–	–	**0.10**	–	–	–	–	–	–	–	–
		M-F	100	0.10	0.10	–	–	–	0.48	–	–	–	–	–	–	–	–
25	**ARTICHOKE, JERUSALEM, SILAGE** *Helianthus tuberosus*	**A-F**	**30**	–	–	–	–	–	–	–	–	–	–	–	–	–	–
		M-F	100	–	–	–	–	–	–	–	–	–	–	–	–	–	–
	ATLAS SORGHUM–SEE SORGHUM, SILAGE																
26	**BARLEY, SILAGE** *Hordeum vulgare*	**A-F**	**31**	**0.11**	**0.09**	**0.00**	–	**0.04**	**0.63**	**0.04**	**0.212**	**1.5**	–	**0.009**	**24.8**	**0.029**	**7.0**
		M-F	100	0.34	0.28	0.01	–	0.13	2.01	0.11	0.674	4.9	–	0.028	78.9	0.092	22.4
27	**BARNYARD GRASS (MILLET, JAPANESE)-SOYBEAN, SILAGE** *Echinochloa crusgalli-Glycine max*	**A-F**	**21**	–	–	–	–	–	–	–	–	–	–	–	–	–	–
		M-F	100	–	–	–	–	–	–	–	–	–	–	–	–	–	–
28	**BEAN, LIMA, SILAGE** *Phaseolus limensis*	**A-F**	**25**	–	–	–	–	–	–	–	–	–	–	–	–	–	–
		M-F	100	–	–	–	–	–	–	–	–	–	–	–	–	–	–
29	**BEAN, MUNG, SILAGE** *Phaseolus aureus*	**A-F**	**27**	–	–	–	–	–	–	–	–	–	–	–	–	–	–
		M-F	100	–	–	–	–	–	–	–	–	–	–	–	–	–	–
	BEET, MANGEL, SILAGE *Beta vulgaris, macrorrhiza*																
30	ROOTS	**A-F**	**10**	–	–	–	–	–	–	–	–	–	–	–	–	–	–
		M-F	100	–	–	–	–	–	–	–	–	–	–	–	–	–	–
31	TOPS WITH CROWNS	**A-F**	**13**	–	–	–	–	–	–	–	–	–	–	–	–	–	–
		M-F	100	–	–	–	–	–	–	–	–	–	–	–	–	–	–
	BEET, SUGAR–SEE SUGAR BEET, SILAGE																

Entry Number	Fat-Soluble Vitamins					Water-Soluble Vitamins								
	A (1 mg Carotene = 1667 IU Vit A)	Carotene (Provitamin A)	D	E	K	B–12	Biotin	Choline	Folacin (Folic Acid)	Niacin	Pantothenic Acid (B–3)	(Pyri-doxine) B–6	Ribo-flavin (B–2)	Thiamin (B–1)
	IU/g	ppm or mg/kg	IU/kg	ppm or mg/kg	ppm or mg/kg	ppm or mg/kg	ppm or mg/kg	ppm or mg/kg	ppm or mg/kg	ppm or mg/kg	ppm or mg/kg	ppm or mg/kg	ppm or mg/kg	ppm or mg/kg
1	**42.2**	**25.3**	–	–	–	–	–	–	–	–	–	–	–	–
	155.0	93.0	–	–	–	–	–	–	–	–	–	–	–	–
2	–	–	–	–	–	–	–	–	–	–	–	–	–	–
	–	–	–	–	–	–	–	–	–	–	–	–	–	–
3	**103.6**	**62.1**	–	–	–	–	–	–	–	–	–	–	–	–
	238.4	143.0	–	–	–	–	–	–	–	–	–	–	–	–
4	**101.7**	**61.0**	–	–	–	–	–	–	–	–	–	–	–	–
	178.0	106.8	–	–	–	–	–	–	–	–	–	–	–	–
5	**41.5**	**24.9**	**0**	–	–	–	–	–	–	–	–	–	–	–
	100.8	60.5	1	–	–	–	–	–	–	–	–	–	–	–
6	–	–	–	–	–	–	–	–	–	–	–	–	–	–
	–	–	–	–	–	–	–	–	–	–	–	–	–	–
7	–	–	–	–	–	–	–	–	–	–	–	–	–	–
	–	–	–	–	–	–	–	–	–	–	–	–	–	–
8	–	–	–	–	–	–	–	–	–	–	–	–	–	–
	–	–	–	–	–	–	–	–	–	–	–	–	–	–
9	–	–	–	–	–	–	–	–	–	–	–	–	–	–
	–	–	–	–	–	–	–	–	–	–	–	–	–	–
10	–	–	–	–	–	–	–	–	–	–	–	–	–	–
	–	–	–	–	–	–	–	–	–	–	–	–	–	–
11	–	–	–	–	–	–	–	–	–	–	–	–	–	–
	–	–	–	–	–	–	–	–	–	–	–	–	–	–
12	–	–	–	–	–	–	–	–	–	–	–	–	–	–
	–	–	–	–	–	–	–	–	–	–	–	–	–	–
13	**53.8**	**32.3**	–	–	–	–	–	–	–	–	–	–	–	–
	180.4	108.2	–	–	–	–	–	–	–	–	–	–	–	–
14	–	–	–	–	–	–	–	–	–	–	–	–	–	–
	–	–	–	–	–	–	–	–	–	–	–	–	–	–
15	–	–	–	–	–	–	–	–	–	–	–	–	–	–
	–	–	–	–	–	–	–	–	–	–	–	–	–	–
16	–	–	–	–	–	–	–	–	–	–	–	–	–	–
	–	–	–	–	–	–	–	–	–	–	–	–	–	–
17	–	–	–	–	–	–	–	–	–	–	–	–	–	–
	–	–	–	–	–	–	–	–	–	–	–	–	–	–
18	–	–	–	–	–	–	–	–	–	–	–	–	–	–
	–	–	–	–	–	–	–	–	–	–	–	–	–	–
19	–	–	–	–	–	–	–	–	–	–	–	–	–	–
	–	–	–	–	–	–	–	–	–	–	–	–	–	–
20	–	–	–	–	–	–	–	–	–	–	–	–	–	–
	–	–	–	–	–	–	–	–	–	–	–	–	–	–
21	–	–	–	–	–	–	–	–	–	–	–	–	–	–
	–	–	–	–	–	–	–	–	–	–	–	–	–	–
22	–	–	–	–	–	–	–	–	–	–	–	–	–	–
	–	–	–	–	–	–	–	–	–	–	–	–	–	–
23	–	–	–	–	–	–	–	–	–	–	–	–	–	–
	–	–	–	–	–	–	–	–	–	–	–	–	–	–
24	–	–	–	–	–	–	–	–	–	–	–	–	–	–
	–	–	–	–	–	–	–	–	–	–	–	–	–	–
25	–	–	–	–	–	–	–	–	–	–	–	–	–	–
	–	–	–	–	–	–	–	–	–	–	–	–	–	–
26	**13.4**	**8.0**	–	–	–	–	–	–	–	**75**	**8.3**	–	**1.7**	**7.2**
	42.6	25.6	–	–	–	–	–	–	–	240	26.5	–	5.3	22.9
27	–	–	–	–	–	–	–	–	–	–	–	–	–	–
	–	–	–	–	–	–	–	–	–	–	–	–	–	–
28	–	–	–	–	–	–	–	–	–	–	–	–	–	–
	–	–	–	–	–	–	–	–	–	–	–	–	–	–
29	–	–	–	–	–	–	–	–	–	–	–	–	–	–
	–	–	–	–	–	–	–	–	–	–	–	–	–	–
30	–	–	–	–	–	–	–	–	–	–	–	–	–	–
	–	–	–	–	–	–	–	–	–	–	–	–	–	–
31	–	–	–	–	–	–	–	–	–	–	–	–	–	–
	–	–	–	–	–	–	–	–	–	–	–	–	–	–

(Continued)

TABLE V-1D SILAGES AND HAYLAGES, COMPOSITION OF FEEDS, DATA EXPRESSED **AS-FED** AND **MOISTURE-FREE**—*(Continued)*

Entry Number	Feed Name Description	International Feed Number	Moisture Basis: A-F (as-fed) or M-F (moisture-free)	Proximate Analysis: Dry Matter	Ash	Crude Fiber	Neutral Det. Fib. (NDF)	Acid Det. Fib. (ADF)	Lignin	Ether Extract (Fat)	N-Free Extract	Crude Protein	Digestible Protein: Ruminant	Swine	Horse
				%	%	%	%	%	%	%	%	%	%	%	%
	BERMUDAGRASS, COASTAL, SILAGE *Cynodon dactylon*														
32	WILTED	3-09-097	A-F	39	3.2	12.5	–	–	–	1.3	16.6	5.7	3.7	–	–
			M-F	100	8.1	31.9	–	–	–	3.4	42.2	14.4	9.5	–	–
33	**BERSEEM (EGYPTIAN) CLOVER, SILAGE**	3-01-353	A-F	24	3.6	7.8	–	–	–	0.9	7.7	4.1	1.9	–	–
	Trifolium alexandrinum		M-F	100	15.0	32.4	–	–	–	3.7	32.0	17.0	7.8	–	–
34	**BLUEGRASS-CLOVER, WHITE, MOLASSES ADDED, SILAGE**	3-08-357	A-F	35	3.6	8.9	–	–	–	1.8	13.9	6.8	4.8	–	–
	Poa spp-Trifolium repens		M-F	100	10.3	25.4	–	–	–	5.1	39.7	19.4	13.6	–	–
	BROOMCORN (HOG MILLET; MILLET, PROSO), SILAGE *Panicum miliaceum*														
35	STOVER WITHOUT HEADS	3-08-460	A-F	27	2.2	9.5	–	–	–	0.4	13.3	1.3	0.5	–	–
			M-F	100	8.2	35.6	–	–	–	1.5	49.8	4.9	1.8	–	–
36	**CANARYGRASS, REED, SILAGE** *Phalaris arundinacea*	3-01-114	A-F	27	2.0	9.0	–	–	–	0.7	13.1	2.4	1.2	–	–
			M-F	100	7.3	33.1	–	–	–	2.5	48.2	8.8	4.4	–	–
	CHINESE LESPEDEZA–SEE LESPEDEZA, SERICEA, SILAGE														
	CITRUS, SILAGE *Citrus spp*														
37	PULP	4-01-234	A-F	21	1.2	3.3	–	4.2	–	2.1	12.8	1.5	0.7	1.0	1.0
			M-F	100	5.6	15.7	–	20.0	–	9.9	61.5	7.3	3.5	4.7	4.8
38	**CLOVER, EGYPTIAN (BERSEEM), SILAGE**	3-01-353	A-F	24	3.6	7.8	–	–	–	0.9	7.7	4.1	1.9	–	–
	Trifolium alexandrinum		M-F	100	15.0	32.4	–	–	–	3.7	32.0	17.0	7.8	–	–
39	**CLOVER, LADINO, SILAGE** *Trifolium repens*	3-01-384	A-F	25	2.5	5.2	–	–	–	1.0	10.2	6.0	4.3	–	–
			M-F	100	10.0	20.8	–	–	–	3.9	41.1	24.2	17.4	–	–
40	**CLOVER, LADINO-GRASS, SILAGE** *Trifolium repens-grass*	3-08-384	A-F	30	2.6	7.5	–	–	–	1.5	12.9	5.4	3.9	–	–
			M-F	100	8.7	25.1	–	–	–	5.0	43.1	18.1	13.0	–	–
	CLOVER, RED, SILAGE *Trifolium pratense*														
41	ALL ANALYSES	3-01-441	A-F	25	2.5	8.4	–	–	–	1.0	9.9	3.4	2.2	–	–
			M-F	100	9.9	33.4	–	–	–	3.8	39.3	13.7	8.5	–	–
42	MOLASSES ADDED	3-01-451	A-F	34	2.8	10.2	–	–	–	1.0	15.1	4.5	2.5	–	–
			M-F	100	8.4	30.3	–	–	–	2.9	45.1	13.4	7.4	–	–
43	**CLOVER, RED-GRASS, SILAGE** *Trifolium pratense-grass*	3-26-122	A-F	30	2.6	9.4	–	–	–	0.9	13.4	3.7	2.2	–	2.2
			M-F	100	8.5	31.3	–	–	–	3.1	44.8	12.3	7.4	–	7.4
44	**CLOVER, SWEET, YELLOW, SILAGE** *Melilotus officinalis*	3-04-771	A-F	28	2.5	8.8	–	–	–	0.9	11.5	3.9	2.6	–	–
			M-F	100	8.9	31.9	–	–	–	3.2	41.7	14.3	9.2	–	–
	CORN, SILAGE *Zea mays*														
45	ALL ANALYSES	3-02-822	A-F	26	1.5	6.6	–	8.9	1.2	0.8	15.2	2.2	1.0	1.3	–
			M-F	100	5.6	25.1	–	34.0	4.4	3.2	57.8	8.3	4.0	5.1	–
46	IMMATURE	3-02-817	A-F	24	1.7	5.8	–	–	–	0.7	13.1	2.3	1.2	–	–
			M-F	100	7.3	24.6	–	–	–	3.1	55.3	9.7	5.2	–	–
47	MILK STAGE	3-02-818	A-F	22	1.1	6.9	–	8.3	–	0.7	11.8	2.0	1.3	–	–
			M-F	100	5.1	30.6	–	37.0	–	3.1	52.3	8.8	5.8	–	–
48	DOUGH STAGE	3-02-819	A-F	27	1.2	6.6	–	8.3	–	0.8	16.2	2.1	1.2	–	–
			M-F	100	4.6	24.5	–	30.9	–	2.8	60.3	7.7	4.5	–	–
49	MATURE[1]	3-02-820	A-F	30	1.4	6.5	–	4.2	–	1.3	18.7	2.5	1.5	–	–
			M-F	100	4.7	21.3	–	14.0	–	4.2	61.6	8.2	4.9	–	–
50	OVERRIPE	3-02-821	A-F	30	1.5	7.0	–	–	–	1.4	17.2	2.3	1.6	–	–
			M-F	100	5.2	23.8	–	–	–	4.8	58.3	7.9	5.3	–	–
51	LESS THAN 30% DRY MATTER	3-20-507	A-F	26	2.0	6.5	14.7	8.8	1.5	0.9	14.1	2.1	1.4	–	–
			M-F	100	7.9	25.4	57.0	34.1	5.7	3.7	54.9	8.3	5.3	–	–
52	30–50% DRY MATTER	3-20-506	A-F	37	2.2	8.1	21.7	9.8	1.7	1.4	22.2	3.1	1.8	–	–
			M-F	100	5.9	21.9	59.0	26.7	4.7	3.8	60.1	8.3	5.0	–	–
53	MORE THAN 50% DRY MATTER	3-20-505	A-F	54	1.2	11.6	–	–	–	1.3	35.7	4.3	2.3	–	–
			M-F	100	2.3	21.4	–	–	–	2.4	65.9	8.0	4.3	–	–
54	MOLASSES ADDED	3-02-834	A-F	29	1.9	6.8	–	–	–	0.8	17.2	2.3	1.1	–	–
			M-F	100	6.4	23.4	–	–	–	2.9	59.3	8.0	3.7	–	–
55	EARS	4-07-740	A-F	70	1.6	7.3	–	–	–	2.5	51.5	7.1	4.2	5.1	5.0
			M-F	100	2.3	10.4	–	–	–	3.6	73.7	10.1	6.0	7.3	7.1
56	EARS WITH HUSKS	4-02-839	A-F	43	1.5	4.8	–	–	–	1.6	31.5	4.0	2.2	2.8	2.7
			M-F	100	3.4	11.1	–	–	–	3.7	72.7	9.2	5.2	6.4	6.3
57	HUSKS (HUSKLAGE)	3-26-074	A-F	78	2.7	26.8	–	–	–	0.7	45.0	2.9	–	–	–
			M-F	100	3.4	34.3	–	–	–	0.9	57.7	3.7	–	–	–
58	**CORN, FLINT, SILAGE** *Zea mays, indurata*	3-02-941	A-F	23	1.5	5.5	–	–	–	0.6	14.2	1.7	1.0	–	–
			M-F	100	6.4	23.2	–	–	–	2.7	60.6	7.1	4.1	–	–
	CORN, SWEET, SILAGE *Zea mays, saccharata*														
59	CANNERY RESIDUE	3-07-955	A-F	31	1.7	10.1	–	10.5	–	1.3	15.4	2.5	1.2	–	–
			M-F	100	5.4	32.7	–	34.0	–	4.3	49.6	8.0	3.7	–	–
	CORN-SORGHUM, SILAGE *Zea mays-Sorghum bicolor*														
60	MATURE	3-03-013	A-F	34	2.1	8.5	–	–	–	1.1	19.7	2.7	1.7	–	–
			M-F	100	6.1	24.9	–	–	–	3.3	57.9	7.9	4.9	–	–

Entry Number	TDN Ruminant	TDN Swine	TDN Horse	Digestible Energy Ruminant		Digestible Energy Swine		Digestible Energy Horse		Metabolizable Energy Ruminant		Metabolizable Energy Swine		Poultry ME_n		Metabolizable Energy Horse		Net Energy Ruminant NE_m		Ruminant NE_g		Lactating Cows NE_{lc}	
	%	%	%	Mcal		kcal		Mcal		Mcal		kcal		kcal		Mcal		Mcal		Mcal		Mcal	
				lb	kg	lb	kg	lb	kg	lb	kg	lb	kg	lb	kg	lb	kg	lb	kg	lb	kg	lb	kg
32	23	–	–	0.46	1.00	–	–	–	–	0.39	0.86	–	–	–	–	–	–	0.22	0.49	0.12	0.26	0.22	0.48
	59	–	–	1.16	2.55	–	–	–	–	0.99	2.19	–	–	–	–	–	–	0.56	1.24	0.30	0.67	0.56	1.23
33	10	–	–	0.22	0.49	–	–	–	–	0.18	0.39	–	–	–	–	–	–	0.12	0.26	0.06	0.13	0.13	0.28
	40	–	–	0.93	2.05	–	–	–	–	0.73	1.62	–	–	–	–	–	–	0.49	1.07	0.24	0.52	0.53	1.18
34	23	–	–	0.46	1.01	–	–	–	–	0.39	0.87	–	–	–	–	–	–	0.23	0.51	0.14	0.30	0.23	0.51
	67	–	–	1.31	2.90	–	–	–	–	1.12	2.48	–	–	–	–	–	–	0.66	1.45	0.39	0.86	0.66	1.46
35	13	–	–	0.27	0.60	–	–	–	–	0.22	0.49	–	–	–	–	–	–	0.13	0.28	0.06	0.13	0.14	0.31
	49	–	–	1.02	2.25	–	–	–	–	0.83	1.82	–	–	–	–	–	–	0.48	1.06	0.23	0.50	0.53	1.17
36	14	–	–	0.29	0.64	–	–	–	–	0.24	0.53	–	–	–	–	–	–	0.15	0.33	0.08	0.18	0.16	0.35
	50	–	–	1.07	2.36	–	–	–	–	0.88	1.93	–	–	–	–	–	–	0.55	1.21	0.29	0.65	0.58	1.28
37	18	18	5	0.37	0.80	354	781	0.11	0.24	0.33	0.72	335	739	–	–	0.09	0.19	0.21	0.45	0.14	0.31	0.19	0.42
	88	85	24	1.76	3.87	1706	3762	0.51	1.13	1.57	3.46	1614	3558	–	–	0.42	0.93	0.99	2.18	0.68	1.50	0.92	2.04
38	10	–	–	0.22	0.49	–	–	–	–	0.18	0.39	–	–	–	–	–	–	0.12	0.26	0.06	0.13	0.13	0.28
	40	–	–	0.93	2.05	–	–	–	–	0.73	1.62	–	–	–	–	–	–	0.49	1.07	0.24	0.52	0.53	1.18
39	17	–	–	0.33	0.74	–	–	–	–	0.29	0.63	–	–	–	–	–	–	0.17	0.37	0.10	0.22	0.17	0.37
	70	–	–	1.34	2.96	–	–	–	–	1.15	2.54	–	–	–	–	–	–	0.67	1.47	0.40	0.88	0.67	1.47
40	21	–	–	0.41	0.90	–	–	–	–	0.35	0.78	–	–	–	–	–	–	0.20	0.45	0.12	0.27	0.20	0.45
	72	–	–	1.37	3.03	–	–	–	–	1.18	2.61	–	–	–	–	–	–	0.68	1.50	0.41	0.90	0.68	1.49
41	14	–	–	0.29	0.63	–	–	–	–	0.24	0.53	–	–	–	–	–	–	0.15	0.33	0.08	0.18	0.15	0.34
	57	–	–	1.14	2.52	–	–	–	–	0.95	2.09	–	–	–	–	–	–	0.59	1.29	0.33	0.72	0.61	1.34
42	20	–	–	0.40	0.89	–	–	–	–	0.34	0.75	–	–	–	–	–	–	0.19	0.41	0.10	0.22	0.20	0.43
	60	–	–	1.20	2.65	–	–	–	–	1.01	2.23	–	–	–	–	–	–	0.56	1.23	0.30	0.66	0.59	1.29
43	18	–	–	0.37	0.81	–	–	–	–	0.30	0.66	–	–	–	–	–	–	–	–	–	–	–	–
	61	–	–	1.23	2.70	–	–	–	–	1.00	2.20	–	–	–	–	–	–	–	–	–	–	–	–
44	15	–	–	0.32	0.70	–	–	–	–	0.26	0.58	–	–	–	–	–	–	0.16	0.34	0.09	0.19	0.16	0.36
	56	–	–	1.14	2.52	–	–	–	–	0.95	2.09	–	–	–	–	–	–	0.56	1.24	0.31	0.68	0.59	1.30
45	18	19	–	0.34	0.75	–	–	–	–	0.30	0.66	–	–	–	–	–	–	0.19	0.42	0.12	0.26	0.17	0.37
	68	72	–	1.30	2.86	–	–	–	–	1.14	2.50	–	–	–	–	–	–	0.73	1.61	0.46	1.01	0.65	1.42
46	15	–	–	0.31	0.68	–	–	–	–	0.26	0.58	–	–	–	–	–	–	0.15	0.33	0.09	0.19	0.15	0.34
	65	–	–	1.30	2.86	–	–	–	–	1.11	2.44	–	–	–	–	–	–	0.64	1.40	0.37	0.82	0.64	1.42
47	15	–	–	0.28	0.63	–	–	–	–	0.24	0.54	–	–	–	–	–	–	0.14	0.32	0.09	0.19	0.14	0.31
	67	–	–	1.26	2.78	–	–	–	–	1.08	2.38	–	–	–	–	–	–	0.64	1.41	0.38	0.83	0.62	1.37
48	19	–	–	0.36	0.79	–	–	–	–	0.30	0.67	–	–	–	–	–	–	0.19	0.43	0.12	0.27	0.18	0.40
	69	–	–	1.34	2.95	–	–	–	–	1.13	2.50	–	–	–	–	–	–	0.72	1.59	0.46	1.01	0.68	1.50
49	22	–	–	0.41	0.91	–	–	–	–	0.35	0.76	–	–	–	–	–	–	0.22	0.48	0.12	0.26	0.20	0.43
	74	–	–	1.36	3.00	–	–	–	–	1.14	2.51	–	–	–	–	–	–	0.72	1.58	0.38	0.84	0.65	1.42
50	20	–	–	0.41	0.90	–	–	–	–	0.35	0.78	–	–	–	–	–	–	0.21	0.47	0.13	0.29	0.21	0.46
	69	–	–	1.39	3.06	–	–	–	–	1.20	2.64	–	–	–	–	–	–	0.72	1.58	0.45	0.98	0.71	1.56
51	18	–	–	0.36	0.79	–	–	–	–	0.31	0.68	–	–	–	–	–	–	0.18	0.40	0.11	0.24	0.18	0.39
	70	–	–	1.39	3.07	–	–	–	–	1.20	2.65	–	–	–	–	–	–	0.70	1.54	0.43	0.95	0.69	1.52
52	25	–	–	0.49	1.07	–	–	–	–	0.42	0.91	–	–	–	–	–	–	0.27	0.59	0.17	0.36	0.24	0.53
	68	–	–	1.32	2.90	–	–	–	–	1.13	2.48	–	–	–	–	–	–	0.72	1.59	0.45	0.99	0.66	1.45
53	38	–	–	0.72	1.58	–	–	–	–	0.66	1.45	–	–	–	–	–	–	0.37	0.82	0.23	0.50	0.36	0.79
	71	–	–	1.32	2.91	–	–	–	–	1.21	2.67	–	–	–	–	–	–	0.68	1.51	0.42	0.92	0.66	1.46
54	19	–	–	0.38	0.85	–	–	–	–	0.33	0.73	–	–	–	–	–	–	0.19	0.42	0.11	0.25	0.19	0.42
	66	–	–	1.33	2.92	–	–	–	–	1.13	2.50	–	–	–	–	–	–	0.65	1.43	0.38	0.85	0.65	1.44
55	55	55	46	1.10	2.42	1105	2436	0.84	1.86	0.97	2.13	1042	2297	–	–	0.69	1.53	0.58	1.28	0.38	0.84	0.55	1.22
	78	79	66	1.57	3.46	1579	3481	1.21	2.66	1.38	3.04	1489	3282	–	–	0.99	2.18	0.83	1.83	0.54	1.20	0.79	1.75
56	31	35	27	0.66	1.45	692	1526	0.49	1.09	0.57	1.27	653	1439	–	–	0.41	0.89	0.33	0.74	0.21	0.47	0.32	0.71
	72	80	62	1.51	3.33	1595	3516	1.14	2.51	1.32	2.92	1504	3317	–	–	0.93	2.06	0.77	1.70	0.49	1.08	0.75	1.64
57	47	–	–	0.92	2.03	–	–	–	–	0.78	1.72	–	–	–	–	–	–	–	–	–	–	–	–
	60	–	–	1.18	2.60	–	–	–	–	1.00	2.20	–	–	–	–	–	–	–	–	–	–	–	–
58	17	–	–	0.31	0.69	–	–	–	–	0.27	0.59	–	–	–	–	–	–	0.15	0.33	0.09	0.20	0.15	0.34
	70	–	–	1.34	2.95	–	–	–	–	1.15	2.53	–	–	–	–	–	–	0.64	1.42	0.38	0.83	0.65	1.43
59	22	–	–	0.42	0.92	–	–	–	–	0.36	0.79	–	–	–	–	–	–	0.24	0.53	0.15	0.34	0.23	0.51
	72	–	–	1.35	2.97	–	–	–	–	1.16	2.55	–	–	–	–	–	–	0.77	1.70	0.49	1.08	0.75	1.64
60	24	–	–	0.45	1.00	–	–	–	–	0.39	0.86	–	–	–	–	–	–	0.22	0.49	0.13	0.29	0.22	0.49
	70	–	–	1.34	2.94	–	–	–	–	1.14	2.52	–	–	–	–	–	–	0.65	1.44	0.39	0.85	0.66	1.45

(Continued)

TABLE V-1D SILAGES AND HAYLAGES, MINERAL AND VITAMIN COMPOSITION OF FEEDS, DATA EXPRESSED **AS-FED** AND **MOISTURE-FREE**—*(Continued)*

Entry Number	Feed Name Description	Moisture Basis: A-F (as-fed) or M-F (moisture-free)	Dry Matter	Macro Minerals							Micro Minerals						
				Calcium (Ca)	Phosphorus (P)	Sodium (Na)	Chlorine (Cl)	Magnesium (Mg)	Potassium (K)	Sulfur (S)	Cobalt (Co)	Copper (Cu)	Iodine (I)	Iron (Fe)	Manganese (Mn)	Selenium (Se)	Zinc (Zn)
			%	%	%	%	%	%	%	%	ppm or mg/kg	ppm or mg/kg	ppm or mg/kg	%	ppm or mg/kg	ppm or mg/kg	ppm or mg/kg
	BERMUDAGRASS, COASTAL, SILAGE *Cynodon dactylon*																
32	WILTED	**A-F**	**39**	–	–	–	–	–	–	–	–	–	–	–	–	–	–
		M-F	100	–	–	–	–	–	–	–	–	–	–	–	–	–	–
33	**BERSEEM (EGYPTIAN) CLOVER, SILAGE** *Trifolium alexandrinum*	**A-F**	**24**	–	–	–	–	–	–	–	–	–	–	–	–	–	–
		M-F	100	–	–	–	–	–	–	–	–	–	–	–	–	–	–
34	**BLUEGRASS-CLOVER, WHITE, MOLASSES ADDED, SILAGE** *Poa spp-Trifolium repens*	**A-F**	**35**	–	–	–	–	–	–	–	–	–	–	–	–	–	–
		M-F	100	–	–	–	–	–	–	–	–	–	–	–	–	–	–
	BROOMCORN (HOG MILLET; MILLET, PROSO), SILAGE *Panicum miliaceum*																
35	STOVER WITHOUT HEADS	**A-F**	**27**	–	–	–	–	–	–	–	–	–	–	–	–	–	–
		M-F	100	–	–	–	–	–	–	–	–	–	–	–	–	–	–
36	**CANARYGRASS, REED, SILAGE** *Phalaris arundinacea*	**A-F**	**27**	–	–	–	–	–	–	–	–	–	–	–	–	–	–
		M-F	100	–	–	–	–	–	–	–	–	–	–	–	–	–	–
	CHINESE LESPEDEZA–SEE LESPEDEZA, SERICEA, SILAGE																
	CITRUS, SILAGE *Citrus spp*																
37	PULP	**A-F**	**21**	**0.42**	**0.03**	**0.02**	–	**0.03**	**0.13**	**0.00**	–	–	–	**0.004**	–	–	**3.3**
		M-F	100	2.04	0.15	0.09	–	0.16	0.62	0.02	–	–	–	0.016	–	–	16.0
38	**CLOVER, EGYPTIAN (BERSEEM), SILAGE** *Trifolium alexandrinum*	**A-F**	**24**	–	–	–	–	–	–	–	–	–	–	–	–	–	–
		M-F	100	–	–	–	–	–	–	–	–	–	–	–	–	–	–
39	**CLOVER, LADINO, SILAGE** *Trifolium repens*	**A-F**	**25**	–	–	–	–	–	–	–	–	–	–	–	–	–	–
		M-F	100	–	–	–	–	–	–	–	–	–	–	–	–	–	–
40	**CLOVER, LADINO-GRASS, SILAGE** *Trifolium repens-grass*	**A-F**	**30**	**0.31**	**0.07**	–	–	**0.09**	–	–	–	–	–	–	–	–	–
		M-F	100	1.04	0.23	–	–	0.30	–	–	–	–	–	–	–	–	–
	CLOVER, RED, SILAGE *Trifolium pratense*																
41	ALL ANALYSES	**A-F**	**25**	**0.40**	**0.05**	**0.06**	**0.21**	**0.10**	**0.43**	**0.04**	–	**2.7**	–	**0.009**	**32.0**	–	–
		M-F	100	1.57	0.20	0.23	0.84	0.39	1.72	0.16	–	10.8	–	0.032	127.0	–	–
42	MOLASSES ADDED	**A-F**	**34**	–	–	–	–	–	–	–	–	–	–	–	–	–	–
		M-F	100	–	–	–	–	–	–	–	–	–	–	–	–	–	–
43	**CLOVER, RED-GRASS, SILAGE** *Trifolium pratense-grass*	**A-F**	**30**	**0.33**	**0.06**	–	–	–	–	–	–	–	–	–	–	–	–
		M-F	100	1.11	0.20	–	–	–	–	–	–	–	–	–	–	–	–
44	**CLOVER, SWEET, YELLOW, SILAGE** *Melilotus officinalis*	**A-F**	**28**	**0.38**	**0.06**	–	–	**0.17**	**0.54**	–	–	–	–	–	**6.6**	–	–
		M-F	100	1.36	0.21	–	–	0.62	1.96	–	–	–	–	–	24.0	–	–
	CORN, SILAGE *Zea mays*																
45	ALL ANALYSES	**A-F**	**26**	**0.08**	**0.07**	**0.01**	**0.05**	**0.06**	**0.32**	**0.03**	**0.026**	**2.4**	–	**0.005**	**10.8**	–	**5.5**
		M-F	100	0.31	0.27	0.03	0.18	0.22	1.22	0.12	0.097	9.2	–	0.018	41.1	–	21.2
46	IMMATURE	**A-F**	**24**	**0.12**	**0.07**	–	–	**0.07**	**0.39**	–	–	–	–	**0.012**	–	–	–
		M-F	100	0.52	0.31	–	–	0.31	1.64	–	–	–	–	0.049	–	–	–
47	MILK STAGE	**A-F**	**22**	**0.06**	**0.05**	**0.00**	–	**0.09**	**0.35**	–	–	–	–	–	–	–	–
		M-F	100	0.28	0.24	0.01	–	0.41	1.57	–	–	–	–	–	–	–	–
48	DOUGH STAGE	**A-F**	**27**	**0.07**	**0.05**	–	–	–	–	–	–	–	–	–	–	–	–
		M-F	100	0.27	0.19	–	–	–	–	–	–	–	–	–	–	–	–
49	MATURE	**A-F**	**30**	**0.10**	**0.24**	–	–	**0.05**	–	–	**0.019**	–	–	–	–	–	–
		M-F	100	0.33	0.79	–	–	0.16	–	–	0.060	–	–	–	–	–	–
50	OVERRIPE	**A-F**	**30**	–	–	–	–	–	–	–	–	–	–	–	–	–	–
		M-F	100	–	–	–	–	–	–	–	–	–	–	–	–	–	–
51	LESS THAN 30% DRY MATTER	**A-F**	**54**	**0.11**	**0.12**	**0.01**	–	**0.09**	**0.54**	**0.08**	–	**1.0**	–	**0.008**	**11.1**	–	**14.9**
		M-F	100	0.20	0.23	0.02	–	0.17	1.00	0.14	–	1.8	–	0.015	20.4	–	27.5
52	30-50% DRY MATTER	**A-F**	**37**	**0.08**	**0.09**	**0.00**	–	**0.07**	**0.40**	**0.04**	**0.395**	**2.5**	–	**0.009**	**22.7**	**0.049**	**9.3**
		M-F	100	0.22	0.24	0.01	–	0.18	1.07	0.10	1.072	6.7	–	0.023	61.5	0.133	25.2
53	MORE THAN 50% DRY MATTER	**A-F**	**26**	**0.07**	**0.06**	**0.00**	–	**0.05**	**0.25**	**0.04**	–	–	–	**0.005**	–	–	–
		M-F	100	0.26	0.23	0.01	–	0.18	0.95	0.14	–	–	–	0.018	–	–	–
54	MOLASSES ADDED	**A-F**	**29**	**0.07**	**0.09**	–	–	–	–	–	–	–	–	–	–	–	–
		M-F	100	0.25	0.30	–	–	–	–	–	–	–	–	–	–	–	–
55	EARS	**A-F**	**70**	–	–	–	–	–	–	–	–	–	–	–	–	–	–
		M-F	100	–	–	–	–	–	–	–	–	–	–	–	–	–	–
56	EARS WITH HUSKS	**A-F**	**43**	**0.04**	**0.12**	**0.00**	–	**0.05**	**0.21**	**0.06**	–	–	–	**0.004**	–	–	–
		M-F	100	0.10	0.29	0.01	–	0.12	0.49	0.13	–	–	–	0.008	–	–	–
57	HUSKS (HUSKLAGE)	**A-F**	**78**	–	–	–	–	–	–	–	–	–	–	–	–	–	–
		M-F	100	–	–	–	–	–	–	–	–	–	–	–	–	–	–
58	**CORN, FLINT, SILAGE** *Zea mays, indurata*	**A-F**	**23**	–	–	–	–	–	–	–	–	–	–	–	–	–	–
		M-F	100	–	–	–	–	–	–	–	–	–	–	–	–	–	–
	CORN, SWEET, SILAGE *Zea mays, saccharata*																
59	CANNERY RESIDUE	**A-F**	**31**	**0.10**	**0.24**	**0.01**	–	**0.07**	**0.36**	**0.03**	–	–	–	**0.007**	–	–	–
		M-F	100	0.32	0.77	0.03	–	0.24	1.15	0.11	–	–	–	0.020	–	–	–
	CORN-SORGHUM, SILAGE *Zea mays-Sorghum bicolor*																
60	MATURE	**A-F**	**34**	**0.15**	**0.08**	–	–	–	**0.35**	–	–	–	–	–	–	–	–
		M-F	100	0.45	0.24	–	–	–	1.03	–	–	–	–	–	–	–	–

Entry Number	Fat-Soluble Vitamins: A (1 mg Carotene = 1667 IU Vit A)	Carotene (Provitamin A)	D	E	K	Water-Soluble Vitamins: B–12	Biotin	Choline	Folacin (Folic Acid)	Niacin	Pantothenic Acid (B–3)	(Pyri-doxine) B–6	Ribo-flavin (B–2)	Thiamin (B–1)
	IU/g	ppm or mg/kg	IU/kg	ppm or mg/kg	ppm or mg/kg	ppm or mg/kg	ppm or mg/kg	ppm or mg/kg	ppm or mg/kg	ppm or mg/kg	ppm or mg/kg	ppm or mg/kg	ppm or mg/kg	ppm or mg/kg
32	–	–	–	–	–	–	–	–	–	–	–	–	–	–
	–	–	–	–	–	–	–	–	–	–	–	–	–	–
33	–	–	–	–	–	–	–	–	–	–	–	–	–	–
	–	–	–	–	–	–	–	–	–	–	–	–	–	–
34	–	–	–	–	–	–	–	–	–	–	–	–	–	–
	–	–	–	–	–	–	–	–	–	–	–	–	–	–
35	–	–	–	–	–	–	–	–	–	–	–	–	–	–
	–	–	–	–	–	–	–	–	–	–	–	–	–	–
36	–	–	–	–	–	–	–	–	–	–	–	–	–	–
	–	–	–	–	–	–	–	–	–	–	–	–	–	–
37	–	–	–	–	–	–	–	–	–	–	–	–	–	–
	–	–	–	–	–	–	–	–	–	–	–	–	–	–
38	–	–	–	–	–	–	–	–	–	–	–	–	–	–
	–	–	–	–	–	–	–	–	–	–	–	–	–	–
39	–	–	–	–	–	–	–	–	–	–	–	–	–	–
	–	–	–	–	–	–	–	–	–	–	–	–	–	–
40	**57.3**	**34.4**	–	–	–	–	–	–	–	–	–	–	–	–
	191.7	115.0	–	–	–	–	–	–	–	–	–	–	–	–
41	**36.5**	**21.9**	–	–	–	–	–	–	–	–	–	–	–	–
	144.9	86.9	–	–	–	–	–	–	–	–	–	–	–	–
42	–	–	–	–	–	–	–	–	–	–	–	–	–	–
	–	–	–	–	–	–	–	–	–	–	–	–	–	–
43	–	–	–	–	–	–	–	–	–	–	–	–	–	–
	–	–	–	–	–	–	–	–	–	–	–	–	–	–
44	**31.9**	**19.2**	–	–	–	–	–	–	–	–	–	–	–	–
	115.6	69.3	–	–	–	–	–	–	–	–	–	–	–	–
45	**15.2**	**9.1**	**0**	–	–	–	–	–	–	**11**	–	–	–	–
	58.1	34.9	0	–	–	–	–	–	–	43	–	–	–	–
46	**43.6**	**26.2**	–	–	–	–	–	–	–	–	–	–	–	–
	184.5	110.7	–	–	–	–	–	–	–	–	–	–	–	–
47	–	–	–	–	–	–	–	–	–	–	–	–	–	–
	–	–	–	–	–	–	–	–	–	–	–	–	–	–
48	**29.1**	**17.4**	–	–	–	–	–	–	–	–	–	–	–	–
	108.5	65.1	–	–	–	–	–	–	–	–	–	–	–	–
49	**7.9**	**4.8**	–	–	–	–	–	–	–	–	–	–	–	–
	26.1	15.7	–	–	–	–	–	–	–	–	–	–	–	–
50	–	–	–	–	–	–	–	–	–	–	–	–	–	–
	–	–	–	–	–	–	–	–	–	–	–	–	–	–
51	–	–	–	–	–	–	–	–	–	–	–	–	–	–
	–	–	–	–	–	–	–	–	–	–	–	–	–	–
52	–	–	–	–	–	–	–	–	–	–	–	–	–	–
	–	–	–	–	–	–	–	–	–	–	–	–	–	–
53	–	–	–	–	–	–	–	–	–	–	–	–	–	–
	–	–	–	–	–	–	–	–	–	–	–	–	–	–
54	–	–	–	–	–	–	–	–	–	–	–	–	–	–
	–	–	–	–	–	–	–	–	–	–	–	–	–	–
55	–	–	–	–	–	–	–	–	–	–	–	–	–	–
	–	–	–	–	–	–	–	–	–	–	–	–	–	–
56	–	–	–	–	–	–	–	–	–	–	–	–	–	–
	–	–	–	–	–	–	–	–	–	–	–	–	–	–
57	–	–	–	–	–	–	–	–	–	–	–	–	–	–
	–	–	–	–	–	–	–	–	–	–	–	–	–	–
58	–	–	–	–	–	–	–	–	–	–	–	–	–	–
	–	–	–	–	–	–	–	–	–	–	–	–	–	–
59	**6.9**	**4.2**	–	–	–	–	–	–	–	–	–	–	–	–
	22.3	13.4	–	–	–	–	–	–	–	–	–	–	–	–
60	**3.1**	**1.9**	–	–	–	–	–	–	–	–	–	–	–	–
	9.2	5.5	–	–	–	–	–	–	–	–	–	–	–	–

(Continued)

TABLE V-1D SILAGES AND HAYLAGES, COMPOSITION OF FEEDS, DATA EXPRESSED **AS-FED** AND **MOISTURE-FREE**—*(Continued)*

Entry Number	Feed Name Description	International Feed Number	Moisture Basis: A-F (as-fed) or M-F (moisture-free)	Proximate Analysis: Dry Matter	Ash	Crude Fiber	Neutral Det. Fib. (NDF)	Acid Det. Fib. (ADF)	Lignin	Ether Extract (Fat)	N-Free Extract	Crude Protein	Digestible Protein: Ruminant	Swine	Horse
				%	%	%	%	%	%	%	%	%	%	%	%
61	**CORN-SOYBEAN, SILAGE** *Zea mays-Glycine max*	3-03-015	A-F	**27**	**1.8**	**7.0**	–	–	–	**0.9**	**14.6**	**2.7**	**1.7**	–	–
			M-F	100	6.6	25.7	–	–	–	3.5	54.1	10.1	6.2	–	–
	COWPEA, COMMON, SILAGE *Vigna sinensis*														
62	WILTED	3-01-659	A-F	**30**	**3.3**	**8.6**	–	–	–	**1.2**	**12.6**	**4.6**	**2.6**	–	–
			M-F	100	11.0	28.3	–	–	–	4.0	41.7	15.0	8.6	–	–
	DARSO SORGHUM–SEE SORGHUM, SILAGE														
	DURRA SORGHUM–SEE SORGHUM, SILAGE														
63	**FESCUE, MEADOW, SILAGE** *Festuca elatior*	3-01-925	A-F	**33**	**3.4**	**9.5**	–	–	–	**1.2**	**14.7**	**4.3**	**2.6**	–	–
			M-F	100	10.3	28.8	–	–	–	3.5	44.5	12.9	7.7	–	–
	FETERITA SORGHUM–SEE SORGHUM, SILAGE														
	GRASS, SILAGE														
64	IMMATURE	3-02-217	A-F	**28**	**2.5**	**9.0**	–	–	–	**1.1**	**11.9**	**3.7**	**2.3**	–	–
			M-F	100	8.7	31.9	–	–	–	4.0	42.2	13.2	8.3	–	–
65	EARLY BLOOM	3-02-218	A-F	**23**	**2.1**	**7.3**	–	–	–	**0.6**	**10.6**	**2.8**	**1.7**	–	–
			M-F	100	9.1	31.0	–	–	–	2.7	45.1	12.1	7.3	–	–
66	STEM CURED	3-02-219	A-F	**21**	**2.3**	**5.4**	–	–	–	**1.1**	**9.6**	**2.7**	**1.7**	–	–
			M-F	100	11.0	25.5	–	–	–	5.0	45.6	12.9	8.0	–	–
67	DRIED WHEY ADDED, LACTIC ACID BACTERIA ADDED	3-02-231	A-F	**24**	**2.3**	**6.1**	–	–	–	**1.0**	**11.6**	**3.2**	**2.3**	–	–
			M-F	100	9.5	25.0	–	–	–	4.3	48.0	13.2	9.4	–	–
68	MOLASSES ADDED	3-02-261	A-F	**27**	**3.8**	**6.4**	–	–	–	**1.4**	**10.6**	**4.3**	**3.1**	–	–
			M-F	100	14.4	24.1	–	–	–	5.3	40.0	16.1	11.5	–	–
	GRASS-LEGUME, SILAGE														
69	ALL ANALYSES	3-02-303	A-F	**31**	**2.5**	**10.1**	**18.1**	**12.4**	**3.1**	**1.1**	**13.6**	**3.7**	**1.9**	–	–
			M-F	100	8.0	32.6	58.4	40.0	10.0	3.6	43.8	11.9	6.1	–	–
70	MOSTLY GRASS	3-08-599	A-F	**28**	**2.5**	**9.7**	–	–	–	**1.1**	**11.1**	**3.2**	–	–	**1.9**
			M-F	100	9.1	35.1	–	–	–	4.0	40.2	11.6	–	–	6.8
71	MOSTLY LEGUME	3-08-598	A-F	**26**	**1.7**	**8.6**	–	–	–	**0.9**	**10.8**	**3.6**	–	–	**2.3**
			M-F	100	6.6	33.6	–	–	–	3.5	42.2	14.1	–	–	9.0
72	BARLEY GRAIN ADDED	3-02-305	A-F	**34**	**2.2**	**8.6**	–	–	–	**1.4**	**16.7**	**5.2**	**3.4**	–	–
			M-F	100	6.4	25.4	–	–	–	4.0	49.0	15.2	10.0	–	–
73	CITRUS PULP FINES ADDED	3-02-267	A-F	**26**	**1.6**	**8.9**	–	–	–	**1.3**	**10.7**	**3.7**	**2.4**	–	–
			M-F	100	6.0	34.1	–	–	–	4.8	40.9	14.2	9.2	–	–
74	MOLASSES ADDED	3-02-309	A-F	**28**	**2.0**	**9.1**	–	–	–	**1.1**	**12.8**	**3.4**	**2.0**	–	–
			M-F	100	7.0	32.0	–	–	–	4.0	45.2	11.8	7.1	–	–
75	WILTED	3-02-304	A-F	**35**	**3.2**	**10.8**	–	–	–	**1.2**	**14.7**	**4.7**	**2.6**	–	–
			M-F	100	9.3	31.1	–	–	–	3.5	42.5	13.5	7.5	–	–
	HEGARI SORGHUM–SEE SORGHUM, SILAGE														
	HOG MILLET (BROOMCORN; MILLET, PROSO), SILAGE *Panicum miliaceum*														
76	STOVER WITHOUT HEADS	3-08-460	A-F	**27**	**2.2**	**9.5**	–	–	–	**0.4**	**13.3**	**1.3**	**0.5**	–	–
			M-F	100	8.2	35.6	–	–	–	1.5	49.8	4.9	1.8	–	–
77	**HORSE BEAN, SILAGE** *Vicia faba, equina*	3-02-406	A-F	**21**	**1.4**	**5.7**	–	–	–	**0.5**	**10.3**	**3.3**	**2.2**	–	–
			M-F	100	6.6	26.9	–	–	–	2.4	48.5	15.6	10.4	–	–
	JOHNSONGRASS SORGHUM–SEE SORGHUM, SILAGE														
	KAFIR SORGHUM–SEE SORGHUM, SILAGE														
	LADINO CLOVER–SEE CLOVER, LADINO, SILAGE														
	LEGUME-GRASS–SEE GRASS-LEGUME, SILAGE														
78	**LESPEDEZA, COMMON-KOREAN, SILAGE** *Lespedeza striata-stipulacea*	3-08-455	A-F	**30**	**1.8**	**9.5**	–	–	–	**0.8**	**13.8**	**4.3**	**2.8**	–	–
			M-F	100	6.0	31.5	–	–	–	2.6	45.7	14.2	9.2	–	–
79	**LESPEDEZA, SERICEA (CHINESE), SILAGE** *Lespedeza cuneata*	3-02-614	A-F	**30**	**1.7**	**9.5**	–	–	–	**0.9**	**14.0**	**4.3**	**2.7**	–	–
			M-F	100	5.5	31.3	–	–	–	3.1	46.1	14.0	9.0	–	–
	LUCERNE–SEE ALFALFA														
	MAIZE–SEE CORN														
	MANGEL BEET, SILAGE *Beta vulgaris, macrorrhiza*														
80	ROOTS	4-00-636	A-F	**10**	**1.9**	**0.8**	–	–	–	**0.1**	**6.4**	**1.0**	**0.5**	**0.8**	**0.7**
			M-F	100	18.8	7.6	–	–	–	1.2	62.5	9.9	4.5	7.5	6.9
81	TOPS WITH CROWNS	3-00-635	A-F	**13**	**2.8**	**1.9**	–	–	–	**0.6**	**5.7**	**1.8**	**1.2**	–	–
			M-F	100	22.1	14.5	–	–	–	4.6	44.9	13.9	9.2	–	–
82	**MEADOW FESCUE, SILAGE** *Festuca elatior*	3-01-925	A-F	**33**	**3.4**	**9.5**	–	–	–	**1.2**	**14.7**	**4.3**	**2.6**	–	–
			M-F	100	10.3	28.8	–	–	–	3.5	44.5	12.9	7.7	–	–
83	**MILLET, JAPANESE (BARNYARD GRASS)-SOYBEAN, SILAGE** *Echinochloa crusgalli-Glycine max*	3-26-066	A-F	**21**	**2.9**	**7.3**	–	–	–	**1.1**	**6.9**	**2.9**	**1.7**	–	–
			M-F	100	13.7	34.6	–	–	–	5.2	32.8	13.7	8.1	–	–

SILAGES AND HAYLAGES

Entry Number	TDN			Digestible Energy						Metabolizable Energy								Net Energy					
	Ruminant	Swine	Horse	Ruminant		Swine		Horse		Ruminant		Swine		Poultry ME_n		Horse		Ruminant NE_m		Ruminant NE_g		Lactating Cows NE_{lc}	
	%	%	%	Mcal		kcal		Mcal		Mcal		kcal		kcal		Mcal		Mcal		Mcal		Mcal	
				lb	*kg*	lb	*kg*	lb	*kg*	lb	*kg*	lb	*kg*	lb	*kg*	lb	*kg*	lb	*kg*	lb	*kg*	lb	*kg*
61	**19**	–	–	**0.37**	***0.82***	–	–	–	–	**0.32**	***0.70***	–	–	–	–	–	–	**0.21**	***0.46***	**0.13**	***0.29***	**0.20**	***0.44***
	69	–	–	1.37	*3.02*	–	–	–	–	1.18	*2.60*	–	–	–	–	–	–	0.77	*1.69*	0.49	*1.08*	0.75	*1.64*
62	**18**	–	–	**0.36**	***0.79***	–	–	–	–	**0.30**	***0.67***	–	–	–	–	–	–	**0.18**	***0.39***	**0.10**	***0.22***	**0.18**	***0.41***
	59	–	–	1.19	*2.62*	–	–	–	–	1.00	*2.20*	–	–	–	–	–	–	0.59	*1.29*	0.33	*0.72*	0.61	*1.34*
63	**20**	–	–	**0.40**	***0.88***	–	–	–	–	**0.33**	***0.74***	–	–	–	–	–	–	**0.19**	***0.42***	**0.10**	***0.23***	**0.20**	***0.44***
	62	–	–	1.21	*2.66*	–	–	–	–	1.01	*2.23*	–	–	–	–	–	–	0.57	*1.27*	0.32	*0.70*	0.60	*1.32*
64	**17**	–	–	**0.34**	***0.75***	–	–	–	–	**0.29**	***0.63***	–	–	–	–	–	–	**0.17**	***0.37***	**0.09**	***0.21***	**0.17**	***0.38***
	60	–	–	1.20	*2.65*	–	–	–	–	1.01	*2.22*	–	–	–	–	–	–	0.59	*1.30*	0.33	*0.73*	0.61	*1.34*
65	**14**	–	–	**0.27**	***0.61***	–	–	–	–	**0.23**	***0.51***	–	–	–	–	–	–	**0.13**	***0.28***	**0.07**	***0.15***	**0.14**	***0.30***
	59	–	–	1.17	*2.58*	–	–	–	–	0.98	*2.16*	–	–	–	–	–	–	0.55	*1.21*	0.29	*0.65*	0.58	*1.28*
66	**11**	–	–	**0.25**	***0.54***	–	–	–	–	**0.21**	***0.45***	–	–	–	–	–	–	**0.13**	***0.29***	**0.08**	***0.17***	**0.14**	***0.30***
	55	–	–	1.17	*2.59*	–	–	–	–	0.98	*2.16*	–	–	–	–	–	–	0.64	*1.40*	0.37	*0.82*	0.64	*1.42*
67	**18**	–	–	**0.33**	***0.74***	–	–	–	–	**0.29**	***0.64***	–	–	–	–	–	–	**0.16**	***0.34***	**0.09**	***0.20***	**0.16**	***0.35***
	75	–	–	1.38	*3.04*	–	–	–	–	1.19	*2.62*	–	–	–	–	–	–	0.64	*1.41*	0.38	*0.83*	0.65	*1.43*
68	**18**	–	–	**0.34**	***0.75***	–	–	–	–	**0.29**	***0.64***	–	–	–	–	–	–	**0.16**	***0.36***	**0.09**	***0.20***	**0.17**	***0.37***
	68	–	–	1.29	*2.83*	–	–	–	–	1.09	*2.41*	–	–	–	–	–	–	0.61	*1.34*	0.35	*0.76*	0.62	*1.37*
69	**18**	–	–	**0.38**	***0.84***	–	–	–	–	**0.31**	***0.68***	–	–	–	–	–	–	**0.19**	***0.42***	**0.11**	***0.24***	**0.19**	***0.41***
	57	–	–	1.23	*2.71*	–	–	–	–	0.99	*2.18*	–	–	–	–	–	–	0.62	*1.36*	0.36	*0.78*	0.60	*1.33*
70	**16**	–	–	**0.31**	***0.69***	–	–	–	–	**0.26**	***0.56***	–	–	–	–	–	–	**0.15**	***0.33***	**0.07**	***0.15***	**0.16**	***0.35***
	57	–	–	1.13	*2.49*	–	–	–	–	0.93	*2.04*	–	–	–	–	–	–	0.55	*1.21*	0.24	*0.53*	0.57	*1.25*
71	**16**	–	–	**0.31**	***0.68***	–	–	–	–	**0.25**	***0.56***	–	–	–	–	–	–	**0.15**	***0.33***	**0.08**	***0.17***	**0.15**	***0.35***
	61	–	–	1.21	*2.67*	–	–	–	–	0.99	*2.19*	–	–	–	–	–	–	0.59	*1.30*	0.31	*0.68*	0.60	*1.36*
72	**23**	–	–	**0.45**	***1.00***	–	–	–	–	**0.39**	***0.86***	–	–	–	–	–	–	**0.23**	***0.51***	**0.14**	***0.31***	**0.23**	***0.50***
	67	–	–	1.34	*2.94*	–	–	–	–	1.14	*2.52*	–	–	–	–	–	–	0.68	*1.49*	0.41	*0.90*	0.67	*1.48*
73	**16**	–	–	**0.33**	***0.72***	–	–	–	–	**0.28**	***0.61***	–	–	–	–	–	–	**0.17**	***0.37***	**0.10**	***0.22***	**0.17**	***0.37***
	62	–	–	1.24	*2.74*	–	–	–	–	1.05	*2.32*	–	–	–	–	–	–	0.64	*1.41*	0.37	*0.82*	0.65	*1.42*
74	**16**	–	–	**0.33**	***0.72***	–	–	–	–	**0.27**	***0.60***	–	–	–	–	–	–	**0.16**	***0.34***	**0.08**	***0.18***	**0.17**	***0.36***
	58	–	–	1.16	*2.55*	–	–	–	–	0.96	*2.12*	–	–	–	–	–	–	0.55	*1.21*	0.29	*0.64*	0.58	*1.28*
75	**20**	–	–	**0.40**	***0.88***	–	–	–	–	**0.33**	***0.73***	–	–	–	–	–	–	**0.20**	***0.44***	**0.11**	***0.24***	**0.21**	***0.45***
	57	–	–	1.16	*2.55*	–	–	–	–	0.96	*2.12*	–	–	–	–	–	–	0.57	*1.26*	0.31	*0.69*	0.60	*1.32*
76	**13**	–	–	**0.27**	***0.60***	–	–	–	–	**0.22**	***0.49***	–	–	–	–	–	–	**0.13**	***0.28***	**0.06**	***0.13***	**0.14**	***0.31***
	49	–	–	1.02	*2.25*	–	–	–	–	0.83	*1.82*	–	–	–	–	–	–	0.48	*1.06*	0.23	*0.50*	0.53	*1.17*
77	**12**	–	–	**0.25**	***0.56***	–	–	–	–	**0.21**	***0.47***	–	–	–	–	–	–	**0.13**	***0.29***	**0.07**	***0.16***	**0.13**	***0.29***
	58	–	–	1.19	*2.63*	–	–	–	–	1.00	*2.21*	–	–	–	–	–	–	0.61	*1.35*	0.35	*0.77*	0.63	*1.38*
78	**15**	–	–	**0.33**	***0.73***	–	–	–	–	**0.27**	***0.60***	–	–	–	–	–	–	**0.18**	***0.40***	**0.10**	***0.22***	**0.19**	***0.41***
	50	–	–	1.10	*2.42*	–	–	–	–	0.91	*2.00*	–	–	–	–	–	–	0.59	*1.31*	0.33	*0.74*	0.61	*1.35*
79	**19**	–	–	**0.38**	***0.84***	–	–	–	–	**0.32**	***0.71***	–	–	–	–	–	–	**0.19**	***0.41***	**0.11**	***0.24***	**0.19**	***0.42***
	62	–	–	1.25	*2.75*	–	–	–	–	1.06	*2.33*	–	–	–	–	–	–	0.62	*1.36*	0.35	*0.78*	0.63	*1.38*
80	**5**	**8**	**6**	**0.12**	***0.27***	**167**	***367***	**0.11**	***0.23***	**0.10**	***0.23***	**157**	***347***	–	–	**0.09**	***0.19***	**0.08**	***0.17***	**0.05**	***0.10***	**0.07**	***0.16***
	51	81	54	1.20	*2.64*	1618	*3567*	1.02	*2.24*	1.00	*2.21*	1527	*3367*	–	–	0.84	*1.84*	0.73	*1.60*	0.45	*1.00*	0.71	*1.57*
81	**7**	–	–	**0.14**	***0.32***	–	–	–	–	**0.12**	***0.26***	–	–	–	–	–	–	**0.07**	***0.15***	**0.04**	***0.08***	**0.07**	***0.16***
	56	–	–	1.13	*2.49*	–	–	–	–	0.94	*2.06*	–	–	–	–	–	–	0.55	*1.21*	0.29	*0.64*	0.58	*1.27*
82	**20**	–	–	**0.40**	***0.88***	–	–	–	–	**0.33**	***0.74***	–	–	–	–	–	–	**0.19**	***0.42***	**0.10**	***0.23***	**0.20**	***0.44***
	62	–	–	1.21	*2.66*	–	–	–	–	1.01	*2.23*	–	–	–	–	–	–	0.57	*1.27*	0.32	*0.70*	0.60	*1.32*
83	–	–	–	–	–	–	–	–	–	–	–	–	–	–	–	–	–	–	–	–	–	–	–
	–	–	–	–	–	–	–	–	–	–	–	–	–	–	–	–	–	–	–	–	–	–	–

(Continued)

SILAGES AND HAYLAGES

TABLE V-1D SILAGES AND HAYLAGES, MINERAL AND VITAMIN COMPOSITION OF FEEDS, DATA EXPRESSED **AS-FED** AND **MOISTURE-FREE**—*(Continued)*

Entry Number	Feed Name Description	Moisture Basis: A-F (as-fed) or M-F (moisture-free)	Dry Matter	Macro Minerals							Micro Minerals						
				Calcium (Ca)	Phosphorus (P)	Sodium (Na)	Chlorine (Cl)	Magnesium (Mg)	Potassium (K)	Sulfur (S)	Cobalt (Co)	Copper (Cu)	Iodine (I)	Iron (Fe)	Manganese (Mn)	Selenium (Se)	Zinc (Zn)
			%	%	%	%	%	%	%	%	ppm or mg/kg	ppm or mg/kg	ppm or mg/kg	%	ppm or mg/kg	ppm or mg/kg	ppm or mg/kg
61	**CORN-SOYBEAN, SILAGE** *Zea mays-Glycine max*	**A-F**	**27**	**0.19**	**0.09**	**0.01**	**–**	**0.08**	**0.29**	**0.05**	**–**	**2.2**	**–**	**0.012**	**12.4**	**–**	**7.6**
		M-F	100	0.70	0.32	0.05	–	0.30	1.06	0.19	–	8.0	–	0.042	46.0	–	28.0
	COWPEA, COMMON, SILAGE *Vigna sinensis*																
62	WILTED	**A-F**	**30**	**0.49**	**0.10**	**0.07**	**0.05**	**0.13**	**0.89**	**0.01**	**–**	**–**	**–**	**0.030**	**–**	**–**	**–**
		M-F	100	1.60	0.33	0.23	0.17	0.43	2.93	0.03	–	–	–	0.097	–	–	–
	DARSO SORGHUM–SEE SORGHUM, SILAGE																
	DURRA SORGHUM–SEE SORGHUM, SILAGE																
63	**FESCUE, MEADOW, SILAGE** *Festuca elatior*	**A-F**	**33**	**–**	**–**	**–**	**–**	**–**	**–**	**–**	**–**	**–**	**–**	**–**	**–**	**–**	**–**
		M-F	100	–	–	–	–	–	–	–	–	–	–	–	–	–	–
	FETERITA SORGHUM–SEE SORGHUM, SILAGE																
	GRASS, SILAGE																
64	IMMATURE	**A-F**	**28**	**–**	**–**	**–**	**–**	**–**	**–**	**–**	**–**	**–**	**–**	**–**	**–**	**–**	**–**
		M-F	100	–	–	–	–	–	–	–	–	–	–	–	–	–	–
65	EARLY BLOOM	**A-F**	**23**	**–**	**–**	**–**	**–**	**–**	**–**	**–**	**–**	**–**	**–**	**–**	**–**	**–**	**–**
		M-F	100	–	–	–	–	–	–	–	–	–	–	–	–	–	–
66	STEM CURED	**A-F**	**21**	**–**	**–**	**–**	**–**	**–**	**–**	**–**	**–**	**–**	**–**	**–**	**–**	**–**	**–**
		M-F	100	–	–	–	–	–	–	–	–	–	–	–	–	–	–
67	DRIED WHEY ADDED, LACTIC ACID BACTERIA ADDED	**A-F**	**24**	**–**	**–**	**–**	**–**	**–**	**–**	**–**	**–**	**–**	**–**	**–**	**–**	**–**	**–**
		M-F	100	–	–	–	–	–	–	–	–	–	–	–	–	–	–
68	MOLASSES ADDED	**A-F**	**27**	**0.28**	**0.07**	**–**	**–**	**–**	**–**	**–**	**–**	**–**	**–**	**–**	**–**	**–**	**–**
		M-F	100	1.04	0.28	–	–	–	–	–	–	–	–	–	–	–	–
	GRASS-LEGUME, SILAGE																
69	ALL ANALYSES	**A-F**	**31**	**0.26**	**0.08**	**0.02**	**0.33**	**0.08**	**0.57**	**0.17**	**0.040**	**1.9**	**–**	**0.013**	**18.0**	**–**	**8.7**
		M-F	100	0.84	0.26	0.07	1.06	0.25	1.83	0.54	0.126	6.0	–	0.041	58.1	–	28.0
70	MOSTLY GRASS	**A-F**	**28**	**–**	**–**	**–**	**–**	**–**	**–**	**–**	**–**	**–**	**–**	**–**	**–**	**–**	**–**
		M-F	100	–	–	–	–	–	–	–	–	–	–	–	–	–	–
71	MOSTLY LEGUME	**A-F**	**26**	**–**	**–**	**–**	**–**	**–**	**–**	**–**	**–**	**–**	**–**	**–**	**–**	**–**	**–**
		M-F	100	–	–	–	–	–	–	–	–	–	–	–	–	–	–
72	BARLEY GRAIN ADDED	**A-F**	**34**	**0.26**	**0.12**	**–**	**–**	**–**	**–**	**–**	**–**	**–**	**–**	**–**	**–**	**–**	**–**
		M-F	100	0.75	0.35	–	–	–	–	–	–	–	–	–	–	–	–
73	CITRUS PULP FINES ADDED	**A-F**	**26**	**–**	**–**	**–**	**–**	**–**	**–**	**–**	**–**	**–**	**–**	**–**	**–**	**–**	**–**
		M-F	100	–	–	–	–	–	–	–	–	–	–	–	–	–	–
74	MOLASSES ADDED	**A-F**	**28**	**0.30**	**0.10**	**0.04**	**–**	**0.09**	**0.55**	**0.07**	**–**	**–**	**–**	**0.015**	**–**	**–**	**–**
		M-F	100	1.07	0.34	0.13	–	0.32	1.92	0.24	–	–	–	0.053	–	–	–
75	WILTED	**A-F**	**35**	**0.27**	**0.10**	**–**	**–**	**–**	**–**	**–**	**–**	**–**	**–**	**–**	**–**	**–**	**–**
		M-F	100	0.79	0.28	–	–	–	–	–	–	–	–	–	–	–	–
	HEGARI SORGHUM–SEE SORGHUM, SILAGE																
	HOG MILLET (BROOMCORN; MILLET, PROSO), SILAGE *Panicum miliaceum*																
76	STOVER WITHOUT HEADS	**A-F**	**27**	**–**	**–**	**–**	**–**	**–**	**–**	**–**	**–**	**–**	**–**	**–**	**–**	**–**	**–**
		M-F	100	–	–	–	–	–	–	–	–	–	–	–	–	–	–
77	**HORSE BEAN, SILAGE** *Vicia faba, equina*	**A-F**	**21**	**0.19**	**0.06**	**–**	**–**	**–**	**0.44**	**–**	**–**	**–**	**–**	**–**	**–**	**–**	**–**
		M-F	100	0.90	0.28	–	–	–	2.08	–	–	–	–	–	–	–	–
	JOHNSONGRASS SORGHUM–SEE SORGHUM, SILAGE																
	KAFIR SORGHUM–SEE SORGHUM, SILAGE																
	LADINO CLOVER–SEE CLOVER, LADINO, SILAGE																
	LEGUME-GRASS–SEE GRASS-LEGUME, SILAGE																
78	**LESPEDEZA, COMMON-KOREAN, SILAGE** *Lespedeza striata-stipulacea*	**A-F**	**30**	**–**	**–**	**–**	**–**	**–**	**–**	**–**	**–**	**–**	**–**	**–**	**–**	**–**	**–**
		M-F	100	–	–	–	–	–	–	–	–	–	–	–	–	–	–
79	**LESPEDEZA, SERICEA (CHINESE), SILAGE** *Lespedeza cuneata*	**A-F**	**30**	**–**	**–**	**–**	**–**	**–**	**–**	**–**	**–**	**–**	**–**	**–**	**–**	**–**	**–**
		M-F	100	–	–	–	–	–	–	–	–	–	–	–	–	–	–
	LUCERNE–SEE ALFALFA																
	MAIZE–SEE CORN																
	MANGEL BEET, SILAGE *Beta vulgaris, macrorrhiza*																
80	ROOTS	**A-F**	**10**	**–**	**–**	**–**	**–**	**–**	**–**	**–**	**–**	**–**	**–**	**–**	**–**	**–**	**–**
		M-F	100	–	–	–	–	–	–	–	–	–	–	–	–	–	–
81	TOPS WITH CROWNS	**A-F**	**13**	**–**	**–**	**–**	**–**	**–**	**–**	**–**	**–**	**–**	**–**	**–**	**–**	**–**	**–**
		M-F	100	–	–	–	–	–	–	–	–	–	–	–	–	–	–
82	**MEADOW FESCUE, SILAGE** *Festuca elatior*	**A-F**	**33**	**–**	**–**	**–**	**–**	**–**	**–**	**–**	**–**	**–**	**–**	**–**	**–**	**–**	**–**
		M-F	100	–	–	–	–	–	–	–	–	–	–	–	–	–	–
83	**MILLET, JAPANESE (BARNYARD GRASS)-SOYBEAN, SILAGE** *Echinochloa crusgalli-Glycine max*	**A-F**	**21**	**–**	**–**	**–**	**–**	**–**	**–**	**–**	**–**	**–**	**–**	**–**	**–**	**–**	**–**
		M-F	100	–	–	–	–	–	–	–	–	–	–	–	–	–	–

Entry Number	Fat-Soluble Vitamins A (1 mg Carotene = 1667 IU Vit A)	Carotene (Provitamin A)	D	E	K	Water-Soluble Vitamins B–12	Biotin	Choline	Folacin (Folic Acid)	Niacin	Pantothenic Acid (B–3)	(Pyri-doxine) B–6	Ribo-flavin (B–2)	Thiamin (B–1)
	IU/g	ppm or mg/kg	IU/kg	ppm or mg/kg	ppm or mg/kg	ppm or mg/kg	ppm or mg/kg	ppm or mg/kg	ppm or mg/kg	ppm or mg/kg	ppm or mg/kg	ppm or mg/kg	ppm or mg/kg	ppm or mg/kg
61	**89.0**	**53.4**	–	–	–	–	–	–	–	–	–	–	–	–
	329.3	197.5	–	–	–	–	–	–	–	–	–	–	–	–
62	–	–	–	–	–	–	–	–	–	–	–	–	–	–
	–	–	–	–	–	–	–	–	–	–	–	–	–	–
63	–	–	–	–	–	–	–	–	–	–	–	–	–	–
	–	–	–	–	–	–	–	–	–	–	–	–	–	–
64	–	–	–	–	–	–	–	–	–	–	–	–	–	–
	–	–	–	–	–	–	–	–	–	–	–	–	–	–
65	–	–	–	–	–	–	–	–	–	–	–	–	–	–
	–	–	–	–	–	–	–	–	–	–	–	–	–	–
66	–	–	–	–	–	–	–	–	–	–	–	–	–	–
	–	–	–	–	–	–	–	–	–	–	–	–	–	–
67	–	–	–	–	–	–	–	–	–	–	–	–	–	–
	–	–	–	–	–	–	–	–	–	–	–	–	–	–
68	–	–	–	–	–	–	–	–	–	–	–	–	–	–
	–	–	–	–	–	–	–	–	–	–	–	–	–	–
69	**102.4**	**61.4**	**0**	–	–	–	–	–	–	**14**	–	–	–	–
	329.5	197.6	0	–	–	–	–	–	–	46	–	–	–	–
70	**76.1**	**45.6**	–	–	–	–	–	–	–	**13**	–	–	–	–
	275.6	165.3	–	–	–	–	–	–	–	46	–	–	–	–
71	**62.8**	**37.7**	–	–	–	–	–	–	–	–	–	–	–	–
	245.5	147.3	–	–	–	–	–	–	–	–	–	–	–	–
72	–	–	–	–	–	–	–	–	–	–	–	–	–	–
	–	–	–	–	–	–	–	–	–	–	–	–	–	–
73	–	–	–	–	–	–	–	–	–	–	–	–	–	–
	–	–	–	–	–	–	–	–	–	–	–	–	–	–
74	–	–	–	–	–	–	–	–	–	–	–	–	–	–
	–	–	–	–	–	–	–	–	–	–	–	–	–	–
75	**21.1**	**12.7**	–	–	–	–	–	–	–	–	–	–	–	–
	61.1	36.6	–	–	–	–	–	–	–	–	–	–	–	–
76	–	–	–	–	–	–	–	–	–	–	–	–	–	–
	–	–	–	–	–	–	–	–	–	–	–	–	–	–
77	–	–	–	–	–	–	–	–	–	–	–	–	–	–
	–	–	–	–	–	–	–	–	–	–	–	–	–	–
78	–	–	–	–	–	–	–	–	–	–	–	–	–	–
	–	–	–	–	–	–	–	–	–	–	–	–	–	–
79	–	–	–	–	–	–	–	–	–	–	–	–	–	–
	–	–	–	–	–	–	–	–	–	–	–	–	–	–
80	–	–	–	–	–	–	–	–	–	–	–	–	–	–
	–	–	–	–	–	–	–	–	–	–	–	–	–	–
81	–	–	–	–	–	–	–	–	–	–	–	–	–	–
	–	–	–	–	–	–	–	–	–	–	–	–	–	–
82	–	–	–	–	–	–	–	–	–	–	–	–	–	–
	–	–	–	–	–	–	–	–	–	–	–	–	–	–
83	–	–	–	–	–	–	–	–	–	–	–	–	–	–
	–	–	–	–	–	–	–	–	–	–	–	–	–	–

(Continued)

TABLE V-1D SILAGES AND HAYLAGES, COMPOSITION OF FEEDS, DATA EXPRESSED **AS-FED** AND **MOISTURE-FREE**—*(Continued)*

Entry Number	Feed Name Description	International Feed Number	Moisture Basis: A-F (as-fed) or M-F (moisture-free)	Proximate Analysis: Dry Matter	Ash	Crude Fiber	Neutral Det. Fib. (NDF)	Acid Det. Fib. (ADF)	Lignin	Ether Extract (Fat)	N-Free Extract	Crude Protein	Digestible Protein: Ruminant	Swine	Horse
				%	%	%	%	%	%	%	%	%	%	%	%
	MILLET, PROSO (BROOMCORN; HOG MILLET), SILAGE *Panicum miliaceum*														
84	STOVER WITHOUT HEADS	3-08-460	A-F	27	2.2	9.5	–	–	–	0.4	13.3	1.3	0.5	–	–
			M-F	100	8.2	35.6	–	–	–	1.5	49.8	4.9	1.8	–	–
	MILO SORGHUM–SEE SORGHUM, SILAGE														
	MUNG BEAN–SEE BEAN, MUNG, SILAGE														
	NAPIERGRASS, SILAGE *Pennisetum purpureum*														
85	IMMATURE	3-08-463	A-F	20	2.6	7.3	–	–	–	0.5	8.4	1.1	0.3	–	–
			M-F	100	13.1	36.7	–	–	–	2.5	42.2	5.5	1.6	–	–
86	MATURE	3-03-169	A-F	26	2.3	10.4	–	–	–	0.6	11.1	1.1	0.1	–	–
			M-F	100	9.1	40.9	–	–	–	2.2	43.5	4.3	0.4	–	–
	OATS, SILAGE *Avena sativa*														
87	DOUGH STAGE	3-03-296	A-F	35	2.4	11.6	–	–	–	1.4	16.1	3.5	2.0	–	–
			M-F	100	6.9	33.0	–	–	–	4.1	46.0	10.0	5.6	–	–
88	MOLASSES ADDED	3-03-300	A-F	33	2.5	10.4	–	–	–	1.2	15.8	2.9	1.5	–	–
			M-F	100	7.5	31.7	–	–	–	3.8	48.3	8.8	4.6	–	–
89	**OATS-COWPEA, SILAGE** *Avena sativa-Vigna* spp	3-03-397	A-F	27	3.0	7.8	–	–	–	1.1	11.1	4.0	2.6	–	–
			M-F	100	11.0	28.8	–	–	–	4.2	41.2	14.8	9.7	–	–
90	**OATS-PEA, SILAGE** *Avena sativa-Pisum* spp	3-03-402	A-F	26	2.5	8.7	–	11.6	1.2	0.9	10.9	2.8	1.5	–	–
			M-F	100	9.7	33.7	–	45.0	4.6	3.4	42.2	10.9	5.9	–	–
	OATS-VETCH, SILAGE *Avena sativa-Vicia* spp														
91	MILK STAGE	3-03-408	A-F	27	2.2	8.0	–	–	–	1.2	12.4	3.4	2.1	–	–
			M-F	100	8.1	29.4	–	–	–	4.3	45.6	12.6	7.8	–	–
92	**ORCHARDGRASS, SILAGE** *Dactylis glomerata*	3-03-457	A-F	30	2.8	10.6	–	–	–	1.2	10.6	4.4	3.0	–	–
			M-F	100	9.6	35.7	–	–	–	4.2	35.6	14.9	10.0	–	–
	PEA, SILAGE *Pisum* spp														
93	VINES (WITHOUT SEEDS, WITH PODS)	3-03-596	A-F	25	2.2	7.3	–	–	–	0.8	11.0	3.2	1.9	–	–
			M-F	100	9.0	29.8	–	–	–	3.3	44.9	13.1	7.7	–	–
94	**PEA, FIELD, SILAGE** *Pisum sativum, arvense*	3-03-609	A-F	27	2.5	7.5	–	–	–	1.2	12.2	3.8	2.5	–	–
			M-F	100	9.3	27.5	–	–	–	4.3	44.8	14.1	9.1	–	–
	PEAR, SILAGE *Pyrus* spp														
95	CANNERY RESIDUE	3-03-659	A-F	28	2.2	9.0	–	–	–	0.9	12.2	3.3	2.0	–	–
			M-F	100	8.0	32.6	–	–	–	3.3	44.1	12.0	7.2	–	–
	PINEAPPLE, SILAGE *Ananas comosus*														
96	LEAVES WITH STEMS	3-08-488	A-F	21	1.5	4.8	–	–	–	0.6	12.8	1.6	0.7	–	–
			M-F	100	7.0	22.5	–	–	–	2.8	60.1	7.5	3.3	–	–
	POTATO, SILAGE *Solanum tuberosum*														
97	TUBERS, ALFALFA HAY ADDED	3-03-770	A-F	35	2.2	7.1	–	–	–	0.5	20.7	4.2	2.6	–	–
			M-F	100	6.3	20.5	–	–	–	1.4	59.7	12.2	7.4	–	–
98	TUBERS, BOILED	4-03-767	A-F	23	1.5	0.7	–	–	–	0.1	19.1	1.9	0.4	1.2	1.3
			M-F	100	6.5	3.2	–	–	–	0.4	81.7	8.2	1.8	5.1	5.6
99	VINES	3-03-765	A-F	15	2.8	3.4	–	–	–	0.5	5.7	2.3	1.7	–	–
			M-F	100	19.1	23.0	–	–	–	3.7	38.6	15.6	11.2	–	–
	RED CLOVER–SEE CLOVER, RED, SILAGE														
100	**REED CANARYGRASS, SILAGE** *Phalaris arundinacea*	3-01-114	A-F	27	2.0	9.0	–	–	–	0.7	13.1	2.4	1.2	–	–
			M-F	100	7.3	33.1	–	–	–	2.5	48.2	8.8	4.4	–	–
101	**RUSSIAN THISTLE (TUMBLING TUMBLEWEED), SILAGE** *Salsola kali, tenuifolia*	3-03-984	A-F	34	6.2	10.3	–	–	–	0.9	14.5	2.5	1.1	–	–
			M-F	100	17.9	29.8	–	–	–	2.8	42.2	7.4	3.1	–	–
	RYE, SILAGE *Secale cereale*														
102	MOLASSES ADDED	3-04-021	A-F	24	1.8	8.4	–	–	–	0.7	10.4	2.4	1.3	–	–
			M-F	100	7.6	35.4	–	–	–	3.0	43.9	10.1	5.6	–	–
103	WILTED	3-08-601	A-F	30	2.5	10.8	–	–	–	1.0	12.5	3.5	2.1	–	–
			M-F	100	8.2	35.6	–	–	–	3.3	41.3	11.6	6.8	–	–
	SORGHUM, SILAGE *Sorghum bicolor*														
104	MILK STAGE	3-09-092	A-F	26	1.7	7.6	18.7	9.9	1.8	0.7	13.2	2.4	1.1	–	–
			M-F	100	6.7	29.8	72.8	38.5	7.0	2.9	51.4	9.2	4.4	–	–
105	DOUGH STAGE	3-04-321	A-F	29	2.5	8.3	19.2	10.9	1.9	0.9	15.1	2.3	0.9	–	–
			M-F	100	8.5	28.6	66.2	37.4	6.5	3.1	52.0	7.9	3.0	–	–
106	MATURE	3-04-322	A-F	32	2.3	7.8	16.3	8.5	2.1	0.9	18.2	2.7	1.0	–	–
			M-F	100	7.1	24.3	51.0	26.6	6.5	2.8	57.1	8.6	3.0	–	–
107	STOVER WITHOUT HEADS	3-04-326	A-F	62	5.6	16.2	–	–	–	1.2	36.3	2.7	–	–	0.1
			M-F	100	9.0	26.1	–	–	–	2.0	58.6	4.3	–	–	0.2
108	ATLAS, STOVER WITHOUT HEADS	3-04-340	A-F	24	1.4	6.8	–	10.8	1.6	0.5	13.7	1.4	0.6	–	–
			M-F	100	5.9	28.4	–	45.5	6.6	2.1	57.6	5.9	2.6	–	–

Entry Number	TDN Ruminant	TDN Swine	TDN Horse	Digestible Energy Ruminant		Digestible Energy Swine		Digestible Energy Horse		Metabolizable Energy Ruminant		Metabolizable Energy Swine		Metabolizable Energy Poultry ME_n		Metabolizable Energy Horse		Net Energy Ruminant NE_m		Net Energy Ruminant NE_g		Net Energy Lactating Cows NE_{lc}	
	%	%	%	Mcal lb	Mcal kg	kcal lb	kcal kg	Mcal lb	Mcal kg	Mcal lb	Mcal kg	kcal lb	kcal kg	kcal lb	kcal kg	Mcal lb	Mcal kg	Mcal lb	Mcal kg	Mcal lb	Mcal kg	Mcal lb	Mcal kg
84	13	–	–	0.27	0.60	–	–	–	–	0.22	0.49	–	–	–	–	–	–	0.13	0.28	0.06	0.13	0.14	0.31
	49	–	–	1.02	2.25	–	–	–	–	0.83	1.82	–	–	–	–	–	–	0.48	1.06	0.23	0.50	0.53	1.17
85	8	–	–	0.18	0.40	–	–	–	–	0.14	0.31	–	–	–	–	–	–	0.09	0.19	0.04	0.08	0.10	0.22
	41	–	–	0.90	1.99	–	–	–	–	0.71	1.56	–	–	–	–	–	–	0.43	0.95	0.18	0.40	0.49	1.09
86	13	–	–	0.26	0.57	–	–	–	–	0.21	0.46	–	–	–	–	–	–	0.11	0.25	0.05	0.11	0.13	0.28
	51	–	–	1.02	2.24	–	–	–	–	0.82	1.81	–	–	–	–	–	–	0.44	0.98	0.20	0.43	0.51	1.11
87	20	–	–	0.39	0.85	–	–	–	–	0.34	0.75	–	–	–	–	–	–	0.18	0.40	0.09	0.20	0.18	0.40
	57	–	–	1.10	2.43	–	–	–	–	0.97	2.13	–	–	–	–	–	–	0.52	1.14	0.26	0.58	0.52	1.15
88	17	–	–	0.37	0.81	–	–	–	–	0.31	0.68	–	–	–	–	–	–	0.20	0.43	0.11	0.24	0.20	0.44
	53	–	–	1.13	2.49	–	–	–	–	0.94	2.07	–	–	–	–	–	–	0.60	1.32	0.34	0.74	0.61	1.35
89	16	–	–	0.33	0.72	–	–	–	–	0.27	0.60	–	–	–	–	–	–	0.16	0.35	0.09	0.20	0.16	0.36
	60	–	–	1.21	2.66	–	–	–	–	1.02	2.24	–	–	–	–	–	–	0.59	1.30	0.33	0.72	0.61	1.34
90	15	–	–	0.29	0.65	–	–	–	–	0.24	0.54	–	–	–	–	–	–	0.14	0.31	0.07	0.16	0.15	0.33
	57	–	–	1.13	2.50	–	–	–	–	0.94	2.07	–	–	–	–	–	–	0.54	1.19	0.28	0.63	0.57	1.26
91	17	–	–	0.34	0.75	–	–	–	–	0.29	0.64	–	–	–	–	–	–	0.17	0.38	0.10	0.22	0.17	0.38
	63	–	–	1.25	2.76	–	–	–	–	1.06	2.34	–	–	–	–	–	–	0.63	1.38	0.36	0.80	0.64	1.40
92	18	–	–	0.35	0.77	–	–	–	–	0.29	0.65	–	–	–	–	–	–	0.17	0.36	0.09	0.20	0.17	0.38
	61	–	–	1.18	2.60	–	–	–	–	0.99	2.18	–	–	–	–	–	–	0.56	1.22	0.30	0.66	0.58	1.29
93	14	–	–	0.28	0.61	–	–	–	–	0.23	0.51	–	–	–	–	–	–	0.13	0.29	0.07	0.15	0.14	0.31
	57	–	–	1.13	2.49	–	–	–	–	0.94	2.07	–	–	–	–	–	–	0.53	1.18	0.28	0.61	0.57	1.25
94	18	–	–	0.35	0.76	–	–	–	–	0.29	0.65	–	–	–	–	–	–	0.17	0.38	0.10	0.22	0.17	0.38
	65	–	–	1.27	2.80	–	–	–	–	1.08	2.38	–	–	–	–	–	–	0.63	1.38	0.36	0.80	0.64	1.40
95	16	–	–	0.33	0.72	–	–	–	–	0.27	0.61	–	–	–	–	–	–	0.16	0.35	0.09	0.19	0.16	0.36
	59	–	–	1.19	2.62	–	–	–	–	0.99	2.19	–	–	–	–	–	–	0.57	1.26	0.31	0.69	0.60	1.31
96	9	–	–	0.22	0.49	–	–	–	–	0.18	0.40	–	–	–	–	–	–	0.14	0.30	0.08	0.18	0.14	0.31
	42	–	–	1.05	2.32	–	–	–	–	0.86	1.89	–	–	–	–	–	–	0.64	1.42	0.38	0.83	0.65	1.43
97	20	–	–	0.42	0.93	–	–	–	–	0.35	0.78	–	–	–	–	–	–	0.22	0.48	0.13	0.28	0.22	0.49
	59	–	–	1.21	2.68	–	–	–	–	1.02	2.25	–	–	–	–	–	–	0.63	1.40	0.37	0.81	0.64	1.41
98	18	21	18	0.35	0.78	416	917	0.33	0.73	0.31	0.68	394	868	–	–	0.27	0.60	0.17	0.38	0.11	0.24	0.17	0.37
	75	89	78	1.51	3.33	1776	3915	1.41	3.11	1.32	2.91	1682	3708	–	–	1.16	2.55	0.73	1.61	0.46	1.01	0.72	1.58
99	8	–	–	0.17	0.36	–	–	–	–	0.14	0.30	–	–	–	–	–	–	0.07	0.16	0.04	0.08	0.08	0.18
	57	–	–	1.11	2.45	–	–	–	–	0.92	2.02	–	–	–	–	–	–	0.50	1.10	0.25	0.54	0.54	1.20
100	14	–	–	0.29	0.64	–	–	–	–	0.24	0.53	–	–	–	–	–	–	0.15	0.33	0.08	0.18	0.16	0.35
	50	–	–	1.07	2.36	–	–	–	–	0.88	1.93	–	–	–	–	–	–	0.55	1.21	0.29	0.65	0.58	1.28
101	14	–	–	0.31	0.69	–	–	–	–	0.24	0.54	–	–	–	–	–	–	0.15	0.32	0.06	0.13	0.17	0.37
	42	–	–	0.90	1.99	–	–	–	–	0.71	1.56	–	–	–	–	–	–	0.42	0.93	0.17	0.38	0.49	1.08
102	11	–	–	0.25	0.55	–	–	–	–	0.20	0.45	–	–	–	–	–	–	0.13	0.28	0.07	0.15	0.14	0.30
	48	–	–	1.05	2.31	–	–	–	–	0.85	1.88	–	–	–	–	–	–	0.54	1.20	0.29	0.63	0.58	1.27
103	14	–	–	0.32	0.70	–	–	–	–	0.26	0.57	–	–	–	–	–	–	0.17	0.36	0.09	0.19	0.17	0.38
	48	–	–	1.05	2.30	–	–	–	–	0.85	1.88	–	–	–	–	–	–	0.54	1.20	0.29	0.63	0.58	1.27
104	15	–	–	0.30	0.65	–	–	–	–	0.24	0.52	–	–	–	–	–	–	0.15	0.32	0.08	0.17	0.14	0.32
	58	–	–	1.15	2.54	–	–	–	–	0.91	2.02	–	–	–	–	–	–	0.56	1.24	0.31	0.68	0.56	1.23
105	16	–	–	0.31	0.68	–	–	–	–	0.24	0.53	–	–	–	–	–	–	0.16	0.35	0.08	0.18	0.16	0.35
	55	–	–	1.06	2.34	–	–	–	–	0.82	1.81	–	–	–	–	–	–	0.54	1.19	0.29	0.63	0.54	1.19
106	17	–	–	0.33	0.72	–	–	–	–	0.27	0.58	–	–	–	–	–	–	0.16	0.35	0.08	0.17	0.16	0.36
	53	–	–	1.03	2.26	–	–	–	–	0.83	1.83	–	–	–	–	–	–	0.50	1.10	0.25	0.54	0.51	1.12
107	29	–	–	0.69	1.53	–	–	–	–	0.57	1.25	–	–	–	–	–	–	0.34	0.74	0.14	0.31	0.35	0.78
	46	–	–	1.12	2.46	–	–	–	–	0.92	2.02	–	–	–	–	–	–	0.54	1.20	0.23	0.50	0.57	1.25
108	13	–	–	0.24	0.52	–	–	–	–	0.21	0.47	–	–	–	–	–	–	0.10	0.23	0.05	0.10	0.11	0.24
	56	–	–	1.00	2.20	–	–	–	–	0.90	1.98	–	–	–	–	–	–	0.44	0.97	0.19	0.42	0.46	1.01

(Continued)

TABLE V-1D SILAGES AND HAYLAGES, MINERAL AND VITAMIN COMPOSITION OF FEEDS, DATA EXPRESSED **AS-FED** AND **MOISTURE-FREE**—*(Continued)*

Entry Number	Feed Name Description	Moisture Basis: A-F (as-fed) or M-F (moisture-free)	Dry Matter	Macro Minerals Calcium (Ca)	Phosphorus (P)	Sodium (Na)	Chlorine (Cl)	Magnesium (Mg)	Potassium (K)	Sulfur (S)	Micro Minerals Cobalt (Co)	Copper (Cu)	Iodine (I)	Iron (Fe)	Manganese (Mn)	Selenium (Se)	Zinc (Zn)
			%	%	%	%	%	%	%	%	ppm or mg/kg	ppm or mg/kg	ppm or mg/kg	%	ppm or mg/kg	ppm or mg/kg	ppm or mg/kg
	MILLET, PROSO (BROOMCORN; HOG MILLET), SILAGE *Panicum miliaceum*																
84	STOVER WITHOUT HEADS	A-F	27	–	–	–	–	–	–	–	–	–	–	–	–	–	–
		M-F	100	–	–	–	–	–	–	–	–	–	–	–	–	–	–
	MILO SORGHUM–SEE SORGHUM, SILAGE																
	MUNG BEAN–SEE BEAN, MUNG, SILAGE																
	NAPIERGRASS, SILAGE *Pennisetum purpureum*																
85	IMMATURE	A-F	20	–	–	–	–	–	–	–	–	–	–	–	–	–	–
		M-F	100	–	–	–	–	–	–	–	–	–	–	–	–	–	–
86	MATURE	A-F	26	–	–	–	–	–	–	–	–	–	–	–	–	–	–
		M-F	100	–	–	–	–	–	–	–	–	–	–	–	–	–	–
	OATS, SILAGE *Avena sativa*																
87	DOUGH STAGE	A-F	35	0.17	0.12	–	–	–	–	–	–	–	–	–	–	–	–
		M-F	100	0.47	0.33	–	–	–	–	–	–	–	–	–	–	–	–
88	MOLASSES ADDED	A-F	33	0.10	0.09	–	–	–	0.31	–	–	–	–	–	–	–	–
		M-F	100	0.31	0.28	–	–	–	0.94	–	–	–	–	–	–	–	–
89	**OATS-COWPEA, SILAGE** *Avena sativa-Vigna* spp	A-F	27	–	–	–	–	–	–	–	–	–	–	–	–	–	–
		M-F	100	–	–	–	–	–	–	–	–	–	–	–	–	–	–
90	**OATS-PEA, SILAGE** *Avena sativa-Pisum* spp	A-F	26	0.16	0.08	–	0.31	0.11	0.48	–	0.063	4.8	–	0.017	8.5	–	–
		M-F	100	0.62	0.32	–	1.19	0.44	1.86	–	0.241	18.5	–	0.062	33.1	–	–
	OATS-VETCH, SILAGE *Avena sativa-Vicia* spp																
91	MILK STAGE	A-F	27	–	–	–	–	–	–	–	–	–	–	–	–	–	–
		M-F	100	–	–	–	–	–	–	–	–	–	–	–	–	–	–
92	**ORCHARDGRASS, SILAGE** *Dactylis glomerata*	A-F	30	–	–	–	–	–	–	0.24	–	–	–	–	–	–	–
		M-F	100	–	–	–	–	–	–	0.82	–	–	–	–	–	–	–
	PEA, SILAGE *Pisum* spp																
93	VINES (WITHOUT SEEDS, WITH PODS)	A-F	25	0.32	0.06	0.00	–	0.10	0.34	0.06	–	–	–	0.003	–	–	–
		M-F	100	1.31	0.24	0.01	–	0.39	1.40	0.25	–	–	–	0.010	–	–	–
94	**PEA, FIELD, SILAGE** *Pisum sativum, arvense*	A-F	27	0.37	0.08	–	–	0.11	0.38	0.07	–	–	–	–	–	–	–
		M-F	100	1.36	0.29	–	–	0.39	1.40	0.25	–	–	–	–	–	–	–
	PEAR, SILAGE *Pyrus* spp																
95	CANNERY RESIDUE	A-F	28	–	–	–	–	–	–	–	–	–	–	–	–	–	–
		M-F	100	–	–	–	–	–	–	–	–	–	–	–	–	–	–
	PINEAPPLE, SILAGE *Ananas comosus*																
96	LEAVES WITH STEMS	A-F	21	–	–	–	–	–	–	–	–	–	–	–	–	–	–
		M-F	100	–	–	–	–	–	–	–	–	–	–	–	–	–	–
	POTATO, SILAGE *Solanum tuberosum*																
97	TUBERS, ALFALFA HAY ADDED	A-F	35	–	–	–	–	–	–	–	–	–	–	–	–	–	–
		M-F	100	–	–	–	–	–	–	–	–	–	–	–	–	–	–
98	TUBERS, BOILED	A-F	23	–	–	–	–	–	–	–	–	–	–	–	–	–	–
		M-F	100	–	–	–	–	–	–	–	–	–	–	–	–	–	–
99	VINES	A-F	15	0.31	0.03	–	0.06	0.02	0.59	0.06	–	–	–	–	–	–	–
		M-F	100	2.12	0.20	–	0.38	0.14	3.95	0.37	–	–	–	–	–	–	–
	RED CLOVER–SEE CLOVER, RED, SILAGE																
100	**REED CANARYGRASS, SILAGE** *Phalaris arundinacea*	A-F	27	–	–	–	–	–	–	–	–	–	–	–	–	–	–
		M-F	100	–	–	–	–	–	–	–	–	–	–	–	–	–	–
101	**RUSSIAN THISTLE (TUMBLING TUMBLEWEED), SILAGE** *Salsola kali, tenuifolia*	A-F	34	–	–	–	–	–	–	–	–	–	–	–	–	–	–
		M-F	100	–	–	–	–	–	–	–	–	–	–	–	–	–	–
	RYE, SILAGE *Secale cereale*																
102	MOLASSES ADDED	A-F	24	–	–	–	–	–	–	–	–	–	–	–	–	–	–
		M-F	100	–	–	–	–	–	–	–	–	–	–	–	–	–	–
103	WILTED	A-F	30	–	0.07	–	–	–	0.56	–	–	–	–	–	–	–	–
		M-F	100	–	0.23	–	–	–	1.85	–	–	–	–	–	–	–	–
	SORGHUM, SILAGE *Sorghum bicolor*																
104	MILK STAGE	A-F	26	–	–	–	–	–	–	–	–	–	–	–	–	–	–
		M-F	100	–	–	–	–	–	–	–	–	–	–	–	–	–	–
105	DOUGH STAGE	A-F	29	–	–	–	–	–	–	–	–	–	–	–	–	–	–
		M-F	100	–	–	–	–	–	–	–	–	–	–	–	–	–	–
106	MATURE	A-F	32	–	–	–	–	–	–	–	–	–	–	–	–	–	–
		M-F	100	–	–	–	–	–	–	–	–	–	–	–	–	–	–
107	STOVER WITHOUT HEADS	A-F	62	0.25	0.07	–	–	–	–	–	–	–	–	–	91.7	–	–
		M-F	100	0.41	0.11	–	–	–	–	–	–	–	–	–	147.9	–	–
108	ATLAS, STOVER WITHOUT HEADS	A-F	24	0.09	0.03	–	–	–	–	–	–	–	–	–	–	–	–
		M-F	100	0.39	0.11	–	–	–	–	–	–	–	–	–	–	–	–

SILAGES AND HAYLAGES

Entry Number	Fat-Soluble Vitamins: A (1 mg Carotene = 1667 IU Vit A)	Carotene (Provitamin A)	D	E	K	Water-Soluble Vitamins: B-12	Biotin	Choline	Folacin (Folic Acid)	Niacin	Pantothenic Acid (B-3)	(Pyridoxine) B-6	Riboflavin (B-2)	Thiamin (B-1)
	IU/g	ppm or mg/kg	IU/kg	ppm or mg/kg	ppm or mg/kg	ppm or mg/kg	ppm or mg/kg	ppm or mg/kg	ppm or mg/kg	ppm or mg/kg	ppm or mg/kg	ppm or mg/kg	ppm or mg/kg	ppm or mg/kg
84	–	–	–	–	–	–	–	–	–	–	–	–	–	–
	–	–	–	–	–	–	–	–	–	–	–	–	–	–
85	–	–	–	–	–	–	–	–	–	–	–	–	–	–
	–	–	–	–	–	–	–	–	–	–	–	–	–	–
86	–	–	–	–	–	–	–	–	–	–	–	–	–	–
	–	–	–	–	–	–	–	–	–	–	–	–	–	–
87	35.1	21.1	–	–	–	–	–	–	–	–	–	–	–	–
	100.0	60.0	–	–	–	–	–	–	–	–	–	–	–	–
88	–	–	–	–	–	–	–	–	–	–	–	–	–	–
	–	–	–	–	–	–	–	–	–	–	–	–	–	–
89	–	–	–	–	–	–	–	–	–	–	–	–	–	–
	–	–	–	–	–	–	–	–	–	–	–	–	–	–
90	33.5	20.1	–	–	–	–	–	–	–	–	–	–	–	–
	129.5	77.7	–	–	–	–	–	–	–	–	–	–	–	–
91	–	–	–	–	–	–	–	–	–	–	–	–	–	–
	–	–	–	–	–	–	–	–	–	–	–	–	–	–
92	83.6	50.1	–	–	–	–	–	–	–	–	–	–	–	–
	281.9	169.1	–	–	–	–	–	–	–	–	–	–	–	–
93	77.2	46.3	–	–	–	–	–	–	–	–	–	–	–	–
	315.0	189.0	–	–	–	–	–	–	–	–	–	–	–	–
94	–	–	–	–	–	–	–	–	–	–	–	–	–	–
	–	–	–	–	–	–	–	–	–	–	–	–	–	–
95	–	–	–	–	–	–	–	–	–	–	–	–	–	–
	–	–	–	–	–	–	–	–	–	–	–	–	–	–
96	–	–	–	–	–	–	–	–	–	–	–	–	–	–
	–	–	–	–	–	–	–	–	–	–	–	–	–	–
97	–	–	–	–	–	–	–	–	–	–	–	–	–	–
	–	–	–	–	–	–	–	–	–	–	–	–	–	–
98	–	–	–	–	–	–	–	–	–	–	–	–	–	–
	–	–	–	–	–	–	–	–	–	–	–	–	–	–
99	–	–	–	–	–	–	–	–	–	–	–	–	–	–
	–	–	–	–	–	–	–	–	–	–	–	–	–	–
100	–	–	–	–	–	–	–	–	–	–	–	–	–	–
	–	–	–	–	–	–	–	–	–	–	–	–	–	–
101	–	–	–	–	–	–	–	–	–	–	–	–	–	–
	–	–	–	–	–	–	–	–	–	–	–	–	–	–
102	–	–	–	–	–	–	–	–	–	–	–	–	–	–
	–	–	–	–	–	–	–	–	–	–	–	–	–	–
103	–	–	–	–	–	–	–	–	–	–	–	–	–	–
	–	–	–	–	–	–	–	–	–	–	–	–	–	–
104	–	–	–	–	–	–	–	–	–	–	–	–	–	–
	–	–	–	–	–	–	–	–	–	–	–	–	–	–
105	–	–	–	–	–	–	–	–	–	–	–	–	–	–
	–	–	–	–	–	–	–	–	–	–	–	–	–	–
106	–	–	–	–	–	–	–	–	–	–	–	–	–	–
	–	–	–	–	–	–	–	–	–	–	–	–	–	–
107	7.1	4.2	–	–	–	–	–	–	–	–	–	–	–	–
	11.4	6.8	–	–	–	–	–	–	–	–	–	–	–	–
108	9.0	5.4	–	–	–	–	–	–	–	–	–	–	–	–
	37.9	22.7	–	–	–	–	–	–	–	–	–	–	–	–

(Continued)

TABLE V-1D SILAGES AND HAYLAGES, COMPOSITION OF FEEDS, DATA EXPRESSED **AS-FED** AND **MOISTURE-FREE**—*(Continued)*

Entry Number	Feed Name Description	International Feed Number	Moisture Basis: A-F (as-fed) or M-F (moisture-free)	Proximate Analysis									Digestible Protein		
				Dry Matter	Ash	Crude Fiber	Neutral Det. Fib. (NDF)	Acid Det. Fib. (ADF)	Lignin	Ether Extract (Fat)	N-Free Extract	Crude Protein	Ruminant	Swine	Horse
				%	%	%	%	%	%	%	%	%	%	%	%
	SORGHUM, SILAGE (Continued)														
109	DARSO	**3-04-356**	**A-F**	**27**	**1.5**	**6.5**	**–**	**–**	**–**	**0.4**	**16.7**	**1.9**	**0.1**	**–**	**–**
			M-F	100	5.7	24.0	–	–	–	1.4	61.9	7.0	0.2	–	–
110	DURRA	**3-04-364**	**A-F**	**29**	**2.8**	**10.2**	**–**	**–**	**–**	**1.0**	**13.8**	**1.8**	**0.6**	**–**	**–**
			M-F	100	9.4	34.4	–	–	–	3.3	46.9	6.0	2.0	–	–
111	FETERITA	**3-04-368**	**A-F**	**32**	**2.4**	**6.9**	**–**	**–**	**–**	**0.7**	**19.5**	**2.7**	**1.3**	**–**	**–**
			M-F	100	7.4	21.4	–	–	–	2.3	60.7	8.3	4.0	–	–
112	HEGARI	**3-04-395**	**A-F**	**35**	**2.9**	**8.9**	**–**	**–**	**–**	**0.8**	**19.4**	**2.5**	**1.1**	**–**	**–**
			M-F	100	8.4	25.9	–	–	–	2.4	56.1	7.2	3.1	–	–
113	HEGARI, STOVER WITHOUT HEADS	**3-04-396**	**A-F**	**29**	**3.4**	**7.9**	**–**	**–**	**–**	**0.6**	**16.3**	**0.9**	**–0.2**	**–**	**–**
			M-F	100	11.7	27.1	–	–	–	2.1	56.0	3.1	–0.6	–	–
114	JOHNSONGRASS	**3-04-413**	**A-F**	**39**	**3.8**	**12.6**	**–**	**–**	**–**	**0.9**	**18.7**	**2.7**	**1.1**	**–**	**–**
			M-F	100	9.9	32.5	–	–	–	2.4	48.2	7.0	2.8	–	–
115	JOHNSONGRASS, MOLASSES ADDED	**3-04-414**	**A-F**	**33**	**3.5**	**10.4**	**–**	**–**	**–**	**0.7**	**16.0**	**2.1**	**0.8**	**–**	**–**
			M-F	100	10.8	31.8	–	–	–	2.2	48.8	6.4	2.3	–	–
116	KAFIR	**3-04-425**	**A-F**	**30**	**2.2**	**8.1**	**–**	**–**	**–**	**1.0**	**16.1**	**2.1**	**0.9**	**–**	**–**
			M-F	100	7.6	27.3	–	–	–	3.3	54.6	7.2	3.0	–	–
117	MILO	**3-04-437**	**A-F**	**31**	**2.5**	**5.5**	**–**	**–**	**–**	**0.5**	**20.0**	**2.3**	**1.0**	**–**	**–**
			M-F	100	8.2	17.9	–	–	–	1.8	64.7	7.4	3.2	–	–
118	SAGRAIN	**3-08-518**	**A-F**	**38**	**2.3**	**9.1**	**–**	**–**	**–**	**1.2**	**22.7**	**2.8**	**1.2**	**–**	**–**
			M-F	100	6.0	23.9	–	–	–	3.2	59.6	7.3	3.1	–	–
119	SORGO	**3-04-468**	**A-F**	**29**	**2.4**	**7.0**	**18.5**	**11.0**	**1.7**	**0.7**	**16.7**	**1.9**	**0.5**	**–**	**–**
			M-F	100	8.3	24.4	64.0	38.0	6.0	2.5	58.1	6.8	1.7	–	–
120	SUDANGRASS	**3-04-499**	**A-F**	**23**	**2.1**	**7.9**	**16.4**	**1.2**	**9.7**	**0.7**	**9.9**	**2.6**	**1.6**	**–**	**–**
			M-F	100	9.2	34.0	71.0	5.0	42.0	2.9	42.9	11.1	6.8	–	–
121	**SORGHUM-COWPEA, SILAGE** *Sorghum bicolor-Vigna* spp	**3-04-521**	**A-F**	**32**	**2.2**	**8.4**	**–**	**–**	**–**	**1.0**	**18.0**	**2.4**	**1.0**	**–**	**–**
			M-F	100	6.8	26.3	–	–	–	3.1	56.4	7.4	3.2	–	–
122	**SOYBEAN, SILAGE**[1] *Glycine max*	**3-04-581**	**A-F**	**30**	**3.0**	**9.0**	**–**	**12.0**	**–**	**0.8**	**12.2**	**5.2**	**3.2**	**–**	**–**
			M-F	100	9.9	29.9	–	40.0	–	2.6	40.5	17.1	10.6	–	–
123	**SOYBEAN-SORGHUM, SUDANGRASS, SILAGE** *Glycine max-Sorghum bicolor, sudanense*	**3-04-644**	**A-F**	**25**	**2.2**	**8.6**	**–**	**–**	**–**	**0.5**	**10.5**	**3.4**	**2.1**	**–**	**–**
			M-F	100	8.6	34.1	–	–	–	2.1	41.9	13.3	8.4	–	–
	SUDANGRASS SORGHUM–SEE SORGHUM, SILAGE														
	SUGAR BEET, SILAGE *Beta vulgaris, altissima*														
124	TOPS WITH CROWNS	**3-00-660**	**A-F**	**25**	**8.6**	**3.2**	**–**	**–**	**–**	**0.7**	**9.3**	**3.4**	**2.2**	**–**	**–**
			M-F	100	34.0	12.8	–	–	–	2.9	37.0	13.4	8.7	–	–
125	PULP	**4-00-662**	**A-F**	**12**	**0.6**	**3.8**	**–**	**4.8**	**–**	**0.2**	**5.8**	**1.5**	**0.9**	**1.1**	**1.1**
			M-F	100	4.8	32.0	–	40.0	–	1.9	48.6	12.7	8.0	9.6	9.1
126	ROOTS	**4-00-665**	**A-F**	**17**	**0.9**	**2.1**	**–**	**–**	**–**	**0.4**	**12.5**	**1.5**	**0.6**	**1.1**	**1.1**
			M-F	100	5.3	12.2	–	–	–	2.4	71.5	8.8	3.4	6.0	6.0
	SUGARCANE, SILAGE *Saccharum officinarum*														
127	ALL ANALYSES	**3-04-693**	**A-F**	**22**	**1.4**	**7.5**	**–**	**–**	**–**	**0.4**	**11.6**	**1.1**	**0.3**	**–**	**–**
			M-F	100	6.4	34.1	–	–	–	2.0	52.4	5.1	1.2	–	–
128	TOPS	**3-08-528**	**A-F**	**30**	**2.8**	**10.6**	**–**	**–**	**–**	**0.6**	**14.1**	**1.5**	**0.8**	**–**	**–**
			M-F	100	9.5	35.8	–	–	–	2.0	47.6	5.1	2.7	–	–
129	**SUMMER CYPRESS, BELVEDERE, SILAGE** *Kochia scoparia*	**3-04-718**	**A-F**	**31**	**3.9**	**9.7**	**–**	**–**	**–**	**0.7**	**11.6**	**5.1**	**3.4**	**–**	**–**
			M-F	100	12.6	31.3	–	–	–	2.3	37.4	16.4	11.1	–	–
	SUNFLOWER, SILAGE *Helianthus* spp														
130	LATE BLOOM	**3-04-732**	**A-F**	**26**	**3.9**	**9.9**	**–**	**–**	**–**	**0.5**	**9.5**	**2.1**	**0.9**	**–**	**–**
			M-F	100	15.0	38.2	–	–	–	2.1	36.5	8.2	3.6	–	–
131	MILK STAGE	**3-04-733**	**A-F**	**21**	**2.1**	**6.2**	**–**	**–**	**–**	**1.3**	**9.5**	**2.1**	**1.0**	**–**	**–**
			M-F	100	10.0	29.4	–	–	–	5.9	45.0	9.7	4.7	–	–
132	MATURE	**3-04-735**	**A-F**	**26**	**2.6**	**10.2**	**9.2**	**8.6**	**–**	**1.1**	**9.9**	**2.2**	**0.5**	**–**	**–**
			M-F	100	10.0	39.1	35.3	33.2	–	4.3	38.0	8.6	1.9	–	–
	SWEET CLOVER, YELLOW–SEE CLOVER, SWEET, SILAGE														
	SWEET CORN, SILAGE *Zea mays, saccharata*														
133	CANNERY RESIDUE	**3-07-955**	**A-F**	**31**	**1.7**	**10.1**	**–**	**10.5**	**–**	**1.3**	**15.4**	**2.5**	**1.2**	**–**	**–**
			M-F	100	5.4	32.7	–	34.0	–	4.3	49.6	8.0	3.7	–	–
	SWEET POTATO, SILAGE *Ipomoea batatas*														
134	VINES	**3-04-785**	**A-F**	**13**	**1.5**	**3.6**	**–**	**–**	**–**	**0.5**	**5.9**	**1.7**	**0.7**	**–**	**–**
			M-F	100	11.1	27.4	–	–	–	4.1	44.6	12.9	5.2	–	–
135	TUBERS	**4-04-787**	**A-F**	**38**	**1.5**	**1.5**	**–**	**–**	**–**	**0.3**	**34.0**	**0.8**	**–**	**–**	**–**
			M-F	100	4.0	3.9	–	–	–	0.7	89.4	2.0	–	–	–
	TIMOTHY, SILAGE *Phleum pratense*														
136	ALL ANALYSES	**3-04-922**	**A-F**	**34**	**2.4**	**12.0**	**–**	**–**	**–**	**1.2**	**15.1**	**3.6**	**2.0**	**–**	**–**
			M-F	100	7.1	35.0	–	–	–	3.4	44.0	10.5	5.8	–	–
137	AIV ADDED	**3-04-934**	**A-F**	**35**	**2.3**	**12.3**	**–**	**–**	**–**	**1.0**	**15.8**	**3.4**	**1.8**	**–**	**–**
			M-F	100	6.7	35.3	–	–	–	2.8	45.5	9.7	5.2	–	–
138	CEREAL GRAIN ADDED	**3-08-543**	**A-F**	**33**	**2.0**	**10.4**	**–**	**–**	**–**	**1.1**	**15.6**	**4.0**	**2.3**	**–**	**–**
			M-F	100	6.0	31.4	–	–	–	3.3	47.1	12.1	7.0	–	–
139	MOLASSES ADDED, WILTED	**3-04-954**	**A-F**	**43**	**2.9**	**14.3**	**–**	**–**	**–**	**1.4**	**20.1**	**4.4**	**2.2**	**–**	**–**
			M-F	100	6.8	33.2	–	–	–	3.2	46.7	10.1	5.2	–	–

Entry Number	TDN Ruminant	TDN Swine	TDN Horse	DE Ruminant		DE Swine		DE Horse		ME Ruminant		ME Swine		Poultry ME_n		ME Horse		Net Energy Ruminant NE_m		Ruminant NE_g		Lactating Cows NE_{lc}	
	%	%	%	Mcal		kcal		Mcal		Mcal		kcal		kcal		Mcal		Mcal		Mcal		Mcal	
				lb	*kg*	lb	*kg*	lb	*kg*	lb	*kg*	lb	*kg*	lb	*kg*	lb	*kg*	lb	*kg*	lb	*kg*	lb	*kg*
109	15	–	–	0.32	0.70	–	–	–	–	0.27	0.59	–	–	–	–	–	–	0.16	0.36	0.09	0.21	0.17	0.37
	56	–	–	1.18	2.59	–	–	–	–	0.98	2.17	–	–	–	–	–	–	0.61	1.34	0.35	0.77	0.62	1.37
110	16	–	–	0.33	0.72	–	–	–	–	0.27	0.60	–	–	–	–	–	–	0.16	0.35	0.08	0.18	0.17	0.37
	56	–	–	1.11	2.45	–	–	–	–	0.92	2.03	–	–	–	–	–	–	0.53	1.17	0.28	0.61	0.57	1.25
111	18	–	–	0.38	0.85	–	–	–	–	0.32	0.71	–	–	–	–	–	–	0.20	0.45	0.12	0.26	0.21	0.45
	57	–	–	1.20	2.64	–	–	–	–	1.00	2.21	–	–	–	–	–	–	0.63	1.39	0.37	0.81	0.64	1.41
112	19	–	–	0.39	0.87	–	–	–	–	0.34	0.74	–	–	–	–	–	–	0.20	0.45	0.11	0.25	0.21	0.46
	56	–	–	1.14	2.51	–	–	–	–	0.97	2.14	–	–	–	–	–	–	0.59	1.29	0.33	0.72	0.61	1.34
113	15	–	–	0.31	0.69	–	–	–	–	0.26	0.56	–	–	–	–	–	–	0.15	0.33	0.08	0.17	0.16	0.35
	53	–	–	1.07	2.36	–	–	–	–	0.88	1.93	–	–	–	–	–	–	0.51	1.12	0.26	0.57	0.55	1.22
114	22	–	–	0.44	0.96	–	–	–	–	0.36	0.80	–	–	–	–	–	–	0.20	0.44	0.10	0.22	0.21	0.47
	56	–	–	1.12	2.48	–	–	–	–	0.93	2.05	–	–	–	–	–	–	0.51	1.12	0.26	0.57	0.55	1.22
115	18	–	–	0.35	0.78	–	–	–	–	0.29	0.64	–	–	–	–	–	–	0.16	0.36	0.08	0.18	0.18	0.39
	54	–	–	1.08	2.37	–	–	–	–	0.88	1.95	–	–	–	–	–	–	0.49	1.09	0.24	0.53	0.54	1.19
116	17	–	–	0.35	0.77	–	–	–	–	0.29	0.64	–	–	–	–	–	–	0.18	0.40	0.10	0.23	0.19	0.41
	57	–	–	1.18	2.61	–	–	–	–	0.99	2.18	–	–	–	–	–	–	0.61	1.35	0.35	0.78	0.63	1.38
117	18	–	–	0.40	0.88	–	–	–	–	0.34	0.75	–	–	–	–	–	–	0.17	0.38	0.10	0.21	0.18	0.40
	58	–	–	1.29	2.85	–	–	–	–	1.10	2.43	–	–	–	–	–	–	0.56	1.24	0.31	0.68	0.59	1.30
118	22	–	–	0.47	1.03	–	–	–	–	0.39	0.87	–	–	–	–	–	–	0.25	0.55	0.15	0.33	0.25	0.55
	58	–	–	1.22	2.70	–	–	–	–	1.03	2.27	–	–	–	–	–	–	0.66	1.45	0.39	0.86	0.66	1.45
119	17	–	–	0.34	0.74	–	–	–	–	0.28	0.62	–	–	–	–	–	–	0.16	0.35	0.09	0.19	0.17	0.37
	58	–	–	1.17	2.58	–	–	–	–	0.98	2.15	–	–	–	–	–	–	0.55	1.22	0.30	0.66	0.58	1.29
120	13	–	–	0.26	0.58	–	–	–	–	0.22	0.48	–	–	–	–	–	–	0.13	0.28	0.07	0.15	0.13	0.30
	57	–	–	1.14	2.51	–	–	–	–	0.94	2.08	–	–	–	–	–	–	0.55	1.21	0.29	0.64	0.58	1.28
121	18	–	–	0.38	0.84	–	–	–	–	0.32	0.71	–	–	–	–	–	–	0.20	0.44	0.12	0.26	0.20	0.45
	57	–	–	1.19	2.63	–	–	–	–	1.00	2.20	–	–	–	–	–	–	0.63	1.38	0.36	0.80	0.64	1.40
122	16	–	–	0.32	0.71	–	–	–	–	0.26	0.58	–	–	–	–	–	–	0.15	0.33	0.07	0.16	0.16	0.36
	53	–	–	1.07	2.35	–	–	–	–	0.87	1.92	–	–	–	–	–	–	0.50	1.10	0.25	0.54	0.54	1.20
123	14	–	–	0.28	0.62	–	–	–	–	0.23	0.51	–	–	–	–	–	–	0.13	0.29	0.07	0.14	0.14	0.31
	57	–	–	1.12	2.47	–	–	–	–	0.93	2.04	–	–	–	–	–	–	0.51	1.13	0.26	0.57	0.55	1.22
124	13	–	–	0.26	0.58	–	–	–	–	0.21	0.47	–	–	–	–	–	–	0.13	0.28	0.06	0.14	0.14	0.30
	52	–	–	1.04	2.29	–	–	–	–	0.84	1.86	–	–	–	–	–	–	0.50	1.11	0.25	0.55	0.55	1.20
125	9	–	–	0.17	0.38	–	–	–	–	0.15	0.33	–	–	–	–	–	–	–	–	–	–	–	–
	73	–	–	1.45	3.20	–	–	–	–	1.26	2.78	–	–	–	–	–	–	–	–	–	–	–	–
126	11	14	11	0.24	0.52	287	632	0.20	0.45	0.20	0.44	271	597	–	–	0.17	0.37	0.14	0.30	0.09	0.20	0.13	0.29
	61	82	63	1.34	2.96	1638	3611	1.17	2.57	1.15	2.54	1547	3410	–	–	0.96	2.11	0.79	1.74	0.51	1.12	0.76	1.68
127	13	–	–	0.26	0.57	–	–	–	–	0.22	0.48	–	–	–	–	–	–	0.12	0.26	0.06	0.14	0.13	0.28
	61	–	–	1.17	2.58	–	–	–	–	0.98	2.15	–	–	–	–	–	–	0.53	1.18	0.28	0.62	0.57	1.25
128	15	–	–	0.31	0.68	–	–	–	–	0.25	0.56	–	–	–	–	–	–	0.14	0.31	0.07	0.15	0.16	0.34
	52	–	–	1.05	2.31	–	–	–	–	0.85	1.88	–	–	–	–	–	–	0.48	1.05	0.23	0.50	0.53	1.16
129	17	–	–	0.34	0.74	–	–	–	–	0.28	0.61	–	–	–	–	–	–	0.15	0.33	0.07	0.16	0.17	0.37
	55	–	–	1.08	2.39	–	–	–	–	0.89	1.96	–	–	–	–	–	–	0.49	1.08	0.24	0.52	0.54	1.18
130	11	–	–	0.24	0.52	–	–	–	–	0.19	0.41	–	–	–	–	–	–	0.09	0.19	0.02	0.05	0.11	0.24
	43	–	–	0.91	2.01	–	–	–	–	0.72	1.58	–	–	–	–	–	–	0.33	0.72	0.09	0.19	0.43	0.94
131	11	–	–	0.24	0.53	–	–	–	–	0.20	0.44	–	–	–	–	–	–	0.10	0.21	0.04	0.09	0.11	0.24
	51	–	–	1.13	2.48	–	–	–	–	0.93	2.06	–	–	–	–	–	–	0.45	0.99	0.20	0.44	0.51	1.12
132	10	–	–	0.24	0.52	–	–	–	–	0.19	0.41	–	–	–	–	–	–	0.06	0.14	0.00	0.00	0.10	0.21
	38	–	–	0.91	2.01	–	–	–	–	0.72	1.58	–	–	–	–	–	–	0.24	0.53	0.00	0.01	0.37	0.81
133	22	–	–	0.42	0.92	–	–	–	–	0.36	0.79	–	–	–	–	–	–	0.24	0.53	0.15	0.34	0.23	0.51
	72	–	–	1.35	2.97	–	–	–	–	1.16	2.55	–	–	–	–	–	–	0.77	1.70	0.49	1.08	0.75	1.64
134	6	–	–	0.14	0.31	–	–	–	–	0.12	0.26	–	–	–	–	–	–	0.08	0.17	0.04	0.10	0.08	0.18
	48	–	–	1.08	2.39	–	–	–	–	0.89	1.96	–	–	–	–	–	–	0.59	1.31	0.33	0.73	0.61	1.35
135	–	–	–	–	–	–	–	–	–	–	–	–	–	–	–	–	–	–	–	–	–	–	–
	–	–	–	–	–	–	–	–	–	–	–	–	–	–	–	–	–	–	–	–	–	–	–
136	20	–	–	0.41	0.90	–	–	–	–	0.34	0.75	–	–	–	–	–	–	0.20	0.44	0.11	0.25	0.21	0.46
	59	–	–	1.19	2.62	–	–	–	–	1.00	2.20	–	–	–	–	–	–	0.58	1.29	0.32	0.71	0.60	1.33
137	19	–	–	0.40	0.88	–	–	–	–	0.33	0.73	–	–	–	–	–	–	0.18	0.40	0.09	0.21	0.20	0.43
	56	–	–	1.14	2.52	–	–	–	–	0.95	2.09	–	–	–	–	–	–	0.53	1.16	0.27	0.60	0.56	1.24
138	19	–	–	0.39	0.87	–	–	–	–	0.33	0.73	–	–	–	–	–	–	0.20	0.45	0.12	0.25	0.21	0.46
	58	–	–	1.19	2.63	–	–	–	–	1.00	2.20	–	–	–	–	–	–	0.61	1.35	0.35	0.77	0.62	1.38
139	26	–	–	0.52	1.15	–	–	–	–	0.44	0.97	–	–	–	–	–	–	0.26	0.58	0.15	0.33	0.27	0.59
	60	–	–	1.21	2.66	–	–	–	–	1.02	2.24	–	–	–	–	–	–	0.61	1.34	0.35	0.76	0.62	1.37

(Continued)

TABLE V-1D SILAGES AND HAYLAGES, MINERAL AND VITAMIN COMPOSITION OF FEEDS, DATA EXPRESSED **AS-FED** AND **MOISTURE-FREE**—*(Continued)*

Entry Number	Feed Name Description	Moisture Basis: A-F (as-fed) or M-F (moisture-free)	Dry Matter	Macro Minerals Calcium (Ca)	Phosphorus (P)	Sodium (Na)	Chlorine (Cl)	Magnesium (Mg)	Potassium (K)	Sulfur (S)	Micro Minerals Cobalt (Co)	Copper (Cu)	Iodine (I)	Iron (Fe)	Manganese (Mn)	Selenium (Se)	Zinc (Zn)
			%	%	%	%	%	%	%	%	ppm or mg/kg	ppm or mg/kg	ppm or mg/kg	%	ppm or mg/kg	ppm or mg/kg	ppm or mg/kg
	SORGHUM (Continued)																
109	DARSO	**A-F**	**27**	–	–	–	–	–	–	–	–	–	–	–	–	–	–
		M-F	100	–	–	–	–	–	–	–	–	–	–	–	–	–	–
110	DURRA	**A-F**	**29**	–	–	–	–	–	–	–	–	–	–	–	–	–	–
		M-F	100	–	–	–	–	–	–	–	–	–	–	–	–	–	–
111	FETERITA	**A-F**	**32**	**0.12**	**0.09**	–	–	–	–	–	–	–	–	–	–	–	–
		M-F	100	0.37	0.27	–	–	–	–	–	–	–	–	–	–	–	–
112	HEGARI	**A-F**	**35**	–	–	–	–	–	–	–	–	–	–	–	–	–	–
		M-F	100	–	–	–	–	–	–	–	–	–	–	–	–	–	–
113	HEGARI, STOVER WITHOUT HEADS	**A-F**	**29**	–	–	–	–	–	–	–	–	–	–	–	–	–	–
		M-F	100	–	–	–	–	–	–	–	–	–	–	–	–	–	–
114	JOHNSONGRASS	**A-F**	**39**	–	–	–	–	–	–	–	–	–	–	–	–	–	–
		M-F	100	–	–	–	–	–	–	–	–	–	–	–	–	–	–
115	JOHNSONGRASS, MOLASSES ADDED	**A-F**	**33**	–	–	–	–	–	–	–	–	–	–	–	–	–	–
		M-F	100	–	–	–	–	–	–	–	–	–	–	–	–	–	–
116	KAFIR	**A-F**	**30**	**0.07**	**0.05**	–	–	**0.08**	**0.50**	–	–	–	–	–	–	–	–
		M-F	100	0.24	0.17	–	–	0.27	1.68	–	–	–	–	–	–	–	–
117	MILO	**A-F**	**31**	**0.11**	**0.06**	–	–	–	–	–	–	–	–	–	–	–	–
		M-F	100	0.34	0.19	–	–	–	–	–	–	–	–	–	–	–	–
118	SAGRAIN	**A-F**	**38**	–	–	–	–	–	–	–	–	–	–	–	–	–	–
		M-F	100	–	–	–	–	–	–	–	–	–	–	–	–	–	–
119	SORGO	**A-F**	**29**	**0.10**	**0.06**	**0.04**	**0.02**	**0.08**	**0.32**	**0.03**	–	**9.0**	–	**0.006**	**17.6**	–	–
		M-F	100	0.35	0.21	0.15	0.06	0.27	1.12	0.10	–	31.1	–	0.020	61.0	–	–
120	SUDANGRASS	**A-F**	**23**	**0.12**	**0.05**	**0.01**	–	**0.10**	**0.60**	**0.01**	**0.063**	**8.5**	–	**0.003**	**22.8**	–	–
		M-F	100	0.50	0.21	0.02	–	0.42	2.61	0.06	0.270	36.6	–	0.012	98.8	–	–
121	**SORGHUM-COWPEA, SILAGE** *Sorghum bicolor-Vigna* spp	**A-F**	**32**	**0.14**	**0.04**	–	–	–	**0.30**	–	–	–	–	–	–	–	–
		M-F	100	0.43	0.12	–	–	–	0.93	–	–	–	–	–	–	–	–
122	**SOYBEAN, SILAGE** *Glycine max*	**A-F**	**30**	**0.40**	**0.13**	**0.03**	–	**0.12**	**0.39**	**0.09**	–	**2.9**	–	**0.010**	**42.7**	–	**10.3**
		M-F	100	1.32	0.44	0.09	–	0.40	1.28	0.31	–	9.6	–	0.033	141.3	–	34.0
123	**SOYBEAN-SORGHUM, SUDANGRASS, SILAGE** *Glycine max-Sorghum bicolor, sudanense*	**A-F**	**25**	–	–	–	–	–	–	–	–	–	–	–	–	–	–
		M-F	100	–	–	–	–	–	–	–	–	–	–	–	–	–	–
	SUDANGRASS SORGHUM–SEE SORGHUM, SILAGE																
	SUGAR BEET, SILAGE *Beta vulgaris, altissima*																
124	TOPS WITH CROWNS	**A-F**	**25**	**0.39**	**0.07**	**0.14**	–	**0.27**	**1.45**	**0.14**	–	–	–	**0.006**	–	–	–
		M-F	100	1.56	0.28	0.54	–	1.07	5.74	0.57	–	–	–	0.020	–	–	–
125	PULP	**A-F**	**12**	–	–	–	–	–	–	–	–	–	–	–	–	–	–
		M-F	100	–	–	–	–	–	–	–	–	–	–	–	–	–	–
126	ROOTS	**A-F**	**17**	–	–	–	–	–	–	–	–	–	–	–	–	–	–
		M-F	100	–	–	–	–	–	–	–	–	–	–	–	–	–	–
	SUGARCANE, SILAGE *Saccharum officinarum*																
127	ALL ANALYSES	**A-F**	**22**	**0.08**	**0.04**	–	–	**0.05**	–	–	–	–	–	–	–	–	–
		M-F	100	0.35	0.18	–	–	0.22	–	–	–	–	–	–	–	–	–
128	TOPS	**A-F**	**30**	–	–	–	–	–	–	–	–	–	–	–	–	–	–
		M-F	100	–	–	–	–	–	–	–	–	–	–	–	–	–	–
129	**SUMMER CYPRESS, BELVEDERE, SILAGE** *Kochia scoparia*	**A-F**	**31**	–	–	–	–	–	–	–	–	–	–	–	–	–	–
		M-F	100	–	–	–	–	–	–	–	–	–	–	–	–	–	–
	SUNFLOWER, SILAGE *Helianthus* spp																
130	LATE BLOOM	**A-F**	**26**	–	–	–	–	–	–	–	–	–	–	–	–	–	–
		M-F	100	–	–	–	–	–	–	–	–	–	–	–	–	–	–
131	MILK STAGE	**A-F**	**21**	–	–	–	–	–	–	–	–	–	–	–	–	–	–
		M-F	100	–	–	–	–	–	–	–	–	–	–	–	–	–	–
132	MATURE	**A-F**	**26**	–	–	–	–	–	–	–	–	–	–	–	–	–	–
		M-F	100	–	–	–	–	–	–	–	–	–	–	–	–	–	–
	SWEET CLOVER, YELLOW–SEE CLOVER, SWEET, SILAGE																
	SWEET CORN, SILAGE *Zea mays, saccharata*																
133	CANNERY RESIDUE	**A-F**	**31**	**0.10**	**0.24**	**0.01**	–	**0.07**	**0.36**	**0.03**	–	–	–	**0.007**	–	–	–
		M-F	100	0.32	0.77	0.03	–	0.24	1.15	0.11	–	–	–	0.020	–	–	–
	SWEET POTATO, SILAGE *Ipomoea batatas*																
134	VINES	**A-F**	**13**	–	–	–	–	–	–	–	–	–	–	–	–	–	–
		M-F	100	–	–	–	–	–	–	–	–	–	–	–	–	–	–
135	TUBERS	**A-F**	**38**	**0.12**	**0.03**	–	–	–	–	–	–	–	–	–	–	–	–
		M-F	100	0.32	0.08	–	–	–	–	–	–	–	–	–	–	–	–
	TIMOTHY, SILAGE *Phleum pratense*																
136	ALL ANALYSES	**A-F**	**34**	**0.19**	**0.10**	**0.04**	–	**0.05**	**0.58**	**0.05**	–	**1.9**	–	**0.004**	**30.9**	–	–
		M-F	100	0.57	0.29	0.11	–	0.15	1.69	0.13	–	5.5	–	0.011	90.2	–	–
137	AIV ADDED	**A-F**	**35**	–	–	–	–	–	–	–	–	–	–	–	–	–	–
		M-F	100	–	–	–	–	–	–	–	–	–	–	–	–	–	–
138	CEREAL GRAIN ADDED	**A-F**	**33**	**0.19**	**0.11**	–	–	–	–	–	–	–	–	–	–	–	–
		M-F	100	0.57	0.33	–	–	–	–	–	–	–	–	–	–	–	–
139	MOLASSES ADDED, WILTED	**A-F**	**43**	**0.22**	**0.16**	–	–	–	–	–	–	–	–	–	–	–	–
		M-F	100	0.52	0.38	–	–	–	–	–	–	–	–	–	–	–	–

Entry Number	Fat-Soluble Vitamins					Water-Soluble Vitamins								
	A (1 mg Carotene = 1667 IU Vit A)	Carotene (Provitamin A)	D	E	K	B-12	Biotin	Choline	Folacin (Folic Acid)	Niacin	Pantothenic Acid (B-3)	(Pyri-doxine) B-6	Ribo-flavin (B-2)	Thiamin (B-1)
	IU/g	ppm or mg/kg	IU/kg	ppm or mg/kg	ppm or mg/kg	ppm or mg/kg	ppm or mg/kg	ppm or mg/kg	ppm or mg/kg	ppm or mg/kg	ppm or mg/kg	ppm or mg/kg	ppm or mg/kg	ppm or mg/kg
109	–	–	–	–	–	–	–	–	–	–	–	–	–	–
	–	–	–	–	–	–	–	–	–	–	–	–	–	–
110	–	–	–	–	–	–	–	–	–	–	–	–	–	–
	–	–	–	–	–	–	–	–	–	–	–	–	–	–
111	–	–	–	–	–	–	–	–	–	–	–	–	–	–
	–	–	–	–	–	–	–	–	–	–	–	–	–	–
112	**10.8**	**6.5**	**0**	–	–	–	–	–	–	–	–	–	–	–
	31.2	18.7	1	–	–	–	–	–	–	–	–	–	–	–
113	–	–	–	–	–	–	–	–	–	–	–	–	–	–
	–	–	–	–	–	–	–	–	–	–	–	–	–	–
114	–	–	–	–	–	–	–	–	–	–	–	–	–	–
	–	–	–	–	–	–	–	–	–	–	–	–	–	–
115	–	–	–	–	–	–	–	–	–	–	–	–	–	–
	–	–	–	–	–	–	–	–	–	–	–	–	–	–
116	**5.3**	**3.2**	–	–	–	–	–	–	–	–	–	–	–	–
	18.0	10.8	–	–	–	–	–	–	–	–	–	–	–	–
117	–	–	–	–	–	–	–	–	–	–	–	–	–	–
	–	–	–	–	–	–	–	–	–	–	–	–	–	–
118	–	–	–	–	–	–	–	–	–	–	–	–	–	–
	–	–	–	–	–	–	–	–	–	–	–	–	–	–
119	**20.4**	**12.2**	–	–	–	–	–	–	–	–	–	–	–	–
	70.7	42.4	–	–	–	–	–	–	–	–	–	–	–	–
120	**40.6**	**24.3**	–	–	–	–	–	–	–	–	–	–	–	–
	175.4	105.2	–	–	–	–	–	–	–	–	–	–	–	–
121	–	–	–	–	–	–	–	–	–	–	–	–	–	–
	–	–	–	–	–	–	–	–	–	–	–	–	–	–
122	**52.2**	**31.3**	–	–	–	–	–	–	–	–	–	–	–	–
	172.9	103.7	–	–	–	–	–	–	–	–	–	–	–	–
123	**15.7**	**9.4**	–	–	–	–	–	–	–	–	–	–	–	–
	62.5	37.5	–	–	–	–	–	–	–	–	–	–	–	–
124	–	–	–	–	–	–	–	–	–	–	–	–	–	–
	–	–	–	–	–	–	–	–	–	–	–	–	–	–
125	–	–	–	–	–	–	–	–	–	–	–	–	–	–
	–	–	–	–	–	–	–	–	–	–	–	–	–	–
126	–	–	–	–	–	–	–	–	–	–	–	–	–	–
	–	–	–	–	–	–	–	–	–	–	–	–	–	–
127	–	–	–	–	–	–	–	–	–	–	–	–	–	–
	–	–	–	–	–	–	–	–	–	–	–	–	–	–
128	–	–	–	–	–	–	–	–	–	–	–	–	–	–
	–	–	–	–	–	–	–	–	–	–	–	–	–	–
129	–	–	–	–	–	–	–	–	–	–	–	–	–	–
	–	–	–	–	–	–	–	–	–	–	–	–	–	–
130	–	–	–	–	–	–	–	–	–	–	–	–	–	–
	–	–	–	–	–	–	–	–	–	–	–	–	–	–
131	–	–	–	–	–	–	–	–	–	–	–	–	–	–
	–	–	–	–	–	–	–	–	–	–	–	–	–	–
132	–	–	–	–	–	–	–	–	–	–	–	–	–	–
	–	–	–	–	–	–	–	–	–	–	–	–	–	–
133	**6.9**	**4.2**	–	–	–	–	–	–	–	–	–	–	–	–
	22.3	13.4	–	–	–	–	–	–	–	–	–	–	–	–
134	–	–	–	–	–	–	–	–	–	–	–	–	–	–
	–	–	–	–	–	–	–	–	–	–	–	–	–	–
135	–	–	–	–	–	–	–	–	–	–	–	–	–	–
	–	–	–	–	–	–	–	–	–	–	–	–	–	–
136	**51.3**	**30.8**	–	–	–	–	–	–	–	–	–	–	–	–
	149.6	89.8	–	–	–	–	–	–	–	–	–	–	–	–
137	–	–	–	–	–	–	–	–	–	–	–	–	–	–
	–	–	–	–	–	–	–	–	–	–	–	–	–	–
138	–	–	–	–	–	–	–	–	–	–	–	–	–	–
	–	–	–	–	–	–	–	–	–	–	–	–	–	–
139	–	–	–	–	–	–	–	–	–	–	–	–	–	–
	–	–	–	–	–	–	–	–	–	–	–	–	–	–

(Continued)

SILAGES AND HAYLAGES

TABLE V-1D SILAGES AND HAYLAGES, COMPOSITION OF FEEDS, DATA EXPRESSED **AS-FED** AND **MOISTURE-FREE**—*(Continued)*

Entry Number	Feed Name Description	International Feed Number	Moisture Basis: A-F (as-fed) or M-F (moisture-free)	Proximate Analysis: Dry Matter	Ash	Crude Fiber	Neutral Det. Fib. (NDF)	Acid Det. Fib. (ADF)	Lignin	Ether Extract (Fat)	N-Free Extract	Crude Protein	Digestible Protein: Ruminant	Swine	Horse
				%	%	%	%	%	%	%	%	%	%	%	%
	TIMOTHY, SILAGE (Continued)														
140	WILTED	**3-04-930**	**A-F**	**42**	**2.9**	**14.2**	–	–	–	**1.3**	**18.7**	**4.4**	**2.2**	–	–
			M-F	100	7.0	34.2	–	–	–	3.2	45.0	10.7	5.4	–	–
141	WILTED, PHOSPHORIC ACID ADDED	**3-08-544**	**A-F**	**37**	**2.6**	**12.6**	–	–	–	**1.1**	**17.1**	**3.2**	**1.4**	–	–
			M-F	100	7.1	34.4	–	–	–	3.0	46.7	8.7	3.8	–	–
142	**TUMBLEWEED, TUMBLING (RUSSIAN THISTLE), SILAGE** *Salsola kali, tenuifolia*	**3-03-984**	**A-F**	**34**	**6.2**	**10.3**	–	–	–	**0.9**	**14.5**	**2.5**	**1.1**	–	–
			M-F	100	17.9	29.8	–	–	–	2.8	42.2	7.4	3.1	–	–
143	**VETCH, SILAGE** *Vicia* spp	**3-05-112**	**A-F**	**30**	**2.4**	**9.8**	–	–	–	**1.0**	**13.4**	**3.5**	**2.0**	–	–
			M-F	100	7.9	32.7	–	–	–	3.3	44.4	11.6	6.5	–	–
	VETCH-OATS–SEE OATS-VETCH, SILAGE														
144	**VETCH-WHEAT, MOLASSES ADDED, SILAGE** *Vicia* spp-*Triticum aestivum*	**3-08-548**	**A-F**	**29**	**2.2**	**7.9**	–	–	–	**1.0**	**15.2**	**2.9**	**1.6**	–	–
			M-F	100	7.5	27.1	–	–	–	3.4	52.1	9.9	5.4	–	–

[1]Neutral Detergent Fiber (NDF) and Acid Detergent Fiber (ADF) values taken from *Nutrient Requirements of Dairy Cattle,* 6th rev. ed., NRC, National Academy Press, 1988, pp. 90–110, Table 7–1.

TABLE V-1E PASTURE AND RANGE PLANTS, COMPOSITION OF FEEDS, DATA EXPRESSED **AS-FED** AND **MOISTURE-FREE** (See footnote at end of table.)

Entry Number	Feed Name Description	International Feed Number	Moisture Basis: A-F (as-fed) or M-F (moisture-free)	Proximate Analysis: Dry Matter	Ash	Crude Fiber	Neutral Det. Fib. (NDF)	Acid Det. Fib. (ADF)	Lignin	Ether Extract (Fat)	N-Free Extract	Crude Protein	Digestable Protein: Ruminant	Swine	Horse
				%	%	%	%	%	%	%	%	%	%	%	%
	ACACIA, PRAIRIE *Acacia angustissima*														
1	BROWSE, FRESH	**2-00-009**	**A-F**	**35**	**1.7**	**5.9**	–	–	–	**1.3**	**16.7**	**9.3**	–	–	**7.0**
			M-F	100	4.9	16.9	–	–	–	3.8	47.8	26.6	–	–	20.1
	ADAM'S NEEDLE *Yucca smalliana*														
2	FRESH	**2-08-561**	**A-F**	**49**	**3.5**	**21.1**	–	–	–	**1.0**	**20.0**	**3.8**	**1.4**	**2.2**	–
			M-F	100	7.1	42.7	–	–	–	2.0	40.5	7.7	2.8	4.5	–
	ALFALFA (LUCERNE) *Medicago sativa*														
3	FRESH, ALL ANALYSES	**2-00-196**	**A-F**	**26**	**2.5**	**6.0**	**11.8**	–	–	**1.0**	**11.2**	**5.3**	**4.0**	**3.8**	–
			M-F	100	9.5	23.0	45.4	–	–	3.8	43.1	20.5	15.5	14.8	–
4	PREBLOOM, FRESH	**2-00-181**	**A-F**	**20**	**2.1**	**4.9**	**6.4**	**5.2**	**1.5**	**0.6**	**8.5**	**4.3**	**3.3**	**3.0**	–
			M-F	100	10.2	24.2	31.7	25.5	7.5	2.9	41.6	21.1	16.3	15.0	–
5	EARLY BLOOM, FRESH	**2-00-184**	**A-F**	**24**	**3.1**	**6.6**	**9.0**	**7.2**	**2.0**	**0.7**	**8.2**	**5.4**	**4.3**	**3.9**	–
			M-F	100	12.7	27.6	37.3	29.9	8.2	3.0	34.1	22.5	17.8	16.4	–
6	MIDBLOOM, FRESH	**2-00-185**	**A-F**	**24**	**2.2**	**7.2**	**9.1**	**8.5**	**2.5**	**0.6**	**8.9**	**4.8**	**3.5**	**3.7**	–
			M-F	100	9.3	30.3	38.5	35.6	10.5	2.6	37.7	20.2	14.9	15.7	–
7	FULL BLOOM, FRESH	**2-00-188**	**A-F**	**24**	**2.4**	**7.2**	**9.6**	**9.0**	**3.2**	**0.6**	**9.0**	**4.6**	**3.4**	**3.3**	–
			M-F	100	10.1	30.4	40.1	37.9	13.6	2.6	37.7	19.2	14.4	13.7	–
	ALFALFA-BROMEGRASS *Medicago sativa-Bromus* spp														
8	FRESH	**2-08-328**	**A-F**	**23**	**2.2**	**5.3**	–	–	–	**0.8**	**9.4**	**4.8**	**3.3**	**3.5**	–
			M-F	100	9.8	23.6	–	–	–	3.6	41.8	21.3	14.7	15.4	–
	ALFALFA-BROMEGRASS, SMOOTH *Medicago sativa-Bromus inermis*														
9	FRESH	**2-00-262**	**A-F**	**22**	**2.1**	**5.5**	–	–	–	**0.8**	**9.0**	**4.2**	**3.2**	**3.0**	–
			M-F	100	9.8	25.3	–	–	–	3.6	41.7	19.6	14.9	14.0	–
	ALFALFA-ORCHARDGRASS *Medicago sativa-Dactylis glomerata*														
10	FRESH	**2-00-323**	**A-F**	**25**	–	–	–	–	–	–	–	–	–	–	–
			M-F	100	–	–	–	–	–	–	–	–	–	–	–
	ALFALFA-TIMOTHY *Medicago sativa-Phleum pratense*														
11	FRESH	**2-08-574**	**A-F**	**22**	**2.2**	**4.7**	–	–	–	**0.8**	**9.6**	**4.6**	**3.5**	**3.3**	–
			M-F	100	10.0	21.5	–	–	–	3.7	43.8	21.0	16.2	15.1	–
	ALFILERIA, REDSTEM (FILAREE) *Erodium cicutarium*														
12	FRESH	**2-00-356**	**A-F**	**18**	**2.4**	**3.6**	–	**5.3**	–	**0.6**	**8.2**	**2.8**	**2.0**	**1.9**	–
			M-F	100	13.7	20.4	–	30.0	–	3.5	46.4	15.9	11.5	11.1	–
	ALKALI SACATON *Sporobolus airoides*														
13	MIDBLOOM, FRESH	**2-05-601**	**A-F**	**40**	**3.7**	**14.5**	–	–	–	**0.7**	**18.6**	**2.5**	**1.1**	**1.3**	–
			M-F	100	9.2	36.2	–	–	–	1.8	46.5	6.3	2.8	3.3	–
	ALSIKE CLOVER–SEE CLOVER, ALSIKE														
	ALTA (TALL) FESCUE *Festuca arundinacea*														
14	FRESH	**2-01-889**	**A-F**	**28**	**2.5**	**7.5**	**19.5**	–	–	**0.9**	**14.3**	**2.7**	**1.7**	**1.7**	–
			M-F	100	9.0	26.7	69.9	–	–	3.3	51.2	9.8	6.0	6.1	–

SILAGES AND HAYLAGES

Entry Number	TDN Ruminant	TDN Swine	TDN Horse	Digestible Energy Ruminant		Digestible Energy Swine		Digestible Energy Horse		Metabolizable Energy Ruminant		Metabolizable Energy Swine		Metabolizable Energy Poultry ME_n		Metabolizable Energy Horse		Net Energy Ruminant NE_m		Net Energy Ruminant NE_g		Net Energy Lactating Cows NE_{lc}	
	%	%	%	Mcal lb	Mcal kg	kcal lb	kcal kg	Mcal lb	Mcal kg	Mcal lb	Mcal kg	kcal lb	kcal kg	kcal lb	kcal kg	Mcal lb	Mcal kg	Mcal lb	Mcal kg	Mcal lb	Mcal kg	Mcal lb	Mcal kg
140	24	–	–	0.49	1.07	–	–	–	–	0.41	0.89	–	–	–	–	–	–	0.24	0.53	0.13	0.29	0.25	0.55
	58	–	–	1.17	2.57	–	–	–	–	0.97	2.15	–	–	–	–	–	–	0.58	1.28	0.32	0.71	0.60	1.33
141	20	–	–	0.41	0.91	–	–	–	–	0.34	0.75	–	–	–	–	–	–	0.20	0.45	0.11	0.24	0.21	0.47
	55	–	–	1.13	2.48	–	–	–	–	0.93	2.06	–	–	–	–	–	–	0.56	1.23	0.30	0.66	0.59	1.29
142	14	–	–	0.31	0.69	–	–	–	–	0.24	0.54	–	–	–	–	–	–	0.15	0.32	0.06	0.13	0.17	0.37
	42	–	–	0.90	1.99	–	–	–	–	0.71	1.56	–	–	–	–	–	–	0.42	0.93	0.17	0.38	0.49	1.08
143	19	–	–	0.38	0.83	–	–	–	–	0.32	0.71	–	–	–	–	–	–	0.19	0.42	0.11	0.25	0.19	0.43
	63	–	–	1.25	2.76	–	–	–	–	1.06	2.34	–	–	–	–	–	–	0.64	1.41	0.37	0.82	0.65	1.42
144	17	–	–	0.36	0.78	–	–	–	–	0.30	0.66	–	–	–	–	–	–	0.18	0.40	0.11	0.23	0.19	0.41
	60	–	–	1.22	2.69	–	–	–	–	1.03	2.26	–	–	–	–	–	–	0.62	1.37	0.36	0.79	0.63	1.40

PASTURE AND RANGE PLANTS

Entry Number	TDN Ruminant	TDN Swine	TDN Horse	Digestible Energy Ruminant		Digestible Energy Swine		Digestible Energy Horse		Metabolizable Energy Ruminant		Metabolizable Energy Swine		Metabolizable Energy Poultry ME_n		Metabolizable Energy Horse		Net Energy Ruminant NE_m		Net Energy Ruminant NE_g		Net Energy Lactating Cows NE_{lc}	
	%	%	%	Mcal lb	Mcal kg	kcal lb	kcal kg	Mcal lb	Mcal kg	Mcal lb	Mcal kg	kcal lb	kcal kg	kcal lb	kcal kg	Mcal lb	Mcal kg	Mcal lb	Mcal kg	Mcal lb	Mcal kg	Mcal lb	Mcal kg
1	26	–	–	0.53	1.16	–	–	–	–	0.43	0.95	–	–	–	–	–	–	0.27	0.60	0.18	0.39	0.28	0.61
	75	–	–	1.51	3.32	–	–	–	–	1.23	2.72	–	–	–	–	–	–	0.78	1.71	0.51	1.12	0.79	1.73
2	27	–	–	0.54	1.20	–	–	–	–	0.45	0.99	–	–	–	–	–	–	0.25	0.55	0.12	0.27	0.27	0.60
	56	–	–	1.10	2.42	–	–	–	–	0.91	1.99	–	–	–	–	–	–	0.50	1.11	0.25	0.55	0.55	1.21
3	16	–	–	0.32	0.70	–	–	–	–	0.27	0.59	–	–	–	–	–	–	0.16	0.35	0.09	0.20	0.16	0.36
	61	–	–	1.22	2.68	–	–	–	–	1.03	2.26	–	–	–	–	–	–	0.61	1.34	0.35	0.77	0.62	1.37
4	12	12	–	0.24	0.54	–	–	–	–	0.20	0.44	–	–	–	–	–	–	0.12	0.27	0.07	0.15	0.13	0.28
	61	58	–	1.20	2.65	–	–	–	–	0.99	2.17	–	–	–	–	–	–	0.60	1.31	0.34	0.74	0.63	1.38
5	15	–	–	0.30	0.66	–	–	–	–	0.25	0.56	–	–	–	–	–	–	0.15	0.34	0.09	0.20	0.15	0.34
	63	–	–	1.25	2.76	–	–	–	–	1.06	2.34	–	–	–	–	–	–	0.63	1.40	0.37	0.82	0.64	1.42
6	14	14	–	0.29	0.63	–	–	–	–	0.24	0.53	–	–	–	–	–	–	0.14	0.31	0.08	0.18	0.14	0.31
	61	59	–	1.20	2.65	–	–	–	–	1.01	2.23	–	–	–	–	–	–	0.60	1.32	0.34	0.74	0.59	1.29
7	13	–	–	0.28	0.62	–	–	–	–	0.23	0.51	–	–	–	–	–	–	0.15	0.33	0.09	0.19	0.15	0.34
	55	–	–	1.17	2.58	–	–	–	–	0.98	2.16	–	–	–	–	–	–	0.63	1.39	0.37	0.81	0.64	1.41
8	14	–	–	0.29	0.64	–	–	–	–	0.25	0.55	–	–	–	–	–	–	0.16	0.35	0.10	0.22	0.16	0.35
	62	–	–	1.29	2.85	–	–	–	–	1.10	2.43	–	–	–	–	–	–	0.71	1.55	0.43	0.96	0.70	1.53
9	14	–	–	0.28	0.62	–	–	–	–	0.24	0.53	–	–	–	–	–	–	0.14	0.30	0.08	0.18	0.14	0.31
	63	–	–	1.30	2.86	–	–	–	–	1.10	2.43	–	–	–	–	–	–	0.64	1.41	0.38	0.83	0.65	1.42
10	–	–	–	–	–	–	–	–	–	–	–	–	–	–	–	–	–	–	–	–	–	–	–
	–	–	–	–	–	–	–	–	–	–	–	–	–	–	–	–	–	–	–	–	–	–	–
11	14	–	–	0.29	0.65	–	–	–	–	0.25	0.56	–	–	–	–	–	–	0.16	0.35	0.10	0.21	0.15	0.34
	66	–	–	1.34	2.95	–	–	–	–	1.15	2.53	–	–	–	–	–	–	0.72	1.58	0.44	0.98	0.70	1.55
12	10	–	–	0.21	0.46	–	–	–	–	0.18	0.39	–	–	–	–	–	–	0.11	0.25	0.07	0.15	0.11	0.25
	55	–	–	1.18	2.61	–	–	–	–	0.99	2.19	–	–	–	–	–	–	0.64	1.41	0.37	0.82	0.65	1.42
13	22	–	–	0.44	0.96	–	–	–	–	0.36	0.79	–	–	–	–	–	–	0.21	0.45	0.10	0.23	0.22	0.49
	55	–	–	1.09	2.41	–	–	–	–	0.90	1.98	–	–	–	–	–	–	0.51	1.13	0.26	0.57	0.55	1.22
14	17	–	–	0.34	0.76	–	–	–	–	0.29	0.64	–	–	–	–	–	–	0.17	0.38	0.10	0.22	0.18	0.39
	61	–	–	1.23	2.71	–	–	–	–	1.04	2.28	–	–	–	–	–	–	0.62	1.37	0.36	0.79	0.63	1.40

(Continued)

TABLE V-1D SILAGES AND HAYLAGES, MINERAL AND VITAMIN COMPOSITION OF FEEDS, DATA EXPRESSED **AS-FED** AND **MOISTURE-FREE**—*(Continued)*

Entry Number	Feed Name Description	Moisture Basis: A-F (as-fed) or M-F (moisture-free)	Dry Matter	Macro Minerals: Calcium (Ca)	Phosphorus (P)	Sodium (Na)	Chlorine (Cl)	Magnesium (Mg)	Potassium (K)	Sulfur (S)	Micro Minerals: Cobalt (Co)	Copper (Cu)	Iodine (I)	Iron (Fe)	Manganese (Mn)	Selenium (Se)	Zinc (Zn)
			%	%	%	%	%	%	%	%	ppm or mg/kg	ppm or mg/kg	ppm or mg/kg	%	ppm or mg/kg	ppm or mg/kg	ppm or mg/kg
	TIMOTHY, SILAGE (Continued)																
140	WILTED	**A-F**	**42**	**0.23**	**0.12**	**0.08**	**0.28**	**0.06**	**0.70**	**0.06**	–	**2.3**	–	**0.006**	**37.5**	–	–
		M-F	100	0.56	0.29	0.20	0.66	0.15	1.69	0.15	–	5.5	–	0.013	90.2	–	–
141	WILTED, PHOSPHORIC ACID ADDED	**A-F**	**37**	–	–	–	–	–	–	–	–	–	–	–	–	–	–
		M-F	100	–	–	–	–	–	–	–	–	–	–	–	–	–	–
142	**TUMBLEWEED, TUMBLING (RUSSIAN THISTLE), SILAGE** *Salsola kali, tenuifolia*	**A-F**	**34**	–	–	–	–	–	–	–	–	–	–	–	–	–	–
		M-F	100	–	–	–	–	–	–	–	–	–	–	–	–	–	–
143	**VETCH, SILAGE** *Vicia* spp	**A-F**	**30**	–	–	–	–	–	–	–	–	–	–	–	–	–	–
		M-F	100	–	–	–	–	–	–	–	–	–	–	–	–	–	–
	VETCH-OATS–SEE OATS-VETCH, SILAGE																
144	**VETCH-WHEAT, MOLASSES ADDED, SILAGE** *Vicia spp-Triticum aestivum*	**A-F**	**29**	–	–	–	–	–	–	–	–	–	–	–	–	–	–
		M-F	100	–	–	–	–	–	–	–	–	–	–	–	–	–	–

TABLE V-1E PASTURE AND RANGE PLANTS, MINERAL AND VITAMIN COMPOSITION OF FEEDS, DATA EXPRESSED **AS-FED** AND **MOISTURE-FREE**

Entry Number	Feed Name Description	Moisture Basis: A-F (as-fed) or M-F (moisture-free)	Dry Matter	Macro Minerals: Calcium (Ca)	Phosphorus (P)	Sodium (Na)	Chlorine (Cl)	Magnesium (Mg)	Potassium (K)	Sulfur (S)	Micro Minerals: Cobalt (Co)	Copper (Cu)	Iodine (I)	Iron (Fe)	Manganese (Mn)	Selenium (Se)	Zinc (Zn)
			%	%	%	%	%	%	%	%	ppm or mg/kg	ppm or mg/kg	ppm or mg/kg	%	ppm or mg/kg	ppm or mg/kg	ppm or mg/kg
	ACACIA, PRAIRIE *Acacia angustissima*																
1	BROWSE, FRESH	**A-F**	**35**	**0.34**	**0.10**	–	–	**0.14**	**0.45**	–	–	–	–	–	–	–	–
		M-F	100	0.97	0.29	–	–	0.41	1.28	–	–	–	–	–	–	–	–
	ADAM'S NEEDLE *Yucca smalliana*																
2	FRESH	**A-F**	**49**	–	–	–	–	–	–	–	–	–	–	–	–	–	–
		M-F	100	–	–	–	–	–	–	–	–	–	–	–	–	–	–
	ALFALFA (LUCERNE) *Medicago sativa*																
3	FRESH, ALL ANALYSES	**A-F**	**26**	**0.40**	**0.07**	**0.05**	**0.12**	**0.09**	**0.83**	**0.10**	**0.092**	**3.2**	–	**0.009**	**24.1**	–	**9.4**
		M-F	100	1.52	0.28	0.17	0.46	0.34	3.18	0.38	0.352	12.4	–	0.032	92.7	–	36.1
4	PREBLOOM, FRESH	**A-F**	**20**	**0.44**	**0.07**	**0.04**	**0.09**	**0.05**	**0.44**	**0.10**	**0.034**	**2.2**	–	**0.003**	**8.3**	–	–
		M-F	100	2.19	0.33	0.21	0.44	0.27	2.14	0.48	0.167	10.8	–	0.012	40.8	–	–
5	EARLY BLOOM, FRESH	**A-F**	**24**	**0.39**	**0.07**	**0.04**	–	**0.12**	**0.88**	–	**0.107**	**4.4**	–	**0.008**	**33.2**	–	**9.6**
		M-F	100	1.61	0.28	0.18	–	0.49	3.64	–	0.443	18.3	–	0.032	138.3	–	40.0
6	MIDBLOOM, FRESH	**A-F**	**24**	**0.40**	**0.07**	**0.03**	**0.11**	**0.10**	**0.79**	**0.07**	**0.091**	**4.8**	–	**0.011**	**37.1**	–	**9.1**
		M-F	100	1.69	0.31	0.14	0.45	0.41	3.31	0.29	0.381	20.2	–	0.044	156.4	–	38.5
7	FULL BLOOM, FRESH	**A-F**	**24**	**0.28**	**0.06**	**0.04**	**0.10**	**0.10**	**0.86**	**0.07**	**0.117**	**3.6**	–	**0.007**	**26.4**	–	**8.5**
		M-F	100	1.19	0.26	0.16	0.43	0.40	3.63	0.31	0.489	14.9	–	0.030	110.9	–	35.9
	ALFALFA-BROMEGRASS *Medicago sativa-Bromus* spp																
8	FRESH	**A-F**	**23**	**0.28**	**0.07**	–	–	–	**0.63**	–	–	–	–	–	–	–	–
		M-F	100	1.24	0.31	–	–	–	2.80	–	–	–	–	–	–	–	–
	ALFALFA-BROMEGRASS, SMOOTH *Medicago sativa-Bromus inermis*																
9	FRESH	**A-F**	**22**	**0.33**	**0.08**	**0.09**	–	**0.08**	**0.84**	**0.05**	–	–	–	**0.003**	–	–	–
		M-F	100	1.52	0.37	0.42	–	0.35	3.87	0.23	–	–	–	0.013	–	–	–
	ALFALFA-ORCHARDGRASS *Medicago sativa-Dactylis glomerata*																
10	FRESH	**A-F**	**25**	**0.10**	**0.13**	–	–	**0.06**	–	–	–	–	–	–	–	–	–
		M-F	100	0.40	0.52	–	–	0.24	–	–	–	–	–	–	–	–	–
	ALFALFA-TIMOTHY *Medicago sativa-Phleum pratense*																
11	FRESH	**A-F**	**22**	**0.30**	**0.08**	–	–	–	**0.49**	–	–	–	–	–	–	–	–
		M-F	100	1.37	0.37	–	–	–	2.24	–	–	–	–	–	–	–	–
	AFILERIA REDSTEM (FILAREE) *Erodium cicutarium*																
12	FRESH	**A-F**	**18**	**0.35**	**0.08**	–	–	–	**0.59**	–	–	–	–	–	–	–	–
		M-F	100	1.99	0.43	–	–	–	3.37	–	–	–	–	–	–	–	–
	ALKALI SACATON *Sporobolus airoides*																
13	MIDBLOOM, FRESH	**A-F**	**40**	**0.23**	**0.07**	**0.02**	–	**0.12**	**0.69**	–	**0.228**	**6.4**	–	**0.019**	**12.0**	–	**10.8**
		M-F	100	0.58	0.17	0.04	–	0.29	1.73	–	0.570	16.0	–	0.048	30.0	–	27.0
	ALSIKE CLOVER–SEE CLOVER, ALSIKE																
	ALTA (TALL) FESCUE *Festuca arundinacea*																
14	FRESH	**A-F**	**28**	**0.13**	**0.05**	**0.03**	–	**0.07**	**0.70**	–	**0.113**	**1.0**	–	**0.003**	**18.0**	–	**5.9**
		M-F	100	0.48	0.19	0.12	–	0.25	2.51	–	0.401	3.4	–	0.011	64.4	–	21.0

SILAGES AND HAYLAGES

Entry Number	Fat-Soluble Vitamins					Water-Soluble Vitamins								
	A (1 mg Carotene = 1667 IU Vit A)	Carotene (Provitamin A)	D	E	K	B-12	Biotin	Choline	Folacin (Folic Acid)	Niacin	Pantothenic Acid (B-3)	(Pyri-doxine) B-6	Ribo-flavin (B-2)	Thiamin (B-1)
	IU/g	ppm or mg/kg	IU/kg	ppm or mg/kg	ppm or mg/kg	ppm or mg/kg	ppm or mg/kg	ppm or mg/kg	ppm or mg/kg	ppm or mg/kg	ppm or mg/kg	ppm or mg/kg	ppm or mg/kg	ppm or mg/kg
140	–	–	–	–	–	–	–	–	–	–	–	–	–	–
	–	–	–	–	–	–	–	–	–	–	–	–	–	–
141	–	–	–	–	–	–	–	–	–	–	–	–	–	–
	–	–	–	–	–	–	–	–	–	–	–	–	–	–
142	–	–	–	–	–	–	–	–	–	–	–	–	–	–
	–	–	–	–	–	–	–	–	–	–	–	–	–	–
143	–	–	–	–	–	–	–	–	–	–	–	–	–	–
	–	–	–	–	–	–	–	–	–	–	–	–	–	–
144	–	–	–	–	–	–	–	–	–	–	–	–	–	–
	–	–	–	–	–	–	–	–	–	–	–	–	–	–

PASTURE AND RANGE PLANTS

Entry Number	Fat-Soluble Vitamins					Water-Soluble Vitamins								
	A (1 mg Carotene = 1667 IU Vit A)	Carotene (Provitamin A)	D	E	K	B-12	Biotin	Choline	Folacin (Folic Acid)	Niacin	Pantothenic Acid (B-3)	(Pyri-doxine) B-6	Ribo-flavin (B-2)	Thiamin (B-1)
	IU/g	ppm or mg/kg	IU/kg	ppm or mg/kg	ppm or mg/kg	ppm or mg/kg	ppm or mg/kg	ppm or mg/kg	ppm or mg/kg	ppm or mg/kg	ppm or mg/kg	ppm or mg/kg	ppm or mg/kg	ppm or mg/kg
1	–	–	–	–	–	–	–	–	–	–	–	–	–	–
	–	–	–	–	–	–	–	–	–	–	–	–	–	–
2	–	–	–	–	–	–	–	–	–	–	–	–	–	–
	–	–	–	–	–	–	–	–	–	–	–	–	–	–
3	**101.3**	**60.8**	**0**	–	–	–	**0.13**	**374**	**0.64**	**15**	**8.9**	**1.66**	**4.6**	**1.7**
	389.3	233.5	0	–	–	–	0.49	1439	2.47	59	34.3	6.38	17.5	6.4
4	–	–	**0**	**34.8**	–	–	–	–	–	–	–	–	–	–
	–	–	0	171.5	–	–	–	–	–	–	–	–	–	–
5	**69.9**	**41.9**	–	–	–	–	–	–	–	–	–	–	–	–
	291.0	174.6	–	–	–	–	–	–	–	–	–	–	–	–
6	**56.1**	**33.7**	–	–	–	–	–	–	–	–	–	–	–	–
	236.4	141.8	–	–	–	–	–	–	–	–	–	–	–	–
7	–	–	–	–	–	–	–	–	–	–	**7.5**	–	–	–
	–	–	–	–	–	–	–	–	–	–	31.3	–	–	–
8	–	–	–	–	–	–	–	–	–	–	–	–	–	–
	–	–	–	–	–	–	–	–	–	–	–	–	–	–
9	–	–	–	–	–	–	–	–	–	–	–	–	–	–
	–	–	–	–	–	–	–	–	–	–	–	–	–	–
10	–	–	–	–	–	–	–	–	–	–	–	–	–	–
	–	–	–	–	–	–	–	–	–	–	–	–	–	–
11	–	–	–	–	–	–	–	–	–	–	–	–	–	–
	–	–	–	–	–	–	–	–	–	–	–	–	–	–
12	–	–	–	–	–	–	–	–	–	–	–	–	–	–
	–	–	–	–	–	–	–	–	–	–	–	–	–	–
13	–	–	–	–	–	–	–	–	–	–	–	–	–	–
	–	–	–	–	–	–	–	–	–	–	–	–	–	–
14	–	–	–	–	–	–	–	–	–	–	–	–	–	–
	–	–	–	–	–	–	–	–	–	–	–	–	–	–

TABLE V-1E PASTURE AND RANGE PLANTS, COMPOSITION OF FEEDS, DATA EXPRESSED **AS-FED** AND **MOISTURE-FREE**—*(Continued)*

Entry Number	Feed Name Description	International Feed Number	Moisture Basis: A-F (as-fed) or M-F (moisture-free)	Proximate Analysis: Dry Matter	Ash	Crude Fiber	Neutral Det. Fib. (NDF)	Acid Det. Fib. (ADF)	Lignin	Ether Extract (Fat)	N-Free Extract	Crude Protein	Digestible Protein: Ruminant	Swine	Horse
				%	%	%	%	%	%	%	%	%	%	%	%
15	**ALYCE CLOVER** *Alysicarpus vaginalis* FRESH	2-00-362	A-F	20	2.2	5.5	–	–	–	0.6	8.6	3.1	–	–	–
			M-F	100	10.8	27.4	–	–	–	3.2	43.2	15.4	–	–	–
16	**ARTICHOKE, JERUSALEM** *Helianthus tuberosus* FRESH	2-20-655	A-F	27	2.1	4.9	–	–	–	0.3	18.4	1.4	–	–	–
			M-F	100	7.7	18.0	–	–	–	1.1	68.0	5.1	–	–	–
17	**ASPARAGUS, GARDEN** *Asparagus officinalis* IMMATURE, FRESH	2-00-435	A-F	8	0.1	0.9	–	–	–	0.2	3.8	2.4	–	–	–
			M-F	100	0.7	11.1	–	–	–	2.6	47.8	30.1	–	–	–
18	**ASPEN, QUAKING** *Populus tremuloides* BROWSE, FRESH	2-00-442	A-F	38	2.8	5.4	–	–	–	2.1	22.3	5.5	–	–	3.7
			M-F	100	7.4	14.3	–	–	–	5.5	58.6	14.4	–	–	9.8
19	**BAHIAGRASS** *Paspalum notatum* FRESH[1]	2-00-464	A-F	30	3.3	9.0	20.4	11.4	–	0.5	14.2	2.6	1.5	1.6	–
			M-F	100	11.1	30.4	68.0	38.0	–	1.6	48.0	8.9	5.1	5.4	–
20	**BALSAMROOT, ARROWLEAF** *Balsamorhiza sagittata* FRESH	2-00-479	A-F	35	4.4	6.5	–	–	–	1.5	16.1	6.4	5.0	–	4.6
			M-F	100	12.7	18.5	–	–	–	4.3	46.1	18.4	14.2	–	13.2
21	**BANANA** *Musa* spp FRESH	2-00-483	A-F	16	2.1	3.8	–	–	–	0.1	9.0	1.0	0.6	0.5	–
			M-F	100	13.1	23.7	–	–	–	0.8	56.0	6.4	3.5	3.4	–
22	**BARLEY** *Hordeum vulgare* FRESH	2-00-511	A-F	21	2.7	4.7	–	–	–	0.8	8.8	4.3	3.1	3.1	–
			M-F	100	12.6	22.1	–	–	–	3.8	41.2	20.4	14.4	14.6	–
23	**BARLEY GRASS** *Hordeum leporinum* FRESH	2-07-758	A-F	29	1.9	–	–	–	–	0.4	–	1.4	0.4	0.6	–
			M-F	100	6.6	–	–	–	–	1.5	–	4.8	1.5	2.1	–
24	**BARNYARD GRASS (MILLET, JAPANESE)** *Echinochloa crusgalli* FRESH	2-03-108	A-F	22	1.6	6.8	–	–	–	0.6	11.0	1.7	1.0	1.0	–
			M-F	100	7.4	31.3	–	–	–	2.8	50.7	7.8	4.7	4.6	–
25	**BEAN, LIMA** *Phaseolus limensis* PODS, FRESH	2-00-607	A-F	28	–	10.9	–	–	–	0.5	–	2.0	1.0	1.1	–
			M-F	100	–	38.6	–	–	–	1.9	–	7.2	3.6	4.1	–
26	**BEAN, MUNG** *Phaseolus aureus* FRESH	2-00-619	A-F	16	2.1	4.1	–	–	–	0.4	6.6	2.8	2.1	2.0	–
			M-F	100	13.1	25.6	–	–	–	2.3	41.4	17.6	13.1	12.4	–
	BEARDGRASS–SEE BLUESTEM														
27	**BEET, MANGEL** *Beta vulgaris, macrorrhiza* TOPS WITH CROWNS, FRESH	2-00-632	A-F	13	2.4	1.4	–	–	–	0.5	6.1	2.1	1.8	1.5	–
			M-F	100	19.2	11.4	–	–	–	4.2	48.2	17.0	13.9	11.9	–
28	**BEET, SUGAR** *Beta vulgaris, altissima* TOPS WITH CROWNS, FRESH	2-00-649	A-F	17	3.4	1.8	–	–	–	0.3	8.6	2.5	1.9	1.7	–
			M-F	100	20.4	10.7	–	–	–	2.1	51.9	15.0	11.3	10.3	–
29	**BEGGARWEED (TICKCLOVER, CHEROKEE)** *Desmodium tortuosum* FRESH	2-08-541	A-F	27	3.2	7.5	–	–	–	0.5	11.7	4.2	3.0	2.9	–
			M-F	100	11.8	27.7	–	–	–	1.8	43.2	15.5	11.2	10.7	–
30	**BENTGRASS, COLONIAL** *Agrostis tenuis* FRESH	2-00-687	A-F	29	3.0	6.6	–	–	–	1.1	13.1	5.4	4.1	3.8	–
			M-F	100	10.2	22.6	–	–	–	3.6	45.0	18.5	13.9	13.2	–
31	**BERMUDAGRASS** *Cynodon dactylon* FRESH	2-00-712	A-F	29	3.3	7.6	–	–	–	0.6	13.0	4.2	3.0	2.9	–
			M-F	100	11.4	26.6	–	–	–	2.1	45.4	14.6	10.4	10.0	–
32	**BERMUDAGRASS, COASTAL** *Cynodon dactylon* FRESH	2-00-719	A-F	29	1.8	8.3	–	–	–	1.1	13.6	4.4	3.2	3.0	–
			M-F	100	6.3	28.4	–	–	–	3.8	46.6	15.0	11.0	10.3	–
	BERSEEM CLOVER (EGYPTIAN CLOVER)–SEE CLOVER, EGYPTIAN														
33	**BIRCH, SWEET** *Betula lenta* BROWSE, IMMATURE, FRESH	2-00-724	A-F	33	2.5	5.6	–	–	–	2.8	12.7	9.2	7.4	6.8	–
			M-F	100	7.7	16.9	–	–	–	8.6	38.7	28.1	22.6	20.8	–
34	**BIRDSFSOOT TREFOIL (DEERVETCH, BIRDSFOOT)** *Lotus corniculatus* BROWSE, FRESH	2-20-786	A-F	19	2.2	4.1	9.5	–	–	0.8	8.5	3.7	2.8	2.7	–
			M-F	100	11.2	21.2	49.4	–	–	4.0	44.3	19.3	14.6	13.8	–

Entry Number	TDN Ruminant	TDN Swine	TDN Horse	Digestible Energy Ruminant		Digestible Energy Swine		Digestible Energy Horse		Metabolizable Energy Ruminant		Metabolizable Energy Swine		Metabolizable Energy Poultry ME_n		Metabolizable Energy Horse		Net Energy Ruminant NE_m		Net Energy Ruminant NE_g		Net Energy Lactating Cows NE_{lc}	
	%	%	%	Mcal		kcal		Mcal		Mcal		kcal		kcal		Mcal		Mcal		Mcal		Mcal	
				lb	*kg*	lb	*kg*	lb	*kg*	lb	*kg*	lb	*kg*	lb	*kg*	lb	*kg*	lb	*kg*	lb	*kg*	lb	*kg*
15	–	–	–	–	–	–	–	–	–	–	–	–	–	–	–	–	–	–	–	–	–	–	–
	–	–	–	–	–	–	–	–	–	–	–	–	–	–	–	–	–	–	–	–	–	–	–
16	–	–	–	–	–	–	–	–	–	–	–	–	–	–	–	–	–	–	–	–	–	–	–
	–	–	–	–	–	–	–	–	–	–	–	–	–	–	–	–	–	–	–	–	–	–	–
17	–	–	–	–	–	–	–	–	–	–	–	–	–	–	–	–	–	–	–	–	–	–	–
	–	–	–	–	–	–	–	–	–	–	–	–	–	–	–	–	–	–	–	–	–	–	–
18	28	–	–	0.56	*1.22*	–	–	–	–	0.46	*1.00*	–	–	–	–	–	–	0.29	*0.64*	0.18	*0.40*	0.29	*0.64*
	73	–	–	1.46	*3.22*	–	–	–	–	1.20	*2.64*	–	–	–	–	–	–	0.76	*1.67*	0.48	*1.05*	0.76	*1.67*
19	16	–	–	0.32	*0.70*	–	–	–	–	0.26	*0.57*	–	–	–	–	–	–	0.15	*0.33*	0.08	*0.17*	0.16	*0.36*
	54	–	–	1.08	*2.37*	–	–	–	–	0.88	*1.95*	–	–	–	–	–	–	0.51	*1.13*	0.26	*0.57*	0.55	*1.22*
20	23	–	–	0.47	*1.04*	–	–	–	–	0.39	*0.85*	–	–	–	–	–	–	0.23	*0.52*	0.14	*0.32*	0.28	*0.61*
	67	–	–	1.35	*2.97*	–	–	–	–	1.10	*2.43*	–	–	–	–	–	–	0.67	*1.47*	0.41	*0.90*	0.79	*1.73*
21	10	–	7	0.20	*0.43*	–	–	0.14	*0.31*	0.17	*0.36*	–	–	–	–	0.12	*0.25*	0.09	*0.20*	0.05	*0.11*	0.09	*0.21*
	65	–	46	1.23	*2.70*	–	–	0.87	*1.93*	1.03	*2.28*	–	–	–	–	0.72	*1.58*	0.56	*1.23*	0.30	*0.66*	0.58	*1.29*
22	13	–	–	0.27	*0.60*	–	–	–	–	0.23	*0.51*	–	–	–	–	–	–	0.14	*0.31*	0.09	*0.19*	0.14	*0.31*
	63	–	–	1.28	*2.83*	–	–	–	–	1.09	*2.41*	–	–	–	–	–	–	0.67	*1.47*	0.40	*0.88*	0.67	*1.47*
23	–	–	–	–	–	–	–	–	–	–	–	–	–	–	–	–	–	–	–	–	–	–	–
	–	–	–	–	–	–	–	–	–	–	–	–	–	–	–	–	–	–	–	–	–	–	–
24	14	–	–	0.27	*0.60*	–	–	–	–	0.23	*0.51*	–	–	–	–	–	–	0.13	*0.29*	0.07	*0.16*	0.13	*0.29*
	65	–	–	1.26	*2.77*	–	–	–	–	1.07	*2.35*	–	–	–	–	–	–	0.60	*1.32*	0.34	*0.74*	0.61	*1.36*
25	–	–	–	–	–	–	–	–	–	–	–	–	–	–	–	–	–	–	–	–	–	–	–
	–	–	–	–	–	–	–	–	–	–	–	–	–	–	–	–	–	–	–	–	–	–	–
26	10	–	–	0.20	*0.44*	–	–	–	–	0.17	*0.38*	–	–	–	–	–	–	0.10	*0.22*	0.06	*0.12*	0.10	*0.22*
	63	–	–	1.26	*2.78*	–	–	–	–	1.07	*2.35*	–	–	–	–	–	–	0.61	*1.35*	0.35	*0.77*	0.63	*1.38*
27	8	–	–	0.16	*0.36*	–	–	–	–	0.14	*0.30*	–	–	–	–	–	–	0.08	*0.18*	0.05	*0.10*	0.08	*0.18*
	65	–	–	1.28	*2.82*	–	–	–	–	1.09	*2.40*	–	–	–	–	–	–	0.64	*1.40*	0.37	*0.82*	0.64	*1.42*
28	11	–	–	0.21	*0.46*	–	–	–	–	0.18	*0.39*	–	–	–	–	–	–	0.10	*0.22*	0.06	*0.13*	0.10	*0.23*
	65	–	–	1.25	*2.76*	–	–	–	–	1.06	*2.34*	–	–	–	–	–	–	0.61	*1.33*	0.34	*0.76*	0.62	*1.37*
29	14	–	–	0.31	*0.68*	–	–	–	–	0.26	*0.56*	–	–	–	–	–	–	0.16	*0.36*	0.09	*0.20*	0.17	*0.37*
	53	–	–	1.13	*2.50*	–	–	–	–	0.94	*2.07*	–	–	–	–	–	–	0.60	*1.33*	0.34	*0.75*	0.62	*1.36*
30	20	–	–	0.39	*0.86*	–	–	–	–	0.33	*0.74*	–	–	–	–	–	–	0.20	*0.44*	0.12	*0.27*	0.20	*0.44*
	67	–	–	1.33	*2.94*	–	–	–	–	1.14	*2.52*	–	–	–	–	–	–	0.69	*1.52*	0.42	*0.93*	0.68	*1.51*
31	17	–	–	0.35	*0.77*	–	–	–	–	0.29	*0.65*	–	–	–	–	–	–	0.18	*0.39*	0.10	*0.22*	0.18	*0.40*
	61	–	–	1.22	*2.68*	–	–	–	–	1.02	*2.26*	–	–	–	–	–	–	0.62	*1.36*	0.36	*0.78*	0.63	*1.39*
32	19	–	–	0.37	*0.82*	–	–	–	–	0.32	*0.70*	–	–	–	–	–	–	0.18	*0.40*	0.11	*0.23*	0.19	*0.41*
	64	–	–	1.28	*2.83*	–	–	–	–	1.09	*2.41*	–	–	–	–	–	–	0.63	*1.39*	0.37	*0.81*	0.64	*1.41*
33	–	–	–	–	–	–	–	–	–	–	–	–	–	–	–	–	–	–	–	–	–	–	–
	–	–	–	–	–	–	–	–	–	–	–	–	–	–	–	–	–	–	–	–	–	–	–
34	13	–	–	0.26	*0.58*	–	–	–	–	0.23	*0.50*	–	–	–	–	–	–	0.16	*0.35*	0.10	*0.22*	0.15	*0.33*
	68	–	–	1.36	*2.99*	–	–	–	–	1.17	*2.57*	–	–	–	–	–	–	0.81	*1.79*	0.53	*1.16*	0.78	*1.72*

(Continued)

TABLE V-1E PASTURE AND RANGE PLANTS, MINERAL AND VITAMIN COMPOSITION OF FEEDS, DATA EXPRESSED **AS-FED** AND **MOISTURE-FREE**—*(Continued)*

Entry Number	Feed Name Description	Moisture Basis: A-F (as-fed) or M-F (moisture-free)	Dry Matter	Macro Minerals							Micro Minerals						
				Calcium (Ca)	Phosphorus (P)	Sodium (Na)	Chlorine (Cl)	Magnesium (Mg)	Potassium (K)	Sulfur (S)	Cobalt (Co)	Copper (Cu)	Iodine (I)	Iron (Fe)	Manganese (Mn)	Selenium (Se)	Zinc (Zn)
			%	%	%	%	%	%	%	%	ppm or mg/kg	ppm or mg/kg	ppm or mg/kg	%	ppm or mg/kg	ppm or mg/kg	ppm or mg/kg
	ALYCE CLOVER *Alysicarpus vaginalis*																
15	FRESH	A-F	20	–	–	–	–	–	–	–	0.018	–	–	–	–	–	–
		M-F	100	–	–	–	–	–	–	–	0.086	–	–	–	–	–	–
	ARTICHOKE, JERUSALEM *Helianthus tuberosus*																
16	FRESH	A-F	27	0.43	0.03	–	–	–	0.37	–	–	–	–	–	–	–	–
		M-F	100	1.60	0.11	–	–	–	1.36	–	–	–	–	–	–	–	–
	ASPARAGUS, GARDEN *Asparagus officinalis*																
17	IMMATURE, FRESH	A-F	8	–	–	–	–	–	–	–	–	–	–	–	–	–	–
		M-F	100	–	–	–	–	–	–	–	–	–	–	–	–	–	–
	ASPEN, QUAKING *Populus tremuloides*																
18	BROWSE, FRESH	A-F	38	1.15	0.11	–	–	0.22	–	0.10	–	–	–	–	–	–	–
		M-F	100	3.02	0.30	–	–	0.58	–	0.26	–	–	–	–	–	–	–
	BAHIAGRASS *Paspalum notatum*																
19	FRESH	A-F	30	0.14	0.06	–	–	0.07	0.43	–	–	–	–	–	–	–	–
		M-F	100	0.46	0.22	–	–	0.25	1.45	–	–	–	–	–	–	–	–
	BALSAMROOT, ARROWLEAF *Balsamorhiza sagittata*																
20	FRESH	A-F	35	–	–	–	–	–	–	–	–	–	–	–	–	–	–
		M-F	100	–	–	–	–	–	–	–	–	–	–	–	–	–	–
	BANANA *Musa* spp																
21	FRESH	A-F	16	–	–	–	–	–	–	–	–	–	–	–	–	–	–
		M-F	100	–	–	–	–	–	–	–	–	–	–	–	–	–	–
	BARLEY *Hordeum vulgare*																
22	FRESH	A-F	21	0.13	0.09	–	–	–	–	–	–	–	–	–	–	–	–
		M-F	100	0.60	0.40	–	–	–	–	–	–	–	–	–	–	–	–
	BARLEY GRASS *Hordeum leporinum*																
23	FRESH	A-F	29	–	–	–	–	–	–	–	–	–	–	–	–	–	–
		M-F	100	–	–	–	–	–	–	–	–	–	–	–	–	–	–
	BARNYARD GRASS (MILLET, JAPANESE) *Echinochloa crusgalli*																
24	FRESH	A-F	22	0.11	0.07	–	–	–	0.52	–	–	–	–	–	–	–	–
		M-F	100	0.51	0.32	–	–	–	2.40	–	–	–	–	–	–	–	–
	BEAN, LIMA *Phaseolus limensis*																
25	PODS, FRESH	A-F	28	–	–	–	–	–	–	–	–	–	–	–	–	–	–
		M-F	100	–	–	–	–	–	–	–	–	–	–	–	–	–	–
	BEAN, MUNG *Phaseolus aureus*																
26	FRESH	A-F	16	–	–	–	–	–	–	–	–	–	–	–	–	–	–
		M-F	100	–	–	–	–	–	–	–	–	–	–	–	–	–	–
	BEARDGRASS–SEE BLUESTEM																
	BEET, MANGEL *Beta vulgaris, macrorrhiza*																
27	TOPS WITH CROWNS, FRESH	A-F	13	–	–	–	–	–	–	–	–	–	–	–	–	–	–
		M-F	100	–	–	–	–	–	–	–	–	–	–	–	–	–	–
	BEET, SUGAR *Beta vulgaris, altissima*																
28	TOPS WITH CROWNS, FRESH	A-F	17	0.17	0.04	0.09	0.09	0.18	0.96	0.09	–	2.3	–	0.003	9.0	–	–
		M-F	100	1.01	0.23	0.54	0.56	1.07	5.79	0.57	–	13.6	–	0.017	54.5	–	–
	BEGGARWEED (TICKCLOVER, CHEROKEE) *Desmodium tortuosum*																
29	FRESH	A-F	27	–	0.12	–	–	–	0.47	–	–	–	–	–	–	–	–
		M-F	100	–	0.44	–	–	–	1.73	–	–	–	–	–	–	–	–
	BENTGRASS, COLONIAL *Agrostis tenuis*																
30	FRESH	A-F	29	0.19	0.12	–	–	–	0.65	–	–	–	–	–	–	–	–
		M-F	100	0.65	0.41	–	–	–	2.21	–	–	–	–	–	–	–	–
	BERMUDAGRASS *Cynodon dactylon*																
31	FRESH	A-F	29	0.16	0.06	0.13	–	0.07	0.55	–	0.022	1.6	–	0.033	28.6	–	–
		M-F	100	0.55	0.21	0.44	–	0.24	1.92	–	0.075	5.7	–	0.112	100.1	–	–
	BERMUDAGRASS, COASTAL *Cynodon dactylon*																
32	FRESH	A-F	29	0.14	0.08	–	–	–	–	–	–	–	–	–	–	–	–
		M-F	100	0.49	0.27	–	–	–	–	–	–	–	–	–	–	–	–
	BERSEEM CLOVER (EGYPTIAN CLOVER)–SEE CLOVER, EGYPTIAN																
	BIRCH, SWEET *Betula lenta*																
33	BROWSE, IMMATURE, FRESH	A-F	33	0.40	0.11	–	–	–	–	–	–	–	–	–	479.9	–	–
		M-F	100	1.21	0.32	–	–	–	–	–	–	–	–	–	1467.5	–	–
	BIRDSFOOT TREFOIL (DEERVETCH, BIRDSFOOT) *Lotus corniculatus*																
34	BROWSE, FRESH	A-F	19	0.34	0.05	0.02	–	0.08	0.63	0.05	0.094	2.5	–	0.006	16.0	–	6.0
		M-F	100	1.74	0.26	0.11	–	0.40	3.26	0.25	0.487	12.8	–	0.031	82.9	–	31.1

PASTURE AND RANGE PLANTS

Entry Number	Fat-Soluble Vitamins: A (1 mg Carotene = 1667 IU Vit A)	Carotene (Provitamin A)	D	E	K	Water-Soluble Vitamins: B–12	Biotin	Choline	Folacin (Folic Acid)	Niacin	Pantothenic Acid (B–3)	(Pyri-doxine) B–6	Ribo-flavin (B–2)	Thiamin (B–1)
	IU/g	ppm or mg/kg	IU/kg	ppm or mg/kg	ppm or mg/kg	ppm or mg/kg	ppm or mg/kg	ppm or mg/kg	ppm or mg/kg	ppm or mg/kg	ppm or mg/kg	ppm or mg/kg	ppm or mg/kg	ppm or mg/kg
15	**35.0** 174.9	**21.0** 104.9	– –	– –	– –	– –	– –	– –	– –	– –	– –	– –	– –	– –
16	– –	– –	– –	– –	– –	– –	– –	– –	– –	– –	– –	– –	– –	– –
17	– –	– –	– –	– –	– –	– –	– –	– –	– –	– –	– –	– –	– –	– –
18	– –	– –	– –	– –	– –	– –	– –	– –	– –	– –	– –	– –	– –	– –
19	**89.7** 304.2	**53.8** 182.5	– –	– –	– –	– –	– –	– –	– –	– –	– –	– –	– –	– –
20	– –	– –	– –	– –	– –	– –	– –	– –	– –	– –	– –	– –	– –	– –
21	– –	– –	– –	– –	– –	– –	– –	– –	– –	– –	– –	– –	– –	– –
22	**81.8** 384.0	**49.1** 230.4	– –	– –	– –	– –	– –	– –	– –	– –	– –	– –	– –	– –
23	– –	– –	– –	– –	– –	– –	– –	– –	– –	– –	– –	– –	– –	– –
24	– –	– –	– –	– –	– –	– –	– –	– –	– –	– –	– –	– –	– –	– –
25	**11.0** 39.0	**6.6** 23.4	– –	– –	– –	– –	– –	– –	– –	– –	– –	– –	**1.1** 4.0	– –
26	– –	– –	– –	– –	– –	– –	– –	– –	– –	– –	– –	– –	– –	– –
27	– –	– –	– –	– –	– –	– –	– –	– –	– –	– –	– –	– –	– –	– –
28	**9.6** 58.1	**5.8** 34.8	– –	– –	– –	– –	– –	– –	– –	– –	– –	– –	**1.1** 6.6	– –
29	– –	– –	– –	– –	– –	– –	– –	– –	– –	– –	– –	– –	– –	– –
30	– –	– –	– –	– –	– –	– –	– –	– –	– –	– –	– –	– –	– –	– –
31	**147.8** 517.3	**88.7** 310.3	– –	– –	– –	– –	– –	– –	– –	– –	– –	– –	– –	– –
32	**160.3** 550.9	**96.1** 330.5	– –	– –	– –	– –	– –	– –	– –	– –	– –	– –	– –	– –
33	– –	– –	– –	– –	– –	– –	– –	– –	– –	– –	– –	– –	– –	– –
34	– –	– –	– –	– –	– –	– –	– –	– –	– –	– –	– –	– –	– –	– –

(Continued)

TABLE V-1E PASTURE AND RANGE PLANTS, COMPOSITION OF FEEDS, DATA EXPRESSED **AS-FED** AND **MOISTURE-FREE**—*(Continued)*

Entry Number	Feed Name Description	International Feed Number	Moisture Basis: A-F (as-fed) or M-F (moisture-free)	Proximate Analysis: Dry Matter	Ash	Crude Fiber	Neutral Det. Fib. (NDF)	Acid Det. Fib. (ADF)	Lignin	Ether Extract (Fat)	N-Free Extract	Crude Protein	Digestible Protein: Ruminant	Swine	Horse
				%	%	%	%	%	%	%	%	%	%	%	%
	BITTERBRUSH, ANTELOPE *Pushia tridentata*														
35	BROWSE, FRESH	2-00-732	A-F	33	1.6	7.1	–	–	–	1.7	18.8	3.9	2.6	2.6	–
			M-F	100	4.9	21.4	–	–	–	5.0	56.9	11.8	7.8	7.7	–
	BLACK SAGE *Salvia mellifera*														
36	BROWSE, STEM CURED, FRESH	2-05-564	A-F	65	3.6	–	–	–	–	7.0	–	5.5	2.9	3.3	–
			M-F	100	5.5	–	–	–	–	10.7	–	8.5	4.5	5.1	–
	BLUEBELL *Mertensia* spp														
37	FRESH	2-00-739	A-F	25	2.9	4.3	–	–	–	0.8	10.7	6.3	5.0	4.6	–
			M-F	100	11.6	17.1	–	–	–	3.3	42.7	25.3	20.1	18.6	–
	BLUEGRASS, CANADA *Poa compressa*														
38	FRESH	2-00-764	A-F	31	2.8	8.3	–	–	–	1.2	13.8	5.3	3.9	3.7	–
			M-F	100	8.9	26.4	–	–	–	3.7	44.0	17.0	12.5	11.9	–
39	IMMATURE, FRESH	2-00-763	A-F	26	2.4	6.6	–	–	–	1.0	11.2	4.9	3.6	3.4	–
			M-F	100	9.1	25.5	–	–	–	3.7	43.0	18.7	14.1	13.3	–
	BLUEGRASS, KENTUCKY *Poa pratensis*														
40	IMMATURE, FRESH[1]	2-00-777	A-F	31	2.9	7.8	17.1	9.0	–	1.1	13.7	5.4	4.0	3.8	–
			M-F	100	9.4	25.2	55.0	29.0	–	3.6	44.4	17.4	12.9	12.2	–
41	EARLY BLOOM, FRESH[1]	2-00-779	A-F	35	2.5	9.6	22.8	11.2	–	1.4	15.7	5.8	4.2	4.1	–
			M-F	100	7.1	27.4	65.0	32.0	–	3.9	44.9	16.6	12.0	11.6	–
42	MILK STAGE, FRESH[1]	2-00-782	A-F	42	3.1	12.7	28.6	16.0	–	1.5	19.8	4.9	3.2	3.2	–
			M-F	100	7.3	30.3	68.0	38.0	–	3.6	47.2	11.6	7.6	7.6	–
	BLUEGRASS, KENTUCKY-CLOVER, WHITE *Poa pratensis-Trifolium repens*														
43	FRESH	2-08-356	A-F	24	2.7	4.5	–	–	–	0.9	11.3	5.0	3.8	3.6	–
			M-F	100	11.0	18.3	–	–	–	3.5	46.5	20.6	15.8	14.8	–
	BLUEGRASS, MUTTON *Poa fendleriana*														
44	FRESH	2-00-794	A-F	27	3.0	9.0	–	–	–	0.6	12.4	1.9	0.9	1.1	–
			M-F	100	11.1	33.5	–	–	–	2.4	46.0	7.0	3.4	3.9	–
	BLUEJOINT REEDGRASS *Calamagrostis canadensis*														
45	FRESH	2-03-904	A-F	45	4.1	15.2	–	–	–	1.2	20.0	4.1	2.4	2.5	–
			M-F	100	9.2	34.1	–	–	–	2.7	44.8	9.2	5.4	5.7	–
	BLUESTEM (BEARDGRASS) *Andropogon* spp														
46	IMMATURE, FRESH	2-00-821	A-F	27	2.4	6.7	–	–	–	0.7	13.6	3.4	2.3	2.3	–
			M-F	100	8.9	24.9	–	–	–	2.8	50.6	12.8	8.7	8.5	–
47	MATURE, FRESH	2-00-825	A-F	59	3.3	20.2	–	–	–	1.4	30.6	3.4	1.4	1.7	–
			M-F	100	5.6	34.2	–	–	–	2.4	51.9	5.8	2.4	2.9	–
	BRISTLEGRASS *Setaria* spp														
48	FRESH	2-00-876	A-F	26	3.2	8.3	–	–	–	0.5	10.9	3.1	2.0	2.0	–
			M-F	100	12.3	31.9	–	–	–	2.1	41.9	11.8	7.8	7.7	–
	BROMEGRASS *Bromus* spp														
49	IMMATURE, FRESH[1]	2-00-892	A-F	34	3.8	7.5	19.0	10.5	–	1.2	15.5	5.8	4.7	4.1	–
			M-F	100	11.4	22.1	56.0	31.0	–	3.7	45.8	17.1	14.0	12.0	–
50	MATURE, FRESH[1]	2-00-898	A-F	56	3.7	18.5	40.3	24.6	–	1.2	29.0	3.6	1.6	1.9	–
			M-F	100	6.6	33.0	72.0	44.0	–	2.2	51.8	6.4	2.9	3.4	–
	BROMEGRASS, CHEAT (CHEATGRASS) *Bromus tectorum*														
51	IMMATURE, FRESH	2-00-908	A-F	22	2.1	5.0	–	–	–	0.6	10.8	3.5	2.5	2.4	–
			M-F	100	9.6	22.9	–	–	–	2.7	49.0	15.8	11.4	11.0	–
52	MATURE, FRESH	2-00-911	A-F	55	4.8	19.1	–	–	–	0.7	28.2	2.2	0.4	0.8	–
			M-F	100	8.7	34.8	–	–	–	1.3	51.3	3.9	0.7	1.4	–
	BROMEGRASS, RESCUE (RESCUEGRASS) *Bromus catharticus*														
53	FRESH	2-08-361	A-F	29	4.0	6.7	–	–	–	1.0	12.2	5.0	3.7	3.5	–
			M-F	100	13.8	23.2	–	–	–	3.5	42.2	17.3	12.8	12.2	–
	BROMEGRASS, SMOOTH *Bromus inermis*														
54	FRESH	2-00-963	A-F	27	–	7.7	–	–	–	0.8	13.1	3.1	2.0	2.0	–
			M-F	100	–	28.4	–	–	–	3.0	48.6	11.4	7.4	7.4	–
55	MATURE, FRESH	2-08-364	A-F	55	3.8	19.1	–	–	–	1.3	27.4	3.3	1.4	1.7	–
			M-F	100	6.9	34.8	–	–	–	2.4	49.9	6.0	2.5	3.1	–
	BROOMCORN (HOG MILLET; MILLET, PROSO) *Panicum miliaceum*														
56	FRESH	2-03-811	A-F	25	1.8	7.4	–	–	–	0.6	13.1	2.1	1.1	1.2	–
			M-F	100	7.4	29.5	–	–	–	2.5	52.4	8.2	4.6	4.9	–
	BRUSSELS SPROUTS *Brassica oleracea, gemmifera*														
57	HEADS, FRESH	2-08-187	A-F	15	1.2	1.6	–	–	–	0.4	6.7	4.9	3.9	4.1	3.7
			M-F	100	8.1	10.8	–	–	–	2.7	45.3	33.1	26.6	28.0	25.1
	BUCKWHEAT, COMMON *Fagopyrum sagittatum*														
58	FRESH	2-00-989	A-F	37	3.6	8.0	–	–	–	0.9	19.5	4.6	2.9	3.1	–
			M-F	100	9.8	21.9	–	–	–	2.5	53.3	12.6	8.0	8.4	–

PASTURE AND RANGE PLANTS

Entry Number	TDN Ruminant	TDN Swine	TDN Horse	Digestible Energy Ruminant		Digestible Energy Swine		Digestible Energy Horse		Metabolizable Energy Ruminant		Metabolizable Energy Swine		Metabolizable Energy Poultry ME_n		Metabolizable Energy Horse		Net Energy Ruminant NE_m		Net Energy Ruminant NE_g		Net Energy Lactating Cows NE_{lc}	
	%	%	%	Mcal		kcal		Mcal		Mcal		kcal		kcal		Mcal		Mcal		Mcal		Mcal	
				lb	kg	lb	kg	lb	kg	lb	kg	lb	kg	lb	kg	lb	kg	lb	kg	lb	kg	lb	kg
35	22	–	16	0.45	0.98	–	–	0.30	0.65	0.38	0.84	–	–	–	–	0.24	0.54	0.25	0.54	0.15	0.34	0.24	0.53
	67	–	47	1.35	2.97	–	–	0.90	1.98	1.16	2.55	–	–	–	–	0.74	1.62	0.74	1.64	0.47	1.03	0.73	1.60
36	32	–	–	0.62	1.38	–	–	–	–	0.31	0.68	–	–	–	–	–	–	–	–	–	–	–	–
	49	–	–	0.96	2.12	–	–	–	–	0.47	1.04	–	–	–	–	–	–	–	–	–	–	–	–
37	18	–	–	0.36	0.79	–	–	–	–	0.31	0.68	–	–	–	–	–	–	0.19	0.41	0.12	0.26	0.18	0.40
	72	–	–	1.43	3.15	–	–	–	–	1.24	2.74	–	–	–	–	–	–	0.75	1.65	0.47	1.05	0.73	1.61
38	20	–	–	0.41	0.90	–	–	–	–	0.35	0.77	–	–	–	–	–	–	0.23	0.51	0.15	0.32	0.23	0.50
	65	–	–	1.30	2.86	–	–	–	–	1.11	2.44	–	–	–	–	–	–	0.74	1.63	0.47	1.03	0.72	1.60
39	18	–	–	0.36	0.79	–	–	–	–	0.31	0.68	–	–	–	–	–	–	0.20	0.43	0.12	0.27	0.19	0.42
	71	–	–	1.37	3.03	–	–	–	–	1.18	2.61	–	–	–	–	–	–	0.76	1.67	0.48	1.06	0.74	1.62
40	22	–	–	0.42	0.93	–	–	–	–	0.37	0.81	–	–	–	–	–	–	0.24	0.52	0.15	0.33	0.23	0.51
	72	–	–	1.38	3.03	–	–	–	–	1.19	2.61	–	–	–	–	–	–	0.77	1.70	0.49	1.08	0.75	1.64
41	24	–	–	0.49	1.07	–	–	–	–	0.42	0.93	–	–	–	–	–	–	0.26	0.57	0.16	0.36	0.25	0.55
	69	–	–	1.39	3.06	–	–	–	–	1.20	2.64	–	–	–	–	–	–	0.73	1.62	0.46	1.01	0.72	1.58
42	26	–	–	0.52	1.15	–	–	–	–	0.44	0.97	–	–	–	–	–	–	0.27	0.59	0.15	0.34	0.27	0.59
	62	–	–	1.24	2.73	–	–	–	–	1.05	2.31	–	–	–	–	–	–	0.63	1.39	0.37	0.81	0.64	1.41
43	17	–	–	0.34	0.74	–	–	–	–	0.29	0.64	–	–	–	–	–	–	0.18	0.39	0.11	0.24	0.17	0.38
	69	–	–	1.38	3.04	–	–	–	–	1.19	2.62	–	–	–	–	–	–	0.72	1.59	0.45	0.99	0.71	1.56
44	15	–	–	0.30	0.66	–	–	–	–	0.25	0.55	–	–	–	–	–	–	0.14	0.31	0.07	0.16	0.15	0.33
	56	–	–	1.11	2.45	–	–	–	–	0.92	2.02	–	–	–	–	–	–	0.51	1.13	0.26	0.57	0.55	1.22
45	25	–	–	0.51	1.12	–	–	–	–	0.42	0.93	–	–	–	–	–	–	0.25	0.55	0.13	0.29	0.26	0.57
	56	–	–	1.14	2.51	–	–	–	–	0.94	2.08	–	–	–	–	–	–	0.55	1.22	0.30	0.66	0.58	1.29
46	18	–	–	0.35	0.78	–	–	–	–	0.30	0.67	–	–	–	–	–	–	0.18	0.39	0.10	0.23	0.18	0.39
	68	–	–	1.32	2.90	–	–	–	–	1.13	2.48	–	–	–	–	–	–	0.65	1.44	0.39	0.86	0.66	1.45
47	31	–	–	0.67	1.47	–	–	–	–	0.55	1.22	–	–	–	–	–	–	0.35	0.76	0.19	0.43	0.36	0.79
	53	–	–	1.13	2.48	–	–	–	–	0.93	2.06	–	–	–	–	–	–	0.59	1.29	0.33	0.72	0.61	1.34
48	15	–	–	0.30	0.66	–	–	–	–	0.25	0.55	–	–	–	–	–	–	0.14	0.31	0.07	0.16	0.15	0.33
	58	–	–	1.16	2.55	–	–	–	–	0.97	2.13	–	–	–	–	–	–	0.54	1.18	0.28	0.62	0.57	1.26
49	25	–	–	0.50	1.11	–	–	–	–	0.44	0.97	–	–	–	–	–	–	0.24	0.53	0.15	0.33	0.24	0.52
	74	–	–	1.48	3.27	–	–	–	–	1.29	2.85	–	–	–	–	–	–	0.71	1.57	0.44	0.97	0.70	1.55
50	36	–	–	0.70	1.54	–	–	–	–	0.59	1.31	–	–	–	–	–	–	0.38	0.83	0.23	0.50	0.37	0.83
	65	–	–	1.25	2.75	–	–	–	–	1.06	2.33	–	–	–	–	–	–	0.67	1.47	0.40	0.88	0.67	1.47
51	15	–	–	0.29	0.64	–	–	–	–	0.25	0.55	–	–	–	–	–	–	0.15	0.33	0.09	0.20	0.15	0.33
	66	–	–	1.32	2.92	–	–	–	–	1.13	2.50	–	–	–	–	–	–	0.68	1.49	0.41	0.90	0.68	1.49
52	31	–	–	0.62	1.36	–	–	–	–	0.51	1.13	–	–	–	–	–	–	0.29	0.63	0.15	0.32	0.31	0.68
	56	–	–	1.13	2.48	–	–	–	–	0.93	2.06	–	–	–	–	–	–	0.52	1.14	0.27	0.58	0.56	1.23
53	20	–	–	0.38	0.84	–	–	–	–	0.33	0.72	–	–	–	–	–	–	0.18	0.40	0.10	0.23	0.18	0.40
	70	–	–	1.32	2.91	–	–	–	–	1.13	2.48	–	–	–	–	–	–	0.62	1.37	0.36	0.79	0.63	1.40
54	17	–	–	0.34	0.74	–	–	–	–	0.28	0.62	–	–	–	–	–	–	0.17	0.37	0.10	0.22	0.17	0.38
	62	–	–	1.24	2.73	–	–	–	–	1.05	2.31	–	–	–	–	–	–	0.62	1.38	0.36	0.80	0.63	1.40
55	29	–	–	0.61	1.34	–	–	–	–	0.50	1.10	–	–	–	–	–	–	0.31	0.68	0.17	0.37	0.32	0.71
	53	–	–	1.11	2.44	–	–	–	–	0.91	2.01	–	–	–	–	–	–	0.56	1.24	0.31	0.67	0.59	1.30
56	16	–	–	0.31	0.68	–	–	–	–	0.26	0.58	–	–	–	–	–	–	0.15	0.34	0.09	0.19	0.16	0.34
	63	–	–	1.24	2.74	–	–	–	–	1.05	2.32	–	–	–	–	–	–	0.61	1.35	0.35	0.78	0.63	1.38
57	–	–	–	–	–	–	–	–	–	–	–	–	–	–	–	–	–	–	–	–	–	–	–
	–	–	–	–	–	–	–	–	–	–	–	–	–	–	–	–	–	–	–	–	–	–	–
58	22	–	–	0.45	1.00	–	–	–	–	0.38	0.85	–	–	–	–	–	–	0.24	0.54	0.15	0.32	0.24	0.54
	59	–	–	1.24	2.73	–	–	–	–	1.05	2.31	–	–	–	–	–	–	0.66	1.46	0.40	0.87	0.66	1.46

(Continued)

TABLE V-1E PASTURE AND RANGE PLANTS, MINERAL AND VITAMIN COMPOSITION OF FEEDS, DATA EXPRESSED **AS-FED** AND **MOISTURE-FREE**—*(Continued)*

Entry Number	Feed Name Description	Moisture Basis: A-F (as-fed) or M-F (moisture-free)	Dry Matter	Macro Minerals							Micro Minerals						
				Calcium (Ca)	Phosphorus (P)	Sodium (Na)	Chlorine (Cl)	Magnesium (Mg)	Potassium (K)	Sulfur (S)	Cobalt (Co)	Copper (Cu)	Iodine (I)	Iron (Fe)	Manganese (Mn)	Selenium (Se)	Zinc (Zn)
			%	%	%	%	%	%	%	%	ppm or mg/kg	ppm or mg/kg	ppm or mg/kg	%	ppm or mg/kg	ppm or mg/kg	ppm or mg/kg
	BITTERBRUSH, ANTELOPE *Pushia tridentata*																
35	BROWSE, FRESH	A-F	33	–	–	–	–	–	–	0.04	–	–	–	–	–	–	–
		M-F	100	–	–	–	–	–	–	0.13	–	–	–	–	–	–	–
	BLACK SAGE *Salvia mellifera*																
36	BROWSE, STEM CURED, FRESH	A-F	65	0.53	0.11	–	–	–	–	–	–	–	–	–	–	–	–
		M-F	100	0.81	0.17	–	–	–	–	–	–	–	–	–	–	–	–
	BLUEBELL *Mertensia* spp																
37	FRESH	A-F	25	–	–	–	–	–	–	–	–	–	–	–	–	–	–
		M-F	100	–	–	–	–	–	–	–	–	–	–	–	–	–	–
	BLUEGRASS, CANADA *Poa compressa*																
38	FRESH	A-F	31	0.12	0.12	0.04	–	0.05	0.64	0.05	–	–	–	0.010	24.8	–	–
		M-F	100	0.39	0.39	0.14	–	0.16	2.04	0.17	–	–	–	0.030	79.1	–	–
39	IMMATURE, FRESH	A-F	26	0.10	0.10	0.04	–	0.04	0.53	0.04	–	–	–	0.008	–	–	–
		M-F	100	0.39	0.39	0.14	–	0.16	2.04	0.17	–	–	–	0.030	–	–	–
	BLUEGRASS, KENTUCKY *Poa pratensis*																
40	IMMATURE, FRESH	A-F	31	0.15	0.14	0.04	–	0.05	0.70	0.05	–	–	–	0.010	–	–	–
		M-F	100	0.50	0.44	0.14	–	0.18	2.27	0.17	–	–	–	0.030	–	–	–
41	EARLY BLOOM, FRESH	A-F	35	0.16	0.14	0.05	–	0.04	0.70	0.06	–	–	–	0.011	–	–	–
		M-F	100	0.46	0.39	0.14	–	0.11	2.01	0.17	–	–	–	0.030	–	–	–
42	MILK STAGE, FRESH	A-F	42	–	–	–	–	–	–	–	–	–	–	–	–	–	–
		M-F	100	–	–	–	–	–	–	–	–	–	–	–	–	–	–
	BLUEGRASS, KENTUCKY-CLOVER, WHITE *Poa pratensis-Trifolium repens*																
43	FRESH	A-F	24	0.31	0.11	–	–	–	–	–	–	–	–	–	–	–	–
		M-F	100	1.29	0.46	–	–	–	–	–	–	–	–	–	–	–	–
	BLUEGRASS, MUTTON *Poa fendleriana*																
44	FRESH	A-F	27	–	–	–	–	–	–	–	–	–	–	–	–	–	–
		M-F	100	–	–	–	–	–	–	–	–	–	–	–	–	–	–
	BLUEJOINT REEDGRASS *Calamagrostis canadensis*																
45	FRESH	A-F	45	–	0.10	–	–	–	–	–	–	–	–	–	–	–	–
		M-F	100	–	0.22	–	–	–	–	–	–	–	–	–	–	–	–
	BLUESTEM (BEARDGRASS) *Andropogon* spp																
46	IMMATURE, FRESH	A-F	27	0.17	0.05	–	–	–	0.46	–	–	12.6	–	0.024	28.5	–	–
		M-F	100	0.63	0.20	–	–	–	1.72	–	–	47.0	–	0.090	106.3	–	–
47	MATURE, FRESH	A-F	59	0.23	0.07	–	–	0.04	0.30	–	–	15.6	–	0.064	35.9	–	–
		M-F	100	0.40	0.12	–	–	0.06	0.51	–	–	26.5	–	0.108	60.9	–	–
	BRISTLEGRASS *Setaria* spp																
48	FRESH	A-F	26	0.10	0.05	–	–	0.07	1.51	–	–	1.7	–	–	10.9	–	–
		M-F	100	0.37	0.19	–	–	0.28	5.82	–	–	6.6	–	–	42.1	–	–
	BROMEGRASS *Bromus* spp																
49	IMMATURE, FRESH	A-F	34	0.20	0.13	0.01	–	0.06	1.46	0.07	–	–	–	0.007	–	–	–
		M-F	100	0.59	0.37	0.02	–	0.18	4.30	0.20	–	–	–	0.020	–	–	–
50	MATURE, FRESH	A-F	56	0.17	0.15	0.01	–	0.10	0.70	0.11	–	–	–	0.012	–	–	–
		M-F	100	0.30	0.26	0.02	–	0.18	1.25	0.20	–	–	–	0.020	–	–	–
	BROMEGRASS, CHEAT (CHEATGRASS) *Bromus tectorum*																
51	IMMATURE, FRESH	A-F	22	0.14	0.06	–	–	–	–	–	–	–	–	–	–	–	–
		M-F	100	0.64	0.28	–	–	–	–	–	–	–	–	–	–	–	–
52	MATURE, FRESH	A-F	55	0.18	0.06	0.01	–	0.04	0.21	0.02	–	2.5	–	–	38.4	–	8.7
		M-F	100	0.33	0.10	0.02	–	0.08	0.39	0.04	–	4.6	–	–	69.8	–	15.8
	BROMEGRASS, RESCUE (RESCUEGRASS) *Bromus catharticus*																
53	FRESH	A-F	29	0.15	0.08	–	–	–	–	–	–	–	–	–	–	–	–
		M-F	100	0.52	0.28	–	–	–	–	–	–	–	–	–	–	–	–
	BROMEGRASS, SMOOTH *Bromus inermis*																
54	FRESH	A-F	27	–	–	–	–	–	–	–	0.022	–	–	–	–	–	–
		M-F	100	–	–	–	–	–	–	–	0.080	–	–	–	–	–	–
55	MATURE, FRESH	A-F	55	0.14	0.09	–	–	–	–	–	–	1.2	–	–	–	–	–
		M-F	100	0.26	0.16	–	–	–	–	–	–	2.2	–	–	–	–	–
	BROOMCORN (HOG MILLET; MILLET, PROSO) *Panicum miliaceum*																
56	FRESH	A-F	25	–	–	–	–	–	–	–	–	–	–	–	–	–	–
		M-F	100	–	–	–	–	–	–	–	–	–	–	–	–	–	–
	BRUSSELS SPROUTS *Brassica oleracea, gemmifera*																
57	HEADS, FRESH	A-F	15	0.04	0.08	0.01	–	–	0.39	–	–	–	–	0.002	–	–	–
		M-F	100	0.27	0.54	0.07	–	–	2.64	–	–	–	–	0.014	–	–	–
	BUCKWHEAT, COMMON *Fagopyrum sagittatum*																
58	FRESH	A-F	37	–	–	–	–	–	–	–	–	–	–	–	–	–	–
		M-F	100	–	–	–	–	–	–	–	–	–	–	–	–	–	–

	Fat-Soluble Vitamins					Water-Soluble Vitamins								
Entry Number	A (1 mg Carotene = 1667 IU Vit A)	Carotene (Provitamin A)	D	E	K	B-12	Biotin	Choline	Folacin (Folic Acid)	Niacin	Pantothenic Acid (B-3)	(Pyri-doxine(B-6	Ribo-flavin (B-2)	Thiamin (B-1)
	IU/g	ppm or mg/kg	IU/kg	ppm or mg/kg	ppm or mg/kg	ppm or mg/kg	ppm or mg/kg	ppm or mg/kg	ppm or mg/kg	ppm or mg/kg	ppm or mg/kg	ppm or mg/kg	ppm or mg/kg	ppm or mg/kg
35	–	–	–	–	–	–	–	–	–	–	–	–	–	–
	–	–	–	–	–	–	–	–	–	–	–	–	–	–
36	–	–	–	–	–	–	–	–	–	–	–	–	–	–
	–	–	–	–	–	–	–	–	–	–	–	–	–	–
37	–	–	–	–	–	–	–	–	–	–	–	–	–	–
	–	–	–	–	–	–	–	–	–	–	–	–	–	–
38	**199.9**	**119.9**	–	–	–	–	–	–	–	–	–	–	–	–
	637.6	382.5	–	–	–	–	–	–	–	–	–	–	–	–
39	**172.6**	**103.5**	–	–	–	–	–	–	–	–	–	–	–	–
	665.1	399.0	–	–	–	–	–	–	–	–	–	–	–	–
40	**247.6**	**148.5**	–	**47.8**	–	–	–	–	–	–	–	–	–	–
	803.4	481.9	–	155.0	–	–	–	–	–	–	–	–	–	–
41	**163.4**	**98.0**	–	–	–	–	–	–	–	–	–	–	–	–
	466.8	280.0	–	–	–	–	–	–	–	–	–	–	–	–
42	–	–	–	–	–	–	–	–	–	–	–	–	–	–
	–	–	–	–	–	–	–	–	–	–	–	–	–	–
43	–	–	–	–	–	–	–	–	–	–	–	–	–	–
	–	–	–	–	–	–	–	–	–	–	–	–	–	–
44	–	–	–	–	–	–	–	–	–	–	–	–	–	–
	–	–	–	–	–	–	–	–	–	–	–	–	–	–
45	–	–	–	–	–	–	–	–	–	–	–	–	–	–
	–	–	–	–	–	–	–	–	–	–	–	–	–	–
46	**97.9**	**58.7**	–	–	–	–	–	–	–	–	–	–	–	–
	365.3	219.1	–	–	–	–	–	–	–	–	–	–	–	–
47	–	–	–	–	–	–	–	–	–	–	–	–	–	–
	–	–	–	–	–	–	–	–	–	–	–	–	–	–
48	–	–	–	–	–	–	–	–	–	–	–	–	–	–
	–	–	–	–	–	–	–	–	–	–	–	–	–	–
49	**259.6**	**155.7**	–	–	–	–	–	–	–	–	–	–	–	–
	765.9	459.4	–	–	–	–	–	–	–	–	–	–	–	–
50	**77.1**	**46.2**	–	–	–	–	–	–	–	–	–	–	–	–
	137.5	82.5	–	–	–	–	–	–	–	–	–	–	–	–
51	–	–	–	–	–	–	–	–	–	–	–	–	–	–
	–	–	–	–	–	–	–	–	–	–	–	–	–	–
52	–	–	–	–	–	–	–	–	–	–	–	–	–	–
	–	–	–	–	–	–	–	–	–	–	–	–	–	–
53	–	–	–	–	–	–	–	–	–	–	–	–	–	–
	–	–	–	–	–	–	–	–	–	–	–	–	–	–
54	**142.0**	**85.2**	**0**	–	–	–	–	–	–	–	–	–	**2.1**	**0.8**
	525.9	315.5	0	–	–	–	–	–	–	–	–	–	7.7	3.1
55	–	–	–	–	–	–	–	–	–	–	–	–	–	–
	–	–	–	–	–	–	–	–	–	–	–	–	–	–
56	–	–	–	–	–	–	–	–	–	–	–	–	–	–
	–	–	–	–	–	–	–	–	–	–	–	–	–	–
57	–	–	–	–	–	–	–	–	–	**9**	–	–	**1.6**	**1.0**
	–	–	–	–	–	–	–	–	–	61	–	–	10.8	6.8
58	–	–	–	–	–	–	–	–	–	–	–	–	–	–
	–	–	–	–	–	–	–	–	–	–	–	–	–	–

(Continued)

TABLE V-1E PASTURE AND RANGE PLANTS, COMPOSITION OF FEEDS, DATA EXPRESSED **AS-FED** AND **MOISTURE-FREE**—*(Continued)*

Entry Number	Feed Name Description	International Feed Number	Moisture Basis: A-F (as-fed) or M-F (moisture-free)	Proximate Analysis									Digestible Protein		
				Dry Matter	Ash	Crude Fiber	Neutral Det. Fib. (NDF)	Acid Det. Fib. (ADF)	Lignin	Ether Extract (Fat)	N-Free Extract	Crude Protein	Ruminant	Swine	Horse
				%	%	%	%	%	%	%	%	%	%	%	%
	BUFFALOGRASS *Buchloe dactyloides*														
59	FRESH	2-01-010	A-F	46	5.6	12.7	33.9	16.7	2.9	0.9	22.0	4.7	2.5	3.0	–
			M-F	100	12.3	27.7	74.0	36.5	6.3	1.9	48.0	10.2	5.5	6.5	–
	BUFFELGRASS *Cenchrus ciliaris*														
60	PREBLOOM, FRESH	2-10-253	A-F	21	2.8	–	15.1	–	–	–	–	1.7	0.9	1.0	–
			M-F	100	13.4	–	73.1	–	–	–	–	8.1	4.4	4.8	–
	BUNDLEFLOWER, RAYADO (DESMANTHUS; DWARF KOA) *Desmanthus virgatus*														
61	FRESH	2-01-024	A-F	41	2.4	17.5	–	–	–	0.9	15.7	4.9	3.0	3.0	–
			M-F	100	5.7	42.2	–	–	–	2.2	38.0	11.9	7.3	7.3	–
	BUR-CLOVER, CALIFORNIA *Medicago hispida*														
62	FRESH	2-01-035	A-F	27	1.8	–	–	–	–	0.6	–	6.2	–	–	–
			M-F	100	6.8	–	–	–	–	2.3	–	23.0	–	–	–
	BUR-CLOVER, TOOTHED *Medicago hispida*														
63	FRESH	2-08-366	A-F	21	2.3	3.9	–	–	–	1.7	7.8	5.1	–	–	3.8
			M-F	100	11.1	18.8	–	–	–	8.2	37.5	24.5	–	–	18.3
	BURROGRASS *Scleropogon brevifolius*														
64	FRESH	2-01-040	A-F	29	2.7	9.1	–	–	–	–	–	2.1	1.0	1.2	–
			M-F	100	9.4	31.4	–	–	–	–	–	7.1	3.5	4.0	–
	CABBAGE *Brassica oleracea, capitata*														
65	HEADS, FRESH	2-01-046	A-F	9	0.9	1.0	–	–	–	0.2	5.3	1.9	1.6	1.6	1.4
			M-F	100	9.8	10.5	–	–	–	2.2	56.9	20.6	17.7	16.7	15.3
66	OUTSIDE LEAVES, FRESH	2-01-047	A-F	14	1.9	1.9	–	–	–	0.4	7.1	2.4	1.7	1.7	–
			M-F	100	13.6	13.8	–	–	–	3.1	52.0	17.6	12.2	12.4	–
	CACTUS, CANE–SEE CHOLLA MEXICALI														
	CACTUS, PRICKLY PEAR *Opuntia* spp														
67	FRESH	2-01-061	A-F	17	3.4	2.2	5.3	–	–	0.3	9.9	0.9	0.4	0.5	–
			M-F	100	20.5	13.2	31.6	–	–	1.9	59.2	5.1	2.3	3.2	–
	CANADA BLUEGRASS–SEE BLUEGRASS, CANADA														
	CANARYGRASS *Phalaris canariensis*														
68	FRESH	2-01-093	A-F	26	2.4	6.9	–	–	–	1.0	12.1	3.4	2.3	2.3	–
			M-F	100	9.4	26.8	–	–	–	3.7	47.0	13.1	9.0	8.8	–
	CANARYGRASS, REED *Phalaris arundinacea*														
69	FRESH	2-01-113	A-F	23	2.3	5.6	10.6	6.5	1.0	0.9	10.1	3.9	2.9	2.7	–
			M-F	100	10.2	24.4	46.4	28.3	4.3	4.1	44.4	17.0	12.5	11.9	–
	CANE CACTUS–SEE CHOLLA MEXICALI														
	CANNA *Canna* spp														
70	IMMATURE, FRESH	2-01-126	A-F	17	2.7	3.2	–	–	–	0.8	8.0	1.7	0.7	1.1	–
			M-F	100	16.5	19.6	–	–	–	5.1	48.6	10.2	4.5	6.5	–
	CARPETGRASS *Axonopus* spp														
71	FRESH	2-01-140	A-F	30	3.1	8.5	–	–	–	0.5	15.0	2.8	1.4	1.7	–
			M-F	100	10.3	28.5	–	–	–	1.7	50.1	9.4	4.7	5.8	–
	CARROT *Daucus* spp														
72	FRESH	2-08-371	A-F	16	2.4	2.9	–	–	–	0.6	8.0	2.1	1.4	1.4	–
			M-F	100	15.0	18.1	–	–	–	3.8	50.0	13.1	9.0	8.8	–
	CAULIFLOWER *Brassica oleracea, botrytis*														
73	HEADS, FRESH	4-08-189	A-F	9	0.9	1.0	–	–	–	0.2	4.2	2.7	2.1	2.3	2.0
			M-F	100	10.0	11.1	–	–	–	2.2	46.7	30.0	23.8	25.2	22.7
74	LEAVES, FRESH	2-01-192	A-F	9	–	0.9	–	–	–	0.4	–	2.4	1.9	1.8	–
			M-F	100	–	9.5	–	–	–	4.1	–	26.6	21.2	19.6	–
	CELERY, WILD *Apium graveolens*														
75	FRESH	2-01-195	A-F	6	1.2	0.7	–	–	–	0.1	2.9	0.9	0.7	0.6	–
			M-F	100	20.1	12.2	–	–	–	1.7	50.3	15.7	11.3	10.9	–
	CENTIPEDEGRASS *Eremochloa ophiuroides*														
76	FRESH	2-01-198	A-F	25	1.4	7.9	–	–	–	0.7	12.2	2.8	1.8	1.8	–
			M-F	100	5.6	31.6	–	–	–	2.8	48.7	11.3	7.3	7.3	–
	CHEATGRASS (BROMEGRASS, CHEAT) *Bromus tectorum*														
77	IMMATURE, FRESH	2-00-908	A-F	22	2.1	5.0	–	–	–	0.6	10.8	3.5	2.5	2.4	–
			M-F	100	9.6	22.9	–	–	–	2.7	49.0	15.8	11.4	11.0	–
78	MATURE, FRESH	2-00-911	A-F	55	4.8	19.1	–	–	–	0.7	28.2	2.2	0.4	0.8	–
			M-F	100	8.7	34.8	–	–	–	1.3	51.3	3.9	0.7	1.4	–
	CHICORY *Cichorium* spp														
79	FRESH	2-01-220	A-F	17	3.1	2.7	–	–	–	0.4	8.7	2.5	1.8	1.7	–
			M-F	100	18.0	15.3	–	–	–	2.2	50.1	14.4	10.2	9.8	–

Entry Number	TDN Ruminant	TDN Swine	TDN Horse	Digestible Energy Ruminant		Digestible Energy Swine		Digestible Energy Horse		Metabolizable Energy Ruminant		Metabolizable Energy Swine		Metabolizable Energy Poultry ME_n		Metabolizable Energy Horse		Net Energy Ruminant NE_m		Net Energy Ruminant NE_g		Net Energy Lactating Cows NE_{lc}	
	%	%	%	Mcal		kcal		Mcal		Mcal		kcal		kcal		Mcal		Mcal		Mcal		Mcal	
				lb	kg	lb	kg	lb	kg	lb	kg	lb	kg	lb	kg	lb	kg	lb	kg	lb	kg	lb	kg
59	26	–	–	0.52	1.15	–	–	–	–	0.43	0.95	–	–	–	–	–	–	0.26	0.57	0.14	0.31	0.27	0.59
	56	–	–	1.14	2.51	–	–	–	–	0.94	2.08	–	–	–	–	–	–	0.56	1.24	0.30	0.67	0.59	1.30
60	–	–	–	–	–	–	–	–	–	–	–	–	–	–	–	–	–	–	–	–	–	–	–
	–	–	–	–	–	–	–	–	–	–	–	–	–	–	–	–	–	–	–	–	–	–	–
61	22	22	–	0.46	1.02	–	–	–	–	0.38	0.85	–	–	–	–	–	–	0.23	0.51	0.12	0.27	0.24	0.53
	54	54	–	1.12	2.47	–	–	–	–	0.93	2.04	–	–	–	–	–	–	0.56	1.23	0.30	0.66	0.59	1.29
62	–	–	–	–	–	–	–	–	–	–	–	–	–	–	–	–	–	–	–	–	–	–	–
	–	–	–	–	–	–	–	–	–	–	–	–	–	–	–	–	–	–	–	–	–	–	–
63	15	–	–	0.30	0.67	–	–	–	–	0.25	0.55	–	–	–	–	–	–	0.16	0.35	0.10	0.22	0.16	0.35
	73	–	–	1.45	3.20	–	–	–	–	1.19	2.62	–	–	–	–	–	–	0.75	1.66	0.47	1.04	0.75	1.66
64	–	–	–	–	–	–	–	–	–	–	–	–	–	–	–	–	–	–	–	–	–	–	–
	–	–	–	–	–	–	–	–	–	–	–	–	–	–	–	–	–	–	–	–	–	–	–
65	8	6	7	0.14	0.32	126	279	0.12	0.27	0.13	0.28	118	261	–	–	0.10	0.22	0.07	0.15	0.04	0.10	0.07	0.15
	84	68	72	1.54	3.40	1356	2989	1.32	2.90	1.35	2.98	1270	2799	–	–	1.08	2.38	0.74	1.62	0.46	1.02	0.72	1.58
66	10	–	–	0.19	0.42	–	–	–	–	0.16	0.36	–	–	–	–	–	–	0.10	0.21	0.06	0.13	0.10	0.21
	70	–	–	1.38	3.04	–	–	–	–	1.19	2.62	–	–	–	–	–	–	0.70	1.55	0.43	0.95	0.69	1.53
67	10	10	6	0.20	0.44	–	–	0.13	0.28	0.16	0.36	–	–	–	–	0.10	0.23	0.10	0.21	0.05	0.12	0.10	0.21
	58	58	38	1.18	2.60	–	–	0.75	1.65	0.98	2.15	–	–	–	–	0.61	1.35	0.58	1.28	0.32	0.71	0.57	1.26
68	16	–	–	0.32	0.71	–	–	–	–	0.27	0.60	–	–	–	–	–	–	0.16	0.36	0.10	0.21	0.17	0.37
	62	–	–	1.25	2.75	–	–	–	–	1.06	2.33	–	–	–	–	–	–	0.64	1.40	0.37	0.82	0.64	1.42
69	14	–	–	0.29	0.64	–	–	–	–	0.24	0.54	–	–	–	–	–	–	0.15	0.34	0.09	0.20	0.15	0.34
	61	–	–	1.26	2.78	–	–	–	–	1.07	2.36	–	–	–	–	–	–	0.67	1.47	0.40	0.88	0.67	1.47
70	8	–	–	0.17	0.38	–	–	–	–	0.14	0.31	–	–	–	–	–	–	0.09	0.21	0.05	0.11	0.10	0.22
	47	–	–	1.05	2.32	–	–	–	–	0.86	1.89	–	–	–	–	–	–	0.57	1.26	0.31	0.69	0.60	1.32
71	19	–	–	0.37	0.81	–	–	–	–	0.31	0.69	–	–	–	–	–	–	0.17	0.38	0.10	0.21	0.18	0.40
	64	–	–	1.23	2.71	–	–	–	–	1.04	2.29	–	–	–	–	–	–	0.58	1.28	0.32	0.71	0.60	1.33
72	12	–	–	0.22	0.49	–	–	–	–	0.19	0.42	–	–	–	–	–	–	0.10	0.22	0.06	0.13	0.10	0.22
	78	–	–	1.39	3.07	–	–	–	–	1.20	2.65	–	–	–	–	–	–	0.62	1.37	0.36	0.79	0.63	1.39
73	–	–	–	–	–	–	–	–	–	–	–	–	–	–	–	–	–	–	–	–	–	–	–
	–	–	–	–	–	–	–	–	–	–	–	–	–	–	–	–	–	–	–	–	–	–	–
74	–	–	–	–	–	–	–	–	–	–	–	–	–	–	–	–	–	–	–	–	–	–	–
	–	–	–	–	–	–	–	–	–	–	–	–	–	–	–	–	–	–	–	–	–	–	–
75	4	–	–	0.07	0.16	–	–	–	–	0.06	0.13	–	–	–	–	–	–	0.03	0.08	0.02	0.04	0.04	0.08
	62	–	–	1.25	2.75	–	–	–	–	1.06	2.33	–	–	–	–	–	–	0.60	1.32	0.34	0.75	0.62	1.36
76	16	–	–	0.32	0.70	–	–	–	–	0.27	0.59	–	–	–	–	–	–	0.16	0.35	0.10	0.21	0.16	0.36
	63	–	–	1.27	2.79	–	–	–	–	1.07	2.37	–	–	–	–	–	–	0.64	1.42	0.38	0.83	0.65	1.43
77	15	–	–	0.29	0.64	–	–	–	–	0.25	0.55	–	–	–	–	–	–	0.15	0.33	0.09	0.20	0.15	0.33
	66	–	–	1.32	2.92	–	–	–	–	1.13	2.50	–	–	–	–	–	–	0.68	1.49	0.41	0.90	0.68	1.49
78	31	–	–	0.62	1.36	–	–	–	–	0.51	1.13	–	–	–	–	–	–	0.29	0.63	0.15	0.32	0.31	0.68
	56	–	–	1.13	2.48	–	–	–	–	0.93	2.06	–	–	–	–	–	–	0.52	1.14	0.27	0.58	0.56	1.23
79	11	–	–	0.22	0.47	–	–	–	–	0.18	0.40	–	–	–	–	–	–	0.11	0.23	0.06	0.13	0.11	0.24
	62	–	–	1.24	2.72	–	–	–	–	1.04	2.30	–	–	–	–	–	–	0.60	1.33	0.34	0.75	0.62	1.36

(Continued)

TABLE V-1E PASTURE AND RANGE PLANTS, MINERAL AND VITAMIN COMPOSITION OF FEEDS, DATA EXPRESSED **AS-FED** AND **MOISTURE-FREE**—*(Continued)*

Entry Number	Feed Name Description	Moisture Basis: A-F (as-fed) or M-F (moisture-free)	Dry Matter	Macro Minerals							Micro Minerals						
				Calcium (Ca)	Phosphorus (P)	Sodium (Na)	Chlorine (Cl)	Magnesium (Mg)	Potassium (K)	Sulfur (S)	Cobalt (Co)	Copper (Cu)	Iodine (I)	Iron (Fe)	Manganese (Mn)	Selenium (Se)	Zinc (Zn)
			%	%	%	%	%	%	%	%	ppm or mg/kg	ppm or mg/kg	ppm or mg/kg	%	ppm or mg/kg	ppm or mg/kg	ppm or mg/kg
	BUFFALOGRASS *Buchloe dactyloides*																
59	FRESH	**A-F**	**46**	**0.26**	**0.09**	–	–	**0.06**	**0.33**	–	–	–	–	–	–	–	–
		M-F	100	0.57	0.21	–	–	0.14	0.71	–	–	–	–	–	–	–	–
	BUFFELGRASS *Cenchrus ciliaris*																
60	PREBLOOM, FRESH	**A-F**	**21**	**0.19**	**0.03**	**0.03**	–	**0.12**	**0.89**	–	**0.063**	**2.0**	–	**0.014**	**27.7**	–	**9.5**
		M-F	100	0.92	0.16	0.15	–	0.57	4.31	–	0.303	9.7	–	0.067	134.0	–	46.0
	BUNDLEFLOWER, RAYADO (DESMANTHUS; DWARF KOA) *Desmanthus virgatus*																
61	FRESH	**A-F**	**41**	**0.60**	**0.13**	–	–	**0.14**	**0.75**	–	–	–	–	–	–	–	–
		M-F	100	1.44	0.31	–	–	0.33	1.81	–	–	–	–	–	–	–	–
	BUR-CLOVER, CALIFORNIA *Medicago hispida*																
62	FRESH	**A-F**	**27**	–	–	–	–	–	–	–	–	–	–	–	–	–	–
		M-F	100	–	–	–	–	–	–	–	–	–	–	–	–	–	–
	BUR-CLOVER, TOOTHED *Medicago hispida*																
63	FRESH	**A-F**	**21**	–	–	–	–	–	–	–	–	–	–	–	–	–	–
		M-F	100	–	–	–	–	–	–	–	–	–	–	–	–	–	–
	BURROGRASS *Scleropogon brevifolius*																
64	FRESH	**A-F**	**29**	**0.15**	**0.03**	–	–	–	–	–	**0.086**	**2.6**	–	–	**20.7**	–	–
		M-F	100	0.51	0.11	–	–	–	–	–	0.296	8.8	–	–	71.4	–	–
	CABBAGE *Brassica oleracea, capitata*																
65	HEADS, FRESH	**A-F**	**9**	**0.06**	**0.03**	**0.01**	**0.05**	**0.02**	**0.26**	**0.11**	–	**1.3**	–	**0.001**	**2.8**	–	–
		M-F	100	0.64	0.35	0.07	0.53	0.21	2.81	1.17	–	14.1	–	0.008	30.5	–	–
66	OUTSIDE LEAVES, FRESH	**A-F**	**14**	**0.09**	**0.03**	–	–	–	–	–	–	–	–	–	–	–	–
		M-F	100	0.63	0.21	–	–	–	–	–	–	–	–	–	–	–	–
	CACTUS, CANE–SEE CHOLLA MEXICALI																
	CACTUS, PRICKLY PEAR *Opuntia* spp																
67	FRESH	**A-F**	**17**	**1.47**	**0.02**	–	–	**0.23**	**0.37**	–	–	–	–	–	–	–	–
		M-F	100	8.78	0.12	–	–	1.38	2.21	–	–	–	–	–	–	–	–
	CANADA BLUEGRASS–SEE BLUEGRASS, CANADA																
	CANARYGRASS *Phalaris canariensis*																
68	FRESH	**A-F**	**26**	**0.11**	**0.08**	–	–	**0.07**	**0.82**	–	–	–	–	–	–	–	–
		M-F	100	0.44	0.29	–	–	0.27	3.17	–	–	–	–	–	–	–	–
	CANARYGRASS, REED *Phalaris arundinacea*																
69	FRESH	**A-F**	**23**	**0.08**	**0.08**	–	–	–	**0.83**	–	–	–	–	–	–	–	–
		M-F	100	0.36	0.33	–	–	–	3.64	–	–	–	–	–	–	–	–
	CANE CACTUS–SEE CHOLLA MEXICALI																
	CANNA *Canna* spp																
70	IMMATURE, FRESH	**A-F**	**17**	–	–	–	–	–	–	–	–	–	–	–	–	–	–
		M-F	100	–	–	–	–	–	–	–	–	–	–	–	–	–	–
	CARPETGRASS *Anoxopus* spp																
71	FRESH	**A-F**	**30**	**0.12**	**0.05**	**0.16**	**0.13**	**0.07**	**0.28**	**0.03**	**0.028**	–	–	**0.009**	–	–	–
		M-F	100	0.40	0.16	0.52	0.42	0.24	0.92	0.09	0.091	–	–	0.028	–	–	–
	CARROT *Daucus* spp																
72	FRESH	**A-F**	**16**	**0.31**	**0.03**	–	–	–	**0.30**	–	–	–	–	–	–	–	–
		M-F	100	1.94	0.19	–	–	–	1.88	–	–	–	–	–	–	–	–
	CAULIFLOWER *Brassica oleracea, botrytis*																
73	HEADS, FRESH	**A-F**	**9**	**0.02**	**0.06**	**0.01**	–	–	**0.30**	–	–	–	–	**0.001**	–	–	–
		M-F	100	0.22	0.67	0.11	–	–	3.33	–	–	–	–	0.012	–	–	–
74	LEAVES, FRESH	**A-F**	**9**	–	–	–	–	–	–	–	–	–	–	–	–	–	–
		M-F	100	–	–	–	–	–	–	–	–	–	–	–	–	–	–
	CELERY, WILD *Apium graveolens*																
75	FRESH	**A-F**	**6**	**0.04**	**0.03**	**0.12**	–	–	**0.33**	–	–	–	–	**0.001**	–	–	–
		M-F	100	0.66	0.48	2.14	–	–	5.78	–	–	–	–	0.006	–	–	–
	CENTIPEDEGRASS *Eremochloa ophiuroides*																
76	FRESH	**A-F**	**25**	**0.15**	**0.07**	–	–	**0.13**	–	–	**0.009**	–	–	**0.004**	–	–	–
		M-F	100	0.58	0.26	–	–	0.50	–	–	0.036	–	–	0.013	–	–	–
	CHEATGRASS (BROMEGRASS, CHEAT) *Bromus tectorum*																
77	IMMATURE, FRESH	**A-F**	**22**	**0.14**	**0.06**	–	–	–	–	–	–	–	–	–	–	–	–
		M-F	100	0.64	0.28	–	–	–	–	–	–	–	–	–	–	–	–
78	MATURE, FRESH	**A-F**	**55**	**0.18**	**0.06**	**0.01**	–	**0.04**	**0.21**	**0.02**	–	**2.5**	–	–	**38.4**	–	**8.7**
		M-F	100	0.33	0.10	0.02	–	0.08	0.39	0.04	–	4.6	–	–	69.8	–	15.8
	CHICORY *Cichorium* spp																
79	FRESH	**A-F**	**17**	**0.33**	**0.04**	–	–	**0.13**	–	–	–	–	–	–	–	–	–
		M-F	100	1.89	0.22	–	–	0.73	–	–	–	–	–	–	–	–	–

PASTURE AND RANGE PLANTS

Entry Number	Fat-Soluble Vitamins: A (1 mg Carotene = 1667 IU Vit A)	Carotene (Provitamin A)	D	E	K	Water-Soluble Vitamins: (B–12)	Biotin	Choline	Folacin (Folic Acid)	Niacin	Pantothenic Acid (B–3)	(Pyridoxine) B–6	Riboflavin (B–2)	Thiamin (B–1)
	IU/g	ppm or mg/kg	IU/kg	ppm or mg/kg	ppm or mg/kg	ppm or mg/kg	ppm or mg/kg	ppm or mg/kg	ppm or mg/kg	ppm or mg/kg	ppm or mg/kg	ppm or mg/kg	ppm or mg/kg	ppm or mg/kg
59	**71.6**	**42.9**	–	–	–	–	–	–	–	–	–	–	–	–
	156.2	93.7	–	–	–	–	–	–	–	–	–	–	–	–
60	–	–	–	–	–	–	–	–	–	–	–	–	–	–
	–	–	–	–	–	–	–	–	–	–	–	–	–	–
61	–	–	–	–	–	–	–	–	–	–	–	–	–	–
	–	–	–	–	–	–	–	–	–	–	–	–	–	–
62	–	–	–	–	–	–	–	–	–	–	–	–	–	–
	–	–	–	–	–	–	–	–	–	–	–	–	–	–
63	–	–	–	–	–	–	–	–	–	–	–	–	–	–
	–	–	–	–	–	–	–	–	–	–	–	–	–	–
64	**9.1**	**5.4**	–	–	–	–	–	–	–	–	–	–	–	–
	31.2	18.7	–	–	–	–	–	–	–	–	–	–	–	–
65	**0.7**	**0.4**	–	–	–	–	–	**249**	**0.63**	**3**	–	–	**0.5**	**0.6**
	7.8	4.7	–	–	–	–	–	2671	6.80	35	–	–	5.6	6.8
66	–	–	–	–	–	–	–	–	–	–	–	–	–	–
	–	–	–	–	–	–	–	–	–	–	–	–	–	–
67	–	–	–	–	–	–	–	–	–	–	–	–	–	–
	–	–	–	–	–	–	–	–	–	–	–	–	–	–
68	–	–	–	–	–	–	–	–	–	–	–	–	–	–
	–	–	–	–	–	–	–	–	–	–	–	–	–	–
69	–	–	–	–	–	–	–	–	–	–	–	–	–	–
	–	–	–	–	–	–	–	–	–	–	–	–	–	–
70	–	–	–	–	–	–	–	–	–	–	–	–	–	–
	–	–	–	–	–	–	–	–	–	–	–	–	–	–
71	**72.4**	**43.4**	–	–	–	–	–	–	–	–	–	–	–	–
	241.5	144.8	–	–	–	–	–	–	–	–	–	–	–	–
72	**176.7**	**106.0**	–	–	–	–	–	**950**	**0.64**	**15**	**2.0**	–	**0.7**	**0.7**
	1104.6	662.6	–	–	–	–	–	5939	4.00	92	12.4	–	4.1	4.1
73	–	–	–	–	–	–	–	–	–	**7**	–	–	**1.0**	**1.1**
	–	–	–	–	–	–	–	–	–	78	–	–	11.1	12.2
74	**27.8**	**16.6**	–	–	–	–	–	–	–	–	–	–	**2.1**	–
	308.3	185.0	–	–	–	–	–	–	–	–	–	–	23.1	–
75	–	–	–	–	–	–	–	–	–	**3**	–	–	**0.3**	**0.3**
	–	–	–	–	–	–	–	–	–	51	–	–	5.1	5.1
76	–	–	–	–	–	–	–	–	–	–	–	–	–	–
	–	–	–	–	–	–	–	–	–	–	–	–	–	–
77	–	–	–	–	–	–	–	–	–	–	–	–	–	–
	–	–	–	–	–	–	–	–	–	–	–	–	–	–
78	–	–	–	–	–	–	–	–	–	–	–	–	–	–
	–	–	–	–	–	–	–	–	–	–	–	–	–	–
79	–	–	–	–	–	–	–	–	–	–	–	–	–	–
	–	–	–	–	–	–	–	–	–	–	–	–	–	–

(Continued)

TABLE V-1E PASTURE AND RANGE PLANTS, COMPOSITION OF FEEDS, DATA EXPRESSED **AS-FED** AND **MOISTURE-FREE**—*(Continued)*

Entry Number	Feed Name Description	International Feed Number	Moisture Basis: A-F (as-fed) or M-F (moisture-free)	Proximate Analysis: Dry Matter	Ash	Crude Fiber	Neutral Det. Fib. (NDF)	Acid Det. Fib. (ADF)	Lignin	Ether Extract (Fat)	N-Free Extract	Crude Protein	Digestible Protein: Ruminant	Swine	Horse
				%	%	%	%	%	%	%	%	%	%	%	%
	CHINESE LESPEDEZA–SEE LESPEDEZA, SERICEA														
	CHOLLA, MEXICALI (CANE CACTUS) *Opuntia cholla*														
80	FRESH	2-08-367	A-F	21	3.4	3.3	–	–	–	0.4	12.4	1.5	0.8	0.8	–
			M-F	100	16.2	15.7	–	–	–	1.9	59.0	7.1	3.6	4.0	–
81	FRUIT, FRESH	2-08-368	A-F	19	2.7	3.2	–	–	–	0.8	10.4	1.5	0.8	0.9	–
			M-F	100	14.5	17.2	–	–	–	4.3	55.9	8.1	4.4	4.8	–
82	STEMS, FRESH	2-08-369	A-F	22	3.8	3.4	–	–	–	0.4	12.6	1.5	0.7	0.8	–
			M-F	100	17.5	15.7	–	–	–	1.8	58.1	6.9	3.4	3.8	–
	CLOVER, ALSIKE *Trifolium hybridum*														
83	FRESH	2-01-316	A-F	22	2.1	5.2	–	–	–	0.8	10.3	4.1	3.0	2.9	–
			M-F	100	9.3	23.3	–	–	–	3.6	45.7	18.1	13.5	12.8	–
	CLOVER, CRIMSON *Trifolium incarnatum*														
84	FRESH	2-01-336	A-F	18	1.7	4.9	–	–	–	0.6	7.5	3.0	2.3	2.1	–
			M-F	100	9.5	27.7	–	–	–	3.3	42.6	17.0	13.1	11.9	–
	CLOVER, CRIMSON-RYEGRASS *Trifolium incarnatum-Lolium* spp														
85	FRESH	2-08-380	A-F	18	1.9	3.4	–	–	–	1.1	8.0	3.9	3.0	2.8	–
			M-F	100	10.4	18.6	–	–	–	6.0	43.7	21.3	16.4	15.4	–
	CLOVER, EGYPTIAN (BERSEEM CLOVER) *Trifolium alexandrinum*														
86	FRESH	2-01-349	A-F	16	2.3	3.7	–	–	–	0.6	6.6	2.7	2.1	1.9	–
			M-F	100	14.7	23.4	–	–	–	3.7	41.4	16.8	13.0	11.8	–
	CLOVER, HOP *Trifolium agrarium*														
87	FRESH	2-01-361	A-F	25	1.6	5.1	–	–	–	1.0	13.3	4.3	2.7	3.0	–
			M-F	100	6.5	20.0	–	–	–	3.8	52.5	17.1	10.8	12.0	–
	CLOVER, LADINO *Trifolium repens*														
88	FRESH	2-01-383	A-F	18	1.9	2.5	–	–	–	0.9	8.1	4.4	3.6	3.2	–
			M-F	100	10.5	14.2	–	–	–	4.8	45.7	24.7	20.2	18.1	–
	CLOVER, LADINO-GRASS *Trifolium repens-grass*														
89	EARLY BLOOM, FRESH	2-01-513	A-F	20	2.2	4.8	–	–	–	0.6	9.4	3.1	2.2	2.1	–
			M-F	100	11.2	23.9	–	–	–	2.8	46.8	15.3	11.0	10.6	–
	CLOVER, RED *Trifolium pratense*														
90	EARLY BLOOM, FRESH[1]	2-01-428	A-F	20	2.0	4.6	8.0	6.2	–	1.0	8.3	3.8	2.8	2.7	–
			M-F	100	10.2	23.3	40.0	31.0	–	5.0	42.3	19.4	13.9	13.8	–
91	FULL BLOOM, FRESH[1]	2-01-429	A-F	26	2.0	6.8	11.2	9.1	–	0.8	12.7	3.8	2.7	2.6	–
			M-F	100	7.8	26.1	43.0	35.0	–	2.9	48.6	14.6	10.4	10.0	–
	CLOVER, STRAWBERRY *Trifolium fragiferum*														
92	FRESH	2-26-067	A-F	20	3.2	3.1	–	–	–	0.8	7.3	5.6	–	–	–
			M-F	100	16.1	15.6	–	–	–	4.0	36.5	27.8	–	–	–
	CLOVER, SUBTERRANEAN *Trifolium subterraneum*														
93	FRESH	2-26-068	A-F	21	2.1	6.0	–	–	–	1.0	9.4	2.6	–	–	–
			M-F	100	9.8	28.7	–	–	–	4.7	44.6	12.5	–	–	–
	CLOVER, WHITE *Trifolium repens*														
94	FRESH	2-01-468	A-F	18	2.1	2.8	–	–	–	0.6	7.2	5.0	4.0	3.7	–
			M-F	100	11.9	15.7	–	–	–	3.3	40.9	28.2	22.7	20.9	–
	COMFREY, PRICKLY *Symphytum asperrimum*														
95	FRESH	2-01-579	A-F	13	2.4	1.8	–	–	–	0.3	6.0	2.6	1.9	1.8	–
			M-F	100	18.2	13.8	–	–	–	2.4	45.8	19.7	14.9	14.1	–
	CORDGRASS *Spartina* spp														
96	FRESH	2-01-588	A-F	28	2.4	9.5	–	–	–	0.6	13.6	1.9	0.9	1.1	–
			M-F	100	8.6	33.8	–	–	–	2.3	48.4	6.9	3.4	3.8	–
	CORN *Zea mays*														
97	FRESH, ALL ANALYSES	2-02-799	A-F	23	1.3	5.7	–	–	–	0.9	12.9	2.4	1.5	1.5	–
			M-F	100	5.6	24.5	–	–	–	3.7	56.0	10.2	6.5	6.5	–
98	MILK STAGE, FRESH	2-02-802	A-F	24	1.2	5.7	–	–	–	0.7	14.2	1.9	1.1	1.1	–
			M-F	100	5.1	24.2	–	–	–	2.9	59.8	8.0	4.7	4.7	–
99	MATURE, FRESH	2-02-804	A-F	34	1.4	8.1	–	–	–	0.9	20.6	2.6	1.6	1.5	–
			M-F	100	4.3	24.1	–	–	–	2.6	61.5	7.6	4.8	4.4	–
100	LEAVES WITH TOPS, FRESH	2-08-391	A-F	16	1.2	4.8	–	–	–	0.5	7.9	1.5	0.9	0.9	–
			M-F	100	7.5	30.4	–	–	–	3.0	49.9	9.2	5.5	5.7	–
101	STOVER WITHOUT EARS, WITHOUT HUSKS, FRESH	2-02-809	A-F	31	3.0	8.2	22.4	14.1	2.1	0.5	17.4	1.8	0.7	0.9	–
			M-F	100	9.6	26.4	72.5	45.7	6.8	1.8	56.4	5.8	2.4	3.0	–
	CORN, SWEET *Zea mays, saccharata*														
102	STOVER WITHOUT EARS, WITHOUT HUSKS, FRESH	2-02-969	A-F	22	1.4	5.7	–	–	–	0.4	12.8	1.6	0.8	0.9	–
			M-F	100	6.2	26.0	–	–	–	1.8	58.6	7.3	3.7	4.2	–
103	EARS WITH HUSKS, FRESH	4-08-408	A-F	38	0.9	4.3	–	–	–	2.6	26.2	3.8	2.3	2.7	2.7
			M-F	100	2.4	11.4	–	–	–	6.9	69.3	10.1	6.0	7.2	7.0
104	CANNERY RESIDUE	2-02-975	A-F	77	3.5	17.0	–	22.3	–	1.9	47.6	6.8	3.9	4.1	–
			M-F	100	4.6	22.2	–	29.0	–	2.5	62.0	8.8	5.1	5.4	–

Entry Number	TDN Ruminant	TDN Swine	TDN Horse	Digestible Energy Ruminant		Digestible Energy Swine		Digestible Energy Horse		Metabolizable Energy Ruminant		Metabolizable Energy Swine		Metabolizable Energy Poultry ME_n		Metabolizable Energy Horse		Net Energy Ruminant NE_m		Net Energy Ruminant NE_g		Net Energy Lactating Cows NE_{lc}	
	%	%	%	Mcal lb	Mcal kg	kcal lb	kcal kg	Mcal lb	Mcal kg	Mcal lb	Mcal kg	kcal lb	kcal kg	kcal lb	kcal kg	Mcal lb	Mcal kg	Mcal lb	Mcal kg	Mcal lb	Mcal kg	Mcal lb	Mcal kg
80	12	–	9	0.25	0.55	–	–	0.18	0.39	0.21	0.46	–	–	–	–	0.15	0.32	0.12	0.27	0.07	0.15	0.13	0.28
	59	–	44	1.18	2.60	–	–	0.85	1.88	0.99	2.17	–	–	–	–	0.70	1.54	0.58	1.28	0.32	0.71	0.60	1.33
81	11	–	7	0.23	0.50	–	–	0.13	0.29	0.19	0.42	–	–	–	–	0.11	0.24	0.11	0.25	0.06	0.14	0.12	0.26
	61	–	36	1.22	2.68	–	–	0.71	1.56	1.02	2.26	–	–	–	–	0.58	1.28	0.61	1.34	0.35	0.76	0.62	1.37
82	13	–	9	0.25	0.55	–	–	0.18	0.39	0.21	0.46	–	–	–	–	0.15	0.32	0.12	0.27	0.07	0.14	0.13	0.28
	58	–	42	1.15	2.54	–	–	0.82	1.81	0.96	2.12	–	–	–	–	0.67	1.48	0.56	1.23	0.30	0.67	0.59	1.29
83	16	–	–	0.32	0.71	–	–	–	–	0.28	0.61	–	–	–	–	–	–	0.17	0.37	0.11	0.24	0.17	0.36
	71	–	–	1.42	3.14	–	–	–	–	1.23	2.72	–	–	–	–	–	–	0.76	1.67	0.48	1.06	0.74	1.62
84	11	–	–	0.23	0.50	–	–	–	–	0.19	0.42	–	–	–	–	–	–	0.11	0.25	0.07	0.15	0.12	0.25
	65	–	–	1.29	2.84	–	–	–	–	1.10	2.42	–	–	–	–	–	–	0.65	1.43	0.38	0.85	0.65	1.44
85	13	–	–	0.26	0.58	–	–	–	–	0.23	0.51	–	–	–	–	–	–	–	–	–	–	–	–
	72	–	–	1.44	3.18	–	–	–	–	1.25	2.76	–	–	–	–	–	–	–	–	–	–	–	–
86	10	–	–	0.20	0.44	–	–	–	–	0.17	0.37	–	–	–	–	–	–	0.10	0.21	0.06	0.12	0.10	0.22
	63	–	–	1.24	2.73	–	–	–	–	1.05	2.31	–	–	–	–	–	–	0.61	1.34	0.34	0.76	0.62	1.37
87	18	–	–	0.36	0.80	–	–	–	–	0.31	0.69	–	–	–	–	–	–	0.19	0.42	0.12	0.27	0.19	0.41
	72	–	–	1.43	3.16	–	–	–	–	1.24	2.74	–	–	–	–	–	–	0.76	1.67	0.48	1.06	0.74	1.62
88	13	–	–	0.27	0.60	–	–	–	–	0.24	0.52	–	–	–	–	–	–	0.15	0.33	0.10	0.22	0.14	0.32
	76	–	–	1.53	3.37	–	–	–	–	1.34	2.95	–	–	–	–	–	–	0.86	1.89	0.57	1.25	0.82	1.80
89	13	–	–	0.26	0.56	–	–	–	–	0.22	0.48	–	–	–	–	–	–	0.13	0.28	0.08	0.17	0.13	0.29
	64	–	–	1.28	2.82	–	–	–	–	1.09	2.40	–	–	–	–	–	–	0.64	1.42	0.38	0.83	0.65	1.43
90	14	–	–	0.27	0.58	–	–	–	–	0.23	0.50	–	–	–	–	–	–	0.14	0.32	0.09	0.20	0.14	0.31
	69	–	–	1.34	2.96	–	–	–	–	1.15	2.54	–	–	–	–	–	–	0.73	1.61	0.46	1.01	0.72	1.58
91	17	–	–	0.34	0.75	–	–	–	–	0.29	0.64	–	–	–	–	–	–	0.17	0.38	0.10	0.22	0.17	0.38
	64	–	–	1.30	2.86	–	–	–	–	1.11	2.44	–	–	–	–	–	–	0.65	1.44	0.39	0.86	0.66	1.45
92	–	–	–	–	–	–	–	–	–	–	–	–	–	–	–	–	–	–	–	–	–	–	–
	–	–	–	–	–	–	–	–	–	–	–	–	–	–	–	–	–	–	–	–	–	–	–
93	–	–	–	–	–	–	–	–	–	–	–	–	–	–	–	–	–	–	–	–	–	–	–
	–	–	–	–	–	–	–	–	–	–	–	–	–	–	–	–	–	–	–	–	–	–	–
94	13	–	–	0.26	0.57	–	–	–	–	0.22	0.49	–	–	–	–	–	–	0.14	0.30	0.09	0.19	0.13	0.29
	72	–	–	1.45	3.19	–	–	–	–	1.26	2.77	–	–	–	–	–	–	0.77	1.70	0.49	1.09	0.75	1.65
95	8	–	–	0.17	0.37	–	–	–	–	0.14	0.31	–	–	–	–	–	–	0.08	0.18	0.05	0.11	0.08	0.19
	64	–	–	1.27	2.81	–	–	–	–	1.08	2.38	–	–	–	–	–	–	0.64	1.41	0.38	0.83	0.65	1.43
96	16	–	–	0.32	0.71	–	–	–	–	0.27	0.59	–	–	–	–	–	–	0.15	0.34	0.08	0.18	0.16	0.36
	58	–	–	1.15	2.54	–	–	–	–	0.96	2.12	–	–	–	–	–	–	0.55	1.21	0.29	0.65	0.58	1.28
97	16	–	11	0.32	0.70	–	–	0.21	0.46	0.28	0.61	–	–	–	–	0.17	0.38	0.16	0.35	0.10	0.22	0.16	0.35
	71	–	47	1.38	3.04	–	–	0.90	1.99	1.19	2.62	–	–	–	–	0.74	1.63	0.70	1.53	0.43	0.94	0.69	1.52
98	17	–	12	0.33	0.72	–	–	0.22	0.49	0.28	0.62	–	–	–	–	0.18	0.40	0.16	0.36	0.10	0.22	0.16	0.36
	71	–	49	1.38	3.04	–	–	0.93	2.05	1.19	2.62	–	–	–	–	0.76	1.68	0.69	1.52	0.42	0.93	0.68	1.51
99	24	–	17	0.49	1.08	–	–	0.33	0.72	0.43	0.94	–	–	–	–	0.27	0.59	0.27	0.59	0.17	0.38	0.26	0.57
	73	–	52	1.46	3.21	–	–	0.98	2.15	1.27	2.79	–	–	–	–	0.80	1.76	0.79	1.75	0.51	1.13	0.77	1.69
100	10	–	–	0.20	0.44	–	–	–	–	0.17	0.38	–	–	–	–	–	–	0.10	0.21	0.06	0.12	0.10	0.22
	65	–	–	1.27	2.79	–	–	–	–	1.07	2.37	–	–	–	–	–	–	0.61	1.35	0.35	0.77	0.63	1.38
101	18	–	–	0.36	0.80	–	–	–	–	0.30	0.67	–	–	–	–	–	–	0.18	0.40	0.10	0.22	0.19	0.41
	57	–	–	1.17	2.58	–	–	–	–	0.98	2.15	–	–	–	–	–	–	0.59	1.30	0.33	0.72	0.61	1.34
102	12	–	–	0.26	0.58	–	–	–	–	0.22	0.49	–	–	–	–	–	–	0.14	0.31	0.08	0.19	0.14	0.32
	57	–	–	1.21	2.66	–	–	–	–	1.01	2.24	–	–	–	–	–	–	0.65	1.44	0.39	0.85	0.66	1.44
103	29	31	19	0.59	1.30	623	1373	0.35	0.78	0.52	1.14	558	1297	–	–	0.29	0.64	0.32	0.71	0.21	0.47	0.31	0.68
	77	82	49	1.55	3.43	1648	3633	0.94	2.07	1.37	3.01	1557	3432	–	–	0.77	1.69	0.86	1.89	0.57	1.25	0.81	1.80
104	54	–	42	1.05	2.31	–	–	0.79	1.74	0.90	1.99	–	–	–	–	0.65	1.43	0.57	1.26	0.36	0.79	0.56	1.23
	70	–	55	1.37	3.01	–	–	1.03	2.27	1.18	2.59	–	–	–	–	0.84	1.86	0.74	1.63	0.47	1.03	0.72	1.60

(Continued)

TABLE V-1E PASTURE AND RANGE PLANTS, MINERAL AND VITAMIN COMPOSITION OF FEEDS, DATA EXPRESSED **AS-FED** AND **MOISTURE-FREE**—*(Continued)*

Entry Number	Feed Name Description	Moisture Basis: A-F (as-fed) or M-F (moisture-free)	Dry Matter	Macro Minerals							Micro Minerals						
				Calcium (Ca)	Phosphorus (P)	Sodium (Na)	Chlorine (Cl)	Magnesium (Mg)	Potassium (K)	Sulfur (S)	Cobalt (Co)	Copper (Cu)	Iodine (I)	Iron (Fe)	Manganese (Mn)	Selenium (Se)	Zinc (Zn)
			%	%	%	%	%	%	%	%	ppm or mg/kg	ppm or mg/kg	ppm or mg/kg	%	ppm or mg/kg	ppm or mg/kg	ppm or mg/kg
	CHINESE LESPEDEZA–SEE LESPEDEZA, SERICEA																
	CHOLLA, MEXICALI (CANE CACTUS) *Opuntia cholla*																
80	FRESH	**A-F**	**21**	**–**	**0.01**	**–**	**–**	**–**	**0.17**	**–**	**–**	**–**	**–**	**–**	**–**	**–**	**–**
		M-F	100	–	0.05	–	–	–	0.81	–	–	–	–	–	–	–	–
81	FRUIT, FRESH	**A-F**	**19**	**–**	**–**	**–**	**–**	**–**	**–**	**–**	**–**	**–**	**–**	**–**	**–**	**–**	**–**
		M-F	100	–	–	–	–	–	–	–	–	–	–	–	–	–	–
82	STEMS, FRESH	**A-F**	**22**	**–**	**0.04**	**–**	**–**	**–**	**0.40**	**–**	**–**	**–**	**–**	**–**	**–**	**–**	**–**
		M-F	100	–	0.18	–	–	–	1.84	–	–	–	–	–	–	–	–
	CLOVER, ALSIKE *Trifolium hybridum*																
83	FRESH	**A-F**	**22**	**0.31**	**0.06**	**0.10**	**0.17**	**0.07**	**0.61**	**0.05**	**–**	**1.3**	**–**	**0.010**	**26.3**	**–**	**–**
		M-F	100	1.36	0.29	0.45	0.77	0.32	2.70	0.22	–	6.0	–	0.044	117.1	–	–
	CLOVER, CRIMSON *Trifolium incarnatum*																
84	FRESH	**A-F**	**18**	**0.24**	**0.05**	**0.07**	**0.11**	**0.05**	**0.55**	**0.05**	**–**	**–**	**–**	**–**	**43.1**	**–**	**–**
		M-F	100	1.38	0.29	0.40	0.61	0.29	3.10	0.28	–	–	–	–	245.8	–	–
	CLOVER, CRIMSON-RYEGRASS *Trifolium incarnatum-Lolium* spp																
85	FRESH	**A-F**	**18**	**0.12**	**0.12**	**–**	**–**	**–**	**–**	**–**	**–**	**–**	**–**	**–**	**–**	**–**	**–**
		M-F	100	0.66	0.66	–	–	–	–	–	–	–	–	–	–	–	–
	CLOVER, EGYPTIAN (BERSEEM CLOVER) *Trifolium alexandrinum*																
86	FRESH	**A-F**	**16**	**0.57**	**0.05**	**–**	**–**	**–**	**0.37**	**–**	**–**	**–**	**–**	**–**	**–**	**–**	**–**
		M-F	100	3.56	0.32	–	–	–	2.31	–	–	–	–	–	–	–	–
	CLOVER, HOP *Trifolium agrarium*																
87	FRESH	**A-F**	**25**	**0.28**	**0.08**	**–**	**–**	**0.05**	**0.49**	**–**	**0.012**	**–**	**–**	**0.007**	**25.0**	**–**	**–**
		M-F	100	1.09	0.33	–	–	0.19	1.94	–	0.045	–	–	0.025	99.2	–	–
	CLOVER, LADINO *Trifolium repens*																
88	FRESH	**A-F**	**18**	**0.22**	**0.07**	**0.02**	**–**	**0.09**	**0.33**	**0.02**	**–**	**–**	**–**	**0.007**	**12.7**	**–**	**–**
		M-F	100	1.27	0.42	0.12	–	0.48	1.87	0.12	–	–	–	0.037	71.7	–	–
	CLOVER, LADINO-GRASS *Trifolium repens-grass*																
89	EARLY BLOOM, FRESH	**A-F**	**20**	**–**	**–**	**–**	**–**	**–**	**–**	**–**	**–**	**–**	**–**	**–**	**–**	**–**	**–**
		M-F	100	–	–	–	–	–	–	–	–	–	–	–	–	–	–
	CLOVER, RED *Trifolium pratense*																
90	EARLY BLOOM, FRESH	**A-F**	**20**	**0.45**	**0.08**	**0.04**	**–**	**0.10**	**0.49**	**0.03**	**–**	**–**	**–**	**0.006**	**–**	**–**	**–**
		M-F	100	2.26	0.38	0.20	–	0.51	2.49	0.17	–	–	–	0.030	–	–	–
91	FULL BLOOM, FRESH	**A-F**	**26**	**0.27**	**0.07**	**0.05**	**–**	**0.13**	**0.51**	**0.05**	**–**	**–**	**–**	**0.008**	**–**	**–**	**–**
		M-F	100	1.01	0.27	0.20	–	0.51	1.96	0.17	–	–	–	0.030	–	–	–
	CLOVER, STRAWBERRY *Trifolium fragiferum*																
92	FRESH	**A-F**	**20**	**0.37**	**0.09**	**–**	**–**	**–**	**–**	**–**	**–**	**–**	**–**	**–**	**–**	**–**	**–**
		M-F	100	1.83	0.43	–	–	–	–	–	–	–	–	–	–	–	–
	CLOVER, SUBTERRANEAN *Trifolium subterraneum*																
93	FRESH	**A-F**	**21**	**0.31**	**0.07**	**–**	**–**	**–**	**–**	**–**	**–**	**–**	**–**	**–**	**–**	**–**	**–**
		M-F	100	1.49	0.33	–	–	–	–	–	–	–	–	–	–	–	–
	CLOVER, WHITE *Trifolium repens*																
94	FRESH	**A-F**	**18**	**0.25**	**0.09**	**0.07**	**0.11**	**0.08**	**0.38**	**0.06**	**–**	**–**	**–**	**0.006**	**54.4**	**–**	**–**
		M-F	100	1.40	0.51	0.39	0.61	0.45	2.14	0.33	–	–	–	0.034	307.1	–	–
	COMFREY, PRICKLY *Symphytum asperrimum*																
95	FRESH	**A-F**	**13**	**–**	**0.07**	**–**	**–**	**–**	**0.59**	**–**	**–**	**–**	**–**	**–**	**–**	**–**	**–**
		M-F	100	–	0.55	–	–	–	4.53	–	–	–	–	–	–	–	–
	CORDGRASS *Spartina* spp																
96	FRESH	**A-F**	**28**	**0.11**	**0.03**	**–**	**–**	**–**	**–**	**–**	**–**	**–**	**–**	**–**	**–**	**–**	**–**
		M-F	100	0.38	0.11	–	–	–	–	–	–	–	–	–	–	–	–
	CORN *Zea mays*																
97	FRESH, ALL ANALYSES	**A-F**	**23**	**0.07**	**–**	**–**	**–**	**0.21**	**–**	**–**	**–**	**1.8**	**–**	**0.008**	**24.6**	**–**	**16.1**
		M-F	100	0.30	–	–	–	0.89	–	–	–	7.6	–	0.034	106.5	–	69.8
98	MILK STAGE, FRESH	**A-F**	**24**	**–**	**–**	**–**	**–**	**–**	**–**	**–**	**–**	**–**	**–**	**–**	**–**	**–**	**–**
		M-F	100	–	–	–	–	–	–	–	–	–	–	–	–	–	–
99	MATURE, FRESH	**A-F**	**34**	**0.09**	**0.08**	**–**	**–**	**0.04**	**0.33**	**–**	**–**	**3.1**	**–**	**0.005**	**10.4**	**–**	**–**
		M-F	100	0.28	0.24	–	–	0.13	0.99	–	–	9.3	–	0.014	31.1	–	–
100	LEAVES WITH TOPS, FRESH	**A-F**	**16**	**–**	**–**	**–**	**–**	**–**	**–**	**–**	**–**	**–**	**–**	**–**	**–**	**–**	**–**
		M-F	100	–	–	–	–	–	–	–	–	–	–	–	–	–	–
101	STOVER WITHOUT EARS, WITHOUT HUSKS, FRESH	**A-F**	**31**	**0.19**	**0.03**	**–**	**–**	**–**	**0.50**	**–**	**–**	**–**	**–**	**–**	**–**	**–**	**–**
		M-F	100	0.62	0.09	–	–	–	1.63	–	–	–	–	–	–	–	–
	CORN, SWEET *Zea mays, saccharata*																
102	STOVER WITHOUT EARS, WITHOUT HUSKS, FRESH	**A-F**	**22**	**–**	**–**	**–**	**–**	**–**	**–**	**–**	**–**	**–**	**–**	**–**	**–**	**–**	**–**
		M-F	100	–	–	–	–	–	–	–	–	–	–	–	–	–	–
103	EARS WITH HUSKS, FRESH	**A-F**	**38**	**–**	**–**	**–**	**–**	**–**	**–**	**–**	**–**	**–**	**–**	**–**	**–**	**–**	**–**
		M-F	100	–	–	–	–	–	–	–	–	–	–	–	–	–	–
104	CANNERY RESIDUE, FRESH	**A-F**	**77**	**0.25**	**0.54**	**0.02**	**–**	**0.18**	**0.88**	**0.10**	**–**	**5.4**	**–**	**0.016**	**–**	**–**	**–**
		M-F	100	0.32	0.70	0.03	–	0.24	1.15	0.13	–	7.0	–	0.020	–	–	–

Entry Number	Fat-Soluble Vitamins: A (1 mg Carotene = 1667 IU Vit A)	Carotene (Provitamin A)	D	E	K	Water-Soluble Vitamins: B–12	Biotin	Choline	Folacin (Folic Acid)	Niacin	Pantothenic Acid (B–3)	(Pyri-doxine(B–6	Ribo-flavin (B–2)	Thiamin (B–1)
	IU/g	ppm or mg/kg	IU/kg	ppm or mg/kg	ppm or mg/kg	ppm or mg/kg	ppm or mg/kg	ppm or mg/kg	ppm or mg/kg	ppm or mg/kg	ppm or mg/kg	ppm or mg/kg	ppm or mg/kg	ppm or mg/kg
80	–	–	–	–	–	–	–	–	–	–	–	–	–	–
	–	–	–	–	–	–	–	–	–	–	–	–	–	–
81	–	–	–	–	–	–	–	–	–	–	–	–	–	–
	–	–	–	–	–	–	–	–	–	–	–	–	–	–
82	–	–	–	–	–	–	–	–	–	–	–	–	–	–
	–	–	–	–	–	–	–	–	–	–	–	–	–	–
83	–	–	–	–	–	–	–	–	–	–	–	–	**4.4**	**2.0**
	–	–	–	–	–	–	–	–	–	–	–	–	19.6	8.8
84	–	–	–	–	–	–	–	–	–	–	–	–	–	–
	–	–	–	–	–	–	–	–	–	–	–	–	–	–
85	**124.9**	**74.9**	–	–	–	–	–	–	–	–	–	–	–	–
	682.8	409.6	–	–	–	–	–	–	–	–	–	–	–	–
86	–	–	–	–	–	–	–	–	–	–	–	–	–	–
	–	–	–	–	–	–	–	–	–	–	–	–	–	–
87	–	–	–	–	–	–	–	–	–	–	–	–	–	–
	–	–	–	–	–	–	–	–	–	–	–	–	–	–
88	**96.2**	**57.7**	–	–	–	–	–	–	–	–	–	–	**4.2**	–
	545.2	327.0	–	–	–	–	–	–	–	–	–	–	24.1	–
89	–	–	–	–	–	–	–	–	–	–	–	–	–	–
	–	–	–	–	–	–	–	–	–	–	–	–	–	–
90	**81.5**	**48.9**	–	–	–	–	–	–	–	–	–	–	–	–
	412.6	247.5	–	–	–	–	–	–	–	–	–	–	–	–
91	**90.6**	**54.4**	–	–	–	–	–	–	–	–	–	–	–	–
	345.9	207.5	–	–	–	–	–	–	–	–	–	–	–	–
92	–	–	–	–	–	–	–	–	–	–	–	–	–	–
	–	–	–	–	–	–	–	–	–	–	–	–	–	–
93	–	–	–	–	–	–	–	–	–	–	–	–	–	–
	–	–	–	–	–	–	–	–	–	–	–	–	–	–
94	**44.0**	**26.4**	–	**54.6**	–	–	–	–	–	**11**	–	–	**16.0**	**2.5**
	248.4	149.0	–	308.6	–	–	–	–	–	63	–	–	90.2	14.1
95	–	–	–	–	–	–	–	–	–	–	–	–	–	–
	–	–	–	–	–	–	–	–	–	–	–	–	–	–
96	–	–	–	–	–	–	–	–	–	–	–	–	–	–
	–	–	–	–	–	–	–	–	–	–	–	–	–	–
97	–	–	–	–	–	–	–	–	–	–	–	–	–	–
	–	–	–	–	–	–	–	–	–	–	–	–	–	–
98	–	–	–	–	–	–	–	–	–	–	–	–	–	–
	–	–	–	–	–	–	–	–	–	–	–	–	–	–
99	**21.1**	**12.7**	–	–	–	–	–	–	–	–	–	–	–	–
	62.8	37.7	–	–	–	–	–	–	–	–	–	–	–	–
100	–	–	–	–	–	–	–	–	–	–	–	–	–	–
	–	–	–	–	–	–	–	–	–	–	–	–	–	–
101	–	–	–	–	–	–	–	–	–	–	–	–	–	–
	–	–	–	–	–	–	–	–	–	–	–	–	–	–
102	–	–	–	–	–	–	–	–	–	–	–	–	–	–
	–	–	–	–	–	–	–	–	–	–	–	–	–	–
103	–	–	–	–	–	–	–	–	–	–	–	–	–	–
	–	–	–	–	–	–	–	–	–	–	–	–	–	–
104	**17.3**	**10.4**	–	–	–	–	–	–	–	–	–	–	–	–
	22.5	13.5	–	–	–	–	–	–	–	–	–	–	–	–

(Continued)

TABLE V-1E PASTURE AND RANGE PLANTS, COMPOSITION OF FEEDS, DATA EXPRESSED **AS-FED** AND **MOISTURE-FREE**—*(Continued)*

Entry Number	Feed Name Description	International Feed Number	Moisture Basis: A-F (as-fed) or M-F (moisture-free)	Proximate Analysis Dry Matter	Ash	Crude Fiber	Neutral Det. Fib. (NDF)	Acid Det. Fib. (ADF)	Lignin	Ether Extract (Fat)	N-Free Extract	Crude Protein	Digestible Protein Ruminant	Swine	Horse
				%	%	%	%	%	%	%	%	%	%	%	%
	CORN-SUNFLOWER *Zea mays-Helianthus spp*														
105	FRESH	2-03-016	A-F	18	1.6	6.1	–	–	–	0.6	8.5	1.4	0.7	0.8	–
			M-F	100	8.8	33.5	–	–	–	3.3	46.7	7.7	4.1	4.5	–
	COTTONTOP, ARIZONA *Trichachne californica*														
106	FRESH	2-01-634	A-F	30	3.5	10.8	–	–	–	0.6	12.5	2.7	1.6	1.6	–
			M-F	100	11.5	36.1	–	–	–	1.9	41.6	8.9	5.2	5.4	–
	COWPEA, COMMON *Vigna sinensis*														
107	FRESH	2-01-655	A-F	25	3.0	6.1	–	–	–	1.0	10.4	4.0	2.9	2.9	–
			M-F	100	12.3	25.0	–	–	–	4.1	42.4	16.2	12.0	11.8	–
	CRABGRASS *Digitaria spp*														
108	FRESH	2-01-668	A-F	30	4.1	8.9	–	–	–	0.9	13.2	2.9	1.7	1.8	–
			M-F	100	13.6	29.7	–	–	–	3.2	44.1	9.5	5.7	5.9	–
	CRESTED WHEATGRASS *Agropyron desertorum*														
109	IMMATURE, FRESH	2-05-420	A-F	28	2.9	6.2	–	–	–	0.6	12.9	6.0	5.1	4.3	–
			M-F	100	10.0	21.6	–	–	–	2.2	45.2	21.0	17.8	15.1	–
110	EARLY BLOOM, FRESH	2-05-422	A-F	41	–	8.9	–	–	–	1.7	–	4.8	3.1	3.1	–
			M-F	100	–	21.7	–	–	–	4.1	–	11.7	7.7	7.6	–
111	FULL BLOOM, FRESH	2-05-424	A-F	45	4.2	13.6	–	–	–	1.6	22.2	3.4	1.8	1.9	–
			M-F	100	9.3	30.3	–	–	–	3.6	49.3	7.5	3.9	4.3	–
112	OVERRIPE, FRESH	2-05-428	A-F	75	3.1	25.2	–	–	–	1.6	42.3	2.8	0.4	1.0	–
			M-F	100	4.1	33.6	–	–	–	2.1	56.4	3.8	0.5	1.3	–
	CRIMSON CLOVER–SEE CLOVER, CRIMSON														
	CRIMSON CLOVER-RYEGRASS–SEE CLOVER, CRIMSON-RYEGRASS														
	CROTALARIA *Crotalaria spp*														
113	MIDBLOOM, FRESH	2-01-678	A-F	27	1.5	10.9	–	–	–	0.6	10.5	3.6	2.5	2.4	–
			M-F	100	5.6	40.3	–	–	–	2.1	38.7	13.3	9.2	9.0	–
	CURLY MESQUITE *Hilaria belangeri*														
114	BROWSE, FRESH	2-01-728	A-F	35	5.1	9.8	–	–	–	0.7	16.4	3.0	1.7	1.8	–
			M-F	100	14.6	27.9	–	–	–	2.1	46.8	8.6	4.9	5.2	–
115	BROWSE, MATURE, FRESH	2-01-729	A-F	50	7.6	14.3	32.0	–	–	1.1	23.9	2.8	1.1	1.4	–
			M-F	100	15.3	28.8	64.4	–	–	2.2	48.1	5.6	2.2	2.8	–
	DALLISGRASS *Paspalum dilatatum*														
116	FRESH	2-01-741	A-F	25	3.0	7.3	–	–	–	0.6	11.1	3.0	2.0	2.0	–
			M-F	100	12.0	29.1	–	–	–	2.5	44.4	12.1	8.0	8.0	–
	DANDELION, COMMON *Taraxacum officinale*														
117	LEAVES, FRESH	2-01-748	A-F	14	1.8	1.6	–	–	–	0.7	7.6	2.7	2.0	1.9	–
			M-F	100	12.5	11.1	–	–	–	4.9	52.8	18.7	14.1	13.3	–
	DARSO SORGHUM–SEE SORGHUM														
	DEERVETCH, BIRDSFOOT (BIRDSFOOT TREFOIL) *Lotus corniculatus*														
118	FRESH	2-20-786	A-F	19	2.2	4.1	9.5	–	–	0.8	8.5	3.7	2.8	2.7	–
			M-F	100	11.2	21.2	49.4	–	–	4.0	44.3	19.3	14.6	13.8	–
	DESERT MOLLY (SUMMER CYPRESS, GRAY) *Kochia vestita*														
119	STEM CURED, FRESH	2-08-843	A-F	85	21.5	18.7	–	–	–	3.1	35.0	6.6	4.3	3.9	–
			M-F	100	25.3	22.0	–	–	–	3.7	41.2	7.8	5.0	4.5	–
	DESMANTHUS (BUNDLEFLOWER, RAYADO; DWARF KOA) *Desmanthus virgatus*														
120	FRESH	2-01-024	A-F	41	2.4	17.5	–	–	–	0.9	15.7	4.9	3.0	3.0	–
			M-F	100	5.7	42.2	–	–	–	2.2	38.0	11.9	7.3	7.3	–
	DIGITARIA–SEE CRABGRASS; PANGOLAGRASS														
	DOCK, CURLY *Rumex crispus*														
121	FRESH	2-01-794	A-F	18	0.6	4.3	–	–	–	0.6	10.2	2.3	1.6	1.6	–
			M-F	100	3.6	23.7	–	–	–	3.2	56.5	13.0	8.9	8.7	–
	DOGWOOD, FLOWERING *Cornus florida*														
122	BROWSE, FRESH	2-26-117	A-F	27	2.5	4.3	–	–	–	1.5	15.5	3.3	–	–	2.1
			M-F	100	9.2	16.0	–	–	–	5.4	57.3	12.1	–	–	7.8
	DROPSEED, SAND *Sporobolus cryptandrus*														
123	STEM CURED, FRESH	2-05-596	A-F	88	7.0	31.6	–	–	5.2	1.1	43.0	5.4	1.3	2.8	–
			M-F	100	7.9	35.9	–	–	5.9	1.2	48.8	6.1	1.5	3.2	–
	DURRA SORGHUM–SEE SORGHUM														
	DWARF KOA (BUNDLEFLOWER, RAYADO; DESMANTHUS) *Desmanthus virgatus*														
124	FRESH	2-01-024	A-F	41	2.4	17.5	–	–	–	0.9	15.7	4.9	3.0	3.0	–
			M-F	100	5.7	42.2	–	–	–	2.2	38.0	11.9	7.3	7.3	–

Entry Number	TDN Ruminant	TDN Swine	TDN Horse	Digestible Energy Ruminant		Digestible Energy Swine		Digestible Energy Horse		Metabolizable Energy Ruminant		Metabolizable Energy Swine		Metabolizable Energy Poultry ME_n		Metabolizable Energy Horse		Net Energy Ruminant NE_m		Net Energy Ruminant NE_g		Net Energy Lactating Cows NE_{lc}	
	%	%	%	Mcal		kcal		Mcal		Mcal		kcal		kcal		Mcal		Mcal		Mcal		Mcal	
				lb	kg	lb	kg	lb	kg	lb	kg	lb	kg	lb	kg	lb	kg	lb	kg	lb	kg	lb	kg
105	10	–	–	0.21	0.46	–	–	–	–	0.17	0.38	–	–	–	–	–	–	0.10	0.22	0.06	0.12	0.11	0.24
	57	–	–	1.15	2.53	–	–	–	–	0.95	2.10	–	–	–	–	–	–	0.56	1.23	0.30	0.66	0.59	1.29
106	17	–	–	0.33	0.73	–	–	–	–	0.27	0.60	–	–	–	–	–	–	0.15	0.33	0.07	0.16	0.16	0.36
	55	–	–	1.11	2.44	–	–	–	–	0.91	2.01	–	–	–	–	–	–	0.50	1.09	0.24	0.54	0.54	1.19
107	16	15	–	0.29	0.65	–	–	–	–	0.26	0.58	–	–	–	–	–	–	0.15	0.32	0.08	0.18	0.14	0.32
	67	62	–	1.20	2.64	–	–	–	–	1.07	2.37	–	–	–	–	–	–	0.59	1.31	0.33	0.73	0.58	1.29
108	17	–	–	0.34	0.74	–	–	–	–	0.28	0.62	–	–	–	–	–	–	0.16	0.35	0.08	0.18	0.17	0.37
	57	–	–	1.12	2.48	–	–	–	–	0.93	2.05	–	–	–	–	–	–	0.53	1.16	0.27	0.60	0.56	1.24
109	21	–	–	0.41	0.90	–	–	–	–	0.35	0.78	–	–	–	–	–	–	0.20	0.45	0.13	0.28	0.20	0.44
	75	–	–	1.43	3.16	–	–	–	–	1.24	2.74	–	–	–	–	–	–	0.71	1.57	0.44	0.97	0.70	1.54
110	–	–	–	–	–	–	–	–	–	–	–	–	–	–	–	–	–	–	–	–	–	–	–
	–	–	–	–	–	–	–	–	–	–	–	–	–	–	–	–	–	–	–	–	–	–	–
111	26	–	–	0.52	1.15	–	–	–	–	0.44	0.96	–	–	–	–	–	–	0.26	0.57	0.14	0.32	0.27	0.59
	58	–	–	1.16	2.56	–	–	–	–	0.97	2.14	–	–	–	–	–	–	0.58	1.27	0.32	0.70	0.60	1.32
112	45	–	–	0.90	1.99	–	–	–	–	0.76	1.67	–	–	–	–	–	–	0.45	1.00	0.26	0.56	0.46	1.02
	60	–	–	1.21	2.66	–	–	–	–	1.01	2.23	–	–	–	–	–	–	0.60	1.33	0.34	0.75	0.62	1.36
113	16	–	–	0.33	0.72	–	–	–	–	0.28	0.61	–	–	–	–	–	–	0.16	0.35	0.09	0.19	0.16	0.36
	61	–	–	1.22	2.68	–	–	–	–	1.02	2.26	–	–	–	–	–	–	0.58	1.28	0.32	0.71	0.60	1.33
114	21	–	–	0.40	0.88	–	–	–	–	0.33	0.73	–	–	–	–	–	–	0.18	0.40	0.09	0.20	0.19	0.43
	59	–	–	1.14	2.51	–	–	–	–	0.94	2.08	–	–	–	–	–	–	0.52	1.14	0.26	0.58	0.56	1.22
115	27	–	–	0.53	1.17	–	–	–	–	0.44	0.96	–	–	–	–	–	–	0.24	0.52	0.11	0.25	0.26	0.58
	53	–	–	1.07	2.36	–	–	–	–	0.88	1.93	–	–	–	–	–	–	0.48	1.05	0.23	0.50	0.53	1.17
116	16	–	–	0.31	0.67	–	–	–	–	0.26	0.57	–	–	–	–	–	–	0.14	0.31	0.08	0.17	0.15	0.33
	64	–	–	1.22	2.69	–	–	–	–	1.03	2.27	–	–	–	–	–	–	0.57	1.25	0.31	0.68	0.59	1.31
117	10	–	–	0.20	0.44	–	–	–	–	0.17	0.38	–	–	–	–	–	–	0.11	0.24	0.07	0.15	0.11	0.23
	69	–	–	1.38	3.04	–	–	–	–	1.19	2.62	–	–	–	–	–	–	0.75	1.65	0.47	1.05	0.73	1.61
118	13	–	–	0.26	0.58	–	–	–	–	0.23	0.50	–	–	–	–	–	–	0.16	0.35	0.10	0.22	0.15	0.33
	68	–	–	1.36	2.99	–	–	–	–	1.17	2.57	–	–	–	–	–	–	0.81	1.79	0.53	1.16	0.78	1.72
119	43	–	–	0.77	1.71	–	–	–	–	0.68	1.51	–	–	–	–	–	–	0.33	0.73	0.12	0.27	0.40	0.88
	50	–	–	0.91	2.01	–	–	–	–	0.80	1.77	–	–	–	–	–	–	0.39	0.86	0.15	0.32	0.47	1.03
120	22	22	–	0.46	1.02	–	–	–	–	0.38	0.85	–	–	–	–	–	–	0.23	0.51	0.12	0.27	0.24	0.53
	54	54	–	1.12	2.47	–	–	–	–	0.93	2.04	–	–	–	–	–	–	0.56	1.23	0.30	0.66	0.59	1.29
121	12	–	10	0.25	0.55	–	–	0.19	0.42	0.22	0.48	–	–	–	–	0.16	0.34	0.13	0.30	0.09	0.19	0.13	0.29
	69	–	56	1.39	3.06	–	–	1.05	2.32	1.20	2.64	–	–	–	–	0.86	1.90	0.75	1.65	0.47	1.04	0.73	1.60
122	19	–	–	0.37	0.82	–	–	–	–	0.31	0.67	–	–	–	–	–	–	0.19	0.42	0.11	0.25	0.19	0.42
	69	–	–	1.37	3.03	–	–	–	–	1.13	2.49	–	–	–	–	–	–	0.70	1.55	0.42	0.93	0.71	1.57
123	52	–	–	0.96	2.12	–	–	–	–	0.83	1.82	–	–	–	–	–	–	0.47	1.04	0.25	0.54	0.50	1.11
	59	–	–	1.09	2.41	–	–	–	–	0.94	2.07	–	–	–	–	–	–	0.54	1.18	0.28	0.62	0.57	1.26
124	22	22	–	0.46	1.02	–	–	–	–	0.38	0.85	–	–	–	–	–	–	0.23	0.51	0.12	0.27	0.24	0.53
	54	54	–	1.12	2.47	–	–	–	–	0.93	2.04	–	–	–	–	–	–	0.56	1.23	0.30	0.66	0.59	1.29

(Continued)

TABLE V-1E PASTURE AND RANGE PLANTS, MINERAL AND VITAMIN COMPOSITION OF FEEDS, DATA EXPRESSED **AS-FED** AND **MOISTURE-FREE**—*(Continued)*

Entry Number	Feed Name Description	Moisture Basis: A-F (as-fed) or M-F (moisture-free)	Dry Matter	Macro Minerals							Micro Minerals						
				Calcium (Ca)	Phosphorus (P)	Sodium (Na)	Chlorine (Cl)	Magnesium (Mg)	Potassium (K)	Sulfur (S)	Cobalt (Co)	Copper (Cu)	Iodine (I)	Iron (Fe)	Manganese (Mn)	Selenium (Se)	Zinc (Zn)
			%	%	%	%	%	%	%	%	ppm or mg/kg	ppm or mg/kg	ppm or mg/kg	%	ppm or mg/kg	ppm or mg/kg	ppm or mg/kg
	CORN-SUNFLOWER *Zea mays-Helianthus spp*																
105	FRESH	A-F	18	–	–	–	–	–	–	–	–	–	–	–	–	–	–
		M-F	100	–	–	–	–	–	–	–	–	–	–	–	–	–	–
	COTTONTOP, ARIZONA *Trichachne californica*																
106	FRESH	A-F	30	0.14	0.05	–	–	–	–	–	–	–	–	–	–	–	–
		M-F	100	0.45	0.17	–	–	–	–	–	–	–	–	–	–	–	–
	COWPEA, COMMON *Vigna simensis*																
107	FRESH	A-F	25	0.38	0.08	0.06	0.05	0.11	0.41	0.08	–	–	–	0.020	–	–	–
		M-F	100	1.53	0.31	0.25	0.18	0.43	1.66	0.31	–	–	–	0.080	–	–	–
	CRABGRASS *Digitaria spp*																
108	FRESH	A-F	30	0.17	0.07	–	–	0.12	1.26	–	–	–	–	–	–	–	–
		M-F	100	0.57	0.23	–	–	0.40	4.22	–	–	–	–	–	–	–	–
	CRESTED WHEATGRASS *Agropyron desertorum*																
109	IMMATURE, FRESH	A-F	28	0.13	0.10	–	–	0.08	–	–	–	–	–	–	–	–	–
		M-F	100	0.44	0.33	–	–	0.28	–	–	–	–	–	–	–	–	–
110	EARLY BLOOM, FRESH	A-F	41	0.09	0.07	–	–	–	–	–	–	–	–	–	–	–	–
		M-F	100	0.22	0.18	–	–	–	–	–	–	–	–	–	–	–	–
111	FULL BLOOM, FRESH	A-F	45	0.17	0.07	0.00	–	0.04	0.47	0.21	–	2.9	–	–	19.5	–	6.1
		M-F	100	0.37	0.15	0.01	–	0.09	1.04	0.47	–	6.5	–	–	43.3	–	13.5
112	OVERRIPE, FRESH	A-F	75	0.20	0.11	–	–	–	–	–	0.187	6.3	–	–	39.7	–	–
		M-F	100	0.27	0.15	–	–	–	–	–	0.250	8.4	–	–	52.9	–	–
	CRIMSON CLOVER–SEE CLOVER, CRIMSON																
	CRIMSON CLOVER-RYEGRASS–SEE CLOVER, CRIMSON-RYEGRASS																
	CROTALARIA *Crotalaria spp*																
113	MIDBLOOM, FRESH	A-F	27	0.51	0.18	–	–	0.20	–	–	–	–	–	–	–	–	–
		M-F	100	1.88	0.67	–	–	0.75	–	–	–	–	–	–	–	–	–
	CURLY MESQUITE *Hilaria belangeri*																
114	BROWSE, FRESH	A-F	35	0.18	0.05	–	–	0.06	0.23	–	–	3.5	–	–	16.4	–	–
		M-F	100	0.52	0.15	–	–	0.16	0.67	–	–	10.1	–	–	47.0	–	–
115	BROWSE, MATURE, FRESH	A-F	50	0.27	0.04	–	–	0.08	0.19	–	–	–	–	–	–	–	–
		M-F	100	0.55	0.08	–	–	0.15	0.39	–	–	–	–	–	–	–	–
	DALLISGRASS *Paspalum dilatatum*																
116	FRESH	A-F	25	0.14	0.05	0.09	–	0.10	0.43	–	0.019	–	–	0.005	–	–	–
		M-F	100	0.56	0.20	0.34	–	0.40	1.72	–	0.073	–	–	0.016	–	–	–
	DANDELION, COMMON *Taraxacum officinale*																
117	LEAVES, FRESH	A-F	14	0.19	0.07	0.08	–	–	0.40	–	–	–	–	0.004	–	–	–
		M-F	100	1.30	0.46	0.53	–	–	2.76	–	–	–	–	0.022	–	–	–
	DARSO SORGHUM–SEE SORGHUM																
	DEERVETCH, BIRDSFOOT (BIRDSFOOT TREFOIL) *Lotus corniculatus*																
118	FRESH	A-F	19	0.34	0.05	0.02	–	0.08	0.63	0.05	0.094	2.5	–	0.006	16.0	–	6.0
		M-F	100	1.74	0.26	0.11	–	0.40	3.26	0.25	0.487	12.8	–	0.031	82.9	–	31.1
	DESERT MOLLY (SUMMERCYPRESS, GRAY) *Kochia vestita*																
119	STEM CURED, FRESH	A-F	85	2.09	0.12	–	–	–	–	–	–	–	–	–	–	–	–
		M-F	100	2.46	0.14	–	–	–	–	–	–	–	–	–	–	–	–
	DESMANTHUS (BUNDLEFLOWER, RAYADO; DWARF KOA) *Desmanthus virgatus*																
120	FRESH	A-F	41	0.60	0.13	–	–	0.14	0.75	–	–	–	–	–	–	–	–
		M-F	100	1.44	0.31	–	–	0.33	1.81	–	–	–	–	–	–	–	–
	DIGITARIA–SEE CRABGRASS; PANGOLAGRASS																
	DOCK, CURLY *Rumex crispus*																
121	FRESH	A-F	18	–	–	–	–	–	–	–	–	–	–	–	–	–	–
		M-F	100	–	–	–	–	–	–	–	–	–	–	–	–	–	–
	DOGWOOD, FLOWERING *Cornus florida*																
122	BROWSE, FRESH	A-F	27	0.70	0.04	–	–	–	–	–	0.027	1.4	–	0.002	11.8	–	–
		M-F	100	2.59	0.15	–	–	–	–	–	0.100	5.1	–	0.007	43.7	–	–
	DROPSEED, SAND *Sporobolus cryptandrus*																
123	STEM CURED, FRESH	A-F	88	0.40	0.07	0.01	–	0.06	0.28	–	0.503	13.5	0.599	0.043	41.4	–	36.8
		M-F	100	0.45	0.08	0.01	–	0.06	0.32	–	0.572	15.3	0.681	0.049	47.0	–	41.8
	DURRA SORGHUM–SEE SORGHUM																
	DWARF KOA (BUNDLEFLOWER, RAYADO; DESMANTHUS) *Desmanthus virgatus*																
124	FRESH	A-F	41	0.60	0.13	–	–	0.14	0.75	–	–	–	–	–	–	–	–
		M-F	100	1.44	0.31	–	–	0.33	1.81	–	–	–	–	–	–	–	–

PASTURE AND RANGE PLANTS

Entry Number	Fat-Soluble Vitamins					Water-Soluble Vitamins								
	A (1 mg Carotene = 1667 IU Vit A)	Carotene (Provitamin A)	D	E	K	B-12	Biotin	Choline	Folacin (Folic Acid)	Niacin	Pantothenic Acid (B-3)	(Pyri-doxine) B-6	Ribo-flavin (B-2)	Thiamin (B-1)
	IU/g	ppm or mg/kg	IU/kg	ppm or mg/kg	ppm or mg/kg	ppm or mg/kg	ppm or mg/kg	ppm or mg/kg	ppm or mg/kg	ppm or mg/kg	ppm or mg/kg	ppm or mg/kg	ppm or mg/kg	ppm or mg/kg
105	–	–	–	–	–	–	–	–	–	–	–	–	–	–
	–	–	–	–	–	–	–	–	–	–	–	–	–	–
106	–	–	–	–	–	–	–	–	–	–	–	–	–	–
	–	–	–	–	–	–	–	–	–	–	–	–	–	–
107	–	–	–	–	–	–	–	–	–	–	–	–	–	–
	–	–	–	–	–	–	–	–	–	–	–	–	–	–
108	–	–	–	–	–	–	–	–	–	–	–	–	–	–
	–	–	–	–	–	–	–	–	–	–	–	–	–	–
109	**205.8**	**123.4**	–	–	–	–	–	–	–	–	–	–	–	–
	722.9	433.6	–	–	–	–	–	–	–	–	–	–	–	–
110	–	–	–	–	–	–	–	–	–	–	–	–	–	–
	–	–	–	–	–	–	–	–	–	–	–	–	–	–
111	**115.1**	**69.0**	–	–	–	–	–	–	–	–	–	–	–	–
	255.8	153.4	–	–	–	–	–	–	–	–	–	–	–	–
112	**0.3**	**0.2**	–	–	–	–	–	–	–	–	–	–	–	–
	0.4	0.2	–	–	–	–	–	–	–	–	–	–	–	–
113	–	–	–	–	–	–	–	–	–	–	–	–	–	–
	–	–	–	–	–	–	–	–	–	–	–	–	–	–
114	–	–	–	–	–	–	–	–	–	–	–	–	–	–
	–	–	–	–	–	–	–	–	–	–	–	–	–	–
115	–	–	–	–	–	–	–	–	–	–	–	–	–	–
	–	–	–	–	–	–	–	–	–	–	–	–	–	–
116	**126.0**	**75.6**	–	–	–	–	–	–	–	–	–	–	–	–
	503.8	302.2	–	–	–	–	–	–	–	–	–	–	–	–
117	**48.1**	**28.8**	–	–	–	–	–	–	–	–	–	–	**2.6**	**1.9**
	332.6	199.5	–	–	–	–	–	–	–	–	–	–	18.1	13.2
118	–	–	–	–	–	–	–	–	–	–	–	–	–	–
	–	–	–	–	–	–	–	–	–	–	–	–	–	–
119	**18.1**	**10.9**	–	–	–	–	–	–	–	–	–	–	–	–
	21.3	12.8	–	–	–	–	–	–	–	–	–	–	–	–
120	–	–	–	–	–	–	–	–	–	–	–	–	–	–
	–	–	–	–	–	–	–	–	–	–	–	–	–	–
121	–	–	–	–	–	–	–	–	–	–	–	–	–	–
	–	–	–	–	–	–	–	–	–	–	–	–	–	–
122	–	–	–	–	–	–	–	–	–	–	–	–	–	–
	–	–	–	–	–	–	–	–	–	–	–	–	–	–
123	**14.0**	**8.4**	–	–	–	–	–	–	–	–	–	–	–	–
	15.9	9.6	–	–	–	–	–	–	–	–	–	–	–	–
124	–	–	–	–	–	–	–	–	–	–	–	–	–	–
	–	–	–	–	–	–	–	–	–	–	–	–	–	–

(Continued)

TABLE V-1E PASTURE AND RANGE PLANTS, COMPOSITION OF FEEDS, DATA EXPRESSED **AS-FED** AND **MOISTURE-FREE**—*(Continued)*

Entry Number	Feed Name Description	International Feed Number	Moisture Basis: A-F (as-fed) or M-F (moisture-free)	Proximate Analysis: Dry Matter	Ash	Crude Fiber	Neutral Det. Fib. (NDF)	Acid Det. Fib. (ADF)	Lignin	Ether Extract (Fat)	N-Free Extract	Crude Protein	Digestible Protein: Ruminant	Swine	Horse
				%	%	%	%	%	%	%	%	%	%	%	%
	EGYPTIAN CLOVER (BERSEEM CLOVER)–SEE CLOVER, EGYPTIAN														
	EPHEDRA *Ephedra* spp														
125	BROWSE, FRESH	2-26-118	A-F	63	3.4	24.5	–	–	–	2.5	27.8	4.7	–	–	2.5
			M-F	100	5.4	38.9	–	–	–	4.0	44.2	7.5	–	–	3.9
	FESCUE, MEADOW *Festuca elatior*														
126	FRESH	2-01-920	A-F	28	2.4	8.6	–	–	–	1.0	12.9	3.5	2.4	2.4	–
			M-F	100	8.4	30.1	–	–	–	3.7	45.3	12.5	8.4	8.3	–
	FESCUE, TALL (ALTA) *Festuca arundinacea*														
127	FRESH	2-01-889	A-F	28	2.5	7.5	19.5	–	–	0.9	14.3	2.7	1.7	1.7	–
			M-F	100	9.0	26.7	69.9	–	–	3.3	51.2	9.8	6.0	6.1	–
	FILAREE (ALFILERIA, REDSTEM) *Erodium cicutarium*														
128	FRESH	2-00-356	A-F	18	2.4	3.6	–	5.3	–	0.6	8.2	2.8	2.0	1.9	–
			M-F	100	13.7	20.4	–	30.0	–	3.5	46.4	15.9	11.5	11.1	–
	FLAT PEA *Lathyrus sylvestris*														
129	FRESH	2-02-033	A-F	23	1.6	6.2	–	–	–	0.8	7.4	6.5	5.7	4.8	–
			M-F	100	7.1	27.6	–	–	–	3.7	32.7	28.9	25.5	21.5	–
	FOXTAIL, MEADOW *Alopecurus pratensis*														
130	IMMATURE, FRESH	2-02-073	A-F	26	2.8	5.6	–	–	–	1.2	12.0	4.5	3.3	3.2	–
			M-F	100	10.7	21.5	–	–	–	4.6	46.0	17.2	12.7	12.1	–
	FRINGED SAGEBRUSH *Artemisia frigida*														
131	BROWSE, MIDBLOOM, FRESH	2-04-129	A-F	43	2.8	14.3	–	–	–	0.9	21.0	4.0	2.4	2.5	–
			M-F	100	6.5	33.2	–	–	–	2.0	48.9	9.4	5.6	5.8	–
132	BROWSE, MATURE, FRESH	2-04-130	A-F	60	10.3	19.1	27.4	21.1	5.9	2.0	24.4	4.3	2.1	2.4	–
			M-F	100	17.1	31.8	45.6	35.1	9.8	3.4	40.6	7.1	3.5	4.0	–
	GALLETA *Hilaria jamesii*														
133	STEM CURED, FRESH	2-05-594	A-F	86	13.3	28.4	–	–	–	1.4	38.5	4.3	1.3	2.0	–
			M-F	100	15.5	33.0	–	–	–	1.7	44.8	5.1	1.5	2.3	–
	GAMAGRASS, EASTERN *Tripsacum dactyloides*														
134	FRESH	2-02-086	A-F	28	2.9	8.5	–	–	–	0.6	13.7	2.4	1.3	1.4	–
			M-F	100	10.2	30.2	–	–	–	2.0	49.1	8.5	4.8	5.1	–
	GOLDENROD *Solidago* spp														
135	FRESH	2-02-132	A-F	39	2.8	4.9	–	–	–	5.6	22.5	3.4	2.0	2.1	–
			M-F	100	7.1	12.6	–	–	–	14.2	57.4	8.7	5.0	5.3	–
	GRAMA *Bouteloua* spp														
136	IMMATURE, FRESH	2-02-163	A-F	41	4.6	11.2	–	–	–	0.8	19.0	5.4	3.7	3.6	–
			M-F	100	11.3	27.2	–	–	–	2.0	46.4	13.1	9.0	8.8	–
137	MATURE, FRESH	2-02-166	A-F	63	7.2	20.7	–	–	–	1.1	30.2	4.1	1.9	2.2	–
			M-F	100	11.4	32.7	–	–	–	1.7	47.7	6.5	3.0	3.5	–
	GRASS-LEGUME														
138	FRESH	2-08-439	A-F	24	2.5	5.6	–	–	–	0.8	10.4	4.2	3.1	2.9	–
			M-F	100	10.5	23.9	–	–	–	3.4	44.4	17.8	13.3	12.5	–
	GREASEWOOD *Sarcobatus* spp														
139	BROWSE, FRESH	2-02-312	A-F	50	7.3	11.7	–	–	–	1.7	18.6	10.7	8.2	7.7	–
			M-F	100	14.6	23.4	–	–	–	3.4	37.3	21.3	16.5	15.4	–
	GROUNDSEL, BUTTERWEED *Senecio serra*														
140	FRESH	2-02-325	A-F	26	3.5	3.5	–	–	–	0.9	12.6	5.4	4.2	3.9	–
			M-F	100	13.6	13.3	–	–	–	3.6	48.6	20.9	16.1	15.1	–
	GUAR *Cyamopsis tetragonoloba*														
141	FRESH	2-02-333	A-F	19	2.3	4.4	–	–	–	0.5	9.2	2.8	2.0	1.9	–
			M-F	100	12.2	22.8	–	–	–	2.5	47.9	14.6	10.3	10.0	–
	GUINEAGRASS *Panicum maximum*														
142	FRESH	2-02-345	A-F	27	2.9	10.5	–	–	–	0.5	12.1	1.6	0.9	0.8	–
			M-F	100	10.6	38.0	–	–	–	1.8	43.8	5.7	3.2	2.9	–
	HOG MILLET (BROOMCORN; MILLET, PROSO) *Panicum miliaceum*														
143	FRESH	2-03-811	A-F	25	1.8	7.4	–	–	–	0.6	13.1	2.1	1.1	1.2	–
			M-F	100	7.4	29.5	–	–	–	2.5	52.4	8.2	4.6	4.9	–
	HONEYSUCKLE, JAPANESE *Lonicera japonica*														
144	BROWSE, FRESH	2-08-447	A-F	55	3.2	9.2	–	–	–	1.9	36.3	4.0	2.7	2.3	–
			M-F	100	5.9	16.9	–	–	–	3.5	66.4	7.4	4.9	4.2	–
	HOP CLOVER–SEE CLOVER, HOP														
	HORSE BEAN *Vicia faba, equina*														
145	FRESH	2-02-405	A-F	18	1.7	4.1	–	–	–	0.5	7.9	3.6	2.8	2.6	–
			M-F	100	9.6	23.1	–	–	–	2.8	44.1	20.3	15.8	14.6	–

Entry Number	TDN Ruminant	TDN Swine	TDN Horse	Digestible Energy Ruminant		Digestible Energy Swine		Digestible Energy Horse		Metabolizable Energy Ruminant		Metabolizable Energy Swine		Metabolizable Energy Poultry ME_n		Metabolizable Energy Horse		Net Energy Ruminant NE_m		Net Energy Ruminant NE_g		Net Energy Lactating Cows NE_{lc}	
	%	%	%	Mcal lb	Mcal *kg*	kcal lb	kcal *kg*	Mcal lb	Mcal *kg*	Mcal lb	Mcal *kg*	kcal lb	kcal *kg*	kcal lb	kcal *kg*	Mcal lb	Mcal *kg*	Mcal lb	Mcal *kg*	Mcal lb	Mcal *kg*	Mcal lb	Mcal *kg*
125	**35**	–	–	**0.69**	***1.53***	–	–	–	–	**0.57**	***1.25***	–	–	–	–	–	–	**0.34**	***0.74***	**0.13**	***0.30***	**0.35**	***0.77***
	55	–	–	1.10	*2.42*	–	–	–	–	0.90	*1.98*	–	–	–	–	–	–	0.54	*1.18*	0.21	*0.47*	0.55	*1.22*
126	**17**	–	–	**0.35**	***0.77***	–	–	–	–	**0.29**	***0.65***	–	–	–	–	–	–	**0.18**	***0.39***	**0.10**	***0.23***	**0.18**	***0.40***
	61	–	–	1.23	*2.70*	–	–	–	–	1.03	*2.28*	–	–	–	–	–	–	0.62	*1.37*	0.36	*0.79*	0.63	*1.39*
127	**17**	–	–	**0.34**	***0.76***	–	–	–	–	**0.29**	***0.64***	–	–	–	–	–	–	**0.17**	***0.38***	**0.10**	***0.22***	**0.18**	***0.39***
	61	–	–	1.23	*2.71*	–	–	–	–	1.04	*2.28*	–	–	–	–	–	–	0.62	*1.37*	0.36	*0.79*	0.63	*1.40*
128	**10**	–	–	**0.21**	***0.46***	–	–	–	–	**0.18**	***0.39***	–	–	–	–	–	–	**0.11**	***0.25***	**0.07**	***0.15***	**0.11**	***0.25***
	55	–	–	1.18	*2.61*	–	–	–	–	0.99	*2.19*	–	–	–	–	–	–	0.64	*1.41*	0.37	*0.82*	0.65	*1.42*
129	**15**	–	–	**0.31**	***0.69***	–	–	–	–	**0.27**	***0.60***	–	–	–	–	–	–	**0.17**	***0.38***	**0.11**	***0.24***	**0.17**	***0.37***
	68	–	–	1.39	*3.07*	–	–	–	–	1.20	*2.65*	–	–	–	–	–	–	0.76	*1.67*	0.48	*1.06*	0.74	*1.62*
130	**17**	–	–	**0.34**	***0.75***	–	–	–	–	**0.29**	***0.64***	–	–	–	–	–	–	**0.18**	***0.40***	**0.11**	***0.24***	**0.18**	***0.39***
	65	–	–	1.30	*2.88*	–	–	–	–	1.11	*2.45*	–	–	–	–	–	–	0.69	*1.51*	0.42	*0.92*	0.68	*1.50*
131	**26**	–	–	**0.53**	***1.16***	–	–	–	–	**0.45**	***0.98***	–	–	–	–	–	–	**0.26**	***0.57***	**0.15**	***0.32***	**0.27**	***0.59***
	61	–	–	1.23	*2.70*	–	–	–	–	1.03	*2.28*	–	–	–	–	–	–	0.60	*1.33*	0.34	*0.75*	0.62	*1.36*
132	**30**	–	–	**0.60**	***1.33***	–	–	–	–	**0.49**	***1.08***	–	–	–	–	–	–	**0.26**	***0.58***	**0.11**	***0.25***	**0.30**	***0.66***
	50	–	–	1.01	*2.22*	–	–	–	–	0.81	*1.79*	–	–	–	–	–	–	0.44	*0.96*	0.19	*0.41*	0.50	*1.10*
133	**44**	–	–	**0.72**	***1.58***	–	–	–	–	**0.59**	***1.29***	–	–	–	–	–	–	**0.51**	***1.13***	**0.29**	***0.63***	**0.53**	***1.16***
	51	–	–	0.83	*1.83*	–	–	–	–	0.68	*1.50*	–	–	–	–	–	–	0.59	*1.31*	0.33	*0.74*	0.61	*1.35*
134	**16**	–	–	**0.33**	***0.73***	–	–	–	–	**0.28**	***0.61***	–	–	–	–	–	–	**0.16**	***0.35***	**0.09**	***0.19***	**0.17**	***0.36***
	59	–	–	1.18	*2.60*	–	–	–	–	0.99	*2.17*	–	–	–	–	–	–	0.56	*1.24*	0.31	*0.68*	0.59	*1.30*
135	–	–	–	–	–	–	–	–	–	–	–	–	–	–	–	–	–	–	–	–	–	–	–
	–	–	–	–	–	–	–	–	–	–	–	–	–	–	–	–	–	–	–	–	–	–	–
136	**25**	–	–	**0.51**	***1.12***	–	–	–	–	**0.43**	***0.94***	–	–	–	–	–	–	**0.25**	***0.54***	**0.14**	***0.31***	**0.25**	***0.56***
	62	–	–	1.23	*2.72*	–	–	–	–	1.04	*2.30*	–	–	–	–	–	–	0.60	*1.32*	0.34	*0.75*	0.62	*1.36*
137	**35**	–	–	**0.71**	***1.56***	–	–	–	–	**0.59**	***1.29***	–	–	–	–	–	–	**0.32**	***0.71***	**0.16**	***0.36***	**0.35**	***0.77***
	56	–	–	1.12	*2.46*	–	–	–	–	0.92	*2.04*	–	–	–	–	–	–	0.51	*1.13*	0.26	*0.57*	0.55	*1.22*
138	**15**	–	–	**0.31**	***0.68***	–	–	–	–	**0.26**	***0.58***	–	–	–	–	–	–	**0.16**	***0.35***	**0.10**	***0.21***	**0.16**	***0.35***
	65	–	–	1.31	*2.88*	–	–	–	–	1.12	*2.46*	–	–	–	–	–	–	0.67	*1.48*	0.40	*0.89*	0.67	*1.48*
139	**23**	–	–	**0.55**	***1.20***	–	–	–	–	**0.45**	***0.99***	–	–	–	–	–	–	**0.32**	***0.70***	**0.18**	***0.41***	**0.32**	***0.71***
	47	–	–	1.09	*2.41*	–	–	–	–	0.90	*1.98*	–	–	–	–	–	–	0.63	*1.39*	0.37	*0.81*	0.64	*1.41*
140	**18**	–	–	**0.36**	***0.79***	–	–	–	–	**0.31**	***0.68***	–	–	–	–	–	–	**0.19**	***0.42***	**0.12**	***0.26***	**0.19**	***0.41***
	69	–	–	1.38	*3.04*	–	–	–	–	1.19	*2.62*	–	–	–	–	–	–	0.73	*1.60*	0.45	*1.00*	0.71	*1.57*
141	**12**	–	–	**0.24**	***0.54***	–	–	–	–	**0.21**	***0.46***	–	–	–	–	–	–	**0.12**	***0.27***	**0.07**	***0.16***	**0.12**	***0.27***
	63	–	–	1.27	*2.79*	–	–	–	–	1.08	*2.37*	–	–	–	–	–	–	0.63	*1.39*	0.37	*0.81*	0.64	*1.41*
142	**14**	–	–	**0.29**	***0.63***	–	–	–	–	**0.23**	***0.52***	–	–	–	–	–	–	**0.13**	***0.29***	**0.06**	***0.14***	**0.15**	***0.32***
	52	–	–	1.04	*2.30*	–	–	–	–	0.85	*1.87*	–	–	–	–	–	–	0.47	*1.04*	0.22	*0.49*	0.53	*1.16*
143	**16**	–	–	**0.31**	***0.68***	–	–	–	–	**0.26**	***0.58***	–	–	–	–	–	–	**0.15**	***0.34***	**0.09**	***0.19***	**0.16**	***0.34***
	63	–	–	1.24	*2.74*	–	–	–	–	1.05	*2.32*	–	–	–	–	–	–	0.61	*1.35*	0.35	*0.78*	0.63	*1.38*
144	**26**	–	**28**	**0.52**	***1.15***	–	–	**0.53**	***1.17***	**0.42**	***0.92***	–	–	–	–	**0.43**	***0.96***	**0.14**	***0.30***	**0.01**	***0.02***	**0.21**	***0.45***
	48	–	51	0.95	*2.10*	–	–	0.97	*2.13*	0.76	*1.67*	–	–	–	–	0.79	*1.75*	0.25	*0.56*	0.01	*0.03*	0.38	*0.83*
145	**12**	–	–	**0.24**	***0.53***	–	–	–	–	**0.20**	***0.45***	–	–	–	–	–	–	**0.13**	***0.28***	**0.08**	***0.17***	**0.12**	***0.27***
	67	–	–	1.34	*2.95*	–	–	–	–	1.15	*2.53*	–	–	–	–	–	–	0.70	*1.55*	0.43	*0.95*	0.69	*1.53*

(Continued)

TABLE V-1E PASTURE AND RANGE PLANTS, MINERAL AND VITAMIN COMPOSITION OF FEEDS, DATA EXPRESSED **AS-FED** AND **MOISTURE-FREE**—*(Continued)*

Entry Number	Feed Name Description	Moisture Basis: A-F (as-fed) or M-F (moisture-free)	Dry Matter	Macro Minerals							Micro Minerals						
				Calcium (Ca)	Phos-phorus (P)	Sodium (Na)	Chlo-rine (Cl)	Mag-nesium (Mg)	Potas-sium (K)	Sulfur (S)	Cobalt (Co)	Copper (Cu)	Iodine (I)	Iron (Fe)	Man-ganese (Mn)	Sele-nium (Se)	Zinc (Zn)
			%	%	%	%	%	%	%	%	ppm or mg/kg	ppm or mg/kg	ppm or mg/kg	%	ppm or mg/kg	ppm or mg/kg	ppm or mg/kg
	EGYPTIAN CLOVER (BERSEEM CLOVER)–SEE CLOVER, EGYPTIAN																
	EPHEDRA *Ephedra spp*																
125	BROWSE, FRESH	A-F	63	–	–	–	–	–	–	–	–	–	–	–	–	–	–
		M-F	100	–	–	–	–	–	–	–	–	–	–	–	–	–	–
	FESCUE, TALL (ALTA) *Festuca arundinacea*																
126	FRESH	A-F	28	0.13	0.05	0.03	–	0.07	0.70	–	0.113	1.0	–	0.003	18.0	–	5.9
		M-F	100	0.48	0.19	0.12	–	0.25	2.51	–	0.401	3.4	–	0.011	64.4	–	21.0
	FESCUE, MEADOW *Festuca elatior*																
127	FRESH	A-F	28	0.15	0.11	–	–	0.11	0.67	–	0.039	1.1	–	–	–	–	–
		M-F	100	0.53	0.39	–	–	0.37	2.34	–	0.135	4.0	–	–	–	–	–
	FILAREE (ALFILERIA, REDSTEM) *Erodium cicutarium*																
128	FRESH	A-F	18	0.35	0.08	–	–	–	0.59	–	–	–	–	–	–	–	–
		M-F	100	1.99	0.43	–	–	–	3.37	–	–	–	–	–	–	–	–
	FLAT PEA *Lathyrus sylvestris*																
129	FRESH	A-F	23	0.11	0.11	–	–	–	–	–	–	–	–	–	–	–	–
		M-F	100	0.49	0.49	–	–	–	–	–	–	–	–	–	–	–	–
	FOXTAIL, MEADOW *Alopecurus pratensis*																
130	IMMATURE, FRESH	A-F	26	0.15	0.12	–	–	–	–	–	–	–	–	–	–	–	–
		M-F	100	0.57	0.46	–	–	–	–	–	–	–	–	–	–	–	–
	FRINGED SAGEBRUSH *Artemisia frigida*																
131	BROWSE, MIDBLOOM, FRESH	A-F	43	–	–	–	–	–	–	–	–	–	–	–	–	–	–
		M-F	100	–	–	–	–	–	–	–	–	–	–	–	–	–	–
132	BROWSE, MATURE, FRESH	A-F	60	–	–	–	–	–	–	–	–	–	–	–	–	–	–
		M-F	100	–	–	–	–	–	–	–	–	–	–	–	–	–	–
	GALLETA *Hilaria jamesii*																
133	STEM CURED, FRESH	A-F	86	0.60	0.06	0.01	–	0.07	0.41	0.09	0.591	16.3	–	0.044	67.7	–	19.5
		M-F	100	0.70	0.07	0.01	–	0.08	0.48	0.10	0.687	19.0	–	0.052	78.7	–	22.7
	GAMAGRASS, EASTERN *Tripsacum dactyloides*																
134	FRESH	A-F	28	0.17	0.09	–	–	–	–	–	–	–	–	–	–	–	–
		M-F	100	0.62	0.31	–	–	–	–	–	–	–	–	–	–	–	–
	GOLDENROD *Solidago spp*																
135	FRESH	A-F	39	0.44	0.20	–	–	–	–	–	–	–	–	–	–	–	–
		M-F	100	1.13	0.52	–	–	–	–	–	–	–	–	–	–	–	–
	GRAMA *Bouteloua spp*																
136	IMMATURE, FRESH	A-F	41	0.22	0.08	–	–	–	–	–	–	2.3	–	–	18.2	–	–
		M-F	100	0.53	0.19	–	–	–	–	–	–	5.5	–	–	44.3	–	–
137	MATURE, FRESH	A-F	63	0.22	0.08	–	–	–	0.22	–	0.115	8.1	–	0.083	30.0	–	–
		M-F	100	0.34	0.12	–	–	–	0.35	–	0.181	12.8	–	0.130	47.4	–	–
	GRASS-LEGUME																
138	FRESH	A-F	24	0.15	0.08	–	–	0.08	0.40	–	–	–	–	–	–	–	–
		M-F	100	0.62	0.34	–	–	0.32	1.70	–	–	–	–	–	–	–	–
	GREASEWOOD *Sarcobatus spp*																
139	BROWSE, FRESH	A-F	50	0.46	0.09	–	–	–	–	–	0.030	7.8	–	–	12.9	–	–
		M-F	100	0.91	0.18	–	–	–	–	–	0.060	15.7	–	–	25.8	–	–
	GROUNDSEL, BUTTERWEED *Senecio serra*																
140	FRESH	A-F	26	–	–	–	–	–	–	–	–	–	–	–	–	–	–
		M-F	100	–	–	–	–	–	–	–	–	–	–	–	–	–	–
	GUAR *Cyamopsis tetragonoloba*																
141	FRESH	A-F	19	–	–	–	–	–	–	–	–	–	–	–	–	–	–
		M-F	100	–	–	–	–	–	–	–	–	–	–	–	–	–	–
	GUINEAGRASS *Panicum maximum*																
142	FRESH	A-F	27	–	–	–	–	–	–	–	–	–	–	–	–	–	–
		M-F	100	–	–	–	–	–	–	–	–	–	–	–	–	–	–
	HOG MILLET (BROOMCORN; MILLET, PROSO) *Panicum miliaceum*																
143	FRESH	A-F	25	–	–	–	–	–	–	–	–	–	–	–	–	–	–
		M-F	100	–	–	–	–	–	–	–	–	–	–	–	–	–	–
	HONEYSUCKLE, JAPANESE *Lonicera japonica*																
144	BROWSE, FRESH	A-F	55	–	–	–	–	–	–	–	–	–	–	–	–	–	–
		M-F	100	–	–	–	–	–	–	–	–	–	–	–	–	–	–
	HOP CLOVER–SEE CLOVER, HOP																
	HORSE BEAN *Vicia faba, equina*																
145	FRESH	A-F	18	0.16	0.05	–	–	–	0.37	–	–	–	–	–	–	–	–
		M-F	100	0.92	0.29	–	–	–	2.07	–	–	–	–	–	–	–	–

	Fat-Soluble Vitamins					Water-Soluble Vitamins								
Entry Number	**A (1 mg Carotene = 1667 IU Vit A)**	**Carotene (Provitamin A)**	**D**	**E**	**K**	**B–12**	**Biotin**	**Choline**	**Folacin (Folic Acid)**	**Niacin**	**Pantothenic Acid (B–3)**	**(Pyri-doxine) B–6**	**Ribo-flavin (B–2)**	**Thiamin (B–1)**
	IU/g	ppm or mg/kg	IU/kg	ppm or mg/kg	ppm or mg/kg	ppm or mg/kg	ppm or mg/kg	ppm or mg/kg	ppm or mg/kg	ppm or mg/kg	ppm or mg/kg	ppm or mg/kg	ppm or mg/kg	ppm or mg/kg
125	–	–	–	–	–	–	–	–	–	–	–	–	–	–
	–	–	–	–	–	–	–	–	–	–	–	–	–	–
126	–	–	–	–	–	–	–	–	–	–	–	–	–	–
	–	–	–	–	–	–	–	–	–	–	–	–	–	–
127	**160.0**	**96.0**	–	**46.9**	–	–	–	–	–	–	–	–	**2.4**	**3.4**
	562.7	337.5	–	165.1	–	–	–	–	–	–	–	–	8.6	11.9
128	–	–	–	–	–	–	–	–	–	–	–	–	–	–
	–	–	–	–	–	–	–	–	–	–	–	–	–	–
129	–	–	–	–	–	–	–	–	–	–	–	–	–	–
	–	–	–	–	–	–	–	–	–	–	–	–	–	–
130	–	–	–	–	–	–	–	–	–	–	–	–	–	–
	–	–	–	–	–	–	–	–	–	–	–	–	–	–
131	–	–	–	–	–	–	–	–	–	–	–	–	–	–
	–	–	–	–	–	–	–	–	–	–	–	–	–	–
132	–	–	–	–	–	–	–	–	–	–	–	–	–	–
	–	–	–	–	–	–	–	–	–	–	–	–	–	–
133	**0.3**	**0.2**	–	–	–	–	–	–	–	–	–	–	–	–
	0.3	0.2	–	–	–	–	–	–	–	–	–	–	–	–
134	–	–	–	–	–	–	–	–	–	–	–	–	–	–
	–	–	–	–	–	–	–	–	–	–	–	–	–	–
135	–	–	–	–	–	–	–	–	–	–	–	–	–	–
	–	–	–	–	–	–	–	–	–	–	–	–	–	–
136	–	–	–	–	–	–	–	–	–	–	–	–	–	–
	–	–	–	–	–	–	–	–	–	–	–	–	–	–
137	**32.2**	**19.3**	–	–	–	–	–	–	–	–	–	–	–	–
	50.7	30.4	–	–	–	–	–	–	–	–	–	–	–	–
138	–	–	–	–	–	–	–	–	–	–	–	–	–	–
	–	–	–	–	–	–	–	–	–	–	–	–	–	–
139	**36.2**	**21.7**	–	–	–	–	–	–	–	–	–	–	–	–
	72.4	43.4	–	–	–	–	–	–	–	–	–	–	–	–
140	–	–	–	–	–	–	–	–	–	–	–	–	–	–
	–	–	–	–	–	–	–	–	–	–	–	–	–	–
141	–	–	–	–	–	–	–	–	–	–	–	–	–	–
	–	–	–	–	–	–	–	–	–	–	–	–	–	–
142	**107.1**	**64.3**	–	–	–	–	–	–	–	–	–	–	–	–
	389.6	233.7	–	–	–	–	–	–	–	–	–	–	–	–
143	–	–	–	–	–	–	–	–	–	–	–	–	–	–
	–	–	–	–	–	–	–	–	–	–	–	–	–	–
144	–	–	–	–	–	–	–	–	–	–	–	–	–	–
	–	–	–	–	–	–	–	–	–	–	–	–	–	–
145	–	–	–	–	–	–	–	–	–	–	–	–	–	–
	–	–	–	–	–	–	–	–	–	–	–	–	–	–

(Continued)

TABLE V-1E PASTURE AND RANGE PLANTS, COMPOSITION OF FEEDS, DATA EXPRESSED **AS-FED** AND **MOISTURE-FREE**—*(Continued)*

Entry Number	Feed Name Description	International Feed Number	Moisture Basis: A-F (as-fed) or M-F (moisture-free)	Proximate Analysis: Dry Matter	Ash	Crude Fiber	Neutral Det. Fib. (NDF)	Acid Det. Fib. (ADF)	Lignin	Ether Extract (Fat)	N-Free Extract	Crude Protein	Digestible Protein: Ruminant	Swine	Horse
				%	%	%	%	%	%	%	%	%	%	%	%
	INDIANGRASS *Sorghastrum spp*														
146	FRESH	2-08-770	A-F	57	4.1	19.4	–	–	–	1.3	29.4	2.8	0.9	1.2	–
			M-F	100	7.2	34.0	–	–	–	2.3	51.6	4.9	1.5	2.2	–
	INDIAN RICEGRASS *Oryzopsis hymenoides*														
147	FRESH	2-03-944	A-F	48	–	–	–	–	–	–	–	–	–	–	–
			M-F	100	–	–	–	–	–	–	–	–	–	–	–
	JACK BEAN, COMMON *Canavalia ensiformis*														
148	FRESH	2-02-434	A-F	23	2.7	6.4	–	–	–	0.5	8.4	5.2	4.0	3.8	–
			M-F	100	11.6	27.6	–	–	–	2.2	36.2	22.4	17.4	16.3	–
	JOHNSONGRASS SORGHUM–SEE SORGHUM														
	JUNEGRASS *Koeleria cristata*														
149	IMMATURE, FRESH	2-02-437	A-F	28	2.2	6.4	–	–	–	0.6	12.8	6.0	4.7	4.4	–
			M-F	100	7.8	22.9	–	–	–	2.1	45.7	21.6	16.7	15.6	–
150	OVERRIPE, FRESH	2-02-440	A-F	70	5.4	22.1	–	–	–	2.3	35.8	4.4	2.0	2.3	–
			M-F	100	7.7	31.5	–	–	–	3.4	51.2	6.3	2.8	3.3	–
	KAFIR SORGHUM–SEE SORGHUM														
	KALE *Brassica oleracea, acephala*														
151	FRESH	2-02-446	A-F	11	1.6	1.5	–	–	–	0.5	5.1	2.6	2.2	1.9	–
			M-F	100	14.5	13.4	–	–	–	4.3	45.0	22.8	19.0	16.6	–
	KENTUCKY BLUEGRASS–SEE BLUEGRASS, KENTUCKY														
	KIKUYU GRASS *Pennisetum clandestinum*														
152	IMMATURE, FRESH	2-05-666	A-F	21	2.4	6.1	–	8.0	1.2	0.6	8.7	2.9	1.8	2.0	–
			M-F	100	11.5	29.3	–	38.8	5.9	3.1	41.9	14.2	8.8	9.7	–
	KOA HAOLE (LEAD TREE, WHITE POPINAC) *Leucaena glauca*														
153	BROWSE, FRESH	2-02-495	A-F	30	1.9	10.7	–	–	–	0.6	11.5	5.5	4.1	3.9	–
			M-F	100	6.4	35.3	–	–	–	2.1	38.0	18.2	13.4	12.9	–
	KOHLRABI *Brassica oleracea, gongylodes*														
154	FRESH	2-02-477	A-F	9	1.3	1.3	–	–	–	0.1	4.3	2.0	1.6	1.5	–
			M-F	100	14.4	14.4	–	–	–	1.1	47.8	22.2	17.3	16.1	–
	KUDZU *Pueraria spp*														
155	FRESH	2-02-482	A-F	28	2.1	8.5	–	–	–	0.7	11.6	4.9	3.6	3.4	–
			M-F	100	7.6	30.5	–	–	–	2.7	41.6	17.6	13.1	12.4	–
	LADINO CLOVER–SEE CLOVER, LADINO														
	LEAD TREE, WHITE POPINAC (KOA HAOLE) *Leucaena glauca*														
156	BROWSE, FRESH	2-02-495	A-F	30	1.9	10.7	–	–	–	0.6	11.5	5.5	4.1	3.9	–
			M-F	100	6.4	35.3	–	–	–	2.1	38.0	18.2	13.4	12.9	–
	LESPEDEZA, COMMON *Lespedeza striata*														
157	IMMATURE, FRESH	2-20-879	A-F	24	3.1	7.7	–	–	–	0.5	8.8	3.9	2.9	2.7	–
			M-F	100	12.8	32.0	–	–	–	2.0	36.8	16.4	12.0	11.4	–
158	EARLY BLOOM, FRESH	2-20-880	A-F	28	–	6.2	–	–	–	–	–	4.6	3.4	–	2.4
			M-F	100	–	22.0	–	–	–	–	–	16.4	12.1	–	8.5
159	MATURE, FRESH	2-02-567	A-F	37	2.9	16.1	–	–	–	0.9	12.2	5.1	3.5	3.5	–
			M-F	100	7.8	43.3	–	–	–	2.4	32.8	13.7	9.5	9.3	–
	LESPEDEZA, COMMON-KOREAN *Lespedeza striata-stipulacea*														
160	PREBLOOM, FRESH	2-26-028	A-F	25	3.2	8.0	–	–	–	0.5	9.2	4.1	–	–	2.9
			M-F	100	12.8	32.0	–	–	–	2.0	36.8	16.4	–	–	11.5
161	MATURE, FRESH	2-26-032	A-F	35	2.6	15.8	–	–	–	0.7	11.8	4.4	3.0	–	2.9
			M-F	100	7.4	45.1	–	–	–	2.1	33.7	12.7	8.7	–	8.3
	LESPEDEZA, SERICEA (CHINESE LESPEDEZA) *Leucaena cuneata*														
162	FRESH	2-02-611	A-F	33	2.0	7.5	–	–	–	1.2	16.2	5.9	4.4	4.2	–
			M-F	100	6.2	22.7	–	–	–	3.8	49.3	18.0	13.4	12.7	–
163	IMMATURE, FRESH	2-02-609	A-F	28	2.2	–	14.5	–	–	–	–	5.7	4.4	4.1	–
			M-F	100	7.8	–	51.7	–	–	–	–	20.4	15.6	14.6	–
164	FULL BLOOM, FRESH	2-02-610	A-F	25	–	–	–	–	–	–	–	–	–	–	–
			M-F	100	–	–	–	–	–	–	–	–	–	–	–
	LETTUCE *Lactuca sativa*														
165	FRESH	2-02-624	A-F	5	0.9	0.6	–	–	–	0.2	2.5	1.2	0.9	0.9	–
			M-F	100	15.9	11.2	–	–	–	4.1	46.9	22.0	17.0	15.9	–
166	OUTSIDE LEAVES, FRESH	2-02-625	A-F	–	–	–	–	–	–	–	–	–	–	–	–
			M-F	100	20.8	19.1	–	–	–	6.1	41.2	12.8	8.7	8.6	–
	LICHEN *Lichen planus*														
167	FRESH	2-02-626	A-F	33	6.8	4.4	–	–	–	1.6	18.1	2.1	1.0	1.1	–
			M-F	100	20.7	13.2	–	–	–	4.8	54.9	6.4	2.9	3.4	–
	LIMA BEAN–SEE BEAN, LIMA														

PASTURE AND RANGE PLANTS

Entry Number	TDN Ruminant	TDN Swine	TDN Horse	DE Ruminant		DE Swine		DE Horse		ME Ruminant		ME Swine		ME Poultry ME_n		ME Horse		NE Ruminant NE_m		NE Ruminant NE_g		NE Lactating Cows NE_{lc}	
	%	%	%	Mcal lb	Mcal kg	kcal lb	kcal kg	Mcal lb	Mcal kg	Mcal lb	Mcal kg	kcal lb	kcal kg	kcal lb	kcal kg	Mcal lb	Mcal kg	Mcal lb	Mcal kg	Mcal lb	Mcal kg	Mcal lb	Mcal kg
146	33	–	–	0.66	1.45	–	–	–	–	0.55	1.21	–	–	–	–	–	–	0.32	0.70	0.17	0.38	0.33	0.74
	58	–	–	1.16	2.55	–	–	–	–	0.96	2.12	–	–	–	–	–	–	0.56	1.23	0.30	0.66	0.59	1.29
147	–	–	–	–	–	–	–	–	–	–	–	–	–	–	–	–	–	–	–	–	–	–	–
	–	–	–	–	–	–	–	–	–	–	–	–	–	–	–	–	–	–	–	–	–	–	–
148	14	–	–	0.29	0.64	–	–	–	–	0.25	0.54	–	–	–	–	–	–	0.15	0.33	0.09	0.20	0.15	0.33
	61	–	–	1.25	2.75	–	–	–	–	1.06	2.33	–	–	–	–	–	–	0.65	1.43	0.38	0.84	0.65	1.44
149	20	–	–	0.40	0.88	–	–	–	–	0.35	0.76	–	–	–	–	–	–	0.21	0.45	0.13	0.29	0.20	0.44
	71	–	–	1.42	3.14	–	–	–	–	1.23	2.72	–	–	–	–	–	–	0.74	1.62	0.46	1.02	0.72	1.59
150	41	–	–	0.82	1.81	–	–	–	–	0.68	1.51	–	–	–	–	–	–	0.41	0.90	0.23	0.50	0.42	0.93
	58	–	–	1.17	2.58	–	–	–	–	0.98	2.15	–	–	–	–	–	–	0.58	1.29	0.32	0.71	0.60	1.33
151	8	–	–	0.15	0.34	–	–	–	–	0.13	0.29	–	–	–	–	–	–	0.08	0.17	0.05	0.10	0.08	0.17
	67	–	–	1.35	2.97	–	–	–	–	1.16	2.55	–	–	–	–	–	–	0.68	1.50	0.41	0.91	0.68	1.49
152	13	–	–	0.25	0.55	–	–	–	–	0.21	0.46	–	–	–	–	–	–	0.12	0.27	0.07	0.15	0.13	0.28
	61	–	–	1.20	2.65	–	–	–	–	1.01	2.23	–	–	–	–	–	–	0.59	1.30	0.33	0.73	0.61	1.34
153	18	–	–	0.37	0.82	–	–	–	–	0.31	0.69	–	–	–	–	–	–	0.19	0.43	0.11	0.25	0.20	0.43
	60	–	–	1.23	2.70	–	–	–	–	1.03	2.28	–	–	–	–	–	–	0.64	1.41	0.37	0.83	0.65	1.42
154	7	–	–	0.13	0.29	–	–	–	–	0.11	0.25	–	–	–	–	–	–	0.06	0.14	0.04	0.09	0.06	0.14
	78	–	–	1.45	3.20	–	–	–	–	1.26	2.78	–	–	–	–	–	–	0.71	1.55	0.43	0.96	0.70	1.53
155	18	–	–	0.36	0.79	–	–	–	–	0.31	0.68	–	–	–	–	–	–	0.18	0.40	0.11	0.24	0.18	0.41
	65	–	–	1.29	2.85	–	–	–	–	1.10	2.43	–	–	–	–	–	–	0.66	1.45	0.39	0.86	0.66	1.46
156	18	–	–	0.37	0.82	–	–	–	–	0.31	0.69	–	–	–	–	–	–	0.19	0.43	0.11	0.25	0.20	0.43
	60	–	–	1.23	2.70	–	–	–	–	1.03	2.28	–	–	–	–	–	–	0.64	1.41	0.37	0.83	0.65	1.42
157	14	–	–	0.29	0.63	–	–	–	–	0.24	0.53	–	–	–	–	–	–	0.13	0.29	0.07	0.16	0.14	0.31
	60	–	–	1.20	2.64	–	–	–	–	1.01	2.22	–	–	–	–	–	–	0.56	1.23	0.30	0.66	0.59	1.29
158	16	–	–	0.33	0.73	–	–	–	–	0.27	0.59	–	–	–	–	–	–	–	–	–	–	–	–
	58	–	–	1.18	2.60	–	–	–	–	0.95	2.09	–	–	–	–	–	–	–	–	–	–	–	–
159	16	–	–	0.36	0.80	–	–	–	–	0.29	0.64	–	–	–	–	–	–	0.20	0.43	0.10	0.22	0.21	0.46
	42	–	–	0.98	2.15	–	–	–	–	0.78	1.72	–	–	–	–	–	–	0.53	1.16	0.27	0.60	0.56	1.24
160	14	–	–	0.28	0.63	–	–	–	–	0.23	0.51	–	–	–	–	–	–	0.14	0.31	0.06	0.14	0.14	0.32
	57	–	–	1.13	2.50	–	–	–	–	0.93	2.05	–	–	–	–	–	–	0.56	1.23	0.25	0.54	0.58	1.27
161	20	–	–	–	–	–	–	–	–	–	–	–	–	–	–	–	–	–	–	–	–	–	–
	58	–	–	–	–	–	–	–	–	–	–	–	–	–	–	–	–	–	–	–	–	–	–
162	21	–	–	0.44	0.96	–	–	–	–	0.37	0.83	–	–	–	–	–	–	0.22	0.47	0.13	0.28	0.22	0.48
	64	–	–	1.33	2.93	–	–	–	–	1.14	2.51	–	–	–	–	–	–	0.65	1.44	0.39	0.86	0.66	1.45
163	–	–	–	–	–	–	–	–	–	–	–	–	–	–	–	–	–	–	–	–	–	–	–
	–	–	–	–	–	–	–	–	–	–	–	–	–	–	–	–	–	–	–	–	–	–	–
164	–	–	–	–	–	–	–	–	–	–	–	–	–	–	–	–	–	–	–	–	–	–	–
	–	–	–	–	–	–	–	–	–	–	–	–	–	–	–	–	–	–	–	–	–	–	–
165	4	–	–	0.07	0.16	–	–	–	–	0.06	0.14	–	–	–	–	–	–	0.04	0.08	0.02	0.05	0.04	0.08
	68	–	–	1.36	3.01	–	–	–	–	1.17	2.59	–	–	–	–	–	–	0.72	1.58	0.44	0.98	0.70	1.55
166	–	–	–	–	–	–	–	–	–	–	–	–	–	–	–	–	–	–	–	–	–	–	–
	–	–	–	–	–	–	–	–	–	–	–	–	–	–	–	–	–	–	–	–	–	–	–
167	18	–	–	0.36	0.79	–	–	–	–	0.30	0.65	–	–	–	–	–	–	0.18	0.39	0.09	0.20	0.19	0.41
	54	–	–	1.09	2.40	–	–	–	–	0.90	1.97	–	–	–	–	–	–	0.54	1.18	0.28	0.62	0.57	1.26

(Continued)

TABLE V-1E PASTURE AND RANGE PLANTS, MINERAL AND VITAMIN COMPOSITION OF FEEDS, DATA EXPRESSED **AS-FED** AND **MOISTURE-FREE**—*(Continued)*

Entry Number	Feed Name Description	Moisture Basis: A-F (as-fed) or M-F (moisture-free)	Dry Matter	Macro Minerals							Micro Minerals						
				Calcium (Ca)	Phosphorus (P)	Sodium (Na)	Chlorine (Cl)	Magnesium (Mg)	Potassium (K)	Sulfur (S)	Cobalt (Co)	Copper (Cu)	Iodine (I)	Iron (Fe)	Manganese (Mn)	Selenium (Se)	Zinc (Zn)
			%	%	%	%	%	%	%	%	ppm or mg/kg	ppm or mg/kg	ppm or mg/kg	%	ppm or mg/kg	ppm or mg/kg	ppm or mg/kg
	INDIANGRASS *Sorghastrum spp*																
146	FRESH	**A-F**	**57**	**0.19**	**0.04**	–	–	–	–	–	–	–	–	–	–	–	–
		M-F	100	0.33	0.08	–	–	–	–	–	–	–	–	–	–	–	–
	INDIAN RICEGRASS *Oryzopsis hymenoides*																
147	FRESH	**A-F**	**48**	**0.28**	**0.02**	–	–	**0.07**	–	**0.07**	–	–	–	–	–	–	–
		M-F	100	0.58	0.05	–	–	0.15	–	0.14	–	–	–	–	–	–	–
	JACK BEAN, COMMON *Canavalia ensiformis*																
148	FRESH	**A-F**	**23**	–	–	–	–	–	–	–	–	–	–	–	–	–	–
		M-F	100	–	–	–	–	–	–	–	–	–	–	–	–	–	–
	JOHNSONGRASS SORGHUM–SEE SORGHUM																
	JUNEGRASS *Koeleria cristata*																
149	IMMATURE, FRESH	**A-F**	**28**	**0.09**	**0.07**	–	–	–	–	–	–	–	–	–	–	–	–
		M-F	100	0.31	0.26	–	–	–	–	–	–	–	–	–	–	–	–
150	OVERRIPE, FRESH	**A-F**	**70**	**0.20**	**0.12**	–	–	–	–	–	–	–	–	–	–	–	–
		M-F	100	0.28	0.17	–	–	–	–	–	–	–	–	–	–	–	–
	KAFIR SORGHUM–SEE SORGHUM																
	KALE *Brassica oleracea, acephala*																
151	FRESH	**A-F**	**11**	**0.17**	**0.07**	**0.06**	–	–	**0.35**	–	–	–	–	**0.002**	–	–	–
		M-F	100	1.52	0.57	0.56	–	–	3.04	–	–	–	–	0.016	–	–	–
	KENTUCKY BLUEGRASS–SEE BLUEGRASS, KENTUCKY																
	KIKUYU GRASS *Pennisetum clandestinum*																
152	IMMATURE, FRESH	**A-F**	**21**	**0.07**	**0.05**	–	–	**0.06**	**0.95**	–	–	–	–	**0.005**	–	–	–
		M-F	100	0.34	0.25	–	–	0.28	4.59	–	–	–	–	0.023	–	–	–
	KOA HAOLE (LEAD TREE, WHITE POPINAC) *Leucaena glauca*																
153	BROWSE, FRESH	**A-F**	**30**	–	–	–	–	–	–	–	–	–	–	–	–	–	–
		M-F	100	–	–	–	–	–	–	–	–	–	–	–	–	–	–
	KOHLRABI *Brassica oleracea, gongylodes*																
154	FRESH	**A-F**	**9**	**0.08**	**0.07**	–	–	–	**0.37**	–	–	–	–	–	–	–	–
		M-F	100	0.89	0.78	–	–	–	4.11	–	–	–	–	–	–	–	–
	KUDZU *Pueraria spp*																
155	FRESH	**A-F**	**28**	**0.87**	**0.06**	–	–	–	–	–	–	–	–	–	–	–	–
		M-F	100	3.14	0.23	–	–	–	–	–	–	–	–	–	–	–	–
	LADINO CLOVER–SEE CLOVER, LADINO																
	LEAD TREE, WHITE POPINAC (KOA HAOLE) *Leucaena glauca*																
156	BROWSE, FRESH	**A-F**	**30**	–	–	–	–	–	–	–	–	–	–	–	–	–	–
		M-F	100	–	–	–	–	–	–	–	–	–	–	–	–	–	–
	LESPEDEZA, COMMON *Lespedeza striata*																
157	IMMATURE, FRESH	**A-F**	**24**	–	–	–	–	–	–	–	–	–	–	–	–	–	–
		M-F	100	–	–	–	–	–	–	–	–	–	–	–	–	–	–
158	EARLY BLOOM, FRESH	**A-F**	**28**	**0.38**	**0.08**	–	–	–	**0.31**	–	–	–	–	**0.008**	**58.4**	–	–
		M-F	100	1.35	0.27	–	–	–	1.12	–	–	–	–	0.025	208.6	–	–
159	MATURE, FRESH	**A-F**	**37**	**0.38**	**0.06**	–	–	–	**0.34**	–	–	–	–	–	–	–	–
		M-F	100	1.02	0.16	–	–	–	0.91	–	–	–	–	–	–	–	–
	LESPEDEZA, COMMON-KOREAN *Lespedeza striata-stipulacea*																
160	PREBLOOM, FRESH	**A-F**	**25**	**0.28**	**0.07**	–	–	–	**0.32**	–	–	–	–	–	–	–	–
		M-F	100	1.12	0.28	–	–	–	1.28	–	–	–	–	–	–	–	–
161	MATURE, FRESH	**A-F**	**35**	**0.35**	**0.07**	–	–	–	–	–	–	–	–	–	–	–	–
		M-F	100	1.00	0.20	–	–	–	–	–	–	–	–	–	–	–	–
	LESPEDEZA, SERICEA (CHINESE LESPEDEZA) *Lespedeza cuneata*																
162	FRESH	**A-F**	**33**	**0.42**	**0.10**	–	–	**0.07**	**0.39**	–	**0.024**	–	–	**0.008**	**34.1**	–	–
		M-F	100	1.27	0.29	–	–	0.22	1.20	–	0.071	–	–	0.024	103.8	–	–
163	IMMATURE, FRESH	**A-F**	**28**	**0.35**	**0.09**	**0.01**	–	**0.13**	**0.63**	–	**0.046**	**3.6**	–	**0.006**	**33.2**	–	**12.2**
		M-F	100	1.24	0.32	0.04	–	0.45	2.27	–	0.164	13.0	–	0.021	118.5	–	43.5
164	FULL BLOOM, FRESH	**A-F**	**25**	**0.29**	**0.07**	–	–	**0.06**	**0.25**	–	–	–	–	**0.007**	**26.2**	–	–
		M-F	100	1.16	0.26	–	–	0.22	0.98	–	–	–	–	0.025	104.9	–	–
	LETTUCE *Lactuca sativa*																
165	FRESH	**A-F**	**5**	**0.05**	**0.03**	**0.01**	–	–	**0.24**	–	–	–	–	**0.002**	–	–	–
		M-F	100	0.86	0.46	0.17	–	–	4.52	–	–	–	–	0.025	–	–	–
166	OUTSIDE LEAVES, FRESH	**A-F**	–	–	–	–	–	–	–	–	–	–	–	–	–	–	–
		M-F	100	–	–	–	–	–	–	–	–	–	–	–	–	–	–
	LICHEN *Lichen planus*																
167	FRESH	**A-F**	**33**	**1.22**	**0.03**	–	–	–	–	–	–	–	–	–	–	–	–
		M-F	100	3.70	0.09	–	–	–	–	–	–	–	–	–	–	–	–
	LIMA BEAN–SEE BEAN, LIMA																

PASTURE AND RANGE PLANTS

Entry Number	Fat-Soluble Vitamins: A (1 mg Carotene = 1667 IU Vit A)	Carotene (Provitamin A)	D	E	K	Water-Soluble Vitamins: B–12	Biotin	Choline	Folacin (Folic Acid)	Niacin	Pantothenic Acid (B–3)	(Pyri-doxine) B–6	Ribo-flavin (B–2)	Thiamin (B–1)
	IU/g	ppm or mg/kg	IU/kg	ppm or mg/kg	ppm or mg/kg	ppm or mg/kg	ppm or mg/kg	ppm or mg/kg	ppm or mg/kg	ppm or mg/kg	ppm or mg/kg	ppm or mg/kg	ppm or mg/kg	ppm or mg/kg
146	**92.5**	**55.5**	–	–	–	–	–	–	–	–	–	–	–	–
	162.6	97.5	–	–	–	–	–	–	–	–	–	–	–	–
147	**0.4**	**0.2**	–	–	–	–	–	–	–	–	–	–	–	–
	0.7	0.4	–	–	–	–	–	–	–	–	–	–	–	–
148	–	–	–	–	–	–	–	–	–	–	–	–	–	–
	–	–	–	–	–	–	–	–	–	–	–	–	–	–
149	–	–	–	–	–	–	–	–	–	–	–	–	–	–
	–	–	–	–	–	–	–	–	–	–	–	–	–	–
150	**38.8**	**23.3**	–	–	–	–	–	–	–	–	–	–	–	–
	55.5	33.3	–	–	–	–	–	–	–	–	–	–	–	–
151	–	–	–	–	–	–	–	–	–	–	–	–	–	–
	–	–	–	–	–	–	–	–	–	–	–	–	–	–
152	**88.7**	**53.2**	–	–	–	–	–	–	–	**10**	–	–	**4.3**	–
	427.8	256.6	–	–	–	–	–	–	–	50	–	–	20.5	–
153	–	–	–	–	–	–	–	–	–	–	–	–	–	–
	–	–	–	–	–	–	–	–	–	–	–	–	–	–
154	–	–	–	–	–	–	–	–	–	–	–	–	–	–
	–	–	–	–	–	–	–	–	–	–	–	–	–	–
155	–	–	–	–	–	–	–	–	–	–	–	–	–	–
	–	–	–	–	–	–	–	–	–	–	–	–	–	–
156	–	–	–	–	–	–	–	–	–	–	–	–	–	–
	–	–	–	–	–	–	–	–	–	–	–	–	–	–
157	–	–	–	–	–	–	–	–	–	–	–	–	–	–
	–	–	–	–	–	–	–	–	–	–	–	–	–	–
158	–	–	–	–	–	–	–	–	–	–	–	–	–	–
	–	–	–	–	–	–	–	–	–	–	–	–	–	–
159	–	–	–	–	–	–	–	–	–	–	–	–	–	–
	–	–	–	–	–	–	–	–	–	–	–	–	–	–
160	–	–	–	–	–	–	–	–	–	–	–	–	–	–
	–	–	–	–	–	–	–	–	–	–	–	–	–	–
161	–	–	–	–	–	–	–	–	–	–	–	–	–	–
	–	–	–	–	–	–	–	–	–	–	–	–	–	–
162	–	–	–	–	–	–	–	–	–	–	–	–	–	–
	–	–	–	–	–	–	–	–	–	–	–	–	–	–
163	–	–	–	–	–	–	–	–	–	–	–	–	–	–
	–	–	–	–	–	–	–	–	–	–	–	–	–	–
164	–	–	–	–	–	–	–	–	–	–	–	–	–	–
	–	–	–	–	–	–	–	–	–	–	–	–	–	–
165	–	–	–	–	–	–	–	–	–	**3**	–	–	**0.7**	**0.6**
	–	–	–	–	–	–	–	–	–	65	–	–	13.1	10.6
166	–	–	–	–	–	–	–	–	–	–	–	–	–	–
	–	–	–	–	–	–	–	–	–	–	–	–	–	–
167	–	–	–	–	–	–	–	–	–	–	–	–	–	–
	–	–	–	–	–	–	–	–	–	–	–	–	–	–

(Continued)

TABLE V-1E PASTURE AND RANGE PLANTS, COMPOSITION OF FEEDS, DATA EXPRESSED **AS-FED** AND **MOISTURE-FREE**—*(Continued)*

Entry Number	Feed Name Description	International Feed Number	Moisture Basis: A-F (as-fed) or M-F (moisture-free)	Proximate Analysis									Digestible Protein		
				Dry Matter	Ash	Crude Fiber	Neutral Det. Fib. (NDF)	Acid Det. Fib. (ADF)	Lignin	Ether Extract (Fat)	N-Free Extract	Crude Protein	Ruminant	Swine	Horse
				%	%	%	%	%	%	%	%	%	%	%	%
	LOVEGRASS *Eragrostis spp*														
168	IMMATURE, FRESH	2-02-647	A-F	43	2.8	13.1	–	–	–	1.3	20.2	5.4	3.6	3.6	–
			M-F	100	6.5	30.6	–	–	–	3.1	47.2	12.6	8.5	8.4	–
	LUCERNE–SEE ALFALFA														
	LUPINE, WHITE *Lupinus albus*														
169	FRESH	2-02-740	A-F	16	1.5	4.8	–	–	–	0.5	6.2	2.8	2.1	2.0	–
			M-F	100	9.4	30.6	–	–	–	3.0	39.4	17.6	13.2	12.4	–
	MAIZE–SEE CORN														
	MANGEL BEET *Beta vulgaris, macrorrhiza*														
170	TOPS WITH CROWNS, FRESH	2-00-632	A-F	13	2.4	1.4	–	–	–	0.5	6.1	2.1	1.8	1.5	–
			M-F	100	19.2	11.4	–	–	–	4.2	48.2	17.0	13.9	11.9	–
	MAPLE, RED *Acer rubrum*														
171	BROWSE, FRESH	2-03-062	A-F	35	1.6	6.2	16.9	15.0	6.2	2.1	19.7	5.4	3.8	3.7	–
			M-F	100	4.7	17.8	48.3	42.9	17.6	6.0	56.2	15.3	11.0	10.6	–
	MEADOW FESCUE *Festuca elatior*														
172	FRESH	2-01-920	A-F	28	2.4	8.6	–	–	–	1.0	12.9	3.5	2.4	2.4	–
			M-F	100	8.4	30.1	–	–	–	3.7	45.3	12.5	8.4	8.3	–
	MEADOW FOXTAIL *Alopecurus pratensis*														
173	IMMATURE, FRESH	2-02-073	A-F	26	2.8	5.6	–	–	–	1.2	12.0	4.5	3.3	3.2	–
			M-F	100	10.7	21.5	–	–	–	4.6	46.0	17.2	12.7	12.1	–
	MEDIC, BLACK (YELLOW TREFOIL) *Medicago lupulina*														
174	FRESH	2-03-070	A-F	23	2.3	5.6	–	–	–	0.8	9.1	4.9	3.8	3.5	–
			M-F	100	10.2	24.7	–	–	–	3.4	40.2	21.6	16.7	15.6	–
	MELIC, COASTRANGE (ONIONGRASS) *Melica imperfecta*														
175	FRESH	2-03-075	A-F	26	1.7	7.9	–	–	–	0.7	12.9	2.9	1.8	1.8	–
			M-F	100	6.5	30.3	–	–	–	2.6	49.6	11.0	7.1	7.1	–
	MESQUITE, COMMON *Prosopis chilensis*														
176	BROWSE, FRESH	2-03-081	A-F	35	2.1	9.6	–	–	–	1.2	14.8	7.4	5.7	5.3	–
			M-F	100	5.9	27.3	–	–	–	3.4	42.3	21.1	16.2	15.2	–
	MILLET, FOXTAIL *Setaria italica*														
177	FRESH	2-03-101	A-F	29	2.5	9.2	–	–	–	0.9	13.4	2.8	1.7	1.7	–
			M-F	100	8.6	32.0	–	–	–	3.1	46.7	9.6	5.9	6.0	–
	MILLET, JAPANESE (BARNYARD GRASS) *Echinochloa crusgalli*														
178	FRESH	2-03-108	A-F	22	1.6	6.8	–	–	–	0.6	11.0	1.7	1.0	1.0	–
			M-F	100	7.4	31.3	–	–	–	2.8	50.7	7.8	4.7	4.6	–
	MILLET, PEARL (PEARL MILLET) *Pennisetum glaucum*														
179	FRESH	2-03-115	A-F	21	1.9	6.5	–	–	–	0.6	9.7	2.1	1.3	1.3	–
			M-F	100	9.2	31.1	–	–	–	2.9	46.8	10.1	6.3	6.4	–
	MILLET, PROSO (BROOMCORN; HOG MILLET) *Panicum miliaceum*														
180	FRESH	2-03-811	A-F	25	1.8	7.4	–	–	–	0.6	13.1	2.1	1.1	1.2	–
			M-F	100	7.4	29.5	–	–	–	2.5	52.4	8.2	4.6	4.9	–
	MILO SORGHUM–SEE SORGHUM														
	MOLASSES GRASS *Melinis minutiflora*														
181	FRESH	2-03-130	A-F	37	2.7	14.6	–	–	–	0.9	17.1	1.5	0.4	0.5	–
			M-F	100	7.4	39.8	–	–	–	2.4	46.4	4.0	1.0	1.5	–
	MUHLY, BUSH *Muhlenbergia porteri*														
182	MIDBLOOM, FRESH	2-05-619	A-F	43	2.4	16.2	–	–	–	0.7	20.9	2.8	1.3	–	1.5
			M-F	100	5.5	37.7	–	–	–	1.7	48.6	6.5	3.0	–	3.4
	MUNG BEAN–SEE BEAN, MUNG														
	NAPIERGRASS *Pennisetum purpureum*														
183	PREBLOOM, FRESH[1]	2-03-158	A-F	20	1.7	6.7	14.0	9.0	–	0.6	9.5	1.8	1.0	1.1	–
			M-F	100	8.6	33.0	70.0	45.0	–	3.0	46.7	8.7	5.0	5.3	–
184	LATE BLOOM, FRESH[1]	2-03-162	A-F	23	1.2	9.0	17.3	10.8	–	0.3	10.8	1.8	0.8	1.0	–
			M-F	100	5.3	39.0	75.0	47.0	–	1.1	46.8	7.8	3.6	4.5	–
	NEEDLE-AND-THREAD *Stipa comata*														
185	STEM CURED, FRESH	2-07-989	A-F	92	19.4	–	–	39.7	5.9	5.0	–	3.7	1.0	1.4	–
			M-F	100	21.1	–	–	43.2	6.4	5.4	–	4.1	1.1	1.5	–
	NEEDLEGRASS *Stipa spp*														
186	MATURE, FRESH	2-03-205	A-F	42	2.9	17.1	–	–	–	0.5	18.4	3.2	1.7	1.8	–
			M-F	100	6.9	40.7	–	–	–	1.1	43.7	7.6	4.0	4.4	–
	NUTTALL SALTBUSH *Atriplex nuttallii*														
187	BROWSE, STEM CURED, FRESH	2-07-993	A-F	55	11.8	–	–	–	–	1.2	–	4.0	1.8	2.2	–
			M-F	100	21.5	–	–	–	–	2.2	–	7.2	3.4	4.1	–

PASTURE AND RANGE PLANTS

Entry Number	TDN Ruminant	TDN Swine	TDN Horse	Digestible Energy Ruminant		Digestible Energy Swine		Digestible Energy Horse		Metabolizable Energy Ruminant		Metabolizable Energy Swine		Metabolizable Energy Poultry ME_n		Metabolizable Energy Horse		Net Energy Ruminant NE_m		Net Energy Ruminant NE_g		Net Energy Lactating Cows NE_{lc}	
	%	%	%	Mcal		kcal		Mcal		Mcal		kcal		kcal		Mcal		Mcal		Mcal		Mcal	
				lb	*kg*	lb	*kg*	lb	*kg*	lb	*kg*	lb	*kg*	lb	*kg*	lb	*kg*	lb	*kg*	lb	*kg*	lb	*kg*
168	27	–	–	0.54	*1.19*	–	–	–	–	0.46	*1.01*	–	–	–	–	–	–	0.28	*0.61*	0.16	*0.36*	0.28	*0.61*
	63	–	–	1.27	*2.80*	–	–	–	–	1.08	*2.37*	–	–	–	–	–	–	0.65	*1.42*	0.38	*0.84*	0.65	*1.44*
169	10	–	–	0.20	*0.45*	–	–	–	–	0.17	*0.38*	–	–	–	–	–	–	0.10	*0.22*	0.06	*0.13*	0.10	*0.22*
	66	–	–	1.29	*2.83*	–	–	–	–	1.09	*2.41*	–	–	–	–	–	–	0.63	*1.39*	0.37	*0.81*	0.64	*1.41*
170	8	–	–	0.16	*0.36*	–	–	–	–	0.14	*0.30*	–	–	–	–	–	–	0.08	*0.18*	0.05	*0.10*	0.08	*0.18*
	65	–	–	1.28	*2.82*	–	–	–	–	1.09	*2.40*	–	–	–	–	–	–	0.64	*1.40*	0.37	*0.82*	0.64	*1.42*
171	–	–	18	–	–	–	–	0.33	*0.73*	–	–	–	–	–	–	0.27	*0.60*	–	–	–	–	–	–
	–	–	50	–	–	–	–	0.95	*2.09*	–	–	–	–	–	–	0.78	*1.71*	–	–	–	–	–	–
172	17	–	–	0.35	*0.77*	–	–	–	–	0.29	*0.65*	–	–	–	–	–	–	0.18	*0.39*	0.10	*0.23*	0.18	*0.40*
	61	–	–	1.23	*2.70*	–	–	–	–	1.03	*2.28*	–	–	–	–	–	–	0.62	*1.37*	0.36	*0.79*	0.63	*1.39*
173	17	–	–	0.34	*0.75*	–	–	–	–	0.29	*0.64*	–	–	–	–	–	–	0.18	*0.40*	0.11	*0.24*	0.18	*0.39*
	65	–	–	1.30	*2.88*	–	–	–	–	1.11	*2.45*	–	–	–	–	–	–	0.69	*1.51*	0.42	*0.92*	0.68	*1.50*
174	14	–	–	0.29	*0.64*	–	–	–	–	0.25	*0.54*	–	–	–	–	–	–	0.16	*0.35*	0.10	*0.21*	0.16	*0.34*
	61	–	–	1.27	*2.81*	–	–	–	–	1.08	*2.39*	–	–	–	–	–	–	0.69	*1.52*	0.42	*0.93*	0.69	*1.51*
175	16	–	–	0.33	*0.72*	–	–	–	–	0.28	*0.61*	–	–	–	–	–	–	0.17	*0.37*	0.10	*0.21*	0.17	*0.37*
	63	–	–	1.26	*2.78*	–	–	–	–	1.07	*2.36*	–	–	–	–	–	–	0.64	*1.41*	0.37	*0.82*	0.64	*1.42*
176	24	–	–	0.49	*1.08*	–	–	–	–	0.42	*0.93*	–	–	–	–	–	–	0.26	*0.56*	0.16	*0.35*	0.25	*0.55*
	70	–	–	1.39	*3.07*	–	–	–	–	1.20	*2.65*	–	–	–	–	–	–	0.73	*1.61*	0.46	*1.01*	0.72	*1.58*
177	18	–	–	0.35	*0.77*	–	–	–	–	0.30	*0.65*	–	–	–	–	–	–	0.17	*0.37*	0.09	*0.21*	0.17	*0.38*
	63	–	–	1.22	*2.69*	–	–	–	–	1.03	*2.27*	–	–	–	–	–	–	0.58	*1.29*	0.33	*0.72*	0.61	*1.33*
178	14	–	–	0.27	*0.60*	–	–	–	–	0.23	*0.51*	–	–	–	–	–	–	0.13	*0.29*	0.07	*0.16*	0.13	*0.29*
	65	–	–	1.26	*2.77*	–	–	–	–	1.07	*2.35*	–	–	–	–	–	–	0.60	*1.32*	0.34	*0.74*	0.61	*1.36*
179	13	–	–	0.25	*0.56*	–	–	–	–	0.21	*0.47*	–	–	–	–	–	–	0.12	*0.27*	0.07	*0.15*	0.13	*0.28*
	62	–	–	1.22	*2.68*	–	–	–	–	1.02	*2.26*	–	–	–	–	–	–	0.59	*1.29*	0.33	*0.72*	0.61	*1.34*
180	16	–	–	0.31	*0.68*	–	–	–	–	0.26	*0.58*	–	–	–	–	–	–	0.15	*0.34*	0.09	*0.19*	0.16	*0.34*
	63	–	–	1.24	*2.74*	–	–	–	–	1.05	*2.32*	–	–	–	–	–	–	0.61	*1.35*	0.35	*0.78*	0.63	*1.38*
181	19	–	–	0.39	*0.86*	–	–	–	–	0.32	*0.70*	–	–	–	–	–	–	0.18	*0.41*	0.09	*0.20*	0.20	*0.44*
	52	–	–	1.06	*2.33*	–	–	–	–	0.86	*1.90*	–	–	–	–	–	–	0.50	*1.10*	0.25	*0.55*	0.55	*1.20*
182	–	–	–	–	–	–	–	–	–	–	–	–	–	–	–	–	–	–	–	–	–	–	–
	–	–	–	–	–	–	–	–	–	–	–	–	–	–	–	–	–	–	–	–	–	–	–
183	11	–	–	0.22	*0.49*	–	–	–	–	0.19	*0.41*	–	–	–	–	–	–	0.11	*0.25*	0.06	*0.13*	0.12	*0.26*
	55	–	–	1.11	*2.44*	–	–	–	–	0.91	*2.01*	–	–	–	–	–	–	0.55	*1.21*	0.29	*0.64*	0.58	*1.28*
184	12	–	–	0.24	*0.54*	–	–	–	–	0.20	*0.44*	–	–	–	–	–	–	0.13	*0.28*	0.07	*0.15*	0.13	*0.29*
	53	–	–	1.06	*2.34*	–	–	–	–	0.87	*1.91*	–	–	–	–	–	–	0.55	*1.21*	0.29	*0.64*	0.58	*1.28*
185	45	–	–	0.87	*1.93*	–	–	–	–	0.76	*1.67*	–	–	–	–	–	–	–	–	–	–	–	–
	49	–	–	0.95	*2.09*	–	–	–	–	0.82	*1.81*	–	–	–	–	–	–	–	–	–	–	–	–
186	24	–	–	0.48	*1.06*	–	–	–	–	0.40	*0.88*	–	–	–	–	–	–	0.22	*0.48*	0.11	*0.25*	0.24	*0.52*
	57	–	–	1.14	*2.52*	–	–	–	–	0.95	*2.10*	–	–	–	–	–	–	0.52	*1.15*	0.27	*0.59*	0.56	*1.23*
187	20	–	–	0.37	*0.82*	–	–	–	–	0.33	*0.73*	–	–	–	–	–	–	–	–	–	–	–	–
	36	–	–	0.68	*1.49*	–	–	–	–	0.60	*1.32*	–	–	–	–	–	–	–	–	–	–	–	–

(Continued)

TABLE V-1E PASTURE AND RANGE PLANTS, MINERAL AND VITAMIN COMPOSITION OF FEEDS, DATA EXPRESSED **AS-FED** AND **MOISTURE-FREE**—*(Continued)*

Entry Number	Feed Name Description	Moisture Basis: A-F (as-fed) or M-F (moisture-free)	Dry Matter	Macro Minerals							Micro Minerals						
				Calcium (Ca)	Phosphorus (P)	Sodium (Na)	Chlorine (Cl)	Magnesium (Mg)	Potassium (K)	Sulfur (S)	Cobalt (Co)	Copper (Cu)	Iodine (I)	Iron (Fe)	Manganese (Mn)	Selenium (Se)	Zinc (Zn)
			%	%	%	%	%	%	%	%	ppm or mg/kg	ppm or mg/kg	ppm or mg/kg	%	ppm or mg/kg	ppm or mg/kg	ppm or mg/kg
	LOVEGRASS *Eragrostis spp*																
168	IMMATURE, FRESH	A-F	43	0.20	0.10	–	–	–	–	–	–	–	–	–	–	–	–
		M-F	100	0.47	0.24	–	–	–	–	–	–	–	–	–	–	–	–
	LUCERNE–SEE ALFALFA																
	LUPINE, WHITE *Lupinus albus*																
169	FRESH	A-F	16	–	0.04	–	–	–	0.38	–	–	–	–	–	–	–	–
		M-F	100	–	0.23	–	–	–	2.41	–	–	–	–	–	–	–	–
	MAIZE–SEE CORN																
	MANGEL BEET *Beta vulgaris, macrorrhiza*																
170	TOPS WITH CROWNS, FRESH	A-F	13	–	–	–	–	–	–	–	–	–	–	–	–	–	–
		M-F	100	–	–	–	–	–	–	–	–	–	–	–	–	–	–
	MAPLE, RED *Acer rubrum*																
171	BROWSE, FRESH	A-F	35	0.30	0.08	–	–	–	–	–	0.039	2.2	–	0.011	271.3	–	–
		M-F	100	0.85	0.22	–	–	–	–	–	0.111	6.4	–	0.029	775.1	–	–
	MEADOW FESCUE *Festuca elatior*																
172	FRESH	A-F	28	0.15	0.11	–	–	0.11	0.67	–	0.039	1.1	–	–	–	–	–
		M-F	100	0.53	0.39	–	–	0.37	2.34	–	0.135	4.0	–	–	–	–	–
	MEADOW FOXTAIL *Alopecurus pratensis*																
173	IMMATURE, FRESH	A-F	26	0.15	0.12	–	–	–	–	–	–	–	–	–	–	–	–
		M-F	100	0.57	0.46	–	–	–	–	–	–	–	–	–	–	–	–
	MEDIC, BLACK (YELLOW TREFOIL) *Medicago lupulina*																
174	FRESH	A-F	23	–	–	–	–	–	–	–	–	–	–	–	–	–	–
		M-F	100	–	–	–	–	–	–	–	–	–	–	–	–	–	–
	MELIC, COASTRANGE (ONIONGRASS) *Melica imperfecta*																
175	FRESH	A-F	26	0.10	0.07	–	–	–	0.50	–	–	–	–	–	–	–	–
		M-F	100	0.39	0.26	–	–	–	1.94	–	–	–	–	–	–	–	–
	MESQUITE, COMMON *Prosopis chilensis*																
176	BROWSE, FRESH	A-F	35	0.68	0.07	–	–	0.08	0.49	–	–	–	–	–	–	–	–
		M-F	100	1.94	0.19	–	–	0.23	1.41	–	–	–	–	–	–	–	–
	MILLET, FOXTAIL *Setaria italica*																
177	FRESH	A-F	29	0.09	0.05	–	–	–	0.56	–	–	–	–	–	–	–	–
		M-F	100	0.32	0.19	–	–	–	1.94	–	–	–	–	–	–	–	–
	MILLET, JAPANESE (BARNYARD GRASS) *Echinochloa crusgalli*																
178	FRESH	A-F	22	0.11	0.07	–	–	–	0.52	–	–	–	–	–	–	–	–
		M-F	100	0.51	0.32	–	–	–	2.40	–	–	–	–	–	–	–	–
	MILLET, PEARL (PEARL MILLET) *Pennisetum glaucum*																
179	FRESH	A-F	21	–	–	–	–	–	–	–	–	–	–	–	–	–	–
		M-F	100	–	–	–	–	–	–	–	–	–	–	–	–	–	–
	MILLET, PROSO (BROOMCORN; HOG MILLET) *Panicum miliaceum*																
180	FRESH	A-F	25	–	–	–	–	–	–	–	–	–	–	–	–	–	–
		M-F	100	–	–	–	–	–	–	–	–	–	–	–	–	–	–
	MILO SORGHUM–SEE SORGHUM																
	MOLASSES GRASS *Melinis minutiflora*																
181	FRESH	A-F	37	–	–	–	–	–	–	–	–	–	–	–	–	–	–
		M-F	100	–	–	–	–	–	–	–	–	–	–	–	–	–	–
	MUHLY BUSH *Muhlenbergia porteri*																
182	MIDBLOOM, FRESH	A-F	43	0.13	0.03	0.00	–	0.02	0.22	–	0.215	0.2	–	0.006	5.2	–	5.2
		M-F	100	0.30	0.07	0.01	–	0.04	0.51	–	0.501	0.4	–	0.012	12.0	–	12.0
	MUNG BEAN–SEE BEAN, MUNG																
	NAPIERGRASS *Pennisetum purpureum*																
183	PREBLOOM, FRESH	A-F	20	0.12	0.08	0.00	–	0.05	0.27	0.02	–	–	–	–	–	–	–
		M-F	100	0.60	0.41	0.01	–	0.26	1.31	0.10	–	–	–	–	–	–	–
184	LATE BLOOM, FRESH	A-F	23	0.08	0.07	0.00	–	0.06	0.30	0.02	–	–	–	–	–	–	–
		M-F	100	0.35	0.30	0.01	–	0.26	1.31	0.10	–	–	–	–	–	–	–
	NEEDLE-AND-THREAD *Stipa comata*																
185	STEM CURED, FRESH	A-F	92	0.99	0.06	–	–	–	–	–	–	–	–	–	–	–	–
		M-F	100	1.08	0.06	–	–	–	–	–	–	–	–	–	–	–	–
	NEEDLEGRASS *Stipa spp*																
186	MATURE, FRESH	A-F	42	–	–	–	–	–	–	–	–	–	–	–	–	–	–
		M-F	100	–	–	–	–	–	–	–	–	–	–	–	–	–	–
	NUTTAL SALTBUSH *Atriplex nuttallii*																
187	BROWSE, STEM CURED, FRESH	A-F	55	1.22	0.06	–	–	–	–	–	–	–	–	–	–	–	–
		M-F	100	2.21	0.12	–	–	–	–	–	–	–	–	–	–	–	–

Entry Number	Fat-Soluble Vitamins					Water-Soluble Vitamins								
	A (1 mg Carotene = 1667 IU Vit A)	Carotene (Provitamin A)	D	E	K	B-12	Biotin	Choline	Folacin (Folic Acid)	Niacin	Pantothenic Acid (B-3)	(Pyridoxine) B-6	Riboflavin (B-2)	Thiamin (B-1)
	IU/g	ppm or mg/kg	IU/kg	ppm or mg/kg	ppm or mg/kg	ppm or mg/kg	ppm or mg/kg	ppm or mg/kg	ppm or mg/kg	ppm or mg/kg	ppm or mg/kg	ppm or mg/kg	ppm or mg/kg	ppm or mg/kg
168	–	–	–	–	–	–	–	–	–	–	–	–	–	–
	–	–	–	–	–	–	–	–	–	–	–	–	–	–
169	–	–	–	–	–	–	–	–	–	–	–	–	–	–
	–	–	–	–	–	–	–	–	–	–	–	–	–	–
170	–	–	–	–	–	–	–	–	–	–	–	–	–	–
	–	–	–	–	–	–	–	–	–	–	–	–	–	–
171	–	–	–	–	–	–	–	–	–	–	–	–	–	–
	–	–	–	–	–	–	–	–	–	–	–	–	–	–
172	**160.0**	**96.0**	–	**46.9**	–	–	–	–	–	–	–	–	**2.4**	**3.4**
	562.7	337.5	–	165.1	–	–	–	–	–	–	–	–	8.6	11.9
173	–	–	–	–	–	–	–	–	–	–	–	–	–	–
	–	–	–	–	–	–	–	–	–	–	–	–	–	–
174	–	–	–	–	–	–	–	–	–	–	–	–	–	–
	–	–	–	–	–	–	–	–	–	–	–	–	–	–
175	–	–	–	–	–	–	–	–	–	–	–	–	–	–
	–	–	–	–	–	–	–	–	–	–	–	–	–	–
176	–	–	–	–	–	–	–	–	–	–	–	–	–	–
	–	–	–	–	–	–	–	–	–	–	–	–	–	–
177	–	–	–	–	–	–	–	–	–	–	–	–	–	–
	–	–	–	–	–	–	–	–	–	–	–	–	–	–
178	–	–	–	–	–	–	–	–	–	–	–	–	–	–
	–	–	–	–	–	–	–	–	–	–	–	–	–	–
179	**63.0**	**37.8**	–	–	–	–	–	–	–	–	–	–	–	–
	304.2	182.5	–	–	–	–	–	–	–	–	–	–	–	–
180	–	–	–	–	–	–	–	–	–	–	–	–	–	–
	–	–	–	–	–	–	–	–	–	–	–	–	–	–
181	–	–	–	–	–	–	–	–	–	–	–	–	–	–
	–	–	–	–	–	–	–	–	–	–	–	–	–	–
182	–	–	–	–	–	–	–	–	–	–	–	–	–	–
	–	–	–	–	–	–	–	–	–	–	–	–	–	–
183	–	–	–	–	–	–	–	–	–	–	–	–	–	–
	–	–	–	–	–	–	–	–	–	–	–	–	–	–
184	–	–	–	–	–	–	–	–	–	–	–	–	–	–
	–	–	–	–	–	–	–	–	–	–	–	–	–	–
185	–	–	–	–	–	–	–	–	–	–	–	–	–	–
	–	–	–	–	–	–	–	–	–	–	–	–	–	–
186	–	–	–	–	–	–	–	–	–	–	–	–	–	–
	–	–	–	–	–	–	–	–	–	–	–	–	–	–
187	**17.4**	**10.5**	–	–	–	–	–	–	–	–	–	–	–	–
	31.7	19.0	–	–	–	–	–	–	–	–	–	–	–	–

(Continued)

TABLE V-1E PASTURE AND RANGE PLANTS, COMPOSITION OF FEEDS, DATA EXPRESSED **AS-FED** AND **MOISTURE-FREE**—*(Continued)*

Entry Number	Feed Name Description	International Feed Number	Moisture Basis: A-F (as-fed) or M-F (moisture-free)	Proximate Analysis									Digestible Protein		
				Dry Matter	Ash	Crude Fiber	Neutral Det. Fib. (NDF)	Acid Det. Fib. (ADF)	Lignin	Ether Extract (Fat)	N-Free Extract	Crude Protein	Ruminant	Swine	Horse
				%	%	%	%	%	%	%	%	%	%	%	%
	OATGRASS, TALL *Arrhenatherum elatius*														
188	FRESH	**2-03-267**	**A-F**	**30**	**2.0**	**10.5**	–	–	–	**0.9**	**14.3**	**2.6**	**1.5**	**1.6**	–
			M-F	100	6.6	34.7	–	–	–	3.0	47.2	8.6	4.9	5.2	–
189	IMMATURE, FRESH	**2-03-261**	**A-F**	**22**	**2.2**	**4.1**	–	–	–	**1.1**	**10.0**	**4.3**	**3.2**	**3.0**	–
			M-F	100	10.1	19.0	–	–	–	4.9	46.3	19.7	15.0	14.1	–
	OATS *Avena sativa*														
190	IMMATURE, FRESH	**2-03-286**	**A-F**	**16**	**1.7**	**4.0**	–	–	–	**0.4**	**7.4**	**2.5**	**1.8**	**1.7**	–
			M-F	100	10.6	24.9	–	–	–	2.6	46.3	15.6	11.3	10.8	–
191	DOUGH STAGE, FRESH	**2-03-289**	**A-F**	**31**	**2.5**	**10.8**	–	–	–	**1.0**	**13.7**	**2.6**	**1.5**	**1.5**	–
			M-F	100	8.3	35.2	–	–	–	3.3	45.0	8.4	4.9	5.0	–
	OATS, WILD *Avena fatua*														
192	FRESH	**2-03-393**	**A-F**	**36**	**2.6**	**8.4**	–	–	–	**1.4**	**20.8**	**2.5**	**1.3**	**1.4**	–
			M-F	100	7.4	23.5	–	–	–	3.8	58.2	7.1	3.5	4.0	–
	OATS-PEA *Avena sativa-Pisum* spp														
193	EARLY BLOOM, FRESH	**2-03-399**	**A-F**	**23**	**2.0**	**7.0**	–	–	–	**0.9**	**9.5**	**3.5**	**2.5**	**2.4**	–
			M-F	100	8.8	30.5	–	–	–	3.9	41.5	15.3	11.0	10.6	–
	OATS-VETCH *Avena sativa-Vicia* spp														
194	MILK STAGE, FRESH	**2-03-407**	**A-F**	**33**	**2.5**	**9.1**	–	–	–	**1.0**	**16.3**	**3.5**	**2.2**	**2.3**	–
			M-F	100	7.8	28.1	–	–	–	3.0	50.3	10.8	6.9	6.9	–
	ONIONGRASS (MELIC, COASTRANGE) *Melica imperfecta*														
195	FRESH	**2-03-075**	**A-F**	**26**	**1.7**	**7.9**	–	–	–	**0.7**	**12.9**	**2.9**	**1.8**	**1.8**	–
			M-F	100	6.5	30.3	–	–	–	2.6	49.6	11.0	7.1	7.1	–
	ORCHARDGRASS *Dactylis glomerata*														
196	FRESH, ALL ANALYSES	**2-03-451**	**A-F**	**26**	**2.6**	**6.4**	**13.9**	–	–	**1.6**	**11.3**	**3.9**	**2.7**	**2.7**	–
			M-F	100	10.0	24.8	54.0	–	–	6.4	43.8	15.0	10.5	10.3	–
197	IMMATURE, FRESH	**2-03-439**	**A-F**	**23**	**2.6**	**5.7**	**9.8**	**5.2**	**0.7**	**1.1**	**7.6**	**5.5**	**4.0**	**4.1**	–
			M-F	100	11.4	25.2	43.2	22.8	3.3	5.0	33.8	24.5	17.6	18.0	–
198	EARLY BLOOM, FRESH	**2-03-442**	**A-F**	**24**	**1.9**	**7.5**	**12.6**	**7.2**	**1.3**	**0.9**	**10.2**	**3.0**	**2.1**	**2.0**	–
			M-F	100	8.1	32.0	53.7	30.7	5.6	3.7	43.3	12.8	8.7	8.6	–
199	MIDBLOOM, FRESH	**2-03-443**	**A-F**	**27**	**2.1**	**9.2**	**15.8**	**9.8**	**2.1**	**1.0**	**12.4**	**2.8**	**1.6**	**1.8**	–
			M-F	100	7.5	33.5	57.6	35.6	7.6	3.5	45.4	10.1	5.9	6.4	–
200	MILK STAGE, FRESH	**2-03-446**	**A-F**	**34**	**2.4**	**12.1**	**22.0**	**13.9**	**3.4**	**1.3**	**15.0**	**3.6**	**2.2**	**2.3**	–
			M-F	100	7.1	35.2	64.0	40.5	9.8	3.7	43.6	10.4	6.5	6.6	–
	PANGOLAGRASS *Digitaria decumbens*														
201	FRESH	**2-03-493**	**A-F**	**20**	**1.5**	**6.6**	–	**7.5**	**1.0**	**0.5**	**9.8**	**1.8**	**0.9**	**0.9**	–
			M-F	100	7.6	32.6	–	36.9	5.0	2.3	48.4	9.1	4.7	4.5	–
	PANICUM *Panicum* spp														
202	FRESH	**2-03-499**	**A-F**	**29**	**4.1**	**8.6**	–	–	–	**0.7**	**11.9**	**3.7**	**2.5**	**2.5**	–
			M-F	100	14.2	29.6	–	–	–	2.3	41.2	12.7	8.6	8.5	–
	PARAGRASS *Brachiaria mutica*														
203	FRESH	**2-03-525**	**A-F**	**26**	**3.0**	**9.1**	–	–	–	**0.4**	**11.9**	**2.0**	**1.2**	**1.1**	–
			M-F	100	11.3	34.6	–	–	–	1.5	45.1	7.4	4.5	4.2	–
	PARSNIP, GARDEN *Pastinaca sativa*														
204	LEAVES, FRESH	**2-03-534**	**A-F**	**18**	–	**1.4**	–	–	–	**0.9**	–	**4.1**	**3.2**	**3.0**	–
			M-F	100	–	8.0	–	–	–	5.0	–	22.9	17.9	16.7	–
	PASPALUM *Paspalum* spp														
205	FRESH	**2-03-543**	**A-F**	**27**	**3.5**	**8.9**	–	–	–	**0.5**	**12.1**	**1.8**	**0.9**	**1.0**	–
			M-F	100	12.9	33.4	–	–	–	1.7	45.2	6.9	3.4	3.8	–
	PEA *Pisum* spp														
206	PODS, FRESH	**2-03-588**	**A-F**	**13**	**0.9**	**3.0**	–	–	–	**0.3**	**6.9**	**1.8**	**1.0**	**1.2**	–
			M-F	100	7.0	23.3	–	–	–	2.3	53.4	14.0	7.4	9.5	–
	PEA, FIELD *Pisum sativum, arvense*														
207	FRESH	**2-03-603**	**A-F**	**18**	**1.7**	**4.6**	–	–	–	**0.6**	**7.6**	**3.6**	**3.0**	**2.6**	–
			M-F	100	9.1	25.4	–	–	–	3.4	42.2	20.0	16.8	14.3	–
	PEA-OATS *Pisum spp-Avena sativa*														
208	FRESH	**2-08-483**	**A-F**	**23**	**1.9**	**6.4**	–	–	–	**0.9**	**10.3**	**3.2**	**2.4**	**2.2**	–
			M-F	100	8.4	28.2	–	–	–	4.0	45.4	14.1	10.4	9.6	–
	PEARL MILLET (MILLET, PEARL) *Pennisetum glaucum*														
209	FRESH	**2-03-115**	**A-F**	**21**	**1.9**	**6.5**	–	–	–	**0.6**	**9.7**	**2.1**	**1.3**	**1.3**	–
			M-F	100	9.2	31.1	–	–	–	2.9	46.8	10.1	6.3	6.4	–
	PEAVINE *Lathyrus* spp														
210	FRESH	**2-03-669**	**A-F**	**17**	**5.0**	**4.2**	–	–	–	**0.6**	**4.4**	**3.2**	**2.5**	**2.3**	–
			M-F	100	28.3	24.2	–	–	–	3.7	25.5	18.3	14.5	13.0	–
	PIGEON PEA *Cajanus cajan*														
211	FRESH	**2-03-715**	**A-F**	**41**	**1.9**	**15.9**	–	–	–	**1.7**	**15.5**	**5.5**	**3.5**	**3.7**	–
			M-F	100	4.8	39.2	–	–	–	4.2	38.3	13.5	8.6	9.1	–

Entry Number	TDN Ruminant	TDN Swine	TDN Horse	Digestible Energy Ruminant		Digestible Energy Swine		Digestible Energy Horse		Metabolizable Energy Ruminant		Metabolizable Energy Swine		Poultry ME_n		Metabolizable Energy Horse		Net Energy Ruminant NE_m		Ruminant NE_g		Lactating Cows NE_{lc}	
	%	%	%	Mcal		kcal		Mcal		Mcal		kcal		kcal		Mcal		Mcal		Mcal		Mcal	
				lb	kg	lb	kg	lb	kg	lb	kg	lb	kg	lb	kg	lb	kg	lb	kg	lb	kg	lb	kg
188	17	–	–	0.35	0.78	–	–	–	–	0.30	0.65	–	–	–	–	–	–	0.18	0.39	0.10	0.22	0.18	0.41
	57	–	–	1.17	2.58	–	–	–	–	0.98	2.15	–	–	–	–	–	–	0.59	1.29	0.33	0.72	0.61	1.34
189	15	–	–	0.29	0.65	–	–	–	–	0.25	0.56	–	–	–	–	–	–	0.16	0.35	0.10	0.22	0.15	0.34
	68	–	–	1.36	3.00	–	–	–	–	1.17	2.58	–	–	–	–	–	–	0.73	1.61	0.46	1.01	0.72	1.58
190	10	–	–	0.21	0.46	–	–	–	–	0.18	0.39	–	–	–	–	–	–	0.10	0.23	0.06	0.14	0.10	0.23
	64	–	–	1.29	2.83	–	–	–	–	1.09	2.41	–	–	–	–	–	–	0.65	1.42	0.38	0.84	0.65	1.43
191	18	–	–	0.36	0.79	–	–	–	–	0.30	0.66	–	–	–	–	–	–	0.18	0.40	0.10	0.23	0.19	0.41
	60	–	–	1.17	2.59	–	–	–	–	0.98	2.16	–	–	–	–	–	–	0.59	1.31	0.33	0.74	0.61	1.35
192	23	–	15	0.46	1.02	–	–	0.29	0.64	0.39	0.87	–	–	–	–	0.24	0.52	0.24	0.52	0.14	0.31	0.24	0.52
	64	–	42	1.29	2.84	–	–	0.81	1.78	1.10	2.42	–	–	–	–	0.66	1.46	0.66	1.45	0.39	0.87	0.66	1.46
193	14	–	–	0.29	0.63	–	–	–	–	0.24	0.53	–	–	–	–	–	–	0.14	0.32	0.08	0.19	0.15	0.32
	62	–	–	1.25	2.74	–	–	–	–	1.05	2.32	–	–	–	–	–	–	0.63	1.39	0.37	0.81	0.64	1.41
194	20	–	–	0.41	0.90	–	–	–	–	0.34	0.76	–	–	–	–	–	–	0.21	0.46	0.12	0.27	0.21	0.46
	63	–	–	1.25	2.76	–	–	–	–	1.06	2.34	–	–	–	–	–	–	0.64	1.40	0.37	0.82	0.64	1.42
195	16	–	–	0.33	0.72	–	–	–	–	0.28	0.61	–	–	–	–	–	–	0.17	0.37	0.10	0.21	0.17	0.37
	63	–	–	1.26	2.78	–	–	–	–	1.07	2.36	–	–	–	–	–	–	0.64	1.41	0.37	0.82	0.64	1.42
196	17	–	–	0.35	0.77	–	–	–	–	0.30	0.66	–	–	–	–	–	–	0.19	0.42	0.12	0.26	0.18	0.41
	67	–	–	1.35	2.98	–	–	–	–	1.16	2.56	–	–	–	–	–	–	0.73	1.61	0.46	1.01	0.72	1.58
197	15	–	–	0.30	0.66	–	–	–	–	0.26	0.57	–	–	–	–	–	–	0.16	0.34	0.10	0.21	0.16	0.34
	67	–	–	1.33	2.93	–	–	–	–	1.14	2.51	–	–	–	–	–	–	0.69	1.52	0.42	0.93	0.68	1.51
198	16	–	–	0.30	0.66	–	–	–	–	0.25	0.56	–	–	–	–	–	–	0.16	0.35	0.10	0.22	0.16	0.35
	66	–	–	1.27	2.79	–	–	–	–	1.08	2.37	–	–	–	–	–	–	0.68	1.51	0.41	0.91	0.68	1.50
199	16	–	–	0.33	0.73	–	–	–	–	0.28	0.61	–	–	–	–	–	–	0.16	0.36	0.09	0.20	0.17	0.37
	60	–	–	1.20	2.64	–	–	–	–	1.01	2.22	–	–	–	–	–	–	0.59	1.31	0.33	0.73	0.61	1.35
200	20	–	–	0.41	0.90	–	–	–	–	0.34	0.75	–	–	–	–	–	–	0.20	0.45	0.11	0.25	0.21	0.46
	59	–	–	1.18	2.61	–	–	–	–	0.99	2.19	–	–	–	–	–	–	0.59	1.30	0.33	0.72	0.61	1.34
201	12	11	–	0.24	0.54	–	–	–	–	0.20	0.45	–	–	–	–	–	–	0.12	0.26	0.07	0.14	0.12	0.26
	60	54	–	1.20	2.65	–	–	–	–	1.01	2.23	–	–	–	–	–	–	0.58	1.28	0.32	0.71	0.57	1.26
202	17	–	–	0.33	0.74	–	–	–	–	0.28	0.61	–	–	–	–	–	–	0.15	0.34	0.08	0.18	0.17	0.36
	58	–	–	1.15	2.54	–	–	–	–	0.96	2.12	–	–	–	–	–	–	0.53	1.17	0.28	0.61	0.57	1.25
203	14	–	–	0.28	0.63	–	–	–	–	0.23	0.51	–	–	–	–	–	–	0.13	0.29	0.07	0.15	0.14	0.32
	54	–	–	1.08	2.37	–	–	–	–	0.88	1.95	–	–	–	–	–	–	0.50	1.11	0.25	0.55	0.55	1.21
204	–	–	–	–	–	–	–	–	–	–	–	–	–	–	–	–	–	–	–	–	–	–	–
	–	–	–	–	–	–	–	–	–	–	–	–	–	–	–	–	–	–	–	–	–	–	–
205	13	–	–	0.28	0.62	–	–	–	–	0.23	0.51	–	–	–	–	–	–	0.11	0.25	0.05	0.10	0.13	0.29
	49	–	–	1.05	2.32	–	–	–	–	0.86	1.89	–	–	–	–	–	–	0.42	0.94	0.18	0.39	0.49	1.08
206	9	–	–	0.18	0.40	–	–	–	–	0.16	0.35	–	–	–	–	–	–	0.09	0.20	0.06	0.12	0.09	0.20
	73	–	–	1.40	3.09	–	–	–	–	1.21	2.67	–	–	–	–	–	–	0.70	1.55	0.43	0.95	0.69	1.53
207	13	–	–	0.25	0.55	–	–	–	–	0.21	0.47	–	–	–	–	–	–	0.13	0.28	0.08	0.17	0.12	0.27
	71	–	–	1.37	3.02	–	–	–	–	1.18	2.60	–	–	–	–	–	–	0.69	1.53	0.42	0.93	0.69	1.51
208	14	–	–	0.29	0.64	–	–	–	–	0.25	0.54	–	–	–	–	–	–	0.15	0.32	0.09	0.19	0.15	0.33
	63	–	–	1.27	2.80	–	–	–	–	1.08	2.38	–	–	–	–	–	–	0.65	1.43	0.38	0.84	0.65	1.44
209	13	–	–	0.25	0.56	–	–	–	–	0.21	0.47	–	–	–	–	–	–	0.12	0.27	0.07	0.15	0.13	0.28
	62	–	–	1.22	2.68	–	–	–	–	1.02	2.26	–	–	–	–	–	–	0.59	1.29	0.33	0.72	0.61	1.34
210	9	–	–	0.19	0.42	–	–	–	–	0.16	0.34	–	–	–	–	–	–	–	–	–	–	–	–
	54	–	–	1.08	2.38	–	–	–	–	0.89	1.95	–	–	–	–	–	–	–	–	–	–	–	–
211	24	–	–	0.49	1.07	–	–	–	–	0.41	0.90	–	–	–	–	–	–	0.25	0.55	0.14	0.31	0.25	0.56
	59	–	–	1.20	2.64	–	–	–	–	1.01	2.22	–	–	–	–	–	–	0.61	1.35	0.35	0.77	0.63	1.38

(Continued)

TABLE V-1E PASTURE AND RANGE PLANTS, MINERAL AND VITAMIN COMPOSITION OF FEEDS, DATA EXPRESSED **AS-FED** AND **MOISTURE-FREE**—*(Continued)*

Entry Number	Feed Name Description	Moisture Basis: A-F (as-fed) or M-F (moisture-free)	Dry Matter	Macro Minerals							Micro Minerals						
				Calcium (Ca)	Phosphorus (P)	Sodium (Na)	Chlorine (Cl)	Magnesium (Mg)	Potassium (K)	Sulfur (S)	Cobalt (Co)	Copper (Cu)	Iodine (I)	Iron (Fe)	Manganese (Mn)	Selenium (Se)	Zinc (Zn)
			%	%	%	%	%	%	%	%	ppm or mg/kg	ppm or mg/kg	ppm or mg/kg	%	ppm or mg/kg	ppm or mg/kg	ppm or mg/kg
	OATGRASS, TALL *Arrhenatherum elatius*																
188	FRESH	**A-F**	**30**	**0.12**	**0.14**	–	–	–	**0.91**	–	–	–	–	–	–	–	–
		M-F	100	0.40	0.46	–	–	–	3.00	–	–	–	–	–	–	–	–
189	IMMATURE, FRESH	**A-F**	**22**	–	–	–	–	–	–	–	–	–	–	–	–	–	–
		M-F	100	–	–	–	–	–	–	–	–	–	–	–	–	–	–
	OATS *Avena sativa*																
190	IMMATURE, FRESH	**A-F**	**16**	–	–	**0.02**	**0.02**	–	–	**0.01**	–	–	–	–	–	–	–
		M-F	100	–	–	0.11	0.10	–	–	0.08	–	–	–	–	–	–	–
191	DOUGH STAGE, FRESH	**A-F**	**31**	**0.10**	**0.07**	–	–	**0.06**	**0.53**	–	–	**1.3**	–	–	**12.7**	–	**7.5**
		M-F	100	0.32	0.23	–	–	0.20	1.73	–	–	4.3	–	–	41.7	–	24.7
	OATS, WILD *Avena fatua*																
192	FRESH	**A-F**	**36**	**0.09**	**0.10**	–	–	–	–	–	–	–	–	–	–	–	–
		M-F	100	0.25	0.27	–	–	–	–	–	–	–	–	–	–	–	–
	OATS-PEA *Avena sativa-Pisum* spp																
193	EARLY BLOOM, FRESH	**A-F**	**23**	–	–	–	–	–	–	–	–	–	–	–	–	–	–
		M-F	100	–	–	–	–	–	–	–	–	–	–	–	–	–	–
	OATS-VETCH *Avena sativa-Vicia* spp																
194	MILK STAGE, FRESH	**A-F**	**33**	–	–	–	–	–	–	–	–	–	–	–	–	–	–
		M-F	100	–	–	–	–	–	–	–	–	–	–	–	–	–	–
	ONIONGRASS (MELIC, COASTRANGE) *Melica imperfecta*																
195	FRESH	**A-F**	**26**	**0.10**	**0.07**	–	–	–	**0.50**	–	–	–	–	–	–	–	–
		M-F	100	0.39	0.26	–	–	–	1.94	–	–	–	–	–	–	–	–
	ORCHARDGRASS *Dactylis glomerata*																
196	FRESH, ALL ANALYSES	**A-F**	**26**	**0.09**	**0.05**	**0.03**	–	**0.06**	**0.74**	–	**0.055**	**2.5**	–	**0.003**	**28.5**	–	**5.3**
		M-F	100	0.37	0.18	0.13	–	0.24	2.88	–	0.212	9.8	–	0.010	110.4	–	20.6
197	IMMATURE, FRESH	**A-F**	**23**	**0.13**	**0.12**	–	**0.02**	**0.06**	**0.73**	–	–	**1.6**	–	–	**7.1**	–	–
		M-F	100	0.57	0.54	–	0.08	0.27	3.21	–	–	7.0	–	–	31.4	–	–
198	EARLY BLOOM, FRESH	**A-F**	**24**	**0.06**	**0.09**	**0.01**	–	**0.07**	**0.80**	**0.06**	–	**7.8**	–	**0.019**	**24.5**	–	–
		M-F	100	0.25	0.39	0.04	–	0.31	3.38	0.26	–	33.1	–	0.079	104.1	–	–
199	MIDBLOOM, FRESH	**A-F**	**27**	**0.17**	**0.05**	**0.07**	–	**0.09**	**0.57**	–	**0.028**	**13.7**	–	**0.002**	**37.2**	–	**6.9**
		M-F	100	0.60	0.17	0.26	–	0.33	2.09	–	0.102	50.1	–	0.007	135.9	–	25.1
200	MILK STAGE, FRESH	**A-F**	**34**	–	–	–	–	–	–	–	–	–	–	–	–	–	–
		M-F	100	–	–	–	–	–	–	–	–	–	–	–	–	–	–
	PANGOLAGRASS *Digitaria decumbens*																
201	FRESH	**A-F**	**20**	**0.08**	**0.05**	–	–	**0.04**	**0.29**	–	–	–	–	–	–	–	–
		M-F	100	0.38	0.22	–	–	0.18	1.43	–	–	–	–	–	–	–	–
	PANICUM *Panicum* spp																
202	FRESH	**A-F**	**29**	**0.14**	**0.05**	–	–	**0.10**	**0.93**	–	–	–	–	–	–	–	–
		M-F	100	0.47	0.18	–	–	0.36	3.19	–	–	–	–	–	–	–	–
	PARAGRASS *Brachiaria mutica*																
203	FRESH	**A-F**	**26**	**0.10**	**0.10**	–	–	–	**0.42**	–	–	–	–	–	–	–	–
		M-F	100	0.40	0.40	–	–	–	1.58	–	–	–	–	–	–	–	–
	PARSNIP, GARDEN *Pastinaca sativa*																
204	LEAVES, FRESH	**A-F**	**18**	–	–	–	–	–	–	–	–	–	–	–	–	–	–
		M-F	100	–	–	–	–	–	–	–	–	–	–	–	–	–	–
	PASPALUM *Paspalum* spp																
205	FRESH	**A-F**	**27**	–	–	–	–	–	–	–	–	–	**0.047**	–	–	–	–
		M-F	100	–	–	–	–	–	–	–	–	–	0.172	–	–	–	–
	PEA *Pisum* spp																
206	PODS, FRESH	**A-F**	**13**	**0.19**	**0.03**	–	–	–	–	–	–	–	–	–	–	–	–
		M-F	100	1.47	0.23	–	–	–	–	–	–	–	–	–	–	–	–
	PEA, FIELD *Pisum sativum, arvense*																
207	FRESH	**A-F**	**18**	**0.22**	**0.04**	–	–	**0.04**	**0.27**	–	**0.028**	–	–	**0.008**	**15.4**	–	–
		M-F	100	1.21	0.23	–	–	0.22	1.50	–	0.150	–	–	0.040	85.1	–	–
	PEA-OATS *Pisum spp-Avena sativa*																
208	FRESH	**A-F**	**23**	**0.17**	**0.07**	–	–	–	**0.38**	–	–	–	–	–	–	–	–
		M-F	100	0.75	0.31	–	–	–	1.67	–	–	–	–	–	–	–	–
	PEARL MILLET (MILLET, PEARL) *Pennisetum glaucum*																
209	FRESH	**A-F**	**21**	–	–	–	–	–	–	–	–	–	–	–	–	–	–
		M-F	100	–	–	–	–	–	–	–	–	–	–	–	–	–	–
	PEAVINE *Lathyrus* spp																
210	FRESH	**A-F**	**17**	–	–	–	–	–	–	–	–	–	–	–	–	–	–
		M-F	100	–	–	–	–	–	–	–	–	–	–	–	–	–	–
	PIGEON PEA *Cajanus cajan*																
211	FRESH	**A-F**	**41**	–	–	–	–	–	–	–	–	–	–	–	–	–	–
		M-F	100	–	–	–	–	–	–	–	–	–	–	–	–	–	–

	Fat-Soluble Vitamins					Water-Soluble Vitamins								
Entry Number	A (1 mg Carotene = 1667 IU Vit A)	Carotene (Provitamin A)	D	E	K	B-12	Biotin	Choline	Folacin Acid Acid)	Niacin	Pantothenic Acid (B-3)	(Pyri-doxine) B-6	Ribo-flavin (B-2)	Thiamin (B-1)
	IU/g	ppm or mg/kg	IU/kg	ppm or mg/kg	ppm or mg/kg	ppm or mg/kg	ppm or mg/kg	ppm or mg/kg	ppm or mg/kg	ppm or mg/kg	ppm or mg/kg	ppm or mg/kg	ppm or mg/kg	ppm or mg/kg
188	–	–	–	–	–	–	–	–	–	–	–	–	–	–
	–	–	–	–	–	–	–	–	–	–	–	–	–	–
189	–	–	–	–	–	–	–	–	–	–	–	–	–	–
	–	–	–	–	–	–	–	–	–	–	–	–	–	–
190	**150.0**	**90.0**	–	–	–	–	–	–	–	–	–	–	–	–
	934.4	560.5	–	–	–	–	–	–	–	–	–	–	–	–
191	–	–	–	–	–	–	–	–	–	–	–	–	–	–
	–	–	–	–	–	–	–	–	–	–	–	–	–	–
192	–	–	–	–	–	–	–	–	–	–	–	–	–	–
	–	–	–	–	–	–	–	–	–	–	–	–	–	–
193	–	–	–	–	–	–	–	–	–	–	–	–	–	–
	–	–	–	–	–	–	–	–	–	–	–	–	–	–
194	–	–	–	–	–	–	–	–	–	–	–	–	–	–
	–	–	–	–	–	–	–	–	–	–	–	–	–	–
195	–	–	–	–	–	–	–	–	–	–	–	–	–	–
	–	–	–	–	–	–	–	–	–	–	–	–	–	–
196	**137.1**	**82.2**	–	**112.3**	–	–	–	–	–	–	–	–	–	**1.9**
	531.8	319.0	–	435.6	–	–	–	–	–	–	–	–	–	7.3
197	**236.1**	**141.6**	–	–	–	–	–	–	–	–	–	–	–	–
	1044.6	626.6	–	–	–	–	–	–	–	–	–	–	–	–
198	–	–	–	–	–	–	–	–	–	–	–	–	–	–
	–	–	–	–	–	–	–	–	–	–	–	–	–	–
199	–	–	–	–	–	–	–	–	–	–	–	–	–	–
	–	–	–	–	–	–	–	–	–	–	–	–	–	–
200	–	–	–	–	–	–	–	–	–	–	–	–	–	–
	–	–	–	–	–	–	–	–	–	–	–	–	–	–
201	–	–	–	–	–	–	–	–	–	–	–	–	–	–
	–	–	–	–	–	–	–	–	–	–	–	–	–	–
202	–	–	–	–	–	–	–	–	–	–	–	–	–	–
	–	–	–	–	–	–	–	–	–	–	–	–	–	–
203	–	–	–	–	–	–	–	–	–	–	–	–	–	–
	–	–	–	–	–	–	–	–	–	–	–	–	–	–
204	**69.6**	**41.7**	–	–	–	–	–	–	–	–	–	–	**2.1**	–
	386.6	231.9	–	–	–	–	–	–	–	–	–	–	11.9	–
205	–	–	–	–	–	–	–	–	–	–	–	–	–	–
	–	–	–	–	–	–	–	–	–	–	–	–	–	–
206	–	–	–	–	–	–	**0.03**	–	–	**4**	**1.3**	–	**0.9**	**0.3**
	–	–	–	–	–	–	0.22	–	–	32	10.1	–	6.8	2.6
207	–	–	–	–	–	–	–	–	–	–	–	–	–	–
	–	–	–	–	–	–	–	–	–	–	–	–	–	–
208	–	–	–	–	–	–	–	–	–	–	–	–	–	–
	–	–	–	–	–	–	–	–	–	–	–	–	–	–
209	**63.0**	**37.8**	–	–	–	–	–	–	–	–	–	–	–	–
	304.2	182.5	–	–	–	–	–	–	–	–	–	–	–	–
210	**29.7**	**17.8**	–	–	–	–	–	–	–	–	–	–	**2.2**	–
	170.1	102.1	–	–	–	–	–	–	–	–	–	–	12.6	–
211	–	–	–	–	–	–	–	–	–	–	–	–	–	–
	–	–	–	–	–	–	–	–	–	–	–	–	–	–

(Continued)

TABLE V-1E PASTURE AND RANGE PLANTS, COMPOSITION OF FEEDS, DATA EXPRESSED **AS-FED** AND **MOISTURE-FREE**—*(Continued)*

Entry Number	Feed Name Description	International Feed Number	Moisture Basis: A-F (as-fed) or M-F (moisture-free)	Proximate Analysis									Digestible Protein		
				Dry Matter	Ash	Crude Fiber	Neutral Det. Fib. (NDF)	Acid Det. Fib. (ADF)	Lignin	Ether Extract (Fat)	N-Free Extract	Crude Protein	Ruminant	Swine	Horse
				%	%	%	%	%	%	%	%	%	%	%	%
	PINEAPPLE *Ananas comosus*														
212	LEAVES WITH STEMS, FRESH	2-03-720	A-F	18	1.3	4.4	–	–	–	0.4	10.2	1.5	0.9	1.1	–
			M-F	100	7.1	24.8	–	–	–	2.3	57.2	8.6	4.9	6.4	–
	PRICKLY PEAR *Opuntia* spp														
213	FRESH	2-01-061	A-F	17	3.4	2.2	5.3	–	–	0.3	9.9	0.9	0.4	0.5	–
			M-F	100	20.5	13.2	31.6	–	–	1.9	59.2	5.1	2.3	3.2	–
	QUACKGRASS *Agropyron repens*														
214	FRESH	2-03-829	A-F	27	2.3	8.2	–	–	–	1.1	11.8	3.6	2.8	2.5	–
			M-F	100	8.6	30.2	–	–	–	4.1	43.6	13.5	10.5	9.1	–
	RABBITBRUSH *Chrysothamnus* spp														
215	BROWSE, FRESH	2-03-835	A-F	38	3.0	9.0	–	–	–	2.1	18.8	5.1	3.5	3.4	–
			M-F	100	8.0	23.8	–	–	–	5.4	49.4	13.4	9.3	9.0	–
	RABBITBRUSH, SMALL (YELLOWBRUSH) *Chrysothamnus stenophyllus*														
216	BROWSE, FRESH	2-03-849	A-F	40	3.2	9.5	–	–	–	2.2	19.8	5.4	–	–	–
			M-F	100	8.0	23.8	–	–	–	5.4	49.4	13.4	–	–	–
217	BROWSE, STEM CURED, FRESH	2-07-997	A-F	85	7.7	–	–	–	–	8.5	–	4.8	1.9	2.4	–
			M-F	100	9.0	–	–	–	–	10.0	–	5.7	2.2	2.9	–
	RAGWEED *Ambrosia* spp														
218	FRESH	2-03-853	A-F	27	2.4	6.7	–	–	–	1.9	10.3	5.2	3.9	3.7	–
			M-F	100	9.2	25.1	–	–	–	7.3	38.9	19.5	14.8	13.9	–
	RAMIE *Boehmeria nivea*														
219	FRESH	2-03-859	A-F	9	1.4	2.4	–	–	–	0.3	3.0	1.8	1.5	1.3	–
			M-F	100	15.3	26.9	–	–	–	3.1	34.1	20.7	17.4	14.9	–
	RAPE *Brassica napus*														
220	FRESH	2-03-867	A-F	17	2.1	2.4	–	–	–	0.6	8.5	2.9	2.4	2.1	–
			M-F	100	12.6	14.7	–	–	–	3.8	51.2	17.6	14.5	12.4	–
	RED CLOVER–SEE CLOVER, RED														
	REDTOP *Agrostis alba*														
221	FULL BLOOM, FRESH[1]	2-03-891	A-F	26	1.8	6.6	16.6	–	–	0.9	14.8	2.1	1.2	1.3	–
			M-F	100	7.0	25.1	64.0	–	–	3.5	56.3	8.1	4.4	4.8	–
	REED CANARYGRASS *Phalaris arundinacea*														
222	FRESH	2-01-113	A-F	23	2.3	5.6	10.6	6.5	1.0	0.9	10.1	3.9	2.9	2.7	–
			M-F	100	10.2	24.4	46.4	28.3	4.3	4.1	44.4	17.0	12.5	11.9	–
	REEDGRASS, BLUEJOINT *Calamagrostis canadensis*														
223	FRESH	2-03-904	A-F	45	4.1	15.2	–	–	–	1.2	20.0	4.1	2.4	2.5	–
			M-F	100	9.2	34.1	–	–	–	2.7	44.8	9.2	5.4	5.7	–
	RESCUEGRASS (BROMEGRASS, RESCUE) *Bromus catharticus*														
224	FRESH	2-08-361	A-F	29	4.0	6.7	–	–	–	1.0	12.2	5.0	3.7	3.5	–
			M-F	100	13.8	23.2	–	–	–	3.5	42.2	17.3	12.8	12.2	–
	RHODESGRASS *Chloris gayana*														
225	FRESH	2-03-916	A-F	26	3.1	9.9	–	–	–	0.4	11.0	2.0	1.2	1.1	–
			M-F	100	11.6	37.4	–	–	–	1.7	41.7	7.6	4.4	4.2	–
	RICEGRASS, INDIAN *Oryzopsis hymenoides*														
226	OVERRIPE, FRESH	2-03-944	A-F	48	–	–	–	–	–	–	–	–	–	–	–
			M-F	100	–	–	–	–	–	–	–	–	–	–	–
	RUSH *Juncus* spp														
227	FRESH	2-03-965	A-F	31	2.2	9.8	–	–	–	0.6	15.1	3.4	2.2	2.2	–
			M-F	100	7.1	31.5	–	–	–	1.9	48.6	10.9	7.0	7.1	–
	RUSSIAN THISTLE (TUMBLING TUMBLEWEED) *Salsola kali, tenuifolia*														
228	FRESH	2-03-990	A-F	30	6.6	9.3	11.1	6.6	1.0	0.6	9.0	4.6	3.3	3.1	–
			M-F	100	21.9	30.9	37.0	22.1	3.4	2.0	30.1	15.2	10.9	10.5	–
	RUTABAGA *Brassica napus, napobrassica*														
229	TOPS WITH CROWNS, FRESH	2-03-996	A-F	11	2.2	1.5	–	–	–	0.5	4.7	2.0	1.7	1.4	–
			M-F	100	19.9	14.0	–	–	–	4.6	42.9	18.6	15.9	13.2	–
	RYE *Secale cereale*														
230	FRESH	2-04-018	A-F	20	1.9	5.9	–	–	–	0.8	8.1	3.6	2.8	2.5	–
			M-F	100	9.3	29.1	–	–	–	3.9	40.0	17.6	13.9	12.4	–
	RYEGRASS *Lolium* spp														
231	FRESH	2-04-062	A-F	24	1.8	7.0	–	–	–	0.7	12.1	2.5	1.5	1.6	–
			M-F	100	7.6	29.2	–	–	–	2.8	50.3	10.2	6.3	6.4	–
	RYEGRASS, ITALIAN *Lolium muliflorum*														
232	FRESH	2-04-073	A-F	23	3.9	4.7	–	–	–	0.9	9.0	4.0	3.0	2.9	–
			M-F	100	17.4	20.9	–	–	–	4.1	39.8	17.9	13.3	12.6	–

Entry Number	TDN Ruminant	TDN Swine	TDN Horse	Digestible Energy Ruminant		Digestible Energy Swine		Digestible Energy Horse		Metabolizable Energy Ruminant		Metabolizable Energy Swine		Metabolizable Energy Poultry MEn		Metabolizable Energy Horse		Net Energy Ruminant NEm		Net Energy Ruminant NEg		Net Energy Lactating Cows NElc	
	%	%	%	Mcal		kcal		Mcal		Mcal		kcal		kcal		Mcal		Mcal		Mcal		Mcal	
				lb	kg	lb	kg	lb	kg	lb	kg	lb	kg	lb	kg	lb	kg	lb	kg	lb	kg	lb	kg
212	9	13	9	0.21	0.46	–	–	0.17	0.37	0.17	0.38	–	–	–	–	0.14	0.30	0.12	0.26	0.07	0.15	0.12	0.26
	53	75	49	1.17	2.57	–	–	0.93	2.06	0.98	2.15	–	–	–	–	0.77	1.69	0.66	1.45	0.39	0.86	0.66	1.45
213	10	10	6	0.20	0.44	–	–	0.13	0.28	0.16	0.36	–	–	–	–	0.10	0.23	0.10	0.21	0.05	0.12	0.10	0.21
	58	58	38	1.18	2.60	–	–	0.75	1.65	0.98	2.15	–	–	–	–	0.61	1.35	0.58	1.28	0.32	0.71	0.57	1.26
214	18	–	–	0.35	0.77	–	–	–	–	0.30	0.66	–	–	–	–	–	–	0.17	0.37	0.10	0.22	0.17	0.38
	68	–	–	1.30	2.87	–	–	–	–	1.11	2.45	–	–	–	–	–	–	0.63	1.38	0.36	0.80	0.64	1.40
215	24	–	–	0.49	1.07	–	–	–	–	0.41	0.91	–	–	–	–	–	–	0.26	0.58	0.16	0.35	0.26	0.57
	64	–	–	1.28	2.82	–	–	–	–	1.09	2.40	–	–	–	–	–	–	0.69	1.52	0.42	0.92	0.68	1.51
216	–	–	–	–	–	–	–	–	–	–	–	–	–	–	–	–	–	–	–	–	–	–	–
	–	–	–	–	–	–	–	–	–	–	–	–	–	–	–	–	–	–	–	–	–	–	–
217	38	–	–	0.82	1.80	–	–	–	–	0.58	1.28	–	–	–	–	–	–	–	–	–	–	–	–
	44	–	–	0.96	2.12	–	–	–	–	0.68	1.50	–	–	–	–	–	–	–	–	–	–	–	–
218	–	–	–	–	–	–	–	–	–	–	–	–	–	–	–	–	–	–	–	–	–	–	–
	–	–	–	–	–	–	–	–	–	–	–	–	–	–	–	–	–	–	–	–	–	–	–
219	6	–	–	0.12	0.25	–	–	–	–	0.10	0.22	–	–	–	–	–	–	0.05	0.12	0.03	0.07	0.05	0.12
	69	–	–	1.29	2.84	–	–	–	–	1.10	2.42	–	–	–	–	–	–	0.59	1.30	0.33	0.73	0.61	1.34
220	13	–	–	0.25	0.54	–	–	–	–	0.21	0.47	–	–	–	–	–	–	0.12	0.26	0.07	0.16	0.12	0.26
	79	–	–	1.47	3.24	–	–	–	–	1.28	2.83	–	–	–	–	–	–	0.71	1.57	0.44	0.97	0.70	1.54
221	16	–	–	0.33	0.72	–	–	–	–	0.28	0.61	–	–	–	–	–	–	0.16	0.36	0.10	0.21	0.17	0.37
	62	–	–	1.24	2.73	–	–	–	–	1.04	2.30	–	–	–	–	–	–	0.62	1.38	0.36	0.80	0.64	1.40
222	14	–	–	0.29	0.64	–	–	–	–	0.24	0.54	–	–	–	–	–	–	0.15	0.34	0.09	0.20	0.15	0.34
	61	–	–	1.26	2.78	–	–	–	–	1.07	2.36	–	–	–	–	–	–	0.67	1.47	0.40	0.88	0.67	1.47
223	25	–	–	0.51	1.12	–	–	–	–	0.42	0.93	–	–	–	–	–	–	0.25	0.55	0.13	0.29	0.26	0.57
	56	–	–	1.14	2.51	–	–	–	–	0.94	2.08	–	–	–	–	–	–	0.55	1.22	0.30	0.66	0.58	1.29
224	20	–	–	0.38	0.84	–	–	–	–	0.33	0.72	–	–	–	–	–	–	0.18	0.40	0.10	0.23	0.18	0.40
	70	–	–	1.32	2.91	–	–	–	–	1.13	2.48	–	–	–	–	–	–	0.62	1.37	0.36	0.79	0.63	1.40
225	15	15	–	0.31	0.68	–	–	–	–	0.26	0.58	–	–	–	–	–	–	0.15	0.33	0.08	0.18	0.15	0.33
	58	57	–	1.16	2.56	–	–	–	–	0.99	2.17	–	–	–	–	–	–	0.57	1.25	0.31	0.68	0.56	1.24
226	–	–	–	–	–	–	–	–	–	–	–	–	–	–	–	–	–	–	–	–	–	–	–
	–	–	–	–	–	–	–	–	–	–	–	–	–	–	–	–	–	–	–	–	–	–	–
227	19	–	–	0.38	0.85	–	–	–	–	0.32	0.71	–	–	–	–	–	–	0.19	0.42	0.11	0.24	0.20	0.43
	62	–	–	1.23	2.72	–	–	–	–	1.04	2.29	–	–	–	–	–	–	0.62	1.36	0.35	0.78	0.63	1.39
228	13	–	–	0.28	0.61	–	–	–	–	0.22	0.48	–	–	–	–	–	–	0.12	0.27	0.05	0.11	0.15	0.32
	43	–	–	0.92	2.02	–	–	–	–	0.72	1.59	–	–	–	–	–	–	0.41	0.91	0.17	0.37	0.48	1.07
229	8	–	–	0.15	0.32	–	–	–	–	0.12	0.27	–	–	–	–	–	–	0.07	0.15	0.04	0.09	0.07	0.15
	72	–	–	1.33	2.94	–	–	–	–	1.14	2.52	–	–	–	–	–	–	0.62	1.36	0.35	0.78	0.63	1.38
230	14	–	–	0.27	0.60	–	–	–	–	0.23	0.51	–	–	–	–	–	–	0.12	0.27	0.07	0.15	0.13	0.28
	67	–	–	1.34	2.96	–	–	–	–	1.15	2.54	–	–	–	–	–	–	0.61	1.34	0.35	0.77	0.62	1.37
231	15	–	–	0.30	0.66	–	–	–	–	0.25	0.56	–	–	–	–	–	–	0.15	0.34	0.09	0.20	0.16	0.34
	63	–	–	1.25	2.75	–	–	–	–	1.06	2.33	–	–	–	–	–	–	0.64	1.40	0.37	0.82	0.64	1.42
232	14	–	–	0.28	0.61	–	–	–	–	0.23	0.51	–	–	–	–	–	–	0.14	0.31	0.08	0.18	0.14	0.32
	61	–	–	1.23	2.70	–	–	–	–	1.03	2.28	–	–	–	–	–	–	0.62	1.38	0.36	0.80	0.64	1.40

(Continued)

TABLE V-1E PASTURE AND RANGE PLANTS, MINERAL AND VITAMIN COMPOSITION OF FEEDS, DATA EXPRESSED **AS-FED** AND **MOISTURE-FREE**—*(Continued)*

Entry Number	Feed Name Description	Moisture Basis: A-F (as-fed) or M-F (moisture-free)	Dry Matter	Macro Minerals							Micro Minerals						
				Calcium (Ca)	Phosphorus (P)	Sodium (Na)	Chlorine (Cl)	Magnesium (Mg)	Potassium (K)	Sulfur (S)	Cobalt (Co)	Copper (Cu)	Iodine (I)	Iron (Fe)	Manganese (Mn)	Selenium (Se)	Zinc (Zn)
			%	%	%	%	%	%	%	%	ppm or mg/kg	ppm or mg/kg	ppm or mg/kg	%	ppm or mg/kg	ppm or mg/kg	ppm or mg/kg
	PINEAPPLE *Ananas comosus*																
212	LEAVES WITH STEMS, FRESH	A-F	18	0.08	0.03	–	–	0.06	0.18	–	–	1.1	–	0.015	136.2	–	4.3
		M-F	100	0.46	0.16	–	–	0.34	1.00	–	–	6.3	–	0.080	765.0	–	23.9
	PRICKLY PEAR *Opuntia* spp																
213	FRESH	A-F	17	1.47	0.02	–	–	0.23	0.37	–	–	–	–	–	–	–	–
		M-F	100	8.78	0.12	–	–	1.38	2.21	–	–	–	–	–	–	–	–
	QUACKGRASS *Agropyron repens*																
214	FRESH	A-F	27	0.09	0.08	–	–	–	–	–	–	–	–	–	–	–	–
		M-F	100	0.35	0.30	–	–	–	–	–	–	–	–	–	–	–	–
	RABBITBRUSH *Chrysothamnus* spp																
215	BROWSE, FRESH	A-F	38	–	–	–	–	–	–	0.12	0.042	–	–	–	–	–	–
		M-F	100	–	–	–	–	–	–	0.32	0.111	–	–	–	–	–	–
	RABBITBRUSH, SMALL (YELLOWBRUSH) *Chrysothamnus stenophyllus*																
216	BROWSE, FRESH	A-F	40	0.88	0.04	–	–	–	–	–	–	–	–	–	–	–	–
		M-F	100	2.21	0.11	–	–	–	–	–	–	–	–	–	–	–	–
217	BROWSE, STEM CURED, FRESH	A-F	85	1.58	0.07	–	–	–	–	–	–	–	–	–	–	–	–
		M-F	100	1.86	0.08	–	–	–	–	–	–	–	–	–	–	–	–
	RAGWEED *Ambrosia* spp																
218	FRESH	A-F	27	–	–	–	–	–	–	–	–	–	–	–	–	–	–
		M-F	100	–	–	–	–	–	–	–	–	–	–	–	–	–	–
	RAMIE *Boehmeria nivea*																
219	FRESH	A-F	9	0.41	0.02	–	–	–	–	–	–	–	–	–	–	–	–
		M-F	100	4.60	0.24	–	–	–	–	–	–	–	–	–	–	–	–
	RAPE *Brassica napus*																
220	FRESH	A-F	17	0.25	0.07	–	–	0.01	0.56	0.11	–	1.4	–	0.004	7.7	–	–
		M-F	100	1.47	0.43	–	–	0.06	3.37	0.68	–	8.1	–	0.019	46.0	–	–
	RED CLOVER–SEE CLOVER, RED																
	REDTOP *Agrostis alba*																
221	FULL BLOOM, FRESH	A-F	26	0.16	0.10	0.01	–	0.07	0.62	0.04	–	–	–	0.006	–	–	–
		M-F	100	0.62	0.37	0.05	–	0.25	2.35	0.16	–	–	–	0.020	–	–	–
	REED CANARYGRASS *Phalaris arundinacea*																
222	FRESH	A-F	23	0.08	0.08	–	–	–	0.83	–	–	–	–	–	–	–	–
		M-F	100	0.36	0.33	–	–	–	3.64	–	–	–	–	–	–	–	–
	REEDGRASS, BLUEJOINT *Calamagrostis canadensis*																
223	FRESH	A-F	45	–	0.10	–	–	–	–	–	–	–	–	–	–	–	–
		M-F	100	–	0.22	–	–	–	–	–	–	–	–	–	–	–	–
	RESCUEGRASS (BROMEGRASS, RESCUE) *Bromus catharticus*																
224	FRESH	A-F	29	0.15	0.08	–	–	–	–	–	–	–	–	–	–	–	–
		M-F	100	0.52	0.28	–	–	–	–	–	–	–	–	–	–	–	–
	RHODESGRASS *Chloris gayana*																
225	FRESH	A-F	26	0.13	0.10	–	–	–	0.61	–	–	–	–	–	–	–	–
		M-F	100	0.51	0.39	–	–	–	2.29	–	–	–	–	–	–	–	–
	RICEGRASS, INDIAN *Oryzopsis hymenoides*																
226	OVERRIPE, FRESH	A-F	48	0.28	0.02	–	–	0.07	–	0.07	–	–	–	–	–	–	–
		M-F	100	0.58	0.05	–	–	0.15	–	0.14	–	–	–	–	–	–	–
	RUSH *Jancus* spp																
227	FRESH	A-F	31	0.09	0.08	–	–	–	0.73	–	–	–	–	–	–	–	–
		M-F	100	0.30	0.25	–	–	–	2.36	–	–	–	–	–	–	–	–
	RUSSIAN THISTLE (TUMBLING TUMBLEWEED) *Salsola kali, tenuifolia*																
228	FRESH	A-F	30	0.74	0.08	–	–	0.26	1.72	0.05	0.051	5.8	–	–	10.0	–	–
		M-F	100	2.47	0.27	–	–	0.88	5.73	0.17	0.170	19.2	–	–	33.3	–	–
	RUTABAGA *Brassica napus, napobrassica*																
229	TOPS WITH CROWNS, FRESH	A-F	11	–	–	–	–	–	–	–	–	–	–	–	–	–	–
		M-F	100	–	–	–	–	–	–	–	–	–	–	–	–	–	–
	RYE *Secale cereale*																
230	FRESH	A-F	20	0.09	0.08	0.01	–	0.06	0.69	–	–	–	–	–	–	–	–
		M-F	100	0.45	0.38	0.07	–	0.31	3.40	–	–	–	–	–	–	–	–
	RYEGRASS *Lolium* spp																
231	FRESH	A-F	24	–	–	–	–	–	–	–	–	–	–	–	–	–	–
		M-F	100	–	–	–	–	–	–	–	–	–	–	–	–	–	–
	RYEGRASS, ITALIAN *Lolium muliflorum*																
232	FRESH	A-F	23	0.15	0.09	0.00	–	0.08	0.45	0.02	–	–	–	0.023	–	–	–
		M-F	100	0.65	0.41	0.01	–	0.35	2.00	0.10	–	–	–	0.101	–	–	–

Entry Number	Fat-Soluble Vitamins					Water-Soluble Vitamins								
	A (1 mg Carotene = 1667 IU Vit A)	Carotene (Provitamin A)	D	E	K	B-12	Biotin	Choline	Folacin (Folic Acid)	Niacin	Pantothenic Acid (B-3)	(Pyridoxine) B-6	Riboflavin (B-2)	Thiamin (B-1)
	IU/g	ppm or mg/kg	IU/kg	ppm or mg/kg	ppm or mg/kg	ppm or mg/kg	ppm or mg/kg	ppm or mg/kg	ppm or mg/kg	ppm or mg/kg	ppm or mg/kg	ppm or mg/kg	ppm or mg/kg	ppm or mg/kg
212	–	–	–	–	–	–	–	–	–	–	–	–	–	–
	–	–	–	–	–	–	–	–	–	–	–	–	–	–
213	–	–	–	–	–	–	–	–	–	–	–	–	–	–
	–	–	–	–	–	–	–	–	–	–	–	–	–	–
214	55.1	33.1	–	–	–	–	–	–	–	–	–	–	–	–
	204.3	122.6	–	–	–	–	–	–	–	–	–	–	–	–
215	–	–	–	–	–	–	–	–	–	–	–	–	–	–
	–	–	–	–	–	–	–	–	–	–	–	–	–	–
216	–	–	–	–	–	–	–	–	–	–	–	–	–	–
	–	–	–	–	–	–	–	–	–	–	–	–	–	–
217	–	–	–	–	–	–	–	–	–	–	–	–	–	–
	–	–	–	–	–	–	–	–	–	–	–	–	–	–
218	–	–	–	–	–	–	–	–	–	–	–	–	–	–
	–	–	–	–	–	–	–	–	–	–	–	–	–	–
219	45.2	27.1	–	–	–	–	–	169	–	6	2.4	–	1.5	0.6
	508.3	304.9	–	–	–	–	–	1900	–	67	26.7	–	16.5	7.3
220	–	–	–	–	–	–	–	–	–	–	–	–	–	–
	–	–	–	–	–	–	–	–	–	–	–	–	–	–
221	66.9	40.1	–	–	–	–	–	–	–	–	–	–	–	–
	254.4	152.6	–	–	–	–	–	–	–	–	–	–	–	–
222	–	–	–	–	–	–	–	–	–	–	–	–	–	–
	–	–	–	–	–	–	–	–	–	–	–	–	–	–
223	–	–	–	–	–	–	–	–	–	–	–	–	–	–
	–	–	–	–	–	–	–	–	–	–	–	–	–	–
224	–	–	–	–	–	–	–	–	–	–	–	–	–	–
	–	–	–	–	–	–	–	–	–	–	–	–	–	–
225	–	–	–	–	–	–	–	–	–	–	–	–	–	–
	–	–	–	–	–	–	–	–	–	–	–	–	–	–
226	0.4	0.2	–	–	–	–	–	–	–	–	–	–	–	–
	0.7	0.4	–	–	–	–	–	–	–	–	–	–	–	–
227	–	–	–	–	–	–	–	–	–	–	–	–	–	–
	–	–	–	–	–	–	–	–	–	–	–	–	–	–
228	2.3	1.4	–	–	–	–	–	–	–	–	–	–	–	–
	7.7	4.6	–	–	–	–	–	–	–	–	–	–	–	–
229	–	–	–	–	–	–	–	–	–	–	–	–	–	–
	–	–	–	–	–	–	–	–	–	–	–	–	–	–
230	115.0	69.0	–	–	–	–	–	–	–	–	–	–	–	–
	571.1	342.6	–	–	–	–	–	–	–	–	–	–	–	–
231	–	–	–	–	–	–	–	–	–	–	–	–	–	–
	–	–	–	–	–	–	–	–	–	–	–	–	–	–
232	–	–	–	–	–	–	–	–	–	–	–	–	–	–
	–	–	–	–	–	–	–	–	–	–	–	–	–	–

(Continued)

TABLE V-1E PASTURE AND RANGE PLANTS, COMPOSITION OF FEEDS, DATA EXPRESSED **AS-FED** AND **MOISTURE-FREE**—*(Continued)*

Entry Number	Feed Name Description	International Feed Number	Moisture Basis: A-F (as-fed) or M-F (moisture-free)	Proximate Analysis: Dry Matter	Ash	Crude Fiber	Neutral Det. Fib. (NDF)	Acid Det. Fib. (ADF)	Lignin	Ether Extract (Fat)	N-Free Extract	Crude Protein	Digestible Protein: Ruminant	Swine	Horse
				%	%	%	%	%	%	%	%	%	%	%	%
	RYEGRASS, ITALIAN (Continued)														
233	EARLY BLOOM, FRESH	2-04-071	A-F	**35**	**2.9**	**10.6**	**–**	**–**	**–**	**0.3**	**19.5**	**2.0**	**0.9**	**1.0**	**–**
			M-F	100	8.1	30.1	–	–	–	0.8	55.2	5.8	2.5	2.9	–
	SACATON (ALKALI) *Sporobolus airoides*														
234	MIDBLOOM, FRESH	2-05-601	A-F	**40**	**3.7**	**14.5**	**–**	**–**	**–**	**0.7**	**18.6**	**2.5**	**1.1**	**1.3**	**–**
			M-F	100	9.2	36.2	–	–	–	1.8	46.5	6.3	2.8	3.3	–
	SAGE, BLACK *Salvia mellifera*														
235	BROWSE, STEM CURED, FRESH	2-05-564	A-F	**65**	**3.6**	**–**	**–**	**–**	**–**	**7.0**	**–**	**5.5**	**2.9**	**3.3**	**–**
			M-F	100	5.5	–	–	–	–	10.7	–	8.5	4.5	5.1	–
	SAGEBRUSH, BIG *Artemisia tridentata*														
236	BROWSE, STEM CURED, FRESH	2-07-992	A-F	**65**	**4.3**	**–**	**–**	**–**	**–**	**6.4**	**–**	**6.1**	**3.2**	**3.7**	**–**
			M-F	100	6.6	–	–	–	–	9.8	–	9.3	4.9	5.8	–
	SAGEBRUSH, BUD *Artemisia spinescens*														
237	BROWSE, IMMATURE, FRESH	2-07-991	A-F	**23**	**4.9**	**–**	**–**	**–**	**–**	**1.1**	**–**	**4.0**	**3.1**	**2.8**	**–**
			M-F	100	21.4	–	–	–	–	4.9	–	17.3	13.7	12.2	–
238	BROWSE, FRESH	2-04-125	A-F	**27**	**5.8**	**6.1**	**–**	**–**	**–**	**0.7**	**9.6**	**4.7**	**–**	**–**	**3.3**
			M-F	100	21.6	22.7	–	–	–	2.5	35.7	17.5	–	–	12.4
	SAGEBRUSH, FRINGED *Artemisia frigida*														
239	BROWSE, MIDBLOOM, FRESH	2-04-129	A-F	**43**	**2.8**	**14.3**	**–**	**–**	**–**	**0.9**	**21.0**	**4.0**	**2.4**	**2.5**	**–**
			M-F	100	6.5	33.2	–	–	–	2.0	48.9	9.4	5.6	5.8	–
240	BROWSE, MATURE, FRESH	2-04-130	A-F	**60**	**10.3**	**19.1**	**27.4**	**21.1**	**5.9**	**2.0**	**24.4**	**4.3**	**2.1**	**2.4**	**–**
			M-F	100	17.1	31.8	45.6	35.1	9.8	3.4	40.6	7.1	3.5	4.0	–
	SAINFOIN *Onobrychis* spp														
241	FRESH	2-04-146	A-F	**23**	**1.9**	**5.8**	**–**	**–**	**–**	**0.7**	**10.3**	**3.9**	**2.9**	**2.8**	**–**
			M-F	100	8.4	25.5	–	–	–	3.3	45.4	17.4	12.9	12.2	–
	SAINT AUGUSTINEGRASS *Stenotaphrum secundatum*														
242	FRESH	2-08-090	A-F	**18**	**1.3**	**5.4**	**–**	**–**	**–**	**0.5**	**8.2**	**2.7**	**1.9**	**1.9**	**–**
			M-F	100	7.2	29.8	–	–	–	2.8	45.3	14.9	10.6	10.3	–
	SALTBUSH, NUTTALL *Atriplex nuttallii*														
243	BROWSE, STEM CURED, FRESH	2-07-993	A-F	**55**	**11.8**	**–**	**–**	**–**	**–**	**1.2**	**–**	**4.0**	**1.8**	**2.2**	**–**
			M-F	100	21.5	–	–	–	–	2.2	–	7.2	3.4	4.1	–
	SALTBUSH, SHADSCALE *Atriplex confertifolia*														
244	BROWSE, STEM CURED, FRESH	2-05-565	A-F	**80**	**18.9**	**–**	**–**	**–**	**–**	**2.0**	**–**	**6.1**	**3.5**	**3.6**	**–**
			M-F	100	23.6	–	–	–	–	2.5	–	7.7	4.4	4.4	–
	SALTGRASS *Distichlis* spp														
245	FRESH	2-04-170	A-F	**74**	**5.6**	**22.5**	**–**	**–**	**–**	**1.3**	**40.1**	**4.8**	**2.2**	**2.6**	**–**
			M-F	100	7.5	30.3	–	–	–	1.8	53.9	6.5	3.0	3.5	–
246	OVERRIPE, FRESH	2-04-169	A-F	**74**	**5.4**	**26.0**	**–**	**–**	**–**	**1.9**	**37.9**	**3.1**	**0.7**	**1.2**	**–**
			M-F	100	7.3	34.9	–	–	–	2.6	51.0	4.2	0.9	1.6	–
	SALTGRASS, DESERT *Distichlis stricta*														
247	FRESH	2-04-171	A-F	**29**	**2.0**	**8.6**	**–**	**–**	**–**	**0.5**	**16.2**	**1.7**	**0.7**	**0.9**	**–**
			M-F	100	6.8	29.7	–	–	–	1.7	55.9	5.9	2.5	3.0	–
	SEDGE *Carex* spp														
248	FRESH	2-04-195	A-F	**25**	**2.2**	**–**	**15.4**	**–**	**1.0**	**–**	**–**	**3.0**	**2.0**	**2.0**	**–**
			M-F	100	8.8	–	61.6	–	4.1	–	–	12.1	8.1	8.0	–
	SERRADELLA, COMMON *Ornithopus sativus*														
249	FRESH	2-04-216	A-F	**21**	**3.1**	**4.8**	**–**	**–**	**–**	**0.7**	**8.9**	**3.0**	**2.1**	**2.1**	**–**
			M-F	100	15.1	23.4	–	–	–	3.6	43.4	14.6	10.3	10.0	–
	SHADSCALE SALTBUSH *Atriplex confertifolia*														
250	BROWSE, STEM CURED, FRESH	2-05-565	A-F	**80**	**18.9**	**–**	**–**	**–**	**–**	**2.0**	**–**	**6.1**	**3.5**	**3.6**	**–**
			M-F	100	23.6	–	–	–	–	2.5	–	7.7	4.4	4.4	–
	SNOWBERRY *Symphoricarpos* spp														
251	BROWSE, MIDBLOOM, FRESH	2-04-250	A-F	**44**	**3.4**	**6.0**	**–**	**–**	**–**	**2.4**	**27.8**	**4.2**	**2.5**	**2.6**	**–**
			M-F	100	7.8	13.7	–	–	–	5.5	63.4	9.6	5.8	6.0	–
	SOAPWEED *Yucca glauca*														
252	FRESH	2-04-268	A-F	**45**	**3.3**	**14.5**	**–**	**–**	**–**	**0.8**	**23.7**	**2.5**	**0.9**	**1.2**	**–**
			M-F	100	7.5	32.3	–	–	–	1.8	52.9	5.5	2.1	2.7	–
	SORGHUM *Sorghum bicolor*														
253	DARSO, FRESH	2-04-355	A-F	**29**	**2.1**	**7.3**	**–**	**–**	**–**	**0.6**	**17.7**	**1.3**	**0.3**	**0.5**	**–**
			M-F	100	7.2	25.2	–	–	–	2.1	61.0	4.5	1.2	1.9	–
254	DURRA, FRESH	2-04-363	A-F	**22**	**1.8**	**6.2**	**–**	**–**	**–**	**0.6**	**11.8**	**2.0**	**1.2**	**1.2**	**–**
			M-F	100	8.0	27.7	–	–	–	2.7	52.7	8.9	5.2	5.4	–
255	JOHNSONGRASS, IMMATURE, FRESH	2-04-409	A-F	**20**	**2.1**	**5.6**	**–**	**–**	**–**	**0.6**	**8.4**	**3.1**	**2.2**	**2.1**	**–**
			M-F	100	10.6	28.5	–	–	–	3.2	42.2	15.5	11.2	10.7	–
256	JOHNSONGRASS, FULL BLOOM, FRESH	2-04-410	A-F	**35**	**3.5**	**11.4**	**–**	**–**	**–**	**0.8**	**16.4**	**2.8**	**1.6**	**1.7**	**–**
			M-F	100	10.0	32.7	–	–	–	2.3	46.9	8.1	4.4	4.8	–
257	KAFIR, FRESH	2-04-424	A-F	**24**	**1.9**	**6.6**	**–**	**–**	**–**	**0.7**	**12.0**	**2.4**	**1.5**	**1.5**	**–**
			M-F	100	8.1	28.0	–	–	–	3.0	50.8	10.2	6.3	6.4	–

PASTURE AND RANGE PLANTS

Entry Number	TDN Ruminant	TDN Swine	TDN Horse	Digestible Energy Ruminant		Digestible Energy Swine		Digestible Energy Horse		Metabolizable Energy Ruminant		Metabolizable Energy Swine		Metabolizable Energy Poultry MEn		Metabolizable Energy Horse		Net Energy Ruminant NEm		Net Energy Ruminant NEg		Net Energy Lactating Cows NElc	
	%	%	%	Mcal		kcal		Mcal		Mcal		kcal		kcal		Mcal		Mcal		Mcal		Mcal	
				lb	kg	lb	kg	lb	kg	lb	kg	lb	kg	lb	kg	lb	kg	lb	kg	lb	kg	lb	kg
233	20	–	–	0.41	0.90	–	–	–	–	0.34	0.75	–	–	–	–	–	–	0.20	0.45	0.11	0.25	0.21	0.47
	57	–	–	1.16	2.55	–	–	–	–	0.96	2.13	–	–	–	–	–	–	0.58	1.27	0.32	0.70	0.60	1.32
234	22	–	–	0.44	0.96	–	–	–	–	0.36	0.79	–	–	–	–	–	–	0.21	0.45	0.10	0.23	0.22	0.49
	55	–	–	1.09	2.41	–	–	–	–	0.90	1.98	–	–	–	–	–	–	0.51	1.13	0.26	0.57	0.55	1.22
235	32	–	–	0.62	1.38	–	–	–	–	0.31	0.68	–	–	–	–	–	–	–	–	–	–	–	–
	49	–	–	0.96	2.12	–	–	–	–	0.47	1.04	–	–	–	–	–	–	–	–	–	–	–	–
236	27	–	–	0.66	1.46	–	–	–	–	0.37	0.81	–	–	–	–	–	–	–	–	–	–	–	–
	42	–	–	1.02	2.25	–	–	–	–	0.56	1.24	–	–	–	–	–	–	–	–	–	–	–	–
237	12	–	–	0.27	0.59	–	–	–	–	0.21	0.46	–	–	–	–	–	–	–	–	–	–	–	–
	51	–	–	1.16	2.56	–	–	–	–	0.91	2.01	–	–	–	–	–	–	–	–	–	–	–	–
238	14	–	–	0.28	0.62	–	–	–	–	0.23	0.51	–	–	–	–	–	–	0.14	0.30	0.04	0.10	0.14	0.31
	52	–	–	1.04	2.30	–	–	–	–	0.86	1.89	–	–	–	–	–	–	0.51	1.12	0.16	0.36	0.53	1.16
239	26	–	–	0.53	1.16	–	–	–	–	0.45	0.98	–	–	–	–	–	–	0.26	0.57	0.15	0.32	0.27	0.59
	61	–	–	1.23	2.70	–	–	–	–	1.03	2.28	–	–	–	–	–	–	0.60	1.33	0.34	0.75	0.62	1.36
240	30	–	–	0.60	1.33	–	–	–	–	0.49	1.08	–	–	–	–	–	–	0.26	0.58	0.11	0.25	0.30	0.66
	50	–	–	1.01	2.22	–	–	–	–	0.81	1.79	–	–	–	–	–	–	0.44	0.96	0.19	0.41	0.50	1.10
241	15	–	–	0.30	0.67	–	–	–	–	0.26	0.57	–	–	–	–	–	–	0.16	0.34	0.10	0.21	0.16	0.34
	66	–	–	1.33	2.93	–	–	–	–	1.14	2.51	–	–	–	–	–	–	0.69	1.51	0.42	0.92	0.68	1.50
242	11	–	–	0.23	0.51	–	–	–	–	0.20	0.43	–	–	–	–	–	–	0.12	0.26	0.07	0.16	0.12	0.26
	63	–	–	1.27	2.80	–	–	–	–	1.08	2.38	–	–	–	–	–	–	0.65	1.44	0.39	0.86	0.66	1.45
243	20	–	–	0.37	0.82	–	–	–	–	0.33	0.73	–	–	–	–	–	–	–	–	–	–	–	–
	36	–	–	0.68	1.49	–	–	–	–	0.60	1.32	–	–	–	–	–	–	–	–	–	–	–	–
244	25	–	–	0.47	1.03	–	–	–	–	0.32	0.70	–	–	–	–	–	–	–	–	–	–	–	–
	31	–	–	0.59	1.29	–	–	–	–	0.40	0.88	–	–	–	–	–	–	–	–	–	–	–	–
245	45	–	–	0.90	1.98	–	–	–	–	0.76	1.67	–	–	–	–	–	–	0.44	0.97	0.25	0.55	0.46	1.00
	60	–	–	1.21	2.66	–	–	–	–	1.02	2.24	–	–	–	–	–	–	0.59	1.31	0.33	0.73	0.61	1.35
246	42	–	–	0.84	1.86	–	–	–	–	0.70	1.54	–	–	–	–	–	–	0.41	0.90	0.22	0.48	0.43	0.95
	57	–	–	1.13	2.50	–	–	–	–	0.94	2.08	–	–	–	–	–	–	0.55	1.20	0.29	0.64	0.58	1.27
247	18	–	–	0.35	0.78	–	–	–	–	0.30	0.66	–	–	–	–	–	–	0.18	0.39	0.10	0.22	0.18	0.40
	61	–	–	1.22	2.69	–	–	–	–	1.03	2.27	–	–	–	–	–	–	0.60	1.33	0.34	0.76	0.62	1.37
248	–	–	–	–	–	–	–	–	–	–	–	–	–	–	–	–	–	–	–	–	–	–	–
	–	–	–	–	–	–	–	–	–	–	–	–	–	–	–	–	–	–	–	–	–	–	–
249	12	–	–	0.25	0.54	–	–	–	–	0.21	0.46	–	–	–	–	–	–	0.12	0.27	0.07	0.15	0.13	0.28
	60	–	–	1.20	2.64	–	–	–	–	1.01	2.22	–	–	–	–	–	–	0.59	1.29	0.33	0.72	0.61	1.34
250	25	–	–	0.47	1.03	–	–	–	–	0.32	0.70	–	–	–	–	–	–	–	–	–	–	–	–
	31	–	–	0.59	1.29	–	–	–	–	0.40	0.88	–	–	–	–	–	–	–	–	–	–	–	–
251	28	–	20	0.56	1.22	–	–	0.38	0.84	0.47	1.04	–	–	–	–	0.31	0.69	–	–	–	–	–	–
	63	–	45	1.26	2.79	–	–	0.86	1.91	1.07	2.36	–	–	–	–	0.71	1.56	–	–	–	–	–	–
252	24	–	–	0.50	1.10	–	–	–	–	0.41	0.91	–	–	–	–	–	–	0.26	0.56	0.14	0.31	0.27	0.59
	52	–	–	1.11	2.45	–	–	–	–	0.92	2.02	–	–	–	–	–	–	0.57	1.26	0.31	0.69	0.60	1.31
253	19	–	–	0.39	0.86	–	–	–	–	0.34	0.74	–	–	–	–	–	–	–	–	–	–	–	–
	67	–	–	1.34	2.96	–	–	–	–	1.15	2.54	–	–	–	–	–	–	–	–	–	–	–	–
254	15	–	–	0.29	0.63	–	–	–	–	0.24	0.54	–	–	–	–	–	–	0.14	0.31	0.08	0.18	0.14	0.31
	66	–	–	1.28	2.82	–	–	–	–	1.09	2.40	–	–	–	–	–	–	0.62	1.37	0.36	0.79	0.63	1.40
255	12	–	–	0.25	0.54	–	–	–	–	0.21	0.46	–	–	–	–	–	–	0.12	0.27	0.07	0.16	0.13	0.28
	62	–	–	1.24	2.74	–	–	–	–	1.05	2.32	–	–	–	–	–	–	0.62	1.36	0.36	0.78	0.63	1.39
256	20	–	–	0.40	0.89	–	–	–	–	0.34	0.74	–	–	–	–	–	–	0.19	0.42	0.10	0.22	0.20	0.45
	58	–	–	1.15	2.54	–	–	–	–	0.96	2.11	–	–	–	–	–	–	0.55	1.20	0.29	0.64	0.58	1.27
257	14	–	–	0.29	0.64	–	–	–	–	0.25	0.54	–	–	–	–	–	–	0.15	0.33	0.09	0.19	0.15	0.33
	61	–	–	1.23	2.72	–	–	–	–	1.04	2.30	–	–	–	–	–	–	0.63	1.39	0.37	0.81	0.64	1.41

(Continued)

TABLE V-1E PASTURE AND RANGE PLANTS, MINERAL AND VITAMIN COMPOSITION OF FEEDS, DATA EXPRESSED **AS-FED** AND **MOISTURE-FREE**—*(Continued)*

Entry Number	Feed Name Description	Moisture Basis: A-F (as-fed) or M-F (moisture-free)	Dry Matter	Macro Minerals							Micro Minerals						
				Calcium (Ca)	Phosphorus (P)	Sodium (Na)	Chlorine (Cl)	Magnesium (Mg)	Potassium (K)	Sulfur (S)	Cobalt (Co)	Copper (Cu)	Iodine (I)	Iron (Fe)	Manganese (Mn)	Selenium (Se)	Zinc (Zn)
			%	%	%	%	%	%	%	%	ppm or mg/kg	ppm or mg/kg	ppm or mg/kg	%	ppm or mg/kg	ppm or mg/kg	ppm or mg/kg
	RYEGRASS, ITALIAN (Continued)																
233	EARLY BLOOM, FRESH	A-F	35	–	–	–	–	–	–	–	–	–	–	–	–	–	–
		M-F	100	–	–	–	–	–	–	–	–	–	–	–	–	–	–
	SACATON (ALKALI) *Sporobolus airoides*																
234	MIDBLOOM, FRESH	A-F	40	0.23	0.07	0.02	–	0.12	0.69	–	0.228	6.4	–	0.019	12.0	–	10.8
		M-F	100	0.58	0.17	0.04	–	0.29	1.73	–	0.570	16.0	–	0.048	30.0	–	27.0
	SAGE, BLACK *Salvia mellifera*																
235	BROWSE, STEM CURED, FRESH	A-F	65	0.53	0.11	–	–	–	–	–	–	–	–	–	–	–	–
		M-F	100	0.81	0.17	–	–	–	–	–	–	–	–	–	–	–	–
	SAGEBRUSH, BIG *Artemisia tridentata*																
236	BROWSE, STEM CURED, FRESH	A-F	65	0.46	0.12	–	–	–	–	–	–	–	–	–	–	–	–
		M-F	100	0.71	0.18	–	–	–	–	–	–	–	–	–	–	–	–
	SAGEBRUSH, BUD *Artemesia spinescens*																
237	BROWSE, IMMATURE, FRESH	A-F	23	0.22	0.08	–	–	–	–	–	–	–	–	–	–	–	–
		M-F	100	0.97	0.33	–	–	–	–	–	–	–	–	–	–	–	–
238	BROWSE, FRESH	A-F	27	0.42	0.11	–	–	0.13	–	0.07	–	–	–	–	–	–	–
		M-F	100	1.57	0.42	–	–	0.49	–	0.26	–	–	–	–	–	–	–
	SAGEBRUSH, FRINGED *Artemesia frigida*																
239	BROWSE, MIDBLOOM, FRESH	A-F	43	–	–	–	–	–	–	–	–	–	–	–	–	–	–
		M-F	100	–	–	–	–	–	–	–	–	–	–	–	–	–	–
240	BROWSE, MATURE, FRESH	A-F	60	–	–	–	–	–	–	–	–	–	–	–	–	–	–
		M-F	100	–	–	–	–	–	–	–	–	–	–	–	–	–	–
	SAINFOIN *Onobrychis* spp																
241	FRESH	A-F	23	–	–	–	–	–	–	–	–	–	–	–	–	–	–
		M-F	100	–	–	–	–	–	–	–	–	–	–	–	–	–	–
	SAINT AUGUSTINEGRASS *Stenotaphrum secundatum*																
242	FRESH	A-F	18	–	–	–	–	–	–	–	–	–	–	–	–	–	–
		M-F	100	–	–	–	–	–	–	–	–	–	–	–	–	–	–
	SALTBUSH, NUTTALL *Atriplex nuttallii*																
243	BROWSE, STEM CURED, FRESH	A-F	55	1.22	0.06	–	–	–	–	–	–	–	–	–	–	–	–
		M-F	100	2.21	0.12	–	–	–	–	–	–	–	–	–	–	–	–
	SALTBUSH, SHADSCALE *Atriplex confertifolia*																
244	BROWSE, STEM CURED, FRESH	A-F	80	1.78	0.07	–	–	–	–	–	–	–	–	–	–	–	–
		M-F	100	2.23	0.08	–	–	–	–	–	–	–	–	–	–	–	–
	SALTGRASS *Distichlis* spp																
245	FRESH	A-F	74	0.16	0.07	–	–	–	0.18	–	–	–	–	0.015	115.1	–	–
		M-F	100	0.21	0.09	–	–	–	0.24	–	–	–	–	0.019	154.8	–	–
246	OVERRIPE, FRESH	A-F	74	0.17	0.05	–	–	0.22	–	–	–	–	–	–	–	–	–
		M-F	100	0.23	0.07	–	–	0.30	–	–	–	–	–	–	–	–	–
	SALTGRASS, DESERT *Distichlis stricta*																
247	FRESH	A-F	29	0.05	0.03	–	–	–	–	–	–	–	–	–	–	–	–
		M-F	100	0.16	0.09	–	–	–	–	–	–	–	–	–	–	–	–
	SEDGE *Carex* spp																
248	FRESH	A-F	25	–	0.05	0.05	0.06	–	–	0.06	–	–	–	–	–	–	–
		M-F	100	–	0.19	0.21	0.24	–	–	0.23	–	–	–	–	–	–	–
	SERRADELLA, COMMON *Ornithopus sativus*																
249	FRESH	A-F	21	0.29	0.09	–	–	–	0.44	–	–	–	–	–	–	–	–
		M-F	100	1.39	0.45	–	–	–	2.13	–	–	–	–	–	–	–	–
	SHADSCALE SALTBUSH *Atriplex convertifolia*																
250	BROWSE, STEM CURED, FRESH	A-F	80	1.78	0.07	–	–	–	–	–	–	–	–	–	–	–	–
		M-F	100	2.23	0.08	–	–	–	–	–	–	–	–	–	–	–	–
	SNOWBERRY *Symphoricarpos* spp																
251	BROWSE, MIDBLOOM, FRESH	A-F	44	0.91	0.18	–	–	0.26	–	–	–	–	–	–	–	–	–
		M-F	100	2.08	0.41	–	–	0.60	–	–	–	–	–	–	–	–	–
	SOAPWEED *Yucca glauca*																
252	FRESH	A-F	45	0.41	0.08	–	–	–	–	–	–	–	–	–	–	–	–
		M-F	100	0.92	0.18	–	–	–	–	–	–	–	–	–	–	–	–
	SORGHUM *Sorghum bicolor*																
253	DARSO, FRESH	A-F	29	–	0.04	–	–	–	–	–	–	–	–	–	–	–	–
		M-F	100	–	0.14	–	–	–	–	–	–	–	–	–	–	–	–
254	DURRA, FRESH	A-F	22	–	–	–	–	–	–	–	–	–	–	–	–	–	–
		M-F	100	–	–	–	–	–	–	–	–	–	–	–	–	–	–
255	JOHNSONGRASS, IMMATURE, FRESH	A-F	20	0.18	0.06	–	–	–	–	–	–	–	–	–	–	–	–
		M-F	100	0.93	0.31	–	–	–	–	–	–	–	–	–	–	–	–
256	JOHNSONGRASS, FULL BLOOM, FRESH	A-F	35	0.29	0.06	–	–	–	–	–	–	–	–	–	–	–	–
		M-F	100	0.83	0.17	–	–	–	–	–	–	–	–	–	–	–	–
257	KAFIR, FRESH	A-F	24	0.09	0.04	–	–	–	0.40	–	–	–	–	–	–	–	–
		M-F	100	0.38	0.17	–	–	–	1.70	–	–	–	–	–	–	–	–

Entry Number	Fat-Soluble Vitamins					Water-Soluble Vitamins								
	A (1 mg Carotene = 1667 IU Vit A)	Carotene (Provitamin A)	D	E	K	B-12	Biotin	Choline	Folacin (Folic Acid)	Niacin	Pantothenic Acid (B-3)	(Pyri-doxine) B-6	Ribo-flavin (B-2)	Thiamin (B-1)
	IU/g	ppm or mg/kg	IU/kg	ppm or mg/kg	ppm or mg/kg	ppm or mg/kg	ppm or mg/kg	ppm or mg/kg	ppm or mg/kg	ppm or mg/kg	ppm or mg/kg	ppm or mg/kg	ppm or mg/kg	ppm or mg/kg
233	–	–	–	–	–	–	–	–	–	–	–	–	–	–
	–	–	–	–	–	–	–	–	–	–	–	–	–	–
234	–	–	–	–	–	–	–	–	–	–	–	–	–	–
	–	–	–	–	–	–	–	–	–	–	–	–	–	–
235	–	–	–	–	–	–	–	–	–	–	–	–	–	–
	–	–	–	–	–	–	–	–	–	–	–	–	–	–
236	17.3	10.4	–	–	–	–	–	–	–	–	–	–	–	–
	26.6	15.9	–	–	–	–	–	–	–	–	–	–	–	–
237	9.1	5.5	–	–	–	–	–	–	–	–	–	–	–	–
	39.7	23.8	–	–	–	–	–	–	–	–	–	–	–	–
238	10.7	6.4	–	–	–	–	–	–	–	–	–	–	–	–
	39.7	23.8	–	–	–	–	–	–	–	–	–	–	–	–
239	–	–	–	–	–	–	–	–	–	–	–	–	–	–
	–	–	–	–	–	–	–	–	–	–	–	–	–	–
240	–	–	–	–	–	–	–	–	–	–	–	–	–	–
	–	–	–	–	–	–	–	–	–	–	–	–	–	–
241	–	–	–	–	–	–	–	–	–	–	–	–	–	–
	–	–	–	–	–	–	–	–	–	–	–	–	–	–
242	–	–	–	–	–	–	–	–	–	–	–	–	–	–
	–	–	–	–	–	–	–	–	–	–	–	–	–	–
243	17.4	10.5	–	–	–	–	–	–	–	–	–	–	–	–
	31.7	19.0	–	–	–	–	–	–	–	–	–	–	–	–
244	24.1	14.5	–	–	–	–	–	–	–	–	–	–	–	–
	30.1	18.1	–	–	–	–	–	–	–	–	–	–	–	–
245	–	–	–	–	–	–	–	–	–	–	–	–	–	–
	–	–	–	–	–	–	–	–	–	–	–	–	–	–
246	–	–	–	–	–	–	–	–	–	–	–	–	–	–
	–	–	–	–	–	–	–	–	–	–	–	–	–	–
247	–	–	–	–	–	–	–	–	–	–	–	–	–	–
	–	–	–	–	–	–	–	–	–	–	–	–	–	–
248	–	–	–	–	–	–	–	–	–	–	–	–	–	–
	–	–	–	–	–	–	–	–	–	–	–	–	–	–
249	–	–	–	–	–	–	–	–	–	–	–	–	–	–
	–	–	–	–	–	–	–	–	–	–	–	–	–	–
250	24.1	14.5	–	–	–	–	–	–	–	–	–	–	–	–
	30.1	18.1	–	–	–	–	–	–	–	–	–	–	–	–
251	–	–	–	–	–	–	–	–	–	–	–	–	–	–
	–	–	–	–	–	–	–	–	–	–	–	–	–	–
252	–	–	–	–	–	–	–	–	–	–	–	–	–	–
	–	–	–	–	–	–	–	–	–	–	–	–	–	–
253	–	–	–	–	–	–	–	–	–	–	–	–	–	–
	–	–	–	–	–	–	–	–	–	–	–	–	–	–
254	–	–	–	–	–	–	–	–	–	–	–	–	–	–
	–	–	–	–	–	–	–	–	–	–	–	–	–	–
255	–	–	–	–	–	–	–	–	–	–	–	–	–	–
	–	–	–	–	–	–	–	–	–	–	–	–	–	–
256	–	–	–	–	–	–	–	–	–	–	–	–	–	–
	–	–	–	–	–	–	–	–	–	–	–	–	–	–
257	6.9	4.2	–	–	–	–	–	–	–	9	3.3	1.41	1.0	–
	29.4	17.6	–	–	–	–	–	–	–	39	14.1	5.95	4.2	–

(Continued)

TABLE V-1E PASTURE AND RANGE PLANTS, COMPOSITION OF FEEDS, DATA EXPRESSED **AS-FED** AND **MOISTURE-FREE**—*(Continued)*

Entry Number	Feed Name Description	International Feed Number	Moisture Basis: A-F (as-fed) or M-F (moisture-free)	Proximate Analysis: Dry Matter	Ash	Crude Fiber	Neutral Det. Fib. (NDF)	Acid Det. Fib. (ADF)	Lignin	Ether Extract (Fat)	N-Free Extract	Crude Protein	Digestible Protein: Ruminant	Swine	Horse
				%	%	%	%	%	%	%	%	%	%	%	%
	SORGHUM (Continued)														
258	MILO, FRESH[1]	**2-04-436**	**A-F**	**30**	**1.8**	**7.6**	**19.5**	**12.0**	**–**	**0.5**	**17.2**	**2.6**	**1.5**	**1.5**	**–**
			M-F	100	6.2	25.7	65.0	40.0	–	1.8	57.8	8.6	4.9	5.2	–
259	SUDANGRASS, FRESH	**2-04-489**	**A-F**	**19**	**1.5**	**5.8**	**–**	**7.9**	**1.0**	**0.6**	**8.6**	**2.3**	**1.5**	**1.5**	**–**
			M-F	100	8.0	30.9	–	42.0	5.3	3.2	45.8	12.1	8.3	8.0	–
260	SUDANGRASS, IMMATURE, FRESH[1]	**2-04-484**	**A-F**	**18**	**1.5**	**6.7**	**9.9**	**5.2**	**–**	**0.6**	**6.8**	**2.6**	**1.8**	**1.8**	**–**
			M-F	100	8.2	36.7	55.0	29.0	–	3.2	37.5	14.4	9.9	9.8	–
261	SUDANGRASS, MATURE, FRESH	**2-04-487**	**A-F**	**30**	**2.4**	**10.6**	**–**	**–**	**–**	**0.5**	**14.5**	**1.6**	**0.6**	**0.8**	**–**
			M-F	100	8.1	35.8	–	–	–	1.7	48.9	5.5	2.1	2.7	–
	SOTOL *Dasylirion* spp														
262	HEADS, FRESH	**2-04-526**	**A-F**	**39**	**1.7**	**10.0**	**–**	**–**	**–**	**0.6**	**24.3**	**2.2**	**0.9**	**1.1**	**–**
			M-F	100	4.4	25.7	–	–	–	1.6	62.6	5.7	2.3	2.8	–
	SOYBEAN *Glycine max*														
263	FRESH	**2-04-574**	**A-F**	**23**	**2.4**	**6.3**	**–**	**–**	**–**	**0.9**	**9.2**	**4.1**	**3.2**	**2.9**	**–**
			M-F	100	10.5	27.3	–	–	–	4.0	40.3	17.9	14.0	12.7	–
264	MILK STAGE, FRESH	**2-04-572**	**A-F**	**21**	**2.0**	**5.7**	**–**	**–**	**–**	**0.7**	**9.0**	**3.4**	**2.5**	**2.3**	**–**
			M-F	100	9.8	27.4	–	–	–	3.4	43.2	16.1	12.0	11.2	–
265	DOUGH STAGE, FRESH	**2-04-573**	**A-F**	**26**	**2.5**	**7.4**	**–**	**–**	**–**	**1.3**	**9.9**	**4.5**	**3.6**	**3.2**	**–**
			M-F	100	9.8	29.0	–	–	–	5.1	38.4	17.7	14.0	12.5	–
	SOYBEAN-MILLET *Glycine max-Setaria* spp														
266	FRESH	**2-04-639**	**A-F**	**25**	**1.8**	**7.3**	**–**	**–**	**–**	**0.6**	**13.1**	**2.3**	**1.4**	**1.4**	**–**
			M-F	100	7.0	29.1	–	–	–	2.4	52.2	9.3	5.5	5.7	–
	SOYBEAN-SORGHUM, SUDANGRASS *Glycine max-Sorghum bicolor, sudanense*														
267	FRESH	**2-04-646**	**A-F**	**24**	**1.6**	**8.3**	**–**	**–**	**–**	**0.5**	**11.1**	**2.7**	**1.7**	**1.8**	**–**
			M-F	100	6.6	34.3	–	–	–	2.1	45.9	11.2	7.2	7.2	–
	SPANISHMOSS *Tillandsia usneoides*														
268	WHOLE, FRESH	**2-04-648**	**A-F**	**41**	**2.4**	**12.3**	**–**	**–**	**–**	**0.8**	**23.0**	**2.6**	**1.2**	**1.4**	**–**
			M-F	100	5.8	29.9	–	–	–	2.0	56.0	6.4	2.9	3.4	–
	SPINACH *Spinacia oleracea*														
269	LEAVES, FRESH	**2-08-125**	**A-F**	**9**	**2.2**	**0.7**	**–**	**–**	**–**	**0.4**	**3.0**	**3.1**	**2.5**	**2.3**	**–**
			M-F	100	23.6	7.5	–	–	–	4.0	32.2	32.7	26.8	24.5	–
	SQUIRREL TAIL *Sitanion* spp														
270	STEM CURED, FRESH	**2-05-566**	**A-F**	**50**	**8.5**	**–**	**–**	**–**	**–**	**1.1**	**–**	**1.6**	**0.0**	**0.4**	**–**
			M-F	100	17.0	–	–	–	–	2.2	–	3.1	0.0	0.8	–
	STARGRASS *Cynodon plectostachyus*														
271	FRESH	**2-09-730**	**A-F**	**63**	**5.0**	**–**	**51.6**	**–**	**–**	**–**	**–**	**–**	**–**	**–**	**–**
			M-F	100	7.9	–	82.2	–	–	–	–	–	–	–	–
	SUBTERRANEAN CLOVER–SEE CLOVER, SUBTERRANEAN														
	SUDANGRASS SORGHUM–SEE SORGHUM														
	SUGARCANE *Saccharum officinarum*														
272	FRESH	**2-04-689**	**A-F**	**28**	**1.6**	**8.8**	**–**	**–**	**–**	**0.6**	**15.2**	**1.4**	**0.8**	**0.7**	**–**
			M-F	100	5.9	31.7	–	–	–	2.3	55.0	5.0	2.8	2.7	–
273	LEAVES AND TOPS, FRESH	**2-04-692**	**A-F**	**26**	**3.0**	**8.5**	**–**	**–**	**–**	**0.4**	**12.3**	**1.3**	**0.6**	**0.6**	**–**
			M-F	100	11.8	33.1	–	–	–	1.7	48.1	5.2	2.2	2.5	–
	SUMAC *Rhus* spp														
274	BROWSE, FRESH	**2-04-707**	**A-F**	**48**	**2.9**	**6.5**	**–**	**–**	**–**	**2.3**	**31.7**	**4.8**	**3.0**	**3.0**	**–**
			M-F	100	6.0	13.5	–	–	–	4.7	65.8	10.0	6.2	6.3	–
	SUMMER CYPRESS, GRAY (DESERT MOLLY) *Kochia vestita*														
275	STEM CURED, FRESH	**2-08-843**	**A-F**	**85**	**21.5**	**18.7**	**–**	**–**	**–**	**3.1**	**35.0**	**6.6**	**4.3**	**3.9**	**–**
			M-F	100	25.3	22.0	–	–	–	3.7	41.2	7.8	5.0	4.5	–
	SUNFLOWER *Helianthus* spp														
276	FRESH	**2-04-723**	**A-F**	**15**	**1.7**	**4.7**	**–**	**–**	**–**	**0.4**	**7.2**	**1.4**	**0.8**	**0.8**	**–**
			M-F	100	11.1	30.3	–	–	–	2.9	46.9	8.8	5.1	5.3	–
	SUNFLOWER, COMMON *Helianthus annuus*														
277	FRESH	**2-10-697**	**A-F**	**17**	**2.3**	**3.5**	**–**	**–**	**–**	**0.9**	**7.8**	**2.4**	**1.4**	**–**	**1.6**
			M-F	100	13.5	20.7	–	–	–	5.5	45.9	14.4	8.5	–	9.7
	SWEET CLOVER, YELLOW *Melilotus officinalis*														
278	FRESH	**2-04-766**	**A-F**	**23**	**1.8**	**6.9**	**–**	**–**	**–**	**0.7**	**9.5**	**4.3**	**3.4**	**3.1**	**–**
			M-F	100	7.7	29.7	–	–	–	3.0	41.0	18.7	14.6	13.3	–
	SWEET CORN *Zea mays, saccharata*														
279	STOVER WITHOUT EARS, WITHOUT HUSKS, FRESH	**2-02-969**	**A-F**	**22**	**1.4**	**5.7**	**–**	**–**	**–**	**0.4**	**12.8**	**1.6**	**0.8**	**0.9**	**–**
			M-F	100	6.2	26.0	–	–	–	1.8	58.6	7.3	3.7	4.2	–
280	EARS WITH HUSKS, FRESH	**4-08-408**	**A-F**	**38**	**0.9**	**4.3**	**–**	**–**	**–**	**2.6**	**26.2**	**3.8**	**2.3**	**2.7**	**2.7**
			M-F	100	2.4	11.4	–	–	–	6.9	69.3	10.1	6.0	7.2	7.0
281	CANNERY RESIDUE, FRESH	**2-02-975**	**A-F**	**77**	**3.5**	**17.0**	**–**	**22.3**	**–**	**1.9**	**47.6**	**6.8**	**3.9**	**4.1**	**–**
			M-F	100	4.6	22.2	–	29.0	–	2.5	62.0	8.8	5.1	5.4	–

PASTURE AND RANGE PLANTS

Entry Number	TDN Ruminant	TDN Swine	TDN Horse	Digestible Energy Ruminant		Digestible Energy Swine		Digestible Energy Horse		Metabolizable Energy Ruminant		Metabolizable Energy Swine		Metabolizable Energy Poultry ME_n		Metabolizable Energy Horse		Net Energy Ruminant NE_m		Net Energy Ruminant NE_g		Net Energy Lactating Cows NE_{lc}	
	%	%	%	Mcal		kcal		Mcal		Mcal		kcal		kcal		Mcal		Mcal		Mcal		Mcal	
				lb	kg	lb	kg	lb	kg	lb	kg	lb	kg	lb	kg	lb	kg	lb	kg	lb	kg	lb	kg
258	17	–	–	0.36	0.80	–	–	–	–	0.31	0.67	–	–	–	–	–	–	0.20	0.43	0.12	0.26	0.20	0.44
	57	–	–	1.22	2.69	–	–	–	–	1.03	2.26	–	–	–	–	–	–	0.66	1.46	0.40	0.87	0.66	1.46
259	13	–	–	0.25	0.56	–	–	–	–	0.22	0.47	–	–	–	–	–	–	0.13	0.29	0.08	0.18	0.13	0.28
	68	–	–	1.35	2.98	–	–	–	–	1.15	2.53	–	–	–	–	–	–	0.71	1.56	0.44	0.96	0.68	1.50
260	12	–	–	0.25	0.54	–	–	–	–	0.21	0.47	–	–	–	–	–	–	0.14	0.30	0.09	0.19	0.13	0.29
	68	–	–	1.35	2.98	–	–	–	–	1.16	2.56	–	–	–	–	–	–	0.74	1.63	0.47	1.03	0.72	1.60
261	19	–	–	0.36	0.79	–	–	–	–	0.30	0.67	–	–	–	–	–	–	0.16	0.35	0.08	0.18	0.17	0.37
	65	–	–	1.21	2.67	–	–	–	–	1.02	2.25	–	–	–	–	–	–	0.53	1.17	0.28	0.61	0.57	1.25
262	24	–	–	0.49	1.07	–	–	–	–	0.41	0.91	–	–	–	–	–	–	–	–	–	–	–	–
	62	–	–	1.25	2.75	–	–	–	–	1.06	2.33	–	–	–	–	–	–	–	–	–	–	–	–
263	14	–	–	0.29	0.64	–	–	–	–	0.25	0.54	–	–	–	–	–	–	0.15	0.33	0.09	0.19	0.15	0.33
	63	–	–	1.27	2.79	–	–	–	–	1.07	2.37	–	–	–	–	–	–	0.65	1.42	0.38	0.84	0.65	1.44
264	13	–	–	0.26	0.57	–	–	–	–	0.22	0.48	–	–	–	–	–	–	0.13	0.29	0.08	0.17	0.14	0.30
	61	–	–	1.25	2.75	–	–	–	–	1.05	2.32	–	–	–	–	–	–	0.64	1.42	0.38	0.83	0.65	1.43
265	17	–	–	0.33	0.73	–	–	–	–	0.28	0.63	–	–	–	–	–	–	0.17	0.37	0.10	0.22	0.17	0.37
	66	–	–	1.29	2.85	–	–	–	–	1.10	2.43	–	–	–	–	–	–	0.65	1.42	0.38	0.84	0.65	1.43
266	17	–	–	0.33	0.72	–	–	–	–	0.28	0.62	–	–	–	–	–	–	0.16	0.35	0.09	0.20	0.16	0.35
	68	–	–	1.30	2.88	–	–	–	–	1.11	2.45	–	–	–	–	–	–	0.63	1.39	0.37	0.80	0.64	1.41
267	16	–	–	0.31	0.67	–	–	–	–	0.26	0.57	–	–	–	–	–	–	0.15	0.32	0.08	0.18	0.15	0.33
	66	–	–	1.26	2.78	–	–	–	–	1.07	2.36	–	–	–	–	–	–	0.60	1.33	0.34	0.75	0.62	1.36
268	24	–	–	0.50	1.10	–	–	–	–	0.42	0.92	–	–	–	–	–	–	0.26	0.56	0.15	0.33	0.26	0.57
	60	–	–	1.21	2.68	–	–	–	–	1.02	2.25	–	–	–	–	–	–	0.62	1.37	0.36	0.79	0.63	1.40
269	–	–	–	–	–	–	–	–	–	–	–	–	–	–	–	–	–	–	–	–	–	–	–
	–	–	–	–	–	–	–	–	–	–	–	–	–	–	–	–	–	–	–	–	–	–	–
270	25	–	–	0.46	1.01	–	–	–	–	0.39	0.85	–	–	–	–	–	–	–	–	–	–	–	–
	50	–	–	0.92	2.02	–	–	–	–	0.77	1.70	–	–	–	–	–	–	–	–	–	–	–	–
271	–	–	–	–	–	–	–	–	–	–	–	–	–	–	–	–	–	–	–	–	–	–	–
	–	–	–	–	–	–	–	–	–	–	–	–	–	–	–	–	–	–	–	–	–	–	–
272	16	16	–	0.33	0.73	–	–	–	–	0.28	0.62	–	–	–	–	–	–	0.17	0.36	0.09	0.21	0.16	0.36
	59	59	–	1.20	2.65	–	–	–	–	1.02	2.24	–	–	–	–	–	–	0.60	1.32	0.34	0.74	0.59	1.29
273	12	–	–	0.26	0.58	–	–	–	–	0.21	0.47	–	–	–	–	–	–	0.13	0.28	0.06	0.14	0.14	0.30
	49	–	–	1.02	2.25	–	–	–	–	0.83	1.82	–	–	–	–	–	–	0.49	1.09	0.24	0.53	0.54	1.19
274	32	–	25	0.64	1.40	–	–	0.47	1.04	0.54	1.20	–	–	–	–	0.39	0.85	–	–	–	–	–	–
	66	–	52	1.32	2.92	–	–	0.98	2.16	1.13	2.49	–	–	–	–	0.80	1.77	–	–	–	–	–	–
275	43	–	–	0.77	1.71	–	–	–	–	0.68	1.51	–	–	–	–	–	–	0.33	0.73	0.12	0.27	0.40	0.88
	50	–	–	0.91	2.01	–	–	–	–	0.80	1.77	–	–	–	–	–	–	0.39	0.86	0.15	0.32	0.47	1.03
276	9	–	–	0.18	0.39	–	–	–	–	0.15	0.33	–	–	–	–	–	–	0.09	0.19	0.05	0.10	0.09	0.20
	58	–	–	1.15	2.54	–	–	–	–	0.96	2.12	–	–	–	–	–	–	0.55	1.22	0.30	0.65	0.58	1.29
277	11	–	–	0.21	0.47	–	–	–	–	0.18	0.39	–	–	–	–	–	–	0.10	0.23	0.06	0.13	0.11	0.24
	63	–	–	1.26	2.77	–	–	–	–	1.03	2.27	–	–	–	–	–	–	0.61	1.35	0.34	0.74	0.64	1.42
278	15	–	–	0.30	0.66	–	–	–	–	0.25	0.56	–	–	–	–	–	–	0.16	0.34	0.09	0.21	0.16	0.34
	64	–	–	1.29	2.85	–	–	–	–	1.10	2.43	–	–	–	–	–	–	0.67	1.48	0.40	0.89	0.67	1.48
279	12	–	–	0.26	0.58	–	–	–	–	0.22	0.49	–	–	–	–	–	–	0.14	0.31	0.08	0.19	0.14	0.32
	57	–	–	1.21	2.66	–	–	–	–	1.01	2.24	–	–	–	–	–	–	0.65	1.44	0.39	0.85	0.66	1.44
280	29	31	19	0.59	1.30	623	1373	0.35	0.78	0.52	1.14	588	1297	–	–	0.29	0.64	0.32	0.71	0.21	0.47	0.31	0.68
	77	82	49	1.55	3.43	1648	3633	0.94	2.07	1.37	3.01	1557	3432	–	–	0.77	1.69	0.86	1.89	0.57	1.25	0.81	1.80
281	54	–	42	1.05	2.31	–	–	0.79	1.74	0.90	1.99	–	–	–	–	0.65	1.43	0.57	1.26	0.36	0.79	0.56	1.23
	70	–	55	1.37	3.01	–	–	1.03	2.27	1.18	2.59	–	–	–	–	0.84	1.86	0.74	1.63	0.47	1.03	0.72	1.60

(Continued)

TABLE V-1E PASTURE AND RANGE PLANTS, MINERAL AND VITAMIN COMPOSITION OF FEEDS, DATA EXPRESSED **AS-FED** AND **MOISTURE-FREE**—*(Continued)*

Entry Number	Feed Name Description	Moisture Basis: A-F (as-fed) or M-F (moisture-free)	Dry Matter	Macro Minerals: Calcium (Ca)	Phosphorus (P)	Sodium (Na)	Chlorine (Cl)	Magnesium (Mg)	Potassium (K)	Sulfur (S)	Micro Minerals: Cobalt (Co)	Copper (Cu)	Iodine (I)	Iron (Fe)	Manganese (Mn)	Selenium (Se)	Zinc (Zn)
			%	%	%	%	%	%	%	%	ppm or mg/kg	ppm or mg/kg	ppm or mg/kg	%	ppm or mg/kg	ppm or mg/kg	ppm or mg/kg
	SORGHUM (Continued)																
258	MILO, FRESH	**A-F**	**30**	**0.09**	**0.05**	–	–	–	**0.81**	–	–	–	–	–	–	–	–
		M-F	100	0.30	0.17	–	–	–	2.73	–	–	–	–	–	–	–	–
259	SUDANGRASS, FRESH	**A-F**	**19**	**0.09**	**0.08**	–	–	**0.07**	**0.40**	**0.02**	**0.025**	**6.7**	–	**0.004**	**15.3**	–	–
		M-F	100	0.49	0.44	–	–	0.35	2.14	0.11	0.133	35.9	–	0.021	81.4	–	–
260	SUDANGRASS, IMMATURE, FRESH	**A-F**	**18**	**0.08**	**0.08**	**0.00**	–	**0.06**	**0.39**	**0.02**	–	–	–	**0.004**	–	–	–
		M-F	100	0.43	0.41	0.01	–	0.35	2.14	0.11	–	–	–	0.020	–	–	–
261	SUDANGRASS, MATURE, FRESH	**A-F**	**30**	**0.09**	**0.06**	–	–	–	–	–	–	–	–	–	–	–	–
		M-F	100	0.32	0.21	–	–	–	–	–	–	–	–	–	–	–	–
	SOTOL *Dasylirion* spp																
262	HEADS, FRESH	**A-F**	**39**	–	–	–	–	–	–	–	–	–	–	–	–	–	–
		M-F	100	–	–	–	–	–	–	–	–	–	–	–	–	–	–
	SOYBEAN *Glycine max*																
263	FRESH	**A-F**	**23**	**0.25**	**0.07**	–	–	**0.12**	**0.21**	–	–	**2.1**	–	**0.005**	**27.3**	–	–
		M-F	100	1.08	0.29	–	–	0.54	0.92	–	–	9.2	–	0.021	119.4	–	–
264	MILK STAGE, FRESH	**A-F**	**21**	**0.36**	**0.06**	–	–	**0.15**	**0.25**	**0.04**	–	–	–	–	–	–	–
		M-F	100	1.71	0.30	–	–	0.72	1.19	0.19	–	–	–	–	–	–	–
265	DOUGH STAGE, FRESH	**A-F**	**26**	**0.34**	**0.08**	–	–	**0.21**	**0.20**	–	–	–	–	–	–	–	–
		M-F	100	1.31	0.31	–	–	0.83	0.79	–	–	–	–	–	–	–	–
	SOYBEAN-MILLET *Glycine max-Setaria* spp																
266	FRESH	**A-F**	**25**	**0.08**	**0.05**	–	–	–	**0.34**	–	–	–	–	–	–	–	–
		M-F	100	0.30	0.19	–	–	–	1.35	–	–	–	–	–	–	–	–
	SOYBEAN-SORGHUM, SUDANGRASS *Glycine max-Sorghum bicolor, sudanese*																
267	FRESH	**A-F**	**24**	–	–	–	–	–	–	–	–	–	–	–	–	–	–
		M-F	100	–	–	–	–	–	–	–	–	–	–	–	–	–	–
	SPANISHMOSS *Tillandsia usneoides*																
268	WHOLE, FRESH	**A-F**	**41**	–	–	–	–	–	–	–	–	–	–	–	–	–	–
		M-F	100	–	–	–	–	–	–	–	–	–	–	–	–	–	–
	SPINACH *Spinacia oleracea*																
269	LEAVES, FRESH	**A-F**	**9**	**0.09**	**0.05**	**0.07**	–	–	**0.48**	–	–	–	–	**0.004**	–	–	–
		M-F	100	0.97	0.54	0.75	–	–	5.05	–	–	–	–	0.033	–	–	–
	SQUIRREL TAIL *Sitanion* spp																
270	STEM CURED, FRESH	**A-F**	**50**	**0.19**	**0.03**	–	–	–	–	–	–	–	–	–	–	–	–
		M-F	100	0.37	0.06	–	–	–	–	–	–	–	–	–	–	–	–
	STARGRASS *Cynodon plectostachyus*																
271	FRESH	**A-F**	**63**	**0.39**	**0.19**	**0.02**	–	**0.20**	**2.07**	–	**0.133**	**6.7**	–	**0.012**	**50.6**	–	**38.1**
		M-F	100	0.61	0.31	0.03	–	0.32	3.30	–	0.211	10.7	–	0.018	80.6	–	60.7
	SUBTERRANEAN CLOVER–SEE CLOVER, SUBTERRANEAN																
	SUDANGRASS SORGHUM–SEE SORGHUM																
	SUGARCANE *Saccharum officinarum*																
272	FRESH	**A-F**	**28**	**0.11**	**0.05**	–	–	**0.11**	**0.29**	–	–	**0.6**	–	**0.005**	**17.3**	–	**6.9**
		M-F	100	0.39	0.18	–	–	0.41	1.06	–	–	2.1	–	0.017	62.6	–	25.0
273	LEAVES AND TOPS, FRESH	**A-F**	**26**	**0.09**	**0.07**	–	–	–	**0.76**	–	–	–	–	–	–	–	–
		M-F	100	0.35	0.27	–	–	–	2.96	–	–	–	–	–	–	–	–
	SUMAC *Rhus* spp																
274	BROWSE, FRESH	**A-F**	**48**	**0.93**	**0.07**	–	–	**0.14**	**0.84**	–	–	–	–	–	–	–	–
		M-F	100	1.94	0.15	–	–	0.30	1.74	–	–	–	–	–	–	–	–
	SUMMER CYPRESS, GRAY (DESERT MOLLY) *Kochia vestita*																
275	STEM CURED, FRESH	**A-F**	**85**	**2.09**	**0.12**	–	–	–	–	–	–	–	–	–	–	–	–
		M-F	100	2.46	0.14	–	–	–	–	–	–	–	–	–	–	–	–
	SUNFLOWER *Helianthus* spp																
276	FRESH	**A-F**	**15**	–	–	–	–	–	–	–	–	–	–	–	–	–	–
		M-F	100	–	–	–	–	–	–	–	–	–	–	–	–	–	–
	SUNFLOWER, COMMON *Helianthus annuus*																
277	FRESH	**A-F**	**17**	–	–	–	–	–	–	–	–	–	–	–	–	–	–
		M-F	100	–	–	–	–	–	–	–	–	–	–	–	–	–	–
	SWEET CLOVER, YELLOW *Melilotus officinalis*																
278	FRESH	**A-F**	**23**	**0.31**	**0.06**	**0.02**	**0.09**	**0.08**	**0.38**	**0.11**	–	**2.3**	–	**0.004**	**29.0**	–	–
		M-F	100	1.32	0.27	0.10	0.38	0.33	1.65	0.49	–	9.9	–	0.014	125.4	–	–
	SWEET CORN *Zea mays, saccharata*																
279	STOVER WITHOUT EARS, WITHOUT HUSKS, FRESH	**A-F**	**22**	–	–	–	–	–	–	–	–	–	–	–	–	–	–
		M-F	100	–	–	–	–	–	–	–	–	–	–	–	–	–	–
280	EARS WITH HUSKS, FRESH	**A-F**	**38**	–	–	–	–	–	–	–	–	–	–	–	–	–	–
		M-F	100	–	–	–	–	–	–	–	–	–	–	–	–	–	–
281	CANNERY RESIDUE, FRESH	**A-F**	**77**	**0.25**	**0.54**	**0.02**	–	**0.18**	**0.88**	**0.10**	–	**5.4**	–	**0.016**	–	–	–
		M-F	100	0.32	0.70	0.03	–	0.24	1.15	0.13	–	7.0	–	0.020	–	–	–

Entry Number	Fat-Soluble Vitamins					Water-Soluble Vitamins								
	A (1 mg Carotene = 1667 IU Vit A)	Carotene (Provitamin A)	D	E	K	B-12	Biotin	Choline	Folacin (Folic Acid)	Niacin	Pantothenic Acid (B-3)	(Pyri-doxine) B-6	Ribo-flavin (B-2)	Thiamin (B-1)
	IU/g	ppm or mg/kg	IU/kg	ppm or mg/kg	ppm or mg/kg	ppm or mg/kg	ppm or mg/kg	ppm or mg/kg	ppm or mg/kg	ppm or mg/kg	ppm or mg/kg	ppm or mg/kg	ppm or mg/kg	ppm or mg/kg
258	–	–	–	–	–	–	–	–	–	–	–	–	–	–
	–	–	–	–	–	–	–	–	–	–	–	–	–	–
259	57.1	34.3	–	–	–	–	–	–	–	–	–	–	–	–
	304.6	182.7	–	–	–	–	–	–	–	–	–	–	–	–
260	59.9	35.9	–	–	–	–	–	–	–	–	–	–	–	–
	329.2	197.5	–	–	–	–	–	–	–	–	–	–	–	–
261	–	–	–	–	–	–	–	–	–	–	–	–	–	–
	–	–	–	–	–	–	–	–	–	–	–	–	–	–
262	–	–	–	–	–	–	–	–	–	–	–	–	–	–
	–	–	–	–	–	–	–	–	–	–	–	–	–	–
263	121.6	73.0	–	64.2	–	–	–	–	–	–	–	–	–	–
	531.2	318.7	–	280.4	–	–	–	–	–	–	–	–	–	–
264	37.0	22.2	–	–	–	–	–	–	–	–	–	–	–	–
	178.2	106.9	–	–	–	–	–	–	–	–	–	–	–	–
265	–	–	–	–	–	–	–	–	–	–	–	–	–	–
	–	–	–	–	–	–	–	–	–	–	–	–	–	–
266	–	–	–	–	–	–	–	–	–	–	–	–	–	–
	–	–	–	–	–	–	–	–	–	–	–	–	–	–
267	–	–	–	–	–	–	–	–	–	–	–	–	–	–
	–	–	–	–	–	–	–	–	–	–	–	–	–	–
268	10.3	6.2	–	–	–	–	–	–	–	–	–	–	–	–
	25.0	15.0	–	–	–	–	–	–	–	–	–	–	–	–
269	56.4	33.8	–	38.5	–	–	–	–	–	6	–	–	2.0	1.0
	600.1	360.0	–	410.0	–	–	–	–	–	65	–	–	21.1	10.8
270	–	–	–	–	–	–	–	–	–	–	–	–	–	–
	–	–	–	–	–	–	–	–	–	–	–	–	–	–
271	–	–	–	–	–	–	–	–	–	–	–	–	–	–
	–	–	–	–	–	–	–	–	–	–	–	–	–	–
272	–	–	–	–	–	–	–	–	–	–	–	–	–	–
	–	–	–	–	–	–	–	–	–	–	–	–	–	–
273	–	–	–	–	–	–	–	–	–	–	–	–	–	–
	–	–	–	–	–	–	–	–	–	–	–	–	–	–
274	–	–	–	–	–	–	–	–	–	–	–	–	–	–
	–	–	–	–	–	–	–	–	–	–	–	–	–	–
275	18.1	10.9	–	–	–	–	–	–	–	–	–	–	–	–
	21.3	12.8	–	–	–	–	–	–	–	–	–	–	–	–
276	–	–	–	–	–	–	–	–	–	–	–	–	–	–
	–	–	–	–	–	–	–	–	–	–	–	–	–	–
277	–	–	–	–	–	–	–	–	–	–	–	–	–	–
	–	–	–	–	–	–	–	–	–	–	–	–	–	–
278	102.5	61.5	–	–	–	–	–	–	–	8	–	–	19.4	1.2
	443.6	266.1	–	–	–	–	–	–	–	36	–	–	84.0	5.3
279	–	–	–	–	–	–	–	–	–	–	–	–	–	–
	–	–	–	–	–	–	–	–	–	–	–	–	–	–
280	–	–	–	–	–	–	–	–	–	–	–	–	–	–
	–	–	–	–	–	–	–	–	–	–	–	–	–	–
281	17.3	10.4	–	–	–	–	–	–	–	–	–	–	–	–
	22.5	13.5	–	–	–	–	–	–	–	–	–	–	–	–

(Continued)

TABLE V-1E PASTURE AND RANGE PLANTS, COMPOSITION OF FEEDS, DATA EXPRESSED **AS-FED** AND **MOISTURE-FREE**—*(Continued)*

Entry Number	Feed Name Description	International Feed Number	Moisture Basis: A-F (as-fed) or M-F (moisture-free)	Proximate Analysis									Digestible Protein		
				Dry Matter	Ash	Crude Fiber	Neutral Det. Fib. (NDF)	Acid Det. Fib. (ADF)	Lignin	Ether Extract (Fat)	N-Free Extract	Crude Protein	Ruminant	Swine	Horse
				%	%	%	%	%	%	%	%	%	%	%	%
	SWEET VERNALGRASS *Anthoxanthum odoratum*														
282	FRESH	2-05-094	A-F	22	2.2	3.9	–	–	–	1.0	10.5	4.4	3.4	3.2	–
			M-F	100	10.0	17.8	–	–	–	4.6	47.5	20.0	15.2	14.3	–
	SWITCHGRASS *Panicum virgatum*														
283	FRESH	2-04-800	A-F	55	3.5	19.2	41.7	–	5.3	1.3	27.8	3.5	1.6	1.9	–
			M-F	100	6.4	34.7	75.4	–	9.7	2.3	50.2	6.4	2.9	3.4	–
	TALL (ALTA) FESCUE *Festuca arundinacea*														
284	FRESH	2-01-889	A-F	28	2.5	7.5	19.5	–	–	0.9	14.3	2.7	1.7	1.7	–
			M-F	100	9.0	26.7	69.9	–	–	3.3	51.2	9.8	6.0	6.1	–
	TALL OATGRASS *Arrhenatherum elatius*														
285	FRESH	2-03-267	A-F	30	2.0	10.5	–	–	–	0.9	14.3	2.6	1.5	1.6	–
			M-F	100	6.6	34.7	–	–	–	3.0	47.2	8.6	4.9	5.2	–
286	IMMATURE, FRESH	2-03-261	A-F	22	2.2	4.1	–	–	–	1.1	10.0	4.3	3.2	3.0	–
			M-F	100	10.1	19.0	–	–	–	4.9	46.3	19.7	15.0	14.1	–
	TEOSINTE *Euchlaena* spp														
287	FRESH	2-04-823	A-F	21	2.0	6.7	–	–	–	0.5	10.4	1.7	1.0	1.0	–
			M-F	100	9.6	31.3	–	–	–	2.3	48.6	8.2	4.5	4.9	–
	THISTLE, RUSSIAN (TUMBLING TUMBLEWEED) *Salsola kali, tenuifolia*														
288	FRESH	2-03-990	A-F	30	6.6	9.3	11.1	6.6	1.0	0.6	9.0	4.6	3.3	3.1	–
			M-F	100	21.9	30.9	37.0	22.1	3.4	2.0	30.1	15.2	10.9	10.5	–
289	STEM CURED, FRESH	2-08-000	A-F	88	10.7	–	56.3	38.8	9.9	–	–	8.1	4.8	5.0	–
			M-F	100	12.2	–	64.0	44.1	11.2	–	–	9.2	5.5	5.7	–
	THREE-AWN (WIREGRASS) *Aristida* spp														
290	FRESH	2-04-838	A-F	39	2.3	13.3	–	–	–	0.9	18.6	3.8	2.3	2.4	–
			M-F	100	5.9	34.2	–	–	–	2.3	47.8	9.8	6.0	6.1	–
	TICKCLOVER, CHEROKEE (BEGGARWEED) *Desmodium tortuosum*														
291	FRESH	2-08-541	A-F	27	3.2	7.5	–	–	–	0.5	11.7	4.2	3.0	2.9	–
			M-F	100	11.8	27.7	–	–	–	1.8	43.2	15.5	11.2	10.7	–
	TIMOTHY *Phleum pratense*														
292	FRESH	2-04-912	A-F	28	2.3	7.5	19.4	–	–	1.1	13.4	3.4	2.1	2.3	–
			M-F	100	8.2	27.0	69.9	–	–	4.1	48.4	12.3	7.5	8.2	–
293	PREBLOOM, FRESH	2-04-903	A-F	26	1.9	8.5	15.1	–	–	1.0	11.8	3.3	1.8	2.2	–
			M-F	100	7.0	32.1	57.0	–	–	3.8	44.5	12.5	6.9	8.3	–
294	MIDBLOOM, FRESH	2-04-905	A-F	29	1.9	9.8	–	–	–	0.9	14.0	2.7	1.4	1.6	–
			M-F	100	6.6	33.5	–	–	–	3.0	47.9	9.1	4.9	5.6	–
	TOBOSA *Hilaria mutica*														
295	IMMATURE, FRESH	2-08-578	A-F	40	4.5	12.6	27.6	–	–	0.5	18.2	3.8	2.8	2.3	–
			M-F	100	11.4	31.8	69.6	–	–	1.4	45.9	9.5	7.1	5.9	–
296	MATURE, FRESH	2-05-036	A-F	50	5.3	17.2	–	–	–	0.7	23.9	2.8	1.5	1.4	–
			M-F	100	10.7	34.4	–	–	–	1.5	47.9	5.6	3.0	2.7	–
	TREFOIL, BIRDSFOOT (DEERVETCH, BIRDSFOOT) *Lotus corniculatus*														
297	FRESH	2-20-786	A-F	19	2.2	4.1	9.5	–	–	0.8	8.5	3.7	2.8	2.7	–
			M-F	100	11.2	21.2	49.4	–	–	4.0	44.3	19.3	14.6	13.8	–
	TREFOIL, YELLOW (MEDIC, BLACK) *Medicago lupulina*														
298	FRESH	2-03-070	A-F	23	2.3	5.6	–	–	–	0.8	9.1	4.9	3.8	3.5	–
			M-F	100	10.2	24.7	–	–	–	3.4	40.2	21.6	16.7	15.6	–
	TUMBLEWEED, TUMBLING (RUSSIAN THISTLE) *Salsola kali, tenuifolia*														
299	FRESH	2-03-990	A-F	30	6.6	9.3	11.1	6.6	1.0	0.6	9.0	4.6	3.3	3.1	–
			M-F	100	21.9	30.9	37.0	22.1	3.4	2.0	30.1	15.2	10.9	10.5	–
300	STEM CURED, FRESH	2-08-000	A-F	88	10.7	–	56.3	38.8	9.9	–	–	8.1	4.8	5.0	–
			M-F	100	12.2	–	64.0	44.1	11.2	–	–	9.2	5.5	5.7	–
	TURNIP *Brassica rapa, rapa*														
301	FRESH	2-05-063	A-F	13	2.1	1.4	–	–	–	0.3	6.8	2.9	1.1	2.1	–
			M-F	100	15.7	10.3	–	–	–	2.6	50.2	21.2	8.1	15.3	–
	VASEY GRASS *Paspalum urvillei*														
302	MATURE, FRESH	2-05-072	A-F	42	4.3	15.8	–	–	–	0.9	17.8	3.2	1.7	1.8	–
			M-F	100	10.2	37.7	–	–	–	2.1	42.5	7.6	4.0	4.4	–
	VELVETBEAN *Mucuna* spp														
303	DOUGH STAGE, FRESH	2-05-084	A-F	22	2.4	5.3	–	–	–	0.5	10.7	3.5	2.5	2.5	–
			M-F	100	10.7	23.6	–	–	–	2.3	47.7	15.7	11.0	10.9	–
	VERNALGRASS, SWEET *Anthoxanthum odoratum*														
304	FRESH	2-05-094	A-F	22	2.2	3.9	–	–	–	1.0	10.5	4.4	3.4	3.2	–
			M-F	100	10.0	17.8	–	–	–	4.6	47.5	20.0	15.2	14.3	–
	VETCH *Vicia* spp														
305	FRESH	2-05-111	A-F	22	2.1	6.2	–	–	–	0.5	8.9	4.7	3.5	3.4	–
			M-F	100	9.4	27.6	–	–	–	2.3	39.7	20.9	15.7	15.0	–

PASTURE AND RANGE PLANTS

Entry Number	TDN Ruminant	TDN Swine	TDN Horse	Digestible Energy Ruminant		Digestible Energy Swine		Digestible Energy Horse		Metabolizable Energy Ruminant		Metabolizable Energy Swine		Metabolizable Energy Poultry ME_n		Metabolizable Energy Horse		Net Energy Ruminant NE_m		Net Energy Ruminant NE_g		Net Energy Lactating Cows NE_{lc}	
	%	%	%	Mcal		kcal		Mcal		Mcal		kcal		kcal		Mcal		Mcal		Mcal		Mcal	
				lb	kg	lb	kg	lb	kg	lb	kg	lb	kg	lb	kg	lb	kg	lb	kg	lb	kg	lb	kg
282	14	–	–	0.29	0.64	–	–	–	–	0.25	0.55	–	–	–	–	–	–	0.16	0.36	0.10	0.23	0.16	0.35
	61	–	–	1.31	2.90	–	–	–	–	1.12	2.48	–	–	–	–	–	–	0.74	1.63	0.47	1.03	0.72	1.60
283	33	–	–	0.65	1.44	–	–	–	–	0.55	1.20	–	–	–	–	–	–	0.32	0.70	0.18	0.39	0.33	0.73
	59	–	–	1.18	2.60	–	–	–	–	0.99	2.17	–	–	–	–	–	–	0.57	1.27	0.32	0.70	0.60	1.32
284	17	–	–	0.34	0.76	–	–	–	–	0.29	0.64	–	–	–	–	–	–	0.17	0.38	0.10	0.22	0.18	0.39
	61	–	–	1.23	2.71	–	–	–	–	1.04	2.28	–	–	–	–	–	–	0.62	1.37	0.36	0.79	0.63	1.40
285	17	–	–	0.35	0.78	–	–	–	–	0.30	0.65	–	–	–	–	–	–	0.18	0.39	0.10	0.22	0.18	0.41
	57	–	–	1.17	2.58	–	–	–	–	0.98	2.15	–	–	–	–	–	–	0.59	1.29	0.33	0.72	0.61	1.34
286	15	–	–	0.29	0.65	–	–	–	–	0.25	0.56	–	–	–	–	–	–	0.16	0.35	0.10	0.22	0.15	0.34
	68	–	–	1.36	3.00	–	–	–	–	1.17	2.58	–	–	–	–	–	–	0.73	1.61	0.46	1.01	0.72	1.58
287	13	–	–	0.26	0.57	–	–	–	–	0.22	0.48	–	–	–	–	–	–	0.12	0.27	0.07	0.14	0.13	0.28
	63	–	–	1.21	2.68	–	–	–	–	1.02	2.25	–	–	–	–	–	–	0.56	1.24	0.31	0.68	0.59	1.30
288	13	–	–	0.28	0.61	–	–	–	–	0.22	0.48	–	–	–	–	–	–	0.12	0.27	0.05	0.11	0.15	0.32
	43	–	–	0.92	2.02	–	–	–	–	0.72	1.59	–	–	–	–	–	–	0.41	0.91	0.17	0.37	0.48	1.07
289	–	–	–	–	–	–	–	–	–	–	–	–	–	–	–	–	–	–	–	–	–	–	–
	–	–	–	–	–	–	–	–	–	–	–	–	–	–	–	–	–	–	–	–	–	–	–
290	23	–	–	0.47	1.04	–	–	–	–	0.40	0.87	–	–	–	–	–	–	0.24	0.52	0.13	0.30	0.24	0.53
	60	–	–	1.21	2.67	–	–	–	–	1.02	2.24	–	–	–	–	–	–	0.61	1.34	0.35	0.76	0.62	1.37
291	14	–	–	0.31	0.68	–	–	–	–	0.26	0.56	–	–	–	–	–	–	0.16	0.36	0.09	0.20	0.17	0.37
	53	–	–	1.13	2.50	–	–	–	–	0.94	2.07	–	–	–	–	–	–	0.60	1.33	0.34	0.75	0.62	1.36
292	18	–	–	0.35	0.78	–	–	–	–	0.30	0.66	–	–	–	–	–	–	0.19	0.42	0.12	0.26	0.19	0.42
	64	–	–	1.27	2.81	–	–	–	–	1.08	2.39	–	–	–	–	–	–	0.69	1.52	0.42	0.92	0.68	1.50
293	17	–	–	0.34	0.75	–	–	–	–	0.29	0.64	–	–	–	–	–	–	0.18	0.41	0.11	0.25	0.18	0.40
	64	–	–	1.28	2.83	–	–	–	–	1.09	2.40	–	–	–	–	–	–	0.70	1.54	0.43	0.94	0.69	1.52
294	19	–	–	0.37	0.82	–	–	–	–	0.32	0.70	–	–	–	–	–	–	0.20	0.44	0.12	0.27	0.20	0.44
	64	–	–	1.28	2.81	–	–	–	–	1.08	2.39	–	–	–	–	–	–	0.68	1.51	0.41	0.91	0.68	1.50
295	22	–	–	0.44	0.97	–	–	–	–	0.36	0.80	–	–	–	–	–	–	0.21	0.47	0.11	0.25	0.23	0.50
	55	–	–	1.11	2.44	–	–	–	–	0.91	2.02	–	–	–	–	–	–	0.54	1.18	0.28	0.62	0.57	1.26
296	22	–	–	0.49	1.08	–	–	–	–	0.39	0.87	–	–	–	–	–	–	0.25	0.55	0.12	0.27	0.27	0.60
	44	–	–	0.98	2.16	–	–	–	–	0.79	1.74	–	–	–	–	–	–	0.50	1.11	0.25	0.55	0.55	1.20
297	13	–	–	0.26	0.58	–	–	–	–	0.23	0.50	–	–	–	–	–	–	0.16	0.35	0.10	0.22	0.15	0.33
	68	–	–	1.36	2.99	–	–	–	–	1.17	2.57	–	–	–	–	–	–	0.81	1.79	0.53	1.16	0.78	1.72
298	14	–	–	0.29	0.64	–	–	–	–	0.25	0.54	–	–	–	–	–	–	0.16	0.35	0.10	0.21	0.16	0.34
	61	–	–	1.27	2.81	–	–	–	–	1.08	2.39	–	–	–	–	–	–	0.69	1.52	0.42	0.93	0.69	1.51
299	13	–	–	0.28	0.61	–	–	–	–	0.22	0.48	–	–	–	–	–	–	0.12	0.27	0.05	0.11	0.15	0.32
	43	–	–	0.92	2.02	–	–	–	–	0.72	1.59	–	–	–	–	–	–	0.41	0.91	0.17	0.37	0.48	1.07
300	–	–	–	–	–	–	–	–	–	–	–	–	–	–	–	–	–	–	–	–	–	–	–
	–	–	–	–	–	–	–	–	–	–	–	–	–	–	–	–	–	–	–	–	–	–	–
301	10	–	–	0.19	0.41	–	–	–	–	0.16	0.36	–	–	–	–	–	–	0.10	0.21	0.06	0.13	0.10	0.21
	71	–	–	1.39	3.07	–	–	–	–	1.20	2.65	–	–	–	–	–	–	0.72	1.58	0.45	0.98	0.71	1.55
302	23	–	–	0.46	1.02	–	–	–	–	0.38	0.84	–	–	–	–	–	–	0.21	0.46	0.10	0.23	0.23	0.50
	55	–	–	1.10	2.42	–	–	–	–	0.91	2.00	–	–	–	–	–	–	0.50	1.09	0.24	0.54	0.54	1.19
303	15	–	–	0.30	0.67	–	–	–	–	0.26	0.58	–	–	–	–	–	–	0.15	0.32	0.09	0.19	0.15	0.33
	68	–	–	1.35	2.98	–	–	–	–	1.16	2.56	–	–	–	–	–	–	0.65	1.44	0.39	0.85	0.66	1.45
304	14	–	–	0.29	0.64	–	–	–	–	0.25	0.55	–	–	–	–	–	–	0.16	0.36	0.10	0.23	0.16	0.35
	61	–	–	1.31	2.90	–	–	–	–	1.12	2.48	–	–	–	–	–	–	0.74	1.63	0.47	1.03	0.72	1.60
305	13	–	–	0.28	0.62	–	–	–	–	0.24	0.52	–	–	–	–	–	–	0.15	0.33	0.09	0.20	0.15	0.33
	60	–	–	1.25	2.75	–	–	–	–	1.06	2.33	–	–	–	–	–	–	0.67	1.48	0.40	0.89	0.67	1.48

(Continued)

TABLE V-1E PASTURE AND RANGE PLANTS, MINERAL AND VITAMIN COMPOSITION OF FEEDS, DATA EXPRESSED **AS-FED** AND **MOISTURE-FREE**—*(Continued)*

Entry Number	Feed Name Description	Moisture Basis: A-F (as-fed) or M-F (moisture-free)	Dry Matter	Macro Minerals							Micro Minerals						
				Calcium (Ca)	Phosphorus (P)	Sodium (Na)	Chlorine (Cl)	Magnesium (Mg)	Potassium (K)	Sulfur (S)	Cobalt (Co)	Copper (Cu)	Iodine (I)	Iron (Fe)	Manganese (Mn)	Selenium (Se)	Zinc (Zn)
			%	%	%	%	%	%	%	%	ppm or mg/kg	ppm or mg/kg	ppm or mg/kg	%	ppm or mg/kg	ppm or mg/kg	ppm or mg/kg
	SWEET VERNALGRASS *Anthoxanthum odoratum*																
282	FRESH	**A-F**	**22**	**0.14**	**0.08**	–	–	–	–	–	–	–	–	–	–	–	–
		M-F	100	0.64	0.36	–	–	–	–	–	–	–	–	–	–	–	–
	SWITCHGRASS *Panicum virgatum*																
283	FRESH	**A-F**	**55**	**0.16**	**0.05**	–	–	–	–	–	–	–	–	–	–	–	–
		M-F	100	0.29	0.10	–	–	–	–	–	–	–	–	–	–	–	–
	TALL (ALTA) FESCUE *Festuca arundinacea*																
284	FRESH	**A-F**	**28**	**0.13**	**0.05**	**0.03**	–	**0.07**	**0.70**	–	**0.113**	**1.0**	–	**0.003**	**18.0**	–	**5.9**
		M-F	100	0.48	0.19	0.12	–	0.25	2.51	–	0.401	3.4	–	0.011	64.4	–	21.0
	TALL OATGRASS *Arrhenatherum elatius*																
285	FRESH	**A-F**	**30**	**0.12**	**0.14**	–	–	–	**0.91**	–	–	–	–	–	–	–	–
		M-F	100	0.40	0.46	–	–	–	3.00	–	–	–	–	–	–	–	–
286	IMMATURE, FRESH	**A-F**	**22**	–	–	–	–	–	–	–	–	–	–	–	–	–	–
		M-F	100	–	–	–	–	–	–	–	–	–	–	–	–	–	–
	TEOSINTE *Euchlaena* spp																
287	FRESH	**A-F**	**21**	–	–	–	–	–	–	–	–	–	–	–	–	–	–
		M-F	100	–	–	–	–	–	–	–	–	–	–	–	–	–	–
	THISTLE, RUSSIAN (TUMBLING TUMBLEWEED) *Salsola kali, tenuifolia*																
288	FRESH	**A-F**	**30**	**0.74**	**0.08**	–	–	**0.26**	**1.72**	**0.05**	**0.051**	**5.8**	–	–	**10.0**	–	–
		M-F	100	2.47	0.27	–	–	0.88	5.73	0.17	0.170	19.2	–	–	33.3	–	–
289	STEM CURED, FRESH	**A-F**	**88**	–	**0.09**	–	–	–	–	–	–	–	–	–	–	–	–
		M-F	100	–	0.10	–	–	–	–	–	–	–	–	–	–	–	–
	THREE-AWN (WIREGRASS) *Aristida* spp																
290	FRESH	**A-F**	**39**	–	–	–	–	–	–	–	–	–	–	–	–	–	–
		M-F	100	–	–	–	–	–	–	–	–	–	–	–	–	–	–
	TICKCLOVER, CHEROKEE (BEGGARWEED) *Desmodium tortuosum*																
291	FRESH	**A-F**	**27**	–	**0.12**	–	–	–	**0.47**	–	–	–	–	–	–	–	–
		M-F	100	–	0.44	–	–	–	1.73	–	–	–	–	–	–	–	–
	TIMOTHY *Phleum pratense*																
292	FRESH	**A-F**	**28**	**0.14**	**0.08**	**0.03**	**0.14**	**0.06**	**0.69**	**0.04**	**0.041**	**2.2**	–	**0.004**	**24.6**	–	**7.5**
		M-F	100	0.51	0.27	0.11	0.51	0.23	2.47	0.13	0.147	8.0	–	0.012	89.0	–	26.9
293	PREBLOOM, FRESH	**A-F**	**26**	**0.11**	**0.07**	**0.03**	–	**0.04**	**0.72**	**0.03**	**0.040**	**2.4**	–	**0.004**	**33.5**	–	**9.4**
		M-F	100	0.40	0.26	0.11	–	0.16	2.73	0.13	0.150	8.9	–	0.014	126.8	–	35.7
294	MIDBLOOM, FRESH	**A-F**	**29**	**0.11**	**0.09**	**0.06**	**0.19**	**0.04**	**0.60**	**0.04**	–	**3.3**	–	**0.006**	**56.2**	–	–
		M-F	100	0.38	0.30	0.19	0.64	0.14	2.06	0.13	–	11.2	–	0.018	192.5	–	–
	TOBOSA *Hilaria mutica*																
295	IMMATURE, FRESH	**A-F**	**40**	**0.18**	**0.05**	**0.01**	–	**0.04**	**0.21**	–	**0.277**	**5.6**	–	**0.023**	**32.3**	–	**12.3**
		M-F	100	0.45	0.14	0.01	–	0.09	0.54	–	0.698	14.1	–	0.059	81.6	–	31.2
296	MATURE, FRESH	**A-F**	**50**	**0.18**	**0.06**	**0.01**	–	**0.06**	**0.38**	–	**0.271**	**10.0**	–	**0.017**	**40.2**	–	**13.0**
		M-F	100	0.35	0.13	0.01	–	0.12	0.76	–	0.542	20.0	–	0.034	80.4	–	26.0
	TREFOIL, BIRDSFOOT (DEERVETCH, BIRDSFOOT) *Lotus corniculatus*																
297	FRESH	**A-F**	**19**	**0.34**	**0.05**	**0.02**	–	**0.08**	**0.63**	**0.05**	**0.094**	**2.5**	–	**0.006**	**16.0**	–	**6.0**
		M-F	100	1.74	0.26	0.11	–	0.40	3.26	0.25	0.487	12.8	–	0.031	82.9	–	31.1
	TREFOIL, YELLOW (MEDIC, BLACK) *Medicago lupulina*																
298	FRESH	**A-F**	**23**	–	–	–	–	–	–	–	–	–	–	–	–	–	–
		M-F	100	–	–	–	–	–	–	–	–	–	–	–	–	–	–
	TUMBLEWEED, TUMBLING (RUSSIAN THISTLE) *Salsola kali, tenuifolia*																
299	FRESH	**A-F**	**30**	**0.74**	**0.08**	–	–	**0.26**	**1.72**	**0.05**	**0.051**	**5.8**	–	–	**10.0**	–	–
		M-F	100	2.47	0.27	–	–	0.88	5.73	0.17	0.170	19.2	–	–	33.3	–	–
300	STEM CURED, FRESH	**A-F**	**88**	–	**0.09**	–	–	–	–	–	–	–	–	–	–	–	–
		M-F	100	–	0.10	–	–	–	–	–	–	–	–	–	–	–	–
	TURNIP *Brassica rapa, rapa*																
301	FRESH	**A-F**	**13**	**0.40**	**0.05**	–	**0.26**	**0.07**	**0.41**	**0.04**	–	**2.4**	–	**0.006**	**55.2**	–	**5.0**
		M-F	100	2.93	0.39	–	1.93	0.54	3.03	0.27	–	17.6	–	0.041	408.6	–	36.8
	VASEY GRASS *Paspalum urvillei*																
302	MATURE, FRESH	**A-F**	**42**	**0.23**	**0.06**	–	–	–	–	–	–	–	–	–	–	–	–
		M-F	100	0.54	0.13	–	–	–	–	–	–	–	–	–	–	–	–
	VELVETBEAN *Mucuna* spp																
303	DOUGH STAGE, FRESH	**A-F**	**22**	–	–	–	–	–	–	–	–	–	–	–	–	–	–
		M-F	100	–	–	–	–	–	–	–	–	–	–	–	–	–	–
	VERNALGRASS, SWEET *Anthoxanthum odoratum*																
304	FRESH	**A-F**	**22**	**0.14**	**0.08**	–	–	–	–	–	–	–	–	–	–	–	–
		M-F	100	0.64	0.36	–	–	–	–	–	–	–	–	–	–	–	–
	VETCH *Vicia* spp																
305	FRESH	**A-F**	**22**	–	–	**0.11**	**0.42**	–	–	**0.03**	**0.068**	–	–	–	–	–	–
		M-F	100	–	–	0.49	1.85	–	–	0.15	0.300	–	–	–	–	–	–

Entry Number	Fat-Soluble Vitamins					Water-Soluble Vitamins								
	A (1 mg Carotene = 1667 IU Vit A)	Carotene (Provitamin A)	D	E	K	B–12	Biotin	Choline	Folacin (Folic Acid)	Niacin	Pantothenic Acid (B–3)	(Pyri-doxine) B–6	Ribo-flavin (B–2)	Thiamin (B–1)
	IU/g	ppm or mg/kg	IU/kg	ppm or mg/kg	ppm or mg/kg	ppm or mg/kg	ppm or mg/kg	ppm or mg/kg	ppm or mg/kg	ppm or mg/kg	ppm or mg/kg	ppm or mg/kg	ppm or mg/kg	ppm or mg/kg
282	**–**	**–**	**–**	**–**	**–**	**–**	**–**	**–**	**–**	**–**	**–**	**–**	**–**	**–**
	–	–	–	–	–	–	–	–	–	–	–	–	–	–
283	**83.0**	**49.8**	**–**	**–**	**–**	**–**	**–**	**–**	**–**	**–**	**–**	**–**	**–**	**–**
	150.1	90.1	–	–	–	–	–	–	–	–	–	–	–	–
284	**–**	**–**	**–**	**–**	**–**	**–**	**–**	**–**	**–**	**–**	**–**	**–**	**–**	**–**
	–	–	–	–	–	–	–	–	–	–	–	–	–	–
285	**–**	**–**	**–**	**–**	**–**	**–**	**–**	**–**	**–**	**–**	**–**	**–**	**–**	**–**
	–	–	–	–	–	–	–	–	–	–	–	–	–	–
286	**–**	**–**	**–**	**–**	**–**	**–**	**–**	**–**	**–**	**–**	**–**	**–**	**–**	**–**
	–	–	–	–	–	–	–	–	–	–	–	–	–	–
287	**–**	**–**	**–**	**–**	**–**	**–**	**–**	**–**	**–**	**–**	**–**	**–**	**–**	**–**
	–	–	–	–	–	–	–	–	–	–	–	–	–	–
288	**2.3**	**1.4**	**–**	**–**	**–**	**–**	**–**	**–**	**–**	**–**	**–**	**–**	**–**	**–**
	7.7	4.6	–	–	–	–	–	–	–	–	–	–	–	–
289	**–**	**–**	**–**	**–**	**–**	**–**	**–**	**–**	**–**	**–**	**–**	**–**	**–**	**–**
	–	–	–	–	–	–	–	–	–	–	–	–	–	–
290	**–**	**–**	**–**	**–**	**–**	**–**	**–**	**–**	**–**	**–**	**–**	**–**	**–**	**–**
	–	–	–	–	–	–	–	–	–	–	–	–	–	–
291	**–**	**–**	**–**	**–**	**–**	**–**	**–**	**–**	**–**	**–**	**–**	**–**	**–**	**–**
	–	–	–	–	–	–	–	–	–	–	–	–	–	–
292	**103.2**	**61.9**	**–**	**42.6**	**–**	**–**	**–**	**–**	**–**	**–**	**–**	**–**	**3.2**	**0.8**
	372.7	223.6	–	153.9	–	–	–	–	–	–	–	–	11.5	2.9
293	**103.1**	**61.8**	**–**	**–**	**–**	**–**	**–**	**–**	**–**	**–**	**–**	**–**	**–**	**–**
	390.1	234.0	–	–	–	–	–	–	–	–	–	–	–	–
294	**94.5**	**56.7**	**–**	**–**	**–**	**–**	**–**	**–**	**–**	**–**	**–**	**–**	**–**	**–**
	323.4	194.0	–	–	–	–	–	–	–	–	–	–	–	–
295	**–**	**–**	**–**	**–**	**–**	**–**	**–**	**–**	**–**	**–**	**–**	**–**	**–**	**–**
	–	–	–	–	–	–	–	–	–	–	–	–	–	–
296	**2.0**	**1.2**	**–**	**–**	**–**	**–**	**–**	**–**	**–**	**–**	**–**	**–**	**–**	**–**
	4.0	2.4	–	–	–	–	–	–	–	–	–	–	–	–
297	**–**	**–**	**–**	**–**	**–**	**–**	**–**	**–**	**–**	**–**	**–**	**–**	**–**	**–**
	–	–	–	–	–	–	–	–	–	–	–	–	–	–
298	**–**	**–**	**–**	**–**	**–**	**–**	**–**	**–**	**–**	**–**	**–**	**–**	**–**	**–**
	–	–	–	–	–	–	–	–	–	–	–	–	–	–
299	**2.3**	**1.4**	**–**	**–**	**–**	**–**	**–**	**–**	**–**	**–**	**–**	**–**	**–**	**–**
	7.7	4.6	–	–	–	–	–	–	–	–	–	–	–	–
300	**–**	**–**	**–**	**–**	**–**	**–**	**–**	**–**	**–**	**–**	**–**	**–**	**–**	**–**
	–	–	–	–	–	–	–	–	–	–	–	–	–	–
301	**–**	**–**	**–**	**–**	**–**	**–**	**–**	**–**	**–**	**11**	**–**	**–**	**5.4**	**2.9**
	–	–	–	–	–	–	–	–	–	82	–	–	40.2	21.6
302	**–**	**–**	**–**	**–**	**–**	**–**	**–**	**–**	**–**	**–**	**–**	**–**	**–**	**–**
	–	–	–	–	–	–	–	–	–	–	–	–	–	–
303	**–**	**–**	**–**	**–**	**–**	**–**	**–**	**–**	**–**	**–**	**–**	**–**	**–**	**–**
	–	–	–	–	–	–	–	–	–	–	–	–	–	–
304	**–**	**–**	**–**	**–**	**–**	**–**	**–**	**–**	**–**	**–**	**–**	**–**	**–**	**–**
	–	–	–	–	–	–	–	–	–	–	–	–	–	–
305	**–**	**–**	**–**	**–**	**–**	**–**	**–**	**–**	**–**	**–**	**–**	**–**	**–**	**–**
	–	–	–	–	–	–	–	–	–	–	–	–	–	–

(Continued)

TABLE V-1E PASTURE AND RANGE PLANTS, COMPOSITION OF FEEDS, DATA EXPRESSED **AS-FED** AND **MOISTURE-FREE**—*(Continued)*

Entry Number	Feed Name Description	International Feed Number	Moisture Basis: A-F (as-fed) or M-F (moisture-free)	Proximate Analysis									Digestible Protein		
				Dry Matter	Ash	Crude Fiber	Neutral Det. Fib. (NDF)	Acid Det. Fib. (ADF)	Lignin	Ether Extract (Fat)	N-Free Extract	Crude Protein	Ruminant	Swine	Horse
				%	%	%	%	%	%	%	%	%	%	%	%
	VETCH, COMMON *Vicia sativa*														
306	FRESH	2-05-123	A-F	20	2.1	5.5	–	–	–	0.5	8.5	3.8	2.8	2.7	–
			M-F	100	10.3	27.0	–	–	–	2.5	41.6	18.6	14.0	13.2	–
	VETCH, HAIRY *Vicia villosa*														
307	FRESH	2-05-124	A-F	18	2.0	5.2	–	–	–	0.6	6.2	4.4	3.6	3.2	–
			M-F	100	10.7	28.3	–	–	–	3.5	33.7	23.8	19.7	17.4	–
	VETCH-OATS *Vicia spp-Avena sativa*														
308	FRESH	2-05-133	A-F	26	2.8	7.0	–	–	–	0.9	11.4	4.3	3.3	3.0	–
			M-F	100	10.4	26.4	–	–	–	3.5	43.2	16.4	12.3	11.5	–
	VINE-MESQUITE *Panicum obtusum*														
309	MATURE, FRESH	2-05-139	A-F	33	3.5	11.8	–	–	–	0.6	15.3	1.8	0.7	0.9	–
			M-F	100	10.5	35.8	–	–	–	1.8	46.5	5.4	2.0	2.6	–
	WHEAT *Triticum aestivum*														
310	IMMATURE, FRESH	2-05-176	A-F	22	3.0	3.9	10.2	6.3	1.0	1.0	8.3	6.1	4.9	4.5	–
			M-F	100	13.3	17.4	46.2	28.4	4.5	4.4	37.5	27.4	22.0	20.3	–
	WHEATGRASS, BLUEBUNCH *Agropyron spicatum*														
311	PREBLOOM, FRESH	2-05-387	A-F	32	2.4	8.0	22.8	–	–	0.8	16.0	4.5	3.2	3.1	–
			M-F	100	7.7	25.2	72.1	–	–	2.4	50.5	14.2	10.0	9.7	–
	WHEATGRASS, CRESTED *Agropyron desertorum*														
312	IMMATURE, FRESH	2-05-420	A-F	28	2.9	6.2	–	–	–	0.6	12.9	6.0	5.1	4.3	–
			M-F	100	10.0	21.6	–	–	–	2.2	45.2	21.0	17.8	15.1	–
313	EARLY BLOOM, FRESH	2-05-422	A-F	41	–	8.9	–	–	–	1.7	–	4.8	3.1	3.1	–
			M-F	100	–	21.7	–	–	–	4.1	–	11.7	7.7	7.6	–
314	FULL BLOOM, FRESH	2-05-424	A-F	45	4.2	13.6	–	–	–	1.6	22.2	3.4	1.8	1.9	–
			M-F	100	9.3	30.3	–	–	–	3.6	49.3	7.5	3.9	4.3	–
315	OVERRIPE, FRESH	2-05-428	A-F	75	3.1	25.2	–	–	–	1.6	42.3	2.8	0.4	1.0	–
			M-F	100	4.1	33.6	–	–	–	2.1	56.4	3.8	0.5	1.3	–
	WHEATGRASS, SLENDER *Agropyron trachycaulum*														
316	FRESH	2-05-439	A-F	32	3.0	10.4	23.1	–	–	1.4	14.5	3.1	1.8	1.9	–
			M-F	100	9.2	32.1	71.4	–	–	4.2	45.0	9.5	5.7	5.9	–
	WHEATGRASS, WESTERN *Agropyron smithii*														
317	FRESH	2-05-410	A-F	39	3.3	–	25.1	15.8	2.0	–	–	4.3	2.8	2.8	–
			M-F	100	8.5	–	63.8	40.0	5.0	–	–	11.0	7.1	7.1	–
	WHITE CLOVER–SEE CLOVER, WHITE														
	WILD OATS *Avena fatua*														
318	FRESH	2-03-393	A-F	36	2.6	8.4	–	–	–	1.4	20.8	2.5	1.3	1.4	–
			M-F	100	7.4	23.5	–	–	–	3.8	58.2	7.1	3.5	4.0	–
	WILD-RYE, RUSSIAN *Elymus junceus*														
319	FRESH	2-05-469	A-F	35	3.1	7.8	25.6	–	–	0.9	18.9	4.2	2.8	2.8	–
			M-F	100	8.8	22.4	73.2	–	–	2.6	54.0	12.1	8.1	8.0	–
	WILLOW *Salix spp*														
320	BROWSE, FRESH	2-05-472	A-F	41	3.0	11.2	–	–	–	2.0	20.8	4.0	1.2	2.5	–
			M-F	100	7.4	27.2	–	–	–	4.9	50.7	9.8	2.8	6.1	–
	WINTERFAT, COMMON *Eurotia lanata*														
321	BROWSE, STEM CURED, FRESH	2-26-142	A-F	80	12.7	–	–	–	–	2.2	–	8.7	5.4	5.6	–
			M-F	100	15.8	–	–	–	–	2.8	–	10.8	6.7	7.0	–
	WIREGRASS (THREE-AWN) *Aristida spp*														
322	FRESH	2-04-838	A-F	39	2.3	13.3	–	–	–	0.9	18.6	3.8	2.3	2.4	–
			M-F	100	5.9	34.2	–	–	–	2.3	47.8	9.8	6.0	6.1	–
	YARROW, COMMON *Achillea millefolium*														
323	FRESH	2-05-511	A-F	23	2.5	4.6	9.4	–	1.5	0.4	12.3	3.2	2.2	2.1	–
			M-F	100	10.8	20.1	40.8	–	6.6	1.8	53.6	13.8	9.6	9.3	–
	YELLOWBRUSH (RABBITBRUSH, SMALL) *Chrysothamus stenophyllus*														
324	BROWSE, FRESH	2-03-849	A-F	40	3.2	9.5	–	–	–	2.2	19.8	5.4	–	–	–
			M-F	100	8.0	23.8	–	–	–	5.4	49.4	13.4	–	–	–
325	BROWSE, STEM CURED, FRESH	2-07-997	A-F	85	7.7	–	–	–	–	8.5	–	4.8	1.9	2.4	–
			M-F	100	9.0	–	–	–	–	10.0	–	5.7	2.2	2.9	–
	YELLOW TREFOIL (MEDIC, BLACK) *Medicago lupulina*														
326	FRESH	2-03-070	A-F	23	2.3	5.6	–	–	–	0.8	9.1	4.9	3.8	3.5	–
			M-F	100	10.2	24.7	–	–	–	3.4	40.2	21.6	16.7	15.6	–
	YUCCA *Yucca smalliana*														
327	FRESH	2-08-561	A-F	49	3.5	21.1	–	–	–	1.0	20.0	3.8	1.4	2.2	–
			M-F	100	7.1	42.7	–	–	–	2.0	40.5	7.7	2.8	4.5	–
	YUCCA, SOAPWEED *Yucca glauca*														
328	FRESH	2-04-268	A-F	45	3.3	14.5	–	–	–	0.8	23.7	2.5	0.9	1.2	–
			M-F	100	7.5	32.3	–	–	–	1.8	52.9	5.5	2.1	2.7	–

[1]Neutral Detergent Fiber (NDF) and Acid Detergent Fiber (ADF) values taken from *Nutrient Requirements of Dairy Cattle,* 6th rev. ed., NRC, National Academy Press, 1988, pp. 90–110, Table 7–1.

Entry Number	TDN Ruminant	TDN Swine	TDN Horse	Digestible Energy Ruminant		Digestible Energy Swine		Digestible Energy Horse		Metabolizable Energy Ruminant		Metabolizable Energy Swine		Metabolizable Energy Poultry ME_n		Metabolizable Energy Horse		Net Energy Ruminant NE_m		Net Energy Ruminant NE_g		Net Energy Lactating Cows NE_{lc}	
	%	%	%	Mcal		kcal		Mcal		Mcal		kcal		kcal		Mcal		Mcal		Mcal		Mcal	
				lb	kg	lb	kg	lb	kg	lb	kg	lb	kg	lb	kg	lb	kg	lb	kg	lb	kg	lb	kg
306	12	–	–	0.25	0.55	–	–	–	–	0.21	0.47	–	–	–	–	–	–	0.13	0.29	0.08	0.17	0.13	0.29
	59	–	–	1.23	2.71	–	–	–	–	1.04	2.29	–	–	–	–	–	–	0.65	1.43	0.39	0.85	0.65	1.44
307	13	–	–	0.25	0.54	–	–	–	–	0.21	0.47	–	–	–	–	–	–	0.12	0.27	0.07	0.16	0.12	0.27
	69	–	–	1.34	2.95	–	–	–	–	1.15	2.53	–	–	–	–	–	–	0.67	1.47	0.40	0.88	0.67	1.47
308	17	–	–	0.34	0.74	–	–	–	–	0.29	0.63	–	–	–	–	–	–	0.17	0.37	0.10	0.22	0.17	0.38
	65	–	–	1.28	2.82	–	–	–	–	1.09	2.40	–	–	–	–	–	–	0.64	1.42	0.38	0.83	0.65	1.43
309	18	–	–	0.36	0.79	–	–	–	–	0.30	0.65	–	–	–	–	–	–	0.16	0.36	0.08	0.18	0.18	0.39
	55	–	–	1.09	2.41	–	–	–	–	0.90	1.98	–	–	–	–	–	–	0.49	1.09	0.24	0.53	0.54	1.19
310	17	–	–	0.33	0.73	–	–	–	–	0.29	0.64	–	–	–	–	–	–	0.19	0.42	0.13	0.28	0.18	0.40
	78	–	–	1.49	3.28	–	–	–	–	1.30	2.86	–	–	–	–	–	–	0.86	1.89	0.57	1.25	0.81	1.80
311	21	–	–	0.42	0.92	–	–	–	–	0.36	0.79	–	–	–	–	–	–	0.21	0.47	0.13	0.29	0.21	0.47
	66	–	–	1.32	2.92	–	–	–	–	1.13	2.50	–	–	–	–	–	–	0.68	1.49	0.41	0.90	0.68	1.49
312	21	–	–	0.41	0.90	–	–	–	–	0.35	0.78	–	–	–	–	–	–	0.20	0.45	0.13	0.28	0.20	0.44
	75	–	–	1.43	3.16	–	–	–	–	1.24	2.74	–	–	–	–	–	–	0.71	1.57	0.44	0.97	0.70	1.54
313	–	–	–	–	–	–	–	–	–	–	–	–	–	–	–	–	–	–	–	–	–	–	–
	–	–	–	–	–	–	–	–	–	–	–	–	–	–	–	–	–	–	–	–	–	–	–
314	26	–	–	0.52	1.15	–	–	–	–	0.44	0.96	–	–	–	–	–	–	0.26	0.57	0.14	0.32	0.27	0.59
	58	–	–	1.16	2.56	–	–	–	–	0.97	2.14	–	–	–	–	–	–	0.58	1.27	0.32	0.70	0.60	1.32
315	45	–	–	0.90	1.99	–	–	–	–	0.76	1.67	–	–	–	–	–	–	0.45	1.00	0.26	0.56	0.46	1.02
	60	–	–	1.21	2.66	–	–	–	–	1.01	2.23	–	–	–	–	–	–	0.60	1.33	0.34	0.75	0.62	1.36
316	19	–	–	0.37	0.83	–	–	–	–	0.31	0.69	–	–	–	–	–	–	0.19	0.41	0.10	0.23	0.19	0.43
	58	–	–	1.16	2.55	–	–	–	–	0.97	2.13	–	–	–	–	–	–	0.58	1.27	0.32	0.70	0.60	1.32
317	–	–	–	–	–	–	–	–	–	–	–	–	–	–	–	–	–	–	–	–	–	–	–
	–	–	–	–	–	–	–	–	–	–	–	–	–	–	–	–	–	–	–	–	–	–	–
318	23	–	15	0.46	1.02	–	–	0.29	0.64	0.39	0.87	–	–	–	–	0.24	0.52	0.24	0.52	0.14	0.31	0.24	0.52
	64	–	42	1.29	2.84	–	–	0.81	1.78	1.10	2.42	–	–	–	–	0.66	1.46	0.66	1.45	0.39	0.87	0.66	1.46
319	23	–	–	0.46	1.00	–	–	–	–	0.39	0.86	–	–	–	–	–	–	0.23	0.52	0.14	0.31	0.23	0.52
	65	–	–	1.30	2.88	–	–	–	–	1.11	2.45	–	–	–	–	–	–	0.67	1.48	0.40	0.89	0.67	1.48
320	23	–	–	0.49	1.08	–	–	–	–	0.41	0.90	–	–	–	–	–	–	0.27	0.59	0.16	0.35	0.27	0.59
	55	–	–	1.19	2.62	–	–	–	–	1.00	2.20	–	–	–	–	–	–	0.65	1.43	0.38	0.84	0.65	1.44
321	28	–	–	0.60	1.33	–	–	–	–	0.48	1.05	–	–	–	–	–	–	–	–	–	–	–	–
	35	–	–	0.76	1.66	–	–	–	–	0.59	1.31	–	–	–	–	–	–	–	–	–	–	–	–
322	23	–	–	0.47	1.04	–	–	–	–	0.40	0.87	–	–	–	–	–	–	0.24	0.52	0.13	0.30	0.24	0.53
	60	–	–	1.21	2.67	–	–	–	–	1.02	2.24	–	–	–	–	–	–	0.61	1.34	0.35	0.76	0.62	1.37
323	15	–	–	0.30	0.67	–	–	–	–	0.26	0.57	–	–	–	–	–	–	0.15	0.34	0.09	0.20	0.15	0.34
	66	–	–	1.32	2.91	–	–	–	–	1.13	2.49	–	–	–	–	–	–	0.67	1.47	0.40	0.88	0.67	1.47
324	–	–	–	–	–	–	–	–	–	–	–	–	–	–	–	–	–	–	–	–	–	–	–
	–	–	–	–	–	–	–	–	–	–	–	–	–	–	–	–	–	–	–	–	–	–	–
325	38	–	–	0.82	1.80	–	–	–	–	0.58	1.28	–	–	–	–	–	–	–	–	–	–	–	–
	44	–	–	0.96	2.12	–	–	–	–	0.68	1.50	–	–	–	–	–	–	–	–	–	–	–	–
326	14	–	–	0.29	0.64	–	–	–	–	0.25	0.54	–	–	–	–	–	–	0.16	0.35	0.10	0.21	0.16	0.34
	61	–	–	1.27	2.81	–	–	–	–	1.08	2.39	–	–	–	–	–	–	0.69	1.52	0.42	0.93	0.69	1.51
327	27	–	–	0.54	1.20	–	–	–	–	0.45	0.99	–	–	–	–	–	–	0.25	0.55	0.12	0.27	0.27	0.60
	56	–	–	1.10	2.42	–	–	–	–	0.91	1.99	–	–	–	–	–	–	0.50	1.11	0.25	0.55	0.55	1.21
328	24	–	–	0.50	1.10	–	–	–	–	0.41	0.91	–	–	–	–	–	–	0.26	0.56	0.14	0.31	0.27	0.59
	52	–	–	1.11	2.45	–	–	–	–	0.92	2.02	–	–	–	–	–	–	0.57	1.26	0.31	0.69	0.60	1.31

TABLE V-1E PASTURE AND RANGE PLANTS, MINERAL AND VITAMIN COMPOSITION OF FEEDS, DATA EXPRESSED **AS-FED** AND **MOISTURE-FREE**—*(Continued)*

Entry Number	Feed Name Description	Moisture Basis: A-F (as-fed) or M-F (moisture-free)	Dry Matter	Macro Minerals: Calcium (Ca)	Phosphorus (P)	Sodium (Na)	Chlorine (Cl)	Magnesium (Mg)	Potassium (K)	Sulfur (S)	Micro Minerals: Cobalt (Co)	Copper (Cu)	Iodine (I)	Iron (Fe)	Manganese (Mn)	Selenium (Se)	Zinc (Zn)
			%	%	%	%	%	%	%	%	ppm or mg/kg	ppm or mg/kg	ppm or mg/kg	%	ppm or mg/kg	ppm or mg/kg	ppm or mg/kg
	VETCH, COMMON *Vicia sativa*																
306	FRESH	**A-F**	**20**	**0.27**	**0.07**	–	–	**0.04**	**0.51**	**0.02**	–	**2.0**	–	**0.008**	**24.5**	–	–
		M-F	100	1.32	0.34	–	–	0.20	2.50	0.10	–	9.7	–	0.040	119.9	–	–
	VETCH, HAIRY *Vicia villosa*																
307	FRESH	**A-F**	**18**	**0.20**	**0.06**	–	–	–	**0.41**	–	–	–	–	–	–	–	–
		M-F	100	1.10	0.33	–	–	–	2.25	–	–	–	–	–	–	–	–
	VETCH-OATS *Vicia spp-Avena sativa*																
308	FRESH	**A-F**	**26**	**0.18**	**0.08**	–	–	**0.06**	**0.45**	–	–	–	–	–	–	–	–
		M-F	100	0.68	0.30	–	–	0.23	1.70	–	–	–	–	–	–	–	–
	VINE-MESQUITE *Panicum obtusum*																
309	MATURE, FRESH	**A-F**	**33**	**0.11**	**0.06**	–	–	–	–	–	–	–	–	–	–	–	–
		M-F	100	0.32	0.19	–	–	–	–	–	–	–	–	–	–	–	–
	WHEAT *Triticum aestivum*																
310	IMMATURE, FRESH	**A-F**	**22**	**0.09**	**0.09**	**0.04**	–	**0.05**	**0.78**	**0.05**	–	–	–	**0.003**	–	–	–
		M-F	100	0.42	0.40	0.18	–	0.21	3.50	0.22	–	–	–	0.010	–	–	–
	WHEATGRASS, BLUEBUNCH *Agropyron spicatum*																
311	PREBLOOM, FRESH	**A-F**	**32**	**0.12**	**0.08**	**0.02**	–	**0.06**	**1.03**	–	**0.063**	**2.6**	–	**0.008**	**15.7**	–	**9.2**
		M-F	100	0.37	0.25	0.07	–	0.19	3.27	–	0.197	8.1	–	0.025	49.7	–	29.2
	WHEATGRASS, CRESTED *Agropyron desertorum*																
312	IMMATURE, FRESH	**A-F**	**28**	**0.13**	**0.10**	–	–	**0.08**	–	–	–	–	–	–	–	–	–
		M-F	100	0.44	0.33	–	–	0.28	–	–	–	–	–	–	–	–	–
313	EARLY BLOOM, FRESH	**A-F**	**41**	**0.09**	**0.07**	–	–	–	–	–	–	–	–	–	–	–	–
		M-F	100	0.22	0.18	–	–	–	–	–	–	–	–	–	–	–	–
314	FULL BLOOM, FRESH	**A-F**	**45**	**0.17**	**0.07**	**0.00**	–	**0.04**	**0.47**	**0.21**	–	**2.9**	–	–	**19.5**	–	**6.1**
		M-F	100	0.37	0.15	0.01	–	0.09	1.04	0.47	–	6.5	–	–	43.3	–	13.5
315	OVERRIPE, FRESH	**A-F**	**75**	**0.20**	**0.11**	–	–	–	–	–	**0.187**	**6.3**	–	–	**39.7**	–	–
		M-F	100	0.27	0.15	–	–	–	–	–	0.250	8.4	–	–	52.9	–	–
	WHEATGRASS, SLENDER *Agropyron trachycaulum*																
316	FRESH	**A-F**	**32**	**0.16**	**0.05**	**0.03**	–	**0.08**	**1.04**	–	**0.067**	**1.5**	–	**0.003**	**19.8**	–	**7.4**
		M-F	100	0.49	0.14	0.09	–	0.24	3.22	–	0.207	4.7	–	0.008	61.2	–	23.0
	WHEATGRASS, WESTERN *Agropyron smithii*																
317	FRESH	**A-F**	**39**	**0.15**	**0.07**	**0.11**	–	**0.06**	**1.02**	**0.06**	**0.052**	**0.4**	–	**0.005**	**15.1**	–	**10.7**
		M-F	100	0.38	0.18	0.29	–	0.14	2.59	0.16	0.130	1.1	–	0.013	38.3	–	27.2
	WHITE CLOVER–SEE CLOVER, WHITE																
	WILD OATS *Avena fatua*																
318	FRESH	**A-F**	**36**	**0.09**	**0.10**	–	–	–	–	–	–	–	–	–	–	–	–
		M-F	100	0.25	0.27	–	–	–	–	–	–	–	–	–	–	–	–
	WILD-RYE, RUSSIAN *Elymus junceus*																
319	FRESH	**A-F**	**35**	**0.11**	**0.06**	**0.08**	–	**0.05**	**1.06**	–	**0.097**	**1.1**	–	**0.004**	**9.5**	–	**5.7**
		M-F	100	0.31	0.17	0.23	–	0.14	3.03	–	0.276	3.2	–	0.011	27.1	–	16.3
	WILLOW *Salix spp*																
320	BROWSE, FRESH	**A-F**	**41**	–	–	–	–	–	–	–	–	–	–	–	–	–	–
		M-F	100	–	–	–	–	–	–	–	–	–	–	–	–	–	–
	WINTERFAT, COMMON *Eurotia lanata*																
321	BROWSE, STEM CURED, FRESH	**A-F**	**80**	**1.58**	**0.09**	–	–	–	–	–	–	–	–	–	–	–	–
		M-F	100	1.98	0.12	–	–	–	–	–	–	–	–	–	–	–	–
	WIREGRASS (THREE-AWN) *Aristida spp*																
322	FRESH	**A-F**	**39**	–	–	–	–	–	–	–	–	–	–	–	–	–	–
		M-F	100	–	–	–	–	–	–	–	–	–	–	–	–	–	–
	YARROW, COMMON *Achillea millefolium*																
323	FRESH	**A-F**	**23**	–	**0.07**	–	–	–	–	–	–	–	–	–	–	–	–
		M-F	100	–	0.28	–	–	–	–	–	–	–	–	–	–	–	–
	YELLOWBRUSH (RABBITBRUSH, SMALL) *Chrysothamus stenophyllus*																
324	BROWSE, FRESH	**A-F**	**40**	**0.88**	**0.04**	–	–	–	–	–	–	–	–	–	–	–	–
		M-F	100	2.21	0.11	–	–	–	–	–	–	–	–	–	–	–	–
325	BROWSE, STEM CURED, FRESH	**A-F**	**85**	**1.58**	**0.07**	–	–	–	–	–	–	–	–	–	–	–	–
		M-F	100	1.86	0.08	–	–	–	–	–	–	–	–	–	–	–	–
	YELLOW TREFOIL (MEDIC, BLACK) *Medicago lupulina*																
326	FRESH	**A-F**	**23**	–	–	–	–	–	–	–	–	–	–	–	–	–	–
		M-F	100	–	–	–	–	–	–	–	–	–	–	–	–	–	–
	YUCCA *Yucca smalliana*																
327	FRESH	**A-F**	**49**	–	–	–	–	–	–	–	–	–	–	–	–	–	–
		M-F	100	–	–	–	–	–	–	–	–	–	–	–	–	–	–
	YUCCA, SOAPWEED *Yucca glauca*																
328	FRESH	**A-F**	**45**	**0.41**	**0.08**	–	–	–	–	–	–	–	–	–	–	–	–
		M-F	100	0.92	0.18	–	–	–	–	–	–	–	–	–	–	–	–

Entry Number	Fat-Soluble Vitamins: A (1 mg Carotene = 1667 IU Vit A)	Carotene (Provitamin A)	D	E	K	Water-Soluble Vitamins: B-12	Biotin	Choline	Folacin (Folic Acid)	Niacin	Pantothenic Acid (B-3)	(Pyri-doxine) B-6	Ribo-flavin (B-2)	Thiamin (B-1)
	IU/g	ppm or mg/kg	IU/kg	ppm or mg/kg	ppm or mg/kg	ppm or mg/kg	ppm or mg/kg	ppm or mg/kg	ppm or mg/kg	ppm or mg/kg	ppm or mg/kg	ppm or mg/kg	ppm or mg/kg	ppm or mg/kg
306	–	–	–	–	–	–	–	–	–	–	–	–	–	–
	–	–	–	–	–	–	–	–	–	–	–	–	–	–
307	–	–	–	–	–	–	–	–	–	–	–	–	–	–
	–	–	–	–	–	–	–	–	–	–	–	–	–	–
308	–	–	–	–	–	–	–	–	–	–	–	–	–	–
	–	–	–	–	–	–	–	–	–	–	–	–	–	–
309	–	–	–	–	–	–	–	–	–	–	–	–	–	–
	–	–	–	–	–	–	–	–	–	–	–	–	–	–
310	**192.5**	**115.4**	–	–	–	–	–	–	–	**13**	**4.7**	–	**6.1**	–
	866.9	520.1	–	–	–	–	–	–	–	57	21.2	–	27.6	–
311	**173.6**	**104.2**	–	–	–	–	–	–	–	–	–	–	–	–
	549.1	329.4	–	–	–	–	–	–	–	–	–	–	–	–
312	**205.8**	**123.4**	–	–	–	–	–	–	–	–	–	–	–	–
	722.9	433.6	–	–	–	–	–	–	–	–	–	–	–	–
313	–	–	–	–	–	–	–	–	–	–	–	–	–	–
	–	–	–	–	–	–	–	–	–	–	–	–	–	–
314	**115.1**	**69.0**	–	–	–	–	–	–	–	–	–	–	–	–
	255.8	153.4	–	–	–	–	–	–	–	–	–	–	–	–
315	**0.3**	**0.2**	–	–	–	–	–	–	–	–	–	–	–	–
	0.4	0.2	–	–	–	–	–	–	–	–	–	–	–	–
316	–	–	–	–	–	–	–	–	–	–	–	–	**3.4**	**1.6**
	–	–	–	–	–	–	–	–	–	–	–	–	10.6	5.1
317	**83.6**	**50.2**	–	–	–	–	–	–	–	–	–	–	–	–
	212.0	127.2	–	–	–	–	–	–	–	–	–	–	–	–
318	–	–	–	–	–	–	–	–	–	–	–	–	–	–
	–	–	–	–	–	–	–	–	–	–	–	–	–	–
319	–	–	–	–	–	–	–	–	–	–	–	–	–	–
	–	–	–	–	–	–	–	–	–	–	–	–	–	–
320	–	–	–	–	–	–	–	–	–	–	–	–	–	–
	–	–	–	–	–	–	–	–	–	–	–	–	–	–
321	**24.1**	**14.5**	–	–	–	–	–	–	–	–	–	–	–	–
	30.2	18.1	–	–	–	–	–	–	–	–	–	–	–	–
322	–	–	–	–	–	–	–	–	–	–	–	–	–	–
	–	–	–	–	–	–	–	–	–	–	–	–	–	–
323	–	–	–	–	–	–	–	–	–	–	–	–	–	–
	–	–	–	–	–	–	–	–	–	–	–	–	–	–
324	–	–	–	–	–	–	–	–	–	–	–	–	–	–
	–	–	–	–	–	–	–	–	–	–	–	–	–	–
325	–	–	–	–	–	–	–	–	–	–	–	–	–	–
	–	–	–	–	–	–	–	–	–	–	–	–	–	–
326	–	–	–	–	–	–	–	–	–	–	–	–	–	–
	–	–	–	–	–	–	–	–	–	–	–	–	–	–
327	–	–	–	–	–	–	–	–	–	–	–	–	–	–
	–	–	–	–	–	–	–	–	–	–	–	–	–	–
328	–	–	–	–	–	–	–	–	–	–	–	–	–	–
	–	–	–	–	–	–	–	–	–	–	–	–	–	–

TABLE V-1F MINERAL SUPPLEMENTS, COMPOSITION, DATA EXPRESSED **AS-FED** AND **MOISTURE-FREE** (See footnotes at end of table.)

Entry Number	Feed Name Description	International Feed Number	Moisture Basis: A-F (as-fed) or M-F (moisture-free)	Proximate Analysis: Dry Matter	Ash	Crude Fiber	Ether Extract (Fat)	N-Free Extract	Crude Protein (6.25 x N)	Digestible Protein: Ruminant	Non-Ruminant	Horse
				%	%	%	%	%	%	%	%	%
1	AMMONIUM PHOSPHATE, MONOBASIC	6-09-338	A-F	98	53.0	–	–	–	69.4	–	–	–
			M-F	100	54.2	–	–	–	71.0	–	–	–
2	AMMONIUM PHOSPHATE, DIBASIC	6-00-370	A-F	98	35.5	–	–	–	112.9	–	–	–
			M-F	100	36.3	–	–	–	115.5	–	–	–
3	AMMONIUM POLYPHOSPHATE SOLUTION	6-08-042	A-F	60	–	–	–	–	54.8	–	–	–
			M-F	100	–	–	–	–	92.0	–	–	–
4	AMMONIUM SULFATE [1]	6-09-339	A-F	100	–	–	–	–	134.1	–	–	–
			M-F	100	–	–	–	–	134.1	–	–	–
5	BONE, CHARCOAL	6-00-402	A-F	94	79.3	3.7	1.1	1.8	–	–	–	–
			M-F	100	84.6	3.9	1.1	1.9	–	–	–	–
6	BONE MEAL	6-00-397	A-F	94	60.5	2.9	6.5	–	24.8	17.1	–	–
			M-F	100	64.1	3.0	6.9	–	26.2	18.1	–	–
7	BONE MEAL, STEAMED*	6-00-400	A-F	95	67.3	1.9	3.6	3.8	18.6	–	–	–
			M-F	100	70.7	2.0	3.8	4.0	19.5	–	–	–
8	CALCIUM CARBONATE*	6-01-069	A-F	100	97.1	–	–	–	–	–	–	–
			M-F	100	97.5	–	–	–	–	–	–	–
9	CALCIUM GLUCONATE	6-01-073	A-F	99	–	–	–	–	–	–	–	–
			M-F	100	–	–	–	–	–	–	–	–
10	CALCIUM HYDROXIDE	6-14-014	A-F	98	–	–	–	–	–	–	–	–
			M-F	100	–	–	–	–	–	–	–	–
11	CALCIUM IODATE	6-01-075	A-F	98	–	–	–	–	–	–	–	–
			M-F	100	–	–	–	–	–	–	–	–
12	CALCIUM OXIDE	6-14-003	A-F	97	–	–	–	–	–	–	–	–
			M-F	100	–	–	–	–	–	–	–	–
13	CALCIUM PERIODATE*	6-09-335	A-F	–	–	–	–	–	–	–	–	–
			M-F	100	–	–	–	–	–	–	–	–
14	CALCIUM PHOSPHATE, MONOBASIC, FROM DEFLUORINATED PHOSPHORIC ACID	6-01-082	A-F	99	87.1	–	–	–	–	–	–	–
			M-F	100	88.3	–	–	–	–	–	–	–
15	CALCIUM PHOSPHATE, MONOBASIC, FROM FURNACED PHOSPHORIC ACID	6-26-334	A-F	96	–	–	–	–	–	–	–	–
			M-F	100	–	–	–	–	–	–	–	–
16	CALCIUM PHOSPHATE, DIBASIC, FROM DEFLUORINATED PHOSPHORIC ACID*	6-01-080	A-F	97	89.7	–	–	–	–	–	–	–
			M-F	100	92.5	–	–	–	–	–	–	–
17	CALCIUM PHOSPHATE, DIBASIC, FROM FURNACED PHOSPHORIC ACID*	6-26-335	A-F	97	85.6	–	–	–	–	–	–	–
			M-F	100	88.2	–	–	–	–	–	–	–
18	CALCIUM PHOSPHATE, DIBASIC, FROM FURNACE PHOSPHORIC ACID (DICALCIUM PHOSPHATE), $CaHPO_4$ [2]	6-28-335	A-F	100	–	–	–	–	–	–	–	–
			M-F	100	–	–	–	–	–	–	–	–
19	CALCIUM PHOSPHATE, TRIBASIC, FROM FURNACED PHOSPHORIC ACID	6-01-084	A-F	98	92.1	–	–	–	–	–	–	–
			M-F	100	93.7	–	–	–	–	–	–	–
20	CALCIUM SULFATE, ANHYDROUS	6-01-087	A-F	–	–	–	–	–	–	–	–	–
			M-F	100	–	–	–	–	–	–	–	–
21	CALCIUM SULFATE (GYPSUM)	6-01-090	A-F	95	–	–	–	–	–	–	–	–
			M-F	100	–	–	–	–	–	–	–	–
22	COBALT ACETATE	6-01-554	A-F	99	–	–	–	–	–	–	–	–
			M-F	100	–	–	–	–	–	–	–	–
23	COBALT CARBONATE*	6-01-566	A-F	99	–	–	–	–	–	–	–	–
			M-F	100	–	–	–	–	–	–	–	–
24	COBALT SULFATE*	6-01-564	A-F	–	–	–	–	–	–	–	–	–
			M-F	100	–	–	–	–	–	–	–	–
25	COBALTOUS CHLORIDE	6-01-558	A-F	98	–	–	–	–	–	–	–	–
			M-F	100	–	–	–	–	–	–	–	–
26	COBALTOUS OXIDE	6-01-560	A-F	99	–	–	–	–	–	–	–	–
			M-F	100	–	–	–	–	–	–	–	–
27	COLLOIDAL CLAY (SOFT ROCK PHOSPHATE)	6-03-947	A-F	100	–	–	–	–	–	–	–	–
			M-F	100	–	–	–	–	–	–	–	–
28	COPPER (CUPRIC) CARBONATE	6-01-703	A-F	98	–	–	–	–	–	–	–	–
			M-F	100	–	–	–	–	–	–	–	–
29	COPPER (CUPRIC) CHLORIDE	6-01-705	A-F	99	–	–	–	–	–	–	–	–
			M-F	100	–	–	–	–	–	–	–	–
30	COPPER CHLORIDE, DIHYDRATE, $CuCl_2 \cdot 2H_2O$ [2]	6-01-706	A-F	100	–	–	–	–	–	–	–	–
			M-F	100	–	–	–	–	–	–	–	–
31	COPPER (CUPRIC) GLUCONATE	6-01-707	A-F	99	–	–	–	–	–	–	–	–
			M-F	100	–	–	–	–	–	–	–	–
32	COPPER (CUPRIC) HYDROXIDE	6-01-709	A-F	98	–	–	–	–	–	–	–	–
			M-F	100	–	–	–	–	–	–	–	–
33	COPPER (CUPRIC) ORTHOPHOSPHATE	6-01-713	A-F	99	–	–	–	–	–	–	–	–
			M-F	100	–	–	–	–	–	–	–	–

*Sources most commonly used in commercial feeds.

Entry Number	Macro Minerals							Micro Minerals							
	Calcium (Ca)	Phosphorus (P)	Sodium (Na)	Chlorine (Cl)	Magnesium (Mg)	Potassium (K)	Sulfur (S)	Cobalt (Co)	Copper (Cu)	Fluorine (F)	Iodine (I)	Iron (Fe)	Manganese (Mn)	Selenium (Se)	Zinc (Zn)
	%	%	%	%	%	%	%	ppm or mg/kg	ppm or mg/kg	ppm or mg/kg	ppm or mg/kg	%	ppm or mg/kg	ppm or mg/kg	ppm or mg/kg
1	**0.38**	**24.42**	**0.08**	–	**0.46**	**0.14**	**0.82**	–	**86**	**1833**	–	**0.991**	**462**	–	**640**
	0.39	24.99	0.08	–	0.47	0.14	0.84	–	88	1876	–	1.014	473	–	655
2	**0.50**	**20.09**	**0.04**	–	**0.45**	–	**2.47**	–	**81**	**1548**	–	**1.514**	**504**	–	**303**
	0.52	20.54	0.04	–	0.46	–	2.53	–	83	1582	–	1.548	516	–	309
3	**0.10**	**13.44**	–	–	–	–	**0.50**	–	–	**1341**	–	**0.505**	–	–	–
	0.17	22.58	–	–	–	–	0.85	–	–	2254	–	0.848	–	–	–
4	–	–	–	–	–	–	**24.10**	–	**1**	–	–	**10.000**	**1**	–	–
	–	–	–	–	–	–	24.10	–	1	–	–	10.000	1	–	–
5	**31.92**	**14.84**	–	–	**0.55**	**0.15**	–	–	–	–	–	–	–	–	–
	34.08	15.85	–	–	0.59	0.16	–	–	–	–	–	–	–	–	–
6	**24.52**	**11.43**	**0.61**	**0.22**	**0.35**	**0.14**	**0.12**	–	**19**	**2014**	–	**0.057**	**9**	–	**377**
	25.98	12.11	0.65	0.23	0.37	0.14	0.13	–	20	2134	–	0.060	9	–	399
7	**25.98**	**11.80**	**0.40**	**0.01**	**0.78**	**0.18**	**0.34**	**0**	**162**	**637**	**29**	**0.085**	**37**	–	**362**
	27.31	12.40	0.42	0.01	0.82	0.19	0.36	0	170	669	31	0.089	39	–	381
8	**37.97**	**0.04**	**0.07**	**0.04**	**0.41**	**0.04**	**0.08**	–	**14**	**0**	–	**0.059**	**159**	**0.07**	**17**
	38.13	0.04	0.07	0.04	0.41	0.04	0.08	–	14	0	–	0.059	160	0.07	17
9	**8.85**	–	–	–	–	–	–	–	–	–	–	–	–	–	–
	8.94	–	–	–	–	–	–	–	–	–	–	–	–	–	–
10	**53.01**	–	–	–	–	–	–	–	–	–	–	–	–	–	–
	54.09	–	–	–	–	–	–	–	–	–	–	–	–	–	–
11	**10.00**	–	–	–	–	–	–	–	–	–	**635040**	–	–	–	–
	10.20	–	–	–	–	–	–	–	–	–	648000	–	–	–	–
12	**69.33**	–	–	–	–	–	–	–	–	–	–	–	–	–	–
	71.47	–	–	–	–	–	–	–	–	–	–	–	–	–	–
13	–	–	–	–	–	–	–	–	–	–	–	–	–	–	–
	31.01	–	–	–	–	–	–	–	–	–	392800	–	–	–	–
14	**18.55**	**20.98**	**0.06**	–	**0.81**	**0.40**	**0.81**	**5**	**5**	**1410**	–	**1.007**	**201**	–	**419**
	18.80	21.27	0.06	–	0.82	0.41	0.82	5	5	1429	–	1.021	204	–	424
15	**22.00**	**23.00**	–	–	–	–	–	–	–	**300**	–	–	–	–	–
	22.92	23.96	–	–	–	–	–	–	–	313	–	–	–	–	–
16	**22.00**	**18.43**	**1.56**	–	**0.51**	**0.10**	**0.69**	**8**	**9**	**940**	–	**0.844**	**253**	–	**122**
	22.67	19.00	1.61	–	0.52	0.10	0.71	9	9	969	–	0.870	261	–	126
17	**23.00**	**18.50**	**0.08**	–	**0.60**	**0.07**	–	–	**80**	**1150**	–	**1.000**	**300**	**0.60**	**220**
	23.71	19.07	0.08	–	0.62	0.07	–	–	82	1186	–	1.031	309	0.62	227
18	**26.30**	**18.07**	–	–	–	**0.07**	–	–	–	–	–	–	**300**	–	–
	26.30	18.07	–	–	–	0.07	–	–	–	–	–	–	300	–	–
19	**30.90**	**17.04**	**0.17**	–	–	–	–	–	–	**501**	–	–	–	–	–
	31.44	17.34	0.17	–	–	–	–	–	–	510	–	–	–	–	–
20	–	–	–	–	–	–	–	–	–	–	–	–	–	–	–
	29.43	–	–	0.00	–	–	23.54	–	–	–	–	0.003	–	–	–
21	**21.86**	–	–	–	**0.46**	–	**16.20**	–	–	**27**	–	–	–	–	–
	23.01	–	–	–	0.48	–	17.05	–	–	29	–	–	–	–	–
22	–	–	–	–	–	–	–	**234234**	–	–	–	–	–	–	–
	–	–	–	–	–	–	–	236600	–	–	–	–	–	–	–
23	–	–	**0.25**	**0.01**	–	–	**0.03**	**465000**	**15**	–	–	**0.020**	**100**	–	**15**
	–	–	0.25	0.01	–	–	0.03	469697	15	–	–	0.021	101	–	15
24	–	–	–	–	–	–	–	–	–	–	–	–	–	–	–
	–	–	–	–	0.04	–	11.46	211055	10	–	–	0.002	20	–	–
25	–	–	–	**29.20**	–	–	**0.07**	**242648**	**20**	–	–	**0.003**	–	–	**196**
	–	–	–	29.80	–	–	0.07	247600	20	–	–	0.004	–	–	200
26	–	–	–	**0.01**	–	–	**0.20**	**703494**	–	–	–	**0.050**	–	–	–
	–	–	–	0.01	–	–	0.20	710600	–	–	–	0.051	–	–	–
27	**16.01**	**9.00**	**0.10**	–	**0.38**	–	–	–	–	**12061**	–	**1.911**	**995**	–	–
	16.09	9.05	0.10	–	0.38	–	–	–	–	12121	–	1.920	1000	–	–
28	–	–	–	–	–	–	**0.17**	–	**530000**	–	–	**0.147**	–	–	**196**
	–	–	–	–	–	–	0.17	–	540816	–	–	0.150	–	–	200
29	–	–	–	**41.17**	–	–	**0.03**	–	**368973**	–	–	**0.006**	–	–	–
	–	–	–	41.59	–	–	0.03	–	372700	–	–	0.007	–	–	–
30	–	–	–	**41.18**	–	–	–	–	**369000**	–	–	–	–	–	–
	–	–	–	41.18	–	–	–	–	369000	–	–	–	–	–	–
31	–	–	–	–	–	–	–	–	**133353**	–	–	–	–	–	–
	–	–	–	–	–	–	–	–	134700	–	–	–	–	–	–
32	–	–	–	–	–	–	–	–	**602994**	–	–	–	–	–	–
	–	–	–	–	–	–	–	–	615300	–	–	–	–	–	–
33	–	**14.11**	–	–	–	–	–	–	**434214**	–	–	–	–	–	–
	–	14.25	–	–	–	–	–	–	438600	–	–	–	–	–	–

(Continued)

TABLE V-1F MINERAL SUPPLEMENTS, COMPOSITION, DATA EXPRESSED **AS-FED** AND **MOISTURE-FREE**—*(Continued)*

				Proximate Analysis						Digestible Protein		
Entry Number	Feed Name Description	International Feed Number	Moisture Basis: A-F (as-fed) or M-F (moisture-free)	Dry Matter	Ash	Crude Fiber	Ether Extract (Fat)	N-Free Extract	Crude Protein (6.25 x N)	Ruminant	Non-Ruminant	Horse
				%	%	%	%	%	%	%	%	%
34	COPPER (CUPRIC) OXIDE	6-01-711	A-F	99	–	–	–	–	–	–	–	–
			M-F	100	–	–	–	–	–	–	–	–
35	COPPER (CUPRIC) SULFATE, PENTAHYDRATE*	6-01-719	A-F	99	–	–	–	–	–	–	–	–
			M-F	100	–	–	–	–	–	–	–	–
36	COPPER (CUPROUS) IODIDE	6-01-721	A-F	–	–	–	–	–	–	–	–	–
			M-F	100	–	–	–	–	–	–	–	–
37	COPPER (CUPROUS) OXIDE*	6-28-224	A-F	99	–	–	–	–	–	–	–	–
			M-F	100	–	–	–	–	–	–	–	–
38	CURACAO PHOSPHATE, GROUND	6-05-586	A-F	99	94.1	–	–	–	–	–	–	–
			M-F	100	95.0	–	–	–	–	–	–	–
39	DIIODOSALICYLIC ACID*	6-01-787	A-F	99	–	–	–	–	–	–	–	–
			M-F	100	–	–	–	–	–	–	–	–
40	ETHYLENEDIAMINE DIHYDROIODIDE*	6-01-842	A-F	98	–	–	–	–	54.3	–	–	–
			M-F	100	–	–	–	–	55.4	–	–	–
41	FERRIC (IRON) AMMONIUM CITRATE	6-01-857	A-F	99	–	–	–	–	42.1	–	–	–
			M-F	100	–	–	–	–	42.5	–	–	–
42	FERRIC (IRON) CHLORIDE	6-01-865	A-F	98	–	–	–	–	–	–	–	–
			M-F	100	–	–	–	–	–	–	–	–
43	FERRIC (IRON) CHLORIDE, HEXAHYDRATE, $FeCl_3 \cdot 6H_2O$[2]	6-28-101	A-F	100	–	–	–	–	–	–	–	–
			M-F	100	–	–	–	–	–	–	–	–
44	FERRIC (IRON) OXIDE*	6-02-431	A-F	97	–	–	–	–	–	–	–	–
			M-F	100	–	–	–	–	–	–	–	–
45	FERROUS (IRON) CARBONATE*	6-01-863	A-F	99	–	–	–	–	–	–	–	–
			M-F	100	–	–	–	–	–	–	–	–
46	FERROUS (IRON) FUMARATE	6-08-097	A-F	99	–	–	–	–	–	–	–	–
			M-F	100	–	–	–	–	–	–	–	–
47	FERROUS (IRON) GLUCONATE	6-01-867	A-F	99	–	–	–	–	–	–	–	–
			M-F	100	–	–	–	–	–	–	–	–
48	FERROUS (IRON) OXIDE	6-20-728	A-F	97	–	–	–	–	–	–	–	–
			M-F	100	–	–	–	–	–	–	–	–
49	FERROUS (IRON) SULFATE, MONOHYDRATE*	6-01-869	A-F	98	98.0	–	–	–	–	–	–	–
			M-F	100	100.0	–	–	–	–	–	–	–
50	FERROUS (IRON) SULFATE, HEPTAHYDRATE	6-20-734	A-F	99	–	–	–	–	–	–	–	–
			M-F	100	–	–	–	–	–	–	–	–
51	KELP (SEAWEED) *Laminariales* (order), *Fucales* (order) WHOLE, DEHY	1-08-073	A-F	91	35.0	6.5	0.5	42.4	6.5	2.9	2.2	3.4
			M-F	100	38.6	7.1	0.5	46.7	7.1	3.2	2.4	3.7
52	LIMESTONE, GROUND*	6-02-632	A-F	100	93.8	–	–	–	–	–	–	–
			M-F	100	94.1	–	–	–	–	–	–	–
53	LIMESTONE, MAGNESIUM (DOLOMITE), GROUND*	6-02-633	A-F	100	–	–	–	–	–	–	–	–
			M-F	100	–	–	–	–	–	–	–	–
54	MAGNESIUM CARBONATE	6-02-754	A-F	98	–	–	–	–	–	–	–	–
			M-F	100	–	–	–	–	–	–	–	–
55	MAGNESIUM HYDROXIDE	6-26-012	A-F	98	–	–	–	–	–	–	–	–
			M-F	100	–	–	–	–	–	–	–	–
56	MAGNESIUM OXIDE*	6-02-756	A-F	98	98.3	–	–	–	–	–	–	–
			M-F	100	100.0	–	–	–	–	–	–	–
57	MAGNESIUM SULFATE (EPSOM SALTS)*	6-02-758	A-F	99	–	–	–	–	–	–	–	–
			M-F	100	–	–	–	–	–	–	–	–
58	MANGANESE CHLORIDE	6-03-038	A-F	99	–	–	–	–	–	–	–	–
			M-F	100	–	–	–	–	–	–	–	–
59	MANGANESE DIOXIDE	6-03-042	A-F	98	–	–	–	–	–	–	–	–
			M-F	100	–	–	–	–	–	–	–	–
60	MANGANOUS (MANGANESE) CARBONATE	6-03-036	A-F	97	–	–	–	–	–	–	–	–
			M-F	100	–	–	–	–	–	–	–	–
61	MANGANOUS (MANGANESE) CITRATE	6-03-040	A-F	99	–	–	–	–	–	–	–	–
			M-F	100	–	–	–	–	–	–	–	–
62	MANGANOUS (MANGANESE) GLUCONATE	6-03-044	A-F	99	–	–	–	–	–	–	–	–
			M-F	100	–	–	–	–	–	–	–	–
63	MANGANOUS (MANGANESE) OXIDE*	6-03-054	A-F	99	–	–	–	–	–	–	–	–
			M-F	100	–	–	–	–	–	–	–	–
64	MANGANOUS (MANGANESE) SULFATE*	6-26-136	A-F	100	–	–	–	–	–	–	–	–
			M-F	100	–	–	–	–	–	–	–	–
65	MANGANOUS (MANGANESE) SULFATE, MONOHYDRATE, $MnSO_4 \cdot H_2O$[2]	6-28-103	A-F	100	–	–	–	–	–	–	–	–
			M-F	100	–	–	–	–	–	–	–	–

*Sources most commonly used in commercial feeds.

Entry Number	Macro Minerals: Calcium (Ca)	Phosphorus (P)	Sodium (Na)	Chlorine (Cl)	Magnesium (Mg)	Potassium (K)	Sulfur (S)	Micro Minerals: Cobalt (Co)	Copper (Cu)	Fluorine (F)	Iodine (I)	Iron (Fe)	Manganese (Mn)	Selenium (Se)	Zinc (Zn)
	%	%	%	%	%	%	%	ppm or mg/kg	ppm or mg/kg	ppm or mg/kg	ppm or mg/kg	%	ppm or mg/kg	ppm or mg/kg	ppm or mg/kg
34	**0.01**	–	–	–	**0.00**	–	–	–	**753827**	–	–	**0.020**	**10**	–	**800**
	0.01	–	–	–	0.00	–	–	–	761441	–	–	0.020	10	–	808
35	–	–	–	–	–	–	**13.25**	–	**250976**	–	–	**0.010**	**2**	–	**9**
	–	–	–	–	–	–	13.32	–	252257	–	–	0.011	2	–	9
36	–	–	–	–	–	–	–	–	–	–	–	–	–	–	–
	–	–	–	–	–	–	–	–	333600	–	666400	–	–	–	–
37	–	–	–	–	–	–	–	–	**879318**	–	–	–	–	–	–
	–	–	–	–	–	–	–	–	888200	–	–	–	–	–	–
38	**35.10**	**14.24**	**0.20**	–	**0.80**	–	–	–	–	**5445**	–	**0.347**	–	–	–
	35.45	14.38	0.20	–	0.81	–	–	–	–	5500	–	0.350	–	–	–
39	–	–	–	–	–	–	–	–	–	–	**644391**	–	–	–	–
	–	–	–	–	–	–	–	–	–	–	650900	–	–	–	–
40	–	–	–	–	–	–	–	–	–	–	**787234**	–	–	–	–
	–	–	–	–	–	–	–	–	–	–	803300	–	–	–	–
41	–	–	–	–	–	–	–	–	–	–	–	**15.840**	–	–	–
	–	–	–	–	–	–	–	–	–	–	–	16.000	–	–	–
42	–	–	–	**64.27**	–	–	–	–	–	–	–	**33.742**	–	–	–
	–	–	–	65.58	–	–	–	–	–	–	–	34.431	–	–	–
43	–	–	–	**39.35**	–	–	–	–	–	–	–	**57.000**	**3000**	–	–
	–	–	–	39.35	–	–	–	–	–	–	–	57.000	3000	–	–
44	**0.36**	**0.10**	–	–	**0.66**	–	–	–	–	–	–	**58.800**	**3600**	–	–
	0.37	0.10	–	–	0.68	–	–	–	–	–	–	60.619	3711	–	–
45	**1.24**	**0.01**	–	–	**0.33**	–	**1.77**	**200**	**3000**	–	–	**40.667**	**9000**	–	–
	1.25	0.01	–	–	0.33	–	1.79	202	3030	–	–	41.077	9091	–	–
46	–	–	–	–	–	–	–	–	–	–	–	**32.542**	–	–	–
	–	–	–	–	–	–	–	–	–	–	–	32.870	–	–	–
47	–	–	–	–	–	–	–	–	–	–	–	**11.465**	–	–	–
	–	–	–	–	–	–	–	–	–	–	–	11.580	–	–	–
48	–	–	–	–	–	–	–	–	–	–	–	**75.369**	–	–	–
	–	–	–	–	–	–	–	–	–	–	–	77.701	–	–	–
49	–	–	–	–	**0.50**	–	**17.80**	–	–	–	–	**31.000**	–	–	–
	–	–	–	–	0.51	–	18.16	–	–	–	–	31.633	–	–	–
50	–	–	–	–	**0.21**	–	**11.00**	–	**100**	–	–	**20.899**	**0**	–	**100**
	–	–	–	–	0.21	–	11.06	–	101	–	–	21.006	0	–	101
51	**2.47**	**0.28**	–	–	**0.85**	–	–	–	–	–	–	–	–	–	–
	2.72	0.31	–	–	0.93	–	–	–	–	–	–	–	–	–	–
52	**37.12**	**0.21**	**0.06**	**0.03**	**1.13**	**0.11**	**0.04**	–	**11**	–	–	**0.357**	**269**	–	**19**
	37.22	0.22	0.06	0.03	1.13	0.11	0.04	–	11	–	–	0.358	270	–	19
53	**20.61**	**0.02**	**0.38**	**0.12**	**10.37**	**0.27**	**0.01**	–	**20**	–	–	**0.053**	–	–	–
	20.65	0.02	0.38	0.12	10.39	0.27	0.01	–	20	–	–	0.053	–	–	–
54	**0.02**	–	–	–	**30.19**	–	–	–	–	–	–	**0.020**	–	–	–
	0.02	–	–	–	30.81	–	–	–	–	–	–	0.020	–	–	–
55	–	–	–	–	**40.86**	–	–	–	–	–	–	–	–	–	–
	–	–	–	–	41.69	–	–	–	–	–	–	–	–	–	–
56	**1.66**	–	–	–	**55.19**	–	**0.10**	**501**	**5**	**251**	–	**1.048**	**80**	**0.35**	**9**
	1.69	–	–	–	56.15	–	0.10	510	5	255	–	1.066	82	0.35	9
57	**0.02**	–	–	**0.01**	**9.60**	–	**13.00**	–	–	–	–	–	–	**10.12**	–
	0.02	–	–	0.01	9.68	–	13.11	–	–	–	–	–	–	10.20	–
58	–	–	–	**35.47**	–	–	–	–	–	–	–	–	**274824**	–	–
	–	–	–	35.83	–	–	–	–	–	–	–	–	277600	–	–
59	–	–	–	–	–	–	–	–	–	–	–	–	**619262**	–	–
	–	–	–	–	–	–	–	–	–	–	–	–	631900	–	–
60	–	–	–	–	–	–	–	–	–	–	–	–	**463660**	–	–
	–	–	–	–	–	–	–	–	–	–	–	–	478000	–	–
61	–	–	–	–	–	–	–	–	–	–	–	–	**300465**	–	–
	–	–	–	–	–	–	–	–	–	–	–	–	303500	–	–
62	–	–	–	–	–	–	–	–	–	–	–	–	**320463**	–	–
	–	–	–	–	–	–	–	–	–	–	–	–	323700	–	–
63	**0.16**	**0.10**	**0.06**	–	**0.70**	**0.58**	**0.01**	**300**	**724**	–	–	**3.436**	**620217**	–	**1349**
	0.16	0.10	0.06	–	0.71	0.59	0.01	303	731	–	–	3.470	626482	–	1363
64	–	–	–	–	**0.30**	–	**19.01**	–	–	–	–	**0.040**	**250000**	–	–
	–	–	–	–	0.30	–	19.10	–	–	–	–	0.041	251256	–	–
65	–	–	–	–	–	–	**18.97**	–	–	–	–	–	**325000**	–	–
	–	–	–	–	–	–	18.97	–	–	–	–	–	325000	–	–

(Continued)

TABLE V-1F MINERAL SUPPLEMENTS, COMPOSITION, DATA EXPRESSED **AS-FED** AND **MOISTURE-FREE**—*(Continued)*

Entry Number	Feed Name Description	International Feed Number	Moisture Basis: A-F (as-fed) or M-F (moisture-free)	Proximate Analysis						Digestible Protein		
				Dry Matter	Ash	Crude Fiber	Ether Extract (Fat)	N-Free Extract	Crude Protein (6.25 x N)	Ruminant	Non-Ruminant	Horse
				%	%	%	%	%	%	%	%	%
66	OYSTER SHELLS, GROUND (FLOUR)*	6-03-481	A-F	99	79.0	1.8	0.3	17.0	0.7	–	–	–
			M-F	100	79.9	1.8	0.3	17.2	0.7	–	–	–
67	PHOSPHATE, DEFLUORINATED	6-01-780	A-F	100	99.3	–	–	–	–	–	–	–
			M-F	100	99.7	–	–	–	–	–	–	–
68	PHOSPHATE ROCK, GROUND (RAW)	6-03-945	A-F	–	–	–	–	–	–	–	–	–
			M-F	100	–	–	–	–	–	–	–	–
69	PHOSPHATE ROCK, LOW FLUORINE*	6-03-946	A-F	–	–	–	–	–	–	–	–	–
			M-F	100	–	–	–	–	–	–	–	–
70	PHOSPHATE SOFT ROCK (COLLOIDAL CLAY)	6-03-947	A-F	100	–	–	–	–	–	–	–	–
			M-F	100	–	–	–	–	–	–	–	–
71	PHOSPHORIC ACID, FEED GRADE (ORTHO)*	6-03-707	A-F	75	–	–	–	–	–	–	–	–
			M-F	100	–	–	–	–	–	–	–	–
72	POTASSIUM BICARBONATE	6-09-337	A-F	99	–	–	–	–	–	–	–	–
			M-F	100	–	–	–	–	–	–	–	–
73	POTASSIUM CARBONATE	6-09-336	A-F	–	–	–	–	–	–	–	–	–
			M-F	100	–	–	–	–	–	–	–	–
74	POTASSIUM CHLORIDE*	6-03-755	A-F	100	98.9	–	–	–	–	–	–	–
			M-F	100	99.0	–	–	–	–	–	–	–
75	POTASSIUM IODIDE*	6-03-759	A-F	–	–	–	–	–	–	–	–	–
			M-F	100	–	–	–	–	–	–	–	–
76	POTASSIUM MAGNESIUM SULFATE	6-06-177	A-F	98	–	–	–	–	–	–	–	–
			M-F	100	–	–	–	–	–	–	–	–
77	POTASSIUM SULFATE	6-08-098	A-F	98	97.0	–	–	–	–	–	–	–
			M-F	100	99.0	–	–	–	–	–	–	–
78	SEAWEED (KELP) *Laminariales* (order), *Fucales* (order) WHOLE, DEHY	1-08-073	A-F	91	35.0	6.5	0.5	42.4	6.5	2.9	2.2	3.4
			M-F	100	38.6	7.1	0.5	46.7	7.1	3.2	2.4	3.7
79	SODIUM BICARBONATE*	6-04-272	A-F	100	–	–	–	–	–	–	–	–
			M-F	100	–	–	–	–	–	–	–	–
80	SODIUM CHLORIDE*	6-04-152	A-F	97	93.0	–	–	–	–	–	–	–
			M-F	100	95.9	–	–	–	–	–	–	–
81	SODIUM FLUORIDE	6-04-275	A-F	–	–	–	–	–	–	–	–	–
			M-F	100	–	–	–	–	–	–	–	–
82	SODIUM IODIDE*	6-04-279	A-F	–	–	–	–	–	–	–	–	–
			M-F	100	–	–	–	–	–	–	–	–
83	SODIUM PHOSPHATE, MONOBASIC*	6-04-288	A-F	97	96.9	–	–	–	–	–	–	–
			M-F	100	99.8	–	–	–	–	–	–	–
84	SODIUM PHOSPHATE, DIBASIC	6-04-286	A-F	97	96.7	–	–	–	–	–	–	–
			M-F	100	99.7	–	–	–	–	–	–	–
85	SODIUM SELENATE*	6-26-014	A-F	99	–	–	–	–	–	–	–	–
			M-F	100	–	–	–	–	–	–	–	–
86	SODIUM SELENITE*	6-26-013	A-F	99	–	–	–	–	–	–	–	–
			M-F	100	–	–	–	–	–	–	–	–
87	SODIUM SULFATE, DECAHYDRATE	6-04-291	A-F	97	–	–	–	–	–	–	–	–
			M-F	100	–	–	–	–	–	–	–	–
88	SODIUM TRIPOLYPHOSPHATE*	6-08-076	A-F	97	89.7	–	–	–	–	–	–	–
			M-F	100	92.8	–	–	–	–	–	–	–
89	SULFUR*	6-04-705	A-F	99	–	–	–	–	–	–	–	–
			M-F	100	–	–	–	–	–	–	–	–
90	ZINC ACETATE	6-05-547	A-F	99	–	–	–	–	–	–	–	–
			M-F	100	–	–	–	–	–	–	–	–
91	ZINC CARBONATE	6-05-549	A-F	99	–	–	–	–	–	–	–	–
			M-F	100	–	–	–	–	–	–	–	–
92	ZINC CARBONATE, TETRAHYDRATE	6-29-585	A-F	98	–	–	–	–	–	–	–	–
			M-F	100	–	–	–	–	–	–	–	–
93	ZINC CHLORIDE	6-05-551	A-F	98	–	–	–	–	–	–	–	–
			M-F	100	–	–	–	–	–	–	–	–
94	ZINC OXIDE*	6-05-553	A-F	–	–	–	–	–	–	–	–	–
			M-F	100	–	–	–	–	–	–	–	–
95	ZINC SULFATE, MONOHYDRATE*	6-05-555	A-F	99	–	–	–	–	–	–	–	–
			M-F	100	–	–	–	–	–	–	–	–
96	ZINC SULFATE, HEPTAHYDRATE	6-20-729	A-F	98	–	–	–	–	–	–	–	–
			M-F	100	–	–	–	–	–	–	–	–

*Sources most commonly used in commercial feeds.

[1]From: *Nutrient Requirements of Dairy Cattle,* 6th rev. ed., NRC, National Academy Press, 1988, pp. 111–112, Table 7-2.

[2]From: *Nutrient Requirements of Swine,* 9th rev. ed., NRC, National Academy Press, 1988, pp. 60–61, Table 6-4.

Entry Number	Macro Minerals							Micro Minerals							
	Calcium (Ca)	Phosphorus (P)	Sodium (Na)	Chlorine (Cl)	Magnesium (Mg)	Potassium (K)	Sulfur (S)	Cobalt (Co)	Copper (Cu)	Fluorine (F)	Iodine (I)	Iron (Fe)	Manganese (Mn)	Selenium (Se)	Zinc (Zn)
	%	%	%	%	%	%	%	ppm or mg/kg	ppm or mg/kg	ppm or mg/kg	ppm or mg/kg	%	ppm or mg/kg	ppm or mg/kg	ppm or mg/kg
66	**35.85**	**0.10**	**0.21**	**0.01**	**0.24**	**0.10**	–	–	**15**	–	–	**0.254**	**178**	–	**7**
	36.27	0.10	0.21	0.01	0.24	0.10	–	–	15	–	–	0.257	180	–	7
67	**31.99**	**17.07**	**3.26**	–	**0.29**	**0.10**	**0.13**	**10**	**40**	**1794**	–	**0.840**	**496**	–	**90**
	32.10	17.13	3.27	–	0.29	0.10	0.13	10	41	1800	–	0.843	498	–	90
68	–	–	–	–	–	–	–	–	–	–	–	–	–	–	–
	35.00	13.00	0.03	–	0.41	0.06	–	10	10	37000	–	1.680	200	–	100
69	–	–	–	–	–	–	–	–	–	–	–	–	–	–	–
	36.00	14.00	–	–	–	–	–	–	–	4500	–	–	–	–	–
70	**16.01**	**9.00**	**0.10**	–	**0.38**	–	–	–	–	**12061**	–	**1.911**	**995**	–	–
	16.09	9.05	0.10	–	0.38	–	–	–	–	12121	–	1.920	1000	–	–
71	**0.14**	**20.88**	**0.18**	–	**0.40**	**0.06**	**1.56**	–	**17**	**1900**	–	**0.913**	**500**	–	**210**
	0.18	27.84	0.23	–	0.53	0.08	2.08	–	22	2533	–	1.217	667	–	280
72	–	–	–	–	–	**38.67**	–	–	–	–	–	–	–	–	–
	–	–	–	–	–	39.06	–	–	–	–	–	–	–	–	–
73	–	–	–	–	–	–	–	–	–	–	–	–	–	–	–
	–	–	–	–	–	56.58	–	–	–	–	–	–	–	–	–
74	**0.05**	–	**1.00**	**46.88**	**0.23**	**51.31**	**0.32**	–	**7**	–	–	**0.061**	**7**	–	**9**
	0.05	–	1.00	46.93	0.23	51.37	0.32	–	7	–	–	0.061	7	–	9
75	–	–	–	–	–	–	–	–	–	–	–	–	–	–	–
	–	–	0.01	–	–	21.00	–	–	–	–	681700	–	–	–	–
76	**0.06**	–	**0.75**	**1.24**	**11.58**	**18.45**	**21.97**	–	**2**	**10**	–	**0.010**	**10**	–	**9**
	0.06	–	0.76	1.27	11.82	18.83	22.42	–	2	10	–	0.011	10	–	9
77	**0.15**	–	**0.09**	**1.50**	**0.59**	**43.04**	**17.64**	–	**3**	–	–	**0.069**	**9**	–	**4**
	0.15	–	0.09	1.53	0.60	43.92	18.00	–	3	–	–	0.071	9	–	4
78	**2.47**	**0.28**	–	–	**0.85**	–	–	–	–	–	–	–	–	–	–
	2.72	0.31	–	–	0.93	–	–	–	–	–	–	–	–	–	–
79	–	–	**26.87**	–	–	**0.01**	–	–	–	**450138**	–	**0.001**	–	–	–
	–	–	27.00	–	–	0.01	–	–	–	452400	–	0.002	–	–	–
80	–	–	**38.17**	**58.46**	–	–	–	–	–	–	–	–	–	–	–
	–	–	39.34	60.26	–	–	–	–	–	–	–	–	–	–	–
81	–	–	–	–	–	–	–	–	–	–	–	–	–	–	–
	–	–	54.75	–	–	–	–	–	–	452400	–	–	–	–	–
82	–	–	–	–	–	–	–	–	–	–	–	–	–	–	–
	–	–	15.33	–	–	–	–	–	–	–	–	–	–	–	–
83	**0.04**	**24.84**	**18.65**	–	–	**0.14**	–	–	**7**	–	–	–	–	–	**5**
	0.04	25.60	19.23	–	–	0.14	–	–	7	–	–	–	–	–	5
84	–	**21.65**	**31.04**	–	–	–	–	–	–	**300**	–	–	–	–	–
	–	22.32	32.00	–	–	–	–	–	–	309	–	–	–	–	–
85	–	–	**24.18**	–	–	–	–	–	–	–	–	–	–	**415898.96**	–
	–	–	24.42	–	–	–	–	–	–	–	–	–	–	420100.00	–
86	–	–	**26.40**	–	**0.01**	–	–	–	**10**	–	–	**0.031**	–	**452927.78**	–
	–	–	26.60	–	0.01	–	–	–	10	–	–	0.031	–	456386.34	–
87	–	–	**31.33**	–	–	–	**9.66**	–	–	–	–	**0.001**	–	–	–
	–	–	32.30	–	–	–	9.96	–	–	–	–	0.002	–	–	–
88	–	**24.53**	**30.18**	–	–	–	–	–	–	**247**	–	**0.004**	–	–	–
	–	25.38	31.23	–	–	–	–	–	–	256	–	0.004	–	–	–
89	–	–	–	–	–	–	**99.00**	–	–	–	–	–	–	–	**291951**
	–	–	–	–	–	–	100.00	–	–	–	–	–	–	–	294900
90	–	–	–	–	–	–	**0.07**	–	–	–	–	**0.001**	–	–	**294822**
	–	–	–	–	–	–	0.07	–	–	–	–	0.002	–	–	297800
91	–	–	–	–	–	–	–	–	–	–	–	–	–	–	**516285**
	–	–	–	–	–	–	–	–	–	–	–	–	–	–	521500
92	–	–	–	–	–	–	–	–	–	–	–	–	–	–	**534100**
	–	–	–	–	–	–	–	–	–	–	–	–	–	–	545000
93	–	–	–	**50.99**	–	–	**0.07**	–	–	–	–	**0.001**	–	–	**470008**
	–	–	–	52.03	–	–	0.07	–	–	–	–	0.002	–	–	479600
94	–	–	–	–	–	–	–	–	–	–	–	–	–	–	–
	4.29	–	–	–	0.30	–	1.00	1500	500	–	–	0.551	800	–	724968
95	**0.05**	–	–	**0.20**	–	–	**17.62**	–	**55**	–	–	**0.053**	**169**	**99.24**	**359073**
	0.05	–	–	0.20	–	–	17.76	–	56	–	–	0.053	171	100.00	361815
96	–	–	–	–	–	–	**10.93**	–	–	–	–	–	–	–	**222460**
	–	–	–	–	–	–	11.15	–	–	–	–	–	–	–	227000

TABLE V-1G VITAMIN SUPPLEMENTS, COMPOSITION OF FEEDS, DATA EXPRESSED **AS-FED** AND **MOISTURE-FREE**

Entry Number	Feed Name Description	International Feed Number	Moisture Basis: A-F (as-fed) or M-F (moisture-free)	Proximate Analysis: Dry Matter	Ash	Crude Fiber	Neutral Det. Fib. (NDF)	Acid Det. Fib. (ADF)	Lignin	Ether Extract (Fat)	N-Free Extract	Crude Protein	Digestible Protein: Ruminant	Swine	Horse
				%	%	%	%	%	%	%	%	%	%	%	%
	ALFALFA (LUCERNE) *Medicago sativa*														
1	IMMATURE, FRESH	2-00-177	A-F	21	2.3	4.3	–	–	–	0.7	8.6	5.4	4.5	4.0	–
			M-F	100	10.7	20.1	–	–	–	3.4	40.4	25.3	21.0	18.6	–
2	HAY, SUN-CURED	1-00-078	A-F	90	8.6	28.2	35.4	30.9	8.9	1.7	35.9	16.0	11.2	7.5	11.9
			M-F	100	9.5	31.2	39.2	34.2	9.8	1.9	39.7	17.7	12.4	8.3	13.2
3	HAY, SUN-CURED, PELLETED	1-00-124	A-F	92	10.2	25.4	–	31.9	6.1	1.9	39.1	15.7	11.5	10.2	10.4
			M-F	100	11.1	27.5	–	34.6	6.6	2.1	42.3	17.0	12.5	11.1	11.3
4	LEAVES, SUN-CURED, GROUND	1-00-146	A-F	88	9.2	15.0	–	–	–	2.7	40.7	20.1	15.8	14.3	13.9
			M-F	100	10.5	17.1	–	–	–	3.1	46.3	22.9	18.0	16.3	15.9
5	MEAL, DEHY, 17% PROTEIN	1-00-023	A-F	92	9.7	24.0	41.3	31.5	9.7	2.8	37.8	17.4	12.6	11.7	11.8
			M-F	100	10.6	26.2	45.0	34.3	10.6	3.0	41.2	18.9	13.7	12.8	12.8
6	MEAL, DEHY, 20% PROTEIN	1-00-024	A-F	92	10.2	20.8	–	27.0	–	3.3	37.1	20.2	15.1	14.4	14.0
			M-F	100	11.1	22.7	–	29.4	–	3.6	40.5	22.1	16.4	15.7	15.2
7	MEAL, DEHY, 22% PROTEIN	1-07-851	A-F	93	10.2	18.3	–	25.3	–	4.1	38.1	22.2	16.4	16.4	15.5
			M-F	100	11.0	19.8	–	27.3	–	4.4	41.0	23.9	17.7	17.7	16.6
	ANIMAL														
8	LIVER-GLANDS, GROUND	5-00-390	A-F	93	5.9	1.8	–	–	–	15.8	3.3	66.5	58.2	64.8	66.5
			M-F	100	6.3	1.9	–	–	–	16.9	3.6	71.3	62.4	69.5	71.2
9	LIVER, MEAL	5-00-389	A-F	93	6.3	1.4	–	–	–	15.7	3.2	66.1	57.9	64.4	66.1
			M-F	100	6.8	1.5	–	–	–	17.0	3.5	71.4	62.5	69.6	71.4
10	MEAT SOLUBLES, DEHY	5-00-393	A-F	90	5.7	–	–	–	–	–	–	80.0	70.6	78.3	82.0
			M-F	100	6.3	–	–	–	–	–	–	88.9	78.5	87.0	91.2
	BLUEGRASS, KENTUCKY *Poa pratensis*														
11	IMMATURE, FRESH	2-00-777	A-F	31	2.9	7.8	–	–	–	1.1	13.7	5.4	4.0	3.8	–
			M-F	100	9.4	25.2	–	–	–	3.6	44.4	17.4	12.9	12.2	–
	BREWERS GRAINS														
12	DEHY	5-02-141	A-F	92	3.6	13.0	38.7	23.9	4.6	6.6	41.6	27.3	20.1	21.8	21.0
			M-F	100	4.0	14.1	42.0	26.0	5.0	7.1	45.2	29.6	21.8	23.7	22.8
	BUTTERMILK, CATTLE *Bos taurus*														
13	CONDENSED	5-01-159	A-F	29	3.6	0.1	–	–	–	2.4	12.4	10.8	9.2	10.0	9.5
			M-F	100	12.2	0.3	–	–	–	8.3	42.3	36.9	31.1	33.9	32.4
	CARROT *Daucus* spp														
14	ROOTS, FRESH	4-01-145	A-F	11	1.0	1.1	–	–	–	0.2	8.1	1.2	0.8	0.8	0.8
			M-F	100	8.4	9.5	–	–	–	1.3	70.7	10.0	7.3	7.2	7.0
	CATTLE *Bos taurus*														
15	BUTTERMILK, CONDENSED	5-01-159	A-F	29	3.6	0.1	–	–	–	2.4	12.4	10.8	9.2	10.0	9.5
			M-F	100	12.2	0.3	–	–	–	8.3	42.3	36.9	31.1	33.9	32.4
16	LIVER, FRESH	5-01-166	A-F	28	1.4	0.2	–	–	–	5.1	1.9	19.5	17.0	19.0	19.4
			M-F	100	4.9	0.6	–	–	–	18.3	6.7	69.6	60.9	67.8	69.4
17	WHEY, DEHY	4-01-182	A-F	93	8.8	0.2	0.3	0.2	–	0.8	70.2	13.3	8.9	13.1	9.6
			M-F	100	9.4	0.2	0.3	0.2	–	0.8	75.3	14.2	9.5	14.1	10.3
	COD, FISH *Gadus morrhua, Gadus macrocephalus*														
18	LIVER, MEAL	5-08-423	A-F	93	2.9	0.7	–	–	–	28.9	9.6	50.4	43.6	48.9	48.4
			M-F	100	3.1	0.8	–	–	–	31.2	10.4	54.5	47.2	52.8	52.3
19	LIVER OIL	7-01-993	A-F	100	–	–	–	–	–	99.5	–	–	–	–	–
			M-F	100	–	–	–	–	–	100.0	–	–	–	–	–
	CORN *Zea mays*														
20	DISTILLERS GRAINS WITH SOLUBLES, DEHY	5-02-843	A-F	92	4.5	9.1	–	–	–	9.2	41.9	27.1	17.2	25.7	22.1
			M-F	100	4.9	9.9	–	–	–	10.1	45.7	29.5	18.7	28.0	24.1
	CRAB *Callinectes sapidus, Cancer* spp, *Paralithodes camschatica*														
21	CANNERY RESIDUE, MEAL (CRAB MEAL)	5-01-663	A-F	92	41.1	10.7	–	–	–	2.2	5.9	32.2	27.1	30.9	27.9
			M-F	100	44.6	11.6	–	–	–	2.4	6.4	35.0	29.4	33.5	30.3
	DISTILLERS PRODUCTS (ALSO SEE CORN)														
22	GRAINS, DEHY	5-02-144	A-F	93	1.5	12.8	–	–	–	7.4	43.5	27.3	18.9	26.0	22.3
			M-F	100	1.6	13.8	–	–	–	8.0	47.0	29.5	20.4	28.1	24.1
23	SOLUBLES, DEHY	5-02-147	A-F	92	6.2	3.4	–	–	–	8.9	44.8	28.8	24.0	27.4	24.0
			M-F	100	6.7	3.7	–	–	–	9.7	48.6	31.3	26.0	29.8	26.1
	FATS AND OILS														
24	GERM OIL (WHEAT)	7-05-207	A-F	100	–	–	–	–	–	99.5	–	–	–	–	–
			M-F	100	–	–	–	–	–	100.0	–	–	–	–	–
25	LIVER OIL (COD)	7-01-993	A-F	100	–	–	–	–	–	99.5	–	–	–	–	–
			M-F	100	–	–	–	–	–	100.0	–	–	–	–	–
	FISH														
26	MEAL MECH EXTD	5-01-977	A-F	92	21.4	0.7	–	–	–	6.0	–	64.3	57.1	59.1	64.1
			M-F	100	23.3	0.8	–	–	–	6.6	–	70.2	62.3	64.5	70.0
27	SOLUBLES, CONDENSED	5-01-969	A-F	50	10.1	0.5	–	–	–	6.1	2.2	31.5	28.0	29.3	30.9
			M-F	100	20.0	1.0	–	–	–	12.2	4.4	62.5	55.6	58.1	61.3
28	SOLUBLES, DEHY	5-01-971	A-F	93	12.7	2.0	–	–	–	9.0	8.7	60.4	52.7	58.8	59.7
			M-F	100	13.7	2.1	–	–	–	9.7	9.4	65.1	56.8	63.4	64.3
	FISH, COD *Gadus morrhua, Gadus macrocephalus*														
29	LIVER, MEAL	5-08-423	A-F	93	2.9	0.7	–	–	–	28.9	9.6	50.4	43.6	48.9	48.4
			M-F	100	3.1	0.8	–	–	–	31.2	10.4	54.5	47.2	52.8	52.3

VITAMIN SUPPLEMENTS

Entry Number	TDN Ruminant	TDN Swine	TDN Horse	Digestible Energy Ruminant		DE Swine		DE Horse		Metabolizable Energy Ruminant		ME Swine		ME Poultry ME_n		ME Horse		Net Energy Ruminant NE_m		NE Ruminant NE_g		NE Lactating Cows NE_{lc}	
	%	%	%	Mcal		kcal		Mcal		Mcal		kcal		kcal		Mcal		Mcal		Mcal		Mcal	
				lb	kg	lb	kg	lb	kg	lb	kg	lb	kg	lb	kg	lb	kg	lb	kg	lb	kg	lb	kg
1	15	–	–	0.30	0.66	–	–	–	–	0.26	0.57	–	–	–	–	–	–	0.16	0.35	0.10	0.22	0.15	0.34
	70	–	–	1.40	3.08	–	–	–	–	1.21	2.66	–	–	–	–	–	–	0.74	1.63	0.47	1.03	0.72	1.59
2	51	32	48	1.03	2.28	–	–	0.90	1.98	0.90	1.99	–	–	–	–	0.74	1.62	0.48	1.06	0.25	0.55	0.49	1.07
	56	35	53	1.14	2.52	–	–	0.99	2.19	1.00	2.20	–	–	–	–	0.81	1.79	0.53	1.18	0.28	0.61	0.54	1.18
3	54	44	48	1.01	2.22	–	–	0.90	1.98	0.84	1.86	–	–	–	–	0.74	1.62	0.49	1.09	0.26	0.57	0.53	1.16
	58	48	52	1.09	2.40	–	–	0.97	2.14	0.91	2.01	–	–	–	–	0.80	1.76	0.53	1.18	0.28	0.62	0.57	1.25
4	56	63	56	1.13	2.49	–	–	1.03	2.27	0.96	2.12	–	–	–	–	0.85	1.86	0.58	1.28	0.35	0.77	0.58	1.28
	64	72	64	1.29	2.84	–	–	1.18	2.59	1.10	2.42	–	–	–	–	0.96	2.13	0.66	1.46	0.40	0.87	0.66	1.46
5	55	44	47	1.12	2.47	643	1418	0.89	1.96	0.96	2.12	601	1326	682	1504	0.73	1.61	0.60	1.32	0.36	0.79	0.57	1.25
	60	48	51	1.22	2.69	701	1546	0.97	2.14	1.05	2.32	656	1446	744	1640	0.80	1.75	0.65	1.44	0.39	0.86	0.62	1.37
6	57	48	50	1.07	2.36	943	2079	0.94	2.08	0.93	2.06	872	1923	737	1625	0.77	1.70	0.58	1.28	0.34	0.75	0.59	1.30
	62	52	55	1.17	2.58	1030	2270	1.03	2.27	1.02	2.25	952	2099	805	1774	0.84	1.86	0.64	1.40	0.37	0.82	0.64	1.42
7	60	49	52	1.20	2.65	991	2186	0.98	2.15	1.02	2.26	841	1855	768	1692	0.80	1.76	0.63	1.39	0.38	0.84	0.63	1.38
	65	53	56	1.29	2.85	1068	2355	1.05	2.32	1.10	2.43	907	1999	827	1823	0.86	1.90	0.68	1.49	0.41	0.90	0.67	1.49
8	90	95	–	1.87	4.11	1711	3771	–	–	1.69	3.73	1503	3314	1323	2917	–	–	1.11	2.44	0.79	1.73	1.02	2.25
	97	102	–	2.00	4.41	1833	4042	–	–	1.82	4.00	1611	3552	1418	3126	–	–	1.19	2.61	0.84	1.86	1.09	2.41
9	89	93	–	1.79	3.94	1867	4116	–	–	1.62	3.57	1645	3627	1306	2878	–	–	1.00	2.21	0.70	1.55	0.93	2.04
	97	101	–	1.93	4.26	2017	4446	–	–	1.75	3.85	1777	3918	1410	3109	–	–	1.08	2.38	0.76	1.67	1.00	2.21
10	–	–	–	–	–	–	–	–	–	–	–	–	–	–	–	–	–	–	–	–	–	–	–
	–	–	–	–	–	–	–	–	–	–	–	–	–	–	–	–	–	–	–	–	–	–	–
11	22	–	–	0.42	0.93	–	–	–	–	0.37	0.81	–	–	–	–	–	–	0.24	0.52	0.15	0.33	0.23	0.51
	72	–	–	1.38	3.03	–	–	–	–	1.19	2.61	–	–	–	–	–	–	0.77	1.70	0.49	1.08	0.75	1.64
12	65	66	48	1.25	2.76	1045	2303	–	–	1.01	2.22	1038	2288	1047	2308	–	–	0.64	1.41	0.39	0.86	0.67	1.48
	71	71	52	1.36	2.99	1134	2500	–	–	1.09	2.41	1127	2484	1137	2506	–	–	0.69	1.53	0.42	0.93	0.73	1.61
13	26	22	–	0.52	1.14	442	974	–	–	0.46	1.02	383	844	–	–	–	–	0.29	0.63	0.20	0.43	0.27	0.59
	88	76	–	1.77	3.89	1503	3314	–	–	1.58	3.48	1303	2872	–	–	–	–	0.98	2.15	0.67	1.48	0.91	2.01
14	10	10	8	0.19	0.43	208	458	0.14	0.31	0.17	0.38	195	430	208	458	0.12	0.26	0.10	0.23	0.07	0.16	0.10	0.22
	84	90	67	1.69	3.72	1805	3979	1.23	2.72	1.50	3.30	1695	3736	1805	3979	1.01	2.23	0.91	2.00	0.61	1.35	0.86	1.89
15	26	22	–	0.52	1.14	442	974	–	–	0.46	1.02	383	844	–	–	–	–	0.29	0.63	0.20	0.43	0.27	0.59
	88	76	–	1.77	3.89	1503	3314	–	–	1.58	3.48	1303	2872	–	–	–	–	0.98	2.15	0.67	1.48	0.91	2.01
16	29	31	–	0.58	1.28	615	1356	–	–	0.53	1.17	544	1200	–	–	–	–	0.34	0.76	0.25	0.54	0.32	0.70
	104	110	–	2.08	4.59	2201	4852	–	–	1.90	4.18	1947	4293	–	–	–	–	1.23	2.71	0.88	1.93	1.13	2.49
17	76	77	–	1.51	3.33	1444	3183	–	–	1.28	2.83	1411	3110	880	1939	–	–	0.87	1.92	0.59	1.30	0.78	1.71
	82	83	–	1.62	3.57	1549	3415	–	–	1.38	3.03	1514	3337	944	2081	–	–	0.93	2.06	0.63	1.40	0.83	1.83
18	109	118	–	2.18	4.81	2357	5196	–	–	2.01	4.44	2098	4625	–	–	–	–	1.27	2.81	0.92	2.03	1.17	2.57
	118	127	–	2.36	5.20	2548	5617	–	–	2.18	4.80	2268	5001	–	–	–	–	1.38	3.04	1.00	2.19	1.26	2.78
19	–	–	–	–	–	–	–	–	–	–	–	–	–	–	–	–	–	–	–	–	–	–	–
	–	–	–	–	–	–	–	–	–	–	–	–	–	–	–	–	–	–	–	–	–	–	–
20	81	79	–	1.55	3.43	1466	3232	–	–	1.34	2.95	1278	2817	1149	2533	–	–	0.92	2.03	0.63	1.40	0.86	1.89
	88	86	–	1.69	3.73	1597	3522	–	–	1.46	3.22	1392	3070	1252	2760	–	–	1.00	2.21	0.69	1.52	0.94	2.06
21	27	–	–	0.54	1.18	686	1511	–	–	0.35	0.78	555	1224	827	1823	–	–	0.07	0.16	–	–	0.25	0.54
	29	–	–	0.58	1.29	744	1640	–	–	0.39	0.85	602	1328	897	1977	–	–	0.08	0.18	–	–	0.27	0.59
22	76	83	–	1.52	3.36	1669	3680	–	–	1.35	2.97	1463	3225	1138	2509	–	–	0.87	1.92	0.59	1.31	0.82	1.80
	82	90	–	1.65	3.63	1804	3977	–	–	1.46	3.21	1581	3485	1230	2711	–	–	0.94	2.08	0.64	1.41	0.88	1.95
23	78	82	–	1.61	3.54	1637	3609	–	–	1.43	3.16	1433	3160	1316	2901	–	–	0.93	2.06	0.65	1.43	0.87	1.92
	84	89	–	1.74	3.85	1779	3922	–	–	1.56	3.43	1558	3434	1430	3153	–	–	1.02	2.24	0.70	1.55	0.95	2.09
24	–	–	–	–	–	–	–	–	–	–	–	–	–	–	–	–	–	–	–	–	–	–	–
	–	–	–	–	–	–	–	–	–	–	–	–	–	–	–	–	–	–	–	–	–	–	–
25	–	–	–	–	–	–	–	–	–	–	–	–	–	–	–	–	–	–	–	–	–	–	–
	–	–	–	–	–	–	–	–	–	–	–	–	–	–	–	–	–	–	–	–	–	–	–
26	67	66	–	1.34	2.95	1317	2903	–	–	1.17	2.57	1138	2508	1174	2587	–	–	0.64	1.40	0.39	0.86	0.63	1.39
	73	72	–	1.46	3.22	1438	3169	–	–	1.27	2.81	1242	2738	1281	2825	–	–	0.69	1.53	0.42	0.93	0.69	1.51
27	41	44	–	0.85	1.87	866	1909	–	–	0.78	1.73	736	1623	755	1665	–	–	0.54	1.20	0.38	0.84	0.44	0.97
	82	87	–	1.68	3.71	1717	3784	–	–	1.55	3.42	1459	3217	1497	3300	–	–	1.08	2.37	0.75	1.66	0.87	1.92
28	77	66	–	1.50	3.30	1467	3234	–	–	1.21	2.66	1278	2818	1322	2915	–	–	0.81	1.78	0.54	1.19	0.77	1.70
	83	71	–	1.61	3.56	1581	3485	–	–	1.30	2.87	1377	3036	1424	3140	–	–	0.87	1.92	0.58	1.28	0.83	1.83
29	109	118	–	2.18	4.81	2357	5196	–	–	2.01	4.44	2098	4625	–	–	–	–	1.27	2.81	0.92	2.03	1.17	2.57
	118	127	–	2.36	5.20	2548	5617	–	–	2.18	4.80	2268	5001	–	–	–	–	1.38	3.04	1.00	2.19	1.26	2.78

(Continued)

TABLE V-1G VITAMIN SUPPLEMENTS, MINERAL AND VITAMIN COMPOSITION OF FEEDS, DATA EXPRESSED AS-FED AND MOISTURE-FREE—*(Continued)*

Entry Number	Feed Name Description	Moisture Basis: A-F (as-fed) or M-F (moisture-free)	Dry Matter	Macro Minerals: Calcium (Ca)	Phosphorus (P)	Sodium (Na)	Chlorine (Cl)	Magnesium (Mg)	Potassium (K)	Sulfur (S)	Micro Minerals: Cobalt (Co)	Copper (Cu)	Iodine (I)	Iron (Fe)	Manganese (Mn)	Selenium (Se)	Zinc (Zn)
			%	%	%	%	%	%	%	%	ppm or mg/kg	ppm or mg/kg	ppm or mg/kg	%	ppm or mg/kg	ppm or mg/kg	ppm or mg/kg
	ALFALFA (LUCERNE) *Medicago sativa*																
1	IMMATURE, FRESH	A-F	21	0.50	0.09	0.04	0.08	0.05	0.48	0.13	–	–	–	0.006	6.7	–	–
		M-F	100	2.33	0.40	0.20	0.36	0.22	2.23	0.61	–	–	–	0.025	31.2	–	–
2	HAY, SUN-CURED	A-F	90	1.28	0.24	0.07	0.33	0.30	1.85	0.25	0.250	10.0	–	0.019	41.4	–	21.9
		M-F	100	1.42	0.26	0.08	0.37	0.33	2.05	0.28	0.277	11.0	–	0.020	45.8	–	24.3
3	HAY, SUN-CURED, PELLETED	A-F	92	1.48	0.20	–	–	0.25	–	0.24	–	0.6	–	–	–	–	–
		M-F	100	1.61	0.22	–	–	0.27	–	0.26	–	0.6	–	–	–	–	–
4	LEAVES, SUN-CURED, GROUND	A-F	88	2.32	0.24	–	–	0.36	1.47	–	–	–	–	0.031	29.8	–	–
		M-F	100	2.64	0.27	–	–	0.41	1.67	–	–	–	–	0.035	33.9	–	–
5	MEAL, DEHY, 17% PROTEIN	A-F	92	1.40	0.23	0.10	0.47	0.29	2.38	0.23	0.302	8.6	0.148	0.041	31.0	0.335	19.3
		M-F	100	1.52	0.25	0.11	0.52	0.32	2.60	0.25	0.329	9.3	0.162	0.045	33.8	0.365	21.1
6	MEAL, DEHY, 20% PROTEIN	A-F	92	1.59	0.28	0.11	0.47	0.33	2.41	0.50	0.259	12.2	0.135	0.036	45.2	0.285	21.8
		M-F	100	1.74	0.31	0.13	0.51	0.36	2.63	0.55	0.283	13.3	0.147	0.039	49.4	0.311	23.8
7	MEAL, DEHY, 22% PROTEIN	A-F	93	1.69	0.30	0.12	0.52	0.31	2.40	0.30	0.311	9.8	0.166	0.036	36.4	0.534	19.5
		M-F	100	1.82	0.33	0.13	0.56	0.33	2.58	0.32	0.336	10.5	0.179	0.039	39.2	0.576	21.0
	ANIMAL																
8	LIVER-GLANDS, GROUND	A-F	93	0.63	1.18	0.10	–	–	0.40	–	0.170	94.2	–	0.056	8.1	–	50.1
		M-F	100	0.67	1.26	0.11	–	–	0.43	–	0.182	101.0	–	0.060	8.7	–	53.6
9	LIVER, MEAL	A-F	93	0.56	1.26	–	–	0.10	–	–	0.135	89.4	–	0.064	8.8	–	61.8
		M-F	100	0.61	1.36	–	–	0.11	–	–	0.146	96.5	–	0.069	9.5	–	66.8
10	MEAT SOLUBLES, DEHY	A-F	90	0.45	0.67	–	–	–	–	–	–	–	–	–	–	–	–
		M-F	100	0.50	0.74	–	–	–	–	–	–	–	–	–	–	–	–
	BLUEGRASS, KENTUCKY *Poa pratensis*																
11	IMMATURE, FRESH	A-F	31	0.15	0.14	0.04	–	0.05	0.70	0.05	–	–	–	0.010	–	–	–
		M-F	100	0.50	0.44	0.14	–	0.18	2.27	0.17	–	–	–	0.030	–	–	–
	BREWERS GRAINS																
12	DEHY	A-F	92	0.30	0.51	0.21	0.15	0.15	0.09	0.30	0.076	21.7	0.066	0.024	37.2	–	27.3
		M-F	100	0.33	0.55	0.23	0.17	0.17	0.09	0.32	0.083	23.6	0.072	0.026	40.4	–	29.6
	BUTTERMILK, CATTLE *Bos taurus*																
13	CONDENSED	A-F	29	0.44	0.26	0.31	0.12	0.19	0.23	0.03	–	–	–	–	–	–	–
		M-F	100	1.51	0.89	1.06	0.41	0.65	0.79	0.10	–	–	–	–	–	–	–
	CARROT *Daucus* spp																
14	ROOTS, FRESH	A-F	11	0.05	0.04	0.06	0.06	0.02	0.32	0.02	–	1.2	–	0.002	3.6	–	–
		M-F	100	0.40	0.35	0.48	0.50	0.20	2.80	0.17	–	10.4	–	0.013	31.5	–	–
	CATTLE *Bos taurus*																
15	BUTTERMILK, CONDENSED	A-F	29	0.44	0.26	0.31	0.12	0.19	0.23	0.03	–	–	–	–	–	–	–
		M-F	100	1.51	0.89	1.06	0.41	0.65	0.79	0.10	–	–	–	–	–	–	–
16	LIVER, FRESH	A-F	28	0.01	0.23	0.10	–	0.01	0.20	–	–	6.1	–	0.005	2.8	–	26.6
		M-F	100	0.04	0.82	0.35	–	0.04	0.72	–	–	21.9	–	0.017	9.9	–	95.0
17	WHEY, DEHY	A-F	93	0.86	0.76	0.62	0.07	0.13	1.11	1.04	0.111	46.5	–	0.017	5.9	–	3.2
		M-F	100	0.92	0.82	0.66	0.08	0.14	1.19	1.11	0.119	49.9	–	0.019	6.3	–	3.4
	COD, FISH *Gadus morrhua, Gadus macrocephalus*																
18	LIVER, MEAL	A-F	93	0.16	0.69	–	–	–	–	–	–	–	–	–	–	–	–
		M-F	100	0.17	0.75	–	–	–	–	–	–	–	–	–	–	–	–
19	LIVER OIL	A-F	100	–	–	–	–	–	–	–	–	–	–	–	–	–	–
		M-F	100	–	–	–	–	–	–	–	–	–	–	–	–	–	–
	CORN *Zea mays*																
20	DISTILLERS GRAINS WITH SOLUBLES, DEHY	A-F	92	0.16	0.69	0.47	0.17	0.18	0.47	0.31	0.152	52.6	0.051	0.024	24.0	0.331	80.7
		M-F	100	0.17	0.76	0.52	0.18	0.20	0.51	0.33	0.165	57.3	0.055	0.026	26.1	0.361	87.9
	CRAB *Callinectes sapidus, Cancer* spp, *Paralithodes camschatica*																
21	CANNERY RESIDUE, MEAL (CRAB MEAL)	A-F	92	14.46	1.58	0.88	1.51	0.94	0.45	0.25	–	32.7	0.557	0.435	132.8	–	–
		M-F	100	15.69	1.72	0.95	1.63	1.02	0.49	0.27	–	35.5	0.605	0.472	144.0	–	–
	DISTILLERS PRODUCTS (ALSO SEE CORN)																
22	GRAINS, DEHY	A-F	93	0.12	0.54	0.05	0.05	0.09	0.20	0.46	0.092	47.9	–	0.027	35.0	–	–
		M-F	100	0.13	0.59	0.05	0.06	0.10	0.21	0.49	0.099	51.7	–	0.029	37.8	–	–
23	SOLUBLES, DEHY	A-F	92	0.24	1.35	0.45	–	0.53	1.97	–	0.196	71.6	–	0.031	64.1	–	138.0
		M-F	100	0.26	1.47	0.49	–	0.58	2.14	–	0.213	77.9	–	0.034	69.7	–	150.0
	FATS AND OILS																
24	GERM OIL (WHEAT)	A-F	100	–	–	–	–	–	–	–	–	–	–	–	–	–	–
		M-F	100	–	–	–	–	–	–	–	–	–	–	–	–	–	–
25	LIVER OIL (COD)	A-F	100	–	–	–	–	–	–	–	–	–	–	–	–	–	–
		M-F	100	–	–	–	–	–	–	–	–	–	–	–	–	–	–
	FISH																
26	MEAL MECH EXTD	A-F	92	6.63	3.61	1.11	1.25	0.21	0.40	0.25	0.110	15.1	–	0.038	23.6	–	99.1
		M-F	100	7.24	3.94	1.21	1.37	0.23	0.44	0.27	0.120	16.5	–	0.042	25.8	–	108.2
27	SOLUBLES, CONDENSED	A-F	50	0.16	0.57	2.45	2.93	0.03	1.64	0.12	0.069	46.6	1.111	0.028	13.2	–	43.2
		M-F	100	0.32	1.14	4.86	5.81	0.06	3.24	0.25	0.137	92.4	2.202	0.055	26.2	–	85.6
28	SOLUBLES, DEHY	A-F	93	0.40	1.27	1.70	–	0.30	2.50	0.45	–	20.0	–	0.095	50.4	2.692	76.7
		M-F	100	0.43	1.37	1.83	–	0.32	2.69	0.48	–	21.5	–	0.102	54.3	2.901	82.6
	FISH, COD *Gadus morrhua, Gadus macrocephalus*																
29	LIVER, MEAL	A-F	93	0.16	0.69	–	–	–	–	–	–	–	–	–	–	–	–
		M-F	100	0.17	0.75	–	–	–	–	–	–	–	–	–	–	–	–

VITAMIN SUPPLEMENTS

Entry Number	Fat-Soluble Vitamins					Water-Soluble Vitamins								
	A (1 mg Carotene = 1667 IU Vit A)	Carotene (Provitamin A)	D	E	K	B-12	Biotin	Choline	Folacin (Folic Acid)	Niacin	Pantothenic Acid (B-3)	(Pyridoxine) B-6	Riboflavin (B-2)	Thiamin (B-1)
	IU/g	ppm or mg/kg	IU/kg	ppm or mg/kg	ppm or mg/kg	ppm or mg/kg	ppm or mg/kg	ppm or mg/kg	ppm or mg/kg	ppm or mg/kg	ppm or mg/kg	ppm or mg/kg	ppm or mg/kg	ppm or mg/kg
1	85.9	51.5	–	–	–	–	–	–	–	–	8.9	–	–	–
	402.8	241.6	–	–	–	–	–	–	–	–	41.7	–	–	–
2	45.0	27.0	2	55.9	–	–	0.18	892	3.07	43	18.1	–	9.5	3.1
	49.8	29.9	2	61.9	–	–	0.20	988	3.40	48	20.1	–	10.5	3.4
3	48.2	28.9	–	–	–	–	–	–	–	–	–	–	–	–
	52.2	31.3	–	–	–	–	–	–	–	–	–	–	–	–
4	97.7	58.6	–	–	–	–	–	–	–	–	–	–	20.9	–
	111.3	66.8	–	–	–	–	–	–	–	–	–	–	23.8	–
5	200.3	120.2	–	105.7	8.24	–	0.33	1369	4.37	37	29.7	7.18	12.9	3.4
	218.5	131.1	–	115.3	8.98	–	0.36	1494	4.77	40	32.4	7.83	14.1	3.7
6	265.4	159.2	–	143.3	14.19	–	0.35	1417	2.96	48	35.5	8.72	15.2	5.4
	289.7	173.8	–	156.4	15.50	–	0.39	1547	3.24	52	38.8	9.52	16.6	5.9
7	391.4	234.8	–	221.3	11.65	–	0.33	1605	5.15	50	39.0	8.28	17.6	5.9
	421.8	253.0	–	238.4	12.55	–	0.36	1729	5.55	54	42.0	8.92	19.0	6.3
8	–	–	–	–	–	440.5	0.41	10610	4.00	172	90.8	5.01	42.4	0.2
	–	–	–	–	–	472.1	0.44	11372	4.29	185	97.3	5.37	45.4	0.2
9	–	–	–	–	–	501.3	0.02	11370	5.56	205	29.2	–	36.2	0.2
	–	–	–	–	–	541.5	0.02	12281	6.01	221	31.5	–	39.1	0.2
10	–	–	–	–	–	881.6	–	–	–	–	–	–	–	–
	–	–	–	–	–	979.7	–	–	–	–	–	–	–	–
11	247.6	148.5	–	47.8	–	–	–	–	–	–	–	–	–	–
	803.4	481.9	–	155.0	–	–	–	–	–	–	–	–	–	–
12	0.8	0.5	–	26.7	–	3.6	0.44	1651	0.22	44	8.2	1.03	1.5	0.6
	0.8	0.5	–	29.0	–	3.9	0.48	1792	0.24	47	8.9	1.11	1.6	0.7
13	25.6	15.4	–	–	–	–	–	–	–	–	–	–	12.6	–
	87.2	52.3	–	–	–	–	–	–	–	–	–	–	42.8	–
14	129.9	77.9	–	6.9	–	–	0.01	–	0.14	7	3.5	1.39	0.6	0.7
	1129.4	677.5	–	60.2	–	–	0.07	–	1.21	58	30.1	12.05	4.9	5.8
15	25.6	15.4	–	–	–	–	–	–	–	–	–	–	12.6	–
	87.2	52.3	–	–	–	–	–	–	–	–	–	–	42.8	–
16	–	–	–	7.1	–	425.8	0.98	1424	2.33	75	46.1	5.03	25.8	1.8
	–	–	–	26.4	–	1523.2	3.51	5092	8.35	269	164.9	18.00	92.2	6.3
17	–	–	–	0.2	–	18.9	0.35	1790	0.85	11	46.2	3.21	27.4	4.0
	–	–	–	0.2	–	20.3	0.38	1921	0.91	11	49.6	3.45	29.4	4.3
18	–	–	–	–	–	–	–	–	–	132	46.1	32.85	33.3	18.1
	–	–	–	–	–	–	–	–	–	143	49.8	35.51	36.0	19.5
19	845.8	–	–	39.5	–	–	–	–	–	–	–	–	–	–
	850.0	–	–	39.7	–	–	–	–	–	–	–	–	–	–
20	6.2	3.7	1	39.8	–	1.5	0.69	2582	0.91	73	13.8	4.74	8.5	3.0
	6.8	4.1	1	43.4	–	1.6	0.75	2813	0.99	80	15.1	5.17	9.2	3.3
21	–	–	–	–	–	437.6	0.07	2008	0.11	45	6.5	6.62	6.1	0.4
	–	–	–	–	–	474.8	0.07	2179	0.12	49	7.0	7.18	6.7	0.5
22	13.0	7.8	–	30.5	–	–	–	2645	–	47	11.9	6.00	6.6	2.5
	14.0	8.4	–	32.9	–	–	–	2858	–	51	12.9	6.48	7.1	2.6
23	1.9	1.1	–	–	–	2.9	2.84	4992	–	143	25.3	8.66	11.3	6.9
	2.0	1.2	–	–	–	3.1	3.09	5425	–	155	27.5	9.42	12.3	7.5
24	–	–	–	–	18.66	–	–	–	–	–	–	–	–	–
	–	–	–	–	18.75	–	–	–	–	–	–	–	–	–
25	845.8	–	–	39.5	–	–	–	–	–	–	–	–	–	–
	850.0	–	–	39.7	–	–	–	–	–	–	–	–	–	–
26	–	–	–	19.2	–	258.6	–	3644	–	75	15.0	14.68	5.6	0.8
	–	–	–	20.9	–	282.3	–	3979	–	82	16.3	16.03	6.1	0.9
27	2.2	1.3	–	–	–	506.6	0.14	3370	0.22	176	35.7	12.20	12.9	5.5
	4.4	2.6	–	–	–	1004.2	0.28	6680	0.44	348	70.7	24.19	25.5	11.0
28	–	–	–	6.1	–	485.9	0.40	5525	0.57	256	50.4	19.71	13.5	7.4
	–	–	–	6.5	–	523.6	0.43	5953	0.62	276	54.3	21.24	14.6	8.0
29	–	–	–	–	–	–	–	–	–	132	46.1	32.85	33.3	18.1
	–	–	–	–	–	–	–	–	–	143	49.8	35.51	36.0	19.5

VITAMIN SUPPLEMENTS

TABLE V-1G VITAMIN SUPPLEMENTS, COMPOSITION OF FEEDS, DATA EXPRESSED **AS-FED** AND **MOISTURE-FREE**—*(Continued)*

Entry Number	Feed Name Description	International Feed Number	Moisture Basis: A-F (as-fed) or M-F (moisture-free)	Proximate Analysis									Digestible Protein		
				Dry Matter	Ash	Crude Fiber	Neutral Det. Fib. (NDF)	Acid Det. Fib. (ADF)	Lignin	Ether Extract (Fat)	N-Free Extract	Crude Protein	Ruminant	Swine	Horse
				%	%	%	%	%	%	%	%	%	%	%	%
	FISH, COD (Continued)														
30	LIVER OIL	7-01-993	A-F	**100**	–	–	–	–	–	**99.5**	–	–	–	–	–
			M-F	100	–	–	–	–	–	100.0	–	–	–	–	–
	FISH, SARDINE *Clupea* spp, *Sardinops* spp														
31	MEAL MECH EXTD	5-02-015	A-F	**93**	**15.8**	**1.0**	–	–	–	**5.0**	**6.1**	**65.2**	**53.5**	**63.6**	**65.1**
			M-F	100	17.0	1.1	–	–	–	5.4	6.5	70.0	57.4	68.2	69.8
32	SOLUBLES, CONDENSED	5-02-014	A-F	**50**	**10.2**	–	–	–	–	**9.4**	**0.6**	**29.5**	**25.6**	**28.7**	**28.7**
			M-F	100	20.5	–	–	–	–	18.9	1.2	59.4	51.6	57.7	57.8
	KENTUCKY BLUEGRASS *Poa pratensis*														
33	IMMATURE, FRESH	2-00-777	A-F	**31**	**2.9**	**7.8**	–	–	–	**1.1**	**13.7**	**5.4**	**4.0**	**3.8**	–
			M-F	100	9.4	25.2	–	–	–	3.6	44.4	17.4	12.9	12.2	–
	LIVER														
34	CATTLE, FRESH	5-01-166	A-F	**28**	**1.4**	**0.2**	–	–	–	**5.1**	**1.9**	**19.5**	**17.0**	**19.0**	**19.4**
			M-F	100	4.9	0.6	–	–	–	18.3	6.7	69.6	60.9	67.8	69.4
35	SHEEP, FRESH	5-08-116	A-F	**29**	**1.4**	–	–	–	–	**3.9**	**2.9**	**21.0**	**18.4**	**20.5**	**21.0**
			M-F	100	4.8	–	–	–	–	13.4	9.9	71.9	63.0	70.1	72.0
36	SWINE, FRESH	5-04-792	A-F	**30**	**1.6**	**0.1**	–	–	–	**5.0**	**2.8**	**20.8**	**18.2**	**20.3**	**20.7**
			M-F	100	5.3	0.3	–	–	–	16.5	9.1	68.8	60.2	67.1	68.5
37	MEAL	5-00-389	A-F	**93**	**6.3**	**1.4**	–	–	–	**15.7**	**3.2**	**66.1**	**57.9**	**64.4**	**66.1**
			M-F	100	6.8	1.5	–	–	–	17.0	3.5	71.4	62.5	69.6	71.4
	LIVER-GLANDS														
38	MEAL	5-00-390	A-F	**93**	**5.9**	**1.8**	–	–	–	**15.8**	**3.3**	**66.5**	**58.2**	**64.8**	**66.5**
			M-F	100	6.3	1.9	–	–	–	16.9	3.6	71.3	62.4	69.5	71.2
	LUCERNE–SEE ALFALFA														
	MAIZE–SEE CORN														
	MEAT														
39	SOLUBLES, DEHY	5-00-393	A-F	**90**	**5.7**	–	–	–	–	–	–	**80.0**	**70.6**	**78.3**	**82.0**
			M-F	100	6.3	–	–	–	–	–	–	88.9	78.5	87.0	91.2
	OATS *Avena sativa*														
40	IMMATURE, FRESH	2-03-286	A-F	**16**	**1.7**	**4.0**	–	–	–	**0.4**	**7.4**	**2.5**	**1.8**	**1.7**	–
			M-F	100	10.6	24.9	–	–	–	2.6	46.3	15.6	11.3	10.8	–
	RICE *Oryza sativa*														
41	BRAN WITH GERMS, MEAL SOLV EXTD (RICE BRAN, SOLV EXTD)	4-03-930	A-F	**91**	**14.5**	**12.9**	–	–	–	**1.5**	**48.1**	**14.0**	**9.8**	**9.1**	**10.2**
			M-F	100	15.9	14.2	–	–	–	1.6	52.9	15.4	10.7	10.0	11.2
42	POLISHINGS	4-03-943	A-F	**90**	**7.6**	**3.2**	–	**3.6**	–	**12.6**	**54.9**	**12.0**	**8.6**	**10.1**	**8.6**
			M-F	100	8.4	3.5	–	4.0	–	13.9	60.9	13.3	9.5	11.2	9.5
	SHEEP *Ovis aries*														
43	LIVER, FRESH	5-08-116	A-F	**29**	**1.4**	–	–	–	–	**3.9**	**2.9**	**21.0**	**18.4**	**20.5**	**21.0**
			M-F	100	4.8	–	–	–	–	13.4	9.9	71.9	63.0	70.1	72.0
	SPINACH *Spinacia oleracea*														
44	LEAVES, FRESH	2-08-125	A-F	**9**	**2.2**	**0.7**	–	–	–	**0.4**	**3.0**	**3.1**	**2.5**	**2.3**	–
			M-F	100	23.6	7.5	–	–	–	4.0	32.2	32.7	26.8	24.5	–
	SWINE *Sus scrofa*														
45	LIVER, FRESH	5-04-792	A-F	**30**	**1.6**	**0.1**	–	–	–	**5.0**	**2.8**	**20.8**	**18.2**	**20.3**	**20.7**
			M-F	100	5.3	0.3	–	–	–	16.5	9.1	68.8	60.2	67.1	68.5
	TURNIP *Brassica rapa, rapa*														
46	FRESH	2-05-603	A-F	**13**	**2.1**	**1.4**	–	–	–	**0.3**	**6.8**	**2.9**	**1.1**	**2.1**	–
			M-F	100	15.7	10.3	–	–	–	2.6	50.2	21.2	8.1	15.3	–
	WHALE *Balaena glacialis, Balaenoptera* spp, *Physeter catodon*														
47	LIVER, DEHY	5-05-157	A-F	**93**	–	–	–	–	–	–	–	–	–	–	–
			M-F	100	–	–	–	–	–	–	–	–	–	–	–
	WHEAT *Triticum aestivum*														
48	GERM MEAL	5-05-218	A-F	**88**	**4.3**	**3.1**	–	**4.4**	–	**8.5**	**48.1**	**24.4**	**22.9**	**23.1**	**19.4**
			M-F	100	4.9	3.5	–	5.0	–	9.6	54.4	27.6	26.0	26.2	21.9
49	GERM OIL	7-05-207	A-F	**100**	–	–	–	–	–	**99.5**	–	–	–	–	–
			M-F	100	–	–	–	–	–	100.0	–	–	–	–	–
	WHEY, CATTLE *Bos taurus*														
50	DEHY	4-01-182	A-F	**93**	**8.8**	**0.2**	**0.3**	**0.2**	–	**0.8**	**70.2**	**13.3**	**8.9**	**13.1**	**9.6**
			M-F	100	9.4	0.2	0.3	0.2	–	0.8	75.3	14.2	9.5	14.1	10.3
	YEAST, BREWERS *Saccharomyces cerevisiae*														
51	DEHY	7-05-527	A-F	**93**	**6.5**	**3.0**	–	**3.7**	–	**0.9**	**38.8**	**43.8**	**39.0**	–	–
			M-F	100	7.0	3.2	–	4.0	–	1.0	41.7	47.1	41.9	–	–
	YEAST, PRIMARY *Saccharomyces cerevisiae*														
52	DEHY	7-05-533	A-F	**93**	**8.0**	**3.1**	–	–	–	**1.0**	**32.5**	**48.0**	–	–	–
			M-F	100	8.6	3.3	–	–	–	1.1	35.1	51.8	–	–	–
	YEAST, TORULA *Torulopsis utilis*														
53	DEHY	7-05-534	A-F	**93**	**8.0**	**2.5**	–	**3.7**	–	**1.6**	**31.5**	**49.6**	**45.1**	**40.6**	–
			M-F	100	8.6	2.7	–	4.0	–	1.7	33.8	53.3	48.5	43.7	–

VITAMIN SUPPLEMENTS

Entry Number	TDN Ruminant	TDN Swine	TDN Horse	Digestible Energy Ruminant		Digestible Energy Swine		Digestible Energy Horse		Metabolizable Energy Ruminant		Metabolizable Energy Swine		Metabolizable Energy Poultry ME_n		Metabolizable Energy Horse		Net Energy Ruminant NE_m		Net Energy Ruminant NE_g		Net Energy Lactating Cows NE_{lc}	
	%	%	%	Mcal lb	Mcal kg	kcal lb	kcal kg	Mcal lb	Mcal kg	Mcal lb	Mcal kg	kcal lb	kcal kg	kcal lb	kcal kg	Mcal lb	Mcal kg	Mcal lb	Mcal kg	Mcal lb	Mcal kg	Mcal lb	Mcal kg
30	–	–	–	–	–	–	–	–	–	–	–	–	–	–	–	–	–	–	–	–	–	–	–
	–	–	–	–	–	–	–	–	–	–	–	–	–	–	–	–	–	–	–	–	–	–	–
31	70	67	–	1.40	3.09	1327	2925	–	–	1.23	2.71	1148	2531	1313	2896	–	–	0.76	1.67	0.49	1.08	0.73	1.60
	75	72	–	1.51	3.32	1425	3141	–	–	1.32	2.90	1233	2717	1410	3109	–	–	0.81	1.79	0.53	1.16	0.78	1.72
32	–	–	–	–	–	–	–	–	–	–	–	–	–	–	–	–	–	–	–	–	–	–	–
	–	–	–	–	–	–	–	–	–	–	–	–	–	–	–	–	–	–	–	–	–	–	–
33	22	–	–	0.42	0.93	–	–	–	–	0.37	0.81	–	–	–	–	–	–	0.24	0.52	0.15	0.33	0.23	0.51
	72	–	–	1.38	3.03	–	–	–	–	1.19	2.61	–	–	–	–	–	–	0.77	1.70	0.49	1.08	0.75	1.64
34	29	31	–	0.58	1.28	615	1356	–	–	0.53	1.17	544	1200	–	–	–	–	0.34	0.76	0.25	0.54	0.32	0.70
	104	110	–	2.08	4.59	2201	4852	–	–	1.90	4.18	1947	4293	–	–	–	–	1.23	2.71	0.88	1.93	1.13	2.49
35	–	–	–	–	–	–	–	–	–	–	–	–	–	–	–	–	–	–	–	–	–	–	–
	–	–	–	–	–	–	–	–	–	–	–	–	–	–	–	–	–	–	–	–	–	–	–
36	31	32	–	0.62	1.37	650	1433	–	–	0.56	1.24	574	1266	–	–	–	–	0.37	0.80	0.26	0.57	0.34	0.74
	102	107	–	2.05	4.52	2148	4736	–	–	1.87	4.11	1899	4186	–	–	–	–	1.21	2.66	0.86	1.89	1.11	2.45
37	89	93	–	1.79	3.94	1867	4116	–	–	1.62	3.57	1645	3627	1306	2878	–	–	1.00	2.21	0.70	1.55	0.93	2.04
	97	101	–	1.93	4.26	2017	4446	–	–	1.75	3.85	1777	3918	1410	3109	–	–	1.08	2.38	0.76	1.67	1.00	2.21
38	90	95	–	1.87	4.11	1711	3771	–	–	1.69	3.73	1503	3314	1323	2917	–	–	1.11	2.44	0.79	1.73	1.02	2.25
	97	102	–	2.00	4.41	1833	4042	–	–	1.82	4.00	1611	3552	1418	3126	–	–	1.19	2.61	0.84	1.86	1.09	2.41
39	–	–	–	–	–	–	–	–	–	–	–	–	–	–	–	–	–	–	–	–	–	–	–
	–	–	–	–	–	–	–	–	–	–	–	–	–	–	–	–	–	–	–	–	–	–	–
40	10	–	–	0.21	0.46	–	–	–	–	0.18	0.39	–	–	–	–	–	–	0.10	0.23	0.06	0.14	0.10	0.23
	64	–	–	1.29	2.83	–	–	–	–	1.09	2.41	–	–	–	–	–	–	0.65	1.42	0.38	0.84	0.65	1.43
41	55	72	–	1.15	2.54	1481	3264	–	–	0.98	2.15	1199	2643	909	2003	–	–	0.62	1.36	0.37	0.82	0.61	1.35
	61	79	–	1.26	2.79	1628	3589	–	–	1.07	2.37	1318	2907	999	2203	–	–	0.68	1.49	0.41	0.90	0.68	1.49
42	81	88	–	1.56	3.45	1684	3713	–	–	1.43	3.16	1555	3428	1367	3015	–	–	0.88	1.93	0.60	1.32	0.83	1.82
	89	97	–	1.73	3.82	1866	4114	–	–	1.59	3.50	1723	3798	1515	3340	–	–	0.97	2.14	0.67	1.47	0.92	2.02
43	–	–	–	–	–	–	–	–	–	–	–	–	–	–	–	–	–	–	–	–	–	–	–
	–	–	–	–	–	–	–	–	–	–	–	–	–	–	–	–	–	–	–	–	–	–	–
44	–	–	–	–	–	–	–	–	–	–	–	–	–	–	–	–	–	–	–	–	–	–	–
	–	–	–	–	–	–	–	–	–	–	–	–	–	–	–	–	–	–	–	–	–	–	–
45	31	32	–	0.62	1.37	650	1433	–	–	0.56	1.24	574	1266	–	–	–	–	0.37	0.80	0.26	0.57	0.34	0.74
	102	107	–	2.05	4.52	2148	4736	–	–	1.87	4.11	1899	4186	–	–	–	–	1.21	2.66	0.86	1.89	1.11	2.45
46	10	–	–	0.19	0.41	–	–	–	–	0.16	0.36	–	–	–	–	–	–	0.10	0.21	0.06	0.13	0.10	0.21
	71	–	–	1.39	3.07	–	–	–	–	1.20	2.65	–	–	–	–	–	–	0.72	1.58	0.45	0.98	0.71	1.55
47	–	–	–	–	–	–	–	–	–	–	–	–	–	–	–	–	–	–	–	–	–	–	–
	–	–	–	–	–	–	–	–	–	–	–	–	–	–	–	–	–	–	–	–	–	–	–
48	83	80	–	1.66	3.66	1727	3807	–	–	1.50	3.30	1522	3357	1223	2696	–	–	0.96	2.11	0.67	1.48	0.89	1.95
	94	91	–	1.88	4.14	1954	4308	–	–	1.69	3.73	1723	3798	1384	3051	–	–	1.08	2.38	0.76	1.67	1.00	2.21
49	–	–	–	–	–	–	–	–	–	–	–	–	–	–	–	–	–	–	–	–	–	–	–
	–	–	–	–	–	–	–	–	–	–	–	–	–	–	–	–	–	–	–	–	–	–	–
50	76	77	–	1.51	3.33	1444	3183	–	–	1.28	2.83	1411	3110	880	1939	–	–	0.87	1.92	0.59	1.30	0.78	1.71
	82	83	–	1.62	3.57	1549	3415	–	–	1.38	3.03	1514	3337	944	2081	–	–	0.93	2.06	0.63	1.40	0.83	1.83
51	73	70	–	1.46	3.21	–	–	–	–	1.28	2.83	1299	2865	928	2047	–	–	0.81	1.79	0.54	1.19	0.77	1.69
	78	76	–	1.57	3.45	–	–	–	–	1.38	3.04	1396	3078	998	2199	–	–	0.87	1.92	0.58	1.28	0.83	1.82
52	–	–	–	–	–	–	–	–	–	–	–	–	–	–	–	–	–	–	–	–	–	–	–
	–	–	–	–	–	–	–	–	–	–	–	–	–	–	–	–	–	–	–	–	–	–	–
53	72	64	–	1.49	3.29	1287	2837	–	–	1.27	2.81	1096	2416	840	1851	–	–	0.82	1.81	0.55	1.21	0.78	1.71
	78	69	–	1.60	3.53	1383	3049	–	–	1.37	3.02	1178	2597	902	1989	–	–	0.88	1.95	0.59	1.30	0.84	1.84

TABLE V-1G VITAMIN SUPPLEMENTS, MINERAL AND VITAMIN COMPOSITION OF FEEDS, DATA EXPRESSED **AS-FED** AND **MOISTURE-FREE**—*(Continued)*

Entry Number	Feed Name Description	Moisture Basis: A-F (as-fed) or M-F (moisture-free)	Dry Matter	Macro Minerals							Micro Minerals						
				Calcium (Ca)	Phos-phorus (P)	Sodium (Na)	Chlo-rine (Cl)	Mag-nesium (Mg)	Potas-sium (K)	Sulfur (S)	Cobalt (Co)	Copper (Cu)	Iodine (I)	Iron (Fe)	Man-ganese (Mn)	Sele-nium (Se)	Zinc (Zn)
			%	%	%	%	%	%	%	%	ppm or mg/kg	ppm or mg/kg	ppm or mg/kg	%	ppm or mg/kg	ppm or mg/kg	ppm or mg/kg
	FISH, COD (Continued)																
30	LIVER OIL	A-F	100	–	–	–	–	–	–	–	–	–	–	–	–	–	–
		M-F	100	–	–	–	–	–	–	–	–	–	–	–	–	–	–
	FISH, SARDINE *Clupea* spp, *Sardinops* spp																
31	MEAL MECH EXTD	A-F	93	4.61	2.68	0.18	0.41	0.10	0.32	–	0.183	20.2	–	0.030	23.2	1.772	–
		M-F	100	4.95	2.88	0.19	0.44	0.11	0.35	–	0.197	21.7	–	0.033	24.9	1.903	–
32	SOLUBLES, CONDENSED	A-F	50	0.14	0.83	0.18	0.28	–	0.18	0.11	–	25.8	4.934	0.002	24.9	–	–
		M-F	100	0.28	1.67	0.36	0.56	–	0.36	0.22	–	51.9	9.928	0.005	50.1	–	–
	KENTUCKY BLUEGRASS *Poa pratensis*																
33	IMMATURE, FRESH	A-F	31	0.15	0.14	0.04	–	0.05	0.70	0.05	–	–	–	0.010	–	–	–
		M-F	100	0.50	0.44	0.14	–	0.18	2.27	0.17	–	–	–	0.030	–	–	–
	LIVER																
34	CATTLE, FRESH	A-F	28	0.01	0.23	0.10	–	0.01	0.20	–	–	6.1	–	0.005	2.8	–	26.6
		M-F	100	0.04	0.82	0.35	–	0.04	0.72	–	–	21.9	–	0.017	9.9	–	95.0
35	SHEEP, FRESH	A-F	29	0.01	0.35	0.06	–	0.02	0.20	–	–	37.5	–	0.007	3.6	–	36.0
		M-F	100	0.03	1.20	0.19	–	0.07	0.69	–	–	128.4	–	0.022	12.3	–	123.3
36	SWINE, FRESH	A-F	30	0.01	0.37	0.07	–	0.01	0.26	–	0.255	56.4	0.340	0.015	1.8	0.340	44.2
		M-F	100	0.04	1.22	0.24	–	0.04	0.85	–	0.842	186.6	1.122	0.049	5.8	1.122	146.1
37	MEAL	A-F	93	0.56	1.26	–	–	0.10	–	–	0.135	89.4	–	0.064	8.8	–	61.8
		M-F	100	0.61	1.36	–	–	0.11	–	–	0.146	96.5	–	0.069	9.5	–	66.8
	LIVER-GLANDS																
38	MEAL	A-F	93	0.63	1.18	0.10	–	–	0.40	–	0.170	94.2	–	0.056	8.1	–	50.1
		M-F	100	0.67	1.26	0.11	–	–	0.43	–	0.182	101.0	–	0.060	8.7	–	53.6
	LUCERNE–SEE ALFALFA																
	MAIZE–SEE CORN																
	MEAT																
39	SOLUBLES, DEHY	A-F	90	0.45	0.67	–	–	–	–	–	–	–	–	–	–	–	–
		M-F	100	0.50	0.74	–	–	–	–	–	–	–	–	–	–	–	–
	OATS *Avena sativa*																
40	IMMATURE, FRESH	A-F	16	–	–	0.02	0.02	–	–	0.01	–	–	–	–	–	–	–
		M-F	100	–	–	0.11	0.10	–	–	0.08	–	–	–	–	–	–	–
	RICE *Oryza sativa*																
41	BRAN WITH GERMS, MEAL SOLV EXTD (RICE BRAN, SOLV EXTD)	A-F	91	0.11	1.37	–	–	–	1.48	0.18	0.111	13.0	0.045	0.019	232.2	–	30.0
		M-F	100	0.12	1.50	–	–	–	1.63	0.20	0.122	14.3	0.049	0.021	255.3	–	33.0
42	POLISHINGS	A-F	90	0.05	1.34	0.04	0.11	0.60	1.28	0.17	3.890	8.0	–	0.009	126.8	–	63.2
		M-F	100	0.05	1.49	0.05	0.12	0.66	1.41	0.19	4.311	8.8	–	0.009	140.5	–	70.0
	SHEEP *Ovis aries*																
43	LIVER, FRESH	A-F	29	0.01	0.35	0.06	–	0.02	0.20	–	–	37.5	–	0.007	3.6	–	36.0
		M-F	100	0.03	1.20	0.19	–	0.07	0.69	–	–	128.4	–	0.022	12.3	–	123.3
	SPINACH *Spinacia oleracea*																
44	LEAVES, FRESH	A-F	9	0.09	0.05	0.07	–	–	0.48	–	–	–	–	0.004	–	–	–
		M-F	100	0.97	0.54	0.75	–	–	5.05	–	–	–	–	0.033	–	–	–
	SWINE *Sus scrofa*																
45	LIVER, FRESH	A-F	30	0.01	0.37	0.07	–	0.01	0.26	–	0.255	56.4	0.340	0.015	1.8	0.340	44.2
		M-F	100	0.04	1.22	0.24	–	0.04	0.85	–	0.842	186.6	1.122	0.049	5.8	1.122	146.1
	TURNIP *Brassica rapa, rapa*																
46	FRESH	A-F	13	0.40	0.05	–	0.26	0.07	0.41	0.04	–	2.4	–	0.006	55.2	–	5.0
		M-F	100	2.93	0.39	–	1.93	0.54	3.03	0.27	–	17.6	–	0.041	408.6	–	36.8
	WHALE *Balaena glacialis, Balaenoptera* spp, *Physeter catodon*																
47	LIVER, DEHY	A-F	93	–	–	–	1.99	–	–	–	–	–	–	–	–	–	–
		M-F	100	–	–	–	2.13	–	–	–	–	–	–	–	–	–	–
	WHEAT *Triticum aestivum*																
48	GERM MEAL	A-F	88	0.06	0.95	0.02	0.06	0.25	0.94	0.27	0.120	9.2	–	0.006	132.5	0.463	119.4
		M-F	100	0.06	1.07	0.03	0.07	0.28	1.06	0.30	0.136	10.4	–	0.007	149.9	0.524	135.1
49	GERM OIL	A-F	100	–	–	–	–	–	–	–	–	–	–	–	–	–	–
		M-F	100	–	–	–	–	–	–	–	–	–	–	–	–	–	–
	WHEY, CATTLE *Bos taurus*																
50	DEHY	A-F	93	0.86	0.76	0.62	0.07	0.13	1.11	1.04	0.111	46.5	–	0.017	5.9	–	3.2
		M-F	100	0.92	0.82	0.66	0.08	0.14	1.19	1.11	0.119	49.9	–	0.019	6.3	–	3.4
	YEAST, BREWERS *Saccharomyces cerevisiae*																
51	DEHY	A-F	93	0.14	1.36	0.07	0.07	0.24	1.69	0.43	0.506	38.4	0.358	0.009	6.7	0.911	39.0
		M-F	100	0.15	1.47	0.08	0.08	0.26	1.82	0.46	0.544	41.3	0.384	0.009	7.2	0.979	41.9
	YEAST, PRIMARY *Saccharomyces cerevisiae*																
52	DEHY	A-F	93	0.36	1.72	–	0.02	0.36	–	0.57	–	–	–	0.030	3.7	–	–
		M-F	100	0.39	1.86	–	0.02	0.39	–	0.62	–	–	–	0.033	4.0	–	–
	YEAST, TORULA *Torulopsis utilis*																
53	DEHY	A-F	93	0.55	1.61	0.01	0.02	0.14	1.92	0.55	0.031	11.9	2.502	0.011	9.3	–	99.5
		M-F	100	0.59	1.73	0.01	0.02	0.15	2.06	0.59	0.033	12.8	2.689	0.012	10.0	–	107.0

VITAMIN SUPPLEMENTS

Entry Number	Fat-Soluble Vitamins A (1 mg Carotene = 1667 IU Vit A)	Carotene (Provitamin A)	D	E	K	Water-Soluble Vitamins B-12	Biotin	Choline	Folacin (Folic Acid)	Niacin	Pantothenic Acid (B-3)	(Pyridoxine) B-6	Riboflavin (B-2)	Thiamin (B-1)
	IU/g	ppm or mg/kg	IU/kg	ppm or mg/kg	ppm or mg/kg	ppm or mg/kg	ppm or mg/kg	ppm or mg/kg	ppm or mg/kg	ppm or mg/kg	ppm or mg/kg	ppm or mg/kg	ppm or mg/kg	ppm or mg/kg
30	**845.8**	–	–	**39.5**	–	–	–	–	–	–	–	–	–	–
	850.0	–	–	39.7	–	–	–	–	–	–	–	–	–	–
31	–	–	–	–	–	**238.0**	**0.10**	**3277**	–	**75**	**11.0**	–	**5.4**	**0.3**
	–	–	–	–	–	255.5	0.11	3518	–	81	11.8	–	5.8	0.3
32	–	–	–	–	–	**1041.0**	**0.13**	**3009**	–	**356**	**41.2**	–	**16.8**	**4.0**
	–	–	–	–	–	2094.6	0.27	6054	–	716	82.9	–	33.7	8.0
33	**247.6**	**148.5**	–	**47.8**	–	–	–	–	–	–	–	–	–	–
	803.4	481.9	–	155.0	–	–	–	–	–	–	–	–	–	–
34	–	–	–	**7.1**	–	**425.8**	**0.98**	**1424**	**2.33**	**75**	**46.1**	**5.03**	**25.8**	**1.8**
	–	–	–	25.4	–	1523.2	3.51	5092	8.35	269	164.9	18.00	92.2	6.3
35	–	–	–	–	–	–	–	–	–	**169**	–	–	**32.8**	**4.0**
	–	–	–	–	–	–	–	–	–	579	–	–	112.3	13.7
36	–	–	–	–	–	**282.7**	**0.75**	–	**2.07**	**165**	**23.6**	**3.02**	**27.3**	**2.3**
	–	–	–	–	–	934.6	2.49	–	6.85	544	77.9	9.97	90.3	7.7
37	–	–	–	–	–	**501.3**	**0.02**	**11370**	**5.56**	**205**	**29.2**	–	**36.2**	**0.2**
	–	–	–	–	–	541.5	0.02	12281	6.01	221	31.5	–	39.1	0.2
38	–	–	–	–	–	**440.5**	**0.41**	**10610**	**4.00**	**172**	**90.8**	**5.01**	**42.4**	**0.2**
	–	–	–	–	–	472.1	0.44	11372	4.29	185	97.3	5.37	45.4	0.2
39	–	–	–	–	–	**881.6**	–	–	–	–	–	–	–	–
	–	–	–	–	–	979.7	–	–	–	–	–	–	–	–
40	**150.0**	**90.0**	–	–	–	–	–	–	–	–	–	–	–	–
	934.4	560.5	–	–	–	–	–	–	–	–	–	–	–	–
41	–	–	–	**60.7**	–	–	**0.42**	**1128**	**2.21**	**284**	**23.0**	**29.11**	**2.9**	**22.6**
	–	–	–	66.8	–	–	0.46	1240	2.43	312	25.3	32.01	3.2	24.9
42	–	–	–	**90.2**	–	–	**0.62**	**1248**	–	**506**	**46.4**	**27.89**	**1.8**	**20.0**
	–	–	–	100.0	–	–	0.68	1383	–	560	51.4	30.90	2.0	22.1
43	–	–	–	–	–	–	–	–	–	**169**	–	–	**32.8**	**4.0**
	–	–	–	–	–	–	–	–	–	579	–	–	112.3	13.7
44	**56.4**	**33.8**	–	**38.5**	–	–	–	–	–	**6**	–	–	**2.0**	**1.0**
	600.1	360.0	–	410.0	–	–	–	–	–	65	–	–	21.1	10.8
45	–	–	–	–	–	**282.7**	**0.75**	–	**2.07**	**165**	**23.6**	**3.02**	**27.3**	**2.3**
	–	–	–	–	–	934.6	2.49	–	6.85	544	77.9	9.97	90.3	7.7
46	–	–	–	–	–	–	–	–	–	**11**	–	–	**5.4**	**2.9**
	–	–	–	–	–	–	–	–	–	82	–	–	40.2	21.6
47	–	–	–	–	–	**499.0**	–	**3351**	–	**200**	**36.4**	**9.04**	**79.1**	**2.6**
	–	–	–	–	–	534.3	–	3588	–	214	38.9	9.68	84.7	2.8
48	–	–	–	**141.2**	–	–	**0.22**	**3062**	**2.12**	**68**	**18.6**	**9.97**	**6.0**	**23.1**
	–	–	–	159.7	–	–	0.24	3465	2.40	77	21.0	11.28	6.8	26.2
49	–	–	–	–	**18.66**	–	–	–	–	–	–	–	–	–
	–	–	–	–	18.75	–	–	–	–	–	–	–	–	–
50	–	–	–	**0.2**	–	**18.9**	**0.35**	**1790**	**0.85**	**11**	**46.2**	**3.21**	**27.4**	**4.0**
	–	–	–	0.2	–	20.3	0.38	1921	0.91	11	49.6	3.45	29.4	4.3
51	–	–	–	**2.1**	–	**1.1**	**1.04**	**3847**	**9.69**	**443**	**81.5**	**36.67**	**34.1**	**85.2**
	–	–	–	2.3	–	1.1	1.12	4134	10.41	476	87.6	39.40	36.6	91.6
52	–	–	–	–	–	**6.2**	**1.61**	–	**31.13**	**301**	**312.0**	–	**38.8**	**6.4**
	–	–	–	–	–	6.7	1.74	–	33.62	325	336.9	–	41.9	6.9
53	–	–	–	–	–	**4.0**	**1.19**	**2981**	**25.66**	**512**	**107.5**	**34.48**	**47.7**	**6.8**
	–	–	–	–	–	4.3	1.27	3203	27.58	550	115.6	37.06	51.3	7.3

TABLE V-2 AMINO ACID, COMPOSITION OF FEEDS, DATA EXPRESSED **AS-FED** AND **MOISTURE-FREE** (See footnote at end of table.)

Entry Number	Feed Name Description	International Feed Number	Moisture Basis: A-F (as-fed) or M-F (moisture-free)	Dry Matter	Crude Protein	AMINO ACIDS													
						Arginine	Cystine	Glycine	Histidine	Isoleucine	Leucine	Lysine	Methionine	Phenylalanine	Serine	Threonine	Tryptophan	Tyrosine	Valine
				%	%	%	%	%	%	%	%	%	%	%	%	%	%	%	%
	ENERGY FEEDS																		
	BAKERY																		
1	WASTE, DEHY (DRIED BAKERY PRODUCT)	4-00-466	A-F	91	10.1	0.47	0.17	0.69	0.16	0.45	0.77	0.31	0.17	0.45	0.65	0.46	0.10	0.36	0.47
			M-F	100	11.1	0.51	0.19	0.76	0.18	0.50	0.85	0.33	0.19	0.49	0.72	0.50	0.11	0.40	0.51
	BARLEY *Hordeum vulgare*																		
2	GRAIN	4-00-549	A-F	88	11.7	0.51	0.20	0.37	0.25	0.46	0.75	0.40	0.16	0.58	0.46	0.36	0.15	0.35	0.57
			M-F	100	13.2	0.58	0.23	0.42	0.28	0.52	0.85	0.45	0.18	0.65	0.52	0.41	0.17	0.39	0.64
3	GRAIN, PACIFIC COAST	4-07-939	A-F	89	9.5	0.44	0.19	0.30	0.21	0.40	0.60	0.26	0.14	0.47	0.32	0.31	0.12	0.31	0.46
			M-F	100	10.8	0.50	0.22	0.34	0.23	0.45	0.67	0.30	0.16	0.53	0.36	0.35	0.14	0.34	0.52
4	GRAIN SCREENINGS	4-00-542	A-F	89	11.5	—	—	—	—	—	—	—	—	—	—	0.36	—	—	—
			M-F	100	13.0	—	—	—	—	—	—	—	—	—	—	0.40	—	—	—
5	MALT SPROUTS, DEHY	5-00-545	A-F	93	22.9	1.05	0.23	0.81	0.43	0.88	1.36	1.12	0.31	0.80	0.47	0.85	0.41	0.46	1.16
			M-F	100	24.6	1.13	0.25	0.87	0.46	0.95	1.47	1.21	0.33	0.87	0.51	0.91	0.44	0.49	1.25
6	**BEAN, BROAD**[1] *Vicia faba*	5-09-262	A-F	—	26.7	2.45	0.32	—	0.67	1.08	1.98	1.68	0.20	1.13	—	0.96	0.23	0.84	1.21
	BEAN, NAVY *Phaseolus vulgaris*																		
7	SEEDS	5-00-623	A-F	89	22.9	1.19	0.23	0.80	—	—	—	1.29	0.23	—	—	—	—	0.24	—
			M-F	100	25.6	1.33	0.26	0.89	—	—	—	1.44	0.26	—	—	—	—	0.27	—
	BEET, SUGAR *Beta vulgaris, altissima*																		
8	PULP, DEHY	4-00-669	A-F	91	8.8	0.30	0.01	—	0.20	0.30	0.60	0.60	0.01	0.30	—	0.40	0.10	0.40	0.40
			M-F	100	9.7	0.33	0.01	—	0.22	0.33	0.66	0.66	0.01	0.33	—	0.44	0.11	0.44	0.44
	BROOMCORN (HOG MILLET; MILLET, PROSO) *Panicum miliaceum*																		
9	GRAIN	4-03-120	A-F	90	11.6	0.34	0.20	0.25	0.20	0.44	1.13	0.22	0.26	0.54	0.63	0.37	0.16	0.23	0.55
			M-F	100	12.9	0.38	0.22	0.28	0.22	0.49	1.26	0.25	0.28	0.60	0.70	0.41	0.17	0.26	0.61
	BUCKWHEAT, COMMON *Fagopyrum sagittatum*																		
10	GRAIN	4-00-994	A-F	88	11.1	0.96	0.17	0.61	0.26	0.37	0.59	0.62	0.19	0.44	0.41	0.44	0.18	0.21	0.53
			M-F	100	12.6	1.09	0.19	0.70	0.30	0.42	0.67	0.71	0.22	0.50	0.46	0.51	0.21	0.24	0.61
	CITRUS *Citrus spp*																		
11	PULP WITHOUT FINES, DEHY (DRIED CITRUS PULP)	4-01-237	A-F	91	6.1	0.25	0.11	—	0.09	0.18	0.31	0.20	0.09	0.18	—	0.18	0.06	—	0.25
			M-F	100	6.7	0.28	0.12	—	0.10	0.20	0.34	0.22	0.09	0.20	—	0.20	0.07	—	0.28
	CORN, DENT WHITE *Zea mays, indentata*																		
12	GRAIN	4-02-928	A-F	90	10.8	0.27	0.09	—	0.18	0.45	0.90	0.27	0.09	0.36	—	0.36	0.09	0.45	0.36
			M-F	100	11.9	0.30	0.10	—	0.20	0.50	1.00	0.30	0.10	0.40	—	0.40	0.10	0.50	0.40
	CORN, DENT YELLOW *Zea mays, indentata*																		
13	GRAIN, ALL ANALYSES	4-02-935	A-F	88	9.9	0.43	0.12	0.37	0.27	0.35	1.19	0.30	0.18	0.46	0.49	0.36	0.09	0.31	0.48
			M-F	100	11.2	0.49	0.13	0.42	0.31	0.40	1.35	0.34	0.20	0.52	0.55	0.41	0.10	0.35	0.54
14	GRAIN, GRADE 2, 54 lb/bu (*69.5 kg/hl*)	4-02-931	A-F	87	8.9	0.45	0.11	0.45	0.20	0.40	1.00	0.19	0.11	0.45	—	0.35	0.09	0.43	0.35
			M-F	100	10.2	0.52	0.13	0.52	0.23	0.46	1.15	0.22	0.13	0.52	—	0.40	0.10	0.49	0.40
15	GRAIN, FLAKED	4-28-244	A-F	89	9.9	0.44	0.25	0.36	0.28	0.34	1.24	0.25	0.15	0.44	0.48	0.35	—	0.39	0.47
			M-F	100	11.2	0.49	0.28	0.40	0.31	0.38	1.40	0.28	0.17	0.50	0.54	0.39	—	0.44	0.53
16	DISTILLERS' GRAIN W/SOLUBLES, DEHY[1]	5-28-236	A-F	—	27.0	0.96	0.29	—	0.64	1.38	2.21	0.70	0.49	1.47	—	0.92	0.17	0.69	1.48
17	DISTILLERS SOLUBLES, DEHY	5-28-237	A-F	93	27.4	0.99	0.44	1.12	0.67	1.32	2.38	0.92	0.55	1.47	1.22	1.01	0.25	0.88	1.53
			M-F	100	29.5	1.06	0.48	1.21	0.73	1.42	2.56	0.99	0.59	1.58	1.32	1.09	0.27	0.95	1.65
18	EARS, GROUND (CORN-AND-COB MEAL)	4-28-238	A-F	87	7.8	0.36	0.12	0.31	0.16	0.35	0.86	0.17	0.14	0.39	—	0.28	0.07	0.32	0.31
			M-F	100	9.0	0.42	0.14	0.36	0.19	0.40	1.00	0.20	0.16	0.45	—	0.33	0.08	0.38	0.36
19	GERM MEAL, WET MILLED, SOLV EXTD	5-28-240	A-F	92	20.7	1.31	0.40	1.10	0.70	0.70	1.81	0.90	0.58	0.90	1.00	1.09	0.20	0.70	1.20
			M-F	100	22.6	1.43	0.44	1.20	0.76	0.76	1.97	0.98	0.64	0.98	1.09	1.19	0.21	0.76	1.31
20	GRITS (HOMINY GRITS)	4-03-011	A-F	90	10.3	0.47	0.15	0.35	0.20	0.39	0.85	0.38	0.16	0.33	—	0.39	0.11	0.50	0.49
			M-F	100	11.4	0.52	0.16	0.38	0.22	0.44	0.94	0.42	0.18	0.36	—	0.44	0.12	0.55	0.55

TABLE V-2 AMINO ACID, COMPOSITION OF FEEDS, DATA EXPRESSED **AS-FED** AND **MOISTURE-FREE**—*(Continued)*

Entry Number	Feed Name Description	International Feed Number	Moisture Basis: A-F (as-fed) or M-F (moisture-free)	Dry Matter	Crude Protein	Arginine	Cystine	Glycine	Histidine	Isoleucine	Leucine	Lysine	Methionine	Phenylalanine	Serine	Threonine	Tryptophan	Tyrosine	Valine
						AMINO ACIDS													
				%	%	%	%	%	%	%	%	%	%	%	%	%	%	%	%
	CORN, OPAQUE 2 (HIGH LYSINE) *Zea mays*																		
21	GRAIN	4-11-445	A-F	90	10.1	0.64	0.19	0.48	0.35	0.33	0.98	0.42	0.16	0.43	0.46	0.37	0.12	0.40	0.48
			M-F	100	11.2	0.71	0.21	0.53	0.38	0.37	1.09	0.46	0.18	0.48	0.51	0.42	0.13	0.44	0.54
	COWPEA, COMMON *Vigna sinensis*																		
22	SEEDS	5-01-661	A-F	89	23.8	1.70	–	–	0.70	1.10	2.31	2.10	0.20	1.30	–	0.80	0.30	1.10	1.20
			M-F	100	26.7	1.91	–	–	0.79	1.24	2.58	2.36	0.23	1.46	–	0.90	0.34	1.24	1.35
	DARSO SORGHUM–SEE SORGHUM																		
	DISTILLERS PRODUCTS (ALSO SEE CORN; WHEAT)																		
23	SOLUBLES, DEHY	5-02-147	A-F	92	28.8	1.06	0.40	1.20	0.66	1.21	2.35	0.95	0.50	1.24	0.93	1.00	0.24	0.93	1.40
			M-F	100	31.3	1.15	0.44	1.30	0.72	1.32	2.55	1.03	0.54	1.35	1.01	1.09	0.26	1.01	1.52
	EMMER *Triticum dicoccum*																		
24	GRAIN	4-01-830	A-F	91	11.7	0.46	–	–	0.20	0.42	0.67	0.29	0.16	0.46	–	0.38	0.12	–	0.47
			M-F	100	12.9	0.51	–	–	0.22	0.46	0.74	0.32	0.18	0.51	–	0.42	0.13	–	0.52
	FETERITA SORGHUM–SEE SORGHUM																		
	GOOSEFOOT, LAMBS QUARTER *Chenopodium album*																		
25	SEEDS	5-08-424	A-F	90	18.8	0.08	–	0.05	0.02	0.03	0.05	0.04	0.02	0.03	0.03	0.03	–	0.02	0.03
			M-F	100	20.9	0.09	–	0.05	0.03	0.03	0.05	0.04	0.02	0.04	0.04	0.03	–	0.03	0.04
	GRAPE *Vitis spp*																		
26	POMACE, DEHY (MARC)	1-02-208	A-F	90	12.1	0.67	0.17	0.89	0.26	0.55	1.63	0.50	0.18	0.55	–	0.38	0.07	0.16	1.09
			M-F	100	13.4	0.74	0.19	0.99	0.29	0.60	1.80	0.55	0.20	0.60	–	0.42	0.08	0.18	1.21
	HEGARI SORGHUM–SEE SORGHUM																		
	HOG MILLET (BROOMCORN; MILLET, PROSO) *Panicum miliaceum*																		
27	GRAIN	4-03-120	A-F	90	11.6	0.34	0.20	0.25	0.20	0.44	1.13	0.22	0.26	0.54	0.63	0.37	0.16	0.23	0.55
			M-F	100	12.9	0.38	0.22	0.28	0.22	0.49	1.26	0.25	0.28	0.60	0.70	0.41	0.17	0.26	0.61
28	**HOMINY GRITS (CORN, DENT YELLOW, GRITS)** *Zea mays, indentata*	4-03-011	A-F	90	10.3	0.47	0.15	0.35	0.20	0.39	0.85	0.38	0.16	0.33	–	0.39	0.11	0.50	0.49
			M-F	100	11.4	0.52	0.16	0.38	0.22	0.44	0.94	0.42	0.18	0.36	–	0.44	0.12	0.55	0.55
	KAFIR SORGHUM–SEE SORGHUM																		
	MAIZE–SEE CORN																		
	MILLET *Setaria spp*																		
29	GRAIN	4-03-098	A-F	90	12.1	0.35	0.12	0.40	0.23	0.49	1.23	0.26	0.30	0.59	–	0.44	0.12	–	0.62
			M-F	100	13.5	0.39	0.13	0.44	0.26	0.54	1.37	0.29	0.33	0.66	–	0.49	0.14	–	0.69
	MILLET, PROSO (BROOMCORN; HOG MILLET) *Panicum miliaceum*																		
30	GRAIN	4-03-120	A-F	90	11.6	0.34	0.20	0.25	0.20	0.44	1.13	0.22	0.26	0.54	0.63	0.37	0.16	0.23	0.55
			M-F	100	12.9	0.38	0.22	0.28	0.22	0.49	1.26	0.25	0.28	0.60	0.70	0.41	0.17	0.26	0.61
	MILO SORGHUM–SEE SORGHUM																		
	MOLASSES[1]																		
31	CANE	4-04-696	A-F	–	4.4	–	–	–	–	–	–	–	–	–	–	–	–	–	–
32	BEET	4-00-668	A-F	–	6.6	–	–	–	–	–	–	–	–	–	–	–	–	–	–
	OATS *Avena sativa*																		
33	GRAIN	4-03-309	A-F	89	11.9	0.71	0.19	0.51	0.17	0.48	0.87	0.40	0.18	0.57	0.50	0.38	0.15	0.45	0.62
			M-F	100	13.3	0.80	0.21	0.57	0.19	0.54	0.97	0.45	0.20	0.64	0.56	0.43	0.17	0.50	0.69

(Continued)

TABLE V-2 AMINO ACID, COMPOSITION OF FEEDS, DATA EXPRESSED **AS-FED** AND **MOISTURE-FREE**—*(Continued)*

Entry Number	Feed Name Description	International Feed Number	Moisture Basis: A-F (as-fed) or M-F (moisture-free)	Dry Matter	Crude Protein	AMINO ACIDS: Arginine	Cystine	Glycine	Histidine	Isoleucine	Leucine	Lysine	Methionine	Phenylalanine	Serine	Threonine	Tryptophan	Tyrosine	Valine
				%	%	%	%	%	%	%	%	%	%	%	%	%	%	%	%
	OATS (Continued)																		
34	GRAIN, GRADE 1, 34 lb/bu *(43.8 kg/hl)*	4-03-313	A-F	88	11.2	0.79	0.22	0.49	0.19	0.52	0.89	0.49	0.18	0.59	–	0.39	0.16	0.52	0.69
			M-F	100	12.7	0.89	0.25	0.56	0.22	0.59	1.01	0.56	0.20	0.67	–	0.45	0.18	0.59	0.78
35	GRAIN, PACIFIC COAST	4-07-999	A-F	91	9.1	0.58	0.17	0.40	0.17	0.38	0.70	0.33	0.13	0.43	0.40	0.30	0.12	0.70	0.49
			M-F	100	10.0	0.63	0.18	0.44	0.18	0.42	0.77	0.37	0.14	0.47	0.44	0.33	0.13	0.77	0.54
36	MIDDLINGS, LESS THAN 4% FIBER (FEEDING OAT MEAL)	4-03-303	A-F	91	14.8	0.81	0.25	0.62	0.30	0.56	1.05	0.53	0.21	0.69	0.70	0.48	0.20	0.72	0.74
			M-F	100	16.3	0.89	0.28	0.69	0.33	0.61	1.16	0.58	0.23	0.76	0.77	0.53	0.22	0.80	0.81
37	GROATS	4-03-331	A-F	90	15.8	0.89	0.21	0.61	0.27	0.54	1.04	0.54	0.20	0.70	0.62	0.44	0.19	0.51	0.74
			M-F	100	17.6	0.99	0.23	0.68	0.31	0.60	1.16	0.61	0.23	0.78	0.69	0.49	0.21	0.57	0.83
	PEA, GARDEN *Pisum sativum, sativum*																		
38	SEEDS	5-08-482	A-F	89	23.8	1.43	–	–	0.63	1.03	1.61	1.47	0.34	1.20	–	0.80	0.23	–	1.18
			M-F	100	26.7	1.60	–	–	0.71	1.16	1.81	1.65	0.38	1.35	–	0.90	0.26	–	1.32
	PUMPKIN *Cucurbita pepo*																		
39	SEEDS	5-03-817	A-F	94	38.3	4.93	0.93	1.96	0.75	1.29	2.29	1.45	0.77	1.67	2.03	0.95	0.47	1.38	1.57
			M-F	100	40.9	5.25	0.99	2.09	0.80	1.38	2.44	1.54	0.82	1.79	2.16	1.02	0.50	1.47	1.68
	RICE *Oryza sativa*																		
40	GRAIN, GROUND (GROUND ROUGH RICE; GROUND PADDY RICE)	4-03-938	A-F	89	7.5	0.54	0.12	0.62	0.16	0.27	0.54	0.25	0.14	0.30	0.50	0.23	0.10	0.63	0.40
			M-F	100	8.4	0.61	0.14	0.69	0.18	0.30	0.60	0.28	0.15	0.34	0.56	0.26	0.12	0.71	0.45
41	GRAIN, POLISHED & BROKEN[1]	4-03-932	A-F	–	7.8	0.49	0.08	–	0.18	0.33	0.68	0.27	0.13	0.39	–	0.24	0.10	0.41	0.47
42	BRAN WITH GERMS (RICE BRAN)	4-03-928	A-F	91	13.0	0.82	0.16	0.81	0.29	0.50	0.84	0.54	0.26	0.53	0.73	0.44	0.10	0.59	0.75
			M-F	100	14.3	0.91	0.18	0.90	0.32	0.56	0.93	0.60	0.29	0.58	0.81	0.49	0.11	0.65	0.83
43	GROATS, POLISHED (RICE, POLISHED)	4-03-942	A-F	89	7.0	0.48	0.09	0.42	0.18	0.28	0.47	0.24	0.17	0.31	0.29	0.25	0.09	0.23	0.40
			M-F	100	7.9	0.54	0.11	0.47	0.20	0.32	0.53	0.27	0.19	0.35	0.33	0.28	0.11	0.26	0.45
44	POLISHINGS	4-03-943	A-F	90	12.0	0.57	0.14	0.65	0.19	0.37	0.73	0.51	0.22	0.43	0.49	0.35	0.11	0.45	0.68
			M-F	100	13.3	0.63	0.16	0.72	0.21	0.41	0.81	0.57	0.24	0.47	0.54	0.39	0.12	0.50	0.76
	RYE *Secale cereale*																		
45	GRAIN	4-04-047	A-F	87	12.0	0.52	0.19	0.44	0.25	0.48	0.68	0.42	0.17	0.56	0.61	0.35	0.12	0.26	0.58
			M-F	100	13.8	0.60	0.21	0.50	0.29	0.55	0.77	0.48	0.20	0.64	0.70	0.40	0.13	0.30	0.66
	SAFFLOWER *Carthamus tinctorius*																		
46	SEEDS	4-07-958	A-F	93	14.9	1.60	0.35	1.00	0.48	0.80	1.20	0.60	0.33	1.00	–	0.64	0.28	–	1.00
			M-F	100	16.0	1.72	0.38	1.07	0.52	0.86	1.29	0.64	0.35	1.07	–	0.69	0.30	–	1.07
	SCREENINGS, GRAIN (CEREAL) (ALSO SEE WHEAT)																		
47	REFUSE	4-02-151	A-F	91	12.6	0.68	–	0.59	0.30	0.52	0.98	0.48	0.15	0.64	0.57	0.46	–	0.32	0.63
			M-F	100	13.8	0.75	–	0.65	0.33	0.58	1.08	0.53	0.16	0.71	0.63	0.51	–	0.35	0.70
48	UNCLEANED	4-02-153	A-F	92	13.7	0.67	–	0.61	0.30	0.45	0.90	0.42	0.19	0.58	0.67	0.44	–	0.58	0.58
			M-F	100	14.9	0.73	–	0.66	0.33	0.49	0.98	0.46	0.21	0.63	0.73	0.48	–	0.63	0.63
	SHALLU SORGHUM–SEE SORGHUM																		
	SORGHUM *Sorghum bicolor*																		
49	GRAIN, ALL ANALYSES	4-04-383	A-F	90	11.5	0.39	0.21	0.34	0.24	0.42	1.47	0.26	0.14	0.56	0.49	0.36	0.09	0.40	0.50
			M-F	100	12.8	0.43	0.23	0.38	0.26	0.47	1.63	0.29	0.15	0.62	0.54	0.40	0.10	0.45	0.56
50	GRAIN, LESS THAN 9% PROTEIN	4-08-138	A-F	89	8.9	0.28	0.14	0.27	0.19	0.46	1.40	0.19	0.12	0.47	–	0.36	0.12	0.60	0.53
			M-F	100	10.1	0.32	0.16	0.31	0.22	0.52	1.58	0.22	0.14	0.53	–	0.41	0.14	0.68	0.60
51	DARSO, GRAIN	4-04-357	A-F	90	10.1	0.36	–	–	0.18	0.45	1.23	0.19	0.11	0.48	–	0.31	0.11	–	0.48
			M-F	100	11.2	0.40	–	–	0.20	0.51	1.38	0.21	0.12	0.53	–	0.34	0.12	–	0.53
52	FETERITA, GRAIN	4-04-369	A-F	89	11.7	0.46	–	–	0.26	0.58	1.78	0.20	0.18	0.67	–	0.46	0.17	–	0.67
			M-F	100	13.0	0.51	–	–	0.29	0.65	1.99	0.23	0.21	0.75	–	0.51	0.19	–	0.75
53	HEGARI, GRAIN	4-04-398	A-F	89	10.4	0.29	–	–	0.18	0.47	1.40	0.17	0.12	0.54	–	0.36	0.11	–	0.55
			M-F	100	11.7	0.33	–	–	0.21	0.52	1.56	0.20	0.13	0.61	–	0.41	0.12	–	0.61
54	KAFIR, GRAIN	4-04-428	A-F	89	10.8	0.38	0.17	0.30	0.27	0.55	1.62	0.26	0.19	0.64	–	0.45	0.15	–	0.62
			M-F	100	12.1	0.43	0.19	0.33	0.30	0.62	1.81	0.29	0.21	0.71	–	0.50	0.17	–	0.69
55	MILO, GRAIN	4-04-444	A-F	89	10.1	0.37	0.13	0.35	0.24	0.44	1.32	0.23	0.16	0.49	0.49	0.35	0.10	0.37	0.53
			M-F	100	11.4	0.42	0.15	0.39	0.27	0.50	1.49	0.26	0.18	0.55	0.56	0.40	0.11	0.41	0.60

TABLE V-2 AMINO ACID, COMPOSITION OF FEEDS, DATA EXPRESSED **AS-FED** AND **MOISTURE-FREE**—*(Continued)*

Entry Number	Feed Name Description	International Feed Number	Moisture Basis: A-F (as-fed) or M-F (moisture-free)	Dry Matter	Crude Protein	AMINO ACIDS Arginine	Cystine	Glycine	Histidine	Isoleucine	Leucine	Lysine	Methionine	Phenylalanine	Serine	Threonine	Tryptophan	Tyrosine	Valine
				%	%	%	%	%	%	%	%	%	%	%	%	%	%	%	%
	SORGHUM (Continued)																		
56	MILO, GLUTEN WITH BRAN, MEAL	5-08-089	A-F	89	23.2	0.90	0.20	0.68	0.60	1.00	2.51	0.70	0.40	1.00	–	0.80	0.20	0.90	1.30
			M-F	100	26.0	1.01	0.23	0.77	0.68	1.13	2.81	0.79	0.45	1.13	–	0.90	0.23	1.01	1.46
57	SHALLU, GRAIN	4-04-456	A-F	90	11.5	0.31	–	–	0.19	0.38	0.97	0.19	0.17	0.40	–	0.30	0.10	–	0.46
			M-F	100	12.7	0.34	–	–	0.21	0.42	1.08	0.21	0.18	0.44	–	0.33	0.11	–	0.51
	SOYBEAN *Glycine max*																		
58	SEEDS	5-04-610	A-F	92	38.4	2.63	0.42	1.42	0.92	1.62	2.72	2.32	0.48	1.76	1.99	1.46	0.56	1.29	1.61
			M-F	100	41.7	2.86	0.45	1.55	1.00	1.76	2.95	2.52	0.52	1.91	2.16	1.58	0.61	1.40	1.75
59	SEEDS, FULL-FAT, COOKED [1]	5-04-597	A-F	–	36.7	2.54	0.55	–	0.87	1.60	2.64	2.25	0.46	1.80	–	1.42	0.54	1.26	1.62
60	SOYBEAN MILL FEED	4-04-594	A-F	90	12.6	0.70	0.13	0.47	0.18	0.41	0.58	0.59	0.12	0.37	–	0.30	0.13	0.23	0.37
			M-F	100	14.1	0.78	0.15	0.52	0.20	0.45	0.64	0.66	0.14	0.42	–	0.34	0.15	0.26	0.42
	SPELT *Triticum spelta*																		
61	GRAIN	4-04-651	A-F	90	12.0	0.45	–	–	0.18	0.36	0.63	0.27	0.18	0.45	–	0.36	0.09	–	0.45
			M-F	100	13.3	0.50	–	–	0.20	0.40	0.70	0.30	0.20	0.50	–	0.40	0.10	–	0.50
	TOMATO *Lycopersicon esculentum*																		
62	POMACE, DEHY	5-05-041	A-F	92	21.0	1.20	–	–	0.40	0.70	1.70	1.60	0.10	0.90	–	0.70	0.20	0.90	1.00
			M-F	100	22.9	1.30	–	–	0.43	0.76	1.85	1.74	0.11	0.98	–	0.76	0.22	0.98	1.09
	TRITICALE *Triticale hexaploide*																		
63	GRAIN	4-20-362	A-F	89	15.4	0.85	0.27	0.68	0.38	0.58	1.11	0.52	0.22	0.77	0.73	0.53	0.18	0.49	0.78
			M-F	100	17.3	0.95	0.30	0.76	0.43	0.65	1.25	0.58	0.24	0.86	0.82	0.60	0.20	0.55	0.88
	WHEAT *Triticum aestivum*																		
64	GRAIN, ALL ANALYSES	4-05-211	A-F	89	13.1	0.61	0.22	0.59	0.30	0.49	0.90	0.39	0.18	0.61	0.63	0.40	0.15	0.37	0.61
			M-F	100	14.7	0.69	0.25	0.66	0.34	0.55	1.01	0.44	0.20	0.69	0.71	0.45	0.17	0.41	0.69
65	GRAIN, HARD RED SPRING	4-05-258	A-F	88	14.2	0.64	0.22	0.60	0.27	0.52	0.91	0.38	0.20	0.66	0.61	0.39	0.15	0.45	0.61
			M-F	100	16.2	0.73	0.25	0.68	0.31	0.60	1.04	0.44	0.23	0.75	0.70	0.44	0.17	0.51	0.69
66	GRAIN, HARD RED WINTER	4-05-268	A-F	89	12.8	0.65	0.30	0.58	0.30	0.53	0.87	0.36	0.22	0.63	0.59	0.37	0.17	0.46	0.58
			M-F	100	14.5	0.74	0.34	0.65	0.34	0.60	0.98	0.41	0.24	0.71	0.66	0.42	0.19	0.52	0.66
67	GRAIN, SOFT RED WINTER	4-05-294	A-F	88	11.4	0.65	0.36	0.55	0.32	0.45	0.90	0.36	0.22	0.64	0.65	0.39	0.27	0.37	0.58
			M-F	100	12.9	0.73	0.41	0.62	0.36	0.51	1.02	0.41	0.24	0.72	0.73	0.44	0.30	0.42	0.65
68	GRAIN, SOFT WHITE WINTER	4-05-337	A-F	90	10.2	0.47	0.27	0.50	0.22	0.42	0.66	0.32	0.16	0.46	0.46	0.32	0.13	0.37	0.45
			M-F	100	11.3	0.52	0.30	0.55	0.24	0.46	0.73	0.35	0.18	0.51	0.51	0.35	0.14	0.41	0.50
69	GRAIN, SOFT WHITE WINTER, PACIFIC COAST	4-08-555	A-F	89	10.0	0.45	0.24	0.50	0.20	0.40	0.75	0.30	0.14	0.48	0.49	0.31	0.12	0.36	0.46
			M-F	100	11.2	0.50	0.27	0.56	0.22	0.45	0.84	0.34	0.16	0.54	0.54	0.34	0.13	0.41	0.52
70	BRAN	4-05-190	A-F	89	15.5	0.85	0.26	0.77	0.33	0.55	0.89	0.54	0.17	0.50	0.68	0.40	0.25	0.38	0.67
			M-F	100	17.5	0.96	0.29	0.86	0.37	0.62	1.00	0.61	0.19	0.56	0.76	0.45	0.28	0.43	0.76
71	DISTILLERS GRAINS, DEHY	5-05-193	A-F	93	31.6	1.10	–	–	0.80	2.01	1.71	0.70	–	1.71	–	0.90	–	0.50	1.71
			M-F	100	33.9	1.18	–	–	0.86	2.15	1.83	0.75	–	1.83	–	0.97	–	0.54	1.83
72	ENDOSPERM	4-05-197	A-F	88	11.1	0.60	0.30	–	0.30	1.10	1.70	0.40	0.20	0.60	–	0.40	0.30	–	0.60
			M-F	100	12.6	0.68	0.34	–	0.34	1.25	1.93	0.46	0.23	0.68	–	0.46	0.34	–	0.68
73	FLOUR, LESS THAN 1.5% FIBER	4-05-199	A-F	88	13.7	0.42	0.25	0.46	0.28	0.56	0.89	0.27	0.13	0.62	0.51	0.30	0.10	0.25	0.51
			M-F	100	15.5	0.47	0.28	0.52	0.32	0.63	1.01	0.30	0.15	0.70	0.58	0.34	0.11	0.28	0.58
74	GRAIN SCREENINGS	4-05-216	A-F	89	13.3	0.68	0.12	0.53	0.30	0.45	0.74	0.43	0.26	0.49	0.40	0.33	0.13	0.23	0.55
			M-F	100	15.0	0.77	0.14	0.60	0.34	0.50	0.83	0.48	0.29	0.55	0.45	0.38	0.14	0.26	0.62
75	MIDDLINGS, LESS THAN 9.5% FIBER	4-05-205	A-F	89	16.4	0.98	0.22	0.96	0.40	0.68	1.11	0.68	0.19	0.66	0.80	0.57	0.19	0.43	0.80
			M-F	100	18.5	1.10	0.24	1.08	0.45	0.77	1.25	0.77	0.21	0.75	0.90	0.64	0.22	0.49	0.90
76	MILL RUN, LESS THAN 9.5% FIBER	4-05-206	A-F	90	15.1	0.94	0.23	0.53	0.40	0.70	1.20	0.57	0.33	–	–	0.50	0.21	0.50	0.80
			M-F	100	16.7	1.04	0.26	0.59	0.44	0.78	1.33	0.64	0.37	–	–	0.56	0.23	0.56	0.89
77	RED DOG, LESS THAN 4% FIBER	4-05-203	A-F	88	15.6	0.96	0.36	0.74	0.38	0.58	1.08	0.60	0.22	0.65	0.76	0.50	0.20	0.46	0.73
			M-F	100	17.6	1.08	0.40	0.84	0.43	0.66	1.22	0.68	0.24	0.74	0.87	0.56	0.22	0.52	0.83
78	SHORTS, LESS THAN 7% FIBER	4-05-201	A-F	88	16.5	1.20	0.38	0.96	0.44	0.57	1.07	0.80	0.28	0.67	0.77	0.60	0.23	0.47	0.82
			M-F	100	18.7	1.35	0.43	1.08	0.50	0.64	1.22	0.90	0.32	0.75	0.87	0.67	0.26	0.53	0.92
	WHEAT, DURUM *Triticum durum*																		
79	GRAIN	4-05-224	A-F	88	13.8	0.58	0.13	0.46	0.27	0.48	1.40	1.05	0.14	0.53	0.45	0.37	0.26	0.29	0.54
			M-F	100	15.7	0.67	0.15	0.52	0.31	0.55	1.60	1.19	0.16	0.61	0.51	0.42	0.30	0.33	0.62

(Continued)

AMINO ACIDS

AMINO ACIDS

TABLE V-2 AMINO ACID, COMPOSITION OF FEEDS, DATA EXPRESSED **AS-FED** AND **MOISTURE-FREE**—*(Continued)*

Entry Number	Feed Name Description	International Feed Number	Moisture Basis: A-F (as-fed) or M-F (moisture-free)	Dry Matter	Crude Protein	Arginine	Cystine	Glycine	Histidine	Isoleucine	Leucine	Lysine	Methionine	Phenylalanine	Serine	Threonine	Tryptophan	Tyrosine	Valine
				%	%	%	%	%	%	%	%	%	%	%	%	%	%	%	%
	PROTEIN FEEDS																		
	ACACIA, SWEET *Acacia farnesiana*																		
80	SEEDS	5-09-110	A-F	87	47.9	4.40	–	1.63	1.10	1.67	3.58	2.25	0.44	1.67	1.97	1.20	–	1.34	1.86
			M-F	100	55.0	5.06	–	1.87	1.26	1.92	4.12	2.58	0.50	1.92	2.26	1.38	–	1.54	2.14
	ANIMAL																		
81	BLOOD, MEAL	5-00-380	A-F	91	80.5	3.23	1.25	3.45	3.93	0.85	10.07	6.43	0.94	5.56	3.95	3.59	1.01	1.94	6.56
			M-F	100	88.2	3.54	1.37	3.78	4.31	0.94	11.03	7.04	1.03	6.09	4.32	3.93	1.10	2.12	7.19
82	BLOOD, SPRAY DEHY	5-00-381	A-F	93	86.0	3.59	1.03	3.83	5.18	0.91	10.97	7.44	1.05	5.89	3.53	3.63	1.05	2.26	7.52
			M-F	100	93.0	3.88	1.11	4.14	5.59	0.98	11.85	8.04	1.14	6.36	3.82	3.93	1.13	2.44	8.13
83	LIVER, MEAL	5-00-389	A-F	93	66.1	4.04	0.94	5.61	1.48	3.11	5.31	5.22	1.22	2.92	2.50	2.50	0.69	1.70	4.15
			M-F	100	71.4	4.37	1.01	6.05	1.60	3.36	5.74	5.63	1.32	3.15	2.70	2.70	0.74	1.84	4.49
84	MEAT AND BONE MEAL, 50% [1]	5-09-322	A-F	–	50.9	3.65	0.46	–	0.96	1.47	3.02	2.89	0.68	1.65	–	1.60	0.28	0.79	2.14
85	MEAT MEAL, 55% [1]	5-09-323	A-F	–	55.6	3.79	0.68	–	1.04	1.84	3.51	3.09	0.73	1.91	–	1.78	0.38	0.96	2.61
86	MEAT, MEAL RENDERED	5-00-385	A-F	94	50.7	3.58	0.60	7.23	0.87	1.63	3.11	3.00	0.69	1.74	2.31	1.67	0.35	1.09	2.42
			M-F	100	54.0	3.81	0.64	7.70	0.93	1.74	3.31	3.19	0.74	1.86	2.46	1.78	0.37	1.17	2.58
87	MEAT WITH BONE, MEAL RENDERED	5-00-388	A-F	93	50.4	3.53	0.53	6.49	1.01	1.64	3.10	2.93	0.67	1.71	1.90	1.66	0.31	0.89	2.44
			M-F	100	54.0	3.78	0.56	6.96	1.08	1.76	3.33	3.14	0.71	1.84	2.03	1.78	0.33	0.96	2.61
88	TANKAGE, MEAL RENDERED	5-00-386	A-F	92	60.5	3.60	0.48	6.45	2.06	1.82	5.10	3.89	0.75	2.56	2.81	2.34	0.65	1.38	3.83
			M-F	100	65.6	3.91	0.52	7.00	2.23	1.97	5.54	4.22	0.81	2.78	3.05	2.54	0.71	1.50	4.16
89	TANKAGE WITH BONE, MEAL RENDERED	5-00-387	A-F	93	46.6	2.82	0.27	6.58	1.76	1.87	5.26	3.32	0.69	2.28	–	2.18	0.62	–	3.42
			M-F	100	50.2	3.03	0.29	7.08	1.90	2.01	5.67	3.57	0.74	2.46	–	2.34	0.67	–	3.68
	BABASSU *Orbignya* spp																		
90	KERNELS WITH COATS, MEAL MECH EXTD (BABASSU OIL MEAL)	5-00-454	A-F	92	22.3	2.87	–	–	0.40	1.04	1.34	0.89	0.30	0.89	–	0.60	0.20	0.40	1.09
			M-F	100	24.2	3.12	–	–	0.43	1.13	1.45	0.97	0.32	0.97	–	0.65	0.22	0.43	1.18
91	KERNELS WITH COATS, MEAL SOLV EXTD (BABASSU OIL MEAL)	5-00-455	A-F	93	21.2	3.19	–	–	0.41	0.88	1.40	0.98	0.53	1.35	–	0.71	0.24	–	1.19
			M-F	100	22.9	3.44	–	–	0.44	0.95	1.51	1.06	0.57	1.46	–	0.77	0.26	–	1.28
	BEAN, PINTO *Phaseolus vulgaris*																		
92	SEEDS	5-00-624	A-F	90	22.7	1.55	–	–	0.64	1.14	1.11	1.60	0.26	1.20	–	1.09	0.32	–	1.23
			M-F	100	25.1	1.72	–	–	0.71	1.26	1.23	1.77	0.29	1.33	–	1.21	0.35	–	1.36
	BLOOD																		
93	MEAL	5-00-380	A-F	91	80.5	3.23	1.25	3.45	3.93	0.85	10.07	6.43	0.94	5.56	3.95	3.59	1.01	1.94	6.56
			M-F	100	88.2	3.54	1.37	3.78	4.31	0.94	11.03	7.04	1.03	6.09	4.32	3.93	1.10	2.12	7.19
94	SPRAY, DEHY (BLOOD FLOUR)	5-00-381	A-F	93	86.0	3.59	1.03	3.83	5.18	0.91	10.97	7.44	1.05	5.89	3.53	3.63	1.05	2.26	7.52
			M-F	100	93.0	3.88	1.11	4.14	5.59	0.98	11.85	8.04	1.14	6.36	3.82	3.93	1.13	2.44	8.13
	BUTTERMILK (CATTLE) *Bos taurus*																		
95	DEHY	5-01-160	A-F	92	31.7	1.08	0.39	0.47	0.85	2.42	3.21	2.28	0.71	1.46	1.50	1.52	0.49	1.00	2.58
			M-F	100	34.4	1.17	0.42	0.51	0.92	2.62	3.48	2.47	0.76	1.58	1.62	1.64	0.53	1.08	2.80
96	**CANOLA SEEDS, MEAL, PREPRESSED, SOLV** [1]	5-06-145	A-F	–	38.0	2.32	0.47	–	1.07	1.51	2.65	2.27	0.68	1.52	–	1.71	0.44	0.93	1.94
	CASEIN																		
97	ACID PRECIPITATED, DEHY	5-01-162	A-F	91	84.0	3.49	0.31	1.61	2.59	5.72	8.80	7.14	2.81	4.81	5.46	3.91	1.08	4.90	6.71
			M-F	100	92.7	3.85	0.34	1.77	2.86	6.32	9.71	7.88	3.10	5.31	6.03	4.32	1.19	5.41	7.40
	CASTOR BEAN *Ricinus communis*																		
98	SEEDS WITHOUT TOXIN, MEAL	5-01-155	A-F	87	26.0	2.77	–	0.86	0.48	1.01	1.40	0.76	0.36	0.82	1.13	0.72	–	0.68	1.27
			M-F	100	30.0	3.20	–	0.99	0.55	1.17	1.61	0.88	0.42	0.94	1.30	0.83	–	0.78	1.46
	CATTLE *Bos taurus*																		
99	BUTTERMILK, DEHY	5-01-160	A-F	92	31.7	1.08	0.39	0.47	0.85	2.42	3.21	2.28	0.71	1.46	1.50	1.52	0.49	1.00	2.58
			M-F	100	34.4	1.17	0.42	0.51	0.92	2.62	3.48	2.47	0.76	1.58	1.62	1.64	0.53	1.08	2.80
100	MILK, FRESH	5-01-168	A-F	12	3.3	–	–	–	–	0.32	0.25	0.28	0.18	0.07	–	0.16	0.05	–	0.25
			M-F	100	26.7	–	–	–	–	2.58	2.03	2.27	1.43	0.55	–	1.33	0.39	–	2.03

TABLE V-2 AMINO ACID, COMPOSITION OF FEEDS, DATA EXPRESSED **AS-FED** AND **MOISTURE-FREE**—*(Continued)*

Entry Number	Feed Name Description	International Feed Number	Moisture Basis: A-F (as-fed) or M-F (moisture-free)	Dry Matter	Crude Protein	AMINO ACIDS Arginine	Cystine	Glycine	Histidine	Isoleucine	Leucine	Lysine	Methionine	Phenylalanine	Serine	Threonine	Tryptophan	Tyrosine	Valine
				%	%	%	%	%	%	%	%	%	%	%	%	%	%	%	%
	CATTLE (Continued)																		
101	MILK, DEHY	5-01-167	A-F	95	25.3	0.92	–	–	0.71	1.32	2.54	2.24	0.61	1.32	–	1.02	0.41	1.32	1.73
			M-F	100	26.6	0.96	–	–	0.75	1.39	2.67	2.35	0.64	1.39	–	1.07	0.43	1.39	1.81
102	SKIM MILK, DEHY	5-01-175	A-F	94	33.3	1.16	0.45	0.29	0.86	2.18	3.33	2.54	0.90	1.57	1.67	1.57	0.43	1.14	2.29
			M-F	100	35.4	1.23	0.48	0.31	0.92	2.32	3.53	2.70	0.96	1.66	1.78	1.66	0.46	1.22	2.43
103	WHEY, DEHY	4-01-182	A-F	93	13.3	0.33	0.30	0.44	0.17	0.78	1.18	0.94	0.19	0.35	0.47	0.90	0.20	0.25	0.67
			M-F	100	14.2	0.36	0.32	0.47	0.18	0.83	1.26	1.00	0.20	0.37	0.50	0.97	0.21	0.26	0.72
104	WHEY, LOW LACTOSE, DEHY	4-01-186	A-F	93	16.7	0.60	0.43	0.72	0.27	0.96	1.54	1.40	0.41	0.55	0.59	0.95	0.27	0.46	0.87
			M-F	100	17.9	0.64	0.46	0.77	0.29	1.03	1.65	1.50	0.43	0.59	0.63	1.01	0.29	0.49	0.93
	CHICKPEA (GARBANZO; GRAM PEA) *Cicer arietinum*																		
105	SEEDS	5-01-218	A-F	89	19.1	1.52	–	0.69	0.40	0.76	1.32	1.25	0.24	1.14	0.90	0.61	–	0.57	0.80
			M-F	100	21.4	1.71	–	0.78	0.45	0.85	1.47	1.40	0.27	1.28	1.01	0.68	–	0.64	0.89
	COCONUT *Cocos nucifera*																		
106	KERNELS WITH COATS, MEAL MECH EXTD (COPRA MEAL)	5-01-572	A-F	92	21.2	2.30	0.21	1.05	0.33	0.90	1.35	0.55	0.31	0.81	–	0.60	0.20	0.58	0.98
			M-F	100	23.1	2.52	0.23	1.15	0.36	0.99	1.48	0.60	0.34	0.88	–	0.66	0.22	0.63	1.07
	CORN *Zea mays*																		
107	DISTILLERS GRAINS, DEHY	5-02-842	A-F	93	27.8	0.99	0.23	0.75	0.62	1.00	3.01	0.76	0.42	0.99	1.01	0.56	0.20	0.84	1.21
			M-F	100	29.7	1.06	0.25	0.80	0.66	1.07	3.23	0.81	0.44	1.06	1.08	0.60	0.21	0.89	1.29
108	DISTILLERS SOLUBLES, DEHY	5-02-844	A-F	93	27.4	0.99	0.44	1.12	0.67	1.32	2.38	0.92	0.55	1.47	1.22	1.01	0.25	0.88	1.53
			M-F	100	29.5	1.06	0.48	1.21	0.73	1.42	2.56	0.99	0.59	1.58	1.32	1.09	0.27	0.95	1.65
109	GLUTEN FEED	5-02-903	A-F	90	23.0	0.78	0.42	0.85	0.61	0.88	2.20	0.64	0.37	0.81	0.85	0.78	0.15	0.72	1.10
			M-F	100	25.6	0.87	0.47	0.94	0.68	0.98	2.44	0.71	0.41	0.90	0.94	0.87	0.17	0.81	1.22
110	GLUTEN FEED [1]	5-28-243	A-F	–	23.3	0.79	0.43	–	0.62	0.89	2.21	0.64	0.37	0.82	–	0.79	0.15	0.74	1.11
111	GLUTEN MEAL	5-02-900	A-F	91	43.2	1.40	0.67	1.51	0.97	2.25	7.38	0.80	1.03	2.85	1.70	1.43	0.21	1.01	2.23
			M-F	100	47.3	1.53	0.73	1.65	1.06	2.46	8.08	0.88	1.13	3.12	1.86	1.56	0.23	1.11	2.44
112	GLUTEN MEAL, 41% [1]	5-12-354	A-F	–	42.1	1.37	0.66	–	0.97	2.25	6.00	0.78	1.07	2.84	–	1.42	0.21	1.01	2.22
113	GLUTEN MEAL, 60% [1]	5-28-242	A-F	–	61.2	2.08	1.01	–	1.40	2.54	10.44	1.03	1.78	4.04	–	2.25	0.30	3.31	3.11
	COTTON *Gossypium* spp																		
114	SEEDS, MEAL MECH EXTD, 36% PROTEIN	5-01-625	A-F	92	37.2	3.55	0.79	1.83	0.91	1.32	–	1.22	0.55	1.88	–	1.12	0.46	–	2.84
			M-F	100	40.5	3.86	0.86	1.99	0.99	1.44	–	1.33	0.60	2.04	–	1.21	0.50	–	3.09
115	SEEDS, MEAL MECH EXTD, 41% PROTEIN	5-01-617	A-F	93	41.0	4.20	0.71	1.87	1.07	1.42	2.30	1.60	0.57	2.19	1.70	1.33	0.52	0.97	1.89
			M-F	100	44.3	4.54	0.77	2.02	1.15	1.54	2.49	1.73	0.62	2.36	1.84	1.44	0.57	1.05	2.04
116	SEEDS, MEAL, PREPRESSED, SOLV EXTD, 41% PROTEIN	5-07-872	A-F	90	41.3	4.32	0.78	1.89	1.14	1.42	2.42	1.80	0.56	2.05	1.80	1.34	0.50	1.14	1.97
			M-F	100	45.7	4.78	0.87	2.09	1.26	1.57	2.67	1.99	0.61	2.27	1.99	1.48	0.56	1.27	2.18
117	SEEDS, MEAL, SOLV EXTD, 41% PROTEIN	5-01-621	A-F	91	41.2	4.24	0.76	1.95	1.10	1.50	2.46	1.69	0.58	2.23	1.76	1.37	0.55	1.04	1.97
			M-F	100	45.4	4.66	0.84	2.14	1.22	1.65	2.70	1.86	0.64	2.46	1.93	1.51	0.60	1.15	2.17
118	SEEDS, MEAL, EXPELLER [1]	5-01-609	A-F	–	36.8	4.26	0.75	–	0.97	1.29	2.10	1.51	0.56	1.98	–	1.18	0.52	1.02	2.08
119	SEEDS, MEAL, SOLVENT [1]	5-01-619	A-F	–	41.7	4.27	0.57	–	1.01	1.22	2.21	1.70	0.49	2.05	–	1.23	0.48	1.04	1.67
120	SEEDS WITHOUT HULLS, MEAL, PREPRESSED, SOLV EXTD, 50% PROTEIN	5-07-874	A-F	93	50.3	4.83	1.05	2.82	1.21	1.86	2.82	1.93	0.76	2.62	–	1.66	0.62	0.81	2.16
			M-F	100	54.0	5.20	1.13	3.03	1.30	2.00	3.03	2.08	0.81	2.81	–	1.78	0.67	0.87	2.32
	CRAB *Callinectes sapidus, Cancer* spp, *Paralithodes camschatica*																		
121	CANNERY RESIDUE, MEAL (CRAB MEAL)	5-01-663	A-F	92	32.2	1.66	0.24	1.74	0.48	1.16	1.54	1.38	0.52	1.16	1.38	1.00	0.29	1.17	1.47
			M-F	100	35.0	1.80	0.26	1.89	0.53	1.26	1.67	1.50	0.57	1.26	1.49	1.09	0.32	1.26	1.59
	CRAMBE, ABYSSINIAN *Crambe abyssinica*																		
122	SEEDS WITHOUT HULLS, MEAL MECH EXTD	5-16-453	A-F	92	45.8	–	–	–	–	–	–	–	–	–	–	–	–	–	–
			M-F	100	49.8	–	–	–	–	–	–	–	–	–	–	–	–	–	–
	DISTILLERS PRODUCTS (ALSO SEE CORN; RYE)																		
123	GRAINS, DEHY	5-02-144	A-F	93	27.3	1.04	0.42	0.56	0.53	1.16	2.66	0.81	0.46	1.03	0.70	0.81	0.21	0.73	1.22
			M-F	100	29.5	1.13	0.45	0.61	0.57	1.25	2.88	0.87	0.50	1.12	0.75	0.88	0.22	0.79	1.32
124	SOLUBLES, DEHY	5-02-147	A-F	92	28.8	1.06	0.40	1.20	0.66	1.21	2.35	0.95	0.50	1.24	0.93	1.00	0.24	0.93	1.40
			M-F	100	31.3	1.15	0.44	1.30	0.72	1.32	2.55	1.03	0.54	1.35	1.01	1.09	0.26	1.01	1.52

(Continued)

TABLE V-2 AMINO ACID, COMPOSITION OF FEEDS, DATA EXPRESSED **AS-FED** AND **MOISTURE-FREE**—*(Continued)*

						AMINO ACIDS													
Entry Number	Feed Name Description	International Feed Number	Moisture Basis: A-F (as-fed) or M-F (moisture-free)	Dry Matter	Crude Protein	Arginine	Cystine	Glycine	Histidine	Iso-leucine	Leucine	Lysine	Methi-onine	Phenyl-alanine	Serine	Threo-nine	Trypto-phan	Tyrosine	Valine
				%	%	%	%	%	%	%	%	%	%	%	%	%	%	%	%
	FISH																		
125	MEAL MECH EXTD	5-01-977	A-F	92	64.3	31.28	0.62	3.99	1.46	3.27	4.90	5.26	1.63	2.60	2.42	2.59	0.75	1.79	3.14
			M-F	100	70.2	34.16	0.68	4.35	1.59	3.57	5.35	5.74	1.78	2.84	2.64	2.83	0.82	1.96	3.43
126	SOLUBLES, CONDENSED	5-01-969	A-F	50	31.5	1.63	0.39	3.87	1.54	1.09	1.94	1.85	0.70	1.07	1.05	0.90	0.33	0.50	1.26
			M-F	100	62.5	3.23	0.77	7.67	3.06	2.17	3.84	3.67	1.39	2.12	2.08	1.79	0.65	0.99	2.50
	FISH, HERRING[1]																		
127	MEAL	5-02-000	A-F	–	72.0	4.65	0.74	–	1.66	3.17	5.23	5.64	2.08	2.73	–	2.89	0.78	2.21	4.36
	FISH, ANCHOVY *Engraulis ringen*																		
128	MEAL MECH EXTD	5-01-985	A-F	92	65.4	3.78	0.60	3.69	1.60	3.11	4.99	5.02	1.99	2.78	2.42	2.76	0.75	2.24	3.50
			M-F	100	71.1	4.10	0.66	4.01	1.74	3.38	5.42	5.46	2.16	3.03	2.63	3.01	0.81	2.44	3.81
	FISH, MENHADEN *Brevoortia tyrannus*																		
129	MEAL MECH EXTD	5-02-009	A-F	92	61.2	3.74	0.58	4.19	1.44	2.85	4.48	4.74	1.75	2.46	2.25	2.51	0.65	1.93	3.19
			M-F	100	66.8	4.08	0.63	4.57	1.58	3.11	4.89	5.17	1.91	2.68	2.45	2.74	0.71	2.11	3.48
	FISH, TUNA *Thunnus thynnus, Thunnus albacares*																		
130	MEAL MECH EXTD	5-02-023	A-F	93	59.0	3.43	0.47	4.09	1.75	2.45	3.79	4.06	1.47	2.15	2.08	2.31	0.57	1.69	2.77
			M-F	100	63.6	3.69	0.50	4.41	1.89	2.64	4.09	4.37	1.58	2.32	2.25	2.49	0.62	1.82	2.98
	FISH, WHITE *Gadidae* (family), *Lophiidae* (family), *Rajidae* (family)																		
131	MEAL MECH EXTD	5-02-025	A-F	91	62.6	4.26	0.77	5.15	1.38	2.85	4.65	4.70	1.79	2.44	3.44	2.56	0.67	2.27	3.25
			M-F	100	68.8	4.68	0.84	5.66	1.52	3.13	5.11	5.16	1.97	2.68	3.78	2.82	0.73	2.49	3.57
	FLAX, COMMON *Linum usitatissimum*																		
132	SEEDS, MEAL MECH EXTD (LINSEED MEAL)	5-02-048	A-F	90	34.6	2.94	0.61	1.74	0.69	1.68	2.02	1.16	0.54	1.46	1.93	1.22	0.51	1.09	1.74
			M-F	100	38.4	3.25	0.67	1.93	0.77	1.87	2.24	1.28	0.60	1.62	2.13	1.35	0.56	1.21	1.93
133	SEEDS, MEAL SOLV EXTD (LINSEED MEAL)	5-02-045	A-F	91	34.3	2.81	0.61	1.64	0.65	1.69	1.92	1.18	0.58	1.38	1.90	1.14	0.51	0.96	1.61
			M-F	100	37.8	3.10	0.67	1.80	0.71	1.86	2.11	1.30	0.64	1.53	2.09	1.25	0.56	1.06	1.77
	GARBANZO (CHICKPEA; GRAM PEA) *Cicer arietinum*																		
134	SEEDS	5-01-218	A-F	89	19.1	1.52	–	0.69	0.40	0.76	1.32	1.25	0.24	1.14	0.90	0.61	–	0.57	0.80
			M-F	100	21.4	1.71	–	0.78	0.45	0.85	1.47	1.40	0.27	1.28	1.01	0.68	–	0.64	0.89
	GRAM PEA (CHICKPEA; GARBANZO) *Cicer arietinum*																		
135	SEEDS	5-01-218	A-F	89	19.1	1.52	–	0.69	0.40	0.76	1.32	1.25	0.24	1.14	0.90	0.61	–	0.57	0.80
			M-F	100	21.4	1.71	–	0.78	0.45	0.85	1.47	1.40	0.27	1.28	1.01	0.68	–	0.64	0.89
	HORSE *Equus caballus*																		
136	MILK, FRESH	5-02-401	A-F	17	4.2	–	–	–	0.11	0.25	0.34	0.25	0.07	0.18	–	0.16	0.05	–	0.29
			M-F	100	24.7	–	–	–	0.64	1.49	2.02	1.49	0.43	1.06	–	0.96	0.32	–	1.70
	HORSE BEAN *Vicia faba, equina*																		
137	SEEDS	5-02-407	A-F	88	25.5	–	–	–	–	–	–	–	–	–	–	–	–	–	–
			M-F	100	29.0	–	–	–	–	–	–	–	–	–	–	–	–	–	–
	LINSEED–SEE FLAX																		
	LIVER																		
138	MEAL	5-00-389	A-F	93	66.1	4.04	0.94	5.61	1.48	3.11	5.31	5.22	1.22	2.92	2.50	2.50	0.69	1.70	4.15
			M-F	100	71.4	4.37	1.01	6.05	1.60	3.36	5.74	5.63	1.32	3.15	2.70	2.70	0.74	1.84	4.49
	LOCUST, NEW MEXICO *Robinia neomexicana*																		
139	SEEDS	5-09-055	A-F	89	36.5	3.01	–	1.32	0.73	0.87	1.75	1.32	0.26	1.02	1.28	0.87	–	0.84	1.17
			M-F	100	41.0	3.38	–	1.48	0.82	0.98	1.97	1.48	0.29	1.15	1.44	0.98	–	0.94	1.31
	MAIZE–SEE CORN																		

TABLE V-2 AMINO ACID, COMPOSITION OF FEEDS, DATA EXPRESSED **AS-FED** AND **MOISTURE-FREE**—*(Continued)*

Entry Number	Feed Name Description	International Feed Number	Moisture Basis: A-F (as-fed) or M-F (moisture-free)	Dry Matter	Crude Protein	AMINO ACIDS Arginine	Cystine	Glycine	Histidine	Isoleucine	Leucine	Lysine	Methionine	Phenylalanine	Serine	Threonine	Tryptophan	Tyrosine	Valine
				%	%	%	%	%	%	%	%	%	%	%	%	%	%	%	%
	MEAT																		
140	MEAL RENDERED	5-00-385	A-F	94	50.7	3.58	0.60	7.23	0.87	1.63	3.11	3.00	0.69	1.74	2.31	1.67	0.35	1.09	2.42
			M-F	100	54.0	3.81	0.64	7.70	0.93	1.74	3.31	3.19	0.74	1.86	2.46	1.78	0.37	1.17	2.58
141	MEAL, 55% [1]	5-09-323	A-F	–	55.6	3.79	0.68	–	1.04	1.84	3.51	3.09	0.73	1.91	–	1.78	0.38	0.96	2.61
142	WITH BLOOD, MEAL RENDERED (TANKAGE)	5-00-386	A-F	92	60.5	3.60	0.48	6.45	2.06	1.82	5.10	3.89	0.75	2.56	2.81	2.34	0.65	1.38	3.83
			M-F	100	65.6	3.91	0.52	7.00	2.23	1.97	5.54	4.22	0.81	2.78	3.05	2.54	0.71	1.50	4.16
143	WITH BLOOD, WITH BONE, MEAL RENDERED (TANKAGE)	5-00-387	A-F	93	46.6	2.82	0.27	6.58	1.76	1.87	5.26	3.32	0.69	2.28	–	2.18	0.62	–	3.42
			M-F	100	50.2	3.03	0.29	7.08	1.90	2.01	5.67	3.57	0.74	2.46	–	2.34	0.67	–	3.68
144	WITH BONE, MEAL RENDERED	5-00-388	A-F	93	50.4	3.53	0.53	6.49	1.01	1.64	3.10	2.93	0.67	1.71	1.90	1.66	0.31	0.89	2.44
			M-F	100	54.0	3.78	0.56	6.96	1.08	1.76	3.33	3.14	0.71	1.84	2.03	1.78	0.33	0.96	2.61
145	WITH BONE, MEAL, 50% [1]	5-09-322	A-F	–	50.9	3.65	0.46	–	0.96	1.47	3.02	2.89	0.68	1.65	–	1.60	0.28	0.79	2.14
	MILK																		
146	FRESH (CATTLE *Bos taurus*)	5-01-168	A-F	12	3.3	–	–	–	–	0.32	0.25	0.28	0.18	0.07	–	0.16	0.05	–	0.25
			M-F	100	26.7	–	–	–	–	2.58	2.03	2.27	1.43	0.55	–	1.33	0.39	–	2.03
147	DEHY (CATTLE *Bos taurus*)	5-01-167	A-F	95	25.3	0.92	–	–	0.71	1.32	2.54	2.24	0.61	1.32	–	1.02	0.41	1.32	1.73
			M-F	100	26.6	0.96	–	–	0.75	1.39	2.67	2.35	0.64	1.39	–	1.07	0.43	1.39	1.81
148	SKIMMED, DEHY (CATTLE *Bos taurus*)	5-01-175	A-F	94	33.3	1.16	0.45	0.29	0.86	2.18	3.33	2.54	0.90	1.57	1.67	1.57	0.43	1.14	2.29
			M-F	100	35.4	1.23	0.48	0.31	0.92	2.32	3.53	2.70	0.96	1.66	1.78	1.66	0.46	1.22	2.43
149	FRESH (HORSE *Equus caballus*)	5-02-401	A-F	17	4.2	–	–	–	0.11	0.25	0.34	0.25	0.07	0.18	–	0.16	0.05	–	0.29
			M-F	100	24.7	–	–	–	0.64	1.49	2.02	1.49	0.43	1.06	–	0.96	0.32	–	1.70
150	FRESH (SHEEP *Ovis aries*)	5-08-510	A-F	19	4.6	–	–	–	0.18	0.39	0.60	0.51	0.17	0.32	–	0.30	0.09	–	0.48
			M-F	100	24.7	–	–	–	0.94	2.08	3.23	2.71	0.89	1.72	–	1.61	0.47	–	2.55
151	FRESH (SWINE *Sus scrofa*)	5-08-537	A-F	20	7.3	–	–	–	0.20	0.42	0.59	0.50	0.14	0.34	–	0.37	0.09	–	0.45
			M-F	100	36.3	–	–	–	1.00	2.09	2.94	2.49	0.70	1.69	–	1.84	0.45	–	2.24
	MILKWEED, COMMON *Asclepias syriaca*																		
152	SEEDS	5-09-137	A-F	86	31.8	3.05	–	1.65	0.73	1.12	1.97	1.56	0.45	1.53	1.31	0.86	–	1.08	1.37
			M-F	100	37.0	3.55	–	1.92	0.85	1.30	2.29	1.81	0.52	1.78	1.52	1.00	–	1.26	1.59
	PALM *Elaeis* spp																		
153	KERNELS WITH COATS, MEAL SOLV EXTD	5-03-486	A-F	90	18.2	2.52	–	–	0.31	0.76	1.22	0.66	0.41	0.81	–	0.60	0.20	–	1.02
			M-F	100	20.3	2.81	–	–	0.34	0.84	1.36	0.73	0.45	0.91	–	0.66	0.22	–	1.14
	PEA *Pisum* spp																		
154	SEEDS	5-03-600	A-F	89	23.2	1.40	0.21	1.09	0.60	1.20	1.81	1.53	0.27	1.25	–	0.94	0.21	–	1.25
			M-F	100	26.0	1.56	0.23	1.22	0.67	1.35	2.02	1.71	0.31	1.39	–	1.05	0.24	–	1.39
	PEA, FIELD *Pisum sativum, arvense*																		
155	SEEDS	5-08-481	A-F	91	23.2	1.86	0.26	1.05	0.51	0.91	1.59	1.44	0.23	1.00	1.05	0.82	0.22	0.77	1.00
			M-F	100	25.4	2.04	0.28	1.15	0.56	1.00	1.74	1.58	0.25	1.10	1.15	0.90	0.24	0.85	1.10
	PEANUT *Arachis hypogaea*																		
156	PODS WITH SEEDS, MEAL SOLV EXTD	5-03-656	A-F	92	47.4	5.19	0.70	2.39	1.10	1.92	3.20	1.75	0.43	2.49	3.05	1.38	0.49	1.68	2.48
			M-F	100	51.3	5.62	0.76	2.59	1.19	2.08	3.47	1.90	0.47	2.70	3.30	1.50	0.53	1.82	2.68
157	SEEDS WITHOUT HULLS, MEAL MECH EXTD (PEANUT MEAL)	5-03-649	A-F	93	49.2	5.08	0.96	2.49	1.03	1.78	3.13	1.69	0.50	2.38	1.44	1.27	0.46	1.59	2.29
			M-F	100	53.1	5.49	1.04	2.69	1.11	1.92	3.38	1.83	0.54	2.56	1.56	1.37	0.49	1.72	2.47
158	SEEDS WITHOUT HULLS, MEAL SOLV EXTD (PEANUT MEAL)	5-03-650	A-F	93	49.0	5.82	0.54	2.88	1.46	1.84	3.27	1.45	0.44	2.12	3.12	1.37	0.48	–	2.16
			M-F	100	52.9	6.29	0.58	3.11	1.58	1.99	3.53	1.57	0.48	2.29	3.37	1.48	0.52	–	2.33
	POULTRY																		
159	BY-PRODUCT, MEAL RENDERED	5-03-798	A-F	94	61.2	4.01	0.85	6.09	1.13	2.35	4.10	3.12	1.14	2.04	2.88	2.10	0.47	1.84	2.94
			M-F	100	65.3	4.27	0.91	6.49	1.20	2.51	4.37	3.33	1.22	2.17	3.07	2.23	0.50	1.96	3.13
160	FEATHERS, HYDROLYZED, MEAL	5-03-795	A-F	93	83.8	5.33	3.21	6.32	0.47	3.51	6.42	1.55	0.54	3.59	9.16	3.63	0.52	2.35	5.85
			M-F	100	90.2	5.74	3.46	6.80	0.50	3.78	6.91	1.67	0.58	3.86	9.85	3.91	0.56	2.53	6.29
	RAPE, SUMMER *Brassica napus, annua*																		
161	SEEDS, MEAL, PREPRESSED, SOLV EXTD	5-08-135	A-F	92	40.5	2.23	–	1.94	1.09	1.46	2.71	2.15	0.77	1.54	1.70	1.70	0.49	0.85	1.94
			M-F	100	44.0	2.42	–	2.11	1.19	1.59	2.95	2.33	0.84	1.67	1.85	1.85	0.53	0.92	2.11

(Continued)

TABLE V-2 AMINO ACID, COMPOSITION OF FEEDS, DATA EXPRESSED AS-FED AND MOISTURE-FREE—*(Continued)*

Entry Number	Feed Name Description	International Feed Number	Moisture Basis: A-F (as-fed) or M-F (moisture-free)	Dry Matter	Crude Protein	Arginine	Cystine	Glycine	Histidine	Isoleucine	Leucine	Lysine	Methionine	Phenylalanine	Serine	Threonine	Tryptophan	Tyrosine	Valine
				%	%	%	%	%	%	%	%	%	%	%	%	%	%	%	%
	RYE *Secale cereale*																		
162	DISTILLERS GRAINS, DEHY	5-04-023	A-F	92	23.0	—	—	—	—	—	—	—	—	—	—	—	—	—	—
			M-F	100	25.1	—	—	—	—	—	—	—	—	—	—	—	—	—	—
163	DISTILLERS GRAINS WITH SOLUBLES, DEHY	5-04-024	A-F	90	27.2	1.00	—	—	0.70	1.50	2.10	1.00	0.40	1.30	1.20	1.10	0.30	0.50	1.60
			M-F	100	30.1	1.11	—	—	0.77	1.66	2.32	1.11	0.44	1.44	1.33	1.22	0.33	0.55	1.77
	SAFFLOWER *Carthamus tinctorius*																		
164	MEAL, SOLVENT[1]	5-04-110	A-F	—	22.9	1.94	0.36	—	0.51	0.28	1.22	0.71	0.34	1.02	—	0.51	0.27	—	1.02
165	SEEDS WITHOUT HULLS, MEAL MECH EXTD	5-08-499	A-F	91	42.0	5.44	—	2.52	—	—	—	1.31	0.71	—	—	0.81	—	—	—
			M-F	100	46.1	5.97	—	2.76	—	—	—	1.44	0.77	—	—	0.88	—	—	—
166	SEEDS WITHOUT HULLS, MEAL SOLV EXTD	5-07-959	A-F	91	42.8	3.67	0.71	2.36	0.97	1.58	2.42	1.26	0.67	1.73	—	1.30	0.59	1.01	2.17
			M-F	100	47.0	4.03	0.78	2.59	1.07	1.73	2.66	1.38	0.74	1.90	—	1.43	0.65	1.11	2.39
	SESAME *Sesamum indicum*																		
167	SEEDS, MEAL MECH EXTD	5-04-220	A-F	93	45.0	4.55	0.59	3.96	1.07	1.96	3.20	1.26	1.37	2.14	2.94	1.60	0.71	1.87	2.32
			M-F	100	48.6	4.91	0.64	4.28	1.16	2.12	3.45	1.36	1.48	2.31	3.18	1.72	0.76	2.02	2.51
	SHEEP *Ovis aries*																		
168	MILK, FRESH	5-08-510	A-F	19	4.6	—	—	—	0.18	0.39	0.60	0.51	0.17	0.32	—	0.30	0.09	—	0.48
			M-F	100	24.7	—	—	—	0.94	2.08	3.23	2.71	0.89	1.72	—	1.61	0.47	—	2.55
	SHRIMP *Pandalus* spp, *Penaeus* spp																		
169	CANNERY RESIDUE, MEAL (SHRIMP MEAL)	5-04-226	A-F	90	38.7	2.33	0.47	1.31	0.87	1.51	2.37	2.05	0.84	1.55	1.25	1.26	0.36	1.10	1.71
			M-F	100	43.0	2.58	0.52	1.45	0.97	1.67	2.63	2.28	0.93	1.72	1.39	1.40	0.40	1.22	1.89
	SORGHUM *Sorghum bicolor*																		
170	GLUTEN MEAL	5-04-388	A-F	90	44.4	1.26	0.73	0.95	1.07	2.39	7.85	0.74	0.71	2.70	—	1.45	0.44	—	2.50
			M-F	100	49.2	1.39	0.81	1.05	1.19	2.65	8.70	0.82	0.78	2.99	—	1.61	0.48	—	2.77
	SOYBEAN *Glycine max*																		
171	FLOUR, SOLV EXTD	5-04-593	A-F	93	51.6	4.27	0.64	1.65	1.26	1.90	3.33	4.48	0.57	2.00	2.09	1.58	0.79	1.44	1.86
			M-F	100	55.3	4.58	0.69	1.77	1.35	2.04	3.57	4.81	0.61	2.14	2.24	1.69	0.85	1.54	1.99
172	MEAL, SOLV EXTD	5-04-612	A-F	90	49.7	3.67	0.70	2.27	1.20	2.13	3.63	3.12	0.71	2.36	2.49	1.90	0.69	1.71	2.47
			M-F	100	55.1	4.07	0.78	2.51	1.33	2.37	4.03	3.46	0.79	2.62	2.76	2.11	0.77	1.89	2.74
173	MEAL, SOLV EXTD, 44% PROTEIN	5-20-637	A-F	89	44.4	3.26	0.67	2.10	1.13	2.12	3.49	2.85	0.59	2.23	2.37	1.81	0.62	1.60	2.37
			M-F	100	49.8	3.65	0.75	2.36	1.27	2.38	3.92	3.20	0.67	2.51	2.66	2.03	0.69	1.80	2.66
174	MEAL, SOLV EXTD, 49% PROTEIN	5-20-638	A-F	90	49.0	3.62	0.75	2.39	1.28	2.34	3.77	3.08	0.66	2.47	2.76	2.00	0.70	1.96	2.49
			M-F	100	54.6	4.03	0.83	2.66	1.43	2.60	4.20	3.44	0.74	2.76	3.08	2.23	0.78	2.18	2.77
175	MEAL, SOLVENT[1]	5-04-604	A-F	—	44.0	3.20	0.66	—	1.12	2.00	3.37	2.90	0.52	2.10	—	1.70	0.64	1.50	2.02
176	**SUNFLOWER MEAL, DEHULLED, SOLVENT**[1]	5-04-739	A-F	—	45.5	3.62	0.73	—	0.96	1.97	2.77	1.68	0.82	2.12	—	1.63	0.60	0.68	2.22
	SWINE *Sus scrofa*																		
177	MILK, FRESH	5-08-537	A-F	20	7.3	—	—	—	0.20	0.42	0.59	0.50	0.14	0.34	—	0.37	0.09	—	0.45
			M-F	100	36.3	—	—	—	1.00	2.09	2.94	2.49	0.70	1.69	—	1.84	0.45	—	2.24
	WHALE *Balaena glacialis, Balaenoptera* spp, *Physeter catodon*																		
178	MEAT, MEAL RENDERED	5-05-160	A-F	91	71.4	2.49	0.63	6.31	1.19	2.72	4.27	3.48	1.01	2.06	—	1.63	0.82	—	2.81
			M-F	100	78.1	2.72	0.69	6.90	1.30	2.97	4.67	3.80	1.10	2.26	—	1.79	0.89	—	3.07
	WHEAT *Triticum aestivum*																		
179	GERM MEAL	5-05-218	A-F	88	24.4	1.83	0.47	1.46	0.62	0.95	1.47	1.53	0.41	0.93	1.12	0.94	0.30	0.74	1.16
			M-F	100	27.6	2.07	0.53	1.65	0.70	1.07	1.67	1.73	0.47	1.05	1.27	1.07	0.34	0.84	1.31
180	GLUTEN	5-05-221	A-F	90	63.4	2.97	1.74	2.77	1.64	3.39	5.54	1.54	1.23	4.21	4.10	2.15	0.72	2.36	3.90
			M-F	100	70.3	3.30	1.93	3.07	1.82	3.75	6.14	1.71	1.36	4.66	4.55	2.39	0.80	2.61	4.32
	WHEY (CATTLE) *Bos taurus*																		
181	DEHY	4-01-182	A-F	93	13.3	0.33	0.30	0.44	0.17	0.78	1.18	0.94	0.19	0.35	0.47	0.90	0.20	0.25	0.67
			M-F	100	14.2	0.36	0.32	0.47	0.18	0.83	1.26	1.00	0.20	0.37	0.50	0.97	0.21	0.26	0.72

TABLE V-2 AMINO ACID, COMPOSITION OF FEEDS, DATA EXPRESSED **AS-FED** AND **MOISTURE-FREE**—*(Continued)*

Entry Number	Feed Name Description	International Feed Number	Moisture Basis: A-F (as-fed) or M-F (moisture-free)	Dry Matter	Crude Protein	AMINO ACIDS Arginine	Cystine	Glycine	Histidine	Isoleucine	Leucine	Lysine	Methionine	Phenylalanine	Serine	Threonine	Tryptophan	Tyrosine	Valine
				%	%	%	%	%	%	%	%	%	%	%	%	%	%	%	%
	WHEY (CATTLE) (Continued)																		
182	LOW LACTOSE, DEHY (DRIED WHEY PRODUCT)	4-01-186	A-F	93	16.7	0.60	0.43	0.72	0.27	0.96	1.54	1.40	0.41	0.55	0.59	0.95	0.27	0.46	0.87
			M-F	100	17.9	0.64	0.46	0.77	0.29	1.03	1.65	1.50	0.43	0.59	0.63	1.01	0.29	0.49	0.93
	YEAST, IRRADIATED *Saccharomyces cerevisiae*																		
183	DEHY	7-05-529	A-F	94	48.1	2.46	–	–	1.00	2.94	3.56	3.70	1.00	2.77	–	2.41	0.73	–	3.06
			M-F	100	51.2	2.62	–	–	1.06	3.13	3.79	3.94	1.06	2.95	–	2.56	0.78	–	3.26
	YEAST, TORULA *Torulopsis utilis*																		
184	DEHY	7-05-534	A-F	93	49.6	2.52	0.59	2.54	1.34	2.69	3.39	3.65	0.76	2.63	2.75	2.67	0.52	1.94	2.88
			M-F	100	53.3	2.71	0.63	2.73	1.44	2.89	3.65	3.93	0.82	2.83	2.96	2.87	0.56	2.08	3.10
	DRY FORAGES																		
	ALFALFA (LUCERNE) *Medicago sativa*																		
185	HAY, SUN-CURED	1-00-078	A-F	90	16.0	0.81	–	–	0.28	0.87	1.12	1.00	0.12	0.71	–	0.62	0.18	0.50	0.69
			M-F	100	17.7	0.89	–	–	0.31	0.96	1.24	1.11	0.13	0.79	–	0.69	0.20	0.55	0.76
186	HAY, SUN-CURED, EARLY BLOOM, MEAL	1-00-108	A-F	92	22.5	–	–	–	–	–	–	–	–	–	–	–	–	–	–
			M-F	100	24.5	–	–	–	–	–	–	–	–	–	–	–	–	–	–
187	LEAVES, SUN-CURED, MEAL	1-00-146	A-F	88	20.1	–	–	–	–	–	–	–	–	–	–	–	–	–	–
			M-F	100	22.9	–	–	–	–	–	–	–	–	–	–	–	–	–	–
188	LEAVES, MEAL, DEHY	1-00-137	A-F	92	20.6	0.97	0.27	0.99	0.39	0.92	1.45	0.95	0.32	0.94	0.89	0.86	0.43	0.60	1.05
			M-F	100	22.3	1.05	0.29	1.08	0.42	1.00	1.57	1.03	0.34	1.02	0.97	0.93	0.47	0.65	1.13
189	MEAL, DEHY, 15% PROTEIN	1-00-022	A-F	90	15.6	0.59	0.17	0.70	0.27	0.64	1.02	0.59	0.22	0.62	0.60	0.56	0.38	0.41	0.75
			M-F	100	17.3	0.65	0.19	0.78	0.30	0.71	1.13	0.66	0.24	0.69	0.67	0.62	0.42	0.45	0.83
190	MEAL, DEHY, 17% PROTEIN	1-00-023	A-F	92	17.4	0.77	0.29	0.84	0.33	0.81	1.28	0.85	0.27	0.80	0.71	0.71	0.34	0.54	0.88
			M-F	100	18.9	0.84	0.31	0.91	0.36	0.88	1.39	0.93	0.29	0.87	0.77	0.77	0.37	0.59	0.96
191	MEAL, DEHY, 20% PROTEIN	1-00-024	A-F	92	20.2	0.95	0.32	0.99	0.38	0.89	1.43	0.89	0.32	0.94	0.90	0.82	0.41	0.60	1.05
			M-F	100	22.1	1.04	0.35	1.08	0.41	0.97	1.56	0.97	0.34	1.03	0.98	0.89	0.45	0.66	1.15
192	MEAL, DEHY, 22% PROTEIN	1-07-851	A-F	93	22.2	0.96	0.30	1.09	0.44	1.06	1.63	0.97	0.34	1.13	0.97	0.97	0.49	0.64	1.29
			M-F	100	23.9	1.04	0.32	1.18	0.47	1.15	1.75	1.05	0.37	1.22	1.05	1.04	0.52	0.69	1.39
	ALFALFA-GRASS *Medicago sativa-grass*																		
193	HAY, SUN-CURED	1-08-331	A-F	91	14.5	–	–	–	–	–	–	–	–	–	–	–	–	–	–
			M-F	100	15.9	–	–	–	–	–	–	–	–	–	–	–	–	–	–
	BEET, SUGAR *Beta vulgaris, altissima*																		
194	LEAVES, SUN-CURED	1-00-641	A-F	91	23.2	1.00	–	–	0.27	1.00	1.55	1.27	0.36	0.91	–	0.91	0.27	–	1.18
			M-F	100	25.5	1.10	–	–	0.30	1.10	1.70	1.40	0.40	1.00	–	1.00	0.30	–	1.30
	CLOVER, LADINO *Trifolium repens*																		
195	HAY, SUN-CURED	1-01-378	A-F	89	20.0	–	–	–	–	–	–	–	–	–	–	–	–	–	–
			M-F	100	22.4	–	–	–	–	–	–	–	–	–	–	–	–	–	–
	COWPEA, COMMON *Vigna sinensis*																		
196	HAY, SUN-CURED	1-01-645	A-F	90	17.7	1.11	–	–	0.45	1.27	2.01	1.08	0.51	1.26	–	1.06	0.52	–	1.44
			M-F	100	19.6	1.23	–	–	0.50	1.40	2.22	1.20	0.57	1.39	–	1.18	0.58	–	1.59
	LESPEDEZA, COMMON *Lespedeza striata*																		
197	HAY, SUN-CURED	1-08-591	A-F	89	13.8	–	–	–	–	–	–	–	–	–	–	–	–	–	–
			M-F	100	15.6	–	–	–	–	–	–	–	–	–	–	–	–	–	–
	OATS *Avena sativa*																		
198	HULLS	1-03-281	A-F	92	3.7	0.15	0.06	0.15	0.08	0.15	0.25	0.17	0.08	0.15	–	0.16	0.09	0.14	0.19
			M-F	100	4.0	0.16	0.07	0.16	0.08	0.16	0.27	0.18	0.09	0.16	–	0.17	0.09	0.15	0.21

(Continued)

TABLE V-2 AMINO ACID, COMPOSITION OF FEEDS, DATA EXPRESSED **AS-FED** AND **MOISTURE-FREE**—*(Continued)*

Entry Number	Feed Name Description	International Feed Number	Moisture Basis: A-F (as-fed) or M-F (moisture-free)	Dry Matter	Crude Protein	AMINO ACIDS													
						Arginine	Cystine	Glycine	Histidine	Isoleucine	Leucine	Lysine	Methionine	Phenylalanine	Serine	Threonine	Tryptophan	Tyrosine	Valine
				%	%	%	%	%	%	%	%	%	%	%	%	%	%	%	%
	SOYBEAN *Glycine max*																		
199	HAY, SUN-CURED	1-04-558	A-F	89	14.1	–	–	–	–	–	–	–	–	–	–	–	–	–	–
			M-F	100	15.8	–	–	–	–	–	–	–	–	–	–	–	–	–	–
	VETCH *Vicia spp*																		
200	HAY, SUN-CURED	1-05-106	A-F	89	18.4	–	–	–	–	–	–	–	–	–	–	–	–	–	–
			M-F	100	20.7	–	–	–	–	–	–	–	–	–	–	–	–	–	–
	PASTURE AND RANGE PLANTS																		
	COWPEA, COMMON *Vigna sinensis*																		
201	FRESH	2-01-655	A-F	25	4.0	–	–	–	–	–	–	–	–	–	–	–	–	–	–
			M-F	100	16.2	–	–	–	–	–	–	–	–	–	–	–	–	–	–
	SPINACH *Spinacia oleracea*																		
202	LEAVES, FRESH	2-08-125	A-F	9	3.1	0.11	–	–	0.04	0.09	0.18	0.12	0.06	0.12	–	0.10	0.03	–	0.13
			M-F	100	32.7	1.20	–	–	0.40	1.00	1.90	1.30	0.60	1.30	–	1.10	0.30	–	1.40
	VITAMIN SUPPLEMENTS																		
	ALFALFA (LUCERNE) *Medicago sativa*																		
203	MEAL, DEHY, 20% PROTEIN	1-00-024	A-F	92	20.2	0.95	0.32	0.99	0.38	0.89	1.43	0.89	0.32	0.94	0.90	0.82	0.41	0.60	1.05
			M-F	100	22.1	1.04	0.35	1.08	0.41	0.97	1.56	0.97	0.34	1.03	0.98	0.89	0.45	0.66	1.15
204	MEAL, DEHY, 22% PROTEIN	1-07-851	A-F	93	22.2	0.96	0.30	1.09	0.44	1.06	1.63	0.97	0.34	1.13	0.97	0.97	0.49	0.64	1.29
			M-F	100	23.9	1.04	0.32	1.18	0.47	1.15	1.75	1.05	0.37	1.22	1.05	1.04	0.52	0.69	1.39
	BREWERS GRAINS																		
205	DEHY	5-02-141	A-F	92	27.3	1.27	0.35	1.09	0.53	1.57	2.53	0.88	0.46	1.46	1.30	0.93	0.37	1.16	1.58
			M-F	100	29.6	1.38	0.38	1.18	0.58	1.71	2.75	0.95	0.50	1.58	1.41	1.01	0.40	1.26	1.72
	CORN *Zea mays*																		
206	DISTILLERS GRAINS WITH SOLUBLES, DEHY	5-02-843	A-F	92	27.1	0.97	0.31	0.59	0.64	1.33	2.31	0.70	0.50	1.47	1.21	0.93	0.18	0.72	1.47
			M-F	100	29.5	1.05	0.34	0.64	0.69	1.45	2.52	0.77	0.54	1.60	1.31	1.01	0.19	0.78	1.60
	FISH																		
207	SOLUBLES, DEHY	5-01-971	A-F	93	60.4	3.06	0.62	5.75	2.10	2.05	2.98	3.52	1.18	1.53	2.03	1.35	0.60	0.85	2.10
			M-F	100	65.1	3.29	0.66	6.20	2.26	2.21	3.21	3.79	1.27	1.65	2.19	1.46	0.64	0.92	2.26
	FISH, SARDINE *Clupea spp, Sardinops spp*																		
208	MEAL MECH EXTD	5-02-015	A-F	93	65.2	2.70	0.80	4.50	1.80	3.34	–	5.91	2.01	2.00	–	2.60	0.50	2.79	4.10
			M-F	100	70.0	2.90	0.86	4.84	1.93	3.59	–	6.34	2.16	2.15	–	2.79	0.54	3.00	4.40
209	SOLUBLES, CONDENSED	5-02-014	A-F	50	29.5	1.50	0.20	–	2.00	0.90	1.60	1.60	0.90	0.80	–	0.80	0.10	–	1.00
			M-F	100	59.4	3.02	0.40	–	4.02	1.81	3.22	3.22	1.81	1.61	–	1.61	0.20	–	2.01
	RICE *Oryza sativa*																		
210	BRAN WITH GERMS, MEAL SOLV EXTD (RICE BRAN, SOLV EXTD)	4-03-930	A-F	91	14.0	0.98	0.21	0.91	0.33	0.52	1.02	0.61	0.26	0.57	0.70	0.53	0.21	0.55	0.76
			M-F	100	15.4	1.07	0.23	1.00	0.37	0.57	1.12	0.67	0.29	0.63	0.77	0.58	0.23	0.60	0.84
	YEAST, BREWERS *Saccharomyces cerevisiae*																		
211	DEHY	7-05-527	A-F	93	43.8	2.26	0.52	1.77	1.13	2.03	2.86	2.98	0.66	1.60	–	2.06	0.51	1.47	2.25
			M-F	100	47.1	2.43	0.56	1.90	1.21	2.18	3.07	3.20	0.71	1.72	–	2.22	0.54	1.58	2.42
	YEAST, PRIMARY *Saccharomyces cerevisiae*																		
212	DEHY	7-05-533	A-F	93	48.0	2.60	0.50	–	5.60	3.60	3.70	3.80	1.00	2.50	–	2.50	0.40	–	3.20
			M-F	100	51.8	2.81	0.54	–	6.05	3.89	4.00	4.10	1.08	2.70	–	2.70	0.43	–	3.46

[1]From: *Nutrient Requirements of Swine,* 9th rev. ed., National Research Council, National Academy Press, 1988, p. 58–59, Table 6-3.

Fig. V-3. Grain sorghum, headed out and nearing harvest. (Courtesy, Northrup King Co., Minneapolis, Minn.)

Fig. V-4. Environmentally controlled hog finishing building, with partially slotted floor and scraping gutter manure-handling system. This unit will accommodate 1,200 hogs. (Courtesy, *National Hog Farmer*, St. Paul, Minn.)

TABLE V-3 APPARENT ILEAL DIGESTIBILITY OF CRUDE PROTEIN AND ESSENTIAL AMINO ACIDS IN FEEDSTUFFS FOR SWINE, AND

FEEDSTUFF (AS FED BASIS)	**Obs	DM %	Crude Protein			Arginine			Cystine			Histidine			Isoleucine		
			Total %	Coefficient %	Digestible %	Total %	Coefficient %	Digestible %	Total %	Coefficient %	Digestible %	Total %	Coefficient %	Digestible %	Total %	Coefficient %	Digestible %
ALFALFA	3	89.85	16.95	39	6.61	0.66	59	0.39	0.18	15	0.03	0.32	46	0.15	0.66	55	0.36
BARLEY	33	87.74	10.59	70	7.41	0.52	75	0.39	0.22	74	0.16	0.24	71	0.17	0.37	73	0.27
*BARLEY, NAKED	1	87.20	11.70	78	9.13	0.64	85	0.54	—	—	—	0.30	90	0.27	0.45	87	0.39
BLOOD MEAL	3	90.57	87.87	87	76.45	3.68	94	3.46	—	—	—	5.44	94	5.11	0.72	70	0.50
BONE MEAL	2	94.11	28.10	72	20.23	1.78	79	1.41	0.20	38	0.08	0.26	71	0.18	0.54	71	0.38
BREWERS GRAINS	3	91.60	26.03	70	18.22	1.82	—	—	—	—	—	0.67	—	—	1.11	79	0.88
CANOLA MEAL	4	90.43	37.28	69	25.72	2.21	82	1.81	0.42	89	0.37	1.20	78	0.94	1.57	75	1.18
CASEIN	4	90.69	86.19	89	76.71	2.88	93	2.68	—	62	—	2.00	94	1.88	4.22	90	3.80
COCONUT OIL MEAL	1	94.70	21.58	52	11.22	2.18	—	—	0.29	—	—	0.42	—	—	0.60	65	0.39
CORN	10	88.34	9.19	77	7.08	0.41	82	0.34	0.17	74	0.13	0.25	82	0.21	0.33	80	0.26
CORN GLUTEN FEED	4	90.30	19.58	54	10.57	1.07	72	0.77	0.47	—	—	0.66	58	0.38	0.66	62	0.41
CORN GLUTEN MEAL	2	89.35	59.00	80	47.20	2.00	86	1.72	1.06	73	0.77	1.31	81	1.06	2.61	84	2.19
COTTONSEED MEAL GLANDED	9	91.60	41.14	72	29.62	4.42	88	3.89	—	—	—	1.10	77	0.85	1.28	67	0.86
COTTONSEED MEAL GLANDLESS	2	94.82	42.40	83	35.19	5.56	95	5.28	—	—	—	1.36	88	1.20	1.45	85	1.23
DRIED SKIM MILK	4	95.78	33.96	85	28.87	1.24	85	1.05	0.28	86	0.24	1.47	93	1.37	1.73	85	1.47
FEATHER MEAL	3	94.30	83.20	71	59.07	5.09	79	4.02	3.85	72	2.77	0.57	45	0.26	3.69	78	2.88
FISH MEAL	9	94.49	61.83	79	48.85	3.91	88	3.44	0.66	62	0.41	1.37	81	1.11	2.60	86	2.24
GROUNDNUT MEAL	2	89.30	41.70	85	35.45	4.92	95	4.67	0.49	78	0.38	0.88	85	0.75	1.38	85	1.17
LUPIN MEAL	2	88.25	33.07	80	26.46	3.50	93	3.26	0.43	70	0.30	0.77	85	0.65	1.47	83	1.22
L-LYSINE HCI	2	98.50	94.40	100	94.40												
MEAT & BONE MEAL	29	93.38	55.07	67	36.90	3.44	77	2.65	0.63	51	0.32	0.93	71	0.66	1.53	68	1.04
OATS	1	87.60	13.00	61	7.93	0.65	—	—	0.29	—	—	0.21	—	—	0.39	—	—
OAT GROATS	1	91.62	16.32	84	13.71	1.15	90	1.04	1.15	90	1.04	0.46	86	0.40	0.52	86	0.45
PEANUT MEAL	4	93.87	44.23	73	32.29	5.20	90	4.68	—	—	—	1.04	73	0.76	1.46	76	1.11
PEAS	3	89.03	20.87	73	15.24	1.83	86	1.57	0.37	61	0.23	0.57	81	0.46	0.90	74	0.67
POULTRY-BY-PRODUCT MEAL	5	92.76	63.71	72	45.87	4.34	88	3.82	2.04	—	—	1.23	81	1.00	2.52	79	1.99
RAPESEED MEAL	7	91.45	35.32	70	24.72	2.05	83	1.70	0.83	73	0.61	0.99	83	0.82	1.48	75	1.11
*RICE	1	85.50	8.80	72	6.34	0.98	90	0.88	—	—	—	0.18	90	0.16	0.32	86	0.28
RYE	3	88.27	10.03	68	6.82	0.48	74	0.36	0.20	72	0.14	0.21	70	0.15	0.33	70	0.23
SAND EEL	1	89.42	72.56	79	57.32	3.45	—	—	0.70	—	—	1.63	54	0.88	2.93	88	2.58
SESAME MEAL	1	92.70	44.04	76	33.47	5.39	—	—	0.86	—	—	1.01	36	0.36	1.40	79	1.11
SORGHUM	2	89.99	9.83	81	7.96	0.45	85	0.38	—	—	—	0.24	81	0.19	0.44	88	0.39
SOY FLOUR	1	94.44	50.68	82	41.56	4.02	91	3.66	0.74	78	0.58	1.39	88	1.22	2.42	83	2.01
SOYBEANS EXTRUDED	2	92.68	35.11	74	25.98	2.72	84	2.28	0.57	64	0.36	0.99	80	0.79	1.69	74	1.25
SOYBEAN MEAL 44%	9	89.62	44.27	80	35.42	3.29	90	2.96	0.69	74	0.51	1.18	86	1.01	2.03	83	1.68
SOYBEAN MEAL 48%	18	89.51	47.29	80	37.83	3.56	90	3.20	0.66	79	0.52	1.28	86	1.10	2.22	84	1.86
SOYFLAKES RAW	1	90.72	48.72	35	17.05	3.91	56	2.19	0.64	35	0.22	1.41	48	0.68	2.42	43	1.04
SOYFLAKES HEATED	1	91.96	49.29	79	38.94	3.91	88	3.44	0.63	74	0.47	1.40	82	1.15	2.48	78	1.93
SUNFLOWER MEAL	10	91.90	35.29	75	26.47	2.93	90	2.64	0.57	74	0.42	0.89	76	0.68	1.45	78	1.13
TRITICALE	5	89.14	13.31	78	10.38	0.55	87	0.48	—	—	—	0.25	84	0.21	0.36	83	0.30
WHEAT	24	87.93	12.68	82	10.40	0.57	83	0.47	0.20	80	0.16	0.25	80	0.20	0.43	83	0.36
WHEAT BRAN 7% FIBER	3	88.30	13.50	67	9.05	0.88	85	0.75	0.37	71	0.26	0.35	79	0.28	0.47	74	0.35
WHEAT FLOUR	1	88.00	13.46	91	12.25	0.43	91	0.39	—	—	—	0.26	91	0.24	0.50	94	0.47
WHEAT MIDDS 9% FIBER	5	89.14	16.63	67	11.14	1.18	84	0.99	—	—	—	0.45	79	0.36	0.55	70	0.39
WHEAT OFFAL 9% FIBER	1	88.00	15.84	70	11.09	0.97	95	0.92	—	—	—	0.39	79	0.31	0.51	73	0.37

*True Ileal Digestibility Values

**The number of observations for tryptophan may differ substantially from the value indicated.

DIGESTIBLE AMINO ACID RECOMMENDATIONS

Heartland Lysine would like to express its appreciation to each of the authors who contributed data to this work. In particular our thanks are extended to Dr. Darrell Knabe (Texas A&M University), Dr. Malcolm Fuller (Rowett Research Institute), and Dr. Michael Taverner (Animal Research Institute) for their contributions.

This chart reflects our interpretation of published literature on digestible amino acids for swine nutrition. It is the responsibility of the purchaser of our products to determine the best application of our products for their needs. Information and recommendations regarding our products and/or nutrient levels for swine feeding are to the best of our knowledge accurate. We do not warrant the accuracy or completeness of this information. Our making this information available does not relieve the purchaser or user of his obligation to verify the suitability of our products and recommendations for their intended application.

Period	Starting		Growing-Finishing[1]	
Weight, lb.	10 - 22	22 - 55	55 - 110	110 - 220
DIGESTIBLE MINIMUMS, %				
CRUDE PROTEIN	16.50	14.50	12.50	11.00
LYSINE/CALORIE (g/Mcal ME)	3.60	3.30	2.50	2.00
LYSINE[2]	1.25	1.00	0.75	0.61
THREONINE	0.81	0.65	0.49	0.40
TRYPTOPHAN	0.23	0.18	0.14	0.11
ISOLEUCINE	0.75	0.60	0.45	0.37
METHIONINE & CYSTINE	0.69	0.55	0.41	0.34
HISTIDINE	0.41	0.33	0.25	0.20
LEUCINE	1.35	1.08	0.81	0.66
PHENYLALANINE & TYROSINE	1.50	1.20	0.90	0.73
VALINE	0.91	0.73	0.55	0.45

[1]Boar requirements are 10 to 15% higher than gilts or barrows

DIGESTIBLE AMINO ACID RECOMMENDATIONS FOR SWINE FEED FORMULATION[1]

Leucine			Lysine			Methionine			Phenylalanine			Threonine			Tryptophan			Valine		
Total %	Coefficient %	Digestible %	Total %	Coefficient %	Digestible %	Total %	Coefficient %	Digestible %	Total %	Coefficient %	Digestible %	Total %	Coefficient %	Digestible %	Total %	Coefficient %	Digestible %	Total %	Coefficient %	Digestible %
1.10	60	0.66	0.69	48	0.33	0.21	64	0.13	0.68	59	0.40	0.62	49	0.30	0.16	—	—	0.83	52	0.43
0.71	75	0.53	0.39	67	0.26	0.18	79	0.14	0.52	78	0.41	0.38	63	0.24	0.12	73	0.09	0.53	71	0.38
0.95	85	0.81	0.54	89	0.48	0.25	87	0.22	0.76	90	0.68	0.47	87	0.41	0.17	—	—	0.62	87	0.54
11.89	93	11.06	9.05	93	8.42	—	—	—	6.65	92	6.12	4.68	87	4.07	0.99	89	0.88	8.61	92	7.92
1.20	74	0.89	1.05	72	0.76	0.23	78	0.18	0.71	75	0.53	0.67	68	0.46	—	50	—	1.00	73	0.73
2.23	—	—	1.05	68	0.71	—	—	—	1.49	—	—	0.99	67	0.66	0.28	71	0.20	1.53	—	—
2.77	78	2.16	2.19	73	1.60	0.98	82	0.80	1.57	77	1.21	1.78	68	1.21	0.43	71	0.31	2.03	67	1.36
7.81	94	7.34	6.09	95	5.79	2.61	96	2.51	4.34	95	4.12	3.22	86	2.77	0.98	91	0.89	5.62	92	5.17
1.20	70	0.84	1.98	53	1.05	0.38	—	—	0.89	76	0.68	0.68	49	0.33	0.22	—	—	0.99	70	0.69
1.12	88	0.99	0.27	68	0.18	0.18	86	0.15	0.43	84	0.36	0.32	71	0.23	0.06	72	0.04	0.44	78	0.34
2.01	77	1.55	0.67	48	0.32	0.37	—	—	0.75	75	0.56	0.74	50	0.37	0.07	33	0.02	1.00	68	0.68
10.71	90	9.64	1.10	74	0.81	1.42	86	1.22	3.76	88	3.31	2.10	80	1.68	0.27	72	0.19	2.85	81	2.31
2.29	70	1.60	1.68	59	0.99	0.64	72	0.46	2.16	80	1.73	1.28	61	0.78	0.47	72	0.34	2.57	71	1.82
2.52	84	2.12	2.04	82	1.67	0.77	84	0.65	2.62	91	2.38	1.47	78	1.15	0.57	81	0.46	2.83	85	2.41
3.21	95	3.05	2.79	95	2.65	0.78	96	0.75	1.72	96	1.65	1.49	89	1.33	0.56	—	—	2.15	88	1.89
6.63	77	5.11	1.59	51	0.81	0.39	71	0.28	4.29	81	3.47	3.49	74	2.58	0.42	60	0.25	6.75	78	5.27
4.51	87	3.92	4.69	87	4.08	1.87	91	1.70	2.43	84	2.04	2.71	81	2.20	0.55	74	0.41	2.95	83	2.45
2.46	87	2.14	1.46	82	1.20	0.38	84	0.32	1.88	89	1.67	1.04	77	0.80	—	—	—	1.58	83	1.31
2.35	83	1.95	1.54	79	1.22	0.21	65	0.14	1.30	82	1.07	1.19	77	0.92	—	—	—	1.41	79	1.11
			78.80	100	78.80															
3.32	71	2.36	2.75	70	1.93	0.75	77	0.58	1.79	72	1.29	1.75	63	1.10	0.26	54	0.14	2.51	70	1.76
0.75	—	—	0.40	70	0.28	0.17	79	0.13	0.48	—	—	0.34	55	0.19	0.10	—	—	0.52	—	—
1.04	85	0.88	0.64	82	0.52	0.23	89	0.20	0.68	90	0.61	0.53	78	0.41	0.17	81	0.14	0.78	85	0.66
2.65	78	2.07	1.55	72	1.12	—	—	—	2.12	85	1.80	1.16	68	0.79	0.35	68	0.24	1.75	76	1.33
1.50	75	1.13	1.54	82	1.26	0.23	74	0.17	0.98	75	0.74	0.79	75	0.59	—	—	—	0.96	71	0.68
4.61	80	3.69	3.57	81	2.89	1.07	—	—	2.37	82	1.94	2.53	71	1.80	0.43	79	0.34	3.39	77	2.61
2.49	79	1.97	2.00	73	1.46	0.71	85	0.60	1.46	80	1.17	1.53	68	1.04	—	—	—	1.91	70	1.34
0.67	86	0.58	0.33	81	0.27	0.19	87	0.17	0.43	88	0.38	0.31	84	0.26	0.11	—	—	0.48	87	0.42
0.59	71	0.42	0.36	65	0.23	0.15	77	0.12	0.44	78	0.34	0.32	59	0.19	—	—	—	0.45	69	0.31
5.40	89	4.81	5.58	91	5.08	2.01	—	—	2.77	87	2.41	3.06	80	2.45	0.91	—	—	3.54	86	3.04
2.75	82	2.26	1.06	71	0.75	1.32	—	—	1.92	85	1.63	1.38	66	0.91	0.65	—	—	1.97	80	1.58
1.47	91	1.34	0.24	75	0.18	0.18	87	0.16	0.65	92	0.60	0.40	78	0.31	0.09	80	0.07	0.53	86	0.46
4.23	81	3.43	3.36	88	2.96	0.79	91	0.72	2.66	87	2.31	2.03	76	1.54	0.64	79	0.51	4.12	81	3.34
2.89	73	2.11	2.39	80	1.91	0.57	78	0.44	1.86	80	1.49	1.44	69	0.99	0.47	70	0.33	2.36	73	1.72
3.40	83	2.82	2.81	86	2.42	0.67	85	0.57	2.22	85	1.89	1.74	76	1.32	0.53	80	0.42	2.83	81	2.29
3.61	83	3.00	3.05	85	2.59	0.67	87	0.58	2.39	85	2.03	1.89	77	1.46	0.58	78	0.45	2.45	80	1.96
4.04	37	1.49	3.35	44	1.47	0.79	47	0.37	2.69	45	1.21	2.04	32	0.65	0.57	25	0.14	4.13	35	1.45
4.15	80	3.32	3.35	85	2.85	0.79	82	0.65	2.67	84	2.24	2.03	72	1.46	0.56	77	0.43	4.15	78	3.24
2.11	77	1.62	1.30	74	0.96	0.78	87	0.68	1.56	80	1.25	1.26	71	0.89	0.42	76	0.32	1.71	75	1.28
0.67	84	0.56	0.40	73	0.29	0.16	84	0.13	0.58	83	0.48	0.34	65	0.22	0.14	—	—	0.47	83	0.39
0.83	83	0.69	0.33	71	0.23	0.20	84	0.17	0.55	86	0.47	0.36	69	0.25	0.15	81	0.12	0.53	77	0.41
0.88	76	0.67	0.52	72	0.37	0.21	79	0.17	0.57	79	0.45	0.44	66	0.29	—	—	—	0.65	72	0.47
0.92	95	0.87	0.23	84	0.19	0.16	94	0.15	0.70	96	0.67	0.33	85	0.28	—	—	—	0.56	93	0.52
1.06	72	0.76	0.68	72	0.49	0.23	79	0.18	0.68	82	0.56	0.55	60	0.33	0.19	72	0.14	0.78	73	0.57
0.95	74	0.70	0.54	66	0.36	0.23	78	0.18	0.60	76	0.46	0.41	54	0.22	—	—	—	0.71	72	0.51

FOR SWINE FEED FORMULATION

	Sows		Mature
	Gestation	Lactation	Boar
DIGESTIBLE MINIMUMS, %			
CRUDE PROTEIN	9.50	12.00	12.00
LYSINE/CALORIE (g/Mcal ME)	1.60	2.00	2.00
LYSINE[2]	0.51	0.67	0.67
THREONINE	0.33	0.44	0.44
TRYPTOPHAN	0.09	0.12	0.12
ISOLEUCINE	0.30	0.40	0.40
METHIONINE & CYSTINE	0.28	0.37	0.37
HISTIDINE	0.17	0.22	0.22
LEUCINE	0.54	0.73	0.73
PHENYLALANINE & TYROSINE	0.60	0.80	0.80
VALINE	0.37	0.49	0.49

[2]For each 1% added fat digestible lysine should increase by .02%.

[1]Table V-3 data was assembled by, and is presented through the courtesy of, Heartland Lysine, Inc., 8430 West Bryn Mawr Avenue, Suite 650, Chicago, IL 60631.

TABLE V-4 TRUE DIGESTIBILITY OF ESSENTIAL AMINO ACIDS FOR POULTRY, AND TRUE DIGESTIBILE AMINO ACID RECOMMENDATIONS

FEEDSTUFF (AS FED BASIS)	Obs	DM %**	CP	Arginine Total %	Arginine Coefficient %***	Arginine Digestible %	Cystine Total %	Cystine Coefficient %***	Cystine Digestible %	Histidine Total %	Histidine Coefficient %***	Histidine Digestible %	Isoleucine Total %	Isoleucine Coefficient %***	Isoleucine Digestible %
ALFALFA MEAL	9	89.86	16.23	0.70	81.3 (6.1)	0.57	0.27	41.2 (10.1)	0.11	0.34	74.5 (5.8)	0.25	0.60	76.6 (7.0)	0.46
ALGAE	2	90.00	33.64	2.10	87.1 (5.0)	1.83	1.06	83.2 (0.6)	0.88	0.53	82.4 (4.5)	0.44	1.33	81.2 (5.0)	1.08
BARLEY	30	89.43	11.43	0.56	83.5 (4.8)	0.47	0.28	81.5 (8.5)	0.22	0.26	86.0 (4.5)	0.22	0.37	80.7 (5.9)	0.30
BLOOD MEAL	22	90.00	87.56	4.05	86.5 (4.2)	3.51	1.27	75.8 (5.3)	0.96	5.57	87.1 (3.4)	4.85	0.91	79.8 (6.2)	0.73
BONE MEAL	1	91.50	39.80	2.38	70.5 (0.0)	1.68	0.20	48.7 (0.0)	0.10	0.43	60.5 (0.0)	0.26	0.83	76.2 (0.0)	0.63
BREWERS GRAINS	1	90.00	22.70	1.04	80.9 (0.0)	0.84	—	— —	—	0.42	73.8 (0.0)	0.31	0.78	80.7 (0.0)	0.63
CANOLA MEAL	25	90.38	35.72	2.21	89.8 (1.6)	1.98	0.93	72.4 (8.4)	0.68	0.97	87.0 (2.2)	0.84	1.29	83.5 (2.6)	1.08
CASEIN	2	90.00	61.00	3.94	98.3 (1.0)	3.87	—	— —	—	1.86	88.9 (9.4)	1.65	4.36	98.4 (0.9)	4.28
COCONUT MEAL	1	90.00	20.93	2.58	86.5 (0.0)	2.23	0.41	52.9 (0.0)	0.22	0.35	74.8 (0.0)	0.26	0.60	83.6 (0.0)	0.50
CORN	12	88.75	8.45	0.37	94.6 (4.8)	0.35	0.17	83.7 (7.8)	0.14	0.23	89.2 (7.3)	0.20	0.30	91.3 (3.0)	0.27
CORN GERM MEAL	1	90.00	21.43	1.47	90.3 (0.0)	1.32	0.40	57.8 (0.0)	0.23	0.67	83.7 (0.0)	0.56	0.67	86.2 (0.0)	0.57
CORN GLUTEN FEED	10	90.00	21.88	1.12	88.8 (2.7)	0.99	0.53	64.4 (6.4)	0.34	0.66	84.6 (2.7)	0.56	0.62	82.9 (4.8)	0.51
CORN GLUTEN MEAL	13	90.03	61.75	2.07	95.8 (2.1)	1.98	1.17	87.5 (3.7)	1.02	1.24	94.7 (1.3)	1.18	2.24	95.5 (1.0)	2.14
COTTONSEED MEAL	1	87.40	34.70	3.75	80.0 (0.0)	3.00	—	— —	—	0.81	82.2 (0.0)	0.67	0.90	68.0 (0.0)	0.61
D L METHIONINE	2	99.00	58.10												
D L MHA CALCIUM	2	99.00	50.47												
FEATHER MEAL	14	90.27	79.58	6.05	82.1 (4.8)	4.96	4.2	54.9 (6.9)	2.33	0.80	69.5 (15.3)	0.56	3.50	83.6 (5.0)	2.92
FISH MEAL	34	90.25	62.87	4.01	92.5 (2.4)	3.71	0.57	77.9 (5.9)	0.44	1.46	89.1 (3.1)	1.30	2.40	93.3 (2.9)	2.24
FISH MEAL ANALOG	16	90.00	62.73	4.02	88.4 (2.2)	3.55	—	— —	—	1.59	75.5 (6.6)	1.20	1.79	87.5 (2.2)	1.57
GROUNDNUT MEAL	4	89.26	43.80	4.90	90.7 (3.7)	4.44	0.50	79.5 (3.0)	0.40	0.86	85.4 (5.0)	0.73	1.37	88.7 (2.4)	1.22
HAIR (HYDROLYZED)	2	90.00	66.07	5.77	62.5 (2.6)	3.60	—	— —	—	0.65	51.9 (1.0)	0.34	2.11	64.0 (5.4)	1.35
L-LYSINE HCl	1	98.50	94.40												
LIVER MEAL	1	90.00	73.80	5.51	76.3 (0.0)	4.20	—	— —	—	2.04	73.1 (0.0)	1.49	3.82	73.8 (0.0)	2.82
LUCERNE MEAL	2	90.70	13.90	0.78	82.6 (5.2)	0.64	0.15	24.8 (0.0)	0.04	0.34	65.6 (19.9)	0.22	0.68	74.7 (8.5)	0.51
LUPINSEED MEAL	3	88.90	32.80	2.86	92.7 (3.5)	2.65	0.43	93.7 (0.0)	0.40	0.75	88.4 (3.9)	0.66	1.49	90.0 (5.0)	1.34
MEAT MEAL	20	90.83	47.70	3.27	86.0 (6.2)	2.81	0.48	55.1 (11.0)	0.27	0.91	84.9 (4.3)	0.77	1.14	84.7 (6.7)	0.97
MEAT AND BONE MEAL	22	90.51	54.02	3.79	85.6 (6.9)	3.25	0.70	62.5 (14.1)	0.44	1.04	79.3 (8.1)	0.83	1.54	84.0 (6.3)	1.29
OATS	14	90.00	10.78	0.72	93.2 (3.9)	0.67	0.47	84.2 (9.5)	0.40	0.24	92.4 (4.3)	0.22	0.38	87.9 (4.3)	0.33
POULTRY BYPRODUCT MEAL	7	90.19	55.10	3.96	87.0 (4.9)	3.44	1.35	60.5 (8.5)	0.82	1.15	70.5 (7.4)	0.81	2.22	83.1 (6.7)	1.84
POULTRY OFFAL MEAL	2	90.00	56.40	4.14	88.9 (0.3)	3.67	—	— —	—	0.90	82.9 (3.8)	0.75	2.46	86.1 (1.0)	2.11
RICE (ROUGH)	1	86.60	6.60	0.56	90.6 (0.0)	0.51	—	— —	—	0.14	91.9 (0.0)	0.13	0.22	78.9 (0.0)	0.17
RICE BRAN (DEFATTED)	6	89.68	14.56	1.15	85.7 (2.9)	0.98	0.32	63.4 (7.0)	0.20	0.39	81.7 (3.9)	0.32	0.47	74.1 (5.4)	0.35
SESAMESEED MEAL	3	91.27	43.37	3.82	77.6 (21.0)	2.97	0.90	81.7 (3.5)	0.74	0.87	71.0 (25.9)	0.62	1.31	72.1 (28.1)	0.94
SHRIMP MEAL	1	90.00	36.28	2.25	93.5 (0.0)	2.10	0.43	78.6 (0.0)	0.34	0.90	91.0 (0.0)	0.82	1.40	95.1 (0.0)	1.33
SINGLE CELL PROTEIN	2	90.00	68.18	2.51	88.6 (6.0)	2.22	0.48	66.9 (0.0)	0.32	1.22	88.7 (1.4)	1.08	2.27	87.8 (4.6)	1.99
SORGHUM	2	89.30	9.30	0.43	88.4 (0.9)	0.38	0.17	76.5 (0.0)	0.13	0.24	92.6 (3.4)	0.22	0.39	87.5 (2.8)	0.34
SOYBEAN MEAL (48%)	30	89.84	48.47	3.77	92.1 (2.1)	3.47	0.76	82.5 (4.2)	0.63	1.31	91.4 (4.5)	1.20	2.05	91.9 (1.7)	1.88
SOYBEAN MEAL (FULL FAT)	1	90.00	34.35	2.62	94.6 (0.0)	2.48	—	— —	—	0.88	94.0 (0.0)	0.83	1.59	93.4 (0.0)	1.49
SUNFLOWER MEAL	5	91.51	33.12	2.79	95.3 (2.4)	2.66	0.53	80.9 (6.8)	0.43	0.86	89.1 (4.3)	0.77	1.41	91.2 (1.2)	1.29
TRITICALE	1	90.00	19.00	0.83	93.2 (0.0)	0.77	—	— —	—	0.51	94.8 (0.0)	0.48	0.56	94.6 (0.0)	0.53
WHEAT	30	89.58	15.77	0.69	87.1 (4.3)	0.61	0.37	87.3 (6.3)	0.32	0.35	90.7 (3.5)	0.32	0.48	87.9 (3.9)	0.42
WHEAT BRAN	3	89.27	14.80	1.03	79.7 (1.0)	0.82	0.29	71.8 (0.4)	0.21	0.38	79.4 (1.1)	0.30	0.47	76.2 (2.5)	0.36
WHEAT MIDDS	2	89.75	17.00	0.58	84.3 (2.7)	0.49	0.37	77.2 (0.0)	0.29	0.44	83.1 (0.2)	0.36	0.46	67.4 (10.3)	0.31
WHEAT POLLARD	2	90.00	16.50	0.95	79.3 (5.0)	0.75	—	— —	—	0.39	84.5 (4.0)	0.33	0.49	75.3 (1.0)	0.37
WHEAT SCREENINGS	5	90.00	13.65	0.89	89.4 (5.0)	0.80	0.35	74.3 (14.5)	0.26	0.31	85.5 (8.1)	0.26	0.47	84.4 (5.9)	0.39
WHEAT SHORTS	15	90.00	17.44	1.29	86.4 (4.0)	1.11	0.41	69.1 (7.5)	0.28	0.48	83.8 (3.5)	0.40	0.50	82.3 (4.0)	0.41

*Combination of data from conventional and caecectomized precision-fed rooster assays. Due to lack of data, tryptophan is not included.

**90% dry matter assumed where data unavailable.

***Values in parentheses represent standard deviations of digestibility coefficients.

TRUE DIGESTIBLE AMINO ACID RECOMMENDATIONS

	BROILER CHICKENS			TURKEYS							DUCKS			
Period	Starting	Growing	Finishing						Holding	Breeding Hens	Starting	Growing	Rearing	Breeding
Age (weeks)	0-21	22-42[2]	43+[2]	0-4	4-8	8-12[2]	12-16[2]	16+[2]			0-3	3-8[2]	8-breeding[2]	
Metabolizable Energy (Kcal/kg)	3200	3250	3300	2900	3000	3100	3200	3300	2750	3000	2900	2950	2900	2400
Crude Protein[1] %	21.00	19.00	17.00	26.00	24.00	21.50	18.50	15.00	12.00	13.50	20.00	16.00	14.00	14.00
Methionine + Cystine %	.86	.75	.65	.97	88	.79	.70	.62	.44	.53	83	66	51	49
Methionine + Cystine %/Mcal	269	.230	.197	.334	293	.255	.219	.188	.160	.177	286	224	176	169
Methionine %	.50	.43	.36	.56	51	.46	.41	.36	.23	.29	.48	.38	28	27
Methionine %/Mcal	.156	.132	.109	.193	170	.148	.128	.109	.084	.097	166	.129	.097	.093
Lysine %	1.14	1.00	.90	1.65	1.49	1.29	1.05	90	.57	.65	1.10	94	.72	.71
Lysine %/Mcal	.356	.308	.273	.569	497	416	328	273	.207	.217	.379	.319	.248	245
Threonine %	.69	.63	.59	.88	.79	70	61	.52	.35	.40	62	.53	42	.41
Threonine %/Mcal	.216	.194	.179	.303	263	226	.191	.158	.127	133	214	.180	.145	.141
Arginine %	1.30	1.12	.96	1.49	1.34	1.19	1.03	.90	.60	.65	1.12	1.01	.77	.75
Arginine %/Mcal	.406	.345	.291	.514	.447	.384	.322	.273	.218	.217	.386	342	266	.259
Tryptophan[1] %	.22	.18	.17	.25	.225	.20	.175	.15	.09	.12	.20	.17	.13	.13
Tryptophan %/Mcal	.069	.055	.052	0.86	0.75	.065	.547	.045	.033	.040	.069	.058	045	.045
Histidine %	.34	.29	.26	.54	.49	.43	.38	.32	.24	.28	.33	.29	.22	.22
Histidine %/Mcal	.106	.089	.079	.186	.163	.139	.119	.097	.087	.093	.114	.098	.076	.076
Isoleucine %	.74	.66	.58	.99	.89	.79	.68	.58	.41	.46	.72	.63	.50	.49
Isoleucine %/Mcal	.231	.203	.176	.341	.297	.255	.213	.176	.149	.153	.248	.214	.172	.169
Leucine %	1.35	1.18	1.02	1.74	1.56	1.39	1.21	1.05	.71	.90	1.31	1.17	.90	.88
Leucine %/Mcal	.422	.363	.309	.600	.520	.448	.378	.318	.258	.300	.452	.397	.310	.303
Phenylalanine %	.67	.60	.52	.93	.84	.74	.65	.56	.37	.53	.65	.59	.45	.44
Phenylalanine %/Mcal	.209	.185	.158	.321	.280	.239	.203	.170	.135	.177	.224	.200	.155	.152
Valine %	.76	.73	.60	1.07	.98	.86	.75	.64	.48	.54	.74	.72	.55	.54
Valine %/Mcal	.238	.225	.182	.369	.327	.277	.234	.194	.175	.180	.255	.244	.190	.186

[1] Minimum Total Content.

[2] 70 F Temperature assumed during growing and finishing periods. For each 10 F increase in temperature, increase amino acid levels by 3% of value.

FOR POULTRY FEED FORMULATION[1]

Leucine				Lysine				Methionine			
Total %	Coefficient %	***	Digestible %	Total %	Coefficient %	***	Digestible %	Total %	Coefficient %	***	Digestible %
1.11	79.5	(6.1)	0.88	0.78	60.1	(9.1)	0.47	0.19	76.3	(9.1)	0.14
2.98	82.9	(5.4)	2.47	1.83	82.5	(4.3)	1.51	0.77	83.4	(5.3)	0.65
0.77	84.8	(4.3)	0.65	0.42	77.8	(5.5)	0.33	0.18	78.3	(10.4)	0.14
11.19	90.8	(4.9)	10.16	7.97	87.1	(4.8)	6.94	1.10	92.5	(3.1)	1.02
1.85	77.9	(0.0)	1.44	1.54	69.2	(0.0)	1.07	0.34	79.8	(0.0)	0.27
1.39	81.6	(0.0)	1.13	0.73	72.7	(0.0)	0.53	0.36	80.4	(0.0)	0.29
2.49	87.3	(2.4)	2.17	1.97	78.3	(4.6)	1.54	0.77	89.7	(2.8)	0.69
8.19	99.1	(0.8)	8.11	5.59	97.1	(2.5)	5.43	2.49	99.2	(0.4)	2.46
1.12	80.0	(0.0)	0.89	0.64	79.7	(0.0)	0.51	0.44	88.1	(0.0)	0.39
1.03	95.6	(1.8)	0.99	0.25	84.8	(6.3)	0.21	0.15	94.5	(2.6)	0.15
1.63	87.2	(0.0)	1.42	0.91	83.0	(0.0)	0.75	0.37	85.6	(0.0)	0.32
2.12	89.8	(2.5)	1.90	0.71	71.9	(4.5)	0.51	0.40	84.8	(3.4)	0.34
10.13	98.0	(0.5)	9.93	1.07	89.1	(3.1)	0.95	1.59	97.2	(0.7)	1.54
1.87	74.8	(0.0)	1.40	1.36	61.0	(0.0)	0.83	0.44	80.8	(0.0)	0.36
								99.0	99.7	(0.0)	98.70
								86.0	97.4	(1.4)	83.76
6.60	81.5	(5.2)	5.38	2.24	62.5	(9.0)	1.40	0.57	73.9	(7.1)	0.42
4.47	94.0	(2.2)	4.20	4.66	90.1	(2.9)	4.20	1.75	92.7	(2.6)	1.62
5.81	89.6	(2.5)	5.20	4.00	87.5	(2.6)	3.50	1.49	94.2	(1.6)	1.41
2.53	90.4	(1.0)	2.29	1.34	75.4	(6.4)	1.01	0.40	87.1	(0.9)	0.35
4.53	76.1	(6.6)	3.45	1.86	61.2	(11.1)	1.14	0.36	73.2	(10.8)	0.26
				78.80	100.0	(0.0)	78.80				
7.31	73.8	(0.0)	5.40	5.40	69.5	(0.0)	3.75	2.02	75.0	(0.0)	1.51
1.16	77.3	(6.2)	0.90	0.78	64.6	(10.5)	0.50	0.23	81.2	(3.8)	0.18
2.61	91.2	(4.5)	2.38	1.76	87.3	(2.9)	1.54	0.45	85.0	(7.4)	0.38
2.78	86.7	(6.4)	2.41	2.38	81.8	(9.3)	1.95	0.63	87.6	(5.1)	0.55
3.44	84.6	(7.6)	2.91	2.84	82.6	(7.6)	2.34	0.72	86.6	(6.6)	0.62
0.81	90.2	(4.5)	0.73	0.47	86.5	(4.1)	0.40	0.17	85.5	(4.7)	0.15
4.16	82.4	(6.6)	3.43	3.03	80.0	(6.7)	2.42	0.93	83.6	(7.0)	0.78
4.44	85.4	(1.3)	3.79	2.46	84.3	(3.4)	2.07	0.80	90.3	(1.2)	0.72
0.48	85.7	(0.0)	0.41	0.23	62.7	(0.0)	0.14	0.14	89.5	(0.0)	0.13
0.98	73.1	(4.8)	0.72	0.66	73.4	(5.2)	0.48	0.30	76.1	(3.1)	0.23
2.62	73.1	(25.4)	1.92	0.79	58.4	(41.4)	0.46	1.15	76.6	(24.8)	0.88
2.18	95.2	(0.0)	2.07	2.09	90.0	(0.0)	1.88	0.80	96.2	(0.0)	0.77
3.99	88.0	(3.8)	3.51	3.15	86.6	(2.4)	2.73	1.14	88.8	(0.0)	1.01
1.31	93.6	(0.6)	1.22	0.24	81.3	(0.0)	0.20	0.18	86.0	(3.2)	0.15
3.75	92.1	(1.6)	3.45	3.09	89.8	(2.0)	2.77	0.70	92.2	(2.2)	0.65
2.68	93.6	(0.0)	2.51	2.19	92.2	(0.0)	2.02	0.39	88.9	(0.0)	0.35
2.05	90.9	(1.5)	1.86	1.23	83.6	(7.8)	1.03	0.78	93.8	(1.3)	0.73
1.13	95.5	(0.0)	1.08	0.57	90.7	(0.0)	0.52	0.24	95.7	(0.0)	0.23
1.00	90.0	(3.7)	0.90	0.42	81.4	(6.4)	0.34	0.23	86.7	(3.6)	0.20
0.89	77.1	(1.2)	0.69	0.61	74.1	(2.1)	0.45	0.21	77.3	(9.7)	0.16
0.99	80.9	(0.6)	0.80	0.63	77.8	(4.3)	0.49	0.24	82.4	(0.5)	0.20
0.98	78.0	(0.2)	0.76	0.62	75.7	(0.5)	0.47	0.22	74.9	(1.5)	0.16
0.96	87.1	(5.4)	0.84	0.56	79.3	(6.7)	0.44	0.24	82.4	(4.2)	0.20
1.09	84.2	(2.7)	0.92	0.75	81.0	(5.9)	0.60	0.27	79.5	(2.5)	0.22

Phenylalanine				Threonine				Valine			
Total %	Coefficient %	***	Digestible %	Total %	Coefficient %	***	Digestible %	Total %	Coefficient %	***	Digestible %
0.72	78.7	(5.6)	0.57	0.68	72.5	(7.0)	0.49	0.77	76.4	(6.0)	0.59
1.76	83.4	(5.0)	1.47	1.89	78.0	(4.3)	1.47	2.01	81.1	(4.8)	1.63
0.55	85.3	(5.5)	0.47	0.38	75.8	(6.0)	0.29	0.50	80.1	(5.1)	0.40
6.07	92.3	(3.5)	5.61	4.24	89.1	(4.4)	3.78	6.99	90.1	(4.8)	6.30
1.09	77.5	(0.0)	0.84	1.01	75.4	(0.0)	0.76	1.38	75.6	(0.0)	1.04
0.99	75.1	(0.0)	0.74	0.69	72.2	(0.0)	0.50	1.02	78.2	(0.0)	0.80
1.39	86.8	(2.9)	1.20	1.56	78.7	(4.1)	1.23	1.65	81.8	(3.1)	1.35
4.51	99.1	(0.7)	4.47	3.63	97.7	(1.6)	3.54	5.63	98.8	(0.8)	5.55
0.87	86.2	(0.0)	0.75	0.57	65.6	(0.0)	0.37	0.93	83.8	(0.0)	0.78
0.39	93.6	(2.9)	0.36	0.29	86.3	(3.5)	0.25	0.41	91.0	(3.8)	0.37
0.91	88.0	(0.0)	0.80	0.78	78.6	(0.0)	0.62	1.13	85.6	(0.0)	0.96
0.93	87.1	(3.1)	0.81	0.79	76.9	(4.2)	0.61	0.96	84.0	(4.3)	0.81
3.95	97.4	(0.6)	3.85	2.12	92.9	(1.1)	1.97	2.64	95.5	(1.1)	2.52
1.68	83.2	(0.0)	1.40	1.09	70.9	(0.0)	0.77	1.23	74.1	(0.0)	0.91
3.83	84.4	(4.7)	3.23	3.77	70.5	(4.8)	2.66	5.40	79.6	(4.4)	4.30
2.35	92.5	(2.4)	2.18	2.62	91.4	(2.4)	2.40	2.85	92.5	(2.9)	2.64
3.11	90.7	(2.2)	2.82	2.84	84.3	(3.0)	2.39	4.32	89.8	(2.0)	3.88
1.93	91.8	(0.6)	1.78	1.09	84.7	(1.4)	0.92	1.65	89.2	(1.1)	1.47
1.47	70.1	(8.7)	1.03	4.21	33.3	(3.9)	1.40	3.41	54.6	(6.6)	1.86
3.84	73.3	(0.0)	2.82	3.84	71.3	(0.0)	2.74	4.90	75.7	(0.0)	3.71
0.73	76.2	(4.6)	0.56	0.69	67.1	(9.1)	0.46	0.85	71.6	(8.3)	0.61
1.56	90.8	(5.4)	1.41	1.35	87.3	(5.2)	1.18	1.64	88.1	(5.1)	1.44
1.50	87.1	(6.3)	1.31	1.54	82.8	(8.6)	1.27	1.84	84.3	(8.2)	1.55
1.85	83.0	(7.5)	1.54	1.88	80.6	(7.5)	1.52	2.42	82.6	(8.0)	2.00
0.54	92.2	(3.8)	0.49	0.37	83.0	(6.4)	0.31	0.53	86.6	(4.9)	0.46
2.26	83.0	(7.0)	1.87	2.27	78.9	(6.7)	1.79	3.02	82.0	(6.9)	2.48
2.50	86.2	(0.2)	2.15	2.57	81.7	(1.9)	2.10	3.52	84.6	(1.1)	2.97
0.30	82.2	(0.0)	0.25	0.22	79.4	(0.0)	0.17	0.33	87.4	(0.0)	0.29
0.64	74.4	(4.8)	0.47	0.54	68.6	(4.9)	0.37	0.74	75.2	(4.1)	0.56
1.81	74.9	(25.9)	1.35	1.16	60.7	(37.2)	0.70	1.74	73.0	(25.5)	1.27
12.02	99.1	(0.0)	11.92	1.64	91.1	(0.0)	1.49	1.79	93.5	(0.0)	1.67
2.19	80.9	(12.0)	1.77	2.57	84.5	(4.1)	2.17	2.98	85.8	(4.2)	2.56
0.53	89.5	(0.7)	0.47	0.36	81.0	(1.8)	0.29	0.51	88.5	(0.2)	0.45
2.39	92.9	(1.8)	2.23	1.92	89.3	(2.0)	1.71	2.13	90.8	(2.4)	1.93
1.68	88.8	(0.0)	1.49	1.38	89.5	(0.0)	1.23	1.72	91.9	(0.0)	1.58
1.50	92.7	(0.9)	1.39	1.21	86.4	(4.3)	1.04	1.65	88.6	(2.8)	1.46
0.81	95.7	(0.0)	0.78	0.52	91.9	(0.0)	0.48	0.69	91.7	(0.0)	0.63
0.71	91.4	(3.1)	0.65	0.41	81.9	(5.7)	0.34	0.58	85.9	(4.1)	0.50
0.56	86.5	(9.6)	0.49	0.49	71.4	(1.0)	0.35	0.70	74.9	(2.1)	0.52
0.61	82.5	(0.5)	0.50	0.50	73.0	(0.6)	0.37	0.64	75.1	(4.0)	0.48
0.60	74.0	(4.5)	0.44	0.54	69.7	(1.6)	0.38	0.74	74.5	(0.1)	0.55
0.59	87.3	(5.4)	0.52	0.48	79.8	(6.5)	0.38	0.59	83.6	(6.2)	0.50
0.68	85.3	(2.7)	0.58	0.57	78.6	(4.1)	0.45	0.74	81.7	(4.1)	0.60

FOR POULTRY FEED FORMULATION

GEESE			EGG-TYPE CHICKENS				
Growing		Breeding	Growing				Laying[3]
0-5	5 +[2]		0-6	6-14[2]	14-20[2]		
2900	2950	2900	2900	2900	2900	Metabolizable Energy (Kcal/kg)	2.900
20.00	15.00	14.00	15.00	12.00	12.00	Crude Protein[1] %	14
						Crude Protein mg/hen/day	15.400
.70	.66	.55	.62	.53	.40	Methionine + Cystine %	.54
.241	.224	.190	.214	.183	.138	% Methionine + Cystine mg/hen/day	594
.40	.38	.32	.33	.28	.23	Methionine %	.32
.138	.129	.110	.114	.097	.079	Methionine mg/hen/day	352
1.10	.82	.71	.87	.70	.60	Lysine %	.69
.379	.278	.245	.300	.241	.207	Lysine mg/hen/day	759
.62	.52	.45	.53	.44	.33	Threonine %	.40
.214	.176	.155	.183	.152	.114	Threonine mg/hen/day	440
1.1	.87	.72	.93	.75	.62	Arginine %	.68
.379	.295	.248	.321	.259	.214	Arginine mg/hen/day	748
.16	.15	.12	.16	.12	.11	Tryptophan %[1]	.13
.055	.051	.041	.055	.041	.038	Tryptophan mg/hen/day	143
.27	.25	.21	.27	.21	.18	Histidine %	.20
.093	.085	.072	.093	.072	.062	Histidine mg/hen/day	220
.59	.54	.45	.57	.46	.38	Isoleucine %	.46
.203	.183	.155	.197	.159	.131	Isoleucine mg/hen/day	506
1.08	1.02	.85	1.08	.85	.71	Leucine %	.89
.372	.346	.293	.372	.293	.245	Leucine mg/hen/day	979
.53	.50	.42	.51	.42	.33	Phenylalanine %	.37
.183	.169	.145	.176	.145	.114	Phenylalanine	407
.61	.58	.48	.58	.48	.38	Valine %	.51
.210	.197	.166	.200	.166	.131	Valine mg/hen/day	561

[3]Assumes an average daily intake of 110 g of feed/hen daily. Protein and Amino Acid levels should be adjusted according to feed intake to result in a constant nutrient intake in mg/hen/day.

Heartland Lysine would like to express its appreciation to each of the authors who contributed data to this work. In particular our thanks are extended to Dr. Ian Sibbald (Agriculture Canada) and Dr. Carl Parsons (University of Illinois) for their contributions to the digestibility of feedstuff data.

This chart reflects our interpretation of published literature on digestible amino acids for poultry nutrition. It is the responsibility of the purchaser of our products to determine the best application of our products for their needs. Information and recommendations regarding our products and/or nutrient levels for poultry feeding are to the best of our knowledge accurate. We do not warrant the accuracy or completeness of this information. Our making this information available does not relieve the purchaser or user of his obligation to verify the suitability of our products and recommendations for their intended application.

[1]Table V–4 data was assembled by, and is presented through the courtesy of, Heartland Lysine, Inc., 8430 West Bryn Mawr Avenue, Suite 650, Chicago, IL 60631.

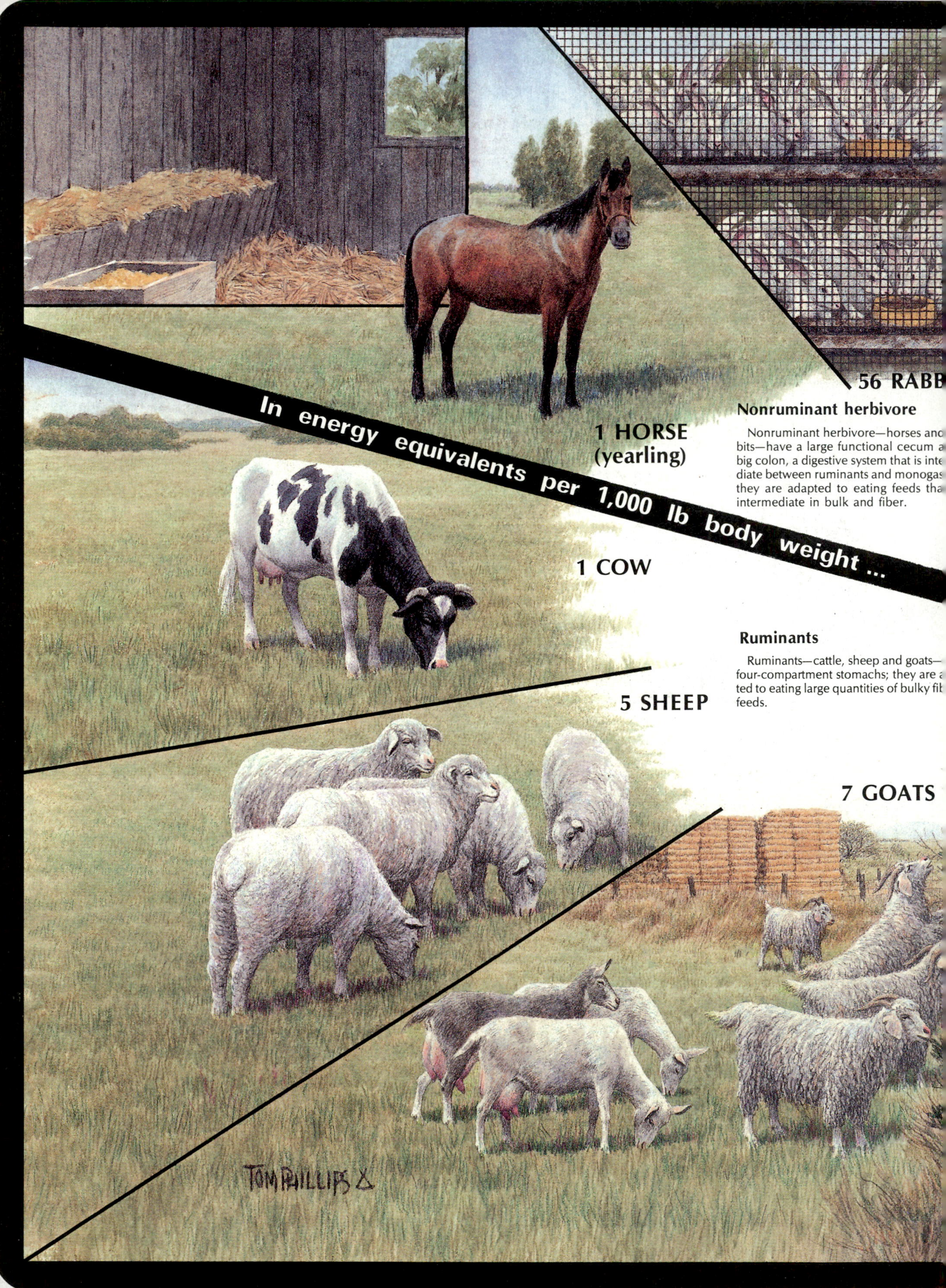

56 RABB
Nonruminant herbivore
Nonruminant herbivore—horses and
bits—have a large functional cecum a
big colon, a digestive system that is inte
diate between ruminants and monogas
they are adapted to eating feeds tha
intermediate in bulk and fiber.
In energy equivalents per 1,000 lb body weight ...
1 HORSE
(yearling)
1 COW
Ruminants
Ruminants—cattle, sheep and goats—
four-compartment stomachs; they are a
ted to eating large quantities of bulky fib
feeds.
5 SHEEP
7 GOATS
TOM PHILLIPS

3 SOWS
116 MINK
Monogastrics
Monogastrics—swine, mink, and fish—have a simple digestive system with limited capacity and limited microbial action and fiber digestion; they are adapted to eating concentrate feeds such as grains.
259 FISH
with each species eating its most common feeds
Poultry
Poultry—chickens and turkeys—have no teeth, store feed in a crop, and crush and grind feed in a gizzard; they are adapted to eating cereal grains and by-product feeds.
71 LAYERS
24 TURKEYS
6
Appendix

SECTION VI

APPENDIX

The Appendix is essential to the completeness of *Feeds & Nutrition*. It provides useful information on weights and measures, animal units, and poison information centers.

ANIMAL UNITS, the preceding two-page spread, is from an original painting by the noted artist, Tom Phillips, prepared specially for this book. It portrays the artist's conception of animal units—the energy equivalents per 1,000 lb body weight, with each species eating its most common feeds.

Section 

Original painting by Tom Phillips

APPENDIX

Contents | Page

Useful supplementary information to all that has gone before is contained in this section.

WEIGHTS AND MEASURES

Weights and measures are the standards employed in arriving at weights, quantities, and volumes. Even among primitive people, such standards were necessary; and with the growing complexity of life, they become of greater and greater importance.

Weights and measures form one of the most important parts of modern agriculture. This section contains pertinent information relative to the most common standards used in the U.S. livestock industry.

Metric System[1]

The United States and a few other countries use standards that belong to the *customary,* or English, system of measurement. This system evolved in England from older measurement standards, beginning about the year 1200. All other countries—including England—now use a system of measurements called the *metric system,* which was created in France in the 1790s. Increasingly, the metric system is being used in the United States. Hence, everyone should have a working knowledge of it.

The basic metric units are the *meter* (length/distance), the *gram* (weight), and the *liter* (capacity). The units are then expanded in multiples of 10 or made smaller by $\frac{1}{10}$. The prefixes, which are used in the same way with all basic metric units, follow:

"milli-"	=	$\frac{1}{1000}$	"deca-"	=	10
"centi-"	=	$\frac{1}{100}$	"hecto-"	=	100
"deci-"	=	$\frac{1}{10}$	"kilo-"	=	1,000

The following tables will facilitate conversion from metric units to U.S. customary, and vice versa:

Table VI–1 Weights and Measures—
Weight
Length
Surface/Area
Volume

Table VI–2 Temperature

[1]For additional conversion factors, or for greater accuracy, see *Misc. Pub. 223,* the National Bureau of Standards.

TABLE VI-1
WEIGHTS AND MEASURES

Weight

Unit	Is Equal To	
Metric system:	**(metric)**	**(U.S. customary)**
1 microgram (mcg)	.001 mg	
1 milligram (mg)	.001 g	.015432356 grain
1 centigram (cg)	.01 g	.15432356 grain
1 decigram (dg)	.1 g	1.5432 grains
1 gram (g)	1,000 mg	.03527396 oz
1 decagram (dkg)	10 g	5.643833 dr
1 hectogram (hg)	100 g	3.527396 oz
1 kilogram (kg)	1,000 g	35.274 oz; 2.2046223 lb
1 ton	1,000 kg	2,204.6 lb; 1.102 tons (short or 0.984 ton (long)
U.S. customary:	**(U.S. customary)**	**(metric)**
1 grain	.037 dr	64.798918 mg; .064798918 g
1 dram (dr)	.063 oz	1.771845 g
1 ounce (oz)	16 dr	28.349527 g
1 pound (lb)	16 oz	453.5924 g or 0.4536 kg
1 hundredweight (cwt)	100 lb	
1 ton (short)	2,000 lb	907.18486 kg or 0.907 (metric) ton
1 ton (long)	2,200 lb	1,016.05 kg or 1.016 (metric) ton
1 part per million (ppm)	1 microgram/gram; 1 mg/l; 1 mg/kg	.4535924 mg/lb; .907 g/ton; .0001%; .00013 oz/gal
1 percent (%) (1 part in 100 parts)	10,000 ppm; 10 g/l	1.28 oz/gal; 8.34 lb/100 gal

Weight Conversions

U.S. Customary to Metric			Metric to U.S. Customary		
To Change		**Multiply By**	**To Change**		**Multiply By**
grains	to milligrams	64.799			
ounces	to grams	28.35	grams	to ounces	0.035
pounds	to grams	453.6			
pounds	to kg	0.454	kg	to pounds	2.205
tons	to metric tons	0.9	metric tons	to tons	1.102

Weight–Unit Conversion Factors

To Change		Multiply By	To Change		Multiply By
mg/lb	to g/ton	2	mg/g	to mg/lb	453.6
g/lb	to g/ton	2,000	mg/kg	to mg/lb	0.4536
lb/ton	to g/ton	453.6	mcg/kg	to g/lb	0.4536
ppm	to mg/lb	0.4536	g/ton	to g/lb	0.0005
ppm	to %	move decimal 4 places to left	g/ton	to lb/ton	0.0022
mg/lb	to ppm	2.2046	g/ton	to %	0.00011
			%	to g/ton	9,072.00
ppm	to g/ton	0.907	g/ton	to ppm	1.1

(Continued)

TABLE VI–1 *(Continued)*

Length

Unit	Is Equal To	
Metric system:	**(metric)**	**(U.S. customary)**
1 millimicron (m)	.000000001 m	.000000039 in.
1 micron ( )	.000001 m	.000039 in.
1 millimeter (mm)	.001 m	.0394 in.
1 centimeter (cm)	.01 m	.3937 in.
1 decimeter (dm)	.1 m	3.937 in.
1 meter (m)	1 m	39.37 in.; 3.281 ft;1.094 yd
1 hectometer (hm)	100 m	328.08 ft; 19.8338 rd
1 kilometer (km)	1,000 m	3,280.8 ft; 0.621 mi
U.S. customary:	**(U.S. customary)**	**(metric)**
1 inch (in.)	1 in.	25 mm; 2.54 cm
1 hand*	4 in.	10.16 cm
1 foot (ft)	12 in.	30.48 cm; .305 m
1 yard (yd)	3 ft	.914 m
1 fathom** (fath)	6.08 ft	1.829 m
1 rod (rd), pole, or perch	16½ ft; 5½ yd	5.029 m
1 chain	792 in.; 66 ft; 22 yd	20.116 m
1 furlong (fur.)	220 yd; 40 rd	201.168 m
1 mile (mi)	5,280 ft; 1,760 yd; 320 rd; 8 fur.	1,609.35 m; 1.609 km
1 knot or nautical mile	6,080 ft; 1.15 land miles	1.85 km
1 league (land)	3 mi (land)	4.827 km
1 league (nautical)	3 mi (nautical)	4.827 km

Length Conversions

U.S. Customary to Metric			Metric to U.S. Customary		
To Change		**Multiply By**	**To Change**		**Multiply By**
inches	to millimeters	25.4	millimeters	to inches	0.04
inches	to centimeters	2.54	centimeters	to inches	0.4
feet	to centimeters	30.5	centimeters	to feet	0.033
feet	to meters	0.305	meters	to feet	3.3
yards	to meters	0.914	meters	to yards	1.1
miles	to kilometers	1.609	kilometers	to miles	0.6

*Used in measuring height of horses.

**Used in measuring depth at sea.

(Continued)

TABLE VI-1 *(Continued)*

Surface/Area

Unit	Is Equal To	
Metric system:	**(metric)**	**(U.S. customary)**
1 square millimeter (mm^2)	.000001 m^2	.00155 in.^{2}
1 square centimeter (cm^2)	.0001 m^2	.155 in.^{2}
1 square decimeter (dm^2)	.01 m^2	15.50 in.^{2}
1 square meter (m^2)	1 centare (ca)	1,550 in.^{2}; 10.76 ft^2; 1.196 yd^2
1 are (a)	100 m^2	119.6 yd^2
1 hectare (ha)	10,000 m^2	2.47 acres
1 square kilometer (km^2)	1,000,000 m^2	247.1 acres; .386 mi^2
U.S. customary:	**(U.S. customary)**	**(metric)**
1 square inch (in.^{2})	1 in. × 1 in.	6.452 cm^2
1 square foot (ft^2)	144 in.^{2}; 0.111 yd^2	.093 m^2
1 square yard (yd^2)	1,296 in.^{2}; 9 ft^2	.836 m^2
1 square rod (rd^2)	272.25 ft^2; 30.25 yd^2	25.29 m^2
1 rood	40 rd^2	10.117 a
1 acre	43,560 ft^2; 4,840 yd^2; 160 rd^2; 4 roods	4,046.87 m^2; 0.405 ha
1 square mile (mi^2)	640 acres; 1 section	2.59 km^2 or 259 ha
1 township	36 sections; 6 miles square	

Surface/Area Conversions

U.S. Customary to Metric			Metric to U.S. Customary		
To Change		**Multiply By**	**To Change**		**Multiply By**
sq in.	to cm^2	6.452	cm^2	to sq in.	0.155
sq ft	to m^2	0.09	m^2	to sq ft	10.764
sq yd	to m^2	0.836	m^2	to sq yd	1.196
sq mi	to km^2	2.6	km^2	to sq mi	0.4
acres	to ha	0.4	ha	to acres	2.5

Weights/Measures/Unit Area

Unit	Is Equal To
Volume per unit area:	
1 liter/hectare	0.107 gal/acre
1 gal/acre	9.354 liter/ha
Weight per unit area:	
1 kilogram/cm^2	14.22 lb/in^2
1 kilogram/hectare	0.892 lb/acre
1 lb/sq in.	0.0703 kg/cm^2
1 lb/acre	1.121 kg/ha
Area per unit weight:	
1 square centimeter/kilogram	0.0703 in.^{2}/lb
1 sq in./lb	14.22 cm^2/kg

(Continued)

TABLE VI–1 *(Continued)*

Volume

Unit		Is Equal To (U.S. customary) (liquid)	(U.S. customary) (dry)
Metric system liquid and dry:			
1 milliliter (ml)	.001 liter	.271 dram (fl)	.061 in.³
1 centiliter (cl)	.01 liter	.338 oz (fl)	.610 in.³
1 deciliter (dl)	.1 liter	3.38 oz (fl)	
1 liter	1,000 cc	1.057 qt or 0.2642 gal (fl)	.908 qt
1 hectoliter (hl)	100 liters	26.418 gal	2.838 bu
1 kiloliter (kl)	1,000 liters	264.18 gal	1,308 yd³

Unit		(ounces)	(cubic inches)	(metric)
U.S. customary liquid:				
1 teaspoon (t)	60 drops	⅛		5 ml
1 dessert spoon	2 t			
1 tablespoon (T)	3 t	½		15 ml
1 fl oz		1	1.805	29.57 ml
1 gill (gi)	½ c	4	7.22	118.29 ml
1 cup (c)	16 T	8	14.44	236.58 ml or 0.24 liter
1 pint (pt)	2 c	16	28.88	.47 liter
1 quart (qt)	2 pt	32	57.75	.95 liter
1 gallon (gal)	4 qt	8.34 lb	231	3.79 liters
1 barrel (bbl)	31½ gal			
1 hogshead (hhd)	2 bbl			
Dry:				
1 pint (pt)	½ qt		33.6	.55 liter
1 quart (qt)	2 pt		67.20	1.10 liters
1 peck (pk)	8 qt		537.61	8.81 liters
1 bushel (bu)	4 pk		2,150.42	35.24 liters

Unit	Is Equal To (metric)	(U.S. customary)
Solid metric system:		
1 cubic millimeter (mm³)	.001 cc	
1 cubic centimeter (cc)	1,000 mm³	.061 cu in.
1 cubic decimeter (dm³)	1,000 cc	61.023 cu in.
1 cubic meter (m³)	1,000 dm³	35.315 ft³; 1.308 yd³
U.S. customary:	**(U.S.customary)**	**(metric)**
1 cubic inch (in.³)		16.387 cc
1 board foot (fbm)	144 in.³	2,359.8 cc
1 cubic foot (ft³)	1,728 in.³	.028 m³
1 cubic yard (yd³)	27 ft³	.765 m³
1 cord	128 ft³	3.625 m³

Volume Conversions

U.S. Customary to Metric			Metric to U.S. Customary		
To Change		**Multiply By**	**To Change**		**Multiply By**
ounces (fluid)	to cc	29.57	cc	to oz (fluid)	0.034
ounces	to ml	29.57	ml	to oz	0.034
qt	to liters	0.946	liters	to qt	1.057
cu in.	to cc	16.387	cc	to cu in.	0.061
cu yd	to cm	0.765	cm	to cu yd	1.308

TABLE VI–2
TEMPERATURE

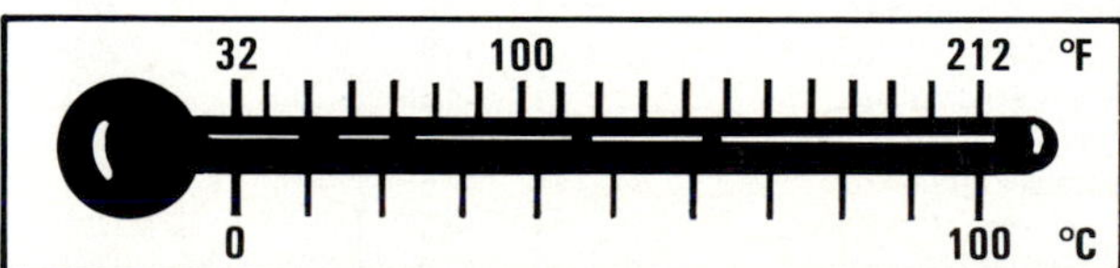

Fig. VI–1. Fahrenheit-Centigrade scale for direct conversion and reading.

One Fahrenheit (F) degree is 1/180 of the difference between the temperature of melting ice and that of water boiling at standard atmospheric pressure. One Fahrenheit degree equals 0.556°C.

One Centigrade (C) degree is 1/100 the difference between the temperature of melting ice and that of water boiling at standard atmospheric pressure. One Centigrade degree equals 1.8°F.

To Change	To	Do This
Degrees Fahrenheit	Degrees Centigrade	Subtract 32, then multiply by .556 (5/9)
Degrees Centigrade	Degrees Fahrenheit	Multiply by 1.8 (9/5) and add 32

Weights and Measures of Common Feeds

In calculating rations and mixing concentrates, it is usually necessary to use weights rather than measures. However, in practical feeding operations it is often more convenient for the farmer or rancher to measure the concentrates by volume. Table VI–3 will serve as a guide in feeding by measure.

TABLE VI–3
WEIGHTS AND MEASURES OF COMMON FEEDS

Feed	Approximate Weight: Lb per Quart[1]	Approximate Weight: Lb per Bushel[1]
Alfalfa meal	0.6	19
Barley	1.5	48
Beet pulp (dried)	0.6	19
Brewers' grain (dried)	0.6	19
Buckwheat	1.6	51
Buckwheat bran	1.0	32
Corn, husked ear	—	70
Corn, cracked	1.6	51
Corn, shelled	1.8	58
Corn meal	1.6	51
Corn-and-cob meal	1.4	45
Cottonseed	0.9–1.0	29–32
Cottonseed meal	1.5	48
Cowpeas	1.9	61
Distillers' grain (dried)	0.6	19
Fish meal	1.0	32
Flax	1.7	54
Gluten feed	1.3	42
Linseed meal (old process)	1.1	35
Linseed meal (new process)	0.9	29
Meat scrap	1.3	42
Milo (grain sorghum)	1.7	54
Molasses feed	0.8	26
Oat middlings	1.5	48
Oats	1.0	32
Oats, ground	0.7	22
Peanut meal	1.0	32
Peas	1.9	61
Rice	1.4	45
Rice bran	0.8	26
Rye	1.7	54
Sorghum (grain)	1.7	54
Soybeans	1.8	58
Sunflower	0.7	22
Tankage	1.6	51
Velvet beans, shelled	1.8	58
Wheat	1.9	61
Wheat bran	0.5	16
Wheat middlings, standard	0.8	26
Wheat screenings	1.0	32

[1]32 qts per bushel.

Storage Space Requirements for Feed and Bedding

The space requirements for feed storage for the livestock enterprise—whether it be for cattle, sheep, hogs, or horses, or, as is more frequently the case, a combination of these—vary so widely that it is difficult to provide a standard method of calculating space requirements applicable to such diverse conditions. The amount of feed to be stored depends primarily upon (1) length of pasture season, (2) method of feeding and management, (3) kind of feed, (4) climate, and (5) the proportion of feeds produced on the farm or ranch in comparison with those purchased. Normally, the storage capacity should be sufficient to handle all feed grain and silage grown on the farm and to hold purchased supplies. Forage and bedding may or may not be stored under cover. In those areas where weather conditions permit, hay and straw are frequently stacked in the fields or near the barns in loose, baled, or chopped form. Sometimes poled, framed sheds or a cheap cover of waterproof paper, grass, or cereal straw grass are used for protection. Other forms of low-cost storage include temporary upright silos, trench silos, temporary grain bins, and open-walled buildings for hay.

Table VI–4 gives the storage space requirements for feed and bedding. This information may be helpful to the individual operator who desires to compute the barn storage space required for a specific livestock enterprise. This table provides a convenient means of estimating the amount of feed or bedding in storage.

TABLE VI–4
STORAGE SPACE REQUIREMENTS FOR FEED AND BEDDING

Kind of Feed or Bedding	Pounds per Cubic Foot	Cubic Feet per Ton	Pounds per Bushel of Grain
Hay-Straw:[1]			
1. Loose			
Alfalfa	4.4–4.0	450– 500	
Nonlegume	4.4–3.3	450– 600	
Straw	3.0–2.0	670–1,000	
2. Baled			
Alfalfa	10.0–6.0	200– 330	
Nonlegume	8.0–6.0	250– 330	
Straw	5.0–4.0	400– 500	
3. Chopped			
Alfalfa	7.0–5.5	285– 360	
Nonlegume	6.7–5.0	300– 400	
Straw	8.0–5.7	250– 350	
Corn:			
15½% moisture:			
Shelled	44.8		56.0
Ear	28.0		70.0
Shelled, ground	38.0		48.0
Ear, ground	36.0		45.0
30% moisture:			
Shelled	54.0		67.5
Ear, ground	35.8		89.6
Barley, 15% moisture	38.4		48.0
Ground	28.0		37.0
Flax, 11% moisture	44.8		56.0
Oats, 16% moisture	25.6		32.0
Ground	18.0		23.0
Rye, 16% moisture	44.8		56.0
Ground	38.0		48.0
Sorghum grain, 15% moisture	44.8		56.0
Soybeans, 14% moisture	48.0		60.0
Wheat, 14% moisture	48.0		60.0
Ground	43.0		50.0

[1]Many factors—other than kind of hay-straw, form (loose, baled, chopped), and period of settling—affect the density of hay-straw in a stack or in a barn, including (a) moisture content at haying time, and (b) texture and foreign material.

Grain Weight in a Bin

Sometimes farmers need to estimate the weight of grain in storage. Such estimates are difficult to make because of differences in moisture content, depth of material stored, and other factors. However, the following procedure will enable one to figure feed quantities fairly closely.

1. **Corn (shelled) or small grain in rectangular cribs or bins.** Multiply the width by the length by the average depth (all in feet) and multiply by 0.8 to get the number of bushels (multiplying by 0.8 is the same as dividing by 1¼, the number of cubic feet in a bushel).

2. **Ear corn in rectangular cribs or bins.** Multiply the width by the length by the average depth (all in feet) and multiply by 0.4 to get the number of bushels (multiplying by 0.4 is the same as dividing by 2½, the number of cubic feet in a bushel of ear corn).

3. **Round bins or cribs.** To find the cubic feet in a cylindrical bin, multiply the squared radius by 3.1416 by the depth.

Thus, the volume of a round bin 20 ft in diameter and 10 ft deep is determined as follows:

a. The radius is half the diameter, or 10 ft

b. 10 × 10 = 100

c. 100 × 3.1416 = 314.16

d. 314.16 × 10 = 3,141.6 cu ft

e. Where shelled corn or small grain is involved, one would multiply 3,141.6 × 0.8, which equals 2,513.28 bu of grain that it would hold if full.

f. Where ear corn is involved, one would multiply 3,141.6 × 0.4 which equals 1,256.64 bu of ear corn that it would hold if full.

Hay Weight in a Barn or Stack

Livestock producers and hay dealers frequently buy and sell large quantities of hay in the stack or in the barn. This practice is especially prevalent in the western and Great Plains states where cattle and sheep are brought into the valleys to be wintered on hay bought from valley hay producers. Under such circumstances, the weight of hay is usually estimated, because (1) no scales are available, and/or (2) it is impractical to weigh the hay due to the time, labor, and wastage involved. In many such instances, the hay is fed directly from the stack or barn, in racks arranged about it. Under these and other circumstances, there is need for a simple and reasonably accurate method of estimating the weight of hay in a stack or in a barn.

In order to estimate the tonnage of hay in a stack or in a barn, it is necessary (1) to compute the volume of hay, and (2) to know the number of cubic feet per ton of hay. Table VI–5 gives the latter information.

TABLE VI–5
CUBIC FEET PER TON OF HAY

Feed	Settled 1–2 Months	Settled Over 3 Months
	(cu ft)	(cu ft)
Alfalfa	485	470
Clover	512	500
Hay, baled (closely stacked)	150–300	150–200
Hay, chopped	225	210
Straw, baled	200	200
Straw, loose	1,000	600–1,000
Timothy	640	625
Wild hay	600	450

In using Table VI–5, it should be recognized that many factors—other than kind of hay, form (loose, chopped, or baled), and period of settling—affect the density of hay in a barn or in a stack, including (1) moisture content at haying time, and (2) texture and foreign material.

It is relatively simple to compute the volume of hay in a mow, but it is more difficult to determine the volume of a stack. Although different rules or formulas may be and are used, the following are recommended by the U.S. Department of Agriculture.[2]

1. **Volume of hay in barns.** Multiply the width by the length by the height, all in feet, and divide by the cubic feet per ton as given in Table VI–6.

2. **Volume of hay in oblong stacks.** Three types of oblong stacks are common, as shown in Fig. VI–2.

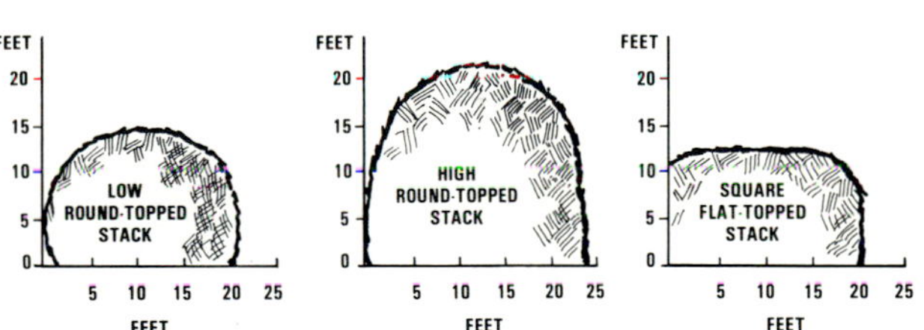

Fig. VI–2. Three common types of oblong stacks.

The volume of each type of oblong stack may be determined as follows:

a. For low, round-topped stacks—

(0.52 × O) − (0.44 × W) × W × L

b. For high, round-topped stacks—

(0.52 × O) − (0.46 × W) × W × L

c. For square, flat-topped stacks—

(0.56 × O) − (0.55 × W) × W × L

In these formulas "O" is the "over" or "overthrow," which is the distance in feet from the ground on one side

[2]*Measuring Hay in Stacks,* USDA Leaflet No. 72.

of the stack, up and over the stack and down to the ground on the other side; W is the width; and L is the length.

The application of this formula is illustrated as follows:

Example. *It is desired to estimate the amount of alfalfa hay in a low, round-topped type of oblong stack that has settled for 4 months. The stack is 20 ft wide, 30 ft long, and has an over of 40 ft.*

The answer is secured as follows:

a. Volume = (0.52 × 40) − (0.44 × 20) × 20 × 30 = 7,200 cu ft.

b. Table VI–6 shows that there are 470 cu ft per ton of settled alfalfa.

c. 7,200 ÷ 470 = 15 tons of hay.

3. **Volume of hay in round stacks.** The rules or formulas used for oblong stacks do not apply to round stacks. The volume of round stacks can be calculated by using the following formula:

$$\text{Volume} = (0.04 \times O) - (0.012 \times C) \times C^2$$

In this formula, C equals the circumference or distance around the stack at the ground, and O equals the over or distance from the ground on one side over the peak to the ground on the other side (usually it is best to take 2 over measurements at right angles to each other, and to average them).

Thus, the computation of the volume of a large round stack may be illustrated by the following example:

Example. *It is desired to determine the amount of alfalfa hay in a round stack that is 100 ft in circumference and has an average over of 60 ft.*

The answer is secured as follows:

a. Volume = (0.04 × 60) − (0.012 × 100) × $(100)^2$ = 12,000 cu ft.

b. Table VI–6 shows that there are 470 cu ft per ton of settled alfalfa.

c. 12,000 ÷ 470 = 25.5 tons of hay.

Animal Weights

Feeders who finish large numbers of animals have scales in their feedyards for use in determining in-weights, out-weights, and interim weight gains of animals while they are on feed. Likewise, both purebred and commercial breeders usually have scales. However, those with only one animal, or a few head—such as 4–H Club and FFA members, and part-time farmers—may not have scales. As a result, rations cannot be accurately evaluated, rate of gain cannot be calculated, and an animal's "weight readiness" for a livestock show or for market cannot be determined. Under such circumstances, a simple but reasonably accurate method of estimating body weight is very useful. Fortunately, animal weights may be determined with reasonable accuracy by taking two body measurements (body length and heart girth), then applying an appropriate formula.

BEEF CATTLE WEIGHTS

Here is how to do it:

Step 1. Measure the circumference (heart girth), from a point slightly behind the shoulder blade, thence down over the foreribs and under the body, behind the elbow (distance C of Fig. VI–3).

Step 2. Measure the length of body, from the point of the shoulder to the point of the rump (pinbone), in inches (distance A-B of Fig. VI–3).

Step 3. Take the values obtained in Steps 1 and 2 and apply the following formula to calculate body weight:

Heart girth × heart girth × body length ÷ 300 = weight in pounds

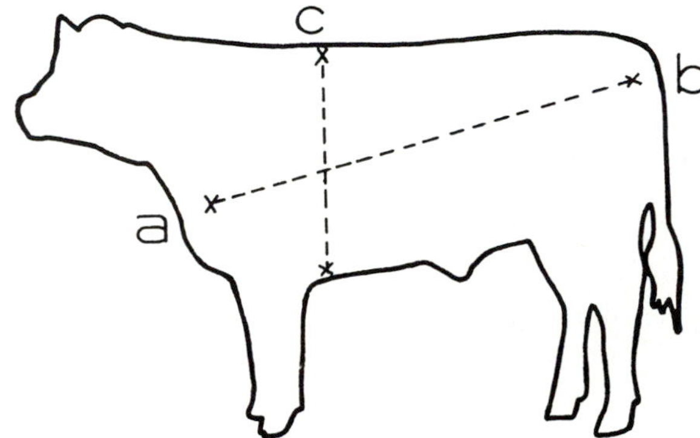

Fig. VI–3. How and where to measure beef cattle.

Example of a beef animal. *Assume that the heart girth measures 76 in. and the body length, 66 in. How much does the animal weigh?*

76 × 76 = 5,776
5,776 × 66 = 381,216
381,216 ÷ 300 = 1,270 lb

DAIRY HEIFER WEIGHTS

Weight for age is important in dairy heifers from the standpoint of determining the growth progress made by herd replacements.

Table 20–24 of Chapter 20, Feeding Dairy Cattle, shows the weight and heart girth measurements of dairy calves or heifers at monthly intervals up to 22 months of age. If producers do not have scales, they can measure the heart girth with a tape (see Fig. VI–4) and use Table 20–24 to estimate weight within 95% accuracy.

Fig. VI-4. How to tape measure a dairy heifer.

SHEEP AND GOAT WEIGHTS

The weight of sheep and goats is estimated in the same way as for beef cattle; hence, it involves making the measurements and applying the formula given for beef cattle. There is one important precaution, however; with unshorn sheep, be sure to part, or compress, the wool to ensure an accurate heart girth measurement.

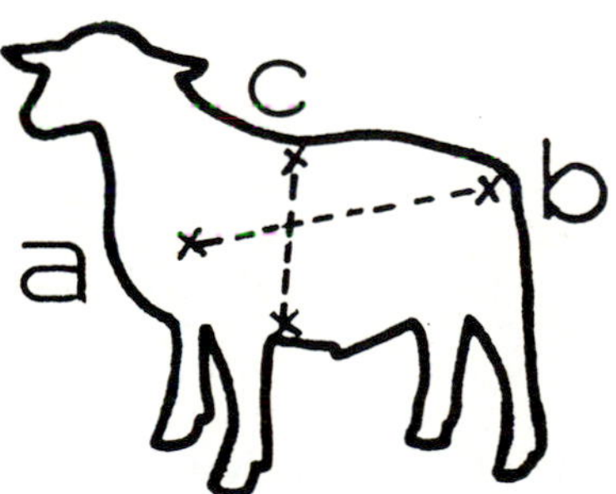

Fig. VI-5. How and where to measure sheep.

SWINE WEIGHTS

Hog weights can be calculated from body measurements, similar to beef cattle, but a different formula must be used. Here is how to estimate the weight of hogs:

Step 1. Measure the circumference (heart girth) of the animal (C in diagram).

Step 2. Measure the length of body (A-B in Fig. VI-6). With the animal standing or restrained in the position shown in Fig. VI-6, measure the distance from the poll (between the ears), over the backbone, to the base of the tail.

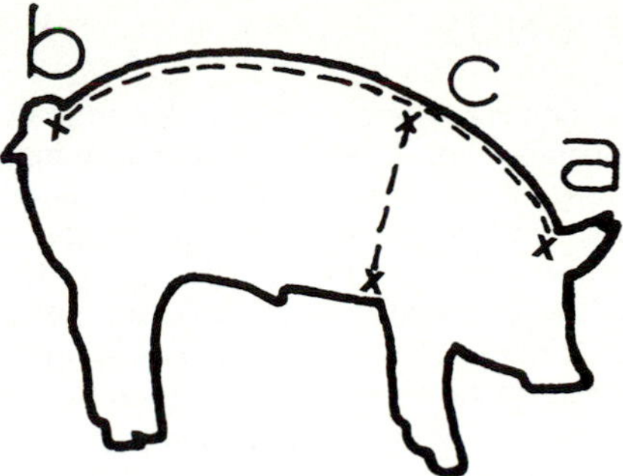

Fig. VI-6. How and where to measure hogs.

Step 3. Apply the following formula:

Heart girth × heart girth × length ÷ 400 = weight in pounds.

Note: For hogs weighing less than 150 lb, add 7 lb to the weight figure obtained from the formula. For animals weighing 151 to 400 lb, no adjustment is necessary.

HORSE WEIGHTS

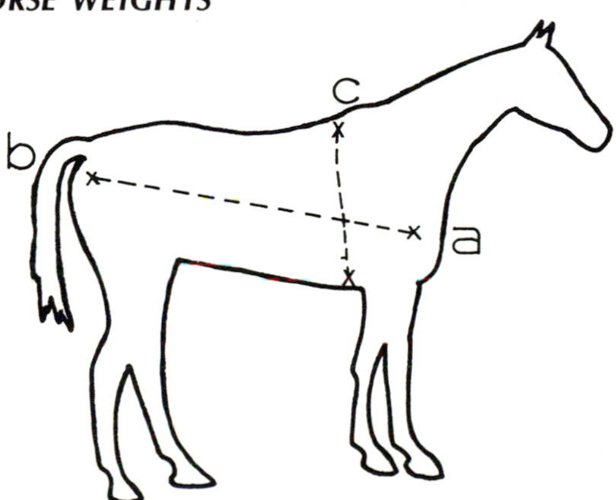

Fig. VI-7. How and where to measure horses.

It is easy to estimate the weight of a horse; and tests have shown that the results obtained this way are accurate within 3% of actual scale weight. This procedure is as follows:

Step 1. Measure the circumference (heart girth) of the body in inches (C in diagram).

Step 2. Measure the length of body from the point of the shoulder to the point of croup (A-B in the diagram).

Step 3. Apply the following formula to calculate the weight of the horse:

Heart girth × heart girth × length ÷ 300 + 50 lb = weight of horse.

Example. *Assume that the heart girth is 70 in. and the body length is 65 in.* How much does the horse weigh?

70 × 70 × 65 ÷ 300 + 50 lb = weight
4,900 × 65 = 318,500
318,500 ÷ 300 = 1,061 lb
1,061 + 50 = 1,111 lb body weight

ANIMAL UNITS

An animal unit is a common animal denominator, based on feed consumption. It is assumed that 1 mature cow represents an animal unit. Then, the comparative (to a mature cow) feed consumption of other age groups or classes of animals determines the proportion of an animal unit which they represent. For example, it is generally estimated that the ration of one mature cow will feed 5 mature ewes, or that 5 mature ewes equal 1.0 animal unit.

The original concept of an animal unit included a weight stipulation—an animal unit referred to a 1,000-lb cow, with or without a calf at side. Unfortunately, in recent years, the 1,000-lb qualification has been dropped. Certainly, there is a wide difference in the daily feed requirements of a 900-lb cow and of a 1,500-lb cow. Both will consume dry matter on a daily basis at a level equivalent to about 2% of their body weight.

Hence, a 1,500-lb cow will consume 50% more feed than a 1,000-lb cow.

Also, the period of time to be grazed has an effect on the total carrying capacity. For example, if an animal is carried for 1 month only, it will take $\frac{1}{12}$ of the total feed required to carry the same animal 1 year. For this reason, the term *animal unit months* is becoming increasingly important. So, in addition to the weight factor, the time factor has a distinct bearing on the ultimate carrying capacity of a tract of land.

Table VI-6 gives the animal units of different classes and ages of livestock.

TABLE VI-6
ANIMAL UNITS

Type of Livestock	Animal Units
Cattle:	
Cow, with or without unweaned calf at side, or heifer 2 years old or older	1.0
Bull, 2 years old or older	1.3
Young cattle, 1 to 2 years	0.8
Weaned calves to yearlings	0.6
Horses:	
Horse, mature	1.3
Horse, yearling	1.0
Weanling colt or filly	0.75
Sheep:	
5 mature ewes, with or without unweaned lambs at side	1.0
5 rams, 2 years old or over	1.3
5 yearlings	0.8
5 weaned lambs to yearlings	0.6
Swine:	
Sow	0.4
Boar	0.5
Pigs to 200 lb	0.2
Chickens:	
75 layers or breeders	1.0
325 replacement pullets to 6 months of age	1.0
650 8-week-old broilers	1.0
Turkeys:	
35 breeders	1.0
40 turkeys raised to maturity	1.0
75 turkeys to 6 months of age	1.0

Fig. VI-8. White turkeys on the range. Forty turkeys raised to maturity equal one animal unit. (Courtesy, J. C. Allen & Son, West Lafayette, Ind.)

POISON INFORMATION CENTERS

With the large number of chemical sprays, dusts, and gases now on the market for use in agriculture, accidents may arise because of operators being careless in their use. Also, there is always the hazard that a child may eat or drink something that may be harmful. Centers have been established in various parts of the country where doctors can obtain prompt and up-to-date information on treatment of such cases, if desired.

Local medical doctors have information relative to the Poison Information Centers of their area, along with some of the names of their directors, telephone numbers, and street numbers. When calling any of these centers, one should ask for the "Poison Information Center." If this information cannot be obtained locally, call the U.S. Public Health Service at Atlanta, Ga., or Wenatchee, Wash.

INDEX

A

C

D

E

F

G

I

J

K

L

S

T

U

V

W

X

Y

Z